P9-CML-814

Concise Encyclopedia Biology

Concise Encyclopedia Biology

Translated and revised by
Thomas A. Scott

Walter de Gruyter Berlin · New York 1996

Title of the original, German language edition
ABC Biologie
Edited by
Friedrich W. Stöcker, Nauen, Germany
Gerhard Dietrich, Berlin/Leipzig, Germany
VEB F. A. Brockhaus Verlag, Leipzig 1986

Translated into English and revised by
Thomas A. Scott, Ph.D.
10, Lidgett Walk
Leeds, LS8 1NW
Great Britain

⊗ Printed on acid-free paper, which falls within the guidelines of the ANSI to ensure permanence and durability.

Library of Congress Cataloging-in-Publication Data

ABC Biologie. English
Concise encyclopedia biology / translated and revised by Thomas A. Scott.
Includes bibliographical references.
ISBN 3-11-010661-2
1. Biology-Dictionaries. I. Scott, Thomas A., 1935– .
II. Title
QH 302.5.12313 1995
574.03—dc 20

Die Deutsche Bibliothek – Cataloging-in-Publication Data

Concise encyclopedia biology / transl. and rev. by Thomas A. Scott. – Berlin ; New York : de Gruyter, 1996
Einheitssacht.: ABC Biologie <engl.>
ISBN 3-11-010661-2
NE: Scott, Thomas [Bearb.]; EST

English language edition

Typesetting and Printing: Buch- und Offsetdruckerei Wagner GmbH, Nördlingen
Binding: Lüderitz & Bauer, Berlin
Cover Design: Hansbernd Lindemann, Berlin. – Printed in Germany.

Translator's preface

Fortunately, science crosses the boundaries of translation with relative ease. This encyclopedia, however, is not simply a translation. The opportunity has been taken to update and revise many entries and to introduce recent material.

Zoological entries have been checked against a wide range of literature sources.

Botanical information has been compared with that in published Floras, as well as being checked against, and sometimes supplemented with, material from botanical gardens at Washington D.C. (USA), Longwood (Pennsylvania, USA), Kew (London, UK), Munich (Germany) and Penang (Malaysia).

Entries concerned with biochemistry and molecular biology have been checked against the manuscript in preparation for the 3rd edition of the Concise Encyclopedia of Biochemistry and Molecular Biology (de Gruyter).

I have included the most notable new animal species described since the publication of the German edition. In recent years, new species of lemurs have been discovered in Madagascar and the sun-tailed monkey was reported in 1988 in Gabon, but the discovery of the Vu Quang ox in 1992 (reported in 1993) marked the first discovery of a really large mammal since the kouprey was found more than 50 years ago. Barely one year later, in 1994, the same team reported a new and larger species of muntjac deer in the Vu Quang Nature Reserve of Vietnam. It appears that Vietnam may have not yielded all its zoological secrets; a pair of horns found on a market stall in Ho Chi Minh has been assigned to a new genus and species of antelope (*Pseudonovibos spiralis*), and an expedition is planned to find a living specimen. "New" species also include *Australopithecus ramidus* (see Anthropogenesis), identified in 1994 from its 4.4 million-year-old remains, and which arguably qualifies as the missing link between the earliest hominids and the last common ancestor of humans and African apes.

German literature references have been omitted or replaced by appropriate English literature, sometimes as an anknowledgement of sources (when I have introduced new material), and otherwise as recommended reading for those who wish to pursue a topic in greater detail.

The use of italics, bold type and bold italics has not been changed. As in the German version, all entry headings are in bold type, including those that are scientific Latin; and in the rest of the text, italics are used for the Latinized names of all taxa (from species to phylum). Although this is no longer universal practice on the part of scientific editors, it has been retained here, because the italicized names represent useful signposts when scanning an entry for specific information.

As a rule, the subject of each entry is abbreviated to its initial letters in the text, but this practice is sometimes abandoned in favor of clarity. It must be emphasized that the cross referencing instruction to "see" is entirely neutral; it does not necessarily infer a synonym, and should be interpreted as "see for more information". Centimeter measurements after the names of bird species refer to the distance from the tip of the bill to the end of the tail with the neck extended and bill in line. Relative molecular mass (formerly called molecular weight) is signified as M_r. For most purposes, the joule (J) is used as the unit of energy (1 calorie = 4.1868 J), and pressure is quoted in pascal (Pa) (1 Pa = 1 N m^{-2}; 1 pound/-square inch = 6 894.76 N m^{-2}). Linear measurements are normally quoted in centimeters (1 cm = 0.3937 inch; 1 inch = 2.54 cm).

Finally, I thank friends and colleagues who provided criticism and advice, in particular Alfred Dillmann (Munich, Germany) who helped me to identify several organisms from their trivial German names, Cicley Shaw and Tonia Mason for their help with preparation of the manuscript, and Wendy, my wife, for keeping my home and my life in a state of negative entropy.

Leeds, 1995 Thomas A. Scott

[illegible]

A

Aardvark, ***ant bear, earth hog***, *Orycteropus afer:* the only member of the order *Tubulidentata* ("tubule-toothed"). The A. is a pig-sized mammal with a very long snout, long ears and a long tail. Dentition consists of five teeth on each side of the upper jaw, and four on each side of the lower jaw. Each tooth is composed of numerous hexagonal prisms of dentine surrounding a tubular pulp cavity, and it is covered externally by a layer of cement, giving the appearance of a flat-crowned column. The A. is a native to Africa, south of the Sahara, mainly in open country and savanna, avoiding thickly forested areas. It is an extremely active and powerful burrower, and sleeps during the day in a self-dug burrow or cave. Its diet consists of ants and termites, whose mounds it pulls apart with its powerful front claws.

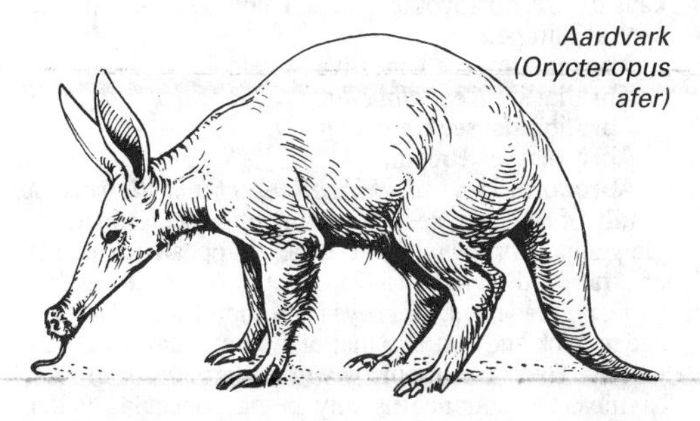
Aardvark (Orycteropus afer)

Aardwolf, *Proteles cristatus:* a terrestrial carnivore (see *Fissipedia),* sometimes placed in the family *Hyaenidae* (see), but more appropriately considered as the sole representative of the family *Protelidae.* Phylogenetic relationships to both the *Viverridae* and the *Hyaenidae* are evident. There is no perineal gland and no anal pocket, and the teeth are reminiscent of those of insectivores. The incisors and canines have typical carnivore structure, but only 2 premolars are present, and these are small and nearly functionless. The single molar is microscopically small. Head plus body length is 55–80 cm, with a tail of 20–30 cm. Superficially, it resembles the striped hyaena, with long legs, black stripes on the flanks, and a posteriorly sloping back like that of the *Hyaenidae.* The black and yellowish dorsal mane extends along the entire back to the base of the tail. The gray to yellowish hair is soft and woolly, but also contains long stiff bristles. Nocturnal in habit, it spends most of the day beneath the ground, often in abandoned wart hog dens, occasionally emerging to sun itself.

Only one species is recognized, but throughout its range (Somalia and Sudan through East Africa to South Africa) there are at least 6 different color varieties. It inhabits dry grasslands, steppes and savannas, where it feeds mainly on the grass-eating termite, *Trinervitermes.* As an adaptation to this dietary habit, the tongue is very mobile.

A-band: see Muscle tissue.

Abbé condenser: see Darkfield illumination.

Abbreviation, ***abbreviated development:*** phylogenetic abbreviation of ontogenesis, resulting from the omission of certain morphological stages. A. may occur by omission of both early and intermediate stages of development (see Baer's law), or by omission of the final stage (see Neoteny). Omission of the last developmental stage is exemplified by the Mexican Axolotl *(Abystoma mexicanum),* which reproduces as a gill-bearing, aquatic larva, and normally does not complete its metamorphosis to the terrestrial form. See Metaboly.

Abdominal reflex: see Protective reflexes.

Abdominal respiration: see Respiratory mechanics.

Abducent nerve: the VI cranial nerve; see Cranial nerves.

Abductors: muscles which move entire limbs away from the midline of the body, or a digit from the plane of the middle finger or the plane of the second toe, e.g. the deltoid muscle raises the arm directly from the side to bring it at right angles to the trunk.

Abele: see *Salicaceae.*

Aberration: see Lens aberrations, Chromatid aberrations, Chromosome aberrations.

Abietic acid: a diterpene resin acid, and the main constituent of colophony, from which it is obtained by distillation. Derivatives of A.a. are present in the alcohol-soluble fraction of amber.

Abietic acid

Abietitol: see Pinitol

Abiogenesis: the development of living organisms from inorganic and organic materials, and not by the reproduction of other living organisms. A. is the subject of several scientific theories. By providing a mechanism for the generation of the simplest life forms from nonliving organic compounds, A. possibly explains the origin of life.

Up to the beginning of the 19th century, the view was widely held that parasitic insects and worms, and especially the microorganisms of mud, intestinal contents and refuse, etc. could arise spontaneously. F. Redi (1649) and J. Swammerdam (1669) had demonstrated that insects, fish and amphibians develop from eggs, and Spallazani (1765) had provided experimental proof that heat sterilization prevented the development of microorganisms in a liquid medium. Nevertheless, the belief in spontaneous generation persisted and was slow to be dispelled.

According to modern concepts, four thousand million (4×10^9) years ago the composition of the primitive atmosphere of the earth was very different from that of the present atmosphere. In particular, it was a reducing atmosphere, lacking oxygen (which did not begin to appear until about 3×10^9 years ago) and containing much nitrogen. Under the highly energetic conditions of the time (high temperatures, UV irradiation, electric discharge), chemical reactions occurred that eventually led to the generation of the precursors of living organisms. Oparin named this process *abiogenic* or *prebiotic organic evolution.* Thus, in the primitive atmosphere, polymeric hydrocarbon derivatives were formed by the action of UV-light on water, methane, ammonia, hydrogen sulfide and carbon monoxide. Hydrogen cyanide was also important in the synthesis of biomolecules. Under simulated primitive earth conditions, three immediate products of HCN (cyanoacetylene, nitriles, cyanamide) can interact with aldehydes, ammonia and water to form various organic compounds, notably amino acids, pyrimidines, purines and porphyrins. In model experiments, Miller synthesized 14 of the 20 proteogenic amino acids by passing an electric discharge through a simulated primitive atmosphere. Von Oró et al. demonstrated the formation of adenine and guanine by the heat polymerization of ammonium cyanide in aqueous solution. Furthermore, formaldehyde is produced readily in simulated primitive atmosphere experiments, and it yields several sugars when heated with calcium carbonate, thus opening the way for the production of nucleosides and nucleotides.

It is assumed that polypeptides were formed by self condensation of abiotically produced amino acids, and polynucleotides by self condensation of abiotically produced nucleotides. Several mechanisms are possible for the promotion of such condensations. In particular, it is known that polyphosphates or polyphosphate esters act as dehydrating agents for the formation of peptides from amino acids when heated or irradiated with UV-light. Such polyphosphates, which are strong candidate forerunners of ATP, could have arisen from phosphate minerals by the action of cyanoacetylene or cyanoguanidine, or under the influence of high temperatures (e.g. near to volcanoes). For the development of a true life form, capable of further development and evolution, with the indispensable coupling of nucleic acids and protein synthesis, it was necessary for prebiotic material to become separated into discrete units. It is therefore thought that the formation of internally organized cells was the next significant step in the evolution of life forms. In this respect, special importance is attached to the formation of precellular structures, such as Oparin's coacervates and Fox's microspheres. It is highly probable that prebiotic membranes were formed by the aggregation of prebiotic lipids.

The direct route from abiotic high molecular mass compounds of the primitive earth to the first true living organisms is, however, still largely unexplained, since it is very difficult to explore experimentally. Essentially, there are two hypotheses for the origin of protobionts, i.e. the metabolic or *protein hypothesis* and the gene or *nucleic acid hypothesis*. In the gene hypothesis, primacy is given to the role of proteins in prebiotic evolution. Thus, membrane-bounded, replicative protocells containing catalytically active proteinoids subsequently acquire a coding system and become primitive living cells. In the gene hypothesis, primacy is given to nucleic acids. Originally proposed by H. Muller in 1929, this hypothesis states that life had its origin in the prebiotic formation of genes, which encoded the potential for metabolism and self-replication. The necessary amino acids were then acquired by random encounter. Further elaboration of this hypothesis became possible when the chemical structure of genetic material was established by Watson and Crick and others.

Abiotic factors: see Environmental factors.

Ablepharus, *lidless skinks:* a genus of the family *Scincidae* (see) containing only 4 species, all native to Europe and western Asia. They have elongated bodies and short, thin, weak legs. Since the eyelids are fused to a transparent spectacle, they are also known as "snake-eyed skinks", but this term is also applied to other genera, such as the Australian *Panaspis*. The basic color is glossy brown with streaking and stippling. The most familiar species is the 10–12 cm-long *A. kitaibelii*. Its subspecies, *A. kitaibelii fitzingeri,* is found from the Balkans to Czechoslovokia, and is the only skink native to central Europe.

Abomasum: see Digestive system.

Abralia: see *Cephalopoda.*

Abraliopsis: see *Cephalopoda.*

Abramis: see Bream.

Abrocomidae, *rat chinchillas*, *chinchilla-rats:* a family of Rodents (see) consisting of 2 species in a single genus *(Abrocoma)*. The general appearance is that of a rat (body length 15–25 cm, tail 6–18 cm) with a large, pointed head, large eyes and short limbs. The pelage is thick and soft like that of a chinchilla, but not as woolly. They live in burrows in the ground or in rock cavities, the entrance usually being concealed under bushes or rocks. *A. bennetti* is found in Chile, while the range of *A. cinerea* includes southern Peru, northern Chile and northwestern Argentia.

Abroscopus: see *Sylviinae*.

Abscisic acid, *ABA, abscisin, dormin:* (S)-(+)-5-(1′-hydroxy-4′-oxo-2′,6′,6′-trimethyl-2-cyclohexen-1-yl)-3-methyl-*cis,trans*-2,4-pentadienoic acid, a sesquiterpene plant hormone of ubiquitous accurrence in higher plants. It is also present in the *Bryophyta,* but was previously reported to be absent from liverworts, where lunularic acid (a dihydro-stilbene) was thought to perform the same function. ABA, in concentrations similar to those in higher plants, has now been unequivocally identified in the gametophyte of the liverwort, *Marchantia polymorpha,* where it probably also functions as a hormone [Xiaoyue Li et al. *Phytochemistry* ***37*** (1994) 625–637].

Two stereoisomeric forms are possible, depending on the *cis* or *trans* orientation of the $\Delta^{2,3}$ double bond; the *cis* isomer predominates in all plants; the *trans* isomer is occasionally found in small quantites, but does not appear to be biologically active. Only the (+)-form occurs naturally, and its concentration depends on the plant organ and its stage of development (average about 100 µg/kg fresh weight); relatively large quantities are present in fruits, dormant seeds, buds and wilting leaves.

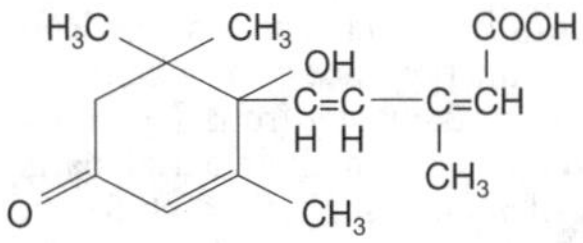

H_3C CH_3 OH C=C—C=CH H H COOH CH_3 O CH_3

Abscisic acid

ABA antagonizes the action of auxin, gibberellins and cytokinins, and in concert with these other phytohormones it is involved in the regulation of important growth and developmental processes such as seed and bud dormancy, stomatal transpiration, flowering, germination and aging. ABA functions in the adaptation of plants to environmental change. Thus, it occurs in the xylem sap where it serves as a chemical signal between root and shoot [A. Bano et al. *Aust. J. Plant Physiol.* ***20*** (1993) 109–115; A. Bano et al. *Phytochemistry* ***37*** (1994) 345–347]. Water stress (water shortage or soil flooding) is first perceived by the roots; the resulting transport of ABA to the shoot induces stomatal closure and sometimes also the synthesis of proteins that possibly protect issues from the effects of desiccation. As seeds develop in the plant, their premature germination is prevented by the presence of ABA, which also appears to induce the synthesis of certain seed proteins. Induction by gibberellic acid of hydrolytic enzymes in the aleurone layer of cereal grains is restrained by the presence of ABA. Stomatal closure by ABA is due mainly to an efflux of K^+. ABA my directly affect K^+ channels, or the loss of K^+ may be caused by an increase in the Ca^{2+} concentration in the guard cell cytoplasm. Many genes encoding stress proteins (proteins induced by desiccation, low temperature, wounding, etc.), storage proteins and proteins conferring desiccation tolerance display an increased rate of expression in the presence of ABA. The molecular basis of ABA action is unknown, and no receptor has been demonstrated.

ABA is synthesized primarily in leaf chloroplasts and in the root. It is a true terpenoid, being derived from mevalonic acid via isopentenyl pyrophosphate. However, it is not derived directly via a C_{15} (sesquiterpenoid) route, but indirectly via a C_{40} (apocarotenoid) by cleavage: mevalonate →→ farnesyl pyrophosphate →→ all-*trans*-violaxanthin → all-*trans*-neoxanthin → xanthoxin → ABA-aldehyde → ABA. The rate-limiting step of biosynthesis appears to be the conversion of all-*trans*-violaxanthin into 9'-*cis*-neoxanthin, and this conversion is promoted by water deficiency. The oxidation of ABA-aldehyde to ABA is catalysed by a Mo-containing aldehyde oxidase [A. D. Parry & R. Horgan *Physiol. Pl.* ***82*** (1991) 320–326]. Direct synthesis of ABA from farnesyl pyrophosphate via the sesquiterpenoid route has been reported in the fungi, *Cercospora rosicola* and *C. cruenta.* In higher plants ABA is accompanied by its glucose esters and its *O*-glucoside, which are thought to be transport and storage forms. It is metabolized and inactivated by 8'-oxidation to phaseic acid.

Abscission: shedding of leaves, fruits and other plant organs, e.g. foliage, flower buds, branches, thorns or flowers. A. serves to remove old, damaged and diseased parts from the plant. An excretory function is also possible, i.e. when organs are shed which contain accumulated metabolic waste products. A. is important in the spread of fruits and asexual reproductive structures. Shedding of the plant part is preceded by characteristic morphological and anatomical changes in the region of separation. A prominent layer of tissue (the *abscission layer)* is formed at the base of the leaf or fruit stalk, consisting of small parenchyma cells with dense cytoplasm. Within this abscission layer, 2–3 layers of cells become considerably weakened by disintegration of the middle lamellae, both middle lamellae and primary walls, or whole cells (depending on the plant species). This is an active, regulated process, requiring oxygen and respiratory substrates (A. is inhibited by respiratory poisons). A. is also accompanied by the synthesis of protein and RNA. In particular, cellulase and pectinase are synthesized, which play a special role in A. Pectinase, which renders propectin water-soluble, is actively secreted into the cell wall from the protoplasm.

Auxin, senescence factors, ethylene and abscisic acid are involved in the regulation of A. Auxin is synthesized in intact leaves which have not yet aged. It migrates through the leaf stalk and inhibits A. In horticulture, premature fruit drop is prevented by spraying auxin solutions (usually nonbiological, chemically synthezised compounds with auxin activity). Senescence factors migrate from aged leaves and flowers, and stimulate A. In many cases of fruit and leaf A. (e.g. in the lupin), there is a correlation between onset of abscisic acid production and A. According to the senescence hypothesis of A., the activity of the abscission layer is regulated by the ratio of auxin to senescence factors. Factors that retard leaf and fruit senescence, e.g. cytokinins, also retard A. Ethylene is directly responsible for the induction of cell changes in the abscission layer leading to A. This primary effect of ethylene is thought to be due to gene activation, resulting, inter alia, in increased synthesis of RNA and cellulase and/or pectinase. Ethylene also inhibits auxin synthesis and transport, thereby decreasing an inhibitory factor of A. Preparations which release ethylene, e.g. Flordimex, are used horticulturally to accelerate or synchronize fruit ripening, which is economically important for mechanized harvesting.

Absinthin: a sesquiterpene bitter principle from wormwood *(Artemisia absinthum).* It is the flavoring agent in absinthe liqueur.

Absinthol: see Thujone.

Absorbance, ***A, extinction, optical density:*** a measure of the quantity of light absorbed by a solution. $A = \log I_0/I$, where I_0 is the intensity of incident light, and I is the intensity of transmitted light. I_0/I is known as the *transmission,* and is usually expressed as a percentage. According to the Lamber-Beer law, A is linearly proportional to the thickness of the solution (i.e. the length of the light path through the solution) and the concentration of dissolved material. The proportionality factor is known as the *absorptivity, absorbance index* or *absorption coefficient,* and it is a characteristic quantity for each molecular species in a given solvent. Expressed on a molar basis, it becomes the *molar absorptivity, molar absorption coefficient* or *molar extinction coefficient,* i.e. $\varepsilon = A/lc$, where l is the length of the light path in centimeters, c is the molar concentration, and ε is the molar absorptivity.

Measurement of A, which is performed with a spectrophotometer, is an important quantitative laboratory procedure in biochemistry and molecular biology, routine clinical biochemistry, food science and biotechnology.

Absorbance index: see Absorbance.

Absorption:

1) The uptake of gases, vapors or dissolved substances by liquids or solids.

2) Absorption of electromagnetic radiation by matter. All atoms and molecules have several possible excited states, corresponding to a series of closely spaced energy levels. Transition from a high to a low energy level is possible by absorption of a discrete quantum of light equal to the difference between two energy states. In molecules, each electronic state is divided into vibrational levels, which in turn are divided into rotational states. These give rise to the line spectra, band spectra and continuous spectra of atoms, molecules and macromolecules, respectively. The singlet and triplet states are especially important biologically; these differ in their orientation of spin, which is antiparallel or parallel, respectively. Transition from the singlet to the triplet state and *vice versa* is forbidden for reasons of quantum mechanics. As a rule, the ground electronic state of a molecule is the singlet state. The excited triplet state of the molecule is therefore long-lived and it can serve as a storage form of light energy, a phenomenon of importance in photosynthesis. Singlet to singlet transition, e.g. $S_1 \rightarrow S_0$, produces fluorescence, whereas the forbidden triplet to singlet transition, e.g. $T_1 \rightarrow S_0$, produces phosphorescence.

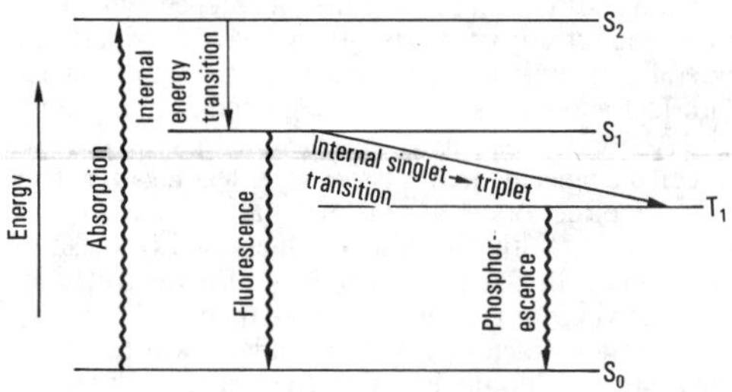

Schematic representation of the states involved in absorption, fluorescence and phosphorescence, and internal energy transfer.
S singlet. *T* triplet.

Absorption coefficient: see Absorbance.

Absorption hairs: see Absorption tissue, Plant hairs, Root.

Absorption spectroscopy: an analytical method for the determination of the wavelength-dependent absorbance *(A)* or extinction *(E)* of a sample. Absorption of the solution under test is compared with that of the pure solvent. A prism, diffraction grating or optical filter is used to generate light of a single wavelength. This monochromatized light beam is split; one of the split beams is passed through the sample, and the other through the pure solvent. The intensity of the transmitted light is measured with a photocell. Modern spectrophotometers are fitted with several light sources, spanning the wavelength range 175–2 000 nm. Fully automatic machines are available, which permit the continuous measurement of absorption changes against time. With the aid of further electronic equipment, derivative spectra can be produced ($dE/d\lambda$), which reveal small differences in the rate of change of absorption at peak wavelengths.

Absorption tissue: plant tissue specialized for the uptake of water and dissolved material; especially the rhizodermis of young roots, which is not cutinized and in which the cuticle is absent or only weakly formed (cf. the epidermis of shoots and leaves, which is heavily cutinized). The water-absorptive surface area of the Root (see) is also increased by the presence of root hairs.

Plants that are unable to take water from the ground possess water-absorptive tissue in their aerial parts. Thus, the aerial roots of many epiphytic orchids and other plant species possess several layers of spongy tissue (the *Velamen radicum*), consisting of cells with large pores which absorb water from the air. Other tropical epiphytes carry epidermal hairs (absorption hairs) on the upper surface of their leaves, which absorb rain water and water from humid air. The leaves and stems of many submerged plants often possess epidermal absorption organs (hydropotes), which absorb mineral salts from the water. The Ligules (see) of *Selaginella* (family *Lycopodiatae*) are also A.t.

Absorptivity: see Absorbance.

Abundance-dominance: see Cover-abundance.

Abundance rule: see Ecological principles, rules and laws.

Aburnus: see Bleak.

Acacia: see *Mimosoceae*.

Acacia gum: see Gum arabic.

Acantharians (Greek *akantha* thorn): marine actinopod amebas placed in the subclass *Acantharia* (class *Actinopodea*, superclass *Sarcodina*), or in the order *Acantharida* (subclass *Actinopedia*, class *Sarcodinea*). The skeleton of A. consists of strontium sulfate. In contrast, the Radiolarians (see) possess siliceous skeletons. Otherwise, A. and radiolarians are biologically and structurally similar; both occur in the same marine habitat, and their reproductive physiology appears to be the same. A. are divided into 2 groups: 1. *Acanthophractida*, which possess a lattice-like skeleton without slender spines, but with 2 to 6 heavy spines (e.g. *Hexaconus serratus*); 2. *Acanthometrida*, which do not have a lattice shield, but possess a skeleton of 10 diametric or 20 radial, slender spines, joined at the center and arranged in a regular pattern (Fig.). A., like the radiolarians, are divided into an ectoplasmic region and a central body. The gelatinous ectoplasmic layer (extracapsular cytoplasm layer or pellicle) contains 3 zones:

1. A periperhal axopodial zone or ectoplasmic cortex, made up of a reticular system of microfibrils, and usually divided into 20 fields, each pierced at its center by a spine, and connected to the spine by a myoneme. The myonemes are microfibrillar and contractile, and are possibly responsible for slight changes of shape.
2. A vacuolated region known as the calymma. Changes in density are generated by collapse or regeneration of the vacuoles, so that the organism sinks or floats. For this reason the calymma is also called the hydrostatic apparatus.
3. The sarcomatrix or phaeodium which adjoins the fibrillar capsular membrane of the central body. This appears to be the digestive region, displaying food vacuoles and often a brownish mass of apparently waste material. The capsular membrane contains pores, presumably permitting movement of digestion products into the central body.

The cellular body harbors yellow, symbiotic cryptomonads or dinoflagellates, known as *zooxanthellae*. The axopods, which are constant in number, arise from a central axoplast, emerging through the capsular membrane. During the brief vegetative state, only one nucleus is present, but most of the time the central body

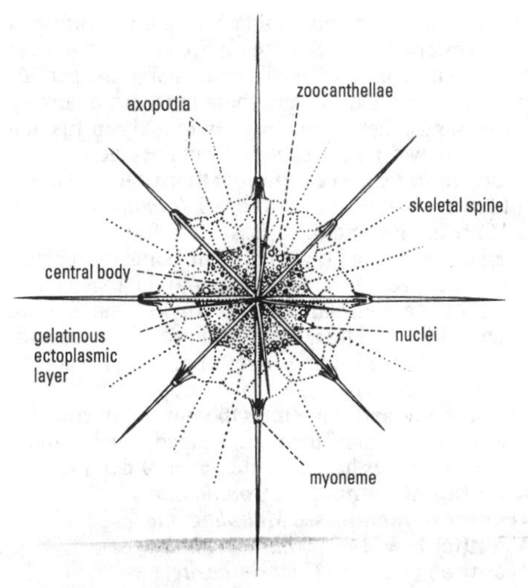

Acanthometra elastica

contains numerous small nuclei. A. are common surface plankton in the Mediterranean in summer.

Acanthella larva: see *Acanthocephala.*

Acanthis: see *Fringillidae.*

Acanthisitta: see *Xenicidae.*

Acanthisittidae: see *Xenicidae.*

Acanthocephala, ***thorny-headed worms, spiny-headed worms:*** a group of bilaterally symmetrical, pseudocoelomate organisms, formerly regarded as a class of aschelminths, but now placed in a separate phylum. The ultrastructure of the body wall, muscles and sperm indicates a close affinity to the aschelminths, but the embryology of the *A.* sets them apart. Some 500 species have been described, all of which are endoparasites of at least two hosts. Adults live exclusively as endoparasites of vertebrates, whereas their larval forms always parasitize an invertebrate. Parasitic *A.* are known for every vertebrate class.

Morphology. Most *A.* are 1.5 mm to 1.5 cm in length, but some species attain 1 m. In many species, the body is ringed or constricted at regular intervals, giving a more or less segmented appearance. They are commonly whitish in color, but a few are yellowish or orange. At the anterior end of the elongate cylindrical body is a retractable, finger-like or spherical proboscis covered with recurved spines. The body wall consists of a cuticle (about 1 mm thick), epidermis, dermis and muscle layer. The entire, thick, fibrous epidermis (sometimes called the hypodermis) forms a syncytium (i.e. it has several nuclei, but no cell walls); it is divided into 3 distinct layers: 1) an outer layer of radially arranged, parallel fibers, 2) a middle of layer of fibers lying in all directions and forming a feltwork, and 3) an inner layer of radially arranged fibers. The inner radial layer of the epidermis contains an elaborate system of lacunae, consisting of channels without definite walls but arranged in a distinct pattern. In most species, the lacunae form two main longitudinal channels with regularly spaced transverse connections. An alimentary canal is lacking in all species. The nervous system consists of a ventral, anterior aggregation of ganglia, with nerves to different body structures. The spacious body cavity is a fluid-filled pseudocoelom, and the most prominent structures within this space are the sex organs. Male sex organs consist of a pair of testes and several cement glands. Females lack true ovaries; eggs are ejected into the pseudocoelom to await fertilization. The body wall invaginates on each side of the neck region to form two fluid-filled *lemnisci,* which project posteriorly into the pseudocoelom, and appear to act as part of the hydraulic system concerned with proboscis eversion and retraction.

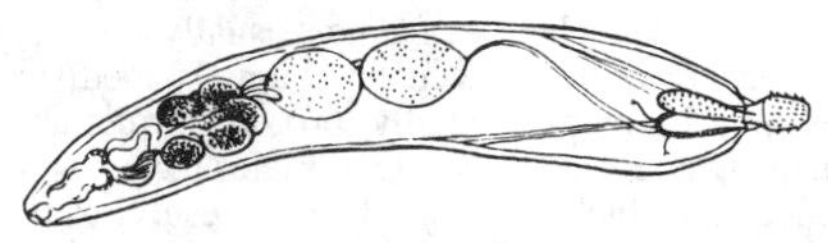

Acanthocephala. Typical thorny-headed worm (male).

Biology. Adults anchor themselves by burying their spinous proboscis in the intestinal mucosa of their vertebrate host, and they feed by absorbing the liquid contents of the host's intestine. Males possess a protrusible penis, and eggs are fertilized inside the body of the female. After fertilization, the eggs develop to a stage possessing an anterior crown of hooks (rostellum), then become enclosed in a shell. A complicated selection mechanism in the oviduct insures that only these mature eggs are laid. Mature eggs are shed with the feces of the vertebrate host, then eaten by an intermediate invertebrate host (e.g. rat feces may be eaten by cockroaches). Inside the invertebrate host, the egg develops into an elongated *acanthor,* which, with the aid of 3 pairs of recurved hooks, bores through the intestinal wall and into the coelom. This intermediate host is almost always an arthropod, in particular a crustacean or an insect. The hooks and associated muscles then degenerate, and the acanthor becomes an *acanthella.* Further development produces the infectious *cystacantha,* which closely resembles the adult form, and which becomes encapsulated in the tissues of the invertebrate host. Further development depends on ingestion of the invertebrate by the final vertebrate host, whereupon the encapsulated cystacantha is released and anchors itself in the vertebrate intestinal mucosa. Often, however, there is a second intermediate host, either a fish, an amphibian or a reptile.

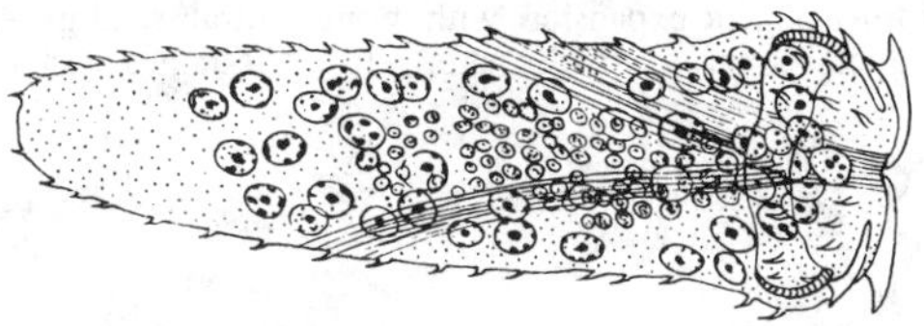

Acanthocephala. Larva of *Macracanthorhynchus hirudinaceus.*

Economic importance. Only a few species attack humans or farm animals, and disease symptoms only occur if the infection is massive. The common spiny-headed worm of pigs *(Macracanthorhynchus hirudinaceus)* lives in the pig intestine, and is found occasionally in humans and dogs. It attains lengths in excess of 60 cm and has a spherical proboscis. In heavy infections, the host may be under-nourished because so much material is absorbed by the parasite, which may also block the intestine.

Classification. Three classes are usually recognized:

Archiacanthocephala: main lacunar channels median; proboscis spines arranged concentrically; protonephridia present; females have 2 persistent ligament sacs; males have 8 cement glands; final host terrestrial.

Palaeacanthocephala: main lacunar channels lateral; proboscis spines in alternating radial rows; protonephridia absent; female ligament sacs ephemeral; males usually with 6 cement glands; final host usually aquatic.

Eocanthocephala: main lacunar channels median; proboscis spines arranged radially; protonephridia absent; female ligament sacs persistent; males have a single cement gland which is syncytial with a reservoir; final host aquatic.

Collection and preservation. *A.* are found in the intestines of vertebrates, especially fishes. Specimens are preferably fixed with the proboscis still embedded in a small piece of intestine wall. Otherwise, care must be taken to insure that the proboscis remains everted during fixation, because it is an important taxonomic character. Fixation is performed in 70% ethanol at about 60 °C. After 10 min, the specimen is placed in fresh 70% ethanol for storage.

Acanthoceras (Greek *akantha* thorn; *keras* horn): an ammonite genus, whose species are worldwide index fossils of the Lower Chalk (Cenomanian). The planispiral, strongly tuberculated shell typically has strong ribs, which are usually straight, less often flexuous and rounded. In the juvenile form, the ribs are alternately long and short, whereas the adult form has only long ribs. The ribs of the inner whorls are often forked, while those of the outer whorls are simple. Inner whorls have umbilicated, inner and outer ventrolateral and siphonal tubercles. On the outer whorls, the two ventrolaterals may fuse into a single, large tubercle, while the siphonal tends to disappear. There is a simple suture with broad, square bifid or trifid elements.

Acanthochitonida: see *Polyplacophora*.

Acanthocinus: see Long-horned beetles.

Acanthocladia: a genus of the *Bryozoa*, distributed from the Upper Carboniferous to the Permian, and largely responsible for the formation of bryozoan reefs in the upper division of the Permian (Zechstein) in the German basin, and the Permian reefs and reef knolls of Texas. The colonies, which resemble broken twigs, form a pinnate expansion, with strong central stems giving off numerous smaller, short, strong branches.

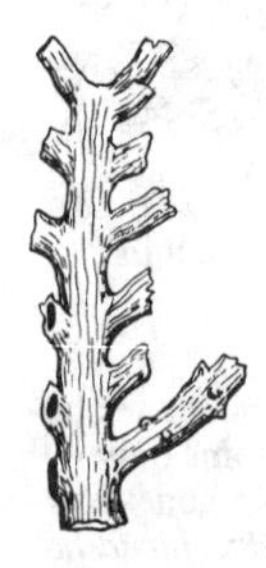

Colony of a reef-forming bryozoan (Acanthocladia anceps v. Schloth.) *from the Lower Zechstein of Thüringen* (x 2).

Acanthodactylus, ***fringe-toed lizards*** or ***fringe-toed lacertids:*** a genus of the family *Lacertidae* (see). Between 15 and 20 cm in length, these ground-dwelling lizards inhabit desert and semidesert areas in Northwest Africa, western Asia and southern Spain. Their toes are bordered with combs of scales that enlarge the surface area of the foot, and prevent the animal from sinking into soft sand. They run extremely quickly in fits and starts, and they dig deep subterranean passages.

Acanthodians (Greek *akantha* thorn), *Acanthodes:* the oldest known order of jawed fish (*Gnathostomata*). The A. are extinct and their fossils are found from the Upper Silurian to the Lower Permian, being most abundant in the Lower Devonian. Owing to their shark-like form, they have been called "spiny sharks", but they are not true sharks, and their scales are not of the elasmobranch type. At the leading edges of all fins, except the caudal, is a prominent spine, which supports the fin like a sail. The body, and sometimes the fins, is covered in a mosaic of very small, diamond-shaped, thick ganoid scales, whose bases have an outer layer of dentine.

Acanthophthalmus: see *Cobitididae*.

Acanthorhynchus: see *Meliphagidae*.

Acanthor larva: see *Acanthocephala*.

Acanthosicyos: see *Cucurbitaceae*.

Acanthuridae, ***surgeonfishes:*** a family of high-backed, marine fishes, which are found in all tropical seas. The family belongs to the order *Perciformes*. Two subfamilies are recognized.

1. Subfamily: *Acanthurinae (Hepatidae, Teuthidae)*, containing about 76 species in 9 genera. They possess one or more sharp lateral spines on the caudal peduncle, which when extended serve as an effective weapon. The largest species attain a maximal length of about 66 cm. *Paracanthurus* has a strikingly colored blue body with black markings and a yellowish tail, and is a popular aquarium fish.

2. Subfamily: *Zanclinae*. This subfamily is represented by a single living species, *Zanclus canescens = cornutus* ***(Moorish idol),*** which is a popular aquarium fish. It inhabits coral reefs, and its extended snout is used for foraging for invertebrates and algae in crevices. The dorsal fin filament is elongated, and the caudal peduncle is unarmed.

Acaracides: plant protection agents for controlling mites and ticks of the order, *Acarina*. Some examples are: ***Dichlorvos*** (2,2-dichlorovinyl dimethyl phosphate), ***Dicofol*** or ***Kelthane*** [2,2,2-trichloro-1,1-bis(4-chlorophenyl)ethanol], ***Naled*** (1,2-dibromo-2,2-dichloroethyl dimethyl phosphate), ***Endosulfan*** [C,C'-(1,4,5,6,7,7-hexachloro-8,9,10-trinorborn-5-en-2,3-ylene) (dimethyl sulfite)]. Most organophosphorus insecticides are also effective against mites and ticks. Some A. are applied as systemic agents. See insecticides. *[The Agrochemicals Handbook.* 2nd. edn. Royal Society of Chemistry, 1987; updated 4 December 1989 and 5 June 1990]

Acari, *Acarina,* ***mites:*** an order of the subphylum *Arachnida*, phylum *Arthropoda*. More than 30 000 species have been described, but it is estimated that 500 000 may exist. Mites are usually only a few millimeters in length, the smallest being only 0.1 mm. Engorged ticks (see *Ixodides)*, however, may attain sizes of up to 30 mm. The body consists of 3 parts, and segmentation is basically different from that of other arachnids. The much reduced, unsegmented (almost always) and sack-like posterior body section *(hysterosoma)* is derived from the gnathobases of the third and fourth pairs of legs, together with the completely fused true abdomen. The gnathobases of the two anterior pairs of legs, together with the bases of the chelicera and pedipalps, form the proterosoma, a structure present in other arachnids. In mites, however, the proterosoma is further divided into the anterior *gnathosoma* and the central *propodosoma*. The gnathosoma is a uniform structure formed from the bases of the pedipalps and the dorsal and ventral plates. The central, uniform propodosoma carries the first and second pairs of legs. The chelicera usually have terminal pincers, but are also often modified to needlelike piercing organs. In predatory forms the pedipalps are large, but otherwise they are usually smaller than the walking legs. In a very few species the most posterior pair of legs is absent.

In some species the male inserts a penis into the sexual opening of the female. In others, a modified appendage (chelicera or leg) is loaded with spermatozoa, which are then introduced into the female. In yet other species the male deposits a spermatophore, which is than taken up by the female into its genital opening. During copulation the male often adheres to the female with the aid of secretions, or attaches itself to the female by suction cups.

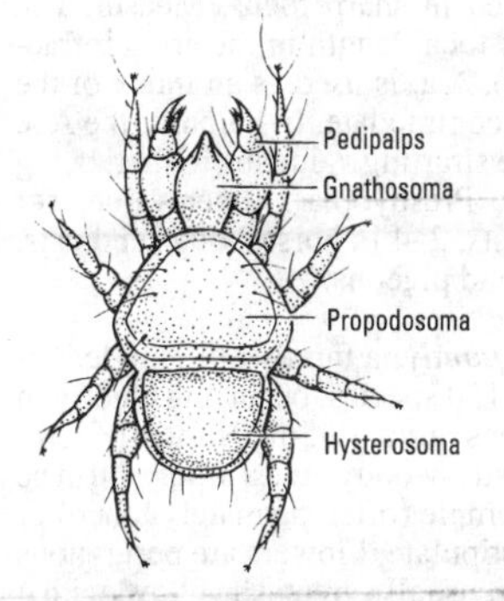

Acari. Dorsal view of the free-living predatory mite, *Cheyletus eruditus,* family *Cheyletidae,* suborder *Trombidiformes.*

There are four stages of postembryonal development, known as the larva, protonymph, deutonymph and tritonymph. One or more nymphal stages may be omitted. All larvae possess only 3 pairs of legs.

Mites occupy an extremely wide variety of environments, from the soil to the tops of trees, in the nests of insects, birds and mammals, in buildings and food stores, in fresh water and in the sea. Many species are parasitic on animals and plants, sometimes causing considerable damage by their chewing and sucking activities, as well as acting as vectors of dangerous diseases.

The following are listed separately: *Ixodides* (ticks), *Oribatei* (beetle mites), *Tetranychidae* (spider mites), *Hydrachnella* (water mites), Scabies mite *(Sarcoptes scabei), Tetrapodili* (gall mites), *Acaridae* (cheese mites and allies), *Trombiculidae*.

[The Acari (Reproduction, development and life history strategies). R. Schuster & P.W. Murphy, Chapman & Hall 1991]

Acaridae, *Tyroglyphidae, **cheese mites and allies:*** a family of the *Acari* (see), suborder *Sarcoptiformes*. The family is represented in all types of habitat from the Arctic tundra to tropical rainforests. Many species have long been notorious for their propensity for infesting stored food products throughout the world. They live in all types of organic substrates, including meats, cured and raw hides, all types of seeds, stored grains and farinaceous products, rotting leaves, tree bark, nests of mammals and birds, pet foods, house plants, decaying bulbs and tubers, fresh and decaying fungi, and cheese. Under unfavorable conditions, many acarids form a resting stage, known as a hypobus, which possesses suckers or claspers for attachment to insects and other animals that might transport them. The ***Meal mite*** *(Acarus siro = Tyroglyphus farinae;* 0.4–0.6 mm) is a destructive cosmopolitan species, occurring in great numbers in grain and flour in granaries, mills and pantries. The grain is eaten, leaving only the husk. Cheese, dried fruit, and vegetables are also attacked. Under favorable conditions of high humidity it reproduces rapidly. In dry environments it forms a resting stage which remains viable for at least 2 years. The ***Cheese mite*** *(Tyrophagus casei;* 0.45–0.7 mm) feeds on cheese and smoked meats, such as ham, bacon and sausages. The ***House mite*** *(Glyciphagus domesticus;* 0.3–0.75 mm) is common in damp houses, where it is a nuisance rather than a pest.

Acarina: see *Acari*.

Acarus: see *Acaridae*.

Accelerated development: see Acceleration.

Acceleration:

1) an increase in the rate of ontogenic development (see Metaboly).

2) the relatively higher rates of growth and development of humans during childhood and youth, compared with those of previous generations (centenary A.). Alternatively, the term can be applied to the accelerated development of an individual, compared with that of other individuals of the same size (individual A.). Centenary A. became apparent in many countries at the end of the nineteenth century, resulting in the advancement of the age of sexual maturity by an average of about 3 years, and an increase in the body height of adults of 7–8 cm. Eruption of the first teeth and appearance of the second teeth also occurred earlier, as well as the incidence of age-specific illnesses. Psychological development is also affected by A.

A. seems to be due to a complex of factors, in which the improvement of living standards (higher protein consumption, improved hygiene, abolition of child labor, etc.) appears to play an important part. An exact analysis of all the factors involved is, however, lacking. There are indications that A. in highly industrialized countries is now slowing down, and has perhaps ended (Table).

Average height (cm) of 14-year-old school children in Jena (Germany)

Year	Boys	Girls
1880	143.6	147.4
1921	149.5	150.2
1932/33	153.5	155.1
1944	153.9	155.3
1954	156.2	156.8
1964/65	158.7	157.9
1975	163.3	162.1
1980	163.7	160.7

Acceptable daily intake, ***ADI:*** the recommended daily limit for ingestion of a potentially harmful chemical. ADI = NOEL/safety factor, where NOEL is the acronym of "no-observed effect level", i.e. the daily intake of a chemical that has no observed effect on the organism, expressed as mg/kg body weight/day. The safety factor (which may be as high as 1000, but is more usually 100) makes allowance for variation between individuals, for interspecies differences (i.e. extrapolation of test results from animals to humans), and the fact that the test conditions (frequency of exposure, route of administration, dose duration) may differ from "real life" conditions. Thus, ADI ultimately becomes: "the daily intake of a chemical throughout an entire lifetime that presents no appreciable risk" ADI and NOEL are clearly applicable to materials that display a threshold of dangerous activity, and they are especially important in establishing safe limits for the intake of dietary contaminants such as pesticide and herbicide residues and veterinary drugs, or food additives such as coloring agents, etc. The risks associated with nonthreshold agents (e.g. many carcinogens) are less easy to quantify, but various mathematical models can be used to extrapolate high dose toxicity data to a relevant low dose region, enabling the assessment of an appropriate ADI.

Accessory nerve: the XI cranial nerve; see Cranial nerves.

Accessory pigments: assimilatory pigments, including carotenoids and chlorophyll *b*, which transfer the energy of absorbed light to the photochemically active chlorophyll *a*. In addition to carotenoids, the A.p. of red algae include protein-bound phycoerythrins, which are structurally similar to bile pigments (4 pyrrole rings in an open chain structure). Phycoerythrins enable the utilization of the predominantly blue-green light that penetrates deep water layers.

Accipitres: see *Falconiformes*.

Accipitridae: see *Falconiformes*.

Accipitrinae: see *Falconiformes*.

Acclimatization: the gradual adaptation of organisms to changed climatic conditions, especially to alterations in the prevailing temperature. A. in warm-blooded animals consists of changes in temperature regulation, e.g. changes in the density of the coat hair or the thickness of the subcutaneous adipose insulating layer, or alterations in physiological heat production (see Temperature regulation of the body). Cold-blooded animals respond to altered climatic temperature by changing their egg laying frequency, by shifting their critical survival temperature, and by adjusting the temperature range in which they are normally and optimally active.

Accommodation:

1) In the physiology of vision, A. enables the formation of a sharp image on the retina (see Light-sensitive organs) by the lens. In most reptiles, birds and mammals (many mammalian eyes do not accommodate), A. occurs by changes in the shape of the lens, but in invertebrates, fish and amphibia, the distance between retina and lens is altered. Changes in lens shape are brought about by changes in the tension of the ciliary muscle acting via the suspensory ligament. Contraction of the ciliary muscle relaxes the suspensory ligament; this permits the lens, through its natural elasticity, to increase its curvature or refractive power, so that objects very near to the eye are brought into focus. If the ciliary muscle is relaxed, the suspensory ligament exerts tension and causes the lens to become flatter, so that distant objects are brought into focus on the retina. Changes in the distance between an inelastic lens and the retina (i.e. changes in the length of the eyeball) are achieved by alterations of muscle tension on the lens (fish, amphibians) or by alterations of muscle pressure on the vitreous humor (cephalopods).

2) A. of nerves. If the applied stimulatory potential is allowed to rise slowly, then the recorded excitatory threshold voltage is higher than that for a suddenly applied potential. The nerve therefore accommodates itself to the slowly changing potential, so that the threshold value increases if the threshold is approached at a slower rate. Nerve fibers are not stimulated by extremely slow increases of potential; thus, very low frequency alternating currents do not stimulate, because they rise and fall too slowly.

Motor nerve fibers of vertebrates show a marked ability to accommodate, whereas the A. of sensory nerves is low. See Excitation, Adaptation.

Accommodation amplitude: the difference between the refractive powers of the eye lens when accommodated to the farthest and shortest distances at which an object is seen in sharp focus. Measured in diopters (reciprocal of focal length in meters = refractive power in diopters), A.a. is used as an index of the ability of the eye to accommodate. In humans, the A.a. decreases with age, the starting value in children being about 14 diopters (see Presbyopia). Other values for A.a. are, e.g. 0 in rabbits, 2–4 in horses, dogs and cats, and 8–12 in chickens and pigeons.

Acer: see *Aceraceae*.

Aceraceae, *maple family:* a family of dicotyledonous plants, containing 150 species, occurring chiefly in nontropical mountainous regions of the Northern Hemisphere. They are mostly woody trees, rarely bushes. Leaves are opposite, simple (often palmately lobed) or rarely pinnate, and exstipulate. Flowers are perigynous or hypogynous, regular, usually greenish yellow, with 4 or 5 sepals, and arranged in umbels, racemes, spikes or panicles. Hermaphroditism is common, but flowers may only be apparently hermaphrodite and functionally monoecious, due to atrophy of the ovaries or stamens. The transition from hermaphrodite to monoecious flowers is correlated with the change from insect- to wind-pollination. The 2-celled, 2-lobed ovary develops into a schizocarp fruit. There are two genera, differing primarily in the structure of their winged fruits. In the larger genus, *Acer* (about 200 species), the seeds possess obovate wings (only one side of each paired fruit is winged). Relatively few species of *Acer* are native to the forests of Central Europe, but several have been introduced as ornamentals. Flowers of the ***Norway maple*** (*Acer platanoides*) are carried in upright umbel-like racemes, while those of the ***Sycamore*** (*Acer pseudoplatanus*) hang in panicles. Both trees attain a height of 20–35 m, and their wood is used for various purposes. ***Common maple*** (*Acer campestris*) is a small tree or shrub (3–15 m), with small, 3-lobed leaves, which is useful for hedging. Native North American species are often introduced in European parks and gardens, e.g. the ***Ash-leaved maple*** or ***Box elder*** (*Acer negundo*) with pinnate and often white variegated leaves, and unisexual, dioecious flowers; and the ***Sugar maple*** (*Acer saccharinum*) with bright green (topside) and silver-white (underside) leaves, whose sap is the source of maple syrup.

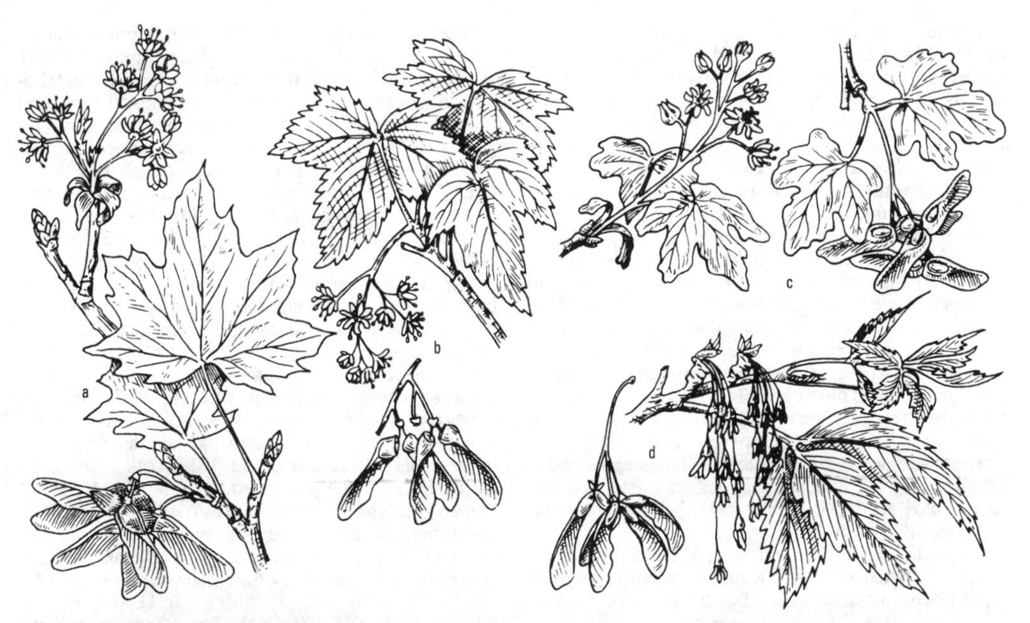

Aceraceae. *a* Norway maple. *b* Sycamore. *c* Common maple. *d* Ash- leaved maple or Box elder.

The native Chinese genus, *Dipteronia*, contains only 2 species, whose fruits are surrounded by a circular membrane.

Acerina: see Perches, Ruffe.

Acetabulum: see *Cestoda*.

Acetes: see Shrimp.

Acetic acid: see Ethanoic acid.

Acetic acid bacteria: bacteria belonging to the genera *Acetobacter* and *Gluconobacter*, which oxidize alcohols to the corresponding acids in the presence of oxygen. In particular they oxidize ethanol to acetic (ethanoic) acid, and are exploited commercially for the production of vinegar from wine, cider, etc. (see Acetic acid fermentation).

Cells of *Gluconobacter* (= *Acetomonas)* are ellipsoid to rod-shaped, 0.6–0.8 by 1.5–2.0 µm, with 3–8 polar flagella (rarely 1 flagellum), and they occur singly, in pairs or in chains. *Gluconobacter* is strictly aerobic, usually produces high levels of ketones as well as acids, and is incapable of the further metabolism of acetic and lactic acid. It is normally Gram-negative, but old cultures may be weakly Gram-positive.

Cells of *Acetobacter* are ellipsoid to rod-shaped, straight or slightly curved, 0.6–0.8 by 1.0–3.0 µm, sometimes with flagella, and occurring singly, in pairs or in chains. *Acetobacter* is also strictly aerobic, but unlike *Gluconobacter* it can oxidize acetate and lactate to carbon dioxide and water, i.e. it is capable of "overoxidation". It is Gram-negative, but old cultures of some strains are Gram-variable.

Both *Acetobacter* and *Gluconobacter* occur naturally on fruits (especially souring fruits) and vegetables, in beer and most other brewed beverages, as well as in garden soil, and admixed with baker's yeast. In liquid medium, they often form a surface pellicle, and the medium does not become turbid. These mats or pellicles of old *Acetobacter* cells are known as "mother of vinegar", since they also form on the surface of alcoholic beverages (e.g. wine or cider) during their conversion into vinegar. Old pellicles contain branching involution forms, and large, swollen cells of *Acetobacter.* See Acetic acid fermentation.

Acetic acid fermentation: an oxidative metabolic process performed by Acetic acid bacteria (see), in which acetic (ethanoic) acid is formed from ethanol in the presence of atmospheric oxygen: $C_2H_5OH + O_2 \rightarrow CH_3$—$COOH + H_2O$. Acetaldehyde is an intermediate in this metabolic conversion. Acetic acid formation can only occur in the presence of oxygen, and the process is strictly aerobic It is therefore not a Fermentation (see), but an incomplete oxidation. Nevertheless, it is customarily (albeit inappropriately) referred to as a fermentation. The A.a.f. is responsible for the spoilage (acidification) of beer and wine.

Technically, the A.a.f. is used for the production of vinegar (which contains about 4% acetic acid) by the oxidation of the ethanol present in grape wine, fruit wine or cider. In a commonly used commercial process, known as the *quick* or *generator* process for acetic acid production, alcohol-containing fluids are allowed to trickle through a tall cylinder (2–6 m-high; known as the Frings generator) filled with shavings of beechwood inoculated with *Acetobacter.* Air is also made to circulate freely through the cylinder and around the wet surfaces of the shavings. Vinegar is drawn off from the bottom collection chamber. More modern methods of vinegar production employ closed containers (see Fermenter), in which the acetic acid bacteria are continually stirred with the ethanolic solution, and oxygen is blown through the mixture.

Preparation of wine vinegar was known to the ancient Babylonians and Egyptians, who used it as a beverage and for medicinal purposes.

Acetoacetic acid, ***β-ketobutyric acid, 2-oxobutyric acid:*** CH_3—CO—CH_2—COOH, a β-ketoacid produced as an intermediate in the β-oxidation of fatty acids. In the healthy organism, A. is degraded to 2 molecules of acetate. When carbohydrate metabolism is disturbed, especially in diabetes mellitus, A. is excreted together with other ketone bodies in the urine.

Acetobacter: see Acetic acid bacteria.

Acetomonas: see Acetic acid bacteria.

Acetylcholine: a vertebrate and invertebrate Neurotransmitter (see) in neuromuscular synapses, and in synapses of the following nerves: all motor nerves to skeletal muscle; all preganglionic nerves, including the nerve supply to the adrenal medulla; all postganglionic parasympathetic nerves; postganglionic sympathetic nerves to sweat glands; and some postganglionic sympathetic nerves to blood vessels in skeletal muscle. After release from the nerve terminal at the synapse, A. binds to, and triggers a response by, receptors in the postsynaptic neuron; it then leaves the receptor and is rapidly degraded by Acetylcholinesterase (see). Nerves that employ acetylcholine for their chemical transmission are called *cholinergic nerves*.

Depending on its concentration, A. exerts two different physiological effects. Injection of small amounts of A. produces the same response as the injection of Muscarin (see), i.e. a fall in blood ressure (due to vasodilation), slowing of the heart beat, increased contraction of smooth muscle in many organs, and copious secretion from exocrine glands; this muscarinic effect of muscarin or A. is abolished by atropne. After administration of atropine, larger amounts of A. cause a rise in blood pressure, similar to that caused by nicotine. Nicotinic cholinergic synapses are found in vertebrate neuromuscular junctions, certain ganglia, central synapses, and the electroplax of *Torpedo*. Muscarinic cholinergic synapses operate in smooth muscle, cardiac muscle, ganglia, and many central brain regions. In the brain and central vervous system, muscarinic synapses outnumber nicotinic synapses by 10–100 fold. In ganglia, nictinic cholinergic receptros are blocked by tetraethylammonium, and in neuromuscular synapses they are blocked by Curare (see), and irreversibly occupied by the snake venom constituent, α-bungarotoxin. Muscarinic cholinergic receptors of the postganglionic prapsympathetic systems are blocked by atropine and scopolamine, which are therefore parasympatholytic agents. Other substances inhibit the activity of acetylcholinesterase, thereby causing nerve paralysis; for example, physostigmine is a reversible inhibtior, whereas certain organic phosphates, which are used as insecticides, are irreversible inhibitors. Other acetylcholinesterase inhibitors, such as the organic fluorophosphates, are among the most potent chemical warfare agents.

Nicotinic and muscarinic effects are mediated by *nicotinic* and *muscarinic receptors,* respectively. These receptors are the products of two distinct gene superfamilies, and their only common property is that they are activated by A. The slower muscrinic receptor response operates via G-protein-coupled receptros (see G-proteins). Depending on the receptor subtype, the subsequent effector mechanism involves inhibition of adenylate cyclase, formation of inositol 1,4,5-trisphosphate from phosphoinositides by a specific phospholipase C, or modulation (opening) of certain K^+ channels [D. Brown *Nature* ***319*** (1986) 358-359].

Muscarinic receptors display a high degree of heterogeneity, but in all forms the primary sequence shows similarities with those of the β_2 adrenoreceptor and rhodopsin, suggesting variations on a common structural theme. Five pharmacologically distinct muscarinic receptor subtypes (m1-m5) have been cloned and sequenced from rat and porcine cerebral cortex [T. Kubo et al. *Nature* ***323*** (1986) 411–426; E. C. Hulme et al. *Annu. Rev. Pharmacol. Toxicol.* ***30*** (1990) 633–673]. m1, m3 and m5 are coupled to inositol 1,4,5-trisphosphate production, m2 and m4 to adenylate cyclase inhibition and modulation of K^+ channels. All are activated by muscarine and oxotremorine, and blocked by atropine. They also show considerable sequence homology, and the sequence of each can be interpreted in terms of seven transmembrane domains.

Occupancy of the nicotinic receptor (by A.) triggers a rapid response (1–2ms) by direct activation of cation-selective ion channels, thereby causing depolarization of the postsynaptic membrane. The nicotinic cholinergic receptor has been isolated from the electroplax of *Torpedo californica* (electric ray) and *Electrophorus electricus* (electric eel) and from vertebrate muscle. From all three tissues, it is a single membrane protein, M_r 250 000, consisting of 4 glycoprotein subunits: M_r 40 000 (50 116) (α), 50 000 (53 681) (β), 60 000 (56 279) (γ) and 65 000 (δ) (the first value is from SDS; the value in brackets is the exact M_r based on amino acid composition) in the ratio 2:1:1:1, with an average 40% amino-acid sequence identity between all 4 chains [B.M. Conti-Troconi et. al. *Science* ***218*** (1982) 1227–1229]. DNA for all 4 subunits has been cloned and sequenced [M. Noda et al. *Nature* ***301*** (1983) 251–254]. The covalent affinity probes, $[^3H]$-bromoacetylcholine nd 4-(N-maleimideo)-^3H benzyl trimethylammonium iodide, label the α-subunit by reacting with cysteine residues 192 and 193. Thus, each of the two α-subunits carries an acetylcholine binding site, and there are 2 sites per oligomeric receptor. When incorporated into liposomes or planar lipid bilayers, the nicotinic receptor permits a flux of $^{22}Na^+$, which is promoted by acetylcholine and blocked by α-bungarotoxin. Nicotinic cholinergic receptor is therefore one of the group of receptors known as *gated ion channels* (others include receptors for gamma-aminobutyric acid, glycine and 5-hydroxytryptamine).

Nicotinic receptors of the vertebrate neuromuscular junction and the electric organs of fishes represent a single subtype, and are known as *peripheral receptors.* Nicotinic receptors are also found in autonomic ganglia and in the central nervous system, where they are collectively known as *neuronal receptors*. The latter are also gated cation channels, but stucturally they are a heterogeneous group, and less well characterized than peripheral receptors.

All 4 subunits of the *Torpedo* and other peripheral nicotinic receptors show a high degree of amino acid sequence identity, with particularly close homologies between the α and β and between the γ and δ polypeptides, suggesting a single ancestral gene with evolutionary branching into the αβ- and γδ-type chains. Hydrophobicity analysis predicts the existence of 4 transmembrane α-helical segments, each containing at least 20 hydrophobic residues. X-ray analysis is not possible because the proteins have not been crystallized, but a structural model has been developed with the aid of electron microsopy of negatively stained preparations and electron image analysis of rapidy frozen receptor proteins. Thus, perpendicular to the plane of the membrane, the receptor is an 80 Å diameter rosette consisting of 5 peaks of electron density (clockwise: $\alpha\beta\alpha\gamma\delta$) around a central pore. This rosette represents the end of a cylinder which spans the membrane, protruding mostly on the extracellular surface, i.e. into the synaptic cleft. [E.S. Deneris et al. *Trends Pharmacol. Sci.* ***12*** (1991) 34–40]

The autoimmune disease, myasthenia gravis, is due to the presence of circulating antibodies to the peripheral nicotinic receptor. Binding of these antibodies to the receptor results in increased receptor degradation and a decreased efficiency of neuromuscular transmission. The condition can be partly relieved by administration of acetylcholinesterase inhibitors. [J. Newsom-Davis et al. in *Clinical Aspects of Immunology,* P.J. Lachmann et al. (eds), Blackwell Scientific publications, Oxford, 1993: pp. 2091–2113]

Histochemical localization of acetylcholinesterase serves to identify cholinergic synapses. It is based on the technique of Koelle and Friedenwald (G.B. Koelle *Handb. Exp. Pharmakol.* ***15*** (1963) 187–298]. The substrate used is acetyl- or butylthiocholine, and the product, thiocholine, is visualized by precipitation with lead or copper salts. A more specific marker for cholinergic neurons (acetylcholinesterase is also present in dopaminergic cells of the substantia nigra) is choline acetylase (EC 2.3.1.6).

A. is a phylogenetically ancient hormone which also appears in protists. It may be an evolutionary precursor of the neurohormones.

Traces of A. also occur in plants, e.g. in the hairs of the stinging nettle. Related compounds, e.g. murexin in the glands of certain gastropods, are probably venoms.

$$H_3C{-}C({=}O){-}O{-}CH_2{-}CH_2{-}N^{\oplus}(CH_3)_3$$

Acetylcholine

Acetylcholinesterase: an enzyme which catalyses the hydrolysis of acetylcholine into choline and acetate, and which plays an essential role in nerve transmission. It is sometimes called "true cholinesterase" in contrast to relatively unspecific acylcholinesterases (see Cholinesterase). Due to the high turnover number of A. (0.5 to 3.0 $\times 10^6$ molecules substrate per molecule enzyme per min), the acetylcholine released at a synapse is hydrolysed within 0.1 ms. The enzyme is found in the central nervous system, in the postsynaptic membranes of striated muscles, parasympathetic ganglia, erythrocytes and the electric organs of fish. It possesses a serine residue in its active center. The activity of organo-phosphate nerve poisons is due to their inhibition of A. See Cholinesterase

Acetyl-CoA: see Coenzyme A.

Acetyl-coenzyme A: see Coenzyme A.

Achatina: see *Gastropoda.*

Achilidae: see *Homoptera.*

Achene: see Fruit.

Achillea: see *Compositae.*

Achondroplasia, ***chondroplasty, osteosclerosis congenita, fetal rickets, Parrot's disease:*** an autosomal dominant inherited disease, thought to arise frequently by mutation. It results in the slow growth of bones derived from cartilage; formation of periosteal bone is practically normal. The primary defect is a deficient conversion of cartilage to bone at the epiphyses of long bones, resulting in disproportionate dwarfism, with short extremities, midfacial hyperplasia, a small foramen magnum and a narrow spinal canal. Intelligence is unaffected. The terms *chondrodystrophia fetalis* and *hypoplastic fetal chondrodystrophy* are also used, but these are considered to be outmoded or inaccurate.

Achromasia: see Color blindness.

Achromycin: see Tetracyclins.

Achroodextrins: see Dextrins.

Achrooglobin: see Blood.

Acidophytes: see Calcifuges.

Acidosis: a decrease in the pH of body fluids. During A., body fluids do not actually become acidic, i.e. the pH remains above 7. If mammalian body fluids become truly acidic, i.e. if the pH falls below 7, then the animal dies. A. is classified as respiratory or metabolic, according to its origin.

1) Respiratory A. can be caused by depression of the respiratory center (e.g. by morphine poisoning) or by paralysis of the respiratory muscles, by an increase in respiratory resistance, or by any condition that impairs gas exchange in, or ventilation of, the lung alveoli. In humans, A. occurs in spinal infantile paralysis, bronchial asthma, lung emphysemia and pneumonia. Decreased respiratory efficiency results in an increase in the carbon dioxide concentration and a decrease in the pH value of the blood. Respiratory A. is compensated metabolically by the formation of bicarbonate (HCO_3^-) ions, and by increased excretion of H^+ by the kidneys.

2) Metabolic A. is caused by a decrease in the bicarbonate fraction, with little or no change in the carbonic acid fraction. There are several possible causes. 1) Vigorous exercise (increase of lactic acid). 2) Severe diarrhea, resulting in the loss of gastrointestinal secretions containing high concentrations of bicarbonate (the resulting acidosis is a contributing factor to infant death in the developing world). 3) Treatment with acetazolamide (Diamox), a drug used to promote diuresis; this inhibits carbonic anhydrase in the brush border of the kidney proximal tubule epithelium, thereby retarding the reabsorption of bicarbonate. 4) Severe renal disease, which may impair the ability of the kidney to remove acids formed in the normal course of metabolism. 5) Vomitting, which usually leads to the loss of bicarbonate from the upper intestine, as well as the acidic stomach contents. Since the loss of alkali exceeds the loss of acid, the net result is acidosis. 6) Diabetes mellitus results in the excessive formation of acetoacetic acid, which accumulates and causes acidosis in the extracellular fluids (as much as 500–1000 mmoles of acid may be excreted per day). Metabolic A. is compensated by an increase in the respiratory minute volume (see Respiratory regulation), which normalizes the quantity of standard bicarbonate and raises the pH value.

Acid rain: rain with a pH lower than 5.6, due to acidification by acidic industrial pollutants. Rain is normally slightly acidic (pH 5.6), because it contains dissolved carbon dioxide, and is therefore a weak solution of carbonic acid. In industrial regions, rain may have a pH of about 4, individual downpours may have a pH less than 3, while the water droplets of industrial fogs may be even more acidic. A major component of A.r. is sulfuric acid (H_2SO_4), produced by atmospheric oxidiation of sulfur dioxide (SO_2) to sulfur trioxide (SO_3), which dissolves in water to form the sulfate ion (SO_4^{2-}) of sulfuric acid. The major part of this conversion occurs in the water droplets of clouds, where it may be catalysed by other industrial pollutants such as metal particles. Ozone (see Greenhouse effect) and ammonia (produced from rotting organic material, and in large quantity from the slurry of factory farms) also accelerate sulfuric acid formation in cloud droplets. In dry air, sulfur dioxide is converted much more slowly, and much of it falls to the ground unconverted. Sulfur dioxide has several sources, such as rotting vegation and volcanoes, but it is estimated that as much as half of the atmospheric SO_2 arises from the burning of fossil fuels, and in industrial areas this proportion may be as high as 85%.

Nitric acid (HNO_3) is also responsible for the production of A.r., and in some areas its contribution may be equal to that of sulfuric acid. Power stations and automobile exhausts both produce nitrogen oxides (nitrogen dioxide, NO_2, and nitric oxide, NO). In the atmosphere, nitric oxide is rapidly oxidized to nitrogen dioxide, which in turn is oxidized and reacts with water to form nitric acid.

Although the acidity in itself is harmful to living organisms, the main deleterious effects of A.r. are due to aluminium, which is released by acid from its bound form in the soil. The death of freshwater fish and of forest trees, as a result of A.r., is due largely to aluminium poisoning. Trees starve to death, because their roots take up aluminium, rather than essential nutrients such as magnesium and calcium. High aluminium concentrations cause fishes' gills to become clogged with mucus.

In addition to A.r., plants and trees are also assaulted by acidic air pollution. Thus, sulfur dioxide blocks stomata, preventing photosynthesis. In the summer, ozone (from automobile exhausts) reaches levels that are toxic to some trees, especially when accompanied by sulfur dioxide.

Acinonyx: see *Felidae.*

Acipenser: see *Acipenseridae.*

Acipenseridae, *sturgeons* (plate 2): a family (26 species; order *Acipenseriformes)* of sluggish, bottom-feeding, mostly marine fishes of the Northern Hemisphere, which ascend large rivers to spawn. They have 5 rows of bony scutes on the body and a row of 4 barbels in front of the mouth, which is protrusible and inferior. The snout is pointed and extended anteriorly. Sturgeons are commercially very important; their flesh is fine-tasting, and the egg roe is processed to make caviar. The ***Common sturgeon*** *(Acipenser sturio)* occurs in the North, Baltic, Mediterranean and Black Seas, and in the Atlantic off the coasts of Norway and Iceland. Up to 3 m in length with a maximal weight of 200 kg, it feeds on small fishes and bivalves and other bottom dwelling animals. The ***Great sturgeon*** or ***Beluga*** *(Huso huso),* of the Black, Caspian and Mediterranean Seas, grows up to 9 m in length, may attain a weight of 1 500 kg, and is the source of Beluga caviar. The freshwater ***Sterlet*** *(Acipenser ruthenus),* up to 120 in length, feeds mainly on aquatic insects; it lives in rivers emptying into the Black and Caspian Seas, as well as in the rivers Dvina and the Ob, and in the river systems of Lakes Ladoga and Onega. *Acipenser güldenstaedti* from the Black and Caspian Seas attains a length of 2 m and yields Malossi caviar.

Acipenseriformes: an order of large, mostly marine fishes, with a partially cartilaginous skeleton and other primitive characteristics. The caudal fin is heterocercal, and the head has a pointed protruding rostrum with a ventral mouth. The skin is either naked, or has 5 longitudinal rows of large, rhomboid bony scutes. Marine species migrate into rivers for spawning. Several extinct families are known from the fossil record (from the Carboniferous onward), and there are two families with living representatives: *Acipenseridae* (sturgeons) (see) and *Polyodontidae* (paddlefishes) (see). The intestine has a spiral valve.

Acochlidiacea: see *Gastropoda.*

Acoela: see *Turbellaria.*

Acoelic or **acoelous vertebra:** see Axial skeleton.

Aconaemys: see *Octodontidae.*

Aconitase, *aconitate hydratase:* a hydratase which catalyses one stage of the tricarboxylic acid cycle, the interconversion of citrate and isocitrate via the enzyme-bound intermediate *cis*-aconitate. At equilibrium, the relative abundances are 90% citrate, 4% *cis*-aconitate and 6% isocitrate. Thus, citrate is favored at equilibrium, but in respiring tissue the reaction proceeds from citrate to isocitrate, because isocitrate is oxidized by isocitrate dehydrogenase. The enzyme consists of a single polypeptide chain (M_r 89 000), which contains Fe(II) and an essential SH-group; the presence of a thiol such as cysteine or glutathione is required for activity. The Fe(II) ion forms a stable chelate with citrate. X-ray analysis of Fe(II) complexes of tricarboxylic acids suggested the "ferrous wheel" hypothesis of A. action. According to this mechanism, three points on the *cis*-aconitate molecule are bound at separate sites on the enzyme surface; in additon, the molecule is also complexed with Fe(II) at the active center. Stereospecific *trans* addition of water to *cis*-aconitate to form either citrate or isocitrate is achieved by rotation of the ferrous wheel, which can add OH to either side of the molecule. A. is inhibited by fluorocitrate. Two isoenzymes are present in animal tissues, one in the cytosol and one in the mitochondria. [Glusker, J.P., in Boyer, P.D. (ed.), *The Enzymes,* ***5,*** 434, Academic Press Inc., 1971]

```
HC—COOH
||
C—COOH
|
H2C—COOH    cis-Aconitic acid
```

cis*-Aconitic acid, *cis-1,2,3-propenetricarboxylic acid: an intermediate in the tricarboxylic acid cycle, and therefore present in all aerobic organisms, at least in traces. The name is derived from the discovery of *cis*-A.a. in monkshood *(Aconitum napellus).* It has also been isolated from sugar cane *(Saccharumm offinarum),* horsetail *(Equisetum)* and beetroot *(Beta vulgaris).*

Aconitum: see *Ranunculaceae.*

Acontium: see Mesenterial filaments.

Acorn: the fruit of an oak tree; see *Fagaceae.*

Acorn barnacles: see *Cirripedia.*

Acorn worms: see *Hemichordata.*

Acorus: see *Araceae.*

Acouchi: see *Dasyproctidae.*

Acoustic nerve: the VIII cranial nerve; see Cranial nerves.

Acquired behavioral response: see Innate releasing mechanism.

Acquired releaser mechanism: see Innate releasing mechanism.

Acrania: see *Acraniata.*

Acraniata, *Acrania:* a subphylum of the *Chordata* (see), comprising the Classes *Hemichordata* (see), *Urochordata* (see) and the *Cephalochordata* (see *Amphioxus lanceolatus).* Also known as protochordates, they are simple or aberrant chordates, differing from more advanced members of the phylum by the absence of a true brain or skull, by the absence of a heart and kidneys, and by their peculiar embryonic development.

Acrasiales: see *Myxomycetes.*

Acridothera: see *Sturnidae.*

Acrobates: see *Phalangeridae.*

Acrocephaly, ***oxycephaly, hypsicephaly, turricephaly, steeple head, steeple skull, tower head, tower skull:*** a condition in humans, in which the top of the head is pointed or conical. It is due to the prevention of saggital skull growth by premature closure (ossification) of the coronal and lambdoid sutures, which is compensated by increased growth in height and width.

Acrochordus: see *Acrochordidae.*

Acrochoridae, ***wart snakes:*** a family of primitive, viviparous, fish-eating snakes, completely adapted to an aquatic existence, and related to the *Boidae* (see). Sluggish and slow swimming, they are found in the brackish estuary zones of rivers in South and Southeast Asia as far as the Indo-Australian archipelago, sometimes swimming short distances out to sea. Nostrils and eyes are situated on top of the snout. Scales are non-overlapping and adjacent, but bear a sharp ridge, so that the skin is very rough and has a warty appearance. The skin is used for the manufacture of "sea snake leather". The family consists of only 2 monotypic genera: *Acrochordus javanicus* ***(Javan wart snake),*** which is up to 2 m in length; and *Chersydrus granulatus* ***(Indian wart snake),*** which has a "keel" consisting of 2 rows of scales extending from the throat to the anus.

Acrocont: a term applied to spores and gametes that possess a single, forward directed, tinsel-type (with flimmer hairs) flagellum.

Acrodont: see Teeth.

Acron: the anterior nonsegmental part of the body of a metameric animal.

Acropora: see *Madreporaria,* Coral reef.

Acrosome: a cap-like structure lying at the anterior of the nucleus in the head of the metazoan sperm (Fig.) (see also Figs. in Spermatogenesis and Fertilization), and which plays a central role in sperm penetration of the egg. During spermatogenesis, the A. arises by the fusion of Golgi vesicles.

The A. is a Lysosome (see), containing enzymes such as hyaluronidase, neuraminidase, glycosidases and acrosin. Contact of the sperm with the primary or secondary egg membrane initiates the A. reaction, in which the A. breaks, releasing its lytic enzymes and undergoing exocytic fusion with the plasma membrane of the sperm. The enzymes digest the membranes surrounding the egg (corona radiata and zona pellucida), thereby opening a pathway to the plasma membrane of the egg cell. The A. membrane, now continuous with the plasma membrane of the sperm, produces a number of tubular projections which make contact with the plasma membrane of the egg cell. The equatorial region of the plasma membrane of the sperm then fuses with the plasma membrane of the egg.

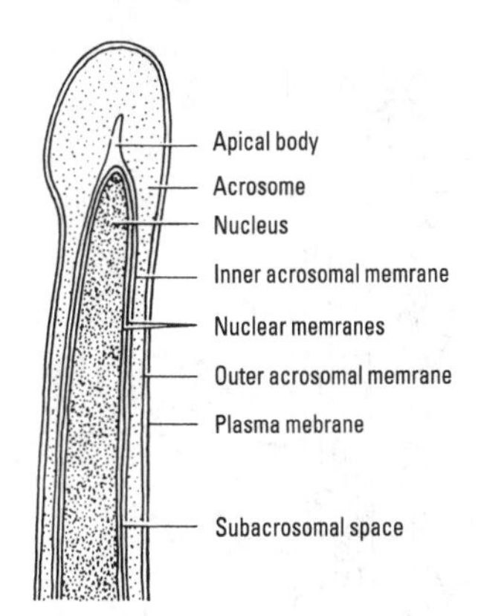

Acrosome. Sagittal section of the anterior region of a ram sperm (schematic).

Recognition molecules appear to be present in the A. tip, and one of these ("bindin") has been tentatively identified in sea urchin sperm.

Acrothoracica: see *Cirripedia.*

Acrotony: stimulation of the growth and development of lateral shoots at the terminus of the main axis of a plant and at the termini of the higher lateral axes, with inhibition of the growth of the lower lateral axes. A. is characteristic of the growth of deciduous trees, where it ensures that new leaves receive adequate light. See Basitony, Mesotony.

Acryllium: see *Numididae.*

Actin: a structural protein, which together with myosin, is responsible for the contraction of muscle. Actin is also present in all eukaryotic nonmuscle cells, and it plays an important part in the movement of, and within, cells, e.g. protoplasmic streaming, migration of cell organelles, ameboid movement of cells, and chromosome migration, as well as the stabilization of cell shape and cell evaginations (see Microvilli). Monomeric, globular G-actin (M_r 43 500) polymerizes with the binding of ATP to form thread-like F-actin, and this transformation is reversible. In nonmuscle cells, G-actin is in equilibrium with F-actin. In the cell, G-actin is not visible under the electron microscope. F-actin is the main component of the thin myofilaments or actin filaments (diam. 5–7 nm), which consist of 2 spirally wound F-actin filaments, as well as tropomyosin and troponin (see Fig.). Nonmuscle cells also contain actin filaments or actin-like filaments, known as contractile microfilaments. In the absence of Myosin (see), actin cannot bring about contraction. Actin filaments interact with myosin oligomers. It is presumed that α-actinin is involved in the binding of actin filaments to the plasma membrane, and to the membranes of secretory granules and phagosomes; actin filaments also bind to microtubules. The cell contains a network of contractile microfilaments, which are particularly numerous in the peripheral protoplasm on the inner side of the plasma membrane. In nerve axons, actin constitutes 10–15% of the cell protein. Together with microtubules and intermediate filaments, actin filaments form part of the cytoskeleton, thereby contributing to the stabilization of cell shape.

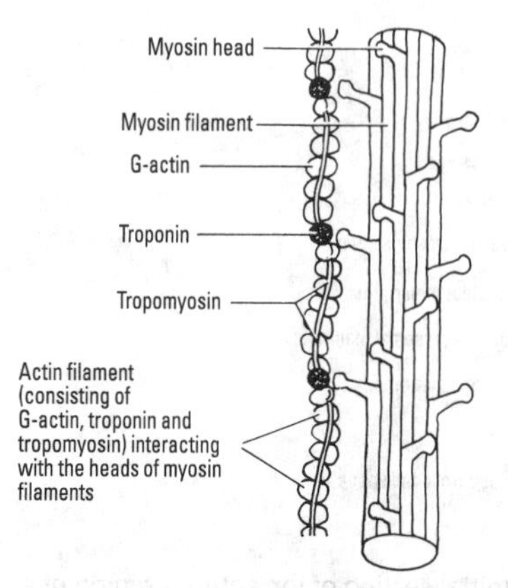

Actin. Section of myosin thick filament, showing globular myosin heads interacting with an actin filament.

Actin filaments can be demonstrated by immunological methods, and they can also be recognized from their specific reaction with HMM (heavy meromyosin, the heavy part of the myosin molecule). The binding sites of myosin fragments to actin filaments can be visualized by negative staining. The controlled aggregation of G-actin to F-actin is inhibited by the antibiotic, cytochalasin B.

Actinia: see *Actiniaria.*

Actiniaria, *sea anemones:* an order of the *Anthozoa,* subclass *Hexacorallia.* Sea anemones are relatively large polyps (the largest may be up to 1.5 m in diameter), and almost all of them are solitary. They are found from the littoral zone to the greatest ocean depths, on the sea floor and often on other animals. Some live in symbiosis with crustaceans or fish. All species are carnivorous, catching relatively large prey with their tentacles, or driving plankton into their mouths by ciliary action. A muscular, ectoderm-lined, flattened pharynx (stomodeum) leads from the mouth into the gastrovascular cavity, and is connected to the body wall by vertical partitions or septa, which increase the digestive surface area of the cavity. A single septum consists of 2 sheets of endoderm separated by an an internal layer of mesoglea. The free edges of the septa are expanded into convoluted septal filamants, containing nematocysts and gland cells, the latter secreting digestive enzymes. At one or both ends of the pharynx is a cilia-lined groove or siphonoglyph.

Over 1 000 species are known, the most familiar genera being *Actinia, Metridium, Sagartia* and *Anemonia. Metridium* is commonly seen attached to rocks in tidepools or to wharf and harbor pilings on the Atlantic coast of USA. It can change its position by a creeping movement. Plankton are stunned by nematocysts, held by mucus glands on the tentacles, carried to the tips of the tentacles by cilia, then pushed toward the mouth; at this point, the lip cilia which normally generate an outward current, reverse their beat and sweep the food into the pharynx. The Great Barrier reef of Australia is naturally home to many species, e.g. *Stoichactis,* which may display several colors (white, green, lilac, yellow) simultaneously.

Actinocamax (Greek *aktis* ray, *kamax* rod, post): characteristic Belemnite genus of the Upper Cretaceous. The rostrum is up to 10 cm long, usually cylindrical, rarely club-shaped, and possesses a short but deep ventral cleft.

Actinoceras (Greek *aktis* ray, *keras* horn): a nautiloid genus from the Ordovician, with an extended straight shell. The siphuncle resembles a string of beads, contains a highly differentiated system of vessels, and lies in the center of the shell.

Actinodont: a hinge type of the *Bivalvia* (see).

Actinomyces: see *Actinomycetales.*

Actinomycetaceae: see *Actinomycetales.*

Actinomycetales, *Actinomycetes:* an order of Gram-positive, nonmotile, predominantly aerobic bacteria, usually growing as branched threads, which readily break down into rod-shaped cells. Some A. form conidia or spores on aerial hyphae or on sporangia. The order contains the following 4 families.

1. ***Mycobacteriaceae.*** The genus *Mycobacterium* is represented by various pathogenic species, which characteristically produce chronic granulomatous lesions. They are nonsporing (but very resistant to drying), noncapsulated, aerobic, and slow-growing. Owing to the presence of waxy constituents, they are more difficult to stain than other bacteria; a strong dye with a mordant is required with prolonged staining or heating. A suitable stain is Ziehl-Neelsen carbol fuchsin, which consists of 1 part of a saturated solution of fuchsin in absolute ethanol and 9 parts of a 5% solution of phenol; mycobacteria appear red on a light blue background. Once stained, decolorization with acid or alcohol is resisted, so that mycobacteria are described as "acid and alcohol-fast". For the same reason, Gram staining is difficult to demonstrate, but all varieties of *M. tuberculosis* (tubercle bacilli) are, in fact, Gram-positive.

Human tuberculosis is caused by *M. tuberculosis* var. *hominis,* which occurs as slender, straight or slightly bowed rods, 0.5–4 μm by 0.3 μm, with rounded, pointed or sometimes expanded ends. The human variety has also been found in monkeys, cattle, pigs and dogs.

M. tuberculosis var. *bovis* (1–2 μm in length) is the causative agent of bovine tuberculosis, and is also the commonest variety responsible for tuberculosis of other domestic animals. The guinea pig is highly susceptible to both the human and bovine varieties.

M. avium (avian tubercle bacillus) causes tuberculosis in birds. It is highly virulent in fowls (which are resistant to the mammalian variety), and it has been reported in pigs, cattle and other domestic animals, but the guinea pig is resistant to it.

M. tuberculosis var. *muris* (vole tubercle bacillus) has been found in a fairly widespread tubercle-like disease of voles. The organism is longer and more slender than other tubercle bacilli, and often slightly curved. It is antigenically similar to the human and bovine varieties, but practically nonpathogenic to humans; it has been used as a vaccine as an alternative to BCG.

M piscium has been isolated from fish, frogs, turtles, etc., and seems to be associated with a turbercle-like disease in these cold-blooded animals; it is not pathogenic to mammals or birds.

M. leprae, the causative agent of human leprosy, takes the Gram stain quite easily. It is otherwise difficult to stain and acid-fast, but somewhat less so than *M. tuberculosis.* Morphologically, it is similar to a tubercle bacillus, with pointed, rounded or club-shaped ends. It is an obligate parasite, i.e. it cannot be cultured on artificial media, but for experimental purposes it can be cultured in armadillos.

M. johnei (= M. paratuberculosis) causes chronic enteritis of cattle and sheep (Johne's disease).

2. ***Actinomycetaceae.*** Species of the genus *Actinomyces* are the causative agents of actinomycoses, e.g. human actinomycoses are caused by *A. israeli,* and actinomycoses in cattle are due to *A. bovis.*

Spore formation in actinomycetes.

3. ***Streptomycetaceae.*** The genus, *Streptomyces,* contains many species of soil-dwelling, aerobic, Gram-positive, non-acid-fast bacteria, which grow as branched, hyphae-like threads, bearing numerous aerial spores (conidia). The hyphae do not break down into individual rods. They are notable for their ability to produce antibiotics, e.g. *Streptomyces griseus* (produces streptomycin), *S. aureofaciens* (chlortetracyclin or aureomycin®), *S. erythreus* (erythromycin), *S. fradii* (neomycin), *S. mediterranei* (rifamycin), *S. noursei* (nystatin), *S. rimosus* (oxytetracyclin or terramycin®), *S. venezuelae* (chloramphenicol).

4. ***Frankiaceae.*** Members of the genus *Frankia* are free-living soil organisms, which can also live in symbiotic association with nonleguminous plants. Their cells grow as threads, but they also display rod-shaped stages. They penetrate the root cells of various plants , e.g. alder *(Alnus),* Sea buckthorn *(Hippophae rhamnoides),* Russian olive *(Elaeagnus),* where they multiply and elicit the formation of root nodules, which may attain the size of tennis balls. Within these nodules. *Frankia* converts atmospheric nitrogen into an organically bound form, which is utilized by the host plant. *Frankia* species are differentiated according to their symbiotic host plant, e.g. *F. alni* is the symbiont of alder.

Actinomycosis: see *Actinomycetales.*

Actinotrocha larva: see *Phoronida.*

Actinula: a tentacled, ciliated, free-swimming larval stage of those coelenterates, which do not have a polyp generation. The A. arises from the Planula (see) and it forms itself into a medusa. In reality, the A. is an unusual, short-stemmed polyp, which can be observed in the gonophore of *Tubularia,* and which eventually leaves the gonophore and becomes attached at its aboral end.

Action potential: the change of electrical potential during the activity of cells, tissues and organs. An unequal distribution of ions between the interior of the cell and its exterior milieu results in a difference of electrical potential between the internal and external surfaces of the cell membrane. In particular, this has been demonstrated and studied for nerve and sensory cells, muscles and gland cells, but a potential difference is doubtless present between the interior and exterior of all physiologically active cells. The potential of the resting cell, in which the interior is negative to the exterior, is known as the Resting potential (see). The change in the value of this resting potential that occurs during cell activity is known as the A.p. The A.p. has the opposite potential to the resting potential, i.e. the generation of an A.p. is equivalent to depolarization or a reversal of polarity. A.p. is used in a restricted sense to mean a nerve impulse. When recorded on an oscillograph, the A.p. of a nerve is seen to consist of a spike, which rapidly rises to a sharp positive maximum then rapidly dies away, followed by a very small negative after-potential (i.e. slightly more negative than the resting potential). The A.p. of a nerve fiber has an amplitude of 60–130 mV, so that the polarization of the internal surface of the cell membrane (resting potential about -75 mV) is reversed, i.e. the internal surface becomes positive (by a value up to +55 mV) relative to the exterior. This change of potential is very brief and dependent on numerous factors. In the large nerve fibers of warm-blooded animals, depolarization may last less than 1 millisecond, whereas in invertebrates it may be longer than 10 milliseconds. Amplitude and duration of an A.p. are independent of the intensity of the excitatory stimulus (see All-or-nothing rule), but the Latent time (see) decreases with the intensity of the stimulus. When the measuring electrodes have a large contact area (see Electroencephalography), they record the response of several or many nerve fibers, rather than a single fiber. In such cases, the measured A.p. represents the summation of many potentials, the number of stimulated nerve fibers increases with the intensity of the stimulus, and the amplitude of the A.p. therefore also increases with the intensity of the stimulus.

See Excitation.

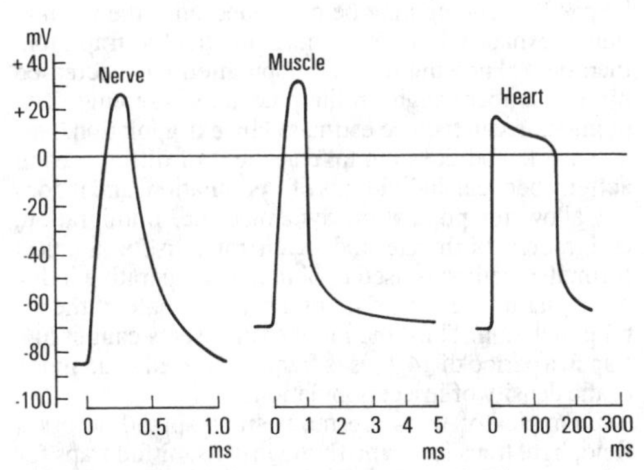

Time course of the action potentials of different cell types of warm- blooded animals.

Action specificity: see Enzymes.

Actitis: see *Scolopacidae.*

Activated sludge: see Biological waste water treatment.

Activation: see Vigilance.

Active dried yeast: see Baker's yeast.

Active resistance: resistance resulting from the reaction of a host against the attack of a parasite. Hypersensitivity (see) is a form of A.r.

Active sulfate: see Sulfur.

Active transport: the transport of certain ions and molecules across biomembranes against a concentration gradient or against an electrochemical potential, with an expenditure of energy (i.e. at the expense of the cleavage of ATP to ADP and inorganic phosphate). For example, amino acids, sugars and nucleotides are actively transported across biomembranes. By A.t., materials that are essential for metabolism and cell function become concentrated in cells, cell compartments or extracellular spaces.

After crossing the membrane, an ion or molecule may immediately become chemically bound or metabolized, or physically immobilized. On the other hand, ions or molecules may be concentrated in a free form, e.g. high intracellular concentrations of K^+ and high extracellular concentrations of Na^+ are maintained in animal and human tissues by the action of membrane Na^+/K^+-ATPase, which actively transports the respective ions in the appropriate directions. See Carrier, Ion pumps.

Activity:

1) A general term for the working intensity of a parameter or functional unit, e.g. enzyme A. or neuronal A.

2) A collective term used in Ethology (see) for movement as an expression of input into the central nervous system. The A. of small rodents in the laboratory is recorded with a running wheel. To determine A. in the wild, a small transmitter is attached to the animal. A. exhibits periodicity during the course of the day (see Biorhythms).

Activity density, ***species activity density:*** an estimate of population density obtained by determining the number of individuals or species that are captured in a trap in a given period of time. If all individuals in a population have the same probability (p) of being captured, and if P is the total number of individuals in the population, then the number of individuals caught in a trap will be p × P. Trapping may be continued until the population is exhausted, but it is more practical to trap more than once, knowing that the population has decreased by the number caught in the preceding trappings. The population can then be estimated by extrapolation.

The method does not take account of differences in activity between individuals of a population, and it does not allow for population dynamics, i.e. immigration, emigration, birth rate and death rate. In its simplest form, the method is used to obtain a comparative index of population density, rather than an estimate of the total population. Thus, the number of insects caught in a trap in a period of 14 days is frequently used as an index of the density of insect populations.

Examples of traps are mousetraps spread across a field, light traps for night-flying insects, pitfall traps for ground beetles, suction traps for aerial insects, adhesive ("flypaper") traps for flying insects, plankton nets, etc.

Activity pattern: see Gene activity pattern.

Activity phases of the heart: see Cardiac cycle.

Activity range, ***operational area:*** the area or space within which an individual or family preferentially performs its essential life functions (feeding, reproduction, rearing of young). The A.r. is usually centered around a structure such as a nest or burrow, and allegiance to this site is often strongly developed. A.rs. represent divisions of space among the members of a population. They often overlap, and members of the same species from other A.rs. are tolerated [cf. Territory (see) in which intruders are not tolerated].

Actomyosin: a protein complex containing 3 myosin molecules and one F-actin molecule. A. possesses ATPase activity, which is activated by calcium or magnesium ions. Muscle contraction is linked to the cleavage of ATP by A.

Actomyosin complex: see Myosin filaments.

Actophilornis: see *Jacanidae*.

Actualism, ***uniformitarianism:*** a principle stating that the evolution of the earth is due to relatively small natural fluctuations and events. By operating throughout entire geological history, these small changes have accumulated to produce changes of great magnitude. Further, the study of present-day geological events provides a key to understanding the processes that have occurred in geological time, i.e. it is assumed that natural geological processes are still operating at essentially the same rate and intensity as they have throughout geological time. A. was founded by Charles Lyell (1797–1875) and slightly altered and extended by Karl Adolf von Hoff (1771–1837). A. has experienced an extension in modern times, with the proposition that processes in geological time that proceeded differently from those of the present, can nevertheless be comprehended actualistically. A. is directly opposed to the Catastrophe theory (see) of Cuvier.

Aculeata: see *Hymenoptera*.

Aculifera: see *Amphineura*.

Acyclic parthenogenesis: see Anholocyclic parthenogenesis.

Acyltransferases: see Transferases.

Adam's apple: see Larynx.

Adaptation:

1) Evolutionary adaptation: interaction of a population of organisms with the environment, resulting in the intensification of specific, heritable characters, so that the chances of survival and of ultimately leaving descendants are increased. Thus, evolutionary adaptation is the natural selection of the most favorable characters and combinations of characters that are available from the existing gene pool (the store of potential genetic variability), each living organism representing a complex of adaptations for a successful existence in its normal habitat. Every specialization thereby represents a loss of plasticity, i.e. greater evolutionary stability, or a decrease in the ability for further adaptation by exploitation of the existing gene pool. Depending on the plasticity of the phenotype (usually greater in plants than in animals), however, environmental changes may result in noninherited alterations of the phenotype, which enable the population to survive for a sufficiently long period for the appearance of favorable mutations. This latter process was called *organic selection* by Baldwin in 1896 and is now known as the *Baldwin effect*. A. is especially important in the adjustment of an organism to changing environmental conditions (see Resistance). The greater the variability of a population, the more rapidly it can adjust to environmental changes, a process usually accompanied by the formation of new species. Thus, according to Schwarz's rule, species of a genus are always more capable of adapting to specific conditions than are the races of a single species (see Preadaptation, Climatic laws, Evolution).

2) Pysiological and sensory adaptation: changes in the performance of tissues, organs and/or individual organisms in response to events in the environment; or changes in the excitability of a sensory organ as a result of continuous stimulation. As a result of A., a more intense stimulus becomes necessary to produce the same activity. Thus, the touch receptors of the skin are excited by contact with an object, but if the contact is maintained, the touch receptors adapt and cease to respond, i.e. the stimulus required to excite them has increased. Similarly, olfactory receptors adapt rapidly, so that many substances cannot be detected by their smell after a relatively brief but sustained action upon the olfactory organ. This A. (i.e. decrease in response with sustained stimulation) occurs by a variety of inadequately characterized mechanisms, but it is not due to tiredness or exhaustion. A. of sensory and nerve cells is due to a blockage of, or decrease in, the number of receptors, or a decrease in the activity of the reactions that link the receptors with the excitatory response. A. can be seen and measured as a decrease in the receptor potential of continuously stimulated sensory cells. The excitation threshold and sensitivity of a sensory organ depend on the degree of A., so that the same brief stimulation of a sensory organ may cause different levels of excitation, depending on whether the organ is strongly or weakly adapted.

Adaptive landscape, ***adaptive surface, adaptive topography, fitness diagram:*** a graphical representation of all the values for total fitness in a population, which is polymorphic (see Polymorphism) at two gene loci that influence Fitness (see). The analysis is restricted to two gene loci, because the results of variabilty at three or more loci can no longer be represented on a single plane. All possible frequencies of genes at one locus are plotted on the abscissa, the possible frequencies of genes at the other locus on the ordinate. If the fitness value of all genotypes is known, then the total fitness of the population can be read off from any point on the surface of the A.l. If all points of equal fitness are joined, an ordered system of *isofitness* lines is obtained. Thus, the fit genotypes able to occupy particular ecological niches are depicted as adaptive peaks, separated by adaptive valleys representing unfit gene combinations. Real populations usually display a fitness peak, or they strive to attain the nearest fitness peak. If a population on a low peak is separated from a high peak by a valley, then it cannot attain the high peak, because selection continually pushes each population in an upward direction. Thus, depending on its starting condition, a population cannot always optimize its gene pool.

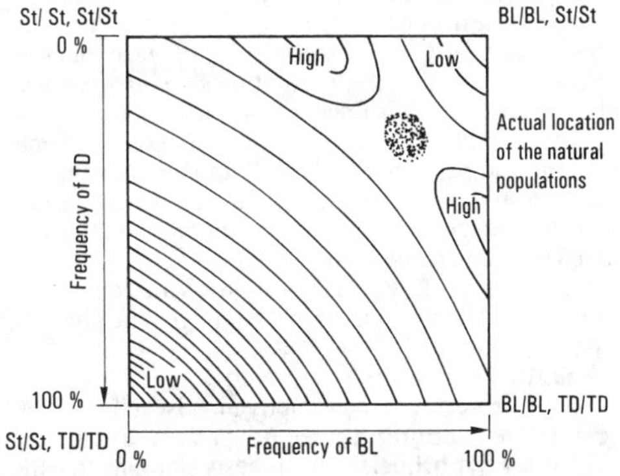

Adaptive landscape. Theoretical fitness diagram for a population of grasshoppers with two polymorphic chromosomes. In this case, the actual position of the population lies in a fitness valley. BL, St and TD are structural types of chromosomes.

Adaptive radiation: the relatively rapid diversification of a group of organisms into separate forms, which are adapted to different environments, or utilize different resources of the same environment. A.r. is favoured by the following conditions:

1. Attainment of a new evolutionary niveau, e.g. the rapid diversification of angiosperms during the Creta-

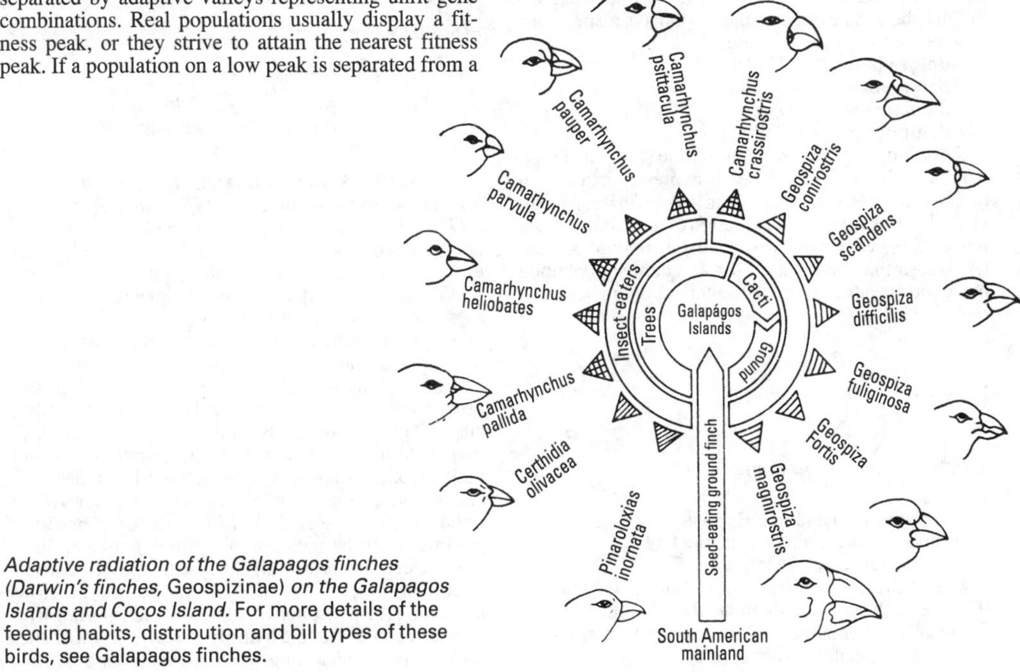

Adaptive radiation of the Galapagos finches (Darwin's finches, Geospizinae) *on the Galapagos Islands and Cocos Island.* For more details of the feeding habits, distribution and bill types of these birds, see Galapagos finches.

ceous; evolution of placental mammals from a primitive insect-eating, five-toed, short-legged creature, so that modern mammals are now represented in all possible environments (e.g. dogs, deers, squirrels, primates, bats, beavers, seals, whales).

2. A severe decrease in the number of species in a taxon, e.g. the decline of the ammonites at the Permian-Triassic boundary, and their almost complete extinction at the end of the Triassic; each of these reductions was followed by re-establishment of an abundance of forms.

3. Availability of new territories and environments free from competition, e.g. the A.R. of Darwin's finches (*Geospinzinae*) on the volcanic Galapagos islands, and of the honeycreeper finches (*Drepanilidae*) on the volcanic Hawaiian archipelago.

A.R. is the most frequent reason for Virescence periods (see), and it often occurs in the typogenetic phase of Typostrophism (see).

Adaptive similarity: see Analogy.

Adaptive surface: see Adaptive landscape.

Adaptive topography: see Adaptive landscape.

Adaptor hypothesis: a hypothesis claiming that the selection and ordering of amino acids during protein synthesis does not depend on their direct interaction with DNA, i.e. other adaptor molecules are present which serve to transfer genetic information from DNA. This is clearly no longer a hypothesis but an experimentally well established fact, the adaptor molecules in question being messenger RNA and transfer RNA. See Protein biosynthesis.

Adaxial petal: see *Papilionaceae.*

Addax: see *Hippotraginae.*

Adductors: muscles that move a limb towards the mid line of the body, e.g. muscles that bring the arms to the side of the body. The meaning is different when applied to the adductor muscles of the fingers and toes, where movement is implied toward an imaginary line through the middle finger or the second toe, respectively.

Adecticous pupa: see Insects.

Adelges laricis: see Galls.

Adelgidae: see *Homoptera.*

Adelidae: see *Lepidoptera.*

Adelphogamy: see Pollination.

Adenine, ***A*** or ***Ade:*** 6-aminopurine (Fig.), a component of nucleic acids, and therefore present in all living organisms. Free A. is found in various plants, especially yeast. It arises from the degradation of nucleic acids, and it is synthesized *de novo* via adenosine monophosphate. (A. is also the abbreviation for an adenine-containing nucleotide residue of a nucleic acid or polynucleotide).

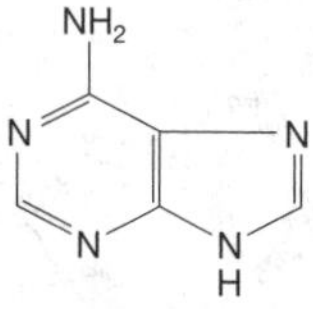

Adenine

Adenohypophysis: see Hypophysis.

Adenoid tissue: see Connective tissues.

Adenophora: see *Nematoda.*

Adenosine, ***Ado:*** 9β-D-ribofuranosyladenine (Fig.), a nucleoside in which D-ribose is linked β-glycosidically to adenine. Adenosine phosphates (see) have special metabolic significance.

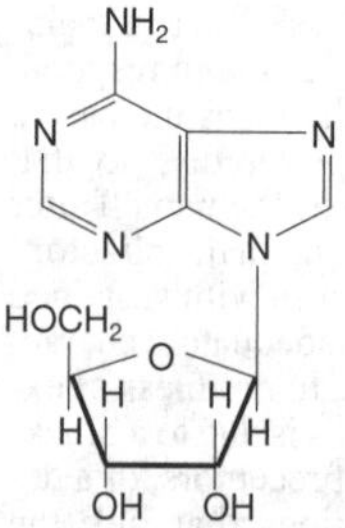

Adenosine

Adenosine 5'-diphosphate, ***ADP:*** a nucleotide containing one labile or energy-rich pyrophosphate bond. ADP becomes phosphorylated to ATP in substrate level phosphorylation, oxidative (respiratory chain) phosphorylation and photophosphorylation.

Adenosine 5'-monophosphate, ***AMP, adenylic acid:*** a nucleotide present in all living organisms, which is formed as an end product of purine biosynthesis, or arises from ATP in reactions involving transfer of the pyrophosphate or AMP moiety. Yeast and muscle serve as ready sources for the laboratory isolation of AMP.

Adenosine phosphates: phosphoric acid esters of adenosine, which are metabolically very important as components of nucleic acids, and as agents for transfer and storage of energy. A.p. also have a regulatory function in glycolysis and the tricarboxylic acid cycle.

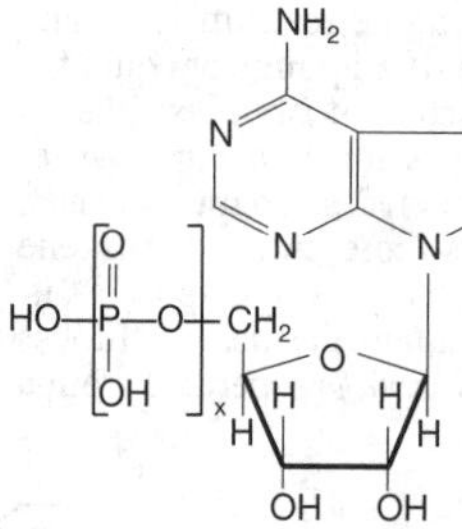

Adenosine phosphate(s)

Adenosine triphosphatase, ***ATP phosphohydrolase, adenylpyrophosphatase, ATP monophosphatase, ATPase:*** an enzyme that catalyses the hydrolytic cleavage of adenosine triphosphate (ATP) into adenosine diphosphate (ADP) and orthophosphate:

ATP + H_2O → ADP + orthophosphate. Some ATPases simply catalyse this hydrolysis, without coupling to another specific cell function. Many ATPases, however, are associated with specific ATP-dependent mechanisms, e.g. myosin ATPase [myosin ATP phosphohydrolase (actin-translocating)] and dynein ATPase [dynein ATP phosphohydrolase (tubulin-translocating)]. Of particular interest is the group of membrane-bound, multisubunit complexes in mitochondria, chloroplasts and bacteria, which couple the reverse reaction (ADP + orthophosphate → ATP + H_2O) with the transport of protons under the pressure of a proton gradient. These enzyme complexes, which catalyse the physiologically important process of ATP synthesis, are generally known as ATP phosphohydrolase (H^+-transporting), i.e. they are named after the hydrolytic reaction, which they also catalyse under nonphysiological conditions; other names are: H^+-transporting ATPase, H^+-transporting ATP synthase, mitochondrial ATPase, coupling factors (F_0, F_1, CF_1), chloroplast ATPase, bacterial Ca^{2+}/Mg^{2+} ATPase. A group of H^+-transporting ATPases from the plasma membranes of yeast and *Neurospora* consist of a single catalytic unit *(M_r* 100 000). H^+/K^+-transporting ATPase is present in the gastric mucosa. The sodium pump of animal cells is a membrane-bound Na^+/K^+-transporting ATPase, which couples the exchange of K^+ and Na^+ with ATP hydrolysis, and is specifically inhibited by ouabain. Similarly, the calcium pump in the plasma membrane of eukaryotes is a Ca^{2+}-transporting ATPase, which couples the transport of Ca^{2+} with ATP hydrolysis.

Adenosine 5'-triphosphate, ***ATP:*** a nucleotide which acts as an energy storage and transfer molecule in all living cells. It is biosynthesized by phosphorylation of ADP, i.e. by substrate level phosphorylation, oxidative (respiratory chain) phosphorylation, or photophosphorylation. Various groups are transferred from ATP, e.g. phosphoryl, ADP, AMP, adenosyl and pyrophosphoryl groups, which activate precursors in the biosynthesis of macromolecules, and activate substrates of metabolic pathways. For example, phosphorylation of glucose by ATP is required for the glycolytic degradation of glucose; β-oxidation of fatty acids commences with an ATP-dependent synthesis of fatty acyl-CoA; urea synthesis requires ATP at two stages (synthesis of carbamoyl phosphate and synthesis of argininosuccinate); ATP is required for the synthesis of aminoacyl-tRNA in protein synthesis; other nucleoside triphosphates (UTP, GTP, TTP, CTP), which are required for nucleic acid synthesis, protein translation (GTP), activation of sugars and phospholipid synthesis, are all synthesized from their mono- and diphosphates by phosphorylation by ATP. Muscle contraction, biological luminescence, current production in electric fish, and substrate transport against a potential gradient all depend on ATP cleavage.

Adenota: see *Reduncinae*.

Adenylate cyclase, ***adenyl cyclase, adenylyl cyclase, ATP pyrophosphate lyase (cyclizing):*** an enzyme localized in the cell membrane, which catalyses the formation of cyclic adenosine 3',5'-monophosphate (cyclic AMP) from adenosine triphosphate (ATP) by elimination of pyrophosphate and esterification of the remaining phosphate residue with the 3'-hydroxyl group: ATP → cyclic AMP + pyrophosphate. Cyclic AMP formation plays an important role as a "second messenger" in the action of Hormones (see). See Cyclic adenosine 3',5'-monophosphate, G-protein.

Adenyl cyclase: see Adenylate cyclase.

Adenylic acid: see Adenosine 5'-monophosphate.

Adenylyl cyclase: see Adenylate cyclase.

Adephage: see *Coleoptera*.

Adherant: see Life form.

Adhesion belt: see Epithelium.

Adhesion organs, ***suction organs, attachment organs, haptors:*** animal organs for attachment to other animals or inanimate objects. They usually have the shape of a cup, basin or disk. Active muscle contraction causes negative air pressure within the organ after it is placed on the attachment surface. A.o. occur in trematodes, tapeworms, leeches, echinoderms, branchiurans, lampreys and remoras. Cephalopods possess particularly powerful suction disks (stalked or arising directly on the tentacles), which consist of a firm muscular ring with a central core of longitudinal muscles arranged to form a hemispherical structure. Contraction of these muscles, after the ring has been pressed against an attachment surface, produces such a powerful suction pressure that less force is required to tear the tentacle off the animal than to break the grip of the suction disk. In octopuses, the ring is thick and powerfully muscular, whereas in cuttlefish and squid the ring usually has an outer edge of keratochitin, armed with teeth or hooks for gripping soft-skinned prey.

ADI: acronym of Acceptable daily intake (see).

Adipocyte: see Connective tissues.

Adipose tissue: see Connective tissues.

Adjutant stork: see *Ciconiiformes*.

Adnate plant: see Life form.

Adonis: see *Ranunculaceae*.

Adrenal corticosteroids, ***adrenocorticoids, corticosteroids, corticoids, cortins:*** an important group of steroid hormones produced in the adrenal cortex in response to adrenocorticotropic hormone (ACTH). A.c. are structurally related to pregnane (see Steroids); they contain a carboxyl group with a neighboring α,β-unsaturated bond in ring A, a ketol side chain in position 17, and other oxygen functions, particularly in positions 11, 17, 21 and 18.

More than 30 steroids have been found in the adrenal cortex; the following show marked corticosteroid activity: Cortisol (see), Cortisone (see), cortexolone, Cortexone (see), 11-dehydrocorticosterone, Corticosterone (see) and Aldosterone (see). Quantitatively the most significant members of this group are cortisol, corticosterone and aldosterone, which are secreted daily into the blood in quantities of 15, 3 and 0.3 mg, respectively.

Production of A.c. is increased in physical and/or psychological stress. Deficiency of A.c., e.g. caused by pathological changes in the adrenal glands, results in Addison's disease (Morbus Addison), characterized by tiredness, emaciation, decrease of blood sugar and dark pigmentation of those areas of the skin exposed to light. Adrenalectomy of experimental animals leads to rapid death unless exogenous A.c. are given. A.c. control mineral metabolism by causing retention of Na^+, Cl^- and water, with a simultaneous K^+ diuresis (mineralocorticoid or mineralotropic action). They also regulate carbohydrate metabolism, in particular glycogen synthesis in the liver (glucocorticoidal or glucotropic action). Depending on the type of activity that predominates, the A.c. are classified as *mineralocorticoids* (aldosterone, cortexone, cortexolone) or *glucocorticoids* (cortisol, cortisone, corticosterone, 11-dehydrocorticosterone).

For therapy of adrenal insufficiency, extracts of adrenal cortex are no longer used and have been replaced by pure A.c. High doses reveal other pharmacological properties, especially anti-inflammatory and antiallergic activities. This discovery led to the development of highly active, synthetic derivatives, which are now used widely for the treatment of rheumatism, asthma, allergies, eczema, etc., e.g. prednisone, dexamethasone and triamcinolone.

Adrenal gland, ***suprarenal gland, glandula suprarenalis:*** an endocrine gland of vertebrates. Phylogenetically, it is derived from two organs (the suprarenal bodies and interrenal bodies of fishes), which in terrestrial vertebrates have become closely associated to form a single body. Many vertebrates possess multiple A.gs., but mammals have just two A.gs., each lying close to and above a kidney. The A.g. consists of a medulla and a cortex. The adrenal cortex secretes steroid hormones that regulate the water economy of the body, influence the salt concentrations of body fluids (mineralocorticoids), and influence carbohydrate metabolism (glucocorticoids), as well as sex hormones that complement the hormones of the gonads. Hormones of the adrenal medulla (adrenalin and noradrenalin) cause an increase in blood glucose in response to stress or excitement.

Adrenalin, *epinephrin:* 4-[1-hydroxy-2-(methylamino)ethyl]-1,2-benzenediol (Fig.), a biogenic amine and a catecholamine. A. has two distinct physiological functions. First, it is a hormone synthesized in the adrenal medulla, where it is stored in chromaffin granules. Second, it is an adrenergic neurotransmitter of the sympathetic nervous system, where is it synthesized and stored in postganglionic neurons. Only the L-isomer is physiologically active. When released into the blood by nervous stimulation of the adrenal medulla, it causes an increase in the beat frequency and contractility of the heart, increased muscular performance by dilation of blood vessels, and an increase in the circulating levels of glucose, lactate and free fatty acids. A. is bound either by an α- or a β-receptor in the membrane of the target organ. Binding to a β-receptor results in activation of the adenylate cyclase system on the cytoplasmic side of the membrane, leading to activation of liver phosphorylase and the lipase of adipose tissues. Biosynthesis of A. proceeds from L-tyrosine via 3,4-dihydroxyphenylalanine (dopa), 3,4-dihydroxyphenylethylamine (dopamine) and noradrenalin, the latter becoming methylated to A. Inactivation of A. occurs by methylation to 3-methoxyadrenalin, which becomes oxidatively deaminated to 3-methoxy-4-hydroxymandelic acid aldehyde. Further oxidation produces the urinary excretion product, 3-methoxy-4-hydroxymandelic acid.

HO–C₆H₃(OH)–CH(OH)–CH_2–NH–CH_3

Adrenalin

Adrenergic: the adjective applied to neurons, nerve fibers and synapses, whose Neurotransmitter (see) is noradrenalin or adrenalin.

Adrenocorticoids: see Adrenal corticosteroids.

Adult stage, *adult period, adult phase:* in humans the period between the 20th and 40th years of life. See Age diagnosis, Ontogenesis, Development.

Adultus: see Age diagnosis.

Adventitious bud: see Bud.

Adventitious embryony, *nucellar embryony:* direct development of embryos from somatic cells of the ovule, e.g. from the nucellus or integuments. A.e. may be apparently spontaneous, or induced by pollination. Alternation of generations and formation of an embryo sac may be completely suppressed by A.e., which can be regarded as a special type of vegetative propagation or asexual reproduction. In some cases (e.g. *Citrus)* adventitious and sexually produced embryos occur together. A.e. is known, e.g. in *Citrus* and *Hosta* and in members of the *Orchidaceae.* See Polyembryony.

Adventitious plant: a plant found in an area to which it is not indigenous, having been introduced, directly or indirectly, by human agency. Depending on their degree of naturalization and their mode and time of introduction, adventitious plants can be classified in various groups.

1. ***Ephemerophytes*** or ***passants*** occur sporadically and have only a transitory existence in a particular area. They may arise, e.g. from imported bird seed, or from seeds and plants carried inadvertently along transport routes.

2. ***Colonizers*** or ***colonists*** are weeds introduced unintentionally with cultivated plants, and which have become established in man-made ruderal and arable habitats, e.g. field poppy (*Papaver rhoeas*), narrow leaved pepperwort (*Lepidium ruderale*).

3. ***Neophytes*** are recent arrivals, which have become naturalized among the native vegetation, e.g. Canadian pondweed (*Elodea canadensis*), groundsel (*Senecio vulgaris*), slender rush (*Juncus tenius*), sweet flag (*Acorus calamus*), small-flowered balsam (*Impatiens parviflora*). Neophytes may displace and supplant native species, and the study of such recent introductions should include a record of the year of their first appearance and their subsequent rate of spread.

4. ***Archeophytes*** are plants that became naturalized in prehistoric or very early historic times. Thus, many plants which occur as garden and agricultural weeds, as well as being found in waste areas, on waysides and even in apparently "neutral" habitats, were brought in with agricultural plants during the early construction of arable fields and the building of extensive settlements in the Neolithic period. The process has continued with each successive wave of human migration, bringing with it not only the economically useful plants favored by the immigrants and invaders, but also their accompanying weeds. European examples are cornflower (*Centaurea cyanus*), buckthorn plantain or ribwort (*Plantago lanceolata*), corn cockle (*Agrostemma githago*), great burdock (*Arctium lappa*). Indigenous plants which move into man-made waste or arable land in the area of their natural habitat are called ***apophytes***. See Ruderal plants, Weeds.

Adventitious root: see Root.

Adventitious shoots: shoots formed at unusual sites, especially on roots, usually following wounding or splitting of the plant. Thus, *stool shoots* often appear on the stumps of felled trees and shrubs. Many plants are propagated commercially and by amateur gardeners from the A.s. produced by *cuttings* (sections of stem, root or leaf).

Adventitious structures: a collective term for adventitious buds (see Bud), adventitious roots (see Root) and adventitious shoots (see).

Aechmophorus: see *Podicipediformes.*

Aecidiospores: unicellular, rust-colored spores of the rust fungi (see *Basidiomycetes,* Fig. 3.1), containing a pair of nuclei. A. are formed in a cup-shaped Aecidium (see). In most of the economically important rust fungi, A.are responsible for infecting the other host, e.g. A. of the cereal black rust, *Puccinia graminis*, are formed on *Berberis*, and they carry the infection to cereals and certain grasses.

Aecidium: the cup-shaped spore container of the rust fungi, in which the rust-colored aecidiospores are formed (see *Basidiomycetes,* Fig. 3.1). Aecidia are formed when the host plant is infected by basidiospores of opposite mating types. The mature basal cells of the respective mycelia conjugate laterally in pairs within a complex of hyphae known as the protoaecidium (a partly plectenchymatous mass of hyphae near the lower epidermis of the host plant). Each of the resulting dikaryotic cells gives rise to a chain of binucleate, rust-coloured aecidiospores, which eventually rupture both the pseudoperidium (outer layer of the A., formed by the adhesion of a layer of spores) and the plant epidermis. Alternatively, the infecting basidiospores may all be of the same mating type. The haploid mycelium then sends out so-called receptive hyphae to the exterior of the plant, each with a single nucleus at its base and no cross walls. Through the activity of insects these may come into contact with pycnospores of the opposite mating type, or the receptive hyphae may send out processes to nearby pycnospores. Fusion then occurs and the pycnospore nucleus passes down the hypha, passing through perforations in the transverse septa of the protoaecidium until it reaches a basal cell of the opposite mating type. An A. and aecidiospores are then formed as already described. Aecidia of *Puccinia graminis*, one of the commonest cereal rust fungi, occur as small, rust-brown dots on the undersurface of the leaves of *Berberis*, which is the intermediate host.

Aegeriidae: see *Lepidoptera*.

Aegithalinae: see *Paridae*.

Aegithina: see *Chloropseidae*.

Aegotheles: see *Caprimulgiformes*.

Aegothelidae: see *Caprimulgiformes*.

Aegypiinae: see *Falconiformes*.

Aeolothripidae: see *Thysanoptera*.

Aepycerini: see *Antilopinae*.

Aepyceros: see *Antilopinae*.

Aepyornithidae, *elephant birds*: an extinct group of flighless, running birds, whose fossilized and subfossilized bones and eggs have been found on the island of Madagascar, dating from the Pleistocene to recent times. All species were tall and massive with very stout legs. In general stature they resembled an ostrich, but they were much bulkier. Weighing 450 kg, *Aepyornis* was up to 3 m tall. Eggs measuring 340 mm by 245 mm, with an estimated weight of 10 kg, have been excavated; with an average capacity of 9 liters, these are the largest known eggs in the animal kingdom.

[*Extinct Birds* by Errol Fuller; Viking/Rainbird (Penguin Books Ltd.) London, Victoria, Aukland, 1987]

Aepypodius: see *Megapodiidae*.

Aerenchyma: see Hydrophytes, Parenchyma.

Aerial hyphae: fungal hyphae that grow upward and away from the growth substrate, as distinct from Substrate hyphae (see).

Aerial mycelium: a layer of fungal mycelium consisting of erect (aerial) hyphae that grow away from the nutritive hyphae in the substrate. An A.m. frequently produces large quantities of asexual spores (conidia), carried on specialized hyphae called conidiophores, or within sporangia at the termini of specialized hyphae called sporangiophores. The A.m. is therefore also known as the *reproductive mycelium*.

Aerial roots: see Epiphytes.

Aerobe, *aerobic organism, aerobiont:* an organism which depends upon oxygen for its energy metabolism. The term is usually applied to microorganisms that are obligate aerobes, i.e. they can only develop in the presence of free oxygen (e.g. many fungi and bacteria). Facultative aerobes can also grow and reproduce in the absence of oxygen (e.g. lactic acid bacteria, *Lactobacillus*). Animals are almost exclusively aerobes, but a few can live anaerobically, e.g. tapeworms, which live as parasites in the anaerobic environment of the host intestine. Microorganisms with a decreased oxygen requirement (see Microaerophil) are also aerobes. The opposite of aerobe is Anaerobe (see).

Aerocharis: see *Vangidae*.

Aerosol: an air or gas mixture containing finely suspended liquid (mist) or solid (smoke) particles. Many pesticides, etc. are applied as aerosols.

Aerotaxis: a special case of Chemotaxis (see), in which the direction of movement of freely motile organisms is influenced by variations in the oxygen concentration of their environment. *Positive A.* is displayed by strongly aerobic bacteria, while anaerobes display *negative A.*

Aerotropism: a special case of Chemotropism (see), in which curvature is induced in plant organs by a decrease in oxygen tension. Thus, roots bend away from regions of low oxygen tension and therefore grow toward better aerated regions of the soil.

Aeschnidae: see *Odontoptera*.

Aesculapian snake, *Elaphe longissima:* a slender, glistening brown, tree-climbing colubrid snake (see *Colubridae*). It is found in open deciduous woodland in southern Europe northward to the Taunus. A further population in the area of Schlangenbad in Germany is probably not truly indigenous, but descended from released specimens. Young snakes feed preferably on lizards. Older snakes (maximal length of 2 m) feed mainly on mice. Between 5 and 8 eggs are usually laid in tree hollows.

Aesthetes: see Esthetes.

Aethalium: see *Myxomycetes*.

Aethopyga: see *Nectariniidae*.

Affect: an emotional state or feeling preceding an action. As. are recognizable in mammals, and they are presumed to be present to a slight extent in other vertebrates. They are manifested in the subsequent action (or inaction), e.g. flight due to fear, or attack due to anger, and by physiological changes, e.g. in pulse rate and/or blood pressure, etc. A. is almost synonymous with emotion, and there is a close relationship between A. and motivation. *Affective disorders* are psychiatric disorders, e.g. schizophrenia, manic depression.

Afference: the totality of nervous impulses leading from a sensory cell, nerve cell or a nerve, which are ultimately processed. In particular, A. is also a collective term for all the excitations from sense organs (including proprioceptors) to the central nervous system. See Efference, Reafference.

Afferent attenuation: see Habituation.

Affinity:

1) A term used to express the strength of receptor-ligand interaction. A. is usually applied to noncovalent bonding between molecules, i.e. ionic interactions, hydrogen bonding, Van der Waals forces, hydrophobic interactions, etc., as illustrated by the following 3 examples.

a) The strength of binding of an Antibody (see) to the epitope of the Antigen (see). Antibodies display different As., depending on their exactness of fit with the binding site of the epitope of the antigen.

b) The A. of an enzyme for its substrate is measured as the Michaelis constant (K_m), i.e. the concentration of substrate that produces one half of the maximal reaction rate (V_{max}).

c) The A. of binding of hormones to their receptors on the cell surface or to their intracellular receptors is expressed as K_d (dissociation constant), which is analogous to K_m (see above) and equivalent to the concentration of hormone causing 50% saturation of the receptor.

2) Genetically different gametes may exhibit different rates of fertilization. Fertilization frequency or *fertilization affinity* is therefore genetically determined, and due to the different attractive forces between male and female gamnetes.

3) An animal species may prefer or be restricted to a habitat, because only that habitat satisfies certain ecological requirements *(ecological affinity,* see Species combination), or the preference may be based on the requirement for certain partner species *(sociological affinity).*

Afghan mole lemming: see *Arvicolinae.*

Aflatoxins: Mycotoxins (see) produced by *Aspergillus flavus, A. parasiticus* and *A. oryzae,* as well as some *Penicillium* strains. A. are among the most potent, orally active, naturally occurring carcinogens. Many liver cancers in the tropics are attributable to the ingestion of A. Unmodified A. are relatively nontoxic *per se,* but they are converted into toxic and carcinogenic derivatives by monofunctional oxygenases (cytochrome P450) in the liver (Fig.). The LD_{50} for aflatoxin B_1 in ducklings is 18.2 μg/50 g body weight. Due to the growth of moulds on stored plant materials, A. occur widely in foodstuffs in the humid tropics, in particular in groundnuts and groundnut oil. There is evidence that Kwashiorkor (see) results from a combination of starvation and aflatoxin poisoning.

African black terrapin: see *Pelomedusidae.*

African buffalo: see *Bovinae.*

African bush pig, ***African water hog, African river hog, red-river hog,*** *Potamochoerus porcus:* a monotypic species of the family *Suidae,* order *Artiodactyla* (even-toed ungulates), found throughout sub-Saharan Africa and on Madagascar, in scrub forest and bush areas. Young animals are longitudinally striped pale yellow on a dark background, but adults are reddish brown to black, often with an admixture of white. The tips of the long pointed ears often carry tufts of long hair. There is a pronounced light-colored mane along the top of the neck and back. The male has a protruding wart in front of each eye. The upper tusks (average length 7.6 cm) are directed downward, and they rub against the lower ones (17–19 cm). During the day, these animals tend to rest in dense vegetation, and they are most active at night, feeding on roots, wild fruit, reptiles, eggs and occasionally young birds.

African cattle sickness: see Trypanosomes.

African clawed toad, *Xenopus laevis:* a member of the *Pipidae* (see), found in Kenya, Zaire, Angola and further south into South Africa. Six subspecies are recognized. Males attain a length of 6 cm, females 12 cm. It is a skilful swimmer, possessing a flattened body and very long hindlegs with long webbed, sideways splayed toes; the inner three toes carry short black claws. It cannot survive for long out of water, and the short, sideways projecting front legs are incapable of supporting the body on land. When moving on land out of neces-

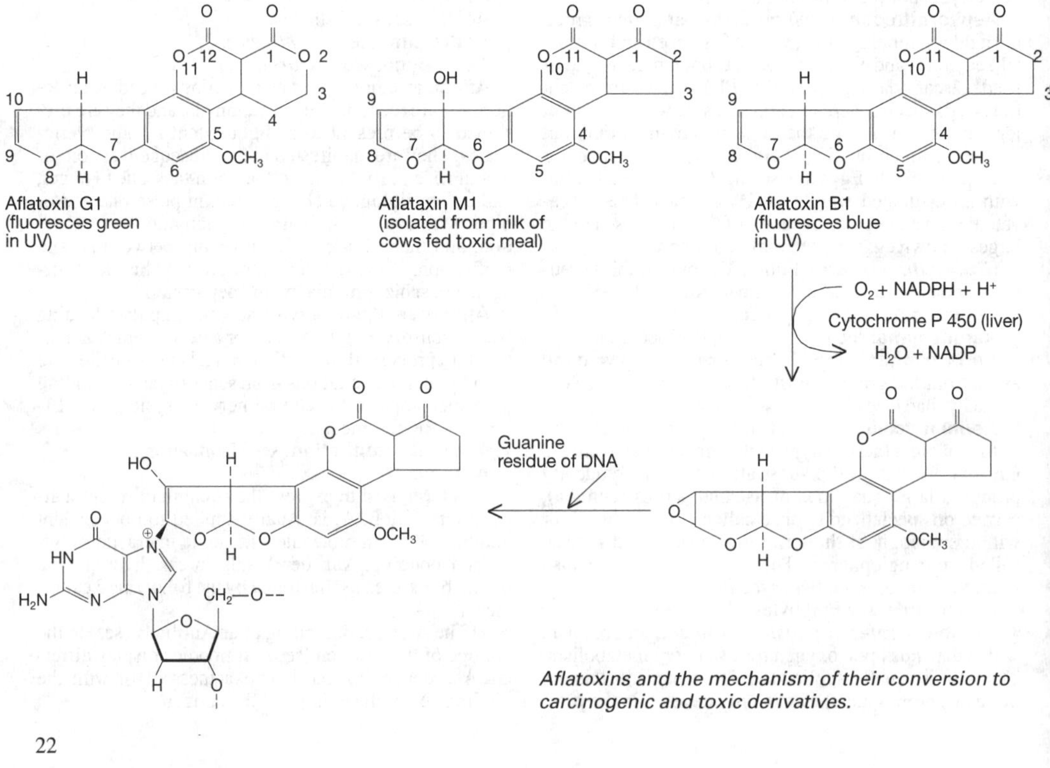

Aflatoxin G1 (fluoresces green in UV)

Aflataxin M1 (isolated from milk of cows fed toxic meal)

Aflatoxin B1 (fluoresces blue in UV)

O_2 + NADPH + H^+
Cytochrome P 450 (liver)
H_2O + NADP

Guanine residue of DNA

Aflatoxins and the mechanism of their conversion to carcinogenic and toxic derivatives.

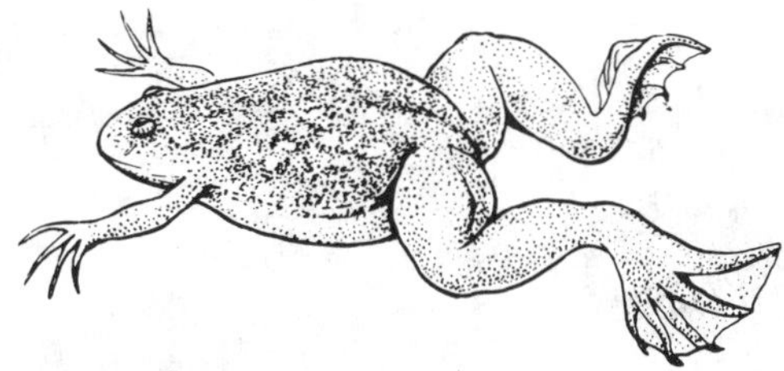

African clawed toad (Xenopus laevis)

sity, the animal leaps clumsily, landing flat on its front and sliding forward. The very smooth skin has a relatively poor blood supply and is covered with a malodorous slime. Respiratory gas exchange via the skin appears to be of minor importance. The lungs are large in proportion to the body, and the animal rises regularly to the surface to gulp air. It typically lives in stagnant, muddy pools, where the oxygen tension is low, and burrows in the bottom mud for food.

Xenopus was the first member of the *Anura* to be used for pregnancy testing. Hormones excreted by the pregnant human initiate egg laying in *Xenopus*, so that injection of human pregnancy urine causes the toad to lay eggs after 5 to 24 hours. Pregnancy testing is now performed by modern chemical methods, and the *Xenopus* test is obsolete. Females lay about 10 000 eggs during the year on water plants.

Members of the related genera, *Hymenochirus* (4 species; Cameroon, Gabon, Zaire) and *Pseudhymenochirus* (1 species; Guinea) are 3–4 cm in length, and often kept in aquaria. Spawning partners lie on their backs with their cloacal openings above the water-level, release eggs and sperm, then return to the bottom.

African dwarf crocodile, *broad-fronted crocodile*, *Osteolaemus tetraspis:* a nearly extinct, very short-snouted member of the *Crocodylidae* (see), attaining a length of only 1.5 m, and found in tropical West and Central Africa. It is found exclusively in fresh water, and is not dangerous to humans.

African eye worm: see Filariasis.

African hunting dog, *Lycaon pictus* (plate 38): a predator of the family, *Canidae* (see). The A.h.d. lives in packs on the savannas of sub-Saharan Africa, where it preys on both small and large mammals, including antelopes and zebras. Prey is either caught by surprise, or it is harried and run down cooperatively by the pack. Head plus body length varies from 76 to 105 cm, and the coat colors (black, yellow and white) occur in very variable proportions with spotting and mottling. There are only 4 toes on each foot, the legs are long and slender, and the jaws broad and powerful. *Lycaon* is a monotypic genus.

African land snail: see *Gastropoda*.

African long-fingered frogs: see *Arthroleptinae*.

African palm squirrels: see *Sciuridae*.

African plated lizard: see *Cordylidae*.

African pygmy squirrel: see *Sciuridae*.

African river martin: see *Hirudinidae*.

African rock martin: see *Hirudinidae*.

African rubber: see *Apocyanaceae*.

African sparmannia: see *Tiliaceae*.

African tree toads: see *Bufonidae*.

Afro-American side-necked turtles: see *Pelomedusidae*.

Afro-American snake-necked turtles: see *Pelomedusidae*.

Afro-European *sapiens*-hypothesis: see Anthropogensis.

Afterbirth: see Placenta.

Afterfeather: see Feathers.

Agamete: a Spore (see).

Agamidae, *agamids, chisel teeth lizards:* a family of lizards containing about 35 genera and more than 300 species. They are mostly 20–30 cm long, and have a typical lizard appearance, with well developed limbs, a moderately long, nonbrittle tail, and a broad head. The scales are mostly keeled or drawn out into spines. Other occasional features are a spiny tail and a scaly crest. The teeth are fixed firmly at their bases to the surface of the major tooth-bearing bones. Most species are day-active. Some arboreal species have laterally compressed bodies, whereas desert forms tend to be dorsoventrally flattened or depressed. *A.* are an exclusively Old World family, found in Africa (not in Madagascar), southeastern Europe, Asia, Polynesia and Australia, and they often display close external similarities to the New World *Iguanidae* (see). These evolutionary convergent features result from similar modes of existence in similar ecological systems. Both families contain representatives with nuchal crests (a crest on the nape of the neck), spiny or prehensile tails, gular pouches (variously known as dewlaps, gular sacs or throat fans); and both contain large tree-dwelling rainforest forms, as well as small, flattened, spiny desert forms. Courtship display is similarly elaborate in *A.* and *Iguanidae*, consisting of head nodding, extension of the gular pouch, tail movements and color changes. The acrodont dentition of *A.*, however, sets them clearly apart from the *Iguanidae*. With few exceptions, *A.* are oviparous. The following examples are listed separately: Rainbow lizard, Rough-tailed agama, Spring-tailed lizards, Toad-headed lizards, Changeable lizards, Water dragons, Sail fin lizard, Bearded dragon, Flying dragon, Thorny devil, Frilled lizard.

Agamogony: Reproduction (see) without fertilization.

Agamous species, *agamo-species:* species which reproduce exclusively by parthenogenesis or other nonsexual means. Prokaryotic organisms appear to be A.s. (e.g. blue-green bacteria or *Cyanophyceae*), although the transfer of plasmids between bacterial cells (see F-plasmid) can be considered as a sexual process. A further example of A.s. is provided by rotifers of the order *Bdelloidea*, whose members are exclusively parthenogenetic females. See Species.

Agaontidae: see *Hymenoptera*.

Agar-agar, *agar:* a gel-forming polysaccharide, obtained from various red algae, mainly the East Asian genera *Gelidium* and *Gracilaria*. It consists of about 70% polygalactans, which in turn consist of about 70% agarose and 30% agaropectin. Agarose is a linear polymer of alternating D-galactose and 3,6-anhydrogalactose. Agaropectin consist of 1–3 glycosidically linked D-galactose units, of which some are sulfated. A. is commercially available as a dry powder or a fibrous preparation. It is used as a gelling agent for microbiological Growth media (see). Growth media containing 1.5–3.0% A. are liquid at 100 °C, and they gel at about 45 °C. As a gelling agent for solid growth media, A. is superior to gelatin because: a) A.-containing media are still solid at 37 °C, b) A. contains no nitrogenous com-

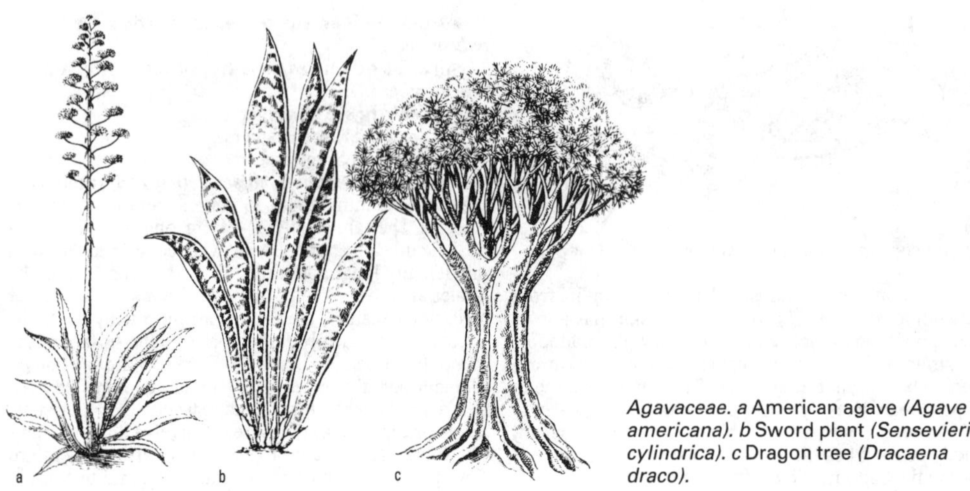

Agavaceae. a American agave *(Agave americana). b* Sword plant *(Sensevieria cylindrica). c* Dragon tree *(Dracaena draco).*

pounds, c) very few microorganisms are able to attack and therefore liquify A. It is also used as a gelling agent in the pharmaceutical and food industries.

Agaricales: see *Basidiomycetes.*

Agarics: see *Basidiomycetes.*

Aga toad: see *Bufonidae.*

Agavaceae, ***agave family:*** a family of monocotyledonous plants, containing 8 genera *(Agave, Furcraea, Yucca, Cordyline, Phormium, Polyanthes, Dracaena, Sansevieria)* and about 560 tropical and subtropical species. The family is regarded as intermediate between the *Liliaceae* and *Arecaceae.* Specimens may be many years old and are often lignified. They possess rhizomes, and some species attain a tree-like form, which is accompanied by anomalous secondary thickening. Their fleshy or leathery leaves are usually crowded at the base of the stem. Pollination is by insects, occasionally by birds. Berries or capsule fruits develop from superior or inferior ovaries.

The most species-rich genus is *Agave,* which is characterized by the rosette arrangement of its leaves. This genus is native to southern North America, Central America and northern South America. ***Mexican sisal grass*** *(Agave sisalanda)* is economically important as a source of very tough sisal fiber, and as a source of steroid saponins which are used commercially for the synthesis of pharmaceutically important steroids. The plant is now widely grown commercially in the tropics. Other agaves also yield commercially important fiber, e.g. ***keratto fiber*** (*Agave morrisii*), ***Bombay aloe fiber*** *(A. vivipara),* ***sisal hemp*** *(A. fourcroydes),* ***istle*** or ***Mexican fiber*** *(A. heteracantha).* ***American agave*** *(A. americana)* is now frequently found growing wild in the Mediterranean region. In Central America, the juice of *A. americana* is collected from the cut flower stem and fermented to yield pulque, a national Mexican drink. Distillation of pulque produces the spirit, mezcal (mescal).

Species of the genus *Yucca* ***(palm lilies)*** are native to southern North America and Central America, and some of these are grown as ornamental plants.

The genus *Dracaena* is mainly native to the Old World tropics, and its members attain tree-like forms, often of considerable size. Best known of these is the ***Dragon tree*** *(Dracaena draco)* of the Canary Islands. *D. draco* and *D. cinnabari* produce a red resin, known as dragon's blood.

Species of *Sansevieria* ***(bowstring hemp),*** mostly native to tropical Africa, are used as fiber plants, e.g. *S. cylindrica* from West Africa. Varieties of *S. trifasciata,* especially those with yellow leaf margins, are cultivated as ornamental sword plants.

New Zealand flax *(Phormium tenax),* native to New Zealand, is also an important fiber plant; its strong bast fibers are used to manufacture coarse matting and cordage. ***Mauritius hemp*** or ***fique*** *(Furcraea foetida),* originally from Central America, is used for the same purpose.

Agave: see *Agavaceae.*

AG-complex: an obsolete term for the totality of factors responsible for the development of male (A) and female (G) characters, once the sex of the organism has been established. Thus, the A and G factors are not themselves sex-determining factors. In a diploid cell, A and G each occur in duplicate as AAGG, while a haploid cell contains only one copy of each. Accordingly, each cell has the potential to develop as male or female (bisexual potential), but it develops in the direction dictated by certain sex-determinants. The AG-complex [now known as male (M) and female (F) factors] is not carried on the sex chromosomes, but on the autosomes (i.e. any chromosomes other than sex chromosomes), or their inheritance may be cytoplasmic. See Sex determination.

Age diagnosis:

1) in humans , the determination of the biological age of an individual, by comparing age-dependent characteristics with those of specific age groups. Such characteristics include the eruption of teeth and their degree of wear through chewing (dental age), the occurrence of secondary sexual characteristics (sexual maturity), bone formation, ossification of the epiphyseal sutures, ossification of the calvarium sutures, age-dependent bone degradation (see Ossification age), etc.

Biological A.d. may be necessary for medical purposes, if it is suspected that the difference between calender age and biological age is abnormally large. A.d. also plays a special role in forensic medicine in the identification of corpses and parts of corpses. In paleodemography, the age-related changes in earlier populations, and the age composition of these populations are determined by the study of skeletal remains. The following age ranges or stages can be differentiated.

a) *Infans I:* from birth until complete eruption of the first permanent molar or incisor (up to 6 years of life).

b) *Infans II:* from the eruption of the first permanent tooth to complete eruption of all definitive teeth, except the wisdom teeth (years 6–14).

c) *Juvenis:* from the end of infans II to the ossification of the sphenobasilar suture between the sphenoid and the basal part of the occipital bone (years 14–20).

d) *Adultus:* from the ossification of the sphenobasilar suture to the beginning a fairly widespread ossification of sutures in the calvarium. The beginning of wear in the masticatory apparatus, and the onset of age-dependent alterations in the postcranial skeleton (years 20–40).

e) *Maturus:* advanced ossification of the skull sutures, marked tooth wear, advanced atrophy of the alveolar border of the jaws, marked structural changes in the region of the pubic symphysis, expansion of the marrow cavity of the humerus and femur (up to 60 years).

f) *Senilis:* extreme, extensive ossification of the sutures of the calvarium, and very marked changes in the masticatory apparatus and postcranial skeleton.

By taking into account further details of dentition and skeletal ossification, a fairly accurate age assessment can be made for the pre-adult. From infans II onward, the age-related changes are different for male and female. Thus, the end of growth, which is marked by ossification of the cartilaginous sutures of the bones, occurs earlier in girls than in boys. With advancing age, there is increasing individual variation in the time of onset and rate of development of age-dependent physiological changes, so that more or less large differences may arise between the biological and the calendar age.

2) Estimation of the age of fossils, using stratigraphic, chemical or physical methods. See Dating methods.

Agemone: see *Papaveraceae.*

Agglutination: clumping of suspended antigenic particles due to interaction with antibodies. A. is used, e.g. in the determination of Blood groups.

Agglutinins: antibodies present in blood serum, which interact with particulate antigens and cause them to form clumps, i.e. they cause agglutination.

Aggregate culture: see Water culture.

Aggregation:

1) See Organism collective

2) Aggregation behavior is the gathering together of individuals of the same species in response to chemical signals released from the body, known as *aggregation pheromones.*

Aggressive behavior: a form of animal behavior defined by Konrad Lorenz (1966) as "the fighting instinct in beast and man which is directed against members of the same species". Other authorities allow the inclusion of interspecific behavior, such as the behavior of a predator toward its prey, but most are strict in the exclusion of such behavior from the definition. A.b. is a special case of Agonistic behavior (see), and in many respects it is almost synonymous with the latter. It involves attack and threat behavior, directed to the driving away of other individuals or the imposition of hierarchy. It symbolizes a readiness to do harm to the recipient, although the dispute may often be resolved without physical harm to either participant, e.g. by retreat of the threatened individual. Alternatively, if the distance between antagonists cannot be increased in any other way, the conflict may become physical rather than symbolic, leading to injury or death, sometimes involving the use of "weapons" such as teeth and claws, which are normally employed for food gathering. A.b. is an evolutionary adaptation, and each pattern and type of A.b. is typical of the species. In some species it is totally absent. It reaches a high and sophisticated level of development in certain social insects like termites and ants, which have evolved aggressive Castes (see) known as soldiers. A.b. has many different functions and many different causes, such as the exclusion of conspecific individuals from an established territory (see Territorial behavior), the hierarchical control of conspecifics by the establishment of dominance, the competition between males for females, disciplining of offspring by parents, and the protection of offspring against intruders. The term "Aggression" should be avoided.

Aggressive mimicry: see Protective adaptations.

Aggressiveness: the degree of pathogenicity as manifested, e.g. by the frequency of lesions or pustules on the host, the quantity of spores produced by a pathogen, or its general rate of development (i.e. the rate at which the host succumbs to the attack). Thus, under suitably humid conditions, the potato blight fungus, *Phytophthora infestans*, can become very aggressive, producing masses of asexual spores that rapidly infect neighboring plants, and sending hyphae throughout the plant, eventually into the tubers themselves, thereby devasating entire potato crops.

Aging: see Lifespan.

Aging of lakes: the gradual infilling of Lakes (see). The process normally starts with the deposition of peat or organic mud derived from the residues of plankton, animals and higher plants. As the water becomes shallower, the lake bottom finally becomes illuminated by penetrating sunlight. Rooted aquatic plants are then able to colonize its entire surface, which leads to an increase in the rate of elevation of the lake bottom. The lake is thereby transformed into a Pond (see), which becomes progressively smaller as its edges rise above the water level, due to the accumulation of the persistent organic remains. The area in question gradually becomes a *raised bog* or *mire,* as it becomes colonized by plants with aerial foliage. If the entire body of water disappears, except for small residual pools, it is called a *low moor bog* (see Bog). As a final stage, the bog may become an *alder swamp*. This develops into an alder wood or *carr,* which becomes progressively drier on its surface. Other pioneer plants also become established, and eventually (after thousands of years) the area becomes dry land.

Agkristrodon: see Halys viper, Water moccasin.

Aglaonema: see *Araceae.*

Aglossa: see *Gastropoda, Anura.*

Aglycon, *aglycone, genin:* the noncarbohydrate part of a glycoside. Aglycons are released by hydrolysis (e.g. by acid or enzymes) of the C-, N-, or S-glycosidic linkage. See Glycoside.

Aglyphic dentition: see Snakes.

Agmatoploidy: an increase in the chromosome number, due to fragmentation of holocentric chromosomes. Since all chromosome segments are capable of migration, the fragments in this special case are not eliminated during nuclear division. A. is a form of Pseudopolyploidy (see).

Agnatha: a class or superclass of primitive, fish-like craniates without jaws, consisting of the Lampreys (see), Hagfishes (see) and their extinct relatives. Lampreys and hagfishes are placed in the subclass or class, *Cyclostomata* (see). With respect to their fossil history, the *A.* are the oldest fish-like vertebrates. They arose from several parallel evolutionary lines, possibly with a common origin in the Ordovician. Cephalization is less complete than in higher forms, and a jaw apparatus is lacking. The skull is, in fact, unique and difficult to homologize with those of other craniates. The vertebral column is feebly developed and the notochord persists. Movement of the round mouth resembles the opening and closing of a photographic lens diaphragm. Some forms have primitive pectoral fins, but paired fins are otherwise lacking. They were most abundant in the Silurian and Lower Devonian. See *Cyclostomi*.

Agnostic behavior: a general term for all behavior associated with conflict between individuals, including attack and escape.

Agnostida: see *Agnostus*.

Agonistic behavior: behavior leading to the resolution of conflict (i.e. regulation of the distance) between competing or rival animals striving for the same objects or resources in their environment, e.g. dwelling holes, egg laying sites, tree perches, females, etc. A.b. is often linked to Territorial behavior (see). It normally occurs between members of the same species, but may rarely occur between members of closely related species with overlapping environmental demands. A.b. represents a complex of attacking, threatening, submissive and fleeing behavior, and within this complex the limitations of Aggressive behavior (see) are difficult to specify. See Threat behavior. See Plate 26.

[Scott, J.P. & Fredericson, E. (1951) *Physiol. Zool.* ***24***, 273–309]

Agnostus (Greek *agnostos* unrecognizable), *Peronopsis:* a genus of small trilobites, 5–10 mm in length. A. is found worldwide in the Upper Cambrian, and is therefore a useful index fossil. Cephalon and pygidium are of approximately equal size and of simple construction, and there are only two thoracic segments. Eyes and facial suture are absent. A. is a member of the trilobite order, *Agnostida*, members of which rarely exceed 1 cm; eyes and facial suture are absent from most genera, and some members possess three thoracic segments. Since the evolutionary trend of trilobites appears to be towards an increase in the number of fused segments in the pygidium and a reduction in the number of thoracic segments, the *Agnostida* are thought to be relatively advanced.

Agouti: see *Dasyproctidae, Agoutidae*.

Agoutidae, ***pacas:*** a family of Rodents (see) comprising a single genus *(Agouti)* with 2 species, inhabiting the banks of streams and rivers in tropical forests from central Mexico to southern Brazil. These large herbivores weigh up to 12 kg, and they are active burrowers and good swimmers. The body is patterned with conspicuous white spots and stripes, the legs are short and the head blunt. Four digits are present on each forefoot, and 5 on each hindfoot. Part of the zygomatic arch is enlarged and contains a large sinus, a feature not found in any other mammal. Sound chambers, formed from the widened zygomatic arch and concavities in the maxillaries, enable the paca to emit a resonant rumble. Dental formula: 1/1, 0/0, 1/1, 3/3 = 20.

Agridiidae: see *Odonatoptera*.

Agriocharis: see *Meleagrididae*.

Agrionemys: see *Testudinidae*.

Agrionidae: see *Odonatoptera*.

Agrobacterium: a genus of motile, Gram-negative, aerobic bacteria, living in the soil or as plant parasites. *A. tumefaciens* is the causative agent of crown gall, a tumor-like growth consisting of undifferentiated, callus-like tissue. Infection of the plant occurs via external wounds; fruit trees and sugar beet are especially susceptible. During infection, DNA from a bacterial plasmid [plasmid Ti (tumor-inducing)] becomes stabily integrated into the plant nuclear DNA. Expression of the bacterial genes within the host plant DNA results in the localized synthesis of excessive amounts of auxin, so that the plant tissue becomes dedifferentiated. Discovery of the Ti plasmid of *A.* has provided plant molecular biologists with an excellent system for the genetic engineering of transgenic plants with novel characters.

Agrocoenosis, ***agrobiocoenosis, agrarian ecosystem:*** a farmland ecosystem of agricultural crops, field weeds and animals. For economic purposes, the plant stock of this coenosis is under human control, through soil management, plant care and protection against diseases and pests, and the regular removal (complete or partial) of crops. The species composition of an A. is altered routinely and at regular intervals, thus imposing a Succession (see) that is unnatural, and restricting the self-regulation of the coenosis. Examples of As. are root crop and cereal fields, fruit orchards, fodder fields, vineyards and agricultural grassland, all of which are subject to varying degrees of human management. Attention to the ecological relationships within As. is important when considering measures for increasing the agricultural yield.

Agrostis: see *Graminae*.

Agrotis: see Cutworms.

Agulus: see *Branchiura*.

Ahaetulla: see Long-nosed tree snake.

AIDS, ***acquired immune deficiency syndrome:*** a syndrome, first identified in 1981 in Los Angeles, and which at the time of writing (1994) has always proved fatal. There is no known cure. The causative virus (HIV: Human Immunodeficiency Virus; a RNA virus or retrovirus) destroys a lymphocyte subgroup, known as T4 lymphocytes, which normally regulate the immune system, protecting the body from disease organisms and from the development of certain tumors. HIV therefore destroys the body's ability to defend itself, and AIDS patients die from opportunistic infections which in healthy individuals are easily combatted by the immune system. Opportunistic infection by the protozoan, *Pneumocystis carinii,* which causes pneumonia, and tumors such as Kaposi's sarcoma, are particulary common in AIDS.

HIV enters T4 lymphocytes by binding to a cell surface protein known as CD4, followed by fusion of the virus with the lymphocyte membrane. Viral RNA is transcribed into DNA with the aid of reverse transcriptase, and this DNA is integrated as a provirus into the host genome. After more or less prolonged periods of dormancy, this provirus reorganizes cell metabolism to synthesize new virus particles. The T4 cell is killed as newly synthesized viruses bud from the cell, leaving a perforated cell membrane. The estimated average incubation period between infection and the manifestation of AIDS is about 10 years, although it may be much shorter (months) in some people and especially in children. It is possible that some infected individuals never manifest AIDS.

Acute infection causes antibody production, and the detection of these antibodies shows that an HIV infection has occurred. This acute phase, involving seroconversion (antibody production) may or may not be followed by the chronic phase of the disease. Onset of the chronic phase is usually marked by the *AIDS-related complex (ARC),* with fever, weight loss, diarrhea, general fatigue and night sweats, followed by fully developed AIDS.

AIDS is essentially a sexually transmitted disease. It is also transferred from mother to fetus in the uterus, or via maternal blood during birth. Infected blood products for transfusion and for the treatment of hemophiliacs have also been responsible for AIDS transmission, but this danger has now been virtually elimated by blood processing techniques. Sharing of syringe needles among drug addicts is a notorious transmission route. Since the virus is fragile and does not survive easily outside the body (except, e.g. in blood and other body fluids), normal social contact with AIDS patients carries no risk of infection.

Ailuroedinae: see *Ptilinorhynchidae*.

Ailuroedus: see *Ptilinorhynchidae*.

Ailurus: see Red panda.

Airbreathing catfishes: see *Clariidae*.

Air sac: see Resonance sac.

Airsac catfishes: see *Heteropneustidae*.

Aix: see *Anseriformes*.

Aizoaceae, *Ficoidaceae, **stone flower family:*** a family of dicotyledonous plants containing about 2 500 species. The main evolutionary dispersal centers are the dry areas of Australia and especially of southern Africa, with a few species in America and Asia. Members of this family are almost entirely leaf-succulent herbs, usually with large and conspicuously colored, insect-pollinated, hermaphrodite, actinomorphic flowers. The numerous petals are derived from the outermost of the many stamens, and they are inserted into the calyx tube in one or more series. The fruit is usually a 2-locular to multilocular, woody capsule, which opens only when moist (hydrochastic). Members of this family show many adaptations to extremely dry habitiats, e.g. reduction of transpiration by the presence of a wax coating or a layer of dense hairs. In "living stones", especially of the genus *Lithops*, the vegetative body is reduced to two spherical leaves that closely resemble stones, practically indistinguishable from the surrounding pebbles, serving both as a water-conserving adaptation and as a protective camouflage against animal predation. Members of the genus *Mesembryanthemum* are popular garden plants for sunny sites, their brilliantly colored flowers opening only in direct sunlight. *Mesembryanthemum cristallinum **(Ice plant)*** is grown both as an ornamental and as a spinach-like vegetable; the entire upper surface of its vegetative body is covered with water-filled papillae resembling beads of ice. *Mesembryanthemum edule **(Hottentot fig),*** whose fruit is fleshy and edible, is native to southern Africa, but it is naturalized in many warmer, temperate regions, including Southwest England. The family is closely related to the *Cactaceae* and the *Portulaceae*.

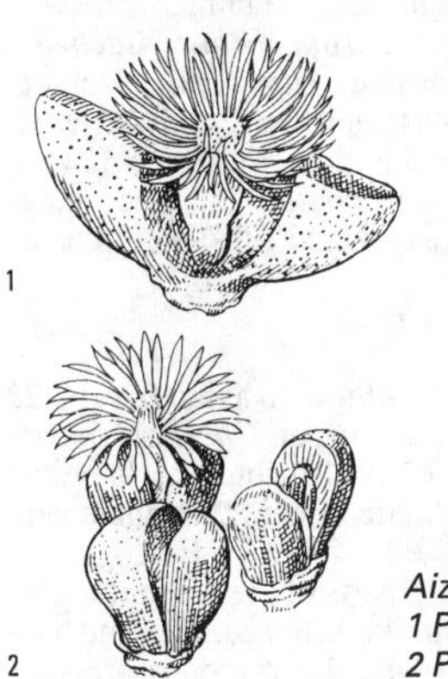

Aizoaceae. "Living stones". *1 Pleiospilos bolusii. 2 Pleiospilos hilmarii.*

Ajaia: see *Ciconiiformes*.

Akepa: see *Drepanididae*.

Akialoa: see *Drepanididae*.

Akinesis, *akinesia:* a reflex inhibition of motility, displayed mainly by arthropods and vertebrates. A. is usually a protective response to perceived danger. Examples are: 1. the total immobility ("freezing") of ground nesting birds, which increases the effectiveness of their existing color and pattern camouflage, and 2. the stiffness or rigidity (feigned death) of insects (especially certain beetles) and spiders in response to shock. See Protective adaptations. Medically, the term is applied to various conditions in which there is an inability to initiate movement.

Ala: the commonly accepted 3-letter abbreviation for the amino acid, alanine.

Alabetidae: see *Gobiesocidae*.

Alanine, *Ala, aminopropionic acid:*

1) L-α-alanine: CH_3—$CH(NH_2)$—COOH, a proteogenic, glucogenic amino acid, found in almost all proteins, and one of the main components of silk fibroin.

2) β-alanine: NH_2—CH_2—CH_2—COOH, a nonproteogenic amino acid, which occurs in the free form in brain, and in combined form in coenzyme A and certain dipeptides (anserine, carnosine). β-A. is produced during the reductive degradation of pyrimidines, and by the decarboxylation of aspartic acid.

Alarm signals: see Protective adaptations.

Alaska blackfish: see *Umbridae*.

Alaskan brown bear: see Brown bear.

Alauda: see *Alaudidae*.

Alaudidae, *larks:* an almost exclusively Old World avian family of 75 ground-nesting species in the order *Passeriformes* (see). Larks are quite uniform in general appearance and habits. They are small (13–23 cm in length), mostly with drab plumage. Most species have a sightly decurved bill and a crest or tuft on the head. Usually the sexes are fairly similar, the female sometimes being slightly smaller and duller. More pronounced sexual dimorphism is shown, e.g. by the African

Finch-larks *(Eremopterix)*. Courting or territory-marking males sing from a high perch or in flight, sometimes extremely high above their territory. Two characteristics set larks apart from all other passerines: a) the back of the shank (tarsus) is scaled and rounded, instead of sharp and unsegmented; b) they lack a pessulus (a cartilaginous or bony bar across the base of the trachea, which separates dorsoventrally the medial tympaniform membranes of the 2 sides). Most larks prefer open, treeless habitats, e.g. beaches, moors, grassy plains, deserts, etc., although the members of the genus, *Mirafra* (containing mainly the African ***Bush larks***), regularly perch on bushes or posts. Larks feed on seeds and other plant materials, larvae, snails, insects, pupae, etc. The ***Skylark*** *(Alauda arvensis)*, immortalized by poets for its song, is found from Britain right across Asia to Kamchatka, and from northern Siberia to India and North Africa. It has been introduced into New Zealand, the Hawaiian Islands, and Vancouver Island off British Columbia. The very widely distributed ***Horned*** or ***Shore lark*** *(Eremophila alpestris;* Eurasia, North Africa, North America, Mexico, Colombia) is the only native member of the family in the New World. Two other well known genera are *Galerida* (e.g. *G. magnirostris,* ***Thick-billed lark;*** 3 races in South Africa extending north to the Transvaal and the Drakensburg mountains) and *Lullula* (e.g. *L. arborea,* ***Wood lark;*** Palaearctic).

Albany pitcher plant: see Carnivorous plants.

Albatrosses: see *Procellariiformes.*

Albinism: the genetically determined absence of pigmentation in the skin, hair, feathers, eyes, etc. of animals. In partial A., pigment is absent only from certain sites of the body. The most common type of A. in humans (oculocutaneous A.) results in white hair, pink skin and an extreme photophobia owing to the lack of pigment in the eyes. A. occurs in all races; in Europeans, it occurs with a frequency of 1:30 000. Total A. is almost always an autosomal recessive character, whereas partial A. is predominantly autosomal dominant. A. is due to the lack of melanin (see Melanins), which in turn is usually due to the absence of the enzyme, tyrosinase, which catalyses the conversion of L-tyrosine into melanin.

Albumin glands: glands in the female genital ducts of e.g. snails, elasmobranch fishes and sauropsids, which contribute to the formation of the egg envelope.

Albuminous glands: glands or glandular cells in the oviducts of amphibians, which secrete the tertiary egg envelope, i.e. the outer layer around the amphibian egg which swells into a gel on contact with water.

Albumins: a group of simple proteins, present in the body fluids and tissues of animals and in some plant seeds. In contrast to the globulins, they have a low molecular mass, they are water-soluble and easily crystallizable, and they have high contents of acidic amino acids. They can be precipitated with neutral salts, but only when these are used in high concentrations. Amino acid compositions show 20–25% glutamate and aspartate, up to 16% leucine and isoleucine, and very little glycine (1%). Important representatives of the group are serum albumin, α-lactalbumin (a milk protein) and ovalbumin (an egg protein) from animals, and the poisonous ricin (from *Ricinus* seeds), leucosin (from seeds of wheat, rye and barley) and legumelin (from legume seeds).

Alburnus: see Bleak.

Alca: see *Alcidae.*

Alcaligenes: a genus of Gram-negative, rod-shaped or coccal bacteria (0.5–1.2 μm × 0.5–2.6 μm), motile with 1–4 (occasionally up to 8) peritrichous or degenerate flagella. Cells usually occur singly, and no resting stages are known. All species are strict aerobes, although some strains are capable of utilizing nitrate or nitrite as a terminal electron acceptor. They never display fermentative metabolism. Casein, gelatin, cellulose, chitin and agar are not hydrolysed. An alkaline reaction is given in the litmus milk test. No nitrogen-fixing members are known. All species give a positive oxidase test (Kovacs' test). Their inability to grow at pH 4.5 can be used diagnostically to distinguish them from *Acetobacter.* The G + C content of the DNA of different species lies in the range 57.9–70 moles%. No pathogenic species are known. They commonly occur as saprophytic inhabitants of the vertebrate intestinal tract, in dairy products and rotting foods, and they are involved in decomposition and mineralization processes in freshwater, marine and terrestrial environments.

The type species is *Alcaligenes faecalis* (Castellani and Chalmers 1919): G + C content of DNA 58.9 moles%; cocci or coccal rods, 0.5 × 0.5–2.0 μm; motile with 1–8 peritrichous flagella; arsenite oxidation is very active in some strains, but it does not support chemolithotrophic growth; simple carbon sources (acetate, propionate, butyrate, etc) and simple nitrogen sources (ammonium salts or nitrates) support growth. *A. faecalis* is a permanent member of the intestinal flora of humans and other vertebrates. Some strains produce a characteristic strawberry-like odor.

Alcedinidae: see *Coraciiformes.*

Alcedininae: see *Coraciiformes.*

Alcedo: see *Coraciiformes.*

Alcelaphinae: subfamily of the *Bovidae* (see), containing 2 tribes.

1. Tribe: *Alcelaphini;* genera: *Alcelaphus* ***(Hartebeests;*** 8 species) and *Damaliscus* ***(Sassabies;*** 8 species).

2. Tribe: *Connochaetini;* genus: *Connochaetes* ***(Wildebeest*** or ***Gnu;*** 2 species).

Damaliscus dorcas (head plus body 140–160 cm) has horns up to 47 cm in length, and very distinct white markings on the legs, chest and abdomen, a white escutcheon, and notably a white blaze on the face. It was nearly exterminated in South Africa by hunting, and the subspecies *D.d. dorcas* ***(Bontebok)*** still exists only in small numbers in protected areas. *D.d. philippsi* ***(Blesbok)*** has, however, made a better recovery.

Gnus have a characteristically broad muzzle and a horse-like tail, and both sexes have heavy, recurved horns. Head plus body length is 150–240 cm. Two species are recognized, both inhabiting open grassy plains in Africa. *Connochaetes gnou* ***(Black wildebeest, White-tailed wildebeest;*** central and eastern South Africa) is generally brown to black in color, with tufts of long black hair protruding from the muzzle, throat and chest. There is an erect mane, and the tail is white with a black base. *Connochaetes taurinus* ***(Blue wildebeest, Brindled gnu, White-bearded wildebeest;*** southern Kenya and southern Angola to northern South Africa) is generally silvery gray in coloration, with brown bands on the neck, shoulders and foreparts. The face, mane, beard and tail are black (Kenyan and Tanzanian subspecies have a white beard).

Alcelaphus: see *Alcelaphinae.*

Alces: see Moose.

Alcidae, ***alcids, auks and allies:*** an avian family (23 species in 14 genera) in the order *Charadriiformes* (see). They are Holarctic in distribution, breeding around the fringes of the Arctic Ocean, their main center of distribution being the Bering Sea. Ecologically, they are the northern counterparts of the Antarctic penguins, both groups being similar in appearance and habits. Phylogenetically, however, the 2 groups are very distant. In addition, all living alcids can fly (a fast whirring flight), whereas penguins cannot. Alcids are heavy-bodied swimming birds, which use their wings underwater in pursuit of fish. On land they appear clumsy, because the legs are set far back on the body. The recently extinct ***Great auk*** *(Pinguinus impennis;* 76 cm; islands of the North Atlantic), although flightless, had well developed flight feathers. Bill shapes vary considerably, e.g. the stubby bill of the ***Little auk*** or ***Dovekie*** *(Plautus alle;* 20 cm; Greenland to Novaya Zemlya), the razor-like bill of the ***Razorbill*** *(Alca torda;* 40.5 cm; North Atlantic coasts), the dagger-shaped bill of the Guillemots, e.g. the ***Common guillemot*** or ***Common murre*** *(Uria aalge;* 43 cm; coasts of North Atlantic and North Pacific) and the familiar red and yellow, parrot-like bills of the Puffins, e.g. the ***Common puffin*** *(Fratercula artica;* 30.5 cm; North Atlantic coasts). Guillemots and Razorbills breed on narrow ledges on seaward facing cliffs. They make no nest, and lay a single egg on bare rock. Puffins nest in burrows, which they dig about 1 m into the turf of grassy slopes near the sea, and they lay a single egg. In the postnuptual molt, the bright red and yellow outer covering of the puffin bill is lost. It is formed again the following year in readiness for the breeding season. The Little auk nests in crevices of rocky cliffs, often at some distance from the sea.

Alcids: see *Alcidae.*

Alcohol dehydrogenase, ***ADH:*** a zinc-containing oxidoreductase, which in the presence of NAD^+ reversibly oxidises primary and secondary alcohols to the corresponding aldehydes and ketones. ADH occurs in bacteria, yeasts, plants and the liver and retina of animals. The ADH from yeast, which is distinguished by its high affinity for ethanol, is of practical significance as the last enzyme in the pathway of Alcoholic fermentation (see). In the liver, ADH converts ethanol into acetaldehyde, representing the first stage in the removal of ethanol from the blood and the body. ADH in the retina, however, serves to convert the vitamin A aldehyde, all-*trans*-retinal, to retinol (see Vitamins). ADH from animal organs and yeast has low substrate specificity, and it dehydrogenates both short (C2 to C6) and long-chain alcohols. Yeast ADH (M_r 145 000) consists of four catalytically active, zinc-containing subunits (M_r 35 000) with four NAD^+ or NADH binding sites per molecule of ADH. The dimeric horse liver enzyme (M_r 80 000) contains two zinc atoms (one is essential for catalysis) and one coenzyme binding site per subunit (M_r 40 000).

Alcoholic fermentation: conversion of sugars to ethyl alcohol (ethanol) by microorganisms, especially yeasts, according to the equation: $C_6H_{12}O_6$ (hexose) + $2P_i$ + 2ADP → $2CH_3CH_2OH$ (ethanol) + $2CO_2$ + 2ATP + $2H_2O$; where P_i is inorganic phosphate. A.f. occurs under anaerobic conditions, and its energy yield of 2 mol of ATP per mol of glucose is lower than for aerobic respiration. A.f. can also be carried out by the tissues of higher plants, e.g. carrots and maize roots. There is no A.f. in animals, which lack pyruvate decarboxylase (see below). Within the cell, the starting point for A.f. is glucose 6-phosphate, which is converted to pyruvate by the reactions of glycolysis. Pyruvate is then decarboxylated by the action of pyruvate decarboxylase, and the resulting acetaldehyde is reduced to ethanol by alcohol dehydrogenase. In the presence of oxygen, the glycolytic pathway still proceeds, but the pyruvate is not decarboxylated by pyruvate decarboxylase. Instead, it undergoes oxidative decarboxylation to CO_2 and acetyl-CoA, the latter entering the tricarboxylic acid cycle. Operation of the tricarboxylic acid cycle and the associated respiratory chain gives a relatively high yield of ATP, which regulates the activity of phosphofructokinase (an early enzyme in glycolysis) and slows down the rate of glycolysis (the important regulatory factor is the ratio of [ATP] to [ADP] + [AMP]). Thus, oxygen not only inhibits A.f., but also decreases the rate of consumption of glucose. This inhibitory effect of oxygen, first described by Pasteur, is known as the Pasteur effect (see).

Human society has used A.f. for thousands of years for the preparation of alcoholic beverages; beer was in fact brewed by the ancient Babylonians. Modern brewing uses carefully selected strains of different yeasts, especially *Saccharomyces cerevisiae* (see Beer yeast, Wine yeast). The most important substrates for A.f. are the monosaccharides D-fructose, D-glucose, D-mannose and sometimes D-galactose. In some cases the disaccharides sucrose and maltose and the polysaccharide starch are employed as substrates. Convenient starting materials for the preparation of alcoholic beverages are e.g. malt, fruit juices, molasses, potatoes, etc. For industrial ethanol production, A.f. is also performed on sulfite wastes from the paper industry and on wood hydrolysate. Bacteria as well as yeasts are involved in the A.f. of Pulque (see).

Historical. The simple equation: 1 glucose → $2CO_2$ + 2 ethanol was established in 1815 by Gay-Lussac. In 1857, Pasteur proposed that A.f. could only be carried out by living organisms (vitalistic theory of fermentation). This was disproved in 1897 by Buchner, who established that a cell-free filtrate of disrupted yeast cells was capable of A.f. This discovery marked the beginning of modern enzymology. The enzyme system responsible for A.f., which was originally thought to be a single enzyme, was called zymase. In 1905, the role of phosphate in A.f. was described by Harden and Young. In 1912, Neuberg proposed the first fermentation scheme, which was revised in 1933 by Embden and Meyerhof.

Alcohols: compounds whose characteristic functional group is hydroxyl (—OH). The simplest A. are hydroxylated derivatives of hydrocarbons. According to the position of the OH-group, the alcohol is a primary alcohol (R—CH_2OH), a secondary alcohol (R_1R_2—CHOH), or a tertiary alcohol ($R_1R_2R_3$—COH). Polyhydric A. contain more than one hydroxyl group, e.g. glycerol is a trihydric alcohol, containing three OH-groups. Waxes are esters of long chain fatty acids with long A. Many essential oils contain volatile esters of A. with fatty acids. Triacylglycerols (triglycerides) are esters of fatty acids with glycerol. A number of lower A., e.g. ethanol, are formed from carbohydrates (themselves polyalcohols) and/or proteins by fermentative processes.

Alcyonacea: see *Alcyonaria.*

Alcyonaria, *Alcyonacea,* ***soft corals:*** an order of the *Octocorallia* (see), containing about 800 species, of which the majority are colonial. Colonies are stiffened by a skeleton of small, separate, calcareous sclerites or spicules. The coenenchyme forms a rubbery mass. Some colonies grow as incrustations, while others have a mushroom shape. An example is *Alcyonium digitatum* ***(Dead mens's fingers),*** the broad lobed masses of which are found attached to stones below the low tide mark.

Alcyonium: see *Alcyonaria.*

Aldabrachelys: see *Testudinidae.*

Aldanophyton antiquissimum: a fossil plant with shoots up to 13 mm wide and 8.5 cm long, covered with microphyllous leaves attaining a length of 9 mm. Discovered in 1963 by Kryshtofovich in the Middle Cambrian of the Aldan mountain range in Siberia, this supposedly lycopodiaceous species is the oldest known land vascular plant plant. The reproductive organs are not known.

Alder: see *Betulaceae.*

Alderflies: see *Megaloptera.*

Aldonic acids: monocarboxylic acids derived from aldoses (sugars containing an aldehyde group) by oxidation of the aldehyde group. The names of these acids are formed by replacing the "-ose" of the parent aldose with "-onic acid". A.a. may form a 1,4 lactone ring (γ-lactones) or the more stable 1,5 lactone (δ-lactones). Some naturally occurring A.a. are L-arabonic acid, xylonic acid, D-gluconic acid, D-mannonic acid and galactonic acid.

Aldoses: monosaccharides characterized by the presence of a terminal aldehyde group (—CHO), which is always given the number 1 in systematic nomenclature. The A. are formally related to their simplest representative, glyceraldehyde, by chain extension. According to the number of carbon atoms in the molecule, A. are classified as trioses, tetroses, etc. The aldopentoses and aldohexoses are particularly important biochemically. Glucose is both an aldose and a hexose (i.e. an aldohexose). D-ribose, which is present in ribonucleic acid (RNA), is an aldopentose.

Aldosterone: 11β, 21-dihydroxy-3,20-dioxopregn-4-en-18-al, the most important mineralocorticoid hormone. It is produced in the adrenal cortex, and it is 1000 times more active than cortisol. In contrast to other adrenal cortex hormones, A. has a carbonyl group on C-18, which forms a hemiacetal with the 11β-hydroxyl group. It promotes increased resorption of sodium ions and increased excretion of potassium ions in the kidney tubule. A. also exhibits low glucocorticoid activity, which is expressed primarily as an influence on liver glycogen storage. Synthesis of A. (about 0.3 mg/day) in the human adrenal cortex is regulated by the renin-angiotensin system, in which the octapeptide, angiotensin II, induces A. synthesis.

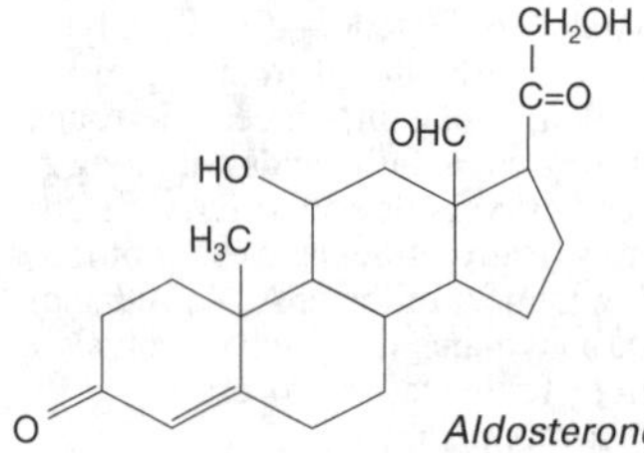

Aldosterone (aldehyde form)

Aldrovanda: see Carnivorous plants, *Droseraceae.*

Alecost: see *Compositae.*

Alectura: see *Megapodiidae.*

Alepas: see *Cirripedia.*

Aleppo galls: see Tannins.

Alethe: see *Turdinae.*

Aleuriospore: one of the spore types found in certain conidial fungi. The A. is a terminal portion of a hypha that becomes separated early by a septum from the parent hypha. After spore maturation and separation from the conidiophore, the vacated conidiophore does not normally produce a new spore immediately. If a new spore is formed directly, then it does not occur at the same level as the preceding spore. The term is also applied to mycelial spores that arise from hyphae and are not released, but remain until the hyphae die.

Aleurites: see *Euphorbiaceae.*

Aleurodoidea: see *Homoptera.*

Alfalfa: see *Papilionaceae.*

Alfalfa mosaic virus: see Multipartite viruses.

Alfaroa: see *Juglandaceae.*

Algae, *Phycophyta:* a phylum of the plant kingdom, containing about 33 000 species. The phylum comprises variously colored, autotrophic, unicellular or multicellular plants. Some members live as plankton or benthos in salt or fresh water, and others are primitive moisture-dependent plants, in which the vegetative body is a thallus; the sexual organs are usually unicellular and never enclosed by structures of sterile cells. The nucleus of each algal cell is well defined and possesses a nuclear membrane; mitochondria are also present. The A. are therefore eukaryotes, and they differ from the prokaryotic blue-green algae (more correctly called blue-green bacteria) and other *Schizophyta,* in which the nuclear material is not enclosed by a membrane, and in which mitochondria are absent. The flagella of A. also have typical eukaryotic structure. Algal chromatophores usually contain Pyrenoids (see).

The many different types of A. probably evolved separately at an early stage from primitive flagellates. Affinities between the different groups are still recognizable.

Evolution of the thallus can be considered as proceeding from the unicellular A., via unbranched then branched filamentous forms, finally leading to the formation of a three-dimensional thallus, in some cases showing tissue differentiation.

An evolutionary sequence can also be traced from the simple, exclusively haploid A., via A. which show an alternation of haploid and diploid plants, finally to reduction of the haploid phase and the existence of a single diploid plant. Sexual reproduction shows a progression from primitive isogamy, via anisogamy, to the more highly evolved oogamy.

Nonsexual reproduction also occurs, usually by flagellated zoospores, and in some groups by nonmotile spores.

Certain A., especially the larger species, are used as vegetables or other types of food, animal feedstuffs, and fertilizer. They also provide raw material for the preparation of medically important and cosmetic products; they have various other technological uses, and in some cases are of considerable economic importance.

A. are separated into a number of classes, mainly according to their pigments and other constituents, then further subdivided on the basis of their level of organization. Classification is problematic and it is treated differently by various authorities.

Strasburger (1978) divides the A. into 10 classes: *Euglenophyceae* (see Phytoflagellates), *Cryptophyceae* (see Phytoflagellates), *Pyrrhophyceae* (see Phytoflagellates), *Haptophyceae* (see Phytoflagellates), *Chrysophyceae* (see), *Bacillariophyceae* (see Diatoms), *Xanthophyceae* (see), *Chlorophyceae* (see), *Phaeophyceae* (see), *Rhodophyceae* (see).

Algal fungi: see *Phycomycetes.*

Algarroba: see *Mimosaceae.*

Algicide: a plant protection agent or a water purifying agent for the destruction of algae. On part per million (1 ppm) of copper sulfate is regularly employed for the control of algae in reservoir waters. In Australia, 5 ppm copper sulfate is added to rice irrigation water, to control *Spirogyra, Lyngbya* and *Phormidium,* which interfere with the growth of rice seedlings. Examples of organic algicides are 2,3-dichloronaphthoqinone, which is toxic to "bloom"-producing *Cyanophyta* at 5 µg/l; and phenanthraquinone, which is effective against bloom-producing species and *Chlorella* at 80 µg/l.

Alginic acid: a polyuronide, M_r in the range 12 000–120 000, which is a chain polymer of 1,4β-glycosidically linked D-mannuronic acid and 1,4α-glycosidically linked L-guluronic acid. A. is a characteristic, structurally important mucilage component of brown algae, from which it is extracted with NaOH. The prepared material can absorb up to 300 times its weight of water. It is used primarily in the food industry (thickening agent in soups, custard powder, ice cream), in dentistry (for making impressions), in medicine (resorbable sutures), in pharmacy (emulsions, dispersions, ointment bases, and appetite supressant), in the cosmetic industry, in the preparation of textiles (finishing agent), and in the manufacture of paper goods and finishing inks.

Algonkium (from Algonkin, a tribe of Canadian Indians): an obsolete term for the early Precambrian. It is more or less equivalent to the Proterozoic or the transition from Middle to Upper Precambrian (about 2 000 million years ago). See Geological time scale.

Alimentary oocyte nutrition: see Oogenesis.

Alimentary tract: see Digestive system.

Alismatidae: see *Monocotyledoneae.*

Alisterus: see *Psittaciformes.*

Alizarin: 1,2-dihydroxyanthraquinone, a red dye. In the root of the madder plant *(Rubia tinctorum* L.) and other *Rubiaceae,* it is present in combination with 2 moles of glucose, forming the compound ruberythic acid. Prepared from the madder plant, A. used to be an important natural dye, but since 1871 it has been prepared by chemical synthesis. Several derivatives are also important as alizarin dyes. A. was isolated in 1826 by Colin and Robiquet, and its structure was determined in 1868 by Graebe and Liebermann. The first technical syntheses were developed independently by Caro and W.H. Perkin.

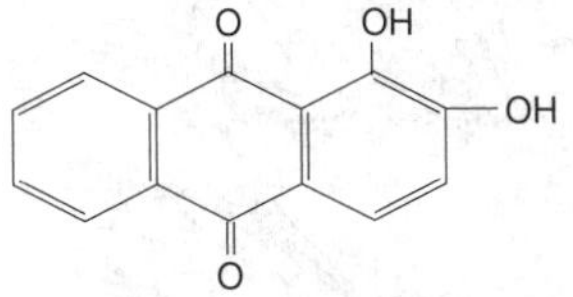

Alizarin

Alkali plants: see Calcicoles.

Alkali reserve: the quantity of alkali present in the blood, which is necessary for maintaining the blood pH. A more exact measure of blood buffering (see) than the A.r. is the Standard bicarbonate (see) of the plasma.

Alkaloids: an important group of secondary natural products, named for their basic (alkali-like) properties. They occur mainly in plants, but a few are also found in animals (e.g. toad and salamander toxins) and in microorganisms. A. contain nitrogen, usually in a heterocyclic ring structure, and they display a variety of physiological properties. The majority of A. are stored in peripheral plant parts (roots, leaves, fruits, bark), as the salts of organic acids. Often a plant contains several structurally related A. Certain plant families are notable for their high content of A., e.g. *Papaveraceae, Solanaceae, Ranunculaceae, Apocynaceae, Papilionaceae.* Closely related plants often contain similar A., so that in certain cases the A. of a plant can be used to clarify taxonomic problems (see Chemotaxonomy)

Isolation of A. involves alkalization of macerated plant material to release the A. from their salts, followed by extraction of the A. with organic solvents. Several further steps are then usually necessary to obtain pure A. (e.g. various chromatographic methods, recrystallization). Liquid A. are often separated by steam distillation.

According to their chemical structure, A. can be classified into several groups; some of these are: ***1) pyridine and piperidine A.,*** e.g. Nicotine (see), coniine, Lobeline (see); ***2) quinoline A.,*** e.g. Quinine (see), quinidine, cinchonine; ***3) isoquinoline A.,*** e.g. Morphine (see), Codeine (see), Papaverine (see), Narcotine (see); ***4) tropane A.,*** e.g. Atropine (see), hyoscymine, Cocaine (see), Scopolamine (see); ***5) purine A.,*** e.g. Caffeine (see), Theobromine (see), Theophylline (see); ***6) indole A.,*** e.g. Ergot A. (see), Reserpine (see), Physostigmine (see); ***7) strychnine A.,*** e.g. Strychnine (see), Brucine (see); ***8) steroid A.,*** e.g. Solanine (see), Tomatine (see), Veratrum A. (see).

A. may also be classified according to their origin (e.g. Tobacco A., *Solanum* A.), or according to their main biosynthetic precursors, e.g. A. derived biosynthetically from anthranilic acid. There is often a close correlation between the chemical structure of A. and their biogenetic precursors, e.g. furoquinoline A. are normally derived from anthranilic acid; indole A. are normally derived from tryptophan, etc.

Alkalosis: an increase in the pH of the blood. A. may be respiratory or metabolic.

a) ***Respiratory A.*** can result from hyperventilation, which may occur in central nervous system diseases affecting the nervous control of breathing, in the early stages of salicylate poisoning, and during exposure to high altitudes. It can also be promoted by voluntary hyperventilation. Hyperventilation results in an increased removal of carbon dioxide from the blood, thereby raising the blood pH-value. Respiratory A. is compensated metabolically, by the excretion of hydrogencarbonate in the urine.

b) ***Metabolic A.*** is relatively rare. The most familiar cause is excessive vomitting, in which only the stomach contents are voided and not the alkaline contents of the intestine (e.g. in pregnancy sickness). This results in the loss of HCl secreted by the gastric mucosa, the lost Cl^- being replaced by HCO_3^-, leading to A., also called hypochloremic A. Metabolic A. can also be caused by the ingestion of alkaline drugs, e.g. $NaHCO_3$ for the treatment of a peptic ulcer. A further known cause is the excess secretion of aldosterone by the adrenal glands, which promotes increased reabsorption of Na^+ from the distal tubules of the nephrons. The latter process is coupled with an increased secretion of H^+, thereby leading to A. Metabolic A. can be compensated by a decreased respiration rate.

Allantoic circulation: see Allantois.

Allantois: an embryonic intestinal appendage, consisting of a ventral diverticulum of the hindgut in the caudal region of the embryo. It arises as a small groove, then develops rapidly into a stalked vesicle, which grows very large in reptiles and birds, but usually remains small in mammals. It is always covered with connective tissue containing a rich network of blood vessels, communicating with the embryonic circulation. Fish and amphibian embryos have no A.

In ***reptiles*** and ***birds,*** development of the A. starts at about the same time as that of the amnion and chorion (see Extraembryonal membranes). Continually increasing in size, it grows around and completely surrounds the embryo and yolk sac, and fills the *extraembryonic coelom*. Throughout the entire development of the embryo, the A. remains attached to the embryonal gut by a double-walled *allantoic stalk*. Initially, the A. is simply an embryonal urinary reservoir filled with a turbid solution of excretory material. Solid crystals of uric acid that accumulate in the allantois during embryonal development are left behind in the egg when a young bird hatches. Eventually, the outer visceral A. wall grows into contact with the inner parietal chorionic wall, and the two layers fuse to form the *chorio-allantois*. This has an outer epithelium of chorionic ectoderm in contact with the shell membrane, an inner epithelium of allantoic endoderm, and between these is connective tissue derived from the fused chorionic (somatic) and allantoic (splanchnic) mesoderm. At the same time, a second extraembryonal vascular system *(umbilical* or *allantoic circulation)* develops, which serves the chorio-allantois. This takes over from the yolk sac vascular system, as the yolk is utilized and the yolk sac disappears. In this way, the chorio-allantois becomes the respiratory organ of the later embryo, and it is positioned immediately beneath the egg shell membrane and the calciferous shell, next to the air chamber. In addition, parts of the chorio-allantois in the region of the yolk sac fold around the egg-white to form the albumen sac, which functions as a protein and water resorption organ. The A. wall facing the embryo becomes loosely associated with the amnion to form the *amnio-allantois,* and parts of the A. wall may also fuse with the yolk sac wall.

In ***mammals,*** the A. is usually a long-stalked, pear-shaped vesicle, and only in pigs, horses, carnivores, etc. does it attain the size of a sauropsid A. The allantoic circulation supplies blood to the placenta, serving not only respiratory gas transport, but also the transport of nutrients and excretory products. While the amnion is forming, the A. grows from the hindgut and spreads out into the extraembryonal coelom, as in birds and reptiles. Its subsequent development, however, differs from that seen in sauropsids. The A. grows into contact with a rather restricted region of the chorion behind the embryo. The splanchnic mesoderm of the A. then fuses with the somatic mesoderm of the chorion to form the chorio-allantois, which in most mammalian species is a much smaller structure than its sauropsid counterpart. The necks of the yolk sac and the stalk of the A., together with the blood vessels of the A. (arising from the chorionic mesoderm) become enclosed in an investment of the amnion to form the umbilical cord. The A. vascular system spreads out under the chorionic capillary system and penetrates the fetal placental villi, thereby establishing a connection between the extra- and intra-embryonal circulatory systems (placental circulation). In most mammals, the proximal part of the allantoic stalk is retained and contributes to formation of the definitive urinary bladder, whereas the distal part eventually atrophies. In humans, however, the entire allantoic stalk is lost, leaving only the connective tissue of its wall. This connective tissue forms a strand, the *Chorda urachii,* which runs from the top of the bladder on the anterior body wall to the umbilicus, and serves to suspend the bladder. In humans, the bladder is derived entirely from the ventral residues of the embryonal cloaca.

Alleles: forms of a gene that occupy the same locus in the coupling group of homologous chromosomes. A. undergo meiotic pairing, and they may be modified (including their mutual interconversion) by mutation. A. may be identical (AA) or different (Aa). Depending on whether A. arise by mutation at nonhomologous or homologous sites of a gene, they are termed nonidentical (heteroalleles) or identical (homoalleles), respectively. See Pseudoalleles, Isoalleles.

Multiple A. are more or less large series of different A., all representing mutations of a single gene. Each of the multiple A. affects the same phenotypic character and modifies it in a characteristic way. Depending on the number of A. in the multiple allele, up to 20 and more different phenotypes of the relevant character can occur.

Allelochemicals: chemical signals or messengers produced by animals, especially insects. In contrast to Pheromones (see), A. serve for communication between individuals of different species. They are called *allomones* when they confer an ecological advantage on the producing organism, e.g. defense secretions, whereas *kairomones* are of benefit to the recipient.

Allelomimetic behavior, ***mimesis***, ***contagious behavior***, ***mutual imitation***: similar behavior within a group, which is initiated by one individual. This may occur in feeding, body grooming, adoption of a resting posture, flight, movement of young animals as a group, following of a mother by offspring, and other forms of contagious behavior. It is assumed that adoption of the appropriate motor component of the behavioral pattern generates the corresponding inner response, e.g. allelometric feeding behavior generates a receptiveness for food. A.b. is probably a widespread and fundamental element of biosocial behavior. It is also known as the *group effect*.

Allelomimetic behavior in pelicans, which has evolved into cooperative fishing.

Allelopathy: a term coined by Molisch in 1937 for inhibitory and stimulatory interactions between all types of plants, including microorganisms. In the modern, narrower definition of A. the following are excluded: 1) interactions between microorganisms (see Antibiotics), 2) effects of higher plants on microorganisms, e.g. by Phytoalexins (see) or by Root secretions (see), and 3) effects of microorganisms on higher plants, e.g. by the Wilt toxins (see) of plant pathogens, by Gibberellins (see), or by Symbiosis (see). It is essential to the definition of A. that it is due to the addition of a chemical compound to the environment. Such compounds leave the producer plant by volatilization, exudation from roots, leaching by rain from plants or their residues, or degradation of plant residues in the soil; these compounds then inhibit or stimulate the growth or development of neighboring species. The substances in question represent a variety of different classes of compounds, and in many cases their structure is not known.

Many fruits, especially apples, produce the gas, Ethylene (see), which is now classified as a Phytohormone (see). It accelerates fruit ripening, e.g. later apple varieties ripen more quickly when stored with early ripeners, due to the effect of the ethylene from the latter. It also promotes development of dormant plant organs, causes various types of growth inhibition and derangement, and causes premature leaf fall. The adverse effects of tobacco smoke and coal gas on plants (e.g. leaf shedding) are probably due ot traces of ethylene. Other volatile compounds from higher plants with allelopathic effects are essential oils, isothiocyanates (produced mainly by the *Cruciferae*), alkyl sulfides (produced by *Allium* spp., see *Liliaceae*), and hydrogen cyanide (HCN) from the hydrolysis of certain glycosides. Volatile compounds in the leaves of wormwood (*Artemisia absinthium*) strongly inhibit the growth of other plants, creating a 1–2 m wide bare zone around wormwood communities.

It has long been known (even to Pliny about 2 000 years ago) that the walnut (*Juglans regia*) prevents growth of plants in its vicinity; this effect is due to Juglone (see) and its oxidation products.

Cultivated sorghum and the weed, Johnsongrass (*Sorghum halepense*), contain dhurrin (a cyanohydrin), which becomes hydrolysed enzymatically to glucose, HCN and *p*-hydroxybenzaldehyde. Both species of sorghum are very allelopathic, due to the presence of the allelochemically potent HCN and *p*-hydroxybenzaldehyde, which prevent the growth of other species.

Other compounds shown to be allelopathic include long chain fatty acids (in decomposing remains of *Polygonum aviculare*), polyacetylenes (e.g. α-terthienyl produced by the roots of the marigold, *Tagetes erecta*, has an LD_{50} value of about 1 ppm for various seedlings), phenols and phenolic acids derived from benzoic and cinnamic acids (e.g. *p*-hydroxybenzoic, vanillic, *p*-coumaric, ferulic acids), coumarins (inhibitors of seed germination produced by legumes and cereal plants) and hydrolysable and condensed tannins (generally inhibitory to growth and seed germination).

A. is often largely responsible for soil tiredness or soil sickness, i.e. the inability of plants to thrive repeatedly and continuously in the same site. Thus, HCN and benzaldehyde, produced by hydrolysis of amygdalin in peach roots, inhibits the growth of peach seedlings, and is thereby responsible for the difficulty of replanting old peach orchards. Similarly, phlorizin (a flavonoid) in apple roots is toxic to young apple trees. Red clover is also allelopathic against itself, resulting in clover soil sickness. Several isoflavonoids have been identified in the aerial parts of red clover (*Trifolium pratense*), which are degraded in the soil to phenolic compounds inhibitory to seed germination and seedling growth.

A. is of considerable ecological importance, affecting the dispersion and patterning of plants, and the establishment of plant communities. It is therefore important in forestry and agriculture. Crop plants may be selected for the ability to counteract weed growth by A., and it is conceivable to breed allelopathic genes into cultivated varieties as an aid to weed control.

[Recent Advances in Phytochemistry, Vol. 19. *Chemically Mediated Interactions Between Plants And Other Organisms*, edit. Gillian A. Cooper-Driver, Tony Swain and Eric E. Conn, Plenum Press, 1985].

Allen-Doisy test: see Estrogens.

Allen's swamp monkey, *Allenopithecus nigroviridis:* the only species of this genus of Old World monkeys (see). It is found sporadically in the central basin of the Zaire river, apparently preferring swamp forest. The body coat is black with yellowing at the ends of the hairs. Eyelids and chin are pale, whereas the rest of the face is dark, with narrow dark bands at the brow extending to each ear. The scrotum is whitish blue, and there is a conspicuous red-brown tuft beneath the tail. It has not been extensively studied.

Allergy: acquired sensitivity to foreign substances, known as *allergens*. An A. arises after a single contact or repeated contact with an allergen. Sensitization of the organism is due to an immunological response, in which the allergen acts as an Antigen (see) and gives rise to an undesirable and harmful immune reaction. Illnesses due to allergies are, e.g. hay fever, bronchial asthma and urticaria.

The antibodies produced in response to allergens are known as Reagins (see), which are stored in blood granulocytes or in the mast cells of the tissues. These cellular antibodies bind the corresponding allergens, resulting in cell damage and release of substances that are harmful to the organism (e.g. histamine).

Different types of A. are recognized: Anaphylaxis (see), Arthus reaction (see), Contact hypersensitivity (see), Delayed hypersensitivity (see), and Serum sickness (see).

Alleröd period, ***Alleröd interstadial:*** a 3-fold division of the late glacial period originally recognized by Herz and Milthers in 1901 at the Danish village of Alleröd northwest of Copenhagen, and subsequently recognized at many other sites in northwest Europe and Ireland. The A.p. was a late glacial temperature change between 10 000 and 9 000 BC, corresponding to the period between Dryas and the late sub-Arctic. It was characterized by the development of open woodlands of birch or pine (see Pollen analysis). Peat and loess profiles of the A.p. contain thin layers of pumice, carried there by the wind over a distance of 500 km from volcanoes in Laacher Sea area (Eifel).

Alliance:

1) Federation: a ranked category in the classification of vegetation, consisting of one or more floristically and sociologically related associations. Alliances are characterized by their typical combinations of Character species (see).

2) See Relationships between living organisms.

Alligator: see *Alligatoridae.*

Alligatoridae, ***alligators:*** a family of the *Crocodylia* (see) containing 7 species in 4 genera. See separate entries for: ***American alligator*** *(Alligator mississippiensis),* ***Chinese alligator*** *(Alligator sinensis),* and ***Spectacled caiman*** *(Caiman crocodilus).* The Smooth-fronted caiman *(Paleosuchus palpebrosus),* Black caiman *(Melanosuchus niger)* and Broad-nosed caiman *(Caiman latirostris)* are found in Central and South America. In contrast to the *Crocodylidae* (see), the 4th mandibular tooth of the *A.* fits into a groove in the upper jaw and cannot be seen when the jaw is closed. The dorsal, and to a lesser extent the ventral plates have dermal ossifications. The snout is relatively short and broad.

Alligator lizards: see *Anguidae.*

Alligator snapping turtle: see Snapping turtles.

Alliin, ***(-)-S-allyl-L-cysteine sulfoxide:*** a compound present in the bulbs (cloves) of garlic *(Allium sativum).* By the action of the enzyme, alliin lyase, A. is converted into allizin, which is bacteriostatic and largely responsible for the smell of garlic.

$$2\,CH_2{=}CH{-}CH_2{-}\underset{\underset{O}{\downarrow}}{S}{-}CH_2{-}\underset{\underset{NH_2}{|}}{CH}{-}COOH \xrightarrow[\text{(Alliin-Lyase)}]{H_2O}$$

Alliin

$$CH_2{=}CH{-}CH_2{-}\underset{\underset{O}{\downarrow}}{S}{-}S{-}CH_2{-}CH{=}CH_2 + CH3{-}CO{-}COOH + 2NH_3$$

Allizin

Allis shad, *Alosa alosa:* a fine tasting fish of the family *Clupeidae* (order Clupeiformes), maximal length 70 cm (usually 30–40 cm). An inhabitant of the Atlantic coasts of Europe and the western Mediterranean, it spawns in river estuaries and further upstream in fresh water, but it is now extinct in many areas. The genus *Alosa* (shads) characteristically possesses very long scales at the base of the caudal fin. The A.s. is flat-sided with a sharp belly ridge. It has a dark green-blue back. The silvery sides have a few dark spots, and there is a conspicuous large dark spot on the body near the edge of each operculum.

Allizin: see Alliin.

Allochthone:

1) A non-indigenous or non-native organism (see Allochthonous).

2) Exogenous food material transported into a cave or cave system from outside.

Allochthonous, ***xenogenous, ectogenous:*** not indigenous; present at the site of occurrence due to active or passive transport from its true origin by wind, water currents or human agency. Animals native to mountain areas are often transported to lowlands by rivers, but are unable to establish themselves there. The converse of A. is Autochthonous (see).

Allogamy: see Pollination.

Allometric growth: see Allometry.

Allometry, ***allometric growth, heterauxesis, heterogony:*** regular alterations in the shape or proportion of an organ or body part in relation to body size, as a result of a differential growth rate. If the development rate of an organ is more rapid than that of general body development, the phenomenon is known as *positive A.,* whereas slower development than the rest of the body is called *negative A.* Preservation of positive A. during a phylogenetic increase in body size can lead to the evolution of disproportionately large organs. An advantageous increase in body size may therefore be accompanied by disadvantageously large organs or structures.

Allomimesis: see Protective adaptations.

Allopatric speciation: see Speciation.

Allopatry: the occurrence of closely related species, taxa, or populations in different and disjunct geographical areas. A. is an important criterion for recognition of the taxonomic status of closely related forms. See Vicariance, Vicarism. The converse of A. is Sympatry (see).

Allopolyploidy, ***alloploidy:*** a polyploid state arising from the combination of genetically different sets of chromosomes.

Allopora: see *Millepora.*

Allopreening: see Grooming.

All or nothing reaction: the behavior of certain systems, in which a stimulus either has no effect, or it causes the maximal response when it equals or exceeds a threshold value. Thus, the response either occurs with its greatest possible intensity or not at all. A further increase in the intensity of the stimulus cannot cause an increase in the intensity of the response. The A.o.n.r. therefore has binary character, and a system displaying this type of reaction can exist in only in a plus (+) or a minus (-) state, i.e. it is either "on" or "off". Irritability phenomena often exhibit an A.o.n.r. response, e.g. the leaf movements of *Mimosa pudica* (the "sensitive plant") and the movement of the stamens of *Berberis.* See also Excitation, Action potential.

Allosomal inheritance: the same as Sex-linked inheritance (see).

Allosomes: see Sex chromosomes.

Allosteric effector: a low molecular mass compound that influences the activity of an enzyme by binding at an allosteric site, i.e. it does not bind at the catalytically active center. Binding of the A.e. alters the conformation of the enzyme protein, so that its catalytic activity is decreased (allosteric inhibition) or increased (allosteric activation).

Allosyndesis: meiotic pairing of Homologous chromosomes (see) in an Allopolyploid (see). Let the genetic composition of an allopolyploid organism be XXYY, where XX and YY are the chromosomes from the respective parents. Presumably due to the common ancestry of the parents, X and Y have some homologous segments, so that X and Y display allosyndetic pairing during the prophase of meiosis. In Autosyndesis (see) X pairs only with X, and Y with Y. In diploid forms, chromosome pairing is always an A. In polyploids, chromosome pairing may be an A. or an autosyndesis, and these two types of pairing may occur with the same (autopolyploidy) or different frequencies. Through the operation of A., *segmental alloploids* (in which many chromosomal segments are similar) form both bivalents and multivalents during meiosis.

Allotetraploidy: see Amphiploidy.

Allochory: see Seed dispersal.

Almond: see *Rosaceae.*

Almoviruses: see Multipartite viruses.

Alocasia: see *Araceae.*

Aloe: see *Agavaceae.*

Alopecurus: see *Graminae.*

Alopex: see Foxes.

Alopiinae: see *Selachimorpha.*

Alosa: see Shad.

Alouatta: see Howler monkeys.

Alpacca, *Lama guanacoë* var. *pacos:* a very long-haired, small, variously colored, domesticated form of the Guanaco (see).

Alpheus: see Shrimps.

Alpine brown bear: see Brown bear.

Alpine flora: bushy and herbaceous and cushion-forming plants, which inhabit the high mountain regions between the timber line and the snow, e.g. dwarf and espalier flowering shrubs, grasses, sedges, herbaceous perennials, mosses, lichens and algae, some rooted in relatively flat, turfed areas, others in rock fissures or scree. The vascular plants of this high altitude have strongly developed root systems, often with a stunted, dwarf or creeping growth habit, and occasionally forming hemispherical cushions. As an adaptation to the mountain climate and the high light intensity, these plants often possess robust boundary tissue, and their surfaces are frequently hairy; leaves in particular often possess a very dense covering of hairs. A.f. are physiologically adapted to the low temperature (see Temperature factor) and the short growth period (1–3 months). Thus, A.p. are naturally slow growing, and most of them are perennials. Their requirements for growth are therefore satisfied by a relatively low annual rate of assimilation of carbon and nitrogen.

Alpine grass heath zone: see Altitudinal zonation.

Alpine newt: see *Salamandridae.*

Alpine salamander: see *Salamandridae.*

Alpine zone: see Altitudinal zonation.

Alternation of generations: the usually regular alternation between a sexually reproductive generation and one or more generations that reproduce asexually. A.o.g. occurs in both plants and animals.

In plants, the asexual generation that reproduces by spore formation is called the sporophyte, while the generation that reproduces sexually by gametes is called the gametophyte. The sporophyte and gametophyte may be similar to one another, or they may differ morphologically. In all higher plants, the A.o.g. is accompanied by an alternation of the nuclear phase, i.e. the sporophyte is almost always diploid (see Diplont, Diplophase), and the gametophyte is haploid (see Haplont, Haplophase); the A.o.g. is then known as *heterophasic* or *antithetic.* In the mosses *(Bryophyta),* the green plant is the haploid gametophyte, while the spore capsule (which always remains attached to the gametophyte) is the diploid sporophyte. In the *Pteridophyta,* the green plant is the diploid sporophyte. The latter produces spores asexually, which germinate to form the gametophye prothallus. Fertilization of a gametophyte egg produces a zygote, which germinates to form the diploid sporophyte, thereby completing a single cycle of the A.o.g. In gymnosperms and angiosperms the gametophyte is reduced even further, consisting merely of male and female gametes and a few surrounding cells; the gametophyte never has an independent existence, and the actual green plant is the diploid sporophyte.

In animals, heterophasic A.o.g. occurs only in unicellular organisms. *Homophasic* A.o.g., in which the chromosome number does not change, occurs in Heterogony (see) and Metagenesis (see). In higher animals there is no A.o.g.

Alternative habitat: see Habitat alternation.

Althaea: see *Malvaceae.*

Altitudinal zonation: the establishment of characteristic flora and fauna in dependence on altitude. With increasing altitude, the decrease of temperature, changes in the type and quantity of precipitation, shortening of the growth period, and increased irradiation intensity all lead to multiple changes in the vegetation and the formation of altitudinal zones or vegetation zones. In general, the following zones can be defined. 1) *Planar zone.* 2) *Hill zone (colline zone).* In central Europe the planar and colline zones consist predominantly of mixed deciduous forests with beech, oak and lime, with an admixture of coniferous forest in the northeast; in central and northeastern Europe most of the agricultural land lies in these two zones. 3) *Intermediate montane forest zone* with beech and mixed beech-fir-spruce forest. The lower part of this zone is known as the *submontane zone;* some areas are cleared for pasture and arable land. 4) *Upper montane forest zone,* characterized by spruce, mixed spruce-larch and mixed *Pinus cembra*-larch forests, with clearings of mountain meadows and pastures. For example, the highest altitudes of the Harz mountains, the Thüringian forest and the Erz mountains correspond to the upper montain forest zone. This zone extends to the limit of the closed forest or timberline. 5) *Alpine zone,* or *alpine grass heath zone.* The lower part of this zone (the *dwarf scrub* or *subalpine zone)* extends to the upper tree limit (often isolated spruces) and includes hardy dwarf shrubs such as *Loiseleuria procumbens (Ericaciae).* The alpine zone proper consists of natural high mountain dwarf shrub heaths, turfs, rocks and screes, with many dwarf perrenial herbs. Many of the grasses and sedges that form the turf and many of the perennial herbs are strictly alpine species, and they differ according to whether the soil is acid or calcareous. Thus, turfs of *Carex curvula* are found on acid soils, whereas *Sesleria albicans* and *Carex firma* are important on calcareous soils. 6) *Nival zone* or permanent snow zone. This begins with the subnival zone which is covered for long periods by snow and ice, with some turfed areas, and some areas covered with cushion plants, together with moss and lichens. The actual nival zone lies above the climatic snow line, with only occasional snow-free patches where plant life is possible.

Altricial: adjective applied to young animals that are born or hatch in an immature and helpless condition, so that they must rely on brood care during completion of their development to a state of independence and self sufficiency. They cannot independently change their location, and they are unable to respond to environmental demands.

Food reserves in the eggs of altricial birds are sufficient only for embryonal and fetal growth and development up to the point of hatching. The hatched chick therefore has no further reserves to draw upon and must be fed.

Animals born in the converse condition, i.e. in a state of independence, are described as Precocial (see). See Gaping.

Aluminium, *Al:* the third most abundant element and the most abundant metal in the earth's crust, where it is present as oxides and more commonly as aluminosilicates. Despite this abundance, significant amounts of Al are not normally found in surface waters and oceans (less than 1 mg Al/l sea water). Higher levels are present in some volcanic regions and in alum springs.

There is now a tendency, however, for Al (as well as certain other metals) to appear in higher concentrations in fresh water, due to its release from mineral sources by the action of acid rain. This is already an ecological problem; for example, fish are poisoned by nanomolar concentrations of Al^{3+}, and the threshold of toxicity has been exceeded in certain freshwater lakes.

Alzheimer's disease, a form of senile dementia, involves an abnormal degeneration of the brain. Brain autopsies of Alzheimer victims reveal neurofibrillary tangles (deposited fibrillar amyloid protein) and senile plaques containing high levels of Al^{3+} and aluminosilicate. Al is also associated with other neurotoxic conditions, e.g. amylotropic lateral sclerosis. The latter is a form of Parkinson's disease with severe dementia, also accompanied by an accumulation of Al^{3+} in neurofibrillary tangles in the brain; moreover, the complaint is common in the indigenous populations of southern Guam, Kii peninsular in Japan and western New Guinea, where the soils are high in Al^{3+} and low in Mg^{2+} and Ca^{2+}. It is not known whether Al causes these dementias, or whether the metal simply has a affinity for the abnormal neurofibrillary tangles.

Long-term hemodialysis patients may suffer Al toxicity, which is manifested as microcytic anemia, bone disease and dementia, leading ultimately to death. This problem is now overcome by meticulous deionization of the dialysis water.

Al appears to have no specific function in the metabolism of higher plants, but it is an essential trace element for ferns and horsetails. Uptake of very small quantities of Al by plants generally has a favorable effect on growth, resulting from an unspecific effect of plasma swelling. The Al content of higher plants is usually about 200 mg/kg dry mass, but the tea plant contains 5000 mg/kg. Al toxicity is one of the most important growth-limiting factors for plants on acid soils, one of the first signs of Al stress being disorganization of the plasma membrane. Cells of the root cap often become vacuolated, with disruption of the Golgi function and plastid development, changes in nuclear structure and release of cytoplasm.

[*Metals in biological systems, vol.4. Aluminium and its role in biology,* edit. Helmut Sigel with Astrid Sigel; Marcel Dekker Inc., New York and Basel, 1988]

Alveolar air: the air in the alveoli of the lung. Oxygen diffuses from the alveolar air into the blood, while carbon dioxide diffuses in the reverse direction. The respective rates of diffusion are determined by the Partial pressure (see) of each gas.

Alveolar hydatid disease: see Dog tapeworm.

Amadou: see *Basidiomycetes (Poriales).*

Amaltheus (Greek *amalteia* nymph): a characteristic fossil of the Upper Pliensbachian (Lias δ). The shell is plani-spiral, discus-like when viewed from the edge, umbilicate, and fairly involute. The ribs have a shallow S-shape, and they are curved forward (toward the aperture). The suture resembles that of the ammonites, and the keel is corded.

Amandibulata: a group of the *Arthropoda* (see) comprising two subphyla: 1)*Trilobitomorpha* [an extinct subphylum containing the single class,*Trilobita*; the subphylum is also referred to as the *Trilobita*], and 2) *Chelicerata.* In contrast to the *Mandibulata*, members of the A. lack mandibles (see Mouthparts).

Amanin: see Amatoxin.

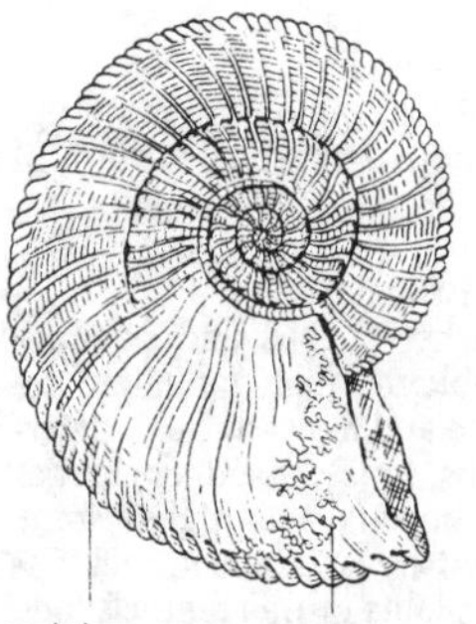

Shell of Amaltheus margaritatus *Montf., showing the deeply indented suture and corded keel.*

Amanita: see *Basidiomycetes (Agaricales).*

Amanitin: see *Amatoxin*

Amanullin: see *Amatoxin.*

Amaryllidaceae: a family of about 850 species of monocotyledonous plants, distributed throughout the world. They are herbaceous plants with leafless flower stems, and bulbs which enable them to survive the nonvegetative period of the year. *A.* are very similar to the *Liliaceae*, and differ from the latter morphologically mainly in their possession of inferior ovaries, and chemically by the absence of saponins and the presence of characteristic alkaloids. Many species have beautiful flowers and are kept as ornamentals. The ***Snowdrop*** (*Galanthus nivalis*), one of the earliest spring flowers, native in Central Europe as far as the Caucasus, is a familiar garden plant. Various species and varieties of ***Narcissus***, especially the white-flowered ***Pheasant's Eye Narcissus*** (*Narcissus poeticus*) and the yellow-flowered ***Daffodil*** (*Narcissus pseudonarcissus*) are cultivated as ornamentals. Narcissi are characterized by a secondary corolla, and they occur naturally mostly in the Mediterranean region. The ***Spring Snowflake*** (*Leucojum vernum*), a native, protected plant of Central Europe, is also a garden flower. Hybridized species of the South American ***Amaryllis*** (genus *Hippeastrum*) and the South African ***Clivia*** (*Clivia miniata*) are important indoor ornamentals. Another favored decorative species is *Haemanthus albiflos*, a plant with broad fleshy leaves, and flowers which carry dense clusters of long, white stamens.

Amaryllidaceae-type stomata: see Stomata.

Amaryllis: see *Amaryllidaceae.*

Amastigote: see Trypanosomes.

Amathes: see Cutworms.

Amatoxins: a group of bicyclic octapeptides which, together with the phallatoxins, are mainly responsible for the toxicity of the death cap fungus, *Amanita phalloides.* These poisons inhibit nucleoplasmic RNA polymerase II of eukaryotic cells, which leads to necrosis of liver and kidney cells. The poisonous effects of the A, and phallatoxins can be inhibited by simultaneous application of antamanid. More than 90% of fatal mushroom poisonings are due to consumption of *Amanita phalloides* and related species.

Amazonian tree boa, *Corallus caninus:* a tree-dwelling snake of the subfamily *Boinae* (boas) in the

Amaryllidaceae. a Spring snowflake *(Leucojum vernum). b* Amaryllis *(Hippeastrum). c* Clivia *(Clivia miniata). d Haemanthus albiflos.*

	R_1	R_2	R_3	R_4
α-Amanitin	OH	OH	NH_2	OH
β-Amanitin	OH	PH	OH	OH
γ-Amanitin	OH	H	NH_2	OH
Amanin	OH	OH	OH	H
Amanulin	H	H	NH_2	OH

Amatoxins

family *Boidae* (see). Native to the northern rainforests of South America, it is up to 2 m long, brilliant green with light transverse bands, and a triangular head. At rest it rolls itself around a branch. As an adaptation to its arboreal existence, the tail is designed for clinging. The diet consists mainly of birds, which are caught with the long front fangs. Like other *Boinae*, it is vivaparous.

Amazons: see *Psittaciformes.*

Ambari hemp: see *Malvaceae.*

Amber, *succinite, succinum:* a yellow to brown, solidified fossil resin, especially common in the coastal regions of the North Sea and the Baltic, and often containing embedded insects and plant parts. It originated from certain extinct coniferous trees of the Tertiary, which probably formed extensive forests over the northern parts of Scandanavia. Chemically, it is a complex mixture of resin acids and alcohols and their oxidation products, the characteristic component being succinic acid.

Ambivalent behavior: behavior displaying different and often opposing types of behavior, e.g. elements of both attack and escape behavior. In many species of animals, "threat behavior" or "threat postures" have evolved from such a combination of opposing elements of communicative behavior. A.b. or threat behavior is a type of agonistic behavior towards other individuals in the close vicinity, i.e. a type of behavior or communicative (signalling) posture oriented toward a competitor or rival that is in the range of sensory perception.

Ambivalent motivation, *ambivalent state:* an inner physiological state containing two opposing vectors, but directed toward a single stimulus, resulting in Ambivalent behavior (see).

Amblyopsidae: see *Percopsiformes.*

Amblyornis: see *Ptilinorhynchidae.*

Amblypterus (Greek *amblys* blunt, *pteron* fin or wing): an extinct genus of the *Holostei,* with rounded ventral fins and flat, almost quadratic scales. Fossils are found from the Permian to the Tiassic, especially in the Rotliegendes of the Lower Permian.

Amblypygi, ***tailless whipscorpions, whipspiders:*** a suborder of the order *Pedipalpi,* class *Arachnida* (see), containing about 60 tropical and subtropical species with several similarities to spiders. The body consists of a prosoma and a sac-like opisthosoma. A telson appendage, venom glands, and spinnerets are all completely absent. The large pedipalps are held flexed and horizontal in front of the body; they are either provided with pincers or they form a basket-like structure for capturing prey, and they are not fused at the base to form a trough (cf. *Uropygi).* The first pair of legs is often unusually long and whip-like with many false segments; they are very sensile and used as feelers, especially in water location and in mating. Two pairs of lung books serve for respiratory gas exchange.

Amblypygids are nocturnal in habit, hiding beneath stones, leaves or loose bark during the day. They walk with the last three pairs of legs, holding the feeler legs spread widely to the sides. Until the eggs hatch, they are carried by the female in a secreted egg case at the sexual opening.

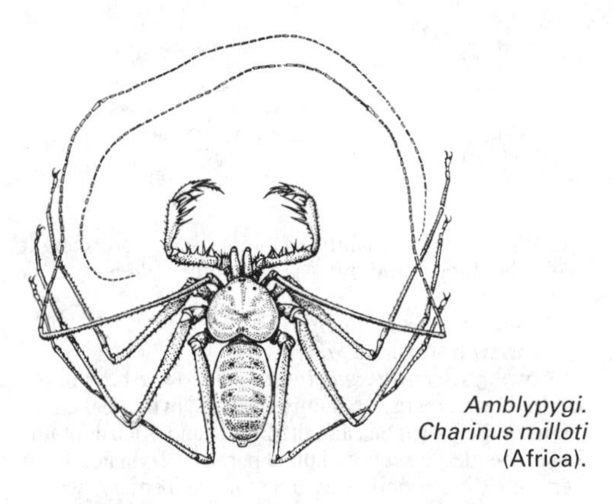

Amblypygi. Charinus milloti (Africa).

Amblyrhynchus: see Marine iguana.

Amboceptor, ***desmon:*** an obsolete term for any complement-binding hemolysin, i.e. a complement-binding antibody, whose reaction with antigen results in lysis of target cells. *Bordet's A.* is the A. in blood which is involved in the fixation of complement . A *hemolytic A.* is an A. involved in the lysis of erythrocytes. Similarly, a *bacteriolytic* or *bacteriocidal A.* is a bacteriolysin or bacteriocidin involved in the lysis or killing of bacteria.

Amboyna pitch tree: see *Araucariaceae.*

Ambramycin: see Tetracyclins.

Ambrette oil: see Ambrettolide

Ambrette seeds: see Ambrettolide.

Ambrettolide: an unsaturated lactone with a musk-like odor. It is the main odorous principle of ambrette oil (musk seed oil) from the seeds of the musk mallow *(Abelmoschus moschatus = Hibiscus abelmoschus),* which are known as ambrette seeds. A. is used as a fixative in perfumes.

$$\begin{array}{l} HC{-}(CH_2){-}CH_2 \\ \| \qquad\qquad\quad \backslash \\ \| \qquad\qquad\quad\; O \\ \| \qquad\qquad\quad / \\ HC{-}(CH_2)_5{-}CO \end{array}$$

Ambrettolide

Ambrosia beetles: see Engraver beetles.

Ambulacral: an adjective applied to one surface of the body of an echinoderm, and to structures arising from that surface, e.g. ambulacral feet, ambulacral groove (see *Echindermata).* Since the ambulacral surface contains the mouth opening, it is also known as the *oral* surface. The other surface is called the *aboral* or *abambulacral* surface. These terms are used in preference to ventral and dorsal, because the two surfaces of an echinoderm do not correspond to the ventral and dorsal sides of the larva, but to the left and righ sides of the larva.

Ambulacral system: see *Echinodermata.*

Ambush bugs: see *Heteroptera.*

Ambush hunters: see Modes of nutrition.

Ambystomatidae, ***mole salamanders:*** an American family (more than 30 species) of the *Urodela* (see). The palatine teeth of mole salamanders are arranged in continuous or interrupted diagonal rows on the roof of the mouth. Most species have squat bodies, wide heads, prominent vertical rib grooves along the sides of the body, and a laterally compressed tail. They vary from 8 to 30 cm in length, and display a variety of skin patterns and colors: black or brown, spotted, mottled or striped with white, yellow, green, pink, orange, red or blue. Lungs are present, and there are 4 fingers and 5 toes. They inhabit damp forests (often burrowing in the forest floor), as well as ponds and mountain streams, although many species seek out water only for mating, egg laying and egg fertilization (fertilization is internal; the male deposits a spermatophore, which is taken up by the female into her cloaca). Eggs are laid in water, singly or in clumps. The family is represented from southern Alaska and Labrador to the southern edge of the Mexican plateau.

A very familiar member is the ***Axolotl*** *(Ambystoma = Siredon mexicanum),* which is the classical example of obligate Neoteny (see). The axolotl often kept in aquaria is the 25 cm-long, albino form with red external gills; it is a non-metamorphosed "adult" larva, which is not capable of reproduction. By feeding thyroid extract and gradually lowering the water level, this larval stage can be induced to metamorphose into the gill-less, lung-breathing terrestrial form. The axolotl also occurs in a neotenic, dark brown, reproductive form in Lake Xochimilco in Mexico, and its metamorphosed form is not known in the wild.

The light- and dark-spotted *Ambystoma tigrinum* ***(Tiger salamander)*** is the longest terrestrial salamander, growing up to 35 cm in length. Water is required only for reproduction, and the female can lay more than 1 000 eggs. It occurs in the lowland regions of North America, and western subspecies often display facultative neoteny, which is attributed to an iodine deficiency in the local waters.

Other representative species are *Ambystoma opacum* ***(Marbled salamander),*** *Ambystoma macrodactylum croceum* ***(Santa Cruz long-toed salamander),*** *Ambystoma maculatum* ***(Spotted salamander).***

Amebas, ***amoebae,*** *Amoebina:* an order of the class *Rhizopoda (Sarcodina),* phylum *Protozoa.* The body of

A. is a mass of protoplasm, which is usually clearly differentiated into endoplasm and an outer boundary layer of ectoplasm. Pseudopodia are used for locomotion and feeding. Reproduction is usually by binary fission, and encystment is a common means of dispersal and survival. A. regularly change their shape, while retaining a distinct outline. When not moving, most A. are more or less rounded, but when active they may be flattened, lens-shaped, ribbon-like, sometimes with fine indentations, star-shaped or aborescent, with entirely irregular folding, extending pseudopodia in different directions at the same time. Such variety of shape is displayed especially by members of the genus *Amoeba,* whereas the more rounded "Limax" A. (most species belong to the genus *Hartmanella)* are rather slug-like, extending only a single pseudopodium. *Radiosa* A., which are found floating in fresh water, have long, streaming pseudopodia. *Verrucosa* A. (0.1 mm diam) are characterized by their wrinkled body surface. *Amoeba proteus* (0.2–0.5 mm diam) (Fig.) is commonly found on the bottom of pools and puddles. The exceptionally large *Pelomyxa palustris* (diam. up to 2 mm) is roughly egg-shaped, the entire body behaving as a single large pseudopodium; in addition to its many nuclei, the endoplasm also contains highly refractive glycogen particles, which are known as "reflecting bodies", as well as numerous particles of grit and other foreign material. *P. palustris* is commonly found in water overlying rotting soil material. Other common environments that harbor A. are sea mud, feces, moist earth, the water trapped in moss cushions, as well as the intestinal tracts of vertebrates and invertebrates. The genus, *Malpighiella,* parasitizes the nephridia of insects. *Entamoeba hystolytica (= E. dysenteriae)* is the causative agent of human amebic dysentery, also occasionally causing abscesses of the liver and other organs.

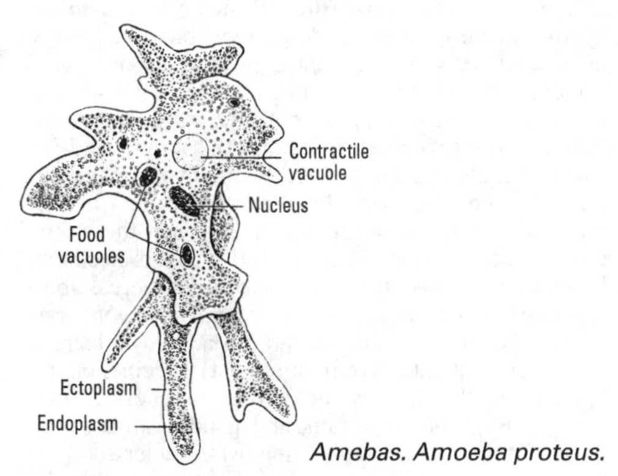

Amebas. Amoeba proteus.

Amebic dysentery: see Amebas.

Amebozygote: see *Myxomycetes.*

Ameiurus: see *Siluriformes.*

Ameiva: see Jungle runners.

Amentiflorae: see *Dicotyledoneae.*

American alligator, *Alligator mississippiensis:* a member of the *Alligatoridae* (see), average length 3 m (maximum 6 m), which inhabits rivers and swamps in southeastern USA. It feeds mainly on fish, together with water birds and mammals up to the size of a large calf, the latter being subjected to a sudden surprise attack, then dragged into the water. For incubation of the eggs, a 1 m-high heap of vegetation is constructed, and a clutch of 40–60 eggs is deposited in the center. The female watches over this nest until the young emerge (after about 10 weeks), having first announced their imminent arrival by loud squeaking within the egg. The skin is used extensively for the manufacture of luxury leather goods. Hunting caused severe depletion of numbers and the disppearance of the A.a. from parts of its natural range. Protected by conservation measures, it has made an excellent recovery. The leather goods industry now meets part of its requirements by breeding and farming alligators and especially certain crocodiles.

American badger, *Taxidea taxus:* a monotypic genus of the *Mustelidae* (see). It is somewhat smaller than the European badger, and ranges from southwest Canada and north central USA, through the western USA to Central Mexico, usually inhabiting dry, open country. The upper parts are gray to reddish, and a white stripe runs from the nose at least to the shoulders. Chin and throat, extending to the midventral region, are whitish, and the remainder of the underside is buffy. There are black patches on the face and cheeks. It is usually solitary, feeding on rodents and other animals and carrion. In the northern part of its range, it sleeps during the period that the ground is frozen.

American blackbirds: see *Icteridae.*

American bullfrog: see *Raninae.*

American harvest mice: see *Cricetinae.*

American leaf-nosed bats, *Phyllostomatidae:* a family of microchiropteran bats (see *Chiroptera),* containing about 140 species in 51 genera, distributed in the American tropics and subtropics. In some species, the nose leaf is long and sharply pointed rather than leaf-like (e.g. in the sword-nosed bat, *Lonchorhina aurita).* The nose leaf does not show the complex development seen in the Old world leaf-nosed bats, and in a few genera it is reduced or absent. Those species lacking a nose leaf have a plate-like outgrowth on the lower lip. Members of the family feed variously on fruit, nectar, pollen and insects. Members of the subfamilies, *Glossophaginae* and *Phyllonycterinae,* bear a resemblance to the *Macrochiroptera,* in that they possess an elongated snout and a long, extensible tongue covered with bristle-like papillae.

American lobster: see Lobsters.

American mud and musk turtles: see *Kinosternidae.*

American orioles: see *Icteridae.*

American potto: see Kinkajou.

American robin: see *Turdinae.*

American rock rat: see *Octodontidae.*

American toad: see *Bufonidae.*

American vultures: see *Falconiformes.*

American warblers: see *Parulidae.*

Amictic lake: see Lake.

Aminergic: an adjective applied to neurons, nerve fibers, nerve endings and synapses that release amine-type Transmitter (see) substances when stimulated. Amine-type transmitters are catecholamines (e.g. noradrenalin, adrenalin, dopamine) and indolamines (e.g. 5-hydroxytryptamine or serotonin).

Amines: organic compounds that can be represented as formal derivatives of ammonia, in which one or more hydrogen atoms of the ammonia are replaced by hydrocarbon groups (R). According to the number of hydrogen atoms attached to the nitrogen, A. are classified as primary, secondary or tertiary (Fig.). A. occur in plants and animals, where they are formed by the decarboxylation of amino acids. See Biogenic amines.

```
            R'            R'
             \             \
R—NH2         NH            N—R'''
             /             /
            R''           R''
primary    secondary     tertiary amine
```

Amino acid decarboxylases: Ligases (see), which catalyse the conversion of α-L-amino acids into the corresponding amine and carbon dioxide. Most A.a.d. are specific for a particular L-amino acid, and pyridoxal phosphate (see Vitamins) is required as a cofactor of the reaction.

Amino acid oxidases, ***amino acid dehydrogenases:*** flavo-enzymes that catalyse the oxidative deamination of amino acids to imino acids in the presence of oxygen. The imino acid is spontaneously and rapidly hydrolysed to the corresponding 2-oxo acid (α-keto acid) and ammonia. The hydrogen removed from the amino acid is transferred to the flavin cofactor of the enzyme, then to molecular oxygen with the formation of hydrogen peroxide. The overall reaction is therefore: amino acid + O_2 + H_2O → oxoacid + NH_3 + H_2O_2. It is irreversible under physiological conditions, and does not contribute to the assimilation of ammonia. A.a.o. are specific for either L- or D-amino acids. In animals, they are present especially in the liver and kidneys. D-A.a.o. (cofactor FAD) may have a protective role in the destruction of potentially toxic D-amino acids, while L-A.a.o. (cofactor FMN) may be important in the metabolism of L-lysine. A.a.o. are not thought to play a major quantitative role in the metabolism of amino groups. Both D- and L-A.a.o. are localized in Peroxisomes (see), which also contain the necessary catalase for destruction of the product hydrogen peroxide.

L-Glutamate dehydrogenase is not a flavoprotein; it is strictly specific for L-glutamate, and its cofactor is NAD. Unlike the Amino acid oxidases, L-glutamate dehydrogenase occupies a central position in the nitrogen metabolism of living organisms (see Nitrogen).

Amino acids: organic acids possessing one or more amino groups (—NH_2). Proteins and peptides are polymers of A.a., and the biological function of A.a. as precursors of proteins is quantitatively and qualitatively very important. A.a. also occur free in all living cells. According to the position of the amino group in relation to the terminal carboxyl group, A.a. are classified as α-, β, γ-A.a.., etc. (in an α-amino acid the amino group is attached to the α-carbon, i.e. to C-2 of the carboxylic acid; in a β-amino acid, the amino group is attached to C-3, etc.). Most naturally occurring A.a. are of the α-type, i.e. with the general formula: R—CH(NH_2)—COOH. With the exception of glycine, all α-A.a. have an asymmetric C-atom (the α-C), so that they are able to exist in the L- or D-configuration. All A.a. found in proteins have the L-configuration, in fact the overwhelming majority of naturally occurring, bound and free A.a. have the L-configuration. Certain D-amino acid residues are present in some peptide antibiotics, and in bacterial cell wall polymers. Since they contain both an acidic (COOH) and a basic (NH_2) group, A.a. are amphoteric, and in solution they behave as ampholytes. In strongly polar solvents (e.g. water), internal salt formation produces a zwitterion (NH_3^+—CHR—COO^-), which is fully formed at a certain pH, known as the isoelectric point. Abolition of the hydrophilicity of the amino group or carboxyl group (e.g. by derivatization) markedly decreases the water solubility of the amino acid. At acidic pH-values, each amino acid exists as a cation (H_3N^+—CHR—COOH), whereas at basic pH-values the anion (H_2N—CHR—COO^-) is present. A.a. are solid, crystalline compounds with relatively high melting points, generally soluble in water, but poorly soluble in organic solvents.

Naturally occurring A.a. can be classified in various ways.

1) According to the position of the isoelectric point, they are classified as *neutral, acidic* or *basic.* A basic amino acid (e.g. lysine) possesses an extra basic group in addition to the α-NH_2; an acidic amino acid (e.g. glutamic acid) possesses an extra acidic group in addition to the COOH; while the only dissociable groups of neutral amino acids (e.g. leucine) are the α-NH_2 and the COOH.

2) According to the structure of the side chain, A.a. may be classified as aliphatic, heterocyclic, aromatic or imino acids.

3) Depending on their polarity, A.a. can be divided into polar (hydrophilic) and nonpolar (hydrophobic) types.

4) A.a. that are metabolized to C4-dicarboxylic acids or pyruvic acid are potential precursors of gluconeogenesis, and are therefore known as *glucogenic* A.a. Those that are metabolized to ketone bodies, especially acetoacetate are known as *ketogenic* A.a.

6) A.a. that participate in protein synthesis, i.e. those that are encoded in the Genetic code (see), are known as *proteogenic* A.a., while all the other A.a. that do not normally occur in proteins (about 200 are known) are nonproteogenic.

Plants are able to synthesize all proteogenic A.a. from simple precursors. Animals, however are able to synthesize only certain of the proteogenic A.a., i.e. glycine, alanine, serine, cysteine, proline, aspartic acid, asparagine, glutamic acid and glutamine. Most animals can also synthesize arginine, but not always at a rate sufficient to meet their entire need for this amino acid. Tyrosine can also be synthesized by the hydroxyaltion of phenylalanine, but phenylalanine itself cannot be synthesized in the animal organism. A.a. that the animal can synthesize in sufficient quantity for its needs are known as *dispensable* or *nonessential,* whereas those that must be supplied in the diet are *indispensable* or *essential.* For humans, the essential A.a. are (in decreasing order of quantitative requirement): leucine, phenylalanine, methionine, lysine, valine, isoleucine, threonine and tryptophan (arginine and histidine are essential for infants but not for adults). It must be understood that the terms essential/nonessential or indispensable/dispensable refer to the dietary requirement; the presence of all the proteogenic A.a., whether from endogenous synthesis or from the diet, is essential for the synthesis of protein. If any single essential amino acid is absent from the diet, the organism quickly goes into negative nitrogen balance (more organic nitrogen excreted than consumed by the organism). A prolonged dietary deficiency of an essential amino acid severely impairs health, and ultimately leads to death.

A.a. are obtained technically by the hydrolysis of proteins, and by chemical and microbial synthesis.

All the A.d. mentioned above are described more fully under separate entries.

Classification of proteogenic amino acids according to polarity

Amino acids with nonpolar (hydrophobic) and neutral side chains	Amino acids with polar (hydrophilic) side chains: Neutral	Basic	Acidic
Glycine, Isoleucine,	Serine	Lysine	Aspartic acid
Alanine, Proline,	Threonine	Arginine	Glutamic acid
Valine, Methionine,	Cysteine	Histidine	Tyrosine
Leucine, Phenylalanine	Asparagine		

p*-Aminobenzoic acid, *PAB: an aromatic carboxylic acid, which is a component of folic acid (see Vitamins). PAB is an essential growth factor for many bacteria. Sulfonamides (which are derivatives of sulfonilamide) exert their bacteriostatic effect by releasing sulfanilic acid, which is an antagonist of PAB (see Antivitamins); this inhibits folic acid synthesis by blocking the enzyme that incorporates PAB.

Aminoethanol: see Ethanolamine.

Aminosugars: monosaccharides, in which a hydroxyl group has been replaced by an amino group. In addition, the amino group is often acylated (usually acetylated). Biologically important A.s. are D-galactosamine, D-glucosamine, neuraminic acid, D-mannosamine and muramic acid. The 2-amino-2-deoxyaldoses are particularly important as components of bacterial cell walls, of some antibiotics, blood group substances, milk oligosaccharides, and high molecular mass natural products such as chitin.

Amitosis, *nuclear fragmentation:* a process of nuclear division, in which the cell nucleus simply separates into 2 parts, i.e. it divides by constriction. During this process, a spindle is not formed, the nuclear membrane persists, the chromosomes do not change their shape or form, and the partitioning of the daughter genomes is haphazard. The daughter nuclei are therefore genetically different. Ring-shaped Microtubules (see) are involved in the constriction process, the microtubule ring being formed from the diplosome (see Centrosome). Usually, A. is not accompanied by cell division, so that the cell becomes bi- or multinucleate. A. results essentially in an enlargement of the nuclear surface area, enabling an increased rate of metabolite exchange between nucleoplasm and cytoplasm. It occurs, e.g. in highly differentiated liver, heart muscle and skeletal muscle cells, as well as in the macronuclei of ciliates (polygenomic protozoa) and certain fungi.

Ammodorcas: see *Antilopinae*.

Ammodorcatini: see *Antilopinae*.

Ammodytidae, *sand eels, sand lances:* a family (3 genera, about 12 species) of small, elongated marine fishes of the order *Perciformes*. Common inshore fishes, they occur in shoals on sandy coasts and can bury themselves with lightning speed in the bottom sand. They form an important part of the diet of predatory fishes, such as cod and salmon. There are no teeth, and the single dorsal fin extends along most of the dorsal surface. Pelvic fins are usually absent.

Ammonia, *NH_3* : a colorless gas with a sharp, characteristic smell. It is very soluble in cold water, and the aqueous solution is weakly basic, due to the formation of ammonium ions ($NH_3 + H_2O \rightleftharpoons NH_4^+ + OH^-$; the equilibrium lies far to the left). Ammonia, especially as ammonium ions, is important in biological nitrogen metabolism, e.g. it is formed by the reduction of nitrate, and by the biological fixation of atmospheric nitrogen, in the deamination of amino acids, and in various catabolic processes like oxidative purine degradation and reductive pyrimidine breakdown. Free ammonia is, however, toxic to the cell, and its accumulation is prevented by assimilation processes. Thus, in mammals, much of the nitrogen from the turnover of amino acids is converted via NH_3 (NH_4^+) into the organically bound nitrogen of urea. Plants in particular have well developed systems for the metabolism of inorganic nitrogen. Many microorganisms can use simple ammonium salts as their sole source of nitrogen for growth. Higher green plants have only a limited ability to take up ammonium ions from the soil; they depend largely on nitrate, which is formed in the soil by the microbial oxidation of ammonium (nitrification). After entering the plant cell cytoplasm, nitrate is reduced to ammonium (see Nitrogen), which is then assimilated. Carbamoyl phosphate and glutamine can be regarded as "activated ammonia", since they participate in the biosynthesis of many nitrogen-containing compounds.

Ammonification: the formation of ammonia by microoganisms, especially in the soil. Soil microorganisms degrade organic substances with the release of ammonia (e.g. by deamination of amino acids; see equation). Release of ammonia into the soil by soil microorganisms is generally known as Mineralization (see).

CH_3—$CH(NH_2)$—COOH + $\frac{1}{2} O_2 \rightarrow$
Alanine

CH_3—CO—COOH + NH_3
Pyruvic acid   Ammonia

Certain bacteria produce ammonia from nitrate by using nitrate as an electron acceptor in respiration (see Nitrate respiration). Whether from the degradation of nitrogenous organic compounds or from nitrate, the resulting ammonia may be released into the atmosphere, or reoxidized to nitrate by other bacteria. Thus, A. is important for the nitrogen economy of the soil and the entire nitrogen cycle of the biosphere.

Ammonites, *Ammonoidea:* an extinct order or subclass of the *Cephalopoda*. In most A., the shell is essentially a cone, which is spirally coiled in one plane. The few groups in which the shell is not spirally coiled are collectively known as heteromorphs. A. possess a small, ventrally situated marginal siphuncle, which starts at the posterior wall of the living chamber and extends back to the nepionic whorls (the first 3–4 whorls formed in infancy), gradually shifting from its initial marginal position to the center of the respective septa. *Clymenia* from the Devonian is unique in having a dorsal siphuncle. The siphuncle ends blindly in a bulbous extension (cecum) of the first septum, which protrudes into the protoconch (embryonic chamber). The protoconch is ovoid to spherical, and calcified. The aperture of the relatively long final chamber (living chamber), which houses the body of the animal, can be closed by an operculum, which is known in some forms as an aptychus (see Aptychia) and in others as an Anaptychus (see). Intersection of the septum with the outer wall forms a line known as the suture. Those parts of the suture curved toward the open end of the shell are called saddles, while those curved in the opposite direction are known as lobes. Three basic suture patterns can be recognized, and these serve for a classification of A., which is descriptive and not necessarily phylogenetic: *goniatitic A.* (relatively simple and undivided lobes; Ordovician to the Triassic), *ceratitic A.* (lobes divided into 2nd order lobes and saddles; Middle Permian to the Triassic), *ammonitic A.* (lobes and saddles contain 2nd, 3rd and sometimes 4th order subdivisions; Liassic to Upper Cretaceous). Without exception, the surface of the shell carries growth rings, and ribbing is also usually present. The ribs may be coarse and strong (costae) or very fine (lirae). Small, round tuberosities (nodes) are also present in some forms, often on the costae. If the tuberosities are high and hollow, they are known as tubercles. Sometimes the nodes are spirally or radially elongated, and tubercles may carry spines. Other forms of ornamentation include spiral ridges and grooves. A smooth or beaded keel may also be present. From the Ordovician to the Permian, A. are usually smooth or have simple ribs. During the Triassic, the first smooth forms appear, followed by forms with divided ribs. In a second evolutionary phase from the Jurassic to the Cretaceous, the same pattern of development is repeated, i.e. smooth forms in the Early Jurassic are later replaced replaced by forms with simple ribs, and these in turn are replaced by forms displaying an ever increasing complexity in the subdivision and branching of their ribs. In the Upper Cretaceous, ribbing is once more reduced, and smooth forms reappear.

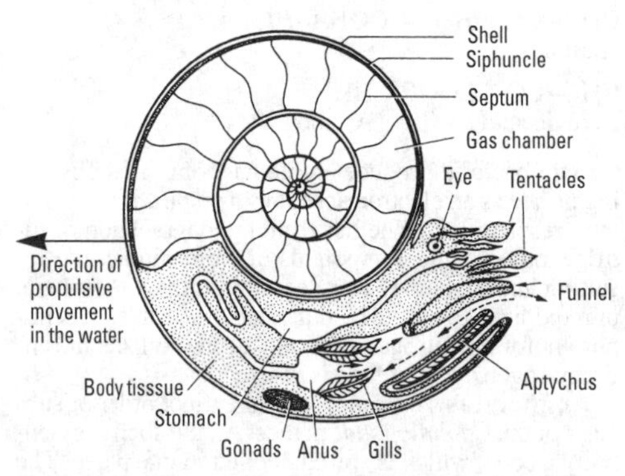

Ammonites. Reconstruction of an ammonite of the genus *Aspidoceras* (after Trauth).

Geological distribution. A. are found from the Lower Ordovician to the Cretaceous with evolutionary peaks in the Early Paleozoic, Triassic and Jurassic. From the Devonian to the Upper Cretaceous, A. are important index fossils.

Ammonitic ammonites: see Ammonites.

Ammotragus: see Barbary sheep.

Amnio-allantois: see Allantois.

Amniocentesis, ***amniotic punction*** (plate 27): sampling of the amniotic fluid during pregnancy, normally between the 12th and 16th week of pregnancy, by insertion of a needle through the abdominal wall into the uterus. By determining the Karyotype (see) of the fetal cells contained within the amniotic fluid, and by the biochemical investigation of the proteins present, it is possible to determine the sex of the fetus and to diagnose the presence of genetic diseases, e.g. Down's syndrome and Spina bifida. Genetic analysis can also be performed on the DNA isolated from the fetal cells. See Prenatal diagnosis.

Amnion: the nonvascularized, thin-walled sac filled with amniotic fluid *(Liquor amnii),* which encloses the developing fetus of reptiles, birds and mammals. The fetus hangs free in the amniotic fluid, protected from physical shock and dehydration. For the embryonic origin and structure of A., see Extraembryonic membranes.

Amniotic fluid: see Extraembryonic membranes, Amnion.

Amoeba: see Amebas.

Amoebina: see Amebas.

Amoebozygote: see *Myxomycetes.*

Amorph, ***isomorph:*** a mutant allele that has little or no detectable effect on the expression of the trait that is controlled by the normal allele. See Antimorph, Hypermorph, Neomorph.

3',5'-AMP: see Cyclic adenosine 3',5'-monophosphate.

cAMP: see Cyclic adenosine 3',5'-monophosphate.

Ampelidae: see *Bombycillidae.*

Amphetamine, ***β-phenylisopropylamine, Benzedrine®:*** an antidepressant that stimulates the circulation and respiration. It produces a euphoric state, and is classified as a narcotic.

$C_6H_5-CH_2-CH(NH_2)-CH_3$

Amphetamine

Amphibia, ***amphibians*** (plate 5): a class of vertebrates comprising newts, salamanders, frogs and toads, with variable body temperature and usually a naked, slippery skin rich in gland cells. Normally there are two pairs of limbs, which are rarely regressed or totally absent. Claws are present in only a few species. The ossified dorsal skin of certain species and the minute calcified scales of the *Apoda* are the only remaining evolutionary relicts of an exoskeleton of bony scales which clothed the extinct giant ancestors of the *A.,* the *Stegocephalia.* Many species are able to change color, due to the presence of displaceable pigment cells in the hypodermis. In the adult, respiratory gas exchange generally occurs via the lungs; gills are always present in the larvae, but only certain aquatic adult forms retain their gills. In many species the skin (which usually has a plentiful blood supply) also serves as a respiratory surface. The mucosa of the buccal cavity serves as a respiratory surface in buccopharyngeal respiration: the floor of the mouth is raised and lowered, often in rapid sequence so that the throat pulsates; this constantly replenishes the mouth with fresh air by driving air in and out of the internal nares, the choanae. The amphibian heart has two separate atria and a single ventricle (see Blood circulation, Heart). The colon, and the

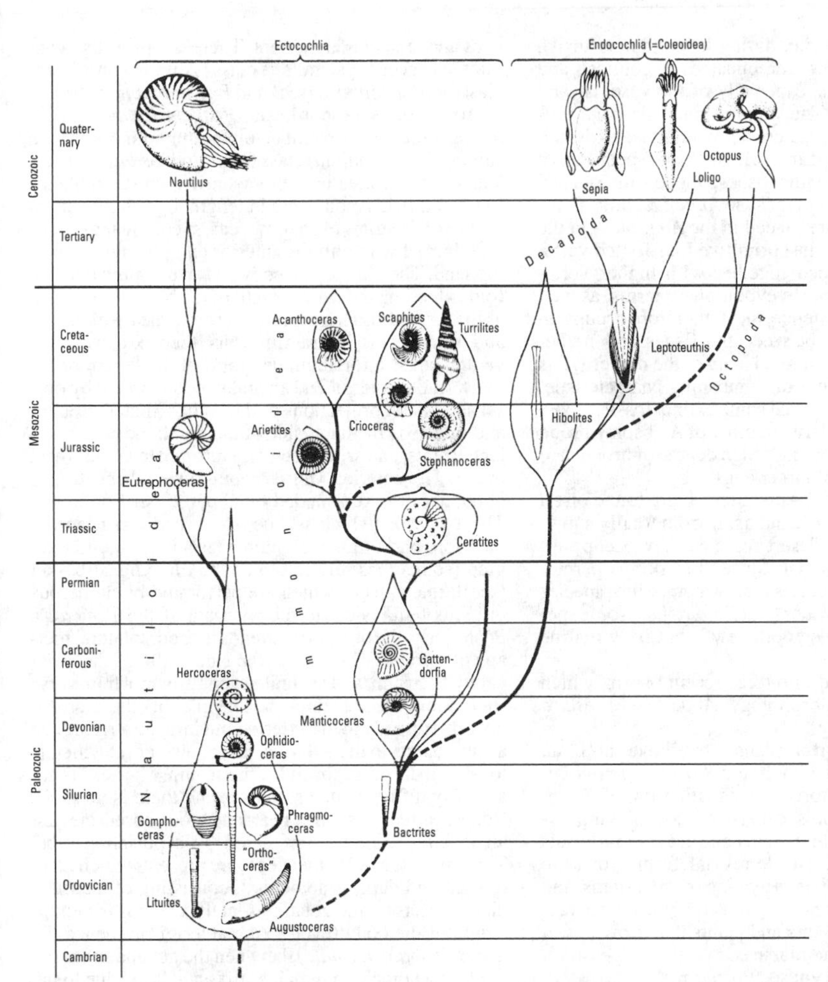

Ammonites. Evolution of the cephalopods.

urinary and reproductive tracts all open into the cloaca. With the exception of the tadpoles of the *Anura*, all *A.* devour live animal prey. Important sense organs are the tactile receptors (which exist as free nerve endings in the skin) and the eyes. The lens of the eye is fixed for relatively distant vision, and accommodation to nearer objects occurs by contraction of small ciliary muscles, which move the lens nearer to the cornea. In aquatic *A.* the eyelids are poorly developed. In terrestrial species, the lower lid is usually more movable than the fleshy, opaque upper lid. Continuous with the lower lid and folded behind it when at rest, is a transparent layer, which performs the same function as the nictating membrane of higher vertebrates, moving upward over the eye to moisturize and protect it. The *Caudata* and *Apoda* lack a middle ear, but this structure is relatively well developed in the *Anura* (frogs and toads). In the *Anura*, a flat ear drum or tympanic membrane is located behind each eye. Sounds are transmitted from the ear drum across the tympanic cavity to the inner ear by a bone called the columella, which is homologous with the hyomandibular element of the gill arches of cartilaginous fishes. Embryologically, the tympanic cavity is derived from the first pharyngeal or gill pouch; pressure equalization is achieved via an eustachian tube connecting the tympanic cavity with the pharynx. A ventral evagination of the sacculus of the inner ear (the lagena) is seen as the evolutionary precursor of the mammalian cochlea (see Auditory organ). The olfactory organ is paired and connected to the buccal cavity via internal choanae. Olfactory epithelium is smooth and restricted to the upper part of the nasal passage. Another olfactory structure is Jacobson's organ (see); this sac-like evagination of the nasal passage is lined with olfactory epithelium, and probably serves both olfaction and taste. In evolutionary terms, Jacobson's organ makes its first appearance in the *A*.

The *A.* are not fully adapted to a terrestrial existence. Adults require a damp environment, and some have become secondarily adapted to a totally aquatic existence. With a few exceptions (e.g. viviparous species) adults must return to the water to breed; the eggs are laid in water and are unable to withstand dry conditions. Fertilization is internal (most of the *Caudata*) or external (most of the *Anura*). Development from the egg to the adult proceeds via an aquatic, gill-respiring larval stage, known as a Tadpole (see). During its transition to a terrestrial existence (i.e. during its metamorphosis), the tadpole undergoes complicated anatomical and physiological changes, especially of the vascular and respiratory systems. The amnion and allantois, embryonic structures characteristic of reptiles, birds and mammals, do not appear in *A*. In this respect the *A*. show an affinity with the fishes; indeed the *A*. and fishes are grouped together as the *Anamia*, while all remaining vertebrates are placed in the *Amniota*. On the other hand, *A*. are the most primitive four-footed vertebrates, and are grouped together with higher vertebrates as the *Tetrapoda*. In evolutionary terms, as well as in their mode of existence, the *A*. therefore occupy an intermediate position between the fishes and higher vertebrates. The four-fingered hand of the modern *A*. is often considered to be a distinguishing characteristic, notwithstanding the fact that some extinct fossil *A*. possessed five fingers. A large number of *A*. display brood care; those species showing a high degree of brood care lay smaller numbers of larger eggs.

With the exception of Antarctica, *A*. are found on all land masses of the word, and apart from totally snowbound, glaciated and desert regions, they occupy almost every type of habitat. Since their skin is permeable, they cannot tolerate saline water. A few species are found in brackish water (e.g. *Bufo viridis*, some species of *Rana*, and the smooth newt), but truly marine species are unknown.

The class is divided into three recent orders, which differ in their basic morphology: *Apoda* (see), *Anura* (see) and *Caudata* (see).

Amphibious plants: plants that have both an aquatic and a terrestrial habit, e.g. *Polygonum amphibium* (amphibious bistort). The aquatic form of *P. amphibium* is found in pools, canals and slow moving rivers, where it has floating stems and leaves, and roots formed at the nodes. The terrestrial form grows on banks near to water; it has ascending or erect stems, and roots only at the lower nodes. A.p. combine the characters of truly aquatic plants and plants that grow in wet places such as bogs and marshes.

Amphiblastula larva: see *Porifera*.

Amphibolurus: see Bearded dragon.

Amphicoela: see *Anura*.

Amphicoelic vertebra: see Axial skeleton.

Amphidiploidy, *allotetraploidy:* a condition in which a diploid hybrid between 2 different species doubles its chromosome number, i.e. it has the diploid chromosome complement of both parents in its somatic cells. Diploid hybrids are usually sterile, because the chromosomes cannot undergo meiotic pairing. In A., however, each chromosome has a homolog, and meiosis takes its normal course, thus resulting in a new species; tobacco probably originated in this way. A. is known only in plants, and the artificial preparation of amphiploids is used for the production of new agricultural and horticultural varieties. See Polyploidy.

Amphigastria: see *Hepaticae*.

Amphigony: sexual reproduction with cross fertilization.

Amphilinidea: see *Cestoda*.

Amphineura, *Aculifera:* a class of elongate, bilaterally symmetrical marine mollusks, containing about 1 240 living species. The dorsal surface is covered either by a spiny cuticle, or by 8 (rarely 7) overlapping calcareous plates. The head is poorly differentiated, and eyes and tentacles are absent. There is a primitive ganglionic nervous system. The class is divided into 2 subclasses: *Aplacophora* (see) and *Polyplacophora* (see).

Amphioxus lanceolatus, *Branchiostoma lanceolatum*, ***lancelet:*** a member of the phylum *Chordata*, subphylum *Acraniata*, class *Cephalochordata*. *A.l.* is a small, slender creature with a superficial resemblance to a small fish. About 6 cm in length, it is an inhabitant of the soft bottoms of shallow seas, spending most of its time buried with only its anterior end protruding from the sand. The transparent body has an unpaired delicate fold extending the entire length of the body, and known as the dorsal fin. This continues around the tip of the tail and becomes the ventral fin, which then extends on the ventral surface for about one third of the length of the body. At the anterior end an oral hood is formed by dorsal and lateral projections of the body. Ciliated grooves and ridges on the underside of the oral hood create water currents, and are collectively referred to as the wheel organ. The anterior, ventral mouth opens beneath this hood, and it is surrounded by about 20 stiff oral cirri. The pharynx (which is lined with endoderm and is therefore part of the mesenteron) is a large sac-like region [sometimes referred to as the Pharyngeal basket (see)], the walls of which are perforated by numerous gill slits (equivalent to the stigmata of the *Tunicata)*. Posteriorly, the oral hood extends as 2 dorsolateral metapleural folds, which form the side walls of the atrial cavity or atrium, and are united ventrally by a transverse shelf, known as the epipleur. The gill slits therefore do not open directly to the exterior, but into the atrium. The atrium opens to the exterior via the atriopore, which is located at the posterior end of the atrium. The head is not a clearly differentiated structure, and there is no skull. Bones, paired eyes, a true heart and blood corpuscles are all lacking. Blood is circulated by a pulsating tube. There is no brain. Pigment dots are present, which react to light and dark. A notochord, consisting of alternate disks of fibrous and gelatinous cells, extends the entire length of the body (even to the anterior tip, hence the name *Cephalochordata)*, between the gut and the nerve cord. The notochord remains stiff and elastic due to the turgor pressure of its cells against the resistance of its connective tissue sheath. Metamerism is marked, and the segmented musculature of *A.l.* resembles that of fishes. A definite coelom is present. Further information and a lateral view of *A.l.* are to be found under Pharynx (see). See also Gills.

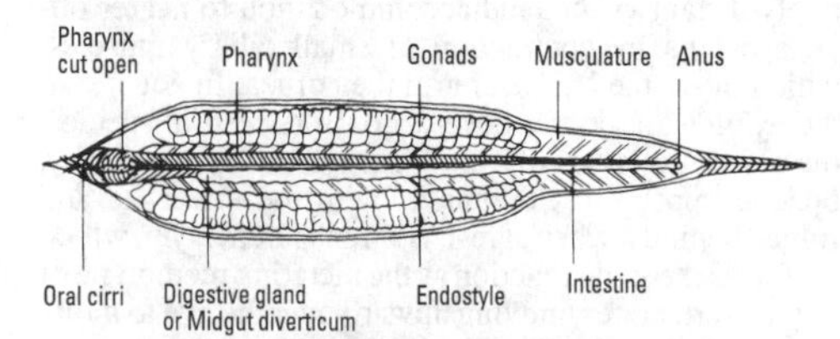

Amphioxus (Branchiostoma lanceolatum). View of the opened underside.

Amphiphotoperiodic plants: see Flower development.

Amphiploidy: a term describing all types of Polyploidy (see) that arise after the crossing of 2 or more diploid species.

Amphipnous: see *Synbranchiformes*.

Amphipoda, *amphipods, water fleas:* an order of crustaceans (subclass *Malacostraca*), without a carapace and usually with a laterally compressed body. The first pair of thoracic limbs are modified as maxillipeds, each of which carries a grasping claw (subchela). Other thoracic limbs are of more than one type, and the second and third pair are usually prehensile. The first three segments of the abdomen carry multiarticulated pleopods, which possess bristles and serve as swimming legs, whereas the posterior three abdominal segments carry uropods. There is no tail fan. The uropods, which possess only a few segments, serve as jumping legs, or to provide thrust when walking. Eggs develop in a marsupium on the ventral side of the female. Most of the 3 600 (approx.) known species are marine. The *Gammaridea* include some freshwater species and some soil-dwelling species; the *Hyperiidea* live pelagically in the high seas, often inhabiting pelagic tunicates and jellyfish; the *Cyamidae* are parasitic whale lice.

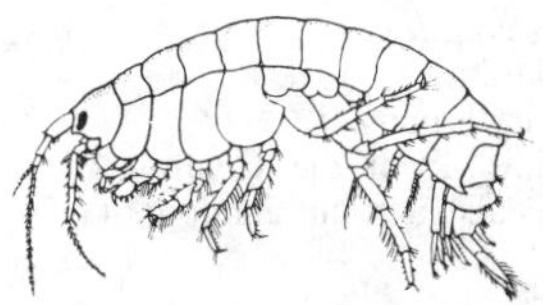

Amphipoda. Freshwater shrimp or water flea *(Gammarus locusta)* (female).

Classification (including the main families):

1. Suborder: *Gammaride.*

Family: *Gammaridae*, e.g. the common water flea (*Gammarus = Rivulogammarus pulex*), which is found amongst water plants throughout Europe in brooks and rivulets, in limpid shallow water and near the shores of lakes; the female is up to 20 mm long, the male somewhat shorter. *Niphargus* ("cave fleas") is a European genus of about 35 species, often found in cave waters and natural springs arising from groundwater; it is also found at deep levels in Alpine lakes.

Family: *Talitridae*, e.g. sand fleas and beach fleas, which often live in wet sand in the intertidal zone, as well as in the sea or in brackish water. Shore-dwelling species secrete themselves during the day and emerge to feed at night. When disturbed they respond with a powerful jump. The common sand flea (*Talitrus saltator*) lives in washed up seaweed above the waterline.

Family: *Corophiidae* (e.g. *Corophium).*

2. Suborder: *Ingolfiellidea.*

3. Suborder: *Laemodipodea.*

Family: *Caprellidae*, e.g. ghost shrimps (*Caprella*), which have very elongate, slender bodies. They are found in the sea attached to plants and hydroids, and they lure their prey. The abdomen is reduced to a vestigial stump. Legs are absent from the 4th and 5th thoracic segments, and the remaining legs carry a subchela.

Family: *Cyamidae*. Whale parasites with broad, flat bodies.

4. Suborder: *Hyperiidea*

Amphiprioninae: see Damselfishes.

Amphisbaenia, *amphisbenians, worm lizards:* a suborder of the *Squamata* (see), containing 130 species in 23 genera. The status of several apparent subspecies or races is not clear. Two families are recognized, the pleurodont (side-toothed) *Amphisbaenidae* and the acrodont (tip-toothed) *Trogonophidae*. Members of the *Amphisbaenidae* occur throughout Central and South America from Patagonia to as far north as Mexico, Lower California and the Florida peninsula. They are present in Trinidad, various coastal islands and the Atlantic island of Fernando da Noronha. They are also present in Africa, extending from the Cape to West Africa, and northeast to Somalia, but are absent from Madagascar. Related forms of one species occur in Asia Minor, Morocco and Spain *(Blanus cinereus,* the only European representative of the *A.)*. Members of the *Trogonophidae* are found in Somalia, the Arabian peninsula, Iran and the island of Socotra. One species occurs in Morocco, Algeria and Tunisia.

The phylogeny of the *A.* is uncertain. They share characters with both snakes and lizards, and they also differ from these two groups in many respects. Between 20 and 60 cm in length, they have no limbs, except for 3 species of the Mexican genus, *Bipes,* which have small, well developed forelimbs. The limbless majority have no pelvic or pectoral girdles, or these may be evident only as small splints or plates of cartilage and bone. Like other members of the *Squamata,* the *A.* possess a hemipenis, a true egg tooth, a transverse cloacal slit, and a squamate skin organization. On the other hand, *A.* show reduction of the right lung, whereas limbless lizards, snakes, and even amphibians show reduction or absence of the left lung. Only the *A.* possess an enlarged medial premaxillary tooth and a heavily ossified and completely reinforced skull, as well as a characteristic enlargement of the stapes, which makes contact with an enlarged extracolumella lateral to the lower jaw. Most *A.* have a cylindrical trunk with blunt head and tail. The skull is adapted as a burrowing organ, and the skin of the trunk is divided into annuli, usually 2 annuli per vertebra, each annulus consisting of many small rectangular segments. The eyes are concealed beneath the skin, and external auditory apertures are absent. Locomotion is by serpentine and wormlike undulatory movements, and by rectilinear progression. The latter is facilitated by the skin, which slips freely backward and forward like a loose sleeve ("concertina" movements). *A.* inhabit subterranean burrows, and they are formidable predators above and below ground. Their rigid skull and jaw construction, and massive temporal musculature permit them to feed not only on annelids and arthropods, but also on relatively large-sized prey, such as reptiles and rodents. Most species lay eggs, but a few are viviparous.

Amphitoky: production of both male and female progeny; see Parthenogenesis.

Amphitretus: see *Cephalopoda.*

Amphitrichous flagella: see Bacterial flagella.

Amphiumidae, *amphiumids, Congo eels, Congo snakes:* a family of the *Urodela* (see), containing 3 species in a single genus. Up to 1 m in length, amphiumids are neither snakes nor eels, and their trivial names of "Congo" must not be confused with conger eels. They are entirely aquatic, inhabiting the coastal plain of southeastern USA from Virginia to Louisiana and the Mississippi drainage northward to Missouri. In the the adult, 1 of the 3 larval gills slits is retained on each side, but external gills are absent, and they also possess lungs. The head is eel-like, eyelids are absent, the diminutive limbs carry 1, 2 or 3 toes, and there are vertical rib grooves along the sides of the body. They feed on water insects, worms, small fish and crabs. During mating, the partners clasp one another, and the male transfers spermatophores directly into the female's cloaca. The female coils herself around the 2 strings of up to 150 eggs each until they hatch. Larvae have both gills and limbs (incomplete metamorphosis). The ***Two-toed amphiuma*** *(Amphiuma means)* is a nocturnal lowland form, commonly found in drainage ditches, muddy polls, swamps and wet, swampy meadows. It often hides in crayfish burrows and beneath bark and debris at the water's edge. The larger ***Three-toed amphiuma*** *(Amphiuma tridactylum)* (Fig.) lives in lakes, open streams, drainage ditches and woodland alluvial swamplands. The ***One-toed amphiuma*** *(Amphiuma pholeter)* is found in parts of Florida.

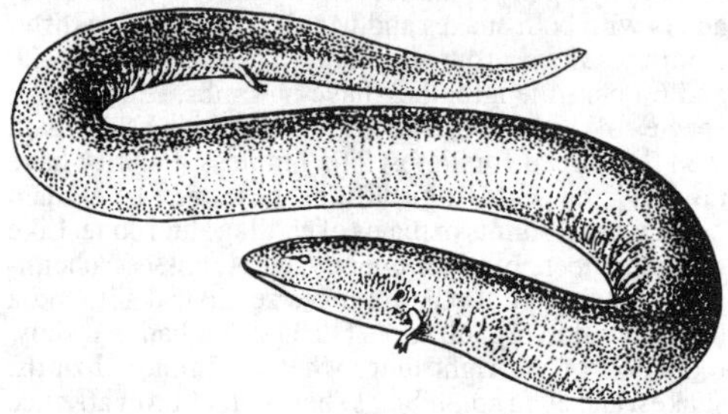

Three-toed amphiuma (Amphiuma tridactylus)

Amphiura: see *Ophiuroidea.*

Ampullae of Lorenzini: see Electroreceptors.

Amygdalin: a glycoside present in almonds and the seeds of other fruit stones, and responsible for their taste and toxicity. By the action of the enzyme, emulsin, A. is hydrolysed to benzaldehyde, hydrogen cyanide and glucose.

CH—O—Gentiobiose
|
CN

Amygdalin

Amylases, ***diastases:*** glycosidases that hydrolyse oligosaccharides, especially polysaccharides, e.g. starch, glycogen and dextrins. They are classified as α-, β- and γ-amylases. ***α-Amylase*** is an endoamylase, which degrades starch to maltose via intermediate dextrins. It is present in saliva and pancreas, and is an important digestive enzyme. ***β-Amylase*** is an exoamylase, which removes maltose units from the terminus of the polysaccharide chain. It is known only in plant seeds. The ***γ-amylase*** of liver and intestine removes single glucose units and catalyses the hydrolysis of both 1→4 and 1→6 linkages, leading to the complete degradation of starch or glycogen.

In animals, A. are important for the digestion of dietary polysaccharides, and in plants for the degradation and mobilization of storage carbohydrates.

Amylodextrins: see Dextrins.

Amylopectin: a component of starch (the other is amylose), M_r 500 000 to 1 000 000. A. is a branched, water-insoluble polysaccharide, consisting of a main chain of α(1→4)-linked D-glucose units, with side chains attached to every 8th or 9th glucose by an α(1→6) bond. The side chains consist of 15 to 25 D-glucose units. With A., iodine forms violet to red-violet inclusion compounds. A. swells in water, and upon heating with water it forms a paste.

β-Amylase degrades A. as far as limit dextrins, while α-amylase degrades it to a mixture containing about 70% maltose, 10% isomaltose and 20% glucose. Hydrolysis of A. with dilute acid yields D-glucose.

Amylose: a component of starch (the other is amylopectin), M_r 10 000 to 50 000. A. is an unbranched, water-soluble polysaccharide, consisting of 100 to 300 α(1→4)-linked D-glucopyranose residues. The fundamental unit of A. is the disaccharide, maltose. The polysaccharide chain is held in a spiral arrangement by hydrogen bonds, with 6 monosaccharide units per turn of the helix. This arrangement differs from that of cellulose. A. forms blue inclusion compounds with iodine. Degradation by α-amylase produces a mixture containing about 90% maltose and 10% glucose. Gentle degradation yields dextrins.

Amyrin: a triterpene pentacyclic alcohol, present in balsams and latexes, e.g. in the latex of the dandelion, *Taraxacum officinale.*

Anabaena: see *Hormogonales.*

Anabantidae: see Gouramies.

Anabas: see Respiratory organs.

Anabasine: an alkaloid from the Central Asian plant, *Anabis aphylla.* It is also a minor alkaloid of *Nicotiana* species.

N
N
H

Anabasine

Anabiosis: see Resting state.

Anablepidae, ***four-eyed fishes:*** a family (1 genus, *Anableps,* 3 species) of live-bearing fishes of the order *Cypriniformes,* found in fresh and brackish water (rarely coastal marine) from southern Mexico to northern South America. They are up to 30 cm-long, slender surface fishes with eyes raised above the top of the head. Each eye is subdivided by a horizontal septum, which corresponds to the water line when the fish swims on the surface. Thus, images can be perceived simultaneously from above and below the water. Most rays of the anal fin of the male are modified into a tubular gonopodium for intromission.

Anableps: see *Anablepidae.*

Anabolic steroids: see Androgens, Testosterone.

Anabolism: the totality of metabolic reactions involving an input of energy, and leading to the synthesis of new intracellular and extracellular materials, which are necessary for growth and development, and the replacement of existing body or cellular constituents. The converse of A. is Catabolism (see).

Anacanthine fishes: see *Gadiformes.*

Anacardiaceae, ***cashew family:*** a family of dicotyledonous plants containing about 600, almost exclusively tropical species, in 60 genera. They are woody plants, usually trees, with alternate, simple or aparipinnate, exstipulate leaves. The small, actinomorphic, hermaphrodite or monosexual, 5-merous flowers have 5–10 stamens and usually a 3-locular ovary, which deveops into various types of fruits. All members of the family characteristically possess resin canals, which contain mastic, latex, or balsams rich in essential oils.

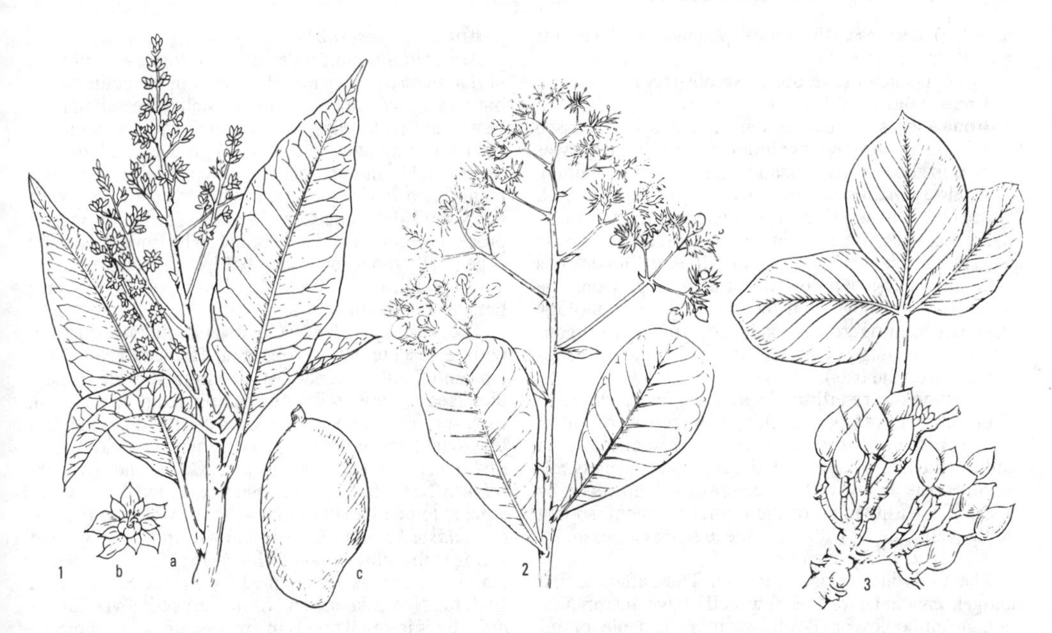

Anacardiaceae. 1 Mango tree *(Mangifera indica): a* flowering branch, *b* flower, *c* fruit. *2* Branch of cashew tree *(Anacardium occidentale)* with flowers and young fruits. *3* Branch of *Pistacia vera* with leaves and fruits.

The ***Mango tree*** *(Mangifera indica)* is widely grown in the tropics for its large, reddish-yellow fruits. The fleshy fruit stalk of *Anacardium occidentale* has the appearance and taste of an apple, and its hard, reniform fruits ***(cashew nuts)*** are eaten and used like almonds. *Pistacia lentiscus,* a shrub or small tree indigenous to the Mediterranean region, yields a resin known as mastic or mastich, formerly employed as a stimulant and in the manufacture of varnishes. The seeds ("pistachio nuts") of *Pistacia vera* are used in confectionary. *Cotinus (= Rhus), coggygria* ***(Smoke tree)*** is distributed from southern Europe to Central Asia; it has undivided leaves and loose feathery panicles of small purple flowers, and during fruiting the flower stalks become greatly elongated and covered with long, silky, reddish hairs. A popular ornamental tree of parks and gardens is the North American ***Stag's horn sumach*** *(Rhus typhina),* with large pinnate leaves and minute hairy, pale red flowers borne in dense conical panicles. Some species of *Rhus* are exceptionally poisonous, causing unpleasant skin reactions simply by touching, e.g. *R. radicans, R. toxicodendron (Toxicodendron quercifolium)* and *R. vernis* from the USA, and *R. coriaria* from Japan. Leaves of *Rhus coriaria* contain about 25% tannins, which are used for leather tanning.

Anacardium: see *Anacardiaceae*.

Anaconda, *Eunectes murenus* (plate 5): a snake of the family *Boidae* (see). It is patterned with round black spots on a yellow-brown background, and it inhabits the tropical rainforests of the Amazon and Orinoco basin. Attaining lengths of up to 11 m, it is the longest living member of the *Squamata,* although specimens greater than 7 m in length are extremely rare. It is an aquatic snake of rivers and swamps, leaving the water only to sun itself on sandbanks or tree branches. It feeds on mammals and birds that come to the water to drink, and on fish. The female gives birth to up to 50 live young, about 70 cm in length.

Anaerobes: microorganisms, especially bacteria, that are able to live in the absence of oxygen. Obligate A. (e.g. Tetanus bacterium) can develop only in the absence of oxygen. They live by Fermentation (see), Nitrate respiration (see) or Desulfurication (see Sulfur). Facultative A. (e.g. *Escherichia coli)* can live in the presence or absence of oxygen. In the presence of oxygen, some facultative species simply continue to metabolize anaerobically, whereas others switch their metabolism to aerobic respiration.

Many methods are used for the laboratory maintenance and propagation of A. ***1) Mechanical removal of oxygen.*** The lower region of a deep layer of growth medium is sufficiently anaerobic to allow the growth and multiplication of most anaerobic organisms. Anaerobic jars are also widely used: air is pumped out of the jar and replaced with hydrogen; in addition, a palladium or platinum catalyst in the jar insures that the last trace of oxygen is removed by reaction with hydrogen. ***2) Chemical removal of oxygen.*** This can be achieved by absorption of oxygen with alkaline pyrogallol, or by the catalytic reduction of oxygen with hydrogen (as in the anaerobic jar, above). ***3) Biological removal of oxygen*** (Fortner method). An aerobic, oxygen-utilizing bacterium, usually *Serratia marcescens,* and the anaerobic organism under investigation are inoculated side by side on the surface of the culture medium in an air-tight culture vessel. The aerobe develops first and uses the oxygen, followed by development of the anaerobe in the resulting anaerobic atmosphere. ***4)*** Many A. can develop in a culture medium with a ***low redox potential.*** Addition to the medium of e.g. sodium thioglycolate, cysteine, or even brain or liver fragments (Tarozzi-bouillon) decreases the redox potential and permits growth of certain A.

The opposite of anaerobe is Aerobe (see).

Anaethalion: see *Leptolepiformes.*

Anagenesis: an expression coined by B. Rensch (1947) for progressive evolution toward higher levels of organization or specialization. See Higher evolution.

Analogy, *adaptive similarity:* a term coined by R. Owen (1848) to describe the existence of structural similarities between organs that arise from their adaptation to a similar function, i.e. the similarities are not due to a common phylogenetic origin. Structures may only be described as analogous, if the existence of Homology (see) has been rigorously excluded. Analogous organs are, e.g. the wings of insects and the wings of vertebrates (birds and bats).

Anamnestic reaction, *immunological memory:* a strong immune reaction resulting from renewed contact with an antigen. The primary immune response, which follows the first contact with the antigen, is usually relatively weak and slow. It is accompanied, however, by a specific sensitization of the immune system, so that subsequent contact with the same antigen causes an intensified secondary response.

The A.r. is due to clonal selection. Thus, after the first antigen encounter only a few cells have membrane-bound antibodies or T-cell receptors capable of responding to the antigen. But these cells are stimulated to divide, resulting in the proliferation and maturation of the B- or T-lymphocytes induced by the first antigen encounter. In addition, large clones of immature B- and T-cells are formed ("memory cells"), and these respond to a second encounter with the same antigen by differentiating into functional B- and T-effector cells.

Anamorpha: see *Chilopoda.*

Anancus: see *Mastodon.*

Anaphase: see Meiosis, Mitosis.

Anaphylactic shock: see Anaphylaxis.

Anaphylaxis: an allergic reaction (see Allergy) due to anaphylactic antibodies. The term was coined as an opposite of prophylaxis, to draw a distinction between the unfavorable action of anaphylactic antibodies and the protective action of antibodies that confer immunity. The most severe form of A. is anaphylactic shock, which can lead to death. A. is due to the occurrence of the antibody-antigen reaction on the surface of cells carrying the antibody, followed by release from the cells of vasoactive substances and substances that promote contraction of smooth muscle. Symptoms are powerful itching of the scalp and tongue, flushing of large areas of the body surface, shortage of breath, vomitting, defecation, perspiration. An abrupt fall of arterial pressure may endanger the liver. There is also an increase in the heart rate with a weak and racing pulse. Fainting may ensue.

Anaplasia: loss of the characteristic structure of a differentiated cell, accompanied by its reversion to a less differentiated, primitive or embryonic type. A. occurs in certain tumors. The term is also applied to the loss of form and function of an organ.

Anaptychus (Greek *anaptychon* unfolded, spread out): the unpaired, radially or concentrically ribbed, horny (chitinous) operculum of the Paleozoic ammonites. See Aptychia.

Anarhicus: see Catfish.

Anarhynchus: see *Charadriidae.*

Anas: see *Anseriformes.*

Anaspidacea: an order of the *Crustacea* (subclass *Malacostraca).* They have long cylindrical bodies without a carapace, and the eyes are stalked, sessile or absent. The first thoracomere is united with the cephalon (which is marked with a cervical groove), and the second to eighth thoracomeres are free. The pleon (abdomen) has 6 free segments, and the pleopods are variously well developed or absent. A secondary incisor process is present on the mandible The first thoracopod is generally modified as a maxilliped. Pereiopod endopods are uniform, and the first 5 or 6 pereiopods carry branchial epipodites.

They display a classic Gondwana relict distribution, with representatives in Australia, Tasmania, New Zealand and South America. All live in fresh water, where they feed mainly on detritus and plankton. In addition, each species is rare, and is known from only one or a few localities. *Parastygocaris* is found only in Argentina, *Oncostygocaris* in southern Chile, and *Stygocarella* in New Zealand. *Psammaspides williamsi* is restricted to one locality in New South Wales, while *Eucrenonaspides oinotheke* is known only in a ground spring in the wine cellar of a house in Northwest Tasmania. *Stygocaris* is represented by 3 species, found in Victoria, New Zealand and Chile, respectively. *Allanaspides* lives in small ponds in grass swamps in remote areas of Southwest Tasmania. Other genera are *Paranaspides, Anaspides, Micraspides* and *Koonunga.* The largest species is *Anaspides tasmaniae,* which is 5 cm in length.

[Knott, B, & Lake, P.S. (1980): Eucrenonaspides oinotheke from Tasmania, and a new taxonomic scheme for Anaspidacea. *Zool. Scripta,* ***9***, 25–33]

Anaspidea (sea hares): see *Gastropoda.*

Anastomosis:

1) A linkage or shunt between 2 blood vessels, lymph vessels, nerves or muscle fibers.

2) In fungi, a linkage between 2 adjacent hyphae of the same species. The A. arises by direct fusion of the hyphae, or by the formation of lateral processes. The protoplasm of the 2 fungal cells becomes continuous, and nuclei can move from one former cell into the other via the A.

Anatidae: see *Anseriformes.*

Anatinae: see *Anseriformes.*

Anatini: see *Anseriformes.*

Anatomy: the study of the organization of living organisms, in particular the positions and structures of their organs and tissues. A. is branch of morphology. For medical purposes, *human anatomy* is often accorded a separate status from *animal anatomy,* whereas the study of *plant anatomy* is yet another discipline. A. is subdivided into: 1) *macroscopic A.,* i.e. the structure of the organism that is visible to the naked eye; 2) *microscopic A.* or *histology,* i.e. the fine structure of tissues; 3) *descriptive A.,* in which the various systems of the body are described separately, e.g. osteology (bone structure and skeletal anatomy), neurology (nervous A.), syndesmology (ligament A.), myology (muscle A.), angiology (A. of the circulatory system), and splanchnology (visceral A.); 4) *topographic A.,* which describes the relative positions of the body parts and organs; 5) *comparative A.,* which compares the individual organs and organ systems of different species; and 6) *pathological A.,* which, in contrast to the A. of the healthy organism, is concerned with the pathological changes in the body's organs caused by disease, physical injury, poisoning, and inherited defects.

Ancestrula: see *Bryozoa.*

Anchovies: see *Engraulidae.*

Ancyclostoma: see Hook worm; *Nematoda.*

Ancyclus: see *Gastropoda.*

Ancyclus lake: see *Gastropoda.*

Andantia: see Movement.

Andigena: see *Ramphastidae.*

Andreaea: see *Musci.*

Andreaeaceae: see *Musci*

Andreaeidae: see *Musci.*

Androgamete: see Reproduction.

Androgenesis: male Parthenogenesis (see). Following penetration of the egg by the sperm, the egg nucleus is eliminated, so that the resulting individual possesses only the chromosome complement of the male parent. Such individuals are said to be androgenetic. The converse of A. is Gynogenesis (see).

Androgenic gland: an endocrine gland in the males of many crustaceans, the hormone of which induces development of the primary and secondary male characteristics.

Androgens: a group of male gonadal hormones, including testosterone, androsterone and androstenolone, which are formed in the intermediary cells of the testes tissue. A number of less active A. are also formed in the adrenal cortex, e.g. androstenedione and andrenosterone. Since A. are the precursors of estrogens, they are also synthesized in the ovaries and the fetoplacental unit. At present more than 30 naturally occurring A. have been discovered, all of them structurally related to the parent hydrocarbon, androstane (see Steroids). They are biosynthesized from cholesterol via pregnenolone and progesterone, or alternatively via pregnenolone and 17α-hydroxpregnenolone. During their transport in the blood they are bound to carrier proteins. They are further metabolized to Androsterone (see) and 3α-hydroxy-5β-androstan-17-one, which are excreted in the urine, together with unchanged A. Urinary A. and their metabolites are often present as conjugates with glucuronic acid, sulfuric acid or protein. A. are also found in sperm.

The biological function of A. is to induce the development of both physical and psychological secondary male sex characteristics; they are also required for the maturation of sperm and the activity of the accesory glands of the genital tract. In addition to these sex-specific effects, A. also stimulate anabolic processes, leading to an increase in muscle protein and increased nitrogen retention. Certain synthetic A., known as ***anabolic steroids,*** display very active anabolic activity with minimal androgenic effects (e.g. methenolone, 4-chlorotestosterone, androstanazole); these are used clinically to increase protein turnover, and are also used (illegally) to increase the performance of athletes and racing animals.

Castration causes the disappearance of secondary sex characteristics, and this can be counteracted by the administration of A. In the capon comb assay of androgenic activity, admistration of A. to a capon causes its degenerated comb to grow. One capon unit is the amount of androgen that produces an increase in the comb surface of 20% (e.g. 15 mg testosterone). One international unit (IU) of androgen activity corresponds to 0.1 mg androsterone. Immunoassays are now used for the clinical determination of A. Highly active, synthetic A. (e.g. methyltestosterone, mesterolone), which are effective when administered orally, are used therapeutically to correct the deficiency symptoms following castration, hypogenitalism, impotency due to lack of hormones, climacterium, mammary carcinoma and peripheral circulatory impairment.

Androgenic gland: an inner secretory gland in the males of many crustaceans, the hormone of which induces the development of the primary and secondary male characteristics.

Androstane: see Steroids.

Androstenedione, ***androst-4-en-3,17-dione:*** a weakly androgenic steroid hormone, m.p. 174 °C, which occurs in the testes and the adrenal cortex, but in much smaller concentrations than testosterone. It is synthesized from pregnenolone via progesterone, and it is the immediate precursor of testosterone. Since the estrogens are synthesized via testosterone, A. is also present as a biosynthetic intermediate in the ovaries and the fetoplacental unit.

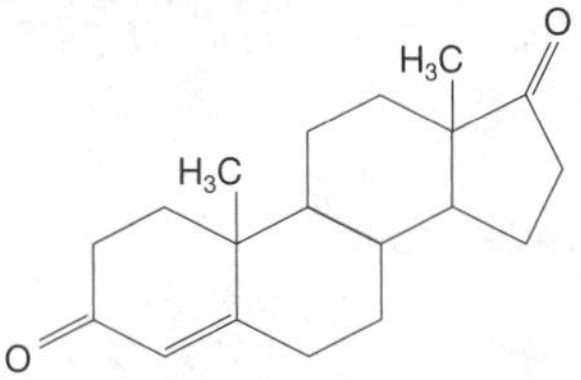

Androstenedione

Androstenolone, ***dehydroepiandrosterone, DHEA, 3β-hydroxyandrost-5-en-17-one:*** an Androgen (see), m.p. 142 °C, less potent than testosterone, but with similar physiological effects m.p. 142 °C. A. is biosynthesized from pregnenolone via 17α-hydroxypregnenolone, and it is an intermediate in the biosynthesis of testosterone in the adrenal cortex.

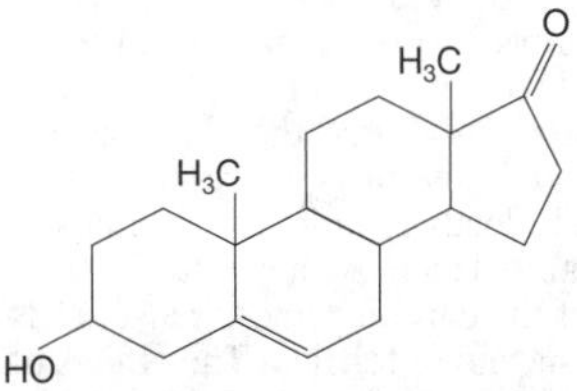

Androstenolone

Androsterone, ***3α-hydroxy-5β-androstan-17-one:*** an Androgen (see), m.p. 183 °C, which is formed in the interstitial cells of the testes. Its androgenic potency is 7 times less than that of testosterone, but it is used as the international reference standard for androgenic activity: 1.0 international unit (IU) corresponds to 0.1 mg A. A. is excreted in urine as a metabolic product of testosterone. It was first isolated and purified in 1931 by Butenandt, who obtained 15 mg A. from 1 500 liters of urine.

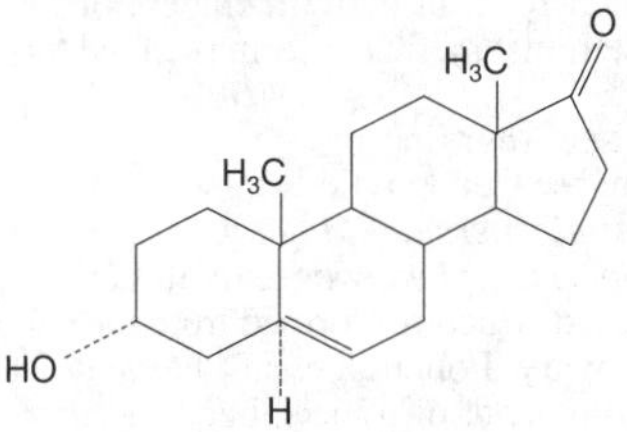

Androsterone

Anelectrotonus: see Electrotonus.

Anelytropsidae: see Mexican blind lizard.

Anelytropsis papillosus: see Mexican blind lizard.

Anemochory: see Seed dispersal.

Anemogamous flower: see Pollination.

Anemogamy: see Pollination.

Anemone: see *Ranunculaceae.*

Anemone fishes: small coral fishes of the genera *Amphiprion* and *Premnas* (subfamily *Amphiprioninae,* family *Percidae),* which display a commensal or close territorial relationship with large sea anemones. They find protection by living about and among the tentacles of the sea anemones. Discharge of the sea anemone nematocysts appears to be inhibited by a factor in the skin of the A.f. See Damselfishes.

Anemonia: see *Actiniaria.*

Anemophilous flower: see Pollination.

Anemophily: see Pollination.

Anemotaxis: see Wind factor.

Anethole, *p-methoxy-4-propenylbenzene:* a major component and aromatic principle of anise *(Pimpinella anisum),* star anise *(Illicium verum)* and fennel *(Foeniculum vulgare).*

$H_3CO-C_6H_4-CH{=}CH-CH_3$

Anethole

Aneuploidy: a type of Heteroploidy (see), in which the chromosome complement of individual cells, tissues or entire organisms is not an exact multiple of the haploid number, i.e. in which there is a loss or gain of single chromosomes. If chromosomes have been lost, the condition is "hypo", e.g. hypodiploidy, hypotriploidy, etc., whereas the gain of chromosomes is represented by the prefix "hyper" (e.g. hyperdiploidy). If the parent chromosome complement is diploid (2n), some possible types of A. are: monosomy (2n-1), trisomy (2n+1), tetrasomy (2n+2), and double trisomy (2n+1+1). A. can arise in diploid or polyploid cells. Since the chromosome complements of diploid individuals represent a balanced system, A. usually leads to more or less marked disturbances of the gene balance. As a result of the altered meiotic pairing of the chromosomes, it also leads to impairment of meiosis and of fertility. The effects of A. on gene balance are less drastic in polyploid systems. A. is usually the result of Nondisjunction (see).

Aneusomy: the occurrence within an individual organism of cells with different chromosome numbers. See Somatic inconstancy.

Angel-fishes: a name applied to fishes of the marine family *Pomacanthidae,* and to freshwater Cichlids (see) of the genus *Pterophyllum.*

Angelica: see *Umbelliferae.*

Angelic acid: an unsaturated carboxylic acid, which is present in esterified form in many members of the *Umbelliferae.* The stereoisomeric *cis*-form is called ***tiglic acid.***

Angiospermae: see Angiosperms.

Angiosperm era: see Cenophytic.

Angiosperms, *Angiospermae, Magnoliophytina:* a subdivision of the *Spermatophya,* whose ovules are always housed in a closed structure (formed from the carpels) known as the ovary. Pollen does not have direct access to the ovules, but lands on a specialized and characteristic receptive structure, known as the stigma. A pollen tube then carries sperm cells from a pollen grain on the stigma to an ovule in the embryo sac, where fertilization occurs. Typical of A. is the predominance of hermaphrodite flowers (see Flower) with a floral envelope or perianth, and a propensity for pollination by animals. Insects, attracted by pollen as a food source, probably played an important part in the evolution of hermaphrodite flowers by becoming pollinators. In primitive spermatophytes, however, such insects would also have damaged the open, exposed, nutrient-rich ovules, so that these became enclosed in carpels for protection.

Almost invariably, the stamens of A. consist of a filament and an anther head, the latter consisting of two thecae, each with two pollen sacs. The stamens probably evolved from the simple microsporophylls of pinnate-leaved gymnosperms *(Cycadophytina),* while the carpels evolved from the macrosporophylls. Ovaries may contain one or several ovules, giving rise to single-seeded, or multi-seeded fruits. In most cases, an ovule consists of two integuments, i.e. the nucellus and the embryo sac. The female gametophyte develops inside the embryo sac, and it generally consists of only a few cells, being even more reduced than in the gymnosperms. At one end of the female gametophyte are a single egg cell and two synergidae. Three antipodal cells lie at the opposite end, and there is a more or less centrally situated embryo sac containing the secondary embryo sac nucleus. A. do not possess archegonia. Male gametophytes are formed by mitosis of the uninucleate pollen grain, producing a vegetative and a generative (or antheridial) cell; the latter divides again into two sperm cells. A. do not form spermatozoids. The triploid secondary endosperm, which serves as a nutritive tissue for the embryo in the seed, arises by fertilization of the secondary endosperm by one of the two sperm cells; this occurs only after the egg cell has been fertilized by the other sperm cell ("double fertilization"). The integuments develop into the seed coat (testa), while the entire ovule becomes the seed. In the A., the seed is always enclosed by an envelope, which is formed from the ovary with the participation of other flower or even leaf parts, the entire structure constituting the fruit. The fruits of individual A. display a wide variety of sizes, shapes and surface modifications, related to their means of dispersal.

In most A., the water-conducting tissue consists of vessels. Tracheids are still present, but they have become modified to fiber tracheids and wood fibers. Sieve tubes are also prominent, and they are accompanied by companion cells. These companion cells, which are physiologically and ontogenetically closely associated with the sieve tubes, make their first appearance in plant evolution in the A. [The genus *Austrobaileya* (family *Austrobaileyaceae,* order *Magnoliales)* is the only angiosperm lacking companion cells in the phloem]. Also characteristic of A. is the variety of form displayed by the vegetative leaves. The original primitive form appears to be the undivided, pinnate-nerved leaf, which in turn has given rise to various other designs (see Leaf). Two types of branching occur in the A., monopodial and sympodial (see Shoot). Ninety percent of the herbaceous and shrub A. show sympodial branching. Among the A., herbs predominate, both with respect to the number of different species, and with respect to the total

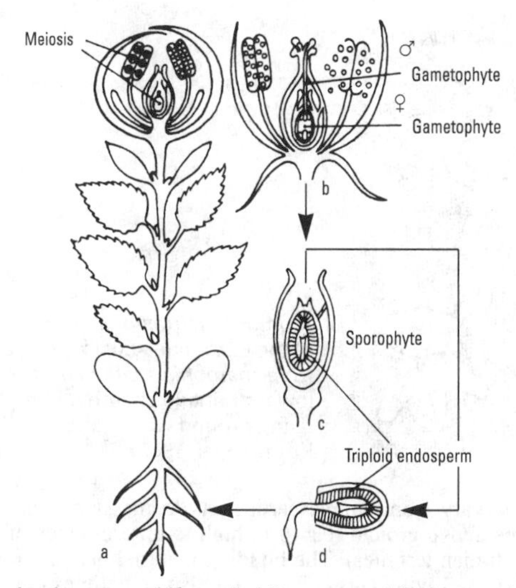

Angiosperms. Life cycle (diagrammatic) of an angiosperm (after Firbas). *a* Entire plant with root, stem, leaves and hermaphrodite flower bud. *b* Open flower with perianth (sepals and petals), stamens (with pollen grains), and carpels (ovary, style, stigma and enclosed ovule); pollination has occurred (pollen tube growing) but egg cell in embryo sac is not yet fertilized. *c* Seed becoming detached from the fruit. *d* Germinating seed.

number of individual plants. Trees are seen as a primary form, evergreen trees evolving into deciduous trees and shrubs, followed by perennial herbs and finally annual herbs. It is thought that this development was primarily the result of a decrease in the functional capability of the cambium. Herbaceous plants reproduce more rapidly than trees, producing more generations in a given period. Herbs are therefore the more adaptable, and are able to invade regions that are unavailable to trees. Some treelike species have, however, evolved secondarily from herbaceous species; this transformation is achieved by development of an active cambium, or development of primary thickening, the latter occurring in the treelike monocotyledonous A. (e.g. palms).

Evolution of A. probably started in the Mesozoic, more than 130 million years ago. Fossils from this time are rare. Early ancestors were probably the fossil seed ferns *(Lyginopteridatae),* from which the *Cycadatae, Bennettitatae, Gnetatae* and *Angiospermae* separated and evolved in parallel. The A. are therefore seen as a natural evolutionary group (monophyletic group). Containing some 230 000 species (new species are continually being described), the A. are the most numerous of all plant groups. Their taxonomic separation into the *Dicotyledoneae* (see) and *Monocotyledoneae* (see) is based on the observation that the germinating seed produces one or two cotyledons. These two classes separated very early (the ancestral forms being dicotyledonous), their distinct fossil representatives being found as early as the Lower Cretaceous.

Angler: see Goosefishes.

Anglerfishes: see *Lophiiformes.*

Anguidae, ***lateral fold lizards and allies:*** a family of the *Lacertilia* (see), containing about 75 species in 8 genera, distributed on every continent except Australia. They have a very limited range in Africa, being restricted to the northwestern corner. They attain their greatest diversity in the Americas, where they are found from southwestern Canada and much of the northern USA to northern Argentina, and in the West Indies. They are absent from Sri Lanka and the Arabian Peninsula, but present in Sumatra and Borneo. Members vary from lizards with 4 well developed limbs to legless snakelike forms. The tail is autotomous and usually very long. Species of the genus,*Gerrhonotus* ***(Alligator lizards*** in the narrow sense), which occur from southern Canada to Central America, have 4 legs, each with 5 toes. Species of the South American genus, *Ophiodes* ***(Worm lizards),*** have no forelimbs, and small vestigial toe-like hindlimbs. The snake-like members of the genus, *Ophisaurus* ***(Glass lizards),*** found in Eurasia, Africa and America, have only minute spurs in place of the hindlimbs. The Blindworm *(Anguis fragilis)* (see) is entirely limbless, although parts of the pelvic and pectoral girdles are still present. Some species have an extensible longitudinal fold on either side of the body. The sides and belly are covered with rounded scales, which are underlain by bony plates (osteoderms). Eyelids are separate and freely movable. The diet consists mainly of slow moving animals, such as snails and worms, but larger species also prey on mice. Some species are egg-laying (oviparous), while others are live bearing (vivi-oviparous), the latter occurring primarily in climatically unfavorable regions, such as high mountains and northern latitudes. The genus, *Gerrhonotus,* contains both oviparous and vivi-oviparous species. The Blindworm is ovi-viviparous, whereas the European glass lizard (see) lays eggs. Three subfamilies are recognized: *Diploglossinae* (found only in the Americas; do not possess a lateral fold); *Gerrhonotinae* ***(Alligator lizards;*** characterized by the lateral fold; found in both the Old World and the Americas; includes the Sheltopusik); *Anguinae* (Old World only; includes the Blindworm).

Anguina tritici: see Eelworms.

Anhole: see *Bovinae.*

Anholocyclic parthenogenesis, ***acyclic parthenogenesis:*** a cycle of generations in aphids, in which the sexual reproduction phase has been lost, and reproduction is entirely by parthenogenesis. The converse of A. is parthenogenesis (cyclic parthenogenesis, monocyclic parthenogenesis), in which parthenogenetic generations alternate with a sexually reproducing generation. See Parthenogenesis.

Anhimae: see *Anseriformes.*

Anhimidae: see *Anseriformes.*

Anhinga: see *Pelicaniformes.*

Anhingidae: see *Pelicaniformes.*

Aniliidae, ***pipe snakes:*** a family of small, short-tailed, ground-living or burrowing Snakes (see) with a nondemarcated head, small eyes and smooth scales. Remnants of the pelvic girdle and of the hindlimbs (which are evident as claw-like spurs) are normally present. Native to the tropics, the A. are anatomically and phylogenetically closely related to the *Boidae* (see). The 80 cm-long South American ***False coral snake*** *(Anilius scytale)* is bright red with black transverse rings.

Anilius: see *Aniliidae.*

Animal: a uni- or multicellular, heterotrophic organism. Among the simplest life forms, the differentiation of A. and plants is often difficult. Alternatively, the *Protozoa* may be classified not as A., but in the separate kingdom, *Protista*.

The following principal properties distinguish A. from plants. The nutrition of A. is independent of light, but dependent on a supply of high molecular mass organic compounds, i.e. A. are heterotrophic, whereas green plants are autotrophic. In A., the exchange of materials with the environment occurs primarily in internal body cavities, and not at the external body surface as in plants. Consequently, A. display extensive internal organization and differentiation, so that internal surface areas for respiratory gas exchange and nutrient uptake are large in comparison with the external surface area and body mass. In contrast, plants display external differentiation. Whereas plant cells have strong, cellulose-containing walls, animal cells are bounded by thin, cellulose-free membranes. The tissues of A. contain intercellular substances, like the ground material of cartilage and bone, elastic and collagen fibers, keratin and chitin, which together with inorganic materials (e.g. calcium calts, silicic acid) form part of the mechanical and support tissue of the endo- or exoskeleton. In contrast to the continuous growth of plants, the growth of A. eventually decreases or stops completely, due to the loss of the ability of young, undifferentiated cells to divide. A. represent a "closed system", i.e. they possess no growth zones or vegetative points such as cambium or buds. With the evolution of nervous tissue, A. have acquired the properties of excitability, sensory perception and coordination of movement, which are required, inter alia, for the often very complicated behavior of A. in relation to their environment. Reproductive organs of A. are usually situated in body cavities, whereas the reproductive organs of plants are developed almost without exception on the body surface. Furthermore, many A. exhibit brood care.

Animal constructions: structures built by animals for protection of the animal or its offspring, for rearing the young, or for food acquisition. They are biologically useful structures produced by the remodelling of materials in the animal's habitat. Shells, cocoons and webs are not A.c. in the true sense. Although they serve for protection of the animal, passages and cavities formed in the substratum during feeding and migration, e.g. the leaf mines of certain phytophagous insects, are also not true A.c. The feeding channels and brooding passages of bark beetles are united in a single system, which is therefore intermediate between a true A.c. and a system of feeding channels.

Dwelling (residential) structures are widespread in the animal kingdom, e.g. tubes of many annelids, the self-excavated subterranean cavities of many crustaceans, living passages and cavities excavated by insects in the earth and in plant tissues, soil cavities of many amphibians. Special structures are made by a great many birds and mammals. Birds' nests are perhaps the most familiar example of intricate A.c. Some birds (e.g. wrens) construct several nests, one of which is reserved for brooding. The living quarters of rodents, some predators, insectivores, etc. are almost exclusively earth excavations, except in those cases where natural cavities are exploited.

Many dwelling constructions serve simultaneously or primarily for reproduction and brood care, especially in the case of colonial insects, whose constructions vary from simple earth nests to massive structures above ground (e.g. 6 m-high termite mounds of Australian termites). The building material is soil, often mixed with saliva. Feces are also sometimes employed. Wasps use chewed wood to construct paper nests. The red wood ant uses small pieces of wood and pine needles. Weaver ants use the silk threads produced by their larvae to "sew" leaves together into a leaf nest. *Halictus* species fabricate nests from a mud-like mortar, which they prepare themselves. Terrestrial snails and some fishes, reptiles and amphibians prepare depressions in the soil or bottom substratum for egg laying. Mammalian subterranean burrows serve as dwelling places and for accommodation of the young, often with special nursery chambers containing additional nesting material. Tree nests are constructed by squirrels, and the beaver constructs a "lodge" within a specially constructed dam.

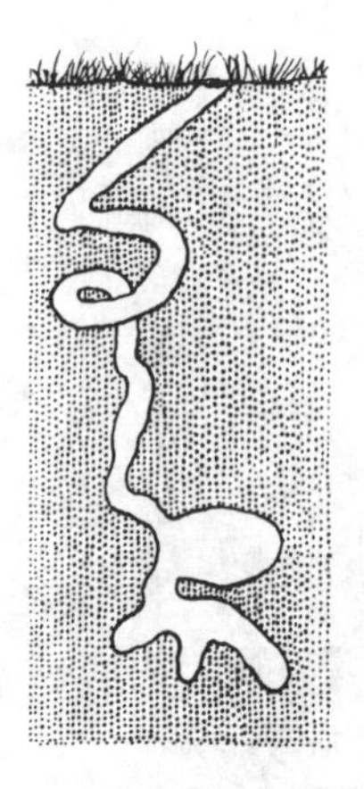

Animal constructions. Burrow of the European souslik *(Spermatophilus citellus)* with lower drainage chambers for rising ground water (after Krumbiegel, 1955).

Animal constructions. Bank burrow and living chamber of the Eurasian river otter *(Lutra lutra)*. The entrance is below the water level, with an emergency exit on the land side of the bank (after Krumbiegel, 1955).

Birds build a wide variety of different nests. Generally, the nests of precocial birds are simpler and usually on the ground, whereas altricial species are more dependent on the nest, which is therefore more intricately and strongly constructed and located at a more inaccessible site. Building materials are mainly plant material, but mud and saliva are also used, as in the nests of swallows and South American ovenbirds. Gray and gray-rumped swiftlets of Malaya and Indochina build their nests en-

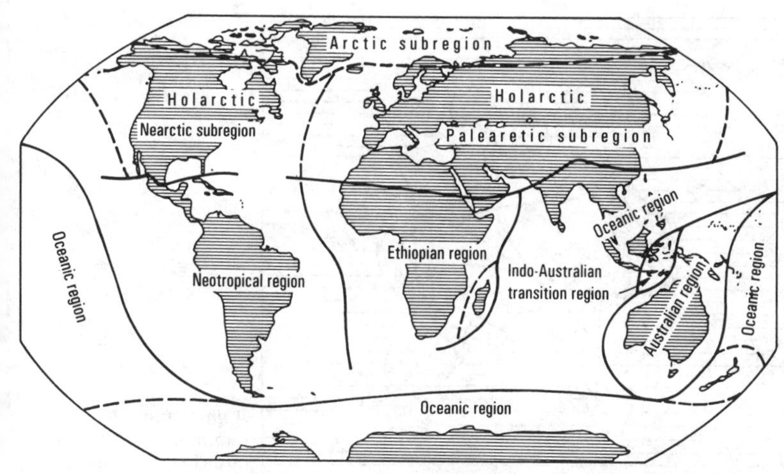

Animal geographical regions (terrestrial)

tirely of hardened saliva. Large mounds of vegetation are built by megapodes purely for incubating their eggs, the incubation temperature being generated by fermentation.

The earth constructions of the *Pompilidae* and *Sphecidae,* as well as the passageways of beetles that display brood care, also serve to insure the undisturbed development of the offspring.

Constructions for the acquisition of food include the sand funnel of the ant-lion and the U-shaped tubes of the marine polychaete, *Arenicola,* which are also used as dwelling places by the animals that construct them.

Animal geographical regions, ***zoogeographical regions, faunal regions:*** regions of the earth's surface, each containing a fauna that is more less peculiar to it. The terrestrial regions are combined into larger regions ("empires") (see Arctogea, Neogea, Notogea), and divided into subregions then provinces, although these terms are sometimes used in a different hierarchical order, or replaced by terms such as district, area and zone. There is a lack of consensus on the number of recognized regions, but the following six (each is listed separately) are generally accepted by all authorities: Australian, Ethiopian, Nearctic, Neotropical, Oriental and Palearctic.

Animal geography, ***zoogeography:*** a branch of biogeography which describes the present and past distribution of animals on the earth *(descriptive A.g.),* and analyses and interprets the factors leading to this distribution *(causal A.g.).* The established pattern of animal distribution is largely interpreted as the result of the matching of ecological conditions and ecological requirements *(ecological A.g.).* A.g. also includes the study of the extent and course of Animal migration (see).

A.g. also contributes to other disciplines. For example the Continental drift (see) hypothesis is now fundamental to the interpretation of earlier animal dispersal routes, but for a long time the students of A.g. were the only biologists to support the views of Alfred Wegener. Such a mutual dependence of disciplines often leads to circular reasoning.

See Areal, Dispersal center, MacArthur-Wilson theory of island biogography.

Animal migration: active change of territory by animal populations or parts of populations, resulting in an alteration of the geographical distribution of the species (see Dispersion). Movement from a hitherto occupied region is called *emigration,* whereas the penetration of new territory is called *immigration.* Movement through already occupied territory is called *permigration.* In the stricter sense, migration is a regular periodic movement to and from a given area, usually via established routes. If the new territory does not become occupied, the movement is an invasion (e.g. invasion of central Europe by the Siberian pine jay *(Nucifraga caryocatactes).*

The migration rate expresses the proportion of migrating animals in a population (e.g. 64% for young great tits); high migration rates often lead to dispersal of the original population (e.g. the early summer migration of lapwings). Massive immigrations and invasions are made by locusts, Migratory lepidoptera (see) and lemmings.

If movement is practically continuous (often determined by the search for food), the condition is called *nomadism,* e.g. many African grazing animals move according to the availability of pasture, and the rosy pastor *(Sturnus roseus)* follows locust swarms.

Periodic translocations may occur daily (e.g. movement to specific water depths by small marine crustaceans, and the movement from feeding to sleeping sites by starlings), or seasonally. Such movements may be determined by the requirements of different developmental stages or generations (e.g. summer and winter hosts of aphids) or by changes of normal diet (e.g. feeding migrations of cod and haddock), or they may be related to mating and reproduction (e.g. Palolo worm, many mollusks and echinoderms, mitten crabs, salmon, eels, amphibians, penguins and seals).

Finally, the overwintering migrations of birds, bats and some insects are also determined by the need for food. Bird migration may take place on a broad front (song thrush) or a narrow front (white stork). Predatory birds are solitary in their migrations, whereas small song birds migrate in flocks. Storks, swallows and swifts migrate in the day, whereas the yellow warbler *(Hippolais icterina),* the European robin and many ducks migrate during the night. The arctic tern holds the

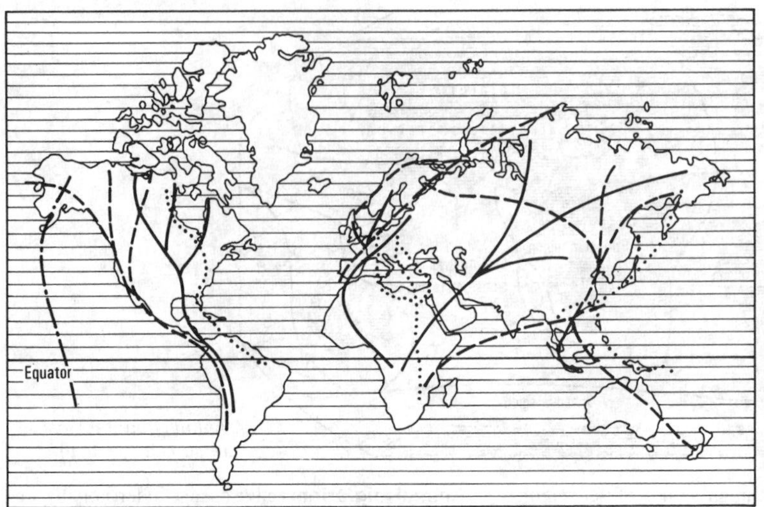

Animal migration. Main migration routes of birds in the Northern Hemisphere (after Creutz, 1951).

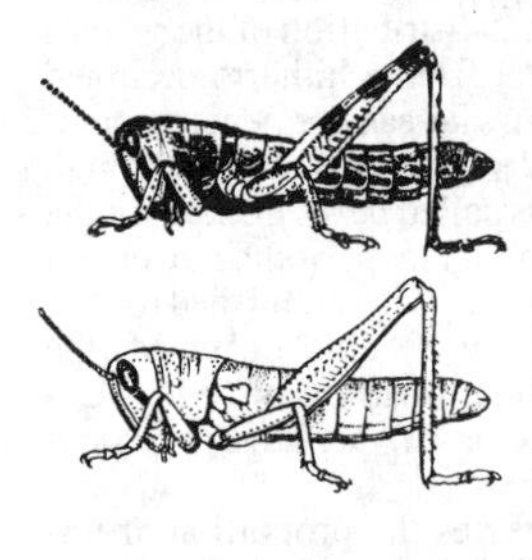

Animal migration. Larvae of the migratory locust, *Schistocerca paranensis. a* Migratory phase. *b* Solitary phase. (After Dampf).

record for the greatest recorded migration distance (17 000 km). In the tropics, the alternation of wet and dry seasons largely determines the time and direction of the migrations of birds, as well as other animals.

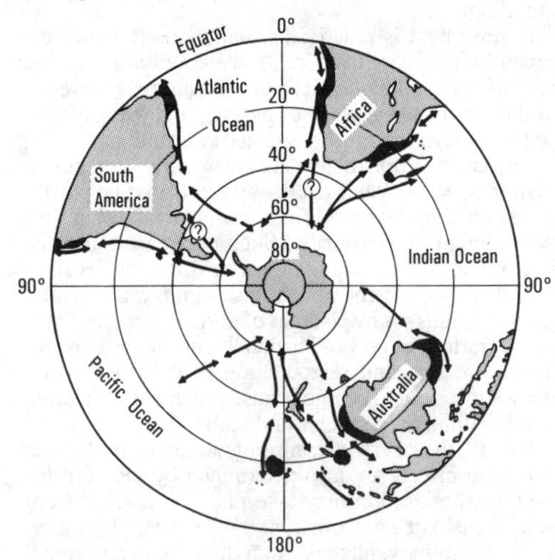

Animal migration. Migratory routes and catching areas (black) of the humpback whale *(Megaptera novaeangliae)* in the Southern Hemisphere (after Slijper. 1958).

Animal migration. Song thrush *(Turdus philomelos),* an example of broad front migration (after Eichler, 1934).

Animal lice: see *Phthiraptera.*

Animal protein factor: a factor present in animal protein, which was ultimately identified as vitamin B_{12} (see Vitamins).

Animal repellants: chemical compounds that repel wild animals, e.g. fish oil, acid-free wood tar, distillation residues of mineral oil, etc.

Animal semiotics: see Zoosemiotics.

Animal sociology: see Sociology of animals.

Anion: a negatively charged atom or molecule, formed by the uptake of electrons.

Anis: see *Cuculiformes.*

Anise: see *Umbelliferae.*

Anisian stage: see Triassic.

Anisogamy: see Reproduction, Fertilization.

Anisognathous dentition: see Dentition.

Anisoptera: see *Odonatoptera.*

Ankylosaurus: see Dinosaurs.

Annelida, ***annelids:*** a phylum of coelomate, segmented animals, containing 17 000 known species. Almost all species are elongate and wormlike, with pronounced metamerism (segmentation) of the trunk. The head or acron, represented by the prostomium, is not a segment, nor is the terminal pygidium. There is a tendency for anterior trunk segments to fuse with the prostomium. Trunk segments are generally internally and externally similar, separated externally by circular grooves, and internally by membranous partitions (dissepiments). Each segment carries bristles (setae or chetae), embedded in, and secreted by, pits of the ectoderm. In the *Polychaeta,* the setae are borne in groups on processes called parapodia. Each segment has a pair of coelomic cavities, each housing the ciliated nephrostome of a metanephridium, whose nephridiophore opens to the exterior on the following segment. The mouth is usually found in the first trunk segment, where it is overlapped by the prostomium, while the anus opens on the pygidum. The alimentary canal consists of pharynx, esophagus, midgut and hindgut. Compared with the brain of platyhelminths, the annelid brain represents a considerable evolutionary development. It lies above the pharynx, and is joined by two circumpharyngeal commissures to the subpharyngeal ganglion; a ventral ganglion chain (two fused ganglia in each segment, linked by a paired ventral nerve cord) then extends to the last segment. Lateral nerves to body wall muscles, blood vessels and excretory organs are segmentally arranged. The usually closed vascular system consists of a longitudinal dorsal vessel linked by looped lateral vessels to a longitudial ventral vessel. Respiratory gas exchange occurs through the skin or by means of gills.

Annelids are primarily aquatic animals, but many species have adapted to existence in moist terrestrial environments. The mode of feeding is very variable; in addition to predators, herbivores and ciliary feeders, there are also blood-sucking parasites. Polychaetes and myzostomids develop via a free-swimming trochophore larva, or via the specially modified trochophore, the mitraria. Other annelids develop directly from the egg without metamorphosis.

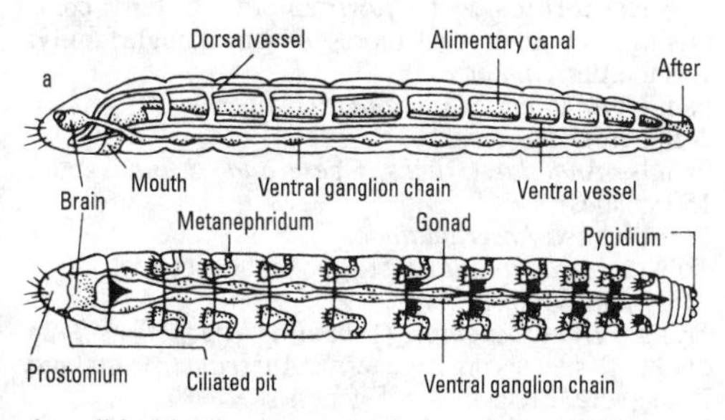

Annelida. Idealized structural plan of an annelid. *a* Lateral view. *b* Ventral view.

Three classes are recognized: *Polychaeta* (see), *Myzostomida* (see), *Clitellata* (see), the latter containing the subclasses, *Oligochaeta* (see) and *Hirudinea* (see).

In addition, there is an enigmatic group, parasitic or commensal on freshwater crayfish, and known as *branchiobdellid annelids.* They are usually included with the *Oligochaeta,* but they probably diverged early from the common oligochaete-hirudean stock. They are all very small (only 14 or 15 segments), and the head is modified to a sucker with a circle of finger-like projections. Setae are absent. The buccal cavity contains two teeth. Posterior segments are also modified to a sucker

Fossil record. Fossils are found from the Cambrian to the present, with possible evidence for annelids in the early Precambrian of Australia. Fossilized remains include the jaw apparatus (see Scolecodonts), scales, setae, living tubes and burrows, and tracks. The serlupids of the Upper Cretaceous and of the Tertiary are used for stratigraphic purposes.

Collection and preservation. Polychaetes are found on the surface of flooded soil. Bright illumination at night attracts mainly sexually mature animals. They are narcotized by dropwise addition of 70% ethanol or 2% formaldehyde, laid on a glass plate, then fixed with a mixture of 40% formaldehyde and ethanol under a cover glass. They are stored in the same mixture. Oligochaetes and leeches are collected as above. Terrestrial forms are killed with chloroform, aquatic forms by the dropwise addition of chloral hydrate or magnesium sulfate solution to their water. For fixing, warm, 4% formaldehyde solution is poured over the extended animal. Specimens are stored in a mixture of 40% formaldehyde and ethanol.

Annidation, ***Ludwig effect:*** a hypothesis that explains the coexistence of mutants or species. If the population size of a species is limited by a minimum factor (e.g. food), then a mutant that exploits a different ecological niche (e.g. a different food source) cannot be displaced, even if it is otherwise markedly inferior to the original genotype. Provided the original genotype is not completely displaced by the mutant, an equilibrium is established between the two. Such equilibria permit the coexistence of closely related species.

Anniella: see *Anniellidae.*

Anniellidae, ***shovel-snouted legless lizards:*** a family of the *Lacertilia* (see), related to the *Anguidae* (see), and containing only 2 species in a single genus. Limbs are absent, and only minute vestiges of the pectoral girdle remain. They are burrowing reptiles, wormlike or snakelike in appearance. The eyelids are separate and movable. As an adaptation to burrowing, the skull is compact and firm (the skull bones are fused to one another), the lower jaw is set back, there are no external ear openings, and the scales are smooth. The diet consists of insect larvae, beetles and spiders. Both species are found in California and on neighboring islands: ***Californian legless lizard*** *(Anniella pulchra);* ***Geronimo legless lizard*** *(Anniella geronimensis).*

Annual cycles: variations in the behavior of living organisms which are synchronized with the annual seasons. A.c. occur particularly in animals that live for several years, and they are associated with metabolic and hormonal changes. In the course of the year, an animal may display changes in the intensity of its activity (quantitative changes), and its daily behavioral pattern may show qualitative changes. Thus, in many long-lived animals reproduction occurs in a yearly cycle, an important environmental cue ("Zeitgeber") being the change in daylength. The timing of reproduction in monoestrus species (i.e. estrus occurs once a year) has evolved so that the offspring develop or are reared during the most favorable part of the year. Occasionally, the timing of mating is not strictly controlled, and the appearance of offspring is timed by an appropriate prolongation of gestation. Territorial behavior is also subject to A.c., in particular in species that regularly migrate. Some social species show annual changes in dominance between the sexes. In some species, A.c. are so firmly and endogenously established that the rhythm remains unchanged when the animal is transferred to a different geographical latitude, e.g. when the Australian black swan *(Chenopis atrata)* is transferred to the Northern Hemisphere it incubates its eggs and produces young in the winter.

Annual plants: plants that complete their life cycle from seed germination to fruit ripening and death of the plant within one year, more precisely within a single vegetative period. *Summer annuals* germinate in the spring, flower in the summer of the same year, set fruit and die (e.g. summer barley). *Winter annuals* germinate in autumn, then flower, set fruit and die in the following year (e.g. winter barley). See Biennial plants, Perennial plants, Life form, Lifespan.

Annual ring: see Wood.

Annular membranes: see Annulated lamellae.

Annulated lamellae, *annular membranes:* cisternae (compressed cavities bounded by a membrane) containing pore complexes. They occur in the nucleus and cytoplasm of some rapidly growing animal cells, e.g. Cancer cells (see) and egg cells. The pore complex is similar in structure to that of the nuclear membrane (see Nucleus), from which A.l. are thought to be derived. The function of A.l. is unknown.

Anoa: see *Bovinae*.

Anodonta: see Freshwater mussels.

Anoles, *Anolis:* the largest genus of the *Iguanidae* (see), with species from the southern USA, throughout Central America and the West Indies, to Paraguay. A. have a pointed head, and they are 10–20 cm long; arboreal forms are green, and terrestrial forms brownish. The middle phalanges are particularly dilated and carry a series of transverse adhesive lamellae, which function like those of Geckos (see). Males have a large gular pouch, which is colored and patterned characteristically according to the species, and which can be distended by the hyloid bones. Most species perform marked color changes. The majority are insectivorous. Eggs (maximum of 2) are buried in the ground. *A.* are especially common in the West Indies, being numerous in both forests and cultivated areas. Their distribution among many islands has led to the evolution of a rich variety of species, e.g. more than 20 species are known in Cuba alone. The largest species, the ***giant Cuban anole*** (*Anolis equestri*) is 45 cm long. The ***great anole*** or ***American chameleon*** (not a true chameleon) (*Anolis carolinensis*) is common in the extreme south of the USA; barely 15 cm long, the whole dorsal surface of this species is golden green, and the gular pouch becomes vermilion when stretched, but is white with occasional red lines and spots when flaccid.

Anomalodesmata: see *Bivalvia*.

Anomocoela: see *Anura*.

Anomocoelous vertebra: see *Anura*.

Anomura: an infraorder of the *Decapoda* (see).

Anoplocephala: see *Oribatei*.

Anopla: see *Nemertini*.

Anostraca (plate 44): a sub-class of the *Crustacea*, containing about 175 species, all 10–20 mm in length. In some systems of classification, *A.* is an order of the *Branchiopoda*. The extended body comprises 19 or more approximately equal segments. The head is free and there is no carapace (Fig.). In most species, only the anterior 11 segments of the body have appendages (phyllopodia), which are all alike. The telson carries unjointed caudal rami (furca). The phyllopodia serve simultaneously for locomotion, respiration, and for collecting food. Between each limb and the one behind it is a space which becomes enlarged to a chamber with each forward stroke, exerting suction which draws water from the ventral median gully between the limbs of the right and left sides. In the backward stroke the chamber becomes smaller, and water is driven backward in two ventrolateral streams, driving the animal forward. During this action, small food particles are strained from the water by the limb bristles, and transported to the mouth in the median ventral groove. Thus, the *A.* are filter feeders. They are incessantly in motion, always with the ventral side upwards. *A.* live exclusively in fresh water, preferably small ponds and puddles, which periodically dry out. *Artemia* (brine shrimp) is found in salt lakes and marshes and in pans in which brine is being concentrated, but never in the sea; it can withstand salt concentrations from 4 to 23%. Other familiar genera are *Branchipus* and *Chirocephalus*. *A.* are often called "fairy shrimps".

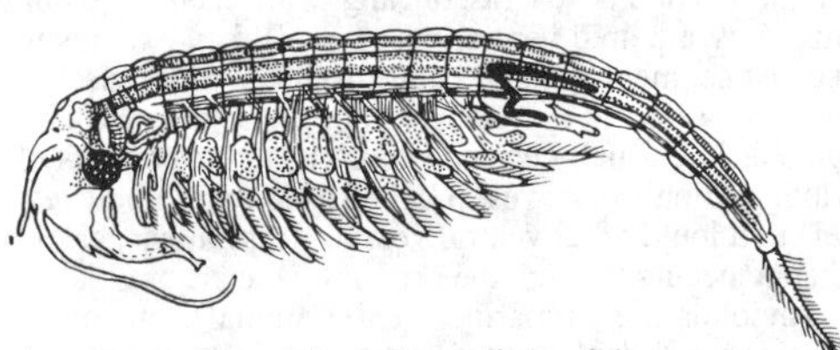

Anostraca. Branchipus stagnalis (male).

Anous: see *Laridae*.

Anser: see *Anseriformes*.

Anseranatinae: see *Anseriformes*.

Anseranatini: see *Anseriformes*.

Anseranus: see *Anseriformes*.

Anseres: see *Anseriformes*.

Anseriformes, *water fowl:* an order of birds comprising 2 suborders, each represented by a single family:
1. Suborder: *Anhimae*.
Family: *Anhimidae* ***(Screamers)*** (3 species in 1 genus).
2. Suborder: *Anseres*.
Family: *Anatidae* ***(Ducks, Geese and Swans)*** (about 150 species).
1. Subfamily: *Anseranatinae*.
Tribe: *Anseranatini* ***(Magpie goose)*** (1 species).
2. Subfamily: *Anserinae*.
Tribes: *Dendrocygnini* ***(Whistling ducks*** or ***Tree ducks;*** 9 species in 2 genera); *Anserini* ***(Swans*** and ***True geese;*** 21 species in 5 genera).
3. Subfamily: *Anatinae*.
Tribes: *Tadornini* ***(Shelducks;*** 15 species in 5 genera); *Anatini* ***(Dabbling ducks;*** 19 species in 4 genera); *Aythyini* ***(Pochards;*** 16 species in 3 genera); *Cairinini* ***(Perching ducks*** and ***geese;*** 13 species in 9 genera); *Somateriini* ***(Eiders;*** 4 species in 2 genera); *Mergini* ***(Scoters, Goldeneyes*** and ***Mergansers;*** 16 species in 6 genera); *Oxyurini* ***(Stifftails;*** 8 species in 3 genera).

Four species have been domesticated: 1. ***Muscovy duck*** *(Cairina moschata;* tribe *Cairinini)* of Central and

South America; 2. ***Gray lag goose*** *(Anser anser;* tribe *Anserini)* of Eurasia and Iceland; 3. ***Mallard*** *(Anas platyrhyncos;* tribe *Anatini)* found throughout the whole of the Northern Hemisphere, except the Arctic; 4. ***Swan goose*** *(Anser cygnoides;* tribe *Anserini)* of Eurasia.

Screamers are widely distributed in tropical and subtropical South America. About the size of large geese, but longer in the leg, they swim well, although the toes are webbed only slightly at the base. Two forward-directed sharp spurs on the leading edge of each wing act as effective weapons. All the bones are hollow, including those of the toes and wings. Natural inhabitants of marshy areas, screamers are also often seen walking among cattle in the pampas, or soaring in circles at considerable heights like birds of prey. Most familiar is the ***Crested screamer*** *(Chauna torquata),* which ranges from eastern Bolivia, Paraguay and southern Brazil to Argentina. Its plumage is gray with a dark neck ruff and a short nuchal crest.

Members of the *Anatidae* are essentially aquatic with relatively long necks, broad bills, and body down which is used to line the nest. The eggs are unspotted, and all members of this family pass through a flightless period during the molt. Swans and geese show little or no sexual dimorphism, and both sexes are involved in rearing the young. In contrast, the males of many ducks display bright breeding plumage, and usually take no part in rearing the young. Horny lamellae on the bill serve as a sieve or fish trap in the ducks, for seizing prey in the mergansers, and for grasping and cutting vegetation in the geese. They nest on the ground or in depressions, and the young are able to swim and seek their own food immediately upon hatching.

The ***Magpie goose*** or ***Semipalmated goose*** *(Anseranus semipalmata)* is an Australian gooselike bird with black and white plumage, long legs, a sturdy bill and slightly webbed feet. It is semiterrestrial and frequently perches, and is probably phylogenetically intermediate between the screamers and the other water fowl.

Swans are large, herbivorous birds with notably long necks and powerful spatulate bills (Fig.1). The Australian ***Black swan*** *(Cygnus atratus)* is unique in being entirely black except for the lighter wing tips and a bright red bill. ***True geese*** (Gray geese, *Anser;* and Black geese, *Branta)* are confined to the Northern Hemisphere, most of them breeding in the Arctic or sub-Arctic. The ***Gray lag goose*** *(Anser anser)* is a widespread Palaearctic species and the ancestor of the domestic or farmyard goose. The most familiar and widespread goose in North America is the ***Canada goose*** *(Branta canadensis)* with its black neck and head and white cheeks.

Shelducks are the most primitive of the *Anatinae,* and the tribe *Tadornini* contains some of the largest ducks. Some are so large that they are called geese, e.g. the ***Abyssinian blue-winged goose*** *(Cyanochen cyanoptera).* Most familiar of the Old World shelducks is the Eurasian ***Common Shelduck*** *(Tadorna tadorna),* a large (61 cm) gooselike duck, boldly patterned in black, white and chestnut, the male carrying a conspicuous knob on his bright red bill.

Most of the *Anatini* ***(Dabbling*** or ***Dipping ducks)*** belong to the genus *Anas.* They include the ***Mallards, Pintails, Wigeon, Shovelers*** and various species of ***Teal*** (plate 6). Many of the northern species display pronounced sexual dimorphism, the breeding plumage of

Fig. 1. *Anseriformes.* Mute swan *(Cygnus olor).*

the male being brighly colored, but becoming dull like the female during the nonbreeding, flightless period. About three quarters of all wild fowl hunted for sport belong to the genus *Anas.*

The aberrant ***Steamer ducks*** are usually classified with the dabbling ducks, although some taxonomists assign them to a separate tribe *(Tachyerini).* Two of the 3 species are flightless, i.e. the ***Megellanic flightless steamer duck*** *(Tachyeres pteneres)* of southern South America, and the ***Falkland flightless steamer duck*** *(T. brachypterus)* . The ***Flying steamer duck*** *(T. patachonicus)* of southern South America and the Falkland Islands has the smallest body of the 3 species (the wings of all three are about the same size). These birds are named for their habit of "steaming", i.e. running along the top of the water propelled by their feet and wings like a paddle steamer.

The 16 species of highly gregarious freshwater ***Pochards*** or ***Bay ducks*** have a rather short, heavy body with large legs placed far apart and further back than those of the dabbling ducks, and they are also characterized by a strong lobe or flap on the hindtoe. They have an almost cosmopolitan distribution, and are some of the best known sporting ducks, especially the North American ***Canvasback*** *(Aythya vallisneria).* Other examples are the European ***Tufted duck*** *(Aythya fuligula)* and the circumpolar ***Greater scaup, Bluebill*** or ***Broadbill*** *(Aythya marila).*

Largest of the ***Perching ducks*** and ***geese*** is the male ***Spurwinged goose*** *(Plectopterus gambensis;* up to 91 cm; Gambia, Sudan and Ethiopia down to the Zambesi) which weighs up to nearly 10 kg, while the smallest representatives are the ***Pygmy geese*** or ***Cotton teal*** *(Nettapus),* which may weigh as little as 225 g. This is a heterogeneous group of forest birds that spend much time in trees. They have long, sharp, strong claws, and a well developed, nonlobed hindtoe The tribe *Cairinini* also includes brilliantly colored and elaborately marked species such as the Chinese ***Mandarin duck*** *(Aix galericulata)* and the North American ***Wood*** or ***Carolina duck*** *(Aix sponsa).*

All 4 species of ***Eiders*** or ***Eider ducks*** are found in the Northern Hemisphere. Especially *Somateria mollissima* is exploited for the manufacture of eiderdowns. The female eider lines the nest with down plucked from her breast, and the best quality eiderdown is obtained from nest linings.

With the exception of the rare ***Brazilian Merganser*** *(Mergus octosetaceus)* and the extinct ***Auckland Island Merganser*** *(M. australis),* the tribe *Mergini (Sea **ducks**)* is confined to the Northern Hemisphere. These are mostly fish-eating ducks with the narrowest bills of any waterfowl. Sometimes called ***Sawbills,*** their bills have serrated edges. They are strong swimmers and divers. Some are truly marine, while others breed in fresh water. The ***Smew*** *(Mergus albellus)* is a small merganser of Europe and Asia, while the largest merganser is the ***Goosander*** or ***Common merganser*** *(Mergus merganser)* (Fig.2) which is widely distributed in Eurasia and North America.

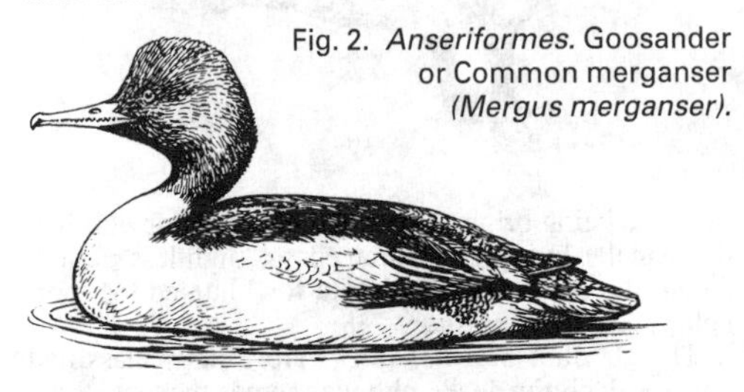

Fig. 2. *Anseriformes.* Goosander or Common merganser *(Mergus merganser).*

The ***Stifftails*** *(Oxyurini)* form a phylogenetically rather isolated tribe of extremely aquatic, largely nocturnal freshwater ducks. They are small, round, long-tailed diving ducks, and their legs are placed so far back on the body that they can only waddle clumsily on land. The genus *Oxyurus* is widely distributed and species are found in all 6 continents; typically the male breeding plumage is chestnut red with a bright blue bill, e.g. the North American ***Ruddy duck*** *(Oxyura jamaicensis),* which breeds from the Great Slave Lake south to the northern USA and winters as far south as northern South America and the West Indies. The ***Blackheaded duck*** *(Heteronetta atricapilla)* of South America lacks the long tail typical of other stifftails, and is a nest parasite. It lays its eggs in the nests of coots, rails, herons and even the hawk, *Polyborus.*

Anserinae: see *Anseriformes.*

Anserini: see *Anseriformes.*

Answering frogs: see *Microhylidae.*

Ant: see Ants.

Antabuse: see *Basidiomycetes (Agaricales; Coprinus).*

Antagonism: inhibition or opposed activity. A. is apparent in numerous metabolic, growth and developmental processes, e.g. the mutual A. of ions in the mineral metabolism of plants; interaction of growth-promoting and growth-inhibiting factors; the opposing action of muscles (flexor and extensor) and nerves. In some formulations of plant protection agents, the activity can be significantly lower than the sum of the activity of individual components, because certain constituents act antagonistically. Ecologically, A. is taken to mean the impairment or inhibition of the growth of a species by environmental conditions generated by another organism, e.g. exhaustion of nutrients, or production of antibiotics. The components or organisms that act against each other are called *antagonists.*

Antagonism factor: an important biotic factor (see Environmental factors), consisting of the totality of predators, parasites and pathogens that directly influence the population density of an animal species in a habitat. The A.f. is an antibiosis factor (see Relationships between living organisms). Biological pest control is based essentially on the promotion of the A.f. as a means of controlling mass outbreaks of pests. The potency of the A.f. is determined largely by the degree of mutual dependence of host (prey) species and their parasites (predators), and their respective population densities.

Antarctica, ***Antarctic:*** the geographical term for the South Polar area, i.e. the land mass surrounding the South Pole, and the adjoining ocean. A. should not be confused with the Antarctic kingdom (see), which is not geographically equivalent.

Antarctic kingdom: a Floristic kingdom (see), including the geographical Antarctic, the southermost part of South America, and islands of the South Atlantic and South Pacific. The South American part of the A.k. still contains permanently wet southern beech wood communities *(Notofagus)* with some deciduous and some evergreen trees, together with rich growths of mosses and ferns. The treeless Antarctic islands are home to cushion plants *(Azorella).* On the periphery of the ice-bound Antarctic continent, numerous mosses, lichens and terrestrial algae are able to thrive, but there are only a few flowering plants. These represent the relicts of an earlier, richer circumpolar flora of vascular plants, which were the southern counterpart of the Holarctic kingdom (see).

Antbirds: see *Formicariidae.*

Ant chats: see *Turdinae.*

Ant cows: see Ants.

Antelope ground squirrel: see *Sciuridae.*

Antelopes: a collective term for a large number of fast running, graceful and nimble ungulates of the family *Bovidae.* It is not a scientific term, and it is applied widely to several different forms, which can be generally defined as all those members of the *Bovidae* that are not cattle, bison, buffalo, sheep or goats, i.e. members of the subfamilies *Cephalophinae* (Duikers), *Neotraginae* (Dwarf antelopes), *Tragelaphinae* (Spiral-horned antelopes, Eland, Nilgais), *Alcelaphinae* (Hartebeests, Gnus), *Hippotraginae* (Roan and Sable antelopes), *Reduncinae* (Reedbucks), *Antilopinae* (Gazelles), *Saiginae* (Tibetan and Saiga antelopes). Even the pronghorn, which is not even a member of the *Bovidae,* has been called an antelope.

Antennae of insects: a single pair of antennae (this number is never exceeded in insects) situated on the head and near to the compound eyes of ectognathous insects. They are absent in the *Protura* and in many larvae of the *Hymenoptera.* Since the nerve centers for olfactory reception, as well as the motor centers for antennal movement are present in the deutocerebrum (see Brain), insect antennae are considered to belong to the preoral region of the head. Each antenna in generally divided into three parts: 1) a basal muscular *scape,* inserted into a membranous socket of the head (antennal socket), and which can be moved in all directions; 2) a short *pedicel* containing Johnston's organ (see); and 3) a nonmuscular, multisegmented flagellum.

Insect antennae display a variety of homologous forms, the greatest differences occurring in the flagellum. Thus, they may be decribed as filiform (threadlike), moniliform (beadlike), setaceous (tapering), geniculate (bent), capitate (with a headlike terminus), lamellate (leaflike), pectinate (comblike), serrate (serrated or sawlike), or clavate (club-shaped).

Functionally, insect antennae may be olfactory (beetles, moths, flies), tactile (grasshoppers and allies), auditory (mosquitos), or even respiratory in aquatic beetles. Olfactory pits are present on various segments of the antennae of beetles and aphids. Antennae often display sex differences, the antennae of the male serving as olfacory receptors (many moths) or sound receptors (mosquitos) for location of the female.

Antennal gland: a paired excretory organ situated at the base of the second antennae of many crustaceans.

Antennariidae: see Frogfishes.

Antennata: see Tracheata.

Antennules: the first (i.e. upper or most anterior) of the crustacean antennae. The A. are primarily uniramous, and they are not derived from biramous appendages. The lower or posterior pair of crustacean antennae are known as the second antennae, and they are usually typical biramous appendages.

Anterior horn, motor cells of: see Motor cells of the anterior horn.

Anterior pituitary: see Hypophysis.

Ant guests: see Myrmecophiles.

Anther: see Flower.

Antheraea: see *Bombycidae.*

Antheridium: see *Bryophyta,* Reproduction.

Antherurus: see *Hystricidae.*

Anthesins: see Flower development.

Anthoceros: see *Hepaticae.*

Anthocerotales: see *Hepaticae.*

Anthocyanins: widely occurring flavonoid plant pigments, responsible for the red, violet, blue or black colors of blossoms, leaves and fruits of higher plants. The variety of colors and patterns is due to the occurrence of A. either alone or in combination with other pigments, usually other flavonoids or tannins (copigmentation). A. are the water-soluble glycosides of hydroxylated 2-phenylbenzopyrylium salts. With the aid of acid or enzymes (glycosidases), the carbohydrate moiety can be removed, leaving the unstable, water-insoluble aglycon (anthocyanidin).

Anthocyanidins are formal derivatives of the C_{15} flavylium cation, and they differ from one another in the number and positions of hydroxyl groups. More than 20 types of A. are known, which differ in the number, type and position of their carbohydrate residues (Fig.). Well over 100 natural A. have been isolated and structurally characterized. A. are amphoteric, and their color depends on the pH of the cell sap. Salts of A. with acids are red, at neutral pH they form colorless pseudobases, while in the alkaline range they form unstable blue salts. Free A. are violet. These color changes depend on the reversible interconversion of the oxonium salt and a quinoid structure. Complexes of A. with metal ions, e.g. iron, magnesium and aluminium, also make an important contribution to flower coloration. Thus, the blue color of the cornflower is not due to a basic reaction of the cell sap, but to complex formation. Structurally, A. are closely related to flavones. Anthocyanidins can be reduced to catechins, and oxidized to flavonols; such conversions also occur naturally in the plant.

Benzopyrilium cation

Ring B and C-atoms 2,3 and 4 of the flavylium ring system are biosynthesized from a C_6-C_3 unit. Ring A is derived from a C_6 unit formed from 3 molecules of acetate.

Native A. have high relative molecular masses (e.g. the M_r of protocyanin from the cornflower is about 20 000). In the cell, A. are usually bound to polyuronides.

Elucidation of the structure of A. is due largely to the intensive research of R. Willstätter, R. Robinson and P. Karrer.

Anthomedusae: see *Hydroida.*

Anthomyiids, *Anthomyiidae:* a family of the *Diptera*, comprising more than 1 000 species. They feed on flower pollen or nectar, and some species prey on other insects. The larvae feed on plant tissues or rotting organic materials, and some species are agricultural and garden pests, e.g. the larva (cabbage maggot) of the ***cabbage root fly*** (*Phorbia floralis* Fall. = *Phorbia brassicae* Bouche = *Erioischia brassicae* = *Hylemyia brassicae*) destroys the roots of vegetables of the *Brassica* tribe and of wild *Cruciferae*. Species of *Pegomyia* are leaf miners; the larva of the ***beet*** or ***spinach leaf miner*** (*Pegomyia hyoscyami* Panz.) destroys the foliage of sugar beets, beets, spinach and other crops. The ***wheat bulb fly*** (*Hylemyia coarctata*) is a serious pest in many parts of Europe, its larva attacking the central heart leaves of wheat shoots. Other examples are *Hylemyia rubivora* (***raspberry cane maggot***) and *Hylemyia antiqua* (***onion maggot***). The ***kelp flies*** (*Fucellia*), whose larvae feed on brown seaweed, are well known on beaches.

Anthomyiidae. Imago of the beet or spinach leaf miner *(Pegomyia hyoscyami* Panz.).

Anthornis: see *Meliphagidae.*

Anthoscopus: see *Remizidae.*

Anthozoa: a class of the phylum, *Cnidaria.* All members occur only as polyps, which are either solitary or colonial, and the medusoid stage is completely absent. They are cylindrical in shape and are often beautifully colored. At the center of the oral disk is a transverse, compressed mouth, which leads into a tubular ectodermal pharynx, extending more than half way into the gastrovascular cavity. At least one longitudinal strip of cilia is present in the pharynx. One or more circles of tentacles are situated on the periphery of the oral disk. The gastrovascular cavity is divided into radial compartments by longitudinal gastral septa or mesenteries, and it also extends into the tentacles. The mesenteries are fused to the basal disk and to the oral disk, and often also to the pharynx. They possess a powerful endodermal longitudinal muscle and carry a Mesenterial filament (see) on their free edge. In addition, the mesenteries contain the gonads which are also of endodermal origin. A cell-containing supporting lamella (mesoglea) separates the endoderm and ectoderm.

Reproduction is sexual, or asexual by budding or cleavage.

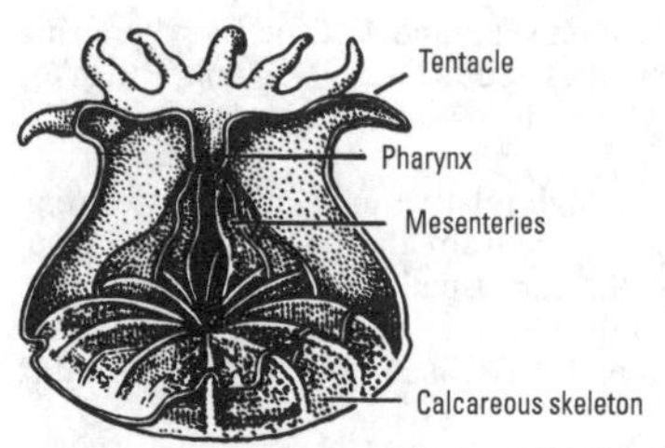

Anthozoa. Structure of the anthozoan *Astroides calycularis.*

About 6 000 species are known, and representatives are found in all the seas of the world, from shallow intertidal zones to depths of 1000 m and more. Many species secrete a skeleton, e.g. the stony corals, which are responsible for the formation of coral reefs.

Classification. Recent forms are divided into 2 subclasses: *Hexacorallia* (see) and *Octocorallia* (see). Extinct forms are divided into the *Tabulata* (see) and the *Tetracorallia* (see).

Collection and preservation. In the coastal zone, specimens are collected by hand by divers, and from deeper levels with a dredge or drag net. The animals are narcotized by adding 10% magnesium sulfate to the sea water. 5% Formaldehyde solution is suitable for fixation and storage.

Anthracosia: an extinct genus of bivalves of the subclass *Palaeoheterodonta,* order *Unionoida* (see *Bivalvia).* The shape, which is variable, is generally compressed and ovate, with a small rounded anterior and an elongate, tumid posterior. The thin, usually small shell is isomyarian and integripalliate, with fine concentric growth lines on its surface. There is often one hinge tooth on each side. Fossils of A. are found in the Upper Carboniferous (Westphalian), and it sometimes occurs in lumachelles (a sedimentary rock consisting mainly of mollusk remains).

Anthrax: see Bacillus.

Anthreptes: see *Nectariniidae.*

Anthropochory: see Seed dispersal.

Anthropocoenosis: an interactive community of living organisms centered around humans, and consisting of humans, domestic animals, their parasites, as well as house plants and synanthropic organisms (flies, food and wood pests, cockroaches, spiders, etc.). In the wider sense, rural human settlements represent a special type of Cultural biocoenosis (see). In contrast, towns and cities are complex systems in which A. (residential houses) combine with Ruderal coenoses (see), Agrocoenoses (see) (i.e. gardens) and woodland to form a functional unit (see Urban ecology).

Anthropogenesis: the process of descent and evolution from the early hominids to *Homo sapiens sapiens.* A. is an historical process, whose elucidation depends almost entirely on indirect evidence. Since the events under investigation occurred in prehistoric time, they cannot be observed directly,and they are not accessible to experimentation. The only direct evidence of human evolution is represented by the fossil record,and this is supplemented by studies on recent living forms, especially the primates. The development of social behavior, and the complicated dialectic interaction of physical and spiritual components in human evolution represent a specific and fundamental problem in the study of A., known as anthroposociogenesis.

That humans evolved within, and belong to, the animal kingdom is supported by abundant evidence from morphology, anatomy, physiology, genetics, serology, immunology, behavioral studies, paleontology and other disciplines. Factors affecting the early phases of human evolution were the same as those generally responible for the evolution of all animals and plants. Genetic changes arose by random mutation and recombination. By natural selection, the frequency of a new genotype increased or decreased in the population, depending on whether it was advantageous or disadvantageous to species survival. The rate of evolution depended on the number of mutations and recombinations, their selection value and the size of the population in question. In the later stages of evolution, humans altered their own environment, exerting direct control over food availability and territorial range. Biological evolution was then influenced by social factors rather than natural selection.

The animal origins of humans are now undisputed. Like all historical processes, human evolution has occurred in space and time. In geological time, the basic features of human evolution were established by the beginning of the Tertiary. The former concept of linear evolution is no longer acceptable. At various stages, branching must have occurred, and some of these branches have subsequently become extinct. Since man belongs to the suborder *Anthopoidea (= Simiae),* the roots of his evolutionary line cannot extend further than the basis of this group. There is general agreement that the Old World monkeys and baboons *(Catarrhina)* and humans have branched from a common evolutionary stem. Hominoid and cercopithecoid lines probably separated during the Oligocene. Existence of a common ancestral stock for all the catarrhini (procatarrhini) is not unequivocally proven by the fossil record. If a procatarrhini group never existed, then the three recent superfamilies of the *Anthropoidea (Ceboidea, Cercopithecoidea, Hominoidea)* must have arisen from a single radiation. The earlier existence of a complex group known as *Parapithecoidea* is evident from the fossil record, and it can be classified as an extict superfamily under the *Catarrhina.* The parapithecine radiation occurred in the lower to middle Oligocene. Within the group, *Propliopithecus* (characterized from lower jaw fragments from the Fayum oasis in Egypt) occupies a special place. *Propliopithecus* is thought to be a generalized hominoid (prothominoid), which qualifies as the ancestor of *Hylobates, Pongo* and possibly even *Homo.* The dental arcade is V-shaped, the canines protrude beyond the other teeth, the two premolars are bicuspid, and the molars show the *Dryopithecus* pattern (see) that is typical only of the *Hominoidea.*

The Miocene and Pleistocene have yielded the pliopithecines *(Pliopithecus, Epipliopithecus, Limnopithecus),* a group that shows an affinity to *Hylobates* (gibbons).*Limnopithecus* provedes the earliest evicence of arm lengthening and shortening of the olecranon process, which are adaptations to an arboreal existence. Pliopethicines became extinct in the Pleistocene. The lower jaw of *Aelopithecus* from the upper Oligocene of the Faym oasis (Egypt) is considered to represent a bridge between the pliopethicines and protohominoids. The upper Oligocene at El Faym has also yielded a rel-

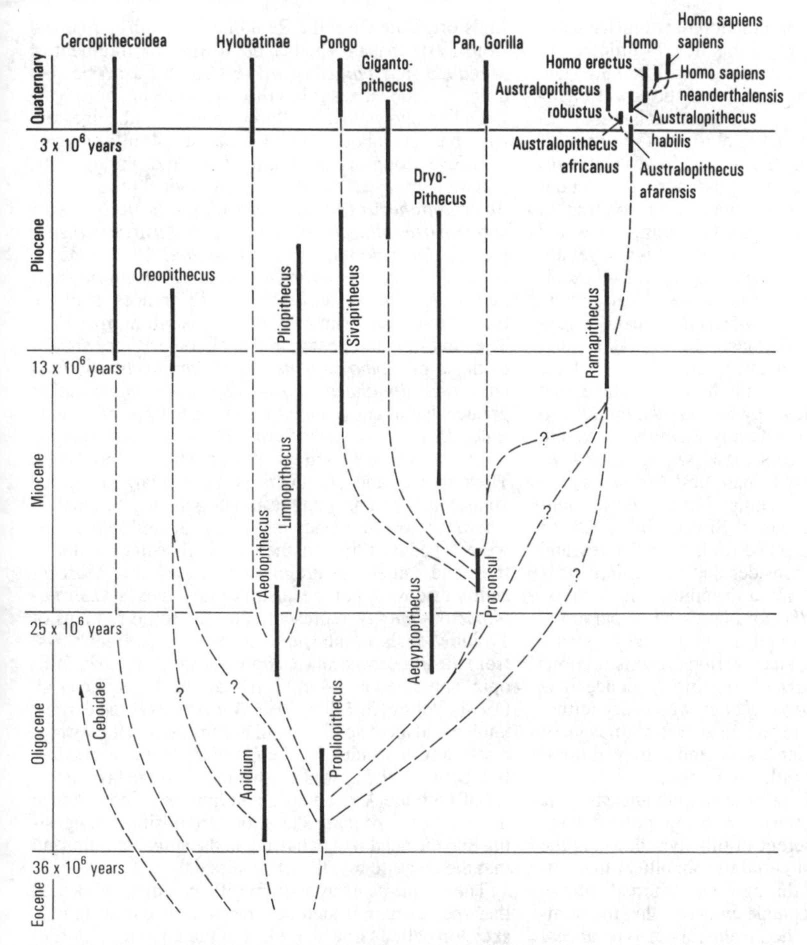

Fig. 1. *Anthropogenesis.* Evolution of the hominoids (after Knuflmann).

atively well preserved skull, lower jaw fragments and remains of the postcranial skeleton of *Aegyptopithecus,* whichin turn could represend a bridge between the protohominoids and the dryopithecines (a pongid-group). Both *Aegyptopithecus* and the pliopithecines possess large orbits, but in constrast to the latter, *Aegyptopithecus* has a long snout with powerful canines and a simian shelf, which are features of the modern great apes.

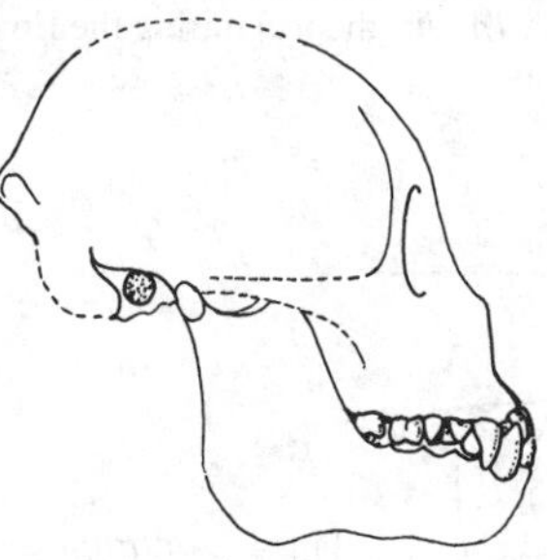

Fig. 3. *Anthropogenesis.* Reconstruction of the skull of *Proconsul africanus* (after Campbell).

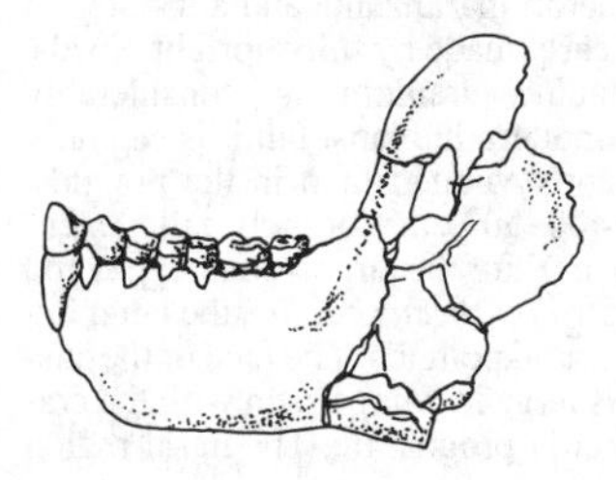

Fig. 2. *Anthropogenesis. Propliopithecus haeckeli:* half of lower jaw with teeth (after Kälin).

In East Africa the dryopithecine radiation and early pongid evolution can be traced through the whole of the Miocene into the late Pleistocene, while in Europe and Southern Asia the record extends from 16 million years ago to the late Pleistocene. Many fossils are included under *Dryopithecus,* and their systematic classification varies considerably with different authors. The genera, *Proconsul, Dryopithecus, Sivapithecus, Gigantophitecus* and *Ramapithecus* are generally acknowledged. *Proconsul* remains from the East African Miocene *(Proconsul fricanus, Proconsul nyanzae,Pro-*

consul major) display a combination of primitive characteristics and genuine pongid features. Prominent supra-orbital ridges (brow ridges) are still lacking, and the dentition shows some features of the lower catarrhins. In the whole, however, *Proconsul* is largely pongid (*Dryopithecus* pattern, parallel rows of molars), although there is no simian shelf on the lower jaw symphysis. Postcranial remains indicate a slender physique, with three size variants corresponding to the sizes of the pygmy chimpanzee, chimpanzee and gorilla. A marked adaptation for climbing is not yet apparent, but the beginnings of specialization for brachiation are evident. Later dryopithecines already show pongid-type specialization. *Proconsul* appears to represent the immediate evolutionary basis fo the main dryopithecine group, but it is separate from it. Thus *Proconsul* could represent a phylogenetic stage that subsequently gave rise not only to pongids, but also to hominids. Members of the family *Ramapithecus* are characterized only from fossil jaw fragments: *Ramapithecus wickeri* from Fort Ternan, East Africa, 14 million years old from K/Ar dating; *Ramapithecus punjabicus* from the India-Pakistan Siwalik hills, 8.5–13 million years old; also disputed finds from Turkey and Hungary. Many authors consider that *Ramapithecus* is a basic hominid, rather than a dryopithecine. The assumptions that *Ramapithecus* possessed a parabolic dental arcade and possibly an intratemporal fossa are not universally accepted, since various reconstructions are possible. On the basis of existing evidence it is equally possible that *Ramapithecus* was a dryopithecine, which gave rise to the human ancestral lineage, or that it arose from an earlier branch and achieved hominid characteristics by parallel evolution.

According to all the theories of human ancestry, the lineage of *Homo* can be traced back to a pongid stage (see Pongid theory). Inherent in this hypothesis is the assumption (which has a certain probability) that human evolution passed through an arboreal phase. Also,there is still no acceptable evidence that the hominid line could have branched from Tarsier-type ancestors (as proposed by the *Tarsius* theory) during the Eocene.

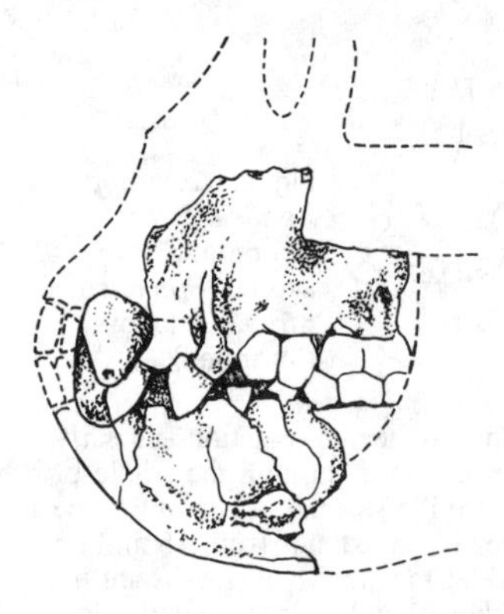

Fig. 4. *Anthropogenesis.* Reconstruction of the jaw region of *Ramapithecus wickeri* (after Walker and Andrews).

The first undoubted hominids are the australopithecines (prehominids, early hominids, ape-man, man-ape, near-man). Fossils of over 2000 individuals are known, and all skeletal parts are more or less well preserved. Jaw fragments and teeth occur most frequently. The finds originate from the Republic of South Africa and from East Africa. Whether the lower jaw fragment of *Meganthropus palaeojavanicus* (South East Asia) belongs to a robust australopithecine is disputed.

Dating shows that the known australopithecines existed between about 1.75 and at least 4.4 million years ago. Five groups can be distinguished morphologically, and they represent a certain chronological progression: *Australopithecus ramidus, Australophithecus afarensis* and *Australopithecus africanus* (the *Australopithecus* group); *Australopithecus robustus and Australopithecus boisei* (the *Paranthropus* group);*Australopithecus habilis (= Homo habilis).* The differences between these groups are primariliy those of skull morphology. The dentition and postcranial skeleton are very similar to those of *Homo sapiens. Australopithecus robustus* and *Australopithecus boisei,* however, show rather greater dental and skeletal differences from other australopithecine groups and from *Homo sapiens. Australopithecus robustus* and *boisei* were large, robust types. Their molars and premolars possess enlarged surface areas, which strongly suggests a largely vegetarian diet. There is strong evidence that all the australopithecines were bipedal, although they may also have climbed trees and sometimes progressed on all fours. Most recently discovered of the australopithecines is *Australopithecus ramidus,* represented by the fossil remains of 17 individuals (dental, cranial and postcranial parts) from the Pliocene strata at Aramis, Middle Awash, Ethiopia, and dated at 4.4 million years [T. D. White et al. (1994) *Nature,* ***371*** 306–312]. *A. ramidus* is the earliest known and most apelike of all hominids, and it may represent a root species for the *Hominidae,* i.e. a possible link between the australopothecines and the last ancestor of both the African apes and humans. The foramen magnum of *A. ramidus* has a forward position, suggesting that the head was balanced on the spinal column and that the creature was therefore bipedal.

The *A. afarensis* hypodigm (all specimens allocated that species name) includes the famous partial female skeleton called "Lucy". Oldest of the known *A. afarensis* remains (3.9 million years) is the Belohdelie frontal from the Middle Awash. The youngest yet discovered ("Son of Lucy", 3 million years) include a male skull, humerus, ulna and other pieces from the Hadar formation of Ethiopia [W. H. Kimbel et al. (1994) *Nature* ***368,*** 449–451]. There is thus evidence that *A. afarensis* remained relatively unchanged for at least 0.9 million years. Curvature of the ulna and its length relative to the well muscled and robust humerus suggest a tree-climbing creature, which may also have walked on two legs (cf. chimpanzee); but there is no proximo-anterior orientation of the trochlear notch and the olecranon is very small, suggesting a terrestrial quadruped. On the other hand, 3.5 million year-old fossil footprints, found in volcanic ash at Laetoli in Tanzania and attributed to *A. afarensis,* were clearly made by fully upright bipeds.

The australopethicine cranium is considerably smaller than that of modern humans, but it is sagitally much more prominently vaulted than in the pongids. Skull capacities are 400–500 cm^3 for the smaller, slenderer types, 500–600 cm^3 for the large, robust types and 600–800 cm^3 for *Australopithecus habilis,* the latter exceeding the average for all pongids. The face of the australopethicine skull is large in comparison with the cranium, and it is markedly prognathic, The nasal region

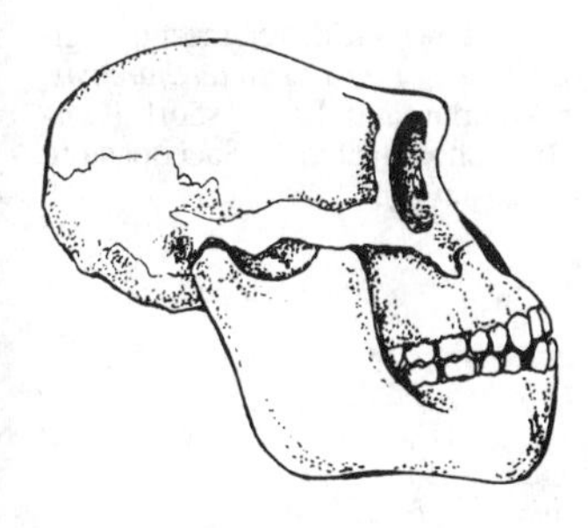

Fig. 5. *Anthropogenesis.* Reconstruction of the skull of *Australopithecus afarensis* (after Matternes).

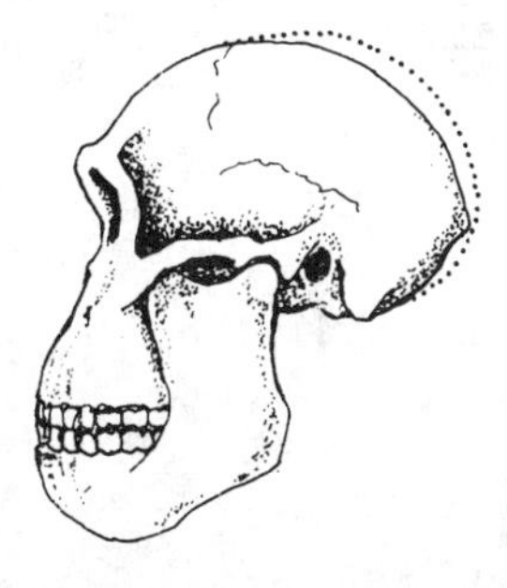

Fig. 6. *Anthropogenesis.* Reconstruction of the skull of *Australopithecus africanus.* For comparison, the skull of *Australopithecus habilis* is shown in dotted outline.

has pongid affinities, but the zygomatic bones are more hominid in character, being frontally flattened with a wide sideways projectory, before bending sharply toward the rear; a shallow infraorbital fossa is also present. In the robust group, the mandible is especially powerful with a large ramus. In contrast, the mandible of the *habilis* group is relatively slender with a smaller ramus and a large submandibular angle. Australopethicine dentition conforms to the pattern of the *Hominidae;* as in humans, the dental arcade is a smooth, regular parabola with no diastematic intervals. On the basis of the complex of numerous hominid-like skeletomorphological features, many authorities place *Australopithecus habilis* in the *Homininae,* renaming it *Homo habilis,* but more evidence is required to fully justify this reclassification. No stones modified as tools have been found associated with australopethicines, but it can be assumed that they used tools and hunted communally.

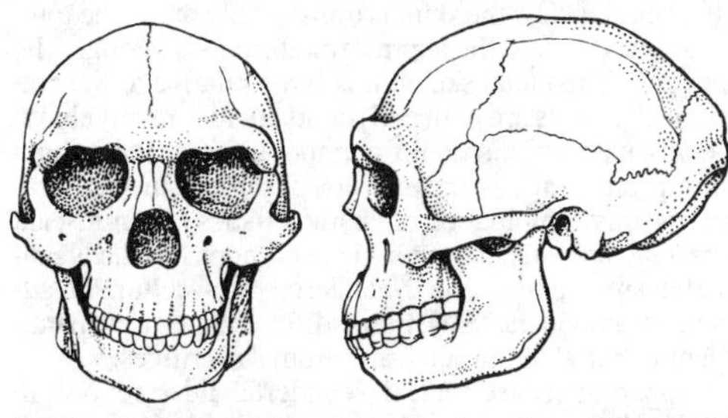

Fig. 7. *Anthropogenesis.* Reconstruction of the skull of *Homo erectus pekinensis* (after Weidenreich).

The earliest known representatives of the subfamily of true humans *(Homininae)* are the pithecanthropines (= archanthropines) (primitive man), which are grouped together as the species *Homo erectus.* Dating of their fossils from the Lower and Middle Peistocene shows that they existed between 1.9 million and 200 000–300 000 years ago. The oldest of these originate from East Africa (Olduvai Bed I and II) and from South-East Asia (Djetis layers at Modjokerto and Sangiran), and they overlap in time with the later australopethicines. In Europe, only the Upper Middle Pleistocene has yielded pithecanthropine fossils. Skull fragments from numerous individuals have been unearthed,but postcranial remains (which are largely identical with those of *Homo sapiens*) are less abundant. Compared with the australopithecines, the pithecanthropines show more advanced features, like a higher skull capacity (750–1200 cm^3), a decrease in the size of the facial skeleton relative to the cranium, centralized position of the occipital foramen, increased folding in the base of the skull,a prominent nasal bone, and a wide nasal opening with a sharply delineated lower edge.On the other hand, the sagittal vaulting of the cranium is less than in *Homo sapiens sapiens,* and even less than in *Australopithecus habilis.* Some further differences from *Homo sapiens sapiens* are: a powerful supra-orbital ridge is present; the widest part of the skull lies directly above the auditory aperture (in occipital view, the skull has a tent shape); the skull possesses a medium sagittal crest; the tabular part of the occipital bone is bent; an occipital crest is present; the jaw section is massive; and there is no chin. Between individuals, the spectra of characteristics show wide variations. The extensive temporal and spatial spread of the erectus group apparently resulted in variations in the rate of development, with trends towards regional differences. Accordingly, classification of the following subspecies is more or less justified: *Homo erectus modjokertensis, H. erectus lantianensis, H. erectus leakeyi, H. erectus mauritanicus, H. erectus heidelbergensis, H. erectus bilzingslebenensis.* Furthermore, some of the fossil pithecanthropines (especially those living most recently) provide evidence for continued independent development of the *Homo erectus* lineage, and as such can be considered as transitional forms between *Homo erectus* and *Homo sapiens.*

At the European, African and Chinese discovery sites, stone tools have been found in obvious association with *Homo erectus.* These include not only crude hands axes, but Acheulean hand axes, Levalloisian flakes and microlith Clactonian scrapers, which are finely worked and retouched, and belong to a relatively high level of cultural development. These peoples hunted large mammals and turned their bones into a variety of tools. By inference, it is assumed that they used fire. There are many different views on the possible beginnings of oral communication, which is thought to have arisen at some time between *Ramapithecus* and *Homo sapiens.* Since skull casts show that the relative proportions of the brain parts are very similar in *Homo erectus* and *Homo sapiens,* and a social structure is archeologically documented for *Homo erectus,* it seems more than probable that *Homo erectus* could communicate by speech.

Towards the end of the subspecies radiation of *Homo erectus,* certain forms appear in Europe, North Africa and the Near East, known as *Homo sapiens praesapiens.* Their skull structure is only slightly different from that of the neanthropines. The forehead is relatively steep, the occiput well rounded with a practically horizontal base. In occipital view, the skull is house-like

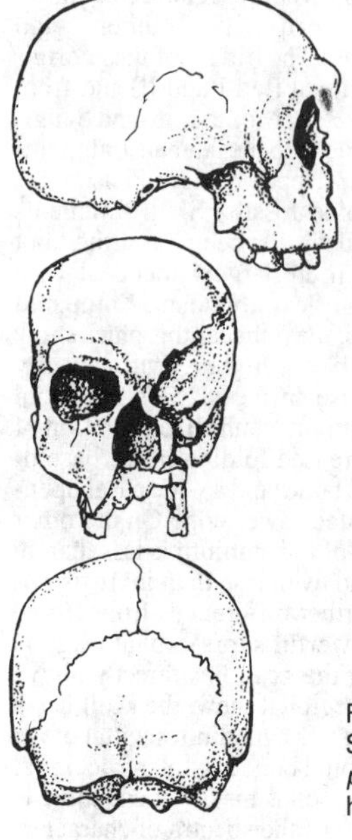

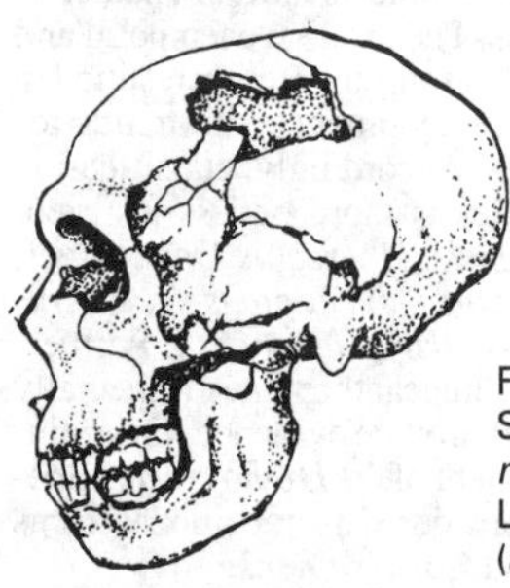

Fig. 8. *Anthropogenesis.* Skull of *Homo sapiens praesapiens* from Steinheim.

Fig. 9. *Anthropogenesis.* Skull of *Homo sapiens neanderthalensis* from La Capelle aux Saints (after Boule).

with steep sides (skull capacity 1200–1450 cm^3). Early representatives of this group still possess a relatively prominent supra-orbital ridge, the orbits are rectangular, a shallow infratemporal fossa is present, and the jaw is only slightly prognathic. Skeletal fossils, morphologically different from *Homo erectus* and generally acknowledged as *Homo sapiens praesapiens,* originate from Steinheim near Stuttgart and Swanscombe near London (dated between 200 000 and 300 000 years old) and from Arago, Montmaurin and Nizza in Southern France (possibly a later date). It seems that *Homo sapiens* arose from a certain evolutionary level within the *Homo erectus* population, while the development of *Homo erectus* continued locally in isolated evolutionary side branches. These different lines, however, were not necessarily genetically isolated.

Sapient forms probably split into two lines at an early stage of their evolution, one leading to *Homo sapiens sapiens,* the other via the pre-Neandertals to the classical Neandertals (*Homo sapiens neanderthalensis;* paleoantropines). [The original German spelling, *Neanderthal,* is also used, and Neandertals are also called Neandert(h)alers]. Fossil remains of pre-neandertals *(Homo sapiens praeneanderthalensis),* dated at 100 000–70 000 years old, which differ morphologically and in age from *Homo sapiens neanderthalensis,* have been found at Weimar and Ehringsdorf (Germany), Granovce (Czechoslovakia), Saccopastore (Italy) and Krapina (Yugoslavia).

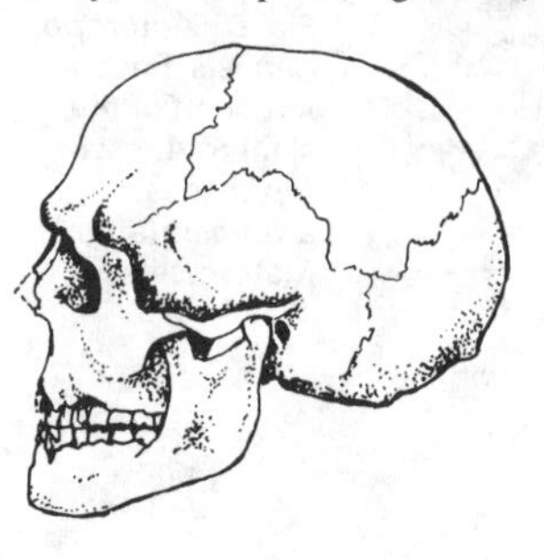

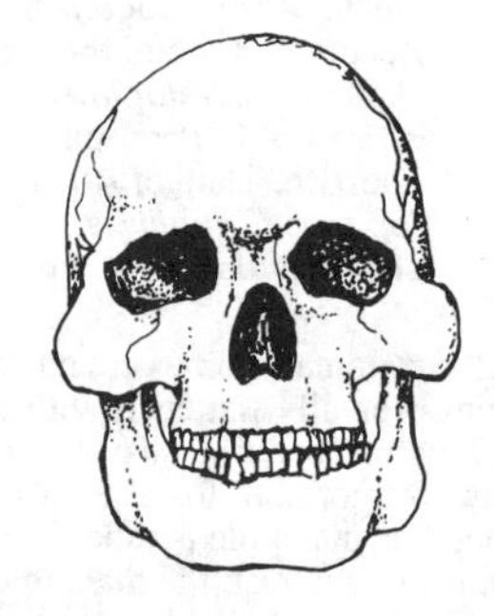

Fig. 10. *Anthropogenesis.* Skull of *Homo sapiens sapiens* of the Crô- Magnon type from the Würm glacial period (after Graham).

Neandertals in the narrow sense (the name is derived from the site of their discovery in the Neander Valley near Düsseldorf) originate from the first half of the Würm glaciation (Upper Pleistocene, 35 000 to 70 000 years ago). Skulls and practically all skeletal parts have been recovered. Neandertals were especially widespread in West, Central and Southern Europe, whereas the "eastern Neandertals", which possess all the Neandertal characteristics but in less prominent form, existed in southeastern Europe and the Near East. The Neandertal skull is characterized by its size (skull capacity 1350 to 1750 cm^3, average about 1550 cm^3); it is long and relatively low, the occiput is rounded with vaulted side walls (in occipital view the skull is transversely oval), the forehead recedes, and the supra-orbital ridge is strongly developed. The facial skeleton is large relative to the cranium, the orbits are high and round, there is relatively little prognathism and no infratemporal fossa, and a slight chin prominence is only rarely observed. In many respects, this combination of characteristics is reminiscent of *Homo erectus,* although the skull bones of the Neandertal are thinner. The Neandertal was culturally advanced, as evidenced by highly differentiated tool manufacture, burial of the dead, and animal sacrifices.

Apparently, the classical Neandertal did not evolve in total isolation from the evolutionary line of *Homo sapiens sapiens.* Fossil remains from israel, Iraq and Yugoslavia display morphological and cultural characteristics from both types, which indicates the existence of intermediate forms, or at least the coexistence of both evolving lines.

According to the presently available fossil record,

Homo sapiens sapiens appeared in its characteristically modern human form 35 000 to 40 000 years ago, in the middle of the Würm glaciation. At this time, Neandertal man disappears from the fossil record. *Homo sapiens sapiens* from the Würm Ice Age existed in many different forms, showing the almost uninterrupted, smooth progression from a robust, tall form with a broad and relatively low face (Cro-magnon type) to a less robust, intermediate sized form with a small high face (Combe-Capelle type). Culturally, these early paleolithic humans were more developed than previous groups. They manufactured a variety of stone, bone and horn implements, produced decorative art work, and left evidence of funeral rites that indicate a form of discriminate cult worship. Similar discoveries have been made in Europa, Asia, Africa and Australia.

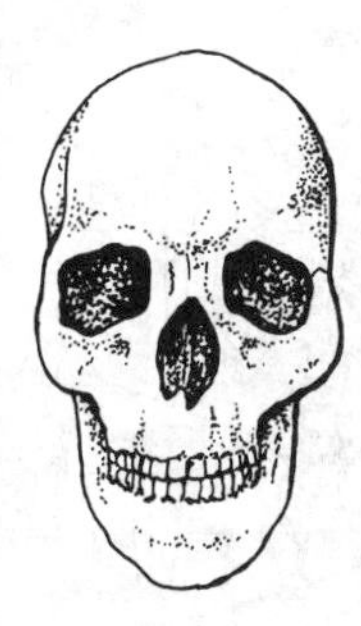

Fig. 11. *Anthropogensis.* A skull of *Homo sapiens sapiens* of the Combe-Capelle type from the Würm glacial period.

Up to the 1970s, the possible existence of two parallel lines of pre-*sapiens* evolution in Europe was accepted, one to the Neandertals, the other (via Fontéchevade) to Cromagnon man. This proposition now seems increasingly unlikely, and the European hominids of the Middle and Upper Pleistocene are regarded as a single line, albeit with considerable variability, leading from ante-Neandertals, via pre-Neandertals (or early Neandertals) to the late Neandertals of the Würm glaciation. This leaves in question the roots of modern *Homo sapiens.* The suggesion that modern man originated in the Middle East has little support, especially since the fossil remains of Neandertaloids and hominids at various Middle Eastern sites show that these two types were morphologically quite separate, with no evidence of intermediate forms. According to the *out-of-Africa model,* modern *Homo sapiens* had a single African origin, and from there progressively supplanted other archaic hominids. In contrast, the *multiregional evolution model* proposes that archaic *Homo sapiens* evolved into modern *Homo sapiens* independently in many different parts of the Old World, and not necessarily at the same time. The most recent hypothesis, known as the "Afro-European *sapiens*-hypothesis",first proposed in 1983 [G. Bräuer, *Early anatomically modern man in Africa and the replacement of the Mediterranean and European Neandertals.* 1er Congr. Intern. Paléont. Hum., p 112 (Abstract), C.N.R.S., Nice] proposes:

1. During the late Middle Pleistocene, early anatomically modern *Homo sapiens* evolved from early archaic forms of *Homo sapiens* via late archaic forms in eastern and/or southern Africa.
2. During the early Upper Pleistocene, early anatomically modern *Homo sapiens* was widespread throughout Africa.
3. These anatomically modern humans expanded northward during the Würm glaciation, moving out of East Africa. As a result of this northward migration, various groups of modern humans mixed in both the Near East and Northern Africa. Following this intermediate period of mixing, modern man eventually supplanted Neandertaloid populations wherever they encountered them in Africa and the Near East. During the subsequent millenia, in which further hybridization occurred, they moved further north and superceded the European Neandertals.

According to archeological findings, America was first settled about 20 000 years ago by *Homo sapiens sapiens.* It must be generally assumed that the hominids at every geological horizon were subject to considerable variability, and that the fossil finds have provided only a few random samples. Thus, the extent to which these finds are truly representative is open to question. Also important is the concept of genetically isolated populations. Fossil finds give no information about reproductive barriers; taxonomic groupings can therefore be inferred only from morphological differences and temporal and geographical data, and from the occurrence of intermediate populations. The available evidence therefore lends itself to different interpretations, thus providing more than one possibility for describing the evolutionary pathway leading to the modern human. Further information is needed for an unequivocal portrayal of hominid evolution.

For the systematic classification of the primates, including some fossil groups see Primates.

[*The Human Evolution Source Book,* Russel L. Ciochon & John G. Fleagle, Prentice Hall Inc., New Jersey, 1993; ISBN 0-13-446097-9. *The Order of Man. A Biomathematical Anatomy of the Primates,* C. E. Oxnard, Yale University Press, New Haven and London, 1984. *Human Variations and Origiins,* W. H. Freeman and Company, San Francisco and London, 1967]

Anthropoid apes: see Hominoids.

Anthropoides: see *Gruiformes.*

Anthropoids: see Hominoids.

Anthropological-morphological assessment: see Forensic anthropology.

Anthropometer: an instrument used in anthropology for measuring the human body. The A. consists of 4 hollow tubes, which fit into one another to form a rigid rod, with a sliding crosspiece. Each rod is 2.1 m in length, and graduated in millimeters. A second crosspiece can be fitted at the upper end of the hollow rod, so that the upper part of the A. can be used as a beam compass or trammel (Fig.). The straight crossbars can be replaced by recurved crosspieces, which can be used as caliper arms for depth measurements, e.g. on the thorax.

Anthropometry: determination of the absolute and relative dimensions of the human body and its parts. Statistically analysed anthropometric data (see Biometrics, Biostatistics) form the basis of most anthropological studies. In addition, A. of children provides a useful index of health, and the corresponding data for adults may be used in the design of clothing, seating, manually operated machinery, etc.

A. is subdivided into cephalometry (measurement of the head), somatometry (measurement of the body), craniometry (measurement of the skull) and osteometry (measurement of the postcranial skeleton). The dimensions of interest are width, depth, circumference, and an-

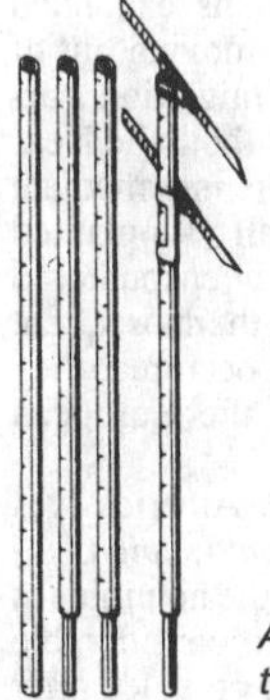

Anthropometer, showing separated tubes and horizontal crosspieces.

gles, together with weight and volume, often necessitating the use of special instruments, e.g. the Anthropometer (see), sliding calipers, spreading calipers with recurved arms, and a goniometer for measuring angles, in particular of the face and skull. To insure that recorded values are comparable in different populations throughout the world, recommended and exactly defined procedures are followed. Measurements are made between anthropological points, chosed to provide the maximal possible information on the natural structure of the relevant region. These points therefore often lie at the extremities of bones, which are relatively easy to locate and have a constant position. Before recording the following normae (outlines) of the cranium, the head is first oriented so that the Frankfort plane (see Cranial planes) is horizontal: norma basilaris or n. ventralis (outline of the inferior aspect of the skull), norma frontalis, n. anterior or n. facialis (outline of skull viewed from the front), norma occipitalis or n. posterior (outline of skull seen from behind), norma sagittalis (outline of sagittal section through skull), norma verticalis or n. superior (outline of skull viewed from above). Since absolute measurements are subject to inter-individual variation, anthropologists use about 170 different ratios or indexes, e.g. the Cranial index (see).

Anthropomorphology: the morphology of humans. See Anthropology.

Anthroponoses: infectious diseases that attack only humans, and can only be transferred between humans.

Anthropos: see Human.

Anthropozoonoses: infectious diseases that can be transferred between humans, between animals, and between humans and animals. Important foci of human diseases are rodents (mice, rats), dogs, cats, cattle, sheep, pigs and birds (chickens, pigeons, game birds).

Anthrus: see *Basidiomycetes (Gasteromycetales).*

Anthuridea: a suborder of the *Isopoda (see).*

Anthurium: see *Araceae.*

Anthus: see *Motacillidae.*

Antiaris: see *Moraceae.*

Antiauxins: competitive inhibitors of auxin activity, which have little or no auxin activity, and which compete with auxins for binding to their receptor sites. Thus, excess antiauxin prevents auxin action by blocking receptors, and this inhibition can be overcome by excess auxin. Under certain conditions, *trans*-cinnamic acid behaves as an antiauxin. Naturally occurring antiauxins probably play a role in the metabolism of auxins and the control of growth. Phenylacetic acid and phenylbutyric

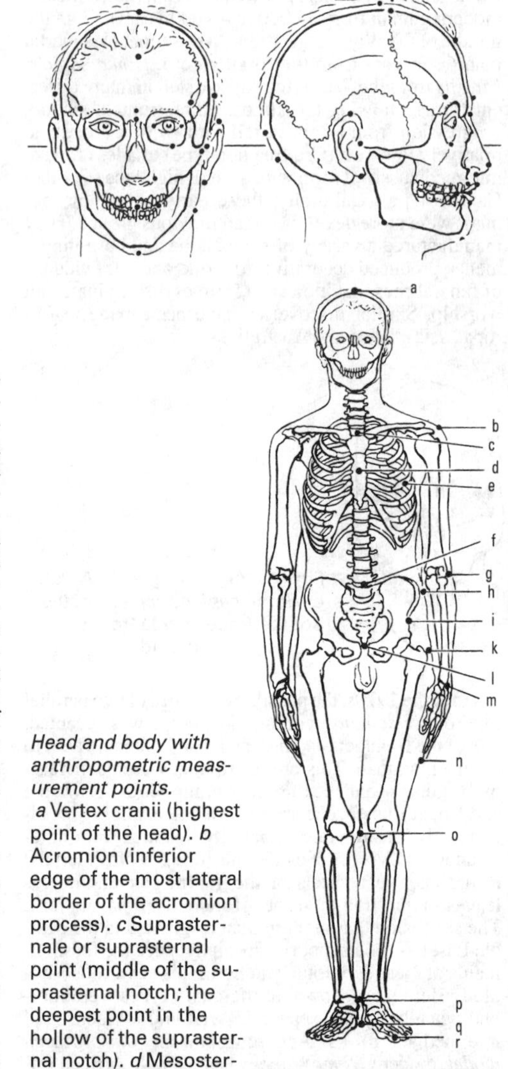

Head and body with anthropometric measurement points.
a Vertex cranii (highest point of the head). *b* Acromion (inferior edge of the most lateral border of the acromion process). *c* Suprasternale or suprasternal point (middle of the suprasternal notch; the deepest point in the hollow of the suprasternal notch). *d* Mesosternale or mesosternal point (mid point of the sternum; the estimated union of the 3rd and 4th sternebrae). *e* Thelion (level of the nipple). *f* Omphalion (level of the umbilicus). *g* Radiale (upper border of the head of the radius. *h* Iliocristale (iliac crest). *i* Iliospinale ant. or anterior superior iliac spine (most prominent medial point of the anterior superior spine of the ilium). *k* Trochanterion (upper border of the great trochnater). *l* Symphysion (pubic symphysis; the term, symphysion, is also used for a craniometric point on the lower jaw, corresponding to the anterior border of the alveolar process). *m* Stylion (tip of the styloid process of the radius). *n* Dactylion (tip of the longest finger). *o* Tibiale (upper point of the inner border of the medial tibial condyle). *p* Sphyrion (distal border of the medial malleolus). *q* Pternion (most posterior part of the heel). *r* Acropodium (end of the longest toe).

acid strongly compete with true auxins and displace them from their receptors, thereby causing a decrease in auxin activity. In the absence of true auxins, however, both of these compounds display weak, auxin-like activity.

Antibiosis: see Relationships between living organisms.

Antibiotics: defined chemical substances (and their laboratory modifications), which are synthesized by (and are isolated from) living cells, and which cause the stasis, degeneration, lysis or death of plant or animal microorganisms, e.g. viruses, rickettsias, bacteria, actinomycetes, fungi, algae or protozoa. They do not act upon their producer organisms. A. are active in very low concentrations, but they do not act catalytically, i.e. they are not enzymes or coenzymes.

Occurrence. The property of A. production occurs sporadically among various families of microorganisms, and it has little taxonomic significance. Nevertheless, the actinomycetes (especially *Streptomyces)* are notable producers of A. The polypeptide A. produced by some *Bacillus* species are clinically important (see Gramicidin, Bacitracin). Among the ascomycetes, the genera, *Penicillium* and *Aspergillus,* are important sources of A. *Cephalosporium* is also a pharmaceutically important genus (see Cephalosporins).

Classification. A. are secondary metabolites, which can be classified in different ways, e.g. according to the taxonomic position of their producing organisms, the spectrum of their antimicrobial activity, or the mechanism of their biological activity. Thus, they can be classified according to whether they are active against Gram-positive or Gram-negative bacteria, or against both (broad spectrum A.). On the basis of their action mechanism, three important classes are recognized: 1) A. that interfere with the biosynthesis of nucleic acids or proteins; 2) A. that interfere with cell wall biosynthesis; and 3) A. that interfere with the function of the cytoplasmic membrane. A. are often classified according to their pathways of biosynthesis or their chemical structures, these two properties being closely related.

Structurally, the A. represent a very heterogeneous group of natural products. Important biosynthetic precursors are e.g. amino acids, acetate, malonate, sugars, sugar derivatives, and cyclic triterpenes (see Polyene antibiotics). A. can also be prepared by chemical synthesis, but technically they are obtained predominantly from natural sources (see Fermentation). Semisynthetic A. (see Penicillin) are prepared by laboratory modification (chemical or enzymatic) of natural A.

Modern research into A. began with the discovery of penicillin by Fleming (1928). About 3 500 A. have now been described, but only about 80 of these are prepared commercially for pharmaceutical purposes. The annual world production of A. for medical use is about 20 000 tonnes. A further 30 000 tonnes are used in animal feeds (they promote an increase in body mass and a general improvement in health), and in plant protection.

Medically and economically, A. are the most important of all natural pharmaceutical products. Their indiscriminate use can lead to the development of resistant strains of pathogenic organisms.

Antibodies: specific defense proteins found in the blood plasma, lymph and many body secretions of all vertebrates. A. are formed in response to contact of the body with Antigens (see). A. specificially bind to the antigens that induce their production. According to the reaction observed in vitro between an antibody and its antigen, A. are classified as Precipitins (see), Agglutinins (see), Lysins (see), or Antitoxins (see). The biological importance of A. lies in their ability to neutralize harmful antigens, such as disease-causing microorganisms or their toxic products.

All A. are immunoglobulins (see). Humans produce 5 different classes of A. (immunoglobulins G, M, A, D and E). Historically, this classification is based on electrophoretic behavior, but these electrophoretically different immunoglobulins also display different biological behavior. Thus, the most investigated antibodies are immunoglubulins G (IgG), also known as gamma-globulins (γ-globulins), and most of these act as a defense against foreign proteins, microorganisms, etc. In contrast, immunoglobulins E (IgE) are formed as a defense against multicellular parasites, and they are also involved in allergic reactions (see Anaphylaxis).

In the antibody-antigen interaction, a specific binding region of the antibody (the paratope), binds to a complementary region of the surface of the antigen (the epitope). The Affinity (see) of this binding is variable.

Antical leaves: see *Hepaticae.*

Anticipatory movements: near field movments which serve as a prelude to physical contact with the object of the behavior. These actions may also serve as signals. As part of behavior preliminary to bodily contact, dogs touch a partner with a raised foreleg; the partner may reciprocate, and the two animals may take turns in this behavior. Humans use many such gestures, e.g. the offering of a helping hand, or holding out both arms in anticipation of a hug or embrace.

Anticlinical cell walls: cell walls aligned approximately at right angles to the outer surface of the plant part in which they are situated.

Anticoagulants: agents that prevent blood coagulation by acting as inhibitors of components of the Blood coagulation (see) cascade, or by complexing and thereby rendering unavailable essential cofactors of the blood coagulation cascade. Thus, ethylenediaminetetraacetic acid (EDTA) acts as an A. by complexing calcium ions, which are essential for several stages of the blood coagulation cascade. Heparin, on the other hand, combines with, and causes a conformational change in, the proteinase inhibitor, antithrombin III, leading to an accelerated inactivation of the serum proteinases involved in blood coagulation.

Anticodon: a sequence of three nucleotide residues in one loop (the anticodon loop) of a transfer RNA molecule. This sequence is complementary to the nucleotides of a codon in messenger RNA, i.e. it forms H-bonded base pairs with (or "recognizes") a specific codon in mRNA. For example, the anticodon UAG binds to the complementary codon AUC.

Since the specificity of each tRNA molecule is at least 2-fold, i.e. it possesses a specific anticodon, as well as binding a specific amino acid, the anticodon-codon interaction insures that the correct amino acid is incorporated into the polypeptide sequence during protein biosynthesis. See Protein biosynthesis, Genetic code.

Antidepressants: drugs with stimulatory and antidepressant action. They decrease fatigue, reduce appetite and decrease sleeping time. These effects are the result of reduced autonomic activity (especially of cholinergic systems) and of central sympathetic activation.

A. are inhibitors of monoamine oxidase. They are all amines, structurally related to adrenalin (epinephrin), and presumably they act as analogs of natural monoamine oxidase substrates; inhibition of monoamine oxidase delays or prevents destruction of the natural catecholamines (adrenalin, dopamine, noradrenalin), which therefore persist at elevated concentrations and cause stimulation. Examples of A. are Amphetamine (see), Methamphetamine (see), ephedrine, harmine, etc. The first A. to be discovered was iproniazid (*N*-isonicotinyl-*N'*-isopropylhydrazide); during the original tests of this compound as an antitubercular drug, it was observed that patients became elated. A. are used for the treatment of mental depression; they have also been used for doping racehorses and athletes. They can be addictive, especially amphetamine and methamphetamine.

Antidorcas: see *Antilopinae*.

Antidorcatini: see *Antilopinae*.

Antienzymes: polypeptides or proteins that act as enzyme inhibitors, including antibodies formed in response to antigenic enzyme proteins or coenzymes in the blood. A. may be formed within the organisms, or they may be encountered as exogenous inhibitors. They include the many animal and plant protease inhibitors, such as soybean trypsin inhibitor and serum antitrypsin, which form tight complexes with the corresponding enzymes. Antibody-type A., raised by injection of enzymes, are important tools in the purification and characterization of enzymes. Injection of a whole enzyme protein induces antibodies against the apoenzyme (which determines the catalytic specificity of the enzyme). Injection of an antigenic coenzyme induces antibodies against the coenzyme, which therefore inactivate all enzymes with that particular cofactor.

Antifeedants: endogenous substances that protect plants against consumption by animals, mainly insects. A. are not necessarily acutely toxic, and their effect is behavioral, causing the animal to largely avoid the plant. Many compounds have been shown to have antifeedant activity for insects, e.g. aristolochic acids, cyclitols, chromenes, cyclopropenoid acids, phenolic compounds, furanocoumarins, polyacetylenes, cardenolides, diterpenes, flavonoids, iridoid glycosides, sesquiterpenes, tannins, triterpenoids, etc. Even the most toxic alkaloids appear to deter rather than to kill phytophagous insects.

The term has also been used for chemical compounds applied externally to control the pests of economically important plants, e.g. fentol acetate for control of the Colorado beetle.

Antifreeze protein: an extracellular glycoprotein found in the blood of certain Arctic and Antarctic fishes. It contains a repeating sequence of alanine-alanine-threonine, in which each threonine residue carries a unit of D-galactosyl-*N*-acetyl-D-galactosamine. It appears to prevent the initiation of ice crystal formation, and together with the sodium chloride in the blood it enables the fish to live at temperatures down to -1.85 °C.

Antigens: substances, organisms or cells that are foreign to the body, and which induce an immune reaction in humans or other vertebrates. Exposure of the body to an antigen leads to the production of antibodies by activated lymph cells. The antibodies bind the antigen, which then undergoes lysis, precipitation, or immobilization, and is finally eliminated by phagocytosis.

Biologically important A. include all disease-causing organisms, i.e. microorganisms and parasites, as well as their products, e.g. bacterial toxins. Neutralization of toxins is an especially important function of A.

Proteins, carbohydrates and nucleoproteins may all act as A. Compounds of low relative molecular mass (e.g. amino acids, dipeptides, monosaccharides, disaccharides) are not antigenically active, whereas the most active A. have a high relative molecular mass (e.g. proteins). The blood group substances are medically important A.

Antigibberellins: growth regulators that inhibit the action of gibberellins by competing with them for their binding sites. Natural A. have been demonstrated in different legumes, but they have not been studied chemically. The Morphactins (see) are synthetic A.

Certain quaternary ammonium salts, e.g. chlorocholine chloride, are more appropriately described as Gibberellin antagonists (see), because they act by inhibiting gibberellin biosynthesis.

Antihemorrhagic vitamin: vitamin K (see Vitamins).

Antilocapra: see *Antilocapridae*.

Antilocapridae, *pronghorns:* a North American family of even-toed ungulates, now represented only by the single species, *Antilocapra americana* (***Pronghorn***). Both sexes of this roe-deer-sized mammal possess curved horns with a single side branch, consisting of a bony core with a sheath of fused hairs; the sheath is shed annually. Dewclaws are absent. Vision and the sense of smell are acute. Dentition: 0/2, 0/1, 3/3, 3/3 = 32; the cheek teeth are selenodont.

Antilocapridae. Pronghorn *(Antilocapra americana).*

Antilope: see *Antilopinae*.

Antilopinae: a subfamily of the *Bovidae* (see). The A. are divided into 6 tribes.

1. Tribe: *Gazellini;* genera: *Gazella* (20 species, e.g. ***Thomson's gazelle,*** *G. thomsoni;* see Gazelles) and *Procapra* (1 species: *P. picticaudata,* ***Tibetan gazelle***).

2. Tribe: *Antilopini;* genus: *Antilope* (1 species: *A. cervicapra,* ***Blackbuck***)

3. Tribe: *Litocraniini;* genus: *Litocranius* (1 species: *L. walleri,* ***Gerenuk***).

4. Tribe: *Ammodorcatini;* genus: *Ammodorcas* (1 species: *A. clarkei, **Dibatag**).*

5. Tribe: *Antidorcatini,* with the genus: *Antidorcas* (1 species: *A. marsupialis, **Springbuck**)*

6. Tribe: *Aepycerini;* genus: *Aepyceros* (1 species: *Ae. melampus, **Impala**).*

The ***Blackbuck*** (head plus body length about 120 cm) lives in herds on the plains of India. The female is light brown and usually hornless. The male is dark gray sometimes almost black, with a white belly and white head markings. Males have horns up to 75 cm in length with 3 and occasionally more spiral twists. Blackbuck are excellent jumpers, covering a distance of 6 m and a height of over 200 cm in a single leap.

Antilopini: see *Antilopinae.*

Antilymphocyte serum: an antiserum produced in animals by immunization with lymphocytes of a different species. The antibodies in the A.s. can destroy lymphocytes, so that the A.s. is an active immunosuppressive agent. A.s. is therefore used to suppress undesired immune reactions, e.g. to prevent rejection after organ transplantation.

Antimitotic drugs: see Microtubules.

Antimorph: a mutant allele that has the opposite effect to that of the normal allele. See Amorph, Hypermorph, Neomorph.

Antimutagens: agents that counteract the action of mutagens. They decrease the rates of spontaneous and induced muitations.

Antimycotic agent: an inhibitor with special activity against fungi, e.g. the antibiotics, griseofulvin and fungicidin.

Antineuritic vitamin: vitamin B_1. See Vitamins.

Antinuclear factors: antibodies that bind to constituents of cell nuclei. They appear in the blood in certain diseases, e.g. autoimmune diseases. Very low concentrations of A.f. are also detectable in the blood of healthy individuals. Their concentration increases with age, and it is a matter of debate whether they are a cause or a result of the aging process.

Antipatharia, *black corals, horny corals:* an order of the *Hexacorallia.* The upright, plantlike colonies are found mainly in deep tropical waters. Polyps are arranged around a thorny, noncalcareous skeleton, which consists of a black, horny material. This material ("black coral") can be distorted with heating, and in many tropical countries it is used for making necklaces, brooches, etc.

Antipernicious anemia factor: a vitamin B_{12}. See Vitamins.

Antiphytopathogenic potential: the sum of all factors in the soil that contribute to the promotion and inhibition of pests and disease-causing organisms, and which lead to the exclusion or the diminution in numbers and potency of such organisms.

Antirachitic vitamin: vitamin D. See Vitamins.

Antisaprophytic conditions: see Saprophytic system.

Antiscorbutic vitamin: vitamin C. See Vitamins.

Antiserum: blood serum of an animal or human, containing antibodies specific for a certain antigen(s). A. raised in animals by immunization with specific antigens are used both diagnostically and therapeutically.

Antisomatogen: an obsolete term for antigen.

Antitoxins: antibodies formed in the organism against a toxin. A. neutralize the harmful activity of the toxin, and therefore have a protective function. A. obtained by immunization of animals are used therapeutically, e.g. against snake venoms.

Antitranspirants: chemicals used for reducing the transpiration rates of plants, to enable them to withstand periods of water stress, e.g. following transplantation or during drought. A distinction is drawn between A. that cause stomatal closure (e.g. mercury phenylactetae, abscisic acid) and A. that simply limit water loss by providing an impermeable film over the surface of the plant (e.g. paraffins and silicones). Materials that are impermeable to both water and carbon dioxide, e.g. paraffins, have the same effect as those that cause stomatal closure, i.e. photosynthesis and biomass production are greatly decreased. Some silicones are more permeable to carbon dioxide, and do not drastically retard photosynthesis when they are applied as A.

Antivitamins, *vitamin antagonists:* analogs of those vitamins that function as coenzymes. An antivitamin or a metabolic derivative of an antivitamin may specifically compete with the vitamin-derived coenzyme for binding to its apoenzyme. For example, the classical competitive inhibitor of vitamin B_6 is 4-deoxypyridixine; the 5'-phosphorylated derivative of this analog competes with pyridoxal 5'-phosphate for binding to a number of B_6-dependent enzymes, such as amino acid decarboxylases, transaminases, etc. Alternatively, the antivitamin may compete in those reactions that convert the vitamin into a coenzyme. For example, the therapeutically important sulfonamides are antimetabolites of *p*-aminobenzoic acid, and therefore act as inhibitors of bacterial folic acid synthesis (see Folic acid in the entry Vitamins). A. inhibit the growth of microorganisms, and in mammals they lead to the appearance of vitamin deficiency symptoms. The effects of A. can be counteracted with high doses of vitamins (competitive inhibition).

Antlers: paired, branched bony processes growing from the frontal bone of cervids. They are covered with a well vascularized integument (bast or velvet), and the nature of the branching is characteristic of the species. When growth of the A. is complete and at the beginning of the mating season, the velvet is removed largely by striking and rubbing against bushes and tree branches. After mating, the A. are shed, and they grow again during the following year, often with more branches. The process of shedding is regulated by sex hormones, and the new growth is promoted by thyroid hormone. Male and female reindeer have well developed A. In all other cervids, the A. of the female are less developed than those of the male, and may even be absent. They may be stemlike with side branches (red deer, roe deer, reindeer) or spatulate (elk, fallow deer).

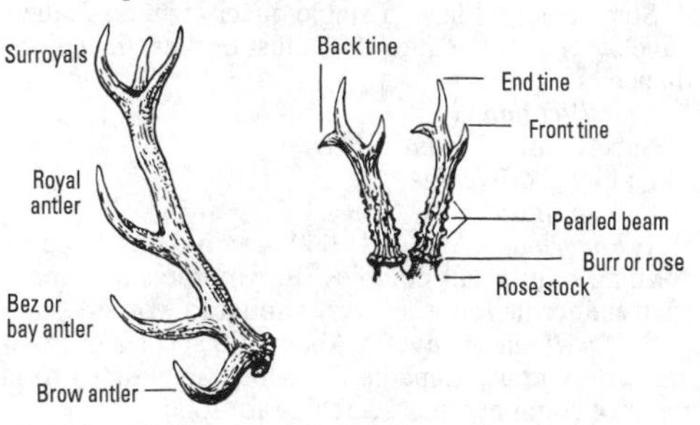

Antlers. a Red deer. *b* Roebuck.

Antlions: see *Neuroptera.*
Antpipits: see *Conopophagidae.*
Ant plants: see *Rubiaceae.*
Ants, *Formicidae:* the only family of the superfamily *Formicoidea*. A. belong to the *Aculeata* of the order, *Hymenoptera* (see). The earliest fossils are found in the Cretaceous. A. originated in the tropics, but are now known in all other regions of the world as far as the Arctic and the Antarctic. At least 10 000 species have been described. Sixty colony-forming species are known in Central Europe. Antennae are characteristically bent, and the abdominal stalk of the adult carries a single narrow, scale-like segment, or 2 nodule-like segments; all such segments are called pedicels. Nearly all A. form communities (a few solitary parasitic forms are the exception), varying in size from barely a dozen to more than a million individuals. Species that form large colonies display highly developed social systems, with division of labor between castes. Winged fertile males and females perform a nuptial flight, after which the male soon dies, whereas the inseminated female founds a colony and becomes its "queen". Most of her eggs develop into workers (wingless, infertile females), which perform all tasks of the colony (nest building, foraging for food, care of the brood), except egg laying. The majority of the inhabitants of any colony belong to the worker caste, which is often subdivided into small, medium and large types. By analogy with termites, the large workers are called soldiers; they have a large head, powerful mouthparts and potent poison glands. Some species of the subfamilies *Formicinae* and *Dolichoderinae* possess a further caste, known as repletes. These are workers which are able to distend themselves and act as storage vessels for sweet fluids brought to the nest by foragers. Their contents are regurgitated when required for use by the colony. Species possessing this caste are found in North America, South Africa and Australia, and are known as honey ants.

A. feed on plant sap, nectar and other insects. Some culture fungi on decaying vegetable matter and ant excreta in subterranean chambers. Many A. have formed a symbiotic association with honeydew-secreting insects (mainly aphids or "ant cows"); in return for honeydew, the A. protect the insects, provide shelter for them, and store their eggs through the winter. In addition to symbiotic insects, ants' nests often contain Myrmecophiles (see) or ant guests.

Nests are built in the earth or above ground, in the trunks of live trees, in rotting tree stumps, in excavated cavities and under stones. Particularly striking are the large hills made by wood A., which are useful predators on insect pests.

Some colonies have a single queen, whereas others have several. If the queen(s) is lost or dies, the colony disperses.

Classification

Superfamily: *Formicoidea*

Family: *Formicidae*

Subfamilies:

1. *Ponerinae*. About 1 000 carnivorous species, forming only small colonies. Polymorphism is minimal, and fertile forms are very similar to workers.

2. *Dorylinae* (***Army A.***). About 200 species of tropical, carnivorous, nomadic A., which sometimes form massive columns when searching for food.

3. *Myrmicinae*. The commonest and largest subfamily, containing over 3 000 known species. Many species collect seeds and are known as ***Harvester A.*** (e.g. *Pogonomyrmex* of North America). ***Leaf cutter A.*** (e.g. *Atta*) of tropical and subtropical America collect vegetable material, which they chew to form a nutrient medium for fungal growth; the A. feed on the fungus and use it to feed the brood. Many myrmecines are social parasites of other A., e.g. the myrmecine, *Strongylognathus testaceus*, has saber-like mandibles and is unable to feed its own larvae, a task performed by the workers of its myrmecine host, *Tetramorium caespitum* (***Pavement ant***); both queens live in the same nest. The genus, *Crematogaster*, nests inside hollow acacia thorns and hollow cecropia stems.

4. *Dolichoderinae*. 300 species distributed throughout the world. In place of the sting, the dolichoderines have a gland at the end of the abdomen, which secretes a foul-smelling defensive substance. The pedicel has a single segment. Species of the genus *Azteca* nest inside hollow acacia thorns and hollow cecropia stems.

5. *Formicinae*. Over 2 500 almost cosmopolitan species. As in the dolichoderines, the sting is regressed. The poison glands, however, are retained, and the poison is either sprayed or dripped into wounds made by the mandibles. Many species have a scale on the pedicel. In the ***Wood A.*** (*Formica*) the poison is a strong solution of formic acid. The red wood ant, whose ant hills may be several meters high, is well known in the temperate and cool latitudes of the Northern Hemisphere. ***Slave-making A.*** (e.g. the Amazonian *Polyergus rufescens*) carry off pupae from the nests of other A.; these hatch into workers, which care for the strange queen and her progeny. Slave-makers are unable to feed themselves or their progeny, because their mouthparts are modified for aggression rather than feeding. A slave-making colony is established when an inseminated female enters the nest of the slave species (usually the genus *Formica*), kills the queen and takes her place. The ***European jet ant*** (*Lasius fuliginosus*) and related species of *Lasius* exploit aphids, which are then popularly known as "ant cows" (see text). ***Carpenter A.*** (*Camponotus* spp.) also feed on aphid honeydew, and are notoriously destructive of wood, chewing out nests in coniferous trees. ***Weaver A.*** (e.g. genus *Oecophylla*) use their own larvae (which produce a silk thread) like a weaver's shuttle to bind leaves together, forming a compartment for rearing the brood.

6. *Myrmecinae* (***Bulldog A.***). About 100 aggressive Australian species, often 20 mm in length, with a painful sting.

7. *Pseudomyrmecinae*. About 200 mainly tropical species, with very long bodies and conspicuously large eyes. They often nest inside the hollow thorns of acacia species and the hollow stems of South American cecropias.

Anulus:

1) The remains of the torn Indusium (see) on the stipe of a basidiomycete fungus.

2) In pteridophytes, a row of cells with greatly thickened radial and inner walls, which extends over the dorsal surface of the top of the sporangium as far as the middle of the ventral surface. It functions in the cohesion mechanism which opens the sporangium and ejects the spores.

3) A narrow circular zone beneath the rim of the lid of the moss sporangium (capsule). The cells of the A. contain mucilage which swells and causes the lid to spring open when the sporangium is ripe.

Anura, *Salienta, Ecaudata,* ***anurans, frogs and toads:*** an order of the *Amphibia.* Adult members lack tails (which are lost at metamorphosis) and have a squat or stocky overall shape, with 4 well developed limbs bearing 4 fingers and 5 toes, the latter often webbed. The hindlimbs are usually longer and adapted for jumping and swimming. The whole body surface is naked, and it may be smooth or warty. Some species lack a tongue, but when present the tongue is usually attached at the front of the floor of the bucco-pharyngeal cavity, and it can be projected outward to catch insects. Practically all species possess ear drums. The presence or absence of various types of teeth (see Dentition) is not normally taken into account in the taxonomy of the *A.*

The skeleton is characterized by the presence of only a small number of vertebrae, and ribs are present only in the most primitive families. There are basically 2 types of frog, according to whether the 2 halves of the pectoral girdle are fused in the midline (firmisternal) or whether they overlap (arciferal). Arcifery is considered to be the more primitive condition. In the arciferal girdle the 2 epicoracoid cartilages are fused in the region between the clavicles, but they are free and overlapping posterior to the clavicles, where they form a pair of posteriorly directed epicoracoid horns, which articulate with the sternum. In the firmisternal girdle, the epicoracoid cartilages are fused to one another, epicoracoid horns are lacking, and the sternum is fused to the pectoral arch. Firmisternal girdles also usually carry an omosternum. In its simplest form the omosternum is a cartilaginous disk lying anterior to the precoracoid bridge or the median articulation of the clavicles. In some anurans, ranids in particular, the omosternum becomes a long style (sometimes ossified) with a terminal cartilaginous disk. In the arciferal girdle the omosternum is usually absent or very small. With the exception of *Rhinophrynus* (family *Rhinophrynidae), Brachycephalus* and *Psyllophryne* (family *Brachycephalidae),* all *A.* possess a sternum. In the firmisternal girdle the sternum is usually long (except in microhylids), forming an ossified stylus (mesosternum) with an expanded cartilaginous end (xiphisternum). In most arciferal girdles the sternum is broad and ovoid, and shorter than that of the firmisternal type. In the genus *Pipa* the sternum is especially large, covering most of the abdominal region. Conditions intermediate between arcifery and fermisterny are also known. Fusion of the epicoracoid cartilages in the region between the clavicles (epicoracoid bridge) may extend posteriorly so that the cartilages no longer overlap freely, representing a condition known as pseudofirmisterny (e.g. the bufonids, *Atelopus, Oreophrynella, Dendrophrynisus;* and the leptodactylids, *Sminthillus, Geobatrachus, Phrynopus).* A few species of ranids and sooglossids display the pseudoarciferal condition, in which the epicoracoids overlap and are partially free, but are fused with one another and with the sternum posteriomedially.

Males usually possess an expandable vocal sac, which amplifies the many types of call used to attract the female at mating time, and for various other types of signalling. Males are usually smaller than females (males and females are about the same size in many populations of dendrobatids); they often have nuptial pads (swellings formed by numerous acinous glands) on the propellex region (or the propellexes themselves may be enlarged as gripping organs), which enable them to cling to the female during mating. The male usually grasps the female so that he is dorsal to her, a temporary association known as amplexus. Fertilization is external in most anurans, and the pair is in amplexus at the time of oviposition and fertilization. Ejaculation usually occurs as the eggs are deposited. Mating generally occurs in water, but a few species mate on land. The eggs, enclosed in a gelatinous material and known as spawn, are laid in clumps, strings or singly. The larvae (tadpoles) possess a caudal fin (tail); they have external gills at first, and internal gills develop later. The larval mouth has horny jaws equipped for scraping or rasping. During metamorphosis to the lung-breathing, terrestrial adult, the intestine becomes shorter as an adaptation to a carnivorous diet. The hindlimbs appear first (cf. the *Urodela,* in which the forelimbs emerge first). When the metamorphosed anuran leaves the water, it has 4 legs and a stump representing the remains of the tail. In some anurans, e.g. species of *Leiopelma* (family *Leiopelmatidae)* and *Sooglossus gardineri* (family *Sooglossidae),* metamorphosis occurs in the egg, and the newly hatched young are already fully formed miniature adults. The viviparous toads *(Nectophrynoides;* family *Bufonidae)* give birth to fully developed young adults. Many species display a highly evolved and often complex form of brood care, in which both sexes may take part. Tadpoles usually feed on plant material, whereas adults are predatory, feeding on insects, worms, snails and even small vertebrates.

Most of the 2 600 species of anurans are active at night or during twilight (the dendrobatins are exceptional in being active only in the daytime). With the exception of deserts and polar regions, they inhabit all regions of the world. Since their skins are permeable, they are unable to tolerate a marine existence. There is no universally accepted system of classification of the *A.*, and the taxonomy of this amphibian order has even been described as chaotic. The system presented below recognizes 23 families in 6 suborders. Like most other classifications of the *A.*, it relies heavily on the structure of the vertebrae and the pectoral girdle, as well as other anatomical features. The terms, frog and toad, are based on superficial appearance and do not represent phylogenetic relationships.

Suborder 1: ***Amphicoela.*** These are primitive members, in which the ends of the bony centrum of each vertebra are flat or amphicoelous (biconcave). Remnants of notochord persist, and the joints between vertebrae, including that between the sacrum and urostyle, are formed from a combination of hyaline and fibrous cartilage. Ribs are present. Members of this suborder, sometimes referred to as "tailed frogs", are unique in their possession of vestigial tail-moving muscles, the pyriformis and the caudaliopuboischiotibialis.

Families: *Leiopelmatidae, Ascaphidae.*

Suborder 2: ***Aglossa.*** Vertebrae are opisthocoelous, i.e. the presacral centra are concave at the rear, and they carry a condyle (formed from intervertebral cartilage) on their front end, which articulates with the centrum of the anteriorly adjacent vertebra. The *Aglossa* are unique among anurans in not possessing a tongue. Ribs are present; these are free in sub-adults, but ankylosed to the transverse processes of presacrals II-IV in adults.

Family: *Pipidae*.

Suborder 3: ***Opisthocoela.*** Vertebrae are opisthocoelous, as in the *Aglossa*, but the tongue is present. Free ribs are present in the *Discoglossidae*, but absent from the *Rhinophrynidae*.

Families: *Discoglossidae, Rhinophrynidae*.

Suborder 4: ***Anomocoela.*** The coccyx and sacrum are either fused or have a monocondylar articulation. The presacral centra display various configurations. They may be biconcave with a free intervertebral element (disk), which may be cartilaginous or bony. The disk sometimes adheres to the rear of the centrum, but is not fused to it, or it may be joined synostotically to the centrum, so that the vertebrae become procoelous. Ribs are absent.

Families: *Pelobatidae* (spade foot toads), *Pelodytidae*.

Suborder 5: ***Diplasiocoela*** . Most members display the diplasiocoelous condition, i.e. all but the last presacral vertebra are procoelous. The last presacral is biconcave, and it therefore has a procoelous relationship with the anteriorly adjacent veretebra, but it articulates at its rear with a condyle on the anterior end of the sacrum. In some families of this suborder, all the vertebrae are procoelous. Ribs are absent.

Families: *Ranidae* (true frogs), *Sooglossidae* (Seychelles frogs), *Dendrobatidae* (poison-arrow frogs), *Platymantidae, Rhacophoridae, Microhylidae* (narrow-mouthed frogs), *Phrynomeridae*.

Suborder 6: ***Procoela*** . These are the most advanced frogs. The presacral centra are uniformly procoelous, i.e. concave at the front, with a condyle at the rear that articulates with the posteriorly adjacent centrum. Occasionally free intermediate elements are present. Ribs are absent.

Families: *Pseudidae, Bufonidae* (toads), *Brachycephalidae, Atelopodidae, Hylidae* (tree frogs), *Leptodactylidae, Rhinodermatidae, Heliophrynidae, Centrolenidae, Myobatrachidae*.

See separate entries for each of the above families.

Anus: see Digestive system.

Anvil of the middle ear: see Auditory organ.

Aoudad: see Barbary sheep.

Apalis: see *Sylviinae*.

Aparity: a rare and abnormal occurrence in oribatids, in which the female dies shortly before laying eggs. The larvae hatch in the body of the mother and eat their way out in the region of the mother's mouthparts. A. is observed in *Notapsis, Coleoptratus* and others. It possibly represents an adaptation to adverse environmental conditions.

Apataelurus: see *Creodonta*.

Aperture: see Resolving power.

Apes: see Simians

Apfelbeckia: see *Diplopoda*.

Aphaniptera: see *Siphonaptera*.

Aphanomyces astaci: see Crayfish.

Aphasmida: see *Nematoda*.

Aphetohyoidea: the most primitive class or subclass of fishes, including several Paleozoic orders, which displays an evolutionary peak in the Devonian. They already possessed paired fins (preceded by a spine) or similar appendages, and they were usually of bizarre appearance. In some groups the bony endoskeleton was supplemented by an armored epidermis, and the edges of the mouth were supported by jawbones. The hyoid arch played no part in suspending the jaws to the cranium, hence the name *Aphetohyoidea* (free hyoid). Both upper and lower jaws consisted of two pieces, and bore a strong resemblance to a branchial arch; a series of bony rays projected back from the hinder end of the jaws to form an operculum over the visceral arches. Elasmobranchs and teleosts both probably evolved from the subclass *Acanthodi*.

Aphid: see *Homoptera*.

Aphididae: see *Homoptera*.

Aphid lions: see *Neuroptera*.

Aphidoidea: see *Homoptera*.

Aphis: see Heterogony, *Homoptera*.

Aphodiinae: see Lamellicorn beetles.

Aphredoderidae: see *Percopsiformes*.

Aphriza: see *Scolopacidae*.

Aphrodite: see *Errantia*.

Aphroditidae: see *Errantia*.

Aphyllophorales: see *Basidiomycetes*.

Apiaceae: see *Umbelliferae*.

Apical cone: see Meristems.

Apical cell: see Meristems.

Apical dominance: hormonal inhibition of the growth of lateral shoots by the apical shoot. A.d. occurs with varying degrees of intensity. Often the growth of lateral shoots is completely suppressed. Lateral shoots then remain smaller than 1 mm in diameter, and they are known as dormant buds or eyes. When the apical shoots are removed, as in the cutting of a hedge, the lateral buds in the leaf axes can then develop and the hedge becomes denser. The threadlike or ropelike aerial roots of certain climbing plants can also be induced to branch by removing the root tip, whereas under normal conditions branching does not occur until the root reaches the ground. In other cases, growth of lateral shoots is only delayed or somewhat retarded by the apical shoot, or the lateral shoots are prevented from growing directly upward, as in many trees; if the apical shoot is removed, however, the nearest lateral shoot becomes stronger and grows upward. Auxin is always involved in A.d. It is synthesized in the apical shoot, then transported downward, where it inhibits the growth of lateral shoots and buds. Accordingly, the application of auxin can mimic the action of an apically dominant shoot. Cytokinins and morphactins cause auxin-inhibited lateral buds to develop by stimulating DNA synthesis. A.d. appears to be regulated by an equilibrium between auxin and cytokinin. Especially in plants that display only weak A.d., it is evident that A.d. is the net result of opposing systems. When such plants are deficient in nitrogen, the A.d. becomes more pronounced, because the nitrogen supply is sufficient only for the growth of the apical shoot.

Apical growth:

1) Growth of plant organs from their tips or apices. A.g. is very pronounced in the roots of higher plants, in which extension growth is restricted to a narrow zone directly behind the root tip. In contrast, the apical extension zone of the shoot may be considerably wider. Intercalary growth zones may also be present, e.g. in the stem nodes of grasses. Unlike most spermatophyte leaves, A.g. of fern fronds continues actively over an extended period. At first, A.g. of a fern frond proceeds from a single apical cell with two dividing faces, and this is replaced later by a group of dividing cells. In the dicotomous branching of liverworts, a new apical cell arises from the abaxial part of a cell recently abstricted from the main apex. Some liverworts and many mosses possess tetrahedral apical cells at their apices. The posteriorly directed 3 abstriction faces produce a regular sequence of basipetal segments, contributing cells to the ground and cortical tissue, as well as leading to the formation of leaflet and lateral shoot primordia. In other lower plants, e.g. filamentous algae, A.g. is due exclusively or largely to division of a single apical cell.

2) Preferential growth of some cells at their ends, e.g. root hair cells, xylem fibers, pollen tubes or fungal hyphae.

Apical meristem: see Meristems, Flower development.

Apicomplexa: see *Sporozoa.*

Apidae: see Bees.

Aplacentalia: mammals that lack a placenta, i.e. marsupials and monotremes.

Aplacophora, ***worm mollusks:*** a subclass of the *Amphineura* (see), containing about 240 living species. They are benthic, wormlike mollusks, usually 0.5–3 cm in length, but some species may attain 30 cm. They lack a shell or an obvious foot. The entire epidermis is covered by a secreted cuticle, which usually contains solid calcareous spicules, often in several layers. Development is direct, or proceeds via a trochophore larva. The eggs hatch into trocophore larvae after a few days incubation in the anal cavity; the ciliated bands around the equator and mouth are soon lost, and the resulting larvae then closely resemble miniature adults. All species are marine, living in the bottom mud or creeping on the surfaces of hydrozoan and coral colonies at depths between 18 m and 6 000 m. They feed on microorganisms and on coral polyps or their coenosarc (material linking individuals in the stems of colonial polyps). Movement is usually wormlike with the aid of the outer sheath of skin muscles. Some forms progress snail-like on their ventral surface, probably with the aid of cilia in the ventral groove.

Aplacophora. The worm mollusk, *Proneomenia.*

Two orders are recognized. The approximately 60 species of the *Chaetodermatida (Caudofoveata)* lack a ventral groove, their simple bell-shaped mantle cavity contains true gills, and individuals are unisexual. The approximately 180 species of the *Neomeniida (Solenogastres)* have no true gills, and respiratory gas exchange occurs via internal folds of the simple mantle cavity. A ventral groove is present, containing a ciliated ridge, and individuals are hermaphrodite.

Aplites: see Sunfishes.

Aplodontidae, ***mountain beavers, boomers, sewellels:*** a family of Rodents (see) represented by a single living species, *Aplodontia rufa,* found in conifer forests along the Pacific coast of USA and Canada. The mountain beaver, a strict vegetarian, is a pest of young timber plantations, partly through eating young shoots, but mainly through its habit of gnawing off small shoots and branches as nesting material. Head plus body length is 30–41 cm. The short tail (2.5–3.5 cm) is hardly noticeable, and the animal appears to be tailless. Adults are black to reddish brown with a lighter underside. Most of the time is spent in burrows, venturing out briefly for foraging. It is the most primitive of all the living rodents.

Aplousiobranchiata: see *Ascidiacea.*

Aplysia: see *Gastropoda.*

Apocrenic acid: see Humus.

Apocrita: see *Hymenoptera.*

Apocyanaceae, ***dogbane family, periwinkle family:*** a family of dicotyledonous plants containing 180 genera and almost 2 000 species, most of which occur in the tropics and subtropics. Some species are herbaceous, but most are woody, often with a climbing habit (lianas); some succulent forms are also known. Leaves are usually opposite, rarely whorled or spiral. The presence of a milky latex is characteristic of the family; the latex canals are nonseptate. Flowers are 4–5-merous, hermaphrodite, actinomorphic, solitary or in cymose inflorescences, arising from the leaf axes, and insect-pollinated. The gynecium (usually a superior ovary) is cenocarpous, and there is a tendency for only the styles and stigma to be free during flowering; the various fruits (resembling multiseeded follicles or capsules) therefore appear to be 2-lobed and secondarily almost choricarpous.

Chemically, the family is characterized by the occurrence of indole alkaloids and cardenolides, and consequently contains many poisonous and pharmaceutical plants. African species of the genus *Strophanthus* are used for the preparation of arrow poisons; their seeds yield the strophanthins, which are used medically as cardiac stimulants. *Rauwolfia serpentina* yields the alkaloid, Reserpine (see), an antihypertensive agent used widely in the treatment of mental illnesses and high blood pressure. The poisonous ***Oleander*** *(Nereum oleander)* from the eastern Mediterranean has been a popular ornamental pot plant since the 15th century, on account of its beautiful foliage and its white, yellow or red flowers. The European, blue-flowered ***Periwinkle*** *(Vinca minor)* favors shady habitats, and is therefore often planted as an ornamental in cemeteries. *Carissa macrocarpa (= C. grandiflora)* ***(Natal plum)*** and *C. carandas* are grown for their edible fruit. The Brazilian *Hancornia speciosa* ***(Mangabeira)*** is also grown for its fruit, but mainly as a source of rubber in South America. *Plumeria rubra* ***(Frangipangi)*** is grown widely in the tropics as an ornamental. *Funtumia elastica* of Africa yields the Lagos silk rubber or African rubber, while lianas of the African genus *Landolphia* yield commercial Landolphia rubber.

Apoda:

1) Gymnophiona, ***caecilians*** (plate 3): an ancient, and at the same time, greatly modified order of the *Amphibia* (see), with members in all tropical regions except Australia. Many species are uniformly blue-gray and others have lighter lateral or segmental stripes. They are difficult to observe in the wild, and their phylogenetic relationships are unknown. The fossil history of the *A.* is almost nonexistent, consisting of a single vertebra from Brazil, dated from the Paleocene (65 million years ago). Varying in length from 7 cm to 1.5 m, they are long-bodied and limbless, lacking even the rudiments of limb girdles. Forward progress is achieved by body undulation. The body surface is encircled by a number of grooves, producing segmental rings. In many species, these rings also carry patches of scales, which resemble fish scales in structure and embryonic origin. The more primitive species have scales for the entire length of the body, whereas scales are totally absent from the more advanced, aquatic typhlonectids. The scales are evolutionary relicts of the armored exoskeleton of the extinct giant ancestors of the amphibians (*Stegocephalia*). *A.* are unique among vertebrates in possessing tentacle sensors, which develop at metamorphosis on the uper jaw, one between each eye and nostril. Tentacle sensors convey chemical information from the environment to the nasal cavity, and they are retracted by muscles, which, in other amphibians, retract the eyes. The eyes are regressed and probably function only by detecting the difference between light and dark. Almost all species retain the optic nerve, but the lens cannot be moved. In some species both lens and retina are reduced, and some species have no eyeball muscles. Skulls are massively bony with fusion of some elements, e.g. maxilla plus palatine into the maxillopalatine; otic plus occipital plus parasphenoid into the basale. The middle ear bone (columella or stapes) is massive or absent. Vertebrae are amphicoelous and ribs are bicapitate. Teeth (2 rows in the upper jaw; 1 or 2 rows in the lower jaw) are pointed and curved inward, with sharp edges and sometimes 2 cusps. The jaw muscles are especially powerful. Only a single lung is present. The inner skin layer contains mucus glands and poison glands, whose toxic secretions can be harmful to predators and to humans.

Most species are subterranean, living in loose earth and ground litter in tropical forests, often near to water. The typhlonectids, however, lead an aquatic existence. Food consists of earthworms, termites and orthopterans. Some appear to be specialist termite feeders (*Afrocaecilia*, *Boulengerula*), while others prefer earthworms (*Dermophis*) or beetle pupae (*Typhlonectes obesus*). Small lizards may even be eaten by *Dermophis*, while *Siphonops* has taken baby mice under laboratory observation.

Fertilization is always internal; the male possesses a single protrusible copulatory organ (phallodeum). Depending on the species, *A.* are oviparous, ovoviparous or viviparous. In viviparous species the young are fed with secretions from glands in the oviducts ("uterine milk"). In oviparous species, the larvae have gills and are aquatic, becoming terrestrial after metamorphosis (except the aquatic typhlonectids). Six families are recognized, comprising about 160 species in 34 genera. Classification is based to a large extent on the bone structure of the skull.

1. Family: *Rhinatrematidae* (rhinatrematids, 9 species in 2 genera). The most primitive family of the *A.*

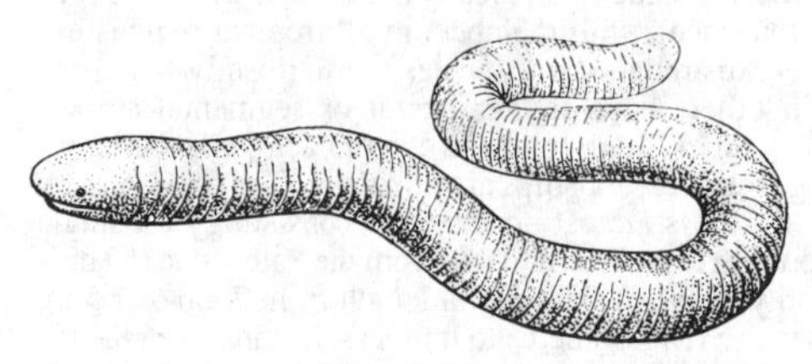

Apoda. Ceylon caecilian *(Ichthyophis glutinosus).*

Adults live in damp earth in the vicinity of water. All members are oviparous, and the gill-bearing, newly hatched, free-living larvae make their way to the water. The skulls of the rhinatrematids have a unique feature, in which the basale of the skull carries paired dorsolateral processes, which fit into notches in the posterior ends of the squamosals. The family is restricted to South America, occurring in the Caribbean, the Guianan region, Amazonia, and Pacific drainages from Venezuela to Peru. The largest member of the family (*Epicrionops petersi*) is 328 mm in length. Other members (genus *Rhinatrema*) are among the smallest *A.* known.

2. Family: *Ichthyophiidae* (ichthyophiids, fish caecilians, 35 species in 2 genera). The family is allied to the rhinatrematids, but more skull bones are present, and there are some differences in musculature. They appear to occupy a status intermediate between the rhinatrematids and more advanced families. Vertebrae extend beyond the anus, i.e. a tail is present. Adults live in damp earth near water. All are oviparous, and the gill-bearing, newly hatched larvae make their way to the water. Females of the Ceylon caecilian (*Ichthyophis glutinosus*) (plate 3) (Fig.) lie coiled around their eggs. The family is represented in both Asia and South America.

3. Family: *Uraeotyphylidae* (uraeotyphlids; a single genus, *Uraeotyphus*, with 4 species). All species are relatively small (about 300 mm in length) and restricted to a small area in the southern part of the Indian peninsula. They share features with the ichthyophiids, but they are cladistically closer to the more advanced families. All are oviparous.

4. Family: *Scolecomorphidae* (scoleomorphids; a single genus, *Scolecomorphus*, with 7 species). Members are of moderate size (e.g. *S. convexus* is 448 mm in length), and restricted to Africa. All are viviparous. Skull bones are reduced in number, and the columella is absent.

5. Family: *Caeciliidae* (caeciliids, worm caecilians, 88 species in 24 genera). The family is represented throughout tropical South and Central America (13 genera), in sub-Saharan Africa (6 genera), in the Indian subcontinent (2 genera) and in the Seychelles (3 genera). Members are oviparous, ovoviparous or viviparous. The young may or may not pass through a free-living larval stage. Skull bones are reduced in number, and there is no tail. *Caecilia thompsoni* of Columbia (1.5 m) is the longest member of the *A.* Other species are shorter and stouter, e.g. *Caecilia nigricans* measures 80 cm, with a body diameter of 4 cm. Examples of other genera are: *Boulengerula*, *Dermophis*, *Afrocaecilia*, *Siphonops*. Smallest of all the *A.* is the 7 cm-long *Idiocranium* from *Nigeria*.

6. Family: *Typhlonectidae* (typhlonectids, aquatic caecilians, 19 species in 4 genera). All are viviparous, and the young are born at an advanced state of development without gills. All adults are aquatic, living in lowland rivers and streams of South America; they are the only members of the *A.* that are not terrestrial burrowers. The body possesses a small to medium-sized dorsal "fin".

[R.A. Nussbaum, Rhinatrematidae: a new family of caecilians (Amphibia: Gymnophiona), *Occ. Pap. Mus. Zool. Univ. Michigan* (1977) ***682***, 1–30. R.A. Nussbaum, The evolution of a unique dual jaw-closing mechanism in caecilians (Amphibia: Gymnophiona) and its bearing on caecilian ancestry, *J. Zool.* (London)

(1983) ***199***, 545–554. R.A. Nussbaum, The taxonomic status of the caecilian genus Ureatryphlus Peters, *Occ. Pap. Mus. Zool. Univ. Michigan* (1979) ***687***, 1–20. R.A. Nussbaum & G. Treisman, Cytotaxonomy of Ichthyophis glutinosus and I. kohtaoensis, two primitive caecilians from Southeast Asia. *J. Herpetol.* (1981) ***15***, 109–113.]

2) See *Cirripedia.*

Apodidae: see *Apodiformes.*

Apodiformes, ***swifts*** and ***hummingbirds:*** an order of birds (see *Avies)* comprising 3 families. Whether it is appropriate to group the swifts with the hummingbirds is debatable, but their wing structure is similar, in that the upper arm bone or humerus is short and strong, and the wing is mainly formed by elongation of the bones beyond the humerus.

1. Family: *Apodidae* ***(Swifts;*** 67 species in 8 genera). Swifts feed entirely on insects, which they catch in flight, usually at considerable heights. They have small bills and wide gapes, and unusually for birds that catch insects in flight they have no rictal bristles. Despite their superficial resemblance to swallows and their similar mode of feeding, the two groups are not closely related. The wings of swifts form a characteristic smooth crescent. Their wing movements are often too rapid to follow with the eye, and they appear to control direction by beating one wing faster than the other. This latter technique is also used for the otherwise difficult process (for swifts) of taking off from flat ground. Some fly very fast, reaching speeds up to 100 miles (160 km) per hour, and air speeds of 150–200 miles (240–320 km) per hour are claimed for the large ***Spine-tailed swifts*** (genus *Chaetura)* of East Asia. Swifts have small weak feet and short legs. They never perch, and some species even mate and sleep in the air; when nest building or roosting they cling to vertical surfaces with the aid of their sharp claws. Plumage is generally drab gray and brown, often with lighter patches on the underparts. The common European ***Swift*** *(Apus apus;* 16.5 cm) (plate 6) breeds throughout the whole of Europe, extending well into Central Asia almost to the Pacific coast; the entire breeding population overwinters in sub-Saharan Africa.

Of particular interest are the small cave ***Swiftlets*** (genus *Collocalia)* of Southeast Asia, Australia and the western Pacific islands. Species differences are only slight and their distribution is complex, so that taxonomy is difficult, but there seem to be between 15 and 20 distinct species. These small birds (9–16.5 cm) do not fly particularly high or fast, and they feed by darting about in forest clearings. The rounded, bracket-shaped nests of cave swiftlets are "glued" with saliva to the vertical rock surface of a cliff or cave, or a tree trunk; they contain an extremely high proportion of saliva, and they are used to make "birds nest soup". Depending on the species, the saliva is mixed with varying proportions of vegetable material or feathers. All swiftlet nests have culinary value, but most favored (and expensive) are those of the ***Gray*** and ***Gray-rumped swiftlets*** of Southeast Asia, whose nests consist entirely of hardened saliva.

2. Family: *Hemiprocnidae* **(Crested swifts).** This small family contains only 3 species, all in the genus, *Hemiprocne,* distributed from India to the Solomon Islands. Less specialized than the *Apodidae,* they can perch, and they are more brightly colored than the other swifts. They also possess a distinctive forehead crest and a long forked tail. The ***Indian crested swift*** *(H. longipennis;* 20.5–23 cm) inhabits mangrove swamps, rubber plantations and forests up to 1000 m in India and Southeast Asia; upper parts, including the crown and pointed vertical crest on the forehead, are green, while the neck is tinged with bronze, and the flight and tail feathers are dark green.

3. Family: *Trochilidae* ***(Hummingbirds;*** 320 species in 80 genera). The deep keel of the sternum serves for the attachment of flight muscles that constitute 25–30% of the body weight. With the aid of such powerful flight musculature, the wings beat extremely fast, conferring the ability to hover, to move straight up and down in the air, to fly sideways, and even to fly backward. When hovering, the ***Broad-tailed hummingbird*** beats its wings 55 times per second, rising to 75 beats per second as it moves away in level flight. The small feet are used only for perching and never for progressing; humming birds use their wings even to move only few centimeters along a perch. Their long, thin, sometimes curved beaks and very long tongues (tubular at the tip) are adapted for taking nectar from tubular and trumpet-shaped flowers, while hovering. They also take insects from inside flowers, hunt insects on the wing, and sometimes forage for insects among leaves. In taking nectar, hummingbirds also perform a cross pollination. Found only in the New World, they have no counterpart in the Old World avifauna, where their particular ecological niche appears to be unfilled. Smallest members of the family are the ***Bee hummingbirds*** of the West Indies. The smallest of all birds is the ***Cuban bee hummingbird,*** which is little more than 5.5 cm from the tip of its bill to the tip of its tail; it weighs less than 2 g, and a single egg weighs only 200 mg. Hummingbird plumage is usually vividly colored and irridescent due to light diffraction. The illustrated ***Swordbilled hummingbird*** (Fig.), which inhabits the forests of the Andes from Venezuela to Peru, is 21.5 cm in length and colored irridescent green, somewhat darker around the head, wings and tail. Names such as ***Ruby and Topaz hummingbird*** *(Chrysolampis mosquitus;* 9 cm; northern and eastern South America), ***Crimson topaz*** *(Topaz pella;* 19 cm; Guianas to Ecuador), ***Ruby-throated hummingbird*** *(Archilocus colubris;* 9 cm; eastern USA and southern Canada) bear witness to the beautiful colors of these birds. Both the male and female of *Archilocus colubris* are dark green above and white below, and the male also has an iridescent ruby throat. *Archilocus* breeds throughout North America east of the Plains from Alberta to Nova Scotia and is the only humming bird found in the eastern USA. It winters in Mexico and Central America, after migrating across the Gulf of Mexico. During the autumn, before migration, it adds fat equivalent to half again of its body

Sword-billed hummingbird (Docimastes ensifer)

weight, then flies a considerable distance (from the coast of Louisiana to the tip of Yucatan is 575 miles or 925 km) at about 30 miles (48 km)/h, with a wing beat of 75 times/second, and a heart beat of 615/min. With no adverse winds, 10 hours are required for this flight; thousands perish on the way, but tens of thousands survive. Average weight before leaving is 0.2 ounce (5.67g), and the weight on arrival is 0.125 ounce (3.54 g). The reverse migratioin is undertaken in the spring.

Apoenzyme: see Enzymes.

Apogamy: a type of Apomixis (see), in which the sporophyte or embryo does not develop from an egg cell, but either asexually from a diploid prothallium (gametophyte) cell (as in pteridophytes), or from the synergidae or antipodal cells [as in some phanerogams, e.g. *Taraxacum* (dandelion) and species of *Hieracium* (hawkweed)]. The synergidae and antipodal cells of the embryo sac are normally haploid, but in cases of A. they become diploid by Apospory (see) or Apomeiosis (see).

Apolonis: see *Sturnidae*.

Apomeiosis: the formation of gametophytes without meiosis. In higher plants A. leads to the production of diploid egg cells.

Apomixis: reproduction without fertilization and/or meiosis, but with the superficial appearance of an ordinary sexual cycle. In plants, A. is the formation of embryos in the ovules without prior fertilization. Different types of A. are Parthenogenesis (see), Apogamy (see) and Adventitious embryony (see). The adjective to A. is apomictic.

Aponeurosis: see Connective tissues.

Apoplastic route: see Water economy.

Aporogamy: see Fertilization.

Apospory:

1) The absence of (meio)spore formation (see Spores) in some pteridophytes, e.g. in *Athyrium filix-femina,* in which a diploid prothallus develops from a diploid cell of the Sporophyte (see). Thus, an ordinary diploid sporophyte cell substitutes for the spore, meiosis being omitted, so that the gametophyte is diploid.

2) In phanerogams, the development, without meiosis, of a vegetative cell of the ovule into an embryo sac containing a diploid egg cell.

Apostematic coloration: see Protective adaptations.

Apostoma: see *Ciliophora*.

Apothecium: the flat, cup- or saucer-shaped fruiting body of certain ascomycete fungi *(Discomycetes* or cup-fungi) and lichens, which is lined with a hymenium of asci and paraphyses. The often brightly colored A. may be sessile or stalked, and it varies in diameter from a few millimeters to over 40 cm.

Apparent photosynthesis: see Net photosynthesis.

Appeasement behavior: behavior that inhibits or neutralizes aggression in a behavioral partner. A.b. also activates behavioral tendencies that are different from, and incompatible with, aggression or attack. It is often difficult to distinguish clearly between A.b. and behavior that signals a state of subdominance.

Appendiculariae, *Copelata, Larvaceae:* a class of tunicates with 60 pelagic species, only a few millimeters in length. The sexually mature form retains the organisation of the larva. The tail, containing the notochord, is attached at right angles to the ventral side of the body, and the notochord is associated with a series of ganglia. There is no cloacal cavity, and the intestine and two gill

Appeasement behavior. a Looking away (herring gull). *b* Offering the throat (red fox). (After Tinbergen and Tembrock)

clefts open directly to the exterior. Gonads are usually large and lie in the hinder region, and nearly all *A.* are hermaphrodite. The only form of reproduction is sexual, and metamorphosis does not occur. All members of the *A.* reside inside a test, known as a house, which is not composed of tunicin. In relation to the animal, the house is very large. Water is driven through the house by movements of the tail, and plankton are filtered from the resulting water stream by a filtering apparatus. If conditions become adverse, e.g. if the filter trap becomes blocked, the animal can escape and form a new house.

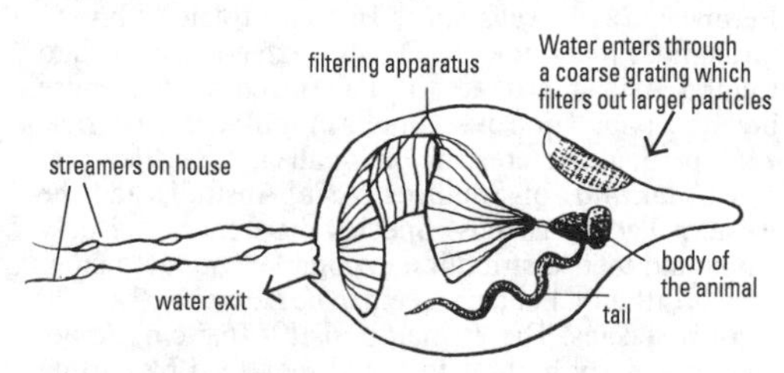

An appendicularian in its house

Appendix: see Cecum.

Appetitive behavior: motivated behavior toward a particular object, which ends when the object is found, i.e. when recognition stimuli become operative.

A.b. includes seeking and decision making The actual search is known as *orientation A.b. (A.b.I),* and this is ended as soon as orientation stimuli come into play, i.e. the sought object (e.g. food) becomes the subject of sensory perception (e.g. can be seen or smelled). Location and identification of the object is followed by *decision A.b. (A.b.II),* which involves acceptance or refusal (or selection in those cases where a choice is offered, e.g. the choice of a food plant by a herbivore). Another example is the search for, and eventual choice of, a reproductive partner. When the decision has been made, the releasing stimulus of the sought or chosen object becomes stronger, and the A.b. is replaced by Cunsummatory action (see). Consummation is followed by a refractory phase, in which A.b. cannot recur.

Apple: see *Rosaceae.*

Apple codlin moth, *Cydia pomonella* L.: a moth of the family *Tortricidae* (see), with a wingspan of about 15 mm. Eggs are laid singly on young fruits, leaves or twigs. The caterpillars (popularly known as "maggots") tunnel into the center of the fruit, where they remain until ready to pupate, feeding on the apple pips and flesh around the core. Fully grown caterpillars leave their food source and pupate nearby, so that larvae and pupae are often found in other foods stored near apples. Pears, plums, quinces and walnuts are also attacked.

Apple fruit moth: see *Yponomeutidae.*

Appleskin worm, *Adoxophyes orana* F.R.: an orchard pest of the family *Tortricidae* (see). The caterpillar attacks fruits of apple, pear, cherry and many other trees and plants. It does not bore into the fruit, but eats along the surface, forming a trough, which provides access for fungal attack.

Apposition: the thickening of plant cell walls by the addition of a layer of new material alongside the older layers. After completion of cell growth, the secondary wall is formed mainly by A. During growth, the irreversibly stretched cell wall is reinforced by Intussusception (see).

Apposition image: see Light sensory organs.

Apricot: see *Rosaceae.*

Apteria: see Feathers.

Apterogasterina: see *Oribatei.*

Apterygiformes, ***kiwis:*** an order of flightless, running birds, comprising 3 recent species in New Zealand. The ***Common, Brown*** or ***Striated kiwi*** *(Apteryx australis)* exists as 3 subspecies: *A.a. mantelli* (North Island), *A.a. australis* (South Island), *A.a. lawryi* (Stewart Island). The ***Great spotted*** or ***Large gray kiwi*** *(Apteryx haasti)* is found in the south and west of the South Island. Although now very rare on the South Island, the ***Little spotted*** or ***Little gray kiwi*** *(Apteryx oweni)* is numerous on Kapiti Island off the southwestern tip of the North Island, where it has been introduced. The plumage of the 2 latter species is softly barred.

In certain areas of their range, kiwis are not uncommon, but they are rarely seen, being nocturnal and unobtrusive. They hunt at night and at twilight for worms and insects, which they locate by smell, and they also eat berries. With approximately the same size range as domestic fowls, they weigh from 1.4 to 4 kg. Their short legs are powerful and muscular, and the claws are sharp and strong. Wings are vestigial and buried in the feathers, and there is no tail. Large ear apertures suggest that hearing is well developed, while facial bristles and the long, flexible bill confer a keen sense of touch. In contrast, the eyes are small and the eyesight poor. The nostrils are situated at the tip of the bill, and the olfactory sense is well developed. Pectoral development is poor, so that the trunk is wedge-shaped, tapering toward the strong neck, which carries a small, compressed skull. Feathers are weakly barbed, lying rather haphazardly, ruffled and hairlike. The female kiwi lays a single egg about one quarter of her own weight; this contains a copious food supply, which sustains the embryo through its long incubation, and provides temporary nourishment (in the form of a yolk sac) for the newly hatched chick. Incubation of the eggs and care of the young (although little attention is paid to them after hatching) are performed entirely by the male. The nearest known phylogenetic allies of the kiwis are the extinct, flightless moas (see *Dinornithae).*

Members of the orders *Struthioniformes, Rheiformes, Casuariiformes* and *Apterygiformes* were formerly grouped together as the "Ratitae": this close relationship is now doubtful, but members of these 4 orders are still collectively known as Ratites. In all ratites, the wing skeleton and pectoral girdle show varying degress of reduction, and the sternum is flat and raft-like with no keel. See *Avies.*

Apterygota: a paraphyletic group of relatively primitive, wingless insects, in which winglessness is a primary (primitive) characteristic and not the result of secondary wing reduction. It includes the probably monophyletic *Entognatha,* the *Archaeognatha* (bristletails) and the *Zygentoma* (silverfish). Members retain a relatively large number of the primitive characters of the *Mandibulata,* e.g. winglessness, abdominal limb primordia and jointed antennae, but they cannot be considered as phylogenetically intermediate between myriapods and true insects. They appear rather to represent a secondarily highly specialized group of soil insects.

In the *Entognatha* (composed of the *Collembola, Protura* and *Diplura),* the labium is fused with the side walls of the head capsule to form a pouch enclosing the eversible mandibles and maxillae (Fig.), enabling the development of biting, tearing or piercing-sucking mouthparts. The *A.* is a heterogeneous group, but its members share characters such as reduction of compound eyes to maximally 8 ocelli, as well as a tendency for regression of the Malpighian tubules.

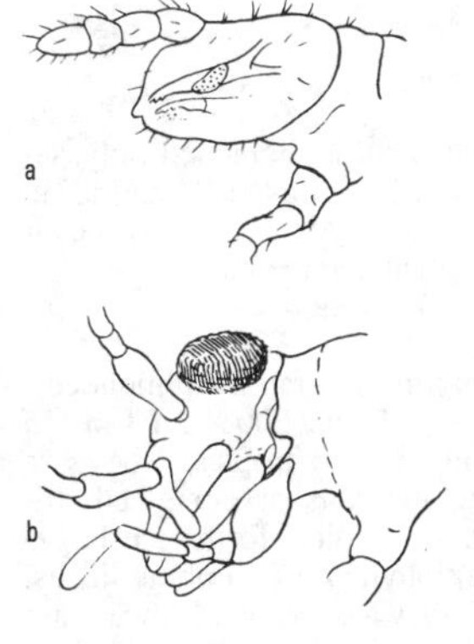

Apterygota. a Entognathous mouthparts *(Collembola). b* Ectognathous mouthparts *(Archaeognatha).*

The *Archaeognatha* and *Zygentoma* (formerly combined as the *Thysanura)* are, however, clearly related phylogenetically. The dicondylar mandibles of the *Zygentoma* and of pterygote (winged) insects represent a crucial evolutionary development, which clearly separates them from the moncondylar ectognathous *Archaeognatha.*

Since the *Collembola, Protura* and *Diplura* are now regarded as distinct subclasses, the *A.* are considered by most authorities to comprise only the *Archaeognatha* and *Zygentoma.*

See Coast bristletails.

Aptychia (sing: **Aptychus)** (Greek *aptychon* a folding panel consisting of two parts): opercula from the apertures of ammonites that occur from the Permian to the Upper Cretaceous. They consist of a symmetrical pair of plates with a superficial resemblance to bivalves; the concave surface is smooth, while the convex surface is granulated, ribbed, punctate or furrowed [Paleozoic ammonites had an unpaired operculum, called an Anaptychus (see)]. A. are chitinous and calcareous, or purely calcareous. The calcareous material of A. is calcitic, whereas that of ammonite shells is aragonitic, and more easily dissolved. A. have therefore sometimes survived to form aptychian chalk (e.g. in the European Alps), while the parent ammonite shells have disappeared. See Ammonites.

Aptychian chalk: see Aptychia.

Aptyctima: see *Oribatei.*

Apus: see *Apodiformes.*

Aquatic bugs: see *Heteroptera.*

Aquatic plants: see Hydrophytes.

Aqueduct of Sylvius: see Ventricle.

Aquila: see *Falconiformes.*

Aquilegia: see *Ranunculaceae.*

Ara: see *Psittaciformes.*

Arabian camel: see Dromedary.

Arabinose: a pentose occurring naturally in both the D- and L-form. β-L-A. is a component of hemicelluloses, e.g. the araban of cherry gum, and it is found in plant mucilages, glycosides and saponins. β-D-A. has been isolated from the glycosides of some bacteria.

```
    _H
  C
  | \\O
HO—C—H
   |
 H—C—OH
   |
 H—C—OH
   |
  CH2OH
```

D-Arabinose

Arabans: high molecular mass, branched polysaccharides composed of L-arabinose linked 1,5 and 1,3 in the furanose form (see Carbohydrates). They occur widely as components of plant hemicelluloses.

Aracaninae: see *Ostraciontidae.*

Araçaris: see *Ramphastidae.*

Araceae, ***aroids, arum family:*** a family of monocotyledonous plants, containing about 2 000 species in 115 genera, distributed mainly in the tropics. They are mostly herbs, often with tuberous or elongated rhizomes, rarely woody. Many tropical forms are large-leaved rosette plants or epiphytic root-climbing lianes. A few species are marsh or water plants. Flowers are without bracts and usually crowded together on a club-shaped spadix which is generally enclosed in a spathe (see plate 14, Fig. 6), the latter often brightly colored. Flowers are hermaphrodite or unisexual. Unisexual flowers are usually monoecious, with males on the upper part of the spadix and females below. The genus, *Arisaema,* is dioecious. A perianth is present in hermaphrodite, but not in unisexual flowers. Stamens (usually fewer than 6) are often united into a synandrium (e.g. *Colocasia, Spathicarpa).* Staminodes are often present and united. The ovary is superior or sunk into the spadix, and often reduced to 1 carpel. The one to many seeded fruit is a fleshy or coriaceous (leathery) berry. Flowers are usually protogynous with an offensive odor. Many species contain a poisonous latex, which can be rendered harmless by heating. Pollination is often by insects attracted by the carrion-like odor of the spadix; for a description of this pollination mechanism in *Arum,* see Pollination.

The large starchy tubers of ***Taro*** *(Colacasia antiquorum = esculenta)* are an important tropical root crop. Species of *Alocasia* (see plate 14, Fig. 6) and *Xanthosoma* are also cultivated in the tropics for their roots or leaves. In the tropics, *Monstera deliciosa* is cultivated for its edible fruit. In temperate regions, *Monstera* is also a popular decorative house plant (Swiss cheese plant), with curiously perforated leaves. Other genera grown for their leaves (often variegated) are *Caladium, Aglaonema, Philodendron, Syngonium* and *Dieffenbachia.* Genera grown for their flowers are *Anthurium, Spathiphyllum, Sauromatum* and *Zantedeschia.* ***Sweet flag*** *(Acorus calamus)* is a familiar wild European, Asian and North American species; its tightly packed yellowish flowers completely cover the spadix, and its leaves emit a tangerine smell when bruised. *Calla palustris* is found in swamps and wet woodlands in northern Asia, Europe and North America; its flat spathe does not enclose the stout, oval, cuspidate spadix. The erect, glabrous *Arum maculatum* ***(Lords and ladies, Cuckoo pint)*** is native from North Africa, through Europe to southern Sweden; the spathe is pale yellow-green, sometimes spotted with purple.

Arachidic acid, ***icosanoic acid:*** CH_3—$(CH_2)_{18}$—COOH, a fatty acid which occurs naturally in the form of its glycerol esters. The latter are widely distributed, but usually in low concentrations (3%) in plant oils, e.g. groundnut, cocoa, olive, sunflower, soybean and rapeseed oils.

Arachidonic acid, ***5,8,11,14-icosatetraenoic acid:*** $CH_3(CH_2)_4$—CH = CH—CH_2—CH = CH—CH_2—CH = CH—CH_2—CH = CH—$(CH_2)_3$—COOH, an unsaturated, higher fatty acid containing 4 double bonds. It occurs as an esterified component of animal fats and phospholipids, being particularly abundant in fish liver oils. Esters of A.a. are also present plant oils. It is an essential dietary component for many mammals, including humans, and is therefore considered to be a vitamin.

Arachis: see *Papilionaceae.*

Arachnida, ***arachnids:*** a class of the phylum *Arthropoda.* The *A.* are almost exclusively terrestrial. They include a wide variety of different forms, all with the same basic body plan. Head and thorax are amalgamated into a single anterior body section, known as the cephalothorax or *prosoma.* The prosoma is composed of six adult segments, each bearing a pair of appendages, and it is usually covered by a uniform dorsal shield (carapace). The fifth and sixth segments are rarely differentiated, so that a proterosoma is formed from the four anterior segments and the acron. The hind body section (abdomen or *opisthosoma)* consists of a maximum of 13 segments, but this number is usually reduced by regression of posterior segments. Boundaries marking the opisthosomal segments are often absent, so that the opisthosoma becomes a uniform sac-like structure (e.g. in spiders). In some cases, a telson is still present, e.g. the venom gland of scorpions and the flagellum of whipscorpions. Prosoma and and opisthosoma may be broadly jointed or connected by a pedicel. The first (preoral) segment of the prosoma carries a pair of prehensile appendages, known as *chelicerae*. Each chelicera has a maximum of 3 segments, and in most orders the terminal segment is chelate (pincer-like or claw-like); in many parasitic mites, the chelicerae are modified to narrow piercing structures. The second pair of appendages, the *pedipalps,* may serve as walking legs, but in arachnids with large chelicerae (spiders, microwhipscorpions, windscorpions, many harvestmen and mites) they have become true palps. In orders with small chelicerae (scorpions, whipspiders, whipscorpions, pseudoscorpions), the pedipalps are modified to large, usually chelate raptorial organs. Thus, according to the order, either the chelicerae or the pedipalps are developed for the performance of a major function, such as the capture and dispatch of prey, for crushing ("chewing") food material, for defense, or for digging. In male spiders, the pedipalps are modified as organs of sexual intromission. The next four arachnid appendages serve as walking legs, but they also have a sensory function, and they are also used for drawing out and stretching threads from the spinnerets, for digging or for swimming. In a very few species, the posterior pair of walking legs is absent. In contrast to the related *Merostomata,* the opisthosoma of the arachnids is totally devoid of ambulatory appendages. Limb rudiments are, however, present in the embryo; in scorpions some of the embryonal mesosomatic appendages are modified during development and become the lung books of the adult, whereas in spiders the spinnerets are derived from embryonal opisthosomal appendages.

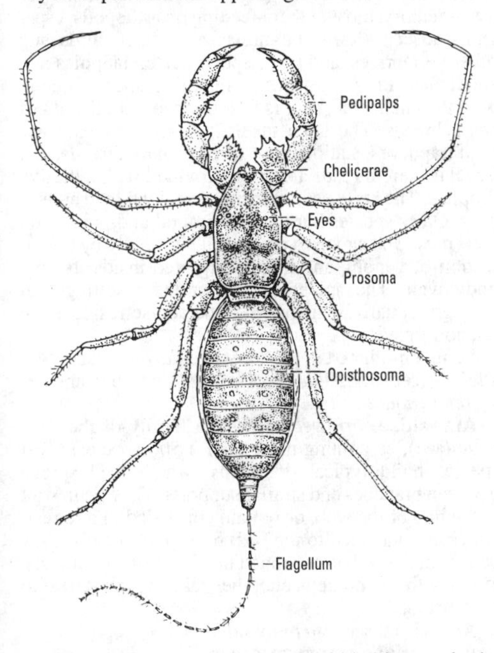

Arachnida. Structure of an arachnid (whipscorpion).

Eyes are always simple (not compound), and sometimes highly sensitive. They vary in number; some scorpions have up to twelve, and some arachnids possess both median and lateral eyes. In most orders, however, vision is relatively poor, especially in the many nocturnal species, which rely on sensory hairs for the detection of prey. Associated with the alimentary canal are paired midgut glands, which are located mainly in the opisthosoma. Excretory Malpighian tubules arise from the midgut (not from the hindgut, as in the *Tracheata),* and they form branches between the midgut glands. The primary excretory product of the Malpighian tubules is guanine, with some uric acid. The prosoma also contains one or two coxal glands, which correspond to the nephridia of other arthropods. Respiratory gas exchange is performed by tracheae, most commonly by tubular tracheae, and sometimes by the type known as lung books. Sexes are separate, and the genital organs exit on the opisthosoma. Many species display brood care, in that they carry the eggs with them in a cocoon or secreted sac.

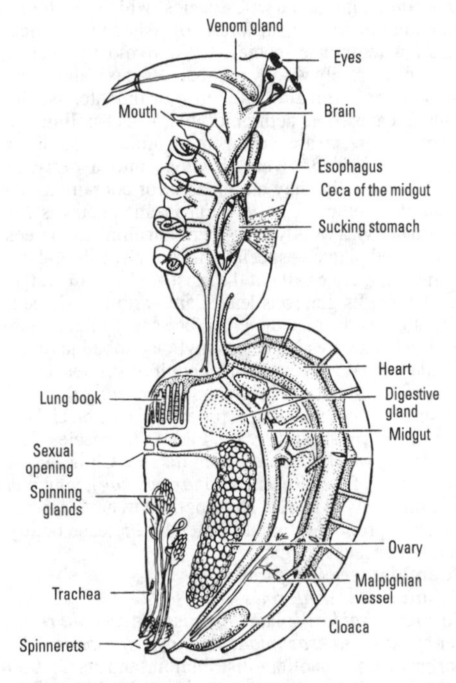

Arachnida. Diagram of a spider.

About 37 000 species are known, practically all of them terrestrial. A few species have become secondarily adapted to an aquatic existence (e.g. the water spider and water mites).

Classification. Nine living orders are recognized: 1. *Scorpiones* (scorpions); 2. *Pedipalpi* [including the suborders *Uropygi* (whipscorpions) and *Amblypygi* (whipspiders)]; 3. *Palpigradi* (microwhipscorpions); 4. *Araneae* (spiders); 5. *Ricinulei* (ricinuleids); 6. *Pseudoscorpiones* (pseudoscorpions); 7. *Solifugae* (windscorpions); 8. *Opiliones* (harvestmen); 9. *Acari* (mites and ticks).

Fossil record. Fossils are found from the Silurian to the present. The earliest Silurian fossils are those of scorpions, and the Silurian *Palaeophonus nuncius* was possibly the first terrestrial animal. Fossil spiders first appeared in the Devonian. Since the *A.* are not easily preserved, their remains are relatively rare. Nevertheless, fossil *A.* display a greater variety of forms than the modern living *A.* Including fossil *A.*, fourteen orders are recognized. Five of these became extinct in the Paleozoic, leaving the present nine orders of living A.

Collection and preservation. Specimens can be caught practically everywhere by hand, or trapped in glass vessels set into the soil. Ectoparasites (mites and ticks) can be picked off the host animal. They are killed in ethyl acetate vapor and stored in 75% ethanol. Many species can be dehydrated by passing through a series of increasing concentrations of ethanol or acetone, then finally dried.

Arachnis: see *Orchidaceae*.

Arachnothera: see *Nectariniidae*.

Araliaceae, *ivy family:* a family of dicotyledonous plants containing about 700 species, which are found predominantly in tropical forests of Asia and America, with a few temperate species. They are mostly trees or shrubs, more rarely woody climbers (lianas), and occasionally herbs, with alternate lobed or pinnate, usually stipulate leaves. The actinomorphic, insect-pollinated, 5-merous flowers are arranged in umbels, heads or spikes. The inferior ovary develops into a berry or drupe. Chemically, they are notable for containing triterpenoid saponins. The only European species is ***Ivy*** *(Hedera helix),* a woody evergreen root climber on trees or rocks, and planted especially in churchyards and cemeteries. It tolerates all but the most acid, dry or waterlogged soils. Its glabrous leaves are variously shaped, with a shiny dark green upper surface and a paler underside. The leaves are also heterophyllous (lobed juvenile or shade leaves, and rhomboidal adult or sun leaves on the flowering shoots). The green flowers are pollinated in autumn by flies and wasps, producing ripe black berries in the following spring. The East Asian species, *Tetrapanax papyrifer,* is used to manufacture Chinese rice paper. Roots of the ***Ginseng*** *(Panax ginseng),* which is native primarily to Korea, are supposed to have universal healing properties, and to generally increase bodily performance and strength.

Aramidae: see *Gruiformes*.

Aramus: see *Gruiformes*.

Araneae, *spiders* (plate 44) (Fig.2, *Arachnida):* an order of the class *Arachnida,* phylum *Arthropoda.* The order has a cosmopolitan distribution, and its 21 000 species have invaded a wide variety of habitats. The body is divided into two main parts, the *prosoma* and the sac-like, extremely manoeuvrable *opisthosoma* (abdomen), which is attached to the prosoma by a narrow, stalk-like pedicel. The prosoma is covered by a uniform dorsal carapace (tergal shield), and also has a nonsegmented ventral plate. Only a groove marks off the head, which is otherwise an integral part of the prosoma. It carries two large, powerful chelicerae, each with a fang-like subchela, which houses the opening of a venom gland. Also on the prosoma are the leg- or feeler-like pedipalps and four pairs of walking legs. Appendages are absent from the opisthosoma (apart from two pairs of spinnerets, which are modified vestigial appendages). The spinnerets arise on body segments 10 and 11, but migrate posteriorly during embryonal development. The familiar elastic threads of the spider's web are formed by drawing out the secretion of the spinning glands, whose opening are housed in the spinnerets. Sensory organs consist of 2–8 (usually 8) eyes, whose axes can point in any direction. Numerous sensory setae and trichobothria are also present, as well as chemical and tactile receptors on the tarsi and pedipalps. In the primitive condition, the respiratory organs consist of two pairs of book lungs, but one or both of these are often replaced by tracheae. Males are usually smaller and more slender than females, and they often allow themselves to be transported by the wind at the end of their own spun thread.

Before pairing, a complicated appendage at the tip of each male pedipalp is filled with a sperm droplet form the sexual opening of the opisthosoma. The male then approaches the female and performs a highly stereotyped pattern of movements, which is different for each species. This is necessary, in order to gain recognition by the female, who would otherwise regard the male as prey. The tips of the pedipalps are then inserted into the sexual opening of the female. In some species, it is normal for the female to kill and feed on the male after copulation. Eggs are fertilized as they are laid. Almost all species enclose their eggs in a spun cocoon, which is then often guarded by the female.

Nearly all species are strictly terrestrial. Only one species lives in water, and a few species hunt or lie in wait on the surface film of water. All species construct living quarters (a lair) from spun threads. This usually tubular or funnel-shaped construction is hung or attached in various ways on or in the soil, beneath loose bark, in passages and fissures. In addition to a lair, many species also build a web (e.g. *Areneidae).* All species are predatory, most of them feeding on arthopods, especially insects. Large tarantulas also catch birds and other vertebrates, and some species catch tadpoles and small fish. Prey is caught at lightning speed directly with the chelicerae, or it is first trapped with the aid of spun threads. The latter take the form of webs, traps with trapdoors, stationary or thrown snare threads, etc. Special snare threads are often provided with adhesive droplets. The lassoo spiders fling a sticky thread at their prey. Other species hunt on the ground and pounce on their prey. The captured prey is killed rapidly by the injection of venom, and largely digested inside its own body wall. The spider sucks out the resulting brei through its narrow mouth, and the coarse residues are not consumed.

Some familiar examples are *Aviculariidae* (see), *Areneidae* (see), *Lycosidae* (see), *Salticidae* (see), and *Argyroneta aquatica* (see).

Araneidae, *orb-web spiders:* a family of the *Araneae* (see), containing more than 2 500 species. Most species build typical orb webs, which they spread between branches and similar supports. They lie in wait at the hub of the web, or remain concealed in a retreat, which is connected to the web by signal threads. Prey caught in the web is cocooned in a layer of silk threads, killed with a venomous bite, then taken to the retreat to be consumed.

Arapaima: see *Osteoglossidae*.

Araucaria: see *Araucariaceae*.

Araucariaceae, *araucaria family:* a family of about 40 species of evergreen conifers, found predominantly in the Southern Hemisphere (South America, Australia, New Guinea, New Caledonia, Malaysia, New Hebrides, Norfolk Island), but with an almost worldwide distribution in the Triassic.

Young trees are clothed in branches from gound level to summit, whereas old trees have bare trunks for most of their height, and flat, untidy heads. The bark of old trees is thick and resinous, ridged with the bases of old leaves, and sometimes peeling into papery scales. Branching is regular, usually in whorls, and the branches are horizontal. The family contains the few conifers with broad leaves. In *Agathis,* the leaves are broad, stalked, and uncrowded, whereas in *Araucaria* they are broad, unstalked and densely crowded. Leaves are often very robust and always spirally arranged.

Male and female inflorescences are usually on different trees, and both are spirally arranged. In the female cone, the bracts and scales (ligules) are fused to form single-seeded bract-ovuliferous scales, each female cone consisting of many of these woody scales. Male strobili are dense, cylindrical and catkin-like. The single ovule is centrally placed and inverted. Usually the seeds are relatively large and thick-walled, adnate to the base of the scale, and winged in most species. The tracheids of the secondary thickening show a characteristic honeycomb arrangement of bordered pits (the primitive "araucarioid" pattern), a feature otherwise found only in fossil gymnosperms.

Araucariaceae. Shoot and cone of *Agathis* sp.

Some species of *Araucaria* produce useful timber, especially *Araucaria araucana* (named for Arauco, a province of Chile where it is native; known as the ***Chile pine*** or ***Monkey puzzle)*** and *Araucaria angustifolia = A. brasiliana* ***(Parana araucaria*** or ***Candelabra tree).*** *Araucaria excelsa = A. heterophylla* ***(Norfolk Island pine)*** is well known in Europe as an indoor ornamental tree. Other representatives include *Araucaria cunninghamii = Cunningham araucaria* ***(Hoop pine, Colonial pine, Moreton Bay pine, Richmond River pine, White pine)*** of southeast and tropical Queensland, and New Guinea. *Araucaria bidwillii* ***(Bunya-bunya)*** of the Queensland coast is very closely allied with South American species; its edible seeds are eaten by the aborigines.

The genus, *Agathus,* also contains useful timber species, e.g. *Agathus australis = Dammara australis* ***(Cowdie pine, New Zealand kauri, Dammar pine, Kauri pine),*** an important forest tree on the North Island of New Zealand, reported to attain a height of 60 m and an age of 2 000 years. *Agathis dammara = A. loranthifolia* ***(Amboyna pitch tree)*** is a variable species in many areas of Malaysia and Polynesia.

Various species of *Agatha* yield an amber-like resin , known as copal, Australian copal or gum kauri.

Araucarioid pattern: see *Araucariaceae.*

Arboretum: a plantation of trees and woody plants, often of foreign origin, in the interests of scientific study.

Arboricide: a plant protection agent for preventing the growth of unwanted woody plants.

Arbor-vitae: see *Cupressaceae.*

Arbutin: the mono-β-D-glucopyranoside of hydroquinone. It occurs widely in leaves of members of the *Saxifragaceae, Ericaceae, Pyrolaceae, Rosaceae,* and other plant families. It is hydrolysed by the enzyme, emulsin, into D-glucose and hydroquinone. Oxidation products of the released hydroquinone are responsible for the autumnal leaf colors of some fruit trees.

CH_2OH H O O OH H OH H HO H H OH

Arbutin

Arbutin cleavage test, ***arbutin splitting test***: hydrolysis of arbutin by β-glucosidases of microorganisms. Stelling-Dekker (1931) first used the cleavage of aesculin to distinguish between the yeast genera *Pichia* and *Hansenula.* Castellani (1937) and Diddens and Lodder (1942) replaced aesculin with the cheaper compound, arbutin. The yeast culture under test is subcultured onto arbutin agar [0.5% (w/v) arbutin, 2% agar in yeast infusion or dilute (1:10) yeast autolysate; immediately after sterilization (15 min autoclaving), 2–3 drops of 1% ferric ammonium citrate are added aseptically]. After incubation, an intense brown zone forms around positive colonies, i.e. colonies possessing β-glucosidase activity. The released aglucon, hydroquinone, is oxidized to tannins, which form brown complexes with Fe (III). The test has been abandoned for diagnostic purposes because it is now known that the genera, *Pichia* and *Hansenula,* both contain arbutin-positive and arbutin-negative species.

Arbutus: see *Ericaceae.*

Arcellinida, *Testacea, Testacida:* shelled, testate or thecate amebas of the order, *Rhizopodea.* The single-chambered, dish-, urn- or ampulla-shaped test has a more or less large aperture (pseudostome) through which the pseudopodia extend to the exterior (Fig.). The pseudostome is the main region of contact between the animal and its test. The test consists of organic material, often with inclusions of silica platelets, which appear during the vegetative phase. Inclusions of foreign bodies may also be present. At cell division, the platelets migrate to the daughter cell and form an ordered surface arrangement. *A.* are found primarily in freshwater habitats, forming species-rich fauna in bogs, peat and moss-covered areas; certain species occur in the mud of ponds and lakes.

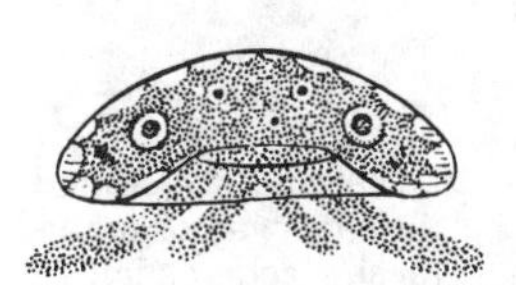

Arcellinida. Arcella vulgaris with chitinous test (lateral view).

Archaebacteria: a group of Prokaryotes (see), which differs from other prokaryotes as well as from eukaryotes. Based on the analysis of 16S ribosomal RNA, it was proposed by Woese that bacteria can be divided into archaebacteria and eubacteria, which differ from one another as well as from eukaryotes. The existence of these 2 evolutionarily different groups of bacteria is now generally accepted, and it is assumed that the separation of A., eubacteria and eukaryotes from common ancestor occurred at a very early stage of evolution.

A. can be divided into 3 main goups: Thermoacidophiles (e.g. *Sulfobolus solfataricus, Thermoproteus tenax),* Halophiles (e.g. *Halobacterium volcanii, Halococcus morrhuae),* Methanogens (see) (e.g. *Methanospirillum hungatei, Methanobacterium formicicum, Methanococcus vaniellii).* Some thermoacidophiles live autotrophically by desulfurication in hot springs up to temperatures of 85 °C. *Halobacteria* occur in salt lakes and even in saturated brine, and one representative harvests light energy for metabolism with the aid of the red pigment, rhodopsin.

There are several fundamental biochemical differences between A. and eubacteria. Thus, the lipids of A. consist of branched chain fatty acids in ether linkage with D-glycerol, whereas the lipids of eubacteria contain D-glycerol and the fatty acids are linear. Elongation factor 2 of A. resembles that of eukaryotes in being sensitive to diphtheria toxin (that of eubacteria is insensitive). Introns are known to be present in the ribosomal and transfer RNA of A., whereas no introns are known in the corresponding RNA of eubacteria. The RNA polymerases of A., like those of eukaryotes, are large, multisubunit proteins, whereas those of eubacteria are relatively simple proteins. Peptidoglycan (murein) is a typical component of the cell walls of all eubacteria, but it is absent from A. Thermoacidophiles possess reverse gyrases that introduce a positive supercoil into DNA, whereas the gyrases of eubacteria introduce negative supercoils.

Archaeoceti: an extinct suborder of Whales (see).

Archaeocyatha: fossil calcareous sponges, cone-shaped and 2–8 cm in length. The calcareous skeleton consists of an inner and an outer porous wall. The space between the two walls is separated into chambers by porous, radial, irregular dividing walls. *A.* occur in particular in the Lower and Middle Cambrian of North America, Australia and Asia. Together with calcareous algae, they were responsible for the formation of massive reefs during the Cambrian.

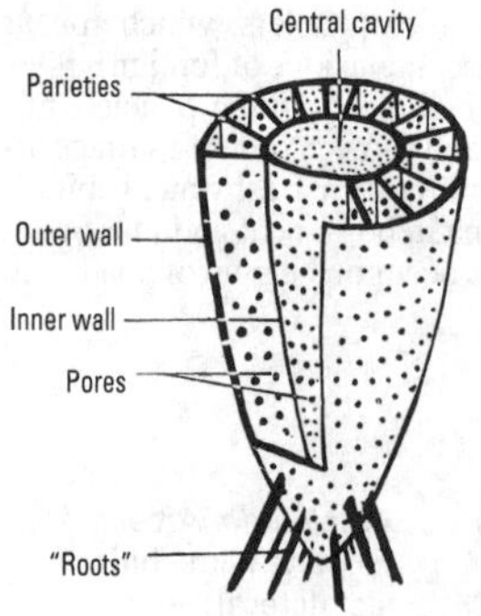

Calcareous skeleton of *Archaeocyathus* (natural size, schematic).

Archaeognatha, *Microcoryphia, **bristletails:*** an order of primarily wingless, ectognathous primitive insects (see *Apterygota),* previously placed with silver fish in the obsolete taxon, *Thysanura.* The more or less spindle-shaped *A.* are between 10 and 23 mm in length, and they are characterized by long, multisegmented, filiform antennae; large, dorsally adjacent or contiguous compound eyes; and three segmented caudal appendages (paired cerci and a terminal filament) (Fig.). The ectognathous mouthparts are designed for chewing; there are conspicuous maxillary palps; the mandibles are elongated and pointed, with a well-defined, projecting molar region, and they have a single articulation with the head. The coxae of the powerful middle and hindlegs often carry a movable unsegmented style. Of the 11 abdominal segments, the second to the ninth carry a ventral pair of styles, and some segments also carry one or two pairs of water-absorbing coxal vesicles. A tracheal system is served by 2 pairs of thoracic and 7 pairs of abdominal spiracles. Excretion occurs via labial nephridia and 12 to 20 Malpighian tubules.

Archaeognatha. Bristletail *(Machilis).*

About 220 recent species are complemented by fossil *A.*, e.g. *Triassomachilis* from the Triassic. The *Monura* (e.g. *Dasyleptus*) from the Permian and Upper Cretaceous are closely related to the *A.* The *A.* contains a single super family, *Machiloidea,* which is divided into 3 families: *Machilidae, Praemachilidae* and *Meinertellidae.* Only members of the *Machilidae* are commonly encountered; they occur widely, but are found predominantly in the Northern Hemisphere.

Life history. Fertilization occurs by the indirect transfer of spermatophores, a process involving elaborate courtship behavior. The total life cycle lasts between 6 and 27 months, with up to 11 months for preadult development. Lifespan is estimated at between 1.5 and 3 years.

A. are active in the evenings and at night, inhabiting primarily warm, variably moist sites, where they feed on the algae of bark and rocks. When disturbed they can jump suddenly and rapidly a distance of 10 to 20 cm.

Archaeolacerta: see Common lizard.

Archaeozoicum: an obsolete term for the geological era, in which the first life forms appeared.

Archanthropines, *Archanthropinae:* see Anthropogenesis, *Homo.*

Archboldia: see *Ptilinorhynchidae.*

Archegoniatae: see Bryophytes.

Archegonium: see Bryophytes, Reproduction.

Archeocyte: see *Porifera.*

Archeogastropoda: see *Gastropoda.*

Archeopteryx, *Archaeopteryx lithographica* von Meyer: an extinct species of primitive bird (order *Archaeopterygiformes),* fossils of which have been found in the lithographic limestone of the Upper Jurassic, Middle Kimmeridgian zone in Bavaria. It is known from the following 4 specimens of fossilized bones and feathers.

1) The London specimen, now in the British museum, was found in 1861 in the Ottmann quarry near Solnhofen in Bavaria.

2) The Berlin specimen, now in the Geological-Palaeontological museum of the Humboldt University in Berlin, was found in 1877 in the Dürr quarry near Eichstätt, 14 km from Sollnhofen.

3) The Erlangen specimen, now in the Geological Institute, University of Erlangen, was found in 1956 in the Opitsch quarry near Solnhofen, 250 m from the discovery site of the London specimen.

4) A specimen originally in the collection of an amateur German geologist, and recognized as that of *Archaeopteryx lithographica* in 1988. The site of its discovery is unknown. Similar to the London specimen (but about 10% larger), it is now in the Bavarian State Collection of Paleontology in Munich. This is the only known fossil of A. containing possible evidence of a furcula (wishbone).

A single fossil flight feather was also found in 1860 at Sollnhofen; the limestone slab carrying the mineralized feather is now in the Deutsches Museum, Munich, while the counter slab containing the impression is in Berlin. The fossil of a single downy contor feather found in 1954 in a limestone quarry (Upper Jurassic Kimmeridgian lithographic limestone) at Rubies near Lerida, Spain, is attributed only tentatively to *A. lithographica*.

A. is an example of mosaic evolution, i.e. intermediate forms of individual body parts are not present, but rather each body part is typical either of the phylogenetically more primitive (in this case reptilian) or of the phylogenetically more advanced (avian) class. During the progress of evolution, the individual parts then undergo complete transformation one after the other.

Reptilian (or dinosaur-like) characteristics of A.: long tail containing 20 free vertebrae; amphicoelous articulation of the vertebrae; short sacrum with only 6 vertebrae; free metacarpal bones in the hand; a claw on each finger of the hand; possession of teeth (thecodont and restricted to the anterior half of the jaw; 6 teeth in the premaxilla, 7 in the maxilla, 3 in the dentary); simple, unjointed ribs; presence of gastralia; simple brain with elongated, smooth cerebral hemispheres and small cerebellum.

Avian characteristics of A.: possession of feathers, which are identical with those of modern birds, those on the forearm being differentiated into primaries on the hand and secondaries on the ulna; presence of a furcula (?); backward direction of the pubes; fusion of the distal tarsals and metatarsals into a tarsometatarsus; opposable hallux; 3 fingers and 4 toes.

A. possessed flat claws, which suggests that it was a ground dweller. It also had solid rather than pneumatic bones, the latter being a feature associated with flight in modern birds. It is calculated that the center of gravity of A. lay above its feet, again suggesting a ground dwelling animal (for flight, the center of gravity ideally lies in the wings). Whether A. could fly is therefore a matter of debate. It lacked a sternum for the attachment of flight muscles, and it may have used its wings to glide rather than to fly. On the other hand, if it possessed the same type of muscles as a modern reptile, these would have been powerful enough to enable short flights, perhaps up to 2 km.

Archerfishes: see *Toxotidae*.

Archer's dart: see Cutworms.

Archesporium: see Bryophytes, Flower.

Archiannelida: a small order of polychaete annelids, containing both primarily primitive and highly degenerate forms. The prolegs are only weakly developed or absent, and only a few chetae are present. *A.* are marine arganisms, living in the sand or on algae, and feeding on detritus and unicellular algae.

Archicoelomata: a group of animal phyla of the *Bilateria,* whose members possess three body compartments (archimetameres), each with its own coelomic cavity, which arises as a fold of the archenteron. The constituent phyla are: *Ctenophora, Hemichordata, Pogonophora, Echinodermata* and *Chaetognatha.* Since, however, the *Bilateria* are subdivided into the *Protostomia* (the mouth develops from the blastopore) and the *Deuterstomia* (mouth develops from a different part of the archenteron), the *Ctenophora* must be assigned to the *Protostomia,* the other archicoelomate phyla to the *Deuterostomia.*

Archidiales: see *Musci.*

Archidium: see *Musci.*

Archilocus: see *Apodiformes.*

Archimetamere: see Mesoblast.

Archipallium: see Brain.

Archiperlaria: see *Plecoptera.*

Architeuthis: see *Cephalopoda.*

Archoophora: see *Turbellaria.*

Archostemata: see *Coleoptera.*

Arcifery: see *Anura.*

Arcoida: see *Bivalvia.*

Arctic, *Arctic subregion:* a zoogeographical subregion of the Holarctic (see). Other subregions of the Holarctic are the Palearctic (see) and Nearctic (see). The separate existence of the A. was recognized in the 19th century, and became important again to students of animal geography in the late 20th century. The A. is more or less delimited by the polar circle, but it also includes the whole of Greenland. The fauna of the A. displays a much greater zonal uniformity than that of more southerly areas, and the ecology of the A. differs from that of the areas surrounding it. However, only a few Arctic species are strictly confined to the A., and most species found in the A. also encroach on the Nearctic and/or Palearctic. In particular, many species, like the permafrost, are also found further south in eastern Siberia and Alaska (see Arctic-alpine distribution). Conversely, there is a slight tendency for some southerly animals to enter the A. by crossing the polar circle. Transition from the surrounding areas into the inhospitable A. is determined by gradual changes in ecological potential (see Ecological valency), rather than by any present or former physical barrier. Nevertheless, the Davis Strait at least acts as an effective barrier to the dispersal of many insects; the insects of Greenland are largely restricted to that land mass, and they are more closely related to those of the Palearctic than the Nearctic.

Typically, relatively few species are found in the A. Terrestrial poikilothermic vertebrates are almost nonexistent. The life cycle of invertebrates is greatly prolonged, due to the brief vegetation period, e.g. butterflies or moths may take 10 years or more from the egg to the imago. Although invertebrates are active for only a short period in the year, they are more numerous than vertebrates. Mass outbreaks of mosquitoes occur in the tundra. On Ellesmere island, below latitude 82 °N, there are 115 species of winged insects. One hundred and fifty eight insect species are known to be present on the colder Devon island at latitude 76 °N, despite its isolated position and bleak aspect (500 km of glaciers and mountains).

The polar bear, Arctic fox, blue hare, reindeer and lemming have a circumpolar distribution, while the musk ox was originally restricted to the American A. and Greenland. The dominant birds are mostly species that breed in colonies and feed from the sea. Most are migratory, and only a few overwinter in the A. (e.g. the willow grouse, *Lagopus lagopus),* and these, like lemmings, live part of the time beneath the snow.

Arctic-alpine distribution: the disjunct Distribution (see) of animals in the Alps and/or other Eurasian (especially Central European) mountains, and the Arctic and sub-Arctic. During the Ice Age, the animals of the A.d. inhabited a continuous, sef-contained area. As the ice receded, however, their environmental requirements were preserved only in the far north and in high mountains. At least, only in these areas could they compete successfully with other species. Examples are the willow grouse *(Lagopus lagopus),* ring ouzel *(Turdus torquatus),* blue hare and numerous insects (especially midges and bumble bees). Some animals of the A.d. are found as far south as the British Isles. A narrower definition admits only those animals living beyond the timber line in the mountains and in the tundra. See Boreo-alpine distribution.

Arctictis: see Binturong.

Arctiidae: a family of the *Lepidoptera* (see) containing the tiger moths, footmen and tussocks. The family, which is related to the *Noctuidae*, contains about 5 000 species, 50 of which occur in Central Europe. Members include *Arctia caja* L. ***(Garden tiger moth)*** and other typically orange or yellow tiger moths with black or brown markings and a wingspan up to 9 cm. Many caterpillars are long and covered with dense hair, and known as woolly bears or tussocks. For pupation, they spin a loose web of silk on the ground. The family also includes New World genera (e.g. *Bertholdia)* with reduced hindwings, often with camouflaged forewings, but with vividly colored hindwings and abdomen. Many of the caterpillars favor poisonous food plants, the toxins of which persist in the caterpillar and the adult, thereby protecting them against predators. Such plants include potato, foxglove and laburnum. The ***Fall webworm*** *(Hyphantria cunea* Drury = *H. textor)* is a native North American species, which was introduced into southern Europe in 1940, and also into Japan. It is a rather plump, white moth (usually with some pale brown spots), with a wingspan of 2–3 cm. The caterpillar has conspicuous, 12 mm-long, lateral white hairs, and it is highly destructive to many broadleaved trees, causing considerable damage in fruit orchards.

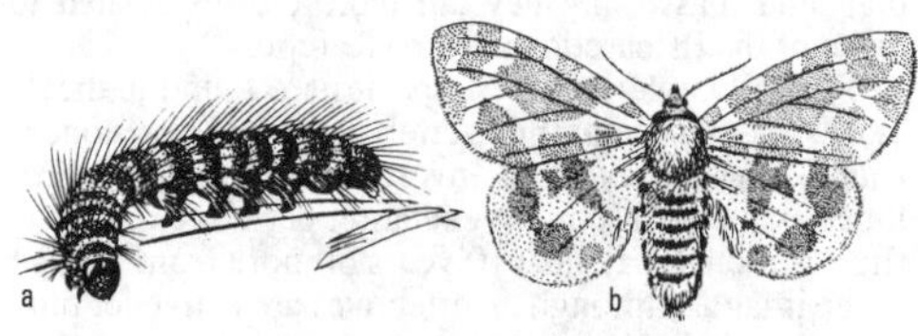

Arctiidae. Caterpillar *(a)* and imago *(b)* of the garden tiger moth *(Arctia caja L.).*

Arctium: see *Compositae.*

Arctocephalus: see *Otariidae.*

Arctogea, *Arctogaea, Megagaea:* a biogeographical area comprising the Palearctic, Nearctic, Ethiopian and Oriental regions. See Notogea.

Arctostaphylos: see *Ericaceae.*

Arctotertiary flora: a species-rich, frost-sensitive, warm climate-dependent flora, which was present in northern polar countries (Greenland, Spitzbergen, Grinnelland) during the early Tertiary, and appeared in Central Europe from the Upper Oligocene onward. The A.f. is known from its fossils, and it formed the basis of the modern Holoarctic flora.

Arctous: see *Ericaceae.*

Arcuate veins: see Kidney functions.

Ardeidae: see *Ciconiiformes.*

Area, ***area of distribution*, *areal:*** the area, space or range populated or colonized, at present or in the past, by a taxon, community or other grouping. The term is used especially with reference to the geographical distribution of individual species. When no physical barriers exist to the spread of a population, the borders of its A. usually fluctuate; the size of the A. is then determined partly by the demand for certain ecological conditions (i.e. environmental conditions suitable for existence), and partly by opportunities for dispersal (e.g. the spread of a plant population is possible only if seeds or propagules can reach a site outside the existing limits of the A.). Stability of the group (see Ecological valency) and an ability for efficient active or passive dispersal, together with a high reproductive rate, favor the formation of large As. Compactness and uniformity of terrestrial and aquatic habitats also favor large As. The As. of euryecious animals and plants are found throughout the world (see Cosmopolitan species). In contrast, the As. of certain stenoecious groups may be no more than a single site consisting of a few square kilometers, e.g. a single cave or a single freshwater source. The strictly defined A. of an animal group comprises only the region where reproduction occurs, any area of migration being excluded. For example, the African wintering range of European migratory birds is not part of their A. This definition is sometimes difficult to apply, especially with respect to certain migratory fish: the A. of the European river eel would thus be the Sargasso Sea, yet it does not become adapted for its breeding migration in the Atlantic until it has spent 9–12 years in freshwater rivers. By the same strict definition, the salmon would have a totally fragmented A.

As. may be continuous (inclusive or compact), i.e. they are coherent and nonfragmented, and they contain no unsuitable habitats; or they may be discontinuous (disjunct), i.e. they consist of several, often widely separated locations, whose populations have no contact with each other, e.g. the wood anemone (*Anemone nemorosa*) occurs in Central Europe, East Asia and North America (see Disjunct distribution).

The study of As. (see Chorology) serves as a basis for all further biogeographical research. Thus, a high frequency of intersection or overlapping of different As. or parts of As. in certain regions is an indication for the presence of Dispersal centers (see). Knowledge of the relationship between the As. of related forms, i.e. the existence of Sympatry (see) or Allopatry (see), is often of crucial importance for their taxonomic ranking as Species (see) or Subspecies (see).

The results of A. distribution studies are recorded as a Distribution map (see).

Area effect, *areal effect:* the apparent dependence

Arecaceae. Growth habits of some palms. *a* Date palm *(Phoenix dactylifera). b* Coconut palm *(Cocos nucifera). c* Washington palm *(Washingtonia). d* Palmetto *(Sabal).*

of the frequency of occurrence of different shell morphologies of the Grove snail *(Cepaea nemoralis)* and the Garden snail *(Ceapaea hortensis)* simply on the site of their occurrence, with no obvious relationship to ecological factors. Certain shell types of *C. nemoralis* and *C. hortensis occur* with similar frequency over large areas, although the character of the vegetation changes markedly within these areas. On the other hand, the frequency of these shell morphologies often alters abruptly between neighboring sites without any recognizable corresponding change in the environment.

Arecaceae, ***palm family,*** *Palmae:* a family of monocotyledonous plants containing about 3 500 species distributed in the tropics and subtropics. They are mainly aborescent plants with an unbranched trunk and a terminal rosette of very large pinnate or flabellate (fan-like) leaves *(feather palms* and *fan palms,* respectively). The stem or trunk usually attains its final diameter before the onset of extension growth, so that the diameter is practically uniform from base to apex. Flowers are generally unisexual (monoecious or dioecious), occasionally hermaphrodite, and contained in an inflorescence subtended by a large, rigid spathe. The 6-merous perianth is usually inconspicuous with similar segments. Pollination is by insects or wind. Fruits are drupes, nuts and berries.

Many palms are economically important. The ***Coconut palm*** *(Cocos nucifera)* has pinnate leaves, and a slender trunk which can attain a length of 30 m. It is planted on all tropical sea coasts for its valuable single seeded drupe, known as a *coconut.* The fibrous mesocarp of the fruit provides coconut fiber. Commercial *copra,* which yields a valuable oil, is the dried endosperm of the coconut. The ***Oil palm*** *(Elais guineensis)* from West Africa, which has pinnate leaves and attains a height of 10–15 m, is an important oil plant. Its seeds contain about 37% of an oil, which is used in the large-scale commercial manufacture of soap, as well as being used widely as an edible and cooking oil. The ***Date palm*** *(Phoenix dactylifera),* which is not known with certainty in the wild, produces berry-like fruits, has a thick, 22 m-high trunk, and is the most important food plant in the oases of the deserts of North Africa and Southwest Asia. The fruit is a unicarpellary berry. The seed, which is very hard due to the presence of reserve cellulose, is surrounded by a sugary, fleshy pericarp. Seeds of the ***Betel nut palm*** *(Areca catechu),* known commercially as *betel nuts,* also possess a hard, brown endosperm. *Areca* is the only palm genus known to contain alkaloids. When betel nuts are chewed with lime (usually wrapped in leaves of the betel pepper, *Piper betle),* the main alkaloid, arecoline, is converted into the stimulant compound, arecaidine. The practice of chewing betel, at least 2 000 years old, is common in East Africa, India and Oceania, and the number of users is estimated at about 200 million. ***Sago palms*** *(Metroxylon rumphii* and *M. sagu)* are grown from Indonesia to Fiji. Their stem pith, which is rich in starch, yields the food material, *sago.* European species of palms are the ***Dwarf palm*** *(Chamaeops humilis)* in the western Mediterranean region, and *Phoenix theophrasti,* which is native to Crete.

Arecidae: see *Monocotyledoneae.*

Arenaria: see *Scolopacidae.*

Arenariinae: see *Scolopacidae.*

Arenicola: see *Sedentaria.*

Areolar tissue: see Connective tissues.

Areole: see *Cactaceae.*

Argasidae: see *Ixodides.*

Arginase: a highly active, highly specific hydrolase, which catalyses one stage of the urea cycle, i.e. the hydrolysis of L-arginine into urea and L-ornithine. A. therefore plays an important part in the detoxification of ammonia in the organism. In terrestrial ureoteles such as mammals, frogs and swamp turtles, A. is found practically only in the liver, with traces in the pancreas, mammary glands, testes and kidneys. In the ureotelic elasmobranch fishes (sharks and rays) the enzyme is not confined to the liver; in these fishes it serves to generate a high urea concentration in the blood (2–2.5%), which is needed to maintain its osmotic pressure. A. is a tetrameric molecule (M_r 118 000) binding 4 Mn(II) ions. Removal of the metal ions causes the enzyme to dissociate into its 4 inactive subunits (M_r per subunit 30 000).

L-Arginine, ***Arg:*** 2-amino-5-guanidovaleric acid, the most strongly basic of the proteogenic amino acids. Arg is half essential for humans (i.e. it is not required by adults) and it is glucogenic. Particularly high concentrations are present in protamines (e.g. 87% in salmine) and in histones. Arg is an important component of the urea cycle, where it is synthesized from carbamoyl phosphate, L-ornithine and the α-amino group of L-aspartate. This pathway is generally responsible for the biosynthesis of the guanido group, all guanidine derivatives (e.g. creatine) being generated from Arg.

$$H_2N{-}\underset{\underset{NH}{\|}}{C}{-}NH{-}(CH_2)_3{-}CH(NH_2){-}COOH,$$

Arginine

Arginine-urea cycle: see Urea cycle.
Argonauta: see *Cephalopoda*.
Argulus: see *Branchiura*.
Argyroneta aquatica, ***water spider:*** the only spider that spends its entire life below water. Like all other arachnids, however, it relies on atmospheric air for respiratory gas exchange. Air is trapped in a dense layer of hairs that covers the body, and the upper layers exchange gases with the surrounding water. The spider also spins a domed underwater bell web, which it fills with air stroked from its body with its legs. Eggs are laid in underwater in air-filled egg bells, and are therefore surrounded by air during their development. The spider walks and swims underwater, and feeds mostly on water slaters *(Asellus aquaticus)*. It is usually placed in the family *Agelenidae* (sheet web spiders), and sometimes in a family of its own.
Arid: the term applied to the climate of steppes and deserts, in which the annual evaporation of water exceeds the water received by precipitation. Arid soils may be very rich in nutrients, due to the accumulation of material by weathering and disintegration, and the relatively slow removal of nutrients by the sparse vegetation. The limiting factor for fertility is water.
Arietites (Latin *aries* ram): an index ammonite genus of the Lias α_3. They are planispiral, with a wide umbilicus, and they attain diameters up to 1 m. The outer edge carries a keel bordered on each side by a groove. Simple ribs on the shell surface extend to the grooves of the outer edge.
Arisaema: see *Araceae*.
Arista: see Awn.
Ark shells: see *Bivalvia*.
Armadillidiidae: a family of the *Isopoda* (see).
Armadillidium: see *Isopoda*.
Armadillos: see *Dasypodidae*.
Armilla: a cuff-like ring of membranous tissue around the stipe of certain fungi of the order *Agaricales*, e.g. Fly agaric *(Amanita muscaria)*. The A. arises at the site of insertion of the stipe into the cap. It consists of the remainder of the inner membrane or partial veil (velum partiale), which in the young ("button") form of the fungus encloses the gills.
Armored scales: see *Homoptera*.
Armor-headed frogs: see *Hylidae*.
Armyworm: see Cutworms.
Arnaudovia: see Carnivorous plants.
Arni: see Barbary sheep.
Arnica: see *Compositae*.
Arnoglossus: see *Pleuronectiformes*.
Aroids: see *Araceae*.
Arolium: a median pad-like lobe of the insect pretarsus, similar in position and origin to an Empodium (see).
Aromorphosis: an expression coined by A.N. Sewertzoff (1931) for evolution characterized by an increase in the degree of integration and organization without marked specialization. See Higher Evolution.
Arousal: see Vigilance.
Arrau: see *Pelomedusidae*.
Arrhenatherum: see *Graminae*.
Arrhenotoky: production of only male progeny; see Parthenogenesis.
Arrow grass: see *Juncaginaceae*.
Arrow worms: see *Chaetognatha*.
Arsenic, ***As:*** a chemical element that occurs in trace amounts in plants. It is not an essential trace element, and it is toxic in relatively small amounts in both plants and animals.
Arses: see *Muscicapinae*.
Artemisia: see *Compositae*.
Artery: a blood vessel that transports blood away from the heart. See Circulation.
Arthrobotrys: see Carnivorous plants.
Arthrodia: see Joints.
Arthrodira, *Coccostei:* an extinct order of fishes containing most of the known placoderms. They were characterized by a heavily armored head and forebody, consisting of separate plates joined by sutures. Paired pectoral, ventral and anal fins were present, as well as dorsal and caudal fins. Free-standing, tooth-like bony plates were inserted at the front of the skull, and these functioned as biting surfaces. Gills opened between the head and body, and the gill-slit was usually covered by a bony plate. A typical genus was *Coccosteus* (Greek: *kokkos* nucleus; *osteon* bone), whose fossils are found from the Middle to the Upper Devonian of Europe, North America and New Zealand, being especially abundant in the Old Red strata of Scotland and Ireland.

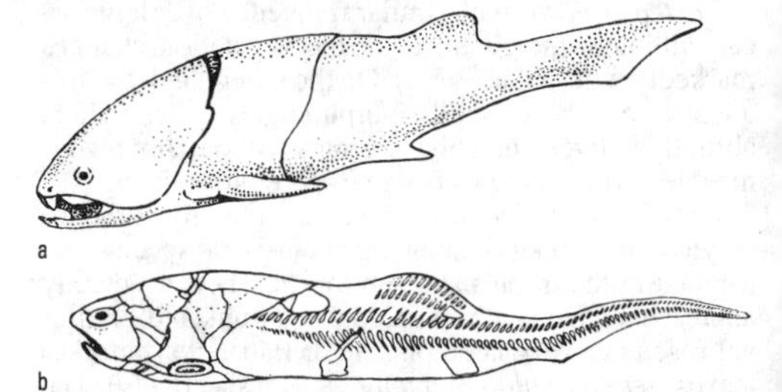

Arthrodira. a Reconstruction of *Coccosteus* (after Colbert). *b* Skeletal structure of *Coccosteus.*

Arthroleptinae, ***arthroleptins:*** a subfamily of the *Ranidae* (see), containing 47 species in 3 genera, all restricted to Africa: *Arthroleptis* (11 species; sub-Saharan Africa); *Cardioglossa* (16 species; West and Central Africa); *Schoutedenella* (20 species; sub-Saharan Africa).
Maxillary and premaxillary teeth are absent in some species. The pupils are horizontal. Some members have only procoelous vertebrae. Except for their small size, species of *Arthroleptis* are almost identical with *Rana*. The most recent addition to the family is *Cardioglossa pulchra* , discovered as recently as 1963 in thick riverside forest in Nigeria. Each genus contains species in which the third finger of the male is greatly elongated, giving rise to the name "Long-fingered frogs": this secondary sexual characteristic is not otherwise recorded in the *Amphibia*.
Arthropoda, ***arthropods:*** the largest phylum of the *Protostomia* (see), containing about 1 million known species. A. are metameric (i.e. they are segmented). In all *A.*, metamerism is evident during embryonal development, and it may still be evident in the adult, especially in primitive groups. An evolutionary reduction of metamerism has occurred in several groups of *A.*, e.g. by loss of segments, or by fusion of segments. In the embryo, each segment corresponds to a coelomic cavity, but during embryonal development these coelomic cavities disappear, their walls giving rise to musculature, connective tissue and other organs. Each body seg-

ment consists of a dorsal plate (tergite or tergum), ventral plate (sternite or sternum) and two flexible side walls (pleura, *sing.* pleuron). The individual segments are joined flexibly to one another by intersegmental membranes. The entire body is covered by a chitin cuticle, which serves simultaneously as a protective and supportive structure (exoskeleton). Usually the body is divided into 3 sections: *head* (prosoma or cephalothorax), *thorax,* and *abdomen* (pleon), which carry differently structured appendages. Head segments are always fused, and, depending on the species, varying degrees of fusion are displayed by the body segments. In the primitive condition, each segment possesses two lateral appendages, which are attached to the respective pleura. These appendages consist of flexibly jointed, tubular sections. In the primitive condition, the segmented appendages are similar, but these show a progressive evolutionary tendency to become structurally and functionally differentiated. Thus, the second antennae and the chelipeds (claws) of the crabs have evolved from similar segmented structures, and are therefore said to be serially homologous. The appendages of the head may be modified to antennae, as well as chewing, gripping or piercing organs. As a rule, the thorax carries locomotory appendages (e.g. walking, swimming, or jumping legs). Insects also possess wings, which are evaginations or folds of the integument. Abdominal appendages may be atrophied or totally absent.

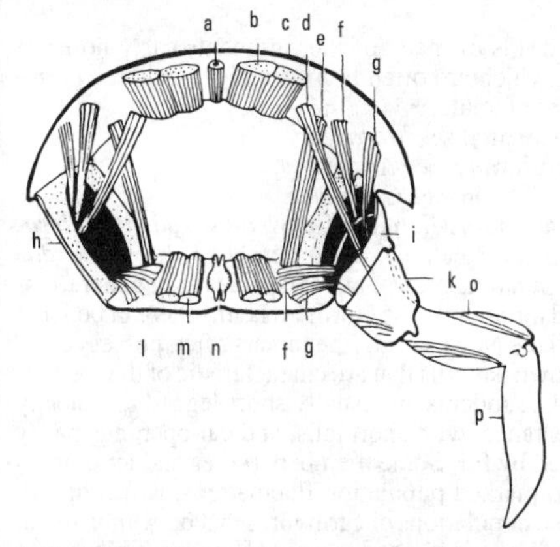

Arthropoda. Schematic transverse section of a segment of an arthropod. *a* Heart. *b* Dorsal longitudinal muscle. *c* Tergum. *d* Trochanter muscle. *e* Dorsoventral muscle. *f* Protractor. *g* Retractor. *h* Pleuron. *i* Subcoxa. *k* Coxa. *l* Sternum. *m* Ventral longitudinal muscle. *n* Ventral nerve cord with ganglion. *o* Extensor muscle. *p* Tibiotarsus.

The brain is dorsal and anterior, and is divided into 3 major regions: *anterior protocerebrum* (receives nerves from the eyes), *median deuterocerebrum* (receives the antennal nerves; this brain region is absent in the chelicerates, which lack antennae), and the *posterior tritocerebrum* (gives rise to nerves to the labium and digestive tract, the chelicerae of chelicerates, and the second antennae of crustaceans). In addition, a ventral nerve cord runs beneath the alimentary tract with a ganglionic swelling in each segment. Eyes vary greatly in complexity. Simple eyes possessing only a few photoreceptors are located centrally on the head, whereas the laterally situated, paired eyes are often large and compound, consisting of many ommatidia. Also present are sensory hairs (sensilla, *sing.* sensillum), which respond to mechanical and chemical stimuli. Both the foregut (stomodeal region) and the hindgut (proctodeal region), which are of ectodermal origin, are lined with a thin layer of chitin. The mid gut, however, is of endodermal origin, does not have a chitin lining, and usually has extensive pouches and/or large digestive glands. Excretion is performed by paired blind saccules, or by Malpighian tubules. The paired blind saccules open by ducts to the exterior of the body, adjacent to an appendage; they are named according to the associated appendage, e.g. coxal glands, maxillary glands, and they are never present in more than two segments. Embryonically, the saccule is derived from the coelom, and its tubule may be an evolutionary relic of a metanephridium. Malpighian tubules (found in centipedes, millipedes, insects, and arachnids) are blind tubules that lie in the hemocoel and open into the gut; soluble waste substances pass from the blood into the tubules and from there into the gut, where they are voided via the anus with fecal material. Aquatic *A.* respire via Gills (see), which are evaginations of the appendages, whereas terrestrial forms possess Tracheae (see), which are invaginations of the body wall; arachnids also possess Book lungs (see). With few exceptions, *A.* are monosexual. Egg cleavage is superficial. Usually, all the body segments are formed during blastogenesis; larvae with fewer than the normal number of segments do, however, hatch from the eggs of many crustaceans and myriapods. *A.* have invaded every type of habitat.

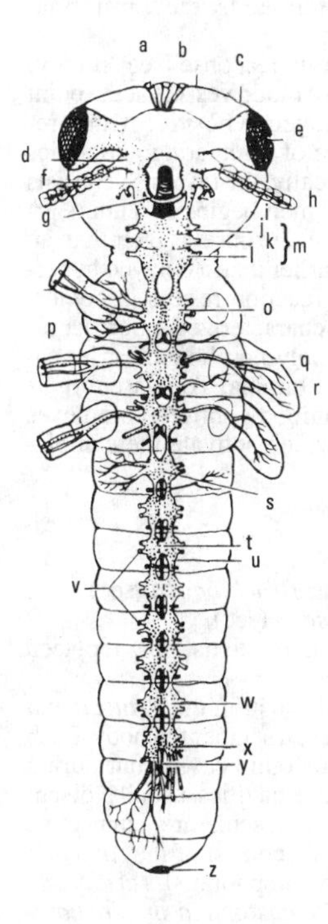

Arthropoda. Plan of the nervous system of an insect. *a* Ocelli. *b* Protocerebrum. *c* Optic lobe. *d* Deuterocerebrum. *e* Compound eye. *f* Tritocerebrum. *g* Commissure of the Tritocerebrum. *h* Antenna. *i* Labial nerve of the tritocerebrum. *j* Mandibular ganglion. *k* Maxillary ganglion. *l* Labial ganglion. *m* Subesophageal ganglion (formed by fusion of the neuromeres of the 3 mouthparts). *n* First ambulatory leg. *o* First thoracic ganglion. *p* Nerve of the second leg. *q* Wing nerve. *r* Forewing. *s* First abdominal ganglion. *t* Commissure. *u* Connective. *v* Abdominal ganglion chain. *w* 7th Abdominal ganglion. *x* 8th Abdominal ganglion. *y* Caudal sympathetic nerve strand. *z* Anus.

Classification. Various classifications are used. Thus, some authorities place the *Onychophora* with the *A*.

1. Group: *Amandibulata*
 1. Subphylum: *Trilobitomorpha*
 Only class: *Trilobita* (trilobites) (extinct)
 2. Subphylum: *Chelicerata*
 1. Class: *Merostomata* (horseshoe crabs with the extinct *Eurypterida)*
 2. Class: *Arachnida* (spiders, scorpions, mites, ticks, etc.)
 3. Class: *Pycnogonida* (sea spiders)
2. Group: *Mandibulata*
 3. Subphylum: *Branchiata*
 Only class: *Crustacea* (crabs, shrimps, lobsters, crayfish, woodlice, etc.)
 4. Subphylum: *Trachiata*
 1. Class: *Myriapoda* (centipedes, millipedes, pauropods, symphylans)
 2. Class: *Insecta* (insects)

Arthroses: see Joints.

Arthrospores: fungal spores formed by the septation and breaking up of simple or branched hyphae. They may remain atached to one another in chains, or become separate. In exogenous A. (e.g. in *Geotrichum candidum),* transverse septa are laid down in hyphae, and splitting occurs at the septa. In endogenous A. (e.g. in *Coremiella ulmariae),* a thinner, internal wall is developed (i.e. the spore wall is synthesized around a delimited volume of cytoplasm), and A. are released by circumscissile rupture of the parent wall. Conceptually, A. are the same as Oidia (see), and they occur, inter alia, in many powdery mildews and dark mildews *(Erysiphales).* Like chlamydospores, endospores and schizospores, A. are by definition formed by the transformation of pre-existing elements.

Arthus reaction: an allergic response (see Allergy) manifested as inflammation of blood vessels at the point of application of a soluble antigen. A positive Arthus reaction indicates the presence of circulating antibodies in the organism. Macroscopically, the response consists of reddening and swelling, which begins 3–4 hours after exposure to the antigen, i.e. the onset is later than an anaphylactic response, but earlier than a delayed hypersensitive reaction. A strong reaction results in hemorrhage and tissue necrosis, characterized microscopically by an accumulation of polymorphonuclear leukocytes in the vessel walls and lumina. The vasculitis is due to the accumulation of antibody-antigen complexes in the vessel endothelium, which activate the complement system.

Artichoke: see *Compositae.*

Articular: see Skull.

Articulata:

1) A class or subclass of the *Brachiopoda* (see).

2) A subclass of the *Crinoidea* (see).

3) An obsolete term used in old texts, now replaced by *Arthropoda.*

4) A group of animal phyla [within the *Protostomia* (see)], whose members possess a segmented body. Each segment corresponds to a coelomic cavity, but during embryonal development these cavities usually disappear, their walls giving rise to musculature, connective tissue and other organs. The A. consist of the *Annelida* (annelids), *Onychophora* (onychophorans), *Tardigrada* (tardigrades, water bears), *Pentastomida* or *Linguatulida* (pentastomids), *Brachiopoda* (lamp shells), and *Arthropoda* (arthropods). Fossil A. are found from the Cambrian to the present, but the greatest diversity and phylogenetic development of A. is displayed among modern living forms, in particular the insects. The *Trilobita* (an extinct subphylum of the *Arthropoda)* and the *Ostracoda* (mussel or seed shrimps; a class of the subphylum *Crustaceae*, phylum *Arthropoda)* provide many important index fossils. Other A. are paleontologically less important.

Articulatae: see *Equisetatae.*

Articulated antenna: see Jointed antenna.

Articulations: see Joints.

Artificial key: see Identification key.

Artiodactyla, ***even toed ungulates:*** a large mammalian order, the members of which possess an even number of toes. Most members are also relatively large in size. The third and fourth toes are strongly developed and carry hooves, the second and fifth toes are greatly reduced, and the first toe is absent. They are classified as follows.

1. Suborder: *Nonruminantia* (Nonruminants)
 Families: *Suidae* (Old World pigs),*Tayassuidae* (Peccaries), *Hippopotamidae* (Hippopotamuses), *Camelidae* (Camels).
2. Suborder: *Ruminantia* (Ruminants)
 Families: *Tragulidae* (Chevrotains), *Cervidae* (Deer), *Giraffidae* (Giraffes), *Antilocapridae* (Pronghorns), *Bovidae* (Antelopes, Cattle, Bison, Buffalo, Goats, Sheep)

The heads of many of the ruminants carry horns or antlers, which are often less strongly developed or absent in the female.

Artocarpus: see *Moraceae.*

Arum family: see *Araceae.*

Arundinaria: see *Graminae.*

Arvicolinae, *Microtinae,* ***arvicoline rodents, microtine rodents*** (plate 7): a subfamily of the *Cricetidae* (see), containing more than 200 species in 21 genera, distributed throughout the Northern hemisphere in both the Old and the New World. The molars often possess complex crown patterns that are characteristic of the species. Arvicoline rodents are usually short-legged and stocky in appearance, with short tails, and ear openings partly protected by fur. Some members (voles and lemmings) undergo marked population fluctuations, which in turn affect the populations of carnivores that prey upon them. The subfamily is divided into the following 3 tribes.

1. *Lemmini* ***(Lemmings,*** 4 genera). Lemmings are plump, especially short-tailed rodents, very common in the northern regions of America and Eurasia. They are proficient excavators, using the long claws on their front feet, and their fur is long, thick and waterproof. Lemmings do not hibernate, and during the winter they live in large groups under the snow cover. In some years, they undergo a population explosion, followed by mass migration. This migration is rarely coordinated and unidirectional; rather, individuals appear to migrate randomly in every direction. They form the main diet of Arctic and red foxes, wolverine, and many birds of prey (e.g. snowy owl). There are several genera of lemmings, all showing more or less marked migratory behavior, but the migrations of the ***Brown lemmings*** *(Lemmus)* are famous. Five species of *Lemmus* are known, distributed throughout northern Scandanavia, Asia and North America. The ***Norway lemming*** *(Lemmus lemmus)* lives in open, flat,

wet areas of the tundra, moving to mountain snowfields during the winter. In "lemming years", migrations occur in both the spring and late summer. Other genera include ***Collared lemmings*** *(Dicrostonyx,* body length 12–15 cm) found in northern Asia, North America and the Arctic as far as Greenland. Five species are recognized, including the ***Arctic lemming*** *(D. torquatus)* and the ***Greenland collared lemming*** *(D. groenlandicus).* ***Bog lemmings*** *(Synaptomys,* body length 10–13 cm) are found from eastern North America to the Pacific coast and north to Alaska. The 2 species, ***Southern bog lemming*** *(S. cooperi)* and ***Northern bog lemming*** *(S. borealis)* prefer bogs or moist grasslands. ***Wood lemmings*** are represented by a single species *(Myopus schisticolor,* body length 8–11 cm), found at high altitudes (up to 2 500 m) in Central Asia; it inhabits marshes during the summer and pine woods in the winter, where it excavates a tunnel system beneath the moss of the forest floor

2\. *Microtini* **(*Voles and Muskrats,*** 16 genera). The term, vole, is widely applied to many genera in this tribe. In the strict sense, voles belong to the genus *Microtus.* The *Microtini* are generally more heavily built than mice, which they otherwise tend to superficially resemble. The tail is always very short with sparse hair, and the head is broad with a blunt snout. Ears are generally short and eyes are small. The ***Common vole*** *(Microtus arvalis),* found throughout Europe and Central Asia, extending into China, is one of the most common European mammals. In causes considerable damage to crops and stored food products, and has a very high reproductive potential, producing 4–7 litters per year, each consisting of 4–8 young. Early springs, followed by long autumns and dry winters, especially when accompanied by dry, warm summers, result in massive population increases and veritable plagues of Common voles. They excavate a branched system of tunnels in the ground, up to 50 cm below the surface. In their search for food, they attack the cereal harvest and clover fields. For overwintering they also invade haystacks, cornstacks and barns. The ***Field vole*** *(Microtus agrestis)* is very similar to the Common vole, but with coarser and rather darker fur. Its range is partly shared with that of the Common vole, but extends further north. Frequently found in the weedy perimeters of forests, it gnaws the bark of young deciduous and pine trees, and bites off their buds and terminal shoots. Of the nearly 50 species of *Microtus,* others are the ***Tundra vole*** *(M. oeconomus;* similar to *M. agrestis* and *arvalis,* but with an especially long tail; found throughout North America and Eurasia), ***Eastern meadow mouse*** *(M. pennsylvanicus;* restricted to, and widely distributed in, North America, being absent from only the most southerly states), ***Brandt's vole*** *(M. brandti;* high steppes of Mongolia), and the Snow vole *(M. nivalis;* considered to be the most characteristic mammal of the European Alps; it is also found in other southern and Central European mountains above 1000 m).

The ***Bank vole*** *(Clethrionomys glareolus)* is a reddish brown European species, body length 8–12 cm. It occurs in deciduous and evergreen forests of northern and Central Europe, where it feeds on roots and shoots. It is therefore regarded as a forestry pest and is controlled by laying poisoned grain and by smoking out its burrows.

The ***European water vole*** *(Arvicola terrestris;* body length 12–19 cm) may indeed be found near to water, but it is just as likely to occur far from water, in meadows, agricultural land and gardens. Numerous subspecies are recognized, but the ***Iberian water vole*** *(A. sapidus)* is now described as a distinct species.

The ***Muskrat*** *(Ondatra zibethica,* body length 30–36 cm, weight 600–1 500 g) is the largest arvicoline rodent; it displays various adaptations to an aquatic existence, such as a laterally compressed tail, fringes of bristles (swim bristles) on the edges of the toes of the hindfeet, and a skin fold that closes the inner ear. It excavates residential and nesting burrows, complete with ventilation shafts, in steep banks, and causes damage by undermining dykes. If the banks shelve gradually, muskrats do not burrow, but build a cone-shaped mound of plant material on the water with the nest in the center. The muskrat does not hibernate and is active throughout the winter, keeping holes open in the ice and feeding (mainly on plant material) on the bottom. It is exploited for its fur, and for its scent glands, which are used in perfumery. In 1906, muskrats were introduced into Europe from North America for fur farming, and later into Asia. They have now become established in many areas of the Old World. One other species of *Ondatra* is known *(O. obscura),* which is restricted to Newfoundland. A close relative is the more terrestrial ***Round-tailed muskrat*** *(Neofiber alleni),* found in the coastal swamps of the Florida peninsula.

3\. *Ellobiini* **(*Mole lemmings,*** 1 genus). This tribe is represented by only 2 species: *Ellobius talpinus* **(*Mole lemming;*** Transurals to Mongolia) and *Ellobius fuscocapillus* **(*Afghan mole lemming;*** Afghanistan to Iran, slightly overlapping on the northern edge of its range with the southern limit of *E. talpinus).* These subterranean rodents use their large, protruding incisors for digging. Body length is 8–15 cm, with a blunt head, yellowish brown velvety fur, small eyes (very poor vision) and no external ears. They are found mainly in dry steppe and semi-desert areas, where they excavate interbranching tunnel systems 20–30 cm below the surface.

Arytenoid cartilages: see Larynx.

Asaphus (Greek *asaphes* indistinct): a large trilobite genus, up to 40 cm in length, with a semicircular cephalon (with no pronounced genal spines) and a semicircular pygidium, which are about the same size. The glabella swells outward in front of the eyes, and the facial suture runs from the posterior border over the semicircular compound eyes to the anterior border of the cephalon. There are no lateral borders on the free cheeks, and the pygidium is also without a border. Clear segmentation is seen on the long axis. The thorax consists of 8 segments with broad articulating facets. The pygidium

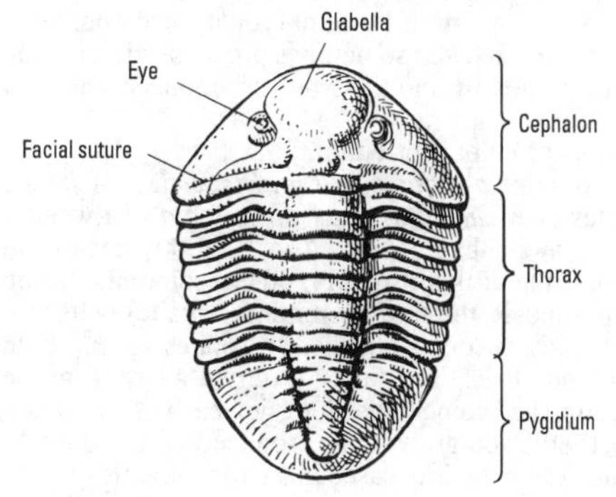

Asaphus expansus Wahlenb. from the Ordovician (x 0.6)

is also clearly segmented, but this is weakly reflected in the pleural folds.

Fossils are found in the Ordovician, and are common in geological debris. The genus provides important index fossils of the Lower and Middle Ordovician.

Ascaphidae, ***ascaphids***, ***tailed frogs***: a primitive family of the *Anura* (see), represented by a single aquatic species, *Ascaphus truei*, 3–5 cm in length, grayish in color, found in the cool mountain streams of Northwest USA and adjacent Canada. Skeletal characteristics are similar to those described for the *Leiopelmatidae* (see), to which it is closely related (in *Ascaphus*, but not in *Leiopelma*, the anterior end of the clavicle is overlain by the scapula). Males possess a cylindrical, 3–10 mm long cloacal appendage (intromittent organ) which is used in copulation, fertilization being internal. The name "tailed frog" is derived from this cloacal appendage. A true tail is absent, but like the leiopelmatids, *Ascaphus* also possesses the vestiges of tail-wagging muscles, the pyriformis and the caudalipuboischiotibialis. Eggs are laid in strings in water, and the tadpole stage lasts 2–3 years.

In some classification systems, *Ascaphus* is placed with the closely allied *Leiopelma* in the family *Leiopelmatidae* (see).

Ascarida: see *Nematoda.*

Ascaris: see *Nematoda.*

Aschelminthes, *Nemathelminthes:* a phylum of about 12 500 species, some of which are free living, aquatic or terrestrial, while others are parasites of animals or plants. The natural relationship of the constituent classes *(Gastrotricha, Rotifera, Nematoda, Nematomorpha, Kinorhynca* and *Acanthocephala)* is questionable, in view of wide differences in their internal and external structures, and in their developmental biology. However, they do share otherwise uncommon characteristics, such as constant cell numbers (eutely) and a tendency to form syncytia. They are worm-shaped and unsegmented Protostomia, all possessing a usually extensive, fluid-filled primary body cavity, which lies between the alimentary canal and the outer muscle layers. This primary body cavity (derived from the blastocoel) has no epithelial lining, and is therefore called the pseudocoel. All A. lack a vascular system. Their outer body layer is a muscular tube, consisting of an outer epidermis (completely or partly covered with cuticle) and an inner layer of longitudinal muscle fibers. A. are usually bisexual, but some species are known only as parthenogenetic females. Egg cleavage is total, and bilateral or spiral. Postembryonal development is sometimes direct, and sometimes proceeds via separate juvenile stages by molting, but pronounced cases of metamorphosis are rare.

Aschoff's rule: Biorhythm.

Ascidiacea, ***ascidians, sea squirts:*** a class of sessile tunicates (see *Tunicata),* varying in length between 1 mm and more than 30 cm. About 2 000 species are known, some of them solitary, others colonial. During metamorphosis, the free-swimming larva loses its tail. As the body becomes attached, it rotates so that both mouth and cloacal opening are turned away from the substratum and come to lie alongside each other. Meanwhile, rootlike outgrowths are produced against the substratum. The body is encased in a strong mantle of tunicin (containing immigrant mesenchymal cells and blood vessels) and by a muscular epidermal layer beneath the mantle. On one side, the pharynx (known as the pharyngeal basket or branchial chamber) opens via numerous clefts (stigmata) into the cloacal space, which also receives the hindgut and gonadal ducts, as well as serving as a brood chamber. Otherwise the stigmata communicate with the atrial or peribranchial space which surrounds the pharyngeal basket. In colony-forming *A.* the individuals have a common cloaca. The heart continually alternates the direction of its beat, driving the red or blue blood alternately to the stigmata of the branchial chamber and to the other organs. All species reproduce sexually, and some also display asexual budding from a stolon (see Stolo prolifer). In Mediterranean countries, the viscera of large species are considered a delicacy.

Three orders are recognized: *Aplousiobranchiata, Phlebobranchiata* and *Stolidobranchiata.*

A. have colonized almost every marine habitat from rocky shores to ocean depths at all latitudes, but the majority of species is found in shallow coastal waters. For collection and preservation, see *Tunicata.*

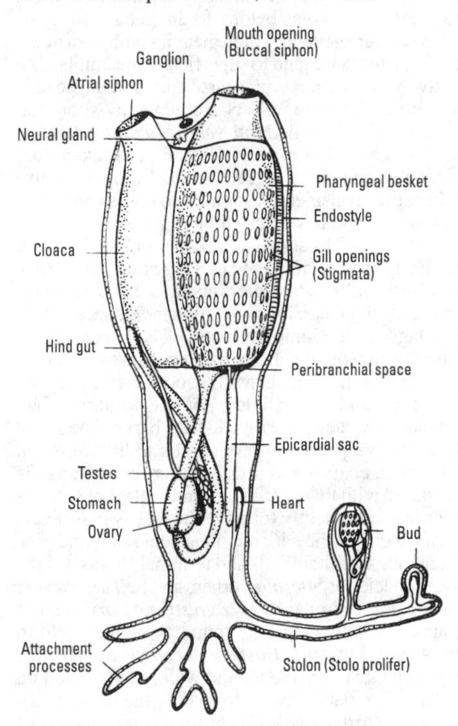

Ascidiacea. Structure of the sea squirt, *Clavelina.*

Ascidiae compositae: see *Synascidiae.*

Ascidiae simplices: see *Monascidiae.*

Ascidiform leaf: see Carnivorous plants.

Asclepiadaceae, ***milkweed family:*** a family of dicotyledonous plants containing about 2 000 species in 130 genera, found principally in the tropics. They are erect or twining shrubs or perennial herbs, with some fleshy succulent forms. A milky sap is usually present. Leaves are simple and generally entire, opposite or whorled, rarely alternate; in some succulent taxa the leaves are caducous or vestigial. Most species possess minute stipules. The inflorescence is usually a cyme, but racemes or umbels occur in some species. Flowers are 5-merous, actinomorphic and hermaphrodite, often with a corolla appendage or paracorolla (an appendage on the outside of the stamens, which forms a nectar-producing corona). At maturity, the pollen grains are united into pollinia. As a rule, two pollinia from adjacent anthers are connected by a bow-shaped translator and a corpusculum formed by the capitate stigma. A groove in the corpusculum is capable of trapping the probiscis or leg of an insect, which then carries the two pollinia to the next flower. [Two subfamilies are recognized: *Periplocoideae* (pollen grains free and united in tetrads) and *Cyanchoideae* (pollen in pollinia)]. The superior (semi-inferior) ovary consists of 2 free carpels, whose style apices are united to form a single, large, 5-lobed stigma. The ovary develops into 2 (often only one develops fully) pod-like, multiseeded fruits, whose seeds carry a terminal plume of long silky hairs (coma).

Most plants of the family cannot withstand a temperate climate, but several are cultivated in glasshouses, e.g. *Hoya* (China, Southeast Asia, Australia, Pacific islands), *Ceropegia* (Canary Islands, tropical and southern Africa, tropical and subtropical Asia, northern Australia, Madagascar), *Caralluma* (Africa, Mediterranean basin and eastward through arid regions of India to Burma), *Huernia* (southern and tropical Africa, extending into southern Arabia), *Stapelia* (tropical and southern Africa), and *Oxypetalum* (Mexico, West Indies, Brazil). The family includes several poisonous and medicinal plants, containing cardiac glycosides and bitter principles, e.g. *Cynanchum vincetoxicum (= Vincetoxicum officinale,* ***White swallow-wort****)*, the roots of which were formerly used as an antidote to poisoning. ***Milkweed*** and ***Butterfly flowers*** *(Asclepius)* are planted as ornamentals. *Asclepius curassavica (****Blood flower****,* native in tropical America and naturalized in southern USA) is often grown in glasshouses in temperate regions for its attractive inflorescences. Some species of *Asclepias* and *Calotropis* are a source of fiber. *Hoya* is a popular genus of tropical climbing plants for cultivation under glass in temperate climates. The ***Wax plant*** *(Hoya carnosa)* is an indoor ornamental, native to China and Australia, whose flesh-colored umbels appear to be molded from wax. The inflorescences of *Hoya sussuela*

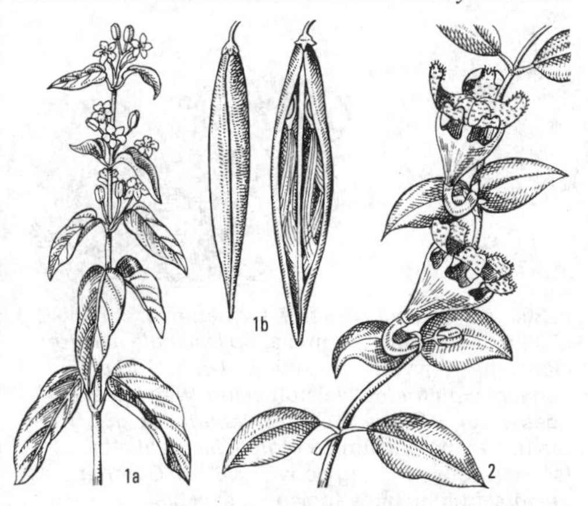

Asclepiadaceae. 1 Cyanchum (swallow wort): *a* flowering branch, *b* open (right) and closed (left) fruits. *2* Flowers of the genus *Ceropegia.*

(native to Borneo and the Moluccas) contain as many as 14 individual fragrant and long-lived flowers, each up to 7.5 cm in diameter. In 1848, *H. sussuela* won the prize for the best new plant at the exhibition in Regent Park Gardens, London. *Cryptostegia grandiflora* has been cultivated commercially as a source of rubber. *Marsdenia tinctoris* yields an indigo-like dye.

In arid areas, various genera of the *Asclepiadaceae* have evolved vegetative organs closely resembling those of cacti or members of the *Crassulaceae* (parallel evolution). Genera displaying these adaptations are placed in a division of the family known as ***Stapeliads,*** after the best known genus, *Stapelia*. The 70 species of *Stapelia* have fleshy, cactus-like stems, sometimes with spines. Flowers of *Stapelia gigantea,* which are clothed in a layer of purple, velvety hairs, are the largest in the family, with a diameters up to 46 cm. The decorative flowers of various *Stapelia* species ***(Carrion flowers)*** emit a carrion-like odor, which attracts pollinating flies; old flowers are often crawling with maggots. The 30 species of *Huernia* also have succulent stems, without hard spines. Flowers of *Huernia hystrix* ***(Porcupine huernia)*** have fleshy, spiny petals, dull yellow in color and banded with red. Other stapeliad genera are *Caralluma* and *Hoodia* (Southwest Africa). *Brachystelma barberiae* is neither spiny nor succulent, but its short stalk and hairy leaves arise from a massive tuber. Discovered in the Transkei, it is found from the eastern cape to Zimbabwe, and it is especially common on the high veld of the Transvaal. When the flowers open, the tips of the petals remain attached to one another, forming a "bird cage" structure.

Ascogonium: see *Ascomycetes*.

Ascolichenes: see Lichens.

Ascomycetes, *Ascomycotina,* ***ascus fungi, cup fungi and allies:*** a class of fungi, containing 20 000 species, characterized by the possession of a tubular sporangium ***(ascus),*** and septate hyphae with mononucleate or multinucleate cells. The transverse walls are perforated by a simple pore, and the cell walls are chitinous. Asexual reproduction is by conidia, which are formed in great abundance, especially in the less advanced forms. Sexual reproduction occurs by the fusion of gametangia. The rounded, multinucleate, female gametangium is called an ***ascogonium,*** and the club-shaped multinucleate male gametangium is known as an ***antheridium.*** The ascogonium has an organ, the ***trichogyne,*** which withdraws the cytoplasm and nuclei from the antheridium and transfers them to the ascogonium (plasmogamy). In a typical ascomycete, the ascogonium then gives rise to ascogenous hyphae (equivalent to the sporophyte), which form asci at their tips. Nuclear fusion (karyogamy) occurs only in the ascus. Thus, plasmogamy and karyogamy are spatially and temporally separated by an intermediate, binucleate stage (dikaryophase). The dikaryophase can be looked upon as a diploid stage, in which the nuclei remain paired and unfused. Haploid ascospores (usually 8, sometimes 4) are produced in the ascus by meiosis, which occurs immediately after the fusion of the two nuclei. In the more advanced forms, the ascospores are ejected from the ascus by osmotic pressure. Asci are usually found side by side and interspersed with sterile hyphae ***(paraphyses),*** forming a palisade-like surface or ***hymenium*** which is present within, or sometimes on, a highly differentiated fructification. The plectenchymatous fructification consists of haploid gametophyte

Ascomycetes

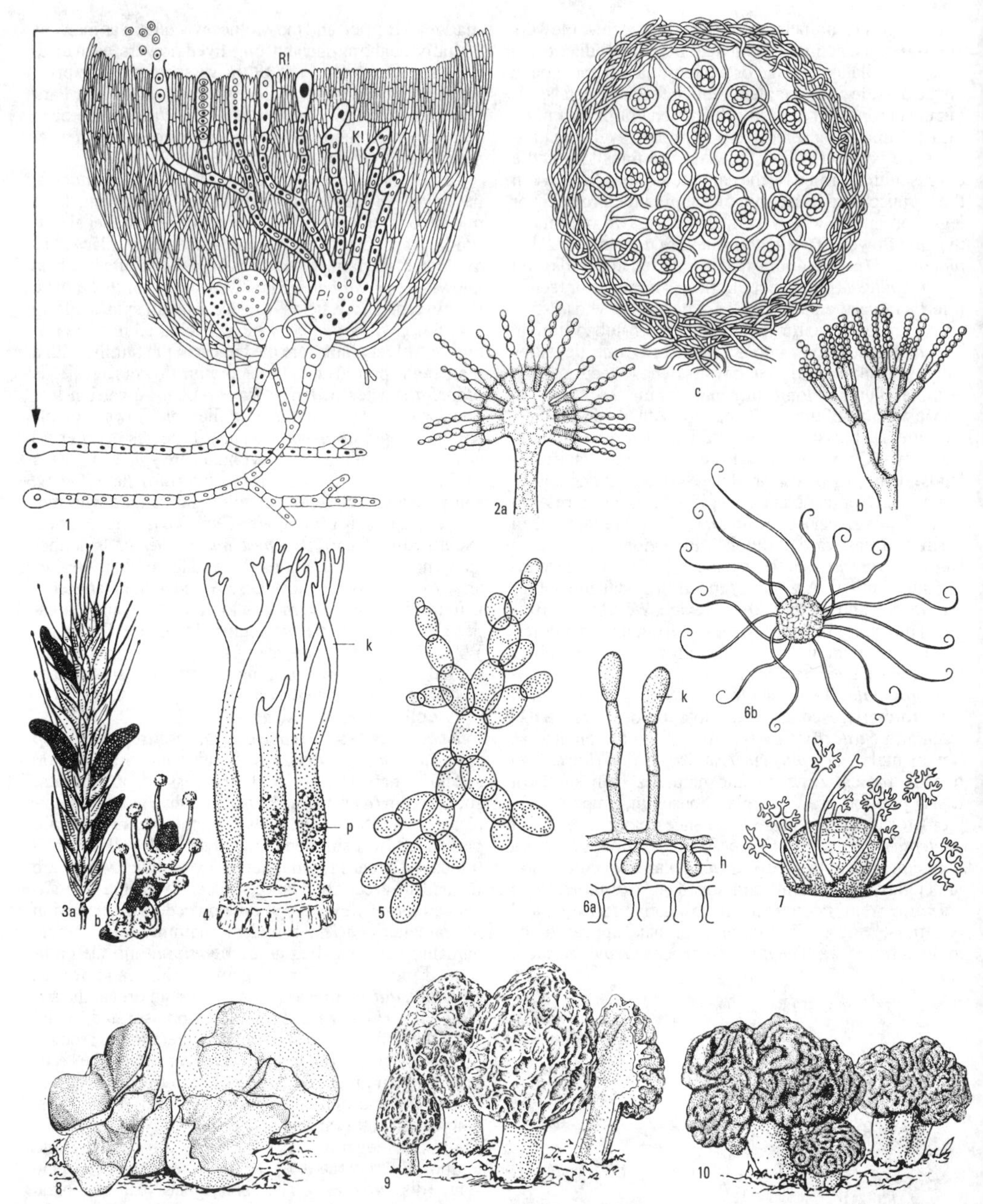

Ascomycetes. 1 Developmental cycle of a dioecious discomycete: *P!* plasmogamy; *K!* karyogamy; *R!* meiosis. *2a* Conidiophore of *Aspergillus. 2b* Conidiophore of *Penicillium. 2c* Cleistothecium of a member of the *Plectascales. 3* Ergot: *a* ear of rye with mature sclerotia (x 2/3), *b* germinated sclerotium with stalked fruiting bodies (enlarged). *4 Xylaria hypoxylon* (natural size): *k* conidial region; *p* perithecial region. *5* Chains of buds of *Saccharomyces cerevisiae* (x 300). *6a Uncinula necator:* formation of conidia (x 100); *h* haustorium; *k* conidia. *6b Uncinula necator:* cleistothecium with appendages (x 30). *7 Microsphaera alphitoides:* cleistothecium with appendages (x 30). *8 Peziza (= Otidea) leporina* (x 2/3): fruiting body (apothecium). *9 Morchella esculenta* (Morel): edible fruiting body (x 3/8). *10 Gyromitra esculenta:* fruiting body (poisonous) (x 3/8).

mycelium. The sporophyte stage is therefore restricted to the fertilized ascogonium, the ascogenous hyphae and the stages immediately before ascus formation.

Five subclasses are recognized, based mainly on the form and mode of development of the fructification, and on the structure of the ascus and its mechanism of dehiscence.

1. Subclass: *Protoascomycetidae*. Members of this subclass do not form fructifications. The asci do not arise from ascogonial hyphae, but directly from the zygote.

1. Order: *Saccharomycetales*. These are the yeasts, and their vegetative cells rarely form hyphae. They usually grow in sugary fluids, such as fruit juices and plant exudates. Most species are unicellular, and the spherical or ovate cells usually multiply by budding, often forming more or less branched chains. The cell wall consists largely of carbohydrate, and the storage material is mainly glycogen.

Sexual reproduction occurs by the conjugation of two cells, and in some species a short conjugation bridge is formed. The resulting zygote develops into an ascus, sometimes directly and sometimes with an intermediate budding phase. After meiosis, the ascus contains 4 or 8 ascospores. Three types of life cycle can be recognized. In the *haplontic* type (e.g. *Schizosaccharomyces)* the haploid phase displays vegetative reproduction; when a zygote is formed by cell conjugation, it immediately undergoes meiosis to form an ascus. In the *haplo-diplontic* type (e.g. *Saccharomyces cerevisiae)* budding occurs in both the haploid cells and the zygote; asci can therefore be formed from budded diploid cells. In the *diploid* type (e.g. *Saccharomyces ludwigii)* ascospores conjugate within the ascus, and only the diploid phase displays vegetative budding. Some more primitive yeasts possess multinucleate hyphae. For example, in *Dipodascus albidus,* adjacent hyphal cells form conjugation tubes while retaining their common cross wall. After the fusion of one "male" with one "female" nucleus, the remaining nuclei disintegrate. By meiosis, the resulting fusion nucleus produces numerous haploid nuclei, each forming an ascospore.

The most familiar representatives of the yeasts are Brewer's yeast (see) (Fig.5) and Baker's yeast (see). These two species of *Saccharomyces,* together with their numerous strains and races, are known only in artificial culture. In contrast, the Wine yeast (see), *Saccharomyces ellipsoideus,* overwinters in the soil and occurs naturally on fruit. Some yeasts cause human skin diseases, especially *Candida albicans*. Dried cell material from the growth of *Candida utilis* on sulfite waste liquor (from the paper and pulp industry) contains 47% protein, 5% fat and 27% carbohydrate, together with vitamins of the B complex. Such material (known as "single cell protein" or SCP) is used for the supplementation of animal feedstuffs. When grown on ethanol as a carbon source, *Candida utilis* yields a SCP known as Torutein, which is reported to improve the texture, flavor and nutritional value of human foods, such as bread and ground meats, etc.

2. Order: *Taphrinales*. These possess asci, which form a hymenium without paraphyses. Some members of the genus, *Taphrina,* are plant parasites, e.g. *Taphrina pruni,* which transforms the ovaries of plums into hollow galls without a stone (hence the name "pocket plums"). The "witches brooms" of many trees are also caused by *Taphrina* species.

2. Subclass: *Plectomycetidae*. In this subclass, the asci usually develop in a closed fructification, known as a cleistothecium (Fig.2c). At first, however, the sex organs are naked, the ascogenous hyphae becoming enclosed later by the developing fructification. The ascospores are not released until the fructification disintegrates or decomposes. Multiplication is predominantly asexual, by the formation of conidia. Large numbers of conidia are abstricted in coherent chains from specialized structures called conidiophores (Figs 2a & 2b).

1. Order: *Plectascales*. Members of this order include the widely occurring and familiar moulds, such as *Aspergillus* and *Penicillium*.

Species of *Aspergillus* occur as saprophytes on foods, fruit, and dead organic materials. *Aspergillus oryzae* has several industrial and domestic applications; it is used in the preparation of saki (saké), and various pastes produced by the growth of *A. oryzae* on soybeans and/or rice are used in Japan and China for seasoning and flavoring, and as whole foods. *A. niger* is used to synthesize citric acid from molasses. *A. fumigatus* can behave as a human pathogen, attacking the ears or the lungs, in the latter case causing a disease resembling tuberculosis. *A. flavus* and other species produce Aflatoxins (see).

Species of *Penicillium* occur as saprophytes on practically all forms of dead organic material. The antibiotic, penicillin, is obtained from special strains of *Penicillium notatum, P. chrysogenum* and others. *Penicillium camemberti* and *P. roqueforti* are used in cheese manufacture.

2. Order: *Erysiphales*. These are the "true" or powdery mildews, which occur as ectoparasites on various agricultural plants, sometimes causing extensive damage. *Erysiphe graminis* attacks grasses and cereals. *Sphaerotheca pannosa* is found on roses and peaches. *Uncinula necator* (Fig.6) attacks the leaves and fruits of the vine, while *Microsphaera quercina* is found on oak, beech and chestnut. The web-like aerial mycelium, which can be wiped from the plant surface, sends haustoria into the epidermal cells of the host. In summer, reproduction occurs by conidia. The asci are found in cleistothecia (Fig.7), which also serve as overwintering organs.

3. Subclass: *Loculomycetidae*. In this subclass, the fructification (pseudothecium) develops early and the gametangia subsequently develop within its plectenchyma. The ascus wall is bitunicate, i.e. it consists of 2 layers. The outer wall is thin but rigid, while the outer wall is thicker but more elastic. At maturity, the inner wall protrudes through the fractured outer wall at the tip of the ascus, and spores are ejected through a pore in the inner wall.

Order 1: *Myriangiales*. These are mostly tropical species, parasitic on plants and insects. The asci are distributed randomly in the plectenchyma.

Order 2: *Dothiorales*. These are plant saprophytes or parasites found on stems, leaves and fruits, or on nonliving materials such as bark. The asci are grouped, and the fructification usually has a large opening, which is formed by breakdown of the surface layer. Often several fructifications are found embedded in the stroma.

Order 3: *Pseudosphaeriales*. In contrast to the *Dothiorales,* the opening of the fructification is a pore or canal (sometimes surrounded by bristles) through the plectenchyma. The *Pseudosphaeriales* include serious plant disease organisms. Thus, *Venturia* (the conidial stage is known as *Fusicladium)* causes the scab disease of apples and pears. The scab is a dark, corky patch, which causes the growing fruit to split. "Sooty moulds" on leaves are caused by *Capnodium*. Conifers are attacked by *Herpotrichia,* which clothes the twigs with a brown hyphal mat and kills the needles.

4. Subclass: *Pyrenomycetidae.* The typical fructification, the perithecium, is more or less flask-shaped, with a preformed aperture, and it develops only after formation of the sex organs. The base of the preformed cavity of the perithecium is covered by a hymenium, containing asci and paraphyses.

1. Order: *Sphaeriales.* This order contains *Neurospora,* which has served several generations of scientists as an organism for genetic and biochemical research. Also of scientific importance is *Gibberella fujikuroi,* the causative agent of a Japanese rice disease, and a source of the growth factors, Gibberellins (see), which are used in research in plant physiology. The red perithecia of various species of *Nectria* are usually found on the dead branches and twigs of various trees and bushes. *Nectria galligena* causes the canker of fruit trees. The causative agent of Dutch elm disease, *Ceratocystis ulmi,* also belongs to this order. In *Xylaria* species the perithecia are embedded in a club-shaped stroma. Thus, in *Xylaria hypoxylon,* found on the stumps of deciduous trees, the black, lower part of the antler-like stroma carries the perithecia, whereas the upper, whitish part carries conidia (Fig.4).

2. Order: *Clavicipitales.* In this group the perithecia are also embedded in a well developed plectenchymatous structure, the stroma. The asci are elongated and contain filamentous ascospores, which are initially unicellular, later becoming septate. This order contains *Claviceps purpurea,* which parasitizes the young ovaries of grasses and cereals, especially rye. When the ovary has been consumed, the mycelium develops into a sclerotium (closely packed hyphae, some with transverse septa, forming a pseudoparenchyma). In the *Claviceps* infestation of rye, the sclerotia are dark violet bodies, called ergot (Fig.3), which project from the ear. These sclerotia contain very toxic alkaloids, especially ergotamine and ergotoxin (see Ergot alkaloids), which are used medicinally to prevent post partum hemorrhage in childbirth, and in the treatment of migraine. The sclerotia are known pharmaceutically as *Secale cornutum,* but in modern practice only the purified alkaloids or their laboratory derivatives are used medicinally. The sclerotia overwinter in the soil, developing subsequently into a stroma of pink, stalked heads with numerous embedded perithecia, carrying tubular asci with filamentous ascospores. Infection of the graminaceous host occurs by wind-borne ascospores, which alight on the stigma of the plant. People eating ergot-infected rye bread suffer from St. Anthony's fire, an affliction well known in medieval Europe: constriction of arteries inhibits the circulation, initially causing a burning sensation, but later leading to gangrenous loss of fingers and other extremities. Some species of *Cordyceps* parasitize insects, while others attack hypogeous fungi (e.g. *Elaphomyces).*

Order 3: *Laboulbeniales.* Represented by about 1 500 species, this group is exceptional in that its members possess virtually no hyphae, consisting mainly of sex organs and perithecia. Free male cells (spermatia) are formed, rather than male gametangia. They are minute, bristle-like insect parasites, and penetration of the chitin of the host is achieved by a short, dark-colored "foot".

5. Subclass: *Discomycetidae.* This subclass is characterized by its fructification (see Apothecium), which carries the hymenium on its upper surface.

1. Order: *Pezizales.* These are saprophytic fungi, whose fructifications may be shaped like a disk, cup, dish, hat or club, and whose asci open by a preformed lid or operculum. Species of *Peziza,* with their often brightly colored, and usually dish-shaped fructifications (Fig.8), are common on woodland paths and the sites of old bonfires. Morel *(Morchella esculenta)* is a delicious, edible fungus, whose fructification is differentiated into a stalk and a honeycombed fertile cap (Fig.9). It is found in the spring in deciduous woods and hedge bottoms. The fructification of the poisonous *Gyromitra esculenta* is also differentiated into a stalk and a cap (Fig.10); the deeply convoluted surface of the cap is colored dark date-brown to sooty brown. It also occurs in the spring, especially on the sandy floors of conifer forests.

2. Order: *Helotiales:* The asci do not possess an operculum, and they open via an apical pore. The order contains both saprophytic and parasitic species. *Sclerotinia fructogena* grows on apples and pears, forming at first concentric circles of conidial pustules (known as *Monilia);* stalked apothecia are formed on the mummified fruit in the spring. *Sclerotinia fuckeliana* (the conidial form is known as *Botrytis cinerea)* is responsible for the gray mold of various fruits. In dry seasons, it is responsible for the "noble decay" of grapes, in which the grapes have a particularly high sugar content, a phenomenon exploited in the fermentation of Beerenauslese hocks. In wet years, the same fungus simply causes the grapes to fall prematurely. This order also contains the Earth tongues *(Geoglossaceae),* which posses a stalk and a more or less flattened or club-shaped head. Several black and sometimes olive colored species are common on grasslands. The olive-green *Geoglossum viride* (2–5 cm high) grows on soil in deciduous woods, often in tufts. *Cudonia circinans* (4 cm high) has a flattened, leather-yellow, hat-shaped head, and it occurs in clusters on the needle cover of conifer plantations.

3. Order: *Phacidiales.* These are predominantly parasitic fungi, e.g. *Rhytisma acerinum,* the causative agent of black spots on sycamore leaves in the autumn. Pine leaf-cast of young conifers is caused by *Lophodermium pinastri.*

4. Order: *Tuberales.* These generally live subterraneously as mycorrhizal fungi, and the fructifications are known as truffles. Fructifications may be dish-shaped with an apical aperture. In most species, however, the fructification is tuberous, containing internal passages (lined with hymenium); in the young fructification these passages are open to the exterior. Species of the genus, *Tuber*, e.g. *Tuber brumale* (beneath oak and beech in northern and Central Europe) and *Tuber melanosporum* (predominantly in France and Italy) have been prized since antiquity as edible fungi.

Ascomycotina: see *Ascomycetes.*

Ascon: see *Porifera,* Digestive system.

Ascorbinogen: a substance originally isolated from *Brassica oleracea.* It is now known to be an artifact of the isolation procedure. During maceration of the plant tissue, the mustard oil glycoside, glucobassicin, is hydrolysed by the action of the enzyme, myrosinase. The resulting 3-hydroxymethylindole then reacts with ascorbic acid to form A. It can also be obtained synthetically in excellent yield simply by reacting ascorbic acid with 3-hydroxymethylindole at moderate temperature.

Ascospores: spores formed in the spore tube (see Ascus) of the *Ascomycetes* (see).

Ascothoracica: see *Cirripedia.*

Ascus: see *Ascomycetes.*

Asellidae: a family of the *Isopoda* (see).

Asellota: a suborder of the *Isopoda* (see).

Asellus: see *Isopoda.*

Ash: see *Oleaceae.*

Ash-grayleaf bugs: see *Heteroptera.*

Ash-leaved maple: see *Aceraceae.*

Asian blind lizards: see *Dibamidae.*

Asian bullfrog: see *Raninae.*

Asian rat snakes, *Ptyas:* a genus of powerful gray to black colubrid snakes (see *Colubridae)* of South and Southeast Asia. In some places they are tolerated, because they keep down rats and mice. They are slender and agile, and attaining a length of 2.5 m or more. The largest species (3.7 m) is the ***Keeled*** or ***Oriental rat snake*** *(P. mucosus)* of eastern India, Sri-Lanka, Malaysia, Indonesia, Afghanistan and southern China. A.r.s. are particularly skilful in trees, as well as on the ground, readily entering human habitations when chasing prey, and climbing to the topmost branches of trees in pursuit of birds.

Asian water buffalo: see *Bovinae.*

Asiatic black bear, *Selenarctos thibetanus*: a black bear with especially dense neck hair and a light V-shaped mark on the chest. It frequents deciduous forests and brushy areas in southern, central and eastern Asia and the Himalayas up to the timber limit (about 4 000 m). It occasionally kills domestic animals and raids beehives.

Asiatic painted frog: see *Microhylidae.*

Asiatic rock python, *Python molurus:* a snake of the subfamily *Boinae*, family *Boidae* (see), native to India, Indochina and Wallacea. It grows up to 7 m in length, and is patterned with red-brown spots. The smaller and lighter nominate form (*Python molorus molorus*) inhabits India and Sri Lanka, whereas the darker form (*Python molorus bivittatus*) is the commonest boid snake of South East Asia and the Sunda islands (islands of the Malay archipelago as far east as the Moluccas and excluding the Philippines). The mode of life and reproduction are similar to those of the Reticulated python (see).

Asiatic salamanders: see *Hynobiidae.*

Asiatic wild ass, ***Equus hemionus:*** a single species of pale yellow to reddish brown asslike members of the family *Equidae* (see). The seven subspecies are or were distributed from western Asia to Mongolia and Tibet. They combine characters of both asses and horses, and they produce infertile hybrids when crossed with asses or horses. Only the distal half of the tail carries hair. They lack the horny warts that are present on the inside of the hindfeet of horses. The ears are long and asslike, and the voice is intermediate between a neigh and a bray. A dorsal stripe is present, and the mane and tail tassel are black-brown.

1. The large ***Kiang*** *(E. h. kiang)* is bright red-brown in summer, brownish gray in winter, with a white belly. Although it is the most numerous of the Asiatic wild asses, it is still an endangered species, found only on the high plateaus of Tibet.

2. The light yellow-brown ***Onager*** *(E. h. onager)* of the plateaus of northern Iran is an endangered species. The hooves have a dark outline, and a dorsal stripe extends to the tail tip.

3. Also an endangered species is the ***Kulan*** *(E. h. kulan)* of Turkmenia and Kazakhstan. It possesses a characteristic white field along the dorsal stripe.

4. The ***Syrian wild ass*** *(E. h. hemippus)* of Syria, Iraq and northern Arabia is possibly extinct. It is sandy to brown-yellow in color, with a white belly. A dorsal stripe extends to the tip of the tail, and there is a dark outline around the hooves.

5. The ***Dzeggetai*** *(E. h. hemionus)* has a narrow and indistinct dorsal stripe. Also an endangered species, only small relict herds exist in Mongolia.

6. In the ***Indian wild ass, Khur*** or ***Ghorkar*** *(E. h. khur)*, the dorsal stripe ends at or near the base of the tail. Relict herds still exist in the desert of the Ran of Kutch.

7. The ***Anatolian wild ass*** *(Equus hemionus anatoliensis)* is now extinct.

Asio: see *Strigiformes.*

Asities: see *Philepittidae.*

Asity: see *Philepittidae.*

Asp:

1) see Cobras.

2) *Aspius aspius:* a predatory freshwater fish, up to 100 cm (usually 50–60 cm) in length, of the family *Cyprinidae*, order *Cypriniformes,* found in rivers and lakes of eastern and Central Europe, and in the Baltic Sea. It feeds on other fish, and also takes insects, frogs, and small mammals such as water shrews. A favorite sporting fish, it is grayish blue-black with silvery sides.

Asparaginase: a hydrolase, widely distributed in plants and animals, which catalyses the hydrolysis of L-asparagine to L-aspartic acid and ammonia.

Asparagus fly: see Fruit flies.

L-Aspartic acid, ***Asp, 2-aminosuccinic acid:*** HOOC—CH_2—$CH(NH_2)$—COOH, a proteogenic, nonessential, glucogenic amino acid. The α-amino group of Asp contributes nitrogen for the synthesis of urea in the urea cycle. In the biosynthesis of purines, Asp provides N-1 of the purine ring system, as well as the 6-amino group of adenine. The entire molecule of Asp is incorporated in the biosynthesis of pyrimidines; thus, Asp gives rise to N-1, C-atoms 4, 5 and 6, and the carboxyl group of orotic acid (cyclic pyrimidine precursor).

Aspen: see *Salicaceae.*

Aspen mushroom: see *Basidiomycetes (Agaricales).*

Aspect succession: the seasonal change in the composition of animal and plant communities. At any time, the current aspect is characterized primarily by the occurrence of season-dependent species or Semaphoronts (see). In temperate zones there are essentially 4 aspects, corresponding to spring, summer, autumn and winter. A true tropical rainforest shows essentially no changes in aspect, while other parts of the tropics alternate between the aspects of the dry season and the wet season.

Aspen: see *Salicaceae.*

Aspergillus: see *Ascomycetes.*

Aspidobranchia: see *Gastropoda.*

Aspidocotylea: see *Trematoda.*

Aspirin: see Salicylic acid.

Aspius: see Asp *(2).*

Aspro: see Zingel.

Ass: see Wild ass.

Assassin bugs: see *Heteroptera.*

Assimilate: see Assimilation.

Assimilation: the uptake of nutrients by organisms, and their conversion into organic body materials. In the *autotrophic A.* of green plants, light energy is used to convert inorganic nutrients (especially CO_2, NO_3^-, NH_4^+, SO_4^{2-} and PO_4^{3-}) into organic substances. The totality of the reactions leading to the synthesis of energy-rich substances that constitute the organism is known as *anabolism.*

In plant physiology, the term A. is sometimes taken to mean only the A. of carbon *(carbon A., carbon dioxide A.),* which is achieved by Photosynthesis (see) or by Chemosynthesis (see).

Nitrogen A. (nitrate A.) involves the conversion of nitrate ions into ammonium ions (to a small extent, ammonium ions may also be taken up from the soil and utilized directly), followed by the conversion of this inorganic form of nitrogen into the numerous nitrogenous organic constituents of the plant, such as amino acids, purines, pyrimidines, etc. The amino acids then serve as the main components of proteins, which may also contain additional constituents, such as carbohydrate, sulfur and phosphorus. In many plants, nitrate A. occurs in all plant parts, and the necessary energy and reducing power are derived from respiration of sugars, the latter having been formed by the A. of carbon dioxide. In many herbaceous plants, however, nitrate A. occurs largely in the leaves, in or in close association with the chloroplasts, and it is directly linked to the provision of energy and reducing power by the light reaction of photosynthesis. In such cases, nitrate A. is truly a form of photosynthesis (see Nitrogen).

Sulfate A. (see Sulfur) also occurs in plants.

In *heterotrophic A.,* which occurs in animals, organic materials derived from other living organisms are partially degraded and/or altered, then reassembled into forms that are specific to the assimilating organism. Thus, purines and pyrimidines enter a network of metabolic reactions and are converted into precursors that can be used for the synthesis of new nucleic acids, as well as being available for degradation. Similarly, amino acids from the digestion of foreign protein are available both for the synthesis of new protein and for degradation.

The products formed by A. are called the *assimilate.* The converse of A. is Dissimilation (see).

Assimilation quotient, ***AQ:*** the ratio of O_2 consumed to CO_2 released in photosynthesis. AQ = O_2/CO_2. In the biosynthesis of carbohydrate, AQ = 1.

Association:

1) The fundamental unit of a plant community. Floristically, it is characterized by a typical combination of species, or by a diagnostically important group of species that forms a significant proportion of the A. Certain species are also specific for a particular A. Specific species also serve to differentiate between the floristically different subdivisions of the A. An A. is usually associated with a particular type of habitat, and it usually has a typical appearance (physiognomy) wherever it occurs. The term was coined by Humboldt (1807) for certain plant groupings. It later became synonymous with Formation (see). See Character species.

2) In animal ecology, the gathering together of animals of different species in confined locations, which offer benefit to the animals in question.

3) A term used in behavioral studies. See Learning.

Association areas of the cerebrum: areas of the lateral cerebral cortex, which lie between, and integrate information from, the motor and sensory areas (e.g. visual cortex, auditory cortex and the somatic-sensory cortex). The A.a. therefore represent an association center that is secondary to the primary centers (e.g. the visual cortex), and therefore more appropriately referred to as the *secondary association cortex.* Parts of the secondary association cortex merge with the speech center. There is strong evidence that this region is concerned with the integrative processing of perception (seeing, hearing and touch, together with the corresponding memory content), as well as the production of abstract thought. This region shows development in animals that are capable of acquiring habits and conditional reflexes. In the great apes, it is smaller than in modern humans.

Assortative mating, ***assortive mating:*** nonrandom pairing and reproduction between individuals in a population. In positive A.m. the paired individuals are more similar (with respect to one or more chosen traits) than predicted from random pairing. In negative A.m. (disassortative mating) the paired individuals are more different than an average pair.

Astacus: see Crayfish.

Astacidea: Crayfish (see) and Lobsters (see), an infraorder of the *Decapoda* (see).

Astaxanthine, ***3,3'dihydroxy-β,β-carotene-4,4'-dione:*** a xanthophyll or oxygen-containing Carotenoid (see). A. occurs widely as a red animal pigment, especially in crustaceans, echinoderms and tunicates. It is also found in the feathers and skin of flamingoes and other birds, in certain fish, in the chicken retina, and in species of *Haematococcus* which are responsible for the phenomenon of "red snow". It is not common in plants. In native form, it can occur free as a red pigment, as an ester such as the dipalmitate, or as a blue, green or brown chromoprotein. The dark, blue-black pigment in the shell of the lobster, *Astacus gammarus,* is a complex of A. with protein, which upon denaturation (e.g. by cooking) releases red A.

Aster: see *Compositae.*

Asteraceae: see *Compositae.*

Asterias: see *Asteroidea.*

Asteridae: see *Dicotyledoneae.*

Asterochelys: see *Testudinidae.*

Asteroidea, ***asteroids, starfishes, sea stars:*** a class (or subclass) of the *Echinodermata* (see) with flattened, star-shaped bodies. The largest members attain an arm length of 45 cm and a span of almost 1 m. Five gradually tapering arms radiate from a central disk; in rare cases, there may be four or more than five arms. On the underside of each arm is an open ambulacral groove bearing rows of large tube feet. The skeleton consists of movable plates, which may give rise to spines and Pedicellariae (see). Respiratory gas exchange occurs via external gills (papulae) on the aboral surface. Starfishes inhabit the sea bottom from shallow coastal waters to the deep oceans. Large individual numbers even occur in polar seas. They are mostly predators. The stomach is often everted through the mouth to digest prey on the exterior; for example a starfish applies its tube feet to the two valves of a clam and gradually pulls them apart until the space is sufficient to allow entry of its stomach, which then digests the soft clam tissues. The larva, known as a *bipinnaria,* either develops directly into the adult animal, or it passes through a second larval stage known as a *brachiolaria* which possesses three attachment arms. About 1 500 living species are known. The most familiar genera are *Astropecten, Solaster* (see Sunstar) and *Asterias.*

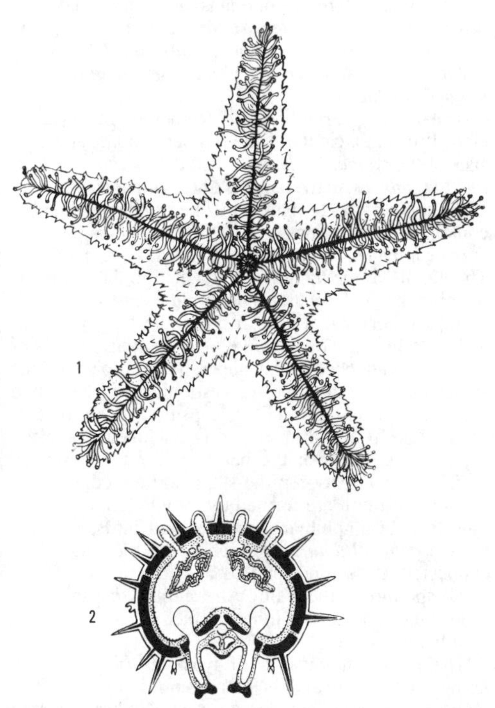

Asteroides. 1 Oral surface of a starfish *(Echinaster). 2* Transverse section of an arm, with skeletal elements shown in black.

Fossil record. Fossils are found from the Ordovician to the present. They have only limited paleontological significance, showing a relatively increased abundance only in the Old Paleozoic. Only about 310 fossil species are known.

Collection and preservation. Specimens are collected from beaches at ebb tide or from the drag nets of fishing boats. They are also easily caught with paternoster lines baited with fish heads. For museum purposes, it is sufficient to preserve specimens in 85% ethanol. Alternatively, they can be dried by dehydration in acetone.

Asteroids: see *Asteroidea.*

Asteroxylaceae: see *Psilophytatae.*

Asteroxylon: see *Psilophytatae.*

Astilbe: see *Saxifragaceae.*

Astmata: see *Ciliophora.*

Astragalus: see *Papilionaceae.*

Astrangia: see *Madreporaria.*

Astrocytes: see Nervous tissue.

Astronesthidae: see *Stomiiformes.*

Astropecten: see *Asteroidea.*

Astrophytum: see *Cactaceae.*

Astylosterninae, *astylosternins:* a subfamily of the *Ranidae* (see), containing 25 species in 5 genera, all restricted to Africa: *Astylosternus (= Gampsosteonyx = Dilobates*) (11 species; West Africa from Sierra Leone to Central African Republic and Zaire); *Nyctibates* (1 species; Southwest Cameroon); *Scotoblebs* (1 species; East Nigeria to Zaire); *Leptodactylon (= Bulua*) (11 species; East Nigeria to Southwest Cameroon).

Astylosternins possess a posteriorly forked, ossified omosternum, and a broad cartilaginous sternum. The coracoid cartilages are usually broad and ossified ("epicoracoid cartilages"). All astylosternins have vertical pupils.

The genus, *Astylosternus,* includes the "Hairy frog" *(A. robustus),* noted for the growth of 10–15 mm long, vascularized villous processes along the thighs and flanks of the male during the mating season. Not all authorities are agreed that these villosities function in respiratory gas exchange; more recent studies suggest that they act as a sex signal. The tadpoles of *Astylosternus* have a large oral sucker.

Asynapsis: a phenomenon in which chromosome pairing in the first meiotic division is absent or incomplete, so that a variable number of unpaired (univalent) chromosomes are present during the metaphase. A. may be genetically based or due to environmental effects. It can also result from inadequate structural complementarity between the chromosomes. Single chromosomes or the entire chromosome complement may be subject to A. Subsequent random distribution of the unpaired chromosomes in the first stage of meiosis results in hypo- and hyperploid gametes. In extreme cases, A. can result in complete sterility. See Desynapsis.

Atavism: the occurrence of characters, which are divergent for the species, but which were originally primary characters of a phylogenetic ancestor. In A., these characters recur in individuals of a species after an interval of several or many generations. *Combination A.* arises occasionally by the mutual encounter of very rare recessive alleles in an individual. If various hereditory factors that determined the original character become distributed in different races, the old character may reappear after the crossing of races *(hybrid A.). Paratypic A.* is caused by unusual environmental conditions. *Mutative A.* occurs when mutations lead to characters resembling those of an early ancestor, but do not necessarily restore the ancestral genetic condition.

Ateles: see Spider monkeys.

Atelura formicaria: see *Zygentoma.*

Athectae hydroids: see *Hydroida.*

Atherurinae: see *Hystericidae.*

Atherurus: see *Hystricidae.*

Atlantic period: a postglacial period with an Atlantic (oceanic) climate, which occurred beteween 5 500 and 2 700 BC. It was characterized by a predominance of mixed oak forest with an undergrowth of hazel. See Pollen analysis.

Atlantic sea-wolf: see Catfish.

Atlantic wreck fish: see Sea basses.

Atlantosaurus: see Dinosaurs.

Atlas: see Axial skeleton.

Atmobios: the totality of organisms that live above ground and are surrounded by the normal atmosphere. The term also embraces inhabitants of the soil surface (epigaion) and vegetation (hypergaion). The A. is separate and different from the Edaphon (see Soil organisms).

Atmungsferment: see Cytochromes.

Atokous: nonreproductive, vegetative, or without offspring. The converse is Epitokous (see).

Atoll: see Coral reef.

ATPase: see Adenosine triphosphatases

ATP pyrophosphate lyase: see Adenylate cyclase.

Atrichornithidae, ***scrub birds:*** an Australian avian family in the order *Passeriformes* (see). Only 2 species of scrub birds are known, and they are very closely related to the lyrebirds *(Menuridae).* They are unique among the *Passeriformes,* in that their collar bones are not fused into a wishbone. The ***Rufous scrub bird*** *(Altrichornis rufescens)* (16.5 cm) is found in Southeast Australia, while the ***Noisy scrub bird*** *(A. clamosus)* (20 cm in length, and at one time thought to be extinct, but rediscovered in 1961) is restricted to a small area of Western Australia. They can run quickly, and they behave more or less like miniature lyre birds.

Atriplex: see *Chenopodiaceae.*

Atrium: an antechamber, especially an antechamber of the Heart (see).

Atropa: see *Solanaceae.*

Atropoda: see *Psocoptera.*

Atropine, ***DL-hyoscyamine:*** the ester of DL-tropic acid with tropine. The racemate is formed during alkaline treatment of the alkaloid, L-hyoscyamine during isolation from its plant source. A. is present in different members of the *Solanaceae,* e.g. *Atropa belladonna* (Deadly nightshade), *Datura stramonium* (Thorn-apple), *Hyoscyamus niger* (Henbane). The water-soluble sulfate of A. is used medicinally. A. inhibits the secretion of saliva, perspiration and gastric juice, relaxes cramps in the smooth muscle of the gastrointestinal tract and bronchi (e.g. in asthma), and causes enlargement of the eye pupil.

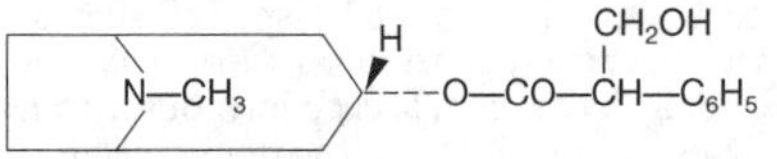

$N{-}CH_3$ (H) $-O-CO-CH(CH_2OH)-C_6H_5$

Atropine

Attached chromosomes: see Isochromosomes.

Attachment organs: see Adhesion organs.

Attachment settlement: see Relationships between living organisms.

Attagis: see *Thinocoridae.*

Attractants: chemical compounds used to attract animal pests, in particular insects. They are used to detect and identifiy pests, or to control pests by attracting them to pesticides or traps. A. include sexual attractants (see Pheromones), mustard oils, terpenes, cantharidine, etc. The converse of A. is Repellents (see).

Atyidae: see Shrimps.

Aubergine: see *Solanaceae.*

Auchenorrhynchi: see *Homoptera.*

Auditory lobes: see Brain.

Auditory nerve: the VIII cranial nerve; see Cranial nerves.

Auditory organ: an animal sensory organ for the perception of sound. Among the ***invertebrates,*** a genuine auditory sense exists only in arthropods, e.g. the Tympanal organ (see) of insects, which responds to sound above a certain amplitude, and Trichobothria (see) of spiders and some insects that lack tympanal organs, which respond to certain sound frequencies. In arthropods, the generation of auditory signals, often advertizing the presence of a potential sexual partner, is a certain indication that the species also possesses a means of detecting such signals. The withdrawal reaction of larvae in response to loud noise is not evidence for their possession of A.os.

In ***vertebrates*** sound waves are perceived by the ***ear,*** which also functions as an equilibrium organ. In lower vertebrates, only the ***inner ear*** is present. The ***middle ear*** evolved with the amphibians, while mammals also possess an ***outer*** or ***external ear,*** consisting of an expanded part called the *pinna* or *auricula* (with associated muscle) and an external *auditory canal* or *external acoustic meatus.*

The middle ear is derived from the first visceral cleft. It lies in a cavity of the tympanic bone, and consists of the *tympanic chamber* and the *auditory* or *Eustachian tube* which connects the tympanic chamber with the pharynx. In mammals, the tympanic chamber contains 3 small bones or *ossicles,* known as the *malleus* or *hammer, incus* or *anvil,* and *stapes* or *stirrup.* The handle of the hammer is attached to the tympanic membrane which separates the middle ear from the external auditory canal. The head of the hammer is attached to the base of the anvil, and the long process of the anvil is attached to the stapes. The footpiece of the stapes occupies the fenestra vestibuli, one of the two small openings in the auditory capsule which separates the middle from the inner ear. [Attachments between the ossicles, between the hammer and the tympanic membrane and between the stape and the edge of the fenestra vestibuli are formed by minute ligaments and muscles]. In amphibians, reptiles and birds, a single slender *columella auris* or *stapes* spans the middle ear, connected at its inner end to a membrane closing a small aperture in the wall of the auditory capsule, and at its outer end to the inner surface of the tympanic membrane.

The inner ear consists of two sets of structures, known as the bony and the membranous labyrinth. The *bony* or *osseus labyrinth* is a series of interconnected channels and cavities hollowed out of the petrous-temporal bone; there are basically three regions, known as the *cochlea, vestibule* and *semicircular canals.* The *membranous labyrinth* is a system of membranous channels and cavities within the osseus labyrinth. Within the cochlea and the semicricular canals, the membranous labyrinth has the same general form as the osseus labyrinth, but it is considerably smaller, being separated from the bony walls by watery perilymph (not floating, but attached by strands of fibrous connective tissue). The vestibule is the central cavity of the osseus labyrinth; its lateral or tympanic wall communicates with the middle ear by means of the fenestra vestibuli. Within the vestibule, the membranous labyrinth consists of two small chambers, the upper *utriculus* or *utricle* and the lower *sacculus* or *saccule.* A blind duct (endolymphatic duct or ductus endolymphaticus) extends from the sacculus. According to the species, utriculus and sacculus may communicate directly, or the connecting channel from the utriculus may move to the endolymphatic duct (e.g. in humans). The vestibule gives rise to three *osseus semicircular canals* at right angles to one another (one horizontal, two vertical, known as the *superior, posterior* and *lateral canals),* which contain *membranous semicircular canals* similar in shape and number to the osseus canals but of smaller diameter, and arising from the utriculus. One end of each osseus canal is distended to form an *ampulla.* In the membranous labyrinth, sensory epithelium or *macula* (groups of elongated cells, each with a tuft of sensory hairs) is located in the region of the ampullae, in the utriculus and in the sacculus, where it responds to the accelerated and decelerated flow of endolymph during head movement, thereby contributing to the sense of equilibrium. Semicircular canals, utriculus and sacculus are filled with endolymph.

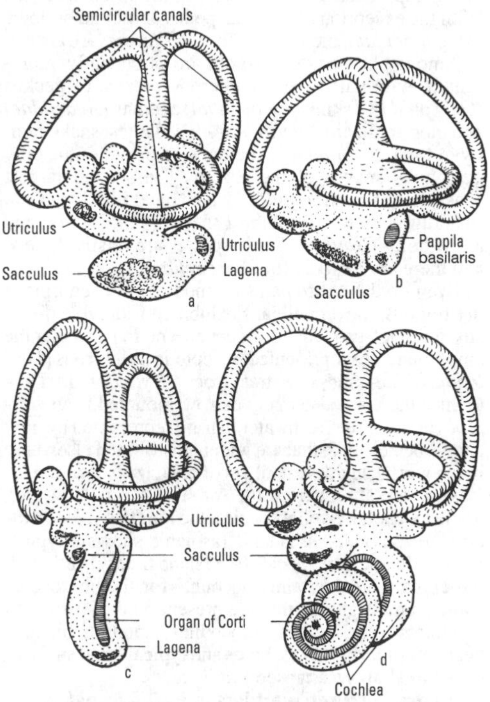

Auditory organ. Labyrinths of vertebrates. *a* Fish. *b* Frog. *c* Bird. *d* Mammal.

In lower vertebrates, the sacculus has a short blind extension called the *lagena,* but in higher vertebrates this has evolved into a spirally coiled tube, which, together with the surrounding auditory capsule, is called the *cochlea.* The cochlea is already evident in the crocodylians, and the actual A.o. of the higher vertebrates, i.e. the *organ of Corti* or *Corti's organ* is located within it. In transverse section the cochlea has the appearance of a bony tube (the cochlear canal), lined with connective tissue and transected by two membranes, which divide it into three chambers (Fig.2); the middle chamber (ductus cochlearis or scala media) is continuous with the sacculus, and it is filled with endolymph. The upper chamber (between Reissner's membrane and the wall of the bony canal) is called the scala vestibuli; this is a perilymphatic space connecting the base of the cochlea with a space below the fenestra ovalis, a small membranous area which abuts the cavity of the middle ear. The scala vestibuli, also filled with perilymph, is actually connected with the lower chamber of the cochlea (scala tympani) by a narrow duct at the tip of the cochlea. At its lower end, the scala tympani ends against the fenestra rotunda, and in this region it also communicates with subarachnoid space via the aqueductus cochleae (an osseus channel with no corresponding component of the membranous labyrinth). The periplymph between the bony and membranous labyrinth performs an important function in the transmission of sound waves to the organ of Corti.

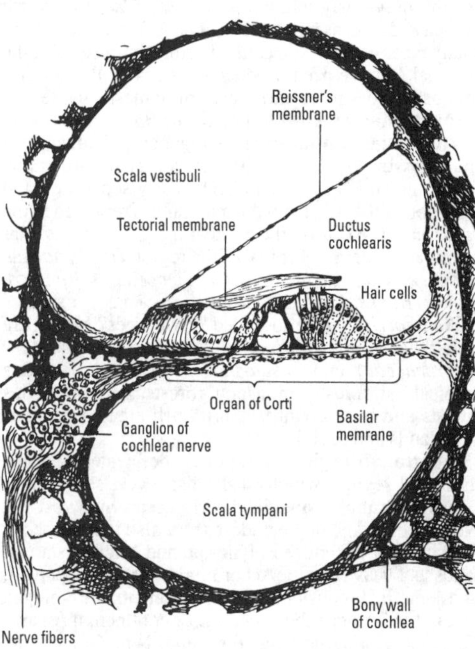

Auditory organ. Transverse section of a mammalian cochlea.

The *acoustic* or *auditory nerve* (VIII) divides into the *cochlear nerve* (ramus cochlearis) and the *vestibular nerve* (ramus vestibularis). The cochlear nerve gives off branches to the sacculus, the ampulla of the posterior semicicular canal, and Corti's organ, while the vestibular nerve serves the utriculus and the ampullae of the superior and external semicircular canals.

Auditory tube: see Eustachian tube.

Auger shells: see *Gastropoda.*

Auks: see *Alcidae.*

Aulacorhynchus: see *Ramphastidae.*

Aulacoceras (Greek *aulax* groove, *keras* horn): an ancestral belemite found in the Alpine Triassic. It consists of a cylindrical, belemite-like rostrum (which is covered with 40 longitudinal grooves) and a phragmacone. The phragmacone, which is usually found separated from the rostrum, is twice as large as the rostrum and resembles the phragmacone of *Orthoceras.*

Aurelia: a scyphomedusa (see *Scyphozoa)* of the order, *Semaeostomae* (see). *Aurelia aurita* is distributed worldwide and is one of the commonest "jellyfish". The medusa has a diameter of 20–40 cm, and its nematocycts are not dangerous to humans. It feeds on planktonic animals and other small creatures such as water fleas, polychaetes, etc. The polypoid form is only a few millimeters in length, and it produces medusae by Strobilation (see).

Aureomycin: see Tetracyclins.

Auricle: an antechamber of the Heart (see).

Auricula: see *Primulaceae.*

Auricula of the ear: see Auditory organ.

Auriculariales: see *Basidiomycetes.*

Auriparus: see *Remizidae.*

Aurochs, *Bos primigenius*: an ox, up to 1.8 m in height, previously inhabiting forests of Europe, North Africa and Asia as far as China. The A. is the ancestor of many races of domestic cattle. Forms reminiscent of the original A. have been produced in zoos by the systematic back-crossing of early races of domestic cattle.

Australian crested bellbird: see *Pachycephalinae.*

Australian kingdom: a phytogeographical term for the Floristic kingdom (see) that comprises Australia, Tasmania and New Zealand. The early separation and isolation of the A.k. from the mainland is reflected in the fact that 80–90% of the more than 12 000 vascular plants are endemic. Typical of the region are *Myrtaceae (Eucalyptus, Melaleauca), Proteaceae (Grevillea, Hakea, Banksia),* phyllode-bearing acacias, 15 species of *Xanthorrhoea* (the so-called "grass trees", which are either placed in the *Liliaceae* or separately in the *Xanthorroeaceae),* and *Casuarinaceae.* The A.k. contains tropical rainforests, monsoon forests, sclerophyllous forests and bush country, thorn bush areas, savannas, grasslands and semideserts.

Australian region: a variously delineated zoogeographical region, which in the narrowest sense comprises Australia, Tasmania, the Moluccas, Sulawesi and New Guinea, and in the widest sense also includes New Zealand and the entire Polynesian and Micronesian islands as far as Hawaii. Accordingly, the A.r. is part of, and sometimes considered as equivalent to, the Notogea (see). It is further subdivided into a continental Australian subregion (Australia, Tasmania, New Guinea, the Moluccas and Sulawesi), a New Zealand subregion, a Micronesian subregion and a Polynesian subregion, the latter possibly including the Hawaiian islands.

In the present account, the A.r. is assumed to comprise only Australia, Tasmania, New Guinea and other nearby islands of the continental shelf. The region contains numerous ancient groups of animals, and it has a high Endemic rate (see). As an original component of the Gondwana Continent, the A.r. was colonized from the Antarctic and from Asia. Transantarctic dispersal accounts for the fact that many invertebrates of the A.r. (and New Zealand) have close relatives in the Neotropical region. It is thought that marsupials originated in Asia, but it is possible that they were also able to exploit the opportunity for dispersal offered by Gondwanaland. On the other hand, the birds and placental mammals of the A.r. certainly originated in Asia.

The A.r. is characterized especially by the monotremes and the high proportion and adaptive diversity of marsupials. It is often overlooked that the fauna of the A.r. also includes a large proportion of placental mammals: the Dingo (introduced by prehistoric man), bats and murid rodents. On the Australian continent, placental mammals account for about 40% of the mammalian species, while in New Guinea they are in the majority. The largest families of marsupials are the *Dasyuridae* (endemic), *Phalangeridae* and *Macropodidae* (both of which are also found in intermediate regions).

Despite the opportunities for dispersal offered by flight (illustrated by the fact that about 20 European birds occur at least temporarily in the A.r.), the bird populations of the A.r. and the Oriental region are markedly different. Surprisingly large differences even exist between the bird populations of the Australian Continent and New Guinea. Thus, out of 531 breeding species (excluding sea birds) in Australia and 566 in New Guinea, only 191 occur in both areas. Endemic or almost endemic birds of the A.r. are, inter alia: *Casuariidae* (cassowaries), *Menuridae* (lyrebirds), *Paradisaeidae* (birds of paradise) and *Ptilonorhynchidae* (bowerbirds). Certain groups, such as the parrots, display a rich diversity. With the exception of the megapodes or mound builders *(Megapodidae),* members of the *Galliformes* are absent.

Among the reptiles, lizards are abundantly represented, with particularly numerous species of geckos *(Gekkonidae),* skinks *(Scincidae),* agamas *(Agamidae)* and monitor lizards *(Varanidae).* The largest snake family of the A.r. *(Elapidae,* cobras) also provides the only poisonous snakes of the region. The second largest snake family is the*Typhlopidae* (typical blind snakes). In addition there are colubrid snakes *(Colubridae)* and a few boids *(Boidae).* Turtle species are relatively few, and there are 3 species of *Crocodylia.*

Even on the Australian Continent with its enormous dry regions, the amphibians exhibit considerable diversity, although urodelic forms are absent. In particular the amphibians are represented by leptodactylid frogs *(Leptodactylidae)* and true tree frogs *(Hylidae).* In New Guinea there are also many narrow-mouthed frogs *(Microhylidae).* True freshwater fish are represented by only a few species. Fish that are widely distributed elsewhere in the world, such as members of the *Cyprinidae* and *Siluridae,* are absent from the A.r. The A.r. is, however, home to a notable relict species, the lungfish *(Neoceratodus forsteri).* As an island, Tasmania shows a typical impoverishment of species (see Island theory), although the climate also certainly accounts for the absence of some animals groups that are present on the Australian Continent (e.g. turtles, crocodiles, monitor lizards, geckos, colubrid snakes, boids and typical blind snakes).

Australian wrens: see *Sylviinae.*

Australian wren warblers: see *Muscicapidae.*

Australopithecines, *Australopithecinae, prehominids:* an extinct subfamily of hominids. See Anthropogenesis.

Austrian briar: see *Rosaceae.*

Austro-American side-necked turtles: see *Chelidae.*

Austro-American snake-necked turtles: *Chelidae.*

Autoantibodies: antibodies that react with material of the organism that produced them. Some A. are responsible for Autoimmune diseases (see). Others serve to destroy nonfuncional cells and unwanted cell material.

Autobasidium: see Holobasidium.

Autochory: see Seed dispersal.

Autochthone:

1) An indigenous or native organism (see Autochthonous).

2) Endogenous food material of a cave or cave system, which is derived within the cave itself.

Autochthonous: indigenous, or native to the site at which it is found.

Autochthonous microflora: soil microflora that actively develop under entirely natural conditions, in contrast to those species of microorganisms that do not make a significant contribution to the total microflora until the soil has been influenced by human activity (application of fertilizers, pesticides, etc.). The A.m. must be differentiated from the *potential microflora,* i.e. the totality of all microorganisms (active or inactive). See Soil organisms.

Autoclave: an airtight, sealable metal container of variable size, which can be heated. It is used to sterilize bandages, instruments, bacteriological growth media, preserved foods, etc. The material to be sterilized is separated from a relatively small quantity of water in the bottom of the A. With the aid of a pressure control valve, the A. can be heated to generate a steam pressure of 15–20 pounds per square inch (103–138 kPa), corresponding to temperatures of 120–135 °C. The first simple A. was used by Louis Pasteur.

Autogamy:

1) In unicellular organisms, the pairing of gametes (autogametes) formed by reduction division from a single parent cell.

2) Nuclear cleavage and refusion within the same cell.

3) In protistans, the division of the two micronuclei to produce eight or more nuclei, followed by fusion of two of these nuclei to form a new macronucleus.

4) Self-fertilization or self-pollination in flowering plants; see Pollination, Self-fertilization.

See also Automixis, P(a)edogamy.

Autoimmune diseases: diseases caused by immune reaction to the body's own cells and materials. Autoantibodies (see) are mainly responsible for some A.d. Examples are A.d. with autoantibodies against red blood cells or against thrombocytes. In other cases, activated lymphocytes that are aggressive toward cells of certain organs are largely responsible for the damage to tissues and organs in A.d. In true A.d., the role of autoimmune processes is clear. Often, however, it is not certain whether autoimmune processes are the cause or the result of the disease, e.g. in rheumatoid diseases.

Autolysis: self-digestion, and therefore the self-destruction of dying cells. During cell death, the lysosomal membranes become leaky and break down. The enzymes of intracellular protein degradation that are normally contained within the lysosomes are therefore released, and they attack the cytoplasmic and structural proteins of the cell.

Automictic parthenogenesis: see Automixis.

Automimicry: see Protective adaptations.

Automixis, ***automictic parthenogenesis:*** obligatory self-fertilization, in which the gametes arise from the same meiosis. Thus, meisosis always occurs and diploidy is always reinstated. See Pedogamy.

Automutagens: substances produced by normal or abnormal metabolic processes, which are able to cause mutations in the same organism. See Mutagens.

Autonomic nervous system, ***vegetative nervous system, visceral nervous system:*** that part of the vertebrate nervous system responsible for innervating the organs of the thorax and abdomen, glands, blood vessels, and the inner musculature of the eye. Functionally, the A.n.s. is divided into two main parts, the *sympathetic* and the *parasympathetic* systems. After leaving the spinal cord, autonomic nerve fibers do not travel directly to their target organs; they are interrupted by synapses, which are located in ganglia outside the spinal cord. On arriving at the synapse, a single preganglionic nerve impulse may generate impulses in up to twenty postganglionic fibers (cf. motor fibers of the somatic outflow, which travel directly to the target muscle).

Fibers of the ***sympathetic system*** arise in all the thoracic segments and in the upper two lumbar segments of the spinal cord. The cell body of each preganglionic sympathetic neuron lies in the intermediolateral horn of the gray matter of the spinal cord; its fiber passes through an anterior root of the cord into a spinal nerve, then immediately passes through a white ramus to a ganglion in the sympathetic trunk (two rows of interconnected sympathetic ganglia lying on either side of the vertebral column). From these ganglia, postganglionic neurons travel to the target organ. Central preganglionic sympathetic neurons receive continuous impulses from higher centers of the central nervous system. They are therefore continually active, and the basal rate of activity is known as *sympathetic tone.*

The central ganglia of the ***parasympathetic system*** lie in the brainstem and in the sacral region of the spinal cord. Preganglionic fibers do not travel to the sympathetic trunk, but to peripheral ganglia, most of which are located in the organ that is to be excited; the postganglionic neurons are therefore relatively short. Parasympathetic nerve fibers from the brainstem centers leave the central nervous system via four cranial nerves, the most important of these being the *vagus* nerve, which innervates practically all the organs of the thorax and abdomen. The parasympathetic system also receives continual impulses from higher centers, the corresponding basal rate of activity being known as *parasympathetic tone (vagus tone* for the vagus nerve alone).

It is relatively easy to block nerve transmission in the peripheral autonomic ganglia with chemical agents, e.g. nicotine, which is used medicinally as a *ganglion blocker* to reduce the activity of the A.n.s.

Almost all organs are innervated by both the sympathetic and parasympathetic systems, and in most cases their actions are potentially antagonistic. However, one of the systems is usually dominant, so that the two systems do not normally actively oppose one another. Both parasympathetic and sympathetic stimulation may be excitatory or inhibitory, depending on the organ in question.

The preganglionic neurons of both the sympathetic and parasympathetic system, as well as the parasympathetic postganglionic neurons are cholinergic neurons, i.e. their nerve endings secrete the neurotransmitter, acetylcholine. Some of the sympathetic postganglionic nerve endings are also cholinergic, but the majority are adrenergic (i.e. they secrete noradrenalin).

Autonomous: due to inner causes; endogenous. Autonomous processes, e.g. certain plant movements, are not induced by external stimuli.

Autoradiography (plate 18): a method for the detection of radioactive isotopes in macroscopic and microscopic preparations. Special photographic emulsions or stripping film are used for the detection. For light microscopy, microscope slides carrying the specimen under investigation are covered with stripping film or immersed in photographic emulsion. For electron microscopy, ultrathin sections are covered with a monograin layer of very fine-grained silver halide emulsion. After an appropriate exposure time, which may vary from a few days to a few months, the resulting silver grains are developed photographically. Thus, superimposed on the morphological picture, the observer sees a pattern of silver grains representing the distribution of radioactivity in the tissue. A. is important in modern physiology and biochemistry for the subcellular localization of compounds and sites of biosynthesis, and for the investigation of transport processes.

The same term is applied to a biochemical technique for detecting radioactive substances on chromatograms and electropherograms. In this case, the chromatogram is placed for an appropriate length of time next to X-ray film, which is then developed. In modern molecular biology, A. is used widely for locating ^{32}P-labeled nucleic acids and fragments thereof, and for detecting the sites of hybridization of labeled probes on electrophoresis gels. Classically, A. was very important in the elucidation of the pathway of carbon in the dark reaction of photosynthesis by Calvin et al.: after brief exposure to $^{14}CO_2$, illuminated algae were rapidly killed in hot ethanol, their cell constituents separated by paper chromatography, and radioactive products of assimilation located on the paper chromatogram by A.

Autoregulation: see Kidney function, Hydrodynamics of blood circulation.

Autosomes: the normal chromosomes in the chromosomal complement of a eukaryote, which form homologous pairs in diplonts, and which differ in structure and function from Heterochromosomes (see Sex chromosomes) and B-chromosomes (see).

Autosynapsis: a type of meiotic pairing that occurs in Isochromosomes (see). The two arms pair with one another by folding downward, and crossing over can occur between them.

Autosyndesis: the meiotic pairing of chromosomes from the same gametes, irrespective of whether the pairing chromosomes are structurally similar or are partly nonhomologous. A. is therefore associated with the occurrence of Aneuploidy (see) or Polyploidy (see). In diploid forms, A. does not occur, and meiotic pairing is always an Allosyndesis (see).

Autotomy, ***self-amputation:*** a Protective reflex (see), in which animals release their body parts or appendages. It occurs in some protozoa, in many arthropods (e.g. the limbs of crustaceans), and in vertebrates (e.g. the tails of certain lizards). A. occurs at predetermined sites, e.g. a particular vertebra in the tail of a lizard is snapped by muscular action when it is seized by a predator. The lost structures can be more or less regenerated.

Autotrophy: a mode of nutrition, in which organic constituents of the organism are synthesized from simple inorganic compounds, like carbon dioxide, ammonia, nitrate and sulfate. Originally, the term was used only for carbon A., i.e. the ability to synthesize organic acids, carbohydrates and other compounds from carbon dioxide and water, the energy required for the synthesis being derived from sunlight (see Photosynthesis) or chemical conversions (see Chemosynthesis). Nitrogen A. of N-autotrophic organisms is the ability to assimilate nitrate, ammonium compounds or atmospheric nitrogen (see Nitrogen fixation). Sulfur A. is the ability of plants and microorganisms to reductively assimilate sulfate (see Sulfur).

Autotropism: the tendency for shoots to return to a normal position after bending in response to a stimulus. Thus, a plant shoot may over-respond to a geotropic stimulus, and undergo excessive curvature. Under the influence of A., acting in concert with a renewed geotropic stimulus, the shoot then normalizes its alignment, often with a pendulum-like swing into a final perpendicular position. See Tropism.

Autozoid: see *Bryozoa.*

Auxanogram: see Auxanography.

Auxanography: a method for studying the growth-promoting or growth-inhibiting effects of chemical compounds on microorganisms. An agar culture medium (see Agar-agar) lacking the substance under investigation is prepared in a Petri dish, and inoculated with the test organism. Samples of the substance under investigation are placed at chosen sites on the surface of the agar medium (see Diffusuion plate test). If the substance promotes growth, the site of its application becomes surrounded by a zone of strong growth, whereas the area around an inhibitor, e.g. an antibiotic, becomes a growth-free zone.

A. is used, e.g. for the identification of yeasts. Each yeast species has a characteristic *auxanogram,* i.e. a characteristic spectrum of utilizable and nonutilizable carbohydrates, which can be determined by A.

Auxesis: see Isometric growth.

Auxiliary intestinal breathing: see *Cobitididae.*

Auxiliary oocyte nutrition: see Oogenesis.

Auxins: a group of natural and synthetic growth regulators, which affect the growth and differentiation of higher plants in a variety of ways. The naturally occurring plant A. are phytohormones. 2-(3-Indolyl)acetic acid (β-indolylacetic acid, IAA) is the most active natural A.; it is an ubiquitous plant constituent, occurring in very low concentrations (1 to 100 µg per kg fresh mass of plant material), and it is also present in lower plants, including bacteria. Phenylacetic acid is also a naturally occurring A., but it has lower biological activity than IAA. Plant tissues contain free IAA and structurally related compounds, some of which may be biosynthetic precursors of IAA, e.g. 4-chloroindolylacetic acid, indolylethanol (tryptophol), indolylacetamide, indolylacetonitrile, indolylacetaldehyde, indolylacetaldoxime. Auxin conjugates are also present, e.g. indolylacetyl-L-aspartate, indolylacetyl-L-alanine, indolylacetyl-L-glutamate, indolylacetylglycine, indolylacetylmyoinositol, indolylacetylmyoinositol-galactoside, indolylacetylmyoinositol-arabinoside,indolylacetyl-1-0-β-D-glucopyranoside, indolyacetyl-S-coenzyme A, indole-3-ethyl-β-D-glucopyranoside (tryptophol glucoside), N-(*p*-coumaryl)-tryptamine, *N*-ferulyl-tryptamine, glucobrassicin (indolylmethylglucosinolate). As storage forms of A. or of auxin precursors, the auxin conjugates play an important part in the regulation of auxin metabolism.

Natural A. are indole derivatives, previously thought to be biosynthesized from L-tryptophan. However, it may be necessary to modify this view, since a tryptophan-independent route of IAA synthesis has been demonstrated in maize and *Arabidopsis*, in which indole-3-acetonitrile is converted into indole-3-acetic acid by the action of nitrilase [J. Normanly et al. *Proc. Natl. Acad. Sci. USA* ***90*** (1993) 10355-10359]. The most active sites of biosynthesis are young, developing and actively growing parts of plant organs, like coleoptiles, tips of shoots, young leaves and active cambium. Transport of A. occurs mainly from the shoot to the root tips. A. are inactivated by enzymatic oxidation or by conjugation.

Several A. receptors have been identified, which may represent the primary elements of more than one auxin signal pathway. Within minutes of A. application to shoot tissue, two responses can be measured: 1. acidification of the cell wall space, which is correlated

CH_2—COOH

N
H

β-Indolylacetic acid

with changes in extensibility of the cell wall, and which initiates turgor-driven cell expansion; and 2. increases in the steady state of several gene transcripts. It is thought that A. alter the Michaelis constant (K_m) of an outwardly directed plasma membrane-associated H^+-ATPase, so that the plasma membrane becomes hyperpolarized and the pH of the cell wall space is lowered. As the cell expands, new cell wall material is added at such a rate that the cell wall never attenuates. Presumably, some of the gene transcripts (mRNA) that increase under the influence of A. encode enzymes for synthesis of cell wall material during expansion. At least one family of auxin-inducible genes has been identified, which is involved in auxin-mediated growth; these have been designated SAUR (small auxin up-regulated). *[Bioessays **14*** No.1 January 1992, pp. 43–48; B.A. McClure and T.J. Guilfoyle, *Science, **243*** (1989), 91–93]

Synthetic A. of practical importance are indolylbutyric acid, indolylpropionic acid, phenyl- and naphthylacetic acids, and phenoxy- and naphthoxyacetic acids.

Many effects of A. are the result of the joint action of A. and other phytohormones. A. are primarily responsible for cell extension, especially in coleoptiles and shoot axes. They also stimulate cambium activity, and are involved in cell division, apical dominance, abscission., photoperiodism, geotropism and other growth and developmental processes. At high dosage levels, A. inhibits growth, a phenomenon which has led to the development of selective herbicides, e.g. 2,4-dichlorophenoxyacetic acid (2,4-D) and 2,4,5-trichlorophenoxyacetic acid (2,4,5-T).

A. are detected and determined mostly with specific biotest systems, e.g. the oat coleoptile curvature test, or by physical chemical methods, e.g. gas chromatography.

IAA and some synthetic A., such as 2,4-D, are used widely as growth regulators in agriculture and horticulture, e.g. in the rooting of cuttings, or as selective herbicides in the growth of crops, such as cereals, cotton, soya beans, sugar beet and others.

Auxin antagonists: inhibitors of the action of auxins in plants. The effects of A.a. can at least be partly reversed by auxins. The term is used irrespective of the mechanism of inhibition. Competitive inhibitors, i.e. inhibitors that compete with auxin for the same active site, are also called Antiauxins (see). Many structurally different compounds behave as A.a., including some synthetic auxins, e.g. 2,3,6-triiodobenzoic acid, phenylacetic acid, phenylbutyric acid.

Auxotroph, ***deficiency mutant:*** a mutant organism (usually a microorganism) which has lost the ability to synthesize an essential metabolite. This metabolite therefore becomes an essential nutrient, which must be provided exogenously to support growth. Gene mutation may affect one or more enzymes of the relevant biosynthetic pathway. See Genetic block.

Auxozygote: see Diatoms.

Avacado: see *Lauraceae.*

Avadavat: see *Ploceidae.*

Avahi: see Lemurs.

Avena: see *Graminae.*

Avenue builders: see *Ptilinorhynchidae.*

Averrhoa: see *Oxalidaceae.*

Aversive conditioning: see Avoidance conditioning.

Avicenna: see Mangrove swamp.

Aviculariidae: a tropical family of the *Araneae* (see), whose members can attain a length of 9 cm. These usually long-haired, nocturnal spiders feed on large insects, and sometimes small vertebrates. Some species habitually jump onto their prey, which they kill with a venomous bite of their chelicerae. The venom of some species is also dangerous to humans. Members of this family are often called Tarantulas (see).

Avicularium: see *Bryozoa.*

Avidin: a basic glycoprotein found in the egg whites of many birds and amphibians, where it protects the egg against bacterial invasion. It has an extremely high binding affinity for the vitamin, biotin. Feeding of A. or raw egg white can therefore lead to experimental biotin deficiency (see Vitamins). A. is denatured and inactivated by heating (cooking).

Chicken A. consists of 4 identical subunits of known primary structure (the single subunit contains 128 amino acid residues, M_r 14 332 without the carbohydrate moiety). A. forms a stoichiometric, noncovalent complex with biotin (4 molecules biotin per tetrameric molecule of A), which is not attacked by proteolytic enzymes, and is therefore not digested.

Avidity: the strength of binding of an antibody to its antigen. It usually results from the multiple binding of the antibody to a multivalent antigen, e.g. a bacterium. A. is essentially the counterpart of Affinity (see).

Avies, *birds* (plate 6): a class of vertebrates. Birds and reptiles, which share a common evolutionary ancestry, are grouped together as the *Sauropsida.* The nearest reptilian relatives of the birds are the archosaurs, represented by the recent *Crocodylia.* Feathers (see) are characteristic of, and unique to birds.

Anatomy and physiology. In contrast to reptiles, birds are homeothermic ("warm blooded"); they accordingly display high levels of bodily activity and a high metabolic rate. The resulting increase in oxygen demand is met by a combination of various anatomical and physiological characters. Thus, the right and left atria are separate, as are the right and left ventricles of the heart, so that the circulation is separated into pulmonary and body circuits. In addition, the fine structure of the lungs is highly developed. Depending on age, sex and seasonal activity, the blood represents 5–13% of the body weight, and the blood concentration of erythrocytes is very high (1.5–7.5 × 10^6/mm^3). Compared with mammals, the avian heart is larger in relation to body size. A high heart rate also results in a relatively high cardiac output. Large running land birds (ostrich, emu, cassowary, nandu, kiwi) cannot fly. Also incapable of flight are the penguins, the Megellanic flightless steamer duck *(Tachyeres pteneres)*, the flightless cormorant *(Nannopterum harrisi)* and the short-winged grebe *(Centropelma micropterum)*. The extinct Drontevogel and the giant auk *(Pinguinus impennis)* were also flightless.

Evolution. The earliest and most primitive birds of which fossils have been found, are placed in the order *Archaeopterygiformes,* which existed 150 million years ago. These primitive ancestors had teeth and a long tail with vertebrae, but only 4 toes and 3 fingers. The fossil remains indicate that the feathers of the *Archaeopterygiformes* were fully developed, and there is no clue as to how feathers may have evolved from reptilian scales. Most recent birds still have 4 toes, some have only the 3 front toes, while the African ostrich has only 2. Flying birds and penguins also retain 3 fingers. The flight or contour feathers are attached to the forearm, the hand and the prolonged second finger. Both phalanges of the second finger possess notches for insertion of the quills of the primary feathers (the largest flight feathers), which propel the bird through the air during the outward wing stroke. The beak or bill is a new evolutionary entity, which has become modified in shape and size in accordance with different modes of feeding. Most birds use their beak for taking animal or plant food. Only members of the *Falconiformes* and owls take prey with their feet, and these birds also transport prey in their talons. The mode of feeding and foraging is reflected in the form of the feet and toes (swimming, climbing, gripping, prehensile), length of the neck, structure of the stomach (seed eaters have a muscular stomach), development of a crop and of a cecum. The flight musculature corresponds to the mode of flight or the inability to fly. Thus, the large pectoral muscles are most strongly developed in birds that fly off rapidly from a standing start and/or fly in still air (e.g. in woodlands).

Birds possess a larynx, but this is much reduced and lacks vocal cords, and contributes little or nothing to voice production. In birds, the organ of voice or song is the Syrinx (see), situated lower than the larynx, at or near the bifurctaion of the trachea into its 2 bronchi.

Mode of life. Birds display brood care. Egg cells ripen in the ovary and are released into the funnel of the oviduct by rupture of the follicle. The egg, floating on a sphere of yolk, must be fertilized soon after release (fertilization is internal, therefore requiring copulation), before it becomes enveloped in egg white and a shell of calcium carbonate further along the oviduct. Eggs are incubated (brooded) by the male, female or both parents [except in the case of the mound builders *(Megapodiidae)*, see *Galliformes]*. Some species do not form pairs [many members of the *Galliformes*, the Ruff *(Philomachus pugnax)* and others], but most species form pairs at least in the mating season, or for an even longer period until the young are fully fledged. Various species pair for life (e.g. nuthatches, magpies, gray lag geese and others).

Not all birds build nests. The number of eggs per clutch differs from species to species, and further depends on the age of the female, the availability of food, the population density and other factors. As a rule, one egg is laid per day, so that it takes several days to lay a large clutch (e.g. tits). All the young, however, hatch at almost the same time. This occurs because brooding is delayed until most of the eggs have been laid, and because hatching is synchronized by sound signals between the unhatched chicks. Young birds are either Precocial (see) *(Galliformes, Anseriformes* and others), or Altricial (see) *(Ciconiiformes, Falconiformes, Piciformes, Passeriformes,* swifts, owls and others); "Platzhocker" (e.g. gulls) (see Precocial) are intermediate between altricial and precocial. After leaving the nest, the young are led by the parent birds, and this serves to supplement instinctive behavior with learning.

Pair bonding and territorial behavior. Relations between older and young birds, between the young of the same brood and between neighbors, etc. are based on specific patterns of behavior, postures and vocal signals, a multiplicity of communication signals which is specific for and understood within each species. Further specific behavioral patterns are also associated with feeding and protection, etc. Migration between breeding grounds and overwintering areas has been intensively studied. The geographical orientation of birds is not fully understood. It has been claimed that bird migration resulted from the occurrence of the Ice Age, but this does not seem to be the case.

Economic importance. The economic importance of birds varies with the species and the geographical location. Laws governing the hunting and legal protection of birds are different in all countries. Birds are a source of meat, feathers (down, ornamentation) and eggs. They eat insect pests, and they damage cereal crops and fruit orchards. Many humans take pleasure in bird song and in the spendidly colored plumage of many species. As a result of hunting by humans or alteration of the environment, many species are threatened with extinction (see Red book), while others multiply strongly in areas under human influence (e.g. gulls and many other species). The bird population (species and numbers) of an area can be used as an environmental indicator (presence or absence of pollution, etc.). As a part of nature conservancy and environmental protection programs, birds are generally protected in all countries of the world.

Classification. The class is divided into the following 28 extant orders (from the most primitive to the most advanced), each of which is listed separately: *Struthioniformes, Rheiformes, Casuariiformes, Apterygiformes, Tinamiformes, Sphenisciformes, Gaviiformes, Podicipediformes, Procellariiformes, Pelicaniformes, Ciconiiformes, Anseriformes, Falconiformes, Galliformes, Opisthocomiformes, Gruiformes, Charadriiformes, Columbiformes, Psittaciformes, Cuculiformes, Strigiformes, Caprimulgiformes, Apodiformes, Coliiformes, Trogoniformes, Coraciiformes, Piciformes, Passeriformes.*

Avocets: see *Recurvirostridae*.

Avoidance behavior, ***aversion:*** motivated behavior involving rejection or avoidance of a particular object or any type of environmental stimulus or pattern of stimuli. It is the opposite of Appetitive behavior (see). In a cycle of motivated behavior, A.b. follows the consummation of appetitive behavior. Thus, A.b. arises in relation to food when the organism is satiated. In this sense, A.b. can only exist toward objects or conditions which at other times are the subject of appetitive behavior.

Avoidance conditioning, ***conditioned aversion, aversive conditioning, learned avoidance behavior, avoidance learning:*** a type of conditioning, in which species-specific behavior becomes linked to stimuli or patterns of stimuli that have negative associations A.c. develops when an environmental factor previously considered to be neutral or positive acquires a negative character, i.e. becomes associated with pain or fear; this particular factor is then subsequently avoided. For example, birds learn to avoid eating Monarch butterflies *(Danaus plexippus),* because they contain emetic poisons. In contrast, if the experience of the environmental factor is positive, an appetite is generated, i.e. positive conditioning occurs.

Avoidance learning: see Avoidance conditioning.

Awn, ***beard, arista:*** a bristle-like structure projecting from the glume of grasses and the fruits of members of the *Geranaceae*. It may be smooth, serrated, angled or twisted, and it may be located at the base, on the back, or at the tip of the glume. In most cereal species, the A. is smooth. The length is variable, and it often exceeds that of the fruit or glume. Some As. undergo hygroscopic movements, which serve to bury the fruit in damp earth.

Axeny: the failure of an organism to serve as the host of a parasite, e.g. to serve as the host of a disease-causing bacterium or fungus. A. may be due to anatomical/morphological factors, or to physiological properties of the nonhost. A. is frequently considered to be a type of Passive resistance (see).

Axial organ, ***axial gland, axial complex:*** see *Echinodermata*.

Axial skeleton: the longitudinal supporting element of the endoskeleton of chordates, lying dorsal to the alimentary tract. *Cephalochordata* and *Tunicata* have no skull, or cartilagenous or bony skeleton, but they possess the rudiment of an endoskeleton, known as the *Chorda dorsalis* or *notochord*. A similar structure, also called the notochord, is formed during the embryonic development of the vertebrates. The notochord is a rod-shaped structure, made up of vacuolated cells surrounded by two layers of connective tissue (the outer elastica externa, and the inner, thicker fibrous sheath), which together form the notochordal sheath. Due to the turgor pressur of the vacuolated cells against the sheath, the notochord is stiff but flexible. In addition to its mechanical function, the notochord is also the site for the subsequent formation of the regularly segmented axial structures, such as vertebrae and dorsal muscles. In vertebrates, this process starts with the formation of metameric blocks of connective tissue, called *sclerotomes*. The sclerotomes give rise to a skeletogenous layer, which eventually forms an axial tube of connective tissue enclosing the notochord on all sides. The skeletogenous layer rapidly becomes cartilagenous and forms the *vertebral column*. In elasmobranch fishes, this process proceeds no further, and the vertebral column remains cartilagenous in the adult. In most vertebrates, the segmented vertebral column becomes ossified, producing a bony A.s., which supplants or encloses the notochord. Residual notochord may remain between the vertebrae (in mammals in the intervertebral disks) or within the vertebrae (e.g. in reptiles).

In the more advanced vertebrates, the A.s. is divided into *cervical*, *thoracic*, *lumbar*, *sacral* and *caudal* regions. The first two cervical vertebrae, the *atlas* and *axis*, are structurally specialized, enabling them to carry the head and permit rotation of the skull on the vertebral column. The sacral vertebrae are usually joined in the pelvis, and they may fuse to form the *sacrum*. The number of vertebrae in a vertebral column varies from 9 in the frog to more than 200 in snakes. Humans have 7 cervical, 12 dorsal and 5 lumbar vertebrae, and the last 9 vertebrae are fused to form the sacrum and coccyx.

Each *vertebra* is a cartilagenous or bony unit of the vertebral column. It consists of a cylindrical centrum, carrying a dorsal neural vertebral arch and various other appendages. Depending on the contors of their end surfaces, vertebrae are classified as *amphic(o)elic* (both ends concave, e.g. fish and certain reptiles), *proc(o)elic* (anteriorly concave and posteriorly convex, e.g. salamanders and some dinosaurs) or *ac(o)elic* (flattened ends, e.g. mammals). With the exception of the atlas and axis, the vertebrae are not interconnected by rotatable joints, but rather by intervertebral disks. The neural arches of the vertebrae form the *neural* or *spinal canal*, which houses the spinal cord. Gaps between neighboring neural arches (nerve foramina, *sing*. foramen) provide exit and entry for spinal nerves. The dorsal appendage of the neural arch is called the *neural spine*. In quadruped vertebrates, the neural arches of neighboring vertebrae are articulated via structures called *zygapophyses*. Each side of the neural arch often carries a transverse appendage called the *diapophysis* or *transverse process*. In the thoracic region of the vertebral column, the vertebrae possess a ventral transverse appendage, the *parapophysis*, which usually only presents a surface for articulation of the capitula of the ribs. A ventral vertebral arch, known as the *hemal arch*, is usually found only in the tail region; in fish it also occurs in the body region, where it encloses large blood vessels. The hemal arch may also carry a ventral spine. The caudal vertebrae of urodeles, birds and higher apes are reduced. In frogs, the caudal region has developed into a single elongated bone, called the *urostyle*, whereas in birds the anterior caudal vertebrae are fused into a *coccyx* or *pygostyle*, which carries the tail feathers.

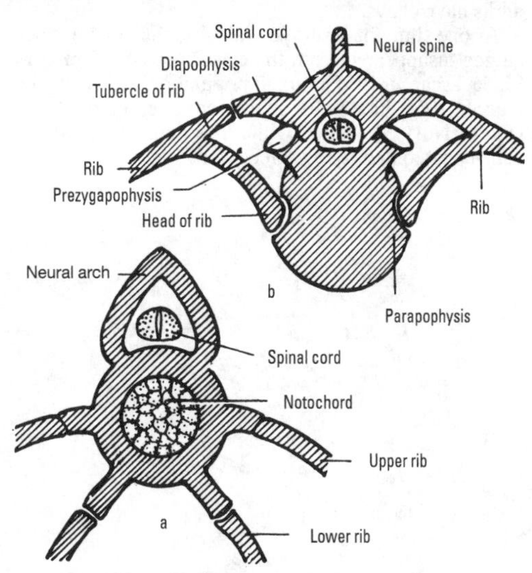

Axial skeleton. Transverse sections of Vertebrae with ribs (schematic). *a* Fish vertebra. *b* Quadruped vertebra.

Axis: see Axial skeleton.

Axis axis: see Chital.

Axis deer: see Chital.

Axocoelom: see Coelom, Mesoblast.

Axolemma: see Neuron.

Axolotl: see *Ambystomatidae*

Axon: see Nerve fiber, Neuron.

Axoneme: see Cilia and Flagella.

Axon hillock: see Nerve impulse frequency, Neuron.

Axon reflex: see Circulatory regulation.

Axopodia: fine, radiating, unbranched extensions of the cytoplasm of Acantharians and Heliozoans. Each axopodium is stiffened by a central axial rod, which consists of packed microtubules.

Axostyle: an organelle providing mechanical support in the cells of certain zooflagellates (e.g. *Trichomonas).* It arises from the basal body, and runs internally and longitudinally the entire length of the cell.

Aye-aye, *Daubentonia madagascariensis:* the only species of the superfamily *Daubentonioidea* (see Primates, Prosimians), and native only to Madagascar. It is cat-sized, with uniformly dark brown shaggy fur, large naked ears, and a body-length extremely bushy tail. The strikingly long, thin middle finger is used to scrape insect grubs from crevices and to remove the pith and flesh of plants and fruits, e.g. the flesh of unripe coconuts is scooped through a hole gnawed in the shell. When the presence of a wood-boring grub is detected by hearing (or smell?), the Aye-aye rapidly gnaws the wood until it can insert its middle finger and squash the grub; the quarry is then removed in pieces, which are licked from the middle finger. The Aye-aye thus appears to fulfil the same ecological role as woodpeckers, which are absent from Madagascar. Only 18 teeth are present (most prosimians have 36), and the typical prosimian dental comb is lacking. The teeth grow continuously, and for this reason the Aye-aye was previously classified as a rodent. The big toes carry nails, but all the other digits have claws.

At one time the Aye-aye was thought to be extinct, having disappeared from the eastern coastal forests of Madagascar. In 1957, a small population was rediscovered. In 1966, nine individuals were released on the island of Nosy Mangabe, which has dense forest cover and appeared to offer a protective habitat.

Aysheaia pedunculata: see *Onychophora.*

Aythya: see *Anseriformes.*

Aythyini: see *Anseriformes.*

Azalea: see *Ericaceae.*

Azoic period: the period in the evolution of the earth before the appearance of living organisms. The first life forms appeared 3 500–4 000 million years ago.

Azonal vegetation: see Zonal vegetation.

Azotobacter: a genus of aerobic Gram-negative bacteria, present in soil and water, usually motile by numerous cilia, and able to fix atmospheric nitrogen. The oval to round, 5 μm diam. cells frequently occur in pairs, and may form capsules and thick-walled cysts. *A.* species are agriculturally important, because they increase the soil content of bound nitrogen available for plant growth. The commonest species is *A. chroococcum.*

Aztec thrush: see *Turdinae.*

Azulenes: blue to violet, unsaturated, bicyclic sesquiterpene hydrocarbons with anti-inflammatory properties, produced nonbiologically from colorless proazulenes, which occur in some essential oils. Thus, camomile and yarrow contain a colorless proazulene precursor, which is converted by steam distillation into chamazulene.

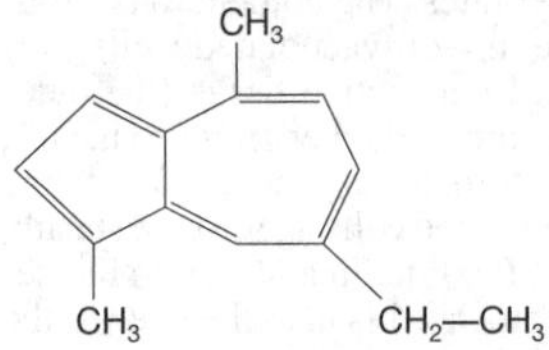

CH_3

CH_3 CH_2—CH_3 *Chamazulene*

B

Babblers: see *Timaliinae.*
Babesian: see Piroplasmids.
Babesidae: see Piroplasmids.

Babirussa, *Babyrousa babyrussa:* a monotypic species of pig (see *Suidae)* found in northern Celebes and the islands of Sulawesi and Buru, where it inhabits moist forests and the shores of lakes and rivers. Between 65 and 80 cm at the shoulder, it has relatively long legs, and the rough, brown-gray skin has a very sparse covering of short, gray-yellow hair. The 4 canine tusks of the male curve upward, and are often so long that they reach the forehead. Dental formula: 2/3, 1/1, 2/2, 3/3 = 34. The B. is hunted for its meat, and it is also tamed and kept in captivity. Nocturnal in habit, it runs swiftly, and swims well, often swimming in the sea to neighboring islands.

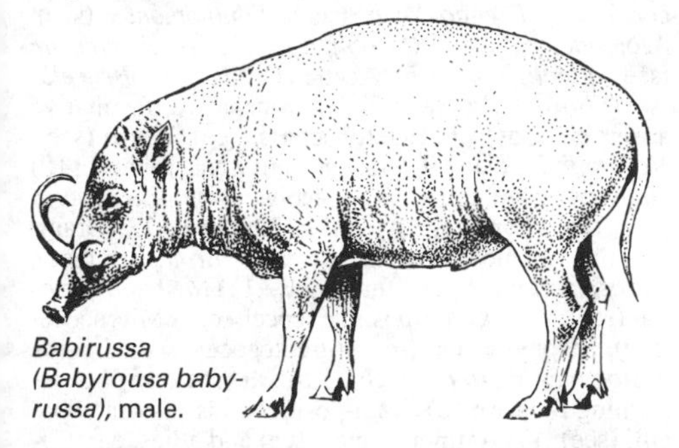

Babirussa (Babyrousa babyrussa), male.

Baboons, *Papio* (plate 41): powerfully built Old World monkeys (see) with a dog-like, long face, prominent muzzle and strong teeth and jaws. Social groups of up to 200, but usually fewer (40–100), inhabit nonforested areas of sub-Saharan Africa and southwestern Arabia. They show no fear of man or even leopards. The diet is very varied, consisting of all forms of vegetation and fruits, insects, small mammals, birds, birds' eggs, roots and even crabs on the sea shore. The genus is represented by 5 species: *P. papio* ***(Guinea baboon;*** reddish coat), *P. anubis* ***(Olive baboon;*** dark olive-gray coat), *P. cynocephalus* ***(Yellow baboon;*** yellowish coat), *P. hamadrayas* ***(Hamadrayas baboon, Sacred baboon),*** *P. ursinus* ***(Chacma baboon;*** dark brown, almost black coat). Hamadrayas B. of Ethiopia live in large herds of up to 750 members, consisting of many groups of up to 12 adult females with a single male. Males are silver-gray with a long-haired shoulder cape; the female is olive brown.

Babyrousa babyrussa: see Babirusa.
Bacillariales: see Diatoms.
Bacillariophyceae: see Diatoms.
Bacillariophyta: see Diatoms.

Bacillus (Latin *bacillus* stick or staff):

1) A genus of spore-forming, aerobic, rod-shaped (about 1.5 μm diameter and up to10 μm in length), usually motile, Gram-positive bacteria. Under appropriate environmental conditions, almost all bacilli form spores (see Spores, Bacterial spores), which are very resistant to high temperatures and desiccation. Most species occur in soil, e.g. the type species of the genus, *Bacillus subtillis* (rods up to 3 μm long), which has a very wide natural distribution, especially in soil. *B. subtilis* is also notably prevalent on hay. The population of organisms in warm water infusions of hay contain a high proportion of *B. subtilis,* which has therefore been called the "hay bacillus". There are, however, several "hay bacilli": *B. subtilis, B.cereus, B. mycoides, B. vulgatus* and *B. mesenteroides,* all of them very prominent in hay infusions. *B. cereus* v. *mycoides* is a common nonmotile soil species (up to 5 μm long), which on solid culture media displays spreading grayish growth resembling a branched root system. *B. anthracis,* the only bacillus pathogenic to man, is the causative agent of anthrax, which is primarily a disease of farm animals, but is also transmissible to humans. Its rod-shaped cells are up to 5 μm in length, and its spores remain viable in soil for decades. When growing parasitically in animals, *B. anthracis* does not form spores, but develops capsules. With his discovery of *B. anthracis,* Robert Koch provided the first proof that bacteria can cause disease. *B. thuringiensis* is pathogenic for insects and is used in biological control. *B. coagulans* is sometimes responsible for spoilage of canned foods. *B. alvei* is one of several organisms associated with foulbrood of bees in Europe. *B. larvae* is associated with American foulbrood. Most baciili are, however, harmless saprophytes. Some yield antibiotics.

2) A general morphological rather than systematic term for rod-shaped or cylindrical bacteria, often simply described as "rods". They occur separately, in pairs, or in chains, and occur in various systematic groupings. The cell wall appears to be a fairly rigid tube or inelastic sac, since growth occurs mainly lengthwise. Some are so short that it is difficult to distinguish them from cocci, but some cells in the same culture are usually somewhat longer, thereby revealing their bacillary nature; such bacilli are often called *cocco-bacilli. Fusiform bacilli* are long and thin, and pointed at both ends.

Bacillus. Different forms of rod-shaped and cylindrical bacteria.

Bacitracins: Six polypeptide antibiotics (A, B, C, D, E and F) produced by various strains of *Bacillus licheniformis*. The most important member of the group is bacitracin A., which contains an unusual thiazoline structure derived from the N-terminal isoleucine and neighboring cysteine. Bacitracin F is a relatively inactive rearrangement product of bactitracin A. B. are used in human and veterinary medicine, primarily against infections due to Gram-positive bacteria. They inhibit cell wall peptidoglycan synthesis in Gram-positive bacteria.

Backswimmers: see *Heteroptera*.

Bacteria: prokaryotic, unicellular microorganisms. Bacterial cells are spherical, rod-shaped, comma-shaped or spirally twisted, rarely branched, and they vary in size from 0.5 to 10 μm. They may exist as separate cells, or adhere to one another in chains or groups, etc. Multiplication occurs by binary fission, and in a minority of B. by budding. Some B. are also motile, possessing flagella (see Bacterial flagella), or contractile, longitudinal elements beneath the cell wall. Some are also capable of gliding movement on a substratum. Most B. are, however, nonmotile. Spore-forming B. (see) produce characteristic Spores (see) in the interior of the cell, some of which are highly resistant resting cells. A few species produce Cysts (see). As prokaryotes, B. do not possess a nuclear membrane. The genetic material (DNA and associated proteins) lies directly in the cytoplasm. There are no mitochondria or plasmids, and the functions performed by these organelles in eukaryotes are performed in B. by the cytoplasmic membrane and related structures (mesosomes and chromatophores). Small ribosomes (70S-ribosomes) are present in the cytoplasm. Most B. have a constant cell shape, which is maintained by a rigid cell wall, consisting almost exclusively of murein (peptidoglycan). Some B. possess an outer capsule of polysaccharide, polypeptide or mucilage (Fig.).

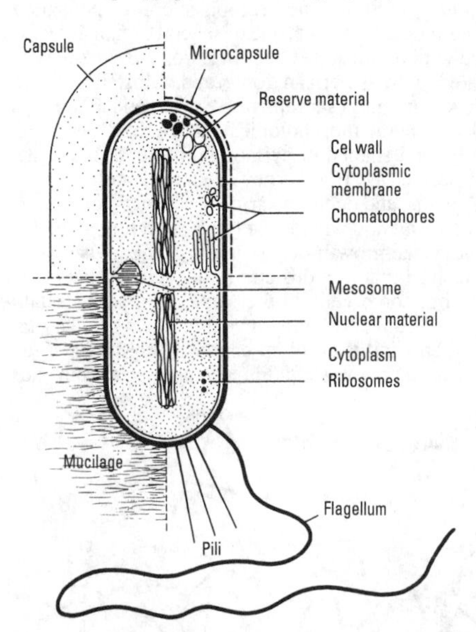

Bacteria. Highly generalized diagram of a bacterial cell, incorporating structures from different types of bacteria.

Most B. are heterotrophic, i.e. they live saprophytically on dead organic material, or as parasites in other living organisms. The parasitic B. include causative agents of disease in humans, animals and plants. Various B. produce highly dangerous toxins (see Bacterial toxins). Autotrophic B. perform chemosynthesis or photosynthesis. Some B. are Anaerobes (see).

B. occur practically everywhere on the planet, especially in soil and in water, but also in the air and in a variety of seemingly very unfavorable habitats. Since they occur naturally in extremely large numbers, and are able to degrade most naturally occurring organic compounds, B. play an essential part in natural recycling processes. Many B. are exploited by humans (see Industrial microbiology).

Since classification on the basis of natural affinities is not possible, B. are grouped according to their morphology and physiological properties. Nineteen groups are recognized: 1) photoautotrophic B., i.e. Rhodobacteria (see) and *Chlorobacteriaceae* (see); 2) gliding B., e.g. *Beggiatoa* (see) and Myxobacteria (see); 3) *Chlamydobacteriaceae* (see); 4) stalked B. and B. that reproduce by budding, e.g. *Hyphomicrobium* (see), *Caulobacter* (see), *Gallionella* (see); 5) Spirochetes (see); 6) Spirally shaped and curved B., e.g. *Spirillum* (see), *Bdellovibrio* (see); 7) Gram-negative aerobic rods and cocci, e.g. *Pseudomonas* (see), *Xanthomonas* (see), *Azotobacter* (see), *Rhizobium* (see), *Agrobacterium* (see), *Alcaligenes* (see), Acetic acid B. (see), *Brucella* (see), *Bordatella* (see); 8) Gram-negative, facultative anaerobic rods, e.g. Enterobacteria (see), *Vibrio* (see), *Haemophilus* (see); 9) Gram-negative anaerobic B.; 10) Gram-negative cocci and coccobacilli, e.g. *Neisseria* (see); 11) Gram-negative anaerobic cocci; 12) Gram-negative chemoautotrophic B., e.g. *Nitrobacter* (see), *Nitrosomonas* (see), sulfurizing B.; 13) Methane bacteria (see); 14) Gram-positive cocci, e.g. Micrococci (see), Staphylococci (see), Streptococci (see), *Leuconostoc mesenteroides* (see), *Sarcina* (see); 15) Spore-forming B. (see); 16) Gram-positive rods, e.g. Lactobacilli (see); 17) Actinomycetes (see) and allies, e.g. *Corynebacterium* (see), Propionic B. (see); 18) Rickettsias (see); 19) Mycoplasmas (see).

Bacterial culture: see Culture of microorganisms.

Bacterial flagella: organelles responsible for the active motility of many bacteria. They are composed of a protein called flagellin, and they are structurally and functionally different from eukaryotic flagella. The filament of the bacterial flagellum is15–20 μm in length, with a diameter of 12–25 nm (120–250 Å). Like eukaryotic tubulin, the flagellin molecules polymerize into strands or subfibrils, and between 3 and 11 of these are arranged helically to form the filament, which is a hollow cylinder (Fig.). Flagellin has no ATPase activity.

The base of the flagellum is known as the hook; it is formed by the aggregation of identical protein subunits, and it is rigid and bent. The hook connects the filament to the basal structure, which is inserted in the cell wall and cytoplasmic membrane. The basal structure rotates (and with it the flagellum), either clockwise or counterclockwise (about 100 revolutions per second). In *Escherichia coli,* counterclockwise rotation causes several flagella to wind together in a rotating bundle, which propels the cell forward. Clockwise rotation unwinds the bundle, and the cell displays erratic tumbling movement. Operating like a propeller, the rotating flagellum

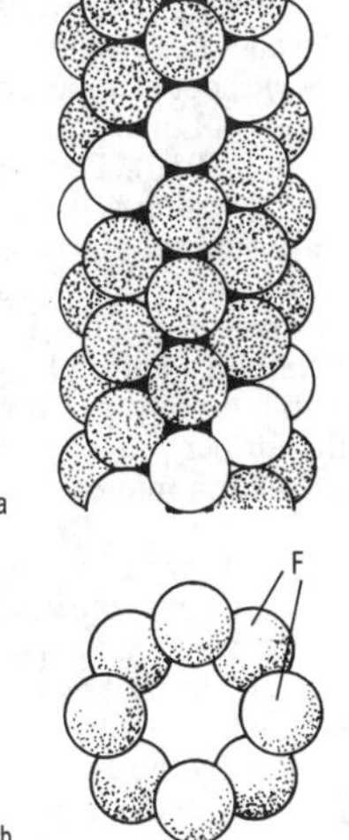

Fig. 1. *Bacterial flagella.* Model of the fine structure: *a* longitudinal, *b* transverse. *F* represents a single flagellin molecule, which is actually ellipsoid rather than spherical.

generates traction or thrust. The basal structure (composed of at least 10 different proteins) acts as a rotary motor. It consists of disks surrounding a thin rod; the L and P disks in the outer membrane and peptidoglycan layer act as a "bearing" for the motor, while the S and M disks act as "stator" and "rotor", repectively. Thus, during rotation of the flagellum, the L, P and S disks remain stationary, while the rod (connected via the hook to the flagellum) and the M disk rotate. Energy for rotation of the basal structure is derived from a proton gradient across the cell membrane, and there appears to be a charge interaction between the M and S disks.

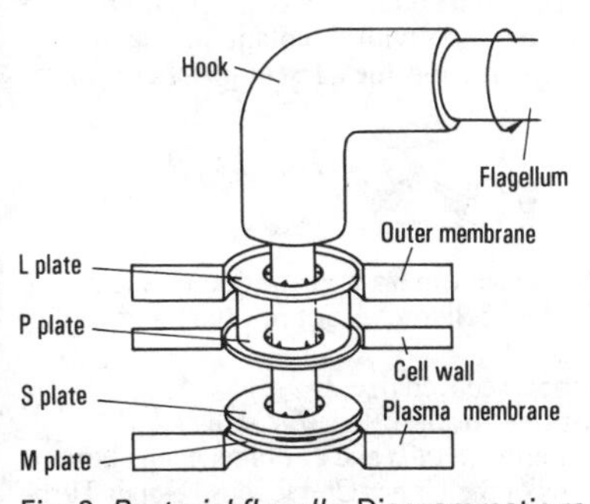

Fig. 2. *Bacterial flagella.* Diagrammatic representation of a flagellum and its mode of insertion into the cell wall and cell membrane.

The number and arrangement of flagella is largely specific for particular bacterial species. Cells may possess a single flagellum *(montrichous cells),* or several flagella *(polytrichous).* In special cases, bundles of flagella may be present *(lophotrichous).* Flagella may be present at only one end or pole of the cell *(polar, monopolar* or *unipolar* flagella), at both ends *(amphitrichous* or *bipolar),* or in circles around the cell *(peritrichous).*

Bacterial oxidation of molecular hydrogen: a property of certain soil bacteria, known in German as "Knallgasbakterien". These organisms can live autotrophically by oxidizing hydrogen as a sole source of energy, while deriving their carbon exclusively from carbon dioxide. They possess hydrogenases, which enable them to oxidize free hydrogen to water: $2H_2O + O_2 \rightarrow 2H_2O$, part of the free energy of this reaction being used for chemosynthesis. These organisms are not obligate autotrophs, and they can live heterotrophically by respiring carbohydrate. The first Knallgasbakterium to be described was *Bacillus pantotrophus,* discovered in 1906 by Kaserer by placing soil inocula in flasks containing an inorganic medium and filled with a mixture of air, carbon dioxide and hydrogen. Pseudomonads are also well known for their ability to oxidize hydrogen.

Bacterial spores, ***endospores:*** highly durable, refractile bodies formed in the cell interior of spore-forming bacteria (primarily by species of *Bacillus* and *Clostridium,* as well as a few cocci and spirilla). Endospores have a low water content, and they contain cytoplasm and nuclear material. Some of them withstand boiling for several hours, and even several minutes at 130 °C. In dry soil, they remain viable for more than 300 years.

Spore formation is initiated by both environmental and endogenous factors. Environmental factors favoring spore formation are low concentrations of nutrients, high concentrations of oxygen, and the presence of certain cations, especially Mn^{2+} and K^+. Usually, only a single endospore is formed within a bacterial cell, which is then called the sporangium. Within this sporangium, the spore genome, together with some sporangial cytoplasm, becomes enclosed by a thin septum. The process therefore resembles an internal cell division, in which the larger of the 2 daughter cells encloses the smaller, the latter being transformed into a spore. As the spore develops, other walls are laid down, their number and thickness varying according to the bacterial species and strain. A membranous germ cell wall (enclosing nuclear material and cytoplasm), followed by a relatively thick and loosely cross-linked layer known as the cortex, are general structural features of all endospores. In some B.s. (e.g. those of *Bacillus coagulans),* the cortex is followed by a laminated inner coat and an electron-dense outer coat. In others (e.g. *Bacillus cereus),* however, the cortex is surrounded by a non-electron-dense, 1- or 2-layered inner coat, and this is enveloped by an outer, thin-walled, loose layer, known as the *exosporium.* Under favorable conditions, the spore germinates with fracture of the cortex and other outer layers, and develops into a vegetative bacterium.

Bacterial toxins: toxins formed by bacteria. They are subclassified into exotoxins and endotoxins. All the B.t. mentioned below are proteins, usually of known amino acid sequence.

Exotoxins, which by definition are secreted into the surrounding medium, are exceptionally poisonous and have very specific modes of action. Five exotoxins are secreted by *Staphylococcus.* In staphylococcal food poisoning, the toxins are secreted in the gastrointestinal tract (they are therefore also called *enterotoxins),* causing diarrhea and vomitting. *Diphtheria toxin* from *Corynebacterium* inactivates peptidyl transferase II in eukaryotic cells by promoting the attachment of ADP-ribose to the enzyme. *Tetanus toxin* from *Clostridium tetani* is a neurotoxin; in the mouse, 0.01 ng/kg body weight is fatal. The five highly toxic *Botulinus toxins* from *Clostridium botulinum* are also neurotoxins. Type A *Botulinus* toxin has been shown to bind specifically to the membranes of peripheral nerve terminals, where it irreversibly inhibits the release of acetylcholine.

Endotoxins are relatively heat-stable, and are released by autolysis after the death of the bacteria. They are produced by various Gram-negative intestinal bacteria, e.g. *Salmonella* and *Shigella*. *Salmonella* is well known as a causative agent of violent food poisoning. *Cholera toxins* are endotoxins released from the Gram-negative *Vibrio cholerae* in the intestine. They are composed of 2 functionally different subunits, L and H. The L-subunit binds with high affinity to gangliosides on the membranes of nerve cells, adipocytes, erythrocytes, etc., permitting entry into the cell of the H-subunit, which is responsible for the toxicity *per se*. The *colicins* are endotoxins produced by *Escherichia coli* and capable of killing other intestinal bacteria. Their toxicity is due to inhibition of cell division, and inhibition of DNA and RNA degradation (colicin E_2), or to inhibition of protein biosynthesis by inactivation of the 30S ribosomal subunit (colicin E_3). See Bacteriocins.

Bacteriochlorophylls: chlorophylls of photosynthetic bacteria. The main chlorophyll of purple sulfur bacteria *(Thiorhodaceae)*, purple nonsulfur bacteria *(Athiorhodaciae)* and green sulfur bacteria *(Chlorobacteriaceae)* is related to chlorophyll *a*, and is known as bacteriochlorophyll *a*. It contains one double bond fewer than chlorophyll *a* and possesses an acetyl group in place of the vinyl group, i.e. it is 2-acetyl-2-devinyl-dihydrochlorophyll *a*. *Thiococcus* and a single strain of *Rhodopseudomonas* (family *Athiorhodaceae)* contain bacteriochlorophyll *b*, while bacteriochlorophylls *c* and *d* (in which the phytol ester is replaced by a farnesol ester) are found in green sulfur bacteria in addition to bacteriochlorophyll *a*. See Chlorophyll.

Bacteriocidal activity: the killing of bacteria. Sterilization procedures with disinfectants, many antibiotics, and ultraviolet irradiation are bacteriocidal. Cf. Bacteriostatic activity.

Bacteriocides, bacteriocidal agents: in the wider sense, all substances that kill bacteria, such as many antibiotics and most disinfectants. In a more restricted sense, the term is applied to plant protection agents for controlling bacterial diseases of plants.

Bacteriocidins: antibodies that bind to bacteria and thereby inhibit or impair their development.

Bacteriocins: proteins produced by bacteria, which are toxic to other bacteria. A single bacteriocin protein is toxic for only a small group of closely related bacteria. B. are therefore much more specific than antibiotics. B. are produced only by bacteriocinogenic bacteria, i.e. bacteria that contain plasmids carrying genes for bacteriocin synthesis. These plasmids, and therefore the ability to synthesize B., can be transferred to other bacteria. Some strains of *Escherichia coli* produce B., which are known as *colicins*. The corresponding genes are known as *colicinogenic factors*. See Bacterial toxins.

Bacteriology: the study of bacteria, and part of the wider field of microbiology. B. is concerned with the structure and physiology of bacteria, their role in natural processes and cycles of the biosphere, and the way in which humans, animals and plants are affected by their useful or deleterious activities. B. is therefore important in human and veterinary medicine, phytopathology, environmental protection, hydrobiology, agriculture, and industrial microbiology.

B. became an independent discipline in the second half of the 19th century. Bacteria were discovered as early as 1675 by Antonie van Leeuwenhoek (1632–1723), but their significance was not recognized until later. B. was founded by the work of Louis Pasteur (1822–1895) and Robert Koch (1843–1910). Pasteur demonstrated that the lactic acid fermentation is due to bacteria, and that bacteria do not arise by spontaneous creation from dead material. Koch identified various bacteria as the causative agents of disease, e.g. the anthrax bacillus, the tuberculosis bacterium and the cholera bacterium. Developments in microscopy made a very significant contribution to the further progress of B. From about 1940 onwards the electron microscope enabled further important advances.

Bacteriolysins: specific antibodies formed by animals after exposure to bacteria. Thereafter, the B. circulate in the blood. Upon re-exposure (i.e. to the same species of bacterium), the bacteria are immobilized by the antibodies, and the infection is curtailed.

Bacteriolysis: lysis of bacteria, e.g. by the enzyme lysozyme, or by Bacteriolysins (see). Immunobacteriolysis is due to specific antibodies, which bind to the bacteria, then activate the Complement system (see) of the serum. The complement system then attacks and destroys the bacteria.

Bacteriostatic activity, *bacteriostasis:* inhibition of the multiplication of bacteria, without killing them. Many antibiotics and preservatives are bacteriostatic. Cf. Bacteriocidal activity (see).

Bacteroids: involution forms of bacteria, i.e. forms that differ in shape and size from the normal cell. B. arise inder the influence of certain environmental conditions. Most familiar are the irregular, swollen or lobed B. of *Rhizobium* (see), found in the Root nodules (see) of leguminous plants. Free-living rhizobia are rod-shaped. Within the root nodule, the B. of *Rhizobium* live in symbiotic association with the plant, and as part of this symbiotic interaction they fix atmospheric nitrogen.

Bacteroids. Stages in the transformation of a rod-shaped *Rhizobium* cell into a root nodule bacteroid.

Bactrian deer: see Red deer.

Bactrian house mouse: see *Muridae*.

Baculites (Latin *baculus* stick or baton): an aberrant ammonite genus of the Upper Cretaceous (Upper Turonian to Maastrichian). The shells are mostly straight and rodlike in shape. The blind end of the shell forms a small planispiral coil of 2 whorls, but in the majority of fossils this is rarely preserved. The surface of the shell is smooth or finely ribbed, and tuberosities may also be present. With regard to the shape of the shell and its type of ribbing, *B.* is a typical degenerate form of ammonite.

Baerman funnel: an apparatus for isolating nematodes. See Soil organisms.

Baer's law: the principle established by Karl Ernst von Baer in 1828 that the common characteristics of the members of large systematic units appear at an early stage of embryonic development, whereas the more specialized features, which distinguish one member from another and permit the construction of ever smaller taxa, appear at a later stage.

B.l. was formulated before the theory of evolution became generally accepted. The biogenetic law of Müller (1864) and Haeckel (1868) represents a reinterpretation of B.l. in the light of evolutionary theory. Thus, the characteristics developing earliest in ontogeny are inherited from the common ancestor of the entire group, whereas the features appearing later are later evolutionary acquisitions, i.e. the embryo undergoes a resumé of its evolutionary history, or ontogeny displays a recapitulation of phylogeny. The biogenetic law explains, for example, the occurrence of gills during the embryonic development of reptiles, birds and mammals. For the reconstruction of phylogenetic relationships, however, the law has only limited value. See Caenogenesis, Palinogenesis.

Bagworm moths: moths of the family *Psychidae;* see *Lepidoptera.*

Bainbridge's reflex: see Cardiac reflexes.

Baker's yeast: a cultivated yeast used for raising dough in baking. It is a selected strain of *Saccharomyces cerevisiae,* characterized by high carbon dioxide production, heat stability, and good keeping properties. When incorporated into a dough, B.y. performs a fermentation in the presence of sugar, producing ethanol and carbon dioxide, the latter serving to raise the dough. B.y. is grown industrially in culture media containing molasses, washed and traditionally pressed into cakes, and sold as *pressed* or *compressed yeast.* Pressed yeast is ready for immediate use; it is perishable and should be stored at 0–4 °C, when it has a shelf life of 3–4 weeks. Alternatively, the yeast may be washed, pressed and carefully dried at 25–30 °C to produce *dried yeast.* Careful drying prevents loss of enzymatic activity and reduces autolysis to a minimum. Dried yeast contains about 30% living cells. *Active dried yeast* is supplied in airtight packages sealed under vacuum or with an inert gas such as nitrogen. It consists of more or less spherical granules with a water content of 6–8%, and in the sealed pack at room temperature it has a shelf life of at least one year. *Instant dried yeast,* prepared by a quick drying process (fluidized bed), contains 4–6% water, and an added emulsifier (e.g. sorbitan monostearate), and has a leavening activity close to that of active dried yeast. It consists of small, highly porous rods, which are easy to rehydrate.

Balaena: see Whales.

Balaeniceps: see *Ciconiiformes.*

Balaenicipitidae: see *Ciconiiformes.*

Balaenidae: see Whales.

Balaenopteridea: see Whales.

Balanced lethal stocks: genotypes of *Drosophila* constructed by H.J. Muller (see Muller's method), involving two homologous chromosomes, each carrying a different homozygously lethal gene, as well as different overlapping inversions (which impede recombination between the homologous chromosomes). For example, a genotype has been constructed in which chromosome II carries the homozygously lethal gene Pm (plum for eye color), while its homologous partner carries the homozygously lethal gene Cy (curly for wing shape), i.e. Pm^+/Cy^+. Crossing within this balanced lethal stock therefore produces only Pm^+/Cy^+ individuals (although Cy^+/Cy^+ and Pm^+/Pm^+ genotypes are formed, they are nonviable).

Balance model of population structure: a proposed model for the effect of selection on the gene pool of natural populations. According to this model, most gene loci in the individuals of a natural population are occupied by two different alleles, because Fitness (see) increases with the degree of heterozygosity. Homozygous loci are nevertheless present, due to segregation according to Mendel's second law. See Classical model of population structure.

Balanomorpha: see *Cirripedia.*

Balanus: see *Cirripedia.*

Balbiani rings: see Giant chromosomes.

Bald-eagle: see *Falconiformes.*

Baldwin effect: an evolutionary principle proposed in 1896 by J.M. Baldwin, designed to reconcile Lamarckism with Weismann's Neo-Darwinism. According to this principle, the phenotype (but initially not the genotype) becomes modified to the new environmental conditions, so that the organism can remain in a favorable environment, thereby allowing time for selection and the genetic fixation of the phenotype. This "organic selection" was considered by Baldwin to be an alternative to natural selection, which he discounted as an evolutionary force. This hypothesis is now largely rejected. In fact, if a phenotype is plastic, this may actually reduce selection pressure, because there is no selective advantage in changing the genotype.

Balearic: see *Gruiformes.*

Baleen: see Whales.

Bali cattle: see *Bovinae.*

Balistidae, *leatherjackets:* a family of marine teleost fishes (order *Tetraodontiformes),* containing about 135 species in about 42 genera, found in the Atlantic, Indian and Pacific Oceans. The body is usually compressed and high-backed, with a large head and small mouth. The first spine of the dorsal fin has a locking mechanism formed by the second spine. Many species are splendidly colored, and are often displayed in marine aquaria. Pelvic fins are absent, but a pelvic spine or tubercle is present in the balistines and some monacanthines. The nonprotractile upper jaw has two rows of protruding incisor-like teeth. The eyes can be rotated independently. Two subfamilies are recognized.

1. Subfamily: *Balistinae* ***(Triggerfishes).*** Three dorsal spines are present, although the third may be minute.

2. Subfamily: *Monacanthinae* ***(Filefishes).*** Two dorsal spines are usually present, the second often much reduced or absent. About 95 species (in 31 genera) are known, and 54 of these are found in Australian waters.

Balistinae: see *Balistidae.*

Ballast material:

1) Superfluous material taken up by plants, which has no nutritional importance. Such material is often eliminated, either by secretion externally or by deposition in the plant in an insoluble form, e.g. Ca^{2+} is deposited as crystals of calcium oxalate. Silica is often deposited in cell walls. B.m. is also secreted into vacuoles.

2) Constituents of the diet that cannot be degraded in the gastrointestinal tract, e.g. cellulose.

Ballistospores: fungal spores that are ejected with the secretion of a moisture droplet. The spore is situated unsymmetrically on the end of a fine sterigma, and very close to its point of attachment to the sterigma it has a small projection, known as the hilar appendix. Immediately before discharge, a droplet appears suddenly at the hilar appendix. This droplet grows rapidly in a few seconds to a definite size, then the spore is discharged. The discharge distance is rarely greater than 0.1 cm, and usually 0.01–0.02 cm. B. include some basidiospores of *Hymenomycetes, Tremallales* and *Uredinales* (rusts), as well as the secondary conidia of *Tilletia* and the aerial spores of the mirror picture yeasts *(Sporobolomycetaceae)*.

Balsa: see *Bombacaceae*

Balsaminaceae, *balsam family:* a family of dicotyledonous plants, containing only 2 genera and about 450 species, most of which are native to moist, tropical, forested regions of Africa and Asia. They are herbs with translucent stems. Leaves are simple, and alternate, opposite, or in whorls of 3. Stipules are absent or represented by glands. The 5-merous flowers are hermaphrodite, strongly zygomorphic and usually brightly colored; the crown-like calyx carries a spur. The 5-celled ovary develops into a 5-valved loculicidal capsule, which dehisces elastically and explosively by coiling when disturbed, scattering the numerous seeds widely. Chemically, the family is characterized by the presence of acetylglycerols in its seed oils.

Balsaminaceae. a Garden balsam *(Impatiens balsaminia). b* Small balsam *(Impatiens parviflora).*

Impatiens noli-tangere ***(Touch-me-not)*** is found from Europe to Japan; its hanging golden yellow flowers have red spots inside. The garden weed, *Impatiens parviflora* ***(Small balsam),*** was introduced into Europe from Siberia 150 years ago. In many parts of Europe it is now completely naturalized in woods and other shady places. *Impatiens balsamina* ***(Garden balsam)*** is a cultivated garden plant from Southeast Asia, which shows variations of flower color. The cultivated red-flowered *Impatiens glandulifera* ***(Policeman's helmet)*** attains a height of 2 m; it was introduced from the Himalayas, and is now completely naturalized in the wild in many parts of Europe, growing especially on river banks.

Balsam of Peru: see Peru balsam.

Balsam pear: see *Cucurbitaceae*.

Balsams, *oleoresins:* solutions of resins in volatile oils. B. are produced by plants, either normally or pathologically. They may be obtained by damaging the plant (e.g. incision of the tissues), causing the B. to be extruded from the intercellular, schizogenous oleoresin ducts of the plant tissue. Thus, ***turpentine,*** one of the more important B., is produced by conifers in response to bark injury in amounts of 1–2 kg per tree per year (Bordeaux turpentine is obtained from *Pinus maritima,* Venice turpentine from *Larix europea)*. Steam distillation of the crude balsam yields turpentine oil, while the residue consists of colophony (rosin). Lower grades of turpentine are obtained by direct steam distillation of the wood. Other important B. are ***Peru balsam*** (extruded from the trunk of *Myroxylon pereiroe* after the bark has been beaten and scorched), ***Canada balsam*** (obtained by incision of the bark of the balsam fir, *Abies balsamea,* and the hemlock spruce, *Abies canadensis),* ***Tolu balsam*** (obtained by incision of the bark of *Myroxylon toluifera)*. B. are variously used in perfumes and/or pharmaceuticals.

Baltic basin: see *Gastropoda*.

Bamboo: see *Graminae*.

Bamboo rats: see *Rhizomidae*.

Bambus: see *Graminae*.

Banana: see *Musaceae*.

Bananaquit: see *Parulidae*.

Banded anteater: see Numbat.

Banded thrips: see *Thysanoptera*.

Bandfishes: a goup of phylogenetically unrelated elongated fishes with strongly laterally compressed bodies. Thus, all members of the marine perciform family, *Cepolidae* (found in the eastern Atlantic and the Indo-West Pacific) are known as B. Other B. are the ***Snake blenny*** *[Lumpenus lampretiformis;* northern seas; family *Stichaeidae* (pricklebacks); order *Perciformes];* and the ***King-of-the-herring*** *[Regalecus glesne;* family *Regalecidae* (oarfishes); order *Lampriformes]*. With a maximal length of 10 m, *R. glesne* is one of the longest fishes, and it is possibly responsible for the sea serpent stories of old mariners.

Bandicoots, *Peramelidae:* a family of marsupials found in Australia, Tasmania, New Guinea and some neighboring islands. B. are fairly large mammals, the smaller species being rat-sized, the larger attaining weights of about 7 kg, and their hindlimbs show a trend towards specialization for running or hopping. They are mostly nocturnal and omnivorous. The dentition resembles that of the insectivores.

Banjo frog: see *Myobatrachidae*.

Banteng: see *Bovinae*.

Banyan tree: see *Moraceae*.

Baobab: see *Bombacaceae*.

B/A-quotient: the ratio of Genetic load (see) in a random-mating population to the genetic load in a fully inbred population. In random mating, relatively few disadvantageous genetic factors are expressed phenotypically. Complete homozygosity can be achieved by extreme inbreeding; all recessive, deleterious genes are then manifested in the population. If deleterious genetic factors are retained in the population by Mutation-selection balance (see), i.e. if the genetic load is largely mutational, then B/A values higher than 10 and probably between 25 and 50 can be predicted. If balanced polymorphism with heterozygote advantage is primarily responsible for the retention of recessive deleterious genes, i.e. if the genetic load is largely segregational, then the B/A-quotient should be much lower, with a value of about 2.

Barasingha deer, *swamp deer,* *Cervus duvauceli:* a large (shoulder height 119–124 cm) deer of northern and central India and southern Nepal. It is found in swampy areas and on grassy plains. The coarse pelage is yellow to brown with yellowish underparts, and in the hot season the hair on the backs of the stags becomes red-brown. Stags have a long, thick mane, and the young have many white spots on the back and sides.

Barb:

1) A structure present in Feathers (see).

2) A fish of the family *Barbinae* (see).

Barbary ape: see Macaques.

Barbary sheep, *arni, aoudad,* *Ammotragus lervia:* the single member of the genus *Ammotragus,* of the tribe *Caprini,* subfamily *Caprinae* (see), family *Bovidae* (see). It is a light brown, powerful animal, with long, mane-like hair on the neck, breast and forelimbs. Horns are almost equally developed in both sexes, and they sweep in a wide semicircle away from the head (Fig.). Phylogenetically it is intermediate between sheep and goats, and in appearance and habits it resembles a goat. It inhabits the high Atlas Mountains of North Africa, and is also found as far east as the Red Sea.

Barbary sheep (Ammotragus lervia)

Barbel:

1) Filiform epidermal organs around the mouths of many fishes. They function as sensory organs of taste and touch.

2) A fish of the family *Barbinae* (see).

Barbel region: see Running water.

Barberry: see *Berberidaceae.*

Barber's trap: see Soil organisms.

Barbets: see *Capitonidae.*

Barbicel: see Feathers.

Barbinae, *barbs* or ***barbels:*** the largest subfamily of the *Cyprinidae* (see). B. are found in fresh water in temperate and warm regions of Africa and Eurasia. They possess 3 rows of pharyngeal teeth. A large number of small species are native to tropical Asia and Africa, and species from southern Asia are especially popular as aquarium fish. The European ***Barbel*** *(Barbus barbus),* which can attain a length of 90 cm, has local importance as a commercial fish, and is also popular as a sporting fish. It is found in clear running water with a sandy or stony bottom. The roe and liver are poisonous, especially during spawning.

Barb ridge: see Feathers.

Barbule: see Feathers.

Barbule plate: see Feathers.

Barbus: see *Barbinae.*

Baribal: see Black bear.

Bark beetles: see Engraver beetles.

Barking deer: see Muntjacs.

Bark lice: see *Psocoptera.*

Barley: see *Graminae.*

Barley stripe mosaic virus: see Virus groups.

Barley yellow dwarf virus: see Virus groups.

Barnacles: see *Cirripedia.*

Baroceptors, *pressure receptors, pressoreceptors:* mechanoreceptor nerve endings in the carotid sinus, left ventricle and aortic arch. These are not B. or pressure receptors in the true sense, but rather "stretch receptors" that are stimulated when they are distended. They act as pressure receptors because the pressure within the blood vessels determines the tension of the vessel walls. Nerve fibers from the B. of the carotid sinus pass to the

Baroceptors. Action of arterial baroceptors.

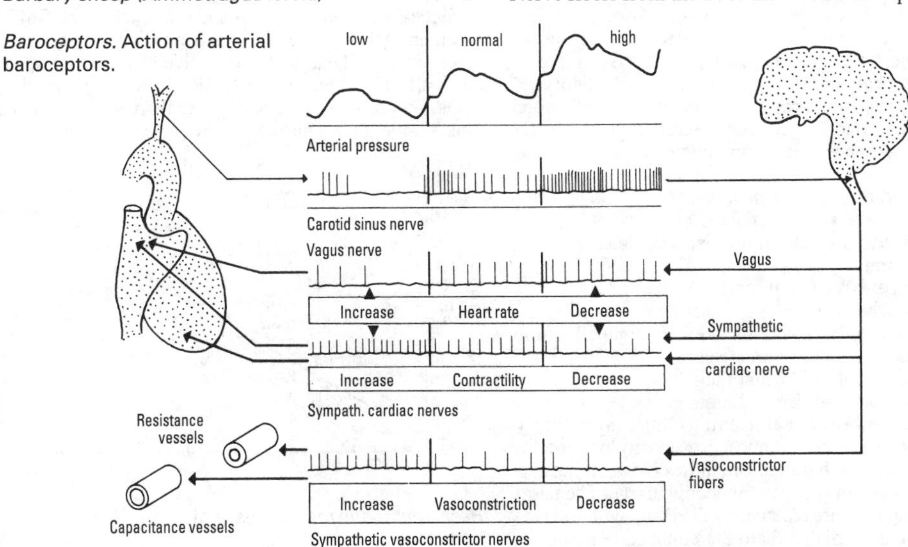

medulla in a branch of the glossopharyngeal nerve. In most animals, including humans, nerve fibers from the B. of the left ventricle and aortic arch ascend in the trunk of the vagus (in the rabbit they form a separate nerve bundle called the depressor nerve). Each pulse wave stretches the walls of the arteries and stimulates the B. Increased pressure (i.e. increased stretch of the B.) causes an increased rate of firing in the afferent nerves that connect the B. with the solitary tract nucleus. This leads to activation of the vagal center and the inhibitory neurons of the vasomotor center. There is therefore an increase in the vagal traffic to the heart, but a decrease in the sympathetic outflow to the heart and blood vessels. A fall in pressure within the carotid sinus and aortic arch results in a reversal of these processes. B. excitation starts at pressures below the normal blood pressure. The B. of the carotid sinus do not display activity below a threshold pressure of 65–70 mm Hg. As the pressure increases above this threshold the rate of firing in the myelinated afferent fibers increases, showing a sigmoidal relationship (not very pronounced and sometimes described as roughly linear) between blood pressure and receptor activity. Maximal excitiation occurs at about 180 mm Hg (an unphysiologically high value). The B. of the aortic arch show a similar relationship, but the threshold is higher, and the pressure/receptor activity curve is displaced with respect to that for the carotid sinus, so that greater pressures are needed to cause the same B. activity. The B. of the aortic arch therefore reinforce the action of the B. in the carotid sinus in opposing pressure increases above the normal.

Arterial B. are excited by each heart beat. Their firing rate increases up to the peak of systolic pressure, then decreases during the diastolic pressure decrease. In this way they inform the vasomotor center about the start, the rate of increase, the magnitude, the decrease, and the end of each pressure pulse.

Excitation of B. acts via the vasomotor center to cause inhibition of the heart and circulation. By activation of the vagus nerve, the heart frequency is decreased, and by the inhibition of the vasoconstrictor fibers of the sympathetic nerve, the blood pressure is decreased (Fig.). However, the system does not just respond to extremes of blood pressure. Precise information about the arterial pulse pressure is relayed at each heart beat, and the intensity of the regulatory response is graded appropriately. Pulse rate and blood pressure are continuously and flexibly adjusted in accordance with the cycle of pressure changes in the circulation.

Barracudas: see *Sphyraenidae*.

Barramunda: see Lungfishes.

Barr body: see Chromosomal sex determination, Sex chromatin.

Barrier reef: see Coral reef.

Basal body:

*1) **Blepharoplast, kinetosome:*** an organelle from which a flagellum or cilium arises.

2) See Centriole, Cilia and Flagella.

Basal lamina: see Epithelium.

Basal membrane: a thin extracellular layer, 50–70 nm in thickness, and composed primarily of proteins and mucopolysaccharides, which lies between the epithelial cells and the deeper connective tissue. The bases of epithelial cells are attached to the B.m., which serves to bind the epithelial cells to the connective tissue beneath. The B.m. represents a very fine network, which probably acts as an ultrafilter for macromolecules.

Basal metabolic rate, *BMR:* that part of energy metabolism necessary for maintenance of an animal completely at rest. It is measured by calorimetry.

Body cells do not naturally disintegrate, because the continuous degradation of their components is balanced by continuous synthesis. The metabolism necessary for the maintenance of this dynamic state in all the cells of the body is the BMR. In addition to the energy turnover of continuously active organs (heart, liver, central nervous system, kidneys), the BMR includes energy consumed in maintaining the tone and viability of muscles and skin. Since the magnitude of the BMR is influenced by many factors (body size, age, sex, time of day, time of year, nutritional state, illumination, environmental temperature, functional state of the nervous and endocrine systems), it must be measured under standard condition; the subject must be fasting (no food for the previous 16 h) and completely at rest (no muscular activity), the temperature must be normal (avoiding the physiological stress of high or low temperatures), there must be no excitement, and comparative measurements must be made at the same time of day.

The nominal value for the BMR of an adult human per day is roughly equivalent to the body mass in kg multiplied by 120. This is about 8 400 kJ for a 70 kg man. For a woman of the same weight, the BMR is about 10% less.

The BMR of animals is subject to continuous nervous and hormonal control. Nerves to the skeletal muscles have only a small influence on the BMR, but the sympathetic nervous system exerts a very large influence. By promoting the secretion of adrenalin and noradrenalin, the sympathetic system increases liver metabolism, raises blood glucose and directly intervenes in cellular metabolism. All endocrine glands, in particular the thyroid and the adrenals, influence the BMR. Removal of the thyroid or adrenals decreases the BMR by 30% and 15%, respectively. It is also decreased by removal of the testes or ovaries. It falls during pregnancy, and increases post partum during suckling.

Since every cell generates and uses energy, the BMR of an animal increases with cell mass, i.e. body size. Also, since the heat generated inside an organism must be lost to the exterior, the BMR per unit volume decreases with increasing body size. Warm-blooded animals smaller than a humming bird or a shrew cannot ex-

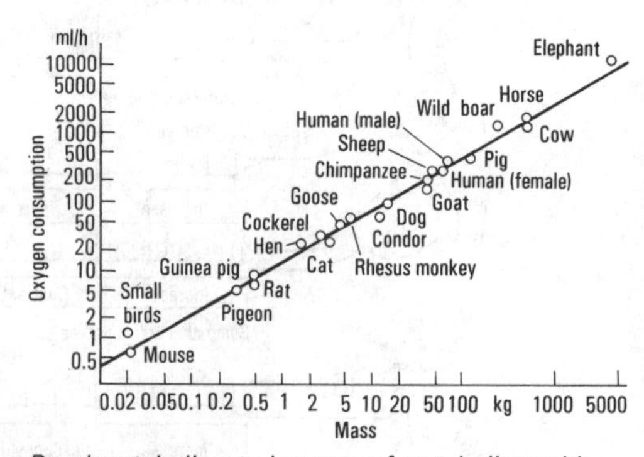

Basal metabolic rate. Increase of metabolism with body mass.

ist, because the rate of heat loss over the relatively large body surface area would be too high. Mammals larger than an elephant would also not be viable, because the rate of heat loss from the relatively small body surface area would be too low. For all warm- and cold-blooded animals, the logarithm of BMR and the logarithm of body mass show a linear relationship, although the slope of the straight line is somewhat different for fishes, reptiles, birds and mammals (Fig.). See Temperature regulation of the body.

Basal nuclei: see Brain.

Base exchange: see Sorption.

Basement membrane: see Epithelium.

Basic group: a group of fossil species that are very similar or related to one another, or a fossil group of higher taxonomic units, which belong to different branches of evolutionary development, but due to their common basal position near the evolutionary branch point have many primitive characteristics in common. E.g. the *Condylarthra* (see).

Basic karyotype: a Karyotype (see) that is specific for an individual group. Its chromosome number respresents the Basic number (see).

Basic number: the smallest haploid chromosome number of a particular group of related organisms. It is represented by the x, so that $x = 4$ indicates a B.n. of 4 chromosomes.

Basiconic sensilla: see Olfactory organ.

Basidiocarp: the three-dimensional fructification of many *Basidiomycetes* (see).

Basidiolichenes: see Lichens.

Basidiomycetes, ***basidium fungi:*** a class of fungi which possess a characteristic sporangium called a basidium. The basidium is the site of meiosis. *B.* usually have a well developed, persistent mycelium, consisting of numerous, septate, chitinous hyphae, whose transverse septa (cross walls) contain a barrel-shaped pore (dolipore) covered on each side by a parenthosome (Fig.1).

Most *B.* have a low rate of asexual reproduction (cf. the *Ascomycetes*, which have a high rate of asexual reprodution), and a high rate of sexual reproduction. The mechanism of sexual reproduction is uniform and characteristic throughout the *B.* Specialized sex organs are typically absent, but sexuality exists in that the Basidiospores (see) are of different mating types (see Diaphoromixis) and the hyphae that arise from them contain mononucleated cells of the same mating type as the parent basidiospores. This haploid phase, which corresponds to the gametophyte, is relatively brief. Whenever two hyphae of complementary mating types meet one another, cell fusion (plasmogamy, somatogamy) occurs, but the nuclei remain as a pair. During the subsequent binucleated phase of more or less prolonged mycelial growth (dikaryotic phase, dikaryophase), each cell division occurs with the formation of a Clamp connection (see) (a structural homolog of the hook of the *Ascomycetes).* Under appropriate conditions (largly undefined) this dikaryotic mycelium produces basidia, which are formed from apical cells of the dikaryotic hyphae. The paired nuclei of the apical cell fuse (karyogamy), followed immediately by meiosis, and formation of 4 haploid basidiospores. Each basidiospore is abstricted exotopically, i.e. it is found on the exterior of the basidium (usually on a peg-like stalk or sterigma) (cf. the ascospores of the *Ascomycetes,* which are found inside the ascus). The basidium is homologous with the ascus of the *Ascomycetes*, but the fructifications of the two groups are not homologus; they differ fundamentally in that they are formed from the dikaryotic mycelium in *B.*, but mainly from the haploid mycelium in the *Ascomycetes*. Basidia-bearing bodies of *B.* vary from the minute pustules of *Puccinia* species and the inconspicuous slimy layers of some *Tremallales* to the large, three dimensional structures of most of the *Holobasidiomycetidae*. The latter are referred to as fruiting bodies, fruit-bodies, fructifications or basidiocarps.

Classification. There is no universally employed classification of the *B.* It is generally agreed that many apparent affinities are the result of parallel evolution, as well as shared heritage. Any scheme therefore tends to combine artificial grades and natural series. This is especially true for the order *Gasteromycetales*. In some systems, the *Uredinales* and *Ustilaginales* are kept apart from the other *B.*, and placed in the subclass *Teliosporaceae (Hemibasidii, "**Brand fungi**")* on the basis that the basidia arise from the thick walled teliospore, they are all obligate parasites, and they never produce fructifications. *Auriculariales, Tremellales* and *Tulasnellales (**Jelly fungi**)* may then be assigned to a separate subclass *(Heterobasidiae),* most members of which are saprophytic, with a few parasites, and with some members that produce fructifications. The system presented below is based mainly on whether the basidia are septate or nonseptate, and on the structure of the fructification and its hymenium.

1. Subclass: *Phragmobasidiomycetidae*. The basidia of most members of this subclass are divided into 4 cells, either by transverse septa *(Uredinales*, and some *Ustilaginales* and *Auriculariales)* or by longitudinal septa *(Tremellales)*. Basidia of the family *Tilletiaceae* lack septa and are therefore unicellular, while basidia of the *Ustilaginaceae* are multicellular. Most of the *Phragmobasidiomycetidae* do not form fructifications, and the members of this subclass are considered to be more primitive than those of the *Holobasidiomycetidae*.

1. Order: *Uredinales* (Figs.3.1–3.5) ***(Rust fungi).*** All members of this order are plant parasites, responsible for the rust diseases of higher plants. They are called rust fungi because the eruption of spores on the surface of the host resembles a deposit of rust. The position, shape and appearance of the spore eruption are typical for each species. The spores characteristically display a great diversity of forms, which develop in regular sequence, and are often associated with an alternation of hosts. During its life cycle, a rust fungus may produce 5 different types of spores: Pycnospores (see), Aecidiospores (see), Uredospores (see), Teleutospores (see) and Basidiospores (see). Infection of the second host is usually accomplished by the aecidiospores. Rust fungi cause considerable damage to agricultural plants. *Puccinia graminis* ***(Black rust*** of cereals) (Fig.3.1) has a cosmopolitan distribution, and its intermediate host is *Berberis*. *Puccinia glumaram* is responsible for the ***Yellow rust*** disease of cereals, and it attacks various other grasses; its intermediate host has not been identified. *Puccinia coronata* causes the ***Brown rust*** of oats, and its intermediate host is buckthorn *(Rhamnus cathartica).* Other species of *Puccinia* parasitize cereals and wild grasses as well as non-graminaceous agricultural crops. Other important genera of rust fungi, whose

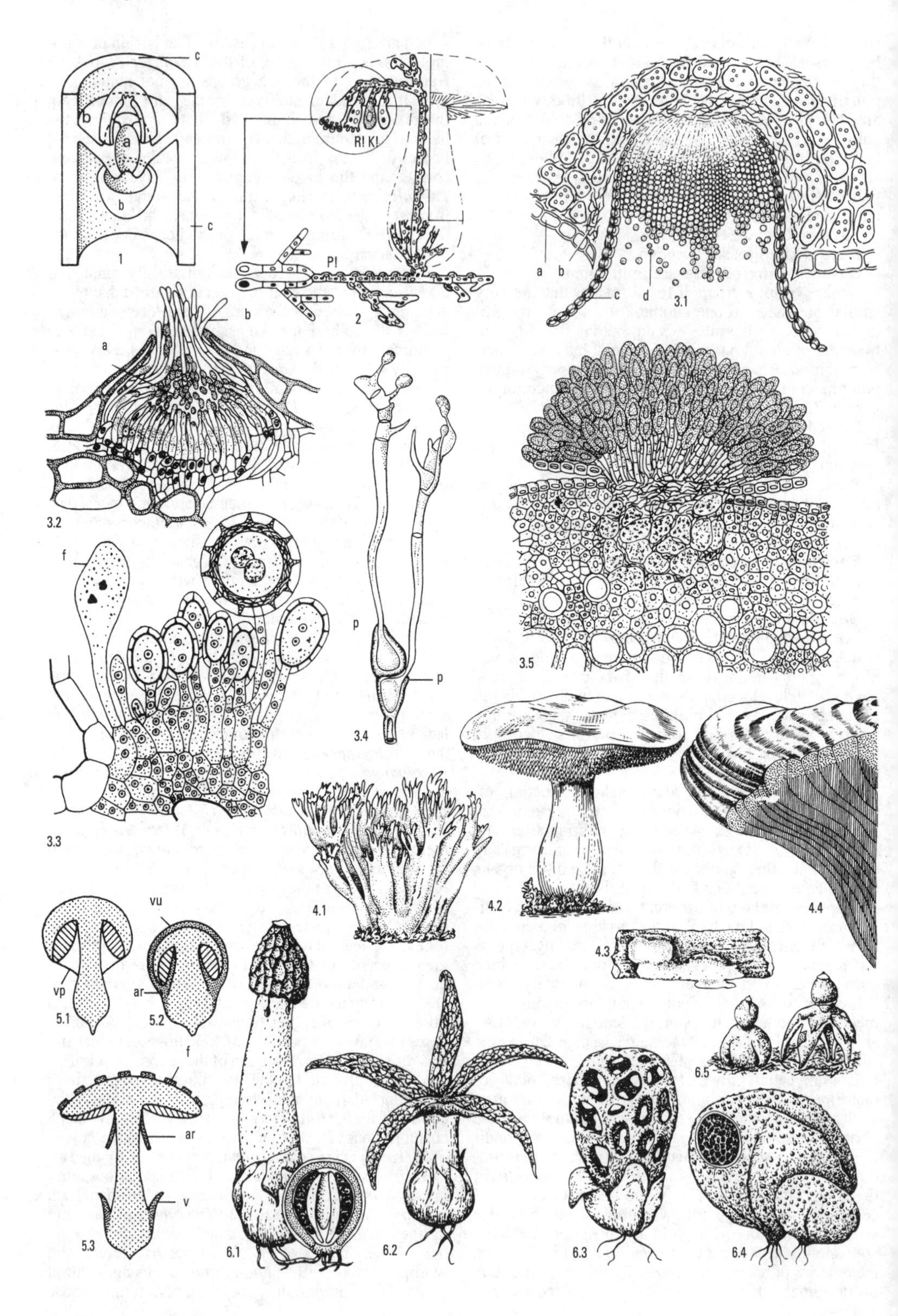
c
b
a
b
c
1
R! K!
P!
2
a
b
c
d
3.1
a
b
3.2
f
p
p
3.4
3.5
3.3
4.1
4.2
4.3
4.4
vu
vp
ar
5.1
5.2
f
ar
v
5.3
6.1
6.2
6.3
6.4
6.5

members attack a wide variety of plant species, are *Uromyces, Gymnosporangium* (Fig.3.2), *Meampsora* and *Cronartium.*

The following differences between the *Uredinales* and other *B.* should be noted: 1) in comparison with other *B.*, the haploid phase is relatively prolonged; 2) pycnospores are not found in other *B.;* 3) the pores of the transverse septa are more like those of *Ascomycetes,* rather than the dolipores of other *B.;* 4) they are all obligate parasites and their rate of asexual reproduction is higher than that of other *B.;* 5) whereas all other *B.* lack male or female reproductive structures (cf. *Ascomycetes,* which have definite male and female structures: antheridia and oogones), the *Uridinales* have female receptive hyphae.

2. Order: *Ustilaginales* ***(Smut fungi).*** All members of this order live parasitically on higher plants, causing the smut diseases. Large masses of spores are formed by fragmentation of the dikaryotic mycelium into single dark colored cells (brand spores), which give a burnt appearance to the infected plant parts. The brand spore (often also called a chlamydospore) behaves as a zygote and is homologous with the teleutospore of the rust fungi. Fusion of the paired nuclei of the brand spore is followed by outgrowth of an elongated basidium, which usually has transverse septa; meiosis then produces basidiospores (gonospores) which lack sterigmata. The germinating basidiospores conjugate to produce, once again, a dikaryotic mycelium. The basidium of the *Ustilaginales* is more primitive than that of other *B.*, and it is often referred to as a Promycelium (see) or metabasidium; similary, the basidiospores are often called sporidia. *Ustilago longissima* (causative agent of "leaf stripe" in *Glyceria* species) does not even produce a distinct promycelium, but merely a short tube, from which sporidia are abstricted successively. Two families of smut fungi are recognized: *Ustilaginaceae* (with multicellular basidia) and *Tilletiaceae* (basidia lack transverse septa). Some smut fungi are economically important, e.g. *Ustilago zeae (= U. maydis*), the causative agent of ***Maize smut,*** whose accumulations of brand spores cause blister-like swellings; other species of *Ustilago* cause ***Loose smuts*** of barely, oats and wheat. *Tilletia tritici,* whose spore masses smell of pickled herrings, is responsible for the ***Stinking smut*** or ***Bunt*** of wheat.

3. Order: *Auriculiales.* This order contains primitive forms with loose hyphal mats, as well as forms with complex fructifications, e.g. *Auricularia auricula-judae* ***(Jew's ear),*** often seen on the stems of old elder *(Sambucus nigra)* trees. The Jew's ear fungus produces ear-shaped, liver-brown fructifications, 3–10 cm broad, with the hymenium on the concave lower surface. In all members of the *Auriculariales,* the basidia are divided into 4 cells by transverse septa, each cell carrying a spore on a lateral sterigma. The basidium of many *Auriculariales* has a basal spherical swelling, the probasidium; this is the remnant of the terminal cell of the dikaryotic hypha in which nuclear fusion occurred prior to basidiospore formation.

4. Order: *Tremallales.* Like the *Auriculariales,* this order contains primitive members which form only hyphal mats, as well as those with complex fructifications. The basidia have 4 longitudinal septa. Examples are *Calocera viscosa,* a small (3–8 cm high), slimy, orange-yellow fungus with the appearance of a stag horn, which grows on dead conifer stumps; and *Tremella mesenterica* , whose gelatinous, contorted, brain-like, folded orange-yellow fructification (2–10 cm wide), is found on dead branches.

2. Subclass: *Holobasidiomycetidae.* Practically all the well known edible and poisonous toadstools and mushrooms belong to this subclass, especially in the order *Agaricales.* The mycelium is almost always perennial, and most species have conspicuous fructifications. *Holobasidiomycetidae* characteristically possess nonseptate basidia. The hymenium consists of the basida, sterile hyphae (pseudoparaphyses, containing degenerate pairs of nuclei), and sterile terminal hyphal cells called cystidia. There are many diverse forms of cystidia, and their shape is an important taxonomic character. Further division of this subclass into several orders is based on the structure and development of the fructification and hymenium.

1. Order: *Exobasidiales.* These are plant parasites without fructifications. Basidia erupt from the surface of the parasitized host, where they form a continuous hymenium, e.g. *Exobasidium,* which produces colored galls on leaves of members of the *Ericaceae.*

2. Order: *Poriales* (Figs.4.1–4.4) *(Aphyllophorales).* From an early stage, the surface of the fructification is totally or partly covered by hymenium, which continues to extend as the fructification increases in size. The hymenophore (the fertile lower surface of the fungus car-

Basidiomycetes. 1 Transverse septum of a basidiomycete hypha: *a* dolipore; *b* parenthosome; *c* cell wall. *2* Developmental cycle of an agaric; haploid phase (light lines), diploid phase (heavy lines); *P!* plasmogamy; *K!* karyogamy; *R!* meiosis. *3.1-3.5* Various spore containers of the rust fungi. *3.1 Puccinia graminis:* aecidium with aecidiospores on *Berberis vulgaris; a* epidermis; *b* intercellular mycelium; *c* pseudoperisium; *d* chains of aecidiospores (x 140). *3.2 Gymnosporangium clavariaeforme:* spermatia in spermogonium on a leaf of *Crataegus,* breaking through the upper epidermis; *a* spermatia; *b* paraphyses (x 450). *3.3 Phragmidium rubi:* margin of a nearly mature sorus of binucleate uredospores (at various stages of development) after rupture of the host epidermis *(e); f* paraphyses (x 565). *3.4* Germinating 2-celled teleutospore with 2 basidia (x 300); *p* germination pore. *3.5* Group of teleutospores on a cereal stem. *4.1-4.4* Order *Poriales. 4.1 Clavaria botrytis* (fairly club) (x 0.5): fruiting body. *4.2 Hydnum repandum* (wood hedgehog) (x 0.5): fruiting body. *4.3 Stereum hirsutum:* fruiting body on small tree branch. *4.4 Phellinus igniarius:* section through an old fruiting body with annual growth zones (x 0.5). *5.1* Order *Agaricales.* Longitudinal section through fruiting body showing partial veil (vp.) *5.2* Young stage of *5.3* with universal veil (vu). *5.3* Mature stage of agaric fruiting body: *ar* annule or armilla; *v* volva; *f* remains of the universal veil on the cap or pileus. *6.1-6.5* Various fruiting bodies of the *Gastromycetales. 6.1 Phallus impudicus* (stinkhorn): mature fruiting body with droplets running from the gleba, and a longitudinal section of an immature fruiting body (x 0.5). *6.2 Anthurus archeri* (x 0.5). *6.3 Clathrus ruber* (x 0.5). *6.4 Scleroderma aurantium* (common earth ball), partly cut away to expose the areolate gleba (x 0.5). *6.5 Geastrum quadrifidum* (earth star) (x 0.5).

rying the hymenium) displays various contors, e.g. wrinkled, folded, warty, spiny or tubular.

Serpula lacrymans ***(House fungus, Weeping merulius)*** forms encrusting fructifications with folded hymenia, 5–50 cm across and bracket-like, on the vertical surface of its substrate. It is responsible for serious "dry rot" in the timbers of buildings.

The hymenium of *Stereum* species (Fig.4.3) is smooth and undivided, and the fructification usually consists of three layers. They are usually found on dead wood, but *Stereum purpureum* parasitizes plum trees, causing "silver leaf disease"; its 1–3 broad purplish to lilac brackets occur in tiers on the vertical surfaces of the host tree.

Cantharellus cibarius ***(Chantarelle)*** is edible and excellent eating; its fructification is obconical with forked ridges on the underside. Fructifications of *Clavaria* species (Fig.4.1) ***(Fairy clubs*** and ***Coral fungi)*** display a variety of erect, fleshy, branched or unbranched forms, covered by a smooth hymenium, except at the base.

In members of the *Hydnum* family (Fig.4.2) *(Hydnaceae,* ***Tooth fungi)***, the spore-forming layer covers individualized teeth or spines on the undersurface of the cap.

The spore-forming layer of the ***Polypores*** *(Polyporacae)* consists of densely packed tubes (unlike the tubes of *Boletus*, the tubes of the polypores do not form an easily detachable layer). Polypore fructifications are often bracket shaped, and known as "bracket fungi". Several genera are recognised in the *Polyporaceae*, e.g. *Fomes,* whose species are parasitic on deciduous trees, and produce woody, perennial, bracket-like fructifications. *Fomes fomentariuds* ***(German tinder fungus, Soft amadou, Punk)*** causes white rot of beech. *Fomes lignosus* is a common, worldwide, tropical species, often found growing at the bases of infected trees on large exposed roots or decaying stumps, especially in the rainy season. The fructification is deep red-brown with a bright yellow margin, and the top often appears green due to the presence of algae; the surface of the hymenium is bright orange. It parasitizes rubber, palm oil and teak, as well as other economically important trees. A similar fungus, *Phellinus igniarius* (Fig. 4.4) ***(Hard amadou, Touchwood)***, grows on apple trees. Since the revision of the taxonomy and nomenclature of the *Polyporaceae,* the genus *Polyporus sensu restricto* includes fewer species, e.g. *Polyporus squamosus* ***(Dryad's saddle, Scaly polyporus)*** parasitizes deciduous trees, especially elm, and the anatomy of its fructification displays some affinities with the next order, *Agaricales*.

3. Order: *Agaricales* (Figs.5.1–5.3) ***(Gill fungi, Gill mushrooms)***. This order contains most of the familiar edible and poisonous toadstools. They have an umbrella- or hat-shaped cap (pileus) with a central stalk (stipe). In the *Agaricales* the whole hymenium develops simultaneously, and not gradually as in the *Poriales*. The hymenophore in most species takes the form of lamella-like, radially arranged vertical gills (gill fungi in the strict sense), which cover the underside of the pileus. In the ***Boleti*** the hymenophore occurs as an easily detachable layer of closely packed tubes, so that the underside of the pileus is covered with small pores. In most young *Agaricales* (i.e. in the "button stage" or unexpanded stage), the margin of the young cap is joined to the stipe by the veil (Fig.5.1). The hymenium develops in the annular space between the pileus and stipe, and it is exposed when enlargement of the pileus causes the partial veil to tear. The remains of the veil may then completely disappear, or form a ring or inferior annulus on the stipe. In some *Agaricales* (e.g. *Amanita muscaria,* ***Fly agaric)*** the young pileus and stipe are enclosed in a common membrane, the universal veil (Fig.5.2). After expansion of the fructification, the remnants of the universal veil remain at the base of the stipe as a volva, and as white patches on the top surface of the pileus (Fig.5.3).

Many of the *Agaricales* form extensive mycorrhizas and they are associated with specific hosts, usually deciduous or coniferous trees.

A) Edible *Agaricales*.

a) *Agaricus* ***(Agarics)***. The cap is more or less white and smooth, or brownish due to scales. The gills are free, and white, gray or pink when young, becoming chocolate brown later. The stipe has a distinct ring. *Agaricus campestris* ***(Common mushroom, Field mushroom)*** occurs in well fertilized meadows. *Agaricus arvensis* ***(Horse mushroom)***, which smells of aniseed, grows in fields and in the undergrowth of hedges and thickets. *Agaricus bisporus* ***(Garden mushroom, Cultivated mushroom)*** is cultivated as an edible fungus.

b) *Macrolepiota procera* ***(Parasol mushroom)*** grows in wooded areas. It has a gray-brown, umbrella-like cap (10–20 cm in diameter) with coarse, dark brown scales. The stipe has a double ring, and a bulbous base.

c) *Amanita rubescens* ***(Blusher, Blushing amanita)*** has a red-brown cap with gray patches. The basal bulb of the stipe is not well differentiated. Older gills have red spots, and the white flesh of the stipe turns red when damaged, e.g. by insect bites. The fungus is often found in deciduous and coniferous forests

d) *Lactarius* species have brittle gills and fleshy stipe and cap, which yield a white or colored milky fluid when broken. Mild tasting species are edible, while astringent tasting species are nonpoisonous but inedible. *Lactarius deliciosus* ***(Delicious lactarius, Saffron milk cup)***, found under coniferous trees, has a reddish orange cap, and the gills and stipe may be a similar color or with greenish patches; its milk is a mild-tasting, and rapidly turns orange after it is released. *Lactarius volemus* ***(Orange-brown lactarius)*** is found in mixed woodlands, has a bright tawny orange cap, cream gills and stipe that is increasingly orange toward the base, and it yields abundant, mild tasting, white milk. Both species are edible and good to eat.

e) *Russula* species have brittle gills and no milky fluid is produced when the fungus is broken. The stipe easily breaks transversely, and becomes bright or dirty salmon pink when rubbed with a crystal or ferrous sulphate. All mild tasting species are edible, especially *Russula vesca* (found in deciduous woodlands, especially under oak; cap brownish red to pinkish buff), and *Russula cyanoxantha* ***(Blue and yellow russula;*** cap bluish green to violet), which is also found in deciduous woodlands, especially under beech.

f) *Kuehneromyces (= Pholiota) mutabilis* ***(Changeable mushroom, Prickly cup)*** has a yellow-brown, smooth cap and a brown, scaly stipe. It grows in clumps on stumps in deciduous woodland.

g) *Armillariella mella* ***(Honey-color Armillariella,***

Mellaarmillaria, ***Honey Agaric, Honey fungus, Bootlace fungus)*** has a yellowish to brownish cap with recurved, fibrillose scales, and a yellow-brown stipe. A common fungus in the late autumn, it is poisonous when raw. It spreads extensively by black rhizomorphs, and is a serious destructive parasite of deciduous and coniferous trees. One of the commonest fungi in both temperate and tropical regions, it also causes serious damage to tropical crops like rubber and tea.

h) *Pleurotus squarrosulus* grows in the tropics and subtropics, where its fructifications develop in groups or tufts on fallen trees and logs. It is common in India, Sri-lanka and Burma, and especially common in tropical West Africa, where it is used locally in soups and stews.

i) *Coprinus* ***(Ink caps)*** have thin and very crowded gills, and most species are characterized by autodigestion of the gills, and sometimes the cap, into a black, inky fluid. Young specimens of *Coprinus comatus* ***(Shaggy cap, Lawyer's wig)*** have a cylindrical, white cap with shaggy scales, later expanding below and becoming brownish, while the gills progressively turn from white through pink to black, finally autodigesting. It is found in fields, on waysides and on refuse tips, sometimes solitary, but often clustered. It is edible and very good, with a delicate taste, and best gathered and cooked before it begins to autodigest. *Coprinus atramentarius* is also edible, but causes nausea if consumed with alcohol. This is due to the presence of bis(diethylthiocarbamoyl) disulfide ("antabuse"), which was discovered independently as an effective treatment for alcoholism, and later found to be the active fungal constituent.

B) Poisonous *Agaricales*

a) *Amanita phalloides* ***(Death cap, Death cap amanita, Deadly amanita, Poison amanita).*** The fructification has a greenish cap, white gills and white stipe, with a ring and bag-like volva, and a large basal bulb. It grows in oakwoods and is one of the most dangerous fungi, accounting for 95% of deaths from fungus poisoning in Europe. Toxicity is due to the presence of phalloidins and amanitins, which cause liver and kidney damage, muscle cramps and respiratory failure. Similar toxicity is displayed by *Amanita verna* ***(Spring amanita)*** and *Amanita virosa* ***(Destroying angel),*** both totally white and far less common than *A. Phalloides.*

b) *Amanita pantherina* ***(Panther cap, Panther amanita)*** possesses a smoky brown cap with white remnants of the universal veil. The young cap is convex, but soon becomes flat. The stipe has 2 or 3 more or less concentric hoops (remnants of the volva) between the ring and the well developed basal bulb. It is somewhat similar in appearance to *Amanita rubescens,* but unlike the latter its flesh remains white when damaged.

c) *Amanita muscaria* ***(Fly agaric, Fly amanita)*** has a red cap with white remnants of the universal veil, white gills, and a white-ringed stipe with a basal bulb. It is found especially under birch and conifers on poor soils.

d) *Inocybe patouillardii* has a whitish cap, which becomes yellowish brown in older specimens, and turns bright pink where split or bruised. The gills are white and the stipe white or reddish. It occurs on calcareous soils in deciduous woodlands, especially under beech, and it is often found on parklands and in meadows. This deadly poisonous fungus is sometimes confused with the edible *Agaricus.*

e) *Paxillus involutus* ***(Involute paxillus)*** grows in masses during the autumn in mixed woodlands, especially under birch. It is supposedly harmless if cooked, but it is of little value and best avoided. The cap is woolly and olive brown, and the gills yellowish, becoming olive-yellow to dark brown in older specimens. The stipe is olive-brown and relatively short.

f) *Boletus satanus* ***(Blood-red boletus, Satan's mushroom)*** is extremely rare and poisonous (but not deadly), with a grayish or gray-green cap with blood-red pores when mature. The spores turn blue when touched, and the white flesh turns pale blue when damaged. *Tylopilus (= Boletus) felleus* ***(Bitter boletus)*** has white to red pores, an ocher to honey-colored cap, and bitter tasting flesh. Most boleti, however, are edible and many are good to eat, e.g. *Boletus edulis* ***(Cepe, Edible boletus;*** deciduous woods, especially beech) has a brown cap, white to green pores and a swollen stipe with a white network of slightly raised veins towards its apex. *Xerocomus (= Boletus) badius* ***(Bay boletus, Polish mushroom;*** common in coniferous woods) has a dark brown cap, green-yellow pores and a brown stipe. *Ixocomus (=Boletus) luteus* ***(Yellow-brown boletus, Butter mushroom;*** often on grassland) has a slimy brown cap, yellow pores and a yellow stipe above a purple-brown ring. *Leccinium (= Boletus) scaber* ***(Birch boletus;*** under birches) has a grayish to dark brown cap, off-white pores and a tall striate stipe with brown to black scales. *Leccinium aurantiacum* ***(Orange cap boletus, Aspen mushroom;*** under aspen and birch) has a brownish red to brownish orange cap, pale, dirty gray-brown pores, and a striate stipe with pallid scales turning to reddish brown.

4. Order: *Gasteromycetales* (Figs. 6.1–6.5). Most members of this order have rounded, closed fructifications, whose outer part (the peridium) dehisces or disintegrates when the spores are ripe. The spore-forming, inner tissue, is called the gleba. Spores are distributed passively, whereas the "ballistospores" of the *Phragmobasidiomycetidae* and *Holobasidiomycetidae* are discharged violently. The order *Gasteromycetales* contains a great diversity of genera, e.g. *Bovista* ***(Bovists, Puff-balls),*** *Lycoperdon* ***(Puff-balls),*** *Geastrum* ***(Earth stars)*** (Fig. 6.5) and the poisonous *Scleroderma auranticum* (Fig. 6.4), which are not as close to each other taxonomically as the genera of other orders.

Even *Phallus impudicus* (Fig. 6.1) ***(Stinkhorn)*** is included in this order, although it possesses a stipe and a structure resembling a pileus. The young fructification is an edible, white, ovoid body, known as a "devil's egg", completely enclosed in a double-walled peridium with an intermediate gelatinous layer. Within this young fructification, the hyphae differentiate into an axial stalk and a bell-shaped cap. The surface membrane of the cap covers the gleba. When mature, the stalk elongates, carrying the cap with it, and ruptures the peridium. The outer membrane of the cap then disintegrates, exposing the blue-green, spore-bearing gleba, which deliquesces and emits a carrion-like smell. Various carrion insects, especially blowflies and dung flies, disperse the spores.

Isolated species of *Clathrus* (Fig. 6.3) ***(Lattice fungus;*** reddish, reticulate receptacle) and *Anthurus* (Fig. 6.2) ***(Cuttle-fish fungus;*** receptacle with a number of reddish arms) are found in temperate climates. These two genera, with their bizarre fructifications, are more numerous in the tropics, where they are known as ***Flower fungi.***

[Basidium and Basidiocarp, K. Wells and E. Wells (eds.), Springer Verlag, New York, Heidelberg, Berlin, 1982]

Basidiospores: spores formed by members of the *Basidiomycetes* (see). B. are abstricted from club-shaped cells, the basidia. They are usually 4 in number, and they occur on the outside of the basidium, usually on peg-like projections (sterigmata). The nuclei of these 4 B. are formed by meiosis, following fusion of the paired haploid nuclei of the basidium. They represent 4 different sexes or mating types. This tetra-polarity does not occur in all *Basidiomycetes,* e.g. species of *Hydnum* produce only 2 types of B. The B. are homologous with the Sporidia (see) of some smut fungi. See Diaphoromixis.

Basidium: see *Basidiomycetes.*

Basidium fungi: see *Basidiomycetes.*

Basihyal: see Skull.

Basihyal throat valve: see *Crocodylia.*

Basil: see *Lamiaceae.*

Basiliscs, *Basiliscus:* large, jungle-dwelling, Central American iguanas (see *Iguanidae)* with high, erectile dorsal and caudal crests, supported in some species by bony struts. There is often a narrow, bizarre, helmet-like protuberance or casque on the head of the adult male. They are excellent climbers and swimmers, and their diet is mainly vegetarian. Five species are recognized, all of which bury their eggs in damp earth. *B. basiliscus* of Columbia, Panama and Costa Rica is 80 cm long. In addition to a 6 cm-high, serrated crest and a large casque toward the back of the head, *B. plumifrons* also has a small protuberance on the front of the head (Fig.).

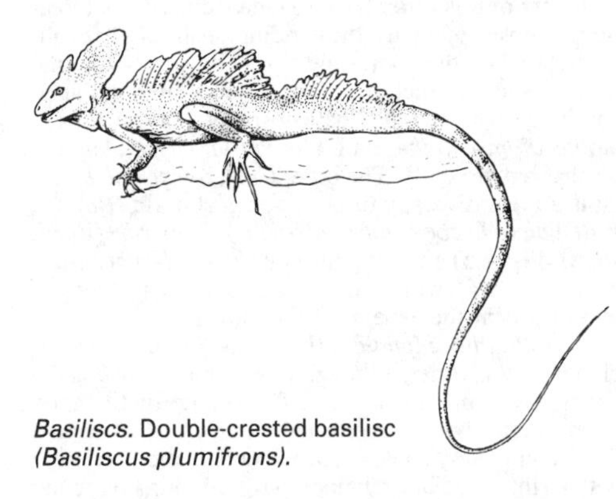

Basiliscs. Double-crested basilisc *(Basiliscus plumifrons).*

Basiliscus: see Basiliscs.

Basitony: development of buds at the lower end of the parent axis, rather than elsewhere. B. is typical of shrubs and perennial herbs. See Acrotony, Mesotony.

Basket stars: see *Ophiuroidea.*

Basomattophora: see *Gastropoda.*

Bass: see Sunfishes.

Bast:

1) The permanent plant tissue that extends outward from, and includes, the cambium. In includes some medullary ray parenchyma, but consists primarily of sieve tubes with companion cells and B. parenchyma. Also included are B. fibers (nonlignified sclerenchyma fibers), which serve as strengthening elements. B. fibers are absent from e.g. beech, birch, alder, spruce and fir. The conducting and storage tissues are collectively known as *soft B.,* and this forms alternate layers with the *hard B.* Several alternate layers of hard and soft B. are formed during a single vegetative period. Unlike the annual rings of the wood, these layers cannot therefore be used to date a tree. B. fiber strips are used in handicrafts and as cordage.

2) The hair-covered skin on the young, growing antlers of deer.

Batch culture: see Discontinuous culture.

Bateleur eagle: see *Falconiformes.*

Batesian mimicry: see Protective adaptations.

Bateson's law: see Regeneration.

Batfish, ***Sea bat:*** a trivial name applied to two systematically unrelated fishes. 1) *Platax orbicularis,* a high-backed, strongly laterally compressed coral fish of the order *Perciformes,* found in the Indo-malayan seas, the Red Sea and off the east coast of Africa. Unlike most coral fishes, *Platax* is not vividly colored, but possesses gray, soiled yellow-white and brown vertical camouflaging stripes. 2) Members of the family *Ogcocephalidae* (order *Lophiiformes),* which are bottom-dwelling fishes of warm seas, with wide, dorsoventrally flattened bodies.

Bat flies: see *Diptera.*

Bath sponges: members of the genera *Spongia* and *Hippospongia* (family *Spongidae,* order *Dictyoceratida,* phylum *Porifera*). All are keratinous sponges without spicules and with only spongiolin fibers, and they attain diameters of up to 50 cm. After removal of soft tissues, drying and bleaching, the remaining network of fibers constitutes the sponge of commerce, which is used as a bath sponge, and for technical applications like filtration, sanding, polishing and painting. They are found and harvested in various warm seas, especially the Mediterannean and Caribbean. Most common are the Dalmation sponge (*Spongia officinalis mollissima*), elephant ear sponge (*Spongia officinalis lamella*), zimocca sponge (*Spongia zimocca*), yellow sponge (*Spongia irregularis*), horse sponge (*Hippospongia communis*), velvet sponge (*Hippospongia communis meandriformis*), grass sponge (*Hippospongia communis cerbriformis*) and the wool sponge (*Hippospongia canaliculata*).

Bat hawk: see *Falconiformes.*

Bathynellacea: an order of the *Crustacea,* subclass *Malacostraca.* They are some of the smallest malacostracans, e.g. *Brasilibathynella florianopolis* from Brazil is little more than 0.5 mm in length. They have worm-like, elongated bodies, and they are eyeless and degenerate, various limbs being reduced or absent. A carapace is lacking. All the thoracic appendages are biramous and similar. All are found in fresh water in the interstitial spaces between sediment particles such as sand grains, where they feed on detritus, bacteria and fungi. Representatives of the order have now been reported from most parts of the world. A well known genus is *Bathynella* from subterranean waters in central and northern Europe.

Bathyteuthis: see *Cephalopoda.*

Batis: see *Muscicapinae.*

Batoids, *Batoidae, Hypotremata:* a superorder of elasmobranch fishes with flattened bodies, in which the olfactory openings, mouth and gill slits are all situated on the light-colored ventral surface. The skin is more or less densely covered with placoid scales. Some B. carry a spine in a whip-like tail, which can inflict considerable damage on an aggressor. Most B. are bottom fishes of warm seas. There are 20 families, distributed in 4 orders:

1. Order: *Rajiformes*

1. Suborder: *Rhinobatoidei*

Families: *Rhinidae, Rhynchobatidae, Rhinobatidae, Platyrhinidae.* Members of all four of these families are ovoviviparous, and generally referred to as ***Guitar fishes.***

2. Suborder: *Rajoidei*

Families: *Arhynchobatidae, Rajidae, Pseudorajidae, Anacanthobatidae.* Most familiar of these families is the *Rajidae,* comprising a nearly cosmopolitan marine group of fishes ranging from estuaries to the depths of warm and cold seas. Examples of *Rajidae* from the northern seas are: *Raja batis* (***Blue*** or ***Gray skate;*** a valuable commercial species); *Raja clavata* (***Thornback ray); Raja radiata (Starry ray).***

2. Order: *Pristiformes*

Family: *Pristidae,* the ovoviviparous ***Sawfishes,*** which live in the shallow waters of tropical and subtropical seas and rivers. They possess an elongate extension of the snout, known as the rostrum, which is a long flat blade armed with teeth set in sockets.

4. Order: *Torpediniformes*

Families: *Torpedinidae, Hypnidae, Narcinidae, Narkidae.* Members of these four families are generally known as ***Electric rays*** and ***Torpedoes.*** All possess electric organs in a disk on each side of the head, which in large specimens can produce a surge of up to 200 volts. The electric discharge is used to stun prey or discourage aggressors. They are found over a wide depth range in tropical and temperate waters.

5. Order: *Myliobatiformes.* All members of this order are warm water fish, usually living close to shore. They are bottom feeders, except for the ***Mantas*** which consume plankton and small schooling fishes. The tail spine is typically barbed with grooved edges. In the ***Sting rays*** the spine is associated with a poison gland.

Families: *Dasyatidae,* ***Sting rays,*** e.g. *Dasyatis pastinaca* found in the North Sea; *Potamotrygonidae,* ***River sting rays*** found in the rivers of South American; *Urolophidae,* ovoviviparous ***Eagle*** and ***Bat rays;*** *Rhinopteridae,* ***Cownose rays;*** *Mobulidae,* ***Mantas*** or ***Devil rays,*** including the gigantic ***Manta ray,*** *Manta birostris,* which can attain weights of 5 000 kg.

Baule unit: see Mitscherlich's law of plant growth factors.

Bay: see *Lauraceae.*

Bayberry: see *Myricaceae.*

Bay ducks: see *Anseriformes.*

B-chromosomes, ***accessory chromosomes:*** heterochromatic, telocentric chromosomes present in some races and species of plants, in addition to the autosomes and sex chromosomes. These additional chromosomes are usually without an apparent genetic function. They occur in varying numbers, and are generally smaller than the autosomes.

Bdellonemertea: see *Nemertini.*

Bdellovibrio: a genus of motile, Gram-negative bacteria, which parasitize other bacteria. After penetrating the cell wall of another bacterium, the small, slightly curved *Bdellovibrio* cell grows within the host into a long, coiled cell, which then divides into several daughter cells. The host cell is destroyed.

Beach flea: see *Amphipoda.*

Beaded lizards: see *Helodermatidae.*

Beaked whales: see Whales.

Bearberry: see *Ericaceae.*

Beard (of grasses): see Awn.

Bearded dragon, *Amphibolurus barbatus:* an Australian agamid (see *Agamidae)* with a wide, spiny skin fold like a frill or ruff at the chin. When the animal is excited, this skin fold is spread to a diameter of up to 18 cm. In comparison, head plus body length is about 20 cm, so that both head and mouth appear to be greatly enlarged, which acts as a deterrant to aggressors. In addition, the animal can run upright very quickly on its hindlegs.

Bearded sakis, *Chiropotes:* a genus of totally arboreal Cebid monkeys (see) in the subfamily *Pitheciinae,* infraorder *Platyrrhini* (see New World monkeys). The genus is represented by only 2 living species: *C. satanus* ***(Black-bearded saki)*** and *C. albinasus* ***(White-nosed bearded saki;*** the animal has in fact a red nose). These medium-sized monkeys inhabit an area approximating to that enclosed by the Orinoco, Negro and Amazon rivers. Head plus body length is 44 cm, and the thick, bushy tail attains 40 cm. The skull bulges at the forehead, where it is decorated by 2 dense whorls of hair. A conspicuous beard gives the genus its name. They feed largely on seeds and fruits, with some insects.

Bearded tit: see *Timaliinae.*

Bears: see *Ursidae.*

Beavers: see *Castoridae.*

Becards: see *Cotingidae.*

Bêche-de-mer: see Trepang.

Bed bugs: see *Heteroptera.*

Bedeguar gall: see Galls.

Bedstraw: see *Rubiaceae.*

Beech: see *Fagaceae.*

Bee-eaters: see *Coraciiformes.*

Bee dance: see Colony formation.

Beef tapeworm: see *Taenia.*

Bee hummingbirds: see *Apodiformes.*

Bee louse: see *Diptera.*

Bees (plate 10), *Apidae:* an insect family of about 20 000 species, order *Hymenoptera,* suborder *Apocrita.* More than 85% are Solitary B. (see), the remainder being social bees (see Honey bee; Bumble bees). B. apparently evolved from predaceous ancestors like the solitary mud-dauber wasps in the middle of the Cretaceous, i.e. when flowering plants became dominant. B. are distributed throughout the world, showing the greatest abundance and species diversity in semi-arid, warm temperate climates. See Colony formation.

Beeswax: a solid secretion from the abdominal glands of the workers in a honeybee colony, which is used to construct the honeycomb. The crude yellow, brown or red wax can be purified by melting and skimming, and by bleaching with light or oxidizing agents. It is a complex mixture containing esters of straight-chain fatty alcohols (even-numbered carbon chains from C_{24} to C_{36}) with straight-chain fatty acids (even-numbered carbon chains up to C_{36}), the most abundant ester being myricyl palmitate. The corresponding unesterified acids and alcohols are also present. The mixture also contains about 20% hydrocarbons with odd-numbered carbon chains, and about 6% propolis (a material of vegetable origin, used as an adhesive in nest construction), as well as pigments and unidentifed substances. B. is very stable, and it is used in the manufacture of cosmetics, pharmaceuticals, polishes, paper, etc. The wax moth *(Galleria)* feeds exclusively on B.

Beetle mites: see *Oribatei.*

Beetles: see *Coleoptera.*

Beet yellows virus: see Virus groups.

Bee toxin: an acidic defense secretion produced in the abdominal venom gland of queen and worker honeybees *(Apis mellifica* L.), and delivered by the sting. The maximal quantity produced by a bee is 0.3 mg. It contains three types of active principle: 1) biogenic amines, including histamine, which cause pain and dilate the blood vessels, allowing the toxin to spread; 2) biologically active peptides such as mellitin and apamin, and 3) enzymes like hyaluronidase and phospholipase A. B.t. causes allergic reactions in humans, although individual sensitivity varies greatly. It is applied intracutaneously or percutaneously for treatment of neuralgias and allergies, and in particular for treatment of rheumatic diseases.

Beggiatoa: a genus of Gram-negative, filamentous bacteria. The cells are arranged in chains within filaments, which range in diameter from 10 to 50 mm and up to several centimeters in length. These filaments display slow creeping and waving movements. They are nonphotosynthetic, sulfur-storing, autotrophic bacteria, which oxidise hydrogen sulfide and sulfur with the aid of atmospheric oxygen (see Chemosynthesis). *B.* forms a white incrustation on stones and plants beneath the water.

Begonia: see *Begoniaceae.*

Begoniaceae, ***begonia family:*** a family of dicotyledonous plants containing 920 species in 5 genera. The family is represented in all tropical and subtropical regions, except Australia. Most members are perennial, monoecious herbs. Leaves are simple, usually unsymmetric, basal or alternate in 2 ranks, with large membranous, deciduous stipules. The unisexual and often irregular flowers have 2 to many petaloid segments. Female flowers have an inferior ovary consisting of 2–5 (usually 3) united carpels, with numerous anatropous ovules on axile placentas. Fruits are usually loculicidal, winged capsules (rarely berries), containing numerous, minute seeds that lack endosperm. Male flowers have numerous stamens, often with united filaments.

Begoniaceae. Begonia x *rexcultorum* (Begonia rex).

The family has no economic importance, but the genus *Begonia* is of considerable horticultural interest and value. Many of its species, hybrids and cultivars are grown as ornamental house plants, and for outdoor summer bedding in temperate climates. There are several separate horticulturally important groups. The ***Tuberous begonias*** include, e.g. *Begonia* x *tuberhybrida,* a moisture-loving, summer-flowering outdoor race of tuberous hybrids with mid-green leaves and large rose-like, often double flowers, the few single female flowers being overshadowed by the double male flowers, 8–15 cm in diameter. Flower colors range from white through pastel shades to deep, rich yellows and crimsons. They are derived from hybrids of wild species found in the South American Andes. ***Lorraine begonias*** are hybrids of *Begonia dregei* from southern Africa and *Begonia socotrana* from the Island of Socotra. They are valued for bearing abundant flowers from November to January. The hybrids known as ***Rex begonias*** *(Begonia* x *rexculturum)* are valued not for their flowers, but for the design and color of their decorative foliage, which ranges from silver patterns on dark green to zoned patterns of cream, red or purple. The parents are all Indian species, the principal parent being *Begonia rex* from Assam. Most popular of all the summer bedding plants, are the ***Semperflorens begonias*** *(Begonia* x *semperflorens cultorum),* which are noted for flowering all the year round; all are hybrids of the Brazilian species, *Begonia semperflorens.*

Behavior, the study of: see Ethology.

Behavioral physiology: the study of physiological processes associated with behavior. B.p. uses all the modern methods of physiology, with the purpose of analysing behavior and interpreting its causes. *Neuroethology* is concerned with the reception and processing of information and signals, the accompanying filtration processes, the selection and identification of relevant information, the conceptualization of space and time, processes of information storage, preconditions for motivation, emotion and its activation niveau, and the central nervous control of behavior. It also studies neuronal processes during ontogenesis.

Behavioral research: see Ethology.

Behavioral sequence: time sequence of behavioral acts; the sequence of defined units or elements of behavior, which are part of an overall pattern of behavior. The simplest analysis determines the probability that two particular elements of behavior will be associated in a given sequence. B.s. can be interpreted as a "Markoff process", so that Markoff chains of the 1st to the nth order can be established. The regularity of a B.s. can also be determined from indices such as the number of different elements in a B.s., the number of times each element is repeated, the relative frequency of each element, and similar parameters. Analyses can be presented as *flow diagrams* or *behavioral algorithms.* For example, the burying of food by a dog: acquisition of meat – carrying the meat in the molar region of the mouth – searching movements – scratching and pawing the ground – laying down the meat – butting the snout against the meat – alternately butting and pushing the meat with the snout until it is covered with earth. Sequences of body care movements also provide a good example for the analysis of B.s. By constructing both an ethogram and a chronogram, the duration of each behavioral element and the time interval between each element are recorded.

Behaviorism: a branch of psychology based on the premise that much behavior is acquired, i.e. it is the result of previous experience. The generalizations of B. display typical reductionist elements, disregarding the evolutionary aspects of behavior and ignoring the existence of genetic elements (i.e. the inheritance) of behavior. In its extreme form, B. gives rise to complex theories of the interrelationships of behavior patterns and the environment, but it has also led to the development of useful experimental procedures.

Beira antelope: see *Neotraginae*.

Belbidae: see *Oribatei*.

Belemnites, *Belemnitida* (Greek *belemnon* lightning, missile): an order of extinct cephalopods. The elongated body was supported by a long, dart-like internal skeleton (endocochlia), which consisted of three parts. The conical, chambered *phragmocone* is traversed by a siphuncle and carries the embryonic vesicle. The cigar-shaped to conical or cylindrical *rostrum,* which receives the pointed end of the phragmocone, consists of radially arranged calcite and organic material. The surface of the rostrum carries furrows (interpreted as attachment sites for fins) and an alveolar depression. At its broad end, the phragmocone carries a thin chitinous plate, the *pro-ostracum,* which is often missing when fossils are recovered. Fossils are found from the Upper Carboniferous to the end of the Cretaceous, extending into the Eocene in one area. They are important index fossils for the classification of strata in the Jurassic and Cretaceous.

Bellbine: see *Convolvulaceae*.

Bellbirds: see *Cotingidae*.

Bell moths: see *Tortricidae*.

Bellerophon: a genus of the extinct superfamily, *Bellerophontacea,* the earliest known representatives of the *Archeogastropoda* (see *Gastropoda).* The usually rounded shell is symmetrical and planospiral with strong growth rings on its surface. A slot, slit or notch for the exhalant water current is present on the outer edge of the shell, and this continues over the entire shell as a slender, anterior, mid-dorsal cleft (slit band or selenizone), which is lined with transverse ridges (Fig.). *B.* is phylogenetically interesting, because the shells of modern gastropods are asymmetrical (in symmetrical forms, the symmetry is secondary). The symmetry of primitive forms like *B.* was superceded by the evolution of asymmetric coiling around a central columella. *B.* is found from the Ordovician to the Lower Triassic, and is especially common in the limestone of the Lower Carboniferous, and the Bellerophon-chalk of the Alpine Upper Permian.

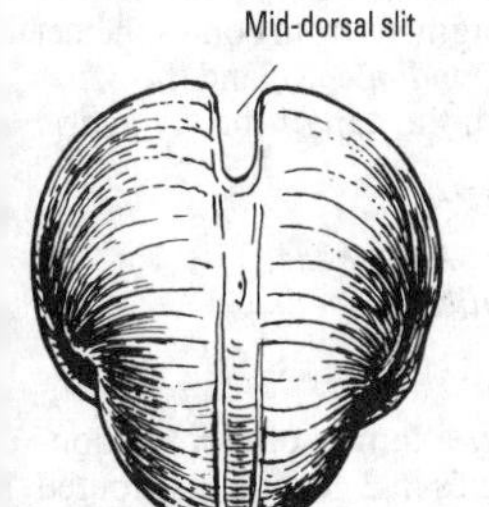

Shell of *Bellerophon* with conspicuous slit band (Lower Carboniferous; natural size).

Bellis: see *Compositae*.

Belone: see Gar pike.

Belontidae: see Gouramies.

Belontiinae: see Gouramies.

Belt desmosome: see Epithelium.

Beluga:

1) See *Acipenseridae*.

2) See Whales.

Bending movements: bending or curving of plant organs or organ parts. B.m. due to unequal growth rates on different sides of the organ are called Nutations (see), whereas B.m. due to changes in turgor pressure are called Variational movements (see). Some B.m. are induced by external stimuli, while others are partly autonomous. If the direction of bending is determined by the position or direction of the stimulus, the movement is called a Tropism (see), whereas a Nasty (see) shows no directional relationship to the inducing stimulus. Most tendril movements are Haptoreactions (see). Autonomous factors play a role in Twining (see) and in certain Flower movements (see).

Bennettitatae: an extinct class of pinnate-leaved gymnosperms, with single stalked ovules borne directly on the floral axis. Members of the order *Bennettitales* had the first truly hermaphrodite flowers with a perianth, which were probably insect-pollinated. Some (e.g. *Williamsonia)* resembled cycads in their pachycaulous habit, whereas others (e.g. *Wielandiella)* were leptocaulous with sympodial branching. Xylem structure was more primitive than that of modern cycads (simple eusteles and mainly scalariform tracheids). As in the cycads, fertilization was by spermatozoids.

The principal genera are *Williamsonia, Wielandiella, Williamsoniella* and *Cycadeoidea*.

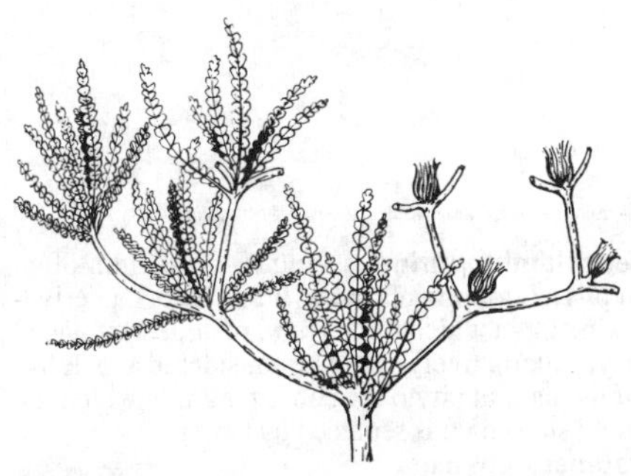

Bennettitatae. Growth habit of *Wielandiella angustifolia.*

Benthal: the bottom or floor of a lake, river, sea, etc. The B. is the habitat of bottom-living organisms (see Benthos), and it is divided iinto the Littoral (see), which extends down to a depth of 200 m, and the deep-water zone or Profundal (see).

Benthos: a Biocoenosis (see) comprising all animal *(zoobenthos)* and plant *(phytobenthos)* inhabitants of the banks and bottom of lakes, rivers, seas, etc., i.e. inhabitants of the Benthal (see). Freely moving animals of this biocoenosis belong to the *vagile, vagrant* or *nonsessile B.,* whereas fixed animals and plants (e.g. brown and red algae on the sea floor) belong to the *sessile* or *sedentary B.*

Benzedrine®: see Amphetamine.

Benzoic acid: C_6H_5COOH, an aromatic carboxylic acid, which occurs both free and esterified in many resins and balsams. It can be obtained from benzoin resin by sublimation.

Benzoquinones: compounds structurally related to *p*-benzoquinone (benzo-1,4-quinone). B. are found widely in microorganisms, plants and animals. *p*-Benzoquinone, and its mono-, di- and trimethyl, ethyl, methoxy and 2-methoxy-3-methyl derivatives are found in the defense secretions of certain arthropods. More than 90 different B. have been isolated from higher plants, fungi and molds. These include the yellow skin irritant, primin, from *Primula obconica,* the yellow perezone from various Mexican *Perezia* species, the orange embelin (which has antihelminthic and antibiotic activities), and the orange rapanon. Embelin and rapanon are found in many members of the *Myrsinaceae.* Fungal B. include the maroon fumigatin, which is excreted in the culture medium of of *Aspergillus fumigatus* and *Penicillium spinulosum,* the bronze-colored polyporinic acid, which is synthesized by the parasitic fungus *Polyporus nidulans,* the bronze-colored atromentin from *Paxillus atromentosus,* and the red spinulosin from *Penicillium* spp. and *Aspergillus fumigatus.* The benzoquinone ring (as well as an isoprenoid side chain) is also an essential feature of ubiquinone and plastoquinone, which are involved in electron transport in the respiratory chain and in the light reactions of photosynthesis, respectively. The ring system of the numerous naturally occurring anthraquinones of lower fungi and higher plants contains a fused benzoquinone ring. The naphthoquinone ring system of phylloquinone (vitamin K) contains a fused benzoquinone ring, and an isoprenoid side chain is also present.

Benzo-l,4 quinone Naphtho-l,4puinone Antrapuinone

6-Benzylaminopurine: a synthetic cytokinin often used in phytohormone research. It represents a derivative of kinetin, in which the furfuryl residue is replaced by benzyl. Alternatively, it can be considered as a derivative of the natural phytohormone, zeatin, in which the isoprenoid side chain is replaced by benzyl.

Beradius: see Whales.

Berberidaceae, ***berberis family:*** a family of about 650 species of dicotyledonous plants, native mostly in the temperate regions of the Northern Hemisphere. They are herbs, bushes or trees (some of them evergreen) with simple or compound leaves. The wind-pollinated flowers are hermaphrodite, single or in racemes, with a double perianth and numerous stems. The superior ovary develops into a berry, rarely into a capsular indehiscent fruit.

The only native European species is ***Barberry*** *(Berberis vulgaris),* a thorny bush growing up to 3 m high, with racemes of yellow flowers, which possess nectar-producing staminodes (honey leaves) and sensitive stamens. Since Barberry is an alternative host of the black stem rust of cereals *(Puccinia graminis),* it has been exterminated in many areas. Its roots yield the isoquinoline alkaloid, berberine. The evergreen North American ***Mahonia*** *(Mahonia aquifolium)* is often planted as a garden ornamental and kept in bush form by trimming

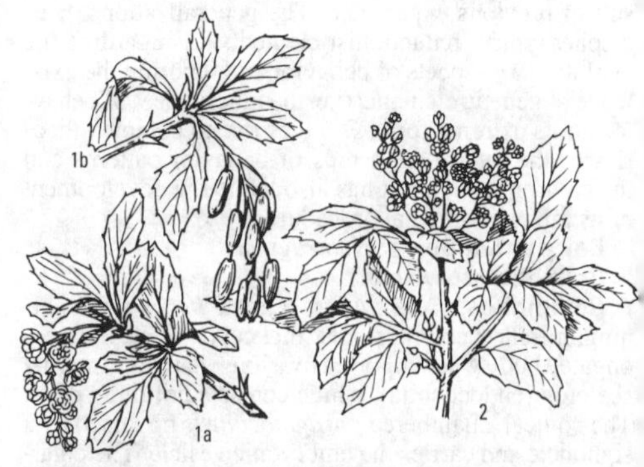

Berberidaceae. 1a and *1b* Barberry *(Berberis vulgaris). 2* Oregon grape *(Mahonia aquifolium).*

Berberine: a yellow isoquinoline-type alkaloid, present in the roots of *Berberis vulgaris* and in some other plants.

Bergapten, ***5-methoxy-furano (2':3'-6:7) coumarin:*** a coumarin derivative present in oil of bergamot *(Citrus bergamia)* and many other plants. It is toxic to fish.

Bergapten

Bergenia: see *Saxifragaceae.*

Berlese fauna: soil fauna, mostly arthropods, isolated from soil with Berlese's apparatus, or Tullgren's funnel. See Soil microorganisms.

Berlese's apparatus: a device for isolating small Soil organisms (see), in particular small arthropods.

Berthelinia limax: see *Gastropoda.*

Beta: see *Chenopodiaceae.*

Betalains: a group of nitrogenous plant pigments found almost exclusively in the order *Centrospermae (Caryophyllales).* They are all derivatives of betalaminic acid. They never occur jointly in the same plant with anthocyanins, and are therefore taxonomically useful. Examples are ***Indicaxanthin,*** a yellow pigment from the prickly pear *(Opuntua ficusindica;* family *Cactaceae);* ***Betanin,*** a red pigment from common beet *(Beta vulgaris;* family *Chenopodiaceae);* and the ***Muscaaurins,*** orange pigments in the cap of the fly agaric fungus *(Amanita muscaria).*

Betanin: see Betalains.

Betel nut (and palm): see *Arecaceae.*

Betsy beetles: see *Lamellicorn beetles.*

Betta: see Gouramies.

Betula: see *Betulaceae.*

Betulaceae, ***birch family:*** a family of dicotyledonous plants containing 95 species in 2 genera, distributed mainly in the Northern Hemisphere. They are deciduous, woody plants, usually trees, with alternate, entire leaves. Stipules are shed at an early stage (caducous stipules). The wind-pollinated flowers are monoecious, with male and female in separate inflorescences. Male inflorescences form a drooping catkin (3 together in the axil of a branch), whereas female inflorescences form erect cylindrical or ovoid catkins (2–3 in the axil of a bract). The 2-celled ovary develops into a winged seed, i.e. a single nutlet carried on the surface of a scale. *Betula (**Birches**)* has 2 stamens which are deeply divided below the anthers, 3 female flowers to each bract, and cylindrical seed-bearing catkins with relatively thin, 3-lobed scales which fall with the fruit. *Alnus (**Alders**)* has 4 stamens which are not deeply divided, 2 female flowers to each bract, and ovoid or cone-like ellipsoid seed-bearing catkins with woody, 5-lobed scales which remain on the cone after the fruit falls.

Birches have a more or less conspicuous white or yellowish bark. The familiar ***Silver birch*** or ***European white birch*** *(Betula pendula* Roth = *B. verrucosa* Ehrh. = *B. alba* L.), which attains a height of 25 m, is known for its silvery white bark, which becomes black and fissured near the base. Widespread in the wild in Europe and Asia Minor, it is also planted as an ornamental and forestry tree in Europe and North America. More common in North America is the native ***American white birch*** or ***Paper-white birch*** *(Betula papyrifera),* which is abundant throughout all the forested regions of northern USA and Canada. Its bark is whiter and less silvery than that of *B. alba,* and it continually peels away in thin papery sheets. Other North American species include *B. nigra* **(*Red* or *River birch*)**, *B. lenta* **(*Cherry birch*)**, *B. alleghaniensis* **(*Yellow birch*)** and *B. populifolia* **(*Gray birch*)**. The ***Brown*** or ***Downy birch*** (*B. pubescens* Ehrh.; height 15 m) is native to Central Europe and Siberia. Both the Silver and Brown birch yield Birch tar (see). *B. pubescens, alba* and *papyrifera* occur as far north as the Arctic.

Alders are moisture-loving trees or bushes, whose winged nutlets are formed in woody, cone-like fructescences. In Europe and the Near East, the ***Common*** or ***Black alder*** *(Alnus glutinosa* (L.) Gaertn. = *A. rotundiflora* Stokes) is found in boggy woodlands and on river banks in low marshy areas, whereas the ***Gray*** or ***European alder*** *(A. incana* (L.) Moench) is found in wet upland habitats. *A. glutinosa* has been introduced into North America. Both the Black and Gray alder possess root nodules (actinorhizas) which harbor symbiotic, nitrogen-fixing actinomycetes of the genus *Frankia.* Since these alders can survive on nitrogen-poor, boggy soils, they were among the pioneer plants that colonized land in the Northern Hemisphere after the last Ice Age. Other species of *Alnus* occur in Southeast Asia and the Andes.

Betulin: a pentacyclic tripterpene diol, found in birch and hazel-nut bark, in rose hips, and in the cactus *Lemaireocereus griseus.*

Beyrichia (named after the paleontologist H.E. Beyrich, 1815–1896): a fossil ostrocod with a semicircular, straight-hinged, 2-valved shell. Each valve has 3 rounded lobes in the dorsal position. It is found from the Silurian to the Devonian.

BFB-cycle: see Break-fusion-bridge cycle.

Biceps muscle: see Musculature.

Bicuspid valve: see Heart valves.

Bidder's organ: see *Bufonidae.*

Biennial plants: a term for Hapaxanthic plants (see), which require two summers and the intervening winter to develop from the seed to the flowering and fruiting stage, after which they die. They develop vegetatively and usually lay down food reserves in the first year, and these reserves contribute to flower and seed formation in the second year. Examples are foxglove, cabbage and beet. See Annual plants, Perennial plants, Life form, Lifespan.

Bifidus bacterium, *Lactobacillus bifidus:* a lactic acid bacterium present in large numbers in the feces of breast-fed infants. It is an essential regulatory component of the intestinal flora of infants. With an average size of 4 × 0.5–0.7 μm, often with expanded ends, it exhibits considerable pleomorphism. It is named from its apparent bifid appearance described by early authors; 3 bacilli can become arranged in a Y-shape. In primary culture it is strictly anaerobic, but may eventually become microaerophilic. It is normally described as Gram-positive, but its reaction to the Gram stain is variable.

Bifurcation: see Far off equilibrium.

Big bud of blackcurrants: see *Tetrapodili.*

Big-headed turtle: see *Platysternidae.*

Bilateral symmetry: two-sided symmetry, i.e. there is only a single plane of symmetry. The bodies of bilaterally symmetrical animals can be divided into two halves, which are mirror images of each other.

Bilateria: all bilaterally symmetrical animals, i.e. all animals belonging to the bilaterally symmetrical eumetazoan phyla of the animal kingdom. *B.* therefore comprise all animal phyla, with the exception of the *Porifera* (sponges), *Cnidaria* (or *Coelenterata;* hydras, jellyfish, sea anemones, etc.), and *Ctenophora* (comb jellies). As a rule, *B.* possess an anterior end with the

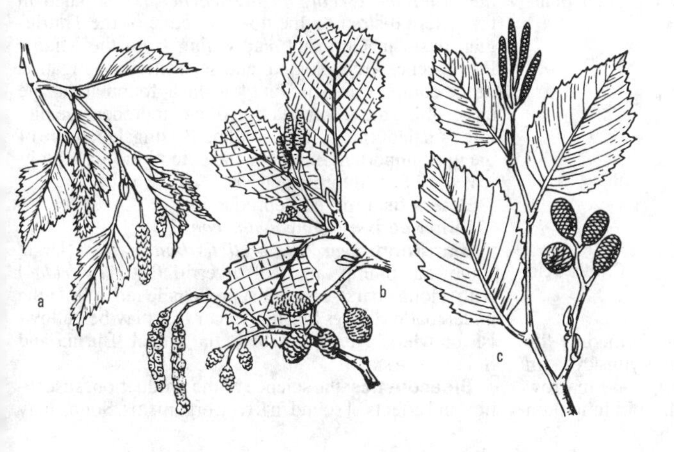

Betulaceae. a Silver birch *(Betula verrucosa* Ehrh, *Betula pendula* Roth). *b* Alder *(Alnus glutinosa). c* Gray alder *(Alnus incana).*

mouth and main sensory organs, and this is differentiated from the posterior end, which contains the anus. The surface of the body carrying the mouth is designated the ventral side, while the opposite surface is known as the dorsal side. In addition to ectoderm and endoderm, *B.* possess a third embryonic germ layer, known as the mesoderm, which makes a large contribution to organ formation. *B.* are divided into 2 branches, the *Protostomia* (see) and the *Deuterostomia* (see).

Bilayer lipid membrane, ***BLM:*** a planar, artificial lipid double layer, which serves as a model of a biological membrane. BLMs are used for physical-chemical studies of e.g. conductivity, capacitance and transport. The BLM technique was introduced in 1960 by Müller and Rudin, and it played an important part in the development of membrane biophysics. As a model membrane system, the BLM has now been largely replaced by Liposomes (see).

Bilberry: see *Ericaceae.*

Bile acids: steroid carboxylic acids conjugated with (bound in amide linkage to) taurine or glycine, and known respectively as *taurocholic* or *glycocholic* acids. Strictly speaking, B.a. are these taurine and glycine derivatives, but the term is also applied to the free steroid carboxylic acids. Glycocholic acids predominate in human and bovine bile [e.g. the glycine conjugate of *chenodeoxycholic acid* (3α, 7α-dihydroxycholanic acid)], whereas taurocholic acids are more common in canine bile. B.a. are are typical components of mammalian bile. The sodium salts of B.a. reduce surface tension and emulsify fats, acting as an aid to their digestion and absorption (in lower vertebrates, the bile alcohols have the same function). B.a. are indispensable for fat absorption in the intestine. Furthermore, B.a. activate the lipases that are responsible for the intestinal hydrolysis of acylglycerols. B.a. are synthesized in the liver from cholesterol, and 90% of the B.a. released in the bile is resorbed in the intestine and returned to the liver in the enterohepatic circulation. One liter of bile contains 30 g B.a. Often the pattern of B.a. is specific for the given species.

For experimental and commercial pruposes, B.a. are usually prepared from bovine or porcine bile. Some of the commonest B.a. are *cholic acid* (3α, 7α, 12α-trihydroxycholanic acid) (Fig.), *deoxycholic acid* (3α, 12a-dihydroxycholanic acid), and *lithocholic acid* (3α-hydroxycholanic acid).

B.a. are used as starting materials for the technical synthesis of adrenocortical hormones and other pharmacologically important steroids.

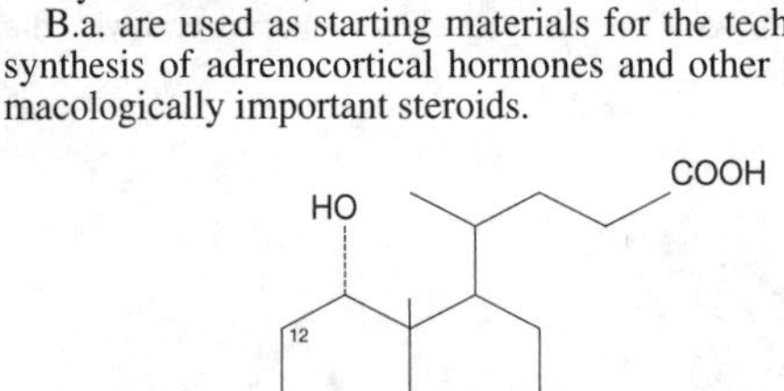

COOH HO 12 3 7 HO H OH

Cholic acid

Bile duct: see Gall bladder.

Bile pigments: a group of pigments formed by the degradation of the hemoglobin of aged erythrocytes in the reticuloendothelial system (spleen), bone marrow and liver. Hydroxylases of the endoplasmic reticulum cleave the α-methine bridge between rings A and B of heme to form CO and blue-green biliverdin IX (an open chain containing 4 pyrrole rings) [an early green-colored product is *verdoglobin (choleglobin),* which still contains globin and iron]. Biliverdin is reduced to orange-red bilirubin and transported to the liver as a complex with serum albumin.

In its free form, bilirubin is highly toxic, especially for newborn infants, in whom it can cross the blood-brain barrier more easily than in adults. In jaundice, bilirubin is increased in the blood, causing yellowing of the skin.

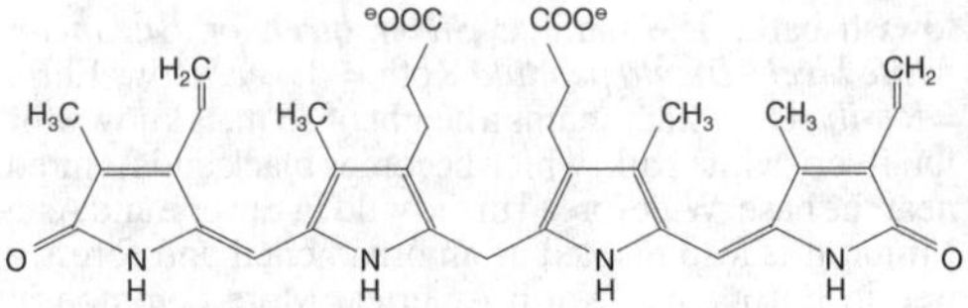

$^{\ominus}$OOC COO$^{\ominus}$ H_2C CH_2 H_3C H_3C CH_3 CH_3 O N H N H N H N H O

Bilirubin is released from serum albumin and concentrated in liver cells (it contributes to the yellow-brown color of this organ), where it becomes conjugated with two glucuronic acid residues by enzymes (glucuronosyltransferases) located mainly in the smooth endoplasmic reticulum. The resulting glucuronide is excreted into the intestinal tract in the bile. In the large intestine, most of the conjugated bilirubin is hydrolysed back to free bilirubin, then reduced by intestinal bacteria to yellow *meso*bilirubin, which in turn is reduced to colorless urobilinogen *(meso*bilirubinogen) and stercobilinogen. The latter two compounds are oxidized by oxygen to orange-yellow urobilin and golden yellow stercobilin, which are responsible for the brown color of feces. In addition to their occurrence as excretory products in feces, B.p. are also found in the egg shells of birds, and they occur as wing and skin pigments of insects. Numerous plants contain phycobiliproteins, i.e. protein conjugates of B.p. containing an ethylidene group. Phycobiliproteins are components of the photosynthetic membranes of red algae and *Cyanophyceae.*

Bilimbi: see *Oxalidaceae.*

Bilirubin: see Bile pigments.

Biliverdin: see Bile pigments.

Billfishes: see *Scombroidei.*

Bilzingsleben: an archeological site with human remains *(Homo erectus bilzingslebenensis),* situated in the Artern district on the northern edge of the Thüringian basin in Germany, and dating from the Middle Pleistocene. Archeological and scientific investigation of the complex association of artifacts found at the site has enabled the reconstruction of a comprehensive picture of paleolithic culture, making Bilzingsleben one of the most important European old paleolithic archeological sites. See Anthropogenesis.

Binary fission: see Reproduction.

Bindweed: see *Convolvulaceae.*

Binturong, ***bear cat, Arctictis binturong:*** a large, somewhat plump, arboreal viverrid (see *Viverridae)* with a long coarse coat and a prehensile tail. While the undercoat is always black, the hair tips may be yellowish or whitish. It occurs in India, Nepal, Burma and Southeast Asia.

Bioacoustics: the science of the production, susception and effects of sound in living organisms. Sound may simply be an accompanying phenomenon (i.e. an unavoidable side effect) of several biological processes, such as limb movement, respiratory activity, alteration of pressure and tension within the body, and the release of gases and other substances from the body. Sound is also generated as a functional signal, in which case specialized organs exist for both its generation and perception. Thus, sound may be produced by rubbing or striking together different body structures (frictional or stridulation organs). Particularly suitable for this purpose are are hard scales (reptiles), teeth, bones (some fish), or parts of a hardened cuticle (arthropods). Also, sound produced by bodily contact with the substratum may be used for signalling, e.g. beating the local substratum with the legs or with parts of the body (some spiders, deathwatch beetle, psocids, cockroaches, stoneflies), and beak "drumming" as performed by woodpeckers. The body itself may also act as a resonance chamber, as in chest beating by gorillas. Respiratory sounds have evolved into true vocalizations; in reptiles and mammals, vocalizations are produced by the larynx, while in birds this is usually the function of the syrinx. In cicadas and certain fishes, sound is produced by vibrating membranes. Various other types of specialized sound generation are known. Sound generation as a means of biocommunication is restricted essentially to vertebrates and arthropods. Sounds are perceived and processed by appropriate receptors, which are collectively known as auditory organs. Sound may also directly influence body functions in the absence of specialized auditory organs. Originally, auditory organs evolved for detecting prey and/or aggressors. By the further adaptive specialization of receptors, signal transfer was improved and refined, and auditory organs evolved into organs of biocommunication. With modern acoustic technology, it is possible to analyse sounds generated for signalling purposes, and to relate these to the intra- and interspecies behavioral responses. Often, acoustic and visual signals are combined as a means of biocommunication. In some cases, the generated sound displays complex patterns, which (theoretically) have a high information content. Signal properties (i.e. the information content) depend on several factors, such as the quality of the sound (the phonetic content), changes of sound intensity, the occurrence of particular sound sequences or "phrases", and other regularities based, e.g. on the tempo of changes in pitch. Often, the sounds produced by organisms lie in the ultrasound range, i.e. partly or entirely outside the frequency range perceived by the human ear. Aided by modern developments in sound technology, the subject of B. has developed since the 1950s into an independent discipline of behavioral biology.

[Arthropod Bioacoustics: Neurobiology and Behaviour, by A.W. Ewing, 1989, Edinburgh University Press]

Biocatalyst: an organic catalysts of high substrate and action specificity. It initiates and regulates the chemical reactions essential to living processes. Bs. are Enzymes (see), Vitamins (see) and Hormones (see). In the strictest sense, a B. must be an enzyme. Vitamins and hormones indeed have a catalytic effect, in that they generate a response that is not proportional to their quantity (they do not provide material for the synthesis of structures, and they do not serve as fuel for energy production), but their actions are ultimately manifested through enzyme action. Thus, vitamins may be the precursors of coenzymes, some hormones trigger the onset of an amplification cascade of enzyme reactions (e.g. epinephrin), while other hormones promote transcription of DNA, leading ultimately to enzyme synthesis (e.g. steroid hormones).

Biochemistry: a largely independent bioscience, which originated toward the end of the 19th century from the interaction of chemistry, biology, medicine and agriculture. In the UK and USA, B. developed from the chemical aspects of physiology in preclinical medical teaching. This physiological chemistry rapidly became a discipline in its own right, and was eventually named B. In France, the physiological studies of Claude Bernard also contributed to the subsequent growth of B. In many countries, notably Germany, organic chemists were involved in isolating, purifying and determining the structures of natural products. This natural product chemistry (i.e. the chemistry of substances of biological interest) is also one of the important precursors of B. Contributions were also made by agriculture and microbiology.

Since the beginning of the 19th century, the growth of B. has been explosive, and it is now probably the most active of all the biological sciences. B. is a biological discipline, in that it attempts to answer biological questions, but its methods are largely those of chemistry and physical chemistry. It investigates 1) the structure of the components of living organisms (this area is known as *descriptive B.,* and it is largely synonymous with natural product chemistry), 2) the chemical reactions involved in the interconversion of substances in living organisms, 3) mechanisms of energy production and utilization, and 4) the chemical basis of information storage and transfer. 2), 3) and 4) are collectively known as *dynamic B.*

Subdivisions of the subject are recognized, based on the area of application, the nature of the experimental material, or the type of problem under investigation. Thus there are several important branches of *applied B.*, e.g. agricultural B., clinical B., industrial B., etc. Since all life forms share a fundamental biochemical unity, *general* or *basic B.* is a necessary first stage in the training of all biologists. Biochemical differences between organisms, however, also justify and necessitate subdivisions such as microbiological B., plant B. (including the biologically fundamental subject of photosynthesis) and animal B. Human B. and clinical B. are, for obvious reasons, intimately related, and they occupy a special position in medical teaching and research. Other biochemical disciplines are, e.g. enzymology, immunology and biochemical genetics. The area of B. called molecular biology, sometimes known as the "new biology", has grown to such an extent that some scientists claim to be molecular biologists rather than biochemists, and some university departments of B. have renamed themselves departments of molecular biology. Important problems at the forefront of modern biochemical research are the B. of differentiation, the B. of biological energy transduction, and the B. of the higher centers of nerve activity.

Biochore, *biochorion:* see Bioc(o)enosis.

Bioclimatology: a branch of climatology, which studies the effects of climate on plants, animals and humans (e.g. the relation between weather and illness), the ability of organisms to adapt to climatic changes (see Acclimatization), and seasonal changes in the physiology and morphology of plants and animals *(Phenology).*

Bioc(o)enology: the differentiation and classification of biotic communities. B. is a branch of synecology (see Ecology), with which it is often considered to be synonymous. It is founded on the existence of the Bioc(o)enosis (see), an empirical unit, which is difficult to classifiy or define.

Bioc(o)enosis, ***biotic community:*** a concept introduced by Möbius (1877) when studying oyster bed communities. He wrote, "Every oyster bed is a community of living beings, a collection of species and a massing of individuals, which find everything necessary for their growth and reproduction … a community whose species and individuals mutually interact, limit and select each other under the prevailing conditions, and continue to occupy a defined territory". He proposed the term, B., for such a community. Members of a B. display a marked degree of interdependence. Any single B. is characterized by the constant presence of certain species; these species exist in defined combinations or groupings, in which the population size of each species is compatible with the balanced composition of the B. (see Dominance).

Balogh defined 4 types of B.: 1) *true* or *planar biocoenoses,* or biocoenoses in the strict sense, which are distributed over a relatively large area, e.g. forest biocoenoses; 2) *border* or *linear biocoenoses,* which are found on the edges or borders of planar biocoenoses; 3) *zonation* or *ribbon biocoenoses,* which are found in gradients between differing environments, e.g. on river banks; 4) *choriocoenoses* or *point biocoenoses,* which occur at isolated sites within the other types of biocoenoses, and differ from these in their characteristic combination of species. Different types of biocoenoses may exist together in a mosaic, e.g. bush-savanna; and zonation complexes are often formed in a very restricted area by successive zonation biocoenoses (e.g. on the sea shore).

In addition to natural biocoenoses, there are also unnatural or man-made communities. These are formed from the components of natural groupings, and they are preserved only by continual human intervention. When left undisturbed, they revert in stages (see Succession) to natural biocoenoses (see Anthopocoenosis, Ruderal coenosis, Agrocoenosis). The degree of human influence is variable. Thus, without doubt, anthropocoenoses represent the ultimate human influence over natural systems, whereas many areas of man-made forest are largely similar to natural forest communities.

All biotic communites are formed from permanent structural elements, and one of the most important of these is the Semaphoront (see) or a population of semaphoronts. Such elements may associate to form complex units or *merocoenoses,* which are characteristic structural components of a B. (e.g. flowers, stem and fruits with their associated microflora and microfauna). Merocoenoses associate in layers to form *stratocoenoses,* i.e. a B. is subdivided into layers, usually based on plant height (e.g. soil horizon, vegetation layer and canopy). Alternatively, the plants of a B. may be grouped in *synusiae,* in which the concept of morphol-

Classification of the most important habitats with their widely varying hiocoenoses

Critical temperature and moisture conditions	Variants	Habitats without continuous vegetation cover	Habitats with grass and herbaceous cover	Forests or bush- and shrub-covered areas
		1. Litorea		2. Hylea
Excess moisture	Warm	Vegetation-free river bank and beach zones of tropics and subtropics (below high water mark)	Tropical and subtropical flood areas. Saline flats and marshland	Tropical and subtropical montane rain forestes and mangrove forests
				3. Silvea
	Cold	Vegetation-free river bank and beach zones of temperate regions (below high water mark)	Reeds, marshy meadows, saline flats and marshland of temperate regions. Subarctic and subalpine tall herbaceaus vegetation	Mesophilic deciduous summer forests. Swamp forests
		4. Tundra		5. Taiga
Low temperature	Dry to damp	Cold deserts (subniveal cushion plant zones, rock outcrops)	Northern and alpine grass meadows. Arctic and maritime dwarf bush heaths	Northern and subalpine conifer forests
	Wet	Glacier and snow border zones (snow fields and snow patches)	Tundra bogs. Treeless high bogs	Bog forests
		6. Desert	7. Steppe	8. Sklerea
Arid conditions	Hot	Stone, clay, loess and sand deserts and semideserts (lomas)	Tree steppe, forb steppe and grass steppe	Dry thorn and savanna forests
	Warm	Drifting sand dunes. Hard, steep rock faces. Stony plains	Steppe meadow. Beach grass heaths	Sclerophyllous forests. Pine heath forests. Deciduous hardwood steppe forests. Dry shrub heaths

ogy is used differently; thus a synusia of climbing plants is distinguished from the synusia of their supporting trees. Particular patterns of existence or microhabitats within a B. are also referred to as *niches,* a term favored by zoologists, but equally applicable in plant ecology. An intricate dynamic relationship or interactive network exists between the various structural elements of a B. Individual components of this complex can be studied in isolation, e.g. the food chain.

The intimate interaction between a B. and the Biotope (see) produces a largely self-regulatory system, which has been variously called a *biogeocoenosis, biotic district* or *biochore,* but is now generally known as an *ecosystem.* The B. is in dynamic equilibrium with its Environment (see), i.e. over a relatively extended period it is able to more or less reproduce itself and remain constant; but it undergoes change, due to Succession (see) and Evolution (see).

According to Tischler (1955), the biotic communities of land animals can be classified on the basis of their habitats, as shown in the Table.

The largest biocoenotic unit is the *biome* or *formation,* i.e. the characteristic biotic community of a bioregion, e.g. deciduous European forest.

Bioc(o)enotic parameters, ***bioc(o)enotic characters, ecological parameters, ecological characters:*** parameters that define the composition and structure of biotic communities. B.p. may be absolute, relative or structural. 1) Parameters of absolute quantity (e.g. abundance, biomass, bioenergy) are related to area or volume. 2) Parameters of relative quantity (e.g. dominance, mass dominance, energy dominance) express the percentage contribution of a species, group or other unit to the totality of individuals, total biomass, etc. 3) Structural parameters or characteristics (e.g. constancy, dispersion, evenness, diversity) provide information on the composition and structure of a community, and they are used primarily to compare the efficiencies of different biotic communities.

Certain comparative parameters [e.g. Jaccard's index (see), Dominance identity (see), Quotient of similarity (see), Divergence (see)] express the relationship between the proportion of species common to the total composition of two biotic communities. All of these parameters or characters can therefore be reduced to the ratio: average number of species common to only one community/number of species common to both communities.

Biocommunication: the transmission of information between organisms. Information is transmitted by means of signals, whose meaning is determined by the sender and decoded by the receiver. A known example is the feeding call of the cock, in which the sequence of sounds indicates food and simultaneously where the food is to be found; the hens react accordingly by running to a certain spot and pecking the ground. B. guides and regulates interaction between members of the same species. Typically, members of the same species use the same innate signals and responses, without the necessity for a prior learning process. In some cases, however, a certain degree of learning is necessary. The following three levels of B. are recognized.

a) Communication of the fact that the sender is prepared for a particular type of behavior. The signal indicates the inner state of the sender, e.g. a readiness for flight, so that the receiver can adapt its own behavior accordingly. b) Communication of the external cause of this inner state, e.g. a warning call to indicate the presence of an aggressor. c) Communication of the fact that the behavior pattern is completed (see Appetitive behavior), e.g. the individual providing the signals flies into the air or runs away, as a signal that other individuals should do likewise.

Many different sensory modalities are exploited in B., but a single species employs only one or two major senses. Thus, B. may occur through chemical, mechanical, electrical or visual channels. Mechanical channels can be subdivided into: 1) tactile-mechanical, 2) vibratory-mechanical, and 3) acoustic or auditory channels. Reptiles and fishes depend mostly on visual signals, birds use both visual and auditory signals, while social insects rely heavily on the release and perception of chemical agents. Chemical signals may be used as markers of territorial boundaries or pathways, or they may be transmitted over a distance as alarm signals or attractants (see pheromones). Included in visual signalling is the coded light emission by various animal species, the light being produced by their own metabolism (e.g. glow worms), or by symbiotic bacteria (see Bioluminescence). The information content of light signals is encoded by their wavelength, the timing of light pulses, their pattern of occurrence on the body surface, and possibly changes of intensity.

Biocybernetics: the study and analysis of the guidance and regulation of biological systems. The boundaries of B. are not clearly defined. According to some authorities, B. includes information theory, communication theory, animal orientation, information processing by sense organs, as well as systems theory.

Bioelements: the chemical elements present in the compounds that constitute living material. The bioelements, carbon (C), oxygen (O), hydrogen (H), nitrogen (N), sulfur (S) and phosphorus (P), account for more than 90% of living material. About 40 bioelements are known.

Bioenergetics: a branch of biophysics, which treats living organisms as physical-chemical systems, and describes them in thermodynamic terms.

Life manifests a high degree of spatial and temporal order, but the formation of biological structures cannot be interpreted as an equilibrium process; in fact thermodynamic equilibrium is synonymous with death in living systems. Since the order displayed by living systems is the result of a nonequilibrium state (i.e. a steady state, rather than an equilibrium), the energy relations of living processes can be described only with the aid of Nonequilibrium thermodynamics (see). In the thermodynamic interpretation of living processes, statements can be made that are independent of the specific molecular structure of the analysed system. Thus, the directional nature of biochemical reactions, the spontaneous formation of ordered structures, and the maintenance of those ordered structures can be explained by continuous entropy production through the dissipation of free energy. The energy input or loss can then be interpreted in terms of molecular structures with the aid of statistical mechanics. Particularly important is the statistical thermodynamic description of evolution, especially molecular evolution. Thus, according to Eigen, the evolution of precellular systems can be explained by nonequilibrium thermodynamics through the formation of dissipative structures known as hypercycles (involving RNA, DNA and proteins). Models of metabolic cycles can be used to predict the levels of components during the steady state, and to identify possible regulatory sites. Such studies have practical importance for the optimal regulation of industrial microbiological processes.

B. is also concerned with processes of energy conservation and transduction in living systems, e.g. the transduction of the redox energy of respiratory substrates to the chemical energy of ATP in oxidative phosphorylation, the transduction of radiant energy to reducing power and ATP in photosynthesis, the storage of chemical energy in the unstable phosphate bonds of ATP and other phosphorylated compounds, and the utilization of this energy to support biosyntheses, heat production, light production, movement, etc.

Biofiltration: see Biological waste water treatment.

Biogas: gas formed by the bacterial fermentation of cellulose-rich organic material, especially sewage sludge and agricultural waste such as straw, dung, etc. The process in conducted in heated decomposition chambers by the action of methane bacteria. B. contains about 60% methane and 40% carbon dioxide, as well as traces of hydrogen sulfide and hydrogen. It is stored in low pressure gas tanks, often installed close to sewage processing plants and large agricultural concerns, where the gas can be used for heating, as a spray propellant, and for generating electricity. B. may also be mixed with town gas (coal gas). It has twice the thermal value of coal gas, but burns less rapidly. The residue from B. is a valuable fertilizer.

Biogenetic law: a law stating that the embryo undergoes a recapitulation of the animal's evolutionary history. It is discussed under Baer's law (see), of which it is a later extension.

Biogenic amines: amines arising from the enzymatic decarboxylation of amino acids. They occur widely in plants and animals, and are formed by bacteria as products of protein decomposition. B.a. often have potent physiological activity (e.g. hypertensive properties of tyramine, and hallucinogenic action of psilocybin). Some are neurotransmitters (e.g. acetylcholine, noradrenalin). Residues of ethanolamine and of choline (derived from L-serine residues) are present in some phospholipids. Some B.a. are precursors of alkaloids (e.g. phenylethylamines are precursors of isoquinoline alkaloids), and some are precursors of vitamins and coenzymes (e.g. 1-amino-2-propanol, derived from L-threonine, is incorporated in the biosynthesis of vitamin B_{12}). Other examples are muscarine, putrescine, cadaverine, spermine, histamine, mescaline, hordenine, dopamine, tryptamine, serotonin and carnitine.

Biogeoc(o)enosis, ***holoc(o)enosis***, ***biosystem:*** terms that are synonymous with Ecosystem (see). The term, B., is used by Russian ecologists. Holoc(o)enosis (Holozön) was introduced by the German author, Friedrich (1930), and biosystem by the German author, Thienemann (1939).

Biogeography: the study of the distribution and dispersal of living forms on the planet. Traditionally, and as the result of objective difficulties, B. is usually divided into Zoogeography (see) and Phytogeography (see), which are treated as separate disciplines. The geographical regions defined in zoogeography differ from those recognized by students of phytogeogrphy. Moreover, identical terms may have different meanings in these two subdivisions of the subject.

Biohelminth: see Geohelminth.

Bioindicators: living organisms, whose established presence reveals the long-term, combined activity of numerous environmental factors (e.g. the typical plants of acidic or basic soils), and can therefore be used to characterize a particular habitat. B. also respond (usually by decreasing or increasing in numbers) to sudden changes in the factors that define their habitat. They therefore also serve as indicators of pollution. See Indicator plants.

Bioliths, ***organogenic sediments:*** sediments and sedimentary rocks formed from the remains of living organisms. *Zooliths* are derived from animals, e.g. radiolarian mud, coral limestone, Muschelkalk (see), echinospherite limestone. *Phytoliths* are of vegetable orgin, e.g. coal, oil shale, algal chalk. Microscopic silicified plant remains are also called Phytoliths (see).

Biological clock: see Biorhythm.

Biological decalcification: the precipitation of calcium carbonate from natural waters as a result of the activities of photosynthetic organisms. In submerged plants, the calcium carbonate is deposited as a crust on the leaves. The activities of phytoplankton lead to the formation of very fine calcium carbonate crystals, which sediment as marine chalk, and are then partly redissolved in the hypolimnion (see Sea).

Biological engineering: in the wider sense, the generation of biological effects by structural alteration of the landscape and/or the use of biological methods to generate specific landscape features.

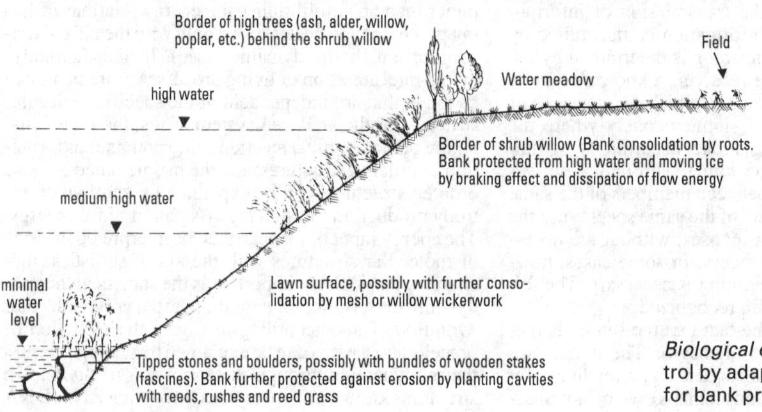

Biological engineering. River control by adapting natural structures for bank protection (after L. Bauer).

Essentially, B.e. consists of the use of suitable plants (in particular trees, bushes and grasses) to consolidate river banks, lake edges and dunes, and to prevent landslip and erosion of waste tips and slag heaps. Biological methods are initially more costly than purely structural engineering methods, but they provide greater and longer lasting protection, are easier to maintain, and blend more satisfactorily with the natural landscape.

Biological filter: see Biological waste water treatment.

Biological law of relativity: a law formulated by Lundegårdh, which describes the relation between the relative effects of an environmental factor on plant growth and the intensity of the factor. The key sentence in Lundegårdh's definition is "The relative activity of a factor is insignificant in the optimal range, but it becomes very great in the minimal and especially in the inhibitory range". As an ecological generalization, it is derived from a law describing the relative effects of factors that influence plant assimilation. Mathematically, it is more or less identical with Mitscherlich's law of plant growth factors (see).

Biological pacemaker: see Biorhythm.

Biological pest control: the use of biological products or living organisms to control or destroy unwanted organisms. This includes the introduction of predators, parasites, or disease organisms, as well as the application of toxins (e.g. pesticides) of biological origin. Also included are techniques such as the introduction of large numbers of artificially sterilized or genetically altered males of a pest polulation , so that subsequent progeny are nonviable, as well as the use of biological attractants (e.g. sex pheromones) to confuse the mating behavior of the pest, or to attract it to a trap or toxin.

Biological control measures have the advantage over the use of chemical control agents (e.g. DDT) that they do not introduce biologically nondegradable toxins into the natural food chain or into harvested food products. Biological agents are also highly specific and do not affect organisms for which they are not intended. In many cases, biological measures are also more enduring than the application of chemicals, and in the long term may therefore be cheaper. Furthermore, once pests have developed resistance to chemical agents, biological methods may be the only effective means of control.

I *Control of mites and insects.*

1) Promotion of antagonists already present, e.g. bird protection measures (nesting aids, winter feed, perches and lairs for buzzards, etc.); protection and conservation of ants; protection of flowering wild plants ("weeds") and sowing of nectar plants to feed flower-visiting parasitic or predatory insects.

2) Naturalization of foreign fauna, i.e. the "classical" method of control, which has been used successfully for the eradication and control of many pests. This approach is particularly successful if the pest in question is also an introduced species, and if the host or prey is accessible to the antagonist for long periods during the year (e.g. scale insects). Appropriate climatic and weather conditions are also important.

3) Mass release of animal antagonists (saturation or inundation method), producing a temporary excessive increase in population density, which is effective against the pest, but which does not persist. A notable example is the mass culture and release of chalcidids of the genus *Trichogramma* as polyphagous egg parasites against a series of lepidopteran pests in the former USSR. Also important is the use of predatory mites *(Phytoseiulus persimilis)* and chalcidids in greenhouses, where many pests are introduced foreign species. The mass release of insect-pathogenic nematodes has been used only to a small extent.

4) The use of pathogens (viruses, bacteria, fungi). Microbiological pest control is technologically similar to the application of pesticides, but it is more or less strictly specific. Biologically, it resembles a saturation method, since the effect is usually not sustained over several generations. Preparations based on *Bacillus thuringiensis,* which act almost exclusively on insect larvae, are used widely (fortunately, during the preparation of the specific toxin, an accompanying toxin with a broader spectrum of activity is lost). More recently, a strain of of *B. thuringiensis* has been isolated whose toxin acts only on mosquito larvae. Fungal preparations are less stable in storage, and fungal spores germinate only under certain favorable weather conditions. The use of virus diseases is restricted, because it is expensive to produce the starting material from living organisms or tissue cultures. Commercial preparations, are, however, already available. Bacteria, fungi and viruses that attack insects are harmless to humans. The use of pathogenic protozoa, such as microsporidia, is still in the experimental stage.

5) Self-extermination. Extermination of the parasitic screw worm fly *(Cochliomyia hominivorax),* first on Curaçao, then in large areas of the USA demonstrated that entirely specific and total self-extermination is possible when all other control methods fail. Sterilized male flies were brought in for several generations in such numbers that they greatly exceeded the number of fertile males in the wild population. In 1988, the screw worm was also inroduced into Libya in an infected consignment of livestock from South America. It killed 12 000 animals in Libya in 1990, and unchecked it could have devastated the cattle of 5 North African countries, with the prospect of eventually spreading into the Middle East, southern Europe and Asia. At the peak of the successful eradication campaign, 40 million sterile males (bred and irradiated in Mexico) were released per week over Libya. The sterile male method is relatively costly, and it must be restricted to isolated areas, where in the long term a fresh immigration of the pest is unlikely. It can be used, however, to maintain a barrier zone, preventing penetration of the pest into a larger area. The sterilization is performed in the laboratory by irradiation or chemical treatment. The use of chemical sterilants in the open environment is not an approved practice, because the compounds in question have a broad spectrum of activity. A further genetic approach is the introduction into the population of mutants containing a genetic defect; this defect is transmitted and gradually spreads in the population, causing high losses, but not complete extermination of the pest. Finally, trials have been conducted with male mosquitoes of geographically distant races, which compete with local males, but are genetically isolated from the native females, and therefore incapable of fathering viable offspring (gene flow between neighboring populations is still possible).

II *Control of other animals.* Control of nematodes (with fungi) or snails (with beetles of the families *Lampyridae* and *Carabidae)* has been used, but only to a small extent. From time to time, mice and rats have been effectively decreased in various countries with *Salmonella,* but on grounds of hygiene there are obvious objections to the use of such agents. In the 1950s, the virus disease, myxomatosis, was successfully used in Australia against wild rabbits, but the rabbit population has risen again due to increasing resistance to the virus. Similar measures were introduced privately on a farm in France, with the result that myxomatosis spread to large areas of Europe, causing the widespread and unwanted deaths of wild and domestic rabbits. Especially in the last century, the naturalization of predators (e.g. mongoose, fox, domestic cat) sometimes led to catastrophic effects on the native fauna.

III *Weed control.* An impressive example of biological weed control is the elimination of *Opuntia* from large areas of Australia (largely by the introduction of a small butterfly) and South Africa (mainly by the cochineal shield bug), allowing the establishment of economic agriculture. In these cases, animal antagonists were imported that were specific for intentionally or carelessly introduced plants. Although this is similar in principle to the use of predators and parasites against insect pests, there is a risk that agricultural plants and native wild plants will also be affected. Thorough preparation and preliminary research are therefore necessary. Fungal plant pathogens have been used to a small extent.

The extent to which biological control has has gained acceptance over chemical control varies widely in different parts of the world. Certainly, the full potential of biological control measures has not yet been realized. The successful application and the choice between the many different types of biological control depend on an understanding of biological relationships and ecology. Such an understanding often seems to be totally absent in the indiscriminate, routine application of pesticides.

Biological relativity: see Biological law of relativity.

Biological self-purification: the oxidative degradation of material, which would otherwise putrify, in waters contaminated with sewage. The oxidative degradation is performed by organisms growing on the bottom surface (biological lawns), and by organisms in the freely suspended biomass (see Seston) of flowing water. The original biological equilibrium is restored stepwise by the mineralization of the organic contaminants. See Saprotrophic system, Biological waste water purification.

Biological value of proteins: the effectiveness of proteins in supporting the growth of young, growing individuals, or in maintaining the nitrogen balance of adult organisms. This value depends on the digestibility of the protein, and on its content of essential amino acids. Since protein synthesis in the body requires the simultaneous presence of all amino acids, efficient protein synthesis depends on the dietary provision of essential amino acids, i.e. those amino acids that the body cannot synthesize for itself. The absence of just one essential amino acid results in a slowing of protein synthesis and a decrease in the utilization (and an increase in the degradation) of all the other amino acids that are present. The most direct means of determining the biological value is to determine the amino acid composition by chemical analysis, but feeding studies are still necessary to account for digestibility. A diet complete with respect to carbohydrate, fats, mineral elements, vitamins, etc., but containing a single protein source (the protein under test), is fed to young animals; their growth rate is measured and compared with that of young animals on an optimal diet. Alternatively, adult animals (e.g. adult humans) are fed an optimal diet, so that they are in nitrogen balance, then all other sources of protein are omitted and replaced with the single protein under test, and the nitrogen balance measured. When the body is in nitrogen balance, the nitrogen excreted (in urine, feces, perspiration, loss of hair, sloughing of skin, etc.) is equal to that ingested in the diet (usually over a 24-hour period), i.e. (weight of N in diet – weight of N excreted) = 0. In negative nitrogen balance, however, this value becomes a negative number, indicating that protein synthesis is slowing down and unused amino acids are being degraded. The biological value of a protein can be expressed as: (N intake – N in feces/N in urine and perspiration) × 100. The nitrogen of the feces is taken to represent undigested material, so that (N intake – N in feces) represents the nitrogen actually absorbed by the body from digestion of the protein. The nitrogen of urine (and perspiration) is mainly derived from the catabolism of amino acids, and in mammals consists mainly of urea. Proteins supplying all the required amino acids in sufficient quantity will therefore have a value of 100 (e.g. whole egg albumin).

Single animal proteins have a higher biological value than single plant proteins, but in mixtures of dietary proteins deficiencies of essential amino acids in one protein are compensated by their presence in others. A diet can therefore have an adequate biological value, although its individual proteins may each have a low biological value.

Biological variability: the existence of random phenotypic differences between individuals of the same species. It is the most significant source of random variation in biology, and it necessitates the use of statistical methods for the analysis and description of biological phenomena. See Biostatistics, Variability.

Biological waste water treatment (plate 29): the purification of sewage and domestic and industrial waste water by the biological degradation of dissolved and finely suspended organic materials. According to the type of waste water, the nature of the land available, and limitations on space, the methods employed for B.w.w.t. are natural, semi-technological or technological.

1. ***Natural methods.*** Sewage water may be applied directly to the soil, where organic constituents are degraded (mineralized) by microorganisms, and the products of degradation contribute to the fertility and agricultural value of the soil. The microorganisms responsible are present only in the upper soil layers, so that relatively large areas of land are needed. This totally natural method, known as *sewage irrigation,* was used in Graeco-Roman times for the watering of gardens, and it is still applied to a greater or lesser extent. Thus, waste water may be applied directly for the improvement of agricultural soil by spraying of standing crops, or by allowing the water to seep into tilled land for the benefit of a subsequent crop, but nowdays the output of such waste is often so great that other means of disposal are also necessary. By the middle of the 16th century, sewerage systems linked to special drainage fields were developed in parts of Europe, and by the middle of the 19th century drainage fields for handling large quan-

tities of waste water had been widely adopted. *Drainage fields* or *sewage lands* are agricultural fields serving more or less exclusively for the purification of waste water, and their agricultural output has no economic importance. After prolonged use, their fertility decreases, probably due to the accumulation of toxic heavy metals. They are also known as *municipal irrigation fields* or *sewage farms*.

2. ***Semi-technological methods.*** Waste water is treated in artificial *sewage* or *waste water ponds,* which provide a biotope for a rich variety of species, and resemble highly eutrophic lakes. Organic materials are degraded by microorganisms, and the resulting nutrients are utilized by plants in the illuminated zone of the pond. Oxygen is derived partly by the photosynthesis of the plant community and partly by uptake from the atmosphere, but in order to prevent putrifaction, forced mechanical aeration is also necessary. In *sewage* or *waste water fish ponds,* fish are the end consumers in the food chain. For this purpose, mechanically clarified sewage must first be diluted 5–10 times with fresh water, and artificial aeration must be employed. Large areas are required for these semi-technological methods. Fish ponds with a total area of 1 hectare are required to treat the sewage from 2 000 inhabitants, and the breeding of edible fish (carp and tench) in such ponds is economically viable.

3. ***Technological methods.*** These intensive methods employ very high concentrations of microorganisms, and they achieve an extensive degradation of organic material in a relatively small space. In the *biofiltration method,* microorganisms live on the surface of solid fragments (rock, slag, pummice or synthetic material), which form the bed or packing of a *biological* or *trickling filter.* The trickling filter is a cylindrical or square structure, 1.8 to 8 m high, open above and below. Waste water is continually sprinkled from above and is exposed to the action of the microbial population as it descends through the bed. Air travels upwards and supplies the necessary oxygen. Bacteria in the microbial population grow into a mucilaginous mass, known as "zoogloea", which forms an adherent film or pellicle of biological slime on the surface of the packing material. The slime is populated by unicellular *(Ciliophora)* and multicellular *(Rotifera, Nematoda* and insect larvae) organisms, which "graze" on the bacteria and on each other. Biological filters are 80–95% efficient in the degradation of the organic constituents of sewage water.

Sludge activation systems are also extremely efficient. *Activated sludge,* which consists mainly of flocculating microorganisms (bacteria and protozoa). It is mixed with waste water under aerobic conditions in an *activation* or *aeration basin,* where constant whirling and agitation due to the forced aeration ensures constant contact between the microbial population and the organic pollutants, and keeps the flocculant bacteria in suspension. The biological processes are similar to those occurring on the surface of biological filters, and there are no fundamental differences between the mineralization organisms in the biological filter slime and the organisms of activated sludge, i.e. the sludge is a biocoenosis or biotope, consisting of a rich flora of aerobic bacteria, together with some fungi and bacteria-consuming protozoa, and a high quality sludge is characterized by a high concentration of protozoa. The organic material is partly respired and partly converted into sedimentary biomass. The mixture is eventually pumped into *secondary clarification basins* or *sedimentation tanks* for the separation of the 2 phases. The amount of sludge necessary for the activation basin is then recirculated. Sludge activation is at present the highest grade technique for B.w.w.t., being more efficient and far less odorous than biological filtration.

Waste water from domestic or industrial sources contains various types of suspended material, which must be removed before the water is purified biologically. This is achieved by screening (filtration through wire mesh) and by sedimentation. In the latter process, sewage flows continuously through a sedimentation tank, which is 3–4 m in depth, and the sedimented sludge is removed continuously or several times a day. Sedimentation (by flocculation) may be aided by the addition of aluminium sulfate, ferric chloride, ferric sulfate and ferrous sulfate, the particular salts and their quantities depending on the nature of the sewage. Such methods increase the speed and efficiency of removal of solid material, but make practically no contribution to the removal of dissolved organic material. The *Imhoff tank* is a funnel-shaped sewage settling tank designed by the German engineer, Karl Imhoff, consisting of a central feed for introduction of the sewage, an upper settling compartment, and a lower compartment for the collection and bacterial decompostion of solids. The collected sludge may be removed regularly *(Dortmund tank),* or the lower decomposition ("fermentation") chamber may be larger and deeper, permitting 2–3 months for completion of the degradation process *(Emscher tank).*

[Waste Water Technology (Origin, Collection, Treatment and Analysis of Waste Waters), edit. W. Fresenius, W. Schneider, B. Böhnke, K. Pöppinghaus, Springer-Verlag, 1989]

Biological water analysis: methods for determining the degree of water pollution, based on the numbers, types and proportions of different living organisms in the water, as well as the presence of certain Indicator organisms (see). See Saprotrophic system.

Biological weapons: live or dead organisms or viruses, or infectious material prepared from them, or their toxic products, intended for causing disease or death of humans, animals or plants for military purposes, together with their means of delivery. Substances of nonbiological origin, which have a biological effect, e.g. nerve gases, plant defoliants, etc., may also be classified as B.w., but are usually referred to as chemical weapons. For military use, the following materials of biological origin are potentially very effective.

Bacteria: *Pasteurella pestis* (plague), *Bacillus anthracis* (anthrax), *Pasteurella (Francisella) tularensis* (tularemia), *Brucella abortus* (brucellosis), *Salmonella typhi* (typhus), etc.

Viruses: various poxes, yellow fever, encephalitis, and other diseases.

Toxins: botulinus toxin from *Clostridium botulinum* (death from respiratory paralysis), *Staphylococcus* enterotoxins, ricin from the fruits of *Ricinus communis,* abrin from the seeds of *Abrus precatorius* (Jequirity seeds or Prayer beads), cicutoxin from the roots of *Cicuta virosa* (Cowbane), pallatoxins from *Amanita phalloides* (Death cap).

The fatal human dose of botulism toxin is 1 μg or much less. Inhalation of less than 1 μg of spores of *Bacillus anthracis* is also fatal.

The first documented use of *Pasteurella pestis* was in 1346 in the capture of the Black Sea port of Feodosija (then called Kaffa). A 3-year siege was brought to a successful conclusion by catapulting plague-infected animal corpses over the city walls.

Both biological and chemical weapons were forbidden by the Geneva Protocol of 17th June, 1925 ("Protocol for the Prohibition of the Use in War of Asphyxiating, Poisonous or Other Gases, and of Bacteriological Methods of Warfare"), which was signed by more than 40 countries. Many signatories, however, reserved the right to use such weapons against nonsignatory countries and/or for defense.

On 10th April, 1972, the Geneva Protocol was replaced by the Biological Weapons convention (Convention on the banning of the development, manufacture and storage of bacteriological (biological) and toxic weapons and their destruction), signed in London, Moscow and Washington. By 31st October, 1981, the Geneva convention had been ratified by 94 countries. These countries thereby undertook that they would: "not develop, manufacture, store, acquire or obtain: 1) microbiological or other biological substances or toxins, irrespective of their origin or means of manufacture, which according to their type and quantity cannot be intended for prophylactic, protective or other peaceful uses, and 2) weapons, equipment and means of transportation, intended for the employment of such substances or toxins for aggressive purposes, or for use in armed conflict". Participants in the Biological Weapons Convention-Review-Conference (Genever, March 1980) also subscribed to the views of the governments of the 3 depository countries of the Convention (USSR, UK, USA) that the formulation "biological substances or toxins, irrespective of origin or means of manufacture" should also include the methods of gene technology (e.g. total synthesis of toxins or infectious nucleic acids). At the same time, the participants stressed the urgent need for a corresponding convention on chemical weapons.

[J. Goldblat: *Agreements for Arms Control,* Taylor and Francis Ltd. London 1982. *The Problem of Chemical and Biological Warfare,* vol I-IV, Stockholm/International Peace Research Institute (SIPRI), Stockholm 1971–1975]

Biologism: in the broadest sense, the more or less wholesale interpretation of human biology and human society in the light of knowledge of the structure and function of biological systems. The biologistic approach was particularly influential in the second half of the 19th century, e.g. in social Darwinism, which asserted that the struggle for existence was also a valid concept in the interpretation of the behavior of human populations. Crass forms of B. are to be found in the origins of national socialist ideology, where they provide the pseudoscientific framework for racism, euthanasia, and the systematic anihilation of life deemed to be inferior. Projects on "human breeding", procedures for discriminating between ethnic or racial groups, as well as proposals for genetic manipulation of the human race, and many more of the numerous postulations of Social biology (see) are actual manifestations of B. As a living organism, the human is a valid object for biological research, but men and women are also social beings. Only by taking both aspects into consideration can students of *Homo sapiens* do justice to their own individuality. B. is therefore not the study of human biology.

Biology: the study of living organisms. B. deals with the complexity and variety of form and function in living organisms, the physiological and biochemical processes that occur during the life cycle of an organism, the laws and principles that govern the behavior and functions of living organisms, the evolution and distribution of living organisms, interrelationships between organisms, and their dependency on, and influence of, the inanimate environment. Numerous areas of research are devoted to the in-depth study of narrow aspects of B. Understanding of the phenomenon of life is then increased step by step, as these painstaking and narrowly based contributions are assimilated into an overall concept. The objective of all research is the discovery of a unifying principle of life. Only by the mutual interaction of all levels of biological research will it be possible to achieve this goal. Thus, at the present time, B. consists of many separate and very different disciplines, but they have in common the fact that they all ultimately ask, and seek to answer, questions about living organisms. Any discussion as to whether B. is an independent discipline is therefore irrelevant. The absence, so far, of a single theoretical basis for B. means that biologists in different branches of the subject may view themselves differently, i.e. have a different sense of scientific identity. Moreover, there is no generally accepted answer to the question "what is life ?" Nevertheless, each advance in our knowledge of living organisms and life processes points to the ultimate discovery of a unifying concept.

Many different areas of biological research are linked to, and depend on, the state of development of nonbiological subjects, e.g physics and chemistry. Knowledge from these other subjects is then used to advantage in the further development of various biological disciplines, e.g. medicine, agriculture, etc. Elementary knowledge of cell structure would not be possible without the advances in engineering and physics that made possible the development of the electron microscope, and our understanding of the chemical structures of biological substances and the enzyme-catalysed reactions responsible for their metabolism depends largely on advances in chemistry and physical chemistry. The important and dynamic role of B. as a modern science can be seen from the continued development of new branches of B. with their accompanying specialization, as well as the increasing number of problems from outside the subject that rely upon B. for an answer.

There are various criteria for the separation and definition of different biological disciplines. *Microbiology* (see) deals with the structure and mode of existence of microorganisms, whereas *Botany* (see) is concerned with plants, *Zoology* (see) with animals, and *Anthropology* (see) with the basis of human existence. These separate areas are the result of historical development, and they correspond only incompletely to a strict classification of organisms.

A distinction between *descriptive* and *experimental* B. is no longer justified, since the boundaries between description and experiment have become very fluid in the repertoire of modern biological research, which is pervaded by the causal-analytical approach.

The customary division into general and specialized B. can be applied to the entire subject, or in a more restricted sense within the subdivisions of microbiology, botany and zoology. A clear division into separate and self-contained generalized and specialized subjects is, however, rarely possible. The very complexity of living phenomena means that different areas of study have overlapping interests, thereby forming an interactive network.

General B. includes, inter alia: *Taxonomy* (see), which seeks to place organisms in a hierarchical system according to their evolutionary relationships; *Morphology* (see), which describes and investigates the form and structure of organisms, and the position and juxtaposition of their parts and organs; and *Anatomy* (see), which investigates the internal structure of organisms. Anatomy proceeds from organ systems to the cellular level, via *Histology* (see), which deals with the fine structure of tissues, and *Cytology* (see), which is concerned with cellular structure and function. Function in the widest sense is the subject of *Physiology* (see), which in turn is subdivided into different areas (e.g. metabolic physiology, physiology of movement, developmental physiology, neurophysiology).

Biophysics (see) was previously regarded as an auxillary subject of medicine, but it has now achieved a much wider significance in the elucidation of physical processes in all living organisms. *Biochemistry* (see) has given an insight into the chemistry of living processes, and made important contributions to the understanding of the structure and function of organisms. *Molecular biology* (see), together with other disciplines, has increased our knowledge of the molecular structures of the living cell and the underlying phenomena of living processes. This, in turn, is closely related to certain aspects of *Genetics* (see), the science of heredity, the importance of which for plant and animal breeding, and especially for problems of human heredity cannot be overestimated. An interaction of numerous disciplines is especially apparent in *Developmental physiology* (see), which is greatly influenced by (and depends upon) genetics, cytology, molecular biology and biochemistry.

Phylogenetics (see) addresses problems of speciation, and in particular Evolution (see) and evolutionary relationships between organisms, as well as the origin of life. It therefore represents the meeting point of the research goals of evolutionary B., biochemistry and molecular biology. *Paleantology* is the study of earlier life forms. This independent discipline makes an important contribution to phylogenetics, as well as representing a link between B. and geology; it makes a practical contribution to the latter in the subject of Biostratigraphy (see).

Biogeography (see) investigates the spread and distribution of life forms in the biosphere (see Plant geography, Animal geography), and it has various subdivisions, according to the emphasis of the research (e.g. Chorology, Sociology of plants). *Ecology* (see) has attained considerable importance in recent times; as the science of the environmental relationships of organisms, it addresses many problems of importance for the maintenance and re-establishment of naturally balanced systems. The complexity of ecology has resulted in the development of several branches of the discipline (e.g. autecology, synecology, population biology, ecophysiology, engineering ecology).

Specialized B. can be described essentially as the B. of specific groups of organisms (e.g. algology, mycology, entomology, herpetology, ornithology, etc.). Clear separation of the various specialized areas is made difficult by the existence of *applied B.* On the one hand, the majority of biological subjects cannot be considered or understood in isolation from their practical applications. On the other hand, subjects such as fishing, agriculture, forestry, veterinary and human medicine belong to applied B. in its widest sense. Plant and animal breeding, phytopathology, pharmacology, certain branches of the food industry, and especially the extensive subject of biotechnology also clearly belong to applied B.

Historical. Prescientific and speculative biological thought was traditional in most ancient human cultures. An early knowledge of anatomy is apparent from the ancient arts of healing and in many religious rites (e.g. embalming of corpses). On a foundation of naive materialism, the ancient Greek philosophers (e.g. Democritus 460–371 BC, Hippocrates 460–377 BC) combined zoological, botanical, anatomical and physiological observations within a framework of natural philosophy. As early as the 4th century BC, the Hippocratic school had acquired an extensive knowledge of the animal and human muscle system, cardiac function and the eye. The doctrine of body fluids (yellow and black bile, blood and mucus) was important in the development of medicine. The first attempt to arrange and classify the variety of living forms is seen in the animal system originating from the Hippocratic school on the island of Cos.

The pinnacle of Greek B. was attained with the teaching of Aristotle (384–322 BC). He brought together the biological knowledge of his time, and enriched it with his own studies on the genesis, life history and structure of animals. Aristotle described and classified approximately 500 animal species and divided them into those with blood and those without blood, then further into genera and species. His view of natural development was teleological; development was determined by a pre-existing, unidirectional force within the organism, and the potentiality of that force became actuality in the developed organism. In this concept (known as *Entelechy),* the material of the organism therefore represented a passive principle. A modified form of this approach is found in all subsequent idealistic theories and doctrines of life (see Vitalism). The work and teachings of Aristotle influenced biological thinking into the 17th century. Lucretius Carus (98–55 BC), in his didactic poem "De rerum natura" transfers the materialism of his teacher, Epicurus, to natural living systems. Plinius Secundus (23–79 AD) summarized the knowledge of the natural world in a 37-volume "Natural history". The development of B. and medicine along these traditional lines was ended by the influence Galen (129–199 AD). Galen was important as a medicial systematist and anatomist. The results of his animal postmortem examinations, and his studies on the determination of drug dosages played a decisive role in medical science for more than a thousand years.

During the early feudal period of medieval Europe, there were no significant advances in B., and thinking was dominated by religion. Meaningful biological thought did not restablish itself until the translation of Aristotle's works from the Arabic into Latin. Jewish, Persian and Arabian physicians and philosophers had inherited Aristotle's theoretical system and had annotated his works [e.g. Avicenna (Abu Ibn Sina) 980–1037, a Persian physician and commentator on Aristotle; and Averroes (Abul Walid Mohammed ben Ahmed ibn Roshd) 1126–1198, a Moslem doctor born at Cordova, and author of a famous commentary on Aristotle]. The church sanctioned the Aristotelian teachings as final and conclusive, and scholastic philosophers simply restricted themselves to the repetition and interpretation of Aristotle (e.g. Albertus Magnus 1207–1280).

The growth of trade with its accompanying journeys and voyages of discovery, and an increased inclination toward empiricism were accompanied by new attitudes, including a desire to investigate the natural world. During the Renaissance, progress occurred primarily in descriptive zoology and botany, and in the study of the anatomy of humans and higher animals. It became clear that the ancient authors had not known all the species of animals and plants. In the 16th century a number of scholars described new animals and plants, and listed these together with the unkown species of the ancient authors. In zoology, the "Historia animalum" (1551–1558) by Konrad Gesner (1516–1565) marked the beginning of a new epoch. Other authors described in particular marine animals, e.g. G. Rondelet (1507–1566) and Pierre Belon (1518–1564). The comprehensive zoological works of Ulisse Aldrovandi (1522–1605) appeared around 1600. Descriptive botany received its impetus from the "fathers of botany": Hieronymus Bock (1489–1554), Otto Brunfels (1489–1534), Leonhart Fuchs (1501–1566). Other "herbals", often extensive folios, describing plants and their uses, originated from: P.A. Matthiolus (1501–1577), Andreas Caesalpinus (1519–1603), J. Th. Tabernaemontanus (1520–1590), Adam Lonicerus (1528–1586), Caspar Bauhin (1560–1624). At this time, it was also recognized that Galen's version of human anatomy contained numerous errors, due to his use of animals (mainly pigs and monkeys) for his autopsies. With the use of human autopsies, a new human anatomy was founded, in particular by Andreas Vesalius (1515–1564) with his excellently illustrated book "De humani corporis fabrica" (1543). For example, Vesalius recognized that the human did not have a 7-membered, but a 3-membered sternum, and that human liver has a different structure from that of the pig, etc. This turning away from ancient models, which had been thought to be adequate, and the turning toward independent research were crucial for the further development of B. During the 16th century, most of the still unknown macroscopic structures of the human body were described, e.g. vein valves and the female sex organs were not described until the 16th century. Some anatomists were also already comparing the skeletal structures of different classes of vertebrates, e.g. P. Belon, B. Eustachius, V. Coiter (1534–1600). Of particular importance, however, was the investigation of the embryonal development of several animals, especially the chicken in the egg. Pioneering work on this subject was carried out by Hieronymus Fabricius ab Aquapendente (1537–1619) and by Coiter. The 17th century was witness to the "scientific revolution", i.e. the emergence of modern science, characterized by new theoretical approaches, new forms of communication between scientists, and by research. Great importance was attached to experimentation, and to quantitation. Two of the foremost representatives of this revolution were Francis Bacon (1561–1626) and Galileo Galilei (1564–1642). Bacon pleaded for the inductive method, in which general conclusions would be drawn from the collection of many individual facts. The influence of Galileo lay especially in his fresh approach to scientific inquiry, his conviction that nature is governed by laws, and the use of experimentation to test these assumptions.

The first significant contribution to the scientific revolution within B. was the discovery of the circulation of the blood (1628) by William Harvey (1578–1657). After Harvey had conceived the idea that blood circulated within the body, he and some of his supporters proceeded to prove the correctness of this theory with meaningful and ingenious experiments, e.g. the ligature of blood vessels to determine the direction of blood flow within them, and the determination of the quantity of blood pumped by the heart per unit time. The latter quantity proved to be much greater than hitherto assumed, thereby excluding the possibility that blood was continually consumed within the body. Following the discovery of lymph vessels by Gasparo Aselli (1581–1626), Jean Pecquet (1622–1674) found that lymph transports the products of food digestion from the intestinal wall directly into the blood system, thereby bypassing the liver. With this discovery, the old physiology of Galen, in which the liver occupies a central position, was superseded. These successful investigations marked the beginning of an expansion in physiological research, notable contributors being: Francis Glisson (1597–1677), Robert Boyle (1627–1691), Richard Lower (1631–1691) and John Mayow (1643–1679).

Influenced by successes in mechanical physics, many 17th century biologists attempted to explain living processes mechanistically, i.e. to consider the body as a complicated machine, a concept known as *iatrophysics*. Other workers devoted their attention primarily to the chemical processes of the body *(iatrochemistry)*, e.g. van Helmont (1577–1644), based, however, on the erroneous chemical concepts current at that time. The iatrophysical, mechanistic view of life found an important proponent in René Descartes (1596–1650). The immortal human soul, however, was supposed to act upon the bodily machine via the pineal gland. G.A. Borelli (1608–1679) attempted in particular to reduce muscle movements to mechanics. S. Sanctorius (1561–1636) was one of the first to apply new physical knowledge and to introduce measurement into medicine; he used a modification of Galileo's thermometer for measuring the temperature of the body, devised an apparatus for comparing pulse rates, and studied changes in the weight of the human body by weighing himself in a balance. Invention of the microscope around 1600 led to the investigation and illustration of the fine structure of organs, and to the discovery of small living organisms not visible to the naked eye. Workers in this field became known as "microscopists", e.g. Antony van Leeuwenhoek (1632–1723; blood corpuscles, capillaries, spermatozoa, bacteria on dental enamel); Marcello Malpighi (1628–1694; capillaries, fine structure of glands and kidneys, fine structure of insects); Nehemiah Grew (1641–1711; fine structure of plants, 1682); Robert Hooke (1635–1703; "Micrographia" 1664). The hope that the investigation of the fine structure of living organisms would lead to a better understanding of living processes was, however, hardly fulfilled. Thus, the versatile physiologist, Regnier de Graaf (1641–1673) discovered the Graafian follicle in the mammalian ovary, but mistakenly believed it to be the mammalian egg.

In the 16th and 17th centuries, the administration and organization of B. were advanced by the establishment of botanical gardens (Padua 1545) and modest zoological gardens, by the formation of natural history collections, by the foundation of scientific academies (Rome 1609, Royal Society in London 1663, Leopoldina 1652/1677, Paris 1666), and by the publication of the first scientific journals (Philosophical Transactions, etc.).

Important botanical discoveries were made by Rudulf Jacob Camerarius (1665–1721), who demonstrated experimentally the sexuality of flowering plants (published 1695), and by Stephen Hales (1677–1761), who investigated the water transport and water economy of plants (1727, "vegetable statics") and thereby qualifies as one of the earliest plant physiologists.

Until the end of the 18th century, embryonal development was explained by many biologists with the theory of *preformation*. According to this theory, the embryo was already preformed in miniature in the sperm (animalculists, e.g. Leeuwenhoek) or in the egg (ovulists, e.g. A. v. Haller, Ch. Bonnet, L. Spallanzani), and embryonal development was the unfolding and enlargement of this miniature form ("evolutio"). Embryonal development was therefore understandable mechanistically, and the preformation theory was in harmony with mechanistic B. By analogy with other objects, living organisms were believed to be preformed from the beginning. For example, Ch. Bonnet thought that all generations of living organisms destined to appear by the end of time were prepacked in the eggs of existing organisms.

During the 18th century, mechanistic B. was increasingly challenged, and by the end of the century most biologists assumed that many living processes were determined by innate factors of the organism ("life force"), thus representing the emergence of a new vitalism in B. As early as the end of the 17th century, Georg Ernst Stahl (1660–1735) opposed the machine theory with a vitalistic approach, claiming that the soul, the actual life principle, conducts and directs the body. In the middle of the 18th century, Albrecht von Haller (1708–1777), the leading physiologist of his time, argued that muscle contraction (irritability) and nervous excitation (sensitivity) cannot be explained mechanistically ("Elementa physiologiae", 1757). A series of discoveries in the 18th century produced increasing difficulties for the mechanistic preformation theory, and they demanded at least the elaboration of auxiliary hypotheses. For example, in 1744, Abraham Trembley (1700–1784) published his sensational discoveries on the regeneration of the freshwater polyp. Starting in 1761, Joseph Gottlieb Kölreuter (1733–1806) published the results of his crossing experiments with flowering plants (tobacco, etc.), which clearly demonstrated that both parents are represented in the offspring. Kaspar Friedrich Wolff (1733–1794) established that the supposed preformed embryos cannot be observed, and that the young embryo develops from unstructured material. Thus, the concept of epigenesis, which was espoused by Aristotle and later by Harvey, became re-established. To explain embryonal development within the framework of epigensis, a special life force was postulated. K.F. Wolff called this the "vis essentialis", while F. Blumenbach referred to the "Bildungstrieb" (developmental drive).

Lazzaro Spallanzani (1729–1799) remained a preformist, but as one of the pioneers of experimental B., he made a number of important discoveries: artificial fertilization of mammals, regeneration, orientation of bats, etc. In numerous experiments, Felice Fontana (1720–1805) studied the action of snake venom. Progress in the chemistry of gases led to the discovery of gas metabolism in plants (carbon dioxide assimilation, oxygen excretion) by J. Priestley and J. Ingenhousz.

By the 18th century there was a need among biologists for greater rationalization in the recording and listing of living organisms. This was satisfied by great advances in taxonomy (systematics), the description of organisms, and in nomenclature. Early contributions to plant and animal systematics were made by A. Qu. Rivinus (1652–1723), John Ray (1627–1705), Joseph Pitton de Tournefort (1656–1708) and others. These were followed by the great systematist, Karl von Linneus (1707–1778). In his work "Systema naturae" (first published 1735), Linneus produced a systematic classification of animals that remained valid for a considerable time, as well as a sexual system for plants, which was however spurious. The Latinized binomial nomenclature introduced by Linneus (accepted in 1753 and current ever since) became the basis for defining and identifying species.

At the end of the 18th century, comparative anatomy received a great impetus from the work of authorities such as Petrus Camper (1722–1789) and Félix Vicq d'Azyr (1748–1794). Even Johann Wolfgang von Goethe (1749–1832), who introduced the term "morphology", made contributions to comparative anatomy, e.g. in his study of intermaxillary bones. Similarities between living organisms were established by studies in comparative morphology, and this aided the attempts to classify and order the natural world. Ch. Bonnet (1720–1793) placed natural objects on the steps of a single ladder from the simplest to the most complicated, whereas Georges Cuvier (1769–1832) and Karl Ernst von Baer (1792–1876) differentiated between 4 different structural types in the animal kingdom. A minority of biologists, including Jean Baptiste de Lamarck (1744–1829), however, deduced from anatomical similarities that species may actually be transformed or remodelled.

A process of evolution was first proposed primarily for heavenly bodies (I. Kant) and the earth's crust. The idea that living organisms might evolve was mentioned only very speculatively in the 18th century (e.g. by Telliamed, and by Georges Louis L. de Buffon, 1708–1788).

From about 1800 onwards, it was increasingly recognized by Cuvier and others that many groups of animals and plants had existed previously and were now extinct. It became clear that in the history of the earth, new groups of organisms had come into existence at different times, and that these represented an ever increasing level of development (progression). Since gradual transitions between the different groups were not apparent, this progression of organisms did not lead to a concept of evolution. At first, the often abrupt alteration of the dominant flora and fauna in each period of the earth's history was not attributed to gaps in the fossil record, but explained by Cuvier and others by catastrophes, which from time to time had almost or completely anihiliated the living world. Paleantology was developed further by William Smith (1769–1839), Alcide d'Orbigny (1802–1857), Richard Owen (1804–1892), Louis Agassiz (fossil fish) (1807–1873) and others.

Spontaneous generation, which was still accepted in the 17th century, even for complex organisms, was refuted experimentally for flies by Francesco Redi (1626–1698), and it became increasingly improbable as the complicated structures of even simple life forms became known (Jan Swammerdam 1637–1680). Spontaneous generation was also in conflict with the preformation theory. With the re-emergence of belief in "life forces" at the end of the 18th century, the concept of spontaneous generation once more asserted itself. Many believed that spontaneous generation was experimentally proven for lower organisms. Furthermore, spontaneous generation was invoked to explain the appearance of new groups of organisms in the fossil record. Not until 1830 was the concept of spontaneous generation finally rejected, when it was disproved for microorganisms by Louis Pasteur (1822–1895).

Before 1859 many biologists believed that the variability of species could lead to new intraspecific forms, but not to new species, i.e. variation could not exceed the boundaries between species.

Between 1800 and 1830, some scientists and philosophers were at least temporarily captivated by "romantic natural philosophy" (F.W. v. Schelling 1775–1854, Lorenz Oken 1779–1851).

In the 18th century, various biologists were seeking the elementary components of living organisms, inspired by the fact that the theory of elementary particles (at that time atoms and molecules) had at last met with significant success in chemistry and physics. In the case of living organisms, fibers and globules, even "infusoria" (L. Oken) were among the elementary components discussed. Following the improvement of the microscope (Amici and others), the cell was recognized as the only elementary component of plants in 1838 by Matthias J. Schleiden (1804–1881), and of animals in 1839 by Theodor Schwann (1810–1882). Even fibers in plants, as well as animal muscle fibers, were shown to be cells. Before the establishment of this cell theory, the cell nucleus had already been discovered in 1833 by Robert Brown (1773–1858). Not until the middle of the 19th century, however, did it become clear that all cells arise from pre-existing cells (R. Remak, Rudolf Virchow 1821–1902: "omnis cellula e cellula"), and that the animal sperm and the egg are single cells (Albert Kölliker 1817–1905, and others). Several biologists developed the concept of the organism as a confederacy of cells (e.g. R. Virchow). According to Virchow's "cellular pathology" (1858), all illnesses were due to the alteration of cells, and their causes were therefore localized.

On the basis of the cell theory, histology was developed and consolidated by A. Kölliker and Franz v. Leydig (1821–1905). Protoplasm was studied by Hugo v. Mohl (1805–1872), Max Schultze (1825–1874) and others.

Great progress was made in the 19th century in the investigation of small living organisms, in both the plant (algae) and animal kingdoms. Christian Gottfried Ehrenberg (1795–1876) studied "infusoria", but C.T. v. Siebold was largely responsible, in 1846, for establishing that these are unicellular organisms. Anton de Berg (1831–1888) made fundamantal contributions to the knowledge of parasitic fungi and their life cycles, e.g. in 1864 he published studies on the causative organism of potato blight, and on the alternation of generations in cereal rusts. Pasteur recognized the role of microbes in fermentation and putrifaction, thereby laying the foundations of microbiology. In 1876, Robert Koch (1843–1910) showed that anthrax is caused by a specific bacterium, leading to the recognition that each infectious disease is due to a specific causative organism. In 1883, Ilya Ilyitsch Metschnikov (1845–1916) discovered phagocytosis. Measures for combating infectious diseases were also discovered, e.g. in 1880 Pasteur reported the technique of active immunization, while serum therapy was discovered by Emil von Behring (1854–1917).

In the 19th century, significant discoveries were also made in embryology. H. Chr. Pander (1794–1865), K.E. v. Baer, M. Heinrich Rathke (1793–1860) and others studied the embryonal development of different classes of animals. In 1842, J. H. Steenstrup (1813–1865) discovered the alternation of generations in the scyphozoan, *Aurelia aureta,* thereby providing a key to the discovery of other complicated reproductive cycles. Wilhelm Hofmeister (1824–1877) showed that all vascular plants share a basically similar type of reproductive mechanism, and he discoverd the alternation between sporophyte and gametophyte generations. Theodor Ludwig Wilhelm Bischoff (1807–1882) clarified some important events in mammalian egg production and fertilization.

Nineteenth century physiology showed various lines of development. Some individual meaningful discoveries were: tropism in plants by Thomas Andrew Knight (1759–1838) and others; osmosis by Henri Dutrochet (1776–1847); the role of the nerves of the spinal cord by Charles Bell (1774–1842). At first, experimentation was not seen as the most important way of obtaining information about life processes. Johann Friedrich Meckel (1781–1833) and others attempted to further knowledge in physiology by comparing the different structures of the same organs in animals with different modes of existence, i.e. by studying comparative anatomy. K. A. Rudolphi (1771–1832) and Johannes Müller (1801–1858) were among those who expected to gain an insight into the physiology of organs by investigating their fine structure. Thus, J. Müller studied the fine structure of glands, and established that gland ducts are separated from the blood stream by a membrane, thereby drawing attention to the mode of formation of glandular secretions. In his handbook of human physiology, J. Müller presented a survey of current physiological knowledge, and he remained for decades one of the most influential biologists; but even he still retained a basic belief in vitalism. Jan Evangelista Purkyne (1787–1863) also made contributions in several different fields of physiology. François Magendie (1783–1855) was responsible for the renewed development of experimental physiology. In exemplary fashion, his famous pupil, Claude Bernard (1813–1878), experimentally demonstrated various fundamental life processes, e.g. the role of glycogen in maintaining the constant blood sugar level in healthy individuals (1843), the importance of pancreatic secretions in this process, and the bases for the toxicity of carbon monoxide and curare. He drew attention to the importance of the "milieu intérieur" (1857), which constantly bathed the organs of the body, and which determined the constancy of living processes, and he also recognized that the various individual functions within the organism were actually part of a system of interrelated and interdependent processes. In Germany, Ernst Brücke (1819–1892), Emil du Bois-Reymond (1818–1896), Hermann Helmholtz (1821–1894) and Carl Ludwig (1816–1896) were responsible for forming the physical school of physiology, which rejected the "life force", and attempted to reduce all life processes to physical processes. They concentrated mainly on isolated processes, e.g. kidney and gland function (Ludwig), and the electrical processes accompanying muscle and nerve action (du Bois-Reymond). The versatile Eduard Pflüger (1829–1910) showed that the tissues are the seat of respiration. Supporters of mechanistic materialism, Ludwig Büchner (1824–1899) and Jacob Moleschott (1822–1893), also became supporters of the purely mechanistic physiology. Other biologists dedicated themselved primarily to studying the chemical processes of living organisms, e.g. Justus von Liebig in his studies on plant metabolism, and Felix Hoppe Seyler (1825–1895), who was one of the pioneers of biochemistry. At the end of the 19th century, the concept of humoral communication (hormones) was successfully established by Ch. E. Brown-Sequard (1817–1894).

Immunology developed from the work of Pasteur, Paul Ehrlich (1854–1915) and Karl Landsteiner.

With his *theory of evolution,* Charles Darwin (1809–1882) was responsible for revolutionary changes in biological thinking. Lamarck had sought to attribute higher levels of development to an inner developmental force. He thought that external factors could modifiy the effects of this inner drive, and that the resulting changes were inherited. In animals, the necessity to perform new tasks (e.g. to exploit new food sources) was though to be a driving force leading to heritable modifications. Darwin sought to explain evolution by processes that are operating continually, even at the present time, and which are therefore amenable to experimental investigation. In this approach, he followed the actualistic principle of Charles Lyell (1797–1875) for the investigation of geological history, according to which all changes in the earth's crust in the course of long periods of time can be explained and understood by reference to current geological phenomena. In animal breeding, Darwin recognized examples of processes that can occur in natural evolution, e.g. the occurrence of heritable alterations, and the selection of those alterations that are favorable. In Nature, the favorable characters are not selected by human intervention (the animal or plant breeder), but by a process called "natural selection". Similar conclusions were reached by Alfred Russel Wallace (1823–1916), but without reference to animal breeding. Darwin's main work, "The Origin of Species" (1859), fundamentally changed all subsequent biological thinking. Many biologists became followers of Darwins's theory of evolution, e.g. Thomas H. Huxley (1825–1895), Ernst Haeckel (1834–1919), August Weismann (1834–1914), K.A. Timirjasew (1843–1920), A.O. Kowalewski (1840–1901). Huxley, Haeckel and others also considered the position of humans in the process of natural evolution. In Germany, the efforts of evolutionary biologists were dominated for a long time by attempts to clarify the phylogenetic interrelationships of taxa (notably Haeckel worked on evolutionary trees), and to determine the evolutionary origin of organs and structures (Karl Gegenbaur 1826–1903). Evolutionary factors were further investigated, especially by August Weismann. Like Nägeli and others, he attempted to identify the hereditary material that could form the basis of inheritance in the widest sense. He drew a distinction between the "germ plasm" and the body substance or "soma". Inheritance of acquired changes was therefore no longer tenable, and this markedly restricted the type of changes that could be accepted as relevant to evolution. Furthermore, Nägeli believed that higher development must be due to an inner evolutionary force. Botanists like Richard von Wettstein (1863–1931) and zoologists like Paul Kammerer (1880–1926) furthermore maintained that the "inheritance of acquired characters" could in fact occur, and was at least a contributing factor to evolution.

One of the leading branches of biological research in the last decades of the 19th century was *cytology,* and particular attention was paid to the cell nucleus. Behavior of the nucleus in fertilization was studied by Oscar Hertwig (1849–1922), chromosomes were described by Walther Flemming (1843–1905), chromosome behavior during fertilization by Edouard van Beneden (1846–1910), and the properties of chromosomes, in particular by Theodor Boveri (1862–1915).

An upsurge in experimental B. after 1880 introduced a new period in the development of B. Information on the factors active in embryonal development, and in the possible reorganization and modification of adult organisms was sought by expermentation. These developments led to the emergence of developmental mechanics, developmental physiology, and causal morphology. Some representatives of this new experimental B. (e.g. Hans Driesch 1867–1941) even expressed doubt about the value of research into evolution, because this could never yield fully conclusive results, and it would not increase man's control over Nature. This led, in turn, to attacks by the partly dogmatic, older evolutionary biologists (notably Haeckel) against developmental physiology.

Pioneers of developmental animal physiology included E. Pflüger (possible action of gravity on the development of frogs' eggs), Wilhelm Roux (1850–1924), H. Driesch, and Curt Herbst (1866–1946). Those in plant physiology included Julius Sachs (1823–1897), Hermann Vöchting (1847–1917), Fritz Noll (1858–1908), and Georg Klebs (1857–1918). Different experimental results were sometimes obtained by different workers, giving rise to conflicting interpretations. Thus, Roux obtained partial embryos from parts of fertilized frog eggs, whereas Driesch obtained entire embryos from parts of fertilized sea urchin eggs. Driesch invoked a special factor, only present in living organisms, called "entelechy", which must be responsible for the completion of a partial embryo (neo-vitalism). Edmund B. Wilson (1856–1939) found that there was no clear difference between *regulation* and *mosaic* eggs (see Regulation). Hans Spemann (1869–1941) reported that specific regions of the embryo ("organizers") determine the formation of other embryonal parts.

From 1890 onwards, research into inheritance, based on the results of hybridization experiments, underwent an expansion. In 1906, this field of study was named "genetics" by William Bateson (1861–1926). In the 1890s, Bateson, Korschinsky, and especially Hugo de Vries (1848–1935) had reported that only certain, often abrupt and marked alterations in living organisms are heritable, and that only these changes can have any importance in evolution. This was the mutation theory, first put forward by de Vries in 1901. In 1903, Wilhelm Johannsen (1857–1927) showed that selection was ineffective in "pure lines", i.e. in genetically pure populations. In 1900, the laws of inheritance of Gregor Mendel (1822–1884), published earlier in 1865, were rediscovered by Carl Correns (1864–1933), de Vries and Erich von Tschermak (1871–1962). These Medelian laws led to a deeper understanding of the nature of inheritance. The organism appeared to be a "mosaic" of individual characters, which were inherited independently of one another, and which to a certain extent could exist in any combination. By comparing the coincidence of the transmission of characters to the offspring and the behavior of chromosomes, W.S. Sutton and Th. Boveri derived the chromosomal theory of inheritance (1902), in which chromosomes were seen as the essential carriers of genetic material (as already suspected in the 19th century) and the site where the hereditary factors ("genes") are linearly arranged (as shown in particular by the school of Thomas Hunt Morgan, 1866–1945). From the 1920s onwards, "formal" or "classical" genetics was transcended by the formulation of new questions. Attempts were made to obtain information on the nature of gene action (Richard Goldschmidt 1878–1958). Gene physiology was studied by A. Kühn, and especially in the USA by G.W. Beadle and Tatum, culminating in the important and experimantally verified "one gene one enzyme" hypothesis; updated in the light of subsequent knowledge of gene structure and gene action, this may now be somewhat modified to "one cistron one polypeptide". An attempt was also made to bring together the theory of evolution and the results of genetics (mutations and their spread in natural populations), resulting at the end of the 1930s in the synthetic theory of evolution (Th. Dobzhansky, J. Huxley, B. Rensch, and others).

During this period, the various areas of physiology also underwent further development, e.g. the study of neurons by Ramon Y. Cajal (1852–1934), the study of conditional reflexes, etc. by Pavlov (1849–1936), and the investigation of autonomous nerve action by E. v. Holst (1908–1962). Regulatory processes within the organism were investigated by, inter alia, W.B. Cannon (1871–1945) and W.D. Bancroft (1867–1953).

In plant physiology, significant advances were made by Wilhelm Pfeffer (1846–1920; osmosis, chemotaxis, etc.) and Gottlieb Haberlandt (1854–1945; stimulus receptors, wound hormones).

Behavioral research was promoted by Oskar Heinroth (1871–1945), Karl von Frisch (1886–1982), Konrad Lorenz (born 1903) and others.

Biochemistry underwent tremendous development after 1900, with the discovery of vitamins and enzymes, discovery of the respiratory chain (Otto Warburg 1883–1970, and others), recognition that alcoholic fermentation and glycolysis in muscle are basically identical processes (Otto Meyerhof 1884–1951), and the discovery of metabolic cycles (Hans Krebs 1900–1981). In addition, many natural products were synthesized chemically (Richard Willstätter 1872–1942, Richard Kuhn 1900–1967, and others). A milestone in B. was the publication in 1953 of the elucidation of the chemical composition of the genetic material, deoxyribonucleic acid (DNA) (Oswald Avery 1877–1955, F.H.C. Crick, J.D. Watson, and others). This meant that heredity, as well as evolution, could now be related to the structure, properties and behavior of a chemically defined substance.

Bioluminescence: the production of light by living organisms. B. is a chemical process, in which the oxidation of Luciferins (see) is catalysed by the enzyme, luciferase. There are different chemical classes of luciferins, e.g. in bacteria luciferin is a complex of a long-chain aldehyde with FMN, whereas fireflies possess a 2-(4-carbothiazole)-6-hydroxybenzothiazole. Luciferins are usually synthesized by the respective bioluminescent organisms.

In the animal kingdom, B. may be due to one of 3 different processes:

1) ***Intracellular luminescence*** is generated by *light cells* in the body of the organism, and the light radiates through the skin. The light effect is often intensified by reflective material, e.g. urate crystals in fireflies, and platelets of guanine in fishes, which cause sparkling and iridescence.

2) ***Extracellular luminescence arises*** from the interaction of luciferin and luciferase outside the animal body. The two components are synthesized and stored in separate gland cells in the outer skin layer, or immediately below the outer skin layer in small pouches. Release of the reactants gives rise to "clouds" of light. This type of B. is found in many crustaceans and in some deep sea cephalopods.

3) ***Symbiosis with luminescent bacteria.*** This type of B. is found only in marine animals, e.g. coelenterates, worms, mollusks, echinoderms, fishes. At various sites of the body, the animals have small pouches which harbor the luminescent bacteria.

Only very rarely is light production continuous. It usually occurs spontaneously in response to a stimulus, e.g. in the organisms responsible for sea phosphorescence (flagellates, medusae, see *Pelagia),* in which the period of the light pulse varies with the species. Hollow light organs containing luminescent bacteria produce continuous light, which in some cases can be screened out or altered in intensity by special structures. The individual light organs are interconnected via the nervous system.

In all the animals investigated so far, the B. lies in the visible range of the spectrum, varying in color from blue to green. Other colors are occasionally observed, and these are attributable to alteration of the original color by lens systems and reflective layers. Bioluminescent radiation consists of 80–90% cold light rays and 10–20% heat energy.

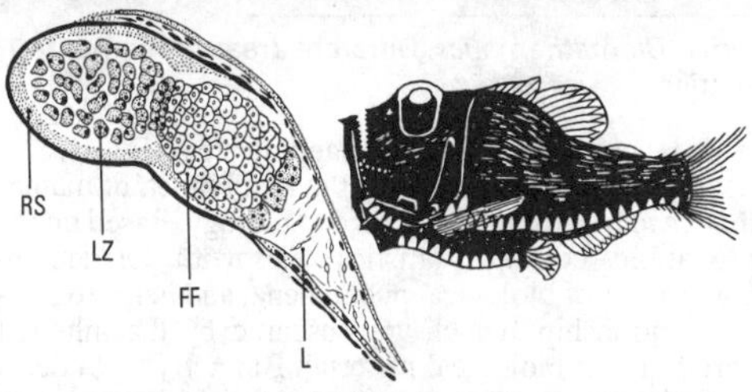

Bioluminescence. Above: deepsea lanternfish *(Diaphus metopoclampus);* light organs shown as white spots. Below: hatchetfish *(Argyropelecus affinis)* and section through a ventral light organ; *LZ* light-generating cells, *RS* reflective layer, *L* lens, *FF* color filter (from Marshall, 1957).

In the plant kingdom, B. occurs in bacteria and certain plants. Some luminescent bacteria occur free in marine waters, e.g. *Bacterium phosphorescens* and the common Baltic sea bacterium, *Vibrio balticum*. Others live parasitically on animals, or in symbiotic association with animals. The light produced by fresh meat or fish is due to bacteria. The luminescence of rotting wood is produced by certain bacteria and fungal mycelia.

The biological significance of B. is not always clear. In many animals, light production is different in the two sexes, and serves as a means of recognizing and locating a sexual partner, e.g. in the glow-worm, *Lampyris noctiluca*. Light organs may be highly specialized, and often act as a bait to attract food. Some crustaceans and cephalopods produce light clouds to confuse predators.

[*Bioluminescence and Chemiluminescence (Proceedings of VIIth International Symposium on Bioluminescence and Chemiluminescence*, Banff, March 1993) edit. A. Szalay, L. Kricka and P. Stanley, John Wiley, 1993]

Biomass: see Standing crop.

Biomathematical modelling: a branch of Biomathematics (see), comprising nonstatistical methods and procedures. Usually biophysical, biochemical or biocybernetic modelling is used to describe and to simulate biological systems and processes.

Nonstatistical mathematical aids are being used increasingly in certain areas of biology and medicine, e.g. biochemistry, animal and plant physiology, morphometry and stereology, chronobiology, developmental biology, agricultural and forestry economics, enzymology, pharmacology, immunology, endocrinology, radiology and molecular biology.

The most important branches of mathematics used for this purpose are: set theory, linear recurrence relations, graph theory, functions, Boolean algebra, formal expressions, differential and integral calculus, automation theory and topology, sets of equations, integral equations, and especially common and partial differential equations. The solution of problems formulated with these mathematical procedures nearly always necessitates the use of numerical methods and the use of electronic calculators or computers.

B.m. is applied mainly to the following problems: growth of organisms and populations, interaction of populations with each other and with the environment, biochemical reaction kinetics, compartmentation kinetics, pharmacokinetics, enzyme kinetics, transport processes (e.g. through biological membranes), diffusion reaction processes, morphogenic, biophysical and biochemical processes of structural differentiation, dynamics and stability of open systems, metabolite transport, information transfer and processing in biological systems, biocommunication, biological oscillators, biological regulation and control systems.

A mathematical relationship or expression, which defines the state of a system or process under investigation, the changes that it undergoes, or the relationship between its dimensions and other parameters, is called a mathematical model. Some mathematical models are extremely representative of the biological situation they attempt to describe, e.g. the Hardy-Weinberg law (see) is a nearly perfect model. Due to the complexity of biological material, however, a biomathematical model is usually an oversimplification. In spite of this, there is often a very close and useful correspondence between the model and reality. A relatively simple model is first devised, then continually refined until it becomes useful.

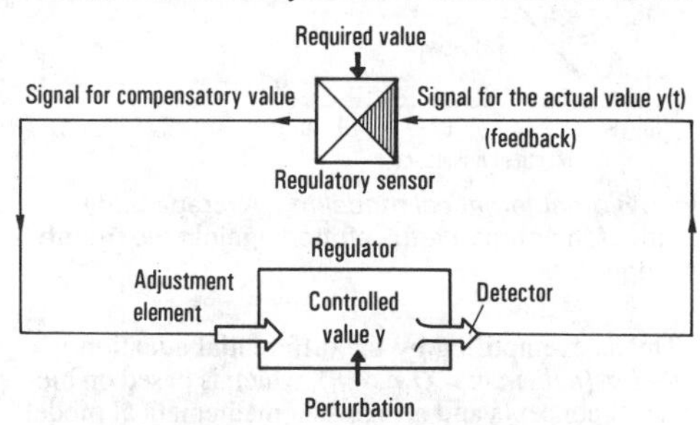

Fig.1. *Biomathematical modelling.* Schematic representation of a closed regulatory circuit, showing the functional elements involved in the regulation.

Particularly important in biology are regulation processes. Fig.1 shows a model of a regulatory loop. As a result of feedback control (i.e. the response of the regulator at all times to the prevailing value of the regulated quantity), a compensatory change is made in response to any perturbation, and the quantity in question is maintained at a nominal, predetermined value. The ability of living systems to regulate the value of physiological quantities, so that they remain within limits necessary for the functioning of the system, is called homeostasis. The normal physiological values themselves are parameters of the highly organized complex of systems that constitute the living organism, and they have been established by evolution. Feedback control systems are active at all levels of biological organization, e.g. at the cellular or molecular level (e.g. in protein biosynthesis), and even at the level of ecological interaction and the social integration of biological systems. Biocybernetic analysis and modelling use a variety of mathematical methods, which have been developed theoretically for describing the function and dynamics of regulatory loops and other regulatory mechanisms.

In the modelling of biological control, and in most of the areas of B.m. listed above, differential equations play an important part. These are derived from assumed correlations in the biophysical, biochemical or biocybernetic model. The differential equations, or their solved functions (provided they can be presented in the form of an analytical expression) contain model mathematical parameters, whose values must still be provided by a real situation. The provision of these values is known as parameter identification. Provided that in principle the mathematical model correctly represents the interrelationship under investigation, any discrepancy between the model and the analytical data can be ascribed to randomness, i.e. the existence of Biological variance (see). Determination of the values of the model parameters is then often performed by the Gaussian method of least squares, i.e. the values assigned to parameters are such that the sum of the squares of the differences between the model and real data are minimal.

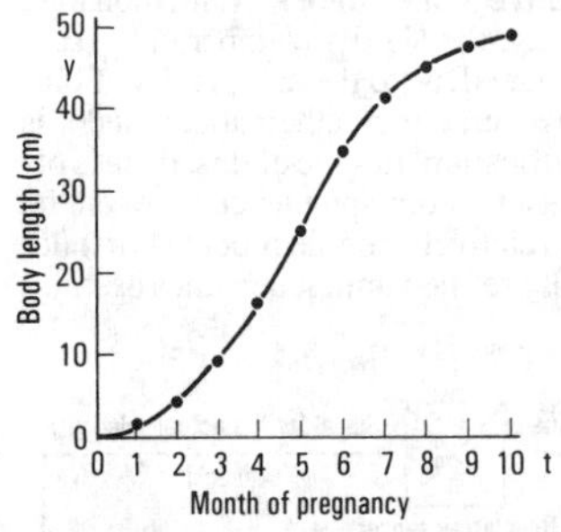

Fig.2. *Biomathematical modelling.* Average body length of the human fetus plotted against the month of pregnancy.

This is exemplified by the differential equation $\bar{y} = ay^m - by^n$ $(a, b, m, n > O, n > m)$, which is based on biophysical concepts and serves as a mathematical model of growth. The first differential coefficient of the growth rate (y) is represented by $\bar{y}$. For many growth processes in living organisms, the value of y follows an S-shaped time course (Fig. 2; see also Fig. 2 in the entry Biomathematics). At first $\bar{y}$ increases with y, as expressed by the term ay^m. As growth continues, however, the attenuation effect of by^n increases until the rate of growth becomes zero. Using the data points in Fig. 2 and the method of least squares, the calculated parameters of the differential equation for this growth process are: $\bar{y} = 1.78y^{0.39} - 0.00018y^{2.93}$. The curve in Fig. 2 is the integral curve of this equation, with the starting value $y(t = 0) = 0.039$. Within the limits of accuracy of the draughtsmanship, the curve is identical with the Bertalanffy growth function, i.e. the line of best fit to the data points; this is expressed as $y = c_1[1 - exp(c_2 - c_3 t)]^{C_4}$, which is the integral of the differential equation $\bar{y} = ay^m - by$ $(0 < m < 1)$.

Biomathematics: the application of mathematics to various biological disciplines. Mathematics is used in biology to overcome problems of analysis and interpretation, which are due to the variability of living organisms (see Biological variance) and the structural and functional complexity of biological systems and phenomena.

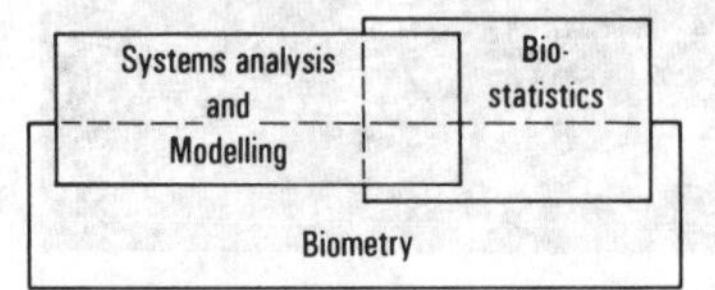

Fig.1. *Biomathematics.* Different areas of biomathematics.

B. is subdivided into 3 overlapping areas (Fig.1). Statistics (see Biostatistics) was the first branch of mathematics to be exploited as an aid to biology. Based on the probability concept, statistics deals with the random component of biological phenomena, and helps to clarify relationships which are obscured by the inherent variability of biological material. Biometry (see) deals with the treatment and presentation of observational and analytical data, so that they can be evaluated biostatistically or serve as a basis for mathematical modelling. Historically, biostatistics and biometry developed together; at first, they were not recognized as different areas, the two subjects were more or less synonymous, and they were often referred to jointly as B. The youngest area of B. is Biomathematical modelling (see) and systems analysis. This is generally closely associated with the biophysical or biochemical modelling of living systems; alternatively, it may be linked to the biocybernetic modelling of biosystems and their environmental interactions, a branch of ecology that arose in the middle of the present century.

In biology, as in many other sciences, the most important source of reproducible data is the planned experiment. Mathematics is an essential aid to the design of, and especially to the evaluation of the results from, such experiments. Thus, mathematical procedures can be used to separate the random and deterministic components of experimental values; these 2 components are present in every phenomenon and process of living systems. In principle, the deterministic component is reproducible when observations and experiments are repeated under the same experimental conditions. Superimposed on the deterministic component is the random component, which usually obscures the underlying, regular behavior of the system. In biological systems, this randomness is largely due to Biological variance (see), i.e. the wide random variation in the degree of expression of properties and processes in otherwise similar biological systems or organisms.

As an example, Fig. 2 shows the result of an experiment, in which a growth inhibitor was applied to wheat seedlings, and the primary leaf lengths of 9 seedlings were measured on 10 successive days. The differences between the 9 leaf lengths for similarly treated seedlings, recorded on a single day, is due to the random variation of biological material, and it is superimposed on the deterministic or actual increase of leaf length over the 10 experimental days. In Fig. 2, the latter is represented by the continuous line, which was determined mathematically (see Biostatistics). Similar experiments can then be performed with different concentrations of growth inhibitor, and the statistically established deterministic value for the leaf growth rate can be used as an index of inhibitor activity.

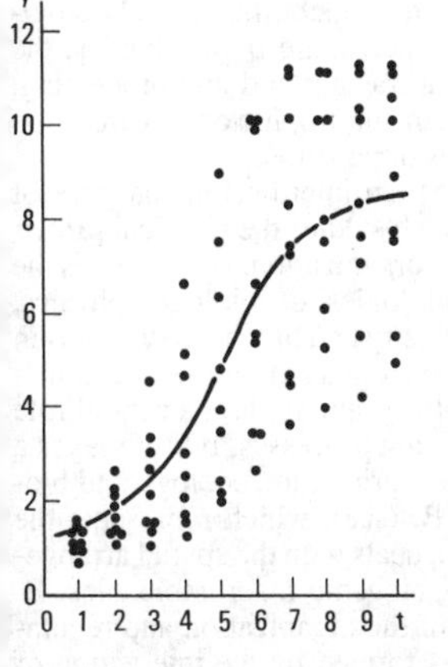

Fig.2. *Biomathematics.* Experimental data for plant growth, showing the statistically determined average (solid line). The time of measurement in days is shown on the abscissa (t). The extent of growth is shown on the ordinate (y).

The particular importance of mathematics to biology lies in the fact that a mathematical interpretation is entirely impartial. This is illustrated by the presentation and evaluation of the experimental data in Fig. 2; the random variability of this experiment would provide licence for numerous subjective values for the deterministic component. Without mathematical criteria for establishing the true deterministic component, subjective analysis could produce several different and even controversial interpretations. Conversely, the application of inappropriate mathematical procedures can lead to false interpretations, e.g. the use of an unsuitable mathematical model for calculation of the regression equation, or an unjustified assumption as to the type of the distribution of the random values.

The main difficulty in applying mathematical concepts and procedures to biological problems lies in the different levels of abstraction of the two subjects. The specific character of a natural object is determined by the totality of numerous properties and interrelationships. In order to formulate a biological problem for mathematical analysis, it is therefore necessary to isolate a single parameter from a multiplicity of parameters, i.e. from the structural and functional complexity of the system under investigation. The abstraction process uses biostatistic, biophysical, biochemical or biocybernetic modelling. It is necessary to test whether and how this abstraction process may place constraints on the quality of the subsequent data, i.e. whether the original formulation of the problem is still valid, or whether it should be modified. Careful attention to this problem is critically important in the interpretation of the results.

Biomechanics: a branch of biophysics, which investigates the mobility and contractility of biological systems. A central problem of B. is the mechanism of chemo-mechanical energy tranduction. At the level of the entire organism, B. also deals with locomotion on land, in the air, and in water. Important branches of B. are Hemorheology (see), and Biostatics (see), which is closely related to bionics. Hemorheology, as well as the study of fluid movement in plants, employ mainly hydrodynamic methods.

Biomembrane: see Membrane.

Biometry: a branch of Biomathematics (see) which attempts to quantify the properties of living systems, biological relationships and natural laws. This abstract process proceeds partly in association with the biostatistic, biophysical or biocybernetic modelling of living phenomena.

Characters; measurement. In order to recognize interrelationships and laws, the sometimes great variety of phenomena and properties available for investigation must be ordered, classified, and/or measured. Such properties are generally known as *characters.* Some examples of characters and their classification are: hair color, type of chemical bonding, sex (type 1); intelligence, condition after an operation, nutritional state (type 2); height above sea level, direction in a plane, electric potential (type 3); body size, body weight, respiration rate (type 4).

A character may be expressed with varying degrees of intensity in different objects, or in the same object at different times, i.e. the character may have different values. Character values are expressed and recorded in different numerical ways, or non-numerically. Characters can be classified (e.g. type 1, type 2, etc, above) according to the method most appropriate for recording their intensity, presence, absence, etc.

The term, measurement, is generally taken to mean the establishment of the intensity of expression of a character in a concrete object. The traditional understanding of measurement as the laying on of a "ruler" and reading from a scale is important in biology, where often qualitative characters are involved.

The distinction between qualitative and quantitative characters is determined by the type of character expression. The character values of *qualitative characters* are denoted by names or symbols, e.g. "sex", which has two values or classes, "male" and "female". *Quantitative characters* are such that their intensity of expression can be recorded numerically as different quantities.

Qualitative characters can usually be expressed by only a limited number of discrete values. In contrast, quantitative characters (also known as variables) may be either discrete or continually changeable. In the latter case, the quantitative character can have any numerical value within its variation interval. Measured values of a continually variable character are always "granulated", i.e. they are represented by measurement intervals, whose size is determined by the accuracy of the measurement (see Biostatistics).

Scale divisions. If the different intensities of character expression are characterized merely by different names or designations, the scale is *nominal* , e.g. the qualitative type 1 characters listed above: hair color (with the values, blond, brown, dark, gray, etc), chemical bonding type (heteropolar, homopolar, metallic, coordinate), and sex.

If the intensities of expression of a qualitaty can be placed in a ranking order, the scale is *ordinal.* Examples of such ordinally scalable characters are the type 2 characters listed above; their values are mostly vague attributes, such as: "seldom", "sometimes", "often", or "good", "bad", etc.

In the case of quantitative characters, the intensity of expression of a character can be expressed quantitatively, and the difference between two values can be expressed numerically. The characters of group 3 (above) can be recorded on a scale with measured intervals. The zero point on an interval scale is an arbitrary reference point, or it is established by convention, and it is not determined by the problem under investigation; the use of ratios or percentages is therefore inappropriate. For characters of type 4 (above), however, a proportional scale is used; the zero of a proportional scale is determined objectively.

The crosses in the table show the appropriate scales for different types of quantitative data. For the statistical evaluation of data on a nominal or ordinal scale, only nonparametric methods (see Biostatistics) are suitable, whereas characters recorded on an interval or proportional scale can be subjected to the entire rannge of statistical treatments.

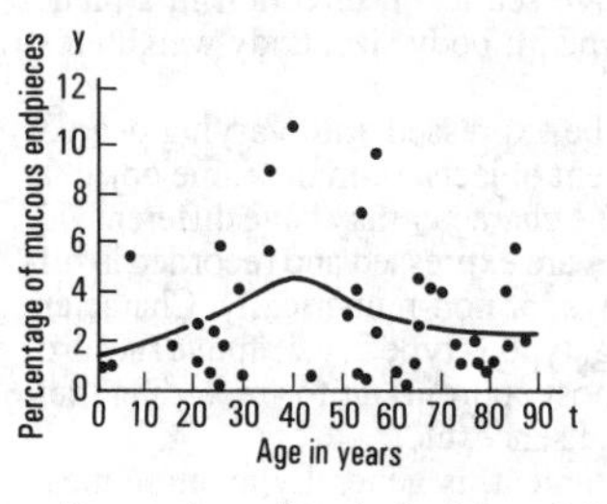

Biometry. Changes in tissue composition as observed in cross sections of human submandibular gland.

Processing of analytical data. Many biological problems are concerned with the dynamic properties of living systems, so that the intensity of expression of characters is determined at different time points. If this is performed on a single individual, the result is a *longitudinal* study, whereas the investigation of a character in different individuals (e.g. of different ages) is known as a *tranverse* study. In the latter case, the deterministic component is subject to random variations (see Biomathematics, Biostatistics), resulting from Biological variability (see).

In Biomathematical modelling (see), the deterministic component is represented in the mathematical model, and the resulting mathematical expression compensates for the random error. In the absence of a mathematical model, the value of the character under investigation can be determined by empirical regression, or other numerical smoothing procedures. An example is shown in the figure, where widely varying individual values have been smoothed into a gentle curve, displaying a gentle increase in the proportion of the investigated tissue up to the 40th year of life, followed by a decrease.

Biomolecule: a general term for any chemical compound that belongs to a biological system, especially a macromolecule like a nucleic acid, protein , or polysaccharide.

Bionics: a branch of science that exploits biological principles in technology. The object of B. is to improve the quality and to extend the capabilities of technological systems. Of central interest are biomechanics, the principles of information transfer and and processing, and energy transduction in autotrophic organisms.

Biophage: see Modes of nutriiton.

Biophysics (plate 28): an interdisciplinary area of biology and physics, which studies the physical principles of life at all levels of organization. B. combines the knowledge and methodologies of biology, physics, physical chemistry, chemistry and biochemistry. B. is in a process of continual development in close interaction with neighboring disciplines, and it plays a central role in the investigation of living processes. It also makes a contribution to medicine, agriculture, ecology and biotechnology. Molecular B. (see), which represents the main focus of modern B., deals with the spatial arrangements and structures of biogenic macromolecules. B. also includes the study of the organization and regulation of complex systems formed by the interaction of macromolecules, organelles and cells, as well as the study of living organisms with respect to mechanics, acoustics and optics.

Biopolymer, ***biomacromolecule:*** a biological molecule, such as a protein, nucleic acid, polysaccharide, or lignin, which can be regarded as a polymer of identical or similar subunits. *Aperiodic* B. (e.g. proteins) are composed of similar but not identical monomers, whereas *periodic* B. (e.g. cellulose) are formed from identical subunits. Bs. are essential for the life processes of the organism, serving as structural material (e.g. cellulose, lignin, keratin, collagen), enzymes (all enzymes are proteins), energy stores (polysaccharides such as starch or glycogen), or for information storage and transfer (nucleic acids). Polyprenols and polyterpenes may also be regarded as Bs.

The monomer sequences of nucleic acids and proteins are determined by nucleic acid templates, and organisms have elaborate systems for synthesizing these B. with very low rates of error. The overall composition of polysaccharides and lignins is determined by the enzymes catalysing their synthesis, but their size and exact monomer sequence may be determined by factors such as the juxtaposition of other structural molecules and rate of growth (e.g. cell wall polysaccharides and lignin) or the rate of degradation and availability of precursor monomers (e.g. storage polysaccharides). Blood

Biometry. Different types of character values and their appropriate scaling systems

Scale	Qualitative characters		Quantitative characters	
	Class or frequency values	Ranking values	Average numerical values	Ratios and percentages
Proportional	X	X	X	X
Interval	X	X	X	
Ordinal	X	X		
Nominal	X			

group antigens and other cell surface recognition oligo- and polysaccharides, however, have precise sizes and monomer sequences.

Biorhiza pallida: see Galls.

Biorhythm: a pattern of biological activity with a characteristic periodicity. A ***circadian*** rhythm is an activity or pattern of behavior that recurs every 24 hours (e.g. the sleep/waking cycle). In ***ultradian*** rhythms the periodicity is markedly shorter, e.g. "bursts" of activity in animals at intervals of 3–4 hours, which are often associated with changes in metabolic activity. Various types of circadian rhythm are synchronized differently with the cycle of light and dark. Thus, animals active in daylight are known as ***diurnal*** species, whereas those active during the night are called ***nocturnal.*** Animals active at dusk or twilight are referred to as ***crepescular.*** An ***annual*** rhythm exhibits a periodicity of 1 year, e.g. animal Hibernation (see) and bird migration, as well as reproductive cycles in many species. In contrast, the ovarian cycle of the laboratory rat is about 4 days, whereas that of the human female is about 28 days. Further examples of rhythmic biological behavior are: the activity of the enzyme, *N*-acetyltransferase, which displays a circadian rhythm in the pineal gland; the short-wave changes in potential of individual nerve fibers recorded by electroencephalography (see); and the periodicity or rhythm displayed by cell division (see Cell cycle).

In the analysis of Bs., a clear distinction must be drawn between: 1) the ***internal, endogenous*** or ***free-running rhythms*** of an organism, which are determined independently of the environment by a mechanism known variously as a ***biological pacemaker, biological clock*** or ***physiological clock;*** and 2) the influence of the environmental factors, which may ***synchronize*** the endogenous biological clock. Such influential environmental factors may be referred to as ***environmental cues,*** but many authors use the German word, ***Zeitgeber*** ("timers or time-givers"). Zeitgeber provide information to animals or plants about the periodicity of environmental variables, and they can influence both the phase and the frequency of an endogenous biological clock. The most familiar and universally active Zeitgeber is the daily light/dark cycle. Zeitgeber may be current or latent. A ***current*** Zeitgeber is currently operating, while a ***latent*** Zeitgeber may become active if the current Zeitgeber ceases to operate, or the animal is removed from its influence. Often there is a hierarchy of Zeitgeber with varying degrees of potency, e.g. three hierarchical Zeitgeber in domestic animals are the feeding timetable, the light/dark cycle and biosocial influences. Zeitgeber serve to continually adjust endogenous rhythms so that they maintain an accurate phasing, a process known as ***synchronization.*** Generation of a rhythm by repeated exposure to Zeitgeber, with subsequent persistence of that rhythm, is known as ***entrainment.***

If unaffected by Zeitgeber, the endogenous biological clock gives rise to a ***free-running rhythm.*** If an animal is kept in constant darkness, its rhythm of daily biological activity continues with a period of 24 hours, but drifts slightly and becomes either shorter or longer, i.e. it becomes free-running. With some exceptions, free-running rhythms conform to Aschoff's rule: "the direction and rate of drift of a free-running rhythm in a constant environment are a function of the light intensity, and whether the animal is normally diurnal or nocturnal". Thus, the free-running rhythm of a nocturnal animal is slightly less than 24 hours, whereas that of a diurnal animal is slightly longer than 24 hours. When isolated in an environment lacking all visual clues about the passage of night and day, most human individuals establish a cyclic pattern of waking, sleeping, physical activity, feeding, etc., based on a free-running rhythm with a period of about 23 hours.

The existence of a truly endogenous clock, which is not influenced by any form of "learning" is shown by ***isolation*** experiments with bird and reptile eggs. Irrespective of the conditions under which the eggs are allowed to hatch and are subsequently maintained (constant light, constant darkness, etc.), the hatched animals display activity cycles of about 24 hours.

Similarly, the existence of a biological clock is shown by ***translocation*** experiments. Isolated under constant conditions, bees forage according to a 24-hour cycle. When such bees are flown overnight from Europe to New York and placed under similar constant conditions, they begin foraging exactly 24 hours after the last foraging. In addition, when bees in outdoor hives (i.e. not under controlled conditions) are transported from the east to the west coast of the USA, they continue to forage according to the 24-hour cycle of the east coast, with a gradual adjustment over a period of days to the west coast diurnal cycle. The subsequent adjustment represents synchronization of the endogenous clock by the west coast Zeitgeber.

It seems that the biological clock(s) of plants and animals is a physiological oscillator, which moves backward and forward between two extreme states at a constant rate. It probably consists of a complicated series of largely temperature-independent reactions. As yet, there is no useful molecular model of such a system. Various findings indicate, however, that nucleic acids may be involved. Thus, in *Acetabularia* and *Gonyaulax* the diurnal rhythm is abolished by the antibiotic, actinomycin, which inhibits transcription. Furthermore, if the nucleus of *Acetabularia* is removed, the customary rhythm carries on as before; but implantation of a nucleus from a cell with a differently phased rhythm, causes the the previously denucleated cell to display the same rhythm as the donor cell.

Under constant conditions of continuous darkness or continuous light, endogenously regulated flower and leaf movements or the cyclic activity of *N*-acetyltransferase in the mammalian pineal gland continue for days, but eventually need to be promoted again by Zeitgeber. Especially in multicellular organisms, there is thought to exist a central, endogenous synchronizer or central clock for all the other possible types of internal pacemakers or endogenous clocks. In cockroaches, the endogenous clock for locomotory activity appears to be localized in the optic centers of the brain and possibly also involves the lateral brain neurosecretory cells. In the rat, a direct neural connection between the retina and the hypothalamus has been shown, which terminates in a specific region of the hypothalamus, known as the suprachiasmatic nuclei. At one time, this was thought to represent the endogenous clock, but evidence now points to the existence of an endogenous clock in the ventromedial nucleus of the hypothalamus, which is a central timekeeper for many rhythms, including, e.g., the rhythms displayed by the pineal gland. It is also evident that in birds and mammals the pineal gland is important in the reception and/or processing of light stimuli, which are among the most obvious Zeitgeber, and that the production of melatonin by the pineal gland is in some way involved in the internal time keeping process.

Biosocial behavior: a term from animal sociology, referring to the interactive behavior of individuals of the same species on a basis of mutual attraction. There are several types of B.b. a) *Prebiosocial behavior:* the meeting of animals at a particular site and at certain times, followed by coordination of behavior, leading to a specific group activity such as sleeping, breeding, hibernation, etc. b) *Aggregation behavior:* the attraction of animals into groups from time to time at certain sites by the secretion of species-specific aggregation hormones. c) *Semibiosocial behavior 1:* the formation (independently of a specific site) of large, anonymous groupings (swarms, herds), which dsiplay coordinated behavior. d) *Sub-biosocial behavior 1:* the grouping of individuals of different ages for the purposes of care and protection, i.e. brood care. e) *Advanced sub-biosocial behavior 1:* this arises when adults and juveniles remain together for a relatively long period. f) *Semibiosocial behavior 2:* the formation of brood care communities containing more than one breeding pair. g) *Eubiosocial behavior 1:* the formation of brood care communities in which special "castes" perform the brood care, i.e. there is a division of labor (e.g. in communities of social insects). h) *Sub-biosocial behavior 2:* individuals come together in search of a sexual partner, followed by courtship and pairing. i) *Advanced sub-biosocial behavior 2:* the members of a successful breeding pair remain together beyond the reproductive period, sometimes until one partner dies. j) *Primitive biosocial behavior:* offspring remain with the breeding pair, thereby forming a family grouping or tribe. k) *Eubiosocial behavior 2:* the formation of tribes and complex, non-anonymous groups with an allocation of roles, i.e. there is a "pecking order" or hierarchy.

These different types of B.b. have arisen by different evolutionary routes. Central to B.b. is the necessity to interact with other members of the same species, whose presence in time and space is nonrandom. It therefore represents a special form of distance regulation, which precedes coordination, cooperation and biocommunication. Eubiosocial behavior is encountered only in arthropods and vertebrates.

Biosphere: the totality of living organisms on the earth. It is often used synonymously with *ecosphere,* which is the totality of habitats occupied by living organisms. The ecosphere includes the upper layer of the earth's crust (lithosphere), with its surface and underground waters (hydrosphere), and the lower layer of the earth's gaseous envelope (atmosphere).

Biostatics: a branch of biomechanics. B. deals with the interaction of internal and external forces and moments in different positions and postures, and during the movement of animal or plants. The support structures of organisms are subject to different forces, such as pressure, tension, bending and torsion. The basic principle in the structure and constitution of support structures is the achievement of maximal resistance to relevant forces at the expense of minimal mass. In this respect, the principles of B. are the same as those applied to technological systems. Thus, the subject of Bionics (see) attempts to use the optimal structural principles of biology in the construction of technical systems.

An interesting area of B. is the investigation of systems with a high ratio of height to width, i.e. systems that are very slender, expressed as $l = \lambda/d$, where λ is the degree of slenderness, h is the height or length, and d is the diameter. For example, the rye halm attains a λ-value of 500, while the λ-values for bamboo and trees are 133 and 10–50, respectively. The λ-values of television towers are 15–20. It is incorrect to suppose that optimal stability with minimal expenditure on material can be achieved by building extremely thin, high towers, based on the dimensions the cereal halm. As the size of a building increases, the volume, and therefore the mass, increase as the cube function of the dimensions. As a first approximation, however, resistance to bending is proportional to the cross sectional area, i.e. to the square function of the dimensions. Strictly proportional enlargement therefore leads to increasing instability. As the size increases, stability can only be maintained by a greater than proportional increase of the diameter ($d \approx h^{3/2}$), a relationship that is found to be widely fulfilled in biology, as well as in technology.

An important principle for resisting bending in the skeletal system is spring loading (Fig.). Bones withstand high pressure and tension, but are susceptible to bending. Counter-bending by muscles and ligaments decreases the bending forces at the expense of increased compression forces. The *moment of resistance* to bending is proportional to the *area moment of inertia.* This is expressed by the integral $I = r^2 dF$, where r is the distance from the axis. The area moment of inertia can therefore be increased by the addition of cylinder-shaped transverse sections. This tube principle is encountered in many systems that are subject to bending forces. The addition of cylindrical sections also increases the moment of resistance to torsion, which is calculated in relation to an axis in the stem center.

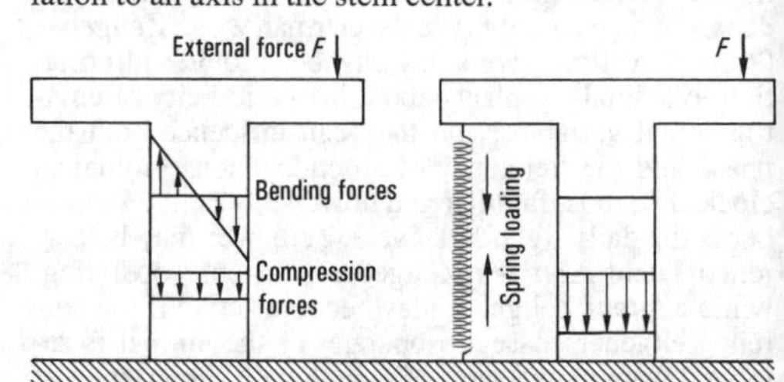

Biostatics. Relief from bending forces at the expense of increased compression forces by the application of spring loading.

Bending forces depend on the distance between the point of application of the force and the point at which the support structure is attached or jointed, i.e. the moment of the force. Support material is used most economically, if the support structure offers the same resistance to bending at all points. Such a structure, with constant resistance to bending, is achieved by alteration of the cross sectional area along its length. In this respect, for example, the human ulna is fully optimized.

Biostatistics: a branch of Biomathematics (see), which describes and analyses the natural variation in data from biological experiments or trials.

Natural variation of data; Influencing factors; Properties, Quantities, Character values. Data for statistical evaluation may be generated by a planned experiment, or they may be derived from a natural process that has occurred without human intervention. Such data are always subject to Biological variability (see), i.e. they consist of values that are subject to chance variation (stochastic variables), due to the influence of factors that are not constant and cannot be made constant. Repetition of an experiment therefore does not always produce the same result. For example, individual plants display unpredictable differences in growth. Therefore, the effect of a fertilizer on the growth of a plant species is not investigated on a single plant, but on a number of plants of the same species. Growth is also affected by factors like the composition, structure and moisture content of the soil, various microclimatic conditions, the presence of pollutants or pests,and many other factors that may be unrecognized, due to the inherent complexity of biological processes and systems. It is assumed that the effect of each of these uncontrolled and uncontrollable factors (known as random factors) is small, and that they do not all affect the result in the same sense or direction, i.e. some cause growth inhibition, while others stimulate growth.

In contrast, the factors under investigation are kept constant or changed by a known amount when the experiment is repeated. In the fertilizer experiment, examples of such constant or controllable experimental factors are the chemical composition of the fertilizer, its mode of application, its quantity, and time and frequency of its application. Thus, in order to investigate the dependency of growth on the concentration of a fertilizer, a range of experiments would be performed, using different concentrations of fertilizer, and maintaining all other factors constant. Especially in field trials, the experimental design is such that several factors and their possible interactions are investigated simultaneously. Usually, the influence of any factor on a system is investigated at more than one level, value or intensity of that factor.

If replicate experiments are performed one after another, the result of one may influence the result of the next. Thus, the effects of treatment on a patient might influence the appraisal of another patient receiving the same therapy. Such interexperimental effects are undesirable; moreover, it is usually impossible to assess their magnitude. For the purposes of statistical evaluation, the replicates of experiments should therefore be mutually independent.

Frequently the question arises as to whether a factor has any influence at all. The experiment (with replicates) is then performed in conjuction with a control experiment (also with replicates). Both experiments are identical with respect to all conditions and factors, except that the factor in question is eliminated from the control.

A change in the value of a property, character or quantity is due to active causative factors in the system. Statistical methods of Factorial analysis (see) are used to determinewhich of a number of factors is most potent in causing a change in the value of a quantity. Statistical procedures that simultaneously take account of several quantities of an experiment are called multivariate methods. For the statistical analysis of only one quantity, univariate methods are used. Owing to the complexity of biological systems and processes, multivariate methods are generally more appropriate for the analysis of biological data.

Random variables; Distribution. Since the change in value of a property is influenced by random factors, it is customary in statistics to also describe such properties as a *Random variables* (see), especially if they are quantitative properties. Random variables are generally represented by large case letters, e.g. X; the value of the property resulting from the j-th performance (the last in a series of j replicates) of an experiment is represented by the corresponding small case letter, x_j. The value of x_j is then known as a realization of the random variable X. If a random variable is always a whole numer (e.g. the number of segments on a worm; the number of flies in a trap), it is known as a *discrete* random variable. If any value in a particular range is possible, then it is a *continuous* random variable; the latter is, in fact, any random variable that is not discrete.

In the fertilizer experiment, let the investigated parameter be the yield. If this is expressed as weight, then X is a continuous random variable, and x can have any positive value up to a biologically plausible upper limit. If "yield" is defined as the number of fruits, only whole numbers are possible, and X is then a discrete random variable.

The *degree of confidence* or *probability (P)* that a random variable will have a certain experimental value is expressed on a scale of 0 to 1, or as a percentage. An impossible result has a probability of 0, whereas a result that is bound to arise in every experiment has a probability of 1. Any single value of a parameter therefore has its own probability. This can be determined theoretically or experimentally. In the experimental determination of probability, the same experiment is repeated many times, and the number of times *(m)* that each value of the quantity appears in *n* experiments is recorded. The *relative frequency, m/n,* approaches closer to the true probability of that value as the number ofexperiments (i.e. the number of experimental values of x) is increased. In other words, the probability of an event is the relative frequency with which it occurs in an infinite (or indefinitely large) number of trials.

If the probablities of all possible experimental values of a parameter are known, then the parameter is statistically fully characterized. Expressed in terms of random variables, this means that the probability that the value of the random variable will lie within any chosen range or interval is known.

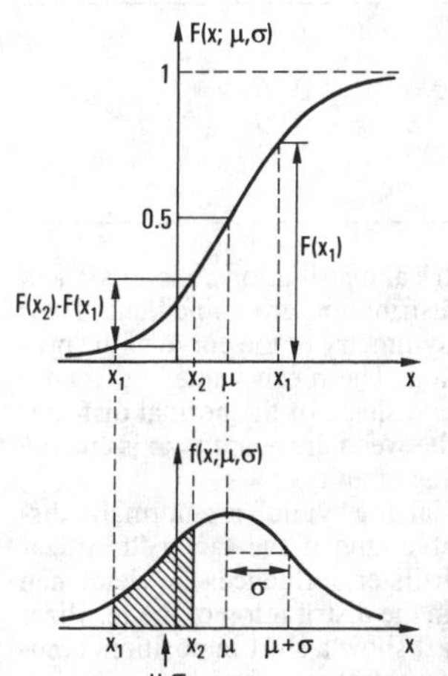

Fig. 1. *Biostatistics.* Distribution function and density of the normal distribution.

The upper part of Fig. 1 shows the curve for the distribution function of the Normal distribution (see). The function value, $F(x_1)$, is the probability that the random variable, X, will have a value in the interval $(-\infty, x_i)$, i.e. that x_j is smaller than x_i. The probability that the value of X will lie between x_1 and x_2 is given by $F(x_2)$-$F(x_1)$.

The function of the normal distributions is continuous; when plotted, its curve shows no discontinuities. Random variables, whose probability distribution is described by such continuous distribution functions, are called *continuous random variables*. The distribution of a continuous random variable can also be represented by the *distribution density* or *probability density*. The probability density for a normally distributed value is shown by the *Gaussian curve* in the lower part of Fig. 1. Mathematically, this is the first derivative of the distribution function. The probability that the normally distributed random variable, X, will possess a value in the interval between x_1 and x_2 is equal to the area of the curve between the abscissa values of x_1 and x_2.

The distribution density of the normal distribution is expressed as the function

$$f(x; \mu, \sigma) = \frac{1}{\sigma\sqrt{2\pi}} \exp[-(x-\mu)^2/2\sigma^2],$$

where μ and σ are the distribution parameters; σ is the *expected value* or *mean value;* σ is the *standard deviation;* and σ^2, the square of the standard deviation, is known as the *variance of scatter.* It can therefore be stated that the random variable, X, is normally distributed with a mean value of μ and a variance of σ^2; this statement is symbolized as $N(\mu,\sigma)$. Table 1 lists the probabilities (P) that the value x for a random variable with a distribution of $N(\mu,\sigma)$ will lie within the symmetrical interval $\mu - k\sigma \leqq x \leqq \mu + k\sigma$, i.e. an area of the Gaussian curve symmetrical around a vertical center line corresponding to μ. For example, the probability that x exceeds the limits $\mu \pm 3\sigma$ is less than 3/1000. There is a 68% probability that an $N(\mu, \sigma)$-distributed variable will lie in the range described by $\mu + \sigma$ in Figs. 1 and 2.

Probabilities (P) for values from a kσ region around the mean value of a normally distributed random variable

k	P	k	P
1	0.6827	1.96	0.95
2	0.9545	2.58	0.99
3	0.9973	3.29	0.999
4	0.999937		

As with all symmetrical distributions, the mode and mean of the normal distribution are coincident. Their ordinate is the axis of symmetry of the curve of the probability density function. The mean value and scatter describe the position and shape of the normal distribution (Fig. 2). The area between the quantity axis and the curve is assigned a value of unity.

If, in the fertilizer trial, the "yield" is a normally distributed random variable, and if the factor "fertilization" (application of fertilizer) influences the yield, then the random variables in the distribution of the fertilizer experiment must at least show a shift in position, compared with those of the control experiment.

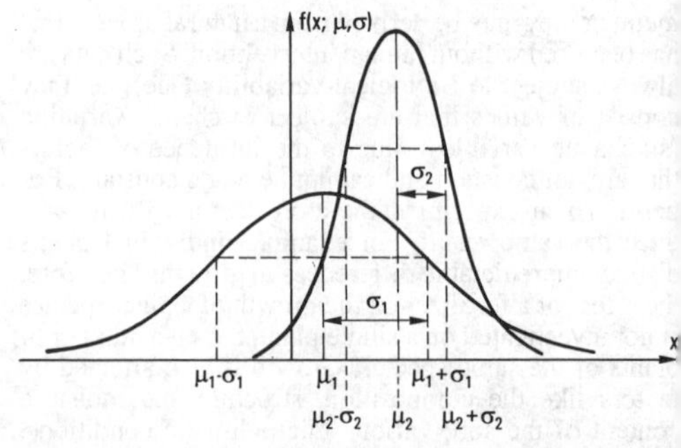

Fig.2. *Biostatistics.* Normal distributions for different values of the mean and scatter.

By scale transformation, any distribution, $N(\mu, \sigma)$, can be converted into the *standard normal distribution*, $N(0; 1)$, with the mean value of 0 and a scatter of 1. The density of scatter with these values is symbolized with the letter φ, so that $f(x; 0; 1) = \varphi(x)$. The distribution function, $F(x; 0; 1) = \Phi(x)$, is listed in published tables. The value, $F(x_1;\mu, \sigma)$, is equal to the value $\Phi(c_1)$, where $c_1 = (x_i - \mu)/\sigma$. The graph of the distribution function of a discrete random value is a stepped curve. If the random value is described by only a few values, the probability distribution can be represented with a bar graph. Fig. 3 shows an example of a binomial distribution, with the stepped plot of the distribution function of probabilities at the top, and the bar graph of probabilities below.

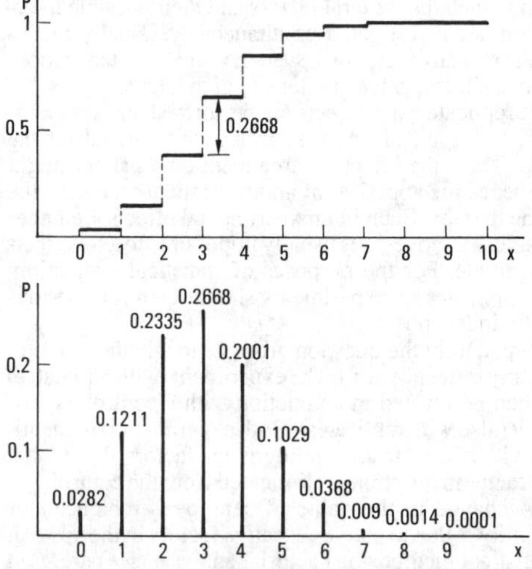

Fig. 3. *Biostatistics.* Distribution function (above) and probability distribution (below) of a binomially distributed random value.

The *binomial distribution* is used in the investigation of characters that have only 2 values, e.g. sex, which can only be male or female. If the probability of the occurrence of one character is p, then the probability for the other ist $(1-p)$. When an experiment has only 2 possible alternative results (plus or minus, male or female, alive or dead,etc.), the frequency, X, of one of the 2 variables determined by n independent experiments, is a disrete, binomally distributed random value. The natural numbers 0, 1, ..., n are its possible realization. Their probabilities, $P(x_i)$ can be calculated from the formula $P(x) = \binom{n}{x_i} p^{x_i}(1-p)^{n-x_i}$. The value $\binom{n}{x_i}$ is called the *binomial coefficient*. The length of the experimental series (n) and the probability (p) are the parameters of the binomial distribution, where $\mu = np$, and σ^2 (variance) = npq. For Fig. 3, $p = 0.3$ and $n = 10$ were chosen; the probability that the same value will be obtained for a parameter in all 10 experiments is less than 0.00001. The heights of the steps of the stepped distribution curves at different points (x_i) on the x axis correspond to the respective probabilities $P(x_i)$. Otherwise, the distribution function has the same meaning as in the case of a constant random variable; its ordinate value at a point x_j gives the probability for the realization $x_i \leqq x_j$ of the random value X. For the discrete random value this is the sum of the individual probabilities $\sum_{x_i \leqq x_j} P(x_i)$.

Statistical universe; Random sample. In theory at least, a random determination can be repeated an infinite number of times. This totality of theoretically possible determinations, in which every one produces an observable or measurable result, is called the *universe* or *statistical universe*, also known as the *target population*. The same term is also used for the quantity of all experimental objects that display the character under observation. From the statistical point of view, the universe is completely characterized by the probability distribution of the considered random variables. If, for example, a random value displays a normal distribution, then the universe must be normally distributed.

In practice, however, a determination can only be repeated a limited number of times. A *random sample* is therefore selected from the universe.

As a rule, the universe is deduced from the results of random sample analysis. In this way, valid general rules or laws are derived from a restricted number of real observations. For example, if fertilized plants give a statistically significant higher yield than those in the control group, one expects higher yields from all fertilized plants in the future.

However, all statements based on the statistical comparison of empirical data, or based on the relationships of empirical data to a universe, i.e. to an underlying theoretical model, are statements of probability, i.e. they are subject to the *probability of error.*

The aim of experimental scientific investigation is to draw conclusions about the universe from the results of analysing or observing random samples. However, the study of random samples only yields true statements of probability, if the random samples are representative of the universe, i.e. if they are an undistorted, scaled down portrayal of the universe. If random samples are "selected", i.e. if objects with certain character values are favored for experimental investigations, or certain character values are favored during the collection and recording of experimental data, then the results do not accurately reflect the distribution of values of the universe. This kind of systematic error, whose magnitude and influence are diffucult to assess, can be avoided by *random selection* of random sample elements. For example, the composition of control and treatment groups for a metabolic experiment on rats should not be determined by reaching into the animal cage and taking the first animals to hand. Although this might appear to be a random process, it is subject to the criticism that the more lethargic and less alert animals will allow themselves to be taken first. Preferably, each animal should be assigned a number, and these numbers should then be chosen at random (if necessary, "taken out of a hat").

Even if several random samples are taken from the same universe, the influence of random factors will insure that these will all have more or less different average character values and different scatters. To decide whether an observed difference is only a chance occurrence, or whether it is real, the limits of that particular character value must be established. If the observed difference is greater than the limiting value, then it is improbable that it is simply due to chance. Synonymous with this in the establishment of a *limiting probability* (α), or a *certainty probability* ($1 - \alpha$). In biology, α is normally assumed to be 5%, unless there is extra information to suggest a more appropriate value.

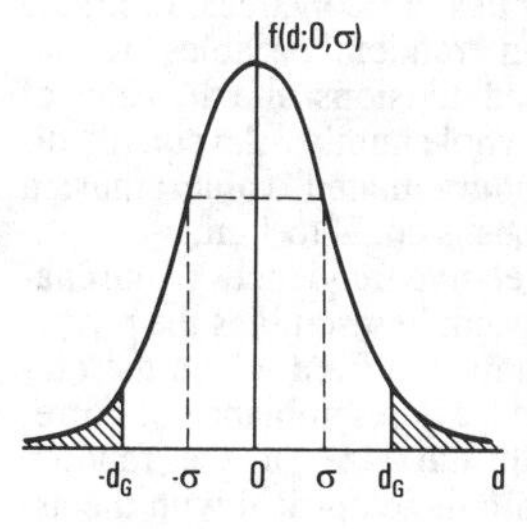

Fig. 4. *Biostatistics.* Symmetrical 95% confidence range of the expected value of a *N(0,s)*-distributed random quantity.

Fig. 4 shows these relationsips for a *confidence interval* that is symmetrical around the expected value of 0, for a random sample with the distribution $N(0, \sigma)$ [under certain conditions the difference, D, between 2 random samples can correspond to this type of distribution]. The hatched areas account for 5% of the total area between the bell-shaped curve and the abscissa. The significance limits are $d_G = \pm 1.96\ \sigma$ (see Table 1). If a mean difference, d, lies outside the limits of the confidence interval, it is considered improbable that it has arisen simply by chance. Limiting probability (α) is often taken as 3σ, i.e. $\alpha = 0.27\%$ (see Table 1). Limiting probabilities of 10%,1% or 0.1% are also sometimes chosen. The choice of these values depends not on mathematical but on practical considerations related to the material under investigations

Frequency distribution. Data from random samples are recorded in the order in which they are obtained. To obtain a view of the frequency distribution of the character values in the random sample, and to facilitate their evaluation, the data are then rearranged in order of their magnitude. In this second list, each value is listed only once, and the frequency of its occurrence is noted. This tabular record of character values and their frequencies is known as a *frequency table*. Table 2 is a frequency table of the random value "body length at birth", measured to the nearest cm in a random sample of 53 male neonates.

Body length at birth **(x_i)** ***of 53 male neonates: measured values with their absolute and relative frequencies***

x_i	41	42	44	45	47	48	49	50
H_i	1	1	2	2	1	1	7	5
h_i	0.019	0.019	0.038	0.038	0,019	0.019	0.132	0.094
x_i	51	52	53	54	55	56	58	61
H_i	4	13	6	4	1	3	1	1
h_i	0.075	0.245	0.113	0.075	0.019	0.057	0.019	0.019

In practice, continuous variates do not exist, since all measured values are rounded values. For example, when the smallest interval that a balance can weigh is a milligram, all weights are rounded off to the nearest milligram, i.e. X changes by discrete intervals of 1 mg. Such a value can be called a *granulated* or *atomic* variable. "Granulated" variables are actually continuous variables which have been converted into discrete variables by rounding off. In the case of discrete variables, the same value may occur two or more times in a sample. This is an almost impossible event in the case of continuous variables, but occurs all the more often, the more "coarsely granulated" they are. If two or more identical values occur in a sample of "granulated" variables, they are known as ties or tied values. From the outset, the values of discrete random variables are placed on a scale of equal-sized divisions, and the value of each variable represents a whole number. An equally divided scale is also used for "granulated" values; thus, in Table 2 every scale divisions is equal to 1 cm.

The distribution of the relative frequences of the character values of a random sample resembles the *probability distribution* of the universe from which the random sample was taken, and this resemblance or correspondence increases with the size of the random sample. In order to compare the empirical with the assumed theoretical distribution, a secondary classification is normally necessary. The variation interval is divided equally, and the number of values within each class or subinterval is reported.

Frequency table of the random sample data from Table 2, with division of the range into 7 equal-sized classes. Cumulative relative frequencies, $kh_i = \sum_{j \leq i} h_j$ ***are shown on line 4.***

x_i	42	45	48	51	54	57	60
H_i	2	4	9	22	11	4	1
h_i	0.038	0.075	0.170	0.415	0.208	0.075	0.019
kh_i	0.038	0.113	0.283	0.698	0.906	0.981	1.000

In Table 3 each group of 3 classes from Table 2 has been combined into a new and wider class; x_i shows the middle value of each new class (e.g. body length 42 cm for the 1st class); H_i represents the frequency of each new class, and h_i the relative frequency. Fig. 5 shows the *histogram* of this frequency distribution. The heights of the columns are proportional to the empirical frequencies.

The frequency distribution of a continuous variable may also be represented graphically by a *frequency polygon,* as shown in Fig. 6, whereas a bar graph (similar to that shown in Fig. 3) is preferred for discrete variables. The summed curve (see Fig. 3), which displays the empirical distribution function, is less informative; its values are the cumulative relative frequencies (see Table 3). If possible, these are obtained from the non-secondary classified distribution, and they are used in the Kolmogorov test of the hypothesis that the empirical distribution function in the statistical sense is the same as the assumed theoretical continuous distribution function. The test of the distribution of the data in Table 2 against a normal distribution, whose parameter values are assessed by the random sample average value and scatter, show that the null hypothesis of similiarty of the two distributions at a confidence level of 5% is not disproved.

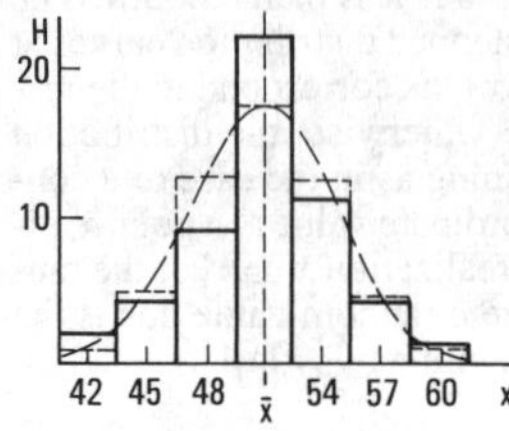

Fig. 5. *Biostatistics.* Histogram of the frequency distribution of body length (cm) of neonates, with the corresponding normal distribution (dashed).

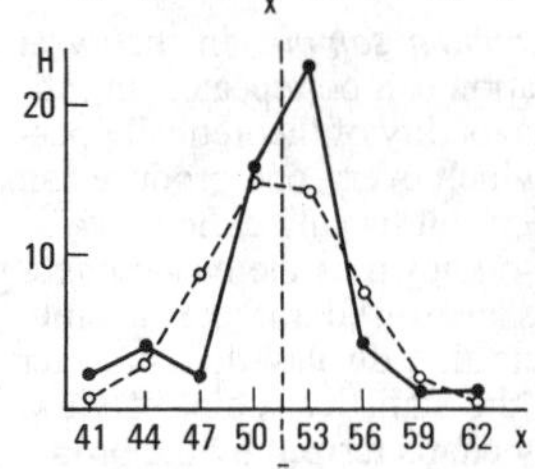

Fig. 6. *Biostatistics.* Frequency polygon for the body length (cm) of 53 male neonates. In each group, the highest and lowest values cover a range of 3 cm. Each axis point corresponds to the smallest value of the group. The polygon of the corresponding normal distribution is also shown (dashed).

The statistical test of an hypothesis for similarity of distribution is frequently performed with the χ^2-fitting test. Here, the differences between the observed and the theoretical distribution frequencies are used. In Figs. 5 and 6, each is represented as columns or the points of a polygon, respectively. If the test is carried out at the 5% level, Fig. 5 does not disprove the hypothesis, whereas Fig. 6 does disprove the hypothesis that the random sample is derived from a normally distributed universe.

Random sample reference values. In most evaluations of random sample results, it is assumed that the universe has a Normal distribution (see). Often the random sample is too small to permit the establishment of the frequency distribution.

Random sample results are characterized by their localization and their dispersion, i.e. by values describing the position and spread of the random sample values on the x-axis.

The most commonly used localization values are the *arithmetical mean* (usually called the *mean value* or *mean*), the *median* and the *mode*. The random sample mean, $\bar{x}$, is calculated from the equation $\bar{x} = \frac{\sum_{i=1}^{n} x_i}{n}$, where n is the size of the random sample, i.e. the number of individual values of x_i. For a nonlinear sample size, the median $\bar{x}_M$, is the most central value, when all the analytical values are arranged in the order of their magnitude. For a linear value of n, the most central va lue is the arithmetic mean of the two most central experimental values. The modal value, $\bar{x}_D$, is the most frequently occurring of the all the analytical values. Taking Table 2 as an example, $\bar{x} = 51.0$, $\bar{x}_M = 52$, and $\bar{x}_D = 52$. The arithmetic mean is calculated from a frequency table according to the formula $\bar{x} = \sum_{j=1}^{k} h_j x_j$, where h_j represents the relative frequencies of k different character values in the random sample. For the data of Table 3, $\bar{x} = 50.9$, $\bar{x}_M = 51$ and $\bar{x}_D = 51$. The most commonly used measures of scatter are *empirical scatter* or *random sample variance*, $s^2 = SQ(n-1)$, and the *extreme range*, which is the difference between the larges and smallest measurement values in the random sample; SQ (abbreviation of "sum of squares") is calculated from $SQ = \sum_{i=1}^{n} (x_i - \bar{x})^2$. If the mean value is unknown, it is calculated from $SQ = \sum x_i^2 - (\sum x_i)^2/n$. Alternatively, it can be calculated from a frequency table, using the equation $SQ = \sum_{j=1}^{k} H_j(x_j - \bar{x})^2$. The empirical standard deviationis $s = \sqrt{s^2}$. For the data in Tables 2 and 3, SQ = 721.9 and 710.8, s^2 = 13.9 and 13.7, s = 3.73 and 3.7. The extreme range = (61–41) = 20, and (60–42) = 18, respectively.

Experimental values from a random sample and the statistical parameters derived from them should be representative of the universe. Thus, the mean and scatter of the random sample should correspond to the expected value and the variance of distribution.

In many medical, psychological or sociological investigations the modal value is often of interest, because medical, hygienic and other measures must be adjusted to the most frequently occurring value. In a symmetrical, single peak distribution (e.g. a normal distribution), median, mode and arithmetic mean all have the same value in the universe (see Fig. 2). The positions of these three values in an unsymmetrical distribution [e.g. a Log-normal distribution (see)] are shown in Fig. 7.

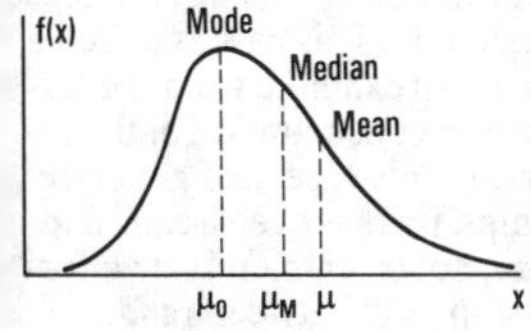

Fig.7. *Biostatistics.* Positions of the three important parameters of a skewed unimodal distribution.

If chance random samples in the same range are repeatedly taken from the same universe, they will all possess different distribution parameters. In this sense, the mean value (for example) is a random variable. The realiziations of these parameteres are the arithmetic means of their values from all the random samples. As the number of random samples increases, the variability decreases. Variability is in turn described by the scatter. The *scatter of the mean value* of a random sample is $s^2_x = s^2/n$, where n is the size of the sample, and s^2 the scatter of individual values x_i (from which $\bar{x}$ is derived). The mean value of the random sample data in Table 2 has a scatter of $s^2_x = 13.9/56 = 0.26$, and its *standard deviation* (also called the *standard error* or *error of the mean*) is $s_x = 0.51$. A random sample result is usually summarized by the expression $\bar{x} \pm s_x$; the body length data in Table 2 are therefore characterized by 51 ± 0.51. The size of the random sample *(n)* is cited to enable further evaluations and comparison with the results of other investigators. For $n > 200$, the interval $\bar{x} \pm s_x$ has the meaning of a confidence interval with a certainty probability of 68.3% (see Table 1), i.e. 68.3% of all random samples of size n contain in this region the unknown mean value (μ) of the universe, while 21.7% of all random samples do not contain it.

Testing hypotheses. An assumption about unknown or incompletely known probability distributions of random variables in called a *statistical hypothesis*. The correctness of any hypothesis formulated in the scientific field can, in principial, not be proved, but merely disproved. A statistical hypothesis is tested with the aid of a *statistical test: On the basis of random sample results, a decision is made as to whether a hypothesis should be rejected or not rejected.* By the very nature of the subject, erroneous decisions are possible. The importance of statistical tests lies, inter alia, in the fact that they permit the probability of an erroneous decision to be expressed objectively.

On the basis of the random sample in table 2, it is possible to test the hypothesis that this sample is from the same normally distributed universe as a random sample of female neonates of sample size $n = 62$, the results of which are summarized in the expression 50.2 ± 0.61. A different hypothesis could be that the data of Table 2 are the result of withdrawing a random sample from a universe with a distribution of $N(\mu, \sigma)$, the values $\mu = 50$ and $\sigma = 3.5$ being known from a large number of earlier investigations. Taking the example of the fertilizer trial, the hypothesis may be formulated that *no increase* in yield occurs as a result of fertilizer application, i.e. that the expected values for the distribution of "fertilized" and "nonfertilized" are the same. If the results of two random samples indicate that the two represented universes are different, then the opposite hypothesis is formulated, i.e. that they are similar. This is known as the null hypothesis, frequently symbolized by H_0. The purpose is to reject this null hpothesis, and rejection occurs if it is too improbable to reconcile random sample results with the valididity of H_0, i.e. a statistically significant difference is established between the two sets of random sample reults. The degree of improbability is determined by a limiting probability, α, which in this connection is the level of significance of the statistical test, the *significance probability* or *error probability*. It is a measure of improbability, and it gives the percentage of erroneous decisions necessary for the rejection of a null hypothesis,although the null hypothesis may be correct. This erroneous rejection is kown as an *error of the first kind*.

If the test result shows that the null hypothesis is not disproved (i.e. H_0 is accepted), it does not mean that its correctness is proven, only that it does not contradict the existing random sample results, and therefore cannot for the time being be rejected. The erroneous decision to retain a null hypothesis, although it is wrong, is called an *error of the second kind*. In effect, it means that the data available so far do not reveal a statistically significant difference. The probability of committing an error of the second kind increases as the probability, α, of committing an error of the first kind decreases. In practice, in establishing the value of α, it must be considered which kind of erroneous decision could have the most adverse effect on the problem in hand.

The counterpart of the null hypothesis is called the *alternative hypothesis*. In the fertilizer trial, "increase of yield" would be the alternative hypothesis to H_0, which is "equal yields".In the comparison of the body lengths of neonatal males and females, $\mu_K \neq \mu_M$ would be the alternative hypothesis to $\mu_K = \mu_M$, which represents H_0 (null hypothesis). In the second example, the statistical test is said to be two-sided, because both $\mu_K < \mu_M$ and $\mu_K > \mu_M$ are included in the alternative hypothesis. In the first example, the test is one-sided. Rejection of a null hypothesis means the acceptance of the alternative hypothesis.

In carrying out a statistical test, a *test value* is calculated from the measurement results. The distribution of the test value is known, and its rejection thresholds, based on the significance level, α, are listed in statistical tables. If the calculated test value exceeds the rejection threshold, H_0 is rejected; otherwise it is accepted. The question of whether the data in Table 2 are derived from a $N(50;3.5)$-distributed universe is answered by testing the null hypothesis "$\mu_K = \mu$", where an estimated value of $x = 51.0$ is used as the estimated random sample value for μ_K. The test value, $D = \sqrt{n}\,|x - \mu|/\sigma$, is distributed standard normally, and the 5% rejection thresholds are $d_{5\%} = \pm 1.96$ (see Table 1). In the present case the value of d, calculated on the basis of the random sample of $Dd = \sqrt{53}\,|51–50|/3.5 = 2.21 > |d_{5\%}| = 1.96$. H_0 is rejected; the difference between μ_K and μ is statistically significant. For the comparison of two mean values, $\bar{x}_1$ and $\bar{x}_2$, *Student's t-test* is used: the $\hat{t}$ value for independent chance random samples from normally distributed universes with the same variances is calculated, using the equation:

$$\hat{t} = \frac{|\bar{x}_1 - \bar{x}_2|}{\sqrt{\left(\frac{n_1 + n_2}{n_1 n_2}\right)\left(\frac{SQ_1 + SQ_2}{n_1 + n_2 - 2}\right)}}.$$

SQ_1 and SQ_2 are the sums of the squares for random samples 1 and 2, respectively, and n_1 and n_2 are their sizes. In the comparison of body length at birth, $\hat{t} = 0.985$, and in a two-sided analysis ($\mu_K \neq \mu_M$), the rejection threshold $(t\alpha)$ is 1.98 (taken from a table of t distributions for an error probability of $\alpha = 0.05$). Since $\hat{t} < t\alpha$, H_0 is not rejected. On the basis of the random sample, therefore, there is no statistically verified difference at the 5% level. The test for similarity of variances carried out beforehand, for which the test value $\hat{F} = s^2_M/s^2_K = 1.66$ is calculated,show that the null hypothesis "similarity of variance" does not contradict the random sample data at the 5% level. In this case, the rejection threshold $F_{0.05} = 1.72$ of the *F-distribution* is not exceeded.

There are also tests, in which it is not necessary to make assumptions about the distribution of the universe. Such tests, known as *distribution-free* or *nonparametric tests,* have great practical importance, and are used relatively widely. They include, inter alia, the Kolmogorov test, the χ^2-fitting test, and the Rank correlation.

Correlation and regression. Especially in biological problems correlations are very often sought between two or more random values, or between random and non-random variables. These are correlations of a functional-stochastic type. Changes in one value are accompanied by functional changes in the other value. The functional relationship is, however, "disturbed" by random factors, i.e. it contains random elements (see Biomathematics).

Relationships of this kind are investigated by correlation and regression analysis Correlation analysis determines the extent of covariation of two random values, Z_1 and Z_2, which have equal status (e.g. body size and body weight), based on the paired realization (z_{1i}, z_{2i}). As a measure of the correlation, the *correlation coefficient,* r, is calculated from the random sample data, and it represents an estimate of the correlation coefficient, ρ, of the universe. Its value can vary only between –1 and +1. If the two random values are not related,in the sense that changes in one are not accompanied by changes in the other, they are *noncorrelated,* i.e. $\rho = 0$. In the random sample, this is expressed in the fact that in the variation interval of one value,the values of the other are grouped randomly around its mean value, and $r \approx 0$. In a negative correlation ($r < 0$), the values of one parameter decrease as the values of the other increase. In a positive correlation ($r > 0$), both values alter in the same sense. For $|r| = 1$, the correlation degenerates into a functional relationship, and the linear correlation between the two values is not blurred by the random variations.

The *rank correlation coefficient* is relatively important as a measure of correlation between two series of analytical values. It can be calculated for ordinally scalable characters (see Biometry), and the statistical test for the existence of a correlation is not tied to the prerequisite of a normal distribution, or the assumption of a linear functional relationship betweenthetwo values.

In *regression analysis,* the primary objective is to estimate the values of one quantity on the basis of the values of the other. Thus, the regression is equivalent to the expected value of the random value, *Y,* provided that the random value, *X* is a realization of x_i. In the mathematical sense, *x* is the independent variable, while *y* (the expected dependent value of *Y*) is the dependent variable of the functional relationship,$\bar{y} = f(x)$.

The nature of the problem under investigation determines which of the two related values is chosen as an independent variable. Often, only the target value, *Y,* is a random value; the values of the independent variables can be established experimentally. This case is called a *modified regression problem.* An example from the fertilizer experiment is the steps in concentration of the influencing factor "fertilizer". In this case, and generally in dose-response relationships, there is even a causal relationship present. This may, however, merely be a formal mathematical relationship, e.g. between randomly influenced growth size and time in the study of growth processes (see Fig. 2 in Biomathematics).

If the distribution of *X* and *Y* in the universe are known, the dependent expected value, $\bar{y} = f(x)$, can be derived. A two-dimensional normal distribution (the plot of its probability density describesa bell-shaped curve above the *x-y axis)* produces a straight line. The estimate of $\bar{y}$ from *n* random sample pairs (x_i, y_i) *is* $\bar{y} = \bar{y} + b_{yx}(x - \bar{x})$, where $\bar{x}$ and $\bar{y}$ are the arithmetic means of x_i and y_i, and b_{yx} is the regression coefficient of *y* against *x*. The latter is calculated from the equation:

$$b_{yx} = \frac{\sum_{i=1}^{n} (x_i - \bar{x})(y_i - \bar{y})}{\sum_{i=1}^{n} (x_i - \bar{x})^2}$$

Fig. 8 shows part of the calculated regression line for the number pairs (0/1), (1/2.5), (2.5/1.5) and (3/2), where $\hat{y} = 1.75 + 0.154\,(x-1.625)$.

An estimate $\hat{y} = f(x)$ of the regression cannot be made, from the random sample data, without assuming the existence of a particular type of distribution. In the overwhelming majority of biological problems, the type of distribution of X and Y is, in any case, not known. Regression estimated in this way is called *empirical regression;* it cannot be incorporated into an analytical expression for the functional relationship $f(x)$. The simplest regression consists of the calculation of the sequence from mean values of $\bar{y}_i$, if several analytical values (y_{ij}), are available for x_i. More modern versions also permit calculation of regression values at chosen points between x_i, so that a smooth empirical regression curve is obtained. Fig. 2 in the entry Biomathematics (see) can be taken as an example of an empirical *regression curve*.

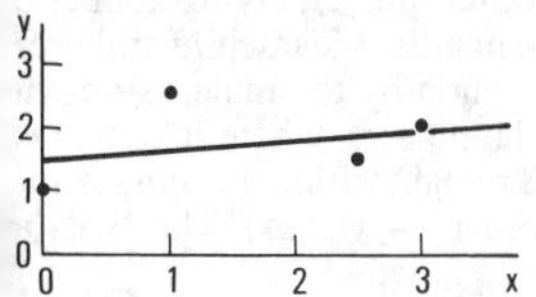

Fig.8. *Biostatistics.* Example of a regression line.

In order to describe the functional relationship, $\hat{y} = f(x)$, with an analytical expression, it must be possible to deduce the type of function from the nature of the experimental problem. If this is not possible, the regression problem is modified, usually according to external or formal mathematical considerations; an appropriate functional expression is chosen, and its parameters are fitted to the random sample data (see Biometry). An expression of this kind is valid only in that interval (usually the variation interval of the independent variables in the random sample) in which the fitting is performed. For the sake of argument,the linear function, $y = a + bx$, is often used for this purpose. Fig. 9 shows results of morphometric measurements; x is not a random value. Since this is a "saturation process" (apparent-from the visible bending of the line as the analytical points progress), the function equation $y = a(1 - \exp(-bx))$ was chosen for the relationship. The calculated paramater values for this type of function, which show the best fit to the analytical values of the curve in Fig. 9, are $a = 0.82$ and $b = 0.27$.

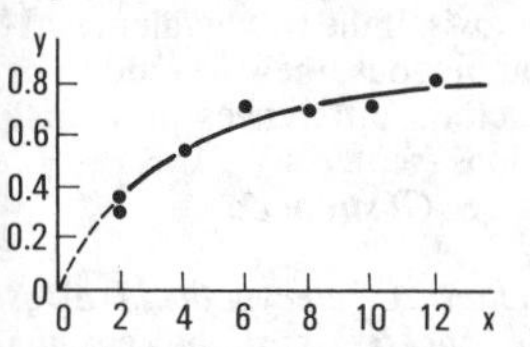

Fig.9. *Biostatistics.* Nonlinear regression. The results are from morphometric measurements, and the functional relationship is described by the nonlinear function that shows the best fit to the measurement values.

Biostratigraphy: a branch of geology. With the aid of index fossils, B. places rocks in their historical sequence and provides a relative biochronological scale for the dating of all processes in the earth's geological history. By using index fossils, all rock formations are classified in relation to different periods in the evolution of living organisms.

Biostratinomy: a method established by J. Weigelt (1933) for determining the specific chemical and physical conditions under which residues of living organisms were deposited and fossilized. It studies the chemical, mechanical and spatial relationships of fossils to one another and to the sedimentary rocks in which they occur. B. is a branch of Paleontology (see).

Biota: all the plants and animals inhabiting a particular biotope, e.g. marine biota, terrestrial biota.

Biotechnology: the use of living organisms or systems derived from them in manufacturing industry. The subject has ancient origins, embracing agriculture, cheese production, fermentation of alcoholic beverages, leather tanning, etc. However, B. is a modern term, referring to a particular type of biological exploitation, and the scope of the subject has been greatly expanded by developments in molecular biology. Important areas of B. are as follows.

1. Genetic engineering, recombinant DNA technology or Gene technology (see).
2. Process engineering (e.g. production of antibiotics by fermentation; microbiological conversion of simple chemical source materials like methanol and ammonia into animal feed; production of fuel alcohol by fermentation).
3. Concentration of minerals, or "microbiological mining".
4. Construction of biosensors.
5. Manufacture of biocatalysts (immobilized enzymes).
6. Waste treatment and biodegradation.
7. Production of monoclonal antibodies for medical research.
8. Study of the relationship between protein structure and function, and the nature of molecular interactions, leading to the designed (nonbiological) synthesis of model catalysts and new drugs.

Biotic community: see Bioc(o)enosis.

Biotic district: see Bioc(o)enosis.

Biotic potential: specific relationship between the reproductive potential and survival potential of a species, as an index of the potential increase in population size of the species. The term is also used for the ability of ecosystems to recover and regenerate.

Biotope: an environmental region, characterized by certain conditions and populated by a characteristic community of living organisms.

Biotraditional behavior: see Tradition.

Biotransformation of information: translation of the German word "Informationswechsel", which refers to biological processes involved in the exchange of information between an organism and its environment. By analogy with metabolism ("Stoffwechsel"), which is concerned with the biotransformation of chemical substances, "Informationswechsel" refers to the biotransformation, processing, storage, etc. of information. It therefore embraces basic phenomenon of living organisms, such as sensitivity, excitability or irritability. In multicellular organisms, information is received and subjected to a certain level of preliminary processing by sensory organs (external receptors). Subsequent *information processing* is linked to conversion (transduction), transport (conduction) and storage, and it finishes with the integration of the processed information with existing stored information. Each of these stages of information biotransformation is accompanied by biochemical and bioelectric changes. The overall process is optimized by movement and complex behavior (see Maintenance activity). Secondary, information-containing signals may also be produced by movements and other efferent nervous impulses (see Biocommunication, Sign stimulus).

Biotype:

1) A term introduced by the geneticist, Johannsen, for all genetically similar individuals in a population, which have arisen by sexual reproduction or by parthenogenesis within the population. B. refers to the totality of phenotypes belonging to a particular genotype.

2) A Physiological race (see).

Bipedalism: a functional-anatomical term for the mode of locomotion of certain vertebrates that progress by the exclusive or almost exclusive use of their hindlimbs. Examples of bipeds (bipedal organisms) are flightless birds (e.g. ostrich, emu, cassowary), kangaroo and human, as well as some dinosaurs like *Tyrannosaurus* and *Iguanodon*. B. arose at a relatively early stage of human evolution.

Bipedal saltation: see Movement.

Biphasic theory of photoperiodic floral induction: see Flower development.

Bipolar flagella: see Bacterial flagella.

Biramous limb, ***stenopodium:*** a typical crustacean limb or appendage, which forks distally into two rami. The two rami, an inner *endopodite* and an outer *exopodite,* are carried on a common stem, the *protopodite*. In many cases, the outer side of the protopodite also carries one or more processes known as *epipodites,* which usually serve as gills. The inner side of the protopodite may carry processes known as *endites,* which in the mouth region are modified to maxillae. The multisegmented expodite is generally adapted for swimming, while the endopodite has fewer segments and is adapted as a walking limb; depending on the mode of locomotion, one of these is degenerate and the other is developed.

Birch: see *Betulaceae*.

Birchmice: see *Zapodidae*.

Birch tar, ***oleum rusci, oleum Betulae empyreumaticum:*** a dark, tarry liquid manufactured in Russia by destructive distillation of birch wood (see *Betulaceae),* and used against skin complaints.

Bird lice: see *Phthiraptera*.

Bird lime trap: see Carnivorous plants.

Bird migration: see Migratory behavior of birds.

Bird of prey, ***raptor:*** a general term for any bird that hunts and kills other animals, especially higher vertebrates, for food. In the strict sense, it is a member of the *Falconiformes* (see). The term may then be extended to cover owls *(Strigiformes)* by calling these "nocturnal birds of prey", the *Falconiformes* then becoming "diurnal birds of prey".

Birds nest soup: see *Apodiformes*.

Birds of paradise: see *Paradisaeidae*.

Birdwing butterflies: butterflies of the family *Papilionidae*. See *Lepidoptera*.

Birgus: see Hermit crabs, Coconut crab.

Birth: see Parturition.

Birth index: see Birth rate.

Birth rate, ***birth index:*** the number of individuals born per thousand members of a population in one year. The population size is taken as that in the middle of the year in question. The B.r. varies between 10 and 60 (1–6%) in the different countries of the world, e.g. 10.7– 17.6 in the former East Germany, 1948 and 1981.

Bishops: see *Ploceidae*.

Bison: see *Bovinae*.

Bistort: see *Polygonaceae*.

Biting lice: see *Phthiraptera*.

Bitis: see Puff adders.

Bittacidae: see *Mecoptera*.

Bitter boletus: see *Basidiomycetes (Agaricales)*.

Bitterling, *Rhodeus sericeus:* a small (6–9 cm) member of the family *Cyprinidae* (minnows or carps) (order *Cypriniformes)*. It lives in lakes and rivers of Central Europe, favoring inlets with sandy or muddy bottoms, where freshwater mussels *(Anodonta* and *Unio)* are also found. Coloration is greenish to silvery gray, with an opalescent turquoise line along the middle of each side. During spawning, however, the male acquires spendidly colored purple sides, and develops islands of tubercles above the eyes and mouth. The female possesses an ovipositor (situated behind the anus, and about 6 cm in length), which she inserts into the excretory siphon of a freshwater mussel, laying about 40 relatively large eggs in the mantle. Meanwhile, milt, released over the inhalant siphon by the male, is drawn into the gill chamber of the mussel, where it fertilizes the eggs. The young fishes remain within the mussel until they are fully developed (3–4 weeks). The B. is a popular fish in cold water aquaria.

Bitterns: see *Ciconiiformes*. See plate 6.

Bitter orange: see *Rutaceae*.

Bitter principles: bitter tasting plant constituents, found especially in members of the *Gentianaceae, Cannabaceae, Labiatae* and *Compositae*. Some known B.p. are absinthin, arnicin, gentiopicrin (gentiopicroside), picrocrocin and lupulone. B.p. are used as flavoring spices, and they are used medicinally to increase the appetite and to promote gastric secretion.

Bittersweet: see *Solanaceae*.

Bittersweet shells: see *Bivalvia*.

Bituminization: see Carbonification.

Bivalent chromosomes: pairs of homologous or partially homologous chromosomes that appear during the first meiotic division. In diploid organisms, the number of bivalent chromosome pairs in meiosis I is normally the same as the haploid chromosome number. The attachment is preserved by crossing over and chiasma formation, so that generally, the bivalent partners remain attached during synapsis until anaphase I of meiosis. Otherwise, the bivalent partners separate prematurely into two single (univalent) chromosomes, a condition known as *desynapsis*. If the two bivalent partners are only partially homologous, they are said to be *heteromorphic*. Such structural differences may arise from Chromosome mutations (see).

Bivalved gastropods: see *Gastropoda*.

Bivalves: see *Bivalvia*.

Bivalvia, *Lamellibranchia(ta), Pelecypoda,* ***bivalves (clams, oysters, mussels, cockles, scallops,*** etc.): a class of the *Mollusca* (see), containing more than 20 000 species, whose shells consist of two valves. The *B.* are distributed in all the seas and oceans of the world, and a small number of species is also found in feshwater.

Morphology and anatomy. There is no head. The body is laterally compressed, usually bilaterally symmetrical along the plane of the junction of the two valves, and completely enclosed by the two valves of the shell (except in the ***Geoduck,*** *Panopea generosa,* order *Myoida,* a giant Californian bivalve, whose body and siphon extend beyond the shell). The two valves are united dorsally by a toothed hinge, and held together by an elastic protein band called the hinge ligament, which is covered on the outside by periostracum. When the valves are closed, the outer part of the hinge ligament is stretched, while the inner part is compressed, so that in the absence of other forces, the elasticity of the hinge keeps the valves open. The valves are closed ventrally (and very tightly) by the contraction of one or two adductor muscles, which act antagonistically to the hinge ligament. The shell is secreted by two lobes of the mantle, whose edges are fused, except for three openings (for the foot and for the inhalant and exhalant water streams). Tubular siphons are often present, formed by elongation of the mantle edges surrounding the inhalant and exhalant apertures. Provided the siphons project beyond the surface, the animals can remain deeply buried in mud, while continuing to circulate water as a source of food and respiratory gases. There is a considerable variation in siphon length and the degree of fusion between the inhalant and exhalant siphon. A spacious mantle cavity is normally divided transversely by the ctenidia. Within the *B.*, all stages of ctenidium (gill) evolution are seen. from protobranch, via filibranch to eulamellibranch (see Gills), while ctenidia have practically disappeared in the *Anomalodesmata* (formerly the *Septibranchia),* which have septibranch gills. Ctenidium structure has therefore been used for the classification of the *B.* The body lies in a compact mass in the upper part of the shell. The nervous system consists of 3 pairs of ganglia and their associated connections, but a circumesophageal commissure is absent. Two cerebropleural ganglia lie on either side of the esophagus (connected by a commissure dorsal to the esophagus), each giving rise to 2 posteriorly directed nerve cords; the upper pair (one from each ganglion) pass through the viscera and terminate in 2 visceral ganglia on the surface of the posterior adductor muscle; the lower pair extend into the foot, where they connect with the 2 pedal ganglia. Apart from sensory epithelia, the only other sense organs generally present are statocysts (usually in the foot near to or within the pedal ganglia) and osphradia. Some species have ocelli along the mantle edge, or even on the siphon. Ocelli are usually simple (pigment spots or cups), but compound ocelli are present in ark shells, while the scallops and thorny oysters possess structures with a cornea and lens. There is no radula, and the stomach contains a crystalline style (for a description of the crystalline style and details of digestion, see *Mollusca).* There is an open circulatory system, and the heart, consisting of a ventricle and two auricles, is usually folded around the intestine. Paired metanephridia are located beneath the pericardium and above the ctenidia, one end of each nephridium opening into the pericardial cavity via the nephrostome, the other into the mantle cavity via the nephridiophore. The foot, which is usually laterally compressed and hatchet-shaped (hence the name *Pelecypoda* "hatchet foot") or wormlike, often carries a Byssus (see) gland; it is normally only capable of slow movement and may also be used for boring into the substratum. Generally, the foot generates locomotory movement by alternately lengthening and shortening (cf. the waves of surface contraction in the gastropod foot). In some cockles, e.g. *Neotrigonia* (order *Trigonioida),* the tip of the foot can be inserted beneath the shell, then quickly flexed so that the animal leaps from the ground.

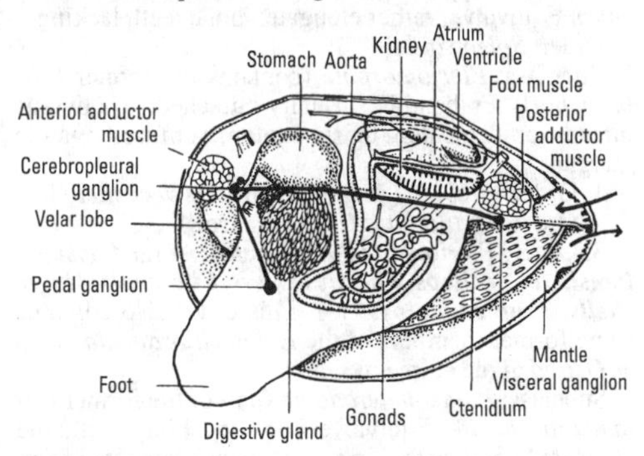

Bivalvia. Structural plan of a bivalve.

Biology. Most *B.* are monosexual, and in most species fertilization occurs externally in the surrounding water. Eggs or sperm are released by the paired gonads into the suprabranchial cavity and transferred to the exterior in the exhalant current. In internal fertilization, only sperm is ejected into the surrounding water, while the eggs are retained by the female, who takes up the sperm in her inhalant stream. Eggs may then be brooded in the suprabranchial cavity (e.g. in some shipworms) or between the gills. In marine *B.*, development typically occurs via a free-swimming trocophore, which develops into a Veliger larva (see). In the freshwater mussels of the families *Margaritiferidae, Unionidae* and *Mutelidae* (order *Unionoida),* the eggs are brooded between the gill lamellae; the highly modified veliger larva, which parasitizes freshwater fishes, is known as a Glochidium (see), haustorium, or lasidium, respectively.

Respiration and feeding operate in parallel. A water stream generated by the cilia of the ctenidia and mantle passes among the ctenidia, which not only exchange respiratory gases, but also filter out food particles (fine plankton and suspended detritus). Trapped in mucus, the food particles are transported by cilia along the margins of the ctenidial filaments, first into ciliated food grooves, then to the mouth. Exhalant water, carrying excretory products, is expelled through the upper mantle opening.

Economic importance. Many *B.* are eaten by humans (e.g. oysters), and some provide decoration (e.g. pearls). They can also be very destructive, e.g. by boring into wooden ships and harbor structures (see Boring bivalves), or by blocking industrial filter plants.

Fossils. Fossil *B.* are known from the Cambrian to the present. They occur with extraordinary frequency, showing a peak of evolutionary development in the Mesozoic. Many have biostratigraphic importance. Thus, *B.* and gastropods provide important index fossils for dating the marine deposits of the Tertiary.

Classification. Earlier classifications based on gill structure (4 orders: *Protobranchia, Filibranchia, Eulamellibranchia, Septibranchia)* are now superceded. The positions of these former orders in relation to the later classification is indicated below.

1. Subclass: *Palaeotaxodonta* (part of the former *Protobrachia).*

Equivalve; taxodont; isomyarian.

Order: *Nuculoida* (**Nut clams**).
2. Subclass: *Cryptodonta* (part of the former *Protobranchia*). Equivalve; rather elongate; hinge teeth lacking.
Order: *Solemyoida.*
3. Subclass: *Pteriomorphia* (contains the former *Filibranchia*). Epibenthic; usually attached by byssus threads or cemented to substratum; unfused mantle margins.
1. Order: *Arcoida* (**Ark shells, Bittersweet shells**).
2. Order: *Mytiloida* (**Marine mussels**).
3. Order: *Pterioidea* (**Purse shells, Winged oysters, Penshells, Scallops, Thorny oysters, File shells, Jingle shells, True edible oysters**). This order also contains some former members of the *Eulamellibranchia,* such as *Ostrea* (True edible oysters).
4. Subclass: *Palaeoheterodonta* (part of the former *Eulamellibranchia*). Equivalve with a few hinge teeth; the elongate lateral teeth, if present, are not separated from the large cardinal teeth.
1. Order: *Trigonioida.*
2. Order: *Unionoida* (**Freshwater mussels** (see); **Pearl mussel** and **River mussels**).
5. Subclass: *Heterodonta* (part of the former *Eulamellibranchia*). Equivalve; a few large cardinal teeth are present and are separated from elongated lateral teeth by a toothless space.
1. Order: *Veneroida* (**Pea mussels, Fingernail clams, Heart cockle, Ocean quahog, Jewel box clams, Cockles, Giant clams, Venus clams, Carpet shell, Qhahog, False angel wing, Tellin clams, Peppery furrow shell, Razor clams, Trough shells, Red-nose clam**). This order also contains the family *Tridacnidae,* with the genus *Tridacna* (**Giant clams**) of the Indo-Pacific. These are the largest living bivalves, measuring up to 135 cm in length, and weighing 200 kg. The valves were formerly used as containers for holy water. The soft body of the animal, weighing up to 10 kg, is highly prized as food.
2. Order: *Myoida.* Thin-shelled burrowers with well developed siphon; 1 or no cardinal teeth. (**Soft-shell clam** or **Sandgaper, Geoduck, Basket clams, Flask shell, Piddocks, Shipworms**). Soft-shelled clams of the genus *Mya* have somewhat unequal valves, their fossils are known from the Tertiary to the present, and they are particularly abundant in the early post-Ice Age waters that developed into the Baltic Sea.
3. Order: *Hippuritoida.* Fossil rudists.
6. Subclass: *Anomalodesmata* (contains the former *Septibranchia*). Equivalve; 1 or no hinge teeth; hinge margin thickened and rolled; isomyarian.
Order: *Pholadomyoida* (**Watering pot shells**), e.g. *Poromya, Cuspidaria, Pandora.*

Hinge and tooth types in the B.:
1) *Heterodont* . Each valve contains a small number of differently shaped teeth and their complementary sockets; the hinge or cardinal teeth lie under the umbones, and the lateral teeth are more or less parallel with the hinge line.
2) *Taxodont.* Numerous parallel teeth (oblique or vertical to the edge of the hinge) fit into sockets in the opposing valve. In the ctenodont or pseudoctenodont type, the teeth converge to the interior, in the actinodont type to the exterior.
3) *Preheterodont* or *diagenodont.* Merely a few ridges converging toward the umbo.
4) *Dysodont.* Very small, weak teeth, or fine fluting of the periphery of the hinge region.
5) *Isodont.* Two teeth and 2 sockets are located symmetrically on either side of the ligament pit.
6) *Desmodont.* Hinge teeth are absent. In some species the ligament is attached to inward projections of the valves.
7) *Schizodont.* The left valve possesses a triangular hinge tooth, which fits into a socket between the teeth of the right valve. The latter usually diverge and are sometimes fluted.
8) *Pachydont.* One to 3 unsymmetrical, peg-like teeth fit into sockets in the opposing valve.

Bixin: the monomethyl ester of a C_{24}-dicarboxylic acid, called norbixin. Norbixin has 4 lateral methyl groups, 2 terminal carboxyl groups and 9 conjugated double bonds. The C_{20}–chain corresponds to the middle part of β-carotene. In naturally occurring B., the Δ^{16} double bond is *cis,* but it easily isomerizes to form the more stable all-*trans* form. B. is a yellow to red-orange pigment, present in the seeds of the tropical tree, *Bixa orellana.* The seed extract is used as a food coloring.

Black arches, *Lymantria monacha* L. (plate 46): a moth of the family *Lymantriidae* (tussock moths) found from Europe though temperate Asia to Japan. Its caterpillar feeds on the young spring shoots, as well as the older needles of pines and firs. It also attacks oak, apple and other deciduous trees, occasionally becoming a minor orchard pest. The gray and brown caterpillars are covered with barbed, poisonous hairs.

Black bear, ***baribal, American black bear,*** *Euarctos americanus:* a North American bear, brown to black, and somewhat smaller than the brown bear, which it resembles in its mode of existence. The B.b originally inhabited practically all wooded areas of North America north of Central Mexico. It is now restricted to regions remote from human settlement, and it is protected in the National Parks of the USA and Canada. It eats vegetation, flesh, fish and carrion. Color variants occur, e.g. cinnamon-brown, blue-black, and white. White variants are usually found on the Pacific coast of North America. Color variants are occasionally found in the same litter. The blue-black variant occurs in the St. Elias range in Alaska. These variants, however, are never in the majority in any population.

Blackberry: see *Rosaceae.*
Blackbird: see *Turdinae.*
Blackbuck: see *Antilopinae.*
Black catechu: see *Mimosaceae.*
Blackcock: see Pheasants.
Black coral: see *Antipatharia.*
Black currant: see *Grossulariaceae.*
Blackcurrant gall mite: see *Tetrapodili.*
Black death: see *Yersinia pestis.*
Black dragonfishes: see *Stomiiformes.*
Blackleg of potatoes: see *Erwinia.*
Black locust: see *Papilionaceae.*
Black-naped flycatchers: see *Muscicapinae.*
Black nightshade: see *Solanaceae.*
Black panther: see *Felidae.*
Black rat: see *Muridae.*
Black slug: see Slugs.
Blacksmith frog: see *Hylidae.*
Black smut: see *Basidiomycetes (Ustilaginales).*
Blackthorn: see *Rosaceae.*
Black-throated blue warbler: see *Parulidae.*
Bladder: see Urinary bladder.
Bladder worm: see *Cestoda,* Dog tapeworm.

Bladderwort: see Carnivorous plants.

Bladder wrack: see *Phaeophyceae.*

Blaeberry: see *Ericaceae.*

Blaniulidae: see *Diplopoda.*

Blaniulus: see *Diplopoda.*

Blastema: see Regeneration.

Blastocladiales: see *Phycomycetes.*

Blastocoel: a cavity formed in the blastula stage of egg cleavage in multicellular organisms (see Cleavage). The B. is enclosed by the blastoderm (usually a single layer of cells) of the blastula. It contains a liquid, which consists partly of the intercellular cement substance that accumulates during cleavage. In lower animals (e.g. flatworms and nematodes), relatively large parts of the B. persist throughout life. In aquatic invertebrates (e.g. annelids and echinoderms), the B. forms the primary body cavity (protocoel, pseudocoel) of the larval stage.

Blastogenesis: see Cleavage, Ontogenesis.

Blastoidea, ***blastoids:*** extinct class of marine echinoderms. Most forms were stalked, with an ovoid (occasionally star-shaped) theca, which carried a boss at the site of attachment of the stalk. Pentamerous symmetry was fully developed. The theca consisted of 13 or 14 plates and 5 lancets, the latter covered with numerous plates to form 5 radial ambulacral fields. The ambulacra were bordered by rows of pinnule-like brachioles. Elongated, internal pouches (hydrospires) ran alongside each ambulacral groove and opened to the exterior at their oral end. Fossils are found from the Middle Ordovician to the Permian, with evolutionary peaks in the Lower Carboniferous of North America and the Permian of Timor Island.

Blastospores: spores of unicellular or multicelluar fungi, which are formed by Budding (see). They frequently remain attached to each other, forming branched chains of cells. B. are particularly common in certain yeast species.

Blastozo(o)id: a modified polyp, lacking a mouth or tentacles, which is formed on a polyp colony. The only function of a B. is to produce medusae by budding. The medusae then reproduce sexually. See Polyp.

Blattariae: see *Blattoptera.*

Blattellidae: see *Blattoptera.*

Blattidae: see *Blattoptera.*

Blattoptera, *Blattariae,* ***cockroaches:*** an insect order containing about 3 500 species, predominantly in the tropics. The earliest fossil *B.* occur in the Carboniferous.

Imagoes. Depending on the species, the body length of the adult varies from 0.2 to 11 cm. The body is flattened, and the prothorax bears a large, dorsal, shield-like protonum, which usually overlaps the head. Antennae are thin and filiform and approximately equal in length to the body. Mouthparts are complete and adapted for biting and chewing. Compound eyes are often well developed, but may be reduced or absent in cavernicolous (cave-dwelling) and myrmecophilous (living in ants' nests) species. Some *B.* possess 2 ocelli, but most species carry two structures which appear to be degenerate ocelli; these are two pale areas known as fenestrae, each with a nerve connexion to the brain. As sense organs, the eyes appear to play a subordinate role (in contrast to the *Manoptera,* in which highly developed compound eyes are used for sighting prey). Important sense organs are the antennae, together with the subgenual organs of the tibiae, which are sensitive to vibration. Walking legs are long, thin and powerful, and cockroaches can run very quickly. At rest, the sturdy, leathery forewings (wing covers) cover the membranous hindwings which lie close to the body. In some species, wings are absent or reduced. In both sexes, the abdomen carries two segmented caudal appendages (cerci). Males also possess a pair of unsegmented styli on the ninth sternite (the subgenital plate). Cockroaches are typically ground-dwelling, secretive insects, preferring dark, warm and humid habitats, often being found in rotting logs and decaying vegetation, under stones, in crevices and in buildings. A few species, however, prefer semidesert or desert conditions, and semiaquatic species are also known. Metamorphosis is incomplete (paurometabolous development).

Eggs. Between 20 and 50 eggs are laid in a leathery or horny capsule (called an ootheca), which can be seen protruding from the genital aperture of the female. It is usually carried in this way for a considerable period, or even until the larvae hatch. In some species it may be deposited in a camouflaged position.

Larvae. The larvae are very similar to adults in structure and mode of existence. The mature adult emerges from the fifth or sixth molt.

Economic importance. Several species are vermin or pests, feeding on plants and on stored products. Throughout the world, the most common cockroach of human dwellings is the ***German cockroach*** *(Blatella germanica;* length 8–13 mm). The second most common species is the ***Oriental cockroach*** *(Blatta orientalis;* length 18–30 mm). Both of these eat and soil foodstuffs, textiles and leather and paper goods during their nocturnal activities. They are also disease vectors (anthrax, tuberculosis) and intermediate hosts of human roundworms. The Oriental cockroach prefers cool habitats (e.g. cellars), but easily tolerates warm conditions

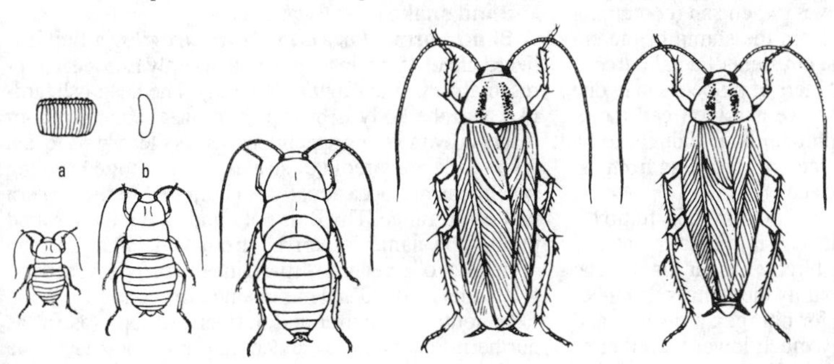

Blattoptera. Metamorphosis of the German cockroach *(Blatella germanica* L.). *a* Egg cocoon. *b* Single egg. *c-e* Second, fourth and sixth larval stages. *f* Adult male. *g* Adult female.

if water is available, whereas the German cockroach thrives in warm habitats, such as bakeries and kitchens. A smaller species, the ***Brown-banded cockroach*** *(Supella supellectilum = S. longipalpa),* is now also becoming established in human dwellings in North America and Europe; it probably originated in tropical Africa, its wider dispersal starting at the beginning of the 20th century. Warmer conditions are required by the large species of *Periplaneta*, e.g. the ***American cockroach*** *(Periplaneta americana*; length 28–45 mm) and the ***Australian cockroach*** *(Periplaneta australasiae),* which, outside the tropics, are encountered in temperate climates in warm greenhouses. An occasional European greenhouse species is the ***Surinan cockroach*** *(Pycnoscelis surinamensis),* which originates from southern Asia and is now also established in Florida and Texas.

Classification. Ten families are recognised. The family *Blattidae* is cosmopolitan and members possess spines on the ventroposterior margin of the femora. It contains the well known *Periplaneta americana*, *Periplaneta australasiae* and *Blatta orientalis*. They probably all originate from Africa, but have become widespread through commerce. *Blattella germanica* and *Supella longipalpa* belong to the family *Blattellidae*, which contains generally small species, usually not more than 1 cm in length.

Blaze: in some animals, a more or less broad and regular white stripe, extending along the top of the muzzle and sometimes as far as the upper lip.

Bleak, *Alburnus alburnus:* a minnow of the family *Cyprinidae* (see), notable for its schooling habit, and found throughout Europe north of the Alps, where it inhabits all standing and flowing waters. The elongated body is quite laterally compressed, and normally 10–13 cm in length (20 cm maximum). It is used as a bait fish, and as a fish and animal food. Guanine crystals from the skin are used in the manufacture of artifical pearls.

Another species, *Alburnus bipunctatus* (body length 2–14 cm), is found throughout most of Europe, except for the southwest and parts of Eurasia. Each lateral line scale has a pair of small black spots. The paired fins are yellowish at the base, turning orange during spawning. It is found exclusively in flowing water, especially in mountainous regions, where it feeds on aquatic invertebrates and flying insects.

Bleeding:

1) In animals and humans the leakage or loss of blood from the circulation.

2) In plants the exit of phloem or xylem sap from wounds. Phloem B. (sieve tube B.) occurs mainly in monocots, e.g. palms or agaves, when sap (containing up to 18% sucrose) exudes from the stump of the cut axis of the inflorescence. The cut stem of a palm tree is capable of releasing up to 19 liters of phloem sap in one day. Since phloem cells are alive and their cell membranes are semipermeable, phloem B. is thought to be due to the osmotic pressure gradient extending from the roots to the wound or cut surface.

Xylem B. (vessel B.) occurs from the root stump after removal of the shoot axis, e.g. in tomato plants, or from the trunks of trees, e.g. birch or sugar maple. The exuded xylem sap contains mainly inorganic substances and compounds responsible for nitrogen transport, and its sugar content is generally much lower than that of phloem sap. In spring, when reserve materials are being mobilized, the exuded sap of tree trunks may, however, contain relatively large quantities of sugars, amino acids and proteins. Root stump B. displays maxima in the morning and at midday, and is hardly detectable at night; it is due to *root pressure,* which arises from the active transport of water into the vessels by endodermis or parenchyma cells of the central cylinder (probably similar to the mechanism operating in glands). Root pressure, which is also partly responsible for Guttation (see), can be measured by attaching a manometer to the stump; it is normally less than 1 atmosphere, but in some cases (e.g. birch) it may exceed 2 atmospheres. The quantity of fluid lost by B. varies considerably according to the plant species. Herbaceous plants lose a few milliliters per 24 h, compared with vines (1 liter) and birch (5 liters). Both the quantity and the chemical composition of the exuded sap are subject to marked seasonal variations. Saps containing high concentrations of organic compounds in the spring are occasionally exploited, e.g. the sap of the sugar maple for the production of maple sugar and maple syrup. Palm wine is prepared by fermentation of palm sap, and pulque is the fermented sap of the agave. The saps of many plants contain secondary products that are synthesized mainly in the root and transported to the aerial organs, e.g. nicotine in the tobacco plant.

Blennioidei: a suborder of fishes, order *Perciformes,* containing about 675 species in 127 genera. Six families are recognized: *Trypterigiidae* (Threefin blennies), *Dactyloscopidae* (Sand stargazers), *Labrisomidae* (Labrisomids), *Clinidiae* (Clinids), *Chaenopsidae* (Pickleblennies, Tubeblennies or Flagblennies), *Blenniidae* (Combtooth blennies). They are mostly small bottom fish, inhabiting primarily the coastal regions of warm seas, only a few species being found in fresh or brackish water. The anal fin has simple, soft rays, and 1 or 2 spines, while the pelvic fins have 2–4 simple, soft rays and a single spine. Cirri are often present at various locations on the head (on the nape, above the eyes, on the nostrils, or the margins of the cephalic sensory pores).

Formerly, the *B.* included the zoarcids, nototheni-oids, trachinoids and blennioids, each of which is now accorded separate subordinal status.

[Gosline, W.A. (1986) *Proc. U.S. Natl. Mus.* ***124,*** 1–78; Gosline, W.A. *Functional morphology and classification of teleostan fishes,* 1971, University press of Hawaii, Honolulu]

Blepharoplast: see Basal body.

Blesbok: see *Alcelaphinae.*

Blicca: see Silver bream.

Blind lizards: see *Dibamidae.*

Blind snakes: see *Typhlopidae.*

Blindworm, ***slowworm,*** *Anguis fragilis:* a limbless lizard, up to 45 cm in length, and the only European representative of the *Anguinidae* (see). The head is lizard-like, and the body is brownish, bronze, copper-red, or (especially in young specimens) glossy lead-gray, often with dark longitudinal stripes. Blue-spotted young specimens are occasionally encountered in the eastern part of the range. The B. is not blind, and it has a round pupil. The name is derived from Old High German "Plintslicho", meaning "sparkling or splendid creeper". It inhabits lowland and mountainous regions up to 2 500 m in southern, central and northern Europe (as far as northern Scandanavia), extending into Southeast Asia and the northwestern corner of Africa. It favors moist, shady habitats. Maximal activity is displayed at dawn and dusk, and the diet consists of worms, slugs and small insects. In the temperate part of its range, the B. hibernates over the winter buried in the earth. Mating occurs from April to May. Between 5 and 20 young are born inside a gelatinous egg membrane, which they leave shortly afterward. The greatest recorded age in captivity is 54 years !

Blister beetles: see *Coleoptera.*

Blockers: substances that occupy molecular Receptors (see), but generate no further response. B. are therefore antagonists of substances that occupy receptors and initiate a response, i.e. *agonists.* B. of cholinergic nerve transmission are, e.g. Ganglion B. (see), and substances acting on the motor endplates, e.g. Curare (see). Adrenergic nerve transmission is blocked by either *alpha-B.* or *beta-B.,* according to the receptor type. The search for new B. and investigations with known B. are important for the elucidation of the mechanisms of nerve transmission (see Transmitters) and other receptor-mediated reactions, for the classification of receptors, and for the design and preparation of pharmaceuticals.

Blood: morphologically, a mesenchymatous tissue, whose cells (see Hemocytes) are suspended in a fluid matrix (see Blood plasma). Physiologically, B. is a heterogeneous fluid, consisting of B. cells and B. plasma.

B. separates and protects the cells of the body from the external environment. The physical-chemical composition of B. is regulated within certain narrow limits, thereby insuring that the body's internal environment is constant. As part of this function, B. also mediates the exchange of material between the environment and the cells and tissues of the body. All vertebrates and many invertebrates depend on this exchange and transport of materials by the B. for the preservation of their life functions.

In animals with open vascular systems (mollusks, arthropods), B. is identical with tissue fluid, and is known as Hemolymph (see). In animals with closed vascular systems (annelids, chordates), B. and Lymph (see) are separate, the former circulating in B. vessels, the latter in lymphatic vessels. In capillary beds, lymph is constantly produced by filtration through the capillary walls. Lymph flows directly into the venous part of the B. circulation via special lymph vessels (in mammals via the thoracic duct). Vertebrate lymph contains only colorless lymphocytes, whereas the B. plasma contains colorless leukocytes, thrombocytes, and specialized red cells (erythrocytes). The erythrocytes contain a red-colored, iron-containing blood pigment called Hemoglobin (see), which serves for oxygen transport. All pigmented blood cells are generally called *chromocytes.* B. cells of invertebrates are normally colorless (leukocytes), pigmented blood cells are absent, and the hemoglobin (or other oxygen-transporting chromoprotein) is dissolved in the plasma. Exceptions are e.g. proboscis worms and sea urchins, whose hemoglobin is present in B. cells. Oxygenated hemoglobin is light red, whereas deoxygenated hemoglobin is dark red. In the B. of freshwater bivalves, higher crustaceans (lobster, crayfish), some snails *(Helix pomatia),* cephalopods, scorpions and some spiders, oxygen is transported by the iron-free, copper-containing pigment, *hemocyanin,* which is always in solution in the plasma. Oxygenated hemocyanin is blue, and the deoxygenated form is colorless. B. cells, eggs, perivisceral fluid and internal organs of some sea urchins contain yellow-red naphthoquinone derivatives known as *echinochromes* (see Spinochromes); echinochrome A was first identified as a red pigment in the eggs of the sea urchin *Arbacia pustulosa.* The body fluids of some polychaetes *(Sabellidae)* contain a dissolved green, iron-containing pigment called *chlorocruorin.* The B. cells of sipuncoloid annelids, brachiopods, and a single polychaete species contain red *hemerythrin.* In some snails and in sea squirts, blood pigments are represented by colorless *achrooglobins.* A *vanadium chromogen* is present in solution in the B. plasma of tunicates.

The numbers and proportions of different B. cell types are characteristic for each species (see Hemogram).

The ratio of B. cell volume to B. plasma volume is expressed as the *hematocrit,* which is determined by centrifugation of anticoagulated fresh blood is a glass capillary tube. Normal hematocrit values are 42% for women and 46% for men.

Functions of B.:

1) Transport of oxygen bound to B. pigments from the respiratory organs to the tissues, and transport of carbon dioxide from the tissues to the respiratory organs.

2) Transport of nutrients absorbed during digestion and nutrients mobilized from reserve materials to various sites in the body, and transport of waste products of metabolism to the excretory organs for removal from the body.

3) Chemical regulation of the entire organism by the transport of hormones from their sites of secretion to their sites of action, and the transport of other active substances such as vitamins.

4) Maintenance of the constant physical-chemical equilibrium, which is necessary for the metabolic activity of all cells. In this respect, the inorganic ions of the internal milieu surrounding all cells are particularly important. Thus, normal cell function is only possible if the total concentration of all ions in the blood is held constant within certain limits. This constancy is achieved essentially by the regulatory activity of the excretory organs, the total salt concentration of human B. being about 300 mmol/l. In addition, acidic compounds like carbon dioxide and carboxylic acids constantly arise from the metabolic oxidation of food materials. These acidic compounds cannot be allowed to decrease the pH of the body fluid, because cells function normally only if they are bathed in medium of a constant pH. This, in turn, means that the pH of the B. must also be controlled within narrow limits. The acid-base equilibrium (balance) of B. is regulated by the activity of the excretory organs (e.g. kidneys) and respiratory organs (carbon dioxide excretion by the lungs), and especially by the buffering capacity of the plasma hydrogen carbonate, which neutralizes the acids that spill over into the B. from the tissues (see Blood buffering). The normal pH of human blood is 7.3–7.4. In insects, certain modified hypodermal cells (enocytes) are responsible, *inter alia,* for maintaining the physical-chemical equilibrium of the hemolymph.

5) To carry defensive substances (antibodies, antitoxins) against toxins and foreign bodies (antigens). Antigens are precipitated or coagulated, then cytosed and digested by colorless blood cells (leukocytes, phagocytes). Febrile infections in humans are overcome by specific types of leukocytes (Fig.). Antibodies are special glycoproteins, produced in a major defense reaction

(immunization) against foreign proteins, such as the protein components of bacteria and viruses, including their endo- and exo-toxins. In addition to antibodies, B. also contains agglutinins, which adsorb to foreign erythrocytes and cause them to clump; these must be taken into consideration when selecting blood for transfusion, and they are used in the clinical laboratory to determine human Blood groups (see).

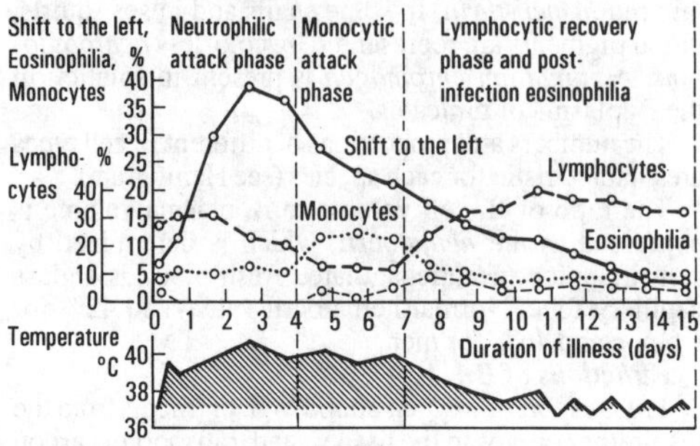

Blood. Changes in the percentages of different leukocytes during a febrile infection. Nuclei of neutrophils have between 1 and 5 lobes. When the number of circulating neutrophils increases as a result of infection (polymorphonuclear leukocytosis), the proportion of cells with 1-, 2-, or 3-lobed nuclei increases at the expense of mature neutrophils with 4-5 lobes. Since development is visualized as proceeding from left to right, this phenomenon is known as a "shift to the left".

6) Closure and sealing of wounds and other perforations in B. vessel walls by Blood staunching (see) and Blood coagulation (see). These processes depend on the thrombocytes, and the several proteins that constitute the coagulation system.

7) Temperature regulation (see). Heat produced by metabolism is transferred to the body surface by the B., where it is lost from the skin by radiation and conduction.

8) In invertebrates, B. may have other functions. Thus, in mollusks, body movements are the result of the interplay of muscular activity and alterations of B. pressure. In arthropods, alterations of B. pressure contribute to ecdysis, pupal molting, and wing expansion of the newly emerged imago. When touched, some beetles (e.g. *Cantharididae)* display a defense reaction known as *reflex bleeding,* in which they release droplets of hemolymph at many different sites (leg joints, wing bases, vicinity of the mouth) containing an irritant (e.g. cantharidine).

Summary of B. components

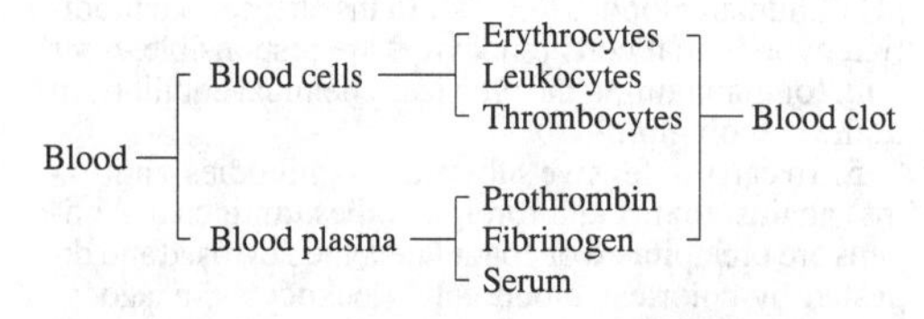

Blood-brain barrier: A barrier to diffusion formed by the total internal surface of the blood capillaries in the brain. This barrier is reinforced by enzymes in the endothelial cells of the capillaries. The B.b.b. displays varying degrees of permeability, depending on the material in question. Thus, the respiratory gases, carbon dioxide and oxygen, cross the capillary walls very easily, whereas proteins and other hydrophilic materials are largely excluded. The B.b.b. clearly serves a protective function, preventing exposure of the central nervous system to a variety of electrolytes and macromolecules which are present in the circulation in health and disease, especially bacterial toxins. It is probably also important in the maintenance of the constant environment of the central nervous system, i.e. for the homeostasis of potassium, calcium, magnesium and hydrogen ions, as well as other materials (e.g. vitamins) in cerebrospinal fluid and cerebral interstitial fluid.

Blood buffering: chemical regulatory mechanisms for maintenance of the normal pH value of blood plasma. Human blood and that of many animals normally has a pH value of 7.4. Marked deviations from this value, constituting Acidosis (see) or Alkalosis (see), are life-threatening.

Two compounds are mainly responsible for the acidifiction of blood, i.e. carbonic acid produced by respiration, and lactic acid formed during muscular contraction. About 80% of the protons of acids entering the circulation are bound by the globin of hemoglobin, which behaves amphoterically (i.e. as an acid toward bases, and as a base toward acids). Oxygenated hemoglobin (HbO_2) is a stronger acid than deoxygenated hemoglobin (Hb). Through the release of oxygen in the capillaries, Hb therefore becomes less acidic and releases part of its bound alkali. Carbon dioxide from the capillary cells then becomes bound in the plasma as sodium hydrogen carbonate (see Blood gases), and is transported to the lungs. The acid burden is released in the lungs, when carbon dioxide leaves the blood and is lost by expiration (the rapid dissociation of H_2CO_3 into carbon dioxide and water is catalysed by carbonic anhydrase).

When organic acids are present (lactic acid, pyruvic acid) they bind part of the standard bicarbonate, which is thereby rendered unavailable for the buffering of carbon dioxide. The carbon dioxide tension of the blood then increases and initiates the Chemical regulation of respiration (see), leading to increased alveolar ventilation and an increased loss of accumulated carbon dioxide by expiration; this effect is observed especially after physical exertion.

Blood circulation: the movement of blood around the body within a system of vessels, contractile organs and body spaces, known collectively as the *vascular system.* Blood may flow simply under the influence of general body movements, or it may be propelled by the rhythmic contraction of elements of the vascular system, or it may be pumped by specially localized hollow muscular organs, known as hearts, which form an integral part of the vascular system. The presence of valves in hearts and blood vessels insures that blood travels in only one direction. Tunicates are an exception; in these organisms the contractile waves of the heart, and therefore the direction of blood flow, are periodically reversed. The vascular system displays varying degrees of evolutionary development. In an *open* vascular system, a heart is present, but its afferent and efferent vessels open into perivisceral cavities (sinuses) which bathe various organs (mollusks, arthropods, tuni-

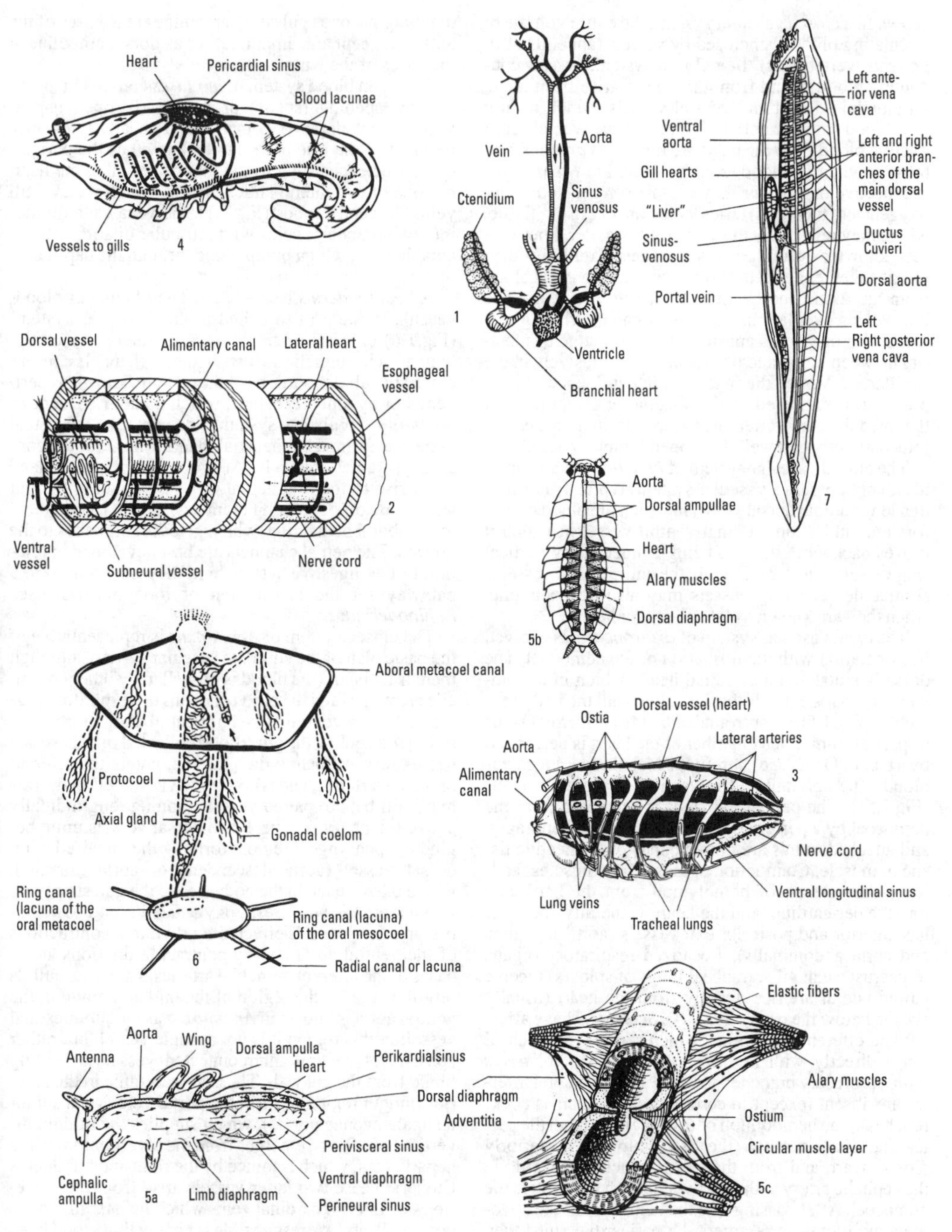

Fig.1. *Blood circulation.* Vascular systems of invertebrates. *1* Open system of a cephalopod. *2* Closed system of an earthworm; left: intestinal region; right; esophagus with pulsating ring or lateral heart connecting the dorsal and ventral vessels. *3* Open system of a scorpion with lung veins and tracheal lungs; the dorsal vessel is developed as tubular, contractile heart. *4* Open system of the river crawfish *(Astacus);* vessels carrying oxygenated blood are shown in black. *5* Open system of insects: *5a* circulation with fully developed pulsating organs (direction of blood flow shown by arrows); *5b* plan of the dorsal vessel (heart) and dorsal diaphragm; *5c* fine structure of the dorsal vessel (ventral view). *6* Blood-lacunar system of echinoderms, which lack a heart or other means of blood propulsion; protocoel, gonads and ring canal of the oral metacoel are shown as a dotted line. *7* Closed system of *Amphioxus (Branchiostoma lanceolatum)* with ventral contractile elements (ventral aorta and bases of the afferent branchial arteries), which propel the blood forward.

cates). In a *closed* vascular system, the entire volume of circulating blood is enclosed by vessels (annelids, *Amphioxus,* vertebrates). In a closed system, vessels conducting blood to and from the organs arise from large, longitudinal dorsal and ventral vessels, which in turn take blood from, or deliver it to, the heart. Vessels conducting blood from the heart are known as *arteries,* and those conducting blood to the heart are known as *veins,* irrespective of whether they contain oxygenated or deoxygenated blood. Arteries enter an organ or tissue, where they repeatedly divide to form smaller, finer vessels, known as Capillaries (see). These capillaries then merge together again to form a vein, which leads blood from the tissue and eventually back to the heart. The intervening capillary bed consists of an interlacing network with numerous anastomoses. Exchange of material between blood and tissues occurs exclusively in the capillaries. Vessel diameter (which influences blood pressure) is regulated by the Vasomotor center (see) in the medulla. In vertebrates, the vascular system is joined directly to a well developed lymph system.

The closed vascular system of *annelids* (small annelids do not possess a vascular system) typically consists of a longitudinal dorsal vessel in which the blood moves forward, and a longitudinal ventral vessel in which it moves backward, the two being connected by vertical ring vessels (Fig.1.2). Usually the entire dorsal vessel is contractile. The ring vessels may also be contracile, when they are known as lateral or ring hearts.

The open vascular system of *arthropods* has evolved in association with the formation of the hemocoel. The dorsally situated, longitudinal heart (which in less advanced groups extends through almost all the body segments) (Fig.1.5b) is surrounded by the pericardial sinus or pericardium. Each chamber of the heart is perforated by a pair of Ostia (see), through which the heart receives blood (hemolymph) from the pericardial sinus (Fig.1.5c). The pericardial sinus is separated from the hemocoel by a perforated dorsal horizontal diaphragm, and attached to this are paired bands of connective tissue or muscle. Contraction of these alary muscles facilitates the passage of hemolymph from the hemocoel into the pericardium and the heart. Generally, the heart has anterior and posterior exit vessels (aorta cephalica and aorta abdominalis). Localized respiratory organs (e.g. crustacean gills, tracheal lungs of spiders) receive paired lateral arteries directly from the heart (usually arising below the ostia) (Fig.1.4). *Scorpions* have afferent and efferent lung veins, which service the tracheal lungs directly with hemolymph (Fig.1.3). In *insects* with a profusely branched tracheal system, lateral arteries are absent (except in certain species, such as cockroaches). The hemolymph of insects flows from the pericardial sinus through the ostia into the posteriorly closed heart, and from there is conducted forward via the cephalic artery to the head region, where it enters the hemocoel. After passing into the hemocoel of the abdomen and thorax, it re-enters the pericardial sinus then the heart. Directional blood flow is insured by the dorsal diaphragm, and by a ventral diaphragm beneath the intestine, which separates the perineural sinus (surrounding the abdominal ganglia) from the perivisceral sinus of the intestinal region (Fig.1.5a). In addition, there are special pulsating organs, which prevent stagnation of hemolymph, particularly in the body appendages. These occur in the limbs as longitudinal muscular diaphragms, or as pulsating ampullae at the bases of the antennae (cephalic ampullae), or as dorsal ampullae at the bases of the wings.

The open blood system of *mollusks* consists of an anteriorly opening (snails) or posteriorly opening (bivalves, cephalopods) dorsal vessel (heart) lying in a remnant of the true body cavity, known as the pericardial space. According to the number of gills, the heart has one or two lateral antichambers, which receive gill veins. In cephalopods (Fig.1.1), the vena cava divides into two branchial veins with muscular dilations (branchial hearts), which pump blood through the capillaries of the ctenidia.

All *echinoderms* lack a heart. Their hemal or blood-vascular system (also called a blood-lacunar system) (Fig.1.6) consists of small sinus channels or lacunae, which lack epithelia. Most of these channels are enclosed by tubular portions of the coelom, known as perihemal or hyponeural canals, which do not represent an accessory circulatory system. The center of the hemal system is formed by the axial gland, whose lumen consists of hemal channels. At one end, the axial gland gives rise to the oral hemal ring (surrounding the gut and giving off 5 radiating branches), and at its other end to the aboral hemal ring, which gives off branches to the gonads. The hemal channels are best developed in relation to the digestive system, and they may represent a pathway for the distribution of food material. See *Echinodermata.*

The closed system of *Amphioxus* is representative of the basic plan of the chordate vascular system, although there is no heart (the blood is propelled by other contracile elements) and the blood contains no respiratory pigment. A main ventral vessel or ventral aorta (aorta ventralis) runs below the gut. Blood collected by this vessel passes forward below the pharynx, enters the afferent branchial arteries, and is conveyed via vessels in the primary gill bars to paired vessels running longitudinally above the pharynx. These two dorsal vessels unite behind the pharyngeal region, forming the unpaired main dorsal vessel (aorta descendens or aorta dorsalis), which passes back in the body, above the gut, supplying various organs with fresh, oxygenated blood. Circulation in this system is effected by rhythmic contractions of the ventral aorta and by contractile dilations at the bases of the afferent branchial arteries, known as bulbils or gill hearts. In the region of the mid and hindgut, the ventral aorta is known in *Amphioxus* as the subintestinal vessel; in this region it is not a single vessel, but rather consists of several interconnected vessels receiving blood from the gut wall. These vessels unite in the anterior midgut region to form the portal vein (vena subintestinalis, venae portae), which supplies the midgut diverticulum. Blood emerges from the diverticulum in the hepatic vein, which is joined by the right and left ductus Cuvieri. These two latter vessels arise from the posterior and anterior cardinal veins which lie laterally in the body wall, and are responsible for collecting blood from the body. As the hepatic vein continues forward, it passes below the pharyngeal region, where it becomes the ventral aorta.

Thus, in the B.c. of craniates, the dorsal aorta is the main vessel for distribution of blood to the body, while the ventral vessel (which in vertebrates contains the heart) has a forward-directed flow and serves as a collecting vessel for blood from the tissues (i.e. deoxygen-

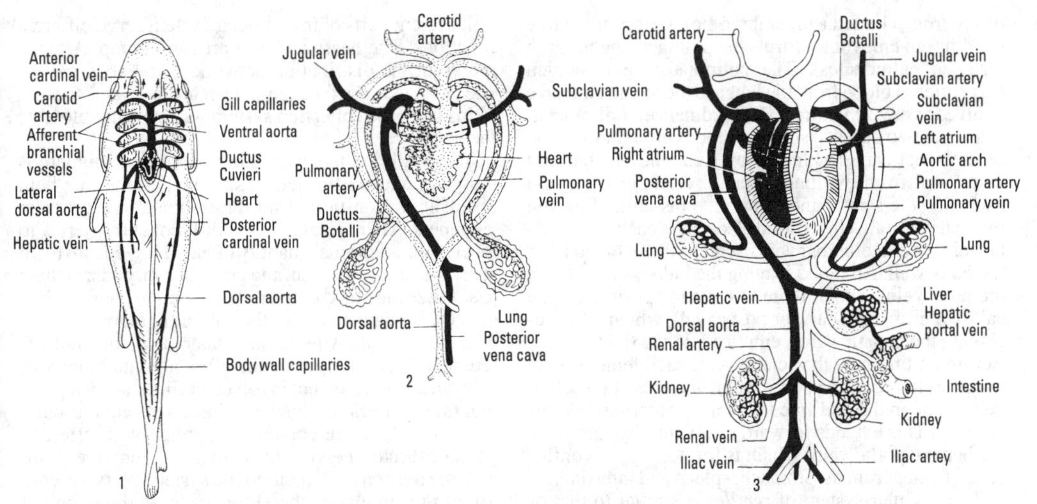

Fig.2. *Blood circulation.* Vascular systems of vertebrates (ventral view). *1* Fish. *2* Frog. *3* Mammal. White: vessels containing arterial blood. Black: venous blood. Shaded: mixed blood.

ated blood carrying waste products). In contrast, in the vascular system of coelomate invertebrates, blood moves forward in a dorsal collecting vessel, and is distributed by a ventral vessel. Subsequent evolution of the vertebrate vascular system has given rise to various departures from the basic plan. Examples are the appearance of *portal systems* and of *double circulation.* A portal system collects blood from one organ or tissue, and conveys it through the capillaries of another organ before it enters the heart. In the *renal portal system,* blood from the caudal region does not pass directly into the posterior vena cava, but is transported to the kidneys by the renal portal veins; after passing through the capillaries of the kidneys, it then enters the posterior vena cava via the renal veins. In the *hepatic portal system,* blood passes via the hepatic portal vein from the gut to the liver, then via the hepatic vein to the posterior vena cava (or directly to the heart).

Although lungs are derived from the posterior end of the pharynx, their blood supply is different from that of gills. There is also a considerable drop in pressure between blood entering and leaving the lungs, due to their extensive capillary systems. Moreover, in active, lung-breathing quadrupeds, especially mammals, the blood must circulate more rapidly than in fish, and a higher aortic pressure is required. After passage through the lungs, oxygenated blood is therefore returned to the heart, whence it is pumped at higher pressure into the arteries supplying the body. In order to receive oxygenated and deoxygenated blood, and to keep the one separate from the other, the Heart (see) has evolved by division of the atrium into two chambers by an interauricular septum, followed by division of the ventricle by an interventricular septum. Stages in this evolution can be seen in the structures of the fish, amphibian, reptilian, bird and mammalian hearts and vascular systems (Fig.2). In the mammalian heart there is complete division into right and left atria and right and left ventricles, and complete separation of venous and arterial blood, so that mammals have a double circulation. In one complete circulation through the body, blood therefore passes twice through the heart, i.e. heart → tissues → heart → lungs → heart.

In *fishes* (Fig.2) an artery (ventral aorta, truncus arteriosus) extends forward from the ventrally situated heart. In the majority of fishes, the heart pumps only venous blood (see Heart). The ventral aorta gives rise to maximally 6 pairs of afferent branchial arteries, which arch around the pharynx and pass to the gills. The first one or two pairs are reduced, so that in adult teleosts, for example, only the last four pairs remain. After passage through the gill capillaries, the now oxygenated blood travels through the epibranchial arteries (one from each gill cleft) to the single median dorsal aorta, which passes backward along the entire length of the body beneath the vertebral column, giving off branches that supply most of the body organs with fresh, oxygenated blood. Anterior branches supply blood to the brain. Deoxygenated blood collects in two ventral veins which transport blood to the sinus venosus of the heart. These veins are mostly large blood spaces or sinuses, rather than the narrow, tubular structures found in higher vertebrates. Nutrient-rich blood from the intestine first passes through the liver (hepatic portal circulation), then travels via hepatic veins directly to the sinus venosus. Two veins from the caudal fin (renal portal veins) pass to the kidneys, then re-emerge as the posterior cardinal veins (postcardinal sinuses; vena cardinalis posterior). These then unite (caudal to the pharyngeal region) with cardinal veins from the head (anterior cardinal veins or sinuses) to form the ductus Cuvieri, whence blood passes into the sinus venosus of the heart.

In *gill-breathing amphibians,* the basic plan of the vascular system is similar to that of fishes. *Lung-breathing quadrupedal amphibians,* however, display considerable differences from this plan, which are due mainly to the presence of a *pulmonary circulation* (Fig.2). Gill vessels are regressed. Venous and arterial blood are mixed in the heart. In the metamorphosis of a tadpole into a frog, the 1st aortic arch of the tadpole (3rd embryonic) becomes the carotid arch system of the frog, while the 2nd (4th embryonic) forms the systemic arch of the frog. The 3rd (5th embryonic) disappears, while the 4th (6th embyonic) forms the pulmocutaneous arch (arteriae pulmonales). The pulmonary veins, which carry arterial blood from the lungs to the left atrium have evolved as new structures. The cardinal veins disappear. There are two portal systems. Blood is brought from the outer aspect of each hindleg by right and left femoral veins, and from the inner aspect by sciatic veins. On each side the sciatic and femoral veins unite, forming two renal portal veins. Blood also enters each renal portal vein via dorsolumbar veins, which collect blood from the body wall. All blood leaving the kidneys travels via the renal veins, which unite to form the posterior vena cava (inferior vena cava or post caval), which, after receiving the hepatic veins, empties into the right atrium. Before uniting with the sciatic vein, each femoral vein branches to produce a pelvic vein, and these two pelvic veins unite in the mid-line to form the anterior abdominal vein. This extends forward mid-ventrally, and joins the hepatic portal vein, which is formed by the confluence of veins from the stomach, spleen and intestine.

The vascular system of *reptiles* is similar to that of lung-breathing amphibia, but the division of the heart is more advanced (see Heart).

Warm-blooded animals display a complete separation of the pulmonary and body circulations. In *birds,* only the right aortic arch remains, and this supplies arteries to the entire body. The hepatic portal system of birds is well developed, whereas the renal portal sytem is of secondary importance.

In contrast to birds, mammals retain only the left aortic arch (Fig.2). This supplies arteries to the head, neck and forelimbs. It continues backward as the dorsal aorta, which lies close to and alongside the vertebral column. It gives off branch arteries to the intestine (mesenteric artery), kidneys (renal arteries) and gonads (genital arteries), ultimately dividing into the iliac arteries, which supply the hindlimbs. The caudal artery to the tail is a direct continuation of the dorsal aorta. Deoxygenated blood from the posterior body collects in the posterior (inferior) vena cava, and that from the head and neck region in the jugular veins, which unite to form the anterior vena cava. Both vena cavae empty into the right atrium. Although a renal portal system is absent, the hepatic portal system is prominent and important. All veins from unpaired organs (stomach, intestine, spleen, pancreas) carry blood to the hepatic portal vein (vena portae), which carries nutrient-rich, but deoxygenated blood to the liver. From there the blood travels via the hepatic vein to the posterior vena cava.

The vascular system of animals has four main physiological functions. 1) Transport. Organs and tissues must be continually supplied with nutrients, including oxygen, and their waste metabolic products must be removed. 2) Humoral coordination. Highly potent agents such as hormones are transported rapidly from their sites of release to their sites of action. 3) Temperature regulation. Heat produced in the tissues by metabolism is transferred to the exterior by the blood. Heat loss is controlled by temperature regulating centers in the brain which modulate blood flow in the skin. 4) Blood distribution. According to their basic function and their current state of activity, different organs have different blood requirements. Hormonal, nervous and local mechanisms insure that blood is distributed to different sections of the circulation according to need.

Different parts of the vascular system serve different functions. The heart acts as a pressure pump. Arteries represent the distribution network. Capillaries are the site of material exchange between blood and tissues. Veins form the collection system for returning blood to the heart.

The walls of the large arteries contain numerous muscular and elastic elements, so that arteries are both extensile and elastic and are able to function as pressure chambers. Their great extensibility enables arteries to receive the additional quantity of blood forced into them at each heart contraction, and their elasticity adapts them to squeeze the blood outward and return to their original diameter in time to receive the volume of blood from the next heart beat. Arteries are innervated, but not efferently. Their diameter is therefore not under nervous control, and they are not involved in Circulatory regulation (see). Arterioles, however, have a strong muscular walls, and they are copiously supplied with efferent, vasoconstrictor nerve fibers, mainly from the sympathetic portion of the autonomic system. Nervous control of the circulation therefore occurs in the region of the arterioles. The capillary system represents the functional part of the B.c. Ultrafiltration of the blood into the tissues occurs under the hydrostatic pressure of the blood, and water is reabsorbed from the tissues into the blood due to the colloidal osmotic pressure of the blood. In the human circulation, about 8 000 liters of blood pass daily through the capillaries. Veins are rich in collagen and have a well developed outer layer of connective tissue. Their walls are therefore very extensible, and they are able to serve as volume reservoirs in the circulation. They can accommodate large volumes of blood without any great increase in wall tension. Since their diameter is controlled by sympathetic nerves, they are also involved in Circulatory regulation (see).

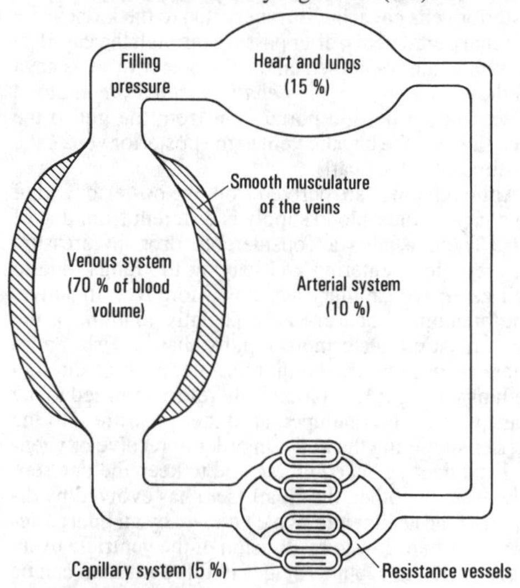

Fig.3. *Blood circulation.* Distribution of blood volume in the circulation.

Physiologically, the B.c. is divided into high- and low-pressure systems. The low-pressure system includes the capillaries, veins, right atrium and right ventricle, the pulmonary circulation, and the left atrium. In this part of the B.c. the blood pressure is relatively low and relatively constant. In the high-pressure system, the blood pressure is much higher, and displays rhythmic variations corresponding to the beating of the heart. This high-pressure system includes the left ventricle, the arteries and the arterioles. High- and low-pressure systems meet in the left half of the heart. During systole, the left side of the heart represents the beginning of the high-pressure system, whereas during diastole it represents the end of the low-pressure system. Flow resistance in the high-pressure system is about ten times greater than in the low-pressure system. Because of its great extensibility, the low-pressure system has a 200 times greater capacity than the high-pressure system. Thus the two systems are equivalent to resistance and capacitance systems, respectively. Due mainly to the different properties of arteries and veins, the total blood volume is distributed unequally in the two systems. In a resting human, the arterial circulation contains about 15% of the total blood volume, and the venous circulation about 85% (Fig.3).

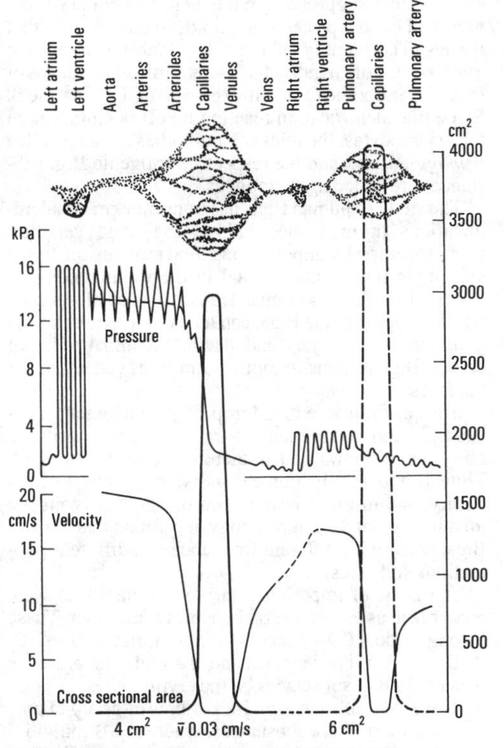

Fig.4. *Blood circulation.* Changes in blood pressure, blood flow rate and cross sectional area of vessels in the human circulation.

Changes in blood pressure and flow rate follow a regular pattern (Fig.4). In the arteries, blood pressure varies continually between the systolic and diastolic value. In the arterioles, these pulse variations disappear, and the pressure falls considerably. Pressure within the capillary system is relatively low, and it falls still further in the venous system. At the entrance of the right atrium, there is even a negative pressure. The transport rate in the arterial system is high, attaining maximal values in systole and minimal values in diastole. In the periphery of the body, the rate of flow is drastically reduced. It is very low in the capillary system, increasing again in the venous system. This is explained by the increase in the total blood space brought about by continual branching of arteries, followed by a decrease in the total blood space as emerging capillaries recombine and finally form the venous system. The low transport rate in the capillaries provides for a relatively long contact period (about 1 second) between blood and tissue, during which materials are exchanged.

Blood circulation, hydrodynamics of: see Hydrodynamics of blood circulation.

Blood clotting: see Blood coagulation.

Blood coagulation, ***blood clotting:*** a process involving a series of sequential enzymatic reactions, which forms a gel (the clot) sufficiently dense to prevent bleeding from a wound. B.c. is the most important process in the Staunching of blood flow (see). It occurs only in arthropods and vertebrates. In most invertebrates, wounding results in the clumping of agglutinized blood cells, but there is no true B.c.

Traditionally, two systems of B.c. are recognized, the intrinsic and extrinsic systems (Fig.). Blood plasma clots slowly (in several minutes) in the presence of a foreign surface like kaolin or glass, and the factors involved are all intrinsic to the plasma, hence the name. The extrinsic system is activated by a specific proteinaceous tissue factor and phospholipids, producing a clot in seconds rather than minutes. The only plasma factor belonging exclusively to the extrinsic system is factor VII, but this may be activated by factor XII_a or kallikrein, both of which belong to the intrinsic pathway. It has also been shown that platelets in the presence of factor V_a are stimulated by low concentrations of prostacyclin to activate factor X. Thus, the distinction between intrinsic and extrinsic pathways may well be artificial.

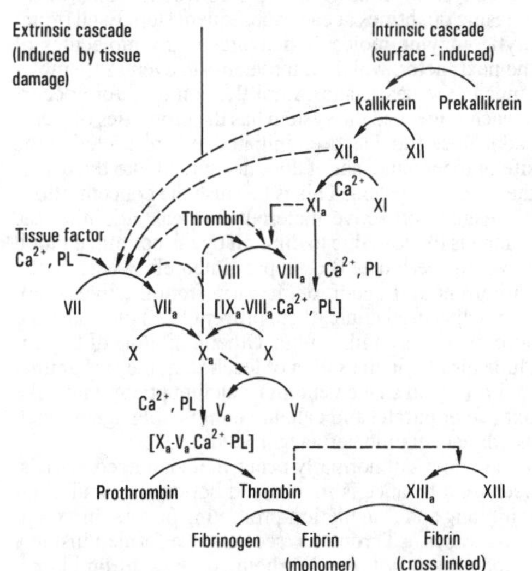

Blood coagulation. The coagulation cascade. PL = phospholipid or cell membrane. Solid arrows indicate reactions; dotted arrows, catalysis. Adapted from L. Lorand, *Methods Enzymol.* 45B (1976) 31-37

Prothrombin and factors V, VII, VIII, IX, X, XI and XII are collectively known as the *prothrombin complex.* Components of the prothrombin complex and other proteins of the B.c. are synthesized in the liver. Prothrombin and factors VII, X and IX contain residues of 4-carboxyglutamic acid, which are formed posttranslationally by carboxylation of glutamate residues, an enzymatic reaction in which vitamin K serves as a cofactor. Vitamin K deficiency or vitamin K antagonists therefore lead to defective B.c., e.g. warfarin is a vitamin K antagonist used as a rodenticide, and dicoumarol in moldy clover hay is a vitamin K antagonist responsible for hemorrhage in cattle. The dicarboxylate side chains of the 4-carboxyglutamate residues enable the respective clotting factors to bind calcium ions (calcium ions are essential for several stages of B.c.).

In the first phase of B.c., prothrombin is converted into thrombin, a process lasting only minutes, and involving several factors (Fig.). In the second phase, fibrinogen is converted into fibrin (by the proteolytic action of thrombin), and the fibrin monomers become cross-linked. In the postcoagulation phase, contractile proteins of the blood platelets pull the fibrin threads together, so that the edges of the wound are drawn closer (this accelerates the healing process). At the same time, all the structured blood components shrink into a scab, with the extrusion of serum.

Each factor of the prothrombin complex is a proteolyic enzyme, and the next factor is its substrate. A factor is activated (converted into an active proteolytic enzyme) when it is partially (but very specifically) hydrolysed by the preceding factor. In the Fig., an "a" subscript (e.g. X_a) indicates that the factor is activated, and can now attack and activate the next inactive factor, which is its specific substrate. Prothrombin (factor II) also conforms to this pattern, so that thrombin is an active protease and may also be designated factor IIa. At the start of B.c., relatively small quantities of substrate are converted, followed by the conversion of ever increasing quantities at each subsequent step. Each proteolytic enzyme molecule activates many molecules of the next factor, which in turn activate even more molecules in the next stage. Amplification therefore occurs at each stage, and the system has the properties of a cascade. Once B.c. has been initiated, it is restricted to the site of the wound, and it does not spread into the rest of the vascular system. This is because the concentrations of coagulation-active materials in areas adjoining the wound is too low, due to dilution by the continual blood flow. Furthermore, B.c. is prevented elsewhere by an antithrombin. In addition, heparin from the liver (heparin is also used clinically to prevent B.c.) increases the activity of this antithrombin. Other inhibitors of B.c. include hirudin in the saliva of leeches, numerous animal toxins (e.g. in snake venoms), calcium precipitants like oxalate or citrate, and calcium complexing agents such as ethylenediaminetetraacetic acid (EDTA).

B.c. does not normally occur in undamaged vessels, because a balance is maintained between coagulation-inhibiting and coagulation-promoting factors. In exceptional cases, a Thrombus (see) may be formed inside a blood vessel. Within a few hours or days, thrombi may removed by fibrinolysis, a process in which fibrin is converted into soluble fibrinopeptides. There is the danger, however, that a thrombus may become dislodged from its site of formation, and lodge in the heart or certain parts of the brain, where it blocks the circulation and causes death. A dislodged thrombus is called an embolus, and the resulting condition is called embolism.

Blood count: see Hemogram.

Blood flow, hydrodynamics of: see Hydrodynamics of blood circulation.

Blood flower: see *Asclepiadaceae.*

Blood gases: the respiratory gases, oxygen and carbon dioxide, which are transported in the blood. In vertebrates, a small quantity of oxygen is dissolved in the plasma, while the vast majority is bound to hemoglobin (see Oxygen binding by blood). A relatively large proportion of the carbon dioxide is dissolved in the plasma, but most is bound by erythrocytes. The blood gases are exchanged in the lungs and tissues. In the lungs, oxygen aids the release of carbon dioxide from the blood (Haldane effect), while in the tissues, carbon dioxide promotes the release of oxygen (see Bohr effect).

Blood group: a serologically identifiable character of animal and human erythrocytes. B.g. substances (B.g. antigens) are specific oligosaccharide structures attached to glycoproteins in the membrane of erythrocytes, and to both proteins and lipids in the cells of other tissues. The identity of these B.g. substances on the erythrocytes of an individual is determined by means of specific antibodies, or with the aid of Lectins (see). Since this identification reaction involves clumping of the erythrocytes, the relevant antibodies are also called *hemagglutinins,* and the respective antigenic B.g. substances are called agglutinogens.

The study of human B.gs. has important practical implications in medicine, anthropology and genetics. B.gs. are crucially important in blood transfusion, tissue and organ tranplantation, and in the immunocompatibility of mother and fetus. Transfusion of blood containing inappropriate B.gs. can lead to intravascular agglutination, hemolysis and death. Incompatibility of certain B.g. systems in mother and fetus can endanger the fetus.

B.gs. are valuable in anthropology and genetics, because, in contrast to many characters, their inheritance obeys simple laws and is therefore easy to interpret. Thus, B.g. determination can assist in questions of paternity, and in the determination of whether twins are identical or not. In anthropology, it is found that various B.gs. occur with different frequencies in different populations and races.

The B.gs. of importance in blood transfusion were recognized as early as 1900 by Karl Landsteiner. These belong to the ***ABO-system,*** which comprises 4 B.gs.: O, A, B, and AB. The letters A and B signify the presence of specific B.g. substances on the erythrocytes, i.e. substances A and substance B. Individuals of group O possess neither of these substances, whereas AB individuals possess both substance A and substance B. This ABO-system has an outstanding peculiarity, in that the serum of each individual contains antibodies (agglutinins) for those group substances that are not present on the erythrocytes (Table). Group AB individuals are an exception, in that their serum does not contain antibodies, so that the erythrocytes of all other groups are unaffected by it. Also, since erythrocytes of group O contain no agglutinogens, they are not agglutinated by any other sera. (If a patient receives repeated transfusions of group O erythrocytes, the production of anti-O agglutinins may be stimulated in the recipient plasma.). There are two A-type B.g. substances, A_1 and A_2. Almost all anti-A sera (from group B-type blood donors) contain two components, anti-A and anti-A_1. Anti-A reacts with both A_1 and A_2, while anti-A_1 reacts only with B.g. substance A_1. By laboratory testing, it is therefore possible to classify AB blood as either A_1B or A_2B; and A-type blood can be more accurately identified as A_1 or A_2.

Agglutination in the ABO-Blood group system.
O = no agglutination of erythrocytes. + = agglutination of erythrocytes.

Serum of group	Erythrocytes of group			
	AB	A	B	O
AB	O	O	O	O
A	+	O	+	O
B	+	+	O	O
O	+	+	+	O

Genetically, the ABO-system is carried on at least 4 alleles (A_1, A_2, B and O). A_1, A_2 and B are each dominant over O; A_1 is dominant over A_2; and B is expressed jointly with A_1 or A_2. Since each parent passes on only one allele of the allele pair to its progeny, only certain phenotypic combinations are possible, as shown in the following table.

Parents (phenotypes)	Possible B.g. phenotypes in the offspring
O × O	O
O × A_1	O, A_1, A_2
O × A_2	O, A_2
O × B	O, B
O × A_1B	A_1, B
O × A_2B	A_2, B
A_1 × A_1	O, A_1, A_2
A_1 × A_2	O, A_1, A_2
A_1 × B	O, A_1, A_2, B, A_1B, A_2B
A_1 × A_1B	A_1, B, A_1B, A_2B
A_1 × A_2B	A_1, A_2, B, A_1B, A_2B
A_2 × A_2	O, A_2
A_2 × B	O, A_2, B, A_2B
A_2 × A_1B	A_1, B, A_2B
A_2 × A_2B	A_2, B, A_2B
B × B	O, B
B × A_1B	A_1, B, A_1B
B × A_2B	A_2, B, A_2B
A_1B × A_1B	A_1, B, A_1B
A_1B × A_2B	A_1, B, A_1B, A_2B
A_2B × A_2B	A_2, B, A_2B

Human blood also contains the ***MN-system,*** which is independent of the ABO-system. Normally, only the B.g. substances of the MN-system are present, and the corresponding antibodies (agglutinins) are absent. There are two alleles (M and N). Homozygotes posses either the M or N B.g. substance, while heterozygotes possess both. The M-gene is rather more abundant than the N-gene in Europens and Asians, whereas American Indians show a very marked predominance of the M-gene. The two genes occur with about equal frequency in Africans, whereas the N-gene is predominant in Australian aborigines.

The ***Rhesus (Rh)-system*** is clinically very important. It appears to consists of several allelic genes. Especially important is the D-antigen. A D-negative pregnant woman with a D-positive fetus can become sensitized by cells from the fetus that may enter her body, so that she produces anti-D antibodies. Since fetal cells are not transferred until the final phase of pregnancy, antibodies are not produced by the mother until the birth, or shortly afterwards. The child is therefore not endangered. In any subsequent pregnancy, however, antibodies from the sensitized mother cross the placenta into the fetus, where they can destroy the fetal red cells (see Erythroblastosis), causing severe damage to the fetus.

There are many other B.g. systems, but they are clinically less important than those already mentioned. Some examples are the P-, Kell-, Lutherean-, Lewis-, Duffy- and Kidd-systems On account of the great variety of different B.g. systems, each person has a very individual B.g. pattern. Very exact information can therefore be obtained from studies of the genetics and inheritance of B.gs.

Blood, oxygen binding by: see Oxygen binding by the blood.

Blood picture: see Hemogram.

Blood plasma: the liquid matrix of the blood without the blood corpuscles. Human B.p. consists of 90% water and 7% protein, the remaining 3% comprising inorganic salts (chloride, hydrogencarbonate, sulfate, phosphate, sodium, potassium, calcium, magnesium, etc.), glucose, lipids, steroids, organic acids, amino acids, urea. Essential nutrients, such as glucose and amino acids, are carried at low concentrations, but are subject to high turnover rates. Vitamins and hormones are also present. Na^+ is a major cation. The concentrations of electrolytes are held within narrow limits, and the H^+-concentration is precisely regulated. Among the plasma proteins are those involved in the blood coagulation cascade (culminating in the quantitatively important fibrinogen: 0.3 g/ 100 ml plasma), specific transport proteins, and numerous antibodies (γ-globulins). The latter include the agglutinins, which interact with blood group substances and must be taken into account in blood transfusion. Both α-globulins and β-globulins function as transport proteins, and the β-globulins also function to a small extent as protective proteins. Specific metal-binding and -transporting proteins include transferrin (iron) and ceruloplasmin (copper). Plasma proteins have a high adsorption capacity, and combine reversibly with many insoluble substances. Plasma albumin, which makes up 55–62% of plasma protein (or 4.5 g/100 ml plasma), has a high binding capacity for water, Ca^{2+}, Na^+, K^+, fatty acids, bilirubin, hormones, drugs and other xenobiotics. Since cell walls are largely impermeable to plasma proteins, the latter (in particular albumin and α-globulin) are important in water exchange between blood and tissues; the major proportion of the protein remaining in the blood exerts a colloidal osmotic pressure, withdrawing water from tissue cells.

Blood pressure: the pressure exerted on the walls of blood vessels, which serves to propel the blood during its circulation in the body. B.p. is highest in the arteries, decreasing in subsequent parts of the circulatory system, and attaining its lowest value in the veins near to the heart. For standardized comparisons, B.p. is therefore always measured at the same site in the circulation. For clinical purposes in humans, the artery of the forearm is used.

Pulsed variations in arterial B.p. are due to rhythmic

beating of the heart. The highest values *(systolic pressure)* occur during systole, i.e. the contraction of the ventricle, which pumps blood into the aorta. The *diastolic pressure* is the lowest value, recorded during relaxation of the heart between beats. Average *arterial pressure* is the arithmetic mean of the diasystolic and systolic pressures in the aorta. Pressure in the peripheral arteries is lower than that recorded in the aorta.

Comparative studies in a large number of healthy humans have shown that B.p. is influenced by the hydrostatic pressure (Fig.), and by a number of biological factors, e.g. age, anxiety, excitement, stressful environmental stimuli, physical work.

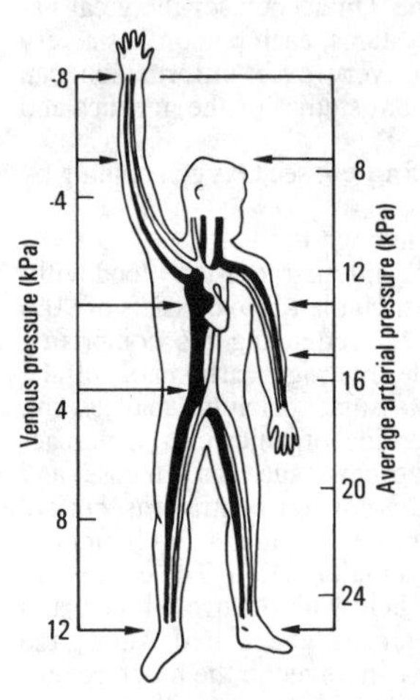

Blood pressure. Influence of hydrostatic pressure on the arterial and venous blood pressures.

B.p. increases with bodily activity, and it can rise as high as 23–24 kPa (systolic B.p.) in extreme physical exertion, compared with about 16 kPa at rest. High performance athletes in regular training usually have low resting values; compared with untrained individuals, their B.p also shows a less pronounced increase in response to physical activity.

Physiologically-based, short-term variations of B.p. must be differentiated from long-term pathological changes. In *hypertension,* the systolic pressure is increased to more than 27 kPa and the diastolic pressure to over 16 kPa. Hypertension may be due to: 1) a decrease in the elasticity of arteries, e.g. in arteriosclerosis; 2) a permanent or prolonged constriction of vessels in the arteriole system, due to nervous activity; or 3) a permanently increased cardiac output, due to impaired Circulatory regulation (see).

Hypotension occurs as an accompanying symptom of poisoning, exhaustion, starvation, adrenal failure, and circulatory collapse. Blood pressure regulation (see) insures that the B.p. is maintained at a normal value, and that it is able to adjust to altered circulatory conditions (exercise, etc.).

Comparison of average values for arterial B.p. in different vertebrates reveals no correlation between B.p. and body size. Fishes, amphibians and reptiles generally have lower values than birds and mammals.

Blood pressure regulation: the complex interplay of cardiac receptors, Baroceptors (see), circulatory centers in the brain and spinal cord, as well as nervous and hormonal regulatory mechanisms, which regulates Blood pressure (see) and adjusts it to the prevailing requirements of the circulatory system. Together with Cardiac regulation (see), B.p.r. is the most important component of Circulatory regulation (see). The interaction of the numerous factors involved in B.p.r. is displayed in the accompanying diagram as a regulatory circuit (Fig.).

Blood reserves: parts of the vascular system, in which the blood flows slowly and forms a reservoir. According to need, this reserve blood can become part of the rapid circulation. Mammalian B.r. are contained in the vessels of the skin and liver; the dog also has B.r. in the spleen.

As a result of nervous control, or in response to an increase in the concentration of circulating adrenalin, B.r.

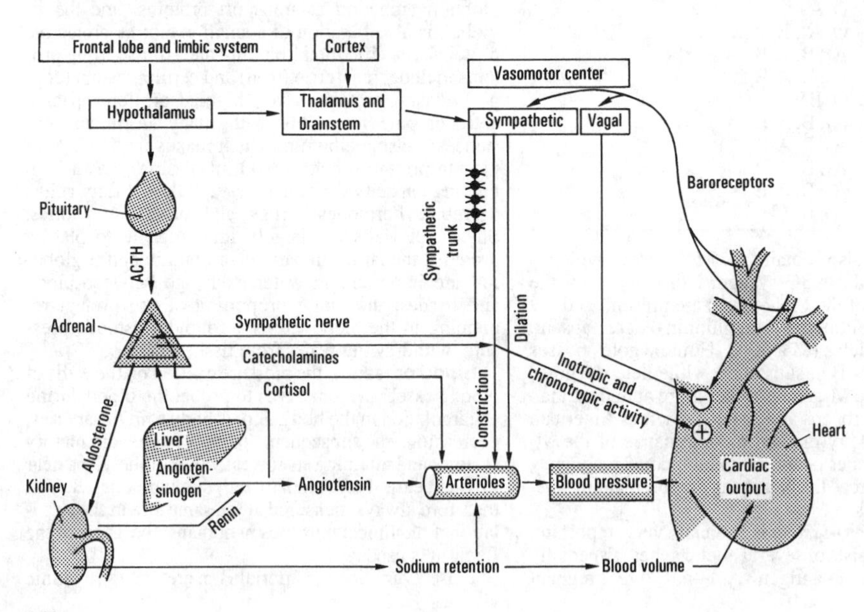

Regulation of blood pressure

are incorporated into the rapid circulation at the beginning of, and during, muscular exercise. The resulting rapid increase in the volume of circulating blood provides for the transport of increased quantities of oxygen, carbon dioxide and metabolic products. Since reserve blood contains relatively high concentrations of carbon dioxide, respiratory ventilation is also intensified, due to the Chemical regulation of respiration (see).

Blood reservoir: see Blood reserves.

Bloodsucker: see Changeable lizards.

Blood sugar level: the concentration of glucose in the blood. The normal fasting glucose concentration in human venous whole blood is 75–91 mg per 100 ml, but it may vary between 70 and 120 mg per 100 ml. Transitory high concentrations occur after ingesting glucose or glucose-containing carbohydrates.

Glucose entering the blood from the intestine is taken up by the liver and converted to glycogen. The latter is stored in the liver, but if the B.s.l. falls (hypoglycemia) as a result of energy-consuming metabolism (e.g. muscle activity), glycogen is reconverted to easily transported glucose, which restores the B.s.l. The brain (hypothalamus) responds to hypoglycemia by nervous stimulation of the adrenal medulla, causing the latter to secrete adrenalin, which in turn promotes the conversion of glycogen to glucose (glycogenolysis). If the B.s.l. rises too high (hyperglycemia), insulin is secreted, and the normal B.s.l. is restored. Disturbance of this homeostatic mechanism leads to diabetes mellitus. See Insulin.

Blood supply to organs, ***vascularization of organs:*** the circulation of blood through individual organs. Blood supply is subject to complicated regulatory mechanisms, which enable it to be adjusted to meet the individual requirements of each organ.

The magnitude of changes in the resting blood supply with changing physiological conditions depends on the organ involved. Whereas the blood supplies of brain, heart and kidneys are relatively constant, the supplies to the alimentary canal, skin and muscles are subject to wide variations. During violent physical exercise, the blood supply of human muscle may increase to 20–30 l/min, representing 40–60% of the cardiac output.

The blood supply is regulated primarily by three factors: 1) the local expansion of capillaries caused by carbon dioxide and certain other metabolites; 2) increased blood pressure, which generally results in a passive increase of blood flow; and 3) nervous control of the diameter of arterioles. The blood supply to any organ is increased by a decrease in Vessel tone (see). See Hydrodynamics of blood circulation.

Blood volume: the volume of blood in an organism. *Total B.v.* represents 6–19% of body mass in reptiles, birds and mammals (8% in humans). The *circulating B.v.* is determined from the degree of dilution of a dye injected into the blood stream; it is always less than the total B.v., indicating that only part of the blood is actively circulating. The other part flows slowly in the low pressure system and represents a Blood reserve (see) or reservoir.

Nervous control of Circulatory regulation (see) insures that mammals can lose up to one quarter of their total B.v. before the blood pressure falls. Humans can tolerate the loss of one third of total B.v., but loss of one half is life-threatening. After heavy blood loss, the normal B.v. is restored in two days. Renewal of B.v. starts with the flow of protein-poor fluid from the extracellular space. The protein content is normalized later, and the lost blood cells are also replaced in 2–3 weeks by an intensification of Hematopoesis (see).

Blood volume regulation: regulation of the water economy of the organism, primarily in response to blood volume. The volume of extracellular fluid is continually monitored by volume receptors, the most important of these being the stretch receptors of the vagus nerve in the

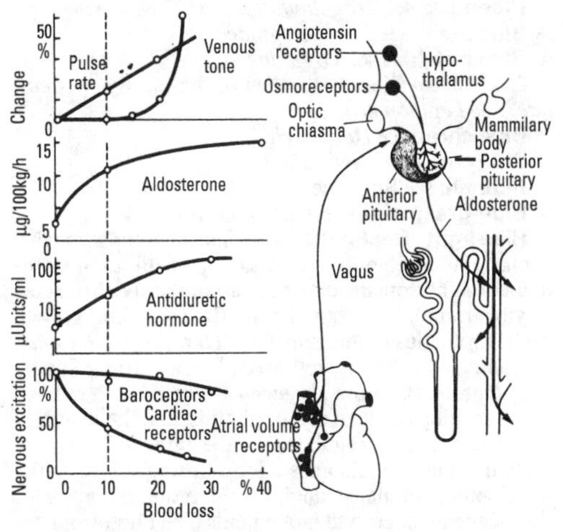

Blood volume regulation. Diuretic cardiac reflex.

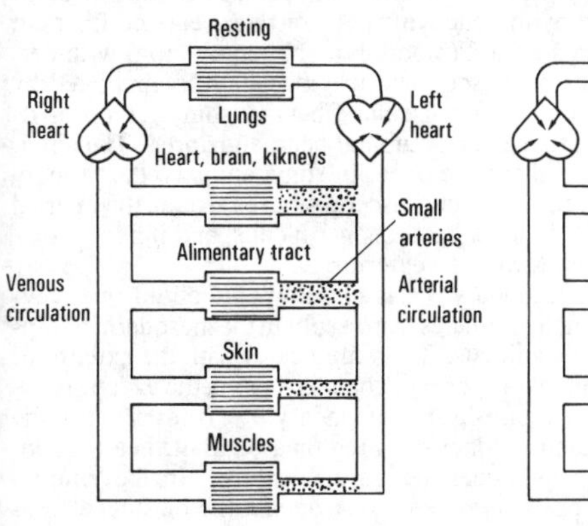

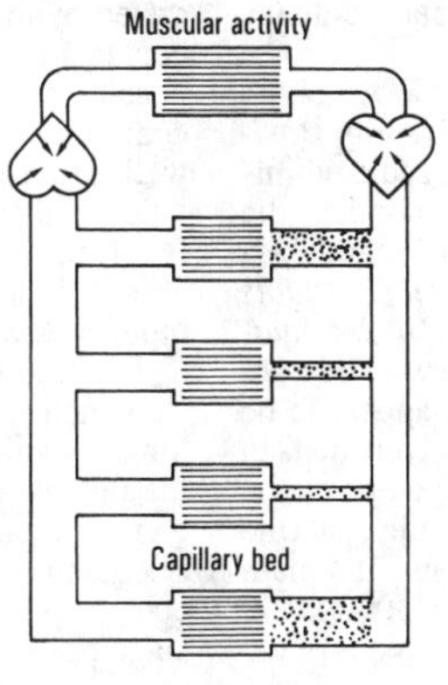

Blood supply to organs. Regulation of blood flow through different organs.

atrial walls of the heart (Fig.). These atrial volume receptors directly monitor the quantity of low pressure blood (see Blood circulation) from the degree of stretch exerted on the atrial walls. This also serves as an indirect measurement of the volume of fluid in the extracellular space, because water and electrolytes diffuse freely from the one space to the other. Thus, stretching of the heart atria by increased vascular volume increases vagal excitation to the diencephalon, which responds by decreasing secretion of antidiuretic hormone by the posterior pituitary, decreasing the secretion of aldosterone by the adrenal cortex, and by inhibiting activity of the sympathetic nerve to the kidneys. Decreased sympathetic activity increases the quantity of primary urine by increasing ultrafiltration in the Malpighian bodies. In response to the decreased concentrations of circulating antidiuretic hormone and aldosterone, the kidney tubule resorbs less water and NaCl, so that more urine is produced and voided, the volume of blood entering the atria is reduced, and excitation of the volume receptors decreases. Secretion of antidiuretic hormone is also inhibited by glucocorticoids, ethyl alcohol, light and cold-stimulation. The action of aldosterone on the kidney tubule is reduced by progesterone, glucocorticoids, and an increase of blood pH (alkalosis).

Bluebirds: see *Turdinae*.

Bluebuck: see *Tragelaphinae*.

Blue buck: see *Hippotraginae*.

Blue butcher: see *Orchidaceae*.

Blue butterflies: butterflies of the family *Lycaenidae*. See *Lepidoptera*.

Blue coral: see *Helioporida*.

Blue crabs: see Swimming crabs.

Bluegill: see Sunfishes.

Blue-green algae: see *Cyanophyceae*.

Blue light effects: effects on plant development, due exclusively to blue light (< 500 nm), with no involvement of light from the dark red part of the spectrum (see Phytochrome). For example, the flat spreading growth of the prothallus of the male fern *(Dryopteris filix-mas)* is promoted by the absorption of blue light (probably by a flavoprotein). In the absence of blue light, a chloronema develops instead of a prothallus, and the length of the chloronema is regulated by the light red-dark red system. Blue light also has a long-lasting effect on the development of many fungi. Furthermore, induction of light-dependent growth movements (see Phototropism) is largely or exclusively due to blue light.

The High energy system (see), which still requires further investigation, has absorption maxima in both the blue and dark red regions of the spectrum.

Blue sheep: members of the genus *Pseudois,* of the tribe *Caprini,* subfamily *Caprinae* (see), family *Bovidae* (see). Only a single species is recognized, and this is divided into 3 subspecies, which are found in the Himalayas and adjacent mountain areas of Central Asia and western China: ***Himalayan blue sheep*** *(Pseudois nayaur nayaur);* ***Szechwan blue sheep*** *(P.n. szechuanensis);* ***Pygmy blue sheep*** *(P.n. schaeferi).* They are neither sheep nor goat and in appearance appear to be intermediate between the two. They differ from goats in the absence of a beard and the absence of glands on the underside of the tail. Only the second juvenile coat (the first winter coat) is blue. They are sure-footed mountain animals, inhabiting canyons and slopes at 3000–5500 m. The habitat of the pygmy subspecies (canyons of the Yangtze-Kiang) is separated from that of the larger subspecies by a belt of forest and brush.

Bluethroat: see *Turdinae*.

Blue tit: see *Paridae*.

Blue-tongued skinks, ***Australian tiliquine skinks,*** *Tiliqua:* a genus of skinks (see *Scincidae),* 20–50 cm in length, found only in the Australian region. They have a cobalt-blue tongue, a broad head and a flattened, plump body with powerful limbs. All species are live-bearing (oviviviparous). Scales are smooth, with the exception of the aberrant ***Stump-tailed skink*** *(T. rugosa),* whose scales are rough with prominent keels, and whose tail is thick and stumpy and can be mistaken for the head. The female of *T. rugosa* gives birth to 2 young, which are about 10 cm in length and therefore already about one third of their final size.

Blusher: see *Basidiomycetes (Agaricales)*.

Blushing amanita: see *Basidiomycetes (Agaricales)*.

Boa constrictor, *Constrictor constrictor:* the most familiar, but not the largest (up to 5.5. m in length) member of the family *Boidae* (see). The B.c. is one of the most beautiful snakes, with a striking pattern of red-brown, pale-centered, yellow-edged spots. It is found from Mexico to central Argentina in a variety of habitats ranging from semi arid deserts to lush, humid tropical forests. It is a proficient climber, and its food is predominantly warm blooded, consisting of birds, small mammals and even predatory animals such as ocelots; lizards are also swallowed. Like all boas, it is viviparous.

Boatbill heron: see *Ciconiiformes*.

Bobac marmot: see *Sciuridae*.

Bobcat: see *Felidae*.

Bobolink: see *Icteridae*.

Bobwhite: see *Phasianidae*.

Body cavity, ***perivisceral space:*** a more or less divided animal body cavity, which contains most of the internal organs, and which in the embryo extends the entire length of the body. It may be a primary B.c. (protocoel, pseudocoel or hemocoel), or a secondary B.c. (coelom, deuterocoel). Certain primitive invertebrates (platyhelminths and aschelminths) are *acoelomate,* i.e. they do not possess a B.c., and their interior is usually occupied by mesenchymal parenchyma. All other animals are *coelomate.*

The primary B.c. is normally an extensive sinusoidal cavity without an epithelial lining, with extensions into the appendages. Morphologically, it represents that part of the embryonic blastocoel that does not become obliterated by mesenchyme cells or their secreted fibers. It contains a fluid (blood, lymph, hemolymph) with corpuscles of mesodermal origin, which is circulated by contractions of muscular fibers in some region of its wall (often a dorsal tubular heart and aorta). The circulation is therefore open. In some animals, the primary B.c. forms large perivisceral sinuses around the internal organs. It never contains germ cells, and it never communicates with the exterior.

The secondary B.c. is a hollow, epithelium-lined cavity, which originates from a split in the mesoderm of triploblastic animals. It communicates with the exterior of the animal via pores or channels, or via the excretory organs. In mollusks, the secondary B.c. is restricted to the sex glands, kidneys and region around the heart (gonocoel, nephrocoel, pericardial cavity). In the embryo stage of *Annelida* and *Arthropoda,* the blastocoel contains a series of small paired *mesodermal sacs* (primordial coelom), lying one behind the other along the length of the embryo between the ectoderm and endoderm. In *annelids,* these mesodermal sacs enlarge so that they fill and almost obliterate the blastocoel. Finally, they meet above and below the alimentary canal, the intervening walls break down, their cavities become continuous around the gut, and the resulting space becomes the perivisceral cavity. The blastocoel is reduced to a series of narrow tubes, which become blood vessels. The outer wall of the perivisceral cavity adheres to the ectoderm and forms the somatic mesoderm, while the inner wall adheres to the endoderm and forms the splanchnic mesoderm. The anterior and posterior walls of successive pairs of sacs meet and fuse to form septa (dissepiments) which separate successive compartments of the coelom. In contrast, in *arthropods,* the mesodermal sacs remain small during embryonal development, so that the blastocoel is not obliterated and persists in the adult as the perivisceral cavity; the mesodermal sacs merely become the cavities of the gonads. Early embryologists claimed the existence of connections between the primary and secondary B.cs. of arthropods, thus producing a tertiary B.c. or mixocoel, but this is now known to be erroneous.

The secondary B.c. of acorn worms and echinoderms consist of 3 paired sections lying one behind the other, the hydrocoel, axocoel and somatocoel.

The secondary B.c. of vertebrates, which usually arises by splitting of the mesoderm, consists of an epithelium-lined cavity, which in the embryo is connected to the exterior by a series of coelomoducts. The coelomoducts are segmentally arranged, but do not have separate segmental openings; they empty into two common lateral ducts, which open externally at the posterior end. In the adult vertebrate, the internal openings of the coelomoducts are closed, and excretory material is extracted by the kidney. During embryonic development, the cranial section of the originally uniform coelom separates as the *pericardial* cavity. The remaining part divides later into the *pleural* and *peritoneal* cavities. Organs project into, and occupy most of the space of these epithelium-lined cavities, so that epithelial surfaces lie close together, separated only by fluid-filled capillary gaps. Thus, the pleural cavity surrounding the lungs is lined by the pleura or pleural membrane. That part of the pleura covering the lung is called the visceral or investing pleura (pleura visceralis), while its continuation which lines the wall of the pleural cavity is called the parietal or lining pleura (pleura parietalis). Similarly, in the abdominal cavity, the peritoneum surrounds most of the organs, in particular the intestine and its auxiliary glands. See Mesoblast.

Body fluids: in the narrow sense, fluids of the animal organism that resemble tissues (and are usually regarded as tissues), in that they serve as functional units. They mediate the intercellular and inter-organ exchange of materials, and they play and important role in the interaction of the organism with its environment. These B.f. are blood, lymph and cerebrospinal fluid, as well as the hemolymph of invertebrates.

Organic components include nutrients like glucose, triacylglycerols and amino acids. Other components are hormones and enzymes. The protein constituents of B.f. include specific and nonspecific carriers, defense proteins (antibodies and complement system), coagulation proteins, etc., and they are essential for the life of the organism. An important organic constituent of mammalian B.f. is urea, which is transported from the liver to the kidneys for excretion. Other excretory products carried by B.f. include uric acid (birds, reptiles, invertebrates) and creatinine. The most important inorganic cations are Na^+, K^+, Ca^{2+} and Mg^{2+}, while the predominant anion is chloride. Bicarbonate is also important, in particular as part of the buffering system of B.f. Since B.f. represent the internal environment of the organism, their constancy with respect to pH, osmolarity, etc. is important. The pH-value of B.f. has a pronounced influence on metabolism, and it is kept relatively constant by buffering systems. Slight differences are associated with the mode of nutrition; thus, the B.f. of herbivores are slightly acidic.

Respiratory gases may also be considered as constituents of body fluids. In vertebrates, carbon dioxide and oxygen are transported free in the blood or reversibly bound to blood pigments.

In the wider sense, B.f. include those with a more restricted and specialized function, e.g. saliva, gastric juice, bile, urine, aqueous humor of the eye, tear fluid, secretions of the testes and vagina, milk, etc.

Body height: linear distance between the standing surface and the highest point of the skull of a human in natural standing posture. It is measured with an anthropometer. The B.h. of modern humans varies between 120 and 200 cm; smaller or greater values (dwarfism or gigantism) usually have a pathological basis. The present world average for men is 168–170 cm, and in all races women are about 7% shorter.

Subdivisions of body height (after Brugsch)

	Men	Women
Small	less than 169 cm	less than 155 cm
Intermediate	169–173 cm	155–161 cm
Large	greater than 173 cm	greater than 161 cm

Women attain their maximal height essentially by the age of 16 to 18 years, whereas men grow until their 20th year. Attainment of maximal height is followed by a period of 20–30 years in which B.h. hardly changes. At a more advanced age, B.h. decreases markedly by about 3%. All races display considerable individual differences of B.h., and group averages are very different in all parts of the world. The majority of individuals, however, are of intermediate or somewhat less than intermediate B.h. Tall individuals are more common in northern Europe, northern India, central and southern Africa, in the middle of North America and in the south of South America, and among some Pacific Islanders. Small races occur in northern Europe, northern Asia and northern North America, in Southeast Asia, Central America and central and southern Africa. Pygmy races, in which the B.h. of men is less than 150 cm, are found only in central Africa, the Andaman Islands of the Indian Ocean, the Philippines, New Guinea and the Admiralty Islands.

During the evolutionary history of *Homo sapiens,* there appear to have been several periods in which B.h. has generally increased or decreased over large areas of the world, independently affecting several races. The average B.h. for many European countries, as well as other parts of the world, currently appears to be undergoing a marked increase. See Acceleration.

Body lice: see *Phthiraptera.*

Body mass: see Body weight.

Body temperature: temperature of the interior of the animal body. In homeothermic animals there is usually a temperature gradient between the body interior and its periphery, so that B.t. is not uniform. A distinction must be made between the constantly maintained interior body temperature and the temperature of the peripheral tissues which varies with the environmental temperature. Only the interior temperature can be considered as a fixed characteristic of the organism, and this approximates to the rectal temperature.

Although relatively constant, the interior temperature does display daily variations, whose periodicity and magnitude are species-specific. In nocturnal animals, B.t. is highest at night, whereas diurnal animals show their highest B.t. during the day (Fig.). The daily periodicity is unaffected by environmental temperature.

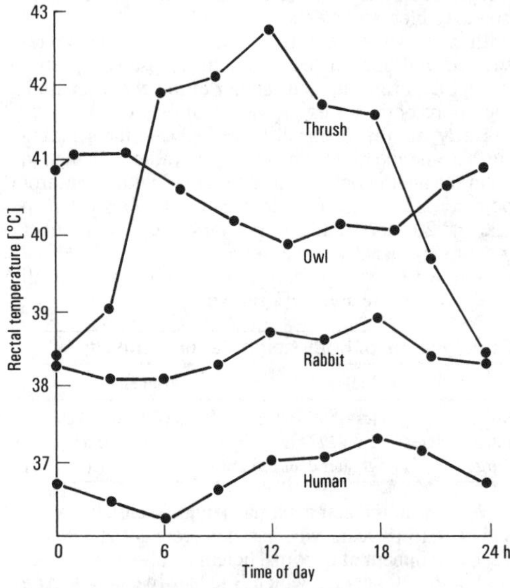

Body temperature. Daily variations in body temperature.

Characteristic changes in B.t. also occur during the human ovarian cycle. It gradually decreases during menstruation and remains low until rupture of the follicle; it then inceases rapidly by about 0.5 °C and remains almost unchanged until onset of the next menstruation. This phenomenon is used to monitor the fertile days during the female cycle, and for the early detection of pregnancy.

During physical work, B.t. increases according to the degree of exertion. In extreme cases in the human this increase may be as much as 2 °C.

See Temperature regulation of the body.

Body temperature, regulation of: see Temperature regulation of the body.

Body weight, ***body mass:*** the total weight of the animal body. It depends on age, height, sex, nutritional state, hormone action and genetic composition. For scientific purposes, the weight of the human body must be determined without clothing and under known conditions (time of day, data on food intake, etc.). The average B.w. of adult Europeans is about 65 kg (men) and 52 kg (women), with wide individual variations. B.w. is also subject to average daily variations of about 2 kg, with an early morning minimum and a late evening maximum. In the first two days after birth, the B.w. of the human neonate decreases by 5%, and the original birthweight is not re-attained until about the 10th day after birth. At the age of 1 year, B.w. is about three times the birth weight, but the rate of increase of B.w. is slower in the second year of life. During the subsequent growth period, B.w. does not increase at a constant rate, but displays several changes related to increases in body height and sexual maturation. Maximal B.w. is attained in about the 50th year.

Bog (plate 47): a region with peat-forming vegetation in a constantly damp to wet location. The water of a B. is highly acidic and deficient in electrolytes, causing a drastic limitation of the flora and fauna capable of colonizing such an environment, and the strict selection of typical B. inhabitants. B. water is also usually very brown due to the presence of humins.

If peat formation occurs near to the ground water, e.g. over marshy mineral soils or over a silted up lake, the product is a *flat B.* with a level surface. Depending on the composition of the ground water, flat Bs. are more or less nutrient-rich and carry a species-rich vegetation, especially of herbaceous plants and bushes that are adapted to a wet soil. According to the vegetation, Bs. are classified as reed, sedge, meadow or forest B. (e.g. alder carr). In regions of high rainfall, permanently wet locations may also be invaded by *Sphagnum* species, which store large quantities of water and form successive layers of growth, gradually raising the surface level of the B. Since rainwater and (if need be) precipitated atmospheric dust provide sufficient nutrients for *Sphagnum,* this moss grows independently of the ground water. It can overgrow older vegetation, even trees, forming a very nutrient-poor, treeless or tree-poor *raised B.*, which arches up in a dome shape above its surroundings. The central surface of a raised B. is usually surrounded by a fairly steep, relatively well drained, often wooded margin, which in turn is surrounded by a lower lying B. or *lagg* (a marginal fen receiving water from the raised B. and its sourroundings). The actual raised B. usually consists of a characteristic mosaic of small hummocks and wet depressions. The hummocks are often colonized by members of the *Ericaceae* and *Cyperaceae,* as well as *Drosera* species.

Only flowering plants with low nutrient requirements (i.e. like *Sphagnum)* and with mechanisms for decreasing transpiration (xeromorphic characters) are capable of colonizing a raised B. Alternatively, some plants (e.g. *Drosera)* may acquire additional nutrients by trapping insects (see Carnivorous plants).

Growth of peat Bs. occurs mainly in the wet depressions. In the course of B. development, hummocks and depressions alternate with each other; the depressions fill with vegetation and rise above the old, non-growing hummocks, so that the latter in turn become wet depressions. At times of low atmospheric precipitation, the ericaceous vegetation of a raised B. quickly becomes dominant, so that the B. becomes a heath or moor, and may also develop tree cover (this stage is called an intermediate B.). Raised Bs. that depend only on aerial precipitation are also known as true or ombrogenic raised Bs., as distinct from vegetatively similar Bs. that do not rise above their surroundings, and which are derived from silted up, nutrient-poor lakes. The latter hardly rise above the level of the former lake, and they occur in regions of relatively low atmospheric precipitation. In humid oceanic climates (e.g. north and west Britain), *Sphagnum* Bs. sometimes form a continuous cover over the entire landscape. Known as *blanket Bs.*, these are often dominated by *Eriophorum vaginatum*.

Bog mosses: see *Musci*.

Bog myrtle: see *Myricaceae*.

Bog turtle: see *Emydidae*.

Bohr effect (after the Danish physiologist, Christian Bohr): a reversible shift in the O_2-binding curve of hemoglobin, which permits O_2-binding and CO_2-release in alveolar capillaries of the lungs, and the reverse process in respiring tissues. Carbonic anhydrase in erythrocytes promotes the rapid formation of carbonic acid from dissolved CO_2 in respiring tissue, and each molecule of H_2CO_3 spontaneously dissociates into bicarbonate and a proton. The protons are absorbed by deoxyhemoglobin, which thereby acts as a buffer. Proton binding by hemoglobin decreases its affinitiy for O_2, so that 4 O_2 molecules are lost for every 2 protons bound. Also, about 15% of CO_2 carried in the blood is bound to hemoglobin as carbamino groups. The process is reversed in the lungs, where O_2 binds to deoxyhemoglobin with release of protons, i.e. binding of O_2 promotes exhalation of CO_2 from the lungs, whereas binding of protons promotes release of O_2 in respiring tissue. See Oxygen binding by blood, Blood gases.

Boidae: a family of mostly large and powerful snakes comprising about 65 species of ***Boas***, ***Pythons*** and ***Anacondas***. Both jaws carry numerous long teeth, but no venom glands. All *B.* are nonvenomous, and they kill their prey by constriction. The palatomaxillary arch is movably attached to the rest of the skull by elastic ligaments. This enables the *B.* to swallow whole carcasses of rodents, wild pigs, young deer and antelopes, as well as domestic animals, which are considerably wider than their own heads. The presence of vestiges of a pelvic girdle (mostly consisting of 3 bony elements) is characteristic of this family, which is a prototype of all modern living snakes. Rudimentary hindlimbs are often visible as claw-like spurs, one on each side and just forward of the cloaca; these can still be actively moved, and play a tactile role in mating, at least in the male. The presence of paired lungs is a further indication of the primitive nature of the *B.* In the more highly evolved snakes, both the pelvic girdle and the left lung have completely disappeared. Another primitive feature of the *B.* is the presence of the coronoid bone in the lower jaw. This bone is always present in lizards, where it is important in locking the other bones together to produce a rigid jaw. In snakes, the coronoid bone has lost its function, and in most it has disappeared entirely, but vestiges persist in the *B.* On the basis of these and other anatomical characteristics, snakes classified as *B.* include not only the large tree, water and land-dwelling forms, but also smaller cryptozoic species with a semi-burrowing existence (see Sand boas). The longest member of the *B.* on record is an anaconda of 1 128 cm. Only 6 species attain lengths greater than 5 m. A maximum of 7 subfamilies has been described, but the classifications are not universally agreed.

Pythons are placed in the separate subfamily, *Pythoninae*, whose members possess supraorbital (postfrontal) bones, premaxillary teeth, and double rows of enlarged ventral caudal scales (gastrosteges). They are the typical *B.* of the old world, and they lay eggs, which are protected and incubated by the female. Examples are the Reticulated python (see), Rock python (see) and the Asiatic rock python (see).

Boinae or ***Boas*** lack the supraorbital bones and premaxillary teeth, and their caudal gastrosteges are arranged in single rows. All are viviparous. Examples are the Anaconda (see), Boa constrictor (see), Amazonian tree boa (see) and the Cuban and Rainbow boas (see).

The subfamily *Bolyerinae* contains only 2 genera, *Bolyerinae* and *Casarea*, which are found only on Round Island off the coast of Mauritius near Madagascar. A rudimentary pelvic girdle and hindlimbs are absent.

The subfamily *Tropidophiinae* contains 4 genera of ***Dwarf boas*** from Central America, northern South America and the West Indies. They have almost lost the left lung, and the female of *Ungaliophus* lacks rudimentary hindlimbs. Best known are the ***Wood snakes*** (*Tropidophis*), which respond to threat by producing a foul smelling anal secretion and exuding blood from their eyes. The *Bolyerinae* and *Tropidophiinae* exhibit a greater number of higher snake features than other boids.

The subfamily *Erycinae* contains 4 genera, *Charina* (***Rubber boa*** of north-west America), *Lichanura* (***Rosy boa*** of north-west America), *Eryx* (see Sand boas) and *Angyrus* (***Pacific boa*** of the East Indies).

The subfamily *Loxoceminae* contains a single primitive burrowing species, *Loxocemus bicolor*, of southern Mexico and Central America. The left lung is large and vestiges of the pelvic girdle are present.

The mainly subterranean ***Calabar rock python*** (*Calabaria reinhardtii*), no more than 1 m long, is placed in a separate subfamily, the *Calabariinae*.

Boiga: see Tree snakes.

Boiginae, ***boigine snakes, boigine vipers:*** a large and varied subfamily of the *Colubridae* (see), containing tree- and ground-dwelling, mostly egg-laying species. All *B.* possess venom glands and opisthoglyphic dentition (see Snakes). They are, however, only weakly venomous, with the exception of the Boomslang (see) and the Twig snake (see), whose bite can be fatal to humans. The subfamily is represented in America (see Banana snakes, Mussurano), Africa (Boomslang, Twig snake), Europe (see Common European grass snake, European smooth snake, *Malpolon monspessulanus*) and the whole of South and Southeast Asia (see Long-nosed tree snake, Tree snakes, Golden snakes).

Bojanus' organs: the excretory organs or kidneys of the *Bivalvia* (clams, oysters, mussels, etc.), discovered by Bojanus (1771–1827). They are situated dorsally immediately below the pericardial cavity, and they are normally paired (one on each side). Each has the structure of a tube bent back on itself, the two arms lying parallel, one above the other. The glandular-walled lower arm is the functional kidney, and it connects at its anterior end with the pericardial cavity. Posteriorly it is continuous with the thin-walled upper arm *(bladder)*, which serves as a duct opening anteriorly into the mantle cavity.

Boletus: see *Basidiomycetes (Agaricales)*.

Bölling period (after Lake Bölling in North Jutland): a period of vegetation development in the late Ice Age. Extending from about 10 750 to 10 350 BC, the B.p. represents the first brief postglacial temperature fluctuation. It led to the development of open park tundra with stands of birch and pine. See Pollen analysis.

Bolting: in plants, the sudden transition of compact or rosette growth to elongate growth, usually associated with flowering. This is due mainly to the stretching or extension growth of cells, sometimes with a contribution from cell division, e.g. in the B. of grasses in the intercalary meristems of the nodes. In winter cereals and other plants, B. will not occur until the plant has been subjected to cold winter temperatures (see Vernalization). In a greenhouse with a constant warm temperature, the same plants can be kept permanently in the rosette habit for several years. In other plants, B. depends on the length of daylight, e.g. in certain long day plants (see Photoperiodism). Under conditions that normally do not lead to B., a number of plants can be induced to bolt by treatment with gibberellins, and usually also made to flower. Gibberellins therefore appear to be critically involved in the induction of B.

Bombacaceae, ***silk-cotton family:*** a family of dicotyledonous plants comprising about 225 species in 31 genera. All are trees, with simple or palmate leaves, and hermaphroditic, 5-merous flowers, which are usually large and vividly colored; the 5 stamens are often fused into a tube. In addition to the calyx and corolla, the whole flower is subtended by a further whorl of sepal-like appendages, called the epicalyx. There are 2–5 fused, superior carpels, with many locules, each containing 2 to numerous erect ovules. Pollination is by insects or birds, and sometimes by bats and other animals. The fruit is a dry or fleshy capsule. The family is restricted to tropical regions, especially South America, with a few representatives in Southeast Asia and Australia, and some unusual ones in Africa and Madagascar. Many species shed their leaves at the end of the rainy season, i.e. they are deciduous. Often the flowers of deciduous species appear after the leaves have been shed, e.g. in the African species, *Bombax costatum* and *Bombax buonopozense*.

One of the best known representatives is *Adansonia digitata* ***(Baobob, Monkey Bread Tree, Monkey Tamarind),*** a curiosity of the African landscape. It is found in dry, sandy habitats throughout tropical and southern Africa, rapidly diminishing in frequency south of a line joining the Olifants River and the Zoutpansberg. It attains heights of 15–21 m, and its massive bole (25–43 m circumference) appears disproportionately large and swollen in comparison with the small crown. Leaves are shed at the end of the rainy season. In addition to insects and birds, pollinators probably include bats and bush babies. The cucumber-shaped fruit is about 30 cm in length, and woody when ripe; it contains a white pulp and oily seeds, both of which are edible.

Fig.1. *Bombacaceae.* Baobab tree *(Adansonia digitata).*

Ceiba pentandra ***(Kapok Tree, Silk-cotton Tree, Kapok Ceiba)*** is a characteristic West African tree (thought by some authorities to be an ancient introduction from South America), possessing a thick, buttressed trunk with thorns, and attaining heights greater than 49 m. It is very common where the natural vegetation has been disturbed, e.g. near to settlements and in regenerating secondary forest. Its fruits contain a white seed wool or kapok, which is harvested as a filling for pillows, saddles and upholstery. The edible seeds are used in soup, and the seed oil can be burned in lamps. The trunk is often used for lightweight dugout canoes.

Fig.2. *Bombacaceae.* Kapok tree *(Ceiba pentandra);* branch with leaves and fruit.

Bombax malabaricum (= Bombax ceiba = Salmalia malabarica) ***(Red Silk-cotton Tree, Semul Tree, Malabar Simal Tree, Simal)*** is native in northern Australia and southern Asia (Indochina, Sri Lanka, India, Thailand, Burma), ascending to about 1 220 m in the Himalayas. Its fruit also contains a silky kapok, which is used for soft upholstery. The wood is also useful for light construction work, matchsticks and dugout canoes.

Durio zibethinus ***(Durian)*** of Malaya and Indonesia is cultivated for its exquisitely flavoured fruit. The aril of the fruit, however, has an offensive odor, which attracts various wild animals. The flowers frequently emerge directly from the base of the trunk.

The South American *Ochroma lagopus (= O. pyramidale)* ***(Balsa Tree)*** is cultivated in the tropics for its light wood (density 0.12–0.30), which has many technical applications, and was formerly used in aircraft construction.

Bombesin: a peptide from frog skin: Pyr-Glu-Arg-Leu-Gly-Asn-Glu-Trp-Ala-Val-Gly-His-Leu-Met-NH_2. It stimulates the release of gastric, pancreatic and adenohypophyseal hormones in mammals, and causes contraction of smooth muscle in the gastric and urinary tracts, and in the uterus. When injected peripherally, B. inhibits the secretion of growth hormone. It also affects body heat production and decreases blood pressure. B.

displays homology with a number of peptides found in mammalian tissues, including the hypothalamus; for example, 9 of the 10 carboxy-terminal amino acids of B. are identical to the 10 carboxy-terminal amino acids of porcine gastric releasing hormone. Similar properties are exhibited by alytensin, litorin and ranatensin.

Bombinae: see Bumble bees.

Bombix: see *Bombycidae.*

Bombus: see Bumble bees.

Bombycidae, ***silkworm moths:*** a lepidopteran family distributed mainly in East and Southeast Asia, whose larvae *(silkworms)* spin a silk pupation cocoon. A white domesticated East Asian species *(Bombyx mori* L.) has been used in China since antiquity for the production of silk. There are no known wild populations, and the wild progenitor is thought to be *Bombyx mandarina* Moore (China, Formosa, Korea, Japan). The closely related *B. huttoni* Moore of northern India may be more correctly regarded as a subspecies of *B. mandarina.* The secret of silkworm culture (earliest record 2 630 BC) was smuggled out of China in 555 AD and reached Europe via India and Turkestan. The larva (caterpillar) normally feeds on black or white mulberry *(Morus),* but it will also feed on osage orange *(Maclura)* and lettuce *(Lactuca).* In Europe, eggs (600–800 from one female) are produced in special breeding centers, stored from Autumn to May under refrigeration, then distributed to silk farmers. The caterpillars are cultured in racks of wooden trays in well ventilated rooms. The culture of 1 g eggs (about 1 500 eggs) requires a surface area of 3 m^2 and about 40 kg of mulberry leaves. After 35 days (average) the caterpillars begin to spin. Between1 000 and 4 000 m of silk thread can be unwound from a single cocoon.

In Asia, species of the family *Saturniidae* (emperor moths) are also used to a small extent for silk production; members of the genus, *Antheraea,* are known as tussah, tussur or tussore silk-moths, or oak silk-moths, and the silk is known commercially as wild silk or tussah silk. The product from the cocoons of *Antheraea pernyi* is also known as Shantung silk. *Antheraea yamamai* (Japanese oak silk moth) is also a source of tussah silk.

Bombycilla: see *Bombycillidae.*

Bombycillidae: a small (8 species) and rather heterogeneous avian family of the Northern Hemisphere, belonging to the order *Passeriformes* (see). They were formerly known as the *Ampelidae.* All are fruit-eating arboreal birds, 15–23 cm in length, with rather broad bills, short legs and 10 primaries. Their affinities are uncertain. The 4 Central American ***Silky flycatchers*** *(Phainopepla nitens, Phainoptila melanoxantha,* and 2 species of *Ptilogonys),* constitute the subfamily, *Ptilogonatinae.* The 3 species of ***Waxwings*** that constitute the subfamily *Bombycillinae* are sleek and strongly crested, with smooth velvety, fawn-brown plumage, with a band of yellow or red at the end of the tail: *Bombycilla garrulus* ***(Bohemian waxwing*** or ***Waxwing;*** conifer and birch forests of northern Eurasia and western North America); *B. cedrorum* ***(Cedar waxwing;*** woodlands of temperate North America); *B. japonica* ***(Japanese waxwing;*** taiga forests of eastern Siberia). Their name is derived from the presence of bright red waxy material (absent in the Japanese waxwing) on the tips of the secondaries. The third subfamily, *Hypocoliinae,* contains a single monotypic species, ***Hypocolius*** *(Hypocolius ampelinus),* a gray bird native to the valleys of the Tigris and Euphrates, which migrates irregularly to Northwest India and occasionally to Northeast Africa.

In some classifications, the ***Palm chat*** *(Dulus dominicus)* is included in the *Bombycillidae,* as the sole member of the subfamily, *Dulinae.* It is more usual, however, to assign the Palm chat to its own monotypic family, *Dulidae.* It is a gregarious, tree-living bird, inhabiting open country with palm trees on the Island of Hispaniola and Gonave Island in the West Indies.

Bombycillinae: see *Bombycillidae.*

Bombykol, ***10-trans-12-cis-hexadecadienol-(1):*** a pheromone exuded by female silk worm moths *(Bombyx mori)* to attract males and predispose them to copulation. It was the first pheromone to be structurally elucidated (Butenandt et al. 1959, using 6 mg of B. isolated from the abdominal glands of 500 000 female moths). The configuration was established by comparison of the biological activities of synthetic compounds.

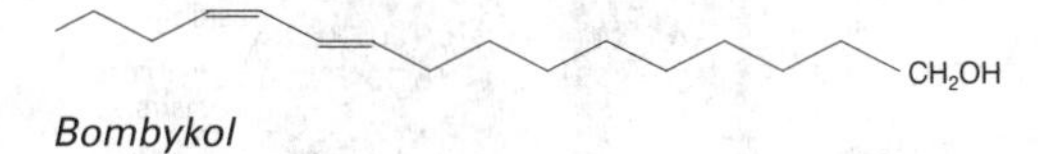

Bombykol

Bone beds: small rock formations consisting of bone breccia in many geological systems (e.g. Triassic, Liassic). They were formed in shallow flowing water, and they consist mostly of bone fragments, teeth, scales and coproliths of fishes and reptiles.

Bone formation, ***osteogenesis, ossification:*** formation of bone tissue by the mineralization of connective tissue. *Osteoblasts (chondroblasts)* (which arise from embryonal connective tissue) secrete intercellular ground substance and collagen fibers. They also secrete calcium salts which harden the intercellular matrix, thereby sealing themselves inside the tissue so that they become *osteocytes (chondrocytes).* Bone can be formed by one of two different processes.

1) Desmal B.f. produces membrane bones, which are derived directly from connective tissue. Connective tissue cells aggregate into a delicate fibrous matrix; differentiated osteoblasts then secrete a three dimensional lattice of small interconnected bars of calcified matrix. The cells remain connected via canaliculi, and the bone has a spongy appearance in cross section.

2) Chondral B.f. produces cartilage bones by the ossification of existing cartilages. Most parts of the vertebrate endoskeleton are laid down as cartilage primordia within the embryo, then later converted and enlarged into cartilage bones by ossification. The process occurs on both the surface and in the interior of the cartilage, giving rise to *perichondral* and *endochondral* bone, respectively. Long bones are formed as follows. Calcified cartilage is invaded by large, multinucleate ameboid cells called *osteoclasts* or *chondroclasts.* The osteoclasts erode the calcified cartilage from the center outward, forming channels for blood vessels and for mesenchyme cells that follow in their wake. Certain of the invading mesenchyme cells (osteoblasts) then secrete calcareous bars (trabeculae), which lie mainly parallel to the long axis of the future bone. As this endochondrial ossification spreads outward toward the periosteum, a central cavity remains to house the marrow. Meanwhile, osteoblasts immediately below the periosteum begin to form a dense layer of bone (perichondral bone) around the more spongy endochondral ossification.

In mammals, ossification zones also appear in the cartilaginous epiphyses of long bones. Between the epiphysis and shaft (diaphysis) there is a narrow, persistent band of cartilage ("epiphyseal joint"). By intercalation of new bone tissue on both sides of this cartilaginous band, the bone is able to grow in length. Growth continues until the end of adolescence, and the maximal length at this time determines the height of the adult skeleton. The epiphyseal joint then becomes ossified, forming a firm bond between the epiphysis and diaphysis.

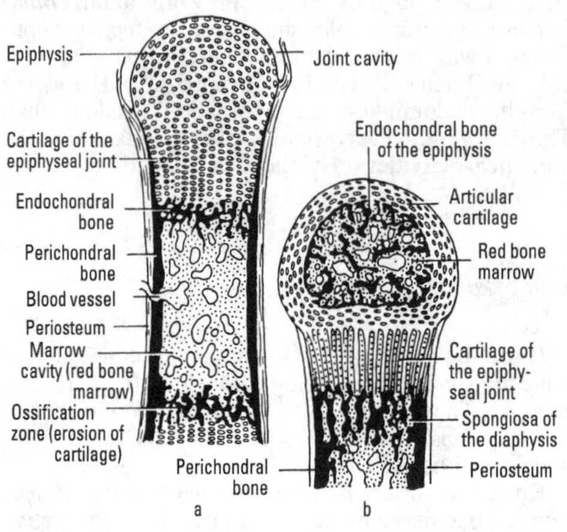

Bone formation. Ossification of a long bone of a mammal (longitudinal section). *a* Endochondral ossification and formation of the vascularized marrow cavity. *b* Ossification of the epiphysis and growth zone ("epiphyseal joint").

Bonellia viridis: see Green boniella.

Bones, ***ossa:*** hard, elastic organs of vertebrates composed of ossified mechanical tissue, which in their totality comprise the skeleton or skeletal system of the body.

1) According to their origin (see Bone formation), B. are classified as *membrane B.* (which arise from specialized connective tissue cells) or as *cartilage B.* (formed by the ossification of embryonic cartilages). Most of the B. of advanced vertebrates are cartilage B. In humans the only membrane B. are the clavicle and certain B. of the cranium.

The particular strength and hardness of bone tissue is due to the chemical composition and complex architecture of its abundant intercellular matrix, which consists of bone minerals (85% calcium phosphate) with embedded bundles of collagen fibrils (the glue prepared from B. by boiling is derived from this collagen). The approximate composition of bone tissue is: organic components 30%, inorganic components 45%, water 25%. The hardness and stability of the tissue are due to the minerals, while elasticity is conferred by the collagen fibers. Enclosed in the intercellular matrix are plum stone-shaped bone cells *(osteocytes* or *chondrocytes).* Osteocytes lie in small cavities or *lacunae* which communicate with neighboring lacunae by fine *caniculi.* Each osteocyte is in contact with its neighbors via numerous branched cellular processes within the caniculi. Thus, the matrix is penetrated in all directions by a protoplasmic network. In lamellar B. (see below) the matrix is also perforated by fine canals *(Haversian canals),* which join with each other, and generally lie more or less parallel to the long axis of the bone. Each Haversian canal contains a small artery and vein, which supply nourishment to the osteocytes, and are ultimately connected with the blood vessels of the periosteum and bone marrow.

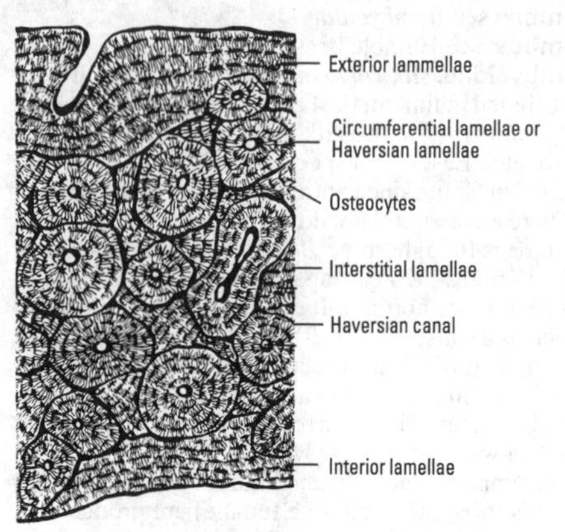

Fig.1. *Bones.* Structure of lamellar bone (section of the compact material of a long bone).

2) According to the arrangement of the intercellular matrix, two types of B. are recognized. a) In *fibrous B.*, the collagen fibers form an irregular network, with osteocytes contained in lacunae. b) In *lamellar B.*, the intercellular matrix forms concentric cylinders *(bone lamellae)* around the Haversian canals, and the collagen fibers are ordered and parallel. There are two types of lamellar bone: dense *cortical bone* found in the shafts of long bones, and the more porous *trabecular* (or *cancellous) bone* at the end of long bones.

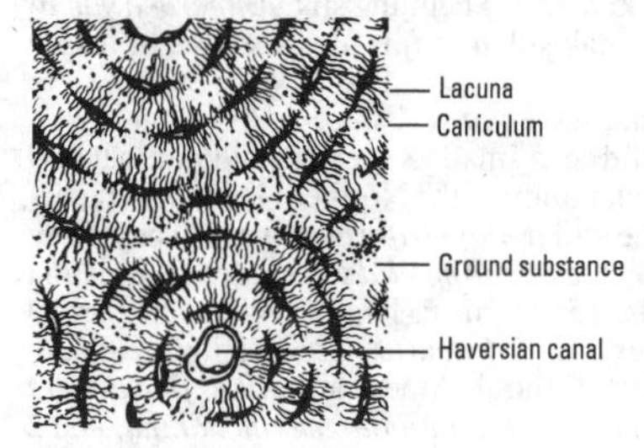

Fig.2. *Bones.* Transverse section of lamellar bone (dried ground section).

All bony skeletal elements arise as fibrous B. In some primitive groups they remain in this form throughout life. In most vertebrates, however, fibrous B. are converted into lamellar B. by complex restructuring processes. In the hollow B. of birds, the collagen fibrils are mostly arranged parallel to the long axis, but without lamellar layering. In the adult human, fibrous B. remain at only a few sites, e.g. in cranial sutures and in the bony labyrinth capsule. B. are surrounded on the outside by a fibrous, vascularized, innervated membrane called the *periosteum.* The inner cavity of the shaft of long B. contains the *bone marrow.* Fully formed B. are not rigid, lead structures; on the contrary, they are metabolically active, and like other living tissues they are in a state of continual degradation and renewal. They may become restructured in response to external forces, and they are capable of self-repair when broken. See Bone formation.

Bone tissue: see Bones.

Bongo: see *Tragelaphinae.*

Bonitation: a measure of the ecological success of a species, scored on a scale of I to V, in which V represents the poorest performance. In plant sociology, it is used to indicate the vitality of trees according to height and age.

Bonito, ***Atlantic bonito,*** *Sarda sarda:* a fish of the family *Scombridae* (mackerels and tunas) (order *Perciformes),* found off the eastern and western shores of the Atlantic, in the North Sea, the Meditteranean, and sometimes in the Black Sea. It is a fast swimmer (often following ships), with a deeply forked tail and fusiform body. Very characteristic are two series of small fins (finlets), dorsally between the dorsal and caudal fin, and ventrally between the anal and caudal fin. A swim bladder is lacking. The light-colored body is patterned with 7–9 oblique dark stripes on the hind part of the body. It feeds on fish, such as anchovies, pilchards, mackerel and flying fish. The Atlantic bonito is caught for its tasty meat, and is economically the most important of the 6 known species of bonito.

Bonobo: see Chimpanzees.

Bontebok: see *Alcelaphinae.*

Bony fishes: see *Osteichthyes.*

Bony labyrinth: see Auditory organ.

Bonytongues: see *Osteoglossidae.*

Boobies: see *Pelicaniformes.*

Booklice: see *Psocoptera.*

Book lungs, ***lung books:*** paired respiratory organs, which are present in many terrestrial arachnids. They open onto the ventral surface of the 2nd and 3rd [*Uropygi* (whip scorpions) and *Araneae* (spiders)] or the 4th to the 7th [*Scorpiones* (scorpions)] segments of the opisthosoma (abdomen). Some species possess both B.l. and tracheae. B.l. arise in close association with the embryonal limb primordia of the abdomen. Thus, in all species that possess B.l., limb primordia are seen on the anterior abdominal segments during development of the embryo; lamellae develop on the posterior borders of these primordia, and at the same time a cavity is formed by invagination, so that the limb primordium becomes part of the floor of the cavity. Thus each lung is a cuticularized pocket in the ventral abdominal wall. On one side, the wall of the pocket is folded into a series of parallel, flat, leaflike lamellae, lying side by side like the pages of a book, and kept apart by cuticular bars, thus ensuring a free circulation of air. The other side of the pocket forms an air chamber *(atrium),* which is continuous with the interlamellar spaces, and which opens to the ventral abdominal surface via a slit-shaped *spiracle (stigma* or *pneumostome).* Respiratory gases are exchanged by diffusion between the blood circulating within the lamellae and the air of the interlamellar spaces. Blood to the lamellae is supplied by a ventral blood sinus, connected directly to the heart via a lung vein. A muscle attached to the dorsal side of the atrium can dilate the atrium and open the spiracle, thereby encouraging air movement, but gas movement in and out of the lung occurs largely by diffusion. The lamellae themseves lack any kind of musculature. B.l. are thought to have evolved by the sinking or invagination of Gill books (see), whereby the appendages or plates disappear or become part of the floor of the lung cavity, and the lamellae or leaflets appear as folds of the lining. See Tracheae, Gill books, Tracheal lungs.

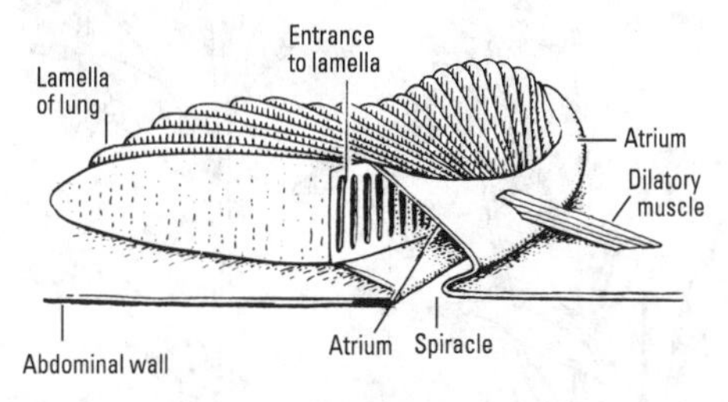

Book lungs. Schematic drawing of a single booklung.

Boomer: see *Aplodontidae.*

Booming ground: see *Tetraonidae.*

Boomslang, *Dispholidus typus:* a green or brown snake (family *Coluibridae,* subfamily *Boiginae)* found on the African savanna. Although it is opisthoglyphic, its grooved venomous teeth are not set as far back in the upper jaw as in most other opisthoglyphic species. Its bite can therefore penetrate human skin and deliver venom into the wound. The B. and the opisthoglyphic twig snake (whose venomous teeth are also placed further forward) are the only two members of the *Boiginae* known to have killed humans. The B. feeds on birds' eggs, and also takes frogs and chameleons. It attains a length of about 1.8 m (2 m maximum), and its scales are very narrow like those of the mambas.

Booted eagles: see *Falconiformes.*

Bootlace worm, *Ligula intestinalis:* a common tapeworm (up to 75 cm in length), order *Pseudophyllidea,* class *Cestoda* (see), which parasitizes the alimentary tract of many diving and wading birds in Europe, but is known only in mergansers *(Mergus)* in North America. The last larval stage is a wide, tape-like plerocercoid, up to 40 cm in length, which lives in the body cavity of carp, pike and perch. It closely resembles the sexually mature form, but its anterior end is blunter, the bothridial grooves are less pronounced, and segmentation is lacking.

Boraginaceae, ***borage family:*** a dicotyledonous plant family of about 2 000 species, mostly native in the temperate zone of the Northern Hemisphere, and especially abundant in the Mediterranean region, Central and East Asia and the Pacific States of North America. They are usually herbaceous plants, sometimes shrubs or trees, mostly hispid (coarsely and stiffly hairy) with alternate, undivided leaves. The flowers are regular, mostly with a 5-lobed, funnel-shaped or campanulate corolla, and arranged in a cymose inflorescence. Pollination is by insects, and the throat of the corolla tube often has scales or hairs. The superior ovary is 2-celled (or 4-celled due to false sepeta), and it develops into a fruit of 4 spiny nutlets, rarely into a drupe. Members of the B. contain pyrrolizidine alkaloids.

Boraginaceae. a Myosotis arvensis (common forget-me-not). *b Pulmonaria longifolia* (lungwort). *c Borago officinalis* (borage). *d Symphytum officinale* (comfrey).

Common borage** (Borago officinalis),* which originated from the Mediterranean region, is now cultivated in America and Europe; its leaves have a cucumber-like taste, and it is useful as a seasoning plant and as a forage plant for bees. Most ornamental B. species are blue, and they include ***Forget-me-not** (Myosotis)* and ***Navelseed or ***Blue-eyed Mary** (Omphalodes).* ***Lungwort*** (*Pulmonaria*) is a former medicinal plant; its flowers are initially pink, turning blue-violet, due to an increase of the pH of the cell sap. ***Prickly comfrey** (Symphytum asperum)* was originally cultivated from Iran to Central Asia as a green fodder plant. In Western Europe, a hybrid *(Symphytum* x *uplandicum)* is used for this purpose.

Bordatella: a genus of very small (up to 1 µm-long), variously shaped (bacillary to oval) bacteria, which live parasitically and cause diseases of the respiratory tract in humans and other mammals. They are nonmotile, or motile with several flagella, aerobic and Gram-negative. *B. pertussis* causes whooping cough, and was first isolated by the Belgian, J Bordet.

Border community, ***border biocoenosis:*** see Biocoenosis.

Bordered white, ***pine looper***, *Bupalus piniarius* L.: a European geometrid moth, wingspan 30 to 40 mm, which displays sexual dimorphism. Males are black and brown with light markings, whereas females are light reddish brown with indistinct markings. The caterpillars are green with three white dorsal stripes and a yellow lateral stripe; they feed on pines, where they are usually found in late summer; the pupa overwinters. This species typically undergoes periodic large increases in population, when it causes considerable damage.

Boreal period (Latin *borealis* northerly): a postglacial climatic and vegetation period, which occurred between 6 800 and 5 500 BC. It is characterized by the rapid spread of hazel and the first appearance of mixed oak woods, with oak, elm, lime and alder among the still predominant pine and birch. The climate was continental and becoming increasingly warmer. See Pollen analysis.

Boreidae: see *Mecoptera.*

Boreo-alpine distribution: the Disjunct distribution (see) of plants and animals in the conifer forests of the north and more southerly montane forests. It arose much later than the similar Arctic-alpine disjunction, because forest development was possible only after the end of the Ice Age. Boreo-alpine flora and fauna consist mainly of elements that advanced westward from a Siberian or Mongolian Refugium (see). Species displaying a B-a.d. are often found in cold nonmountainous areas, such as high moors. In early animal-geographical studies, the term, Boreo-alpine, was also applied to Arctic-alpine animals. Furthermore, the term is widely used for the Paleoarctic, where the Boreo-alpine type of distribution is very common.

Boring thrips: see *Thysanoptera.*

Borneol: a crystallizable monoterpene alcohol, with a camphor-like odor, which occurs naturally as both the dextrorotatory (borneo camphor, D(+)-borneol) and levorotatory (Ngai camphor, L(-)-borneol) form. Both are present in numerous essential oils and resins, and the (+)-form usually predominates. A particularly abundant source of borneo camphor is *Dryobalanops camphora (= D. aromatica* Gaertn.) (family *Dipterocarpaceae),* a tree growing in Sumatra and Borneo. The levorotatory isomer is obtained from valerian oil and from *Blumea balsamifera* L. *(Compositae).*

D-bornyl acetate (borneol acetate) is present in various gymnosperms, and is particularly abundant in Siberian spruce oil. The male American cockroach *(Periplaneta americana)* is sexually excited by D-bornyl acetate (at a concentration of 0.07 mg/cm^2), which appears to have the same effect as the natural female pheromone; the L-isomer has only on hundredth of the activity of the D-isomer.

L(-)-Borneol

Bornyl acetate: see Borneol.

Boron, *B:* a nonmetallic element, and an essential micronutrient for plants. It is taken up by the roots as the borate anion, and this uptake is inhibited by the presence of excess calcium. Within the plant, there is also an interesting and incompletely understood interaction of calcium and borate. Both B and calcium are required for meristem growth and for normal root development. The B content of plants is generally in the range 5–60 mg/kg dry mass. The physiological activity of B differs fundamentally from that of the other micronutrients, but it bears a certain similarity to that of phosphorus. B is an indispensable structural component of plants, and fine, differentiated cell wall structure cannot develop without the prior incorporation of B. Carbohydrate metabolism also displays a striking dependence on the supply of B, which promotes sugar transport and assimilatory processes. The influence of B on the formation of differentiated structures and on the distribution of carbohydrates is undoubtedly associated with its other known requirements for e.g. pollen germination, pollen tube growth, flowering and fruit setting, and its influence on the water economy of the plant.

B deficiency results in various symptoms, such as loss of color via gray-green to yellow, followed finally by leaf drop. The apical meristem dies and the root tips become necrotic. B deficiency symptoms are exacerbated by drought. The most familiar B deficiency disease is heart rot and dry rot of beet, in which the heart leaves become brown and die, and the beet putrifies.

Borrelia: see Spirochetes.

Bos: see *Bovinae.*

Boselaphini: see *Tragelaphinae.*

Boselaphus: see *Tragelaphinae.*

Bosmia: see *Branchiopoda.*

Bostrychoceras (Greek *bostrychos* hair curl, *keras* horn): a genus of ammonites in the family *Nostoceratidae.* The shell is spirally coiled to form a cone or turret; coiling continues in the U-shaped living chamber, but this extends freely and is no longer in contact with the coil below. The shell surface is decorated with simple but large ribs, and there is a single row of tubercles on each side of the venter. *B.* dispays the typical degenerative shell and rib structure of the later ammonites. It is found in the Upper Cretaceous, and certain species are important index fossils of the later strata.

Bostryx: see Flower, section 5) Inflorescence.

Botanical symbols: symbols or signs for expressions frequently used in the description of plants and their properties, in particular in gardening and horticulture. The following list contains the most commonly used B.ss.

♂	male
♀	female
⚥	hermaphrodite
×	hybrid (placed after the name of the genus, this sign indicates a species hybrid, before the genus a genus hybrid); it also means "crossed with"
+	granft or chimaera
⊙	annual
⊙	biennial
♃	perrenial herbaceous plant
♄	semi-shrub (only the lower parts are woody)
♄	shrub
♄	tree
∞	outdoor plant
∞	outdoor plant with winter protection
(K)	cool greenhouse plant
(W)	warm greenhous plant
	pot plant
○	sun plant
◐	semi-shade plant
●	shade plant
≋	aquatic plant
∼	water bank and marsh plant
	bog plant
△	rockery plant
	climbing plant
	hanging plant
	creeping plant
◔	spring flowering
	summer flowering
	autumn flowering
	winter flowering

Botany: the study of the structure and function of plants. Study of the dynamic behavior of plant systems and the associated causal analytical approach are becoming increasingly prominent in B. ***General B.*** deals primarily with the principles of structure and function that are generally applicable throughout the plant kingdom, or throughout large groups of plants, i.e. it concentrates on the unifying principles of structure and life processes. ***Systematics*** is concerned with variations from the basic pattern, the occurrence and distribution of specialized forms, and the extent to which the evolutionary relatedness of plants can be interpreted by studying the differences and similarities in their structure and function. An important branch of general B. is *morphology,* the study of plant structure; this is often subdivided into organography and anatomy. *Organography* (morphology in the narrow sense) deals with external plant organs, while *anatomy,* relying heavily on the microscope, deals with internal structure. Since *histology* (study of tissues) and *cytology* (study of the fine structure and function of cells) are concerned with internal structure, they are also part of anatomy. Studies in cytology are often pursued at a molecular level, so that they overlap and interface with molecular biology. Plant *genetics* is largely an independent botanical discipline, which, in addition to its central themes of replication and mutation, also makes considerable contributions to the understanding of morphogenesis and evolution; at the molecular level, plant genetics (and indeed the genetics of all other organisms) is identical with areas of modern molecular biology. A further area of general B. is represented by plant *Physiology* (see), which deals with functional processes in metabolism, morphogenesis (including growth and development) and movement, all of which are ultimately interpreted at the molecular level, i.e. once again there is an overlap or interfacing with modern molecular biology. *Plant biochemistry* or *Phytochemistry* is the study of the chemical constituents and metabolic processes of plants.

Systematics contains the subject of *Taxonomy* (see), which systematically describes, names and classifies the species of the plant kingdom (more than 500 000 species are known), and has as its goal the erection of a system that is based entirely on natural affinities. Systematics also embraces *Phylogeny* (see), which seeks to elucidate lines of evolution in the plant kingdom. This is aided by the study of evolution, which explores the causes of evolutionary separation and the basis of phylogenetic affinities, and seeks to establish principles governing evolutionary processes. An important contribution is also made by *Paleobotany* (see), which studies plants of earlier epochs of geological history. *Geobotany (Plant geography)* embraces plant distribution; structures of, and interactions within, plant assemblages (i.e. plant sociology); and the analysis of plant habitats. It also studies the histories of floras and plant communities, and seeks to elucidate the principles and causes of the spatial and temporal spread and association of plants on the Earth's surface.

Plant *ecology* is concerned with the relationship of plants and their environment. In this context, *autecology* deals with the adaptation of individual species to their environment, while *synecology* deals with the relationships of entire communities (e.g. plant societies) with environmental factors, e.g. climate and soil. Subjects such as Microbiology (see) or Mycology (see) are devoted to specific groups of organisms within the plant kingdom.

Certain fields of B. are concerned with the economic importance of plants, and are collectively referred to as ***applied B.*** These include branches of agriculture and medicine, such as horticulture, phytopathology and pharmacognosy.

Bothridium: see *Cestoda.*

Bothrium: see *Cestoda.*

Bothrops: see Lance-head snakes.

Botryoidal tissue: mesenchymatous tissue in the musculature of leeches, which is associated with the blood vessels. In particular, in the *Gnathobdellidae,* this tissue is pigmented and its cells are arranged end-to-end and contain intracellular capillaries. It is related to the chlorogogenous tissue of other annelids, and serves for the excretion of waste metabolic products. See Chlorogogenous cells.

Bottom yeasts: see Brewer's yeasts.

Botulism: a fatal form of food poisoning caused by infection of fish, meat, canned vegetables, sausages, paté, etc. with *Clostridium botulinum,* which is a Gram-positive, anaerobic, motile (peritrichous flagella) bacterium, 4–6 μm (maximally 9 μm) × 0.9–1.2 μm, with oval, subterminal spores. The toxins act mainly on the parasympathetic system (e.g. oculomotor paralysis, pharyngreal paralysis, aphonia, etc.). Six main types of *Cl. botulinum* have been differentiated on the basis of their antigenically distinct toxins: A, B, C, D, E and F. Type A toxin has been crystallized, and is arguably the most potent natural toxin known, the estimated lethal dose for mice being 33×10^{-12} g.

Boulengerochromis: see Cichlids.

Boundary tissue: all boundary layers that protect plant tissues externally or separate one tissue from another within the plant. Primary B.t. is derived from primary meristem, and secondary B.t. from secondary mersitem.

The most important B.t. of plants is the *epidermis,* which forms the protective external layer of shoots and roots, and mediates exchange of material with the environment. The epidermis is typically a monolayer of cells, containing leukoplasts rather than chloroplasts. In surface view, the epidermis is seen as an unbroken layer of interlocking cells with wavy or indented contors. Outer walls of epidermal cells are almost always thickened and covered with a delicate but firmly attached film of free cutin, called the *cuticle.* Cutin may also enter the interfibrillar capillary spaces of the network of cellulose fibrils in the outer epidermal cell wall, which is then said to be cutinized. The cuticle and cutinization render the epidermis rather impermeable to water. Wax may also be deposited in the cutinized layer, which then becomes especially impermeable to water. Wax sometimes extends beyond the cuticle, forming a light gray bloom which is easily wiped off.

The epidermis contains typical structures called Stomata (see), which permit the controlled exchange of gases between the plant and the atmosphere. Emergences (see) may also be present.

The outer surface of mature roots consists of a primary B.t. formed by suberization of differentiated cells, sometimes epidermal cells, but more frequently the unbroken subepidermal layer of parenchyma cells (see Root). Within the plant, regions of different tissues are always separated by the endodermis, e.g. in the root the endodermis surrounds the entire central vascular cylinder

The epidermis cannot stretch indefinitely, and epidermal cells are not meristematic. If the epidermis is unable to extend sufficiently to accommodate an increase in the girth of a plant organ, e.g. in the growth of woody plants, then it is destroyed and replaced by secondary B.t. Usually a new meristem appears, called the *cork cambium* or *phellogen,* which is formed in the subepidermal cell layer when a ring of cortical cells, one cell thick, recovers the power to divide. The phellogen cells divide, producing daughter cells on their outer and inner surfaces. The closely packed cells formed on the outer face rapidly develop thick walls heavily impregnated with suberin, forming a suberized tissue called *cork,* which has no intercellular spaces. The smaller number of cells produced on the inner surface of the phellogen layer form the *phelloderm* or *secondary cortex,* a parenchymatous tissue containing chloroplasts. The layers of cork, phellogen and phelloderm are collectively known as *periderm* or *cork tissue.* Gas exchange (not possible through unbroken cork) occurs through lenticels; these narrowly bounded regions of tissue can be seen with the naked eye as brownish, slightly swollen spots or lines on the stems of woody plants. In place of cork, lenticels contain large parenchymatous cells with large intercellular spaces, through which the internal plant tissues communicate with the atmosphere.

In plants of very wide girth, as in most trees, the first phellogen layer is soon expended. It is replaced by a second cork cambium, formed in a deeper layer of tissue. This also is active for only a limited period, and is replaced by a third cork cambium, and so on. The totality of the resulting cork layers is called *bark.* Since bark formation soon extends into the phloem, the later cork layers are separated by phloem cells, resulting in macroscopically visible layers. Since they have no water or nutrient supply, these individual cork layers rapidly die and slough off.

Bouqueria: see *Plantaginaceae.*

Boutu: see Whales.

Bovidae, *bovids, horned ungulates:* a large and varied family of ruminant even-toed ungulates (see *Artiodactyla),* comprising antelopes, cattle, bison, buffalo, goats and sheep: They generally possess horns, which are usually less developed or absent in the female. Incisors and canines are absent from the upper jaw. B. are largely vegetarian, and the family contains a large number of economically important animals. In the present account, the following subfamilies are recognized, and each is listed separately: *Cephalophinae* (Duikers), *Neotraginae* (Dwarf antelopes), *Tragelaphinae* (Spiral-horned antelopes, Eland, Nilgais), *Bovinae* (Cattle, Bison, Buffalo), *Alcelaphinae* (Hartebeests, Gnus), *Hippotraginae* (Roan and Sable antelopes), *Reduncinae* (Reedbucks), *Antilopinae* (Gazelles), *Saiginae* (Tibetan and Saiga antelopes), *Caprinae* (Sheep, Goats and Musk ox).

The phylogeny of the B. is controversial. Some classifications place *Aepyceros* and *Pelea* in their own subfamilies *(Aepycerotinae* and *Peleinae,* respectively), while some authorites do not recognize the *Reduncinae, Alcelophinae, Aepycerotinae* and *Peleinae.*

Bovids: see *Bovidae.*

Bovinae, *bovines, cattle:* large even-toed, horned ungulates comprising a subfamily of the family *Bovidae* (see). They have a massive skull and a broad snout with a wet muzzle. Both sexes possess round to oval, smooth horns. The 5 species (with several subspecies and varieties) of the genus, *Bos,* are generally known as ***oxen.*** They have characteristically large, massive bodies, with stout limbs and a long tail, which usually has a tufted tip. Both sexes have horns, which are larger in males, more or less oval in cross section, and inserted relatively far apart on each extremity of the top of the skull. The occipital area of the skull forms an acute angle with the face, and not a right angle as in *Bubalus* (buffaloes). The Aurochs (see) is the ancestor of domestic European cattle *(Bos primigenius* v. *taurus). Bos javanicus* ***(Banteng)*** (head plus body 180 -225 cm) is found in Burma, Thailand, Indochina, Peninsula Malaysia, Java and Borneo. It appears to favor rather dry, open country, but also requires treed areas for concealment and shelter. An extremely wary animal, it has adopted a nocturnal habit in some areas, presumably to avoid humans. It possesses white stockings and a white escutcheon. Cows are red-brown, and bulls are dark brown to blue-black. ***Bali cattle*** are a domesticated form *(Bos javanicus* v. *domesticus)* found on the islands of Bali and Java, where they are used as draught animals and for meat and milk. *Bos javanicus* may also be an ancestor of the ***Zebu,*** a domestic ox, used widely in India and East Africa as a draught animal and beast of burden, and for meat and milk. The zebu is characterized by its erect horns, fatty hump and large dewlap. One African savanna form of the Zebu is the ***Anhole*** or ***Watussi ox,*** bred by tribes in Uganda and Ruanda. Usually brown in color, it has very long, spreading, only slightly curved horns, which may measure up to 150 cm in length. The zebu has been used in Africa and North America for cross breeding with other domestic cattle.

The ***Gaur*** or ***Seladang*** *(Bos gaurus)* occurs from Nepal and India to Indochina and Peninsula Malaysia, where it inhabits forested hills and associated grassy clearings. It is dark reddish brown to almost black-brown in color with white stockings and escutcheon. Head plus body length is 250–330 cm, and males are about one quarter larger and heavier than females. Adult males have a large hump over the shoulders and a high swollen head crest. Like the Banteng, the Gaur is basically diurnal, but in some areas has adopted a nocturnal existence to avoid humans. In the wild it is an endangered species. A smaller, domesticated form, the ***Gayal*** *(Bos gaurus* v. *frontalis)* lives in a semitame or feral state in eastern India and Burma, where it is used as a working animal, as well as a favored source of meat.

The ***Kouprey*** *(Bos sauveli)* of northern Cambodia and adjoining parts of Thailand, Laos and Vietnam is one of the world's most critically endangered species. Since it became known to science in 1937, its natural range has been a scene of almost continual human conflict. Head plus body length is 210–222 cm. Females and young are generally gray with lighter underparts, while adult males are dark brown. Both sexes have white or grayish stockings. A very long dewlap hangs from the neck, and in old males it may nearly reach the ground. The horns of the female spiral upward in a lyre shape, while the larger horns of the male spread outward and downward at the base, then upward and forward.

The ***Yak*** *(Bos grunniens)* inhabits desolate steppes up to 6 100 m in Tibet and adjacent highland regions. Wild males grow up to 325 cm (head plus body) in length and may weigh about 1 000 kg, while females attain only about one third of this weight and size. The color is generally black-brown, and in the males the large black horns curve upward and forward. Long hair forms a fringe around the lower part of the shoulders, sides of body, flanks and thighs, and the tail is also long-haired. Females gather in large herds, which in former times numberd thousands of individuals. Adult males are solitary or occur in small groups, and they join the herds only for mating (in September). The wild form is an endangered species. Domestication probably dates from the first millenium BC. The domestic Yak is smaller than the wild form, and is red, mottled, brown or black in color; the horns are also more weakly developed and may be absent. Domestic Yaks are found throughout the high plateaus and mountains of Central Asia, where they are used as mounts and beasts of burden, and as a source of milk, meat and wool.

The genus, *Bubalus,* contains the 4 species of ***Asian water buffaloes*** or ***Anoas.*** Three of these are rare animals, inhabiting remote localities, and forming pairs rather than herds: *Bubalus mindorensis* ***(Tamaraw*** or ***Tamarao;*** Mindoro Island of the Philippines; one of the rarest animals, with perhaps less than 200 individuals in existence); *B. depressicornis* ***(Lowland anoa;*** lowlands of Celebes); *B. quarlesi* ***(Mountain anoa;*** highlands of Celebes). Although the fourth species, *B. bubalus* ***(Asian water buffalo*** or ***Carabao)*** is vulnerable in the wild, an estimated 75 million domesticated individuals are in existence. The wild form was originally found from Nepal and India to Vietnam and Malaysia. Domestication probably dates from the third millenium BC. Found mostly in India and Southeast Asia, the domesticated form is indispensable in many communities for tilling rice fields, as well as being an important beast of burden, and a source of milk and leather. It is also used in southern Europe, Asia Minor, North and East Africa, Madagascar, Mauritius, Australia, Japan, Hawaii, and Central and South America. A very docile animal, it can even be controlled by children.

In temperament, the monospecific ***African buffalo*** *(Syncerus caffer)* differs markedly from its Asian relative. It is a deadly and ill-tempered fighter, capable of speeds up to 57 km/h., and one of the most feared of all wild African animals. It originally occurred in most of sub-Saharan Africa, and subspecies display much variation in size. Thus, *S.c. caffer* of the eastern savannas may grow twice as large as *S.c. nana* (head plus body 210–300 cm) of the equatorial forests. Adults have only sparse hair, although young animals have a thick coat. Color varies from brown to black. Head and limbs are massive, and the chest is very broad. In males, the horns are joined by a large shield (boss), which covers the entire top surface of the head. *Syncerus* likes water and enjoys wallowing.

The 2 species of *Bison* (see plate 37) have suffered disastrous reductions in numbers in historical times. The ***European bison*** or ***Wisent*** *(Bison bonasus)* was originally found in much of western, central and eastern Europe, probably extending into Asia. It is now known only in captivity and in reserves. A forest inhabitant, *Bison bonasus* feeds mainly on leaves and twigs. Before colonization of North America by Europeans, the population of ***American bison*** *(Bison bison)* was an estimated 50 million. By 1890, fewer than 1000 individuals existed on the entire continent. By conservation from captive species, this number has been increased to about 50 000, almost all of which are the plains subspecies, *B.b. bison*. The conserved bison of Yellowstone National Park represent a continuity from a truly wild population of the woodland subspecies, *B.b. athabascae,* but these have now become greatly admixed with the plains subspecies. A small, apparently pure herd of woodland bison was discovered in 1957 in Wood Buffalo Park, and these are listed as an endangered subspecies.

Head plus body length of both species is 210–350 cm, males being on average larger than females. A chin beard is usually present, and the hair of the head, neck, shoulders and forelegs is characteristically long and shaggy, and brown-black in color. The remainder of the body is lighter in color and has shorter hair. The forehead is short and broad, the shoulders have a high hump, and the horns of both sexes are short, curved upward and sharp. *B. bison* has longer and thicker hair on the neck, head and forequarters, and therefore appears to be larger; its body is in fact rather more massively built, but the pelvis is smaller and the hindquarters less powerful than those of *B. bonasus*. Some authorities consider that *Bison* is a subgenus of *Bos,* and that *Bison bison* and *Bison bonasus* are conspecific.

The most recently discovered member of this subfamily is the Vu Quang ox (see), which was first reported in 1993.

Bovines: see *Bovinae*.

Bovine spongiform encephalopathy: see Prion.

Bovist: see *Basidiomycetes (Gasteromycetales)*.

Bowdleria: see *Sylviinae*.

Bowerbirds: see *Ptilinorhynchidae*.

Bowhead whale: see Whales.

Bowman's capsule: see Kidney functions.

Box crabs, *Calappidae:* a family of true crabs (see *Brachyura),* with a wide carapace that completely covers the abdomen. They conceal their faces behind large flattened chelae, which are also used for opening gastropod shells. When the crab is buried, the flattened chelae enclose a sand-free, water-filled space, also borderd by the ventral side of the body. In this position, the chelae also cover the water channel on the median part of the branchiostegites that conducts the water from the anterior of the carapace to the openings of the branchial chamber next to the coxae of the large chelipeds. The sand is visibly distubed by the strong current of expiratory water expelled by the scaphognathites into the median excurrent channel. They run as if on stilts, holding the carapace vertically.

Box elder: see *Aceraceae*.

Box family: see *Buxaceae*.

Boxfishes: see *Ostraciontidae*.

Box turtles: see *Emydidae*.

Brachial index: the length of the forearm expressed as a percentage of the length of the upper arm (forearm length/upper arm length × 100). Of the hominoids, the gibbon has on average the largest B.i., and the human has the smallest. There is some overlap between the values for humans and gorillas, i.e. gorillas may occasionally have a relatively shorter forearm than humans.

Brachiation: see Movement.

Brachiator theory: the proposal that the phylogenetic development of humans involved a stage that was adapted to a swinging and climbing arboreal existence, i.e. progress by brachiation, and that this ability was lost by subsequent evolution. Forms displaying extreme adaptation to brachiation have not yet been found in the human fossil record. See Anthropogenesis.

Brachiopoda, ***brachiopods, lamp shells:*** a lophophorate phylum of attached, coelomate, unsegmented, filter-feeding invertebrates, represented by about 325 living species, varying in size from 0.1 to 8 cm. [The other 3 lophophorate phyla are the *Bryozoa, Entoprocta* and *Phoronida]*. All *B.* are exclusively marine, and are found at all depths from intertidal zones to the floors of deep oceans. Most species attach themselves to rocks or other firm substrata, but some forms, e.g. *Lingula,* live in vertical burrows in sand and mud bottoms.

Morphology. The body is dorsoventrally flattened and enclosed by two valves, one on the dorsal, the other on the ventral surface. On acccount of these two valves, the *B.* bear a strong but superficial resemblance to the lamellibranchs *(Mollusca),* but in the lamellibranchs the two valves enclose the body laterally and not dorsoventrally as in the *B*. The valves are secreted by two mantle lobes, which enclose most of the body. Both opening and closing of the valves is controlled by a muscle system, which is more elaborate than that of the lamellibranchs. The body is divided into 3 segments: the *prosoma* or *protosoma* comprises the small upper lip; the *mesosoma* is the central segment with the mouth opening and the double spiral of the lophophore arms; the *metasoma* comprises the rest of the body with the viscera, mantle folds and pedicle (when present). The anterior stalk, penduncle or pedicle attaches the animal to the substratum (usually rock). In those species lacking a pedicle, the valves are attached directly to the substratum. There are 2 spirally coiled tentacular arms (lophophore), which probably correspond to the bryozoan lophophore. A ciliated groove runs along each arm, and a row of ciliated tentacles runs along one side of each groove. A simple gut empties laterally, or caudally, or is blind. A simple, open circulatory system is present, containing a dorsal contractile muscular vesicle or "heart". The blood is colorless, containing a few coelomocytes. There is no specialized organ for the exchange of respiratory gases, which probably occurs mainly via the mantle lobes and lophophore. The nervous system consists of an esophageal ring, as well as a cerebral and a subesophageal ganglion. A true coelom is present, lined with mesoderm.

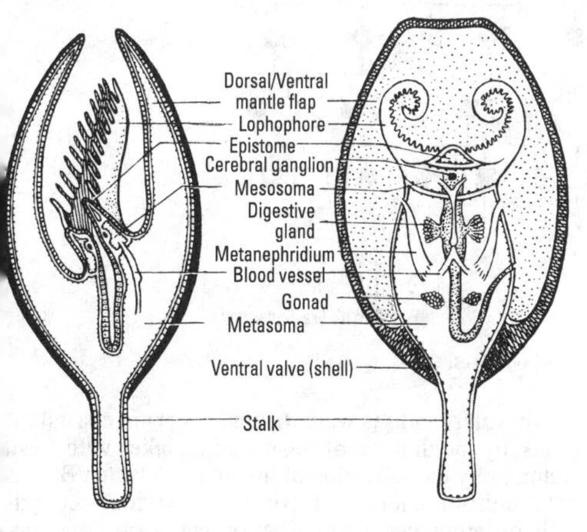

Brachiopoda. Anatomy of a brachiopod (class *Ecardines). a* Side view. *b* Dorsal view.

Biology. In practically all *B.* the sexes are separate. Sex organs lie near the intestine; eggs or spermatozoa are discharged into the coelom, and they reach the exterior through the excretory tubes. Development occurs by metamorphosis, at first in special brood chambers, then via a swimming larva with an apical ciliated plate, eye spots and statocysts. After a few days the larva becomes attached to the substratum. Food consists of microorganisms and detritus, which are swept toward the mouth (situated between the bases of the tentacular arms) by the beating of the tentacular cilia, then transported to the mouth by a mucilage filter.

Classification. *B.* are divided into 2 subphyla.

Subphylum 1: *Inarticulata, Ecardines* or *Gastrocaulia.* The two valves are very similar in shape and size, and not joined by a hinge. There is no brachial skeleton. The shell material is chitin-like, usually a mixture of chitin and calcium phosphate (chitinophosphate), with a horny texture. Most species of the *Inarticulata* possess an attachment stalk, which is usually very long and contains an extension of the coelom.

The two phosphate-infiltrated chitinous valves of the genus *Lingula* are held together by 6 pairs of muscles. *Lingula* valves are elongate, four-sided to oval in shape with a pointed crown; in most species the valves have concentric growth lines. Numerous fossil species of *Lingula* occur from the Ordovician to the present.

Subphylum 2: *Articulata, Testicardines* or *Pygocaulia.* This is a larger group than the *Inarticulata,* and its members have a blind intestine. The ventral valve is larger and more bulbous than the dorsal one. The valves, which consist of calcium carbonate in the form of calcite, are always joined by a hinge. A calcareous *brachium* or *brachial skeleton* is always present on the inner surface of the dorsal valve, where it acts as a support for the lophophore. The diverse forms of the brachial skeleton, ranging from simple hooks to complicated spirals, are diagnostically useful. In stalked members of the *Articulata,* the stalk is noncontractile, and it does not contain an extension of the coelom.

Fossils (see *Obolus, Terebratula, Stringocephalus).* Fossil *B.* occur from the Cambrian to the present, and disputed *B.* fossils may also exist in the early Precambrian. In contrast to the relatively small number of recent forms, fossil *B.* are numerous, representing 1 400 genera and a disputed number (7 000–25 000) of species. Recent *B.* are the last of a line that flourished in the Paleozoic and Mezozoic eras. As early as the Cambrian they represent one third of the total fossil fauna. Following the Mezozoic, the *B.* receded in favor of the lamellibranchs and gastropods. The genus *Lingula,* however, has survived from the Ordovician to the present, although its modern species have evolved more recently and are less numerous.

Collection and preservation. As for the *Tentaculata* (see).

Brachioradial muscle: see Musculature.

Brachiosaurus (Greek *brachion* arm, *sauros* lizard): a genus of quadruped, vegetarian sauropods with 4 columnar limbs, the forelegs longer than the hindlegs. They measured 25–30 m in length, attained a height of up to 13 m, and weighed 80–100 tonnes. They probably lived in swamps and lakes, because without the buoyancy of water they could have hardly moved their own weight. Fossils are found in the Upper Jurassic of North America, and the Upper Jurassic and Lower Cretaceous of East Africa (Tendaguru).

Brachycephalidae, *brachycephalids, gold frog family:* a family of the *Anura* (see) containing 2 monotypic genera: *Brachycephalus ephippium **(Gold frog)*** and *Psyllophryne didactyla,* which are found only in a small, humid coastal area of southeastern Brazil.

There are 7 holochordal, procoelous presacral vertebrae with imbricate (overlapping) neural arches. Presacrals I and II are fused, and the atlantal cotyles of presacral I are widely separated. Ribs are absent. The coccyx has no transverse process, and it has a bicondylar articulation with the sacrum, which has dilated diapophyses. The pectoral girdle is arciferal and totally ossified, and it lacks an omosternum or sternum. There are no teeth and the pupils are horizontal.

Brachycephalus is golden with brown markings on the back and head. Body length of both species is less than 16 mm. There are only 2 functional digits on the hands and 3 on the feet. Only a few, large-yolked eggs are laid on land, and development is direct. *Brachycephalus* was formerly placed in the *Bufonidae,* but the absence of a Bidder's organ justifies its transfer to the *B.;* it has a dermal bony shield dorsal to, and fused with, the vertebrae.

Brachycephaly: "short headedness", a condition in which the Cephalic index (see) lies in the range 81.0–85.4, a characteristic of American Indians, Malays and Burmese. In the early Middle Ages, an increase in the cephalic index or *brachycephalization* (i.e. a general rounding of the head) was apparent in many European populations. Since the middle of the 19th century there has been a general decrease in the cephalic index *(debrachycephalization)*. The reasons for these historical changes in the cephalic index are not known.

Brachydactyly: in humans, the shortening of all or individual bones of the fingers and/or hands. B. was the first inherited human characteristic shown to be autosomally dominant (Farabee, 1905) (Fig.).

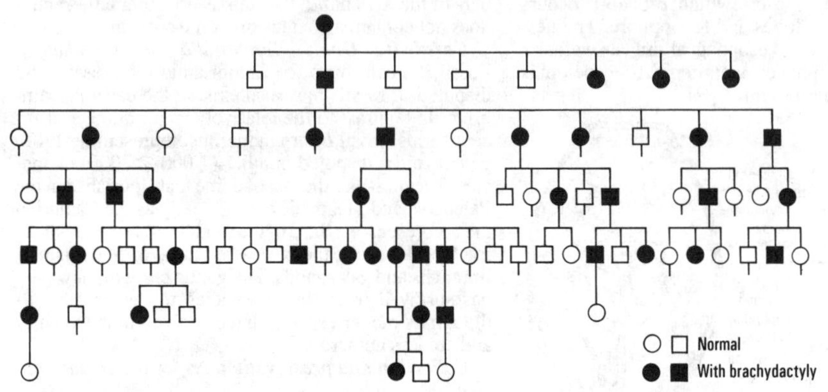

Family tree of a family with brachydactyly (after Farabee and McKusik).

Brachygnatha: a section of the infraorder *Brachyura* (true crabs) of the *Decapoda* (see).

Brachykerkic: having a short forearm or radius, i.e. a radiohumeral index of 75 or less.

Brachyknemic: having a short foreleg or tibia, i.e. a tibiofemoral index of 82 or less.

Brachylagus: see *Leporidae*.

Brachypteraciinae: see *Coraciiformes*.

Brachypteryx: see *Turdinae*.

Brachyskelia: in humans, the ocurrence of short legs in relation to body length. It is more common in women than in men. The converse is Macroskelia. Adjective: brachyskelous.

Brachyteles arachnoides: see Woolly spider monkey.

Brachythecium: see *Musci*.

Brachyura, *true crabs:* a suborder or infraorder (depending on the classification system) of the *Decapoda* (see), containing more than 4 400 species, most of them marine. They have a short, square or triangular body, with a markedly reduced symmetrical abdomen, which is not used in locomotion, and which is tightly folded beneath the anterior cephalthorax. In the majority of species the antennae are very small and lack a scale; alternatively, the scale is immovable fused to the second article. Chelae (pincers) (usually large) are always present on the foremost pair of pereopods, the other legs serving for locomotion (often sideways). The third pair of pereopods is never chelate. Uropods are absent, and males also lack the third, fourth and fifth pleopods. Gills are phyllobranchiate. The zoea and metazoea larvae have long spines.

The most primitive representatives are the dromid crabs (family *Dromiidae)*, which are known from the Jurassic. Dromid crabs often use sponges or seaweed as a form of camouflage, even cutting material to size and planting on their backs. Many other true crabs also display this habit of covering themselves with dead or living material. The following *B.* are listed separately: Spider crabs *(Majidae)*, Box crabs *(Calappidae)*, Swimming crabs *(Portunidae)*, Rock crab *(Cancer pagurus)*, Freshwater crabs *(Potamonidae)*, Fiddler crabs *(Uca)*, Chinese mitten crab *(Eriocheir sinensis)*, Land crabs *(Geocarcinidae)*.

Bracket fungus: see *Basidiomycetes (Poriales)*.

Brackish water (Dutch *brak* salty): rather salty, unpalatable water, formed by the mixing of river water (fresh water) and sea water (saline water) in coastal regions, by the dilution of inland saline lakes with fresh water, or by the salination of inland fresh water. B.w. is not a uniform habitat, but consists of various ecologically differentiated zones. The following system of ecological zones for marine B.w. was established in 1958 in Venice at a meeting of the International Society of Limnology.

Zone	Salinity (%)
Hyperhaline	4
Euhaline	4–3
Mixohaline	4–0.05
Mixoeuhaline	3–0.05
Mixopolyhaline	3–1.8
Mixomesohaline	1.8–0.5
α-Mesohaline	1.8–1.0
β-Mesohaline	1.0–0.5
Mixo-oligohaline	0.5–0.05
α-Oligohaline	0.5–0.3
β-Oligohaline	0.3–0.05
Limnic (fresh water)	< 0.05

Inland saline and brackish waters are not included in this classification, since their ionic composition differs from that of sea water. The salt concentration of *poikilohaline* B.w. is subject to frequent alterations, whereas that of *homeohaline* water is relatively constant.

The animal and plant life of B.w. consists of marine organisms, freshwater organisms, and true B.w.-species. Some β-mesohaline regions of the Baltic contain marine euryhaline species, such as the mussel *Mytilis edulis*, the annelid *Nereis diversicolor*, the small crab *Evadne nordmanni*, the shore crab *Carcinus maenas*, the herring, common cod and flounder and others, together with euryhaline freshwater forms, such as fennel-leaved pondweed *(Potamogeton pectinatus),* the mud snail *Limnaea ovata,* the small crab *Chydorus sphaericus,* the polychaete worm *Nais elinguis,* the pike, perch, roach and others. True B.w.-forms also occur in the same habitat, e.g. the turbellarian *Procerodes ulvae,* the rotifer *Keratella eichwaldi,* the amphipod *Corophium lacustre,* the annelid *Streblospio shrubsoli,* the snail *Hydrobia jenkinsii,* the bryozoan *Membranipora crustulenta,* and others. Of the fish caught commercially in Greifswald Bay (salt concentration 0.7%), the herring is the most abundant (it also breeds there), while the pike is the second most common.

In general, B.w. does not support a great diversity of species, although each species may be abundant. Thus, there are more than 6 000 animal species in the euhaline Mediterranean (3.5% salt), and only 1 033 in the Black Sea (1.86% salt). 476 different animal species have been recorded in the Caspian Sea (which contains 29% salt and has a different ionic composition from that of the open sea), while only 98 are known in the Aral Sea (salt concentration 1.66%). In contrast, Lake Baikal (a freshwater lake) contains 1 800 different species.

As the salt concentration of B.w. decreases, many species undergo a reduction in size. The mussel attains a maximal length of 70 mm in sea water containing 3.5% salt, but never exceeds 20 mm in B.w. containing 0.5% salt. In the case of mussels, this decrease in size is due mainly to calcium deficiency. Oceanic water contains about 430 mg/l Ca, whereas B.w. contains only a fraction of this value.

See Saline lakes.

Braconidae: see *Hymenoptera.*

Bradykinin: Arg-Pro-Pro-Gly-Phe-Ser-Pro-Phe-Arg, one of a group of peptide tissue hormones, known as kinins. Like the structurally and functionally related decapeptide, *kallidin (lysylbradykinin),* it is formed by the enzymatic cleavage of kininogen (an α_2-globulin). It causes dilation of blood vessels and thus a reduction of blood pressure, increases blood vessel permeability, and causes contraction of the smooth muscles of the bronchia, intestines and uterus. It is also a potent pain-producing agent.

Bradypodidae: see Sloths.

Bradypterus: see *Sylviinae.*

Bradytely: see Evolutionary rate.

Brain (plates 24 & 25): the anteriorly situated organization center of the nervous system, also known in certain invertebrates as the ***cerebral ganglion, supraesophageal ganglion*** or ***suprapharyngeal ganglion.***

In evolutionary terms, the cephalic end of the central nervous system shows ever increasing dominance, finally evolving into a morphologically and physiologically complex structure called the B. The anterior location of the B. is correlated with unidirectional movement of most animals, which are therefore bilaterally symmetrical and carry their major receptor systems at the anterior end.

Coelenterates do not possess a central nervous system, or any aggregation of nerve cells suggestive of an organization center; in these organisms, nervous control is served by a continuous, irregular nerve net or plexus. During the evolution of more advanced nervous systems (association of fibers into tracts and eventually true nerves; the appearance of a central nervous system with a differentiated brain), the primitive nerve net loses its importance, but it may still make a minor contribution.

The most primitive invertebrate B. is the cerebral ganglion of the *Turbellaria;* this is simply a concentration of nervous material at the anterior end of the body, below the statocyst and beneath the muscles of the body wall. It can be regarded as a sensory relay which receives stimuli from the anterior epidermal sensory cells and the paired nerves of the eyes and chemoreceptors, then passes them to the primitive nerve net via radial longitudinal tracts *(connectives).* The B. of *Trematoda* (flukes), *Cestoda* (tapeworms), *Acanthocephala* (thorny headed worms), *Rotifera* (wheel animalcules) and *Gastrotricha* corresponds to that of the *Turbellaria.* In contrast, the B. of *Nematoda* (roundworms), *Kinorhyncha, Priapulids* (priapus worms) and *Nemertini* (proboscis worms) is a thickening of the anterior ring commmissure of the nerve net. The *Polyplacophora* (chitons) lack an actual B., but they possess a cerebral commissure which contains a high concentration of ganglion cells. All other mollusks have a well developed B. (cerebral ganglion) which gives rise to the optic, olfactory and statocyst nerves. The tendency to concentrate all ganglia around the B. is most highly developed in cephalopods and opisthobranchs.

Most annelids (e.g. oligochaetes and leeches) are largely devoid of specialized cephalic sense organs, and their B. is a simply organized, uniform structure. In the polychaetes, however, the B. is subdivided into a forebrain (which innervates the palps), a midbrain (which innervates the eyes and tentacles) and a hind B. (which innervates the nuchal organs).

The insect B., which is typical of the arthropod condition, is differentiated into three distinct parts: the protocerebrum (the most anterior of the three), deutocerebrum and tritocerebrum, each being a distinctly paired structure.

1) The primary sensory center or *protocerebrum* receives nerves from the eyes and frontal organs. Each of the two lateral optic lobes contains three centers associated with the compound eyes: the lamina ganglionaris (nearest to the back of the compound eye), followed by the medulla externa and the medulla interna, these three centers being separated from each other by two chiasmata. The protocerebrum also contains medially situated visual centers of the ocelli. Association centers of the protocerebrum are the protocerebral bridge, the Corpora penduculata (see) and the central body. The protocerebral bridge is a median mass of neuropile across the front of the brain; in insects it has axon connections with widely separate parts of the brain, but not with the corpora penduculata (cf. *Chelicerata,* where the protocerebral bridge has connections with the corpora penduculata). The central body consists of a mass of neuropile lying centrally in the protocerebrum with no separate ganglion cell layer; it is a meeting point of axons from different parts of the brain, and it is the source of premotor outflow from the brain to the ventral cord.

2) The *deutocerebrum* contains the antennal lobes, with dorsal sensory (olfactory) ganglia and ventral motor ganglia. The two sides are connected by a commissure. Sensory and motor fiber tracts connect the antennal lobes with the corpora penduculata, while other fibers pass to the tritocerebrum. Most *Chelicerata* lack a deutocerebrum.

3) The *tritocerebrum* is the first postoral ganglion of the ventral ganglion chain, and consists of two lobes beneath the deutocerebrum. Circumesophageal connectives pass from the tritocerebrum to the subesophageal ganglion. The two tritocerebral lobes are connected by a commissure which passes behind the esophagus. In the more advanced condition it fuses with the primary B. (protocerebrum plus deutocerebrum) and becomes a secondary B. It gives rise to the nervus tegumentarius, the labial nerve and the sympathetic nervous system (stomatogastric nervous system) of arthropods.

In the hemichordate *Enteropneusta* (acorn worms), the B. is situated in the collar (mesosomal) region. The subepidermal neural tube (derived from the dorsal nerve strand) is lined internally with nerve cells and is suggestive of the vertebrate spinal cord. Sea squirt larvae also possess a neural tube, which expands at its anterior end, probably representing a forerunner of the vertebrate B.; after metamorphosis, however, all that remains of the entire anterior neural tube is a small ganglion.

The vertebrate B. is an expansion of the cranial section of the neural tube, enclosed and protected by the skull, and therefore also known as the *encephalon*. During embryogenesis, the B. develops almost more rapidly than any other organ. At first, it is simply an expanded area of the neural tube. The anterior end then folds downward, forming a *cephalic flexure,* and defining the terminal saclike primitive *prosencephalon* (forebrain). A second and more posteriorly located flexure then differentiates the *mesencephalon* (midbrain)from the *rhombencephalon* (hindbrain), giving a total of three primitive compartments. Further differentiation of the prosencephalon produces the *telencephalon* and *diencephalon (thalmamencephalon),* while the rhombencephalon differentiates into the *metencephalon* (with *cerebellum* and *pons)* and the *myelencephalon*. The mesencephalon develops with no further subdivisions, giving a total of five compartments. The myelencephalon, also called the *medulla oblongata* or *spinal bulb,* merges with the spinal cord. Within this

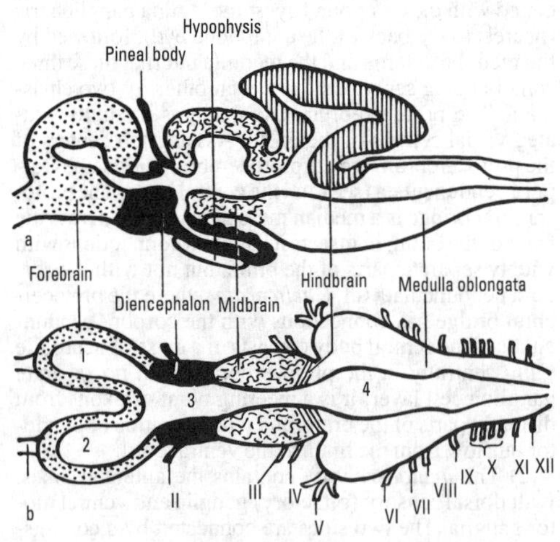

Fig.1. *Brain*. Sections of the vertebrate brain (schematic). *a* Sagittal section. *b* Horizontal section. *I* to *XII* are cranial nerves. *1* to *4* are ventricles.

transition region, fiber tracts (white matter) become relocated, with the result that the gray matter (which is centrally located and compact in the spinal cord) becomes dispersed as isolated nuclei in the medulla. The original cavity of the embryonic neural tube persists in the adult brain as a series of cavities or Ventricles (see).

In the medulla oblongata, the central canal of the spinal cord expands to form the floor of the losenge- or diamond-shaped fourth ventricle. Anteriorly, the roof of the fourth ventricle is formed by the cerebellum; for most of its length, however, the roof is formed by the thin *tela choroidea,* which contains an arterial network, but no nerve elements, and part of which is infolded (corrugated) as the *posterior choroid plexus*. The wall of the floor of the fourth ventricle contains motor nuclei, which give rise to motor nerve fibers, and it receives sensory nerves. Most of the cranial nerves arise from the medulla oblongata, which also contains important centers of respiratory, cardiac, circulatory and metabolic function.

Anterior to the medulla oblongata is the metencephalon with its dorsal evagination known as the *cerebellum,* a high-order coordination center for the reflex control of voluntary muscle in all vertebrates from fishes to humans; impulses from its motor neurons are responsible for muscle tonus and posture, and it acts in the maintenance of equilibrium and body orientation. In addition, the cerebellum processes information from skin sensory areas, from the optic centers and, in lower vertebrates, from the olfactory organs. Unlike most other parts of the brain, the gray matter of the cerebellum is found on the surface. A conspicuous feature of the ventral wall of the mammalian metencephalon is the *pons* or *pons varoli,* situated in front of the cerebellum between the midbrain and the medulla oblongata; it contains decussating fibers (fibers that connect the two halves of the cerebellum), as well as serving as an important relay site for fibers descending from the cerebral hemisphere to the cortex of the cerebellum. The fifth (trigemminal) nerve emerges from the side of the pons near its upper border.

The mesencephalon is primarily of topographic and descriptive value, since its dorsal section functionally and ontogenetically is assigned to the diencephalon and its ventral part to the metencephalon. In all vertebrates, the roof (tectum opticum) of the midbrain has a pair of prominent *optic lobes* (colliculi rostralis, colliculi inferiores, inferior quadrigeminal bodies, anterior tubercula quadrigemmina, nates), bulging from the dorsal surface of the midbrain. The optic lobes receive fibers from the retina and serve partly as optic reflex centers; they are important in the regulation and coordination of visually controlled behavior, and they are especially well developed in birds and many teleost fishes, where they are also known as the corpora bigemina. A pair of *auditory lobes* (colliculi caudalis, colliculi superiores, superior quadrigeminal bodies, posterior tubercula quadrigemmina, testes) is also present in the midbrain roof of all vertebrates, but in fishes and amphibians they are not large enough to bulge from the surface. The auditory lobes receive stimuli from the membranous labyrinth, and they are important in the regulation and coordination of acoustically controlled behavior. Collectively, the optic and auditory lobes are referred to as the corpora quadrigemina or *quadrigeminal bodies*. With the evolutionary increase in the size of the cochlea, the auditory lobes also enlarge and develop afferent connections. Ventrally, the midbrain consists of nuclear masses and fiber tracts connecting higher and lower levels of the B. In the mammalian B. these tracts are massive and visible as surface structures known as *cerebral peduncles*. A prominent transverse fiber pathway is the *trapezoid body*, which enters the brain via the eighth cranial nerve and relays impulses from the cochlea; it first appears on the hindbrain ventral surface of the frog and is largest in humans. In fishes and amphibians, the large midbrain ventricle also extends into the optic lobes to form the optic ventricles. More advanced vertebrates have no optic ventricles, and their midbrain ventricle is confined to a narrow *cerebral aqueduct;* thickened gray matter above the aqueduct forms the mesencephalic *tectum*, while thinner areas of gray matter in the side walls are known as the *tegmentum*.

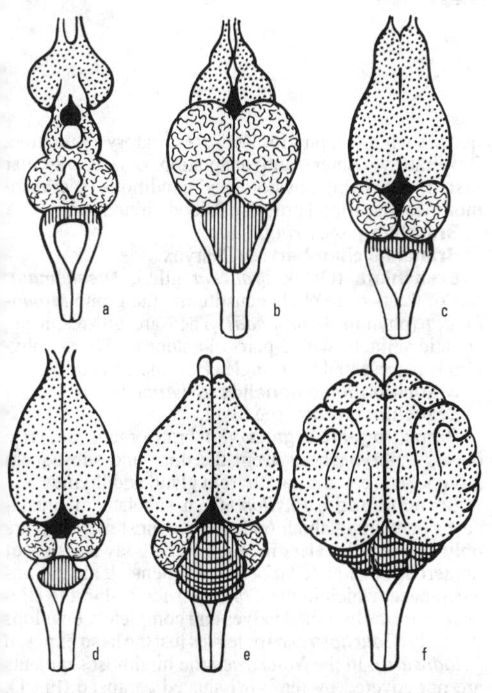

Fig.2. *Brain*. Brains of different vertebrates. *a* Lamprey. *b* Perch. *c* Frog. *d* Alligator. *e* Pigeon. *f* Cat.

The diencephalon is closely associated with visual reception and the endocrine system. On its ventral surface, the optic chiasma (the superficial emergence of the optic nerves from the brain) serves as a ventral landmark. Ventrally, the diencephalon gives rise to the infundibulum, which carries the higher center of the hormonal system, the Hypophysis (see) (also see below). In fish (except cyclostomes) there is also a thin-walled *saccus vasculosus*, caudal to, and often larger than, the hypophysis; it is connected by fiber tracts with the cerebellum.

Largest of the diencephalic subdivisions is the *thalamus*, consisting of a mass of nuclei surrounding the third ventricle (i.e. it is equivalent to the side walls of the diencephalon). All sensory pathways ascend to the telencephalic synapse in one of the thalamic nuclei before they are transmitted further. With vertebrate evolution the thalamus has increased in size in proportion to the rest of the B., relaying an inceasing number of sensory impulses to the cerebral hemispheres. In mammals, its sides bulge into the cavity of the diencephalon (the third ventricle), where they meet to form a commissure.

Located in the floor and ventrolateral walls of the third ventricle, the *hypothalamus* contains the highest centers of the autonomic nervous system, and it is important in homeostasis, e.g. it houses centers for the regulation of water economy, temperature, NaCl and glucose levels; it is also the regulatory center of the entire endocrine system. Structurally, it is linked to the *pituitary gland* via a bridge called the *pituitary stalk*. Functionally, it is linked to the pituitary by cells that function as both nerve and endocrine cells, known as *neuroendocrine* or *neurosecretory* cells. Most neuroendocrine cells respond to stimulation by other brain cells by secreting specific peptide hormones into the blood vessels of the pituitary stalk; these hormones in turn stimulate or suppress the secretion of a second hormone by the pituitary (usually the anterior pituitary or adenohypophysis) into the main circulation. Other hypothalamic neuroendocrine cells extend their axons along the pituitary stalk into the neurohypophysis (posterior pituitary); hormones synthesized in the cell body in the hypothalamus are then transported along the axon and released in the neurohypophysis, e.g. oxytocin is synthesized in the nucleus paraventricularis and transported to the neurohypophysis via the tractus paraventriculo-hypophyseus. In mammals, with the exception of the olfactory nerves, all sensory nerve fibers, including the retinal fibers (optic nerve) are received by the dorsal thalamus, which relays them to the cerebral cortex.

Several protuberances *(habenulae)* on the dorsal surface of the *epithalamus* (the roof of the diencephalon) mark the position of underlying nuclei. In addition, the epithalamus of nearly all vertebrates exhibits evaginations, e.g. the *pineal organ* or *pineal body* (see Epiphysis). In higher vertebrates, the pineal body is not composed of nervous tissue, apart from the sympathetic filaments that enter it with the blood vessels. In other vertebrates, e.g. snakes and lizards, it possesses a long, innervated stalk, resembles structurally an invertebrate eye, and it is carried through an aperture in the cranial wall so that it lies close to the exterior surface of the head. A *parapineal organ* occurs less frequently. A thin-walled evagination known as the *paraphysis* is also evident during embryogenesis, but it is absent from most adult vertebrates.

The telencephalon consists of the paired *cerebral hemispheres* and the *rhinencephalon*. In the most primitive living vertebrates, the cyclostomes, the telencephalon is divided into two hemispheres, each carrying an anterior terminal swelling known as an olfactory bulb. Each bulb receives olfactory nerve fibers, and it contains the primary olfactory nucleus which sends fibers back to the hemisphere. At this level of development, each hemisphere is simply an olfactory lobe, and the entire telencephalon is known as the *rhinencephalon*. At higher evolutionary levels, the hemispheres display differentiated regions, marking the appearance of the *cerebral hemispheres*. Up to the level of primitive reptiles, three regions are recognizable: *paleopallium* (corre-

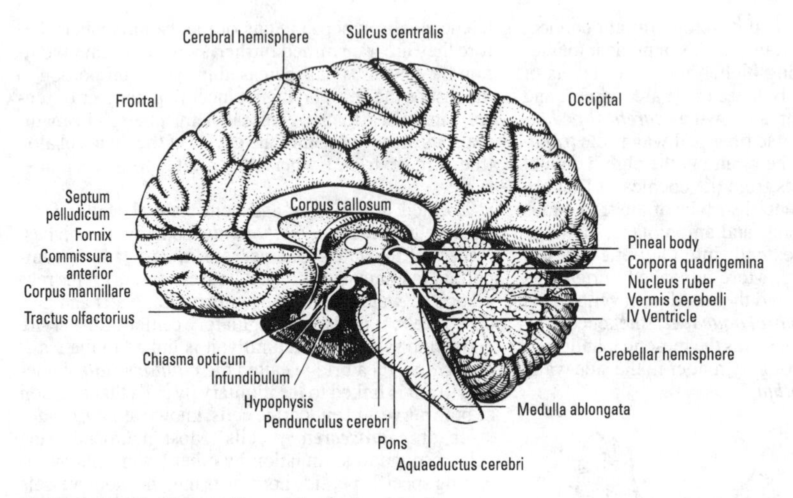

Fig.3. *Brain.* Median section of the human brain.

sponding to the original olfactory lobe, and evolving into the pyriform lobe), a dorsomedial *archipallium* (evolving into the mammalian hippocampus) and the ventral *basal nuclei (= corpus striatum),* the latter dominating the cerebral hemispheres of birds. Advanced reptiles show the first appearance of a *neopallium* between the archipallium and paleopallium. This neopallium becomes a dominant structure in mammals (the archi- and paleopallium decrease in relative size and are displaced toward the base), and it encloses the remaining parts of the hemispheres in a relatively thin layer of gray matter known as the *cerebral cortex,* which consists predominantly of ganglion cells and support tissue. In some mammals the cerebral cortex is thrown into irregular convulutions or *gyri,* separated from each other by furrows, termed *sulci* or *fissures.*

The cerebral cortex is therefore a relatively recent evolutionary acquisition, and it has become the most conspicuous part of the entire mammalian brain. During evolution, the cerebral hemispheres enlarged and extended forward to cover the rhinencephalon, which then became an inconspicuous anteroventral structure. This evolutionary increase in the size of the cerebral hemispheres reflects their increased role in animal behavior.

In birds and mammals, the hindbrain and midbrain show close homologies in their motor and sensory nuclei and in their main structural subdivisions. In contrast, the telencephalon and diencephalon appear to have evolved differently in birds and mammals, after branching from the reptilian type. In birds, neurons homologous to those of the mammalian cerebral cortex are found deep within the cerebral hemispheres, rather than on the surface. Although the avian cerebral cortex is better developed than that of reptiles, its almost complete ablation has no effect on stereotyped behavior (courtship, nest building, egg incubation, care of young), which is associated with the hyperstriatum (additional strata of nuclei superimposed on the corpus striatum).

Brain coral: see *Madreporaria.*

Brainstem centers: groups of nerve cells in the pons, medulla oblongata and midbrain, which are responsible for the nervous regulation of vital body functions, such as respiration (see Respiratory regulation, Respiratory center) and adaptation of the vascular system to existing physiological conditions (see Vasomotor center, Blood pressure, Blood circulation).

Brambling: see *Fringillidae.*

Branchial chamber: see Pharynx.

Branchiata (Greek *branchia* gills), *Dientennata:* one of the two subphyla constituting the group *Mandibulata* (phylum *Arthropoda).* They are gill-respiring, aquatic animals, with 2 pairs of antennae. The subphylum is represented by a single class, the *Crustacea.*

Branchiobdellid annelids: see *Annelida.*

Branchiocranium: see Skull.

Branchiogenic organs: vertebrate organs that arise during embryogenesis from the third and subsequent gill pockets, e.g. the parathyroid gland and thymus

Branchiopoda, *Phyllopoda:* a subclass of crustaceans containing about 600 species, most of which are only a few millimeters in length. The body consists of numerous or a few *(Cladocera)* segments. It always has a carapace, which in the *Conchostraca* and in the *Cladocera* forms two shell valves and completely envelops the body *(Conchostraca)* or leaves just the head exposed *(Cladocera).* In the *Notostraca* the hindmost segments are not covered by the bowl-shaped carapace (Fig.1). Appendages are present only on the forepart of the trunk, and these are almost always phyllopodia, which serve for both respiratory gas exchange and for trapping food material; only rarely do they act as locomotory appendages. Locomotion is generally the function of the large antennae. The telson carries a caudal furca. Reproduction is often parthenogenetic. Members of the *Cladocera* exhibit heterogony. Eggs are usually deposited in the carapace cavity, where they hatch into a nauplius; direct development occurs only in the *Cladocera* (most species). Some species of the *Cladocera* exhibit cyclomorphosis, i.e. a change in body shape from one generation to the next during the course of a year.

Almost all species are freshwater inhabitants, and only a few are marine. They crawl on the bottom and on water plants, although almost all species can also swim. *Cladocera* (water fleas) move forward with a jerky motion (hence the name "flea") by beating their large an-

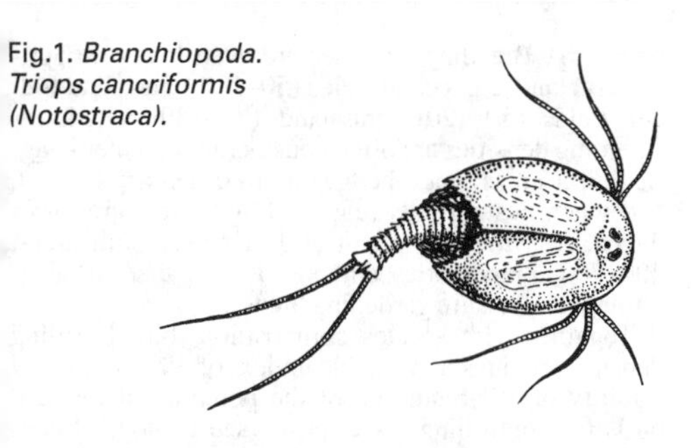
Fig.1. *Branchiopoda. Triops cancriformis (Notostraca).*

tennae. They feed on suspended particles of detritus and plankton, which are filtered from the water by the phyllopodia. The Great water flea *(Daphnia magna)* is particularly common in the small volumes of standing water of small ponds, puddles, drainage gutters, etc. *B.* are ecologically and economically important as fish food (especially water fleas).

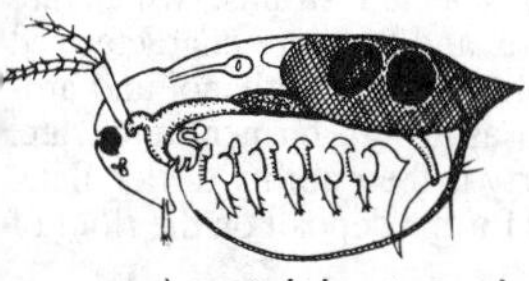
Fig.2. *Branchiopoda.* Anatomy of the water flea, *Cladocera,* showing ephippium (consisting of thickened cuticle of the carapace) containing two winter eggs.

Classification.

1. Order: *Notostraca* (e.g. *Triops* (Fig.1), *Lepidurus);* the carapace forms a broad shallow cover over the back;*Triops* first occurs in the Permian.

2. Order: *Onychura* or *Diplostraca;* the carapace is compressed and usually encloses the trunk and limbs.

1. Suborder: *Conchostraca* (e.g. *Lynceus, Limnadia* and others); the carapace has a hinge and adductor muscle; fossils occur from the Devonian to the Holocene.

2. Suborder: *Cladocera* (Fig.2) (water fleas, e.g. *Sida, Daphnia, Bosmia, Polyphemus* and others); the carapace has no hinge and no adductor muscle; in most groups *(Calyptomera)* the carapace extends downward at the sides and encloses the trunk as a shell, forming a brood pouch over the back; in two abberant groups, the *Gymnomera,* the trunk is partly or entirely exposed, and the shell merely forms a dorsal brood pouch; fossils are rare, and are first encountered in the Cenozoic.

On account of their frequency, some fossil forms are used to characterize certain strata and complexes, e.g. the numerous fossil representatives of the genus *Estheria (Conchostraca)* first encountered in the Devonian; *E. ovata* is abundant in the Triassic of North America, e.g. in the Newark beds of the Atlantic coast and in the Shinarump of Arizona.

Branchiosaurus: a genus of labyrinthodonts, with a small, salamander-like appearance, a short tail, broad skull, and numerous small, smooth, hollow, conical teeth in a wide mouth. The ventral surface of the body had an armor of bony scales. Several species possessed gills. It is now thought that they may be juvenile forms of rachitomous labyrinthodonts. Fossils are found in freshwater strata of the European Upper Carboniferous and Permian, especially in the Rotleigendes of Saxony, Thüringia and Bohemia.

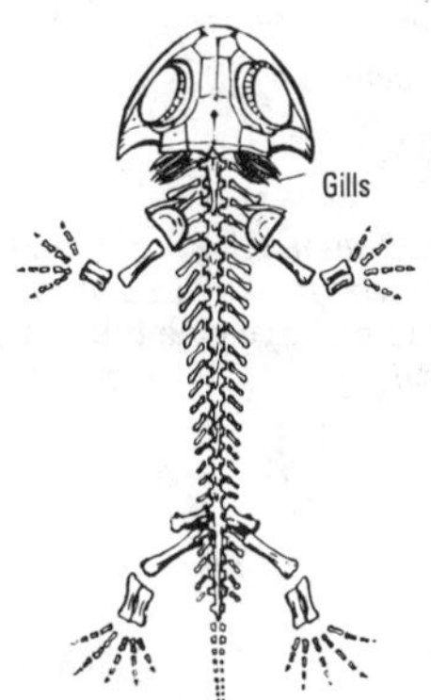

Branchiosaurus from the Rotliegendes (Middle and Lower Permian) (x 0.75).

Branchiostoma lanceolatum: see *Amphioxus lanceolatus.*

Branchiura:

1) Gill tails, fish lice: a subclass of the *Crustacea.* Although they feed exclusively as blood sucking ectoparasites, they do not remain permanently attached to their host. They swim freely, attaching only occasionally to the skin or in the gill cavities of a marine or freshwater fish (a few amphibians are also attacked).

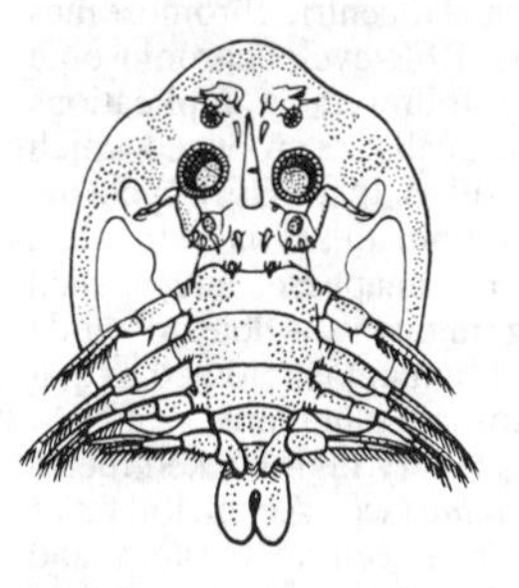
Branchiura. European carp louse *(Argulus foliaceus);* male.

The midgut possesses multibranched diverticula, which serve for food storage between feeding episodes. A shieldlike carapace (formed from the head cuticle and lateral folds of the first thoracic segment) covers the markedly flattened body. A pair of sessile compound eyes is present. The small, unsegmented, degenerate abdomen ends in two well vascularized lobes, which were formerly thought to be gills (hence the name "gill tails" or *B.).* As in the copepods, the first thoracic appendages are included in the head, but in the *B.* they are not modified to maxillipeds. The parasite attaches to its host by means of hooks that are carried on the two short pairs of antennae. In some species, the first maxillipeds also carry attachment hooks. In the European carp louse, *Argulus foliaceus,* however, each of the first maxillipeds terminates in a powerful suction disk. The four pairs of biramous thoracic legs serve for swimming. Fishes are the most common hosts, but amphibians are also attacked. Very unusually for crustaceans, they lay their eggs on stones or plants, and there is no brood care. About 75 species are known, and the largest genus is *Argulus. Argulus foliaceus* (up to 6 mm in length) parasitizes not only carp but several other central European fish species. The largest species is *Argulus scutiformis,* which can attain a length of 30 mm.

2) A genus of the *Oligochaeta.*

Brand fungi: see *Basidiomycetes.*

Branta: see *Anseriformes.*

Brant's vole: see *Arvicolinae.*

Brassia: see *Orchidaceae.*

Brassicasterol, ***ergosta-5,22-dien-3β-ol:*** a plant sterol first isolated from rapeseed *(Brassica campestris)* oil. Together with campesterol, large quantities of B. have also been found in oysters.

Brassicasterol

Brazilian horned toad: see *Leptodactylidae.*

Breadfruit: see *Moraceae.*

Breakage-rejoining hypothesis: see Chromosome mutations.

Break-fusion-bridge cycle, ***BFB-cycle:*** a mechanism involving the formation of dicentric chromosomes or chromatids. During the BFB-cycle chromosome breakage occurs repeatedly, followed by duplications and deletions, so that daughter cells are produced which differ in their genetic composition. In the BFB-cycle the dicentric chromosome (chromatids) is stretched between the cell poles in the anaphase, and is then broken; one of the resulting fragments is deleted, while the broken end of the other fragment is rejoined to that of its duplicate. See Chromosome mutations.

Bream, *Abramis brama:* a very high-backed member of the fish family *Cyprinidae* (see). An inhabitant of standing and flowing water in central, eastern and northern Europe, it is a popular eating and sporting fish. It thrives especially in the deeper lakes of the Baltic region and in large dammed lakes. It grows up to 75 cm in length, and has a single row of pharyngeal teeth. The male B. is notable for the copious production of sperm during egg fertilization.

There are 2 other species of *Abramis.* In the former USSR, *A. ballerus* is fished commercially. It has a long anal fin and a pointed snout, and is found from the Rhine to the Danube in rivers entering the North Sea and the Baltic and in the Ural River. *A. sapa,* which has a long anal fin and a blunt snout, is found in the basin of the Aral, Caspian and Black Seas, including the Danube.

Bream region: see Running water.

Breeding density of birds, ***nesting density:*** the number of breeding pairs or nesting territories of a single species, or of all species, expressed per unit surface area. Breeding density depends on the environmental capacity, which in turn depends on the physical structure of the habitat, availability of food, climate, etc.

In Central Europe, the greatest breeding densities (800–1 600 breeding pairs/km^2) are found in structured deciduous and mixed woodlands. Similar densities also occur in wooded cemeteries and parks. Relatively uniform forests contain 200–800 pairs/km^2, while alder wetland woods are often occupied by only 120–300 pairs/km^2. Breeding densities are even lower on agricultural land, e.g. cereal fields (50–120), rape and fodder fields (50–120), grassland (70–130 pairs/km^2). Breeding densities are often considerably greater in agricultural areas when hedges and copses are present, and when nesting aids (e.g. nesting boxes) are introduced. Densities are about 300 in newly built areas, 400–700 in inner cities, and 600–1 100 pairs/km^2 in allotment and leisure gardening areas.

Together with species composition, bird breeding density provides a valuable index of the ecological stability of a habitat, and of the potential of resident birds for controlling insect pests (see Biological control).

Examples of breeding densities of individual species or groups are: skylarks on grassland (30–90 pairs/km^2), birds of prey in wooded areas in the southwestern part of eastern Germany (3–4 pairs/km^2), Cranes in the northern part of eastern Germany (0.5–5 pairs/100 km^2).

Brewer's yeasts: cultured yeasts used for brewing beer. They are specially selected strains, which are maintained in pure culture, and they are characterized by their relatively high production of ethanol and aromatic substances, as well as a high fermentation rate. Another important property is their ability to floculate, i.e. to clump together and form a deposit on the floor of the fermentation vat.

Low fermentation yeasts are distinguished from *high fermentation* or *distillery yeasts.* Distillery yeasts are the better alcohol producers. Low alcohol beers are brewed with low fermentation yeasts.

Most B.y. are *bottom yeasts,* i.e. they lie on the bottom of the fermentation vat when the fermentation is complete. Bottom yeasts are strains of *Saccharomyces carlsbergensis.* In contrast, *top yeasts* do not settle as readily as bottom yeasts, and they are found mostly in the upper part of the fermentation vat. Top yeasts are strains of *Saccharomyces cerevisiae,* and they are used, inter alia, for the production of ale, porter, and German "Weißbier".

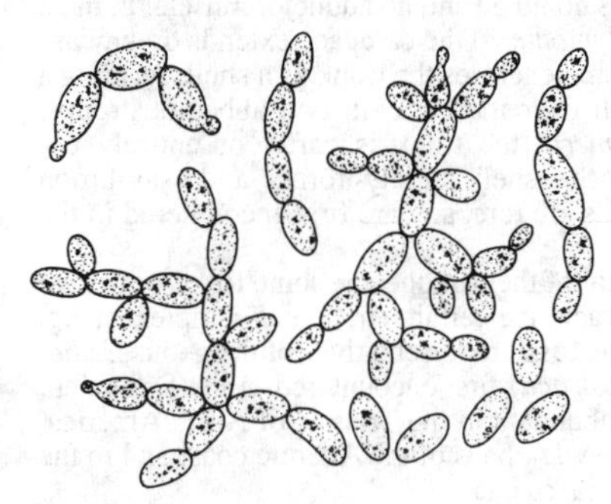

Brewer's yeasts. The bottom yeast, *Saccharomyces carlsbergensis.*

Briar: see *Rosaceae.*

Brill: see *Pleuronectiformes.*

Bristlecone pine: see *Pinaceae.*

Bristlemouths: see *Stomiiformes.*

Bristles: Setae (see) and Chaetae (see).

Bristletails: see *Apterygota.*

Bristleworms: see *Polychaeta.*

British gum: see Dextrins.

British toad: see *Bufonidae.*

Brittle stars: see *Ophiuroidea.*

Broad bean: see *Papilionaceae.*

Broadbills: see *Eurylaimidae.*

Broad fish tapeworm, *Diphyllobothrium latum, Dibothriocephalus latus* (plate 43): a tapeworm of the order *Pseudophyllidea* (see *Cestoda),* over 10 m in length, which lives parasitically in the intestines of predatory animals and humans. It has a lanceolate, flattened scolex, with 2 shallow, slit-like adhesive grooves (bothria) (see *Cestoda,* Fig.1) and a strobila of 3 000–4 000 proglottids. Most of the proglottids mature simultaneously, and the tapeworm may shed a million (10^6) eggs per day within the host intestine. A thick, tanned capsule protects the egg against digestion by the host enzymes and against subsequent desiccation. If the egg reaches water, it develops into a 6-hooked larva, which hatches as a ciliated coracidium. This swims until it is eaten by an appropriate copepod, which serves as the first intermediate host. Within the copepod gut the larva loses its cilia and penetrates the copepod gut wall into the body cavity, where it metamorphoses into a minute, spindle-shaped worm (procercoid) with 6 hooks and a posterior disk. No further development occurs, unless the copepod is eaten by a fish. The procercoid then burrows through the gut wall into the tissues of the fish, where it develops into a plerocercoid, which lacks hooks and possesses a scolex at the anterior end. The plerocercoid may pass from small fish to larger fish in the food chain, simply re-establishing itself in each new host without further development. If a suitable mammal (e.g. human or bear) eats the fish, the plerocercoid attaches to the intestinal wall and grows into a mature tapeworm. Humans become infected by eating raw, undercooked or inadequately smoked freshwater fish. Human *Diphyllobothrium* infection is sometimes accompanied by megablastic anemia. See *Cestoda* (Fig.3).

Broad-fronted crocodile: see African dwarf crocodile.

Broca weight: the desirable body weight of an adult human, based on height. The desirable weight in kilograms is equal to the height in centimeters minus 100.

Brolga: see *Gruiformes.*

Bromeliaceae, ***pineapple family, bromeliads:*** a family of monocotyledonous plants, containing about 1 700 species in 45 genera, native chiefly in tropical and subtropical America, with 2 species in West Africa. Most members are epiphytes, whose roots serve only for attachment. Nutrients are absorbed through the leaf bases from rain water that collects in the cistern or funnel formed by the rosette of fleshy leaves. Flowers are usually actinomorphic, 3-merous, hermaphrodite, in spikes, racemes or panicles. Pollination is by insects or birds. Since the bracts and the foliage leaves adjacent to the inflorescence are usually strikingly colored, many species are popular ornamentals, in particular those of the genera *Vriesea, Billbergia* and *Aechmea.*

The principal economic plant is the ***Pineapple*** *(Ananas comosus),* which grows in the ground. It originated from Brazil, and is now cultivated in most tropical regions of the New and Old World and Oceana. The fleshy infructescence is enjoyed raw as an aromatic fruit. Pineapple is also a source of bromelin, a proteolytic enzyme used pharmaceutically as a digestive aid.

In the warmer parts of the Americas, the universally

Bromeliaceae. Ananas comosus (pineapple): inflorescence and fructescence.

common ***Spanish moss*** or *Old man's beard* ***(Tillandsia usneoides)*** is used as a padding and filling material. Superficially, it resembles beard-moss, has no roots, and absorbs water and nutrients through scale hairs that cover the plant surface.

Brome mosaic virus: see Multipartite viruses, Virus groups.

Bromoviruses: see Multipartite viruses, Virus groups.

Bromus: see *Graminae.*

Bronchidesmus: see Syrinx.

Brontosaurus: see Dinosaurs.

Brood care, ***parental care*** (plate 3): the totality of parental behavior for the protection and rearing of offspring. Unlike brood provision (see), B.c. commences with the appearance of eggs or young. One or both parents display a direct interaction with the offspring. B.c. occurs in many forms, and is especially widespread in more advanced animals.

B.c. serves for *protection,* in that the parents guard their offspring (e.g. sticklebacks, many birds and mammals), or carry them on their bodies (some echinoderms, crustaceans, spiders, sea spiders, midwife toads, rhesus monkeys). Eggs or young are sometimes transported in body cavities, e.g. eggs are carried inside the shell of the oyster, in a ventral brood pocket of the male seahorse, in depressions between the warty skin protuberances of the Surinam toad, in a throat sac of the mouth-brooding frogs, and young are carried in the mouth cavity of mouth-brooding fish.

B.c. also involves the *creation of favorable conditions for growth and development* of the young, e.g. providing a warm environment with temperature regulation, providing oxygen, and providing food. Examples are: the movement of pupae by ants to different parts of the nest, so that they are always at an appropriate temperature; the direct and indirect incubation of eggs by birds, with temperature regulation; the warming of young animals by bodily contact by birds and mammals; measures for increasing the oxygen supply to the young of sticklebacks and some crustaceans; the feeding of larvae by colonial insects; the collection of food and feeding it to their young by birds and by some predators (e.g. fox); direct transfer of food generated in the parent's own body (feeding of crop milk to young pigeons, suckling of young mammals). Care of young animals also includes *cleaning and grooming* them and attending to their *hygiene;* the fox and hedgehog eat the feces of their very young offspring; birds remove the feces of their young from the nest.

Mammals and birds increase the ability of their young to survive after eventual separation from their parents by constantly *leading* them (e.g. ducks and hoofed animals), and *teaching* them *by example,* in particular how to find food or hunt prey (e.g. farmyard fowls, foxes, stoats, cats).

Involvement of parents in B.c. varies both qualitatively and quantitatively. The female parent generally carries the main burden. In some species, several individuals cooperate in a form of *communal B.c.* (e.g. penguins).

B.c. is a form of inherited instinctive behavior.

Brooding, ***incubation:*** a form or stage of brood care by birds, consisting of the continuous incubation of the eggs by the female bird, and in a few species by the male or both parents, until the hatching of the young animals. Depending on the species, incubation temperatures range between +35 ° and +41 °C, and the incubation period between 11 days and 8 weeks.

In poultry breeding, the female bird may incubate the eggs in a brood nest under conditions similar to those in the wild, or the eggs may be hatched in a thermostatically controlled incubator.

As early as 3 000 BC, the ancient Egyptians used artificial brood chambers to hatch large numbers of chickens' eggs. Artificial incubation was introduced into Europe in 1745.

Brood parasitism: a pattern of behavior, in which the incubation of eggs and care of offspring are assigned to a different animal species that has a similar mode of existence. Parasitic bumble bees are brood parasitic insects, while the cuckoo is a well known example among birds. Often the young of the host are incapacitated or killed by the newly hatched young of the parasite For this purpose the young cuckoo has an inborn behavior pattern for levering the host young from the nest.

Brood provision, ***parental provision:*** provision by parents for their young, which ends with the laying of eggs or the appearance of the young. It therefore differs from Brood care (see), which commences with the laying of eggs or appearance of the young. Although the parents do not interact directly with their offspring, B.p. may effect the lives of young animals for a considerable time. In its simplest form, B.p. increases the chances of survival of the young by the laying of eggs at particular sites at particular times of the year.

B.p. serves to *protect the brood,* e.g. eggs may be hidden (Roman snail, many arthropods, turtles), provided with a protective envelope (e.g. the egg-case of cockroaches, the spun cocoon of spiders, the protective secreted layers around the eggs of crickets and amphibia), or remain inside the body of the dead mother (e.g. cyst-forming nematodes and aphids). Finally, eggs may also be laid, for protection, in the body of another animal (eggs of the bitterling in the mantle cavity of a freshwater mussel).

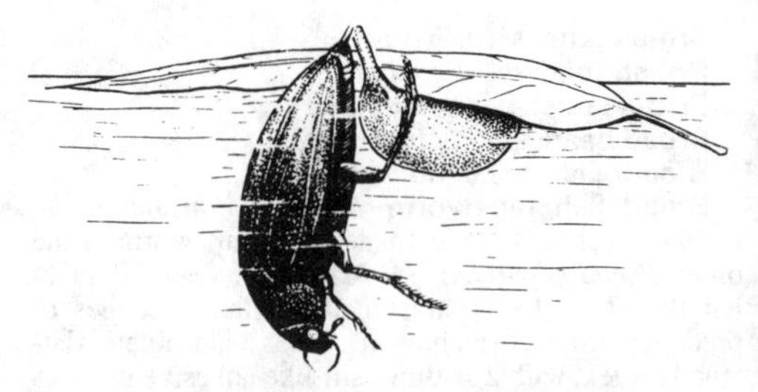

Fig.1. *Brood provision.* Female of the great silver beetle *(Hydrous = Hydrophilus piceus)* attaching a silky cocoon of eggs to the underside of a floating leaf (after v. Lengerken).

To *insure a food supply,* eggs are simply laid on the food (butterflies, flies, carrion beetles), or laid in or on the food which is preprepared in some way (e.g. leaf-rolling beetles, gall-forming insects, paralysis by stinging of live insect larvae by ichneumon flies), or laid in, on, or near the food which is brought from another location (e.g. the provisions laid down by dung beetles, digger wasps and spider wasps). Most of these measures also serve to protect the brood.

Fig.2. *Brood provision.* Leaf roll of the birch leaf roller *(Deporaus betulae;* family *Attelabidae)* (after v. Lengerken).

In some cases, it is also necessary to transfer intestinal symbionts that are essential for digestion, e.g. some bugs smear their eggs with an inoculum.

B.p. is a form of inherited of instinctive behavior.

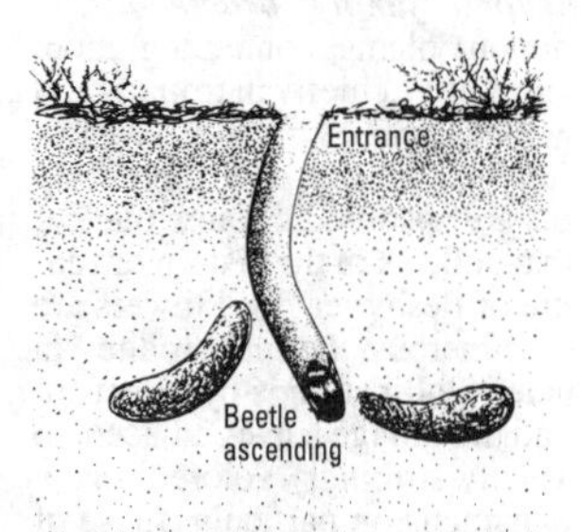

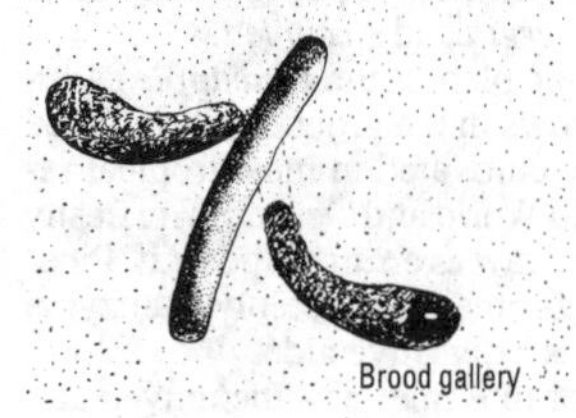

Fig.3. *Brood provision.* Main subterranean passage and brood galleries of the great dor beetle *(Geotrupes silvaticus = stercorosus)* (after v. Lengerken).

Brookesia: see *Chamaeleonidae.*
Brook trout: see *Salvelinus.*
Broom: see *Papilionaceae.*
Broomrape: see *Orobanchaceae.*
Broussonetia: see *Moraceae.*
Brown bat: see *Vespertilionidae.*

Brown bear, *Ursus arctos:* a yellow-brown to black-brown bear with numerous subspecies distributed throughout the temperate and colder regions of the Northern Hemisphere. During the winter, the male "hibernates" (winter sleep) in a natural shelter or self-dug den, while the female gives birth. Newborn young are about the size of guinea pigs. The numbers of Old World (Eurasian) B.bs. have become considerably reduced; they include the Kamchatka, Tibetan, European, Alpine and Syrian races, and their sizes (length about 2 m, weight 150–250 kg) depend on locality and available nutrition. The grizzly bear *(Ursus arctos horribilis)* is now rare in the USA, except under protection in Yellowstone and Glacier National Parks, and it is still found wild in Canada and Alaska; lengths of 2.5 m and weights of 360 kg have been recorded. Alaskan B.bs. (Kodiak bears) are the largest living carnivores, attaining lengths greater than 2.8 m and weighing up to 780 kg.

After their winter sleep, B.bs. feed on roots, moss and ants. Large quantities of vegetable material are generally consumed. Adults are not quick enough to catch wild hoofed animals, except bison, and they sometimes prey on domestic livestock; carrion is also eaten.

Brown birch: see *Betulaceae.*

Brown earth, ***brown forest soil:*** an intrazonal, well weathered, slightly leached, freely draining soil, with a surface horizon of mull humus. Deeper horizons show little differentiation. B.es. are derived from calcareous parent material, and their natural climax vegetation is humid, temperate, deciduous forest. They are used extensively for agriculture.

Brown forest soil: see Brown earth.
Brown frogs: see *Raninae.*
Brown noddy: see *Laridae.*
Brown rat: see *Muridae.*

Browns: butterflies of the family *Satyridae.* See *Lepidoptera.*

Brown smut: see *Basidiomycetes (Ustilaginales).*

Brown soil: a zonal soil formed in temperate and cool climates over calcareous material. There is a dark to gray-brown, slightly alkaline A-horizon and a ligher colored B-horizon.

Brown-tail, *Euproctis chrysorrhoea* L.: a member of the lepidopteran family, *Lymantriidae* (see). It is a white moth with a brown tail tuft. Caterpillars overwinter in communal webs, and complete their growth the following summer. It is an orchard pest, attacking in particular pear, apple, plum and cherry trees. Native to North Africa and Europe, and has now been introduced into North America.

Brucella: a genus of small, nonmotile, aerobic, Gram-negative, usually round to oval coccobacilli (0.4 µm diameter), occasionally also appearing as rods (about 1.5 µm in length). They live parasitically, and are the causative agents of disease in humans and other mammals. *B. melitensis* causes brucellosis (undulant fever, Malta fever) in humans, and also causes abortion in sheep and goats. *B. abortus* causes abortion mainly in cattle, while *B. suis* attacks mainly pigs. The host-parasite relationship is not strict, and both man and domestic animals are susceptible to all 3 species.

Brucine, ***2,3-dimethoxystrychnine:*** a very toxic alkaloid (but with only one tenth of the toxicity of strychnine), which together with strychnine constitutes between 2 and 5% of the seeds of the nux vomica tree, *Strychnos nux-vomica.* B. and strychnine are present in approximately equal quantities. The nux vomica tree grows widely in India, and occurs in other parts of Southeast Asia and northern Australia, and has been introduced into Europe. The main physiological effect of B. is paralysis of the smooth musculature. It is used in preparative chemistry to separate racemic acids into their optical isomers.

Brugia: see Filariasis; *Nematoda.*
Bruguiera: see Mangrove swamp.
Brush-tailed porcupine: see *Hystricidae.*
Brush turkeys: see *Megapodiidae.*
Bryidae: see *Musci.*
Bryony: see *Cucurbitaceae.*
Bryophyllum: see *Crassulaceae.*

Bryophytes, *Bryophyta,* ***mosses*** and ***liverworts*** (plate 32): a plant phylum containing about 26 000 autotrophic terrestrial species. The vegetative body of the more primitive members is thallose, while in all the other members the shoot is differentiated into an axis and leaves. The stem contains only primitive vascular elements, and the leaves almost always consist of a single layer of cells. B. do not possess true roots, only rhizoids, which serve predominantly to anchor the plant in its substratum. Sporangia and gametangia are always multicellular and enclosed by a sheath of sterile cells. B. display an alternation of generations. The unicellular, haploid spore germinates to form a haploid protonema, which is a green, threadlike structure. In some groups the protonema is a highly developed, persistent structure, whereas in other groups it is small and only slightly developed. The plant (the haploid gametophyte) arises from the protonema (sometimes from specialized buds), either as a thallus, or as a structure differentiated into stems and leaves. Gametangia are formed on the haploid plant. The female gametangium, known as an archegonium, has a characteristic structure, and it makes its earliest evolutionary appearance in the B. Archegonia are also present in the pteridophytes, so that B. and pteridophytes may be grouped together as *Archegoniatae.* Reduced archegonia can still be recognized, however, in primitive spermatophytes. The archegonium is a bottle-shaped organ consisting of a long neck and a venter, generally formed by a single layer of cells. The neck contains a row of neck canal cells, while the venter contains a single large central cell. Shortly before maturity, this large central cell divides to form the egg cell and the ventral canal cell, the latter lying at the base of the neck below the last neck canal cell. The male gametangium, the antheridium, is usually a spherical or club-shaped, short-stalked organ; its wall consists of a single layer of cells, which encloses spermatogenous tissue. Each of the usually numerous spermatogenous cells divides to form two spermatozoids, which are spirally twisted structures, consisting mainly of a nucleus with two flagella.

Fertilization can only occur in the presence of water, because the spermatozoids must swim to the archegonia, attracted there chemotactically by specific compounds secreted by the mucilaginous neck canal cells.

After fertilization, the egg develops without pause into the diploid sporophyte. The sporophyte of the B. is a more or less stalked, oval or spherical spore capsule, which in a certain sense lives "parasitically" on the gametophyte. In the mosses, this is covered by a calyptra, which consists of the remains of the haploid archegonial tissue. The entire sporophyte, including its haustorium-like foot, is called the sporogonium.

Spores are produced in groups of four (tetrads), which arise by the meiotic division of spore mother cells in the archesporium (the internal tissue of the capsule). The spore wall consists of inner delicate endospore and an outer resistant exospore.

Several mosses and many liverworts regularly reproduce asexually (vegetatively) by means of clusters of cells (gemmae) which detach from the parent plant and germinate into new plants.

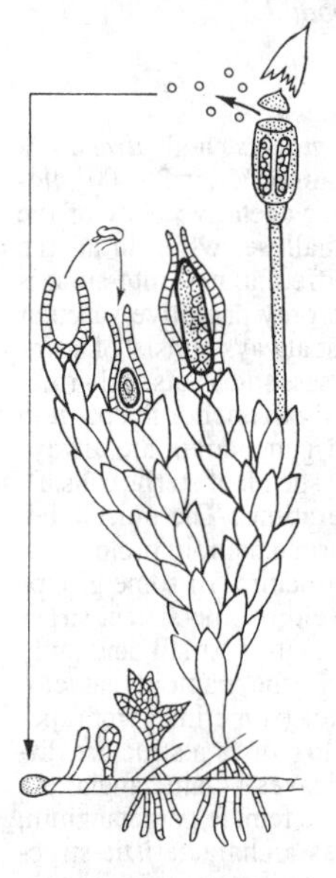

Life cycle of a monoecious moss (schematic). *R!* Meiosis, leading to spore formation in the mature diploid sporophyte. A spore germinates to form a haploid protonema, which in turn develops into a monoecious moss plant, bearing archegonia and antheridia. Fertilization of the ovum in the archegonium by a spermatozoid from the antheridium (water must be present for the passage of the spermatozoid) results in a diploid zygote, which develops into a sporophyte. Left to right on the apices of the 4 branches of the moss plant: antheridium, archegonium, young sporogonium, mature sporogonium.

B. are predominantly terrestrial plants, but on account of their delicate cuticle they are unable to withstand desiccation, and are therefore restricted to wet habitats. For the same reason they are absent from many ecological systems. Crowded growth in the form of cushions also helps to conserve moisture by protecting against evaporation. Some species have reverted to an aquatic existence. Certain species are characteristic of specific habitats, and they are also good bioindicators, often growing together in characteristic bryophyte communities. Fossil B. first appear in the Middle Carboniferous, but they provide no evidence for the evolutionary origin of the phylum. Nevertheless, it seems certain that they evolved from organisms resembling green algae, which displayed an alternation of generations. At one time, the bryophyte protonema was mistakenly thought to be a green alga.

Evolutionary development of the B. has occurred in the water-dependent gametophyte, and not in the sporophyte. In contrast, the pteridophytes have advanced by evolution of their sporophyte generation. Since the latter is less dependent on a wet environment, evolution of the pteridophytes has proceeded further, whereas the B. represent only a side branch in plant evolution.

B. are classified into two sharply distinct classes, the *Musci* (see) and the *Hepaticae* (see). There are no transitional forms between the two classes.

Bryozoa, *Polyzoa, Ectoprocta,* ***moss animals:*** a lophophorate phylum (see *Lophophorates*) containing about 4 000 species.

Morphology. All *B.* exist as colonies of small (about 0.5 mm) individual animals. Colonies may be microscopic, containing only a few zooids; or they may form extensive mat-like or moss-like incrustations, or large foliaceous growths up to 30 cm in length, often with a superficial resemblance to seaweed or coral. Most *B.* are marine and inhabit shallow coastal and estuarine waters, but some species occur at 6 000 m and deeper.

The single animal is called a *zooid,* while the colony is known as a *cormus, zoarium* or *polyzoarium.* Exoskeleton *(zoecium)* plus body wall constitute the *cystid,* while the contents of the zooid within the body wall (lophophore, digestive tract, muscles, etc.) are called the *polypide.*

The coelom of each zooid is divided by a septum, forming an anterior part (the mesocoel, which comprises the internal space of the lophophore and its tentacles) and a larger posterior part (the metacoel, which is traversed by muscle fibers and one or more strings of tissue that constitute the funiculus); these 2 divisions communicate via a pore. In most species, each individual animal is encased in a chitinous or calcareous protective covering *(zoecium),* into which it can withdraw completely, and which possesses an anterior opening for protrusion of the lophophore. The digestive tract is U-shaped and attached to the body wall by a band of cells known as the *funiculus.* The lophophore tentacles are protruded by an increase in the hydraulic pressure of the coelomic fluid; they then fan out, forming a funnel with the mouth or orifice at its base. When retracted by the conspicuous internal retractor muscles, the tentacles become bunched together within a sheath. There is no circulatory system, and specific structures for respiratory gas exchange are lacking. Nephridia are only rarely present.

Biology. All freshwater and most marine *B.* are hermaphrodite. Self-fertilization is common, but protandry also occurs. Ovaries (1 or 2 in number) and testes (1 or many) are simply masses of gametes covered by a peritoneum, which bulges into the coelom. Eggs and spermatozoa are released directly into the coelom by rupture of the peritoneum. In some species the walls of adjacent individuals form a common external brood chamber, the *oecium* or *ovicell.* In the *Stenolaemata,* the parent animal or its intestine becomes the brood chamber. Larvae display a wide variety of forms, but all possess a girdle of cilia *(corona),* as well as long anterior cilia, and a posterior adhesive sac. In those gymnolaemates that do not brood their eggs *(Electra* and *Membranipora,* both in the subclass *Cheilostomata),* the larva is called a *cyphonautes;* it is highly compressed, triangular in shape, and usually carries a chitinous valve on each lateral surface. When a larva comes to rest, the adhesive sac everts and attaches to the substratum. The larval tissues then histolyse, and the first zooid (the *ancestrula)* develops anew from the products of histolysis; a new colony forms by budding from the ancestrula.

Colonies of most marine *B.* are formed by the attachment of adjacent zooids to each other (non-stoloniferous colonies). Attachment occurs at the dorsal surface, so that the ventral surface is exposed (it is then normally referred to as the frontal surface). A more primitive type of colony, the stoloniferous colony, is formed by genera such as *Bowerbankia* and *Amathia* (both in the subclass *Ctenostomata*). The erect or surface-layered stolons consist of specialized zooids *(kenozooids)*. Attached to the stolon by their posterior ends are unmodified feeding zooids *(autozooids)*, which are often totally separate from each other.

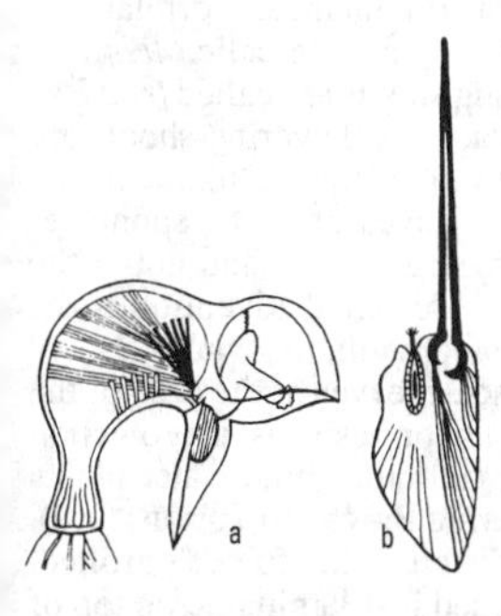

Bryozoa. a Avicularium. *b* Vibracularium.

In many *B.* species, the zooids display polymorphism. As a group, the various structurally and functionally modified zooids are called *heterozooids*. Thus, in addition to the feeding zooids (autozooids), a colony may also possess *avicularia*, which resemble small bird heads. The internal structure of the avicularium is much reduced, and the operculum is modified to form a movable jaw. Avicularia defend the colony against small animals and their larvae. Another type of zooid is the *vibracularium*, whose operculum is modified to a long bristle or seta, which sweeps away falling debris and settling larvae. Kenozooids are very reduced, often consisting of little more than the body wall and interior strands of the funiculus; these heterozooids serve to form stolons, as well as disks and rootlike structures for attachment of the colony to the substratum, and they also act as packing cells in the vegetative part of the colony.

Although the vast majority of species form sessile colonies, the freshwater *Cristatella* can progress slowly (10 cm/day) with the aid of a contractile foot, and the marine *Selenaria* uses the setae of its avicularia to progress at a rate of 1 m/h. Since colonies of *Selenaria* harbor symbiotic zooxanthellae, their observed movement toward light is probably relevant to photosynthesis.

Classification. Three classes are recognized.

1. Class: *Phylactolaemata, Lophopoda.* A class of 50 exclusively freshwater species, whose tentacles (up to 120 in number) emerge from a large, horseshoe-shaped, membranous base. Colonies consist of relatively unspecialized, homomorphic (monomorphic), incompletely separated individuals, whose walls are not calcified. The zooarium is gelatinous or leathery (chitinous). An ancestral-type, tripartite body organization has been retained. Overhanging the orifice (mouth) is a prosomal flap of tissue (the epistome), which contains a hydrostatic protocoel. Well developed circular and longitudinal muscles are present in the body wall. Hydraulic pressure for extrusion of the lophophore is generated by contraction of the circular muscles. The *Phylactolaemata* produce specialized resting cells, known as statoblasts, which develop from aggregations of cells on the funiculus. Highly resistant to freezing and desiccation, statoblasts serve for dispersal and for overwintering

2. Class: *Gymnolaemata.* A class of over 3 000 species. The majority are marine, but a few species are found in brackish and fresh water. The lophophore is relatively small and circular, and many species possess an operculum that closes over the orifice after retraction of the lophophore. Pressure for protrusion of the lophophore is generated by muscular deformation of the body wall by parietal muscles; longitudinal and circular muscles are absent from the body wall itself. Individual zooids are separate, at least by the presence of a wall. The body wall may not be calcified, and when calcification does occur it is only partial. The division of labor or polymorphism of zooids is greatest in this class; avicularia, vibracula, kenozooids and ovicells are all present

1. Subclass: *Ctenostomata* **(Comb-mouth *B.*).** Members have a circular and usually terminal orifice. An operculum is lacking. Walls are not calcified, and the body is often cyclindrical.

2. Subclass: *Cheilstomata.* The orifice is frontal, and an operculum is present (except in *Bugula*). Body walls are partially and often largely calcified. Zooids are typically box-shaped. Adjacent zooids have separate walls. Eggs are commonly brooded in ovicells. Both avicularia and vibracula are present in some species. In *Cribrilina,* arched spines form a vault over the frontal membrane. In some genera, the zoecia are packed closely together in a monolayer, forming encrusting colonies known as "sea mats", e.g. *Membranipora, Flustra.*

3. Class: *Stenolaemata* (in some taxonomies regarded as a subclass of the *Gymnolaemata,* and called the *Cyclostomata* or *Stenostomata*). A class of about 900 living marine species. Fossils of many extinct species are known. After lophophore retraction, the orifice is not closed by an operculum, but by a membrane. The lophophore is circular and carries no more than 30 tentacles; pressure for its protrusion is generated by a hydraulic apparatus, in which dilator muscles exert pressure on the fluid present in both the metacoel and an exosaccal, pseudocoelomic body cavity. The parent zooid or its intestinal tract acts as a brood chamber. Egg development is polyembryonic, i.e. the zygote cleaves to produce a ball of blastomeres, which buds off secondary embryos, which in turn may produce tertiary embryos. In this way, more than 100 blastulas may arise from a single zygote. Polymorphism is not extensive.

Fossil history. Apart from statoblasts in deposits of the Late Tertiary and Quaternary, the *Phylactolaemata* are unknown as fossils. Only those *B.* with calcareous zoecia have left fossils of their vegetative colonies, and these occur from the Cambrian to the present. Evolutionary peaks occur between the Ordovician and the Permian, and between the Cretaceous and the present. In many Middle and Upper Ordovician strata, bryozoan fossils are extremely abundant, forming a large proportion of the sedimentary rocks, such as marine limestones, calcareous shales and mudstones. Cheilostomes do not appear until the Jurassic, being abundant in Jurassic, Cretaceous and Cenozoic formations in many parts of the world. Bryozoan fossils have some value as index fossils for the characterization of geological strata.

Collection and preservation. Large colonies can be brought to the surface with a rake and laid out to dry. To keep colonies moist and supple, they are best stored in 70% ethanol. For the microscopic observation of zooids with extended lophophores, colonies are allowed to thrive in an aquarium, then narcotized by addition of chloral hydrate to the water. They can be preserved in an extended state in 2% formaldehyde or 70% ethanol.

Bryssetaeres: see *Gobiesocidae.*

Bryum: see *Musci.*

BSE: acronym of bovine spongiform encephalopathy (see Prion).

Bubalornis: see *Ploceidae.*

Bubalornithinae: see *Ploceidae.*

Bubalus: see *Bovinae.*

Bubo: see *Strigiformes.*

Bubonic plague: see *Yersinia pestis.*

Bubulcus: see *Ciconiiformes.*

Buccal incubation: see Mouth brooding.

Buccinium: see *Gastropoda.*

Bucconidae, *puffbirds:* an avian family (order *Piciformes),* containing about 30 species (in 10 genera) of small to medium sized birds (14–27 cm), found in continental tropical America, usually confined to the warm lowlands. Their lax, dull-colored plumage gives them a puffy appearance. The medium-sized, often stout and broadly flattened bill is decurved or hooked at the end, and surrounded by conspicuous bristles. The tail is short and the feet are zygodactylous. They feed on insects, which are caught on the wing, and they also take many invertebrates from ground litter. The nest is built in a burrow, usually excavated in a bank. Some species excavate downward from the surface of level ground, while others dig nest burrows in termite nests. One of the largest and most widespread members is the ***White-necked puffbird*** *(Notharchus macrorhynchus;* 24 cm; southern Mexico to Amazonia); less drab than most other puffbirds, both sexes have a thick tapering black bill and a largely black dorsal surface; the forehead, sides of the head, nuchal collar, and underparts are white, with a broad black band across the breast. Members of the genus *Monasa* are called ***Nunbirds.***

Buccopharyngeal respiration: see *Amphibia.*

Buckeye: see *Hippocastanaceae.*

Buckwheat: see *Polygonaceae.*

Bud:

1) In higher plants, the growing point of a shoot, enclosed by leaf primordia and often also by young leaves. A B. is formed when leaf primordia grow more rapidly than the shoot apex. At the same time, the primordia grow more rapidly on their lower surface, so that older leaves come together over the vegetative apex and the younger leaf primordia, providing protection against desiccation, physical damage, etc. Bs. at the end of a shoot axis are called *terminal Bs.* These are normally only a few millimeters in diameter, but in red and white cabbage (where the bud is the head of the cabbage) they may attain 40–50 cm. Bs. formed as outgrowths on the periphery of the shoot are called *lateral Bs.* or *axillary Bs.,* since they are always formed in a leaf axis, the corresponding leaf being the subtending leaf or bract. Axillary Bs. are important overwintering organs of perennial palnts. Brussel sprouts are particularly large axillary Bs., which would form inflorescences in the subsequent year, but are usually picked as a vegetable. Further lateral Bs. *(accessory Bs.)* may be formed by the first axillary B.; these may arise side by side (collateral Bs., e.g. many *Liliaceae,* inflorescence of the banana) or one above the other (serial Bs., e.g. *Lonicera, Robinia, Forsythia).* Axillary Bs. may remain dormant for years (resting Bs.). If their dormancy is broken (e.g. by injury, light conditions, drought, certain chemicals), these previously resting Bs. become the dominant sites of leaf formation for the entire vegetative season. During growth, Bs. show an increase in water content, enzyme content and respiration rate. Normally, the sequence in which Bs. open is determined by correlative, mostly hormone-regulated relationships between the Bs. themselves (see Apical dominance, Correlation). Bs. that give rise to foliage shoots are called *leaf Bs.,* those giving rise to flowering shoots are called *fruit Bs.,* while those producing foliaceous flowering shoots are called *mixed Bs.* A closed single flower is known as a *flower B.* Bs. that arise on leaves or roots, spontaneously or in response to physical injury, and not at the shoot apex or in a leaf axis, are called adventitious Bs. In trees that exhibit seasonal growth, the winter Bs. are enclosed in *B. scales,* i.e. scale leaves at the base of the B. axis, which secrete gum or resin as a protection against desiccation. In the autumn, some water plants form detachable overwintering Bs. called Turions (see). Within the B., leaf primordia are either folded or rolled. If the two halves of the folded leaf lamina lie on top of each other, the arrangment is said to be *conduplicative* (e.g. elm, hazel); multiple folding of the leaf is known as the *plicative* condition (e.g. beech). Folding may be transverse, fan-like, or longitudinal. If the two halves of the leaf primordium are rolled separately, the bud is said to be *involutive,* whereas in *convolutive* Bs., the two halves are rolled around each other.

2) In microorganisms, a small, vesicular evagination of the cell formed in the asexual reproductive process known as budding. Usually, the bud grows to the size of the parent cell, then abstricts to become an independent cell. For example, *Sacccharomyces* grows and multiplies by budding.

Budding: a type of vegetative Reproduction (see), in which a new cell is formed by abstriction of an extrusion or outgrowth of the parent cell. B. occurs widely in yeasts. Basidiospores and blastospores are also formed by B. In *bipolar B.,* abstriction occurs at the ends of oval or elongate cells, whereas in *multilateral B.,* abstriction occurs at any site on the cell. Incomplete abstriction of daughter cells results in loose cell associations. In animal B., a new individual may develop in a bud-like swelling of the parental body wall or cell mebrane, then detach itself.

Budgerigar: see *Psittaciformes.*

Bud moths: two species of the family, *Tortricidae* (order *Lepidoptera): Hedya nubiferana* Haw. ***(Gray bud moth)*** and *Spilonata ocellana* ***(Eye-spotted bud moth).*** Their caterpillars cause severe damage to fruit trees by eating into the fruit and leaf buds, especially the terminal buds.

Bud mutant: see Chimera.

Budorcas: see Takin.

Budorcatini: see *Caprinae.*

Buffalo: see *Bovinae.*

Buffalo currant: see *Grossulariaceae.*

Buffalo weavers: see *Ploceidae.*

Buffer: a solution whose pH-value is only slightly affected by the addition of hydrogen ions (H^+) or hydroxyl ions (OH^-). The buffering action is due to the consumption of the added H^+ or OH^- ions by the formation of weak acids or weak bases, respectively. Bs. designed to prevent an increase in acidity consist of a mixture of a weak acid and a strong base, e.g. sodium hydroxide and acetic acid (sodium acetate) or sodium hydroxide and phosphoric acid (sodium phosphate). Conversely, the effect of OH^- ions is buffered by a mixture of a strong acid and a weak base, e.g. ammonium hydroxide and hydrochloric acid (ammonium chloride). Laboratory Bs. for biochemical studies are usually mixtures of different buffering agents.

Physiologically, Bs. stabilize the pH of cells and body fluids during metabolic under- or overproduction of acids and bases, or during environmental changes of hydrogen ion concentration. Buffering is crucial for all living processes, because enzymes are active only in a certain range of pH, and for most enzymes this range is relatively narrow. Plant cell sap is efficiently buffered, while in animals the blood and other body fluids exhibit strong buffering activity.

Bufonidae, *bufonids, true toads* (plate 5): a family of the *Anura* (see) containing 335 living species in 25 genera, distributed throughout temperate, subtropical and tropical regions from sea level up to 4 000 m, but not represented in Australia or Madagascar. Most members are plump with stocky or squat bodies, broad snouts and relatively short legs that permit only short, clumsy jumps or steps. Ribs are absent. Apart from some species of *Dendrophryniscus,* all males possess a Bidder's organ (a rudimentary, nonfunctional ovary at the anterior end of the testis). Most bufonids lead a concealed existence, and many have horny digging protuberances on the hind legs. Although they cannot respond rapidly to threat or danger, their poisonous skin secretions represent an active means of defense (see Toad poisons). Most are terrestrial or arboreal in habit, entering water only to breed. During amplexus, the male attains a powerful and often prolonged grip on the female. Spawn is deposited in 2 long strings. A few members develop independently of standing water, and one genus is viviparous. Many species feed on insects and snails and are therefore useful as controllers of pests. Their spawning waters, especially in Europe, are threatened by drainage and pollution.

Bufonids have 5–8 holochordal, procoelous presacral vertebrae with imbricate (overlapping) neural arches. Intervertebral bodies are usually ossified, but they remain cartilaginous in *Pseudobufo*. Presacrals I and II are fused in *Atelopus, Leptophryne, Oreophrynella* and *Pelophryne,* but not in other genera. The atlantal cotyles of presacral I are juxtaposed. The coccyx lacks transverse processes, and it usually has a bicondylar articulation with the sacrum. In some taxa, however, the sacral articulation is shifted forward and the original sacral vertebra is incorporated into the coccyx; articulation is then monocondylar, or the coccyx becomes fused with the functional sacral vertebra. The pectoral girdle is usually arciferal, but in some genera *(Atelopus, Osornophryne)* the epicoracoids are fused to produce a pseudofirmisternal condition. An omosternum is usually lacking, but present in *Nectophrynoides, Werneria* and some species of *Bufo*. A bony sternum is always present. Pupils are horizontal. Maxillary and premaxillary teeth are absent.

Bufo represents the true toads in the strict sense. It is the most numerous genus (250 species), and it has an almost cosmopolitan distribution, being absent only from Australia and Madagascar. Species of *Bufo* characteristically have squat bodies, large parotid glands and a very warty skin, and they enter water only to breed. The fingers are free and the toes are more or less webbed. Males usually possess an internal vocal sac. During the winter, temperate species often bury themselves, and in areas of human habitation they may also overwinter in cellars, outhouses, etc.

Bufo bufo ***(Common European toad)*** is the commonest European toad, found from damp plains up to altitudes above the tree limit, and represented by subspecies in Asia and North Africa. Females attain a maximal length of 12 cm, and males are somewhat smaller. The dorsal surface is dirty brown with a dense covering of warts. A vocal sac is lacking. Behind each eye is a parotid gland, which secretes a white, poisonous fluid. Like all European toads, it feeds on slugs, toads and insect pests. At mating time, the male develops dark nuptual pads on the 3 inner fingers. Overwintering animals are often found in European cellars. Mating occurs from March to April in standing water. They return to the same water each year, resulting in sometimes impressive spawning migrations. Each 5 m-long string of spawn contains 6 000 or more eggs. Tadpoles hatch after 2–3 weeks, leaving the water as miniature adults in high summer, often forming swarms of hopping toadlets.

Bufo viridis ***(Green toad),*** which is more agile than *Bufo bufo,* inhabits dry hilly and mountainous areas of Europe, Asia and North Africa. The skin is yellowish gray with olive or brilliant green spots on the back. It is active both in the day and at night, and the male uses its large internal vocal sac to produce a drawn-out, trilling call. Mating occurs from April to May, and each of the 2 strings of spawn contains 10 000 to 12 000 eggs.

Bufo calamita ***(British toad*** or ***Natterjack toad)*** is a 6–8 cm-long, squat toad, olive gray in color with a distinctive sulfur yellow longitudinal stripe down the center of its back. Its range extends from northern and western Europe into central Europe, and it prefers relatively dry areas, where it can overwinter by burying itself up to 1 m deep in a sandy soil. It conceals itself among stones and in holes in the daytime, and it feeds on snails, worms and insects. With its short hindlegs, it does not jump, but progresses quickly like a mouse. Mating occurs from February to March in small areas of standing water that are rich in vegetation. Each short string of spawn contains 3 000 to 4 000 eggs. *B. calamita* and *B. viridis* are closely related and have been known to hybridize in parts of Europe where they share the same breeding water.

Bufo melanosticus is the commonest species of *Bufo* in India and Indo-China, with a mode of existence very similar to that of *Bufo bufo*. It is generally dull brown with black-headed warts, and the epidermis of the fingers and toes is stained black-brown. The parotid glands are very prominent.

Bufo regularis ***(Panther toad)*** is the commonest African *Bufo,* occurring from Cairo to Capetown. Up to 14 cm in length, its back is patterned, panther-like, with square dark spots. Spawn is deposited from August to January, a single batch containing more than 24 000 eggs.

Bufo blomberg ***(Colombian giant toad),*** a chestnut

brown giant toad (up to 25 cm in length), was discovered in 1951 in the wet cloud forests of Colombia. Its diet consists of slugs, worms, small rodents and young birds.

Bufo marinus ***(Aga toad, Giant toad, Neotropical toad, Cane toad),*** occurring in Central and South America, is darkly colored with black spots, and is usually 15–20 cm long, although lengths of 25 cm have been recorded. The parotid glands are very large. This very adaptable toad has been introduced into many areas outside its normal range, e.g. Florida, Solomon Islands, Hawaii and Australia, often with the intention of counteracting insect pests of sugar cane (hence "cane toad"). Notably in Australia, this planned biological control was a failure, but *Bufo marinus* adapted well and is now very common in Queensland, regarded by some with affection and by others as vermin. Males emit a loud barking sound. Poison from the enormous parotid gland can be squirted over a distance of 30 cm. The toxin is powerful and normally proves fatal to dogs and cats that attempt to bite one of these toads. During the hot season they remain concealed, emerging, often in large numbers, during cooler, wet weather.

Bufo mauritanicus ***(Berber toad*** or ***Mauritanian toad)*** is common in North Africa. Up to 12 cm in length, its skin is mottled reddish brown, and it inhabits steppe, river valleys, oases and areas of human settlement.

Bufo is well represented over the entire continent of North America. Thus, *Bufo americanus* ***(American toad),*** with its high trilling voice, ranges over the eastern half of North America, whereas *Bufo cognatus* ***(Great Plains toad)*** occupies the Great Plains east of the Rocky Mountains. *B. cognatus* often spawns in cattle drinking troughs; in North Dakota and Minnesota it has been known to undertake massive migrations, in which millions migrate over hundreds of kilometers, usually northward. In 1941 in Dakota they were so numerous that their squashed bodies on the highway caused skidding and became a traffic hazard. *B. boreas* ***(Western toad)*** is native to northwestern North America, ranging far to the north and up to altitudes of 3 000 m, where it spawns in the snow. A subspecies *(Bufo boreas halophilus),* sometimes called the ***Salt toad,*** is found in California, where it reproduces in salty swamps.

Pseudobufo is a genus of mostly aquatic toads, found in Southeast Asia, e.g. *Pseudobufo subasper,* which grows up to 15 cm in length, typically has large webs on its feet, and its upward-directed nostrils lie on top on the snout. It inhabits slow flowing waters in Sumatra, Borneo and the Malay peninsula.

Nectophryne ***(African tree toads)*** are small, arboreal toads found in Africa and southern Asia. Eggs are laid on land, but the tadpoles have a free swimming stage and must make their own way to water. Adults live mostly on ants and termites.

Nectophrynoides ***(Live-bearing toads)*** is an African genus, with a unique (for the *Anura)* mode of reproduction. Fertilization is internal. Mating occurs on land by pressing together of the cloacae. Although the male lacks a specialized copulatory organ, sperm passes directly into the cloaca of the female. Young, fully developed toads are extruded by the female. A larval tail is present before birth, and this probably serves as an embryonal repiratory organ. During development the larvae lack gills. *Nectophrynoides occidentalis* is found at an altitude of 1 300–1 600 m on Mt. Nimba in Guinea (West Africa), restricted to only a few square meters of the entire mountain. Within this small area, however, they are very numerous, concealing themselves in rock crevices in the dry season and emerging in swarms in the rainy season. *N. vivipara* and *N. tornieri* occur in Tanzania.

Some classifications of the *Anura* place *Atelopus* in a separate family *(Atelopodidae),* but here they are recognized as a genus of the *Bufonidae.* Forty three species have been described, most with rudimentary inner toes, and outer toes that form long claws. Their striking coloration of black, yellow and red rivals that of the *Dendrobatidae.* Both sexes have a bright, bell-like call. When threatened, the Argentinian *Atelopus stelzneri* turns the bright orange-red undersides of its hands and feet toward the attacker; otherwise, its coloration consists of irregular yellow blotches on a dark background.

Bugeranus: see *Gruiformes.*

Bugs: see *Heteroptera.*

Bugula larva: the larval form of the polyzoan genus, *Bugula.* The larva lacks an alimentary canal and has an elongated axis.

Bulb:
1) See Shoot.
2) See *Liliaceae.*

Bulbil: see *Reproduction.*

Bulbogastrone: see Gastrointestinal hormones.

Bulbuls: see *Pycnonotidae.*

Bulking: see *Chlamydobacteriaceae.*

Bulldog ants: see Ants.

Bullfinch: see *Fringillidae.* See plate 6.

Bullfrog: see *Raninae.*

Bullhead: see *Cottidae.*

Bullheads: see *Siluriformes.*

Bumble bees, *Bombinae:* a subfamily of Bees (see) containing about 200 species, mostly in temperate and northern regions, with some in mountainous regions further south. The imago is somewhat plumper and more hairy than other bees, and has a longer proboscis. They often become active in the early spring when climatic temperatures are still relatively low, and when other bees are still inactive. They generate body heat by the hydrolysis of fructose-1,6-bisphosphate in a so-called "futile metabolic cycle"; the hairy body also aids heat conservation. Like honey bees, they live socially (except for *Psithyrus* species) in nests, with a division of labor among queen, workers and males, but the colony comprises no more than 400–500 individuals. Only young fertilized females hibernate and survive the winter, all other members of the colony dying in the autumn. New nests are therefore founded every year by these young queens (cf. honey bee colonies, which survive the winter). Female workers (incompletely developed females) have 12–segmented antennae and a sting. Male workers have 13-segmented antennae and no sting. Nests are usually built below ground in earth cavities, but some species prefer concealed sites above but close to the ground. *Psithyrus* species do not build a nest, do not have workers, and their progeny are reared by the workers of various species of *Bombus.* B.b. feed on nectar and pollen, and they are important pollinators of many wild and agricultural plants, especially fruit trees and soft fruit.

Bunodont: see Teeth.

Bunolagus: see *Leporidae.*

Bunolophodon: see *Mastodon.*

Bunt: see *Basidiomycetes (Ustilaginales).*

Bunter deposits: see Muschelkalk, Geological time scale.

Buntings: see *Fringillidae.*

Bunya-bunya: see *Araucariaceae.*

Bupalis: see Bordered white.

Buphaginae: see *Sturnidae.*

Buphagus: see *Sturnidae.*

Burbot, *Lota lota:* a fish of the order *Gadiformes,* and the only freshwater member of the family, *Gadidae* (cods) [subfamily *Lotinae,* tribe *Lotini* (hakes and burbot)]. It occurs in northern parts of Eurasia and North America.

Burdo(n): a true graft hybrid, in which cell fusion occurs between the graft partners, leading to changes in the chomosomes. The existence of Bs. is disputed. Certain chimaeras of black nightshade *(Solanum nigrum)* and tomato are thought have the partial character of a B.

Burdock: see *Compositae.*

Burhinidae, ***thick-knees*** or ***stone curlews:*** a small avian family (8 species), order *Charadriiformes* (see). Plumage is generally cryptic with black and white wing bars that are visible in flight. The bill is short and plover-like, and there is a general superficial resemblance to bustards. Special distinguishing characteristics are a broad head with enormous yellow eyes (an adaptation to the nocturnal habits of these birds) and enlarged intertarsal joints (hence "thick-knees"). The name, stone curlew, indicates their general preference for dry, stony habitats. *Esacus (= Orthorhamphus) magnirostris* ***(Beach curlew;*** 53 cm) is a monotypic genus inhabiting reefs, beaches and coastal and river mud banks. It exists as 2 races: *Esacus magnirostris recurvirostris* ***(Great stone curlew)*** inhabiting the Indian subcontinent and Burma; and *E.m. magnirostris* of Indonesia, the Philippines, New Guinea, northern Australia, Solomon Islands and the Andamans. All remaining 7 species constitute the genus, *Burhinus.* They are generally gray-brown, streaked and barred with darker browns and blacks. The ***Stone curlew*** *(Burhinus oedicnemus;* 40.5 cm) inhabits dry river beds, sand dunes and deserts, heaths, dry salt marshes and even cultivated land in Europe, North Africa and southwestern Asia. *Burhinus vermiculatus* ***(Water dikkop*** or ***Thick-knee;*** 38 cm) exists as several races in sub-Saharan Africa, where it inhabits river beaches, banks and islets.

Burhinus: see *Burhinidae.*

Burkitt's lymphoma: see Tumor viruses.

Burnet: see *Rosaceae.*

Burnet moths: moths of the family *Zygaenidae.* See *Lepidoptera.*

Burning bush: see *Chenopodiaceae.*

Bursa of Fabricus, ***cloacal bursa:*** a lymphoid organ on the rectum of birds. It consists of a dorsal median diverticulum of the proctodeum, and often contains traces of fecal material. It appears to be unique to birds, and it is not clear whether a corresponding organ is present in mammals. In most birds it is pear-shaped, e.g. in all examined passeriforms and in the domestic fowl. In others (e.g. anseriforms) it is spindle-shaped. The Bursa is the site of differentiation of lymphatic stem cells into immunologically competent bursal lymphocytes (B-lymphocytes). A few T-lymphocytes are also found in the Bursa.

Bursera: see Linalool.

Bushbaby: see Galagos.

Bushbuck: see *Tragelaphinae.*

Bush cricket: see *Saltoptera.*

Bush frogs: see *Rhacophoridae.*

Bush larks: see *Alaudidae.*

Bushman's apron: see Hottentot apron.

Bushmaster, *Lachesis mutus:* a monotypic genus of pit viper (see *Crotalidae)* found in mountainous rainforest regions of Central and South America. It grows up to 3.75 m in length, and is the largest and most powerful of the pit vipers, as well as being the second largest venomous snake in the world after the King cobra. Its background color of gray-brown is patterned with dark diamond shapes with light yellow borders. The venom fangs may be up to 3.5 cm in length, but the venom is somewhat less potent than that of most rattlesnakes and lance-head snakes. Unlike most other pit vipers, it lays eggs.

Bush pig: see African bush pig.

Bush wren: see *Xenicidae.*

Bustard quails: see *Gruiformes.*

Bustards: see *Gruiformes.*

Butastor: see *Falconiformes.*

Butcherbirds: see *Laniidae.*

Buteo: see *Falconiformes.*

Buteogallus: see *Falconiformes.*

Buteoninae: see *Falconiformes.*

Butter crab: a newly molted and therefore soft-skinned crab, lobster or crawfish. The term is applied in particular to crawfish.

Buttercup: see *Ranunculaceae.*

Butterfish, *Pholis gunellus:* a fine-tasting bottom fish of the western and eastern Atlantic. It has a laterally compressed, ribbon-like body, and attains a length of 30 cm. It is a member of the blennylike perciform fish family, *Pholididae* (gunnels), which is now placed in the suborder *Zoarcoidei.* Formerly, this family belonged to the older, expanded form of the *Blennioidei* (see).

Butterflies, *Diurna, Rhopalocera:* a collective term for members of the superfamilies, *Papilionoidea* and *Hesperioidea,* i.e. families *Hesperidae* through *Papilionidae* in the classification list under *Lepidoptera* (see). The evolutionary relatedness of these two superfamilies is disputed. B. fly mainly in daylight, and at rest they fold their wings together vertically above the body (plate 1, brimstone butterfly). They generally lack a bristle coupling mechanism for the wings (frenulum), and the antennae are usually clubbed. Fore- and hindwings overlap, which helps them to beat in unison. See Moths.

Butterflyfishes, *Chaetodontidae:* a family (10 genera, 114 species) of fishes in the order *Perciformes.* They have high-set, strongly laterally compressed bodies, occasionally with a pointed snout (the 2 species of *Forcipiger* of the Indo-Pacific have a very elongated snout), and are often very colorful. Many species have a dark band across the eye, and many have a dorsal or posterior eyespot. B. are common around the coral reefs of tropical seas (where they feed on coral polyps), especially the regions between and including Australia and Taiwan. Thirteen species occur in the Atlantic, and 4 in the eastern Pacific. Many are popular marine aquarium species.

Butterfly flower: see *Asclepiadaceae.*

Butternut: see *Juglandaceae.*

Butterwort: see Carnivorous plants.

Butyric acid, ***butanoic acid:*** CH_3—CH_2—CH_2—COOH, a fatty acid constituting 3–5% of the acids esterified to glycerol in butterfat, and which contributes to the objectionable odor of rancid butter. Esters of B.a. are present in many essential oils. Perspiration contains traces of B.a. It is also produced in the putrifaction of protein, and in the butyric acid Fermentation (see) of carbohydrates. The even more unpleasant-smelling *iso*-butyric acid [$(CH_3)_2$—CH—COOH] also occurs in perspiration, as well as in plants, e.g. fruits of *Ceratonia siliqua* (carob tree).

Butyric acid bacteria: anaerobic bacteria that produce butyric acid from carbohydrate (butyric acid fermentation, see Fermentation). B.a.b. occur widely in the soil, and they are functionally important among the rumen microflora of ruminants. Most species of *Clostridium* ferment carbohydrate with the production of butanol, ethanol, acetic acid, acetone and butyric acid, and may therefore be classified as B.a.b. Butyic acid production, however, is highest in *Cl. butyricum,* and the term B.a.b. is normally reserved for strains of this species. It is a slightly curved, motile rod (0.7 × 5–7 μm) with oval eccentric spores that swell the sporangium. It is normally Gram-positive, but older cultures may become Gram-negative. The organism is used industrially to produce butyric acid, as well as being one of the organisms involved in the anaerobic retting of flax and hemp (soluble plant material and pectic substances are removed by fermentation, leaving the fibers intact).

Buttonquails: see *Gruiformes.*

Buxaceae, ***box family:*** a family of dicotyledonous plants, containing about 60 species in 6 genera, found in tropical and temperate regions. They are evergreen shrubs or small trees, rarely herbs, with leathery leaves. Flowers are regular, usually monoecious, with 4, 6 or numerous stamens, or a superior 3-celled ovary (rarely 2- or 4-celled) ovary with 3 pistils. The calyx usually consists of 4 sepals; petals are absent. Pollination is by insects, and the fruit is a loculicidal capsule or drupe. The seeds have a fleshy endosperm, and the embryo is straight.

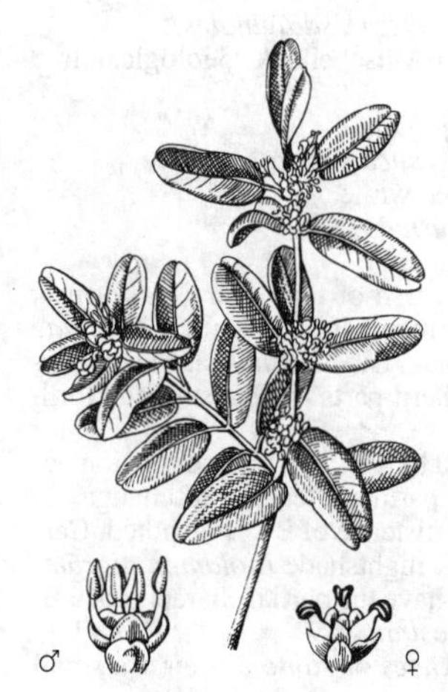

Buxaceae. Flowering branch of *Buxus sempervirens* (box).

The most familiar representative is the Mediterranean ***Box tree*** ***(Buxus sempervirens),*** a shrub or small tree that is relatively tolerant of atmospheric pollution. It is a popular plant of parks and gardens, used for enclosing borders and for topiary.

Buzzards: see *Falconiformes.*

Byblidaceae: see Carnivorous plants.

Byblis: see *Carnivorous plants.*

Byssus: a threadlike, viscous secretion from the byssus gland, which lies ventrally or at the base of the foot of many bivalves. In water, this secretion hardens into bundles of silky, tensile threads of tanned protein (byssus threads), which serve to anchor the bivalve permanently or temporarily to its substratum.

C

Caatinga: light forests of thorn bush communities, containing a high proportion of mimosa and cactus species. C. are found in Brazil, chiefly in the northern provinces of Ceara, Rio Grande do Norte, Pernambuco, Piauhy, Goyaz and Bahia, where the rivers dry out in the summer, and where the soils are sandy, primary granite or Jurassic limestone, and without natural springs.

Cabbage leaf miner: see *Yponomeutidae.*

Cabbage root fly: see Anthomyids.

Cabbage web moth: see *Yponomeutidae.*

Cabbage white, *large white,* *Pieris brassicae* L.: a common butterfly (family *Pieridae),* occurring throughout the entire Palearctic region. It is a well known migrant, occasionally forming large swarms. The yellowish-green caterpillar (plate 45) feeds on various members of the *Cruciferae,* and is a serious pest of cabbages. The male butterfly has black-tipped wings, which are otherwise unmarked. The forewings of the female (Fig.) carry additional black spots.

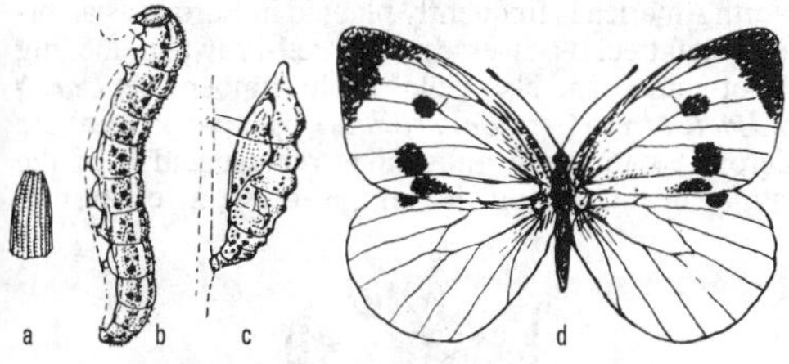

Large cabbage white (Pieris brassicae). a Egg. *b* Caterpillar. *c* Pupa. *d* Butterfly (female).

Cable theory: a theoretical model describing the spread and conduction of an action potential in a nerve fiber or muscle cell. The membrane is assumed to have a resistance r_m and a parallel capacitance c_m. The resistances of the interior solution and the exterior medium are designated r_i and r_e, respectively (all values based on 1 cm length of fiber). Since r_i is small relative to r_m, the only appreciable flow of current occurs longitudinally inside the fiber. The current through the membrane consists of a capacitative current and an ionic current. If r_m remains constant, the capacitative component causes small depolarizations, which die away. Decrease of the electrotonic potential (see Electrotonus) is described by the *membrane length constant:* $\lambda = \sqrt{r_m (r_e + r_i)^{-1}}$. The membrane time constant is equal to $r_m \times c_m$. In nerve fibers, λ is usually several millimeters, so that electrotonic potentials decay to immeasurably small values over distances of only a few centimeters. The conduction velocity of the action potential is determined by the amplitude of the depolarizing inward current of sodium ions, but also by the conditions for the electrotonic spread of membrane currents, i.e. by r_m, r_i and c_m. These conditions are more favorable in thick fibers than in thin ones, so that conduction velocity increases with fiber diameter. Thus, the giant axon of certain cephalpods attains conduction velocities of up to 20 m.s^{-1}. In cephalopods, this rapid nervous response, with its accompanying natural advantages, has been achieved by the development of nerve fibers with diameters up to 1 µm. Vertebrates have achieved the same effect and better by insulating sections of their nerve fibers with myelin; myelinated fibers of 20 mm diameter can achieve conductance velocities of 100 m.s^{-1}. Conductance velocities calculated from the cable theory agree very closely with experimentally determined values. The experimental values can therefore used to calculate r_m and c_m.

Cablin patchouli: see *Lamiaceae.*

Cabomba: see *Nymphaeaceae.*

Cacatua: see *Psittaciformes.*

Cachalotes: see *Furnariidae.*

Caciques: see *Icteridae.*

Cactaceae, *cactus family:* a family of dicotyledonous plants containing about 2 000 species. The family is indigenous to the American tropics and subtropics, especially in desert and semidesert regions. Some species, however, have been introduced and have become established in the tropics and subtropics outside the American Continent, e.g. in the Mediterranean region and South Africa.

Most cacti are markedly xeromorphic plants with flattened, columnar or conical shoots (stem succulents), which may be smooth, longitudinally ribbed or divided into globose sections. Leaves are absent (except in the primitive genus, *Peireska),* thereby effecting a considerable reduction in the surface area of the plant. Spines (generally considered to be modified leaves) and sometimes barbed bristles (glochids) are present. Spines and glochids arise from cushion-like areoles, which are considered to be condensed lateral branches. Water loss is also reduced by the presence of a thick cuticle. The usually large, striking, sessile flowers are multimerous, with numerous, epipetalous stamens. The spirally arranged perianth segments display a gradual transition from sepaloid to petaloid structures, and they are often fused at the base, forming (with the bases of the filaments) an elongate hypanthium. The inferior ovary, which usually develops into a berry, consists of 2 to many united carpels.

Many cacti are grown as ornamental plants. Familiar species are the ***Queen of the night*** *(Selenicereus = Cereus grandiflorus)* and the ***Christmas cactus,*** which is cross between *Schlumbergera truncata (= Epiphyllum truncatum = Zygocactus trancatus)* and *Schlumbergia russelliana (= Epiphyllum russellianum)* made in about 1860 by W. Buckley. Also popular are the ***Bishop's cap*** or ***Star cactus*** *(Astrophytum myriostigma)* and various species of *Opuntia, Cereus* and *Mammillaria.* Flowers of *Nopalxochia* develop slowly from an attractive bud to the open flower, a feature exploited in the attractive

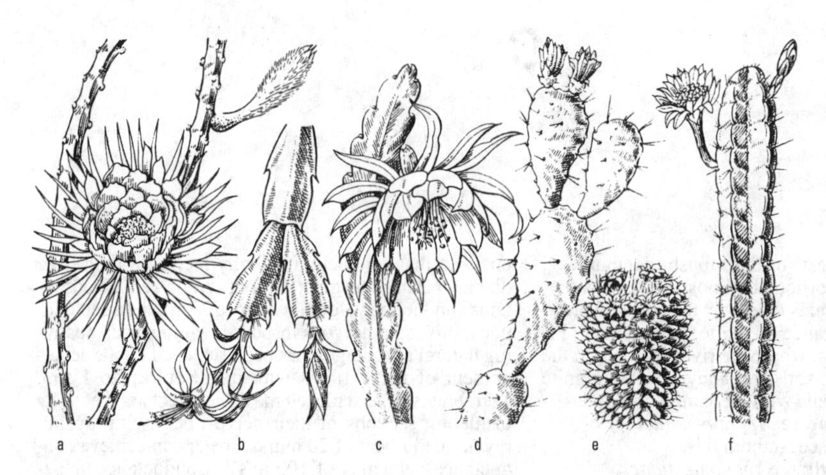

Cactaceae. a Queen of the night *(Selenicus grandiflorus).* b Christmas cactus *(Zygocactus truncatus).* c Leaf cactus *(Nopalxochia* hybrid). d Indian fig *(Opuntia ficus-indica).* e Mammillaria *(Mammillaria* sp.). f Echinocereus *(Echinocereus* sp.).

ornamental hybrids of *N. phyllanthoides* ***(Leaf cacti).*** The ***Indian fig*** *(Opuntia ficus-indica),* which grows wild in the Mediterranean region, produces edible fruits. Some cactus species contain alkaloids, e.g. the Mexican *Lophophora williamsii,* whose dried shoots (known as peyotl) have long been used as an hallucinogenic intoxicant by the Indians living near the Mexico-USA border; the main active constituent is the protoalkaloid, mescaline (3,4,5-trimethoxyphenethylamine).

Cactus: see *Cactaceae.*

Cactus wren: see *Troglodytidae.*

Cadaverine, ***pentamethylenediamine:*** $H_2N—(CH_2)_5—NH_2$, a Biogenic amine (see), formed by the enzymatic decarboxylation of the amino acid lysine. It possesses an unpleasant smell, and it arises during the putrifaction of proteins. Together with putrescine, it was previously considered to be a dangerous product from corpses, but it is now known to be relatively harmless. C. occurs widely in living organisms, and has been found, e.g. in bacteria, porcine pancreas, ergot, soybeans, potatoes and peas.

Caddis flies: see *Trichoptera.*

Cadherins: see Epithelium.

Caecilians: see *Apoda.*

Caeciliidae: see *Apoda.*

Caecophagy: British spelling of Cecophagy (see Cecotrophy).

Caecotrophy: British spelling of Cecotrophy (see).

Caecum (plural **Caeca**): British spelling of Cecum (plural Ceca).

Caenogenesis: a term introduced by E. Haeckel in 1874 for the occurrence of transitory ontogenetic peculiarities (e.g. larval adaptations), which are phylogenetically irrelevant. C. can mislead the interpretation of phylogenesis, in that caenogenetic characters do not represent phylogenetically recapitulated characters. Thus, veliger larvae may be present or absent during the development of aquatic snails *(Prosobanchia),* without affecting the organization of the adult. See Palingenesis.

Caenophidia: see Snakes.

Caesalpiniaceae, ***carob family:*** a dicotyledonous plant family of about 2 000 species, occurring mainly in the tropics and subtropics. They are mostly trees, rarely bushes, with alternate, paripinnate, pinnate or bipinnate leaves, and hermaphrodite, zygomorphic, 5-merous flowers, which are pollinated by insects or birds. Together with the *Papilionaceae* and *Mimosaceae*, the *C.* formerly constituted the family *Leguminosae.* Flowers of the *C.* display an early stage in the evolution of papilionate flower structure (see *Papilionaceae),* but they differ from the flowers of the *Papilionaceae* in that the uppermost petal is overlapped by the 2 lateral petals, which in turn are overlapped by the 2 lower petals. Stamens are usually free. The superior ovary develops into a pod.

The ***Honey locust*** *(Gleditsia triacanthos)* from North America is frequently planted in Europe as an ornamental tree. It possesses shiny, red-brown, branching shoot-spines, and also yields useful timber. The ***Carob*** or ***Locust tree*** *(Ceratonia siliqua),* a native of Southern Europe, is the only indigenous representative of the family in that region; its unripe, dried, sweet-tasting

Caesalpiniaceae. Carob or locust tree *(Ceratonia siliqua).* Branch with flowers and fruits (pods).

fruits (carob beans) are eaten, used in the tobacco industry for aromatization, fed to cattle and sometimes used as a coffee substitute.

Dried pods of various tropical species of *Cassia (C. senna, C. angustifolia, C. acutifolia),* known commercially as senna pods, contain laxative compounds (glycosides of anthraquinones). The dried leaves are also official as a purgative. Often cultivated as an ornamental plant in the tropics, ***Indian laburnum*** *(Cassia fistula)* is also the source of cassia pods, which yield a laxative pulp. The interior of the pod of *C. fistula* is divided into compartments by transverse dissepiments, about 6 mm apart, each compartment containing a single seed; the pulp occurs as a thin, very dark layer, which is soft and fills the compartments of the fresh fruit, but is much contracted in dry pods so that the seeds are loose and rattle when shaken. The ***Tamarind tree*** *(Tamarindus indica)* is a large, handsome tree, native to tropical Africa, and cultivated throughout the tropics, especially in India and the West Indies, where the acid-sweet fruit pulp provides a valuable food and is especially used for beverages. Official tamarinds are the fruits preserved in hot syrup. Many species of *Copaifera* are official, yielding copal resin (e.g. West African copal from *Copaifera guibourtiana),* which is used for the manufacture of laquers and varnish, and for various other purposes in chemical industry and in pharmacy. Copal resin rapidly solidifies to a hard material resembling amber, which is sometimes worked into small ornaments. Copaiba or copaiba balsam, an oleo-resin from the trunk of *Copaifera landsdorfi* and certain other species, was used by Brazilian Indians and introduced into Europe in the 17th century. The constituent volatile oils stimulate the entire genito-urinary tract and the bronchi, increasing mucous secretion and promoting expectoration.

Caffeic acid: an aromatic carboxylic acid, which occurs widely in plants, either free or as a component of chlorogenic acid. It is present, e.g. in coffee, poppy, melissa (lemon balm), dandelion and hemlock.

Caffeic acid

Caffeine, ***thein***, ***1,3,7-trimethylxanthine:*** a purine alkaloid. C. monohydrate crystallizes as white, bitter-tasting felted needles, the anhydrous form as hexagonal prisms. It occurs in coffee beans (1 to 1.5%), leaves of the tea bush (up to 5% of the dried leaves), and some other tropical plants (cola nuts, maté leaves, guaranta paste). C. stimulates the central nervous system, and has a mildly euphoric action. It stimulates the heart, basal metabolic rate, and respiration, and it is the active constituent of coffee and other stimulants. Excessive intake causes excitement, palpitations of the heart, and sleeplessness.

Caffeine

Caiman: see *Alligatoridae,* Spectacled caiman.
Cainophytic: see Cenophytic.
Cainozoic: see Cenozoic.
Cairina: see *Anseriformes.*
Cairinini: see *Anseriformes.*
Cajanus: see *Papilionaceae.*
Calabash: see *Cucurbitaceae.*
Caladium: see *Araceae.*
Calamites: see *Equisetatae.*
Calamocichla: see *Sylviinae.*
Calamus: see Feathers.
Calanoida: see *Copepoda.*
Calanus: see *Copepoda.*
Calappidae: see Box crabs.
Calcarea: see *Porifera.*
Calcareous sponges: see *Porifera.*

Calceola, ***slipper corals:*** an extinct genus of noncolonial, operculate Devonian tetracorals, family *Calceolidae (Cystiphyllidae).* Solitary individuals lived in the mud of the sea bottom. One side of the slipper-shaped calix was flat and triangular, and the calix was marked internally with fine septal striae. A very thick, semicircular, movable cover (operculum) could be swung across the calix like a lid. The undersuface of the operculum was marked with prominent median and fainter lateral septal ridges. Various species occur from the Lower to the Middle Devonian. *C. sandalina* is very common in the Middle Devonian of Europe.

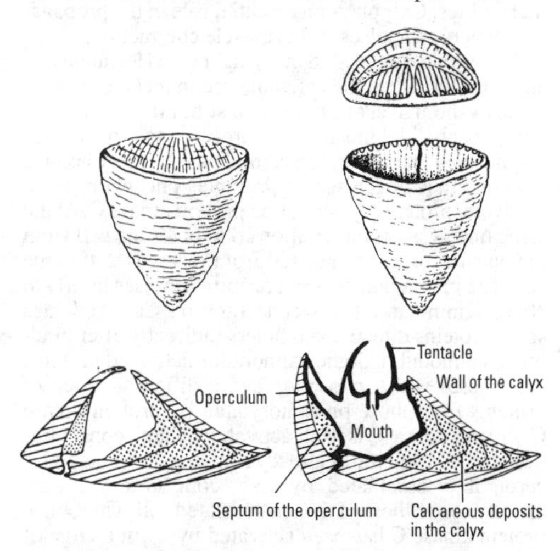

Calceola sandalina (slipper coral), an index fossil of the Middle Devonian.

Calcicoles, ***alkali plants:*** plants that grow preferentially or exclusively on soils with a high calcium carbonate (chalk or limestone) content, e.g. *Adonis flammea, Bupleurum rotundifolium, Ajuga chamaepitys, Coronilla vaginalis.* The presence of chalk or limestone improves soil crumb structure, drainage and warming. In addition, hydrolytic cleavage of the calcium carbonate produces a slightly basic reaction, and helps to bind humic acids.

Calcifuges, ***acidophytes:*** plants which grow preferentially or exclusively on silicate soils that are deficient in calcium carbonate and have a weakly acidic reaction. See Calcicoles.

Calcispongiae: see *Porifera.*

Calcitonin, ***thyreocalcitonin:*** a polypeptide hormone formed in the parafollicular cells (C-cells) of the mammalian thyroid gland, and in the ultimobranchial gland of nonmammalian species (both glands are derived from the fifth gill pocket of the embryonic gut). It contains 32 amino acid residues, M_r (human) 3 420; the amino acid sequence is species-dependent. It causes a rapid but short-lived drop in the level of calcium and phosphate in the blood, by promoting the incorporation of these ions into the bones, thereby preventing hypercalcemia after the dietary intake of calcium. It is released in response to increased levels of Ca^{2+}, and it is a direct antagonist of Parathormone (see).

Calcium, ***Ca:*** an alkaline earth metal, whose divalent cation (Ca^{2+}) has many vital biological functions. As $CaPO_4$ and $CaCO_3$ it provides rigidity and hardness to shells and bones. By acting as a chelating agent, Ca^{2+} stabilizes the structures of proteins and lipids in cell membranes, cytoplasm and organelles, and in chromosomes. It is a cofactor of a number of extracellular enzymes, including prokaryotic and eukaryotic digestive enzymes, factors II, VII, IX and X in blood coagulation, and of complement activation by antigen-antibody complexes. Ca^{2+} binds to tubulin with high affinity, and is required for the entry of a cell into the S-phase of the cell cycle. Contactility, secretion, chemotaxis and aggregation of cells are regulated by Ca^{2+} and arachidonic acid metabolites. Ca^{2+} plays an essential role in the propagation of nerve impulses and in muscle contraction.

Ca^{2+} serves as a second messenger (see Hormones) in animal cells. When the calcium system receptor is activated by the arrival of a molecular stimulus, 4,5-diphospho-phosphatidyl inositol is hydrolysed to inositol 4,5-bisphosphate and diacylglycerol. Diacylglycerol is also a second messenger, which promotes phosphorylation of several proteins by activating protein kinase C. At the same time, Ca^{2+} is either allowed to enter the cell from the outside, or it is mobilized from reserves within the cell. The intracellular Ca^{2+} concentration rises briefly to about 1 mmolar. At this concentration Ca^{2+} activates some proteins directly, and others indirectly after binding to calmodulin. Some calmodulin-activated proteins are kinases which phosphorylate a different group of proteins from those phosphorylated by protein kinase C. Protein kinase C is itself activated by high concentrations of Ca^{2+}, but if it is already sensitized by diacylglycerol, it is stimulated by Ca^{2+} concentrations only slightly above those of the unstimulated cell. Thus, once protein kinase C has been activated by a brief surge of Ca^{2+}, the activation of the cell may persist at much lower Ca^{2+} concentrations.

In many cases, a given stimulus will activate both cAMP and Ca^{2+}/diacylglycerol as second messengers. H. Rasmussen has proposed the term "synarchic" to describe the interactions which then occur. In some cases, artifical stimulation of just one of the two systems (e.g. Ca^{2+} influx though an ionophore, or injection of cAMP) leads to phosphorylation of the same proteins. Often the influx of Ca^{2+} leads to a brief stimulation, whereas the injection of cAMP produces a slow but long-term stimulation; normal stimulation of both systems simultaneously produces a response that is both rapid and long-term. The presence of two mutually regulatory second messenger systems allows considerable plasticity in the response of the cell to the primary stimulus.

There is much evidence that Ca^{2+} serves as a second messenger in plants, which contain calmodulin (cAMP is not a second messenger in plants). Both gravitropic and phototropic responses in plants appear to depend on Ca^{2+}.

Excess Ca^{2+} is toxic to cells, which therefore have efficient mechanisms for removing it. The calcium pump, which is activated by high intracellular Ca^{2+} concentrations, removes Ca^{2+} to the cell exterior. Under normal conditions, Ca^{2+} may be exchanged for Na^+ leaking into the cell. Large amounts of Ca^{2+} may also be sequestrated in mitochondria. The normal intracellular concentration of Ca^{2+} is 0.1 mmolar; blood is about 1.5 mmolar with respect to free Ca^{2+}.

Compared with other plant nutrients, the Ca content of soil is generally high, although a large proportion is present in the lattice structure of basic minerals, or as poorly soluble salts (carbonates and phosphates). By the action of water and CO_2 (produced by soil microorganisms), $CaCO_3$ is converted into soluble $Ca(HCO_3)_2$, which serves as a ready source of Ca^{2+}, but is also easily lost by leaching. For high soil fertility, the soil sorption complexes must be largely saturated with Ca^{2+}. Not only is this adsorbed Ca^{2+} readily available to the plant, the sorption complexes also stabilize soil crumb structure.

Ca^{2+} is taken up by plant roots, and this uptake is competitively inhibited by certain other cations, especially, Sr^{2+}, NH_4^+, K^+, Na^+ and Mg^{2+}. Conversely, the uptake of these ions is inhibited by Ca^{2+}. In particular, competition between Ca^{2+}, K^+ and Mg^{2+} can lead to poor availabilty of one ion or the other, even when the soil appears to contain an adequate concentration of the ion in question. Thus, conifers suffer from K^+ deficiency on calcareous soils.

After entering the plant root, Ca^{2+} is transported in only one direction (acropetally), i.e. to the shoot apex. When administered directly to a leaf, Ca^{2+} ions, penetrate the leaf tissue, but are then transported only to the leaf tip. There is hardly any reverse transport to branches or roots. Leaves and bark therefore accumulate Ca^{2+}, and this accumulation may be considerable in old leaves and old regions of bark. Accumulated Ca^{2+} is often deposited as insoluble Ca salts of organic acids, e.g. calcium oxalate. Other salts occur in the cell sap or as incrustations of the cell wall, e.g. Ca phosphates and Ca carbonate. Free and adsorbed Ca^{2+} also occur in the plant. The Ca content of dicotyledonous plants is generally 10–30 mg/kg dry mass. Grasses contain much less than 10 mg/kg dry mass.

Ca^{2+} promotes the dehydration of negatively charged plasma colloids, i.e. it has the opposite effect to K^+, which promotes plasma colloid hydration. The actual hydration state of the colloidal cytoplasm depends largely on the interplay of the effects of these two cations. As in animal cells, Ca^{2+} is also an important chelating agent in plant cells. The Ca salt of pectin is a structurally important component of the middle lamella of the cell wall. Accordingly, cell multiplication in meristematic tissue and cell extension growth are promoted by Ca^{2+}. It also promotes root growth. Ca^{2+} cannot be replaced physiologically by the chemically closely related Sr^{2+}.

Application of Ca fertilizers promotes plant growth, not only by provision of an essential plant nutrient, but also indirectly by favoring the growth of microorganisms that perform beneficial degradation and synthetic processes, by improving soil crumb structure, and not least by raising the pH value of the soil. For a general improvement of soil quality, much larger quantities of Ca salts are required than are simply needed for plant nutrition. Application of Ca fertilizers, otherwise known as "liming", is therefore primarily a form of soil treatment. It is sufficient to treat light soils with Ca carbonate, which releases Ca^{2+} slowly but persistently, whereas heavy clay soils benefit more readily from the application of Ca oxide (quicklime) or Ca hydroxide (slaked lime), which act very rapidly. Gypsum ($CaSO_4.2H_2O$), a neutral Ca salt, is suitable for improving already alkaline soils when an increase of pH is unnecessary.

Absolute Ca deficiency in plants is rare. The symptoms are not always externally recognizable, and they are rather unspecific. There is general depression of growth, inhibition of root growth, and dying back of the apical mersitem. More frequently, plants on Ca-deficient soils, i.e. strongly acidic soils, display symptoms of acid damage, or toxicity due to aluminium or manganese, which are more available under acidic conditions.

Calcium-magnesium ATPase: see Ion pumps.

Calcium-potassium rule: a rule discovered empirically during the early development of plant nutritional physiology. It states that when the availability of K^+ is increased the uptake of Ca^{2+} is decreased, and vice versa. In the light of present knowledge, the C-p.r. is interpreted as competition between ions in an uptake mechanism.

Caleana: see *Orchidaceae.*

Calendula: see *Compositae.*

Calicalicus: see *Vangidae.*

Calidris: see *Scolopacidae.*

Californian legless lizard: see *Anniellidae.*

California nutmeg: see *Taxaceae.*

Caligus: see *Copepoda.*

Calla: see *Araceae.*

Callaeidae, ***wattlebirds:*** a small avian family in the order *Passeriformes* (see). Only 3 species are known, all of them native New Zealand forest birds, and one of them now extinct. All have wattles drooping from their gapes. The feet are particularly strong with long hind-claws, and their long tails reveal projecting feather shafts. The extinct ***Huia*** *(Heteralocha acutirostris;* 48–54 cm; dark to black in color, with orange wattles), last seen alive in 1907, is famous for the difference between the male and female bills. The male's bill was stout, straight and chisel-like, and used for attacking hard wood to extract grubs. The female's bill was longer, more slender, and decurved, and more suitable for probing rotten wood or narrow crevices for grubs. The remaining 2 living species are the ***Saddleback*** *(Creadon carunculatus;* about 25 cm; dark to black in color, with small yellow wattles) and the ***Kokako*** or ***Wattled crow*** *(Callaeus cinerea;* 43–46 cm; blue-gray in color, with blue wattles and a heavy decurved beak). Both species are rare, and each exists as 2 different races on the North and South Island, respectively.

Callaeus: see *Callaeidae.*

Callinecta: see Swimming crabs.

Calling hares: see *Ochotonidae.*

Callipepla: see *Phasianidae.*

Callipharixenidae: see *Strepsiptera.*

Callistephus: see *Compositae.*

Callitrichaceae, ***water starwort family:*** a cosmopolitan family of dicotyledonous plants, containing about 25 species in a single genus *(Callitriche,* water starwort). Members are fully aquatic or inhabitants of wet and marshy areas on the edges of standing or slow-flowing waters. They are monoecious herbs with undivided opposite leaves. Flowers lack a perianth, but are enveloped by two bracteoles; the family is placed in the sympetalous order, *Lamiales,* on the basis of embryological evidence. Male flowers usually posses only a single stamen, which has a long, slender filament and a reniform anther. Female flowers have a basically 2-celled ovary, which is 4-celled by secondary septation. The 4-lobed fruit separates at maturity into 4 single drupelets. The commonest Central European species is *Callitriche palustris* (= *C. verna),* which has an aquatic and a terrestrial growth form.

Callitriche: see *Callitrichaceae.*

Callorhinus: see *Otariidae.*

Callyodontidae: see *Scaridae.*

Calobryales: see *Hepaticae.*

Caloenas: see *Columbiformes.*

Calopterygidae: see *Odonatoptera.*

Caloric: the same as Calorific, e.g. Caloric value of dietary components.

Calorie: a unit of energy, formerly used widely for quoting heats of combustion and for general thermodynamic measurements. It is still used popularly for stating the energy values of foods. For scientific purposes, it is replaced by the Joule: 1 calorie = 4.184 Joules.

Calorific value of dietary components, ***heat of combustion:*** the energy content of carbohydrates, fats and proteins, determined by measuring the heat produced during their total combustion in a calorimeter. Carbohydrate yields 17.5 kJ/g, fat 38.5 kJ/g, and protein 24 kJ/g. Carbohydrates and fats yield the same quantity of energy during metabolism as they yield during combustion in a calorimeter, i.e. the *physical-chemical* and *physiological* calorific values are the same, because these substances are totally oxidized by the organism. Part of the energy of proteins, however, is present in excreted compounds, such as urea, uric acid and creatinine. The physiological calorific value of protein is consequently 18 kJ/g.

Calorigenic effect: see Specific dynamic action.

Calorimetry: in physiology, a method for the determination of energy turnover in an organism. In *direct C.,* the quantity of heat released by an organism is measured in a thermally insulated chamber. In *indirect C.,* the heat equivalent is calculated from the quantity of carbon dioxide produced and the quantity of oxygen consumed. The energy obtained from the consumption of 1 liter of oxygen varies according to the respired substrate (i.e. each food material has a different calorific equivalent). The Respiratory quotient (see), which is proportional to the calorific equivalent of the respired food material, is therefore used to convert the oxygen consumption into its heat equivalent.

Calotes: see Changeable lizards.

Calotropis: see *Asclepiadaceae.*

Caltha: see *Ranunculaceae.*

Calyptomera: see *Branchiopoda.*

Calyptrogen: see Histogen concept.

Calystegia: see *Convolvulaceae.*

Camarhynchus: see Galapagos finches, Adaptive radiation.

Cambrian (from the Roman, *Cambria,* for North Wales): the earliest of the six periods of the Paleozoic Era. The C. lasted about 70 million years, and its fauna include 11 invertebrate phyla. Complex life forms make their first appearance in the sediments of the C., including the first organisms with mineralized skeletons or other hard parts. The C. therefore marks the beginning of the "fossil age" and the sudden appearance of life forms that are qualitatively much more advanced than those of previous eras. The most important index fossils are the trilobites, with the aid of which the C. is subdivided into the Upper C. *(Olenus* stage), Middle C. *(Paradoxides* stage) and Lower C. *(Olenellus* stage). This classification reflects the evolution of the trilobites; the most primitive forms have a long, cylindrical, segmented glabella, which is fused with the relatively large crescent-shaped eyes; more advanced forms have a shorter, less segmented glabella, widening into a club shape, and the eyes are smaller. In parallel with these changes, there is also reduction in the number of body segments, and the pygidium becomes modified from a small plate to a semicircular segmented element.

Brachiopods were among the most abundant and widespread marine organisms of the C. shelf seas, and are therefore useful for geological dating of C. marine sediments. Most brachiopods of the C. are typically nonarticulated and lack a lophophore, and their shells are chitinous-phosphatic. A few articulated forms with calcareous shells do appear in the C. The extinct *Archaeocyatha* (see) are known only from the C., where they form the oldest known reef communities; they are placed systematically between the sponges and the coelenterates, and their fossils are also important in the fine classification of strata. The oldest nautiloids are represented by the genus *Plectronoceras* of the Upper C. As in previous eras, the dominant plants of the C. were thallophytes (algae and fungi) and bacteria. Calcareous rocks were formed by the activities of calcium carbonate-secreting cyanobacteria and green algae.

Camel, *Bactrian camel,* *Camelus bactrianus:* a member of the *Camelidae* (see), standing about 1.8 m at the shoulder, with a thick, ruddy brown, woolly coat and two humps on its back. There is a deep fringe of hair under the neck, and the upper parts of the forelegs are shaggy with hair. The humps, which stand erect in well nourished animals, contain fat reserves. The C. can go without water for a week, during which time it may lose water equivalent to 30% of its body weight. As in other mammals, the musculature also stores water. Body temperature can be varied by 6 °C, decreasing the temperature differential with the environment, and thereby controlling water loss by evaporation. Throughout the whole of Central Asia, as well as in parts of western Asia and the Near East, it serves as a beast of burden and as a riding animal, as well as providing milk, meat and leather. A few original wild herds still exist in the Gobi desert. It can be crossed with the Dromedary (see).

Camelidae, *camelids, camels:* long-necked, long-legged members of the order *Artiodactyla,* and the only family of the infraorder *Tylopoda.* They possess wide pads on the underside of the two digits of each foot. Hoofs were present in earlier forms, but these now exist only as small, nail-like structures. *C.* ruminate, but their 3-chambered stomach is rather differently structured from that of other ruminants. All members of the family possess a characteristic vacuity between the lachrymal and nasal bones. *C.* spit when provoked or irritated, first projecting saliva and finally stomach contents at the perceived adversary. The only living members of the family are the Alpacca, Camel, Dromedary, Guanaco, Llama and Vicuña, all of which are listed separately.

The earliest known fossil *C.* occur in the Upper Eocene of North America, and the family remained exclusively North American until the end of the Tertiary. In the Pliocene, fossils occur in South America and Asia, spreading later into Africa.

Camellia: see *Theaceae.*

Ca^{2+}/Mg^{2+}-ATPase: see Ion pumps.

Camomile: see *Compositae.*

Camouflage: see Protective adaptations.

cAMP: see Cyclic adenosine 3'5'-monophosphate.

Campanulaceae, *bellflower family:* a family of dicotyledonous plants, containing 60–70 genera and about 2 000 species, distributed throughout most of the world, especially in temperate and subtropical regions and in tropical mountainous areas. Most species are herbs, rarely small shrubs to trees, nearly always with latex. Leaves are alternate (rarely opposite), simple and exstipulate. Flowers are usually showy and blue, nearly always actinomorphic and bell- or funnel-shaped. Stamens are free or united, with pollen sacs opening toward the middle of the flower (introrse anthers). The ovary is inferior (rarely superior), 2-10-locular, with axile (rarely basal or apical) placentas. The fruit is a multiseeded capsule, dehiscing in various ways, depending on the genus, or it may be a berry.

Three subfamilies are recognized:

1. *Campanuloideae*: flowers actinomorphic, rarely slightly zygomorphic; anthers usually free. Typical genera: *Campanula* (north temperate region), *Wahlenbergia* (south temperate region), *Phyteuma* (Europe, Mediterranean, Asia), *Jasione* (Europe, Mediterranean,

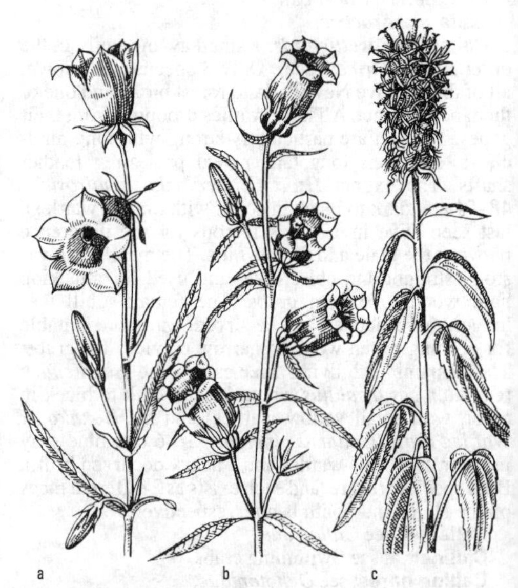

Campanulaceae. a Campanula persicifolia (peach-leaved bellflower). *b Campanula medium* (Canterbury bell). *c Phyteuma spicatum* (spiked rampion).

western Asia), *Canarina* (Africa, Atlantic islands). Various members of this subfamily are arable weeds.

2. *Cyphioideae*: flowers zygomorphic; stamens sometimes united, anthers free. Typical genera: *Cyphia* (Africa), *Nemacladus* (southest USA and Mexico).

3. *Lobelioideae*: flowers zygomorphic; rarely almost actinomorphic; anthers united. Typical genera: *Lobelia* (cosmopolitan), *Centropogon* (tropical America, Caribbean), *Siphocampylus* (tropical America, Caribbean). Genera of the *Lobelioideae* are placed by some authorities in a separate family, the *Lobeliaceae*.

The most numerous genus is *Campanula* ***(Bellflowers)*** with over 300 species. The well known *Campanula rotundiflora* ***(Harebell)*** grows in dry meadows, heaths and rock clefts in Europe, and is common throughout the Prairie Provinces of North America. *Campanula persicifolia* ***(Peach-leaved bellflower)*** is a European species of warm, dry forests. *Campanula pusilla (=C.cochlearifolia)* ***(Fairy's thimble)*** is found in European mountains and Lower Alps, in gravelly, moist sites and shifting screes, especially on limestone, and is a popular garden plant for rockeries and borders. *Campanula patula* ***(Spreading bellflower)*** possesses erect violet flowers, and grows in shady woods and hedge banks. *Campanula medium* ***(Canterbury bell),*** one of the best known garden species, is a biennial native to Southern Europe, which may be double or single, and occurs in color varieites of white, rose and purple. Another garden species is the Chinese *Campanula nobilis*, which occurs in reddish violet or creamy white varieties. Roots of the European *Campanula rapunculus* ***(Creeping campanula)*** are used in salads; this species is therefore cultivated, but rapidly becomes a weed, and in North America it is now established as an escape. *Campanula glomerata* ***(Clustered bellflower)*** is a European alpine species, used as an ornamental, and now established as an escape in North America. *Campanula lasiocarpa* Cham. **(*Alaska bellflower*)** occurs in the mountain meadows of the southern Rocky Mountains of North America. *Campanula uniflora* ***(Alpine bellflower)*** occurs in the Arctic meadows of the northeastern boreal forest and the Rocky Mountains of North America, and in the European Alps. The term "bluebell" is often used in place of "bellflower", but this must not be confused with the monocotyledonous bluebell (*Endymion nonscriptus*) of the *Liliaceae*.

Flowers of the genus *Phyteuma* ***(Rampion)*** occur in dense heads or spikes, surrounded by an involucer of bracts, usually blue. The corolla has 5 long, strap-like lobes, which are joined near the base, and the stigma is long and protruding. The genus is small, and spread over Europe and Western Asia, mainly in mountainous areas. *Phyteuma spicatum* ***(Spiked rampion)*** has yellowish-white flowers and occurs in mountainous pastures and woods. The very similar *Phyteuma nigrum* ***(Black rampion)*** has blackish violet, rarely blue or white flowers. *Phyteuma comosum* ***(Devil's claw)*** is an alpine species (1 000–1 700 m) found only on limestone and dolomite in Austria, where it is a protected species, and in northern Italy. In contrast to the dense heads of sessile flowers of the other 30 *Phyteuma* species, *P. comosum* has large umbrellate, short-stalked flowers, and it is sometimes placed in a separate genus as *Physoplexis comosa*. The decorative, blue-flowered *Jasione montana* ***(Mountain sheep's bit)*** is found on sandy soils in sunny positions. *Canarina abyssinica*, with pinkish orange, elongated corollas streaked with red, is a climbing herb found on high ground from southern Ethiopia to Tanzania. *Canarina canariensis*, on the other hand, is a yellow-orange-flowered herbaceous climber found in shady ravines and woodlands of the Canary Islands. This remarkable disjunct distribution is presumably a reflection of an ancient, pre-Ice Age continuity. *C. canariensis* produces edible black berries, and is now cultivated in Europe. Several species of *Lobelia* are horticulturally important as ornamentals, and medically important as a drug source (see Lobeline).

Campanularia: see *Hydroida*.

Campesterol, ***(24R)-ergost-5-en-3β-ol:*** a phytosterol, which occurs in the oils of rapeseed *(Brassica campestris),* soybean and wheat germ, and in some mollusks.

Camphene: a solid (melting point about 50 °C), unsaturated, cyclic monoterpene hydrocarbon (Fig.), smelling of turpentine and camphor, and found as the (+)-form in ginger, rosemary and spike oils, and as the (-)-form in citronella and valerian oils.

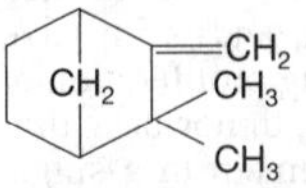

CH_2 $=CH_2$ CH_3 CH_3

Camphene

Camphor: a crystallizable, bicyclic, monoterpene ketone with a characteristic odor. Both optical isomers occur naturally: (+)-C. (Japan C.) is obtained in large quantities by steam distillation of the wood of the East Asian camphor tree, *Cinnamomum camphora*; (-)-C. (Matricaria C.) is present in the essential oils of certain plants, e.g. in valerian *(Valeriana),* sage *(Salvia)* and mint *(Mentha)*. C. is used internally as a cardiac stimulant, and externally in salves. Large quantities of C. are used industrially for the manufacture of plastics.

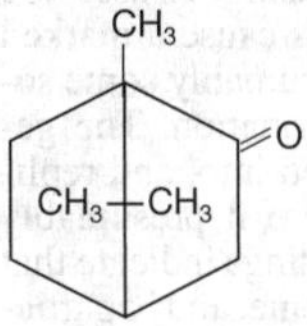

CH_3 $=O$ CH_3–CH_3

Camphor

Campodeiform larva: an insect larval form, which has a long abdomen, possesses well developed legs, biting mouthparts, antennae and cerci. It resembles certain adult forms of primitive, ametabolous, apterygote (wingless) insects, and is named for its resemblance to the genus *Campodea* [family *Campodeidae,* order *Diplura* (2-pronged bristle-tails)]. Campodeiform larvae are often predaceous, and often have a chitinised exoskeleton. Certain beetles have a C.l., e.g. *Staphylinidae* (rove beetles), *Cicindelidae* (tiger beetles) and *Carabidae* (ground beetles).

Campylorhynchus: see *Troglodytidae*.

Canada balsam: see Balsams.

Canadian hemlock: see *Pinaceae*.

Canary: see *Fringillidae*.

L-Canavanine, ***2-amino-4-guanidinohydroxybutyric acid:*** a nonproteogenic amino acid found only in certain legumes. The presence or absence of C. is a useful trait in the chemical taxonomy of *Papilionaceae*. C. is a structural analog of arginine and a competitive inhibitor of arginine metabolism. It is poisonous to most animals and to C.-free plants grafted onto C.-containing plants. In C.-containing plants it serves as a soluble nitrogenous storage material (it accounts for 55% of the nitrogen in seeds of *Dioclea megacarpa)*. Production of poisonous C. may also be an adaptive advantage, e.g. by preventing attack by insects. Toxicity is due in part to the incorporation of C. into proteins in place of arginine. C.-containing plants avoid the toxic effects of C. by producing very specific arginyl-tRNA synthetases, which distinguish between arginine and C.

$$H_2N{-}\underset{\displaystyle \overset{\|}{NH}}{C}{-}NH{-}O{-}CH_2{-}CH_2{-}CH(NH_2){-}COOH,$$

Canavanine

Canbarus: see Crayfish.

Cancellous bone: see Bone.

Cancer cell, *transformed cell, malignant cell:* a cell that differs morphologically, biochemically and/or physiologically from its parent cell; it is autonomous with respect to its host, exhibits seemingly uncontrolled growth and excessive multiplication, and is invasive (i.e. displays metastasis). The resulting proliferation of cells is known as a cancer, malignant tumor or malignant neoplasm. If neoplastic cells remain in a single mass and do not display invasive behavior, they are said to be benign. It is mainly the property of invasiveness (the ability to break loose, enter the circulation and form metastases or secondary tumors at other sites in the organism) that characterizes a malignant neoplasm.

Genesis. C.cs. arise by the transformation of normal dividing cells of higher plants, invertebrates and vertebrates. Transformation is promoted in particular by tumor viruses, numerous carcinogenic chemicals, ionizing radiation, and abnormal gene interaction. Human C.cs. also arise under these influences, although very few examples of human cancer-causing viruses are known. Some carcinogenic chemicals cause a marked increase in the mutation rate, and presumably some somatic mutations lead to cell transformation. The genome of tumor viruses is incorporated into, and replicated in, the host cell nucleus, where it presumably causes errors of regulation. Many findings indicate that C.cs. arise from alterations of the genome, and in particular from disturbances of the functional interaction of genome and cytoplasm. Mitotic division of C.cs. produces only C.cs, and reversion of C.cs. to normal, nonautonomous cells does not occur. The properties of C.cs. are therefore genetically based and inherited. It is now generally accepted that C.cs. arise by somatic mutations, i.e. mutations in cells not destined to become sex cells. Thus, many C.cs. have one or more abnormal chromosomes, usually from the breakage and rejoining of other chromosomes. Every cell in the cancer or tumor possesses these same aberrant chromosome(s), which are not present in the normal cells of the body. There is also evidence that a single somatic mutation is usually insuffient to cause a transformation to malignancy, and that several somatic mutations are necessary to cause cancer.

Biological properties. The cell surface of many C.cs. differs chemically from that of the stem cell, so that cell-cell interactions are affected. Virus-transformed cells display a number of new surface antigens. In addition, the cholesterol and phospholipid contents of the plasma membrane may be increased, and the contents of some proteins decreased or increased. Proteolytic activity is increased. Also, the activity of extracellular glycosyltransferases is usually elevated in C.cs., leading to alterations in the composition of the glycocalyx carbohydrates, and in turn to changes in the antigenic and adhesion properties of the cell surface. Animal C.cs. have a decreased content of a particular glycoprotein (*M_r* 250 000), known variously as *Component I* or *Z*, as *CSP* as *Galactoprotein*, or as *LETS P* (Large External Transformation Sensitive Protein). C.cs. often show poor adhesion properties (decreased adhesion to each other and to a substratum), due mainly to an altered deposition of contractile proteins (i.e. alterations in the cytoskeleton of transformed cells; see Actin, Myosin), and possibly to the absence or low level of fibronectin in transformed cells. Mitotic activity is also usually increased. Compared with normal cells, transformed cells are more easily aggregated by lectins (proteins or glycoproteins which bind specific sugar residues); lectin receptors are unevenly distributed on the surface of C.cs. Normal fibroblasts possess microfibril bundles consisting of actin, myosin, α-actinin, tropomyosin and other proteins, which help to maintain cell shape. In contrast, the actin of virus-transformed fibroblasts is diffusely distributed, and virus-transformed fibroblasts are rounded with pseudopodium-like, actin-containing regions, which serve for ameboid movement. These chemical changes at the cell surface, together with a low requirement for mitogenic substances and other growth factors (some of which are synthesized by C.cs.), lead to the abolition of cell growth control by cell density, i.e. the abolition of contact inhibition. Thus, in cell culture, the maximal density of normal cells is $10^4/cm^2$, whereas C.cs. can attain a density of $10^6/cm^2$. Differentiated normal cells grow into a simple monolayer, whereas the unregulated growth of C.cs. results in the formation of multilayers. Compared with normal cells, the Cell cycle (see) of C.cs. is usually accelerated. On the other hand, some C.cs. live longer than their stem cells. Growth of normal cells can be stopped, e.g. by removal of an essential growth factor, at any stage of the cell cycle, but the cells always proceed to early G_1 phase before becoming quiescent. In contrast, C.cs. have lost the ability to proceed to G_1 quiescence, so that a growth-inhibited culture or colony of C.cs. contains cells at all stages of the cell cycle. Many animal and human C.cs. display invasive behavior, i.e. an ability to migrate into other tissues and organs: individual cells or groups of C.cs. may travel by ameboid movement, or they may be carried by the blood or lymph to adjacent or distant organs, resulting in the formation of secondary tumors or *metastases*.

C.cs. retain some of the characters (expressed to varying degrees) of their differentiated stem cells, so that is usually possible to identify their tissue of origin. For example, cells of an epidermal basal cell carcinoma, derived from a keratinocyte stem cell in the skin usually continue to synthesize cytokeratin intermediate filaments; melanoma cells (from skin pigment cells) often still produce pigment granules; C.cs derived from nervous tissue have excitable membranes. Cancers derived from epithelial cells are called *carcinomas,* those from connective tissue or muscle cells are called *sarcomas,* those from hemopoietic cells are *leukemias,* while those derived from cells of the nervous system are called *neuroblastomas*. In many types of C.c., differentiation appears to be blocked at an early stage, and this blockage is associated with a loss of specific cell function, e.g. most hepatomas (liver-derived cancers) no longer synthesize many normal liver enzymes. Many C.cs. contain lower concentrations of cAMP than their differentiated stem cells. Compared with their stem cells, C.cs. also show altered and often stimulated metabolism. Many tumors show a typical decrease of aerobic cell respiration, correlated with a decrease in the number of mitochondria, and a marked increase in the rate of glycolysis. Loss of specific enzymes results in an accumulation of metabolic intermediates and accompanying structural alterations. The nucleus may be increased in size relative to the cytoplasm, and the proportion of heterochromatin may be increased. Many C.cs. have more Microvilli (see) than their stem cells.

Cancer crab, *rock crab*, *Cancer pagurus:* a true crab (see *Brachyura,* family *Cancridae).* Up to 30 cm wide, the C.c. is common on North Sea coasts, and it makes good eating.

Cancer viruses: see Tumor viruses.

Cancridae: see Cancer crab.

Cancridea: Cancer crabs (see), a section of the infraorder *Brachyura* (true crabs) of the *Decapoda* (see).

Candelabra tree: see *Araucariaceae.*

Canegrass warblers: see *Sylviinae.*

Canidae, *dogs and allies:* a family of Terrestrial carnivores (see), containing about 35 species in 14 genera. They are long-snouted mammals, which walk on their toes and possess nonretractable claws. The carnassial teeth are strongly developed. With the exception of 3 genera, the dental formula is 3/3, 1/1, 4/4, 2/3 = 42. Other dental formulas are *Speothos* (Bush dog of South and Central America): 3/3, 1/1, 4/4, 1/2 = 38; *Cuon* (Indian dholes): 3/3, 1/1, 4/4, 2/2 = 40; *Otocyon* (Big-eared African fox): 3/3, 1/1, 4/4, 4/4 = 48.

The following representatives are listed separately: Wolf, Coyote, Jackals, Foxes, African hunting dog, Japanese fox, Maned wolf.

The domestic dog *(Canis lupus* v. *familiaris)* also belongs to this family. It arose by domestication from the wolf, possibly with contributions from other *Canis* species. The Dingo (see) is a feral form of the domestic dog.

Canine teeth: see Dentition.

Canis: a genus of the family, *Canidae* (see). See Wolf, Coyote, Jackals, Dingo.

Cannabaceae, *Cannabinaceae, **hemp family:*** a family of dicotyledonous plants containing only 4 species in 2 genera, native to Europe and Asia. They are all herbaceous plants with stipulate, lobed or hand-shaped, divided leaves. The dioecious, 5-merous flowers are wind-pollinated. Male flowers are aggregated into a panicle, while the female inflorescence may be bushy *(Cannabis)* or a peduncled cone-like spike *(Humulus).* The fruit is an achene enclosed in the persistent perianth.

The ***Hop plant*** *(Humulus lupulus)* is a climbing (3–6 m) perennial, cultivated for its cone-like inflorescences (hops), which are added during the brewing of beer to impart a bitter taste. Hops have conspicuous bracts, covered with glands containing resins and bitter principles. ***Hemp*** *(Cannabius sativa),* an annual plant from southern Asia, has long been exploited as a source of fiber and to a lesser extent for its oil-containing seeds. Its long (1–2 m), strong phloem fibers are used in the manufacture of ropes, sacking and carpets. The narcotic

Cannabaceae. 1 Hemp *(Cannabis sativa), a* branch with male flowers, *b* male flower, *c* female flower, *d* ovary, *e* opened fruit, *f* seed. *2* Hop *(Humulus lupulus):* shoot of female plant.

drug known as hashish or cannabis resin consists of the resin exuded from multicellular glandular hairs on the stem and leaves of the female plant of *Cannabis sativa.* Marihuana is the flowering or seed-carrying, dried shoot tips of the female plant.

Cannabalism: a specialized mode of nutrition in animals, which feed upon members of their own species or their early developmental stages. C. often results from overpopulation or food shortage, when it serves for the regulation of population density. It occurs in many arthropods, e.g. silverfish, cockchafer larvae, praying mantids, myriapods, spiders, some crustaceans, and many vertebrates such as pike, trout, true lizards, birds of prey, storks, field mice, rats and wolves.

Cannabis: see *Cannabaceae.*

Cantharellus: see *Basidiomycetes (Porialis).*

Cantharidine: a cyclic acid anhydride derivative (Fig.) present (about 0.4%) in the beetle *Cantharis (= Lytta) vesicatoria* (Spanish fly, blistering beetle), which is native to southern and central Europe. C. is a toxic agent which causes blistering and local inflammation, and its use as an aphrodisiac has caused fatalities.

O
CH_3
O
O
CH_3
O *Cantharidine*

Canvasback: see *Anseriformes.*

Caoutchouc: elastic, high molecular mass polyterpenes, which can be converted into rubber by vulcanization. Natural C. is a mixture of polyisoprenes with varying molecular masses, ranging from 300 000 to 700 000. According to X-ray and IR data, the double bonds are *cis* oriented, whereas in the C.-like polyterpenes, gutta and balata, they are *trans*. Hundreds of species of plants contain C. in their latex, but it can only be obtained on a large scale from a few representatives of the *Euphorbiaceae, Apocyanaceae, Asclepiadaceae* and *Moraceae*. The rubber tree, *Hevea brasiliensis,* which is native to the Amazon region, but is also grown in India and Indonesia, is quantitatively the most important world source of C. *Parthenium argentatum* of Mexico and California is the source of guayule C., which is similar to that from *Hevea,* but has a lower molecular mass. C. is harvested by cutting the bark of the trees without injuring the cambium. Latex flows from the cut latex canals, yielding 40–80 ml/tree/ harvest. The latex contains 25–35% C., 60–75% water, 2% protein, 2% resin, 1.5% sugar and 1% ash. Coagulation is prevented by the protein. The latter is precipitated with dilute acid or sodium fluorosilicate, and the coagulated C. pressed into sheets to remove the water. The resulting light-weight, very elastic and flexible material is processed to rubber by vulcanization.

Since 1935, kok-saghys C. has been produced in the former USSR from roots of the dandelion species *Taraxacum kok-saghys*.

Cape gooseberry: see *Solanaceae*.

Capella: see *Scolopacidae*.

Capensis: see Cape region.

Cape pigeon: see *Procellariiformes*.

Caper: see *Capparidaceae*.

Capercailie: see *Tetraonidae*.

Cape region, *capensis, South African kingdom:* the smallest in area of all the floral kingdoms, situated on the southern extremity of Africa. This warm, temperate region, with dry summers and winter rains, is represented by about 6 000 species, including many endemics. Trees are almost absent, and the ground is typically covered by evergreen, sclerophyllous, dwarf shrubs and heaths, with numerous geophytes and annuals. Characteristic are members of the *Proteaceae* and *Liliiflorae,* more than 450 species of *Erica,* and many species of *Pelargonium.* The drier marginal areas are characterized by succulent species of *Mesembryanthemum, Stapelia* and *Euphorbia*.

Caperkailzie: see *Tetraonidae*.

Cape sugarbird: see *Meliphagidae*.

Capillaries: the finest branches of the blood and lymph vessels, with an average diameter (in humans) of 8 μm, which connect the arterioles (smallest arteries) with the venules (smallest veins). The C. wall consists of a single layer of endothelial cells surrounded externally by a thin network of connective tissue. C. communicate freely with one another and form interlacing networks of variable form and size in different tissues. Blood cells must usually pass through C. in single file, and since they are often larger than the caliber of the C., they frequently become distorted during passage. Blood flow through organs is regulated by the dilation and constriction of C. Metabolites and respiratory gases are exchanged through C., e.g. they absorb oxygen and release carbon dioxide in the lungs, discharge waste products in the kidney, take up digestion products in the alimentary canal, etc. In some organs (e.g. liver), blood cells migrate into the intercellular fluid of the tissue through gaps in the C. wall.

Capillary circulation: see Microcirculation.

Capillitium: see *Myxomycetes*.

Capitate antennae: see Antennae of insects.

Capitate feeding tentacles: see Captaculae.

Capitonidae, *barbets:* an avian family (76 species in 13 genera) in the order *Piciformes*. Barbets are stocky, strongly built, small birds (3–30 cm). Representatives are found in the tropical forests of America, Asia and Africa, but the family is absent from the West Indies, Madagascar, Australia, Borneo and Bali. They are most abundant in Africa, where some species are also found in dry bush country. The head is large, with a stout, heavy, conical, sharp bill. Rictal and chin bristles are well developed in most species. The feet are zygodactylous. Plumage is coarse and often brightly colored, and especially in some American species it is spendidly multicolored. The head and foreparts of many Asian barbets are decorated with areas of red, yellow, blue and white. Sexes are usually alike, but some American species display sexual dimorphism. Nesting holes are dug in partly rotted tree trunks. Some feed entirely on insects, ants etc., which are taken from the surfaces of tree branches (cf. the related jacamars and puffbirds, which catch insects on the wing). Other species are vegetarian, but feed their young on insects. On account of their metallic bell-like calls, the Asian *Megalaima haemacephala* is known as the ***Coppersmith,*** while species of *Pogoniulus* are known as ***Tinkerbirds.***

Capitulum: see *Compositae,* Flower (section 5, Inflorescence).

Capon comb assay: see Androgens.

Capparidaceae, *caper family:* a family of dicotyledonous plants, containing about 800 mostly tropical and subtropical species in about 45 genera. They are mostly shrubs or small trees with alternate simple or compound leaves. Flowers are radial or actinomorphic, 4-merous with a superior ovary of 2 carpels; they are usually borne in racemes and are usually large and pleasantly scented. The floral axis may carry a stalklike extension (gynophore, androgynophore), which raises the ovaries (more rarely the stamens) above the perianth. Pollination is by insects or birds, and the fruits are berries, nuts or drupes.

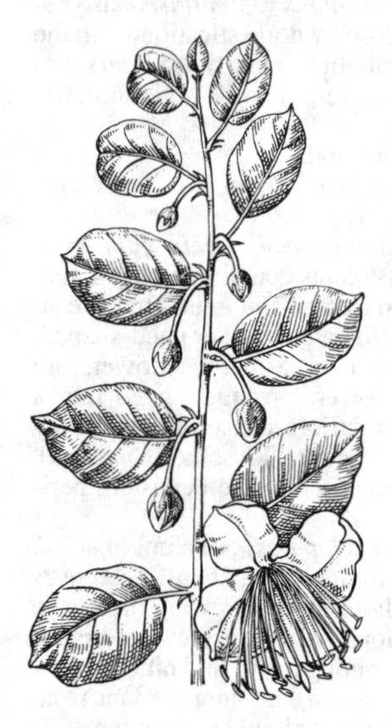

Capparidaceae. Caper *(Capparis spinosa);* branch with flowers and flower buds.

A well known species is *Capparis spinosa* ***(Caper),*** which grows wild in the Mediterranean region, western Asia and India. It is also cultivated for its flower buds (true capers) which are cooked and pickled as a flavoring spice. The characteristic taste of capers is due to the presence of the alkaloid capparidine and the glycoside cutin, as well as caproic acid and an oil. In India, the root bark is used medicinally.

Capparis: see *Capparidaceae.*

Capra: see Goats.

Caprellidae: see *Amphipoda.*

Capreolus: see Roe deer.

Capricornis: see Serows.

Caprifoliaceae, ***honeysuckle family:*** a family of dicotyledonous plants containing about 450 species in 13 genera, with a cosmopolitan distribution but mainly in northern temperate regions. They are woody lianas or shrubs, rarely trees or herbs, with opposite leaves, which are often toothed, lobed, or pinnate. Stipules are usually absent, but when present they are small and adnate to the petiole, and very rarely conspicuous. The hermaphrodite 5-merous flowers may be actinomorphic and radially symmetrical, and aggregated into dense, umbel-like thyrses (e.g. *Viburnum, Sambucus).* In contrast, the flowers of *Lonicera* are zygomorphic with a single plane of symmetry. The superior ovary develops into a berry, drupe, capsule or achene.

The most familiar European representative is the ***Elder*** *(Sambucus nigra),* which bears clusters of edible black drupes. The related species, *Sambucus racemosa* ***(Mountain elder*** or ***Red-berried elder),*** which occurs on mountainous areas, bears red drupes, which are also edible. Naturalized species of *Viburnum* are grown as ornamentals in gardens and parks in Europe, North America and Asia. The marginal flowers of the inflorescence of *Viburnum opulus* ***(Guelder rose)*** are sterile and enlarged as a conspicuous display; the fruits are bright red drupes. The double, garden variety ***(Snowball tree)*** has a globular infloresence containing only sterile flowers. The North American ***Snowberry*** *(Symphoricarpos albus)* is often used as a hedging plant; its fruit is white berry. Various species of ***Honeysuckle*** *(Lonicera)* are also used for hedging and in particular as fragrant climbing plants on walls and trellises; the fruit is a few-seeded berry. Species and hybrids of the East Asian *Weigela* are cultivated as ornamentals in Europe and North America; their striking trumpet-shaped flowers range in color from white to red. *Linnaea borealis* ***(Linnaea)*** is a northern alpine creeping perennial of northern coniferous forests; it has a pink and often beautifully marked corolla, and its fruit is an achene.

Caprimulgidae: see *Caprimulgiformes.*

Caprimulgiformes, ***nightjars and allies:*** an avian order containing nearly 100 species in 5 families. Plumage is generally gray, brown, black and white, and patterned like tree bark. All possess wide gapes, and since they were once erroneously thought to milk goats, they are also known as ***Goatsuckers.*** They hide by day and are active mainly at night (nocturnal) or at twilight (crepuscular).

1. Family: *Steatornithidae.* This is represented by a single species ***(Oilbird,*** *Steatornis caripensis;* Trinidad and northern South America), which broods in large colonies in mountain caves, coming out to feed only at night. Body length is about 33 cm, but the span of the long pointed wings amounts to 76 cm. It is the only vegetarian member of the *C.,* and it plucks fruit on the wing (mainly the oily fruit of palm trees). The young are also fed on oily fruits, so that they become grotesquely fat, weighing almost twice as much as their parents. Venezuelan indians obtain cooking oil by rendering down the young birds, hence the name. The oilbird is thus a unique aberrant member of the *C.*, also with affinities to

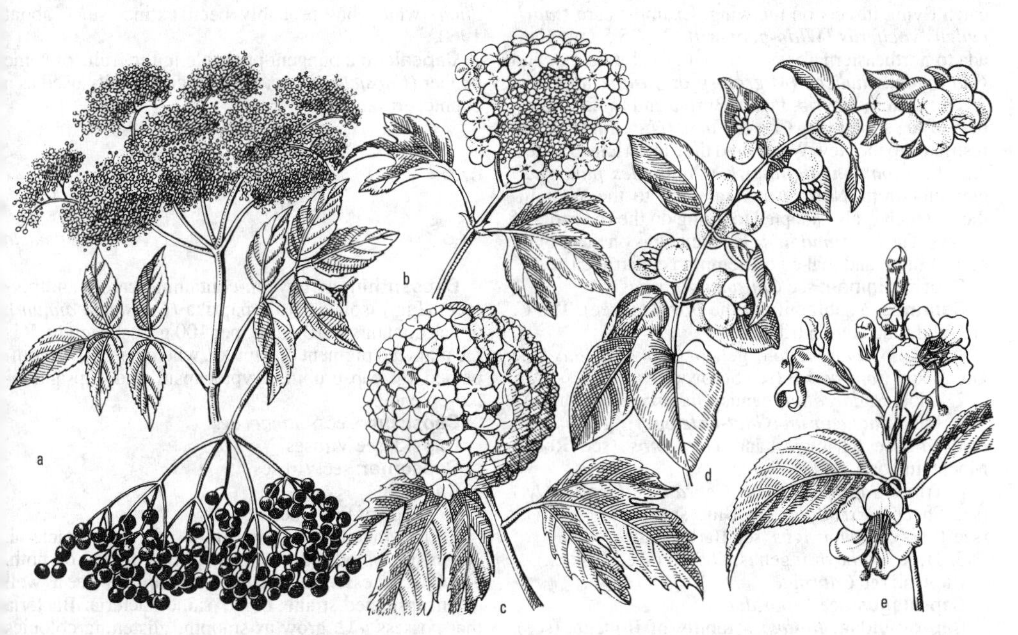

Caprifoliaceae. a Elder *(Sambucus nigra). b* Guelder rose *(Viburnum opulus). c* Snowball tree *(Viburnum opulus* var. *roseum). d* Snowberry *(Symphoricarpos albus). e Weigela.*

the owls. Its well developed olfactory organ suggests that it differs from most other birds in having an active sense of smell.

2. Family: *Nyctibiidae* (***Potoos;*** 5 species; Jamaica, Hispaniola, Central and South America). The potoos catch flying insects on the wing. They sit immobile and upright on the ends of broken branches, so that they look like part of the tree. The ***Common potoo*** *(Nyctibius griseus;* 35.5 cm) is distributed from Mexico to Argentina. A single spotted egg is laid on top of a broken tree stump.

3. Family: *Podargidae* (***Frogmouths;*** 13 species in 2 genera). These are large nightjars with large, flattened, triangular, sharply hooked bills and very wide gapes; the 10 species of *Batrachostomus* and the 3 species of *Podargus* are distributed from India to New Guinea and Australia, and various other Oceanic islands farther east, but not New Zealand. They nest in trees, usually in the fork of a horizontal branch, which is supplemented with a frail construction of twigs, or a cushion of feathers and vegetation.

4. Family: *Aegothelidae* (***Owlet frogmouths;*** 15.5–30.5 cm; 8 species in 1 genus, *Aegotheles).* These plump little birds have been described as resembling small, long-tailed owls. They catch insects in flight, but also take insects, millipedes, etc. from the ground. Most members are found in New Guinea, with others in the Moluccas and New Caledonia, while the ***Moth owl*** *(A. cristata)* is an Australian bird. They nest in hollow trees.

5. Family: *Caprimulgidae* (70 species in 18 genera). Typical ***Nightjars*** belong to the subfamily, *Caprimulginae,* which has an almost cosmopolitan distribution; they are mostly crepuscular in habit, with small bills and wide gapes fringed with bristles. The middle toe of the small feet has a comb claw, which seems to be used for cleaning the rictal bristles around the gape. They catch flying insects on the wing. Examples are *Caprimulgus vociferus* (***Whip-poor-will,*** 23–25.5 cm; Canada to northeastern Texas and south to Honduras), and *Caprimulgus indicus* (***Migratory*** or ***Jungle nightjar;*** 28 cm; eastern Siberia, Japan, Burma and India). Members of the subfamily, *Chordeilinae* (***Nighthawks),*** are restricted to the New World, and they lack rictal bristles, e.g. the ***Common nighthawk*** *(Chordeiles minor;* 23 cm; sub-Arctic North America south to the West Indies), which is also adapted to living on the flat roofs of houses. The *Caprimulgidae* lay their eggs on the ground on leaf litter, and make no attempt to construct a nest.

Caprimulginae: see *Caprimulgiformes.*

Caprinae: a subfamily of the *Bovidae* (see). The *C.* are subdivided into 5 tribes.

1. Tribe: *Nemorhaedini;* genera: *Nemorhaedus* (see Goral) and *Capricornis* (see Serows).

2. Tribe: *Budorcatini;* genus: *Budorcas* (see Takin).

3. Tribe: *Rupicaprini* (***Goat-antelopes);*** genera: *Rupicapra* (see Chamois) and *Oreamnos* (see Rocky Mountain goat).

4. Tribe: *Caprini;* genera: *Capra* (see Goats), *Ovis* (see Sheep), *Pseudois* (see Blue sheep), *Hemitragus* (see Tahrs), *Ammotragus* (see Barbary sheep).

5. Tribe: *Ovibovini;* genus: *Ovibos* (see Musk ox).

Caprini: see *Caprinae.*

Caprolagus: see *Leporidae.*

Capromyidae, ***hutias:*** a family of Rodents (see), somewhat similar in appearance to the coypu. The skull is flattened and long, and the paraoccipital processes are usually separate from the bullae. Fore and hindlimbs are short with 5 digits on each. Up to and shortly after the Spanish conquest, many genera and species of hutias lived in the West Indies. The cats, dogs and mongoose introduced by the Europeans caused the extinction of most Caribbean hutias. The remaining species, all of which are threatened with extinction, are divided into 3 genera.

Capromys has a more or less prehensile tail, and is usually arboreal. Unusually for a rodent, the stomach is divided into 3 compartments by 2 strictures. The following 4 species are recognized: ***Prehensile-tailed hutia*** *(Capromys prehensilis),* ***Black-tailed hutia*** *(C. melanurus),* ***Dwarf hutia*** *(C. nana)* and the ***Cuban hutia*** *(C. pilorides).* The largest of these 4 species, the ***Cuban hutia,*** enjoys sunning itself in the high tree canopy. It is protected in the Cienaga de Zapata reserve in Southwest Cuba, and it is kept in some zoos, where it occasionally breeds in captivity.

Two surviving species of *Geocapromys* are known, i.e. the ***Jamaican hutia*** *(G. brownii)* and the ***Bahaman hutia*** *(G. ingrahami).* A third species, *G. columbianus,* is now extinct. These hutias have heavy bodies (33–35 cm in length) with short limbs and tail. They are largely nocturnal, living in tree holes or under rocks. *G. brownii is* found only in inaccessible rocky areas of the Blue Mountains in Jamaica, and on a small island (Swan Island) more than 700 km south of Jamaica.

There is only one surviving species of *Plagiodontia* (***Hispaniolan hutias).*** Two other species were known in modern times *(P. ipnaeum and P. spelaeum),* but are now extinct. The body length of these rodents is 31–40 cm, and the hair is short and thick. Hair is almost absent from the tail, which is scaly. The surviving species is known as 2 subspecies, the very rare ***Dominican hutia*** *(P. aedium hylaeum)* and ***Cuvier's hutia*** *(P. aedium aedium)* which has probably been extinct since about 1960.

Capsaicin: a pungent principle in the fruits of some pepper *(Capsicum)* species. It is occasionally used as a counter-irritant.

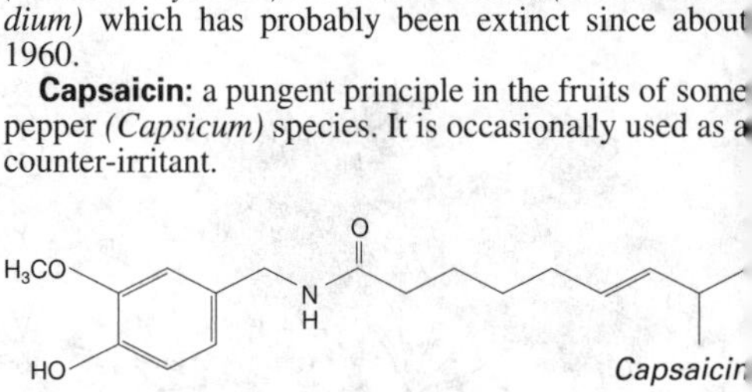

H_3CO … O … N H … HO

Capsaicin

Capsanthin: an oxygen-containing carotenoid pigment (Fig.) isolated from paprika *(Capsicum annum),* which contains 9–10 mg C. per 100 g fresh weight. It is the main red pigment of paprika, where it is also accompanied by capsorubin, cryptocapsin and capsanthin-5,6-epoxide.

Capsicum: see *Solanaceae.*

Capsid: see Viruses.

Capsomer: see Viruses.

Capsule:

1) A type of Fruit (see).

2) The mucilaginous envelope of certain bacteria, consisting of polysaccharides, polypeptides, or both. The C. is not essential, and there are capsule-free as well as encapsulated strains of the same bacteria. Bacteria that possess a C. grow as smooth, glistening colonies (the S-form), whereas colonies of strains lacking a C. have a rough, dull surface (the R-form). The C. protects parasitic bacteria from the immunological defenses of their host, e.g. R-type *Pneumococcus* is nonpathogenic, whereas S-type *Pneumococcus* is pathogenic. The C. can be demonstrated microscopically by preparing the bacteria in Indian ink; against the dark background, the Cs. then appear as light regions around the bacterial cells (Fig.).

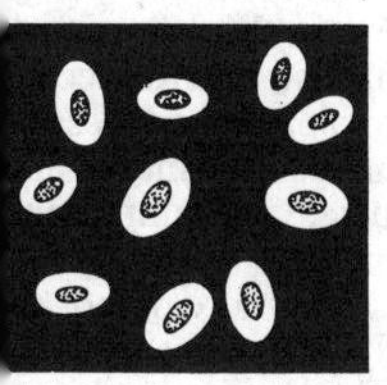

Capsule. Bacteria with capsules (prepared in Indian ink).

In now classical studies, Avery (USA) showed that nonpathogenic R-type *Pneumococcus* can be transformed into pathogenic S-type *Pneumococcus* by treatment with killed S-type cells. In 1944, he showed that the transforming principle is DNA, i.e. "naked DNA" simply penetrates the cells of the R-type and confers the genes for C. synthesis. These experiments provided the first proof that the genetic material of the cell is DNA.

The *microcapsule* is a very thin polysaccharide layer on the exterior of the bacterial cell wall. It can be detected by serological methods, but it is not visible under the light microscope.

Captaculae, *capitate feeding tentacles:* bundles of contractile tentacular threads extending from the head of scaphopods (see *Scaphopoda).* They serve for the capture of food, and they are provided with muscles, nerves and adhesive glands. Food particles are held by the adhesive glands while the C. transfer them to the mouth.

Capture-recapture method: see Lincoln index.

Capuchins, *Cebus:* a large genus of medium sized (head plus body 450 mm; tail 450 mm) Cebid monkeys (see) of the infraorder *Playrrhini* (see New World monkeys). They inhabit the canopy of forested areas of South and Central America, feeding on fruit, insects and snails, and displaying an elaborate social structure. The semi-prehensile tail is often carried rolled downward, and the head hair is formed into a dark skullcap.

Capybara: see *Hydrochoeridae.*

Carabao: see *Bovinae.*

Carabidae: see Ground beetles.

Caracal: see *Felidae.*

Caracara: see *Falconiformes.*

Caralluma: see *Asclepiadaceae.*

Carambola: see *Oxalidaceae.*

Carapace:

1) A structure present in many crustaceans (also known as the shell), consisting of a dorsal, chitinized fold of skin arising from the hind border of the head. It varies greatly in size, and extends backward for a greater or lesser distance over the trunk. In the *Ostracoda* and most conchostracans it encloses the entire body, also extending forward to enclose the head. In other crustaceans, part or all of the head is uncovered. In typical members of the *Malacostraca* it covers the entire thorax, while in others several thoracic somites are left exposed. In the *Syncardia, Isopoda* and *Amphipoda* the C. has disappeared. In *Apus* (see *Branchiopoda)* it forms a flat dorsal shield, but in most crustaceans it is compressed, and some cases it becomes a bivalve with a dorsal hinge *(Conchostraca* and *Ostracoda).* The chamber enclosed by the C. (known as the gill chamber, mantle cavity, etc.) serves to protect the gills, and sometimes serves as a brood pouch for developing embryos.

2) The dorsal shield (also known as the shell) of turtles and tortoises.

Carapace gland: see Y-organ.

Carapidae, *carapids:* a family of marine, elongate, pencil-shaped teleost fishes, in the order *Ophidiiformes.* Scales are absent, the gill openings are wide and extend far forward, and the anus and origin of the anal fin are behind the head and usually beneath the pectoral fin. They thread themselves into cavities and fissures, as well as openings provided by other living organisms. There are 2 distinct larval stages; the vexillifer is pelagic and planktonic, and possesses a long threadlike process anterior to the dorsal fin, known as a vexillum. The vexillifer develops into a tenuis larva, which is benthic, has no vexillum, and has a relatively small head. In parasitic forms (see below), the fish may enter its host at the tenuis stage. Two subfamilies are recognized. The *Carapinae* (formerly *Fierasferidae,* ***Pearlfishes*** or ***Fierasfers)*** are found in tropical and temperate seas, and many species live parasitically inside sea cucumbers, eating their internal organs. Others live commensally with starfishes, sea cucumbers, clams or tunicates, and a few species are apparently free-living. The 2 genera (3 species) of the *Pyramonodontinae* have a circumtropical distribution.

Carapinae: see *Carapidae.*

Carassius: see Goldfish, Crucian carp.

Caraway: see *Umbelliferae.*

Carbohydrates: a class of natural substances, many of plant origin, with the general composition $C_n(H_2O)_n$. Structurally, they are polyhydroxycarbonyl compounds. They were originally characterized as hydrated forms of carbon and were named C. in 1844 by K. Schmidt. The name has been retained, although it is inaccurate from a chemical point of view, since compounds are now included which have other proportions of C, H, and O, e.g. the aldonic acids, uronic acid and deoxy sugars, or which contain additional elements, e.g. aminosugars, mucopolysaccharides.

Conventions for the structural representation of carbohydrates according to *Fischer (a,* D-glucose), *Tollens (b,* b-D-glucose) and *Haworth (c,* b-D-glucose).

Cs. are present in every plant and animal cell, and make up the largest proportion, in terms of mass, of organic compounds present on the earth. They are produced during photosynthesis, and stored in roots and tubers. As one of the main components of the animal diet, they are particularly important as an energy source. According to their molecular size, they are subdivided into mono-, oligo-, and polysaccharides.

1) ***Monosaccharides*** (simple Cs.) cannot be further hydrolysed into simpler types of C. According to the number of constituent carbon atoms, monosaccharides are classified as trioses, tetroses, pentoses, hexoses, etc. They can be regarded formally as the primary oxidation products of aliphatic polyalcohols. Oxidation of the terminal primary alcohol group produces a polyhydroxyaldehyde called an Aldose (see). Oxidation of a secondary hydroxyl, usually that on C2, produces a polyhydroxy-ketone, or Ketose (see).

Nearly all natually occurring monosaccharides have unbranched carbon chains; exceptions are hamamelose, apiose, streptose, etc. Configuration is indicated by the prefix D or L, while optical rotation is indicated by (+) and (-). In D-sugars, the asymmetric center furthest removed from the carbonyl function has the same configuration as the arbitrarily chosen reference substance, D-glyceraldehyde. L-sugars are similarly related to L-glyceraldehyde. The theoretically possible number of stereoisomers of a monosaccharide is 2^n, where n is the number of asymmetric centers.

There are various systems for presenting the formula of a monosaccharide, which emphasize the structure of the molecule or its chemical reactivity. The original straight-chain Fischer projection was modified by Tollens to show the cyclic form of the monosaccharide, in which the carbonyl group forms a half-acetal bond with one of the hydroxyl groups. This was further refined by Haworth (Fig.). If the oxygen-containing ring is five-membered (i.e. it is formed by half-acetal formation between C1 and C4), the sugar is a *furanose,* whereas six-membered ring formation between C1 and C5 produces a *pyranose.* Pyranoses are more common than furanoses.

The spatial configuration of monosaccharides is expressed more clearly by the Haworth projection. The furanose ring is nearly planar, but the pyranose ring is puckered, because the C—O—C angle in the pyranose ring (111°) is very close to the tetrahedral angle (109 °28') of the sp^3-hybridized C-atom. The asymmetry caused by the oxygen atom makes possible two chair and six boat forms. However, pyranoses spend most time in the energetically favorable chair form.

Ring formation creates a new asymmetric center on the atom originally carrying the carbonyl group, and thus two new series of isomers, called the α- and β-forms. In solution, the α- and β-isomers exist in equilibrium via the open chain form (oxo-cyclo tautomerism). Thus when α-D-glucose is dissolved in water, its starting specific optical rotation of +113° decreases over a period of a few hours to +52.3°, representing an equilibrium of 37% α-D-glucose and 63% β-D-glucose (the specific optical rotation of β-D-glucose is +19.7°). This process is known as *mutarotation.*

2) ***Oligosaccharides*** consist of 2 to 10, α- or β-glycosidically linked monosaccharide units. They can be hydrolysed into their subunits by acids or enzymes. According to the number of linked units, they are known as di-, tri-, tetrasaccharides, etc. Disaccharides are particularly important biologically. In the *trehalose type* (e.g. sucrose and trehalose) the two monosaccharide units are joined by a C1-C1 glycosidic linkage (i.e. by the formal exclusion of water from the OH-groups on each of the C-1 atoms). Both monosaccharide units are then full acetals and do not display the typical sugar reactions, such as reduction, hydrazone and oxime formation, or mutarotation. In the *maltose type* (e.g. maltose, cellobiose, gentiobiose, melibiose) the glycosidic hydroxyl of one monosaccharide is linked to an alcoholic hydroxyl of a second monosaccharide (e.g. 1,4- or 1,6-glycosidic linkage). This type of structure permits the existence of a free reducing group, so that maltose-type disaccharides show reductive and mutorotatory properties, as well as forming hydrazones and oximes. All mono- and oligosaccharides are water-soluble, crystalline and sweet-tasting.

3) ***Polysaccharides*** consist of 10 or more monosaccharide units, linked α- or β-glycosidically to form branched or unbranched chains. The chains may be arranged linearly, spirally or a spherically. Polysaccharides generally contain hundreds or thousands of monosaccharide units. They therefore differ in their degree of polymerization, type of component monosaccharide and type of glycosidic bonding. Their chemical and physical properties differ from those of the component monosaccharides. Water solubility, reductive capacity and sweetness all decrease with increasing molecular size. They include the structural Cs., Cellulose (see) and Chitin (see), as well as the storage Cs., Starch (see) Lichenin (see) and Glycogen (see). They are not fermented by yeasts.

Carbol fuchsin stain: see *Actinomycetales.*

Carbon, *C:* an element present in all living material The vast majority of substances which constitute living organisms are carbon compounds. In contrast, only 0.2% of the earth's crust consists of carbon. The infinite variety of carbon conpounds is due not only to the ability of carbon to combine with other elements, but in particular to its ability to combine with itself, forming chains, branched structures and ring-shaped molecules. Carbon-containing residues of prehistoric organisms have accumulated as various grades of coal, and as solid, liquid and gaseous hydrocarbons, such as asphalt and oil and gas deposits.

Common inorganic carbon compounds are Carbon dioxide (see) and carbonates like marble and limestone.

Carbon cycle (Fig.). Atmospheric carbon dioxide assimilated by green plants (see Photosynthesis) is used to synthesize all the carbon-containing compounds of the plant organism. Part of this plant material is consumed by animals, then respired as a source of energy or incorporated into their own body compounds. In the form of carbon dioxide, carbon is returned again to the atmosphere during aerobic energy metabolism, or it may be released as incompletely metabolized products of fermentation or anaerobic respiration, or decomposition products of dead organism (see Dissimilation), which are converted to carbon dioxide by further microbial action. In the form of coal and other geological carbonaceous deposits, carbon is removed for long periods from the natural carbon cycle, but it may eventually be converted into atmospheric carbon dioxide by combustion. Wood also represents a long-term immobilized form of carbon; if this is not converted into coal, its constituent carbon is eventually returned as carbon dioxide to the carbon cycle by decomposition.

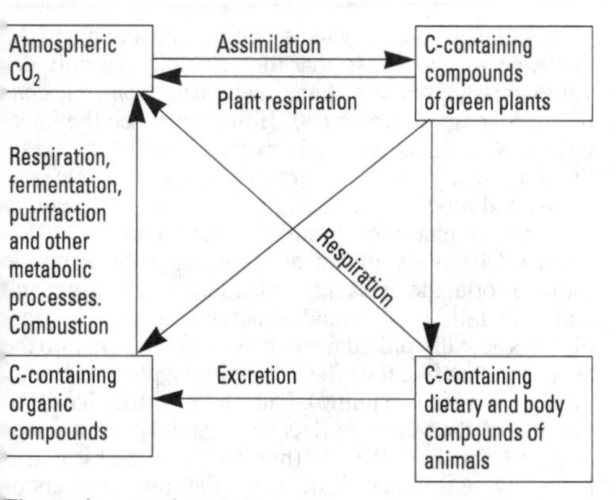

The carbon cycle

Large quantities of carbon dioxide are continually emitted by volcanoes and carbonic acid springs. On the other hand, carbon is immobilized and lost from the carbon cycle as vast deposits of naturally formed inorganic carbonates, often taking the form of entire mountain ranges.

Carbon dioxide: a colorless, nontoxic gas, formula CO_2, which tastes and smells weakly acidic. Large quantities of CO_2 are bound in the earth's crust as calcium and magnesium carbonates. The atmosphere consists of 0.03% CO_2. An atmospheric concentration of 4–5% causes anesthesia and numbness, while 8% causes death by suffocation.

As an end product of biological oxidation, CO_2 is produced in glycolysis, and in the tricarboxylic acid cycle by the decarboxylation of ketoacids. It is utilized in various biological carboxylation reactions, in particular in the dark reaction of Photosynthesis (see). Each year, 5×10^{10} metric tonnes of CO_2 are assimilated by terrestrial plants, and 46×10^{10} metric tonnes by marine plants.

Carbon disulfide, *CS_2*: an unpleasantly smelling, poisonous, strongly refractive liquid, which readily evaporates at room temperature and is very inflammable. It is a good solvent for sulfur, phosphorus, iodine, fats, vegetable oils and resins. It is used to extract sulfur from sulfur ores and the residues of coal gas purification, as well as fats and oils from a variety of sources, such as seeds, bones, etc. Chemically, it was a starting material in the preparation of potassium xanthate, formerly used for destroying the vine louse.

Carboniferous: the system or period of the Paleozoic era, lasting about 65 million years, which was preceded by the Devonian and followed by the Permian. It contains more that 50% of the world's coal reserves. In Europe, the C. is divided into the *Lower C.* or *Dinantian* (consisting of the stages Tournaisian and Visean) and the *Upper C.* or *Silesian* (consisting of the stages Namurian, Westfalian and Stephanian). In North America, these subperiods are known as the Mississippian (equivalent ot the Dinantian) and the Pennsylvanian (equivalent to the Silesian). See Geological time scale.

Terrestrial sequences are classified primarily on the basis of the evolution of higher vascular spore-bearing plants (pteridophytes): clubmosses *(Sigillariaceae, Lepidodenraceae)*, ferns and seed ferns, horsetails *(Calamitaceae)* and the oldest conifers *(Lebachia = Walchia)*, as well as the treelike gymnospermous *Cordaitidae*.

Freshwater habitats were home to the stegocephalic reptiles *(Labyrinthodontia, Lepospondyli)*. The 2.5 m-long cotylosaurians of the C., which are thought to be ancestral to modern reptiles, apparently diverged from the labyrinthodonts.

Marine sequences are based mainly on the evolution of foraminiferans, corals, brachiopods, goniatites, ostracods and conodonts. Foraminiferans attained their first evolutionary peak in the C. The Lower C. contains fossils of small coiled to spiral, twisted, helical and planispiral forms, while rolled fusiform varieties appear in the Upper C. In addition to increasing specialization of the structure of the chamber partitions, the foraminiferans also underwent a general increase in size. Foraminiferans and brachiopods are found predominantly in calcareous strata, and the productid and spirifer brachiopods are important zonal index fossils. Zonal stratigraphy of the C. is based on the goniatites. In contrast to the goniatites of the Upper Devonian, the majority of C. goniatites posses a suture line with a single outer saddle. At the end of the C., there were more than 1 300 insect species, including 170 apterygotes. Dominant fishes were elasmobranchs, ganoids and lungfishes.

Carbonification, *bituminization, coalification, incarbonization, incoalation:* formation of coal from plant material in the absence of air and under pressure. See Fossils. During C. the proportion of carbon increases relative to hydrogen, oxygen and nitrogen. The rate of C. and the nature of the carbonaceous end product are critically dependent on the pressure and temperature of the process.

Carbonization: a process of fossilization in which organic material is reduced to a residual carbon film.

Carbonyl compounds: organic compounds containing the carbonyl group, C=O, e.g. aldehydes and ketones.

Carboxylic acids: organic compounds containing one or more carboxyl groups (—COOH). The carboxyl group is the highest possible oxidation state of a carbon atom, and it formally represents a combination of a carboxyl and an alcohol function. The hydrogen of the carbonyl group can be replaced by metals (salt formation) or organic residues (esterification). C.a. may be classified according to the number of carboxyl groups, e.g. monobasic *monocarboxylic acids* (e.g. formic acid), dibasic *dicarboxylic acids* (e.g. oxalic acid), tribasic *tricarboxylic acids* (e.g. citric acid), etc. According to the nature of the organic residues carrying the carboxyl group(s), C.a. may be unsaturated, aliphatic, aromatic, heterocyclic, etc. They are mostly weakly dissociating acids. A large number of different C.a. occur naturally in the free form, esterified, or as salts.

Carboxypeptidases: exopeptidases which consist of a single polypeptide chain containing zinc. They remove successive amino acids from the *C*-terminal ends of proteins and polypeptides. Animal C. are important digestive enzymes. They are synthesized and secreted by the pancreas as inactive precursors (zymogens or proproteins), then converted into their active forms in the small intestine by the action of trypsin. They are classified according to their substrate specificity into two groups. C.A. preferentially remove terminal amino acids with aromatic or branched aliphatic side chains, whereas C.B remove basic amino acids (arginine and lysine) from the C-terminus of the polypeptide.

With respect to structure and mechanism, bovine C.A (307 amino acid residues; M_r 34 409) is one of the most investigated metalloproteases. The zinc atom is located in a pocket-like active center and participates in the catalytic process.

Carcharhinidae: see *Selachimorpha.*

Carcharhininae: see *Selachimorpha.*

Carchariidae: see *Selachimorpha.*

Carcharodon: see *Selachimorpha.*

Carcinogenic agents: chemicals that cause cancer in experimental animals or in humans, e.g. arsenicals, benzidine, benzene, β-naphthylamine, vinyl chloride and many others. The carcinogenic activity of a compound is not necessarily related to its general toxicity. Some C.a. are mutagens, acting at random on the genome, occasionally altering genes that are responsible for maintenance of growth arrest. Others, called tumor promotors or cocarcinogens, do not interact directly with chromosomes, but make them more susceptible to the effects of carcinogens.

Carcinoma: see Cancer cell.

Carcinus: see Swimming crabs.

Cardamom: see *Zingiberaceae*.

Cardia, *esophageal orifice, cardiac orifice:* the aperture at which the esophagus opens into the stomach, representing the transition zone from the 2-layered esophageal musculature to the 3-layered stomach musculature. In the region of the C. is a valvular arrangement, known as the cardiac sphincter. See Digestive system.

Cardiac: belonging to the heart, or to the upper end of the stomach.

Cardiac cycle: the phases of contraction and relaxation of the heart during a single cycle of activity, i.e. the phases of activity that constitute a heart beat. The contraction phase is called *systole,* and the relaxation phase is called *diastole.* In mammals, the right and left sides of the heart contract and relax simultaneously. In summary, the right and left atria contract simultaneously (atrial systole) and expel their contents into the respective ventricles. After a short pause (diastole), the ventricles contract simultaneously (ventricular systole), expelling their contents into the pulmonary artery and ascending aorta, respectively. The timing of the separate phases of the C.c. is determined by the opening and closing of the heart valves. Since the C.c. consists of similar, simultaneous events on both sides of the heart, the following account describes the sequence of events on one side.

During diastole, the atrio-ventricular valve is open, and the relaxed ventricle fills with blood; the subsequent atrial contraction adds a little more blood to the ventricle, causing a small rise of pressure within the ventricle. As the atrial contraction dies away, the ventricle contracts. The resulting rapid increase in pressure closes the atrio-ventricular valve, and the flaps of this valve are pressed together more tightly by contraction of the fibrous ring to which they are attached. In addition, the papillary muscles shorten, maintaining tension on the chordae tendinae, and preventing eversion of the valve cusps into the atrium. During this first phase of ventricular systole, all valves leading into and out of the ventricle are closed, so that the ventricle can only develop pressure without changing its size *(isometric contraction* or *isometric phase).* However, when the internal pressure of the ventricle exceeds that of the aorta, the semilunar valves are forced open, the ventricle contracts, and blood is ejected into the aorta, representing the second phase of systole *(ejection* or *expulsion phase).* During expulsion of blood from the ventricle into the aorta, the pressure gradient between atrium and ventricle falls rapidly and generates suction, so that blood is actually pulled from the venous system into the atrium (rather like the effect of pulling a plunger from a syringe or a bicycle pump). During this third phase, the volume of the ventricle decreases rapidly. Ventricular systole lasts about 0.3 sec (human heart) and it is followed by ventricular diastole, i.e. the muscle fibers of the ventricle relax (decrease their tension) without altering their length (isometric relaxation), the internal pressure of the ventricle falls, and the higher pressure in the aorta closes the aortic (semilunar) valve. The fourth phase begins with the opening of the atrio-ventricular valve, i.e. the filling phase of the ventricle during ventricular diastole. At first, the ventricle fills rapidly, because its internal pressure is less than that of the atrium, then more slowly in accordance with normal venous blood flow.

The different phases of the C.c. can be demonstrated by pressure measurements in the atrium and ventricle, as well as in the aorta or in the carotid artery, and they are also recognizable in the sounds of the heart beat. Thus, a plot of the pressure within the carotid artery shows a rapid increase, marking the beginning of the ejection or expulsion phase. The pressure then decreases somewhat during the last stages of ventricular systole, and continues to decrease to the baseline during ventricular diastole. At the beginning of ventricular diastole, however, there is a rapid backward movement of blood toward the heart, which is quickly stopped by closure of the aortic valve. This sudden decrease and re-establishment of pressure is shown by a V-shaped cut *(incisura)* in the plot of decreasing pressure.

Each side of the heart contributes to the two main sounds of the heart beat. The first sound is the dull tone of vibrating muscle tissue, generated during the phase of increasing pressure, when the valves of the ventricle are closed and the surging of blood against the ventricle wall causes it to vibrate or shudder. The second, shorter and brighter sound is generated at the end of ventricular systole by the sudden closure and brief vibration of the aortic (semilunar) valve.

For the positions and identities of the valves and heart chambers, see Heart (Fig.).

Cardiac nerves: nerves of the vertebrate autonomic system, which innervate the heart and regulate its autonomic activity. They are not involved in the production of excitation in the automatic centers of the heart (see Cardiac stimulation), but they modulate the activity of these centers, thereby increasing the range of response of cardiac function.

The heart is innervated by both parasympathetic vagus nerves and by the sympathetic system. Different regions of the heart receive different nerve fibers. Several functional quantities of the heart are altered by parasympathetic or sympathetic excitation, which are mutually antagonistic :

Functional quantity	Increased sympathetic stimulation	Increased vagus stimulation
Pulse rate	increased	decreased
Force of a single contraction	increased	decreased, but only in the atria.
Conduction time between atria and ventricles	decreased	increased
Excitability	increased	decreased, but only in the atria

Decreased sympathetic or vagus stimulation has the opposite effect to that listed in the above table. For example, the pulse rate falls if sympathetic excitation is decreased, and it increases with decreased vagal activity. Under physiological conditions, the vagal and sympathetic C.n. are normally transmitting impulses to the heart at an intermediate level of excitation. The functional quantities of the heart can therefore be regulated by either an increase or a decrease in the levels of excitation of the C.n. A simultaneous increase in vagal and sympathetic excitation does not alter the pulse rate or the conduction time, because these two nervous systems act antagonistically. The force of contraction is, however, increased, because the ventricular muscle is innervated only by the sympathetic system. The state of the heart muscles resulting from continuous activity of the C.n. is known as *tonus;* predominance of vagal activity results in *vagal tonus,* whereas the sympathetic system causes *sympathetic tonus*.

Vagal nerve endings release acetylcholine, which inhibits cardiac function, whereas sympathetic nerve endings release adrenalin and noradrenalin, both of which are stimulatory. These chemical transmitters bind to receptor molecules of the heart muscle cells, leading to changes in the ion permeability of the muscle cell membrane. Their effects are terminated enzymatically, i.e. by the hydrolysis of acetylcholine and the oxidation of adrenalin and noradrenalin.

Vagal stimulation causes a decrease in heart beat frequency, because acetylcholine selectively permits the flow of only potassium ions from the cell interior. The resulting increase in the resting potential of the pacemaker muscle cells leads to a decrease in their rate of depolarization, so that excitation occurs later in the sinus node. In contrast, the catecholamines released by the sympathetic nerve endings decrease the pacemaker resting potential, leading to an earlier onset of excitation in the sinus node. The power of contraction is decreased by vagal activity, because acetylcholine decreases the duration of the action potential; increased potassium conductivity leads to more rapid repolarization, so that fewer calcium ions have time to enter the muscle cell and bind to myosin and actin filaments. Conversely, the catecholamines of the sympathetic system increase the inflow of calcium, and the resulting increased binding of calcium to myosin and actin leads to a more powerful muscular contraction.

Cardiac orifice: see Cardia.

Cardiac output, *heart minute volume:* the volume of blood ejected by the heart in one minute. It is the product of the frequency of the heart beat and the Stroke volume (see). For a resting adult human with a pulse rate of 70/min and a stroke volume of 70 ml, the C.o. is 4 900 ml or about 5 l.

Cardiac reflexes: reflex responses of the heart and circulation initiated by cardiac receptors. The mammalian heart possesses *mechanoreceptors,* which are connected, via afferent fibers of the vagus and sympathetic nerves, to the brainstem cardiovascular centers. The atria have more sensory nerve endings than the ventricles.

Afferent fibers arising from the right side of the heart are known as *pressor fibers*. At each heart beat, vagal receptors register the increase in pressure in the atria and great veins, the vagus nerves are stimulated (especially the right vagus nerve) and the heart is reflexly accelerated (right atrial or *Bainbridge's reflex)*.

Afferent fibers arising from the nerve endings in the aortic arch and left ventricle are known as *depressor fibers*. These cardioaortic nerve endings are baroceptors which are stimulated only when pressure in the left ventricle and aorta is high. Such stimulation always causes reflex inhibition (due to impulses passing down the vagus nerve to the heart) (aortic or *Marey's reflex),* manifested as a slowing of the pulse rate and a fall in blood pressure.

Also sensitive to stretching are the nerve endings of the carotid sinus. Increase of pressure within the sinus produces a reflex slowing of the heart and a fall of blood pressure exactly as observed for the cardio-aortic nerve. The carotid sinus nerve and cardio-aortic nerve represent two afferents of a single regulatory mechanism, preventing undue rises of blood pressure and avoiding overloading of the heart. They are therefore also jointly known as "buffer nerves".

Cardiac regulation: the local and nervous control of cardiac activity (Fig.).

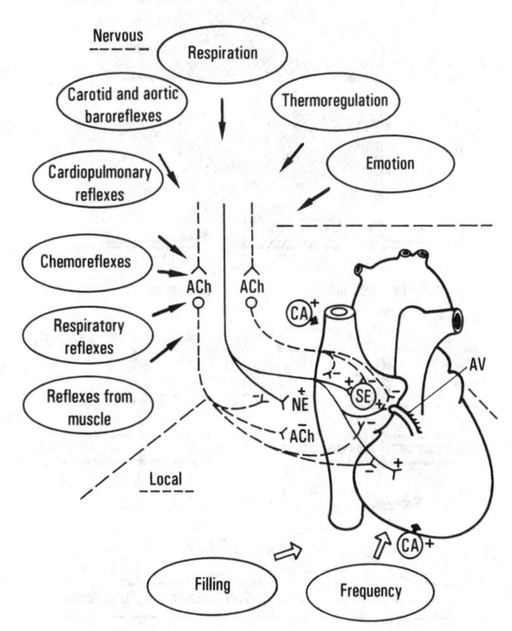

Local and nervous mechanisms of cardiac regulation. NE Noradrenalin. *SA* Sinus node. *AV* Atrioventricular node. *CA* Catecholamines. *ACh* Acetylcholine.

Cardiac stimulation, ***heart stimulation:*** changes in resting potential, caused by ion currents, which lead to Heart contraction (see).

The heart exhibits *automaticity,* i.e. the stimuli that excite it to activity are generated within the tissue itself. An automatic center consists of a group of heart muscle cells that are capable of spontaneous impulse generation. If more than one such center is present, the center with the highest automatic frequency functions as the *pacemaker,* which dominates the frequencies of the other centers. In the mammalian heart, the atrioventricular bundle (A-V bundle) of the ventricles has the lowest automatic frequency, followed by the atrioventricular node (A-V node) at the junction of the right atrium and right ventricle. The sinuatrial node (S-A node) (near the base of the superior vena cava in the right atrium) has the highest frequency and serves as the pacemaker. If the pacemaker fails, automaticity is assumed by the A-V node, and in the event of A-V node failure by the A-V bundle; atrial and ventricular contraction are then, however, uncoordinated. [The A-V bundle, 1–2 mm in cross section and consisting of primitive muscle fibers, begins at the A-V node and passes along the top of the interventricular septum; it then divides into two branches (fasciculi) which pass on each side of the interventricular septum to the right and left ventricles; each branch produces numerous smaller branches which form a reticulated layer of tissue just under the endocardium of both ventricles, called Purkinje tissue.]

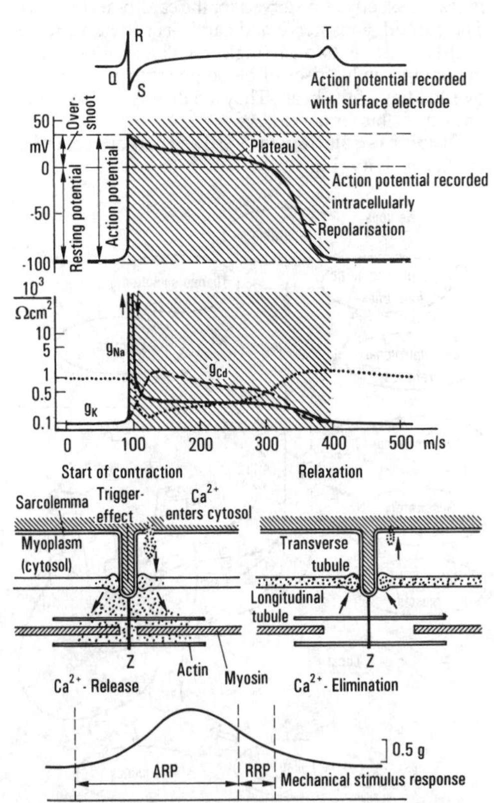

Cardiac stimulation. Progress of cardiac stimulation and contraction. *ARP* Absolute refractory phase. *RRP* Relative refractory phase.

In the mammalian heart, a wave of contraction initiated by the S-A node spreads rapidly through the muscles of both atria. It slows down in the A-V node, so that atrial contraction occurs ahead of ventricle stimulation. Excitation then spreads rapidly via the conducting system of the A-V bundle in the interventricular septum, and contraction occurs from the apex to the base of the ventricle; this contraction is therefore directional, pumping blood from the apex of the ventricles into the aorta and the pulmonary artery.

As the wave of excitation and contraction passes through the heart, different groups of heart muscle cells are depolarized at different times. This pattern of depolarization can be registered by an Electrocardiogram (see). In mammals, the P-wave of the electrocardiogram represents the excitation of the atria. The delay in excitation between atria and ventricles corresponds to the P-Q interval. Peaks Q, R, S and T arise from the overlap of the action potentials of all the stimulated ventricular muscle fibers. The beginning of the Q peak corresponds to the onset of ventricular excitation, Q, R and S to excitation of different regions of the ventricles, and T to repolarization in the ventricles.

Excitation and contraction of heart muscle occur simultaneously. While depolarized, a heart muscle fiber is nonexcitable or refractory. Therefore, during contraction, a second excitation or a second systole cannot be initiated. This protective mechanism insures that each contraction ends normally and that each beat is of the same power. C.s. therefore obeys an all-or-nothing rule, i.e. contraction either does not occur, or it occurs with maximal intensity.

Cardiac work: work performed during systole, consisting of the work done in discharging blood from the heart against the pressure of the aorta, and the work done to impart kinetic energy to the blood. At each systole, the left ventricle of the heart of a person at rest discharges about 70 ml blood. In order to overcome the average aortic pressure (more accurately, the mean pressure operating during the time that the aortic valves are open) of about 13 kPa, it must perform work of Q × R = 0.9 Joule, where Q is the volume of blood expelled at each contraction, and R is the average arterial resistance. To impart a minimal velocity of 40 cm/s, additional work of $mV^2/2g$ = 0.006 Joule must be added, where m is the weight of blood expelled at each contraction, V is the velocity, and g is the specific gravity of the blood. The work required for acceleration is therefore negligible compared with that required for overcoming arterial pressure, although these two factors may become approximately equal if the elasticity of the arteries decreases, e.g. in old age. Also, the kinetic factor may increases to about a quarter of the total work of the heart during strenuous exercise. The work of the right side of the heart is much less, because the mean pressure of the pulmonary artery is only 2 kPa (the kinetic energy imparted to the blood is the same for both the right and left ventricle). The total work of the heart of a person at rest is therefore (0.9 + 0.006) + (0.14 + 0.006) = 1.052 Joule, i.e. about 1 Joule. At a normal heart rate of 70 beats per minute, the daily work of the heart is therefore equivalent to almost 100 kJoules. For clinical purposes, a satisfactory approximation is cardiac work = cardiac output × mean arterial pressure.

Cardinal values: critical limiting values for the influence of environmental factors, i.e. pessimum, minimum, optimum, maximum.

Cardioid condenser: see Darkfield illumination.

Cardisoma: see Land crabs.

Cardoon: see *Compositae*.

Carduelis: see *Fringillidae*.

Carduus: see *Compositae*.

Caretta: see *Chelonidae*.

Carettochelyidae: a family of the order *Chelonia* (see), represented by the single living species, *Carettochelys insculpta* ***(Pig-nosed softshell turtle, New Guinea plateless turtle),*** found in rivers, lakes and lagoons in southern New Guinea and northern Australia. The bony carapace is not covered with horny scutes, but with a leathery integument. The legs are modified to wide flippers, and the snout is elongated. It leaves the water only to lay eggs. *Carettochelys* appears to represent an intermediate form between the *Emydidae* and the *Trionychidae*.

Carettochelys: see *Carettochelyidae*.

Cariama: see *Gruiformes*.

Cariamidae: see *Gruiformes*.

Caribou, *reindeer,* *Rangifer tarandus:* a large deer (head plus body 120–220 cm, shoulder height 87–140 cm), represented by many subspecies in the Arctic tundra and adjoining boreal coniferous forests of Eurasia and North America. The name "caribou" is used in North America, while "reindeer" is strictly a European term for the same animal. It is the only genus of the *Cervidae,* in which the female also has antlers. Relatively large, spreading hooves enable the animal to progress safely over swampy and frozen surfaces. The diet consists mainly of lichens and mosses. The commonest coloration is brown to gray above with paler or white underparts, the winter coat being somewhat paler. There are, however, wide color variations. Some individuals are dark brown to black, while the subspecies *R. t. pearyi* of northern Greenland and Ellesmere Island is almost pure white. European animals are kept in large, semidomesticated herds for their meat and hides. The total world population of domesticated *Rangifer* is estimated at about 3 million, many of those in the former USSR. The domestic form has been introduced into Iceland, Scotland, Orkneys, Canada, and even into South Georgia in the South Atlantic. See Deer.

Caricaceae, *pawpaw family:* a family of dicotyledonous plants, containing 45 species, all restricted to the tropics. All species are relatively small, latex-producing trees, with alternate long-stalked, exstipulate, petiolate leaves, which are simple or deeply palmately lobed or palmately 5-12-foliolate. The trunk is rarely branched and carries a terminal crown of leaves. Most species have separate male and female flowers, and plants are monoecious or dioecious. The actinomorphic and 5-merous flowers are yellow, whitish or greenish and aggregated in axillary inflorescences. The ovary, which is free and unilocular (rarely 5-locular) with 3–5 stigmas and numerous ovules, develops into a large, fleshy berry. The genus *Carica* is typical of the family. Male flowers of *Carica* have 10 stamens, 5 of which alternate with the corolla lobes and have short filaments, while 5 lie opposite the corolla lobes and are sessile. The corolla lobes are oblong or linear, and the rudiment of the ovary is awl-shaped. In the female flower, the petals are linear-oblong.

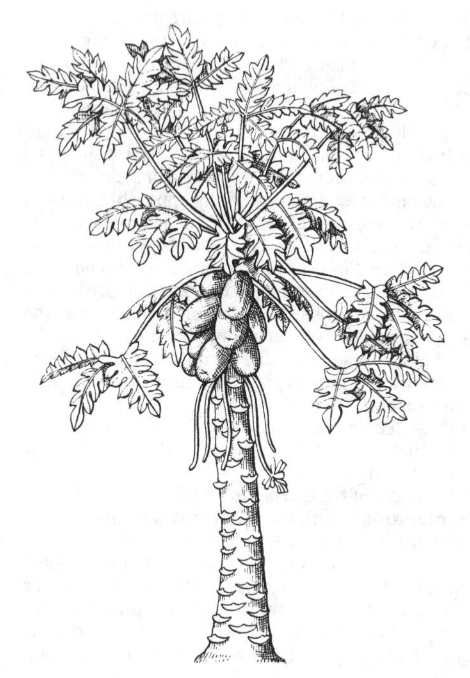

Caricaceae. Pawpaw tree *(Carica papaya).*

Pawpaw or ***Papaw*** *(Carica papaya)* has a soft, hollow, 2–6 m-long trunk, marked with the scars of fallen leaves (Fig.). The ovary is unilocular, and the ripe fruit is short-stalked and pendulous, about 400 cm in length, colored green to yellow, often with a purple tinge and red markings. Ripe fruits are eaten raw, while the unripe fruits may be pickled like gherkins. The seeds have antihelminthic properties. Juice from the pawpaw tree, especially from the fruit, contains a proteolytic enzyme called papain, which is used pharmaceutically to aid digestion, in the leather industry for softening hides, in cheese manufacture, and as a specific proteolyic agent in biochemical research. Cooks sometimes tenderize meat by wrapping it in pawpaw leaves or treating it with juice from the fruit. Although the pawpaw may occasionally exist as an escape, it is not otherwise known in the wild. It is cultivated all over the tropics, and probably originated from Central America.

The related species, *Carica pubescens, C. pentagona* and *C. chrysopetala,* are also cultivated for their fruit.

Caridea: Shrimps (see) and prawns, an infraorder of the *Decapoda* (see).

Carinaria: see *Gastropoda*.

Carissa: see *Apocyanaceae*.

Carlaviruses: see Virus groups.

Carline thistle: see *Compositae*.

Carmine mite: see *Tetranychidae*.

Carnassial teeth: see Dentition.

Carnation latent virus: see Virus groups.

Carnian stage: see Triassic.

Carnivora, *carnivores* (plates 38 & 39): an order of mammals, containing about 238 species in 92 genera and 7 families. Representatives are naturally distributed throughout the world, except in Australia, New Guinea and New Zealand. C. are usually flesh eaters with appropriate dentition, i.e. small incisors; large, strong, elongate, recurved canines, which are round to oval in cross section; premolars which are usually adapted for cutting; and molars usually with 4 or more sharp, pointed cusps. The last upper premolar and first lower molar are called carnassials, and they often work together in a special shearing mechanism. C. have 4 or 5 clawed digits on each limb; the first digit is not apposable, and it is sometimes reduced or absent. Recent suborders are the terrestrial carnivores (see *Fissipedia)* and the aquatic carnivores (see *Pinnipedia).* Fossil C. are found from the Tertiary to the present, e.g. dogs and cats are known from the Eocene onwards, bears and hyenas from the Miocene onwards. The dentition of primitive C. in the Lower Tertiary resembles that of the insectivores. Typical C. dentition appeared gradually during evolution.

Carnivores:

1) Flesh-feeding animals, as well as plants that supplement their diet by capturing animals. See Mode of nutrition, Carnivorous plants.

2) Members of the *Carnivora* (see).

Carnivorous fungi: see Carnivorous plants.

Carnivorous marsupials: see *Dasyuridae.*

Carnivorous plants, ***zoophagous plants:*** plants which are able to live autotrophically (green plants) or saprophytically (fungi), but have the additional capacity to supplement their nutrition by trapping and digesting small animals. About 500 species of C.p. are known.

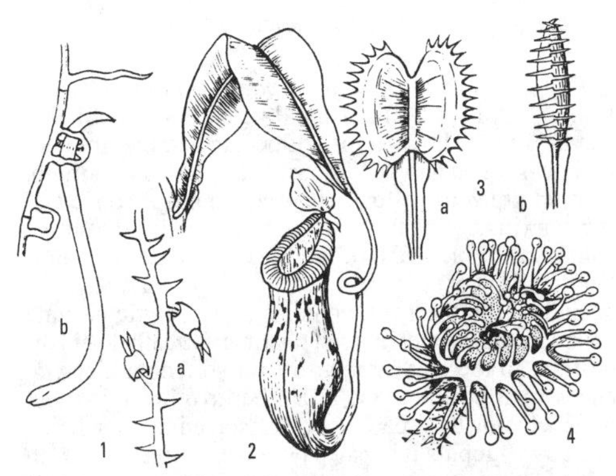

Carnivorous plants. 1 Carnivorous fungi: *a* adhesive hyphae of *Zoophagus* with entrapped rotifers; *b Arthrobotrys* with hyphal nooses and an entrapped nematode. *2* Pitcher of the pitcher plant, *Nepenthes. 3* Venus' fly trap: *a* open leaf lamina; *b* closed trap. *4* Leaf of sundew with adhesive tentacles and an entrapped insect.

Carnivorous fungi employ adhesive hyphae, rhizopodial processes, or hyphal nooses. For example, *Arnaudovia,* a constituent of the Neuston (see), traps unicellular organisms with the aid of 6 long, fine, adhesive hyphae. *Zoophagus* (Fig.) also traps unicellular organisms (e.g. rotifers) by adhesion. *Arthrobotrys* (Fig.) employs hyphal nooses to trap nematodes. Most *Arthrobotrys* noose systems consist of loops on loops, often at 90 ° to each other, forming an elaborate net, aided also by an adhesive secretion. In *Dactylaria candida* the hyphae form closed rings, and nematodes become trapped by trying to force their way through. In *Dactylaria gracilis* the noose structure is similar, but it actively contracts on the nematode. *Stylopage grandis* relies entirely on adhesion, nematodes being retained by a sticky secretion covering the mycelium. In every case a hypha then penetrates and grows into the trapped nematode.

Carnivorous species occur sporadically in various dicotyledonous families. A variety of trapping mechanisms is employed, such as suction traps (e.g. *Utricularia),* spring traps (e.g. *Dionaea),* "flypaper" or "birdlime" traps (e.g. *Pinguicula)* and pitfall traps (e.g. pitcher plants).

About 90 species of *Drosera* (sundew; see *Droseraceae)* are variously distributed in almost all parts of the world, the genus being especially developed in Australia and well represented in South Africa. The modified leaves form a ground rosette. Each leaf lamina carries vascularized emergences (tentacles) and the glandular head of each tentacle secretes a glistening, adhesive liquid with the fragrance of honey. Small animals stick to the tentacles, and the ensuing struggle for freedom stimulates the other tentacles to fold over onto the prey. A glandular secretion containing exoenzymes (mainly proteinases) then digests most of the animal (chitin is not attacked) and the products are absorbed by absorption hairs.

Pinguicula (butterwort; family *Lentibulariaceae)* possesses a ground rosette of more or less elongate leaves, each covered with a pale sticky layer (hence the name butterwort) produced by glandular hairs. The leaf margins are slightly incurved, and they roll over still further when an insect is trapped. About 30 species of *Pinguicula* are distributed throughout the temperate and cool regions of the Northern Hemisphere. The adhesive or "flypaper" principle is also employed by *Byblis* (rainbow plant; family *Byblidaceae),* which is confined to Western Australia. Victims are caught in a mass of sticky glands covering the surface of the stems and leaves; shorter digestive glands are also present. The trapping process is purely passive, since the glands do not move to enfold the prey. A surface layer of sticky glands is also found in *Drosophyllum* (Morocco, Spain, Portugal; family *Droseraceae),* which is not a herbaceous plant, but grows as a bush.

The leaves of the North American *Dionaea muscipula* (venus' fly trap; family *Droseraceae)* are also arranged in a ground rosette. The leaf lamina is rather rounded and the two halves fold upward and together when an insect stimulates the sensitive trigger hairs (6 on each leaf) on the upper leaf surface. Since the lamina margins are toothed and interdigitate with each other, the leaf closes rather like a gin trap and escape is impossible (Fig.). The insect is then digested by glandular secretions and the products are absorbed. *Aldrovanda vesiculosa* (waterwheel plant; family *Droseraceae)* is a rare, submerged, rootless plant, whose folding leaf traps are very similar to, but much smaller than those of *Dionaea;* it ranges from southern France to Japan, and south to Australia, Africa and India, probably as a single species with local varieties.

About 7 species of *Sarracenia* (sun pitchers; family *Sarraceniaceae)* are native to the eastern USA, growing in damp and swampy habitats. The whole leaf is modified to a tube with a lid at the mouth. It appears to be more efficient than the pitcher of *Nepenthes* for trapping prey, and *Sarracenia* pitchers are often almost full of insects. About 6 species of *Heliamphora* are known (rare South American sun pitchers; family *Sarraceniaceae),* all restricted to inaccessible jungle plateaus in Venezuela. The cobra lily of North America *(Darlingtonia = Chrysamphora californica)* is also a pitcher plant of the family *Sarraceniaceae.*

Species of *Utricularia* (bladderwort; family *Lentibulariaceae)* occur on every continent, including Greenland, but appear to be absent from Oceana. *U. nelumbifolia* is found in the water-containing leaf rosettes of large bromeliads, sending out runners into the urns of neighboring rosettes. Some tips of the finely divided leaves are modified to bladders, with a small mouth and an inward opening lid. When the bladder is closed, its walls are under tension and the contents are under negative pressure. Bristles on the outside of the lid act as levers, so that small animals knocking against them cause the valve-like lid to open, and are sucked into the bladder by the water stream. The lid closes and the animal is digested by secreted enzymes. The prey consists of water fleas, small larvae and even very young fish and tadpoles; the latter may be partly trapped rather than completely engulfed, but they are nevertheless digested. The family *Lentibulariaceae* contains 4 genera; the other 2 are *Polypompholyx* (South and Southwest Australia; 2 species, with bladders resembling those of *Utricularia;* the flowers of *Polypompholyx* have 4 calyx lobes, whereas those of *Utricularia* have 2) and *Genlisea* (15 species; South and Central America and Africa; the plants have bottle-like traps with an internal "lobster pot" structure of bands of backward facing hairs, as well as digestive glands).

Pitfall traps are especially well developed in *Nepenthes* (hanging pitchers; family *Nepenthaceae)* of the Old World tropics; the center of distribution of the genus is in the region of Borneo, extending to Northern Australia, New Guinea, Sri-Lanka and Madagascar. The tip of the leaf lamina is modified to an urn-like structure (ascidiform leaf), containing a small quantity of a solution of digestive enzymes (Fig.). Insects are attracted by nectaries on the upper edge of the pitcher, and they slip into the interior on a "slide zone" of wax coating the upper inner surface of the pitcher. A lid over the pitcher prevents the entry of rain, which would dilute the digestive secretion. Pitchers of *Cephalotus follicularis* (fly-catcher plant, Albany pitcher plant, found only in wet areas of extreme Southwest Australia; the only species of the family *Cephalotaceae)* are arranged as a ground rosette, and bear a superficial resemblance to the hanging pitchers of *Nepenthes.*

By culturing C.p. under sterile conditions, it is possible to determine the extent to which an animal food source is necessary for their normal dveelopment. Thus, the sundew is capable of complete development without a supply of organic material. In contrast, the bladderwort flowers only after receiving animal material. Since the natural habitats of C.p., e.g. acidic moorlands, are deficient in nitrogenous compounds and other nutrients, the carnivorous habit is clearly beneficial to the plant under natural conditions.

Carnivorous water beetles: see *Coleoptera.*

Carob (bean, family, tree): see *Caesalpiniaceae.*

Carotenes: in the narrow sense, a group of three yellow-red, isomeric, unsaturated hydrocarbon Carotenoids (see), empirical formula $C_{40}H_{56}$, with nine conjugated *trans* double bonds. They are widely distributed plant pigments, occurring e.g. in green leaves, many flowers, fruits and algae (0.02–1.15% fresh weight), usually accompanied by chlorophyll and xanthophylls. They enter the animal organism in the diet, and are regularly found in serum, milk and animal fat.

In carrots, the three isomers, α-, β- and γ-carotene (Fig.), occur in the ratio 15:85:0.1. α-Carotene differs from β-carotene (provitamin A) in the position of a double bond in one of the two terminal β-ionone rings. γ-Carotene lacks one of the β-ionone rings which is replaced by an open chain. In animals, the C. are nutritionally very important as precursors of vitamin A (i.e. they are provitamins; see Vitamins).

In the wider sense, the term is applied to most hydrocarbon carotenoids. Thus, phytoene is a colorless hydrocarbon carotenoid, which occurs widely in plants and is particularly abundant in tomatoes and carrot oil. It contains 6 branch methyl groups, 2 terminal isopropylidene groups and 9 double bonds, 3 of them conjugated. It is biosynthesized from two molecules of geranylgeranyl pyrophosphate, and it serves as a C_{40} starter molecule in the biosynthesis of other C., e.g. phytofluene, neurosporene, Lycopene (see) and the C. in sensu stricto.

Carotenoids: a large class of naturally occurring yellow and red pigments, consisting of highly unsaturated aliphatic and alicyclic hydrocarbons and their oxidation products. Since they are also nonsaponifiable lipids, they were formerly known as *lipochromes.* C. are biosynthesized from isoprene units (C_5H_8), and therefore

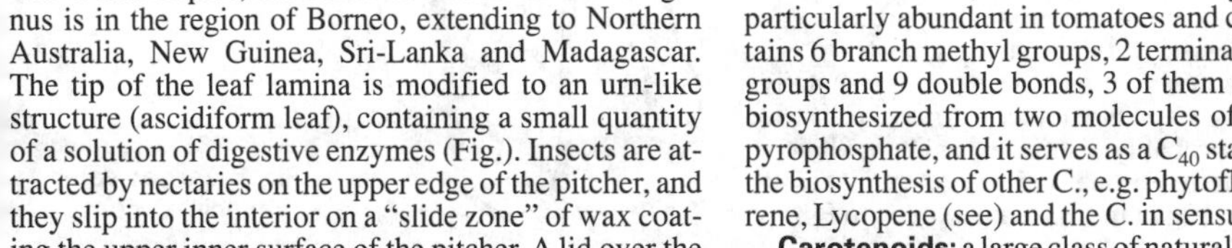

α-Carotene

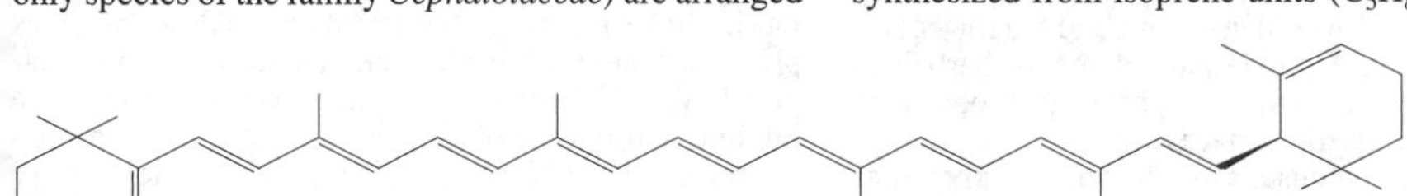

β-Carotene

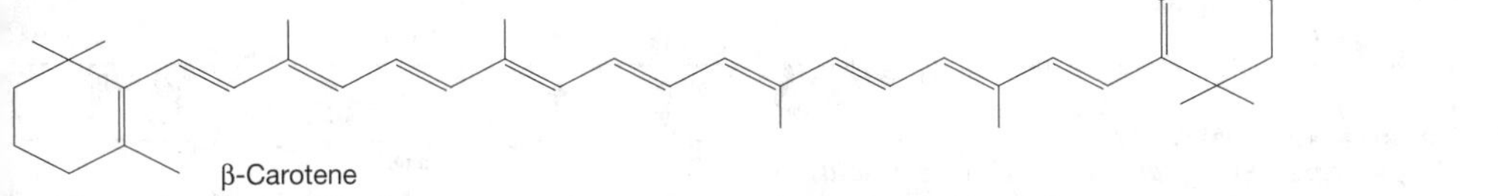

γ-Carotene

have the methyl branches typical of isoprenoid compounds. Most C. have 40 carbon atoms and are therefore tetraterpenes, which are formed from 8 isoprene units. In addition, there are a few C. with 45 and 50 carbon atoms, especially in nonphotosynthetic bacteria.

The C. are divided into the Carotenes (see), which are pure hydrocarbons (e.g. lycopene, carotene, neurosporene, phytofluene and phytoene) and the oxygen-containing yellow xanthophylls (e.g. violaxanthin, zeaxanthin, fucoxanthin, lutein, neoxanthin, cryptoxanthin, astaxanthin, capsanthin, capsorubin, rubixanthin, rhodoxanthin). In the native state, xanthophylls often occur as carotenoproteins (in which the carotenoid is bound to a protein, forming a soluble chromoprotein which is stabilized against the effects of air and light), or esterified to fatty acids, or linked glycosodically to glucose. C. contain 9 to 15 (usually 9 or 11) conjugated double bonds, usually *all-trans,* and are therefore planar polyenes. Their usually intense yellow to red color is due to this chromophoric system.

C. occur in both plants and animals, and are one of the commonest and most widespread groups of natural pigments. In animals, they occur mainly in surface tissues, such as skin, shells, arthropod exoskeleton (e.g. of lobster), scales, feathers and beaks, but also in egg yolk and milk (and therefore butter), and as visual pigments. In plants they occur in green leaves, many flowers, fruits and algae, usually accompanied by chlorophyll. Animals are incapable of the de novo synthesis of C., and all C. found in animals are of plant or bacterial origin. Animals, however, do perform certain conversions of preformed C. (e.g. carotenes are cleaved to form vitamin A).

C. are important in photosynthesis, accompanying chlorophyll in chloroplasts, where they function in energy transfer. They are also important as precursors of vitamin A and therefore of the visual pigments.

Carp, *Cyprinus carpio:* an important representative of the fish family *Cyprinidae* (see). After the last Ice Age, the C. was probably distributed as 2 populations, one in southeastern Europe and the other in eastern Asia. C. breeding began in China and moved to Europe during Roman times. As a commercial fish, the C. is now very widely distributed, especially in Eurasia, as well as North America, Australia, South Africa, New Zealand, and to a lesser extent in South America. Selective breeding has produced several different domestic races. Body length is 120 cm. The back is grayish black or brownish, the sides golden to rust, and the lower surface bright yellow. A single pair of mouth barbels is present, and there are 3 rows of pharyngeal teeth with well developed chewing surfaces.

Members of the genus *Carassius* are also known as C. They differ from *Cyprinus* by the absence of barbs and the presence of a single row of greatly compressed throat teeth. See Crucian carp, Godfish.

Carpel: see Flower.

Carpenter moths: moths of the family *Cossidae.* See *Lepidoptera.*

Caperea: see Whales.

Carpet viper, ***Saw-scaled viper,*** *Echis carinatus:* a snake of the family *Viperidae* (see), found in semidesert regions from West Africa to western India and Sri-Lanka. It is about 60 cm in length, colored sandy brown to dark brown with light dorsal stripes, and not as stout as the Common viper. The scales have prominent keels. Dorsal scales are arranged in straight rows, whereas the lateral scales are arranged obliquely. In addition, the keels of the lateral scales are serrated, so that as the snake progresses forwards by serpentine undulations, the scales rub on one another to produce a dry rasping sound. The C.v. is very aggressive, and its strike is so rapid that the eye can barely perceive it. The extremely potent, hemotoxic venom is arguably the most dangerous in the world. 10–15 young are born live.

Carp louse: see *Branchiura.*

Carpogonium: the female gametangium of the *Rhodophyceae* (see).

Carposis: see Relationships between living organisms.

Carrageen: see *Rhodophyceae.*

Carrier:

1) an organic molecule that acts as a vehicle for the transport of substances through membranes. Each C. is substrate-specific, i.e. it promotes the transport of only a limited number of chemically related substances. Membrane proteins, which are responsible for the substrate specificity of C. transport, may catalyse the binding of the substrate to its low molecular mass C. Alternatively a membrane protein may be the actual C., i.e. a transport protein. The existence of endogenous low molecular mass Cs. is uncertain, but low molecular mass substances are known which serve as Cs., e.g. the antibiotic, Valinomycin, which serves as a C. for K^+ (Fig.1).

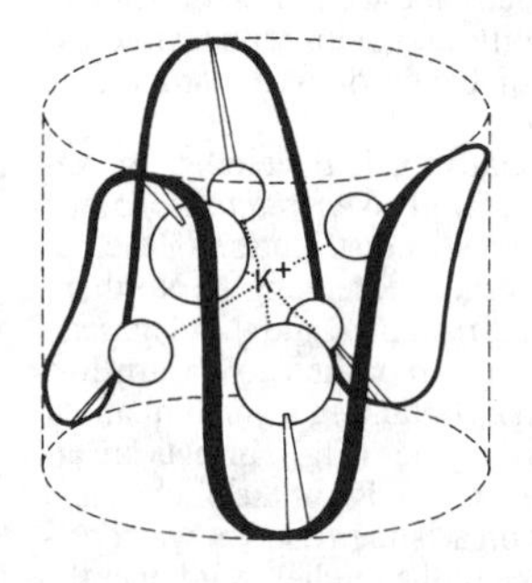

Fig.1. *Carrier.* Valinomycin-K^+ complex.

Several membrane transport proteins have been isolated, most of them large lipoproteins with M_r 200 000 to 700 000, which span the membrane. Since the energy requirement of transport is provided by ATP cleavage, these proteins are also known as *transport ATPases.* During ATP cleavage, the transport ATPase becomes phosphorylated. Phosphorylation and dephosphorylation of the ATPase, which operate in parallel with the binding and release of the substrate, are accompanied by reversible conformational changes in the ATPase that serve to move the substrate molecule through the membrane (Fig.2).

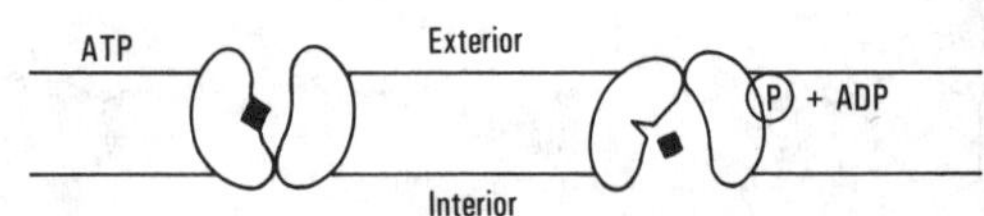

Fig.2. *Carrier.* Carrier protein model. The transport protein exists in two conformations, which are interconvertible by phosphorylation and dephosphorylation. The conformation on the left has a high affinity for the transported substrate, whereas that on the right has none.

Active transport mediated by Cs. is responsible, inter alia, for: 1) the uptake of inorganic ions by plant roots (see Mineral metabolism), 2) the maintenance of specific Na^+- and K^+-concentrations in organisms or cells in media of variable ionic concentration (see Ion pumps), e.g. marine organisms and erythrocytes in blood plasma, 3) the establishment of the bioelectric membrane potential of nerves (see Excitation).

Transport across membranes can also occur by *catalysed* or *facilitated diffusion,* a process independent of metabolism and not requiring energy in the form of ATP. It serves for the transport of, e.g. sugars and some proteins. Facilitated diffusion can only proceed down a concentration gradient and never against such a gradient. Like active C. transport, it is substrate-specific, and related substances may competitively inhibit the transport of each other. Membrane proteins catalysing facilitated diffusion, often called permeases, have comparatively low M_r values between 10 000 and 45 000. Some permeases also catalyse active transport. Conversely, if energy is deficient, transport ATPases may behave as permeases.

2) In human genetics, the vector of abnormal genetic material, who does not display the corresponding abnormal phenotype. The abnormal genetic material is usually in the form of a recessive allele in the heterozygous state, or exists as balanced chromosome aberrations.

Carrion beetles, *Silphidae:* a family of Beetles (see) of about 230 known species. Both adults and larvae feed predominantly on carrion. The genus, *Necrophorus* contains the well known burying or sexton beetles, which excavate and bury the bodies of small vertebrates. Eggs are laid in a gallery leading from the buried corpse. In their earliest instars, the larvae are fed by the female, and the later instars consume the carrion independently. Some C.b. are predacious, e.g. *Phosphuga atrata* preys on snails, and *Xylodrepa punctata* preys on lepidopterous larvae. Some are vegetarian and are occasionally harmful to plants, e.g. *Aclypea (= Blithophaga)* often damages beet and other root crops.

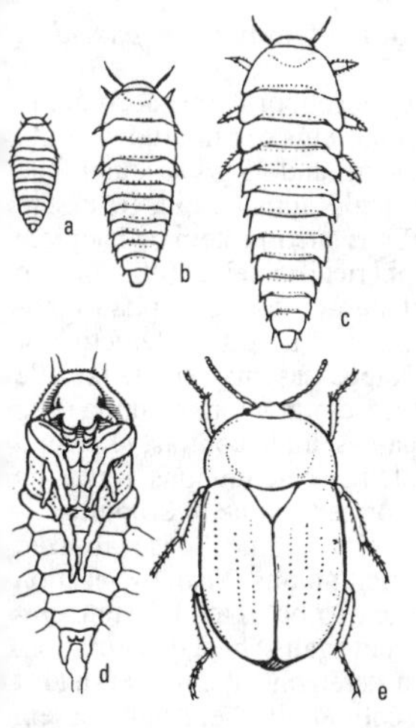

Carrion beetles. A vegetarian member of the *Silphidae (Aclypea undata* Mu. *= Blithophaga undata = Silphia reticulata). a* to *c* Larval stages 1 to 3. *d* Pupa. *e* Imago.

Carrion flowers: see *Asclepiadaceae.*

Carrot: see *Umbelliferae.*

Carrot fly, ***carrot rust fly,*** *Psila rosae* Fabricus: a psilid fly (order *Diptera),* about 5 mm in length, with a yellow head and green body. It lays eggs at the bases of suitable food plants, which are then attacked by the slender, yellowish white larva (about 0.8 cm in length when fully grown). Food plants include carrot, parsnip, celery, parsley and other related species. The entire root becomes scarred and riddled with larval burrows, which are colored rust-red (Fig.). It was imported from Europe into America in 1885 and has now spread over eastern Canada and many parts of northern USA.

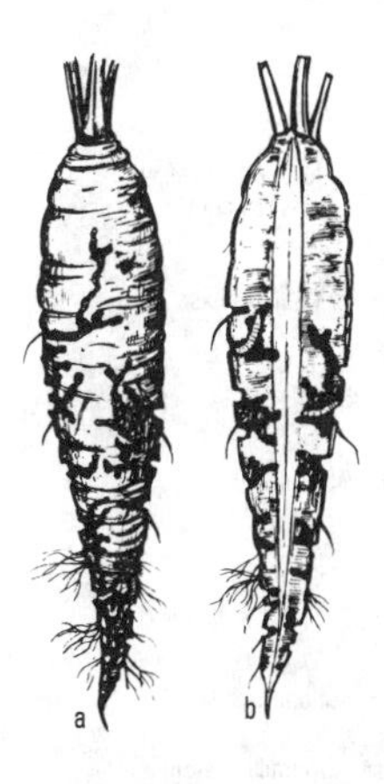

Carrot fly (Psila rosea). a Carrot attacked by larvae. *b* Longitudinal section of *a.*

Carrot rust fly: see Carrot fly.

Carthamus: see *Compositae.*

Cartilage, ***chondros, cartilago:*** an animal skeletal tissue (the other is bone). It occurs rarely in invertebrates (e.g. the cartilaginous radula support of snails and the head capsule of cephalopods), but is found in all vertebrates. C. is derived from embryonal connective tissue, and it forms the cartilaginous skeleton of the vertebrate embryo. In advanced vertebrates, the cartilaginous skeleton becomes ossified to form the bony skeleton of the adult, only a few skeletal elements being retained as C. Structurally, C. consists of *chondroblasts* and surrounding Intercellular substance (see). The latter consists of gelatinous ground substance, which has a characteristically high content of chondroitin sulfate and contains embedded collagen fibers. Chondroblasts occur singly or in groups, in fluid-filled spaces of the intercellular substance called *lacunae.* Except where it forms an articular surface, C. is surrounded by a layer of moderately vascularized, fibrous connective tissue, called the *perichondrium.* C. is nonvascular, and it is nourished by diffusion of nutrients from the perichondrium. C. has only low tensile strength, but high resistance to compression, and high elasticity.

There are 3 types of vertebrate C. 1) *Hyaline C.* (familiarly known as "gristle") is the most widely distributed type. Fresh, living materal is slightly blue in color, with a glassy, macroscopically clear appearance. In fact, it contains a delicate, felt-like network of collagen fibers, which can be revealed by a special staining technique. The skeleton of lower vertebrates consists exclusively of hyaline C. throughout life, whereas in higher vertebrates only the embryonal skeleton consists of hyaline C. In adult mammals, hyaline C. is retained as the costal Cs., as the cartilaginous skeleton of the trachea, and as the articular C. which covers the surfaces

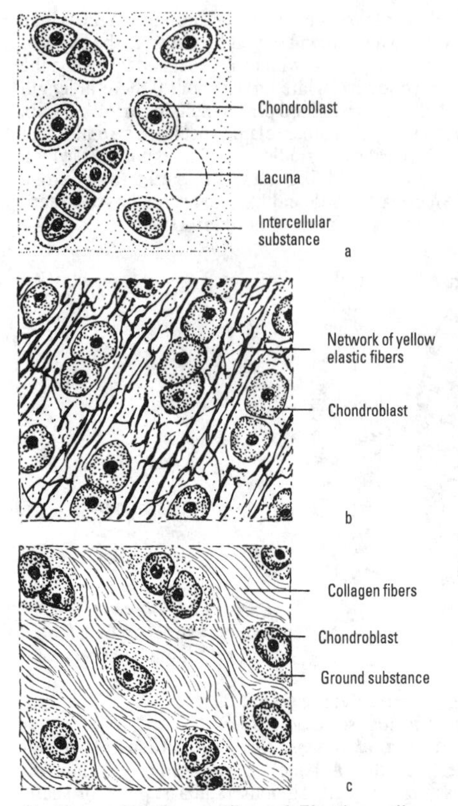

Cartilage. a Hyaline cartilage. *b* Elastic cartilage. *c* Fibrocartilage.

of joints. 2) *Elastic C.* occurs in the epiglottis, larynx, auditory tube, ear pinna and end of the nose. Its intercellular substance is pervaded by a network of yellow elastic fibers, to which it owes its high flexibility. 3) *Fibrocartilage* is not very widely distributed, being found in intervertebral disks and in the symphysis pubis between the pubic bones. Its intercellular substance is pervaded by abundant bundles of white collagen fibers, between which are scattered a relatively small number of chondroblasts. It can be regarded as a transitional structure between hyaline C. and tendon tissue.

Cartilages of Santorini: see Larynx.

Cartilaginous fishes: see *Chondrichthyes.*

Carvone: a monoterpene with the structure of a cyclic, unsaturated ketone. It smells like caraway, and it occurs in certain essential oils. Dextrorotatory C. is present in oil of caraway (*Carum carvi*) and in dill *(Anethum graveolens).* The levorotatory form constitutes 70% of the essential oil of spearmint.

CH3 O H_3C CH_2 Carvone

Carya: see *Juglandaceae.*

Carybdea: see *Cubomedusae.*

Carybdeida: see *Cubomedusae.*

Caryophyliidae: see *Dicotyledoneae.*

Caryophyllaceae: a family of dicotyledonous annual to perennial herbs, comprising about 2 000 species, distributed throughout the world, but occurring chiefly in the temperate regions of the Northern Hemisphere. The leaves are always simple entire, and usually in opposite and decussate pairs. The flowers are hermaphrodite, possess 4 or 5 petals, and are mostly pollinated by insects. The fruit is usually a multiseeded capsule, sometimes a berry *(Cucubalus)* and rarely an indehiscent one-seeded nutlet *(Corrigiola, Scleranthus).* The best known members are various species of *Dianthus,* e.g. the frequently cultivated, variegated ***Carnation,*** *Dianthus caryophyllus,* which originates from southern Europe; more than 100 varieties of this plant were known as early as 1671. *Dianthus barbatus* ***(Sweet William)*** is also widely cultivated; its flowers form dense, compact heads at the end of the stem, characteristically enclosed by very pointed, linear lanceolate or linear acute rough-edged bracts.

Various members of the family have poisonous or pharmacological properties, due to their constituent saponins. Thus, *Agrostemma githago* ***(Corn Cockle),*** formerly widely encountered as an agricultural weed, produces poisonous seeds. *Saponaria officinalis* ***(Soapwort)*** and various species of *Gypsophila* ***(White Soapwort)*** and *Herniaria* ***(Rupture-wort)*** are used as medicinal plants. One of the most widespread weeds is *Stellaria media* ***(Chickweed),*** which flowers from March to December. Another member of the family, *Spergula arvensis* ***(Corn Spurrey),*** is an almost cosmopolitan weed of arable land, and it is sometimes grown on sandy soils as a fodder plant and as green manure.

Caryophyllenes: cyclic, unsaturated sesquiterpene hydrocarbons (Fig.), existing in several isomeric forms, and present in numerous essential oils.

Caryophyllidea: see *Cestoda.*

Caryopsis: see *Graminae,* Fruit.

Cascaval: see Rattlesnakes.

Casebearers: moths of the family *Coleophoridae.* See *Lepidoptera.*

Caseins: a mixture of phosphoproteins, accounting for about 80% of the total proteins of milk (cow's milk contains 31.5 g/l total proteins and 25 g/l caseins). The mixture consists of α-, β-, and Kappa-caseins (α-casein is a group of 6 structurally related proteins). They are present as large stable spherical micelles (40–300 nm diam.), consisting of a calcium-protein-phosphate complex. The micelles are formed by specific interaction of monomers of α-, β- and Kappa-casein; a single micelle consists of 4 000–10 000 casein monomers in the ratio α:β:Kappa = 3:2:1. Kappa-casein, in contrast to α- and β-caseins, is not precipitated by calcium ions. Since the surface of the micelle is formed by the hydrophilic C-terminal part of Kappa-casein, the micelle is stabilized against precipitation by calcium ions. Also, by reaction of Kappa-casein with β-lactoglobulin at high temperatures, the micelle is protected against heat denaturation. Thus Kappa-casein has a detergent-like action, and it functions as protective colloid. If the Kappa-casein structure is disturbed or destroyed by acidification, or by the action of rennin, all the caseins are precipitated, i.e. the milk is coagulated.

Kappa-casein consists of 169 amino acid residues (M_r 19 032). Its *N*-terminal 105 residues are predominantly hydrophobic and located inside the micelle. The carbohydrate rich C-terminal part (64 residues) has a high content of serine, threonine, asparagine and glutamine, with no aromatic amino acids, and it lies on the surface of the micelle. Rennin hydrolyses one peptide bond in Kappa-casein (-Phe_{105}-Met_{106}-), separating the *N*-terminal hydrophobic part (para-Kappa-casein) from the carbohydrate-rich *C*-terminal part (known as macropeptide).

The primary structures of the α-caseins (M_r 33 000) and β-casein (M_r 24 400) are also known.

Cashew nuts: see *Anacardiaceae.*

Casiragua: see *Echimyidae.*

β-Casomorphines: peptides isolated in 1979 from enzymatic digests of bovine casein, and shown to have opioid activity in the guinea pig ileum longitudinal muscle-myenteric plexus preparation. For example, β-casomorphine-7 has the structure: Tyr-Pro-Phe-Pro-Gly-Pro-Ile.

[V. Brantl et al. (1979) *Hoppe-Seyler's Z. Physiol. Chem.* ***350,*** 1211–1216; A. Henschen et al. *ibid.* 1217–1224]

Caspialosa: see Shad.

Cassava: see *Euphorbiaceae.*

Cassia: see *Caesalpiniaceae.*

Cassidinae: see Leaf beetles.

Cassiope: see *Ericaceae.*

Cassowaries: see *Casuariiformes.*

Castanea: see *Fagaceae.*

Caste:

1) Among eubiosocial insects, the existence of functional polymorphism within the same group, e.g. female wasps and bees may be queens or workers. Hive bees have three castes: nonreproductive, usually sterile female workers which forage and provision the hive; reproductive males (drones); and fertile females (queens). In termite colonies, sterile workers and soldiers may be either male or female, whereas all the different worker Cs. of ant colonies are sterile females. In the wider sense, all morphotypes within a species may be termed Cs., e.g. males, females, different stages of development (infant, juvenile), provided that they differ with respect to body structure and function, and behavior.

2) A position in a hierarchical system of some human societies; membership is determined by birth and cannot be changed.

Castor bean tick: see *Ixodides.*

Castoridae, ***beavers:*** a family of Rodents (see), comprising only 2 species in a single genus. Beavers are large (head plus body length 80–120 cm, tail 25–50 cm), plump rodents with a flattened, scaly tail, webbed hindfeet and waterproof fur. Colors vary from yellowish brown to almost black, a reddish brown being the most common. They are expert swimmers and divers. A large conical lodge or house is built from mud and gnawed off sticks, either on the shore or in the middle of a pond. This structure dams the flow of water, and as the water rises, the beaver burrows upward in the mound, extending its living quarters and nest above the water level.

The ***European*** or ***Asiatic beaver*** *(Castor fiber)* is a protected species. It survives only in isolated populations in the Rhone valley in France, the Elbe in Germany, various areas of Scandanavia, Central Russia (Lake Baikal) and Mongolia; many of these isolated groups are classed as subspecies. The ***North American*** or ***Canadian beaver*** *(Castor canadensis)* is common in North America from Alaska, eastward to Labrador, southward to northern Florida, and in Tamaulipas (Mexico); it has also been introduced into Europe and Asia.

Castor oil, ***Ricinus oil, oleum ricini:*** a fatty oil (see Fats) from the seeds of the tropical and subtropical *Ricinus communis* L. (family *Euphorbiaceae).* It is a pale yellow, oily liquid consisting mainly of the glycerol esters of Ricinoleic acid (see), together with gycerol esters of Oleic acid (see), Stearic acid (see) and Linoleic acid (see). C.o. is used medicinally as a mild purgative, in the textile industry as a proofing agent and mordant, in the manufacture of soaps and dyes, and as a lubricant.

Casparian strip: a water-impermeable, cork-like strip in the radial walls of young endodermal cells. The C.s. apparently forms a barrier, thereby compelling water and all its dissolved constituents migrating into the interfibrillar spaces (apoplasts) of the root cortex to pass into the protoplasts of the endodermis cells.

Castilloa: see *Moraceae.*

Casuariiformes: an order of flightless, running birds, comprising a single living species of ***Emu*** *(Dromaius novaehollandiae)* and 3 living species of ***Cassowaries*** *(Casuarius).*

The ***Emu*** is the only surviving member of the family *Dromeidae.* Certain island races were lost soon after white settlement (from Tasmania, King Island, Kangaroo Island). Although the emu population of mainland Australia is excluded from the main areas of industrialization and urban development, this bird has otherwise thrived, and it occurs all over Australia from sclerophyll forest to open plains and grasslands. It even became a pest of marginal wheat-growing areas in Western Australia, leading to the "Emu war" of 1932, in which an attempt was made to control emus in a military style operation using machine guns ! This was a costly failure, and the seasonal movements of emus are now controlled by the erection of emu-proof fences. Sexual dimorphism is only slight. Of the ratites, they are second in height only to the ostrich, standing 150–185 cm tall and weighing as much as 55 kg (more usually about 45 kg). Plumage is brown to black-brown, and the feathers are double, the aftershaft being the same size as the main shaft. The feather barbs are not hooked together, so that the feather is loose and "hairy". When running they can attain speeds up to 50 km/h, and they can also swim. Between 8 and 16 massive, green eggs are laid in a flat bed of trampled vegetation. Incubation of the eggs (58–61 days) and care of the young is performed entirely by the male. In the reduced bone structure of the wing, the manus has only one digit (with 3 phalanges) and a long terminal claw.

Cassowaries (family *Casuariidae)* inhabit rainforests in northeastern Australia, New Guinea and neighboring islands. Three recent species are known: *Casuarius unappendiculatus* ***(One-wattled cassowary;*** a single wattle is present; New Guinea); *Casuarius casuaris* ***(Australian cassowary;*** 2 wattles; northeastern Australia); and *Casuaris bennetti* (***Bennett's Cassowary;*** wattles absent; New Britain Island). The wing quills are reduced to bare, black spines, and the drooping plumage is coarse and bristle-like. The top of the skull of the adult carries a bony helmet or casque; head and neck are naked. Incubation of the eggs and care of the young are performed entirely by the male. Adult cassowaries weigh up to 80 kg, with a body length of 130–165 cm. The legs are powerful, and the innermost of the 3 toes is armed with a long and deadly claw. Even humans have been killed by cassowaries. New Guinea natives keep cassowaries in captivity as food.

All *Casuariformes* are mainly vegetarian, feeding on seeds and soft fruits, and sometimes other vegetable material; insects are also occasionally taken.

The orders *Struthioniformes, Rheiformes, Casuariiformes* and *Apterygiformes* were formerly grouped together as the "Ratitae"; this close relationship is now doubted, but members of these 4 orders are still collectively known as Ratites. In all ratites, the wing skeleton and pectoral girdle show varying degrees of reduction, and the sternum is flat and raft-like with no keel. See *Aves.*

Cat: see *Felidae.*

Catabolism, ***dissimilation:*** the totality of energy-yielding degradative metabolic processes. The chemical energy from catabolic reactions is exploited for the synthesis of new cell material in processes collectively known as Anabolism (see). The organic materials that form the substrates of C. are produced by Assimilation (see) and anabolism.

Cataclysmic theory: see Catastrophe theory.

Catalase: an oxidoreductase of high catalytic activity, which catalyses the decomposition of toxic hydrogen peroxide in living cells: $H_2O_2 \rightarrow H_2O + 1/2O_2$. It is a tetrameric enzyme (M_r 254 000), each subunit (M_r 60 000) with a heme prosthetic group (ferriporoporphyrin IX). C. is found in all animal organs, particularly the liver (where it is contained in special organelles, the peroxisomes) and erythrocytes, all plant organs and nearly all aerobic microorganisms. Its turnover number of 5×10^6 molecules H_2O_2 per min per C. molecule is one of the highest known for an enzyme.

Catalepsy: see Protective adaptations.

Catalysed diffusion: see Carrier.

Catalyst: a material that triggers or accelerates a thermodynamically possible reaction, without itself being altered or consumed. The position of equilibrium of the catalysed reaction is not changed, only the rate at which equilibrium is attained. Cs. act by lowering the activation energy of a reaction. In reversible reactions (e.g. cleavage of an organic ester), a catalyst can be used to accelerate either the cleavage of the starting material, or its synthesis from the cleavage products, until equilibrium is attained. Cs. are widely used in synthetic chemistry, especially for hydrogenations. Most processes that occur in living cells depend on catalysis by Enzymes (see).

Catapulting mechanisms, ***discharge mechanisms:*** rapid, explosive movements in plants caused by the equalization of turgor differences between different layers of tissue (see Explosion mechanisms). They serve for the catapulting, and therefore the dispersal, of seeds, pollen and spores. C.m. are irreversible, and they normally occur as the result of natural aging and maturation. Thus, the middle lamella may break down, so that tissue tensions can express themselves. For example, the outer part of the wall of the fruit of *Impatiens* species consists of thin walled parenchyma, which, as a result of cell shape and a turgor pressure of about 20 atmospheres, is attempting to expand longitudinally. The two innermost layers of the wall are elongated fiber cells, which resist the elongation of the outer tissue. The five carpels forming the ovary are united into a cylinder, which is therefore in a state of tension. At maturation, dissolution of the middle lamellae along the sutures of the carpels renders the fruit susceptible to the slightest disturbance. Each valve then rolls in upon itself like a clock spring, ejecting the adherent seeds over a considerable distance. A similar explosive mechanism, normally triggered by insects, occurs in the ejection of pollen in orchid flowers. See Squirting mechanisms

Catasetum: see *Orchidaceae.*

Catastrophe theory, ***cataclysmic theory:*** the doctrine of G. Cuvier (1769–1832), according to which all life on earth has been periodically exterminated by cataclysmic geological or climatic events, followed by its re-establishment. It was proposed that life was re-established after each catstrophe, either by colonization from outside the earth, or (according the Cuvier's successors) by an act of creation. Proponents of the theory of evolution (Lamarck, Darwin) opposed this doctrine (also known as catastrophism or convulsionism), which finally gave way to Actualism (see).

Catastrophism: see Catastrophe theory.

Catbirds: see *Ptilinorhynchidae.*

Catchweed: see *Rubiaceae.*

Catechins, ***flavan-3-ols:*** a group of plant constituents, which can be considered as phenolic derivatives of flavone, or as hydrogenated anthocyanins. (+)-Catechin and (-)-epicatechin (Fig.) are widespread in the plant kingdom, whereas other C. have a more limited distribution. C. have been implicated in the biosynthesis of condensed tannins (see Tannins). Insoluble, tannin-like oxidation and polymerization products of C., known as phlobaphenes, are deposited in the membrane systems of leaves, where they contribute to brown coloration of leaves in the autumn.

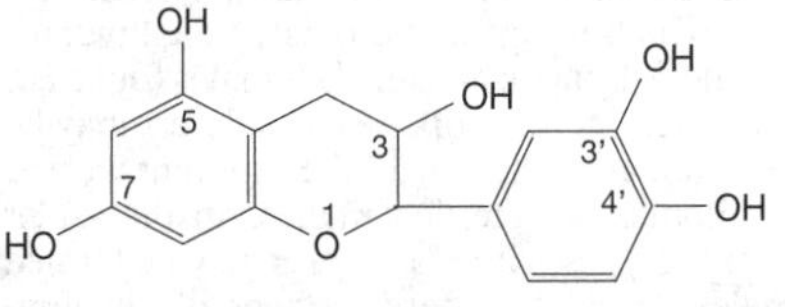

OH, OH, OH, 5, 3, 3', 7, 4', OH, HO, 1, O

Catechins. 5,7,3',4'-Tetrahydroxyflavon-3-ol. Depending on the configuration at position 2 and 3, this is (+)-catechin (2R,3S) or (-)- epicatechin (2R,3R).

Catelectrotonus: see Electrotonus.

Catenulida: see *Turbellaria.*

Caterpillar: a trivial name for eruciform (polypod) insect larvae, which possess thoracic legs and a variable number of abdominal prolegs. The *Lepidoptera, Mecoptera* and some *Hymeoptera* have eruciform larvae.

Catfish, ***common catfish, common wolf-fish, Atlantic sea-wolf,*** *Anarhichus lupus:* an elongate, up to 125 cm-long marine bottom fish of the family *Anarhichadidae* (Wolffishes), order *Perciformes.* The broad head carries a very large mouth and the bluntly conical teeth of a predator. Catfishes swim by eel-like undulating movements of the body. The dorsal fin extends the entire length of the body, from behind the head practically to the caudal fin; the anal fin is also very long. It is largely dark yellowish green in color, with some vertical stripes and a pale yellow belly. It is caught in the North Atlantic with bottom trawls, and sold without the head as rockfish. In Scandanavia, the liver is a delicacy, and the strong skin is used for the manufacture of light leatherwear.

Catfishes: see *Siluriformes.*

Catharacta: see *Stercoraridae.*

Cathepsins: proteases with endopeptidase specificity, localized intracellularly in the lysosomes. C. from protease-rich animal organs, like liver, spleen and kidney, can be separated into C. A, B, C, D and E (pH optima from 2.5 to 6) and C.L. (active at neutral pH). M_r of C. are between 25 000 (C.B.) and 100 000 (C.E).

Catherine's moss: see *Musci.*

Catharobionts: inhabitants of pure water containing little or no organic material, e.g. high mountain streams.

Catharosaprophytic conditions: see Saprophytic system.

Catharsus: see *Turdinae.*

Cathartes: see *Falconiformes.*

Cathartidae: see *Falconiformes.*

Cation: a positively charged atom or molecule, formed by the loss of electrons. See Ions.

Cation exchange: see Sorption.

Cation exchange capacity: see Sorption.

Catkin: see Flower (section 5, Inflorescence).

Cats: see *Felidae.*

Cattle: see *Bovinae.*

Caudofoveata: see *Aplacophora.*

Caularchus: see *Gobiesocidae.*

Cauliflory, ***stem flowering:*** flowering of formerly dormant buds on woody parts of a stem or trunk, especially on tropical trees, e.g. the cocoa tree.

Cauliflower mosaic virus: see Virus groups.

Caulimoviruses: see Virus groups.

Caulobacter: a genus of stalked water and soil bacteria. The cells are Gram-negative, rod-shaped, fusiform or vibrioid, 0.4–0.5 by 1–2 μm. Extending from one pole of the cell (aligned with the long axis of the cell) is a stalk of varying length, diameter about 0.15 μm. The stalk consists of layers of the cell wall and cell membrane, and carries at its end a small mass of adhesive material ("attachment plate"). In crowded populations of *C.*, the cells may adhere to each other and form rosettes, the distal ends of their stalks embedded in a common mass of adhesive material. Cell division occurs by the asymmetrical binary fission of a stalked cell, one daughter cell remaining attached by the original stalk, the other being freely motile and possessing a single polar flagellum. The flagellated cell secretes adhesive material at the base of its flagellum, develops a stalk at this site, and becomes a typical nonmotile, stalked cell.

Cavefishes: see *Percopsiformes.*

Cave flea: see *Amphipoda.*

Cavernicolous animals: see Troglobionts.

Cavia: see *Caviidae.*

Cavies: see *Caviidae.*

Caviidae: a family of Rodents (see), consisting of 16 recent species in 5 genera and 2 subfamilies, comprising the ***guinea pigs*** or ***cavies*** and the ***Patagonian "hares".*** The family is represented throughout South America, except in Chile and parts of eastern Brazil. Habitats include pampas and savanna, as well as marshy, tropical areas and dry, stony plateaus up to 4 000 m, but not usually thick jungle. All wild forms have a coarse pelage. The cheek teeth are rootless (i.e. they grow continually). Dental formula: 1/1, 0/0, 1/1, 3/3 = 20. All caviids display cursorial adaptations (adaptations for running); thus, the clavicle is vestigial, the tibia and fibula are partly fused, and the number of digits is reduced to 4 on each forefoot and 3 on each hindfoot.

Members of the subfamily, *Caviinae* ***(Guinea pig-like cavies)*** are moderately short-limbed with a chubby appearance (head plus body 20–45 cm, tail vestigial), and despite the cursorial adaptations of the skeleton, these animals move with plantigrade feet, scurrying about like rats or mice. The *Caviinae* comprise 4 genera: *Microcavia* ***(Mountain cavies),*** *Galea* ***(Yellow-toothed cavies*** or ***Cuis),*** *Cavia* ***(Cavies*** or ***True guinea pigs)*** and *Kerodon* (monospecific genus, *K. rupestris,* ***Rocky cavy*** or ***Moco).*** Guinea pigs kept as pets and used in research are varieties of *Cavia aperea* from Peru; selection and breeding have given rise to several soft-haired and long-haired varieties, as well as different coat colors.

The subfamily, *Dolichotinae,* is represented by the single genus, *Dolichotis* ***(Patagonian cavies, Patagonian "hares", Pampas "hares",*** or ***Maras),*** containing 2 species *(D. patagonium,* head plus body 69–75 cm; *D. salinicola,* head plus body 45 cm; in both species the tail may be up to 4.5 cm in length). In contrast to the *Caviinae,* Patagonian cavies have long slender legs, and they are able to progress by bounding and running, during which movement the feet are digitigrade. Specialized pads (absent in the *Caviinae)* beneath the digits act as shock absorbers when the feet strike the ground. The general appearance is that of a rabbit, with a laterally compressed skull and long ears; they have even been described as antelope-like.

Caviinae: see *Caviidae.*

Caviar: see *Acipenseridae.*

Cavity-nesting birds: a term for birds that nest in cavities, rather than in open nests. They are essentially Tree-nesting birds (see), which exploit cavities in tree trunks, but they also include birds that nest in earth cavities. C-n.b. are especially numerous (up to 50%) in relatively old deciduous forests, whose tree trunks provide the appropriate nesting cavities. Some important European species are woodpeckers, tits, nuthatches, Old World flycatchers, redstarts and doves. Only the woodpeckers excavate their own nesting cavities, and these may be used later by other species. By providing nesting boxes (i.e. artificial nesting cavities), the breeding density of economically important C-n.b. can be considerably increased. See Biological control.

Cavy: see *Caviidae.*

Cayenne pepper: see *Solanaceae.*

Caytoniales: see *Caytoniopsida, Lyginopteridatae.*

Caytoniopsida: a class of fossil gymnosperms with palmate leaves and simply pinnate megasporophylls, found in the Uper Triassic and the Jurassic. Alternatively, the members of this class are considered to constitute the *Caytoniales,* a Mesozoic order of the seed ferns. The best known genus, *Caytonia,* is named from its first discovery site at Cayton Bay on the Yorkshire coast of England. A complete reconstruction has not been achieved, the plant being known chiefly from its ovary and fruits. The simple pinnae of the megasporophylls formed rounded receptacles enclosing several macrosporangia, with a small micropyle, thus representing a type of ovary (Fig.). The fruits were berry-like little closed sacs, formed from the receptacle and enclosing small seeds. Although the *C.* display some angiosperm affinities, they are not thought to be direct ancestors of the angiosperms, but rather an isolated evolutionary group.

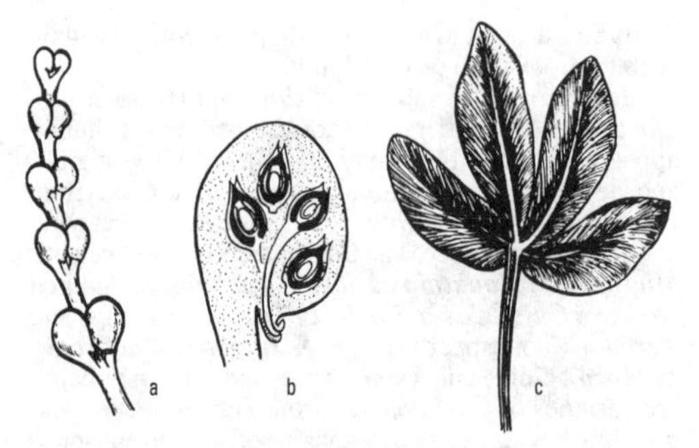

Caytoniopsida. a Macrosporophyll of *Gristhorpia nathorsti. b* Longitudinal section of the fruit of *Caytonia thomasi. c* Leaf of *Sagenopteris rhoifolia.*

CCC: acronym of chlorocholine choride. See Gibberellin antagonists.

cDNA: acronym of complementary DNA (often also referred to as "copy" DNA). cDNA is complementary to a mRNA. It is synthesized in the laboratory as a probe for hybridization, by incubating mRNA with dATP, dGTP, dTTP and dCTP in the presence of reverse transcriptase.

Ceara rubber plant: see *Euphorbiaceae.*

Cebid monkeys, ***cebids***, *Cebidae:* a family of New World monkeys (see). Dental formula: 2/2, 1/1, 3/3, 3/3 = 36. The constituent subfamilies are listed under Primates (see). See the following separate entries: Capuchins, Owl monkeys, Titis, Squirrel monkeys, Sakis, Uakaris, Howler monkeys, Spider monkeys, Woolly monkeys.

CEC: Cation exchange capacity. See Sorption.

Cecidiophyte: see Galls.

Cecidiozoa: see Galls.

Cecidium: see Galls.

Cecidology (Latin *cecidium* gall, Greek *logos* study): the study of plant galls and the processes of gall formation.

Cecidomyiidae: see Gall midges.

Cecophagy: see Cecotrophy.

Cecotrophy, ***cecophagy:*** the passing of food twice through the alimentary canal. C. is practised by the *Lagomorpha* and by some rodents, which take soft fecal pellets directly from the anus at night and store them in the stomach to be mixed with food taken during the day. This may be an aid to digestion, and the animal probably benefits from certain products of the gut flora and fauna, such as vitamins. C. appears to be essential for survival, for if it is prevented the animal dies.

Cecropines: peptides with antibacterial activity, found, inter alia, in the moth *Hyalophora cecropia.* They are chemically unrelated to any other antibiotics.

Cecum, ***caecum:*** a blind, lateral process of the large intestine of vertebrates, located near the junction with the small intestine. Reptiles and mammals (except for armadillos, sloths, toothed whales and some carnivores) possess a single C., whereas birds have two. In herbivores the C. is particularly long, e.g. 50 to 60 cm in the cow, whereas in carnivores it is markedly shorter. In rodents the C. is the largest single section of the gut; it contains bacteria which break down cellulose, and it is also a site of vitamin synthesis. The C. of humans and anthropoid apes is a thin, vermiform structure, known as the appendix or vermiform appendix.

Cedar: see *Pinaceae.*

Celastraceae, ***spindle-tree family:*** a family of dicotyledonous plants containing about 850 species in 55 genera, distributed in the tropics and subtropics. They are shrubs or trees with simple, opposite or alternate leaves, with or without small stipules. Flowers are 4- to 5-merous and actinomorphic, with a superior ovary of 2–5 united carpels. The fruit is a loculicidal capsule, samara or berry, and the seeds often have a brightly colored aril.

The genus *Euonymous* (176 species) is distributed primarily in Southeast Asia, with some representatives in Europe, including *Euonymous europaea* ***(Spindle tree).*** This is a small (up to 7 m high) tree, with 4-angled, green branches and dark red fruit capsules, whose seeds are completely enclosed by an orange-colored aril. The fine-grained wood of certain *Euonymous* species is used for carving; the seeds also yield an oil used in soap manufacture, as well as a yellow dye used for coloring butter. *Euonymous hians* (a Japanese shrub) has a heavy, durable wood which is used for carving and the manufacture of printing blocks.

Celastraceae. a Spindle tree *(Euonymus europaea),* fruit. *b* Khat tree *(Catha edulis),* branch with flower and opened fruit.

Species of *Celastrus* ***(Staff tree*** or ***Bittersweet),*** all of which are vigorous climbers, may wind so tightly round a support plant that the latter eventually dies. Many members of the family contain alkaloids. Thus, the leaves of *Catha edulis* ***(Khat tree),*** which grows in Northeast Africa and Arabia, contain the stimulatory alkaloid, cathine. The leaves are chewed by natives, or used to prepare an infusion known as Arabian tea. In Sri-Lanka, the seeds of *Kokoona zeylanica* ***(Kokoon tree)*** are a source of oil.

Celebes black ape: see Macaques.

Celery: see *Umbelliferae.*

Celery fly: see Fruit flies.

Cell (plates 20 & 21): the smallest structural and functional unit of living organisms, that still displays all the properties of life.

1) All the following properties are associated with Cs.: the ability a) to synthesize proteins and to multiply (replicate) (see Cell multiplication) using the information stored in their nucleic acids, b) to continuously exchange material and energy with the environment, c) to regulate their physiological state and to adapt to changing environmental conditions, d) to store and metabolize material for the provision of energy, material replacement and growth, e) to grow and differentiate, f) to perceive, process and respond to stimuli, and g) to move. All these living phenomena are associated with the synthesis and action of *biomolecules*. Large biomolecules differ essentially from inorganic material by their complicated and sometimes changeable 3-dimensional structure. The large biomolecules of the living C. exist in a highly ordered and organized state, in particular in structural elements such as the membrane systems (lipid/protein complexes) of the C. and its organelles. Organelles are the intracellular sites of partial C. functions (e.g. the tricarboxylic acid cycle occurs in mitochondria; translation of mRNA occurs on ribosomes), and they display quite specific morphological, biochemical and physical properties. Prokaryotic organisms possess only some of the organelles that are present in eukaryotic Cs. (e.g. prokaryotes do not have mitochondria). All membranes, C. organelles, microtubules and most other C. components are in a continual state of synthesis and degradation. Numerous reactions can be studied in isolation from the whole C., and they provide some insight into C. function (e.g. the reactions of glycolysis can be studied in C. extracts from which all organelles and C. wall material have been removed).

Since viruses (see) do not possess all the properties of living Cs. (no metabolism of their own, nonexcitable, no autonomous growth) they are rather equivalent to relatively independent C. constituents.

Considering the extraordinarily wide variety of unicellular and multicellular living organisms and their very different C. types, the essential structural and functional elements and main metabolic pathways of all living Cs. are surprisingly similar.

Cs. are the end result of an evolutionary process that began more than 3×10^9 years ago, in which the interaction of proteins and nucleic acids was crucial. The essential molecular and supramolecular structures of modern pro- and eukaryotic Cs. and the metabolic processes associated with them were all present in the earliest primitive prokaryotic Cs.

Living Cs. now arise only from other living Cs. The living C. is an open, dynamic and highly organized system. It interacts continually with its external environment. Thus, energy and material from the environment are taken into the C., where they are utilized to maintain a steady state, and the waste products of this utilization are passed back to the environment. The C. also responds to information from the environment. Concentrations of C. components remain relatively constant (steady state) and energy-yielding reactions are coupled with energy-consuming reactions. The Plasma membrane (see) is a selective and regulatory barrier to the passage of material, energy and excitatory stimuli into and out of the C., thus separating the C. contents and insuring that they do not equilibrate with the environment.

All the life functions of a unicellular organism proceed in a single C. In terms of visible structures, the most highly specialized unicellular organisms are the ciliated protozoa, which possess, inter alia, a C. mouth (cytostome) and a gullet. However, the cooexistence of all life processes within a single C. implies a high degree of functional organization in all unicellular organisms. The numerous Cs. of a multicellular organism generally arise by C. division from a single fertilized egg C. Thus, the 1 mm-long nematode, *Caenorhabditis elegans,* consists of 600 somatic Cs. (plus some germ Cs.), while an adult human contains some 6×10^{13} Cs., differentiated (during ontogenesis) into various types, e.g. muscle Cs., nerve Cs. (see Neuron), etc.

2) During C. differentiation (see) in multicellular organisms, Cs. acquire specific morphological, biochemical and physiological properties, which optimize or maximize their efficiency for one or more particular functions. This allocation of different functions to differently specialized Cs. is far more effective than the differentiation of a single C. into differently specialized regions (e.g. in ciliated protozoa). Associations of similarly differentiated Cs. are called tissues. Certain fundamental structural and functional properties are common to the Cs. of both unicellular and multicellular organisms. The Cs. of multicellular organisms, however, display additional properties that arise from the requirements for cooperative interaction and coordination of Cs. and tissues, e.g. Cs. may be specialized for the conduction of material and energy, or as components of mechanical tissue. Mitosis and meiosis occur only in eukaryotes. However, the bacterium, *Escherichia coli,* does very infrequently undergo conjugation and recombination (parasexual processes). Each C. type of a multicellular organism has a characteristic C. cycle (see), i.e. timing of the onset and duration of each phase of the C. cycle varies according to the type of C. The lifespan of most Cs. is genetically determined (see C. death). Many cancer Cs. are not greatly differentiated (morphologically, biochemically and/or physiologically) from their parent Cs.

3) Structural and functional elements. Most eukaryotic Cs. possess a Nucleus (see), which serves as a control and information center. C. fusion produces multi- or polynucleated Cs. (syncytia; e.g. skeletal muscle Cs.). Multinucleated Cs. are also formed when nuclear division is not accompanied by C. division (e.g. in *Plasmodium).*

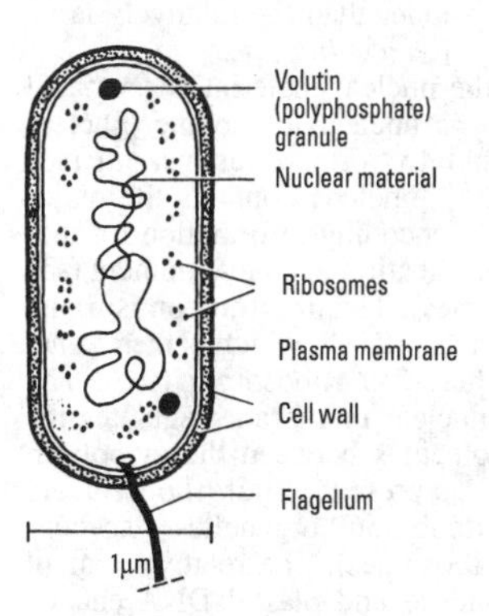

Fig.1. *Cell.* Bacterial cell (schematic).

Prokaryotic Cs. (bacteria and cyanobacteria) (C. volume 1–30 μm³) are smaller than eukaryotic Cs. and have a simpler structure. (Fig.1). Mycoplasmas (diam. about 100 nm) are the smallest known organsms and the smallest known Cs. Possibly their small dimensions represent the lower size limit for a living C. The bacterium, *Escherichia coli,* is probably currently the best studied of all living organisms. Its short generation time (about 20 min under optimal conditions) is possible because of its small size and its relatively large specific surface area. Under optimal conditions, one C. of *Es-*

cherichia coli can produce 8.6×10^9 Cs. in 11 hours. After 43 hours at this growth rate, the C. volume would equal that of the earth (just 1.1×10^{12} km^3). Prokaryotes lack a nuclear membrane (or nuclear envelope; see Nucleus), and the DNA (located more or less centrally in the C.) is known as as the *nuclear equivalent* or *nucleoid*. Prokarytic Cs. contain about 1 000 times less DNA than do eukaryotic C. nuclei. In all bacteria so far investigated, the DNA double helix exists as a circular molecule, which is rather coiled and twisted on itself. Histones are absent. In prokaryotes, the C. organelles of protein synthesis, the Ribosomes (see), are only 15 nm in diameter, and they are densely packed throughout the C., except in the central region of nuclear material. These 70S-ribosomes share many properties with the ribosomes of mitochondria and plastids.

Intracellular Membranes (see) are present in cyanobacteria (thylakoids) and certain bacteria. They are derived from, and remain connected with, the plasma membrane, and their chemical composition differs from that of the eukaryotic plasma membrane. Mitochondria and plastids are absent from prokarytoic Cs. Invaginations of the plasma membrane known as Mesosomes (see) are present in some bacterial Cs., sometimes with additional membranous lamellae inside; they may have respiratory or photosynthetic functions. The shape of most prokaryotes is stabilized by a *C. wall;* this contains peptidoglycan, which is not present in the eukaryotic plant C. wall. Microtubules and C. organelles formed from them (spindle apparatus, centrioles) are not present in prokaryotes. Bacterial flagella (see), only 20 nm in diameter, differ structurally and functionally from the flagella and cilia of eukaryotes. Many prokaryotes are able to fix atmospheric nitrogen, a property unknown in eukaryotic Cs.

Animal eukaryotic Cs. have an average mass of 2 ng, i.e. about 1000 times heavier than the relatively large prokaryotic C. of *Escherichia coli*.

In eukarytoic Cs., the nuclear material is separated from the cytoplasm by a nuclear membrane (nuclear envelope; see Nucleus) to form a distinct *nucleus* (Fig.2; plate 21). The C. nucleus contains the major proportion of the DNA, encoding information for metabolism, growth, multiplication and development (see Chromatin, Chromosomes). This information is transferred to the cytoplasm as mRNA, which is then translated by protein synthesis on Ribosomes (see). The pore complex of the nuclear membrane regulates the exchange of macromolecules between the cytoplasm and nucleus. DNA is also present in mitochondria and plastids; these semiautonomous organelles also contain ribosomes (70S ribosomes). The relatively small quantities of mitochondrial and plastid DNA encode proteins, which are synthesized within the organelle. The Plasma membrane (see) forms a complex outer C. surface, which regulates exchange of material with the environment, mediates or inhibits C. contact, receives, processes, transmits and releases specific information, as well as protecting the C. The C. surface may be modified in various ways, e.g. its surface area may be greatly increased by the development of Microvilli (see). On the internal surface of the plasma membrane, especially in animal Cs., there is a *cytoplasmic skeleton,* which consists of Actin (see), Myosin (see), intermediate filaments (consisting of structural proteins, diam. about 10 nm) and Microtubules (see). Plant Cs. are stabilized mainly by their C. wall (see), but during interphase, microtubules aggregate on the internal face of the plasma membrane.

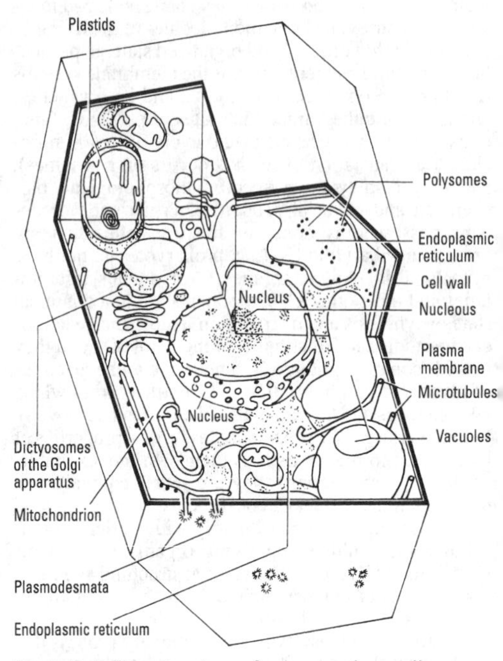

Fig.2. *Cell.* Fine structure of a young plant cell.

The cytoplasm is divided by membranes into numerous *compartments,* each with a different composition and a different complement of metabolic reactions. It is therefore possible for degradative and synthetic processes to operate simultaneously within the C. Membrane-bound C. compartments (organelles) are, e.g. Mitochondria (see), Plastids (see), Golgi apparatus (see), Endoplasmic reticulum (see), Lysosomes (see), Peroxysomes (see). Mitochondria provide energy from the aerobic respiration of substrates, while Photosynthesis (see) occurs in plastids known as chloroplasts. In the Golgi apparatus, proteins are processed for export, in particular by the attachment of carbohydrate residues. The endoplasmic reticulum (ER) extends widely throughout the cytoplasm. Membranes of the rough ER are occupied by ribosomes, representing synthesis sites for enzymes destined for export from the C., whereas the smooth ER lacks ribosomes and is responsible for certain stages of lipid and steroid metabolism and for the modification of xenobiotics. Lysosomes are the site of intracellular digestion. Peroxysomes house several reactions of oxidative metabolism, and they destroy any toxic hydrogen peroxide produced by these reactions.

The ground material of animal Cs. is usually in the gel phase, partly reinforced by an extensive cytoskeleton, consisting of microtubules, intermediate filaments and actin. Actin filaments are also involved in plasma streaming, ameboid movement of Cs. and the shifting of organelles. At the beginning of mitosis, the Spindle apparatus (see) is formed by microtubules, which are also structural and functional elements of the Centriole (see) and of Cilia and Flagella (see), as well as being involved in intracellular movement. The widespread occurrence of the characteristic 9+2 pattern of microtubules in cilia and flagella suggests a common evolutionary origin for all modern eukarytoic Cs. Muscle Cs. contain both actin and myosin filaments, combined reversibly to form contractile *actomyosin*. Nonmuscle Cs. contain an oligomeric, nonfilamentous form of myosin.

Vacuoles and *vesicles* serve as storage and reaction compartments for reserve materials (e.g. glycogen, fat droplets). Some materials are deposited in the cytoplasm (see Paraplasmatic inclusions).

Cell biology, *cytology* (plates 20, 21 & 22): a branch of general biology concerned with the structure, function and life history of the cell. Advances in C.b. are increasingly based on research in molecular biology. Since the 1960s, the most significant contributions to C.b. have been made in molecular biology, immunology, gene technology, electron microscopy, isotopic tracing techniques and cell fusion.

Cell count: the number of microorganisms in a specific quantity of substrate, e.g. in 1 ml of water or 1 g of air-dried soil. The C.c. is often used as an indication of quality and of the degree of contamination, e.g. in the determination of water purity. In microbiological research, the C.c. is an important quantity and its determination is a standard laboratory procedure.

City air contains up to 10 000 dust-borne cells/m^3, whereas sea air usually contains fewer than 50/m^3. Ideally, drinking water should contain no microorganisms, but certainly no more then 100 cells/ml. Fertile soils contain up to 10^{10} microorganisms/g.

The C.c. can be determined *directly* by counting cells under the microscope in a counting chamber, by using an electronic cell counter, or by spreading a cell suspension on nutrient agar and counting the colonies that develop. *Indirect* methods include photometric determination of the turbidity of cell suspensions, and the measurement of metabolic activity, e.g. lactic acid production by lactobacilli. Plating on nutrient agar or measurement of metabolic activity detect only living cells, and the result is known as the *viable C.c.* All other procedures produce the *total C.c.*, i.e. living and dead microorganisms.

Cell culture: maintenance of cells, tissues, organ parts, organ primordia or organs in vitro, in a live state, in an artificial medium for longer than 24 hours.

Animal cell culture. The earliest culture methods consisted of the maintenance of fresh explants of animal or human *tissue* in a suitable nutrient medium, either derived from the organism (blood plasma, embryonal extract) or prepared artificially. During incubation, cells resembling fibroblasts and/or epithelial cells migrate to the exterior and form a layer around the explant. After a few days, oxygen deficiency and accumulation of metabolic products make it necessary to transfer the explants to fresh medium. The classical method is known as the "hanging drop" procedure, in which the tissue is cultured in a drop of medium hanging on the underside of a cover glass over a central concavity of a glass slide. It is still used, mainly for morphological studies.

In contrast to tissue culture, the aim of *organ culture* is to preserve the structure and function of tissues, organ parts or organs at a particular developmental stage, or to study their further development under conditions approximating as nearly as possible to in vivo conditions. The techniques of organ culture are different from those of tissue culture, and some of them are highly specialized. Both procedures, however, generally employ the same synthetic and semisynthetic culture media.

In recent decades, the greatest advances have been made in *cell culture* in the narrow sense. Unlike tissue culture, cell culture starts from isolated single cells, which are released from tissues or organ fragments by the action of proteolytic enzymes (trypsin, elastase, pronase). In specially designed culture vessels, these cells attach to glass or plastic surfaces and multiply to form a monolayer or lawn. The nutrient culture media contain amino acids, vitamins, salts and serum components. Using trypsin or other agents, the cells can be loosened from their substratum and transferred to new culture vessels containing fresh medium.

Since the development of techniques for production of large quantities of cells in a short time, cell culture has been exploited to great advantage by many scientific disciplines, e.g. virology. It is also used for the large scale production of certain biological products, e.g. hormones, interferon, etc., and it has become a routine tool in cancer research. Cell culture made possible the development of somatic cell genetics, and it is becoming increasingly important in medical genetics. Research in the molecular biology of eukaryotic organisms depends largely on cell culture, and new applications of the technique are continually being recognized.

Plant cell culture. Cells, tissues or organs of plants can also be cultured in vitro. In certain respects, the culture methods resemble those for animals cells, with adaptations to the specific requirements of plant cells. Thus, in accordance with plant metabolism, the liquid or semisolid nutrient media contain mostly inorganic salts, glucose and vitamins, with a few natural additives (e.g. yeast extract); individual tissue types may also have specific nutritional requirements.

Classical methods include the culture of isolated roots and callus culture. In *root culture,* excised root tips from seedlings grown under sterile conditions are placed in a suitable liquid nutrient medium. The roots grow, and sections or root tips are periodically transferred to fresh medium, so that in theory such cultures can be maintained indefinitely. Stem buds only rarely develop in root cultures, which are therefore true organ cultures. In contrast, cultured shoot tips regularly form root primordia, and the culture eventually differentiates into an entire plant. The culture of isolated embryos (embryo culture), which has been tried experimentally for many plant species, also inevitably leads to the development of an entire plant.

Tissue culture in the narrow sense refers to *callus culture*. Callus tissue is usually the starting point for the establishment of a cell culture. The callus tissue is mechanically or enzymatically disintegrated to produce separate cells, and these are transferred more or less individually to semisolid media (nutrient media solidified with agar). Alternatively, cells can be cultured in large numbers in suspension in liquid nutrient media.

Undifferentiated callus tissue will redifferentiate into a whole plant, irrespective of whether it was originally derived from a single cell of a stem, leaf, root, petals, etc., thus demonstrating the totipotency of these cells. In contrast, cultured animal tissues retain their original identity, and cannot differentiate into an animal.

Plant cell and tissue culture has rapidly become an

important method of botanical research, particularly in plant genetics and plant pathology. It is now used widely for cloning plants. A callus culture of a new variety can be grown and subdivided many times during a year, eventually serving as a source of thousands of new, completely identical plants. This technique not only has obvious economic advantages in horticulture, it is also being used to preserve and reintroduce threatened plants in some parts of the world. For example, the re-establishment of dwindling teak forests in India will be facilitated by cloning the healthiest and strongest of the remaining trees.

Cell cycle: a sequence of irreversible phases in the life of a eukaryotic cell, following mitosis and leading to the next mitosis. Occurring in the following order, they are known as the M-, G_1-, S- and G_2-phases (G stands for "gap"). The M-phase is mitosis (30 min – several hours), culminating in cell division. DNA replication occurs during the S-phase (synthesis phase). G_1 and G_2 are the postmitotic and premitotic phases, respectively, and DNA synthesis (replication) is absent from both. The C.c. can last 12 hours or several weeks, depending on the species, cell line and environment (e.g. in the root meristem of broad bean, G_1 = 12 h, S = 6 h, G_2 = 8 h, M = 4 h). The widest variations occur in the duration of G_1, which may be almost too brief to measure (e.g. in the early frog embryo) or so extended that the cell appears to be quiescent. In the latter case, the cell is said to be in the G_0-phase. After mitosis, some cells (e.g. neurons) undergo such a marked differentiation that they can no longer divide, but remain viable for several years in G_0. Other cells (e.g. hepatocytes, lymphocytes) undergo considerable differentiation, but prepare again for cell division (i.e. they pass from G_0 to G_1) after a few weeks or months. Unfertilized eggs are arrested at some stage of the C.c., e.g. sea urchin eggs in G_1, frog eggs in M and clam eggs in G_2. Other cells divide regularly, i.e. they continually pass through the C.c. without pausing in any phase. Cell growth usually occurs during G_1. Also during the G_1-phase, RNA, tubulin, etc. are synthesized, together with enzymes of DNA and protein biosynthesis (e.g. DNA polymerase) which are required in the next phase (S-phase). The S-phase can be divided into S_1 (replication of euchromatin) and S_2 (replication of heterochromatin), which overlap one another. The quantity of chromatin is doubled during the S-phase, but the number of chromosomes is unchanged. After G_2, the nuclear membrane disintegrates and the chromosomes condense, marking the beginning of the prophase of mitosis. The numbers and specificities of cell receptors vary during the different phases of the C.c.

The C.c. of neoplastic (cancer) cells is no longer subject to control by the organism. Such cells divide autonomously, and usually pass through their C.c. more rapidly than normal cells; in some cases, however, they may take longer to divide than comparable normal cells.

Advances in the understanding of molecular events during G_1 and G_2, which prepare the cell for entry into S and M, were made possible by the study of temperature-sensitive mutants of yeasts for DNA, RNA, or protein synthesis. Some of these mutations result in blockage at specific points in the cell cycle, and many such *cdc* (cell *d*ivision *c*ycle) mutants have been isolated. A structurally highly conserved protein kinase, which promotes trainsition from G_2 to M, has been found in all investigated eukaryotes, and is considered to be themaster regulator of the C.c. Known as *maturation-promoting factor* (MPF), this protein kinase was first described as a protein fraction promoting the M-phase in amphibian eggs. Purfied MPF consists of 2 proteins in equimolar amounts: a protein kinase, $p34^{cdc2}$ (product of gene *cdc2*$^+$), and a B-cyclin, whose potential biochemical role includes substrate specification and/or subcellular localization of the heterodimer. During interphase, $p34^{cdc2}$ is marked by dephosphorylation of $p34^{cdc2}$, which then becomes an active phosphorylase. Substrates phosphorylated during M-phase include nuclear lamins (contributing to nuclear disassembly), vimentin and caldesmon (potential substrates of cytoskeletal rearrangement).

Commitment to enter S-phase is thought to occur during G_1, in a molecular event known as Start. A gene (*rum1*$^+$), which regulates G_1 progression and the onset of S-phase, has been identified in the fission yeast, *Schizosaccharomyces pombe*. Overexpression of *rum1*$^+$ leads to giant cells containing many times the haploid complement of DNA, whereas deletion of *rum1*$^+$ abolishes the interval between the end of mitosis and Start. $p34^{cdc2}$ possibly exists in two forms, an S form that allows progress through Start with subsequent replication of DNA, and an M form that allows progress through mitosis with subsequent chromosome segregation. The gene product of *rum1*$^+$ may play a critical role in the interconversion of these two forms, e.g. by maintaining $p34^{cdc2}$ in the inhibited G_1/S form until the critical cell mass for Start is attained.

An alternative explanation for the role of *rum1*$^+$ invokes the existing theory of a *licensing factor* that binds to DNA during mitosis, and is consumed during S-phase. A subsequent round of replication is not possible until the nucleus has disintegrated, enabling the chromosomes to bind fresh licensing factor. G_1 cannot be followed by mitosis,because licensing factor activates the gene product of *rum1*$^+$, which in turn inhibits mitotic MPF. During DNA synthesis, licensing factor is consumed and cytoplasmic factor is excluded from the nucleus; the resulting activation of MPF programs the cell for entry into mitosis.

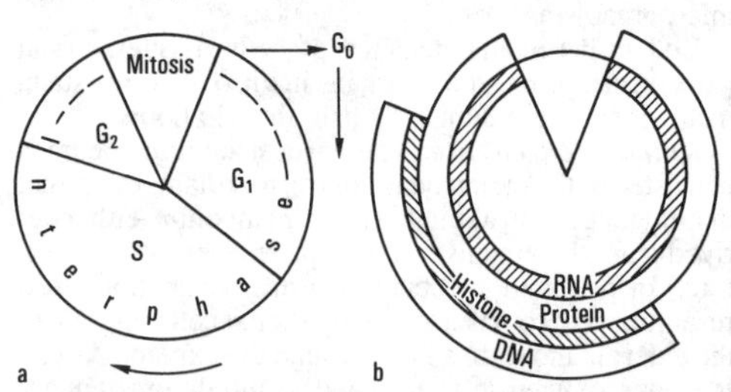

Cell cycle. a Schematic representation of the sequential phases of the cell cycle; the sections of the clock face represent the relative duration of each phase. *b* Periods of synthesis of different cytoplasmic and nuclear components during the cell cycle.

It should be noted that replication and chromosome segregation are not mutually exclusive in bacteria, especially in rapidly growing forms, where DNA synthesis and chromosome segregation may occur at the same time. Thus, when applied to prokaryotes, the term C.c. has a different meaning from that discussed above. [C. N. Norbury & P. Nurse, *Ann. Rev. Biochem.* **61** (1992) 441–470; A. W. Murray, *Nature* **367** (1994) 219–220; S. Moreno & P. Nurse, *Nature* **367** (1994) 236–242]

Cell death: see Programmed cell death.

Cell differentiation: structural and functional differences between cells that arise during development of multicellular organisms. C.d. normally starts during cell division, and results in the specialization of cells for specific functions. This specialization consists of the acquisition of specific morphological, biochemical and physiological properties, and it is the basis of the division of labor among the cells of the organism. The differentiated state is characterized by differential gene activity, in which a different combination of genes is active in each cell type. This is reflected in the fact that each differentiated cell type has a specific complement of proteins.

It can be shown experimentally that the quantities of DNA and the total gene complement are the same in all differentiated cells and the fertilized egg cell from which they are derived. In the development of *Ascaris,* chromosome elimination (loss of chromosomes and parts of chromosomes) occurs in those blastomeres destined to become somatic cells. This type of gene loss is not generally observed, but must also be listed as one of the mechanisms of C.d., in addition to differential gene activity. A further mechanism of C.d. is gene amplification, in which multiple copies of certain genes are produced. The central problem of developmental biology is the initiation and regulation of differential gene activity.

Cell fusion, *cell hybridization* (plate 28) (the entry entitled Fusion should also be consulted): an experimental technique for fusion of somatic cells from the same or different species in cell culture, so that nuclei and cytoplasm from two (or more) cells are enclosed by a single cell wall. The nuclei of these *hybrid cells* may remain separate or they may fuse. In the latter case, the mononuclear cell hybrid is called a *synkaryon.* If the nuclei remain separate (as occurs naturally for certain molds and fungi), the hybrid cell is a *heterokaryon.* A *homokaryon* contains separate but genetically identical nuclei. There are many examples of experimental C.f. between cells of very different species, e.g. human and rodent cells. For fusion of plant cells the cell wall must first be removed enzymatically, and the negative charge removed from the exterior exposed surface of the plasmalemma (see Protoplast). Experimental C.f. can be performed with normal cells, cancer cells, embryonal cells or fully differentiated cells. Spontaneous fusion of animal cells in culture is a rare event, but it is promoted by certain viruses that possess lipoprotein envelopes similar to the plasma membranes of animal cells. Virus-treated cells form numerous microvilli which form intercellular contacts. Cytoplasmic bridges arise at these contact sites, which enlarge until the cells effectively fuse. C.f. is also promoted by addition of polyethylene glycol, which facilitates cell-cell contact, leading to fusion of cell membranes. After completion of cytoplasmic and nuclear fusion, a single spindle apparatus forms during the subsequent cell division, and the nuclei of the daughter cells contain both genomes. Cell division is not impaired, and such cell hybrids divide rapidly, but one or more chromosomes may be lost in the process. Thus, as human/mouse hybrid cells grow and divide, human chromosomes are lost at each division and in random order. Even if the culture medium is capable of supporting both mouse and human cells, all the human chromosomes are eventually lost. Conversely, the medium may support the growth of human cells, but not mouse cells (which require extra genetic information); in this case human chromosomes will again be lost, but not the chromosome carrying the necessary genetic information. In this way, panels of hybrid cell lines have been established, each containing a full complement of mouse chromosomes and different, selected human chromosomes. Individual human chromosomes isolated in this way can be probed for the presence of particular genes, i.e. human genes can be mapped to specific chromosomes.

Somatic hybrids of plants can be been cultured from hybrid cells. This is even possible when the species concerned cannot be sexually crossed either naturally or experimentally.

Cell hybridization: see Cell fusion.

Cell-mediated immunity, cellular immunity: a special type of nonhumoral Immunity (see), due to the presence of activated, thymus-derived lymphocytes (see Thymus). These T-lymphocytes do not release antibodies, but carry cell-bound antibodies. They attack invading microorganisms, intracellular viruses, cancer cells and foreign tissues. The most familiar type of activated T-lymphocyte is the killer cell, which plays an important role in allograft rejection and the killing of tumor cells.

On contact with a specific antigen, T-lymphocytes release numerous factors known as lymphokines, which act on various motile leukocytes, e.g. they attract macrophages and granulocytes, thereby promoting phagocytosis and destruction of antigens such as invading bacteria.

Cell multiplication: a process in which division of a single cell, followed by division of each successive generation of daughter cells, etc, gives rise to a potentially unlimited number of cell progeny. Normally, cell division is preceded by cell growth. In eukaryotes, cell division is associated with Mitosis (see). C.m. is the basis of the growth and multiplication of organisms and for the replacement of dead cells (regeneration and tissue repair). C.m. of the animal embryo is known as Cleavage (see), and in this case mitosis is not followed or preceded by cell growth. Each cell type shows a characteristic division rate (number of cell divisions per unit time) and a charcateristic generation time (interval between two cell divisions) (see Cell cycle).

Bacteria have a short generation time(as short as 20 min under optimal conditions for certain species), so that the dynamics of bacterial C.m. (increase of cell number and mass) are particularly easy to follow and analyse. Some protozoa divide into several or many equal-sized cells. In this type of multiple division, the daughter cells may be produced successively following a succession of nuclear divisions (e.g. the 32 cells of the *Eudorina* colony), or multiple nuclear division may be completed, followed by multiple division of the cell (e.g. in *Sporozoa).*

Tissues or organs of an organism may consist of a fixed number of cells. This *cell constancy* occurs widely in aschelminths.

Cell nucleus: see Nucleus.

Cellobiose: a reducing disaccharide consisting of

two molecules of β-1,4 glycosidically linked D-glucose. C. is the disaccharide unit of some polysaccharides, e.g. cellulose and lichenin, but it does not occur free in nature. It is occasionally found in glycosides. It is not fermented by yeasts or hydrolysed by maltase.

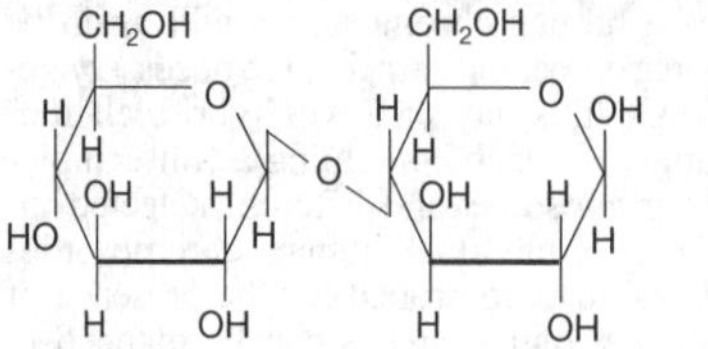

β-Cellobiose

Cellulases: glycosidases that catalyse the hydrolysis of cellulose to cellobiose. They are present in lower plants, wood-destroying fungi and some bacteria. Although the diet of most herbivores consists predominantly of cellulose, they do not produce C. Termites, ruminants and some rodents harbor symbiotic, cellulase-producing bacteria in their digestive tracts which enable them to ultilize cellulose, and the saliva of certain snails is a rich source of C. Snail saliva and the fungus, *Trichoderma viride,* are used as sources of C. for hydrolysis of plant cell walls in the preparation of protoplasts.

Cellulose: a plant polysaccharide, empirical formula $(C_6H_{10}O_5)_n$, M_r between 300 000 and 500 000, consisting of unbranched chains of β-1,4 glycosidically linked D-glucopyranoside residues. It can be hydrolysed to D-glucose by treatment with concentrated acids, such as 40% HCl or 60–70% H_2SO_4 at high temperatures. When applied to wood this process is known as saccharification, and it is used to produce fermentable sugar from wood. C. is the main component of plant cell walls. Cotton, hemp, flax and jute are almost pure C. Wood contains 40–60% C. In the plant cell wall, C. is arranged in microfibrils, which are 100–300 Å wide and about half as thick. The C. chains are parallel, stabilized by interchain hydrogen bonding. The microfibrils are embedded in a matrix of other polysacccharides, including pectins, hemicelluloses and lignin, and small amounts of a protein, extensin.

C. is the most abundant natural product in the world, about 10^{12} tonnes being produced annually by plant metabolism.

It is biodegraded by organisms that possess Cellulases (see).

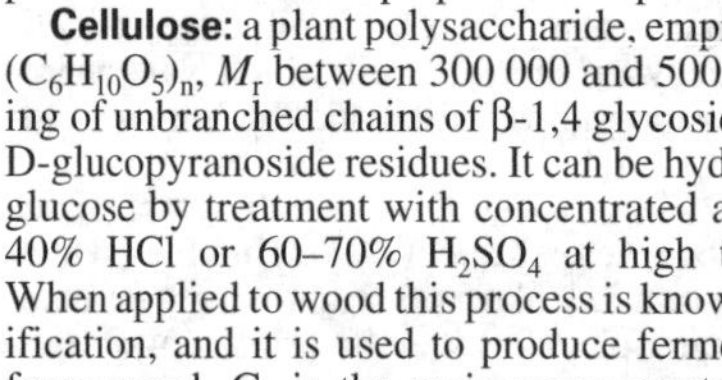

Cellulose

Cell wall: a rigid organic structure external to the plasma membrane of prokaryotes, plant cells, fungi and some protists, and synthesized by the protoplasm. Animal cells do not have a C.w. It functions mainly to protect the cell and to stabilize its structure.

Classification of bacteria as Gram-positive or Gram-negative on the basis of their reaction to the Gram stain corresponds to a fundamental difference in the structures of their C.ws. Gram-positive bacteria have relatively simple walls, usually consisting of two layers; the outer layer is often a teichoic acid, although in some species it may be a neutral polysaccharide or an acidic one called teichuronic acid. The inner layer is peptidoglycan. Gram-negative C.ws. are more complicated. Under the electron microscope they appear to consist of at least five layers, comprising lipoproteins, lipopolysaccharides, proteins and peptidoglycan; the peptidoglycan again forms the innermost layer, or it may be separated from the cell membrane by an extra layer of protein. The bacterial C.w. can be removed by gentle treatment with lysozyme to produce a Protoplast (see).

The plant C.w., which consists almost entirely of polysaccharide, is permeable to water and dissolved substances. It resists the high turgor pressure of the central vacuole, and it is responsible for the elasticity and strength of the plant cell. Elastic cellulose microfibrils are embedded in an amorphous matrix. Microfibrils (diam. 10–25 nm) are a few μm in length, consisting of cellulose molecules linked by hydrogen bonds. Each microfibril contains one or more microcrystalline regions or micelles (3–7 nm wide; Fig.1), in which the cellulose molecules are arranged with parallel long axes and the planes of their sugar rings also in parallel. In other regions of the microfibril, the cellulose molecules are distributed more irregularly.

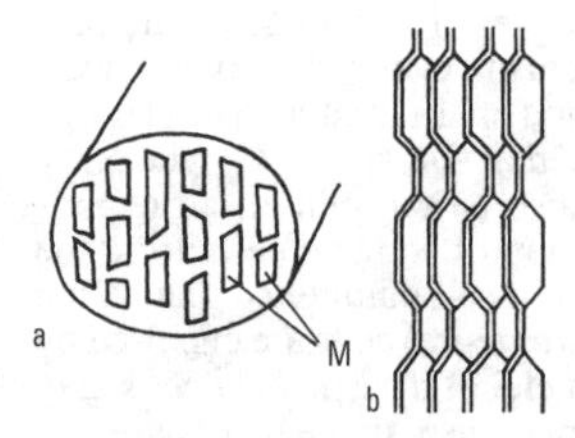

Fig.1. *Cell wall.* Structural model of a microfibril. *a* Transverse section of a microfibril. *b* Arrangement of cellulose molecules in the micelles *(M),* showing parallel long axes and parallel alignment of the planes of the glucopyranose rings.

During cell division in higher plants, Golgi vesicles fuse to form a cellulose-free *cell plate (primordial C.w.)* between the two daughter cells. Precursor material of the cell plate (low molecular mass polygalacturonic acid) is contained in the vesicles. The membranes of the Golgi vesicles fuse to form the new plasma membranes on either side of the cell plate (Fig.2). Layers of cellulose are deposited on each side of the cell plate, which then becomes the *middle lamella.* The latter never contains cellulose.

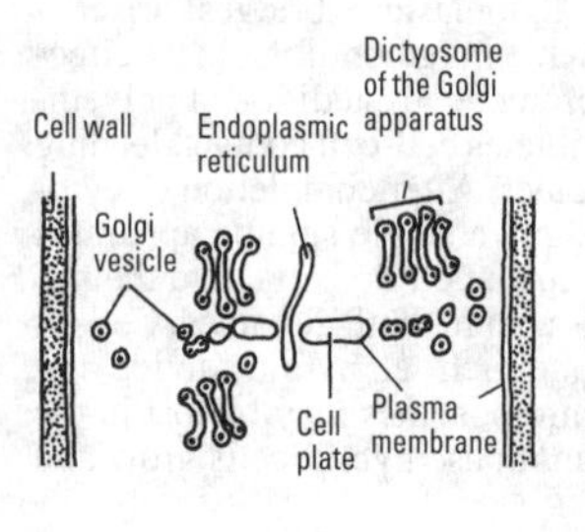

Fig.2. *Cell wall.* Formation of the primordial cell wall (cell plate).

Cellulose microfibrils that are laid down against the middle lamella to form the *primary C.w.* At first these microfibrils show a scattered arrangement with very little ordered alignment. At the completion of extension growth, however, they become aligned almost parallel to the long axis of the cell. The primary C.w. consists of an extremely complicated array of carbohydrates, the cellulose of the primary wall being embedded in an amorphous matrix of hemicelluloses and pectin. Depending on the plant, other matrix components are plant gums, mucilages (e.g. fucoidin, laminarin, alginates, agar, carrageenans) and reserve carbohydrates (e.g. arabans, xylans, mannans and galactoarabans). In certain plants, such as date palms and vegetable ivory, the thick cell wall consists mainly of these primary wall carbohydrates

In most plant cells, the main structural material of the secondary and tertiary C.ws. is cellulose, unlike the primary C.w., where cellulose may be a relatively minor constituent. The *secondary C.w.* shows a parallel arrangement of cellulose microfibrils, which change their direction in each subsequent layer. The resulting mechanical strength is considerable; thus, the tensile strength of steel is 150 kg/mm^2, while that of flax fibers is 110 kg/mm^2. The interstices of the cellulose network contain incrustation material, such as lignin (most higher plants), silicic acid (*Equisetum* and diatoms) or calcium oxalate (cypress). Fungal C.ws. contain microfibrils of chitin, together with glucan and glycoprotein The *tertiary C.w.* is a thin layer bordering the plasma membrane; it consists of a matrix of interwoven microfibrils which becomes completely impregnated with lignin.

In the synthesis of the primary and secondary wall, Golgi vesicles transport part of the C.w. material to the plasma membrane. After exocytosis of their contents, the vesicle membranes are incorporated into the plasma membrane (*membrane flux;* see Membrane). Cellulose synthesis is thought to occur mainly on the outside face of the plasma membrane. Microtubules appear to be involved in the orientation of microfibrils, since they are densely aggregated on the plasma membrane, usually lying parallel to the microfibrils. If microtubules are eliminated by treatment of cells with colchicine, cellulose microfibrils continue to be synthesized, but they are no longer correctly orientated. It has been shown that microfibrils "grow" from their ends.

Electron microscopy of freeze-etched preparations reveals granules (diam. about 10 nm) in the plasma membrane. Some of these are arranged in rows at the ends of microfibrils, and they are thought to represent active enzyme complexes of C.w. synthesis. Using [6-^{3}H]glucose and autoradiography of electron microscope preparations, it is possible to follow the transport of C.w. precursors in Golgi vesicles to the plasma membrane, and to show their incorporation into the C.w. after exocytosis. Presumably microtubules are involved in the transport of organelles and materials (Golgi vesicles, cellulose synthetase, etc.) required for the synthesis of the C.w. Adjacent plant cells are connected via perforations in the C.w. known as plasmodesmata.

At the beginning of the 1960s, Cocking (University of Nottingham, UK) introduced an enzyme preparation for the removal of the C.ws. in the root tip of tomato, and for isolation of the resulting Protoplasts (see). The method was rapidly developed for large quantities of cells from many plant parts (roots, leaves, petals, fruits, coleoptiles, etc.). Basically, a pectinase is used to hydrolyse the middle lamellae (which link all neighboring cells), and a cellulase is used to hydrolyse the cellulose of the C.w. Crude extracts of the bacterium *Rhizopus* can be used as a source of pectinase, while fungi such as *Trichoderma viride* or snail saliva (e.g. *Helix pomatia)* are a rich source of cellulase. The resulting protoplasts show a remarkable capacity for rapid regeneration of a C.w. They are also experimentally important for studies on Cell fusion (see).

The walls of pollen grains and cryptogam (e.g. fern) spores consists of sporopollenin.

See Intussusception, Apposition growth.

Celtis: see *Ulmaceae.*

Cenophytic (Greek *kainos* new), ***Cainophytic, Kainophytic, Neophytic, Angiosperm era:*** the latest era of plant evolution, preceded by the Mesophytic. It extends from the Apt or Gault (late Lower Cretaceous) to the present. The beginning of the C. is marked by the apparently sudden appearance of angiosperms and their dominance of the world's flora.

In the Upper Cretaceous (see Geological time scale), there was little geographical differentiation of plants; the vegetation of Europe and North America was uniformly subtropical. In the subsequent Tertiary, the European vegetation showed an evolutionary divergence; its Paleocene flora was still very similar to that of the Upper Cretaceous, whereas new forms appeared in the tropical to subtropical conditions of the Eocene. From the Oligocene onward, the temperature of the Northern Hemisphere became increasingly cooler, leading to the eradication of tropical forms progressively toward the south. This process continued in the Miocene and into the Pliocene, when there were no longer any tropical plants in Central Europe. Subtropical species also continued to decrease, and the arctotertiary flora became dominant. The climate was then probably similar to that of the present day. The effects of this cooling were most noticeable on the European Upper Pliocene flora; the onset of the Ice Age was already apparent. The Pleistocene is characterized by several glacial periods, with more or less lengthy, warmer intervening periods (interglacials). These cold periods had their greatest effect on the European vegetation, leading to the extinction of many plant groups, which were trapped between the mountains of Central Europe and the advancing glaciers. In North America, which lacks such an east-west barrier, fewer plant species became extinct, because they could disperse to the south without hindrance. Ice ages exerted their least effect on the flora of East Asia.

The end of the last Ice Age (10 000–12 500 years ago) marked the beginning of a period of floral development leading to the present vegetation structure of the Northern Hemisphere. The history of this development has been studied in particular by Pollen analysis (see).

Cenosarc, ***sarcosome:*** material within the stems of colonial polyps that links individuals, and which secretes the coenenchyma. It contains the gastrovascular cavity, and it is surrounded by the perisarc.

Cenozoic, ***Cainozoic, Kainozoic, Neozoic:*** the latest era of animal evolution. It lasted about 65 million years, and is divided into the Tertiary and Quaternary sub-eras, or the Paleogene and Neogene periods (see Geological time scale). The lower boundary of the C. is marked by the extinction of the dinosaurs, and the final large accumulations of ammonites and belemnites (they then became extinct). Brachiopods and crinoids became less common. Gastropods and bivalves, however, underwent a great evolutionary expansion, producing numerous new forms. Particularly important is the explosive and widely varied evolution of mammals, culminating in the evolution of humans. Birds attained their modern form in the Late Tertiary.

Centauria: see *Compositae*.

Center of diversity, ***gene center:*** a geographical location or region, in which certain cultivated plant taxa or their wild counterparts display a greater genetic deversity than elsewhere. It is considered by many authorities to be equivalent to the evolutionary Center of origin (see) of the taxa in question. Many centers of diversity are ice-age refugia, while others typically exist in the tropics. In view of their rich genetic diversity, these centers are extremely important for the breeding and development of useful species and varieties.

Center of origin: a geographical location thought to be the site of evolutionary origin of a taxon. In some cases, a center of origin may be equivalent to a Center of diversity (see), but this interpretation is in conflict with the concept of allopatric Speciation (see).

Centipedes: see *Chilopoda*.

Central American river turtle: see *Dermatemydidae*.

Central coordination: pattern of movement of a freely motile animal, which involves the entire motor system, and therefore excludes other movements. C.c. involves reflex contraction of groups of synergistic muscles, thereby achieving a whole body movement, rather than a single muscle contraction. Variable motoric patterns coupled with coordinated movement are known as *peripheral coordinations*.

Central dogma of molecular biology: see Molecular genetics.

Centrales: see Diatoms.

Central vacuole: see Endoplasmic reticulum, Lysosome.

Centricae: see Diatoms.

Centric fusion, ***fusion translocation:*** a form of Translocation (see), in the course of which the long arms of two acrocentric chromosomes become united to form a metacentric chromosome. The cell number is thereby reduced by one. C.f. is frequently responsible for evolutionary changes of the karyotype.

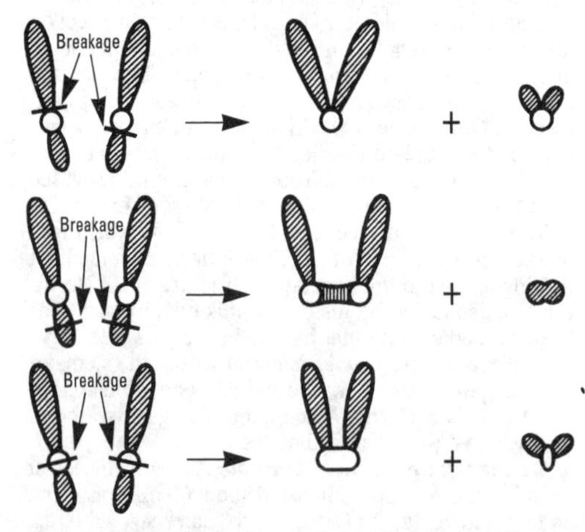

Centric fusion. Three types of centric fusion, all leading to the formation of a V-shaped chromosome and a small fragment from two linear chromosomes.

Centriole: a small cylindrical organelle 150–200 nm wide and 300–600 nm long in the cytoplasm of eukaryotoic cells. The cylindrical wall of the C. consists of a region of low density cytoplasm, containing 9 microtubule triplets, each triplet tilted inward toward the central axis at about 45° to the circumference, like the blades of a turbine (Fig.). DNA and RNA are also probably present, the DNA as a particle (diameter 70 nm) encoding information for replication of the C., and perhaps for cilium and flagellum formation (see below). In the S-phase of the Cell cycle (see), a daughter C. is formed perpendicular to each original C. The early immature C. is smaller than the original C., and it contains a 9-fold symmetrical array of single microtubules. The C. then matures to full size, and each microtubule serves as a template for assembly of the triplet microtubule. Cs. perform two distinct functions in eukaryotic cells: 1) as the site of formation of cilia or flagella (see below), 2) as part of the centrosome, which organizes the microtubules of the spindle apparatus during interphase (see below).

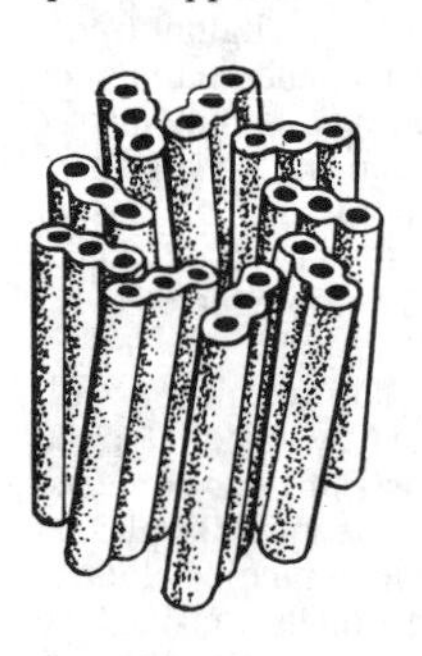

Centriole. Schematic drawing of a centriole, consisting of 9 sets of triplet microtubules.

In ciliated or flagellated cells, the axoneme of each cilium and flagellum arises from a C., which in this case is also known as a basal body (for further details see Cilia and Flagella).

Nearly all animals cells and the cells of some lower plants possess a pair of Cs. in a region of dense cytoplasm known as the *Centrosome* (see) or *cell center.* During the S-phase of the cell cycle (see above), each C. of the pair is replicated, producing two centrosomes. In prophase, the two centrosomes (each containing a mother and a daughter C.) separate and migrate to opposite cell poles, where they act as the focal point for production of a radial array of microtubules, known as the *aster.* The astral fibers radiate outward from the poles toward the periphery of the cell. Other types of microtubules are then synthesized between the poles and in alignment with the astral fibers, leading to the complete mitotic spindle: *polar fibers* extend from the two poles of the spindle to the equator, and *kinetochore fibers* attach to the centromere of each mitotic chromosome and extend toward the poles.

Cs. are absent from higher plants, but they still form fully functional spindles. Such spindles are usually *anastral*, i.e. they have no astral fibers.

Centroacinar cells: see Pancreas.

Centrolenidae, ***centrolenids***, ***glass frogs:*** a family of the *Anura* (see) containing 65 tree-dwelling species in 2 genera, found in the cloud forests of Central and South America, especially in the Andes in northwestern South America.

There are 8 holochordal, procoelous presacral vertebrae with non-imbricate (non-overlapping) neural arches. Presacrals I and II are not fused, and the atlantal cotyles of presacral I are widely separated. Ribs are absent and the pupils are horizontal. The coccyx lacks transverse processes, and it displays a bicondylar articulation with the sacrum. An omosternum is absent but a sternum is present, and the pectoral girdle is arciferal. Maxillary and premaxillary teeth are present. A short cartilaginous intercalary element is present between the terminal and penultimate phalanges, and there is a single bone (tarsus) in the ankle and wrist instead of two. Adhesive disks are present on fingers and toes. Most species are bright green or yellowish, and the skin of the belly is nearly transparent. The eyes are small and set almost on top of the head. Eggs are laid on leaves above running water, and the hatching tadpoles fall directly into the water. Most of the 64 species of *Centrolenella* are small (3 cm), but the single species of *Centrolene (C. geckoideum)* is larger (7–8 cm).

Centromere: the region located in the primary constriction of the chromosome, consisting of a specific DNA sequence required for chromosome segregation. During late prophase, a mature *kinetochore* develops on the centromere of each chromatid. The two kinetochores of sister chromatids face toward opposite poles of the cell, and each becomes attached to a chromosome fiber from opposite poles of the Spindle (see). In higher eukaryotes, the kinetochore is structurally conspicuous, and the attached chromosome fiber of the spindle consists of many microtubules. In most animals and lower plants, the kinetochore has a laminated structure consisting of 3 layers. The kinetochore of most higher plants appears to be hemispherical. In some lower eukaryotes that possess small chromosomes, e.g. yeast, there is no obvious kinetochore structure; each chromosome fiber of the spindle consists of a single microtubule, and its site of attachment within the C. is composed of specific kinetochore proteins, flanked by regions of heterochromatin.

Centromere co-orientation: mutual alignment of the centromeres of a pairing group of two or more homologous chromosomes (bivalents or multivalents, respectively) in the metaphase of the first meiotic division. The aligned centromeres lie in the equatorial plane of the cell, and this arrangement determines the distribution of pairing partners to opposite poles in anaphase. Alignment of chromatid centromeres in mitosis or meiosis II, which leads to the controlled distribution of the chromatids of each chromosome, is called centromere *auto-orientation.*

Centromere distance: the distance between a gene and the centromere, as determined from the Recombination (see) frequency between these two sites.

Centromere interference: the inhibition of crossing-over and chiasma formation in the neighborhood of the centromere. See Differential distance, Interference.

Centromere misdivision, *kinetochore breakage:* an anomolous transverse rather than longitudinal division of the centromere region. The consequences vary, depending on the site of breakage within the centromere. Thus, isochromosomes are formed if the breakage occurs in the innermost zone of the centromere.

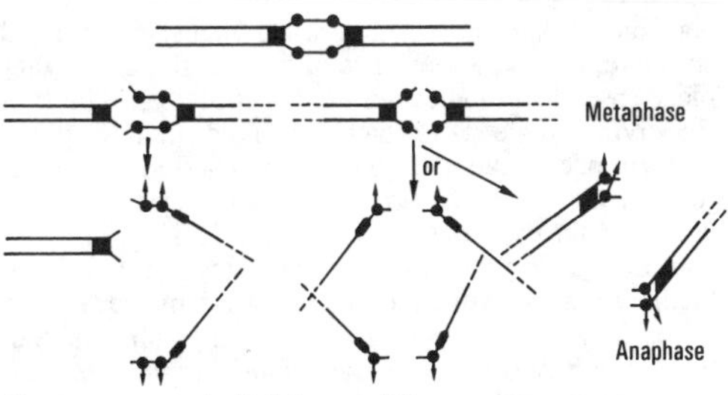

Centromere misdivision and its consequences.

Centropelma: see *Podicipediformes.*

Centropodinae: see *Cuculiformes.*

Centropus: see *Cuculiformes.*

Centrosome: a dense region of cytosol containing the centrioles. Usually (in the G_1-phase of the cell cylce) two Cs. are present *(diplosome).* During nuclear division the C. also divides. The daughter Cs., each containing a pair of centrioles, migrate to opposite poles of the cell. Centriole replication occurs during the preceding S-phase of the cell cycle.

Cepaea: see *Gastropoda.*

Cepe: see *Basidiomycetes (Agaricales).*

Cephalaspidae: see *Cephalaspidiformes.*

Cephalaspidia: see *Gastropoda.*

Cephalaspidiformes (Greek *kephale* head, *aspis* shield), *Osteotraci:* an extinct order of the *Agnatha,* and the best known of the fossil agnathans. They had 10 pairs of gill chambers and 10 pairs of external ventral gill openings. The front of the body was covered with a bony armor, known as the cephalic shield. In *Cephalaspis* (family *Cephalaspidae),* the head and forebody were protected by an anteriorly rounded bony shield, which was prolonged into spines at its posterior corners, and ornamented with stellate tubercles and nodules. Paired pectoral fins were present. The hind body was covered with dorsoventrally elongated scales, and it carried a dorsal fin and an asymmetrical (heterocercal) tail. Fossils are known from the Lower Silurian to the Lower Devonian.

Cephalaspis: see *Cephalaspidiformes.*

Cephalic flexure: see Brain.

Cephalic index: see Cranial index.

Cephalization, *cephalogenesis:* formation of a head. The head represents a concentration of sense organs and nerve ganglia at the anterior end of the body of motile animals. It is also the beginning of the alimentary canal, and often also the respiratory passage. C. occurs independently by different processes of embryonic development in different animal phyla *(Mollusca, Articulata, Vertebrata).*

Cephalobaenida: see *Pentastomida.*

Cephalocarida: a subclass of the *Crustacea,* discovered as recently as 1955 and containing only 4 known species. They are the most primitive of the recent crustaceans. There is no carapace and eyes are absent. The elongated, worm-like body consists of a head, 19 body segments and the telson, which bears a caudal furca (Fig.). Each branch of the furca has a seta sometimes half as long as the animal. Only the thoracic segments (1–8) bear appendages, which are all largely similar and not very different from the second pair of maxillae. Limb rudiments are, however, present on the first pleomere, and these are used for carrying the eggs. Each appendage has a triramous rather than the usual biramous structure, due to the presence of a large, flattened outer piece (pseudoepipodite) on the basal section. The first free-living stage after hatching is a metanauplius. Known species live in the mud of the sea floor in coastal regions. *Hutschinsoniella macracantha* (up to 3.7 mm) occurs on the east coast of North America, e.g. Long Island Sound. *Lightiella incisa* (2.6 mm) is found from the intertidal zone down to depths of 2.5 m near Puerto Rico. *Lightiella serendipita* (3.2 mm) inhabits San Francisco Bay, California. *Sandersiella acuminata* (2.4 mm) is found in Japanese waters.

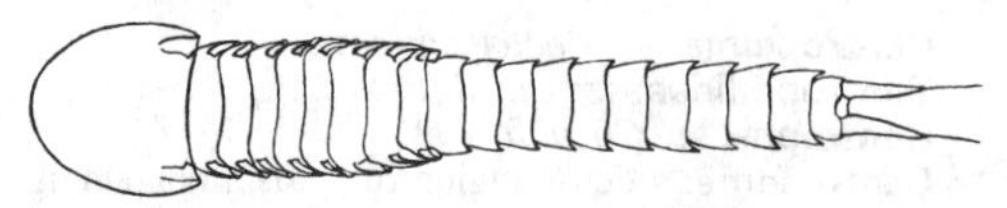

Cephalocarida. Hutchinsoniella macracantha, dorsal surface.

Cephalogenesis: see Cephalization.

Cephalometry: measurement of animal or human head. See Anthropometry.

Cephalophinae, ***duikers:*** an African subfamily of the *Bovidae* (see), containing 15 species of short-horned antelopes, varying in size from that of a hare to that of a deer. Most forms have a dorsal stripe, and the upper parts vary from light brown to almost black. The ***Gray duiker*** *(Sylvicapra grimmia)* is a monotypic species, while the remaining 14 species belong to the genus *Cephalophus.*

Cephalophus: see *Cephalophinae.*

Cephalopoda, *Siphonopoda,* ***cephalopods:*** a class of the phylum *Mollusca* (see), and the most highly organized of all the mollusks. The foot is divided into a number of tentacles, which are located around the head, as well as being modified to form the siphon *(cephalopod* means "head-footed"). Horny, beak-like jaws in the buccal cavity are used to dispatch prey. In the *Dibranchia,* 2 pairs of salivary glands empty into the buccal cavity. In *Sepia* and in the octopods, the posterior salivary glands are modified as poison glands for killing prey. A radula is also present, which functions as a tongue, taking pieces of fragmented tissue from the beak. Various degrees of reduction of the shell are found: *Nautilus* has an external, calcareous shell, which houses its body, and into which it can retract its head and tentacles; *Sepia* has an internal horny plate embedded in the mantle; *Octopus* has no shell at all. Among the approximately 750 living species of *C.,* lengths range from a few centimeters to several meters. The largest recent invertebrates are *C.,* and the largest so far recorded was a 22 meter-long giant squid. All *C.* are marine animals, living in the various oceans and seas of the world.

Morphology. According to the mode of existence, the bilaterally symmetrical body is plump and sac-like *(Octopus),* longitudinally flattened *(Sepia)* or spindle-shaped *(Loligo).* The head is well developed. There is a pocket-like *mantle,* which opens out to form a cavity ventral to the visceral hump. Water is expelled from this *mantle cavity* to the exterior via an anteriorly directed muscular tube *(funnel* or *siphon).* The mantle cavity contains *ctenidia* (gills) (2 pairs in *Nautilus,* 1 pair in all other genera). Contraction and expansion of the mantle provide a steady circulation of water through the mantle cavity; the volume of the mantle cavity is increased by relaxation of the circular muscles and contraction of the radial muscles, causing water to enter dorsally, laterally, and ventrally between the anterior margin of the mantle and the head; muscle action is then reversed, sealing the edges of the mantle tightly around the head, so that water must leave via the siphon. Surrounding the mouth are mobile, often prehensile arms or tentacles, usually armed with suckers (numerous tentacles in *Nautilus,* 8 or 10 in other genera), which serve for catching prey and for locomotion. In the male, one tentacle is often modified for the transfer of spermatophores (see Hectocotylus). Two eyes are present. In *Nautilus,* the eyes have no lens and they consist of a pit (ectodermal in origin), the lining of which forms the retina. All other genera possess camera-type eyes (see Light-sensitive organs), i.e. eyes with a lens and a more highly evolved retina. Camera-type *C.* eyes measuring 40 cm in diameter, found in the stomachs of sperm whales, are the largest known eyes in the animal kingdom. The existence of an image-perceiving camera-type eye with lens, retina, cornea, etc in both vertebrates and C. is a striking example of convergent evolution.

Biology. In all *C.* the sexes are separate. *C.* either swim or crawl on the sea bottom. A moderate rate of travel is achieved with the aid of the tentacles (crawling) and by undulation of the lateral fins (swimming). Spasmodic, powerful contraction of the mantle forces water through the siphon, generating a jet that propels the animal suddenly backward. Rapid forward movement is equally possible by directing the siphon backward. In some species (e.g. certain deep water octopods like *Cirroteuthis,* and some squids, including *Vampyroteuthis),* the arms are webbed like an umbrella, and the animal swims like a jellyfish by opening and closing the web. Some species have an ink gland or sac, whose contents they eject when fleeing from danger; the resulting dense brown cloud acts as a "smoke screen" and conceals the escaping animal; this cloud may equally well appear as an object in itself and divert the attention of a predator. Bottom dwelling species (e.g. *Octopus)* are usually solitary, whereas swimming species (many decapods) live in groups, sometimes forming immense shoals. Copulation is head-to-head, during which one of the male arms transfers spermatophores to the female. Encased eggs are either deposited or shed directly into the sea water. Development is direct, with no trocophore or veliger larva. Their highly evolved sense organs enable *C.* to react very rapidly to external stimuli. Various members of the *C.* are also used widely as human food.

Fossils. About 7 500 different fossil forms are known, first appearing in the Cambrian, and displaying periods of great evolutionary diversity in the Paleozoic and the Mesozoic. Most of the important index fossils are those of *C.* Fossil *C.* have chambered shells (like *Nautilus),* straight shells (e.g. belemnites), or spiral shells (e.g. ammonites). *Nautilus* is the sole survivor of the *Nautiloidea,* which flourished between the Early Cambrian and the Late Cretaceous. The totally extinct *Ammonoidea* lived from the Lower Ordovician until the Cretaceous. Fossils of the *Belemnoidea* occur from the Carboniferous to the Cretaceous. The *Sepioidea* and *Octopodia* are still represented by living species, and their extinct representatives are known from the Jurassic and Cretaceous, respectively. With the exception of the *Nautiloidea,* all these groups belong to the *Dibranchiata. C.* index fossils are important for the differentiation of the Upper Marine Paleozoic from the Mesozoic.

Classification. The class is divided into 2 subclasses: *Dibranchiata* and *Tetrabranchiata.*

1. Subclass: *DIBRANCHIATA (Coleoidea).* A reduced, chalky or horny, internal shell (enveloped by the mantle) may be present or absent. The key characteristic of this subclass is the presence of a single pair of ctenidia. Skin chromatophores are often present, conferring the ability to change color. Eyes are of the camera-type (see above). Fossil *Dibranchiata* occur from the Upper Carboniferous to the present, with their greatest taxonomic diversity in the Lower Jurassic. There are 3 orders: *Decapoda, Octopoda,* and *Vampyromorpha.*

1. Order: *Decapoda, Decabrachia* (**Squids** and **Cuttlefish**). Ten arms are present; 8 of these are short and stout and are usually referred to as "arms", whereas the 4th pair are much longer and called "tentacles". The coelom is well developed. In some squids, the suckers are replaced by claws. The internal shell consists of phragmocome, rostrum, and proostracum, or it is greatly reduced. Squids are generally long and posteriorly tapered, with a pair of posterior fins that act as stabilizers. "Flying squids" (e.g. *Dosidius gigas)* have well developed fin vanes and a powerful siphon, which enable them to shoot out of the water and glide for some distance. Compared with the streamlined squids, cuttlefish are relatively slow swimmers. Cuttlefish and squids tear their prey with their jaws, then swallow the pieces. There are 4 tribes or suborders: *Belemnoidea, Sepioidea, Oegopsida,* and *Myopsida.*

1. Suborder: *Belemnoidea.* These are known only as fossils, and they appear to be the ancestors of the following 3 extant tribes.

2. Suborder: *Sepioidea* (**Cuttlefish** and **Sepiolas**). Shore and bottom living forms. Typical genera are *Sepia, Spirula, Idiosepius, Rossia,* and *Sepiola.* The phragmocome of the internal shell is sometimes bent ventrally. Each eye possesses a cornea. The fins are not united posteriorly. Eight of the 10 tentacles (1st, 2nd, 3rd and 5th pairs) are short and stout and covered with suckers on their inner surfaces. The 4th pair (counting from the dorsal surface), are long and can be retracted into large pits at their base; they carry a sucker at their free end, which is enlarged and club-like, and they serve primarily for catching food (fish and crustaceans). The suckers have a muscular stalk, and they are lined with a toothed, horny ring. In the male, the suckers of the left hand member of the 5th pair of tentacles are reduced.

Sepia (**Cuttlefish**) (Fig. 1) lives in the coastal waters of warm seas, swimming with the aid of the siphon ("jet propulsion") or by undulations of the lateral fins. When the tip of the siphon is directed forward, the animal shoots backward (usual escape behavior), but by directing the siphon backward, it can dart quickly forward to seize prey. It possesses a defensive ink sac (see above), which opens into the siphon near the anus. The artist's brown pigment, known as sepia, was originally prepared by treating the contents of the *Sepia* ink sac with alkali.

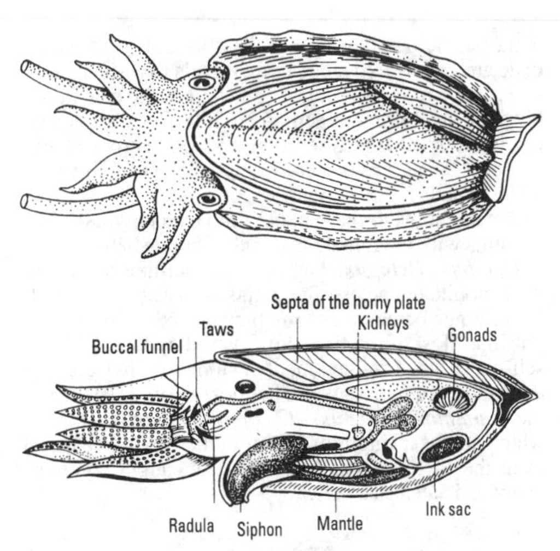

Fig.1. *Cephalopoda.* Cuttlefish *(Sepia).* Above: dorsal view of mounted specimen. Below: body plan in longitudinal section.

3. Suborder: *Oegopsida* (**Squids**). The anterior chamber of the eye is open, i.e. it lacks a transparent cornea, but eyelids are present. All tentacles are usually alike, and the suckers are often modified to hooks. The shell is a vestigial horny structure, known as a *gladius* or *pen.* All members are strong swimmers. The tribe includes many abyssal forms with phosphorescent organs, and some extremely large forms, e.g. *Architeuthis* (up to 18 m in length). Other genera are *Abralia, Abraliopsis, Gonatus, Onychoteuthis, Ctenopteryx, Histioteuthis, Bathyteuthis, Illex, Ommastrephes, Chiroteuthis,* and *Cranchia.*

4. Suborder: *Myopsida* (**Squids**). Shore-living forms, e.g. *Loligo, Lolliguncula, Sepioteuthis.* The eye has a cornea, and the shell is a simple gladius. The 4th pair of tentacles is elongated but not retractile. The fins are united posteriorly. They swim continually, rapidly following fish shoals upon which they prey. In many countries they are eaten as a delicacy.

2. Order: *Octopoda, Octobrachia.* Eight prehensile tentacles are present, the coelom is reduced, and the body is globular or sac-like. Octopods are generally smaller than decapods, maximal body length being about 3 meters. An internal shell is rarely present. They are predominantly bottom dwellers in shallow seas, but some species occur in deep oceans. In contrast to decapods, octopods inject their prey with poison, then with proteolytic enzymes, finally taking in the partially digested tissues.

1. Suborder: *Cirrata.* These are cirrate forms, i.e. the arms carry finger-like cirri and are connected by broad webs, forming and umbrella-like structure. They swim by opening and closing the umbrella. Most are deep sea species. The mantle bears a pair of fins. Typical genera are *Cirrothauma, Opisthoteuthis,* and *Cirroteuthis.*

2. Suborder: *Incirrata.* Fins are absent, and broad webbing between the arms is lacking, although a small degree of webbing often extends from the body between the proximal ends of the tentacles. Typical genera are *Octopus, Argonauta, Eledone, Eledonella, Vitreledonella,* and *Amphitretus.*

Argonauta argo (**Paper nautilus**) is a bottom dweller

in the Mediterranean. Two of the dorsal arms of the female are greatly expanded at their ends to form a membrane, each of which secretes one half of the calcareous bivalved shell (up to 20 cm in length). Eggs are deposited in this shell, which serves not only as a brood chamber, but also as a retreat for the female. The 1 cm-long male (which loses a modified arm when transferring spermatophores to the female; see Hectocotylus) often cohabits with the female inside the brood shell.

Octopus ***(Octopus)*** has a short, sac-like body and very mobile tentacles. The genus is distributed worldwide. When octopuses swim, they do so by siphon propulsion. Most of the time, however, they lurk behind a self-constructed embankment of stones, or they conceal themselves in rock caves. The most familiar species is the ***Common octopus*** *(Octopus vulgaris)* (Fig.2), which attains a length of 70 cm in the North Sea and 3 m in the Mediterranean (where it is caught commercially and eaten as a delicacy).

Fig.2. *Cephalopoda.* Common octopus *(Octopus vulgaris).*

3. Order: *Vampyromorpha* ***(Vampire squids).*** These are small, deep-water forms, superficially resembling octopods. They have 8 arms united by a web, as well as 2 small, coiled, retractile tentacles. Genus: *Vampyroteuthis.*

2. Subclass: *TETRABRANCHIATA (Nautiloidea, Protocephalopoda).* A calcareous, spiral shell is present, separated into internal chambers by transverse septa. The body is housed in the outer chamber of the shell; as the animal grows, it moves forward and the posterior part of the mantle secretes a new septum. A cord of body tissue (the *siphuncle)* extends from the visceral mass through a perforation in each septum. The shell is made buoyant by gas secreted by the siphuncle into each empty chamber. The tentacles arise from lobes arranged in 2 concentric circles around the head; they are adhesive, although they lack suckers or disks, and each can be retracted into a sheath. Above the head is a tough protective hood, representing a fold of one of the tentacular lobes. This hood serves as an operculum, which seals the entrance when the animal withdraws into the shell. The typical cephalopod siphon has not yet fully evolved into a complete tube, so that the functional siphon consists of unfused, overlapping, lateral flaps of the foot. The eyes are open, water-filled vesicles. There is no ink gland. Swimming is by siphon propulsion, the chambered shell acting as a hydrostatic apparatus. Fossil nautiloids occur from the Upper Cambrian to the present, with their greatest evolutionary diversity in the Ordovician. Since the Ordovician, the number of genera has continually decreased, so that the only living genus is now *Nautilus* (Fig.3), represented by 6 species which are bottom-living forms at depths greater than 100 m, e.g. *Nautilus pompilius* ***(Pearly Nautilus).***

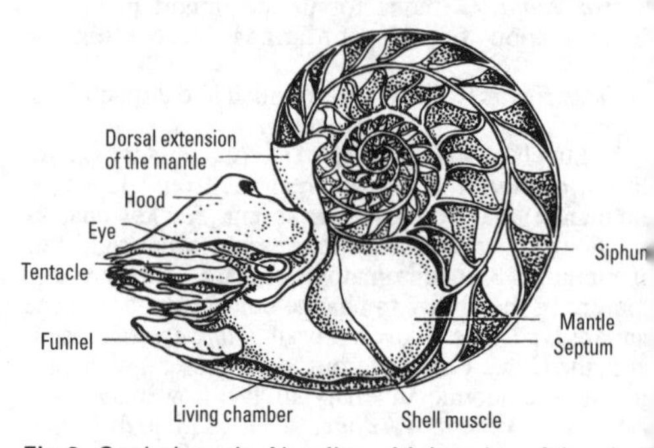

Fig.3. *Cephalopoda. Nautilus* with interior of the shell exposed.

Cephalopterus: see *Cotingidae.*

Cephalorhynchinae: see Whales.

Cephalosporins: an important group of β-lactam antibiotics, chemically closely related to Penicillin (see). The prototype organism for the isolation of cephalosporin C is *Cephalosporium acremonium.* Members of the C-series of cephalosporins are formal derivatives of 7-aminocephalosporanic acid, in which the amino nitrogen is acylated by D-α-aminoadipic acid. C. are active against both Gram-negative and Gram-positive organisms, and are used, e.g. for the treatment of severe bacterial infections and for infections of the urinary system.

R—HN, S, N, O, CH_2—O—$COCH_3$, COOH

Cephalosporin C : R = HOOC—CH(NH_2)—$(CH_2)_3$—CO—

7-Aminocephalosporanic acid: R = H

Cephalotaxaceae: a family of the *Pinidae* (fork- and needle-leaved gymnosperms) containing a single relict genus with 6 species, confined to the Himalayas and East Asia. They are evergreen trees or shrubs with spirally inserted needles and dioecious flowers, the female flowers occurring in small numbers in lateral inflorescences (cones). An oil is obtained from the large seeds of *Cephalotaxus harringtonia* var. *drupaceae.* Otherwise the family has little commercial importance. They were formerly included in the *Taxidae,* which they closely resemble.

Cephalothorax: the anterior body section of some members of the *Arthropoda,* consisting of the head fused with a variable number (sometimes all) of the thoracic segments. It is found in the *Merostomata, Arachnida, Pycnogonida* and most *Crustacea.*

Cephalotus: see Carnivorous plants.

Cephidae: see Stem sawflies.

Cepolidae: see Bandfishes.

Cerambycidae: see Long-horned beetles.

Ceramides: see Glycolipids.

Cerastes: see Horned vipers.

Cerata (sing. **Ceratum):** projection of the body wall f a nudibranch (naked sea slug). See Cnidosac, *Gastro-oda.*

Cerathotherium: see *Rhinocerotidae.*

Ceratida (Greek *keras* horn): the predominant Trias-ic ammonoids, represented mainly by the genus *Cera-tes.* The shell of *Ceratites* is planispiral, relatively ompressed, and displays the typical ceratitic suture saddles smoothly rounded and lobes developed into ubordinate small folds). In a succession of horizons, *'eratites* is represented by numerous variable species isplaying an evolutionary progression from small ork-ribbed forms to large, 36 cm-diameter forms trongly ornamented with coarse ribs, as well as smooth orms. Fossils are found in the Middle Alpine Triassic nd the Muschelkalk of the German Basin.

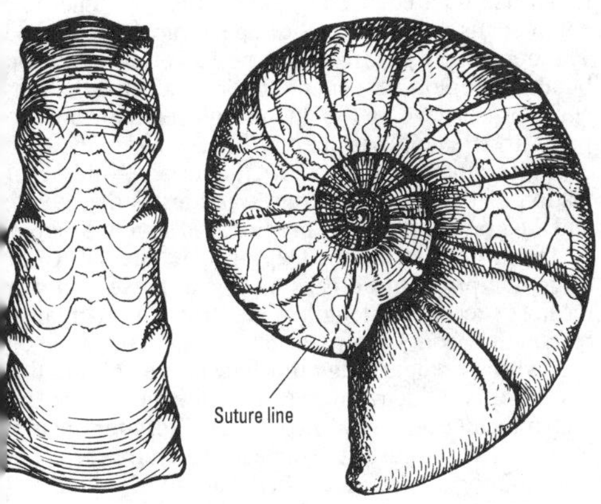

Ceratida. Stone cast of *Ceratites nodosus* Brug. with eratitic suture line (x 0.25).

Ceratiomyxales: see *Myxomycetes.*

Ceratitic ammonites: see Ammonites.

Ceratodus: an extinct genus of lungfishes with ounded scales and paired fins, which contained a long, ointed, cartilaginous axis. The vertebral column ex-ended to the hind end of the body, where it was sur-ounded above and below by the caudal fin. Large cre-aceous tooth plates were present on the palate and ower jaw. Fossils are found from the Lower Triassic to he Upper Cretaceous. A living related form, discovered n 1870 in Queensland, is sometimes placed in the same genus, but is preferably separated as *Epiceratodus* or *Neoceratodus.*

Ceratopsians: see Dinosaurs.

Ceratopyge (Greek *pyge* hind part): a trilobite ge-nus, in which the cephalon (head shield) and pygidium (tail shield) are about the same size. The glabella is smooth, and the facial sutures extend from the posterior border of the cephalon. The pygidium possesses 2 long, lateral spines, and is distinctly separated into pleural lobes laterally adjoining the 5–6 axial rings of the lon-gitudinal axis. Fossils occur in the Lower Ordovician (Tremadoc series).

Ceratotrichia: see Fins.

Ceratum: see Cerata.

Cercaria: see Liver flukes.

Cercopithecine theory: an anthropogenetic theory, proposing that the human evolutionary line diverged from the cercopithecoid stage before the evolution of the other primates. Numerous similarities between hu-mans and the great apes, however, do not support the C.t., but rather provide evidence for the more probable and more modern Pongid theory (see). See Anthropo-genesis, Dryopithecus pattern.

Cercopithecus: see Guenons.

Cerebellum: see Brain.

Cerebral aqueduct: see Brain.

Cerebral commissure: see Brain.

Cerebral cortex: see Brain.

Cerebral ganglion: see Brain.

Cerebral hemispheres: see Brain.

Cerebralization: in phylogenetic evolution, the in-creasing volume and differentiation of individual parts and fine structure of the brain, and the increasing prom-inence of the brain as the control center of the central nervous system. The highest degree of C. is shown in the relatively recent evolutionary development of the human brain.

Cerebral penduncles: see Brain.

Cerebrosides: Glycolipids (see) in which sphingo-sine is combined with a fatty amide or hydroxy fatty amide and a single monosaccharide unit (galactose or glucose). Some examples which have been obtained in pure form are kerosin (containing lignoceric acid), phrenosine (containing cerebronic acid or 2-hydroxy-lignoceric acid: $C_{24}H_{48}O_3$), nervone (containing ner-vonic acid) and hydroxynervone (containing 2-hydro-xynervonic acid). C. occur primarily in the white matter of brain and nerve cells, and they account for 11% of the dry matter of the brain. They are found in smaller amounts in liver, thymus, kidneys, adrenals, lungs and egg yolk.

Cerebrospinal fluid, *liquor cerebrospinalis:* a col-orless, lymph-like, clear liquid, containing only small white blood corpuscles (lymphocytes) and low concen-trations of glucose and proteins. It is produced in the plexus chorioidei which lines the brain ventricles. Ma-terial pass between the blood and the C.f. via special-ized villi which extend into the C.f.

The main function of the C.f. is mechanical protec-tion of the brain and spinal cord. Like the lymph system, it penetrates the intercellular spaces. Together with the blood system, it therefore also provides the nerve cells of the brain with nutrients. Analysis of C.f. is used in the diagnosis of diseases of the nervous system.

Cereus: see *Cactaceae.*

Ceriantharia: a small order of anemone-like antho-zoans, belonging to the *Hexacorallia.* They are solitary forms with no skeleton. The order includes the large, solitary "burrowing sea anemones", which live in a se-creted mucous tube, buried in the soft sea bottom. Oral disk and tentacles are projected when feeding, but they withdraw very rapidly into their tubes, so that even dense populations of *C.* may be overlooked.

Ceriops: see Mangrove swamp.

Ceropegia: see *Asclepiadaceae.*

Cerophagy: the utilization of wax as a food source, as in the birds known as honeyguides (see *Indicatoridae).*

Certation: Nonrandom fertilization by male- and fe-male-specifying gametes, which influences the sex ra-tio of progeny. In seed plants, C. can arise from the dif-ferent growth rates of pollen tubes.

Certhidia: see Galapagos finches; Adaptive radiation.

Certhiidae, ***treecreepers:*** an avian family (5 species in a single genus) in the order *Passeriformes* (see). They are small and brownish in color, streaked above and much paler (almost white) below. As an adaptation to tree climbing, they have long, sharp, slender, curved claws and long hindtoes. In addition, the stiff, pointed tail is pressed against the bark as a support. They creep up the vertical surfaces of trees and along the undersides of branches, and they may also be seen climbing on the vertical surfaces of rocks. The long, slender decurved beak is used for pulling insects from crevices in the bark. *Certhia familiaris* (***Treecreeper*** in Britain, ***Brown creeper*** in USA) is found all over the Northern Hemisphere between latitudes 70 ° and 15 °, whereas the other 4 species have more localized and restricted ranges. The ***Himalayan treecreeper*** *(C. himalayana)* is found in Turkestan, Afghanistan and west to China and Burma; the ***Short toed treecreeper*** *(C. brachydactyla)* is restricted to Europe, southern Scandanavia, Morocco and Tunisia; *C. discolor* (***Brown throated treecreeper***) and *C. nipalensis* (***Stoliczka's treecreeper***) are both found in the mountains of Nepal, *C. discolor* also extending into semitropical Indochina.

Cervus: see Barasingha, Red deer, Sambar, Sika deer.

Cestoda, ***cestodes, tapeworms*** (plate 43): a class of flatworms *(Platyhelminthes)*. The adult worm always lives parasitically in a vertebrate, most species in the intestine of the host, but some in the body cavity [see below, however, *Biacetabulum sieboldii* (order *Caryophyllidea*)]. The life cycle involves 2 to 3 larval stages. Some 3 500 species have been described.

Morphology. Adults of the subclass, *Eucestoda* (merozoic cestodes) consist of a very small (1–2 mm diam.) spherical or egg-shaped "head" (***scolex*** or ***holdfast***), and a flattened, ribbon-like "body" (***strobila***). The latter varies in length from a few millimeters to more than 10 meters, and it may consist of only 3, or up to several thousand segments, called ***proglottids*** or ***proglottides*** (sing. ***proglottis*** or ***proglottid***). The scolex is usually buried in the host's intestinal mucosa, where it is attached by various organs, such as hooks, suckers, adhesive folds (bothridia or bothria) and protrusible spined "tentacles".

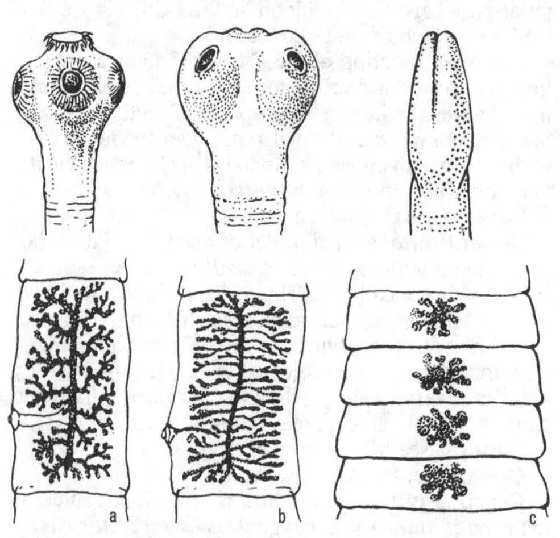

Fig. 1. *Cestoda.* Scolices (heads) and proglottids of tapeworms. *a* Pork tapeworm *(Taenia solum)*. *b* Beef tapeworm *(Taenia saginata)*. *c* Broad fish tapeworm *(Diphyllobothrium latum)*.

Proglottids are formed by budding in orderly succession from a short, unsegmented region behind the scolex, known as the "neck" or growth zone. The oldest proglottid is therefore the one most distant from the scolex. Each proglottid is a self-contained functional unit; apart from the fact that they occur at different stages of development along the length of the strobila all proglottids are identical.

Each proglottid contains 1 or 2 sets of hermaphrodite reproductive apparatus, each set consisting of numerous testes, a single ovary and a yolk gland (vitellarium or vitelline gland). Each testis gives off a vas efferens and all the vasa efferentia unite to form the vas deferens. The duct formed by the confluence of the vitelline duct and uterus receives the oviduct, which arises from a transverse duct connecting the ovaries; this duct then continues as the vagina, which opens into the genital pit. The genital pit also receives the vas efferens. Newly budded proglottids contain no reproductive organs. Male organs develop first in the younger, more anterior proglottids. Female structures then appear in the more centrally situated proglottids of the strobila. After fertilization of the eggs, the ovaries and testes degenerate and the gravid (or "ripe") proglottid is largely occupied by the profusely branched, egg-packed uterus. At this stage, the proglottid breaks away and passes out with the host's feces. The length of the strobila is maintained by budding from the growth zone. Self-fertilization occurs within a single proglottid and between proglottids but there is evidence that self-fertilization leads to a higher incidence of abnormal larvae. Cross-fertilization between two worms in the same host therefore probably leads to progeny with increased viability.

Running through each proglottid and traversing the length of the strobila are 2 nerve cords and typically 4 osmoregulatory canals (often referred to erroneously as "excretory" canals), lying in the peripheral zone of the medullary parenchyma. The 2 ventrolateral canals carry water away from the scolex, while the 2 dorsolateral canals carry water toward it. At the posterior end of each proglottid the two ventrolateral canals are connected by a single transverse canal. In each proglottid, secondary tubules arise from the ventrolateral canals, and these may branch further into tertiary tubules. The free ends of these tubules carry flame cells, which collect water and transfer it to the ventrolateral canals. Thus, the osmoregulatory system is essentially the same as the protonephritic osmoregulatory system of the trematodes.

An alimentary canal is totally lacking, and soluble nutrients are absorbed through the surface of the proglottids. Carbohydrate is stored largely as glycogen, and metabolism is anaerobic.

Life cycle. In most *C.*, segmentation, gastrulation and embryogeny occur while the egg is in the uterus. In many members of the order, *Pseudophyllidea,* however, segmentation occurs when the egg has left the gravid uterus. The resulting embryo or ***onc(h)osphere*** is sometimes surrounded by a ciliated membrane, and it is then known as a ***coracidium.*** The oncosphere or coracidium possesses 6 hooks (hexacanth embryo or larva of the *Eucestoda*) or 10 hooks (decacanth embryo or larva of the *Cestodaria*). It may hatch into the free environment (e.g. the coracidium of *Diphylloboth-*

ium latum), but as a rule it remains inside the egg until the egg is ingested.

In the family *Taeniidae* (order *Cyclophyllidea),* whose members infect warm blooded vertebrates, the oncosphere penetrates the intestinal wall of the secondary host, enters the circulation, and migrates to specific organs, e.g. liver, muscle, brain, and develops there into a ***cysticercus*** (also known as a ***metacestode*** or ***bladder worm).*** Cysticerci are pea or cherry-sized, fluid-filled, translucent bladders, and they are formed only by members of the family *Taeniidae.* Inside the cysticercus, the future scolex develops within an invagination from the outer surface in such a way that pressure on the bladder would cause it to be everted with its holdfast structures correctly positioned on the outside surface (Fig.2). In some species (e.g. *Taenia multiceps, Echinococcus granulosus)* the cysticercus is polycephalic and it attains the size of an egg or fist; it is then known as a ***coenurus*** (see Reproduction Fig.2) (see *Taenia* and Dog tapeworm). If the cysticercus or coenurus is ingested by a suitable carnivore or omnivore, the head evaginates and attaches to the intestinal wall, followed by formation of a segmented strobila.

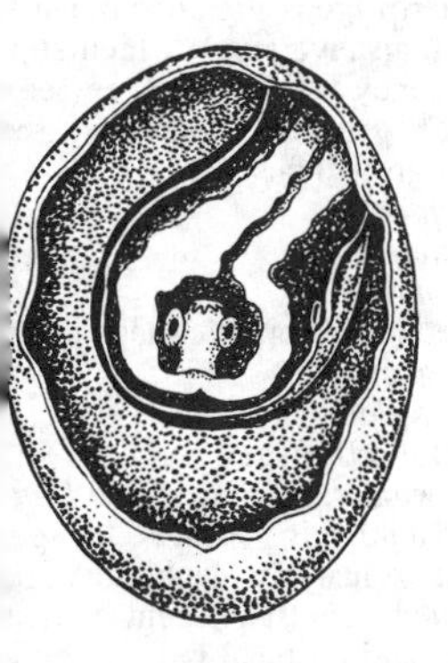

Fig.2. *Cestoda.* Cysticercus of the pork tapeworm *(Taenia solum)* with invaginated scolex.

In some species of the order *Cyclophyllidea* the intermediate host is an invertebrate (insect, crustacean, snail or annelid). In these species, the oncosphere develops into a ***cysticercoid*** larva. The cysticercoid consists of an anterior vesicle containing an invaginateed scolex, and a tail-like posterior region carrying the larval hooks. In contrast to a cysticercus, the cysticercoid contains no fluid-filled cavity. If the cysticercoid is ingested by a suitable primary host, the scolex evaginates and attaches to the intestinal wall, and the body forms a segmented strobila. In other species [e.g. Broad fish tapeworm (see), Bootlace worm (see)], the oncosphere (coracidium) (Fig.3) develops inside a copepod or amphipod host into a ***procercoid*** larva (Fig.3). The procercoid is an elongated, solid-bodied larva, in which the future holdfast has not yet differentiated; oncospheric hooks are still present at the posterior end, and several boring glands are present at the anterior end.

In the second intermediate host (usually a fish) the procercoid develops into a ***plerocercoid*** larva (Fig.3). The plerocercoid is a spindle-shaped, solid larval form with an invaginated holdfast, and no embryonic hooks. Apart from the absence of segmentation and genitalia, the plerocercoid closely resembles the adult form, and if it is ingested by a suitable predatory animal, it rapidly develops into a sexually mature adult.

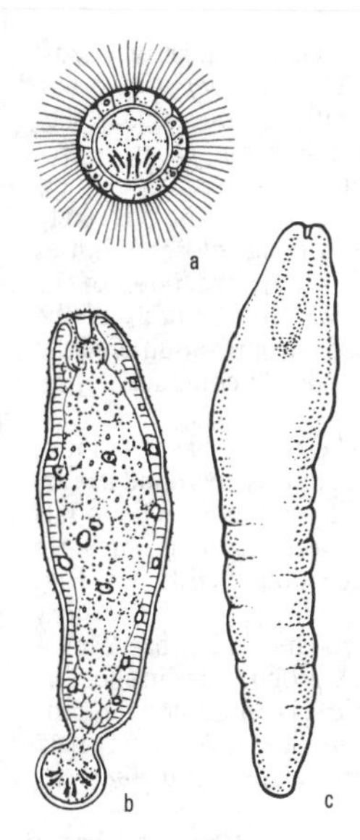

Fig.3. *Cestoda.* Developmental stages of the broad fish tapeworm *(Diphyllobothrium latum). a* Coracidium. *b* Procercoid. *c* Plerocercoid.

Economic importance. A large number of *C.* infect humans, as well as domestic and agricultural animals (e.g. several species of *Taenia,* the dog tapeworm and the broad fish tapeworm). Adult tapeworms represent a nutritional burden on their hosts, as well as causing harm by their toxic secretions and by blockage of the intestines. In addition, contamination of meat products and fish with cysticerci renders them unfit for human consumption. (See *Taenia*, Dog tapeworm, Broad fish tapeworm and plate 43)

Classification. Cestodes are usually placed in 2 subclasses *(Cestodaria* and *Eucestoda)*

1. Subclass: *Cestodaria.* These share characters with both the cestodes and other platyhelminths, especially rhabdocoel turbellarians and monogeneans. According to some authorities, the *Cestodaria* should not be included with the *Cestoda,* and the 2 orders of this group may not even be related to each other. Segmentation (strobilization) is absent, and only a single reproductive system is present, consisting of a single ovary, many testes and a single copulatory organ. *Cestodaria* have no mouth or gut, and the larva (known as a lycophore) has 10 (rather than 6) hooks. A true scolex is absent, but some members have holdfast organs, such as a cup-like acetabulum, or bothria. Most species invade the coelomic spaces of the host, rather than the alimentary canal.

1. Order: *Gyrocotylidea.* These are small, thick, aberrant, unsegmented worms (e.g. *Gyrocotyle),* found only in the large intestines of primitive cartilaginous fishes known as chimaeroids *(Chimera, Callorhynchus, Hydrolagus).* A muscular, cup-like acetabulum is present. This order is sometimes grouped with the monogeneans.

2. Order: *Amphilinidea.* The anterior end has a boring structure for penetrating the body wall of the host. Holdfast organs are absent. They are found in the body cavities of sturgeons (Europe and North America), siluroid fishes (India) and tortoises (Australia). The intermediate hosts are crustaceans.

2. Subclass: *Eucestoda.* This subclass contains about

12 orders, mostly specific for elasmobranch or teleost fishes as the primary hosts. The *Caryophyllidae* and *Spathebothridea* are placed with the eucestodes, because they probably represent interesting cases of progenic development, i.e. the adults are precociously reproductive juvenile forms (see Neoteny), and it is likely that the unknown adult ancestor had close affinities with the present eucestodes. With the exception of the *Caryophyllidea* and *Spathebothridea,* the body of the *Eucestoda* is usually segmented (strobilization) and the embryo has 6 hooks. Each proglottid contains a complete reproductive system.

1. Order: *Tetraphyllidea.* The scolex has 4 bothria, and it also often has hooks. Proglottids are found in various stages of development. The yolk glands lie in lateral bands in the proglottids, and the genital pores are lateral. Members parasitize elasmobranch fishes. E.g. *Phyllobothrium.*

2. Order: *Trypanorhynchea.* The scolex has 2 or 4 bothria, and 4 protrusible, spiny proboscides in sheaths. The yolk glands lie in the cortical layer around the proglottids, and the genital pores are lateral. Members parasitize elasmobranchs and some teleosts. E.g. *Tentacularia.*

3. Order: *Pseudophyllidea.* The anterior end has a single terminal bothrium, or 2 lateral bothria. Most proglottids in the strobila are at the same stage of development. The yolk glands lie in dorsal and ventral layers across the proglottids. The uterus opens at a uterine pore, and the genital pores are usually midventral. Members parasitize teleost fishes and terrestrial vertebrates. See Broad fish tapeworm.

4. Order: *Proteocephaloidea.* The scolex has 4 cup-shaped suckers, as well as an apical sucker or glandular organ. The yolk bands lie in lateral bands in each proglottid. The uterus ruptures to form midventral uterine pores, and the genital pores are lateral. Members parasitize cold-blooded vertebrates. E.g. *Ophiotaenia.*

5. Order: *Cyclophyllidea.* The scolex has 4 cup-shaped suckers, and usually an apical rostellum. Proglottids are found at different stages of development. The yolk gland is compact, and the uterus lies posteriorly without a uterine pore. The genital pores are usually lateral. Members typically parasitize warm-blooded vertebrates. See *Taenia,* Dog tapeworm.

6. Order: *Caryophyllidea.* Bothria are sometimes present, the body does not form a strobila, and only a single hermaphrodite sexual apparatus is present. Members are found in the body cavities of oligochaete worms (family *Tubificidae)* and in the intestines of freshwater fishes (e.g. *Catostomidae, Clariidae, Cyrinidae).* Adults of this order are thought to be pedomorphic larvae, i.e. prematurely reproductive proceroid larvae of ancestral protocestodes. *Caryophyllaeus laticeps* (length 4 cm) is a serious parasite of natural and cultured carp in many areas of the world, especially Europe and the former USSR. Its proceroid larva is found in redworms (genus *Tubifex);* if an infected redworm is eaten by a fish, the proceroid larva becomes sexually mature within the fish and produces fertilized eggs. The most extreme form of neoteny is found in the closely related *Biacetabulum sieboldii* (length up to 3 mm), which does not need a fish host in order to attain sexual maturity; fertilized eggs ripen in the proceroid larva while it is still in the *Tubifex.*

7. Order: *Spathebothridea.* These are parasites of salmonid and coregonoid fishes in Iceland, Greenland and sub-Arctic Canada. Members of the family *Diplocotyliae* possess 2 completely symmetrical pseudobothria The single species *Spathebothrium simplex,* which parasitizes the teleost fish *Liparis liparis,* has a bluntly tapering anterior, with no type of holdfast structure. Members of the family *Cyathocephalidae* have an acetabulum at their anterior end; this is a chitinous, funnel-like or cup-shaped structure with a slightly outward rolled margin. Adult cyathocephalids appear to be pedomorphic larvae, which parasitize whitefish *(Coregonus)* and trout *(Salmo);* although there is no external division into segments, the body contains a series of reproductive systems, suggesting evolutionarily primitive tapeworms.

[Advances in the Zoology of Tapeworms 1950–1970 R.A. Wardle, J.A. McLeod, S. Radinovsky, 1974, University of Minnesota Press, Minneapolis]

Cestodaria: see *Cestoda.*

Cetacea (Cetaceans): see Whales.

Cetoniinae: see Lamellicorn beetles.

Cetraria: see Lichens.

Cettia: see *Sylviinae.*

Cetti's warbler: see *Sylviinae.*

C-factor: a gene or a heterozygous alteration of chromosome structure (usually an inversion), which suppresses or reduces the frequency of Crossing over (see)

C-factors of plant growth factors: see Mitscherlich's law of plant growth factors.

Chachalacas: see *Cracidae.*

Chaemaea: see *Timaliinae.*

Chaenopsidae: see *Blennioidei.*

Chaetae: hair-like structures of variable shape in the skin of *Polychaeta* and *Oligochaeta.*

Chaetodermatida: see *Aplacophora.*

Chaetodontidae: see Butterflyfishes.

Chaetognatha, ***arrow worms:*** phylum of the *Deuterostomia,* containing about 80 living species. All species are marine, and almost exclusively planktonic. The usually arrow-like, elongated, semitransparent body is divided into a head, trunk, and postanal tail region; it has two horizontal fins that flank the trunk, and a tail fin Each body section has its own well developed coelom

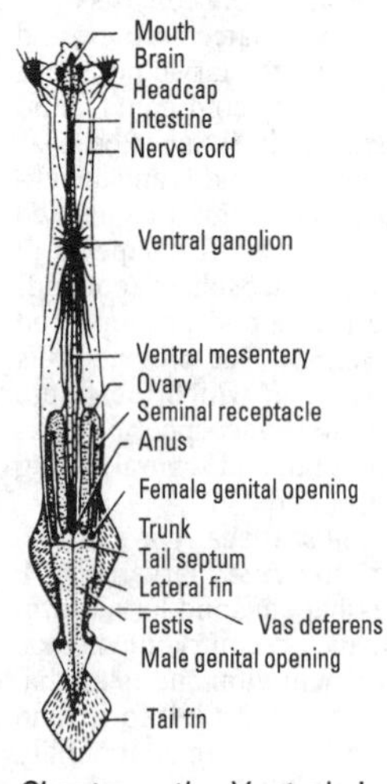

Chaetognatha. Ventral view of an arrow worm.

with a distinct epithelial lining), at least during larval evelopment. The head carries two eyes and 4–14 chinous food-catching spines, which curve outward then oint inward; during swimming, these spines are concealed beneath a cap-like skin fold. Respiratory and excretory organs are absent. All species are hermaphroite, the ovaries being located in the central body secon, the testes in the posterior section. Eggs are usually eleased into the sea. Total cleavage results in an invagnation gastrula, the blastopore of which closes and beomes the posterior end of the body.

The phylogeny of the C. is largely unknown. They ppear to be a very early offshoot of coelomate stock, irst appearing in the fossil record in the Carboniferous. Most species occur in large swarms on the sea surface. They are primarily drifters, but they can swim actively or short distances by jerky movements of the tail. All pecies are predatory, taking in particular small crustaeans and fish larvae. The most familiar genus is *Saitta,* which contains more than 20 species, up to 10 cm n length, and present in all the oceans of the world.

Collection and preservation. C. are caught with arge plankton nets and narcotized by scattering a few menthol crystals on the sea water. Narcotized specimens are smoothed out with a small paint brush in a ish, then fixed with Bouin's fixative. After washing with 50% ethanol, they are preserved in 2% formaldehyde in sea water.

Chaetomys: see *Erethizontidae.*

Chaetonotoidea: see *Gastrotricha.*

Chaetura: see *Apodiformes.*

Chaffinch: see *Fringillidae.*

Chagas' disease: see Trypanosomes.

Chain reaction: in biology, a sequence of (usually eleterious) phenomena, arising from the serious disturance of a natural system. A C.r. may be triggered natually, e.g. by a flood, earthquake or forest fire. Many C.rs. are due to human intervention in the natural order. For example, at the beginning of the 19th century in La Plata (South America), enormous areas of wild land were converted to cereal cultivation. As a result, the natural enemies of the locust were markedly reduced. The locust population increased dramatically and destroyed the cereal crops. It is feared that the present worldwide estruction of forests may lead to dangerous C.rs. Inexpert use of insecticides can also initiate undesirable C.rs. Thus, any intervention in the natural order, e.g. biological methods of environmental protection, should be appropriately designed to avoid negative C.rs.

Chalcides, *cylindrical skinks:* a genus of the *Scincidae* (see), distributed from the Mediterranean through East Africa, and as far as India. They have glossy scales, and display verying degrees of limb regression. The African *C. ocellatus* has normal, short, five-toed limbs, whereas the 40 cm-long *C. chalcides* of the western Mediterranean has the external appearance of a slowworm with only minute, functionless, three-toed limbs.

Chalcididae: see *Hymenoptera.*

Chalinasterol, *ostreasterol, 24-methylenyl cholesterol, ergosta-5,24(28)-dien-3β-ol:* a characteristic sterol of pollen, also found in sponges, oysters, mussels and honeybees.

Chalk: a fine-grained, biogenic, soft, friable rock, derived entirely or largely from the fossil shells of aquatic organisms, such as foraminifera, bryozoans, ostracods, residues of coccolithophorids and diatoms and other marine animals (radiolarians, sponges, corals, sea urchins, belemnites, etc.). Broken or ground shells form a paste or matrix in which more perfect ones are bedded. The color is commonly white or gray, but blue, brown and even black chalks are known. Chert and flint nodules are often present as inclusions; these are derived from siliceous sponge spicules, radiolarians and diatom shells. The Cretaceous (see) is sometimes called the Chalk.

Chalones, *antitemplate substances, tissue-specific endogenous mitosis inhibitors:* proteins produced by mature or differentiated cells, which inhibit cell division in primordial cells by a kind of negative feedback. The effect of C. is reversible and not species-specific. They have been isolated from epidermis, lymphocytes, granulocytes, fibroblasts, erythrocytes and hepatocytes. C. are active in the late presynthetic phase G_1 and the premitotic phase G_2 of the Cell cycle (see). They also inhibit the decision phase E and the first postmitotic maturation phase A_1, which precedes the aging phase A_2. Therefore C. not only inhibit cell division, they also retard the process of aging and thus increase the life expectancy of cells.

Chamaecyparis: see *Cupressaceae.*

Chamaeleo: see *Chamaeleonidae.*

Chamaeleonidae, *chameleons:* a family of the *Lacertilia* (see). Chameleons are highly specialized, mostly aboreal reptiles. On account of their many aberrant characters, they were earlier assigned to a separate suborder, but they are now thought to be phylogenetically related to the *Agamidae.* The body is laterally compressed. The high, slender skull often caries a bony helmet, as well as horns, snout appendages or combs. As an adaptation to their arboreal existence, the feet are zygodactylous; on the forefeet the 2 outer and 3 inner toes are fused; on the hindfeet the 3 outer and 2 inner toes are fused. The muscular prehensile tail serves as an additional appendage. The large, hemispherical, turret-like eyes have scaled lids and can be moved independently. Chameleons are relatively plump, and they climb slowly and deliberately. Insect prey is caught with the aid of the rapidly extensible tongue, which is as long as the body, and which holds the prey by suction (not by a sticky secretion as previously thought). Of all reptiles, chameleons display the most marked faculty for color change; color is influenced by the color of the background, temperature, illumination and the excitatory state of the animal. Average length is 20–30 cm. Some large species attain 60 cm including the tail, while the smallest are only 4 cm in length. Most species lay eggs, and for this purpose the female ascends from the trees and buries 20–30 eggs in the ground. Some southern African species and some montane species are live-bearing. Two genera and about 85 species are recognized, all restricted to the Old World, and most of them in Africa and Madagascar.

Chamaeleontidae. European chameleon *(Chamaeleo chamaeleon).*

The ***European chameleon*** *(Chamaeleo chamaeleon)* has the largest areal of all, inhabiting southern Spain, North Africa, Crete, Cyprus and the Near East, with several subspecies in Arabia and the Indian subcontinent; it has also been introduced into the Canary Islands. One of the most common African species is the ***Flap-necked chameleon*** *(Chamaeleo dilepis)* of tropical and southern Africa, with movable, scaled, membranous lobes on both sides at the back of the head. The ***African chameleon*** *(Chamaeleo africanus)* is found in tropical Africa. All three of these species lay eggs. Live-bearing species include: the East African ***Jackson's chameleon*** *(Chamaeleo jacksoni),* whose males have 3 conspicuous, head-length, anteriorly directed, soft "horns" on the snout (females have a single, greatly reduced horn); the 60 cm-long (with tail) ***Oustalet's chameleon*** *(Chamaeleo oustaleti)* from Madagascar, which also captures young mice as well as insects with its tongue; and the ***Stump-tailed chameleons*** *(Brookesia),* which are found in Central Africa and Madagascar, and which spend more time on the ground than in the trees.

Chamaenerion: see *Onagraceae.*

Chamaephyte: see Life form.

Chamaesaura: see *Cordylidae.*

Chameleons: see *Chamaeleonidae.*

Chamois, *Rupicapra rupicapra:* a medium-sized European horned ungulate (see *Bovidae)* of the subfamily *Caprinae* (see). Both sexes have short, steeply erect, black horns, which bend abruptly backward at their ends to form hooks. An inhabitant of high, rocky, mountainous regions, it is an extremely sure-footed, nimble climber. It is naturally distributed in the mountains of central and southern Europe, Asia Minor and the Caucasus, and it has been introduced into the Black Forest and the high sandstone areas of the River Elbe in Saxony. The summer coat is tawny brown, while the winter coat is much darker with thick woolly underfur. The underparts are pale and there is a white throat patch. The gemsbart or gamsbart of Tyrolean hats is made from the stiff winter hair from the back of the animal. Chamois or "shammy" leather, used for polishing glass, is made from the skin.

Changeable lizards, ***variable lizards***, *Calotes:* a tropical Asiatic genus of spiny, insectivorous, arboreal agamids (see *Agamidae),* with laterally compressed bodies, crested necks and backs, extremely long tails, and usually green coloration. The male of some species has a throat fan, which is extended by the hyoid bone and muscles. These agamids are remarkable for their ability to rapidly change color, excelling most chameleons in this behavior. *Calotes versicolor* ***(Common*** or ***Indian bloodsucker)*** is one of the most common lizards of the Indian subcontinent, and it is widespread from Iran to southern China. Dominant or territory-holding males display magnificent scarlet head coloration (hence the name bloodsucker), and they become drab if they lose a territorial dispute.

Channallabes: see *Clariidae.*

Chantarelle: see *Basidiomycetes (Poriales).*

Chaparral: see Sclerophylous vegetation.

Chaparral cock: see *Cuculiformes.*

Characiformes: an order of fishes considered by some authorities to be a suborder *(Characoidei)* of the *Cypriniformes.* It contains at least 1 335 species, of which only 176 are native to Africa, the remainder being found in Central and South America. The approximately 252 genera are allocated to 10 familes. They are all tropical freshwater fishes, and most species possess an adipose fin. Pharyngeal teeth are usually present, but not specialized as in the *Cypriniformes.* The majority have well developed jaw teeth and are carnivores, whereas the popular aquarium fishes known as headstanders (see below) and their relatives have few if any teeth, and are herbivores. Most characiforms are small and some are spendidly colored, popular aquarium fishes. Some of the larger predatory species are important locally as a food source.

The family *Characidae* ***(characins)*** is large and diverse and divided into several subfamilies, e.g. *Serrasalminae* ***(pacus, silver dollars, piranhas;*** 9 genera, 5? species). Serrasalmins are laterally compressed and disk-shaped, and are notable for their extraordinarily sharp teeth. They feed mainly on sickly or wounded animals that enter the water. For example, ***piranhas*** gather in large schools, which can attack and overcome relatively large terrestrial animals, usually mammals, that venture into the water, reducing them rapidly to a skeleton. They are even dangerous to humans, but normally do not attack unless a person enters the water with an open wound, or is very near during a frenzied attack on another victim. Well known piranhas are *Serrasalmus nattereri,* and *S. piraya.*

Members of the family, *Gasteropelecidae* ***(freshwater hatchet fishes;*** 3 genera, 9 species) have a very strongly laterally compressed head and body, so that the entire ventral region is keel-like and strongly arched. By beating the pectoral fins, they can truly fly (i.e. propel themselves in the air, rather than gliding) for a few meters out of the water.

Members of the South American family, *Anostomidae* (10 genera, about 105 species, maximal length 1? cm) often swim in an oblique head-downward position, feeding from the bottom. Known popularly as headstanders, they have a small, nonprotracile mouth and very few or no teeth. The two genera (3 species, maximal length 40 cm) that constitute the subfamily *Chilodontidae* (family *Curimatidae;* South America) adopt a similar oblique downward swimming position, and are also known as headstanders.

Characins: see *Characiformes.*

Characoidei: see *Characiformes.*

Character:

1) A genetically regulated morphological, physiological or biochemical property, which is the end product of gene expression. Expression of a character depends on the interaction of one or more genes with the rest of the genotype and with environmental factors, so that a C. is qualitatively and quantitatively variable. It is not the C. but rather the genetically determined reaction of the organism to its environment that is inherited. The relative importance of genotype and environment varies greatly for different Cs. A distinction is drawn between 1) *qualitative* or *oligogenic Cs.* which are controlled by only a few genes, each possessing a marked individual action (see Oligogenes), and 2) *quantitative* or *polygenic Cs.* The latter are determined by the combined action of numerous genes (see Polygenes), which usually act cumulatively, each making only a small contribution.

Cs. that are strongly influenced by the environment are known as *environmentally labile Cs.,* whereas those that are hardly influenced by the environment are *environmentally stable Cs.*

2) See Biometry.

Character dominance: the phenotypic expression f only one of two characters, which are determined by ıdependent genes, i.e. the expression of one character s suppressed when both genes are present in the same enome. The expressed gene is said to be epistatic (see pistasis). The normal F_2 segregation of 9:3:3:1 (see Mendelian segregation) is shifted to 12:3:1.

Characteristic species combination: the combiation of permanently established species that occur nore or less uniformly in all subunits of a vegetative nit (see Association). They include diagnostically imortant species, which permit easy identification of that articular type of plant community. If the frequency of ccurrence of a species varies by more than 30% in the ndividual subunits of an association, it is classed as a ifferential species, not a characteristic species. C.s.cs. re used to characterize higher ranks of vegetative units.

See Characteristic species group combination, Charcter species.

Characteristic species group combination: a ystem developed by Schubert, in which all the permaently established species of a region are divided into roups, based on ecological and sociological considertions. Such groupings define and characterize different ypes of vegetative association. They are not based on a mited number of "high fidelity" species, but on the enre complement of species in the association.

See Characteristic species combination, Character pecies.

Character pairs: more or less clearly differing charcters, which in the simplest case are determined by the ction of two different alleles of one gene.

Character species, *characteristic species, faithful pecies:* plant species that are mainly limited to, or ocur optimally in, a given community type. The C.s. of a lant community are ascertained by a statistical floristic nalysis, and comparison with the statistical floristic omposition of other communities, as well as by the tudy of phytosociological progression within the comnunity. A Syntaxon (see) is characterized by the occurence of C.s. in phytoc(o)enoses (see Plant community) f that syntaxon, and by the absence or rarity of these C.s. in the phytoc(o)enoses of other syntaxa. Closely reated syntaxa may be differentiated by the presence or bsence of certain species, which are then known as Differential species (see) (e.g. they may differentiate etween two subassociations). The concept of differenal species is not related to fidelity (see below); differential species define syntaxa on the basis of the distribtional boundaries of species.

According to Braun-Blanquet, syntaxa are defined y diagnostic species, which consist of 3 types: *C.s., differential species* and *constant companions*. The diagostic value of a species for the purpose of defining or dentifying a syntaxon depends on its fidelity.

Plant species have a fidelity rating of 5 and are called *xclusive,* if they are completely or almost completely estricted to one type of community. *Selective* species fidelity rating 4) occur most frequently in one kind of ommunity, but may also occasionally appear in others. *Preferential* species (fidelity rating 3) have their optinum in one type of community, but occur in several thers. *Indifferent* species (fidelity rating 2) show no ronounced affinity or preference for any type of comnunity. *Strange* species (fidelity rating 1) are rare or occasional intruders. Species of fidelity ratings 3 to 5 are C.s., whereas those of ratings 1 and 2 are diagnosticlly unimportant companion species.

In the vegetation classification system based on C.s., the fundamental unit is the Association (see). According to Braun-Blanquet, the association is the smallest unit that still displays its own C.s. In addition, it is also characterized by differential species. Associations can be further subdivided into Subassociations (see) and Variants (see), etc. The latter contain only differential species. Floristically related associations form alliances, which are united in orders. The latter are combined into classes. Within the phytosociological system of Braun-Blanquet, these higher levels of organization are characterized by alliance C.s., order C.s. and class C.s., respectively. They form a hierarchy based on phytosociological progression (level of organization). Phytosociological progression is assessed from such factors as the way in which individuals are grouped, consistency of individuals, degree of interaction between members of the community, extent of sociological differentiation, and stability and constancy, etc. of the community.

Rank within this system is shown by specific endings given to the community names. Rank ending is added to the genitive stem of the second genus in the name, or to the first, if there is only one genus. The species name may be included in the genitive, if the generic name is insufficient for naming the community. Endings: ***-etea*** (denotes *class,* e.g. Festuco-Brometea); ***-etalia*** *(order,* e.g. Brometalia erecti); ***-ion*** *(alliance,* e.g. Bromion); ***-etum*** *(association,* e.g. Cirsio-Brometum); ***-etosum*** *(subassociation,* e.g. Cirsio-Brometum caricetosum); variants are known by their unchanged specific names.

The increasing number of classes of vegetation, as well as the application of Braun-Blanquet's system to floristically widely different parts of the world, led to the development of a grouping above the level of class. This is known as the Class group (see), which is practically identical to the phytogeographical concept of the floral region (see Floristic kingdom).

This phytosociological concept, developed in the 20th century by Braun-Blanquet, was first applied to species-rich Alpine regions. It differs from Sociation (see), which was developed by botanists in the relatively species-poor northern European countries. Study of C.s. is associated with useful methods of vegetation analysis, and especially for this reason, it has a strong following among European botanists. It can, however, be justifiably criticized on the basis that the total complement of species in a syntaxon becomes neglected, due to over-emphasis on C.s. The system developed by Schubert meets this criticism by classifying vegetation according to the Characteristic species group combination (see), which takes into account the ecology of all plant species.

Higher ranks of vegetation units (after Schubert):

Deciduous forest communities

Alnetea glutinosae: species-poor alder woodlands of oligotrophic marshes.

Alnetalia glutinosae.

Carpino Fagetea: mesophilic mixed deciduous forests.

Fraxinetalia: mixed deciduous hardwood forests.

Carpino-Fagetalia: eutrophic beech and hornbeam forests.

Luzulo-Fagetalia: mesophilic beech and hornbeam forests.

Character species

Quercetea robori-petraeae: acidophilous mixed deciduous forests.
Quercetalia robori-petraeae: acidophilous mixed oak forests.
Myrtillo-Fagetalia: acidophilous mixed beech forests.
Quercetea pubescenti-petraeae: dry oak forests.
Quercetalia pubescentis.

Evergreen pine forest communities.

Uliginosi-Betulo-Pinetea: birch, pine and fir swamp woodlands.
Uliginosi-Pinetalia: subcontinental pine swamp woodlands.
Sphagno-Betuletalia: subcontinental-atlantic birch woodlands.
Eriophoro-Betuletalia: cotton grass and deciduous woodland vegetation.
Vaccinio-Piceetea: Eurosiberian fir and pine forests with associated juniper scrub.
Vaccinio-Piceetalia: Eurosiberian continental fir forests.
Vaccinio-Pinetalia: moss-dominated for forests.
Erico-Pinetea: Eurosiberian dry fir forests.
Erico-Pinetalia.
Pulsatillo-Pinetia: subcontinental dry fir forests.
Pulsatillo-Pinetalia.

Deciduous scrub communities.

Carici-Salicetea cinereae: sedge and dwarf willow.
Eriophoro-Salicion cinereae: cotton grass and dwarf willow scrub.
Calamagrostio-Salicetalia cinereae: reed grass and dwarf willow scrub.
Salicetea purpureae: willow scrub and trees on river banks.
Salicetalia purpureae.
Crataego-Prunetea: hawthorn and slow bushes.
Prunetalia: slow hedges and bushes.
Cotinetalia coggyriae: east Mediterranean/submontane thornbush scrub.
Urtico-Sambucetea: stinging nettle and elder scrub.
Rubo-Sambucetalia: elder copse and scrub.
Betulo-Franguletea: acidophilous deciduous shrubs.
Rubo-Franguletalia: blackberry and alder buckthorn scrub.
Betulo-Adenostyletea: Arctic + temperate/alpine tall herb communities and scrub.
Adenostyletalia.

Evergreen scrub communities.

Loiseleurio-Vaccinietea: Arctic + temperate/alpine dwarf bush and grass heaths above the tree line.
Empetretalia hermaphroditi: alpine crowberry scrub.
Vaccinio-Juniperetea: whortleberry and juniper scrub.
Vaccinio-Juniperetalia.

Evergreen dwarf scrub heath communities.

Calluno-Ulicetea: heather and gorse heaths.
Vaccinio-Genistetalia: subatlantic, central European blueberry and gorse heaths.
Erico-Sphagnetalia: heath bogs.

Woodland edge scrub communities.

Galio-Uricetea: mesophilic bush vegetation.
Galio-Alliarietalia: scrub vegetation of forest edges and clearings.
Petasito-Chaerophylletalia: butterbur- and chervil-dominated tall herbaceous vegetation.
Epilobietea angustifolii: acid-nitrophilous woodland edge and clearing communities.
Epilobietalia angustifolii.
Trifolio-Geranietea sanguinei: thermophilous, species-rich communities of scrub grassland boundary zones.
Origanetalia.

Alpine meadow communities.

Salicetea herbaceae: snowfield communities.
Salicetalia herbaceae: snowfield communities on acidic soil.
Arabitetalia caeruleae: snowfield communities on calcareous soil.
Caricetea curvulae: alpine swards on acidic soil.
Caricetalia curvulae.
Elyno-Seslerietea: alpine swards on calcareous soil.
Seslerietalia variae: alpine swards on calcareous soil.
Oxytropido-Elynetalia: alpine bare rock swards.

Saltwater and saline soil communities.

Zosteretea marinae: sea grass-dominated mud flat and estuarine communities of lower intertidal zone.
Zosteretalia marinae.
Ruppietea maritimae: beaked tassel pondweed communities; submerged vegetation of brackish ditches and saltmarsh pools.
Ruppietalia matitimae.
Thero-Salicornietea strictae: stands of annual *Salicornia* on saline mud flats in eu- and supralittoral zones.
Thero-Salicornietalia.
Spartinetea: *Spartina* salt marshes.
Spartinetalia.
Saginetea maritimae: sea pearlwort-dominated small winter-annual and rosette communities of slightly saline dune slacks.
Saginetalia maritimae.
Juncetea maritimi: *Juncus*-dominated halophyte communities.
Juncetalia maritimi.

Pioneer vegetation on rocks and stones.

Asplenietea rupestris: small fern-dominated communities of rocks and walls..
Potentilletalia caulescentis: communities in limestone and chalk crevices and mortar lines of walls.
Androsacetalia: communities in crevices of silicate rocks.
Cymbalario-Parietarietea diffusae: thermophilous nitrophilous communities in mortar lines of walls.
Parietarietalia muralis.
Thlaspietea rotundifolii: alpine and subalpine scree vegetation.
Thlaspietalia rotundifolii: limestone scree communites.
Drabetalia hoppeanae.
Androsacetalia alpinae: silicate rock debris communities.
Epilobietalia fleischeri: alluvial rock debris vegetation.

Freshwater, river bank, spring and alluvial communities.

Lemnetea: free-floating vegetation of eutrophic waters.
Lemnetalia: duckweed communities.
Hydrocharitetalia: permanent frog-bit communities in ponds and ditches.
Potamogetonetea: perennial rooted pondweed communities.
Potamogetonetalia.
Urticularietea intermedio-minoris: bladderwort communities of bog and fen pools.

Urticularietalia intermedio-minoris.
ittorelletea: rooted aquatic vegetation, consisting of osette species in oligotrophic and dystrophic waters.
Littorelletalia.
Montio-Cardaminetea: vegetation of oligotrophic prings and flushes, mainly in montane zones.
Montio-Cardaminetalia.
hragmitetea: tall sedge and reed swamps.
Phragmitetalia.

Vegetation of marshes and bogs.

soeto-Nanojuncetea: short-lived small rush-dominaed communities of sandy habitats, inundated in winter.
Cyperetalia fusci.
cheuchzerio-Caricetea fuscae: swamp communities lominated by marsh scheuzeria and common sedge.
Scheuchzerietalia palustris.
Caricetalia fuscae: common sedge mires.
Tofieldietalia: base-rich low moor bog and sewage communities.
xycco-Sphagnetea: high-moor peat bog communities.
Sphagnetalia: high-moor peat hummock communities.

Vegetation of fields and pastures.

mmophiletea: beach dune communities.
Elymetalia arenarii.
orynephoretea: club awn grass-dominated pioneer egetation.
Corynephoretalia.
edo-Scleranthetea: stonecrop-dominated pioneer egetation.
Sedo-Scleranthetalia: stonecrop-dominated pioneer vegetation of walls and rock surfaces.
Festuco-Sedetalia: pioneer vegetation on high-mineral sands and gravel soils.
estuco-Brometea: vegetation of calcareous grassands.
Bromeria erecti: sub-Mediterranean dry and semidry meadows.
Festucetalia valesiacae: continental dry and semidry swards.
Violetalia calaminariae: open communities of heavy metal mine spoil heaps.
Iolinio-Arrhenatheretea: vegetation of hay meadws, permanent pastures and adjacent footpaths.
Arrhenatheretalia: unused meadows and pastures.
Molinietalia: damp meadows.
Nardetalia: mat-grass meadows.

Weeds and ruderal communities.

akiletea maritimae: open communities of annual halnitrophile species (sea rocket) on coastline and coastal etsam.
Cakiletalia maritimae.
ideutetea tripartitae: bur marigold nitrophilous weed ommunities.
Bideutetalia tripartitae.
henopodietea: nitrophilous ruderal orache and weed ommunities of rootcrop fields and gardens.
Polygono-Chenopodietalia: nitrophilous weed communities of rootcrop fields and gardens.
Sisymbrietalia: ruderal hedge mustard and orache vegetation.
Onopordietalia: thermophilous Scotch thistle communities.
ecalietea: weed and escape communities.
Aperetalia: bent grass-dominated weed communities of arable land.
Secalietalia: basophilic corn buttercup communities.
Artemisietea: nitrophilous mugwort communities of hedgerows and lakesides.
Artemisietalia: tall herb weed communities.
Agropyretea repentis: couch grass pioneer vegetation.
Agropyretalia repentis.
Plantaginetea: pioneer communities of nutrient-rich grass verges and fields.
Agrostietalia stoloniferae: bent grass communities of water meadows and wet fallow land and pastures.
Plantaginetalia: plantain communities of waysides and roadsides.

Charadriidae, *plovers, lapwings and allies:* an avian family in the order *Charadriiformes* (see), containing 56 small to medium-sized species (15–39 cm) with a cosmopolitan but largely tropical distribution. These thick-necked, compact birds inhabit open ground, often near water. They can run quickly and they are also strong fliers. The bill is usually straight, stout and of medium length. The hallux is either vestigial or absent. Although the plumage is boldly patterned (brown, olive, gray, black, white), the disruptive effect affords an effective camouflage. All 24 species of ***Lapwings*** (in the strict sense) constitute the genus *Vanellus*, e.g. the ***Lapwing, Peewit*** or ***Green plover*** (*Vanellus vanellus;* 30.5 cm) is a crested Palaearctic species occurring in a broad band across the temperate zone of Eurasia from the Atlantic to the Pacific, migrating south within its breeding range and sometimes further into North Africa and northern India. The ***Sociable plover*** (*Vanellus gregarius*) has a more restricted breeding range on both sides of the Urals. The 4 species of ***Golden plovers*** constitute the genus *Pluvialis*, e.g. the ***Golden plover*** (*Pluvialis apricaria;* 28 cm), which is found from Iceland to northwestern Siberia, tending to move further south out of the breeding season; its upper parts are spangled black and gold. Members of the genus *Charadrius* are generally known as ***Sand plovers,*** e.g. the ***Ringed plover*** (*Charadrius hiaticula;* 19 cm), which occurs from Iceland and Ireland across northern Asia to the Bering Straits, also in Greenland and Baffin Island; it is essentially a coastal bird, but often moves far inland where there is water. The head is patterned black and white and the bill is short; it possesses a black neck collar, yellow legs and a prominent bar on each wing. The ***Little ringed plover*** (*Charadrus dubius;* 15 cm) (plate 6) is widely distributed over the whole of Europe, most of central, northern and southern Asia and Southeast mainland Asia, as well as Papua New Guinea. In appearance, it is a smaller version of the Ringed plover, but lacks the wing bars. *Anarhynchus frontalis* (***Wrybill plover;*** 20 cm) breeds on the South Island of New Zealand and overwinters on the coastline of the North Island. It runs quickly with its head tucked in, and probes for insects under stones. It is unique among birds in that the distal quarter of its bill (total length about 2.5 cm) is turned to the right. The ***Dotterel*** (*Eudromias morinellus;* 21.5 cm) breeds in the extreme north of Europe and of western Siberia and in mountain ranges further south (Scotland, Tirol, etc.), overwintering in the Mediterranean region and Southeast Asia. It has, however, become a rare bird, and nowadays its distribution is very discontinuous. The female is similarly but more brightly patterned than the male: the white eye stripes meet in a characteristic V at the nape of the neck, and it has a white breast band and chestnut underparts; a pale wing bar becomes visible in flight. Its congener, *E. ruficollis,* breeds on high ground in the most southerly parts of south America, overwintering near the equator. *Phegornis mitchelli* ***(Mitchell's plover*** or ***Diadem sandpiper plover;*** 18 cm) is a Neotropical species breeding by mountain bogs and the rocky borders of mountain streams at 3 000 to 5 000 m on the slopes of the Andes, and overwintering at somewhat lower altitudes. Its true taxonomic position is uncertain.

Charadriiformes: an order of birds (see *Avies)* containing 306 species in 16 families, with a practically worldwide distribution including the polar regions. Most inhabit coastal waters, beaches and marshes. A few are pelagic (inhabiting the open sea) and some remain inland. With the exception of the seedsnipe, their diet consists mainly of animal material. All have similar types of palate bones, and similar syringeal structure (see Syrinx). They also share similarities in the mode of insertion of the leg tendons connecting the muscles to the toes, and in the arrangment of the secondary flight feathers. Most species have 11 primaries, a few only 10. All have a tufted oil gland and a small aftershaft on their body feathers (see Feathers). The sexes are usually alike, and bright plumage is rare. Pronounced sexual dimorphism is found only in the painted snipes, phalaropes and the ruff. Phylogenetically, the *Charadriiformes* are close to the *Gruiformes,* and this is most apparent from the similarities between the jacanas, seedsnipe and thick-knees. The following families are listed separately: *Jacanidae* ***(Jacanas*** or ***Lily-trotters),*** *Rostratulidae* ***(Painted snipe),*** *Haematopodidae* ***(Oystercatchers),*** *Charadriidae* ***(Plovers*** and ***Lapwings),*** *Scolopacidae* ***(Long-billed waders, Sandpipers),*** *Recurvirostridae* ***(Stilts*** and ***Ovocets),*** *Phalaropodidae* ***(Phalaropes),*** *Dromadidae* ***(Crab plover),*** *Burhinidae* ***(Thick-knees*** or ***Stone curlews),*** *Glareolidae* ***(Coursers*** and ***Pratincoles),*** *Thinocoridae* ***(Seedsnipe),*** *Chionididae* ***(Sheathbills),*** *Stercorariidae* ***(Skuas*** or ***Jaegers),*** *Laridae* ***(Gulls*** and ***Terns),*** *Rhynchopidae* ***(Skimmers),*** *Alcidae* ***(Auks, Guillemots*** or ***Murrers, Puffins).***

Charadrius: see *Charadriidae.*

Charr: see *Salvelinus.*

Chasiempis: see *Muscicapinae.*

Chasmogamy: see Pollination.

Chats: see *Turdinae.*

Chaudhuria: see *Mastacembeloidei.*

Chaudhuriidae: see *Mastacembeloidei.*

Chauliodontidae: see *Stomiiformes.*

Chaulmoogra oil: a soft fat from the seeds of the southern Asian tree *Taractogenes kurzii* (family *Flacourtiaceae),* which has been used by the Hindus and Chinese for centuries for the treatment of leprosy. Similar oils are also obtained from the seeds of various species of *Hydnocarpus* (Chaulmoogra tree or Spongeberry tree) of the same family *(H. alcalae,* Philippines; *H. anthelmintica,* China; *H. wightiana,* India). The active component, *chaulmoogric acid,* is present as glycerides. The oil is administered orally, and the free acid is released during digestion. Chaulmoogric acid contains an asymmetric carbon atom, and the natural compound is dextrorotatory. Acids containing the cyclopentenyl ring system are characteristic of the seed fats and oils of the *Flacourtiaceae.*

$(CH_2)_{12}$—COOH

Chaulmoogric acid

Chauna: see *Anseriformes.*

Chavica: see *Piperaceae.*

Chazogamy: see Fertilization.

Cheek teeth: see Dentition.

Cheese mites: see *Acaridae.*

Cheetah: see *Felidae.*

Cheilostomata: see *Bryozoa.*

Cheirogaleidae (Mouse lemurs and **Dwarf lemurs):** see Lemurs.

Chelates, ***internal complexes:*** compounds containing a chelate ring, in which hydrogen or a metal are involved in ring formation. Physiologically important *chelating agents* or *chelators* are compounds that form a chelate ring with metal ions. Natural chelators are nitrogen- or sulfur-containing organic compounds and dicarboxylic acids (see Carboxylic acids), e.g. certain amino acids, peptides and humus compounds. In the earth, many plant mineral nutrients are transported and rendered available in the form of C. (see Mineral metabolism). At normal soil pH-values, some free metal ions would precipitate as insoluble salts, which would be unavailable to plants, e.g. iron (III) precipitates as ferric hydroxide at weakly basic pH-values. The formation of soluble C. of such ions is therefore important for plant nutrition. In plant physiological experiments, iron is often applied as its chelate with ethylenediaminetatraacetic acid (EDTA). Many important cellular compounds are also C., e.g. various metal-containing enzymes (metalloenzymes), chlorophyll (containing chelated magnesium), cytochromes (chelated iron), vitamin B_{12} (chelated cobalt).

Cheleutoptera, *Phasmidae*, *Phasmodea*, *Phasmatodea*, *Gressoria*, ***phasmids***, ***stick*** and ***leaf insects:*** an insect order containing more than 2 500 species, distributed mainly in the tropics. The Mediterannean stick insect *(Bacillus rosii)* is found in southern Europe, and 2 New Zealand stick insects *(Acanthoxyla prasina* and *Clitarchus laeviusculus)* are established in the Scilly Isles off the south-west coast of Britain. Some species are popular terrarium pets and laboratory animals.

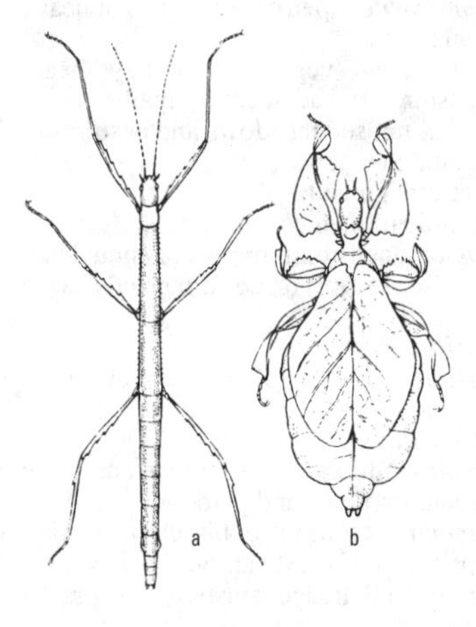

Cheleutoptera. a Stick insect *(Argosarchus* sp.). *b* Moving leaf *(Phyllium* sp.)

Imagoes. The body is either elongate, cylindrical and sticklike (stick insects) or flattened and leaflike (leaf insects), both types varying in length from 5 to 35 cm (see Fig.). The 2 pairs of wings, when present, may be fully developed or more or less reduced. Some species are wingless. The head carries strong, biting mandibulate mouthparts. Antennae may be filiform or moniliform, with between 8 and more than 100 segments. Compound eyes are always present, but ocelli are found only in certain winged species. The prothorax is short, the meso-and metathorax long. All legs are largely similar, possessing small, widely separated coxae; the tarsi usually have 5 segments. Metamorphosis is incomplete (paurometabolous development).

Eggs. Egg production is not always preceded by mating, and parthenogenesis is common. Eggs are laid individually on the larval food plant.

Larvae. In their structure and mode of existence, the larvae are very similar to adults, and they are fully grown after 5 to 8 molts.

Classification. According to Sharov (1968), the fossil evidence suggests that the *C.* are most closely allied with the suborder *Caelifera* of the *Orthoptera*, whereas Kemp (1973) associates them more closely with the *Dermaptera* and *Grylloblattodea*. The order is divided by Günther (1953) into 2 families: *Phyllidae* (leaf insects and related sticklike forms) and *Phasmatidae* (all remaining stick insects).

[K. Günther (1953) *Beitr. Entomol.* ***3,*** 541–563; J.W. Kamp (1973) *Can. Entomol.* ***105,*** 1235–1249; A.G. Sharov (1968) *Tr. Paleontol. Inst. Akad. Nauk. SSSR,* ***118,*** 1–213 (English translation 1971)].

Chelicerata: a subphylum of the *Arthropoda* (or of the *Amandibulata).* The subphylum contains a wide variety of forms, in which the body always consists of an anterior cephalothorax or prosoma and a posterior abdomen or opisthosoma. The prosoma always consists of 6 segments, all with legs. There is also an acron (head lobe), and the prosoma is often covered by a uniform dorsal plate (tergum). Sometimes, however, the acron and the 4 anterior segments form a body section known as the proterosoma, a structure which had already evolved in the trilobites. The opisthosoma does not contain a constant number of segments, since its posterior end is usually degenerate. Antennae and chewing maxillae are always absent. The first pair of appendages are chelicerae, which serve for grasping and tearing prey, and they are often chelate (armed with a pincer). The next 5 pairs of appendages are walking legs, although the first of these may be modified to pedipalps, which are also often chelate. Opisthosomal extremities or limbs are present only in the *Xiphosura* (horseshoe crabs), where they take the form of phyllopodia. In the *Arachnida,* the opisthosoma carries only the modified rudiments of extremites, such as lung books and spinnerets. The opisthosoma often carries an appendage (telson), which takes the form of a caudal spine (horseshoe crab), venom gland (scorpion) or segmented flagellum (whipscorpion). Sense organs include central and lateral eyes, which are usually simple ocelli, only rarely compound eyes. The midgut gives rise to extensive diverticulae with a dimorphic epithelium (glandular and absorptive cells). Excretion is performed by modified nephridia, the coxal glands and the Malpighian vessels. In horseshoe crabs, respiratory gas exchange occurs in the gills of the phyllopodia, whereas arachnids possess tracheae and lung books.

About 38 000 species are known, most of them terrestrial. Only a few forms have colonized the sea, and a few species have become secondarily adapted to an aquatic existence (e.g. water mites).

Classification. The subphylum contains 3 classes: *Merostomata* (see), *Arachnida* (see) and *Pantopoda* (see).

Chelictina: see *Falconiformes.*

Chelidae, ***Austro-American side-necked*** or ***snake-necked turtles:*** a family of the order *Chelonia* (see) containing 37 species in 9 genera, distributed in South America, Australia and Papua New Guinea. They are aquatic or semiaquatic, most species going onto land only to lay eggs. The neck is usually very long and mobile, and the head and neck may be longer than the carapace. The neck vertebrae fold only sideways, and the head can be folded completely beneath the carapace by a lateral S-shaped bending of the neck. The plastron is commonly yellow or white, sometimes orange or red. In some species, the darker, olive or black carapace is decorated with yellow, red or orange markings.

In the South American genus, *Hydromedusa,* the cervical scute is displaced back from the ring of marginal plates, so that is resembles a 6th vertebral scute.

Chelidae. Argentinian snake-necked turtle *(Hydromedusa tectifera).*

Probably the most divergent form of all turtles is the ***Matamata*** *(Chelus fimbriatus),* found in stagnant freshwater in tropical South America. Its flat head ends anteriorly in a slender, flexible proboscis, and appears triangular in outline, due to extensions of skin on either side. Lobes of skin hang from the underside of the head and the long, broad, flat neck. There are three longitudinal nodular ridges on the relatively flat carapace. It feeds on fish and other small aquatic prey by sucking them suddenly into its enormous gape. As in many nocturnal reptiles, the backs of the eyes of the Matamata are lined with a crystalline layer, known as a tapetum lucidum.

The Australian genus, *Chelodonia,* is represented by 8 species, the most familiar of which is the agile *C. longicollis.* The unpaired anterior scute of the plastron is displaced backward, so that it usually no longer reaches the anterior edge.

The 9 species of *Emydura* have relatively short necks. Found only in Papua New Guinea, some are brightly colored, e.g. *E. albertsii,* which has a red plastron and a red and yellow head.

Chelodonia: see *Chelidae.*

Chelonia, *Testudines, Testudinata,* ***turtles, terrapins*** and ***tortoises:*** an order of reptiles, distributed on all continents (excluding Antarctica) and many islands, and in warm seas, the greater number of species being found in the tropics. They possess a relatively short, broad, dorsiventrally compressed body protected by a dorsal bony shield (carapace) and a ventral bony shield (plastron). Carapace and plastron are joined laterally by a bridge. In most species, carapace and plastron are overlain by horny epidermal scutes, and in some species by a leathery integument. The totality of this external bony covering with its horny overlay is known as the "shell". Many species possess additional shell articulations, enabling the anterior and posterior openings to be completely closed. Head and limbs can be withdrawn inside the shell. The bones of the carapace arise from dermal ossifications, and these are rigidly fused with the widely spread ribs (there is no sternum) and with the flattened and expanded neural spines of the vertebral column. The clavicles and the interclavicles are separated from the pectoral girdle, and they form three bones of the plastron. The pattern of scutes does not correspond to the arrangement of bony plates beneath them. Skull bones are firmly fused to one another. Unlike all other reptiles, there are no temporal vacuities in the skull. Teeth are absent and functionally replaced by a sharp horny bill or beak. The limbs and the tail are covered with scales. The pupil is round, and the eyes have 2 lids and a nictating membrane. All species possess four well developed limbs. In terrestrial forms, the limbs are plump and club-like, while in freshwater forms they are laterally compressed like oars, and are usually webbed. The limbs of marine species are modified to paddle-shaped flippers. In aquatic forms, the lungs also function as swim bladders, allowing alterations in buoyancy. Aquatic species also exchange respiratory gases through the skin, in particular through the buccal mucosa. Visual and olfactory senses are well developed. Hearing is probably absent, but vibrations are perceived.

The cloacal opening is longitudinal and the penis is unpaired. All C., including marine forms, lay their hard- or parchment-shelled, white eggs out of water in a self-excavated hollow in the ground. With the exception of the vegetarian Green turtle *(Chelonia mydas),* most freshwater and marine species are predatory, feeding on fish and aquatic invertebrates, as well as worms and snails that they find on land. Terrestrial species, however, are mainly vegetarian.

Classification. The order contains 244 species, 75 genera and 13 (or 12) families. There are 2 suborders:

1. Suborder: *Cryptodira.* A series of inframarginal plates separates the carapace from the plastron. The head is withdrawn into the shell by a S-shaped vertical bending of the vertebral column. The pelvic girdle is not fused with the carapace or plastron. The following families are listed separately: *Chelydridae, Kinosternidae, Staurotypidae, Dermatemydidae, Platysternidae, Emydidae, Testudinidae, Cheloniidae, Dermochelyidae, Carettochelyidae, Trionychidae.*

2. Suborder: *Pleurodira.* Members of this order are more advanced than those of the *Cryptodira.* The inframarginal plates are reduced to an anterior (axial) plate immediately behind each foreleg, and a posterior (inguinal) plate at each groin, or they are entirely absent. The neck is turned sideways into the protection of the shell, and the pelvic girdle is fused with the carapace and the plastron. The subfamily contains 2 families, *Pelomedusidae* and *Chelidae,* which are listed separately.

Chelonia is also a genus of the family *Chelonidae.*

Chelidonichthys: see *Triglidae.*

Cheliped: a type of arthropod leg, present in many members of the *Arthropoda,* in which the terminal segments are modified to a chela (prehensile claw or pincer). Cs. are particularly well developed in the *Decapoda.*

Chelonidae, ***sea turtles, marine turtles:*** a family of the order *Chelonia* (see), containing 6 aquatic marine species in 4 genera, inhabiting the warm parts of all the large oceans and their associated seas The limbs are modified as flippers. The skull is completely roofed and cannot be retracted into the shell. Sea turtles only go to land to lay eggs, undertaking long sea migrations to specific nesting places on the coasts of tropical islands and continents. The body plates of the flat carapace are covered with large horny scutes, with the primitive feature of a complete row of inframarginal plates between the carapace and the plastron. The diet comprises fish and marine invertebrates, as well as eel grass and seaweed. Numbers have been severely depleted, because their flesh and their eggs make good eating. Sea turtles and their eggs are now protected.

Chelonia mydas ***(Green turtle)*** is a threatened species. It lives exclusively on seaweed and other marine plants, attains a length of 1.5 m and a weight of 200 kg. Turtle soup is prepared from the cartilage, and its meat and eggs are considered delicacies. Females congregate only every 2–3 years to lay eggs. A total of about 200 eggs are laid by each female in 2 or 3 clutches in coastal sand. Known egg laying sites are Kalimantan and other islands of Southeast Asia, as well as islands in the Caribbean.

The ***Loggerhead*** and the ***Hawksbill*** are large predatory turtles with transparent horny scutes on the bony plates of the flat, broad carapace. The posteriorly overlapping horny scutes of the 90 cm-long *Eretmochelys imbricata* ***(Hawksbill)*** were the source of tortoiseshell for the manufacture of combs, etc., but since the advent of modern synthetic plastics the animal is no longer exploited on a large scale for its shell. The ***Loggerhead sea turtle*** *(Caretta caretta;* 1 m-long, 150 kg) is found not only in tropical seas, but is also the commonest sea turtle of the Mediterranean and Adriatic. Both the Hawksbill and the Loggerhead are endangered species. They prefer relatively shallow water. Fifty to one hundred eggs are laid in the sand above the tidal zone. The young hatch after 2–3 months and enter the water immediately.

Chelonoidis: see *Testudinidae.*

Chelus: see *Chelidae.*

Chelydra: see Snapping turtles.

Chelydridae: see Snapping turtles.

Chemical weapons: see Biological weapons.

Chemoautotrophy: an autotrophic mode of nutrition based on Chemosynthesis (see), and not on the utilization of light energy as in photosynthesis.

Chemocline: see Lake.

Chemodinesis: see Dinesis.

Chemomorphosis: changes in the form and shape of a plant caused by external chemical factors. See Morphoses.

Chemonasty: a nastic movement induced by a chemical stimulus, especially in carnivorous (insectivorous) plants. The leaves of the sundew *(Drosera)* have tentacles on their surface, whose glandular heads are sensitive to proteins, peptides, amino acids, ammonium salts or phosphate; such a cocktail of compounds probably arises from insects trapped in the adhesive secretion of the glandular heads, and signals the presence of utilizable prey. The hairs are also sensitive to touch. Once the glandular head has been stimulated, the excitation is transferred to the base of the tentacle at the rate of 8 mm/min, where it causes a bending growth movement that moves the tentacle toward the center of the leaf and encloses the insect prey to await eventual digestion. The excitation is transferred to the bases of other tentacles, which therefore also contribute to enclosure of the prey. Chemonastic reactions are also involved in the opening and closing of stomata (see Stomatal movements).

Chemoreception: the generation of a primary electrical response when a sensory cell recognizes a signal molecule. The signal molecule gains access to the sensory cell by Diffusion (see) or Convection (see), where it becomes bound to the specific binding site of a specific receptor protein. Signal transduction occurs when the bound signal molecule causes the receptor protein to undergo an allosteric change, which in turn causes an ion channel to open.

Olfactory cells of insects can produce a nerve impulse in response to a single molecule of olfacory stimulant (e.g. the antenna of a female insect may respond to a single molecule of a male attractant pheromone). This impulse arises from the opening of a single sodium channel. Likewise, on the muscle endplate, one or two molecules of acetylcholine are sufficient to open an ion channel. The change in the receptor potential of the olfactory cell corresponds to an energy of 10^{-17} J. The binding energy of the signal molecule, however, is about 10^{-20} J, thus representing an amplification factor of about 10^3.

Chemoreceptors: nerve endings that are excited by chemical stimuli, in particular those C. that are active in chemical Respiratory regulation (see). C. are located in the arterial circulation, in the medulla oblongata, and in skeletal muscle. Of these, the *carotid* and *aortic bodies* of the arterial circulation are the most extensively investigated. The two carotid bodies are located bilaterally in the bifurcations of the two common carotid arteries. They contain blood spaces (sinusoids) and chromaffin cells (glomus cells), and they are richly supplied with afferent nerve fibers, which pass into the glossopharyngeal nerves (carotid sinus nerves) and from there to the medulla. The aortic bodies are located along the arch of the aorta, and their afferent fibers pass to the medulla via the vagi. The receptor fields of the carotid and aortic bodies appear to be identical or very similar. They react primarily to chemical stimuli, but their sensitivity to chemical stimuli can be increased by other stimuli. Since the threshold for the oxygen receptors lies at a partial oxygen pressure (see Partial pressure) of about 20 kPa, the carotid and aortic bodies are continually stimulated at the normal oxygen tension of the blood of 13 kPa, and they act as antennae of oxygen deficiency. They report only the concentration of dissolved oxygen, and are insensitive to the major fraction that is bound to hemoglobin, and furthermore they react to an increase in hydrogen ion concentration. Oxygen deficiency intensifies the firing of active receptors, and activates previously inactive receptors. The threshold of the carbon dioxide receptors lies at about 3 kPa, which means that these are also continually activated at the normal carbon dioxide tension of more than 5 kPa. The relationship between the firing rate of carbon dioxide receptors and the carbon dioxide tension is linear (Fig.); the stimulus is carbon dioxide itself (i.e. *not* hydrogen ions), and the activity of the receptors is increased by oxygen deficiency. Arterial C. exert a marked effect on the circulation. Their excitation activates the sympathetic nervous system, which then increases blood pressure by a general constriction of arterioles. If the blood pressure falls below the threshold of the Baroreceptors (see), survival of the organism may depend on the activity of the arterial C.

The activities of C. in the respiratory center and in skeletal muscle are described more fully under Respiratory regulation (see).

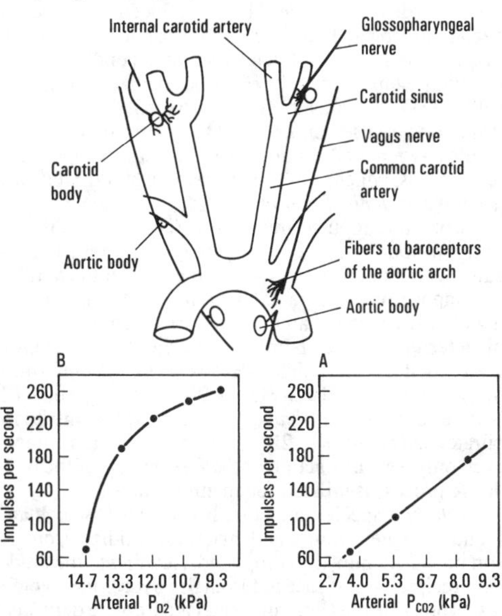

Chemoreceptors. Position of chemoreceptors and graphs showing their response to partial pressures of oxygen and carbon dioxide.

Chemosynthesis: a special form of Autotrophism (see), occurring primarily in certain nonpigmented bacteria, and based on the assimilation of carbon dioxide (CO_2) with the aid of energy derived from the oxidation of inorganic compounds. [In contrast, photosynthesis utilizes the energy of sunlight for the assimilation of CO_2]. As in photosynthesis, the process of C. can be divided into 2 phases: 1) the phase of energy generation and 2) the phase in which CO_2 is reduced to carbohydrate (formation of organic assimilate). The latter phase is essentially the same as the dark reaction of Photosynthesis (see). In C., the reducing power (NADH + H^+) and energy (ATP) required for the assimilatory phase are produced by the oxidation of inorganic materials. Electrons withdrawn from inorganic compounds are used partly for the reduction of NAD^+ to NADH + H^+ and partly for the production of ATP. In the latter process, electrons are transported to oxygen via an electron transport chain similar to the respiratory chain. If the oxidized substrate provides only electrons and has no available hydrogen (e.g. NO_2^-), then electrons are transferred to protons derived from the dissociation of water.

The simplest example of C. is provided by green algae of the genus *Scendesmus,* which are adapted to the utilization of hydrogen gas. In light, with the aid of chlorophyll, electrons are transferred from H_2 via ferredoxin to NAD^+. In the dark, however, *Scendesmus* lives chemoautotrophically by exploiting the "Knallgas" reaction, i.e. electrons are transferred from hydrogen to oxygen: $2H_2 + O_2 \rightarrow 2H_2O$. Among the nonpigmented bacteria, the Knallgas reaction is used chemoautotrophically by species of *Hydrogenomonas.*

Autotrophic C. is also performed by sulfur bacteria, e.g. *Beggiatoa,* which occurs in sulfur springs and marshes. Sulfur bacteria obtain energy from the oxidation of hydrogen sulfide (H_2S) (produced by bacterial decomposition and putrifaction of organic substances): $2H_2S + O_2 \rightarrow S_2 + H_2O$. Under suitable conditions, the resulting elementary sulfur can be oxidized further to sulfate (SO_4^{2-}). Other sulfur bacteria, e.g. *Thiobacillus denitrificans,* use nitrate (KNO_3) rather than atmospheric oxygen for the oxidation of of sulfur compounds; this process, which results in the release of nitrogen gas, is known as *denitrification.*

Chemoautotrophic, free living (i.e. nonsymbiotic) *nitrifying bacteria* are ecologically important organisms, which oxidize ammonia (produced in the soil by decomposition) to nitrate. Nitrifying bacteria, consisting of *nitrite formers* and *nitrate formers,* are abundant in all fertile soils (see Nitrogen). Nitrite formers *(Nitrosomonas* species) oxidize NH_3 (NH_4^+ in aqueous solution) to nitrite: $2NH_4^+ + 3O_2 \rightarrow 2NO_2^- + 2H_2O + 4H^+$. Nitrate formers *(Nitrobacter* species) oxidize nitrite to nitrate: $2NO_2^- + O_2 \rightarrow 2NO_3^-$. In these reactions, however, only a small fraction of the free energy of the oxidative process is utilized for chemosynthesis.

Leptothrix species and other iron bacteria, which are found in water sumps and marshes, oxidize iron(II) compounds to iron(III) compounds, leading to the gradual precipitation of rust-red iron(III) hydroxide, which forms bog iron ore (limonite) and lake iron ore deposits. These conversions have a very low energy yield, so that in order to obtain sufficient energy for an autotrophic existence, very large quantities of iron compounds are often oxidized.

Chemotaxis: determination of the direction of translatory movement of an organism, cell, or organelle by a chemical stimulus (see Taxis). C. occurs primarily in heterotrophic organisms, and causes them to approach food sources. Sex cells, such as spermatozoids, are guided toward a fusion partner by C. Chloroplasts also display C. (see Chloroplast movements). Many different substances act as stimuli. Thus, putrifying bacteria show a positive chemotactic response to numerous organic compounds, like asparagine, dextrin, glucose, urea, creatine, mannitol, peptone, etc., as well as to various inorganic salts. Autotrophic flagellates are attracted by carbonic acid, phosphates and nitrates, while the swarmers of *Myxomycetes* adjust their direction of movement according to pH gradients within the environment; low H^+ concentrations act as an attractant, whereas high H^+ concentrations are chemotactically negative. Highly specific attractants (see Gamones) are often involved in the encounter between sex cells; spermatozoids of various ferns, *Equisitales* (horse-tails) and *Selaginalla* react positively to citric acid; moss spermatozoids respond positively to sucrose, while a specific protein has been identified as the attractant for spermatozoids of the liverwort, *Marchantia.* Presumably, these chemotactic agents are also secreted by the respective female reproductive organs.

The stimulatory threshold, i.e. the lowest substance concentration to cause chemotactic activity, is often extremely low. For example, fern spermatozoids respond to 0.001% malic acid. Of greater importance, however, is the concentration gradient, which must be steep enough to stimulate translational movement. Thus, spermatozoids in 0.001% malic acid solution are attracted by a concentration of 0.03%, whereas spermatozoids in 0.01% malic acid are stimulated only by concentrations of 0.03% and higher, i.e. the concentration difference must exceed a threshold factor of 30. When allowed to stand in homogeneous solutions of malic acid, fern spermatozoids gradually lose their ability to display a chemotactic response. A special case of C. is Aerotaxis (see).

Chemotaxonomy: the use of the occurrence and distribution of chemical constituents to determine the taxonomic position and phylogenetic relationships of organisms. It is applied largely to plants, and is a valuable aid to systematic botany. The biosynthetic mechanisms of natural products may also be used in C. Evidence from C. often reinforces the conclusions reached from morphological studies. Conversely, C. has enabled corrections in the assignment of taxa, whose systematic position was previously uncertain. The expectation, however, that C. might transcend morphological studies and provide a more reliable and fundamental basis of taxonomy, has not been fulfilled.

Some examples of natural products employed in C. are alkaloids, flavonoids, proteins and nonprotein amino acids. Serology is also used in C., i.e. antiserum raised against a purified protein or protein extract from one plant is tested for cross reactivity with the proteins of other plants.

[Chemotaxonomy of the Leguminosae, edit. Harbourne, Boulter and Turner, Academic press, 1971; *The Biology and Chemistry of the Umbelliferae,* edit. Heywood, Academic Press, 1971; *The Biology and Chemistry of the Compositae,* edit. Heywood, Harbourne and Turner, Academic Press, 1977]

Chemotherapy: the use of synthetic compounds to selectively kill infectious organisms, parasites and tumor cells in animals and humans, without causing damage to host cells. C. is based on the principle of selective toxicity, and was pioneered by Paul Ehrlich. The first example of a synthetic therapeutic agent was trypan red, which kills trypanosomes in infected mice; unfortunately, it is inactive in humans. The earliest human therapeutic agent was arsphenamine ("Salvarsan"), which was used in the treatment of syphilis (caused by *Treponemia pallidum).* The investigation of more than 1 000 azo-dyes for antistreptococcal action in the blood stream, led to the discovery of the chemotherapeutic properties of "Prontosil" (Domagk, 1935). Prontosil is reduced in the body to a sulfonamide, and the azo-dye structure *per se* is not concerned in the therapeutic activity. Thus, the sulfonamides were discovered as important antibacterial chemotherapeutic agents. By 1942, penicillin was in general clinical use, marking the beginning of an important new era in C. Although antibiotics are not synthesized in the laboratory (some, however, are semisynthetic; see Antibiotics), they are generally classified as chemotherapeutic agents.

Chemotherapeutic agents, such as fungicides and insecticides, have also been developed against plant diseases and pests. Probably the best known insecticide, DDT is very effective against insect pests, with no deleterious action on the protected plant. Its persistence in the biosphere, its accumulation in the lipids of animals, and its toxicity to animals such as fish has led in many parts of the world to its replacement by phosphate and carbamate insecticides, which are readily degraded in the soil.

See Systemic agents.

[Selective Toxicity, Adrien Albert, Chapman and Hall, London, 1973; Chapter 5: Chemotherapy: history and principles]

Chemotropism: adjustment or regulation of the direction of growth of a plant part by a concentration gradient of a dissolved substance in water, or gaseous material in the air. Bending within the gradient toward the higher concentration is called positive C., bending toward the lower concentration negative C. Some plants display positive C. in the low concentration region of the gradient, but negative C. when the concentration is higher, thus enabling plant parts to grow into regions of optimal concentration. C. is important for the nutrition of many heterotrophic plants (see Heterotrophy). Thus, fungal hyphae display positive C. in concentration gradients of glucose, sucrose, peptone, asparagine, etc., but negative C. toward inorganic and organic acids. Sexual fusion of hyphae, e.g. the (+) and (-) hyphae of the *Mucoraceae* (molds) and of gametangia is made possible by C. Pollen tubes react positively to substances from the stigma tissue, from the inner glandular tissue of the style, and from the ovule, and in this way are attracted and guided to the ovule. Pollen tube attractants are mixtures of many substances, the main active components being calcium ions, together with borate ions, various sugars and proteins. The parasitic dodder plant *(Cuscuta)* finds its host plant by positive C. Movements of *Drosera* tentacles toward a trapped insect are determined partly by C. and partly by chemonasty. Roots display positive C. to oxygen and even to carbon dioxide. Special cases of C. are Aerotropism (see) and Hydrotropism (see).

Chenodeoxycholic acid: see Bile acids.

Chenopodiaceae, *Salsolaceae, **goosefoot family:*** a cosmopolitan family of dicotyledonous plants containing about 1 500 species in about 75 genera. They generally prefer saline soils and soils with a high nitrogen content, being common in salt deserts, along coasts, and as ruderals associated with human activity. They are herbs or shrubs, rarely bushes or trees, with usually alternate leaves which are regressed in succulent forms. The small, inconspicuous, predominantly wind-pollinated flowers are arranged in glomerular, cymose or racemose inflorescences. They have only a single perianth (usually green, sometimes with a reddish tinge), a single whorl of epipetalous stamens, and a one-seeded ovary of 2 or 3 carpels. In most species the flowers are hermaphrodite, but the family displays an evolutionary tendency to a reduction of perianth segments and stamens, so that species of certain genera (e.g. *Atriplex)* have dioecious flowers and no perianth. The superior ovary develops into a single-seeded nutlet or achene. Like other families of the order *Centrospermae (Caryophylalles),* the C. are characterized chemically by the presence of pigments known as betalains.

Economically important, cultivated varieties of *Beta vulgaris* include ***Sugar beet, Beetroot, Spinach beet, Chard*** and ***Mangold.*** Despite their high content of oxalic acid, the long-stalked leaves of ***Spinach*** *(Spinacia oleracea),* which has hermaphrodite flowers, are also important as a culinary vegetable. The genera, *Chenopodium* ***(Goosefoot)*** and *Atriplex* ***(Orache),*** include many ruderal plants and agricultural weeds. *Chenopodium bonus-henricus* ***(Good King Henry)*** is sometimes cultivated as a leaf vegetable. Long used as a vegetable by Indogermanic tribes, the ***Orache*** or ***Garden orache*** *(Atriplex hortensis)* has been cultivated in Europe since the Renaissance as a vegetable and ornamental; about one quarter of its dimorphic flowers have a perianth but lack bracteoles, while the remainder lack a perianth.

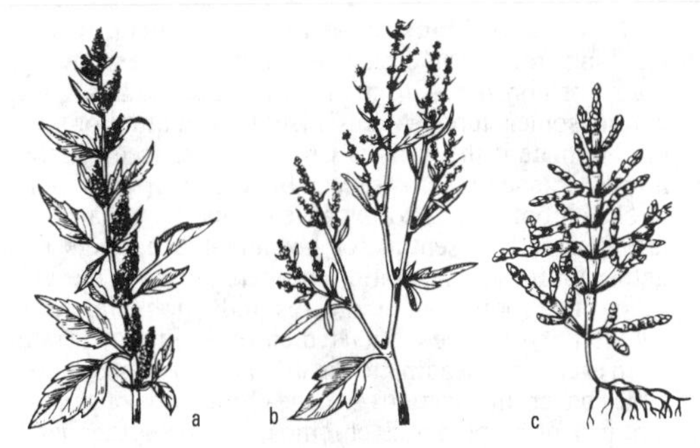

Chenopodiaceae. a Fat hen *(Chenopodium album). b* Common orache *(Atriplex patula). c* Glasswort or marsh samphire *(Salicornia* sp.).

Quinoa or ***Pigweed*** *(Chenopodium quinoa)* is cultivated in the South American Andes, where it is an important food source for the indian population; the seeds are ground to flour, and the plant is used as animal fodder. ***Summer cypress*** or ***Burning bush*** *(Kochia scoparia)* is a popular ornamental of parks and gardens; its leaves change from bright green in the spring to dark red in the autumn. ***Marsh samphire*** *(Salicornia europaea),* a halophytic stem succulent, is a pioneer plant in saline habitats such as silted up salt marshes. Various species of *Haloxylon* ***(Saxaul),*** found in desert areas from North Africa to Central Asia, are known as "Trees of the desert"; they are exploited for a variety of purposes, in particular as firewood.

Chenopodium: see *Chenopodiaceae.*

Cherry: see *Rosaceae.*

Cherry fly: see Fruit flies.

Cherry fruit moth: see *Yponomeutidae.*

Cherry-laurel: see *Rosaceae.*

Chersydrus: see *Acrochordidae.*

Chervil: see *Umbelliferae.*

Chestnut:

1) Sweet chestnut. See *Fagaceae.*

2) Horse chestnut. See *Hippocastanaceae.*

Chevrotains: see *Tragulidae.*

Chewing reflex, *mastication reflex:* nervous regulation of the mechanical disintegration of food in the buccal cavity. Pressure on the incisors or the gum below the incisors activates the *gnawing reflex,* involving rapid movement of the muscles of mastication. Initially, pressure on the teeth causes reflex jaw opening. Rebound of the motor neurons of the jaw-closing muscles then causes the mouth to close, which once more triggers the opening reflex, etc. Pressure on the buccal mucosa in the region of the premolars causes reflex rhythmic vertical movements of the lower jaw *(vertical chewing),* which serve to crush the food. Pressure on the buccal mucosa in the region of the molars activates lateral movements of the lower jaw, which grind the food *(ruminant reflex).* All three reflexes are coordinated by a center in the medulla oblongata.

Chiasma (pl. **Chiasmata**): a crossed structure (Fig.) resulting from an interchange between corresponding points of nonsister chromatids of paired homologous chromosomes during Meiosis (see). Formation of a C. or chiasmata is the basis of Crossing over (see). In the normal course of meiosis, each bivalent pair of homologous chromosomes forms at least one C. Chiasmata are apparently essential for keeping the partners together until their separation in anaphase I. In the absence of C. formation, or in cases of the premature dissolution of a C., the associated chromosomes separate from each other, leading to disturbances of meiosis (see Chiasma terminalization). Let the chromatids of a bivalent pair of homologous chromosomes be symbolized as A, A', B, B'. If there are 2 chiasmata, both between A and B, or between A' and B', they are termed *reciprocal*. If one C. occurs between A and B, the other between A' and B', the condition is *complementary*. If chiasmata are formed between A and B', as well as A' and B, the condition is *diagonal*.

Chiasma frequency is the average number of chiasmata formed per pair of homologous chromosomes, per cell, or per karyotype under given physiological and environmental conditions.

Chiasmata do not occur randomly throughout the chromosome. They show a greater frequency of occurrence in certain chromosomal segments, and in some cases are localized entirely in these segments.

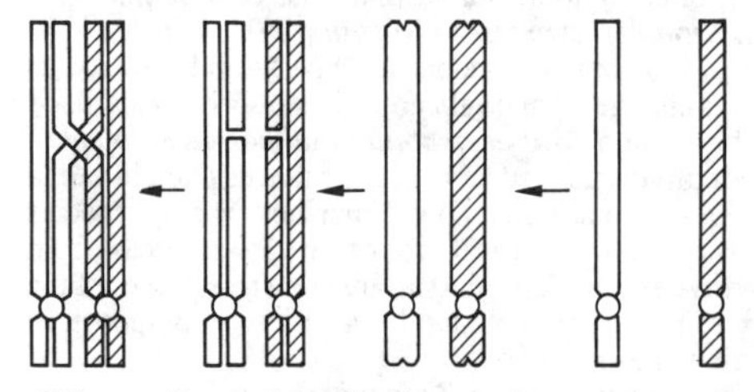

Chiasma. Formation of a chiasma by crossing over.

Chiasma terminalization: the movement of chiasmata to the ends of paired, nonsister chromatids. This movement starts in the diplotene (see Meiosis) and may continue until metaphase I (Fig.). As C.t. reaches completion, the total number of chiasmata among the paired chromosomes decreases, and those that remain become concentrated near to, or at the ends of, each bivalent.

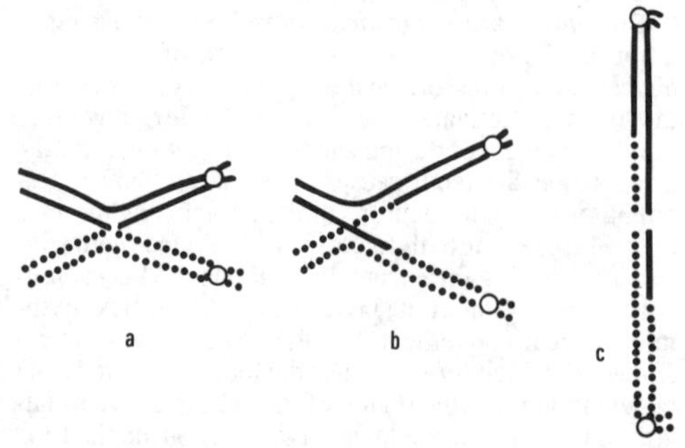

Chiasma terminalization in meiosis I. a Diplotene. *b* Diakinesis. *c* Metaphase I.

Within a single set of chromosomes, the rates of movement of chiasmata may be very different. The ratio of the number of terminal chiasmata to the total number of chiasmata at any stage of meiosis (the stage at which the cell is chemically fixed and investigated microscopically) is called the terminalization coefficient. A value of 1 for the terminalization coefficient means that all chiasmata are terminal; this represents the extreme case where the rate of chiasmata movement is similar in all chromosome pairs, and the coefficient is determined at the completion of terminalization.

Chiastoneury: see *Gastropoda*.

Chickaree: see *Sciuridae*.

Chick pea: see *Papilionaceae*.

Chicory: see *Compositae*.

Chiffchaff: see *Sylviinae*.

Chigger: see *Trombiculidae*.

Chigoe flea: see *Siphonaptera*.

Chile pine: see *Araucariaceae*.

Chilli pepper: see *Solanaceae*.

Chilognatha: see *Diplopoda*.

Chilomonas: see *Phytomastigophorea*.

Chilopoda, *Opisthogoneata*, ***centipedes:*** a class of myriapodous arthropods (see *Myriapoda*) with a terminal genital opening. About 2 800 species are known, and all are thought to be predatory. The oldest fossil remains are found in the Upper Carboniferous. The body consists of at least 19 and as many as 181 segments. With the exception of the first segment and the last 3 segments, each segment carries a pair of walking legs. The first trunk segment carries a large pair of maxillipeds (forcipules or poison claws), each with a terminal pointed fang containing the outlet duct of a poison gland. The sternite of the first trunk segment and the large coxae of the forcipules form a plate beneath the mouthparts. Prey (mostly insects and earthworms) is paralysed by the poison fangs, then ingested suctorially after extraoral digestion. The head also carries a pair of long, multisegmented antennae. Especially in males, the last pair of appendages is modified to a pair of long sensory and grasping structures, which are often used as weapons. Except in the *Scutigeromorpha* (see below), the spiracles of the tracheal system lie in the membranous pleural region above and just behind the coxae. There is basically one pair of spiracles per segment, but they are absent from some segments, depending on the taxon. All species are terrestrial, living under stones, bark, etc., or in the soil. A few species are found near the sea in stony coastal habitats. The great majority of species live in the tropics and subtropics. Lengths vary from 3 mm to 27 cm.

Classification.

1.Subclass: *Epimorpha*. The female broods the eggs. Young animals hatch with the same number of segments and legs as the adult.

Order: *Geophilomorpha*. The body is slender, wormlike, 9–200 mm in length, with 35–175 segments, and very short legs. Eyes are absent. Whitish to yellowish in color, they burrow wormlike through the soil, and feed almost exclusively on earthworms. When threatened, they roll up with the ventral side outward, and produce a protective secretion from their ventral glands. They are widely distributed in temperate and tropical regions throughout the world. *Geophilus longicornis* (20–40 mm) is well known in European soils (Fig.1).

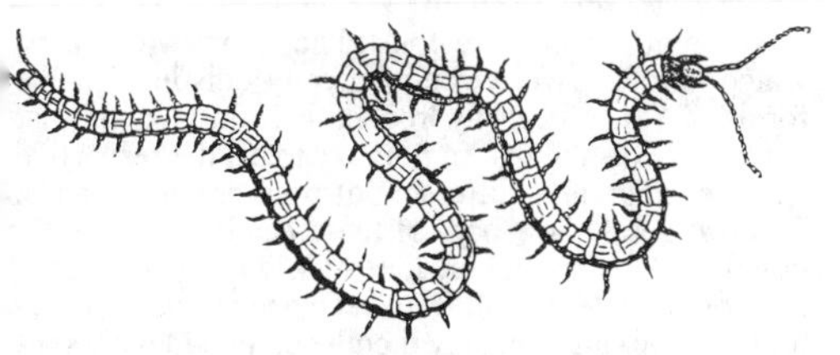

Fig.1. *Chilopoda. Geophilus* (order *Geophilomorpha).*

Order: *Scolopendromorpha.* The body (which is not wormlike) is 15–265 mm in length, with 25–27 segments and 21 or 23 pairs of legs. The predominantly tropical *Scolopendridae* hunt arthropods and small vertebrates at night. The bite of the larger species (e.g. *Scolopendra gigantea,* South America) is even dangerous to humans. The order is well represented in the tropics. Central European representatives are restricted to a few blind, earth-dwelling, 1–2 cm-long species of the family, *Cryptopidae*.

2.Subclass: *Anamorpha.* Development is hemianamorphotic. Young animals hatch with 7 pairs of legs, and attain the complete number of segments and legs after 4–5 molts. The adult has 15 pairs of legs. Eggs are not brooded.

Order: *Lithobiomorpha.* The order has a worldwide distribution, but most genera and species occur in temperate and subtropical regions. The *Lithobiidae* and *Henicopidae* are represented in Central Europe. They are fast running, dorsiventrally compressed animals, occurring mostly in the leaf litter of forest floors, where they hunt insects. The coxae of the 4 or 5 posterior pairs of legs carry conspicuous glands. *Lithobius forficatus* can attain a length of 30 mm.

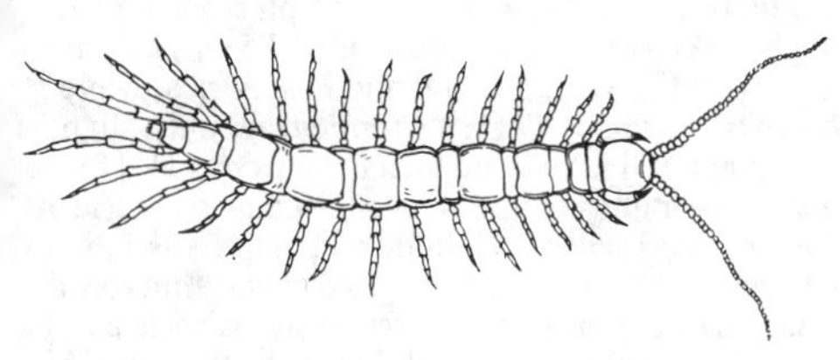

Fig.2. *Chilopoda. Lithobius* (order *Lithobiomorpha).*

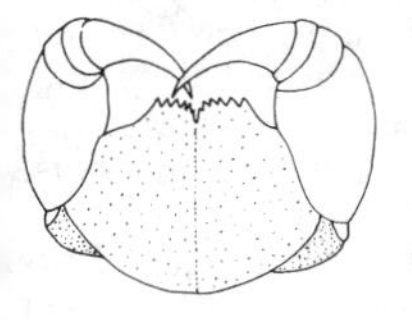

Fig.3. *Chilopoda.* Ventral view of the forcipules (poison glands) of *Lithobius forficatus.*

Order: *Scutigeromorpha.* The order is represented by a single family, the *Scutigeridae,* which has a worldwide distribution, but is represented mainly in the tropics. The 15 pairs of legs are very long (up to 30 cm) and thin. The eyes are large and compound, and the antennae are also very long. With lightning speed, they hunt on rocks and trees for flying insects, which they "lassoo" with their legs. The range of *Scutigera coleoptrata* extends from the Mediterranean region to southern Slovakia and southern Moravia. In contrast to the other orders, the spiracles of the *Scutigeromorpha* are situated middorsally, near the posterior margins of the tergal plates that cover the leg-bearing segments. Each spiracle opens into an atrium, which gives rise to two wide fans of tracheal tubes. The blood bathing these tracheal tubes contains hemocyanin, which is otherwise absent from other uniramians (myriapods and insects).

On account of their different tracheal and stigmatal system, the *Scutigeromorpha* are placed alone in the *Notostigmophora,* while all other orders are placed in the *Pleurostigmophora.*

Chimaera: see *Holocephali.*

Chimaeriformes: see *Holocephali.*

Chimera (Greek *chimaira* a mythical being, with separate parts derived from lion a goat and a dragon): an organism consisting of genetically (idotypically) different cells or tissues (see Idiotype).

1) Plant Cs. A plant C. can be made by grafting, or it may arise by spontaneous mutation.

a) A *graft C.* is produced by grafting, and it arises when callus at the graft site produces adventitious shoots containing tissues from both partners. As in all other types of C., the common observation that the characters of a graft C. represent a blending or mutual approximation of the characters of the two partners has no genetic basis. Breeding of graft Cs. is therefore impossible, and their properties can only be propagated vegetatively. Very rarely, grafting is followed by genuine cell fusion to produce a true hybrid (see Burdon); even then, its existence is disputed.

b) A *mutation C.* is the result of a gene or genome mutation. Initially, the mutation affects only one cell in the apical meristem. All cells derived by mitosis from this single mutated cell then contain the same mutation. This type of mutation gives rise to *bud mutants, shoot mutants* or *sports.* They are particularly important in agricultural species that are propagated vegetatively, e.g. fruit and potatoes; a number of commercially important varieties of apples and potatoes were discovered as sports. In sexual reproduction, the new properties are transmitted to the offspring only if the altered genotype is present in at least two outer cell layers of the meristem, because plant gametes subsequently arise from the subepidermal layer.

Gene mutations caused by spindle poisons, e.g. colchicine, may lead to the formation of *ploidy Cs.,* which posses different degrees of ploidy in their somatic cells.

Plant Cs. may also be classified according the arrangement and mixing of tissues: a) *Sectorial C.,* in which a more or less broad sector is formed by tissues of the scion or the stock; b) *Periclinal* or *coat C.,* in which the epidermis and possibly other outer cell layers originate from one partner, the inner tissues from the other partner; c) *Mericlinal C.,* in which the outer coat does not originate entirely from one partner, and does not completely surround the inner tissues.

C. are frequently formed from the graft callus between members of the *Solanaceae,* e.g. tomato and black nightshade *(Solanum nigrum).* Other known examples are *Laburnocystis adamii,* derived from *Cytisus laburnum* and *Cytisus purpureus,* as well as forms of *Crataegomespilus* which are Cs. of medlar *(Mespilus germanica)* and hawthorn *(Crataegus).*

2) Animal Cs. Animal Cs. (sometimes also plant Cs.) are also called ***genetic mosaics.*** Animal Cs. can be formed experimentally by the fusion of two early embryos. They can also arise by somatic segregation. The exchange of blood cells in the placentas of twin cattle can produce natural Cs. or mosaics with two blood types. Somatic disjunction in humans can produce an individual whose cells contain different numbers of X and Y chromosomes. Every female mammal can be considered as a mosaic with respect to the functional state of her two X chromosomes. Humans containing transplanted organs might also be thought of as Cs. or mosaics, and certainly an irradiated recipient of reconstituted bone marrow may be defined as a C. or mosaic.

Somatic segregation due to *mitotic crossing over* was discovered in 1936 by Curt Stern in studies on sex-linked genes of *Drosophila*. In the cross: +sn/+sn × y+/Y, the female progeny displayed, as expected, a predominantly wild phenotype [y is the sex-linked gene for yellow tissue; sn is the sex-linked gene for short, curly (singed) bristles]. Some individual females, however, possessed adjacent sectors of yellow tissue and singed bristles. These sectors are reciprocal products of the same event, and they arise by crossing over between the sn/+ locus and the centromere during mitotic division. Two such adjacent sectors of mutant phenotypes in a background of wild-type tissue are known as *twin spots*. Segregation of somatic tissue in *Drosophila* also occurs as a result of *mitotic chromosome loss*, e.g. females of genotype M/+ may possess a patch or *sector* of wild-type bristles on a body of otherwise M-phenotype (M is a dominant allele for slender bristles); mitotic chromosome loss is due to *somatic* or *mitotic disjunction*, which causes the dominant allele to be left behind when the daughter nuclei reform after mitosis; thus, in the present case, disjunction would give rise to two types of somatic cell: M/M/+ (phenotype M) and +/0 (wild phenotype). Spontaneously formed genetic mosaics are rare in mammals, and are usually due to somatic mutation.

Experimental fusion of early embryonic tissue. Early experiments on embryo fusion were performed by Tarkowski (Poland 1961) and Mintz (Philadephia 1962). Morula stage embryos were isolated from two inbred strains of mice (A and B), which were easily differentiated by their coat color and electrophoretically identifiable allozymes. The embryos were treated with pronase to remove the outer membrane (zona pellicula), then brought into contact with one another. The resulting cell aggregate was cultured for one day in vitro, then implanted into the uterus of a pseudopregnant mouse (a female mouse with a uterus receptive to embryo implantation, following mating with a sterile male). Such embryos developed into mice consisting of cells of different genetic origin (A and B), and displaying characters of both phenotypes. In analogy with the term allotype, this phenomenon is called *allopheny*, because different phenotypes are combined in a single individual. Similar Cs. have been constructed experimentally between different respective strains of rabbits, rats, and sheep. Animals constructed in this way are known as *tetraparental* animals, since they are derived from four genetic parents.

Experimentally, tetraparental animals are useful for following the destinies of cells throughout embryonic development. To give just one example, cells of striated muscle contain multiple nuclei, and the question arises of whether these arise by the fusion of mononucleate myoblasts, or whether they are formed by nuclear division within myoblasts without division of cytoplasm. Tetraparental mice were constructed from two strains, which produce electrophoretically distinguishable forms of isocitrate dehydrogenase-1, an enzyme that contains two subunits. In homozygotes, the respective enzymes migrate to different positions as single bands. Heterozygotes show both of these bands, and in addition they show an intermediate hybrid band containing one polypeptide subunit from each phenotype. This hybrid enzyme can form only if both polypeptides are synthesized in the same cell. Muscle cells of the tetraparental mice were found to contain the hybrid enzyme, showing that cells of striated muscle arise by the fusion of different myoblasts.

In a different experimental procedure, cells (or a single cell) of one genotype are injected into the blastocyst embryo of another genotype. The injected cells are incorporated into the inner cell mass of the host blastocyst, which eventually develops into a chimeric animal. Experiments of this kind reveal that the cells of the early mammalian embryo (up to the 8-cell stage) initially have an identical developmental potential, i.e. they are totipotent. Important results have been obtained by injecting teratocarcinoma cells into the isolated mouse blastula (teratocarcinoma is a type of malignant tumor of the gonads). The tumor cells become fully integrated and, moreover, they become normal, i.e. the resulting chimeric mouse is normal, has no defects (no teratocarcinoma), and is capable of normal reproduction, even when a clone of tumor-derived cells is included in the gonadal region. This shows that teratocarcinoma cells do not have an altered genome, but have become neoplastic through false programming.

Chimpanzees, *Pan* (plate 41): lively and especially intelligent Primates (see), inhabiting the African rainforest. Two species are recognized: *P. troglodytes* (***Common chimpanzee***; male head plus body 85 cm, weight 40 kg; female head plus body 77.5 cm, weight 30 kg) is found in West, Central and East Africa, while *P. paniscus* (***Bonobo***, ***Pygmy chimpanzee***; only slightly smaller but of lighter build than *P. troglodytes*) is found only in the rainforest between the Zaire river and its tributary, the Lualaba. C. build sleeping nests in trees and are partly arboreal, but spend much time on the ground. Groups of *P. troglodytes* have also been observed in grassland areas of Senegal with very few trees. C. eat mainly fruits with some nuts and leaves. They are also noted for inserting sticks into termite mounds to withdraw termites which they eat ("termite fishing"). Monkeys and young baboons are also sometimes eaten. *P. troglodytes* has been observed to form cooperative hunting groups for the capture of monkeys.

Chinchilla: see *Chinchillidae*.

Chinchillidae, ***chinchillas and viscachas (vizcachas)***: a family of Rodents (see) comprising 6 species in 3 genera. With the exception of the Viscacha, they inhabit the highlands of Argentina, Bolivia, Chile and Peru. They are fairly large rodents, with a compact body, large head and blunt, wide snout. The forelegs are relatively short and the forefeet are small with 4 digits. In contrast, the hindlegs are long and muscular, and the hindfeet are long with either 3 or 4 digits. Dental formula: 1/1, 0/0, 1/1, 3/3 = 20.

The ***Viscacha*** *(Lagostomus maximus)* is probably the only species of its genus; a second species, known only from its skeleton, is assumed to be extinct. The head of the Viscacha is especially large and bulky. Fully grown animals weigh about 7 kg, with a body length of 47–66 cm and a tail length of 15–20 cm. Viscacha burrows often attract "sub-tenants", which inhabit the same burrow, or excavate their own burrows alongside, e.g. the burrowing owl *(Speotyto)* and various snakes and lizards. Vicachas live in colonies in the plains of Argentina (pampas), where they display an unexplained habit of collecting unlikely objects, which they build into mounds outside their burrows, e.g. horns, bones, plant stalks and even human artifacts like pieces of metal.

Mountain chinchillas *(Lagidium)* are smaller and more delicate than the viscacha. Three species are known, e.g. the ***Common mountain chinchilla*** *(L. viscacia).*

Chinchillas *(Chinchilla),* originally native to the Andes, are now extinct in many areas. On account of their unusually thick, soft, silver-gray fur, they have been relentlessly trapped, until their continued existence in the wild is in jeopardy. The ***Short-tailed chinchilla*** *(Ch. chinchilla)* is largely extinct in the wild and is also rare in captivity. The ***Long-tailed chinchilla*** or ***Chinchilla*** *(Ch. laniger)* is also largely unknown in the wild, but is quite common in captivity.

Chinemys: see *Emydidae.*

Chinese alligator, *Alligator sinensis:* the only Asian representative of the *Alligatoridae* (see). It inhabits the lower reaches of the Yangtse River and its tributaries in China. Only 1.5 m in length, it digs itself into the river bank in the cold season. It is not commonly kept in zoos, and little is known about its mode of existence. Only a few hundred individuals exist, and the species is under severe threat of extinction.

Chinese crab: see Chinese mitten crab.

Chinese galls: see Tannins.

Chinese giant fire-bellied toad: see *Discoglossidae.*

Chinese giant salamander: see *Cryptobranchidae.*

Chinese lantern: see *Solanaceae.*

Chinese mitten crab, ***Chinese crab, mitten crab,*** *Eriocheir sinensis:* a true crab (see *Brachyura),* in which the chelae of the male are covered with woolly hairs. It originated in China, but in 1912 was transported in ships to river estuaries (notably the Weser and the Elbe) on the northern coast of Europe. It has since extended its range and is now found widely in rivers and canals of Scandanavia and Europe, penetrating as far inland as Dresden and Prague, and becoming a pest in some freshwater fisheries. For reproduction, however, the crabs must return to the sea.

Chinese mitten crab (Eriocheir sinensis)

Chinese mountain viper: see Lance-head snakes.

Chinese nest frog: see *Raninae.*

Chinese water deer, *Hydropotes inermis:* a monotypic genus of small deer (head plus body 77–100 cm, shoulder height 45–55 cm), which lack antlers. Both sexes have elongated upper canines, which are somewhat shorter in the female. Its natural range is from the Yangtze Basin of east-central China to Korea. It inhabits tall reeds and rushes near rivers, and is also found among tall grass in hilly and mountainous areas. Back and sides of the body are uniform yellow-brown with fine black stippling. The underparts are white, and the face is gray-brown with a white chin and throat. *Hydropotes* is the only member of the *Cervidae* that possesses inguinal glands (one on each side in both sexes). It is bred extensively in captivity in Europe, and escaped animals have given rise to wild European populations. See Deer.

Chionididae, ***sheathbills:*** a small avian family (2 species) in the order *Charadriiformes* (see). These are rather stout birds, about 40.5 cm in length, with short necks, bills and legs. They have vestigial webs between the front toes, and a well developed hallux. Phylogenetically they are thought to represent a link between waders and gulls. As in some plovers, there is rudimentary spur at the joint of the wing. Plumage is entirely white, and the face is wattled. The name, Sheathbill, refers to the horny sheath, which extends over the base of the powerful bill and partly shields the round nostrils. Pugnacious little scavengers, they steal eggs and even chicks from sea bird colonies. They scavenge the shore line and the garbage tips of whaling stations and Antarctic survey bases, and also feed on the afterbirths and stillborn pups of seals. Although largely terrestrial in habit, they are able to fly strongly and swim well. *Chionis alba* ***(Snowy sheathbill)*** has a yellow or pinkish bill and ranges from the southern tip of South America to the Falkland islands and the Palmer Peninsula of Antarctica. *Chionis minor* ***(Black-faced sheathbill)*** has a black bill and inhabits the Crozets, Kerguelen and other neighboring island groups.

Chionis: see *Chionididae.*

Chipmunk: see *Sciuridae.*

Chironex: see *Cubomedusae.*

Chiropotes: see Bearded sakis.

Chiropsalmus: see *Cubomedusae.*

Chiroptera, ***bats, flying foxes*** and ***fruit bats:*** an order of mammals, whose members are adapted for flight. The elongated fingers of the forelimb act as supports for two thin, elastic membranes that extend from the sides of the body and legs, and in some species also include the tail. The membrane consists of two layers of skin, separated only by a thin layer of connective tissue, housing blood vessels and nerves. In most species, a cartilaginous spur (calcar or calcaneum) on the inner side of the ankle joint helps to spread the tail membrane. The third finger is very elongated, and usually equal in length to the head, body and legs. The legs are rotated to support the wing membrane, so that the knee is directed outward and backward. As an aid to clinging to supports, the short thumb (which does not help to extend the wing membranes) carries a hooked claw (except in the *Furipteridae). C.* sleep during the day, hanging head-down in trees or caves. During their darting flight at twilight and at night, the *Microchiroptera* (see below) navigate mainly by emitting and detecting the echos of ultrasonic pulses (30–70 kHz), and are capable of detecting the smallest and finest obstacles. Species native to colder regions hibernate during the winter in a frost-free location, often in large communities. The usually single infant is carried by the mother during flight.

C. are subdivided into *Megachiroptera* (Old world fruit bats and flying foxes) and *Microchiroptera.*

Megachiroptera (about 130 species in 39 genera) have a dog-like face, with large eyes that provide efficient vision in twilight; unlike the *Microchiroptera,* they do not have an ultrasonic apparatus. A tragus (see below) is absent. The nose is not compressed, and it carries no extensions or appendages, and the tongue is covered with bristle-like papillae. The tail is short, vestigial or absent, except the tail of *Notopteris,* which has 10 vertebrae. In addition to the thumb, the second finger also carries a claw. They feed on pollen, nectar and fruits, and some species are flower pollinators. *Megachiroptera* are found throughout the tropics and subtropics of Africa, Asia, Australia and Oceana. They are represented by a single family *(Pteropodidae),* which contains the largest living bats. The largest fruit-eating species weigh more than 900 g, whereas some adult nectar- and pollen-feeding species attain only 15 g and are smaller than some species of *Microchiroptera.* The dental formula varies from 2/2, 1/1, 3/3, 2/3 = 34 *(Pteropus* and *Rousettus)* to 1/0, 1/1, 3/3, 1/2 = 24 *(Nyctimene* and *Paranyctimene).*

Microchiroptera (bats in the narrow sense; about 140 genera in 17 families) have small eyes and a compressed nose, the latter often provided with curious appendages and extensions that serve as directional antennae for the emitted ultrasound. They have large ears, each with a tragus (a lobe in front of the ear orifice). Most species are insect feeders, and only a few feed on fruit or blood (vampire bats). The following families of the *Microchiroptera* are listed separately: Horseshoe bats *(Rhinolophidae* and *Hipposideridae),* American leaf-nosed bats *(Phyllostomatidae),* Vampire bats *(Desmodontidae),* Bulldog bats *(Mollossidae),* and Vespertilionids *(Vespertilionidae).*

Chiropterogamy: see Pollination.

Chiropterophily: see Pollination.

Chiroteuthis: see *Cephalopoda.*

Chirotherium (Greek *cheir* hand, *therion* animal), "***hand animal***": a large extinct genus of reptiles, originally known only from tracks of footprints in the Germanic Triassic, especially the variegated sandstone (Bundsandstein) of Thüringia. The footprints bear a striking resemblance to a human hand, the mark of the "thumb" corresponding to the 5th toe. Similar prints have been found in South America and elsewhere, and are now thought to belong to the pseudosuchian *Ticinosuchus.*

Chiru: see *Saiginae.*

Chisel teeth lizards: see *Agamidae.*

Chital, ***axis deer, spotted deer,*** *Axis axis:* a deer that inhabits forest edges and woodlands in India and Sri Lanka, and has also been introduced into Australia. Its coat is rufous fawn in color, with prominent white spots. The average shoulder height is about 91 cm, and the weight 86 kg. The beam of the antlers curves back and outward in a lyre shape.

Chitin: a nitrogen-containing polysaccharide consisting of straight chains of β-1-4-linked *N*-acetylglucosamine residues. It is a major component of the exoskeletons of invertebrates especially arthropods, and it is also found in the cell walls of diatoms, fungi and higher plants. It usually occurs together with other polysaccharides and proteins, and may be impregnated with calcium salts.

Chitin

Chitinases: glycosidase enzymes that aid the breakdown of chitin by catalysing its hydrolysis to *N*-acetylglucosamine. C. are present in the snail stomach, and are also produced by some molds and bacteria.

Chitobiose: a nitrogen-containing disaccharide, consisting of two β-1-4-glycosidically linked *N*-acetylglucosamine molecules. C. is a structural unit of Chitin (see) and can be isolated from hydrolysates of the latter.

Chitons: see *Polyplacophora.*

Chlamydera: see *Ptilinorhynchidae.*

Chlamydobacteriaceae, ***chlamydobacteria, sheath-forming bacteria:*** rod-shaped, chain-forming, aquatic bacteria, which live inside a thread-like, cylindrical, mucilaginous membrane. Within the cylindrical thread, the rod-shaped bacteria are aligned end-to-end. Individual cells are pushed from the end of the thread by the dividing cells below, or "creep" from the end without apparent pressure from below.

Leptothrix ochracea is a familiar, worldwide distributed species, occurring in natural waters that contain iron combined with organic material (e.g. decomposing plants). Compounds of iron or manganese are present in the sheath, and old filaments carry a thick incrustation of ferric hydroxide. The yellow to red-colored slime on the beds of iron-containing waters is usually due to ferric hydroxide-impregnated and -incrusted filaments of *Leptothrix.* Individual cells released from the sheath may form a new filament. Fragmentation of filaments may also occur. *Leptothrix* also produces protuberances or "swarmers" on the outer surface of the filament; these elongate, break away, and form a new filament. Incrustation of old filaments with ferric hydroxide prevents the formation of swarmers.

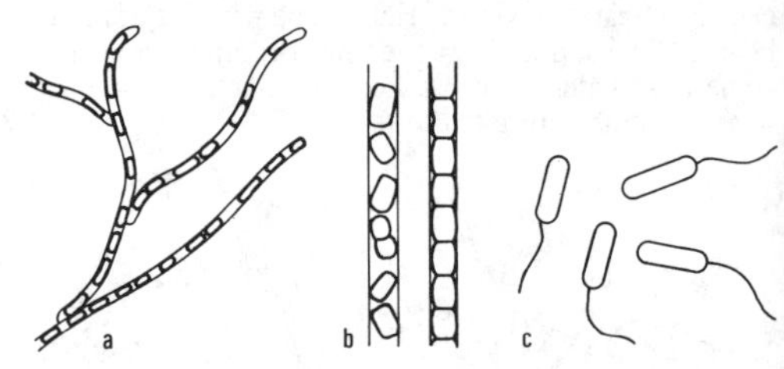

Chlamydobacteriaceae. Sphaerotilus natans. a Threads of ensheathed bacterial cells. *b* Higher magnification of ensheathed cells. *c* Freely motile cells.

Sphaerotilus differs from other *C.* by the presence in its cells of oil droplets and glycogen granules. The sheath of *Sphaerotilus* also contains only minimal quantities of iron salts. *Sphaerotilus natans* often occurs in activated sludge, and it may develop sufficiently to clog the tanks, causing a condition known as "bulking".

The sheaths of *Crenothrix* are hard and impregnated with ferric hydroxide. Topmost cells are continuously pushed from the mouth of the sheath (which is expanded like a trumpet) by the dividing cells below. *Crenothrix polyspora* can live inside water supply pipes, provided the oxygen concentration is low and iron is present. It forms gelatinous masses that clog the pipes. See Iron bacteria.

Chlamydosaurus: see Frilled lizard.

Chlamydoselachidae: see *Selachimorpha.*

Chlamydospermae: see *Gnetatae.*

Chlamydospores: asexual fungal spores formed from a hypha by thickening of the cell wall and separation into individual cells. C. are formed when nutrients are in short supply. Under favorable conditions they germinate by forming a single germination tube. Functionally, individual C. are identical to Gemmae (see), but the latter are multicellular structures. Many fungal species produce C., e.g. the brand spores of the *Ustilaginaceae* (see *Basidiomycetes)* are C.

Chloramphenicol: a broad spectrum antibiotic synthesized by *Streptomyces venezuelae.* It is active against Gram-negative and Gram-positive bacteria, rickettsias and large viruses. It is unusual, in that nitro-groups and dichloromethyl-groups are relatively uncommon in natural products. C. is prepared synthetically, and it is used successfully against typhus, pneumonia and whooping cough. It is also used to suppress the growth of unwanted bacteria in industrial fermentations, e.g. in the yeast fermentation of molasses.

$O_2N-C_6H_4-CH(OH)-CH(NH-CO-CHCl_2)-CH_2OH$

Chloramphenicol

Chlorine, ***Cl:*** a halogen. In the form of the chloride anion (Cl^-) it is an essential micronutrient for plants, and it occurs in relatively high concentrations in plant material, e.g. 20 mg/kg dry mass of cereals. The chloride ion can be taken up via the leaves as well as the roots. In hydroponic culture, the rate of uptake is strongly influenced by the chloride concentration of the medium, and it may greatly exceed the physiological requirement. The metabolic role of chloride is incompletely understood. Through nonspecific physical-chemical interactions with proteins, it exerts an influence on various enzyme reactions, oxygen uptake, cell water content and growth. Its involvement in oxygen evolution in photosynthesis is probably more specific. Plant species differ markedly in their chloride requirement and tolerance. Species requiring and tolerating high chloride levels (chlorophilic species) are e.g. beet, radish, spinach and celery. Chloride-sensitive (chlorophobic) species are e.g. potatoes, tomatoes, beans, gherkins, grape vine, fruit bushes and fruit trees.

Chlormadinone acetate: a synthetic progestin used in oral contaceptives. See Hormonal contraception.

Chlorobacteriaceae, ***green bacteria, chlorobacteria:*** a group of mostly nonmotile, anaerobic, photosynthetic bacteria. They are colored green to brown by the presence of bacteriochlorophylls and carotenoids. *C.* occur in oxygen-poor, sulfur-containing water, where they assimilate carbon dioxide with the aid of light energy. In contrast to the cyanobacteria and green plants, which obtain reducing power for carbon dioxide assimilation by the photolysis of water, the C. derive the necessary reducing equivalents by the photolysis of hydrogen sulfide, and to some extent from molecular hydrogen. The photosynthetic pigments are contained in Chromatophores (see).

Chlorocichla: see *Pycnonotidae.*

Chlorocruorin: see Blood.

Chlorogenic acid: a depside of caffeic acid and quinic acid, which is found in higher plants, and is particularly abundant in coffee beans. It forms a complex with iron, which is responsible for the darkening of cut potatoes.

$(HO)_2C_6H_3-CH=CH-CO-O-C_6H_7(OH)_3(COOH)$

Caffeic acid — Quinic acid

Chlorogenic acid

Chlorogenic acid

Chlorogogenous cells (Greek *chloros* yellow-green *agein* to lead or conduct): brown, yellow or greenish storage and excretory cells from the intestine of polychaete worms. Thus, in the earthworm, where the coelomic epithelium covers the intestine (the splanchnic layer) it gives rise to masses of large yellow C.c., forming a layer that covers the intestine and its blood vessels.

Chlorhemin: see Hemoglobin.

Chlorolab: see Color vision.

Chlorophyceae, ***green algae:*** a large, phylogenetically important class of algae, containing about 10 000 species. The great majority (90%) are freshwater species, living in the plankton or benthos. A few larger marine species are found in coastal waters. Other species are found on rocks, tree bark and in moist soils. Some live symbiotically with other plants (as lichens) or inside the cells of primitive animals. *C.* include free living, unicellular forms and colony-forming unicells, as well as multicellular species that form filaments or leaf-like thalli. Often the cells of multicellular forms are not all identical, but the thallus lacks the high level of differentiation displayed by vascular plants. All species are photosynthetic, and their content of photosynthetic and accessory pigments is qualitatively and quantitatively similar to that of higher green plants (chlorophylls *a* and *b,* carotenes, lutein and other xanthophylls), which is regarded as evidence that higher green plants and *C.* share a common evolutionary origin. Their assimilation product is starch, and their cell wall material, like that of higher plants, is mainly cellulose. On account of their photosynthetic pigments, practically all species are green, but the green color may be masked by hematochrome, which gives a red coloration, e.g. *Haematococcus pluvialis,* which imparts a red color to small patches of standing water, and *Chlamydomonas nivalis,* which is responsible for the "red snow" of Alpine and Arctic regions.

Asexual reproduction occurs by the formation of motile zoospores or nonmotile aplanospores, or by fragmentation of the vegetative organism. Sexual reproduction occurs by iso-, aniso-, or oogamy. With few exceptions, the gametes and zoospores of all members of the

C. bear a close resemblance to the vegetative cells of *Chlamydomonas.* This latter organism is thought to be representative of the primitive stock from which all other groups of *C.* have evolved. As a rule, both zoospores and gametes are pear-shaped with 2 or 4 equal-length, whiplash flagella at the pointed anterior end, and at the posterior end a curved or cup-shaped parietal chloroplast. The flagella are typical eukaryotic flagella, with an internal 9 + 2 arrangement (see Cilia and Flagella), and they lack the lateral, hairlike appendages (flimmer hairs or mastigonemes) found on some other types of flagella. Many *C.* undergo an alternation of generations, in which the gametophyte and sporophyte may have a similar or different overall shape and appearance.

Under adverse environmental conditions, resting cysts (cystozygotes) may be formed.

Fossil filamentous and coccoid organisms, assumed to be green algae, are known from the Precambrian Bitter Springs deposits in Australia (about 10^9 years old). The calcified thalli of some marine *Siphonales* (especially *Dasycladaceae)* have been identified in Cambrian deposits. *Rhabdoporella,* a fossilized, unbranched green alga from the Ordovician, is thought to be close to the ancestral prototype of the *Dasycladales.*

Classification. Some authorities place all the green algae in a division known as the *Chlorophyta,* which consists of 2 classes: *Chlorophyceae* ***(True green algae)*** and *Charophyceae* ***(Stoneworts).*** This separation of the *Charophyceae* is justified by their distinct morphology and reproductive process.

Modern ultrastructural studies have revealed the existence of the following 4 lines of evolution from the primitive unicellular condition.

1. *Chlorophyceae:* primarily freshwater forms related to *Chlamydomonas* and *Volvox.*
2. *Ulvophyceae:* mostly marine species found in the tropics and subtropics, including large forms, but lacking plasmodesmata, e.g. *Ulva, Codium, Valonia, Halimeda.*
3. *Charophyceae:* common freshwater forms like *Spirogyra,* ornate desmids and charophytes; probably the ancestors of green land plants.
4. *Pleurastrophyceae:* a small, recently recognized class containing the flagellate *Tetraselmis (= Platymonas),* as well as the lichen symbionts, *Trebouxia* and *Pseudotrebouxia.*

The following detailed and earlier classification is due to Strasburger (1978). It does not place the Stoneworts in a separate class, and all green algae, according to their level of organization, are assigned to one of 9 orders in the single class, *Chlorophyceae.*

1. Order: *Volvocales.* These occur widely as freshwater plankton, sometimes in such abundance that they turn the water green. There are no marine species. Cells are rounded and radially symmetrical, and in the vegetative state they always have flagella. A starch-producing pyrenoid is generally present at the base of the cell, together with a red eye spot, and the chloroplast is usually cup- or pot-shaped. Some live as individual unicells (e.g. *Chlamydomonas)* and others form colonies, usually with a constant number of unicells (e.g. *Oltmannsiella* colonies contain 4 unicells, *Gonium* 4 to 16, *Pandorina* 16, and *Eudorina* 32).

Members of the primitive, small family, *Polyblepharidaceae,* lack a cellulose wall, and possess 2–8 flagella. Sexual reproduction has not been observed in *Polyblepharides,* and asexual reproduction in this genus consists of longitudinal division into two. More advanced forms of this family display sexual reproduction, e.g. *Dunaliella salina,* which is found in strong brine in salt works, and is colored red due to the presence of hematochrome.

Chlamydomonas is a widely distributed and familiar genus of the family *Chlamydomonadaceae,* all members of which possess a cellulose wall. Some species of *Chlamydomonas* display the simplest mode of sexual reproduction found among the *C.,* in which ordinary, flagellated, vegetative cells function as gametes; they fuse to form a zygote, which then undergoes meiosis to produce 4 spores, which develop directly into vegetative, unicellular individuals. In other species of *Chlamydomonas,* vegetative cells undergo one or more mitotic divisions to form 2,4,8,16 or 32 isogametes, each one similar to the mother cell, but smaller and naked (i.e. no cellulose wall). According to the species, there may be larger gametes (female) and smaller gametes (male) (i.e. a condition of heterogamy), or all gametes may be similar in size. After swimming freely for a time, the gametes conjugate in pairs at their apices. The resulting zygote initially has 4 flagella. It swims briefly, then settles, loses its flagella, and forms a zygospore. Asexual reproduction is by zoospores; between 2, 4, 8 or 16 zoospores are produced by repeated longitudinal division of the contents of the mother cell (sporangium).

Under certain conditions, *Chlamydomonmas* reverts to a nonmotile form, loses its flagella, and becomes embedded in a gelatinous matrix. It then divides vegetatively to form a dense, amorphous mass, known as the

Chlorophyceae. 1 Chlamydomonas angulosa (x 1100): *c* chloroplast; *g* flagellum; *k* nucleus; *p* pyrenoid; *s* eye spot; *v* contractile vacuole. *2* Colony of *Pandorina morum* (x 160). *3 Volvox:* individual with 6 daughters (x 50). *4.1* Disk-shaped association of *Pediastrum granulatum* (x 50). *4.2* Vegetative cell of *Chlorella vulgaris* (x 500). *4.3* A 4-celled aggregate of *Scendesmus acutus* (x 1 000). *5 Closterium moniliferum* (x 400). *6 Micrasterias rotata* (x 100). *7* Single cell in the vegetative strand of *Spirogyra jugalis* (x 150). *8.1 Ulva lactuca,* showing colorless marginal cells after the liberation of zoospores (x 0.5). *8.2 Ulothrix zonata,* showing young filament with rhizoid cell, *(r)* (x 300). *9.1* Fertilization of *Oedogonium ciliatum: z* dwarf male; *o* oogonium (x 300). *9.2* Zoospore of *Oedogonium concatenatum. 10 Fritschiella tuberosa: b* soil surface; *pa* erect, branching, aerial filament; *pr* creeping rows of cells buried in the soil; *r* rhizoid; *sf* secondary aerial filament. *11 Cladophora* (habit) (x 0.3). *12.1* Side view of *Chara fragilis: a* antheridium; *o* oogonium with its enveloping filaments and corona *(c)* (x 50). *12.2* Spermatozoid of *Chara fragilis: g* flagellum; *k* spirally twisted, elongate nucleus; *p* cytoplasm with reserve starch (x 540). *12.3 Nitella flexilis:* manubrium with head cell and spermatogenous filaments. *12.4 Nitella flexilis:* cells of the spermatogenous filaments, each with one spermatozoid. *12.5 Nitella flexilis:* longitudinal section through a young antheridium; *k* head cell; *m* manubrium; *w* wall. *13 Acetabularia mediterranea:* mature thallus (natural size).

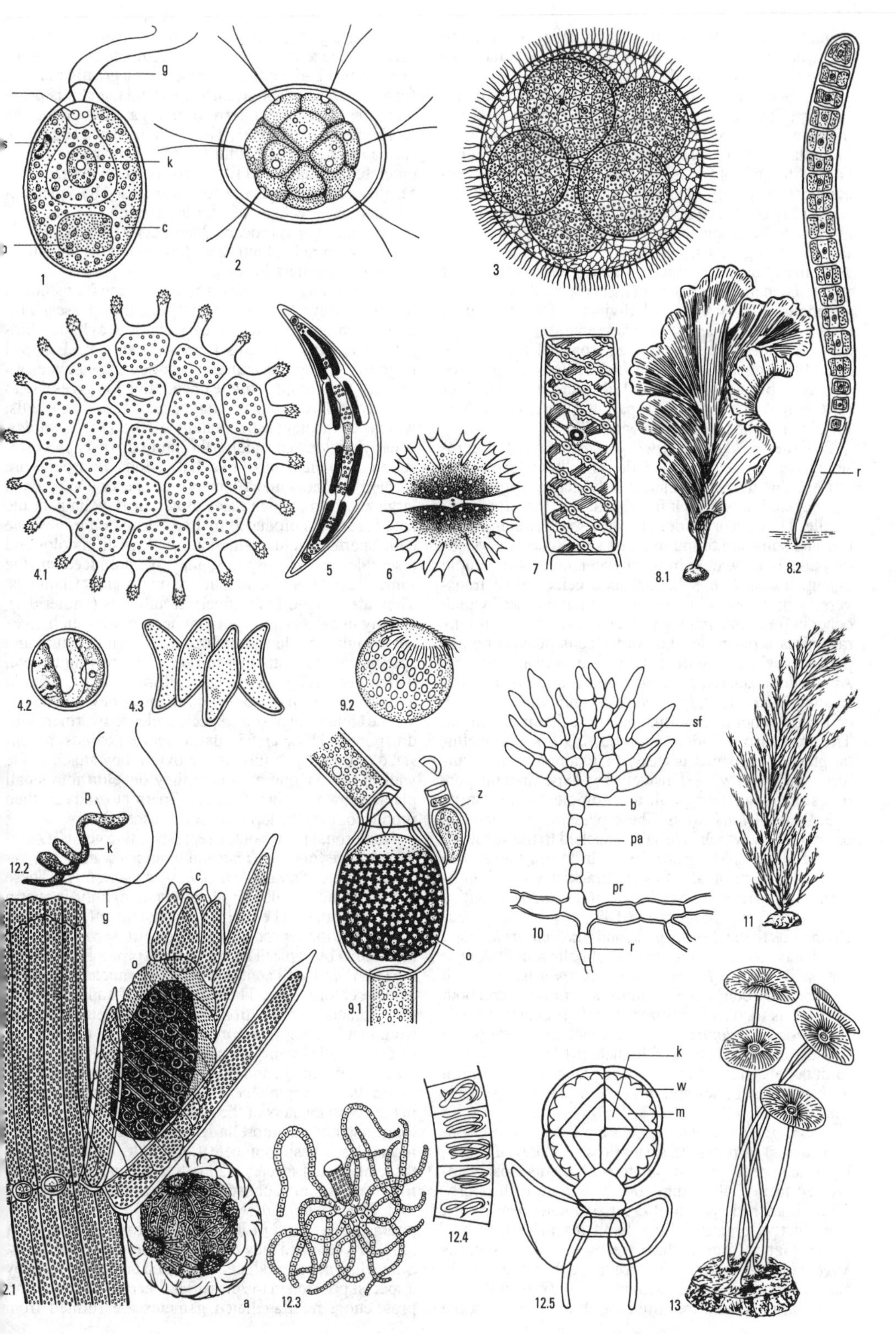
g
s
k
c
1
2
3
4.1
5
6
7
8.1
r
8.2
4.2
4.3
9.2
sf
b
pa
pr
r
10
11
p
k
12.2
g
c
o
z
o
9.1
k
w
m
2.1
a
12.3
12.4
12.5
13

palmelloid state. Within this aggregation, there is no particular pattern or arrangement of cells, and the cells are never interconnected.

Members of the family *Volvocaceae* are colony-formers. The *Volvox* colony is a hollow sphere, consisting of up to 20 000 cells, all interconnected by broad protoplasmic processes. There is also a division of labor among the cells of the sphere. Most of the cells are vegetative (i.e. purely photosynthetic and locomotory), while only certain cells on the hinder part of the sphere are able to form spermatozoids and ova. Sexual reproduction is oogamous, and between 6 and 8 green ova arise singly by enlargement of a vegetative cell. Other vegetative cells give rise to numerous, small, yellowish spermatozoids by repeated division. The vegetative cells display a tendency to differentiation, e.g. the eye spot gradually decreases in size from the front to the rear of the sphere. Various species of *Volvox* reproduce asexually by invagination and abstriction of daughter spheres into the center of the parent sphere, where they remain until the parent disintegrates.

2. Order: *Chlorococcales (Protococcales).* Representatives of this order include unicellular forms, and unicells that form characteristically shaped aggregates. The vegetative cells lack flagella and are therefore nonmotile. They occur predominantly as freshwater plankton, and some are found in moist soil. Species of *Scendesmus,* found widely in fresh water, consist of simple aggregates of 4 to 8 boat-shaped cells united transversely in a row, whereas *Pediastrum* species, which occur in fresh and brackish water, consist of 16 cells arranged as a flat plate, the border cells possessing appendages which give the colony a star-like appearance. Another characteristic member of this order is the free-floating ***Water net*** *(Hydrodictyon reticulatum),* whose cells form a hollow, sac-like net up to 50 cm in length. The order also includes *Chlorella vulgaris,* a unicellular green alga, which is easily maintained in pure culture, and is grown extensively for experimental purposes and on a factory scale as a commercial feedstuff.

3. Order: *Ulotrichales.* These possess a filamentous or flat thallus, which is usually attached to the substrate by rhizoid cells. Many species occur in coastal sea waters, while some inhabit fresh or brackish water. Familiar marine species are ***Sea lettuce*** *(Ulva lactuca)* with a thallus consisting of a large leaf-like expanse of parenchymatous tissue 2 cells thick, and *Enteromorpha* spp., which have a filamentous thallus of hollow cylinders or flat bands; both *Ulva* and *Enteromorpha* have a high content of carbohydrate, vitamins and protein, and both are used as food and fertilizer in various coastal areas.

4. Order: *Cladophorales.* The vegetative body of the *Cladophorales* consists of branched filaments, which form tassels and feel rough to the touch. They occur in the benthos of the sea and fresh water. The type genus is *Cladophora.*

5. Order: *Chaetophorales.* As a rule these consist of two parts: 1) a prostrate system (basal plate) of creeping, often pseudoparenchymatous filaments closely appressed to the substratum, and 2) a more or less profusely branched system of erect filaments bearing the reproductive organs, with a habit resembling that of higher plants. They include freshwater algae such as *Stigeoclonium* and *Coleochaete.* Others are fully adapted to a terrestrial existence, e.g. *Trentepohlia* is found on rocks and tree trunks in Europe, and on the leaves of higher plants in the tropics, and is often colored red due to the presence of hematochrome; *Pleurococcus* (tree bark and rock surfaces) is a greatly reduced form which does not produce sporangia, and it is probably the commonest terrestrial alga throughout the world; *Fritschiella,* which is found in Africa and India, forms rows of cells within the soil, which give rise to erect, branching, aerial filaments. In *Coleochaete,* after fertilization, the zygote becomes enclosed by a plectenchymatous envelope of sterile filaments, forming a fructification or sporocarp. Members of the *Chaetophorales* are pure haplonts, and they do not display an alternation of generations.

6. Order: *Oedogoniales.* The cells of the filamentous, unbranched thalli are uninucleate and they possess a reticulate chloroplast shaped like a hollow cylinder. Single cells of the filaments form swollen, barrel-shaped oogonia, giving the thallus a knotted appearance. An unusual and complex mode of cell division gives rise to cells with slightly projecting caps. These cap cells, which are a diagnostic feature of the *Oedogoniales,* usually divide repeatedly, the number of caps corresponding to the number of cell divisions. Zoospores are formed on the same filaments as the oogonia. Each large zoospore, which has a crown of flagella near the anterior end, is produced by transformation of a whole cell. Spermatozoids, which are yellowish in color and resemble small zoospores, may be produced on the same filaments as the oogonia, or on different filaments. They are formed in relatively small cells (antheridia), usually in pairs. The oogonium has a pore, which gives entry to the motile spermatozoid. Fusion of ovum and spermatozoid results in a thick-walled, resting zygote, which eventually germinates to form 4 large, haploid meiospores, each one developing into a new filament.

The antheridia of some species release swarmers (androspores). These are similar to spermatozoids, but instead of directly fertilizing the ovum they attach to the wall of the oogonium, where they develop into small plants known as "dwarf males". Spermatozoids are then produced from the terminal cells of the latter.

Numerous species of the representative genus, *Oedogonium,* are found in fresh and brackish water.

7. Order: *Siphonales (Chlorosiphonales).* These siphonaceous or tubular algae occur mainly in warm marine habitats. The thallus is nonseptate (lacks cross walls), but the reproductive organs are separated from the thallus by septa. The continuous protoplasm within the single cell wall contains numerous nuclei and small discoid chloroplasts. The proportion of xanthophyll in the pigment composition of the chloroplasts differs from that in other *Chlorophyceae.* Despite the absence of cross walls, many members of the *Siphonales* form macroscopic thalli, and they display a wide diversity of forms. For example, *Acetabularia mediterranea,* found in the coastal waters of the Mediterranean, has a finger-long stalk and an umbrella-like cap, and it superficially resembles a basidiomycete fungus. Its cell walls and those of related species readily become calcified. Fossil forms are common constituents of Alpine limestones.

8. Order: *Conjugales (Conjugatae).* This is a cosmopolitan order. Most of its very numerous species occur in fresh water, and they display a wide variety of forms. Since they never produce flagellated cells, they occupy a special position in the classification of *C.* In sexual reproduction, nonflagellated gametes are formed from

vegetative cells. Two gametes fuse to form a zygote, which after a prolonged period of dormancy, germinates and undergoes meiosis. All members of the order *Conjugales* are therefore haploid organisms. Asexual reproduction occurs by cell division. The family *Desmidiaceae* ***(Placoderm desmids),*** particularly common in the acidic waters of peat bogs, consists of unicellular organisms, which are notable for their complexity of cell outline and beautiful symmetry. Thus, species of the genus *Closterium* are crescent-shaped, while those of the genus *Micrasterias* are stellate. The most familiar genus is *Spirogyra* (family *Zygnemaceae),* with filaments of cylindrical cells arranged end to end, and containing a spiral chloroplast; it forms a free floating tangled mass or green "scum" on the surface of stagnant pools, especially in the spring. In Burma, *Spirogyra* is traditionally harvested as a food.

9. Order: *Charales* ***(Stoneworts).*** A small order of *C.* occupying an isolated taxonomic position and not closely related to the other green algae. The order is distributed worldwide and represented by about 80 species, the majority occurring in fresh water, the others in brackish water. They often form deep "meadows" in ponds and streams. The thallus characteristically has the appearance of a chandelier, with whorls of branches separated by long internodal cells. Short shoots arise from the nodes, and these in turn consist of nodes and internodes. These algae are achored to the substratum (usually sand or mud) by branched, filamentous rhizoids, which arise from the basal nodes. Sexual reproduction is oogamous. The structure of the macroscopically visible sex organs, situated in the nodes of the lateral branches, is characteristic of the order. The ovate oogonia (also called nucules) are surrounded by spirally wound, green filaments, while the orange antheridiophores (also called globules or "antheridia") are hollow spheres containing spermatogenous filaments, the latter consisting of strings of disk-shaped, reduced antheridia. Each antheridium releases a spirally twisted, biflagellate spermatozoid, with an eye spot but no plastids. The fertilized oogonium or zygote becomes surrounded by a thick, colorless membrane. In addition, the enveloping filaments thicken, and often become encrusted with calcium carbonate.

The encrusted cells walls of many members of the *Charales* have made a major contribution to the formation of calcareous tufas. Fossil *Charales* are known from the Devonian.

[*The* Chlamydomonas *Source Book - A Comprehensive Guide to Biology and Laboratory Use,* by Elizabeth Harris, Academic Press, Inc., 1989]

Chlorophylls: green plant pigments, which are usually accompanied in the plant by yellow xanthophylls and carotenes. The singular "chlorophyll" is also used collectively for the mixture of different C. in a plant. Chemically, C. are magnesium (MgII) complexes of tetrahydropyrroles, and can be considered as derivatives of protoporphyrins (see Porphyrins). They occur as protein complexes in the thylakoid membranes of chloroplasts. In contrast to the porphyrins, C. do not possess free acidic side chains; one side chain carboxyl group is present as its methyl ester, while the other is esterified with the C_{20} diterpene alcohol, phytol. The presence of the phytol residue is responsible for the lipid or waxy properties of C., which are therefore difficult to crystallize. The phytol residue also provides a lipophilic anchor by which the molecule is held in the thylakoid membrane. The presence of the alicyclic ring E with a carboxyl group at C9 is characteristic of the porphyrin ring system of C., and this structure is not present in other naturally occurring porphyrins, such as hemoglobin and cytochromes, etc. Removal of the magnesium by gentle acid hydrolysis produces *pheophytin.* Hydrolytic removal of the phytol residue generates a water-soluble *chlorophyllide.* The most widely distributed representative is chlorophyll *a,* which is present in all plants that generate oxygen photosynthetically. In higher green plants and green algae, chlorophyll *a* is accompanied by chlorophyll *b* (see Fig. for structures). Other C. include c_1, c_2, *d* and Bacteriochlorophyll (see). C. and their derivatives are identified and quantified by their characteristic absorption spectra. Less than 1% of the C. present in a plant is directly active in the primary events of Photosynthesis (see), and this small proportion consists of chlorophyll *a* or bacteriochlorophyll *a.* More than 99% of the chlorophyll molecules, together with carotenoids, serve an accessory function, trapping light and transferring the electronic excitation energy to the active centers of photosystems I and II.

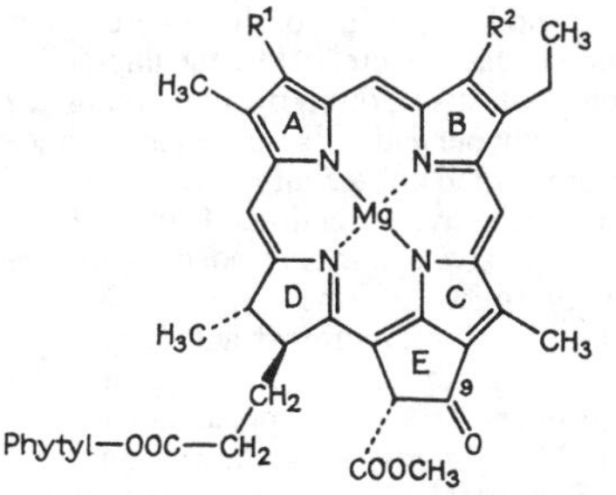

	R^1	R^2
Chlorophyll a	$CH{=}CH_2$	CH_3
Chlorophyll b	$CH{=}CH_2$	CHO

Up to the level of protoporphyrin IX, C. and Porphyrins (see) share a common biosynthetic pathway. A number of alternative pathways are available from protoporphyrin IX to chlorophyll *a.* Other C. are then derived from chlorophyll *a.* The final steps of biosynthesis occur in situ in the thylakoid membrane.

Chloropidae: see Frit flies.

Chloroplast movements: changes in the orientation, position and shape of chloroplasts within the plant.

In the narrow sense, C.m. are positional changes, which are induced and regulated by light. In weak light, many chloroplasts achieve an optimal utilization of the available radiant energy by orienting their greatest possible photosynthesizing surface area toward the light source. This usually occurs by migration of chloroplasts to cell walls that are aligned at right angles to the direction of the light. In strong light, chloroplasts offer the smallest possible photosynthetically active surface to the light source, apparently as a protection against photo-oxidative damage. This is achieved partly by alignment with cell walls that are parallel to the direction of the light, and partly by contracting into a spherical shape. Rotatory movements are rare, but are displayed, e.g. by the platelet-like chromatophores of the alga, *Mougeotia.* Usually, C.m. occur only during the period of light exposure. *Mougeotia* chloroplasts, however, continue to respond for 30 min after the removal of a weak light source. Light susception for C.m. occurs primarily in the peripheral protoplasm of the cell. Blue light is active and red light is inactive, and the action spectra for weak and strong light show that the photoreceptor is probably a flavin or flavoprotein. In some cases there is evidence that chlorophyll functions as a second photoreceptor. In *Mougeotia,* the light red-dark red system (see Phytochrome) is involved. The events that occur between light susception and C.m. are unclear. It has been suggested that translational movements of chloroplasts are only apparent, and that they result from the inhibition or modification of the passive movement that occurs continually by protoplasmic streaming.

Chemical influences can also cause C.m. Positive Chemotaxis (see) is induced by carbon dioxide, glucose, fructose, malic acid and asparagine. In leaves containing numerous intercellular spaces, chloroplasts often aggregate along the lateral walls at night; this is probably due to chemotaxis of chloroplasts in those cells surrounding and adjacent to the intercellular spaces, in response to gases excreted into the intercellular spaces. Chloroplasts also exhibit positive traumatotaxis in the vicinity of wounded cells, as well as thermotaxis when cells are cooled unilaterally.

In addition to these active, directional C.m., chloroplasts also undergo passive changes of position and orientation by protoplasmic streaming.

Chloropseidae, *Irenidae,* ***leafbirds*** and ***fairy bluebirds:*** a small avian family (14 species, 3 genera) in the order *Passeriformes* (see). Representatives are found in the Indo-Malayan region, southern China and the Philippines. They share several characters with the related bulbuls, including hair-like feathers on the nape of the neck. They are tree-dwelling birds, feeding primarily on fruit. The subfamily, *Chloropseinae,* comprises the ***Leafbirds*** *(Chloropsis)* and the ***Ioras*** *(Aegithina).* The 8 largely light green species of *Chloropsis* (17–20 cm) have rather curved, pointed bills, and they spend most of the time in the high crowns of jungle trees. Their beaks and tongues are adapted for nectar feeding, and many species of plants depend on them for pollination. The 4 species of *Aegithina* (15–18 cm) are yellow, or olive green with black and some white. Their bills are straight and fairly strong, and unlike the other members of the family they are mainly insectivorous. As its many subspecies, the ***Common Iora*** *(A. tiphia)* is found throughout the range of the entire family, not only in forests, but also as a very common bird of parks and gardens. The subfamily, *Ireninae,* contains only 2 species of *Irena* (***Fairy bluebirds,*** 28 cm), both of which inhabit evergreen lowland forest. The ***Blue-backed fairy bluebird*** *(Irena puella)* is fairly widespread in Southeast Asia, where it is often seen in small fruit-eating flocks; the male is light blue and velvet black, the female greenish blue. The ***Philippine fairy bluebird*** *(Irena cyanogaster)* is restricted to the Philippines, and both sexes are similarly colored blue and black.

Chloropseinae: see *Chloropseidae.*

Chloropsis: see *Chloropseidae.*

Chlorosis: the yellowing or failure to become green of plant parts that are normally green, resulting from a deficiency or absence of chlorophyll synthesis, or premature chlorophyll degradation. C. does not include the natural yellowing of leaves in autumn. The following types of C. are recognized: 1) genetically based C., arising from defects in the chloroplasts or inhibition of chlorophyll synthesis; 2) physiological C., due to disturbances in mineral metabolism, e.g. deficiency of calcium, iron, magnesium, manganese or copper, or excessive fertilization with potassium; 3) C. due to environmental influences, e.g. light deficiency, drought, low temperature, and smoke damage; 4) C. caused by virus diseases. In its early stages, the C. caused by pathogens is usually manifested as mottled outbreaks of C. over the plant surface. As the severity of C. increases, a stage is reached where only the main veins remain green, followed by total and uniform yellowing of the entire leaf.

Chlortetracyclin: see Tetracyclins.

Choanocyte: see *Porifera,* Digestive system.

Choanoflagellates, *Choanoflagellata, Choanoflagellidae* (Greek *choane* funnel): small, nonpigmented flagellates of the order *Protomonadina.* They are uniflagellate with a funnel-shaped protoplasmic collar around the base of the flagellum. In this and other features, they bear a very close resemblance to the collared flagellate cells (choanocytes) of the sponges. Both marine and freshwater forms are known. Some are free swimming, but most forms are fixed to the substratum by a stalk of nonliving material.

Cholecystokinin: see Pancreozymin.

Choleglobin: see Bile pigments.

Choeropsis: see *Hippopotamidae.*

Cholane: see Steroids.

Cholera: see *Vibrio.*

Cholestane: see Steroids.

Cholestanol, ***5α-cholestan-3β-ol:*** a zoosterol, and the main sterol of some sponges. It occurs in small quantities in animal cells, where it is accompanied by cholesterol. It differs from the stereoisomeric coprostanol in the configuration at C-5 and the *trans*-configuration of rings A and B.

Cholesterol, ***cholest-5-en-3β-ol:*** the most abundant sterol in higher animals. It occurs free or esterified with fatty acids in all mammalian tissues, often together with phospholipids. It occurs in the brain (about 10% of dry matter), adrenals, egg yolk and wool fat. Blood contains about 2 mg/ml, bound to lipoproteins. C. is a component of biomembranes and of the myelin sheaths around nerve axons. It has a detoxifying effect on hemolytically active saponins, with which it forms insoluble complexes, thus protecting red blood cells from lysis. Human skin excretes up to 300 mg C. daily. Deposition of C. on the interior surfaces of artery walls results in arteriosclerosis. It is also deposited in the form of gallstones, from which it was first isolated by Green in 1788. Small amounts of C. have also been isolated from potato plants, from many pollens, from isolated chloroplasts, and from bacteria. C. is a vitamin for many insects, which require it as a precursor of ecdysone and related molting hormones. C. is biosynthesized from acetyl-coenzyme A, via the triterpenes lanosterol and zymosterol. In turn, it is a key intermediate in the biosynthesis of many other steroids, including steroid hormones, steroid sapogenins and steroid alkaloids.

HO Cholesterol

Cholic acid: see Bile acids.

Cholinesterase, ***pseudocholinesterase:*** an unspecific acylcholinesterase which hydrolyses butyroyl and propionoyl choline much faster than acetylcholine. It is found primarily in serum (human serum enzyme M_r 348 000, 4 subunits of M_r 86 000), liver and pancreas. A cholinesterase is also present in cobra venom. Cholinesterases possesses a catalytically important serine residue in the active center (Gly-Gly-Asp-Ser-Gly), so that they are stoichiometrically and irreversibly inhibited by organic phosphate esters (diisopropylfluorophosphate, phenylmethanesulfonylfluoride, E600, E605 and other nerve and tissue poisons). See Acetylcholinesterase.

Cholinergic: the adjective applied to neurons, nerve fibers and synapses, whose Neurotransmitter (see) is acetylcholine.

Chondrichthyes, *Elasmobranchii,* cartilaginous fishes: a subclass of primitive, variously structured Fishes (see). The endoskeleton consists entirely of cartilage; it is usually rigid due to impregnation with calcium salts, but it is seldom ossified. The skin is covered with tooth-like placoid scales. Separate gill clefts open to the exterior, and an operculum is absent (except in the chamaeras). The first (hyomandibular) gill cleft is usually reduced to a spiracle. External nares and mouth are located on the ventral surface of the head, due to anterior elongation of the snout. The absorptive surface of the intestine is greatly increased by a fold, known as the "spiral valve". Fertilization is internal, and the male possesses pelvic claspers (the intromittent or copulatory organ; modified pelvic fins), which are inserted into the female cloaca and oviduct(s). Eggs are large-yolked and enclosed in a horny capsule. Most forms are marine, and recent forms include the sharks, dogfishes, skates, rays and chimaeras.

Fossil elasmobrachs are found from the Lower Devonian to the present, and the oldest forms are particularly well represented in the Middle Devonian and Carboniferous of North America.

Accepting the *C.* as a class, it consists of 2 subclasses:

Holocephali (chimaeras and fossil relatives): an operculum is present, which covers 4 gill openings, leaving one open on each side. The branchial basket is usually beneath the neurocranium. There is no spiracle, and tooth replacement is slow. Teeth are grinding plates. Anal and urogenital openings are separate, and there is no cloaca. Adult skin is naked, and ribs are absent. The palatoquadrate bar is fused to the cranium.

Elasmobranchii (sharks, skates and rays): between 5 and 7 separate gill openings are present on each side, and there is no operculum. The branchial basket is usually behind the neurocranium. A spiracle (remains of the hyoidean gill slit) is usually present, and tooth replacement is relatively rapid. The teeth are sharp, pointed derivatives of the placoid scales. Some ribs are usually present. The palatoquadrate (upper jaw) is not fused to the cranium.

Chondriome: see Mitochondria.

Chondriosomes: see Mitochondria.

Chondroblast: see Bone formation.

Chondroclast: see Bone formation.

Chondrocyte: see Bones.

Chondrodystrophia fetalis: see Achondroplasia.

Chondroitin sulfates: water-soluble mucopolysaccharides, M_r about 250 000, which comprise 40% of the dry mass of cartilage. They are also found in skin, tendons, umbilical cord, heart valves and other connective tissues. Chondroitin sulfate A and C are composed of equimolar amounts of D-glucuronic acid and *N*-acetyl-D-galactosamine linked by alternating β-1,3 and β-1,4 bonds; they differ in the position of the sulfate ester (position 4 of the aminosugar in A, and position 6 in C). Chondroitin sulfate B (dermatan sulfate) contains L-iduronic acid in place of D-glucuronic acid, and sulfate is present on position 4 of the aminosugar. In vivo, chondrotin sulfates are generally bound to protein.

Chondroplasty: see Achondroplasia.

Chondrostei: an infraclass of fishes of the class *Osteichthyes* (bony fishes), with numerous primitive characterisics. Recent *C.* are represented by the *Acipenseridae* (sturgeons) (see) and *Polyodontidae* (paddlefishes), both in the order *Acipenseriformes.* Many extinct orders are recognized. There is no interoperculum, and the premaxilla and maxilla are rigidly attached to the ectopterygoid and dermopalatine. Most species possess a spiracle.

Chondrostoma: see Chrondrostome.

Chondrostome, *Chondrostoma nasus:* a member of the fish family *Cyprinidae* (see), found almost exclusively in flowing waters between the trout zone and the sea in Central and Eastern Europe, from the Rhine and the Rhône to the Ural River and northern Turkey, but not in the Elbe basin. It differs from other members of the family, in that the mouth is diagonal and located on the underside of the body, and it has a thickened, cartilaginous, pointed, protruding lower lip. There is a single row of pharyngeal teeth with very long chewing surfaces. The elongated, fusiform body shows slight lateral compression, and the underside is black. It feeds mainly on vegetation, especially algae which it grazes from stones; invertebrates are also occasionally taken.

Chonotricha: see *Ciliophora.*

Chorda: see *Phaeophyceae.*

Chorda dorsalis: see Axial skeleton.

Chordata, ***chordate animals:*** a phylum of the *Deuterostomia.* It contains the vertebrates and therefore the most highly evolved forms of the animal kingdom. The most important characteristics are: 1) the presence of an endodermal notochord at least during some stage of the life history, which is derived from the roof of the archenteron; 2) a dorsally situated neural tube of ectodermal origin, with a central canal; 3) a foregut, perforated (at least in the embryo) by visceral clefts. The coelom forms by folding from the lateral roof of the archenteron; the two folds grow downward on either side of the enteron, meet ventrally and form a single continuous coelom. Dorsally, the mesoderm may become segmented. In vertebrates, the anterior end of the neural tube is developed into a brain, which is enclosed in a cartilaginous or bony capsule, the skull. The vascular system is always closed, and displays a basically similar plan in all forms. Limbs are present only in vertebrates. Tunicates are aberrant forms, displaying reduction and specialization, and they do not represent the primitively simple chordate type.

The approximately 70 000 living chordate animals are subdivided into 3 subphyla: the exclusively marine *Hemichordata* and *Urochordata* (tunicates), and the *Craniata* (vertebrates). Alternatively, the hemichordates and urochordates may be considered as classes of a single subphylum, the *Acrania*. The vertebrates include marine, terrestrial and freshwater forms, as well as those with the power of flight.

Chorda urachii: see Allantois.

Chordeiles: see *Caprimulgiformes*.

Chordeilinae: see *Caprimulgiformes*.

Chordeumida: see *Diplopoda*.

Chordeumidae: see *Diplopoda*.

Chordotonal organ, ***scolo(po)phorus organ:*** a typical insect sensory organ, consisting of a single Scolopidium (see), or groups of scolopidia. C.os. serve for the susception of internal pressure and changes in the position of individual body parts, as well as the susception of vibrations and sound waves. They are subcuticular, often with no exocuticular component, and they are found in the legs, antennae, mouthparts or thorax.

Johnston's organ is a modified C.o. found only in the second antennal segment of all adult insects, with the exception of the *Collembola* and *Diplura*. It consists of a hollow cylinder containing a single mass or several groups of scolopidia, in the wall of the antennal segment. The most advanced type of Johnston's organ is found in the *Culicidae* and *Chironomidae,* where it is housed in the enlarged pedicel. It serves for the susception of antennal movements, vibrations, and perhaps sound waves.

Subgenual organs are C.os. consisting of 10–40 scolopidia, located in the proximal part of the tibia, and not associated with a joint. Processes from the accessory cells at the distal end of the scolopidia are packed together to form an attachment body, which is fixed to the cuticle with the proximal ends supported by a trachea.

The ***Tympanal organ*** (see) of insects is also a modified C.o.

Chordotonal sensillum: see Scolopidium.

Choricarpous condition: see Fruit.

Chorio-allantois: see Allantois.

Chorioc(o)enosis: see Bioc(o)enosis, Stratum.

Choriogonadotropin, ***placental gonadotropin, human chorionic gonadotropin, hCG:*** a proteohormone and one of the gonadotropins. It is formed in the placenta during pregnancy, and excreted in the urine. It is a glycoprotein (M_r 30 000), containing about 30% carbohydrate, and two polypeptide chains (subunits). The α-subunit contains 92 amino acid residues and is identical with the subunit of follicle-stimulating hormone, luteinizing hormone and thyrotropin. The hormone-specific β-chain contains 139 amino acid residues, and displays 82% sequence homology with the β-chain of luteinizing hormone. hCG stimulates the ovaries to biosynthesize and secrete estrogens and progesterone, thereby maintaining the course of the pregnancy, and indirectly promoting growth of the uterus. Using monoclonal antibodies, hCG can be detected in maternal urine or plasma as early as one week after implantation, and it is therefore used for the early detection of pregnancy.

Chorion frondosum: see Placenta.

Chorionic villi: see Placenta.

Choriotis: see *Gruiformes*.

Choristidae: see *Mecoptera*.

Chorites: see Sorption.

Choroid: see Light sensory organs.

Choroid plexus: see Brain.

Chorological species: species with a similar distribution, i.e. a similar areal. Phytosociological units can be characterized by their combinations of C.s.

Chorology: a branch of biogeography, involving the study of the spatial distribution of plant and animal taxa, and factors influencing their distribution. By analysing and comparing areas occupied by different taxa, it is possible to interpret the complex interrelationships between the environmental conditions of a particular area and the organisms that live there. The same analysis can be applied retrospectively to former environmental conditions and earlier plant and animal populations.

Chorus frogs: see *Hylidae*.

Corticoids: see Adrenal corticosteroids.

Corticosteroids: see Adrenal corticosteroids.

Cortins: see Adrenal corticosteroids.

Chough: see *Corvidae*.

Christmas rose: see *Ranunculaceae*.

Chromaffin tissue: an association of polygonal cells, which are interspersed in a blood sinus of the adrenal medulla, and which possess variously sized, chromatin-poor nuclei. The cells are stained brown by treatment with potassium dichromate (the chromaffin reaction), due to the reduction of potassium dichromate by adrenalin and noradrenalin in the chromaffin cell granules. Following the heavy loss of these hormones by normal physiological secretion, the chromaffin reaction is weaker.

Chromatic adaptation: the adaptation of certain plants to the prevailing illumination wavelengths of their environment, by a shift in the relative proportions of photosynthetic pigments or by the possession of additional pigments. The upper layers of the sea absorb the longer wavelengths of light, so that light reaching lower levels is predominantly in the blue-green region of the spectrum. As an adaptation to living in deep water and exploiting this short wavelength light, red algae possess phycoerythrins and phycocyanins. These pigments absorb energy in the green and yellow region of the spectrum, and transfer it with relatively high efficiency to chlorophyll.

Chromatid aberrations: chromosome mutations that occur after replication of the chromosome in interphase. C.a. give rise to Chromosome mutations (see) by breakage and rejoining.

Chromatid interference: the observation that, during meiosis and multiple crossing over of paired chromosomes, the exchange of chromatid segments is nonrandom. In positive C.i., the probability that a second crossing over will involve different chromatids from those involved in the first crossing over is greater than that predicted for a totally random process. It is negative, if probability favors the involvement of the same two chromatids in a second crossing over.

Chromatids, ***daughter chromosomes:*** the two subunits of a eukaryotic chromosome. C. arise by replication during interphase, and are therefore completely identical with one another. They do not become visible until late metaphase. Each chromatid contains only one DNA duplex During mitosis and meiosis II, the C. separate and migrate to opposite cell poles.

Chromatin: the stainable material of the interphase nucleus, consisting of uncoiled or "relaxed" DNA, as well as RNA and several specialized proteins. During interphase C. is dispersed randomly in the nucleus. Immediately prior to cell division, it condenses into dense bodies (chromosomes), which can be intensely stained with basic dyes. *Heterochromatin* (see) represents chromosomal regions that remain condensed during interphase, and which display an intense but abnormal pattern of staining. *Euchromatin* represents more relaxed or uncoiled chromosomal regions, stains more weakly than heterochromatin, is diffuse throughout interphase, and becomes condensed at the time of nuclear division. Thus, heterochromatin and euchromatin are merely different functional states of C. Transcription is possible only on relaxed, noncondensed genetic material, i.e. on euchromatin, but not on heterochromatin. Filaments of C., known as *nucleofilaments* or *nucleohistone strands,* consist of a DNA double helix with histones and nonhistone proteins. In addition to smooth strands (diameter 20–30 nm) and very small numbers of filaments (diameter 3 nm), certain preparations of C. also contains structures resembling strings of beads. The "beads" (see Nucleosomes) consist of histone complexes. The DNA double helix winds 1–2.5 times around each histone complex, and forms the "string" between the beads. The small diameter filaments represent transcriptionally active, histone-free C. The smooth strands are probably densely packed strings of nucleosomes.

Chromatium: a genus of purple bacteria. *Chromatium* species are red in color, motile, anaerobic and Gram-negative. They live autotrophically by utilizing light energy to cleave hydrogen sulfide. Sulfur is deposited in the cells.

Chromatophores:

1) see Color change.

2) Membranous structures in the cytoplasm of cyanobacteria and certain other bacteria. They contain pigments (e.g. chlorophylls, carotenoids), which absorb radiant energy for photosynthetic carbon dioxide assimilation. C. may be simple vesicular structures derived from the cytoplasmic membrane (e.g. in green sulfur bacteria). The C. of cyanobacteria are complicated systems of stacked membranes.

Chromobacteria: a genus of rod-shaped, Gram-negative bacteria, up to 6 mm in length, with polar or lateral flagella. They produce a violet pigment, violacein. They are found in soil and water, and may also infect animals and humans.

Chromomere: a dense, granular, heterochromatic region of a eukaryotic chromosome, visible under the light microscope during the prophase of meiosis. The term is also applied to the condensed regions at the base of loops on lampbrush chromosomes, and to the condensed bands in the polytene chromosomes of *Diptera.* Cs. represent sites of especially dense packing of genetic material, and they are very rich in DNA.

Chromomonadina: see *Mastigophorea.*

Chromonema: light microscopically visible filamentous structures in giant and lampbrush chromosomes. C. are transcriptionally competent chromatids or nucleohistone strands, and some of them can be seen to be carrying transcription products.

Chromoproteins: proteins containing a covalently or noncovalently bound, colored prosthetic group. They include, e.g. hemoproteins, iron-porphyrin enzymes, flavoproteins, chlorophyll-protein complexes, and non-porphyrin iron and copper proteins in the blood of vertebrates (e.g. transferrin and ceruloplasmin) and invertebrates (e.g. hemerythrin and hemocyanin).

Chromosomal analysis: investigation of the chromosomes of an organism for 1) the presence of aberrations (see Chromosome aberrations), and 2) the pattern of sex chromosomes for the determination of chromosomal sex. C.a. is important for the Hereditary prognosis (see) of numerous diseases. See Human genetic counselling.

Chromosomal homeology: the partial identity of formerly homologous chromosomes. Loss of complete identity is usually due to chromosome mutations. In contrast, homologous chromosomes are genetically and structurally completely identical.

Chromosomal sex determination: microscopic detection of cell structures associated with X- or Y-chromosomes (X-chromatin, Barr-bodies, Y-chromatin). C.s.d. reveals whether a particular cell originates from an organism that is genetically male or one that is genetically female.

Chromosome: see Chromosomes.

Chromosome aberrations, ***chromosome abnormalities:***

1) In the narrow sense, changes in chromosome structure that are present during interphase, before chromosome replication, so that the longitudinal chromosome structures involved in breakage and reunion are functionally single stranded. In the wider sense, Chromosome mutations (see).

2) Deviations from the normal chromosome number (numerical C.a.), caused by abnormal behavior of chromosomes during meiosis, and usually leading to pathological changes in the organism (see Gene mutation).

With improved cytological techniques, considerable progress has been made in the investigation of human C.a. Using appropriate procedures, the chromosomes in metaphase, which are normally closely packed, can be made to move apart. There are also numerous methods for selective staining of specific chromosomal regions *(banding techniques),* which have greatly facilitated investigation of the morphology of individual C.a. So far, in humans, a ploidy mutation with a 3-fold chromosome number (triploidy = 69 chromosomes) has been observed in a physically and mentally retarded child, who also had various deformities. More common human conditions are those in which one member of a pair of homologous chromosomes is lost (monosomy), or in which an extra chromosome is present (trisomy).

The Ullrich-Turner syndrome is associated with the presence of only a single X-chromosome, so that the total chromosome number is only 45. In such cases, the internal and external sex organs are formed normally but the gonads are nonfunctional. Other features include pterygium colli, pectus excavatum, often retarded growth and other defects.

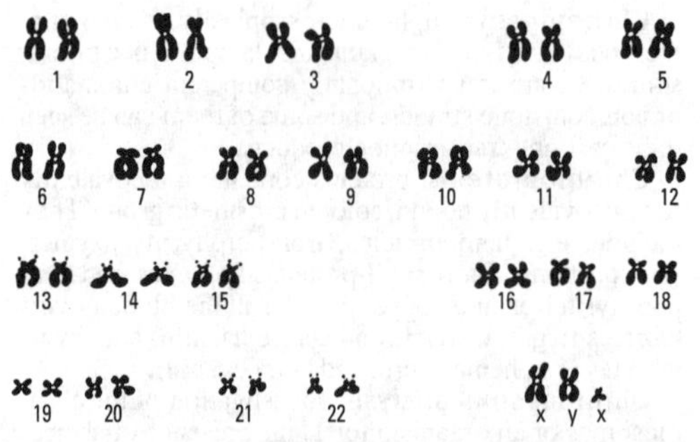

Fig.1. *Chromosome aberrations.* Karyogram of a patient with Klinefelter's syndrome, showing the supernumerary X-chromosome.

In males, the presence of supernumery X-chromosomes leads to Klinefelter syndrome (Fig.1). XXY-males have abnormal gonadal tissue, aspermia, gynecomastia, a eunuchoid disposition and low intelligence. The syndrome is relatively common, whereas XXX-women are rarely encountered.

The most common form of mental deficiency, *Morbus Langdon-Down* (Down's syndrome, mongolism), is due an extra copy of chromosome 21 resulting from nondisjunction during meiotic division I or II (Fig.2).

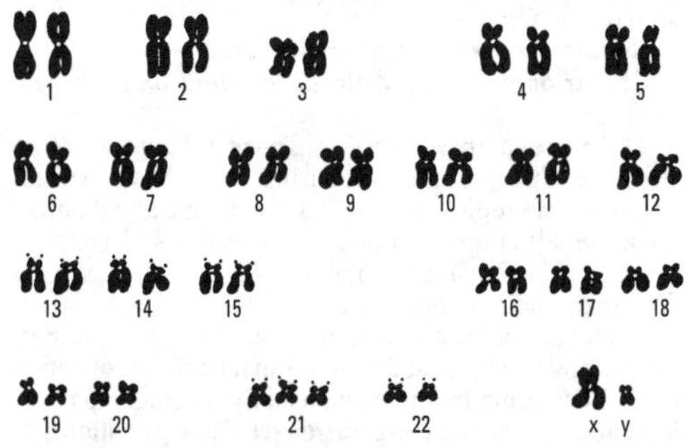

Fig.2. *Chromosome aberrations.* Karyogram of a patient with Down's syndrome or mongolism, showing the supernumerary chromosome No. 21.

The chromosome number may differ in different organs of an individual, a phenomenon known as mosaic formation.

C.a. occur in the newborn individuals of all human populations with a frequency of about 0.5%, and about one half of all spontaneous abortions in early pregnancy (up to the 12th-14th week) display fetal chromosomal abnormalities. In addition to chemicals and ionizing radiation, the ages of the mother and father are also important factors in the etiology of C.a. In mothers older than 38, the statistical probability of bearing a child with a C.a. is 15 times greater than that for the average female population. For pregnant older women, a Prenatal diagnosis (see) is therefore strongly recommended; this involves cytogenic analysis of fetal cells in the amnitoic fluid.

Chromosome interference: a phenomenon in which the occurrence of a Crossing over (see) influences the statistical probability that a further crossing over will occur in its neighborhood.

Chromosome map: a graphical representation showing the positions of genes within a chromosome. In a *genetic* or *theoretical map,* the genes are shown in a linear arrangement, in which the relative distances between genes or sites of mutation within a gene are deduced from genetic recombination experiments, i.e. the distances are proportional to the recombination frequency. In a *cytogenetic* or *real map,* the true loci of genes on a eukaryotic chromosome are shown accurately, based on cytological findings. A relatively inaccurate method of gene mapping is based on the tendency for certain mutagens to promote mutations in restricted regions of the chromosome, so that genes that mutate under the influence of the same mutagen are assumed to be relatively close to one another.

Chromosome mutations: spontaneous or experimentally induced (see Mutagens) structural alterations of chromosomes, representing a displacement of segments within individual chromosomes (intrachromosomal mutation), or between different chromosomes (interchromosomal mutation), or the loss of segments. According to which stage in the cell cycle the chromosome becomes structurally altered, C.m. are truly chromosomal (the alteration occurs before replication), chromatidal, or subchromatidal (occurring after replication). C.m. may be due to Insertion (see), Inversion (see), Translocation (see), Deletion (see) or Duplication (see).

According to the *breakage and rejoining hypothesis,* C.m. arise from breakages in the longitudinal structure of the chromosome. Restitution of the original structure by recombination of the broken ends is inhibited or prevented. Several new breakages then occur, and the structure is stabilized by the formation of new combinations.

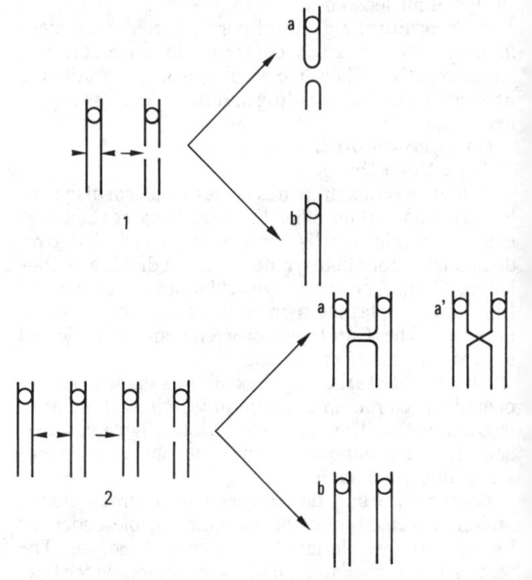

Chromosome mutations that arise from the breakage and rejoining of chromatids. 1a Deletion. *2a* Asymmetrical chromatid translocation. *2a'* Symmetrical chromatid translocation. *1b* and *2b* Restoration of the structure that existed prior to breakage.

Chromosome puffs: see Giant chromosomes.

Chromosomes (plate 27): nucleoprotein structures which are the main carriers of genetic information in prokaryotes and eukaryotes. They become well defined and microscopically visible during nuclear division in eukaryotes and during division of the nucleoid in prokaryotes. They carry genetic information in the form of linear sequences of genes, and they are capable of self replication. The genetic complement of prokaryotes is contained in a single C., whereas the greater informational requirement of the eukaryotic cell is distributed among a species-specific number of C. Special mechanisms insure the faithful transfer of genetic information to daughter cells and from generation to generation (see Mitosis, Meiosis).

I) ***Prokaryotic C.*** In all investigated species, the genetic material of bacteria and cyanobacteria consists of a circular, supercoiled, relatively short DNA double helix. This is usually called a chromosome, although histones and other chromosomal proteins, and the typical structure displayed by eukaryotic chromosomes are absent. The negative charge of the acidic DNA is neutralized in particular by amines and Mg^{2+}, and in this respect it resembles the DNA of eukaryotic plastids and mitochondria. Under suitable growth conditions, bacterial DNA replication starts at a specific initiation site. In contrast, well over 100 replication initiation sites are present in each eukaryotic chromosome or nucleohistone strand. The genetic material of cyanobacteria is microscopically visible as a region of diffuse centroplasm (chromidial apparatus). Eukaryotic and prokaryotic C. also differ in size (quantity of DNA, length of the DNA molecule), and the arrangement of specific nucleotide sequences. Thus, the bacterial C. contains an average of 0.007 pg DNA, compared with 3–6 pg DNA in the mammalian haploid genome. Eukaryotic DNA contains a large number of repetitive nucleotide sequences (100– to a million-fold repetition of certain sequences). These repetitive sequences are, however, part of the structure of the constitutive Heterochromatin (see), and they are not transcribed; they may be important in the maintenance of C. structure. The circular DNA molecule of *Escherichia coli* contains information for about 3 000 diferent proteins.

With regard to nuclear organization, the eukaryotic dinoflagellates differ from both pro- and eukaryotes. The nuclear membrane does not disintegrate during cell division, the DNA is condensed to fibrils in the C., and the C. remain visible as separate structures even during interphase.

II) ***Eukaryotic C.*** are contained within the cell nucleus, and during interphase they are separated from the cytoplasm by the nuclear membrane. Replication and transcription occur on relaxed (uncoiled) chromosomal material in the interphase nucleus. The main chemical components of eukaryotic C. are: 35% DNA, 40% histones (arginine- or lysine-rich, basic proteins), 10% nonhistone proteins (acidic proteins, phospho- and lipoproteins) and transcribed RNA. In the dividing nucleus, the C. can be made visible for light microscopy by specific DNA stains, e.g. Feulgen reagent and other basic dyes. A DNA double helix with bound histones (nucleofilament, nucleohistone strand) runs the entire length, and forms the ground structure of each chromatid. The negative charge of the acidic DNA is neutralized by the basic histones. Histones serve, inter alia, to stabilize the double helix and to nonspecifically inhibit transcription and replication. During gene activation, histones are acetylated and presumably no longer bind to DNA. Nonhistone chromosomal proteins possibly abolish the blocking action of histones.

Human metaphase chromosomes have an average length of 10 mm, each containing a nucleohistone strand with an average length of 7.3 mm, which is densely packed by multiple folding and coiling in the metaphase C. The successful isolation of nucleohistone strands with lengths in the mm to 1 cm range lends support to the *single strand hypothesis,* and is in accordance with the semiconservative replication of C.

Folding and coiling are reversible and are responsible for changes in the shape of eukaryotic C. In the interphase nucleus, C. are largely uncoiled, i.e. they are in a functional state and able to transfer information by transcription, and to undergo replication. The totality of the individual C. of the interphase nucleus represent the Chromatin (see).

Early C. preparations did not permit all human C. to be identified, since the only distinguishing characteristics were length and centromere location. It was particularly difficult to differentiate C. 4 and 5, 8 to 11, and 13 to15, which are otherwise similar in appearance. In 1970, banding techniques were developed, which allowed every C. and even regions within the C. to be distinguished. The first methods used fluorescent derivatives of quinacrine, which bind preferentially to some regions of C. Later, the modified Giemsa stain was developed, which gives banding patterns similar to those produced by quinacrine derivatives (Q-bands), but is easier to perform. It is now possible to apply banding techniques to C. at the less condensed prophase stage, revealing even more bands (as many as 3 000 can be identified). After pretreatment (e.g. with trypsin), followed by staining with adenine/thymine (AT)-specific or guanine/cytosine (GC)-specific dyes, distinct patterns of bands are discernable, which indicate the distribution of GC- and AT-rich regions. Using other methods of cell biology, it is possible to idenify the sites of individual genes on sections of human C. (e.g. the gene loci for the β- and δ-chains of hemoglobin).

The Centromere (see) serves as the attachment point to the spindle microtubules, and it is possibly also involved in the formation of the latter. Depending on the position of the centromere, the two arms of the C. are about the same length *(metacentric C.),* unequal in length *(submetacentric C.)* or have very different lengths *(acrocentric* or *telocentric C.).* In addition to the primary constriction, some C. also have a secondary constriction, known as the *nucleolus organizer,* with an attached short section of C. known as the *satellite.* It is here that the nucleous is reformed after completion of nuclear division. Such C. are known as satellite C. (see). The sections of C. lying directly next to the centromere and to the satellites consist of Heterochromatin (see); they do not relax during interphase and are therefore transcriptionally inactive. In contrast, Euchromatin (see) becomes unfolded and uncoiled during interphase, and serves as a template for transcription.

Following replication, the C. consists of 2 subunits, the Chromatids (see). During mitosis and meiosis II, the chromatids become sister C., which separate and migrate to opposite cell poles. Somatic cells have a diploid set of C. (2n), whereas germ cells possess a haploid number of C. (n). All the haploid C. carry different genetic information, and have different shapes and sizes. In the haploid nucleus, however, each C. has a similar or identical partner (homologous C., autosomes), with the exception of the sex C. The characteristic chromosomal content of the nuclei of an organism, recorded on the basis of the number and morphology of the C., is known as the *karyotype*.

Certain severe inherited illnesses in humans are due to alterations in the chromosomal complement, and can be recognized by prenatal analysis of the karyotype. For example, a single C. may be absent *(monosomy)* or a single C. may occur as 3 copies *(trisomy)*. This condition of *aneuploidy* is caused by Nondisjunction (see); two homologous C. fail to separate during meiosis or the chromatids of an individual C. fail to separate during mitosis.

See Giant chromosomes, Lampbrush chromosomes, Chromosome aberrations, Karyotype, Meiosis, Mitosis.

Chromosome transformation: reversible alterations in the morphology of eukaryotic chromosomes during the cell cycle. During the interphase of mitosis the chromosomes have an open, noncondensed structure, and are functional in transcription, whereas each chromosome of the dividing nucleus (interphase of mitosis) has a closely packed, highly condensed structure, due to complex coiling and folding of the nucleohistone strand.

Chronaxie: the minimal time required for the generation of an action potential by a current that is twice the intensity of the Rheobase (see). For myelinated nerves and vertebrate skeletal muscle C. is 0.3–0.8 of a millisecond. In contrast, the C. for the smooth muscle of *Helix pomatia* is almost 1 second. Chronaxie values are determined in the clinical diagnosis of nerve and muscle degeneration.

Chronobiology: the study of the space-time organization of life processes. See Biorhythms. For example, C. addresses the problem of the synchronization of the body rhythms in humans performing shift work.

Chronoelement: see Floral element.

Chronogram: a record of animal behavior, based on the time of its occurrence, duration, and rhythm. A first order C. (periodicity C.) lists the times at which activities are observed, their absolute and relative frequencies of occurrence, and their duration. The second order C. records the activity dynamics, and determines the interdependence of the timing and duration of types of behavior.

Chronotropic effect: an effect on the heart rate. Factors that increase the heart rate (e.g. noradrenalin) are said to have a positive C.e.; those that decrease the heart rate have a negative C.e.

Chroococcales: an order of the *Cyanophyceae* (see). They are either unicellular organisms (e.g. *Anacystis, Synechococcus),* or they exist as small colonies surrounded by a gelatinous envelope. Members of the genera, *Aphanothece, Chroococcus* and *Gloeocapsa,* form mucilaginous layers on the wet surfaces of rocks and walls. *Merismopedia* forms characteristic free-floating, plate-like colonies in water.

Chrysalid: see Insects.

Chrysalis: see Insects.

Chrysamphora: see Carnivorous plants.

Chrysanthemum: see *Compositae.*

Chrysemys: see *Emydidae.*

Chrysididae: see *Hymenoptera.*

Chrysochloridae, ***golden moles:*** a family of the order *Insectivora* (see) containing 17 species in 5 genera, widely distributed in southern and equatorial Africa. Golden moles are mole-like, subterranean mammals, related to the tenrecs. They have a glossy greenish golden coat, the eyes are vestigial and overgrown by skin, and two large claws on the forefeet are used for digging. The dental formula is 3/3, 1/1, 3/3, 3/3 = 40, except in *Ablysomas* which has the formula: 3/3, 1/1, 3/3, 2/2 = 36.

Chrysocyon: see Maned wolf.

Chrysolampis: see *Apodiformes.*

Chrysomelidae: see Leaf beetles.

Chrysomma: see *Timaliinae.*

Chrysomonadina: see *Phytomastigophorea.*

Chrysopelea: see Golden snakes.

Chrysophyceae, ***"golden algae":*** a class of mostly unicellular brown to gold-brown algae, containing chlorophylls *a* and *c*, β-carotene and various xanthophylls, which are also characteristic constituents of the *Chrysomonadina* and the *Dinoflagellata* (see *Phytomastigophorea)*. Reserve materials are chrysolaminarin and oil.

Familiar freshwater genera are *Synura* and *Uroglena,* which form spherical colonies (cenobia) by the radial arrangment of numerous individual cells in a gelatinous matrix. The freshwater genus, *Mallomonas,* carries on its surface delicate silicified scales with long processes. Marine and freshwater species of the genus, *Dinobyron,* form a cone-shaped cellulose envelope (lorica). After division, the daughter cells become attached to the margin of the lorica, each forming a new lorica to produce a tree-like branched structure.

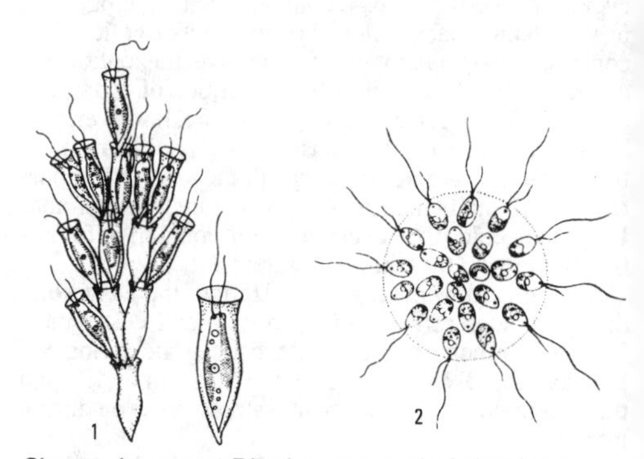

Chrysophyceae. 1 Dinobryon sertularia (x 180). *2 Uroglena americana* (x 300).

Chub, *Leuciscus cephalus:* a member of the fish family *Cyprinidae* (see). Up to 40 cm in length, it has a gray-brown back with a greenish sheen, and whitish gray sides. Dorsal and caudal fins are pale gray, the other fins slightly reddish. A popular sporting fish, it is found in Europe and Turkey, in fresh, flowing water from the trout zone to the river mouth.

Chunga: see *Gruiformes.*

Chyle: the lymph that circulates in the lymph vessels of the intestine and stomach. In the fasting state, it differs only slightly from lymph in the rest of the lymphatic system. During digestion, it contains large quantities of emulsified fat.

Chylophagy, *halmophagy:* the provision of root tissues with mineral nutrients (mainly nitrates and phosphates) by the fluid secretions of mycorrhizal fungi. The salt solution from the fungal hyphae is taken up by the host cells partly by special tubular protrusions. C. often occurs in the ectotrophic mycorrhizas of trees (e.g. spruce and oak), and the ectotrophic associations of a phycomycete fungi with the prothalli of clubmosses.

Chyme: an acidic fluid brei of food material, produced by mechanical reduction of the food by chewing, predigestion by saliva, and admixture with gastric juice in the stomach.

Chymosin: see Rennin.

Chymotrypsin: a family of serine protease (see Proteases), which are important digestive enzymes of vertebrates, with activity optima in the range pH 7–8. They are formed and stored in the pancreas as precursors (proenzymes) known as ***Chymotrypsinogen,*** which are converted in the intestine into proteolytically active C. by the action of trypsin. C. catalyses the hydrolysis of dietary proteins and polypeptides to oligopeptides, with a preference for peptide bonds derived from the carboxylic acid groups of phenylalanine, tyrosine, tryptophan and leucine.

C.A (245 amino acids, M_r 25 670) is cationic at pH 8. and C.B (248 amino acids, M_r 25 760) is anionic at pH 8. The latter is lacking in pig pancreas, from which C.C (281 amino acids, M_r 31 800) has been isolated. C. is structurally related to Trypsin (see).

Chytridiales: see *Phycomycetes.*

Chytridiomycetes: see *Phycomycetes.*

Cicada: see *Homoptera.*

Cicer: see *Papilionaceae.*

Cichlherminia: see *Turdinae.*

Cichlidae: see Cichlids.

Cichlids, *Cichlidae:* a large fish family in the order *Perciformes.* C. are distributed in South and Central America, Africa and Asia. They are small to relatively large (maximal length 80 cm attained by *Boulengerochromis microlepis* of Lake Tanganyika), often colorful fishes. Some are important economically as food sources in tropical countries. Others are popular aquarium fishes, many color patterns having been developed by selective breeding for the aquarium trade.

Endemic C. account for most of the fishes in Lake Malawi (more than 200 species), Lake Victoria (at least 170 species) and Lake Tanganyika (126 species), which contain more different fish species than any other world lake. Study of these species flocks has contributed considerably to the understanding of adaptive radiation and speciation.

Most species have a moderately deep, compressed body, but some are disk-shaped with wide, high fins (e.g. *Pterophyllum,* ***Angel-fishes)*** or low fins (e.g. *Symphysodon,* ***Discus fishes).*** In the genus *Crenicichla* ***(Pike C.),*** the body is elongated.

Cichoriaceae: see *Compositae.*

Cichorium: see *Compositae.*

Ciconia: see *Ciconiiformes.*

Ciconiidae: see *Ciconiiformes.*

Ciconiiformes: an order of birds (see *Avies)* consisting of 114 species in 7 families. All are long-legged wading birds, living mostly in marshes and shallow waters, but a few species live on drier, higher ground. They have long bills. long necks, relatively short tails and rounded wings. Most temperate species are migratory and all members of the order are strong fliers. They feed mainly on fish, water crustaceans and amphibians, also taking insects and reptiles. All have 4 long spreading toes, the 3 front ones slightly webbed. All families except the *Ciconiidae* have powder downs and serrated middle claws. The sexes are generally alike, except for some species of the *Ardeidae* that show sexual dimorphism.

1. Family: *Ardeidae* ***(Herons*** and ***Bitterns;*** 62 species with a cosmopolitan distribution). Members of this family differ from all other *C.* in that the head is completely feathered and the neck vertebrae are of unequal length, so that the neck is carried in an S-shape when in flight. Herons nest in colonies on trees or rocks, while bitterns are more solitary, nesting and feeding among reeds. Bitterns have streaked and speckled brownish plumage, and they depend upon camouflage for protection, reinforcing this by freezing into a reed-like pose with body, neck and bill stretched upward (Fig.1). The ***Little bittern*** *(Ixobrychus minutus;* 35.5 cm; Eurasia, Africa, Australia, New Zealand) is also illustrated on plate 6. Herons and bitterns are notable for their powder downs, which grow in prominent patches on the breast, rump and flanks. Oil and slime are cleaned from the feathers by dressing them with powder down, which is then removed by scratching. The family also includes the ***Egrets,*** e.g. the ***Cattle egret*** *(Bubulcus ibis),* originally of southern Eurasia and North Africa, but through its own ability to migrate and extend its range it is now more widely distributed in most of Africa, and in the New World and Australia. It feeds by foraging for insects at the feet of large ruminants.

Fig.1. *Ciconiiformes.* Little bittern *(Ixobrychus minutus).*

2. Family: *Ciconiidae* ***(Storks;*** 17 species). Storks are widely distributed in the warm and temperate regions, but are absent from northern North America, New Zealand and Oceana. Due to the absence of syringeal musculature, adult storks are mute, and they express themselves by rattling their bill. The ***Marabou*** or ***Adjutant stork*** *(Leptoptilus crumeniferus;* 150 cm; tropical Africa) is the largest of the family and is largely a carrion feeder. The ***European white stork*** *(Ciconia alba)* breeds throughout temperate Eurasia, overwintering in Africa, Arabia, India and Northeast Asia. The European population has abandoned natural nesting sites, and invariably nests on the roofs of buildings.

3. Family: *Threskiornithidae* (***Ibises*** and ***Spoonbills;*** 28 species). Ibises have slender decurved bills, whereas the bills of the spoonbills are straight with spatulate ends. Both fly with their neck and legs extended. Examples are ***Sacred ibis*** *(Threskiornis aethiopica;* Africa and Madagascar), the ***White ibis*** *(Eudocimus albus;* southern USA, West Indies, central and northern South America) and the ***Glossy ibis*** *(Plegadis falcinellus;* cosmopolitan in warmer regions). The ***Waldrapp*** *(Geronticus eremita)* occurs locally in North Africa and the Middle East. All 5 Old World spoonbills are placed in the genus *Platalea*. The only American spoonbill is the rare ***Roseate spoonbill*** *(Ajaia ajaja)*.

4. Family: *Balaenicipitidae*. A monotypic family represented by the ***Shoebill*** or ***Whale-headed stork*** *(Balaeniceps rex),* which is restricted to papyrus swamps of the upper White Nile of southern Sudan, extending into Uganda and Zaire. It stands 107–122 cm tall and possesses a very wide (shoe-shaped), 20 cm-long beak which is used for probing muddy waters for lungfish and gars, and it also feeds on frogs and other small animals. It rattles its bill like a stork and flies with its neck in an S-shape like a heron. As in herons, the hindtoe is on the same level as the 3 front ones, and the central claw carries comb-like serrations. It also has powder down patches on the rump.

5. Family: *Scopidae*. A monotypic family represented by the ***Hammerkop*** or ***Hammerhead*** *(Scopus umbretta),* distributed from Southwest Arabia through East Africa to Madagascar, with a dwarf race in West Africa. The East African race is about 50 cm in length. Its long crest is directed backward and overhangs the back of the head. Powder downs are absent, and it flies with extended neck like a stork. Its vocalizations resemble those of herons, all 4 toes are on the same level, and the middle claw is serrated. The wide flat bill is quite atypical of the order, and the neck is atypically short and thick.

6. Family: *Cochleariidae*. A monotypic family represented by the ***Boatbill heron*** *(Cochlearius cochlearius;* 56 cm), which inhabits freshwater mangrove swamps from Mexico to Peru and southern Brazil. The broad boat-like bill is 7.7 cm long and 5 cm wide, and unlike that of any other of the *Ciconiiformes*. In size and color it resembles the night herons *(Nycticorax),* and it is also nocturnal in habit. Like the herons, it has a serrated middle claw and powder down patches, whereas the distensible gular pouch is more stork-like.

7. Family: *Phoenicopteridae* (***Flamingos;*** Fig.2). Flamingos are sometimes placed in a separate order. In relation to the size of the body, they have the longest necks and legs of any bird, and these are held fully extended in flight. They are also good swimmers. Their plumage is notably pale to rosy pink. The 3 front toes are webbed. When feeding, the large bill is placed upsidedown in the water, and both vegetable and animal material is filtered off between platelets on the lower mandible and the large tongue. The family displays a discontinuous distribution. The ***Greater flamingo*** *(Phoenicopterus ruber)* exists as 3 races, found respectively in the Old World (Kenya, Mediterranean, Kirghiz steppes and India); the West Indies and the Galapagos; and in South America (Peru to Tierra del Fuego). The ***Lesser flamingo*** *(Phoeniconaias minor)* exists in very large numbers on East African alkaline lakes. Two species of *Phoenicoparrus* live on high lakes in the Andes. Flamingos breed in large colonies. They build a mud nest, shaped like a flattened cone and slightly depressed on the top. The young are precocial, and they gather together in "kindergarten" groups.

Fig.2. *Ciconiiformes.* Greater flamingo *(Phoenicopterus ruber).*

Cidaris: see *Echinoidea.*

Cilia (sing. **Cilium**) and **Flagella** (sing. **Flagellum**): *1)* mobile processes present in many eukaryotic cells. They are surrounded by the plasma membrane, and possess a characteristic internal structure, which is referred to as the "9 + 2 arrangement". They are responsible for the locomotion of cells in aqueous media, or for the generation of water currents over the surface of fixed, nonmotile cells. Cilia and flagella have essentially the same fundamental structure. Cilia are usually short (5–10 mm) and present in relatively large numbers on the surface of a single cell, e.g. on ciliated protozoans, or on the ciliated epithelium of bronchi and oviducts (the bronchial epithelium carries 10^9 cilia per cm^2). Flagella are elongated, and usually restricted to one per cell, e.g. sperm "tails" and the flagella of fern spermatozoids. With few exceptions (e.g. some insect sperm flagella), a transverse section of a cilium or flagellum shows a core structure containing a characteristic number and arrangement of structural elements (Fig.).

In all eukaryotic organisms, the core or *axoneme* has a diameter of 200 nm, and is surrounded by the plasma membrane. In transverse section, the axoneme is seen to consist of 9 peripheral microtubule doublets and 2 centrally located individual Microtubules (see). In transverse section the microtubules appear as circles. At very high magnification under the electron microscope, it can be seen that each of the peripheral doublets is composed of one complete and one partial microtubule. The complete member of each doublet contains 13 tubulin protofilaments, and is known as the *A tubule,* while the partial member contains 10 protofilaments, and is known as the *B tubule*. The common shared wall between the A and B tubules consists of 3 tubulin protofilaments of the A tubule. Attached to each A tubule are 2 "arms" (an inner dynein arm and an outer dynein

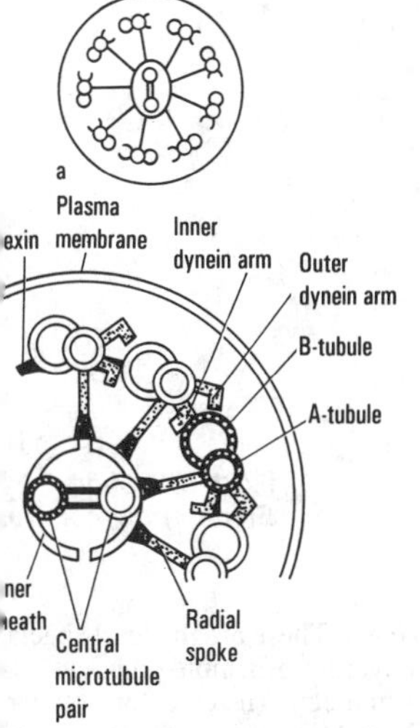

Cila and flagella. a Schematic transverse section of a cilium or flagellum showing the 9 + 2 arrangement of microtubules. The axoneme contains 9 peripheral microtubule doublets and 2 separate central microtubules. *b* Magnified section of *a.*

rm), which consist of the protein, *dynein* . Dynein (M_r bout 2 × 10^6), possesses ATPase activity and is funcionally comparable to myosin. From the clearest elecron micrographs available, dynein consists of 3 globuar heads, each of which is attached by a filament to a ommon root-like base. The dynein arms are spaced long each A tubule at regular 24 nm intervals. They nteract with the B tubules of neighboring doublets and, t the expense of ATP hydrolysis, they push the doublets ast each other, causing the entire structure to bend, enerating a wave that emanates from the base and preads outward to the tip of the cilium or flagellum. This *sliding microtubule mechanism* resembles the nechanism of muscle contraction, dynein being analoous to myosin and the doublets to actin. The direction f sliding is such that the dynein arms of one doublet ush the neighboring doublet toward the tip of the cilum or flagellum. The plane of the stroke is perpendiclar to the paired protein bridges that link the central tuules, the latter probably serving as a flexible counterupport. Adjacent doublets are also linked by the rotein, *nexin* (M_r 165 000), which occurs at wider ntervals along the microtubule than the dynein arms. Nexin is thought to provide elastic resistance to the sliding mechanism. Surrounding the central pair of microubules is an *inner sheath,* and a *radial spoke* extends rom each A tubule to this inner sheath.

Isolated flagella continue to propagate bending movements, indicating that the mechanism for motility s a property of the axoneme. Even an isolated axoneme ropagates bending movements when placed in an appropriate medium with added ATP. The mechanism is herefore clearly different from that of the bacterial flaellum, which depends on a motor at its base.

By treatment of isolated axonemes with proteolyic enzymes, it is possible to destroy the nexin links and the adial spokes, leaving the microtubules and their dynein rms intact. When exposed to ATP, these treated axoemes do not bend, but they elongate up to nine times heir orginal length, i.e. the component microtubule oublets slide past each other without restriction.

The beating of cilia in one plane is due to the rhythmic activity of doublets on opposite sides of the axoneme. The cilium performs a rapid power stroke in one direction, followed by a slower recovery stroke in the opposite direction, attaining a beat frequency of 30 Hz, while the tip of the cilium attains a speed of up to 2.5 mm per sec. Rotary movements are due to cyclic activity of the doublets (in the sequence 1 to 9, etc.).

The rate and pattern of movement is probably controlled by the *kinetosome (basal body),* but the participation of some auto-excitable component of the sliding mechanism cannot be excluded. In ciliated protozoa and ciliated epithelia, the beating of individual cilia is synchronized. In certain organisms, synchronization is imposed by the fusion of cilia, e.g. the cirri of the *Hypotrichida,* the walking cirri of *Stylonychia,* and the combs of the *Ctenophora.*

The occurrence of the 9 + 2 arrangement in many widely differing eukaryotic organisms, from protophytes and protozoa to mammals, is indicative of the common phylogenetic origin of all modern eukaryotes.

Deviations from the 9 + 2 arrangement are found in the sperm tail flagella of certain insects and a few representatives of other animal groups, but these spermatozoa are nonmotile.

Most sensory cilia lack the central tubule pair, and they are immobile. They are found in many sensory cells, e.g. in the secondary sensory cells in the lateral line of fishes. The outer segment of the rod and cone cells of the vertebrate eye appears to be a modified cilium, with a characteristic cilium-like arrangment of microtubules in the neck or "connecting cilium", which connects the outer segment to the rest of the cell; a pair of central tubules is lacking, and the peripheral tubules make contact with the densely packed membrane complexes of the outer segment, which are derived from the plasma membrane.

Cilia and flagella develop from the *basal body,* which is in fact a Centriole (see). The basal body is composed of 9 sets of triplet microtubules, each triplet containing one complete microtubule (the A tubule) fused to 2 incomplete microtubules (the B and C tubules). The 9 triplets form a cylinder, in which they are laterally linked by other proteins [see Centriole (Fig.)]. During development or regeneration of a cilium, each microtubule doublet of the axoneme grows from two of the microtubules in a triplet of the centriole. The origin of the central microtubule pair is not known. The basal body is anchored in the cytoplasm by *striated rootlets,* which link it to other cytoplasmic components.

2) Terminal structures of the appendages of arthropods, consisting of numerous short sections, are also known as Cilia.

Ciliata: see *Ciliophora.*

Ciliatea: see *Ciliophora.*

Ciliated protozoa: see *Ciliophora.*

Ciliophora, ***ciliated protozoa:*** one of the four subphyla of the phylum *Protozoa,* containing a single class, *Ciliata* or *Ciliatea.* They are the most advanced protozoa, often displaying a high degree of differentiation. All possess cilia or ciliary organelles in at least the young stage of the life cycle. They are unique in possessing two distinct types of nucleus, a macronucleus (meganucleus) and a micronucleus. All reproduce by binary fission, and nearly all members are also capable of conjugation with the fusion of gametic nuclei. *C.* are relatively large (mostly in the range 100–250 μm), and in the primitive form the entire cell surface is covered with uniformly distributed, longitudinal rows of cilia. The simplest forms have an anterior cleft-shaped mouth, which expands widely when prey is ingested. A spiral groove (peristome) leads along the ventral surface to the opening (cytostome) of the gullet. Cilia are sometimes modified to ciliary organelles known as cirri, consisting of bundles of united cilia which function as a single unit. Higher forms have a food-gathering row of cirri or membranelles along the outer edge of the peristome. Mouth cilia may also be greatly modified, in particlar to undulating membranes and membranelles.

The highly polyploid macronucleus, which has no recognizable chromosomal organization, is essential for continuous metabolic activity, and its size is roughly proportional to that of the organism. The micronucleus is the source of gametic nuclei during conjugation (sexual reproduction). Individual ciliates may possess from one to many of each type of nucleus.

Binary fission always occurs by a simple or modified form of transverse cleavage, in which the micronucleus divides mitotically, the macronucleus amitotically. Anterior and posterior body halves of the parent may then undergo considerable morphogenic differentiation, so that each becomes identical and possesses an identical pattern of cilia.

Sexual reproduction occurs by conjugation. Two organisms make close contact, usually aligned longitudinally, and fusion occurs in the region of contact. The micronucleus enlarges and undergoes reduction division. Three of the resulting haploid nuclei are resorbed, and the remaining one divides into a stationary and a migratory nucleus. Migratory nuclei are then exchanged: each migratory nucleus passes into the conjugation partner, where it fuses with the resident stationary nucleus to form a synkaryon. Conjugation partners then separate and the synkaryon divides into a new micronucleus and the precursor of a new macronucleus. The old macronucleus disintegrates and is resorbed.

C. are found in all kinds of standing water, even in thin water films on soil or mosses, and in temporary bodies of water such as rain puddles. Species of temporary water usually survive as resting cysts, formed in response to drought, which rapidly transform to free living organisms in a moist environment. Planktonic forms, as well as inhabitants of the bottom mud, occur in permanent freshwater and in the sea. Certain cellulose-degrading *C.* live in the rumen of ruminants and in the cecum and large intestine of other herbivorous mammals.

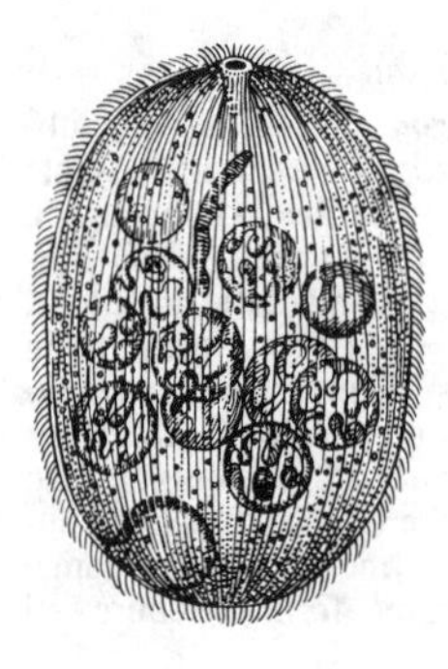

Fig.1. *Ciliophora. Prorodon teres.*

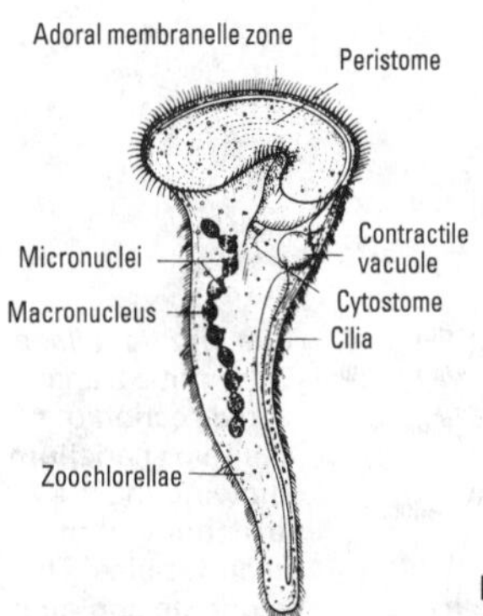

Fig.2. *Ciliophora. Stentor polymorphus.*

Classification.

1. Subclass: *Holotrichia.* These are the least specialized, with a uniform covering of simple cilia and relatively simple axes of symmetry. The cytostome may be shifted to a ventral or lateral location, and there is no adoral zone of membranelles. Orders include the *Gymnostomata, Trichostomata, Apostoma, Hymenostomata, Thigmotricha* and *Astmata.*

The hymenostomes (prominent cytostome accompanied by accessory ciliary organelles, e.g. *Paramecium* and *Tetrahymena)* appear to be a stem group of the holotrichs, as well as being ancestral to all other subclasses.

2. Subclass: *Spirotricha.* These are the most specialized of the *C.* and of the *Protozoa* as a whole. Symmetry is complex, with sharply polarized dorso-ventral and antero-posterior axes. The cytostome is located in a depression, and the oral ciliature includes both an undulating membrane and a densely packed row of membranelles. The peristome spirals clockwise as it approaches the cytostome. Body ciliation is reduced or replaced by compound ciliary organelles.

1.Order: *Oligotricha.* These include free-living freshwater and marine forms. There is an anterior peristome with very strong membranelles, but body cilia are sparse or absent.

2. Order: *Entodiniomorpha.* Close relatives of the *Oligotricha* found in the rumen and reticulum of ruminants and in the cecum and large intestine of other herbivorous mammals.

3. Order: *Heterotricha.* The gullet is permanently open and provided with an undulating membrane. The body is not compressed and it usually has a uniform covering of cilia.

4. Order: *Ctenostomata.* The body is laterally compressed and possesses only a few cilia.

5. Order: *Hypotricha.* Dorsoventrally compressed body, usually with an elaborate system of cirri and other ciliary organelles on the ventral surface.

3. Subclass: *Peritricha.* Highly specialized *C.;* usually sessile and stalked, with a permanently open gullet and an undulating membrane which generates a feeding current. The peristome spirals counter-clockwise. Body cilia are absent from sessile forms, but free-swimming species and stages have an aboral ring of cilia. *Vorticella* has the shape of a solid, inverted bell, the handle or stalk consisting of a cuticle-covered extension of the body.

4. Subclass: *Chonotricha.* Permanently sessile, attached by the posterior end to the bodies of crustaceans

There is a clockwise coiled, internally ciliated peritome, and the rest of the body is naked.

5. Subclass: *Suctoria*. The sessile adult organisms feed by predation with the aid of suctorial tentacles. Cilia are present only in the free-swimming juvenile stages.

Peritricha, Chonotricha and *Suctoria* may be subsumed as orders of the *Spirotricha,* producing just 2 subclasses: *Holotricha* and *Spirotricha.*

Cimbicidae: see Sawflies.

Cinchona: see Quinine, *Rubiaceae.*

Cinchonidine: see Quinine.

Cinchonine: see Quinine.

Cincinus: see Flower, section 5) Inflorescence.

Cinclidae, ***dippers:*** an avian family of the order *Passeriformes* (see), and containing only 4 species distributed in Eurasia and America. Ranging from 18 to 21 cm in length, they resemble thrush-sized wrens, with strong legs, round bodies and short, uptilted tails. They live by mountain streams, and they feed on water insects (mayflies, stoneflies, etc.), aquatic larvae, and various freshwater crustaceans such as amphipods and hog slaters. Small amphibians and fish are also taken. Dippers are the only truly aquatic passerines, and they swim well under water. The plumage is soft and filmy with a thick undercoat of down, and the feathers are kept waterproof with the aid of a large preen gland, which is ten times larger than that of any other passerine. A movable flap over the nostrils and a highly developed nictitating membrane (third eyelid) are the only other obvious adaptations to their exceptional mode of existence. The feet are not webbed. The 4 species are: ***Dipper*** *(Cinclus cinclus;* Europe, western Asia including the Himalayas, and North Africa); ***Brown dipper*** *(C. pallasi;* northern Asia and Japan); ***North American dipper*** *(C. mexicanus;* western North America from Alaska to Panama); ***White headed dipper*** *(C. leucocephalus;* Andes).

Cinclocerthia: see *Mimidae.*

Cinclosomatinae: see *Muscicapidae.*

Cinclus: see *Cinclidae.*

Cineol: see Eucalyptol.

Cingulum: one of the ciliated bands constituting the "wheel organ" of the *Rotifera* (see).

Cinnamic aldehyde: C_6H_5-CH=CH-CHO, an aromatic aldehyde found in balsams and essential oils. It is a pale yellow oil, smelling pleasantly of cinnamon. In particular, it occurs in oil of cinnamon and oil of cassia, to which it imparts the odor of cinnamon.

Cinnamon: see *Lauraceae.*

Cinnamomum: see *Lauraceae.*

Circadian rhythms: variations of metabolism, growth, development and movement according to a 24-hour cycle, which are displayed by numerous plants and animals. Closing and opening of flowers in the evening and morning, accompanied by the drooping and raising of leaves, are visibly striking and therefore familiar examples of C.r. The time course of these processes is often directly regulated by environmental factors. For example, *Crocus* flowers open in response to increased temperature, and the inflorescences of *Taraxacum* open when illuminated. In other cases, the timing is regulated endogenously (see Physiological clock). The leaves of bean plants kept in constant darkness since germination display circadian movements, once these movements have been triggered by a single brief exposure to light. Even when the parent plant is compelled experimentally to respond to a different periodicity (e.g. a 16-hour rhythm), its seedlings still display an endogenous 24-hour cycle of leaf movements. After C.r. of plants have been set in motion by an initial exposure to the natural 24-hour cycle of light and dark, they may continue for days or even months in the absence of an environmental stimulus, i.e. in constant light or dark, with only a gradual decrease in oscillatory amplitude, e.g. the growth rate of *Avena* coleoptiles, conidia production by the fungus, *Neurospora.*

The activity of most animals is subject to C.r. See Biorhythm, Orientation.

Circaea: see *Onagraceae.*

Circaerus: see *Falconiformes.*

Circinae: see *Falconiformes.*

Circular dichroism, ***CD:*** the elliptical polarization of light which occurs when linearly polarized light (regarded as consisting of superposed equal left and right circularly polarized components) passes through an optically active medium. The absorption coefficient of the optically active medium is different for the right and left circularly polarized light, i.e. the medium displays CD. Both optical rotation and the ellipticity of CD are dependent upon the wavelength. A plot of the angle of rotation against wavelength is known as the optical rotation dispersion (ORD) spectrum, while the plot of CD or ellipticity against wavelength is called a CD spectrum. Depending on the relative values for the absorption coefficients of the left and right circularly polarized component, optical rotation and ellipticity are negative or positive. These spectral effects are known as Cotton effects (named for their discoverer). CD spectrocopy is used to investigate the conformation of macromolecules.

Circulation time: the time taken for the blood to pass from one point to another in the circulation. C.t. is measured by injection of various substances at easily accessible sites in the circulatory system, e.g. the total C.t. can be determined by injecting fluorescein into a vein and timing its appearance in the corresponding vein on the opposite side of the body, i.e. the time taken to pass through the systemic and pulmonary circulation. The jugular to jugular time in humans is about 22 seconds, while the time from the veins of one foot to those of the other foot is about 28 seconds. As a close approximation, the total C.t. for all species is the time of 28 heart beats.

Circulatory regulation: the adjustment of cardiac activity, blood pressure, and vascularization to meet the functional requirements of the organs.

C.r. is effected primarily by the regulation of blood vessel diameter. Dilation of blood vessels causes the blood pressure to decrease, and increases the quantity of blood in that section of the circulation (i.e. increases the blood supply), whereas constriction of blood vessels has the opposite effects. Cardiac activity also contributes to C.r., in that increased heart contraction increases the blood pressure and the rate of circulation.

1) C.r. is initiated either by local factors or by the processing of stimuli in the periperal receptor system.

a) Local factors. The blood supply is decreased by mechanical stimuli, such as incision or puncture, brief chilling, and by endogenous effectors like adrenalin, noradrenalin, angiotensin, serotonin and vasopressin. The blood supply is increased by carbon dioxide, acids (e.g. lactic acid), oxygen deficiency, warming, and by endogenous effectors like histamine, acetylcholine, bradykinin and prostaglandins.

b) Some centrally processed receptor systems for C.r. are Baroceptors (see), arterial Chemoreceptors (see), and cardiac receptors (see Cardiac reflexes).

2) Centers for C.r. are seldom anatomically well defined structures, but rather widely branched functional circuits, present in several regions of the central nervous system. Spinal cord centers are present in the cervical region, and the medulla oblongata of the brainstem is the site of the actual *Vasomotor center* (see). The latter adjusts blood pressure and blood flow by regulating Vessel tone (see). As shown in Fig.1, the vasomotor center specifically intensifies or decreases the circulation according to the blood requirement in different parts of the circulation. Close to the vasomotor center is a *cardiac control center.* Centers affecting the circulation are also present in the hypothalamus and cerebral cortex.

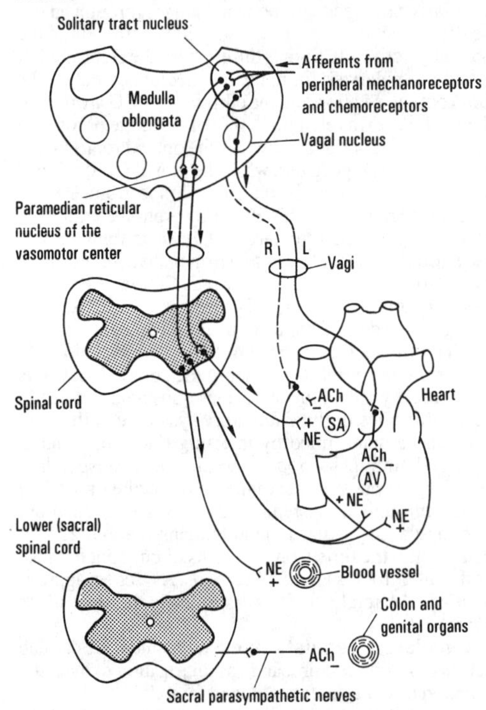

Fig.1. *Circulatory regulation.* Nervous control of cardiac activity and blood vessel diameter. *NE* Noradrenalin. *SA* Sinus node. *AV* Atrioventricular node. *ACh* Acetylcholine.

3) C.r. may be local, nervous, or hormonal.

a) Vessel tone (see) is adjusted by local changes in blood vessels.

b) *Circulatory reflexes.* The main mechanism of nervous C.r. is the adjustment of the diameter of arterioles via sympathetic nerves, known as *vasoconstrictors*. Increased vasoconstrictor activity causes constriction of arterioles and venules, whereas decreased activity results in their dilation (increase in internal diameter) (Fig.2). *Axon reflexes* may be regionally involved in nervous C.r. [an axon is the long extension (neurite) of a neuron]. In an axon reflex, chemical stimuli (e.g. metabolic products in the blood) excite sensory endings in the vessel walls. Their action potentials are not transmitted via ganglion cells and synapses to another neuron, but return to the blood vessel in the same axon via a colateral. Because of their short conduction path, axon reflexes operate rapidly, and because of the absence of synaptic transmission, they operate with minimal energy expenditure. Axon reflexes operate independently of, and are not influenced by, centers in the central nervous system. An example of an axon reflex is *dermographism:* after hard rubbing of the skin with a blunt instrument, which stimulates the pain receptors, the skin becomes locally flushed, due to increased blood flow.

c) The most important hormones of C.r. are the two catecholamines, adrenalin and noradrenalin, which cause constriction of arterioles and venules. Vessel diameter is continually adjusted by their graded release.

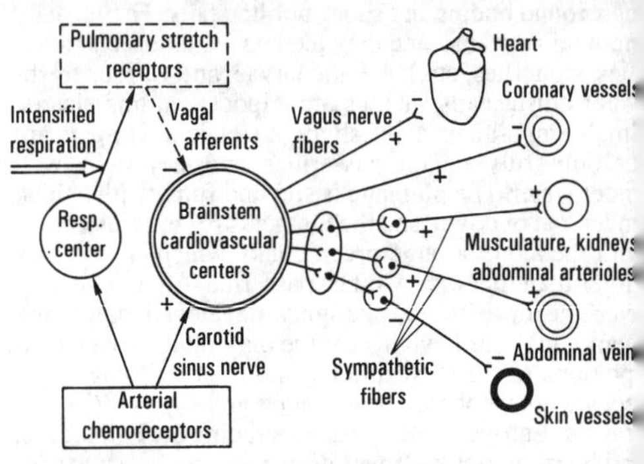

Fig.2. *Circulatory regulation.* The influence of respiration on cardiac activity and the blood supply to important areas of the circulation.

Circulatory system: see Blood circulation.

Circumnutation: a nutation movement of growin plant organs, in which the longitudinal growth c shoots, roots, leaves, etc. does not occur strictly in straight line, because one side of the plant organ grow more rapidly than the other. The zone of rapid growt shifts at random to different sides of the growing orga so that the organ tip describes an irregular and compl cated path. In contrast, Cyclonasty (see) is a regular cy clic movement.

Circumscissile capsule: see Fruit.

Circus: see *Falconiformes.*

Cirphus: see Cutworms.

Cirripedia, ***barnacles:*** a subclass of the *Crustace* Adults are always attached to a substratum, or they liv as parasites of other animals. The body is short ar compresssed, the pleon strongly reduced and alway without appendages. The thorax consists originally six segments, but in most species there is practically r evidence of external segmentation. Each segment ca ries a pair of long, biramous appendages (cirri), eac consisting of many segments, densely covered with s tae. Quite unsuitable for locomotion, the cirri serve feeding appendages by scooping particles from the w ter on their setae. In parasitic forms, the cirri are dege erate or absent. The larval body is enclosed in a tw valved carapace, which persists in the adult where may also be called the mantle; the mantle cavity is th the space between the carapace and the body. In nonpa asitic forms, characteristic calcareous plates may form a solid shell on the outside of the mantle. The majority of species is hermaphrodite. In dioecious species the male is extremely small and is attached to the mantle of the female.

The egg hatches to a Nauplius (see) larva, which remains for some time in the mantle cavity of the mother. Nauplius larvae then swarm into open water, where they metamorphose via the metanauplius to the Cypris (see) larva, which has a two-valved carapace and closely resembles an ostracod. The cypris larva always becomes attached by its cephalic region to the substratum, where it develops into the adult.

About 800 species are known, practically all of them marine. Five orders are recognized.

1. Order: *Thoracica.* These attach to stones, wooden pilings and ships' hulls, as well as crabs, turtles and whales. The order contains the ***Goose barnacles*** *(Lepas),* e.g. the common goose barnacle *(Lepas anatifera),* in which the foremost part of the head forms a stalk (peduncle) attached to the substratum. Attached to the stalk is the capitulum which consists of the rest of the head, the thorax (with 6 pairs of cirri), the remains of the abdomen and the shell lobes with their calcareous plates. They are found attached to seaweed, driftwood and ships' hulls. In the *Balanomorpha,* the body is enclosed by an outer wall of skeletal plates. This ring of plates is firmly attached to the substratum, and it is closed at the top by two movable plates, the scutum and the tergum. Most familiar of the *Balanomorpha* is the ***Common acorn barnacle,*** *Balanus.* Acorn barnacles cause considerable damage by forming incrustations on the bottoms of ships.

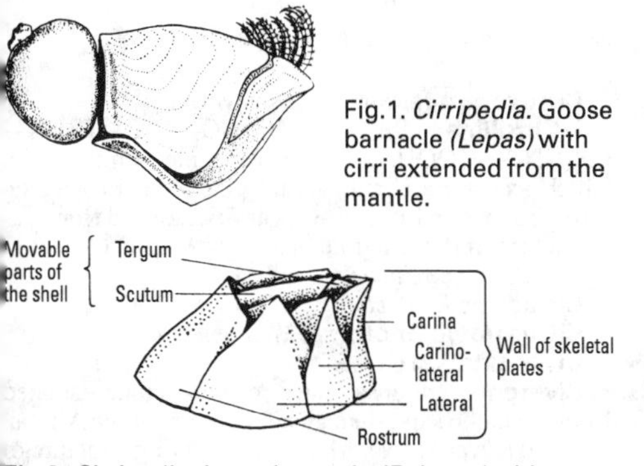

Fig.1. *Cirripedia.* Goose barnacle *(Lepas)* with cirri extended from the mantle.

Fig.2. *Cirripedia.* Acorn barnacle *(Balanus)* with closed mantle.

2. Order: *Acrothoracica,* with the genus *Trypetesa.* The sexes are separate and there are fewer than 6 pairs of thoracic appendages. Females are extremely small, and they live in hollows excavated in the shells of mollusks. Males are regressed.

3. Order: *Apoda.* Contains only a single member, *Proteolepas,* a small, maggot-like creature first described by Darwin in the mantle cavity of the stalked barnacle, *Alepas.* There is no mantle and there are no thoracic appendages.

4. Order: *Rhizocephala,* with the genera *Peltogaster* and *Sacculina.* These are endoparasites in other crustaceans, usually decapods. Most familiar is *Sacculina,* which is a parasite of crabs. It sends out a highly branched network of roots within its host. The externally visible part of the parasite, which hangs on a short stalk, consists of practically only the sex organs enclosed by the mantle (carapace). Although highly modified, it is recognizable as a crustacean from its nauplius and cypris larvae.

5. Order: *Ascothoracica.* Parasites of corals and echinoderms, often embedded in the host tissues, e.g. *Laura* parasitizes the black coral, *Gerardia.* They represent an early evolutionary branch of of the cirripeds, which has retained the abdomen; the latter is sometimes well segmented and provided with movable caudal rami.

Cirroteuthis: see *Cephalopoda.*

Cirrothauma: see *Cephalopoda.*

Cirrus (plural **Cirri):**

1) In some platyhelminths and trematodes, an eversible male copulatory organ (penis), often provided with barbs.

2) In the wider sense it is a tendril-like body processes or correspondingly modified limb, which serves as a tactile organ, locomotory organelle, organ for circulating water, or for attachment to the substratum, e.g. the water circulating cirri of barnacles; the locomotory organelle of some ciliate protozoans which is formed by fusion of a group of cilia; the slender ciliated appendage of polychaetes.

3) A tuft of skin on the jaw of certain fishes (e.g. *Creediidae).*

Cirsium: see *Compositae.*

Cisco: see *Coregonus.*

Cis-configuration: a term apllied to the configuration of the alleles of two coupled genes (see Coupling), or the Heteroalleles (see) of a cistron. In heterozygous or heterogenic cells, the two normal alleles are localized in one and the same homologous coupling structure (chromosome or genophore), while the two mutant alleles are localized in the other, so that the genotype can be expressed as: ++/ab. See Trans-configuration, Cis-trans test.

Cisticola: see *Sylviinae.*

Cistothorus: see *Troglodytidae.*

Cis-trans test: a genetic test to determine whether two mutations (m^1 and m^2) occur in the same functional gene (cistron), and to establish the limits of this genetically active region. Two types of heterozygote (or heterogenote) are prepared, one containing the mutations under investigation in the same chromosome (cis-configuration: $++/m^1m^2$), the other containing them in different chromosomes (trans-configuration: $+m^2/+m^1$). If two recessive mutations are present on the same cistron, the heterozygous trans-configuration displays the mutant phenotype, whereas the cis-configuration displays the normal (wild-type) phenotype. If one or both mutations is dominant, and the cis- and trans-heterozygotes are phenotypically different, then both mutations are present in the same cistron. Conversely, if the cis- and trans-heterozygotes are phenotypically identical, this is taken as evidence that the mutations are present in different cistrons. Results of a C-t.t. must be interpreted with caution, because mutations within a cistron can exhibit intra-cistron complementation (see Complementation). Thus, a mutant in which two mutations are present in the trans-configuration may display a normal phenotype. But even in cases of intra-cistron complementation, the normal phenotype of the cis-heterozygote is more strongly developed than that of the trans-heterozygote.

Cistugo: see *Vespertilionidae.*

CITES, ***Convention on International Trade in Endangered Species:*** an agreement signed in Washington in 1973 by 57 sovereign states, which came into power in 1975. By the end of 1980 it had been signed by a further 4 states. The convention protects free living animal and plant species that are threatened with extinction by controlling the international trade in those species. Trade in especially threatened species is totally prohibited, while trade in other species is subject to stringent restrictions and dependent on permission from the country of origin. The corresponding lists, which also include products from endangered species (furs, ivory, rhinoceros horn, etc.), are periodically reviewed and usually extended. Signatory states, however, have the right to omit measures for the implementation of CITES from their national legislature. CITES has been instrumental in insuring that, should the need arise, some traditional zoo animals can be procured from breeding colonies. On the other hand, it has caused such a rise in the black market price of many species or their products that poaching, smuggling and the falsification of official documents is far more profitable than previously. Illegal international organizations trading in CITES-listed species and their products have been formed, which are comparable to those responsible for the illicit international trade in narcotics. A special problem is that control agencies must be skilled in identification, and these skills are sometimes very demanding, especially in the identification of products (e.g. different crocodile skins).

Citral, ***3,7-dimethyl-2,6-octadienal***: a doubly unsaturated monoterpene aldehyde, smelling of lemons. It is present in many essential oils, e.g. verbena, lemon and orange oils, and it is the main constituent (about 80%) of lemon grass oil (from *Cymbopogon citratus* or *C. flexuosus).* C. occurs naturally as a mixture of 2 isomers (Fig.): citral A (geranial) and citral B (neral). It is used in the manufacture of cosmetics, perfumes and soaps, and as a precursor in the commercial synthesis of vitamin A.

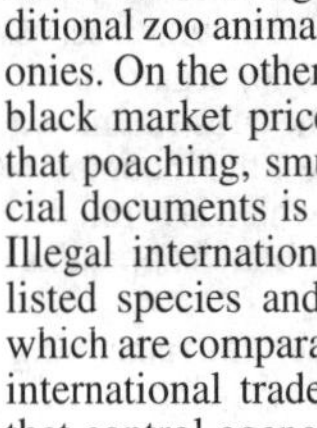

CHO OHC CH3 CH3 H3C CH3 H3C CH3

Citral A (Geranial) Citral B (Neral)

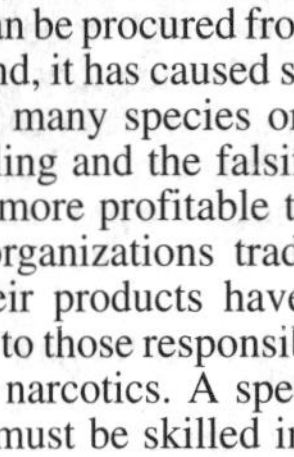

Citric acid: a tricarboxylic acid (see Carboxylic acids) found widely in the plant kingdom, e.g. in lemon juice (5 to 7%), from which it was first isolated by Scheele in 1784. Other sources are apples, pears, raspberries, black currants, milk, conifers, fungi, tobacco leaves, wine. Since C.a. is an intermediate in the Tricarboxylic acid cycle (see), at least small quantities are present in all aerobic living organisms. It is prepared commercially by the fermentation of carbohydrates by certain molds, e.g. *Aspergillus niger.* C.a. is used in the pharmaceutical, food and textile industries. For structure, see Tricarboxylic acid cycle.

Citric acid cycle: see Tricarboxylic acid cycle.

Citron: see *Rutaceae.*

Citronellal, ***3,7-dimethyl-6-octenal***: a mono-unsaturated, monoterpene aldehyde, present in many essential oils, e.g. lemon, lemon grass and melissa oils. C. is the main constituent of citronella oil (from *Cymbopogon nardus,* family *Graminaceae)* and it is an important constituent of the essential oils of various *Eucalyptus* species. It is also an alarm pheromone for ants of the genus *Lasius.*

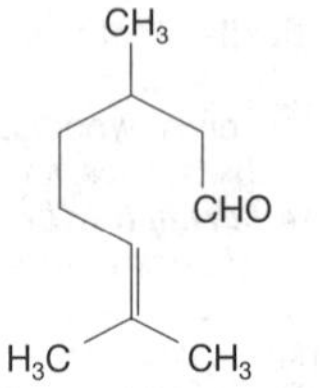

CH3 CHO H3C CH3 Citronellal

Citronellol, ***3,7-dimethyl-6-octen-1-ol***, or ***2,6-dimethyl-2-octen-8-ol***: an unsaturated monterpene alcohol, with an odor of roses. The levorotatory form occurs in rose and geranium oils, and the dextrorotatory form occurs in lemon and citronella oils.

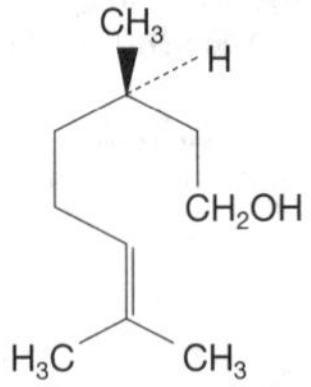

CH3 H CH2OH H3C CH3 (-)-β-Citronellol

Citrovorum factor: see Leucovorin.

L-Citrulline: $H_2N—CO—NH—(CH_2)_3—CH(NH_2)—COOH$, a nonproteogenic amino acid, which occurs free in animals and plants, e.g. in bleeding sap of birches and alder. L-C. was first isolated from water melons. It is an intermediate in the Urea cycle (see).

Citrullus: see *Cucurbitaceae.*

Citrus: see *Rutaceae.*

Citrus flower moth: see *Yponomeutidae.*

Civet cats: see Civets.

Civetone: an unsaturated cyclic ketone isolated from civet, the salve-like gland secretion of the African civet cat *(Viverra civetta).* It contains a ring structure of 17 carbon atoms, and forms colorless crystals. C. smells strongly of musk and civet cats, and in the concentrated form the smell is offensive. Highly diluted, it is used as a fixation agent in perfumery.

Civets, ***civet cats,*** *Viverra:* a genus of viverrids (see *Viverridae)* represented by 5 species, which are native to sub-Saharan Africa and southern Asia. Some authorities place the 5 species in 2 genera: *Viverra* and *Civettictis.* They are primarily ground dwellers and do not burrow, and they have sharp, arched, semi-retractile claws on each of their 5 fingers and toes. C. have large anal pouches, the contents of which have a powerful musk-like odor. The contents of the anal glands of the ***African civet*** *(Viverra civetta)* , which are known as "civet", are used in greatly diluted form in the preparation of perfume (see Civetone).

Civets. Asiatic civet cat.

Civettictis: see Civets.

Cladocera: see *Branchiopoda.*

Cladistic distance, ***cladistic separation:*** the number of branching points between any two points on a phylogenetic tree.

Cladogenesis: a term coined in 1947 by B. Rensch for the branching of an ancestral lineage to form equal sister taxa (species, genera, families,etc.) in the process of evolution.

Cladonia: see Lichens.

Clamatores: an alternative name for the suborder, *Tyranni,* of the *Passeriformes* (see). See Syrinx.

Clamp connection: a looped structure on the hyphae of basidiomycete fungi, which serve for the transfer of a nucleus of the appropriate mating type to each successive dicaryotic cell (Fig.).

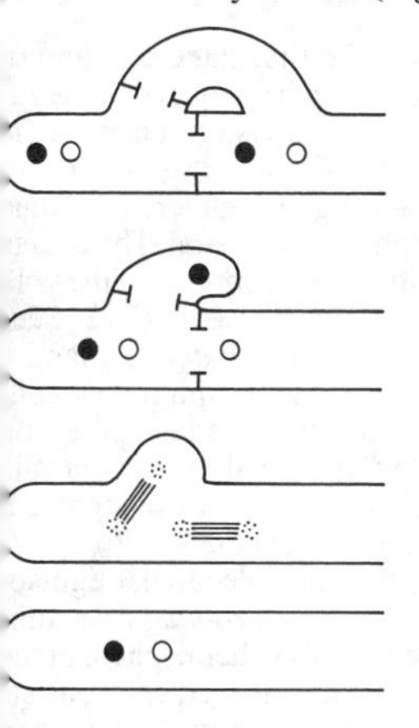

Clamp connection. Progressive stages (bottom to top) in the formation of a clamp connection on a fungal hypha.

Mycelia from germinating basidiospores contain unnucleate cells. When mycelia of opposite mating types make contact, they fuse (somatogamy) and the nuclei pair off but do not fuse. The fusion cell then develops a lateral, hook-shaped outgrowth, which represents the beginning of a C.c. One of the nuclei passes into this structure, then divides; one daughter nucleus remains in the hook, while the other moves to the apex of the cell. Meanwhile, the other nucleus also divides, one daughter nucleus migrating to the cell apex, the other to the cell base. A transverse wall develops immediately below the hook, to form a terminal and a subterminal cell. A developing transverse wall separates the hook from the terminal cell, which therefore now contains two nuclei of opposite mating types. The hook continues to grow, forming a complete loop (C.c.) and fusing with the basal cell. The nucleus that was temporarily isolated in the hook now enters the subterminal cell via the C.c. With continued growth of the terminal cell, C.cs. are formed at each cell division, so that each transverse wall is associated with C.c.

Clams: see *Bivalvia.*

Clarias: see *Clariidae.*

Clariidae, ***airbreathing catfishes, labyrinthine catfishes:*** a family of catfishes (see *Siluriformes),* containing 13 genera and about 100 species, inhabiting fresh water in Africa, Syria, and southern and western Asia. They have long, plump bodies. The dorsal fin is very long, usually with more than 30 rays, not preceded by a spine, and separate or continuous with the dorsal fin. The gill openings are wide, and there are 4 pairs of barbels. An opening between the 2nd and 3rd gill arches leads to 2 sacs, which have highly vascularized (from all 4 afferent branchial arteries) aborescent extensions, with cartilaginous supports originating from gills II and IV. The walls of this labyrinthic structure function in aerial respiration. Some members move overland between water sources. *Clarias batrachus,* a species of walking catfish, is distributed naturally throughout most of the range of the family, and has also been introduced into southern Florida. The burrowing habit is well developed in the 3 African genera, *Gymnallabes, Channallabes* and *Dolichallabes,* which also have small eyes and reduced or absent pectoral and pelvic fins. Blind forms are *Uegitglanis* (Somalia), *Horaglanis* (India), and a single species of *Clarias* from Southwest Africa.

Clarkia: see *Onagraceae.*

Clasping antennae: modified antennae of insects, used for clinging to a sexual partner. For example, the male of *Sminthurides aquaticus* (see *Collembola)* uses its C.a. to cling to mature females, and it is often carried by a female for days.

Class:

1) A systematic unit (see Taxonomy).

2) The highest vegetation unit of the plant sociological system of Braun-Blanquet (see Character species).

Class group of vegetation: the highest and largest unit of vegetation in the phytosociological system of Braun-Blanquet (see Character species). The class group is defined by plant species and genera, and by plant communities (usually classes and order) which are specific to, or characteristic of it, or show their optimal development within it. Class groups embrace large regions, and they are delineated partly by macroclimate and partly floristically. They correspond closely to the phytogeographical floral regions (see Floristic kingdom). European class groups are the Eurosiberian-North American, the Alpine-Nordic, the Mediterannean and the Irano-Caspian class groups.

Classical model of population structure: the assumption that, in natural populations, all the gene loci of favorable alleles become monomorphic through selection. The persistence of polymorphism is explained by the generation of deleterious factors by mutation. See Balance model of population structure.

Clathrus: see *Basidiomycetes (Gasteromycetales).*

Clavaria: see *Basidiomycetes (Poriales).*

Clavate antennae: see Antennae of insects.

Clavicipitales: see *Ascomycetes.*

Clavicle: see Pectoral girdle.

Clawed toads: see *Pipidae,* African clawed toad.

Claws, hoofs and nails: thickened, keratinized epidermal structures covering the ends of the fingers and toes, which grow continually from a germinative layer near the base. Claws (faculae) are curved and pointed; they occur in some amphibians, most reptiles, birds and mammals, and are used variously for burrowing, cleaning, offense and defense, and for climbing and gripping supports. True hoofs (ungulae) occur in the *Ungulata*. Most primates have nails, some have both nails and claws, while the most primitive have only claws. In hyraxes, claws and hoof-like claws are found in the same individual. Agouitis and pacas have hoof-like claws. The claw can be considered as the basic type of structure, and nails and hoofs as broadened modifications.

The underlying corium (nail bed) of the human nail is highly vascularized, giving a pink appearance. At the base, the epidermis is invaginated to form a nail root, where a short covering projection of the stratum corneum forms a pale-colored fold (eponychium). Beneath the nail root the corium is less vascularized, producing a whitish lunula. Claws, hoofs and nails all have a hard upper unguis (e.g. the nail plate) and a softer subunguis. The subunguis or sole plate is least prominent in the human nail; it consists of somewhat less modified epidermis on the opposite side of the terminal phalanx from the unguis, and it is particularly well developed in the hoof, where the weight is carried on the tip of the digit. Beyond the sole plate lies the pad (e.g. the ball of the primate finger), and in the horse's hoof this is also known as the frog or cuneo. The strong, transversely arched, shoelike horny unguis of the hoof completely envelops the digit, whereas nails and claws occur only on the upper surface.

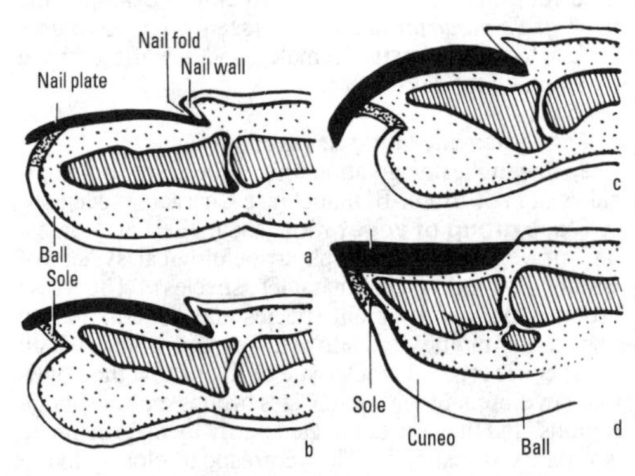

Claws, hoofs and nails. a Nail (human). *b* Nail (monkey). *c* Claw (cat). *d* Hoof (horse).

Cleaning behavior: see Grooming.

Cleaning shrimps: see Shrimps.

Clearance: the volume of plasma that is cleared of a specified material in one minute. In most cases, plasma clearance is only partial, and C. is a theoretical value. C. is an important kidney function test, because it shows whether excretory material is simply filtered, or whether it is also resorbed or secreted, and to what extent its rate of excretion may be pathologically altered. $C = UV/P$, where C = clearance value in ml/min, U = urinary concentration of test substance in mg/ml, V = rate of urine production in ml/min, and P = plasma concentration of test substance in mg/ml.

The reference material is inulin, a plant polysaccharide found in members of the genus *Inula*. Inulin is excreted almost exclusively by glomerular filtration, and in humans shows a C. of 120 ml/min. C. values lower than 120 ml/min are observed when a substance is resorbed after filtration, whereas higher C. values indicate that secretion has occurred in addition to filtration. *p*-Aminohippuric acid is excreted by both glomerular filtration and tubular secretion, and is removed almost completely by a single passage through the kidneys, thus providing a means of measuring the true plasma flow rate through the kidneys. The ratio of the Cs. of inulin and *p*-aminohippuric acid therefore represents the filtration fraction: $C_{In}/C_{pAH} = FF$. Radioactive materials are now often used for the measurement of C., e.g. derivatives of inulin containing radioactive iodine or chromium, vitamin B_{12} containing radioactive cobalt, etc., thus providing a more rapid and exact means of measurement.

If oxidative phosphorylation is inhibited, the C. of materials normally subject to tubular resorption and secretion becomes identical with that of inulin. This shows that glomerular ultrafiltration is a passive physical process, whereas resorption and secretion are active biochemical processes.

Clear water stage: a stage in the development of natural waters, in which filtrable zooplankton become abundant at the expense of phytoplankton, and the latter almost totally disappears.

Clearwing moths: moths of the family *Aegeriidae*. See *Lepidoptera*.

Cleavage, ***blastogenesis:*** the first stage of embryogenesis (see Ontogenesis), in which a sequence of regular and mostly geometric mitotic divisions occur in the fertilized and unfertilized egg. Growth does not occur during C., so that the resulting blastomeres become smaller at each successive phase of division. The course of C. depends on the quantity and distribution of the yolk (deutoplasm), which can present a barrier to C. (See entry under Egg for definition of terms related to quantity and distribution of yolk). In eggs with a high yolk content, the yolk accumulates more at the vegetal pole, and C. tends to be restricted to the animal or cytoplasmic pole. Thus, two families of C. types can be recognized: 1. *holoblastic* (complete) C., in which the whole egg becomes subdivided into blastomeres (alecithal, oligoisolecithal, mesotelolecithal eggs); 2. *meroblastic* (incomplete) C., in which the vegetal (yolky) hemisphere of the egg remains uncleaved, so that the cleaved egg consists of blastomeres at the animal pole and a residual mass of yolky cytoplasm with scattered nuclei at the vegetal pole (polytelolecithal and centrolecithal eggs).

The most regular type of C. is *complete-equal C* (Fig. 1), which occurs in the iso-oligolecithal eggs of many sponges, many coelenterates, some echinoderms and crustaceans, mammals, and exceptionally in the low-yolk, centrolecithal eggs of some insects (e.g. springtails).

In complete-equal C., the egg divides into equal sized blastomeres, a process initiated by two meridional and one equatorial C., and two further Cs. parallel to the equator at the tropics of Cancer and Capricorn. Early in this process, the mass of cells acquires a superficial resemblance to a mulberry, and is therefore called a morula. As C. continues, a massive structure may be formed with blastomeres filling all the space once occu

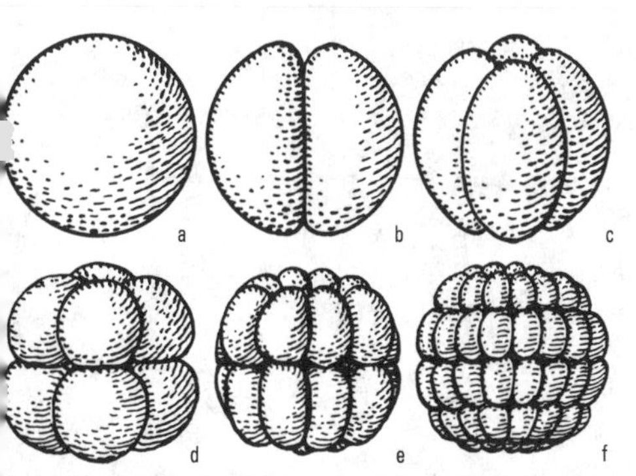

Fig.1. *Cleavage.* Complete-equal cleavage (e.g. sea urchin). *a* Egg. *b* 2-Cell stage. *c* 4-Cell stage. *d* 8-Cell stage. *e* 16-Cell stage. *f* Morula.

pied by the uncleaved egg. This structure is also referred to as a morula. More often, as C. progresses, the blastomeres become distributed in a single layer, all of them contributing to the external surface of the embryo, and enclosing a central cavity known as the blastocoel. The blastocoel appears at first as a narrow space in the center of the group of blastomeres usually at the third division (i.e. the 8-cell stage). As C. continues, the blastocoel increases in size, and the resulting hollow sphere is called a blastula. There are several variations of complete C. In *radial C.,* the blastomeres lie exactly one above the other after the first sequence of division, i.e. during divisions perpendicular to the egg axis, the mitotic spindles lie parallel to this axis (e.g. coelenterates and nematodes). In the *spiral C.* of marine turbellaria, polychaetes, snails and some nematodes, the blastomere layers are shifted by 45 ° in relation to one another. This arises because each C. plane deviates from the horizontal position and is inclined at an angle. As C. continues, the angled divisions occur alternately clockwise (dexio-division) and anticlockwise (leiotropic division). At the third C., the four spindles are arranged in a rough spiral, which determines the final direction of spiraling as observed from above (i.e. clockwise or dextral; anticlockwise or sinistral). *Bilateral C.* (in rotifers, tunicates, lancet fishes, vertebrates, some nematodes) approximates to radial C., but two of the first four blastomeres are larger than the other two, thereby establishing a plane of bilateral symmetry, which may become more distinct with subsequent Cs.

Ctenophores or comb jellies show a very rare type of C., called *disymmetrical C.,* in which the morula possesses two planes of symmetry, one above the other. The protoplasmic layer of the fertilized egg accumulates on the side which will become dorsal, and C. occurs along a vertical plane. The resulting two blastomeres therefore each possess a protoplasmic cap. A second division at 90 ° to the first results in four blastomeres. This is followed by further division into daughter cells of unequal size, producing an 8-blastomere embryo, each blastomere with a protoplasmic cap at its dorsal end. A horizontal C. then divides off the protoplasmic caps to form 16 distinct blastomeres, consisting of 8 large ventral yolk-containing macromeres and 8 small, dorsal, protoplasmic micromeres.

In *determinate* or *mosaic C.* (in rotifers, nematodes, mollusks, tunicates and others), specific blastomeres are predestined to give rise to specific parts of the embryo.

Even relatively low levels of yolk, as in the eggs of lancet fishes and echinoderms, lead to *complete-unequal C.,* i.e. the equatorial C. plane is shifted toward the animal pole. The subsequent mono- or multilayered blastula therefore consists of micromeres and macromeres, as in some coelenterates, echinoderms, tunicates, lung fishes and amphibians (Fig. 2). As in the case of complete-equal C., there are also different subtypes of complete-unequal C.

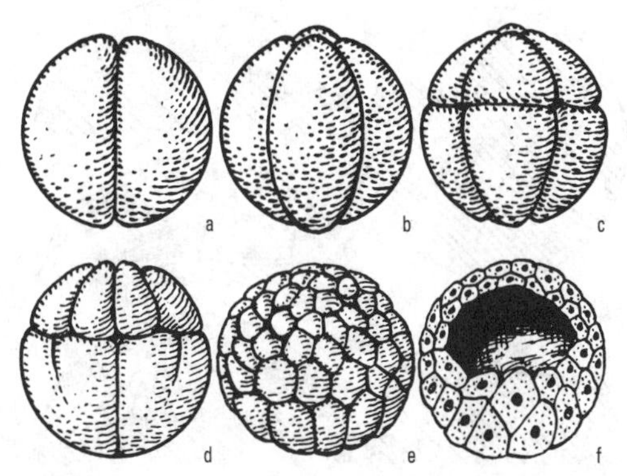

Fig.2. *Cleavage.* Complete-unequal cleavage (e.g. frog egg). *a* 2-Cell stage. *b* 4-Cell stage. *c* 8-Cell stage. *d* Several cell stage. *e* Morula. *f* Blastula (longitudinal section).

A special type of C. (*asynchronous-complete-unequal*) occurs in the eggs of humans and other mammals (excluding monotremes) (Fig. 3). Within the compact morula, crevices appear between the inner cell mass and the cells of the trophoblast (*trophoblast C.).* These crevices enlarge to a fluid-filled cavity, and the trophoblast and inner cell mass (embryoblast) become separated over most of the inner surface, leaving the embryoblast attached on one side only. At this stage the embryo is called a blastocyst (see Gastrulation).

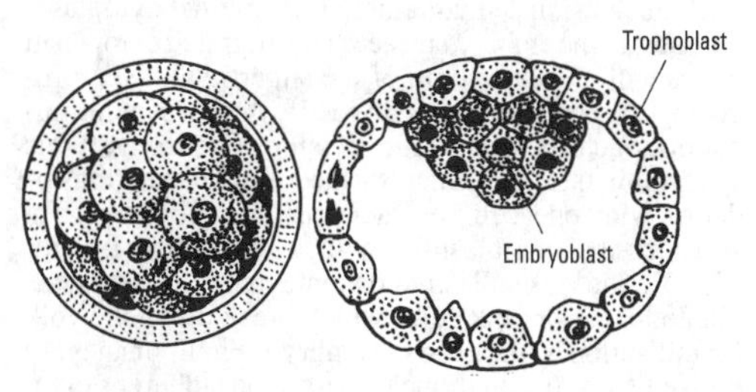

Fig.3. *Cleavage. a* Morula (human). *b* Median section through a blastocyst (human).

Holosteans and lung fishes show various types of C., which are transitional between total and partial.

In the extremely telolecithal eggs of cephalopods, sharks, rays, teleosts (Fig. 4), monotremes, reptiles and birds, C. occurs only in the blastodisc at the animal pole, and is known as *discoidal C.* The resulting discoblast consists of three or more layers of blastomeres. Beneath the discoblast and adhering to the yolk is a thin syncytial layer, called the periblast, which is not destined to participate in the formation of the later embryo. Nuclei from peripheral cells of the discoblast migrate into the periblast, which then appears to function by breaking down the yolk to provide nutrients for the growing embryo. The discoblast and periblast separate from one another, leaving a cavity called the blastocoel. In birds an extension of the blastocoel, the subgerminal cavity, is formed by liquifaction of the yolk. The eggs of some crustaceans and spiders also show discoidal C.

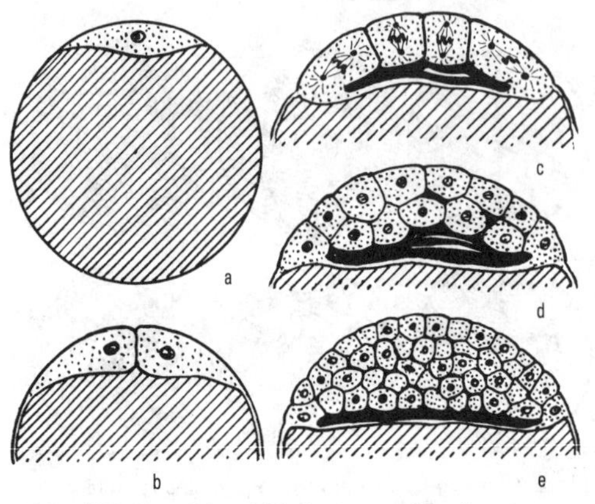

Fig.4. *Cleavage.* Discoidal cleavage of a teleost egg. *a* Fertilized egg. *b* to *e* Subsequent stages of cleavage, all restricted to the embryonal disk.

Superficial C. (a type of incomplete C.) is characteristic of the centrolecithal eggs of arthropods (spiders, crustaceans, myriapods, insects). Certain transitional types of C. between complete and superficial C. are exhibited by many primitive insects and certain spiders. At the onset of superficial C., the nucleus (synkaryon) of the centrolecithal egg is located at the center of the yolk mass, surrounded by a small amount of cytoplasm. Mitosis then produces several nuclei embedded in a central mass of cytoplasm (the cytoplasm does not divide). Each nucleus, in a small amount of original cytoplasm migrates to the embryo surface. The migrated cytoplasm unites with the cytoplasm of the superficial layer, producing a syncytium of undivided cytoplasm with numerous nuclei. Cytoplasmic divisions then convert this syncytium into a cellular surface epithelium (blastoderm) with one cell for each nucleus. The resulting structure is a periblastula, with a yolk-filled interior (Fig. 5). The yolk still contains some free cells, known as vitellophages or yolk cells, which break down the yolk for utilization by the growing embryo. Vitellophages are derived from residual nuclei with their adherent cytoplasm, or from cells migrating in from the blastoderm (primary or secondary vitellophages, respectively).

Cleavers: see *Rubiaceae*.

Cleistocarpy: fruit formation in flowers that never open and are self-pollinated (cleistogamous flowers). See Pollination.

Cleistogamy: see Pollination.

Cleistothecium: see *Ascomycetes*.

Cleithrum: see Pectoral girdle.

Clelia: see Mussurano.

Clematis: see *Ranunculaceae*.

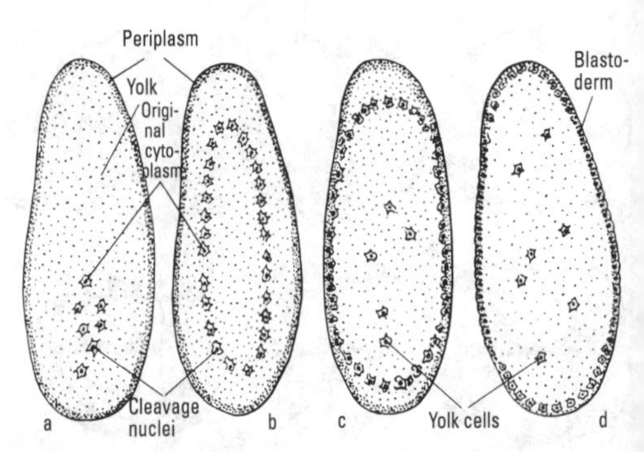

Fig.5. *Cleavage.* Superficial cleavage with blastoderm formation in the insect egg *(Hydrophilus piceus).*

Clibanarius: see Hermit crabs.

Climacograptus: a genus of diplograptid Graptolites(see), found from the Lower Ordovician to the Lower Silurian. The rhabdosome is elongate with two rows of thecae. The thecae do not overlap and stand nearly vertically from the rhabdosome; they are suboval in section, with more or less slit-like, narrow apertures, and they are distally somewhat contracted.

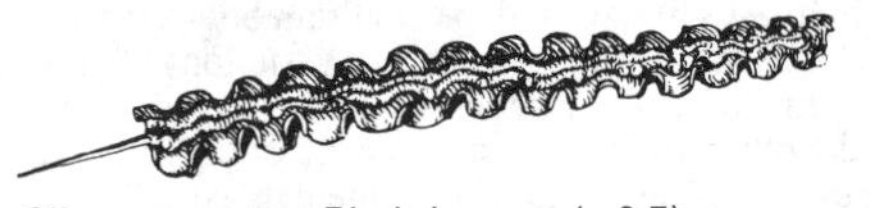

Climacograptus. Rhabdosome (x 0.5).

Climate: average annual weather conditions recorded over a long period (usually more than 30 years). C. is determined by the interaction of climatic factors: intensity of solar irradiation, temperature, level of precipitation, rate of evaporation, air humidity, air movement, atmospheric pressure, etc., representing an important complex of abiotic environmental factors. The interdependence of biomes, C. and soil group is shown in Fig.1. According to its distance from the equator, a C. may be tropical, subtropical, temperate or polar. In many respects, this horizontal zonation of Cs. corresponds to the vertical zonation of mountain Cs. (see Altitudinal zonation). Maritime (oceanic) Cs. are characterized by high relative humidity and small annual temperature variations,

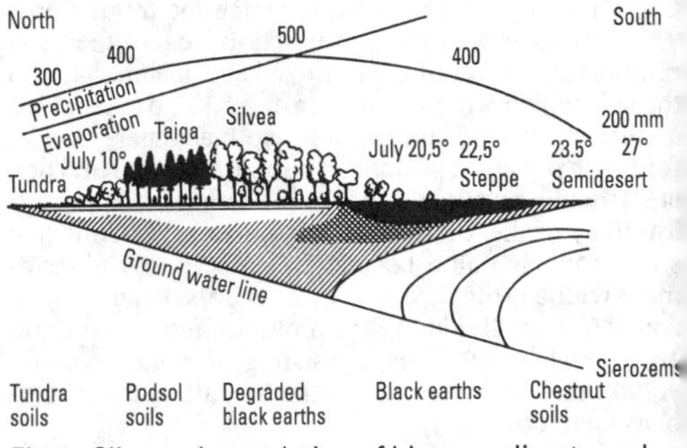

Fig.1. *Climate.* Interrelation of biomes, climate and soil groups.

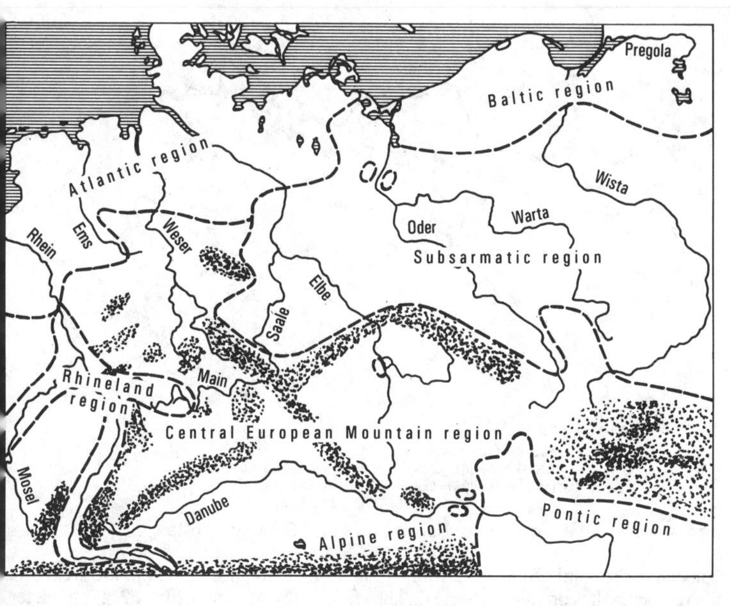

Fig.2. *Climate.* Climatic regions of Central Europe.

whereas a continental C. typically displays extreme seasonal variations. Within each macroclimate (tropical, temperate, etc.), there are different climatic regions (Fig.2). The general C. of a particular ecosystem is known as the ecoclimate. Each organism in an ecosystem is generally associated with a particular microclimate. Thus, the microclimates of mammalian nests and dens, birds' nests, wind-protected sites, sun-exposed sites, soil interstices, tree canopy, etc. may differ considerably from the general climate of the ecosystem that contains them.

Climatic chamber: a chamber designed for generating specific microclimates. Generally, only temperature, illumination and relative humidity are regulated. With appropriate relay systems, daily variations of these factors can also be simulated. A battery of C.cs. can be used to study the response of organisms simultaneously to a range of factor intensities, and to determine the optimal value of each factor.

Climatic laws: laws which attribute morphological and physiological differences between closely related animal races or species to the effects of climate on evolution. According to Schwarz's law, these differences are greater between climatically separated species than between climatically separated races, because racial differences arise from smaller evolutionary changes.

Some important C.l. are stated briefly below. All are based on average or frequently observed tendencies. Within observed trends there may be a considerable variation due to other influences on the evolutionary process. A Lamarkian interpretation of climatic factors as "causes" of change is not acceptable.

1. *Bergmann's law.* Of closely related warm-blooded animals, the larger live in colder climates. For a threefold increase in body volume, the surface area increases only twofold. Thus, if two organisms of the same shape are compared, the larger possesses a relatively smaller surface area and is therefore subject to relatively less heat loss. Deviations from the law have been reported, e.g. wolves (*Canis lupus*) become progressively larger from the mountains of northern Mexico northwards; but at the 60th parallel the trend is reversed, so that Arctic wolves are somewhat smaller than their temperate counterparts [V. Geist (1987) *Canadian Journal of Zoology*, ***65***, 1035–1038].

2. *Allen's law.* The feet, ears and tails of mammals tend to be shorter in colder climates when closely allied forms are compared, thereby reducing heat loss and the possibility of freezing.

3. *Rensch's hair law.* The density and length of mammalian hair is less in warm climates.

4. *Hesse's law.* In warm-blooded animals the relative heart weight is greater in cold climates, representing a higher metabolic and circulatory demand.

5. *Rensch's egg number law.* Under long day conditions (temperate zone) song birds lay more eggs per clutch than under short day conditions. This correlates with the longer available period for food collection for the brood.

6. *Gloger's law.* Damp, warm climates (humidity seems to be important) favor eumelanin formation, so that southern races tend to be black, brown, gray and especially rust-red. *Kleinschmidt's law* is similar, and it states that northern races of birds are relatively large and lightly pigmented, whereas southern races are smaller and darker.

Climax: hypothetical final stage in the development of an ecological system of vegetation. Under natural conditions and with sufficient time, vegetation and the soil are thought to pass through intermediate stages of development, eventually achieving a more or less stable state, representing an equilibrium with the existing environmental conditions. The final stage or C. is a rather specialized phytosociological unit, e.g. oak forest. The concept of C. is now only occasionally used, and then it is more or less equivalent to Zonal vegetation (see).

Climbing perch: see Respiratory organs.

Climbing plants, ***lianes:*** ground-rooted, thin-stemmed herbs or shrubs, rarely trees, which by climbing on other plants, rockfaces, walls, etc., bring their

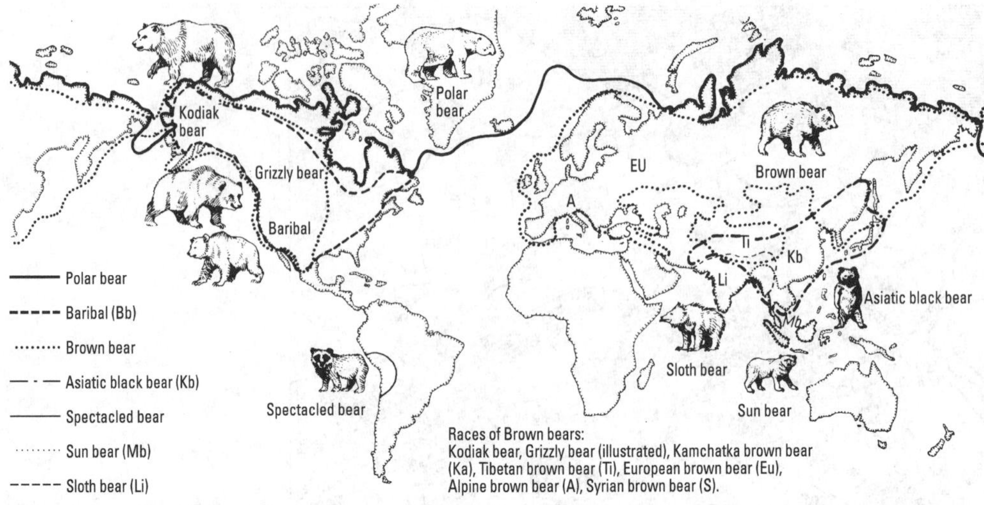

Climatic laws. Distribution of large bears as an illustration of Bergmann's and Schwarz's rules.

foliage from lower shaded regions into the sunlight. C.p. display intense, longitudinal shoot growth with little or no secondary thickening, and their shoots contain very long (up to 5 m), wide vessels, which enable water to be conducted as high as 300 m. They do not develop self-supporting stems or trunks.

Some C.ps. climb by means of ***hooked lateral shoots*** [e.g. Woody nightshade *(Solanum dulcamara)*], while others use ***stiff hairs*** [e.g. Cleavers *(Galium aparine)*], ***prickles*** [e.g. Climbing roses *(Rosa* spp.) and Brambles *(Rubus* spp.)] or ***thorns*** [e.g. Duke of Argyll's tea plant *(Lycium)* and *Bougainvillea*]. ***Root climbers*** employ adventitious roots which serve as attachment organs [e.g. Ivy *(Hedera)* and many members of the *Araceae*]. ***Tendril climbers*** attach themselves with the aid of strongly haptotropic tendrils, which wrap around supports and anchor the shoot [e.g. Grape vine *(Vitis)* and Bryony *(Bryonia)*]. In ***twining plants,*** the stem has long internodes and ascends by winding around a support [e.g. Hops *(Humulus)* and Runner bean *(Phaseolus)*].

Climbing plants. Adaptations for climbing. *1* Hooked hairs (goosegrass). *2* Leaf thorns (ratan palm). *3* Twining shoot axis (runner bean). *4* Tendrils (bryony). *5* Adventitious attachment roots (ivy). *6* Attachment and feeding roots (strangling fig).

Cline: a gradation of a character within the areal of a species. The frequency or intensity of a character changes more or less regularly along a C. Conversely, an *isophene* is a line along which the frequency or intensity of a character is more or less constant. A C. runs at right angles to an isophene (Fig.).

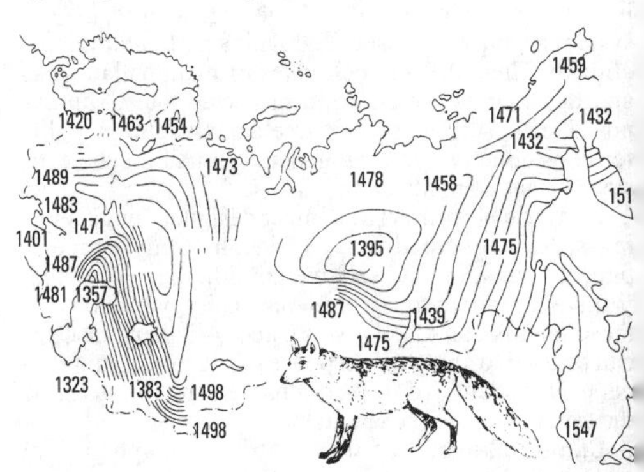

Clinal variations in the skull length of the fox (Vulpes vulpes). The lines are isophenes, and the numbers are the skull lengths in millimeters.

Clingfishes: see *Gobiesocidae*.

Clinical picture: in clinical medicine, the totality of externally recognizable signs of a disease, disorder or injury. See Symptom.

Clinidae: see *Blennioidei*.

Clinodont: see Dentition.

Clionidae: see *Porifera*.

Clitellata: a group of annelids comprising the *Oligochaeta* and the *Hirudinea*. The segments bearing the sex organs become swollen to form a glandular clitellum (at least during the reproductive period), which secretes a coccoon for the eggs.

Clitellum: a circular glandular thickening of epidermis at the anterior end of oligochaete worms and leeches, which are sometimes placed together in a single class, the *Clitellata*. The C. is a secondary reproductive structure, which secretes material for cocoon formation, as well as food reserves for nourishment of the embryo.

Clivia: see *Amaryllidaceae*.

Cloaca: terminal part of the alimentary canal of some invertebrates (e.g. aschelminths) and of most vertebrates, which receives the openings of the alimentary, urinary and reproductive systems, and which has a single aperture to the exterior. In most forms, the C. is differentiated into a dorsal diverticulum or urinogenital pouch (which receives the ureters and reproductive ducts) and a ventral chamber or diverticulum lying posterior to the urinogenital pouch (which receives the opening of the rectum). Among mammals, only the monotremes have a C. In all other mammals, the C. becomes partitioned during embryonal development into a dorsal section (which forms the end of the rectum) and a ventral section. The latter houses the genital duct, as well as giving rise dorsocranially to the urinary bladder, with which it remains in communication via the urethra.

Clonal reproduction: formation of progeny that are genetically identical to the parents. In prokaryotes and unicellular organisms, C.r. occurs, inter alia, by binary fission. In multicellular organisms, it occurs by layering or formation of bulbils, by ameiotic parthenogenesis, and in rare cases by Hybridogenesis (see).

Clone: descendants of a single individual, produced by asexual (vegetative) multiplication. All individuals of a C. are genetically identical, i.e. they have the same genotype (assuming the absence of mutations). A C. can therefore also be described as a vegetatively produced, genetically pure line. *Cloning* is very important in those areas of plant breeding where subsequent propagation is vegetative. Thus every potato variety is a C. Cloning is also achieved by dividing plant tissue cultures and allowing each piece of callus to differentiate into a new plant. Animals can be cloned by nucleus transplantation. The replication of genes and other DNA sequences by the methods of gene technology is also called cloning; the "cloning of DNA" has become one of the most powerful tools of modern molecular biology.

Clone selection theory: the most soundly based theory of antibody formation, first put forward in 1959 by the Australian biologist, Burnet. It represents a development of the side-chain theory of Paul Ehrlich, published in 1898, which was also based on a selection mechanism. It opposes the instructional hypotheses of antibody formation. According to the C.s.t. the organism possesses a large number of lymphocytes which differ in the specificity of their antigen receptors. An antigen entering the organism is bound by the receptors of certain lymphocytes, and it stimulates them to multiply and differentiate. Each of these lymphocytes gives rise to a clone of antibody-synthesizing cells. The antibody produced by each clone has the same specificity as the antigen-binding receptor, and therefore specifically binds the antigen.

Clonorchis sinensis: see Liver flukes.

Closed plant community: a plant community at a particular site, which has reached the last natural stage of its development with respect to constituent species and their relative numbers. See Succession, Climax.

Closing movements: flower movements, usually the result of thermo- or photo-nastic growth of the upper surface of the petals (see Flower movements, Hygroscopic movements).

Closteroviruses: see Virus groups.

Clostridium: a genus of spore-forming, anaerobic, rod-shaped, Gram-positive, usually motile bacteria. They occur in soil, water and in the alimentary canal of humans and animals. Some species are important as nitrogen-fixing organisms, and several are Butyric acid bacteria (see). Important disease organisms are Tetanus bacterium (see), Botilinus bacterium (see) and the agent of gas gangrene, *C. perfringens (= C. welchii).*

Clostridium botulinum: see Botulism.

Clostridium tetani: see Tetanus bacterium.

Clothes moths: Two moths of the family *Tineidae*.

1) Tineola bisselliella Hummel (common or webbing clothes moth) with straw-yellow wings, wingspan 8 to 15 mm. The caterpillars feed on wool, feathers or furs, and damage textiles and clothes.

2) Tinea pellionella L. (case-bearing or case-making clothes moth) with glossy mud-yellow forewings patterned with three dark spots, wingspan 12 mm. The caterpillars feed on furs, feathers or wool, and damage textiles, clothing, carpets and soft furnishings. The head and foreparts of the caterpillar protrude from a small silken case, which it spins for itself after hatching.

Clouded leopard: see *Felidae*.

Cloud forest: see Rainforest.

Clover: see *Papilionaceae*.

Clubmosses: see *Lycopodiatae*

Clubmoss trees: see *Lycopodiatae*.

Club root: see *Myxomycetes*.

Clupea: see Herring, Shad.

Clupeiformes, *herrings and allies:* a relatively primitive order of fishes, most of which are marine and occur in large shoals. A few species occur in freshwater. They include some of the commercially most important fishes, e.g. herrings, sardines, anchovies. The order is characterized by the presence of the *recessus lateralis*, in which the infraorbital canal merges with the preopercular canal in a chamber of the neurocranium, which is not known in any other order of fishes. Most species are plankton feeders. Four families are recognized: *Denticipitidae* (***Dentical herring,*** a single species), *Clupeidae* (***Herrings,*** including ***Shads, Sardines*** and ***Menhadens;*** 50 genera, 190 species), *Engraulididae* or *Stolephoridae* (***Anchovies;*** 16 genera, about 139 species), *Chirocentridae* (***Wolf herring,*** a single species).

Clymenia (Greek *klymene* daughter of the sea god Okeanos): a genus of ammonites with a smooth, disk-like, planispiral shell. The suture line shows simple waves, and the siphuncle lies on the inner side of the coil. Index fossils of the higher strata of the Upper Devonian (Famenium) are primarily *C*.

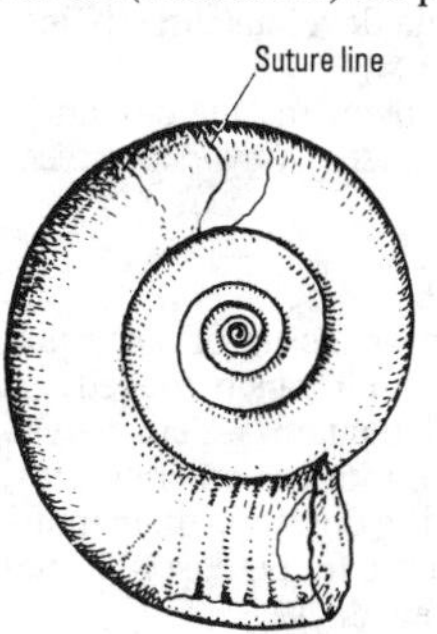

Clymenia. Stone core (x 0.5) from the later strata of the Upper Devonian.

Clypeaster: see *Echinoidea.*

C-meiosis: meisosis altered by the action of agents that cause Polyploidy (see), in particular by the action of colchicine. Microtubule assembly is inhibited, so that a functional spindle is not formed. In addition, Chiasma (see) formation is affected. In normal meiosis, the somatic chromosome number is decreased, but in C-m. this does not occur, leading to gametes with more than the normal complement of chromosomes.

C-mitosis: a characteristically modified form of mitosis, in which spindle formation is partially or completely inhibited, so that chromosomes do not migrate to opposite cell poles and daughter cells are not formed. It is typically caused by colchicine and a number of other agents. In parallel with the absence of chromosome movements, the chromosomes show increased contraction, and the separation of the centromeres is delayed. The chromatids repel one another, but at first remain attached at the centromere, producing a cruciform structure (see Fig., Mitosis). After separation of the chromatids, a typical C-m. finally results in a new cell containing a doubled chromosome number, i.e. polyploidy occurs. See Restitution nucleus.

CMU: acronym of the systemic herbicide, 3-(4-chlorophenyl)-1,1-dimethylurea. See Herbicides.

Cnemidophora: see Whiptail lizards.

Cnidaria: a phylum of the *Coelenterata,* whose members are characterized by the possession of Nematocysts (see) (cf. *Ctenophora,* which do not possess nematocysts). The majority of the *C.* are marine, with only a few freshwater species. They occur in two structural forms, i.e. a Polyp (see) or a Medusa (see), both of which display radial symmetry.

In sexual reproduction, the egg forms a blastula, which develops into a free swimming, 2-layered larva, known as a Planula (see). This later attaches itself to a solid substrate, develops a mouth and tentacles at its free end, and becomes a polyp. In some forms, the planula develops into a medusa. As a rule, polyps only reproduce asexually. This can occur by the budding of daughter polyps, often leading to the formation of a colony. Alternatively, budding of the polyp may produce medusas, which detach themselves, become free swimming, and reproduce sexually. This gives rise to the alternation of generations that is characteristic of many *C.,* i.e. alternation between the sessile, nonsexual polyp generation and the free swimming sexual medusa generation. This condition is also an example of metagenesis. In some forms (e.g. *Hydra* and corals) the medusoid form is absent, and the polyps reproduce both asexually and sexually. In certain other species the polyp generation is absent, so that the planula develops directly into a medusa.

About 10 000 species are known, and the phylum is divided into 3 classes: *Hydrozoa* (see), *Scyphozoa* (see) and *Anthozoa* (see).

Cnidocil: see Nematocyst.

Cnidocyte: see Nematocyst.

Cnidosac: an internal sac near the tip of a ceratum (dorsal body projection) of a nudibranch (naked sea slug; see *Gastropoda).* Each C. contains a nematocyst, which the sea slug obtains from its coelenterate prey. Nematocysts are separated during digestion, then transported to the Cs., where they serve for defense. The nematocysts are replaced every few days.

Cnidosporidia: a subclass of the *Sporozoa* (see) in which the spores contain 1–4 *polar capsules.* Each polar capsule contains a coiled polar filament, which under the action of digestive juices discharges by eversion and fastens the spore to the wall of the host digestive tract. Some authorities consider that the *C.* are degenerate metazoans; certainly, there are striking structural and functional similarities between the polar capsule and the coelenterate nematocyst. There are 2 principal orders.

Members of the *Myxosporidia* are principally parasites of fish, and there is no intermediate host. Two nuclei of the sporoplasm fuse before the spore hatches to invade the host tissue. Subsequent growth results in a vegetative multinucleate mass; this mass consists of several discrete parts, known as pansporoblasts, each containing several nuclei and giving rise to a new spore. Two nuclei are gametic, while other nuclei are responsible for formation of the polar capsules and valves.

Members of the *Microsporidia* are insect parasites, e.g. *Nosema apis* causes nosema disease in honeybees; *Nosema bombycis* causes a disease known as pebrine in silkworms.

Cnemidophora: see Whiptail lizards.

Coacervate: a droplet of liquid colloidal aggregate, containing a high concentration of organic compounds, which separates spontaneously from a colloidal solution. It is thought that Cs., which separated from the primeval soup, may have been precursors of the first living cells.

Coadaptive gene system: an accumulation of genetic factors in a single population, whose interaction confers a high degree of fitness on members of the population. In a natural population, certain genes become concentrated which interact favorably with other genes in the population. In another population, however, these same genes may even decrease the fitness of their carriers. Crossing of individuals from different populations therefore often produces offspring with decreased viability.

There is no evidence that human populations or races form C.g.ss., since people of mixed race and their offspring exhibit no decrease in fitness.

Coalification: see Carbonification.

Coal tit: see *Paridae.*

Coarctate larva: see Insects.

Coarctate pupa: see Insects.

Coast bristletails, *Halomachilis:* a genus of bristletails found on rock coasts, where they feed on algae as the tide goes out, and are particularly common under heaps of the larger species of seaweed. They jump with the aid of their legs and increase speed by using their elastic hindquarters.

Coatis, *Nasua:* a genus of the *Procyonidae* (see), represented by 3 species: *N. narica* (Southwest USA to South America), *N. nelsoni* (only on Cozumel Island of the Yucatan Peninsula), *N. nasua* (most of South America). All species are reddish brown or reddish brown-gray above, and yellowish to dark brown below, with whitish muzzle, chin and throat, and blackish feet. The tail is typically banded, and the muzzle is characteristically long and pointed with a very mobile tip (Fig.). They live in groups, varying in number from 5 or 6 to as many as 40.

Coat of mail shells: see *Polyplacophora.*

Coati

Cobalt, *Co:* a heavy metal and an important bioelement, present in traces in higher plants (0.01–0.4 mg/kg dry mass), animals and microorganisms. It is a component of vitamin B_{12}, and it also occurs in various sugar complexes. Unlike most animals, ruminants have a high nutritional requirement for Co, probably because it is needed by Vitamin B_{12}-synthesizing bacteria in the intestinal tract. Co is also needed for the efficient nodulation of leguminous plants, and is therefore indirectly required for symbiotic nitrogen fixation.

Cobitididae, ***loaches:*** a family (at least 175 species in 21 genera) of freshwater fishes (order *Cypriniformes),* found in lakes and rivers of Eurasia and Africa. The body is wormlike to fusiform with a subterminal mouth. They are capable of intestinal respiration, also known as "auxiliary intestinal breathing", in which air is swallowed and oxygen is absorbed by the intestine. The mouth carries 3–6 pairs of barbels, and many species carry a spine under, or in front of, each eye. Members of the genus *Acanthophthalmus* (coolie loaches) and *Misgurnis* (weatherfishes) are popular aquarium fishes, including a color form of the Japanese weatherfish, called the golden dojo. See Spined loach, Stone loach, Weatherfish.

Cobitis: see Spined loach.

Cobra lily: see Carnivorous plants.

Cobras, *Naja:* a genus of the *Elapidae* (see), whose several species are distributed over Africa and the whole of tropical Asia. When excited, they place the anterior part of the body in an upright position, and are then able to spread their neck skin sideways with the aid of elongated neck ribs, producing a plate-like disk or "hood", which may be several times wider than the normal diameter of the animal. In the Asiatic *Naja naja* and its subspecies, a light-colored design appears between the parted dorsal scales when the hood is extended. In the Indian cobra *(Naja naja naja,* the ***Spectacled cobra),*** one of the commonest poisonous snakes of India, this design has the appearance of spectacles. In the Southeast Asian subspecies *(Naja Naja kaouthia,* ***Monocled cobra)*** it has the appearance of a monocle. The neurotoxic venom of C. (see Snake venoms) is responsible for more than 10 000 deaths annually in India. Snake charmers in India use *Naja naja naja,* whereas in Egypt they use the legendary "Asp", *Naja haje.* The Egyptian *Naja haje* is usually jellowish brown with bars on the underside; it is found over most of the semidesert country of Africa, often spending the day in rodent burrows. The "dancing" of the snake has no connection with the music (snakes are deaf), rather the animal follows the movements of the snake charmer in order to keep him in view. The large African species occur in many habitats, and have no spectacle pattern. The ***Spitting cobra*** *(Naja nigricollis)* of Central and West Africa attains a length of 2 m. As a means of defense, it can eject venom from its front fangs into the eyes of its perceived attacker, causing considerable pain and injury. Having disabled its attacker in this way, it may also follow up with a venomous bite. C. feed mainly on rodents, and all species lay eggs (cf. King cobra, *Ophiophagus hannah).*

Cocaine: a bitter-tasting alkaloid, which is chemically the methyl ester of benzoyl-ecgonine (Fig.). Leaves of the coca plant *(Erythroxylum coca)* contain 1–2% C. To prepare C. commercially, an extract of leaves of the coca plant is hydrolysed, converting the C. into ecgonine. The latter is isolated and easily reconverted into C. by esterification with methanol and benzoic acid. C. was formerly widely used as a local anesthetic, but it is now largely replaced by less toxic, synthetic compounds. It also causes vasoconstriction and pupil dilation. At higher doses, it has euphoric or hallucinogenic effects, and it is an addictive narcotic. The use of C. as a stimulant can be traced back to the Incas before 1000 AD. See *Erythroxylaceae.*

COOH
$N{-}CH_3$
OH
Ecgonine

H
$COOCH_3$
$N{-}CH_3$
$OCOC_6H_5$
H
Cocaine

Coca plant: see *Erythroxylaceae.*

Cocci (sing. **Coccus):** spherical, Gram-positive bacteria, which are mostly nonmotile and largely incapable of spore formation. The spherical shape may be distorted, so that the cells appear oval, elliptical or conical. They are classified according to the groupings they form after cell division. Diplococci divide in one plane and remain in pairs. Streptococci (see) divide in one plane and remain together in chains. Micrococci (see) divide irregularly in 3 planes, and form a mass or cluster. Species of *Sarcina* (see) divide regularly in 3 planes at right angles to each other, forming cubical groupings or packets of 8 cells. *Gaffkya* divide at right angles but in a single plane, forming tetrads. See Staphylococci.

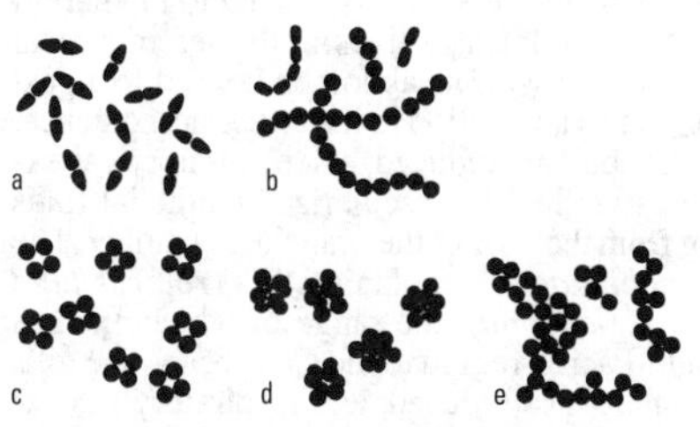

Cocci. a Diplococci. *b* Streptococci. *c* Tetrads *(Gaffkya). d* Cubical groupings of 8 cells *(Sarcina). e* Staphylococci.

Coccidia: see *Telosporida.*

Coccidiomorpha: see *Telosporidia.*

Coccidioses: see *Telosporidia.*

Cocco-bacilli: see Bacillus.

Coccoidea: see *Homoptera.*

Coccolith: see Coccolithophorids.

Coccolithophorids, *cocclithophores,* *Coccolithophorida, Coccolithineae:* a family of marine, planktonic flagellates, placed either in the algal class *Prymnesiophyceae,* or in the protozoan subclass *Phytomastigophorea* (see). They have a spherical or oval shape, are less than 20 mm in diameter, and are covered in calcareous plates (coccoliths) embedded in a gelatinous sheath. Fossils are known from the Jurassic to the present, and they have made a large contribution to chalk formation in the Tertiary and in recent times. Recently, colonies of C. have proved useful as index fossils.

Coccothraustes: see *Fringillidae.*

Cocculin: see Picrotoxin.

Coccyx: see Axial skeleton, Pelvic girdle.

Cochlea: see Auditory organ.

Cochleariidae: see *Ciconiiformes.*

Cochlearius: see *Ciconiiformes.*

Cochylidae: see *Phaloniidae.*

Cockatoos: see *Psittaciformes.*

Cockchafers: see Lamellicorn beetles.

Cockles: see *Bivalvia.*

Cock-of-the-rock: see *Cotingidae.*

Cockroaches: see *Blattoptera.*

Cock's foot: see *Graminae.*

Cocoa butter, *oil of theobroma:* a solid fat representing 40–50% by weight of the seeds of *Theobroma cacao* (Cocoa tree), from which it is expressed. At room temperature, it is hard, brittle and pale yellow, softening at 25 °C and melting at 30–33 °C. It is used pharmaceutically and in the preparation of semiluxury foods.

Cocoa swollen shoot virus: see Virus diseases.

Cocoa tree: see *Sterculiaceae.*

Coconut: see *Arecaceae.*

Coconut crab, *robber crab,* *Birgus latro:* a land crab of the family *Coenobitidae,* infraorder *Anomura,* order *Decapoda* (see) found on various islands in the Indian and Pacific oceans. It feeds on carrion and on both decaying and fresh vegetation. Mainly nocturnal in habit, it is often also active in full daylight. Normally it lives close to the ocean, but it is sometimes encountered high above the shore line; for example, it has been found on the Solomon Islands some distance inland and 100 m above sea level. Using the sharp ends of its legs as climbing spikes, it climbs palms and mangroves to heights of 20 m to obtain fruit. This powerful crab can open fallen coconuts. It is closely related to the hermit crabs, and its juvenile stage lives in the sea in typical hermit crab fashion with its abdomen housed in a gastropod shell. As an adaptation to its terrestrial existence, the gills of the adult are reduced, and respiratory gas exchange occurs via highly vascularized epithelial folds which hang from the roof of the branchial chamber. The family *Coenobitidae* also contains the ***Tropical land hermit crab*** *(Coenobita),* the range of which extends from the Indo-Pacific region to the Caribbean. The gills of *Coenobita* are also reduced, and respiratory gas exchange occurs via the highly vascularized floor of the branchial chamber and via a vascularized region on the anterodorsal surface of the abdomen.

Cod, *common cod,* *Gadus morrhua:* a predatory fish found in the deeper waters of the continental shelf of the North Atlantic, White Sea, Barents Sea, Baltic and northern Pacific; family *Gadidae,* order *Gadiformes.* It is one of the largest gadiform fishes, attaining a maximal length of 150–180 cm (usually 40–80 cm). The upper jaw protrudes slightly, and the lower jaw carries a long barbel. Back and sides are olive-colored and decorated with numerous yellow-brown spots; the belly is off-white. The prominent, pale lateral line travels upward in a shallow curve above the pectoral fin. It is one of the most important commercial fishes.

Codehydrogenase I: see Nicotinamide-adenine-dinucleotide.

Codehydrogenase II: see Nicotinamide-adenine-dinucleotide phosphate.

Codeine, *methylmorphine:* an opium alkaloid (see Opium). Chemically, it is closely related to Morphine (see) and differs from the latter in possessing a methoxy group in place of the phenolic OH-group. Although C. can be prepared from opium, most of the C. used for medical purposes is prepared by the chemical methylation of morphine. It is a narcotic analgesic, but much milder in action than morphine and less addictive. C. is used medicinally as its sulfate or phosphate, mainly as a centrally acting antitussive.

Coding ratio:

1) Average number of nucleotides needed to unequivocally encode a particular amino acid. If 4 nucleotides are available for 20 amino acids, the C.r. must be greater than 2.

2) Number of nucleotides in a codon.

3) Ratio of the number of nucleotides in an mRNA molecule to the number of amino acids in the polypeptide chain encoded by the mRNA.

Codlin moths: see *Tortricidae.*

Codlins and cream: see *Onagraceae.*

Codominance: full expression of both alleles at the same gene locus. C. is displayed by certain alleles that apparently behave independently, and do not show mutual recessivity and dominance.

Codon: see Genetic code.

Coefficient of yield: see Economic coefficient.

Coelacanthini: see Coelacanths.

Coelacanths, *Coelacanthini, Crossopterygii:* a group of very primitive, plump teleost fishes (see *Osteichthyes).* They include the extinct *Osteolepiformes,* which are thought to be ancestral to the amphibians and therefore to terrestrial animals. Their brush-like fins are very narrow at the base and have a more or less stalked appearance. The skeleton is somewhat reduced, and internal nares are absent.

Fossil remains of C. are found from the Lower Devonian to the present. In UK, Scandanavia and North America, they serve as index fossils of the Devonian, which was their period of greatest diversity and abundance.

Until 1938, C. were thought to have become extinct in the Cretaceous. In 1939, a live specimen *(Latimera chalumnae)* was caught off the coast of East Africa. It attains a length of about 150 cm and lives as a predator at depths of 70–400 m. The paired fins are used for locomotion on the sea bottom. See plate 4.

Coelenterata: a subdivision of the *Eumetazoa,* containing more than 10 000 species. In contrast to the *Bilateria,* members of the *C.* display radial symmetry. They can exist in 2 different forms, i.e. a Polyp (see) or a Medusa (see). The axis of symmetry always passes through the mouth and the opposite pole. There are at least 4 identical copies of each organ, arranged symmetrically around the axis, e.g. tentacles, gonads, gastral pouches. The very mobile tentacles, which serve to capture prey or to ward off aggressors, are arranged in a circle around the oral cone of the polyp or around the rim of the medusa. Nematocysts (see) serve as organs of defense and capture in the *Cnidaria,* but are largely absent from the *Ctenophora.* The body consists of only two cellular layers, i.e. ectoderm and endoderm, separated by a gelatinous mesoglea. The mesoglea may be totally noncellular, or it may contain wandering amebocytes, but a true mesoderm is never present. *C.* possess a single internal space (gastrovascular cavity), which can be a simple sac, but in many forms it is divided into compartments (gastral pouches) by projecting septa. There is no anus, and the mouth is the only opening connecting the gastrovascular cavity to the exterior. A very primitive muscular system is present, represented merely by epithelial muscle cells in both body layers. Epithelial sensory cells are interconnected by a primitive nerve network.

The *C.* are divided into 2 phyla: *Cnidaria* (see) and *Ctenophora* (see).

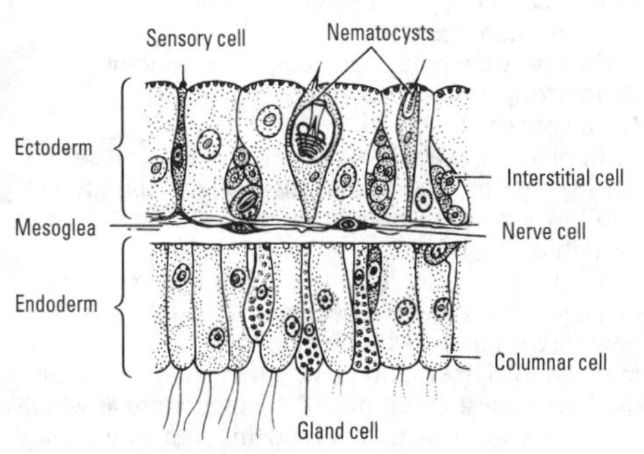

Coelenterata. Schematic longitudinal section through the body wall of *Hydra.*

Coeloconic sensilla: see Olfactory organ.
Coelodonta: see *Rhinocerotidae.*
Coelom: see Body cavity.
Coelomoducts: see Nephridia.
Coenagriidae: see *Odonatoptera.*
Coendou: see *Erethizontidae.*
Coenenchyme: see *Octocorallia.*

Coenobia: very loose associations of unicellular organisms. After cell division, daughter cells remain in a shared, jointly secreted, swollen, gelatinous cell wall, or in a mucilaginous envelope, or cellulose shell. The individual cells do not interact to form a functional unit. According to whether all the cells divide in the same single plane, or whether they divide in 2 or 3 planes, the coenobium is a filament, a plate, or a three-dimensional clump. Filamentous forms may display false branching when one of two parallel filaments emerges through a break in the gelatinous sheath. C. reproduce vegetatively by releasing fragments, i.e. smaller C. They can also break down completely into viable individual cells. C. are formed, inter alia, by bacteria, cyanobacteria and diatoms.

Coenobita: see Hermit crabs, Coconut crab.
Coenocarpous condition: see Fruit.
Coenoelement: see Floral element.
Coenosarc: British spelling of Cenosarc (see).
Coenosis: see Biocoenosis.
Coenothecalia: see *Helioporida.*
Coenurus: see *Cestoda,* Dog tapeworm, *Taenia.*
Coenzyme: see Enzymes.

Coenzyme A, ***CoA, CoA-SH:*** a group-transferring coenzyme in many reactions of intermediary metabolism. CoA is present in all living cells. Chemically, it is a compound of adenosine, pantothenic acid (see Vitamins), cysteamine and three molecules of phosphoric acid. The free thiol group (—CH_2—SH) of the cysteamine residue is responsible for the biological transfer activity of CoA. Thus, the thiol group interacts with acetic (ethanoic) or other carboxylic acids, forming an energy-rich (labile) thioester bond, and activating both the carboxyl group and the α-position of the acid. This activated carboxylic acid residue can participate in a variety of biochemical reactions, e.g. reduction, transacylation, carboxylation and condensation. The free energy of hydrolysis of the thioester bond is relatively high.

OC—NH—CH_2—CH_2—CO—NH—CH_2—CH_2—SH
H—C—OH
H_3C—C—CH_3
H_2C—O—Ⓟ—Ⓟ—O—CH_2
NH_2
N N N N
O
O OH
Ⓟ

Coenzyme A

Acetyl-CoA ("active acetate") (CH_3CO~S—CoA) is an especially important CoA derivative, in which the thiol group carries an acetyl residue (CH_3CO—). It occupies a key position in intermediary metabolism, on the one hand representing an endproduct of carbohydrate, fatty acid and amino acid metabolism, and on the other hand serving as a precursor for many syntheses and acetylation reactions. Of particular importance is the condensation of acetyl-CoA with oxaloacetate to form citrate, a reaction which channels the acetyl group of acetyl-CoA into the tricarboxylic acid cycle. Acyl derivatives of CoA occur as intermediates in the β-oxidation of fatty acids.

Coenzyme I: see Nicotinamide-adenine-dinucleotide.
Coenzyme II: see Nicotinamide-adenine-dinucleotide phosphate.
Coereba: see *Parulidae.*
Coffee: see *Rubiaceae.*
Cohesion theory: see Water economy.

Coincidence: occurrence of two partners (e.g. sexual partners) at the same place and the same time. C. is crucial for the predator-prey relationship, for parasite-host interaction, and for the coexistence of symbiotic partners. An exact knowlege of the seasonal occurrence of C. is important in the control of agricultural pests.

Coincidence index: ratio of the observed number of double crossovers to the expected number of random double crossovers in a pair of homologous chromosomes during meiosis (see Interference). The C.i. is a measure of crossover interference. A value of 1 indicates the absence of interference, values greater than 1 indicate negative interference, and values less than 1 indicate positive crossover interference.

Coitus: see Copulation.

Colacium: see *Phytomastigophorea.*

Colaptes: see *Picidae.*

Cola tree: see *Sterculiaceae.*

Colcemid: see Microtubules.

Colchicine: a very poisonous alkaloid from the meadow saffron *(Colchicum autumnale).* In very small quantities, C. relieves pain and suppresses inflammation. It is used against gout and to inhibit neoplastic growth, but as little as 20 mg (the content of 5 seeds of the meadow saffron) is a fatal human dose, leading to paralysis of the central nervous system and respiratory failure. Because it binds specifically to tubulin, C. causes depolymerization of Microtubules (see), including those of the mitotic and meiotic spindles. C. is therefore a mitotic poison. In plants, C. induces Polyploidy (see) and often gigantism by preventing the separation of chromosomes during cell division.

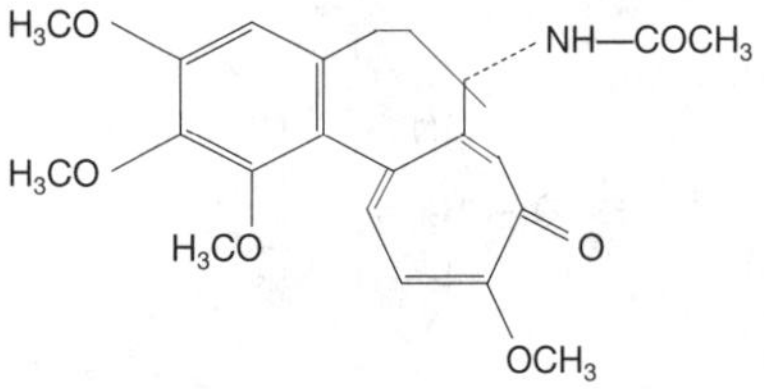

H_3CO, H_3CO, H_3CO, NH—$COCH_3$, O, OCH_3

Colchicine

Cold breeders: animals that reproduce in the winter. Examples of cold breeding insects are the Winter moth *(Operophthera brumata)* and the Mottled umber moth *(Erannis defoliaria).* Bears are cold breeding mammals.

In contrast, in *warm breeders,* reproduction occurs from spring onwards, and ceases with the onset of the next cold season. The majority of insects, amphibians, reptiles, birds and mammals of temperate zones are warm breeders. Fish can also be classified as cold or warm spawners.

Continuous breeders reproduce throughout the year, irrespective of the season, e.g. Field slug *(Deroceras)* and Crossbill *(Loxia curvirostra).*

Cold death: death resulting from irreversible changes in essential life processes, caused by low temperatures. As in Cold torpor (see), the temperature of onset of C.d. is species-specific, and it also depends on the stage of development of the organism, its sex, and its physiological state (e.g. the temperature of C.d. can be decreased by a preliminary period of cold adaptation). Insects are less able to tolerate spring frosts than the sub-zero temperatures they encounter during hibernation. A low temperature may be tolerated if applied gradually, but becomes lethal when applied suddenly. The duration of exposure is also important.

Homeothermic animals suffer C.d. when the body temperature falls to 15–20 °C. Even with a temperature difference of 100 °C between the body and the environment, body temperature can be maintained. This is achieved mainly by efficient body heat regulation and insulation against heat loss (feathers, hair, subcutaneous adipose layers); also, large animals lose relatively less heat, owing to their favorable volume to surface area ratio. Cold tolerance is particularly marked in Arctic and Antarctic homeothermic animals (seals, penguins).

Cold injury: adverse effects on plants of temperatures that are above freezing but below those normally encountered. Such chilling may cause reversible damage or death. In tropical plants, algae of warm seas, and some fungi, the effect occurs at relatively high tempertaures. C.i. occurs, inter alia, because the rates of different metabolic reactions and entire metabolic pathways within the same plant are affected differently by the temperature decrease. For example, respiration generally has a higher temperature optimum than photosynthesis. Also, in some plants, the optimal temperature for chlorophyll biosynthesis is higher than that for growth, so that C.i. causes these plants to turn a pale yellowish color. Other symptoms are general stunting of growth, structural deformities, flower and fruit drop, empty seed heads, and reversible or irreversible wilting. Examples of C.i. due to disturbed metabolism are the sweetening of potato tubers and winter vegetables.

Injury caused by temperatures below freezing is called Frost damage (see).

Cold-stenothermal: see Temperature factor.

Cold steppe mammoth: see Mammoth.

Cold torpor:

1) In plants, the abolition of sensitivity to all stimuli and a slowing or cessation of metabolism and growth, due to low environmental temperatures. It may lead to irreversible cold injury.

2) In animals, a decrease or abolition of movement, associated with a marked decrease of metabolism, due to low environmental temperatures. It is an essential component of Hibernation (see) in homeothermic animals. The highest environmental temperature at which C.t. occurs is species-specific, and this temperature may also vary according to the stage of development of the organism. Thus, the winter fly, *Chionea araneoides,* is still active at -6 °C, the caterpillar of the moth, *Lymantria monacha* (Black arches), displays C.t. at about 0 °C, whereas the cricket ceases to be active at +16 °C. C.t. is generally reversible, but further cooling leads eventually to Cold death (see).

The converse of C.t. is known as *heat torpor.* In poikilothermic animals, the upper temperature limit for activity is usually between +40 °C and +50 °C. It is restricted to a relatively narrow temperature range, and any further increase rapidly leads to heat death. Inhabitants of hot springs or dunes characteristically display a high temperature limit for heat torpor and a correspondingly high temperature for heat death (e.g. some protozoans, nematodes, insects).

Cold water bacteria: bacteria with optimal growth temperatures between 0 ° and 10 °C. They generally grow and multiply relatively slowly. In winter, they are responsible for the degradation and conversion of material in all natural freshwaters, but in the summer they are active only at lower depths in thermally layered lakes. They are also continually active in many ground water courses.

Coleoidea: see *Cephalopoda.*

Coleophoridae: see *Lepidoptera.*

Coleoptera, *Eleutherata*, *Elytroptera*, ***beetles:*** an insect order containing 350 000 described species. The earliest fossil *C.* are found in the Permian.

Imagoes. Adult body length varies between 0.3 and 155 cm, i.e. the *C.* include some of the smallest and the largest recent insects. The head is usually smaller and narrower than the first abdominal segment, and partly sunk into the latter. Antennae are very variable, usually 11-segmented. Mouthparts are adapted for chewing. The dorsal surface of the first thoracic segment (prothorax) is usually modified to a more or less large plate (protonum). The dorsal surface of the second thoracic segment (mesothorax) is reduced to a small triangular shield (scutellum) between the inner angles of the wing covers, while the third thoracic segment (metathorax) is completely covered by the wings. More heavily sclerotized than in any other insect family, the vein-less, robust forewings (known as wing covers, wing cases or elytra) are laid protectively over the soft dorsal surface of the mesothorax, metathorax and abdomen. The hindwings, folded beneath the elytra, are membranous and veined. Only the hindwings are used for flying, and only a few species are strong fliers. There is a wide variety of leg shapes, reflecting the habits of walking, digging, jumping or swimming. Externally, 8 abdominal segments are visible. The relatively short-lived adults feed on a variety of plant and animal materials. Many *C.* make special provision for their subsequent offspring, and also display brood care and protection. Many species live as tenants, guests or parasites of ants (myrmecophilous species); others live in the nests of mammals and birds (nidicolous species). Most *C.* produce a single generation per year. Metamorphosis is complete (holometabolous development).

Eggs. These are usually elongate-cylindrical, but in some families they are rounded. Eggs are usually laid in several batches, either individually, or in rows, plates or heaps. When laid in a concealed position, the eggs are usually whitish, whereas exposed eggs are often yellow, orange or red. As in all insect orders, the number of eggs depends on the size of the species, the size of the eggs and on the statistical chances of hatching and survival; e.g. the June beetle (*Melolontha*) lays 60–80 eggs concealed in the soil, whereas the Colorado beetle (*Leptinotarsa*) lays 500–2 500 on leaf surfaces.

Larvae. Larvae display a variety of external forms, e.g. elongate-flattened with long walking legs (capodeiform larvae; carabids, rove beetles), long-cylindrical and uniformly segmented (eruciform larvae; click beetles), scarabaeiform (cockchafers), or grub-like (apodous larvae; weevils, scolytid beetles). Hypermetamorphosis occurs in a few families, e.g. *Meloidae*, in which the larva passes through all 4 forms. The head capsule is always well developed and sclerotized. Each of the 3 thoracic segments carries a pair of legs, which may be regressed, especially in endophagous species. There are 9 abdominal rings, lacking appendages. Most *C.* larvae molt twice, e.g. carabids, carnivorous water beetles, weevils, rove beetles, but in some species the larvae may molt up to 6 times, e.g. mealworms. Generally, the separate instars are very similar. Larvae feed on plant material (click beetles, leaf beetles, weevils, long-horned beetles); some species are predators on other insects or small animals (carabids, carnivorous water beetles). Adult and larva usually have similar feeding habits.

Pupae. Many larvae prepare a pupal cell, consisting of a depression or cocoon-like covering, lined with salivary secretion or feces, in wood, humus or earth. All pupae are adecticous and most are exarate (pupa libera) with freely movable extremities. In a few species of *Staphylinidae*, the pupa is obtect. The pupa is usually concealed in cracks in bark, in bored holes in wood, under stones or in the earth. In temperate and cold regions, it is normally the overwintering form of the beetle.

Economic importance. The order is extremely numerous and therefore, not surprisingly, contains many pest species. For example, many cockchafers, click beetles, sap beetles, weevils and leaf beetles are agricultural pests, while long-horned and scolytid beetles are forestry pests. Stored food products and materials are attached by drugstore beetles, hide beetles, seed beetles and weevils. On the other hand, the predatory carabids (e.g. caterpillar hunters), checkered beetles (which prey on scolytids) and ladybirds (which prey on aphids) play a significant role in controlling other insect pests. Some are also useful flower pollinators. Cantharidin from blister beetles, e.g. the "Spanish fly" (*Lytta vesicatoria*) and *Mylabris* spp., is used pharmaceutically.

Classification. About 150 families are recognized. These are grouped differently into superfamilies and suborders, depending on the authority. A selection is given below.

1. Suborder: *Archostemata.* Fewer than 30 species in two families: *Cupedidae* (North America, Asia, Southeast Africa); found under bark in rotting wood; *Micromalthidae* (a single species found in South America, South Africa and Gibraltar). The distal part of the wings is coiled when at rest. Hind coxae are not immovably fixed to the metasternum. A notopleural suture is usually present in the prothorax. Larval mandibles possess a molar area. Cerci are absent. Larval legs are 5-segmented with 1 or 2 claws. The hindwing cross veins define a wing cell, called an oblongonum.

2. Suborder: *Adephaga.* A notopleural suture is present in the prothorax. Hind coxae are immovably fixed to the metasternum. Larval legs with tarsus and 2 (rarely 1) claws. Larval mandibles lack a molar area. Hindwings usually with an oblongonum.

Families: *Cicindelidae* ***(Tiger beetles)****, Carabdae* ***[Ground beetles*** (see)], *Dytiscidae* ***(Carnivorous water beetles)****, Gyrinidae* ***(Whirligig beetles)****, Haliplidae* ***(Crawling water beetles)****, Rhysodidae* ***(Wrinkled bark beetles).***

3. Suborder: *Myxophaga.* Minute beetles with clubbed antennae. The prothorax has a notopleural suture. Hindwings have an oblongonum and a fringe of long hairs, and are coiled apically at rest. Larvae are aquatic and have mandibles with a molar area.

Families: *Lepiceridae, Sphaeriidae, Hydroscaphidae, Calyptomeridae.*

4. Suborder: *Polyphaga.* The hindwings lack an oblongonum and are not coiled distally. The prothorax has no notopleural suture, and hind coxae are movable. Larval legs (if not vestigial or absent) are 4-segmented with a single claw and lacking a tarsus. Larval mandibles have a molar area.

Section 1. *Hologastra.*

Families: *Canthariddae* ***(Soldier beetles)****, Lampyridae* ***(Fireflies).***

Section 2. *Haplogastra.*

Families: *Silphidae* [***Carrion beetles*** (see)], *Staphylinidae* (***Rove beetles***), *Lucanidae* (***Stag beetles***), *Geotrupidae* (***Dung beetles***), *Scarabaeidae* (see ***Lamellicorn beetles***).

Section 3. *Cryptogastra*.

Families: *Meloidae* (***Blister beetles***), *Tenebrionidae* (***Darkling beetles***), *Anobiidae* (***Deathwatch beetles***), *Elateridae* [***Click beetles*** (see); larvae are called wireworms], *Buprestidae* (***Metallic woodborers*** or ***Jewel beetles***), *Cleridae* (***Checkered beetles***), *Dermestidae* (***Skin beetles, Carpet beetles***), *Nitidulidae* (***Sap-feeding beetles***), *Coccinellidae* (***Lady beetles*** or ***Lady bugs***), *Cerambycidae* [***Long-horned beetles*** (see)], *Chrysomelidae* [***Leaf beetles*** (see)], *Curculionidae* [***Snout beetles*** or ***Weevils*** (see)], *Scolytidae* [***Bark, Cone*** and ***Engraver beetles*** (see)].

Coley, *Pollachius virens, Gadus virens:* a commercially important marine teleost of the family *Gadidae* (order *Gadiformes*). It is a predatory, pelagic fish of the European Atlantic and North Sea, extending north as far as the Barents Sea, the southern Greenland coast, and along the western Atlantic shores from the Hudson Bay the New York. Its southern limit is the Bay of Biscay. Back and head are dark olive green, brown or black. The sides are yellow-gray, and the belly silver-white. It possesses a tiny chin barbel and a distinctive black spot in the region of the pectoral fin.

Colicinogenic factors: see Bacteriocins.

Colicins: see Bacterial toxins, Bacteriocins.

Colies: see *Coliiformes*.

Coliidae: see *Coliiformes*.

Coliiformes, ***colies, mousebirds:*** an order of birds containing 6 finch-sized species in a single family (*Coliidae*). They are mainly vegetarian, feeding on seeds and buds, and sometimes causing considerable damage to crops. All have crests, and in some species the tapering tail is almost twice as long as the body. The first and fourth toes can be turned either forward or backward. Some species hang upside down like titmice. All are confined to sub-Saharan Africa, where they live gregariously in open areas. The usual feather tracts are absent, and the soft, loose, gray plumage grows all over the body rather like fur. In addition, they move busily on the ground like mice, hence the name, mousebird. They lay a clutch of 2–3 eggs.

Colins: old collective term for all metabolic products of higher plants that inhibit the growth of other plants. See Allelopathy.

Colinus: see *Phasianidae*.

Collagen: a tough and only slightly extensible, extracellular protein, containing an unusually high proportion of proline and hydroxyproline. It is reponsible for the strength and flexibility of animal connective tissue, and it accounts for 25–30% of the protein in an animal. Denaturation of C. by heating or by salts produces gelatin.

Collagen fibers: slightly corrugated or spirally twisted bundles of collagen, which account for most of the intercellular substance in animal support tissue. The individual fibers consist of fine *collagen fibrils*, held together by organic cement material. Fibrils are visible under the light microscope; under the electron microscope, they are seen to be composed of microfibrils. Due to end-to-end alignment of the macromolecular structural unit, tropocollagen, the fibrils have a characteristic striation with a repeat distance of about 670 Å. C.f. have extremely high tensile strength, but are nonelastic. They are disintegrated by boiling or by the action of acid (softening of meat by cooking; digestion of meat by pepsin and hydrochloric acid). Treatment with tannic acid causes C.f. to shrink and become resistant to bacterial attack (tanning of animal hides).

Collar cell: see *Porifera*, Digestive system.

Collared dove: see *Columbiformes*.

Collared hemipode: see *Gruiformes*.

Collared flycatcher: see *Muscicapinae*.

Collective of organisms: see Organism collective.

Collective types: see Intermediate forms.

Collembola, ***springtails:*** an order of insects, which in terms of individual numbers is probably the most abundant insect order. They have a thin, soft skin, and vary in length from 0.3 to about 9 mm in length, most species being between 1 and 2 mm. About two thousand species have been described, but it is estimated that there may be as many as 5 000 different species, with high population densities worldwide, including polar and alpine regions. The oldest known fossil insect, *Rhyniella praecursor*, from the Middle Devonian, belongs to the still recent family, *Neanuridae*.

Morphology. Winglessness is a primary characteristic and not a secondary adaptation. They posses chewing, tearing or piercing-sucking mouthparts, recessed within the head capsule; they therefore belong to the *Entognatha* (see). Antennae usually have 4 segments, and the abdomen characteristically has only 6 segments (Fig.1).

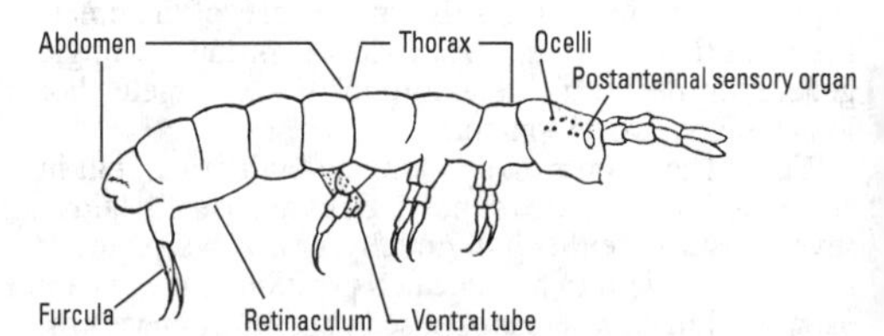

Fig.1. *Collembola.* Springtail (genus *Isotoma*).

Compound eyes are absent. On each side of the head are 8 to 10 ocelli and a paired circular or rosette-shaped postantennal sensory organ, which serves as a humidity, chemo- and thermo-receptor. Dorsally, the prothorax is very poorly developed. There are three pairs of legs, each carrying a claw (empodium) on the tibia (the tarsus is absent) and often an empodial appendage. The first abdominal segment carries a single ventral tube (collophore), from which either a membranous double sac or two single hoselike tubules can be everted by blood pressure. This appears to be a multipurpose organ, used for attachment to smooth surfaces, for respiratory gas exchange, and for water uptake under dry conditions. In the *Sminthuridae*, the collophore tubules are particularly long and appear to be used for cleaning the body. The fourth abdominal segment carries a 2-pronged projection called the *furcula*. In the resting position, the furcula is folded against the underside of the body with the tip pointing forward, where it is held by another smaller 2-pronged appendage called the *retinaculum*. Powerful abdominal muscles cause the furcula to unfold suddenly downward and backward, striking against the ground and propelling the insect in an aimless somersault (backward or forward, depending on the center of gravity) over a distance of 5–20 cm. The common basal piece of the furcula (the *manubrium)* has a pair of distal arms or *dentes*, each *dens* carrying a variably shaped claw or *mucro*. In some genera the furcula is well developed and extends beyond the ventral tube when folded beneath the abdomen (e.g. *Entomobrya);* in others (e.g. *Hypogastrura)* it is short, and in some it is absent *(Neanura* and *Anurida)*. Most springtails lack special organs of external respiration, and respiratory gas exchange is cutaneous. Exceptions include *Sminthurus,* which has a reduced tracheal system and a single pair of spiracles between the head and prothorax. There is no excretory system, and labial nephridia are also absent. Excretory products are stored in the fat body and in the intestinal epithelium, which is regularly shed and renewed. Malpighian tubules are likewise absent.

Reproduction. A sperm droplet (usually stalked) is deposited by the male, which rarely shows courtship behavior or has any sex-related contact with the female. Females independently locate the sperm droplets and take them into their slit-shaped genital opening. Some species (e.g. *Dicyrtomina labelli)* show slight pairing behavior, in that the male plants several sperm droplets in the vicinity of the female, increasing the probability that she will encounter one of them. Eggs are laid in small clusters of 5–20. Development occurs with little change in external form (epimetabolous development). Although the insect is sexually mature after 6 to 10 molts, growth may continue until 30–45 molts have occurred. Some species *(Folsomia candida)* live for 6–9 months with a generation time of about 4 weeks; others may produce only one generation per year, or even every 2 or 3 years. Facultative and obligatory parthenogenesis are common. Extremes of temperature and aridity or humidity during development often lead to ecologically determined morphological changes (see Ecomorphology).

Ecology. Springtails display many different life forms, representing adaptations to life at different soil depths (Fig.2). Surface species (e.g. many *Entomobryomorpha* and *Symphypleona)* are sometimes even found in tree canopies (epedaphon). These species are large with dense hairs or scales, and strongly pigmented (brown, violet, greenish). Their antennae (secondarily 5- or multisegmented) may be longer than the body; they also have long legs, and a long furcula. Eyes and sensory hairs are well developed, but the postantennal sensory organ is absent. In contrast, species of the upper soil layer (hemiedaphon) and especially of the soil interior (euedaphon) are small, short, sparsely haired and hardly pigmented. They have short legs and antennae, and furcula and ocelli may be absent. Such species usually have very elaborate postantennal sensory organs (e.g. many *Poduromorpha)*.

Most spingtails feed on fungal hyphae, spores, algae, dead organic material (fallen leaves, rotting wood), and very rarely on animal material. Species with tearing-sucking mouthparts *(Friesea* and *Anurida)* appear to regularly feed on rotifers, nematodes and unicellular organisms. Those with piercing-sucking mouthparts feed on liquid decomposition products and bacterial slime (e.g. *Neanura* and *Pseudachorutes,* Fig.3). Some of the few species that feed on living plants can become pests of seedlings, especially in Atlantic climates (e.g. *Bourletiella, Sminthurus, Onychiurus).* The colored, rounded species, *Sminthurus viridis,* is a well known pest of alfalfa.

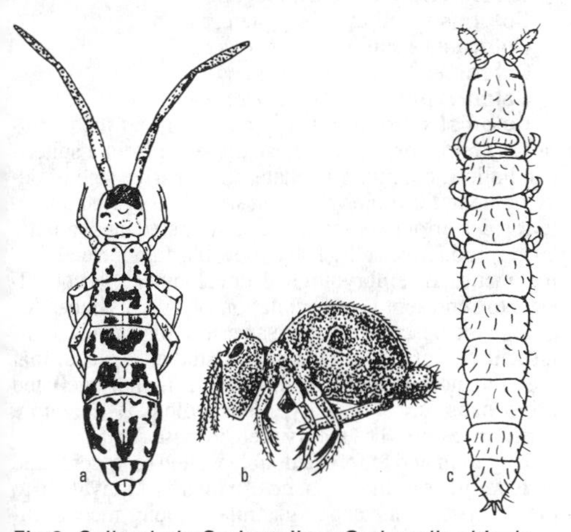

Fig.2. *Collembola.* Springtails. *a Orchesella alticola* (4 mm). *b Bourletiella hortensis* (1 mm). *c Tullbergia quadrispina* (1 mm). *a* and *b* are epiedaphic (surface-dwelling); *c* is eudaphic (an inhabitant of the soil interior).

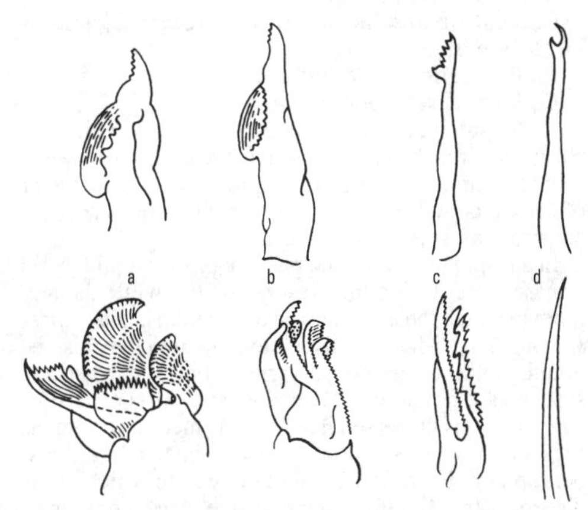

Fig.3. *Collembola.* Mouthparts. Top row: mandibles. Bottom row: maxillae. *a* Chewing (hard food). *b* Chewing (soft food). *c* Tearing- sucking. *d* Piercing-sucking.

Distribution. Springtails are present in all soils, attaining maximal densities (700 000 individuals under each square meter of soil surface) in the cavity-rich fresh humus and leaf mold of forest floors. Densities are between 10 000 and 40 000 individuals per square meter. By preparing organic material for the activity of microorganisms, springtails make an important contribution to soil biology. Some species live on the surface of standing water *(Podura aquatica,* species of *Sminthurides),* but subaquatic forms are not known. They often survive spring flooding in an air pocket surrounding their water-repellant skin. Flooding in the warmer seasons is often survived only by the eggs. Only a few species can withstand a decrease in atmospheric humidity for more than a few hours. Most species prefer a temperature between 10 and 15 °C, and remain active above -2 °C and below +28 °C, but the glacier "flea" *(Isotoma saltans)* has an optimal temperature range of +5 °C to -5 °C. Practically all species, including those without ocelli, retreat from light. Inhabitants of deeper soil levels, e.g. *Onychiurus,* can respire in the soil atmosphere of 1% O_2 and up to 3.5% CO_2; surface species are more exacting in their requrements.

Classification. Forms with a segmented abdomen are placed in the suborder, *Arthropleona.* Its superfamily, *Poduromorpha,* is dominated by members with a well differentiated prothorax (e.g. the hemiedaphic, pigmented *Hypogastruridae* and the euedaphic, white and blind *Onychiuridae),* as well as containing the *Neanuridae,* whose fossil record extends back to the Devonian. The superfamily, *Entomobryomorpha,* includes in particular the epidaphic and atmobiotic families with a reduced prothorax *(Isotomidae, Entomobryidae, Tomoceridae,* etc.). Members of the second suborder, *Symphypleona* (rounded springtails), have a rounded, fused abdomen, and are divided into 7 families. These are mainly medium to large, well pigmented inhabitants of the vegetation layer. Some minute, colorless inhabitants of the soil interior, which possess a rounded abdomen, are placed in a third suborder, *Neelipleona.*

Collenchyma: see Mechanical tissues.

Collencyte: see *Porifera.*

Colliculi (caudalis, inferiores, rostralis, superiores): see Brain.

Colline zone: see Altitudinal zonation.

Collocalia: see *Apodiformes.*

Colloidal systems: disperse systems, in which the particles of the disperse phase have a diameter between 1 and 500 nm, the upper limit representing the diameter of certain complex molecules of high molecular mass, e.g. proteins, starch, rubber, etc.

In a *disperse system,* the particle phase (liquid, solid or gaseous) is called the *disperse phase,* while the medium in which the particles are distributed is the *dispersion medium.* Since the dispersion medium may also be liquid, solid or gaseous, various types of disperse systems are possible, e.g. smokes and dusts consist of solid particles dispersed in a gaseous medium, whereas fogs and mists (also called aerosols) have a liquid disperse phase. Foam is a 2-phase system, in which gas is dispersed in a liquid. If the disperse phase consists of relatively large (i.e. relatively highly aggregated) solid particles, the system is called a suspension; if it consists of relatively large liquid droplets it is called an emulsion.

Below a certain size of disperse particle (see above), *colloidal systems* are formed. Colloidal particles can only be visualized with the aid of the electron microscope. When a light beam is passed through a colloidal solution, the light is scattered sideways (Tyndall effect), whereas a light beam is hardly scattered when it passes through an optically clear solution of a truly dissolved crystalloid substance. Colloidal particles can be separated from the dispersion medium by ultrafiltration through an animal, plant or artificial membrane. Liquid colloids are known as *sols,* whereas the occlusion of water by a network of fibrous particles produces a colloid known as a *gel.* Colloidal systems with water as dispersion medium can be categorized as hydrophilic or hydrophobic systems. In hydrophilic sols, the disperse phase has an affinity for water, but may be unable to form a true, crystalloid solution (e.g. on account of its size), whereas the disperse phase of a hydrophobic sol is often an uncharged, unhydrated, water-repelling type of molecule, such as a metal or lipid.

Colloidal systems of macromolecules, in particular proteins, nucleic acids and polysaccharides, occur widely in living cells. From the physical-chemical point of view, such compounds are dispersed as single molecules and not as aggregates, although functional interaction between different types of colloidal macromolecules is crucial for cell metabolism, growth and replication.

Collum:

1) Neck.

2) The neck segment (first segment) of the *Diplopoda* (millipedes). It is particularly well developed in the orders, *Julida, Cambolida, Spirobolida* and *Spirostreptida* (previously combined as the *Opisthospermophora* or *Juliforma),* and is used as a battering ram when the animal burrows in the earth.

Colobine monkeys, ***colobid monkeys***, ***colobines***, ***colobes***, *Colobinae:* a subfamily of the *Cercopithecidae.* All colobines are arboreal. With the exception of the Pig-tailed leaf monkey, all have long tails. As an adaptation to a diet of foliage, with some buds and fruits, the stomachs of colobines are large and capable of holding a quantity of leaves equal to one third of the animal's weight. The stomach is also subdivided into three, semi-independent sacs, which function like those of a ruminant. As an adaptation to the mastication of leaves, the incisors are small and sharp, and the molars are proportionally much larger. Leaves are ground by side-to-side jaw movements, resulting in pronounced development of the masseters and the temporal muscles of the cranium. Colobines are Old World monkeys (see) with genera in Africa and southern Asia. See the following entries for the individual genera: Guerezas, Langurs, Douc, Proboscis monkey, Snub-nosed monkeys, Olive colobus and Pig-tailed monkey.

Colobognatha: see *Diplopoda.*

Colobus monkeys: see Guerezas.

Colocasia esculenta: see *Araceae.*

Colon: see Digestive system.

Colonial pine: see *Araucariaceae.*

Colonial theory: a theory claiming that the evolutionary ancestor of the metazoans arose from a spherical, hollow, colonial flagellate. At an early stage of development, this colony (the *blastaea)* possessed an anterior-posterior axis and was differentiated into somatic and reproductive cells. The hollow blastula formed during metazoan embryological development is considered to represent a recapitulation of this blastaea. Although the description of the ancestral blastaea is reminiscent of *Volvox,* ultrastructural studies indicate that the choanoflagellates (animal-like, monoflagellated protozoans, some solitary, some colonial) are more likely ancestors. The theory then proposes that the blastaea invaginated to form a double walled (ectoderm and endoderm), sac-like *gastraea,* which is equivalent to the embryonic metazoan gastrula. Among recent animals, the polyps of the hydrozoan *Cnidaria* bear a close similarity to this hypothetical, primitive gastraea. The colonial theory was first eleborated by Haeckel (1874), later modified by Metschnikoff (1887) and revived in 1940 by Hyman. Metschnikoff pointed out that primitive gastrulation in cnidarians occurs by ingression, i.e. cells are proliferated from the blastula wall into the blastocoel to produce a solid gastrula. Accordingly, the gastraea may have been a solid structure with no opening or internal space. In this hypothetical, solid gastraea (known as the *planuloid ancestor)*, the interior cells were active in nutrition and reproduction, and food was engulfed anywhere on the exterior surface, thus resembling the planula larva of cnidarians. According to Grell (1981), the gastraea may have evolved from a *Placozoa*-like organism (see *Placozoa).*

Colony:

1) a society of animals which transcends the family unit, but retains certain family structures. It is often formed for a particular purpose (e.g. nesting Cs. formed for brood care; settlement Cs. for building living quarters), and it is often temporary.

2) An accumulation of microorganisms produced by vigorous cell multiplication under favorable conditions. Cs. are usually visible to the naked eye, and may attain a diameter of several centimeters. Such large microorganism Cs. are produced by inoculating the center of a plate of nutrient medium with only a few cells, or even a single cell, of the relevent organism, and incubating for a few days or weeks. Under standardized laboratory conditions (e.g. on nutrient agar medium; see plate 16), the color, size and shape of the C. are typical of the organism (e.g. different yeast species) and are therefore used as taxonomic characters. Encapsulated bacteria (S-forms; S for smooth) produce smooth, glistening Cs., whereas colonies of noncapsulated forms (R-forms; R for rough) of the same species have a dull, rough appearance. The S-form can be converted into the R-form by mutation (loss of the ability to synthesize the polysaccharide of the bacterial capsule). In 1944, the transforming principle that converted the R- to the S-form of *Pneumococcus* (i.e. restored of the ability to synthesize the capsule polysaccharide) was identified as DNA, thereby providing proof that DNA is the true genetic material.

Cs. also develop under nonstandardized conditons, e.g. fungal Cs. on moldy bread and other foodstuffs.

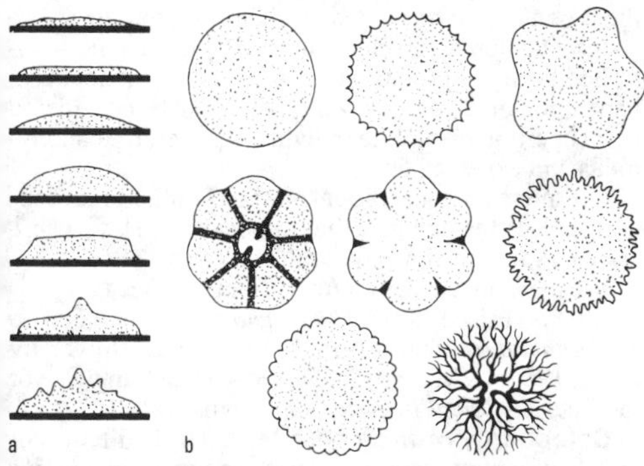

Colony. Bacterial colonies. *a* Transverse section from top to bottom of the colony. *b* Surface of colony viewed from above.

Colony formation: the most highly evolved form of association in animals, based on a family structure (see Sociology of animals). Members display a division of labor, often with corresponding morphological and physiological differentiation (caste formation). A colony remains active and viable for several years, and it can be looked upon as an organism of a higher order.

C.f. is observed only in insects (termites, ants, bees). All termites and ants form colonies, whereas bees display a range of behavior from solitary existence to true C.f. These precursors of colonial behavior provide an interesting insight into its evolution.

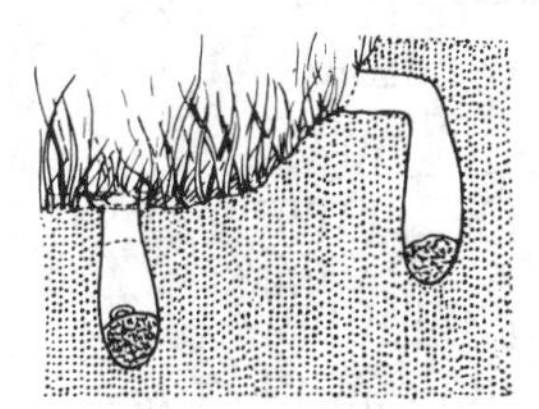

Fig. 1. *Colony formation.* Nest burrows of the solitary bee, *Osmia papaveris.* The cells are lined with poppy petals (after Goetsch, 1953).

Many solitary *Hymenoptera* provide a food store for the next generation, and this is placed in a specially constructed food burrow (e.g. *Osmia papaveris*). Often the older daughter bees contribute to the efforts of the mother bee by constructing more burrows (e.g. the mining bee, *Halictus malachurus*); the mother lays more eggs, while the daughter bees care exclusively for the hatched larvae. This is also true of the bumble bee, which in addition uses wax from its own body to complete the cell structure. Wasps and hornets actually build honeycombs of six-sided cells. The family associations of the tropical stingless bees (*Meliponini*) are stable over long periods; they also contain true workers, which are characterized by special adaptations (e.g. wax glands) (in contrast to bumble bees, where "workers" are simply regressed females). True C.f. and colonial behavior are characterized as follows:

1. *An inhabited structure.* Brood care, food storage and protection of the colony are served by a variety of subterranean and aerial structures, which are built by the insects, using foreign materials (earth, wood, plant material) and their own body products (saliva, feces, wax). See Animal constructions.

2. *Collective feeding behavior.* The acquisition of food for so many individuals (bee population 35 000 to 50 000, migratory ants up to 20 million) living in a confined space creates special problems. Bees forage within their flight range and store honey in their nest or hive. Termites and ants also carry provisions into the colony (e.g. harvester ants). Many termites and ants culture ectosymbiotic fungi (*Hypomyces*, *Rhozites*) as a food source, using a compost of feces or chewed plant material; in the nests of leaf cutting ants, these fungal gardens may support over a million inhabitants (see Symbiosis). Many ants form a trophobiotic association (see Relationships between living organisms) with *Homoptera* and *Psocoptera*; the ants feed and protect the aphids and "milk" them for their sugary exudate (honeydew). Driver ants (*Dorylidae*) live as predators, and at times they lead a nomadic existence.

3. *Division of labor.* Acquisition of food, construction of the habitation, brood care and protection of the colony are performed by functionally specialized colony members. The greatest specialization occurs in reproduction, which is sharply separated from all other functions, and is restricted to one or a few sexually ac-

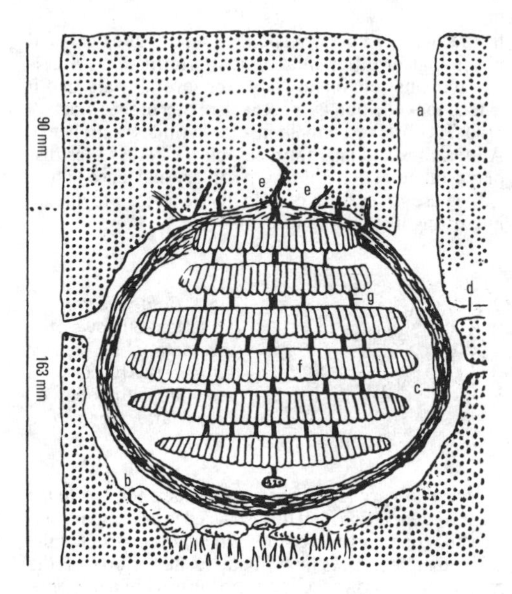

Fig.2. *Colony formation.* Section through a wasp nest *(Vespa germanica). a* Entrance. *b* Excavated cavity. *c* Nest outer wall of paper pulp. *d* Side galleries. *e* Roof attachment. *f* Combs. *g* Support columns. (After Janet)

tive animals. All other individuals in the colony are sexually regressed and they function as sterile female or male workers, or as soldiers. The function of a colony member often shows successive changes; in the case of the honey bee, "duties" change in the order: cell cleaning, larval feeding, food processing and cell construction, guard duty and finally foraging for food.

4. *Communication.* The complex interaction of numerous individuals is only possible with an efficient communication system, which depends on highly evolved sense organs and their associated centers in the brain. Mutual recognition and understanding are achieved by optical, chemical and mechanical signals, e.g. to indicate the location of food sources, to identify pathways, and to raise an alarm. Termites communicate alarm by tapping signals; ants communicate by touching each other, and fellow individuals are recognized from the nest smell.

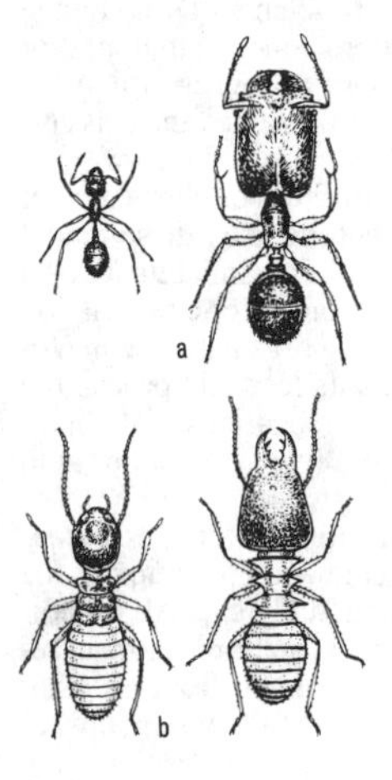

Fig.3. *Colony formation.* Different castes of colony-forming insects. *a* Ant worker and soldier. *b* Termite worker and soldier.

Bees communicate the distance and direction of forage plants by ritualized movements (dancing). The position of the sun or the polarized light from the sky serves as an orientation reference. The dancing-communication behavior of different bees represents an evolutionary sequence. Thus, the stingless bee (*Trigona irridipennis*) does not perform a patterned dance, but simply moves around in an agitated fashion on the comb, bumping into other bees. This primitive behavior appears to indicate to the rest of the colony that a food source has been found, but it provides no information on direction and distance. Other bees experience the odor of the nectar on the informant's body, then fly out to locate its source, a process which may take a long time. The dance of the dwarf bee (*Apis florea*) is more advanced; it contains a straight run at the same angle to the sun as the direction of the food source, and it is performed in the light on the horizontal surface of the single comb in the upper branches of a tree. Further evolution of the dance is shown by the giant bee (*Apis dorsata*) which performs a complete round dance on the vertical surface of the comb, containing angular components related to the direction of gravity. Nevertheless, the comb of the giant bee is exposed to daylight, and the dance appears to contain direct information on the position of the sun. The ultimate in this evolutionary sequence is found in the European honey bee (*Apis melifica*) and the Indian honey bee (*Apis indica*), which communicate entirely in the darkness of their hives. Here, the relative positions of sun and food source are first communicated by the angle of the dance movements to the direction of gravity (perpendicular to the hive). An accurate "internal clock" allows the bees to adjust continually to the changing position of the sun.

5. *Social regulation.* Crucial functions of the animal colony are regulated by Pheromones (see). In bee colonies, development of the sexually mature, reproductive female is determined by the different type of food given to her by the workers during her larval development. Sexual development of the numerous other females in the brood is also prevented by chemical inhibitors.

Insect colonies are characterized by the instinct-controlled behavior of individuals within the functional organization of the colony, which itself can be considered as a superorganism. The individual is therefore anonymous and expendable.

It is interesting that the ant and bee families both contain species that are parasites, predators or slave-keepers, or exploit the broods of other species.

See separate entries on Ants, Termites, Bees.

[K. Von Frisch, *Bees, Their Vision, Chemical Senses and Language*. Ithaca, New York, Cornell University Press, 1950; K. Von Frisch, Dialects in the Language of the Bees, in *Scientific American* August 1962].

Colophony, ***rosin:*** residue left after distillation of oil of turpentine from the crude oleo-resin of various species of *Pinus*. It consists mainly of abietic acid and related resin acids, and it is used technically for the manufacture of laquers, dyes, soaps, and linoleum, as an ingredient of ointments and plasters, and as violin rosin. See Balsam.

Colorado beetle: see Leaf beetles.

Coloration, ***coloring:*** the typical color or color pattern of the body or body parts. In animals C. is due to the deposition of pigments in the skin, hairs, feathers or scales, to the structure of the body surface, or to the color of internal organs or gut contents visible through the transparent body wall. Various physical principles and combinations thereof are involved in color generation, e.g. light diffraction by crystals within cells, color effects of clouded or turbid bodies, diffraction by thin layers, total reflection by air-filled cells, scales and hairs.

Ecologically, F. is an important environmental adaptation, providing protection against irradiation and temperature, and camouflage against predators, etc. (see Protective adaptation). Alternatively, conspicuous colors serve to attract sexual partners (e.g. breeding plumage of birds), while exteme C. and patterning of certain body parts serve to threaten, frighten, warn or confuse potential predators.

Soil and cave animals and endoparasites are often colorless, whereas the body suface of diurnal insects in sun-exposed habitats often has an extremely reflective metallic luster. In addition to light irradiation, C. also displays adaptation to the environmental temperature and humidity, and is also influenced by the diet (see Climatic rules). Animals with little endogenous coloration often acquire the color of the animals in their diet (e.g. fish, mollusks); the colors of some phytophagous insects are derived from the pigments (e.g. carotenoids) of their food plants.

Certain animal species periodically alter their C. (e.g. summer and winter plumage and coats of birds and mammals), or they may exist as differently colored morphological forms (e.g. seasonal dimorphism in butterflies). Some species are capable of rapid changes of C. (see Color change).

Color blindness: strictly speaking, the inability to differentiate any colors (see Color vision), but usually used for conditions in which the ability to recognize red, green or blue (violet) is absent or weak. The deficiency may be genetic, or it may be caused by damage to the retina, optic nerve, or central nervous system. Total color blindness *(achromasia, monochromasia)* is very rare. The inability to perceive red (red blindness) is known as *protanopy,* green blindness as *deuteranopy,* and blue blindness as *tritanopy.* The commonest color anomaly (poor color vision rather than color blindness), affecting 8% of men and 0.4% of women, is the inability to distinguish between red and green.

Color change: an alteration of body color which occurs in many animals. *Morphological C.c.* involves the gradual change of coloration by the formation or degradation of certain pigments, which may be present in pigment-containing cells known as chromatophores. In *physiological C.c.,* pigments are aggregated within chromatophores, making little contribution to color, or they confer color by becoming dispersed throughout the cell. Physiological C.c. is regulated by the nervous system or by hormones, or by a combination of both. In cephalopods, the chromatophores are associated with smooth muscle fibers, which are supplied by neurons. Numerous teleost fishes possess sympathetically and parasympathetically innervated chromatophores. C.c. in the chameleon occurs by sympathetic control of chromatophores. C.c. in crustaceans involves hormones e.g. peptide hormones which have been isolated and characterized from the eye stalk region. *Hormonal C.c.* in vertebrates is determined by melanophore stimulating hormone (MSH) (see Melanotropin) from the opiomelanotropinergic cells of the pars intermedia of the pituitary and by neurons of the central nervous system. In species lacking the pars intermedia (e.g. chickens, porpoises, whales), MSH is produced by the neurohypophysis. MSH, a peptide hormone of known sequence and conformation, was first recognized and named for its ability to cause dispersion of melanin granules within melanin-containing melanophores in the skin of cold-blooded vertebrates. It causes increased deposition of melanin in mammals, but its role in human pigmentation is doubtful. It also causes dispersion of pigment granules within the yellow (xanthophores) and red chromatophores (erythrophores). In fishes and amphibians, the response of skin pigmentation to light is mediated by Melatonin (see) (*N*-acetyl-5-methoxytryptamine), which is produced in the pineal gland and retina. Melatonin reverses the darkening effect of melanotropin by causing the melanin granules within the melanocytes to aggregate. Other endogenous factors are involved in the regulation of C.c., e.g. factors regulating hormone release from the hypophysis, pineal gland and adrenal medulla. Light intensity is a particularly important exogenous factor in the control of C.c.

Color vision: the ability to distinguish between different wavelengths of electromagnetic radiation. C.v. is widespread in invertebrates, especially in crustaceans and many insects. According to their retinal structure, cyclostomes, elasmobranchs and lungfishes are color-blind. Among the amphibians, some are color-blind (midwife toad), some can differentiate between red and blue (frogs and various newts), or blue and violet (common toad, *Bufo bufo)* whereas others can recognize all colors (salamanders). Reptiles are generally color-competent, some showing particular efficiency in specific spectral regions. Diurnal birds and mammals see colors, the most competent being monkeys and humans. In contrast, rats, hamsters, the golden hamster, rabbits and lemurs are color-blind. Other species differentiate only certain colors, e.g. the domestic cat in the long wave region of the visible spectrum. Bees see numerous colors in the range 300–650 nm, whereas humans can distinguish about 180 gradations of color between 380 and 760 nm. In humans, C.v. starts when the light intensity exceeds the threshold of the photopic system *(color threshold);* at this point, three colors can be distinguished: red (760–570 nm), green (570–480 nm) and blue-violet (480–380 nm). The basis of this phenomenon appears to be the excitation of three different types of cone cells, containing photosensitive pigments for red (erythrolab), green (chlorolab) or blue (cyanolab), respectively. The sensitivity maxima of these three receptors, determined from the receptor potentials of individual cones of carp, monkeys, humans, etc. correspond to the absorption maxima of the three visual pigments, which accords with the *trichromatic principle* of C.v. elaborated by Young (1807) and Helmholtz (1867). This principle states that the *color sensation* is determined by the relative excitation of each receptor type, i.e. either by the excitation of a single receptor for red, green or blue-violet, or by various mixtures of these components. Equal excitation of all three receptor types gives the impression of colorlessness (white). This is in accordance with the *principle of opponent colors* (Hering, 1874), and it is due to the interaction of output signals from the different cones before they arrive in the central nervous system. For example, some spectral signals pass through the corpus geniculatum laterale, which contains the magnocellular nucleus and the parvocellular nucleus. Parvocellular neurons consist of four types of spectral filters; two of these have the properties of wide band (425–600 nm), short wavelength filters with maximal transmission at 525 nm, while two have wide band (500–700 nm), long wavelength sensitivity with maximal transmission at 600–625 nm. In each case, monochromatic light outside the excitatory sensitivity range hyperpolarizes the neurons, whereas wavelengths within the excitatory range cause depolarization. Parvocellular neurons are therefore also called color opponent cells, because they are stimulated by one part of the spectrum and inhibited by another. In addition to these two chromatically polar systems, there is a black-white system which signals only brightness. The trichromatic (receptor) principle and the opponent color (neuron) principle are combined in the *zonal theory* of C.v. Receptors sensitive to different regions of the spectrum have also been demonstrated in the ommatidia of insects, e.g. two in *Diptera* (green- and blue-receptors), three in bees (yellow-, blue- and ultraviolet-receptors).

Colostrum: the first milk produced by the mother mammal and fed to the newborn animal. C. of many mammals contains large quantities of antibodies, which are absorbed through the intestinal wall and into the circulation, conferring immunity on the newborn animal.

Coltsfoot: see *Compositae.*

Colubridae, ***harmless snakes, colubrid snakes:*** the largest and most varied snake family, containing more than 1 800 species in 280 genera, and represented in all parts of the world. Many different types of environment have been invaded by colubrid snakes. In addition to aquatic and terrestrial species, there are burrowing species, as well as numerous bush- and tree-dwelling forms, which are often colored green. Longitudinal ridges on the ventral surface represent a special adaptation for climbing. Some species can even project themselves from branches and glide downward in a brief flight. Colubrid snakes are most common in the tropics, but the range of some species extends as far as latitude 65 °N, e.g. the Common European grass snake, *Natrix natrix.*

They are generally slender and long-tailed, and the head is not strongly demarcated from the neck. There are no vestiges of the pelvic girdle. The highly specialized jaws are provided with numerous teeth, and only the intermaxillary bone lacks teeth. Some species have aglyphic (solid and ungrooved) teeth that are not associated with venom glands (which may be present or absent). Other species possess opisthoglyphic teeth, i.e. the rear upper jaw teeth are grooved and are associated with venom glands. Snakes with this latter type of dentition (primarily members of the subfamily *Boiginae)* can deliver a poisonous bite, but on account of the inaccessible position of the venom teeth, this can only be achieved by "chewing" on their prey. In general, they therefore do not represent a serious danger to humans, with the exception of the Boomslang (see) and the Twig snake (see), whose bites can prove fatal. Forward-situated venom fangs in the upper jaw, like those found in the *Elapidae* (see), *Hydrophiidae* (see), *Viperidae* (see) and *Crotalidae* (see), never occur in colubrid snakes. In a few species the teeth are regressed as an adaptation for feeding on birds' eggs. Colubrids also differ from more primitive families in lacking a coronoid bone in the jaw, and in the presence of only the left carotid artery. Depending on their ecological adaptations, colubrid snakes feed on small rodents, birds, birds' eggs, lizards, frogs, toads, fish, crustaceans, snails, earthworms, etc. Most colubrid snakes are oviparous, but some give birth to live young (vivi-oviparous), and a small minority is truly viviparous.

The earlier classification of the C. according to tooth structure is now abandoned, since grooved teeth have evolved in several genera that are not closely related, and nongrooved and grooved teeth can occur in closely related species in a single genus. Even aglyphic, nonvenomous snakes may possess functional venom glands (e.g. *Natrix natrix),* which are not discharged until the prey has passed along the pharynx, and which are not associated with teeth.

Eleven subfamilies are recognized. Less advanced members belong to the *Xenodontinae,* e.g. Hog-nosed snakes (see). Two other important subfamilies are: the ***Water snakes*** *(Natricinae),* e.g. Garter snakes (see), Common European grass snake (see), Tessellated snake (see), Dice snake (see); and the ***True terrestrial and arboreal colubrids*** *(Colubrinae* , the largest subfamily), e.g. Rat snakes (see), European smooth snake (see), Asian rat snakes (see), King snakes (see), Whip snakes (see), Egg-eating snakes (see), Snail-eating snakes (see), *Boiginae* (see).

Columba: see *Columbiformes.*

Columbidae: see *Columbiformes.*

Columbiformes, ***pigeons*** or ***doves, sandgrouse*** and ***dodos:*** an order of birds (see *Avies),* containing about 300 living species, most of which belong to the family *Columbidae.* Wing and palate structure are similar to those of the *Charadriiformes,* but other characters, like the structure of the vertebrae, sternum and feet, justify their allocation to a separate order. The family has a cosmopolitan distribution, excluding the polar regions. For the first few days after hatching, the young birds are fed by both parents with regurgitated "pigeon milk", which is produced in the adult crop by secretion and sloughing of the epithelial lining. Production of pigeon milk is stimulated by the hormone prolactin, the same hormone that stimulates milk production in mammals. After the first few days, the parents provide other food items, but feeding with pigeon milk continues until the young are well grown. *Columbiformes* are almost unique among birds in being able to drink by sucking up water through their beaks without raising their heads; the Buttonquail (family *Turnicidae,* order *Gruiformes)* is the only noncolumbiform species that also takes water without raising its head, with a "bibbling" action of the mandibles. None of the *Columbiformes* is aquatic or particularly associated with water. The order is divided into 3 families: 2 living and 1 extinct. The extinct family is listed separately (see *Raphidae).*

1. Family: *Pteroclidae* ***(Sandgrouse;*** 23–40.5 cm; 16 species in 2 genera). These gregarious birds live in open dry areas in southern Europe, Asia and Africa. They are strong fliers, and they make at least one daily flight, sometimes over a considerable distance (e.g. 10 miles or 16 km), to obtain water. With their very short, feathered legs, running is impossible. They forage on the ground for berries, seeds, buds and occasionally in-

sects, walking with short, mincing steps. Plumage is attractively patterned in orange, chestnut, black, white, grays and browns, affording an effective camouflage in desert surroundings. Males are brighter than females, and both sexes incubate, the male by night, the female by day; both sexes feed the young by regurgitation. The cryptically colored and patterned eggs are laid on the bare ground, sometimes in a slight scrape. Some species are migratory, while others are irruptive, i.e. they undergo sporadic population movements, e.g. ***Pallas' sandgrouse*** *(Syrrhaptes paradoxus;* 33–40.5 cm) of central Asia, which invades western Europe after hard winters. Most members of the family (14 species) belong to the genus *Pterocles,* e.g. ***Chestnut-bellied sandgrouse*** *(P. exustus;* 32 cm), which inhabits the African desert north of the equator.

2. Family: *Columbidae* **(*Pigeons*** or ***Doves;*** about 280 species in 40–43 genera). The names, pigeon and dove, are interchangeable, although there is a tendency to call the smaller species doves. All *Columbidae* resemble one another closely, and are immediately recognizable as members of the same family. They have short legs, a stout body, a small head, soft plumage, a slender bill, which is constricted in the middle, and a fleshy cere (housing the nostrils) at the base of the bill. In addition they have a large crop, which has 2 side pockets. Representatives are found in all temperate and tropical zones of the world, the main center of distribution being Southeast Asia and Australia. In the present account, 6 subfamilies are presented, but these divisions are sometimes rather artificial.

1. Subfamily: *Columbinae.* Members of this large subfamily (about 180 species) are primarily seed eaters. Most have a strong, muscular gizzard and a long intestine, and they ingest grit to help grind the seeds in the gizzard. In contrast to the sometimes brilliantly colored fruit pigeons (see below), the *Columbinae* are generally quietly colored in various soft shades of brown, gray, wine-red, often with some iridescent white or black areas on the neck, wings and tail. This subfamily contains the 50 species of *Columba,* which are most "typical" of all pigeons. The ***Wood pigeon*** *(Columba palumbus;* 40.5 cm) is generally blue-gray, with a white patch on its underparts, and glossy green and purple on the neck; flanks and belly are paler, and a white wing band is visible in flight; it has red-purple legs, and a golden bill, which is red at the base with a white cere. *C. palumbus* breeds from Europe to Iran and northern India, and from North Africa to the Azores. It is very adaptable, inhabiting wooded country of any kind, suburbs and towns, and nesting (the nest is a twig platform) in trees or on the ledges of buildings (it is known to nest on the ground in Orkney). Food consists of grains, fruits, nuts, leaves, buds, flowers, but mainly cereals and clover; insects are taken only occasionally. This remarkable ability to establish a dependence on humans, and to adapt to life in agricultural areas and even in towns and cities, is displayed by several other species. Thus, the domestic pigeon and its feral descendants are familiar in most towns and cities throughout the world. Ancestor of many breeds of domestic pigeon is the ***Rock pigeon*** *or* ***dove*** *(Columba livia;* 33 cm; blue-gray with white wing bars and iridescent neck), which is native to southern Eurasia and North Africa. After the domestic or feral pigeon, the pigeon most closely associated with humans in Europe is the ***Collared dove*** or ***Collared turtledove*** *(Streptopelia decaocto;* 32 cm) (plate 6), which is widely distributed throughout Asia and Europe from the Atlantic to the Pacific. Formerly common in dry country bordering cultivated areas, it has now spread almost explosively into the agricultural heartland, as well as into towns and villages. Members of the genus, *Streptopelia,* are generally known as ***Turtledoves.*** In North and South America, the ***Inca dove*** *(Scardafella inca;* 20 cm; southwest USA to Costa Rica) is often found near human habitations, particularly farms, and in the winter it becomes completely dependent on these for food and shelter on the northern edge of its range. In the last century, it is claimed that the ***Passenger pigeon*** *(Ectopistes migratorius;* 43 cm) was the most abundant species of bird, not only in its native habitat of temperate central and eastern North America, but in the world. Their vast nesting colonies extended for 20 miles (32 km) or more in the North American oak and beech forests, each tree containing sometimes hundreds of nests. A nesting in Wisconsin in 1878 covered 850 square miles (2 200 square km), and the number of birds was estimated conservatively at 136 million. Profligate shooting and trapping for food reduced their numbers, so that the species was nearly extinct by the end of the last century. The last wild bird was recorded in 1899, and the last living specimen died in 1914 in a Cincinnati zoo. Commonest pigeon of North America is the ***Mourning dove*** *(Zenaidura macroura;* 30.5 cm), which occurs in every state of the Union except Hawaii. Protected as a songbird in the north, it is a sporting bird in the Southern States.

2. Subfamily: *Treroninae* (20 species). A group of arboreal, fruit-eating pigeons of sub-Saharan Africa and Southeast Asia, also known as ***Green pigeons.*** All are similarly colored pale yellowish green, often with extra markings of yellow, orange or mauve, as well as brilliantly colored eyes and cere. Like the *Columbinae,* they have muscular gizzards, and they digest the seeds of the fruit (mainly wild figs) that they eat.

3. Subfamily: *Duculinae* (37 species). Also known as ***Imperial pigeons*** or "Nutmeg pigeons", many of these fruit pigeons live on hard nuts and fruits that are often larger than their heads. Their jaws have elastic sockets and can stretch to accommodate large objects. They are widespread in the Indo-Malayan and Pacific areas, a few species reaching India and Australia. They digest only the pulp of their dietary fruit and void the stones, thereby performing a service in seed dispersal. Accordingly, they have a strong stomach and a short wide gut.

4. Subfamily: *Ptilinopinae* (39 species). Most fruit pigeons are gorgeously patterned and colored, but the *Ptilinopinae* are arguably the most beautiful of all birds. Also known as ***Painted pigeons,*** these fruit-eating pigeons inhabit the South Pacific islands, a few reaching Southeast Asia. *Ptilinopus victor* **(*Flame dove*)** of Fiji displays unusual sexual dimorphism, in which the male is vivid orange with an olive-yellow head, and the female is dark green with an olive-yellow head. The ***Nicobar pigeon*** *(Caloenas nicobaria;* 40 cm) is heavier for its size than most pigeons. Its plumage is iridescent dark green, with a white tail, red legs and a black bill. Each feather of the neck is elongated, the whole forming a display ruff, which is colored green with coppery and blue-green glints. Found from the Nicobar Islands to New Guinea, and as far north as the Pilippines, *C. nicobaria* spends much of its time on the ground feeding on fallen fruit and seeds on the forest floor.

5. Subfamily: *Gourinae*. A small subfamily of 3 closely related ***Crowned pigeons*** (genus *Goura)*, all native to New Guinea. They are the largest pigeons (66–84 cm), and they are characterized by a large, laterally flattened, erect crest of lacy feathers. *Goura cristata* (northwestern New Guinea and neighboring islands) is slaty bue, with a chestnut band across its back and the upper wing coverts, a white wing speculum, black face mask, pink feet and red eyes.

6. Subfamily: *Didunculinae*. A monotypic family consisting of the ***Tooth-billed pigeon*** *(Didunculus strigirostris;* 33 cm) of Samoa. It is greenish black and chocolate brown, with bright red bill and feet, and feeds mainly on the fruits of the banyan tree. The sharply decurved bill has 2 serrations near the tip.

Columbine: see *Ranunculaceae*.

Columella

1) A central, sterile, column-shaped structure: a) in the sporangium of certain fungi, e.g. *Mucor;* b) in the capsule of the sporangium of mosses, where it is surrounded by spore-forming tissue (archesporium) to which it supplies nutrients and water for the developing spores.

2) A columnar, calcareous structure in the skeleton of corals.

3) A spindle or axial rod inside a coiled snail shell, formed from the inner walls of the convolutions that lie close to the ideal shell axis.

4) One of the sound-transmitting bones of the middle ear. See Auditory organ, Skull.

Columnar epithelium: see Epithelium.

Columnar rigidity: see Mechanical tissues, Biostatics.

Combination atavism: see Atavism.

Combtooth blennies: see *Blennioidei*.

Cometabolism: degradation by microorganisms of certain compounds, which cannot serve as a sole source of energy or carbon, i.e. the microorganisms cannot grow at the expense of these compounds alone. Degradation of such compounds is only possible if true nutrients for the microorganisms are also present. C. is particularly important for the removal of xenobiotics, which are produced by human agency and released into the environment, e.g. agrochemicals and industrial waste products.

Comfort behavior: see Grooming.

Commensalism: cohabitation of two species, in which one consumes the excess food of the other, and neither partner is disadvantaged. Thus, small fishes participate in the meals of large predatory fishes by consuming those parts of the prey that are lost by the predator and sink while it is feeding. There are transitional states between extreme C. and associations in which one partner is definitely disadvantaged.

Common adder: see Common viper.

Common catfish: see Catfish.

Common European grass snake, *Natrix natrix:* the most common Central European colubrid snake (see *Colubridae)*, usually gray with small black spots. On both sides of the head is a conspicuous white or yellowish crescent-shaped spot. Females grow up to 1.8 m, males up to 1 m in length. It is a water-loving, diurnal snake, found on level lowland plains, as well as in mountainous areas up to 2 000 m, and in agricultural areas. Several subspecies are distributed from Northwest Africa over the whole of Europe to western Asia. It feeds mainly on frogs, with some toads and fish. When threatened, it retreats into the water, and is a very skilful swimmer and diver. Mating occurs in April to May, and 20–50 eggs are often laid communally at a single site among rotting vegetation or loose earth; garden compost heaps are favored egg-laying sites. The 20 cm-long young snakes hatch after 10 weeks. From October to March, *Natrix natrix* remains buried in the ground in a cold torpor. In Europe it is a protected species.

Common gecko, *Tarentola mauritanica:* a 6–8 cm-long gecko (see *Gekkonidae)*, which occurs from the western islands of the Mediterranean to Egypt. The entire underside of the toes is expanded into an oval lamellate cushion or attachment disk. It is commonly found on stone walls and the outer walls of houses. Like most other geckos, its eyelids are fused to a transparent spectacle.

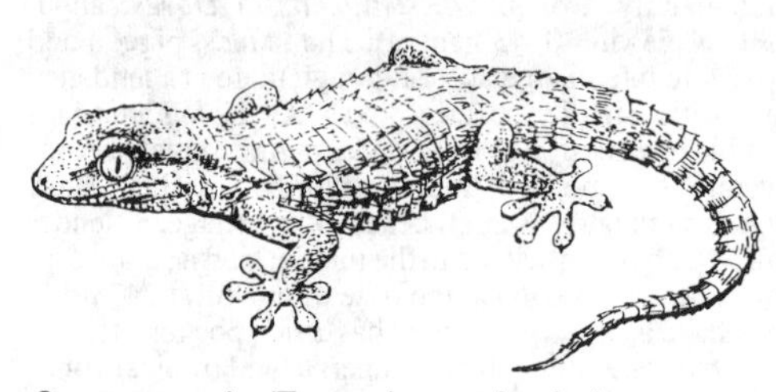

Common gecko (Tarentola mauritanica)

Common iguana: see Green iguana.

Common lizard:

1) Lacerta (Archaeolacerta) saxicola: a slender lacertid (see *Lacertidae)*, greenish brown in color, often spotted or striped, up to 25 cm in length, and found in rocky areas of the Caucasus. What was previously thought to be a collective of subspecies and varieties, has been shown by recent research to contain a number of independent species. Some of these (e.g. the Armenian *Lacerta armenica, Lacerta unisexualis,* etc.) reproduce parthenogenetically. Cf. also *Teiidae* (see).

2) Lacerta vivipara: a member of the family *Lacertidae* (see). It attains a maximal length of 18 cm, but most specimens are 12–16 cm in length. The back is reddish brown with black spots, and the belly is yellow to orange. An inhabitant of mountain forests and wet meadows (up to 300 m) in Europe and Asia (as far as Sachalin), the range of *L. vivipara* also extends northward to the Siberian tundra, and it is the only Eurasian lizard to cross the Arctic circle. Mating occurs in May to June, and it is the only member of the *Lacertidae* to bear live young. It is actually oviviparous (i.e. there is no placenta), and the young are born inside a transparent membrane, which is broken during or shortly after birth. The diet consists of slugs, worms and insects. *L. vivipara* often lives with the blindworm in the same biotope.

Common name: the accepted international, nonsystematic, nonproprietary name of a chemical agent, in particular a pharmaceutical compound or plant protection agent.

Common skink, *Scincus scincus:* a viviparous, light brown skink (see *Scincidae)*, about 20 cm in length, patterned with dark, transverse bands. It has a strurdy, cylindrical body with 4 short limbs, each with 5 toes. The tip of the snout is shaped like an elongated wedge, and the tail (which is shorter than the rest of the body) is

:onical. It inhabits the hot North African deserts, and on account of its faculty for "swimming" through the loose sand with serpentine movements, it is also known as the 'sandfish". In the Middle Ages its flesh was considered o possess medical and aphrodisiac properties. There are several geographical subspecies.

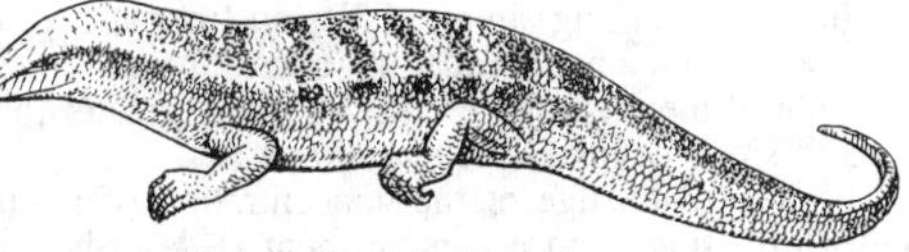

Common skink (Scincus scincus)

Common viper, ***common adder,*** *Vipera bereus:* one of the most widely distributed Old World species of snake, found from western Europe over temperate Asia o Sachalin, and occurring as far north as latitude 67 °. This member of the *Viperidae* (see) is normally 50–60 cm in length, although females may occasionally attain 80 cm. The dorsal surface of males is predominantly gray, while females are brown with a black or dark brown zig-zag band. The head carries a dark, cruciform pattern. Some specimens, especially in damp habitats, are completely black. Red-brown variants are also known. It inhabits lowlands and mountainous areas up o 3 000 m, and it is common in the central mountains of Europe and on the Baltic coast, but absent from large agricultural and industrial areas. Mating occurs in April o May, starting with courtship fights between the males, in which they raise the front part of the body and strike at their rival without biting. Birth is vivi-oviparous, and 5–18 young are born in August or September. The new-born C.v. are 15–20 cm in length and possess functional venom fangs. C.v. overwinter in groups in earth cavities. Young snakes feed on small lizards and frogs, but the adults feed mainly on mice. They do not attack humans unless provoked. The bite can be dangerous to humans, especially children, but its venomous properties are greatly exaggerated. Over periods of many years, this snake causes very few human fatalities in Central Europe.

Communication: transfer of information between two dynamic systems. The different stages of C. form a *C. chain* (Fig.), the minimal components of which are a transmitter, a channel of C. and a receiver. The *transmitter* generates, encodes and transmits *signals,* which are conveyed via the C. channel, then received, decoded and evaluated by the *receiver.*

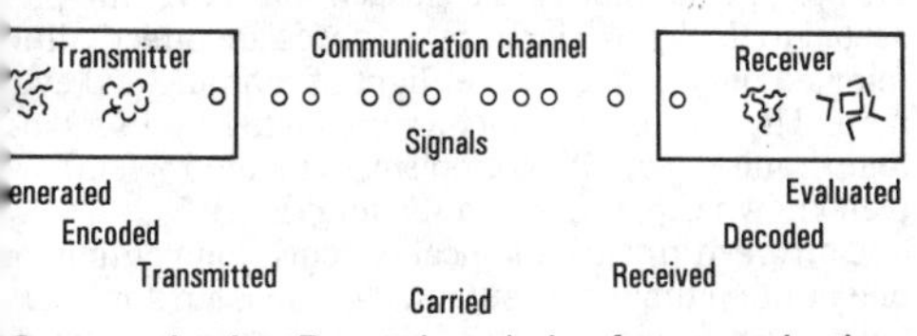

Communication. Events in a chain of communication.

Signals have the dimensions of quantity and intensity, and may take the form of spatial patterns or impulse frequencies. C. has important implications in endocrinology, neurophysiology and ethology (see Biocommunication). The cells of endocrine glands serve as transmitters, which synthesise and release signals (hormones; familiarly known as "chemical messengers"). The C. channel for these hormone signals is the vascular system, which carries the hormones to cell populations with specific hormone receptors (e.g. in the cell membrane), i.e. a cell population serves as the receiver of the hormone signal. Signal reception (hormone-receptor contact) triggers a physiological response (i.e. an evaluation of the signal). In neurophysiology, neurons or parts of neurons can be considered as transmitters, and the signals are either Neurotransmitters (see) or electrical impulses. The synaptic cleft is the C. channel of the released neurotransmitter, and the postsynaptic membrane is the receiver of the transmitted signal; the resulting potential difference represents the evaluation of the signal.

Ethology is concerned with C. between organisms (interorganismal C.), in which signals generated by one organisms are received and evaluated by another organism (with the aid of sensory organs), followed by a behavioral response. Fundamental animal behavior, such as feeding, protection and reproduction (e.g. courtship behavior) involves C.

Information (e.g. about a current state of affairs, or the condition of the transmitter) can be conveyed or transmitted only if it is *encoded* by specific signals. Thus, the impulse frequency of a conductile nerve membrane (specific signal) encodes the excitatory state of the neuron (condition of the transmitter).

Communication theory: see Information theory.

Communicative display, ***behavioral signalling***, ***display signalling:*** Signal behavior (see) based on movement and posture, which elicits specific behavior in the recipient (Figs. 1 and 2).

Fig.1. *Communicative display in the wolf. a* Threat. *b* Threat with uncertainty. *c* Very mild threat. *d* Weak threat with strong element of uncertainty. *e* Nervousness and anguish. *f* Defensive response to an antagonist, indicating uncertainty and subordination. (After Schenkel. 1947)

Fig.2. *Communicative display.* Typical behavioral postures in song birds. *a* Intimidatory stretching and *b* submission (great tit). *c* Attacking threat against a rival (bullfinch). *d* Courtship display by male to female (canary). *e* Lateral courtship display by male, performed side-on to the female (chaffinch). *f* Side-on courtship display by male to female (bullfinch). *g* Invitation gesture of the female chaffinch. *h* Posture of male chaffinch before copulation. (After Hinde, 1961)

Comoviruses: see Multipartite viruses, Virus groups.

Companion cell: see Conducting tissue.

Comparative behavioral research: the application of ethological methods and the formulation of ethological problems with respect to evolutionary relationships between of organisms. C.b.r. is therefore concerned with the variability of behavior between species, and it attempts to understand the causal relationships of those variations.

Compartment:

1) A selected functional unit of an ecosystem. The identity of a C. depends on the level of inquiry into the system, and it cannot be functionally subdivided, i.e. all the elements of a C. have the same function. Thus, in studying the stability of a freshwater ecological system, producers, consumers and destruents may each be considered as a C. In a different line of investigation, the relevant Cs. may be species, populations, individual organisms, and even parts of organisms (e.g. leaves, stems, roots).

This collective reduction of complicated structural elements to their functional aspects is important in the mathematical modelling of ecosystems.

2) A subcellular compartment, such as the nucleus mitochondria, cytosol, chloroplasts, etc., which serv specific functions in cell metabolism.

3) In physical chemistry, part of an exchange system which is isolated from adjacent Cs. either spatially by membrane, or by a chemical process. Since the content of a C. must be homogeneous and contain no gradients it follows that during simple diffusion from one C. to another through a membrane, the diffusion rates on either side of the membrane must be much greater than the diffusion rate inside the membrane.

The rate of change of the concentration of a substance diffusing from a compartment is described by first order linear differential equations. These equation can be solved in relation to the duration of diffusion, using exponential expressions, in which the time constants are combinations of the rate constants of the exchange processes. In analogy with electrical circuits Cs. may be connected in series or in parallel. An exchange system finally attains a state in which there is no longer a net flux from one C. to another, i.e. net exchange (e.g. net simple diffusion) has ceased. The system is then said to be in dynamic equilibrium. This final state of a dynamic system is independent of its starting state, and depends only on the rate constants of each exchange process. Therefore, the concentration of components in a system in dynamic equilibrium (i.e the quantity of substance in each C.) can be regulated by alterating the reaction rates. Exchange of material between the Cs. of biological systems is often subject to nonlinear regulatory mechanisms, so that the above theoretical analysis has only limited application. Under laboratory conditions, however, biological systems may be isolated so that such an analysis is possible.

Compartmentation, ***compartmentalization:*** division of the living cell, e.g. by biomembranes, into structurally and biochemically distinct reaction compartments. Many metabolic conversions in the cell, which would otherwise interfere with one another, are able to proceed in orderly sequence, because they occur in separate compartments. C. is also important in regulating the complex interactions of different metabolic processes. An example of conversions involving cytoplasmic and noncytoplasmic compartments is Crassulacean acid metabolism (see).

Compass plants: plant species which, as an adaptation to strong solar irradiation, hold their foliage leaves vertically and oriented in a north-south direction, by torsion of the leaf base. C.ps. therefore have a compressed appearance. When the irradiation intensity is greatest, i.e. at midday, the edges of the leaves are presented to the light. The upper surfaces are directly illuminated only by the weaker light of morning and evening. The leaf movements are regulated by light and temperature (see Phototropism, Thermotropism). A well known C.p. is *Lactuca serriola* (Fig.).

Competition: in biological systems, competition for one or more limited resources, e.g. food, territory, sexual partner. As the population density of competitors increases, resources become increasingly scarce, so that C. is an important factor in the regulation of population size (see Environmental factors, Abundance). C. between members of the same species *(intraspecific C.)* is a crucial factor in the selection of suitable genotypes (see Selection). In intraspecific C., however, some members must perish, decreasing the average chance of

Compass plants. Prickly lettuce *(Lactuca serriola).* *a* View from north or south. *b* View from east or west.

urvival for all individuals in the population. Lessening f intraspecific C. by subdivision of a habitat into iches (see Niche) is therefore an important ecological trategy, leading to the evolution of new species. Within particular niche, intraspecific C. is total (i.e. it applies o every resource). On the other hand, *interspecific C.* C. between members of different species) is usually restricted to only one or a few resources. Thus, in the presence of a superior competitior (see k-Strategy), a predator can resort to alternative prey, or avoid C. by hunting in an area that cannot be exploited by the ompetitor.

Competition-exclusion principle, *Gause's hypothesis, Grinnell's axiom:* a rule stating that the reater the similarity in the environmental requirements f 2 competing species, the less is the likelihood that hat they will both form a permanent association with he same environment. If the 2 species have identical requirements, they cannot coexist indefinitely. One species will always prove to be the stronger K-Selected pecies (see), and suppress the other.

Complementary genes: nonallelic genes which nust be present together in the genome to produce a articular phenotypic trait.

Complementation: functional complementation etween two mutants. C. is revealed by comparing heterozygous, heterokaryotic or heterogenotic cells (i.e. cells ontaining both mutant genes) with cells containing only ne of the mutations in question. In the first case, the partially or totally blocked biochemical function is restored e.g. the catalytic concentration of a crucial enzyme may e much greater in the heterozygote than the sum of its atalytic concentrations in the two mutants). There are wo fundamentally different types of C.

a) *Intercistron C. (intergene C.)* occurs when the two nutations are present in different functional genes (cistrons). In this case, C. always occurs when the mutaions are present in the same cell. Each cistron deternines the structure of a polypeptide. The cell is thereore fully functional if each polypeptide is present in its ormal configuration, and this is possible if at least one unctional gene is present for each polypeptide. Such olypeptides may be parts of a single enzyme, they may e separate enzymes in the same metaboilic pathway, or they may catalyse reactions in quite different areas of metabolism.

b) *Intracistron C. (intragene C., intra-allele C.)* occurs when both mutations are present in the same cistron, and affect the same polypeptide chain. In this case, restoration of enzyme activity only occurs between certain mutant pairs, and the quantity of active enzyme (usually very much less than in intercistron C.) varies according to the types and sites of mutation within the cistron.

On the basis that noncomplementary mutant alleles contain overlapping defects in the same region of the gene, whereas complementary mutations occur in different regions of the gene, these relationships can be represented by a linear C. map. The latter shows complementary allele pairs as nonoverlapping segments, and noncomplementary allele pairs as overlapping segments. The C. map frequently correlates with the linear fine structure of the cistron, which can be determined by recombination analysis. C. is not due to the physical interaction of genes, but to the action of gene products. Thus, C. can be demonstrated in vitro between purified enzyme preparations from appropriate mutants. If, for example, a functional enzyme normally consists of two identical polypeptide chains (aa), mutation may give rise to nonfunctional enzymes such as $\alpha_1\alpha_1$ or $\alpha_2\alpha_2$. In some cases, $\alpha_1\alpha_2$ (although different from the normal aa) expresses partial or full biological activity, and it can be produced simply by mixing enzyme preparations from the two mutants.

Complement system: a nonspecific defense mechanism in the sera of all vertebrates. It consists of a heat-labile (100% inactivation after 30 min at 56 °C) cascade system. In mammals it involves at least 20 glycoproteins, and its action is interpreted on the basis of 9 main factors. Each activated component (or complex) is a highly specific protease acting only on the next component of the cascade. Two pathways are involved, the *classical* and the *alternative*. The classical pathway is activated by the binding of C1 to immune complexes containing IgG or IgM. C1 consists of 3 subunits: C1q, C1r and C1s. The binding sites for IgG and IgM reside on C1q, and occupation of these sites confers proteolytic activty, which cleaves a single peptide bond of C1r. The

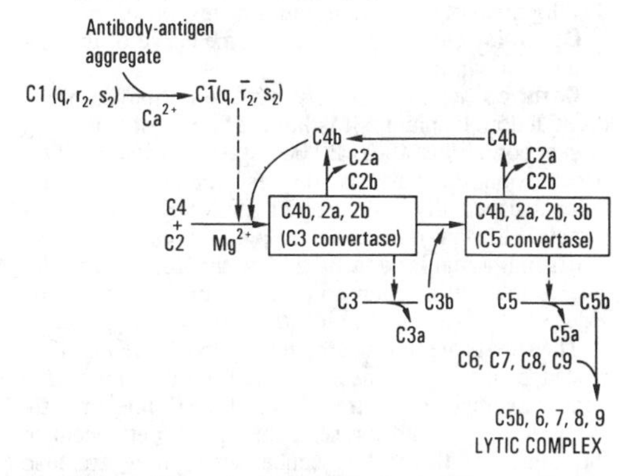

Fig. 1. *Complement system.* Classical pathway of complement activation. A bar over the number or letter of a factor indicates that the factor is activated (i.e. proteolytically active). Activation by proteolysis occurs near to the *N*-terminus, producing a small (a) and a large (b) cleavage product, e.g. C3 → C3a + C3b.

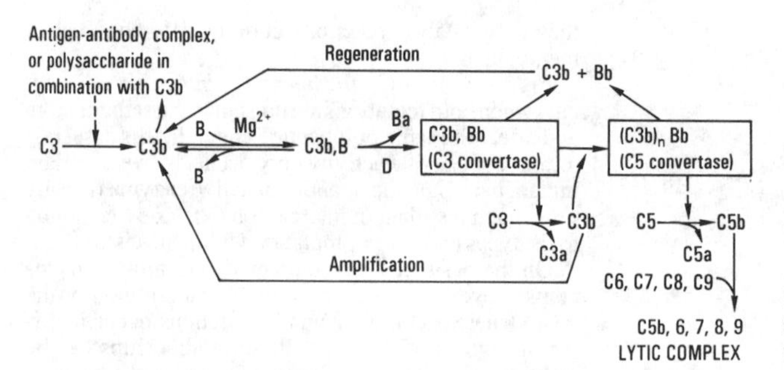

Fig.2. *Complement system.* Alternative pathway of complement activation. See legend to Fig.1.

resulting "activated" C1r in turn hydrolyses a peptide bond in C1s, thus yielding the fully active C1 complex, which initiates the sequential assembly of circulating components into a surface-bound protein complex (Fig.1). The alternative pathway is initiated by a repertoire of activators, including antibodies (IgA and IgE, which do not activate the classical pathway when complexed with antigens), large polysaccharides of bacteria and yeasts, fragments of plant cells walls, and protozoa. This pathway is thought to provide the initial response to bacterial invasion, since it can be activated in the absence of antibodies. Factor D of the alternative pathway is already proteolytically active in nonactivated serum, i.e. its formation from an inactive precursor is not part of the amplification system of the cascade.

Early components of each pathway are primarily concerned with the formation of two protein complexes, which function respectively as C3 and C5 convertases. Proteolytic activation of C5 is the final enzymatic event which triggers the spontaneous association of the late components (C6–C9) to form a nonenzymatic lytic complex capable of puncturing cell membranes (Figs.1 & 2). Thus, the main function of the C.s. is lysis of foreign invasive cells. Also, through the anaphylatoxic and chemotactic properties of some components (notably C3a, C4a and C5a), foreign cells are rendered susceptible to phagocytosis. Particulate antigens bound to C3b and IgG are engulfed by phagocytosis. The C.s. is also involved in solubilization of immune complexes, and development of the cellular immune response.

Compliance: see Pressure-volume curve of the respiratory apparatus.

Compositae, ***daisy family:*** a cosmopolitan family of dicotyledonous plants. It is largest family of flowering plants, containing about 20 000 species. Most are herbaceous, but there are also bushes, small trees and succulents, especially in the tropics. Leaves are usually alternate. Flowers *(florets)* have 5 petals, which are fused, either into a regular corolla tube with 5 teeth *(tube-floret),* or into an irregular corolla tube prolonged on one side as a strap-shaped ligule *(strap-floret)* (Fig.1).

The florets are aggregated into a characteristic inflorescence (composite head or capitulum), which may consist exclusively of strap florets, tube florets, or both. When both types are present, the tube florets occur in the center of the inflorescence where they are also known as *disk-florets;* the outer rim of the inflorescence then consists of strap-florets, also known as *ray-florets.* Disk-florets are mostly hermaphrodite, whereas ray-florets are female or sterile. The calyx is often totally regressed, and sometimes modified to a wreath of hairs or scales (pappus), which remain attached to the fruit. The

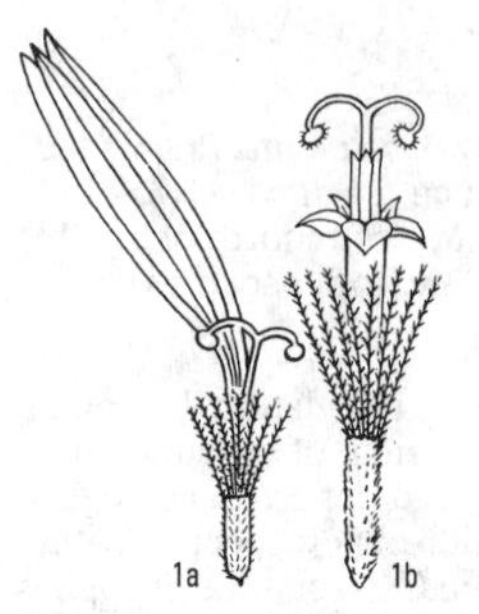

Fig.1. *Compositae. a* Strap floret. *b* Tube floret.

pappus may form a light parachute which aids wind dispersal, or it may be barbed as an aid to dispersal by animals. Typically, the stamens are united laterally, forming a closed cylinder around the style. The upper part of the style is divided and there are two stigmas. Pollination is predominantly by insects, rarely by wind. Self pollination is common. The inferior, 1-celled ovary develops into an indehiscent fruit (nut or nutlet) called an achene, in which the testa and pericarp are fused into a single layer. Inulin is universally present as a storage polysaccharide, especially in roots and tubers.

Fig.2. *Compositae* Safflower *(Carthamus tinctorius). a* Capitulum. *b* Single floret. *c* Fruit (achene).

Taxonomy of the C. is problematic. Subdivisions may be based not only on flower structure and other morphological characteristics, but also on chemical affinities Three large groups are usually recognized: 1) *C.* with strap-florets and abundant septate latex vessels; 2) *C* with tube-florets; 3) *C.* with both tube- and strap-florets

Members of groups 2 and 3 also characterically possess schizogenous resin or oil ducts and typical glandular hairs. When individual species are mentioned below, their group number is indicated in parentheses. Group 1 constitutes the *Liguliflorae* or *Cichoriaceae*. Groups 2 and 3 constitute the *Tubuliflorae* or *Asteraceae*.

The family contains many economically important and ornamental plants. The ***Sunflower*** (3) *(Helianthus annus)*, originally from northern Central America, bears oil-rich seeds; it is now the most important oil plant in the world, as well as being widely cultivated as an ornamental. In contrast, cultivation of the oil plant, ***Chilean tarweed*** (3) *(Madia sativa)*, is restricted almost exclusively to Chile. The thistle-like ***Safflower*** or ***Dyer's saffron*** (2) *(Carthamus tinctorius)*, native in the Near East, is now cultivated mainly for its oil, but was formerly a commercial source of dye. Fruits of the East African ***Ethiopian niger seed*** (3) *(Guizotia abyssinica)* are the source of niger-seed oil, which is used as a fuel oil, as an edible oil, and for the preparation of dyes. ***Jerusalem artichoke*** (3) *(Helianthus tuberosus)*, which originates from North America, is cultivated as a vegetable and as animal fodder. Roots of ***European black salsify*** (1) *(Scorzonera hispanica)* are rich in inulin and are also eaten as a vegetable. The ***Globe artichoke*** (2) *(Cynara scolymus)* is unknown in the wild; it is related to the thistles, and the fleshy parts of its inflorescence are eaten. The ***Spanish artichoke*** or ***Cardoon*** (2) *(Cynara cardunculus)* is cultivated especially in Spain and Portugal, the blanched leaf stalks and ribs being eaten as a vegetable. Leaves of the ***Garden lettuce*** (1) *(Lactuca sativa)* and ***Endive*** (1) *(Cichorium endiva)* are eaten as salad; both plants are unknown in the wild. The blanched, forced leaf buds of ***Chicory*** or ***Wild succory*** (1) *(Cichorium intybus)* are used as winter salad, and the roots are used as a coffee substitute and supplement. In the wild, chicory is a common North American, blue-flowered, wayside plant. The fragrant leaves of the Asian ***Costmary*** or ***Alecost*** (3) *(Chrysanthemum balsamita)* were once used as a seasoning and to flavor ale.

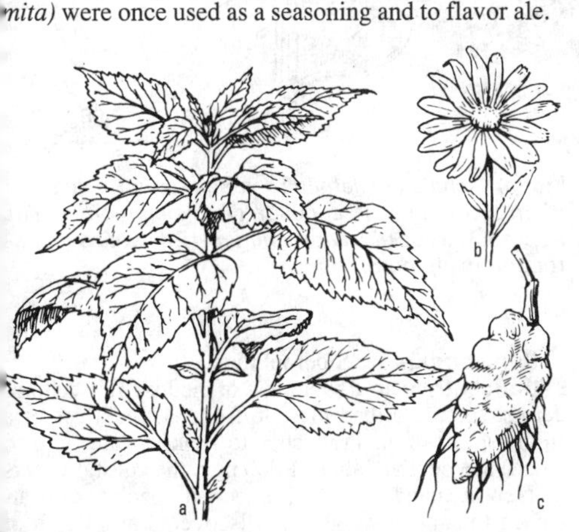

Fig.3. *Compositae.* Jerusalem artichoke *(Helianthus tuberosus)*. *a* Shoot. *b* Flower head (capitulum). *c* Tuber.

Several species are pharmaceutically important, on account of their content of essential oils or bitter principles, in particular sesquiterpene lactones which are characteristic components of the *C.* Examples are ***Wild camomile*** (3) *(Matricaria chamomilla)*, a native agricultural weed of central and southern Europe; ***Wormwood*** (2) *(Artemisia absinthum)* of ruderal communities; ***Coltsfoot*** (3) *(Tussilago farfara)*, which favors loam-clay soils; ***Arnica*** (3) *(Arnica montana)* of alpine meadows; and the Central Asiatic ***Levant wormwood*** (2) *(Artemisia cina)*. Examples of spice and seasoning plants are ***Mugwort*** (2) *(Artemisia vulgaris;* a common ruderal) and ***Tarragon*** (2) *(Artemisia dracunculus)* from southern France and western Asia.

Fig.4. *Compositae.* Globe artichoke *(Cynara scolymus)*. *a* Flowering head. *b* Young head ready for harvesting. *c* Longitudinal section of *b*.

The following is only a small selection of the very numerous ornamental *C.* Ornamental annuals: ***Chinese aster*** (3) *(Callistephus chinensis)*, ***Mexican zinnia*** (3) *(Zinnia elegens)*, ***Mexican cosmos*** (3) *(Cosmos bipinnatus)*, ***African marigold*** (3) *(Tagetes erecta)*, ***French marigold*** (3) *(Tagetes patula)*, ***Mexican marigold*** (3) *(Tagetes lucida)*, and the ***Mediterranean pot marigold*** (3) *(Calendula officinalis)*. North American ornamental perennials: ***Canada goldenrod*** (3) *(Solidago canadensis)* which often escapes and grows as a ruderal, ***Michaelmas daisy*** (3) *(Aster novae-angliae)*, ***Rudbeckia*** or ***Conflower*** (3) *(Rudbeckia)*, ***Garden dahlia*** (3) *(Dahlia variabilis)*, ***Sneezewort*** (3) *(Helenium)*, and ***Gaillardia*** (3) *(Gaillardia)*. The florist's autumn-flowering, half-hardy ***Chrysanthemums*** (3) are of uncertain origin, but were probably native in East Asia. Most of these were formerly referred to as *Chrysanthemum morifolium*, now renamed *Dendranthema morifolium*. Finally, ***Gerbera*** (3) *(Gerbera jamesonii)* was introduced from South Africa.

Native European *C.* of fields and pastures are the ***Daisy*** or ***English daisy*** (3) *(Bellis perennis)*, ***Moon daisy, Oxeye daisy*** or ***Marguerite*** (3) *(Chrysanthemum leucanthemum)*, ***Yarrow*** or ***Milfoil*** (3) *(Achillea millefolium)*, and ***Dandelion*** (1) *(Taraxacum officinale)*. Common weeds are ***Thistles*** of the genera *Carduus* (2) (simple, non-feathery pappus) and *Cirsium* (2) (feathery pinnate pappus); ***Burdock*** (2) *(Arctium)*; and ***Cornflower*** (2) *(Centaurea cyanus)*. The wild *Achillea ptarmica* or ***Sneezewort*** (3) is not to be confused with the cultivated Sneezewort (3) *(Helenia autumnale)*.

Fig.5. *Compositae. a* Coltsfoot *(Tussilago farfara). b* Arnica *(Arnica montana). c* Chinese aster *(Callistephus chinensis). d* Mexican cosmos *(Cosmos bipinnalis). e* Creeping thistle *(Cirsium arvense). f* Michaelmas daisy *(Aster novae-angliae). g* Yarrow or Milfoil *(Achillea millefolium). h* Sneezewort *(Helenium autumnale). i* Dandelion *(Taraxacum officinale) k* Lesser burdock *(Arctium minus). l* Edelweiss *(Leontopodium alpinum).*

European protected species include ***Arnica,*** the stemless ***Carline thistle*** (2) *(Carlina acaulis),* and ***Edelweiss*** (2) *(Leontopodium alpinum).*

Commercial sources of caoutchouc include the ***Mexican bush guayule*** (3) *(Parthenium argentatum),* and the roots of the ***Russian dandelion*** (1) *(Taraxacum koksaghyz)* from Central Asia. The dried flowers of ***Dalmatian pyrethrum*** (3) *(Chrysanthemum cinerariifolium)* are a commercial source of the insecticidal pyrethrins.

Compound chromosomes: see Isochromosomes.

Compound epithelium: see Epithelium.

Compound latex duct: see Excretory tissues of plants.

Compressed yeast: see Baker's yeast.

Conalbumin: see Siderophilins.

Concentricycloidea: a class of medusa-like echinoderms, first described in 1986 from 9 specimens of a single species found (attached to sunken water-logged wood) at a depth of about 1 000 m in the coastal waters of New Zealand. In 1988, a second species, also attached to sunken wood, was discovered at a depth of about 2 000 m in the coastal waters of the Bahamas. Both species are ascribed to the single genus, *Xyloplax*. The body, which is less than 15 mm in diameter, consists of a flat, armless disk encircled by spines. The upper surface is covered with scale-like plates, while the lower surface carries two concentric ambulacral ring canals, which give rise to a single ring of marginal be feet. In both species there are five pairs of gonads, e sexes are separate, and there is sexual dimorphism. he New Zealand species lacks a mouth or gut, and ere is a membranous velum on the central part of the wer surface. The Bahamian species possesses a wide, entral mouth, leading to a shallow stomach, but the omach ends blindly and there is no anus.

Conceptacles: flask-shaped cavities in the swollen nds of the branched thallus of the brown algal genus, *ucus*. They contain antheridia and oogonia interersed with sterile hairs (paraphyses).

Conch: a name for various marine gastropods, and r their shells. See *Gastropoda*.

Conchifera: subphylum of the *Mollusca*, containing ome 129 000 species, and consisting of the 5 classes at possess a shell (single shell or 2 valves): *Monoplacphora, Gastropoda, Scaphopoda, Bivalvia, Cephalooda*.

Conchology: the study of mollusks, in particular eir shells. Malacology is similar but without the emhasis on shells.

Conchostraca: see *Branchiopoda*.

Conditioned aversion: see Avoidance conditioning.

Conditioned reflex: the association of a neutral timulus with an unconditioned (inborn) reflex. A *neual stimulus* is one that normally does not give rise to ny kind of reflex activity. An *unconditioned stimulus* is stimulus or constellation of stimuli that releases an unonditioned or inborn reflex. An *unconditioned reflex* is herited and independent of the environmental condions of the animal; examples are the knee jerk, the steping reflex, saliva secretion in response to food, etc. If an nimal is repeatedly and simultaneously subjected to oth a neutral and an unconditioned stimulus, the neutral timulus loses its neutral character and becomes capable f stimulating the unconditioned reflex in the absence of e unconditioned sitmulus, i.e. the neutral stimulus beomes a *conditioned stimulus*. This process of condioning occurs in 3 stages: 1) the unconditioned stimulus eleases the unconditioned reflex, but the neutral stimus does not; 2) the unconditioned becomes associated ith the neutral stimulus; 3) the neutral stimulus beomes a conditioned stimulus, and is able to release the nconditioned reflex on its own. This particular process f conditioning is also known as *classical conditioning*, s opposed to Operant conditioning (see).

For example, the presence of food in the mouth auses the reflex secretion of saliva, and this secretion an unconditioned (inborn) reflex. If a neutral stimulus e.g. a particular sound) is applied simultaneously with eeding, and this combination of stimuli is repeated sevral times, the sound alone eventually acquires the roperty of stimulating salivary secretion in the absence f food. This example is taken from the classical work f Pavlov, who established the separate existence of onditioned and unconditioned reflexes in his work on he salivation of dogs.

In order to maintain the conditioned reflex, it is usully necessary to periodically repeat the conditioning rocess; otherwise the intensity of expression will deay. The reflex is usually restored to its original value by nly brief reconditioning (e.g. only a few exposures to particular sound when an animal is given food), a process known as *reinforcement*.

Conditioned reflexes have been imposed experimenally on skeletal muscles, respiratory activity, the vasomotor system, milk secretion, bladder contraction, gastric secretion, etc. and effective neutral (conditional) stimuli include visual, auditory, tactile, thermal, kinesthetic and olfactory stimuli.

Conditioned stimulus: see Conditioned reflex.

Condors: see *Falconiformes*.

Conducting tissue: permanent plant tissue, whose cells serve to transport water and dissolved metabolites and nutrients. C.t. became necessary in the course of evolution as terrestrial plants developed more aerial parts extending beyond the immediate source of water and soil nutrients. Essential elements of C.t. are generally elongated tubular cells with tapered end-walls lying close to each other, often also with pores to aid the intercellular transfer of material.

1) ***Sieve cells*** and ***Sieve tubes*** transport organic substances. They always consist of living cells. The watery protoplasm, which contains numerous mitochondria and leukoplasts (often with starch) and a loose meshwork of proteinaceous fibrils, fills the entire cell lumen. Nucleus and tonoplast disintegrate and are lost early in the differentiation of the sieve cell. Sieve cells are named for the sieve-like perforations of their nonlignified cellulose walls, which permit free contact between the protoplasm of adjacent cells in the conducting system. These pores are usually very fine and grouped into large *sieve fields*. Often the entire transverse wall forms a *sieve plate*. In some plant species, sieve fields are also present in the longitudinal walls of the sieve tubes. Toward the end of the growth period, the sieve tubes become blocked with a water-insoluble polysaccharide, called callose, which is already present in young cells as a thin lining on the internal walls of the pores. Further exchange of materials between the sieve tubes is therefore prevented. The sieve tubes of angiosperms are associated with *companion cells*, which are in intimate contact with the sieve tubes via plasmodesmata. Companion cells arise by unequal division from the same mother cells as the sieve tubes; they are narrower than sieve tubes, and their dense cytoplasm contains a large (often polyploid) nucleus, numerous mitochondria, but no plastids.

2) ***Tracheids*** and ***vessels***. These are always dead cells, serving mainly for the long-distance transport of water and dissolved salts from the root to the shoot. Ferns, gymnosperms and primitive angiosperms possess tracheids, which are tubular, short, thick-walled, lignified single cells with a relatively narrow lumen, whose transverse and longitudinal walls contain numerous *bordered pits* (see Pits). In ferns and gymnosperms, tracheids serve for both water conduction and mechanical strengthening. Exclusively mechanical tissue has evolved only in the angiosperms, which with few exceptions possess water-conducting *vessels*. Vessels have a wide lumen and are formed by the joining of cells with extensive loss of transverse walls, forming a closed, tubular system, often of considerable length. Transverse walls remain at certain intervals, so that the length of the tubular system between these walls is typical of the plant species. In climbing plants, vessels attain lengths of 5 m and more, but in most plants lengths vary between 10 cm and 1 m.

Tracheids and vessels are reinforced by lignified cell walls, so that they do not collapse under the reduced pressure caused by water loss. Wall thickening may consist of rings, a spiral, or a network, and the corresponding vessels are referred to accordingly as annular, spiral or reticulate vessels. In pitted vessels, however, most of the cell wall is lignified, but many pits and their lamellae remain unlignified, so that water can still be supplied to adjacent cells.

3) ***Vascular bundles.*** In the primary tissues of higher plants, the conducting tissues are usually aggregated into vascular bundles, in which the sieve tubes constitute the *phloem,* and the vessels constitute the *xylem.* The phloem is also known as the *leptome* and the xylem as the *hadrome.* According to the arrangement and development of phloem and xylem within the vascular bundle (which is characteristic of the species), it is possible to distinguish between collateral, radial and concentric bundles.

Collateral bundles are the most common, being present in the shoots of most gymnosperms and angiosperms. These contain a single strand of xylem and usually only one of phloem, with the phloem facing the periphery of the stem and the xylem facing the center. In the *closed* collateral bundles of monocotyledons, the two conducting tissues are directly adjacent to each other, whereas in the *open* collateral bundles of gymnosperms and dicotyledons the xylem and phloem are separated by a layer of *fascicular cambium.* In *bicollateral bundles (Solanaceae, Cucurbitaceae),* phloem occurs on both the peripheral and adaxial sides of the xylem.

Radial bundles are circular in cross section, and the alternating strands of xlem and phloem are arranged as radii of the circle. Radial bundles occur in roots. They are rare in stems, but occur singly in the stems of many species of *Lycopodium.*

Concentric bundles represent a very primitive type; either the phloem is completely surrounded by a sheath-like strand of xylem, or the xylem is similarly surrounded by phloem. Concentric bundles with internal xylem are found in most ferns, while those with external xylem are found in various monocotyledons.

In all types of vascular bundles, sieve tubes and vessels are embedded in parenchyma. Furthermore, the bundles are always surrounded by a bundle sheath, which may consist partly of closely packed parenchyma and partly of mechanical tissue (sclerenchyma). Sclerenchyma is especially common in that part of the bundle sheath facing the periphery of the stem (on the outside of the phloem) and on the adaxial side (adjacent to the xylem), so that a transverse section shows two caps of sclerenchyma. The sides of the bundle sheath (i.e. opposite the junction of xylem and phloem) usually contain unthickened or less heavily thickened parenchymatous cells *(transfusion tissue),* which facilitates the exchange of water and nutrients between the bundle and the surrounding tissue.

Condylarthra: an extinct order of mammals, considered to be the primitive ancestors of the ungulates. Some forms had claws, suggesting that they were arboreal (family *Hyopsodontidae).* Other families were highly specialized, but the beginnings of hoof development and ungulate dentition can be seen in this order. Fossils are found from the Upper Cretaceous to the European and American Old Tertiary.

Condylarthrosis: a condyloid joint (see Joints).

Condylura cristata: see Star-nosed mole.

Cone beetles: see Engraver beetles.

Conebills: see *Parulidae.*

Conepatus: see Skunks.

Cone shells: see *Gastropoda.*

Configurational specificity: see Enzymes.

Conflower: see *Compositae.*

Conger eel: see *Congridae.*

Conglobation: see Organism collective.

Congo eels: see *Amphiumidae.*

Congo snakes: see *Amphiumidae.*

Congridae, *conger eels:* a family of scaleless, exclusively marine fishes of the order *Anguilliformes,* containing about 109 species in 42 genera. The ***Conger eel (Conger conger)*** is found at depths of 40–100 m on rocky bottoms in the northeastern Atlantic and adjoining seas, where it feeds on herrings, flatfishes, crabs etc. Males attain a length of 1.5 m, while females are larger, attaining lengths of up to 3 m. It has local economic importance, and its flesh is good to eat.

Conidia (sing. **Conidium), *ectospores:*** thin-walled asexual spores produced on fungal mycelia by abstriction.

Coniferae: see *Pinidae.*

Coniferophytina: see Gymnosperms.

Conifers: see *Pinidae.*

Coniine, *2-propylpiperidine:* a piperidine derivative and the main alkaloid of poison hemlock *(Conium maculatum).* A dose of 0.5 to 1 g can cause death in humans. It paralyses the spinal cord and the motor nerve endings of smooth muscle. Death results from paralysis of the thoracic muscles without prior loss of consciousness. In ancient Greece, the death sentence was carried out by the administration of poison hemlock.

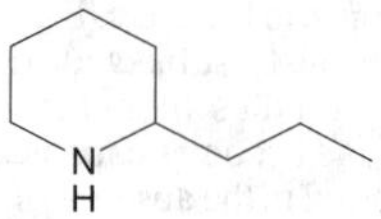

N
H

Coniine

Conjugation:

1) See Fertilization, Reproduction, *Ciliophora.*

2) Pairing of certain bacteria, accompanied by the transfer of genetic factors, e.g. in *Escherichia coli.* The *donor* or F^+ *cells* form F-pili (see Pili), which make contact with *recipient* or F^-*-cells.* Donor behavior is conferred by a plasmid (see) known as the F-factor (F = fertility). The F-factor replicates in the donor cell and one copy is transferred to the recipient, which then also becomes a potential donor.

In some cells the F-factor is integrated into the bacterial chromosome, so that parts of the bacterial chromosome may be transferred into the recipient, resulting in recombination of the genetic characters of donor and recipient. This type of donor cell is therefore known as an Hfr cell (Hfr = high frequency of recombination). C. has been widely used to determine the position and sequence of different genes on the bacterial chromosome. Such determinations are based on the time that it takes for a genetic factor to be transferred by C. from the donor to the recipient. After mixing suspensions of donor and recipient under strictly standardized conditions, C. is interrupted at different times by mechanical homogenization. In such "time of entry" experiments, it is necessary to distinguish between donor and recipient strains in the subsequent plating out procedure. Thus the recipient may be penicillin resistant and the donor pencillin sensitive, so that treatment with penicillin allows only recipient cells to grow. For example, in the C.

of a penicillin-sensitive, wild-type donor with a penicillin-resistant histidine auxotroph, the time is determined at which penicillin-resistant cells able to grow in the absence of histidine first appear in the conjugation mixture. A similar experiment using a tryptophan auxotroph shows whether the operon for histidine biosynthesis, carried by the F-factor, enters the recipient earlier or later than that for tryptophan biosynthesis (i.e. whether it is nearer or further from the site of attachment of the F-factor on the bacterial chromosome).

Connectance, ***connectedness:*** the degree of interrelationship and interdependence among organisms of an ecosystem. C. is usually based on Nutritional interrelationships (see), when it is equivalent to the number of food links in a food web as a fraction of the number of topologically possible links.

Connecting link: a modern term for Darwins's "missing link", i.e. a transitional form between two more or less phylogenetically well separated groups. Examples are *Xenusion* (worms/arthropods), *Ichthyostega* (fishes/amphibians), *Gephyrostegus* (amphibians/reptiles), *Archaeopteryx* (reptiles/birds).

Connective tissues: tissues that surround animal organs, and form connections between them. They serve especially as support tissues, and also fulfil certain metabolic and defense functions. C.t. characteristically contain a high proportion of intercellular matrix. Some cells are relatively sedentary, while other ameboid cells (wandering cells) are capable of limited migration and have a defense function. The various types of C.t. display wide structural differences.

1) ***Embryonal connective tissue, mesenchyme*** (Fig.1.). The first type of connective tissue to appear in the embryo, where it serves as filling or packing material. It consists of a syncytium of stellate cells arranged in a 3-dimensional lattice, whose interstices are filled with a fluid matrix. All other support tissues, including the notochord, arise from mesenchyme.

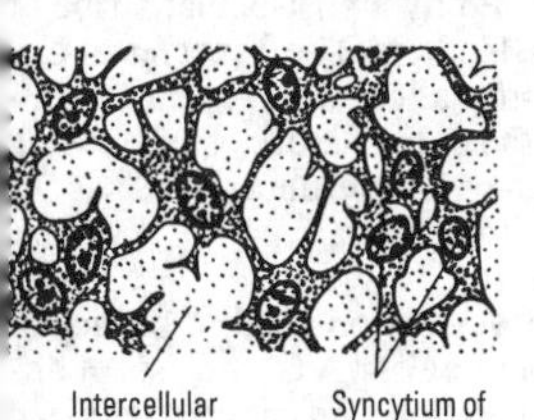

Fig.1. *Connective tissues.* Embryonal connective tissue.

2) ***Mucous connective tissue.*** A gelatinous or mucous ground substance (matrix) containing a network of irregularly branched (stellate) cells and small bundles of white fibrils. It occurs in the vitreous body of the eye, and in the umbilical cord, where it is also known as Wharton's jelly.

3) ***Reticular connective tissue.*** A mesh of stellate cells in a fluid intercellular matrix, the whole being reinforced by a reticulum of fibers. The main function of the cells is to synthesize the fibers, but both the reticulum cells and the free cells derived from them are also capable of phagocytosis. Reticular connective tissue provides a supporting framework in bone marrow, the lymph nodes and muscular tissue. It is also present in the spleen, in the mucous membrane of the gastrointestinal tract, in the lungs, liver and kidneys. The most common variety of reticular connective tissue is *lymphoid* or *adenoid* tissue, in which the meshes of the network are occupied by lymph cells.

4) ***Fat tissue, adipose tissue*** (Fig.2). Fat-storing connective tissue, with numerous fat cells (adipocytes), derived from reticular connective tissue. Two types are recognized, i.e. white and brown adipose tissue. Within the cells of white adipose tissue, small fat droplets coalesce to form large fat globules, which displace the nucleus and cytoplasm to the cell periphery. Adipose tissue contains storage fat, which is important in the energy metabolism of the body. The tissue also serves as insulation and padding. Insects possess *fat bodies,* which are sometimes relatively massive.

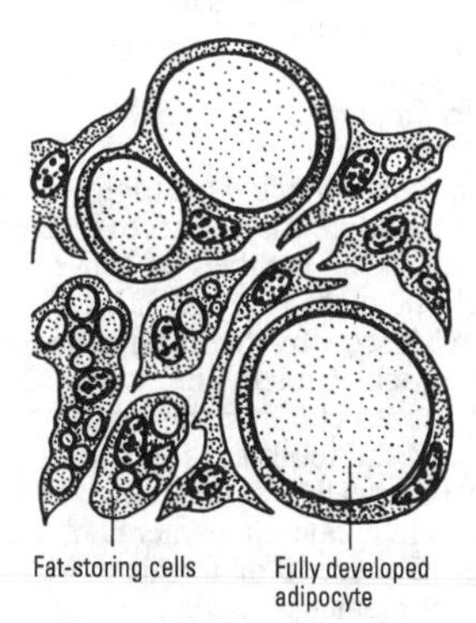

Fig.2. *Connective tissues.* Adipose tissue.

5) ***Diffuse, interstitial,*** or ***areolar connective tissue.*** This tissue is continuous throughout the body. It forms a packing material and flexible connection between organs, and is also found interspersed with organ parenchyma. The large, flat, stellate fibroblasts (fibrocytes) may be solitary or part of an interlacing network; free lymphocytes are also present, e.g. macrophages and mast cells. The aqueous phase (often called tissue fluid) serves as a transport medium between capillaries and parenchyma cells. The intercellular substance consists of tissue fluid with collagen and elastin fibers. In addition to its support function, this tissue also temporarily stores water, salts, nutrients, and other metabolites, and it also has a defensive role. It can become locally inflamed, accompanied by an increase in the blood supply and cell numbers; the macrophages become phagocytic and destroy invading organisms.

6) ***Elastic connective tissue.*** An extensile and elastic tissue, resembling areolar tissue, but with a greater abundance of elastin fibers. It is present in the lungs, and it serves to link the cartilages of the larynx. It is also found in the walls of arteries, the trachea, bronchial tubes, and vocal folds. It is present in some elastic ligaments and between the laminae of adjacent vertebrae.

7) ***Fibrous connective tissue*** (Fig.3). This is characterized by an abundance of structural intercellular material (mostly collagen and some elastin) and a relatively small number of cells. The Collagen fibers (see) cohere very closely and are arranged side by side in dense bundles. The Fibroblasts are arranged in rows between the fibers. Fibrous connective tissue forms: *ligaments* (bands or capsules that hold bones together at joints), *tendons* or *sinews* (attachment of muscles to bones), *aponeuroses* (bands of tissue connecting one

muscle with another or with the periosteum of bone), *membranes* investing and protecting organs such as the heart and kidneys, and *fasciae* (sheets of tissue wrapped around muscles, serving to hold them in place).

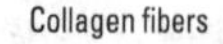

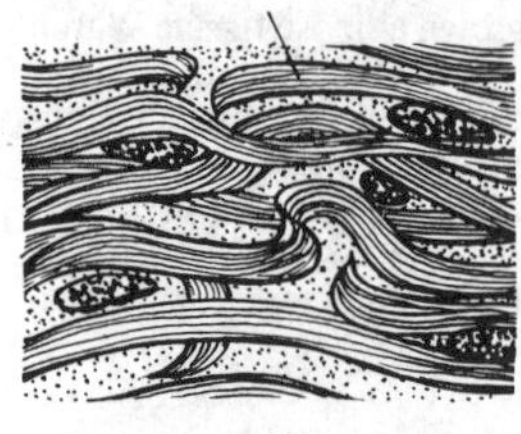

Fig.3. *Connective tissues.* Fibrous connective tissue.

Conocephalum: see *Hepaticae.*

Connochaetes: see *Alcelaphinae.*

Connochaetini: see *Alcelaphinae.*

Conodonts (Latin *conus* cone, Greek *odon* tooth): a name originally applied to toothlike fossils discovered between 1833 and 1844 in washed residues of Lower Ordovician and Silurian clastic rocks from Estonia and in carboniferous rock from Moscow by the Russian paleontologist, Christian Heinrich Pander. The animals that originally possessed these structures are now customarily referred to as C., and the toothlike fossils are called *conodont elements.* Practically any Paleozoic or Triassic marine rock will yield conodont elements if it is dissolved slowly in 10–15% acetic or formic acid. Pander classified condodont elements as simple (single) and compound (composite) types. Compound conodont elements are subclassified into bars, blades and plates (platforms). Alternatively, the simple elements are represented largely by coniform crowns, while compound elements are represented mainly by raniform (ray-shaped) crowns and pectiniform (comb-shaped) crowns. Internally, elements display a lamellar and fibrous structure, and they were previously interpreted as support elements in the gripping and filter organs of a worm-like animal.

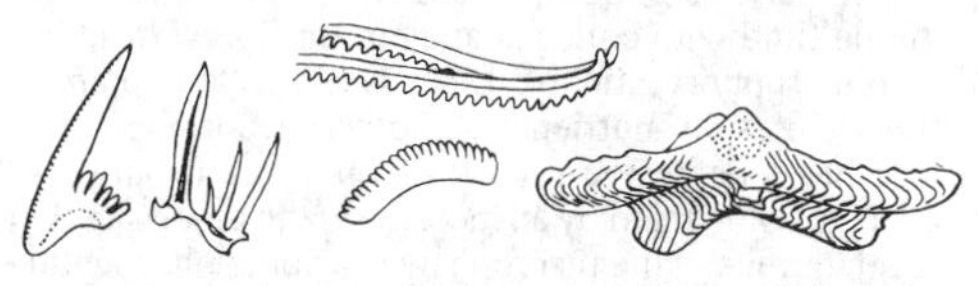

Conodont elements. Some serve as index fossils in the correlation of sedimentary sequences of the Late Cambrian and Triassic.

Until 1983, the rest of the organism was completely unknown, and its true nature was the subject of debate. Fossils revealing soft tissue morphology were then discovered in Lower Carboniferous shales in the Edinburgh district of Scotland. These showed that the C. were laterally compressed, eel-like animals, about 5 cm long. They possessed up to 20 teeth (each 2 mm long) in a bilaterally symmetrical feeding apparatus in the head, flanked by a pair of lobes that possibly represent eyes. A series of V-shaped structures in the trunk represents muscle blocks separated by myosepta. Two parallel lines extending the length of the trunk are interpreted as the margins of the notochord. The tail region has short, closely spaced rays. These morphological characters suggest that the C. were primitive craniates. Final proof came in 1992, with the investigation of polished slices of conodont elements, using interference contrast microscopy and scanning electron microscopy. These techniques revealed the presence of lamellar crystallites typical of the structure of enamel in vertebrate teeth, as well as features histologically identical to the lacunae and canaliculi of cellular bone, and structures suggesting the globular calcified cartilage of vertebrates. Dentine was absent. Since cellular bone is unique to vertebrates, its presence, together with enamel and probably globular cartilage, confirms the conodonts as the earliest known craniates. Cellular bone is found essentially only in the *Eugaleaspida, Osteostraci* and *Gnathostomata;* its discovery in C. supports the claim that it evolved earlier than acellular bone.

The presence of a bilaterally operating feeding mechanism suggests that C. are related to the *Myxinoidea* (hagfishes). If it is assumed, however, that the tissue of conodont elements is homologous with the mineralized dermal tissue of heterostracans, the C. may be considered as primitive relatives of higher craniates, excluding the *Myxinoidea.*

[The Conodonta, Walter C. Sweet, Oxford Monographs on Geology and Geophysics No. 10, Clarendon Press, Oxford, 1988; *Conodonts: A Major Extinct Group Added to the Vertebrates,* D.E.G. Briggs, *Science, **256*** (1992) 1285–1286; *Presence of the Earliest Vertebrate Hard Tissues in Conodonts,* I.J. Sansom et al., *Science, **256*** (1992) 1308–1311]

Conolophus: see Land iguana.

Conopophaga: see *Conopophagidae.*

Conopophagidae, *antpipits* or ***gnateaters:*** an insectivorous Neotropical avian family (11 species in 2 genera) in the order *Passeriformes* (see), found in forest habitats in South America. Members of the genus, *Conophaga* (9 species), are small (about 15 cm), long-legged, short-tailed, round-winged ground birds with thick plumage, characterized by a post-ocular stripe or tuft of silky, usually white feathers. The 2 species of *Corythopus* are pipit-like, long-tailed, walking birds. Antpipits and Tapaculos are the only passerines possessing 4 notches to the back edge of the sternum.

Conopophila: see *Meliphagidae.*

Conostoma: see *Timaliinae.*

Consociation: unit of vegetation used in particular by Scandanavian plant sociologists. A C. consists of different levels (layers) of vegetation, and it is characterized by a single dominant species in only one of the levels.

Constancy: percentage of reference samples in a single biocoenosis that contain a particular plant species, i.e. the percentage of different stands which contain that particular species. It is usually expressed on the scale: r (less then 1%), I (1–20%), II (21–40%), III (41–60%), IV (61–80%) and V (81–100%). In contrast, *frequency* is the percentage of equal sized areas within one uniform stand that contain a particular species, expressed on the scale: 0–25% (sporadic), 25–50% (scattered), 50–75% (dense), 75–100% (very dense). The constancy terms for these same percentages are: 0–25% (accidental), 25–50% (accessory), 50–75% (constant), 75–100% (euconstant). *Fidelity* is the degree of restriction of a plant species to a particular situation, community or association, expressed on a 5-point scale: 1 (strange), 2 (indifferent), 3 (preferential), 4 (selective), 5 (exclusive). *Presence,* as defined by by Braun-Blanquet, is the persistent occurrence of a species in all stands of a plant community. Presence is also used in the sense of constancy, i.e. as an expression of the frequency of occurrence of a species in all samples taken from a community.

Constant field approximation: see Diffusion potential.

Constant field theory: see Diffusion potential.

Consummatory action, ***consummatory reaction:*** the final active phase of motivated behavior, involving contact with the behavioral object. In metabolically-dependent behavior, the ingestion of food or liquid is the C.a. In sexual behavior, this phase is represented by copulation. Normally, C.a. represents the conclusion of a pattern of motivated behavior, and it temporarily inhibits the recurrence of the same motivated behavior, i.e. C.a. is followed by a more or less extended refractory period, after which a new cycle of Appetitive behavior (see) can start.

Contact allergy: see Contact hypersensitivity.

Contact community: see Substitute plant community.

Contact field: the event field of behavior which becomes defined after the operation of a releasing stimulus. It is characterized by physical contact with the reference object. The resulting behavior, known as Consummatory action (see), temporarily abolishes the preceding Appetitive behavior (see). A typical example is feeding, which must involve contact with the food.

Contact preparation: a preparation for the microscopic investigation of microorganisms, prepared by pressing a glass slide or cover slip directly onto the material of interest, e.g. foodstuffs. C.ps. are studied directly under the microscope, usually with prior staining and fixing.

Contact hypersensitivity, ***contact allergy:*** a type of Allergy (see), which is usually caused by allergens that penetrate the skin. These low molecular mass substances (e.g. chromate, nickel, picrylchloride) are not antigenic *per se,* but they become antigenic by binding to skin proteins, and stimulate the organism to produce antibodies. The low molecular mass substance is thus a Hapten (see). In any subsequent encounter with the body, the low molecular mass substance has the properties of an antigen and it reacts with activated lymphocytes. The lymphocytes in turn release lymphokines, which cause reactions of varying intesity from skin inflammation to destruction of entire areas of skin. C.h. represents an occupational hazard in some branches of industry where workers are exposed to skin contact with various chemicals, e.g. the leather dyeing industries.

Contact stimulus, ***touch stimulus, haptic stimulus, thigmostimulus:*** a stimulus resulting from sustained contact with a solid body. Such stimuli are perceived and responded to in particular by plant tendrils, but also by other plant organs (see Haptonasty, Haptotropism). For an effective stimulus, immediately neighboring sites of the organ must be subject to rapid changes of pressure or tension, rather than static pressure or a Seismic stimulus (see). Thus, being struck by liquid (e.g. rain) or pressed with a glass rod coated with gelatine does not elicit a response in touch-sensitive plant organs. On the other hand, a plant tendril responds, sometimes in less than 30 seconds, if a fine thread of wool is drawn across it, causing local rapid changes of pressure equivalent to "tickling". The most sensitive part of the tendril is its tip (approximately the distal third of the tendril). The mechanism of susception is not understood, but it is thought that the stimulus is perceived by sensory pits, papillae or other excitable elements. Rapid growth occurs on the side of the tendril opposite the site of contact causing marked curvature of the tendril, whereas the site of contact itself shows no growth and may even contract. Before curvature occurs, an Action potential (see) passes transversely across the tendril.

Contact types: animal species, which even in the absence of functional requirements (e.g. mating) decrease the distance between individuals to the point of bodily contact. The proximity of individuals during rest or sleep is often a good indicator of whether they are C.t. or Distance types (see).

Contamination: impurities, consisting of e.g. radioactive materials or microorganisms. C. by microorganisms occurs when sterile equipment becomes infected with bacteria, or when a pure culture is invaded by a different organism or a different strain of the same organism.

Continental drift: a geophysical phenomenon

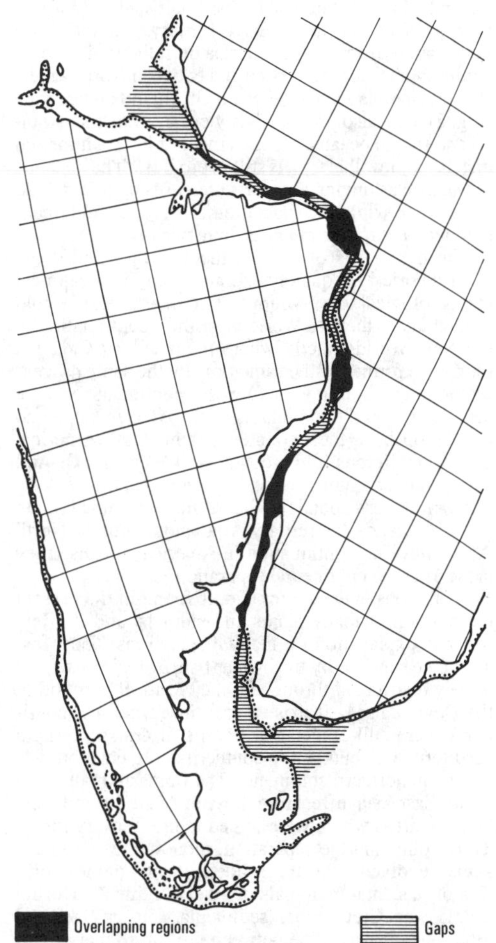

Fig. 1. *Continental drift.* Positions of the continents before the opening of the Atlantic (computer prediction).

which accounts for the present distribution of the continents and oceans, and various other facts of geology and natural history. The existence of C.d. was first suggested in 1915 by Wegener, who noted the striking match between the profiles of the west coasts of the Old World and the east coasts of the New World (Fig. 1). According to Wegener, the land masses of the earth were not always in their present positions, but were originally united in a single primitive continent called Pang(a)ea. Between 200 and 250 million years ago (during the Lower Jurassic) the land mass began gradually to split into various shapes and sizes. The first cleavage separated Australia and Antarctica (still attached to South America) from the Africa/Asia mass.

The united southern continents, including Antarctica, are thought to have formed a distinct land mass, which is called Gondwanaland, while Europe, Asia and North America were associated in a mass known as Laurasia. Later, Africa and South America started to move apart by a north-south cleavage, which by the middle of the Cretaceous had completely separated the two continents. At this time, India also separated from Africa, and Australia from Antarctica. An especially active phase of C.d. occurred during the Early Tertiary. Antarctica did not separate from South America until the Pleistocene; similarly late cleavages occurred between North America and Greenland, and between Greenland and Europe (Fig. 2). C.d. also produced new oceans, like the Atlantic and Indian oceans. Measurement of intercontinental distances show that C.d. is still in progress. The distance between Washington and Paris increases by 0.32 ± 0.08 m per year, while the separation of Europe and Greenland increases by as much as 36 m per year.

The hypothesis of C.d., without further help from other theories, adequately explains many geographical and geological observations. For example, the fold mountains of the Old World have their continuation in the New World exactly where predicted by C.d., i.e. they were formed at the same time by the same process of mountain building, before the continents became separated (e.g. the mountains of the African Cape and Cameroon have their counterparts in South America; the Carboniferous fold mountains of West and Central Europe are continued in the Appalachians; the geology of Norway corresponds to that of Eastern Canada). On the other hand, the Tertiary Atlas mountains of North Africa have no continuation in America, because they arose after the continental separation.

Analogous evidence for C.d. is found in the present distribution of many plants and animals, and it offers the only explanation for many disjunctions. Thus, fossils of the supposedly freshwater fish, *Bothriolepsis*, are widely distributed throughout the world. It is found in the Devonian, i.e. it existed at a time when the continents were still associated. Fossils of *Glossopteris*, a seed fern, are abundant in southern continents, but unknown in northern continents. The floras and faunas of South East Asia differ from those of Australia and New Guinea, areas which were joined in pre-Tertiary times. Thus, following the separation, monotremes and marsupials evolved in Australia independently of the evolution of placental mammals in other continents. During the Lower Cretaceous, sedimentary basins formed along the cleavage line between the future margins of Brazil and West Africa (Fig. 1). Sandstones and shales deposited at that time contain fossils of freshwater fish and ostracods which are almost identical in the basins of eastern Brazil and the Gabon basin in Africa.

According to the mechanism of C.d., the drifting continental layers or plates are about 100 km thick, consisting of relatively light gneis and granite rocks with high contents of silica and aluminium. They are therefore said to consist of sial, and they float on a denser sima (main constituents silica and magnesium) like "ice floes on water". The sial protrudes only 4.8 km above the surface of the sima, which is in accordance with the two statistical maxima for the height of the earth's crust at +100 m and at -4700 m.

Wegener also recognized that a shift had occurred in the position of the earth's magnetic poles; the south pole moved closer to the southern cape of Africa during the Carboniferous, and the north pole lay in the Pacific ocean. This pole shift explains the global distribution of coal deposits, which require a tropical climate for their formation. It also explains the evidence of glaciation in South Africa. At the end of the Carboniferous, the magnetic north pole migrated eastwards, then turned north and crossed the north-east part of the American continent during the Miocene. In the Late Tertiary, it crossed the north polar sea to its present position. Well documented changes of vegetation that occurred in the late Tertiary were the result of this pole migration.

Certain geological and geophysical objections have been raised to the theory of C.d. It provides, however the only simple explanation for many otherwise puzzling observations of animal and plant distribution and of paleobotany.

Continuous culture: a culture of microorganisms in which the cells are maintained under constant conditions in the logarithmic growth phase over a long time period. C.c. is required for certain physiological studies, and it is used in production processes in industrial microbiology. It is achieved by continually adding fresh nutrient medium to the stirred growing bacterial suspension, and removing an equivalent volume of used nutrient medium containing suspended bacteria. The *chemostat* is a culture vessel into which fresh nutrient medium flows at a constant rate; one of the nutrients is growth-limiting, so that the rate of medium addition regulates the growth of the culture. The *turbidostat* is a self-regulating system, in which the rate of addition of fresh medium is controlled by continually monitoring the turbidity of the liquid culture; the cell density (turbidity) of the culture therefore remains constant.

Contor feathers: see Feathers.

Contractile vacuole: a cell organelle of all freshwater protists that lack a cell wall. Its main function appears to be osmoregulation; due to the high osmotic pressure of the cell contents, water continually enters the cell, and is then pumped out by the C.v. While in communication with cisternae of the smooth endoplasmic reticulum and other smaller vacuoles, the C.v. withdraws water from the cytoplasm (diastole). In the subsequent systole, water is forced to the exterior by a contractile system. During systole, connections between the C.v., endoplasmic reticulum and other smaller vacuoles are closed, then re-established during diastole.

Contracture: see Muscle contracture.

Contrast in electron microscopy: see Fixation.

Conularia: see *Conulata*.

Conulata (Latin *conus* skittle): a group of fossil animals often placed with the *Scyphozoa*. Some authorities

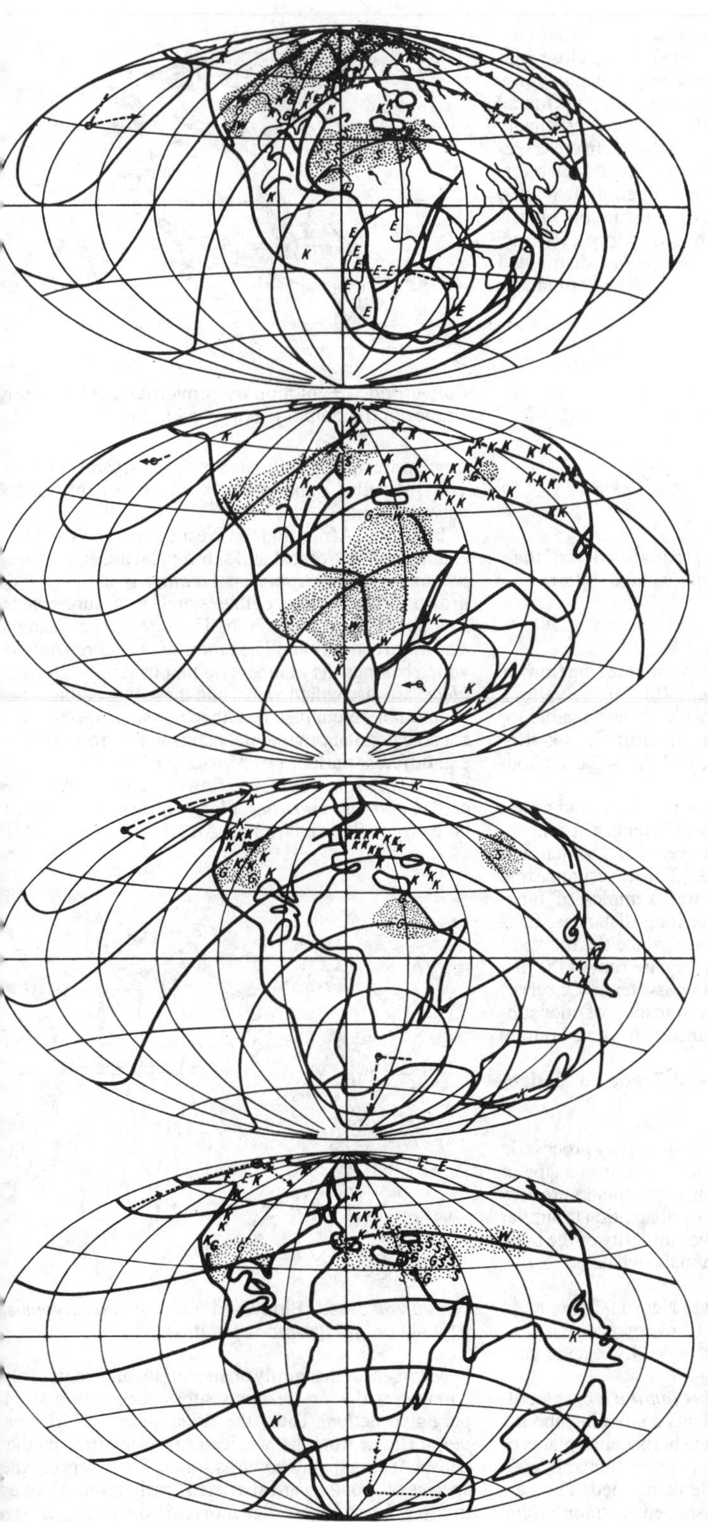

Fig.2. *Continental drift.* Positions of the continents and poles at different times in the earth's history according to Wegener (from Köppen and Wegener). *1* Carboniferous. *2* Jurassic. *3* Eocene. *4* Miocene. *E* Evidence of glaciation. *G* Gypsum. *K* coal. *S* Salt. *W* Desert sandstone. Dotted areas = dry regions.

however, claim that similarities to the scyphozoans are superficial, and that *C.* have no known living close relatives and should not be assigned to any extant phylum. Members of the *C.* possessed a steep sided pyrimidal chitinophosphatidic test or shell, characteristically marked with fine, transverse growth ridges. In cross section it was square or rhomboidal with four internal bifurcating septa, and surrounded by numerous tentacles. Juvenile stages were attached to the substratum. Fossils occur from the Upper Cambrian to the Upper Triassic, and fossils of the genus, *Conularia,* are distributed worldwide from the Upper Cambrian to the Permian.

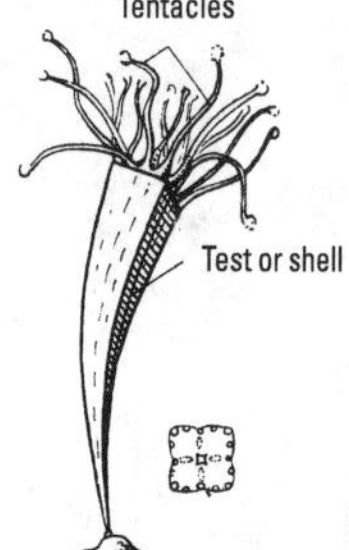

Conulata. A reconstruction of the genus *Conularia* (x 0.3) (after Kiderlen).

Conures: see *Psittaciformes.*

Convallatoxin: a cardiac glycoside from the flowers and leaves of *Convallaria majalis* (lily of the valley), which together with other glycosides is responsible for the toxicity of the plant. The aglycon of C. is k-strophanthidin (see Strophanthins), and the sugar component is L-rhamnose.

Convection: transport by flow in liquids and gases. In the presence of relatively large differences in concentration or temperature, a flow is generated which will transport dissolved substances. Thus, protoplasmic streaming and axon transport are examples of functional C. in the living cell. Over long distances, C. is more effective than Diffusion (see), since the transport time of diffusion increases as the square root of the distance. In aquatic and terrestrial ecosystems, C. serves for the long distance (kilometers) transport of odor substances, pheromones and substances affecting homing behavior (e.g. of fishes).

Convention on International Trade in Endangered Species: see CITES.

Convergence:

*1) **Convergent evolution:*** An evolutionary process, in which organs or organisms become similar in shape or structure, although their evolutionary precursors are very dissimilar. C. is due to evolutionary adaptation to similar environmental conditions, e.g. the similarity of the body shapes of fishes and aquatic mammals (whales).

2) See Neuronal circuit.

Convergence of neurons: see Neuronal circuit.

Convergent evolution: see Convergence.

Conversion: see Gene conversion, Lysogenic conversion.

Convolvulaceae, *convolvulus family:* a cosmopolitan family of dicotyledonous plants containing about 1 600 species in 55 genera. They are herbaceous plants or shrubs (rarely trees), often with twining stems. Leaves are alternate, exstipulate, simple or divided. The hermaphrodite, 5-merous, funnel-shaped, actinomorphic

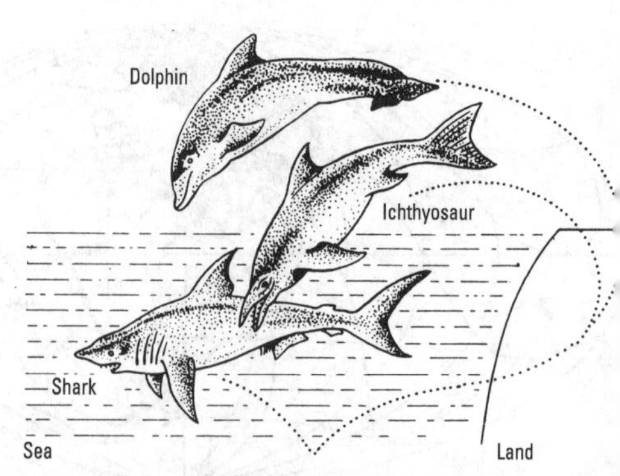

Convergence. Evolutionary convergence of the bod shapes of the dolphin, ichthyosaur and shark.

flowers are mostly pollinated by insects, rarely by bird The 1-4-celled superior ovary develops into a 2- valved capsule.

In tropical Africa and the West Indies, and to a less extent in other tropical and subtropical areas, the ***Swe potato*** or ***Batata*** *(Ipomoea batatas)* is a staple foo grown for its swollen, edible, starch-rich, subterrane root tubers. Especially in the USA, the sweet potato also known as a ***Yam.*** [The tubers of *Dioscorea batata* which belongs the monocotyledonous family, *Diosco ideae,* are also called yams, and these two edible tube should not be confused]. Other *Ipomoea* species hav become popular ornamental plants in nontropical area e.g. different varieties of ***Morning glory*** *(Ipomoea tr color),* which has sky-blue flowers, and the red-flow ered ***Common morning glory*** *(Ipomoea purpurea* both originally from tropical America.

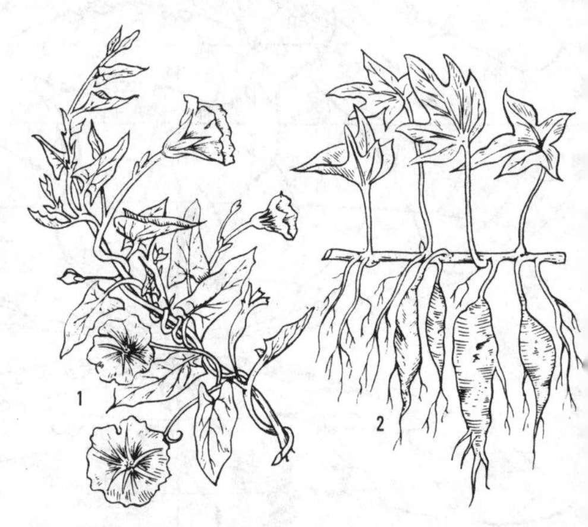

Convolvulaceae. 1 Bindweed *(Convolvulus arvensis 2* Sweet potato *(Ipomoea batatas).*

Members of the family characteristically contain gly coretins, which are resinous substances with a stron purgative action. Thus, the dried tubers of *Ipomoe purga* Hayne from the Mexican Andes are used medic nally as a purgative, known as *Jalap.* The tubers of oth species of *Ipomoea* are also used for their purgative a tion, e.g. *I. orizabensis* Ledanois (Mexican jalap), *I. sim lans* Hanbury (Tampico Jalap) and *I. tuberosa* Linn. Brazilian jalap). *Scammony root (Convolvulus scammonia* Linn.) from Asia Minor was also formerly used or the same purpose, but is now seldom employed. *Kal-lana* or *Pharbis* seeds from the Indian plant, *I. hederacea* Jacq., are also used for their purgative action.

The white-flowered ***Larger bindweed*** or ***Bellbine*** *Calystegia sepium)* grows in the temperate regions of both hemispheres, rarely in the tropics, often becoming a troublesome weed. Similar in distribution and also a persistent weed is the white- or pink-flowered ***Bindweed*** or ***Cornbine*** *(Convolvulus arvensis),* which is common in fields and waste areas and on waysides.

Convolvulus: see *Convolvulaceae.*

Convulsionism: see Catastrophe theory.

Cooperation: a concept of behavioral evolution: by cooperating, an individual acquires increased fitness, as well as increasing the fitness of the cooperation partner.

Cooperative self assembly: see Self assembly.

Coordination: see Central coordination.

Coots: see *Gruiformes.*

Copaiba: see *Caesalpiniaceae.*

Copals, ***copal resins:*** a collective term for very hard, amber-like resins from a variety of plants *(Agathis* species, various leguminous trees, etc.). They are entirely soluble in ethanol, but only partially soluble in benzene or chloroform. Genuine C. are mostly obtained from fossil deposits, but recent trees are also used as a source. Some important C. are ***Zanzibar copal*** (gum animi) from *Trachylobium hornemannianum,* a fossil resin dug up on the east coast of Africa; ***Manilla copal*** (East Indian dammar) from *Dammorca orientalis;* ***Australian copal*** (gum Kauri) from *Agathis australis* (mostly fossil); ***West African copal*** from *Copaifera guibourtiana;* and ***American copal*** from *Hymenoea courbaril* (Brazil). C. are used in the manufacture of cables and linoleum, and in chemical industry; they have no medical applications. See *Caesalpiniaceae.*

Copeognatha: see *Psocoptera.*

Copepoda, ***copepods:*** a diverse subclass of crustaceans (or a class if the crustaceans are considered as a subphylum) containing about 7 500 species. The majority of species is marine, but representatives also occur in all possible freshwater habitats, including ground water and the moisture of moss cushions. Many species are planktonic and often occur in vast numbers, forming an important link between phytoplankton and higher trophic levels of marine habitats. Thus, species of the genus *Calanus* are important as a food source for fishes and baleen whales (see Krill). Numerous species are ecto- or endoparasites, in particular of fishes, but also of whales, as well as mollusks and other marine invertebrates. Some parasitic species have become so modified and reduced that they have lost entirely their characteristic crustacean appearance; the females, however, always produce the typical copepod egg cases, and development always proceeds via the typical copepodid larva. In free living forms the body is only 1–2 mm in length, but the females of some parasitic species can attain 32 cm *(Penellus,* an ectoparasite of fishes and whales). Body form varies greatly, from the segmented, free swimming genera with thoracic appendages to the unsegmented and sometimes limbless parasitic genera. The head (with or without a pointed anterior rostrum) is posteriorly fused with the first thoracic segment and sometimes also with the second thoracic segment. Compound eyes are always absent, but a nauplius eye is typically present. In planktonic forms, the mouthparts serve as filters, whereas in many parasites they are modified to piercing structures. When segmentation is complete, the thorax consists of six segments and the abdomen of five segments. The telson carries two caudal rami. The first pair of thoracic appendages is modified to maxillipeds. With the exception of the last one or two pairs, the other thoracic appendages are similar and symmetrically biramous. The biramous limbs are responsible for the characteristic jerky swimming movements of copepods. Appendages are never present on the narrow, cylindrical abdomen. A carapace is always absent. The sexes are always separate. In both males and females, the sex ducts open on the first abdominal segment. Sexual dimorphism is usually extremely pronounced, especially in parasitic species. Males of all species produce a spermatophore, and the female transports her eggs in strings or sacs on the first abdominal segment. Each egg hatches into a nauplius larva, which becomes a metanauplius, followed by the copepodid stage. The latter develops into the adult. All larvae are free living aquatic forms, and parasites become attached to the host during the copepodid stage.

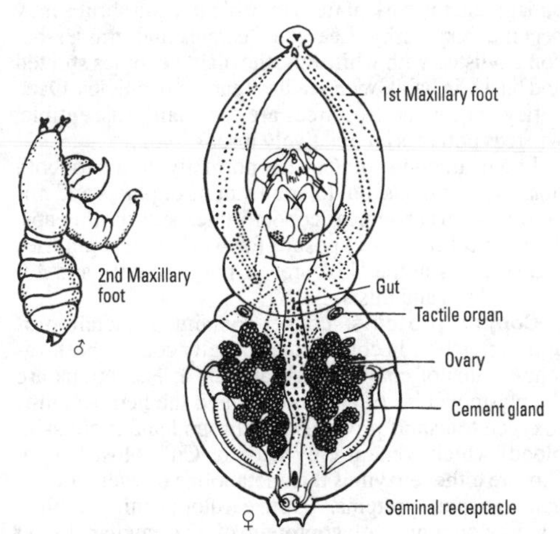

Copepoda. Achtheres percarum.

Classification. Seven orders are recognized: ***1.*** *Calanoida* (free living, largely planktonic; first antennae very long with 16–26 articles), e.g. *Calanus, Diaptomus.* ***2.*** *Misophrioida* (free living on or above bottom surface; first antennae shorter than in calanoids, with 11–16 articles), e.g. *Misophria.* ***3.*** *Harpacticoida* (mostly free living marine and freshwater bottom dwellers; some are planktonic and many are terrestrial; the first antennae are short, with fewer than 10 articles), e.g. *Harpacticus.* ***4.*** *Monstrilloida* (marine species, whose larval stages are parasitic on polychaetes and gastropods; adults are nonfeeding and planktonic, lacking second antennae and mouthparts), e.g. *Monstrilla.* ***5.*** *Siphonostomatoida* (freshwater and marine species, the adults of which parasitize fish and invertebrates), e.g. *Caligus, Penella.* ***6.*** *Cyclopoida* (marine and freshwater, planktonic and benthic species, including some commensals and parasites; first antennae with 10–16 articles; second antennae uniramous), e.g. *Cyclops,*

Lernaeopoda. 7. *Poecilostomatoida* (marine species, of which the adults parasitize invertebrates and fishes), e.g. *Ergasilus*.

Copepodite stage: typical larval form of the *Copepoda*, which develops from the metanauplius. Greatly modified parasitic copepods still pass through the typical C.s.

Copepods: see *Copepoda*.

Cope's giant salamander: see *Dicamptodontidae*.

Cope's rule: see Ecological principles, rules and laws.

Cope's spea: see *Pelobatidae*.

Copper, *Cu:* a bioelement needed by plants and animals. Plants contain 2–20 mg Cu per kg dry mass. The daily requirement of the human is about 2 mg, and the entire body contains 100–150 mg. Cu is a component of various enzymes (see Copper proteins), and it is involved in electron transport in mitochondrial membranes. In plants, it is necessary for the biosynthesis of chlorophyll and in animals for the biosynthesis of hemoglobin, although it is not part of these molecules. Cu deficiency in animals leads to "licking sickness". In plants, Cu deficiency is manifested in various ways; thus, leaves may become chlorotic or, in contrast, they may become dark green; the bark of trees may blister, and shrubs may become very bushy. Leaves of cereals and grasses become twisted with white tips, the plant becomes stunted and bushy in habit, with seedless ears or pannicles. Oats, barley, wheat and fruit trees are especially susceptible, whereas potatoes are relatively insensitive.

Large amounts of Cu are normally toxic to both plants and animals, but tolerant strains of plants and microorganisms arise near copper mines and spoil heaps. Threshold limits of exposure for workers in foundries and smelters in the USA are 0.1 mg/m^3 fumes and 1.0 mg/m^3 dusts and mists.

Copper proteins: copper-containing metalloproteins, often blue in color, which usually contain both valency states of copper (Cu^{2+} and Cu^+); exceptions are the plastocyanins of chloroplasts and the hemocyanins (oxygen transport proteins of arthropod and molluskan blood) which contain only Cu^{2+} or Cu^+. Most known C.p. are either enzymes that metabolize oxygen or oxygen transport proteins. Even ceruloplasmin (a blue, copper-containing glycoprotein of mammalian blood plasma), which functions as a copper transport protein, also has oxidase activity for certain unsaturated compounds. Examples of copper-containing oxidases are galactose oxidase, amine oxidase, dopamine β-monooxygenase, monophenol monooxygenases, laccases, tyrosinase, cytochrome *c* oxidase, ascorbate oxidase and superoxide dismutase.

Coppers: butterflies of the family *Lycaenidae*.

Coppersmith: see *Capitonidae*.

Copra: see *Arecaceae*.

Coprinae: see Lamellicorn beetles.

Coprinus: see *Basidiomycetes (Agaricales)*.

Coproliths: pieces of fossil excrement, which sometimes provide information on the diet and digestion of their producer. C. of fishes, reptiles, birds and echinoderms are particularly common, but mammalian C. are rare. Masses of fossil excrement may contribute to the formation of geological sediments, e.g. bird guano on islands off the west coasts of Chile and Peru, and chiropterite (bat feces) in the dragon cave of Mixnitz (Steiermark, Austria).

Coprophage: an organism that temporarily or continually feeds on feces, (usually the feces of herbivores), e.g. beetles, mites, aschelminths. Small Cs. (aschelminths) actually feed on the microorganisms of the feces. See Modes of nutrition.

Coprophagy must be distinguished from *Cecophagy* (see).

Coprophagy: see Coprophage.

Coprostane: see Steroids.

Coprostanol, *5β-cholestan-3β-ol:* a zoosterol formed in the gut by the microbial reduction of cholesterol. It is the main sterol of feces, and differs from its stereoisomer, cholestanol, in the configuration at C-5 and the *cis*-configuration of rings A and B.

Copsychus: see *Turdinae*.

Copulation: the union of two individuals of different sexes (male and female). In humans, C. is also known as *coitus*. C. serves for the transfer of spermatozoa into the female genital tract, and it is not synonymous with insemination, impregnation, or fertilization. Transfer of spermatozoa is usually achieved with the aid of specialized Copulatory organs (see).

Copulatory organs: variously designed external organs that serve for the transfer of sperm into the female genital tract. They are particularly developed in males, usually consisting of the eversible or extended final unpaired section of the male sexual apparatus (see Cirrus, Penis), often enclosed in erectile tissue (vertebrate penis). In addition to structures directly involved in internal fertilization, many invertebrates possess secondary male C.o., which are developed on or from various parts of the body, and which are not a physical extension of the internal sexual apparatus. In *cephalopods*, a modified tentacle called a Hectocotylus (see) is used to transfer a spermatophore into the mantle cavity of the female. In *Argonauta* and related forms, the hectocotylus detaches from the male during copulation, remains in the female mantle cavity, and continues to display movement for a considerable time.

In *crustaceans*, various pairs of limbs are modified to Gonopods (see). For example, in the higher crustaceans *(Malacostraca)*, the first and second pairs of abdominal limbs are modified to a tubular structure *(petasma)* which can be laid against or inserted into the female sex aperture. In copepods, the fifth or sixth thoracic limb is modified to a small pincer-like structure, which transfers a spermatophore into the female aperture.

In *diplopods*, the limbs on the 7th body segment are modified to complicated gonopods, which are inserted directly into the female aperture. In *spiders*, the final segment of the pedipalps is developed as a C.o., which the male fills with sperm then inserts into the female. *Mites* often use their chelicera to transfer a spermatophore to the female genital aperture. Male arthropods often employ other modified body appendages (legs, antennae) for gripping the female during copulation.

Copy-choice model of crossing over: see Crossing over.

Coracias: see *Coraciiformes*.

Coracidium: see *Cestoda*.

Coraciidae: see *Coraciiformes*.

Coraciiformes: an order of birds (see *Avies)* containing nearly 200 species. The young are altricial and they gape for food. The sexes are generally alike or very similar. It is diagnostic of the order that the front toes are joined. Eight related families are recognized.

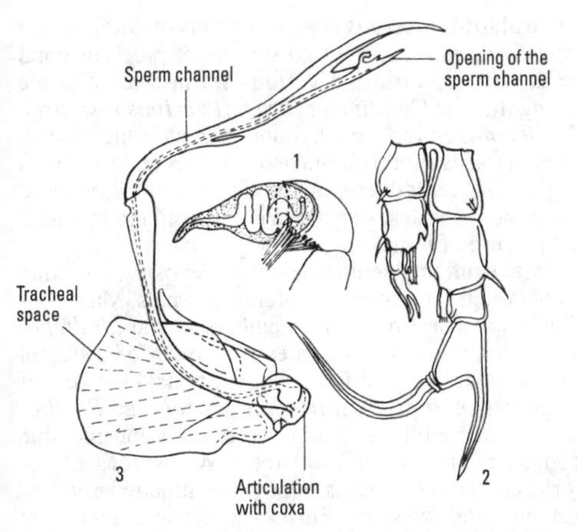

Copulatory organs of some arthropods. 1 Pedipalp of a spider *(Philistata testacea);* the tarsus is modified to a copulatory organ. *2* Fifth biramous limb of a male copepod *(Diaptomus gracilis);* the right ramus serves as a clasping organ, the left for the transfer of the spermatophore. *3* Gonopodium of a male millipede *(Bollmannia nodifrons),* derived from a modified leg of the seventh double segment.

1. Family: *Alcedinidae* ***(Kingfishers;*** 10.5–46 cm; 84 species in 12 genera). Kingfishers have a practically worldwide distribution (absent from the polar regions and some Oceanic islands) but are mainly tropical, 15 species being found in Africa, and 60 species from Southeast Asia to Australasia. Plumage colors are usually brilliant with iridescent greens and blues. The 2nd and 3rd toes are joined at the base, the 3rd and 4th toes for more than a third of their lengh. When perching, the 1st toe points backward. The familiar fishing kingfishers (with strong, narrow, sharp-pointed bills) are placed in the subfamily *Alcedininae,* while the less familiar, larger and more primitive forest kingfishers (broader, flatter and sometimes hooked bills) belong to the *Daceloninae.* Fishing kingfishers live by streams and lakes and are well known for their method of fishing by plunging from an exposed perch above the water. Some forest kingfishers live by streams and feed on fish, and some are coastal birds favoring mangrove swamps; others feed on insects, reptiles, amphibians, even small mammals and birds, in forests or on savanna. One of the largest members of the family is the Australian ***Kookaburra*** or ***Laughing jackass*** *(Dacelo gigas,* subfamily *Daceloninae;* 43 cm), which feeds on snakes and lizards. Other well known species are the ***Common kingfisher*** *(Alcedo atthis;*16.5 cm) of Eurasia and Africa, and the ***Belted kingfisher*** *(Megaceryle alcyon;* 33 cm) of temperate North America, both in the *Alcedininae.* All kingfishers nest in holes, usually dug in the bank of a stream. Some of the *Daceloninae* nest in hollow trees, and many tropical species excavate burrows in termite mounds.

2. Family: *Todidae* ***(Todies).*** This is the only avian family confined to the West Indies. It contains only 5 species in a single genus *(Todus),* all very similar, about 10 cm in length, with long, peculiarly flattened, serrated bills, prominent rictal bristles and short tails. The front toes are united like those of the kingfishers. Plumage is green above and white below with a red throat. They nest in ground burrows.

3. Family: *Momotidae* ***(Motmots;*** 16.5–25 cm). The 8 species (in 6 genera) are confined to continental tropical America, mainly at low altitudes. The bill, which is almost as long as the head, is broad with serrated edges, and decurved at the tip. Colors tend to be subdued browns and greens, often with some bright blue coloration in the head plumage, and occasionally with blue feathers in the tail and wings. One striking feature is the tail, which is long and strongly graduated. The 2 central retrices are far longer than the others in length, and although they grow with vanes along the entire shaft, the vanes along the subterminal section gradually break away leaving small fans at the tip. Some nest in rock crevices, but most dig long tunnels in a bank with a nesting chamber at the end.

4. Family: *Meropidae* ***(Bee-eaters;*** 24 species in 7 genera). The members of this family have the usual coraciid features, such as united front toes, small feet and legs, etc., but their wings are relatively longer and adapted for flight after aerial insects. Tails are also long, sometimes with elongated retrices and sometimes forked. Many kinds of insects are taken, but especially members of the bee and wasp family. Among the most brilliantly colored of all birds, bee-eaters are predominantly green with a black line throught the eye, and patches of reds, yellows and blues around the head and rump. An Old World family, 18 species are found only in Africa and 6 only in Asia, including the ***Rainbow bird*** *(Merops ornatus)* of Australia.

5. Family: *Leptosomatidae* ***(Cuckoo roller).*** The family contains a single living species *(Leptosomus discolor),* confined to Madagascar and the Comoro Islands. It is sometimes classified with the other rollers, but it differs from these in that the sexes are not alike, and it is the only coraciid with well marked powder downs (2 patches on the rump). The male is glossy green above and white below, while the slightly larger female is brownish, with bars above and spots below. Both sexes have small crests.

6. Family: *Coraciidae* ***(Rollers*** and ***Ground rollers).*** The family comprises 16 pigeon-sized Old World species, mainly blue, green and chestnut in color, with large heads,and strong legs (toes syndactylous), bills and wings. They seize insects in flight, and also take small lizards on the ground. They are named from the aerial courtship display of the males of the 7 species of *Coracias,* which include the ***European roller*** *(C. garrulus)* and the ***Indian "Blue jay"*** or ***Indian roller*** *(C. benghalansis),* the remaining 5 species being found in Africa. The 4 species of *Eurystomas* (2 in Africa, 1 in Madagascar, 1 in Asia extending south to Australia and the Solomon Islands) swoop rather than roll in display. The 5 ***Ground rollers*** are placed in the separate subfamily, *Brachypteraciinae.* Found only in Madagascar, these are stout birds, 30–46 cm in length, and more protectively colored (mottled yellows and greens) than the true rollers. They resemble other rollers in their behavior of sitting quietly on a high perch in wait of prey, and of a tumbling flight when excited.

7. Family: *Upupidae* ***(Hoopoe).*** Comprising the single monotypic species, *Upupa epops,* with 9 subspecies ranging over the Palaearctic, Africa, Madagascar, India, and the larger islands of Southeast Asia. Only the inner 2 front toes are fused at the bases. It is a delicate fawn

color, with black and white bars across the wings; the feathers of its striking crest (extending from bill to nape) are tipped with black. The long, curved bill is used for probing for worms and insects. The preen gland of the female emits an objectionable musty smell.

8. Family: *Phoeniculidae* ***(Wood hoopoes).*** The family comprises 6 species (in 3 genera), all confined to sub-Saharan Africa. All 6 have metallic blue, green and purple plumage. They have no crest, and there is a long graduated tail of 10 feathers. They feed mostly on insects and other invertebrates, which they obtain by probing the crevices of tree bark with their long curved bills. They prefer open country with trees rather than continuous forest. Like the hoopoe, female wood hoopoes emit a musty smell in the breeding season.

9. Family: *Bucerotidae* ***(Hornbills).*** The family is divided into 2 subfamilies: *Bucoracinae* ***(African ground hornbills;*** only 2 species in the genus *Bucorus)* and *Bucerotinae* (43 species in sub-Saharan Africa, India and Southeast Asia). Hornbills display the coraciiform characteristic of united 2nd, 3rd and 4th toes, but in many other respects they are unique. Notably, they possess an ungainly looking, variously shaped casque or helmet surmounting the curved, serrated and often grooved bill. Both the bill and casque are usually hollow, or filled with a spongelike structure of small cells. Hornbills also possess eyelashes, a rare avian feature present in only a few other birds (ostrich, hoatzin, and a few cuckoos and hawks). Plumage is mostly black, brown and white. In the *Bucerotinae* the nesting female is walled up in her tree hole with a mixture of mud and regurgitated material. In some species the wall is built by the male, but usually the female works from within and the male from without, leaving a narrow central slit for the female to poke her bill through. She is then fed by the male for the duration of incubation and for a short time after the eggs have hatched. The wall material sets hard and acts as an effective protection against predators like monkeys and tree snakes. After the female breaks out (by diligent chipping of the hard wall), the young may wall themselves up again until they are ready to fly. Most hornbills feed on fruit, insects and small animals. Large insects, small reptiles and amphibians are killed by beating with the bill before they are swallowed. Food is usually swallowed whole, then indigestible material (fruit pips, insect carapaces, animal bones) is regurgitated.

Coracoid bone: see Pectoral girdle.

Coragyps: see *Falconiformes.*

Coral: see *Anthozoa.*

Coral bank: see Coral reef.

Coral fishes: telost fishes of various taxa that inhabit coral reefs. Most are splendidly colored, and many species have the ability to change color. The coloring and patterning of juveniles and adults may be very different. On account of their generally slender or small bodies, they can retreat from danger into fissures and holes. Other C.f. are armored with spines, some can inflate themselves, while certain species produce toxic secrtetions. Powerful, beaklike mouths enable some species to graze coral polyps. Other have forcep-like, elongated jaws, with which they reach concealed worms and other small animals. Many C.f. are particularly attractive in marine aquaria, e.g. butterfly fishes and anemone fishes.

Coral fungus: see *Basidiomycetes (Poriales).*

Corallium, ***precious corals:*** a genus of corals of the order *Gorgonaria.* About 20 species of precious coral are known, occurring in various warm seas. Notable among these is *Corallium rubrum* ***(Precious red coral, Mediterranean red coral),*** found in the Meditteranean, where it forms slightly branched colonies up to 40 cm in height. The central axis is a hollow cylinder of fused calcareous sclerites, varying in color from red to white, and it is used for making jewellery.

Coral reef: unlayered calcareous deposits consisting of the skeletal structures of colonial animals. Main contributors to coral reefs are members of the *Madreporaria* (stony corals). The rates of growth of different stony corals is very variable. For example, members of the genera, *Acropora* ***(Stag's horn coral)*** and *Pocillopora,* grow rapidly, and they represent a considerable proportion of tropical coral reefs. Many seas of the world contain coral reefs. On the continental shelf of northern and western Europe, extensive reefs are formed by *Lophelia* and *Amphihelia* at depths of 60 to 2 000 m. They are, however, particularly widespread and evident in the clear, illuminated coastal waters of the tropics and subtropics, which provide optimal conditions for the growth of reef-forming organisms, i.e. water temperatures greater than 20 °C and maximal penetration of sunlight. Symbiotic zooxanthellae in the reef-forming corals require light for photosynthesis. Many reef inhabitants, e.g. sea anemones and hydrozoan corals, also contain zooxanthellae and are similarly dependent on the penetration of sunlight.

Coral banks or *fringing reefs* are broad expanses of coral that extend from the shore, forming the shallows before descending to greater depths. *Barrier reefs,* which are separated from the coastal land, have a platform top, and they form a wall enclosing a relatively shallow coastal lagoon. *Atolls,* which lie on the summits of submerged volcanos, are more or less circular, and they enclose a central, lagoon. In addition, there are *table reefs* (small ocean reefs with no central lagoon) and *patch reefs* (small reefs arising from the floor of atolls or behind barrier reefs). The great barrier reef off the northeastern coast of Australia is the longest in the world (more than 1000 miles or 1 600 km). The coral reef harbors a rich and diverse ecosystem of animals and plants, and in this respect it may be regarded as the marine equivalent of the tropical rainforest.

Enormous reefs were formed in earlier geological epochs. Thus, the Dolomites are the remains of Paleozoic and Mesozoic reefs.

Corals: see *Anthozoa.*

Coral snakes: members of 3 large snake genera of Central and South America: *Micrurus, Micruroides* and *Leptomicrurus,* all in the family *Elapidae* (see). They have conspicuous warning coloration and patterning, consisting of brilliant red, white or yellow, and blue-black or shiny black transverse rings. They feed mainly on lizards and small snakes, and to some extent on birds, frogs and insects. The largest species is *Micrurus spixii,* which attains a length of 1.5 m. The ***Eastern coral snake*** *(Micrurus fulvius)* has a black snout and often inhabits rodent dens. The ranges of three species, including that of the Eastern coral snake, extend via Mexico into southwestern USA. Despite their very small venom teeth, the bite of a coral snake can be fatal to a human.

Cordaitidae: an extinct subclass of fork- and needle-leaved Gymnosperms (see), known only from their fossils and named after the Prague botanist, A.J. Corda (1808–1849). They represent the direct precursors of the conifers. Uncertain forms are known from the Upper Devonian, and the period of their greatest evolutionary diversity ranges from the Upper Carboniferous into the Middle and Lower Permian (Rotliegendes). They then died out and were replaced by the conifers. Together with the *Lepidodendrales* ("clubmoss trees"), Horsetails and Seed ferns, the *C.* were among the most numerous plants of the prehistoric carboniferous forests. They were tall trees with copiously branched crowns, and trunks up to 30 m in length. Their wood was largely similar to that of the recent conifers, possessing secondary thickening, tracheids with araucarioid bordered pits as in the *Araucariaceae,* and a transversely septate medulla. The leaves, which are common fossils of the carboniferous, were broad, strap-like, linear or lanceolate, with dichotomously branched, parallel veins, sometimes attaining lengths of 1 m. Spikelike catkins (genus *Cordaianthus)* were borne in the axils of bracts, and in some forms the heart-shaped or ovoid seeds were carried on long stalks. Isolated fossil seeds have been attributed to the genus *Cardiocarpus.* Different genera are recognized according to the venation of the leaves, e.g. *Dorycordaites, Eucordaites, Poacordaites, Cordaites,* etc. (Fig.).

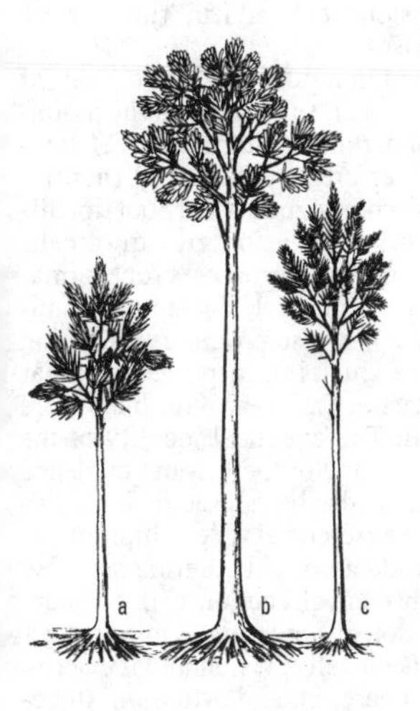

Cordaitidae from the Carboniferous (reconstruction). *a* Dorycordaites. *b* Eucordaites. *c* Poacordites.

Cordon-bleu: see *Ploceidae.*

Cordylidae, *girdle-tailed lizards, plated lizards:* a family of the *Lacertilia* (see) containing about 50 species, native to southern Africa and Madagascar. The head is broad, and in most species the body is covered with longitudinal and transverse rows of rectangular scales, which are underlain by bony plates (osteoderms). Dorsal scales are keeled. Transverse rows of large, ossified tail plates, which may be elongated into spines, are present in many species. The majority of species are rock dwellers, inconspicuous gray or brown in color, and often able to conceal themselves in rock crevices on account of their flattened body. Species of the genus, *Chamaesaura **(Snake lizards),*** are snake-like with regressed limbs. They eat all types of arthropods, as well as snails, flowers and fruits. Two subfamilies are recognized. Members of the subfamily, *Cordylinae,* are vivi-oviparous, and they usually have large spines protecting the nape of the neck. The largest species is the 40 cm-long *Cordylus giganteus **(Sungazer, Lord Derby's zonure,*** or ***Giant girdled lizard)*** of southern Africa, with a triangular head, powerful limbs, and large curved spines along the neck, back and tail. Members of the subfamily, *Gerrhosaurinae,* occur in both Africa (east and southeast) and Madagascar; they have relatively large bony plates beneath their scale covering, and lateral skin folds on the body. In some species the number of toes is reduced. The black, yellow-spotted *Gerrhosaurus major **(African plated lizard)*** attains a length of 45 cm, and is distributed from Ethiopia to South Africa.

Cordyline: see *Agavaceae.*

Cordylophora caspia: a cnidarian of the order, *Hydroida,* occurring in fresh water (e.g. in the lakes near Berlin) and in brackish water (Baltic, and the estuaries of the Elbe and Weser). The polyp colonies are up to 8 cm in height. The degenerate medusoid stage remains attached to the colony, but is still capable of sexual reproduction.

Coregonus, *whitefishes* and ***ciscoes:*** a genus of commercially important, freshwater (occasionally entering the sea), nonpredatory, plankton-feeding teleost fishes of the family *Coregonidae* (order *Salmoniformes),* found mainly in the deep oceans of the Northern Hemisphere. Eurasian species include ***Vendace** (C. albula),* ***Tshiir** (C. nasus),* ***Freshwater houting** (C. lavaretus),* and ***Arctic cisco** (C. autumnalis).*

Coremia: spinous, erect structures of the fungal order *Moniliales (Fungi imperfecti),* consisting of parallel hyphae united into bundles, and usually carrying conidia.

Coriander: see *Umbelliferae.*

Coriandrol: see Linalool.

Corioxenidae: see *Strepsiptera.*

Corium:

1) see Skin.

2) see *Heteroptera.*

Cork oak: see *Fagaceae.*

Cormo-epiphyte: see Life form.

Cormophyta, *cormophytes:* plants that are differentiated into shoot and root, as distinct from the *Thallophyta* (see) which have not attained this level of organization. Most representatives of the *Pteridophyta* (see) and the *Spermatophyta* (see) are cormophytes.

Cormorants: see *Pelicaniformes.*

Cormus: a term applied to the bodies of those plants that possess a distinct shoot axis, leaves and true roots. The tissues of a C. are characteristically well differentiated, e.g. into boundary, vascular, mechanical, absorption and secretory tissues.

Cornbine: see *Convolvulaceae.*

Corn borer: see European corn borer.

Cornea: see Light sensory organs.

Cornflower: see *Compositae.*

Corniculate cartilages: see Larynx.

Coro-coro: see *Echimyidae.*

Corona radiata: see Oogenesis.

Coronata: see *Scyphozoa.*

Coronella: see European smooth snake.

Corophiidae: see *Amphipoda.*

Corpora allata (sing. **Corpus allatum):** endocrine organs that form part of the insect Retrocerebral complex (see). In almost all insects they are typically paired, lying separately and behind the corpora cardiaca, but in cyclorrhaphous *Diptera* they are fused with the latter to form a single structure, the *ring gland* or *Weissmann's ring*. They are innervated by the nervi corporis allati, whose fibers generally pass through the corpus cardiaca as an extension of the nervi corporis cardiaci. Axons from neurosecretory cells of the brain transport neurosecretory products downward to the C.a. In addition to storing and eventually releasing these neurohormones (neurohemal function), the C.a. also secrete material synthesized by their own intrinsic glandular cells (endocrine function). In the larva, the C.a. synthesize juvenile hormone. In a classical experiment, Wigglesworth showed that implantation of active C.a. into the 5th instar larva of *Rhodnius* prevents the molt to the adult and results in a giant 6th instar larva. Conversely, removal of the C.a. (allatectomy) from the 3rd instar causes the larva to molt to a precocious adult. In the imago, the C.a. store neurosecretory substances that are mainly involved in the control of metabolism, ion balance and heart rate.

Corpora bigemina: see Brain.

Corpora cardiaca (sing. **Corpus cardiacum**): organs with both an endocrine and neurohemal function that form part of the insect Retrocerebral complex (see). In almost all insects they are typically paired, lying separately and in front of the corpora allata, but in cyclorrhaphous *Diptera* they are fused with the latter to form a single structure, the *ring gland* or *Weissmann's ring*. They are innervated by the nervi corporis cardiaca. Axons from neurosecretory cells of the brain transport neurosecretory products downward to the C.c. In addition to storing and eventually releasing these neurohormones (neurohemal function), the C.c. also secrete material synthesized by their own intrinsic glandular cells (endocrine function).

Corpora quadrigemina: see Brain.

Corpora penduculata, ***penduculate bodies, stalked bodies, mushroom bodies:*** conspicuous paired regions in the protocerebrum of most annelids and arthropods. They consist of dense neuropile associated with groups of small darkly staining unipolar neurons with relatively large nuclei (globuli cells). Histologically similar regions are found in the higher centers of other groups, such as polyclads, nemertineans, gastropods and cephalopods. In insects, a single mushroom body consists of one or two (usually two) stalked calyces of neuropile overlain by a dense accumulation of globuli cells. The single short stalks of each calyx fuse to a single long stalk, which enters deeply into the medulla and divides ventrally into two lobes (α-lobe directed outward, β-lobe directed toward the center of the protocerebrum). C.p. are widely thought to be associated with complex behavior, but this interpretation must be treated with caution. Thus, among different castes and species of bees and termites, those with the more highly developed activities have larger C.p. relative to total brain size. In worker ants the C.p. constitute about one fifth of brain volume. On the other hand the C.p. of the apterygote *Japyx* are well developed, but the animal is blind and exhibits no particularly complex behavior. Similarly, no functional significance can be attributed the excessively developed C.p. of *Limulus* or to the very well developed C.p. of cockroaches. In many other cases (e.g. polychaetes and onychophorans), well developed C.p. cannot be correlated with any known complex behavior. Certainly in spiders, the size of the C.p. is not correlated with the complexity of the web.

Corpulence index: see Rohrer index.

Corpus: see Meristems.

Corpus pineale: see Epiphysis.

Corpus striatum: see Brain.

Correlation:

1) Interactions of cells, tissues and organs, resulting in the harmonized operation of the individual processes of the living organism. Thus, although numerous plant cells are totipotent (i.e. capable of differentiating into any other type of cell, and performing many different functions), only some of this potential is realized in the contribution of any particular cell to the overall plan of development and the final organized plant structure; other potentialities are inhibited, a process known as *correlative inhibition*. Removal of this inhibition restores the totipotency of the original embryonic state. Repair processes following tissue damage are also regulated by C. An important and familiar example of C. is Apical dominance (see). Transition from the juvenile to adult stage of plant development, as well as the onset of senescence are also subject to C. The most active, senescence-causing inhibitory centers are the reproductive organs. Annual and biannual plants that flower only after a cold shock, become perrenial if they are prevented from flowering, e.g. by growth in a greenhouse or in the tropics. Agaves as a rule live vegetatively for 8–10 years, then flower, fruit and die in the same year, but they can live for up to 100 years if maintained in a vegetative state. Abscission (see) and fruit ripening are also a correlative processes.

The state of nutrition of individual cells or regions of cells, i.e. their supply of water, salts and organic assimilate, plays an important part in C. (nutritional C.). Phytohormones (see), however, are essential for C. (hormonal C.). Auxin, which is synthesized at the shoot tip, diffuses downward and causes extension growth of cells along the axis of the shoot; it also promotes root formation, and in particular is responsible for apical dominance. It is possible that auxin also promotes formation of a correlative inhibitor, which is transported in apolar fashion and therefore reaches the bases of the buds more readily than auxin itself. The chemical identity of the correlative inhibitor is unknown; there some evidence that it may be identical with abscisic acid. It is also thought that auxin can cause correlative inhibition by stimulating ethylene production. Gibberellins are also involved in C., through their effects on cell division, cell extension and initiation of flowering. Kinins are active through their effects on cell division and the activity of seeds and buds. Gibberellins and cytokinins therefore also inhibit senescence. Initiation of senescence by reproductive organs is thought to be due to transported *senescence factors,* but their identity is unknown.

2) See Biostatistics.

Correlation theory: see Cuvier's principle.

Corrodentia: see *Psocoptera.*

Cortex:

1) In certain animal organs, the outer layer of tissue which is differentiated from the central tissue, e.g. adrenal cortex, kidney cortex.

2) In plants, the tissue immediately below the epidermis, but outside the vascular tissue. See Shoot, Root.

3) For cerebral cortex of vertebrates, see Brain.

4) In the *Heliozoa,* the outer ectoplasm, often highly vacuolated.

5) In some members of the *Porifera,* an outer syncytial layer of ameboid pseudopodia.

6) In some aquatic invertebrates, an outer fibrous layer of the periderm.

Cortexone, ***Reichstein's substance Q:*** 11-deoxycorticosterone (abb. DOC), 21-hydroxypregn-4-ene-3,20-dione, a mineralocorticoid hormone from the adrenal cortex. It is an intermediate in the biosynthesis of corticosterone and aldosterone from progesterone. The acetate or glucoside is used in the treatment of Addison's disease and of shock.

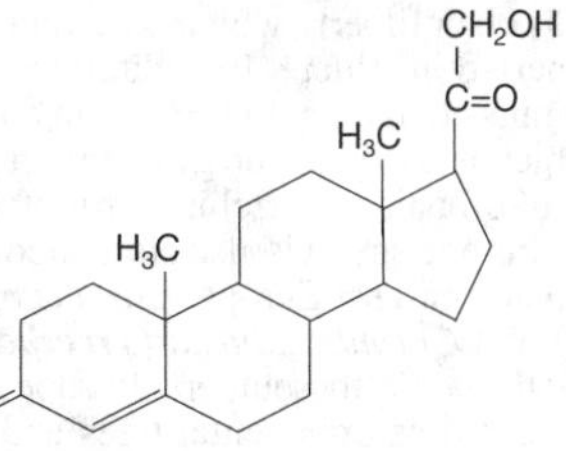

Cortexone

Corticosterone, ***Reichstein's substance H, Kendall's substance B:*** 11β,21-dihydroxypregn-4-ene-3,20-dione, a glucocorticoid hormone from the adrenal cortex. It is biosynthesized from progesterone in two stages: hydroxylation of C21 to form cortexone, which is then hydroxylated in position 11 to form C. It is inactivated in the liver by reduction to 3α,5β-tetrahydrocorticosterone, which is excreted in the urine as its glucuronide.

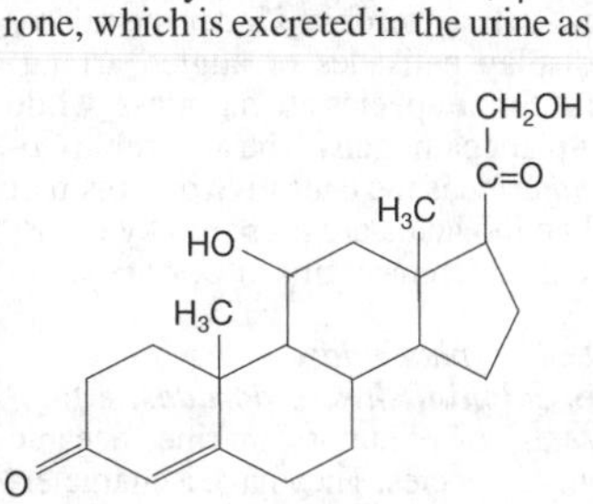

Corticosterone

Corticotropin, ***adrenocorticotropin, adrenocorticotropic hormone, ACTH:*** a basic polypeptide hormone secreted by the anterior pituitary. The unbranched polypeptide chain consists of 39 amino acid residues (M_r 4 541 for human ACTH). Biological activity is determined by the first 24 amino acids, which are constant in ACTH from all investigated species, whereas residues 25–33 display small species-specific differences. Sequence 1–13 is identical with that of α-melanotropin. ACTH is made in the γ-cells (basophilic cells) of the anterior pituitary when they are stimulated by corticotropin releasing hormone. It is stored in secretory granules of the γ-cells. The target organ of ACTH is the adrenal cortex, which is stimulated to growth and increased production of glucocorticoids (in particular cortisol and corticosterone); this stimulatory activity is mediated by the adenylate cyclase system.

Cortisol, ***Reichstein's substance M, Kendall's substance F:*** 17α-hydroxycorticosterone, 11β,17α,21-trihydroxypregn-4-ene-3,20-dione, a glucocorticoid hormone from the adrenal cortex, which is released directly into the circulation without prior storage. In the blood C. is bound to α-globulin, which transports it to the target organ. C. is biosynthesized from cholesterol via pregnenolone, 17α-hydroxyprogesterone and 11-deoxycortisol. Production of C. is regulated by the pituitary hormone, corticotropin. Adult humans produce 10–20 mg C. daily, and the blood concentration of C. displays a diurnal rhythm, with a minimum at night and a maximum in the morning. C. is inactivated in the liver by conversion to tetrahydrocortisol (urocortisol), which is excreted in the urine as its glucuronide.

As a glucocorticotropic hormone, C. stimulates carbohydrate synthesis, promoting the degradation of protein to amino acids, which are utilized by the liver for gluconeogenesis. It also promotes glycogen storage in the liver, and it increases the blood sugar concentration.

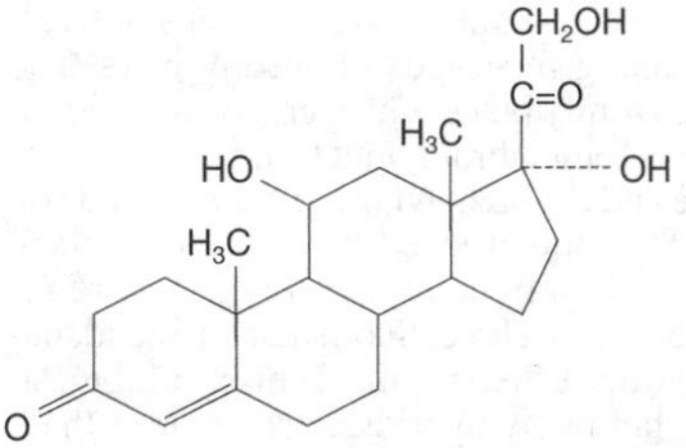

Cortisol

Cortisone, ***Reichstein's substance F, Kendall's substance E:*** 11-dehydro-17α-hydroxycorticosterone, 17α,21-dihydroxypregn-4-ene-3,11,20-trione, a glucocorticoid hormone from the adrenal cortex. It is biosynthesized from cortisol by dehydrogenation, and differs from the latter by the presence of a keto group on C11. Like cortisol, it stimulates the formation of carbohydrate from proteins, promotes glycogen storage in the liver and raises the blood sugar level. C. is inactivated by reduction to cortolone (pregnane-3α,17α,20α,21-tetraol-11-one), which is excreted in the urine as its glucuronide. C. acetate is used as a drug for rheumatoid arthritis and allergic skin reactions, but it is surpassed in these properties by synthetic derivatives such as prednisolone (11β,17α,21-trihydroxypregna-1,4-diene-3,20-dione) and prednisone (17α,21-dihydropregn-1,4-diene-3,11,20-trione).

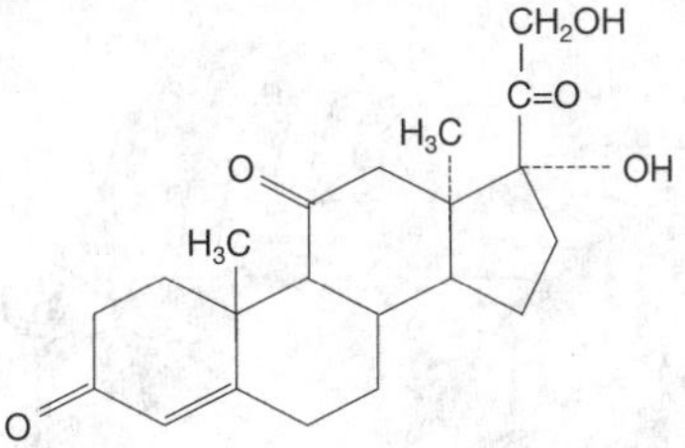

Cortisone

Corti's organ: see Auditory organ.

Corvidae, ***crows, magpies*** and ***jays:*** a family containing about 100 species, and belonging to the order, *Passeriformes* (see). Members of the *Corvidae* are apparently inquisitive about their environment. They pick up and sometimes collect and hide inedible objects. They also display a marked ability to learn and to exploit new situations, and many are notorious egg thieves, especially the carrion crow. The beak is stout and fairly long, often with a small terminal hook. The round nostrils do not have an operculum, and they are protected by forward directed bristles. Smaller members of the family are about the size of a thrush. The almost cosmopolitan family is well represented in the temperate regions of the Northern Hemisphere, and is represented in South America only by the jays. It is absent from New Zealand and most islands of the South Pacific. Examples are the ***Raven*** *(Corvus corax)*, ***Carrion crow*** *(Corvus corone corone)*, ***Hooded crow*** *(Corvus corone cornix)*, ***Rook*** *(Corvus frugilegus)*, ***Jackdaw*** *(Corvus monedula)*, ***Fish crow*** *(Corvus ossifragus)*, ***Magpie*** *(Pica pica)* (plate 6), ***Jay*** *(Garrulus glandularis)*, ***Nutcracker*** *(Nucifraga caryocatactes)*, ***Chough*** *(Pyrrhocorax pyrrhocorax)*, ***White-necked raven*** *(Corvultur albicollis)*, ***Blue jay*** *(Cyanocitta cristata)*.

Rooks are notable for their colonial breeding habits, forming "rookeries" in small woods or groups of trees, sometimes containing thousands of nesting pairs. The carrion crow (uniformly black with a green gloss; about 47 cm in length) is found throughout Central and northern Asia, Europe and Scandanavia, and it also breeds on the Nile delta. The similar sized hooded crow (black head and throat with gray body) has a rather more restricted distribution, in Ireland, Scotland and Scandanavia, and between the Elbe and the Yenisea in the Palaearctic region, but many hybrids exist between the 2 subspecies.

The blue jay is a familiar North American bird from Canada to the Gulf of Mexico. Its upper parts are bright blue with white patches on the wings and tail, and it has a pointed crest.

Corvultur: see *Corvidae*.

Corvus: see *Corvidae*.

Corylaceae, *hazel family*: a family of dicotyledonous plants, containing 62 species in 4 genera, distributed mainly in the northern temperate region. They are deciduous woody plants with alternate, simple leaves and usually caducous stipules. The wind-pollinated flowers are monoecious, and the sexes occur in different inflorescences. Male inflorescences are pendulous catkins. Female flowers occur in pairs in the axil of a bract, sometimes isolated and resembling a bud, sometimes aggregated into a pendulous catkin. The tips of the anthers carry a tuft of hairs. The inferior, 2-locular ovary develops into a single nut, surrounded or subtended by an involucre formed by fusion of the accrescent bract and bracteoles.

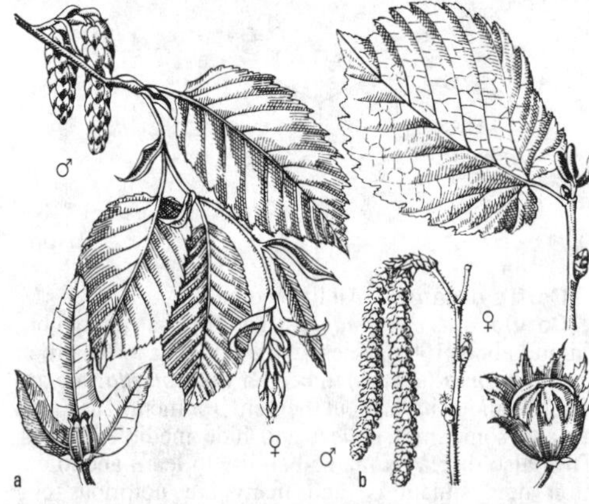

Corylaceae. a Hornbeam *(Carpinus betulus);* branch with male and female inflorescences; *left:* winglike fused bracts with fruit. *b* Hazel *(Corylus avellana),* branch with flower buds; *left;* branch with male and female inflorescences; *right:* fruit.

Hornbeam *(Carpinus betulus)* is an important native European tree (up to 30 m in height, but usually less) frequently planted in woodlands and parks. The tree has a smooth bark, and the fruit has an open, 3-lobed involucre. Since it has a marked ability to sprout and regenerate when cut, it is also common as a coppiced shrub and as a hedging plant. ***Hazel*** *(Corylus avellana)* forms a 5 m-high shrub or small tree. It produces the edible hazel nut (also called a cob or filbert), which is surrounded by a cup-shaped, incised involucre. In additon to vitamins C and B_1, the nuts contain up to 60% of an edible saponifiable oil, which is used in food preparation and in the manufacture of oil paints. Hazel is distributed in the wild throughout Europe, and it is also cultivated, especially in southern and eastern Europe. *Corylus maxima* ***(Lambert nut)*** and *Corylus colurna* ***(Tree hazel, Turkish filbert)***, both native to southern Europe and western Asia, are planted as ornamental trees and for their nuts, which have a high fat content.

Other genera are *Ostrya* ***(Hop hornbeam,*** 10 species; northern temperate region, South to Central America) and *Ostryopsis* (2 species, eastern Mongolia and Southwest China).

Corynebacterium: a genus of Gram-positive rod-shaped bacteria, often with club-shaped swellings at the poles. Most species are aerobic, but some are microaerophilic or facultatively anaerobic. Under the microscope they often display palisades or angled arrangments of paired cells. Some species are harmless, while others are animal or plant pathogens. The anaerobic *Corynebacterium diphtheriae* is the causative organism of diphtheria. Several amino acids are prepared by industrial fermentation with different strains of *Corynebacterium glutamicum*.

Coryphaena: see *Coryphaenidae*.

Coryphaenidae, *dolphinfishes* or *dorados:* a family (order *Perciformes)* of predatory marine, oceanic fishes containing only 2 species. They have a characteristically high forehead, which is almost vertical in the males, the body is very tapered, the caudal fin is deeply forked, and the long dorsal fin orginates on the head. They are beautifully colored, popular sporting fishes, feeding mainly on flying fish and squid. Their flesh is highly prized. *Coryphaena hippurus* attains a length of 150 cm and a weight of 30 kg. Swimming speeds greater then 60 km/h have been recorded.

Coryphaenoididae: see *Macrouridae*.

Corythopus: see *Conopophagidae*.

Cosmopolitan species: a plant or animal species found worldwide in appropriate habitats, i.e. on all inhabitable continents, or in all the oceans. Early C.ss. were mainly passively distributed small animals and plants. Many other species or genera have become cosmopolitan by active extension of their territory, and many have achieved a cosmopolitan distribution after being transported by humans. See Ubiquitous species.

Cosmos: see *Compositae*.

Cossidae: see *Lepidoptera*.

Cossypha: see *Turdinae*.

Costal respiration: see Respiratory mechanics.

Costmary: see *Compositae*.

Cosubstrate: see Enzymes.

Cotingas: see *Cotingidae*.

Cotingidae, *cotingas, becards, tityras* and ***bellbirds:***

a Neotropical avian family in the order *Passeriformes* (see), containing 90 fruit and insect eating species. Most build open nests, and the eggs are speckled for camouflage. All have compact bodies (7.5–50 cm in length), short legs, rounded wings and slightly hooked bills. They are found in Central America and South America as far as Argentina. Two ***Becards*** *(Platypsaris)* are found as far north as southern USA and Jamaica. Many species have spendidly colored plumage with decorative plumes and crests, and other specialized structures such as inflatable throat sacs, and strands of skin or bare lappets on the forehead or at the angle of the beak.

Umbrellabirds *(Cephalopterus)* have a large canopy-like, uniformly glossy blue-black crest of silk filoplumous feathers, which projects over the end of the heavy beak. They also possess a feathered wattle or throat sac (about 40 cm in length in the long-wattled umbrellabird), which hangs down from the breast. During the courtship display, the inflated wattle (resembling a pine cone with spread scales) swings like a pendulum, then forward as the head is thrown back for the loud courtship call. Female umbrellabirds are smaller and inconspicuously colored. The three umbrellabirds may be conspecific: *Cephalopterus ornatus galbricollis* Gould 1851 ***(Bare-necked umbrellabird;*** humid forests at 800–2 000 m in Costa Rica and western Panama, moving to the lowlands in the nonbreeding season); *Cephalopterus ornatus penduliger* Sclater 1959 ***(Long-wattled umbrellabird;*** humid forests on western slopes of the Andes at 700–1800 m in Colombia and Ecuador); *Cephalopterus ornatus ornatus* Geoffroy Saint-Hilaire 1809 ***(Amazonian umbrellabird*** or ***Ornate umbrellabird;*** humid forests in eastern foothills of the Andes up to 1 200 m in Colombia, Ecuador and Peru, and from southern Venezuela and Guyana to northern Amazonian Brazil).

Probably the most familiar members of the family are the 2 ground-dwelling, ledge-nesting Cocks-of-the-rock, which are known for their communal "lek"displays ***(Guianan cock-of-the-rock,*** *Rupicola rupicola;* ***Andean cock-of-the-rock,*** *Rupicola peruviana).*

Bellbirds are named for their bell-like mating call. The ***Three-wattled bellbird*** *(Procnias triarunculatus)* has 3 lappets from the angle of its beak, which are 3–4 times longer than the beak itself. The male ***White bellbird*** *(Procnias alba)* has a cylindrical, sparsely feathered wattle hanging from its forehead.

Cotinus: see *Anacardiaceae.*

Cotoneaster: see *Rosaceae.*

Cottidae, ***sculpins*** and ***bullheads:*** a family of marine and freshwater bottom fishes of the order *Scorpaeniformes* (see), containing about 300 species in 70 genera, which are widespread in nontropical regions in the Northern Hemisphere and near New Zealand. They have a broad, flat head and wide mouth, and the preopercula (fore-gill covers) and fins are armed with spines. Eyes are placed high on the head, and the body is more or less elongate. There are two dorsal fins, which may be separate, or linked by a membrane. The anterior spiny dorsal fin is always shorter and higher than the nonspiny posterior dorsal fin. Pelvic fins are either absent, or represented by a spiny ray and 2–5 soft rays. Scales are either degenerate or absent, and there is no swim bladder. Pectoral fins are large, prominent and fan-shaped. The natural center of radiation appears to be the northern Pacific Ocean. The ***Bullhead, Sculpin*** or ***Miller's thumb*** *(Cottus gobio)* occurs in freshwater in Europe, western Asia and Siberia, and is also found in brackish inlets of the Baltic. The ***Shorthorn sculpin*** or ***Father lasher*** *(Myoxocephalus scorpius)* is found in the Baltic and off the stony coasts of the North Atlantic; it is a voracious predator, but is nevertheless used as a bait fish, as well as being exploited as a source of fish meal. The ***Sea scorpion*** or ***Long-spined bullhead*** *(Taurulus bubalis = Cottus bubalis = Cottus scorpius)* is found in inshore and in intertidal waters off the European and Scandanavian shores of the North Atlantic and in the Baltic, among rocks and seaweed at a depth of 30 m.

Cotton: see *Malvaceae.*

Cottonmouth: see Water moccasin.

Cotton rose: see *Malvaceae.*

Cottontails: see *Leporidae.*

Cotton teal: see *Anseriformes.*

Cottus: see *Cottidae.*

Cotylosauria (Greek *cotule* cup, hollow, *sauros* lizard): extinct order of stem reptiles. Up to 3 m in length, these primitive, plump animals possessed a strongly ossified skull roof with no temporal fossae. The earliest amphibians also lacked temporal fossae, but cotylosaurs differed from amphibians in their possession of only a single occiptal condyle, as well as a fifth digit (thumb) on the forelimbs. Jaws and teeth were poorly developed. The dentition indicates a vegetarian diet, although some forms had fangs. The limbs projected laterally from the body, the upper arm and thigh being practically parallel to the ground. Fossils first appear in the Carboniferous, when cotylosaurs diverged from the *Labyrinthodontia.* They flourished in the Paleozoic, with an evolutionary peak in the Permian, then became extinct in the Triassic. Important fossil sites are located in South Africa, Texas and in the north of the former USSR.

Coua: see *Cuculiformes.*

Coucals: see *Cuculiformes.*

Couch's spadefoot: see *Pelobatidae.*

Couinae: see *Cuculiformes.*

Coumarin, o-*hydroxycinnamic acid lactone,* cis-o-*cumarinic acid lactone, 2H-1-benzopyran-2-one, 1,2-benzopyrone:* a pleasantly smelling compound which occurs in *Galium odoratum,* grass and clover species, tonka beans and many other plants. It is used as an aromatic agent, but its use is now very restricted in view of its possible toxicity. Several derivatives of C. occur as secondary plant products, e.g. umbelliferone (7-hydroxycoumarin; *Ferula galbaniflua, Hydrangea macrophylla),* scopoletin (6-methoxy-7-hydroxycoumarin; *Avena* roots), Esculin (see).

Coumarin

Countess bark: see Quinine.

Counting chamber: an aid for the microscopic counting of microorganisms, blood cells, etc. It consists of a special microscope slide, the surface of which is ruled in a fine grid, with two slightly raised borders for the support of a cover slip. Cells are counted in a known volume, which is determined by the area of the grid and the depth of cell suspension above it (i.e. the distance between the cover slip and the surface of the slide).

Thoma's C.c. is subdivided into squares of 0.0025 mm², and has a chamber depth of 0.1 mm.

Coursers: see *Glareolidae*.

Courtship behavior, ***reproductive behavior***: A species-specific pattern of behavior preceding mating, or intended to lead to mating. In certain vertebrates, it may not lead directly to copulation, but to pair formation or the reinforcement of a pair bond. It is especially marked in arthropods, some mollusks, and in vertebrates. The C.b. of vertebrates often involves Ambivalent behavior (see) and often contains elements of developmental behavior ("infantile elements"). In accordance with the crucial biological function of C.b., it is usually genus-typical, species-specific and multifacetted, consisting of a complicated and coordinated pattern of different elements. It is regulated by biocommunicative signalling between the male and female.

Courtship behavior in the three-spined stickleback (after Tinbergen, 1952). The male (top left) performs a backward and forward dance in front of the female (top right with swollen body) then leads her to the nest. The male points himself toward the nest opening, and the female enters (botton left).

Couvinian stage: see Devonian.

Cover-abundance, ***abundance-dominance:*** an analytical parameter for the characterization of a stand of vegetation. C-a. refers to individual species, and it represents a combination of two factors: abundance (number of individuals; density of a species) and degree of coverage (dominance or area occupied) (see Relevé). It is commonly used in phytosociology as an analytical quantity of a relevé.

C-a. is usually estimated on a 7-point scale according to Braun-Blanquet (1927). ***C-a. 5:*** more than 75% of a sample area covered by the species in question, irrespective of the number of individuals of that species. ***C-a. 4:*** 50–75% coverage, irrespective of numbers. ***C-a. 3:*** 25–50% coverage, irrespective of numbers. ***C-a. 2:*** 5–25% coverage, irrespective of numbers, or very high density of individuals providing low degree of coverage. ***C-a. 1:*** moderate numbers of individuals and less than 5% coverage, or small numbers with rather high coverage. In addition, two symbols are used: **+** indicates sparse C-a., i.e. individuals of a species are few in number and the degree of coverage is very low; ***r*** (rare) indicates that only 1 or 2 individuals of a species are present and they provide extremely small coverage.

The Domin scale employs eleven categories: + (single individual), 1 (very few individuals), 2 (sparsely distributed, less than 1% cover), 3 (frequent, but less than 4% cover), 4 (4–10% cover), 5 (11–25% cover), 6 (26–33% cover), 7 (34–50% cover), 8 (51–75% cover), 9 (76–90% cover), 10 (91–100% cover).

Cover glass: a thin glass plate, 0.15–0.18 mm in thickness, for covering a preparation on a microscope slide. At numerical apertures greater than 0.3, a sharp image is obtained only if the thickness of the G.g. does not exceed a certain limit. Microscope objectives are usually corrected for a C.g. thickness of 0.17 mm.

Cowbane: see *Umbelliferae*.

Cowberry: see *Ericaceae*

Cowbird: see *Icteridae*.

Cowdie pine: see *Araucariaceae*.

Cow lily: see *Nymphaeaceae*.

Cowpea mosaic virus: see Multipartite viruses, Virus groups.

Cowrie shell: see *Gastropoda*.

Cowslip: see *Primulaceae*.

Coxal gland: primary, segmentally arranged excretory organ of arthropods, which has evolved from the nephridia of annelids. Functional C.gs. are present in the *Arachnida,* whereas the *Crustacea* possess modified forms, such as Antennal glands (see) and Maxillary glands (see). Structurally, a C.g. consists of a terminal vesicle (sacculus) (formed from a secondary coelom residue), emptying into a spiral exit at the base of the coxa.

Coxal vesicles: paired, vesicular appendages on the abdominal sternites of primitive insects (e.g. *Diplura, Archaeognatha, Zygentoma).* They are everted by hemolymph pressure, and retracted by muscles.

Coyote, *Canis latrans:* a smaller (head plus body about 90 cm) relative of the wolf (see *Canidae).* The C. is distributed from Alaska to Costa Rica, and on account of its great adaptability and intelligence, it is very common in certain parts of its range. It lives in pairs or small packs and feeds on small animals (jack rabbits, rabbits, rodents) and carrion. Cooperative hunting is performed in relays, rather than by the onslaught of the entire pack. Sheep and goats are also killed occasionally. The coat color is brown sprinkled with black and gray.

Coypu: see *Myocastoridae*.

Cozymase: see Nicotinamide-adenine-dinucleotide.

C3-plants: plants in which the first products of photosynthetic CO_2 assimilation are compounds containing 3 carbon atoms, which arise from the carboxylation of ribulosebisphosphate (ribulose diphosphate), accompanied by cleavage. In contrast, C4 plants (see) and plants operating Crassulacean acid metabolism (see) have an auxiliary pathway for the initial fixation of CO_2, which later releases CO_2 for assimilation by the Calvin cycle (i.e. by the C3-pathway). The majority of photosynthetic green plant species are C3-plants.

C4-plants: plants in which the first products of photosynthetic CO_2-fixation are dicarboxylic acids containing 4 carbon atoms. The C4-cycle or pathway of photosynthesis occurs in many grasses that grow in dry, warm environments, including sugar cane, maize and various species of millet, as well as a number of dicotyledonous plants, e.g. some members of the *Amaranthaceae* and *Chenopodiaceae*. The genus, *Atriplex (Chenopodiaceae)* contains species that operate the C4-pathway, as well as species that incorporate CO_2 solely by the C3-pathway (see Figs. 4 and 5 in Photosynthesis, section: Calvin cycle). In all C4-plants, the C3-pathway is also present. The C4-pathway can be regarded as an auxiliary pathway with an especially high affinity for CO_2, which transfers CO_2 to the C3-pathway for further assimilation.

In the C4-pathway, CO_2 is incorporated into oxaloacetate by the action of phospho*enol*pyruvate carboxylase, which is still active at extremely low concentrations of CO_2 (see Fig.4 in Photosynthesis). In certain plant species, this oxaloacetate is converted into aspartate by the action of oxaloacetate-aspartate transaminase. Other species convert the oxaloactetate into malate by the action of $NADP^+$-dependent malate dehydrogenase. In yet other species, malate appears to arise in a single step by the fixation CO_2 (not shown in Fig.4 of Photosynthesis). These processes occur in the chloroplasts of the mesophyll cells, where phospho*enol*pyruvate carboxylase and malate dehydrogenase are also localized. Mesophyll cells export the C4-acids to the vascular bundle cells, probably via the numerous plasmodesmata. In C4-plants, the vascular bundle sheath is characteristically surrounded by cells ("Kranz" cells), which contain much larger chloroplasts than the mesophyll cells. In malate-forming species, these large chloroplasts also lack grana. The organic acids are decarboxylated in the vascular bundle cells, and the released CO_2 is incorporated into ribulosebisphosphate (ribulose diphosphate) in the large chloroplasts of the Kranz cells. Photosynthesis then continues by operation of the Calvin cycle (see Photosynthesis). The pyruvate, formed in the oxidation and decarboxylation of malate in the vascular bundle cells (or alanine from the decarboxylation of aspartate) returns to the mesophyll cells, where it serves as a substrate for the regeneration of phospho*enol*pyruvate, thereby completing a catalytic cycle for incorporating CO_2 from the atmosphere and transferring it to the vascular bundle cells. Conversion of metabolites in the mesophyll and vascular bundle cells, and the transfer of metabolites between these two types of cell are summarized diagrammatically in Fig.5 of Photosynthesis. Thus, the mesophyll cells of C4-plants contain a CO_2 trapping system, which is capable of utilizing low concentrations of CO_2. Decarboxylation in the vascular bundle cells then generates a very high local concentration of CO_2, which promotes the optimal incorporation of CO_2 into ribulosebisphosphate. This is important in situations where the CO_2 concentration is the limiting factor in photosynthesis, e.g. under light saturation, or during closure of stomata to prevent water loss. Thus, C4-plants are better adapted than C3-plants to warm, dry habitats. This superiority can be demonstrated experimentally by placing C4-plants and C3-plants together in an illuminated chamber, which is closed with respect to the loss or entry of CO_2. Under these conditions, the C4-plants eventually starve out the C3-plants, which require a relatively high atmospheric concentration of CO_2. C4-plants compete more effectively for the available CO_2, because the affinity of phospho*enol*pyruvate carboxylase for CO_2 is much higher than that of ribulosebisphosphate carboxylase. In addition, photosystem 2 does not operate in C4-type vascular bundles, where CO_2 is generated from malate, so that very little O_2 is formed at this site. The resulting low O_2 concentration suppresses Photorespiration (see), a process in which some photosynthesized material is normally catabolized. C4-plants are therefore more economic then C3-plants in their use of the photoassimilate and in their use of transpiration water. For each gram dry mass of photoassimilate, C4-plants use only 300 g water, compared with 610 g required by C3-plants.

The C4-cycle also occurs in Crassulacean acid metabolism (see), where the different stages are separated temporally within the cell, rather than spatially as in other C4-plants. The C4-cycle is also known as the Hatch-Slack-Kortschak cycle, or Hatch-Slack cycle.

Crab apple: see *Rosaceae.*

Crab lice: see *Phthiraptera.*

Crab plague: see Crayfish.

Crab plover: see *Dromadidae.*

Crabs: see *Brachyura, Decapoda.*

Cracidae, *curassows, guans* and ***chachalacas:*** an avian family of Central and South America in the order *Galliformes* (see). All 38 species (in 11 genera) have long tails, short, rounded wings and strong feet. Like the *Megapodiidae,* they differ from other *Galliformes* in having large hindtoes on the same level as the 3 front toes. Although mostly ground birds, living on vegetation, fruit, insects and worms, they nest in trees rather than on the ground.

Largest members of the family are the 7 ***Curassows,*** which are forest birds, similar in size to, but less heavy than turkeys. The bright yellow or orange bill often has a fleshy frontal protuberance above the base of the upper mandible, and there is a strongly recurved, erectile crest. The male of the ***Yellow-knobbed curassow*** *(Crax daubentoni;* 84 cm; Venezuela and Colombia) is glossy black above with white underparts and a white tip to the long tail; there is yellow knob at the base of the black bill, and a prominent recurved crest. The female lacks the bill knob, and its crest, wing coverts, breast and flanks are barred white.

Guans are brownish to olive green and the sexes are alike. The face is bare of feathers, and some species have throat wattles. They live mostly in treetops, where they feed on fruit. Most guans belong to the genus *Penelope* (11 species).

Smallest members of the family are the ***Chachalacas,*** the largest of these attaining a length of 63 cm. Thirteen species are placed in the genus *Ortalis.* They are slender, long-tailed, graceful birds, usually brownish green to olive in color. Most species have naked, bright red throat skin, and there is often a bare grayish area of skin around the eyes.

Crago: see Shrimps.

Crakes: see *Gruiformes.*

Cranchia: see *Cephalopoda.*

Cranes: see *Gruiformes.*

Cranesbill: see *Geranaceae.*

Crangon: see Shrimps.

Crangonidae: see Shrimps.

Cranial capacity: see Skull capacity.

Cranial index: the maximal breadth of the skull expressed as a percentage of its maximal length (B/L x 100). Maximal breadth is measured at 90 ° to the sagittal plane (plane dividing the skull into right and left halves), and it is generally the distance between the parietal eminences. Maximal length is measured from the glabella (bony prominence between the eyebrows and above the nasal depression) and the most posterior point of the occiput (opisthocranion). C.i. values of male skulls are classified as follows: 75% and less (long or dolichocranial), 75–79.9% (medium length or mesocranial), 80–84.4% (short, rounded or brachycranial), 84.5% and greater (very short or hyperbrachycranial). For the female skull, the limits of the index classes are one unit higher.

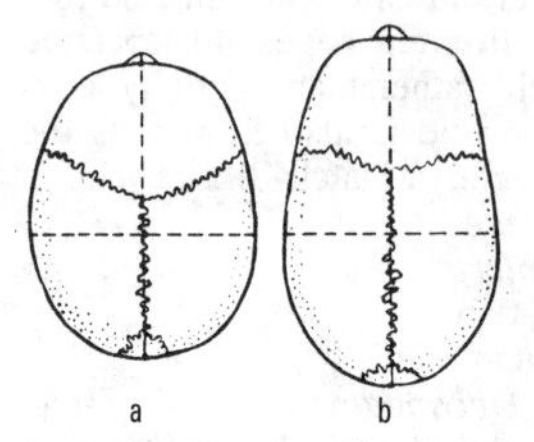

Cranial index. Human skulls (after Bach): *a* with a high cranial index; *b* with a low cranial index.

The C.i. should not be confused with the *cephalic index,* which is the maximal breadth of the head of a living person expressed as a percentage of its maximal length. In this case, the maximal breadth is the greatest distance between the two sides of the head, measured above the ears and at 90 ° to the sagittal plane; the maximal length is the distance from the most prominent point in the median sagittal plane between the eyebrows to the most distant point in the same plane at the back of the head. The cephalic index of the living subject is about 2 units higher than the corresponding C.i. of the skull.

Cranial nerves, ***nervi craniales:*** nerves of vertebrates emerging from various regions on the undersurface of the brain, mainly from the brainstem. They are classified as sensory, motor, and mixed nerves. There are 12 pairs of C.n., which are assigned Roman numerals (in order of appearance from the most anterior or cranial to the most posterior or caudal), as well as being named according to their function or distribution: I or olfactory nerve (sensory), II or optic nerve (sensory), III or oculomotor (motor), IV or trochlear (motor), V or trigeminal (mixed), VI or abducent (motor), VII or facial (mixed), VIII or acoustic (sensory), IX or glossopharyngeal (mixed), X or vagus (mixed), XI or accessory (motor), XII or hypoglossal (motor). Fishes and amphibians have only 10 C.n. (I to X). Amniotes also possess the accessory and hypoglossal nerves. In fishes, the optic nerve emerges from the midbrain. The lateral line organs of fishes and amphibians are innervated by the facial nerve. The most important peripheral innervation regions and the main functions of the Human C.n. are as follows.

I. The ***olfactory nerve (nervus olfactorius)*** is concerned exclusively with the sense of smell. Arising from processes of the olfactory cells of the nasal mucous membrane, its fibers collect into about 20 branches, which pass through the cribriform plate of the ethmoid bone and synapse with the cells of the olfactory bulb. From the olfactory bulb other fibers extend inward to centers in the cerebrum.

II. The ***optic nerve (nervus opticus)*** serves the sense of sight. Its fibers are derived from ganglion cells in the retina and extend to the diencephalon

III. The ***oculomotor nerve (nervus oculomotorius)*** arises from a nucleus in the floor of the cerebral aqueduct. Motor fibers to four extrinsic muscles of the eyeball (superior rectus, inferior rectus, medial rectus, inferior oblique) and to two intrinsic muscles (ciliaris and sphincter pupillae) are responsible for eye movements, pupil contraction and accommodation.

IV. The ***trochlear nerve (nervus trochlearis)*** arises from a nucleus in the floor of the cerebral aqueduct. Motor fibers to the superior oblique muscle of the eye are reponsible for eye movement.

V. The ***trigeminal*** or ***trifacial nerve (nervus trigeminus)*** is the largest cranial nerve and the main sensory nerve of the face and head. Motor fibers to the jaw muscles are responsible for chewing. At the point of emergence from the brain, the V cranial nerve consists of a small motor and a large sensory root. Motor fibers arise from a nucleus in the cerebral aqueduct and a nucleus in the upper pons. Sensory fibers arise from cells in the trigeminal ganglion. The two roots combine into a single trunk, which then divides into three branches:

1. The small sensory *ophthalmic nerve,* which divides again into three branches (lacrimal, frontal and nasociliary), as well as communicating with the oculomotor, trochlear and abducent. Branches extend to the cornea, ciliary body and iris; to the lacrimal gland and conjunctiva; to part of the mucous membrane of the nasal cavity; and to the skin of the eyelid, eyebrow, forehead and nose.

2. The sensory *maxillary nerve,* which sends numerous branches to the dura mater, forehead, lower eyelid, lateral angle of the orbit, upper lip, gums and teeth of the upper jaw, and the mucous membrane and skin of the cheek and nose.

3. The *mandibular nerve* (sensory and motor) sends branches to the temple, ear pinna, lower lip, lower part of the face, teeth and gums of the mandible, and the chewing muscles. It also supplies the lingual nerve to the front of the tongue.

VI. The ***abducent nerve (nervus abducens)*** arises in a small nucleus beneath the floor of the fourth ventricle and sends motor fibers to the lateral rectus muscle of the eye.

VII. The ***facial nerve (nervus facialis).*** Motor fibers arise from a nucleus in the lower part of the pons, while sensory fibers arise from the geniculate ganglion on the facial nerve. The single process of the ganglionic cells divides into central and peripheral fibers. Central fibers pass into the medulla oblongata and end in the terminal nucleus of the glossopharyngeal nerve. Sensory root (peripheral fibers) and motor root emerge together from the brain. Motor fibers are supplied to the muscles of the face, part of the scalp, the pinna, and muscles of the neck (facial expression and movement). Vasodilator fibers are supplied to the submaxillary and sublingual glands (salivary secretion). Sensory fibers are supplied to the anterior two thirds of the tongue (taste) and a few to the region of the middle ear.

VIII. The ***acoustic*** or ***auditory nerve (nervus statoacusticus),*** which emerges from the medulla oblongata, contains two sets of fibers that differ in origin, destination and function.

1. The *cochlear nerve* or nerve of hearing originates in the spiral ganglion of the cochlea.

2. The *vestibular nerve* or nerve for the maintenance of equilibrium originates in the vestibular ganglion of the internal auditory meatus (semicircular canals, utriculus and sacculus).

IX. The ***glossopharyngeal nerve (nervus glossopha-yngicus)*** emerges from the medulla oblongata. Sen-ory fibers arise from the superior and petrous ganglia, notor fibers from the nucleus ambiguus. Sensory fibers re supplied to the fauces (aperture between mouth and harynx), tonsils, pharynx and the posterior third of the ongue (taste). Motor fibers extend to the muscles of the harynx (swallowing, coughing, sneezing), and secre-ory fibers to the parotid gland (salivary ecretion).

X. The ***vagus*** or ***pneumogastric nerve (nervus va-us)*** has the most extensive distribution of all the cranial erves. Motor fibers from the nucleus ambiguus supply he muscles of the pharynx, larynx, trachea, heart, nouths of the large arteries and veins, aortic arch, sophagus, stomach, small intestine, pancreas, liver, pleen, ascending colon, kidneys and visceral blood ves-els. Inhibitory fibers are supplied to the heart, and se-retory fibers to the gastric and pancreatic glands. Sen-ory fibers from the jugular ganglion and the ganglion odosum are disributed to the mucous membrane of the arynx, trachea, lungs, esophagus, stomach, intestines nd gall bladder. The vagus is involved in swallowing, oughing, sneezing and vomitting, and regulation of vis-eral activity (see Autonomic nervous system).

XI. The ***accessory nerve (nervus accessorius)*** arises rom the nucleus ambiguus in the medulla oblongata, nd distributes fibers to the pharyngeal and superior la-yngeal branches of the vagus (swallowing, coughing neezing). It also contains fibers from the spinal cord hat arise at the level of the fifth cervical nerve; these en-er the skull through the foramen magnum, pass through he jugular foramen, then descend to the sternocleido-nastoid and trapezius muscles.

XII. The ***hypoglossal nerve (nervus hypoglossus)*** rises from the hypoglossal nucleus in the medulla blongata, and supplies motor fibers to the muscles of he tongue and the hyoid bone at the root of the tongue movement of the tongue, sucking).

Craniata: see Vertebrates.

Craniates: see Vertebrates.

Craniology: study of the morphology of human and nimal skulls.

Craniometry: measurement of animal or human kull. See Anthropometry.

Cranium: see Skull.

Crassulaceae, ***stonecrop family:*** a family of dicot-ledonous plants containing 1 500 species in 35 genera, osmopolitan in distribution, but mainly in dry habitats n South Africa, Mexico and the Mediterranean region. hey are mostly herbaceous, succulent plants, rarely ow bushes, usually with xeromorphic characters like hick, fleshy leaves and stems, sunken stomata, waxy urfaces and closely packed leaves. Vegetative repro-luction by rhizomes or offsets is common. Flowers are ctinomorphic, hermaphrodite (rarely dioecious) and 3-2-merous (usually 5-merous). Stamens are often obdi-lostemonous. The number of stamens is either equal o, or twice the number of, petals. Pollination is by in-ects. The superior carpels are equal in number to the etals, free or united at the base, each base often carry-ng a nectar-secreting scale. The fruit is a group of folli-les containing numerous minute seeds. Chemically the amily is characterized by an abundance of piperidine lkaloids, in particular sedamine.

The large South African genus, *Crassula*, has pro-ided many ornamental plants, e.g. the familiar indoor plant, *Crassula portulacacea*, with ovoid, opposite leaves; and *Crassula lycopodioides*, with small, scale-like, dark green leaves.

The genus, *Kalanchoe*, is especially well represented in Madagascar and South Africa. *Kalanchoe blossfeldi-ana*, used experimentally in the study of photoperiodic-ity, has bright red flowers in terminal cymose inflores-cences.

Bryophyllum is a tropical genus, containing species which reproduce vegetatively by forming self-rooting adventitious buds on the edges of their leaves.

The Central and South American genus, *Echeveria*, also contains many ornamental plants, with colored, hoary leaf rosettes and majestic, usually multiflowered inflorescences.

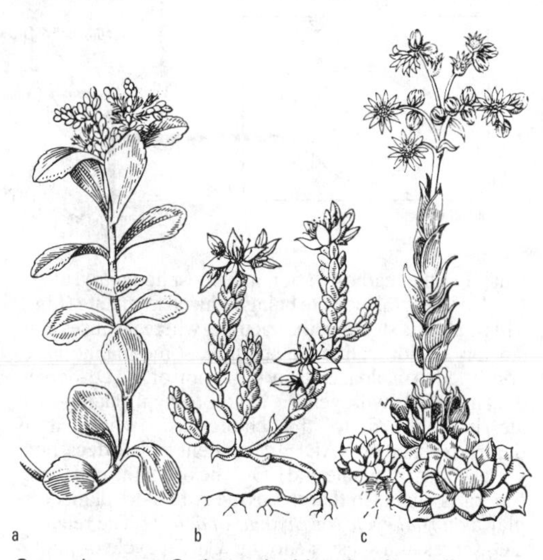

Crassulaceae. a Sedum telephium (orpine or live-long). *b Sedum acre* (wall-pepper). *c Sempervivum tectorum* (houseleek).

Various well known and cultivated species are con-tained in the genus, *Sedum* ***(Stonecrops),*** e.g. *S. teleph-ium* ***(Orpine*** or ***Livelong)*** with greenish-yellow to pur-ple-red flowers, which grow in walls and rocky out-crops and in dry woodlands; and the yellow-flowered, sharp-tasting *S. acre* ***(Wall-pepper),*** which also favors stone walls and rocky ground.

The thick, fleshy leaves of *Sempervivum* characteris-tically form a basal rosette; several species, native in the high mountains of Central and South Europe, are culti-vated in rock gardens, e.g. *Sempervivum tectorum* ***(Houseleek).***

Crassulacean acid metabolism, ***CAM:*** the tempo-ral separation of two types of photosynthetic CO_2 fixa-tion: carboxylation of phospho*enol*pyruvate (see C4-plants) and carboxylation of ribulosebisphosphate in the Calvin cycle (see Photosynthesis). CAM occurs in chlo-rophyll-containing tissues of succulent plants, espe-cially in cells containing both chloroplasts and large vacuoles, e.g. in members of the *Cactaceae* and *Crassu-laceae.* It is called "crassulacean" because it was first observed in members of the *Crassulaceae.* Plants pos-sessing this type of metabolism are known as CAM-plants. In the dark, they incorporate CO_2 into oxaloac-

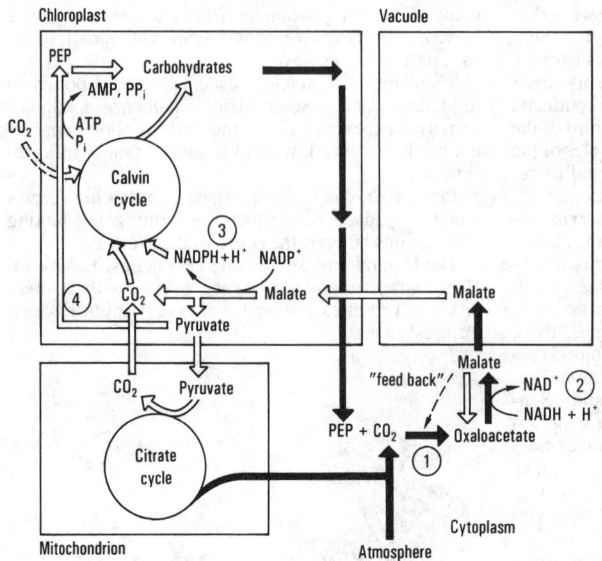

Crassulacean acid metabolism. The figure shows the intracellular compartmentalization of the different reactions of crassulacean acid metabolism; those occurring in the dark are shown as thick black arrows, whereas those occurring in daylight are shown an open arrows. *1* Phospho*enol*pyruvate carboxylase. *2* NAD+-dependent malate dehydrogenase. *3* Malic enzyme (NADP+-dependent malate decarboxylase). *4* Pyruvate phosphate dikinase. *PEP* = Phospho*enol*pyruvate.

teate by the carboxylation of phospho*enol*pyruvate, much of the oxaloacetate being reduced to malate (Fig.). This malate is stored in the vacuole, where it causes considerable acidification. In sunlight, some malate is reconverted to oxaloacetate by the action of NAD+-dependent malate dehydrogenase, but more significantly malate is exported to the chloroplast, where it is decarboxylated by NADP+-dependent malate decarboxylase to form pyruvate and CO_2. The pH of the cell therefore rises again, so that the tissues of CAM-plants display a *diurnal cycle* (or *rhythm) of acidity.* The released CO_2 serves as a substrate for the Calvin cycle of photosynthesis in the chloroplast. The enzymatic reactions correspond closely to those of C4-plants, i.e. initial fixation of CO_2 by carboxylation of phospho*enol*pyruvate, and the final fixation of CO_2 by the Calvin cycle. In C4-plants, however, these two processes are separated spatially, whereas in CAM-plants they are separated in time.

CAM represents an adaptation to extremely dry habitats. An ecological advantage is gained by the ability to fix CO_2 at night, when the humidity is relatively high and water loss by transpiration is low, so that CAM-plants can afford to have open stomata. At the same time, CO_2 released by respiration is also immediately fixed again. In daylight, the stomata are closed early as a result of water stress, but by daybreak the CO_2 needed for the Calvin cycle of photosynthesis is already in the plant, trapped as a carboxyl group of malate. The malate is then decarboxylated and assimilated by the Calvin cycle. Thus, photosynthesis is able to proceed during the day, although the rate of transpiration is extraordinarily low. Consequently, the water economy of CAM-plants is very efficient. On average, they need only 240 g water (compared with 610 g needed by C3-plants) to support the production of 1 g dry mass of photoassimilate. CAM-plants are therefore highly competitive in dry habitats with cool nights, where occasional precipitation permits replenishment of water storage organs.

Crawfish: see Spiny lobsters.

Crax: see *Cracidae.*

Crayfish: large crustaceans of the family *Astacidae* (containing about 100 species), infraorder *Astacidea*, order *Decapoda* (see). The first 3 pairs of legs are chelate and the first pair are greatly enlarged as chelipeds (claws). Almost all species are found in fresh water and in the Northern Hemisphere. *Astacus astacus,* once widely distributed in Central Europe, was almost anihilated in 1870 by a fungus disease known as "crab plague", caused by *Aphanomyces astaci.* The American crayfish, *Canbarus affinis,* introduced in 1890, is immune to the disease as well as being more tolerant of pollution; it has now largely replaced *A. astacus* in wide areas of Europe. C. are caught for their meat. They are closely related to Spiny lobsters (see) and Lobsters (see).

Creadon: see *Callaeidae.*

Cream cups: see *Papaveraceae.*

Creatophora: see *Sturnidae.*

Creeping movements: forms of free locomotion in lower organisms in contact with a surface. Some C.m are perhaps better described as *gliding* movements. Organisms lacking a rigid cell wall display *ameboid* movement. Members of the *Desmidiaceae* move by the unilateral protrusion of mucilaginous threads from the pores of their rigid cell membrane. C.m. of cyanobacteria are also associated with the secretion of mucilaginous carbohydrate. Diatoms of the order *Pennales* display gliding movements, brought about by movement of protoplasm through the raphe, which functions like the caterpillar tread of a tractor or tank.

Cremaster: the hooked caudal extremity of lepidopteran pupae.

Crenic acid: see Humus.

Crenicichla: see Cichlids.

Crenilabrus: see Wrasses.

Crenoplankton: see Plankton.

Crenothrix: see *Chlamydobacteriaceae.*

Creodonta: an extinct suborder of carnivores. They lacked canine teeth, and their carnassial teeth were developed from the first and/or second molars, and no from the upper premolar and lower molar as in modern carnivores. Other primitive features were 5 toes and a small brain with hardly any cerebral sulci. Two families are recognized.

The *Oxyaenidae* were rather mustelid in appearance, although *Paleonictis* resembled a cat. Fossils appear in the Late Paleocene, and the family appears to have flourished in the Eocene. The broad skull carried a large deep jaw. *Oxyaena* (fossils from North America and East Africa) was an exceptionally powerful predator, and it was probably the ancestor of the bear-sized *Patriofelis* and the even larger *Sarkastodon.* Fossils of the *Hyaenodontidae* appear early in the Eocene, and the family appears to have flourished throughout the Tertiary. The skull was narrower, legs longer and carnassials better developed than in the *Oxyaenidae.* They displayed a wide range of sizes from very small, weasel-like forms to lion- or leopard-sized predators. The jaws of *Machaeroides* and *Apataelurus* are reminiscent of those of the later saber-toothed cats.

Representative fossils of the *C.* are well known from the Germanic Tertiary (Lower Eocene to Lower Miocene), especially from the Eocene brown coal measures of Geiseltal near Halle, Germany.

Crepescular species: see Biorhythm.

Crepidula: see *Gastropoda.*

Crested bellbird: see *Pachycephalinae.*

Crested cardinal: see *Fringillidae.*

Crested keeled lizards, *Leiocephalus:* iguanas (see *Iguanidae*) found on open ground and in rocky areas on the Caribbean islands from Cuba to Trinidad, and on the mainland from Columbia to Bolivia. These insectivorous iguanas are about 20 cm long with a flattened head and a somewhat laterally compressed body. The dorsal crest extends to the tail, and the spined, keeled scales of the body and tail overlap like roof shingles. *Leiocephalus carinatus coryi* ***(Curl-tailed lizard)*** of Cuba rolls its tail spirally over its back when excited.

Crested tit: see *Paridae.*

Cretaceous, ***chalk:*** latest of the three systems of the Mesozoic. Facies of the upper C., consist of white to light gray, brittle, friable chalk and chalk marl ("blackboard chalk") (see Chalk). The C. lasted for 75 million years and is divided into Lower and Upper C. Division of marine deposits is based primarily on the evolution of foraminifera, ammonites, belemnites, certain bivalve genera (e.g. *Inoceramus)* and echinoderns. Ammonites of the Lower C. include those with bifurcated ribs, as well as forms with degenerate shell structures, ribbing and suture lines. The Lower C. and the later strata of the Upper C. are classified largely on the basis of their belemnite fossils. Separation of the lower and middle strata of the Upper C. is based essentially on the evolution of the bivalve genus, *Inoceramus.* Classification of terrestrial deposits relies primarily on the wide variety of different reptilian fossils. The extremely large dinosaurs, e.g. *Brachiosaurus* (13 m high, 28 m long, 80 tonnes) and *Brontosaurus* (20–30 m long), appear in the C. In contrast, the mammals of the C. are very small. The transition from the C. to the Tertiary marked the extinction of the giant dinosaurs and pterodactyls, ammonites, belemnites, rudists, inoceramids and a number of foraminifera.

In the vegetable kingdom, the C. was a time of many crucial evolutionary changes. The Lower C. belongs to the Mesophytic (see), containing a little changed Jurassic flora dominated by gymnosperms. In contrast, the Upper C. contains an ever increasing variety of angiosperms that are characteristic of the Cenophytic.

Creutzfeldt-Jacob disease: see Prion.

Cribriform plate: see Skull.

Cricetidae, ***cricetid rodents (hamsters, gerbils, voles, lemmings*** and ***New World rats and mice)*** (plate 7): a family of rodents (see) containing about 570 species in 97 genera, representing a wide variety of forms. Most are terrestrial. Some are fossorial (mole rats), semiaquatic (muskrat) or saltatorial (gerbils). With the exception of the Austro-Malayan region, the family has a cosmopolitan distribution. Dental formula: 1/1, 0/0, 0/0, 3/3 = 16. The cusps of the molars are arranged in 2 rows (cf. the *Muridae,* in which the cusps are arranged in 3 rows). The 5 subfamilies of the C. are listed separately: *Arvicolinae* ***(Lemmings, Voles, Mole lemmings, Muskrat);*** *Gerbillinae* ***(Gerbils);*** *Cricetinae* ***(New World mice, Hamsters, White-tailed rats, Mole mice);*** *Nesomyinae* ***(Malagasy rats);*** *Lophiomyinae* ***(Maned rats).***

Cricetinae: a subfamily of the *Cricetidae* (see), containing about 400 species in 76 genera, and divided into the following 4 tribes.

1. *Hesperomyini* ***(New World mice,*** 366 species in 69 genera). New World mice are distributed from Alaska to Patagonia. Very familiar are the ***White-footed mice*** *(Peromyscus),* which resemble Old World wood mice *(Apodemus)* in both appearance and behavior. All species are nocturnal. Some are arboreal, but most species are terrestrial, living in a variety of habitats such as forests, prairies, rocky outcrops, and near to human settlements. Examples are the ***Golden mouse*** *(Peromyscus nuttalli),* the ***White-footed mouse*** *(P. leucopus)* and the ***Deer mouse*** *(P. maniculatus).* The subspecies, *Peromyscus leucopus tornilla,* lives entirely in darkness, deep in the Carlsbad cave of New Mexico, feeding almost exclusively on the cave cricket, *Ceutophilus.* Other representatives of this new World tribe are *Baiomys taylori* ***(Pygmy mouse),*** *Onychomys* ***(Grasshopper mice),*** *Reithrodontomys* ***(American harvest mice)*** and *Oryzomys* ***(Rice rats).***

2. *Cricetini* ***(Hamsters,*** 24 species in 5 genera). Hamsters are small to medium-sized rodents, most species possessing cheek pouches in which they store food. They are restricted to the Old World, from Central Europe to Asia. The surfaces of the molars always have 2 parallel raised ridges. The ***Common*** or ***Black-bellied hamster*** *(Cricetus cricetus)* is found on arable land and steppes from Central Europe to Asia Minor; it has a body length of 24–34 cm (tail 4–6 cm), weighs up to and exceeds 500 g, and has rather short legs, so that the body nearly touches the ground. In contrast to other lighter areas, the belly coat is dark and plays a role in threat posture. Albinos are quite common, as well as totally black varieties. The fur is used commercially by furriers. The eyes are relatively large and very dark, the ears membranous, and the short tail only sparsely haired. The Common hamster builds an underground structure, where it lays in a store of seeds (carried there in its cheek pouches), and where it hibernates in the winter.

Dwarf hamsters *(Phodopus,* body length 5–10 cm, tail 0.5–1.8 cm) have a dense covering of hair on the soles of their hindfeet, and the tail is hidden in the depth of fur. Three species are known in Siberia, Mongolia and North China, including the ***Striped hairy-footed hamster*** or ***Dzungarian hamster*** *(P. sungorus;* gray-brown to yellow-brown, underside almost white, and a dark line along the back; northern populations become completely white in winter) and ***Roborovsk's dwarf hamster*** *(P. roborovskii;* underside and feet white, ears dark with white edges, upper coat light yellow with reddish tinge, white spot above each eye).

Four species of ***Golden hamsters*** *(Mesocricetus)* are known from eastern Europe, the Middle East and the Caucasus. The well known ***Golden hamster*** *(M. auratus,* body length 17–18 cm, tail about 1 cm, weight about 130 g) was formerly little known, and was represented by a single specimen collected in 1839. In 1930, a female and litter were found in Syria and kept as a breeding colony. Their descendants are now numerous in Europe and USA, and widely favored as pets and laboratory animals. Of all nonmarsupial mammals, the Golden hamster has the shortest gestation period (16 days), producing 7 or 8 litters per year, each of 6–12 young, becoming sexually mature at 2 1/2 months.

The ***Mouselike hamster*** *(Calomyscus bailwardi,* body length 7–8.5 cm, tail 8.5–10 cm; Afghanistan, southern USSR, Parkistan) is the only species of its genus, has a rather long tail for a hamster, and lacks cheek pouches. The tip of the tail carries a brush of densely packed hairs.

Ratlike hamsters *(Cricetulus)* are found in Southeast Europe, Asia Minor and northern Asia. There are 11 species, including the ***Korean gray rat*** *(C. triton).*

3. *Mystromyini* (***White-tailed rat,*** 1 species). The ***White-tailed rat*** *(Mystromys albicaudatus)* is found in South Africa and Lesotho, where it occupies the same ecological niche as hamsters in Europe and Asia. It inhabits flat grassland and dry sandy areas, were it is preyed upon by the civet, *Suricata suricata.*

4. *Myospalacini* (***Zokors, Mole rats, Central Asiatic mole rats,*** 6 species in one genus). Zokors are greatly adapted to an underground existence and are very similar to moles. The forefeet, which are used for digging, carry a powerful long claw on the third digit, and the 2 neighboring claws are also long. Examples are the ***Manchurian Zokor*** *(Myospalax psilurus)* and the ***Transbaikal Zokor*** *(M. aspalax).*

Cricetini: see *Cricetinae.*

Cricket frogs: see *Hylidae.*

Crickets: see *Saltatoptera.*

Cricoid cartilage: see Larynx.

Cricosaura: see Night lizard.

Crinifera: see *Cuculifromes.*

Criniger: see *Pycnonotidae.*

Crinoidea (Greek *krinos* lily), ***crinoids, sea lilies*** and ***feather stars:*** a class of the *Echinodermata* (see), containing about 650 known living species. Modern and fossil *C.* display the greatest diversity of form, and are the most highly evolved taxon within the *Echinodermata.* The body consists of a calyx to which are attached five long pinnate or branched arms (Fig.). Calyx and arms form the crown. The calyx extends into a stalk, which carries whorls of slender, jointed appendages known as cirri. Most species lose the stalk during early development and attach themselves by means of cirri on the calyx to supporting structures such as rock or seaweed; a small amount of local motility is then possible. Calyx, arms, stalk and cirri consist largely of skeletal ossicles (plate or ring-shaped), which lie in the dermis. The arms attain lengths up to 1.2 m, while the stalk of some species may be several meters in length, e.g. up to 18 m in *Seirocrinus.* Food consists of plankton, which is trapped by lateral appendages of the arms (pinnulae and tentacles) and conducted to the mouth via a food channel. The larval form is called a Doliolaria.

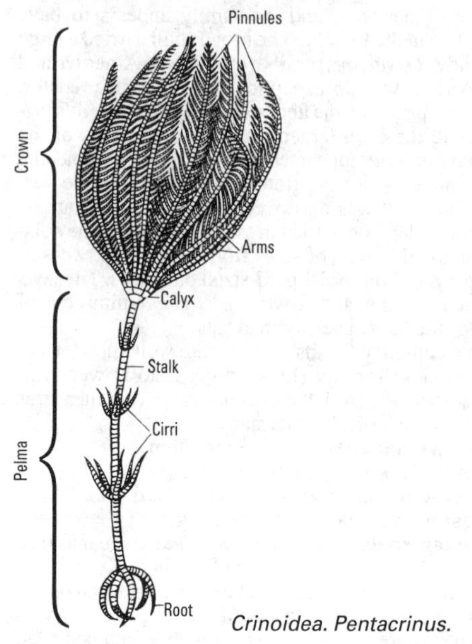

Crinoidea. Pentacrinus.

Fossils. About 5 000 fossil *C.* are known from the Cambrian to the present. In contrast to the 650 recent species, more than 4 300 fossil species have been recorded, all the main groups appearing as early as the Ordovician. The class reached its evolutionary climax in the Mississippian (Lower Carboniferous), with a second, somewhat lesser development in the Permian. There is noticeable evolutionary decline in the transition from the Permian to the Triassic, followed by a further increase in the Mesozoic, when the modern orders of *C.* appeared. Fossil *C.* were mostly sessile inhibitants of shallow seas, and the extension of this habitat to greater depths is not apparent until the beginning of the Tertiary. Complete fossils of sea lilies are rare, e.g. in the Posidonia shale (Liassic) of Holzminden, Germany, and in the Muschelkalk (Triassic) of Freyburg, Germany.

Collection and preservation. Sessile *C.* are collected with a dredge or sea floor grab. Free swimming *C.* are collected in shallow water with a fine net. Specimens are narcotized by adding 30% magnesium chloride solution dropwise to the sea water. After extending the arms, the *C.* are fixed in 70% ethanol, then after a few hours placed in 85% ethanol for storage.

Classification. The extinct subclasses, *Inadunata, Flexibilia* and *Camerata,* were all stalked Paleozoic forms, with or without cirri, some lacking pinnules. All living forms and some fossil forms belong to the subclass *Articulata,* which is divided into the following 5 orders.

1. Order: *Millericrinida.* Sea lilies without cirri, e.g. *Hyocrinus, Calamocrinus.*

2. Order: *Cyrtocrinida.* Two living species from the Caribbean and mid Atlantic; the aboral end of the crown is attached directly to the substratum: *Holopus.*

3. Order: *Bourgueticrinida.* Mostly small sea lilies with a slender stalk lacking cirri: *Rhizocrinus, Bathycrinus.*

4. Order: *Isocrinida.* Sea lilies with cirri: *Metacrinus, Cenocrinus.*

5. Order: *Comatulida.* Feather stars; stalkless, unattached, free swimming or crawling *C.*

Crioceratites (Greek *krios* ram, *kera* horn): a characteristic ammonite of the Lower Cretaceous (Barrême). The shell is coiled like a clock spring, usually oval or subquadratic in cross section, and has simple ribs. Shell shape and ribbing are of the degenerate type often displayed by the later ammonites.

Crioceratities (stone core; x 0.3)

Crista: a ridge or crest-like process especially on bones, e.g. *Crista sagittalis,* a crest of bone along the sagittal suture of the skull of many mammals, some prehominids, and the great apes. For *Crista sterni,* see Thorax. Mitochondrial cristae or *Cristae mitochondriales* are invaginations of the inner mitochondrial membrane.

Crithidial stage of trypanosomes: see Trypanosomes.

Critical phase, ***critical period, imprinting phase:*** see Imprinting.

Croaker: see *Sciaenidae.*

Croaking bag: see Resonance sac.

Crocetin: a brick-red C_{20} carotenoid dicarboxylic acid (see Carotenoids) with 7 conjugated *trans* double bonds, 4 methyl branches and 2 terminal carboxylic acid groups (Fig.). Its digentiobiose ester, crocin, occurs as a yellow pigment in crocuses and a few other higher plants.

HOOC … COOH

Crocetin

Crocin: see Crocetin.

Crocodile: see *Crocodylidae.*

Crocodile lizard: see *Xanosauridae.*

Crocodylia, ***crocodylians, crocodiles, alligators*** and ***gharials*** (plate 5): an order of large (some species nearly 7 m in length), predatory, aquatic or amphibious reptiles with a lizard-like appearance. Propulsion through the water is achieved by serpentine body movements, and by lashing of the powerful, body-length, laterally compressed tail. A single comb or crest of scales extends from the tip to about midway along the tail, then continues as a double row from the middle to the base of the tail. Five toes are present on the forelimbs and 4 on the hindlimbs. In contrast to other reptiles, the limbs lift the belly clear of the ground when walking on land. During swimming, the 4 powerful limbs are folded back along the body and take no part in propulsion. The skin of the head is firmly fused to the skull. Back and tail are covered by thick, rectangular, ridged, horny plates, which are partly ossified on their undersides (dermal ossifications). On the ventral surface of the animal, the plates are smaller, and in the alligators and some crocodiles they are also ossified. The number and arrangement of large, sharply ridged bosses on the neck are taxonomically useful for the identification of some species. Set high in the head, the eyes have a vertical pupil, with 2 lids and a transparent nictitating membrane ("third eyelid"). The nostrils can be closed, and they are also set high, as well as being directed upward. *C.* can therefore see and breathe while lying or floating almost totally submerged. The eardrums are also situated behind a closeable skin fold. Conical in shape, the powerful teeth are set in deep hollows or alveoli. The choanae (internal nares) are set well back in the buccal region. A wide flap of skin at the back of the mouth (false palate, or basihyal throat valve) closes the glottis (the fleshy tongue also contributes to this closure), so that the pharynx can be sealed off when the jaws are open, allowing the animal to feed while submerged. Ribs are present on all the trunk vertebrae, the sacral vertebrae, the first 5–10 tail vertebrae, and the 2 cervical vertebrae. Eight pairs of true ribs are present in the abdominal region. In addition, there are 7–8 pairs of abdominal ribs or gastralia lying free in the musculature. The stomach is very muscular, and like that of a bird of prey, it forms pellets of undigested residues. *C.* feed on fish, water birds and small animals that come to the water's edge to drink. Larger species also attack large animals and even humans. The cloacal aperture is longitudinal, and the penis is unpaired. On the basis of their tooth structure, the possession of a 4-chambered heart, the presence of a diaphragm, and the development of the sense organs, *C.* are anatomically the most advanced reptiles. Young animals communicate by squeaking, while adults have a loud roar or bellow. All species lay white, hard-shelled eggs, which are buried in a specially prepared site in the sand, or in a heap of vegetation.

All 22 species are found in the tropics or subtropics, but especially the tropics. The American alligator and the Chinese alligator can withstand colder temperatures than most *C.*, but the Mugger *(Crocodylis palustris)* and the Gharial also inhabit areas with cold winters in northern India and Nepal. No species is known to hibernate. Most species avoid cold night-time temperatures by sleeping at the bottom of rivers or ponds; during this sleep, metabolism is slowed and oxygen is witheld from most tissues. Periodically, as the heart and brain become threatened with an oxygen deficit, the animal surfaces to breathe. Most species are restricted to freshwater, while the minority of species are tolerant of, or even prefer saltwater. Saltwater-tolerant species can travel further, and therefore have a much wider distribution, e.g. the Saltwater crocodile (see). In contrast, the Cuban crocodile *(Crocodylis rhombifer)* is restricted to 2 swamps in Cuba. Nearly all species are threatened with extinction due to hunting for "crocodile leather", and all species are CITES-listed. On the basis of skull structure and the presence of dermal ossifications, they are divided into 3 families: *Alligatoridae* (see), *Crocodylidae* (see), and *Gavialidae* (see).

Crocodylidae, ***crocodiles:*** a family of the *Crocodylia* (see) containing 14 species in 3 genera. They are found in all tropical regions of the world where there is water. The teeth of the lower and upper jaws bite in the same plane, and in contrast to the *Alligatoridae* (see), the 4th mandibular tooth rests in an exterior groove of the upper jaw, so that it is visible when the mouth is closed. Dermal ossifications are usually found only on the dorsal surface and neck. The non-ossified belly skin yields the valued "Crocodile leather" of commerce. Snouts range from short and broad to long and slender. Some species display brood care, in that the mother carries the newly hatched young to the water in her mouth or throat sac, a form of behavior first observed in the ***Nile crocodile*** (see). Other representatives are: ***Saltwater crocodile*** (see), ***African dwarf crocodile*** (see), and the ***False gharial*** (see).

Crocodylis: see *Crocodylia.*

Crocodylus: see Nile crocodile, Saltwater crocodile.

Crocosmia: see *Iridaceae.*

Crocus: see *Iridaceae.*

Crocuta: see *Hyaenidae.*

Crop dusting agents: plant protection agents prepared as very fine powders with a particle size of 0.02–0.06 mm, and intended for application as a dust (e.g. crop dusting from low-flying aircraft).

Crop protection agents: see Plant protection agents.

Crossbills: see *Fringillidae.*

Cross feeding: see Syntrophism.

Crossing over: a Recombination (see) process in the prophase of the first meiotic division, in which segments are exchanged between homologous, nonsister chromatids of paired chromosomes (see Meiosis). Two reciprocal recombination products are formed simultaneously and with equal frequency. The site of C.o. is recognizable under the microscope as a chiasma, and it is detectable genetically from the recombination of coupled genes (see Coupling). Processes analogous to meiotic C.o. in higher organisms also occur in bacteria and viruses, in which the chromosome is usually a simple DNA molecule displaying neither true mitosis nor meiosis.

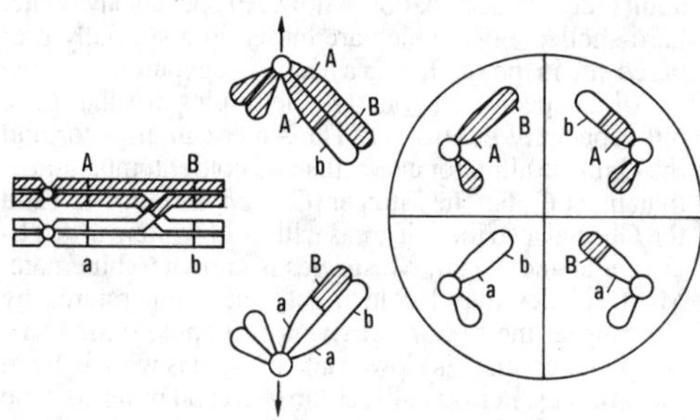

Fig.1. *Crossing over.* By breakage and rejoining, genetic material is exchanged between homologous chromatids or paired chromosomes. All recombinant material was physically present in the parent chromosomes.

According to the classical concept each C.o. occurs by breakage and rejoining of the chromatids (Fig.1), in which the initial breakage occurs at identical sites on each chromatid and the segment exchange is truly reciprocal. The copy-choice model (Fig.2) assumes a temporal relationship between C.o. and chromosome replication, and envisages C.o. as a template change during duplication. Studies on bacteria and viruses show that both processes may be sometimes be involved in the same recombination event, but the weight of evidence supports the classical model of breakage-reunion.

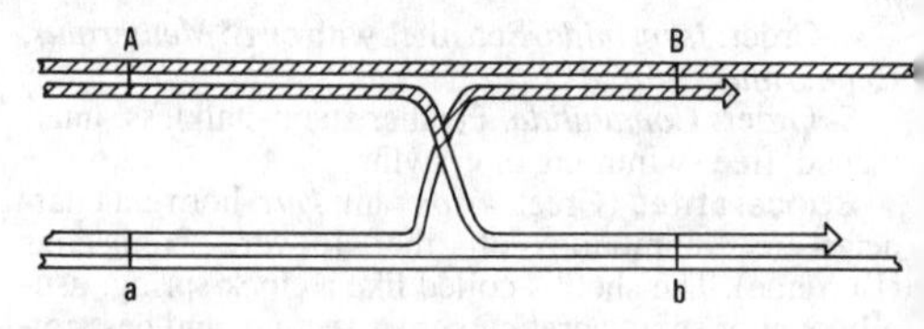

Fig.2. *Crossing over.* Copy-choice model. The recombinant chromatids produced by this mechanism are synthesized from new material during replication, and crossing over represents a change in the choice of template. The arrows indicate the progress of DNA synthesis.

Formation of several C.o. sites between paired chromosomes results in a multiple exchange of segments. Double C.o. (i.e. 2 sites of C.o. present) can occur in several different ways, each with different genetic consequences (Fig.3). C.o. can occur between genes, as well as between the functional subunits of a functional gene (cistron).

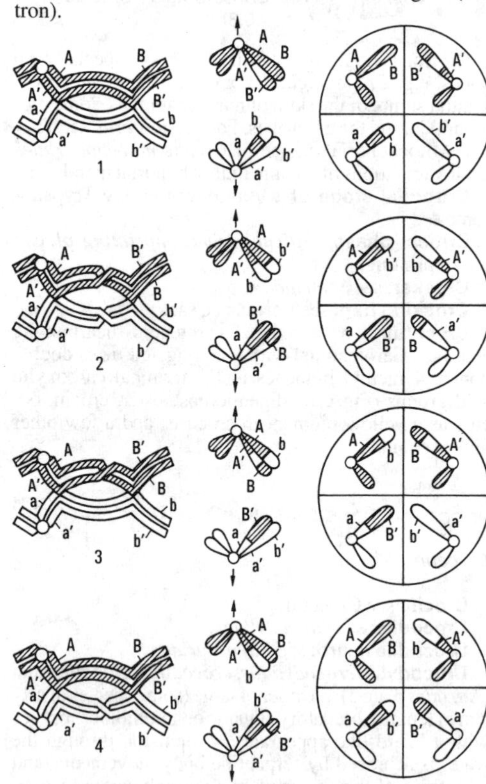

Fig.3. *Crossing over.* Double crossing over involving two chromatids *(1),* four chromatids *(2),* three chromatids (a'b' not involved) *(3),* or three chromatids (AB not involved) *(4).*

C.o. may also occur in the somatic cells of higher organisms (somatic C.o.). In the special case of unequal C.o. the two exchanged segments are different in size.

Crossing over suppressor: a gene or heterozygous structural change in a chromosome, which suppresses or reduces the exchange of genetic material between chromatids that occurs by crossing over.

Crossing over value: the percentage of gametes, in which it can be shown genetically that crossing over has occurred between 2 allele pairs during the preceding meiosis.

Crossopterygii: see Coelacanths.

Cross-pollination: see Pollination.

Cross reactivity: reaction of an antibody, not only with the antigen to which it was raised, but also to different but related antigens. C.r. is exploited in the use of specific antibodies to detect related disease organisms. It is also used in the detection of mutant enzymes, which have lost their catalytic activity, but retain sufficient structural similarity with the native enzyme to cross react with the specific antibody; such mutant, cross-reacting proteins are called CRiM proteins (cross-reacting material; the "i" is added to make a pronouncable word).

Cross resistance: a phenomenon sometimes observed after a pest organism has been exposed to a chemical Plant protection agent (see). Strains or races of the organism then develop which are resistant to other active chemical substances, with which they have had no previous contact.

Crotalidae, ***pit vipers:*** phylogenetically the most advanced and most recent snake family (see Snakes). It contains 130 species in 6 genera, and it is represented in North and South America, and in South and Southeast Asia. Pit vipers posses a deep groove or pit between the eyes and nose, which houses a sunken, richly innervated membrane. This serves as an infrared detector, i.e. a temperature sensor, which responds to the low heat radiation of the prey (mainly rodents), permitting its detection at night. Like their close relatives, the *Viperidae* (see), pit vipers have solenoglyphic dentition, they fold and erect their venom fangs, and their venom is hemotoxic. Most pit vipers are ground-dwelling, and they are the commonest and most dangerous poisonous snakes of North and South America. With few exceptions they give birth to live young, which leave the gelatinous egg covering during or shortly after birth. The following examples are listed separately: Rattlesnakes, Water moccassin, Halys viper, Lance-head snakes, Bushmaster.

Crotalus: see Rattlesnakes.

Croton oil, ***oleum crotonis:*** a fatty oil from the seeds of *Croton tiglium,* an East Asian species of the family *Euphorbiaceae.* It is brownish yellow to dark reddish brown in color, viscous, and slightly fluorescent, with a disagreeable odor and intensely acrid taste. It is used as a drastic laxative.

Crotophaginae: see *Cuculiformes.*

Crowned pigeons: see *Columbiformes.*

Crown roots: roots that arise from the tiller nodes of grasses. See Tiller.

Crows: see *Corvidae.*

Crucian carp, *Carassius carassius:* one of the 2 species of carp in the genus *Carassius* (the other is the goldfish). It is a high-backed, fine-tasting carp with a straight lateral line. Under favorable conditions, it may attain a length of up to 50 cm, but 20–30 cm is more common. It was originally found in rivers flowing into the North, Baltic and Black Seas and into the Arctic Ocean, but it has now been introduced widely as an angling and commercial fish. A very hardy fish, it can hibernate if the water becomes depleted of oxygen, and it can survive for long periods without food. For differences between the C.c. and goldfish, see Goldfish.

Cruciferae, *Brassicaceae,* ***crucifers, brassicas, mustard family:*** a family of dicotyledonous plants containing 3 200 species in 375 genera. The family is cosmopolitan in distribution, but occurs mainly in the northern temperate region, where species are even found at high altitudes and up to the vegetation limits of the Arctic. Crucifers are especially well represented in the Mediterranean region. They are mostly herbaceous plants with alternate (usually), exstipulate leaves, which may be entire, pinnate, digitate, pinnatifid or deeply incised.

Flowers are usually aggregated in a raceme or corymb, nearly always without bracts or bracteoles. The bilaterally symmetrical flowers have a very characteristic structure. As a rule, each flower consists of 4 free sepals in 2 whorls of 2, and 4 petals alternating with the sepals and usually arranged as a cross. There are 6 stamens (an outer whorl of 2 short stamens and an inner whorl of 4 long ones), and a single superior ovary of 2 united carpels. Within the ovary, outgrowths from the placentas unite to form a partition called a replum or false septum, so that the ovary is effectively 2-locular. The ovules are attached to the fused suture of the two carpels. The fruit is usually a specialized, pod-like capsule. If this capsule is at least 3 times as long as broad, it is called a ***siliqua,*** but a ***silicula*** if it is shorter. Rarely the fruit is a single-seeded or multi-seeded achene. The seed contains an oily embryo and little or no endosperm. In a few genera, flower structure deviates from the norm, i.e. petals are of unequal size, and there may be more or less than 6 stamens and more than 2 carpels. Most species are insect-pollinated, and they accordingly possess characteristic nectaries, which usually arise as annular growths at the bases of the stamens.

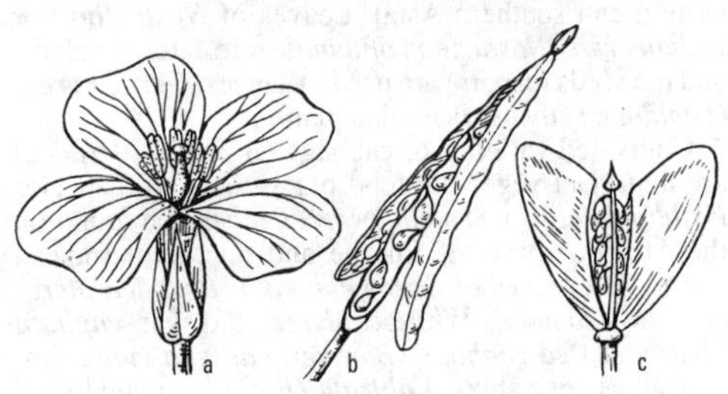

Fig.1. *Cruciferae. a* Flower of lady's smock *(Cardamine pratensis). b* Siliqua of the common wallflower *(Erysimum cheiri). c* Silicule of shepherd's purse *(Capsella bursa-pastoris).*

Mustard oil thioglycosides or Glucosinolates (see) are characteristic constituents of practically all crucifers (these natural products are also especially abundant in the *Capparidaceae* and *Resedaceae).* Damage to the plant results in hydrolysis of the glucosinolates with the release of volatile, pungent, lacrimatory isothiocyanates, also known as mustard oils. This ability to form mustard oils has led to the adoption of several crucifers as spices and condiments (see below). The hydrolysis is

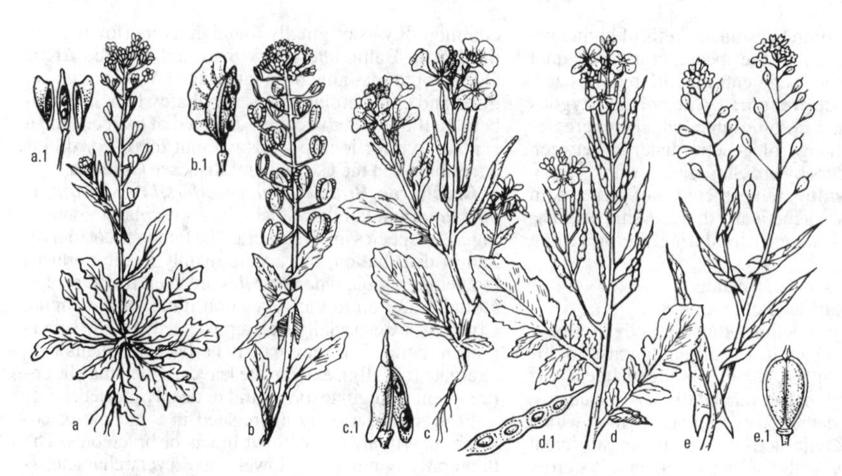

Fig.2. *Cruciferae.* *a* Shepherd's purse *(Capsella bursa-pastoris).* *b* Field penny cress *(Thlaspi arvense).* *c* Charlock or wild mustard *(Sinapis arvensis).* *d* Wild radish or runch *(Raphanus raphanistrum).* *e* Gold of pleasure *(Camelina sativa).* *a1* to *e1* are the fruits of the respective plants.

catalysed by myrosinase (thioglucoside glucohydrolase), which is released during tissue damage from "myrosin cells" or idioblasts.

The taxonomy of this very extensive family is made difficult by the extreme uniformity of floral structure. Certain morphological features such as the position of the embryo in the seed, and the presence of nectaries and myrosin cells are therefore also used in classification.

Many species are exploited as spice and condiment plants. The roots of ***Horse radish*** *(Armoracia rusticana;* originally from southern Europe or Asia) are used in sauces and relishes. Seeds of various species are used to prepare mustard, and to a lesser extent for the extraction of oil, e.g. ***Black mustard*** *(Brassica nigra* , cultivated mainly in Central Europe), ***White mustard*** *(Sinapis alba,* cultivated from the Mediterranean region to eastern India) and ***Indian*** or ***Chinese mustard, Leaf mustard*** or ***Mustard greens*** (*Brassica juncea,* from eastern and southern Asia). Leaves of ***Nasturtium*** or ***Indian cress*** *(Nasturtium officinale)* are eaten in salads, and the seeds (capers) are used in sauces. ***Garden cress*** *(Lepidium sativum)* is a salad plant.

Cultivated varieties of cabbage *(Brassica oleracea)* are important vegetable food plants. The wild species ***(Wild cabbage*** or ***Sea Colewort)*** can still be found on the Atlantic coasts of Europe and on Mediterranean coasts. The leaves of ***Savoy cabbage*** *(Brassica oleracea* var. *sabauda),* ***White cabbage*** *(B.o.* var. *capitata alba)* and ***Red cabbage*** *(B.o.* var. *capitata rubra)* are cooked as vegetables. ***Kohlrabi*** *(B.o.* var. *gongylodes)* is grown for its thick, fleshy stem, which is cooked and eaten. Brussels sprouts are the tight, miniature cabbage-like axial buds of *B.o. var. gemmifera.* The fleshy, thicked, more or less deformed inflorescences of ***Broccoli*** *(B.o. var. cymosa)* and ***Cauliflower*** *(B.o. var. botrytis)* are popular vegetables. Other vegetable or fodder plants are: ***Kale, Borecole*** or ***Collard*** *(B.o. var. acephala);* ***Pe-tsai, Celery cabbage, Peking cabbage*** or ***Chinese cabbage*** *(Brassica pekinensis)* from East Asia; ***Swede*** *(Brassica napus var. napobrassica);* ***Turnip*** *(Brassica rapa var. rapa);* ***Seakale*** *(Crambe maritima),* whose wild ancestor is a protected European plant; ***Garden radish*** *(Raphanus sativus);* and ***Small radish*** *(Raphanus sativus* var. *sativus).* Well known and widely cultivated oil plants are ***Rape*** *(Brassica napus var. napus)* and ***Bird rape*** *(Brassica rapa var. silvestris).* ***Big seeded false flax*** or ***Gold of pleasure*** *(Camelina sativa = Myagrum sativum)* is still cultivated to a small extent as an oil plant. ***Rocket salad*** *(Eruca sativa = Brassica eruca)* was formerly popular as a salad and seasoning plant, but is now grown less widely. Into the 18th century, ***Dyer's woad*** *(Isatis tinctoria)* was economically important, but was later replaced by *Indigofera tinctoria (Papilionaceae)* and by synthetic dyes. Many crucifers are arable weeds, e.g. ***Shepherd's purse*** *(Capsella bursa-pastoris),* which flowers throughout the growth season; ***Field pennycress*** *(Thlaspi arvense),* which is notable for its round siliculas; ***Charlock*** or ***Wild mustard*** *(Sinapis arvensis = Brassica kaber);* and ***Wild radish, White charlock*** or ***Runch*** *(Raphanus raphanistrum).*

Decorative and ornamental crucifers are: the sweetly scented ***Common wallflower*** *(Cheiranthus cheiri)* from southern and western Europe; ***Stock*** or ***Gilliflower*** *(Matthiola incana)* which has delicately colored flowers and is native to the Mediterranean region; ***Aubretia*** *(Aubrietia deltoidea),* a popular bedding plant originating from Southeast Europe; and various species of ***Candytuft*** *(Iberis),* an important garden plant originating from the Mediterranean region. The deserts of North Africa and parts of the Middle East are home to a curious plant known as ***Jericho resurrection mustard, Rose of Jericho*** or ***Mary's flower*** *(Anastatica hierochuntica),* which displays hygroscopic movements. After flowering and setting fruit, the plant turns into a skeletonized hollow ball, i.e. the branches curve in and the stem snaps off at ground level, leaving a ball-like structure which blows around in the desert wind. If it lands in a moist area or is moistened by rain, the ball absorbs water, the branches spread and the seeds are shed. When conditions become dry again, the ball is reformed and rolls around as before.

Crude fiber: collective term for all nitrogen-free components, which remain undissolved after boiling plant material for 30 min with 1.25% sulfuric acid, followed by boiling with 1.25% potassium hydroxide, and finally washing with warm water. This procedure removes proteins and storage carbohydrates (starch, sugars, pectins, etc.), and the residue consists of the structual components, cellulose, lignin and pentosans. The :.f. content is often used instead of dry mass as a refernce value for physiological and biochemical studies, .g. on storage organs or fully grown leaves. The C.f. ontent is also important in straw analysis, and in the deermination of the food value of agricultural products.

Crustacea, ***crustaceans:*** a class of the *Arthropoda,* nd the only class of the subphylum, *Branchiata.* Alnost all of the 31 400 species are aquatic, and respiraory gas exchange occurs via gills. Two pairs of antenae and 3 pairs of mouth appendages (chewing moutharts in the primitive condition) are present. Some pecies are less than 1 mm in length, whereas other nembers of the class attain body lengths of up to 60 cm.

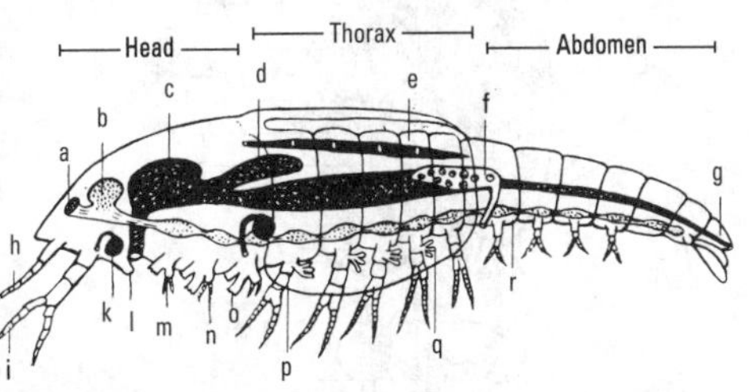

ig.1. *Crustacea.* Body organization of a crustacean. a Eye. *b* Brain. *c* Foregut. *d* Hepatopancreas. *e* Heart. Gonad. *g* Telson. *h* Antennule. *i* Antenna. *k* Nephridum. *l* Labrum. *m* Mandible. *n* Maxillule. *o* Maxilla. *p* Carapace. *q* Gill. *r* Ventral ganglion chain.

The body is divided into head, thorax and abdomen or pleon), and each of these may carry appendages. The ead is a fused, nonsegmented structure comprising a resegmental region (acron), 2 former antennal segnents and 3 former mouthpart segments. Usually one or nore former thoracic segments are included in the head, which is then known as a cephalothorax. The remaining egments comprise the thorax and abdomen, which are clearly differentiated by the form of their appendages. In some groups, the thoracic segments (pleomeres) are fused. At the end of the abdomen is a post-segmental region or telson, which may carry a pair of jointed appendages or caudal rami, forming the caudal furca. Many crustaceans possess a shell or carapace, formed by a fold from the posterior rim of the head, which extends backward for a greater or lesser distance. The carapace either forms a loose shield over the dorsal surface or it is fused with the segments that it covers. In the extreme condition (e.g. in the *Ostracoda),* the carapace envelops the entire body. Beginning anteriorly, the head appendages are: 1) the first antennae or antennules, 2) the second antennae or simply the antennae, 3) the mandibles, 4) the first maxillae or maxillules, 5) the second maxillae or simply the maxillae. With the exception of the antennules, all appendages are primitively and typically biramous, i.e. forking distally into 2 jointed rami, the endopodite and exopodite (Fig.2). Depending on the mode of existence and nutrition, limbs and mouthparts show various specializations. In some orders the limbs are flattened without true joints, and are known as phyllopodia (Fig.3); these include the trunk limbs of the *Branchiopoda,* most maxillules and maxillae, and notably the decapod maxillae. The outer layer or exoskeleton consists of chitin, which may be hardened by calcium carbonate, and serves for outer protection and muscle attachment. Since chitin forms a rigid cover, it must be periodically shed to permit growth.

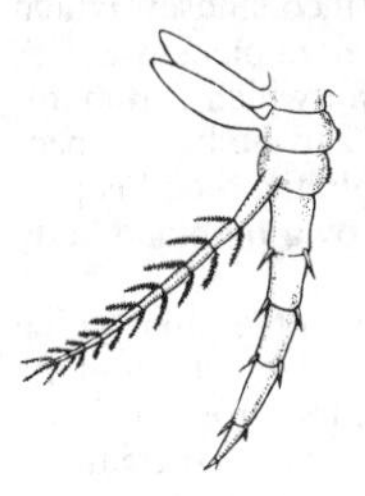

Fig.2. *Crustacea.* Typical biramous limb or stenopodium.

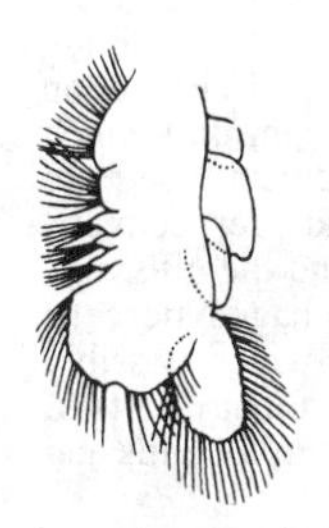

Fig.3. *Crustacea.* Phyllopodium of a member of the *Branchiopoda.*

In general plan, the nervous system consists of a brain or cerebral ganglion, lying above the esophagus and connected to a ventral chain of thoracic and abdominal ganglia which run beneath the alimentary canal. In the primitive condition, each segment contains a ganglion pair, but in most forms the more posterior of these have migrated anteriorly. The ganglia of the mouthpart appendages are normally fused into a single subesophageal ganglion, connected to the brain by 2 circumesophageal commissures. Sense organs include: the single median eye (the eye of the nauplius larva, which persists in most adults); paired and often stalked compound eyes; statocyts for the sense of balance; chemoreceptors or olfactory hairs (esthetascs) on the antennules; sensory cells (modified bristles) on the surface of the body and limbs.

The alimentary canal is divided into 3 main regions, the foregut (stomodeum), midgut (mesenteron) and hindgut (proctodeum). Various lateral expansions or diverticula of the midgut produce digestive enzymes and contribute to the absorption of digestion products; most prominent of these structures is the hepatopancreas. In the subclass, *Malacostraca* (see), in particular in the suborder, *Decapoda,* the foregut is subdivided into an esophagus, cardiac stomach and pyloric stomach. The decapod cardiac stomach has a thick chitinous lining, with calcified ossicles, representing the masticatory apparatus of the gastric mill. As food leaves the cardiac stomach, large particles are held back by the cardiopyloric valve.

The vascular system is open. In the primitive condition, the heart is an extended hollow tube, but in most species it has evolved into a shorter structure, restricted to the region of the body containing the respiratory organs. The heart lies dorsally beneath the exoskeleton in a cavity separated from the rest of the body by a membrane called the pericardium. Valves or ostia permit entry, but not exit, of blood to the heart from the pericardial cavity. Contraction of the heart forces blood through various apertures, which may or may not be continuous with an arterial system. In the *Malacostraca,* the heart has a system of true vessels (arteries), leading to rather diffuse and ill-defined sinuses, which may be considered to represent the venous system. A large sternal sinus is usually present, which communicates with the cavities of the gills and limbs. Harpacticid copepods lack a heart; probably the blood is propelled sufficiently by the animal's own limb and body movements.

Respiratory gas exchange occurs in the gills, which are expansions or specifically modified structures of the appendages. Gas exchange also occurs across the inner surface of the carapace. In small species, specialized respiratory organs may be absent, and gas exchange occurs over the entire body surface. Organs resembling trachea are found in many terrestrial isopods. Excretion occurs via glandular coelomoducts (nephridia) which lie in the antennary or maxillary segments, and are known as antennary or maxillary glands, respectively. In addition, parts of the mesenteron and the gills may have an excretory role. Specialized phagocytic cells (nephrocytes and nephrophagocytes), which possibly consume waste materials, are present in many parts of the body, but are particularly abundant in the gills and bases of the legs.

Hermaphroditism is rare, but is found in certain species of *Isopoda, Notostraca, Cephalocaria* and *Cirripedia.* Parthenogenesis is encountered sporadically. The great majority of species reproduces sexually and the sexes are separate. Gonads are usually paired. Ovaries often have glandular structures, which secrete a secondary covering for the eggs and a cement substance. Limbs in the region of the male sexual opening are often modified for copulation (gonopods). Release of free eggs is not usual. Most species exhibit brood care, in which the eggs are attached to the body or appendages of the female, or laid in a special brood pouch. Cleavage is superficial. In most groups, the eggs hatch into larvae which metamorphose to adults, but in some groups the eggs hatch into young adults. The main larval types are the nauplius, metanauplius, cypris and zoea.

Many crustaceans have evolved into parasites, with such extreme modification that they are hardly recognizable as belonging to the class. In all such cases, however, the larva is still typically crustacean.

Practically every mode of existence is represented, including permanently mobile aquatic forms, motile terrestrial forms, inhabitants of sand holes, semi-sessile tube builders and species that remain permanently fixed to a substratum. Most species are marine, with some freshwater and terrestrial forms. Only the terrestrial isopods are totally independent of free-standing water. There are also predators, filter feeders and those that ingest mud, plants or carrion, as well as sucking parasites. Species are usually dispersed as their free-swimming larvae. Small planktonic forms are ecologically and economically very important as a food source for fishes and whales. Prawns, lobsters, shrimps, crayfish and crabs are commercially important as human food sources.

Fossil record. Fossils are found from the Cambrian to the present. Crustaceans readily become fossilized by impregnation of their outer chitin layer with calcium carbonate. Their remains are therefore known in many geological deposits. After the appearance of the earliest

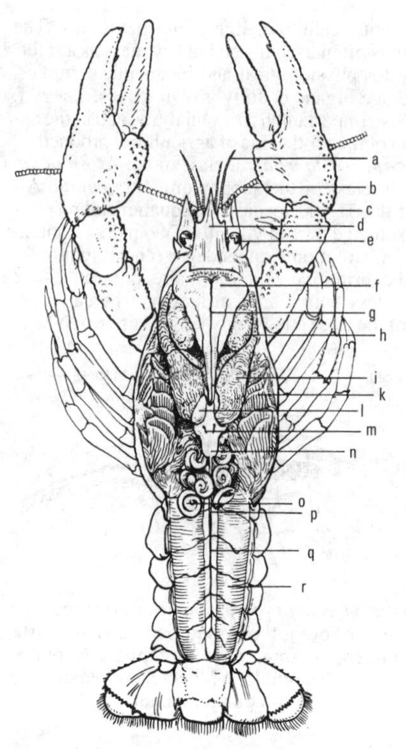

Fig.4. *Crustacea.* Anatomy of a crayfish (male). *a* Firs antenna or antennule. *b* Second antenna. *c* Scale of the antenna. *d* Rostrum. *e* Eye. *f* "Stomach". *g* Ophthalmic artery (anterior aorta). *h* Mandibular muscle. *i* Hepatopancreas. *k* Gills. *l* Gonad. *m* Ostia of the heart. *n* Heart. *o* Vas deferens. *p* Abdominal artery. *q* Hindgut. *r* Tail musculature.

fossils, there is a gradual increase in the variety of forms, culminating in the Upper Cretaceous in a peak of evolutionary development, which persists to the present day. Only the *Ostracoda* (see), however, are used as index fossils for typing and dating strata.

Classification. Further subdivisions will be found under each of the following subclasses: *Cephalocarida, Anostraca, Branchiopoda, Ostracoda, Mystacocarida, Copepoda, Branchiura, Ascothoracida, Cirripedia, Malacostraca.*

Collection and preservation. Small species are caught with a plankton net, or sieved from sediment. Larger forms are most conveniently caught at night using a lamp, or in traps baited with fish. Marine species may be caught with a trawl. Large decapods are killed with choloroform vapor. Specimens are preserved in 75% ethanol containing 5% glycerol, or in neutralized 2% formaldehyde.

Crustecdysone: see Ecdysterone.

Cryobionts: inhabitants of snow and ice. True C. are found primarily among the phytoflagellates, green algae and cyanobacteria. About 70 snow-inhabiting algal and bacterial species are known. Large accumulations of C. may color the snow surface, e.g. red by *Chlamydomonas nivalis,* brownish by *Ancylonema nordenskiöldii,* and green by various green algae. These organsms spend most of the year in a frozen state, but in the pring they live during the day in the melt water on the now surface, then freeze again at night. There are very ew species of animal C. In addition to one nematode pecies, one rotifer species and a single dipteran *(Trihocera hiemalis),* the commonest animal C. are pringtails like the glacier flea *(Isotoma saltans),* and ome 15 species of snow-scorpion flies (family *Boreiae),* e.g. the snow flea *(Boreus hiemalis).* C. are disributed mainly on snow and ice surfaces of lower latiudes and high mountains.

Cryomethods: methods for preparing specimens or light and electron microscopy, in which the specinen is fixed by freezing. Formation of ice crystals, which would damage the tissue, is prevented by addiion of freeze-protection agents and by very rapid freezng. The frozen tissue is sectioned directly (see Freeze ectioning), or the ice is replaced with solvent or embedding medium. In *freeze-substitution* the ice is relaced at about 190 °K with an organic solvent (diethyl ther, acetone), then embedded in paraffin wax or resin. n *freeze-drying,* the ice is removed by sublimation at a ressure of 0.13–0.00013 Pa, and the dry tissue is embedded in the normal way.

Cryoscopic method: in plant physiology, a method or determining the osmolarity and thereby the molar oncentration of plant cell sap (see Osmosis) from the lepression of freezing point.

Crypsis: see Protective adaptations.

Cryptobranchidae, ***giant salamanders:*** a family of he *Urodela* (see), containing 3 species in 2 genera. These are the largest modern urodeles, possessing a nassive, squat body and a broad, flat head with a large nouth. Folds of skin occur along the sides of the body, nd the tail is short. There are 4 fingers and 5 toes. Metmorphosis is incomplete (see Neoteny). Adults lack yelids; they retain one pair of gill slits, and they have a arallel arrangement of maxillary and prevomerine eeth. All members are wholly aquatic. They lie in wait or prey (fish, frogs, aquatic rodents, worms) on the river ed, and rise to the surface to breathe. Fertilization is exernal; spawn is deposited in rosary-like strings in a deression (nest) excavated in the river bed by the male, which then fertilizes the eggs with a cloud of sperm.

Cryptobranchidae. Chinese giant salamander *(Andrias davidianus).*

Two species occur in Asia: *Andrias davidianus* ***(Chinese Giant Salamander***, over 1 m in length) (Fig.) and *Andrias japonicus* ***(Japanese Giant Salamander***, over .5 m in length, the largest living amphibian). Both species inhabit cold mountain rivers and streams. The nales exhibit brood care, in that they keep watch over ne string of spawn.

The single North American species, *Cryptobranchus llaganiensis* **(Hellbender**, up to 70 cm in length) is ound in the fast-flowing waters of the Mississippi and Missouri regions. Several females often spawn in the ame nest, which is guarded by the male for 60–80 days ntil the 3 cm long larvae have hatched.

Cryptocecidium: see Galls.

Cryptodira: a suborder of the *Chelonia* (see).

Cryptodonta: see *Bivalvia.*

Cryptogams, ***nonflowering plants, spore plants:*** all lower plants that do not form flowers, and which generally reproduce by forming spores, e.g. algae, bryophytes and pteridophytes. Flowering plants are called phanerogams. See Spermatophytes.

Cryptogastra: see *Coleoptera.*

Cryptomery, ***latency:*** a term coined by Tschermak for the behavior of certain genes that are not expressed unless accompanied by another specific gene (complementary gene). Thus, by bringing together latent and complementary genes, a genetic cross can result in a completely unexpected phenotypic character, which is unknown in the ancestral organisms. See Polygeny.

Cryptomonadina: an order of usually inflexible or rigid flagellates, with 2 chromatophores (sometimes absent) and 2 flagella. A ventral invagination (gullet), or ventral longitudinal groove is present, which renders the animal asymmetrical.

Cryptomonas: see *Phytomastigophorea.*

Cryptomonadina: see *Phytomastigophorea.*

Cryptophyte: see Life form.

Cryptopidae: see *Chilopoda.*

Cryptostegia: see *Asclepiadaceae.*

Crypotozoa: species of epedaphic soil fauna which spend the day concealed beneath stones or bark, or in soil cavities.

Crystalline cone: see Light sensory organs.

Crystalline style: see *Mollusca.*

Ctenacanthiformes: see *Selachimorpha.*

Ctenidia, ***comb gills:*** the gills of mollusks. They are paired extensions of the body wall extending into the mantle cavity. In gastropods, primitive bivalves and cephalopods, each ctenidium is a single filament with a comb-like row of branches on each side *(dipectinate* gill or ctenidium). In the bivalves, the C. display an evolutionary sequence, the simplest (protobranchiate) ctenidium being short, simple and dipectinate. In more advanced forms (filibranchiate and lamellibranchiate), the filament has elongated and turned back on itself to form an ascending and a descending limb. The limbs of adjacent filaments are connected by ciliary junctions or by growths of tissue, forming a gill plate. By fusion of the ctenidial axis with the body wall, the ctenidium becomes monopectinate (a single row of comblike branches). Some specialized gastropods *(Monotocardia,* e.g. periwinkle and whelk) have lost the right ctenidium (the primitive left ctenidium before torsion), while in others both C. may be of equal size or the right ctenidium may be smaller.

Ctenidium: see *Cephalopoda.*

Ctenodactylidae, ***gundis:*** a family of Rodents (see) consisting of 5 species in 4 genera. They have short legs, short tails, flat ears, big eyes and long whiskers, and they inhabit rock outcrops in desert and semidesert areas in various parts of Africa (see below). In some species the tail is fan-shaped and used as a balancing organ. They do not build nests, and they shelter in rock piles and rock fissures. Unlike many small desert mammals, they are active during the day. The earliest known and most primitive fossil rodents, from the lower Eocene of China, resemble ctenodactylids. The fossil record of the true ctenodactylids begins in the Oligocene of Asia. The 2 inner toes of each hindfoot carry rows of stiff bristles (ctenodactylid = "comb toe"), which are used for combing and scratching; presumably their very sharp claws would damage their coat if used for this purpose. All *C.* are herbivores. The incisors lack the hard orange enamel typical of most rodents, and *C.* do not often gnaw their food. Dental formula: 1/1, 0/0, 1–2/1–2, 3/3 = 20–24.

There are 5 species. *Ctenodactylus gundi* ***(North African gundi)*** has a small, whispy tail, and is found in southeastern Morocco, northern Algeria, Tunisia and Libya. *Ctenodactylus vali* ***(Sahara gundi)*** has a small, whispy tail, and is found in southeastern Morocco, northwestern Algeria and Libya. *Massoutiera mzabi* ***(Mzab gundi*** or ***Lataste's gundi)*** has flat, immovable ears and a fan tail, and is found in Algeria, Niger and Chad. *Felovia vae* ***(Felou gundi)*** has a fan tail, and is found in southwestern Mali and Mauritania. *Pectinator spekei* ***(Speke's gundi*** or ***East African gundi)*** has a fan tail (which is also used for social display), and is found in Ethiopia, Somalia and northern Kenya. All species are about the same size, the largest being *Felovia* (head plus body 17–18 cm, tail 2.8–3.2 cm), the smallest *Pectinator* (head plus body 17.2–17.8 cm, tail 5.2–5.6 cm).

Ctenodactylus: see *Ctenodactylidae.*

Ctenodont: a hinge type of the *Bivalvia* (see).

Ctenolabrus: see Wrasses.

Ctenomyidae, ***tuco-tucos:*** a family of rodents (see) consisting of more than 20 species in a single genus *(Ctenomys),* found on wastelands and areas of sparse vegetation from the Altiplano (plateau between the mountain chains of the Andes) as far as Patagonia and Tierra del Fuego. The head is large with sturdy, protruding incisors. Dental formula: 1/1, 0/0, 1/1, 3/3 = 20. The eyes and ears are small, and all digits carry strong digging claws and a hair comb. In general appearance, habits and bodily adaptations for digging, they resemble the North and Central American pocket gophers *(Geomyidae),* but they lack cheek pouches. The diet consists of roots, bulbs and plant stalks.

Ctenomys: see *Ctenomyidae.*

Ctenophora, ***comb jellies:*** the only class of the coelenterate phylum of the same name (*Ctenophora*), containing only about 80 known species, 1 to 10 cm in length. *C.* have a disymmetric structure; the body can be divided into symmetrical halves by two vertical planes, one above the other. Most *C.* have a spherical or pear-like globular shape, but in certain aberrant forms this basic shape may be extremely modified. The lower end (oral pole) contains a transverse oval mouth. The upper pole (aboral pole) carries a gravity-sensitive organ (statocyst). On the surface of the body are 8 meridional rows of comb plates, which consist of strong cilia carried by modified ectodermal cells. These cilia serve for locomotion, and the remaining body surface is otherwise not ciliated. Most forms belong to the order (or subclass) *Tentaculifera,* which possess two long tentacles set in deep pouches on opposite sides of the body. The tentacles have muscular bases, which enable them to be withdrawn and extended. They are armed with "lasso cells" or colloblasts, and nematocysts are absent. Each colloblast consists of a sticky head, carrying at its base a spiral thread wound around a stiff central filament. Prey is caught by the tentacles, becoming trapped by the sticky heads of the colloblasts. There are only two orders (or subclasses), the other being the *Nuda* or *Atentaculata,* which lack tentacles and swallow their prey by greatly enlarging the mouth opening. The *Atentaculat*[...] are represented by a single genus (*Beroe*). The gastri[...] cavity of *C.* consists of a vertical pharynx, the centra[...] stomach, and eight subcostal (meridional) vessels beneath the rows of comb plates. The subcostal vessel[...] and central stomach are linked to one another b[...] branched canals. Between the ectoderm and endoderm is a large quantity of gelatinous support material (mesogl(o)ea), containing cells and muscle strands.

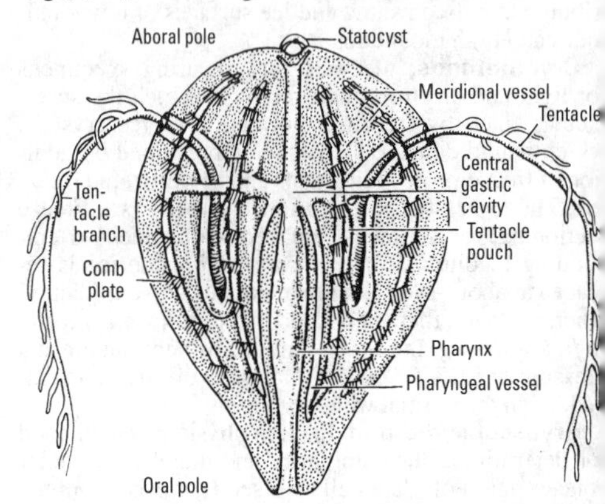

Ctenophora. Body plan of a comb jelly (subclass *Ten*[...]*taculifera).*

C. are hermaphrodite and reproduction is exclusivel[...] sexual. Male and female gonads lie close to each othe[...] in the subcostal canals, and it is assumed that self-fertil[...]ization occurs. Many species are sexually matur[...] shortly after emergence from the egg, when they reproduce for the first time. The gonads then regress, but develop again as the organism grows to full size and sexual maturity and reproduces a second time; this phenomenon is known as *dissogeny.*

All species are marine, swimming free in the water c[...] crawling on the substrate. In calm seas, they appear i[...] spring and summer, usually in swarms near the surface[...] In stormy weather, they are found at deeper levels. Mos[...] *C.* are completely transparent, and they consist of 99[...] water. The largest species, *Cestus Veneris* ***(Venus' girdle),*** from the surface waters of the Mediterranean, i[...] flattened laterally and its body is a 5 cm-wide band, a[...]taining exceptional lengths up to 1.5 m. Many specie[...] are luminescent.

Collection and preservation. Due to their large water content, *C.* easily burst when handled. It is therefo[...] advisable to collect them in containers of sea water,an[...] to narcotize them by addition of magnesium chlorid[...] They can be fixed in a 4% solution of formaldehyde i[...] sea water, then prepared for storage by passage throug[...] increasing concentrations of ethanol from 30% to 70%

Ctenopteryx: see *Cephalopoda.*

Ctenostomata: see *Bryozoa, Ciliophora.*

Cuban boa: a snake of the genus *Epicrates,* closel[...] related to the Rainbow boa (see).

Cubebs: see *Piperaceae.*

Cuboidal epithelium: see Epithelium.

Cubomedusae, ***Carybdeida, sea wasps:*** an order c[...] the *Scyphozoa,* occurring predominantly in warm sea[...] The bells have 4 flattened sides and simple margin[...] With their tentacles they catch fish which may be large an themselves. Some species of the Indo-Australian gion are among the most feared and dangerous marine nimals, the sting from their nematocysts being capable f killing a human within a few minutes. Examples are *arybdea, Chironex* and *Chiropsalmus.*

Cuckoo pint: see *Araceae.*

Cuckoo roller: see *Coraciiformes.*

Cuckoos: see *Cuculiformes.*

Cuckoo spit: see *Homoptera.*

Cuckoo wasps: see *Hymenoptera.*

Cuculidae: see *Cuculiformes.*

Cuculiformes: an order of birds (see *Avies)* with afnities to the parrots. The feet are zygodactylous, but ss prehensile than those of parrots, and with a differt musculature. There is no cere at the nostrils (cf. parts). Sexes are generally alike. The order is divided to 2 families.

1. Family: *Musophagidae* ***(Touracos, Turacos*** or ***lantain eaters;*** 38–63 cm; 19 species). The members this African family are mostly forest birds, feeding ainly on fruit and also taking insects and larvae. They ive soft feathers, long tails, rounded, rather ineffective ings, stubby, often brightly colored bills, and usually prominent mop-like or narrow crest. The outer toe can directed either backward or forward, as in the owls d ospreys. The young are born with wing claws. Toucos are generally more brightly colored than the cuckos, with green and blue bodies and red wing markings, e latter being seen in flight and during the courtship splay, e.g. the ***Knysna touraco*** *(Tauraco corythaix;* 5.5 cm) of South Africa. These colors are not due to ght diffraction, but to pigments unique to this avian mily, i.e. red *turacin* and green *turacoverdin.* Memrs of the genus, *Crinifer,* are, however, gray in color, g. the ***Gray plantain eater*** *(Crinifer zonurus;* 51 cm; rk gray with white wing bar) of East Africa, which is otably unafraid of humans and makes itself conspicus by its noisiness and frenzied activity.

2. Family: *Cuculidae* (127 species in 38 genera). hese are relatively slender, long-tailed species, inhabng mainly forests or scrubland. Compared with tourass, they feed more on animal prey and less on fruit and getable material. Six subfamilies are recognized.

All 47 species of the *Cuculinae* ***(Cuckoos)*** are brood rasites. The subfamily is confined to the Old World. ost temperate species migrate in the winter. One of e most remarkable migrations undertaken by a land rd is that of the ***Shining cuckoo*** *(Chalcites lucidus),* hich flies 2 000 miles (3 200 km) over the ocean from ew Zealand to the Solomon Islands. Most familiar is e ***Common cuckoo*** *(Cuculus canorus;* 33 cm; Eurasia d Africa) (plate 6), which normally lays 4 or 5 eggs ometimes many more), one every 48 hours, each in a fferent nest, but always of the same species. Individl birds produce eggs of different colors, and this apars to determine the chosen dupe, e.g. a cuckoo layg blue eggs usually selects the nests of hedge sparws, which also lay blue eggs. The female cuckoo moves one egg of the prospective foster parent and ys one in its place. The stolen egg may be eaten or opped. The newly hatched, naked cuckoo chick innctively ejects other eggs or nestlings from the nest working its way underneath them then pushing upard until they fall over the edge.

Members of the nonparasitic subfamily, *Phaenicoaeinae* (32 species), represented in Asia, Africa and the New World, differ from the *Cuculinae* mainly in that they are not nest parasites. This subfamily contains the ***Yellow-billed*** and ***Black-billed cuckoos*** that inhabit the woodlands and parks of North America, as well as the handsome Old World, tropical forest-dwelling ***Malcohas*** (or ***Malkohas).***

The 10 species of *Coua* constitute the subfamily *Couinae* ***(Couas),*** which is restricted to the Malagasy region.

The subfamily, *Centropodinae* ***(Coucals),*** contains 27 medium-sized, slow flying, mostly terrestrial cuckoos, all in the genus *Centropus,* represented from Africa to Australia and as far east as the Solomon islands, e.g. the ***Senegal coucal*** *(C. senegalensis),* a familiar bird of the savanna in tropical West Africa.

The subfamily, *Neomorphinae* ***(Ground cuckoos*** and ***Roadrunners;*** 13 species), is largely confined to the Americas, but 2 terrestrial Ground cuckoos are native to Southeast Asia. Three of the South American terrestrial Ground cuckoos are also brood parasites. Most familiar of this subfamily is the long-legged ***Roadrunner*** or ***Chaparral cock*** *(Geococcyx californianus;* 58.5 cm; deserts of southwest USA and Mexico), which feeds mostly on small snakes and lizards; it is a weak flier, but it can run at up to 23 miles (37 km) per hour.

The small subfamily, *Crotophaginae,* contains the ***Guira cuckoo*** (a crested, slender-billed bird of the Argentine pampas) and 3 glossy black ***Anis,*** e.g. the ***Greater ani*** *(Crotophaga major;* 48 cm), which inhabits swamps and mangroves in tropical South America; it has a long tail (28 cm) and a large, black, vertically compressed bill with a prominent dorsal keel. Anis are notoriously gregarious birds, foraging in flocks and nesting in colonies.

Cuculinae: see *Cuculiformes.*

Cuculus: see *Cuculiformes.*

Cucumaria: see *Holothuroidea.*

Cucumber mosaic virus: see Multipartite viruses, Virus groups.

Cucumis: see *Cucurbitaceae.*

Cucumoviruses: see Multipartite viruses, Virus groups.

Cucurbita: see *Cucurbitaceae.*

Cucurbitaceae, *gourd family:* a family of dicotyledonous, mainly tropical plants, comprising 640 species in 110 genera. Most are rapidly growing, climbing herbs with short tendrils and abundant sap. Leaves are alternate, palmately or pinnately lobed (rarely simple, roundish and entire) and exstipulate. Plants are commonly monoecious or dioecious, and the unisexual (usually), actinomorphic, 5-merous flowers are pollinated by insects. The 5 petals of the corolla are usually united, as are the 5 sepals of the calyx. There are typically 5 stamens, which are very variable in structure and more or less united into a column. The inferior ovary consists of 3–5, usually unilocular carpels, with an equal number of usually forked stigmas. Each carpel has between 1 and several anatropous ovules on parietal placentas, often projecting deep into the cavity. Interpretation of the placentation is often complicated by intrusion of the carpel margins. The ovary usually develops into a large, multiseeded, firm-walled, waterrich, berry-like fruit, called a pepo. More rarely, the fruit is a capsule. With the exception of cultured varieties, the fruits have a bitter taste due to the presence of triterpenes.

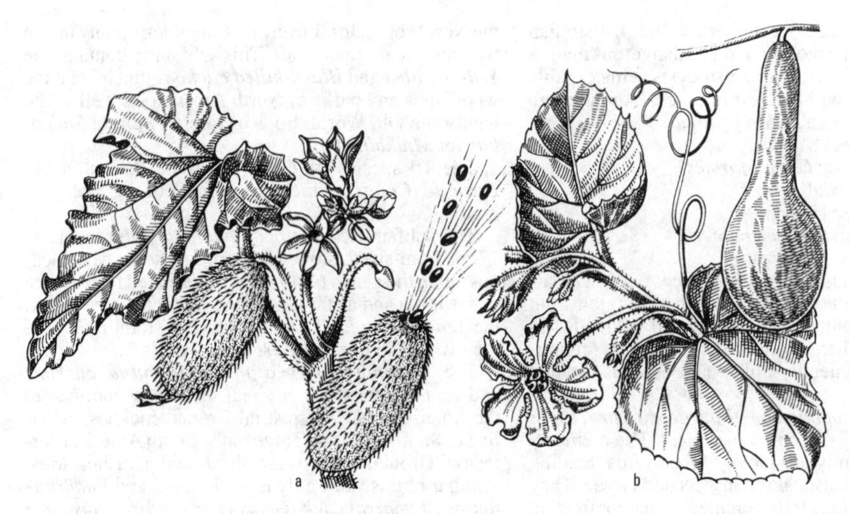

Cucurbitaceae *a* Squirting cucumber *(Ecbal lium elaterium),* branch with flowers and fruits. *b* Calabash or bottle gourd *(Lagenaria siceraria),* flowering branch an fruit.

Pumpkin *(Cucurbita pepo)* and ***Winter squash*** *(Cucurbita maxima),* which originate from Central and South America, have large fruits, and are cultivated as vegetables and ornamentals, and for their oil. Other well known members are the ***Cucumber*** *(Cucumis sativus)* and the ***Musk melon*** *(Cucumis melo),* both from tropical Asia. The network of vascular bundles of the ***Loofah, Vegetable sponge*** or ***Suakwa vegetable sponge*** *(Luffa cylindrica)* form the loofah sponge of commerce. Poisonous European species of ***Bryony*** *(Bryonia)* are grown for pharmaceutical purposes and for ornamentation. Fruits of the ***Squirting cucumber*** *(Ecballium elaterium),* which is found throughout the Mediterranean region, undero explosive dehiscence when the fruit is separated from the plant, ejecting the seeds and surrounding liquid a considerable distance from the parent plant (Fig.). Ripe fruits of the ***Calabash*** or ***Bottle gourd*** *(Lagenaria siceraria)* have a hard, durable wall, and are used in the tropics as vessels, especially as liquid containers (Fig.). Other food plants are the ***Balsam pear*** *(Momordica charantia)* of the Old World tropics, and the familiar, refreshingly juicy ***Water melon*** *(Citrullus lanatus).* A curious xerophytic and dioecious member of the family is *Aconthosicyos horrida* of the Namib Desert. It is a densely branched shrub with photosynthetic spines, and reduced, minute scale leaves. Seeds of the 15 cm-diameter, pale orange fruits are collected (about 30 tonnes annually) by nomadic Hottentots of the Topnaar tribe, and exported to Cape Town as an almond-flavor substitute in the baking industry.

Culicidae: see Mosquitoes.

Culm measures: heterogeneous, predominantly clastic (sand/clay) formations of the marine Lower Carboniferous of Europe and Southeast England.

Culture: see Cell culture, Culture of microorganisms.

Culture biocoenosis: see Culture community.

Culture collection: a collection of live, pure cultures of different species and strains of microorganisms, established and maintained for scientific and industrial purposes. Sloped cultures (see) are stored at low temperatures, sometimes under liquid paraffin to prevent desiccation. Liquid cultures may also be stored, as well as spore samples. Some liquid cultures (usually prepared in very rich media, and sometimes containing high concentrations of glycerol) are stored above liqui nitrogen at -196 °C. Dry lyophilizates are also com monly used (see Freeze drying). Organisms maintaine in liquid media or on agar media must be routinely sub cultured, usually every 2 to 6 months, before they be come nonviable.

Culture community, ***culture biocoenosis:*** an community established or modified from its natur condition by human agency, then maintained by th continuation of human influence. When the human in fluence is removed, a Succession (see) of changes oc curs, leading to reestablishment of the natural commu nity. Categories of C.cs. include Anthropocoenosi (see), Ruderal community (see), Agrocoenosis (see forest (woodlands) and coppice. The greatest human in fluence is evident in anthropocoenoses, whereas som managed forests are largely natural in composition.

Culture-indifferent species: see Hemerophobi species.

Culture media: see Growth media.

Culture of microorganisms: microorganism grown or maintained in or on a Nutrient medium (see The nutrient medium may be prepared as a gel by the in clusion of agar or gelatin (in special cases media ar gelled with silicic acid); such media are referred to a solid media, and the respective cultures are known a agar or gelatin cultures. A liquid culture is grown in o on a liquid medium. Thus, some microorganisms for a coherent layer on the surface of a liquid mediun which is therefore known as a *surface culture,* e.g. th fungus, *Aspergillus niger,* is cultured for the industri production of citric acid as a thick lawn on the surfac of a molasses-containing liquid medium. In a *sub merged culture,* microorganisms multiply within th liquid nutrient medium, which is usually continuall agitated, either by the passage of sterile air, by stirring or by shaking (see Shaking platform).

Cultures on solid media differ according to th method of inoculation. Plate cultures consist of micro organisms growing on the surface of, or within, a thi layer of gelled nutrient medium in a Petri dish (see Pla ing, Streaking, Sloped culture). A *stab culture* is pre pared by stabbing an inculation needle into a deep laye of solid culture medium in a culture tube; the inoculate microorganisms develop along the path of the needle. Stab cultures are used for storing microorganisms, for testing their mobility, and for testing their oxygen requirements.

See also Pure culture, Synchronous culture, Continuous culture, Single cell culture.

Culture slope: see Sloped culture.

Culture tube: a thick-walled test tube without a lip, used in microbiology for culturing and storing microorganisms under sterile conditions. The C.t. is closed with cellulose wadding, cotton wool, heat-resistant plastic foam, a slip-on metal cap, or more rarely a glass cap. Screw-capped bottles are sometimes used instead of culture tubes. A mat area on the upper end of the tube serves for labeling.

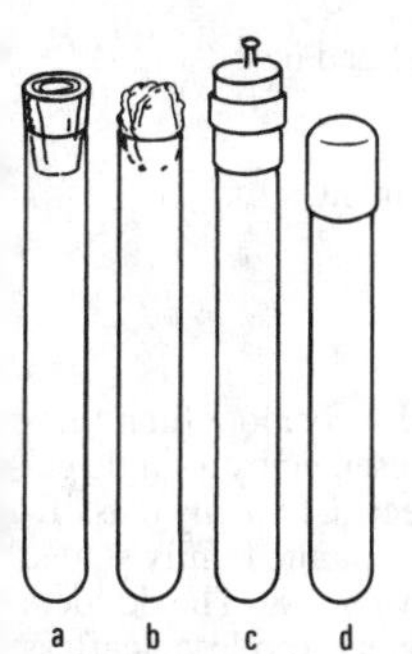

Culture tube. Various types of material are used for closure, and these must be heat-resistant for sterilization purposes. *a* Cellulose wadding. *b* Cotton wool. *c* Slip-on metal cap. *d* Ground glass cap.

Culture yeasts (plate 17): varieties of yeast, obtained mostly by selection and culture methods, which are used for industrial processes, on account, e.g. of their high fermentation rate, low mean generation time, or high protein content. All are derived originally from naturally occurring wild yeasts (class *Ascomycetes).* Wine yeasts (see), Brewer's yeasts (see) and distillery yeasts are used for the industrial production of alcoholic beverages and of ethyl alcohol. Baker's yeasts (see) are used for the preparation of bread and risen pastries, while protein yeasts (see Yeast protein) are used for production of edible protein.

Cumacea, ***cumacids:*** an order of the *Crustacea,* subclass *Malacostraca.* They have a very distinctive body shape, in which the head and thorax are greatly enlarged, and the abdomen is long and thin, terminating in slender, elongated uropods. The short carapace may unite with up to 6 thoracic segments, but it is always united with at least the first 3; it is expanded laterally on each side into a branchial chamber. The carapace also possesses 2 anterior-lateral extensions (lappets) that come together in front of the animal to form a ventral false rostrum (pseudorostrum) [In *Eudorella* and *Eudorellopis* the lappets extend upward and the pseudorostrum is absent]. The anterior 3 pairs of thoracopods are modified as maxillipeds. A series of elaborately lobed filamentous gills are located on the epipodites of the first pair of thoracopods; the latter also possess expanded exopodites, which are housed under the false rostrum. Together with the lappets of the pseudorostrum, the first thoracic exopodites form a valved exhalant siphon, an adaptation that accommodates the animal to its buried position in the bottom mud. Eggs develop in a ventral brood pouch (marsupium) of the female. Most of the 1 000 (approx.) recognized species live on the sea bottom, but they may also rise to higher levels or to the surface, especially at night and when they swarm for breeding. Population densities can be high, e.g. the bottom mud of European coastal waters may contain more than 1 200 individuals per m^2 of the filter-feeding *Diastylis rathkei* (Fig.). Most species are between 0.5 and 1 mm in length, but the largest species *(Diastylis goodsiri)* attains a length of 3.5 cm.

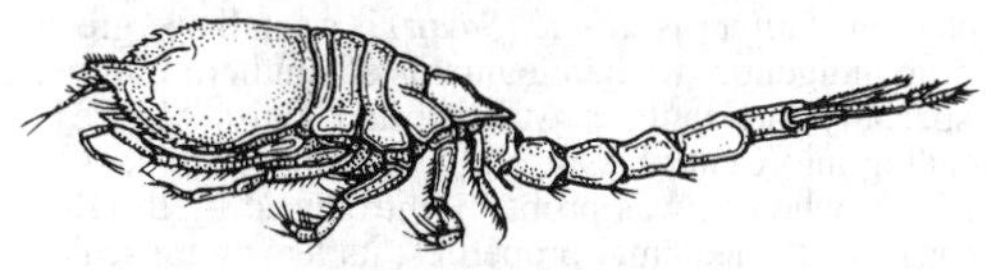

Cumacea. Diastylis rathkei (female).

Cumin: see *Umbelliferae.*

Cumulus oophorus: see Oogenesis.

Cuneiform cartilages: see Larynx.

Cuneo of horse's hoof: see Claws, hoofs and nails.

Cuon: see *Canidae.*

Cupressaceae, ***cypress family***: a cosmopolitan family of the *Coniferae,* containing about 130 species in 15 genera. Members are evergreen trees or shrubs. Young plants have needle-like leaves, but in most species the adult plant has scale-like leaves appressed to and hiding the surface of the stem (see plate 35, No.7). Flowers are monoecious, sometimes dioecious, and arranged in small cones of opposite or whorled scales. Male cones contain 4–8 whorls of microsporophylls, each with 3–5 pollen sacs on the undersurface. Female cones have woody (rarely fleshy when ripe) scales, bearing 2 to many erect ovules on the upper surface. *Juniperus*, a genus of the Northern Hemisphere, differs from other genera in that it forms fleshy, berry-like cones: as the seeds mature, the 3 uppermost scale leaves enlarge, forming a fleshy structure with a distinct tri-radiate scar at its apex indicating the sutures of the 3 bracts enclosing the seeds.

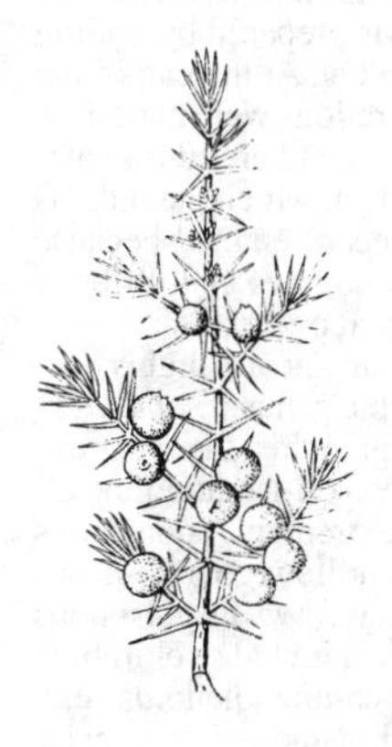

Cupressaceae. Juniperus communis (common juniper). Shoot of female plant with female flowers and one- and two-year old berry-like cones.

Juniperus communis ***(Common juniper)*** grows characteristically on pastures and dwarf shrub heaths; it has needle-like leaves in whorls of 3, which persist in the adult plant; its dark purplish juniper "berries" (about 8 mm diameter) contain about 1% essential oil, which imparts a distinctive flavor to the alcoholic drink, gin. The essential oil has also been used medicinally as a diuretic, stomachic and stimulant. After distillation of the oil, an aqueous extract of the residue is evaporated to give a soft, sweet preparation known as Succus juniperi or Roob juniperi. Above the Alpine forest limit and in the Arctic, the erect and spreading *J. communis* ssp. *communis* is replaced by the procumbent subspecies, *J. communis* ssp. *nana*, which is also found on moors and mountains in the British Isles, and whose leaves are only slightly spiny with a tendency to be loosely appressed to the stem. *Juniperus sabina* ***(Savin)*** is a small evergreen shrub indigenous to the mountains of southern Europe, especially the southern Austrian and Swiss Alps, extending into Central Asia, and frequently cultivated in Britain, where it was probably introduced by the Romans for its medicinal properties; its leaves are scale-like and appressed to the stem. *Cupressus sempervirens* ***(Mediterranean cypress)*** is indigenous to the mediterranean region, where it grows to a height of 25–50 m. Native to southern California, *Cupressus macrocarpa* ***(Monterey cypress)***, grows to a height of 25 m, usually with a pyramidal shape, but sometimes becoming flat-topped, and it is commonly planted in the Southwest British Isles as an ornamental windbreak or hedge plant. The genus *Thuja* ***(Arbor-vitae)*** contains 6 species, native in North America and East Asia, all with scaly bark and a narrow, pyramidal habit. *Thuja plicata* ***(Western red cedar, Giant Arbor-vitae)***, which is native from Alaska to northern California and Montana, grows to a height of 60 m with a buttress at the base of the trunk. Several species of *Thuja* are grown as garden ornamentals. The 6 species of *Chamaecyparis* from North America, Japan and Formosa, also display a narrow pyrimidal habit, and are often planted for decorative purposes, e.g. *Chamaecyparis lawsoniana* ***(Lawson's cypress)***, a native of Southwest Oregon and Northwest California, which attains a height of 60 m (see plate 35, No.7).

Cupressus: see *Cupressaceae.*

Cupule: see *Fagaceae.*

Cupuliferae: see *Fagaceae.*

Curare: a collective term for several poisonous plant extracts used by South American Indians as arrow and bait poisons. The crude material is prepared by boiling down an aqueous extract of the plant. At the end of the 19th century, these various preparations were classified by Boehm as pot, tube (or bamboo) and gourd (or calabash) Cs., according to the vessels in which the indians stored them. These terms have been retained because the poisons come from different sources and differ in their chemical and pharmacological properties.

Pot and *tube* C. are rather similar, and not highly poisonous. Obtained from plants of the genus *Chondodendron,* they contain isoquinoline alkaloids, the most important being the quaternary bisbenzyl-isoquinoline alkaloid, *tubocurarine.* The extremely poisonous *calabash* C., which is packed in hollow gourds, is obtained from plants of the genus *Strychnos.* Its poisonous constituents (all indole derivatives) include yohimbine alkaloids (e.g. mavacurine), strychnine alkaloids (e.g. Wieland-Gumlich aldehyde) and bisindoles (e.g. calabash toxiferin-I).

In current scientific usage, C. refers only to those C. alkaloids with muscle-relaxing (paralysing) effects, namely the dimeric indole alkaloids with a strychnine skeleton and two quaternary nitrogen atoms. Calabash alkaloids E and G are among the most toxic compounds known, causing death by respiratory paralysis.

By displacing Acetylcholine (see), C. alkaloids interrupt nervous impulses at the end plates of motor nerves, leading to paralysis of voluntary muscle. Because the alkaloids are absorbed very slowly from the intestine, animals killed by arrow poisons can be eaten with impunity. For clinical use, natural preparations are now replaced by the pure alkaloids or synthetic or semisynthetic analogs (e.g. alloferin). They are used as Muscle relaxants (see) in surgical operations, severe tetanus and nervous muscle cramps.

Curassows: see *Cracidae.*

Curculionidae: see Weevils.

Curcuma: see *Zingiberaceae.*

Curiosity behavior: a term applied to Exploratory behavior (see) in humans.

Curlews: see *Scolopacidae.*

Curl-tail lizard: see Crested keeled lizards.

Currants: see *Vitaceae.*

Currentia: see Movement.

Cursorial animals: see Movement.

Cursoriinae: see *Glareolidae.*

Cururo: see *Octodontidae.*

Cuscus: see *Phalangeridae.*

Cuscuta: see *Cuscutaceae.*

Cuscutaceae, ***dodder family:*** a cosmopolitan family of dicotyledonous plants, represented by a single genus *(Cuscuta)* with about 170 species. In many classifications, *Cuscuta* is not accorded separate family status, but incorporated into the *Convolvulaceae.* The dodders are exclusively chlorophyll-deficient, rootless, leafless (small scale-like leaf vestiges are present) parasites, which wind their threadlike stems around the host plant, extracting nutrients with the aid of suckers (haustoria). The hermaphrodite, actinomorphic, 4–5-merous flowers usually occur in stalked or sessile ebracteate clusters, and are pollinated by insects. Between 4 and 5 stamens are inserted on the corolla tube. The 2-celled ovary develops into a capsule or berry-like fruit. Many agricultural plants serve as hosts, in particular hops, various members of the *Papilionaceae,* flax and willow. The common *C. europaea* ***(Large dodder)*** attacks hops and a wide variety of other plant species.

Cutch: see *Mimosaceae.*

Cuticle: a more or less structured external layer secreted by the epidermal cells of animals and aerial plant organs (see Boundary tissue), which is relatively impermeable to water and gases, and which serves as a protection against evaporation. Different animal phyla have differently structured Cs. In some coelenterates it is an elastic, chitinous envelope (periderm), whereas in many platyhelminths and aschelminths it is an elastic, nonchitinous, proteinaceous layer. In most mollusks and brachiopods it takes the form of a calcareous exoskeleton, and in the arthropods it serves as a chitinous exoskeleton. Cuticular structures in vertebrates are the oolemma (zona pellucida), the lens capsule and tooth enamel. In arthropods the C. may form an extremely thin layer (e.g. terminal branches of tracheae), or a thick, hard protective layer. A typical arthropod cuticle consists of the following 3 layers. 1) The outermost *epicuticle* or *boundary lamella* is only a few mm thick; it consists of a lower layer of cuticulin (lipoprotein), a middle waxy layer and an outer layer of cement material; chitin is absent. 2) The *exocuticle, primary C.,* or *pigment layer* is a tough, chitinous, amber-colored to black layer, which accounts for between 1/12 and nearly one half of the total thickness of the C. 3) The *endocuticle, secondary C.,* or *inner layer* is chitinous, elastic and colorless, and always accounts for more than 50% of the total C. thickness.

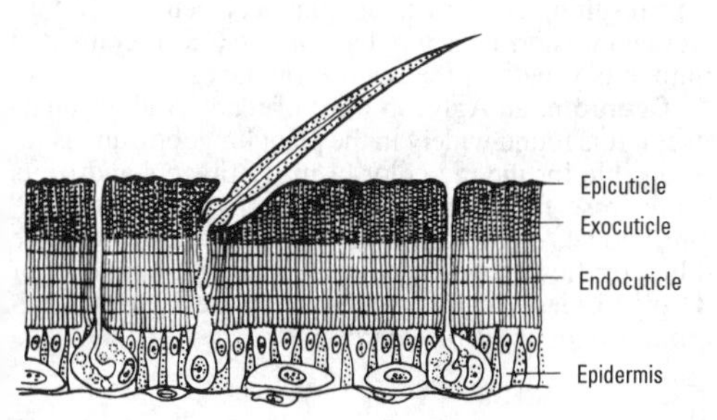

Cuticle. Transverse section of an insect cuticle.

Cutin: a waxy substance consisting of the esters of saturated and unsaturated fatty acids. It is the main component of plant cuticle and is responsible for the waterproof properties of the latter. Pure C. is prepared from *Agave americana.*

Cutis: see Skin.

Cutlassfishes: see *Scombroidei.*

Cutleriales: see *Phaeophyceae.*

Cutting: see *Adventitious shoots.*

Cuttlefish: see *Cephalopoda.*

Cuttlefish bone, ***os sepioe:*** the inner dorsal shell of the cephalopod genus *Sepia.* It is used as a very fine polishing material for metal, wood, enamel and varnish coatings and veneers, and is also used as an ingredient of dentrifices. Large quantities are washed up on Mediterranean shores, and it is also obtained from captured animals. Formerly, it had limited medicinal use against sprue and dysentery. Its chief constituents are calcium carbonate (80–85%), calcium phosphate, sodium chloride, and 10–15% organic material.

Cuttlefish fungus: see *Basidiomycetes (Gasteromycetales).*

Cutworms: the naked larvae or caterpillars of many species of noctuid moths (see *Noctuidae).* Many are very destructive agricultural or forestry pests. Numbers vary greatly from year to year, sometimes attaining epidemic proportions. Most species remain concealed in the earth during the day. Their destructive feeding may occur above ground at night, or below ground at any time. Solitary surface feeders attack the stems of plants more or less at ground level. They eat enough of the stem to cause the plant to fall, and no more, so that their potential for crop destruction is very high. The ***Dark sword-grass moth*** and its larva the ***Black cutworm*** *(Agrotis ypsilon),* named for the Y-shaped marking on each forewing of the imago, is found in North, Central and South America, Europe, Asia, Australia and Hawaii, and is one of the most destructive species of its genus in the eastern USA. It attacks a wide variety of crops including cotton, tobacco, grape vines and brassicas. Other solitary surface feeders are e.g. the ***Bronzed cutworm*** *(Nephelodes emmedonia* Cramer), and ***Dingy cutworm*** *(Feltia subgothica* Haworth).

Some cutworms climb the stems of herbaceous plants and even trees, eating the buds, leaves and fruits. Examples are the cosmopolitan ***Pearly underwing moth*** and its larva the ***Variegated cutworm*** *(Peridroma caucia* Hübner = *margaritosa* Haworth); the ***Spotted cutworm*** *(Amathes = Agrotis c-nigrum* Linné); the ***Southern armyworm*** *(Prodenia eridania* Cramer), whose caterpillars (gray to black with yellow stripes) attack vegetable crops in the southern USA; and the ***Beet armyworm*** *(Spodoptera = Laphygma exigua* Hübner), which favors beets but also attacks many other plants. These cutworms also sometimes appear in large numbers and behave as army worms (see below). The ***White-line dart*** *(Euxoa tritici)* is one of the most variable European moths, with whitish hindwings and forewings ranging from pale yellow-brown through red-brown or gray to deep brown or almost black. Its yellow to red-brown caterpillars, which resemble those of *Agrostis vestigialis,* feed on various small herbaceous plants *(Stellaria, Spergula, Cerastrum, Galium),* and they sometimes attack wheat and buckwheat *(Fagopyrum esculentum).*

Army cutworms are notorious for their appearance in massive numbers, advancing from one field to the next by crawling along the ground, and destroying all crops in their path. They prefer the tops of plants, but sometimes consume entire plants. One notorious example is the ***Armyworm*** *(Leucania = Pseudaletia = Cirphus unipuncta),* which during its worse epidemics is capable of destroying most of the vegetation over several hundred square kilometers. The up to 5 cm-long, white-striped, dark green caterpillars attack all graminaceous crops, as well as many other plants. It is almost cosmopolitan in distribution and a particularly serious pest in North America east of the Rockies. Another well known army cutworm is the ***Fall armyworm*** *(Laphygma frugiperda* Smith), which becomes a serious pest when it adopts the army habit, attacking maize, sorghum, alfalfa, peanuts, tobacco, cotton and other crops. It is common in the southern states of USA (sometimes migrating as far north as Michigan), in Central and South America, and on some of the Caribbean islands. The ***Army cutworm*** *(Chorizagrotis auxiliaris* Grote), found in western North America, is adapted to dry conditions. It is a typical surface feeder, burrowing little if at all, which periodically undergoes a massive population increase and adopts the army habit. It is claimed that it destroyed 100 000 acres of winter wheat in one year in Montana.

Some cutworms remain permanently in the soil, feeding on roots and the underground parts of stems, e.g. the ***Pale western cutworm*** *(Agrotis orthogonia* Morrison), which attacks graminaceous crops, beets, alfalfa, etc., causing several million dollars worth of agri-

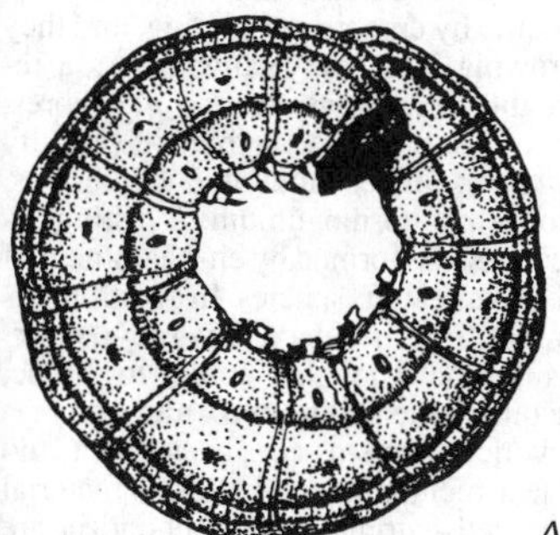

A typical cutworm

cultural damage in western North America; the ***Glassy cutworm*** *(Crymodes devestator),* also strictly subterranean, and widespread in North America except for the most southerly states; and the ***Cotton cutworm*** or ***Yellow-striped armyworm*** *(Prodenia ornithogalli* Guenée), a day-feeding species very destructive to cotton. The ***Turnip moth*** or ***Turnip dart*** *(Agrotis = Scotia segetum)* is somewhat smaller than, but otherwise very similar in appearance to *Agrotis ypsilon*. In Europe and Asia, its subterranean caterpillars bore into and excavate large cavities in the roots of young winter turnips, swedes, carrots and beets, while in Africa it causes damage to the bases of young coffee bushes.

The ***Archer's dart*** *(Agrotis vestigialis* Hfn.) is a European noctuid moth whose large caterpillars (cutworms) are found in light sandy soils, where they feed on the roots of several bedstraws *(Galium)* and a number of common grasses during the day, and on the aerial parts at night. The caterpillars, which remain dormant in the ground during the winter, sometimes also cause damage in young pine nurseries.

Cuvier, ducts of: short veins, present in lower vertebrates, which open into the sinus venosus. They are formed by union of the anterior and posterior cardinal veins.

Cuvierian organs, ***slime tubes:*** glandular tubes extending from the cloaca of holothurians.

Cuvier's principle: the principle, first recognized by Georges L. C. F. D. Cuvier (1769–1832), that the parts of an animal are not only adapted to the environment, but also to each other. This is also known as the "correlation theory". Thus, the natural classification of animals should reflect this correlation of their parts, and it should be possible to predict the nature of various structures of the body from the examination of others. With the aid of this principle, he was able to deduce the structures of incomplete fossils. For example, from the jaw and dentition of an incompletely exposed skeleton of a marsupial, he predicted the presence of marsupial bones in the pelvis, and was susequently proved correct. Cuvier was one of the founders of paleontology and comparative anatomy. Solely on the basis of comparative anatomy, he removed the hyrax from the rodents and placed it with the elephants.

Cyamidae: see *Amphipoda.*

Cyanide: the acid radicle or anion of hydrocyanic acid (old name: prussic acid), which is formed when hydrogen cyanide (HCN) is dissolved in water. HCN has the characteristic odor of bitter almonds. It occurs naturally in the widely distributed cyanogenic glycosides of plants [α-hydroxynitriles (cyanohydrins), in which the hydroxyl group is glycosylated]. HCN is generated from cyanogenic glycosides by the action of β-glycosidases (e.g. emulsin) and oxinitrilases. Examples of cyanogenic glycosides are linamarin, lotaustralin, prunasin, Amygdalin (see) and dhurrin. Naturally occurring nitriles (R-C≡N) are rare, although the nitrile group does occur in the alkaloid, ricinine, and in the rare amino acid, β-cyanoalanine. HCN can be assimilated by plants by combination with L-serine to form β-cyanoalanine; the latter is then hydrated to L-asparagine.

The cyanide ion (CN^-) is a powerful and quick-acting toxin. The average lethal oral dose of HCN for humans is 60–90 mg (equivalent to 200 mg KCN). CN^- blocks electron transport and ATP synthesis by combining with cytochrome a_3 (cytochrome oxidase) of the respiratory chain. CN^- does not react with hemoglobin. It does, however, combine with methemoglobin, which can be used as an antidote to cyanide poisoning, since its affinity for CN^- is higher than that for cytochrome oxidase. The resulting cyanomethemoglobin is then slowly converted into normal hemoglobin as the CN^- is converted into thiocyanate by the enzyme rhodanese.

Cyanidin: an Aglycon (see) of many Anthocyanins (see). It is found widely in the plant kingdom, and is responsible for the red color of many flowers and fruits (tulip, rose, field poppy). More than 20 glycosides of C. are known, e.g. *cyanin* (3,5-di-β-glucoside) from the blue cornflower, violet and other flowers. Complexes of C. glycosides with iron or aluminium are deep blue, e.g. *protocyanin* of the blue cornflower.

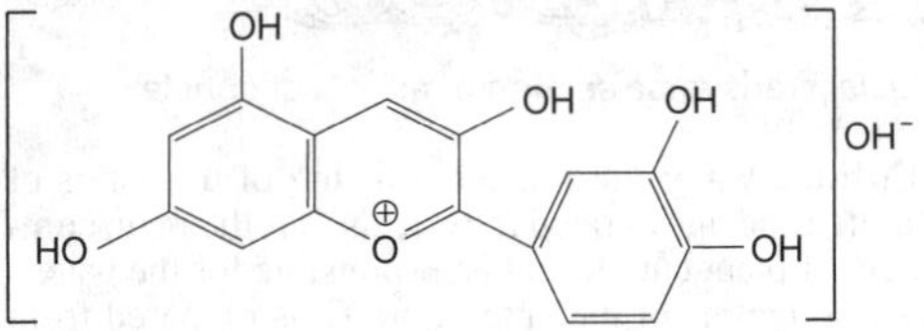

Cyanidin

Cyanobacteria: see *Cyanophyceae.*

Cyanochen: see *Anseriformes.*

Cyanchoideae: see *Asclepiadaceae.*

Cyanocitta: see *Corvidae.*

Cyanogenic glycosides: see Cyanide.

Cyanolab: see Color vision.

Cyanophyceae, *Schizophyceae, Mixophyceae,* ***cyanobacteria, blue-green bacteria:*** a division of the prokaryotes, containing about 2 000 species. They are autotrophic microorgansims, which use light energy to photolyse water (with release of oxygen) and use the resulting chemical energy and reducing power to assimilate carbon dioxide (i.e. photosynthetic assimilation of carbon dioxide energetically similar to that in higher green plants). Glycogen-like cyanophyte starch is deposited in the cytoplasm as cyanophycin granules. *C.* are unicellular, or they form cell associations or filaments. Sexual reproduction is unknown, and flagella have not been observed. Some filamentous forms are motile by gliding or creeping on a moist surface. Reproduction is always vegetative by cell division. Most unicellular forms divide into two by constriction. In a few unicellular species, the parent cell enlarges and the contents divide successively to form numerous naked endospores, which then develop into new individuals with cells walls. Some filamentous species release short, motile pieces *(hormogonia),* consisting of a few cells; hormogonia are motile by creeping or gliding, and they are capable of growing into new filaments. As a response to unfavorable conditions, resting exospores, known as *akinetes* or cysts, are also formed (especially in the *Hormogonales);* under favorable conditions, these germinate to form a hormogonium. A resting organ *(hormocyst)* can also be formed by ensheathment of a lateral branch of a filamentous species. In addition, filamentous forms may also produce *heterocysts,* i.e. thick-walled cells that have lost the ability to divide, and which have acquired the ability to fix nitrogen.

Like all prokaryotic organisms (i.e. bacteria), *C.* do not possess a nuclear membrane, and nuclear material in dispersed in the cell cytoplasm. Mitochondria are also absent. Light-absorbing pigments, chlorophyll *a,* carotenoids and phycobilins (chromoproteins, in which the prosthetic groups are blue phycocyanin or red phycoerythrin) are located on vesicles, or more commonly on stacked membrane systems. Cells are enclosed by a rigid, multilayered cell wall, whose inner layer consists of murein (peptidoglycan). Externally, there is often an additional gelatinous or mucilaginous envelope.

C. are found mostly in fresh water and in wet soils, but they also occur in the sea, on tree bark and on rock surfaces. The mass development of aquatic forms gives rise to water "blooms". *C.* have an ecological role as nitrogen-fixing organisms, and they are especially important for the nitrogen economy of rice paddies. Some *C.* live symbiotically with fungi (see Lichens).

C. are classified into two major groups: the unicellular *Chroococcales* (see) and the filamentous *Hormogonales* (see).

C. are also often called "blue-green algae". This term should be avoided, since algae are eukaryotic organisms, and in no way related to the *C.*

Cyathium: see *Euphorbiaceae,* Flower (section 5, Inflorescence).

Cyathocrinites, *Cyathocrinus* (Greek *kyathos* ladle or bucket, *krinos* lily): a genus of the order *Crinoidea* (sea lilies). Species of *C.* are stalked, with a rounded to cup-shaped calyx, which possesses a dicyclic base. The long, delicate arms, which are composed of elongate, cylindrical joints, give off numerous branches, most of which divide again. The column (stalk) is round in cross section. Fossils are distributed from the Silurian to the Carboniferous of Europe and North America.

Cyathocrinus: see *Cyathocrinites.*

Cybrid: hybrid cell produced by fusion of the protoplast of an enucleated cell with the protoplast of a normal cell (see Cell fusion). It therefore consists of the genome of one parent cell and the cytoplasm of both parent cells.

Cycadatae, ***cycads:*** a class of pinnate-leaved gymnosperms, represented by numerous fossil forms. Today, only 10 genera survive, each in turn represented by only a few very disjunctly distributed species in the tropics and subtropics. In habit they resemble palms. The stem is usually unbranched, bearing a terminal tuft of large, spirally arranged, pinnate leaves. Cycads are dioecious. Male flowers are usually cone-like, with stamens arranged spirally on an axis, carrying numerous groups of pollen sacs on their undersides. The terminal portion of each stamen is sterile. Among the cycads, the most primitive female reproductive organs are those of *Cycas,* in which the growth terminus periodically produces carpels instead of leaves. It is obvious that these yellow-brown, densely haired carpels are homologous with leaves, since their terminal portion is rather large, sterile, and often pinnate. The basal part carries several ovules. In the evolution of the cycads, the sterile terminal portion on the carpel has been suppressed, and the number of ovules has been continually reduced, culminating in more advanced species possessing only simple peltate carpels with only 2 ovules. The carpels are arranged spirally on an extended axis, forming a cone. Fertilization is effected by remarkably large spermatozoids, which have a spiral band of flagella and are freely motile; with a diameter of 0.3 mm, these spermatozoids are the largest in the plant or animal kingdoms. Fossil fragments of plants attributed to the cycads are known from various epochs of the Mesozoic.

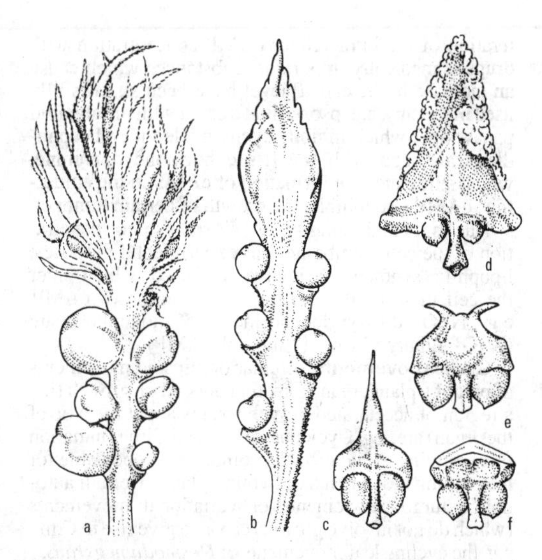

Cycadatae. Carpels (megasporophylls) of cycads, showing the reduction in the number of ovules. *a Cycas revoluta. b Cycas circinalis. c Macrozamia* sp. *d Dioon edule. e Ceratozamia mexicana. f Zamia skinneri.*

The most important living genera are *Cycas* (Madagascar to East Africa), *Encephalartos* (Africa), *Lepidozamia, Macrozamia* and *Bowenia* (all 3 from Australia), and *Dion, Microcycas, Ceratozamia* and *Zamia* (all 4 from the Americas). Sago is prepared from the starchy pith of some species. The leaves of *Cycas* are used for making wreaths, and they are favored in southern Europe as "palm leaves" in religious festivals.

Cycadeoidea: see *Bennettitatae.*

Cycadofilicales: see *Lyginopteridatae.*

Cycadophytina: see Gymnosperms.

Cycads: see *Cycadatae.*

Cyclamen: see *Primulaceae.*

Cyclamen mite: see Strawberry crown mite.

Cyclic adenosine 3′,5′-monophosphate, ***3′,5′-AMP, cyclo-AMP, cAMP:*** a cyclic adenosine monophosphate generated from ATP by the action of Adenylate cyclase (see). Hydrolysis of cAMP to AMP is catalysed by 3′,5′-cyclic nucleotide phosphodiesterase, which is specific for cyclic nucleotides. The intracellular concentration of cAMP is determined by the relative activities of these two enzymes. In many cells, adenylate cyclase is located just inside the plasma membrane, while the receptors for hormones and other chemical signals are on the outside of the plasma membrane. Binding of a hormone or other activator to its receptor leads to an activation of the adenylate cyclase and an increase in the cAMP concentration. The cAMP often serves as an effector molecule, which increases the activity of a protein kinase (or other enzyme) that in turn regulates a cellular process by covalent modification (often phosphorylation) of another enzyme. In this way, cAMP serves as a "second messenger" for a variety of hormones. In addition, cAMP affects the production and release of hormones, e.g. acetylcholine, glucagon, insulin, melanotropin, parathyrin, vasopressin and corticotropin. In many cases the physiological effects of cAMP are only seen in the presence of calcium ions. Alteration of the intracellular cAMP concentration with drugs is medically important. Substances which cause an increase in the cAMP level have been successfully used in therapy, e.g. psoriasis is treated with the alkaloid papaverine, which inhibits cyclic nucleotide phosphodiesterase, and with the tissue hormone dopamine, which stimulates the formation of cAMP in the epidermis. cAMP also inhibits the growth of certain tumors.

Due to its high polarity, cAMP shows poor penetration of the cell membrane. For experimental purposes, lipophilic synthetic derivatives are used, which enter the cell more easily and mimic the action of cAMP, e.g. $N^6,O^{2'}$-dibutyryladenosine 3′,5′-monophosphate ($N^6,O^{2'}$-dibutyryl-cAMP, abb. DBcAMP).

Cyclic movements: circular or elipsoid movements of part of a plant organ. C.m. that depend on growth (i.e. a region of accelerated growth rotates about the axis of the organ) are, e.g. Cyclonasty (see) and Circumnutation (see). Similar cyclic growth promotion is reponsible for the twining movements of twining plants. Certain autonomic turgor movements or variational movements (which do not involve growth) can also give rise to C.m., e.g. the cycling leaf movements of *Desmodium gyrans.*

Cyclic parthenogenesis: see Anholocyclic parthenogenesis.

Cyclitols: compounds derived from 1,2,3,4,5,6-hexahydroxycyclohexane. They have the same empirical formula ($C_6H_{12}O_6$) as the hexoses, and are biosynthetically related to them, but they have an isocyclic rather than a heterocyclic ring, and the hexane ring has the chair configuration. The hydroxyl groups can lie either in the equatorial or the axial position, which makes possible 8 *cis,trans* isomers, of which 7 (the *meso* forms) are optically inactive. Another isomer pair results from the D- and L-forms of the optically active 8th *cis,trans* isomer. All 9 isomers are found in nature, and Myoinositol (see) is very common. In plants, the most common C. are inositol and scyllitol, and 2 methyl esters of D-inositol (pinitol and quebrachitol).

Cycloma: see *Monoplacophora.*

Cyclonasty: in plants, the conspicuous and regular elliptical or circular cycling movements ("searching movements") of growing shoot tips or tendrils, caused by the constant migration of a rapid growth zone around the plant organ (Fig.), in contrast to the irregular movements of Circumnutation (see). Cycling time varies between 40 min and several hours, according to the plant species. C. is autonomic, i.e. determined by internal factors, but to a certain extent it can be influenced by external factors, especially gravity.

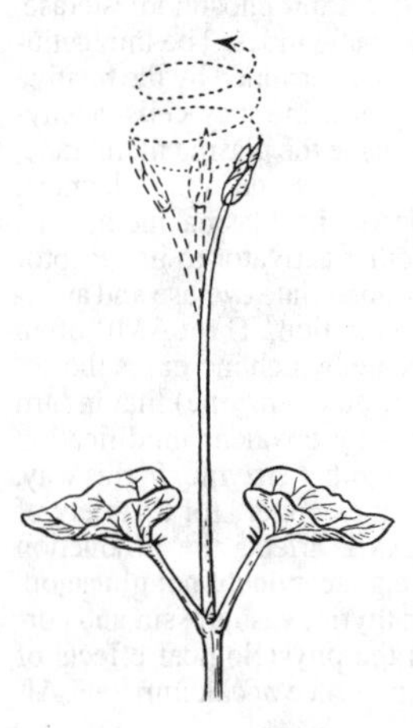

Cyclonasty. Left hand rotation of the growing tip of a bean plant.

Cyclophyllidea: see *Cestoda,* Dog tapeworm, *Taenia.*

Cycloplerus: see *Cyclopteridae.*

Cyclopoida: see *Copepoda.*

Cyclops: see *Copepoda.*

Cyclopteridae, *lumpfishes* and *snailfishes:* a family marine fishes of the order *Scorpaeniformes* (see), containing about 140 species, distributed in the North Atlantic and the Antarctic. Characteristic of the family is a broad, round, ventral suctorial disk formed by fusion of the pelvic fins. Two subfamilies are recognized. Members of the *Liparinae* ***(Sea smails*** or ***snailfishes)*** are sometimes placed in a separate family, the *Liparidae;* they have a naked skin, pink to brownish yellow in color, and a few species also possess small spines. The suction disk is sometimes degenerate in deep sea species, when it is replaced by elongated lower pectoral fin rays. All species inhabit the sea bottom down to depths greater than 600 m. Most species are found in the North Pacific. The ***Sea smail*** *(Liparis liparis)* occurs in the North and Baltic Seas. Members of the *Cyclopterinae* are found mainly in cold seas. They have a particularly powerful suction disk, and the skin is covered with bony plates, which may be spiny or warty. Plump with a small tail, they give the impression of being inflated or swollen. The skelton is largely cartilaginous. The roe of the 50 cm-long *Cyclopterus lumpus* ***(Atlantic lumpsucker),*** found on both sides of the North Atlantic, is used as a caviar substitute.

Cyclopterinae: see *Cyclopteridae.*

Cyclopterus: see *Cyclopteridae.*

Cyclorhagida: see *Kinorhyncha*

Cyclostomata:

1) ***Cyclostomes,*** *Marsipobranchii,* ***lampreys and hagfishes:*** a subclass or class, comprising the living representatives of the *Agnatha* (see). They are aquatic, eel-like animals predatory on fishes. The endoskeleton is cartilaginous and rudimentary; the vertebrae are reduced, the skull consists only of a cartilaginous capsule, and a jaw apparatus is lacking. The mouth is round and suctorial, and the buccal cavity has horny (epidermal) teeth. A single nasohypophyseal aperture opens medially to the exterior on the dorsal surface of the head. The heart is composed of an auricle and a ventricle, but there is no bulbus arteriosus. See Lampreys, Hagfishes. For collection and preservation, see Fishes.

2) See *Bryozoa.*

Cyclura: see West Indian rock inguanas.

Cygnus: see *Anseriformes.*

Cylisticus: see *Isopoda.*

Cylindrical skinks: see *Chalcides.*

Cymbopogon: see *Graminae.*

Cyme: see Flower, section 5) Inflorescence.

Cymene, p-*cymene,* p-*methyl-isopropylbenzene:* a pleasant smelling aromatic hydrocarbon, present in various essential oils, e.g. oils of caraway, eucalyptus, and thyme.

Cymose inflorescence: see Flower, section 5) Inflorescence.

Cymothoidae: see *Isopoda.*

Cynanchum: see *Asclepiadaceae.*

Cynara: see *Compositae.*

Cynocephalidae: see Flying lemurs.
Cynocephalus: see Flying lemurs.
Cynomolgus monkey: see Macaques.
Cynipinae: see Gall wasps.
Cynipidae: see Gall wasps.
Cyperaceae, ***sedge family:*** a family of monocotyledonous plants containing about 4 000 species in 90 genera, found in all parts of the world, but the majority occurring in the temperate and cold regions. They are herbaceous, grass-like plants, with solid (i.e. not hollow), often trigonous stems (culms), which lack nodal thickenings. The leaves are narrow and linear and occur in 3 ranks. Leaf sheaths are usually entire. Inflorescences vary from branched panicles to simple spikes, whose spikelets contain 1 to many hermaphrodite or unisexual, wind-pollinated flowers, arising from the axil of a glume. The perianth is strongly reduced, consisting of bristles or scales, or entirely absent. Female flowers are sometimes enclosed by a modified glume (bracteole, perigynium or utricle). There are usually 1–3 stamens. A superior ovary of 2–3 united carpels develops into a trigonous or biconvex achene (except in *Scirpodendron* and allies).

Frequently the epidermal cell walls are heavily silicified, and therefore very hard. Other chemical characteristics are the occurrence of volatile oils, and the presence of various polyphenols together with tannins derived from them.

Three subfamilies are recognized:

1. *Cyperoideae* (flowers usually hermaphrodite in multi-flowered spikelets, or single, unisexual flowers with or without a perianth). The genus, *Cyperus,* is represented by about 600 species in the tropics and warm temperate regions. *Cyperus papyrus* ***(Papyrus)*** from Central Africa was formerly used to manufacture papyrus paper, and it is now sometimes grown in hothouses in temperate regions as an aquatic ornamental. The closely related *C. alternifolius* ***(Umbrella plant)*** is also a popular aquatic ornamental, which can be grown indoors with copious watering. The rhizome of *C. esculentus* ***(Chufa*** or ***Tiger nut)*** is rich in starch and oil, and was formerly important as a food and for its extracted oil, but it is now barely exploited. *Eriophorum* ***(Cotton sedge*** or ***Cotton grass),*** occurs in northern temperate and Arctic regions; its perianth consists of numerous bristles, which elongate and form a cottony tuft after flowering. *Eliocharis* ***(Spike sedge*** or ***Spike rush)*** is a genus of cosmopolitan distribution, characterized by a terminal spike. The perianth of *Scirpus* ***(Bulrush*** and ***Clubrush)*** usually consists of bristles. Many species of *Scirpus* are economically useful. For example, *Scirpus californicus* ssp. *tatora* ***(Totora*** or ***Tatora reed)*** on Lake Titicaca in the Peruvian Andes is a fundamental commodity in the lives of various indian tribes in that area. It serves as a building material for boats, for basket making and weaving, as animal fodder, and to a lesser extent as a human food material. In pre-Columbian times, the coastal peoples of Peru cultivated it in irrigated swamps and used it for making reed boats.

2. *Caricoideae* (flowers unisexual, naked, usually in multi-flowered spikes; female flowers enclosed by a utricle which persists in the fruit). The largest genus is *Carex* ***(Sedges),*** represented by about 1 100 species, cosmopolitan in distribution, but especially numerous in temperate regions. Characteristic of all species of *Carex* is the open-topped tube or utriculus, formed by the bracts of the female flowers, from which the style and stigma project. Of the approximately 90 European species of sedge, *Carex hirta (= pilosa)* ***(Hairy Sedge)*** is very common, occurring on meadows, waysides and embankments.

3. *Rhynchosporoideae* (flowers hermaphrodite or unisexual, with or without perianth; spike-like cymes containing only a few flowers are aggregated into spikes or heads). The genus *Rhynchospora* ***(Beak-sedges)*** is cosmopolitan and especially numerous in the tropics. *R. alba* ***(White beak-sedge)*** is found in wet, usually peaty places with acid soils throughout Europe, UK and Siberia.

Cyperaceae. a Hairy sedge *(Carex hirta). b* Hare's tail cotton grass *(Eriophorum vaginatum). c* Bullrush *(Scirpus = Schoenoplectus lacustris). d* Common spike rush *(Eleocharis palustris). e* Chufa or tiger nut *(Cyperus esculentus)* showing rhizomes.

Cyphonautes: see *Bryozoa.*

Cypraea: see *Gastropoda.*

Cypress: see *Cupressaceae.*

Cyprinidae, ***minnows*** or ***carps:*** a large family of fish (about 2 070 species in about 194 genera), order *Cypriniformes* (see), suborder *Cyprinoidae.* They are nonpredatory, and their mouths lack teeth. Between 1 and 3 rows of pharyngeal teeth are present, with no more than 8 teeth in each row. They are distributed in freshwater in North America, Europe, Asia and Africa. Examples are Carp (see), Chondrostome (see), Barbel (see *Barbinae),* Bream (see), European roach (see), Chub (see), Dace (see), Rudd (see), Minnow (see) and Bleak (see). The term, ***Minnow,*** is applied specifically to certain species (e.g. see Minnow) and it is also applied generally and somewhat indiscriminately to species that are small, especially if they have shining silvery sides. Strictly speaking, the members of the subfamily *Leuciscinae* are the true minnows (e.g. Roach, Rudd, Chub, Dace, Orfe, Bleak, Bream, Tench, etc.). Most members of the *Cyprinidae* are palatable, but bony.

Cypriniformes, ***carps and allies:*** an order of fish containing about 2 422 species in 256 genera and 6 familes, showing its greatest diversity in Southeast Asia They live predominantly in fresh water. Characteristic of most cypriniform fish is a chain of small bones (Weberian ossicles) connecting the anterior end of the swim bladder to a thin patch in the wall of the auditory capsule (which contains the membranous labyrinth). The mouth is usually protractile and always toothless. There is no adipose fin (except in some cobitidids), the head is scaleless, and the teeth on the 5th ceratobranchial are ankylosed to the bone. There are 3 branchiostegal rays. Some species have spinelike rays in the dorsal fin. Many species are fished commercially, and many members, especially minnows and loaches, are popular aquarium fishes. See also *Gymnotiformes* and *Characiformes.*

Cyprinodontiformes: an order of fishes, containing about 845 species in 120 genera, distributed in 3 suborders. All have a single dorsal fin. The second circumorbital bone is absent, and fin spines are rarely present.

1. Suborder: *Exocoetoidei,* comprising the ***Flying fishes*** *(Exocoetidae;* about 48 species in 8 genera), ***Needlefishes*** *(Belonidae;* 32 species in 9 genera), ***Halfbeaks*** *(Hemiramphidae;* about 80 species in 12 genera), and ***Sauries*** *(Scomberesocidae;* 4 species in 4 genera).

2. Suborder: *Adrianichthyoidei,* comprising the ***Medakas*** or ***Ricefishes*** *(Oryziidae;* 7 species in 1 genus), the *Adrianichthyidae* (3 species in 2 genera), and the *Horaichthyidae* (1 species).

3. Suborder: *Cyprinodontoidei* (see).

Cyprinodontoidei, *Microcyprini:* a suborder of the *Cyprinodontiformes* (see). These small freshwater fishes are found in all tropical and subtropical regions, with the exception of Australia. The lateral line is mainly on the head, rather than the body. Six families are recognized, containing about 670 species in 82 genera: *Aplocheilidae* ***(Rivulines),*** *Cyprinodontidae* ***(Killifishes*** or ***Toothcarps),*** *Goodeidae* ***(Goodeids),*** *Anablepidae* ***(Four-eyed fishes)*** (see), *Jenynsiidae* ***(Jenynsiids),*** *Poeciliidae* ***(Livebearers)*** (see Guppy).

Cyprinus: see Carp.

Cypris larva: a larval form of the *Cirripedia,* which arises from the late nauplius, and which possesses many adult features. The C.l. is enclosed in a bivalve shell with an adductor muscle. There are 6 pairs of biramous thoracic appendages, a small abdomen comprising 4 somites, and a pair of compound eyes.

Cyrtoceratites (Greek *kyrtos* bent or bowed, *keras* horn): a fossil nautiloid genus, found from the Silurian to the Devonian. The shell was not coiled, but extended and shallowly curved like a horn (arcuate), and it had an elliptical or circular cross section. The siphuncle lay along the wall of the shell, and it displayed a beaded structure, due to restriction at each septum (known as a cyrtochoanitic siphuncle).

Cyrtograptus: the type genus of the Upper Silurian monograptid Graptolites (see), with a cosmopolitan distribution. The rhabdosome was coiled, with straight side branches. The thecae were markedly hooked proximally, less so distally, and not hooked on the side branches.

Rhabdosome of *Cyrtograptus* (natural size)

Cyst: a structure secreted under certain conditions by certain animals, which envelops the organism completely and protects it from the external environment. Cysts are commonly produced by freshwater and parasitic protozoans, by other primitive animals, and by certain bacteria (genus *Azotobacter).* ***Resistance cysts*** are often formed in response to unfavorable environmental conditions, e.g. by the drying out of inhabited water. ***Resting cysts*** enable the organism to continue its metabolism protected from the external environment, or to conserve its energy by quiescence during starvation. Thus, digestion ***(digestion cyst)*** or cell division ***(multiplication cyst)*** may continue within a resting cyst. Fusion of gametes occurs within ***gamocysts*** (e.g. of the *Gregarinidea).* After gamete fusion, the zygote-containing cyst becomes an ***oocyst;*** further development within the cyst produces small individual organisms, and the structure is then called a ***sporocyst.*** The cysts of parasites serve as agents of transmission to another host. At excystment, the cyst is ruptured and left behind as a dead structure. In certain agriculturally important phytopathogenic aschelminths (in particular species of the genus *Heterodera),* the female forms a spherical resting cyst beneath the outer epidermis of the root. Eggs and later the hatched larvae are all produced within the cyst.

Cystacantha larva: see *Acanthocephala.*

Cysteamine: $H_2N—CH_2—CH_2—SH$, a Biogenic amine (see) produced by decarboxylation of the amino acid, cysteine. It is a structural component of Coenzyme A (see).

L-Cysteine, *Cys:* $HS—CH_2—CH(NH_2)COOH$, a sulfur-containing, proteogenic amino acid. It plays a central role in sulfur metabolism. Through its reversible oxidation to cystine $[HOOC(NH_2)CH—CH_2—S—S—CH_2—CH(NH_2)COOH]$ it functions as an important biological redox system. The SH-groups of Cys residues are often functionally important in the active sites of enzymes, whereas the S-S bonds of oxidized Cys residues (i.e. cystine residues) serve to link linearly distant parts of the polypeptide chain and maintain the tertiary structure of the protein. Cys can be prepared by hydrolysis of proteins; since keratins have a high Cys content, human hair collected from the hairdresser is a cheap and appropriate starting material for the laboratory isolation of Cys. It is biosynthesized from methionine, and the end products of its biological degradation are usually oxidized sulfur compounds like taurine, NH_2—CH_2—CH_2—SO_3H.

Cystic duct: see Gall bladder.

Cysticercoid: see *Cestoda*.

Cysticercosis: see *Taenia*.

Cysticercus: see *Cestoda*, Dog tapeworm.

Cystic fibrosis: see Inborn illnesses.

Cystid: see *Bryozoa*.

Cystidium: see *Basidiomycetes*.

Cystine: see L-cysteine.

Cystoidea, *cystoids:* an extinct class of echinoderms with an ovoid or globular shell (theca) consisting of numerous tightly fitting, jointed, often hexagonal, calcareous plates, which are organized in an irregular or regular pattern. These thecal plates are perforated by pores, which occur either in pairs (diplopores) or in a diamond pattern (pore-rhomb). Most species were permanently attached to the sea bottom by a jointed stalk or column. Fossils show between 2 and 5 ambulacral fields at the upper end of the shell, which were occupied by rudimentary arms. A mouth, anus and genital pore can be identified in the plates at the vertex of the shell. Ciliated grooves were present, either on the surface of the shell (epithecal) or on special ossicles (exothecal). Cystoids are the oldest and least specialized of the echinoderms, and their evolurionary peak occurred in the Ordovician. They are classified into 2 groups: ***Diploporida*** (with diplopores and epithecal grooves, e.g. *Eucystis* and *Pleurocystis* of the Ordovician); and ***Rhombifera*** (with pore-rhombs and exothecal grooves, e.g. *Echinosphaera* of the Ordovician, *Lepadocrinus* of the Silurian).

Fossils are found from the Ordovician to the Upper Devonian; their occurrence in the Cambrian is doubtful.

Cystophora cristata: see *Phocidae*.

Cytisus: see *Papilionaceae*.

Cytochrome oxidae: see Cytochromes.

Cytochromes: a group of heme proteins that serve as redox catalysts in respiration, photosynthesis, and certain oxygenations. They also transfer electrons to terminal electron acceptors other than oxygen (e.g. nitrate) in the anaerobic respiration of certain bacteria. They function as electron donors and acceptors by a reversible valency change in the iron atom at the center of their heme prosthetic group, i.e. two electrons are gained by the conversion of 2Fe(III) to 2Fe(II), and lost again by the conversion of 2Fe(II) to 2Fe(III).

The importance of C. is indicated by the fact that they are very old, having evolved more than 2 billion (2 x 10^9) years ago, during which time their structures have undergone only relatively minor alterations by point mutations. They are found in all organisms. On the basis of the structures of their heme groups and their spectra (especially the α, β and γ bands in the visible spectrum of the reduced C.), the C. fall into three main groups, *a, b* and *c*. All three types are found in the mitochondria of higher plants and animals, where they are essential components of the respiratory chain.

The *cytochrome* a/a_3 complex is identical with *cytochrome oxidase* (Warburg's respiratory enzyme, or "Atmungsferment"). Cytochrome oxidase is the last member of the electron transport chain (respiratory chain) of mitchondria. It normally transfers electrons to molecular oxygen, and it can be inhibited by reaction with cyanide or carbon monoxide. Its prosthetic group is hemin *a* (cytohemin). Copper is also bound to the enzyme, and the transfer of electrons to oxygen involves valency changes in both the heme iron and the copper. It appears to be a single heme protein, rather than a complex of two proteins, as previously thought.

The prosthetic group of *cytochrome b* is the same as that of hemoglobin (i.e. iron(II)-protoporphyrin IX). Of all the C. in the respiratory chain, it has the lowest redox potential, and is accordingly located between ubiquinone and cytochrome *c* in the mitochondrial membrane, where it serves to transfer electrons from the former to the latter. Cytochrome *c* has a redox potential intermediate between those of cytochrome *b* and cytochrome oxidase, and is appropriately situated between the former and the latter in the mitochrondrial membrane.

Cytochrome c is the most widespread and most thoroughly studied of the respiratory C. It is easy to extract and purify from mitochondria, being water soluble and relatively small. Vertebrate cytochrome *c* has M_r 12 400 (104 amino acid residues), while plant cytochrome *c* has M_r 13 100 (111 amino acid residues). In contrast, cytochrome *b* (M_r 60 000) and cytochrome oxidase (M_r 440 000 from heart muscle) are much larger, more tightly embedded in the mitochondrial membrane, and extracted only with difficulty (e.g. by using detergents).

Bacterial species of cytochrome *c* are similar to the eukaryotic protein, e.g. cytochrome c_2 from *Rhodospirillum rubrum* (112 amino acids, M_r 13 500), cytochrome c_5 from *Pseudomonas* (87 amino acids, M_r 10 000). Cytochrome c_{551} (82 amino acids, M_r 9600) is an electron carrier for the nitrite reductase of *Pseudomonas*. Cytochrome c_{553} and c_{555} (also called cytochrome c_6 and cytochrome *f*, respectively) are components of the sulfide reductases of green sulfur bacteria. Cytochrome *f* is also found in the chloroplast membranes of *Euglena gracilis* and in higher plants, where it is part of the photosynthetic electron transport chain, together with cytochrome b_{559}.

Comparison of the amino acid sequences of more than 50 species of cytochrome *c* from phylogenetically distant species shows that the primary structure has been extremely conserved. The configuration of the chains is also similar in all species. In addition to cytochrome *c*, eukaryotes also have a cytochrome c_1, which is larger (M_r 37 000), insoluble, of a different amino acid composition, and has a redox potential intermediate between cytochrome *c* and cytochrome oxidase.

Cytochrome P450 is an oxygenase, which is also a cytochrome *b*. It is named for the unusual absorption maximum at 450 nm of the reduced form when complexed with carbon monoxide. Prokaryotic cytochrome P450s are soluble, whereas the eukaryotic enzymes are membrane-bound within the endoplasmic reticulum (e.g. in liver), or in the inner mitochronrial membrane. In the mitochondria of the mammalian adrenal cortex, various species of cytochrome P450 are responsible for hydroxylations in corticosteroid biosynthesis. In steroidogenic tissues, e.g. adrenals and placenta, cytochrome P450 in the inner mitochondrial membrane is responsible for cholesterol side chain cleavge, the first stage in the conversion of cholesterol into a variety of steroid hormones. Mammalian liver contains a superfamily of cytochrome P450s with overlapping substrate specificity, which is responsible for the oxidative converison of various endogenous metabolites (e.g. in bile acid biosynthesis) and xenobiotics into more polar (and therefore more easily excretable and conjugable) compounds.

Cytoecology: a branch of ecology that studies the relation between a cell of a multicellular organism and its intraorganismal environment.

Cytogenetics: a branch of genetics that studies the relation between the genetic complement of the cell and its structure, function and life history. It combines the methods of cytology and genetics, and many studies are concerned with the effects of mutations.

Cytokinesis: see Mitosis.

Cytokinins: phytohormones, present in all higher plants, where they exert multiple effects on growth and development. They are also found as nonhormonal compounds in animal tissues, bacteria, fungi and algae. In plant tissues, C. are present either in free form, or as part of specific tRNA molecules. Their concentration in plant tissue is extremely low, and it varies according to the plant organ and its stage of development.

The most important natural C. are *cis-* and *trans*-zeatin, zeatin riboside, zeatin ribotide, dehydrozeatin, and isopentenyl adenine [N^6-(Δ^2-isopentenyl)adenine; abb. i^6Ade], i.e. derivatives of adenine with an isoprenoid C_5 side chain on the exocyclic nitrogen atom (N^6). The first compound with cytokinin activity to be discovered was kinetin, which was isolated from hydrolysed herring sperm DNA by Skoog and Miller in 1955. The first cytokinin obtained from a plant was zeatin, which was isolated in 1964 from unripe maize kernels, and subsequently shown to be widely distributed in higher plants. In addition to the basic C., plants also contain the corresponding ribonucleosides (9-ribosides) and ribonucleotides (9-riboside-5'-monophosphates), as well as N-7-, N-9- and O-4-glucosides. Important synthetic compounds with cytokinin activity are kinetin (6-furfurylaminopurine), 6-benzylaminoadenine and SD 8 339 (Fig.), as well as certain derivatives of *N,N*-diphenylurea. Kinetin is used as a model compound in phytohormone research.

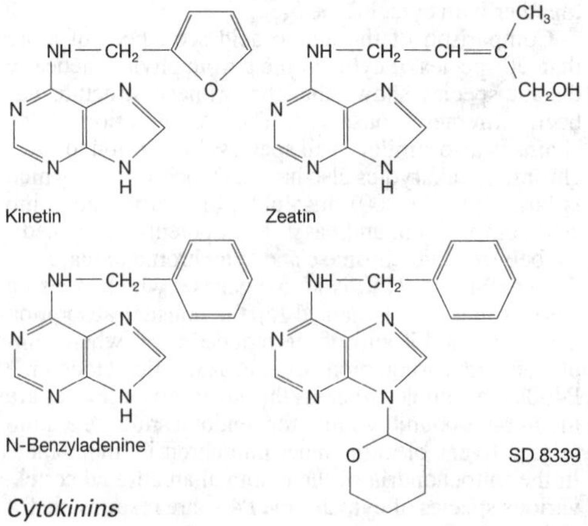

Kinetin

Zeatin

N-Benzyladenine

SD 8339

Cytokinins

Biosynthesis of natural C. is closely linked to RNA metabolism, either proceeding directly from low molecular mass comopounds via purine ring synthesis, or by the release of C. by degradation of cytokinin-containing tRNA. Thus 0.05–0.1% of the purine bases in tRNA have cytokinin activity, e.g. in *Escherichia coli* the tRNA species for phenylalanine, leucine, serine, tyrosine and tryptophan contain various purine residues with cytokinin activity. In higher plants the main sites of cytokinin biosynthesis are the root tips, young fruits, and seeds. C. are transported in the xylem. They have a broad activity spectrum, which overlaps those of other phytohormones. Characteristic responses to C. are, inter alia, stimulation of cell division in callus tissue and intact plants, and the stimulation of extension growth. Biotests for the detection and quantitation of C. activity measure one of the following. 1) stimulation of cell division in tissue culture, e.g. in soybean or tobacco callus tissue; 2) enlargement of cells in the leaf disk test (often performed with bean or radish leaves); 3) inhibition of the loss of chlorophyll from detached leaves; 4) promotion of the germination of seeds in the dark. Interacting closely with other phytohormones (especially auxins and abscisic acid) and certain environmental factors (e.g. light), C. regulate a variety of plant growth and differentiation processes, e.g. they are active in delaying senescence. The action mechanism and molecular site of action of C. are still largely unknown.

Cytology: see Cell biology.

Cytomixis: the movement of usually amorphous beads of chromatin from one cell to another.

Cytophilic antibodies: antibodies that adhere specifically to macrophages. The macrophage cell membrane contains appropriate receptors.

Cytoplasm: that part of the cell plasm of eukaryotic cells that lies outside the nucleus. It consists of Cytosol (see) (which appears amorphous under the electron microscope), the cell organelles embedded in it, and paraplasmatic inclusions. The *hyaloplasm* is that part of the C. which appears amorphous under the light microscope, but it contains numerous structural elements that can be visualized by electron microscopy.

Cytoplasmic ground substance: see Cytosol.

Cytoplasmic inheritance, ***extrachromosomal inheritance, extranuclear inheritance:*** inheritance due to the transfer of genetic material localized outside the nucleus. The demonstration of C.i. is based mainly on 3 criteria: 1) the inherited characters are not segregated according the Mendelian laws; 2) reciprocal crosses produce different offspring, because the cytoplasm is transmitted essentially from the maternal parent; 3) irregular distribution of different cytoplasmic factors in cell division can lead to a demixing of cytoplasmic genes during ontogenesis. The totality of *cytoplasmic genes* is known as the *plasmotype* or *plasmon.* Genes carried in the plastids of an organism are known as the *plaston,* while those localized in the mitochodria are called the *chondriome.* It is still debatable whether certain submicroscopic particles are involved in C.i. All cytoplasmic genes decribed so far replicate autonomously, and are present in the DNA of either mitochondria or plastids. Mitochondrial and plastid DNA are now generally referred to as the mitochondrial or plastid "chromosome".

Cytoplasmon: the totality of cytoplasmic (extranuclear) genetic information of a eukaryotic cell, excluding that in the mitochondria and plastids. See Plastome, Plasmon.

Cytoskeleton: a complex network of protein filaments extending through the cytoplasm of eukaryotic cells. It is directly responsible for the ability of cells to move over a substratum, for muscle contraction, and for the active migration of organelles within the cytoplasm. It enables cells to alter their shape. Changes in shape that occur during embryonic development also depend on the presence of the C. The C. consists of 3 main types of protein filaments: Actin (see), Microtubules (see) and Intermediate filaments (see). Linkage of filaments with each other or with other cell structures is mediated by associated proteins. Other associated proteins control the site and rate of formation of actin filaments, microtubules or intermediate filaments. Muscle contraction and the beating of cilia depend on the interaction of associated proteins with actin and microtubules, respectively. The term, "cytoskeleton", was first applied to the network of intermediate filaments remaining after the extraction of cells with high or low ionic strength salt solutions, or with nonionic detergents, but this definition is now obsolete.

Cytosol, ***cytoplasmic ground substance:*** that part of the cytoplasm which appears amorphous even under the electron microscope. Cell organelles and other inclusions are embedded in the C., which acts as the medium of communication and intracellular transport between all cell components. Many important biochemical reactions occur in the C., which therefore plays an important role in cytoplasmic and cellular dynamics. It exists either as a sol or a gel, and is able to change from one state to the other. The liquid plasma-sol is 2–10 times more viscous than water, whereas the plasma-gel has a highly viscous and elastic structure. Cytosolic viscosity can be determined by special methods, e.g. measurement of the rate of movement of minute steel particles in the C. under the influence of a magnetic field. In the region of the plasma membrane the C. is usually a gel, becoming a sol toward the cell interior.

The C. of plant cells is usually less viscous than that of animal cells. Whereas the shape of plant cells is largely determined by the Cell wall (see), the shape of animal cells is determined by the Cytoskeleton (see), helped by the high viscosity of the cytosolic gel.

After sedimentation of all particulate cell material by ultracentrifugation (e.g. $100\,000 \times g$ for 30 min), the C. remains as the supernatant. It consists of 60–80% water, with proteins (including many enzymes), lipids, glycogen and other storage materials, amino acids, sugars, tRNA, nucleotides, and various inorganic ions. The colloidal properties of C. are due to its macromolecular components, in particular enzymes and other proteins, which exhibit inter- and intramolecular bonding (e.g. hydrogen bonding, ionic interactions). Sol-gel interconversion depends on the breaking and reformation of these bonds. Through these reversible interactions, the macromolecules form a dynamic three-dimensional netwok. The viscosity of the C. is also related to the degree of hydration of its proteins.

Many reactions of intermediary metabolism occur only in the C., e.g. the reactions of glycolysis, as well as several stages in the biosynthesis of proteins and lipids. The enzymes catalysing these reactions serve experimentally as marker enzymes. Thus, the presence of hexokinase activity in a laboratory preparation of mitochondria shows that the mitochondria are contaminated with C.

In addition to being a site of active intermediary metabolism, the C. also contains a pool of materials, especially proteins and lipids, which serve as precursors for the synthesis of membranes, filaments, microtubules, and other structural components of the cell. This pool is maintained by de novo synthesis of its components, and by recycling of the degradation products of cell structures. Examples of pool components are the G-actin of nonmuscle cells, and tubulin dimers (see Microtubules).

Contractile microfilaments play an essential part in cytoplasmic streaming (e.g. in slime molds or amebas) and in the migration of organelles in all cells. These consist of F-actin, which is formed reversibly by the polymerization of G-actin in the presence of ATP (see Actin).

The nucleoplasm (see Nucleus) also contains an amorphous fraction that is structurally and functionally analogous to the extranuclear C.

Cytostatic agents: substances that inhibit cell growth. C.a. are classified as follows, according to their action mechanism and site of action. 1) Mitotic poisons (see). 2) Toxins that attack the resting nucleus; the effect of these agents is not manifested until the commencement of cell division. 3) Antimetabolites of nucleic acid metabolism. 4) Cytotoxic agents that generally act at any stage of cell metabolism, excluding nucleic acid metabolism.

Cytotoxicity: a form of cell destruction. It can result from the activity of antibodies or activated T-lymphocytes. Cytotoxic antibodies damage antigen-bearing cells, especially in the presence of complement. Cytotoxic T-lymphocytes are a subset of the lymphocytes responsible for lysing target cells and for killing virus-infected cells.

Cytotrophoblast: see Extraembryonic membranes, Gastrulation, Placenta.

D

Dab, *Limanda limanda, Pleuronectes limanda:* a commercially important, small flatfish (maximal length 45 cm) of the family *Pleuronectidae,* order *Pleuronectiformes* (see). The upper surface is very rough, due the presence of toothed scales. The D. lives in coastal waters on sandy bottoms at depths of 20–50 m. It is very common on all sandy European shores, and it is caught off the European North Atlantic coast, in the North Sea and in the Baltic. The flesh makes extremely good eating.

Dabbling ducks: see *Anseriformes.*

Daboia: see Russell's viper.

Dace, *Leuciscus leuciscus:* a member of the fish family *Cyprinidae* (see), closely resembling the chub, but smaller (up to 30 cm in length). The anal fin is slightly indented and the scales are smaller than in the chub. It is found throughout Siberia and northern Europe.

Dacelo: see *Coraciiformes.*

Daceloninae: see *Coraciiformes.*

Dacrydium: see *Podocarpaceae.*

Dactylaria: see Carnivorous plants.

Dactylis: see *Graminae.*

Dactylogram: see Finger print pattern.

Dactylogyrus: see *Trematoda.*

Dactylomyinae: see *Echimyidae.*

Dactylomys: see *Echimyidae.*

Dactylortyx: see *Phasianidae.*

Dactyloscopidae: see *Blennioidei.*

Dactylozooid: see Polyp.

Daffodil: see *Amaryllidaceae.*

Dahlia: see *Compositae.*

Daily requirement: the daily food intake of an organism that is necessary to satisfy energy requirements. To determine the D.r., both the basal metabolic rate and the food composition must be known. After taking account of the Specific dynamic action (see) of the dietary constituents, the necessary food intake can be calculated from the Calorific value of dietary components (see). As an empirical rule, a human at rest requires sufficient energy to satisfy the needs of basal metabolism plus a further 30%. Of this extra 30%, one third compensates for the average specific dynamic action of a mixed diet, and two thirds supply the extra energy needed for daily muscular activity.

Daisy: see *Compositae.*

Dalatiinae: see *Selachimorpha.*

Dallia: see *Umbridae.*

Dalmatian sponge: see Bath sponges.

Dama: see Fallow deer.

Damage threshold: see Economic damage threshold.

Damaliscus: see *Alcelaphinae.*

Dammar pine: see *Araucariciae*

Damselbugs: see *Heteroptera.*

Damselfishes, *Pomacentridae:* a family of fishes (about 235 species in 25 genera) in the order *Perciformes,* found in all tropical seas, usually as reef inhabitants. Many species are very colorful, and are popular marine aquarium fishes. The body is usually high set and laterally compressed. Members of the subfamily *Amphiprioninae* ***(Anemonefishes)*** have a commensal relationship with large sea anemones, living near to them and among their tentacles for protection; they appear to possess a skin factor that inhibits, or renders the skin incapable of stimulating, nematocyst discharge.

Damselflies: see *Odonatoptera.*

Dandelion: see *Compositae.*

Danube salmon, *huchen,* *Hucho hucho, Salmo hucho:* a large predatory fish of the family *Salmonidae,* order *Salmoniformes.* It is found only in the Danube and its tributaries, where it attains a maximal length of 1.5 m (usually 50–80 cm); attempts to introduce it into other areas have so far failed. The flesh is white and makes excellent eating.

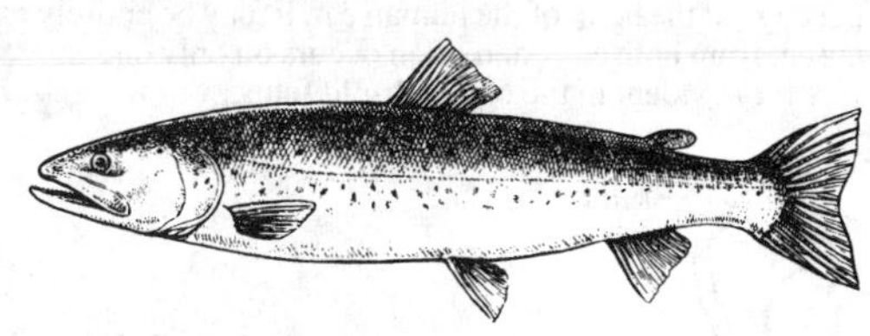

Danube salmon (Salmo hucho)

Daonella: an anisomyarian genus of bivalves, whose fossils are found in the Alpine Triassic. The valves were equal, shallowly compressed, and radially ribbed.

Daphne: see *Thymelaeaceae.*

Daphnia: see *Branchiopoda.*

Daption: see *Procellariiformes.*

Daptriinae: see *Falconiformes.*

Darkfield illumination: a method of illumination for the microscopic examination of contrast-poor and ordered periodic structures. Only light from the specimen enters the objective, so that the specimen appears light on a dark background. This can be achieved by cutting out the central illumination rays with a central or star diaphragm. More intense D.f.i. is achieved with a darkfield condenser, which contains an opaque stop to prevent entrance of direct light directly upward into the tube of the microscope. All peripheral rays are reflected obliquely to the center of the upper surface of the microscope slide, and they cannot enter the objective unless an object on the slide reflects them upward. The empty field therefore appears dark. The Abbé, Paraboloid and Cardioid darkfield condensers are all based on this principle.

Darling effect: see Fraser-Darling effect.

Dartford warbler: see *Sylviinae.*

Darkling beetles: see *Coleoptera.*

Dark sword-grass: see Cutworms.

Darlingtonia: see Carnivorous plants.

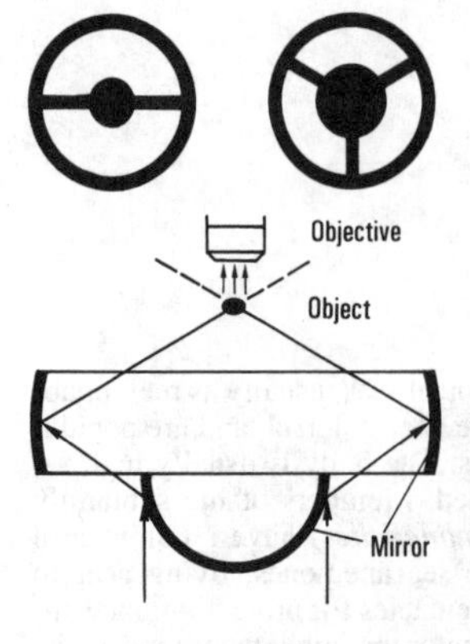

Darkfield illumination. Above: central and star diaphragms for darkfield microscopy. Below: light path in the cardioid condenser.

Darters: see *Pelicaniformes.*

Darwin, Charles Robert (1809–82): an English naturalist and founder of the theory of Evolution (see). The name, Darwin, is also given to the unit of Evolutionary rate (see).

Darwinism: the theory of Evolution (see) by Natural selection (see), founded by Charles Darwin (1809–1882).

Darwin's finches: see Galapagos finches.

Darwin's tubercle: a knob-like thickening on the inner edge of the helix of the human ear. It may be entirely absent from both ears, and often occurs on only one ear. It is very evident in the 6-month-old fetus.

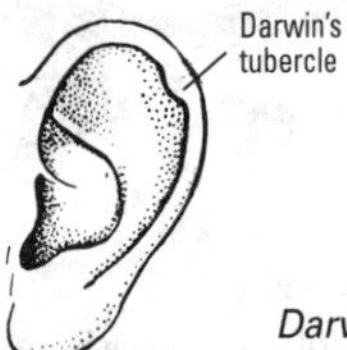

Darwin's tubercle. Human ear with Darwin's tubercle (after Schwalbe).

Dasypeltinae: see Egg-eating snakes.

Dasypeltis: see Egg-eating snakes.

Dasypodidae, *armadillos* (plate 38, No. 3): an American family (9 genera, 21 species) of the *Edentata* (see). All species are found in Central and/or South America, and some have extended their territory recently into North America. Armadillos are terrestrial burrowing, omnivorous animals with a dorsoventrally flattened head. Usually there are 14–50 homodont teeth (up to 90 have been reported). The hindlimbs possess five digits, while the forelimbs have 3–5 claws which are used for digging. The top and sides of the body, and the top of the head are encased in bony armor. In primitive members the armor is a series of transverse movable bands of bony scutes beneath horny plates, along the length of the body; flexible skin between the plates permits some species to roll into a ball. In more advanced forms the front and back armor form solid partitions over the hips and shoulders, with movable hoops between. The short tail is usually encased in bony rings. Hair is present on the lower surface and between the plates.

Dasyproctidae: a family of Rodents (see) consisting of 13 species in 2 genera [11 species of *Dasyprocta* ***(Agoutis)***, 2 species of *Myoprocta* ***(Acouchis)]*** inhabiting the jungles of Central and South America. They are nimble, herbivorous animals with a hare- or guinea pig-like habit. The forefeet carry 4 digits, the hindfeet 3 digits, and the claws are sharp and hoof-like. The skull is robust and the incisors relatively thin. Dental formula: 1/1, 0/0, 1/1, 3/3 = 20.

Some authorities include the ***Pacas*** (genus *Agouti)* in this family, but on account of their enlarged zygomatic arch they are usually placed in the separate family, *Agoutidae* (see). The nomenclature is unfortunate, since the trivial name, Agouti, and the Latin genus, *Agouti,* apply to different animals.

Dasyprocta: see *Dasyproctidae.*

Dasyuridae, *Dasyurids:* a family of carnivorous marsupials distributed in Australia, New Guinea, Tasmania, the Aru islands and Normanby island. The dentition shows adaptation to carnivorous and insectivorous diets; the upper molars have three sharp cusps, the canines are large with a sharp edge, and the incisors are pointed or blade-like. Dental formula: 4/3, 1/1, 2–3/2–3, 4/4 = 42–46. In some species the tail stores fat. *D.* occupy a variety of terrestrial habitats, a few species being arboreal. Smaller species resemble the shrews of North America and Eurasia, and they occupy a similar feeding niche. The larger, rat-sized *D.* prey on insects and small vertebrates. Even larger are the marsupial "cats", which are fierce predators. Most powerful of these marsupial carnivores is the Tasmanian devil (*Sarcophilus harrisi*), a stocky (70 cm total length), short-limbed animal weighing 4.5–9.5 kg; now found only in Tasmania, it feeds on poultry and mammals up to the size of small sheep. Depending on the authority, there are maximally 4 subfamilies: *Dasyurinae* (including the native "cats", e.g. *Sarcophilus harrisi*), *Phascogalinae* (marsupial mice or squirrels, e.g. *Antechinus*, a small, nocturnal, shrew-like marsupial), *Thylacininae* (see below) and *Myrmecobiinae* (see below). The Tasmanian wolf (see) may be classified as the only representative of the subfamily *Thylacininae*, but it is usually regarded as the sole member of a separate family, *Thylacinidae*. Similarly, the most divergent of the *D.,* the Numbat (see) may be classified as the only member of the subfamily *Myrmecobiinae*, but it is usually regarded as the sole species of the family, *Myrmecobiidae*.

Dasyurinae: see *Dasyuridae.*

Date palm: see *Arecaceae.*

Dating methods: methods for the dating of fossils and archeological remains. Relative dating determines the relative ages of fossils, and makes it possible to arrange them in a time sequence. Absolute dating determines the true age of fossils by chronometric and chronographic methods.

Relative D.m.

1) *Stratigraphic D.m.* In the undisturbed sediments of the earth's crust, the geological age of strata always increases with depth. The thickness of a stratum, however, is not necessarily an indication of the time required for its formation. In addition, folding, faulting and displacement can disturb the sequence of strata. Furthermore, it must be determined whether a fossil was incorporated during the formation of the investigated stratum, or whether it was introduced later. With the aid of *index fossils* (fossils whose geological time is known, and which occur in large numbers in a single, restricted time horizon), geographically separate sites

can often be compared and their fossils assigned to a specific time horizon.

The archeological differentiation of stages of human culture is also a form of stratigraphic D. Thus, it is possible to observe a progression from techniques for the manufacture of stone implements, the production of ceramics and decorative articles, the exploitation of metals, etc.

2) *Fluorine test.* This is a physical-chemical method, based on the observation that buried bones exchange small quantities of fluoride ions with the ground water. In this process, hydroxyapatite (the main component of bones and teeth) is converted into fluoroapatite. In addition, calcium ions are replaced by uranium ions, so that information can also be obtained from the analysis of uranium. Nitrogen is also determined as representative of the quantity of remaining organic material. The combined determination of Fluorine, Uranium and Nitrogen is known as the FUN test. The ratio of fluorine to uranium plus nitrogen, however, depends strongly on the local geohydrological conditions, so that the test is only suitable for the determination of relative ages at a single site.

Absolute D.m.

1) *Radiocarbon test, or ^{14}C-test.* Cosmic irradiation in the outer atmosphere converts nitrogen into the radioactive carbon isotope, ^{14}C. In the form of CO_2, all three isotopes of carbon (unstable ^{14}C, and stable ^{12}C and ^{13}C) are assimilated by plants, then transferred to animals in the food chain. After the death of a living organism, no more ^{14}C is incorporated, the ^{14}C that is present decays at a constant rate, and the ratio of the three C-isotopes gradually changes. Based on a half life of 5 570 years for ^{14}C, the age of a fossil (bone, wood, etc.) can be calculated from the ratio of the three isotopes. Owing to the relatively short half-life of ^{14}C, the method can only be used for dating within the last 50 000–75 000 years. New spectrographic methods for ^{14}C, may extend the dating period to 100 000 years.

2) *Potassium-argon test.* In analogy with the radiocarbon method, the measurement of K/Ar depends on the radioactive decay of ^{40}K to ^{40}Ca and ^{40}Ar. Since the half-life of ^{40}K is 1.3×10^9 years, the test can be used to date deposits older than 500 000 years, but it can only be applied to volcanic rocks.

3) *Thorium-uranium test and the Protactinium-thorium test.* Thorium 230 and protactinium 231 are intermediates in the decay series of uranium, which is found in sediments. The presence of thorium 230 depends on the original concentration of ^{234}U and its half-life. With a half-life of 75 000 years for the radioactive decay of ^{230}Th, the optimal dating period for the thorium-uranium method lies between 200 000 and 300 000 years. The protactinium-thorium method uses the ratio of protactinium 231 (half-life 32 500 years) to thorium 230, which also depends on the original concentration of the U-isotope in the sediment; the useful dating period extends to about 140 000 years.

4) *Racemization method, or Amino acid method.* The amino acids of living organisms exist predominantly in the L-configuration. After the death of an organism, there is a gradual racemization to an equilibrium mixture of the D- and L-forms. Dating by determination of the racemization of amino acids can be applied directly to bone material, and requires only small quantities of bone. The age is determined from the L/D ratio of its amino acids. Since the half-life for the conversion (e.g. L-isoleucine into D-alloisoleucine) is about 110 000 years, the method is suitable for the age range 100 000–500 000 years, thereby bridging the gap between the radiocarbon and the potassium-argon methods. Racemization is, however, temperature-dependent, so that the results are influenced by any geological temperature changes that may have occurred.

5) *Fission track method.* This depends on the very slow but constant decay of the radioactive isotope, ^{238}U, to lead. ^{238}U decays even more slowly than ^{40}K, with a half-life of many billions (10^9) of years. Whereas the decay of potassium to argon is "quiet", the decay of the uranium nucleus to form lead is accompanied by the release of small quantities of energy. Fragments of the ^{238}U nucleus are projected for a distance of a few microns through the mineral crystalline layer, leaving fission tracks, which can be observed under the light microscope. The age of a sample can be determined from the ratio of the number of fission tracks to the uranium content. The feasibility of the determination depends, however, on the uranium content of the sample. If too much uranium is present, the fission tracks are so numerous that they cannot be counted. Conversely, the method cannot be applied if the uranium content is less than 50 parts per million. With appropriate samples, dating is possible between 100 000 and 5×10^9 years.

In this and other *chronometric D.m.,* the accuracy of dating can be greatly influenced by a variety of factors. In practice, all possible D.m. should be used. Close or approximate agreement of data from different methods increases the reliability that can be placed on the result.

6) *Varve chronology.* This is a chronographic method, which enables the assignment of an absolute time scale to geological strata. A varve is a layer of fine clay sediment (varved clay) deposited annually in melt-water lakes during the course of a year in periglacial areas; it comprises a coarse, thick, light-colored layer deposited in spring and summer, and a finer, darker layer in autumn and winter. The resulting pattern of bands can be dated for the last 16 000 years in northern Eurasia and North America.

7) *Dendrochronology.* Counting the growth rings of trees (annual ring analysis). A Californian pine *(Pinus arista)* attains an age of more than 4 000 years. Its wood is especially resistant, so that with the additional aid of fossil remains, it has been possible to construct an annual ring sequence of more than 7 000 years. Dendrochronological measurements have now been performed on trees in all parts of the world. Tree ring age has been used to correct ^{14}C dating. In certain periods, the ^{14}C years deviate from the calendar years, because past environmental levels of ^{14}C were sometimes different from modern levels.

8) see *Pollen analysis.*

Datura: see *Solanaceae.*

Dauer-modification: see Modification.

Dawn redwood: see *Taxodiaceae.*

Day geckos: see *Phelsuma.*

Day-neutral plants: plants that are insensitive to the photoperiodicity of their environment, and in which flowering is independent of day-length (or length of the dark period). Examples are *Poa annua, Senecio vulgaris, Stellaria media.* See Photoperiodicity, Flower development.

DCMU: acronym of the systemic herbicide, dichlorophenyldimethylurea. See Herbicides.

Deadly nightshade: see *Solanaceae.*

Dead man's ropes: see *Phaeophyceae.*

Dead men's fingers: see *Alcyonaria.*

Deamination: removal of the amino group (-NH_2) from chemical compounds, in particular from amino acids. D. may occur by Oxidative deamination (see), or by Transamination (see).

Death cap: see *Basidiomycetes (Agaricales).*

Deathwatch beetles: see *Coleoptera.*

Debrachycephalization: see Brachycephaly.

Decapoda:

1) An order of the *Crustacea,* subclass *Malacostraca.* The 8 500 described species represent almost one third of of all the known species of crustaceans. They can be divided into 3 different anatomical types. *Macrurous forms* (members of the suborder *Dendrobranchiata),* i.e. shrimps prawns, lobsters and crayfish, have large, posteriorly extended, elongated abdomens, and compressed or cylindrical bodies. *Anomurous forms* (members of the infraorder *Anomura),* i.e. hermit crabs, porcellanid crabs, etc., have reduced abdomens (which show some evidence of spiralling), less than 4 pairs of thoracic legs for locomotion, and cylindrical or depressed bodies. *Brachyurous forms* (members of the infraorder *Brachyura)* are the true crabs, with reduced and flattened abdomens folded beneath the depressed cephalothorax, and usually 4 pairs of thoracic legs for locomotion. All decapods, however, have a similar functional anatomy. Head and thorax are fused to form a cephalothorax, covered by a large, well-developed carapace, which is almost always united with all thoracic segments. The abdomen is free, and its segments are rarely fused. There is a pair of large compound eyes on movable stalks. A spacious gill cavity, containing the thoracic gills, is formed by the overhang of the carapace. Gills are present on the coxae of the thoracic limbs (podobranchiae), between each limb and the thorax (arthrobranchiae), and on the wall of the thorax (pleurobranchiae). Normally, in any one group, only some of these gills are developed. The first 3 pairs of thoracic appendages are modified as maxillipeds, and the next 5 pairs are uniramous walking legs (or pereiopods) (hence the name, decapod). In many species, the first pair of pereiopods are much larger and heavier, and modified as large pinching claws or chelae, which serve for defense and/or feeding, and in some species for mating; a limb carrying a chela is said to be chelate, and it is known as a cheliped. In lobsters, the abdominal legs (pleopods) are well developed and serve for swimming, but in the crabs they are regressed. The female carries the eggs on her pleopods, and these usually hatch into a zoea, less often (e.g. in primitive shrimps) into a nauplius. Most forms are marine, but a few species inhabit fresh water (freshwater crabs, crayfish), and the land crabs can be considered as terrestrial, although they must return to the water for mating, and their larval forms are marine. Most forms are omnivorous, and only a few are filter feeders. Many edible species are fished on a commercial basis (lobsters, crayfish, many crabs). Fossil decapods are found from the Muschelkalk to the present, but the phylogenetic relationship between extinct and recent forms is unclear.

Classification. Formerly, classification was based on the division of members into *Natantia* ("swimmers") (comprising mainly the shrimps and prawns) and the *Repantia* ("crawlers"). This is now replaced by the following.

1. Suborder: *Dendrobranchiata.*
   Infraorder: *Penaeidea* (primitive shrimps).
2. Suborder: *Pleocyemata.*
   1. Infraorder: *Stenopodidea* [Shrimps (see), e.g. *Stenopus].*
   2. Infraorder: *Caridea* [prawns and Shrimps (see)].
   3. Infraorder: *Astacidea* [Crayfish (see) and Lobsters (see) with large chelipids].
   4. Infraorder: *Palinura* [Spiny lobsters (see), shovel-nosed lobsters].
   5. Infraorder: *Anomura.*
      1. Superfamily: *Thalassinoidea* (marine burrowing shrimps).
      2. Superfamily: *Paguroidea* [Coconut crab (see), Hermit crabs (see), Stone crabs (see)].
      3. Superfamily: *Galatheoidea* (porcellain crabs, South American freshwater *Aegla).*
      4. Superfamily: *Hippoidea* (sand or mole crabs).
   6. Infraorder: *Brachyura* (true crabs); see Crabs.
      1. Section: *Dromiacea* .
      2. Section: *Oxystomata* [Box crabs (see)].
      3. Section: *Oxyrhyncha* [decorator and Spider crabs (see)].
      4. Section: *Cancridea* [Cancer crabs (see)].
      5. Section: *Brachygnatha.*
         Infrasection: *Brachyrhyncha* [Chinese mitten crab (see), Fiddler crabs (see), Freshwater crabs (see), Land crabs (see), Swimming crabs (see), mud crabs, pea crabs, ghost crabs, soldier crabs, coral gall crabs].

2) See *Cephalopoda.*

Decabrachia: see *Cephalopoda.*

Decarboxylation: removal of the carboxyl group of a carboxylic acid, with the release of carbon dioxide. D. of keto acids (oxo acids) and amino acids is metabolically important. See Amino acid decarboxylases, Oxidative decarboxyaltion.

Decay: a special type of Decomposition (see), in which organic material is degraded under warm, dry, aerobic conditions. Aerobic bacteria bring about complete oxidative mineralization, with the release of ammonia and carbon dioxide. Nitrogen-rich material with a high initial moisture content (animal corpses) decays rapidly, whereas nitrogen-poor, dry material (wood) decays slowly. D. of wood is accomplished partly by wood-rotting fungi which, depending on their specificity, attack only the cellulose (brown rot), or the lignin (white rot). D. is promoted by animals such as ants and long-horned beetles, which tolerate dry conditions, and which attack dry vegetable matter, such as wood. It is prevented by animals which bury corpses in order to promote putrifaction (e.g. sexton beetles). D. is finished when all the organic material has disappeared.

Deccan hemp: see *Malvaceae.*

Decidua: see Placenta.

Deciduate placenta: see Placenta.

Deciduous trees: various angiosperm trees, shrubs and dwarf shrubs, whose leaves have a wide lamina, and are shed at the end of each vegetative period. In contrast, the leaves of conifers (referred to as "evergreen") are needlelike and are retained for several vegetative periods. Some angiosperm families have no members that are D.t. (e.g. *Ranunculaceae, Cruciferae),* whereas other families consist only of D.t. (e.g. *Ulmaceae, Betulaceae, Fagaceae).*

Decomposition: a change in the state of dead organic material, involving mineralization and the release of energy. In terrestrial ecosystems, D. is the alternative to consumption by feeding, i.e. the utilisation of primary plant products by phytophagous animals. Thus, a high proportion of the chemically bound energy of green plants leaves the respective ecosystem via the chain of organisms involved in D. (decomposition chain, detritus chain): about 60% in meadows and about 95% in mixed woodlands in temperate climates.

Important stages of D. are: 1. physical disintegration (diminution) of plant and animal remains, which exposes a greater surface area for further D. by microflora and microfauna; 2. biochemical catabolism, i.e. energy-releasing enzymatic reactions; and 3. leaching of organic substances from the ecosystem. D. can be interrupted by immobilization, e.g. the conversion of organic material into humins (see Humus). D. is complete when all the organic material has been converted into inorganic compounds of relatively low molecular mass (= mineralization).

Heterotrophic organisms participating in D. are all saprophytes, and are also known as saprotrophs, saprobionts, saprophages, detritovores, humivores or humiphages; with reference to their ecological role, they are also called destruents. The saprotrophs include microorganisms, protozoa and representatives of almost all groups of animals, including vertebrates. When D. occurs slowly (due to the nature of the substrate), there is a clear differentiation between primary decomposers, which initiate the D. process, and secondary decomposers, which become established only at a later stage. Thus, in the D. of wood or leaf litter, a succession of processes is observed. At the molecular level, the stepwise depolymerization of structural polysaccharides is critical for the subsequent route and rate of D. Furthermore, the protein content of the decomposing material and the abiotic environmental conditions are very important determining factors for the course of D. In the D. of protein-rich material, a restricted oxygen supply leads to Putrifaction (see), while aeration, warmth and dry conditions lead to Decay (see). Wet, cold and acidic conditions prevent complete D. and lead to Peat formation (see).

The particular course of D. is characteristic for each ecosystem, and it determines the potential usefulness of that ecosystem to man. In tropical rainforests, the rate of D. is at least 100 times greater than in tundra soils. Since D. in tropical forest soils is rapid and almost total (mineralization), the humus content of such soils is very low. In contrast, the D. of temperate soils is slower and interrupted by humus formation, so that these soils are rich in organic material.

Decticous pupa: see Insects.

Dedifferentiation: the reversion of a differentiated cell to an embryonal state, i.e. the reestablishment of meristematic behavior. D. restores the previously suppressed ability to differentiate into a variety of cell types (see Totipotency), by the reactivation of blocked genes for the synthesis of certain proteins. It is also associated with cytoplasmic growth and cell division. Examples of D. in plants are the formation of wound callus from parenchyma cells, and the development of secondary meristems, e.g. cork cambium, interfascicular cambium and root cambium. Tissues that readily undergo D. are parenchyma, epidermal tissue, vascular tissue and glandular cells, whereas D. occurs only with difficulty in morphologically strongly modified cells, e.g. collenchyma cells, and immature vessel and sclerenchyma cells. After extreme differentiation (e.g. of sieve tubes) D. is no longer possible.

The formation of cancer and tumor cells in plants and animals is a pathological manifestation of D.

Deep sea: deep zone of the oceans, extending beyond 1 000 m. At these depths there is total darkness, no or extremely slow water movement, a uniform low temperature of 1–2 °C and high hydrostatic pressure. Deep sea fauna (see) are adapted to the extreme conditions. There are no plants at these depths.

Deep sea fauna: animals of the nonilluminated ocean depths. Almost all animal phyla possess representative species which have adapted to life in the deep ocean. Dredge nets at 1 000 m in the north Pacific have collected 120 different species. Three catches at 8 500 m contained 6, 8 and 17 species respectively. The quantity by weight of bottom animals on the continental shelf has been determined at between 500 g/m² and 4 000 g/m², whereas at 5 000 m this quantity is 2 to 5 g/m², and at 10 000 m it is 20 to 30 mg/m². The following table shows the maximal depths of various groups of animals, based on results from deep sea catches by the Soviet research ship Witjas, and the Danish Galathea expedition.

Group	Depth in meters
Foraminifera	10 687
Sponges	8 660
Hydrozoa	8 300
Octocoralla	8 660
Hexacoralla	10 710
Proboscis worms	7 230
Nematodes	10 687
Annelids	10 710
Ostracods	7 657
Copepods	10 002
Cirripedia	7 000
Amphipod	10 687
Isopods	10 710
Shrimps	7 230
Decapod crustaceans	5 300
Caprellids	6 860
Moss animals	5 850
Brachiopods	5 457
Gastropod	10 687
Bivalves	10 687
Cephalopods	8 100
Sea stars	7 614
Brittle stars	8 006
Sea urchins	7 290
Sea cucumber	10 710
Sea lilies	9 735
Sea squirts	7 230
Fish	7 579

In accordance with the constant low temperature of the ocean depths, the D.s.f. consists mainly of cold water forms. The light-sensitive organs of these animals are either regressed, or extremely large and efficient, often taking the form of telescopic eyes. Blind forms possess long feelers for sensory contact with the environment. Light-emitting organs, arranged in species-specific patterns, are a common feature; these aid recognition between members of the same species, and are often used to lure prey. Since light is absent, there are no primary producers. All D.s.f. feed on detritus, or they are predators. The fish have especially wide gapes and long, pointed teeth. Despite the high hydrostatic pressures (greater than 100 000 kPa at 10 000 m), D.s.f. are mostly delicate and fragile; the external and internal pressures are the same, so that the organism is not endangered by a large pressure difference. The swim bladders of deep sea fish are regressed or filled with lipids. If a deep sea fish possessing a gas-filled swim bladder is brought rapidly to the surface, the swim bladder is forced out through the mouth by the high internal pressure ("tympanites").

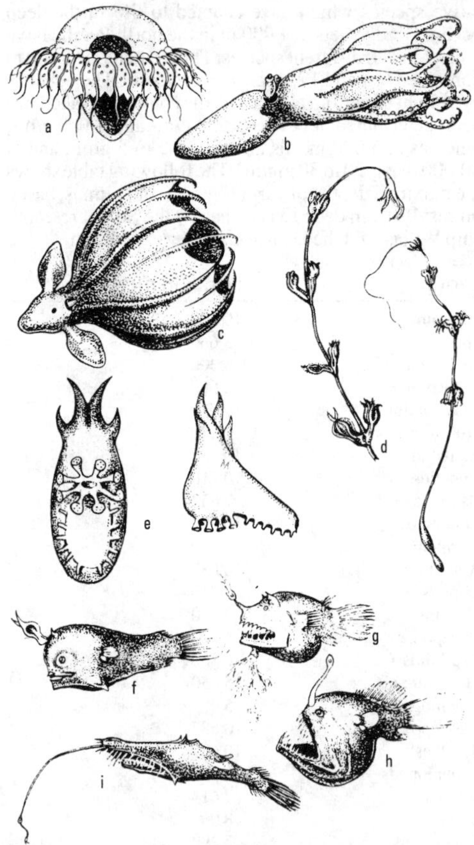

Deep sea animals. a Atolla (medusa, diameter about 7 cm). *b Amphitretus pelagicus* (a squid with telescopic eyes, length of animal 12 cm). *c Cirrothauma murrayi* (a squid with degenerate eyes, length of animal 15 cm). *d Umbellula* (a coral): *right,* the complete animal; *left,* detail. *e Peniagone wyvillei* (a sea cucumber), ventral and lateral views, length 9 cm. *f* to *i* Females of the family, *Ceratiidae,* length 1.5 to 8 cm.

Deer, *Cervidae:* a family of slim, long-legged ungulates, containing 38 species in 17 recent genera. They are naturally distributed in Eurasia, North Africa and North America. Many species have been introduced and are now established well beyond the natural range of the family. The males of most species have Antlers (see). In 2 less advanced genera (*Hydropetes* and *Moschus,* Musk deer) the males lack antlers, and possess elongated canines. Elongated canines are also found in the Muntjacs *(Muntiacus).* The Caribou *(Rangifer)* is unique among the deer family in that both sexes carry antlers. The following representatives are listed separately: Axis deer, Barasingha, Caribou, Chinese water deer, Fallow deer, Moose, Mule deer, Muntjacs, Musk deer, Pére David's deer, Pudu, Red deer, Roe deer, Sambar, Sika deer, White-tailed deer.

Deer mouse: see *Cricetinae.*

Defecation: the reflex extrusion of feces. When feces enter the rectum, the resulting stretching of the rectum wall stimulates the fibers of the nervus pudendus and nervus pelvicus. The center of the D. reflex lies partly in the lumbar and mostly in the sacral region of the spinal cord. It stimulates motor fibers, particularly those of the nervus pelvicus, causing powerful peristalsis in the colon descendens, contraction of the rectum, and relaxation of the inner anal sphincter. Thus, the section of the alimentary canal extending from the colon transversus to the anus is emptied. Voluntary relaxation of the outer anal sphincter is possible via the nervus pudendus. D. is aided by simultaneous contractions of the diaphragm and abdominal muscles.

Defective viruses: viruses that lack genes for the synthesis of functional capsid or envelope proteins. For example, the Rous sarcoma virus is unable to synthesize one of its envelope proteins. In order to complete its replication cycle and form new virus particles, the missing gene must be provided by a *helper virus.* The helper virus is a member of a group of viruses often found in association wth the Rous sarcoma virus, and known as Rous-associated viruses (RAV). In the absence of an RAV, the normal host cell is transformed into a tumor-forming cell (see Tumor virus). Most acute transforming *Retroviridae* are D.v. Deficient viruses (see) are often included with the D.v.

Deficiency diseases: diseases of plants and animals resulting from inadequate provision or availability of essential nutrients, or from a wrongly balanced diet.

D.d. of plants arise when individual nutrients in the soil are not present in sufficient quantity in a form that can be taken up by the plant (i.e. a nutrient may be present, but in an unavailable form), or when the soil is simply deficient in an essential nutrient.

Symptoms of D.d. are abnormal or stunted growth and/ or development, abnormal coloration (e.g. chlorosis), localized tissue necrosis, wilting, etc. An abnormal deposition of anthocyanin pigments is characteristic of certain deficiencies. Thus, the "purpling" of leaves often occurs in phosphorus deficiency. Purpling also occurs in Mg deficiency in cotton, and in boron deficiency in clover. In nitrogen deficiency in dicots, the older, lower leaves turn yellow (i.e. become chlorosed), whereas in monocots yellowing proceeds from the tips of the leaves to the bases. The earliest effects of Ca deficiency are seen in the roots, the tips of which become slimy and turn black; new roots may form near the base of the stem, but these then suffer the same fate. In K deficiency, the tips and margins of the most recently matured leaves turn yellow, then brown, finally becoming dry and brittle ("scorched"), and appearing ragged due to the loss of dead tissue. Some species behave differently, however, and the first signs of K deficiency in leguminous plants is a white speckling or freckling of the leaf blades. Chlorosis, followed by necrosis, and especially wilting are typical of Cl deficiency. Other familiar manifestations of D.d. in plants are heart rot, and a discoloration known as "marsh spot", which occurs in plants growing on waterlogged, acidic soils.

D.d. in animals are usually the result of badly balanced diets or inadequate diets. Young, growing animals are most at risk. Mineral deficiency leads to "licking sickness" (Cu deficiency in cattle), rickets and other skeletal abnormalities (Ca deficiency), anemia (Fe, Cu and iodine deficiencies), neonatal mortality (iodine deficiency), thyroid enlargement or goiter (iodine deficiency). Vitamin deficiencies lead to a variety of symptoms (see Vitamins). Often several factors are deficient, resulting in the general impairment of growth, bodily weakness, skeletal changes, inhibition of mating behavior, sterility, decreased competitiveness. Typical manifestations of D.d. in animals are wool and feather eating, torpor and cannabalism.

Human D.d. have essentially the same origin as animal D.d. They are manifested as general poor health, loss of weight, tiredness, decrease in bodily performance and susceptibility to certain infectious diseases, as well as the specific clinical pictures of vitamin D.d., e.g. beriberi, scurvy, pellagra, rickets, etc.

Deficient viruses: viruses which, through mutation, have lost the ability to synthesize certain essential enzymes, e.g. RNA-dependent RNA replicase (see Viruses). The replication cycle can be completed only if the cell is superinfected with a *helper virus,* which carries the gene for the missing protein. D.v. are often so dependent on their helper virus that they are only found in association with it. In such cases, the deficient virus is known as a satellite virus, e.g. the tobacco necrosis satellite virus (TNSV), which requires the RNA replicase of the tobacco necrosis virus (TNV) in order to complete its replication cycle. As a result, the replication of the TNV is retarded. The TNSV is therefore a virus parasite, because it parasitizes not only the host cell, but also the TNV. A similar relationship exists between the adeno-associated satellite viruses and certain adenoviruses.

Under certain conditions, especially in cell cultures, the missing enzyme can be supplied by the cells of a second organism, which are then known as *helper cells.* D.v. are sometimes grouped with the Defective viruses (see).

Definition: see Resolving power.

Defoliant: an agent for removing leaves before the harvest. It is used to accelerate fruit ripening and to facilitate mechanized harvesting.

Deformities: see Malformations.

Degeneration: the anamolous development or regression of cells, tissues or organs within a living organism.

Nerve D. In the human nervous system, the following types of D. are recognized. a) *Wallerian D.:* in a cut nerve, the axons and dendrites separated from the cell body are degraded and resorbed. The part left connected to the parent cell, usually in the central nervous system, remains healthy. b) *Retrograde D.:* progressive degeneration from a lesion or site of damage toward the cell body. This is a reversible process. c) *Transneuronal D.:* degeneration of apparently healthy nerves, due to spread of degenerative processes from neighboring nerves undergoing Wallerian or retrograde D. d) *Physiological D.:* the natural degeneration of nerves with increasing age; this often becomes apparent in humans around the age of 40, with decreased skin sensitivity due to the loss of neurons in the spinal ganglia. The total number of physiologically degenerating neurons can only be estimated, since it is not yet possible to count them directly. Their significance in the aging process is probably exaggerated. On the other hand, certain forms of *pathological D.* (e.g. cerebral atrophy) represent very grave medical conditions.

Deglutition: see Swallowing.

Degu: see *Octodontidae.*

Dehiscent fruits: see Fruit.

Dehydration: removal of the elements of water from chemical compounds, e.g. concentrated sulfuric acid removes the elements of water from ethanol to form ethene. In living cells, D. is catalysed by dehydratases, e.g. Aconitase (see), Enolase (see) (see also Lyases).

7-Dehydrocholesterol, ***provitamin D_3, cholesta-5,7-dien-3β-ol:*** a zoosterol which occurs in relatively high concentrations in animal and human skin. Prevention and cure of rickets by sunlight are due to the conversion of 7-D. into vitamin D_3 by ultraviolet irradiation. Derivatives of 7-D. with a hydroxyl group in position 1α, 24, or 25 are converted into more potent antirachitic D-vitamins.

H

HO

7-Dehydrocholesterol

Dehydroepiandrosterone: see Androstenolone.

Dehydrogenases: oxidoreductase enzymes which remove hydrogen from their substrate (hydrogen donor) and transfer it to a second substrate (hydrogen acceptor). Anaerobic D. employ NAD^+ or $NADP^+$ as their coenzyme, and transfer hydrogen and electrons to another acceptor that is not oxygen; they catalyse redox reactions in fundamental metabolic pathways, such as glycolysis, the tricarboxylic acid cycle, the respiratory chain, the pentose phosphate cycle, and various biosyntheses. D. that contain the prosthetic groups FMN or FAD, e.g. NADH-dehydrogenase and succinate dehydrogenase, are important in the respiratory chain and other areas of metabolism.

Aerobic D. include xanthine oxidase, aldehyde dehydrogenase, and the L- and D-amino acid oxidases, which employ FMN or FAD as coenzymes, and transfer hydrogen and electrons to molecular oxygen.

Dehydrogenation: removal of hydrogen, more correctly the removal of two electrons and two protons, from a chemical compound (see Oxidation). In living cells, D. involves removal of electrons and protons from respiratory substrates. By a series of coupled, enzymatically catalysed redox reactions, these electrons and protons are then transferred to molecular oxygen to form water (see Respiratory chain).

Deinotherium (Greek *deinos* fruitful, *therion* animal): an extinct genus of the *Proboscida.* The elephant-sized D. had a long trunk, and tusks in the lower jaw, which were directed downward and curved slightly backward. In contrast to all other proboscids, tusks were absent from the upper jaw. Fossils are found in Eurasia from the Miocene to the Pliocene, and in Africa up to the Pleistocene.

Delayed hypersensitivity: a type of Allergy (see) due to cellular immune reactions. Sensitization occurs through contact with the allergen, and the D.h. is manifested on renewed contact with the allergen. Activated lymphocytes (see Thymus) are specifically involved in D.h. On contact with the antigen they cause inflammatory reactions, which can lead to the destruction of entire areas of tissue, e.g. the rejection reaction of organ transplants (see Transplantation). A delayed hypersensitive reaction develops slowly and may not attain maximal intensity for several days.

Deletion: a chromosome mutation, in which a centrally (intercalary) or terminally localized part of a chromosome or chromatid is lost. Loss of a terminal part of a chromosome is called a chromosome *deficiency.* The genetic consequences of a D. depend largely on the number and function of the lost genes. Extensive homozygous Ds. are almost always lethal, whereas extensive heterozygous Ds. are often lethal.

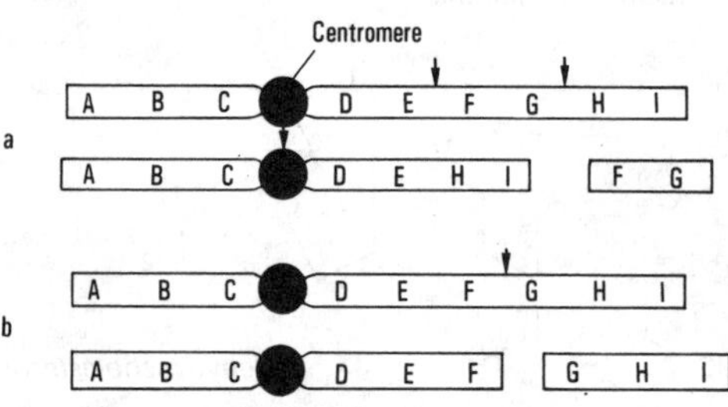

Deletion. a Intercalary deletion. *b* Terminal deletion.

Delicious Lactarius: see *Basidiomycetes (Agaricales).*

Delphinapterus: see Whales.

Delphinidae: see Whales.

Delphinidin: 3,3′,4′,5,5′,7-hexahydroxyflavylium cation (Fig.), the aglycon of many Anthocyanins (see). Glycosides of D. are widespread among plants and are responsible for the mauve and blue colors of many flowers and fruits, e.g. tulipanin from tulips is the 3-rhamnoglucoside of D. Hydrolysis of delphinin from the violet flowers of delphinium *(Delphinium consolida)* yields 2 molecules of glucose, 2 molecules of *p*-hydroxybenzoic acid and 1 molecule of D. *Malvidin* is a dimethyl ether of D. A monoglucoside of malvidin (oenin) occurs in blue grapes, while a diglucoside of malvidin (malvin) occurs in the flowers of the wild mallow.

Delphinidin

Delphininae: see Whales.

Delphinium: see *Ranunculaceae.*

Delphinus: see Whales.

Demissine: one of the *Solanum* alkaloids, found in wild potatoes *(Solanum demissum).* It is repellant to the larvae of potato beetles, and therefore acts as a natural protection against these pests. D. is a glycoalkaloid, which on acid hydrolysis yields the steroid aglycon, demissine (5α-solanidan-3β-ol), and the tetrasaccharide, β-lycotetraose (which contains D-galactose, D-xylose and 2 molecules of D-glucose).

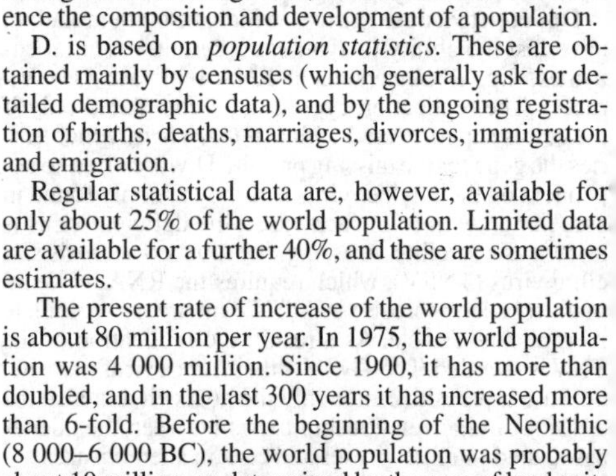

Demissine

Demography: a branch of science that considers the population of a village, town, region, state, nation, or economic area as a unit, and investigates its numerical size, density, age composition, birth and death rate, sex ratio, immigration and emigration, and other factors that influence the composition and development of a population.

D. is based on *population statistics.* These are obtained mainly by censuses (which generally ask for detailed demographic data), and by the ongoing registration of births, deaths, marriages, divorces, immigration and emigration.

Regular statistical data are, however, available for only about 25% of the world population. Limited data are available for a further 40%, and these are sometimes estimates.

The present rate of increase of the world population is about 80 million per year. In 1975, the world population was 4 000 million. Since 1900, it has more than doubled, and in the last 300 years it has increased more than 6-fold. Before the beginning of the Neolithic (8 000–6 000 BC), the world population was probably about 10 million, as determined by the area of land suitable for a hunting and gathering economy. Even as the human race began to organize its own food supply, a further 2 500 to 1 000 years elapsed before the population had doubled itself.

The population increase of the 20th century was brought about primarily by advances in medicine, which led to a considerable decrease in infant and child mortality, and to widespread control of infectious diseases. Furthermore, humans have continued to open up new areas, and have exploited even more intensively the food-producing potential of older inhabited areas, leading to marked increases in population densities (individuals per km^2). The aboriginal hunters of Australia required 110 km^2 per person, whereas about 10 persons per km^2 can be supported by primitive agriculture and cattle herding. Intensive agriculture can support an even greater population density (127/km^2 in Korea). The greatest population densities are attained in highly developed industrial countries, e.g. 154/km^2 in former East Germany, 320/km^2 in Belgium. Under- and Overpopulation are therefore relative terms which must be interpreted on the basis of the local economy.

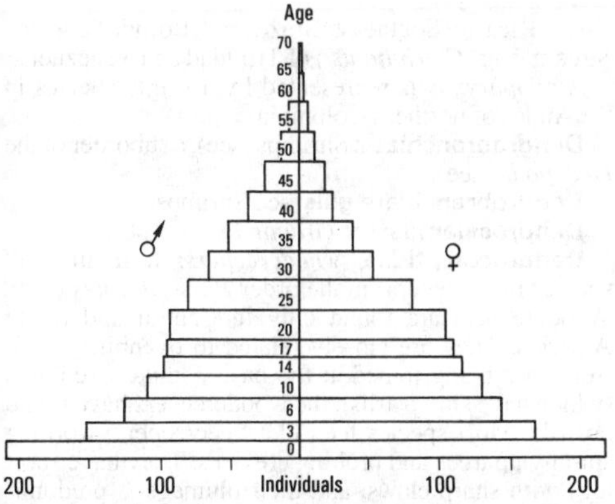

Demography. A population pyramid of the population of Espenfeld (Arnstadt, Germany) in the Middle Ages, based on the analysis of skeletal remains (after H. & A. Bach).

The biological condition of the population is represented by the *population composition,* i.e. the proportion of different age groups and the distribution of the sexes. The population composition can be represented graphically by a *population pyramid* (Fig.), which also provides further information on population dynamics. Thus, if the proportion of children and young adults is relatively high, the pyramid has a broad base, and it can be concluded that the population is growing. A stationary population is reflected by a narrower base, while the age structure of a shrinking population causes the population pyramid to have an inverted bell or vase shape. Average life expectancy can also be derived from the shape of the pyramid. Nonsymmetry of the population pyramid is due to unequal distributions of the sexes in the individual age groups.

Further important demographic factors are natality and mortality. The Birth rate (see) is expressed as the number of births per 1 000 members of the population per year, whereas the general Fertility index (see) is the number of births per year divided by the number of women of reproductive age. In highly industrialized nations, both of these values have decreased markedly in recent years. This has not, however, been accompanied by a population decrease, because at the same time the average life expectancy for each age group has undergone a marked increase. In many countries the average life expectancy for new-born children of both sexes in now more than 70 years, a value that is greatly influenced by the decrease in infant and child mortality. Women usually have a somewhat greater life expectancy than men. Toward the end of the Middle Ages in Germany, 50% of the population died before attaining adulthood. At that time, the life expectancy of women was much less than that of men.

Demography and population dynamics are also politically important, because they provide information required for the forward planning of public services.

Demoiselle: see *Gruiformes.*

Demospongiae: see *Porifera.*

Denaturation: usually irreversible structural changes of proteins or other biopolymers, caused by the action of acids, bases, chemical agents such as organic solvents, concentrated solutions of urea or dodecylsulfate, heat, or irradiation. D. destroys the biological properties of biopolymers (e.g. enzyme activity), as well as altering their physical properties, e.g. denatured (coagulated) proteins are no longer soluble. D. is due to a transition from a highly ordered native configuration to a disordered state.

D. of DNA is widely used in experimants on hybridization and in recombinant DNA technology. In these situations, the term has a specific meaning, i.e. separation of the two strands of the DNA double helix by heating or treatment with alkali. The process is reversible, and the temperature at which the double stranded nucleic acid is 50% denatured is known as its "melting point" (T_m).

Dendranthema: see *Compositae.*

Dendrite: see Neuron.

Dendroaspis: see Mambas.

Dendrobatidae, ***dendrobatids, poison-arrow frogs*** (plate 3): a family of the *Anura* (see), containing 130 species in 4 genera, found in the New World tropics (Central and South America from Costa Rica to South Brazil), and sometimes classified as a subfamily *(Dendrobatinae)* of the *Ranidae* .

Members are characterized by a pair of plate-like scutes (glandular muscular pads) on the upper side of each toe and finger tip. Dendrobatids possess 8 holochordal, procoelous presacral vertebrae with nonimbricate (nonoverlapping) neural arches; presacrals I and II are not fused, and the atlantal cotyles of presacral I are widely separated; in certain species of *Dendrobates,* presacral VIII is synostotically united with the sacrum. The sacrum has rounded diapophyses and it displays a dicondylar articulation with the coccyx. Proximal transverse processes are usually present on the coccyx. The pectoral girdle is firmisternal. Maxillary and premaxillary teeth may be present or absent.

Eggs are laid in moist places on land, such as leaf litter and rock crevices, or in plants like bromeliads and arums. Compared with the hundreds or thousands of eggs laid by water-breeding frogs, dendrobatids lay a small numbers of eggs, sometimes only 1 or 2, and rarely more than 40. In many populations of dendrobatids, the disadvantage of small egg clutches is offset by the fact that breeding is continuous throughout the year. The large-yolked eggs are often guarded and kept moist by an adult until they hatch, whereupon the adult (male or female, depending on the species) carries the tadpoles on its back to a suitable source of water; this may be a stream or pond, or even the small pool of water between the leaf and stem of bromeliads or other plants. Tadpoles adhere to a patch of mucus on the back of the adult; in some species this forms a powerful glue, and the tadpoles are released into their new environment by soaking. This form of brood care closely resembles that of the *Sooglossidae* (see), but the dendrobatid tadpoles (unlike those of the *Sooglossidae)* possess horny jaws and teeth. The adult that carries the tadpoles is known as the "nurse frog".

Many species produce powerful skin toxins which were formerly (and to some extent are still) used by indians in Western Colombia as arrow poisons (east of the Andes, Colombian indians use curare as an arrow and blow dart poison). An early report (1824) describes the use of a frog, three inches long with a yellow back and large black eyes: "... *they take one of the unfortunate reptiles and pass a pointed piece of wood down his throat, and out at one of his legs. This torture makes the poor frog perspire very much, especially on the back, which becomes covered with white froth; this is the most powerful poison that he yields, and in this they dip or roll the points of their arrows, which will preserve their destructive power for a year. Afterwards, below this white substance appears a yellow oil, which is carefully scraped off, and retains its deadly influence for four or six months, according to the goodness (as they say) of the frog. By this means, from one frog sufficient poison is obtained for about fifty arrows.*" There seems little doubt that the species described was *Phyllobates bicolor* (Kokoi poison arrow frog), which always has an orange or yellow back, with pale green to black legs. The precise method of preparation appears to vary, as later observers report that the impaled frogs are held over a fire to sweat out the poison, which is fermented in vessels before application to the arrow heads. Three species have been identified as main sources of arrow poison in western Colombia: *Phyllobates aurotaenia* (Pacific drainage of Northwest Colombia; dark skin with 2 parallel longitudinal yellow stripes on the back), *P. bicolor* (Pacific drainage of Northwest Colombia), and *Phyllobates terriblis* (usually uniformly yellow, orange or pale green; discovered in the early 1970s on the Pacific lowlands of Southwest Colombia). *P. terriblis* yields about 20 times more toxin than *P. aurotaenia* or *bicolor* . So powerful is the poisonous secretion of *P. terriblis* that it is dangerous even to handle, and the indians poison their darts simply by wiping them across the back of a live specimen. The earlier description of *Dendrobates tinctorius* as a main source of Colombian arrow poison was due to an error of identification. The remaining two species of *Phyllobates* are found in a small area on the Pacific side of Costa Rica *(P. vittatus* ; dark skin with ventral green mottling and 2 parallel dorsal orange stripes) and on the Caribbean side of Costa Rica extending into adjoining Panama *(P. lugubris* ; smaller than *P. vittatus*, with similar color and patterning, but orange mottling on the limbs); these are less toxic than the Colombian species, and arrow and dart poisons are not (and are not known to have been) used in these areas.

The genus, *Dendrobates (= Hylaplesia = Dendromedusa*), is represented by 48 usually strikingly colored species, distributed from Nicaragua to Southeast Brazil and Bolivia. *Dendrobates tinctorius* has a glossy black or maroon body with large metallic blue spots. Other species are green, red, pink or golden, often with darker spots and stripes. Remarkable variations of size, color, pattern, habitat, behavior and even poisonous skin secretions occur within single species. Thus, different populations of *D. pumilio* (Red-and-blue poison-arrow frog), which inhabit the rainforest on the Caribbean side of Nicaragua, Costa Rica and western Panama, may be red, red with black or bright blue legs, blue, green, yellow, uniformly colored or spotted or striped. A total of 80 alkaloids has been identified from *D. pumilio* in western Panama, but only between 6 and 24 different alkaloids are found in any single population. Behavior toward potential predators ranges from apparently fearless to very wary. Different preferences are also displayed with respect to habitat, some populations living on the ground, others in trees.

The genus, *Colostethus (= Prostherapis = Hyloxalus = Phyllodromus*) is represented by 63 brown-colored and mostly nonpoisonous species, distributed from Costa Rica to Southeast Brazil and Bolivia, e.g. the Stream frog *(C. trinitatus*) of Trinidad and Venezuela.

Atopophrynus is represented by a single species in the Andes of northern Colombia.

Dendrobranchiata: Shrimps (see), a suborder of the *Decapoda* (see).

Dendrobranchiate gills: see Shrimps.

Dendrocalamus: see *Graminae.*

Dendrocolaptidae, *woodcreepers:* a family (59 species in 13 genera) in the order *Passeriformes* (see). Woodcreepers are found only in Central and South America. They are closely related to ovenbirds, their front toes being joined at the base. Unlike ovenbirds (which have slit nostrils), the woodcreepers have round nostrils. Most species feed like treecreepers, working spirally up trees and probing crevices. They have short legs with sharp claws, and their plumage is predominantly brown with paler undersides.

Dendrocopos: see *Picidae.*

Dendrocygnini: see *Anseriformes.*

Dendrochronology: see Dating methods

Dendrogaea: see Neogea.

Dendroica: see *Parulidae.*

Dendroidea: see Graptolites.

Dendrology: a field of applied botany concerned mainly with the breeding and cultivation of economically important and ornamental trees and shrubs.

Dendronanthus: see *Motacillidae.*

Dendrordyx: see *Phasianidae.*

Dens:

1) Tooth.

2) A tooth-like process, e.g. on a vertebra.

3) Part of the caudal furcula of the *Collembola.*

Dentary: see Skull.

Dentition: all the teeth used for biting and chewing. In the primitive condition, practically all the bones of the buccal cavity carry teeth, e.g. in fish, whereas in more advanced forms such as mammals, teeth arise only from the maxilla, mandible and the intermaxillary bones (premaxillae). Vertebrate teeth may serve not only for seizing and masticating food, but also as weapons, tools, or locomotory aids (e.g. in the walrus).

The teeth of nonmammals are generally uniform in shape. For example the teeth of the frog, which occur only in the premaxilla and maxilla of the upper jaw and in the vomers, are all alike, and are therefore known as *homodont teeth.* In mammals, however, the teeth are differentiated *(heterodont teeth)* into incisors, canines, premolars and molars. If the upper and lower incisors bite directly onto each other, the condition is known as *orthodont*. Incisors that meet obliquely are *clinodont.* If the upper and lower incisors pass over one another, the condition is *psalidont*. In *isognathous* dentition. the upper and lower rows of teeth span an equal distance, whereas in the *anisognathous* condition the upper rows of teeth extend further than the lower.

Mammalian dentition has evolved from a basic pattern of 44 teeth, which can be expressed in the dental formula: $3/3i$, $1/1c$, $4/4p$, $3/3m = 44$, where *i* represents the incisors, *c* the canines, *p* the premolars, and *m* the molars. The upper numbers of each fraction represent the upper jaw, the lower numbers the lower jaw. Only the left side of the symmetrical dentition is represented, so that the total number of teeth is given by doubling the number of teeth in the formula. Some examples of mammalian dentition are:

Hedgehog 3/2, 1/1, 3/3, 3/3 = 38
Rat 1/1, 0/0, 0/0, 3/3 = 16
Dog 3/3, 1/1, 4/4, 2/3 = 42
Seal 3/2, 1/1, 4/4, 1/1 = 34
Cow 0/3, 0/1, 3/3, 3/3 = 32
Chimpanzee 2/2, 1/1, 2/2, 3/3 = 32
Human 2(2)/2(2), 1(1)/1(1), 2(2)/2(2), 3(0)/3(0) = 32 (20); the milk teeth are shown in brackets.

In many mammals, the *milk dentition* of the young animal is shed and replaced in the adult by *permanent dentition.* This is known as *diphyodont dentition,* in contrast to *monophyodont,* in which there is only a single set of teeth. In certain species the permanent teeth do not grow once they have completely erupted; cessation of growth is due to constriction of the blood flow at the base of the root, so that the pulp remains alive, but is insufficently nourished to permit growth. On the other hand, the pulp cavity may not constrict, so that the blood supply is sufficient to sustain continuous growth, e.g. the incisors of a rodent grow throughout life; such teeth are said to have a *persistent pulp.* Premolars and molars are referred to collectively as *cheek teeth.* Molars and premolars do not always differ greatly in form, but premolars are the only cheek teeth in the milk dentition. The upper incisors are always borne on the premaxillae. In the dog, the last pair of premolars in the upper jaw and the first premolars in the lower jaw have pointed ridges and sharp edges, and are called the *carnassial* teeth.

With respect to function, the dentition may be designed for seizing (toothed whales and pinnipeds), grazing (horses and cattle), gnawing (rodents and rabbits), chewing (pigs and monkeys), crushing and grinding (hippopotamus), shearing (insectivores), or tearing and shearing (carnivores).

See Teeth, Tooth development.

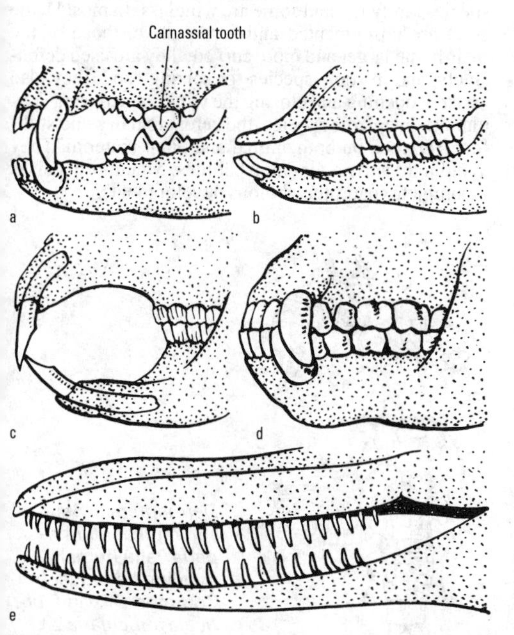

Dentition. a Carnivore. *b* Ruminant. *c* Rodent. *d* Great ape. *e* Toothed whale.

Deoxycholic acid: see Bile acids.

Deoxycorticosterone: see Cortexone.

Deoxyribonucleases, ***DNases:*** phosphodiesterases that catalyse the hydrolytic cleavage of phosphodiester bonds of deoxyribonucleic acids (DNA), with the formation of oligonucleotides. Intestinal DNase (DNase II; pH optimum 7–8; M_r 321 000) is secreted into the duodenum by the pancreas, and serves for the digestion of DNA in the food. Intracellular DNases are also known, e.g. DNase I, which preferentially attacks double-stranded DNA. DNase I is also an *endodeoxyribonuclease,* i.e. it attacks phosphodiester bonds within the DNA molecule to produce oligonucleotides. *Exodeoxyribonucleases* remove successive mononucleotides from the end of the polynucleotide chain. Some S. are specific for single-, others for double-stranded DNA.

Deoxyribonucleic acid, ***DNA:*** a high molecule mass polynucleotide (see Nucleic acids) which is the genetic material of prokaryotes, eukaryotes and DNA-viruses. It is therefore present in chromosomes, and in cell organelles that contain extranuclear genetic material (see Cytoplasmic inheritance). DNA is a polymer of nucleotides, which are joined by phosphodiester linkages, i.e. the respective 3'- and 5'-hydroxyl groups of neighboring carbohydrate moieties (see 2-Deoxy-D-ribose) are esterified with a single molecule of phosphoric acid.

DNA contains 4 different bases: the purines, adenine and guanine, and the pyrimidines, cytosine and thymine. Two polynucleotide strands are attached along their length by complementary base pairing, i.e. by hydrogen bonding between adenine and thymine, and between guanine and cytosine. The base composition of DNA from different organisms varies widely (e.g. 18% adenine in tuberculosis bacteria to 30% adenine in calf thymus), but the amount of adenine is always equal to the amount of thymine, and the amount of guanine is always equal to the amount of cytosine. This fact and the results of X-ray diffraction studies by Wilkins and Franklin led Watson and Crick to propose the double-helix model of DNA structure [*Nature* ***171*** (1953) 737], which introduced a new era of biology. According to the Watson and Crick model, the DNA molecule consists of two complementary, but not identical strands which spiral around an imaginary common axis. The two spiral bands are made up of sugar phosphate chains, from which the bases protrude at regular intervals into the interior of the helix. The two strands are held together by hydrogen bonds between bases on opposite strands. In order for the strands to fit together in a helix, a purine on one strand must always be opposed by a pyrimidine on the other. Hydrogen bonds can only form (within the constraints of the double helix) between adenine and thymine (A-T) or guanine and cytosine (G-C), so that the sequence of bases along one strand determines the sequence of the other. The genetic information is encoded in the sequence of bases (see genetic code). The two strands are antiparallel, meaning that the phosphate diesters between the deoxyribose units read 3' to 5' on one chain and 5' to 3' on the other. In most organisms, only one of the two strands is transcribed. That DNA is the genetic material was already known from earlier studies on bacterial transformation. The elucidation of its structure, however, provided a basis for understanding how genetic information is faithfully replicated in succeeding generations (see Replication).

The right-handed double helix described by Watson

and Crick has 10 base pairs per helical turn, and is known as B-DNA. It probably approximates closely to the structure of relaxed (i.e. unstrained) DNA. It is generally accepted, however, that DNA is a dynamic molecule with different conformations in equilibrium with one another. This equilibrium is affected by the nucleotide sequence, ionic strength of the environment, presence of proteins (e.g. histones and other DNA binding proteins), and the extent to which the molecule is under topological strain. Two other recognized conformations are those of Z-DNA and A-DNA. In Z-DNA, the helix is left-handed, there are 12 residues per turn, and an imaginary line joining the phosphate groups around the outer surface of the molecule describes a zig-zag course (in B-DNA the phosphate groups follow a smooth spiral). A-DNA is observed in X-ray studies of DNA made under conditions of low (75%) humidity; it is right-handed and has 10.9 residues per turn.

Carefully prepared DNA may have a molecular mass greater than 100 million. Viral DNA consists of at least a few thousand nucleotide residues. The DNA that constitutes the 23 chromosome pairs of the human cell consists of about 5 thousand million (5 x 10^9) nucleotide pairs, encoding about 50 000 genes (see Gene).

2-Deoxy-D-ribose: a deoxy pentose sugar, which forms the cabohydrate part of deoxyribonucleic acids. The free sugar exhibits cytostatic activity and inhibits glycolysis.

$HOCH_2$ OH
O
H H
H H
OH H

β-2-Deoxy-D-ribose

Deoxy-sugars: monosaccharides in which a hydroxyl group(s) is replaced by hydrogen. D. often occur as the sugar components of glycosides. Examples are L-rhamnose, D-digitose, and deoxy-D-ribose, the latter being a component of DNA.

Depigmentation: a decrease in the number of pigment granules in the skin, hair and eyes during the phylogenetic development of humans. It is assumed that human ancestors displayed an intermediate degree of pigmentation. As a selective adaptation to intense ultraviolet irradiation, pigmentation then increased in the tropics and subtropics. On the other hand, D. occurred in regions of weak UV-irradiation. See Skin color, Compexion.

Depolarization: see Excitation.

Depsides: organic compounds consisting of hydroxycarboxylic acids (usually phenolcarboxylic acids), in which the carboxyl group of one acid is esterified with the hydroxyl group of the other. Many tannins are depside derivatives. Chlorogenic acid (see) is a simple didepside, which is present in large amounts in coffee beans.

Depsipeptides: heterodetic peptides that contain ester bonds in addition to peptide bonds. Many naturally occurring D. are cyclic peptides, also called *peptolides,* which generally have α- or β-hydroxyacids as heterocomponents. In the wider sense, this class also includes O-peptides and peptide lactones. D. occur as the metabolic products of microorganisms, and they often possess very high antibiotic activity, e.g. Actinomycins from different stains of *Streptomyces.*

Derby: see *Tragelaphinae.*

Derivative: in developmental physiology, a term for organs that have evolved from other structures. Thus, the nervous system is a D. of the ectoderm.

Derived character: a phylogenetically advanced character derived from a more primitive character by evolution. See Apomorphic, Plesiomorphic.

Dermal muscles, ***skin musculature:*** muscles of the skin, which are not associated with bones or joints. In reptiles, e.g. snakes, they serve to erect the scales, thereby preventing the body from sliding backward. In gliding mammals (flying squirrels, flying lemurs) they provide tension to the patagium. Birds can erect or "fluff out" their feathers with the aid of D.m., which also provide tension to the skin of the extended wings. Mammalian hair and fur are erected by D.m., and mammals can roll their bodies by the action of D.m., an ability which is particularly marked in e.g. the hedgehog. Independent movement of separate regions of skin is observed in most mammals, and is particularly obvious in the generation of facial expressions (see Facial muscles).

Dermaptera, ***earwigs:*** an insect order, containing 1 250 species, mainly in the tropics. The earliest fossil D. are found in the Lower Miocene.

Imagoes. These are generally elongate, body length 1 to 3 cm, maximum 5 cm. The prognathous head carries long, multisegmented, filiform antennae, and biting mouthparts. Compound eyes are normally present, but they are absent or reduced in epizoic species. Ocelli are absent. The forewings are modified to short, smooth, robust tegmina. When at rest, the radially veined, semicircular, membranous hindwings are folded longitudinally and transversely beneath the tegmina. European species rarely fly, and some are wingless. In most D. the cerci are unsegmented and forceps-like, those of the male being larger and more curved; they are used defensively, and in some species for seizing food; they also assist in opening and closing the wings, and during copulation. In epizoic species, the cerci are hairy and styliform. D. feed on both animal and plant material. They avoid light and prefer damp habitats. Metamorphosis is incomplete (paurometabolous development).

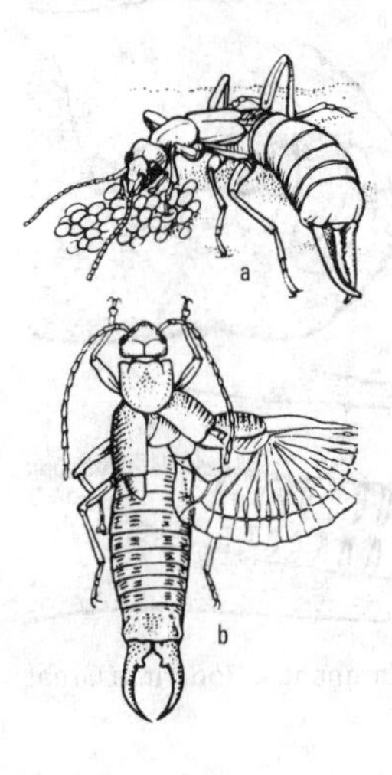

Dermaptera. a Sand earwig *(Labidura riparia* Pall.) (female) minding her eggs. *b* Common earwig *(Forficula auricularia* L.) (male) with right wing unfolded.

Eggs. The female of the common European earwig (*Forficula auricularia*) lays on average 40–50 (maximally 80) eggs. She rests over the eggs and exhibits care for the newly hatched larvae.

Larvae. These are similar to the adult, but wingless. The newly hatched larva of *Forficula auricularia* reaches adulthood in 8 weeks.

Economic importance. D. damage plants and are especially destructive to buds, flowers and fruits of fruit trees; they are sometimes responsible for considerable losses of maize and vegetables. In addition, they are vectors of plant diseases, e.g. cort smut.

Classification. 3 suborders are recognized.

1. Suborder: *Forficulina* (free-living forms)

1. Superfamily: *Pygidicranoidea.* These are the most primitive D. They are found in Asia, Australia, South Africa and South America.

2. Superfamily: *Karschielloidea.* Members of this group are found only in South Africa. They are large, carnivorous species, specially adapted for feeding on ants.

3. Superfamily: *Labioidea*

Family: *Labiidae* (e.g. *Labidura riparia*, the ***Sand earwig)***

Family: *Carcinophoridae*

4. Superfamily: *Forficuloidea*

Family: *Chelisochidae*

Family: *Labiduridae* (e.g. *Anisolabis maritima*, the ***Seaside earwig,*** found on the Atlantic and Pacific coasts of North America).

Family: *Forficulidae* (e.g. *Forficula auricularia*, the common ***European earwig,*** now introduced throughout temperate regions of the world, where it is sometimes a pest).

2. Suborder: *Arixeniina* This suborder contains just 2 species of *Arixenia*, which are epizoic on cave bats in eastern India. Some authorities deny suborder ranking to this group and consider it to be a family (*Arixeniidae*) of the *Labioidea.*

3. Suborder: *Hemimerina.* This suborder comprises about 10 species of the genus, *Hemimerus*, which are epizoic on African giant rats of the genus *Cricetomys.*

Dermatan sulfate: see Chondroitin sulfates.

Dermatemydidae: a family of the order *Chelonia* (see), represented by the single living species, *Dermatemys mawei* ***(Central American river turtle).*** It inhabits rivers and lakes from Vera Cruz (Mexico) to Honduras. It attains a length of 40 cm or more, and displays the primitive characteristic of a complete row of inframarginal plates between the carapace and plastron. Juveniles are carnivorous and adults are herbivorous. The snout is tubular and slightly upturned. Fossil evidence shows that the family was once much more numerous and widely distributed.

Dermatemys: see *Dermatemydidae.*

Dermato-calyptrogen: see Histogen concept.

Dermatocranium: see Skull.

Dermatogen: see Histogen concept, Meristems.

Dermatoglyphics, *dactyloscopy:* the study of the skin relief of the fingers, inner hand surface, toes and sole of the foot. D. are important in anthropolgy for the study of heredity and family relationships, in the diagnosis of twins, in the determination of paternity, and in criminology for the identification of individuals. Many genetic diseases are accompanied by characteristic skin relief patterns, so that D. plays in increasing role in early diagnosis. Every human has a different skin relief pattern, which remains unchanged throughout life. See Fingerprints.

Dermis: see Skin.

Dermochelyidae: a family of the order *Chelonia* (see), represented by the single living species, *Dermochelys coriacea* ***(Leatherback sea turtle),*** of the Pacific, Indian and Atlantic Oceans and the Mediterranean. Measuring up to 2.2 m in length and weighing 600 kg, it is the largest and heaviest of all living turtles. The original shell has almost disappeared (it is represented by small areas of residual bone) and is replaced by a covering of mosaic-like bony platelets embedded in a thick leathery skin. There are 7 longitudinal ridges on the back and 5 on the ventral surface. The forelimbs are modified into wide paddles with a span of 3 m. They leave the water only to lay eggs, but unlike the Sea turtles (see *Chelonidae)* they do not undertake mass migrations. Their egg laying sites on tropical islands are still largely undiscovered. *Dermochelys coriacea* is a threatened species.

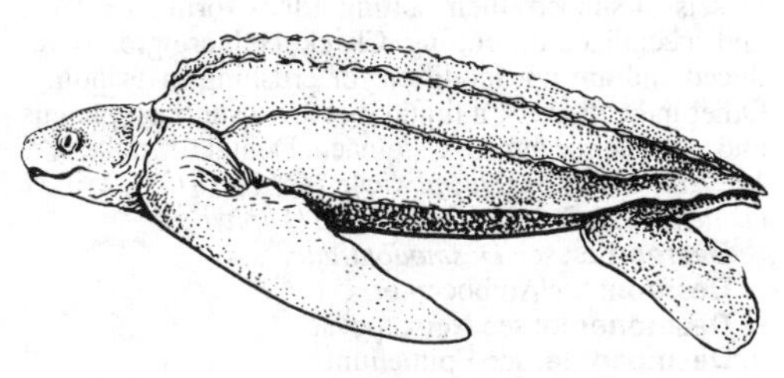

Dermochelyidae. Leatherback sea turtle *(Dermochelys coriacea).*

Dermochelys: see *Dermochelyidae.*

Dermogenys: see *Hemiramphidae.*

Dermographism: see Circulatory regulation.

Dermoptera: see Flying lemurs.

Descensus testiculorum: see Descent of the gonads.

Descent of gonads: in mammals, the migration of the gonads during ontogenesis from their site of development to their final location. Thus, the ovaries move into the pelvic region. Depending on the species, the testes may migrate a considerable distance *(Descensus testiculorum)* or hardly at all. In a relatively large number of mostly primitive mammals, the testes remain in the abdomen: *Insectivora, Edentata, Hyracoidea,* elephants and whales. In many *Rodentia, Insectivora, Pinnipedia, Tapiridae and Rhinocerotidae,* the testes descend to a position directly beneath the skin, whereas in marsupials, some rodents, bats, ruminants, horses, most carnivores and most primates the testes descend from their position near the kidneys into a scrotum, which suspends the testes outside the surface of the pubic region. In insectivores, bats and rodents, the testes may descend into the scrotum only during the mating season, and are then raised into the abdomen until the next mating season.

Desert: see Biocoenosis.

Desmans: large, rat-like, semiaquatic members of the *Talpidae* (see), with webbed forefeet. The ***Russian desman*** *(Desmana moschata)* occurs in and near to rivers in southeastern Europe and central western Asia. The diet of water insects, leeches, snails and spawn is taken while swimming. An earth den is built in the river bank with the entrance below the waterline. *Desmana* has a musky scent and flavor, which deter predators. The smaller ***Spanish desman*** *(Galemys pyrenaicus)* is represented by two geographical races, one in the Pyrenes, the other in the Portugese mountains. They also lead a semiaquatic existence, feeding on aquatic insects, small fish and crustaceans.

Desmodactylae: a suborder of the *Passeriformes* (see).

Desmodont: a hinge type of the *Bivalvia* (see).

Desmodontidae, ***vampire bats:*** a family of bats (see *Chiroptera),* comprising 3 species in 3 genera, distributed in Central and South America. The tip of the short, conical muzzle bears naked pads with U-shaped grooves, and a true nose-leaf is lacking. They feed by drinking the blood of vertebrates, after making an incision with their sharp front teeth. This is done quietly and painlessly while the animal is asleep. The loss of blood is not usually excessive, although frequent visits by vampire bats can be debilitating. They are more feared as vectors of disease. The incisors and canines are sickle or scissor-shaped, their cutting edges forming a "V", and specialized for cutting. Cheek teeth are greatly reduced and are not modified for crushing or grinding. Other indications of a liquid diet are a short esophagus and a slender cecum-like stomach. Dental formula: 1/2, 1/1, 2/3, 0/0 = 20 *(Desmodus);* 1/2, 1/1, 1/2, 2/1 = 22 *(Diaemus);* 2/2, 1/1, 1/2, 2/2 = 26 *(Diphylla).*

Desmodus: see *Desmodontidae.*

Desmon: see Amboceptor.

Desmoneme: see Nematocyst.

Desmosome: see Epithelium.

Desogestrel: a synthetic progestin used in oral contaceptives; see Hormonal contraception.

Destroying angel: see *Basidiomycetes (Agaricales).*

Destruent: see Soil organisms.

Desulfovibrio: a genus of comma-shaped, Gram-negative, obligate anaerobic bacteria with polar flagella. Species of *D.* occur in waters and soils that are rich in organic material, but poor in oxygen. They reduce sulfates, thiosulfate and elemental sulfur to hydrogen sulfide (H_2S), deriving the necessary reducing power from the dehydrogenation of organic materials.

Desulfurication: see Sulfur.

Desynapsis: a phenomenon in meiosis, in which paired chromosomes do not persist as far as anaphase I, but separate due to the absence of chiasma formation or premature chiasma separation (see Chiasma). As in Asynapsis (see), D. usually leads to a disturbance of meiosis, due to the inappropriate distribution of the desynaptic chromosomes, which behave as Univalent chromosomes (see).

Detector center, ***detector mechanism:*** a center in the brain or nervous system, which helps to register and pass on information and signals received from the environment via sense organs. D.cs. were invoked in 1964 by R. Jander, who proposed that the nerve circuits in the insect eye and brain are similar to those in, e.g. a cat. As a result of their smaller size, however, insects have fewer D.cs. Insects can therefore distinguish between certain shapes for the purposes of homing, finding flowers, finding the entrance to a hive or nest, etc., but they do not perceive objects in fine definition. The eyes of ants possess a dark D.c., a light D.c., a D.c. for vertical movement, and a D.c. for certain organized patterns. An ant faced with the choice of a black dot or a flower-like shape chooses the dot, presumably because it resembles the entrance to an ant nest. Faced with a similar choice, a foraging bee would choose the flower shape as a potential source of nectar. In the perception of sound by cats, D.cs. have been demonstrated for vowels and consonants, as well as for brief changes in frequency (transients).

Determination: the commitment of one or more undifferentiated cells to a particular differentiation pathway. D. can be compared to the receiving of orders while the subsequent differentiation (see Cell differentiation) is analogous to the execution of those orders. In D., a commitment is made to one of a number of possible developmental pathways. This commitment is made by the selection of certain genes from the genetic complement of the cell, i.e. a specific pathway of *differential gene expression* is selected by allowing certain genes to remain or become active in information transfer, while simultaneously repressing other genes. In some cases, this progressive restriction of the Totipotency (see) of the original egg cell may itself be genetically programmed, e.g. in the mosaic of determined cells in the animal embryo, and in the fertilized egg cell in the plant ovule. After a single division of the fertilized plant egg cell, one daughter cell is destined to develop into the shoot, the other into the primary root and embryo stalk (suspensor); subsequent divisions of these cells are also programmed. D. may also be induced by neighboring cells, by hormones or phytohormones, and by various external factors. In *homeogenetic induction,* a differentiated permanent tissue induces its own type differentiation pathway in adjacent and neighboring cells. For example, in plant wound callus, vascular bundles arise in those parts of the callus in direct contact with the old damaged vascular bundles. Similar effects occur in the grafting of plant parts. During the development of undamaged plants the fascicular cambium activates the development of interfascicular cambium. On the other hand, *polarity,* i.e. the physiological inequality of opposite poles, is often induced by external factors, e.g. in the spores of many mosses and pteridophytes, and in the egg cells of certain algae, polarity is induced by light. Once established, D. may be final and stable, or it may be labile, i.e. capable of being redirected. As a rule, D. is not morphologically recognizable, and its presence can only be shown experimentally. Thus, if a piece of determined embryo is transplanted to a foreign environment (a different site in the same embryo), it develops in accordance with its origin. If, however, it is not already determined, then it develops in accordance with its new surroundings. For example, a piece of an early gastrula of the salamander is taken from the region that will eventually become the neural plate, then transplanted to the presumptive abdominal region, and vice versa. Each transplant develops in accordance with its new site, i.e. a piece of presumptive medullary plate develops into abdominal epidermis, and presumptive abdominal epidermis eventually becomes nerve tissue. It has also been shown that, at the early gastrula stage, presumptive medullary plate cells can be redirected to develop into mesodermal tissues (notochord, sclerotome, lateral plate, pronephros), or into endodermal intestinal epithelium. However, if such transplants are made at a later stage of embryo development, in the early neurula, this adaptability is no longer observed. It is then found, for example, that a transplant from the region of the presumptive eye always develops into an eye, irrespective of where it s placed in the neurula. D. therefore occurs between the early gastrula and the early neurula (see Organiser). D. does not occur simultaneously in all parts of the same embryo.

Detrital pathway: see Detritus.

Detritivores: see Decomposition, Detritus.

Detritus, ***tripton:*** in a terrestrial habitat, the remains of dead plants (primary producers), i.e. leaf litter and humus. In aquatic environments, D. is the bottom sediment, consisting of dead plant material and sedimented minerals (clay, sand, etc.). A small proportion of D. may also be derived from the dead remains of organisms at a higher trophic level than the primary producers (i.e. small animals). Ecologically, D. represents plant material that is not consumed by grazing herbivores, and is therefore recycled by the action of *detritivores* (e.g. earthworms). D. may first be attacked by saprophytes (fungi and bacteria) which in turn are consumed by detritivores such as nematodes and small insects. Aquatic detritivores are mostly bottom animals like bivalves and worms. This stepwise transfer of the energy content of D. via detritivores to carnivores is known as the *detrital pathway* or *detritus food chain* (e.g. leaf litter → fungal decomposition → nematodes → fish → otter). Through decomposition and mineralization, part of the D. is also converted into soluble and assimilable nutrients (especially nitrate), which are taken up directly by plants. *Allochthonous D.* of aquatic habitats is washed into the water from the surroundings. *Autochthonous D.* is formed in the water where it is deposited.

Detritus food chain: see Detritus.

Deuteranopy: see Color blindness.

Deuterocerebral segment: see *Mesoblast.*

Deuterometameres: see Deuterometamerism.

Deuteromyces: see *Fungi imperfecti.*

Deuterometamerism: formation of the first segments (deuterometameres) from primordial mesoderm cells during larval development of the *Articulata.* This occurs by division of the original uniform coelomic cavity, the somatocoel, into metameres, the number of which varies according to the animal group. D. is then usually followed by Tritometamerism (see) (see Mesoblast). In the development of certain animals (e.g. molluska), D. is not followed by tritometamerism, so that the metameres are exclusively deuterometameres.

Deuterostomia, ***deuterostomes,*** *Notoneuralia:* a branch of the *Bilateria* (see) of the animal kingdom, in which the mouth arises at a considerable distance from the site of the blastopore. The blastopore itself forms the anus, or the anus is formed near to the site of the closed blastopore (hence *deuterostome,* "second mouth"). The central nervous strand of the *D.* lies on the ventral side of the body. In contrast to the *Protostomia* (see), cleavage in *D.* is radial, and it is not determinate. In *D.,* the mesoderm arises by a process known as enterocoelic pouching, i.e. the wall of the archenteron evaginates to form pouches. These pouches separate from the archenteron as a pair, or (in metameric *D.)* as a series of lateral pairs; the pouch cavities become the coelom, and the wall becomes the mesoderm. *D.* are therefore known as *enterocoelous coelomates* or *enterocoelomates.* Chitin is not common in *D.,* and the phosphagen of *D.* is typically creatine phosphate. *D.* comprise the *Hemichordata* (hemichordates), *Echinodermata* (echinoderms), *Pogonophora* (pogonophorans or "beard bearers"), *Chaetognatha* (chaetognaths or arrow worms) and the *Chordata* (chordates). The other main branch of the *Bilateria* is represented by the *Protostomia* (see), which differ from the *D.* embryologically and in other ways.

Deuterotoky: production of both male and female progeny; see Parthenogenesis.

Deutocephalon: see Mesoblast.

Deutocerebrum: see Brain.

Deutoplasm: see Cleavage, Yolk.

Development: directional change; transition from lower to higher, from simple to complex.

The final form of every organism is the result of the following two types of D. 1) Individual D. or ontogenesis, which embraces all the changes in form that occur between the fertilized egg and the adult, senile organism. 2) phylogenetic D. or evolution.

Species-specific D. is genetically controlled, so that developmental physiology and genetics, in particular molecular genetics, are closely interactive. The entire developmental potential of any organism is encoded in the genome. The realization of this genetic information, i.e. establishment of the species-specific metabolism, growth and organ formation, can, however, be modified by environmental factors. In plants, the D. of the organism is particularly influenced by light (photomorphosis) and temperature, but also by the supply of water and nutrient salts, as well as by gravity (geomorphosis).

In higher plants, individual D. is made possible by fertilization, and it starts with embryogenesis and seed D. Seed germination marks the beginning of the active independent existenceof the individual plant. The entire remaining process of D. then consists of the growth of individual cells, differentiation and organ formation. New individual developmental cycles continually arise through reproduction, but each individual is eventually removed from its own developmental cycle by senility and death.

The D. of a plant is not continuous, but occurs in a usually regular succession of discrete stages. Vegetative stages are followed by flower formation, i.e. the plant passes into its reproductive phase. Frequently, primary or juvenile stages differ from the adult forms (see Heteroblastic development). Active periods of growth and/or D. are often followed by dormant or resting stages. An extraordinary variety of individual processes is involved in plant D. In plants, the essentially important phase of ontogentic D. consists organ formation. This occurs either autonomously, or under the influence and control of Correlation (see) and the environment. See the following entries: Leaf, Flower formation, Tuber formation, Shoot, Root formation.

In animals, individual D. can be divided into the following four periods.

1) *Embryogenesis* or *embryonal D.* This includes all stages of D. leading to the independent, viable young animal, which then hatches from the egg or leaves the parent by parturition. It includes: a) blastogenesis, b) formation of the germ layers, with *gastrulation* and *mesoderm formation* (see Mesoblast), c) separation of organ primordia, followed by organ formation *(organogenesis* or *morphogenesis),* d) histological differentiation of organs *(histogenesis),* as well as e) all the growth processes of the embryo or fetus, and f) hatching or parturition.

2) *Postembryonal D.* or *juvenile D.* In direct D., this consists of the continued and progressive formation of organs, in particular the sex organs and secondary sex

structures, so that finally the young animal has all the essential characteristics of an adult. Indirect postembryonal D. proceeds by the complicated processes of metamorphosis via one or more larval stages.

3) *Adult phase.* The period of high bodily performance, sexual maturity, and the generation of new individuals by reproduction. This phase merges imperceptibly into:

4) *Senescence.*

Developmental cycle: the regular sequence of developmental stages during the life of an individual organism, in particular the totality of the developmental processes of a parasite from the egg to the sexually mature animal. The D.c. may be direct (without an intermediate host) (e.g. see Geohelminths), or it may take place in one or more intermediate hosts. If, in the latter case, a suitable intermediate host is not available, the parasite dies.

Developmental physiology, ***causal morphology, experimental biology, developmental mechanics*** (plate 23): a branch of physiology which attempts to describe and causally explain the development of genetically determined primordia, in combination with the environment, to produce the organism in its final form. According to Roux, developmental mechanics is the study of the causality of organically based form and structure.

The same approach and methodology are used for investigating the D.p. of both plants and animals. In both cases, the purpose is to establish the causes of the developmental process, and to derive laws governing the generation of form and structure. The methodology and results of molecular biology and molecular genetics are particularly important in D.p., and they have led to the identification of the causes of physiological and morphological differences that occur during development. Thus, a succession of changes in RNA and protein composition have been demonstrated during embryonal development, while the DNA composition remains unaltered. This biochemical differentiation, also known as *differential gene activation,* is the basis of physiological and morphological differentiation. Differential gene activation can also be observed microscopically on the giant chromosomes of *Diptera* and of certain members of the plant family *Papilionaceae.* During larval metamorphosis or flower formation, regions of intense RNA synthesis appear on the chromosome. These are recognizable as distended loops or rings, known as Puffs (see), around the chromosomal axis, and they occur one after the other at different sites along the chromosome, i.e. they arise on different genes. The molecular switching mechanisms that initiate these changes, are, however, largely unexplained. Regulatory processes analogous to those in the Jacob-Monod model may contribute, but histones and nonhistone chromosomal proteins, as well as translation control mechanisms also appear to be involved.

Due to its particular structure and size (up to 8 cm in length), the unicellular green alga, *Acetabularia,* is a useful organism for experimental nucleus transplantation and grafting. It has been used to show convincingly that differential gene activation can lead to the production of RNA for morphogenesis.

In the experimental history of D.p., the early work of Spemann on the Transplantation (see) of parts or individual cells of amphibian embryos is particularly important. In particular, this work showed that after a certain stage of development cells behave according ot their origin, i.e. they are "determined" or "programmed" to develop into a particular type of tissue or organ, irrespective of their environment (see Induction, Organizer) Transplantation has now been refined to the point where cell nuclei can be transplanted from one cell to another.

Transplantation methodology led to the development of the technique of Explantation (see), in which a piece of living tissue is not transferred to another embryo, but is cultured in a suitable nutrient solution. Using this approach, it is possible to investigate the influence of particular substances or extracts on the further development of the embryo fragment, or to determine the developmental potential of the explant when it is not under the influence of neighboring cells in the embryo. The sandwich method of Holtfreter is especially useful for studying Induction (see) in the amphibian embryo: two pieces of ectoderm are isolated from the early gastrula, and the tissue to be tested for inductor activity (or a small agar block saturated with an extract to be tested) is sandwiched between them; the rims of the ectodermal pieces fuse with one another and completely enclose the central tissue or agar block.

The experimental objects of D.p. are potentially as many and varied as plant and animal forms themselves. There are, however, standard experimental organisms which offer a number of advantages. Thus, any required number of sea urchin eggs can easily be obtained from marine biological stations. They can be artificially fertilized, and since they are transparent their subsequent development is easily followed, even at high magnification. The amphibian egg is also a popular experimental object; for this purpose, *Xenopus laevis* is a popular organism, but eggs of several other amphibians such as newts are also used, Amphibian eggs are also easy to obtain, as well as being relatively large and robust, so they lend themselves readily to transplantation and similar manipulations. Since amphibians are vertebrates, their eggs are suitable for the investigation of special developmental problems associated with vertebrate embryos, and they can even contribute to an understanding of certain aspects of human D.p.

Since research in D.p. is designed to gain an exact understanding of the normal developmental process *(normogenesis),* the developmental physiologist often wishes to know the ultimate fate of each individual part of the uncleaved egg.

The fate of individual regions of the newt *(Triton)* embryo was first determined by using the familiar color marking technique of the German anatomist, Walter Vogt. Each region was traced during the extraordinarily complicated morphogenetic movements of gastrulation, and thereafter until organ differentiation. A small agar block saturated with a nontoxic dye (neutral red, or Nile blue sulfate) was brought into close contact with the embryo surface. At the contact point with the agar, dye diffused into the cells and provided an effective marker, so that the cells could be distinguished for several weeks. The results of numerous marking experiments provide a fate map, showing which regions of the early gastrula will develop into endoderm, mesoderm and ectoderm, and subsequently into the neural tube, the notochord, somites, etc. Since invagination (gastrulation) has not yet occurred, regions of all three germ layers are still on the surface. An elegant modern method for labeling the vertebrate embryo consists of the injection of silicone droplets, e.g. into the early mammalian embryo, to determine the fate of cells during formation of the tropho- or embryoblast. A droplet injected into a blastomere is divided between the daughter cells at each cell division. See Cleavage, Gastrulation, Mesoblast.

Historical. The founders of D.p. were the Halle anatomist Wilhelm Roux (1850–1924) (a pupil of Haeckel), and the Leipzig zoologist and later natural philosopher Hans Driesch (1867–1941). The direction of D.p. was greatly influenced by the work of the Freiburg zoologist, Hans Spemann (1869–1941), who received the Nobel prize in 1935 for his contribution. D.p. has now developed into an extensive discipline, which has provided a useful insight into human normo- and pathogenesis.

Micromethods of biochemistry, metabolic physiology and histochemistry are used increasingly in D.p., creating an area of research known as biochemical embryology. A further area, known as developmental genetics, has developed from the study of the effects of mutations on development, and the effects of inserting cloned genetic material into developing embryos.

Devensian: see Würm.

Devernalization: the decrease or abolition of the effect of Vernalization (see). Plants can be devernalized by several days of heat treatment (20–40 °C), applied immediately after vernalization. Devernalized plants can be revernalized.

Devil's fig: see *Papaveraceae.*

Devonian: a geological period of about 50 million years, representing the first period of the Upper Paleozoic sub-Era, and the fourth of the six periods of the Paleozoic Era (see Geological time scale). All the invertebrate phyla are represented, as well as jawed fishes (gnathostomes) and the oldest labyrinthodont amphibia which still possessed some fish-like characters. Most of the fishes lived in brackish water, or in freshwater lagoons and shallow lakes of the Old Red Continent. Freshwater bivalves and terrestrial snails appear for the first time in the D. The remaining fauna were entirely marine. Subdivisions of the brackish and freshwater D. are based on the development of fishes, important indices being the *Placodermi,* lung fishes and primitive sharks. Stratigraphic subdivisions of the marine D. are based mainly on the corals, brachiopods, goniatites and clymenians, ostracods and conodonts. Also stratigraphically important are the trilobites, tentaculites and crinoids. Graptoliths became extinct at the end of the Lower D. The Lower D. is characterized by brachiopods, the Middle D. by corals, brachiopods and goniatites, and the Upper D. by goniatites, clymenians, ostracods and conodonts. Based on fossiliferous deposits in the Ardennes in Belgium, the marine D. is divided into the following stages: Gedinnian (408–401) + Siegennian (401–394) + Emsian (394–387) = Lower D.; Eifelian or Couvinian (387–380) + Givetian (380–374) = Middle D.; Frasnian (374–367) + Famennian (367–360) = Upper D. (numbers in brackets refer to millions of years ago)

Psilophytes (the oldest plants with vascular bundles and stomata) of the Upper Silurian and Lower and Middle D. were restricted to the junction of land and water (water was required for egg fertilization in the archegonium on the prothallus). This dependence on water was then broken during the D. by the development of the seed and retention of the megaprothallus on the mother plant. The D. is therefore an important period of plant evolution, marking the appearance of the first truly terrestrial plants and the evolution of the first spermatophytes. Botanically, it is divided into three strata: 1) Lower D., characterized by the *Psilophyton* flora, 2) Middle D., characterized by the *Hyeniales* (see) flora; and 3) Upper D., characterized by the *Archaeopteris* flora. Whereas the later psilophytes probably gave rise to the pteridophytes and lycopods (their earliest representatives appear in the Upper D.), the more primitive psilophytes (consisting of telomes with no clear differentiation of stem, leaf and root) probably gave rise to the progymnosperms, which in turn gave rise to the younger spermatophytes *(Coniferophytina* and *Cycadophytina).* These progymnosperms were woody plants occurring from the Middle D. (possibly also the Lower D.) to the Lower Carboniferous. Those in the Middle D. were still homosporous, but heterosporous forms are often found in the Upper D. (e.g. *Archaeopteris).* The remains of seeds in the Upper D. represent the earliest unequivocal fossil evidence of spermatophyes *(Archaeosperma).*

Dextrans: large polysaccharides consisting of D-glucose linked α-glycosidically, mainly by 1,6 bonds, with some 1,3 and 1,4 linkages. D. can be isolated from cultures of lactic acid bacteria, *Leuconostoc mesenteroides* and *Leuconostoc dextranicum* grown anaerobically in sucrose-containing media. The M_r of this microbial product is several million, but smaller molecules can be produced by controlled microbial synthesis. Alternatively, smaller D. can be made by partial hydrolysis of the larger forms. The colloidal pressure of D. with M_r about 75 000 corresponds to that of blood, so they are used in the manufacture of blood plasma substitute. Dextran gels obtained by cross-linking the polysaccharide chain (sold under the name of Sephadex) are used widely as molecular sieves for the experimental and preparative separation of macromolecules.

Dextrins: water-soluble degradation products of starch, produced from the latter by the action of heat, acid, or amylases. According to the color of the reaction products that they form with iodine, they are classified as *amylodextrins* (blue), *erythrodextrins* (red), and low molecular mass *achroodextrins* (no colored reaction product). Heating starch with 3% HCl or HNO_3 produces *acid D. Schardinger D.* are produced by the action of *Bacillus macerans* on starch. The molecule of a Schardinger D. consists of a ring of 6–8 glucose units, linked α-1,4. Depending on the size of the ring, they are called α-, β-, or γ-D. D. (in particular British gum, starch gum or gommelin, produced by dry heating starch at 160 °–200 °C) are used in the manufacture of various commercial products, e.g. thickeners for fabric dyes, sizes for paper and textiles, glues, matches, emulsifying agents, printing inks, etc.

Dhole: see *Canidae.*

Diacetyl, ***butane-2,3-dione:*** CH_3—CO—CO—CH_3, a yellow liquid, which in high concentrations is a respiratory irritant, and in low concentrations smells of butter. D. is a characteristic component of butter aroma, and it has also been found in roast coffee, essential oils, tobacco smoke, human urine, and among the metabolic products of plants.

Diaemus: see *Desmodontidae.*

Diagenodont: a hinge type of the *Bivalvia* (see).

Diageotropism: see Geotropism.

Diagnosis:
1) The recognition and naming of an illness, disease or injury, based on typical symptoms or patterns of symptoms; or the determination of the cause of an illness or injury.
2) See Nomenclature.

Diakinesis: see Meiosis.

Diamond-back moth: see *Yponomeutidae.*

Diamondback terrapin: see *Emydidae.*

Diamond birds: see *Dicaeidae.*

Diapause: in invertebrates (insects in the narrow sense), a normal (endogenously determined) phase of markedly decreased development and metabolism. D. may be initiated by light (decrease in daylength), temperature, humidity and a decrease in the food supply. In insects, D. may occur in the embryonal, larval, pupal, or imaginal stage. It is often part of the Hibernation (see) behavior of the species.

Diapedesis: see Phagocytes.

Diaphoromixis: a type of heterothallism, in which there are many different genetically determined mating types. D. occurs only in the *Basidiomycetes,* which possess allelomorphic series of mating type factors. There are therefore many kinds of genetically determined individuals, and any two may conjugate, provided they carry different mating type factors. These factors may be at 1 locus or at 2 loci. Thus, meiosis can produce 2 or 4 types of genetically differentiated basidiospores on a single basidium. With a single gene locus (A), 2 mating types are produced, i.e. A1 and A2 (dipolar D.). If 2 gene loci (A and B) are involved, then 4 mating types are produced (tetrapolar D.), i.e. A1B1 and A2B2 (parentals), A1B2 and A2B1 (recombinants). Conjugation is possible between A1B2 and A2B2, and between A1B2 and A2B1 (in each case forming A1A2B1B2), but not between any other pairs. Tetrapolar spores are usually produced in equal proportions, i.e. loci A and B are unlinked (on separate chromosomes). See Basidiospore.

Diaphorophyia: see *Muscicapinae.*

Diaphototropism: see Phototropism.

Diaphragm: an important respiratory muscle, which is present only in mammals, and which forms a musculofibrous partition between the thoracic and abdominal cavities. Muscle tissue is less concentrated toward the center of the D., which is therefore more fibrous. When the D. contracts it arches toward the stomach, thereby enlarging the thoracic cavity and promoting inspiration. Relaxation of the D. is accompanied by expiration. In whales, the D. is aligned very obliquely, so that the lungs extend a considerable distance posteriorly and function as hydrostatic organs.

The D. is perforated by three openings: 1. the esophageal opening for passage of the esophagus, certain esophageal arteries and the vagus nerve, 2. the aortic opening for passage of the aorta, the azygos vein and the thoracic duct, and 3. the vena caval opening for passage of the inferior vena cava and some branches of the right phrenic nerves.

Diaphragm respiration: see Respiratory mechanics.

Diapophysis: see Axial skeleton.

Diaptomus: see *Copepoda.*

Diastema: in many mammals, a naturally occurring gap in the tooth row, usually between the incisors and the first premolar. It may be due to the evolutionary loss of one or more teeth (rodents) or to elongation of the jaw (horse, ruminants). In some species (e.g. monkeys), the canine fits into the D. of the opposite jaw. All apes possess a D., but it is absent from fossil and recent humans.

Diastole: see Cardiac cycle.

Diatomaceous earth: see Diatoms.

Diatomales: see Diatoms.

Diatomeae: see Diatoms.

Diatoms, *Bacillariophyceae, Diatomeae:* a large class of algae containing over 10 000 known species. Alternatively, it may be regarded as an algal division *(Bacillariophyta)* or order *(Bacillariales, Diatomales).* They occur in large individual numbers in fresh and marine waters. Some occur in moist soils and other wet substrates. They are solitary or colonial, brown, unicellular organisms, with pigments and reserve materials similar to those of the *Chrysophyceae,* with which they were formerly combined in a single class. They differ from the *Chrysophyceae* by the possession of two silicified walls *(frustules),* which overlap each other like a box with a lid (the larger overlapping lid is called the *epitheca,* the smaller component the *hypotheca).* The frustules are also known as valves, and both are loosely referred to as the shell. The cell has a very different appearance when viewed from above and below (valve view) or from the sides (girdle view). Intercalary bands may be present between valve and girdle, and in some species these carry septa projecting into the cell. The siliceous wall is delicately ornamented with pores, ridges or nodules, often in rows, which under the electron microscope are seen to consist of chambers communicating with the cell interior. In motile species, a longitudinal groove (raphe) runs along the line of symmetry of the valves. It is thought that protoplasm, flowing from the cell and along the raphe, propels the cell by friction with the substratum. Assimilatory products are the polysaccharide, chrysolaminarin, and a copious oil.

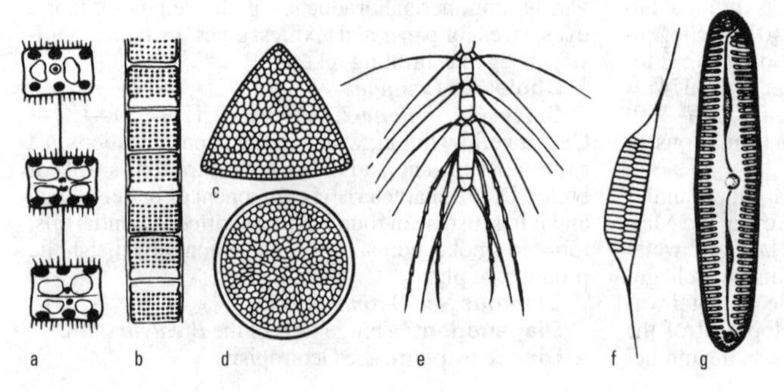

Diatoms. a Coscinodiscus polychordus (floating colony with gelatinous threads). *b Melosira varians* (cells associated into a chain). *c Triceratium favus. d Coscinodiscus sp.* (valve view). *e Chaetoceros coarctatus* (cells associated into a chain). *f Rhizosolenia eriensis. g Pinnularia viridis* (valve view).

Both asexual and sexual reproduction are observed. Asexual reproduction occurs by longitudinal fission, in which the two frustules are pushed apart at the girdle. The frustule of each daughter cell serves as the epitheca, and each cell synthesizes a new hypotheca. One daughter cell is therefore the same size as the parent cell, whereas the other is smaller, leading to a decrease in size at each fission. When a certain minimal size is attained, sexual reproduction occurs; the resulting zygote (auxozygote) grows to the original size, then forms two new frustules.

Two orders are recognized, according to the type of sexual reproduction and the bilateral or radial symmetry of the frustules.

1. Order: *Centrales* (or suborder *Centricae).* These are radially symmetrical, with round or bluntly triangular frustules, which sometimes have bizarre appendages. They are nonmotile, and many are marine plankton. Sexual reproduction is by oogamy; the haploid ova and spermatozoids are formed by meiosis, and the spermatozoids carry flagella with flimmer hairs. The diploid auxozygote grows to full size, then develops into a new individual. Examples are *Biddulphia* and *Melosira,* which form long chains.

2. Order: *Pennales* (or suborder *Pennatae).* These bilaterally symmetrical, elongated, rod- or boat-shaped cells display gliding motility in contact with a surface. Sexual reproduction is by isogamy; two nonflagellated isogametes conjugate to form an auxozygote. The *Pennales* are predominantly bottom dwellers, found on aquatic plants and in the bottom mud. Some species associate into long chains or form star-shaped colonies. Examples are *Navicula, Pinnularia, Pleurosigma, Tabellaria, Aterionella, Synedra.*

Fossil D. occur as early as the Jurassic. In the Tertiary and Interglacial periods they formed massive deposits of diatomaceous earth (Kieselguhr), which is mined (e.g. at Lompoc, California) as a white, friable, highly porous rock. In its crushed form, it is used technically as an insulator and abrasive, as well as serving as a highly insoluble, inert material for filtration and chromatography.

D. are used as indicator organisms in hydrobiology, and they contribute to the bioloigal purification of waste water.

Diauxie: the biphasic growth of a microorganism in a culture medium containing two different growth substrates. Utilization by the organism of one substrate inhibits the formation of the enzymes necessary for utilization of the second substrate. Only when the first substrate is exhausted, can the organism adapt its metabolism to the utilization of the second substrate. The resulting growth curve shows two logarithmic phases separated by a shoulder (Fig.).

Logarithm of the number of living cells

Time

Diauxie. Diauxic growth of a culture of microorganisms. At ↓ the culture begins to utilize the second substrate.

Dibamidae, ***flap-legged skinks, Asian blind lizards:*** a small family of the *Lacertilia* (see), containing only 3 species in a single genus *(Dibamus),* which occur in Indochina and throughout the Indo-Australian archipelago as far as Papua New Guinea. Some authorites also include the Mexican blind lizard (see) in this family. They are 15–30 cm-long, wormlike, short-tailed reptiles. The eyes are lidless and covered with skin, and the external ear openings are covered with scales. Females lack all 4 limbs, while males lack forelimbs but retain small, flaplike vestiges of the hindlimbs. They live in rotting wood or in the earth, and the skull is modified to a compact excavating organ. The head scales are large and platelike, especially on the snout and lower jaw. Body and tail scales are smooth and overlapping, and without a bony underlay. The teeth (absent from the palate) are small and curved backward. They feed mainly on ants and termites. *Dibamus* species are are thought to lay eggs; at least the remnants of egg shells have been found and attributed to *Dibamus.* Most familiar of the 3 species is *D. novaeguineae,* found in the humus layer of tropical forests in Papua New Guinea and the Sunda Islands.

Dibamus: see *Dibamidae.*

Dibatag: see *Antilopinae.*

Dibothriocephalus: see Broad fish tapeworm.

Dibranchiata: see *Cephalopoda.*

Dicaeidae, ***flowerpeckers*** and ***diamond birds:*** an avian family containing 54 species in the order, *Passeriformes* (see). With the exception of 2 species of *Dicaeum* that reach the Palaearctic, they are all inhabitants of the Old World tropical rainforests. They show their greatest diversity in the Papuan region, which is thought to be the center of their evolutionary radiation. Family characteristics are a short, stumpy tail, fine serrations on the terminal third of the bill (which has many different shapes), a semi-tubular tip to the tongue, and reduction in the length of the outermost primary. They feed on insects, fruit and flowers. ***Dicaeum hirundinaceum (Mistletoe bird;*** 9.5 cm; dense tropical forests of Australasia) feeds on mistletoe berries and is held responsible for the spread of this parasitic plant.

The Australian and Tasmanian ***Diamond birds*** or ***Pardalotes*** (members of the genus *Pardalotus),* which have characteristic bright spots on their plumage, are rather aberrant members of the family. Thus, they have shorter, deeper bills than the flowerpeckers, and these are not serrated but notched behind the tip. In addition, most diamond birds build a nest at the end of a tunnel, which they dig into the earth, whereas most flowerpeckers weave a deep, bag-shaped nest from plant down and fibers, which they suspend from a high branch.

Dicaeum: see *Dicaeidae.*

Dicamptodontidae, ***Pacific mole salamanders:*** a family of the *Urodela* (see) comprising 3 species in 2 genera. Pacific mole salamanders are closely related to the *Ambystomatidae* (see), and they are restricted to the Pacific coast of North America. They are brown or gray in color, mottled with brown or black above, with an orange or yellow belly. The body is sturdy with vertical rib grooves along the sides, the head is broad, and the tail is laterally compressed. Eyelids are present and movable. There are 4 fingers and 5 toes, and the lungs are small. Breeding takes place in water, and fertilization is internal. The adults are terrestrial.

Rhyacotriton olympicus ***(Olympic salamander)*** is a 10 cm-long, white-spotted species found in mountain streams in the coastal forests of Oregon and Carolina. It is an accomplished swimmer, using lateral body undulations. The lungs are degenerate, but still functional.

The other 2 species are *Dicamptodon copei* ***(Cope's giant salamander)***, which is paedomorphic, and *D. ensatus* ***(Pacific giant salamander)***.

Dicentric: a term applied to chromosomes or chromatids, which have two Centromeres (see) as a result of mutation. Normally only one centromere is present, and the chromosome is therefore normally monocentric.

Diceros: see *Rhinocerotidae*.

Dice snake, *Natrix piscator:* the commonest aquatic colubrid snake (see *Colubridae)* of South and Southeast Asia. It lies in the water, snout tip at the surface, waiting for fish and frogs. It is especially well known in flooded rice fields, and it is given to biting.

Dichasium: see Stem, Flower.

2,4-Dichlorophenoxyacetic acid, *2,4-D:* a synthetic auxin of great practical importance. It is used primarily as a selective weedkiller for the protection of cereal crops, as well as a growth-stimulant for the generation of seedless fruits (e.g. lemons, oranges). See Parthenocarpy.

Cl
Cl—(benzene ring)—O—CH_2—COOH

2,4-Dichlorophenoxyacetic acid

Dichogamous flower: see Pollination.

Dichogamy: see Pollination.

Dichopatric, ***macrodichopatric:*** used of populations or species whose geographical ranges are separate to the extent that individuals from the two populations never meet and gene flow does not occur.

Dicondylia: the main group of ectognathous insects *(Zygentoma* plus *Pterygota).* Their mandibles are doubly articulated, i.e. they have two condyles, in contrast to the monocondylar mandibles of the *Archaeognatha.* See Insects.

Dicots: see *Dicotyledoneae*.

Dicotyledoneae, *Magnoliatae,* ***dicotyledonous plants, dicotyledons, dicots*** (plate 15): a class of angiosperms. With few exceptions, they are characterized by the presence of two lateral cotyledons in the embryo. The open vascular bundles are arranged cylindrically, and appear as a circle in the transverse section of the stem. Parallel strands of vascular bundles travel the length of the shoot axis and its branches, eventually opening into the leaves. Secondary thickening is an almost universal character. Venation of foliage leaves is generally reticulate. Leaves display a wide variety of shapes, and often have a petiole. Flowers with pentamerous whorls (including calyx and corolla) predominate, but dimerous, trimerous and tetramerous flowers also occur. The primary root is usually richly branched and long-lived (allorhizy). Chemically, dicots are characterized by the widespread occurrence of essential oils, alkaloids and tannins. Ellagic acid is a common constituent, and most of the saponins found in dicots have a triterpenoid structure. Calcium oxalate occurs in different crystalline forms.

The approximately 172 000 species of monocots are assigned to 6 subclasses. About 10% of these species are found in the subclasses *Magnoliidae* and *Caryophylliidae,* the remaining 90% being distributed among the *Dilleniidae, Hamamelididae, Rosidae* and *Asteridae.*

The *Magnoliidae (Polycarpiaceae)* (e.g. families *Magnoliaceae, Piperaceae, Ranunculaceae, Papaveraceae)* exhibit the most primitive characteristics; they are not only considered to represent the phylogenetic base of the dicots, but are also thought to have close affinities with the ancestors of the *Monocotyledoneae*. Their gynaecium is predominantly choricarpous (numerous free carpels); in addition, the carpels are spirally arranged, as are the numerous stamens.

The *Caryophyllidae* (e.g. families *Cactaceae, Polygonaceae, Caryophyllaceae)* appear to represent an early direct and separate branch of the *Magnoliidae*, there is a tendency to the reduction of stamens to a single whorl. Gynaecia tend to be paracarpous, with a central placenta and a reduction in the number of ovules. They are strongly specialized with respect to habitat and chemical constituents.

The *Dilleniidae* (e.g. families *Paeoniaceae, Brassicaceae, Ericaceae)* and *Rosidae (Rosiflorae)* (e.g. families *Crassulaceae, Rosaceae, Papilionaceae, Rutaceae, Euphorbiaceae)* represent separate but related lines of development from the *Magnoliidae.* They are not especially characterized by any primitive or derived characters. All parts of the corolla are usually free and arranged in 5 whorls. There is a secondary increase in the number of stamens, which are arranged centrifugally in the *Dilleniidae* and centripetally in the *Rosidae.*

The *Hamamelididae (Amentiflorae)* (e.g. families *Fagaceae, Platanaceae, Betulaceae, Urticaceae)* are derived from the *Magnoliidae* or related primitive members of the *Rosidae.* They are adapted to wind pollination with accompanying flower reduction, and they are almost always woody. Flowers are often unisexual and ovaries are generally coenocarpous. The *Hamamelididae* represent a very ancient assembly of forms which are perhaps not closely related, but have evolved in parallel.

The *Asteridae (Sympetalae tetracyclicae)* (e.g. families *Gentianaceae, Rubiaceae, Convolvulaceae, Solan-*

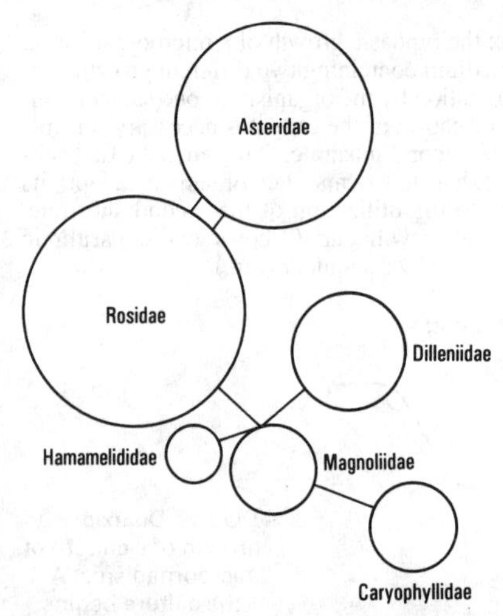

Dicotyledoneae. Presumed evolutionary relationships of the six subclasses of dicotyledonous plants.

ceae, Compositae), considered to be the most advanced subclass of dicots, probably evolved from the *Rosidae.* Members of the *Asteridae* show the greatest number of derived characters, e.g. fusion of perianth segments, arrangement of flower parts in 4 whorls, a maximum of 5 stamens, and usually only 2 carpels.

Dicotyledonous plants: see *Dicotyledoneae.*

Dicotyledons: see *Dicotyledoneae.*

Dicotyles: see *Tayassuidae.*

Dicranales: see *Musci.*

Dicranium: see *Musci.*

Dicrocoelium: see Liver flukes.

Dicruridae, *drongos:* an avian family of about 50 Old World tropical species in the order *Passeriformes* (see). They occur in Africa and Madagascar, and are particularly concentrated in the Indo-Malayan region. They feed on insects and small vertebrates, and are able to catch insects from the air. All species belong to the genus *Dicrurus,* with the exception of the monospecific ***Mountain drongo*** *(Chaetorhynchus papuensis)* of New Guinea. Plumage is generally metallic black; some species have crests and often a red iris. The tail of 10 to 12 feathers tends to be long, and it is especially well developed in the southern Asian ***Greater racket-tailed drongo*** *(D. paradiseus).* Two other southern Asian species, the ***Crow-billed drongo*** *(D. annectans)* and the ***Black drongo*** or ***King crow*** *(D. macrocercus),* are difficult to distinguish in the field from *D. paradiseus.* Drongos are solitary and aggressive, and they fiercely attack birds of prey.

Dictyonema (Greek *diktyon* net, *nema* thread): a genus of the graptolite order, *Dendroidea,* containing more than 200 known species (see Graptolites). The rhabdosome has a rigid basket- or funnel-shape, consisting of practically parallel branches, joined at intervals by tiny transverse bars, called dissepiments. Fossils are cosmopolitan, and found from the Upper Cambrian to the Lower Carboniferous. It is important as an index fossil of the Tremadoc series and *Dictyonema* slate of the Ordovician.

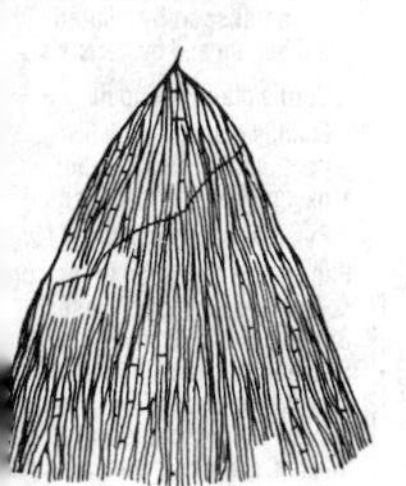

Rhabdosome of *Dictyonema* (x 0.4)

Dictyoptera: according to some authorities, an order of insects consisting of two suborders: *Blattodea* (cockroaches) and *Mantodea* (mantids). In the present work, cockroaches and mantids are accorded ordinal status, and referred to as *Blattoptera* and *Mantoptera,* respectively.

Dictyosome: see Golgi apparatus.

Dictyostelium: see *Myxomycetes.*

Dictyotales: see *Phaeophyceae.*

Didelphia: see *Metatheria.*

Didelphidae: see Opossums.

Didermocerus: see *Rhinocerotidae.*

Didunculinae: see *Columbiformes.*

Didunculus: see *Columbiformes.*

Dieffenbachia: see *Araceae.*

Diencephalon: see Brain.

Differential affinity: a phenomenon of meiotic chromosome pairing, in which homologous chromosomes preferentially or exclusively pair with one another, rather than with partly homologous chromosomes that are also present. The extent of D.a. is generally determined by the size of the homologous and nonhomologous regions in the partly homologous chromosomes. In addition, D.a. is influenced by absolute chromosome size, chiasma frequency and chiasma distribution in the respective chromosomes.

Differential blood count: see Hemogram.

Differential distance: see Interference *(1).*

Differential sensitivity: the ability of a sensory system to distinguish between different intensities of the same stimulus. See Weber's law.

Differentiation center: a term applied in 1924 by Seidel (University of Marburg) to the site in the blastoderm of the insect egg, where differentiation begins and the first segments become visible. Differentiation commences midventrally in the region of the presumptive thorax. The central plate first separates from the two lateral plates, then the primitive pit begins to sink and later to close, the boundaries of the first segments become apparent, and the first leg buds appear. Thus, the D.c. is the region of maximal differentiation activity in a differentiation gradient, which extends away from the D.c. in all directions.

Diffusion: the movement of particles from one site to another under the influence of thermal agitation. In solution, a molecule is subject to 10^{13} to 10^{15} impacts per second, which result in random movement. On average, the square of the distance moved is proportional to time. According to Fick's law of D., the D. coefficient corresponds to the mass of material diffusing across a plane of area 1 cm^2 in unit time under a concentration gradient of unity. The D. coefficient is also related to the relative molecular mass, and may be used experimentally to assess the sizes of biological macromolecules. D. of a particle in a phase of identical particles is called *self-diffusion.* D. in a single plane, e.g. in a membrane, is known as *lateral D.* Random changes in rotary moment with time are called *rotary D.* Lateral and rotary D. provide information on membrane fluidity and molecular interaction within membranes,

Within cell organelles and cytoplasm, D. is a very important transport process. Over larger distances within living organisms, D. is less efficient and responds less rapidly to alterations of concentration, because the diffusion distance increases with the square of time.

Diffusion plate test: a test for determining the inhibitory or stimulatory properties of chemical compounds on the growth of bacteria, e.g. for determining the drug sensitivity and drug resistance of different strains. The entire surface of a layer of nutrient agar in a Petri dish is inoculated uniformly with test organism. The substances under investigation are placed at specific sites on the surface of the agar. From their sites of application, the test substances diffuse outward into the growth medium. Any influence on the growth of the test organism is seen as a concentric zone of increased or decreased growth in the carpet or lawn of bacteria. The diameter of these zones is related to the concentration and the activity of the test substance. Round holes may be punched in the agar surface and filled with a solution of the substance under test. Alternatively small circles of filter paper impregnated with the test substance may be placed on the agar surface. In a third method, glass or metal rings are placed on the agar surface and filled with a solution of the test substance. Standard strains of microorganisms are used for quantitative determinations, e.g. determination of the concentration of a solution of an antibiotic. The method is also used to determine the sensitivity of unknown strains, e.g. to different antibiotics.

Diffusion potential, ***liquid junction potential:*** the potential difference at the boundary of two solutions of the same electrolyte separated by a permeable membrane. The more concentrated solution will tend to diffuse into the more dilute, the rate of diffusion of each ion being roughly proportional to its velocity in an electric field. If the cation moves more rapidly than the anion, the former will tend to diffuse ahead of the latter into the dilute solution, which will therefore become positively charged with respect to the concentrated solution. If the anion is faster moving, then the dilute solution will acquire a negative charge. In both cases an electrical transition layer (each membrane surface acquires a thin layer of excess positive or negative charge) is formed at the junction of the solutions, in which the attraction between opposite charges limits the diffusion rate of the fast moving ion and accelerates the diffusion of the slower ion, producing a steady state in which both ions diffuse at the same rate. The magnitude of the D.p. depends on the relative speeds of the ions, and its direction is such that the more dilute solution assumes the charge of the faster moving ion.

If both ions are permeable, the D.p. is a nonequilibrium potential, generated by the continual transduction of chemical energy into electrical energy. If one ion is permeable and the other impermeable, an equilibrium is established. The D.p. is then an equilibrium potential, given by the Nernst equation: $E = (RT/zF) \ln (c_0/c_1)$, where z is the valency of the permeant ion ($z < 0$ for anions), and c_0 and c_1 are the bulk concentrations of this ion on either side of the membrane. For the calculation of the D.p. in biological membranes, it is necessary to make simplifying assumptions. Thus, the assumption that the drop of electrical potential occurs only in the membrane, and that it describes a uniform gradient across the membrane, so that the electric field strength is constant in the membrane phase, is known as the "constant field approximation" or "constant field theory".

Digenea: see *Trematoda.*

Digestion: the degradation of food into absorbable substances. All the chemical reactions of D. are hydrolyses, initiated and maintained by very specific digestive enzymes celled hydrolases, e.g. cleavage of proteins to peptones and eventually to amino acids, cleavage of triacylglycerols (neutral fats) to glycerol and free fatty acids, cleavage of polysaccharides to smaller saccharides and eventually to monosaccharides.

Processes of D. in the gastrointestinal tract of vertebrates are described in the accompanying diagram. D. begins in the buccal cavity by the action of carbohydrate-cleaving salivary amylases (collectively known as ptyalin). The mixture of masticated food and saliva enters the esophagus by the act of deglutition (swallowing), and is transported by peristalsis to the stomach. Carbohydrate cleavage continues in the weakly alkaline food brei until the pH is lowered by the hydrochlori acid of the gastric secretion. Protein cleavage then com mences, catalysed by the enzyme pepsin. In some mam mals (pig, hamster ruminants) the stomach is function ally divided into different sections, allowing the spatia separation of carbohydrate and protein D. In ruminants hippopotamuses, sloths, and some leaf-eating monke species, the stomach has a spacious antechamber wher carbohydrates (in particular cellulose) are fermented b bacteria. In ruminants, this antechamber (the rumen) i particularly well differentiated, and it contains both cil iated protozoa and bacteria which contribute to foo degradation. In those herbivores that lack separat stomach compartments, the colon and cecum serve a sites of bacterial carbohydrate fermentation.

Fatty and/or protein-rich foods tend to remain longe in the stomach than carbohydrate-rich foods and thos containing a high content of indigestible fiber. After few hours, the stomach contents are delivered batch wise into the duodenum, where they are made weakl alkaline by secretions of the pancreas and walls of th intestine. This increase in pH inhibits the action of th gastric enzymes and promotes the activity of the pan creatic and intestinal enzymes. Bile is necessary for th D. of fats and the absorption of the products. The solu ble cleavage products from all food materials are ab sorbed throughout the intestinal tract, and especially i the small intestine. In the large intestine (colon) the bre of food residues is concentrated by water absorption with the production of feces. Food residues are attacke further by bacteria in the colon and cecum.

Digestion. Structural and functional divisions of the gastrointestinal tract (schematic and largely representative of vertebrates).

Digestion cyst: see Cyst.

Digestive system: the totality of organs and orga nelles that contribute to the intake of food, its degrada tion and absorption, and the excretion of indigestibl residues. Many different types of digestive system are found in the animal kingdom, the structural plan depending largely on the mode of nutrition (Figs.1 & 2).

Apart from specialized structures for capturing or ingesting food, the D.ss. of unicellular animals consist simply of *food vacuoles,* which are formed transitorily in the cytoplasm. In contrast, metazoans have evolved specialized organ systems for digestion. The highly evolved D.s. of higher animals is known as the *alimentary* or *intestinal system* (gastroma), which is basically a hollow cavity (gastrocoel) lined with epithelium. Phylogenetically and during embryonic development, the D.s. arises from the archenteron, which is formed from the endoderm of the gastrula. Only certain parasites (e.g. *Mesozoa* and *Cestoda)* actually lack a D.s.

Sponges do not possess a true D.s. Simple sponges have a uniform cavity *(spongiocoel* or *paragaster),* lined by the *gastral layer,* which is formed from collared flagellate cells, known as *choanocytes.* The paragaster opens to the exterior via the *osculum,* which serves as an exit rather than an entrance. Water enters the paragaster via numerous intra- or intercellular pores on the body surface, which are continuous with a system of canals within the dermal layer. The structure of this canal system and the arrangement of the choanocytes differ according to the sponge species.

The ***coelenterate*** D.s. typically has a single opening to the exterior, which serves for both ingestion and egestion. The enteron serves for both the digestion of food and the distribution of nutrients in the body. Accordingly, it branches into smaller vessels, and is known as the *gastrovascular system.* In medusae, this gastrovascular system consists of a *peripheral ring canal,* connected to the central gastric cavity by 4 *radial canals.* In the *Ctenophora* (comb jellies), the gastric cavity consists essentially of a central canal with branches to 8 peripheral longitudinal canals. The gastric cavity of the *Anthozoa* is divided radially into *gastric pouches* by 6 or 8 vertical folds of the body wall *(gastric septa).* The central free edge of each septum is folded and enlarged to form the *mesenteric filament,* whose gland cells produce a digestive secretion. *Hydropolyps* possess a single opening or mouth at the apex of the anteriorly situated oral cone (hypostome), which communicates with the uniform gastric cavity; the tentacles arise from around the base of the oral cone.

A single aperture, serving both for the intake of food and expulsion of waste, is also found in the *Turbellaria* (free-living flatworms) and *Trematodes* (flukes), both being classes of the *Platyhelminthes.* As an adaptation to their parastic mode of existence, the *Cestoda,* which constitute the third class of *Platyhelminthes,* lack any type of alimentary canal. In the simplest case, the platyhelminth D.s. consists of a single anteriorly directed and 2 posteriorly directed alimentary branches, e.g. in the *Turbellaria.* These branches arise directly from the muscular *pharynx,* which is a wide, ventrally located, often eversible sac-like structure. In larger forms, e.g. *Planaria* and *Fasciola,* the D.s. consists of a richly branched *gastrovascular system.*

All remaining animals have evolved an anus, which is derived from from an ectodermal invagination. Secondary loss of the anus has sometimes occurred, e.g. in ant lion larvae and some mites. In the *Protostomia* the blastopore (or part thereof) of the gastrula develops into the mouth, whereas in the *Deuterostomia* the blastopore becomes the anus, and a new mouth is formed ventrally at the other end of the enteron. With the evolution of the anus, the alimentary canal (gut) then consists of an ectodermal *foregut* or *stomodeum* (in vertebrates represented only by the anterior part of the buccal cavity), an endodermal *midgut, mesenteron* or *mesodeum,* and an ectodermal *hindgut, end gut* or *proctodeum.* In arthropods, those gut sections derived from the ectoderm are lined with a layer of chitin (intima), which is discarded at every molt. Each of the 3 gut divisions serves a special function. Food is captured, disintegrated and softened in the foregut, which also serves as a site for some early reactions of digestion. Cleavage of food material into simpler, soluble and absorbable compounds is performed by the midgut, the necessary enzymes being secreted by the associated glands. Then hindgut reabsorbs water, and receives and expels undigested food residues and waste material.

In all animals that possess a coelom, the gut wall consists of several layers. In arthropods, the intestinal wall from the interior to the exterior consists of: 1) a single cell layer known as the *intestinal epithelium,* 2) a layer of connective tissue *(tunica propria),* and 3) an outer muscular layer *(tunica muscularia)* consisting of both circular and longitudinal muscles. The two inner layers constitute the *intestinal mucosa.* Under the control of the autonomic nervous system, wave-like contractions (peristalsis) of the wall musculature pass backward along the gut, and generally serve to carry the gut contents from the mouth to the anus. In those animals with a secondary coelom, the gut is attached by a dorsal and ventral Mesentery (see) to the wall of the abdominal cavity.

Invertebrates. The invertebrate foregut is variously structured, according to the animal's mode of existence and nutrition. It starts with a mouth, which is provided with more or less complicated structures for the capture and mechanical disintegration of food or prey, e.g. the chitinous teeth and spines of the *Aschelminthes,* the chitinous plates or jaws of blood-sucking leeches, the horny upper and lower jaws (beak) of cephalopods, etc. In arthropods, these functions are performed by the Mouthparts (see), which do not arise from the foregut, but are in fact modified appendages, arranged around the mouth in the form of an antechamber.

Food passes through the mouth into the more or less spacious mouth cavity, which receives the ducts of the salivary glands. In the primitive condition, salivary glands provide lubrication for the passage of food, and they are absent from most aquatic animals, e.g. from all bivalves and aquatic crustaceans. In many terrestrial animals, e.g. herbivorous snails and insects, the salivary secretion also contains carbohydrases which initiate the enzymatic degradation of carbohydrates in the mouth cavity. Salivary glands may be modified for other functions, e.g. the venom glands of many spiders and insects, and the spinnerets of many lepidopterous larvae are derived from salivary glands. The mouth cavity is followed by the muscular *pharynx,* which may be provided with powerful jaws (e.g. in polychaete worms) or with a Radula (see) (e.g. in mollusks). From the pharynx, food is transported by the *esophagus* to the subsequent sections of the alimentary tract for digestion. In some animals, the esophagas is dilated or has a lateral chamber, which retains food for some time and assists in the mechanical and enzymatic degradation of resist-

Digestive system

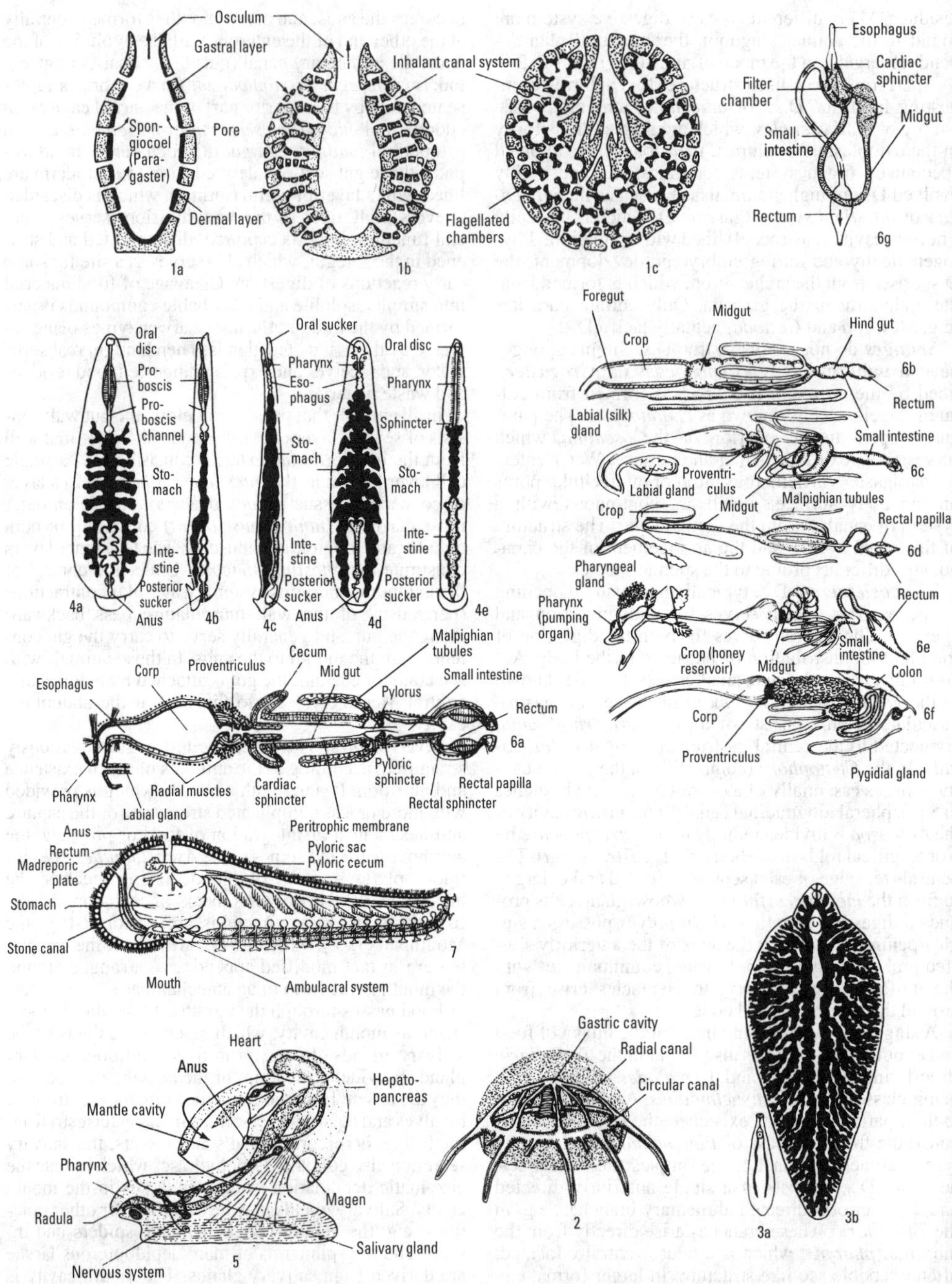

Fig.1. *Digestive system.* Invertebrate digestive systems. *1* Schematic longitudinal sections of the three types of structure found among sponges: *1a* ascon, *1b* sycon, *1c* leucon. *2* Gastrovascular system of a medusa *(Coelenterata)*. *3* Alimentary canal of trematodes: *3a* small liver fluke *(Dicrocoelium dendriticum)*, *3b* large liver fluke *(Fasciola hepatica)*. *4* Alimentary canal of leeches: *4a* and *4b Brachiobdella paludosa* (not a true leech, but a greatly modified oligochaete worm strongly resembling a leech), *4c* and *4d* Medicinal leech *(Hirudo medicinalis)*, *4e* common dog leech *(Erpobdella octoculata)*. *5* Schematic longitudinal section through a snail of the subclass *Prosobranchiata*. *6* Digestive systems of insects: *6a* schematic longitudinal section through the alimentary tract of an idealized insect; alimentary tracts of a silkworm caterpillar *(6b)*, a cockroach *(6c)*, a mosquito *(6d)*, a bee *(6e)*, and a carnivorous beetle *(6f)*; the simple filter chamber of a homopteran insect *(6g)*. *7* Digestive tract of a starfish.

nt material. This structure, known as a *crop,* occurs primarily in herbivorous invertebrates (snails, annelids, insects). In sucking insects, the crop (ingluvies) is a stalked, sac-like appendage of the esophagus. As an adaptation to a fluid diet, spiders possess a complicated *sucking stomach,* which is actually part of the esophagus. In many insects, the crop is followed by a *proventriculus* or *gizzard,* which contains chitinous teeth or ridges. A circular fold of the foregut epithelium extends from the end of the proventriculus into the midgut, forming a nonreturn valve (valvula cardiaca).

Generally, the midgut is differentiated into two main parts; these are the *stomach* (ventriculus, gaster, stomachus) where digestion occurs, and the *small intestine* (mesogaster) where degradation products are absorbed from the fluid emulsion (chyme). Sometimes, the midgut is not differentiated and consists of a simple tube, e.g. in the *Aschelminthes.* In many species, however, the internal surface area of the midgut is enlarged. This is achieved in flatworms by regular symmetrical evaginations of the intestinal wall, while in annelids (e.g. earthworm) the absorptive surface of the intestine is increased by an internal dorsal ridge, the *typhlosole.* In mollusks, the anterior part of the midgut is expanded into a stomach (usually tubular). A diverticulum of the lamellibranch intestine secretes a Crystalline style (see), which represents a special provision for the digestion of carbohydrates; it is also present in some gastropods. Variously shaped and sized glandular evaginations (diverticula) of the intestinal wall empty their contents at the junction of the stomach and the small intestine. Thus, the midgut of arachnids and crustaceans is a simple, often relatively short tube, but it gives rise to large tubular or branched midgut diverticula, known as "hepatic ceca". In crustaceans (e.g. crawfish), digestion and absorption occur exclusively in these hepatic ceca, which produce all the digestive enzymes. Although these ceca are called "hepatic" and are also sometimes referred to as a "liver", they are obviously not homologous with the vertebrate liver. In insects, *pyloric ceca* arise from the junction of the midgut and foregut; these vary greatly in number and may be numerous and arranged in a complete circle around the gut (the primitive condition), or they may be paired. Otherwise, the insect midgut is usually a more or less wide, uniform tube. The midgut epithelium in insects has no mucus glands, and although it also lacks a chitinous lining, special cells of the proventricular region (which may be ectodermal in origin) secrete a chitinous tube (peritrophic membrane), which lies free in the midgut lumen; the food brei is contained within this membrane, thereby preventing the food from making contact with the epithelium. The midgut of the *Asteroidea* (starfishes) consists of a sac-shaped stomach, the upper part of which is five-sided, each angle giving rise (separately or via a common duct) to a pair of pyloric ceca or digestive glands (one pair in each arm). In contrast, sea urchins have a long, tubular, coiled intestine, accompanied by a small cylindrical tube, the *siphon,* which opens into either end of the midgut. This siphon arises by abstriction of the ciliated longitudinal groove of the intestine; it never contains food material, and in nonfeeding sea urchins it generates a water current toward the anus (partly by waves of contraction, partly by the action of cilia), which serves for respiratory gas exchange.

The last section of the alimentary canal of invertebrates is ectodermal in origin. It is known as the hindgut or end gut, and it exits to the exterior via the anus. In the *Aschelminthes,* it is a simple, somewhat expanded section of the intestine, often terminally enlarged to form a cloaca, which also receives the ducts of the sex organs. In blood-sucking leeches, the mucosa of its anterior section is thrown into transverse folds, and its posterior section forms a smooth rectum. In arthropods the hindgut, like the ectodermal foregut, is lined with chitin, and it is usually much longer than the endodermal midgut.

In spiders, the hindgut is short and expanded into a *stercoral pocket* where feces accumulate; this pocket also receives the Malpighian tubules (see Excretory organs). In scorpions, the simple, tubular hindgut begins where the Malpighian tubules enter.

Three clearly differentiated sections are usually recognizable in the insect hindgut: pylorus, intestinum and rectum. The *pylorus* marks the beginning of the hindgut and is the site of entry of the Malpighian tubules; it consists of raised epithelial cells and circular muscles, and serves as a nonreturn valve between the midgut and the hindgut. A large circular fold (valvula pylorica) is often present where the pylorus merges into the intestinum. The usually tubular *intestinum* (or "small intestine") conducts food residues and secretions from the Malpighian tubules into the rectum. Within the normally short, expanded, globular *rectum,* food residues are formed into fecal balls or pellets. Often the intestinum empties laterally into the rectum, which is then called a *rectal sac, rectal pouch,* or *rectal ampulla.* Especially large rectal ampullae are found in the larvae of *Scarabaeidae* (chafers and dung beetles), where they house symbiotic microorganisms and serve as fermentation chambers for the degradation of cellulose-containing dietary plant material. Rectal ampullae also serve as fermentation chambers in wood-eating insects, e.g. *Cerambycidae* (longhorn beetles) and termites. In aquatic forms (e.g. dragonfly larvae), the rectum may serve for respiratory gas exchange (intestinal respiration). In many insects, the rectum epithelium contains thickened patches known as rectal glands or rectal papillae, which are sites of active water resorption. Rectal glands are, however, absent from most *Coleoptera, Hemiptera* and *Ephemeroptera,* and from the larvae of many of the more advanced insects. As an adaptation to the intake of large volumes of liquid, the digestive tract of many sap-sucking insetcs *(Homoptera)* is modified to a filter chamber. Filter chambers are formed by looping of the intestine, so that separate parts are brought into contact with one another. As a rule, the final section of the midgut or the beginning of the intestinum lies against the anterior of the midgut, so that excess dietary water and dissolved carbohydrates diffuse directly across the gut walls from the beginning of the midgut into the hindgut.

Echinoderms characteristically possess a short hindgut. In sea urchins it is provided with a few irregularly branched tubes, the rectal diverticula. In holothuroids (sea cucumbers) it is enlarged to a cloaca, which receives the long, branched, usually paired Respiratory trees (see).

Vertebrates. In vertebrates, food is taken in through the mouth, which is bordered by Lips (see). The mouth opens into the buccal cavity, which is derived from the embryonal stomodeum and partly from the embryonal pharynx. In mammals, the buccal cavity contains char-

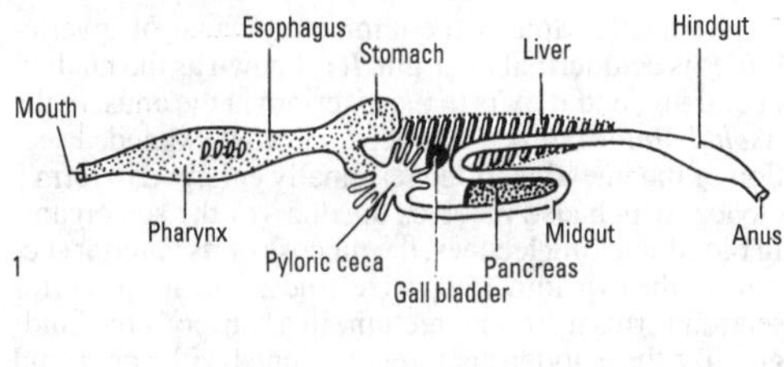

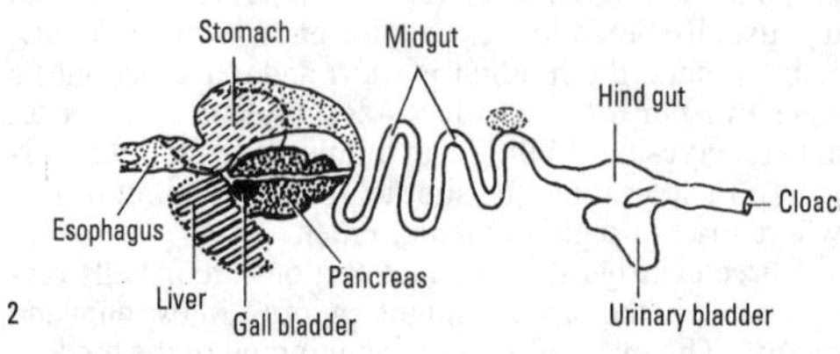

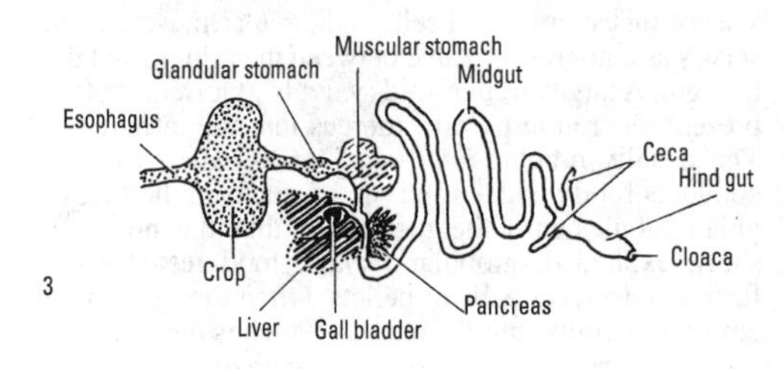

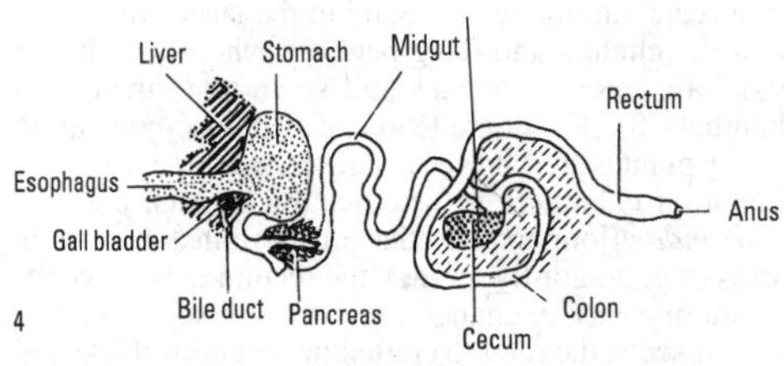

Fig.2. *Digestive system.* Digestive tracts of some vertebrates. *1* Teleost fish. *2* Frog. *3* Pigeon. *4* Guinea pig.

acteristic structures, i.e. Teeth (see), Tongue (see) and salivary glands, any of which may be absent from other vertebrates. The roof of the mouth is formed by a hard palate, which in mammals often has transverse horny ridges. In baleen whales, these horny ridges hang down into the buccal cavity as meter-long baleen plates of baleen or "whalebone". They serve as a filter for trapping small marine organisms, which are removed with the tongue and passed into the pharynx (gullet). Some rodents and monkeys have cheek pouches for the brief storage of food. Salivary glands occur in terrestrial animals; saliva serves as a lubricant, and it contains amylases that initiate the digestion of carbohydrates. The venom glands of snakes are modified salivary glands. Swallows and swifts use an air-drying secretion of their salivary glands as a nest building cement.

After passage through the pharynx (crossing point of the alimentary and respiratory tracts), the food is transported by the *esophagus*. In birds of prey *(Falconiformes)*, fowl-like birds *(Galliformes)*, waterfowl *(Anseriformes)* and others, the esophagus is expanded into a *crop*, which softens and disintegrates hard food. In pigeons, this is the site of production of "crop milk" for feeding the young. The esophagus joins the *stomach* (ventriculus, gaster, stomachus), which in birds is divided into a glandular stomach (for the enzymatic degradation of food) and a muscular stomach (for mechanical disintegration). The mammalian stomach has three regions: *cardia* (a valvular arrangement around the entrance of the esophagus), *fundus* (containing the peptic cells that produce pepsinogen, and the parietal or oxyntic cells that produce hydrochloric acid), and the *pylorus* (the exit from the stomach where the food brei is released into the small intestine). Some division of the stomach into separate chambers occurs in several animals, e.g. in rodents and bats, but the most extreme condition is found in ruminants (Fig.3), where the 4 divisions are known as the *rumen (paunch), reticulum (honeycomb* or *tripe), omasum (psalterium)* and *abomasum*. Contractions of the reticulum wash fluid back into the rumen, followed by contraction of the rumen, which returns material to the reticulum, then to the omasum. The rumen and reticulum serve as fermentation chamber and holding space for cellulose degradation, the omasum is the site of water absorption, while only the abomasum secretes acid and pepsinogen and is homologous with the stomach of other mammals. Generally in mammals, most of the hydrolytic degradations that constitute digestion occur in the *small intestine*, catalysed by enzymes secreted by the intestinal wall and the pancreas; the emulsifying activity of bile from the gall bladder aids the digestion of fats. Generally, the intestine is structurally modified to increase its absorptive surface. Thus, the elasmobranch intestine has a *spiral valve*, the intestine of telost fishes is provided with *ceca*, and the mucosa of the mammalian small intestine is carpeted with *villi*. In addition, the intestine is long, and is packed into the abdominal cavity by coiling and looping. Carnivores have a shorter intestine than herbivores. At the junction of the small intestine and large intestine (colon) is the Cecum (see). The primary function of the *colon* is to reabsorb water. In most vertebrates, the *rectum* opens into the cloaca, but in mammals it opens separately to the exterior via the *anus*.

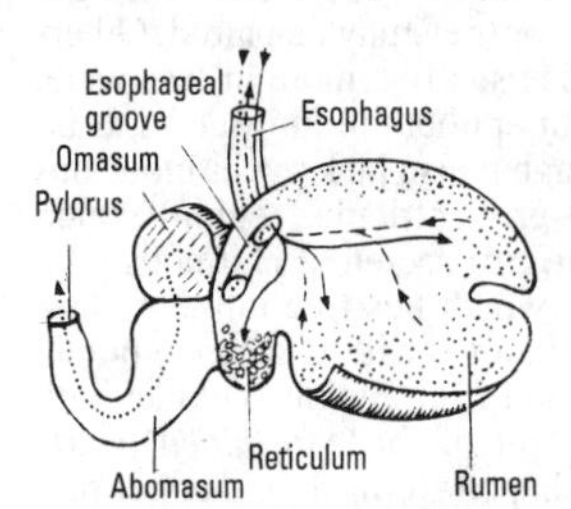

Fig.3. *Digestive system.* Stomach of a ruminant (sheep).

Digitalis glycosides: cardiac glycosides of the cardenolide group found in the leaves of *Digitalis* species, especially *Digitalis purpurea* (purple foxglove) and *Digitalis lanata* (white foxglove), and responsible for the toxicity of these plants. D.g. are formed from naturally occurring *lanatosides* during the preparation of *Digitalis* leaves. In the lanatosides, the digoxose is acetylated at C-3, and this acetylated digoxose also carries a β1,4-glycosidically linked D-glucose residue. The most important D.g. are *digitoxin, digoxin* and *gitoxin* and their aglyca are called *digitoxigenin, digoxigenin* and *gitoxigenin,* respectively. The latter 2 compounds differ from digitoxigenin in the possession of an extra hydroxyl group at C-12 or C-16, respectively. The sugar component is always 3 molecules of D-digitoxose.

The D.g. are important cardiotonic agents, used for long-term treatment of chronic heart weakness and defective heart valves. The pure glycosides are now used instead of leaf powders or extracts.

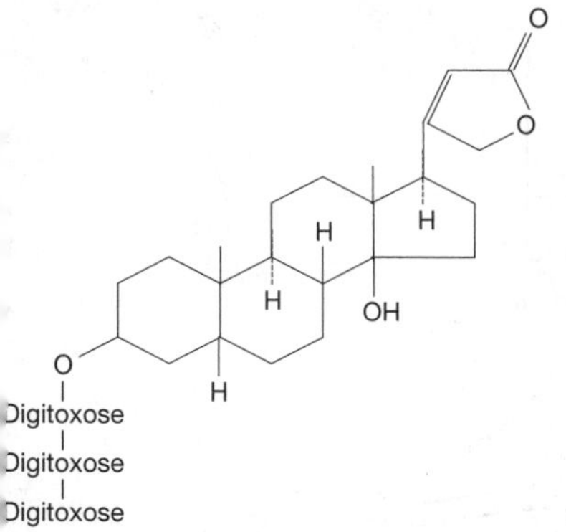

Digitoxose — Digitoxose — Digitoxose

Digitoxin

Digitigrade: walking on the toes, and not on the whole foot, i.e. on the ventral surface of the digits only, and not touching the gound with the heel, e.g. marten, cat, dog. See Plantigrade, Unguligrade.

Digitonigenin: see Digitonin.

Digitonin: a glycoside from the seeds and leaves of the purple foxglove *(Digitalis purpurea)* and other *Digitalis* species. The aglycon is the steroid, digitogenin [(25R)-5α-spirostan-2α,3β,15β-triol], and the sugar component is a pentasaccharide (Fig.). D. is a powerful hemolytic poison, due to its affinity for cholesterol, with which it forms a poorly soluble molecular compound. In the laboratory, it is used as a precipitating agent for cholesterol and other sterols.

The name, digitonin, may also be applied to a mixture of 4 different steroid saponins from *Digitalis* species, which contains 70–80% of the illustrated compound (Fig.).

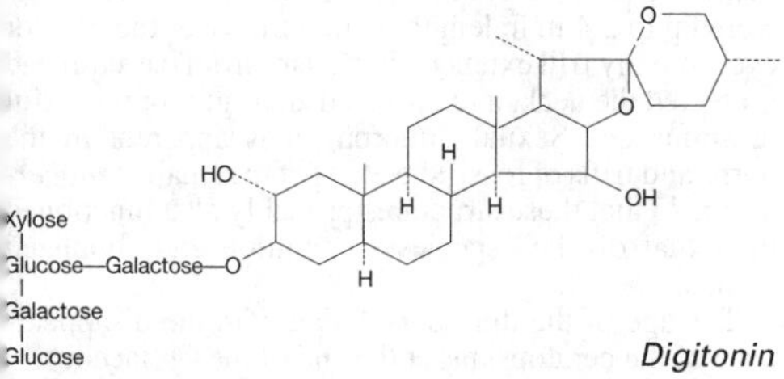

Xylose — Glucose—Galactose—O — Galactose — Glucose

Digitonin

Dihydroxyacetone: a triose and the simplest of the ketoses. Dihydroxyacetone phosphate is an important carbohydrate metabolite, serving as an intermediate in fermenatation and glycolysis.

1α, 25-Dihydroxycholecalciferol: the active form of vitamin D_3. See Vitamins.

22-Dihydroergocalciferol: the same as vitamin D_4. See Vitamins.

Dik-diks: see *Neotraginae.*

Dill: see *Umbelliferae.*

Dilleniidae: see *Dicotyledoneae.*

Diluvium (Latin *diluvium* biblical flood): an old term for the lower division of the Quaternary, i.e. the Pleistocene.

Dimethisterone: a synthetic progestin used in oral contaceptives; see Hormonal contraception.

Diminishing yield, law of: see Mitscherlich's law of plant growth factors.

Dimyarian condition: synonymous with homomyarian or isomyarian condition, terms applied to members of the *Bivalvia* (see) that possess 2 adductor muscles of about the same size.

Dinantian subperiod: see Carboniferous.

Dinesis: excitation or acceleration of plasma streaming in plant cells, as the result of various stimuli. In *chemodinesis,* the stimulus is a chemical agent, e.g. histidine or methylhistidine promotes plasma streaming in the pondweed, *Elodea,* and auxin causes chemodinesis in *Avena* coleoptiles. In *photodinesis,* light is the stimulant, and in *traumatodinesis,* plasma streaming is promoted by cell wounding.

Dingo, *Canis familiaris* v. *dingo:* a wild dog of Australia. It is descended from early domesticated dogs brought by the aboriginies to Australia, a continent that lacked indigenous representative of the *Carnivora.* It is generally similar in appearance to a German shepherd dog, but smaller and red-brown in color. Since it attacks sheep, the D. was almost exterminated, but its numbers are now relatively stable. The smaller Jungle dingo or Hallstrom dog *(Canis lupus* v. *hallstromi)* is found in New Guinea.

Dinobyron: see *Chrysophyceae.*

Dinoflagellata: see *Phytomastigophorea.*

Dinoflagellatae: see *Phytomastigophorea.*

Dinoflagellates: see *Phytomastigophorea.*

Dinornithidae, ***moas:*** a species-rich group of extinct, flightless birds in New Zealand. The largest species was *Dinornis maximus,* standing more than 3 m tall, and taller than the largest known elephant bird (see *Aepyornithidae);* the smallest species were about 1 m tall. The last moa was exterminated on the North Island of New Zealand in the 17th. century, and on the South Island in the 19th. century. They were well known to the Maoris.

[*Extinct Birds* by Errol Fuller; Viking/Rainbird (Penguin Books Ltd.) London, Victoria, Aukland, 1987]

Dinosaurs (Greek: *deinos* fearful, *sauros* lizard): the members of 2 extinct orders of saurians, the *Saurischia* ("reptile hipped") and the *Ornithischia* ("bird hipped"). They possessed a long tail and 4 legs, and locomotion was bipedal or quadrupedal. The forelegs were shorter than the hindlegs, the brain was small, and the skin was naked or armored. *Saurischia* and *Ornithischia* differed systematically in the structures of their pelvises. The saurischian pelvis was triradiate, i.e. essentially reptilian and similar to that of their ancestral thecodonts. In ornithischians the pubis was rotated backward, so that it lay parallel to the ischium, i.e. essentially like that of a bird.

Saurischians evolved in 2 directions, resulting in: ***1.*** The herbivorous sauropods (Greek *sauros* lizard, *podes* feet) (Jurassic to Cretaceous), which were 5-toed, swamp-dwelling animals and the largest 4-legged terrestrial vertebrates of all time; thus, *Brontosaurus* (Greek *bronte* thunder, *sauros* lizard) (syn. *Atlantosaurus*), from the Morrison formation in North America, was about 22 m in length and 10 m high with a markedly

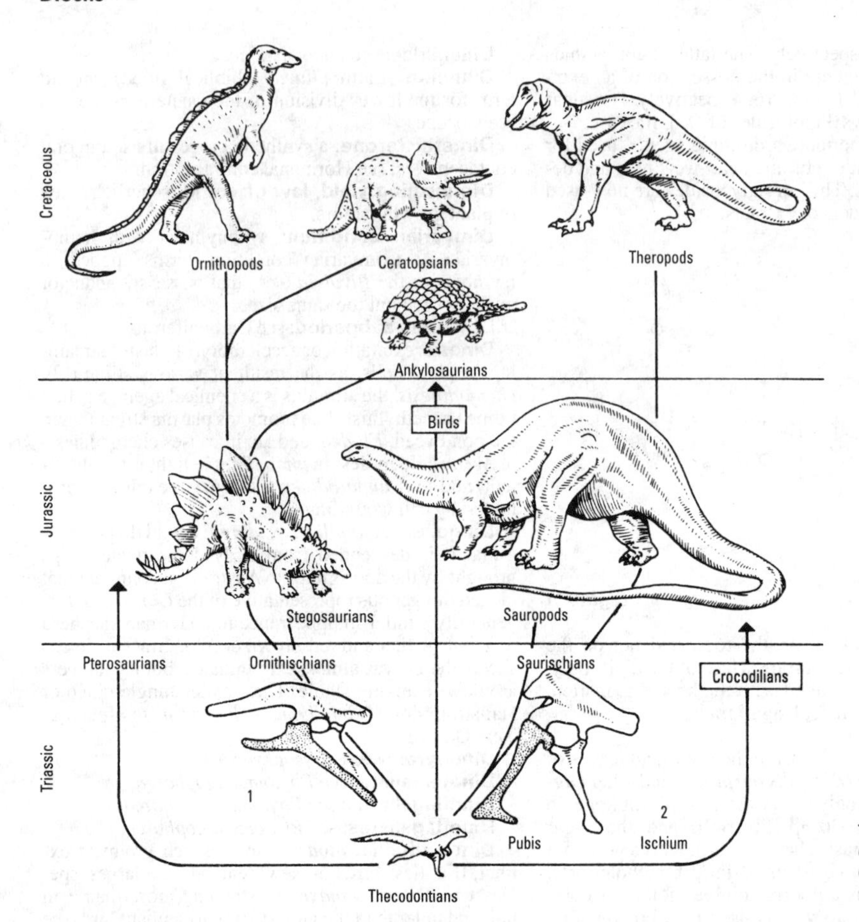

Dinosaurs. Evolutionary tree of the dinosaurs (after Colbert and Kuhn-Schnyder).

small head and tiny brain, and it weighed about 30 tonnes. The forelimbs of *Brontosaurus* were somewhat shorter than its hindlimbs, and it moved quadrupedally. ***2.*** The carnivorous theropods (Triassic to Cretaceous), including small forms no larger than a dog, as well as the fearsome *Tyrannosaurus*.

Ornithischians evolved into 4 basic types, all of them herbivorous: ***1.*** Small bipedal ornithopods of the Late Triassic, culminating in the Late Cretaceous with large, duck-billed ornithopods possessing up to 400 continually replaced teeth, long narial passages (suggesting they could trumpet and/or bellow), and often tall crests (suggesting species and sexual recognition). ***2.*** Large, quadrupedal stegosaurs of the Jurassic, ungainly in appearence, with a very small skull, a tiny brain, and armor plating. Thus, *Stegosaurus* (Greek *stegos* roof, *sauros* lizard), from the Dogger and Malm strata of Europe, East Africa, China and North America, had an alternating double row of upright, triangular bony plates and spines down the center of its back. These plates had a rich blood supply and may have functioned in heat exchange; they possibly also served for display and/or defense. The forelimbs were shorter than the hindlimbs, so that the animals were lower to the ground anteriorly, and the back formed a huge curve. ***3.*** Ankylosaurs, which appeared in the Cretaceous and probably occupied the same ecological niche as the earlier stegosaurs. They were large, broad, short-legged, with a heavy armor of bony plates covering the back, and laterally projecting stout spikes. The skull was triangular, narrowing to the front, and the teeth were small and apparently weak. ***4.*** Horned dinosaurs or ceratopsians, whose remains are restricted to the Late Cretaceous. Their skulls were up to 2.4 m in length, with horns over the nose or eyes. A bony frill extended backward from the skull and protected the neck, and provided an origin for powerful jaw muscles. Sexual dimorphism is apparent in the horns and frills of fossil skeletons of the small *Protoceratops*, so that these structures probably also functioned in sexual display, species recognition and ritualized combat.

The age of the dinosaurs ended with the disappearance of the ceratopsians at the end of the Cretaceous.

See *Pterosauria, Thecodontia, Archosauria.*

Diochs: see *Ploceidae.*

Dioctophyme: see *Nematoda.*

Dioctophymida: see *Nematoda.*

Diodontidae: see *Tetraodontiformes.*

Diogenes: see Hermit crabs.

Diomedeidae: see *Procellariiformes.*

Diomedia: see *Procellariiformes.*

Dionaea: see Carnivorous plants, *Droseraceae.*

Diotocardia: see *Gastropoda.*

Diphycercal fin: see Fins.

Diphycercous fin: see Fins.

Diphylla: see *Desmodontidae.*

Diphyllobothrium: see Broad fish tapeworm.

Diphylodes: see *Paradisaeidae.*

Diphyodont: see Dentition.

Diplany: the successive occurrence of two morphologically distinct zoospore generations. D. occurs in lower fungi, e.g. in the genus *Saprolegnia.*

Diplasiocoela: see *Anura.*

Diplasiocoelous vertebra: see *Anura.*

Diplocaulous: an adjective applied to plants that produce flowers on axes of the first order, i.e. on the first lateral branches, and are unable to form reproductive organs on the main axis. See Haplocaulous, Triplocaulous.

Diplococci: see Cocci.

Diplodocus (Greek *diploos* doubled, *dokos* beam or rafter): a sauropod genus, whose members attained a length of almost 30 m and a height of 5 m. They were vegetarian quadrupeds, which progressed on all 4 feet. The skull was only 60 cm in length, and carried at the end of a long, slender neck. *D.* possessed a long, whip-like, pointed tail (70 vertebrae). The cervical vertebrae were characteristically double-yolked, with forked, pointed processes. It is the largest terrestrial animal known to have existed, and its fossilized remains are found in the Upper Jurassic of North America.

Diplograptus (Greek *diploos* doubled, *graptos* written): a genus of graptoliths with a straight to sigmoid rhabdosome, carrying 2 rows of thecae. The simple or pipe bowl-shaped thecae are densely packed. Fossil *D.* are found from the Middle Ordovician to the Lower Silurian.

Diplohaplonts: organisms in which (in contrast to haplonts and diplonts) the Haplophase (see) and the Diplophase (see) are both vegetative forms, with an intervening meiosis. The haploid, gamete-forming generation is known as the gametophyte, while the diploid, spore-forming generation is the sporophyte. There is therefore a marked alternation of generations between a sexually and an asexually reproducing generation.

Diploidy: the presence of two complete sets of native chromosomes in the cells of an organism, which is then said to be *diploid.* D. is the result of the fusion of two haploid gametes in fertilization.

Diplolepis rosae: see Galls.

Diplont: see Meiosis, Alternation of generations.

Diplophase: the developmental phase of an organism between fertilization (zygote formation) and the onset of meiosis, i.e. the phase during which the cells contain two sets of chromosomes. The converse is Haplophase (see).

Diploporida: see *Cystoidea.*

Diploid: see Diploidy.

Diplopoda, *millipedes:* a phylogenetically ancient class of myriapodous arthropods (see *Myriapoda)* represented by early fossils in the Upper Silurian. The sides of the head are covered by the bases of the large mandibles, the distal ends of which carry a movable gnathal lobe with teeth and a rasping surface. Attached to the posterior ventral surface of the head is a broad flattened plate, which forms the floor of the preoral chamber. Known as the *gnathochilarium,* this plate is formed mainly by fusion of the first maxillae, together with other elements such as the vestiges of the appendages of the last head somite and vestiges of the second maxillae. The third postoral segment, i.e. the first trunk segment after the head, is therefore limbless, and it forms a conspicuous collar known as the collum. Each of the next 3 trunk segments carries one pair of legs. These first 4 segments after the head are considered to represent the thorax. In adult millipedes, all subsequent segments (i.e. those of the abdomen) are double (diplosegments or diplosomites); they are derived by the fusion of 2 originally separate somites, and each carries 2 pairs of legs (Fig.1). Trachaea open to the exterior via lateral spiracles (4 per diplosegment) located anterior to each coxa. Antennae have 7 or 8 segments, sometimes 9. The genital opening (gonopore) is situated on the ventral surface of the third segment (progoneate). More than 7 500 species have been officially listed, and there are probably hundreds more awaiting recognition. Millipedes are distributed worldwide and are exclusively terrestrial, although in rare cases they have penetrated as far as the tidal zones of sea coasts (e.g. *Hydroschendyla).* They live in the upper layers of the soil, under leaf litter, under stones and under tree bark. They shun light and prefer damp habitats, and many are also cave inhabitants. The majority are elongated and cylindrical, while a few species are dorsally or ventrally compressed. Some European species grow up to 5 cm in length, while some tropical species attain 30 cm. The number of leg pairs varies between 13 and about 250. Most species possess a hard exoskeleton, which is strengthened by a calcareous deposit, and made waterproof by a lipid layer; it provides protection and stability while burrowing in the soil, as well as protecting against desiccation and attack by predators. In many millipedes defense is also provided by repugnatorial glands or stink glands (Fig.1). Normally, there is a single pair of glands per segment, which usually open on the sides of the tergal plates, or (in the flat-backed millipedes) on the margins of the tergal lobes. Hydrogen cyanide has been identified in the repugnatorial secretion of many millipedes; in addition, secretions of the *Julidae* have been shown to contain irritant quinones, which have a burning taste and produce allergies, while the *Glomeridae* produce a secretion containing an alkaloid even more bitter than quinine. With the exception of a few carnivorous cave-dwelling forms (e.g. *Apfelbeckia),* the millipedes are exclusively vegetarian.

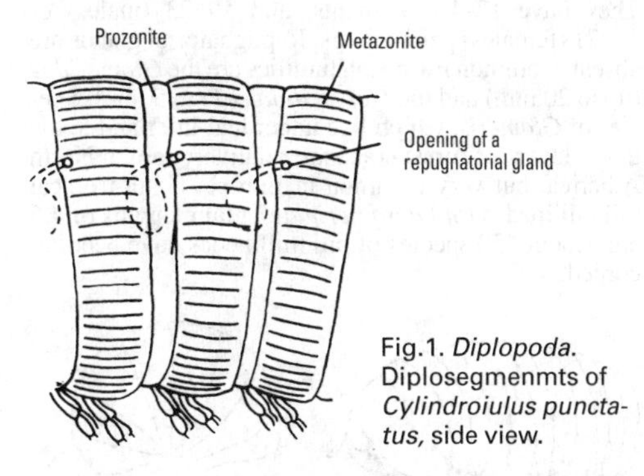

Fig.1. *Diplopoda.* Diplosegmenmts of *Cylindroiulus punctatus,* side view.

Development. After internal fertilization, the eggs are laid in the soil, under bark, etc., often in a protective covering of soil particles and secreted material (members of the *Nematophora* construct a protective layer of spun threads), or in special egg chambers. Larvae hatch after 2–4 weeks, possessing only 3 pairs of legs (4 pairs in the *Colobognatha).* Development of the *Glomeridae* lasts 3 years, while that of the *Julidae* requires 1–3 years, and it is estimated that members of these 2 groups live for about 7 years after hatching. The *Nematophora* live for 6–7 months, and the *Polydesdae* about 2 years. Molting may continue after sexual maturity and mating.

Importance. Millipedes, especially the *Glomeridae* and *Julidae,* make a significant contribution to humus formation. They feed mainly on decomposing plant material, soft plant parts and fungi. Some species may be harmful to cultivated plants, e.g. *Blaniulus guttulatus* ***(spotted snake millipede)*** is a greenhouse and agricultural field pest.

Classification.

1. Superorder: *Pselaphognatha* or *Penicillata.*

Order: *Schizicephala.* These small (2–3 mm) forms have no calcified exoskeleton, and the tergites and pleurites of their soft integument carry dorsal and lateral tufts and rows of hairy bristles (setae). They have 11–13 segments and 13–17 pairs of legs. Repugnatorial glands are absent. The anus is located in the penultimate trunk segment. About 90 species have so far been described. *Polyxenus lagurus* (Fig.2) lives mainly under bark.

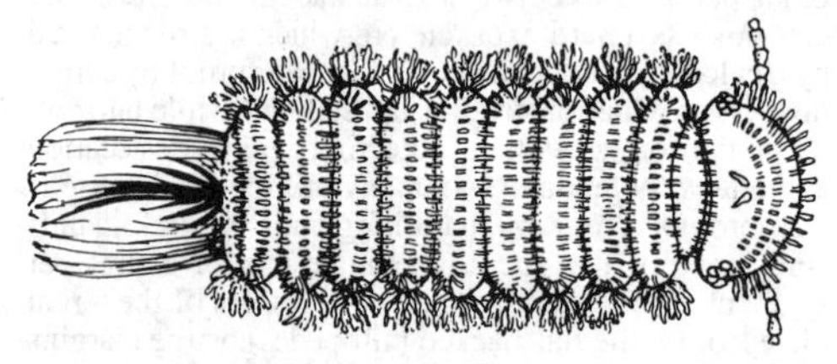

Fig.2. *Diplopoda. Polyxenus lagurus.*

2. Superorder: *Chilognatha.* These are the millipedes in the narrow sense, possessing a hard, calcified exoskeleton and 17 or more pairs of legs.

Order: *Oniscomorpha* or *Glomerida* ***(pill millipedes).*** These are relatively broad millipedes, which are flattened on their ventral surface, and able to roll the body into a tight ball (Fig.3). Most species are tropical. They have 12–13 segments, and 19–23 (males) or 17–21 (females) pairs of legs. Repugnatorial glands are absent. Common European families are the *Glomeridae* (up to 20 mm) and the *Gervaisiidae* (up to 5 mm). Species of *Glomeris,* which live under leaf litter and in the upper layers of forest soils, are relatively rare in North America, but very common in Europe. Giant tropical pill millipedes *(Sphaerotheriidae)* attain lengths of 4.5 cm. About 450 species of pill millipedes have been recorded.

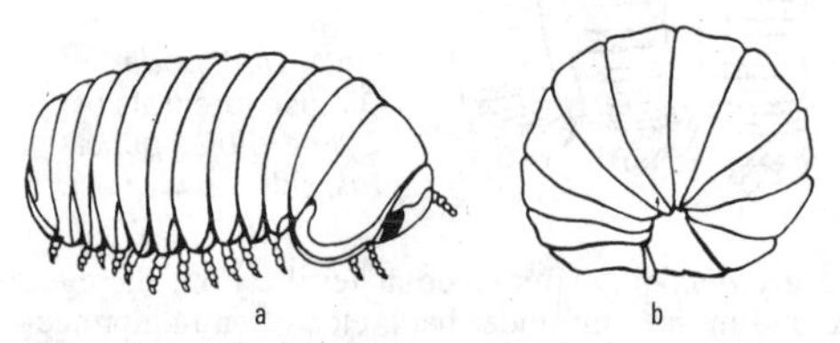

Fig.3. *Diplopoda.* Pill millipede. *a* Progressing. *b* Rolled.

Order: *Nematophora* or *Chordeumida.* These long millipedes are either cylindrical, or they have weak, lateral, tergal keels. Most species have eyes, and all species have one or 3 pairs of spinnerets on the last tergal plate. The trunk is composed of 22–60 segments. As in other orders, both pairs of legs of the seventh segment of the male are modified to gonopods, which are complex structures for sperm transfer, and the most important taxonomic feature of the millipedes. Repugnatorial glands are present. Members of the *Chordeumidae* are 4–25 mm in length and possess 28–30 segments. Like the members of the following orders, they live in the upper soil layers, between the soil and the leaf litter, and under stones. About 700 species are known.

Order: *Proterospermophora* or *Polydesmidea.* The dorsally flattened, elongated body (19–22 segments) often has prominent lateral tergal keels. Repugnatorial glands are present. There are more than 2 700 recorded species, every one of them without eyes. European species of *Polydesmus* are 7–28 mm in length.

Order: *Opisthospermophora* or *Juliformia.* These long, cylindrical millipedes are circular in cross section, with at least 30 body segments. Eyes are usually present, and repugnatorial glands are always present. European members of the *Julidae* and *Blaniulidae* are 5–47 mm in length (Fig.4). They burrow through the soil, using the head and collum as a ram rod. In hard soils, they eat a pathway with their mouthparts. When resting or withdrawing from danger they roll up spirally. Species of the tropical *Spirostreptidae* (mainly Africa) and *Spirobolidae* may attain lengths of 28 cm or more. *Narceus* is a common millipede of eastern USA. About 2 800 juliform species are known.

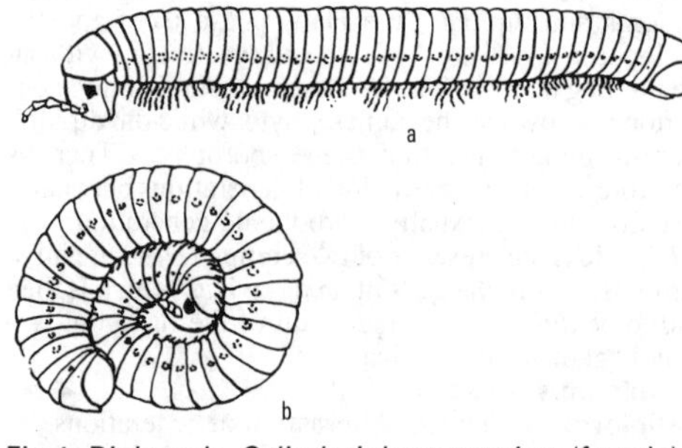

Fig.4. *Diplopoda. Cylindroiulus teutonicus* (female). *a* Progressing. *b* Rolled.

Order: *Colobognatha.* The compressed body has wide, lateral tergal keels, and the head is extended and trunk-like. Repugnatorial glands are present. Eyes are absent. Tropical species attain lengths up to 40 cm, and have 130 segments. The single European species, *Polyzonium germanicum,* has up to 55 segments and is 10–15 cm in length. About 250 species are recognized.

Diplosegment: see *Diplopoda.*

Diplostraca: see *Branchiopoda.*

Diplotene: see Meiosis.

Diplozoon: see *Trematoda.*

Diplura, ***two-pronged bristletails:*** an order of wingless, entognathous *Apterygota* (see). The absence of wings is a primary characteristic, not a secondary adaptation. They have long-jointed antennae, caudal appendages (cerci), and rasping or piercing mouthparts which are sunk into the head. The body is elongate (2 to 58 mm), the largest genus being *Heterojapyx.* Compound eyes and ocelli are absent. Scales occur only in a few members of the *Campodeidae,* and the integument of all other forms is thin and pale. The sterna of abdominal segments 2 to 7 carry lateral styliform appendages, and a variable number of segments carry eversible vesicles. The abdominal coxites are fused with the sterna, forming coxosternites. Sperm transfer is indirect; a stalked sperm droplet is deposited by the male, then taken up by the female. Newly hatched nymphs have the full number of 10 abdominal segments. Excretion occurs via labial nephridia and Malpighian tubules.

D. live in the cavity-rich surface of permanently damp soils. They are especially abundant in the tropics and subtropics, and they also occur in the temperate regions of Europe and North America. The approximately 500 known species are placed in 3 families. Members of the *Campodeidae* have long, fragile, filiform cerci, which are used like posterior antennae. Species of *Campodea* are the commonest *D.* of Central European soils (Fig.). The *Japygidae* (rare in Central Europe, but common in the tropics and subtropics) are characterized by unsegmented, pincer-like cerci, which are used to grasp prey (soft insects, small worms) (Fig.). The *Projapygidae* (found predominantly in the tropics and subtropics) also possess pincer-like cerci, but each cercus is segmented. Some members of the *Campodeidae* are also predators, but most feed saprophagously on dead plant remains, soil fungi, etc. The soil population density of *D.* is low (a few hundred below each square meter of soil surface), and they do not appear to be very important ecologically.

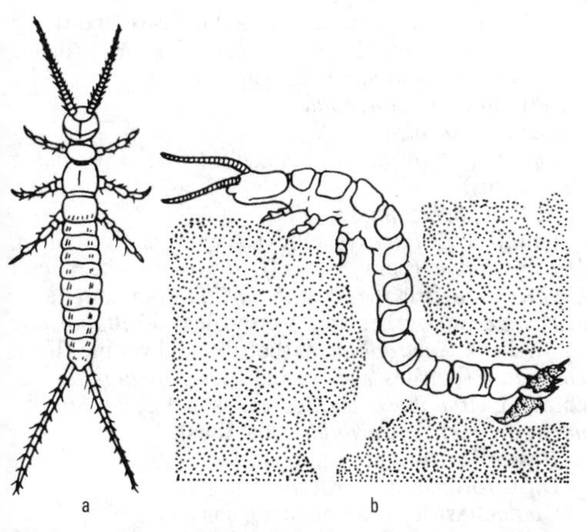

Diplura. a Campodea staphylinus. b Japyx solifugus capturing a springtail.

Dipnoi: see Lungfishes.

Dipodidae, ***jerboas:*** a family of Rodents (see) containing 30 species in 11 genera and 3 subfamilies. Jerboas (head plus body 4–26 cm, tail 7–30 cm) have very long hindlegs and a very long tail (often with a white tuft at the tip), which is used for support when the animal is standing upright and for balancing when jumping. They have large eyes and are nocturnal in habit. In most species (i.e. in the subfamiles *Dipodinae* and *Euchoreutinae),* the three elongated metacarpal bones are fused into a single metatarsus or "cannon bone". Depending on the species, the hindfeet have 3,4 or 5 digits. They inhabit the steppes and deserts of Asia and North Africa and do not drink, relying on the water produced during metabolism. By leaping and springing they can progress extremely quickly.

Examples of the subfamily *Cardiocraniinae* (3 main foot bones not fused) are the ***Five-toed dwarf jerboa*** *(Cardiocranius paradoxus* from western China and Mongolia), and the ***Three-toed dwarf jerboas*** *(Salpingotus* from the Asian deserts). The subfamily *Dipodinae* contains the archetypal species, *Jaculus jaculus* ***(Desert jerboa,*** North Africa and Southwest Asia), as well as 17 other species, e.g. *Allactaga sibirica* ***(Mongolian five-toed jerboa,*** Russian steppes to northeastern and northern Central China). The subfamily *Euchoreutinae* is represented by *Euchoreutes naso* ***(Long-eared jerboa,*** China and Mongolia).

Dippers: see *Cinclidae.*

Dipping ducks: see *Anseriformes.*

Diprionidae: see Sawflies.

Dipsadinae: see Snail-eating snakes.

Dipsas: see Snail-eating snakes.

Diptera, ***mosquitoes*** and ***flies:*** a cosmopolitan order of insects containing 85 000 species. The earliest fossil *D.* are found in the Jurassic (Lias, about 190 million years ago).

Imagoes. Adults are small to moderate-sized (1 to 55 mm). The head is freely mobile and joined to the thorax by a thin neck. Antennae of mosquitoes *(Nematocera)* are long, consisting of 6 to 140 uniform segments, aligned like beads on a string. Antennae of flies *(Brachycera)* are short with only 3 segments. Mouthparts are adapted for piercing and sucking, or for licking and sucking. Of the 3 thoracic segments, the anterior one is always small, whereas the middle one is large and highly articulated. *D.* are so-called because the imago has only 2 wings (forewings); these are normally transparent, and in mosquitoes they are richly veined. In all groups the hindwings are regressed to halteres which serve as flight stabilizers. In some families the forewings are also regressed, e.g. bee lice *(Braulidae)* lack both wings and halteres;the sheep ked *(Hippoboscidae)* is wingless and haltereless; deer keds *(Hippoboscidae)* are fully winged, but the wings of both sexes break off at the base after mating and finding a suitable host; bat lice of the family *Nycteribiidae* are all wingless. Each leg usually carries a foot of 5 subsegments, with paired claws and 2 pad-like lobes (pulvilli; sing. pulvillus). In some families a median pad-like organ (arolium) or a median bristle (empodium) is found between the pulvilli. Midges and mosquitoes have an elongate abdomen, whereas the abdomen of flies is usually short and fat. The terminal abdominal segments of the female form an ovipositor, which in some fly families can be telescopically withdrawn or extended. All dipteran imagoes are truly terrestrial. They feed on feces, plant juices or blood, or they prey on smaller insects. The imago's diet is often different from that of the larva.

Eggs. Most females lay eggs, but some species give birth to larvae, i.e. the larvae hatch from eggs within the female (vivipary). Vivipary is especially common in flies of the infraorder *Cyclorrhapha.* The shape and number of eggs, as well as the way in which they are deposited, differ considerably from species to species. Blowflies lay several separate batches on fresh meat, animal corpses or the wounds of injured animals; horseflies cement their eggs, usually as egg packets, to the aerial parts of plants; frit flies inject their eggs into plant tissues; the eggs of mosquitoes float on the surface of standing water; members of the *Tachinidae* lay their eggs (viviparous species deposit a larva) directly on the body of the host (e.g. a butterfly caterpillar).

Larvae. Legless larvae (maggots) hatch from the eggs, usually very shortly after the eggs are laid. After 3 to 8 molts the larvae are ready for pupation; they are usually found in decaying animal or plant material, in the earth, or in mud. Larvae of mosquitoes, blackflies, nonbiting midges *(Nematocera)* and certain flies, e.g. soldier flies *(Brachycera)* live free in water. The head of the dipteran larva displays varying degrees of reduction. *Nematocera* larvae still possess a well developed head capsule (eucephalic larvae) with chewing mandibles moving in a horizontal plane. *Brachycera* larvae are hemicephalic (head capsule greatly reduced) or hemicephalic (head capsule absent), and they have a maggot-like appearance. The latter condition is found in particular in the infraorder, *Cyclorrhapha.* Hemicephalic larvae have sickle-shaped mandibles moving in a verticle plane. Acephalic larvae lack eyes, and most mouthparts have been lost, leaving only two hooks. The outer integument is usually soft and practically unpigmented. Locomotion is by wriggling, and aquatic larvae swim by lashing their posterior end. Larvae of louseflies, keds and batflies leave the female only when they are ready to pupate; for this reason they were previously classified in a separate group called the *Pupipara,* but this is now thought to be an unnatural division.

Pupae. Eucephalic larvae develop into exarate pupae, from which the adult emerges by a longitudinal split in the thorax. The acephalic larvae of the *Cyclorrhapha* form coarctate pupae, the last larval skin being retained as a protective puparium. The emerging fly must therefore break through both the pupal skin and the puparium. During emergence, a transverse or circular split is formed in the puparium; in the *Schizophora,* the top is pushed away by an eversible head sac, the ptilinum; in the *Aschiza,* there is no ptilinum, and the top is pushed away by the lower face of the emerging imago. Members of the *Orthorrapha* possess no ptilinum and they are liberated by a loingitudinal split in the mid-dorsal line of the pupal case.

Economic importance. Larvae of hover flies (plate 1) attack plant lice. Larve of the *Tachinidae* parasitize butterfly caterpillars, while tachinid imagoes and those of certain other dipteran families play a part in flower pollination. Set against these beneficial effects are the widespread damage and inconvenience caused by other members of the family. Mosquitoes (see) represent an especial menace to humans, since they transmit malaria and other diseases. The African tsetse fly prevents the farming of cattle in parts of Africa, in addition to transmitting human sleeping sickness. The common housefly is not only a nuisance, but also carries typhus, tuberculosis, poliomyelitis and other diseases. Botflies and warbleflies of the *Oestridae* include some of the most economically important and severe parasites of domestic and wild animals; larvae occur in the nasal and pharyngeal cavities, where they form "bots", or beneath the skin, where they form "warbles". Botfly larvae of the family, *Gastrophilidae,* infect the alimentary canal of horses, elephants and rhinoceroses. There are also numerous plant pests, in particular Gall midges (see), Fruit flies (see), Frit flies (see) and Anthomyiids (see). Blowflies, which lay their eggs on stored food products, are well known household pests.

Classification. Two suborders are recognized: *Nematocera* (mosquitoes, midges and allies; about 53 families) and *Brachycera* (flies; about 100 familes). In addition to the divisions shown below, the *Cyclorrhapha* may be separated into 1) acalyptrate families which lack a calypter or lobe covering the halteres at the base of the forewing *(Psilidae* through *Anthomyiidae* in the list below) and 2) calyptrate families which have a calypter *(Muscidae* through *Tachinidae).* Due to the high specialization and degree of regression of the remaining families *(Hippoboscidae* through *Nycteribiidae),* which were formerly assigned to the *Pupipara* (see above), it is difficult to identify them as capyptrate or acalyptrate.

1. Suborder: *Nematocera* (mosquitoes, midges and allies)

Familes: *Tipulidae* (craneflies), *Culicidae* (mosquitoes and lake flies), *Chironomidae* (nonbiting midges), *Ceratopogonidae* (biting midges), *Simuliidae* (blackflies), *Mycetophilidae* (fungus gnats), *Cecidomyiidae* (gall midges), *Anisopodidae* (wood gnats), *Bibionidae* (march flies), *Sciaridae* (dark-winged fungus gnats), *Trichoceridae* (winter craneflies), *Ptychopteridae* (phantom craneflies), *Psychodidae* (moth flies and sandflies), *Blepharoceridae* (net-winged midges)

2. Suborder: *Brachycera* (flies)

Infraorder: *Orthorrapha*

Familes: *Tabanidae* (horseflies and clegs), *Stratiomyidae* (soldier flies), *Asilidae* (robber flies), *Bombyliidae* (bee flies), *Xylophagidae* (xylophagids), *Rhagionidae* (snipe flies), *Therevidae* (stiletto flies), *Omphralidae* (omphralids), *Nemestrinidae* (tangle-veined flies), *Acroceridae* (small-headed flies), *Empidae* (dance flies), *Dolichopodidae* (long-legged flies)

Infraorder: *Cyclorrapha*

Section: *Aschiza*

Families: *Phoridae* (humpbacked flies), *Syrphidae* (hoverflies)

Section: *Schizophora*

Families: *Psilidae* (rust flies), *Trypetidae* (fruit flies), *Drosophilidae* (pomace flies), *Braulidae* (bee lice), *Chloropidae* (frit flies), *Anthomyiidae* (anthomyiids), *Muscidae* (houseflies and tstetse flies), *Calliphoridae* (blowflies), *Gastrophilidae* (botflies and warble flies), *Oestridae* (botflies and warble flies), *Tachinidae* (tachinids), *Hippoboscidae* (louseflies and keds), *Streblidae* (bat flies), *Nycteribiidae* (bat flies).

Dipteriformes: see Lungfishes.

Dipteronia: see *Aceraceae.*

Disaccharide: a carbohydrate consisting of two glycosidically linked monosaccharides. Important examples are sucrose, cellobiose, maltose, lactose, all with the empirical formula: $C_{12}H_{22}O_{11}$.

Disassortative mating, *disassortive mating:* the same as negative Assortative mating (see).

Discharge mechanisms: see Catapulting mechanisms.

Discoglossidae, *discoglossids, disc-tongued toads:* a family of the *Anura* (see) containing 14 living species in 5 genera. The round tongue is almost totally fused with the floor of the buccal cavity. Eight stegochordal, episthocoelous presacral vertebrae are present, with imbricate (overlapping) neural arches. Presacrals I and II are not fused, and the atlantal cotyles of presacral I are closely juxtaposed. Free ribs persist after metamorphosis and are present on presacrals II-IV. The coccyx has a bicondylar articulation (monocondylar in *Barbourula*) with the sacrum, which has expanded diapophyses. The coccyx carries proximal transverse processes. The pectoral girdle is arciferal, with a cartilaginous omosternum (reduced in *Bombina* and *Discoglossus*) and a cartilaginous sternum. Maxillary and premaxillary teeth are present. All discoglossids have aquatic larvae with beaks and denticles, and a single median spiracle formed by a hiatus in the operculum.

Bombina ***(Fire-bellied toads)*** is a Eurasian genus containing 6 living species distributed in Europe, Turkey, western and eastern former USSR and East Asia. They are predominantly aquatic with warty backs and triangular pupils. Eardrums are absent. The belly is vividly marked red, orange, yellow, white and black. When threatened, they throw themselves onto their backs and display their bright ventral coloration. Their call is a hollow bell sound. During mating, the eggs are deposited in small clumps on water plants. *Bombina bombina* is a European species found in central and eastern Europe as far north as Denmark and southern Sweden, east to the Urals, Caucasus and Turkey, and west to Romania and central Germany; the male possesses internal vocal sacs. *Bombina variegata*, which lacks vocal sacs, is found in central Europe, France, Italy, Greece and Germany but not in Spain. Its range overlaps that of *Bombina bombina* in eastern Europe, and populations are decreasing markedly due to alterations in their habitat. The largest species of *Bombina* is *B. maxima* ***(Chinese giant fire-bellied toad***; up to 8 cm in length) native to southern China and eastern Siberia.

The genus *Alytes* ***(Midwife toads)*** contains only 2 living species. *Alytes obstetricans* ***(Common midwife toad***; plate 4) is toad-like in appearance and nocturnal in habit, and has a bell-like mating call. It is found in western parts of Central Europe, extending eastwards as far as Thüringen in Germany. Mating occurs on land and the eggs are exuded in 2 strings. These are immediately fertilized by the amplectant male, who then carries the eggs adhering to his back and thighs for 3 to 4 weeks (during this period he becomes silent and does not emit the mating call) until they are ready to hatch. He then wipes the spawn off into the water, where the tadpoles hatch.

The other discoglossid genera are: *Baleaphyrne* (1 species; Mallorca, Balearic Islands, Spain); *Barbourula* (2 species; Palawan, Philippines, Borneo and Sarawak); *Discoglossus* (3 species; southern Europe, Northwest Africa, Israel, Syria).

Discomycetidae: see *Ascomycetes.*

Discontinuous culture, *batch culture:* a method for culturing microorganisms, in which a culture medium is inoculated, the organism is allowed to grow, and the product is then harvested. If larger quantities of material are required, the entire process must be repeated. In industrial microbiology, D.c. is usually less efficient and therefore less economic than Continuous culture (see). Nevertheless, certain industrial fermentations are performed by D.c., e.g. the production of antibiotics and baker's yeast.

Discontinuous distribution: see Disjunct distribution.

Discordance analysis: an investigative method of human genetics, which attempts to determine the zygosity of twins by studying dissimiliarities (discordances) in certain genetically determined characters. See Zygosity analysis.

Discovery site: the geographical site where a plant is first discovered, or recorded in a survey This is not necessarily a typical growth site where the plant is permanently established.

Disc-tongued toads: see *Discoglossidae.*

Discus fish, *Symphysodon discus:* a cichlid (see Cichlids) from the Amazonian region, with a high-set, disk-shaped body, which can attain the size of a small plate. On account of its splendid coloration and interesting brood care, it is a very popular aquarium fish, but it is not easy to maintain.

Dishabituation: see Habituation.

Disinfection: the killing of disease-causing microorganisms and viruses on equipment, clothing and body surfaces, and in rooms and buildings, etc. D. prevents the transfer of disease organisms and removes the danger of infection, and is particularly important in the control of epidemics. It is achieved: 1) By the application of chemical disinfectants, such as phenolic compounds [e.g. phenol (carbolic acid)], mercuric chloride (corrosive sublimate), calcium hypochlorite (bleaching powder), formaldehyde, ethanol, iodine, oxidizing agents (e.g. hydrogen peroxide, potassium permanganate solution, etc.). 2) By physical methods, such as the application of wet or dry heat, ultraviolet irradiation (sunlight, quartz lamp), etc. 3) By a combination of chemical agents and physical methods.

Disjunct distribution, *discontinuous distribution:* the occurrence of plant or animal taxa in two or more geographically separated areas with no connecting population. If the areas are very different in size, the larger is called an *exclave.*

The term D.d. should not be applied to residual areas following environmental destruction or alteration of intervening areas by humans. Neither does it apply to small outposts beyond the obvious borders of the main area, or to islands off the coast of a main area, since these represent situations in which population contact can still conceivably occur. Disjunct areas must be sep-

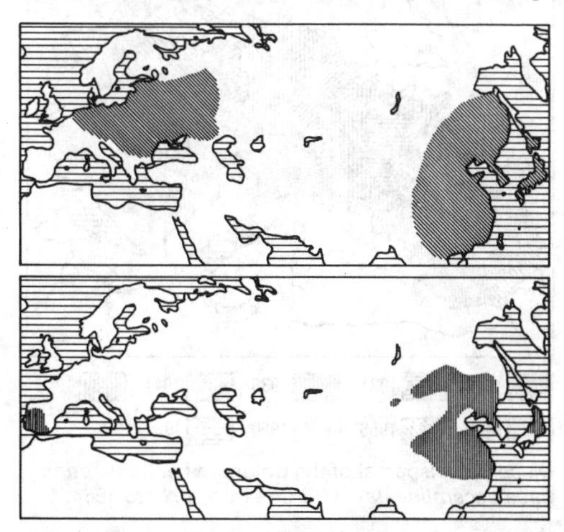

Disjunct distribution (two known cases). Above: the weatherfish *(Misgurnus fossilis).* Below: the azure-winged magpie *(Cyanopica cyanus).*

arated by a relatively large geographical distance. Such disjunctions are often important indicators of earlier land bridges; e.g. the D.d. of many animal and plant species in North America and Eurasia can only be explained in this way. Other D.ds. are due to climatic changes; e.g. in the Palearctic this is illustrated in particular by the Arctic-alpine distribution (see) and the Boreo-alpine distribution (see).

Disjunction: see Zonal vegetation.

Disk-floret: see *Compositae.*

Disk-tongued toads: see *Discoglossidae.*

Disomy: the presence of two complete, homologous sets of chromosomes in a cell, i.e. the diploid state. See Monosomy, Nullisomy, Polysomy, Trisomy, Tetrasomy.

Dispensable amino acid: see Amino acids.

Dispersal: active or passive extension of territory by populations, species or biotic communities. Organisms display a natural tendency to spread from the area in which they live, and to fully exploit the potential of other colonizable sites. Active D. is often prevented, however, by natural barriers. Passive D. (e.g. transport by wind or water) or deportation by humans may overcome these barriers, thereby opening up new colonizable areas. D. sometimes has great economic consequences, e.g. the spread of the Colorado beetle (Fig.), and the introduction of mice and rabbits into Australia. D. is often achieved by an appropriately equipped form or developmental stage of a species, e.g. winged forms of insects, or the dormant seeds of plants which may be carried great distances by wind or water, or on the feet of birds. Encroachment on, or transport to, a new geographical area does not guarantee D., unless the organism survives, reproduces and becomes established. See also Adventitious plants.

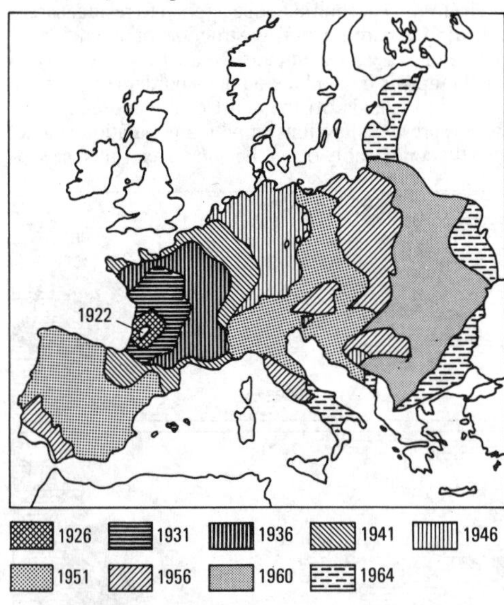

Dispersal. Dispersal of the Colorado beetle *(Leptinotarsa decemlineata)* in Europe from 1922 to 1964.

Dispersal boundaries: environmental factors that limit the dispersal of organisms. A factor is limiting when it represents a pessimum (see Ecological valency). For example, the distribution of the beetle, *Cillenus lateralis,* which inhabits Atlantic coasts, is limited to the coasts of North Africa and western Europe by its requirements for salinity, while the northerly and southerly limits of its range are determined by climatic temperature. For many organisms, dispersion is prevented by unsurmountable physiographic factors (mountain ranges, river systems, oceans, etc.). Such geomorphological boundaries are naturally accompanied by extreme alterations in the intensity of environmental factors.

Under climatically favorable conditions, organisms can often considerably extend their D.b. Thus, certain species present in the British Isles during warm interglacial periods are today only found in the Mediterranean region. Species present in the same region of the British Isles during the colder periods of the Ice Age are now found only in Scandanavia, Siberia or the Himalayas.

Dispersal center, ***expansion center:*** an area from which plant or animal taxa have expanded to colonize a larger area. The D.c. is usually the meeting point of numerous different areals (see Area) and therefore contains a rich variety of species.

A D.c. may be an *evolutionary center,* but most D.cs. of recent historical times are *centers of survival* or Refugia (see), e.g. many species survived in refugia during the Ice Age, when they were driven from their original pre-Ice Age areas by adverse climatic conditions. Upon return of favorable conditions (e.g. the ice receded), a new expansion occurred from the refugia, which therefore represent D.cs. Alternating phases of regression and expansion have also occurred in subtropical and tropical regions due to the alternation of wet and dry periods, e.g. the expansion and contraction of forested areas in the Neotropical region and elsewhere.

Disperse systems: see Colloidal systems.

Dispersion: the distribution of the individuals of a species in space or on a surface, usually measured as the *dispersion number,* which gives the relation between the actual and the ideal (uniform) distribution. A species with an average population density of 1000 individuals per hectare would have an ideal D. (D. number) of 1.0, with a single individual in every 10 m^2 (constant minimal area). If only 80% of these 10 m^2 areas are inhabited, the D. number is 0.8. The basis of the calculation is always the constant minimal area.

Dispholidus: see Boomslang.

Displacement behavior, ***displacement activity, substitute behavior:***

1) Behavior that is inappropriate to the current situation, but which would be functional in a different context. Examples are the displacement pecking of roosters during a fight or threat display, or the brief adoption of a sleeping posture during antagonistic behavior. D.b. is usually due to an internal conflict between stimuli to contrasting patterns of behavior (e.g. attack and flight). There may, however, be other reasons, and there is as yet no uniform hypothesis to explain D.b. See Ambivalent behavior.

2) Behavior of an animal directed to an inadequate or irrelevant object. For example, a male red deer kept in a paddock during the rut will "fork" tree branches lying on the ground with its antlers. This behavior relieves the "frustration" or tension arising from the absence of female deer and male rivals.

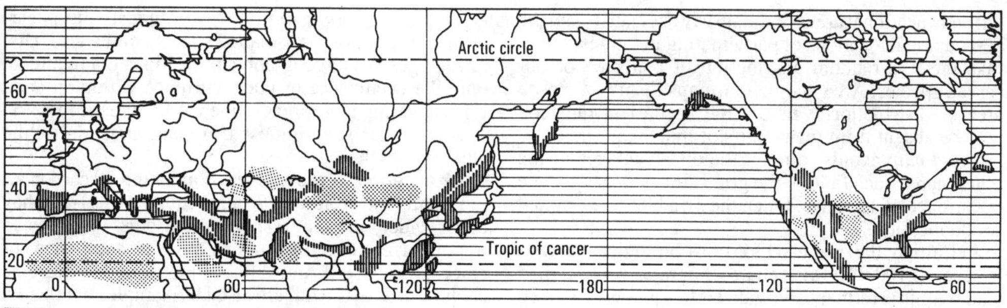

Dispersal centers in the Holarctic

Displacement behavior in birds. a Ringed plover *(Charadrius hiaticula)* seeking imaginary food, and thereby giving an alarm signal. *b* Displacement sleeping by an avocet *(Recurvirostris avosetta);* one eye is kept open and fixed on the adversary. (After Simmons, 1955)

Dissepiment:

1) In corals, thin calcareous plates, which transversely link the radial sclerosepta of the calcareous skeleton.

2) In brachiopods and arrow worms, two transverse walls which separate the secondary body cavity into two successive smaller cavities.

3) In segmented animals (in particular annelids), transverse walls of the secondary body cavity, which subdivide the latter into as many successive cavites as there are body segemnts.

4) In graptoliths, a strand of chitinous material connecting adjacent branches in the dendroid colony.

Dissimilation: see Catabolism.

Dissipative structure: a spontaneous spatial or temporal inhomogeneity in a thermodynamic system (see Far off equilibrium). Prerequisite for the occurrence of D.ss. is a sufficient input of energy and material. In principle, all biological systems are D.ss. in the thermodynamic sense.

Dissogony, ***dissogeny:*** a form of multiple reproduction found in hermaphrodite *Ctenophora* (comb jellies) and some related coelenterates, as well as in bisexual species of *Nereis.* The animal becomes sexually mature at separate stages of the life cycle, with an intervening period in which no gametes are produced. Thus, both larvae and adults become sexually mature and produce fertilized eggs. Sometimes more than two separate stages of the life cycle may produce fertilized eggs.

Distance field behavior: motivated behavior, characterized by the absence of information about the object of the behavior (e.g. food). This event field is replaced by Near field behavior (see) as soon as information on the object in question is available. D.f.b. consists principally of searching for information about the object in question and its position in space and time; it includes Appetitive behavior (see).

Distance maintaining animals, ***distance types:*** animal species that always maintain a certain distance from other members of the same species (known as the *individual distance).* Physical contact with conspecifics is not normally tolerated, but is permitted for copulation and brood care.

Distance types: see Distance maintaining animals.

Distribution: a biogeographical term for the area in which an animal or plant taxon occurs on the earth's surface, or an ecological term for the taxa occurring in specific biotopes or climatic zones. Biogeographically, the area of distribution of a plant taxon grows (and is eventually limited by environmental factors) by the distribution of propagules (see Seed dispersal); disjunct areas may be the remnants of an orginally larger and continuous distribution. In animal geography, almost all authorities distinguish clearly between distribution and dispersal, whereas in plant geography and ecology the dynamic process of Dispersal (see), which gives rise to the distribution, is often used as a synonym of the latter.

Distribution map: a cartographical representation of the area occupied by a taxon. For small, well known areas *point maps* are suitable, i.e. the geographical location of each sighting or recorded occurrence is represented by a point, providing at the same time a visual representation of the frequency or density of distribution. For larger regions, an *area distribution map* is more appropriate, in which individual areas are patterned or shaded; or a *contor distribution map,* marking the outer limits of each area with a line. The contor map is especially useful for recording the occurrence of several taxa on the same map. Point maps often suffer from the disadvantage that they tend to represent the residential and vacational areas of the survey specialists and the geographical locations of participating institutes, rather than the true random distribution of collected or observed species. Area and contor maps display a closed area of distribution, whereas the species in question may be absent from many parts of that area. The modern *grid map* avoids some of the disadvantages of the other maps. Each field of the grid is investigated for the occurrence of a species. Often the grid lines are omitted for publication.

In Europe, the grid used for most biological surveys consists of 100 x 100 km squares. For relatively detailed studies, this is further subdivided into squares of 10 x 10 km. Advantages of a grid map of uniform squares are: 1) a statement regarding species occurrence is cumpulsory for each square (nonoccurrence, occurrence, not investigated); 2) observers and collecters are compelled to investigate less attractive biotopes; 3) it facilitates comparison of the distributions of different animal and plant taxa, and reveals possible correlations; 4) it provides a statistical evaluation and a basis for computer analysis; 5) in large areas affected by human activity, it is very useful to be able to plot the decrease of a species before its total extinction by recording its disappearance from previously occupied squares of the grid. Disadvantages: exact geographical locations not recorded, only the occurrence or nonoccurrence within an area; grids covering large areas of the world, like the maps on which they are superimposed, are inaccurate due to the curvature of the earth.

Diterpenes: Terpenes (see) containing 20 carbon atoms, and built from four isoprene units ($C_{20}H_{32}$). They occur widely in plant resins and balsams. Phytol is an important D., which occurs as an ester component of chlorophyll, and as part of vitamins K and E. Resenic acids (e.g. abietic acid) are also D. Examples of other D. are gibberellins, retinol and vitamin A.

Ditylenchus: see Eelworms.

Diuresis: increased production of urine. According to its cause, D. may be classified as 1) *water D.*, following increased water intake, 2) *pressure D.* caused by increased blood pressure, 3) *osmotic D.* following increased salt intake, and 4) *pharmacological* or *pathological D.* The most familiar pathological D. is Diabetes insipidus, in which individuals produce up to 30 liters of urine per day; this is due to a deficiency of antidiuretic

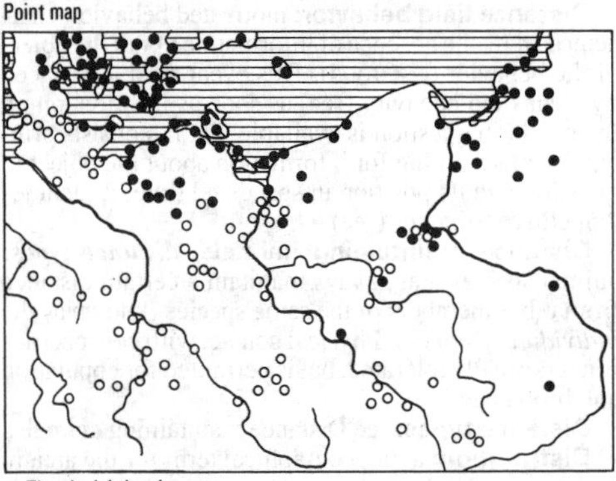

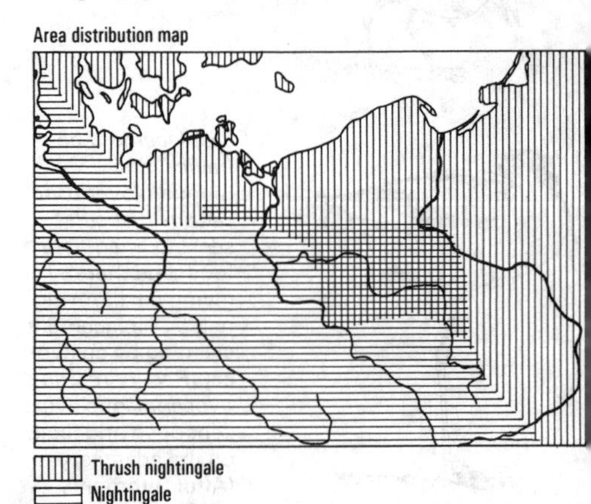

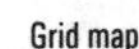

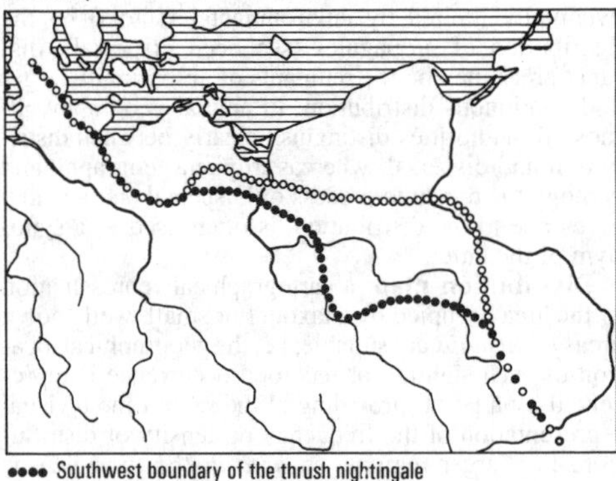

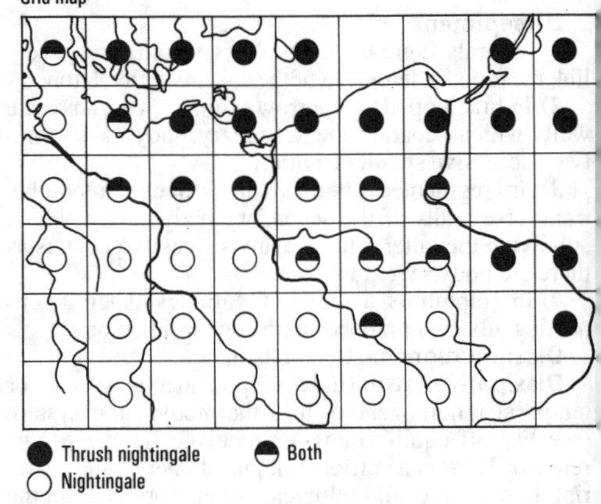

Distribution maps. Different types of distribution maps showing the distribution of the nightingale *(Luscinia megarhynchos)* and the thrush nightingale *(Luscinia luscinia)* in part of their areal. The large 100 x 100 km squares of the grid do not show whether the areals of these two species meet at their boundaries, or overlap.

hormone, which in turn leads to a decrease in facultative water reabsorption in the kidney.

Diuresis, pressure: see Hairpin loop countercurrent pinciple.

Diurna: see Butterflies, *Lepidoptera.*

Diurnal species: see Biorhythm.

Diuron: see Herbicides.

Divergence:

1) A measure of the difference in species composition of two communities. Dominance D. is inversely related to Dominance identity (see). The D. rate (i.e. D. related to time) serves as an index of change in biotic communities (see Succession).

2) See Leaf.

3) See Nerve junction.

Divergence index, ***patristic distance:*** a measure of the genetically determined change that has occurred between two points on a phylogenetic tree.

Divergence of neurons: see Neuronal circuit.

Divergent evolution: evolution of different types of organisms from a common ancestral stock. As a result of competition for food and living space, organisms evolve to occupy as many different habitats as possible. Thus, the wide variety of placental mammals has evolved from a primitive insect-eating, 5-toed, short legged ancestor. D.e. is more or less synonymous with Adaptive radiation (see).

Divers: see *Gaviiformes.*

Diversity index: a measure of the number of different species in a biotic community, and their relative abundance. The Shannon-Wiener D.i. (H_s) is loosely based on information theory, and is calculated from the formula:

$$H_s = -\sum_{r=1}^{s} p_r \log p_r,$$

which expresses the average uncertainty that the r-th species out of a total of s species will be encountered in a random sample; p_r represents the probability that any individual will belong to the r-th species ($r = 1,2, \dots ,s$). Thus, $p_r = N_r/N$, where N is the total number of individuals belonging to species r, and N is the total number of individuals in the sample. Values of p_r therefore vary beetween approaching zero (each individual belongs to a different species) and 1 (all individuals belong to the same species). s represents the number of different species present. The lowest value for diversity is realized when all individuals in a sample belong to to a single species ($s = 1$; $H_s = \log s$). The calculated H_s value, however, does not indicate whether the sample in question is species-rich or species-poor H_s is also influenced by the evenness of distribution of species. The maximal possible value for s would be realized only if all species were equally abundant, i. e. $H_{s\,max} = \log s$. In practice, this maximum is never achieved. Diversity might be increased by the introduction of new species, or by adjustment of the population sizes of rarer existing species, to that they become more common. There are therefore 2 components of diversity, i.e. the number of species and evenness. Since numerical equality among species is highly improbable, the term "equitability" has been proposed in place of "evenness" (Lloyd & Ghelardi, 1964).

In the model proposed by Mac Arthur (1957), individuals are apportioned among species as equitably as can be expected in nature:

$$\pi_r = \frac{1}{s} \sum_{i=1}^{r} 1/(s-i+1),$$

where π_r represents the theoretical proportion of individuals in the r-th most abundant species ($r = 1,2, \dots ,s$), in which each theoretical proportion is obtained by summing over r terms ($I = 1,2, \dots ,r$). The equation generates a complete set of s proportions, π_r ($r = 1,2, \dots ,s$) for each possible number of species. The Shannon-Wiener function can be applied to MacArthur's model to give the following equation:

$$M_s = -\sum_{r=1}^{s} \pi_r \log \pi_r.$$

Since the different species in a biotic community cannot be expected to even approach equality of numbers, log s (the theoretical mathematical maximum) is not a realistic measure of equitability. According to Lloyd and Ghelardi, experience suggests that M_s is at least approximately equivalent to the "ecological maximum" of diversity, i.e. that which is attainable in the field among a given number of species.

Equitability can be represented by the ratio of the observed H_s to the theoretical M_s for the same number of species. It is more appropriate to set $H_s = M_{s'}$ and to calculate the number of hypothetical "equitably distributed species" *(s')* needed to produce a species diversity equivalent to that observed. Thus, equitability (ε) is equivalent to s'/s. A table of s' values (1–1000) (the number of hypothetical species) and the corresponding $M_{s'}$ values (diversity of species based on Mac Arthur's model) has been published [M. Lloyd & R. J. Ghelardi (1964) *Journal of Animal Ecology* **33,** 217–225] for calculating the equitability component of species diversity.

In general, stable ecosystems posses the highest diversity indices.

Divided genome viruses: see Multipartite viruses.

Diving petrels: see *Procellariiformes.*

Division of labor: see Colony formation, Ants, Termites, Bees.

DNA replication: faithful duplication of deoxyribonucleic acid (DNA, the principal genetic material), thereby insuring the continual transfer of encoded genetic information from generation to generation. It is primarily this property of faithful duplication that qualifies DNA (and the RNA of RNA viruses) as the genetic material of living organisms. The ability of DNA (and viral RNA) to replicate is based on its double stranded nature. DNA may not be constantly in a double stranded state, and the double strandedness may only be partial, but only double stranded DNA (duplex DNA) is present in the replication cycle. The two strands of duplex DNA are complementary, in the sense that they are held together by hydrogen bonding between complementary nucleotides (more precisely, complementary bases of nucleotide residues) in opposite strands (in addition to hydrogen bonding between bases, other weaker forces also make a contribution to the stability of the DNA duplex, e.g. the van der Waals forces of base stacking). Hydrogen bonding between complementary bases is commonly known as *base pairing.* The discovery that DNA exists as a duplex of two strands, that these two strands are twisted to form a double helix, and that base pairing is the main interactive force for establishing and maintaining this structure was made in 1953 by Watson and Crick; it ranks as the most important advance in biology since Darwin's evolutionary theory. The base sequence of a DNA strand (e.g. ACTTGAGTAT) determines the complementary sequence of the other strand (in this example, TGAACTCATA). Replication is therefore semiconservative, i.e. each single strand of a DNA duplex serves as a template for the synthesis of a complementary daughter strand, so that a parent DNA duplex gives rise to two identical daughter duplexes (Fig.1). Detailed studies, mostly in bacteria and viruses, have revealed two different mechanisms of semiconservative replication. In the *Y-mechanism,* replication may be initiated at one site (unidirectional replication) or two sites (bidirectional replication), each site being known as a *replication fork* and envisaged as a "Y"; as the two strands separate, each serves as template for the synthesis of its complementary partners (Fig.2). The two strands of the DNA duplex are always antiparallel, i.e. the 3' → 5' alignment of one strand corresponds to the 5' → 3' alignment of its complementary strand. Synthesis of new DNA always occurs in the 5' → 3' direction (by reading the template strand in the 3' → 5' direction), so that synthesis occurs in different directions on the separating parent strands. Replication depends on the action of more than 20 enzymes and proteins. The several sequential steps include recognition of the origin or starting point of the process, unwinding of the parent duplex, maintenance of strand separation in the replicating region, initiation of daughter strand synthesis, elongation of daughter strands, rewinding of the double helix, and termination of the process. The complex of factors involved in replication is called the *DNA replicase system* or *replisome.*

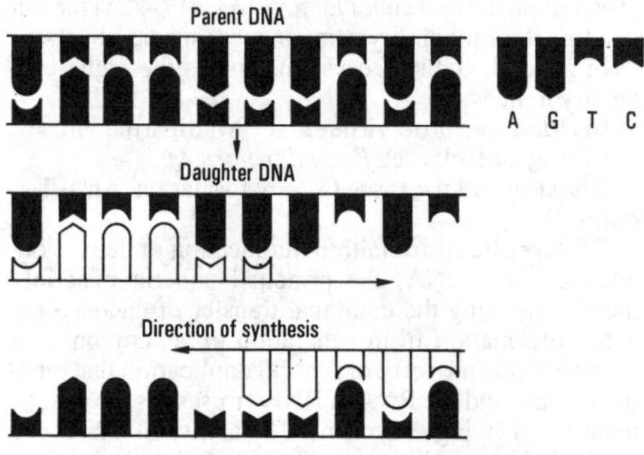

Fig.1. *DNA replication.* Schematic representation of the mechanism of semiconservative DNA replication.

It is clear from autoradiographic studies that synthesis occurs simultaneously from the 3' end of one parent strand and the 5' end of the other, but all three of the known DNA polymerases work only in the 5' → 3' direction. Replication occurs in short pieces (about 2 000 nucleotides in bacteria, less than 200 nucleotides in animal cells). These short pieces, known as *Okazaki fragments,* are later linked by a DNA ligase. Each Okazaki fragment is synthesized (by the action of DNA polymerase III) as an extension of a short RNA primer (about 10 nucleotides). The RNA primer is synthesized in the 5' → 3' direction of the template strand by the action of a specialized RNA polymerase, called a *primase.* The primase is associated with other proteins in a *primosome complex,* which may also be considered as a part of the replisome complex. DNA synthesis proceeds from the 3' end of the primer. The RNA primer is then removed, one nucleotide at a time, by the 5' → 3' exonuclease activity of DNA polymerase I. As each ribonucleotide is removed, it is replaced by a corresponding deoxyribonucleotide by the polymerase activity of the same enzyme. Final splicing of the extended Okazaki fragments is due to DNA ligase, which catalyses phosphodiester bond synthesis between the 3' phosphate group of the elongating DNA and the 5' hydroxyl group of the newly synthesized Okazaki fragment. Ligation is coupled to pyrophosphate bond cleavage (in bacteria NAD is cleaved to nicotinamide mononucleotide and AMP; in animal cells ATP is cleaved to AMP and pyrophosphate). Both strands of DNA are replicated by the above discontinuous mechanism, although discontinuous synthesis is not theoretically necessary for the progress of replication along the 3' → 5' replicating parent strand. The Y mechanism of DNA replication also operates in so-called "unscheduled" DNA replication, which occurs in the *excision repair* of damaged DNA.

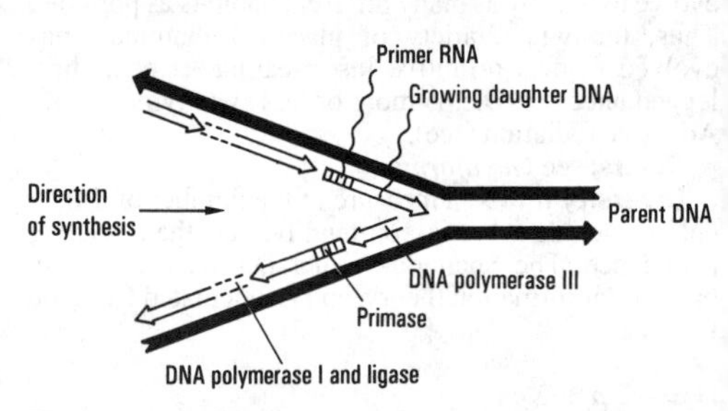

Fig.2. *DNA replication.* Replication fork of DNA. The arrows on the daughter DNA indicate the direction of synthesis.

The second mechanism of DNA replication, the *rolling circle mechanism,* operates for some circular DNA molecules, such as certain viral DNAs and the DNA of bacterial plasmids (Fig.3). In this mechanism, one strand of the circular DNA duplex is first cleaved ("nicked') enzymatically. New nucleotide units are added to the 3' terminus of the broken strand by the action of DNA polymerase III, and the continuous growth of the new strand displaces the 5' tail of the broken strand from the rolling circular template. Thus, the 5' tail becomes a linear template for the synthesis of a new

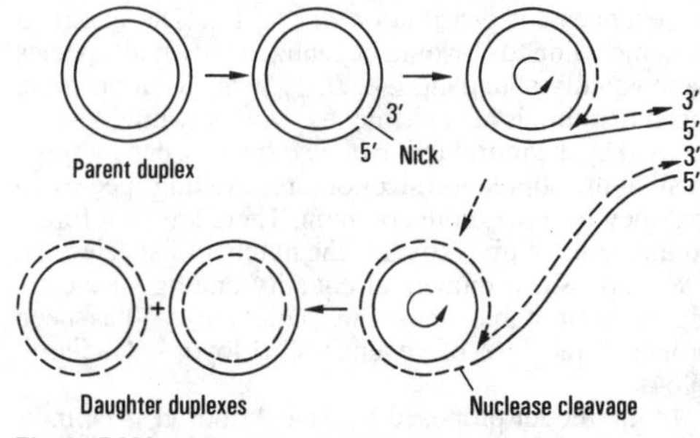

Fig.3. *DNA replication.* Rolling circle mechanism of replication.

complementary strand. The duplex originating from the 5' tail is then cleaved from the other daughter duplex by a nuclease.

Any DNA molecule capable of autonomous replication is called a *replicon.*

DNA: the acronym of Deoxyribonucleic acid.

Dobsonflies: see *Megaloptera.*

DOC: acronym of 11-deoxycorticosterone (see Cortexone).

Docimastes: see *Apodiformes.*

Dock: see *Polygonaceae.*

Dodder: see *Cuscutaceae.*

Dodo: see *Raphidae.*

Dogbane family: see *Apocyanaceae.*

Dogfishes: see *Selachimorpha.*

Dogger: the middle series of the Jurassic.

Dogs: see *Canidae.*

Dog's mercury: see *Euphorbiaceae.*

Dog tapeworm, *Echinococcus granulosus*: a cosmopolitan tapeworm of the order *Cyclophyllidea,* family *Taeniidae,* class *Cestoda* (see). The adult worm, which parasitizes the intestinal tract of dogs, other canids and cats, is only 3–6 mm long, and consists of only 3 oval proglottids. The cysticercus (hydatid cyst or cystic metacestode) is a fluid-filled bladder, containing multiple proscoleces produced by asexual multiplication. It may attain the size of child's head, and it is found in the liver and sometimes the lungs of the secondary host (especially ungulates and less often primates, humans and marsupials) (see Reproduction, Fig.2), causing the condition known as ***hydatid disease*** or ***hydatidosis.*** The disease is also sometimes called ***echinococcosis,*** but this term should be reserved for infection of the primary host with the adult tapeworm. *Echinococcus granulosus* is responsible for ***unilocular*** hydatid disease, i.e. single, large cysts are present. The definitive or primary host is always a carnivore. Other types of hydatid disease are:

Polycystic hydatid disease. This is caused by clusters of relatively small cysts, e.g. *Echinococcus oligarthrus* (primary host: wild felids) is responsible for polycystic hydatid disease in the muscles or viscera of agoutis *(Dasyprocta),* paca *(Cuniculus paca)* and spiny rats *(Proechimys). Echinococcus vogeli* (primary host: bush dog, *Speothos venaticus)* is responsible for polycystic hydatid disease in the liver of agoutis, paca, spiny rats and humans. Both of these species of *Echinococcus* are found in Central and South America.

Alveolar or ***Multivesicular*** hydatid disease. This is caused by a proliferating, invasive mass of hydatid germinal layer tissue. Metastases are formed, which cause secondary alveolar cyst formation, usually in the lungs or brain. *Echinococcus multilocularis* (primary host: foxes, other canids and cats) is responsible for alveolar hydatid disease in the liver of arvicolid rodents and humans in the Holarctic region.

[*The Biology of Echinococcus and Hydatid Disease*, R.C.A. Thompson (edit.), George Allen & Unwin, London, Boston, Sydney, 1986]

Dolichallabes: see *Clariidae.*

Dolichocephaly: a condition in which the head is relatively long, i.e. the skull has a low Cranial index (see).

Dolichonyx: see *Icteridae.*

Dolichotinae: see *Caviidae.*

Dolichotis: see *Caviidae.*

Doliolida, *Cyclomyaria:* an order of the *Thaliacea* (see) with ring-shaped, entire muscle bands around the body, numerous gill slits, and a tailed larva whose tail is reduced during metamorphosis. *D.* display an interesting alternation of generations. The tailed larva develops directly into a fully formed, free-swimming oozoid. The oozoid then begins to degenerate internally, and buds are formed on a ventral appendage (stolon or stolo prolifer). These buds migrate to a long dorsal appendage (cadophore) on the opposite side of the body, where they settle and become attached in three rows as lateral and median buds. The lateral buds serve as gasterozooids, which gather and respire food for the colony, whereas the median row of buds forms phorozooids. The latter act as nurses and form more buds, including gonozooids, which produce eggs and spermatozoa. Larvae hatch from the fertilized eggs.

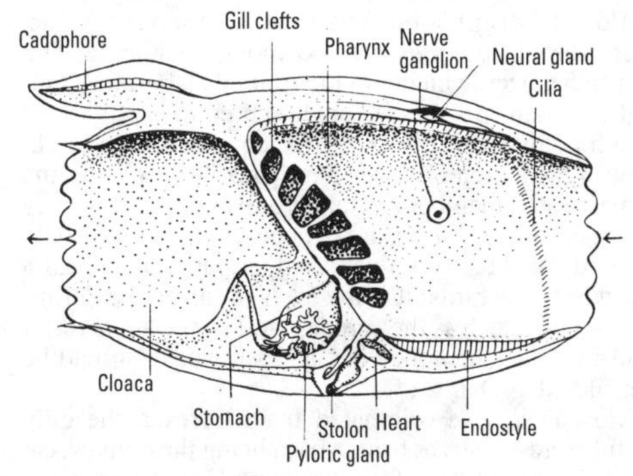

Doliolida. Structure of the adult asexual individual (oozoid) of *Doliolum.*

Doliolum: a genus of the order *Doliolida* (see), in which the complex metagenesis of the *Tunicata* was discovered.

Dollo's law: a law named after the paleontologist, Louis Dollo, stating that evolution cannot be reversed. Due to the complexity of the genetic structure of organisms, evolution can never give rise to ancestral forms, even in the presence of the appropriate environmental conditions. This is especially true of highly specialized organisms, whose potential for further useful mutation is probably limited.

In the Jurassic, some turtles took to the sea, and this change of environment was accompanied by regression of the carapace and plastron. At the beginning of the Tertiary their descendants reinvaded the land and developed a new bony shell. This, however, was not identical with the original shell, but lay above it in the leathery integument, and consisted of numerous bony plates. As these animals at the end of the Tertiary once again colonized the sea, this shell also degenerated. Modern leatherback sea turtles possess only traces of these two shells.

Whether D.l. applies without exceptions depends on what is understood by identity with an ancestral form. The uniform dentition of toothed whales is considered by some authorities to be equivalent to the homodont dentition of reptiles, and therefore an exception to D.r. Other workers interpret this as a confirmation of D.r., because the structures of whale and reptile teeth are different.

Dolphinfishes: see *Coryphaenidae.*

Dolphins: see Whales.

Domestic animals, *domesticated animals:* animals which have lived for many generations under special conditions determined by man. D.a. have existed for at least 10 000 years, those with the longest histories of domestication being the sheep, goat, cow, pig and dog. These were first used by humans as sources of food and raw materials; even the dog was first domesticated as a source of meat. Exploitation of animals for carrying and pulling loads arose later, as did the use of the dog for hunting and guarding.

Of the 600 known species of mammals, relatively few, i.e. about 20, have been domesticated: sheep, goat, cow, gayal, banteng, buffalo, yak, dromedary, camel, llama, alpaca, pig, horse, donkey, reindeer, rabbit, guinea pig, dog, cat, ferret. Only some of these have a worldwide distribution and/or real economic value. More recent additions are the mouse, white rat and golden hamster, which are important as laboratory animals. Certain species are kept and bred in captivity for their fur (silver fox, racoon, mink, coypu, chinchilla). In view of the extremely large number of bird species, the domesticated members of this group are also relatively few (hen, guinea fowl, turkey, peacock, goose, mute swan, duck, Muscovy duck, budgerigar, canary and a few other cage birds). On the basis of the above definition, species such as the carp, as well as certain invertebrates such as the honey bee and silkworm, should be considered as D.a.

Most D.a. have originated from Eurasia, the only useful representatives from Africa being the donkey, cat and guinea fowl. D.a. from precolumbian America are the llama, alpaca, guinea pig, turkey and Muscovy duck. Australia has not produced any D.a.

Elucidation of the origin and history of each domestic species has been considerably aided not only by research in the biological sciences, but also by research into early and pre-history, folk lore and linguistics. The marked alteration of form sometimes brought about by domestication, and the often considerable differences between races formerly led to the conclusion that each domestic species had evolved from several different wild species (polyphyletic D.a.). It is now known that most domestic species are derived from a single wild species (monophyletic D.a.). See Domestication.

Domestication: the conversion of wild animals into Domestic animals (see). The dog was domesticated very early in the Mesolithic period in the region of the Baltic sea; the other oldest domestic species (sheep, goat, cow) were first domesticated during the Neolithic period in the Middle East. Prerequisites for D. were the establishment of human settlements (i.e. the change from a hunting to a pastoral existence) and a certain level of cultural development. Early attempts at D. would require particular care and attention in the rearing and maintenance of young, captured animals. Since the oldest domestic animals have originated from herd or pack species, it appears that social and gregarious animals proved to be especially amenable to D. It must also be assumed that most domestic animals arose from several geographically separate D. centers.

D. signifies a sustained alteration in the mode of existence. Individuals are removed from natural populations and subjected to altered conditions, including an altered choice of sexual partners (selective breeding). This results in changes in genetic constitution and a great variation in the variety of form. The result of thousands of years of artificial selection is that domesticated species show a much wider variation of form than the corresponding wild species. This is abundantly illustrated by the dog, ranging from the Chihuahua to the Saint Bernard. At the same time, certain parallel alterations occur preferentially in all forms, and these are known as domestication characters, e.g. shortening of the skull and decrease of the brain mass of domestic mammals. It can therefore be concluded that those domestic animals showing a wide variety of form are characterized by genetic plasticity, i.e. the total population contains a great variety of genes for each character. In spite of the strong morphological, physiological and behavioral changes, and the considerably increased variability, all races of a domestic species and the wild population can still interbreed and produce fertile offspring. All members therefore belong to a single species in the zoological sense.

Dominance:

1) Biosocial status, which guarantees "privilege" for a dominant individual in a defined environment. Other individuals are then subordinate or subdominant.

2) A bioc(o)enotic character. It expresses the percentage of individuals of a species in the total number of individuals in a sample (see Relevé). Group dominance expresses the percentage of individuals. Mass or weight dominance expresses the relative biomass of a species or taxon in the total biomass of a sample. See Cover-abundance.

3) A term used in genetics to indicate the action of a dominant allele, which is expressed phenotypically, and thereby more or less completely masks or prevents the phenotypic expression of another allele (i.e. a recessive allele) of the same gene. In addition to complete dominance and recessiveness of alleles, intermediate behavior is also encountered (e.g. semidominance, weak and strong recessiveness). The degree of dominance, i.e. the extent to which the phenotype of an allele is expressed, may be altered by environmental conditions, and by the genetic environment of the allele in question. Genes that are able to influence the degree of dominance of a different, nonallelic gene are known as dominance modifiers.

Dominance hierarchy: see Rank order.

Dominance identity, *Renkonen index:* a bioc(o)enotic parameter for the percentage of dominant species that are common to two biotic communities. D.i. is calculated by summing the lowest dominance values of all species occurring in both communities.

Dominance reversal: see Dominance transfer.

Dominance theories: theories explaining the causes of dominance and recessivity, especially the frequent recessivity of mutants. R.A. Fischer's theory of 1928 is still important today. Accordingly, disadvantageous alleles were originally expressed in an intermediate form in their heterozygous carriers. Their disadvantageous action was then often gradually suppressed by modification genes, until heterozygotes became phenotypically identical to the advantaged. homozygotes. According to a more modern theory, the high frequency of recessive mutants is the result of the complicated integrated network of enzyme-catalysed reactions in living organisms, within which there alternative pathways to the same product. This means that large alterations in the activity of a single enzyme (due to mutation) may lead to only small changes in the rate of formation of the gene product, and a relatively large decrease in enzyme activity may not be manifested in the phenotype of many heterozygous mutants.

Dominance transfer, ***dominance reversal:*** transfer of dominance from one allele to the other ($A_1a_2 \rightarrow a_1A_2$) during the ontogenic development of a hybrid, so that both alleles are expressed phenotypically at different developmental stages of the organism.

Donkey: see Wild ass.

Donnan equilibrium: see Donnan principle.

Donnan potential: the electric potential difference between two phases in equilibrium, when one phase contains polyions that cannot diffuse into the other phase. For example, a protein solution may be separated by a membrane from an inorganic electrolyte solution. The protein molecules carry a fixed, nondiffusible charge, whereas the electrolyte ions are mobile and diffusible. The mobile counterion (its sign is opposite to that of the net charge of the protein molecule) therefore diffuses into, and accumulates in, the protein phase. In the protein phase, the cation and anion concentrations of the inorganic electrolyte are no longer balanced, because one ion (the counterion) is in excess. This difference in ion activity is compensated by an electric potential difference between the two phases, so that the electrochemical potential for all permeable ions is the same in both phases. The D.p. is an equilibrium potential and therefore cannot be exploited for the performance of electrical work without destroying the system. Simultaneously, an osmotic pressure difference is also generated, which leads to the uptake of water. If the membrane cannot compensate this pressure, the water uptake leads to swelling and subsequent destruction of the system. See Donnan principle.

Donnan principle: a law described in 1911 by Donnan to explain the distributions of ions that occur when two phases are separated by a membrane, one phase containing diffusible ions, the other containing nondiffusible polyions. If the free diffusion of anions is prevented (e.g. carboxyl groups in protopectin, or in plasma or membrane proteins), they attract diffusible cations from the environment until the potential gradient on one side is balanced by the resulting concentration gradient on the other. This equilibrium is called the *Donnan equilibrium.* It is characterized by the fact that the *Donnan phase,* which contains the nondiffusible trapped or fixed ion, contains a higher total concentration than the external phase, resulting in a potential gradient known as the Donnan potential (see). The polarity of the Donnan potential is determined by the charge of the nondiffusible ion. Donnan potentials influence, inter alia, the sorption of ions on soil colloids, and the uptake of ions and water by the protoplasm. Two generalizations can be applied to the Donnan equilibrium: 1) The tendency of an ion to migrate into the region containing the fixed counterion increases with the number of charges that it carries (the valency effect); 2) This effect is more pronounced, the higher the concentration of the fixed ion compared with that of the diffusible ion (concentration effect).

Dorados: see *Coryphaenidae.*

Dorcotragini: see *Neotraginae.*

Dorcotragus: see *Neotraginae.*

Dormancy:

1) In plants, the resting state of buds and seeds (see Seed dormancy, Resting periods) between two active growth periods. Often D. occurs during unfavorable environmental conditions, e.g. during a dry or cold period, and its onset is usually coordinated with the appearance of Senescence (see) of the entire plant (in annuals) or parts of the plant (e.g. the leaves or all the aerial part of perennials). Obvious coordination of D. with environmental conditions is seen in the shedding of leaves by deciduous trees in the autumn, a lack of obvious growth or developmental activity in the winter, then renewed growth and leaf production in the spring; this is followed by flower formation and the development of buds for the next year's growth. It should be noted, however, that growth and development are not uniform, even under constant conditions. In the unchanging environment of the humid tropics, trees still periodically shed their leaves and pass through a period of D., albeit not always strictly according to an annual rhythm, and perhaps involving only part of the plant. It seems that an alternation of growth and dormancy is normal in plants.

2) In animals, a slowing down of metabolism and suspension of individual development in response to changes in one or more environmental factors. D. takes various forms. *Quiescence* is an immediate response to a current deterioration of environmental conditions (e.g. Black bean aphid, *Doralis fabae* Scop.). *Diapause,* which is especially widespread among insects, is always associated with a particular developmental stage; it may be obligatory or facultative, and it can be classified as: a) *oligopause,* whose onset and completion occur in direct response (with a slight delay) to certain factors (temperature, light) (e.g. Dried fruit moth, *Ephestia cautella* Walker); b) *parapause,* which is a prospective form of D., i.e. it anticipates future changes in environmental factors by occurring while conditions are still favorable; it is ended by further changes in the same factors (e.g. Field cricket); c) *eudiapause,* which is also triggered prospectively, usually by the day length, and is ended by environmental temperature changes. The sensitive phase, during which the animal is programmed for eudiapause, usually occurs long before onset of the latter (e.g. Cabbage white butterfly). By prolonging eudiapause until the following year, some insect species maintain a reserve population.

A special form of animal D. is Hibernation (see).

Dormouse: see *Gliridae.*

Dorsal sac: see Parietal organs.

Dorsal stripe: a dark, dorsal, longitudinal stripe in the coat color of certain wild and domestic mammals, especially horses and goats. It occurs more frequently and is more pronounced in wild forms. Thus, the D.s. of the South Russian plains tarpan *(Equus przewalski gmelini)* is a broad, black stripe extending from the mane to the base of the tail. It may also be present or absent in the Kinkajou *(Potos flavus),* cats *(Felis silvestris),* African climbing mice *(Dendromus),* as well as several other mammals.

Dorsiventral, ***monosymmetrical, zygomorphic:*** a term applied to organisms with only a single plane of symmetry. Many foliage leaves have a D. structure. Flowers of the *Papilionaceae* and *Lamiaceae* are D. (the term zygomorphic is usually applied to flowers).

Dorstenia: see *Moraceae.*

Dortmund tank: see Biological waste water treatment.

Dorylaimida: see *Nematoda.*

Dorylinae: see Ants.

Dosage effect: the quantitative effect on the expression of phenotypic characters exerted by the alleles of a gene, depending on their frequency in the genotype. To investigate a D.e., genotypes are constructed containing different frequencies of the allele in question. This is possible with the aid of deletions, duplications, aneuploidy and polyploidy.

Dose:

1) The administered quantity of a therapeutic agent, plant protection agent, toxin, nutrient, hormone, vitamin, etc., usually expressed as quantity (weight) per volume or mass of the treated organism.

2) The quantity of absorbed radiation energy per unit mass of the irradiated organism or material. The statutory unit of D. is the Gray (Gy). 1 Gy = 1 $J.kg^{-1}$. Since the radiation effect depends not only on the dose but also on the time of exposure, it is also necessary to introduce the concept of the dose rate, which is defined as the dose per unit time ($Gy.s^{-1}$). The older unit, the Rad (rd), corresponds to 10^{-2} Gy. Identical energy doses from different types of radiation (e.g. X-rays, γ-rays, heavy charged particles) may have differing biological effects. The ratio between the dose of a standard radiation (D_S) (200 kV X-rays) and the dose of another radiation (D_O) that causes the same biological effect is called the *relative biological effectiveness* (RBE). Thus, RBE = D_S/D_O. The standard radiation dose is often given as the *equivalent dose* in Sieverts (Sv,), i.e. D_S (in Sv) = RBE x D_O (in Gy). The older unit of equivalent dose is the rem (Roentgen-equivalent-man); 1 rem = 0.01 Sv. The RBE is sometimes known as the quality factor (QF) for the type of radiation, e.g. QF for X-rays, γ-rays, β-particles and positrons = 1; QF for α-particles = 10.

Dose-effect curve: see Target theory.

Dosidius: see *Cephalopoda.*

Dothiorales: see *Ascomycetes.*

Dotterel: see *Charadriidae.*

Douc, *douc langur, variegated langur,* *Pygathrix nemaeus:* a member of the *Colobinae* (see Colobine monkeys). The genus has only one species, which is strikingly patterned. A bright yellow face, with white whiskers and oblique, almond-shaped eyes, is set in a brown head with light chestnut bands behind the ears. The mottled grayish trunk is lighter below, while the tail and rump are white. Upper arms and legs, and hands and feet are black, while the forearms are white and the lower legs are chestnut. Habitat: primary and secondary forests in Laos and Vietnam, and possibly Hainan island.

Dourine: see Trypanosomes.

Dovekie: see *Alcidae.*

Doves: see *Columbiformes.*

Dowitchers: see *Scolopacidae.*

Down: see Feathers.

Downy mildew: see *Phycomycetes.*

Doxycyclin: see Tetracyclins.

DPN: see Nicotinamide-adenine-dinucleotide.

Dracaena: see *Agavaceae.*

Draco: see Flying dragon.

Dracontiasis: see Guinea worm.

Dracunculida: see *Nematoda.*

Dracunculus: see Guinea worm, *Nematoda.*

Dragonflies: see *Odonatoptera.*

Dragon tree: see *Agavaceae.*

Dragon weever: see *Trachinidae.*

Drainage fields: see Biological waste water treatment.

Drepanidae: see *Lepidoptera.*

Drepanididae, *Hawaiian honeycreepers:* an avian family in the order *Passeriformes* (see), found exclusively on the Hawaiian islands. The family contains only 22 known species (11–23 cm in length), although 9 of these have become extinct in recent times. They provide a dramatic example of adaptive radiation from a common ancestor. Thus, their bills are crossed, parrot-like, short and pointed, or long and decurved, representing specializations for different food sources, such as seeds, fruit, insects, nectar, etc. For example, the rare ***Akepa*** *(Loxops coccinea)* has a short, sideways crossed bill, possibly as an adaptation for opening leaf buds, whereas the ***Akialoa*** *(Hemignathus wilsoni = H. procerus)* has a long curved bill, which is used for probing flowers for nectar and insects. They are divided into 2 subfamilies. The more primitive *Drepanidinae* ***(Thick-skinned honeycreepers)*** are mainly red and black, with stiff and pointed head feathers, and sharp edged primaries which produce a whirring sound in flight. They inhabit forests in small flocks, some species extracting nectar from flowers by piercing a hole in the base, and others feeding on soft-bodied insects. The extinct ***Mamo,*** which was used by the Hawaiians to make feather cloaks, belonged to this group; it was last reported in 1898. The more advanced *Psittirostrinae* ***(Green honeycreepers)*** are insect, fruit and seed eaters. They have mostly greenish or yellowish plumage, with soft and fluffy head feathers, e.g. the ***Palila*** *(Psittirostra bailleui;* short parrot-like bill). All drepanidids have a musky smell.

Drepanidinae: see *Drepanididae.*

Depanium: see Flower, section 5) Inflorescence.

Dressing of seeds: see Seed dressing.

Dried yeast: see Baker's yeast.

Drigalski spatula: a glass rod, with one end bent into a triangle. Named for the bacteriologist, W. von Drigalski, it is used for spreading suspensions of viruses, bacteria or other cells uniformly on the surface of solid culture media.

Drill: see Mandrill.

Drill shells: see *Gastropoda.*

Drive: see Vigilance.

Dromadidae: a monotypic avian family in the order *Charadriiformes* (see). Sole representative of the family is the ***Crab plover*** *(Dromas ardeola;* 38 cm), found on the coasts of East Africa and the Indian Ocean. Its partly webbed front toes resemble those of avocets, but the hallux is much larger. Crabs and other crustaceans are pounded to pieces with the long, heavy, black, pointed bill, which resembles that of a tern. It is white, except for a black mantle and the black primaries of its long pointed wings. Its long legs are blue-gray and the eyes dark brown.

Dromas: see *Dromadidae.*

Dromedary, *Arabian camel,* *Camelus dromedarius:* a fawn to beige colored member of the *Camelidae* (see), with a single hump. It is kept as a domestic animal in North Africa, Arabia and western Asia, and is unknown in the wild. Like the Camel (see), it is an easily kept and undemanding animal. With aid of its fat reserves in the hump, it can survive for long periods without water. It has broad, flat feet, its nostrils can be closed, and its thick lips enable it to browse very coarse vegetation. There are two principal domestic breeds of D., a relatively heavy, slow animal used as a beast of burden, and a lighter, long-legged, faster breed for riding. An especially fine, North African breed of riding D. is known as the *Mehari.* The D. and the camel can be cross-bred to form a hybrid known as a *Tulu,* which is often larger and stronger than either parent. Since it is sterile or produces weak offspring, the tulu is always produced by hybridizing the original pure species or by crossing a hybrid with a pure species.

Dromiacea: a section of the infraorder *Brachyura* (true crabs) of the *Decapoda* (see).

Droseraceae, *sundew family:* a cosmopolitan family of dicotyledonous plants, containing about 90 species. They are herbaceous, carnivorous plants, with variously modified leaves for trapping and digesting small animals, chiefly insects.

The flowers are actinomorphic, hermaphrodite, and usually 5-merous, with a superior 1-celled ovary, which develops into a loculicidal capsule with numerous small seeds. The family contains 4 genera: *Dionaea, Aldrovanda, Drosera* and *Drosophylla,* which are described under Carnivorous plants (see). Three species of *Drosera* occur in Europe: *D. rotundifolia* ***(Sundew), D. anglica (Great sundew)*** and *D. intermedia* ***(Long-leaved sundew).***

Drosophyllum: see Carnivorous plants.

Drought avoidance: see Drought resistance.

Drought damage: injury to plant tissues caused by an insufficiency of water. The earliest signs are limpness and wilting of herbaceous plant parts. If this wilting damage is not too extreme, it can be reversed by the reintroduction of an adequate water supply. Extreme water loss, however, leads to cell death, and the plant withers. D.d. resulting from a sudden transition to extreme drought conditions closely resembles Frost damage (see), which is also due to water deprivation. Slow dehydration, which occurs naturally with onset of a dry season, first affects the growth rate, because stretching growth only occurs in fully turgid cells. This is followed by the degradation of proteins and nucleic acids, a decrease in the rate of photosynthesis, and other metabolic disturbances, leading, inter alia, to an accumulation of toxins such as ammonia. Cell death may then insue. See Drought resistance.

Drought resistance: in plants, the ability to withstand and survive periods of drought. D.r. is species-specific and inherited, although the intensity of its expression can be modified by environmental conditions.

Plants that can dry out without injury are well protected against drought. This type of D.r. is called *true D.r.* or *drought tolerance.* It occurs in Poikilohydric (see) plants, whose water potential largely corresponds to that of the environment. Many bacteria, algae and lichens. some fungal hyphae and various mosses are able to survive in a desiccated state. Most spores, pollen grains and cormophyte seeds can survive in a desiccated condition for more or less lengthy periods without losing the ability to germinate. In some fern species, e.g. the rusty-back fern *(Ceterach officinarum)* and some seed plants, e.g. the native Southwest African species *Myrothamnus flabellifolia,* as well as some native South African *Graminae* and *Cyperaceae,* the vegetative organs (e.g. leaves and shoots) are able to survive desiccation. Other plants are not fully resistant to desiccation, but are able to tolerate a considerable loss of water. Thus, the olive tree *(Olea europaea)* can tolerate the loss of up to 70% of its water. The fig tree *(Ficus carica)* tolerates the loss of 25% of its water. Physiological characteristics of resistance include an increased ability to bind water by the protoplasm (increased plasma hydration) and an increased osmolarity of the cytoplasm.

Plants displaying *drought resistance* limit their water loss, thereby avoiding desiccation. They display various adaptations that decrease transpiration, e.g. folding and rolling of leaves, closure of the often numerous and small stomata, strong development of the cuticle and

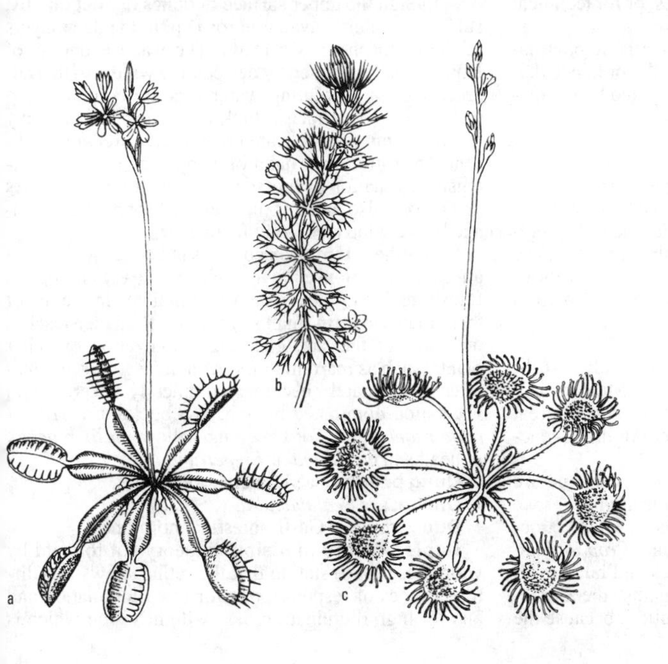

Droseraceae. a Venus' fly-trap *(Dionaea muscipula). b* Water-wheel plant *(Aldrovanda vesiculosa). c* Sundew *(Drosera rotundiflora).*

wax layer, a decrease in the area of transpiring surfaces, as well as other characteristic xeromorphic adaptations (see Xerophytes). A further means of decreasing water loss is the shedding of leaves, e.g. by xeromorphic trees during summer drought, and by deciduous trees in temperate regions as an adaptation to frost drought. Water stores are often developed, succulence (see Xerophytes) being an adaptation to dry conditions. Other plants avoid drought by compensating increased water loss with increased water uptake, with the aid of an extensive root system. For example, the roots of field holly *(Eryngium campestre)* penetrate to a depth of 6 m, whereas the maximal shoot height is 1 m.

The D.r. of plants can be weakened by the continual optimal availability of water, but it can be restored by a process of *hardening*. In the hardening process, the plant is subjected to intermediate intensities of drought, or brief periods of water deprivation, leading to reversible wilting. Alternatively, the seeds are allowed to imbibe water, then dried again. D.r. is also induced by short daylength. According to recent theory, this restoration of D.r. in response to wilting is due to gene activation; it is associated with RNA and protein synthesis, resulting, inter alia, in the synthesis of proteins with a different isoelectric point and an increased level of hydration.

Since D.r. is largely determined by genetic factors, plant breeders attempt to produce strains and varieties of agricultural plants that maintain their yield under conditions of water shortage. Plants can be tested directly for D.r. by planting in dry geographical areas or in controlled artificial dry environmnets, or indirectly by determination of their absorptive capacity and relative rate of transpiration, and by evaluation of root development.

Drought tolerance: see Drought resistance.

Drugs:

1) In the older literature, preparations of plant or animal material, preserved by drying, and used for medicinal purposes (official drugs), as spices, or for technical purposes.

2) The term is now applied to all synthetic pharmaceuticals, especially those taken internally or injected. It is also a blanket term, used colloquially and by the media for addictive narcotics.

Drum: see *Sciaenidae.*

Drupe: see Fruit.

Dryad's saddle: see *Basidiomycetes (Poriales).*

Dryas (from *Dryas octopetala,* Mountain avens): a late Ice Age period of vegetation development. The Pollen analysis (see) of the D. indicates the predominance of steppe-tundra. In addition to tree pollen from willow, birch and pine, there is an abundance of pollen from smaller woody and herbaceous plants.

Drymodes: see *Turdinae.*

Dryopithecines, *Dryopithecinae:* a widely distributed group of primates of the Old World Miocene and Pleiocene. They were the ancestors of the modern great apes, and probably also of humans. See Anthropogenesis.

Dryopithecus pattern: the 5-cusp crown structure of the lower molars of Dryopithecines. The groove between the cusps form a 5-fold Y shape. The D.p. occurs in the very early Oligocene hominid, *Propliopithecus,* as well as in the recent great apes and humans. It represents an important argument against the Cercopithecine theory (see) of human evolution, because the genetic determination of tooth structure is very complex, and the proposal that the same structure has arisen several times independently in unacceptable.

Dryotriorchis: see *Falconiformes.*

Dry rot: see *Basidiomycetes (Poriales).*

Duck-billed platypus: see *Monotremata.*

Duck mussel: see Freshwater mussels.

Ducks: see *Anseriformes.*

Duck's-meat: see *Lemnaceae.*

Duckweed: see *Lemnaceae.*

Ducts of Cuvier: see Cuvier, ducts of.

Ductus choledochus: see Gall bladder.

Ductus cysticus: see Gall bladder.

Ductus hepaticus: see Gall bladder.

Duculinae: see *Columbiformes.*

Dugong: see *Sirenia.*

Duiker: see *Cephalophinae.*

Dulidae: see *Bombycillidae.*

Dulinae: see *Bombycillidae.*

Dulus: see *Bombycillidae.*

Dumetella: see *Mimidae.*

Dune plants: advanced vascular plants, as well as more primitive plants that inhabit coastal and inland sand dunes, which characteristically consist of nutrient-poor, loosely packed sand grains. Examples are Marram grass *(Ammophila arenaria),* which is abundant and often dominant on European coastal sand dunes, and Sand couch grass *(Agropyron junceiforme),* also common on western and northern European coastal dunes. An essential characteristic of the habitat of D.p. is the mobility of the sand, with the accompanying danger that roots may be exposed, or that aerial parts may be buried. Many D.p., however, have extremely long roots, often branched and displaying vigorous growth, which penetrate to considerable depths.

Depending on their density, D.p. decrease or prevent sand movement; they therefore stabilize dunes and help to protect coastlines from erosion.

Although the upper surface of dunes dry out rapidly, sufficient water is available for D.p. in the deep layers of the rhizosphere (see Dew). The water economy of D.p. is greatly stressed by desiccating winds, so that adaptations for preventing water loss are common, e.g. rolling of leaf surfaces, thick epidermal layers, waxy layers. Plants with delicate green leaves are largely absent. The rigid, spiny habit of many D.p., due to the extensive formation of protective tissue in the shoot axes and leaves, also protects them against mechanical damage by the wind (see Wind factor).

On northern European coasts, sand couch grass often appears as a pioneer plant on the salty, waterlogged foreshore. Wind-blown sand accumulates in the lee of these plants to form small sand dunes. Rain then leaches out most of the salt, enabling colonization by other plants, such as marram grass. As a dune grows and matures, it ultimately becomes completely covered with vegetation dominated by, e.g. Sea buckthorn *(Hippophae rhamnoides),* or Creeping willow *(Salix repens),* or heaths like *Calluna* or *Empetrum.*

Dung beetles: see Lamellicorn beetles.

Dunlin: see *Scolopacidae.*

Duocrinin: see Gastrointestinal hormones.

Duplex theory of vision: a theory put forward by von Kries (1896) stating that the retina shows two different types of response, one for low illumination and one for high illumination, i.e. twilight vision which is color blind, and day vision which responds to color. Rods, which contain visual purple, are thought to be responsible for twilight vision, while cones are active in day vision.

Duplication: a Chromosome mutation (see), in which a chromosome segment with all its constituent genes occurs twice in the haploid chromosome. D. may be intra- or inter-chromosomal, i.e. the duplicated segment may occur in the same or in two different chromosomes, respectively. During diplophase, a D. may be homozygous or heterozygous. D. may be advantageous or nonadvantageous for the carrier organism, depending on the size the duplicated segment, and the resulting qualitative and quantitative effects on the genotype and phenotype.

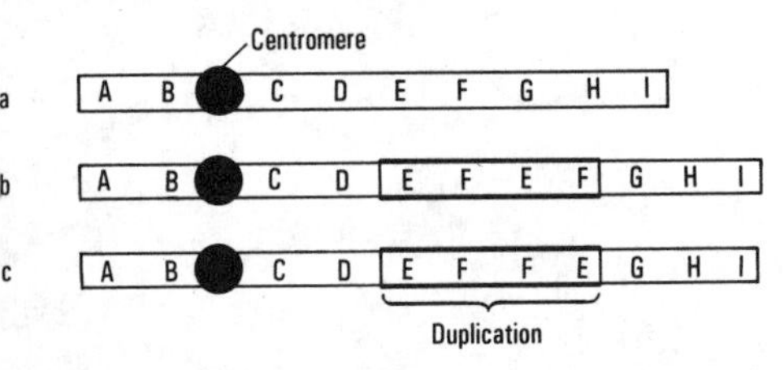

Duplication. a Normal chromosome. *b* Non-inverted intrachromosomal duplication. *c* Inverted intrachromosomal duplication.

Duplicidentata: see *Lagomorpha.*

Duplicidentate condition: see *Lagomorpha.*

Duplicity theory of vision: a theory attributed to Hecht, namely that retinal receptors consist of both rods and cones. These differ in the quantity of light required to stimulate them, rods being on average more sensitive than cones. However, the sensitivities of rod and the cone populations overlap, so that as illumination intensity is increased from zero, an increasing number of rods is brought into play, followed by an increasing number of cones.

Duramen: heartwood. See Wood.

Durchbrenner: a German term for a genotype, which despite the presence of a lethal dose of a lethal factor, survives the resulting developmental crisis and continues its development. The crisis occurs at a specific time during development, which is determined by the factor in question, and it normally ends in death.

Durham tube: a glass tube named after the English physician, A.E. Durham, and used for the demonstration of gas production by microorganisms. The D.t. is filled with culture medium and placed upright in the same medium, with its sealed end upward and its open end downward. If the inoculated microorganism produces gas, this collects in the D.t.

Durian: see *Bombacaceae.*

Durra: see *Graminae.*

Dusting agents: see Crop dusting agents.

Dutch elm disease: see *Ascomycetes.*

Dynastinae: see Lamellicorn beetles.

Dwale: see *Solanaceae.*

Dwarf African otter shrew: see Water shrews.

Dwarf antelopes: see *Neotraginae.*

Dwarfism: see Nanism.

Dwarf pond snail *(Galba truncatula):* see *Gastropoda.*

Dwarf scrub zone: see Altitudinal zonation.

Dy: see Humus.

Dyer's saffron: see *Compositae.*

Dynein: see Cilia and Flagella.

Dysentery: see Amebas.

Dysodont: a hinge type of the *Bivalvia* (see).

Dytiscus: see Respiratory organs.

Dzeggetai: see Asiatic wild ass.

E

Eagles: see *Falconiformes.*
Ear: see Auditory organ.
Eared seals: see *Otariidae.*
Earless monitor: see *Lanthanotidae.*
Early wood: see Spring wood.
Early woodland plants: see Woodland spring plants.
Earth-ball: see *Basidiomycetes (Gasteromycetales).*
Earthcreepers: see *Furnariidae.*
Earth stars: see *Basidiomycetes (Gasteromycetales).*
Earth tongues: see *Ascomycetes.*
Earthworms: see *Oligochaeta.*
Earwigs: see *Dermaptera.*
East coast fever: see Piroplasmids.
Easter-ledges: see *Polygonaceae.*
Eastern bluebird: see *Turdinae.*
Eastern hemlock: see *Pinaceae.*
Eastern house mouse: see *Muridae.*
Eastern meadow mouse: see *Arvicolinae.*
Ecballium: see *Cucurbitaceae.*
Ecdemic: foreign; non-native.
Ecdysis: see Molting.

Ecdysone: a steroidal insect hormone, which is produced in the insect prothoracic gland in response to prothoracotropic hormone. It is converted into the active molting hormone, Ecdysterone (see), by hydroxylation at C20. E. is biosynthesized from cholesterol or phytosterols, which the insect obtains as essential dietary components. E. and structurally related steroids, known as phytoecdysones, occur in certain higher plants.

The isolation and crystallization of E. by Butenandt and Karlson in 1954, from pupae of the silkworm *(Bobyx mori),* was the first reported crystallization of an insect hormone. It was subsequently identified in other insects.

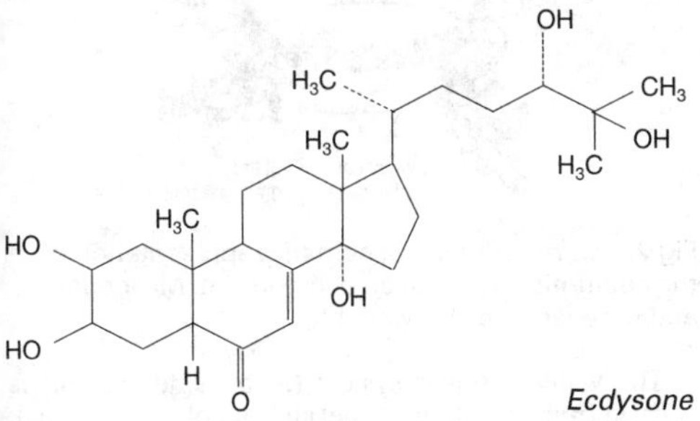

Ecdysone

Ecdysterone, *20-hydroxyecdysone:* a steroidal insect hormone, responsible for the larval molt prior to pupation, and later for the pupal molt that releases the imago. In concert with juvenile hormone, E. is responsible for larval molting (larva to larva). It is formed in the peripheral fat body tissues by the C20-hydroxylation of ecdysone.

E. is identical with *Crustecdysone,* the molting hormone of crustaceans. It is also present in certain higher plants, where its function is unknown.

ECG: acronym of electrocardiogram. See Electrocardiography.

Echeneidae, *remoras:* a family (8 species in 7 genera) of marine fishes, normally placed in the order *Perciformes,* although some authorities place them in a separate order. They grow up to 60 cm in length, and the first dorsal fin is modified to a suction disk on the head. With the aid of the latter, they attach themselves to larger fishes and to ships, and allow themselves to be transported over considerable distances. In some regions turtle fishermen attach a line to the tail, and pull the fish in after it has attached itself to the carapace of a turtle.

Echeveria: see *Crassulaceae.*
Echidnas: see *Monotremata.*

Echimyidae, *spiny rats:* a family of Rodents (see) consisting of 58 species in 16 recent genera and 2 subfamilies, widely distributed throughout South America. The body is rat-sized and rat-like in appearance, with medium sized ears and eyes and a somewhat truncated, pointed nose. The first digit of the forefoot is stunted. Body length ranges from 10 to 48 cm, and the tail (which in some species is longer than the body) from 4.5 to 43 cm. The tail easily breaks, due to a weak point at the centrum of the 5th caudal vertebra, and it is often lost. Dental formula: 1/1, 0/0, 1/1, 3/3 = 20. Skeletal remains of 2 extinct genera have been found in Indian kitchen middens in Cuba and Haiti. One of these *(Heteropsomys)* apparently became extinct as late as the 19th century. With the exceptions of *Trichomys* (a single species known as the ***Punaré)*** and *Isothrix* (3 species known as ***Toros),*** members of the subfamily *Echimyinae* have a bristly or spiny pelage, consisting of oblate, sharp-pointed, furrowed quills, attached to the flesh by narrow basal stalks. Members of the *Echimyinae* (11 genera) include the ***True spiny rats*** or ***Casiragua*** *(Proechimys,* e.g. the ***Cayenne spiny rat,*** *P. guyannensis),* the ***Arboreal spiny rats*** *(Echimys,* e.g. the ***Armored spiny rat,*** *E. armatus)*. Members of the smaller subfamily, *Dactylomyinae,* have soft hair without quills, e.g. the ***Coro-coros*** *(Dactylomys)* and the ***Rato de Taquara*** *(Kannabateomys amblyonyx),* the latter being found in bamboo thickets along river banks in Brazil.

Echimyinae: see *Echimyidae.*
Echimys: see *Echimyidae.*
Echinocardium: see *Echinoidea.*
Echinochrome: see Blood, Spinochromes.
Echinococcus: see *Dog tapeworm.*
Echinodera: see *Kinorhyncha.*

Echinodermata, *echinoderms:* a phylum of the *Deuterostomia.* The *E.* have a most unusual structure, and an aberrant (or unique) appearance compared with other members of the animal kingdom. In the adult form they are radially symmetrical, usually with five radii *(pentam-*

eral symmetry), and may be shaped like a cup, star, sphere, disk or cylinder (Fig.1). All species are marine, living almost exclusively on the sea bottom. All members of the phylum possess characteristic calcareous skeletal elements, which arise in the subdermal tissue. In the *Holothuroidea* (sea cucumbers), these skeletal elements remain as minute sclerites or spicules, whereas in all other classes they are developed into large plates (often carrying spinous appendages) that cover the entire body. Externally, the five-fold symmetry is readily apparent from the arrangement of the body appendages (arms) and the skeletal plates. Internally, the coelomic canals and nervous system also clearly conform to a five-fold symmetrical pattern. The mouth opening lies at the center of the radial structure, and it usually faces the sea bottom. This surface of the body is called the *oral* or *ambulacral* surface, whereas the opposite surface, containing the anus, is said to be *aboral* or *abambulacral.* The terms "ventral" and "dorsal" should not be applied to these surfaces, because they correspond to the left and right sides, rather than the ventral and dorsal sides of the larva.

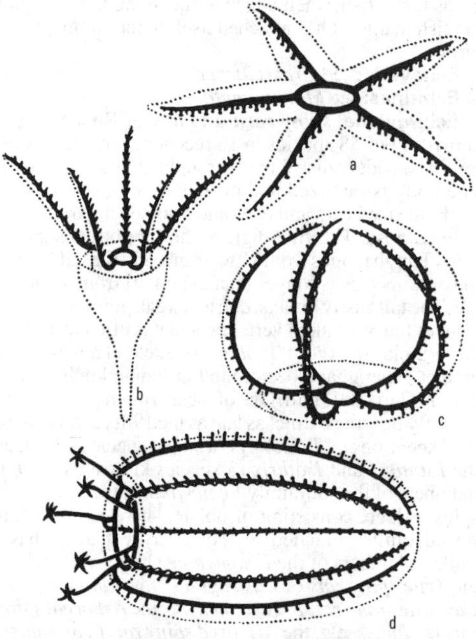

Fig.1. *Echinodermata.* Body forms of echinoderms. *a* Starfish *(Asteroidea). b* Feather star or sea lily *(Crinoidea). c* Sea urchin *(Echinoidea). d* Sea cucumber *(Holothuroidea).* Mesocoel canals (in tube feet and tentacles) are shown in black.

The symmetry of the *E.*, which differs from that of all other animal phyla, arises secondarily during metamorphosis. Larvae are bilaterally symmetrical, and like the acorn worms *(Enteropneusta)* and beard worms *(Pogonophora),* they possess three coelomic cavities (protocoel, mesocoel and metacoel), usually referred to in the *E.* as the axocoel, hydrocoel, and somatocoel (anterior, middle and posterior, respectively). Each coelomic cavity of the E. consists of paired regions (one region to each side of the body), and even in the early larva these regions are asymmetrical, the axocoel and hydrocoel being developed more on the left than on the right. Pentameral symmetry first arises as part of a fundamental reorganization during metamorphosis. This transformation starts with the extension of the left mesocoel to form the beginning of the water vascular system with its five radii (see below). All the other organs then become arranged according to the pattern of these five radii. This abolishes the bilateral symmetry of the larva, and the original three coelomic cavities are no longer recognizable in the adult animal. Through the development of the different coelomic cavities, the individual parts of the larval mesoderm therefore develop into a variety of different structures (Fig.2). The left protocoel becomes the unpaired axial organ (axial gland, axial complex), extending from the oral to the aboral body pole, where it expands into an ampulla which opens to the exterior via the madreporite; the right protocoel forms a small dorsal vesicle. In the *Crinoidea* (feather stars and sea lilies) and *Holothuroidea* (sea cucumbers), the protocoel is markedly regressed. The main body cavity is derived from the metacoel ("somatocoel").

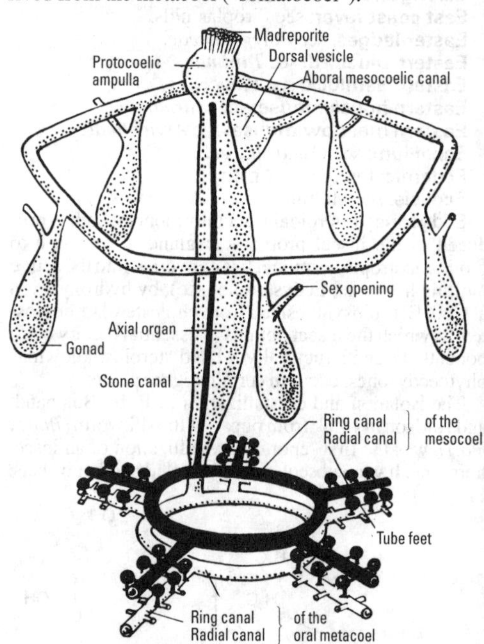

Fig.2. *Echinodermata.* Schematic representation of the coelomic spaces of an echinoderm. Mesocoel and tube feet are shown in black.

The water vascular system (ambulacral system) is derived from the left mesocoel and part of the protocoel. It contains sea water, and is in indirect communication with the surrounding sea. Confined to the oral side of the animal, the ambulacral system encircles the foregut and gives rise to five radial canals. These radial canals have numerous lateral branches, each ending in a small ampulla, with a connected papilla or tentacle-like structure which projects from the body as a small appendage. On the surface of the body facing the substratum, these appendages are modified to small tube-feet, which erve for locomotion, and often also act as attachment isks. The distribution of fluid and hydrostatic pressure vithin the tube feet and the radial water canals is conrolled by muscular valves and constrictions, so that ach tube foot can be operated separately by the conraction of muscles in the adjacent part of the system. In ddition, a rigid, calcified stone canal extends from the ing canal of the water vascular system to the aboral ide, where it empties into the protocoel ampulla; the nesocoel is therefore connected indirectly with the exerior. The right mesocoel is lost during larval developnent. The posterior paired coelom (metacoel) gives rise o the general body cavity, as well as an oral canal ystem (hyponeural system, which encloses part of the emal system: see below); the latter also consists of a ing canal with five radial canals, which lie parallel to hose of the water vascular system. On the aboral side of he body, the metacoel forms another ring canal, the enital canal. At inter-radial positions, five gonads arise s pendulous, sac-like evaginations of the genital canal. The hyponeural ring communicates with the axial organ, and the aboral genital ring communicates with the orsal vesicle of the protocoel.

The hemal, blood-vascular, or blood lacunar system s well developed in holothuroids and echinoids, and oorly developed in other classes. It consists of small luid-filled sinus channels without a distinct lining. Most of the channels are enclosed within tubular porions of the coelom, which are often referred to as the erihemal system, but to avoid the impression that they epresent an accessory circulatory system, they are referably called hyponeural canals; their relation to the ervous system appears to be functional, possibly servng as a cushion against shock. The principal hemal hannels include oral and aboral hemal rings and a raial hemal sinus extending into each arm, all running arallel to the coelom rings and canals, and to the alinentary canal. The hemal channels are best developed n relation to the alimentary canal, suggesting that they lay a role in the uptake and distribution of digestion roducts. From the oral hemal ring, a channel ascends hrough the dark elongated spongy axial organ or gland. There is no heart., but some authorities consider that the xial gland is an evolutionary remnant of a heart. Cerainly, in the asteroids, echinoids and ophiuroids, the xial gland is histologically similar to other parts of the emal system (its lumen consists of hemal channels), nd it appears to represent the functional and structural enter of the hemal system.

Excretory organs are also absent. The straight or ooped intestine passes from one body pole to the other from the oral to the aboral surface). In the *Crinoidea,* owever, the mouth and anus lie close to each other on he upper surface of the chalice. The *Ophiuroidea* (brite stars) lack a hindgut and anus. The nervous apparaus consists of three epithelial systems: the ectoneural nd hyponeural systems both lie on the oral side of the ody, and follow essentially the same course as the waer vascular system; the aboral nervous system lies in he walls of the genital canal, and it is absent from the *Asteroidea* (starfishes) and *Holothuroidea* (sea cucumers).

With few exceptions, the sexes are separate, but sexal dimorphism is rarely observed. Each of the usually ive gonads exits to the exterior via a canal. Holothuroids possess only a single gonad. Development occurs with total cleavage to an invagination gastrula, whose blastopore subsequently develops into the anus, while the adult mouth arises anew on the opposite side of the animal. The coelom forms from outpocketing of the archenteron (primitive gut). Larvae conform to the basic dipleurula type. Depending on the class of echinoderm, various structures are then developed, such as various types of ciliated bands, pointed appendages and calcareous support rods. Thus, the holothuroid larva develops into an auricularia, the asteroid larva becomes a bipinnaria, while the ophiuroid and echinoid larvae become pluteus larvae (ophiopluteus and echinopluteus, respectively).

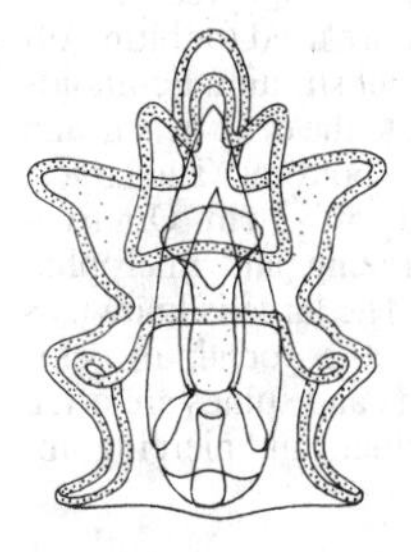

Fig.3. *Echinodermata.* Bipinnaria larva of a starfish.

Echinoderms have affinities with the hemichordates. The dipleurula-type larva closely resembles the tornaria larva of acorn worms. Hemichordates also develop a tripartite coelom, which in the adult is associated with a corresponding division of the body into prosoma, mesosoma and metasoma.

Sea cucumbers have economic importance as food (see Trepang). Sea urchins and starfishes also have a limited commercial value. About 6 000 living species of echinoderms are known.

Classification. The following 6 classes are recognized (see separate entries for each): *Crinoidea* (feather stars and sea lilies); *Holothuroidea* (sea cucumbers); *Echinoidea* (sea urchins); *Asteroidea* (star fishes); *Ophiuroidea* (brittle stars); *Concentricycloidea.*

The *Asteroidea* and *Ophiuroidea* may also be considered as subclasses of the single class, *Stellaroidea.*

Fossil record. Fossils are found from the lower Cambrian to the present. Paleozoic representatives are primarily sessile forms [sea lilies and various extinct classes, e.g. the *Cystoidea* (see), *Blastoidea* (see)] with a spherical or cup-shaped body attached aborally to the substratum by a stalk, the mouth being directed upward and encircled by five arms. Derived, freely motile, nonstalked forms are also present in the Paleozoic, but do not become particularly abundant until the Mesozoic and Cenozoic, while today they account for the majority of the *E.*

Echinoderm saponins: see Echinoderm toxins.

Echinoderm toxins, *echinoderm saponins:* low molecular mass steroid toxins synthesized in the glands of echinoderms. Sea cucumbers of the genus *Holothuria* secrete highly toxic sulfated steroid glycosides called holothurins. Starfishes *(Asteroidea)* produce asteriotoxins or asteriosaponins, in which the main aglycon is pregnene diolone.

Echinoidea, *echinoids, sea urchins, sand dollars, heart urchins:* a class of the *Echinodermata* (see), containing about 860 known living species. The body is enclosed in a hard test, which has the shape of a compressed sphere or heart, or is discoidal. The test consists of large, interlocking, calcareous skeletal plates, which carry movable spines and defensive Pedicellariae (see). On the exterior surface of the test, 20 vertical rows of plates are arranged in 5 double rows of ambulacral (perforate) plates and 5 double radial rows of interambulacral (imperforate) plates. The tube feet of the water vascular system project through the pores in the ambulacral plates. The aboral body surface, which contains the anus, is very small. The foregut is encircled by a jaw apparatus (Aristotle's lantern), consisting of five jaws, each ending in a tooth. These teeth form a conical covering to the mouth opening and can be spread out or closed by attached muscles. They are used for biting off pieces of food or for gnawing the substratum. Echinoids are found from the intertidal zone to the deep ocean, and they inhabit both hard and soft sea bottoms. The largest known species attains a diameter of 32 cm. Omnivorous, predatory, herbivorous, grazing and microphagous species are all represented. The larva is known as an *echinopluteus.* In some parts, the gonads of echinoids are eaten as a delicacy. Tests and spines are often used for decoration, and as containers and piercing implements.

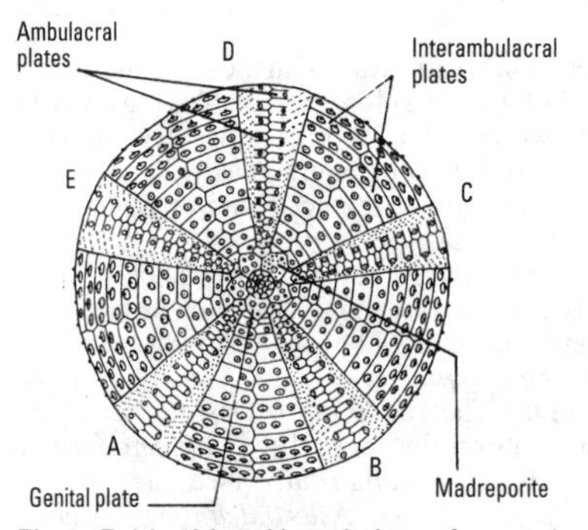

Fig.1. *Echinoidea.* Aboral view of a regular sea urchin *(Echinus). A-E* are radial double rows of ambulacral plates. The madreporite lies on the inter-radius *CD.*

The *E.* are classified as regular or irregular, according to the position of mouth and anus. In the regular *E.* (subclass *Regularia* or *Endocyclica),* the mouth and anus lie opposite to one another at the ends of the main axis, and the outline of the test is circular. Genera of this group include *Echinus, Psammechinus, Strongylocentrotus* and *Cidaris.* In the irregular *E.* (subclass *Irregularia* or *Exocyclica),* the body is flattened and usually has an oval outline; the anus is laterally displaced. Irregular genera include *Spatangus, Echinocardium* and *Clypeaster* (see).

The greenish colored *Psammechinus miliaris* is one of the most common regular sea urchins of western Europe, occurring from Norway and Iceland along the north European coasts, along West Africa as far as the Cape Verde Islands, but not in the Mediterranean, where the genus is represented by the greenish brown *P. microtuberculatus. P. miliaris* is found mainly in the surf zone, the largest specimens attaining a diameter of 47 mm. It is also a well known rock borer or "digger", living among kelp and other seaweeds, and under rocks

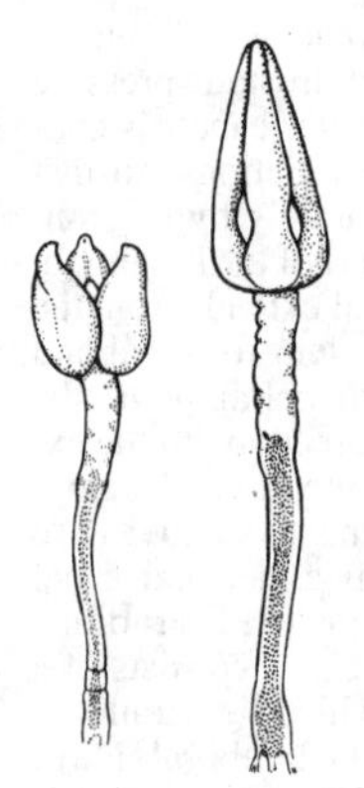

Fig.2. *Echinoidea.* Stalked pedicellariae of a sea urchin, showing the three jaw-like valves.

and stones. It is mainly carnivorous, feeding on hydroids and tubiculous polychaetes, as well as other animal and detritus.

The dirty yellow-colored *Cidaris cidaris* is also a very familiar representative of the *Regularia,* with a maximal diameter of 6.5 cm, height 4 cm. Thick primary spines, up to 13 cm in length, emerge from the broad field of interambulacral plates. Each primary spine is surrounded by a ring of small, flat secondary spines. The ambulacral fields are relatively slender. It is common in the Mediterranean and eastern Atlantic at depths of 30–1000 m. Fossils are found from the Jurassic to the present, and certain species in the Upper Cretaceous have stratigraphic value.

One of the best studied sea urchins is the irregular heart-shaped ***Common Heart urchin*** *(Echinocardium cordatum).* It has an almost cosmopolitan distribution in coastal waters, occurring from Norway and Iceland along western Europe to the Mediterranean and as far as South Africa, from Carolina to Brazil, and off Australia, New Zealand and Japan. It dwells in a cavity several centimeters deep in the sand, connected by a canal to the sand surface.

Clypeaster is a genus of irregular sea urchins which have a 5-sided, flattened, disk-like shell with a radially vaulted upper surface. On the internal surface of the shell is a secondary calcareous layer, which is especially well developed around the rim of the mouth. Processes from this secondary layer form internal connections between the upper to the lower sides of the shell. Some species of *Clypeaster* are extremely large. *C. subdepressus* is common on the Florida coast.

Fossil record. Fossils are found from the Ordovician to the present, and are important index fossils of the Early Tertiary. The wide variety of fossil species greatly excedes that of recent forms. The exclusively Paleozoic *Palechinoidea* are distinct from the more recent *Euechinoidea,* which occur from the Permian onward. The euechinoids include several important index fossil species of the genus *Micraster,* an irregular, oval to heart shaped sea urchin, found from the Upper Cretaceous to the Tertiary; it has a star-like arrangement of five ambulacral plates, and a shallow depression runs from the center of the top of the shell to the anterior edge (Fig.3).

On account of their calcareous skeleton, the *E.* have left an excellent, unbroken fossil record of complete forms in all geological systems. Even frequently occurring incomplete remains and spines can be used confi-

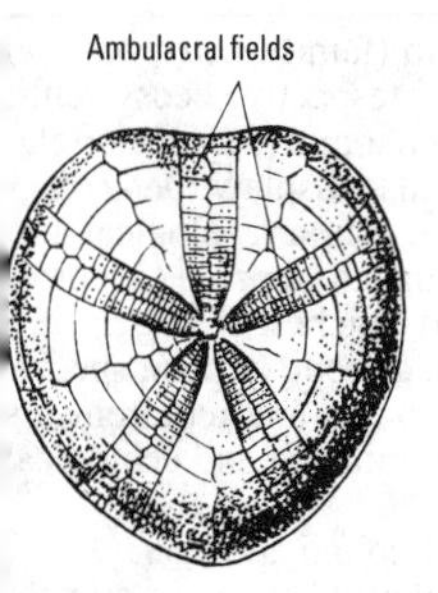

Fig.3. *Echinoidea.* Fossil sea urchin, *Micraster cortestudinarium* (x 0.5).

dently for taxonomic purposes, because the pattern of the test is regular and repetitive, and can therefore be derived from fragments.

Echinoids: see *Echinoidea.*

Echinoprocta: see *Erethizontidae.*

Echinorhininae: see *Selachimorpha.*

Echinosphaerites (Greek *echinos* hedgehog, *sphaira* sphere): a genus of primitive, spherical, Ordovician cystoids, in which the calyx is made up of more than 300 irregular, 5- to 6-sided individual plates. The stem is rudimentary or absent. The anal opening is surmounted by a paneled pyramid. Genital and oral openings are also present. Fossils are found in the Ordovician of Europe and North America.

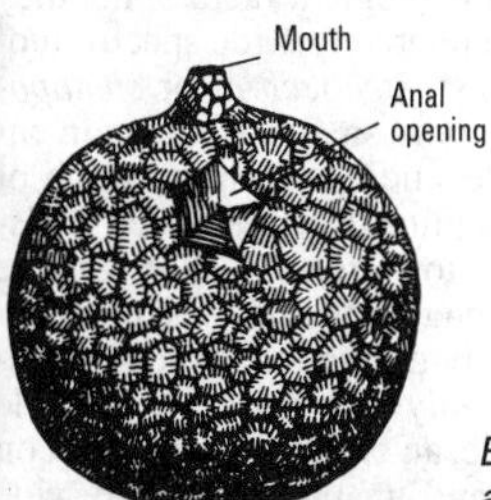

Echinosphaerites aurantium His (x 0.75)

Echinus: see *Echinoidea.*

Echis: see Carpet viper.

Echiurida, ***spoon worms:*** a phylum of the *Protostomia* containing about 140 species.

Morphology. *E.* are 2 to 20 cm in length, one species attaining a length of 40 cm. The body consists of a cylindrical or sac-shaped trunk, with a single pair of ventral chaetae, and an anterior, nonretractable, muscular proboscis (terminally bifurcated in some species), which is often several times the length of the trunk. Mature animals lack metameric segmentation. The mouth opening lies in a funnel-shaped depression at the base of the proboscis. In some species, the posterior end of the trunk is armed with rings of spines. The body cavity is a true epithelium-lined coelom. There is a very simple, closed vascular system. The sexes are always separate. Sperms or eggs are produced by testes or ovaries located internally on the coelom wall, and they pass to the exterior via the ducts of the metanephridia (usually 1–4 pairs).

Biology. *E.* inhabit the ocean floor, living in self-excavated burrows, rock clefts, or the discarded shells of other animals. They feed on microorganisms and detritus, grazed from the sustratum with the proboscis. Development proceeds through a swimming larval stage, which resembles a polychaete trochophore larva. The males of some species are extremely small, and live in the excretory ducts of the female. In *Bonellia viridis* (see), sex is determined after the larval stage by external factors.

Taxonomy. Three orders are recognized: *Echiuroinea, Xenopneusta* and *Heteromyota.* The order *Xenopneusta* is represented by a single genus, whose members have a characteristically degenerate proboscis. The order *Heteromyota* consists of a single species, *Ikeda taenioides,* with a body length of about 40 cm, and a proboscis of up to 100 cm.

Collection and preservation. Since most representatives live at great depths, their planned collection is very time consuming. Most specimens are encountered fortuitously in bottom drag nets. Extended *E.* are killed in boiling water and preserved in sea water containing 2% formaldehyde. Alternatively, specimens are narcotized while in sea water by the dropwise addition of 70% ethanol.

Echiuroinea: see *Echiurida.*

Eclectus: see *Psittaciformes.*

EC number: the number allocated to an enzyme in the Enzyme Commission system of nomenclature. See Enzymes.

Ecoelement: a component of an Ecosystem (see). Es. are defined according to their level or status within the ecosystem under consideration. Es. with the same function form a Compartment (see).

Eco-ethology, ***etho-ecology:*** a discipline of both behavioral biology and ecology. It is concerned with those aspects of behavior that regulate the relationships between organisms, and those which are determined by, or related to, the environment. Evolutionary theory is central to E-e., which attempts to elucidate the evolutionary mechanisms of environmentally related behavior. According to the modern *optimization model* of E-e., the search for food, the territorial area, group size, mating and other forms of interaction are optimized with respect to the environment, so that the reproduction rate of the species is high, i.e. the probability that fertile offspring will be produced is increased. Such an optimized state is also known as "inclusive fitness". In this respect, Evolutionary stable strategies (see) are particularly important, e.g. the relative frequencies of males and females in a population. Important principles of E-e. are revealed by comparative studies of different animal species.

ECoG: acronym of electrocorticogram. See Electroencephalogram.

Ecological amplitude: see Ecological valency.

Ecological characters: see Bioc(o)enotic parameters.

Ecological equilibrium: the long-term, stable condition of a self-regulated ecosystem, in which the degradative and synthetic processes balance one another (see Dynamic equilibrium). The frequency of occurrence and identity of the different species that constitute the system are stable characteristics, which are subject to only minor variations. If environmental influences exceed the self-regulatory capacity of the ecosystem, then directional changes occur (see Succession), leading to the establishment of a new equilibrium with new species combinations (see Climax).

Ecological factors: see Environmental factors.

Ecological niche: see Niche.

Ecological parameters: see Bioc(o)enotic parameters.

Ecological principles, rules and laws: ecologically important relationships between organisms and their environment, which have been discovered in field studies, or established by planned laboratory experiments. They may be relevant only under certain circumstances or in certain ecological systems, or they may be universally applicable; accordingly, they are known as principles, rules, or laws.

1. Autecology

1.1. Laws governing the activity of environmental factors (see Law of minima, Ecological valency)

1.2. See van't Hoff's rule

1.3. See Climatic rules.

2. Demecology

2.1. Growth law of Verhulst and Pearl (see Population ecology)

2.2. Fluctuation laws (see Volterra's rules).

3. Synecology

3.1. *Biocoenotic rules (Thienemann's rules)*

These state: a) The number of different species increases with the diversity of the habitat. b) As conditions become nonoptimal for most species, the species diversity decreases, but each remaining species is more numerous. c) Habitats that develop progressively and without sudden alterations, then remain unchanged for a prolonged period, support a relatively wide variety of species and a relatively balanced and stable ecosystem.

3.2. *Biogeogrphical rules.* a) *Synanthropy rule:* animals and plants that migrate to the northern limit of their distribution for reasons of climate or nutrition are most numerous in the proximity of human settlements and activities; the number of synanthropic organisms increases toward the north. b) *Phytophage rule (-Remmert's rule):* oceanic habitats of temperate regions are characterized by relatively low species diversity, because the exploitation of plant and saprobiotic food sources is not possible in the winter; animals in oceanic habitats remain active in winter after food becomes scarce, and they therefore suffer high energy losses. c) *Oceanic rule (Heydemann's rule):* as the oceanic nature of a region increases, so does the number of carabid beetles with overwintering larvae.

3.3. *Evolutionary strategy rules.* a) *Abundance rule:* in diverse habitats, eurypotent organisms (ubiquists) are more abundant, whereas in highly specialized habitats that lack diversity, stenopotent organisms (specialists) show the greater abundance. b) see *Competition-exclusion principle.* c) *Cope's rule:* the final members of a phyolgenetic series increase in size before they become extinct (see K-selection).

Ecological relationships: see Environmental relationships.

Ecological site equivalence, *site equivalence*: the existence of similar ecological niches in similarly structured ecosystems. Thus, in the geographically separated Eurasian steppes and North American prairies , the specific occupants of analogous niches are functionally similar, but usually genetically distinct species, which to some extent have arisen by convergent evolution (see Nutritional interrelationships, Fig.2). Such species are known as *ecological equivalents.* For example, the role of fossorial zoophagous animal is played by the European mole *(Talpa europaea)* in Eurasia, by the American garden mole *(Scalopus aquaticus)* in North America, by the golden mole (family: *Chrysochloridae)* in Africa, by the marsupial mole (family: *Notoryctidae)* in Australia, and by the armadillo (family *Dasypodidae*) in South America. Within their respective ecosystems, all of these animals display similar nutritional interrelationships. Often, such species also display true *ecological equivalence,* i.e. they have similar ecological amplitudes (ranges of tolerance) and are capable of replacing each other in their respective niches.

Ecological species group: a group of plant species with practically the same responses and requirements with respect to the main environmental and ecological factors of a particular habitat..

Ecological tolerance: see Ecological valency.

Ecological valency: the intensity range of an ecological factor, within which a given species is able to survive. The tolerated intensity range is part of the total range of variation (amplitude) of the intensity of the factor. The lowest intensity of an ecological factor that is still effective in supporting survival is known as the *minimum,* while the highest tolerable intensity is called the *maximum.* Using quantifiable parameters of living processes, the most favorable intensity *(optimum)* can be distinguished from the least favorable *(pessimum)* Between the optimum and pessimum lies the region known as the *pejus.* The valency of the factor has its counterpart in the potency of the organism, i.e. the specific tolerance of the organism for the intensity of the factor. *Euryecious* or *eurypotent* organisms are very tolerant of a large number of ecological factors, i.e. they are capable of exploiting a relatively wide spectrum of factor intensities. In contrast, *stenoecious* or *stenopotent* organisms can tolerate only small variations in environmental conditions. Depending on the position of the optimum within the amplitude of the factor, a distinction is drawn between low (oligo-), intermediate (meso-) and upper (poly-) ranges of potency. These concepts can be used to roughly define the ecological demands of an organism, but only with respect to tangible factors (Table). For example, an organism dependent on a high salt concentration and unable to tolerate wide variations in this factor, would be classified as stenopolyhaline (Fig.).

Many authors apply the term, valency, to the species rather than the factor, i.e. valency is synonymous with potency. Thus, eurytopic species have a high valency and are widely distributed. Conversely, stenotopic species (species with limited distribution) are often also stenoecious. On the other hand, a species is both eurytopic and stenoecious, if it is restricted to a small specialized habitat, which is widely distributed, e.g. *Artemia saline* (a phyllopod crustacean) is found only in waters of a very high salt concentration; such habitats are generally small in size, but they are found scattered throughout the world.

Environmental parameter	Word roots denoting the specific potency or valency of an organism
Water	-hydric
Atmospheric humidity	-hygric
Water pressure	-bathic
Salt concentration	-haline
Barometric pressure	-baric
Temperature	-thermic
Light	-photic
Food	-phagous (-trophic)

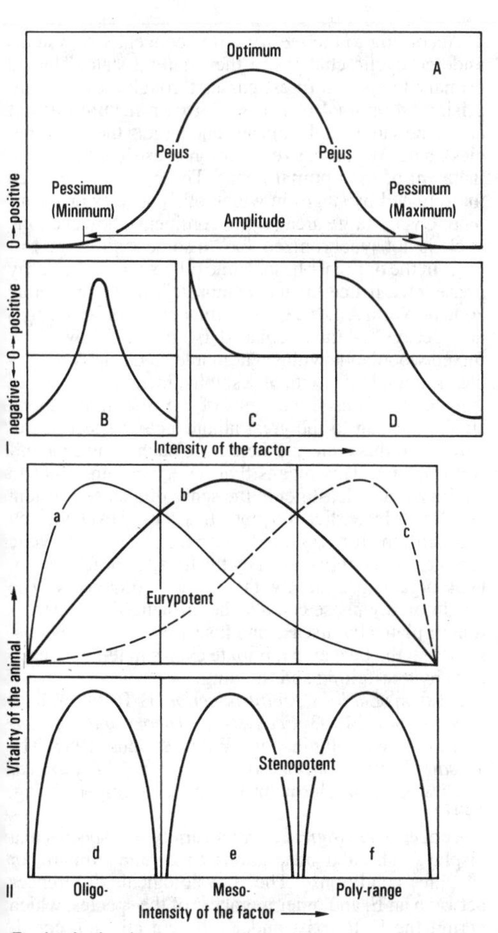

Ecological valency.
I Types of valency. *A* The effect of the factor on the organism is only positive. *B* High and low intensities of the factor have a negative effect on the organism. *C* Only low intensities of the factor have a positive effect. *D* As the factor increases in intensity, its negative action on the organism also increases.
II Types of potency. Euryecious or eurypotent organisms can be classified according to the position of the optimum: *a* oligoeurypotent, *b* mesoeurypotent, *c* polyeurypotent. Similarly, stenoecious or stenopotent organisms can be classified as: *d* oligostenopotent, *e* mesostenopotent, *f* polystenopotent.

Ecology: a term coined by Haeckel (1866) for "the total science of the relationships of organisms to the surrounding outside world".

As a biological discipline, E. is concerned with the interaction of organisms and their environment, and the establishment of laws and principles that govern these interactions. *Autecology* or *physiological E.* studies the interaction of individual organisms with specific environmental factors. Reactions of homotypic populations to environmental changes, and their development and alteration in dependence on environmental conditions are the subject of *demecology* (see Population ecology). *Synecology* (see), which is concerned with communities of living organisms (see Biocoenosis), studies the structure and dynamics of communities and seeks bases for their delineation and classification (see Biocoenology); it also designs theoretical models which can be used to investigate the modification and regulation of ecosystems by human intervention.

As the science of "Nature's economic system", E. can be considered on different levels. In its most integrated form, it transcends biology and encroaches on sociology, but it is ultimately based on geophysical laws.

Two separate approaches can be recognized in modern E., the theoretical and the empirical. Both are considerably dependent on cybernetics and mathematics.

In addition to the classical divisions of *marine, limnic* and *terrestrial E.*, the institutional divisions of *plant* and *animal E.* are still encountered.

Microbial E. studies the role of microorganisms in the economies of natural systems and the interaction of microorganisms with other organisms and with the abiotic environent. Other specialized areas are Human E. (see), Urban E. (see), Paleoecology (see) and Cytoecology (see).

Great importance is attached to *applied E.*, in particular in agriculture, forestry, pest control, fisheries management, landscape management and nature conservation.

Important tasks of modern E. are the recognition of essential ecological interrelationships and their incorporation into appropriate models, with the aim of regulating ecostems and enabling the maintenance and planned use of natural resources (soil, water, air, living organisms).

Ecomorphology, *ecological morphology:* study of the influence of environment on the morphology of living organisms. It is an independent discipline, particularly in relation to the E. of green plants. E. is concerned with all external characters that contribute to *ecomorphosis*, i.e. to the development of a morphology that is adapted to the environment (see Epitoky, Climatic laws, Convergence).

Within a species, environmentally relevant polymorphism can be attributed either to polyphenism (see Modification) or to genetic differences (see Polymorphism).

Some organisms *(Daphnia, Cyclops)* display an annual cycle of morphological changes between generations, which is repeated each year (see Cyclomorphosis). Other species display adaptation to the time of year, known as seasonal dimorphism (e.g. wet and dry season forms of butterflies).

Economic coeficient, coefficient of yield: the ratio of mass produced to substrate consumed. For example, in green plants it is the ratio of material produced by photosynthesis to that consumed by respiration; in the growth of microorganisms in a fermentor, the E.c. is the ratio of total mass of harvested microorganisms to the mass of substrate consumed during growth. It can also be defined as the energy yield coefficient, by comparing the actual yield of ATP with the theoretically possible yield. A high E.c. means a high "yield", i.e. a high excess of assimilated material.

Economic damage threshold: the population density of a pest, above which 1) the resulting damage justifies control measures, or 2) noticeable damage is caused to agricultural crops.

Ecosphere: see Biosphere, Habitat.

Ecosystem: a term coined in 1935 by A.G. Tansley. who considered the E. to be an association of plants and animals and the physical factors of their environment. In emphasizing the influence of environmental factors, Tansley improved on previous concepts of natural communities, which up to the 1920s were simply described in terms of their species composition.

In 1927, Charles Elton introduced the concept of feeding relationships, and coined the terms "food chain" and "food cycle" (later renamed "food web").

In 1925, A.J. Lotka published a thermodynamic approach, in which he recognized the flow of energy and transformation of mass among the members of a community, and compared a community to a machine (i.e. the quantity of energy transformed is proportional to the size of the community). For its time, Lotka's concept represented a considerable advance on previous thinking, as well as a physical and mathematical basis for Tansley's subsequent definition of an ecosystem. Lotka's work, however, remained largely unacknowledged.

It was not until 1942 that the concept of an ecosystem as an energy-transforming system reached a wide audience of ecologists, due to the work of R. Lindemann, who introduced the concepts of trophic levels (primary producers, herbivores, carnivores). At each trophic level, work is performed in the maintenance of organisms. In addition, biological energy transformations are relatively inefficient, so that less energy is available to each higher trophic level. When successive trophic levels are arranged as layers (whose widths are proportional to their energy contents), the result is therefore a pyramid (see Nutritional interrelationships). In addition to this thermodynamic treatment of the food web, Lindemann's work also incorporated Tansley's view of the ecosystem.

By the 1950s, the ecosystem, defined as the totality of interactions of its living members with each other and with the environment, was firmly established as a part of biological thinking. Defined in this way, an ecosystem, can be subdivided into smaller systems, each representing a different hierarchical level with its own scale of time and space. Among these subsystems, some will be strongly interrelated and interdependent, others will show less stringent coupling, and others may be largely independent of each other; the whole ecosystem can then be subjected to systems analysis (systems ecology) (Fig.).

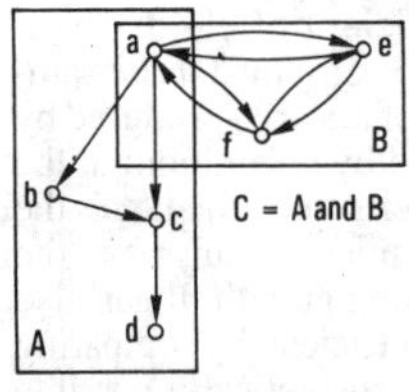

Ecosystem. An ecosystem consisting of elements *a* to *f* can be divided into smaller systems representing different types of ecological relationship. *A* is based on the trophic relationships between the primary producers *(a)*, the phytophages *(b)*, the omnivores *(c)* and the higher predators *(d)*. *B* is based on other important relationships (e.g. competition) between the primary producers *(a, e* and *f)* which belong to the same hierarchical level. Integration of part-systems *A* and *B* produces the total system *C*.

According to the *mosaic-cycle* concept, ecosystems undergo cyclic changes in their natural state. Thus, a primary temperate forest passes through a cycle comprising an optimal phase (well grown mature trees of about the same height and age and largely the same species), a decay phase, a rejuvention phase, then re-establishment of the optimal phase. These stages are often interpolated by stages in which other species predominate. Over a large area (e.g. a continent), however, the cycle is not synchronized, e.g. in some regions the forest is in the optimal phase, while others it is in the decay phase, etc., hence the term "mosaic". In the temperate zone of North America, an optimal phase of beech ages and decays, is then replaced by birch, followed by mixed woodlands with sugar maple, and finally the cycle is completed with re-establishment of beech. In tropical rainforests, the areas of the mosaic are much smaller, owing to the great abundance of species. According to this concept, there is no such thing as a uniform habitat. This means that an ecosystem is not in equilibrium, at least not in the sense of a closed system as inferred by earlier concepts. It is caused to change by competition for essential nutrients, and its cyclic changes are simply driven by the lifespan of its component organisms, mainly the primary producers. The length of any phase can also be influenced by destruction of plants by disease and insect pests, and it is possible that the beaver can initiate cycles in temperate forests by tree-felling and flooding.

[Systems Ecology, An Introduction, H.T. Odum, John Wiley & Sons, 1983; *Ecosystems, Theory and Applications,* edit. N. Polunin, John Wiley & Sons, 1986; The *Mosaic-Cycle Concept of Ecosystems (Ecological Studies 85)* edit. Hermann Remmert, Springer-Verlag, 1991]

Ecotype, ***ecological race:*** a variety of a species that displays inherited adaptations to certain climatic or edaphic conditions. The physiological differences between an E. and other members of the species, which permit the E. to exist under different climatic conditions, or to exploit a new food source, are often very slight. An understanding of Es. is of practical importance, especially in agriculture and forestry.

Ectocarpales: see *Phaeophyceae.*

Ectocarpene: see Gamones.

Ectoderm (Greek *ectos* outer, *derma* skin): the outer germ layer of the metazoan gastrula. The E. gives rise mainly to the epidermis, the nervous system and the sense organs.

Ectogenous: see Allochthonous.

Ectognatha, *Ectotropha:* a superorder of insects with mouthparts projecting outward from the head. It comprises the primitive groups, *Pterygota* and *Thysanura.*

Ectoparasite: see Parasite.

Ectoparasitic lice: see *Phthiraptera.*

Ectopistes: see *Columbiformes.*

Ectospores: see Conidia.

Ectotropha: see *Ectognatha.*

Edaphon: see Soil organisms.

Edaphobionts: see Soil organisms.

Edaphophytes: plants of the microscopic soil flora. See Life form.

Edelweiss: see *Compositae.*

Edentata: an ancient order of 3 highly specialized mammalian families, whose members are toothless (see *Myrmecophagidae),* have only a few teeth (see *Brady-*

podidae), or have numerous peg-like teeth (see *Dasypodidae).*

Ediacara fauna: the earliest known Precambrian metazoans, whose fossils were first discovered in the hills of southern Australia. Similar fauna were later found in Namibia *(Nama* formation), Siberia *(Vend* formation) and Canada. More than 1 600 individual fossils are known, representing 30 species. They comprise coelenterates (especially medusoids and sea pens; 67%), annelids (25%), arthropods (5%) and possibly an echinoderm. Only soft-bodied organisms are found. Trace fossils are also present. The sediments indicate a biotope in strongly flowing water, up to 25 m in depth. The organisms represented are considered to be ancestors of various phyla of the Paleozoic fauna.

Edible snail: see *Gastropoda.*

EDTA: see Ethylenediaminetetraacetic acid.

EEG: acronym of electroencephalogram. See Electroencephalography.

Eel: see Eels.

Eelpout, ***viviparous blenny,*** *Zoarces viviparus:* a 45 cm-long, viviparous, edible, bottom-dwelling fish. It is found inshore and offshore in the north-eastern Atlantic, in the Baltic, and up to the White Sea. It is abundant in Scandanavian and Scottish waters. It is occasionally also known as an "eel-mother".

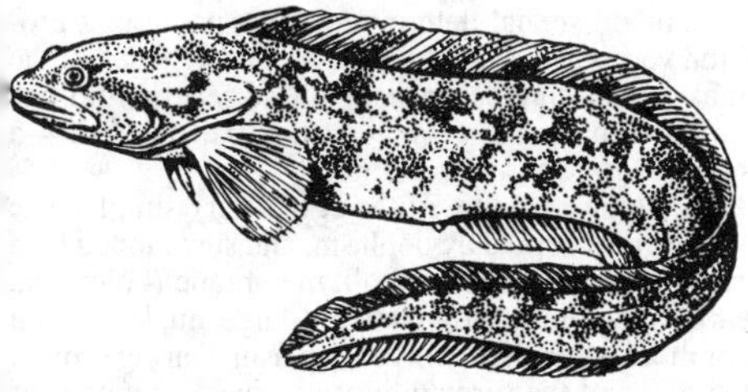

Eelpout or *viviparous blenny (Zoarces viviparus)*

Eels, *Anguilliformes, Apodes:* a species-rich order of teleost fish. The body is long, slender and snake-like, and the mouth is deeply cleft. Dorsal, caudal and anal fins form a single continuous fin; ventral fins are absent and pelvic fins are usually absent. The skin has no scales, except in the *Anguillidae* (see River eels). With the exception of the river E., all have an entirely marine habitat.

Families: ***River E. (Anguillidae), Moray E.*** *(Muraenidae),* ***Conger E.*** *(Congridae),* ***Snake E.*** *(Ophichthyidae),* ***Snubnosed*** or ***Parasitic E.*** *(Simenchelidae),* ***Snipe E.*** *(Nemichthyidae, Serivomeridae* and *Cyemidae).* The ***Spiny*** or ***Spiny-headed E.*** constitute a separate order *(Notocanthiformes).* The ***Gulper E.***, previously grouped in a separate order *(Saccopharyngiformes* or *Lyomeri)* are now included in the suborder *Anguilloidei,* which contains most of the known families.

Eelworms: a term applied to many of the smaller nematodes, especially soil inhabitants and plant parasites. Plant nematodes cause considerable economic damage to crop plants by feeding on the tissues of the host plant, by transmitting viruses, and by providing entry for bacteria and fungi. With the exception of *Trichodorus,* which belongs to the order *Dorylaimida,* all the nematodes mentioned below are in the order *Tylenchida* (see *Nematoda).*

The ***Beet cyst eelworm*** or ***Beet eelworm*** *(Heterodera schachtii)* is notorious for its deleterious effects on sugar beet. Control is primarily through crop rotation, although this is difficult, because the host range of this nematode is actually very wide, including many major crops and weeds, e.g. sugar and fodder beet, mangolds, red and spinach beet, brassicas, mustard, cress, rhubarb, and various members of the *Caryophyllaceae, Amaranthaceae, Lamiaceae, Portulacaea* and *Scrophulariaceae.* By feeding on the sap of young roots, *H. schachtii* retards the growth of beet. The root tissue undergoes marked proliferation, and the leaves become limp and flaccid, a condition known as beet-sickness. The sugar content of the beet is decreased. Males are vermiform and free-living (0.4–1.6 mm in length), whereas the females are sedentary in, or attached to the root. The female body forms a resistant, pinhead-sized brown cyst, containing many eggs, each with a second-stage larva. Toward the end of the vegetation period, the roots of the host are found to contain large numbers of these cysts, which can remain dormant for several years.

The ***Potato cyst eelworm*** or ***Potato eelworm*** *(Heterodera (= Globodera) rostochiensis)* penetrates the roots of young potato plants, and continues to feed and develop within the host root tissue, causing stunted growth. Crop yields can be decreased by up to 70%. Mature females form lemon-shaped, yellow-brown cysts, which are seen in the roots when the potatoes are harvested. Each cyst contains up to 300 eggs, and it can remain viable in the soil for up to 12 years. Occurrence of *H. rostochiensis* is a notifiable agricultural disease. Infestation can be prevented by appropriate crop rotation (4 or 6 course rotations, with or without the use of nematicides) and the planting of resistant potato cultivars.

The ***Stem eelworm*** *(Ditylenchus dipsaci)* has a host spectrum of more than 450 plant species. It lives in the intercellular spaces of the aerial portions of many crop plants, or in subterranean corms and bulbs, causing atrophy and stunted growth. Economically important host plants are rye, oats, buckwheat, maize, rape, clover, field beans, strawberries, onions and garlic, as well as flower bulbs such as narcissi, tulips and hyacynths. Both the aerial and the subterranean parts of the teasel plant are attacked, leading to total plant destruction. Control is by extensive rotation of crops.

The various species of *Meloidogyne* are known as ***Root-knot eelworms*** or ***Root-gall eelworms.*** They do not form cysts, but cause the production of galls on the attacked roots. They are especially harmful to greenhouse crops, such as cucumbers and tomatoes. The main means of control is extensive crop rotation. Root-knot eelworms seem to be present in most soils throughout the world, but they represent a potent source of crop damage and crop loss especially in the tropics and subtropics.

The ***Wheat gall eelworm*** *(Anguina tritici)* causes the formation of galls in the ear of the wheat plant. It is usually first detected by the presence of brown or black, rounded galls (ear cockles) in the threshed grain. These mature galls contain up to 40 adults or their dead remains, together with as many as 30 000 eggs or hatched larvae. The latter develop only to the second stage, which is resistant and infective. Larvae feed ectoparasitically near the growing point of young leaves, causing stunted growth, and abnormal twisting and crinkling of the leaves. Development is completed in the infloresence, causing undersized ears with abnormally spreading glumes, in which many seeds are replaced by galls. The disease is now much less widespread than previously, due to seed-cleaning methods, which insure that the infective galls are not retained and sown with the next crop of seed. *Anguina tritici* interacts with *Corynebacterium tritici* to produce the serious disease of wheat known as ***Yellow ear-rot.*** This is mainly a disease of subtropical areas, especially in India, where it is known as ***Tundu*** disease. Before the ears emerge, a bright yellow bacterial slime is seen on the abortive ears and leaf sheaths. Few, if any galls are formed in infected heads.

In addition to the well recognized, specific disease-causing plant nematodes, there are many species present in the soil that do not form cysts within plant tissue, and do not cause gall formation, but migrate from one plant to another, feeding largely as ectoparasites on epidermal tissue, sometimes penetrating into the root cortex. Many previously unexplained growth defects in fruit trees, cereals and vegetables are now attributed to root damage by these organisms, which are generally known as ***Root nematodes.*** Examples are species of the genera, *Paratylenchus, Pratylenchus, Helicotylenchus, Rotylenchus,* and *Trichodorus.* They are controlled by crop rotation, well fertilized and weed-free conditions, and the use of recommended nematocides.

Effective cation exchange capactity: see Sorption.

Effective dose, *ED:* the dose range within which a biological agent produces a desired effect. The term is often applied to the dosages of Plant protection agents (see). ED50 is the dose that produces an effect in 50% of the organisms treated.

Effective population size: the size of a real population, adjusted mathematically to account for varying sex ratios, degrees of inbreeding, etc., so that different populations may be compared. Certain changes in populations, especially those due to genetic drift, depend strongly on the population size. However, changes do not occur in the same way or to the same extent in all populations of the same size. Natural populations may differ greatly, e.g. in the ratio of different age groups, in the ratio of the two sexes, in the number of progeny per individual, in the proportion of young members involved in reproduction, etc.

In the *idealized* population, there are only adult organisms, the sex ratio is unity, all individuals have the same prospect of contributing offspring to the next generation, and the number of individuals per generation remains constant.

Deviations of real populations from this model are corrected mathematically to give the E.p.s.

If the sex ratio differs from unity, $Ne = 4Nm \times Nf/Nm + Nf$, where Ne is the E.p.s., Nm the number of men, and Nf the number of women.

If individuals have a different potential for producing offspring, then $Ne = 4N - 2/\sigma^2_k + 2$, where N is the number of parents, and σ^2_k the variance of the offspring per parent (K = number of offspring per parent).

If the population size varies between succeeding generations, then $Ne = t/\Sigma^1_{Ni}$, where t is the number of generations, and the Ni-values are the population sizes of the different generations

Efference: the totality of excitations from the central nervous system to effectors or effector organs. The nervous pathways conducting these impulses are called efferent or motoric nerves. See Afference, Reafference.

Efference copy: see Reafference principle.

Egg, *ovum:* the female reproductive cell or Gamete (see) of metazoan (multicellular) animals. It contains at least one complete set of chromosomes. After fertilization by the male gamete, or without fertilization (see Parthenogenesis), the egg cell develops into a new individual.

Eggs are normally formed in the ovary (see Oogenesis). During its formation, the egg takes up reserve material (see Yolk), and may therefore attain a considerable size.

Eggs are usually rounded, but in some species they may be flattened or elongate. Insect eggs are frequently elongate-ovoid (Fig.). The egg case or shell may also display wide variation of shape. Eggs of sponges and coelenterates are capable of ameboid movement. The size of an egg is largely determined by its content of reserve material, and it bears no relationship to the size of the organism that produces it. Sizes range from a few mm (trematodes 12–17 μm, human 120–200 μm) to a few cm (domestic hen 3.0 cm, ostrich 10.5 cm, giant shark 22 cm). Sometimes eggs of different sizes are produced by the same species, e.g. the resting, latent or winter eggs of *Daphnia* and rotifers have a long developmental period and are larger than the summer eggs. The vine phylloxera lays large eggs that hatch into females, and smaller eggs that always produce males (programmed sexual determination). Annual egg production varies according to the mode of existence of the animal, from a single egg (penguin) to several millions *(Ascaris* 64 m, *Taenia* 42 m, termites 10 m, codfish 4–5 m, oyster 1 m, where m = million).

The cellular structure of the egg is very simple. The cell space is occupied by ooplasm, and surrounded by a secreted egg membrane or yolk membrane (oolemma, vitelline membrane). A relatively large nucleus, with one or more nucleoli, lies in the ooplasm. During growth of the egg and the formation of reserves, a number of characteristic changes occur on and within the nucleus; these are accompanied by high respiratory activity and active metabolism in both the ooplasm and the nucleus. The totality of reserve materials stored in the ooplasm (proteins, lipoproteins, neutral fats, phospholipids and glycogen) is known as *yolk* or *deutoplasm.* Yolk is deposited as granules, spheres and platelets; for example, the yolk of insect eggs is formed in three successive phases (fatty yolk, protein yolk, glycogen). The Golgi apparatus probably acts as a yolk-forming organelle for fatty yolk, while mitochondria serve for the accumulation of protein yolk. Completely yolk-free (alecithal) eggs are unknown. In eggs containing only a small amount of yolk *(oligolecithal* eggs) the deutoplasm is uniformly distributed, so the egg is also said to be *isolecithal;* such eggs occur in sea urchins, some worms, *Amphioxus* and mammals (including humans), and sporadically in insects *(Ichneumonidae, Collembola).*

In eggs containing large quantities of yolk *(mesolecithal* to *polylecithal* eggs), the yolk accumulates at one pole of the cell, which is then known as the *vegetal* or *vegetative* pole. The opposite pole is kown as the *animal* pole. The nucleus lies nearer to the animal pole. In vertebrates, the animal pole is the end attached to the follicle cells of the ovary, whereas in invertebrates, the animal pole is that end of the egg nearest the coelomic fluid. Such *telolecithal* eggs are found in many animal groups, from sponges to vertebrates (especially am-

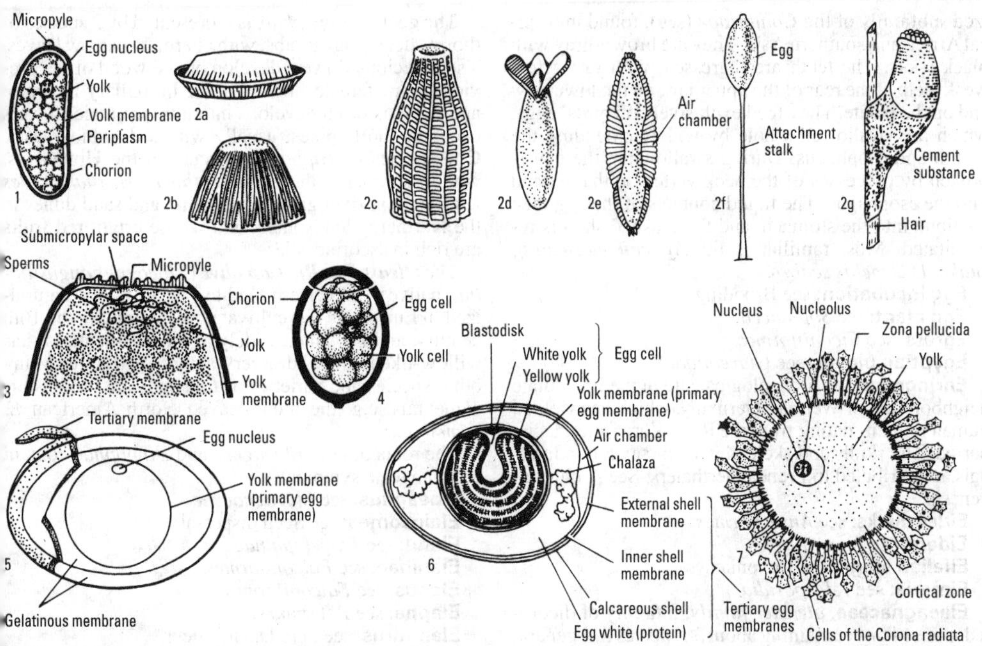

Egg. 1 Insect egg in longitudinal section. *2* Insect eggs, showing various chorion structures: *2a* bell moth *(Tortrix* sp.), *2b* pine beauty *(Panolis flammae), 2c* cabbage white butterfly *(Pieris brassicae), 2d* dung fly *(Scatophaga* sp.), *2e* mosquito *(Anopheles maculipennis), 2f* lacewing *(Chrysopa* sp.) *2g* head louse *(Pediculus captis). 3* Longitudinal section through the upper pole and micropyle of a butterfly egg *(Pieris brassicae). 4* Compound egg of a tapeworm *(Dibothriocephalus latus). 5* Salamander egg with membranes displayed. *6* Longitudinal section through a hen's egg. *7* Mature human egg cell with corona radiata.

phibians). In extremely telolecithal eggs, like those of cephalopods, spiders, fishes, reptiles, birds and some crustaceans, the ooplasm is restricted a small cap at the animal pole. *Centrolecithal* eggs are found almost exclusively in arthropods, i.e. the deutoplasm is located centrally and surrounded by a thin surface layer of ooplasm. Quantity and distribution of yolk are important in determining the type of Cleavage (see). Basically, the yolk acts as an inert material, whose presence slows the rate of formation of cleavage planes. Thus, in telolecithal eggs, cleavage is impeded in the vegetative hemisphere relative to the animal hemisphere. Egg polarity may also be reflected in the distribution of ooplasm, pigments, arginine and granules, and in the egg shape.

Eggs are generally surrounded by egg membranes, which are absent only in sponges and some coelenterates. The primary egg membrane is the above-mentioned yolk membrane (vitelline membrane), which is synthesized by the egg cell itself. It is usually very thin and structureless, rarely strong and multilayered (e.g. in *Ascaris).* The thick, multilayered vitelline membrane of echinoderms, mollusks and others, know as the *zona radiata,* is perforated by fine pores. In mammals (including humans) (Fig., No.7), it is called the *zona pellucida,* and as a rule it contains a single entry point (micropyle) for the spermatozoon. Secondary egg membranes *(chorion* of insects; Fig., 2a–g) are secreted within the ovary from surrounding follicle cells; they are usually tough and multilayered, perforated by pores, and provided with appendages, a sculptured surface, micropyles, operculum, aeration devices, means of attachment (egg stigmata) and cement substances. The insect chorion and the zona radiata consist of chorionin, a chitin-like, sulfur-containing material. The tertiary egg membranes are produced by glands of the female genital tract (albumin, albuminous, shell and nidamental glands), e.g. the horny egg capsules of cephalopods and sharks, the parchment-like outer coverings of reptile eggs, the gelatinous coat of the spawn of mollusks , water insects and amphibians (Fig., No.5), the egg white, the chalazae, the two-layered shell membrane and the porous calcareous shell of birds' eggs (Fig., No.6).

If several or many egg cells are enclosed, with or without separate yolk cells, in a common membrane, they are called ectolecithal eggs, compound eggs, egg capsules and egg cocoons (Fig., No.4). The common membrane is formed either by skin glands (earthworm, leech), or by glands of the oviduct (cockroach, many water beetles, *Turbellaria,* trematodes, snails).

Some animals *(Turbellaria, Rotifera, Branchiopoda)* produce hard-shelled, resistant, resting eggs ("winter eggs") in response to unfavorable environmental conditions (e.g. periodic drought or freezing), whereas they produce thin-shelled "summer eggs" with a low yolk content (usually in large numbers) under normal conditions.

Eggars: moths of the family *Lasiocampidae* (see).
Egg cell: see Reproduction.
Egg cell formation: see Ogenesis.
Egg-eating snakes, *Dasypeltinae:* a highly special-

ized subfamily of the *Colubridae* (see), found in tropical Africa and southern Asia. They are brown-gray with black spots. The teeth are regressed, with just a few weak teeth at the rear of the upper jaw, on the lower jaw and on the palate. They feed exclusively on birds' eggs, which are swallowed whole by widely extending the mouth and esophagus. During swallowing, the egg is broken by processes of the neck vertebrae that extend into the esophagus. The liquid contents of the egg then continue into the stomach, and the crushed shell is regurgitated. Most familiar is the ***African egg-eating snake,*** *Dasypeltis scabra.*

Egg incubation: see Brooding.

Egg plant: see *Solanaceae.*

Egrets: see *Ciconiiformes.*

Egyptian plover: see *Glareolidae.*

Ehringsdorf: an archeological site in the immediate neighborhood of Weimar, Germany, which has yielded human remains dating from the Riss-Würm interglacial period. The skulls and skeletal parts of several individuals are attributed to Preneanderthalers. See Anthropogenesis.

Eider ducks: see *Anseriformes.*

Eiders: see *Anseriformes.*

Eifelian stage: see Devonian.

Eimeria: see *Telosporidia.*

Elaeagnaceae, *oleaster family:* a family of dicotyledonous plants containing about 30 species in 3 genera, distributed in the northern temperate region, southern Asia and Australia. It is the only family of the order *Elaeagnales.* They are almost exclusively trees or shrubs, with entire, exstipulate leaves. The 4-merous flowers consist of a calyx cup without petals. Flowers are solitary or in small clusters, actinomorphic, hermaphrodite or variously unisexual. After wind- or insect-pollination and fertilization, the central, superior, one-celled ovary (with orthotropous ovule) develops into a dry fruit surrounded by the lower half of the receptacle, which becomes fleshy. Plants often have a silvery appearance, due to a dense covering of peltate or stellate, silvery-brown, scale-like hairs.

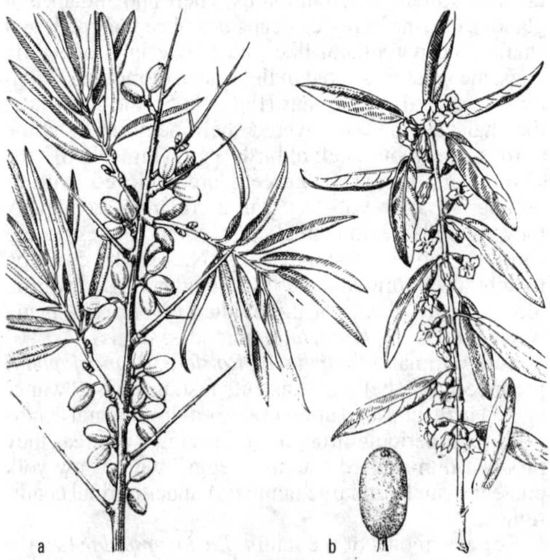

Elaeagnaceae. a Sea-buckthorn *(Hippophaë rhamnoides);* shoot with fruits. *b* Oleaster *(Elaeagnus angustifolia);* flowering shoot and single fruit.

The genus, *Hippophaë,* is represented by 2 species of thorny, deciduous shrubs, with alternate, narrow leaves. The dioecious flowers develop on the wood of the previous year; female flowers occur in axillary racemes, and the axis often develops into a thorn; male flowers occur in short spikes, usually with a deciduous axis. One species of *Hippophaë* occurs in the Himalayas, while the other is the ***Sea-buckthorn*** *(H. rhamnoides* L.) found on river gravels, sea cliffs and sand dunes in the Northern Hemisphere, and whose bright red fruits are rich in ascorbic acid.

The ***Oleaster*** or ***Russian olive*** *(Elaeagnus angustifolia),* native to the eastern Mediterranean, is now naturalized in Europe. It was cultivated in England in the 16th century, and is planted widely as an ornamental; it has willow-like leaves and silvery-amber, oval fruits. Many other species and varieties of *Elaeagus* are used as ornamentals, e.g. the broad-leaved North American *E. commutata* .

The roots of both *Elaeagnus* and *Hippophaë* contain actinomycete symbionts.

Elaeagnus: see *Elaeagnaceae.*

Elaiosomes: see Seed dispersal.

Eland: see *Tragelaphinae.*

Elaninae: see *Falconiformes.*

Elanus: see *Falconiformes.*

Elaphe: see Rat snakes.

Elaphurus: see Père David's deer.

Elapidae, *front-fanged snakes, cobras and allies:* a family of mostly slender, agile Snakes (see) containing more than 180 species in about 40 genera. The family is represented in America, Africa, Asia and Australia, but not in Europe and Madagascar. They bear a close external resemblance to the *Colubridae* (see), especially in the shape of the head and the types of scales, but their dentition is proteroglyphic, and without exception they are very poisonous. The deeply furrowed venom fangs stand erect at the front of the upper jaw. More primitive species have several other grooved teeth behind the functional pair of venom fangs, whereas the most advanced species (which include the cobras) have only 2 fangs in the entire mouth. The venoms are mainly neurotoxic in action (see Snake venoms), and they include the most deadly snake venoms on earth. Coloration varies from the somber gray of some desert inhabitants to the brilliant leaf-green of some tree-dwelling species, and the conspicuous warning coloration of black, yellow and red transverse rings of the coral snakes. They lay eggs or are vivi-oviparous. In addition to small rodents, many species also prey on other snakes. They display their greatest diversity in Australia, where about 80% of all the snakes belong to the *Elapidae.* The following examples are listed separately: Coral snakes, Cobras, King cobra, Kraits, Taipan, Mambas, Long-glanded coral snakes.

Elasis: the assessment of distance and space. It is usually combined with direction orientation (see Taxis). Thus, a tree-dwelling animal about to jump to another branch must assess both the direction and the distance, in order to establish both the starting angle and the necessary mechanical force of the jump.

Elasmobranch: see *Chondrichthyes.*

Elasmobranchii: see *Chondrichthyes.*

Elassoma: see Sunfishes.

Elastase: a proteinase, which catalyses the hydrolytic cleavage of elastin in animal elastic fibers. It behaves as an endopeptidase (attacks bonds within the substrate molecule to form oligopeptides), and displays some homology of primary and tertiary structure with the digestive enzymes, trypsin and chymotrypsin. E. is synthesized in the vertebrate pancreas as an inactive proenzyme (proelastase), which is then converted into active E. by the action of trypsin in the duodenum.

Elastic fibers, ***yellow elastic fibers:*** highly extensile and elastic fibers that form part of the intercellular substance of animal support tissue. They occur in vertebrates, arranged in parallel in elastic ligaments, or as networks in lung tissue, in elastic cartilage and in blood vessels. They consist mainly of elastin and glycoprotein microfibrils, the latter lying mainly on the fiber surface. In the formation of E.f., a bundle of glycoprotein microfibrils *(oxital fibers)* is first laid down. A partial deposition of elastin then produces *elaun fibers,* which finally mature to E.f. by the incorporation of more elastin.

Elastin: a structural protein that forms the main component of the elastic fibers of tendons, ligaments, bronchi and arterial walls. The elasticity of these tissues is due to the presence of E. It owes its elasticity to a high content of glycine, alanine, proline, valine, leucine and isoleucine, as well as to cross-linkages by two unusual amino acids, desmosine and isodesmosine.

Elaters: elongate, fiber-like cells in the capsule of liverworts. Each archesporial cell divides into one narrow and one broad daughter cell. The narrow cell develops without division into the thin-walled fiber-like elaters with spiral, ribbon-like wall thickenings. After dehiscence of the capsule, movement of the elaters by a hygroscopic adhesion mechanism helps to loosen and scatter the spores. See *Hepaticae.*

Elder: see *Caprifoliaceae.*

Electrical double layer: a layer at the boundary of a solid and a liquid phase, in which the condition of electrical neutrality is abolished. By dissociation and adsorption, a transition zone forms at the phase boundaries, in which positive and negative charges are separated. H. von Helmholtz (1879) first suggested that an electrical double layer is generally formed at the interface of two phases, and assumed that the double layer is virtually an electrical condenser with parallel plates no more than a molecule apart. Later development of the concept by G. Gouy (1909) and others supported the existence of a diffuse double layer. Subsequently, O. Stern (1924) showed that neither the sharp nor the diffuse layer adequately explains the phenomenon, and developed a view combining the essential characters of both previous theories. Thus, the double layer has two parts. The first part of the layer is about one ion in depth and it remains almost fixed at the solid surface. Within this layer there is therefore a sharp fall of potential. The second part of the layer extends some distance into the liquid phase and is diffuse; in this region, thermal agitation causes free movement of the particles, but the distribution of positive and negative ions is not uniform, because the electrostatic field at the surface preferentially attracts particles of the opposite sign. The potential reponsible for electro-osmosis and allied phenomena is that between the fixed and freely mobile parts of the E.d.l., and it is known as the *zeta potential.*

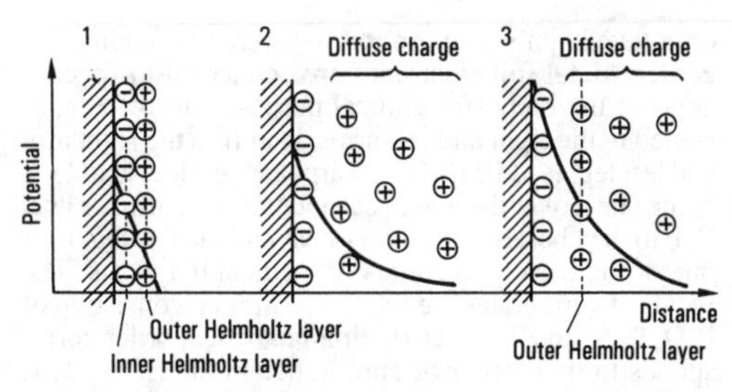

Models of the electrical double layer. 1 Helmholtz's model. *2* Diffuse double layer proposed by Gouy and Chapman. *3* Stern' model. The fall in potential is represented schematically by the plotted curves.

Electrical potential differences: differences of electrical potential between two separate sites, which in principle can be used for performing electrical work. As a rule, E.p.d. occur at phase boundaries, including all such boundaries inside and outside living cells. Electrochemically, it is necessary to distinguish between E.p.d. within a phase (voltaic potential differences) and those between different phases (galvanic potential differences). By measuring E.p.d., it is possible to determine the work that must be done to move a positive charge. If this measurement is attempted between different phases, chemical interactions also occur, and only the sum of electrical and chemical work can be measured, so that in principle a galvanic potential difference between two phases cannot be measured directly. Particularly important are the Membrane potential (see) (E.p.d. between cell interior and liquid of the cell exterior) and the Surface potential (see) (E.p.d. between the bulk of an aqueous solution and a point very near to the surface).

Electric fishes: fishes with electric organs, consisting of a linked series of numerous plates, which are derived from modified skeletal muscle. The electric discharge serves for defense, for stunning prey, and for delineating territory. According to more recent research, it may also represent a form of intraspecies communication. Examples are: Elephant fishes (family *Mormyridae,* order *Osteoglossiformes;* electric organ derived from caudal muscles); Electric knifefish (the single species of the family *Electrophoridae,* order *Salmoniformes);* Electric rays (family *Torpedinidae,* order*Rajiformes;* electric organ derived from branchial muscles in the head region); Electric catfishes (family *Malapteruridae,* order *Siluriformes).*

Electric knife fish: see *Gymnotiformes.*

Electrocardiogram: see Electrocardiography.

Electrocardiography: a method for recording the time course and nature, of the action potentials of the heart. The recording is known as the *electrocardiogram* (ECG). The first human electrocardiogram was obtained in 1889 by Waller, using a capillary electrometer. The first accurate recordings were made possible by the string galvanometer (Einthoven, 1901). Development of valve amplifiers subsequently enabled the use of less sensitive but more robust galvanometers, such as the moving-iron oscillograph and the cathode-ray oscillograph. Earlier methods of amplification have now been transcended by modern electronic systems based on semiconductors. The ECG can be displayed continually on a television screen or/and registered on a chart recorder. Metal foil electrodes are connected to specific areas of the body (for clinical purposes, lead I is connected to the right and left arms, lead II to the right arm and left leg, lead III to the left arm and left leg, lead IVR to the chest over the apex beat and the right arm, or lead IVF to the chest over the apex beat and the left leg. E. is one of the basic procedures of medical diagnosis. The ECGs of vertebrates are relatively similar, consisting of P, Q, R, S, and T waves (their alphabetical order corresponds to their order of appearance in the ECG). P, R and T are negative, while A and S are positive. The P wave begins as excitation spreads from the sinu-atrial node to the atria. The QRS complex marks the beginning of depolarization of the ventricles and it ends when all the ventricular muscle has been depolarized. The T wave signals repolarization in the ventricular muscle, and the U wave is thought to be due to the change in position of the heart in diastole. Fishes and amphibians have a V wave due to the spread of excitation in the sinus.

Electrochemical potential: the sum of chemical and electrical potentials of an ion (it is not experimentally possible to measure separately the chemical potential and the electrical potential of a charged particle): E.p. = $\mu + z\Phi F$, where μ is the chemical potential, Φ is the electrical potential, F is one Faraday of charge (96 485 coulombs/mole), and z is the valency of the ion. The E.p. characterizes the ability of one gram ion to perform work, with reference to a standard electrode. It is an important value for calculating the equilibrium conditions of electrochemical systems. In particular, the E.p. of a charged species is the same at all points in the system when the system is in equilibrium. Differences in E.p. induce fluxes that lead to the establishment of equilibrium. These fluxes constitute the process of *electrodiffusion,* i.e. the diffusion of charged particles in the electric field; the diffusion coefficient (D) can be calculated with the aid of the Nernst-Planck equation: $J = -D\,[dc/dx + zc(F/RT)\,(d\Phi/dx)]$, where c is the concentration of the ion species, and J (the net flux of the ion species) is composed of two parts: the movement of ions along their concentration gradient dc/dx, and the driving force experienced in the electrical potential gradient $d\Phi/dx$.

Electrochromic effect: see Electrochromicity.

Electrochromicity: changes in the absorption of a pigment caused by electrostatic interaction. Light-induced alterations in the absorption of chlorophyll and carotenoid systems can be demonstrated in thylakoid membranes. In this case, the membrane potential is altered by the light-induced energy conduction and subsequent vectorial transport of charge on the thylakoid membrane. This alteration of the electric field of the membrane causes a slight shift in the absorption bands of the membrane-bound pigments. The *electrochromic effect* can therefore be used as an internal voltmeter to determine the membrane potential.

Electrocommunication: biocommunication by electrical signals. This type of signalling was discovered relatively recently (1970s) in various fishes, whose electric organs produce only weak electrical discharges. The signals consist of electrical waves or pulses, and they differ markedly between species (Fig.).

Electrocorticogram: see Electroencephalogram.

Electrodiffusion: see Electrochemical potential.

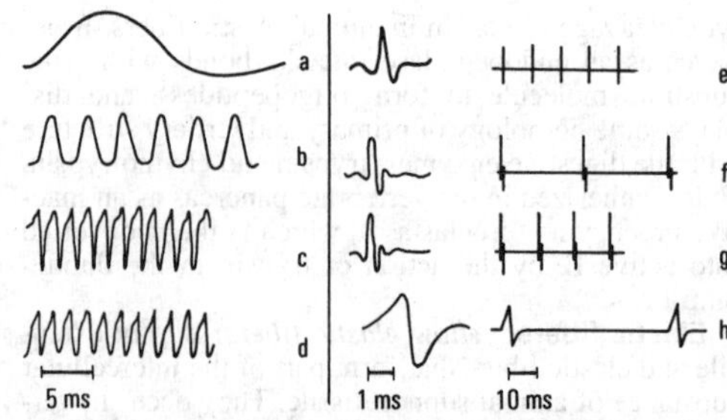

Electrocommunication. Discharge patterns from the electrical organs of fishes. *a Sternopygus macrurus. b Eigenmannia virescens. c Stenarchorhamphus macrostomus. d Apteronotus albifrons. e Rhamphichthys rostratus. f Gymnorhamphichthys hypostomus. g Hypopygus lepturus. h Hypopomus artedi.*

Electroencephalogram: see Electroencephalography.

Electroencephalography: registration of the variations in electrical potential from the brain, usually performed by placing several pairs of electrodes on the scalp. These rhythmic variations in potentials are recorded on a multichannel chart recorder, the resulting traces (a series of waves that are not quite regular) being known collectively as an *electroencephalogram* (EEG). Potentials may also be recorded by placing electrodes directly in or on the brain. An *electrocorticogram* (ECoG) is obtained by placing electrodes directly on the pial surface of the cortex. An EEG may be *bipolar,* i.e. variations in potential are measured between two electrodes placed on the scalp or cortex, or *unipolar,* i.e. one electrode is placed on the scalp or cortex, the other on a part of the body distant from the brain cortex. The background electrical activity of the brain shows wave patterns with the following frequencies: 0.5–3.5 Herz (= oscillations per second) (δ-waves); 4–8 Hz (θ-waves or intermediate waves); 8–13 Hz (α-waves); 13–30 Hz (β-waves). When the subject is relaxed with eyes closed, the α-rhythm is the most prominent, showing a fairly regular pattern of waves of amplitude 50 μV. When the eyes are opened and attention paid to visual objects, or mental tasks are undertaken, the α-rhythm disappears and is replaced by smaller, rapid β-waves. δ-Waves are recorded from subjects in deep sleep. During differing types of sensory stimulation, changes in the EEG can be located to specific fields of the cerebral cortex. To a limited extent, such *evoked* potentials provide information on processing mechanisms in the brain, and they are used experimentally to study sensory processing and brain function in animals. The overall pattern of the EEG provides clinical information on the functional state of the brain. Pathological changes generally cause a slowing of wave frequencies, irregularities in the slower wave patterns, and the erratic appearence of abnormal wave groupings. Thus, in epilepsy, massive discharges of cortical neurons cause much larger, slower waves (2 Hz) in combination with spikes of potential. The existence of pathological changes, as well as the actual site of those changes in the brain, e.g. the presence of a tumor, can be determined by E. In brain death potentials are no longer detectable.

Electrofusion: see Fusion.

Electromyogram: see Electromyography.

Electromyography: a method for recording the action potentials of muscles. The recording or *electromyogram* (EMG) provides information on possible pathological functional disturbances of the muscle system itself and of that part of the nervous system that supplies the group of muscles under investigation. E. is therefore an important tool in medical diagnosis.

Electronasty, *galvanonasty:* movement of plant parts in response to the passage of an electric current. All plant parts that display Seismonasty (see) or Haptonasty (see) are also stimulated by the passage of an electric current. This is exploited for research purposes, because an electric current is more easily quantifiable than the physical processes of striking or touching.

Electron microscope (plate 18): an apparatus for the magnification of very small objects with the aid of an electron beam. Its basic structure is analogous to that of the light microscope. The electron source is a V-shaped tungsten element at a temperature of ~ 2 000 °C. As the emitted electrons travel to the anode, they are accelerated by a high voltage of 50 to 100 kV, then focussed by two condenser lenses (rotation symmetric ecapsulated magnetic coils) onto the surface of the preparation. Scattering of electrons by the atomic nuclei in the preparation generates the necessary contrast for its visualization. The contrast increases in proportion to the atomic number. Since biological objects consists of elements of low atomic number, it is necessary to introduce elements of high atomic number into preparations during Fixation (see). For visualization of the preparation, the electron beam that emerges from it is focussed onto a fluoresent screen with the aid of 3 or 4 additional lenses. The sharpness of this projected image can be controlled by varying the current passing through the objective lens. High current stability is important for the operation of the lenses. An automatic camera is also incorporated into the design, and a reflex arrangement is used to direct the image onto the film for recording purposes (in older designs, the viewing screen is folded back and replaced by a photographic plate). In the most modern instruments, the image can be amplified electronically and transferred to a television camera by fiber optics, so that it can be conveniently studied on a monitor. The tube of the E.m. is evacuated to a very low pressure of 0.0013 Pa, with the aid of rotary and diffusion pumps. Special air locks are incorporated, so that the specimen or photographic plates can be inserted without admitting air to the tube. In the *transmission electron microscope,* the electrons pass through the specimen. In the *Scanning electron microscope* (see), the surface of the specimen is viewed by detecting secondary and back-scattered electrons. If the electron beam between anode and cathode is accelerated with much higher voltages (up to 1 000 kV), the technique is known as high voltage electron microscopy. Technologically, the production and stabilization of such high voltages is considerably more demanding and expensive, but it has the advantage that thicker specimens (up to 3 µm) can be studied; this is mainly necessary for the physical investigation of solid materials, with a few special applications in biology. See Scanning electron microscope.

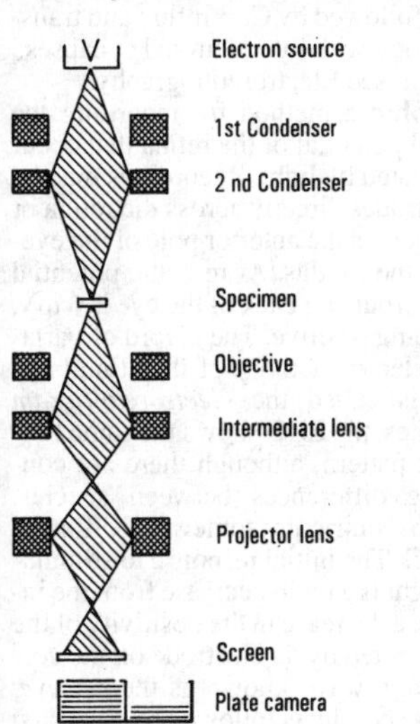

Pathway of the electron beam in the electron microscope (schematic)

Electron-spin resonance spectroscopy, *ESR-spectroscopy:* a spectroscopic method for the investigation of paramagnetic substances. It depends essentially on the interaction of the magnetic moment of an unpaired electron with an external magnetic field. There are only two possible spin states of the electron, corresponding to the magnetic quantum numbers +1/2 and –1/2. In the absence of an external magnetic field the energy of both states is equal and degenerate. Under the influence of an external magnetic field, however, the energy is split into two levels, corresponding to different orientations of the magnetic moment of the electron with respect to the magnetic field. Electromagnetic radiation in the gigahertz (GHz) range (i.e. radiowaves or microwaves) will then induce transitions between these two energy levels if the energy *(hv)* of the radiation (frequency *v)* is exactly equal to the energy level difference: $\Delta E = g\beta H$, where *H* is the magnetic field, β is the Bohr megneton, and *g* is the *Landé g-value* (a dimensionless quantity derived from the contributions of orbital and spin angular momentum to the angular momentum of an electron; for a free electron, $g = 2.0023$).

Normally resonance occurs, because in accordance with the Bolzmann distribution there is a difference in population of the two energy levels. By spin lattice interaction the electrons fall back to the lower energy level, so that continuous ESR-spectroscopy is possible. If the irradiation is too powerful, the population difference between the two levels decreases, producing a state known as *saturation,* i.e. energy cannot be lost at an adequate rate from the lattice. As a result of saturation, the resonance lines become broader.

In biological studies, ESR-spectroscopy is applied to molecules and radicals, so that all possible electric and magnetic interactions of the unpaired electron must be considered. As a result of these interactions, the energy levels undergo further splitting. Of particular importance are: 1) interaction of orbital and spin angular momentum with the external field *(Zeeman interaction);* 2) interaction of electron spin and nuclear spin *(hyperfine interaction);* 3) interaction of the spins of several unpaired electrons *(spin-spin dipole interaction);* and 4) *spin-orbit interaction.* Due to these many possible interactions, ESR-spectra can be very complicated, especially those of biological molecules.

Application. ESR-spectroscopy can be used to study complexes of transitional metals, e.g. the electronic structure of Cu^{2+}, Fe^{3+}, etc. in metalloproteins. The orientation of the ligands can be determined, and electron transfer mechanisms can be investigated. A further important application is the study of radiation-induced radicals. With the aid of ERS-spectroscopy it is possible to identify radicals and to determine their temporal and spatial distribution. Such studies are very important for the design of radiation protection measures and the development of radiotherapy.

Construction of an ESP-spectrometer. Since it very difficult to generate and transmit tunable frequencies in the microwave range, a fixed microwave frequency is used and the magnetic field is altered linearly. The microwave irradiation enters a resonator containing the sample. A powerful magnetic field is generated by an electromagnet. If electron spin resonance occurs, the detector registers a change in intensity of the microwave beam. Since the changes are often very small, a very high frequency magnetic field is used. The signal is then a modulation of the amplitude of the microwave, which can be electronically amplified and more easily recorded.

Electron transport chain: see Respiratory chain.

Electrophoresis: the migration of electrically charged particles in an electric field. *Electrophoretic mobility (u/E,* where *u* is the uniform velocity and *E* is the potential gradient) is the velocity in a potential gradient of 1 volt/cm. Electrophotetic mobility is influenced by the charge density of the particle under investigation, the distribution of charge on the particle surface, and by the shape of the particle. E. may be carrier-free, or it may be performed in the presence of a carrier. In *carrier-free E.* or *free boundary E.,* the particles are simply present in an electrolyte solution. Although no longer widely used, the free boundary procedure of Tiselius for the separation and identification of proteins in biological fluids made an important contribution to biochemical and biophysical studies on proteins well into the 1950s. If the boundary of the migrating molecules can be visualized directly (e.g. if the boundary between a colloidal sol and the pure liquid is sharp and visible by refraction, if boundaries between solute and liquid can be revealed with the aid of ultraviolet light, or if the solute is actually colored) the method is said to be *macroscopic.* If a microscope is used to observe the migration of individual particles, the method is said to be *microscopic.* This must not be confused with the term, microelectrophoresis, which is simply used for electrophoresis on a very small scale. In *carrier E.,* the particles migrate in or on a layer of paper, cellulose powder, starch gel, agar gel, polyacrylamide gel, or other carrier, which also contains a solution of the electrolyte. Gel and paper electrophoresis are used widely in biochemistry for the analytical and preparative separation of macromolecules.

Electrophoridae: see *Gymnotiformes.*

Electrophorus: see *Gymnotiformes.*

Electroreceptors: sensory receptors that respond to external changes in potential in the microvolt range. E. are always found in fishes that possess electric organs, but also in a large number of species without electric organs, especially phylogenetically old fish groups, e.g. lampreys, sharks, rays, sturgeons, ratfishes, bichirs, lungfishes, many freshwater catfishes, as well as several aquatic amphibians. E. have evolved from the so-called acousticolateralis system, which includes the lateral line system of fishes and amphibians, the auditory system and the labyrinth system. Two morphologically and physiologically different E. have been described. *Tuberous receptors,* which occur in gymnotids, mormyrids and *Gymnarchus,* are located in closed skin cavities, and they are excited by high frequency (50–500 Herz) alternating current. *Ampullary receptors* are found in both electric fishes and in those nonelectric species listed above. They are located at the bottom of skin pores, which are filled with a high conductivity gelatinous material. In sharks and rays, these pores are long canals, known as *ampullae of Lorenzini.* They respond to direct current or low frequency alternating current. The absolute sensitivity of ampullary receptors is between one and two orders of magnitude higher (1 μV/cm) than that of tuberous receptors (10–70 μV/cm). E. respond to the discharges from the fish's own electric organ, or from the electric organs of other fishes. They respond to direct current potentials of other aquatic animals (>10 mV/cm) and their muscle potentials (>1mV/cm), and are therefore of use, inter alia, in the detection of prey. They also respond to geophysical phenomena, e.g. lightning and alterations in geomagnetism. When water moves at 1 m/s parallel to the equator (e.g. the Gulf Stream), the vertical component of the gradient induced by the earth's magnetic field can attain 0.4 μV/cm. Since the ampullary receptors (ampullae of Lorenzini) of sharks are known to respond to 0.01 μV/cm, sharks should be able to orientate themselves in the earth's magnetic field. Fish with electric organs have evolved a system of spatial orientation. Fields (distributions of potential) created by electric organs have certain spatial properties around the fish, which depend on the conductivities of the surrounding water, the fish's skin and the fish's internal tissues (more precisely the ratios of these 3 conductivities) and the shape of the fish. In high conductivity water (e.g. sea water), the field is detectably disturbed only by objects within few cm of the fish, but in low conductivity water (e.g. 10–100 KΩcm^2 in some freshwater tropical rivers) small objects (a few mm) can be detected over relatively large distances.

The unusually high sensitivity and the molecular mechanism that actuates the excitation of E. are incompletely understood. Probably the receptor potential opens Ca^{2+} channels, followed by Ca^{2+} influx and transmitter release, in analogy with interneuronal synapses.

Electroretinogram: see Electroretinography.

Electroretinography: a method for recording the variations in electrical potential of the retina that occur when the eye is stimulated by light. Recordings may be made by placing electrodes directly across the retina or by placing one electrode on the anterior pole of the eyeball and the other on the fundus. At rest, the potential diffrence between the front and back of the eye is 6 mV, the front of the eye being positive. The record of variations in the total bioelectric activity of the stimulated retina against time is called the *electroretinogram* (ERG). In all vertebrates, the ERG shows the same fundamental multiphasic pattern, although there are considerable quantitative differences between species. Rods and cones also contribute somewhat different waveforms to the ERG. The initial response to stimulation of the retina by light is a rapid decrease from the initial value of 6 mV, i.e. a decrease in the positivity of the front of the eye (as detected by an electrode on the cornea), to form a negative wave known as the *a-wave.* This lasts less than 0.1 sec, and is followed by a marked reversal of polarity (the cornea becomes strongle positive), producing a sharp positive *b-wave.* The b-wave is followed by a slower *c-wave.* When the light stimulus is removed, there is a small negative "off" deflection, followed by the gradual rise of the *d-wave* to the starting potential of 6 mV. Each wave is composite and the result of more than one neural or retinal factor, but some progress has been made toward a more precise explanation of their origin by the use of penetrating electrodes, which measure the response from a small, defined area.

The a-wave represents the rise of ionically generated receptor potentials of the rods and/or cones, and it attains its maximal amplitude near the distal tips of the photoreceptors. In the ERG, however, it is abruptly terminated by the onset of the large b-wave, which is of opposite polarity. Although it is not immediately apparent from the ERG, there is a further component, known as the *dc component* (it displays the time course of a pulse of direct current), which also increases the positivity of the corneal electrode and contributes to the termination of the a-wave. The type of cell generating the dc-component has not been identified.

The b-wave is generated across the membranes of Müller cells; these glial cells respond to the ionic composition of the extracellular fluid, e.g. a change in the concentration of potassium ions due to the activity of neighboring nerve cells. Thus, the b-wave appears to be a secondary phenomenon of the activity of nerve cells in the inner nuclear layer. Nevertheless, the b-wave is large and sharp and readily recorded, and since it is closely related to ganglion cell activity it is a useful indicator of retinal sensitivity.

The c-wave is initiated by light absorption in the outer segments of the photoreceptors, and it is generated by the transient hyperpolarization of the pigment epithelium cells (after a single bright flash of light, there is a transient decrease in the concentration of extracellular potasium which shows a similar time course to that of the c-wave). The c-wave therefore does not appear to lie on the direct chain of events that link nervous activity to the visual sensation, and it might be decribed as an "epiphenomenon".

Clinically, E. is important for the diagnosis of pathological conditions of the retina.

Electrotonic current: see Electrotonus.

Electrotonic potential: see Electrotonus.

Electrotonus: changes in the membrane potential of electrically excitable systems (nerves or muscles), due to the presence of subthreshold polarizing currents *(electrotonic currents).* Changes of membrane potential are equal to the product of the electrotonic current and the resistance of the membrane Many dendrites of motor neurons do not transmit an action potential because their stimulation thresholds are very high, but they do transmit an electrotonic current to the soma, i.e. a current conducted by the fluids of the dendrites. E. is manifested physiologically when the electrotonic current approaches the stimulation threshold for excitation. E. can be demonstrated experimentally by passing a constant polarizing current through part of a nerve in a nerve-muscle preparation by means of nonpolarizable electrodes placed 2–3 cm apart. Excitability of the nerve is tested by applying a submaximal induction shock. Irrespective of whether the polarizing current is descending (cathode nearer to the muscle) or ascending (anode nearer to the muscle), the excitability of the nerve always increases in the region of the cathode (a formerly submaximal shock becomes maximal), while excitability decreases in the region of the anode. If a stimulus is applied between the anode and the cathode, the resulting impulses must pass a region of *anelectrotonus* (if the polarizing current is ascending) or *catelectrotonus* (if the polarizing current is ascending) before reaching the muscle. If the polarizing current is sufficiently strong, the decrease in excitability due to anelectrotonus may completely block the impulse (an electrotonic block). When the polarizing current is removed, the nerve returns to normal, and any stimulus that was formerly submaximal once again becomes submaximal throughout the excitable system. Catelectrotonus causes depolarization, while anelectrotonus causes hyperpolarization. Electrotonic potentials as well as electrotonic currents can be measured for some distance outside the sites of application of the electrodes (i.e. in the extrapolar portions of the nerve). Both electrotonic currents (extra-anodic and extra-cathodic) flow in the same direction as the polarizing current. Extrapolar electrotonic potentials decrease from their point of origin (in this case one of the electrodes supplying the polarization voltage), and the rate of decrease depends on the membrane length constant (see Cable theory).

Electrotropism, *galvanotropism:* the bending of plant organs in an electric field, or when a direct current is passed transversely through the plant. Shoots always curve toward the anode, roots to the cathode. It is possible that some electrotropic reactions occur in response to electrolytically generated hydrolysis products, so that they are actually a form of Chemotropism (see).

Eledone: see *Cephalopoda.*

Eledonella: see *Cephalopoda.*

Elephant: see Elephants.

Elephant beetles: see Lamellicorn beetles.

Elephant birds: see *Aepyornithidae.*

Elephant ear sponge: see Bath sponge.

Elephantfishes: see *Mormyridae.*

Elephantiasis: see Filariasis; plate 43.

Elephants, *Elephantidae:* the only extant family of Proboscidians (see), and the largest living terrestrial animals. They are herbivorous mammals, possessing a trunk, which is a very mobile extended nose, used for smelling, gripping, drinking, trumpeting and striking. The upper incisors are modified to a pair of long tusks, which are used as weapons and as crowbars. Water is sucked into the trunk then sprayed into the mouth. The trunk is also used to transfer food to the mouth. E. are highly intelligent animals, and they can be trained to work. Especially the Indian elephant has been long used throughout Asia for transport and for lifting heavy goods, like timber. In the wild, E. live in small herds. Gestation lasts 20 to 24 months.

The ***African elephant*** (*Loxodonta africana*) is native to central, eastern and southern Africa south of the Sahara in a variety of habitats, like savanna, river valleys, thornbush, dense forest and scrub. Head and body, including the trunk, measure 6–7.5 m; tail length 1–1.3 m; shoulder height 3–4 m; weight 5–7.5 metric tons. The ears are large and 3 hooves are present on the hindfeet. There is a maximum of 26 caudal vertebrae and 21 pairs of ribs. The forehead is convex, the back slopes down, and the highest point is at the shoulders. At the tip of the trunk are 2 finger-like processes.

The ***Asiatic*** or ***Indian elephant*** (*Elaphas maximus*) is native to India and Southeast Asia, inhabiting thick jungle to open grassy plains. Head and body, including the trunk, measure 5.5–6.4 m; tail length 1.2–1.5 m; shoulder height 2.5–3 m; weight about 5 metric tonnes. The ears are smaller than those of the African elephant, and 4 hooves are present on the hindfeet. There are 33 caudal vertebrae and 19 pairs of ribs. The forehead is flat, the highest point being the top of the head, and there is only one finger-like process at the tip of the trunk.

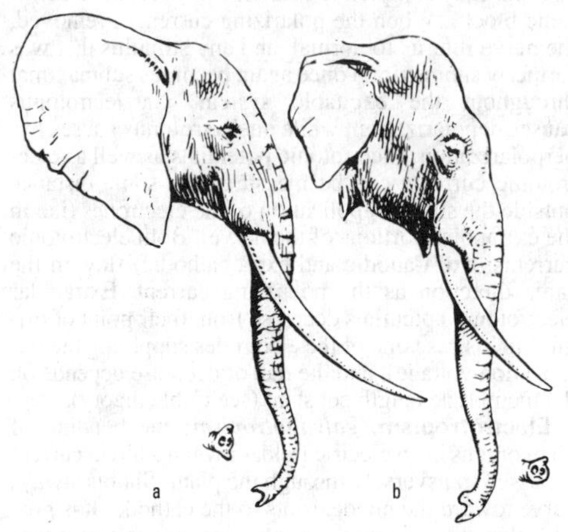

Elephants. Head and trunk tip. *a* African elephant *(Loxodonta africana). b* Indian elephant *(Elaphas maximum).*

Elephant seals: see *Phocidae.*

Elepidote species: see *Ericaceae.*

Elettaria: see *Zingiberaceae.*

Eleusine: see *Graminae.*

Eleutherata: see *Coleoptera.*

Elicitor: a substance produced by pathogenic organisms during infection of a host, which causes the host to produce defense substances [e.g. Phytoalexins (see)], or to otherwise alter its metabolism. The term E. is often used specifically for a glucan fraction, released by heat treatment from the cell wall of a phytopathogenic fungus. In the wider sense, an E. is any factor, biotic or abiotic, which induces formation of Phytoalexins in plant tissues. Heavy metal salts like $HgCl_2$ and $CuCl_2$ are examples of abiotic Es.

Elimination: a term introduced by W. Reinig (1938) for the chance or random loss of an allele from a population, as distinct from its eradication by selection. E. occurs more readily with the decreasing size of the population.

Elimination coefficient: the frequency with which carriers of a specific deleterious gene die prematurely or are prevented from reproducing, and are therefore genetically eliminated from the population. For example, if a certain deleterious gene has an average E.c. of 5%, then 5% of all individuals whose genotype carries this gene, perish before they can transmit the gene to offspring.

Elimination hypothesis: a hypothesis presented in 1938 by W. Reinig, but subsequently disproved, to explain the occurrence of racial differences in animals of the Northern Hemisphere. According to the E.h., after the end of the Ice Age, ever increasing numbers of genes were lost as species dispersed from their previous areas of confinement. This led to infraspecific differences between old and younger populations, with respect to color, shape and body size.

Reinig considered the elimination of genes with an increase in the geographical area of a species to be a separate evolutionary factor. It is now accepted that even if such processes actually occur, they are only a special case of Genetic drift (see). The racial differences that appeared after the Ice Age were due to other factors.

Elimination rate: see Mutation-selection balance.

Elk: see Moose, Red deer.

Ellagic acid: see Tannins.

Ellagitannins: see Tannins.

Ellipura: a hypothetical phylogenetic group of insects, comprising the orders *Protura* and *Collembola.*

Ellobiini: see *Arvicolinae.*

Elm family: see *Ulmaceae.*

Elytroptera: see *Coleoptera.*

Embedding: a method for immersing biological specimens in paraffin wax or resin, prior to the preparation of sections for microscopy. The cells must first be fixed (see Fixation) then dehydrated stepwise by treatment with alcohol or acetone. For light microscopy, this solvent is then replaced with paraffin wax. The very thin sections required for electron microscopy can only be prepared from material embedded in resin. Solvents used for water removal are then in turn replaced stepwise by monomeric epoxy resins or polyesters. A hard block is then formed by polymerization, and this is sectioned with an ultramicrotome (see Microtome).

Emberiza: see *Fringillidae.*

Emberizidae: see *Fringillidae.*

Emberizinae: see *Fringillidae.*

Embiids: see *Embioptera.*

Embioptera, ***embiids, webspinners:*** an insect order of about 300 species, found in the tropics and subtropics of the New and Old World, extending as far north as the Mediterranean and southern Russia, and to Florida, Texas and southern California. Imagoes are yellow or brown, with a slim body usually less than 12 mm in length, and a large head with biting mouthparts, protruding compound eyes and multisegmented filiform antennae. Ocelli are absent. In all species the females are apterous (wingless), and in nearly all species the males have 2 pairs of nearly identical wings, resembling the *Plectoptera* (stoneflies). Larvae and imagoes of both sexes possess up to 100 silk glands on the enlarged first tarsal segment ("heel") of each foreleg. *E.* are gregarious, and live together in tubular shelters of silk threads, which they spin communally under stones, among leaf litter and between sparse root systems. The insects hardly ever leave these shelters. *E.* are vegetarian, feeding on fresh and dry plant material. Larval metamorphosis is incomplete (paurometabolous development), and may last 10 months at more northerly latitudes. Males die soon after pairing, and may be eaten by the female. Parthenogenesis has been observed in some species. Females guard their eggs and care for the young larvae. The earliest fossil *E.* are found in the Permian. Taxonomy of the *E.* is incomplete, and because of the neotenous nature of the females, classification must be based on mature males.

[E.S. Ross, Biosystematics of the Embioptera, *Annu. Ref. Entomol.* ***15*** (1970) 157–172]

Embolism: see Thrombus.

Embolus: see Thrombus, Blood coagulation.

Embryo:
1) Botanically, the young plant that is formed (either immediately or after a certain period of dormancy) from the fertilized (or parthenogenetically from the nonfertilized) ovum of mosses, ferns and seed plants. In mosses and ferns, the E. arises within the archegonium. In seed plants, it is formed in the ovule and is part of the seed. The E. consists of an axis (stem) bearing a plumule (bud), a radicle (root), and one or more cotyledons (seed leaves). As the seed germinates, the E. develops into an Embryo plant (see). See Fertilization.

2) Zoologically, an animal in process of development from the zygote (or parthenogenetically activated ovum). It is known as an E. as long as it is contained within the egg membranes, in the egg shell, or within a parent animal.

Embryology (plate 23): the study of embryonal development. See Development, Ontogenesis.

Embryonal development: the development of a living organism from the fertilized egg cell to the time at which it leaves the egg or is born. See Ontogenesis.

Embryonal growth of plant cells: see Plasmatic growth.

Embryonal organs: organs and structures that have a transient existence, being functionally important only during the embryonal stage of many organisms. They are degraded and reabsorbed before hatching from the egg or before parturition, or they may be discarded after the completion of embryonal development. Notable examples are the Yolk sac (see), Allantois (see) and Extraembryonic membranes (see).

Embryonal type: an adult fossil organism, which in comparison with its living relatives, displays embryonal or juvenile characters.

Embryonic chamber (Protoconch): see Ammonites.

Embryopathy: pathological embryonal development. The term is mostly used for abnormal development resulting from environmental influences.

Embryo plant: the young plant which develops from the embryo, and which represent the transition of the embryo to the mature plant. In pteridophytes the embryo develops into an E.p. as soon as it is formed, whereas in spermatophytes the embryo usually passes through a relatively long period of dormancy in the seed. Seed germination marks the onset of E.p. development; the root emerges first from the seed, takes up water and nutrients, and anchors the plant in the ground. See Germination, Seed germination.

Embryo sac: see Flower, section 3) Carpels or megasporophylls.

Emerald lizard: see Green lizard.

Emergences: multicellular outgrowths from plant epidermis. In contrast to Plant hairs (see), not only the epidermis but also fairly deep-lying parts of the subepidermal tissue participate in the formation of E. Typical E. are the thorns of rose and blackberry, the glandular teeth of the stipules of the pansy, and the tentacles of the sundew. The flesh of citrus fruits is formed by internal E. full of sap which grow into the loculi of the ovary.

EMG: acronym of electromyogram. See Electromyography.

Emitter: see Expedient.

Emperor moths: moths of the family *Saturniidae*. See *Lepidoptera.*

Empodium: a median, unpaired extension of the pretarsus of the insect leg. It lies between the claws and between the pulvilli, as an unpaired prolongation of the planar sclerite distal to the unguitractor plate. In springtails the total pretarsus is prolonged into a single claw, also called an E., usually accompanied by an opposing E. appendage.

Empusa: see *Phycomycetes.*

Emscher tank: see Biological waste water treatment.

Emsian stage: see Devonian.

Emulsin: a mixture of glycosidases from bitter almonds. The main component of the mixture is β-D-glucosidase. E. catalyses the hydrolysis of the cyanogenic glucoside, Amygdalin (see), into glucose, benzaldehyde and hydrocyanic acid. Other natural and synthetic β-glucosides are also attacked.

Emus: see *Casuariiformes.*

Emu wrens: see *Sylviinae.*

Emydidae, ***freshwater turtles, terrapins, pond and river turtles:*** the largest and most varied family of the order *Chelonia* (see), containing 85 species (more than 140 subspecies) in 31 genera, distributed worldwide, but mainly in the tropics and subtropics of America and Asia. The carapace is generally oval and only weakly arched, often with ridges, and not hemispherical as in the *Testudinidae.* Ventral and dorsal scutes meet one another. The intermediate row of scutes present in more primitive families is absent or represented by individual scutes on the shoulders or flanks. The feet are flattened as an adaptation to locomotion in water. There is usually some webbing between the fingers and toes. In many species, the head is strikingly patterned in yellow or red. Fully aquatic forms live in estuaries, coastal waters, or freshwater rivers and lakes, but many species are semi-aquatic, inhabiting forested ponds, streams, marshes and bogs. A few species are fully terrestrial, and have a highly arched carapace. In many species, the plastron is hinged and movable. They are largely omnivorous, feeding on live and dead fish, as well as worms and other small aquatic animals, and vegetation. Some forms have powerful hooked jaws as an adaptation to feeding on snails. The elongated eggs are always buried in the ground. Pond and river turtles are eaten in many parts of the world.

Emys orbicularis ***(European pond*** or ***swamp turtle, European pond terrapin, European pond tortoise)*** is distributed in the Near East, North Africa and central and southern Europe, but is now rare in the northern part of its range, and probably no longer exists in central Europe west of the Elbe. It is strictly protected in Europe. The shell is up to 35 cm in length, and the carapace and soft parts are patterned with yellow stripes and spots on a black background. The plastron is jointed across the middle and joined to the carapace by elastic connective tissue. An inhabitant of large standing and slowly flowing waters and marshy areas, it suns itself on the bank, and rapidly enters the water and submerges if it perceives danger. It overwinters in the mud bottom. Mating usually occurs in April, and the female lays up to 15 eggs in self-excavated holes in the ground; the holes are then filled and the site smoothed over with the underside of the shell. The young animals, which hatch after 8–10 weeks, are quite spherical in shape and about 2.5 cm in diameter. In cold years, the eggs remain unhatched until the spring.

The genus *Clemmys* ***(Pond turtles)*** displays a variety

of forms, which are assigned to different subgroups. Pond turtles are found in both the Americas and the Old World, ranging from southern and southeastern Europe to East Asia. *Clemmys caspica* inhabits the Balkans, the Caucasus and the Near East. It has a rather flat shell and yellow head stripes, and favors shallow areas of water. When threatened it buries itself in the bottom mud. The ***Bog turtle*** or ***Muhlenberg's turtle*** *(Clemmys muhlenbergii;* 11 cm) is the smallest of the *Clemmys* species; it usually stays on land where it feeds on plants, but sometimes also hunts in the water.

Chinemys reevesi ***("Chinese 3-ridged turtle")*** is one of commonest members of the family in Southeast Asia. It has yellow markings on the sides of the head, and the carapace has 3 longitudinal, raised ridges.

Roofed turtles (genus *Kachuga)* are characterized by a carapace that falls away steeply at the sides and carries a high, tuberculate longitudinal ridge. The 6 species are distributed from India and Pakistan to Indochina. In addition to small aquatic animals, they consume much vegetable material.

The many subspecies of the ***Diamondback terrapin*** *(Malaclemys terrapin)* inhabit mainly the brackish coastal waters of much of North America, and their flesh is considered a delicacy. The carapace has a raised relief, consisting of facets of concentric furrows. They feed largely on hard-shelled mollusks, and excess salt from the diet is secreted via glands at the sides of the eyes.

The name, ***Terrapin,*** is given to several different members of the *Chelonia,* but it is also specifically associated with a large group of polymorphic species and subspecies, distributed in North, Central and South America, and displaying wide geographical variability. They are mostly aquatic in habit, but even the more terrestrial species are associated with standing or slowly flowing water. Their phylogenetic relationships are unclear. In particular, the young animals are very strikingly colored, often with eyespots on the carapace and yellow or red patterning on the head. The common North American *Chrysemys scripta elegans* ***(Red-eared turtle)*** has bright red temporal stripes on each side of the head. *Chrysemys picta* ***(Painted turtle)*** has bright red spots on the lateral scutes and yellow head patterning. In the genus *Graptemys* ***(Map turtles),*** the carapace carries a ridge of tuberosities, and is ornamented with a pattern of lines. Terrapins lay their eggs in a self-excavated hole in the ground near to water.

Box turtles (genus *Terrapene)* are represented by 6 species, up to 16 cm in length, and distributed from North America to Mexico. They have a highly arched carapace and are largely adapted to a terrestrial existence, but there is some residual webbing on the feet. The plastron has anterior and posterior transverse joints, which are hinged with connective tissue. Thus, the animal can withdraw entirely into its shell and close the front and rear entrances by folding up the flaps of the plastron. They also resemble terrestrial turtles in their feeding habits, eating fruit, berries, fungi and other vegetation, in addition to worms and snails. Females excavate a deep hole (up to 12 cm) for their eggs.

Emydura: see *Chelidae.*

Emys: see *Emydidae.*

Enarthrosis: a ball-and-socket joint (see Joints).

Enation: a wing-like or ridge-shaped outgrowth on the surface of a plant organ (stem, leaf, fruit, etc.). For example, the small outgrowths or Es. from the stem of *Asteroxylon* are feebly supplied with vascular tissue by a leaf trace from the central vascular column to the base of the E. This is thought to represent an early evolutionary stage of the leaves of the *Lycopsida.* Leaves that evolved from Es. are termed microphylls (one of the distinguishing features of the *Lycopsida),* in contrast to leaves that arose as parts of the branching system. Outgrowths caused by virus infections are also called Es.

Encasement theory: see Preformation theory.

Encephalon: see Brain.

Encephalopathy: see Prion.

Enchanter's nightshade: see *Onagraceae.*

Enchytraeidae, ***enchytraeids:*** a family of small oligochaetes, rarely longer than 3 cm. The largest member is *Fridericia gigantea,* measuring 3–4.5 cm. Other typical genera are *Enchytraeus, Lumbricillus* and *Aspidodrillus.* The *E.* are relatively uncommon in tropical regions and in the Southern Hemisphere, but widespread in the northern temperate zone, especially abundant in European countries and fairly common in Arctic regions. Many genera are littoral, some in fresh water, and they predominate in moist soils, where they represent an important component of the soil fauna. In large numbers they damage cultivated plants in pots or open soil by feeding on the root hairs, which they remove by a sucking action of the pharynx. Some species are used for feeding aquaria fish.

Encinar: see Sclerophylous vegetation.

Encrinus (Greek *enkrinon* closed lily): an extinct, long-stalked genus of the *Crinoidea* (see) with a shallow, dish-shaped calyx. The calyx base consists of two rings of plates, i.e. 5 basal plates and 5 interbasal plates, the latter being overlapped by the uppermost section of the stalk. The single stranded, later two-stranded arms carry long appendages. The stalk consists of many individual Trochites (see). Fossils of E. occur in the Triassic, and *E. liliiformis* is an index fossil of the upper Muschelkalk, where its stalk trochites make a major contribution to the formation of massive reefs, known as trochite chalk.

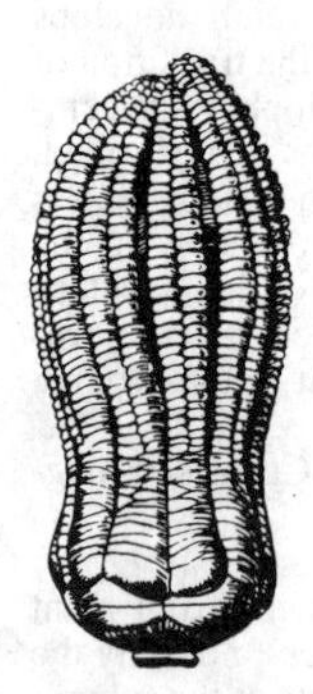

Encrinus liliiformis from the Upper Muschelkalk (x 0.8)

Endemic: native to, and restricted to, a certain geographical area. An adjective applied to infectious diseases, which display a low and more or less constant incidence in a defined locality.

Endemic rate: the proportion of endemic species in the fauna or flora of a particular geographical area, or in a particular taxon (family, order, etc.) in that area. The E.r. is an important index of the duration and degree of isolation of the flora and fauna during their evoluitonary history, and of their evolutionary age. See Endemism.

Endemism, ***endemicity:*** the restriction of a species of living organism to a particular geographical area. Such organisms are said to be *endemic* to that area. Endemic species confined to a small area may be of two types.

Epibionts are survivors from a preceding epoch. Earlier in their evolutionary history, their geographical area was larger, but their endemic area has now contracted, and they are threatened with extinction. This type of E. is known as *conservative E.* or *relict E.* Examples are the Californian redwoods *(Sequoia)* and the New Zealand kauri *(Agathis australis).*

Alternatively, an endemic species may have evolved relatively recently, and its area may be small because it has not yet expanded. This is known as *progessive E.* or *neo-E.* Examples are the Illinois ironweed *(Vernonia illinoiensis)* and Laurentian speedwell *(Veronica peregrina* var. *laurantiana).*

E. is commonly encountered on oceanic islands, where the surrounding sea acts as a barrier to dispersal and to the introduction of new species. Thus, all the land birds on Tristan de Cunha are endemic; the palm genus, *Pritchardia,* is endemic to the Hawaiian Islands, where some species are even restricted to a single valley. High mountains and inland lakes also show high rates of E., e.g. 171 of the 214 species of cichlid fishes in Lake Tanganyika are found nowhere else, i.e. the cichlid species of Lake Tanganyika display an endemic rate of 80%.

The geographical region occupied by an endemic organism is not necessarily the region in which it originated or evolved. Thus, glaciation may drive organisms from their areas of origin, at the same time leaving refugia, which permit the development of new regions of E. See Endemic rate.

Endive: see *Compositae.*

Endo-agar: a selective bacteriological culture medium developed by the Japanese bacteriologist, S. Endo, for the identification of coliform bacteria *(Escherichia coli),* e.g. in drinking water. Additions include lactose and acid fuchsin. Acetaldehyde formed by fermentation of the lactose by the bacteria releases the fuchsin, which stains the bacterial colonies a deep red color.

Endobiosis: a type of coexistence of living organisms, in which one organism lives within the other, e.g. the cellulose-degrading ciliates in the rumen of cattle. See Endoparasite, Symbiosis.

Endocarp: see Fruit.

Endoceras (Greek *endon* within, *keras* horn): an extinct nautiloid genus, in which the shell was an elongated straight cone, up to 2 m in length. In transverse section the shell was circular to elliptical, with the broad siphuncle along one side. Fossils of *E.* are found from the Ordovician to the Silurian.

Endocranium: see Skull.

Endocyclica: see *Echinoidea.*

Endocytosis: a process whereby macromolecules, particles, bacteria, etc. are ingested by the cell. Solid, relatively large particles (e.g. cell debris, bacteria, viruses) are only taken up by certain cells (e.g. monocytes). In this process of *phagocytosis* ("cellular eating"), elucidated by E. Metschnikoff, the particles are enveloped (engulfed) by cytoplasmic processes of the phagocyte (see Macrophages) then enclosed within a vacuole *(phagosome,* diameter >250 nm) in the interior of the cell. The phagocyte may be attracted chemically to the particle, or it may be stimulated immunologically. The phagosome fuses with a primary lysosome to form a *phagolysosome (secondary lysosome),* so that the ingested material is then exposed to lysosomal degradative enzymes (e.g. acidic hydrolases). Indigestible residues are removed to the exterior of the cell by a reversal of E., known as Exocytosis (see). *Pinacytosis* ("cellular drinking") is in principle the same as phagocytosis, but material is taken up in solution, and the resulting pinacytosis vesicles are smaller than phagosomes.

Whereas phagocytosis is limited to certain cells, all cells are capable of *micropinacytosis.* Owing to the small size of the micropinacytosis vesicle (about 40 nm), micropinacytosis can only be demonstrated by electron microscopy. It is thought to be initiated by the binding of dissolved macromolecules to membrane receptors. At the binding site the plasma membrane invaginates, then a fluid-containing vesicle is abstricted and transported into the cytoplasm. The vesicles either fuse with membranes of the endoplasmic reticulum then release their contents into cisternae, or they are incorporated into vesicular bodies (see Multivesicular bodies). They can also probably release their contents into the cytoplasm after disintegration of the membrane. Some endothelial cells lining small blood vessels are thought to transport material from the blood stream to the surrounding extracellular fluid by *transcytosis,* i.e. microvesicles are transported through the cytoplasm, and their contents are released on the other side of the cell by exocytosis. Transcytosis occurs in other cells, but is then always mediated by coated vesicles (see below). The only essential difference between phagosomes and the various sizes of pinacytosis vesicles is their size; they are known collectively as endosomes (diameter 40–5 000 nm). Coated vesicles (diameter 60–100 nm) also serve a very important function in E. as well as in intracellular transport. Their conspicuous surface structure clearly differentiates them from endosomes. In the electron microscopy of freeze-etched cell sections, the surface of coated vesicles is revealed as a network of polygons resembling a honeycomb. The protein coat contains several major structural proteins, the best characterized of which is clathrin (M_r 180 000). The invagination or pit in the cell membrane that precedes the formation of a coated vesicle is known as a *coated pit.* It is thought that invagination is caused by forces generated by the interactions of clathrin and other coat proteins on the cytoplasmic side of the membrane. Finally, the coated vesicle separates from the cell membrane and migrates to the cell interior, e.g. to a primary lysosome. Coated vesicles are also often produced from the membranes of the Golgi apparatus and the Endoplasmic reticulum (see).

All endosomes and coated vesicles are transported by the action of *contractile microfilaments.* Usually, Microtubules (see) are also involved. Transport is therefore prevented by colchicine (which causes disintegration of microtubules) and by cytochalasin B (which blocks development of contractile microfilaments). Material may also be taken into the cell for the purposes of storage, e.g. yolk material.

Endoderm: the innermost cell layer of the two-layered gastrula embryo of multicellular animals. The E. forms the archenteron, which in the course of embryonic development then gives rise to the epithelium of the midgut and in many cases to the mesoderm and notochord.

Endodyogeny: see *Toxoplasmida.*

Endogamy:
1) Inbreeding (see).
2) Pollination of a flower by pollen from a different flower on the same plant.

Endogenote: that part of the bacterial chromosome, which in the course of transformation or conjugation (see F-plasmid) is homologous with and combines with the genetic fragment that enters the cell from the donor. This genetic fragment from the donor cell is called the exogenote.

Endogenous: arising internally; belonging to the interior; occurring inside the body or cells of a living organism; synthesized by a living organism, as opposed to being derived by acquisition from the environment. Occasionally, the term is used as a synonym of Autonomous (see).

Endogenous electric current: a flow of positive charges resulting from a stationary electric field in a growing cell. E.e.cs. have been found in all plant and animal cells investigated so far. They are associated with the local growth processes that precede spatial differentiation. Thus, in the zygote of brown algae, the rhizoid always grows out from the site where a stable current flows into the cell (Fig.). The current is about 100 pA per cell. E.e.c. is generated by laterally asymmetric active transport within the cell, especially of Ca^{2+}-ions and Na^{+}-ions. Fertilization appears to be associated with a powerful, spreading E.e.c., which commences immediately after sperm penetration, and which is also the result of a Ca^{2+} flux. Suppression of the E.e.c. leads to growth inhibition. Growth in the absence of an accompanying E.e.c. has not been observed. The regulatory function of E.e.c. is possibly due to a lateral electrophoretic effect on membrane proteins. It has also been suggested that the function of the E.e.c. is to cause local alterations in ion concentrations.

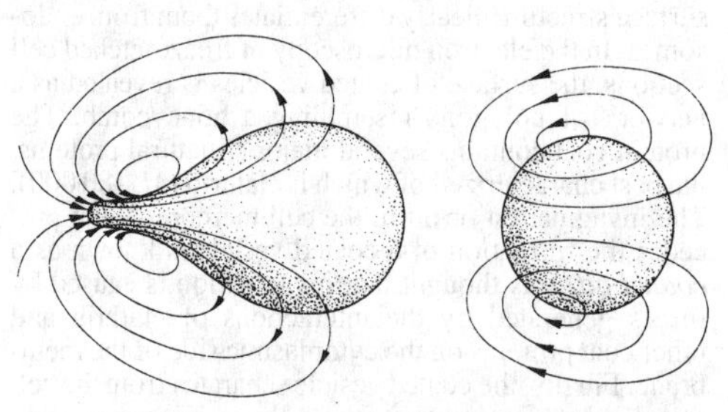

Endogenous electrical current. Endogenous currents in the zygote of a brown alga, before (right) and after (left) formation of the rhizoid primordium. The flow of positive charge is shown by the arrows.

Endogenous opiates: see Endorphins.

Endoliths: see Epiliths.

Endolymphatic duct: see Auditory organ.

Endometrium: the uterine mucosa or epithelium. See Uterus, Placenta.

Endomitosis: in eukaryotes, one or more replications of the chromosome number, in the absence of nuclear division. E. gives rise to somatic polyploidy *(endopolyploidy).* A spindle is not formed during E. The giant polytene chromosomes of dipteran salivary glands and other insect larval tissue are the result of E., although in such cases the daughter chromosomes remain synapsed.

Endomixis: a type of automictic fertilization in *Paramecium caudatum,* reported by Woodruff and Erdmann in 1914. As described, E. consisted of the fusion of two micronuclei, which had been produced by meiosis, all within a single, nondividing cell. Thus, a new macronuclear apparatus would be produced without synkaryon formation. The existence of E. is now doubted. In 1936 Diller failed to observe E., even using the original organism stocks of Woodruff and Erdmann. Genetic evidence for E. is also lacking (Sonneborn, 1954). Diller's investigations, however, led to the description of two other processes: Autogamy (see) and Hemixis (see).

Endoparasite: see Parasite.

Endoplasm: the internal cytoplasm of unicellular organisms, which contains the nucleus and feeding vacuoles.

Endoplasmic reticulum, *ER:* a system of laminar cavities (cisternae) and canals (tubules), present in all eukaryotic cells. It is organized into a reticulate mesh extending throughout the cytoplasm. The extent and the nature of the ER depend on the degree of differentiation, type and physiological state of the cell. The membrane of the ER (a typical lipid bilayer) is a highly convoluted single membrane separating the ER lumen or cisternal space from the cytosol (plates 20 and 21).

Structurally and functionally, there are 2 types of ER. The exterior surface (cytoplasmic side) of the membrane of *rough ER* is occupied by Ribosomes (see) or polysomes. Rough ER is a site of protein synthesis, and many of the synthesized proteins are exported from the cell. The surface of *smooth ER* is not occupied by ribosomes; it serves a number of different functions, but it is not a site of protein synthesis. These 2 types of ER (Fig.1) are continuous with one another and interconvertible within the cell.

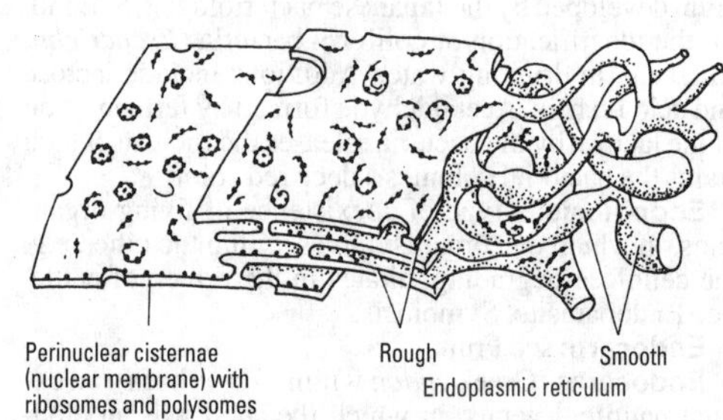

Fig.1. *Endoplasmic reticulum* (schematic)

The nuclear envelope (see Nucleus) consists of two concentric membranes. The outer nuclear membrane (often coated with ribosomes) closely resembles, and is continuous with, the membrane of the ER, so that the space between the outer and inner nuclear membranes can be regarded as a perinuclear cisterna of the ER. Ribosomes are either free in the cytosol or associated with the ER (or outer membrane of the nuclear envelope), and they are never found attached to any other type of membrane. In the membrane-ribosome association, the large ribosomal subunit lies against the membrane surface of the ER that faces into the cytosol. Many of the proteins synthesized on membrane-bound polysomes are secretory proteins for export from the cell. As a polypeptide chain grows, it passes through a pore in the membrane into the lumen of the ER (see Ribosomes).

When synthetic activity is high, proteins accumulate in the lumen and the cisternae expand. Under these conditions (e.g. in the cells of the exocrine pancreas) electron-dense protein granules may be demonstrable in the ER lumen. Membrane proteins for other interacellular structures (but not for the inner membranes of mitochondria or plastids) are also synthesized on the rough ER. From the lumen of the rough ER, proteins are transported to the Golgi apparatus, usually in small vesicles, more rarely via small tubular extensions of the smooth ER (Fig.2). The oligo- and polysacccharide moieties of glycoproteins are added in the Golgi apparatus. The enzymes required for this conversion are also synthesized on the rough ER and transported to their functional site in the Golgi apparatus.

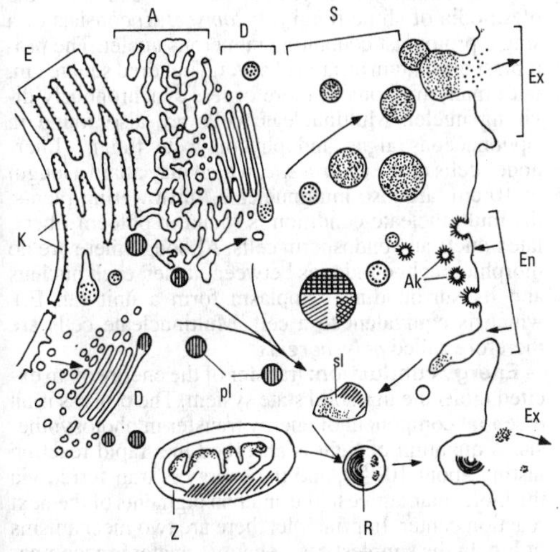

Fig.2. *Endoplasmic reticulum.* The role of membranes in intracellular transport (after Sitte). *A* Smooth endoplasmic reticulum. *AK* Coated vesicles. *D* Dictyosome of the Golgi apparatus. *E* Rough endoplasmic reticulum. *En* Endocytosis. *Ex* Exocytosis. *K* Nuclear membrane. *pL* Primary lysosomes. *sL* Secondary lysosomes. *R* Residual body. *S* Secretory granules. *Z* Cytolysosome.

Structural and functional proteins of Lysosomes (see) and Peroxysomes (see) are also synthesized on the rough ER. Lysosomes are then abstricted from cisternae of the ER or cisternae of the Golgi apparatus. In cells with a high rate of protein synthesis (e.g. plasma cells, neurons, protein-synthesizing exocrine cells), the rough ER is particularly well developed, and is recognizable even under the light microscope as a basophilic region of the cytoplasm. In neurons, it is arranged in discrete clumps, known as Nissl granules. In protein-synthesizing exocrine glands (e.g. exocrine pancreas) the cisternae of the rough ER are densely packed in a largely parallel arrangement. On account of their high biosynthetic activity, such regions were formerly known as *ergoplasm.* Studies with ^{3}H-leucine have shown that antibody molecules synthesised on the rough ER of plasma cells pass into the lumen of the ER and from there are exported from the cell after about 90 min.

During embryonal development, rough ER first appears within the cells of the embryo, followed later by *smooth ER.* Lipid and steroid biosynthesis occur in the smooth ER, which is therefore particularly well developed in steroid-synthesizing cells, e.g. the interstitial cells of the testes. In accordance with the varied functions of hepatocytes, both the smooth and rough ER of these cells are well developed. In the smooth hepatic ER, lipids are synthesized for the synthesis of plasma lipoproteins, glycogen is degraded, and the phosphate residue is removed from the resulting glucose 6-phosphate. In addition, the smooth ER of the liver contains a family of enzymes, known as cytochrome P450, which serve an important function in the detoxication (mainly by hydroxylation) of xenobiotics (foreign substances, such as pharmaceuticals and environmental pollutants) and the modification of endogenous hormones, rendering both types of compound more hydrophilic, more able to form conjugates (glucuronides, sulfates), and more readily excreted. If the body is heavily burdened with pharmaceuticals or other foreign substances, the quanitty of smooth ER in the liver cells increases, and the rough ER decreases.

The central vacuole of plant cells is formed by local vesiculation of the smooth ER, followed by abstriction and fusion of the vesicles.

The smooth ER of striated muscle fibers, known as the sarcoplasmic reticulum (SR), forms a regular envelope around the myofibrils, and it sequesters Ca^{2+} from the cytosol. It is continuous with the plasma membrane (sarcolemma), from which tubular invaginations extend into the cell interior (T-system). Ca^{2+} is pumped into the lumen of the sarcoplasmic reticulum by a Ca^{2+}-ATPase in the reticulum membrane. In the cytosol of the resting muscle fiber, the concentration of Ca^{2+} is 10^{-8} M. When the muscle is stimulated, Ca^{2+} is released from the sarcoplasmic reticulum, briefly increasing the Ca^{2+} concentration to 10^{-5} M. By causing the tropin-tropomyosin system to form cross bridges between Actin (see) and myosin, this concentration increase acts as the signal for the onset of muscle contraction.

When the cell is homogenized in the laboratory, the ER becomes fragmented. The fragments then reseal to form small vesicles known a *microsomes.* The so-called *microsomal fraction,* prepared by differential centrifugation of the cell homogenate, does not therefore contain true intracellular organelles, but it is representative of the ER.

Endorphins: peptide neurohormones that block the pain receptors of the central nervous system, in the same way as the analgesic alkaloid, morphine. Due to their morphine-like action, they are also known as *opioid peptides* or *endogenous opiates.* The name is a contraction of "endogenous morphine". They include the pentapeptide *enkephalins.* All E. contain a common sequence of 4 amino acids, and they bind to the same cell-surface receptors as morphine, i.e. they are the natural ligands of the opiate receptors [opiate receptors have been found in the brains of all vertebrates, from elasmobranch fishes to primates]. Unlike morphine, E. are rapidly degraded, so that they cannot accumulate and induce tolerance. They occur naturally in vertebrate brain and intestine, as well as other parts of the body such as the adrenal medulla, blood, etc.

Endospore: see Bacterial spore.

Endostyle, hypobranchial groove: a glandular, ciliated channel on the floor of the pharynx of tunicates,

Amphioxus and the larvae of lampreys. It secretes mucus and transports trapped particles of food material toward the digestive region of the alimentary canal (see Pharynx). During the metamorphosis of lamprey larvae, the E. becomes abstricted and forms the adult thyroid gland. In all higher vertebrates, the thyroid gland is also derived from the embryonic E.

Endosymbiont hypothesis: the hyphothesis that mitochondria and plastids evolved from prokaryotic intracellular symbionts. Accordingly, mitochondria evolved from bacterial-like symbionts, while plastids evolved from symbionts resembling cyanobacteria. A primitive "eucyte" was supposedly engulfed by phagocytosis, then integerated into the life of the cell. The hypothesis is supported by the fact that both mitochondria and plastids have two boundary membranes which differ structurally and functionally, and that both have their own circular DNA, RNA and ribosomes, and synthesize proteins independently of the rest of the cell. Moreover, the DNA and protein synthesizing systems of mitochondria and plastids closely resemble those of prokaryotes (e.g. size of ribosomes, effect of inhibitors, etc.).

The existence of numerous recent eukaryotes with intracellular prokaryotic symbionts also supports the E.h. For example, the ameba, *Pelomyxa palustris,* has no mitochondria, but contains symbiotic bacteria which perform the functions of mitochondria. The fungus, *Geosiphon pyriforme,* is photosynthetically active, due to the presence of intracellular, symbiotic cyanobacteria. Most plastid-specific proteins are encoded by nuclear DNA, suggesting that establishment of the primitive endosymbiont as a permanent organelle involved gene transfer from the endosymbiont to the cell nucleus.

Strong support for the E.h. is provided by sequence homologies between the proteins and nucleic acids of plastids and mitochondria and those of recent prokaryotes, as well as the marked differences between the plastid or mitochrondrial matrix and the rest of the cell cytoplasm.

Endotheliochorial placenta: see Placenta.

Endothelium: the inner lining of blood and lymph vessels. It consists of a monolayer of squamous epithelium. All capillaries are lined with E. Within the capillaries, the basal membrane of the E. is generally very fine and elastic, and it functions as a selectively permeable ultrafilter. Under the electron microscope it is possible to distinguish between fenestrated and nonfenestrated capillaries, i.e. those with fenestrated or nonfenestrated E.

Fenestrated capillaries have a markedly flattened E. with round pores, which are, however, covered by a membrane. They occur in the kidneys, intestinal mucosa and endocrine glands, where the fenestrations facilitate the passage of dissolved substances. *Nonfenestrated capillaries* occur widely, especially in muscles and lungs, and they possess an uninterrupted endothelial lining. Hepatic sinusoids have a thin E. with intercellular gaps.

Endothelial cells contain only a few mitochondria, only a slightly developed rough endoplasmic reticulum, and only a few cytofilaments and microtubules. On the other hand, they contain abundant micropinocytic vesicles. Transendothelial channels, formed by the fusion of vesicles, permit the passage of relatively small macromolecules. Histamine increases the permeability of the E. by causing it to dilate. See Epithelium.

Endotoxins: see Bacterial toxins.

Endozoochory: see Seed dispersal.

Endplate potential: a brief, localized membrane depolarization in the motor endplate region of a muscle fiber, which follows the release of acetylcholine in the neuromuscular junction. If a sufficient number of acetylcholine molecules makes contact with the receptors on the postsynaptic membrane, the E.p. gives rise to an Action potential (see), which is conducted through the muscle fibers and causes contraction. Substances that block the acetylcholine receptor, e.g. curare, cause muscle relaxation by preventing depolarization of the membrane.

Endproduct inhibition: see Metabolic regulation.

Energid: a single unit (nucleus plus cytoplasm) of an uncompartmentalized protoplast. For example, the plasmodia of slime fungi *(Myxomycetes)* consists of a single protoplast containing numerous nuclei. The protoplast may attain an area of several hundred square cm, with many millions of more or less synchronously dividing nuclei. Multinucleate cells are also found in siphonaceous algae and phycomycete fungi. "Internode" cells of the *Characeae,* which may attain a length of 10 cm, are also multinucleate. In flowering plants, the multinucleate condition is found in phloem fibers, latex ducts and endosperm cells. Although there are no morphlogical boundaries between nuclei, each nucleus and its surrounding cytoplasm form a unit (an E.), which is equivalent to a cell. Multinucleate cells are therefore called *polyenergids.*

Energy conduction: transfer of the energy of an excited molecule in a solid state system. The process is an essential component of energy transfer in photosynthesis. A quantum of light is absorbed in a rapid reaction lasting about 10^{-15} s, and the energy is transferred via the molecular lattice to the antenna pigments of the next reaction center. In principle, there are two mechanisms of E.c. In the simplest case, energy transfer is accompanied by a corresponding mass and charge transport. Thus, electron excitation leads to delocalized electrons in valence bands, and the process is coupled to photoconductivity. However, this mechanism of direct conduction is very improbable, because the high quantum yield of photosynthesis would only be explained by practically resistance-free electron conduction. It is therefore generally accepted that E.c. occurs by *exciton transfer* (excitons are electronically excited states in molecular crystals). In this case, the electronically excited state migrates by resonance transfer, i.e. it migrates through the molecular crystal lattice as a spin-coupled electron-hole pair. This type of E.c. proceeds without mass and charge transfer, and is analogous to the conduction of oscillatory energy in a system of coupled oscillators.

At the photochemical reaction center, the exciton energy is converted into chemical energy, i.e. an electron is released from a *primary donor* molecule and taken up by a *primary acceptor,* and within the thylakoid membrane this electron transfer is coupled to photophosphorylation.

Energy flow: the conversion, storage and distribution of energy in an ecosystem, coupled with the cycling of material in that system (energy flows though an ecological system, whereas elements cycle within the system). Most of the energy consumed by the earth's ecosystems is derived from solar radiation. A relatively small proportion (0.5–1.5%) of the sun's energy reaching the earth is absorbed by green plants (autotrophic producers), which convert it into chemical bond energy for utilization by, and circulation in, the ecosystem. For this energy to flow, appropriate gradients of entropy and temperature must be present. In the ecosystem, this is achieved by the entropy gradients within the Food chain (see Nutritional interrelationships). Each energy conversion is, however, accompanied by the production of waste heat, which is lost from the system. In this way, the high grade starting energy (of low specific entropy) becomes degraded by acquiring a higher specific entropy. Living organisms, however, are able to exploit steep entropy gradients, by abstracting some of the energy and storing it as high grade free energy, i.e. they can use the energy to drive the "entropy pump" and decrease the entropy. It is thought that food chains are relatively short, because the inefficient transfer of energy between trophic levels rapidly leads to a situation where there is insufficient energy to support another level. It is calculated, however, that most ecosystems have sufficient energy to support further trophic levels, i.e. food chains could theoretically be longer. It seems, therefore, that the length of food chains is determined by more than just E.f.

According to Mende (1979), a system must use a certain part of this free energy stored within the system to maintain the following states or processes (regimes):

1) Energy influx (energy must be expended by the system to regulate the inflow of further energy, acquisition of resources, etc.); 2) Internal organization (energy for maintaining internal processes, growth, life functions); 3) Energy efflux (energy for the removal of degradation products, homeostasis, support of energy circulation); 4) Energy export (mostly high grade energy must be exported in order to maintain the organizational level of higher order systems).

Optimal E.f. through the system is determined by the distribution of energy among these regimes. Whenever energy is transferred to the next higher trophic level, it is upgraded, but some is also lost. As energy is transferred from lower to higher niveaux in the food chain, the efficacy of energy conversion also increases, so that even higher and more efficient trophic levels become possible, which explains the great diversity of consumer species. Optimal diversity is a function of the quality and quantity of E.f. In ecosystems highly subsidized by a high quality auxillary E.f. and/or large nutrient inputs, a relatively low diversity may be optimal. Systems limited by the quality of energy input and dependent on internal nutrient cycling attain an optimal performance at relatively high levels of diversity.

An imbalance of energy distribution among the individual regimes frequently leads to excessive demands on the efficiency of energy circulation, and therefore, e.g. to an accumulation of degradation products. This congestion in the system can only be removed by regulatory intervention, i.e. to an ever increasing extent, further portions of the circulating energy are integrated into the system. This expansion of the system increases the possibility of self-induction, i.e. perturbations in the system have repercussions on the system itself, which can only be countered by evolution.

Energy metabolism: the totality of metabolic processes that provide energy for the maintenance of living processes, including heat production for maintenance of body temperature, mechanical energy for movement, as well as energy for the generation of electrical potential, for cell transport, and for the synthesis of cell and body material. In heterotrophic organisms, energy is generated mainly during Respiration (see) or Fermentation (see). In autotrophic, photosynthetic organisms, some of the energy derived from the light reaction of photosynthesis (ATP and NADPH) is used directly in the synthesis of cell material, and some is used in the assimilation of carbon dioxide to storage material (e.g. sucrose, starch), which is respired later to produce energy. Adenosine 5'-triphosphate (ATP) is one of the main products of energy-yielding (exergonic) processes, and it acts as one of the main sources of energy for energy-consuming (endogonic) processes, thereby representing an important energy currency of living organisms.

Energy migration: see Energy conduction.

Energy transfer: see Energy conduction.

Energy turnover: see Metabolic rate.

Engelhardtia: see *Juglandaceae.*

Engraulidae, *Stolephoridae,* ***anchovies:*** a family (16 genera, about 139 species) of marine (occasionally freshwater) fishes of the order *Clupeiformes.* The family has a worldwide distribution, and its members are small (most species less then 20 cm), silvery, shoaling, herring-like fishes with elongated bodies that are almost circular in cross section. Most species feed on zooplankton, but a few larger species (up to 50 cm) take other fishes. Most of the freshwater species occur in South America. The ***European anchovy*** *(Engraulis encrasicolus),* an important commercial fish, occurs off the European and West African coasts, as well as in the Black Sea, the Sea of Asov and the Mediterranean. The annual European catch is about 300 000 tonnes.

Engraulis: see *Engraulidae.*

Engraver beetles, ***bark, cone*** and ***ambrosia beetles,*** *Scolytidae:* a family of the *Coleoptera* (see) containing about 4 600 very small (rarely greater than 5 mm) cylindrical beetles, whose elytra often slope steeply downward at the rear and carry serrations on their apices. In many species the head is more or less covered by the large protonum, while in others it extends to a snout, resembling that of a weevil. Beetles and larvae live and develop beneath the bark in the bast or sapwood (bark breeders), or in wood (wood breeders), especially in diseased, damaged or felled trees. A few species are also found in shrubs and herbaceous plants. The female excavates a maternal gallery, and the larvae tunnel their own passages from it. The pattern of galleries and passages is species-specific. Some wood breeders establish fungal cultures as food for the larvae. The family includes many forestry pests, which undergo mass proliferation in certain years, e.g. the Eurasian *Ips typographus* (Engraver beetle or Spruce bark beetle).

Enhydra lutris: see Sea otter.

Enicurus: see *Turdinae.*

Enkephalins: see Endorphins.

Enocytes, ***oenocytes:*** modified cells of the abdominal epidermis of insects. Some E. remain associated with the epidermis and its pattern of segmentation, either sinking inward as hypodermal invaginations, or remaining in the peripheral tissue as enlarged epidermal cells. Frequently, however, E. dissociate from the epidermis, assume a rounded shape, and form segmented cell aggregates, held in place by tracheae. Single enocytic cells, or *enocytoids,* are found among the fat cells, and floating free in the hemolymph. The functions of E. are not fully understood. They are involved in the synthesis of the water-repellant wax layer (epicuticle) of the cuticle and egg shell, as well as the wax-gland secretion of bees. E. periodically undergo a change of shape, which is synchronized with molting. In insects with incomplete metamorphosis, a new generation of E. arises after each molt. It is also thought that E. have a homeostatic function in the hemolymph, and a role in intermediary metabolism. In aphids, which lack Malpighian tubules, the E. store waste material.

Enocytoids: see Enocytes.

Enolase: a lyase and an enzyme of glycolysis, which catalyses the reversible dehydration of 2-phosphoglyceric acid to form the high energy compound, phospho*enol*pyruvate. It is a dimeric metalloenzyme, requiring 2 magnesium ions per dimer for maintenance of its structure, or 4 magnesium ions for activity. It is an inhibited by fluoride. The M_r of the dimer varies between 80 000 and 100 000, depending on its origin.

Enopla: see *Nemertini.*

Enoplida: see *Nematoda.*

Entamoeba dysenteriae: see Amebas.

Entamoeba hystolytica: see Amebas.

Entelodontidae: see *Suidae.*

Enterobacteria, *Enterobacteriaceae:* anaerobic or facultatively anaerobic, peritrichously flagellated or nonmotile, Gram-negative bacteria with respiratory or fermentative metabolism. Many E. are intestinal inhabitants, the most familiar representative being *Escherichia coli* (see). Important disease-causing E. of humans and animals are *Salmonella* (see), *Shigella* (see) and *Yersinia pestis* (see). Members of the genus *Erwinia* (see) are plant pathogens. Nonpathogenic E. are, e.g. *Serratia* (see) and *Proteus* (see).

Enterobiasis: a disease of humans, especially children, throughout the world. It is caused by infection with the small, whitish nematode, *Enterobius (= Oxyuris) vermicularis* (order *Oxyurida;* see *Nematoda),* known as the Pinworm, Seatworm or Threadworm. The male is only 2–5 mm in length. Females (about 1 cm in length) emerge from the host's anus at night and lay their eggs on the skin in the immediate neighborhood. This produces intense irritation, inducing the host to scratch, resulting in the transfer of eggs to the fingers. Re-infection occurs by ingestion of the eggs from the fingers. Alternatively the eggs, which remain viable for several weeks, may be blown about in dust and eventually become ingested by a new host. The eggs hatch and develop in the lower part of the small intestine and the large intestine. Unlike many other parasitic nematodes, *Enterobius* remains in the intestine and does not migrate into other tissues or organs.

Enterobius: see Enterobiasis; *Nematoda.*

Enterocrinin: see Gastrointestinal hormones.

Enterogastrone: see Gastrointestinal hormones.

Enteroglucagon: see Gastrointestinal hormones.

Enteroponeusta: see *Hemichordata.*

Enterotoxins: see Bacterial toxins.

Entodiniomorpha: see *Ciliophora.*

Entognatha, *Entotropha:* a superorder of insects, whose eversible mouthparts are contained within a small pouch or pocket. The superorder is contained within the *Apterygota,* and it comprises the *Collembola, Protura* and *Diplura.*

Entomogamous flower: see Pollination.

Entomogamy: see Pollination.

Entomology: the study of insects, a branch of zoology that has long been an independent discipline. There are two main divisions of E., general and applied E. *General E.* embraces ontogenic development, mode of existence, body structure, organ function, phylogenetic development, taxonomy, behavior, ecology and distribution. *Applied E.* is concerned with insects that are directly or indirectly useful or deleterious in agriculture, forestry and food storage, and in human and veterinary medicine. One of the main tasks of applied E. is the monitoring of habitats for the early appearance of insect pests, especially the beginnings of mass outbreaks, accompanied by prognosis of the likely severity of the outbreak, and the provision of advanced warning. It also includes the development of specific methods for counteracting economically deleterious insects (chemical, biological and integrated pest control), as well as research on useful organisms, such as pest predators and parasites, and on insect diseases (insect pathology). Applied E. is is also part of plant pathology.

Entomophage: see Modes of nutrition.

Entomophilous flower: see Pollination.

Entomophily: see Pollination.

Entomophthora: see *Phycomycetes.*

Entomophthorales: see *Phycomycetes.*

Entoprocta, *Kamptozoa:* a phylum of the *Protostomia.* On the basis of their external similarity to bryozoans, they were formerly classified in one phylum with the latter, but later work has shown that the two groups are not related.

Morphology. Entoprocts consist of a bell-shaped calyx and a slender stalk. The calyx contains all the internal organs, and its free end carries a circlet of 8–30 tentacles, which surround the mouth, anus and gonopore. The stalk (containing only longitudinal muscles and mesenchyme) is attached to the substratum by a glandular disk or a network of roots. There is single pair of protonephridia, whose blind ends are flame bulbs as in the *Turbellaria.* A few species are hermaphrodite, but most are dioecious. Tentacles and body wall sensory cells are innervated by a ganglion. There is also a nerve plexus, which displays particular development at the bases of the calyx and stalk. Gonads are simple, bladderlike and paired. The mesenchyme-filled body cavity is considered by some authorities to be a pseudocoel. There is a U-shaped alimentary canal, expanded in the middle to form a stomach. The largest species is *Pedicellaria cernua,* which can attain a length of about 5 mm.

Biology. Only a few species are solitary, and most are colonial. They are found in coastal waters, attached to other aquatic animals or plants. They are filter feeders on detritus, microorganisms, or small aquatic animals, which are swirled into the mouth by the tentacles. Development of the fertilized egg starts by spiral cleavage in the funnel-shaped brood chamber; eventually a free-swimming larva is produced, which resembles a polychaete trochophore larva. All E. can also reproduce asexually by budding.

Collection and preservation. At least a binocular lens is needed for the detection and recognition of E. Specimens are anesthetized with 0.5% cocaine, fixed in Bouin's fluid, and stored in 80% ethanol.

Entoprocta. Side view of an entoproct.

Classification. About 100 aquatic species are known, distributed in three families (there is no category higher than the family): *Loxosomatidae* (solitary, marine, epizoic species; the stalk and calyx are not clearly separated); *Pedicellinidae* (calyx and stalk separated by a groove; colonial, marine, and often epizoic); *Urnatellidae* (calyx and stalk separated by a groove; colonial, freshwater; sometimes combined with the *Pedicellinidae* on account of close similarity of larvae).

Entropy, ***S:*** the degree of disorder or chaos of a system, or the sum of all transformations that occur in order to bring the system into its present state. It is the unavailable energy of a system, i.e. the energy that cannot perform useful work. According to the 3rd law of thermodynamics, a perfect crystal of any compound at absolute zero will possess zero entropy. Thermodynamically, changes in the entropy of systems are usually considered, rather than absolute entropy values. The converse of entropy is Information (see), which is the degree of nonrandom order of a system. Entropy cannot be released from a system, and in a closed system it strives to attain its maximum. In an open system, the entropy change is always greater than 0, but it strives to attain a minimum.

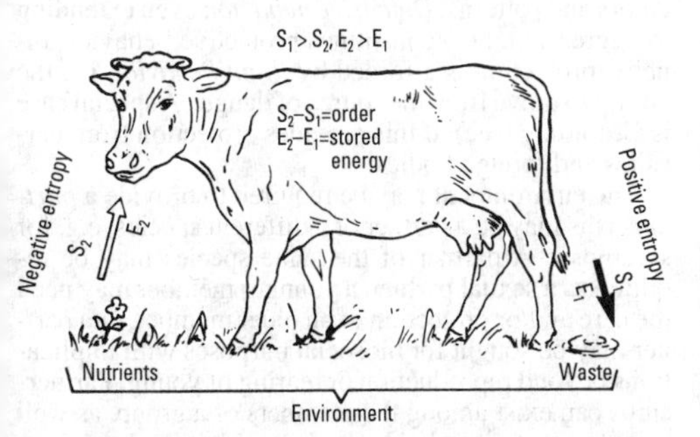

Entropy. The living organism as an entropy pump. Pictorial representation of the Schrödinger-Brillouin model.

The entropy relationships of an individual organism can be appreciated with the aid of the Schrödinger-Brillouin model (Fig.): the organism feeds on negative entropy, i.e. the entropy of the material lost from the organism is greater than the entropy of the dietary material; the organism therefore functions as an "entropy pump". Difficulties are encountered, however, in applying the concept of entropy to complex ecosystems. This is because the increasing entropy on the level of system-building elements (see Diversity) may be linked to a high degree of association and interdependence of those elements (high information content). Without doubt, a field of potatoes has a higher degree of order than the natural woodland that would develop from that field by natural Succession (see), if all human intervention ceased. The stability and the degree of association of the elements (i.e. the information content) of the system is, however, greater in the final stages of succession, but so is the entropy of the distribution of the elements of the ecosystem.

Environment:

1) That part of the external surroundings which influences an organism directly or indirectly. The *informational E.* acts on the organism via its sense organs or exterior receptors. The *noniformational E.* influences the life of an organism, but not via sensory receptors. According to the sources of environmental signals or information and the action field of the associated response, it is possible to differentiate between three types of E.: 1) the *interior E.* of the body itself; 2) the surrounding *ecosystem* or biogeocoenosis; 3) the living *population* of which the organism is a part.

2) The totality of internal and external factors and conditions that influence the realization and phenotypic expression of genetic information. The phenotype of an organism is the product of the genotype and E. The genotype determines the reaction norm, whereas the E. determines which part of the reaction norm is realized. The extent of environmental influence on the realization of specific characters varies considerably, i.e. some characters are environmentally stable, while others are environmentally labile.

Environmental capacity: the maximal number of individuals of a population that can be carried by a specified environment. It is determined essentially by the quantity and availability of limiting resources.

Environmental cue: see Biorhythm.

Environmental demands: see Environmental resistance.

Environmental factors, ***ecological factors:*** environmental quantities that act upon individuals, populations or societies. *Stimulatory factors* excite a response, such as direct flight toward a light source (phototaxis) or the search for a preferred temperature within a temperature gradient. *Modifying factors* actually alter the Life form (see) of the organism (see Climatic rules, Ecomorphology). *Limiting factors* place limits on the existence of a life form (e.g. reef corals can only survive in waters that do not fall below 22 °C). The reaction of an organism to an environmental factor is either all-or-nothing, or its response is graded according to the intensity of the factor (see Ecological valency).

Extensive factors (e.g. temperature) are nonutilizable and nonexploitable. *Intensive factors* (e.g. food) are utilizable and exploitable, and may become, e.g. a factor of competition between organisms.

Abiotic factors are all the nonliving environmental quantities, such as climatic factors (light, temperature, atmospheric humidity), as well as the inorganic components of food and the substratum, e.g. edaphic factors in the soil. Light, as the energy source for autotrophic primary producers, is the most critical abiotic factor for the existence of life on the planet. Animals are also primarily influenced by hydrothermal conditions. Water and air movement, water and air pressure, osmotic pressure, hydrogen ion concentration, and the chemical composition of the substratum are other important abiotic factors.

Biotic factors include all environmental quantities arising from the activities of living organisms, and the responses to them (psychological, sexual, etc.), as well as interactions such as Competition (see) and Interference (see).

Substratum factors and *trophic factors (nutritional factors)* can be both biotic and abiotic.

If an environmental factor is not influenced by a change in population density, it is *density-independent* (e.g. climatic factors). Conversely, changes in the intensity of a density-independent factor causes changes in population density. In contrast, *density-dependent factors* (e.g. competition, interference) have the opposite effect, and prevent or even out changes in population density.

Environmental relationships, ***ecological relationships:*** relationships of organisms, populations and societies to their environment. Such relationships may be trophic, chemical, mechanical, energetic, psychological, etc. Each of these separate components is part of the totality of E.r. According to the direction of influence, a relationship is a nonreciprocal *relation* ($A \rightarrow B$), or a reciprocal *correlation* ($A \rightleftharpoons B$). A correlation that is positive in both directions is a symbiosis; negative in both directions it is a system of competition. If one direction is positive and the other negative, the relationship is that of consumer/consumed. For examples of *interrelation* ($B \rightleftharpoons C \rightleftharpoons A$) see Relationships between living organisms.

Environmental resistance: the totality of physical and biological environmental forces that place constraints on the survival rate of a species and its population size, i.e. negative or positive forces that increase the mortality rate and/or decrease the reproduction rate. Population growth is accompanied by an increase in the total demands made by the species upon the environment, so that a population creates its own resistance to further growth. This is expressed in the equation $R_C = k_C N^2$, where R_C is the rate at which the population growth is checked, N is the number of individuals in the population, and k_C is a constant. The environmental demands made by an organism may be related to its body, its ecosystem, or its population system

Every living organism requires *space* in which to exist, so that the limitation of available space is a source of E.r. Demands on space are made at the following three levels. 1) Body volume requires local space, as expressed, e.g. in the diameter of the subterranean mouse burrow. 2) Freely motile organisms require more space than that occupied by their body volume. For example, the mouse requires the volume of its burrow plus pathways and foraging areas outside. 3) Members of the same species may establish territory, the boundaries of which limit their own available space and that of others.

E.r. also has a *temporal* dimension. Thus, the body may show changes of behavior with time, e.g. young animals and mature animals make different demands on the environment. At a different level, the demand for matter and energy may occur at different times within the ecosystem (e.g. diurnal, nocturnal and crepescular animals). At a third level, interactions between members of the same species depend on the time of year, day, etc. The requirements of space and time constitute a complex *space-time system* for each organism.

An organism or population also requires *information* from its environment. At the organismal level, this ensures body control, the elementary orientation of body functions in space and time, and the maintenance of basic life functions. At the ecological level, it enables the organism to find and fit into its ecological niche, and to interact appropriately with other organisms within the biogeocoenosis. At the population level, information enables the organism to integrate with its own species. The information is demanded from other members of the species, and it becomes the basis of biocommunication.

The most general E.r. comes from the universal requirements of all living organisms for energy and matter, i.e. their *metabolic* demands on the environment. At the organismal level, food is required for the continued physical existence of the body. At the ecological level, the appropriate food must be available, while at the population level the exploitation of food resources must be optimized. Numerous ecologically and environmentally relevant patterns of behavior and physiological functions are associated with these requirements, e.g. respiratory gas exchange, thermoregulation, exchange of matter and energy with the environment (e.g. food cycles), regulation of muscle metabolism and other physiological functions by yawning and stretching, dormancy and sleep.

Organisms also require *protection,* which is provided by a variety of behavioral mechanisms, depending on the species. At the level of the individual organism, the body is protected from immediate harm or damage. Ecologically, protection ensures freedom from influences that disturb normal behavior. At the population level, protection extends to the *collective defensive behavior* of animal groups. Protective behavior takes many forms. *Concealment* may consist of actively seeking a hiding place, or the exploitation of camouflage body colors and patterns. *Defensive behavior,* even extending to aggression, also constitutes protective behavior. Finally, protection is afforded by *flight behavior,* i.e. the ability to move from the source of danger. A special case is Grooming (see), if this provides protection from parasites and foreign bodies.

The environment may be required to provide a *partner.* This may be a partner of a different species, e.g. for symbiosis. A partner of the same species may be required as a sexual partner, a younger member may need the care and/or protection of an older member, or a partner may be sought for biosocial purposes with implications beyond reproduction or rearing of young. Partnerships can exist among the members of a group, as well as between two individuals. In some animal communities partnership is sought with a specific individual or group.

All the above factors may be interpreted as environmental demands, i.e. the demands that an organism must make on the environment, in order to guarantee its own survival and the continued existence of the population. Inasmuch as the organism makes a demand on the environment, the object of that demand can in turn become a factor of E.r.

Enzyme Commission Nomenclature: see Enzymes.

Enzyme induction: stimulation of the synthesis of enzymes by inducers. In bacteria, many catabolic enzymes are only synthesized when the appropriate substrate is available. The classical example is the induction of the enzymes of the lac operon in *Escherichia coli* by their substrate, lactose (Fig.) (the true inducer is allolactose, derived in small quantities by the rearrangement of lactose). The regulator gene R codes for a specific repressor protein which, in the absence of an inducer, binds to the corresponding operator O and blocks transcription of the structural genes *z, y* and *a* (negative control). Since the mRNA for these genes is not produced, neither are the proteins. If lactose (or an artificial inducer) is added to the medium, the inducer binds to the repressor and inhibits its binding to the operator. Now transcription of the structural genes is possible, and the enzymes are synthesized.

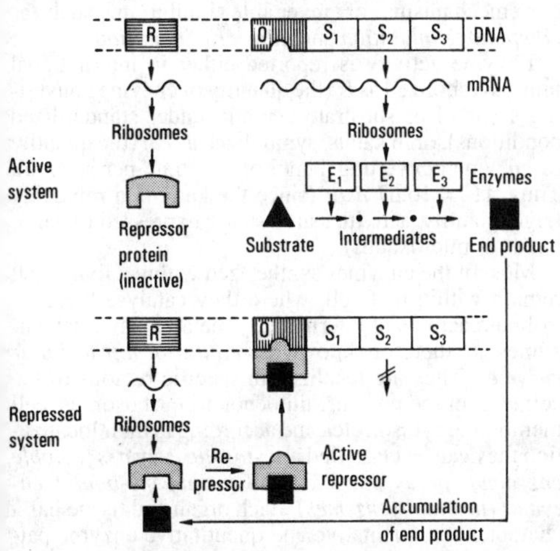

Enzyme induction. Induction of the lac operon in *Escherichia coli.* The top circle represents the chromosome map with the lactose and other regions.

Enzyme polymorphism: the coexistence of enzymes in a population, which catalyse the same reaction but display different electrophoretic mobilities, and are encoded by the same gene locus. In 1966, Lewontin and Hubby discovered that enzyme variants, which are separable by gel electrophoresis, are inherited according to Mendelian laws, i.e. they segregate like different alleles of a single gene. Using other genetic methods, it is only possible to detect the existence in a population of genes with two or more alleles, whereas electrophoretic methods also detect monomorphic genes (see Polymorphism). E.p. enables the determination of the proportion of polymorphic gene loci in a population, as well as the relative proportion of heterozygous genes per individual.

In natural populations of invertebrates, 30–50% of all enzymes are polymorphic. The E.p. of vertebrates is usually less. Under the most favorable circumstances, 85–90% of polymorphic differences between enzymes are detected by electrophoresis. The real extent of E.p. is therefore greater than indicated by protein electrophoresis.

The functional significance of E.p. is a subject of debate (see Non-Darwinian evolution).

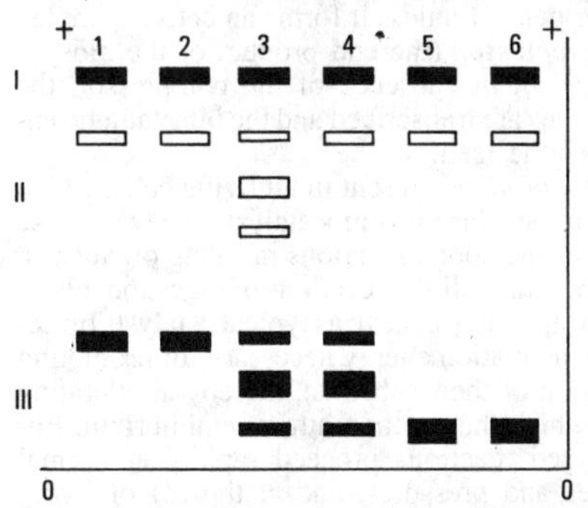

Enzyme polymorphism. Electropherogram of glutamate-oxaloacetate transaminase, which is encoded by genes I, II and III. Enzyme II is heterozygous in animal 3. Enzyme III is heterozygous in animals 3 and 4. The middle bands of the heterozygous enzymes are the mixed proteins from the gene products of the two alleles at that gene locus.

Enzyme-product-complex: see Enzymes.

Enzyme repression: a type of metabolic regulation found in prokaryotes, in which the synthesis of anabolic enzymes is blocked by the end product of the biosynthetic pathway to which they contribute. For example, if an amino acid is present in the growth medium, synthe-

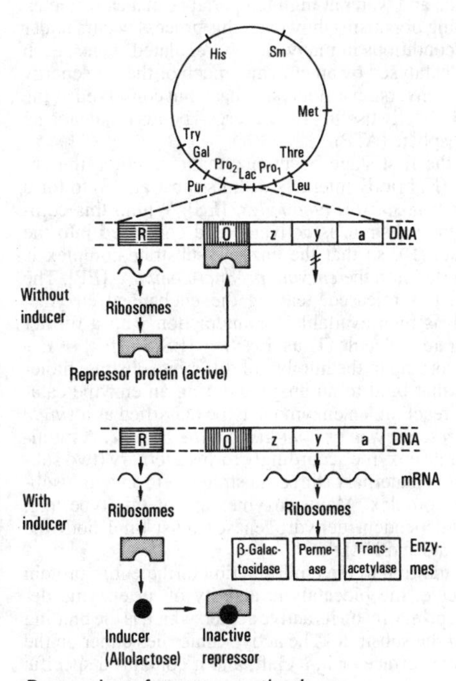

Repression of enzyme synthesis

sis of all the enzymes in the relevant operon is prevented, but if the amino acid is in short supply the operon is derepressed and enzyme synthesis is maximal.

The mechanism of E. is shown in the Fig. Regulator gene R codes for an *aporepressor* which cannot bind to the operator gene O unless it forms an active complex with the *corepressor* (the end product of the biosynthetic chain). In the absence of the corepressor, the structural genes are transcribed and the biosynthetic enzymes are synthesized.

Enzymes: proteins present in all living cells, which act as high molecular mass biocatalysts. They catalyse the totality of metabolic reactions in living organisms, as well as the extracellular reactions of digestion, blood coagulation, etc. They behave as typical catalysts by decreasing the activation energy necessary for promoting the conversion of their substrate, thereby accelerating the rate at which the reaction attains equilibrium. Enzyme-catalysed reactions proceed rapidly at normal temperatures and pressures, and at the pH of living cells, i.e. under relatively mild conditions. Moreover, only one reaction is catalysed, so that there are no side products. Many or most enzyme-catalysed reactions are difficult to replicate in the laboratory in the absence of the specific enzyme. Replication of such reactions requires extreme conditions, often giving a low yield, and numerous side products. By converting a substrate in several discrete enzyme-catalysed stages, the cell and the organism excercise control over the quantity of substrate converted, and the rate and intracellular site of the conversion. If the conversion is exergonic, the cell is then able to control the release of energy and trap it in a chemically useful form, as well as exploiting intermediates for other purposes such as biosynthesis. For example, glucose may be totally and rapidly burned to carbon dioxide and water at high temperature in a calorimeter. In living organisms, however, this process occurs under mild conditions in many closely regulated stages, each stage catalysed by an enzyme; much of the free energy of the conversion is not lost as heat, but conserved as the metabolically useful "high energy" bonds of adenosine triphosphate (ATP).

In the first stage of an enzymatic reaction, the enzyme (E) briefly interacts with its substrate (S) to form an *enzyme-substrate-complex,* [ES]. Within this complex the substrate is activated and converted into the product (P), so that the enzyme-substrate-complex is converted into the *enzyme-product-complex,* [EP]. The product is released, leaving the unchanged enzyme, which is then available for interaction with a further substrate molecule. Thus, $E + S \rightleftharpoons [ES] \rightleftharpoons [EP] \rightleftharpoons E + P$. According to the number of different substrate molecules that bind to an enzyme during an enzyme-catalysed reaction, an enzyme may be classified as a *single substrate enzyme* or a *multisubstrate enzyme.* A multisubstrate enzyme accordingly forms a ternary (two substrates), quaternary (three substrates), etc. enzyme-substrate complex. Many enzymes are of this type, e.g. NAD-dependent dehydrogenases must bind both the substrate and NAD^+.

In addition to the conformation of the entire protein molecule, the biocatalytic activity of an enzyme depends primarily on its active center, which is the binding site of the substrate. The active center lies either on the enzyme surface or in a cleft, and it contains a specific amino acid sequence. Some important components of active centers are the imidazole side chain of histidine, the SH-group of cysteine, and the OH-group of serine. The flexible spatial configuration of the active center enables it to interact with, bind and activate the chemical conversion of the substrate molecule. The active centers of many enzymes also contain a nonprotein group (see below), which is actively involved in binding the substrate and especially in its catalytic conversion.

The rate of an enzyme-catalysed reaction depends on the concentration and activity of the enzyme, the substrate concentration, temperature (the optimum is usually about 40 °C), pH-value, and the presence of effectors. The latter may be activators (e.g. certain metal ions), or inhibitors (competitive, noncompetitve or allosteric inhibitors). All the enzymatic reactions of metabolism are subject to strict systematic control, e.g. by the spatial separation of the enzymes involved in a chain of sequential reactions *(compartmentalization),* by increased enzyme degradation *(enzyme conversion),* by *enzyme induction* (e.g. by hormones), and by *enzyme repression* (decreased enzyme synthesis), as well as various mechanisms of reversible inhibition, such as *competitive inhibition* and *allosteric inhibition.*

Enzyme activity is reported either in international units, symbolized as *U* (the quantity of enzyme converting 1 mmol of substrate per min under standardized conditions), or in katals, symbolized as *kat* (the quantity of enzyme converting 1 mol of substrate per second). Thus, $1U = 16.67$ *nkat* (since the katal is a relatively large quantity, activities are usually expressed in nanokatals or microkatals).

Most of the enzymes synthesized within a living cell remain within that cell, where they catalyse the controlled reactions of intermediary metabolism. Such enzymes are therefore known as *cellular* or *intracellular enzymes.* They are localized in specific regions of the cell, e.g. in the nucleus, mitochondria, ribosomes, cell membrane, cytosol, etc., and according to their localization they can be classified as *cytosolic enzymes (soluble enzymes),* or as *particle- and membrane-bound enzymes (insoluble enzymes).* Each organ and tissue has a characteristic qualitative and quantitative enzyme pattern.

Unlike cellular enzymes, *secretory enzymes* are synthesized in an inactive form in the cells of specific secretory organs, then released into the blood (e.g. enzymes of blood coagulation), or into the digestive tract, where they are converted into an active form; activation of blood coagulation enzymes is, of course, very strictly regulated in response to tissue injury.

More than 2 000 enzymes have been isolated from animals, plants and microorganisms. The chemical reaction catalysed by each of these enzymes is known. In most cases, the primary amino acid sequence of the enzyme protein has also been elucidated. The complete primary, secondary and tertiary structures of several enzymes are also known, and the number of enzymes in this category is constantly increasing. Relative molecular masses of enzymes range between 13 000 and several millions. Many enzymes consist of a high molecular mass protein component *(apoenzyme)* and a low molecular mass nonprotein component *(active group).* When combined, these two components form the holoenzyme.

If the active group is bound firmly to the enzyme protein, remains bound during the enzymatic reaction, and s unchaged at the completion of the reaction, it is :nown as a *prosthetic group* (e.g. the heme group of an ›xidase). If it is chemically altered after the reaction, it s known as a *cosubstrate* or *coenzyme*. After taking part n the first reaction, the altered coenzyme becomes the ubstrate of a second enzyme, which converts it back to ts original state (e.g. NAD^+ is converted into NADH + I^+ as the coenzyme of a dehydrogenase; NADH + H^+ hen serves as the coenzyme of a reductase and is con- 'erted to NAD^+). The terms, coenzyme and prosthetic group are often used synonymously.

Coenzymes are relatively low molecular mass or- ;anic compounds, often chemically related to Vitamins see). The following coenzymes or prosthetic groups ıre responsible for hydrogen and electron transport: ıicotinamide-adenine-dinucleotide, nicotinamide-ade- ıine-dinucleotide phosphate, flavin mononucleotide, 'lavin-adenine-dinucleotide, ubiquinone, heme, lipoic ıcid. The following coenzymes or prosthetic groups ;erve as carriers in the enzyme-catalysed transfer of 'unctional groups (e.g. phosphate, methyl, amino and ıcetyl groups, etc.): adenosine triphosphate, pyridoxal ›hosphate, guanidine triphosphate, cytidine triphos- ›hate, uridine triphosphate, uridine triphosphate, tetra- ıydrofolic acid, biotin, thiamine pyrophosphate, coen- zyme A.

All enzymes are highly specific, i.e. they catalyse ›nly one type of biochemical reaction *(action specificity)*, and they convert only one specific compound or group of compounds *(substrate specificity)*. *Group specificity* is encountered when an enzyme attacks a particular structure or bond in different substrates, e.g. glycosidases attack glycosidic bonds in a variety of oligo- and polysaccharides. With the exception of epimerases, enzymes also display *configurational* or *optical specificity*, i.e. if the substrate exhibits chirality (usually due to the presence of an asymmetrically substituted carbon atom), the enzyme attacks only the D- or L-form.

With the exception of some old trivial names that are still in use (e.g. trypsin, chymotrypsin, emulsin), the names of enzymes end in -ase. The first syllable indicates the nature of the catalysed reaction, or the nature of the substrate, e.g. proteases or proteinases catalyse the cleavage of proteins, while amylase catalyses the cleavage of starch.

The modern classification of enzymes is based on their action specificity, and all enzymes are named and numbered according to a system established by the International Commission on Enzyme Nomenclature (an official commission of the International Union of Biochemistry); thus, each enzyme is listed as the abbreviation EC ("enzyme commission") followed by four numbers, indicating the main class or division of the enzyme, its subclass, sub-subclass, and its serial number in the sub-subclass. There are six main classes or divi-

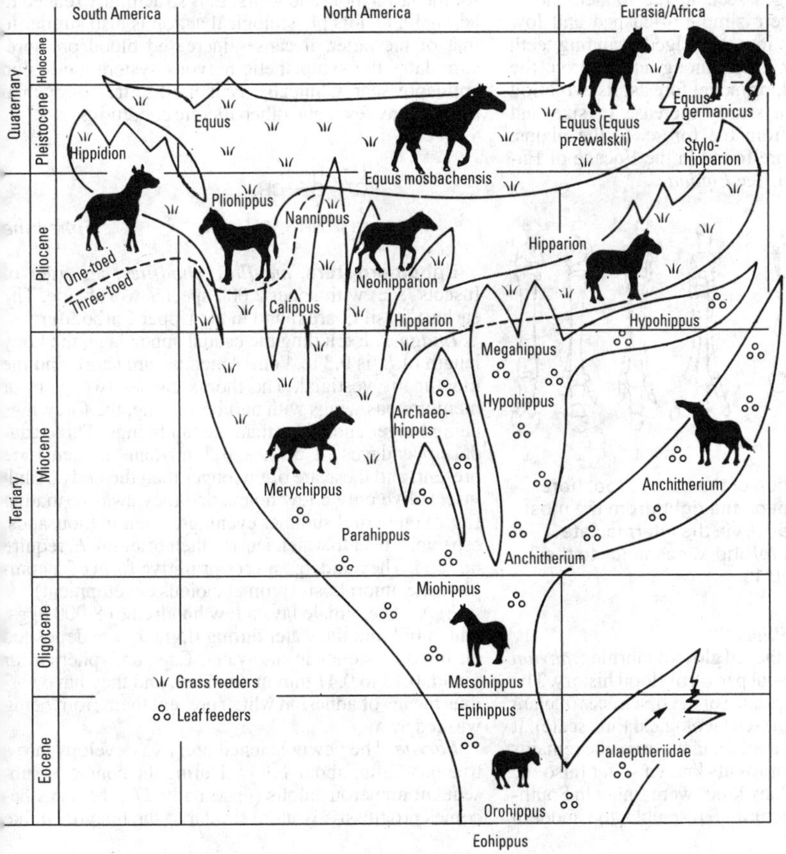

Fig.1. *Eohippus.* Evolutionary tree of the horse (after Simpson and Thenius).

sions: 1) oxidoreductases, 2) transferases, 3) hydrolases, 4) lyases, 5) isomerases, and 6) ligases. In addition to this strict and systematic classification, enzymes may also be classified on the basis of other general properties, e.g. as acidic, neutral, or basic according to their physical properties, or as metallo- or flavo-enzymes according to their prosthetic groups.

Some enzymes are prepared on an industrial scale from producer microorganisms, in particular for use in the food and pharmaceutical industries. Certain microorganisms, with a high enzyme content and therefore high metabolic activity, are exploited industrially for their ability to synthesize, interconvert and degrade certain materials; in particular, they are used to perform specific syntheses and partial syntheses on a commercial scale. Several enzymes are also used as analytical tools in assays performed in clinical chemical laboratories for diagnosis and prognosis. A few enzymes are also used directly as therapeutic agents.

Enzyme-substrate-complex: see Enzymes.

Eocene (Greek *eos* red glow of morning): the second oldest stage of the Tertiary. See Geological time scale.

Eohippus, *Hyracotherium*, ***primitive horse:*** a primitive equine genus, now extinct. These cat to fox-sized ungulates characteristically had 4 functional toes on the forefeet and 3 on the hindfeet, each toe bearing a small hoof. The fossil record of the subsequent evolution of *Eohippus* (Fig.1) shows one toe on each foot becoming larger, while the others shorten and are finally reduced to the splint-like vestiges seen in the modern horse (Fig.2). Cheek teeth were simple, 4-cusped and low crowned, later evolving into the ridged, grinding teeth of the horse. *Eohippus* lived among, and fed on, the undergrowth of tropical, primeval forests. Its evolution was accompanied by a steady increase in size and speed, and migration from the forests to the plains. Their fossilized bones are found in the Eocene of Europe and North America. See *Equidae.*

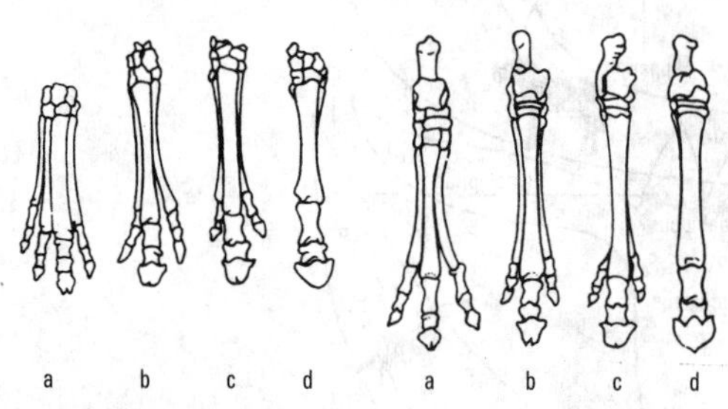

Fig.2. *Eohippus.* Evolution of the horse's foot (forefoot on the left, hindfoot on the right) from the most ancient form *(a, Eohippus)*, via the intermediate forms of the Oligocene *(b)* and Miocene *(c)*, to the modern horse *(d)* (all x 0.15).

Eophona: see *Fringillidae.*

Eophytic (Greek *eos* the red glow of morning, *phyton* plant): the oldest geological period of floral history. The E. extends from the appearance of the first traces of plant life to the Lower Silurian (see Geological time scale). It contains very few plant fossils, and these are the remains of algae. The oldest plant fossils known so far (also the oldest organic fossils of any kind) were found in Southwest Africa. They structurally resemble the modern *Cyanophyceae* (blue-green bacteria), and have been assigned an age of 3.35 billion (3.35 x 10^9) years. Other findings (questionable blue-green bacteria and green algae) from the American lower Middle Precambrian are thought to be 1.3–1.4 billion years old. *Corycium enigmaticum* from the upper Middle Precambrian of Finland is 1.7–1.8 billion years old. The end of the E. probably marks the appearance of the first terrestrial plants.

Eozapus: see *Zapodidae.*

Eozoic: obsolete term for the period of geological history in which the first animal life appeared. It corresponds roughly to the Upper Precambrian. See Geological time scale.

Eozostrodontidae: see Triassic.

Epectal bones: see Wormian bones.

Epedaphic organisms: organisms that live on the soil surface. The term is applied to soil fauna which spend the day concealed in cavities in the upper soil, or under stones (cryptozoa), and emerge at night to feed.

Ependymal cells: see Nervous tissue.

Ephedra: see *Gnetatae.*

Ephedrine, ***1-phenyl-2-methylaminopropane-1-ol,*** an alkaloid from species of *Ephedra* [largest genus of the family *Gnetaceae (Gymnospermae)].* It is also prepared synthetically. Species of *Ephedra* occur in temperate and subtropical regions of Asia, America and the Mediterranean, although most American species are devoid of alkaloids. E. was first isolated in 1887 from the herb "Ma Huang", a species of *Ephedra* used medicinally in China for the last 5 thousand years. E. is structurally related to adrenalin and its physiological action is also similar to that of the latter. It causes increased blood pressure, stimulates the sympathetic nervous system, and is an antidepressant. Clinically, E. is used to treat bronchial asthma, hay fever and other allergic conditions.

$$C_6H_5-CH(OH)-CH(CH_3)-NH-CH_3$$

Ephedrine

Ephemeroptera, ***mayflies, shadflies:*** an order of Insects (see) with about 2 000 species worldwide. The earliest fossil *E.* are found in the Upper Carboniferous.

Imagoes. Excluding the caudal appendage, the body length of *E.* is 0.3 to 4 cm. Antennae are short, and the mouthparts vestigial. The thorax carries two pairs of membranous wings with netlike veining, the forewings being longer and wider than the hindwings. Three caudal appendages (2 cerci and 1 median filament) are present, and these are often longer than the body. Adult insects live only a few hours, and they swarm (mating flight) on humid summer evenings, often in thousands, especially near to water. During their brief life *E.* require no food. They undergo a very primitive form of incomplete metamorphosis (prometabolous development).

Eggs. The female lays a few hundred to 8 000 eggs, which fall into the water during flight, or are deposited on moss or stones in the water. Eggs are spherical or ovoid, 0.15 to 0.41 mm in diameter, and they have various means of adhesion which prevent them from being washed away.

Larvae. The newly hatched prelarva develops into a true larva after about 4 days. During the course of subsequent numerous molts (maximally 27), the larva becomes progressively more similar to the imago. Unlike

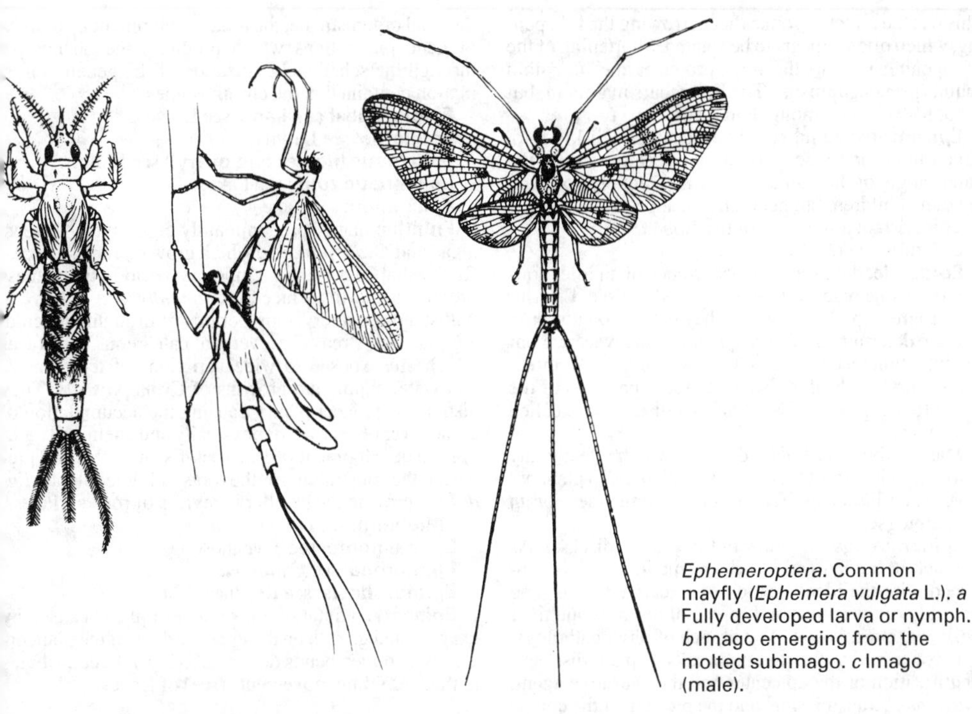

Ephemeroptera. Common mayfly *(Ephemera vulgata* L.). *a* Fully developed larva or nymph. *b* Imago emerging from the molted subimago. *c* Imago (male).

he larvae of the *Plecoptera,* the larvae of *E.* possess (like their imagoes) 3 caudal appendages; the feet have only a single claw. Antennae, compound eyes and ocelli differ only slightly from those of the imago, but in contrast to the imago the larvae possess well developed biting mouthparts. The larvae feed on plant and animal material, and they respire via tracheal gills on the abdomen. The sexually mature imago is preceded by a sexually immature flying stage (subimago). Complete development lasts one to three years.

Economic significance. Dead *E.* serve as angling bait and as food for aquarium fish and cage birds. When they occur in exceptionally large numbers, they are also used as fertilizer and pig feed. The larvae are a natural food source of freshwater fish, and their presence is an indication of the absence of pollution.

Classification. The order is at present separated into 19 families on the basis of variations in the patterns of the wing veins. A comprehensive treatment of the taxonomy of the *E.* worldwide is lacking, but detailed lists exist for separate regions. A specialized, non-Latinized nomenclature is used by anglers (e.g. Claret spinner = imago of *Leptophlebia vespertina;* Blue-winged pale water dun = subimago of *Centroptilum pennulatum),* but common names otherwise hardly exist in English.

[G.F. Edmunds and R.K. Allen, A Checklist of the Ephemeroptera of North America north of Mexico, *Ann. Ent. Soc. Am.*, ***50*** (1957), 317–324.]

Ephestia kuehniella: see Mediterranean flour moth.

Ephippium: a protective covering of the "winter" eggs of many water fleas *(Cladocera).* The E. consists of a preformed, markedly thickened, saddle-shaped part of the cuticle of the carapace. The fertilized winter eggs in the brood pouch remain inside this thickened part of the cuticle when it is shed.

Ephyra: the larva of the medusa of many scyphozoans, formed during asexual reproduction by transverse cleavage of the polypoid generation (see Strobilation). After release from the strobila, the E. swims with the aid of 8 long lappets formed by deep incision of the bell margin. These lappets also serve to catch the small crustaceans that form the diet. New lobes (velar lobes) later arise between the peripheral lobes, and these fuse with the lappets to form a closed medusal umbrella. Development to the sexually mature adult medusa may take a few days or more than a year, according to the species.

Epibiosis: the coexistence of organisms, in which one partner exists on the surface of the other. The converse is Endobiosis (see).

Epicanthic fold, *mongoloid eye:* in humans, a wide fold of the upper eyelid, which extends downward, covering the edge of the eyelid and partly covering the eyelashes. In the inner angle of the eye, this covering fold fuses with the nasal skin to form a crescent-shaped arch.

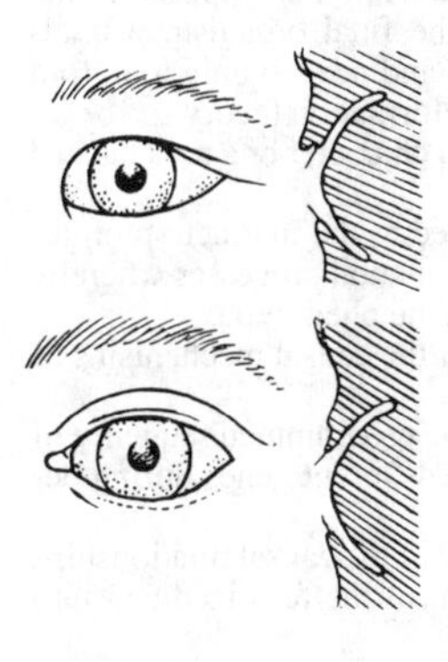

Epicanthic fold. Human eye with epicanthic fold (above), compared with European eye (below). Left: front view. Right: section through the middle of the eye.

This has the effect of somewhat narrowing the lid opening, which often appears to be slanted. Shortening of the lid opening next to the nose produces the mogoloid squint (pseudostrabism). The E.f. occurs mainly in, but is not restricted to, mongoloid races.

Epicanthus: an inherited vertical, curved skin fold, which more or less covers the inner (more rarely the outer) angle of the human eye. It is relatively common in young children, but generally disappears with age.

Epicaridea: a suborder of the *Isopoda* (see).

Epichile: see *Orchidaceae.*

Epidemic: the excessive occurrence of an infectious disease in one place during one period of time. Es. usually progress in phases, and if they depend on a vector, they are determined by the frquency of that vector. Contact infection results in a slow growth of the E. If infection occurs via drinking water or food, progress of the E. is often explosive. Es. of animal diseases are called epizootics.

Mass outbreaks of plant diseases which affect many individuals in a short time are also called Es. In plant pathology, an E. is classified as an Esodemic (see) or an Exodemic (see).

Epidemiology: the study of transmitted diseases. As a branch of human and veterinary medicine, E. is concerned with the identification of infective agents, and the study of the course and spread of diseases and their means of transmission. As a branch of phytopathology, E. is concerned with mass outbreaks of plant diseases, identification of the epicenter and the causative agent, its means of transmission, and the plotting of the course and spread of the disease.

The ultimate purpose of E. is to aid the active prevention and control of transmitted diseases.

Epidermis: see see Boundary tissue, Epithelium, Skin.

Epidermis: see Boundary tissue, Epithelium, Skin.

Epigaea: see *Ericaceae.*

Epigaiic organisms: the same as Edaphic erganisms (see). The term is also sometimes applied to species of the atmobios.

Epigastric reflex: see Protective reflexes.

Epigenesis:

1) The theory, developed by C.F. Wolff (1734–1794), that the development of an embryo from an egg and the subsequent development of an embryo into an individual organism consist of a series of successive gradual changes, starting in the amorphous zygote and proceeding through the formation of new structures and progressive differentiation. Thus, development is a process of a gradual increase in complexity, as opposed to the preformationist view that the final organism was already preformed in the egg, and needed only to unfold and increase in size. See Preformation theory.

2) A change in the mineral character of a rock caused by external influences.

Epigenetic: a term applied to the interaction of genetic factors and the developmental processes whereby the genotype is expressed in the phenotype.

Epigenetics: the study of the causal mechanisms of development.

Epigenetic sequence: the programmed sequence of activation and repression of different genes during development.

Epigenotype: the totality of the causal relationships involved in the development of the fertilized egg into the final organism, i.e. the total developmental comple: of gene interactions which produces the phenotype through the scheduled realization of the genetic infor mation contained in the chromosomes.

Epiglottideal cartilage: see Larynx.

Epiglottis: see Larynx.

Epigynous (flower and ovary): see Flower.

Epilimnetic zone: see Lake.

Epilimnion: see Lake.

Epiliths: plants, predominantly bryophytes, lichens algae and *Cyanophyceae,* which grow on rock surfaces Rocks that experience temporary periods of wetnes provide a growth surface for *Cyanophyceae,* which ca withstand the intervening periods of drought and cold e.g. the "ink streaks" on vertical calcareous rock face which are exposed to wide variations of temperatur and water supply are growths of *Cyanophyceae.* They also act as pioneer plants, causing the accumulation o a fine layer of "earth", or physically and chemically aid ing the development of fine rock fissures, thereby pre paring the substratum for the roots of higher plants. *Endoliths* grow in the capillary crevices of rock surfaces.

Epilobium: see *Onagraceae.*

Epimastigote: see Trypanosomes.

Epimorpha: see *Chilopoda.*

Epimorphosis: see Regeneration.

Epinasty: a nutation movement of plants caused by exaggerated growth on the upper surface of the plant organ, which then bends downward (Fig.). E. contribute to the Unfolding movements (see) of leaves.

Epinasty. 1 Shoot tip of *Parthenocissus* sp. (Virginia creeper). *2* Bud of *Papaver rhoeas* (field poppy).

Epinephelus: see Sea basses.

Epineuston: see Neuston

Epipactis: see *Orchidaceae.*

Epiphragma: a porous calcareous [mostly $Ca_3(PO_4)_2$] operculum secreted by the rim of the mantle in many pulmonate snails. It serves to seal the shell during winter dormancy, protecting the animal against cold and greatly reducing water evaporation, while permitting a low rate of exchange of respiratory gases.

Epiphyll: a plant that grows on a leaf surface. Es. are mostly fungi, algae, lichens and bryophytes. Some Es.

ıay be parasitic, and certain fungal Es. may be mycor-hizal symbionts, but often there is no exchange of ma-erial between the E. and the leaf.

Epiphyllum: see *Cactaceae.*

Epiphysis, ***pineal body, pineal gland, corpus pine-le:*** an organ that arises from the roof of the diencepha-on (thalamencephalon). In the *Petromyzonidae* (lam-reys) and *Anura* (tailless amphibians), it exhibits the haracters of a primitive eye. In mammals it is known as he pineal gland, but it is absent from the sloths. It syn-hesizes melatonin, and with increasing age it accumu-ates a calcareous concretion (acervulus).

Epiphytes, ***aerophytes, air plants:*** plants which are ot rooted in the earth and which usually grow some istance above ground level on trees, or on jutting rocks petrophytes). The host tree serves only as a physical upport, and to provide an elevated and therefore better lluminated position. Epiphytic species are found mong higher plants (cormo-E. or vascular E.) and mong the lower plants or thallophytes (thallo-E.). Growth of E. depends on a constantly high atmospheric umidity. They are therefore especially abundant in ainforests in the tropics and subtropics. In temperate egions, the only E. are usually mosses, lichens and al-gae, i.e. plants which can withstand temporary desicca-ion. Various adaptations are associated with the epi-hytic mode of existence. The seeds of E. germinate di-ectly on the growth support, carried there by animals or vind currents. In some cases, vegetative reproduction lso occurs. The problem of water availability has led to he development of many xeromorphic characters. Thus, numerous epiphytic orchids possess water-stor-ng stem tubers.

Many adaptations exist for trapping water and dis-olved nutrients. The hanging aerial roots of epiphytic rchids and some epiphytic members of the *Araceae* are overed in spongy tissue (velamen), consisting of dead ells, which soaks up rain water. Other species produce dense network of aerial roots, which traps moisture nd humus. In some bromeliads, e.g. *Tillandsia* spp., vater is absorbed by dead scales which clothe the eaves. *Tillandsia usneoides* is Spanish moss or old nan's beard, which festoons trees and telephone wires hroughout the tropical and subtropical Americas. In heir arrangement and shape, leaves may also display daptations for trapping water and humus. Epiphytic romeliads are usually "cistern" plants, i.e. the stem is losely invested by leaf bases, and the leaves form ro-settes, which collect water (Fig.). A similar arrangement s found in the tropical bird's nest ferns, e.g. *Asplenium idus.* In addition to assimilatory leaves (sporotropho-hylls), some epiphytic ferns, especially the tropical stag's horn ferns *(Platycerium),* possess specially mod-fied humus-retaining "nest leaves", which fold round he supporting tree trunk, leaving an opening above for he entry of water and nutrients.

Species of *Myrmecodia* (family *Rubiaceae)* and *Dis-chidia rafflesiana* (family *Asclepiadaceae)* live symbi-otically with ants. *Myrmecodia* species have large hy-pocotyl tubers with interor chambers inhabited by ants. The roots of *Dischidia* grow into the interior of urn-shaped leaves, also inhabited by ants, which bring in earth and humus.

Roots of E. usually press closely against the support-ng tree and serve for attachment. In some cases, as in the epiphytic bromeliads, they are purely attachment organs

Epiphytes. 1 Stag's horn fern *(Platycerium). 2 Bilbergia* (family *Bromeliaceae);* the stem is invested by leaf bases to form a cistern. *3* Orchid *(Aërides)* with aerial roots.

and are incapable of absorbing water and nutrients. In some epiphytic species, the roots are totally regressed.

Some E. have hanging, rope-like aerial roots, which eventually penetrate the soil below and provide support, as well as taking up nutrients (semi- or hemi-E.); these include *Ficus* species (e.g. the rubber tree, *Ficus elasticus)* and many tropical members of the *Araceae.*

Episite: see Relationships between living organisms.

Episitism: see Relationships between living organisms.

Epistasis: interference with, or prevention of, the expression of a gene by another nonallelic gene (the *epistatic gene)* in the same genome. The gene whose expression is affected is called the *hypostatic gene.* The phenomenon of E. displays a characteristically modified Mendelian segregation. See Hypostasis.

Epithecium: a thin, usually brightly colored layer covering the fruiting bodies of the fungal subclass *Discomycetidae* (class *Ascomycetes).* It is formed from Paraphyses (see) (the ends of sterile hyphae). See Apothecium.

Epitheliochorial placenta: see Placenta.

Epithelium, ***epithelial tissue:*** a sheetlike tissue that clothes the external and internal surfaces of the animal body. Epithelial cells are closely apposed to one another as in a mosaic, and intercellular material is almost totally absent. The tissue is strengthened by the presence of special contact zones between the cytoplasmic membranes of neigboring cells. In vertebrate E. there are 3 important types of cell contact: 1) *zonulae occludentes* or *tight junctions,* 2) *zonulae adhaerens* or *adhesion belt,* and 3) *maculae adhaerentes* or *desmosomes.* The basal surface of epithelial cells is connected to the basal lamina by *hemidesmosomes,* which are chemically distinct from, but morphologically similar to, desmosomes.

Tight junctions consist of an anastomosing reticulum of strands, which encircle the apical end of each epithelial cell. Under the electron microscope these appear as a series of point connections between the outer surfaces of the two adjacent plasma membranes. Tight junctions make an important contribution to the nonpermeability properties of the E. They serve as a seal between neighboring cells, preventing the leakage of water and dissolved compounds, as well as serving as barriers to the diffusion of membrane proteins between the apical and basolateral regions of the cytoplasmic membrane.

Several types of cells possess *cell-cell adherens junctions,* i.e. small attachment sites that connect actin filaments in the cortical cytoplasm of adjacent cells. In epithelial tissue these form a continuous adhesion belt or zonula adhaerens around each interacting cell. The adhesion belts of adjacent cells are directly apposed to one another, near the apex of the cell and below the tight junction. Interaction is mediated by calcium-dependent transmembrane linker glycoproteins, which belong to a family of calcium-dependent cell-cell adhesion molecules called *cadherins.* Adhesion belts are sometimes referred to as *belt desmosomes,* but they are chemically quite distinct from other desmosomes.

Desmosomes are small, dense, cytoplasmic plaques, containing transmembrane linker glycoproteins, as well as intracellular attachment proteins. The latter are responsible for attachment to the cytoskeleton. Extracellular domains of the linker glycoproteins interact to form an attachment between adjacent cytoplasmic membranes. The importance of this glycoprotein-mediated linkage is exemplified by the potentially fatal skin disease known as *pemphigus,* in which the body synthesizes antibodies to its own desmosomal linker glycoproteins; the demosomes in the epithelial tissues of the skin are disrupted, so that body fluids leak into the epidermis. Since this only occurs in the epidermis, the desmosomes of other epithelial tissues must be antigenically different.

Three following types of E. are recognized.

1) *Surface E.* or *protective E.* This is the basic type of E., which covers the body surface and serves as the lining of body cavities. *Simple epithelia* consist of only a single layer of cells. In *simple columnar epithelia,* the cells are taller than they are wide, resembling a palisade fence in vertical section, and appearing as sheet of polygons when viewed from above. *Simple cuboïdal epithelia* are similar to the simple columnar type, but the cells are about as long as they are wide. In a *simple squamous* or *pavement E.,* the cells are flattened, sometimes so flattened that they appear in vertical section under the low-power microscope as lines with a hump containing the nucleus.

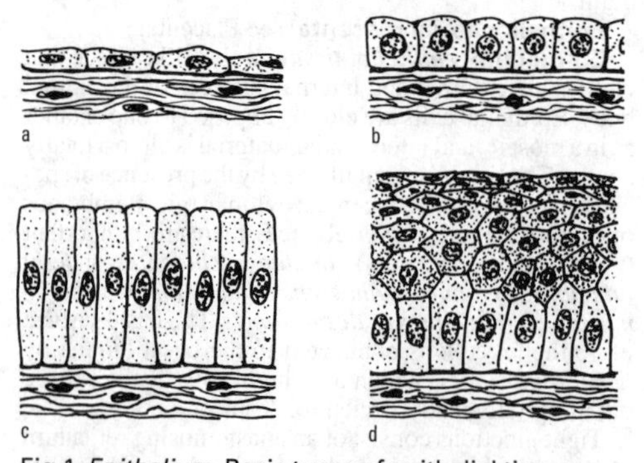

Fig.1. *Epithelium.* Basic types of epithelial tissue. *a* Simple squamous or pavement epithelium. *b* Simple cuboidal epithelium. *c* Simple columnar epithelium. *d* Stratified or compound epithelium.

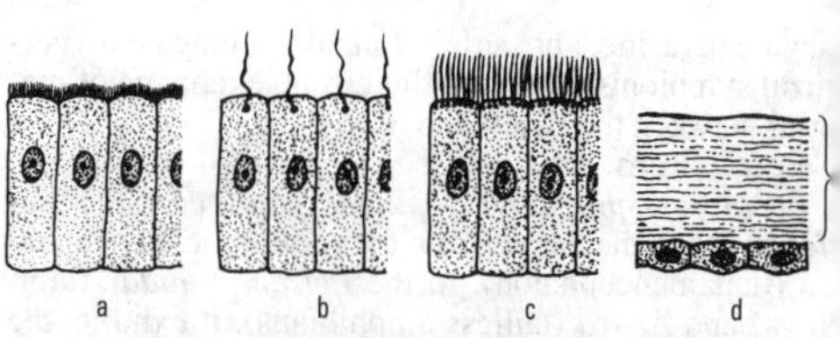

Fig.2. *Epithelium.* Epithelia with special surface structures. *a* Epithelium of brush border cells. *b* Flagellated epithelium. *c* Ciliated epithelium. *d* Epithelium with cuticle.

Stratified or *compound epithelia,* consist of several layers of cells. Of particular importance is *stratified squamous E.,* which may contain many layers of cells the cells of the superficial layers being very flattened like those of a simple squamous E. These superficial cells are continually sloughed off and replaced from below, and the wastage is compensated by cell division in the cuboidal cells of the deepest layers. This deep zone of cell division is known as the *Malpighian* or *germinative layer.*

Surface epithelia lie upon a supporting layer of connective tissue, and are separated from the latter by a specialized mat of extracellular matrix known as the *basement membrane* or *basal lamina.* The exterior or apical surface of an E. may carry various structures. For example, Cilia (see) are often present, appearing most commonly in columnar epithelia, e.g. as the linings of respiratory passages and oviducts. Ciliated epithelia are particularly widespread among invertebrates. Epithelial cells in the *brush border* of the mammalian small intestine possesses, on their apical surfaces, several thousand fine, densely packed processes, called Microvilli (see) which increase the absorptive area. In many invertebrates, the E. of the exterior body surface secretes a layer of organic material, known as the Cuticle (see), which in many species serves as an exoskeleton. In arthropods the cuticle consists of chitin, and this may be strengthened by calcareous inclusions. In all invertebrates, the protective E. of the body surface, the *epidermis,* consists of a monolayer of cuboidal or columnar E. Only vertebrates possess a multilayered epidermis. In most animals, the alimentary canal is lined with a simple columnar E. The walls of coelomic cavities, which are derived from the mesoderm, consist of simple squamous E.

2) *Glandular E.* consists of specialized epithelial cells, which have a secretory function. Glandular epithelia form the inner lining of the distal parts of Glands (see).

3) *Sensory E.* is a specialized type of animal epithelial tissue formed from Sensory cells (see).

Epithem: see Excretory tissues of plants.

Epithemal hydathode: see Excretory tissues of plants.

Epitokous: capable of reproduction, actively reproducing, or possessing offspring. The converse is Atokous (see).

Epitoky: The occurrence of anatomically and morphologically aberrant forms during the sexually active phase of certain animals. E. may be initiated by unfavorable environmental conditions, e.g. in springtails, in which several sexually active stages are separated by inactive periods (see Ecomorphology); or it may occur regularly as in many polychaetes, e.g. it is especially well known in nereids, syllids and eunicids. Thus, the reproductive individual (epitoke) of *Nereis* becomes adapted for a pelagic existence and leaves the bottom mud. It develops large eyes and reduced prostomial palps and tentacles. The anterior 15–20 trunk segments are slightly modified, but the remaining epitokal region is packed with gametes, its segments are elongated, and the parapodia of these segments contain long spatulate swimming setae.

Epithalamus: see Brain.

Epitony, *epitrophy:* promotion of the growth on the upper side of dorsiventral plant organs, e.g. as in the branching of many deciduous trees.

Epitope: the surface structure of an antigen, against which the specificity of the antibody is directed. See Antibodies, Paratope.

Epitrophy: see Epitony.

Epizoic: living on or attached to the body of an animal, not necessarily as a parasite.

Epizoochory: see Seed dispersal.

Epizoon: an animal living on the exterior of another.

Epoecia: see Relationships between living organisms.

Eponychium: see Claws, hoofs and nails.

Epoophoron (Greek *epi* upon, *oon* egg, *pherein* to bear), ***parovarium, Rosenmüller's organ:*** a rudimentary organ of mammals, representing the remains of the Wolffian body of the embryo. It lies in the mesosalpinx between the ovary and the Fallopian tube. In the adult human it consists of a number of small rudimentary blind tubules. One of these tubules, the duct of Gärtner, is a persistent part of the Wolffian duct and it represents the canal of the epididymis in the male.

EPSP: acronym of Excitatory postsynaptic potential. See Synaptic potential.

Epstein-Barr virus: see Tumor viruses.

Equidae, *horses, zebras* and ***asses:*** a family of odd-toed ungulates (see *Perissodactyla).* These long-legged mammals walk only on the tip of the middle toe, which is encased in a broad hoof (see Unguligrade). They are fast, high-endurance runners, inhabiting dry grassy plains and steppes. They are herbivorous, but differ from ruminants in having a relatively small, single-chambered stomach. Members include the Wild horse (see), Asiatic wild asses (see), Wild ass (see) and Zebras (see).

The domestic horse, *Equus przewalskii caballus,* is largely derived from the Wild horse, *Equus przewalskii,* while the domestic donkey or ass, *Equus asinus asinus,* is the domesticated form of the Wild ass, *Equus asinus.*

The ears of horses are short in comparison with those of asses. Horses and asses also differ with respect to their tails, which are entirely covered with hair in the horses, but only haired distally in the asses. Horses (but not asses) also bear characteristic wartlike structures on their hindlegs. A further point of difference is the voice, described as a neigh in the horse, and as a bray in the ass.

Horses and asses can hybridize to produce nonfertile offspring, which are generally and widely used as draft animals and beasts of burden. The offspring of a male horse and a female ass (or jenny) is known as a ***hinny,*** while that of a female horse and a male ass (jackass) is called a ***mule.***

Fossil record. Fossil remains from the Eocene to the present reveal a complete phylogenetic series. The dog-sized *Eohippus* (see) of the Eocene evolved via intermediate forms to the large, long-legged equines of the present. Fossils are found, e.g. in the brown coal measures of the Middle Eocene in the Geisel valley near Halle, Germany.

Equilenin, *3-hydroxyestra-1,3,5(10),6,8-pentaen-17-one:* an estrogen, which occurs with equilin in the urine of pregnant mares. Its total chemical synthesis, in 1939, was the first total synthesis of a natural steroid. It has 1/25 of the biological activity of estrone.

Equilibrium density: stage in the development of a population when factors promoting population growth are balanced by those that limit population growth. Under the conditions obtaining at the time, the E.d. represents a long term, stable state of average Abundance (see) of the population.

Equilibrium reflexes: see Labyrinthine reflexes.

Equilin, *3-hydroxyestra-1,3,5(10),7-tetraen-17-one:* an estrogen, which occurs with equilenin in the urine of pregnant mares. It has 1/10 of the biological activity of estrone.

Equisetales: see *Equisetatae.*

Equisetatae, *Articulatae, Sphenopsida,* ***horsetails:*** a class of the *Pteridophyta* (see). The few living representatives are herbaceous plants, but fossil treelike forms are known. The shoot is hollow and sharply demarcated into nodes and internodes, At each node there is a whorl of scale leaves, united at the base and forming a sheath around the stem. Cone-like structures at the ends of the stems consist of many alternating whorls of sporophylls separated by short internodes. The sporophylls are peltate, with the shape of a one-legged table with 5–10 sac-like sporangia on the underside. The gametophyte, which always develops outside the spore, is a monoecious or dioecious, green prothallus with branched, dorsiventral, curling lobes.

1. Order: *Sphenopsidales, Sphenophyllales.* Paleozoic fossil forms with whorls of forked or wedge-shaped, united leaves with many dichotomous veins. They were herbaceous, possessing long internodes and displaying only minimal branching. Vascular bundles were triarch with secondary thickening, and the tracheids had both reticulate and bordered pits.

Fig.1. *Equisetatae. Sphenophyllum cuneifolium,* showing ears or "cones" of sporophylls and two different leaf forms (heterophylly) (reconstruction).

2. Order: *Equisetales.* A plant group already known from the Carboniferous. The characteristic central medullary cavity was surrounded by a ring of collateral vascular bundles. The order is now represented by only a single living genus *(Equisetum)* with 32 widely distributed species. Some species are deciduous and form tuberous rhizomes. In many species, there is a marked dimorphism of sterile and fertile shoots. Most familiar is the widely distributed agrarian weed, *Equisetum arvense* ***(Common horsetail).*** On account of the heavy impregnation of the outer wall of the stem epidermis with silica, it was formerly used for polishing pewter. It inhabits fields, dune-slacks, waste ground, etc., from Greenland to southern Spain, being found also in Central China, Virginia, Alabama and California. The poisonous *Equisetum palustre* ***(Marsh horsetail)*** is found in bogs, marshes, wet meadows, etc. thoughout Europe into temperate Asia, and in North America from Newfoundland and Alaska to Connecticut and Oregon.

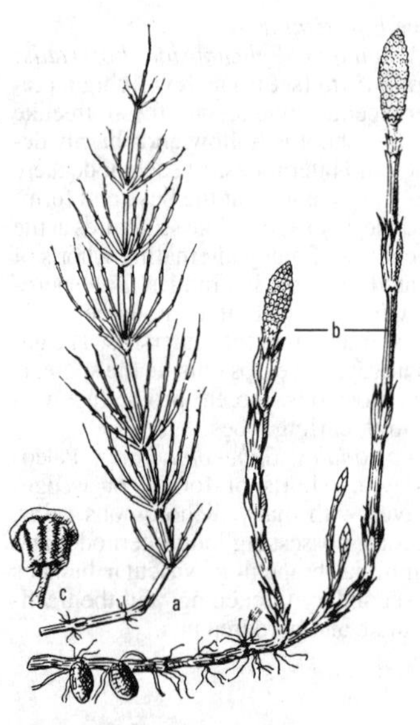

Fig.2. *Equisetatae. Equisetum arvense* (common horsetail). *a* Sterile vegetative shoot. *b* Fertile shoots arising from a rhizome, which has tubers. *c* Sporophyll with dehisced sporangia.

Recent horsetails are similar to the fossil *Equisetites,* which were present in the Upper Carboniferous and sometimes very abundant in the Mesozoic. *Equisetites arenaceus,* from the Bundsandstein and the Keuper, attained a height of 6–10 m, and its shoot was the thickness of a man's arm. The important genus, *Calamites,* which is common in the Carboniferous, had tree-like (arborescent) representatives with secondary thickening.

Equisetites: see *Equisetatae.*

Equisetum: see *Equisetatae.*

Equitability: see Diversity index.

Equus: a genus of the family *Equidae* (see), which includes horses, asses and zebras.

ER: acronym of Endoplasmic reticulum (see).

Eragrostis: see *Graminae.*

Erannis: see Winter moths.

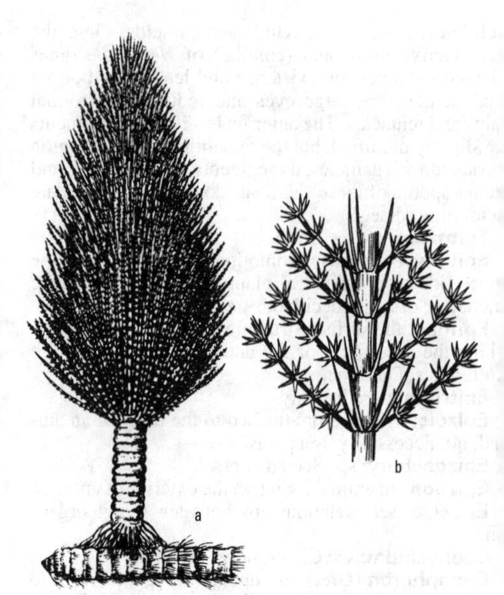

Fig.3. *Equisetatae. a* Reconstruction of a *Eucalamites. b* Stem of *Annularia* showing typical calamite leaf form.

Eremias, ***racerunners:*** a genus of the family *Lacer tidae* (see). Many species, 15–25 cm in length, inhabi the deserts and steppes of North Africa, and especiall Central Asia. They have long tails and pointed heads and they are usually spotted or longitudinally striped The dorsal scales are fine and granular. *E. arguta* range from Central Asia to the southern Ukraine, and is foun as far west as Roumania. *E. velox* ranges from Centra Asia to the north coast of the Caspian Sea.

Eremomela: see *Sylviinae.*

Eremophila: see *Alaudidae.*

Eremopteryx: see *Alaudidae.*

Erethizon: see *Erethizontidae.*

Eretmochelys: see *Chelonidae.*

ERG: acronym of electroretinogram. See Electroret inography.

Ergasilus: see *Copepoda.*

Ergates: see Long-horned beetles.

Ergenia, ***spiny-tailed skinks:*** a genus of powerfull built, 20–40 cm-long Australian skinks (see *Scincidae*) Mostly red-brown in color, they have a spiny tail an keeled scales. They inhabit dry, stony regions, wher they conceal themselves in rock fissures, with the tai turned toward any would-be agressor.

Ergochromes, ***secalonic acids:*** a group of weakl acidic, bright yellow, poisonous natural pigments, iso lated from a number of molds and lichens. Chemically they are dimeric 5-hydroxychromanone derivatives.

Ergostane: see Steroids.

Ergosterol, ***provitamin*** D_2, ***ergosta-5,7,22-trien 3β-ol:*** the most abundant mycosterol (fungal sterol) The main commercial source of E. is yeast. It is con verted by UV irradiation into vitamin D_2 (see Vitamins cholecalciferol). E. is an important component of cell c mycelium membranes, and it is also synthesized b some protozoa (notably *Trypanosomatidae* and soi mebas), *Chlorella,* and the primitive tracheophyte, *Ly-copodium complanatum.* It is not synthesized by all ungi, and is, for example, absent from *Pythium* and *Phytophthera,* which do not synthesize any sterols.

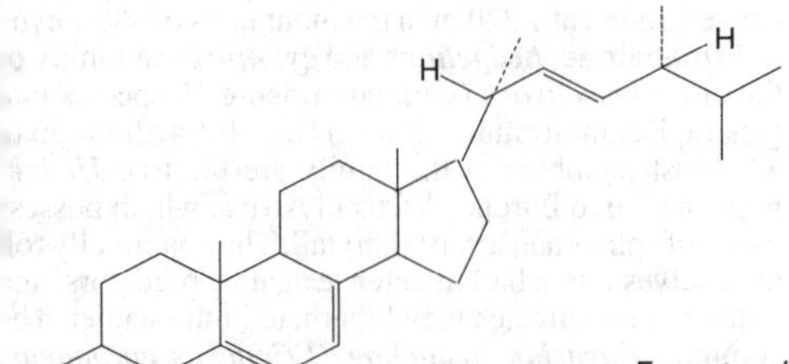

Ergosterol

Ergot: see *Ascomycetes,* Ergot alkaloids.

Ergot alkaloids, ***Claviceps alkaloids, ergoline alkaloids:*** a group of more than 30 indole alkaloids, possessing the ergoline ring system. They are produced by various species of the fungal genus, *Claviceps* (class *Ascomycetes),* which parasitize rye and wild grasses. Following infection by spores of *Claviceps purpurea,* ears of rye develop 1–3 cm long, dark violet (almost black), highly poisonous sclerotia, containing up to 1% E.a. The dug, *ergot,* consists of these poisonous sclerotia. Lysergic acid (Fig.), a typical component of the E.a., occurs in combined form, i.e. with various peptides or amines linked to its carboxyl group. Some of the main derivatives of lysergic acid are *ergotamine, ergocryptine* and *ergocornine.* Another group of E.a. (clavine alkaloids) is derived from 6,8-dimethylergoline, e.g. *agroclavine* and elymoclavine.

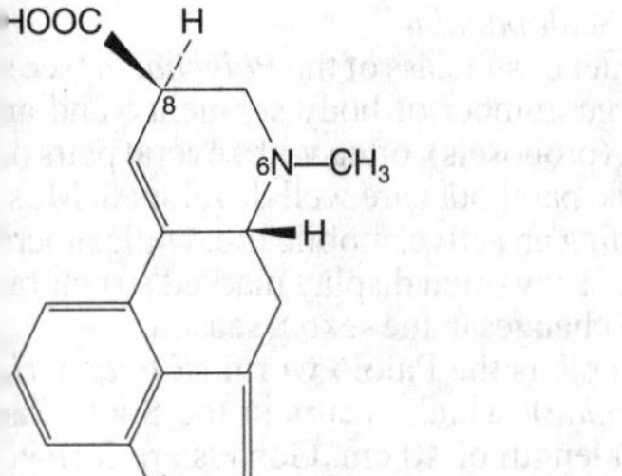

D-Lysergic acid

The lysergic acid derivatives, but not the clavine alkaloids, are physiologically active, exhibiting a wide variety of useful pharmacological properties. Extracts of ergot were used earlier, but owing to their variable alkaloid content, they have now been totally replaced by pure alkaloids (especially ergotamine), or semisynthetic analogs. E.a. stimulate contraction of smooth muscle, especially of the uterus and of arterioles in the peripheral parts of the body; they are used in gynecology for the control of hemorrhage after childbirth. Owing to their wide spectrum of pharmacological activity, they are also used in combination preparations, e.g. with tropane alkaloids for suppression of the autonomic nervous system.

The first description of the drug, ergot, with directions for its use, dates from 1582. In the 19th century, the drug was introduced into many pharmacopeas. For many centruies, it was not realized that ergot was poisonous, and in Europe contamination of rye flour with ergot led to periodic widespread outbreaks of poisoning, known as "St. Anthony's fire", now called ergotism. Toxic symptoms are convulsions, permanent mental damage, limb gangrene, often followed by death.

Ericaceae, ***heath family:*** a family of dicotyledonous plants, containing about 2 500 species in 80 genera. The family is widely distributed, usually on acidic soils, and it contributes important components of the sub-Arctic tundra, the tree line in montane areas, Atlantic dwarf shrub heaths, raised bogs, coniferous forests with abundant raw humus, the Mediterranean maquis and the heaths of the African Cape. They are mostly small, woody (often suffruticose) plants or shrubs, rarely trees, usually with evergreen, simple, often needle- or scale-like, xeromorphic leaves. Their roots harbor symbiotic fungi (see Mycorrhiza), which enable them to colonize soils that are poor in minerals.

Flowers are usually hermaphrodite *(Epigaea* is dioecious) and usually actinomorphic *(Rhododendron* is zygomorphic). The calyx usually consists of 4 or 5 united sepals, the corolla of 4 or 5 united petals. The 8 or 10 stamens (5 in *Loiseleuria)* are obdiplostemonous, and their 2-locular anthers dehisce via terminal pores (longitudinal slits in *Loiseleuria* and *Epigaea).* The ovary is usually superior, sometimes inferior, developing after fertilization (pollination is by insects) into a capsule, drupe or berry. Members of the *Ericaceae* contain various characteristic polyphenols.

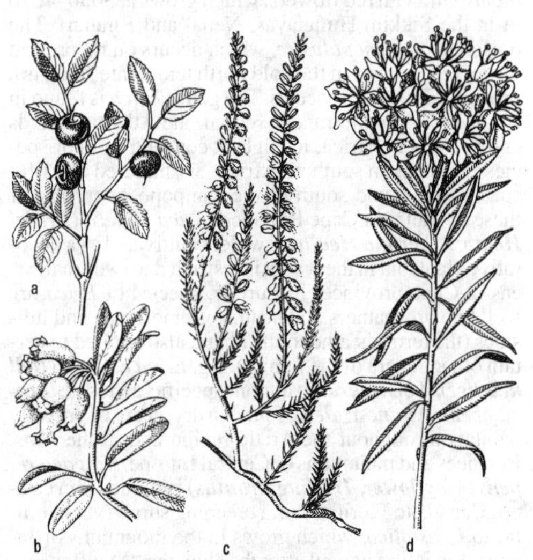

Ericaceae. a Vaccinium myrtillus (bilberry), fruiting branch. *b Arctostaphylos uva-ursi* (bearberry), flowering branch. *c Calluna vulgaris* (ling heather), flowering plant. *d Ledum palustre,* flowering plant.

Economic and ornamental plants include *Vaccinium (= Oxycoccus) myrtillus* ***(Bilberry, Blaeberry, Whortleberry, Huckleberry),*** *V. vitis-idaea* ***(Cowberry, Red whortleberry),*** *V. palustris* ***(Cranberry)*** and *V. macrocarpum* ***(Large cranberry),*** which bear edible berries. Species of the Mediterranean and North American genus, *Arbutus* ***(Strawberry tree),*** are also eaten. Leaves of

Arctostaphylos (= Arctous) uva-ursi ***(Bearberry)*** possess a relatively high content of the phenol heteroside, arbutin, and they are used medicinally against kidney and bladder infections. *Arctostaphylos alpina* ***(Alpine*** or ***Black Bearberry),*** whose edible, ripe, black berries are favored by bears, is widely distributed in the Arctic and in northern temperate alpine regions; its procumbent woody stems are covered with peeling bark entangled with dead vegetation, forming mats up to 90 cm in diameter. The genus *Rhododendron* ***(Rhododendron;*** trees or bushes with large, leathery leaves) is represented by almost 600 species, concentrated mainly in the Himalayas, China and Japan; about 250 species occur in Malaysia, Indonesia and New Guinea, and a few occur in the northern temperate zone in North America and Europe. Cultivated varieties of ***Azalea*** (derived from *Rhododendron simsii,* native to East Asia) are especially popular as decorative plants, and they differ from "true" rhododendrons by the fact that they lose their leaves each year, but this is not a strict botanical division. Species of *Rhododendron,* even from different continents, hybridize very readily in cultivation, resulting in the many hybrids and cultivated varieties that are now planted for decorative purposes. However, the genus can be divided into two groups whose members never hybridize with each other, i.e. species with scales on their leaves, stems and flower parts *(lepidote* species, e.g. *R. fastigiatum,* with pale to intense purple black flowers, which grows at 3350–3960 m in Yunnan) and species lacking these scales *(elepidote* species, e.g. *R. thomsonii*, with red flowers, which grows at 3050–4270 m in the Sikkim Himalayas, Nepal and Bhutan). The poisonous *Ledum palustre,* which occurs on moors and in wet pine forests in the cold north temperate zone, is a protected European species. The genus *Erica* is found in Europe, the Mediterranean region, the Atlantic islands and in tropical Africa, its highest concentration of species occurring in southern Africa. Six hundred and fifty species are found south of the Limpopo river, 580 of these in southern Cape Province. *Erica bauera* ***(Bridal Heath, Albertina Heath)*** is widely cultivated as a decorative plant, but in the wild it is restricted to two small areas in Cape province. In Europe, species of *Erica* are well known heathers and heaths of moorlands and hillsides (the terms heather or heath are also applied to certain other genera of the family), e.g. *Erica cinerea* ***(Bell heather).*** *Loiseleuria* is a monospecific genus *(L. procumbens,* ***Alpine azalea),*** found in dry, peaty or gravelly habitats throughout the Arctic region and in the Alps, Pyrenees and mountains of Central Europe. *Epigaea repens* ***(Mayflower, Trailing Arbutus)*** is found from eastern Canada to Florida. It is a creeping shrub (very similar to *E. asiatica,* which grows in the mountains of Japan), and it was named after the ship, the "Mayflower", by the early pilgrims who found it in flower when they landed at Plymouth, Massachusets. The genus *Gaultheria* is represented by about 200 species distributed from southern India through the western Himalayas, around the Pacific basin into eastern North America and eastern Brazil, with a prominent center of species diversity in the Andes. *G. shallon,* growing to about 1.8 m, is often planted in the USA and elsewhere as game cover. *Cassiope lycopodioides* is a small heather, delicate in appearance, but able to withstand the most adverse habitats, ranging from the Aleutian islands across Alaska, and south to British Columbia at altitudes of 1220 m and west to eastern regions of the former USSR and to Japan. About 15 species of *Paphia* are known from New Guinea, Queensland and Fiji. The first species to be described was the Fijian *Paphia vitiensis*, a shrub or tree (1.2–7 m) with tubular yellow-red flowers and purplish berries, found at 1 220 m in the mountains of Viti Levu.

Erinaceidae, ***hedgehogs*** and ***gymnures:*** a family of the order *Insectivora* (see), comprising 16 species in 8 genera. Dental formula: 2-3/3, 1/1, 3-4/2-4, 3/3 = 36 to 44. Most members of the family are the true ***Hedgehogs***, native to Europe, Asia and Africa, which possess a coat of spines and a very short tail. They habitually roll themselves into a ball in defense against predators, and in temperate climates they hibernate in the winter. The common ***Eurasian hedgehog*** *(Erinaceus europaeus)* has been introduced into Japan and New Zealand. Members of the family found in South-East Asia have normal body hair and a long tail, e.g. the ***Moon rat*** or ***Gymnure*** *(Echinoserex gymnurus).*

Eriocheir: see Chinese mitten crab.

Eriocraniidae: see *Lepidoptera.*

Eriophyes: see *Tetrapodili.*

Eriophyes nervisequens: see Galls.

Erithacus: see *Turdinae.*

Ermine, ***stoat,*** *Mustela erminia:* a mustelid (see *Mustelidae)* found in all of northern Europe and across Asia to Kamchatka and Kuriles to Japan, and in northern North America. Summer coloration is brown above and white below. In winter the fur is entirely white, except for the black tip of the tail. It inhabits hedgerows and the margins of woods, and in winter it may also enter buildings. It feeds almost exclusively on mice, but also takes birds, including domestic poultry.

Ermine moths: see *Yponomeutidae.*

Erodium: see *Geranaceae.*

Eroliinae: see *Scolopacidae.*

Errantia: an order or subclass of the *Polychaeta* (see). They posses a large number of body segments, and an eversible pharynx (proboscis), often with several pairs of chitinous jaws. The parapodia are well developed. Most are predators, leading an active, mobile life, while others often live in tubes. They often display marked structural and physiological changes in the sexual season.

A notable example is the Palolo worm *(Eunice viridis,* family *Eunicidae),* which occurs in the South Pacific and attains a length of 40 cm. Gonads are formed in the posterior part of the body. In the sexual season in October or November (especially at dawn one week after the November full moon), the posterior body is shed. It rises to the surface and swims about, releasing eggs or sperms. Natives of Samoa and other islands net the worms in large quantities and bake or fry them as a delicacy. The anterior body remains among the coral rocks where it regenerates its posterior part.

Sea mice (family *Aphroditidae)* live as predators on the soft bottom mud. In addition to the usual bristles they also possess a fur-like covering of soft bristles which often imparts a metallic irridescence, especially to the sides of the animal. The Common sea mouse *Aphrodite aculeata,* attains a length of 10–20 cm, and occurs in the North Sea. *Neiris* (see Fig.1, *Polychaeta*) also belongs to the *E.*

Errants: plants that are freely motile in various media; motility may be active or passive. E. include, e.g. floating plants, ice and snow inhabitants, edaphophytes and freely motile parasitic plants. See Life form.

Erwinia: a genus of rod-shaped (0.5–1.0 x 1.0–3.0 μm), Gram-negative, motile (peritrichous flagella), facultative anaerobic Enterobacteria (see). They are commonly associated with plants, as constituents of the epiphytic flora, as saprophytes, or as pathogens. As the causative agent of wet rot in stored carrots, potatoes and similar crops, *E. caratovora* causes extensive commercial losses. The same organism is responsible for vascular and parenchymatous disease (blackleg) of potato. *E. amylovora* causes fireblight in fruit trees.

Erysiphales: see *Ascomycetes.*

Erythizontidae, ***New World porcupines:*** a family of Rodents (see) containing between 8 and 23 species in 4 genera. They are large, heavily built mammals with a large, bulky head and blunt muzzle. Except for the possession of quills or spines, they have few affinities with the Old World porcupines *(Hystricidae).* The sharp tips of the quills often have overlapping barbs. The heavily built skull has a deep rostrum and a greatly enlarged infraorbital foramen, which accommodates a large medial masseter. Dental formula: 1/1, 0/0, 1/1, 3/3 = 20. Unlike the *Hystricidae,* they do not dig burrows, and they shelter in rock piles, hollow logs, etc. As arboreal adaptations, the pads of the broad-soled feet are covered with tubercles (which increase traction), the first digit (hallux) in some species is replaced by a broad movable pad, and the toes carry long, curved, climbing claws. Each foot has 4 digits. *Erethizon (E. dorsatum;* ***North American porcupine)*** is a monospecific genus and the largest of the New World porcupines (head plus body 64.5–86 cm, tail 14.5–30 cm). It is found in forested areas (especially conifers and poplar) from Alaska southwards to Mexico, but spends much time on the ground and is semi-arboreal. During the winter it feeds on pine needles and the cambium layer of conifers, sometimes causing considerable destruction. During the spring and summer it feeds on buds and leaves, and grazes other plants, especially sedges. It has a short, stubby tail, and its thick, pointed, barbed quills are hidden between long, stiff hairs. Several subspecies are recognized. The other 3 genera occur in tropical forests from southern Mexico through Central America and into the northern part of South America. *Chaetomys subspinosus* ***(Thin spined porcupine)*** is a monospecific genus, and unique among rodents, in that its orbit is almost ringed by bone. Its quills are short and somewhat wavy, and those on the back are more like bristles. The Thin spined porcupine is found over a wide area of the Amazon basin east of the Andes, usually in dense undergrowth near to swamps and cultivated areas. It is an excellent climber. *Echinoprocta rufescens* ***(Upper Amazonian porcupine)*** is also a monospecific genus, restricted to a small, mountainous region of northern South America. It has short quills, and the short tail is covered with hair. The smallest species belong to the genus *Coendou (= Sphiggurus)* ***(Prehensile-tailed porcupines;*** head plus body 30–60 cm, tail 33–45 cm). Between 5 and 20 species of *Coendu* are recognized, depending on whether certain populations are accorded the status of species or subspecies. They all have a long, downward curved, prehensile tail, and short, thick quills, and may be regarded as truly arboreal, descending to the ground only to eat. Examples are *C. mexicanus* ***(Mexican tree porcupine);*** *C. spinosus* ***(South American tree porcupine);*** and *C. insidosus* ***(Woolly prehensile-tailed porcupine,*** whose quills are hidden in long, soft fur).

Erythroblastosis: the destruction of an infant's red blood cells by maternal antibodies. Most cases of E. are due to the rhesus factor, and they occur when the fetus is Rh-positive and the mother is Rh-negative. In rarer cases, other blood groups may be responsible, e.g. the Diego factor. See Blood groups.

Erythrodextrins: see Dextrins.

Erythrolab: see Color vision.

Erythromycins: a group of antibiotics synthesized by *Streptomyces erythreus.* They are macrolide glycosides, often containing aminosugars and branched sugars. They act primarily against Gram-positive microorganisms, and are used, e.g. against scarlet fever, syphilis and gonorrhea.

Erythropoesis: see Hematopoesis.

Erythropoetin: a glycoprotein hormone which stimulates production and release of erythrocytes in response to lack of oxygen. E. is produced in the kidney and liver of adults, and in the liver of fetuses and neonates. E. increases the rate of mRNA formation in bone marrow, resulting in increased heme synthesis and erythroblast production, and ultimately in large numbers of circulating reticulocytes and erythrocytes (see Hematopoesis). The hormone is quickly degraded and secreted by the kidneys. Human E. has 166 amino acid residues (M_r 18 398 without carbohydrate). The glycosylated protein has M_r 34–39 000, so that about one half of the mass of the native hormone is due to carbohydrate.

Erythropygia: see *Turdinae.*

D-Erythrose: an aldotetrose monosaccharide, the 4-phosphate of which is an intermediate in carbohydrate metabolism. D-E. can be reduced to the optically inactive sugar alcohol, D-erythritol.

Erythroxylaceae, ***coca family:*** a family of dicotyledonous plants containing about 200 species, distributed exclusively in the tropics, particularly in tropical America. They are trees, shrubs or undershrubs, with al-

Erythroxylaceae. Coca plant *(Erythroxylum coca). a* Flowering stem. *b* Flower. *c* Bud. *d* Fruit (longitudinal section).

ternate (rarely opposite), stipulate (stipules often shed early), simple leaves. The inconspicuous, 5-merous flowers are actinomorphic, hermaphrodite (rarely subdioecious) and hypogynous. The ovary develops into a single drupe. The ***Coca plant** (Erythroxylum coca),* native to Peru and Bolivia and cultivated in other tropical regions, is the main commercial source of Cocaine (see). The alkaloid is also prepared from certain other species of *Erythroxylum.* Leaves of the Coca plant are chewed with lime or plant ash by the Andean Indians as a stimulant. They also drink an extract prepared by boiling the leaves.

Erythroxylum: see *Erythroxylaceae.*

Esca: see *Lophiiformes.*

Escargot: see *Gastropoda.*

Escherichia coli: a rod-shaped member of the Enterobacteria (see), up to 6 µm in length, rarely nonmotile and usually with peritrichous flagella. It can form pili, which enable 2 cells to conjugate (see Conjugation). *E. coli* is a predominant and constant inhabitant of the healthy large intestine. In the clinical investigation of urine, or in the testing of water supplies, the the presence of *E. coli* is therefore an indication of fecal contamination. Certain strains of *E. coli* are pathogenic, causing soft tissue inflammation, septicemia and gastroenteritis.

E. coli is widely used in bacteriological, genetic and biochemical research, and it is the most extensively studied of all microorganisms.

Eschrictiidae: see Whales.

Eschrictius: see Whales.

Eschscholzia: see *Papaveraceae.*

Esculin: 6-β-D-glucopyranosyl-esculetin, a bitter tasting glycoside from the bark and leaves of horse chestnut *(Aesculus hippocastanum).* It also occurs in many other plants, e.g. in the bark of hawthorn *(Crataegus).* E. is hydrolysed into glucose and the aglycon, esculetin (6,7-dihydroxycoumarin).

Glucose—O
HO O O *Esculin*

Eserine: see Physostigmine.

Esodemic: a type of Epidemic (see) which is restricted to genetically uniform host tissues or host individuals, which therefore have the same degree of resistance.

Esophageal orifice: see Cardia.

Esophagus: see Digestive system.

ESR-spectroscopy: see Electron-spin resonance spectroscopy.

Essen-Möller method: a procedure for determining the probability of paternity (see Forensic anthropology) by the comparison of numerous individual characteristics. It is based on the observation that the frequency of occurrence of any character displayed by a number of children is higher in the fathers of those children than in the average population. If X is the frequency of the character in true fathers, and Y the average frequency of the character in the male population, then the probabilty (W) of paternity, based on characters 1, 2, 3, … n is: $W = 1/1 + (Y_1Y_2Y_3 \ldots Y_n/X_1X_2X_3 \ldots X_n)$, where W may lie between 0 and 100%.

Essential amino acid: see Amino acids.

Essential nutrients, *indispensable nutrients:* elements or compounds essential for life, which the organism is unable to elaborate for itself, and which must therefore be provided in the diet. Examples of E.n. of mammals are essential Amino acids (see), certain unsaturated Fatty acids (see), Vitamins (see), as well as certain inorganic trace elements such as iron, copper, molybdenum, cobalt, zinc, manganese, chromium, iodine tin, selenium, fluorine and others.

Essential oils: an extremely heterogeneous group of volatile, lipophilic plant products with characteristic odors, occurring in cells, superficial glands and internal cavities of plants. They are complex mixtures of alcohols, aldehydes, ketones, carboxylic acids, esters most of which are also terpenes. They also include aromatics (e.g. eugenol), aliphatics (e.g. 3-*cis*-hexenal) cyclic aliphatics (e.g. cyclopentane), heterocyclics (e.g. coumarins), nitrogenous compounds (e.g. methyl anthranilate), and sulfur-containing compounds (e.g. mustard oils). More than 100 compounds have been isolated and structurally identified from E.o. They occur both in free form and as odorless glycosides, from which they are released by enzymatic action during their isolation.

In contrast to fatty oils (e.g. ground nut oil, olive oil, coconut oil), the E.o. (e.g. aniseed oil, camphor oil, lemon oil, lavender oil, clove oil, peppermint oil, rose oil) evaporate completely and do not leave a greasy spot on paper. E.o. are responsible for the perfume of flowers, and they also occur in many other plant parts. Members of the families *Pinaceae, Rutaceae, Lauraceae, Lamiaceae* and *Umbelliferae* are especially rich in E.o. The chemical composition and odor of the E.o. is very variable, and it is species-specific.

The International Standards Organization (ISO) has defined E.o. in the strict sense as the steam distillates of plants, or of the oils obtained by pressing out rinds of certain citrus fruits. However, in practice, the products extracted with organic solvents, hot oil or melted fat, or by enfleurage or maceration of blossoms (flower oils), are all included by the term E.o. In enfleurage, flowers are spread on glass plates coated with purified fat, followed by extraction with alcohol. About 3 000 E.o. are known and 150 of these are commercially important. They are used mainly in the perfume and food industries, and in pharmacy.

Little is known about the ecological significance of E.o. They serve as insect attractants in flowers. In the vegetative parts of the plant they retard transpiration, discourage consumption of the plant by herbivores (including insects) and inhibit the growth of plant pathogens.

Esterases: widely distributed, naturally occurring hydrolytic enzymes, which catalyse the hydrolysis of esters to the corresponding acids and alcohols. They include *carboxyesterases* [e.g. Lipases (see) and Cholinesterase (see)], *thiolesterases* [e.g. acetyl-CoA thiolesterase], Phosphodiesterases (see), Phosphatases (see) and Sulfatases (see).

Esterification: formation of an *ester* from an acid and an alcohol by removal of the elements of water: R-COOH + R'—OH ⇌ R—COOR' + H_2O. Many metabolic reactions involve E. (see Transferases). The reversal of E. is called Saponification (see), i.e. hydrolysis of esters to alcohols and acids.

Estheria, *Cyzicus, Isaura:* a common genus of freshwater *Branchiopoda* with a thin bivalvular shell. The shell, 2–4 mm in diameter, is round with concentric lines. *E.* has a worldwide distribution, and is found from the Devonian to the present, being especially abundant in the Estheria strata in the Lower Keuper of the Upper Triassic.

Esthetasc: a type of chemoreceptor on the body surface of crustaceans. In aquatic crustaceans, Es. take the form of long, delicate, sensory hairs, whereas in terrestrial forms they are small plates. Es. are found especially on the appendages, and in particular on the first antennae, where they are usually present in rows. Dendritic processes from a group of bipolar sensory neurons extend through the entire length of each E.

Esthetes, ***aesthetes:*** sense organs of the *Polyplacophora* (chitons; phylum *Mollusca).* The valve surfaces of chitons carry numerous "pores" arranged in a definite pattern (some species have more than 10 000). These are not actually pores, but the end caps of strands of cells that ascend vertically through the tegmentum. The larger structures are known as megalopores, the smaller as micropores. The strands leading to the pores were named esthetes by Moseley (1885); the larger esthetes are associated with megalpores, and are known as megalesthetes, while micresthetes are associated with micropores. Esthetes are the terminal parts of long strands known as fiber cords, a bundle of fibers covered with epidermis of the adjacent mantle. They are innervated from the lateral nerve cords, whose branches pass through the pores in the slit rays of the articulamentum. In some species the esthetes are provided with light refracting structures and pigment cups, and therefore appear to represent light receptors on the surface of the shell. Some esthetes are thought to have a hydrostatic function.

Estradiol, ***oestradiol, 17β-estradiol, estra-1,3,5(10)-triene-3,17β-diol:*** a steroid hormone containing 18 C-atoms, m.p. 178 °C, and the most potent natural estrogen. It is biosynthesized from testosterone by a process of aromatization (see Estrogens). E. and estrone are interconvertible in the organism. E. occurs in high concentration in pregnancy urine, as well as in Graafian follicles and placenta. It induces proliferation of the uterine mucosa and development of the mammary glands. Together with progestins (especially progesterone), and certain pituitary hormones [follicle-stimulating hormone (FSH) and luteinizing hormone (LH)], it is also responsible for the regulation of the menstrual cycle. See Estrogens.

OH H_3C HO *Estradiol*

Estrane: see Steroids.

Estrilda: see *Ploceidae.*

Estrildinae: see *Ploceidae.*

Estriol, ***oestriol, estra-1,3,5(10)-trien-3,16α,17β-triol:*** a steroid hormone and an estrogen, containing 18 C-atoms, m.p. 280 °C. It differs from estradiol by the presence of an additional hydroxyl group (α-conformation) at position 16. Within the organism, it is formed from estrone and estradiol. It has been isolated from human female urine (especially during pregnancy) and placenta, from mare's urine and from willow catkins.

OH H_3C OH HO *Estriol*

Estrogens, ***oestrogens:*** a group of female sex hormones, structurally related to the parent hydrocarbon, estrane (see Steroids). They have an aromatic A-ring with a phenolic 3-hydroxy group, and an oxygen function on C-17. The main E. are Estrone (see), Estradiol (see) and Estriol (see); see also Equilenin and Equilin.

E. are produced in the Graafian follicles of the ovary, and in the corpus luteum. During pregnancy they are also formed by the placenta. Small quantities of E. occur in male gonads. In the female, they control the course of the menstrual cycle (in humans and apes), or the estrus or mating cycle (other animals). E., acting together with Progesterone (see) and the Gonadotropins (see), are responsible for the proliferation of the uterine mucosa, growth of the mammary glands, and the appearance of secondary female sex characteristics. E. are metabolized in the liver and kidneys. They are excreted in the urine in free form and as their sulfate and glucuronate esters.

E. are biosynthesized from androgens by aromatization of ring-A and loss of C-19. Three sequential hydroxylations are involved. The first two occur at the C-19 methyl group, and the final and rate-limiting hydroxylation occurs at position 2β, followed by rapid nonenzymatic rearrangement to estradiol. The aromatase system responsible for this conversion is located in the endoplasmic reticulum.

In the Allen-Doisy test for the assay of E., the response of ovarectomized mice to urinary or plasma E. is determined. In clinical practice, this has been replaced by a colorimetric test, the Korber method, in which the E. are treated with H_2SO_4 and hydroquinone to produce intensely fluorescent derivatives. In the clinical routine, E. are now determined by immunoassay, and both earlier methods are obsolete.

E. are used therapeutically for the treatment of menstrual disorders and menopausal problems, and as Ovulation inhibitors (see). Some synthetic E., such as ethinylestradiol, mestranol, and the nonsteroidal diethyl stilbestrol, are more active than natural E., since they are degraded only slowly by the liver.

E. have also been isolated from plants, e.g. estrone from pomegranate seeds and date stones, and estriol from willow catkins. The roots of a Burmese tree, *Pueraria mirifica,* contain mirestrol, which has the same estrogenic potency as 17β-estradiol.

Estrone, ***oestrone, 3-hydroxyestra-1,3,5(10)-trien-17-one:*** an estrogen, which is formed metabolically from estradiol (E. and estradiol are interconvertible within the organism), and it occurs in human urine, ovaries and placenta, as well pomegranate seeds (17 mg/kg) and palm oil. E. was isolated independently by Doisy and by Butenandt in 1929 from pregnany urine. Nowdays some E. is obtained from mare's urine, but most is prepared by the large scale pyrolysis of androsta-1,4-diene-3,17-dione (obtained from cholestrol or sitosterol by microbial oxidation), or by the chemical conversion of diosgenin (isolated from yams). E. is used therapeutically, and as the starting material for the synthesis of other pharmacologically important steroids, e.g. estradiol and ethinylestradiol.

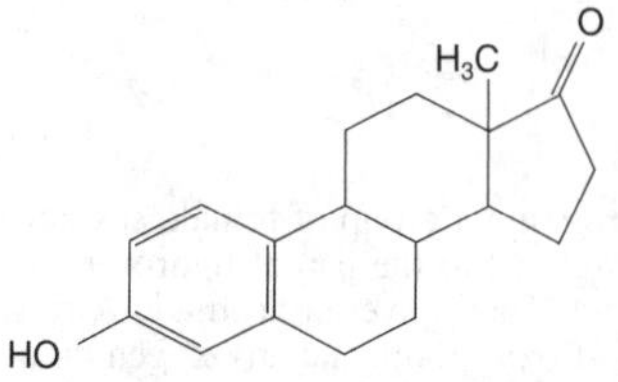

O
H_3C
HO

Estrone

Estuary: the mouth of a river strongly affected by the tide. Estuarine habitats contain distinctive flora and fauna, e.g. the Baltic perch *(Acerina cernua)* is characteristically found in the Elbe E., while flounders *(Platichthys = Pleuronectes flesus)* are commonly caught in the Thames E.

Eté forest: see Rainforest.

Ethanoic acid, *acetic acid, CH_3COOH:* a colorless, sharply smelling liquid which feezes at -16 °C (hence *glacial* acetic acid). It frequently occurs as a stable end product of natural fermentation, putrifaction and oxidation processes. Thus, it can be obtained by the Acetic acid fermentation (see) of ethanol-containing solutions, or by dry distillation of wood. "Active acetate" or acetyl-coenzyme A is an important metabolic intermediate (see Coenzyme A).

Ethanol, *ethyl alcohol, C_2H_5OH:* the end product of the alcoholic fermentation of sugars and starch (see Alcoholic fermentation). It is widely distributed in low concentrations in living organisms (e.g. 0.02–0.03% in mammalian blood). In human liver it is oxidized by the action of NAD-dependent alcohol dehydrogenase, and the resulting acetaldehyde is oxidized to acetate by the action of NAD-dependent aldehyde dehydrogenase.

Ethanolamine, *aminoethanol:* $H_2N—CH_2—CH_2—OH$, a biogenic amine produced by decarboxylation of L-serine. E. is a common constituent of phospholipids in which it is esterified to an acylglycerol phosphate moiety to form phosphatidyl ethanolamine.

Ethephon: see Ethrel.

Ethinylestradiol: a synthetic estrogen used in oral contaceptives; see Hormonal contraception.

Ethiopian niger seed: see *Compositae.*

Ethiopian region: a zoogeographical region comprising sub-Saharan Africa and southern Arabia. Many authors also include Madagascar as a special subregion (see map in entry: Animal geographical regions). The northern boundary of the E.r. is uncertain, for although the Sahara serves as a dispersal barrier, represents the climatic border, and possesses its own desert fauna, characteristic Ethiopian fauna nevertheless lived in North Africa in historical times.

Compared with other regions, the large mammals of the E.r. are unusually diverse and numerous, although bears, deer and sheep are totally absent, and goats (and the Barbary sheep) are only represented peripherally. Voles are also notably absent. The E.r. has retained very old members of its fauna, e.g. mammals such as the golden mole *(Chrysochloridae),* otter shrews *(Potamogalidae),* prosimians *(Lemuroidae)* and hyraxes *(Procaviidae);* and fishes such as the lungfish *(Protopterus)* and bichirs *(Polypteridae).* The great majority of animals, however, belong to more recently evolved groups that originated in the Oriental and Palearctic regions.

In the early Miocene, the first immigrants were probably forest animals from the Oriental region, but the main body of immigrants did not appear until the later Miocene with the development of the relatively dry grasslands. The Red Sea was not formed until the Pliocene. Even today, the A.r. and the Oriental region have numerous species in common, e.g. leopard, lion, cheetah, striped hyena, honey badger, birds of the family *Anhingidae* (darters) and several storks. As late as the Pleistocene, there was ample opportunity for some fauna to migrate from the Palearctic to the E.r. (most of the present Sahara was still not desert), so that the E.r. became a Refugium (see) for numerous groups that are now extinct in the Palearctic region.

Due to the relatively small area covered by forest, the majority of animals are those of semidesert and grassland, mainly ungulates. The viverrids are the most numerous of the predators. Also significant is the species diversity of monkeys and apes. Endemism (see) is marked in the mammals, endemic families of the mainland being: *Chrysochloridae, Potamogalidae, Anomaluridae, Pedetidae, Orycteropidae, Hippopotamidae, Giraffidae,* and an additional 4 rodent familes. The *Macroscelidae* and *Procaviidae* are almost endemic. The number of bird species (1 700) is less than that of the Neotropical region, although the number of families (about 67) is approximately the same for the 2 regions. Most of the bird families have a wide global distribution, but there are some endemic or nearly endemic families: *Struthionidae, Scopiidae, Balaenicipitidae, Sagittariidae, Musophagidae* and *Cliidae.* In contrast to the Neotropical region, the *Passeriformes* predominate (the *Passeriformes* also predominate in other regions). The E.r. has a great variety of lizards and venomous snakes. Urodelic amphibians are absent, while frogs display great species diversity. The fishes are very characteristic, displaying close relationships to the fishes of the Neotropical region, and forming several endemic families.

In the case of insects, endemism is significant at the generic level, and hardly any of the higher taxa are endemic. Many African genera are also distributed in the Palearctic and Oriental regions, and to some extent in the Australian region, whereas only a few genera are common to both the E.r and the Neotropical region.

The species-poor Madagascan subregion, which includes the Mascarene islands and the Seychelles, commands a special position, partly because of its close relationship to the Oriental region, but mainly because of its high proportion of endemic species. Some authors propose the existence of a former land bridge (Lemuria) with India, but Madagascar was probably linked earlier to Africa, or at least was even nearer to the mainland than at present. There are many different species of prosimians and of the endemic tenrecs *(Tenrecidae).* Viverrids are also relatively well represented. Remarkable flightless birds have become extinct or been exterminated. Thus, between 500 and 700 years ago, the elephant birds *(Aepyornithidae)* lived on Madagascar, while the Mascarenes were home to the famous dodo *(Raphidae)* which became extinct in modern times.

Among the reptiles, the occurrence of iguanas and boid snakes is noteworthy, since these groups are otherwise practically restricted to the Neotropical region. The giant turtle of the Indian Ocean *(Testudo gigantea)* survives on Aldabra Island. Chameleons have diversified greatly on Madagascar. Anurans (frogs and toads) exhibit great species diversity in the Madagascan subregion, while primary freshwater fishes, i.e. fishes that are strictly limited to fresh water, are conspicuously absent.

Ethmoid: see Skull.

Etho-ecology: see Eco-ethology.

Ethogram: an inventory of behavior displayed by a species. It includes the absolute and relative frequencies of behavioral events, the frequency with which one type of behavior is regularly followed by another, as well as other observable regularities in the linkage of different types of behavior. The individual behavioral events are described without using any kind of psychological or interpretative terminology (i.e. they are not classified according to the modality of their interaction with the environment). On the other hand, behavioral events may be classified according to their constituent motor elements. The E. is a record of all the external behavioral events inacted in a given environment, i.e. it is a record of the total "behavioral output". The construction of an E. normally starts with observation of sexually mature individuals, and is therefore a record of sex-specific behavior. In addition, behavior specific to different stages of development may also be recorded.

Ethology, *behavioral research* (plate 26): the study of behavior. E. is concerned with the evolved behavior of organisms, which depends on the processing of environmental information by the sense organs, and which governs the relationship between an organism and its environment. The basic approach of ethology is to record observable behavior and to present it as an Ethogram (see). Like other aspects of form and function, the behavior of an organism is seen as the result of environmental adaptation, the limits of which are genetically predetermined. In this sense, E. represents an advance on the old concept of instinct, and it interprets Instinctive movements (see) as genetically determined normal behavior. Nowdays, the central theme of E. is Comparative behavioral research (see). Instinct has been abandoned as a concept, and E. has the task of causally explaining and measuring behavior with scientific methods. For this purpose, however, behavior has a rather resticted definition, being limited to all types of movement, sound production, adoption of postures, color changes, and the release of olfactory substances.

Ethometry: a term for all quantitative methods of behavioral research. Thus, the frequency distribution of particular behavioral events (e.g. warning vocaliza-

Ethogram. Analysis of the courtship behavior of a mallard drake. *a* Individual movements and postures. *b* and *c* Obligatory sequences. *b* Tail wagging – head shaking and stretching – tail wagging – head shaking – grunt whistling – tail wagging. *c* Head up, tail up – turning head to female – nodding and swimming – displaying back of head (after Lorenz, 1960).

tions) can be used as a reference value for determining the strength of response to a stimuli that release that behavior. Complex methods are used to quantify the subcomponents and sequences of behavioral events.

[B.A. Hazlett, *Quantitative Methods in the Study of Animal Behavior,* New York 1976]

Ethomimicry: see Protective adaptations.

Ethoparasites: organisms that disadvantage their host by benefitting unilaterally from its behavior. For example, different species of insects have exploited the ecological niche provided by mouth-to-mouth feeding (trophallaxis). Trophallaxis is characteristic of biosocial insects, among which it serves for the transfer and exchange of food. Thus, different beetle species provide the appropriate stimulus for the release of food by ants. The larvae of other beetle species feed on ant larvae, at the same time providing the stimulus that releases care behavior in their host.

Brood parasitism (see) is a special case of ethoparasitism.

Ethrel, ***ethephon, 2-chloroethylphosphonic acid,*** $ClCH_2CH_2PO_3H_2$**:** a preparation that releases the phytohormone, ethylene, at pH values higher than 4. It is used as a plant growth regulator.

Ethylene, $H_2C{=}CH_2$**:** the simplest unsaturated hydrocarbon of the alkene series. It is an important starting material for many different chemical syntheses. In plant physiology, E. is an important, ubiquitous growth regulatory hormone of higher plants. As a phytohormone, it is necessary for normal plant development, and it differs from other phytohormones in being a gas. In higher plants, E. governs the activity of many biosynthetic pathways; it inhibits or promotes certain enzyme activities, stimulates protein synthesis and affects membrane permeability. In close cooperation with other phytohormones, it intervenes in the entire metabolism of the plant and its morphogenesis from seed germination to fruit ripening and aging. In particular, it affects extension growth, abscission of leaves, flowers and fruits, aging and ripening. The particular physiological action exerted by E. depends on the type of plant tissue and its physiological state.

In the plant, endogenous E. is derived biosynthetically from C-atoms 3 and 4 of the amino acid methionine. Important intermediates are *S*-adenosylmethionine and its cyclization product, 1-aminocyclopropane-1-carboxylic acid (ACC). The latter serves as the transport form of E. (since E. is a gas, control of its transport would be problematic, and it would be lost to the exterior). Within the plant, the endogenous concentration of E. is regulated by the synthesis and degradation of ACC. Average E. production is 0.5–5 nl/g/h, and it is much higher in ripening fruits. E. production is strongly dependent on the type and stage of development of the plant tissue, and on external factors like light and temperature. E. has a variety of applications in agriculture and horticulture, where it is applied in the form of ethylene-producing compounds, e.g. chloroethylphosphonic acid (ethephon, ethrel, CEPA), $ClCH_2CH_2PO_3H_2$. In the aqueous milieu of the plant cell, such compounds are spontaneously hydrolysed to form E. These *ethylene generators* are used to accelerate the ripening of tobacco, sugar cane and fruits (bananas, tomatoes, apples), for the induction of synchronous flowering (e.g. in pineapples and ornamental plants), to promote abscission (e.g. in cherries), to stimulate the latex flow of *Hevea brasiliensis,* and to inhibit the extension growth of cereal crops.

Ethylene chlorohydrin, $HO{-}CH_2{-}CH_2{-}Cl$**:** a compound formed by reaction of ethylene with an alkaline solution of chlorine. The reaction was formerly thought to involve addition of hypochlorous acid (formed by the interaction of water and chlorine), but kinetic evidence reveals a two-step process in which a chlorine cation attacks one of the carbons of ethylene to produce an ionic intermediate, which then combines with a hydroxide ion in the alkaline medium. E.c. is an important component of the germination-promoting preparation, Rindite.

Ethylenediaminetetraacetic acid, ***EDTA, ethylenedinitrolotetraacetic acid:*** $(HOOC{-}CH_2)_2N{-}CH_2{-}CH_2{-}N(CH_2{-}COOH)_2$, a chelating agent, widely used in biochemical systems in vitro for chelating divalent metal ions. It may be used to prevent inactivation of enzymes by chelating toxic metal ions (Hg^{2+}, Cu^{2+}, etc). Alternatively, in plant hydroponic media and bacterial culture media, EDTA can be used to avoid the precipitation of ferric hydroxide, and to keep iron in solution for use by the growing organism. It is commonly used as its disodium salt (Na_2EDTA). Oral administration of EDTA may be used for the chelation of lead in cases of lead poisoning. In the older literature, it was known by its trade name, Versene.

Ethyleneglycol, ***1,2-glycol, ethane-1,2-diol:*** the simplest dihydric alcohol. It is miscible with water and ethanol, and is used as a raw material in the manufacture of polyester fibers, as a laboratory solvent, and as an antifreeze.

Ethynodiol diacetate: a synthetic progestin used in oral contaceptives; see Hormonal contraception.

Etiocholane: see Steroids.

Etiolation: deviation of plant development and growth due to light deficiency. In dicotyledons, E. characteristically consists of a marked elongation of internodes and petioles and the development of abnormally small leaf laminas. The resulting slender stem contains practically no mechanical tissue or vascular bundles. Chlorophyll, carotenoids and anthocyanins are very deficient or absent, so that the etiolated plant is pale yellow in color. Asparagus and chicory are intentionally earthed up to exclude light so that their stems and leaves become etiolated. Shoots of seed potatoes kept in the dark are etioloated and practically colorless. In monocotyledons, etiolation causes abnormal extension of the entire shoot. During E., the plant utilizes available nutrients while they last, in order to bring its assimilatory organs into the light. A few minutes of illumination each day are sufficient to prevent or end E. and to initiate normal growth and development.

Euanthium theory: a theory of plant evolution, in which the angiosperm flower is regarded as a simple, uniaxial aggregation of sporophylls, so that stamens and carpels are homologous with the micro- and megasporphylls of heterosporous pteridophytes. An opposing theory is the Pseudanthium theory (see).

Eubacteria: true bacteria. An order of bacteria in earlier systematic classifications, including essentially all unicellular bacteria with a rigid cell wall.

Eubalaena: see Whales.

Eubryales: see *Musci.*

Eucalyptol, ***cineol:*** a monoterpene, and the main component of the essential oils of *Eucalyptus* species. It is present in smaller quantities in other essential oils. It smells like camphor, and has strong antiseptic properties.

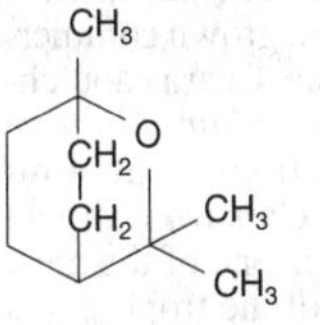

Eucalyptol

Eucestoda: see *Cestoda.*
Euchromatin: see Chromatin, Heterochromatin.
Euedaphic organisms: see Soil organisms (see) which live in the soil interior.
Eudiapause: see Dormancy.
Eudiscopus: see *Vespertilionidae.*
Eudocimus: see *Ciconiiformes.*
Eudromias: see *Charadriidae.*
Euechinoidea: see *Echinoidea.*
Euglena: see *Phytomastigophorea.*
Eugenics: racial improvement by selected mating, so that the better stock prevails. E. aims to increase the prevalence of advantageous or positive genes in a population *(positive* or *progressive E.)* and/or limit the distribution of undesirable genes *(negative* or *preventative E.).*
Euglenida: see *Phytomastigophorea.*
Euglenoids: see *Phytomastigophorea.*
Eugenol: a colorless, oily liquid with a strong smell of cloves. Clove oil contains 80–95% E., together with other essential oils. E. is the starting material for the commercial preparation of vanillin. It is also used in dentistry and perfumery.

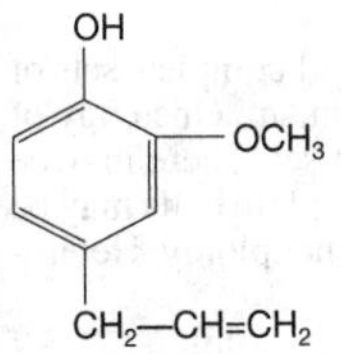

Eugenol

Euglenophyceae: see *Phytomastigophorea.*
Eugregarinaria: see *Telosporidia.*
Euhaline: see Brackish water.
Eukaryotes, *eukaryotic organisms:* organisms whose cells possess a differentiated nucleus, i.e. they possess a nuclear membrane, in contrast to Prokaryotes (see), which have no nuclear membrane. E. possess true chromosomes which become microscopically visible during nuclear division (meiosis and mitosis), when their form and number is characteristic of the organism. The genes of E. are mostly split and divided into exons and introns, so that the processes of gene expression are more complicated than in prokaryotes.

Aerobic E. also possess mitochondria, which are the sites of electron transport and oxidative phosphorylation, whereas in prokaryotes these processes occur in the cell membrane. Chloramphenicol inhibits protein synthesis in prokaryotes but not in E., whereas cycloheximide inhibits protein synthesis in E. but not in prokaryotes. In E., protein biosynthesis is initiated by methionyl tRNA, whereas in prokaryotes it is initiated by formylmethionyl tRNA. The ribosomes of E. (at least in mammals) sediment with a value of 80 Svedberg units (i.e. 80S ribosomes), whereas prokarytoes possess 70S ribosomes.
Eukaryotic organism: see Eukaryote.
Eulamellibranchia: the largest order (about 17 500 species) in the earlier classification of the *Bivalvia* (see). Members of the *E.* possess 2 pairs of eumellibranchiate gills, in which the gill filaments are united by membranes, forming sheets of tissue which partition the gill space. Each gill plate consists of 2 perforated lamellae, formed from the ascending limbs and from the descending limbs, respectively. This gill structure is considered to represent the latest stage in an evolutionary process that started with simple comb gills, gradually becoming complicated respiratory and filter organs.
Eulamellibranchiate gills: see *Eulamellibranchia,* Gills, *Bivalvia.*
Eulimnoplankton: see Plankton.
Eulipoa: see *Megapodiidae.*
Eumecoptera: see *Mecoptera.*
Eumelanins: see Melanins.
Eumetazoa: the totality of multicellular animals, excluding sponges, placozoans and mesozoans. Coelenterates possess two epithelia or germ layers (ectoderm and endoderm), which lie against each other. All other E. possess a third germ layer (mesoderm), located between the ectoderm and endoderm. Other tissues (muscles, nerves, connective tissue, cartilage and bone) and organs are formed by differentiation of one or more of these germ layers. Thus, the cells that associate to form structurally and functionally different tissues may have the same embryonic origin. In all E., development proceeds at least to the stage of the gastrula.
Eumetopias: see *Otariidae.*
Eunectes: see Anaconda.
Eunuchoidism: incomplete sexual development in male humans, characterized by incompletely formed external genitalia and the underdevelopment or absence of secondary sexual characters. There is also increased growth, especially of the distal parts of the extremities, frequently accompanied by gynecomastia and a typically female distribution of subcutaneous fat. Some cases of E. are due to inherited or acquired hypofunction of the gonads, while others are the secondary result of disease of the diencephalon-hypophyseal system.
Euoticus: see Galagos.
Eupagurus: a rejected name, synonymous with *Pagurus.* See Hermit crabs.
Euphagus: see *Icteridae.*
Euphausiacea: an order of the *Crustacea* (subclass *Malacostraca),* approximately 3 cm in length, and superficially resembling a shrimp or prawn. The carapace covers the entire thorax, and is fused with the underlying thoracic segments and with the head to form a cephalothorax. Each thoracic appendage is biramous and carries an exopodite, as well as a large tufted filamentous epipodite gill, known as a podobranch. Most of each thoracic appendage, including the gill, is fully exposed (i.e. not covered by the carapace). There is no specialization of thoracic appendages as maxillipeds, and the last 1 or 2 pairs are often reduced. Special brooding structures are absent, and the eggs are shed freely into the water. Development proceeds through a series of larval stages, the first being a nauplius.

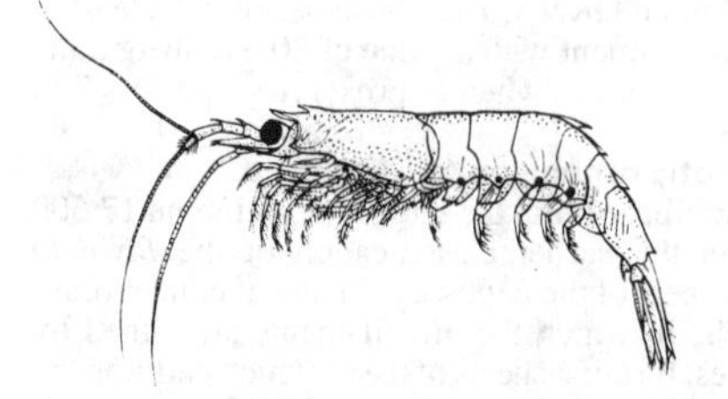

Euphausiacea. Thysanoptera tricuspidata.

All of the 83 recognized species are marine, and with the exception of a few species they are strictly pelagic and rarely found on the bottom. Most species are filter feeders and live on plankton, while a few are predatory, e.g. *Stylocheiron* captures arrow worms and other relatively large pelagic animals. Nearly all species have intracellular light-producing organs called photophores, each consisting of a cluster of light-producing cells, a reflector and a lens. One photophore is usually situated on the upper end of each ocular peduncle, one on the coxae of the seventh thoracic appendages, and one in the middle of each of the sternites of the first four abdominal segments. Luminescence is probably an adaptation for swarming (enormous shoals are often formed) and reproduction. They have great ecological and economic importance as food for baleen whales (see Krill). Well known genera are *Thysanopoda, Meganyctiphanes* and *Euphausia.*

Euphorbia: see *Euphorbiaceae.*

Euphorbiaceae, ***spurge family:*** a family of dicotyledonous plants containing about 8 000 species. Most species are found only in the tropics, but a few are cosmopolitan. They are herbs or shrubs, usually with alternate, simple or compound, stipulate leaves. There are also various succulent forms (especially in the genus *Euphorbia),* with or without very reduced, caducous leaves. Flowers, which display a variety of forms, are always unisexual, and plants are monoecious or dioecious. In many species the perianth is absent, and the flowers usually occur in a complex and varied florescence, which may be spiciform, panicle-shaped or glomerular. In the genus *Euphorbia* (Spurges) the flowers are very small, and a number of male flowers and a single female flower are found within a single cup-shaped, perianth-like involucre, known as a cyathium, in which 4 or 5 small toothed bracts alternate with 4 or 5 crescent-shaped nectary glands (Fig.1.).

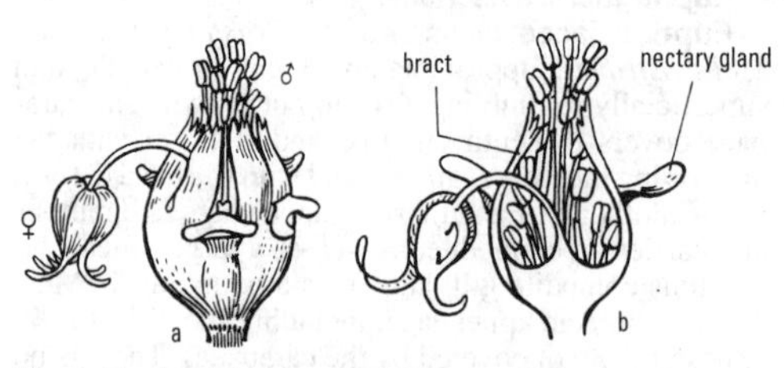

Fig.1. *Euphorbiaceae. a* Cyathium of spurge *(Euphorbia). b* Longitudional section of *a.*

Pollination is predominantly by insects. The fruit is usually a capsule (rarely a drupe), which separates into 2 or 3 parts. Many species of *E.* contain a milky juice, which often contains caoutchouc, and in some species is poisonous. The most important caoutchouc plant of the family is the ***Rubber tree,*** *Hevea brasiliensis,* which is native to the Amazon region, but is also grown commercially in India, Indonesia and Malaysia. Ceara caoutchouc is obtained from the ***Ceara rubber plant,*** *Manihot glaziovii.* Tapioca starch is prepared from the root tubers of ***Manioc*** (American name) or ***Cassava*** (African name) *(Manihot utilissima).* Manioc is one of the most important starchy agricultural crops of the tropics, and it forms the main part of the diet in many tropical areas of South and Central America and Africa. The root tubers, which contain up to 40% starch, are processed into flour, porridge and flat bread cakes. Glycosidically bound cyanide is also present in Manioc tubers, and this must be removed before consumption. The ***Tung oil tree*** *(Aleurites fordii)* is grown for its oil-rich seeds, which yield technically useful tung oil.

The most important oil and pharmaceutical plant of the family is the ***Castor oil plant*** *(Ricinus communis)* from tropical Africa. The seeds contain about 45% of an oil (castor oil, *Ricinus* oil, Oleum ricini), in which ricinoleic acid accounts for 80–85% of the esterified fatty acids. Castor oil is widely used as a mild purgative, and on account of its viscosity, it is also used as a lubricant, especially in warm climates.

The ornamental plant, ***Poinsettia*** *(Euphorbia pulcherrima),* has conspicuous red bracts, blooms in December, and is native to Mexico.

Native European species are, e.g. ***Dog's mercury*** *(Mercurialis perennis),* and various species of of *Euphorbia,* which occur as woodland plants and as garden and field weeds.

Euphysoclistic fishes: see Swim bladder.

Euplectella: see *Porifera.*

Euplectes: see *Ploceidae.*

Euploidy: the exclusive presence of complete sets of chromosomes in the cells of an organism. One copy of each chromosome is present in each set. There may be one or more chromosome sets, i.e. euploid cells may be haploid, diploid or polyploid. See Aneuploidy, Heteroploidy.

Euproctis chrysorrhoea: see Brown-tail

Euproctus: see *Salamandridae.*

Eurasian robin: see *Turdinae.*

European badger, *Meles meles:* a monotypic genus and the largest European member of the *Mustelidae* (see). It is widely distributed in Europe and Asia, south to Tibet, northern Burma and southern China, usually inhabiting wooded regions. The body is relatively plump, with short legs and short tail (head plus body 56–81 cm, tail 11.5–20 cm, weight 10–16 kg). The upper parts are gray, and the underparts, legs and feet are black. The face is striped black and white. Nocturnal in habit, the E.b. remains below ground in its system of burrows and chambers (badger sett) during the day, emerging at night to feed. The diet consists of vegetable matter and small animals, such as snails, beetles and especially earthworms. In the norther part of its range (Scandanavia, Poland, northern Russia) it hibernates in the winter.

European corn-borer, *Ostrinia nubilalis:* a pyralid moth of Europe, USA and Canada, whose caterpillars feed on the stems of millet, hops, hemp, etc. The caterpillars also often feed on tomatoes in the field, and

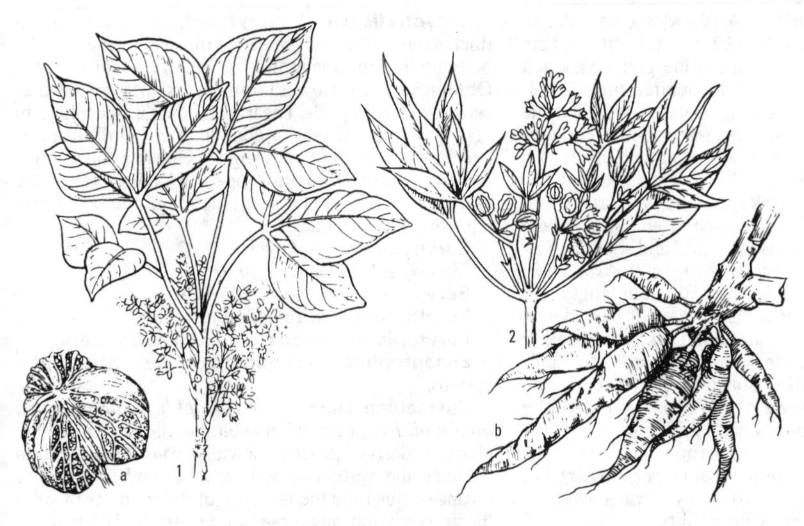

Fig.2. *Euphorbiaceae. 1* Flowering branch of rubber tree *(Hevea brasiliensis), a* fruit capsule. *2* Flowering branch of cassava or manioc *(Manihot utilissima), b* root tubers.

Fig.3. *Euphorbiaceae.* Castor oil plant *(Ricinus communis). a* Shoot. *b* Inflorescence, female flowers above, male flowers below. *c* Fruit. *d* Seed.

sometimes get "canned" with the tomatoes. Above all, however, it is a serious pest of corn (maize), destroying the inflorescence, and sometimes causing damage valued at millions of dollars to the US corn crop (e.g. 183 million dollars in 1969). The female is straw-yellow, the male cinnamon-brown; wingspan 28 to 30 mm.

European grain moth, *Nemopogon granellus* L: a moth of the family *Tineidae.* Through introduction, it now occurs worldwide. The yellowish white caterpillars feed especially on cereal grains and pulses, which they spin together in small clumps, but a great variety of other food sources have been recorded, including nuts, fungi and cork (it has been reported as a pest of wine cellars, where it eats into the corks of wine bottles). Yellow feces are conspicuous among the spun mass of food material.

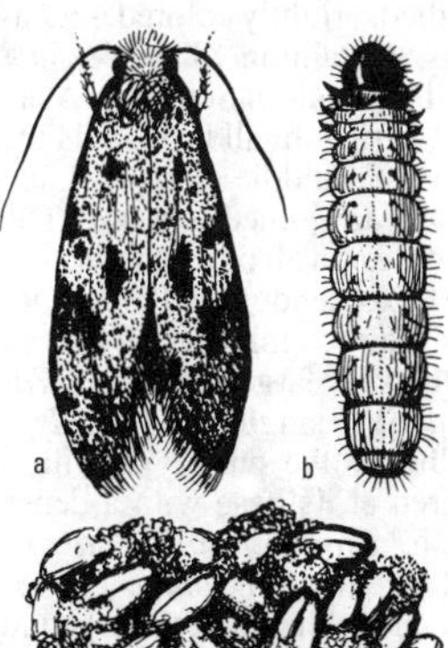

European grain moth (Nemopogon granellus L.). *a* Imago. *b* Caterpillar. *c* Spun mass of corn grains with feces (enlarged).

European grain thrips: see *Thysanoptera.*
European lobster: see Lobsters.
European pond terrapin: see *Emydidae.*
European pond tortoise: see *Emydidae.*
European pond turtle: see *Emydidae.*
European relapsing fever: see Spirochetes.

European roach, *Rutilus rutilus:* a common freshwater minnow of central and northern Europe, belonging to the family *Cyprinidae* (see), and a representative member of the subfamily *Leucistinae.* It is up to 40 cm in length, with an olive-brown to blue-green back and matte silver to silvery sides. The unpaired fins are yellow to bright red, the other fins usually gray, and the eyes are reddish. The mouth opening is almost horizontal. It has local importance as an eating fish, and it is a popular sporting species.

European smooth snake, *Coronella austriaca:* a snake of the family *Colubridae* (see), found in dry areas on the plains and mountains of central and southern Europe, extending eastward to the border with Asia. Diurnal in habit, it is rarely more than 50 cm in length, but may attain 75 cm. There are 2–4 rows of dark dorsal spots, which are sometimes fused into a zig-zag band. It is a good climber, and feeds mainly on lizards and blindworms, as well as small rodents; the prey is entwined and crushed after the style of the constrictors. Young leave the gelatinous egg covering during or shortly after birth, i.e. the birth is vivi-oviparous. The E.s.s. is often mistaken for the Common viper *(Vipera berus),* but like all the *Colubridae* it has round eye pupils, whereas the pupils of the vipers are vertical slits.

European swamp turtle: see *Emydidae.*

Europharyngidae: see *Saccopharyngoidei.*

Europharynx pelecanoides: see *Saccopharyngoidei.*

Eury-: a prefix denoting a broad tolerance of many environmental factors, e.g. euryhaline species occur in estuarine waters, which are subject to wide fluctuations in salinity, See Steno-.

Euryale ferox: see *Nymphaeaceae.*

Euryalaceae: see *Nymphaeaceae.*

Euryceros: see *Vangidae.*

Euryecious conditions: see Ecological valency.

Eurylaimi: a suborder of the *Passeriformes* (see).

Eurylaimidae, ***broadbills***: a passerine family (see *Passeriformes)* of stout-bodied, brightly colored, sedentary, gregarious forest birds, containing 18 species in 8 genera, centered mainly in the islands of Southeast Asia, also extending to India and Africa. In all broadbills the flexor tendons of the third toe (middle front toe) and those of the rear toe (hallux) are joined together. The front toes are fused at their bases. In all other passerines, and in most other birds, the flexor tendons of the toes are separate. Two subfamilies are recognized: *Eurylaiminae* ***(Typical broadbills)*** with very large beaks, and *Calyptomeninae* (***Green broadbills;*** single genus, *Calyptomena).* The green broadbill is the only type with a small beak, which is covered at its base with a dense hood of feathers (rictal brush). *Broadbills in the narrow sense* are represented by the genus *Eurylaimus* (3 species with mainly red, black and white coloration, found in Malaysia and the Sunda islands). Most species are insectivorous. The ***Black-and-red broadbill*** (monotypic genus: *Cymbirhynchus macrorhyncus)* of the Indomalayan region (common along rivers in Borneo) eats berries, small fish and water crustaceans as well as insects. ***Green broadbills*** prefer fruit. Most recently discovered is the monotypic *Pseudocalyptomena graueri* ***(Gray broadbill)*** of eastern Zaire. There are 2 other African broadbills: *Smithornis capensis* ***(Cape broadbill),*** which utters a very unbirdlike explosive grunt, and *Smithornis rufolateralis* ***(Red-flanked broadbill).***

Eurymylidae: see *Lagomorpha.*

Eurypauropus: see *Pauropoda.*

Eurypterus (Greek *eurys* broad, *pteron* fin): an extinct genus of the *Eurypterida* with an elongate, slender body up to 3 m in length. Fossil *E.* are found from the Ordovician to the Lower Permian, and they are the largest known arthropods. The trapezoid-shaped cephalothorax has rounded edges, contains two reniform eyes, and carries 4 pairs of chewing appendages and a single massive pair of swimming appendages. Posteriorly, the body also carries a characteristic long, pointed, terminal, stinglike appendage.

Eurypyga: see *Gruiformes.*

Eurypygidae: see *Gruiformes.*

Eurystomas: see *Coraciiformes.*

Eurythermal: see Temperature factor.

Eurytopic conditions: see Ecological valency.

Eusaprophytic conditions: see Saprophytic system.

Eustachian tube, ***tuba eustachii, auditory tube, tuba auditiva:*** a canal between the middle ear and the pharynx, derived phylogenetically from the connection between the vertebrate gill chamber and foregut. By means of this tube the pressure of the air on both sides of the tympanic membrane is equalized. The pharyngeal opening of the tube is closed except when swallowing, yawning and sneezing. See Auditory organ.

Eutardigrada: see *Tardigrada.*

Eutheria, *Monodelphia, Placentalia,* ***placental mammals:*** an infraclass of the mammals (subclass *Theria)* containing all recent mammals, with the exception of the monotremes and the marsupials. In contrast to the *Didelphia* (see), the vagina of the M. is a single structure, and the end of the penis is not divided. The two uteri are also more or less fused (see Uterus). Urinary and sex ducts are entirely separate, and there is no cloaca. The embryo develops within the uterus, where it is nourished by an allantoic Placenta (see), and it is born in an advanced state of development. The tympanic bone is ringlike or forms a bulla, and the alisphenoid never forms part of the bulla. See *Metatheria, Prototheria, Theria.*

Euthyneura: a subclass of snails, which according to the modern view subsumes the subclasses *Opisthobranchia* and *Pulmonata.* In the *E.,* the rotation of the visceral sack caused by streptoneural crossing over of the pleurovisceral connectives has been reversed. About 50 000 species are classified in 8 orders.

Eutroglobionts: see Troglobionts.

Eutrophic: nutrient-rich. The term is used to characterize the productivity of habitats (e.g. eutrophic lakes are nutrient-rich, highly productive lakes; see Lakes, types of). It is also used of plants which make high demands on the nutrient content of the soil. The converse is oligotrophic.

Eutrophication: an increase in the concentration of plant nutrients, first noticed in the latter half of the 20th century in lakes of Europe and North America. The resulting decrease in the usefulness of the water is a matter of concern. Symptoms of E. include: a) increased biomass of the phyoplankton; b) coloration and clouding of the water by phytoplankton (see Water bloom); c) decreased concentration of dissolved oxygen and increased concentrations of hydrogen sulfide, carbon dioxide, iron and manganese in the deeper layers; d) powerful methane formation in the sediment, accompanied by release of nutrients into the water; e) mass development of encrusting algae and herbaceous plants on the banks.

Euxoa: see Cutworms.

Euzonosoma (Greek *eu* whole or genuine, *soma* body), *Aspidosoma:* an extinct genus of starfish. The borders of the arms carried large plates ornamented with spines or protuberances. Fossils are found in the Devonian. Index forms occur in the Lower Devonian (Bundenbach slate formations of the European Rhineland).

Evening primrose: see *Onagraceae.*

Evenness: a measure of the relative abundance of a species. $E = H/\log_2 S$, where S is the number of species and H is the Shannon-Wiener index of diversity (see).

Evoked potential: changes of potential in the central nervous system, due to the peripheral stimulation of afferent nerves or receptors, e.g. due to stimulation of the eye by flashes of light. See Electroencephalography.

Evolution: see Evolutionary theory, Natural selection, Synthetic theory of evolution.

Evolution of function: see Function evolution.

Evolution, rate of: see Evolutionary rate.

Evolutionary niche: a way of life. This is different from an ecological niche (see Niche).

Evolutionary rate: the rate of occurrence of evolutionary changes. Different criteria can be used for the assessment of E.r., e.g. the appearance of new species, new genera or higher systematic units within lines of phylogenetic development, the frequency of formation of species or higher taxa from a single ancestral species or taxon, and the change in shape and size of individual structures or of the entire body.

A relative index of E.r. is the *Darwin.* One Darwin corresponds to a change of 1/1 000 in an original value in 1 000 years. Rates in the order of a few *milli-Darwin* have been determined for very slow evolutionary processes. In contrast, the rates of alteration of domesticated animals and plants are in the order of *kilo-Darwin.*

A comprehensive index of E.r. would be the totality of genetic changes in a given time in an evolutionary line. Such data can be obtained by the electrophoretic investigation of the proteins of different species (see Polymorphism). If the time of separation of two species from their last common ancestor is known, then their electrophoretically determined genetic differences provide an index of the *genetic E.r.*

Many large systematic units have differing *group-specific E.r.,*e.g. mammals evolve more rapidly than mollusks (Table).

Rates of appearance of new genera within evolutionary lines (after Simpson, 1951)

Group or evolutionary line	No. of genera	Average No. of genera per line per million yrs.
Hyracotherium – Equus	8	0.18
Chalicotheridae (mammals)	5	0.17
Triassic and earlier ammonites	8	0.05

Most genera within such groups develop at approximately the same rate, a relatively large number develop more slowly than the modal value of the taxon, and only a few show a more rapid rate of change. An E.r. lying within this rate distribution is *horotelic.* Unusually slow development lying outside this distribution is *bradytelic.* The bivalve genus with the longest horotelic development existed for 275 000 years. The recent bradytelic genera, *Nucula, Leda, Lima* and *Ostrea,* appeared as long ago as 400 000 years.

Tachytelic development of the snail genus, *Valenciennesia,* from its ancestral form, *Linnaea,* occurred very rapidly during the early Pliocene. The resulting two snail genera are so different that they are placed in separate families; in contrast, certain horse species showed no change at all during this same period.

The reasons for horotely, bradytely and tachytely are unknown. Population size appears to be an important influence; large populations evolve relatively slowly.

Evolutionary stable strategy, *ESS:* a strategy displayed by individuals or species, which if followed by the majority cannot be surpassed by any other alternative strategy. Thus, the frequently occurring sex ratio of 1:1 represents an ESS. In populations with a 1:1 sex ratio, a newly arising allele conferring one particular sex on all its carriers cannot spread within the population. As the frequency of such an allele increases, it suffers a selective disadvantage, because the less numerous sex has a better prospect of finding a copulation partner.

Optimal strategies for populations are often not evolutionarily stable. The rate of increase of many animal populations would be much higher if they contained only a small minority of males. Such a state is, however, unstable, because any new allele conferring the male sex on its carriers would have an enormous selective advantage.

In particular, the theoretical concept of ESS enables an understanding of the behavior that characterizes intra-species conflict. A form of behavior, which is advantageous if followed by all members of the population, is often not evolutionarily stable, because any individual violating that behavior would attain a considerable short-term advantage. Individuals in a population in which conflicts are resolved by threat and not by action have a selective advantage over animals in a population in which fighting actually occurs. Nevertheless, communities that resolve conflict only by threat cannot develop, because aggressive individuals would have an advantage. Therefore, members of the evolutionarily stable population would both threaten and fight.

ESSs also develop between competing species, e.g. between hosts and their parasites.

Evolutionary theory: a theory founded in particular by Charles Darwin (1809–1882), that the variety of living forms on the Earth is the result of a lengthy and complicated process known as *evolution,* and that this process still continues today. In the course of evolution, the numerous and various modern living organisms descended from a few primitive forms with a low level of organization. The E.t. is now accepted by all serious scientists. Strong supporting evidence is provided by Taxonomy (see), Paleontology (see), Comparative anatomy (see Homologies), Embryology (see Biogenetic law) and Biogeography (see).

The driving forces of evolution are explained by Darwin's theory of Natural selection (see). and by the Synthetic theory of evolution (see).

Exarate adecticous pupa: see Insects

Excalfactoria: see *Phasianidae.*

Exchange adsorption: adsorption of ions onto the surface of an adsorbent with simultaneous displacement of a less strongly adsorbed ion. For example, application of Ca^{2+}-containing fertilizer to the soil causes the release of K^+ and other cations from soil colloids. E.a. also makes an important contribution to the Apparent free space (see) of the root in the first stage of ion uptake (see Mineral metabolism).

Exchange of respiratory gases: see Gas exchange.

Excimer: see Fluorescence spectroscopy.

Excision repair of DNA: see DNA replication.

Excitation:

1) Alteration of certain physical and chemical quantities of cell membranes and the fluids on their interior and exterior surfaces.

2) Increased activity of animals, resulting from E. as defined under ***1)***. Different systems (e.g. nerve cells, gland cells, muscles) are affected, so that E. is ultimately the result of complex interactions of numerous elements. For example, E., may be manifested as emotion (see Affect). As a primary process, E. is either *induced* (by an external stimulus), or it arises *spontaneously* in the absence of any recognizable external influence. It is detectable from accompanying variations of bioelectric potential. These fluctuations of potential may be restricted to their site of origin and its immediate vicinity *(local E.)*, or they may spread from this site *(conducted E.)*. The E. intensity is graduated, i.e. it is proportional to the magnitude of the stimulus. E. does not occur, however, until the stimulus attains a certain threshold value; an Action potential (see) is then produced according to the All or nothing rule (see).

The primary process of E. generation has been studied in particular in the giant axons of cephalopods *(Loligo)*, then confirmed in the nerve fibers, ganglion cells, muscles, etc. of other species. Fundamental to the production and conduction of E. is the generation of electrical potential differences by the distribution of ions accross a selectively permeable cell membrane.

Nerve intracellular fluid contains relatively high concentrations of K^+ and low concentrations of Na^+ and Cl^-, whereas this situation is reversed in the extracellular solution. In the extracellular solution the positive and negative charges of Na^+, K^+ and Cl^- are more or less balanced, whereas the charges of Na^+ and K^+ in the intracellular fluid are not neutralized by the small number of Cl^- ions. This balance is accompished by organic anions, which are partially fixed in cytoplasmic proteins. The resting membrane is highly permeable to K^+, but has only a low permeability to Na^+. These permeability differences are due to the presence of specific channels for K^+ and Na^+ (the accessibility of these channels varies according to the state of the membrane). K^+ ions migrate out of the cell under the influence of the concentration gradient. This results in a negative electric potential on the inner side of the membrane, because anions (in particular Cl^- and negatively charged proteins) cannot accompany the K^+ ions.

The relation between membrane potential and K^+ concentration is expressed by the *Nernst equation* (see). The membrane Resting potential (see) is described by the *Goldman equation* (see), which takes into account the relative permeabilities of the main participating ions. This resting potential may change, so that the interior membrane surface becomes more negative *(hyperpolarization)*, or the negative charge of the inner surface is decreased or even becomes positive *(depolarization)*.

When the resting potential is decreased experimentally (e.g. by electrical stimulation) or naturally (e.g. by the activity of synapses), the permeability of the membrane for Na^+ increases exponentially with increasing depolarization. If the depolarization is only slight it remains localized *(local potential)*, and the resting potential is restored by the outward migration of K^+. Otherwise, the rapid inward migration of positive charge destroys the membrane potential, resulting in an Action potential (see).

During an action potential, the Na^+ permeability (or conductance) of the membrane increases 500-fold, Na^+ flows into the cell and the inner side of the membrane becomes positively charged. Increased Na^+ conductance is due to the opening of Na^+ channels (see), which can be inhibited with tetrodotoxin. The number of Na^+ channels per mm^2 of membrane surface varies between about 50 (rabbit vagus nerve, crustacean axons) and 400 (cephalopod axon). However, increased Na^+ conductance is accompanied (albeit with a slight delay) by the opening of more K^+ channels (which can be inhibited with tetramethylammonium), promoting the migration of K^+ out of the cell. There is resulting overshoot, the Na^+ channels close (Na^+ inactivation) as a result of the K^+ activation, and the resting potential is restored.

Depolarization and repolarization are not active transport processes. The sodium-potassium pump represents an active transport system (see Ion pumps), the "pump" being an enzyme, which is sensitive to inhibition by strophanthin (ouabain). The energy for this active transport is derived from hydrolysis of ATP, so that the enzyme is known as a Na^+/K^+-ATPase. Small quantities of ions are transported actively after the passage of the action potential. During the resting state, the concentration gradient is also continually maintained by the action of ion pumps, which counteract the tendency for Na^+ to enter the cell in exchange for K^+. The sodium-potassium pump therefore transports Na^+ out of, and K^+ into, the cell. The nerve cell membrane with its intracellular and extracellular liquid layers therefore resembles a battery which is charged by metabolism. Closer investigation of the mechanisms of ion transport became possible with the introduction of the "voltage clamp" method.

3) In the sensory physiology of plants, changes in the cytoplasm caused by stimulus reception, which may finally lead to movement. The onset of E. is marked by the appearance of an electric potential, known as the Action potential (see), which temporarily abolishes or even reverses the resting potential of -50 to -200 mV of the cell outer surface. The action potential arises from the exit of Cl^- from the cell, often accompanied by the simultaneous entry of Ca^{2+}, due to a stimulus-dependent increase in the permeability of the cell membrane for these ions, each ion migrating along an electrochemical potential gradient. After a certain delay, this displacement of Cl^- and Ca^{2+} is followed by a slow exit of K^+ from the cell. The action potential is therefore quenched, and the original resting potential (but not the original distribution of ions) is restored within about 20 seconds (in contrast, the same process occurs in animal nerves in a fraction of a second). The original ion distribution is finally re-established by active (carrier) transport, in which Cl^- and K^+ are transported into the cell, and Ca^{2+} is transported out.

Excitation conductance, ***excitation propagation:*** the spread of excitation from its site of origin to neighboring regions. Animal nerves are specialized structures for E.c., which can propagate an excitation from its site of origin. Excitation travels in only one direction, due to the presence of synapses between the conducting fibers of the nervous system. Chemical synapses generally allow the passage of excitation in only one direction, although there are possibly exceptions to this rule.

Excitation may occur with or without a decrement. In decremental conduction, which is characteristic of subthreshold excitation, the amplitude of the propagated changes in potential gradually decreases with time and distance from their site of origin. This type of E.c. can also be demonstrated in the dendrites of neurons. Normally, however, excitation in nerves is always propagated without decrement.

In nonmyelinated nerve fibers and in muscle fibers, an action potential arising in one fiber is able to generate excitation in the immediately neighboring fibers. Due to their insulated structure, myelinated nerve fibers cannot perform this type of E.c. Owing to the myelin insulation of the internodal regions of the nerve fiber, an action potential arising in a node of Ranvier (see Neuron) does not cause excitation in that region, but it does produce an excitation in the next and relatively distant node by jumping across the intervening myelinated section *(saltatory E.c.)*. The resulting discontinuous E.c. occurs at very high speed. The rate of propagation of excitation depends on the structure of the conducting system (Table).

Species	Type of nerve fiber	Velocity of propagation (m/s)
Cat	Motor nf. (myelinated)	60–120
Cat	Aff. nf. of pressure receptors (myelinated)	30–45
Cat	Vasomotor nf., slow pain fibers (poorly myelinated)	0.5–2.5
Frog	Motor nf. (myelinated)	8–40
Frog	Aff. nf. of pressure receptors (myelinated)	8–15
Frog	Vasomotor nf., slow pain fibers (poorly myelinated)	0.3–0.8
Shore crab *(Carcinus)*	Leg nerve (poorly myelinated)	0.35
Cephalopod *(Sepia)*	Mantle nerve (poorly myelinated)	0.2–0.4
Mussel *(Anodonta)*	Pedal nerve (poorly myelinated)	0.05
Roman snail *(Helix)*	Intestinal nerve (poorly myelinated)	0.05–0.4
Earthworm	Ventral nerve cord (poorly myelinated)	0.6
Freshwater polyp	capture threads	0.1–0.25

Abbreviations: aff. = afferent. nf. = nerve fiber.

Theoretical explanations of E.c. are based on the small circuit theory of Hermann (1872). In its modern form, this states that at any given point in a nerve fiber, an action potential causes rapid depolarization of the neighboring region of the fiber by setting up small catelectronic circuits or loops. The affected neighboring region then becomes a site of further excitation. The action current produced at the origin of the excitation therefore simultaneously acts as an electrical stimulus to a neighboring region of the nerve.

Excitatory postsynaptic potential: see Synaptic potential.

Exciton transfer: see Energy conduction.

Exclave: a small distribution area of a taxon, which is isolated from the main area of distribution. See Disjunct distribution.

Excretion:

1) In animals, the removal of excretory materials, i.e. removal of carbon dioxide via the gills or lungs; loss of excess water as urine and by evaporation from the body surface and mucous membranes; and removal of nonutilizable end products of protein metabolism, short chain organic acids, salts and toxins in solution in the urine via Excretory organs (see). E. serves to maintain a constant internal milieu, and is the most important factor in the regulation of the osmolarity, volume and pH of body fluids.

In vertebrates, *threshold substances* appear in the urine only if their blood concentration exceeds a certain threshold value (e.g. glucose), whereas *nonthreshold substances* are totally excreted irrespective of their blood concentration.

Removal of water by the vertebrate kidney is regulated by the hormone vasopressin (antidiuretic hormone). Salt excretion is controlled by aldosterone, a hormone of the adrenal cortex.

2) In plants, the expulsion of certain metabolic products to the exterior of the plant or to different sites within the plant where they no longer participate in metaboilism. Nutrients taken up by plants are utilized very rationally and economically, so that plants produce less waste material than do animals. Nitrogenous end products of protein metabolism are not excreted by plants, or only in extremely small quantities. Some of the material produced by ongoing metabolism is not removed to the plant exterior, but transported then stored or encapsulated at certain sites within the plant. Numerous degradation products are removed from the plant by leaf loss during the autumn. There is, however, no plant equivalent of animal feces. A distinction is drawn between *recretion* (removal of unchanged inorganic material which does not appear to have participated in metabolism, e.g. water, calcium salts, potassium salts), *excretion* (release of degradation products which have no further use for the organism), and *secretion* (formation and secretion of functionally important substances). It is often difficult to differentiate between these three types of expulsion, especially since the physiological role of many secondary plant products is unknown. Furthermore, secreted or stored substances may be ecologically important, e.g. by their effects on other organisms (see Allopathy).

In higher plants, substances can be actively secreted or excreted via gland cells and glandular tissue, in which secretions and excretions pass through the cell wall, either into intercellular cavities, or to the exterior of the plant. Known examples are guttation, secretion of sugary fluids by nectaries, the glandular products of carnivorous plants, as well as the formation and storage of essential oils, resins, gums and mucilages, etc. Root secretions (see) are important for the microflora of the rhizosphere.

Different substances are also exported or expelled by microorganisms, e.g. fermentation products like ethanol or ethanoic acid. Many parasitic and saprophytic microorganisms secrete enzymes; others secrete antibiotics, while disease-causing microorganisms secrete toxins.

Within the plant, metabolic products are released by excretory or secretory cells and tissues, e.g. tannins, essential oils, latex.

Excretory materials: end products of metabolism which are no longer utilizable by the body, and which are removed from the body by excretory organs. E.m. also include foreign substances which are taken up by the body but cannot be utilized.

Excretory organs: organs that serve the function of Excretion (see). Specialized E.o. are absent from marine and parasitic unicellular organisms, sponges, coelenterates, *Turbellaria,* echinoderms and tunicates. All the remaining metazoans have specialized E.o. Freshwater unicellular organisms excrete waste products via pulsating vacuoles. In the other invertebrates, the E.o. consist of either simple or branched canals. In the protonephridia of flatworms, proboscis worms and *Rotifera,* and in the solenocytes of some annelids, these canals start blind in the body. In the metanephridia of annelids, mollusks and crustaceans, they start as open, ciliated funnels in the coelom. In mollusks they originate in the pericardium, while in crustaceans they originate as maxillary and antennal glands in the residues of the coelomic space.

Insects, myriapods and arachnids possess specialized E.o. known as *Malpighian vessels* or *tubules,* which are thin, tubular evaginations of the gut. Their function has been investigated in detail in the Indian stick insect *(Carausius = Dixippus morosus),* commonly known as the laboratory stick insect (Fig.1). Under the influence of the relatively high circulatory pressure, a clear fluid is formed inside the blind end of the Malpighian tubule. This fluid is approximately iso-osmotic with the hemolymph, and it contains dissolved sodium and potassium urate. The flow of water from the hemolymph into the Malpighian tubule is coupled to the flow of sodium and potassium ions in the same direction. Uric acid is normally insoluble, but it is transported through the tubule wall as its soluble salts. In the lower section of the tubule, and in the gut and rectum, the sodum and potassium ions are reabsorbed, together with hydrogen carbonate ions, sugars, amino acids and water. As a result of the reabsorption of sodium and potassium, the initially neutral or weakly alkaline fluid ("urine") becomes acidic, and the uric acid crystallizes out.

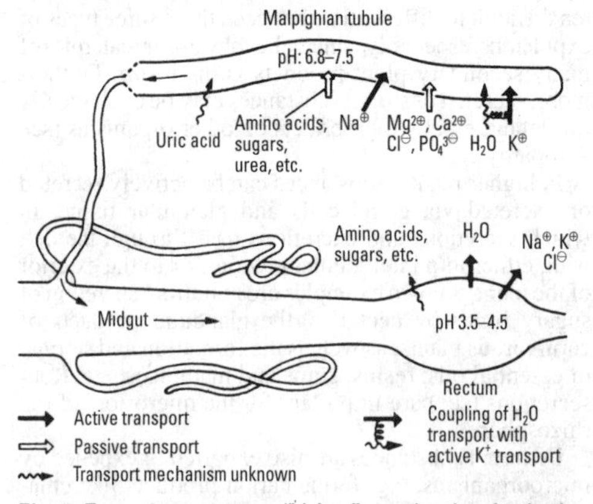

Fig.1. *Excretory organs.* "Urine" production in the laboratory stick insect.

The primary E.o. of vertebrates, including humans, are the paired, dorsally situated *kidneys* (Fig.2). In more primitive vertebrates, these are elongated, lobed structures. In mammals, they are usually compact and bean-shaped. Within the kidney, small arteries form loops *(glomerulus),* in which the material for excretion are transferred from the blood into a capsule *(Bowman's capsule).* Glomerulus and capsule form the *Malpighian body,* which continues as a *uriniferous tubule.* Malpighian body and uriniferous tubule together form the *nephron,* which opens into the renal pelvis, from which the ureter conducts urine to the urinary bladder or cloaca. In particular, the kidneys are responsible for excreting the nitrogenous end products of protein and amino acid metabolism, in addition to playing an important role in regulating the water and salt economy of the body.

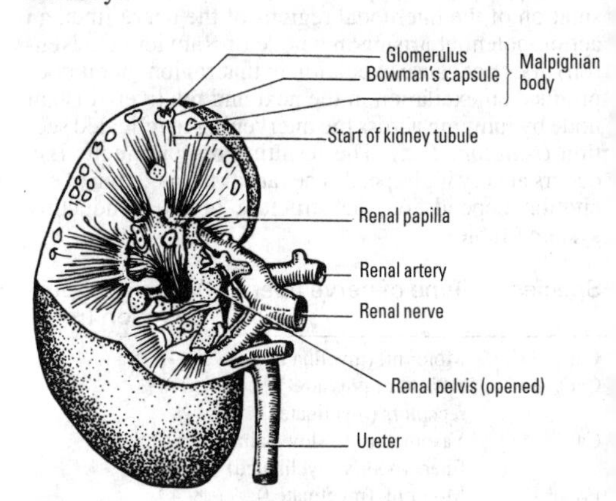

Fig.2. *Excretory organs.* Section through a human kidney.

Ontogenetically and phylogenetically, the kidney passes through three generations or developmental stages. The early vertebrate embryo possesses a pair of structures called *pronephroi* or *protonephroi* (singular *pronephros* or *protonephros),* which are formed dorsolaterally at the anterior end of the trunk. Each pronephros consists of two or three segmentally arranged tubules, which open separately into the coelom by a funnel-shaped *nephrostome,* and at the other end into a *segmental duct,* which empties into the cloaca. Malpighian bodies are absent, but there is a small knot of capillaries on a neighboring part of the coelomic epithelium. The pronephros is generally only a transitory organ; in fishes and amphibians, it may persist into the larval form, but then rapidly degenerates, together with the anterior part of the segmental duct. The pronephroi are succeeded by the *mesonephroi.* Each *mesonephros* consists of segmentally arranged tubules, each with a nephrostome and a Malpighian body, and all opening into the posterior part of the segmental duct, now known as the *mesonephric* or *Wolffian duct.* Association with body segmentation is then lost, the number of mesonephric tubules increases considerably, and a compact kidney is formed. In fishes and amphibians, this mesonephros is the adult kidney; the mesonephric duct becomes the ureter, and also serves as the vas deferens. The mesonephros is therefore the earliest form of kidney to be associated with the sex ducts. In the development of amniotes (reptiles, birds and mammals), pronephroi and mesonephroi are formed, but only transitorily. Before the mesonephroi disappear, a third group of excretory tubules is formed on each side, posterior to the mesonephros, and called the *metanephros,* which becomes the final kidney of the adult amniote.

In the wider sense, CO_2-releasing gills and lungs, as well as glands that secrete perspiration, sebum, mucus and salt, are all E.o.

The principle of urine formation is very similar in all types of E.o., despite their different structures. It commences with filtration of the extracellular fluid (coelomic fluid, hemolymph or blood). This may involve cilia or flagella, but it usually driven by a local hydrostatic pressure difference, resulting in the formation of the primary urine, which is approximately iso-osmotic with the extracellular fluid. Diffusion and transport (the latter is an energy-dependent process) are then responsible for reabsorption of water and dissolved materials from the lumen of the renal tubule, as well as the secretion of further materials into the primary urine from the cells lining the renal tubule. For a detailed account of these processes, see Kidney function.

Excretory tissues of plants:

1) In the narrow sense, all unicellular and multicellular, interior and epidermal systems of plants, whose metabolism is entirely or largely directed to the extensive production of specific metabolites, which in contrast to reserve or storage materials are then excluded from further metabolism.

E.ts. in the narrow sense include:

a) *Glandular cells* and *glandular tissues,* which exist individually or in groups in the epidermis and parenchyma. These are always living cells with abundant cytoplasm, a large nucleus, numerous mitochondria, many Golgi apparatuses, and often a porous cuticle. Their products, many of which are ecologically important, are secreted to the exterior of the plant, or into intercellular spaces. A continuous layer of glandular cells is known as a *glandular epithelium.* Glandular cells and epithelia frequently occur in epidermal tissue, i.e. on the surface of the plant, and they are classified according to their secreted products, e.g. mucilage, resin, salt, oil, and digestive glands (the latter are found, e.g. in the carnivorous plant, *Pinguicula).* Epidermal glandular cells often occur as glandular hairs, e.g. capitate hairs, in which the head or terminal cell is glandular. Glandular cells may also be scale-like, while multicellular glandular papillae are found on some plants. The gland cell secretes its contents (essential oil, resin, mucilage, etc) into the space between the cell wall and the cuticle. This raises the cuticle, which in many plants (e.g. *Primula, Pelargonium,* many *Lamiaceae)* finally ruptures and releases the secreted material. Nectaries, which produce sugary secretions attractive to insects, also release their secretions in this way; they may take the form of glandular epithelia or of glandular hairs, and are naturally regarded as secretory rather than excretory in function. Nectaries occur primarily within flowers as *floral nectaries,* more rarely as *extrafloral nectaries* on petioles (e.g. *Prunus, Acacia),* stipules (e.g. *Vicia),* or in the angle between the veins of the leaf (e.g. *Catalpa). Internal E.ts.* are always enclosed by parenchyma or other tissue. These are also classified according to their products (e.g. oil, resin, latex and mucilage glands), and they usually release their secretions into intercellular spaces or branched or unbranched intercellular canals, formed by glandular cells moving apart from each other. Such channels, which usually permeate the entire plant, are known as *schizogenic secretory reservoirs.* These include, e.g. the resin ducts of many coniferous trees, the oil canals of the *Umbelliferae,* the mucilage and gum channels in the *Cycadaceae,* and the schizogenous sacs of essential oils in species of *Hypericum.*

A *hydathode* is a specialized type of water-secreting gland cell (see Guttation), which enables the plant, under very humid conditions, to secrete droplets of water (which often contain mineral salts). In many mono- and dicotyledonous plants, water-secreting cells are present at the leaf tips (e.g. *Graminae)* or the serrations of the leaf margin (e.g. *Alchemilla, Fuchsia),* or at the ends of the main veins of the leaf (e.g. *Tropaeolium).* These groups of small, water-secreting, chlorophyll-free parenchyma cells (usually with abundant cytoplasm and a large nucleus) are found beneath special water stomata, and are known as *epithems.* A group of these cells inclusive of the water stoma is called an *epithemal hydathode.* Water can also be secreted by epidermal hydathodes, either by groups of modified epidermal cells, or by multicellular hairs known as *trichome hydathodes.*

b) *Excretory cells* and *tissues* are found scattered and embedded in all kinds of primary and secondary plant tissues. They may occur as idioblasts, or they may form an entire tissue. Their secretory products, e.g. resins, mucilages, latex, tannins, alkaloids, and sometimes essential oils are formed by the endoplasmic reticulum or Golgi apparatus; calcium oxalate crystals are also very common. These end products of metabolism normally remain inside the cells that produce them, accumulating in the vacuoles until they fill the entire cell. Often the walls of these cells are suberized.

Unsegmented or *simple* latex ducts are single secretory or excretory cells found in many *Euphorbiaceae, Moraceae, Apocynaceae* and *Esclepiadaceae.* These are branched tubes, up to several meters in length, with a cellulose wall lined with multinucleate cytoplasm. They develop from multinucleate, cylindrical meristematic cells in the young embryo. During plant growth, they branch frequently, grow more or less parallel to the longitudinal axis of the plant organs, and thereby penetrate all parts of the plant. In place of normal cell sap, the ducts are filled with a milky white (occasionally otherwise colored) watery latex that coagulates on exposure to air. In contrast, *segmented* or *compound latex ducts* represent a cell complex formed by dissolution of the transverse walls of neighboring cells, and which forms a reticulum or network within the plant. Compound latex ducts are restricted to certain families: *Euphorbiaceae* (including the economically important rubber tree, *Hevea), Papaveraceae (Papaver* and *Chelidonium,* the latter having an orange latex), *Campanulaceae* and the liguliflorous *Compositae* (e.g. *Cichorium, Taraxacum, Lactuca,* etc.). The latex of both simple and compound ducts contains dissolved sugars, tannins, glycosides, sometimes poisonous alkaloids, and especially calcium malate. The latexes of *Ficus carica* and *Carica papaya* also contain proteolytic enzymes. Some latexes also contain emulsified droplets of essential oils, gum-resins, caoutchouc, guttapercha, etc., as well as solid constituents such as starch and protein granules. *Lysigenic secretory reservoirs* arise when the cell walls of a group of secretory cells become degraded, and the contents (usually an essential oil) form a single deposit. This type of reservoir occurs, e.g. in the skins of many citrus fruits.

2) In the wider sense, all tissues that release material (even oxygen, carbon dioxide and water vapor to the atmosphere) can be regarded as E.ts. of the plant. By this definition, green spongy mesophyll with its many intercellular spaces, as well as green primary cortical parenchyma, which is also well aerated via intercellular spaces, lenticels and stoma, may be considered as E.t.

Exine: the exterior, tough, cutinized layer of spores of the *Archegoniatae* and the pollen of spermatophytes. See Flower.

Exobasidiales: see *Basidiomycetes.*

Exocarp: see Fruit.

Exocoelom: see Allantois, Extraembryonic membranes, Gastrulation.

Exocranium: see Skull.

Exocyclica: see *Echinoidea.*

Exocytosis: the emptying of Golgi vesicles or undigestible residual bodies at the cell surface by fusion of vesicular membranes with the plasma membrane. After releasing their contents to the exterior, the vesicular membranes or the membranes of the residual bodies are integrated into the plasma membrane.

Exodemic: a type of Epidemic (see) which affects genetically different host individuals, whose tissues have different degrees of resistance.

Exogenote: see Endogenote.

Exogenous: arising from outside or the exterior, e.g. environmental conditions are E. factors that act on living organisms.

Exogyra (Greek *exo* outside, *gyros* rounded): an old collective genus of oysters, comprising the modern genera *Aetostreon* and *Ceratostreon.* Each valve possesses a whirl which is turned spirally to the side and bent somewhat inward. In young specimens it is fused with the left vaulted valve, but later becomes free.

Exon: a section of a eukaryotic gene containing genetic information for the synthesis of the partial sequence of a gene product (polypeptide). Many eukaryotic genes are composed of *exons* and *introns,* the latter playing no part in encoding the amino acid sequence of the gene product. In the course of Transcription (see), the total DNA of the gene is first transcribed into pre-mRNA. In a further stage, known as "processing" or "post-transcriptional modification", the mature mRNA required for Translation (see) is formed, i.e. the introns are excised from the pre-mRNA, and the exons containing the genetic information are spiced together in correct sequence.

Exoskeleton: a more or less hard body covering of multicellular animals, which is derived from ectoderm. In arthropods, the E. takes the form of a chitinous cuticle, whereas in mollusks and brachiopods it is a simple or bivalve calcareous shell. Echinoderms possess an internal skeleton (endoskeleton), which is formed by cells beneath the epidermis, and is therefore of mesodermal origin.

Exospore: the outer layer of the spore cell wall. See Endospore.

Exosporium: see Bacterial spore.

Exotoxins: see Bacterial toxins.

Expanded tip receptors: see Tactile stimulus reception.

Expedient, ***emitter:*** the transmitter of news, information or signals in communication between organisms. See Biocommunication.

Expiratory reserve volume of the lungs: see Lung volume.

Explantation: removal of a piece of tissue from an organism, followed by its culture in a suitable nutrient solution. See Transplantation, Experimental biology.

Exploratory behavior, ***"curiosity behavior":*** behavior which serves to exchange information with the environment (i.e. to receive signals from, and emit signals to, the environment). By E.b., animals learn to recognize new objects and situations. E.b. leads to the (partial) space-time orientation of the animal with respect to its surroundings. It may be directed to a particular functional object (e.g. food, or any new object or situation within already familiar surroundings), or it may be autonomous, i.e. unrelated to any specific object.

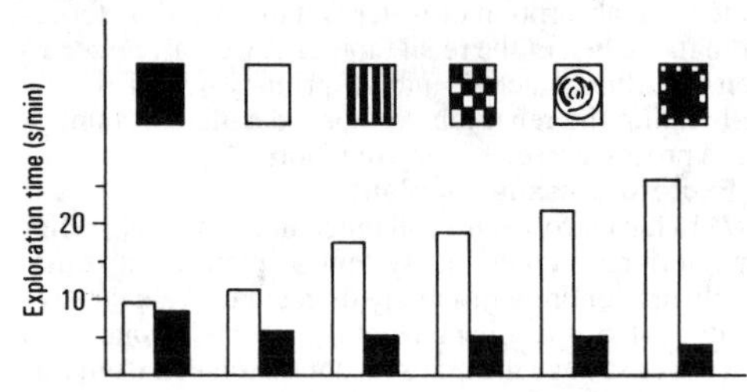

Exploratory behavior. Time spent in the exploration of different patterns by rhesus monkeys reared under different conditions. White columns: reared in the wild. Black columns: reared in captivity and socially isolated from the 9th to 12th month of life. Upper row: different patterns presented to arouse "curiosity". Ordinate: the duration of exploratory behavior in seconds per minute.

Explosive mechanisms, ***explosive movements:*** in plants, mechanisms that serve for the distribution of seeds, pollen or spores. A stable resistant tissue suddenly gives way at a predetermined site, due to the strain exerted on it from within by other tissue under high turgor pressure. The latter, together with the seeds, etc. are sometimes ejected over distances of several meters (see Squirting mechanisms). In other cases, the erectile tissue is external to the resistant tissue, so that when the latter gives way, the seeds, etc. are catapulted (see Catapulting mechanisms).

Expressivity: the degree of expression of a gene or gene combination in the phenotype. E. is influenced by Modifiers (see) and environmental conditions.

Extension growth: an irreversible increase in the volume of plant cells, due to cell extension, associated with a marked uptake of water, vacuole formation, and an increase in the surface area of the cell wall. E.g. generally occurs without an increase in the protein content of the protoplasm. It may occur with surprising speed, e.g. the sprouting of buds in spring, the intercalary growth of grasses and cereals, and the opening of flowers, etc. The stamens of grasses can elongate at the rate of 2 mm/min by E.g. The main root of the broad bean, *Vicia faba,* extends at a rate of 5–15 mm/day. Normally, the average growth rate of plant organs ranges from 0.003 to 0.01 mm/min. E.g. is the exclusive property of young, undifferentiated cells, which still possess only a primary cell wall. Such cells are found in the *extension zones* or *growth zones* of plants. In the shoot of higher plants, *shoot extension growth* is due to more or less elongated apical growth zones (e.g. up to 50 cm long in *Asparagus officinalis)* behind the vegetative point. In plants where the axis is divided into nodes and internodes, growth hardly occurs at the nodes, and each internode has a zone of E.g., usually in its lower part. In the *Graminae,* this *Intercalary growth* (see) continues over an extended period in the life of the plant, and just preceding each zone of E.g. is a mersitematic zone of cell division and plasmatic growth. Similar intercalary growth zones occur in the leaves of conifers, monocotyledons and some dicotyledons. For example, the petiole behaves as an intercalary zone between the lamina and the leaf base. Longitudinal growth of roots occurs exclusively in a narrow extension zone directly behind the root tip. This can be demonstrated by marking the root tip of a young broad bean seedling with Indian ink at fixed intervals. One day later, the ink lines show different spacing on different parts of the root, due to the unequal growth rates of individual root zones. Similar procedures can be used to determine the growth rate, e.g. by simple observation of markings on different parts of the root under a horizontal microscope. There are several other more or less complex devices and methods for measuring the E.g. of plant parts. The E.g. of each growing cell follows a characteristic pattern: immediately behind the meristematic root tip there is only slight growth. E.g. then increases to an optimum and falls away again.

E.g. starts with an increase in the plasticity and deformability (i.e. a decrease in the elasticity) of the cell wall. These changes, which are initiated by auxin, are due to a loosening of the microfibril skeleton. This appears to be the result of the dissolution of the hydrogen bonds that bind cellulose molecules to hemicellulose molecules and anchor them in the wall matrix. Synthesis of a number of enzymes required for this loosening process is known to be activated when the cell is exposed to auxin. Loosening of the cell wall decreases the wall pressure, which is an important osmotic quantity (see Osmosis). If the wall elasticity of a turgid cell suddenly decreases (i.e. it becomes softer and more plastic), the suction pressure equation demands that the resulting decrease in wall pressure leads to an increase in suction pressure. The inflowing water then stretches the cell wall, representing the primary driving force of E.g. As part of osmoregulation, the inflowing water contains salts and sugars so that the osmotic pressure of the cell is not decreased. It is doubtful, however, that the sudden water uptake at the beginning of E.g. is entirely explicable in these simple terms.

The thickness of the original primary lamella is decreased by this initial rapid cell extension. This is compensated by the secretion of new matrix material and the formation of new cellulose microfibrils. At first, these microfibrils are incorporated into the existing meshwork (see Intussusception), and subsequently they form successive layers on the inner surface of the cell wall (see Apposition growth). Microfibrils in each layer tend to be ordered in parallel, and the direction often changes in neighboring layers. Microfibrils in the primary wall are oriented at random, but with E.g. they tend to become parallel to the long axis of the cell, indicating that they undergo gliding movement in the amorphous matrix. During E.g., the orientation of microfibrils, especially those laid down most recently, continues to be changed. As E.g. slows down, the orientation of deposited microfibrils is less affected. The cell wall then increases in thickness by apposition of parallel layers of microfibrils. The oldest microfibrils (i.e. those furthest removed from the cell cytoplasm) therefore display the greatest degree of orientation with the long axis of the cell. Such growth, with ever decreasing displacement of the later microfibrils is called *multi-net growth.* Material for cell wall synthesis is provided by the Glogi vesicles, which are often found concentrated near to the growing cell wall. Auxin stimulates wall synthesis, with the induction of cellulose synthetase and other enzymes involved in the synthesis of cellulose, hemicellulose and pectic acid. Osmoregulation and cell wall synthesis are dependent on a source of energy, and E.g. is prevented by the absence of ATP.

Extensors: muscles that serve to straighten out or extend a limb. See Flexors, Abductor, Adductor.

Extermination: elimination of an animal or plant species in a certain geographical area or throughout its entire range. The process is sometimes intentional, usually unintentional, but largely the result of human activity. See Extinction.

External acoustic meatus: see Auditory organ.

External auditory canal: see Auditory organ.

Extinction:

1) (plates 36 & 37): the discontinuation of the existence of an animal or plant species or taxon. Of all the species that have evolved during the history of life on this planet, only a small proportion is alive today. For animals it is estimated that this proportion is less than 1%. For all types of living organisms, probably about one in every thousand species survives. The estimated total evolutionary progeny is 5 000–50 000 million, whereas only 5–50 million species are represented today. This means that approximately the same number of species has become extinct as has evolved. The process of E., although poorly understood, should therefore be considered as seriously as evolution itself, as a factor determining the quality and quantity of species on the planet.

Many species were anihilated by more adaptable competitors, or were unable to compensate for losses caused by new predators, parasites or diseases. The physical environment has also influenced the process of, and has sometimes been the main cause of E., e.g. adverse climatic changes have often made a significant contribution to E. Moreover, the often proposed E. of species or higher taxa by environmental catastrophes cannot be excluded. Occasionally less specialised species or genera have avoided extinction, whereas related but less adaptable species have died out; the surviving species then become "living fossils". Many animals and plants, however, do not become extinct in the true sense; they undergo *pseudo-E.,* i.e. they disappear from the fossil record by evolving into something else (the genome is not lost but altered). *True E.* occurs when all the representatives of a species die outright without issue. Some groups of organisms, e.g. the ammonites, once displayed great diversity, but nevertheless died out completely. In the ammonites and in certain other animal groups, E. was preceded by the appearance of degenerative and aberrant forms, which has led to the unproven and unexplained hypothesis that animal groups, like individuals, attain maturity, then age, and eventually die.

E. due to human agency. It is probable that stone age hunters at the end of the last Ice Age exterminated some of the animals they hunted. In present times, the E. of animals and plants has reached the scale of a catastrophe, which is hardly matched by the environmental crises of the geological past. Numbers quoted for animals (e.g. almost 500 species have become extinct in the last 2000 years) refer largely to vertebrates. For the much greater loss of invertebrates, the data are only fragmentary. This paucity of data for invertebrates is apparent from the Red Data Book (see), the IUCN (see) and regional red lists. The number of threatened plant species has been estimated at 25 000. Statistical data on the decline of species in relatively large areas conceal the extent of the decrease in separate smaller areas. It is unusual for a single species in an area to become extinct or be threatened with extinction (all other species in the same area being unaffected), and for this process to be due to a single causative factor. For example, excessive exploitation often only becomes dangerous after contraction of the living area or loss of habitats has already led to a weakening of the population, or when commercial hunting and hunting for sport supplement one another in their devastating effects. The following human-related factors are known to contribute to the E. of vertebrates. At the moment, factors 7 and 8 represent added dangers, without being mainly responsible for the E. of any species.

1. Biotope alterations
1.1 Direct: deforestation, drainage, cultivation, etc.
1.2 Indirect: escaped or introduced animals that become established in the wild, e.g. rabbits, sheep, goats, etc.
2. Overexploitation of the stock
2.1 Animals taken for meat and fat.
2.2 Collection of eggs.
2.3 Animals taken for fur, skin or feathers.
2.4 Capture for trade (zoos, pets, etc).
2.5 Animals taken for preparation of traditional remedies.
3. Luxury hunting for trophies and souvenirs
4. Introduction of carnivorous (in the widest sense) animals
4.1 Introduction of free living predators for biological control (foxes, martens, mongooses).
4.2 Naturalization and establishment in the wild of cats, dogs and pigs.
4.3 Involuntary introduction and spread of rats.
5. Control as supposed pests.
6. Involuntary introduction of diseases.
7.Use of plant protection agents.
8. Excessive losses in road traffic.
9. Contamination of water.
10. Introduction of superior competitors.

Geography of E. Island inhabitants are especially susceptible to E. Island populations are naturally small, especially those of larger birds and mammals. Extreme and unfavorable environmental conditions may therefore readily decrease the population to a size from which it cannot recover. In addition, the genetic adaptability of island populations is limited, because the genetic potential of the population is derived from only a few founder individuals. Many island birds have evolved into forms that have lost the ability to fly and/or have lost their protective and defensive instincts against predators. A strikingly high proportion of extinct birds were island inhabitants, whereas mammals have tended to become extinct in continental areas. This is explained mainly by the fact that (with the exception of a few bats) mammals are usually absent from the ecosystems of those islands most greatly affected by man. Furthermore, mainland mammals are often hunted or are claimed to be pests. Due to the widespread destructioin of rainforests, the most threated plant species are those of the tropics and subtropics.

2) see Absorbance.

Extracellular digestion: the commonest form of digestion, in which digestive enzymes are secreted from cells to the exterior (e.g. into the intestinal lumen), where the food is digested.

Extracellular matrix, ***intracellular substance:*** an intricate network of macromolecules occupying the spaces between the cells of multicellular animals. The E.m. is pàrticularly important in the reinforcement of support tissues. During development it is fluid, but becomes stiff in mature tissues. Under the light microscope, it appears as a structureless mass of ground substance with embedded connective tissue fibers. The ground substance consists primarily of a mixture of proteoglycans and glycoproteins, whereas the fibers consist mainly of fibrous proteins. E.m. is formed by the rough endoplasmic reticulum and Golgi apparatus of fibroblasts, then secreted into the intercellular space. E.m. forms an intimate functional connection with neighboring cells; at its point of contact with a cell epithelium the E.m. forms a thin tough layer known as the basal lamina. Depending on the relative quantities and types of its constituent macromolecules, the E.m. displays a variety of forms, e.g. it becomes calcified to the hard structures of bone or teeth, it forms the transparent matrix of the cornea, and it becomes organized into the highly tensile structure of tendons.

Extraembryonic coelom: see Extraembryonic membranes; Allantois.

Extraembryonic membranes, ***embryonal membranes:*** membranous structures formed by the embryo from embryonal cells (see Embryonal organs) and related to the embryonal germ layers, but not participating as differentiating structures of the embryo. After the completion of embryonal development, the E.m. are discarded, either remaining as shrivelled residues in the egg case, or ejected separately during parturition as in mammals. E.m. occur only sporadically in lower vertebrate embryos, but they are typically found in the embryos of insects, scorpions, sauropsids and mammals.

Insects and scorpions form two E.m., an inner Amnion (see) and an outer serosa, both derived from the large-celled membrane primordium, which is part of the blastoderm (see Gastrulation). There are two mechanisms of formation.

1. In insects, in which the embryonic primordium is relatively long (e.g. *Lepidoptera,* many *Diptera* and *Coleoptera),* the margins of the embryonic primordium are composed of presumptive amnion. These fold over the embryonic ectoderm, and carry the extraembryonic

ectoderm with them. The folds meet and merge along the ventral midline, so that the germ band becomes covered by a double layer of extraembryonic ectoderm. The inner layer is the amnion and the outer layer (which extends around the yolk mass) is the serosa.

2. In short and medium length primordia (e.g. in the *Hemiptera, Phthiraptera,* some *Orthoptera),* the germ bands sinks into the yolk. As this occurs, each germ band margin proliferates a sheet of smaller cells between itself and the corresponding margin of the superficial extraembryonic ectoderm. These two sheets of cells form the walls of a cavity as the germ band sinks. The opening may close completely, or a small pore may be retained. As in the long primordia, the extraembryonic ectoderm is called the serosa, and the cavity above the germ band is the amniotic cavity, filled with amniotic fluid. The newly proliferated wall of the cavity is the amnion.

The amnion is always contiguous with the ectoderm of the germ band. The serosa is contiguous with the blastoderm, which itself becomes serosa and covers the yolk. Individual variations of this process are found in different insect orders. It should be noted that these two membranes are not homologous with cellular monolayer structures of the same name in vertebrates.

Vertebrates can be divided into two groups: the *Amniota* (or *Allantoidica),* comprising the reptiles, birds and mammals, which form an allantois (see Embryonal organs) as well as E.m.; and the *Anamniota* (or *Anallantoidica),* which lack these structures. In reptiles and birds, membrane formation is simply a folding process, which occurs in the somatopleure (see Embryonal organs) during neurulation, i.e. after the formation of notochord and mesoderm is finished. This amnionic fold produces two membranes, the Amnion (see) and the chorion (see Serosa), each arising from two layers, the parietal mesoderm (somatopleure) and the ectoderm. Between these two is exocoelom, i.e. residues of the extraembryonal coelom. Folding commences almost simultaneously at the cranial and caudal ends of the embryo, represented by the proamniotic fold (head fold) and caudal fold, respectively. Both rise up as ectodermal folds, and temporarily take the endoderm with them, but this is later displaced and detached by the advancing somatopleure. Shortly afterward, the lateral folds disappear. All the folds grow over each other then fuse into the amniotic navel. The folding phase is seen as a necessary accompaniment of the embryo sinking into the yolk sac, which occurs because the egg shell prevents it from developing outwards.

E.m. formation in mammals, including humans, occurs by two routes. Amniotic folding in rabbits, carnivores, ungulates and lemurs is basically similar to that in birds and reptiles. It commences with the onset of neurulation, i.e. after the appearance of the mesoderm and its differentiation into somites and mesodermal lateral plates (somatopleure and splanchnopleure). Ectoderm and somatopleure fold upward on all sides of the embryo, then unite above it. The chorionic ectoderm becomes blastodermal wall. The somatopleures spread laterally in all directions beneath the blastocyst wall until they grow together under the yolk sac, so that the blastocyst wall is also part of the chorion. In contrast the amnion of insectivores, rodents, primates and humans is formed by cavitation of the inner cell mass of the early embryo (i.e. formation of an embryocyst, see Gastrulation). In most mammals, (except humans) the somatopleure then arches over on all sides, and comes to lie between the amnion and the blastocyst wall. The inner part of this mesodermal fold forms a layer against the internal surface of the blastocyst wall, which thereby becomes the chorion. Here again, the somatopleure spreads in all directions until its layers meet and fuse under the intraembryonic gut.

Formation of the human chorion is much more complicated. Human chorion arises from two different embryonic components: the inner part of the trophoblast (the cytotrophoblast) and the outer peripheral residue of the morula mesoderm (the chorionic mesoderm). The latter lies within the cytotrophoblast; through a process of extraembryonic coelom formation, it becomes separated earlier from the inner amnion and yolk sac mesoblast layers, which remain on the amnion and the yolk sac. The central mass of the morula mesoderm, however, separates as magma reticulare, and its cleavage spaces open into the extraembryonic coelom. For the detailed mechanism of chorion formation, see Placenta.

In monotremes, marsupials, reptiles and birds, the chorion is a smooth membrane. In all placental mammals, however, the surface of the chorion develops villi. It is then known as the shaggy chorion or chorion frondosum, and the region bearing the villi constitutes the placental area. Not all of the chorion is concerned in placenta formation; as the amniotic cavity expands, villi disappear from that part of the chorion in contact with the decidua capsularis, producing a smooth membranous surface, known as the chorion laeve. Some authors consider the chorion of placental mammals to be the true chorion, the name serosa being reserved for the smooth membrane of other vertebrate embryos. Placen-

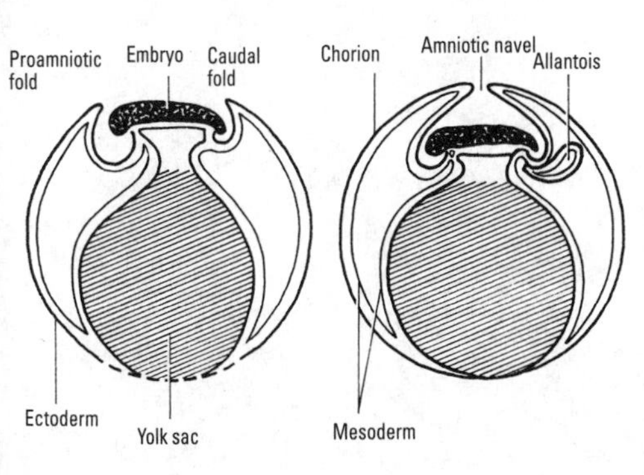

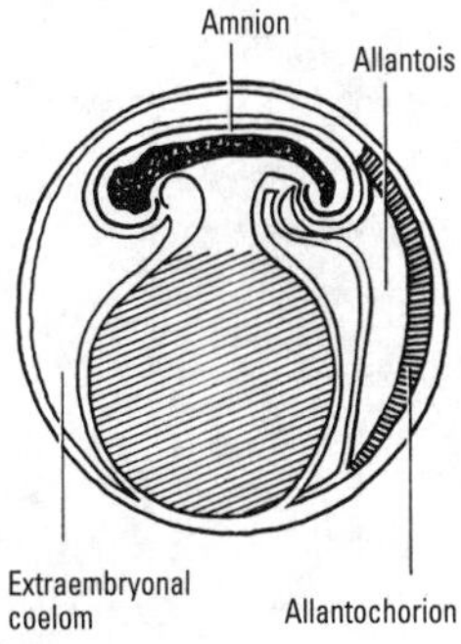

Fig.1. *Extraembryonic membranes.* Schematic representation of the development of the extraembryonic membranes of sauropsids.

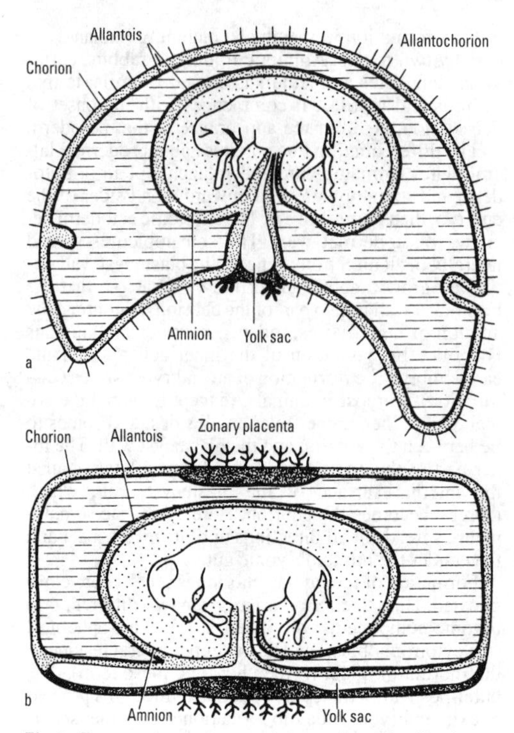

Fig.2. *Extraembryonic membranes.* Longitudinal sections of the mature uteri of a horse *(a)* and a carnivore *(b).*

tal mammals are therefore also called choriates. In addition, in mammals and humans, the amnion and allantois are closely related. Thus, bovines form an extensive amniochorion (amniogenic chorion), whereas horses and carnivores produce a massive allantochorion (see Embryonal organs). In the human embryo, as the allantois and yolk sac decrease in size, the amnion epithelium of the ever increasing amniotic cavity comes to lie firmly against the inside surface of the chorion; the amniotic fluid is weakly alkaline, and contains 1% of dissolved material (proteins, urea, glucose).

Extrazonal vegetation: see Zonal vegetation.

Extreme resistance: a term used in phytopathology for resistance to viruses that displays no specificity for the pathotype. Under outdoor conditions, crop plants with E.r. are free from all types of virus infection. The term is used especially with respect to virus diseases of potatoes.

Exuviae: see Molting.

Exuvial glands: see Molting glands.

Exuviation: see Molting.

Eyebrow ridge, *torus supraorbitalis:* a ridge of bone which lies across the os frontale (frontal bone) of the skull, above and parallel to the upper margin of the orbits. It is present in many apes and in fossil hominids.

Eyed coral: see *Madreporaria.*

Eye-ear plane: see Cranial planes.

Eyed lizard: see Ocellated green lizard.

Eyelash viper: see Lance-head snakes.

Eyes: see Light sensory organs.

Eyespot: see Light sensory organ.

Eye stalks: stalk-like, mobile outgrowths on the heads of many crustaceans. Each E.s. carries a compound eye at its tip.

F

F_1, F_2, F_3 ... F_n: symbols representing successive generations from a genetic cross. F_1 is the first filial generation resulting from the initial cross. F_2 is the second filial generation, and it is the result of self-fertilization of a F_1 individual, or a cross between siblings of the F_1 generation, etc. See Mendel's laws.

Fabales: see Leguminous plants.

Face muscles: muscles of the eyelids, nose and mouth, which by their action or inaction, or by the action of antagonistic muscles, are responsible for the relaxation, tightening, smoothing and creasing of the facial skin, i.e. they are responsible for the wide variety of facial expressions associated with, e.g. joy, laughter, sorrow, anger, etc.

Facial expression: signal behavior of mammals based on the musculature of the head and neck. The signaling system is derived from the primary functions of the corresponding musculature: a) movements of the orofacial region derived from chewing or masticatory movements; b) movements in the region of the eyes, originally to protect the eyes or reinforce vision; c) movement in the aural region, in particular the outer ears, originally to identify the direction of sound; d) movement of the forehead and crown of the head, probably originating from movements to dislodge parasites; e) head movements with changes of tension in the neck and nape region. All of these movements have developed in different ways into secondary movements of F.e., reinforced in primates by further muscle development, in humans by muscles that express laughter. F.e. serves a special "para-acoustic" (in humans paralinguistic) function, i.e. it adds a supplementary visual signal to movements accompanied by the emission of sounds. Since sound emission is also a derived function, F.e. in this case is a tertiary function.

Facial expression. Expression of social contact in the chimpanzee (after Wolf).

Facial nerve: the VII cranial nerve; see Cranial nerves.

Facial profiles: in anthropology, a classification of facial shape based on the degree of protrusion of various facial regions. When a straight line joining the root of the nose (nasion) and the foremost point of the alveolar border of upper jaw (prosthion) is practically perpendicular to the ear-eye plane, the profile is said to be *orthognathic* or *orthognathous.* If the resulting angle is less than 80 °, the profile is *prognathic* or *prognathous.* Protrusion of the entire upper and lower jaw region is *total facial prognathy.* Protrusion of only the alveolar region of the jaws is *alveolar prognathy,* whereas *nasal* or *middle facial prognathy* applies to protrusion of the nasal region. Mongoloid types show slight prognathy of the entire face, whereas negroid and australoid types are markedly prognathic. Alveolar prognathy is the commonest form of prognathy, especially in negroid types, and a less pronounced form of prognathy is fairly common in European types. Pronounced nasal prognathy is generally an individual rather than a racial variant, occuring in particular in humans with a leptosome-type physique.

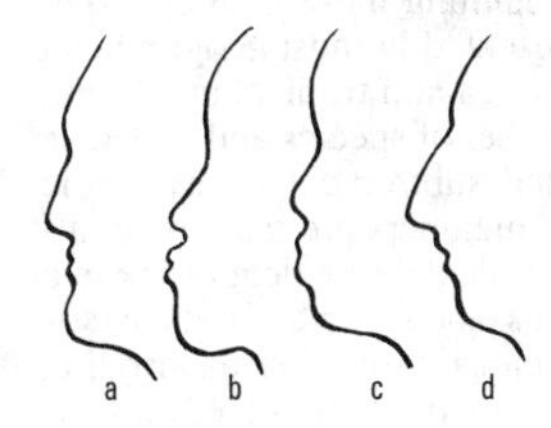

Facial profiles (after H. Bach). *a* Orthognathy. *b* Total facial prognathy. *c* Alveolar prognathy. *d* Middle facial prognathy.

Facial shapes: ten different outline types of the human face used in anthropology (Fig.).

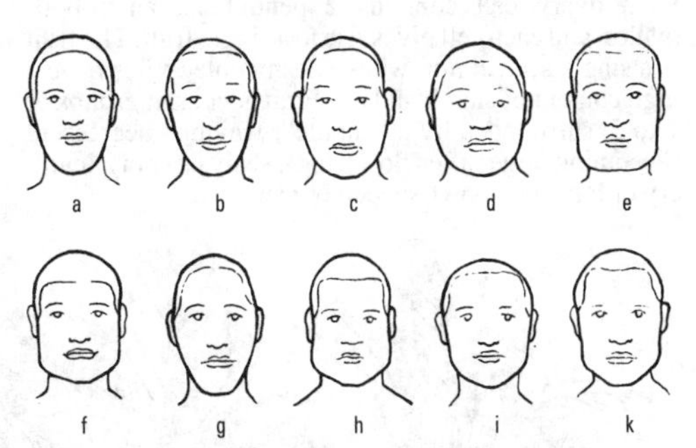

Facial shapes (frontal view, after Pöch). *a* Elliptical. *b* Oval. *c* Inverted oval. *d* Round. *e* Rectangular. *f* Square. *g* Rhombic. *h* Trapezoid. *i* Inverted trapezoid. *k* Pentagonal.

Facies:

1) In geology, the totality of the primary characteristics of a geological sediment or deposit, e.g. the petrographic (petrofacies), lithologic (lithofacies) and fossil content (biofacies). Deposits or strata laid down during the same geological period do not necessarily present the same F., since they may have formed at different rates, or been subject to different modifying influences such as erosion. Accordingly, the following are differentiated: terrestrial or continental F. (land F.), limnic F. (freshwater F., including river, lake and lagoon F.), marine F. (classified into beach, reef, shallow sea and deep sea F.). Formations with the same F. are termed *isopic,* while those with different F. are *heteropic.*

2) In plant ecology, the smallest distinguishable unit of vegetation, characterized by the predominance of one or more particular species.

Facilitated diffusion: see Carrier.

Factor analysis: multivariate statistical analysis, which attempts to relate the expression of a relatively large number of characters of an organism to the activity of a minimal number of factors, when the expression of those characters is subject to random variation. F.a. is used particularly in psychological research, and to determine the relation between morphological and physiological characters.

Factor exchange: recombination of coupled factors or genes in meiosis or mitosis by Crossing over (see).

Facula: claw; see Claws, hoofs and nails.

Facultative aerobe: see Aerobe.

Facultative anaerobe: see Anaerobe.

FAD: acronym of Flavin adenine dinucleotide (see).

Fagaceae, *Cupuliferae, **beech family:*** a family of dicotyledonous plants, containing more than 800 species in 6 genera. It is represented in most geographical regions, except South America and tropical and South Africa, but the greatest number of species and genera is found in the temperate and subtropical zones of the Northern Hemisphere. All members are trees, with alternate, simple and usually dentate or deeply crenate leaves, and usually caducous (shed at an early stage) stipules. Flowers are monoecious, rarely hermaphrodite. Male flowers form catkins or multi-flowered, tassel-like inflorescences (in *Nothofagus* they are solitary or in threes). Female flowers are solitary or in groups of 3, usually with a 3- or 6-celled ovary with 3 or 6 styles. Each ovary cell contains 2 pendulous, anatropous ovules, and each cell gives rise to a single fruit. The fruit is a single seeded nut, whose fleshy coleoptiles have a high content of starch and oil. Fruits occur in groups of 1 to 3, surrounded by a "cupule", which is accrescent (becoming larger after flowering), scaly or spiny, leathery or lignified, bowl-shaped or capsular.

The ***Sweet Chestnut*** or ***Spanish Chestnut*** *(Castanea sativa)* is native to the Mediterranean region and was probably introduced elsewhere by the Romans. Its spiny fruit cupule contains 1 to 3 edible nuts (sweet chestnuts). The tree is planted widely, often in pure stands as a coppice crop for the production of small poles, and as an ornamental, attaining heights of 10–35 m. ***Beech*** *(Fagus sylvatica)* is one of the most important deciduous forest trees of the northern temperate zone. It forms high (30 m), close-canopy, deeply shaded forests, in which the trunks are smooth and free of branches. Free standing trees are shorter with larger crowns. Beech timber is hard and of high commercial value, being used for furniture, parquetry, wooden toys and a variety of small wooden articles. It is used in parts of Europe for railway sleepers, and it is also valued as firewood. The female inflorescence consists of two short-stalked flowers enclosed in a 4-lobed spiny cupule. Beech nuts have a high content of edible oil. ***Copper beech***, a mutant of *Fagus sylvatica*, possesses conspicuous copper-red foliage, which is colored by the presence of anthocyanins in the leaf epidermal cells. The genus *Nothofagus **(Southern beech)*** is found mainly in the temperate zones of the Southern Hemisphere, where it is also an important forest tree.

The most numerous genus is *Quercus **(Oaks)*** with about 300 species, whose fruits are commonly called acorns. ***Common*** or ***"Penduncúlate" oak*** *(Quercus robor* L.), native in Europe, North Africa and West Asia, has stalked acorns and almost sessile (nonstalked) leaves. It is cultivated as a forestry tree, attaining heights up to 50 m. Older trees are notable for their large circumference, which in exceptional cases has been known to attain 12 m. The tree crown has characteristically "crooked" branches and it forms an open canopy, allowing considerable passage of light. Oakwoods therefore permit the growth of many other herbs and bushes. ***Durmast oak*** or ***"Sessile oak"*** *[Quercus petraea* (Mattuschka) Liebl.], with sessile (nonstalked) acorns and stalked leaves, is native in Europe and West Asia, and it is also cultivated as a forest tree. It displays a straighter growth habit than the sessile oak, with more regular branches, and it prefers lighter, more porous

Fagaceae. a Castanea sativa (sweet or Spanish chestnut). *b Fagus sativa* (beech). *c Quercus robur* (common or penduncúlate oak). *d Quercus petrae* (durmast or sessile oak).

soils, often occurring in mountainous regions. The wood of both penduculate and sessile oak is dense, heavy, hard, rich in tannins and very durable; it is widely used for small boat building, for furniture, high quality joinery, parquetry, cooperage and fencing; its exploitation through the ages, especially for large ship building, led to the disappearance of many natural oak forests. Oak bark contains a high proportion of tannins, and was formerly used in leather tanneries. The acorns were once widely used for feeding pigs. The ***Cork oak*** *(Quercus suber* L.) is found in Europe from Spain to Italy, and in Sicily and North Africa. It has thick, fissured, spongy bark, which yields cork; the bark is stripped every 6–12 years, leaving young reddish bark exposed, which subsequently thickens and turns gray. Commercially grown *Qercus suber* is the basis of an important bottle cork industry in the western Mediterranean. The ***Turkey oak*** *(Quercus cerris* L.) is native in Central Europe and Turkey, and widely planted and naturalized in other parts; it is easily identified by its persistent linear stipules surrounding the bud. ***Evergreen oak*** (***Holm oak, Ilex***, *Quercus ilex* L. Costa) is native to the Mediterranean region, extending north in Western Europe to Britanny. The leaves and young shoots of the ***European white oak*** *(Quercus pubescens* Wlld.) are covered on their undersides with a felt of hairs; this oak usually displays the habit of a shrub or bush, and it is found on south facing hillsides and rocky slopes in Central Europe, especially in the Balkans, Greece and Aegean islands. About 65 species of *Quercus* are native to North America. The ***North American white oak*** *(Quercus alba* L.) has a hard, strong, close-grained wood, which is used commercially; it is a forest tree with light gray, scaly bark, found from Maine to Ontario, Minnesota, Florida and Texas. The ***Red oak*** *(Quercus rubra* L.) is native to eastern North America, and it has been introduced into Europe. It prefers gravelly and sandy soils, and is suitable for planting in exposed sites. The name derives from the red autumnal color of the large, tooth-lobed leaves, which gives the tree an ornamental value; the wood is more porous than that of other timber oaks, but it is sometimes used for furniture.

Fagus: see *Fagaceae.*

Fairy bluebirds: see *Chloropseidae.*

Fairy clubs: see *Basidiomycetes (Poriales).*

Fairy ring: a concentric ring of fungal fruiting bodies. A F.r. arises because the mycelium grows outward in the soil from the fungal spore in an ever widening circle, while the center of the mycelial mass dies off.

Fairy shrimps: see *Anostraca.*

Falco: see *Falconiformes.*

Falconets: see *Falconiformes.*

Falcons: see *Falconiformes.*

Falconidae: see *Falconiformes.*

Falconiformes, *Accipitres,* ***diurnal birds of prey, diurnal raptors:*** an order of birds (see *Avies)* represented throughout the world except in Antarctica, and containing more than 260 species of birds of prey divided into 5 families. All *F.* are powerful fliers. They capture their prey with their powerful talons, which carry long, curved, pointed claws; the hindtoe is opposable. The order also contains carrion feeders (notably the vultures), which do not usually capture prey. The upper mandible of all *F.* is sharply hooked and curved downward, and across its base it carries a fleshy cere, which houses the openings of the nostrils. The bill is used for removing feathers, hair, etc. from the prey, and for tearing the meat of the prey. [Many of these characteristics are shared with the owls, but the latter are classified separately in the order, *Strigiformes,* on account of their different skeletal structure.] Many species have been or are hunted by man. As predators, they occupy a position at the top of the food chain, and therefore suffer from the accumulation of environmental pollutants such as herbicides and pesticides. In addition, they naturally lay small clutches of eggs (small and medium sized species lay 3–4 eggs, large species 1–2) and their reproduction rate is slow. Consequently, many species are threatened with extinction.

The common English nomenclature of the *F.* is quite undisciplined. Terms like "hawk", "buzzard", "eagle", etc. are applied indiscriminately to a range of various species, whereas these terms have a more specific meaning in the strict sense, often being synonymous with a genus or family.

1. Family: *Cathartidae* ***(American vultures).*** These large carrion eaters exploit thermal updrafts to soar or hover from just above tree height up to several hundred meters, seeking food with their keen eyesight. They have rather weak bills and can deal only with partly rotted meat. Phylogenetically they show affinities to storks and to cormorants, and they are less related to the Old World vultures than their superficial appearance would suggest. Unlike all other *F.,* their long toes are not strongly hooked. The hing toe is set a little higher than the 3 front toes, and the latter have rudimentary webs at the base. A syrinx is absent, and they can only hiss. The nasal passages within the bill are only partly separated, and the nostrils are are longitudinal rather than round. According to their rich fossil record, they are a very ancient family, formerly most abundant in the Old World 40–50 million years ago. There are now only 6 living species, all restricted to the New World. *Teratornis incredibilis* (wingspan 485–515 cm), which is known from its fossil remains in the Pleistocene of Nevada, is the largest flying bird known to have lived. Living members of the family are the ***Californian condor*** *(Gymnogyps californianus;* 127 cm; coastal mountains of California) and the ***Andean condor*** *(Vultur gryphus;* 132 cm; Andes from Colombia to Megellan Strait); these are the largest of living flying birds, each with a wingspan of about 300 cm. Other species are the black and white ***King vulture*** *(Sarcorhamphus papa;* 81 cm; southern Mexico to Argentina), with a naked, brightly colored purple, green and yellow head; the ***Turkey vulture*** (also incorrectly called the Turkey buzzard) *(Cathartes aura;* 73.5 cm; southern Canada to Megellan Strait); ***Black vulture*** *(Coragyps atratus;* 63.5 cm; central USA to Argentina and Chile); and the ***Yellow-headed vulture*** *(Cathartes burrovianus;* Mexico to northern Argentina and Uruguay).

2. Family: *Accipitridae* ***(Old World vultures, Eagles, Hawks, Buzzards, Kites, Harriers).*** Most of the diurnal birds of prey are included in this family, which contains over 200 species in more than 60 genera. It is represented throughout the world, except in Antarctica, and in all types of habitat. Members differ from falcons by not possessing a notched upper mandible. Females are generally larger than males. Nine subfamilies are recognized.

1. Subfamily: *Aegypiinae* ***(Old World vultures).*** These have strongly hooked feet, round nostrils and a

syrinx. They evolved from an eagle-like ancestor. Like American vultures, they have bare heads, but this is a secondary adaptation to carrion feeding. The ***Egyptian vulture*** *(Neophron peronopterus;* 66 cm; southern Europe, Southwest Asia, Africa) does not have a bare neck and only the face is devoid of feathers. This dirty-white bird is illustrated on ancient Egyptian tombs, and is also known as "Pharoah's chicken". The ***Eared vulture*** *(Torgos tracheliotus;* 109 cm) is found in temperate and tropical Africa. Other species are the ***White-headed vulture*** *(Trigonoceps occipitalis;* 84 cm; open country in central Africa), the ***Griffon vulture*** *(Gyps fulvus;* 104 cm; mountainous regions in southern Europe, Southwest Asia, South Africa), and the ***Hooded vulture*** *(Necrosyrtes monachus;* 58–63 cm, female larger than the male), which replaces the Egyptian vulture in sub-Saharan Africa, and is one of the commonest vultures throughout eastern Africa from southern Egypt to Natal, including the coast; the back of the head and the nape are covered with a hood of down. Some authorities consider that the ***Palm nut vulture*** *(Gyphohierax angolensis;* 51–60 cm, female larger than the male) belongs to the Sea eagles, whereas others classify it as the most primitive of the Old World vultures. It inhabits the mangrove and rainforest regions of Africa, and feeds mainly on the fleshy pericarp of palm nuts of the oil palm *(Elaeis guineensis)* and the raphia palm *(Raffia ruffia),* also taking fish, mollusks and crabs.

2. Subfamily: *Elaninae* **(White-tailed kites).** These small and very small birds of prey have a nearly cosmopolitan distribution, the genus, *Elanus,* being represented in America, the warmer parts of Europe, southern Asia, Africa and Australia. Examples are the American ***White-tailed kite*** *(Elanus leucurus),* and the Old World ***Black-shouldered kite*** *(Elanus caeruleus),* which hovers like a kestrel before dropping onto small mammals and insects. The subfamily also contains the graceful little ***Swallow-tailed kite*** *(Chelictina riocourii;* northern tropical Africa); this monotypic species has gray and white plumage and a swallow-like forked tail, and rather unusually for birds of prey it breeds in colonies. Also classified with the *Elaninae* is the monotypic ***Pearl kite*** *(Gampsonyx swainsoni;* America), which is one of the smallest birds of prey. The Pearl kite is blackish above, rufous on the head and back, and white below, and has well developed powder down tracts.

3. Subfamily: *Perninae* **(Swallow-tailed** and **Hook-billed kites).** This rather heterogeneous subfamily contains such species as the American ***Swallow-tailed kite*** *(Elanoides forficatus),* a handsome bird whose black back, tail and wings contrast with the rest of its white plumage. It also contains the ***Honey buzzard*** *(Pernis apivorus;* Europe and Asia), which feeds on bees and wasps, and their larvae.

4. Subfamily: *Milvinae* **(True kites** and **Fish eagles).** Largest of these is the ***Lammergeyer*** or ***Lämmergeier*** (or ***Bearded vulture)*** *(Gypaetus barbatus;* 102–114 cm), which inhabits high mountain ranges in North Africa, southern Europe, Afghanistan, India and Tibet, and is characterized by a beard of black bristly feathers that hangs down on either side of bill; it is a carrion feeder and is sometimes classified with the *Aegypiinae.* The rare ***Everglade kite*** *(Rostrhamus sociabilis;* 46 cm; Florida Evergaldes, Cuba and southern Mexico) lives almost entirely on a single species of freshwater snail *(Pomacea,* Apple snail). The ***Black kite*** *(Milvus migrans),* a familiar scavenger in towns and villages of the east, is one of the commonest birds of prey in the warmer parts of the world. A number of large species in the genus, *Haliaeetus,* are known as ***Sea eagles*** or ***Fish eagles;*** generally similar in appearance; most possess a white head and tail; illustrated is the ***White-tailed eagle*** *(Haliaeetus albicilla;* 69–91.5 cm; female larger than male); found from Norway and the Mediterranean to Siberia and Japan, and to Greenland and Iceland, overwintering in the Oriental region and sometimes in Africa; it lives on sea coasts or by river valleys or lakes, and snatches fish from the surface of the water. Also in this genus is the famous ***Bald eagle*** *(Haliaeetus leucocephalus;* 86.5 cm; North America and Northeast Siberia), which is the national bird of the USA.

Falconiformes. White-tailed eagle *(Haliaeetus albicilla).*

5. Subfamily:*Accipitrinae.* The subfamily contains 40 species, all in the genus *Accipiter.* These are hunting hawks, with fast, dashing flight, feeding on small mammals and birds. The ***Goshawk*** *(A. gentilis;* 51–61 cm) is found throughout Eurasia and North America. The Old World ***Sparrowhawk*** *(A. nisus;* 28–38 cm) (plate 6) occurs across Europe and Asia from Spain to Japan, in North Africa and Burma.

6. Subfamily: *Buteoninae* **(Buzzards** and **Eagles).** Twenty six species belong to the genus *Buteo* **(*Buzzards*** in the strict sense), e.g. the ***Augur buzzard*** *(Buteo rufofuscus),* which is found in mountainous parts of Africa up to 5 500 m above sea level. The ***Mexican black hawk*** *(Buteogallus anthracinus)* is a New World species found at low altitudes in swampy areas or tropical forest. Species of the genus, *Butastor,* display some of the habits of harriers, e.g. the African ***Grasshopper buzzard*** *(B. rufipennis).* The second division of the *Buteoninae* consists of the large and very large birds of prey known as ***Eagles,*** and also referred to as ***Booted eagles,*** which distinguishes them from the "eagles" in other taxa of the *F.* The ***Golden eagle*** *(Aquila chrysaetos;* 76–89 cm) is found throughout Eurasia to North Africa and Arabia, and from northern North America to Mexico, in mountainous and moorland areas, montane forests and sometimes sea cliffs. The

golden eagle is dark tawny to chocolate brown with a golden crown and nape; the legs are feathered and the toes and cere are yellow. Also in this group is the ***Indian black eagle*** *(Ictinaetus malayenesis)*. Largest and most powerful of all these eagles is the ***Harpy eagle*** *(Harpia harpyja;* 96.5 cm; southern Mexico to northern Argentina); a jungle-dwelling, crested eagle with extremely strong talons; it eats macaws and displays a marked preference for the 3-toed sloth or ai.

7. Subfamily: *Circinae* ***(Harriers*** or ***Marsh hawks;*** 17 species). Most of these are placed in the genus *Circus.* They course over open grasslands and marshes in search of small mammals, birds, amphibians and reptiles. They are relatively long-legged with long square tails, and they have an owl-like facial disk. The ***Hen Harrier*** or ***Marsh hawk*** *(Circus cyaneus;* 61 cm) breeds throughout the temperate Northern Hemisphere.

8. Subfamily: *Circaetinae* ***(Serpent eagles, Snake eagles*** or ***Harrier eagles).*** These soar over flat marshy or grassland areas, feeding mainly on snakes and reptiles. The monotypic ***Bateleur eagle*** *(Terathopius ecaudatus;* 61 cm) has exceptionally long wings and a very short tail; it inhabits open savanna in sub-Saharan Africa, sometimes travelling as far as Arabia. The ***Short-toed eagle*** *(Circaetus gallicus)* is found in the warmer parts of Europe, and in Asia and Africa The montypic ***Serpent eagle*** *(Dryotriorchis spectabilis)*, which inhabits dense tropical forests in West Africa, is notable for its exceptionally large eyes.

9. Subfamily: *Machaerhamphinae.* This monotypic family consists of the ***Bat hawk*** *(Machaerhamphus alcinus;* Africa and Far East), which catches bats at dusk as they emerge from caves and buildings. It also catches swallows and martins, and the prey is swallowed in flight.

3. Family: *Pandionidae.* This monotypic family consists of the ***Osprey*** or ***Fish hawk*** *(Pandion haliatus;* 61 cm), which is distributed throughout the Northern Hemisphere, migrating south in the winter; it also breeds in Australasia and South Africa. It is a large soaring hawk, resembling the accipitrine hawks in its internal anatomy, but with feather tracts arranged more like those of the New World vultures. The thighs are heavily feathered and the legs are bare. All 4 toes are equal in length (unequal in all other birds of prey), and the outer toe is reversible as in owls. These leg and foot structures are the main basis for its separate familial status. It feeds entirely on fish, which it catches alive by diving from a height of 20–70 meters. In Eurasia it favors pine trees for nesting, but in North America the large nest of sticks is usually built on the top of a tall dead tree stump or telegraph pole. It also readily accepts artificial platforms. The same nest is used year after year.

4. Family: *Falconidae* ***(Falcons*** and ***Caracaras).*** Members of this family usually have a characteristic notch or "tooth" in the upper mandible. Tarsi and feet are bare, and the thighs are covered with loose feathers, giving the appearance of pantaloons. The 58 species are divided into 4 subfamilies.

1. Subfamily: *Herpetotherinae.* This small group consists of the monotypic ***Laughing falcon*** *(Herpetotheres cachinnans;* 40 cm), which feeds on snakes, and the 5 ***Forest falcons*** (genus *Micrastur;* 30–60 cm including tail) of Central and South America.

2. Subfamily: *Polyborinae* or *Daptriinae* ***(Caracaras).*** The 9 species of Caracaras are insectivorous or omnivorous, with a strong preference for carrion. They are long-legged, are able to run quickly, and spend much time on the ground. The ***Caracara*** *(Caracara cheriway;* 61 cm) is the national bird of Mexico. The ***Guadeloupe caraca*** *(Polyborus lutosus)* has become extinct in recent times.

3. Subfamily: *Poliohieracinae* ***(Falconets*** or ***Pygmy falcons;*** about 15 cm). The 8 species are distributed in South America, Malaysia, Philippines and Africa. Most species live on insects, which they hunt like flycatchers, e.g. the ***Philippine falconet*** *(Microhierax erythrogonys).*

4. Subfamily: *Falconinae* ***(True falcons).*** The 35 species are all in the genus *Falco.* It is customary to call the female a Falcon, while the male is known as a Tiercel. The ***Peregrine falcon*** *(F. peregrinus;* 46 cm) is one of the most widely distributed of all bird species (entire Northern Hemisphere, Australia, South Africa, Patagonia), and is the most popular bird for falconry. Diving at speeds up to 175 miles (280 km) per hour, it kills its prey with a single strike of its talons. The ***Merlin*** *(F. columbarius;* 27–33 cm, female larger than male), an inhabitant of upland moors, is a Holarctic species found in the British Isles, Iceland, Scandanavia, northern Asia and North America. It chases small birds (its main prey) near to ground level, following every twist and turn of their flight. In the winter, they move to the southern part of the breeding range, some populations moving even further to the tropics; especially during the winter they may supplement the diet with small mammals, lizards and insects. ***Kestrels*** are small falcons with a cosmopolitan distribution; they prey mostly on small mammals and insects, and characteristrically hover over their hunting ground. The ***Kestrel*** *(F. tinnunculus* ; 34.5 cm) is found all over Europe, Asia and Africa, where it seems to be equally at home on high moorlands, low farmlands and sea cliffs; it is a close relative of the unfortunately named American ***Sparrow hawk*** (also known as the ***American kestrel;*** *F. sparverius;* 30.5 cm; New World from northern treeline to Tierra del Fuego). ***Hobbies*** are small, long-winged, very fast flying falcons, feeding mostly on insects, but also taking small birds in flight. The ***Hobby*** *(F. subbuteo;* 33 cm) breeds right across Asia and Europe from the Atlantic to the Pacific, overwintering mainly in Africa. The ***Saker*** *(F. cherrug;* 45.5 cm) breeds in Central Europe to eastern Tibet and Manchuria, and overwinters in Africa and southern Asia. It inhabits steppes and grassy plateaus up to 1 500 m, where it feeds on small mammals and lizards, sometimes insects; it tends to hover like a kestrel. The ***Lanner*** *(F. biarmicus;* 40.5–46 cm) inhabits savannas and deserts from the Mediterranean region to South Africa, the northern populations moving south within the breeding range in winter. It feeds on large birds, bats, lizards and desert locusts. The larger female is generally browner than the male, which has blue-gray upper parts and a white-tipped tail, pinkish underparts, black and white forehead shading into a rufous crown and nape, yellow legs, cere and eye-ring, and brown eyes. There are several races, the northerly ones tending to be browner and paler. Largest of the falcons is the ***Gyrfalcon*** *(F. rusticolus;* 61 cm) of Arctic America and Europe, overwintering in temperate regions.

5. Family: *Sagittariidae* ***(Secretary bird;*** 117 cm) A monotypic family. The Secretary bird *(Sagittarius serpentarius)* is confined to sub-Saharan Africa. It has a

long crest of plumes. The 3 front toes are webbed, and the hindtoe is very small. It has very long legs and hunts on foot, feeding on snakes, other reptiles, small mammals, insects (especially locusts), young birds and eggs.

Falcons: see *Falconiformes.*

Falculea: see *Vangidae.*

Fall army worm: see Cutworms.

Fallow deer, *Dama dama:* a medium-sized deer, originally native to the Mediterranean region. It is now widely established by introduction, e.g. in Europe as far north as Sweden, in western Russia, and in USA, Canada, South America, West Indies, Madagascar, Japan, Fiji, Australia, Tasmania and New Zealand. Antlers of mature males have a wide, flat surface and numerous points. The summer coat is red-brown with white spots, while the winter coat is gray-brown. Black as well as white individuals also occur. *Dama* is a monotypic genus. See Deer.

Fall webworm: see *Arctiidae.*

False acacia: see *Papilionaceae.*

False angel wing: see *Bivalvia.*

False coral snakes:

1) Anilius scytale, see *Aniliidae.*

2) see King snakes.

False gharial, *Malayan gharial,* *Tomistoma schlegelii:* a member of the *Crocodylidae* (see), 5 m in length (7 m specimens have been claimed), native to Malaysia, Sumatra and Borneo. It resembles the gharials (see *Gavialidae)* in that it possesses a long, slender snout, and feeds mainly on fish. Like many crocodilians the F.g. is a threatened species.

False mildew: see *Phycomycetes.*

False scorpions, *Pseudoscorpionida:* an order of the phylum *Arthropoda,* class *Arachnida.* Only a few millimeters long, F.s. bear a certain external resemblance to true scorpions. The flat body consists of a prosoma (carapace), which is entire but with transverse furrows, and a segmented opisthosoma. Appendages consist of the small, 2-jointed chelicerae; large, 6-jointed pedipalps; and 4 pairs of 6- or 7-jointed legs. A silk gland (homologous with the poison gland of the *Aranea)* opens through a galea or spinneret on the tip of the movable finger of each chelicera. The pedipalps sometimes possess poison glands which open through teeth in the chelae.

Biology. F.s. live in vegetation, soil, beneath tree bark, in the nests of birds, mammals and insects, in human dwellings, herbaria and libraries. They capture smaller insects. As with true scorpions, copulation does not occur. The male deposits a stalked spermatophore. He then draws the female forward until her genital aperture is in contact with the spermatophore and closes upon it. In many species, the male performs a courtship dance. Eggs develop inside a gelatinous brood pouch filled with nutritive fluid, which is attached beneath the female genital aperture. The female may shelter in a silk nest during egg development, and the brood pouch may be eventually deposited in the brood nest. Nests are also built before winter hibernation and before molting. Nests are lined with silk spun from chelicerae, and covered externally with fragments of vegetation and other foreign bodies.

About 15 000 species have been recorded, mostly from the tropics and subtropics. The well known book scorpion *(Chelifer cancroides)* is found in books and old warehouses.

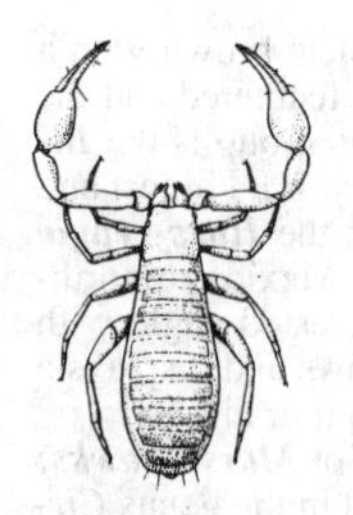

False scorpion (Neobisium muscorum Leach*)*

False sunbird: see *Philepittidae.*

Family, *familia:*

1) A systematic category above that of genus. See Nomenclature, Taxonomy.

2) A group of individuals whose members are genetically related by descent from a common parent.

Family anthropology: a branch of anthropology concerned with biological and social factors in the family, such as birth rate, sex ratio, inbreeding, hereditary composition and social relaitonships.

Fammenian stage: see Devonian.

Fan palms: see *Arecaceae.*

Fantails: see *Muscicapinae.*

Farewell to spring: see *Onagraceae.*

Farnesol: an acyclic sesquiterpene alcohol. There are four isomeric forms, due to *cis-trans* isomerism of the double bonds. *Trans-trans* F. is present in essential oils, especially in lily of the valley and lime tree flowers, to which it imparts its characteristic scent. It is also a bumblebee pheromone. F. is used in the manufacture of perfumes and soaps.

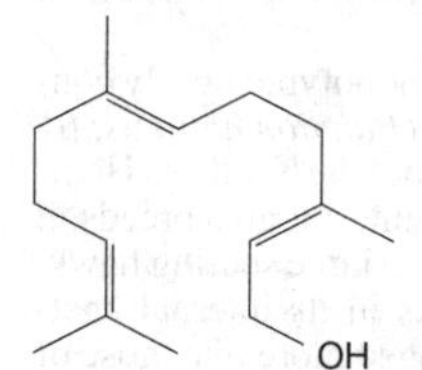

trans-trans-*Farnesol*

Far off equilibrium: a term applied to thermodynamic systems. in which the conditions are far from thermodynamic equilibrium. In most biological systems, the linear flow-force relation is no longer valid, and the laws of linear, irreversible thermodynamics do not apply. Valid generalizations for dealing with such systems have not yet been found. Linearity is violated in particular by autocatalytic chemical reactions, whereas diffusion transport in biological systems usually falls in the linear range.

Systems F.o.e. with nonlinear kinetics characteristically display *bifurcations,* i.e. critical points at which thermodynamic branches become unstable, characterized by sudden changes in the properties of an apparently stable system. A series of bifurcations can lead to spatial and temporal patterns (inhomogeneities), and such pattern-forming processes provide interesting models of morphogenesis. In chemical and biochemical systems, bifurcations cause temporal inhomogeneities, giving rise to order in time. Known as *Hopf bifurcations,* these cause undamped oscillations (provided sufficient energy and material is supplied to maintain a large distance from equilibrium). In biological systems, oscillations occur at all levels, e.g. molecular oscillations apparently form elements of the biological clock. The combination of spatial and temporal oscillations, i.e. the combination of periodic changes in concentration and the breakdown of spatial symmetry, gives rise to chemical waves, which travel through solutions. At a still greater distance from equilibrium, additional Hopf bifurcations occur, leading eventually to the loss of periodicity; the changes become irregular and the behavior becomes *chaotic.*

All of these phenomena have been demonstrated in chemical reactions. Prerequisite for their occurrence are an autocatalytic process and a sufficient input of material and energy. With suitable autocatalytic reactions, periodic spatial inhomogeneities of the reaction partners can be generated in the test tube, and these can be observed through changes in color, pH, etc (Fig.). By virtue of their self-reproduction, biological systems are autocatalytic. The irreversible thermodynamics of F.o.e. systems therefore provides a model of living systems based on physical laws.

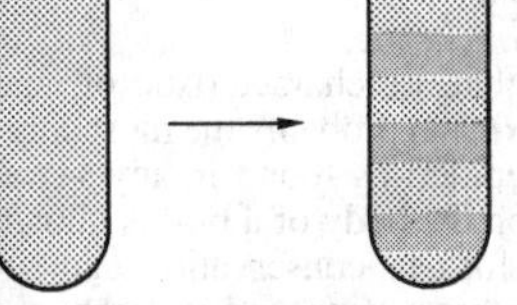

Far off equilibrium. The Belousov-Zhabotinsky reaction in aqueous solution, demonstrating periodic spatial inhomogeneities, leading to ordered zones which differ in composition and physical properties.

Fascia: see Connective tissues.

Fasciation: a form of abnormal development. Developmental processes in plants are usually regular, but they are sometimes seriously disturbed due to unilateral extension of the apical meristem. Fasciated organs often display bizarre bending and twisting. Fasciated stems are flat and strap-shaped instead of round, while fasciated ovaries may open and expose their ovules directly to the outside instead of remaining closed. The horticultural practice of bonsai (abnormal dwarfing) by the control of environmental factors is also a type of F.

Fasciola: see Liver flukes.

Fascioliasis: a disease mainly of herbivorous ungulates, usually caused by trematodes (liver flukes) of the genus *Fasciola,* and associated with chronic irritation of the bile duct epithelia and liver inflammation. The liver may also swell, and there may be digestive disturbances. The condition is diagnosed by investigation of the feces for eggs of the parasite. Other animals are also affected, and other genera of liver flukes may be involved (e.g. *Clonorchis sinensis* commonly infects humans in the Far East). Humans can occasionally become infected with *Fasciola* by taking in metacercaria with fallen fruit on infected meadows, by chewing grass halms, and in particular by eating water cress *(Nasturtium officinale)* carrying larval stages of the parasite. Therapeutic drugs are salicylanilides and nitro- and chorophenols. Emetine and chloroquine are used in the treatment of humans. Heavy infections cause extensive liver damage, known as *liver rot.* See Liver flukes.

Fat cell: see Connective tissues.

Fat hen: see *Chenopodiaceae.*

Father lasher: see *Cottidae.*

Fatigue, ***tiredness:***

1) In humans and animals, a decreased ability to function resulting from continuous demands in the range of maximal physiological perfomance. Organs tire at different rates, e.g. nerves tire more rapidly then muscles. F. partly depends on the organ, and is related, inter alia, to organ-specific metabolism. F. is reversible, and normal performance is restored after a rest or restitution period. F. is different from exhaustion or Stress (see).

2) In the sensory physiology of plants, a decrease in excitability resulting from continuous stimulation, or from repeated stimulation at short intervals.

Fats: esters of glycerol with saturated and unsaturared fatty acids. They occur widely in the animal and plant kingdoms. F. are less dense than water. They are also insoluble in water, but soluble in organic solvents. According to the number of fatty acid residues esterified with the glycerol, F. are called *monacylglycerols* (formerly *monoglycerides), diacylglycerols (diglycerides)* or *triacylglycerols (triglycerides).* Triacylglycerols (all three hydroxyl groups of the glycerol esterified) are also known as *neutral fats.* Many naturally occurring triacylglycerols contain two or three different fatty acid residues, and are called mixed triacylglycerols Their constituent fatty acid resides have an unbranched chain consisting of an even number of carbon atoms, biosynthesized by the successive linkage of two-carbon units (in the form of acetyl-CoA). [Fatty acids (see) with odd numbers of C-atoms and branched carbon chains do occur naturally, especially in certain bacteria, but probably not as constituents of F.] The most common saturated fatty acid components of F. are palmitic and stearic acid. The most common unsaturated components are oleic, linoleic and linolenic acid.

According to their origin, F. are classified as animal or plant F. In plant F., the hydroxyl groups on C-1 and C-3 are usually esterified with saturated fatty acids, whereas C-2 is usually occupied by an unsaturated residue. In the plant, they are accompanied primarily by phytosterols, together with other compounds such as triterpenes, alcohols and free fatty acids. In animal fats, the substitution positions are reversed (C-1 and C-3 esterified with unsaturated fatty acids, C-2 with a saturated residue), and they are often accompanied by cholesterol.

The consistency of a F. is determined mainly by the relative contents of unsaturated and saturated fatty acid residues. Solid F. contain a high proportion of saturated residues, whereas the liquid nature of F. increases with the proportion of unsaturated residues. Liquid F. occur more commonly in plants. They can be solidified by catalytic hydrogenation, a process used industrially for the manufacture of margarine.

F. which are liquid at room temperature are known as *oils* or *fatty oils.* These are further subdivided, according to their tendency to undergo autocatalytic oxidation in the presence of oxygen, into nondrying, semi-drying and drying oils. Drying depends on polymerization and cross-linking by oxy-, peroxy- and hydrocarbon bridges between the fatty acids. Only polyunsaturated acids can undergo these reactions. Linseed and poppy seed oils are are examples of drying oils; peanut and rapeseed oils are half-drying and olive oil is a nondrying oil. Animal fatty oils occur in particular in fish, e.g. cod liver oil.

To prepare animal F., the F.-containing tissue is rendered by heat or extracted with organic solvents, e.g. benzene. Plant fats and oils are most abundant in seeds (40–45% in rapeseed, poppy seed and linseed), from which they are usually prepared commercially by pressing. Olives contain up to 25% F. Other commercially important plant fatty oils are obtained from groundnuts, walnuts, cotton seeds and castor seeds. The most important fruit oils are palm and olive oils; the most important solid seed fats are coconut and palm seed fats and cocoa butter. In higher animals, F. are stored in subcutaneous and omentum tissue, in the peritoneum, the region around the kidneys, and in the bone marrow.

In the pure state, neutral F. are colorless, odorless and tasteless, but they spoil easly. Hydrolysis (e.g. by bacterial attack) releases fatty acids and if unsaturated these are easily oxidized to aldehydes and ketones (i.e. the fat becomes rancid), a process prevented by addition of antioxidants.

F. can be hydrolysed (saponified) with alkali or by fat-hydrolysing enzymes (lipases). Enzymatic hydrolysis produces glycerol and fatty acids, whereas alkali hydrolysis produces glycerol and the alkali salts of fatty acids (soaps). F. provide more metabolic energy (38.5 kJ/g) than proteins (17.1 kJ/g) or carbohydrate (17.5 kJ/g).

Both plant and animal F. and fatty oils are essential for life. Certain F. (containing essential unsaturated fatty acids) are essential dietary constituents for animals. F. also serve as body insulation and cushioning material for organs, and as components of cell membranes.

Commercially, F. are used for the manufacture of fatty acids, glycerol, soaps, salves, candles, fuel and lubricants. Drying oils are used in paints, varnishes and textile dyes.

Fat-tailed mouse: see *Gerbellinae.*

Fatty acids: aliphatic carboxylic acids, which may be saturated or unsaturated. Saturated F.a. have the general formula C_nH_{2n+1} COOH. Unsaturated or olefinic F.a. have double bonds between two or more C-atoms, and correspondingly fewer H-atoms. Simplest representatives of saturated F.a. are formic (methanoic), acetic (ethanoic), propionic (propanoic) and butyric (butanoic) acids, which can be prepared by laboratory synthesis, or by industrial fermentation. F.a. with 12 to 18 C-atoms (e.g. the saturated F.a., palmitic acid and stearic acids, and the unsaturated F.a., oleic, linoleic and linolenic acids) occur naturally as esters of glycerol, i.e. as the acid components of Fats (see) and fatty oils. The great majority of naturally occurring F.a. contain an even number of unbranched C-atoms, in accordance with their biosynthesis from 2-carbon units (acetyl-CoA). Branched-chain F.a. and F.a. with an odd number of C-atoms do, however, occur naturally, e.g. tuberculostearic acid (10-methylstearic acid) found in mycobacteria, *Nocardia, Streptomyces* and *Brevibacterium;* and the mycocerosic acids, a family of multimethyl-branched acids produced by mycobacteria such as *Mycobacterium tuberculosis* var. *bovis.* Degradation of leucine produces isovaleryl-CoA (the CoA derivative of the odd-numbered, branched chain isovaleric acid) and degradation of isoleucine produces 2-methylbutyryl-CoA (the CoA derivative of the odd-numbered, branched chain 2-methylbutyric acid), while propionic bacteria synthesize propionic acid (3 C-atoms) from pyruvate.

Unsaturated F.a. are especially abundant in plant oils, and they are essential dietary constituents for certain mammals which cannot synthesize them (see Vitamins). F.a. are obtained from fats and fatty oils by hydrolysis with alkali or acid, or by enzymatic hydrolysis with lipases (see Fats).

Fatty oils: see Fats.

Fauna: the totality of animal species occurring in a particular habitat. The F. of a geographical region (e.g. F. of Western Australia, F. of the Outer Hebrides, F. of Alaska) is a classified list of all animals in that region, i.e. all the phyla, orders, genera and species, and other taxa where appropriate.

Faunal elements:

1) Components of a local fauna which display certain similarities within their recent areals (see Areal), e.g. they may occur predominantly in the north, south, east or west of their areals.

2) Animal species with different areals, which originate from the same dispersal center, e.g. Holomediterranean, Caspian and Sinopacific F.es.

Faunal regions: see Animal geographical regions.

Favosites: see *Tabulata.*

Feather palms: see *Arecaceae.*

Feathers: structures that are characteristic of, and unique to, birds (see *Aves),* and probably the most complex derivatives of the epidermis found in any vertebrate. The totality of F. on the body of a bird is known as the ***plumage.*** F. serve for heat conservation, camouflage, display, flight, waterproofing and water buoyancy. They are replaced regularly ***(molting*** or ***ecdysis)*** and at least once a year, because they are subject to wear and because they change their function during the life of the bird (e.g. the natal down of chicks, breeding plumage, plain nonbreeding plumage, winter plumage, summer plumage). Complete ecdysis involves the replacement of all F. during one period. Alternatively, small and large F. may be replaced at different times, each representing a partial ecdysis or molt. Phylogenetically, F. are derived from reptilian scales, and transitional forms are present on the legs of birds, where scales give way to F.

The first F. to be produced are the ***embryonic*** or ***natal down F.*** that clothe newly hatched chicks. ***Later*** or ***adult F.*** develop later within the same follicles, pushing out and replacing the earlier F.

During development within the egg, small disklike clusters (placodes) of epidermal cells appear on the skin surface of the embryo bird (in the domestic fowl, placodes first appear on day 5 of incubation). Mesenchymal cells (presumptive dermis) aggregate beneath each placode, thus forming the *feather rudiment, primordium* or *germ.* The epidermis then grows outward to form a conical cylinder containing a central core of mesenchymal pulp ("a thimble of epidermis covering a finger of dermal mesenchyme") (Figs.1a & 1b). The feather is derived entirely by growth of this epidermal cylinder, which then becomes cornified (keratinized), so that the mature feather consists almost entirely of keratin. During these early stages, the projecting feather germ (sometimes called a *pinfeather)* sinks into the skin, forming a tubular invagination, known as the *follicle.* The follicle is slightly dilated at its base, where the *dermal papilla* projects into the *inferior umbilicus* of the feather. The proliferating zone of epidermis at the base of the follicle is known as the *epidermal collar,* and practically all subsequent growth occurs by mitosis from this collar. A single axial artery and a vein extend through the axis of the dermal core. The cornified layer of epidermal tissue, enclosing the developing feather, forms the *feather sheath.*

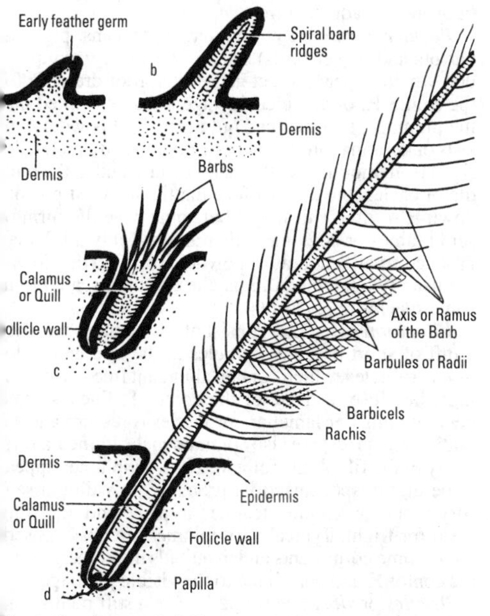

Fig.1. *Feathers. a* & *b* Embryonic feather development. Mesenchymal cells aggregate beneath each placode, forming the feather primordium. Epidermis grows outward to form a conical cylinder containing a central core of mesenchymal pulp; epidermal cells then become grouped internally into a series of spiral barbed ridges (barb primordia). This projection, known as a pinfeather, sinks inward (tubular invagination) to form the follicle. *c* The embryonic or natal down feather. The follicle is fully formed by tubular invagination of the pinfeather, and development has proceeded to the point of eruption of the barbs. There is no rachis, and the plumaceous barbs (with barbules) radiate directly from the calamus. *d* Diagram of a mature adult flight feather.

The presence of a central dermal core differentiates this embryonic feather from a hair, since a hair consists entirely of epidermal cells. At this early stage, the structure has no characteristic featherlike qualities, and in its possession of both an ectoderm (epidermis) and mesoderm or mesenchyme (dermal core) it can be compared to a mammalian horn.

While the outer surface of the epidermal cells remains smooth and covered with a cornified layer, the epidermal cells become grouped internally into 2 series of spiral *barb ridges.* Along the side of each barb ridge, cells become aligned in tangential rows to form *barbule plates.* As the epidermis continues to proliferate from the collar, the tip of the epidermal cylinder becomes further removed from the nutrient supply of the axial artery. Growth therefore ceases in the distal portion of the developing feather, where the dermal core shrinks away and the epidermal cylinder becomes keratinized (cornified), a process that extends back toward the base of the follicle. At hatching, the down feather dries, its sheath splits, and the barbs (each carrying 2 rows of side branches or *barbules)* form a whorl radiating from a small *quill* or *calamus,* which is set in the follicle (Fig.1c).

Later F. or ***adult F.*** At the base of the follicle beneath an epidermal covering lies a *dermal papilla,* which is the germ of the juvenile contor feather. Under the influence of this dermal papilla, the epidermis again grows outward from the follicle as a cylinder, and is filled with dermal mesenchyme proliferated from the dermal papilla. Again, growth occurs above the proliferative zone or collar, the outer cells cornify to form a sheath, and barb ridges become differentiated within the epidermis. Two to four of the most dorsal barb ridges fuse at their bases and the collar proliferates a triangular protrusion beneath them, which continues upward with the extension of the epidermal cylinder, becoming the *rachis* or *shaft.* New barb ridges continue to form on the ventral side, as the bases of existing barb ridges become tilted toward the dorsal side of the epidermal cylinder. As growth continues, the base of each barb ridge shifts from its origin in the collar and roots itself along the side of the rachis, destined to become part of the vane of the feather. Barbules develop from rows of cells at the sides of the barb ridges, and specialized *barbicels* (hooklets) may form on the barbules.

As growth nears completion, the dermal core at the base of the epidermal cylinder is resorbed, leaving only the dermal papilla, which remains quiescent until the next molt. The sheath splits, and the *barbs* unfurl from their curved position to form the *vane* or *vexillum* of the mature feather (a barb consists of an axis or *ramus,* with closely spaced branches known as *barbules* or *radii).* Well developed *barbicels* may link the highly differentiated ("pennaceous") barbules of adjoining barbs to form a surface with a *pennaceous* texture, i.e. a coherent, wind-resistant surface (e.g. in mature flight F.). A pennaceous texture is also important in waterproofing, and the pennaceous exposed distal portion of vanes may also have a visual function in ornamentation and display, or in camouflage. If the smooth, pennaceous surface of a vane is broken, the bird has only to run the barbs through its bill to join them up again. Less differentiated barbules are known as *stylet barbules.* In the absence of barbicels, the barbs remain separate, the feather has a downy (soft, loose and often fluffy) texture, and is referred to as *plumaceous,* a condition that is pronounced in the various types of down F.

F. often carry outgrowths on the underside, attached to the rim of the superior umbilicus. These may constitute a cluster of barbs, an auxillary shaft with lateral barbs, or an intermediate structure. All such structures are known as ***afterfeathers*** or ***hypopennae*** (Fig. 2a); they are highly variable in form, and usually downy in texture. In cassowaries and emus, however, hypopennae are coarse and they are as long as the main feather. Hypopennae are generally not associated with retrices or the larger remiges.

Down F. (Fig.2b) include not only the natal downs of newly hatched chicks, but also the definitive adult downs ***(plumulae).*** These are always entirely plumaceous, and the rachis is either shorter than the longest barb or absent. Plumulae, which are not normally visible, are abundant in ducks, geese and swans, which use them for nest building.

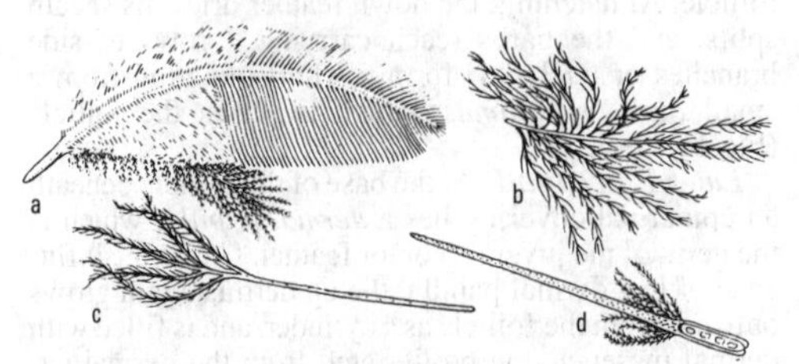

Fig.2. *Feathers. a* Typical body feather, showing pennaceous distal vane and plumaceous proximal region, as well as an afterfeather or hypopenna, which in this view is partly concealed by the plumaceous barbs. *b* Down feather. *c* Filoplume. *d* Bristle or vibrissa.

Contor F. or ***vane F.*** include the ***typical body F.,*** the ***remiges*** (wing flight F.), the ***retrices*** (tail flight F.), and the ***ear coverts*** (surrounding the external opening of the ear). Most of the visible plumage of the adult bird consists of contor F. Members of the *Passeriformes* may possess between 1 500 and 3 000 contor F., while a humming bird may have as few as 1 000. In most birds, the contor F. are distributed over the body in well defined tracts or *pterylae* (sing. *pteryla),* which are clearly visible on plucked birds. The bare areas between the tracts are called *apteria* (sing. *apterium),* most obvious of these being the brood patches used when incubating eggs. The distribution of F. in pterylae is known as *pterylosis,* and the study of pterylosis is known as *pterylography* (used, e.g. in the analysis of molting, description of plumage, and age determination). Most of the body of the bird is covered with relatively unspecialized ***Typical body F.*** Distally, typical body F. are generally pennaceous, but downy (plumaceous) in the proximal region (Fig.2a). Like down F., this downy portion of the typical body F. contributes to thermal insulation. Typical body F. have a relatively short, tubular shaft (calamus), which is seated in the feather follicle; the rachis is relatively long and tapered. ***Flight F.*** (remiges and retrices) are adapted principally for locomotion, but in some birds they are modified for display or even sound production. They are larger and stronger than typical body F., with a relatively longer (and usually asymmetrical) pennaceous vane, and only very few plumaceous barbs in the proximal region (the diagrammatic mature feather in Fig.1 (d) is representative of a flight feather). The most pronounced adaptations for flight are seen in the remiges of the manus ***(primary remiges*** or ***primaries)*** (in many avian families, the number of primaries has taxonomic significance). Often the distal portion of primaries and retrices is distinctly narrowed (notched, incised, emarginate), so that the tips of these F. do not overlap in the spread wing, but are separated by gaps or slots, permitting a propeller-like twist. The pennaceous barbules of remiges and retrices are normally larger, and their barbicels are more developed than those of typical body F. Those areas of flight F. that overlap in the spread wing or tail often contain friction barbules (distal barbules modified by lobate dorsal barbicels), which prevent the vanes of overlapping F. from slipping too far apart.

Semiplumes are characterized by a wholly plumaceous vane, and a rachis that is always longer than the longest barb. They usually lie along the margins of the contor feather tracts, and some are found singly within the tracts. In addition to filling out the contors of the body, they provide thermal insulation, and increase the buoyancy of aquatic birds.

Powder downs (found on herons, toucans, pigeons, parrots and bower birds) shed a fine white waxy powder, which appears to act as a waterproof dressing for the contor F., or as a cleaning agent that is worked into the plumage, then scraped out again. The powder consists of keratin granules, about 1 micron in diameter, derived from modified cells on the surface and in the middle of each ridge of barb-forming tissue. Most powder down F. resemble down F., but some have the form of semiplumes or contor F. Other and possibly all F. may produce small amounts of powder. True powder downs are usually aggregated in patches, but are scattered in some species (e.g. parrots).

Filoplumes (Fig. 2c) consist of a long fine shaft with a tuft of short barbs or barbules at the free end. Filoplume follicles, which contain abundant free nerve endings, lie close to those of the contor F. Encapsulated nerve endings or lamellar corpuscles (pressure and vibration receptors) are also common in the immediate vicinity of the filoplume follicle. Thus, filoplumes appear to be highly specialized F., probably providing a sensory input for keeping feathers in place and adjusting them for flight. Typically, filoplumes are not exposed, but in some cormorants and in bulbuls they emerge over the contor F. and contribute to the visible plumage.

Bristles or ***vibrissae*** (Fig.2d) have a stiff rachis, and may have no barbs or just a few at the proximal end. They are usually found around the gape, especially in species that hunt insects on the wing (e.g. nightjars and flycatchers). The eyelashes of the ostrich consist of bristles. They may also occur around the nostrils, or on the head, and in certain owls thay are also found on the toes. Since the bristle follicle is surrounded by numerous encapsulated sensory corpuscles, it is assumed that bristles have a tactile function. ***Semibristles*** have barbs along most of the length of the rachis.

Plumage colors are largely genetically determined, and they often vary with age and sex. Nongenetic factors such as the nutritional state, may also influence the coloration of F. Red and black colors are normally due to the deposition of pigments, such as porphyrins and melanins, which are synthesized endogenously. Yellow and orange are often due to the deposition of dietary carotenoids. Structural colors (see) arise from the physical nature of the feather surface.

Feather-tailed glider: see *Phalangeridae.*

Feces: a mixture of indigestible food residues, secretions of the gastrointestinal tract, live enteric bacteria and shed dead cells of the intestinal mucosa. It is partially solidified by water absorption in the colon, then voided via the anus by the reflex act of Defecation (see).

Feedback control: see Metabolic regulation.

Feedback inhibition: see Metabolic regulation.

Feedforward control: see Metabolic regulation.

Feeding, types of: see Modes of nutrition.

Felidae, ***cats:*** a family of Terrestrial carnivores (see), containing 36 species in 6 genera, plus the domestic cat. The number of genera is a point of disagreement and some authorities place all species in the genus *Fe* *lis,* except the cheetah. Through introduction the domestic cat is cosmopolitan. Otherwise, the family is represented widely, but is absent from Madagascar, West Indies, Antarctica, the Australian region, and some Oceanic islands. They are muscular, deep chested, lithe mammals, which are able to spring. In daylight their vision is similar to that of a human, but it is far superior in dim light, because the image is intesified by a reflecting layer (tapetum lucidum) that lies outside the receptor layer of the retina. Vision is binocular, and the eyes are relatively large. The forefeet have 5, the hindfeet 4 digits. Cats walk on their toes. Their claws (large, compressed,sharp and strongly curved) can be retracted and thereby protected from wear when not in use, with the exception of the cheetah, whose claws are semiretractile and relatively poorly developed. Dental formula: 3/3, 1/1, 2–3/2, 1/1 = 28 or 30, with small, unspecialized incisors, long, sharp and slightly recurved canines, and large and well developed carnassials. The tongue also assists in the laceration of food, being covered with sharp, bony papillae. The 7 larger species, which can roar (lion, tiger, leopard, jaguar, snow leopard, clouded leopard, cheetah), are known as the "big cats", whereas the smaller species that only "purr" are known as "small cats". Roaring is made possible by replacement of the hyoid bone at the base of the tongue with elastic cartilage, which is absent in the small cats. Since they are extreme predators, feeding only on the flesh of other animals caught by themselves, cats are the final members of many food chains.

Numerous differently colored and patterned subspecies of *Felis sylvestris* (***Wild cat;*** head plus body 50–80 cm, tail 28–35 cm) are distributed throughout Europe, Africa and wide areas of Asia, although the species is now extinct in many parts of Europe. The European type species is medium brown in color, with black stripes. The ***African wild cat*** *(Felis sylvestris catus)* is thought to be the ancestor of the ***Domestic cat*** *(Felis sylvestris catus),* which was domesticated in ancient Egypt, probably around 2 000 BC. Closely related is the ***Jungle cat*** *(Felis chaus;* head plus body 60–75 cm, tail 25–35 cm), which inhabits dry forested areas and reed beds, often near to human settlements, from Egypt to Southeast Asia. It is sandy brown to yellow gray in color, sometimes with dark stripes on the face and legs. ***Pallas's cat, Steppe cat*** or ***Manual*** *[Otocolobus (Felis) manual]* inhabits montainous steppe regions of Central Asia (Iran, Tibet, Kashmir, western China), and is thought to be the ancestor of the domestic Persian cat. Sturdily built and about the size of a large domestic cat, it lacks front premolars, making a total of 28 teeth. Its dense, soft coat is pale yellow to almost white with indistinct black markings, and the fur of the underparts is almost twice as long as that on the back and sides. The head is broad and flat with low set ears and high set eyes, which enable the animal to peer over rocks without exposing itself. Its diet consists mainly of pikas (mouse hares, *Ochotona).* The ***Serval*** *[Leptailurus (Felis) serval]* is a long-legged, spotted, medium sized cat (head plus body about 1 m, tail about 41 cm), which occurs widely in Africa.

Species of the genus *Lynx* are medium sized (head plus body 76–100 cm), long legged cats with ear tufts and a short tail. The feet are furred, and the eye pupil is vertical. The range of the ***Northern lynx*** *(Lynx lynx)* originally extended over the whole of Europe, northern Asia and northern North America. In Europe it is now restricted to Scandanavia, Finland, eastern and southeastern Europe, and parts of Czechoslovakia and Poland. The populations of Alaska, Canada and northern USA are sometimes classified as a separate species *(Lynx canadensis),* but this is probably synonymous with *L. lynx.* It is plain browish gray with a completely black tip to its tail. The related ***Bobcat*** *(Lynx rufus),* which is found in southern Canada and most of USA southward to Mexico, has a black spotted brown coat, relatively small ear tufts, and black bars on the end of its tail. Both the lynx and the bobcat have quite long fur. Also closely related is the ***Caracal*** or ***Desert lynx*** *[Lynx (Caracal) caracal]* of Africa and the Near East Its reddish fur is smooth and short, and like the other lynxes it has tufted ears.

The ***Ocelot*** *[Leopardus (Felis) pardalis;* head plus body 50–97 cm, tail 30–50 cm] is a spotted small cat, which is common from the southern states of USA to South America. It is an expert climber, and preys on animals up to the size of monkeys and small deer. The ***Puma*** or ***Mountain lion*** *[Puma (Felis) concolor]* is a large (head plus body 103–197 cm, tail 53–82 cm) silver-gray to dark brown or yellow-red cat, found from British Columbia to Patagonia, but absent from the eastern part of North America. It feeds on small mammals and birds. The ***Clouded leopard*** *(Neofelis nebulosa)* is a medium sized cat (head plus body 60–106 cm, tail 60–91 cm) found in the jungles, shrublands and swampy areas of South and Southeast Asia. A skilful climber, very much at home in trees, it preys on monkeys and birds, as well as rodents, pigs, deer, cattle, etc. It has short, stout legs and broad paws, and its canines are exceptionally long. The coat is gray-yellow, patterned with various shapes, mostly circles and rosettes. The ***Snow leopard*** or ***Ounce*** *(Unica unica),* which is notable for its short skull, inhabits the rocky mountain sides of Central Asia, above the timber line and near the snow. Its coat of long thick hair with a dense woolly underfur is white to grayish yellow with a pattern of dark and indistinct rosettes or rings on the upper parts. It preys on wild sheep, deer, rodents, etc.

An especially long-legged cat is the ***Cheetah*** or ***Hunting leopard*** *(Acinonyx jubatus;* head plus body 1.4–1.5 m. tail 0.6–0.75 m). It stalks antelope and other animals, then catches and overpowers them in a brief spurt over a short distance, during which speeds of up to 110 km/h are achieved. Originally distributed from Africa to India, it is now extinct in most parts of its original range.

The ***Jaguar*** *(Panthera onca;* head plus body 1.5–1.8 m, tail 70–91 cm) bears a close resemblance to a leopard (see below), but is somewhat stockier, and the rosettes of its coat pattern have central dark spots. Completely black individuals are also known. It is found in wooded country, as well as dry shrubby areas from southwestern USA to Patagonia. It preys on many small animals, including domestic stock, as well as large freshwater fish, alligators and turtles.

The ***Tiger*** *(Panthera tigris;* head plus body 1.8–2.8 m, tail about 90 cm) is a solitary inhabitant of the forests, jungles and reed thickets of Asia from Iran to Siberia and south to Sumatra and Java. Red-brown in color with black transverse stripes, it is fond of swimming and feeds on wild pigs, antelopes, etc. The subspecies *P. tigris longipilis* (***Siberian tiger***) inhabits the cold forests of Manchuria and Siberia. Toward the end of 1993, it was revealed by the WWF that tiger populations are considerably smaller than indicated by official figures. Human pressure on the tiger's natural habitat accounts only partly for this decrease in numbers. Tiger bones are used in traditional medicine in South Korea, China, Hong Kong, Yemen and Oman, and the market is supplied by poachers. Thousands of Indian (Bengal) tigers have been slaughtered by poachers in recent years, and the animal may not survive into the 21st century. A concerted effort by the WWF to increase public awareness, to prevent poaching, and to implement CITES regulations may yet enable the tiger to survive in its allocated reserves.

Probably best known of the big cats is the yellow-brown ***Lion*** *(Panthera leo;* head plus body 1.8–2.4 m, tail 61–91 cm), which inhabits the savannas and steppes of Africa, where it preys on mainly hoofed animals like zebras and antelopes. A small population also exists in the Gir forest on the Kathiawar Peninsula in India. Mature males usually have a more or less dark, dense neck mane, which in some subspecies continues as an abdominal mane. It is polygamnous, breeding throughout the year, and it is the only cat that lives in a group, which is known as a "pride". The range of the ***Leopard*** *(Panthera pardus;* head plus body 92–150 cm, tail about 90 cm) extends throughout Africa and from Mongolia to Java, including both thickly forested regions and more open rocky areas. Its yellow to red-brown coat is patterned with black ring spots, which do not have a black center (cf. Jaguar). Black varieties ("black panthers") also occur. A skilful climber, the leopard spends much of its time resting on tree limbs, and often takes its prey (antelope, deer, pigs, etc.) into a tree.

Felis: see *Felidae.*

Felovia: see *Ctenodactylidae.*

Feltia: see Cutworms.

Fence lizards: see Spiny lizards.

Fennel: see *Umbelliferae.*

Female carrier: in human genetics, a woman who carries an altered recessive gene on her X-chromosomes, which is not, or only slightly, expressed. If this gene is passed to a son, however, it is fully expressed as an inherited disorder. See Hemophilia.

Fenchone: a cyclic monoterpene ketone, isomeric with camphor. The D-form is present in oil of fennel, the L-form in oil of *Thuja* (arbor vitae).

CH_3, O, CH_2, CH_3, CH_3

Fenchone

Fenestrated capillaries: see Endothelium.

Fenestella (Latin *fenestella* little window): an extinct genus of *Bryozoa,* which formed funnel-shaped, reticulate or bush-like colonies. Fossils are found worldwide from the Ordovician to the Permian. *Fenestella retiformis* is an index fossil responsible for the formation of bryozoan reefs in the Germanic Zechstein.

Fennec: see Foxes.

Fennecus: see Foxes.

Feral pigeon: see *Columbiformes.*

Fer-de-Lance: see Lance-head snakes.

Fermentation: a type of catabolism, widespread among microorganisms, in which the organic substrate (in particular carbohydrate) is degraded in the absence of atmospheric oxygen. Hydrogen removed from substrates during F. does not combine with oxygen to produce water (as in aerobic respiration), but is used to reduce organic compounds. Production of reduced organic compounds is therefore an essential feature of F., and different types of F. are classified according to the identity of these endproducts. A ***heterofermentation*** produces several different reduced organic compounds, whereas a ***homofermentation*** yields essentially a single reduced product. Per molecule of substrate, the energy obtained by F. is less than that obtained by aerobic respiration. Therefore, more substrate must be metabolized under fermentative (anaerobic) conditions than in the presence of oxygen, in order to release the same amount of energy (see Pasteur Effect).

Alcoholic F. (see), in which ethanol and carbon dioxide are formed from sugars, is typically associated with yeasts, but is also performed by certain bacteria. ***Lactic acid F.,*** performed by species of *Lactobacillus,* is also economically important; hexoses, pentoses and related disaccharides are converted into lactic acid. Heterofermentative lactic acid Fs. produce ethanol or acetic acid in addition to lactic acid:

$$\underset{\text{Glucose}}{C_2H_{12}O_6} \rightarrow \underset{\text{Lactic acid}}{2CH_3\text{-}CHOH\text{-}COOH}, \text{ or}$$

$$\underset{\text{Glucose}}{C_6H_{12}O_6} \rightarrow \underset{\text{Lactic acid}}{CH_3\text{-}CHOH\text{-}COOH} + \underset{\text{Ethanol}}{C_2H_5OH} + CO_2, \text{ or}$$

$$\underset{\text{Glucose}}{2C_6H_{12}O_6} \rightarrow \underset{\text{Lactic acid}}{2CH_3\text{-}CHOH\text{-}COOH} + \underset{\text{Acetic acid}}{3CH_3\text{-}COOH}.$$

The lactic acid fermentation is used in the manufacture of curds and related products, and in the preparation of silage and sauerkraut.

Butyric acid F., performed by bacteria of the genus *Clostridium,* is usually heterofermentative; hexoses are fermented to butyric acid, sometimes accompanied by acetone and isopropanol. The butyric acid F. is important in the manufacture of butyric acid by the F. of potassium lactate, and in the retting of flax. It may also occur as an unwanted process in the preparation of silage and in putrifaction of stored potatoes. In the ***butanol-acetone F.,*** butyric acid is an intermediate, but it is further reduced to butanol. For example, *Clostridium acetobutylicum* ferments carbohydrates into butanol, acetone carbon dioxide and water. The process is of considerable industrial importance for the manufacture of butanol and acetone, which are widely used as solvents. Maize liquor and sulfite residues from the paper industry are used as raw materials for the butanol-acetone F.

Species of the genus *Propionibacterium* actively ferment glucose and other substrates to form considerable amounts of acid, mainly propionic acid; other byproducts are acetic acid and carbon dioxide. This ***propionic acid F.*** is important in producing the flavors and the holes in Emmenthal cheese, and in the industrial preparation of propionic acid.

The processes known as ***aerobic*** or ***oxidative F.*** can only proceed in the presence of atmospheric oxygen and are therefore not true Fs. Thus, the Acetic acid fermentation (see) is actually an incomplete oxidation, i.e the substrate is degraded aerobically, not to the final respiratory endproducts (carbon dioxide and water), but to the incompletely oxidized product, acetic acid. Other aerobic "Fs." are used for the production of various organic acids. Thus, large quantities of citric acid are produced industrially by the aerobic F. of sucrose or molasses by *Aspergillus niger.* About 60% of the carbohydrate substrate is converted into citric acid, and the product is cheaper than that prepared from citrus fruits.

Fermentative organisms may use other organic substrates besides carbohydrates. The ***fermentation of proteins,*** known as Putrifaction (see), is particularly important for the circulation of elements in the biosphere; putrifaction products include ammonia and hydrogen sulfide.

The term, F., or *industrial F.* is generally employed in industrial microbiology for the conversion or preparation of certain products by microbial action. In this context, F. serves: 1) for the preparation of different chemical compounds, e.g. penicillin by the action of *Penicillium* species, citric acid by the action of *Aspergillus niger;* 2) for the preparation of feedstock from cheap raw materials, e.g. the growth of yeasts for the supplementation of animal feed; 3) for the preparation of plants products, e.g. in Flax retting (see). In the F. of coffee, cocoa, tea and tobacco, undesirable substances are degraded, and smell, taste and color principles are generated by the action of the plant's own enzymes and the enzymes of microorganisms.

Fermentation tube: a glass apparatus for detecting and investigating gas production in fermentation processes. In particular it is used to demonstrate the fermentability of different sugars in the identification of yeasts and bacteria. Various types of F.t. are used, e.g. gas is allowed to collect in the top of a small inverted tube, filled with liquid medium and placed in the main fermentation tube; U-tubes sealed at one end are also used. See Durham tube.

Fermenter: a large-scale vessel for the culture of microorganisms, and for the preparation or conversion of products with the aid of microorganisms (see Fermentation). The main vessel is usually cyclindrical. The requirements of the particular microorganism and the desired product dictate the accessories required for stirring, cooling, aeration, input of medium and harvesting. The simplest F. is a large tank like the vats originally used for wine and beer fermentation (known as a *fermentation tank).* A large vat or tank must be cooled by a water jacket or heat exchanger, so that the organism's metabolic heat does not raise the temperature above its level of tolerance. Provision is also made for the control of pH. Stirring is necessary in large Fs. to keep the organisms from settling and to provide a uniform supply of nutrients, including oxygen in aerobic "fermentations". Alternatively, the microorganism grows as a surface culture or it forms a layer on numerous baffles immersed in, or washed by, the liquid medium. There are many types of stirrers, turbines, baffles and recirculating pumps. In an *air-lift* or *bubble F.,* stirring and aeration are provided by a stream of gas entering the bottom of the tank. In a *cyclone column F.,* a pump continually circulates the culture from the bottom to the top of the vessel. The medium is ejected from a cyclone head tangentially to the wall of the F., so that it spirals down the wall as thin film. Air or gas is injected at the bottom of the vessel and rises in a counter current to the culture flow. *Tower Fs.* are used for beer and other cultures in which the desired products are not produced continuously during the growth cycle. Fs. range in size from a few liters for laboratory use to more than 500 m^3 for industrial applications. The working volume of medium is usually about 2/3 of the total capacity of the F.

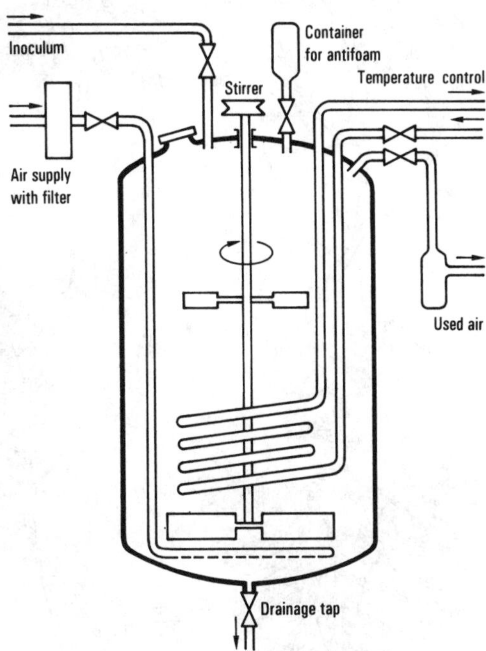

Fermenter. Schematic structure of an aerated fermentation tank with a mechanical stirrer.

Ferminia: see *Troglodytidae.*

Fernbird: see *Sylviinae.*

Ferns (plates 32 and 33), *Filicatae, Filicopsida:* a class of pteridophytes. In contrast to other pteridophytes, the conspicuous sporophytes of F. have monopodial, usually branched stems, mostly with large, abundantly veined, pinnate leaves (megaphylls) which are also known as fronds. In young plants the points of the fronds are inrolled, resembling croziers. Sporangia, usually in large numbers, are found either on the underside of the frond, or enclosed in certain segments of the frond. They often form characteristic aggregates called sori (sing. sorus), which frequently possess a thin membranous covering called the indusium. Sprophylls (leaves carrying sporangia), and trophophylls (assimilatory leaves) may be identical, or the sporophylls may be separate and usually much simpler in structure, but any differentiated structure that could be interpreted as a flower is totally lacking. Close relatives of the 10 000 (approx.) modern fern species are found as fossils in the Lower Carboniferous. More primitive forms *(Primofilices),* closely related to the *Psilophytales,* are present in the Devonian. Members of this homosporous group, known only as fossils, possess terminal sporangia, and in some cases the pinnae are not confined to a single plane (i.e. the fronds are 3-dimensional). There was a gradual evolutionary transition from this primitive form of megaphyll to a typical planar fern frond.

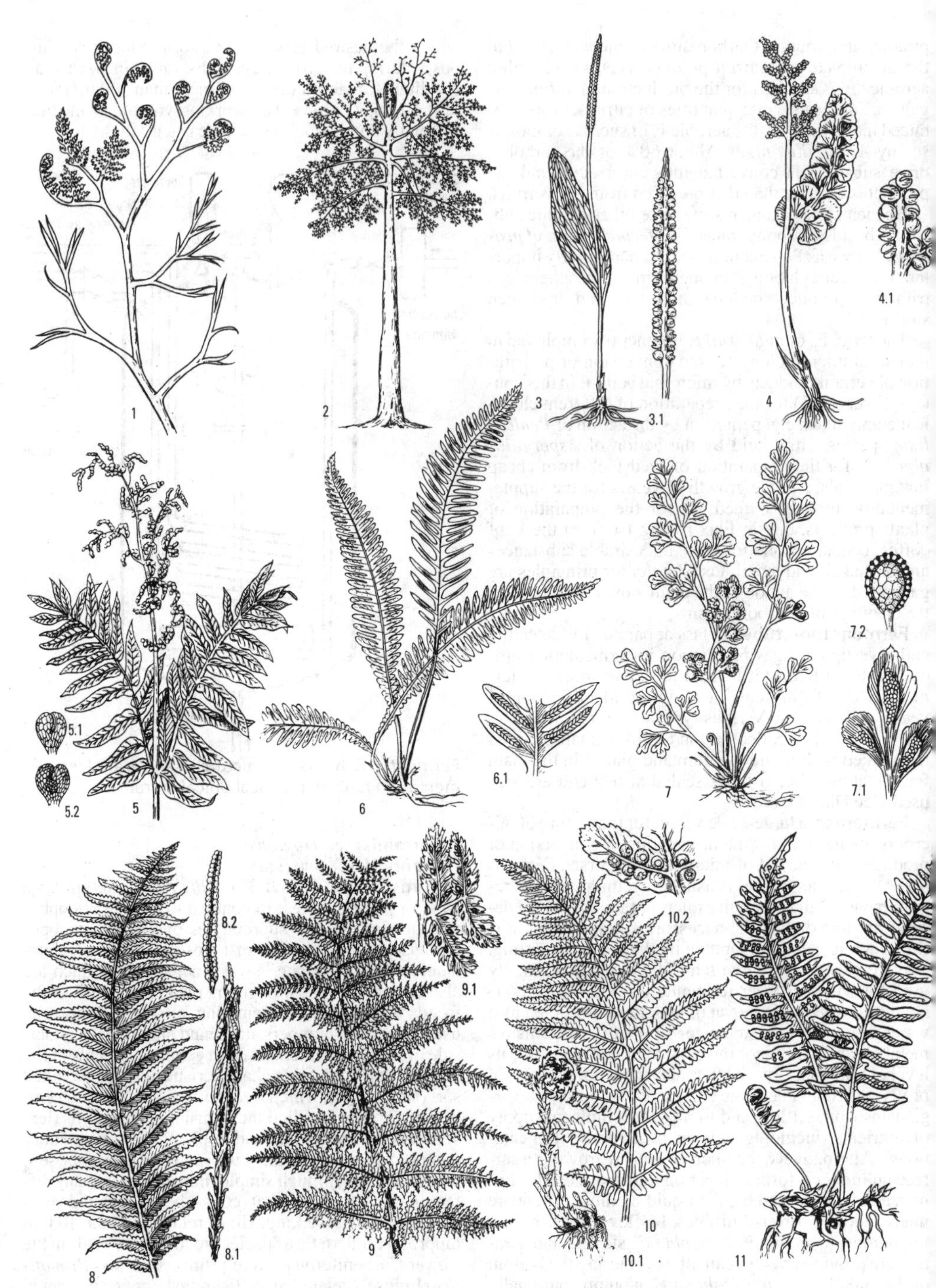

Fig.1. *Ferns. 1 Protopteridium hostimense. 2 Aneurophyton germanicum (1* & *2* are reconstructions from fossils of the Middle Devonian). *3* Adder's tongue *(Ophioglossum vulgatum)* (whole plant); *3.1* sporophyll. *4* Moonwort *(Botrychium lunaria)* (whole plant); *4.1* detail of sporangia. *5* Royal fern *(Osmundia regalis* (whole aerial part of plant); *5.1* & *5.2* sporangia. *6* Hardfern *(Blechnum spicant); 6.1* frond lobes with sori. *7* Wall rue *(Asplenum ruta-muraria); 7.1* underside of frond with sori; *7.2* sporangium. *8* Ostrich fern *(Mattleuccia struthiopteris); 8.1* trophophyll; *8.2* sporophyll. *9* Lady fern *(Athyrium filix-femina); 9.1* frond pinnae with spores. *10* Male fern *(Dryopteris filix-mas); 10.1* lower part of plant; *10.2* frond pinna with sori. *11* Common polypody *(Polypodium vulgare).*

Three orders of *Primofilices* are recognised.

Order 1: *Protopteridiales.* One of the simplest fossil genera is *Protopteridium,* whose several species occur in the Lower and Middle Devonian. They have many affinities with the *Rhyniacae,* but differ from them in having forked, flattened side branches. More highly evolved forms are also included in the *Protopteridiales,* e.g. *Aneurophyton germanicum,* whose fossils from the Middle Devonian are a few meters high; but also in this species the large fronds were spaciously branched shoot systems, which became wider and leaf-like only at the extremities of the branching system.

Order 2: *Cladoxylales.* Members of this order from the Middle Devonian to the Lower Carboniferous represent an independent evolutionary branch. The main genus is *Cladoxylon,* which also possessed nonplanar, 3-dimensional fronds.

Order 3: *Coenopteridales.* This order is represented by several genera from the Upper Devonian to the Lower New Red Sandstone, e.g. *Clepsydropsis, Stauropteris, Etapteris, Rhacophyton.*

Their evolutionary climax was reached in the Carboniferous, and they display a variety of different nonplanar frond structures.

Recent F. are divided into 3 subclasses: the primitive homosporous *Eusporangiatae*, with thick, multilayered sporangia walls and without indusia; the more advanced, but also homosporous *Leptosporangiatae* with single layered sporangia walls and usually indusia; and the highly specialized heterosporous *Hydropterides* (***Water ferns***), with single layered spongia walls, and a sperical covering or sporocarp, which envelops the sori of sporangia. The subterranean prothalli of the *Eusporangiatae* display a variety of shapes; they lack chlorophyll and harbor an endotrophic mycorrhizal fungus.

Prothalli of the *Leptosporangiatae* are always green and above ground. In the *Hydropterides* the megaspore remains inside the sporangium, and the prothalli develop in the micro-and megaspores.

Further classification and important member species are as follows:

1. Subclass: *Eusporangiatae.*

1. Order: *Ophioglossales.* This order contains about 80 cosmopolitan species of small F. with a short unbranched underground stem or rhizome, which bears a single leaf divided into a fertile and a sterile portion. European species are ***Adder's tongue*** *(Ophioglossum vulgatum)* and ***Moonwort*** *(Botrychium lunaria),* which are native in meadows and pastures.

2. Order: *Marattiales.* The order is represented by about 200 extant tropical species. Its members were much more numerous in earlier geological times, especially the Carboniferous and Lower New Red Sandstone. Extant forms resemble living fossils. Most familiar of these is the ***Vessel fern*** *(Angiopteris evecta)* from tropical East Asia.

3. Order: *Archaeopteridales.* A heterosporous fossil order, sharing affinities with both ferns and gymnosperms. Together with the *Protopteridiales*, they are known collectively as "*Progymnospermae*", which are considered to be the evolutionary precursors of the gymnosperms.

2. Subclass: *Leptospongiatae.*

1. Order: *Osmundales.* The order contains about 20 widely distributed species, which show certain affinities with the *Eusporangiatae,* i.e. multilayered sporan-

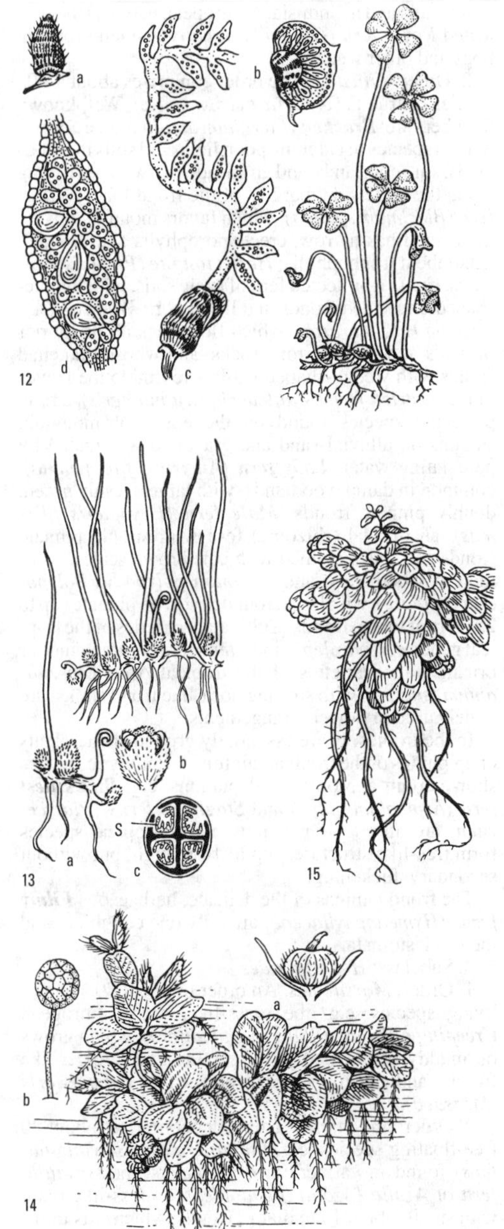

Fig.2. *Ferns. 12 Pepperwort (Marsilea)* (plant and sporocarps); *a* closed sporocarp, *b* opening sporocarp, *c* dehisced sporocarp with ring of mucilage and sporangial sacs, *d* sorus with macro- and microsporangia. *13* Creeping pillwort *(Pilularia globulifera), a* plant with sporocarps, *b* opening sporocarp, *c* transverse section of sporocarp with sporangia, *S* sporangium. *14* Floating moss *(Salvinia natans); a* sporocarp, *b* microsporangium. *15* Azolla *(Azolla filiculoides).*

gium walls and no indusia. A European native is the protected ***Royal fern*** *(Osmundia regalis),* which grows in bogs and other wet areas.

2. Order: *Filicales*. The order comprises about 9 000 species of true F. (F. in the narrow sense). Well known members are: ***Bracken*** *(Pteridium aquilinum),* a cosmopolitan species of calcium-poor, light acid soils in woodlands, on moorlands and in meadows, with sori lying along the underside margins of the frond lobes; ***Hard-fern*** *(Blechnum spicant),* which favors mountainous regions, and has narrow, erect sporophylls and broader, decumbent trophophylls; ***Hart's tongue*** *(Phyllitis scolopendrium),* a protected fern of basic soils, with tongue-shaped undivided fronds and linear sori; ***Wall rue*** *(Asplenium ruta-muraria),* which flourishes in the mortar of walls and on limestone rocks, and whose branched fronds with wedge-shaped leaflets resemble the leaves of rue; ***Ostrich fern*** *(Mattleuccia struthiopteris),* a rare, protected species found on the edges of mountain streams on alluvial sand and gravel soils irrigated by percolating water; ***Lady fern*** *(Athyrium filix-femina),* common in damp woodlands, with large, freshly green, doubly pinnate fronds; ***Male fern*** *(Dryopteris filix-mas),* an official (rhizome) fern with simple pinnate fronds and stalks clothed with pale brown scales, common in woodlands; and ***Common polypody*** *(Polypodium vulgare),* with large, rounded sori on pinnate (up to 25 pairs of lobes) fronds. Cultivated varieties of the tropical genera, *Nephrolepis* and *Pteris,* are common indoor ornamentals. Varieties of the tropical American *Adiantum tenerum,* with striking horn-like frond stalks, are widely used in flower arrangements.

In the tropics, F. are essentially ground shade plants or epiphytes of the tropical rainforest. Epiphytic species show certain appropriate adaptations, e.g. ***Bird's nest fern*** (*Asplenium nidus*) and ***Staghorn ferns*** *(Platycerium)* are able to trap humus. Some tropical species form tree-like structures, up to 18 m high, but without secondary thickening.

The frond laminas of the delicate, herbaceous ***Filmy ferns*** *(Hymenophyllaceae)* are only one cell thick, and they lack stomata.

3. Subclass: *Hydropterides.*

1. Order: *Marsileales*. An order of about 80 swamp-living species, e.g. the grass-like, native European ***Creeping pillwort*** *(Pilularia globulifera),* which grows on muddy banks, and possesses inrolled, crozier-like fronds; and the four-leaved, clover-like ***Pepperworts*** *(Marsilea).*

2. Order: *Salviniales.* This order comprises about 20 free-floating species, e.g. ***Floating moss*** *(Salvinia natans),* found on warmer European waters; and ***Mosquito fern*** or ***Azolla*** *(Azolla filiculoides),* a moss-like plant with small (about 1 cm diam.) fronds, which was introduced from America and is now native in Europe.

Ferret: see Polecat.

Ferritin: an important iron storage protein in mammals. Together with the related hemosiderin it contains 25% of the iron in the body. F. is found in the spleen, liver, bone marrow and reticulocytes, where excess iron is stored intracellularly and can be mobilized when needed. It has also been found in mollusks, plants and fungi.

Fertility:

1) The abiltiy of plants and animals to produce offspring. The absence of fertilility is known as Sterility (see).

2) An index of reproductive capacity, expressed in the number of embryos produced. The F. of plants depends on their lifespan and age, the sizes of the seeds and fruits, the mode of pollination, the available nutrition and other environmental factors. In lower animals, especially parasites, F. is often very high, as an adaptation to the generally unfavorable prospects for further development (i.e. many eggs and embryos perish). In higher forms, which practice brood care, F. is lower. F. is subject to many conditions and limitations. A high F. is a precondition for maintenance of the viability of a species by natural selection. A permanent decrease in F. can lead to the extinction of a species.

Fertility index: the annual birth rate, in relation to number of fertile women (generally in the age range 15–45 years) in the population. For detailed analyses, the F.i. is determined separately for the age, marital status, length of marriage and social conditions of the women.

Fertilization: fusion of a female and a male germ cell (female and male gametes) to form a zygote. First the cytoplasm of the two cells becomes mixed *(plasmogamy),* followed by fusion of the two nuclei *(karyogamy, amphimixis).* F. always involves the combination of male and female genetic material.

1) In ***flowering plants,*** F. is preceded by Pollination (see), i.e. transfer of pollen to the stigma. A distinction is drawn between Cross-fertilization (see) and Self-fertilization (see). Not every pollination leads to F., because the efficacy of the process is influenced by factors such as atmospheric humidity, temperature, light, age of the pollen, maturity of the stigma, etc. The time that elapses between F. and the deposition of pollen on the stigma varies from a few minutes to days (most agricultural plants), and it can also take several weeks, months or years (many deciduous trees and some ornamental plants). As soon as pollen comes into contact with a receptive stigma, it germinates and forms a *pollen tube.* This pollen tube grows through the tissues of the style and ovary as far as the embryo sac, which houses the egg apparatus with the egg cell (ovum). Pollen germination and pollen tube growth are stimulated by chemotropic substances secreted by the stigma and ovule. The pollen tube may take the shortest route to the embryo sac through the micropyle *(porogamy),* or it may arrive at the embryo sac by a less direct route *(aporogamy),* e.g. via the placenta or chalaza *(chalazogamy).* In the early stages of pollen tube development, two nuclei are present, i.e. a vegetative or tube nucleus, and a generative nucleus. During pollen tube growth, the generative nucleus divides once to produce two male nuclei. When the pollen tube reaches the egg apparatus, both male nuclei are released into the embryo sac, while the vegetative nucleus disintegrates and is lost. One of the male nuclei then penetrates the egg cell and fuses with the egg nucleus. The other male nucleus migrates to, and fuses with, the secondary embryo sac nucleus (endosperm nucleus) in the center of the embryo sac. The final result of this *double F.* is a fertilized egg cell (i.e. a zygote) which will develop into the embryo, and a triploid *endosperm nucleus,* which begins to divide to form the endosperm of the Seed (see).

2) In ***animals,*** F. initiates cleavage of the egg cell and development of the embryo (embryogenesis). Many different forms of F. occur in the animal kingdom. In the *Protozoa,* an entire individual can function as a gamete

(hologamy). In such cases the gametes differ only slightly in appearance from vegetative cells. Individuals undergoing sexual fusion may be morphologically similar *(isogamy)* or different *(anisogamy* or *heterogamy).* Often the differentiation is quite marked, resulting in egg-like macrogametes (gynogametes) and sperm-like microgametes (androgametes), a condition known as *oogamy.* Gametes may also be derived from a single protozoan cell *(paedogamy). Conjugation* in ciliates represents a special type of F., which is of interest as a possible link between the reproductive processes of higher protozoa and the metazoa. Thus, in *Paramecium,* two individuals (known as conjugants) become temporarily fused at their oral surfaces. The conjugants must be from different "strains", i.e. mating types. First the micronucleus detaches from the meganucleus (the latter is eventually broken down and absorbed). The micronucleus of each individual increases in size and divides twice to form four haploid nuclei, three of which are aborted. The remaining haploid micronucleus now divides to form two gametic nuclei, known as the migratory nucleus and the stationary nucleus (if these are different in size, the smaller one is always the migratory nucleus) . Each migratory nucleus passes to the other cell, where it fuses with the other stationary nucleus to form the fusion or zygotic nucleus. The conjugants then separate and undergo a series of divisions to produce new individuals, each with a mega- and a micronucleus.

In ***metazoans,*** F. involves the fusion of a true sperm cell (spermatozoon or sperm) with a true egg cell. It begins with the active movement of spermatozoa toward the egg. In mammals, spermatozoa attain a speed of 2–3 mm/s, driven forward by their flagella. When a sperm enters an egg (impregnation), the latter may be in various stages of development. Thus, in *Saccocirrus (Annelida)* and *Ascaris,* the sperm enters when the egg in its growth phase, whereas in insects sperm penetration occurs when the egg is in the metaphase of the first maturation division. In mammals and echinoderms, the sperm enters the egg when the latter is at the stage of the second maturation division, while in other animals the egg has completed meisosis before it receives the sperm. In some cases, therefore, impregnation can occur long before F. Some eggs are receptive to sperm at any point on their surface, while others possess a special *micropyle.* Under the microscope, the micropyle is usually discernable only with difficulty in unstained preparations, but it can be revealed by the addition of Indian ink, toluidine blue or other stains. Arrival of a sperm at the egg surface usually stimulates the formation of a receptive site with protoplasmic processes. By means of its Acrosome (see) the sperm penetrates the *vitelline membrane.* The tail is usually cast off, and the head and middle piece enter deep into the egg cytoplasm. Sperm tails entering the egg, either accidentally or as a normal event, are degraded and absorbed. Between one and two minutes after contact with the first sperm, the surface of many eggs *(Ascaris, Nereis,* sea urchin, mammals) becomes covered with multiple vesicles; these rapidly coalesce to a tough *fertilization membrane* which prevents the entry of other spermatozoa (see Polyspermy). A fertilization membrane can be stimulated to form artificially, e.g. by cytolytic agents (ethyl acetate, formic, acetic, butyric and valeric acids, and saponin) and other substances, as well as increased temperature. Appearance of the fertilization membrane is accompanied by changes in the surface tension of the egg, leading to shrinkage. According to electron microscope studies, in particular those on *Urechis caupo* (an echiurid), the vitelline membrane is attacked and dissolved by lysozyme from the acrosome of the sperm. The resulting channel brings into contact the respective cytoplasmic complements of the egg and sperm. The sperm head and middle piece are drawn into the egg. At the same time, cortical vesicles beneath the plasma membrane of the egg burst and release their contents into the space between the internal plasma membrane and the external vitelline membrane, thereby raising the latter and causing it to become attached to the fertilization membrane. Sperm head and middle piece then rotate through 180 ° and migrate into the interior. In amphibia, this migration through the egg cytoplasm can be followed for some considerable time as a sperm track. After separation from the middle piece, the head swells into the male pronucleus, which moves actively toward the egg nucleus. The male pronucleus at first forms a cap on the female nucleus, and the two nuclei remain morphologically distinct. Meanwhile, the *centriole,* which entered the egg in the middle piece of the sperm, becomes the center of an extensive aster, which then disappears, followed by division of the centriole into two. Daughter centrioles migrate to the nuclear poles. A new aster forms around each daughter centriole, and an achromatic mitotic spindle appears between them. This is normally the point at which karyogamy occurs, followed rapidly by prophase and the first cleavage division. In some animals, however, the two nuclei may remain morphologically separate until the first cleavage division *(gonomery).* The fate of the acrosome is unclear. In the sea urchin and mammals, the male mitochondria, which are brought into the egg in the middle piece of the sperm, disintegrate and are absorbed. The egg cell centriole becomes inactive and is destroyed.

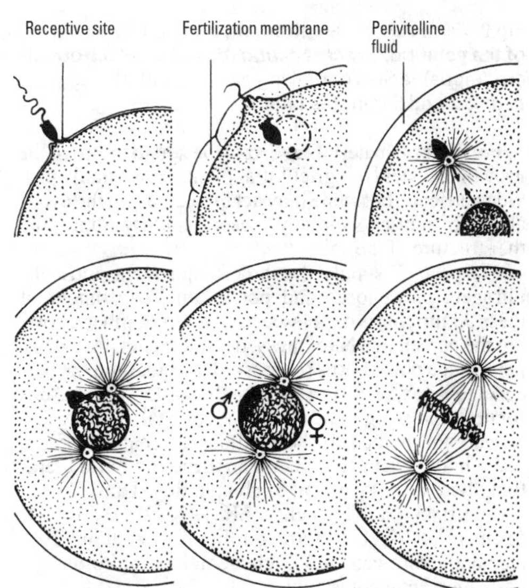

Fig.1. *Fertilization.* Stages in the fertilization of a sea urchin egg.

Generally, spermatozoa are nonmotile or show only slight activity within the testis. They do not become motile and therefore capable of inseminating an egg until placed in an appropriate medium, e.g. the spermatozoa of aquatic organisms become active in water. Mammalian (including human) spermatozoa move actively from the tubules of the testis into the epididymis, where they are apparently rendered immobile by secretions of the epididymal epithelium, which inhibit their metabolism. Subsequently, they become fully activated by secretions of the accessory sex glands, in particular the prostate gland. The vigor of sperm movement is also increased by mild alkalinity, e.g. in mammals by the alkaline reaction of the secretions of the cervix and oviduct (positive chemotaxis). In contrast, motility is inhibited by the acidic vaginal mucus (negative chemotaxis). Mammalian spermatozoa also display negative rheotaxis, in that they orientate themselves against a cilia-generated fluid flow in the uterus and oviduct.

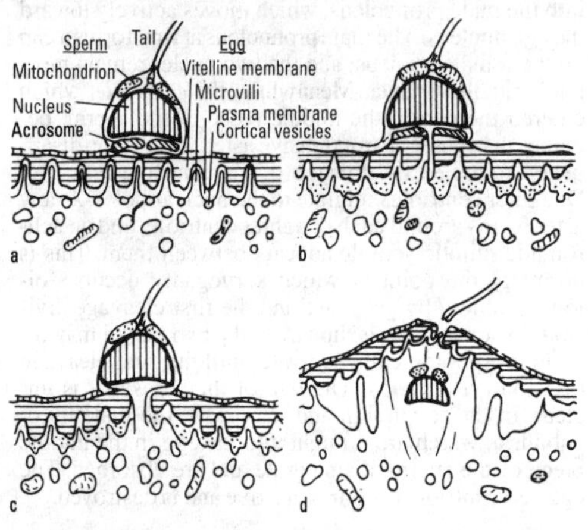

Fig.2. *Fertilization.* Penetration of an egg by a sperm of the echiurid, *Urechis caupo* (based on electron micrographs). *a* Sperm making contact with the egg. *b, c, d:* 1, 2 and 5 minutes later.

Fertilizers: materials added to the soil to improve the growth and yield of agricultural plants. F. normally provide nutrients, but some may act by increasing the availability of nutrients already present, and by improving the structure of the soil.

Complete F. supply the three major plant nutrients, nitrogen, phosphorus and potassium The content of these 3 nutrients is expressed as a 3-part number, showing the percentage N (expressed as elemental nitrogen), percentage phosphorus (expressed as P_2O_5) and the percentage K (expressed as K_2O) in that order. The reasons for not expressing P and K on an elemental basis are historical. and related to the methods of analysis. Especially in the scientific rather than the agricultural literature, there is a movement to change this. Thus, %P (as P_2O_5) x 0.43 = %P (element); %K (as K_2O) x 0.83 = %K (element). A 10/10/10 fertilizer therefore contains 10% elemental nitrogen, 4.3% elemental phosphorus and 8.3% elemental potassium.

F. may be *natural* and *organic,* e.g. animal dung (manure), blood, fish meal, cottonseed meal. Alternatively, they may be *chemical* or *inorganic,* i.e. synthesized from inorganic materials, e.g. ammonium nitrate. Most commercial inorganic nitrogenous fertilizers originate from the industrial Haber process, in which hydrogen and atmospheric nitrogen are reacted under pressure at a high temperature in the presence of a catalyst to form ammonia. Ammonium salts may used directly, or the ammonia may serve for the synthesis of other compounds such as urea or nitrates.

In order to ensure maximal utilization by the plant, to avoid damage by fertilizer excess, and to avoid pollution of ground waters and local rivers, ponds, etc., it is important that F. be applied only where and when they are needed by the plant, and in appropriate quantities. Correct placement and timing are influenced by the stage of plant development, and environmental factors such as weather conditions, efficiency of soil drainage, etc.

Ferulic acid: an aromatic carboxylic acid found in plants. Convenient preparative sources are e.g. the straw of rye, wheat and barley, from which it can be extracted with water. It inhibits germination. In the biosynthesis of F.a., phenylalanine is converted into *trans*-cinnamic acid by the action of phenylalanine ammonia lyase. *Trans*-cinnamic acid is oxidized to *p*-coumaric acid (4-hydroxycinnamic acid); this latter compound may also arise directly from tyrosine. Further oxidation produces caffeic acid (3,4-dihydroxycinnamic acid), which is then methylated to F.a. By further metabolism and polymerization, F.a. also serves as a precursor of lignin.

CH=CH—COOH

$-OCH_3$

OH

Ferulic acid

Fescue: see *Graminae.*

Festuca: see *Graminae.*

Fetalization: occurrence of embryonal or juvenile characters of an ancestral form in phylogenetically more recent forms.

Fetal rickets: see Achondroplasia.

Fetography: X-ray portrayal of the unborn child, made after rendering the body surface of the fetus opaque to X-rays by introducing a fat-soluble contrast agent into the amniotic fluid. F. serves to reveal major malformations in cases considered to be at risk.

Fetoscopy: direct viewing of the fetus with a rod-shaped optical instrument, which is introduced into the uterus via the abdominal wall of the mother. F. serves to reveal major malformations in cases considered to be at risk.

Fever: an increase in body temperature caused by infection. Lipopolysaccharides of the bacterial membrane stimulate leukocytes to produce leukocyte pyrogen. The latter causes fever when it is transported in the circulation to the hypothalamus, or when it is injected experimentally into the hypothalamus. It alters the response of the temperature regulation center, increasing the value at which the body temperature is regulated. Injection of the pyrogen into other regions of the brain is without effect on body temperature.

The febrile condition shows two phases (Fig.). In the first phase, the fever develops, i.e. body temperature in-

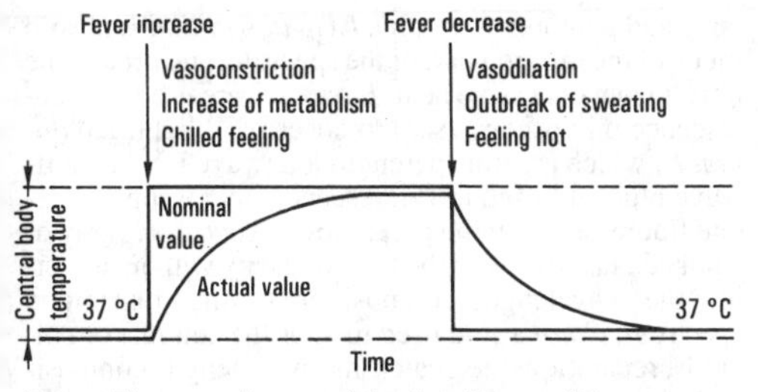

Fever. The two phases of the febrile condition.

creases and approaches the new nominal value. The affected person complains of a subjective cold feeling, and reacts by developing goose flesh (equivalent to the raising of mammalian fur), and by constricting the cutaneous blood vessels. The skin therefore appears pale, and it also displays increased muscle tone with occasional shivering. During this phase, the loss of body heat is reduced and its production is increased. Usually, the body temperature rises above the new nominal value. This is followed by the second phase, in which the patient experiences a subjective feeling of warmth. Cutaneous vessels are now dilated, heat loss is increased, and the skin becomes flushed; extra heat may also be lost by sweating. In both fever phases, Temperature regulation of the body (see) remains intact, so that the response to environmental temperature stimuli is normal, e.g. cold temperatures cause increased heat production.

Fiber plants: plants that yield economically useful fibers. ***Cotton*** is manufactured from the seed hairs of the cotton plant *(Gossypium herbaceum).* ***Flax*** *(Linum usitatissimum),* ***hemp*** *(Cannabis sativa),* ***jute*** *(Corchorus capsularis* and *olitorius),* and ***ramie*** *(Boehmeria vivea)* are all derived from stem fibers. ***Sisal*** is manufactured from the leaf fibers of Mexican sisal grass *(Agave sisalana),* and ***manilla hemp*** from the leaf sheath fibers of abaca *(Musa textilis).*

Fibrin: an insoluble fibrous protein, which is the end product of the Blood coagulation (see) cascade. It is formed from fibrinogen by the action of thrombin. Together with adhering blood cells, it provides a stable wound closure.

Fibroblasts: nonmotile, branched connective tissue cells, spindle-like in shape with an elongated nucleus. They synthesize ground substance and fibrous proteins (see Connective tissue).

Ficedula: see *Muscicapinae.*

Ficoidaceae: see *Aizoaceae.*

Ficus: see *Moraceae.*

Fiddler crabs, *Uca:* a genus of true crabs of the family *Ocypodidae* (Ghost and Fiddler crabs), with representatives on the coasts of all warm seas. They usually live in large communities, each individual excavating its own burrow in the mud of the intertidal zone. Females have equal-sized chelae (pincers), which they use to transfer mud to the mouthparts. After sorting the mud into edible and inedible material, the inedible particles are picked off by the chelae and laid on the ground in geometric patterns as the animal progresses. Males have only one feeding chela, the other having evolved into a greatly enlarged structure (representing up to half the weight of the entire crab), which, in combination with other species-specific body and leg movements, is used in courtship signalling. Thus, to attract a female, *Uca annulipes* waves its outstretched chela from side to side, whereas *Uca beebei* waves it up and down.

Fidelity: a concept which expresses the degree of exclusiveness shown by a species toward a particular habitat (habitat F.) or bioc(o)enosis (community F.). The F. of a species is an indication of its special tolerance of, or preference for, a specific combination of environmental factors. Stenotopic species are found only in a certain type of habitat or a narrowly delimited group of similar habitats. In contrast, suitable conditions for the growth and reproduction of eurytopic or Indifferent species (see) are provided by a variety of different habitats. As a rule, the degree of F. of a species increases toward the boundaries of its area of distribution, i.e. the geographical region is an important factor in the determination of F. Thus, plant species of warm, southern regions often show a strong dependence on limestone soils on the northern boundary of their distribution, or they may be found only on south-facing dry slopes. These specialized habitats may actually be avoided in the center of the normal area, where species are able to occupy a much broader spectrum of other different habitats. In extreme cases, species in different geographical regions are found in totally different habitats (see Habitat alternation). In addition to physiographic factors, which are very important in the determination of F., a strong influence is also exerted by competitors. Thus, weakly competitive species are often obliged to withdraw into extreme habitats that are not occupied by competitors (see Niche).

The following categories of F. are distinghished.

1. ***Euc(o)enic species*** have a low Ecological amplitude (see). These are subdivided into a) c(o)enobionts or characteristic species, which occur only in a certain habitat (e.g. tyrphobionts, which occur only in peat bogs), and b) c(o)enophilic species or selective species, which show a marked preference for one particular habitat, but are also found in other similar, neighboring habitats, e.g. halophilic organisms prefer saline conditions, but are not restricted to such habitats.

2. ***Tychoc(o)enic (preferential) species*** display a relatively strict dependence on a specific combination of ecological factors which occurs in various, but similar habitats.

3. ***Ac(o)enic (indifferent or ubiquitous) species*** display no recognizable requirement for a specific habitat, and they possess a high Ecological amplitude (see).

4. ***Xenoc(o)enic (strange*** or ***accidental) species*** are rare and accidental intruders from other habitats.

See Character species.

Field cricket: see *Saltoptera.*

Fieldfare: see *Turdinae.*

Field mouse: see *Muridae*

Field mushroom: see *Basidiomycetes (Agaricales).*

Field pea: see *Papilionaceae.*

Field resistance, ***relative resistance:*** the resistance (see) to infection under natural outdoor conditions, which is independent of race or variety, and is therefore stable. It is largely identical to *horizontal resistance,* which is defined as pathotype-unspecific resistance with polygenic inheritance.

Field vole: see *Arvicolinae.*

Fierasferidae: see *Carapidae.*

Fig: see *Moraceae.*

Fig birds: see *Oriolidae.*

Fig wasps as specific pollinators: see *Hymenoptera.*

Figwort: see *Scrophulariaceae.*

Filariasis: a complex of diseases of vertebrates (especially birds and mammals) caused by parasitic nematodes known as filarioids (order *Filarioida;* see *Nematoda).* They inhabit the lymph glands and coelom and other sites of the vertebrate host. The females are ovoviviparous and the larvae are known as microfilariae. Transmission occurs by the bite of a blood-sucking insect, which is an intermediate host. At least 7 different species of filarioids are pathogenic to humans, all with different vectors, but the greatest morbidity is associated with developmental and adult stages of endolymphatic-dwelling *Wuchereria bancrofti, W. lewisi, Brugia malayi* and *B. timori,* and with the subcutaneous-dwelling *Onchocerca volvulus.* Various mosquito species of the genera *Culex, Aedes, Mansonia* and *Anopheles* are responsible for transmitting *Wucheria* and *Brugia,* while Onchoceriasis (see) is transmitted by blackflies of the genus *Simulium.*

Loa loa, Loiasis or ***African eye worm,*** which is restricted to African tropical rainforests, affects humans and baboons. It is caused by the filarioid, *Filaria loa,* a migratory worm (male 0.35 x 30 mm; female 0.4–0.5 x 50–60 mm) found mostly in subcutaneous connective tissue and sometimes in subconjunctival tissue. In its migration about the tissues of the host, the worm sometimes passes across the eyeball. Transmission is by mangrove flies of the genus *Chrysops.* Symptoms are due largely to alergic reactions, e.g. localized irritation and swellings, known as calabar swellings.

Adult worms of *Wuchereria bancrofti* (males 0.1 x 40 mm; females 0.24 x 90 mm) live in the lymph glands, and the microfilariae are found in the peripheral blood stream. Migration of microfilariae into the peripheral circulation is triggered when the host is bitten by a mosquito of the appropriate species In severe infections, lymph vessels are blocked by the large number of worms, resulting in grotesque swelling of the body parts (especially the extremities), a condition known as ***Elephantiasis*** (see plate 43).

Filarioida: see *Nematoda.*

Filefishes: see *Balistidae.*

File shells: see *Bivalvia.*

Filial generation: the product of the crossing of members of a previous generation. Thus, the first generation of offspring from a new cross is known as the F_1 generation. The F_2 generation results from the mating of members of the F_1 generation, and so on.

Filibranchia: see *Bivalvia.*

Filicatae: see Ferns.

Filicopsida: see Ferns.

Filiform antennae: see Antennae of insects.

Filipalpia: see *Plectoptera.*

Filoplume: see Feathers.

Filopodium: see *Rhizipodea.*

Filter feeders: see Modes of nutrition.

Filtering: the selection of Sign stimuli (see) from all other simuli impinging from the environment, i.e. the selection of stimuli requiring a behavioral response that is essential for species survival. F. occurs in the sense organs (peripheral F.), as well as in the central nervous system (central F.).

Filters: glass disks which are transparent to only part of the optical spectrum. F. have various uses in microscopy and photomicrography. *Heat absorption F.* absorb much of the infrared part of the spectrum and protect the object from excessive heat. *Barrier F.* are used in Fluorescence microscopy (see) to absorb UV-light. *Excitation F.,* which are transparent to longwave UV or shortwave blue, are used in fluorescence microscopy to excite fluorescence in the preparation. *Neutral F.* (gray or smoked glass) weaken the light intensity without affecting the wavelength composition of the transmitted spectrum. *Color F.* are used to alter the contrast of colored preparations, to match the wavelength composition of the light to that required for correction by an objective, and to adjust the spectral composition to the sensitivity of photographic films. *Interference F.* generate a very narrow wavelength band, and they have special applications in Interference microscopy (see).

Filtration sterilization: filtration of liquids and gases through filters that retain bacterial cells. These bacterial filters consist of unglazed porcellain (Chamberland filter), compressed kieselguhr (Berkefeld filter), sintered glass, or thin membranes (see Membrane filter). F.s. is used primarily for sterilization of heat-sensitive solutions and of air. Most bacterial filters, however, do not retain viruses.

Fimmenite (named for the Oldenburg peat officer, Fimmen): a mass accumulation of fossil pollen in peat and in tertiary brown coal, especially in high bituminous brown coal, e.g. in the Eocene brown coal measures in Geiseltal, Halle, Germany.

Finches: see *Fringillidae.*

Finch-larks: see *Alaudidae.*

Fine spraying: treatment of plants with Plant protection agents (see) by using sprays with a droplet size of 0.025–0.125 mm.

Finfoots: see *Gruiformes.*

Finger and toe disease: see *Myxomycetes.*

Fingerprint pattern, *dactylogram:* a patterned system of raised ridges separated by grooves, present on the lower skin surface of the fingers. The ridges are formed by dermal papillae and they contain touch receptors. Patterns of dermal ridges are also present on the toes, the palm of the hand and the sole of the foot. The detailed F.p.p. is different for every human being, and it even differs for identical twins. Since the F.p.p. remains unchanged throughout life, it is useful in forensic science as an aid to personal identification.

General characteristics of the F.p.p. are determined genetically, and it is thought that three independent genes are involved. They have a certain importance in human genetics, because they occur with different frequencies in individual human populations, and the frequency of their occurrence shows a correlation with certain genetically determined illnesses. The basic components can be recognized: *arcs,* in which the dermal papillae form continuous lines from one side of the finger to the other, usually rising somewhat in the center to form an arc or arch; *loops,* in which several ridges emerge from the side of the finger, bend back on themselves to form more or less concentric loops, then return to the finger edge; *whirls,* in which several ridges form circular or elliptical spirals, double spirals or concentric rings, more or less in the middle of the finger tip.

In European populations, whirls occur with relatively low frequency, whereas loops, and to a lesser extent arcs, are more common. In Europe, the frequency of whirls increases from north to south and from west to east. F.p.p. also show geographical and racial correlations in other parts of the world. Notably, both white and black Americans display the F.p.p. frequencies typical of their ancestral populations. See Fig.

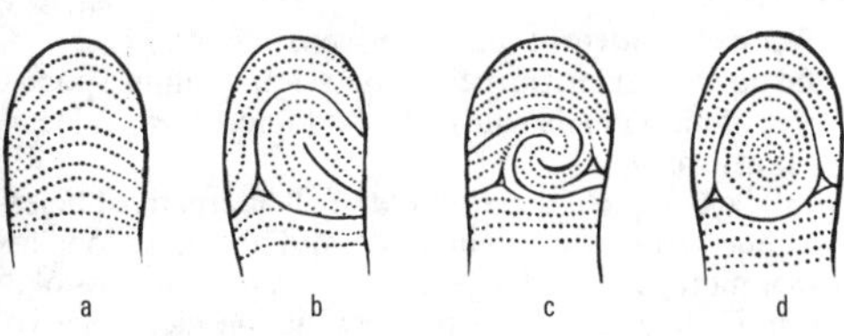

Fingerprint pattern. a Arcs. *b* Loops. *c* Whirls. *c* Concentric rings.

[Dermatoglyphics Today. B. Mohan Reddy, S.B. Roy & B.N. Sarkar. Indian Institute of Biosocial Research and Development Anthropologial Survey of India. Indian Statistical Institute, Calcutta 1991]

Finner: see Whales.

Fins: extremities of fishes and other aquatic vertebrates, as well as the swimming organs of mollusks. The F. of fishes are classified as:

1) *Unpaired fins,* which comprise 1, 2 or 3 *dorsal fins* and a single *anal fin* behind the anus, all of which serve for stabilization. Salmon also have a characteristic *adipose fin* behind the dorsal fin. The *caudal* or *tail fin* is used primarily for propulsion and steering; 3 types are recognized: the primitive *heterocercal (heterocercous)* caudal fin of e.g. the shark and sturgeon, in which the vertebral column extends, with curvature, into the tip of the fin (**a** in Fig.); the *homocercal (homocercous)* caudal fin of most teleost fishes (e.g. salmon), which is externally symmetrical, but in which the verebral column curves upward, ending in the urostyle (**b** in Fig.); the

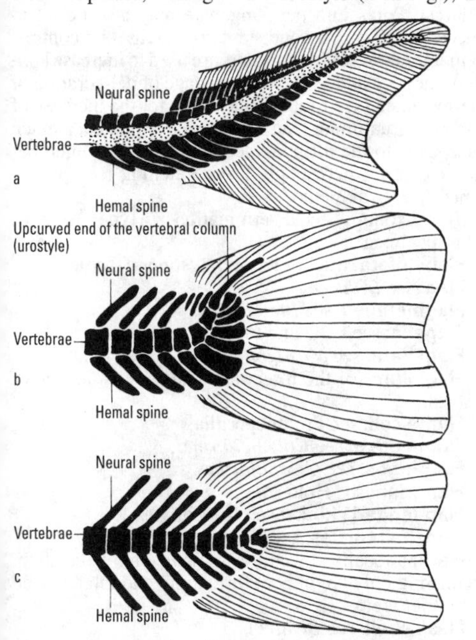

Types of caudal fin. a Heterocercal (sturgeon). *b* Homocercal (salmon). *c* Diphyceral (cod).

symmetrical *diphycercal (diphycercous)* caudal fin of lungfishes and cod (**c** in Fig.). Diphycercal and homocercal caudal fins are stiffened by fin rays (lepidotrichia), while the heterocercal caudal fin is stiffened by horny rays (ceratotrichia).

Whales and sea cows have a horizontally spread tail fin.

2) *Paired fins,* which are homologous with the extremities of terrestrial animals. The usually large *pectoral fins* are fairly constant in position, whereas the location of the *ventral fins* varies greatly, depending on the species, and they may be displaced as far forward as the pectoral region. The ventral fins of teleosts are frequently reduced. In some cartilaginous fishes the ventral fins of the male are modified to a copulatory organ.

Fiorin: see *Graminae.*

Fique: see *Agavaceae.*

Fir: see *Pinaceae.*

Fire-bellied toads: see *Discoglossidae.*

Fireblight of fruit trees: see *Erwinia.*

Firebrats: see *Zygentoma.*

Firecrest: see *Sylviinae.*

Fire-crest aletha: see *Turdinae.*

Fireflies: see *Coleoptera.*

Fireweed: see *Onagraceae.*

Firing rate of nerves: see Nerve impulse frequency.

Fish bird: see *Ichthyornis.*

Fish eagles: see *Falconiformes.*

Fishes, *Pisces* (plate 4): a class of aquatic vertebrates (phylum *Chordata,* subphylum *Craniata),* containing 3 subclasses: 1) *Aphetohyoidea,* 2) *Chondrichthyes* or *Elasmobranchii* (elasmobranchs, or cartilaginous F.), and 3) *Osteichthyes* (teleosts or bony F.). These 3 systematic groups display such great differences, however, that some authorities regard them as separate classes.

The body is differentiated into head, gill, trunk and tail regions, and it is supported by a cartilaginous or bony endoskeleton. In most cases the external form is streamlined to facilitate passage through the water; species that swim in open water are more or less torpedo-shaped, whereas bottom fish are frequently plump, eel-like or disk-shaped. Propulsion is achieved by lateral beating of the muscular tail. Median fins (including the caudal fin) act as keels for the maintenance of stability. The paired pectoral and pelvic (ventral) fins, which represent the fore and hindlimbs of higher vertebrates, are concerned with steering and regulation of the swimming level. There is a true skull attached to a movable lower jaw (for a more detailed description, see Skull). The exoskeleton (sometimes absent) consists mostly of overlapping scales, sometimes with the addition of bony plates, tubercles and spines. Teeth are continually replaced. The brain is well developed, with lobes equivalent to those of reptiles, birds and mammals, although the cerebral hemispheres are small, and the cerebellum is also normally of moderate size (see Brain). Other parts of the nervous system are also relatively well developed. There is no middle or outer ear. The inner ear serves for both equilibration and sound reception. The well developed lateral line organs may consist of free neuromasts on the skin or in pits, or they may constitute a series of receptors in mucus-filled canals or grooves on the head and body; they are primarily responsible for sensing water flow and pressure waves (vibrations) in the surrounding water. Some F. produce sounds. With few exceptions (e.g. lungfishes), the heart pumps only deoxygenated blood, and it is not divided into right and left halves (see Blood circulation, Heart). The gut is usually much shorter than in air-breathing vertebrates, and the vent is situated relatively far forward, thereby separating the trunk into thoracic and caudal regions. The eggs (called spawn) float, sink to the bottom, or are attached to submerged structures. Some species are viviparous, and different F. have characteristic larval stages.

Fossil F. Fossils are found from the Upper Silurian to the present. The oldest F. are the extinct *Aphetohyoidea* (see), which are found almost only in the Devonian. Cartilaginous and bony F. first appeared in the Devonian, and evolved independently.

Collection and preservation. F. are caught with lines, nets, weir baskets or trawls. For preservation they are attached to a board, the fins are extended and fixed with needles or fine nails, and the board (with the specimen beneath) is placed in a dish containing 2–4% formaldehyde or 70% alcohol. For larger specimens, it is desirable to also inject some of this solution into the body cavity. After a few days, the animal can be mounted and stored in a well-sealed glass cylinder in 4% formaldehyde. Saturated sodium chloride solution can also be used for preservation. Small, dead aquarium F. can be placed directly in preservative solution without mounting. See *Chondrichthyes, Osteichthyes.*

Fish lice: see *Branchiura.*

Fish lizards: see *Ichthyosauria.*

Fish tapeworm: see Broad fish tapeworm.

Fissile fruits: see Fruit.

Fission track method: see Dating methods.

Fissipedia, ***terrestrial carnivores*** (plates 38 & 39):a suborder of the *Carnivora* (see Carnivores). F. are nimble, powerful mammals with characteristic dentition, in which the rear upper premolars and the frontal lower molars are modified as cutting and grinding teeth (carnassials). The digits carry sharp claws. Most F. possess anal glands, the strongly odiferous secretions of which are used for territory marking. F. are distributed naturally throughout most regions of the world. Australia has no indigenous F., but the Dingo *(Canis familiaris)* was introduced by prehistoric man and has since reverted to the wild. See the following familes: *Mustelidae* (Weasels, Badgers, Skunks, Otters, etc.), *Procyonidae* (Racoons and allies), *Ursidae* (Bears), *Viverridae* (Civets, Genets, Linsangs, Mongooses and Fossas), *Hyaenidae* (Aardwolf and Hyenas), *Canidae* (Dogs, Wolves, Coyotes, Jackals, Foxes), *Felidae* (Cats).

Fitness: the relative contribution of a genotype to the subsequent generation. The F. of a genotype corresponds to the ratio of the number of its progeny to the number of progeny produced by the optimal genotype. F. is usually denoted by **W,** and optimal F. is expressed as **W** = 1. If an individual of the optimal genotype, ***A,*** produces on average 100 progeny, and an individual of genotype ***a*** produces only 90 progeny, then the F. of ***A*** is by definition $\mathbf{W}_A = 1$, while the F. of ***a*** is $\mathbf{W}_a = 90/100 = 0.9$.

Furthermore, **W** = 1-s, or s = 1-**W,** where s is the *selection disadvantage* or *selection coefficient.* Thus, for genotype ***A:*** s = 1–1 = 0, while for genotype ***a:*** s = 1–0.9 = 0.1.

The F. of a genotype is made up of many components, e.g. the rate of hatching of eggs, the survival rate of juvenile stages, success of pairing, and the number of progeny. In addition to **W,** the Malthus parameter (see) is also used as an index of F.

The F. of a population is calculated from the relative contribution of the F. values of its different constituent genotypes.

Fitness diagram: see Adaptive landscape.

Fitness distribution: the proportions, within a population, of individuals with different Fitness (see) values.

Fixation:

1) The complete replacement of all alternative alleles at a gene locus by a recent successful allele. As long as two or more alleles of a gene or different structures of a chromosome are present in a population their relative frequency can be rapidly altered by selection and genetic drift. Thus, F. is the result of a trend toward homozygosity. When this homogygosity is attained, i.e. only a single allele is present, the population is monomorphic for that gene locus, so that selection and drift then have no relevance until new alleles arise by mutation.

2) Methods for stabilizing and retaining the structural integrity of cells, subcellular structures and macromolecules, so that further preparative procedures (e.g. staining, coating, sectioning) can proceed without structural alteration of the specimen.

Fixatives denature and cross-link proteins. Cell membranes are made permeable, enabling the rapid penetration of solvents and stains. F. may be chemical or physical. Chemical fixatives for the preparation of specimens for light microscopy include ethanol, glacial acetic acid, various mixtures of ethanol and glacial acetic acid, picric acid, chromic acid, etc., depending on the purpose of the investigation. For electron microscopy, F. is usually performed in two stages. The first stage consists of treatment with glutaraldehyde, formaldehyde, or acrolein in various buffers, while osmium tetroxide is introduced in the second stage to increase contrast. Potassium permanganate may also be introduced during this second stage to increase the contrast of membranes. If uranium salts are used to increase contrast, they are added at the early stage of dehydration of the specimen (see Embedding). Also for the increase of contrast, uranium or lead salts can be applied at a much later stage to the ultrathin section of the specimen before electron microscopy. Freeze etching (see) represents a physical method of F.

3) In taxonomy, the determination of a type species or type specimen.

Flabellifera: a suborder of the *Isopoda* (see).

Flag: see *Iridaceae.*

Flagblennies: see *Blennioidei.*

Flagella: see Cilia and Flagella.

Flagellata: see *Mastigophorea.*

Flagellum (of the insect antenna): see Antennae of insects.

Flame cell: see Protonephridia.

Flame dove: see *Columbiformes.*

Flamingos: see *Ciconiiformes.*

Flandrian: see Holocene.

Flap-legged skinks: see *Dibamidae.*

Flapshell turtles: see *Trionychidae.*

Flark: a localized wet area of sparse fen vegetation, displaying only slight peat formation, and interspersed with drier areas.

Flask shell: see *Bivalvia.*

Flat bugs: see *Heteroptera.*

Flatfish: see *Pleuronectiformes.*

Flavin adenine dinucleotide, ***FAD, riboflavin adenosine diphosphate:*** a flavin nucleotide which serves as a hydrogen-transferring prosthetic group of many Flavoproteins (see). FAD and $FADH_2$ constitute a reversible redox system, in which hydrogen is transferred to, or removed from, nitrogen atoms 1 and 10 of the isoalloxazine ring system of the flavin (see Flavins). Reduction converts the yellow flavoquinone (of FAD) into the colorless flavohydroquinone (of $FADH_2$). FAD is biosynthesized from flavin mononucleotide (FMN) and adenosine 5'-triphosphate (ATP) by the action of the enzyme, FAD pyrophosphatase: FMN + ATP $\rightleftharpoons$ FAD + inorganic pyrophosphate.

For formula, see Flavins. Unlike other dinucleotide coenzymes (e.g. NAD), one of the sugar residues of FAD is the sugar alcohol, ribitol, rather than glycosidically linked ribose. Since C-1 of this ribitol forms a direct covalent bond with N-10 of the isoalloxazine ring system (i.e. it is not linked glycosidically), FAD is not a dinucleotide in the strict sense. See also Riboflavin in the entry, Vitamins.

Flavin enzymes: see Flavoproteins.

Flavin mononucleotide, ***FMN, riboflavin 5'-phosphate:*** a flavin nucleotide which serves as the hydrogen-transferring prosthetic group of various flavoproteins. The system FMN/$FMNH_2$ thereby acts as a reversible redox pair. FMN is a component of cytochrome oxidase, cytochrome *c* reductase, L-amino acid oxidase and other enzymes. It is biosynthesized by phosphorylation of the ribityl residue of riboflavin (see Vitamins) by ATP. For formula see Flavins. Unlike the majority of nucleotides, FMN contains the sugar alcohol, ribitol, rather than D-ribose. Since C-1 of this ribitol forms a direct covalent bond with N-10 of the isoalloxazine (i.e. it is not linked glycosidically), FMN is not a nucleotide in the strict sense.

Flavin nucleotides: prosthetic groups of flavoproteins. See Flavin mononucleotide, Flavin-adenine-dinucleotide, Flavins.

Flavins (Latin *flavus* yellow): a group of naturally occurring, nitrogenous pigments, containing the isoalloxazine ring system [a pteridine ring (see Pterins) condensed with a benzene ring]. In the nonreduced form, all F. are yellow, and they produce an intense yellow-green fluorescence in UV-light. They are water-soluble and are therefore also known as *lyochromes* [cf. lipochromes (see Carotenoids), which are lipid soluble]. *Riboflavin* is a vitamin (see Vitamins) and a biosynthetic precursor of *Flavin mononucleotide* (see) and *Flavin adenine dinucleotide* (see), which are prosthetic groups of Flavoproteins (see). The isoalloxazine ring system undergoes reversible oxidation and reduction, so that both flavin mononucleotide (FMN) and flavin adenine dinucleotide (FAD) behave as reversible redox systems (see Redox potential) for the transfer of hydrogen (Fig.).

Flavones: a group of widely distributed Flavonoids (see), which commonly occur (usually as their glycosides) as yellow flower pigments. All contain the flavone ring system. Flavone itself (Fig.) is actually colorless. F. with a hydroxyl group in position 3 are also known as *flavonols.* Familiar F. are luteolin and quercetin. Structurally and biosynthetically, the F. are closely related to the anthocyanidins (see Anthocyanins).

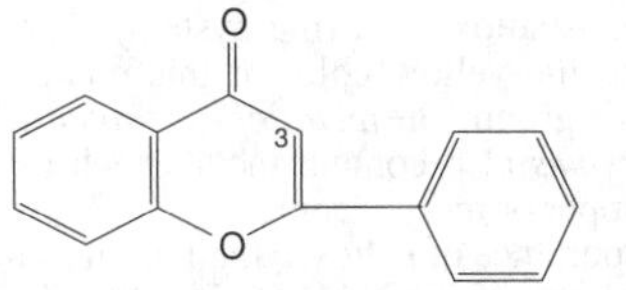

Flavone

Flavonoids: a large group of widely distributed, water-soluble, phenolic plant products, which are not found in bacteria or fungi. They are responsible for the colors of flowers, leaves, fruits and other plant parts. They characteristically contain a $C_6C_3C_6$ skeleton, and in most F. this takes the form of a phenylchroman ring

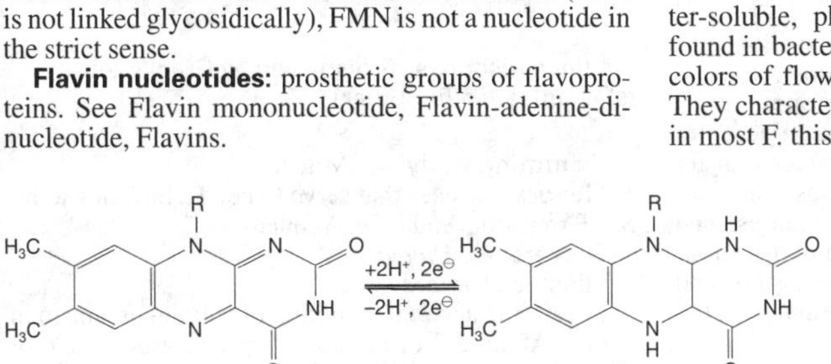

+2H⁺, 2e⊖ / −2H⁺, 2e⊖

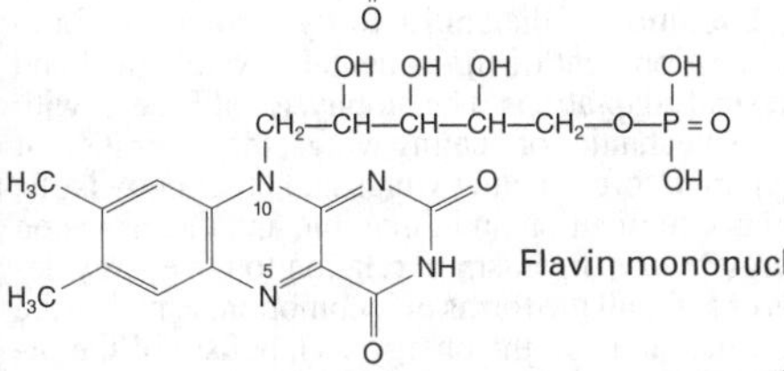

Flavin mononucleotide (FMN)

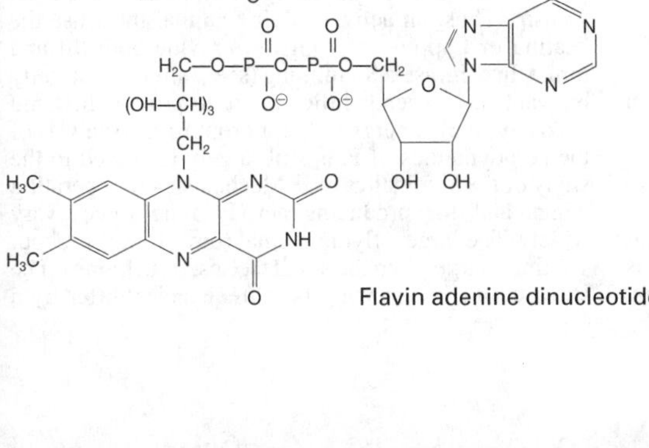

Flavin adenine dinucleotide (FAD)

system. According to the position of the phenyl residue, the F. are classified as *flavans* (2-phenyl), *isoflavans* (3-phenyl) and *neoflavans* (4-phenyl). Further classification, into Anthocyanidins (see Anthocyanins), Flavones (see), isoflavones and flavanones, is based on the degree of oxidation of the pyran ring. They usually occur as their water-soluble *O*-glycosides. In the aurones the oxygen-containing ring is contracted, and in the chalcones it is absent.

Flavan

Flavonols: see Flavones.

Flavoproteins, ***flavin enzymes, yellow enzymes:*** a diverse group of more than 70 oxidoreductases found in animals, plants and microorganisms. All F. possess either Flavin adenine dinucleotide (see) or Flavin mononucleotide as a prosthetic group, which is usually tightly but noncovalently bound to the protein. In some cases, however, the prosthetic group is covalently bound, e.g. the FAD of succinate dehydrogenase (linkage between N-3 of a histidine residue and the 8a methyl group of the isoalloxazine ring system). The group name refers to the yellow color of this nonreduced flavin prosthetic group. The *metalloflavoproteins* or *metalloflavin enzymes* also contain metals such as iron, magnesium, copper or molybdenum.

The biological importance of F. lies in their ability to reversibly accept hydrogen from NAD(P)H or directly from a substrate (e.g. from succinate when it is dehydrogenated by the F., succinate dehydrogenase). In all cases, the hydrogen is accepted by the isoalloxazine ring system of the FAD or FMN (see Flavins).

F. can be classified into the following groups.

1. *Oxidases* can use dioxygen as electron acceptor and transfer two or four electrons. Those which transfer two electrons form H_2O_2, e.g. glucose oxidase, the iron- and molybdenum-containing xanthine oxidases, and the amino acid oxidases. Oxidases which transfer four electrons contain copper; they oxidize the substrate and form water, e.g. laccase, ascorbate oxidase and *p*-diphenol oxidase.

2. *Reductases* react primarily with cytochromes. They include cytochrome b_5 and *c* reductases and glutathione reductase (all containing FAD), GMP reductase and the molybdenum-containing nitrate reductase.

3. *Dehydrogenases.* A well known example is succinate dehydrogenase (containing FAD) of the tricarboxylic acid cycle, as well as the respiratory chain NADH and NADPH dehydrogenases (containing FMN) and the acyl-Co dehydrogenases.

There are also more complex F., such as the hemoflavin enzymes (e.g. formic acid dehydrogenase from *Escherichia coli),* which contain metals, sulfhydryl-disulfide systems and heme groups in addition to the flavin component.

Flax fly: see *Thysanoptera.*

Fleas: see *Siphonaptera.*

Flea seeds: see *Plantaginaceae.*

Flehmen: a form of behavior displayed by many placental mammals, and not observed in marsupials. It is usually observed in males as a response to female odors, especially female urine during estrus. Typically, the head is raised, lips retracted and the nostrils wrinkled and closed. Breathing is momentarily stopped. All species known to flehm possess well developed Jacobson's organs. F. probably opens the orifices of the ductus incisivus, bringing Jacobson's organ into contact with odors. The flehm response can also be evoked by strong, unnatural odors, but it is normally only observed in a sexual context. In the *Bovidae* and *Camelidae*, the response is universal and may also be observed in females. It has also been reported in the black rhinoceros, a bat and some carnivores. Flehmen is a German word, adopted into the English vocabulary of animal behavior without translation.

Flehmen response, as displayed by Grant's zebra (stallion) (after Tembrock).

Flemming body: see Mitosis.

Flexors: muscles that serve to bend a limb at a joint. See Extensors, Abductor, Adductor.

Flickers: see *Picidae.*

Flies: see *Diptera.*

Flight: the active locomotion of a living organism in the air. About 2/3 of all animal species are capable of flight. Evolution of the ability to fly involves the integrated development of fundamental physiological and biophysical adaptations. The biophysics of F. deals with the drive mechanics of beating wings, the generation of aerodynamic forces on the wings and other body parts, the control of position and direction, and the provision of energy by the organism in relation to the energy demands of F. Of all the forms of locomotion, F. makes the greatest demands on the energy metabolism of the organism. Thus, an actively flying animal must use the beating or flapping of wings to provide both lift and thrust. In contrast, a Swimming (see) animal is naturally buoyant (i.e. expends little energy to provide lift) and needs to supply energy only for propulsion. Analysis of the aerodynamics of F. is still largely restricted to the study of the kinematics of F. Mechanisms for operating wings and for producing aerodynamic forces vary widely. The largest flying animal (bird, 15 kg) is about 10^8 times larger than the smallest (insect, 0.15 mg). The slowest and fastest wing beat frequencies differ by a factor of 10^3, and they are inversely related to animal size. Wing surface area increases with the 2/3 power of body mass. During hovering by large species, the power output of the F. muscles greatly exceeds the performance achievable even in a brief burst by a trained human athlete. An important problem of F. aerodynamics is the avoidance of the generation of air turbulence, which would decrease F. stability. This has led to the evolution of many adaptations of flight organs, such as scales and feathers, which prevent air turbulence. See Reynolds number.

Flight membranes: various types of skin folds that serve for aerial locomotion or gliding, e.g. in gliding marsupials *(Petauridae),* flying (gliding) squirrels and flying lemurs *(Cynocephalidae).* In these gliding mammalian species, the skin fold extends between the fore and hindlimbs, and is known as a *patagium.* Among the reptiles, the flying lizard *(Draco volans)* possesses F.m. Bats also have F.m., which are stretched between their extended digits to form true wings.

Flight musculature: the musculature responsible for moving the wings. In vertebrates it is attached to the bony endoskeleton, and in insects to the chitinous exoskeleton.

In birds, the flight muscles are responsible for actively alternately extending and flexing the wings, and for holding them in a resting position or fully outstretched. The large breast muscle moves the humerus downward and backward, and during this movement the humerus turns forward on its long axis, so that the radius, ulna and digits are deflected downward. The small breast muscle lifts the humerus up and forward to its starting position. The biceps and large delta muscle flex the elbow joint, and the triceps extends it. Appropriate muscles are also present for the flexion, extension and rotation of the digits.

In insects, invaginations of the exoskeleton serve for the origin and insertion of muscles. In winged insects, the middle and hind segment of the thorax each carries a pair of wings. The indirect flight muscles raise and lower the dorsal part of the thorax (tergum), and they are not directly attached to the wings. Contraction of the dorsal longitudinal muscles arches the tergum, raising the wing bases, and causing the wings to strike downward. Contraction of the dorsoventral muscles flattens the tergum, so that the wings strike upward. Other wing movements, such as cambering of the wings during flight, or wing folding to a position of rest, are caused by the direct flight muscles, which are attached to the wing bases.

Floating plants: plants which float freely on (rarely below) the water surface, and are not rooted in the bottom substratum, e.g. *Ceratophyllum, Utricularia, Lemna trisulca, Wolffia arrhiza.* See Hydrophytes.

Flora: a classified list of plant taxa (species, etc.) occurring in a habitat or geographical area, often accompanied by a key (which may be artificial and not based on natural affinities) to their identification. On the other hand, the term Vegetation (see) refers to the plant communities that provide the vegetative cover of an area. F. is also used in a more general sense, almost as an alternative plural for the word plant.

Floral diagram: see Flower.

Floral differentiation: see Flower development.

Floral element: a component of the flora of an area (species, group of species, etc.). *Geographical F.es.* or *geoelements* are characteristic species of a region; they have the same distribution within the same or similar areals. *Genetic F.es.* or *genoelements* have a common geographical origin, but different areals. *Historical F.es.* or *chronoelements* are species that arose at the same time. *Immigrant F.es.* or *migroelements* are those species that migrated into a region from the same direction. *Coenoelements* are associated with particular plant communities and therefore correspond to the higher order of Character species (see). *Ecoelements* are species that make the same demands on the environment. Floral kingdoms, regions, provinces, etc. are characterized by geographical F.es.

Floral formula: see Flower.

Floral initiation: see Flower development.

Floral meristem: see Flower development.

Floral primordium: see Flower development.

Floral symmetry: see Flower, Symmetry.

Flor de San Miguel: see *Orchidaceae.*

Floret: see *Compositae.*

Florigen: see Flower development.

Florigenic acid: see Flower development.

Floristic kingdom, ***floristic realm, floristic division:*** an extensive phytogeographical unit of the biosphere. Each F.k. is characterized by predominant higher systematic units, usually genera and families, as well as by the occurrence of endemics. Neighboring F.ks. are delineated from each other by distinct Floristic gradients (see). The Floristic elements (see) of the F.ks. have evolved more or less independently, becoming adapted to the prevailing environmental conditions in different areas of the world. The F.ks. are subdivided latitudinally into climatic zones, which are further subdivided into floristic regions. The latter contain floristic provinces, which are subdivided into floristic districts.

Six F.ks. are generally recognized: Holarctic kingdom (see), Paleotropical kingdom (see), Neotropical kingdom (see), South African kingdom (see), Australian kingdom, Antarctic kingdom (see) (Fig.). An oceanic F.k. of the world's oceans is characterized by only a few vascular plants, e.g. *Zosteraceae,* and by algae. See Phytogeography.

Floristic plant geography, ***floristic geobotany:*** study of the distribution of individual plant taxonomic units in a geographical area, their distribution within that area, their growth conditions, their demands on the environment, and their relationships with other plants and vegetation units. See Plant geography.

Flotation: see Soil organisms.

Flounder, *Platichthys flesus, Pleuronectes flesus* (-plate 4): a flatfish of European coastal waters. Young Fs. also inhabit the brackish and fresh water of river estuaries. It is commercially important, especially in the Baltic, where it is caught in large numbers. It resembles the plaice *(Pleuronectes platessa),* but is smaller, has a rougher surface, and usually a darker color. Both sides of the body are covered with cycloid scales, which are deeply embedded in the skin. A few spiny bony plates are present on the upper surface of the head and along the lateral line. They are found from the intertidal zone to a depth of about 60 m, living on the sand or clay bottom, where they often bury themselves for long periods. They are most active night, feeding on polychaete worms and different mollusks, and they can be a pest on cockle beds. In fresh or brackish water, they may also feed on mosquito larvae.

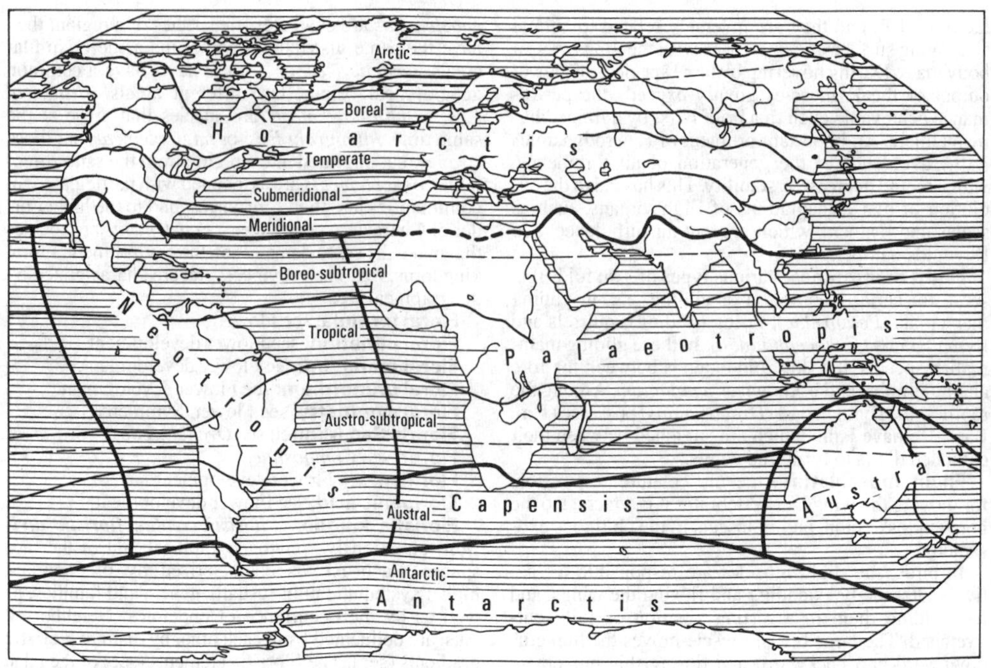

Floristic kingdom. Floristic kingdoms and zones of the world (after Mattick, Meusel, Jäger, Weinert): kingdoms in large type, zones in small type.

Flour moth: see Mediterranean flour moth.

Flow, ***flux:***

1) A quantity of irreversible or nonequilibrium thermodynamics. Vectorial F. (denoted by ***J***) is defined as the quantity per unit area passing per unit time through a plane perpendicular to the vector. Chemical reactions occur in homogeneous solution and are therefore spatially homogeneous, so that the F. of a chemical reaction is not a vectorial quantity. Otherwise, the F. of matter, heat, charge, entropy, etc. is vectorial. The divergence of a vectorial F. ***(div J)*** is the difference between the influx and outflux within an infinitesimal volume element. The F. ***(J)*** may be larger than zero (there is a source of F.), smaller than zero (there is a F. sink), or equal to zero (no sources or sinks of F. in the considered volume element). If ***div J*** is zero throughout, the system is said to be *conservative*. Biological systems are often *nonconservative,* because local changes of concentration occur by diffusion, as well as by chemical reactions, e.g. the Fs. of glucose and oxygen in tissues decrease by consumption and diffusion.

2) For fluid flow, hydrodynamics of flow, etc., see Flow hydrodynamics.

Flower (Plates 10, 34 & 35): a shoot carrying modified leaves, mostly on crowded nodes with very short internodes, whose parts are adapted for sexual reproduction. Vegetative growth and elongation of the floral apex do not occur, and the axis is short. The sporangia-bearing leaves (sporophylls) of pteridophytes, like those of clubmosses and horsetails, are aggregated into structures called strobili; like true Fs., these are borne on an axis of limited growth, i.e. their formation leads to the arrest of vegetative growth.

In the evolution of seed-bearing plants, the primitive condition of a large and indeterminate number of floral organs (microsporphylls and macrosporophylls) arranged spirally on an elongated axis has changed to one of a smaller, fixed number of organs arranged in whorls on a shortened axis. The markedly compressed axis has become the receptacle. A perianth is also formed, especially in hermaphroditic flowers. However, the spiral arrangement of a large, indeterminate number of floral parts is still evident in less advanced angiosperms, e.g. peony.

In dicotyledonous plants, each whorl of the perianth typically possesses 5, 4 or 2 components, while 3 components per whorl are found in monocotyledonous plants. The fully developed hermaphroditic F., which is

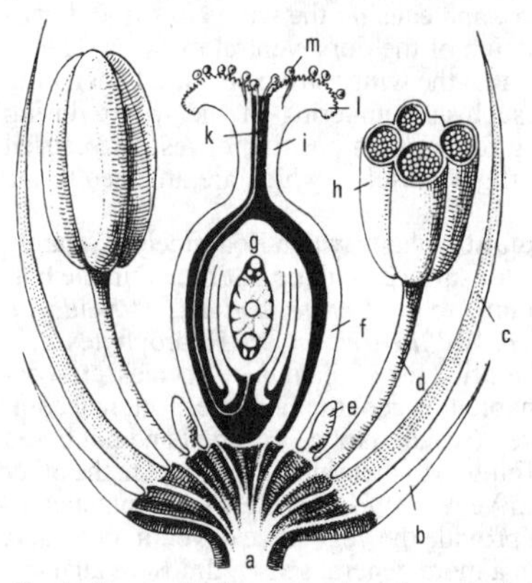

Fig.1. *Flower.* Schematic drawing of a hermaphrodite angiosperm flower. *a* Receptacle or floral axis. *b* Sepal. *c* Petal. *d* Stamen filament. *e* Nectary. *f* Ovary wall. *g* Ovule. *h* Anther head. *i* Style. *k* Pollen tube. *l* Lobes of the stigma. *m* Pollen grain.

typical of many angiosperms, consists of: 1) perianth, 2) stamens (male organs) and 3) carpels (female organs). There are often two whorls of petals or sepals (collectively known as tepals, although some authors reserve this term for the two unchanged petals of orchids), two of stamens and one of carpels; such a F. is referred to as *pentacyclic.* In gymnosperms the perianth is absent, and it is regressed in some angiosperms. Fs. possessing both stamens and carpels are referred to as *hermaphrodite (hermaphroditic)* or *perfect,* whereas *imperfect* or *monosexual* Fs. possess only stamens (staminate or male Fs.) or carpels (pistillate or female Fs.). If male and female Fs. are carried on separate plants, the species is *dioecious,* but if they are on the same plant it is *monoecious.* If hermaphroditic and imperfect Fs. are present on the same plant, it is *polygamous.* Fs. of all recent gymnosperms are imperfect; their stamens or carpels are usually arranged behind each other on a slightly shortened axis, forming a cone. For the derivation of angiosperm Fs., see Angiosperms.

1. ***Perianth.*** The perianth protects the male and female reproductive organs, and often serves to attract pollinating insects. Wind-pollinated Fs. therefore do not need a conspicuous perianth. The perianth arises from modified bracts or from modified stamens. Horticultural "doubling" of ornamental Fs. is due to the conversion of stamens into petals. Development of petals from both leaves and stamens is evident in some primitive angiosperms, especially the *Ranunculaceae.* Angiosperms display two types of perianth structure: a) In *homochlamydeous* (*homochlamydeic*) Fs. the perianth segments (or F. leaves) are similar in size, shape and color, all being inconspicuous or vividly colored. The perianth is then sometimes called a *perigon,* and the individual perianth segments are called *tepals.* If there is only one series or whorl of perianth segments, the F. is simple and the perianth is said to be *haplochlamydeous* or *monochlamydeous.* b) In *heterochlamydeous* Fs. the perianth segments occur in two distinct whorls; the outer whorl or *calyx* consists of *sepals,* which are usually green, while the inner whorl or *corolla* consists of petals, which are often conspicuously colored. If the perianth is strongly reduced, it is referred to as *achlamydeous.* Tepals, sepals and petals may be free, or more or less fused.

2. ***Stamens*** or ***microsporophylls.*** In the primitive condition, numerous stamens are arranged helically or in whorls on the F. axis. In more highly evolved Fs., the number of stamens is usually decreased, sometimes even to a single or even a half-stamen (e.g. *Canna).* The totality of the stamens in a single F. is called the *andrecium.* A single *stamen* usually consists of a *filament* and a *head* or *anther.* The anther is made up of two *thecae* and a central *connective,* which is continuous with the filament and contains vascular tissue. Each theca is divided longitudinally into two united pollen sacs or microsporangia, each containing pollen-forming tissue *(archesporial tissue* or *archesporium).* The archesporium is surrounded by 4 cellular layers: 1) The Tapetum (see) or nutritive layer next to the archesporium, which provides nutrients for the developing pollen grains (microspores), 2) A transitory middle cell layer, 3) The fibrous layer *(endothecum)* whose cells possess thickened bands, which are united and stouter towards the inner wall, but taper outwards. These bands cause opening (dehiscence) of the ripe pollen sacs by a cohesive mechanism; they tend to contract on the outer side, generating tension which ruptures the wall, usually as a longitudinal split alongside the septum between neighboring pollen sacs.

The archesporium differentiates into *pollen mother cells,* each of which gives rise to 4 pollen grains by meiosis. Pollen grains display various shapes and designs, which are characteristic for the plant species (see Pollen analysis). The contents of the pollen grain are nearly always enclosed by two membranes. These are the delicate *intine,* which grows out as the pollen tube; and the thicker, highly resistant *exine,* which forms the outer patterned and sculptured pollen grain surface and contains a small pore for the exit of the pollen tube.

In wind-pollinated (anemophilous) plants, the pollen grains are usually powdery and separate easily from one another. Those of insect- and animal-pollinated (entomophilous and zoophilous) plants usually adhere to each other, due to the presence of an oily *pollen cement,* so that large numbers are transported as crumb-like aggregates. In orchids and members of the *Asclepiadaceae,* the whole pollen mass from one half-anther is transferred as the *pollinium.* Regressed or functionally modified stamens, which no longer produce pollen, are called *staminodes.* They may become nectaries or petals.

3. ***Carpels*** or ***megasporophylls.*** The totality of the female organs or *carpels* is known as the *gynecium.* In the angiosperms the carpels are fused to form one or more ovaries. An ovary together with its *stigma* and *style* is known as a *pistil.* The stigma receives the pollen grains, and the extending pollen tube passes down the interior of the style. If the carpels are separate, each forming an ovary, the gynecium is *apocarpous;* this is considered to be a primitive condition. If individual carpels are fused to form the ovary, the gynecium is *syncarpous.*

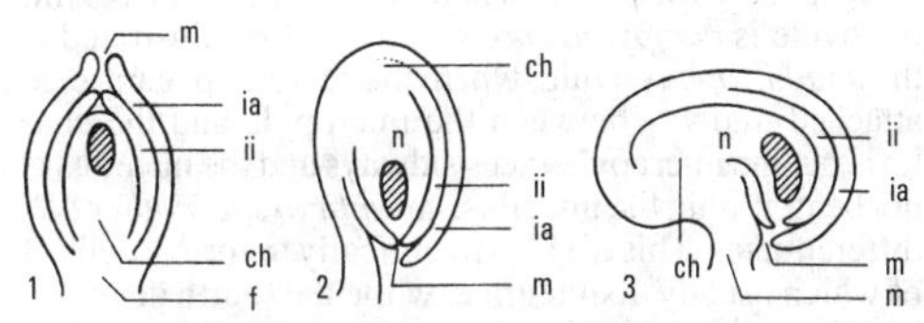

Fig.2. *Flower.* Types of angiosperm ovule. *1* Orthotropous. *2* Anatropous. *3* Campylotropous. *f* Funicle. *ch* Chalaza. *ia* Outer integument. *ii* Inner integument. *m* Micropyle. *n* Nucellus.

The position of the ovary on the F. axis is characteristic of the plant species. If it lies on a conical axis beyond all the other flower parts, it is *superior* and the F. is *hypogynous* (e.g. *Ranunculaceae* and *Papilionaceae).* A free-standing ovary in a depression of the receptacle is *half-inferior,* the F. being *perigynous.* If the ovary lies below the other F. parts, and is encased by and fused with the receptacle, it is *inferior* (e.g. roses) and the F. is *epigynous* (Fig. 3). The ovary encloses the ovules, which are usually attached to a longitudinal thickening called the *placenta.* This thickening corresponds to the ventral suture of the leaf from which the carpel has evolved, whereas the dorsal suture of the carpel corresponds to the original midrib. The arrangement of the ovules within the syncarpous ovary is referred to as *placentation.* Fusion of the carpels only at their edges results in an unchambered compound ovary, with longitudinal placentas on the inside wall; the ovules then line the ovary wall in longitudinal rows and placentation is said to be *parietal,* e.g. in *Viola, Papaver, Ribes.* Where carpellary margins are fused all together centrally, the placentas lie on the longitudinal axis in the center of the ovary and placentation is said to be *axile,* e.g. tomato *(Lycopersicum)* and tulip *(Tulipa).* Parietal and axile placentation are both forms of marginal placentation. In a modified form of axile placentation, known as central placentation, the central axis does not reach the roof of the ovary, so that the ovules appear to arise from a mound in the center, e.g. *Stellaria, Arenia, Spergularia.* Although dividing septa are absent, septation may be evident in the young ovary. In laminar placentation, ovules are inserted all over the inner walls of the carpels; this relatively uncommon, nonmarginal arrangement is found only in the *Nymphaeales* (a dicotyledonous order) and in four monocotyledonous families *(Butomaceae, Hydrocharitaceae, Limnocharitaceae, Alismataceae)* belonging to the order *Alismatales.* The ovule is joined to the placenta by a short stalk or *funicle,* which contains vascular tissue. The thickened upper part of the ovule, which contains the embryo sac, is called the *nucellus.* This is usually invested by two cellular coats or integuments, but in the *Sympetalae* only one integument is normally present. The integuments are derived from the base of the ovule, the *chalaza.* A small opening or *micropyle* at the opposite pole allows access to the nucellus. If the ovule is straight and the nucellus lies directly in line with the funicle, so that the micropyle is opposite the latter, the ovule is said to be *orthotropous*. Where the ovule is bent over against the funicle, so that the micropyle faces the placenta, the ovule is said to be *anatropous*. If the ovule is bent so that the funicle appears to be attached midway between the micropyle and the chalaza, the micropyle faces the placenta and the nucellus is also bent, the ovule is *campylotropous* (Fig. 2). Not illustrated is the *amphitropous* ovule, where the funicle appears to be attached midway between the micropyle and the chalaza, but the micropyle faces sidways and the nucellus is not bent. Within the nucellus, an *embryo sac mother cell* differentiates. This divides meiotically to form 4 cells, 3 of which usually disintegrate, while the fourth develops into the embryo sac. The latter increases in size and its nucleus undergoes three successive divisions. From the resulting 8 nuclei, groups of 3 come to lie at each end of the embryo sac cell, where each nucleus becomes surrounded by cytoplasm and finally enclosed by a membrane. The resulting 3 cells furthest from the micropyle are known as the *antipodal cells,* while the 3 at the micropylar end form the *egg apparatus.* The middle cell of the egg apparatus becomes the egg cell, and the other two are called *synergids.* The remaining 2 nuclei of the original 8 fuse in the center of the embryo sac cell to form the *secondary embryo sac nucleus* or *primary endosperm nucleus.* Development of the female angiosperm gametophyte is then complete, and the embryo sac is ready for fertilization.

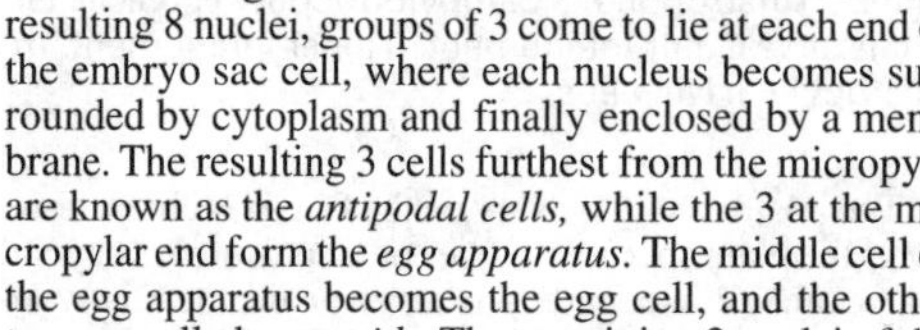

Fig.3. *Flower.* Position of the ovary: *a* superior, *b* half-inferior, *c* inferior.

4. ***Floral symmetry.*** There are 4 main types of floral symmetry. Radial (polysymmetric or actinomorphic) Fs. possess more than two planes of symmetry, e.g. *Sedum.* Dorsiventral (monosymmetric or zygomorphic) Fs. possess only a single plane of symmetry, e.g. *Lamium.* Bilateral (disymmetric) Fs. have two planes of symmetry, e.g. *Dicentra.* Asymmetric Fs. lack any plane of symmetry, e.g. *Canna.* Most Fs. are actinomorphic or zygomorphic. Disymmetry and asymmetry are less common, and are thought to be secondary characteristics. Radial symmetry represents the original condition, and dorsiventral symmetry is thought to be derived from it as an adaptation to the dorsiventral symmetry of pollinating insects. Some dorsiventrally symmetric Fs. have evolved further to achieve a new form of radial symmetry, known as the *peloric condition,* e.g. the terminal Fs. of *Digitalis.*

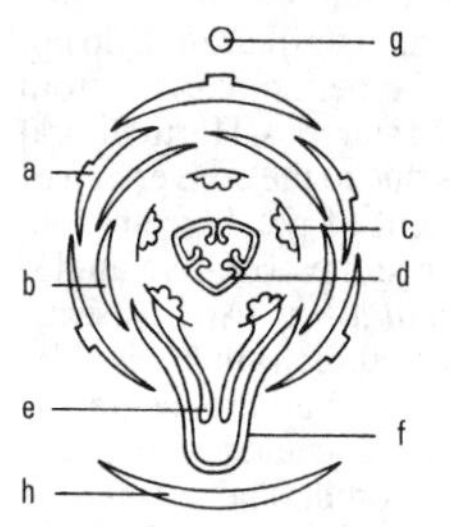

Fig.4. *Flower.* Floral diagram of a violet. *a* Sepals. *b* Petals. *c* Stamens. *d* Carpels. *e* Nectaries. *f* Spur. *g* Flower stalk. *h* Bracts.

Floral symmetry and the relative positions of all parts of the F. can be clearly represented by a *floral diagram,* which is a plan of the F. with certain parts shown in section for clarity (Fig. 4). F. structure can also be briefly expressed in *floral formula,* which uses a letter for each type of F. part, followed by the appropriate number; it also gives the position of the ovary by a line above (superior) or below (inferior) the number for the gynecium, or the line is omitted if the ovary is perigynous. Radial Fs. are denoted by a star ★, dorsiventral Fs. by an arrow ↓. For example, the formula of *Sedum* is: ★ K5, C5, A5 + 5, G5, which is interpreted as 5 sepals, 5 petals, 10 stamens (2 whorls of 5 each), 5 carpels, perigynous ovary, radial F.

5. ***Inflorescence.*** The inflorescence is a specialized part of the shoot carrying one or several Fs., and clearly differentiated from the foliage of the plant. More or less conspicuous bracts may be present, but the inflorescence otherwise possesses no leaves. There is usually an inverse relationship between the number of Fs. in an inflorescence and the individual F. size. When a large number of small Fs. is present, the whole inflorescence is often encircled by a perianth-like involucre of bracts, and the outermost Fs. or F. parts may become modified to form a striking display; the resulting inflorescence often has the appearance of a single F. Such F.-like inflorescences which function biologically as single Fs. are called *pseudanthia* or *false Fs.* (see Pseudanthium theory, Euanthium theory). Examples are the composite heads or *capitula* of the *Compositae,* the *cyathium* of *Euphorbia,* and the *spadix* of the *Araceae,* which is often enclosed in a brightly colored bract or spathe.

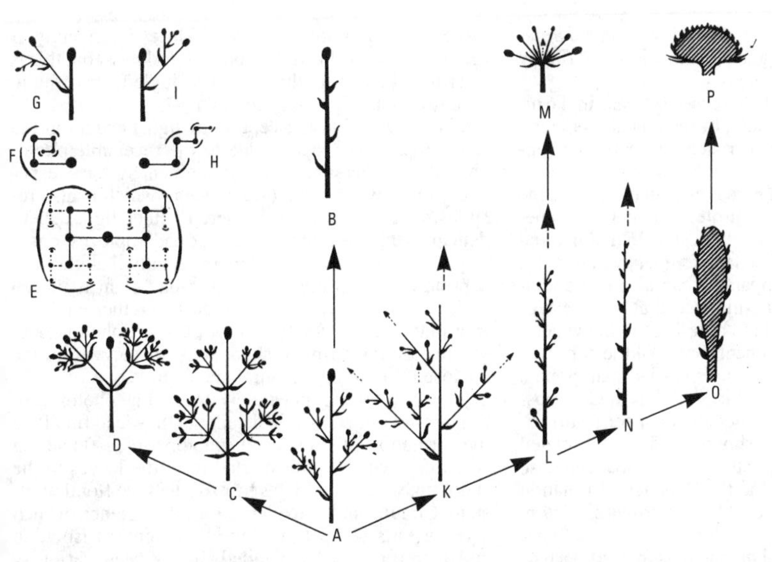

Fig.5. *Flower.* Derivation of the main types of inflorescence from the cymose panicle. For key, see text.

In the simple inflorescence, all the side shoots of the main axis are unbranched and they each terminate in a single F. The *compound inflorescence* is made up of a greater or lesser number of repeatedly branched side axes; the total inflorescence then consists of several partial inflorescences. Inflorescences are classified into 2 main types, depending on the mode of termination of the main axis (or the partial axes of a compound inflorescence). 1) The *cymose* inflorescence is usually obconical in outline, and its growth points all terminate in a flower; continued growth of the inflorescence can proceed only from new lateral shoots, and the oldest branches and Fs. usually occur at the apex. 2) The *racemose* inflorescence, is usually conical in outline, and its growth points continually add to the inflorescence; a terminal F. is usually lacking, and the youngest and smallest branches and Fs. are nearest the apex.

The *cymost panicle* (Fig. 5, A) [e.g. grape *(Vitis)]* is especially significant, because it probably represents the phylogenetically oldest type of inflorescence, from which all other inflorescences are derived. An umbel-like panicle is produced by basitonal development of all side branches, so that the Fs. form a parasol [e.g. European elder *(Sambucus nigra)* and *Viburnum].* Intensified basitonal development may produce lateral branches which extend beyond the terminal F. of the main axis, resulting in a *compound panicle* [e.g. meadowsweet *(Filipendula ulmaria)].* Side branches may be sparse, and in extreme cases they may be totally absent, leaving a solitary F. (e.g. poppy or tulip, Fig. 5, B). Side axes often develop only from the uppermost nodes; if all the internodes of the main axis also remain shortened, the result is a *pleiochasium* [a multi-forked cyme, e.g. houseleek *(Sempervivum),* and many species of *Euphorbia].* If the plant has decussate leaves, side branches of the inflorescence often arise only from two lateral buds of the node below the terminal F.; the result is then a *dichasium* [a doubly forked cyme, e.g. chickweed *(Cerastium),* Fig. 5, D and E]. If a single side branch is formed from only one of these lateral buds, the inflorescence is a *monochasium.* There are 4 types of monochasia, depending on whether branching always occurs on the same side (e.g. branches only formed from the ventral side) or from both sides, and whether branching occurs in a single plane, or revolves in relation to the axis. These are illustrated by the *drepanium* (various rushes, e.g. *Juncus bufonius,* Fig. 5, G), the *rhipidium* (e.g. *Iris,* Fig. 5, I), the *bostryx* or *helicoid cyme* [e.g. day lily *(Hemerocallis)* Fig. 5, F], and the *cincinnus* or *scorpoid cyme* [e.g. comfrey *(Symphytum)* Fig. 5, H]. A compound cymose inflorescence composed of dichasial or monochasial partial inflorescences is known as a *cymose thyrsus* [e.g. calamint *(Calamintha)* Fig. 5, C].

All racemose inflorescences are derived from the *racemose panicle* (e.g. the panicle of grasses, Fig. 5, K). If the side axes of the racemose panicle are reduced to single, stalked Fs., the inflorescence is called a *raceme* (e.g. lupin, Fig. 5, L). When the F. stalk is also absent, the result is an *ear* or *spike* [e.g. plantain *(Plantago),* Fig. 5, N]. A spike of unstalked (sessile) Fs. may be flexible so that it hangs down and is easily disturbed by the wind; it is then called a *catkin* (e.g. the male inflorescence of hazel and oak). An inflorescence with a thickened main axis and sessile Fs. is known as a *spadix* (e.g. the female inflorescence of maize, Fig. 5, O). In the *composite head* or *capitulum* (e.g. sunflower and other members of the *Compositae),* individual sessile Fs. arise from the surface of the receptacle, which is formed by widening of the axis. The *umbel* originates by extreme shortening of the internodes, so that the stalks of the long-stalked individual Fs. appear to arise at the same level (e.g. various species of *Primula,* Fig. 5, M).

Flower beetles: see Lamellicorn beetles.

Flower clock: a collection of plants, first assembled by Linneus, in which the flowers display characteristic and different daily rhythms of opening and closing. From such an assembly of plants, it is therefore possible, within certain limits, to deduce the time.

Flower development: in higher plants the transition from the vegetative phase to sexual reproduction. It includes *floral initiation,* i.e. modification of an apical meristem with leaf primordia to a *floral mersitem* with primordia for carpels, stamens, sepals and petals. This is followed by differentiation of the flower *(floral differentiation),* flower opening and fertilization, and finally development of fruits and seeds.

In the early stages of F.d. there is a decrease in the histone content of the cell nuclei of the apical meristem. The RNA content of the meristematic cells then increases (i.e. RNA is synthesized) and the base composition of the RNA changes. There is also an increase in the number of ribosomes and the protein content of the meristematic cells, accompanied by a sudden and marked increase in DNA synthesis and the frequency of mitosis. Activation of the Golgi apparatus can also be demonstrated. These observations indicate that a variety of genes are activated, which are quiescent in the vegetative phase of plant development. Some of the genes directing F.d. have recently been identified and cloned. Although in its early stages, this work has made it possible to construct molecular models for specification of the floral meristem, and to show that distantly related plants of the *Cruciferae* and the *Scrophulariaceae* use homologous mechanisms of floral pattern formation [Enrico S. Coen and Elliot M. Meyerowitz, *Nature* (1991) ***353,*** 31–37].

In many plants, the floral meristem is formed at a particular stage of plant growth and development, irrespective of environmental conditions, i.e. F.d. is autonomous or endogenously determined. In other plants, special environmental conditions are required, such as a period of low temperature (see Vernalization) or a certain day length (see Photoperiodism). Induction of F.d. by temperature or light is known respectively as *thermoinduction* and *photoinduction.* Much detailed research has been done on the photoperiodic control of F.d., and less is known about the physiological basis of temperature effects. The inductive photoperiod varies according to the plant species. Thus, there are Long-day plants (see), Short-day plants (see), Long-short-day plants (see), Short-long-day plants (see), and Day-neutral plants (see). The existence of Intermediate day-length plants (see) is disputed. Long-short-day and Short-long-day plants are *amphiphotoperiodic plants,* in which F.d. is induced by two different photoperiods.

Often only one cycle of induction is necessary. For example, in certain short-day plants, a single short day (i.e. one long dark phase) is sufficient, while in some long-day plants, 2–3 days of continuous light leads to F.d. After this induction has been achieved, F.d. proceeds and is completed, even if the plant is returned immediately afterward to noninductive environmental conditions.

It has become customary to consider photoperiodic F.d. in terms of the duration of illumination. It is now known, however, that most plants respond not to the length of illumination, but to the duration of uninterrupted darkness. A short-day plant is really a "long-night" plant, requiring a period of uninterrupted darkness that must exceed a certain critical minimum. Conversely, a long-day plant is really a "short-night" plant, which produces a floral meristem only when subjected to a dark period not longer than a certain critical maximum. Cocklebur *(Xanthium)* is a short-day plant, which can be induced to flower by a single cycle of 15 hours light and 9 hours dark. Once this regime has been experienced, flowering will occur, irrespective of subsequent changes in the duration of light/dark exposure. If the plant never experiences a dark period of 9 hours or more, flowering does not occur. More significantly, if the cocklebur is allowed 9 hours of darkness, but this is interrupted half way through by a single flash of light, then flowering does not occur.

Photoperception (susception of light) occurs in the vegetative leaves, and is due to the reversible photochemical conversion of certain pigment systems, especially the Phytochrome (see) system, which is also responsible for regulating numerous other light-dependent growth, developmental, nastic and tropic responses of plants. To demonstrate experimentally that photoreception occurs in the leaf, a short-day plant is placed under long-day conditions; a single leaf is then enclosed in a black bag and subjecting to a short photoperiod, whereupon the terminal shoot some distance from the enclosed leaf initiates a floral mersitem.

The processes immediately following photoreception are not understood in detail, but they lead finally to the formation of a factor(s) (photoperiodic flowering stimulus), which is transported from the leaves to the shoot tip, where it initiates formation of the floral meristem. Circumstantial evidence for the existence of such a factor(s) is compelling, but all attempts at isolation and identification have failed. It has been variously called *flowering hormone, florigen, florigenic acid* and *vernalin.*

If the leaf of a short-day plant (e.g. *Nicotiana tabaccum)* is grafted onto a long-day plant (e.g. *Nicotiana silvestris),* the latter also produces flowers in response to short days. The converse experiment is also successful. Long-day plants also produce flowers under short-day conditions, when grafted onto short-day plants already in flower. Furthermore, both short- and long-day plants produce flowers in the absence of an appropriate inductive photoperiod, if they are grafted onto a flowering

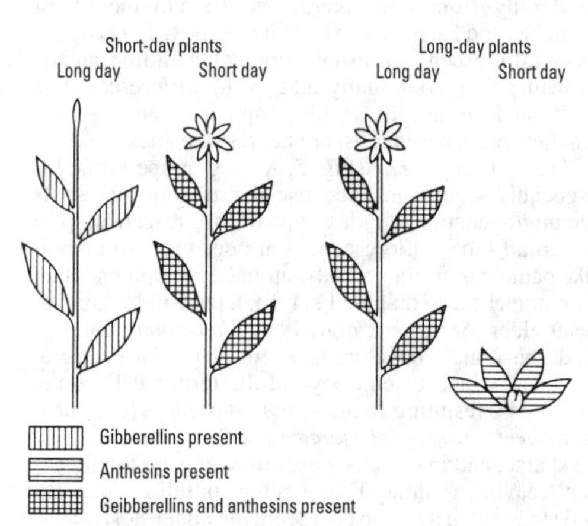

Flower development. Biphasic theory of photoperiodic floral induction (after Chailakhyan). Short-day plants produce sufficient gibberellins for shoot extension under all light conditions, whereas long-day plants produce sufficient gibberellins for shoot extension only under long day conditions. In long-day plants, anthesin production occurs under all conditions, whereas short-day plants produce sufficient anthesin for flower initiation only under short day conditions, and therefore have a rosette habit before flowering.

day-neutral plant. Species of the day-neutral parasitic *Cuscuta* (dodder) flower at the same time as their host, irrespective of whether this is a long- or short-day plant. It is therefore clear that long-day, short-day and day-neutral plants all respond to the same photoperiodic flowering stimulus, whatever this might be.

In some short-day plants, e.g. *Pharbitis (Convolvulaceae),* F.d. can be induced under long-day conditions by the application of Gibberellins (see), while in others (e.g. *Chenopodium rubrum)* F.d. is initiated by the inhibitor, Abscisic acid (see). In the pineapple plant *(Ananas)* F.d. is stimulated by Auxin (see) and by Ethylene (see). It therefore appears that the photoperiodic flowering stimulus or "flowering hormone" may be a complex of all phytohormones. This is somewhat reminiscent of the *biphasic theory of photoperiodic floral induction,* which claims the existence of two hormonal complexes. According to this theory, flowering requires both gibberellins and the nitrogenous, chemically unidentified *anthesins.* Gibberellins are required for extension of the shoot and formation of the floral axis, while anthesins are required for formation of the floral primordia (Fig.).

F.d. may also be controlled by *flowering inhibitors.* If the shoot of a late-flowering pea variety is grafted onto an early flowering variety, flowering of the latter is delayed. The identity of the inhibitor is unknown.

Flower eater: see Lamellicorn beetles.

Flower fungus: see *Basidiomycetes (Gasteromycetales).*

Flowering cherry: see *Rosaceae.*

Flowering currant: see *Grossulariaceae.*

Flowering inhibitors: see Flower development.

Flowering phase of a plant: the phase following development of the flower in the bud. It therefore involves the unfolding or opening of flower buds (see Unfolding movements), subsequent opening and closing movements (see Flower movements), Pollination (see), pollen germination and ovum Fertilization (see), all of which occur during the life of the flower.

Flower movements: various types of movement displayed by flowers or their constituent parts.

1) *Unfolding movements.* The opening of flower buds by unilateral growth processes.

2) *Opening and closing movements.* Thermonastc or photonastic responses of tepals or petals (see Thermonasty, Photonasty), i.e. opening or closing of flowers in response to changes in temperature or light intensity. For example, when tulips or crocuses are transferred to a warm room, the flowers open within a few minutes, and close again if they are cooled. Increased temperature results in faster growth on the upper surface than on the undersurface of the perianth members, and this relationship is reversed at lower temperatures. Temperature-induced opening is therefore accompanied by enlargement of the petals, e.g. the size of a tulip petal increases by about 7% in a single thermonastic movement, while the repeated thermonastic responses that occur during the entire flowering period may cause an increase of up to 100%. Some species are extremely temperature-sensitive; thus, crocus flowers react to temperature changes of 0.2 °C, and tulips to changes of 1 °C. Photonastic opening and closing movements are displayed in particular by flowers of water lilies, cacti, wood sorrel, and the flower heads of many *Compositae* (in which the peripheral strap or ray florets behave like individual petals). Many day-flowering plants open and close their flowers at specific times of the day, and these times vary considerably from species to species. Thus, pumpkins and gourds generally open their flowers between about 5.00 and 15.00. The marsh sow-thistle *(Sonchus arvensis)* blooms between 6.00 and 12.00, the marsh marigold *(Caltha palustris)* from 8.00 to 21.00, upright yellow sorrel *(Oxalis stricta)* from 10.00 to 16.00, and the germander-speedwell *(Veronica chamaedrys)* from 10.00 to 21.00. These differences in daily rhythm were used by Linneus in his concept of a Flower clock (see). Night-flowering plants, e.g. white campion *(Melandrium album)* and thorn-apple *(Datura stramonium),* open their flowers in the evening and keep them closed during the day. Photonastic and thermonastic F.m. are both Nutational movements (see). Under natural conditions, these opening and closing movements are induced primarily by light and temperature variations resulting from the alternation of night and day. Endogenous factors are also probably involved.

3) *Endogenous rhythmic or periodic movements.* These are predominantly or exclusively autonomous, e.g. the 12-hourly cycle of opening and closing of the flower buds of the pot marigold *(Calendula officinalis).* This type of movement is due to different growth rates on the upper and undersides of the peripheral ray florets. In this respect it resembles thermo- and photonastic F.m., but it differs from the latter, in that it operates independently of the day/night cycle. For example, the pot marigold continues to open and close its flowers according to a 12-hour cycle, even when it is kept in continual darkness. Periodic movements that correspond more or less to the periods of night and day are called Nyctinasties (see).

The first three types of movement are due to differential growth. In contrast, type 4) and 5) movements (below) are due to alterations in turgor pressure.

4) *Stamen movements.* These are known to occur in several plants. In the unstimulated condition, the stamens of *Berberis* are spread outward and close to the inside surfaces of the petals. If the base of the inner surface of the filament is touched, e.g. by an insect, the turgor pressure within the stamen immediately decreases, causing the stamen to move rapidly upward and inward, so that it becomes adpressed to the ovary. Turgor is rapidly restored, and the stamens are capable of undergoing repeated cycles of stimulation and recovery. Similar stamen movements occur in cacti flowers (e.g. *Opuntia).* In *Sparmania africana* (family *Tiliaceae)* and Rockroses (*Helianthemum* sp., family *Cistaceae),* the stamens bend outward when stimulated. In the knapweeds *(Centaurea* sp., family *Compositae),* the stamens contract rapidly when touched (also due to a fall in turgor pressure), drawing the closed anther tube downward against the stigmatic head of the style. Pollen is thereby pressed up and out of the top of the anther tube, where it can adhere to a visiting insect. Such seismonastic movements serve the function of pollination.

5) *Stigma movements.* Flowers of various plants species possess touch-sensitive stigmas, e.g. *Mimulus* sp. (family *Scrophulariaceae), Incarvillea* sp. (family *Bignoniaceae), Catalpa* sp. (family *Bignoniaceae)* and others. The usually 2-lobed stigmas are normally widely spread, but the two lobes snap together when touched, trapping between them any pollen that may have rubbed

off an insect. This seismonastic movement is due to a sudden loss of turgor pressure in certain groups of cells. Like stamen movements, stigma movements are known as turgor movements, i.e. they are not due to differential growth, but to reversible changes in turgor pressure in certain cells, which result in the lever-like movements in the structures of organs that contain them.

Flowerpeckers: see *Dicaeidae.*

Flower pigments: Anthocyanins (see), Flavones (see) and Carotenoids (see) responsible for the colors of flowers. Yellow flower colors are due to flavones dissolved in the cell sap, or to carotene-containing chromoplasts in the petal cells. Red and blue colors arise from dissolved anthocyanins. The joint occurrence of different pigment types gives rise to a great variety of colors and color shades.

Flowers of tan: see *Myxomycetes.*

Flow hydrodynamics: movement of liquids and gases. When a liquid moves over a wettable surface, the tangential velocity is zero directly at the surface, where some molecules of the liquid are fixed (adhesive effect). Flow velocity increases asymptotically with distance from the surface. Next to the fixed layer is the boundary layer, in which the rate of flow is a function of the distance from the surface. If flow in the boundary layer is entirely parallel to the surface, it is called *laminar flow.* The rate of flow as a function of distance from the surface is the *shear rate.* In Newtonian (homogeneous) liquids, friction is proportional to the shear rate. In non-Newtonian fluids, Viscosity (see) itself is a function of the shear rate. At a critical rate, the flow becomes turbulent. This critical value depends on the geometry of the body and on the density and viscosity of the flowing medium. It is of considerable interest to compare the flow behavior of differently sized bodies that have the same geometrical shape (see Model system). A dimensionless quantity called the Reynolds number (see) can be calculated from the flow velocity, length of body, density and viscosity of the liquid. Geometrically similar, but differently sized bodies show the same flow behavior in different media, if the rate of flow is adjusted to give the same Reynolds number.

In the flow of a liquid though a tube, the critical Reynolds number for onset of turbulence is about 1 160. Application of a pressure gradient causes a parabolic profile of velocity to form in the tube. According to Poiseuille's law, the quantity of liquid flowing is proportional to the fourth power of the radius. In blood flow, however, the preconditions of Poisseuille's law are not met, because vessel wall structure leads to the development of a totally different type of boundary layer. Due to the orientation and deformation of the formed blood components, the flow behavior of blood is markedly non-Newtonian. In capillary flow in biological systems, flow is greatly influenced by the adhesive effect. The walls of biological transport systems are often elastic, so that the radius is a function of pressure. Flow mechanics are important for an understanding of the biomechanmics of Flight (see), Swimming (see), Hemorheology (see) and the Hydrodynamics of blood circulation (see).

Fluctuation: variations in the density of a population in a particular region during the course of several generations. Annual increases and decreases of density (influenced by a combination of endogenous and exogenous factors, and known as *oscillations*, see Fig.1) are superimposed on the underlying F. A conspicuous rise in the F. curve (e.g. of forest pests) in a relatively short time (a few years) is known as a *mass increase* or *gradation.*

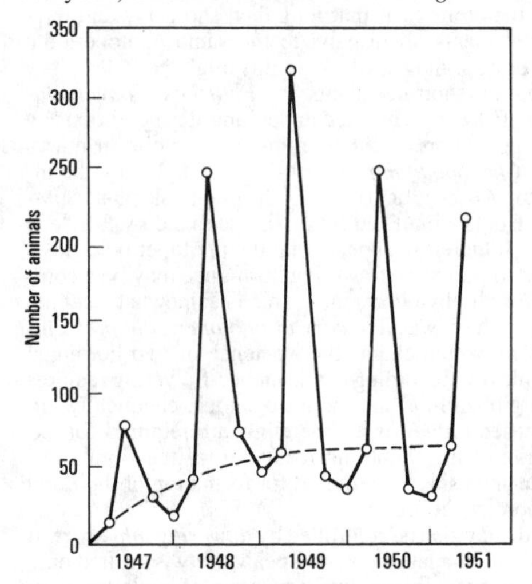

Fig.1. *Fluctuation.* Oscillations (continuous line) and fluctuations (dashed) of a population of great tits in a forest area (from Scherdtfeger, 1978).

A mass increase of animal pests often results in serious economic damage, or a *calamity* (e.g. pine looper moth, *Bupalis pinaria;* pine beauty moth, *Panolis flammea;* fieldmouse). Before a mass increase, the population density is not particularly high, but in a state of latency. The increase to the *culmination point* is known as *progradation,* and the decrease to the next latent state is called *retrogradation.* At the peak of a gradation, the effects of overpopulation are evident, i.e. the *crowding* or *collision factor* shows a marked increase (e.g. competition and interference between members of the same species in feeding, reproduction, etc.). In fieldmouse populations, this may have a shock effect, leading to the elimination of males. Predators, parasites, disease

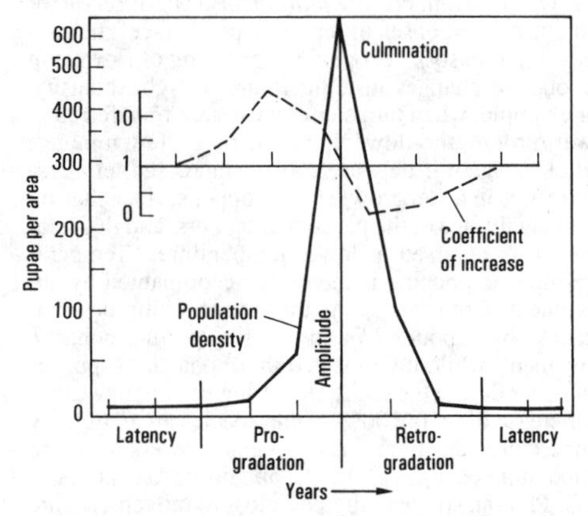

Fig.2. *Fluctuation.* Gradation (mass increase) of a pine beauty *(Panolis flammea)* population in a pine forest (after Scherdtfeger, 1933).

agents and climatic conditions make a considerable contribution to the collapse of a mass increase, and to F. in general. In *cyclic F.,* the time periods between latency and culmination are about the same. F. periodicity for field mice and birds of prey is usually 3–4 years, whereas populations of the American hare *(Lepus americanus)* often show a F. periodicity of 9–10 years.

Monitoring of F. for the prognosis and control of animal pests is of considerable economic importance.

Fluidity: a quantity that describes the viscosity of membranes. F. is the reciprocal of viscosity. The rate of diffusion of molecules in a membrane increases with F. The F. of a membrane depends on the membrane structure, in particular on the phase state of the lipids. Lipid phase states change very sharply in response to temperature or chemical influences. During physiological membrane processes, e.g. Fusion (see), it is assumed that membrane F. is subject to endogenous regulation.

F. is evaluated by measuring the lateral diffusion of membrane components. Lateral diffusion constants of membrane lipids are in the order of 10^{-8} cm^2/s. The mobility of proteins is 1 to 4 orders of magnitude less.

Flukes: trematodes of the order *Digenea* which parasitize the blood vessels of vertebrates. All species are dioecious, and the female resides permanently in a tubular invagination on the underside of the male. The tropical and subtropical species, *Schistosoma haematobium* (plate 42), is the causative agent of Bilharzia. It parasitizes the veins of the intestine and bladder of humans, and is responsible for the appearance of blood in the urine. The eggs are voided from the host and require water for their further development. Miracidia emerge from the eggs and attack water snails. Within water snails the miracidia develop into sporocysts and finally into cercaria (see Liver fluke), which are released into the water, where they reinfect mammalian hosts, e.g. rice farmers. See *Trematoda,* Liver fluke, Schistosomes.

Fluorescence: emission of light when an electron falls from a higher state of excitation to a lower state of excition, whereby the spin of the electron remains unchanged. F. involves singlet-singlet, as well as triplet-triplet transitions. See Absorption.

Fluorescence microscopy: a microscopic method which employs the fluorescence of materials for their visualization. When certain compounds are irradiated with UV or short wavelength visible light, part of the absorbed energy is re-emitted as long wavelength irradiation, a phenomenon known as Fluorescence (see). The light source for F.m. is a high pressure mercury lamp. The excitation light is passed through appropriate excitation filters between the light source and the preparation. A barrier filter between the eyepiece and the preparation insures that the eye receives only fluorescent light and not the short wavelength excitation light. F.m. can be performed by surface illumination or by passing the excitation light through the preparation from below. Modern fluorescence microscopes usually have facilities for other simultaneous procedures, such as Dark field microscopy (see) and Phase contrast microscopy (see). A number of cellular structures fluoresce naturally in UV light *(primary fluorescence).* Many tissues, in the living or fixed state, can be specifically stained with fluorescent dyes *(fluorochromes).* Since only very low concentrations of fluorescent dyes are sufficient, the method is very sensitive and very suitable for the investigation of living cells. F.m. is also widely used for detecting the binding of fluorescent antibodies (treated with fluorescent dyes) in immunocytochemical studies.

Fluorescence quenching: decrease of the quantum yield of fluorescence due to the presence of competing processes. Bimolecular reactions compete with the monomolecular deactivation processes of an excited molecule, so that various bimolecular reactions are exploited for the purposes of quenching, e.g. in Fluorescence spectroscopy (see).

Fluorescence spectroscopy: spectroscopic method for studying the fluorescence of excited molecules. Often a Label (see) is introduced, whose subsequent fluorescence behavior provides information on the molecular and physical properties of the sample. If the label is attached to specific sites on the sample, then exactly localized regions of the sample can be investigated, since the only source of fluorescence is the label itself. These advantages of F.s. are often exploited in the study of cells, organelles, membranes and liposomes. For example, the membrane potential difference can be measured with the aid of F.s. A precondition for such measurements is that the label is distributed in proportion to the membrane potential passively between the cell interior and cell exterior. If a fluorescence quencher is added to the outer solution, then the fluorescence intensities before and after quenching represent the direct ratio of the inner to the total label concentration. Using the Nernst equation, the membrane potential is then easily calculated. There are also labels for the determination of surface potential. F.s. can also be used to determine lateral and rotational diffusion constants of membranes. For this purpose, lipid labels are used, whose diffusion behavior is the same as that of the natural lipids. Lateral diffusion is determined by *excimer fluorescence.* An excimer is a complex between an excited molecule and a molecule in the ground state; the position of the fluorescence maximum of the dimer differs from that of the monomer. Excimer formation depends directly on the probability of an encounter between the two monomers, thereby serving as an index of lateral diffusion. Rotational diffusion is measured from the change in polarization of the emitted fluorescence.

Fluorine dating: see Dating methods.

Flux: see Flow.

Fly agaric: see *Basidiomycetes (Agaricales).*

Flycatcher plant: see Carnivorous plants.

Flycatchers: see *Muscicapinae.*

Flying dragon, *Draco volans:* a fully arboreal, Southeast Asian agamid (see *Agamidae).* The male has a small nuchal crest and an orange gular sac. The gular sac of the female is blue. On each side of the body, a large, brightly colored, thin fold of skin extends between the fore- and hindlimbs. With the aid of several elongated bony ribs, these skin folds can be spread as "wings", enabling the animal to glide passively over distances of 100 m from higher to lower levels in the tree canopy, or to the ground. The genus *Draco* (16 species) is the only group of recent reptiles showing any adaptation whatsoever to exploitation of their air space.

Flying ducks: see *Orchidaceae.*

Flying fishes: fishes that swim near the suface, which can eject themselves from the water and glide through the air with the aid of their long, outspread pectoral fins, assisted by the flapping of the caudal fin and upward currents of air. One example is illustrated in plate 4. Other F.f. eject themselves to a height of 2-3 m above the water, and actually propel themselves by whirring their pectoral fins, e.g. Freshwater hatchetfishes *(Gasteropelecidae,* order *Characiformes).*

Flying frog: see *Rhacophoridae.*

Flying lemurs, *gliding lemurs,* *Dermoptera:* a mammalian order consisting of only 2 species in a single genus *(Cynocephalus = Galeopithecus)* in a single family *(Cynocephalidae). C. volans* is found on the southern islands of the Philippine archipelago, whereas *C. variegatus* occurs in southern Burma, southern Indochina, Malaya, Sumatra, Borneo and other adjacent islands. They are about 50 cm in length, and their faces resemble those of lemurs or Old World fruit bats. Attached to the neck and sides of the body is a large, furry membrane, extending along the limbs to the tips of the fingers, toes and tail. With the aid of this membrane they can glide over distances greater than 100 m when travelling between trees in search of food. F.l. are arboreal and nocturnal, feeding on leaves, fruit, buds and flowers. Dental formula: 2/3, 2/2, 2/2, 3/3 = 38.

Flying lemur

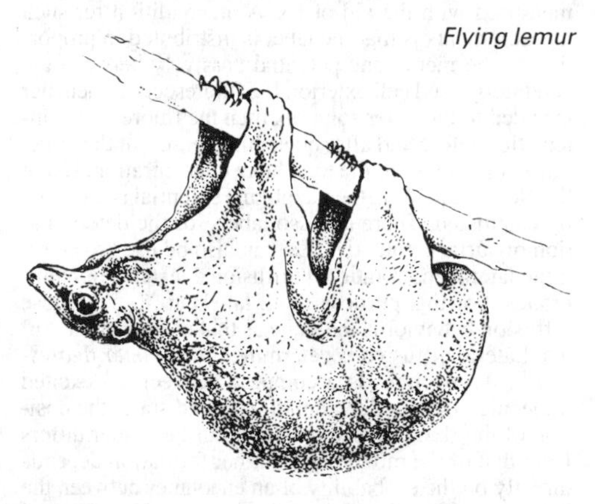

Flying snakes: see Golden snakes.

Flying squid: see *Cephalopoda.*

Flying squirrels: see *Sciuridae.*

Flypaper trap: see Carnivorous plants.

FMN: acronym of Flavin mononucleotide (see).

Fodies: see *Ploceidae.*

Foliage gleaners: see *Furnariidae.*

Follicle: see Fruit.

Follicle cells: cells of the ovary which form an epithelium around the ovum and provide nutrients for the latter. In insects, the F.c. also secrete the egg shell (chorion). In mammals, the *Graafian follicle* is composed of F.c. After release of the mature egg, the Graafian follicle becomes modified to an endocrine gland, the *corpus luteum.*

Follicle stimulating hormone, *FSH, follitropin:* a gonadotropin secreted by the hypophysis. FSH is a glycoprotein (M_r 32 000) containing 27% carbohydrate which consists chiefly of sialic acid, galactose, mannose and glucosamine, with smaller amounts of galactosamine and fucose. The protein consists of an α-subunit (92 amino acids with carbohydrate attached to asparagines 52 and 78) and a β-subunit, which displays microheterogeneity at the *N*- and *C*-terminals. The β-subunit contains between 108 and 115 amino acid residues and is unique to FSH. The α-subunit, however, shows close homologies with the a-chains of human chorionic gonadotropin, luteinizing hormone and thyreotropin.

FSH is synthesized in the adenohypophysis, and its secretion is regulated by follicle stimulating hormone releasing factor or folliliberin (which is identical with luteinizing hormone releasing factor) (see Releasing hormones). FSH is responsible for development and function of the gonads in both men and women. It stimulates spermatogenesis in the male testes, and it controls maturation of the female follicle. Together with luteinizing hormone, estradiol and progesterone, FSH is involved in the hormonal regulation of the monthly cycle. In women, the secretion of FSH is inhibited by estradiol, in men by testosterone.

Follicular oocyte nutrition: see Oogenesis.

Fölling's syndrome: see Phenylketonuria.

Follitropin: see Follicle stimulating hormone.

Fomes: see *Basidiomycetes (Poriales).*

Fontanelle, *fontanel:* a membranous interval between the edges and angles of the bones of the cranial vault of young mammals. In humans, these membranous gaps in the bones of the skull remain for up to two years after birth. Normally they are six in number and correspond in the adult to the position of the bregma and lambda in the mid-line and the pterion and asterion on each side. The diamond-shaped *anterior* or *bregmatic F.* corresponds to the converging angles of the parietals and two halves of the frontal bone. The triangular *posterior F.* lies between the two parietals and the summit of the occipital squama. The *antero-lateral F.* lies between the contiguous margins of the frontal, parietal, squamous temporal, and great wing of the sphenoid, while the *postero-lateral F.* is situated between the adjacent borders of the parietal occipital, and mastoid portion of the temporal.

Fontinalis: see *Musci.*

Food chain: see Nutritional interrelationships.

Food pyramid: see Nutritional interrelationships.

Food web: see Nutritional interrelationships.

Food yeasts: see Yeast protein.

Fool's parsley: see *Umbelliferae.*

Footmen moths: see *Arctiidae.*

Foraging symbiosis: see *Indicatoridae.*

Foraminifera: see *Foraminiferida.*

Foraminiferida, *Foraminifera,* ***foraminiferids:*** an order of the *Rhizopodea* (see), consisting of relatively large, ameboid protozoa with a perforated *(Perforata)* or smooth *(Imperforata)* test (shell). The great majority of species is marine, occurring in the sea and in saltwater lakes, but a few are found in freshwater, e.g. *Allogromia.* The average size is about 0.5 mm, but giant fossil forms are known with a diameter of 19 cm. Species of the fossil genus, *Nummulites,* attain a diameter of 11 cm and a thickness of 1 cm. The test may contain a single chamber (suborder *Monothalamia)* or more than one chamber *(Polythalamia).* In polythalamic forms, the chambers are arranged in a regular pattern, which is typical of the genus (Fig.1). In the most primitive forms, the test is composed of organic material known as tectin. In the *agglutinating F.,* the tectin test is overlain by a layer of particles of foreign sedimentary material (usually sand) bound together with cement secreted by the organism. In the most advanced forms, the tectin layer is overlain by an inorganic test of calcium carbo-

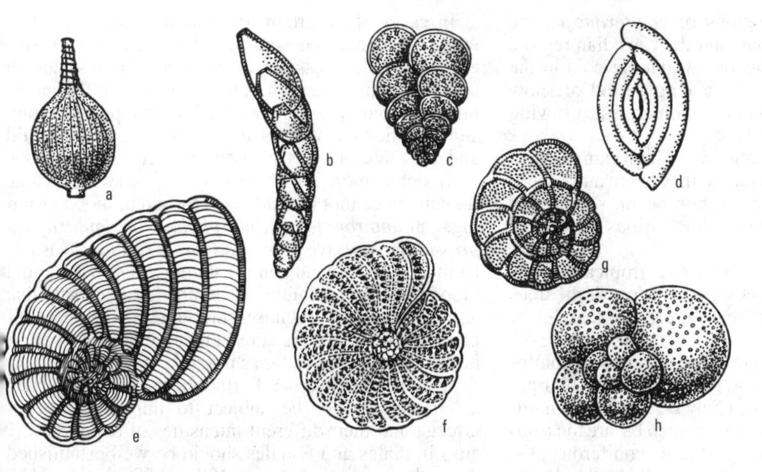

Fig.1. *Foraminiferida.* Tests (shells) of different genera. *a Lagena. b Nodosaria. c Textularia. d Miliola. e Pemeroplis. f Polystomella. g Rotalia. h Globigerina.*

nate, most commonly aragonite or calcite. New chambers are secreted by reticulopods (Fig.2). The first chamber in which the animal starts life is called the *proloculus.* There are 2 types of polythalamous shell, i.e. *megascleric* (with a large proloculus), and *microscleric* (with a small proloculus).

In some *F.* only asexual reproduction has been observed, but many species display an alternation of generations, in which a sexual generation (gamont) alternates with an asexually reproducing generation (agamont or schizont). In the simpler monothalamous *F.*, the two generations are morphologically similar (isomorphic alternation of generations), whereas in most polythalamous genera the schizonts have microsclerotic shells and gamonts have megasclerotic shells (heteromorphic alternation of generations). As schizonts (microscleric individuals) grow they become multinucleate, then undergo multiple fission into mononuclear megascleric individuals, which develop into gamonts. After nuclear reorganization, the gamonts divide to form numerous biflagellate isogametes. The latter fuse in pairs, the resulting zygote developing into a schizont.

F. are predominantly benthic, living on the sea bed and the ocean floor. Warm seas in particular contain a rich diversity of species. On the continental shelf, species tend to be large and contain zooxanthellae. Below 4 000 m, only agglutinating forms with sand shells are found, because calcium carbonate dissolves at this depth. The only pelagic and planktonic members of the *F.* belong to the genus, *Globigerina* (see); their sedimented shells cover a large area of the deep sea floor, forming a layer known as the Globigerina ooze.

Fossil deposits are found from the Cambrian to the present. Only a few sparse forms are found from the Cambrian to the Lower Carboniferous, exclusively with shells of agglutinated sand particles. The first evolutionary peak is observed in the Upper Carboniferous and Permian, where calcareous *F.* shells have formed rock strata (see Fusulinids). Different forms began to develop in the Jurassic, extending through the rest of the Mesozoic and through the Cenozoic, and displaying an extraordinary variety especially in the Tertiary, where individual genera or families were responsible for the formation of entire geological strata (see *Nummulites, Globigerina).* Some *F.* are important index fossils, in particular for the dating of bore samples.

Forest: a natural or partly natural plant formation, in which trees predominate. The area covered by trees must be sufficiently large to create a typical forest floor and forest soil, and to generate its own internal climate which is distinctly different from that of the open land surrounding the F. Although much F. has been cleared, almost 30% of the earth's surface is climatically suitable for F. growth, including central Europe.

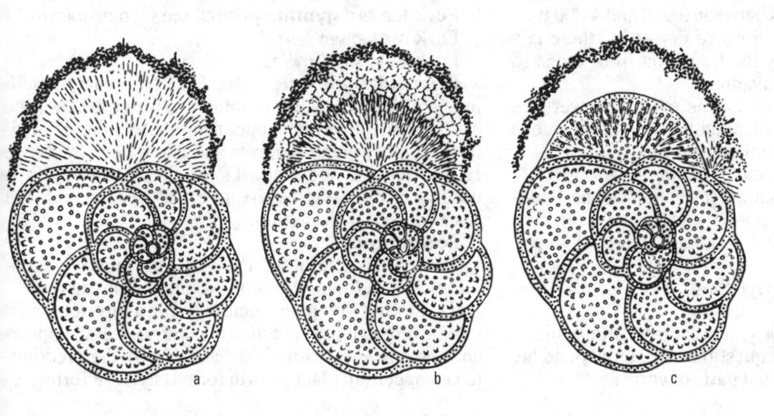

Fig.2. *Foraminiferida.* Formation of a new chamber in *Discorbis bertheloti. a* Fan-like arrangement of the reticulopodia. *b* Withdrawal of the reticulopodia. *c* Secretion of the new chamber wall.

There are two main regions of *coniferous F.,* the northeast European-Siberian and the Canadian region. Both have continental climates, which are cold in the winter with severe frosts and a long period of snow cover, and have warm temperate summers. The growing season is less than half the year.

Deciduous Fs. grow in regions of high rainfall with warm summers and cold winters. In western and central Europe they consist of beech, hornbeam, sessile and pendunculate oak, sycamore, elm, various species of birch, ash, alder, etc.

Monsoon Fs. are found in frost-free, tropical regions with alternating dry and wet seasons. Most of the trees lose their leaves in the dry season, e.g. the teak Fs. of Burma.

In *evergreen Fs.,* leaves are shed and renewed continually throughout the year, and there is no particular period of leaflessness. The trees may be conifers or hardwoods. Most extensive of the evergreen Fs. are the *tropical rainforests* and the northern coniferous Fs. Broad-leaved evergreen F. occurs in regions that lack a prolonged dry season. The evergreen Fs. of the Mediterranean region contain evergreen oak species, olive trees and carob. Evergreen *laurel Fs.* occur in regions of warm, dry summers and mild, wet winters (see Holarctic).

Forest dormouse: see *Gliridae.*

Forest limit, *timber line:* the limit of closed woodland, determined by the occurrence of unfavorable conditions, such as low temperature and the resulting low water availablity (e.g. polar and alpine F.l.), or by high temperature and arid conditions.

In mountainous areas, the tree limit extends beyond the F.l. The tree limit is represented by isolated trees, often killed on the weather side by storms and snow blast, giving a characteristic "flag shape". Thus, many trees between the F.l. and the tree line have a contorted shape, and they often display stunted, shrub-like growth. In different regions, the F.l. (and the tree limit) are represented by different species, e.g. in northern Scandanavia by pine and birch, in Siberia by larch, in the central Alps by larch and Swiss stone pine *(Pinus cembra),* and in the northern Alps by spruce and rarely by Swiss stone pine. Owing to the temperature decrease from the equator to the poles, the *upper* tree limit (height above sea level) increases with nearness to the equator; this upper limit lies at 500–550 m in the Urals, 700–1 800 m in the Bavarian Alps, and 2 400 m in the Caucasus (Elbrus). In the Himalayas (Nanga Parbat) the limit is at 3 800 m, while on Mount Kilimanjaro it lies between 3 900 and 4 000 m.

In the mountainous regions of dry areas, there is a *lower* F.l., determined by the transition from forest to steppe or from forest to savanna.

In densely populated areas, the existing lower tree limit and F.l. are higher than potentially possible, due to timber exploitation and cattle farming.

Forest salamanders: see *Plethodontidae.*

Forest steppe: vegetation formation in the boundary region between Forest (see) and Steppe (see). Tree coverage is less dense and contains islands of steppe vegetation.

Forcipiger: see Butterflyfishes.

Forebrain: see Brain.

Forensic anthropology: a branch of anthropology concerned primarily with questions of paternity and the identification of corpses and parts of corpses.

In cases of uncertain or unknown paternity, blood factors are analysed, and morphological comparisons are made with possible or suspected individuals. In practice, appraisment of paternity based on the inheritance of blood groupings and serum proteins only shows which individuals are *not* the parents of a child, and can therefore only be used for exclusion purposes.

To obtain more information, or in cases where a legal judgement cannot be made with the aid of blood groupings, an *anthropological-morphological similarity appraisal* or a *polysymptomatic similarity analysis* is performed. This depends on the common knowledge that blood relatives resemble one another more closely than any other randomly chosen individuals. For the comparison, characters are chosen that are known to be inherited, and which as far as possible are genetically independent of each other. Furthermore, the chosen characters should not be subject to unpredictable age effects, and their different intensities of expression (if any) in males and females should be well established. According to need, between 100 and 200 distinct characteristics are used, which are mainly concerned with the structure of the head and face [particulaly in the eye region (including iris structure), mouth and ears and other extremities], fingerprints, type of hair, pigmentation of skin, hair and eyes. In additon, physiological characters and a number of head and body measurements are used. If blood analysis does not exclude paternity, the study of morphological similarities permits one of the following conclusions: not possible to decide; paternity probable or improbable; paternity very probable or very improbable; paternity apparent (i.e. to be assumed within certain limits of probability) or apparently impossible (i.e. can be excluded within certain limits of probability). Similar studies have occasionally been necessary when babies have been inadvertently exchanged. All methods for the determination of paternity (indeed all methods for the identification of any individual) are now transcended by the technique of *genetic fingerprinting,* which is based on an analysis of DNA.

Anthropological methods for the identification of corpses or their parts are based primarily on the determination of age (see Age diagnosis), sex (see Sexual dimorphism) and Body height (see). In special cases, it may be necessary to reconstruct the face, using empirical norms for the relationship between bone relief and face shape.

Forest thrushes: see *Turdinae.*

Fork-leaved gymnosperms: see Gymnosperms.

Fork moss: see *Musci.*

Forktails: see *Turdinae.*

Formaldehyde, *methanal, HCHO:* a sharp smelling, irritant gas. It is easily soluble in water, ethanol and diethyl ether, and the vapors from these solutions irritate the mucosae of the eyes and respiratory tract, etc. HCHO is toxic and it attacks metals. In solution, it is used as a preservative, fixative and disinfectant, but must be used with caution.

Formation:

1) Formerly a term for a geologically and/or paleontologically defined period of evolution or history.

2) A term from plant sociology for a physiognomic unit of vegetation *(vegetation F.),* whose overall appearance, shape and profile is determined by the predominance of certain plant growth forms (see Life form), regardless of its floristic composition. Since growth forms of plants represent adaptations to particular environments, similar Fs. occur in similar habitats in different regions, even when the regions have different floristic compositions. The concept of F. is particularly useful for defining the major vegetation types of the world such as tropical rainforest, savanna, steppe, dwarf shrub heath, coniferous forest, evergreen sclerophyllous scrub, etc.

Formation science: a branch of the Sociology of plants (see), which uses the concept of the Formation (see) to describe vegetation, i.e. based on the shape and profile of the vegetation.

Formenkreis: a concept coined in 1900 by Otto Kleinschmidt (1870–1954) to replace the variously interpreted concepts of a species as a zoogeographical unit. He drew attention to the geographical Vicariance (see) of subspecies, which he called forms. These forms were not considered to be nascent species, but the ultimate forms of old species. Later, Formenkreis also became synonymous with superspecific unit (see Species). The concept is difficult to translate from the German into other languages, and together with its suggested nomenclature, has found little acceptance elsewhere. It has nevertheless helped to clarify certain zoogeographical problems of taxonomy.

In the classification system founded by Kleinschmidt, the form replaced the species, and the Formenkreis consisted of more or less different forms. Criteria for the assignment of two or more forms to a Formenkreis are the mutual geographical exclusion of these forms, their correspondence in essential anatomical characters, and the presence of transitional forms.

Kleinschmidt also placed very different forms in the same Formenkreis, and justified this with the argument that similarity is no index of relatedness. Furthermore, basically different forms (e.g. the sympatric marsh and willow tits) can be very similar, whereas very closely related animals can sometimes be very different in appearance (e.g. the sexual dimorphism of the male and female of a mating pair).

According to Kleinschmidt, evolutionary changes occur primarily within a Formenkreis. Whether one Formenkreis can evolve from another, he left as an open question.

In modern systematics, the Formenkreis corresponds most closely to a superspecies.

Form genus: a pseudosystematic artificial unit for the temporary classification of fossil plant remains. See Form species.

Formic acid, *methanoic acid:* HCOOH, the simplest fatty acid. It occurs in the venom glands of ants and bees, as well as in caterpillar hairs, stinging nettles and human urine. F.a. is a colorless, bacteriocidal liquid, b.p. 100.8 °C, with a sharp, irritant smell. It raises blisters when applied to the skin.

Formicariidae, *antbirds:* a large Neotropical avian family (220 species in more than 50 genera) in the order *Passeriformes* (see). Antbirds are small, rather subfuse birds found in the forests of South and Central America. All species are insectivorous, and some species actually follow and feed upon columns of army ants. Sexual dimorphism is pronounced, the males being more strikingly colored. The tip of the upper mandible is hooked and notched.

Formicidae: see Ants.

Formicoidea: see Ants.

Form series: a succession of recent or fossil structures or entire organisms displaying graded modifications. Various evolutionary trends have been recognized from F.ss. An F.s. does not necessarily represent an ancestral line. It may sometimes consist entirely or partly of organisms that are more or less related to the members of a true ancestral line that has not yet been discovered. If the members of a F.s. do not belong to an ancestral line, but nevertheless reveal a particular evolutionary trend, the F.s. is also known as a *stepped series.*

Form species: a pseudosystematic artificial unit formed by temporarily placing together all morphologically similar plant parts. Fossil plants are not usually found in their entirety, but mostly as fragments, especially when the original plant was relatively large. Often these fragments cannot be assigned unequivocally to a particular taxonomic unit. Therefore, as a makeshift but practical meaasure, the F.s. is used. The F.s. contains all the fossil remains of a particular organ that are similar in appearance, despite the fact that thay may be derived from different systematic groups. Thus, there are F.ss. for leaves, flowers, fruits, seeds, pollen, spores and wood structures. Morphologically similar F.ss. can be combined into form genera, and these into form families. Large fossilized remains, however, are classified together no further than the level of the form genus.

Although *organ species* and *organ genera* are also temporary, artificial units, sufficient data are available to establish that they also represent natural units belonging to some single natural familiy. With the aid of additional data, therefore, form genera can sometimes attain the status of organ-genera.

Forsythia: see *Oleaceae.*

Fortner method: see Anaerobes.

Fossilization, *fossil formation:* the totality of processes leading to the production of fossils in earlier geological periods. When living organisms died, they were subjected to the processes of F., leading to their incorporation into the earth's crust, with the total or partial preservation of their three-dimensional structure. Organic material was usually completely degraded, so that the fossilized remains consist primarily of hard parts, such as shells, bones and scales. These were frequently physically and chemically modified after becoming embedded in sediment, a process known as *fossilization diagenesis.* See Fig.

Fossils (Latin *fodere* dug out): modified remains, impressions and other traces (e.g. animal tracks) of earlier living organisms, which provide evidence for the process of evolution. Of particular importance are *index F.* or *characteristic F.,* which have a wide horizontal distribution, but only a restricted vertical distribution, thereby serving to characterize a particular time period in the sequence of geological layers. In addition, there are definite *facies F.* (see Facies), which are not associated with particular time periods, but with particular geophysical and geological conditions. Recent research has shown, however, that there is no clear difference between index and facies F. Many index F. are also facies F., and certain F. associated with specific facies serve as index F. for the corresponding facies region. *Continuous F.* are F. of organisms or systematic groups of organisms (e.g. families, genera) that have remained unchanged or only slightly changed over long periods

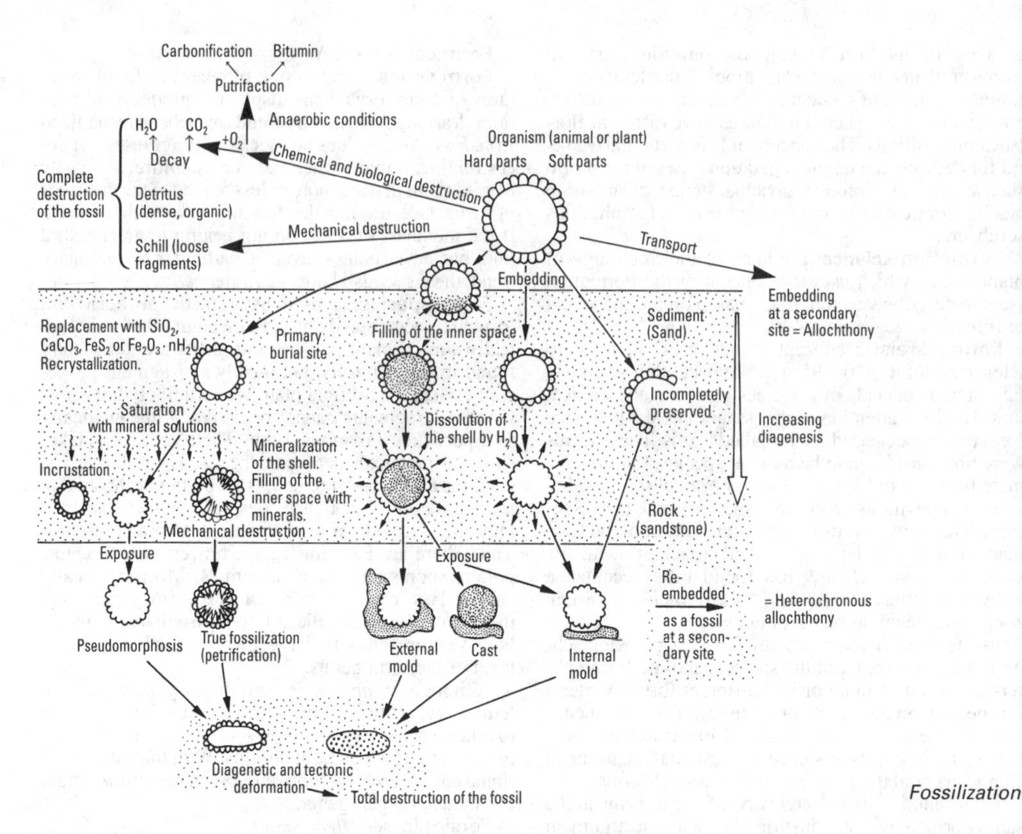

Fossilization

of geological time, sometimes to the present. *Pseudo F.* are not true F., but inorganic structures that resemble animals or plants, which have been formed by the physical and chemical processes of rock formation, compression and weathering.

Fossores: see Movement.

Fossorial animals: see Movement.

Foudia: see *Ploceidae.*

Foulbrood: see *Bacillus.*

Founder effect: derivation of a new, geographically isolated population by only a few individuals, in the extreme case by a single fertilized female. This gives rise to a Genetic revolution (see) and sometimes to the evolution of a new species.

Four-eyed fishes: see *Anablepidae.*

Four-horned antelope: see *Tragelaphinae.*

Four-lined rat snake, *Elaphe quatuorlineata:* a snake of the family *Colubridae* (see), whose several subspecies inhabit dry, bushy country in southern and southeastern Europe. It attains a length of over 2 m, and its laterally compressed body is as wide as a human forearm. Young specimens are gray with black spots, whereas adults are brown with 4 black longitudinal stripes. They lay eggs, and feed mainly on mice and rats, also taking larger animals like rabbits. The prey is constricted in typical rat snake fashion.

Four-strand crossing over, ***four-strand exchange:*** a reciprocal exchange of segments between paired chromosomes during meiosis. Two crossing over events are involved. The second crossing over occurs between the two chromatids of the bivalent that were not involved in the first event (Fig.). See Crossing over, Three-strand crossing over, Two-strand crossing over, Recombination.

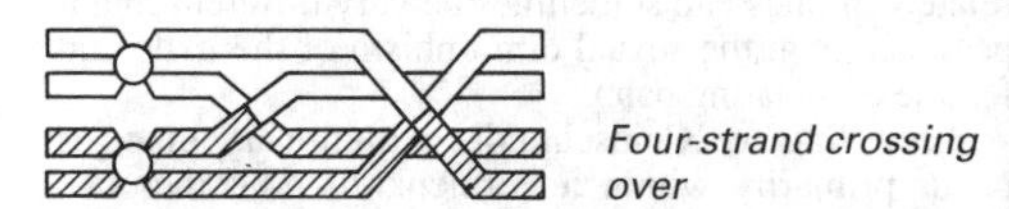

Four-strand crossing over

Foxes: a systematically nonuniform group of the *Canidae* (see). The 9 species of the genus, *Vulpes,* are generally known as ***Red foxes. Vulpes vulpes (Red fox)*** is distributed over the entire Northern Hemisphere. This solitary animal excavates a subterranean "earth" with several openings, and it feeds on vertebrates and insects, occasionally also eating fruit and fungi. A color variant of *Vulpes vulpes,* known as the ***Silver fox,*** occurs in North America, where it is also farmed for its fur. Smaller, lighter colored relatives of the Red fox are the African *Vulpes rüpelli* and the Asian *Vulpes corsac.* ***Kit foxes*** or ***Swift foxes*** *(Vulpes velox* and *Vulpes macrotis)* inhabit the Great Plains of North America, and are found from Canada to Texas and northern Mexico.

The ***Arctic*** or ***Polar fox*** (monotypic genus, *Alopex lagopus;* head plus body 46–67 cm) has a circumpolar distribution in the Arctic; it exists as 2 color variants: "white" (white in winter, brown in summer) and "blue" (pale blue gray in winter, dark blue gray in summer).

The smallest fox and the smallest predator of the *Canidae* is the ***Fennec*** or ***Fennec fox*** (monotypic genus, *Fennecus zerda;* head plus body 36–41 cm), which inhabits dry regions in North Africa and the Sinai and Arabian Peninsulas. It is creamy yellow to almost white in color, and its ears are large (15 cm or longer) in proportion to its body.

F-Plasmid: a circular bacterial Plasmid (see), which behaves as a sex factor in certain microorganisms (F-episome, F-factor). Cells containing an F-plasmid are designated F^+, and during conjugation in *Escherichia coli* they behave as "males". Cells lacking an F-plasmid (F^-) behave as "females". During conjugation, an F^- and an F^+ cell make contact and the F-plasmid is transferred from the F^+ into the F^- cell, the latter thereby becoming F^+. In cell division of F^+ cells, both daughter cells receive a copy of the F-plasmid, which may exist in one of two states: either as a free and independently replicating plasmid, or integrated into the bacterial chromosome. Cells with an integrated F-factor are designated *Hfr* (high frequency of recombination). During conjugation, Hfr cells characteristically transfer fragments of their genome into the F^- cell, leading to the production of Merozygotes (see), a process known as *F-duction* or *sex-duction.* New genotypes then arise from recombination between the complete chromosome and the introduced chromosome fragment. This parasexual recombination can be contrasted with the recombination that occurs in gamete formation during meiosis in higher organisms. See Conjugation.

Fragmentation: see Reproduction.

Frameshift mutation: see Gene mutation.

Francolins: see *Phasianidae.*

Frangipangi: see *Apocyanaceae.*

Frankfort plane: see Skull planes.

Frankia: see *Actinomycetales.*

Frankiaceae: see *Actinomycetales.*

Frank-Starling law: see Starling's law of the heart.

Fraser-Darling effect, ***Darling effect:*** a decrease in threat or danger to a species and its individuals by the synchronization of reproductive behavior, so that the entire breeding period is restricted to a minimum. Synchronization is achieved by conspicuous courtship behavior between mating pairs (especially involving optical and acoustic signals) in the presence of other members of the same species.

Frasnian stage: see Devonian.

Fratercula: see *Alcidae.*

Fraxinella: see *Rutaceae.*

Fraxinus: see *Oleaceae.*

Free-running rhythm: see Biorhythm.

Freesia: see *Iridaceae.*

Freeze-drying:

1) See Cryomethods.

2) ***Lyophilization:*** a protective method for the removal of water, which is used, inter alia, for preserving living microorganisms. It is performed under high vacuum at temperatures between –25 °C and –70 °C, with continual removal of the sublimed water vapor. To achieve a high survival rate, some microorganisms are lyophilized as a suspension in serum or milk. Lyophilized bacteria, sealed in evacuated or nitrogen-filled glass ampules, remain viable for many years. Cultures are easily reestablished by adding an appropriate growth medium to the lyophilizate.

Freeze etching (plate 18): a surface imaging method used in electron microscopy, in particular for the investigation of membranes. Living cells or small pieces of tissue are rapidly frozen to about –150 °C; rapid freezing *(cryofixation)* converts the water into amorphous frozen water, and avoids deformation of the tissues by the growth of ice crystals. The preparation is then fractured with a microtome knife in a high vacuum. The resulting plain of freeze-fracture usually travels within membranes or over their surface, because the lipids concentrated in these regions contain hardly any ice. Some frozen water is then allowed to sublime from the surface of the preparation (etching), leaving exposed structures in relief. At this stage, protoplasm displays a coarse texture, while faces of fractured (split) membranes are smooth with projecting membrane proteins, the latter appearing as small elevations or particles (Fig.). A replica of the surface is then prepared by the deposition of an extremely thin polymer film, and this replica is studied under the electron microscope, usually after emphasizing the relief by shadowing with metal or carbon. Shadowing is performed by subliming metal or carbon at an angle onto the replica in a high vacuum. The result is a faithful three-dimensional representation of the fractured surface of the cell, which was still alive at the time of cryofixation. F.e. avoids artifacts which can arise by drying and by the action of chemical fixatives on organic structures.

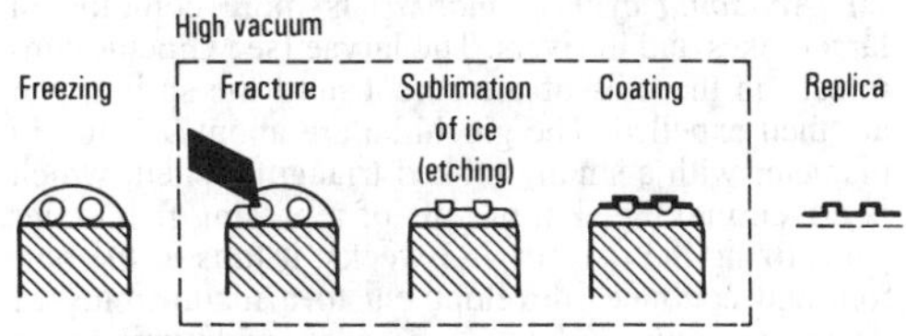

Freeze etching (schematic)

Freeze sectioning: a method for preparing unfixed tissue for microscopy. For certain types of investigation, to save time or to avoid fixation artifacts, the lengthy procedures of fixation and embedding must be bypassed. The tissue is therefore frozen, sectioned with a cold knife, and viewed immediately under the microscope.

Freezing behavior: cessation of all movement, usually caused by stimuli that signal danger or threat. F.b. belongs to the species-specific repertoire of protective behavior of various animal species, e.g. guinea pig.

Fregata: see *Pelicaniformes.*

Fregatidae: see *Pelicaniformes.*

French bean: see *Papilionaceae.*

French purple: see Lichens, Lichen pigments.

Frenulum: see Moths.

Frequency: see Constancy.

Frequency-dependent selection: selection in which Fitness (see) depends on the relative frequency of the genotype. If the fitness of two genotypes changes inversely with their relatively frequencies (each genotype has a high degree of fitness at low frequency and a low fitness at high frequency) then a stable equilibrium is established between them. Under these conditions, both genotypes are retained in the population. In this way, F-d.s. serves as a mechanism for the maintenance of polymorphism. If the two equibrated genotypes have the same fitness, then the population fitness lies below its potential optimum.

One reason for the occurrence of F-d.s. is that predators prey on the most abundant form of a species. This may then become so depleted that it is less common than the formerly less numerous forms. It is then heeded less by predators, and its numbers increase again. It has been shown experimentally that females of *Drosophila* often prefer the male type that happens to be in the minority. Thus, with its decreasing frequency, each male type increases its chances of a successful mating, and its genotype is retained in the population.

Freshwater crabs, *Potamonidae:* a family of true crabs (see *Brachyura),* distributed in freshwater habitats in the tropics and subtropics. There is no larval stage of development; fully developed but miniature crabs hatch from the eggs. They are therefore permanent inhabitants of fresh water, and do not need to return to the sea for reproduction.

Freshwater hatchet fishes: see *Characiformes.*

Freshwater mussels: freshwater bivalves (see *Bivalvia),* all of them important as filter organisms in the biological purification of water. *Anodonta* has thin, blue-green valves lacking hinged teeth. The ***Swan mussel*** *(Anodonta cygnea)* attains a maximal length of 20 cm, and is found in smaller European lakes and ponds. A smaller subspecies known as the ***Duck mussel*** *(Anodonta cygnea anatina)* is more common in larger lakes and in rivers. The larvae (see Glochidium) remain in the gills of the parent until the spring, and are then expelled. The glochidia are about 0.3 mm in diameter with a small, hooked triangular shell, which is driven into the skin or gills of a passing fish. After parasitizing the fish for 2–3 weeks, it falls to the bottom and continues development to a mature mussel. *Anodonta minima* (about 7 cm) is less widespread than *A. cygnea,* and it prefers rivers and canals with sandy bottoms. The ***Painter's mussel*** *(Unio pictorum)* of lakes and rivers grows up to 12 cm in length but is usually smaller. Unlike *Anodonta,* it has a thick, robust shell with strong hinge teeth. Eggs are hatched in the spring. The glochidia are expelled in the summer and they parasitize the gills of fishes. Similar but smaller species are *Unio tumidus* and *Unio crassus.* The ***Pearl mussel*** *(Margaritana margaritifera)* grows up to 12 cm in length and is found in streams and in lakes associated with flowing rivers (see Pearls). Other examples are the ***Horny orb shell*** *(Sphaerium corneum,* up to 13 mm, common in most types of fresh water) and the ***River pea shell*** *(Pisidium amnicum,* up to 10 mm, prefers shallow water in lakes and rivers). There are several smaller species of *Pisidium,* some of them very common. See also Zebra mussel.

Freycinetia: see *Pandanaceae.*

Frigate birds: see *Pelicaniformes.*

Frilled lizard, *Chlamydosaurus kingii*: an arboreal agamid (see *Agamidae)* native to Australia and New Guinea, where it favors damp habitats. It attains a length of 90 cm and has no dorsal crest, but below and at either side of the head it possesses an enormous "frill" or fold of skin supported by cartilaginous extensions ("ribs") of the hyoid bone and covered with very large keeled scales. When excited, the animal spreads out this frill like a wide collar. When fleeing danger, the F.l. always runs upright on its hindlegs. It eats large insects, small rodents and birds' eggs.

Fringillidae, ***finches:*** an avian family of the order *Passeriformes* (see). The family occurs naturally in most parts of the world, excluding Madagascar, Australia and Oceana, although members have been introduced to these regions. It is divided into 3 subfamilies.

Frilled lizard (Chlamydosaurus kingii)

1. Subfamily: *Fringillinae.* This large subfamily (266 species in about 75 genera) is the largest and most widespread of all seedeater groups. The evolutionary center of the subfamily appears to be the American tropics and subtropics where members are the most numerous and best developed. An early migration into the Eastern Hemisphere (probably in the late Miocene) gave rise to 3 Old World genera (40 species), which are represented in Eurasia and Africa.

Most fringillines are terrestrial, inhabiting grasslands or dwelling in bushes and thickets. Adults feed mainly on seeds, which are secured by ground foraging. Every type of climate has been invaded, from the hottest and most arid to the coldest and most humid, from the seashore to barren mountain plateaux, and from the high Arctic to Tierra del Fuego. The ***Snow bunting*** *(Plectrophenax nivalis)* is completely circumpolar and breeds at the tip of northern Greenland, further north than any other land bird. On the whole, they are small birds (11–20 cm) with big feet equipped for scratching. The tail is long, sometimes graduated or slightly forked. Largest members are the American ***Towhees*** *(Pipilo).*

The genus, *Fringilla,* is represented by 3 Old World species, all of which are about 15 cm in length, and lack a crop. Considering that the adults are seed eaters, their bills are rather fine. The young are fed insects. The ***Chaffinch*** *(Fringilla coelebs)* is well known throughout Europe, Scandanavia and parts of North Africa, western Asia and Asia Minor, and appears to be constantly extending its range further eastwards. It breeds in deciduous and conifer woods, parks, gardens and hedgerows. The rare ***Canary Island chaffinch*** *(Fringilla teyda)* is confined to the mountain pine forests of the Canary islands. The ***Brambling*** *(Fringilla montifringilla)* is a migratory bird of the sub-Arctic birchwoods of northern Scandanavia and the northern taiga.

The largest group of Old World fringillines consists of about 30 species of *Emberiza,* commonly known as ***Buntings,*** e.g. the Pine, Cirl, Reed and Little buntings of Eurasia, as well as the common British and European ***Yellowhammer*** *(Emberiza citrinella;* see Plate 6). These were formerly placed in a separate subfamily, the *Emberizinae,* and by some authors in a separate family, the *Emberizidae.*

About 50 species of fringillines occur regularly as familiar garden birds in temperate North America, where they are commonly known as sparrows, e.g. Song sparrow and Chipping sparrow. The decorative and melodius ***White-crowned, Golden-crowned, White-throated*** nd ***Harris sparrows,*** which breed in northern Ameri-an woodlands, belong to the genus, *Zonotrichias.* The our species of *Tiaris **(Grassquits)*** are found from Mex-o to Brazil, including the West Indies. Ten ***Ground inches** (Sicalis)* inhabit open brushland from Mexico Argentina, and are sometimes known as "wild canar-es". The popular cage bird known as the ***Crested cardi-al** (Richmondena cardinalis;* USA to southern Mex-o; 20 cm) is one of the larger and more brightly col-red members of the subfamily.

2. Subfamily: *Carduelinae.* The 128 species (28 gen-ra) of this subfamily are typical seed eaters with strong onical beaks and strong jaw muscles, and a powerful izzard. The gullet is readily stretched, and they possess crop in which seeds are stored. The young are fed on regurgitated paste of ground seeds. Some members re the ***Serin finches** (Serinus), **Goldfinches, Green-inch*** and ***Siskins** (Carduelis), **Redpolls** (Acanthis), **Rosy finches** (Leucosticte), **Trumpeter bullfinches** Rhodopechys), **Rose*** and ***Purple finches** (Carpoda-us),* the monotypic ***Bonin finch** (Chaunoproctus fer-eirostris), **Pine grosbeaks** (Pinicola), **Crossbills** Loxia), **Bullfiches** (Pyrrhula)* (plate 6), ***Hawfinches*** or ***Grosbeaks** (Coccothraustes, Hesperiphona, Mycero-as, Eophona).* Serin finches include the canaries, e.g. ***Canary** (Serinus canaria),* native to the western Canary slands, Madeira and the Azores (males yellow breasted vith ash-gray back; females brown). The bright yellow ***Yellow canary** (Serinus flaviventris)* is found mainly in South Africa. The genus, *Acanthis,* includes the ***Linnet*** *A. cannabina),* the ***Twite** (A. flavirostris)* and the ***Red-poll** (A. flammea).*

3. Subfamily: *Geospizinae,* also known as Darwin's inches or Galapagos finches (see).

Fringing reef: see Coral reef.

Frings generator: see Acetic acid fermentation.

Frit flies, *Chloropidae:* a family of the *Diptera,* con-aining about 1 200 small hairless black, yellow or reen species. Larvae of the yellow *Chlorops pumil-onis* Bjerk. attack the halms of wheat and other cereals, ausing the ear to wither within its leaf sheaf. The ***Frit fly**, Oscinella frit* L., is a European cereal pest. First eneration flies lay eggs on the leaves or stems of ce-eals or grasses, and the larvae migrate to the shoots vhere they eat away the central leaves (yellow heart lisease). In July, the second generation flies lay eggs in he ears of the plant, and the larvae feed on the spikelets nd immature soft milky grains. In September, the third eneration flies lay eggs on winter cereals and grasses, and the larvae spend the winter in the shoot bases, which they destroy (Fig.). Some species are human dis-ease vectors, e.g. in India, *Siphunculina funiculi* trans-mits conjunctivitis and other eye diseases, and species of *Hippolates* transmit yaws (caused by a spirochaete) in the tropics.

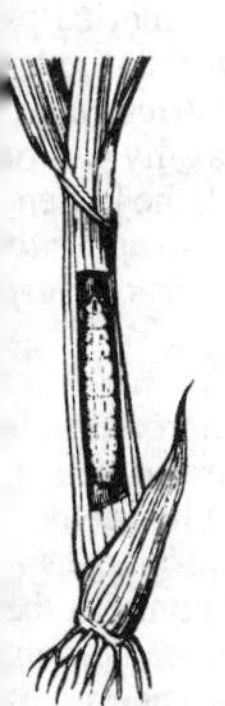

Frit flies. The frit fly, *Oscinella frit* L. Part of wheat halm removed to show larva.

Frog: see Anura, *Ranidae.*

Frogfish: see Goosefishes.

Frogfishes, *Antennariidae:* a fish family of the order *Lophiiformes* (see), containing about 60 species in 15 genera, found in all tropical and subtropical seas. They are bottom fish with plump bodies and very large mouths. Like most anglerfishes they have an antenna-like line and bait (esca) structure on the head, which at-tracts prey to the mouth. This fishing pole device, known as the illicium, is the modified first ray of the first dorsal fin.

Froghoppers: see *Homoptera.*

Frogmouths: see *Caprimulgiformes.*

Frog of horse's hoof: see Claws, hoofs and nails.

Frogspawn alga: see *Rhodophyceae.*

Front-fanged snakes: see *Elapidae.*

Fructification: a term used generally in mycology for any fungal structure that contains or carries spores.

Fructivore: see Modes of nutrition.

D-Fructose, ***fruit sugar, levulose:*** a monsaccharide ketohexose. It tastes sweeter than any other carbohy-drate, and it is fermented by yeast. It crystallizes as its β-pyranose, but in compounds it exists as the furanose form. Together with glucose and sucrose, D-F. occurs in honey and in most sweet fruits. It also occurs in certain pollens and plant roots. It is a component of many oligo- and disaccharides, including sucrose, raffinose, sta-chyose and gentianose, and of various polysaccharides such as inulin and levan. D-F. is used as a sweetener by diabetics, because its consumption, even in large amounts, does not raise the blood glucose level.

β-D-Fructofuranose

β-D-Fructopyranose

Metabolism of D-F. is initiated by phosphorylation to fructose 1-phosphate (only a small amount is phosphor-ylated in position 6). Fructose 1-phosphate is then split into dihydroxyacetone phosphate and glyceraldehyde (in mammals this occurs in the liver). The former enters Glycolysis (see) directly, and the latter is phosphory-lated either to 2-phosphoglyceric acid or glyceralde-hyde 3-phosphate, both of which can enter general car-bohydrate metabolism.

Fructose 6-phosphate (Neuberg ester) and fructose 1,6-bisphosphate (Harden-Young ester) are important intermediates of glycolysis and anaerobic fermentation; they are not major metabolic products of D-F., but are derived from D-glucose by phosphorylation and isom-erization.

Fruit (plates 34 & 35): the organ of higher plants that encloses the seed, and often assists in seed dispersal. F. formation and seed development occur simultaneously. The ovary (see Flower) always participates in F. formation. Other parts of the flower may also contribute to the F., e.g. the floral axis, perianth, bracts and sometimes the style, the latter being modified to form an organ that aids F. dispersal. In many cases, however, the perianth segments, stamens, styles and stigmata wither and are lost as the F. matures.

The ovary wall together with parts of the carpel become modified to form the *pericarp* or *F. wall.* This is normally differentiated into a single-layered *exocarp* (outer) and a single-layered *endocarp* (inner), which are separated by a multilayered *mesocarp.* One or more seeds are located at specific sites on the internal surface of the pericarp. The pericarp may be hard or fleshy, and it may carry various structures for attachment to animals, or it may possess wings or flight hairs for wind dispersal. Alternatively, it may operate a release mechanism for seed scattering.

Strictly speaking, Fs. are formed only from the closed ovaries of angiosperms. Even the primitive *Charales* (stoneworts), however, possess F.-like structures. The cones of gymnosperms can also be considered as Fs., since they contain seeds and aid their dispersal. The seed of *Taxus* with its fleshy aril is sometimes called a F., as are the exposed seeds of *Cycas* and *Ginkgo,* which resemble drupes.

F. formation. After fertilization the style usually withers, leaving a suberized stigma. As a rule, the remaining floral organs are also soon lost. The ovary wall, however, grows by cell division and develops into the pericarp. During this period of intense *F. development,* the cells often accumulate carbohydrates and organic acids, and respiration is normally elevated. The developing F. withdraws nutrients and biosynthetic precursors from the entire plant. *F. ripening* (see) is often accompanied by a color change, especially in fleshy Fs. Thus, chlorophyll is degraded and replaced by yellow or red pigments, a process similar to the color change of foliage leaves in the autumn. A layer of abscission tissue then forms in the elongated F. stalk, leading finally to separation of the F. from the mother plant.

F. formation is regulated by the coordinated action of phytohormones. The earliest developmental stimulus is provided by auxin from the pollen grains. In many plants, subsequent F. growth is positively influenced by the number of pollen grains that originally pollinated the stigma. In addition, the style tissue is stimulated to produce auxin by the pollen tube. After fertilization, auxin is synthesized in both the embryo and the endosperm of the ovule, and this auxin helps to stimulate the rapid development of the pericarp. In many plants, the positive influence on pericarp development and growth normally provided by the developing seed can be simulated by the application of auxin, and in other plants by the application of gibberellin. Thus, in the absence of pollination, seedless Fs. (see Parthenocarpy) can be induced to develop by the application of auxin (tomato, tobacco, fig, gherkin) or gibberellin (cherry, apple, grape). In many cases, however, there is no quantitative correlation between the auxin or gibberellin content of the growing F. and its rate of growth. This is probably because other factors, in particular cytokinins, are also involved in the regulation of F. growth.

Classification of Fs. On the basis of their structures (Fig.), angiosperm Fs. can be classified as follows.

A) *Simple Fs.* These are derived from a single ovary. Flowers with an apocarpous gynaecium each produce a number of simple Fs. According to their mode of disper sal, they can be further subdivided into:

1. *Dehiscent Fs.* The seeds are surrounded by a rigi woody or papery shell, which opens when ripe to re lease the seeds. These include:

1a) *Follicle.* Unicarpellary Fs. (derived from a singl carpel), which open ventricidally (along the ventral su ture) when ripe (e.g. *Paeonia, Delphinium).* This i probably the most primitive form. In particular, follicle are characteristic of the apocarpous members of the or der *Polycarpicae (Ranales),* e.g. the family *Magnolia ceae.*

1b) *Legume or pod.* This is also unicarpellary, but i dehisces along both the ventral and dorsal (abaxial) su tures. Examples are *Magnolia* (several pods per flower and members of the *Papilionaceae* (one pod per flower)

1c) *Siliqua.* Derived from 2 united carpels. When th F. is ripe, the 2 carpels separate as valves from thei marginal placentas and release the seeds. In some spe cies the siliqua is divided into 2 compartments by a fals septum (e.g. many members of the *Brassicaceae).*

1d) *Capsule.* A capsule consists of several united car pels with seeds attached at the sutures (e.g. poppy, to bacco, tulip). It opens either by cleavage along th united margins of the carpels *(septicidal capsules),* o longitudinally along a line of dehiscence on the dorsa side of the carpel *(loculicidal* or *dorsicidal capsule).* I the capsule forms restricted openings to the exterior a specific sites, it is called a *poricidal* or *porose* capsul (e.g. poppy). A *circumscissile capsule, pyxidium* o *pyxis* opens by transverse dehiscence across the entire F (i.e. a "lid" is formed), e.g. *Anagalis arvensis, Hyoscya mus niger.*

2. *Indehiscent Fs.* The pericarp enclosing the seed does not open when ripe. Dry indehiscent Fs. are:

2a) *Nut.* A nut usually contains a single seed, whic is surrounded by a woody, leathery or membranous per icarp (e.g. hazelnut, acorn, beechnut, etc). Very smal nuts are called nutlets (e.g. *Ranunculus,* birch, alder). A nut in which the seed coat (testa) and pericarp are fuse is called a *caryopsis* (e.g. the superior Fs. of grasses which are often also enclosed in bracts), or an achen (e.g. the inferior F. of the *Compositae,* in which the ca lyx is often modified as a flight organ known as *pap pus).*

2b) *Fissile Fs.* There are broadly 2 types, i.e. th *schizocarp* or *separating F.,* and the *lomentum, lomen* or *jointed F.* These are multiseeded Fs., which separat into single-seeded partial Fs. (mericarps) when ripe. I the schizocarp, separation is septicidal, and there ma be several mericarps (e.g. *Malva)* or only 2 mericarp (e.g. *Acer* and members of the *Umbelliferae).* In the lomentum, separation is by septation or abstriction, s that each resulting single-seeded segment is only part o one carpel (e.g. *Ornithopus sativus,* in which the lomentum is a legume that breaks transversely; *Raphanu raphanistrum,* in which the siliqua breaks transversel and is known as a *lomentose siliqua).*

Fleshy indehiscent Fs. are:

2c) *Berry.* A berry usually consists of numerous seed enclosed by the fleshy pericarp (e.g. red currant, grape tomato, paprika, gherkin, cucumber, pumpkin, melon) The banana, too, is a berry; the commercial product usually has no seeds, but the center of the fruit contains th remains of the ovules. Citrus fruits (e.g. orange, lemon grapefruit) are berries in which the mesocarp as well a the epicarp has become firm and leathery; the fleshy edible part consists of the carpels fused in a compound ovary. Berries are usually multiseeded, with the exception of the date, which is a unicarpellary berry, and is therefore considered to be more primitive than other berries. From a phylogenetic point of view, the date is better included with the unicarpellary follicles and legumes.

2d) *Drupe (stone F.).* Only the outer part of the pericarp is fleshy, whereas the inner part is sclerenchymatous and hard, forming one and sometimes more single-seeded stones, which only break open during germination. Cherry, peach and walnut contain a single stone, whereas *Mespilus* (medlar) and *Sambucus* (elder) each contain several stones. The tropical coconut *(Cocos nucifera)* has a fibrous, air-containing mesocarp, which makes the F. buoyant; the hard "shell" of the coconut is the endocarp. Similarly, the walnut *(Juglans regia)* is actually a drupe; the walnut sold in shops is the hard endocarp containing the seed, the epicarp and mesocarp having been stripped away.

B) *Aggregate or compound Fs.* These develop from a single apocarpous gynaecium. The fruitlets are derived from separate ovaries, but they are united directly by their pericarps, or indirectly (e.g. by tissue of the floral axis). Examples are the aggregate drupelets of raspberry, which form an aggregate F. directly. In the strawberry, numerous nutlets are united on a dilated fleshy floral axis. Similarly, the rose hip consists of aggregate nutlets on a fleshy floral axis, but the latter is concave and cup-shaped.

In the *pome* or *core F.* of apple and pear, the parchment-like pericarp of the individual Fs. forms a housing that is fused with, and surrounded by, the fleshy receptacle of the flower. The F. of the medlar also consists of a capsular, seed-containing cavity surrounded by fleshy receptacle, but the pericarp is hard and stony, so that the F. may also be considered as a drupe (see above).

C) *Infructescence, or multiple or collective F.* This highly advanced dispersal organ arises from the inflorescence, and not from a single flower. The fruitlets are united by fleshy axial tissue or by parts of the flower, so that the infructescence has the superficial appearance of a solitary F. Examples are the *berry infructescence* (e.g. pineapple, which consists of a mass of berries and their fleshy bracts united on the fleshy axis of the inflorescence; *nutlet infructescence* (e.g. mulberry, in which all the nutlets arising from an inflorescence are united by fleshy perianth segments; *drupelet inflorescence* (e.g. fig, in which the drupelets of an entire florescence are embedded in a hollow, fleshy axial receptacle).

In the phylogenetic development of the angiosperms, unicarpellary Fs. derived from choricarpous flowers are considered to be the most primitive, i.e. each carpel functions separately and each forms a single F. Such Fs. include follicles, legumes, unicarpellary berries (e.g. members of the *Annonaceae* and the date), unicarpellary drupes (e.g. cherry), and the many unicarpellary nuts and nutlets. Aggregate Fs. are more advanced, but the flowers are still choricarpous, fusion of the fruitlets occurring during F. development. The coenocarpous condition evolves from the choricarpous condition by fusion of carpels. Examples of coenocarpous Fs. are dry capsules, siliqua, fleshy capsules (e.g. *Euonymous),* coenocarpous berries (e.g. red currant, grape, deadly nightshade, Fs. of the *Cucurbitaceae),* coenocarpous drupes *(Juglans, Olea, Sambucus, Cocos),* fissile Fs. (see above), and coenocarpous nuts such as those of *Betula, Ulmus, Fraxinus,* etc. The pome F. (see above), whose

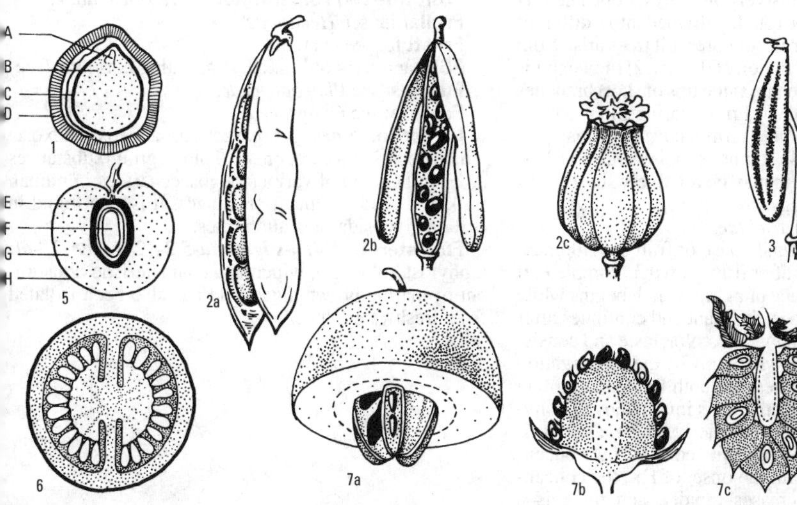

Types of fruit. 1 Indehiscent fruit: a hazelnut *(Corylus avellana)* in schematic longitudinal section: *A* radicle, *B* cotyledon, *C* testa, *D* pericarp. *2a-2c* Dehiscent fruits. *2a* Pod of scarlet runner bean *(Phaseolus). 2b* Cabbage *(Brassica)* siliqua. *2c* Porose capsule of the poppy *(Papaver). 3* A schizocarp: the double achene of caraway *(Petroselenium segetum). 4* Lomentose siliqua of charlock *(Raphanus raphanistrum). 5* Drupe: schematic longitudinal section of a cherry *(Prunus cerasus): E* endocarp, *F* seed with embryo, *G* mesocarp, *H* exocarp. *6* Berry: schematic transverse section of a tomato *(Lycopersicum esculentum). 7a-7c* Compound fruits. *7a* Pome fruit of the apple *(Malus sylvestris),* exposing half of the core (seed housing). *7b* compound nut: schematic longitudinal section of a strawberry *(Fragaria). 7c* Compound drupe: schematic longitudinal section of a raspberry *(Rubus idaeus). 8* Nutlet inflorescence of mulberry *(Morus).*

multiseeded core arises from the carpels, represents an approach to the coenocarpous condition. The most advanced condition is considered to be the infructescence.

Plants that form different types of Fs. are said to be *heterocarpous* (see Heterocarpy).

Fruit flies, *Trypetidae:* a family of the *Diptera* containing about 2 000 species, all of which are only 5mm or less in length. The wings are conspicuously banded or spotted, and the female has a long ovipositor, which it uses to lay eggs in plant tissue. Larvae live in fruits (e.g. the ***Cherry fly***, *Rhagoletis cerasi* L.), in plant stems (e.g. the ***Asparagus fly***, *Platyparea poeciloptera* Schr.), in leaves (e.g. the ***Celery fly***, *Philophylla heraclei* L.) or in flowers. A most serious pest is the ***Mediterranean fruit fly*** (*Ceratitis capitata* L.), whose larvae attack peaches, pears, oranges and other fruits in Africa, Australia, South America, Bermuda and Hawaii. It was introduced relatively recently into the Mediterranean region and has now spread into Central Europe, where it has devastated peach crops. The name, fruit fly, is also erroneously applied to the *Drosophilidae* (pomace flies), but the larvae of the latter feed on the yeasts and fungi of decomposing fruit, rather than the fruit itself.

Fruiting body, ***fruit body, fructification:*** the macroscopic, three-dimensional structure of higher fungi, which consists of plectenchyma, and which carries the hymenophore (fungal hymenium) or spore-bearing layer on either its internal or external surface. "Fungi" of nonscientific speech are the F.bs. of Basidiomycetes (see) and Ascomycetes (see), and they have a variety of shapes and sizes.

Fruiting wood: the branches of fruit trees that bear flowers and fruit. In the *Prunoideae* (producers of stone fruits or drupes) and *Maloideae* (core fruits), flowers and fruits are borne exclusively on short shoots. The various forms of F.w. can be roughly divided into 5 different types. 1) Flowers and fruits of core fruit trees arise from a thickened "placental" region of the F.w. 2) Flower buds of stone fruit trees aggregate at the tips of short branches in a rosette or "bouquet". 3) Fruit-bearing shoots, about 5 cm in length and carrying a terminal fruit bud may persist for several years. 4) Annual shoots (spurs) of F.w. carrying terminal flowers may be relatively short (6–12 cm); or 5) relatively long (up to 25 cm).

Fruit moths: see *Tortricidae.*

Fruit ripening: the final phase of fruit development in plants, which occurs after fruit growth is complete. It is a process of senescence or aging, which begins while the fruit is still attached to the plant and continues after abscission, involving changes in color, taste and consistency (softening). As shown mainly by the investigation of fleshy fruits, the synthesis of both RNA and protein is necessary for F.r. The process is initiated by the phytohormone, Ethylene (see) (formerly known as fruit ripening hormone). Growing fruits contain low concentrations of ethylene. With the onset of F.r., the concentration of ethylene increases rapidly, and there is a simultaneous increase in the responsiveness (sensitivity) of the fruit tissue to ethylene. The onset of increased ethylene synthesis is followed by an intense, temporary increase in respiration, known as the *climacterium.* Timing of the climacterium is unaffected by harvesting, and it may occur before or after the fruit is gathered. During the climacterium, carbohydrate metabolism switches from the pentose phosphate cycle to glycolysis, and there in an increase in the synthesis of ATP, reflecting the increased energy demand by the ripening processes. Respiratory inhibitors inhibit both the climacterium and F.r.

Certain chemical processes of F.r. are species-specific. As a rule, starch is rapidly degraded, whereas the concentration of sugars remains constant or increases. There is also usually a simultaneous decrease in the concentration of organic acids, so that the overall effect is that *the fruit becomes sweet* to the taste (a notable exception is the lemon, in which organic acids increase during F.r.). Cell walls are greatly modified: protopectin decreases, while water-soluble pectin first increases then falls again. The individual cells of the fruit flesh are then more easily separated from each other, i.e. *the fruit softens.* Many proteins are synthesized during F.r., and the synthesis of *odor substances* and *waxes* is also frequently observed. These chemical processes are often accompanied by considerable changes in the permeability of cells and organelles.

The *color changes* of ripening fruits are regulated more or less separately, and they are not directly correlated with the above chemical processes. Very often chlorophyll is degraded and carotenoids are synthesized (e.g. in the tomato). Anthocyanin synthesis, e.g. in the ripening of stawberries and apples, is often controlled by light via Phytochrome (see).

Undesired premature F.r. during storage can be avoided by decreasing the oxygen content of the atmosphere to 3–5% (which leads to suppression of ethylene production), or by increasing the CO_2 content of the air (the CO_2 apparently competes with ethylene at its site of action).

Fruit ripening hormone: old name for the phytohormone, Ethylene (see).

Fruit tree red spider mite: see *Tetranychidae.*

Frullania: see *Hepaticae.*

Frustule: see Diatoms.

FSH: acronym of Follicle stimulating hormone (see)

Fucales: see *Phaeophyceae.*

Fuchsia: see *Onagraceae.*

L-Fucose, ***6-deoxy-L-galactose:*** a deoxy hexose, which occurs as a component of blood group substances A, B and O, and of various oligosaccharides in human milk, seaweed and plant mucilages. It is also found in certain glycosides and antibiotics.

Fucosterol, ***(24E)-stigmasta-5,24(28)-dien-3β-ol:*** a phytosterol (Fig.), which is a characteristic constituent of marine brown algae, and has also been isolated from freshwater algae.

H

HO

Fucosterol

Fucoxanthin: a deep red carotenoid pigment found in many freshwater and marine algae, especially in marine brown algae *(Phaeophyta).* It is the most abundant of the naturally occurring carotenoids.

Fucus: see *Phaeophyceae.*

Fulgoridae (lantern flies): see *Homoptera.*

Fulgoroidea (plant hoppers): see *Homoptera.*

Fulica: see *Gruiformes.*

Fuligo: see *Myxomycetes.*

Fulmars: see *Procellariiformes.*

Fulmarus: see *Procellariiformes.*

Fulvic acid: see Humus.

Fumarase, ***fumarate hydratase:*** a Lyase (see) present in all living cells. It catalyses one of the reactions of the tricarboxylic acid cycle, i.e. the reversible conversion of fumarate into malate by addition of water to the double bond of fumarate. F. (M_r 194 000; 1784 amino acid residues) is a tetramer of 4 identical subunits (M_r 48 500), and it notably contains no disulfide bridges. Unlike many other hydrolyases, F. has no cofactor requirements.

Fumaric acid: an unsaturated dicarboxylic acid, which has the *trans* configuration (see Carboxylic acid). It can be isolated in the free form from many plants, e.g. iceland moss *(Cetraria islandica),* common fumitory *(Fumaria officinalis),* edible field mushrooms *(Agaricus)* and chantarelle *(Cantharellus cibarius).* Since F.a. is an intermediate in the Tricarboxylic acid cycle (see for formula), it occurs in all aerobic living organisms, at least in small quantities.

```
HOOC      H
    \    /
     C=C
    /    \
   H      COOH
```

Fumaric acid

Fumigation: the treatment of spaces, apparatus, plants and soil with gases to control pests and other unwanted organisms.

Funaria: see *Musci.*

Funariales: see *Musci.*

Functional behavior: modalities of behavioral interaction with the environment, determined principally by Environmental resistance (see). F.b. is therefore related to requirements for space (topical F.b.), time (temporal F.b.), food (trophic F.b.), protection (protective F.b.), sexual partners (sexual F.b.), supporting and caring partners (etepimeletic F.b.), and biosocial partners. Further subdivisions are possible.

Functional diversification, ***extension of function:*** the acquisition of other functions by an organ in addition to its main function. For example, the fins of a fish may also serve as support organs, copulatory organs or brood care organs, in addition to serving for locomotion.

Functional residual capacity of the lungs: see Lung volume.

Function communication: an elementary process of biocommunication, in which particular forms of behavior are communicated to members of the same species by exaggerated signals. It is related to Allelomimetic behavior (see), which, however, does not require this signal component. The boundaries between the two are indistinct.

Function evolution: a change in function of an organ within an evolutionary line. Thus, the hyomandibula (upper segment of the hyoid arch) of the dogfish has decreased in size and evolved into an auditory ossicle (the columnella) of the frog. During mammalian development, the stapes of the inner ear is derived from part of the upper end of the hyoid arch, i.e. like the columnella of the frog, it is derived from the hyomandibula. Similarly, the mammalian auditory ossicles known as the malleus and incus are derived from the articular and quadrate, respectively, of lower vertebrates. Evidence for F.e. comes from the study of fossils, which sometimes reveal intermediate stages in F.e., and from embryology.

Fundatrix, ***stem mother:*** the female that hatches from a fertilized aphid egg, and which produces female progeny by parthenogenesis. See Homoptera.

Fundus: see Digestive system.

Fungi, *Mycophyta:* a polyphyletic group, containing about 55 000 widely differing thallophyte species, which are classified as a division of the plant kingdom. Plastids are universally absent, so that F. are incapable of photosynthesis. They are therefore heterotrophic, living either as parasites, symbionts or saprophytes. Most are terrestrial, fewer than 2% being aquatic, the majority of these in fresh water. Most saprophytic F. can be cultured on artificial media. Numerous groups are economically useful, e.g. as food and in the production of antibiotics, whereas many are economically disadvantageous, e.g. as the agents of plant diseases and the rotting of wooden structures.

Storage substances of F. are polyglycans (structurally similar to animal glycogen) and triacylglycerols (neutral fats), as well as soluble carbohydrates like trehalose and mannitol. Starch and sucrose are never synthesized by F. In addition to triterpenes and sterols, F. also produce many phenolic compounds and several species also form organic acids

The fungal thallus is unicellular or multicellular. In the simplest groups it is naked and exhibits ameboid-like motility. More advanced forms have an outer membrane of chitin (rarely cellulose). The vegetative body usually consists of threadlike, branched (rarely unbranched) *hyphae.* The entire system of hyphae is known as the *mycelium.*

Asexual reproduction is common. Examples of the various types of spores are the naked, flagellated *swarmers* or *zoospores* of aquatic fungi, while the nonmotile spores of terrestrial forms arise endogenously *(endospores)* or exogenously by abstriction *(conidia).* Individual cells, groups of cells or an entire mycelium can develop into a resistant resting stage (hard, tuberous masses of resting hyphae are known as sclerotia).

Many different forms of sexual reproduction are found among the F. Thus, Isogamy (see), Anisogamy (see) and Oogamy (see) are well represented, while gametangiogamy (fusion of entire sex organs without gamete formation) and somatogamy (fusion of apparently undifferentiated hyphal cells) are found in more advanced forms. Mycelia may be dioecious or monoecious. A dioecious individual can only function as a nucleus donor (male) or as a nucleus acceptor (female). The dioecious condition may be expressed morphologically as well as physiologically, but in some forms (many zygomycetes and certain yeasts) morphological differentiation is not apparent. In the monoecious condition, each individual organism can donate as well as accept nuclei; here again the sex organs may be morphologically differentiated, or the donor and acceptor functions may only be expressed physiologically. Many monoecious F. display a genetically determined incompatibility, which prevents the fusion of nuclei from the same individual. In the simplest case, the two genetically different mating types are determined by a single allele pair (designated + and –). Fusion can only occur between a + and – individuals. This condition is known as homogenetic incompatibility, in contrast to the heterogenetic incompatibility of animals, higher plants and certain other fungi. After cytoplasmic fusion (plasmogamy) in monoecious or dioecious F., fusion of the respective nuclei (karyogamy) may be preceded by a more or less extended phase in which both nuclei exist side by side in the cell and divide independently of each other. This paired nucleus phase (dikaryophase) is known only in F. The alternation between different nuclear phases can be considered as equivalent to an alternation of generations, in which the mononuclear phase represents the gametophyte, the binuclear phase the sporophyte. In primitive F. the entire vegetative structure consists of fruiting bodies and spores, whereas in more advanced forms, fruiting bodies and spores are formed from only part of the organism. The familiar toadstools, bracket fungi, etc. are the highly differentiated fruiting bodies of ascomycetes and basidiomycetes.

F. are phylogenetically ancient. There is fossil evidence for the existence of *Chytridiales* in the Cambrian, nonspetate hyphae have been found among the remains of terrestrial plants in the Devonian, while *Uredinales* and perhaps *Ascomycetes,* as well as tree root mycorrhizae are well documented in the Carboniferous. Primitive forms without cell walls display affinities with the flagellates, whereas some higher F. have certain similarities to higher algae.

The evolutionary relationships of F. to other organisms and classification within the F. are uncertain. The system used here is the synthetic system described in Strasburger's Textbook of Botany *(Lehrbuch der Botanik,* 1983, Gustav Fischer Verlag, Stuttgart). Accordingly, the *Mycophyta* are divided into 6 classes based on their life cycles and type of reproductive organs: *Myxomycetes* (see), *Phycomycetes* (see) (comprising the *Chytridiomycetes, Oomycetes* and *Zygomycetes), Ascomycetes* (see) and *Basidiomycetes* (see).

In addition, there is a large group of F., of which only the vegetative form exists or is known. These form an appendix to the fungal classification, and are known as the *Fungi imperfecti* (see) or *Deuteromycetes.*

Fungicides: plant protection agents for combatting fungi. Inorganic compounds containing sulfur, copper, or mercury are among the oldest commercial F. Other important F. are organophosphorous compounds, dithiocarbamates, quinomethionate, benzimidazole, derivatives of chlorobenzene, phthalimide, thiourea, pyrimidine or oxathiine, as well as organic compounds of mercury or tin. They are sprayed, sprinkled or dusted, or used as seed dressings.

Fungicidin: see Nystatin.

Fungi imperfecti, *Deuteromyces:* a nonsystematic group of some 20 000 fungi, for which no main organs of reproduction have yet been discovered. Some forms have possibly lost the ability to form such organs. Many are probably ascomycetes, with only a few basidiomycetes or lower fungi. They include a number of important plant pests, e.g. *Septoria apii* (the agent of celery rust), *Fusarium oxysporum* (tomato wilt disease) and many others. Certain human hair and skin diseases are also caused by representatives of this group (e.g. *Trichophyton, Epidermophython, Microsporium). Geotrichum candidum,* which readily metabolizes lactic acid commonly occurs on soured dairy products (sour milk cheese) as well as sauerkraut and silage. Since F.i. are only known to reproduce vegetatively by means of conidia, the conidial form and arrangement are used for classification.

1) *Sphaeropsidales.* Conidia borne on structures resembling perithecia or in cavities.

2) *Melanconiales.* Conidia borne on stroma-like layers.

3) *Moniliales.* Conidia borne on typical conidiophores.

4) *Mycelia sterilia.* Mycelium lacks conidia or any other means of reproduction.

Fungistatic agent: a substance which prevents the growth and development of a fungus without killing it. In appropriate dilutions, fungicides may also be fungistatic.

Furcraea: see *Agavaceae.*

Furmastix: see *Synbranchiformes.*

Furnariidae, ***ovenbirds*** **:** An avian family (220 species in nearly 60 genera) in the order *Passeriformes* (see). Most species are forest dwellers, ranging from Mexico to southern Argentina. The family is named for the nests built by *Furnarius rufus.* Constructed of mud and other materials, the spherical, oven-like nest has a side entrance, and it is divided internally by a high threshold into a nesting chamber and an anteroom. The predominant plumage color is brown. Some species have adapted to open country and mountain areas, e.g. members of the genus *Cinclodes* have become ecologically equivalent to the dippers. One cinclode *(Cinclodes nigrofumosus)* actually feeds offshore on floating seaweed.

Five subfamilies are recognized: *Furnariinae* ***(Miners, Plain runners, Earthcreepers, Shaketails);*** *Synallaxinae* ***(Spinetails,*** which build large stick nests); *Philydorinae* ***(Treerunners, Treehunters, Cachalotes, Foliage gleaners);*** *Margarornithinae* (resembling woodcreepers); *Scerurinae* ***[Leaf scrapers*** *(Scererus* and the ***Sharp-tailed streamcreeper*** *(Lochmias nematura)]* .

Furnariinae: a subfamily of the *Furnariidae* (see).

Furnarius: see *Furnariidae.*

Fungia: see *Madreporaria.*

FUN test: see Dating methods.

Funtumia: see *Apocyanaceae.*

Furca:

1) A sometimes multisegmented appendage on the telson of many crustaceans. It is a paired appendage consisting of two caudal rami, and it is also called the *caudal furca.*

2) The paired structure forming the spring mechanism of the *Collembola* (see). In this case, it is more usually called a *furcula.*

Fur seals: see *Otariidae.*

Fusaric acid, ***5-n-butylpyridine-2-carboxylic acid.*** a phytotoxin produced by fungi of the genera *Fusarium* and *Gibberella.* It is a Wilt toxin (see), e.g. it is responsible for the wilting of tomato plants caused by *Fusarium lycopersici.* It displays powerful fungistatic activity against smut fungi *(Ustilaginales),* and decreases blood pressure in humans.

COOH
N

Fusaric acid

Fuselle: see Graptolites.

Fusel oil: an oily, unpleasantly smelling byproduct of alcoholic fermentation, consisting mainly of *n*-pentanol (amyl alcohol), isopentanol (isoamyl alcohol), isobutanol and *n*-propanol. F.o. arises by fermentation from the yeast proteins and plant products, in particular by deamination and decarboxylation of amino acids.

Fusiform bacilli: see Bacillus.

Fusion (the entry entitled Cell fusion should also be consulted): a physiological process, in which two membranes fuse and their contents are combined. Examples are the fusion of a sperm with an egg cell, membrane fusion during exocytosis, fusion of synaptic vesicles with the presynaptic membrane. F. depends on a controlled local destabilization of the membrane. For this to occur, the membranes must be in close contact, and the area of contact must contain a protein-free region. In this protein-free region an interaction occurs between Ca^{2+} and phosphatidylserine or phosphatidic acid, which induces a phase transition in the lipid phase.

Laboratory methods also exist for promoting F. between cells. Three such artificial methods are used: virus-induced F., chemical F. and electrofusion. For ***virus-induced F.,*** hemagglutinizing viruses (e.g. Sendai virus) are used, which promote F. in a few minutes following the aggregation of the membranous particles. ***Chemical F.*** employs *fusogens* or *fusogenic agents,* e.g. macromolecules like polyethyleneglycol. Fusogens cause aggregation, which is followed by F. when the fusogen is removed by washing. In ***electrofusion,*** the particles are first aggregated, then F. is triggered by a brief impulse of about 1 kV/cm.

Artificial F. is of considerable practical importance. In principle, the vegetative reproduction of mammals is possible by F. of egg cells. "Monster plants" have already been generated by the F. of plant cells. Procedures for the F. of lymphocytes and tumor cells are well established and standardized; the resulting hybridomas are the source of Monoclonal antibodies (see), an extremely important tool in medical research and diagnosis.

Fusion translocation: see Centric fusion.

Fusogen: see Fusion.

Fusogenic agent: see Fusion.

Fusulinidea: see Fusulinids.

Fusulinids, *Fusulinidea* (Latin *fusus* spindle): an extinct suborder of large *Foraminiferida* (see) with fusiform shells, ranging in size from 0.5 to a few centimeters. The numerous chambers are arranged planispirally around the central axis, and further subdivided by septa. Deposits of F. on ancient ocean floors have given rise to fusulina limestones.

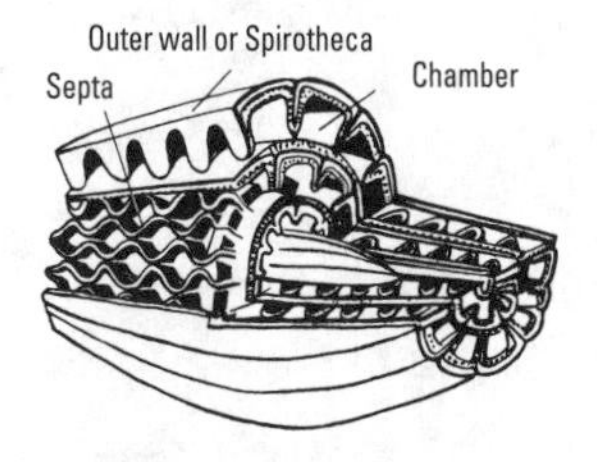

Fusulinids. Structure of the shell of the genus *Fusulina* (schematic, x 10).

Fynbos: see Sclerophylous vegetation.

G

Gaboon viper: see Puff adders.

Gadiformes, *anacanthine fishes:* an order of marine fishes, containing about 700 species in 76 genera and 7 families. With the single exception of the freshwater ***Burbot** (Lota lota),* all species are marine. The majority are free-swimming, predatory fishes, occurring primarily in northern seas. Many species are commercially important, e.g. cod and haddock. Most species have long dorsal and anal fins. The scales are cycloid. Fin spines are absent, and the fins have soft rays. The ventral fins are nearer to the head than the pectoral fins.

Gadus: see Cod.

Gaillardia: see *Compositae.*

Gait: mode of locomotion of terrestrial animals, involving contact with the ground. The term is used primarily for mammals, but is also applied to other animals that have umbulatory legs. Change of G. is associated chiefly with the regulation of speed. Transitional types of G. do not normally occur, and a change of G. involves a change of motor coordination, e.g. walking, trotting, galloping. See also Digitigrade, Plantigrade, Unguligrade.

Galactans: plant polysaccharides (hemicelluloses) consisting mainly of D-galactose. They are usually unbranched and have high molecular masses. Examples are agar-agar and carrageenan

D-Galactosamine, *chondrosamine, 2-amino-2-deoxy-D-galactose:* an aminosugar which usually occurs naturally as its *N*-acetyl derivative, e.g. in certain mucopolysaccharides such as chondroitin sulfate, blood group substance A, etc., as well as in mucoproteins. Structurally, D-G. is a derivative of Galactose (see) in which the OH-group on C2 is replaced by an amino group.

Galactose: an aldohexose monosaccharide, which occurs naturally as both the D- and the L-form. It is a component of many oligosaccharides, e.g. lactose, raffinose, melibiose and stachyose, the cerebrosides and gangliosides of animal nervous tissue, the sugar components of some glycosides (e.g. tomatin and digitonin), as well as galactans. It is not fermented by top fermenting yeasts.

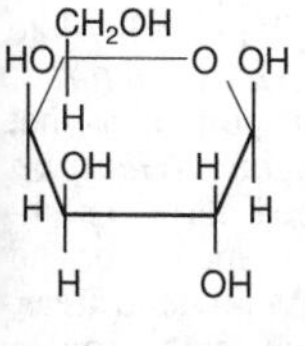

β-D-Galactose

β-Galactosidase, *lactase:* a widely found disaccharidase enzyme which catalyses the hydrolysis of lactose to D-glucose and D-galactose. β-G. activity is conveniently measured with the colorless artificial substrate, *o*-nitrophenyl-β-D-galactoside, which is hydrolysed to galactose and the colored product, *o*-nitrophenol. Classical studies on the induction of β-G. led Jacob and Monod to propose the Operon (see) model for the regulation of protein synthesis. The lactose operon *(lac* operon) of *Escherichia coli* contains structural genes for β-G., galactoside permease and thiogalactoside transacetylase.

D-Galacturonic acid: an uronic acid derived from D-galactose. Pectins contain 40–60% D-G.a., which is present partly as its methyl ester. It is also found in a few other plant polysaccharides.

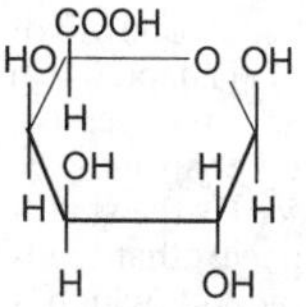

β-D-Galacturonic acid

Galagoides: see Galagos.

Galagos, *bushbabies:* members of the subfamily *Galaginae,* family *Loridae* (see Primates). They are all nocturnal, possessing large reflective eyes and large, mobile, membranous ears. The ears can be folded for protection and can be moved independently to scan for insect prey. G. are found only in Africa, occurring in most wooded savanna or forested areas south of the Sahara to the Orange River in the south of the Continent. They progress by clinging and leaping. The smallest member is *Gallagoides* (head plus body 125 mm, tail 175 mm, weight 60 g) and the largest is *Otolemur* (head plus body 325 mm, weight 180 g). G. comprise the following genera and species: *Galago senegalensis* ***(Senegal bushbaby, Lesser bushbaby, True bushbaby),*** *Galago alleni* ***(Allen's bushbaby),*** *Otolemur crassicaudatus* ***(Greater bushbaby;*** also sometimes listed as *Galago crassicaudatus), Euoticus eleganthus* ***(Western needle-nailed bushbaby),*** *Euoticus inustus* ***(Eastern needle-nailed bushbaby),*** *Galagoides demidoff* ***(Demidoff's galago*** or ***Dwarf galago).*** New species are now being discovered by analysis of their night-time vocalizations.

Galanthus: see *Amaryllidaceae.*

Galapagos finches, *Darwin's finches,* *Geospizinae* (see also Adaptive radiation): a subfamily of the *Fringillidae* (see), consisting of 14 species, 9.5–13.5 cm in length, with a short tail and densely feathered lower back, and largely grayish brown in color. Thirteen species in 3 genera are found on the Galapagos Islands, and 1 monotypic genus inhabits Cocos Island 600 miles (960 km) to the northwest. Observation of these finches provided an important stimulus for Darwin's theory of evolution. Nowdays, they are regarded by biologists as a classic example of adaptive radiation, in which the Galapagos were colonized by an ancestral bunting-like form from Central America, related to the present-day grassquits, followed by evolutionary radiation into various ecological niches. Even the most insectivorous members of the subfamily feed their young by regurgitation (like seed-eaters), rather than from the beak like other insectivorous birds. The Cocos finch is probably very close to the ancestral form, and this ancestral form was in turn probably phylogenetically very close to the ***Blue-black grassquit*** *(Volatina jacarina),* which is widespread in the tropical lowlands from Mexico through much of South America, including the Pacific coast of South America from Colombia to northern Chile.

The adaptive radiation occurred relatively recently in geological time, so that some of the intermediate and transitional stages are extant. Hybrids between species and even genera are frequent. The most obvious morphological variation is in the bill, which in accordance with the type of diet and the way in which it is obtained (probing, biting, grasping or crushing) ranges from stout and finch-like through intermediate stages to long, thin and warbler-like. The various forms of bill are illustrated in the entry "Adaptive radiation" (see). The bill of the ***Cocos finch*** *(Pinaroloxius inornata)* is adapted to a variety of food sources (insects, nectar, small seeds), and it is quite unlike that of any of the other species.

The ***Warbler finch*** *(Certhidia olivacea)* is the smallest of the *Geospizinae.* It is the only species that feeds entirely on animal food (insects and spiders), which it picks from leaves and twigs. It has a thin, pointed, warbler-like, probing bill, and it is the only species found on all the major islands. On the larger islands, it usually nests in the humid highlands, but outside the nesting season it may be found anywhere. It also thrives on the small islands that have only dry habitats. *Camarhynchus pallida* **(Woodpecker finch)** has an elongated and relatively stout, tanager-like probing bill with a biting tip, while the bill of *Camarynchus heliobates* **(Mangrove finch)** is similar but not quite so heavy. Both the Woodpecker finch and the Mangrove finch feed on fairly large insects, but may take some vegetable matter. The Mangrove finch has also been observed to use a twig to poke insects out of crevices, but this habit does not appear to be as well developed as in the Woodpecker finch (see below). All three live in trees, and they differ from the other *Geospizinae* in that the male and female have similar plumage. Moreover, their pale plumage is similar to that of the female tree finches. The Woodpecker finch excavates wood with its bill, and also has the unique habit of taking a cactus spine or small twig to probe bark crevices and its own excavations, in order to drive out insects. This represents a rare example of tool use by animals.

All 3 ***Tree finches*** are largely insectivorous, but also take some plant food. They breed mainly in the transition zone or higher, often descending into the arid lowlands outside the nesting season. They have similarly shaped bills (grasping-type, with a biting tip) and resemble one another very closely in all respects except size. The ***Small tree finch*** *(Camarynchus parvula)* feeds on small insects, fruits and soft seeds, and has a relatively small, conical bill. It is found on all the major islands except Hood, Tower, Bindloe and Culpepper. The ***Large tree finch*** *(Camarynchus psittacula)* has a similar distribution, but is also absent from Wenman. It feeds on fairly large insects and larvae, often obtaining these by breaking up soft plants and twigs with a twisting action of the bill. The bill is of medium length and stout, with the upper surface curved sharply downward and the lower surface upward, giving a parrot-like appearance. Surprisingly, the ***Medium tree finch*** *(Camarynchus pauper)* is found only on Charles Island. It feeds on insects and soft seeds, which are often picked from fruit heads; its bill is rather smaller and less parrot like than that of the Large tree finch.

The ***Cactus finches*** spend much of their time in trees or cacti, and probably evolved directly from the Ground finches. The ***Vegetarian finch*** *[Platyspiza (= Camarynchus) crassirostris],* the only species of *Platyspiza,* feeds mainly on soft seeds, fruits and leaves; its crushing-type bill (a bullfinch-like "fruit squeezer") with a biting tip is noticeably short, deep and broad, with a very convex upper surface. The Vegetarian finch occurs mainly in the humid highlands of Chatham, Charles, Indefatigable, Albemarle, Narborough and Abingdon. The ***Small cactus finch*** *(Geospiza scandens)* has a relatively thick, elongated and slightly decurved probing bill with a crushing edge, used primarily for probing the flowers of the prickly pear cactus for nectar, and for eating the soft, pulpy cactus fruit and its moderately hard seeds. It occurs on all the major islands except Hood, Narborough, Tower, Culpepper and Wenman. The ***Large cactus finch*** *(Geospiza conirostris)* feeds on a variety of seeds and large insects. Its bill is generally of a stout crushing and probing type, but varies in the 2 distinct subspecies; on Hood the subspecies has a very heavy bill, rather similar to that of the Large ground finch, but slightly more elongated and laterally compressed; on Tower, Culpepper and Wenman the subspecies is slightly smaller with a relatively smaller, narrower bill.

Pronounced sexual dimorphism is found in the stout-billed, largely vegetarian, largely seed-eating ***Ground finches,*** the males being black and the females streaked brown. *Geospiza nebulosa (= difficilis)* ***(Sharp-beaked ground finch)*** lives mainly in the humid highlands of Chatham, Charles, Indefatigable, Albemarle, Narborough and Abingdon, also living of necessity in drier habitats on the smaller, lower islands of Tower, Culpepper and Wenman. The subspecies of Tower island *(Geospiza difficilis acutirostris)* is almost entirely vegetarian, taking seeds and flowers, and some insects. The subspecies on Wenman island *(G. d. septentrionalis)* is almost twice as heavy and feeds mainly on animal material, such as the flesh of dead animals. It also breaks open the eggs of sea birds and drinks the contents, and it pecks at the growing wing and tail feathers of molting boobies and drinks the blood. The bill of the Sharp-beaked ground finch is similar to that of the Small ground finch, but slightly longer and more pointed. The ***Small ground finch*** *(Geospiza fuliginosa;* sparrow-like crushing bill), ***Medium ground finch*** *(Geospiza fortis;* a crushing bill that is not as small or as pointed as that of the *G. fuliginosa)* and ***Large ground finch*** *(Geospiza magnirostris;* especially well developed, large, crushing bill) live on all the major islands except Tower and Culpepper. They are usually found in dry lowland areas, where they forage on the ground or in low vegetation. In addition to seeds, they also feed on buds, flowers and young leaves, and even some small insects may be taken, especially by *Geospiza fortis.*

The greatest concentration of species occurs on the central islands of the Galapagos group, with fewer species (especially those displaying intermediate bill forms) on the small outlying islands.

Galapagos giant tortoise: see *Testudinidae.*

Galatheoidea: porcellain crabs and allies, a superfamily of the *Decapoda* (see).

Galba truncatula (dwarf pond snail): see *Gastropoda.*

Galbula: see *Galbulidae.*

Galbulidae, ***jacamars:*** an avian family in the order *Piciformes,* containing about 15 species in 5 genera, found in the less dense areas and edges of the forests of Central and South America. They are small to medium sized birds (13–28 cm) with slender, tapered bodies, short legs and long, curved, pointed and usually thin bills. Most species have four toes, 2 directed backward and 2 forward (i.e. zygodactylous toes), but in *Jacamaralcyon tridactyla* ***(Three-toed jacamar;*** 16cm; southeastern Brazil) the inner hindtoe has been lost. The more characteristic species have iridescent metallic plumage, and females usually differ from males in having brown instead of white throats. They feed exclusively on insects, mostly large butterflies and dragonflies, which are caught in flight. Jacamars nest in burrows, which are usually excavated into a hillside or bank. A widespread and familiar member is the ***Rufous-tailed jacamar*** *(Galbula ruficauda;* 23–28 cm; Central and South America east and west of the Andes, Trinidad and Tobago); the upper plumage, including the wings and central feathers of the long graduated tail, are glittering metallic green with golden glints. The throat is white (male) or buff (female) and separated from the bright chestnut to buff underparts by a green band across the breast.

Galeoidea: see *Selachimorpha.*

Galerida: see *Alaudidae.*

Galidiinae: see Mongooses.

Gall: see Galls.

Gall bladder, ***vesica fellea:*** a thin-walled mucosal sac of vertebrates, which serves as a reservoir for bile formed in the liver. It is absent from certain birds (pigeons, parrots, humming birds and ostriches) and mammals (horse, deer, rat, hippopotamus and whale). Liver bile may bypass the G.b. and flow directly into the duodenum via the hepatic duct (ductus hepaticus) and common bile duct (ductus choledochus). Alternatively, it may pass into the G.b. from the hepatic duct via a lateral branch known as the cystic duct (ductus cysticus). When leaving the G.b., bile returns along cystic duct.

Gall gnats: see Gall midges.

Gallic acid: a common aromatic carboxylic acid, which is particularly abundant in oak bark, oak galls and tea. It is a component of gallotannins, from which it can be obtained by hydrolysis.

COOH
HO OH
OH

Gallic acid

Galliformes, ***gallinaceous birds, fowl-like birds:*** an order of birds (see *Aves)* containing more than 250 species. Most are ground birds, but they usually roost above ground level. All have short, stout, decurved bills and large heavy feet with 3 toes in front and a shorter one behind. They feed largely on berries, seeds and vegetation, and some species also take insects and snails and even small reptiles and amphibians. Food is obtained by scratting, picking and plucking. All have light, succulent breast meast, and many have been domesticated or are hunted by man for food.

The order is usually divided into the following 6 families, which are listed separately: *Megapodidae* ***(Megapodes*** or ***Mound builders),*** *Cracidae* ***(Curassows, Guans*** and ***Chachalacas),*** *Tetraonidae* ***(Grouse*** and ***Ptarmigan),*** *Phasianidae* ***(Pheasants, Peacocks, Partridges, Quails*** and ***Francolins),*** *Numididae* ***(Guinea fowl),*** *Meleagrididae* ***(Turkeys).***

In some classifications, the ***Hoazin*** is also placed in the *Galliformes,* but here it is assigned to its own order, *Opisthocomiformes* (see).

Gallinago: see *Scolopacidae.*

Gallinula: see *Gruiformes.*

Gallinules: see *Gruiformes.*

Gallirallus: see *Gruiformes.*

Gallito: see *Rhinocryptidae.*

Gallionella: a genus of Gram-negative bacteria, found in iron-containing water (see Iron bacteria). It is neither sheath-forming (cf. *Chlamydobacteriaceae)* nor filamentous (cf. *Beggiatoa),* but each cell does form a stalk, which contains iron compounds. The cells are bean-shaped, about 0.5 µm x 2 µm. As each cell divides, the stalk branches, ultimately forming a tangle of stalks, each 0.2 to 0.3 mm in length. The characteristically flat and twisted stalk resembles a loosely coiled rubber band, and is diagnostic of *G.* It also generates deposits of ferric hydroxide, and when it grows in water pipes these deposits may be responsible for blockages.

Gall midges, ***gall gnats,*** *Cecidomyiidae:* a family of the *Diptera,* containing 4 000 species, body length 2–5 mm. Larvae may consume rotting organic material, but most feed on fresh plants, often causing gall formation. Some species are agricultural and garden pests, e.g. wheat midge larvae feed on pollen and the milky juice of immature cereal seeds, causing them to shrivel. The wheat midge, *Thecodiplosis mosellana,* is of European orgin, but is now also a pest in North America. The wheat midge, *Contarinia tritici* Kby. (= *Sitodiplosis mosellana* Geh.) is found only in Europe. Larvae of the Cabbage midge *(Perrisia brassicae* Winn.) destroy the seeds of vegetables of the *Brassica* tribe. Also a serious European fruit pest is the pear midge *(Contarinia pyrivora).* One of the most severe cereal pests is the Hessian fly *(Phytophaga destructor = Mayetioloa destructor),* originating in Europe and now introduced into North America and New Zealand; autumn larvae infest wheat

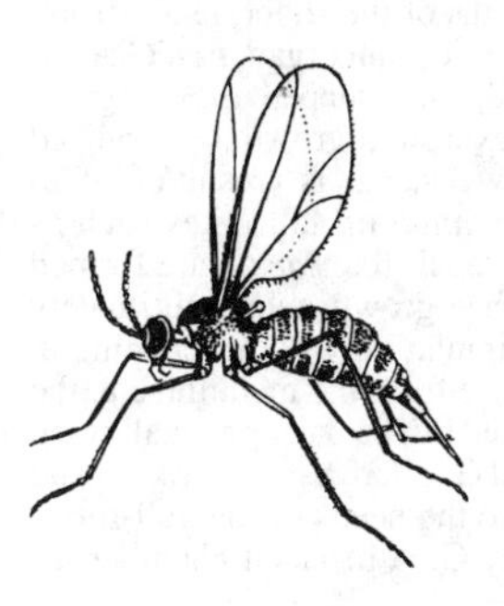

Gall midges. Cabbage midge *(Perrissia brassicae* Winn.).

plants near ground level between the leaf base and the main stalk, whereas adults emerging in the spring deposit eggs in the leaf sheaths well above ground level.

Gall mites: see *Tetrapodili.*

Gallotannins: see Tannins.

Gallotia: see Ocellated green lizard.

Galls, ***plant galls, cecidia*** (plates 12 & 13): localized growth anomalies of plant organs, resulting from an increase in cell numbers (hyperplasia) or enlargement of cells (hypertrophy), often accompanied by endopolyploidy, cell fusion and differentiation into unusual temporal and spatial patterns. G. are tumors with restricted growth, which attain a specific size, and often display organ polarity and symmetry. They arise from the action of animal parasites *(cecidozoa)* or plant parasites *(cecidophytes).* G. caused by animals are known as *zoocecidia.* Gall development is caused by numerous animals, their eggs or their larvae, e.g. nematodes, mites, thrips, plant bugs, cicadas, aphids, psyllids, coccids, butterflies, beetles, sawflies, gall wasps, gall midges and flies. The gall serves as a food source and protective housing for the parasite or its offspring. Animal inhabitants of G. usually have reduced extremities and sense organs, and they are often restricted to a particular plant species. G. caused by plant parasites (mostly bacteria and fungi) are called phytocecidia, e.g. *Agrobacterium tumefaciens* (crown gall of many flowering plants), *Rhizobium* (root nodules of *Leguminoseae),* actinomycetes (root nodules of many nonlegume families), *Plasmodiophora brassicae* (club root of cabbage).

G. display an extraordinary variety of shape and structure. G. which arise inside an organ or tissue and are not externally visible are called *cryptocecidia,* e.g. those caused by potato nematodes. "Physiological G." are morphologically inconspicuous, but are represented by a site of abnormal metabolism. G. that form distinct external structures can be classified according to their morphology and anatomy. 1) *Organoid G.* consist of strongly modified but more or less clearly recognizable basic organs of the host plant. Examples are abnormal branching (see Witches' broom), aggregations of simplified or at least reduced leaves, particularly at the tips of shoots [e.g. the bedeguar gall, robin's pincushion, or rose apple, which is common everywhere on *Rosa canina* (dog rose), where it is caused by the rose gall wasp, *Diplolepis rosae],* doubling and greening and other leaf-like abnormalities of flowers. G. may be described as *simple* or *compound,* according to whether one more plant organs are involved. The compound G. formed on larch by the red fir gall louse *(Adelges laricis)* and on spruce by the yellow fir gall louse *(Sacchiphantes abietis)* bear a superficial resemblance to small pineapples. 2) *Histoid G.* are outgrowths of the shoot, leaf or root, with no resemblance to a basic plant organ. *Felt G.,* consisting of closely packed, club-shaped or elongated hairs, arise by increased extension growth of localized regions of epidermal cell walls, e.g. as a result of infection by *Synchytrium papillatum* or gall mites such as *Eriophyes nervisequens.* Pouch-like *sac G.* are formed by pronounced local surface growth which ultimately encloses the source of stimulation with an opening at the base. *Medullary G.* arise from the medulla, e.g. the oak apple or oak gall caused by the oak apple gall wasp *(Biorhiza pallida).* Histoid G. often display striking and complicated adaptations to the needs of the gall-forming animal. Thus, in many G., a resistant chamber for the developing animal is formed by secondary thickening and differentiation of sclerenchymatous elements. On the interior surface, numerous hairs and thin-walled cells serve for the nutrition of the parasite.

G. can only be formed in plant tissues that are capable of cell division. They arise from the specific localized and transitory action of material from the gall-forming organism. Phytohormones apparently play in important part in gall formation. Thus, *Corynebacterium fascians* produces a cytokinin, which is probably responsible for fasciation. *Rhizobium* produces lipo-oligosaccharides which elicit nodule mersitems in leguminous plants [H.P.Spink et al. *Nature* ***345*** (1991) 125–130]. The salivary gland secretion of many gall insects contain indole acetic acid, as well as amino acids, enzymes and other unknown proliferation-promoting factors, which collectively initiate and program gall formation.

Gallus: see *Phasianidae.*

Gall wasps, *Cynipinae:* a subfamily of the *Cynipidae,* order *Hymenoptera* (see). About 1 600 species of G.w. have been described, 300 in central Europe. In most species the adult is only 1 to 5 mm in length. Larval development and pupation occur within a plant gall. Larvae of most European species produce galls on the leaves, stems or flowers of oak, maple or wild roses, e.g. the common ***Oak gall wasp*** *(Cynips quercusfolii)* causes soft yellow galls (turning red in the autumn) on the underside of oak leaves. In Europe, the familiar oak apples are bud galls formed by the European ***Oak apple gall wasp*** *(Biorhiza pallida).* In the USA, oak galls are caused by species of the genera, *Amphibolips* and *Philonix.* Many G.w. display an alternation between a sexual and a parthenogenetic generation, which form galls on different plant species. Some species undergo their larval development in the galls of other G.w. (see Inquiline).

Galumna: see *Oribatei.*

Galvanonasty: see Electronasty.

Galvanotaxis: directional movement of freely motile organisms (see Taxis) in relation to an electric field. Movement toward the cathode is known as *negative* G., toward the anode as *positive* G. Both forms occur in free-swimming lower plants. The response of a species may be reversed experimentally. Myxomycetes, which progress by ameboid movement, display a negative (cathodic) G.

Galvanotropism: see Electrotropism.

Gambo hemp: see *Malvaceae.*

Gambusia: see Guppy.

Game birds: various members of the order *Galliformes* (see) that are hunted for food.

Gametangiogamy: see Reproduction.

Gametangium: see Reproduction.

Gamete, ***sex cell:*** a haploid eukaryotic cell, which serves for sexual reproduction. A female and a male G. unite, with fusion of their nuclei and cytoplasmic contents, to form a diploid zygote. Male and female G. are biochemically and functionally different, and they also usually differ in size *(anisogametes).* The cell surface of a G. plays an important role in the recognition and establishment of contact between the male and female partners. In particular, the chemical composition of the glycocalyx (see Plasma membrane) is sex- and species-specific, so that fusion is normally only possible between Gs. of the same species. The male G. is usually smaller and more motile than the female G. If both Gs.

are flagellated, the smaller (male) is called a *microgamete,* the larger (female) a *macrogamete.* In the absence of morphological differences *(isogametes),* Gs. are designated + and – in accordance with their behavior. Metazoan animals and ferns and mosses produce small flagellated male Gs. *(sperms* or *spermatozoa,* and *spermatozoids,* respectively), and large nonmotile female Gs. *(eggs).* In most metazoans, egg cells arise in *ovaries,* spermatozoa in the *testes.*

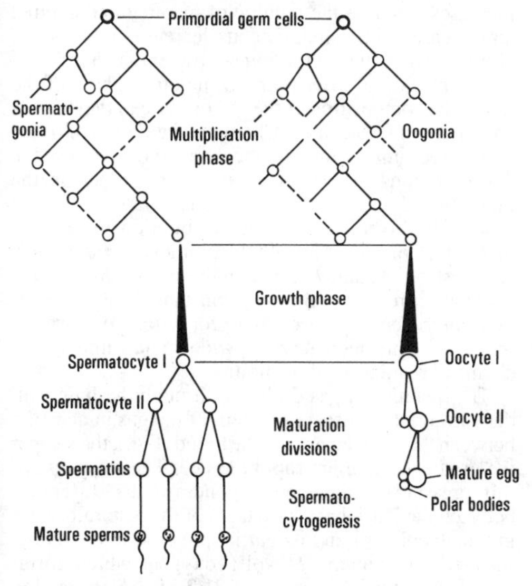

Gamete. Oogenesis and spermatogenesis in animals.

In metazoan *gametogenesis,* Gs. develop from primordial germ cells. By mitosis, primordial germ cells give rise to small, sexually undifferentiated *oogonia* or *spermatogonia,* which continue to multiply by mitosis *(multiplication phase)* until they enter the *growth phase.* During the growth phase, spermatogonia increase only slightly in size to become *primary spermatocytes.* In the *maturation phase,* each of these undergoes meiosis to produce two *secondary spermatocytes,* and finally four haploid *spermatids,* which differentiate *(spermiocytogenesis)* into functional *spermatozoa (sperms)* (plates 22 & 23). The sperm head carries a cap-like Acrosome (see) in front of the nucleus. Acrosomal enzymes aid penetration of the egg cell membrane *(oolemma)* by the sperm. During the usually prolonged growth phase of oogonia, the cells increase considerably in size and accumulate reserve material (yolk), and the cell nucleus is in the prophase of meiosis. After the completion of growth, the resulting primary oocyte undergoes two meiotic divisions, but almost all of the large quantity of cytoplasm *(ooplasm)* is allocated to only one of the final 3 to 4 haploid cells, and this cell, without further differentiation, is the fertile *egg.* The 2 to 3 remaining small cells become *polar bodies,* marking the animal pole of the *egg* (the point on the egg surface nearest to its nucleus, and containing the most cytoplasm and the least yolk). The secondary oocyte (first polar body) usually does not divide again, so that only two polar bodies are normally produced. For the significance of egg polarity and the distribution of yolk and organelles in the ooplasm, see Cleavage.

See Oogenesis, Spermatogenesis.

Gametocyte: see Gamont.

Gametogamy: see Reproduction.

Gametogenesis: formation of functional gametes. G. includes all processes of gamete differentiation, multiplication, growth, formation of reserve material, maturation, envelope synthesis, development of specialized organelles (micropyles, locomotory organelles). It is sometimes referred to as *progenesis,* to distinguish it from the developmental processes that occur after fertilization. In most multicellular organisms (with the exception of a few invertebrates), it occurs in specialized organs known as gonads. The male and female gonads are either carried by separate individuals (gonochorism, monosexuality, unisexuality), or they occur together in the same individual (hermaphroditism, bisexuality). Pulmonate snails represent an extreme form of hermaphroditism, since they possess a true hermaphrodite gland or ovotestis.

See Gamete, Oogenesis, Spermatogenesis.

Gammaridae (Gammarids): see *Amphipoda.*

Gamocyst: see Cyst.

Gamogony, ***gametogony:*** formation of gametes from gamonts in the protozoa. As a rule, a female gamont produces only a single macrogamete, whereas numerous microgametes are produced by male gamonts. For G. in Metazoa, see Reproduction.

Gamones: substances produced in small quantities by the sex cells of plants when at least one of the gametes involved in fertilization has a free existence, i.e. in many algae, lower fungi, mosses and ferns. The gamones that have been investigated, e.g. sirenin and ectocarpene, are produced by the female gametes and act as chemotactic agents, which attract and stimulate the activity of male gametes. In contrast to pheromones, G. influence only the gametes (i.e. they act at the cellular level), and not the behavior of the whole organism.

Sirenin is a sesquiterpene produced by the female gametes of the fungus *Allomyces.* It attracts male gametes at a concentration of 10^{-10} M.

Ectocarpine *[all-cis-*(1-cyclohepta-2',5'-dienyl)-but-1-ene] is secreted by the female gametes of the brown alga *Ectocarpus stiliculosus.*

Gamont, ***gametocyte:*** a cell which forms gametes in the developmental cycle of the protozoa. See Reproduction.

Gamontogamy: see Reproduction.

Ganglion: a region of the nervous system consisting of an accumulation the cell bodies of neurons. In the vertebrate brain, a G. is also known as a nucleus or lamina. Histologically, a lamina appears as a granular layer (cell bodies densely packed side by side) surrounded by a felt-like region (the outwardly extending dendrites). Ganglia are interconnected by *axons,* and the fiber formed by several parallel axons is called a *nerve.* Ganglia also contain glial cells (see Neuron). Numerous ganglia are found in the peripheral nervous system, e.g. in sense organs and in the autonomic nervous system. Through the interaction of their constituent neurons (see Neuron circuit), ganglia serve as centers for the processing, integration and relay of incoming stimuli. The most primitive organisms with recognizable ganglia are certain coelenterates. The complexity and extent of G. development has increased with evolution, leading to the centralized nervous systems of many groups of organisms.

Ganglion blockers: substances that specifically block the transmission of excitatory stimuli (i.e. cholinergic transmission; see Acetylcholine) in ganglia of the Autonomic nervous system (see).

Gangliosides: see Glycolipids.

Gannets: see *Pelicaniformes.*

Gaping: the act of widely opening the bill by young Altricial (see) birds (e.g. young birds of the *Passeriformes,* young cuckoos, etc.), which is part of their reflex feeding behavior. In the first few days after hatching, naked and blind young altricial birds respond to various external (shadow, gentle vibration of the nest, voice of a parent) by stretching their necks steeply upward with widely opened bills, i.e. they gape vertically or nearly so. Later, they gape toward the visual stimulus of the food-bringing parent. The throats of gaping young birds are brightly colored, and often carry a well-marked, species-specific pattern, which in turn acts as a visual stimulus to the feeding parent. The gapes of young nest parasites have the same color and patterning as those of the young of the host bird.

Gar: see *Lepisosteiformes.*

Garden dormouse: see *Gliridae.*

Gardeners: see *Ptilinorhynchidae.*

Garden pea: see *Papilionaceae.*

Garden snail: see *Gastropoda.*

Garden tiger moth: see *Arctiidae.*

Garigue: see Sclerophylous vegetation.

Gar pike, *Belone belone:* an elongate, slender marine fish (maximal length 90 cm, usual length 30–60 cm) of the family *Belonidae* ***(Needlefishes),*** order *Cyprinodontiformes.* It has long jaws with visible teeth, a dark green back, silvery, yellow-spotted sides. Young G.ps. are plankton feeders, and the adults also take small fish. The bones contain iron compounds, causing them to turn green when cooked. The flesh is very tasty but bony, and the fish is not commercially important.

Garrulus: see *Corvidae.*

Garrulux: see *Timaliinae.*

Garter snakes, *Thamnophis:* the commonest North American genus of colubrid snakes (see *Colubridae),* containing many more or less moisture loving species, distributed from Canada to Mexico. They are 0.5–1.5 m in length, usually with light longitudinal stripes on a dark background. Their preferred diet is earthworms. Those that live in rather close association with water also eat frogs and fishes. All give birth to live young and are truly viviparous (i.e. they form a placenta). The ***Common garter snake*** *(T. sirtalis)* occurs from southern Canada to eastern USA.

Gas bladder of fishes: see Swim bladder.

Gas exchange, ***respiratory gas exchange:*** exchange of respiratory gases between the interior of the organism and its environment.

In all animals with a vascular system, in which the blood transports respiratory gases, there is a two-fold G.e., i.e. in the respiratory organs and in the tissues.

1) G.e. in the lungs. In the lungs, atmospheric oxygen passes into the blood, while blood carbon dioxide is released into the pulmonary air space. The exchange occurs entirely by diffusion.

The lungs are not completely emptied by expiration, so that in each cycle of inspiration and expiration the total lung content is not exchanged with the exterior. In quiet respiration, about 500 ml are taken in and expelled with each breath. About 150 ml of this inspired air does not take part in G.e., but remains in the respiratory passages and alveolar space. The remaining 350 ml mix with the expiratory reserve volume (2 000 ml) and the residual air (1 500 ml). The ratio of the volume of inspired involved in G.e. to the volume of air remaining in the lung is known as the *ventilation coefficient.* In quiet respiration, this value is about 350 ml/(2 000 ml + 1 500 ml) = 0.1. With increased respiratory activity, this value increases, because the volume of inspired and expired air increases with a simultaneous decrease in the reserve volume. Thus, the ventilation coefficient is an index of the proportion of "fresh air" in the atmosphere of the lung. The composition of the gas mixture in the lungs is an important factor in the Chemical regulation of respiration (see); this regulation mechanism responds to the concentrations of carbon dioxide and oxygen in the blood, which in turn reflect the concentrations of these gases in the alveolar air. In quiet respiration, exterior air mixes at a constant rate with the pulmonary gases and a steady state is attained in which the composition of the alveolar air remains relatively constant. As the ventilation coefficient increases, the proportions of gases in the alveolar air fluctuate more widely and stimulate the chemical regulation of respiration.

The quantity of gas diffusing through the lung wall depends on the pressure gradient of the gas in question between the alveolar air and the blood (Fig.); the steeper the gradient, the more rapid the G.e. The steepness of this pressure gradient corresponds to the difference between the Partial pressure (see) of the respiratory gas in the alveolar air and its partial pressure in the blood. Inspired air contains 21 vol% oxygen, which corresponds to a partial pressure of 21.3 kPa. After mixing with the air in the lungs, this partial pressure falls to about 13 kPa . In contrast, the oxygen partial pressure of the venous blood delivered to the lung capillaries by the lung arteries is about 5 kPa. After passage through the lung capillaries, the oxygen partial pressure of the blood rises to about 13 kPa. Since the total volume of oxygen in the lung is large compared with the volume that diffuses into the blood, the oxygen partial pressure of the alveolar air decreases only slightly. In the pulmonary vein to the heart, the oxygen concentration decreases somewhat, because it is mixed with "short circuited" blood, which passed through the lungs, but did not take part in G.e. The final oxygen partial pressure of arterial blood is about 12 kPa.

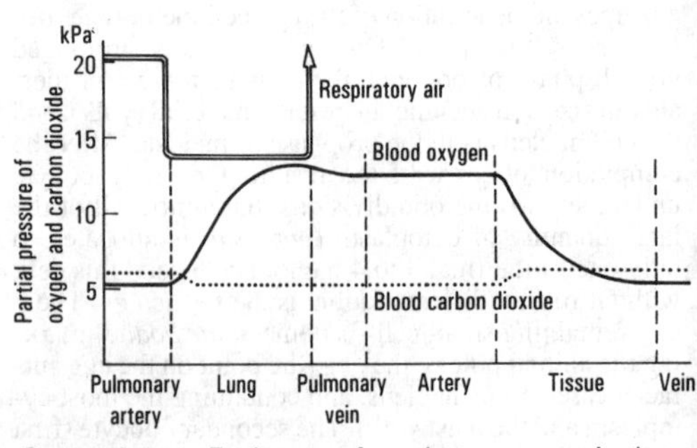

Gas exchange. Exchange of respiratory gases in the human.

In the organism, carbon dioxide diffuses in the opposite direction to oxygen. The carbon dioxide partial pressure of inspired air increases from 0.03 to 5.3 kPa on mixing with the pulmonary air. By G.e., the carbon dioxide partial pressure of the blood decreases from 6 to 5 kPa during passage through the lung capillaries. Carbon dioxide diffuses much more rapidly than oxygen, so that the carbon dioxide exchange is completed before the oxygen exchange.

Diffusion of respiratory gases is governed essentially by the following three factors. a) The pressure gradient of the atmospheric gases between the alveolar air and the blood. This can be changed by two mechanisms. The partial pressures of oxygen and carbon dioxide in the alveolar air depend on the volumes of inspired and expired air and on the frequency of the cycle of inspiration and expiration. Increased ventilation causes an increase in the oxygen partial pressure and a decrease in the carbon dioxide content of the alveolar air; the pressure gradients are therefore increased for each gas and the rate of G.e. increases. Alternatively, the partial pressures of respiratory gases may be altered in the circulation; for example, a higher rate of metabolism increases the oxygen consumption and the carbon dioxide production of tissues, leading to a fall in the oxygen partial pressure and a rise in the carbon dioxide content of the venous blood. This also increases the pressure gradients of the respective gases and leads to an increased rate of G.e. b) The *diffusion factor* or *diffusion capacity*, i.e. the rate of diffusion from the alveolar air into the blood, or vice versa for a given pressure gradient (expressed as volume per min along a pressure gradient of 1.0 kPa). The diffusion factor can also vary. It depends on the surface area available for diffusion. In the resting organism, some alveoli are not ventilated and the capillaries of others do not receive circulating blood. The proportion of such alveoli that do not participate in G.e. is decreased by muscular activity, with a corresponding increase in the diffusion factor. In healthy adult humans at rest, the diffusion factor for oxygen is 2–2.7 ml/min/kPa. Since the oxygen partial pressure difference between alveolar air and capillary blood is 3.3–4.7 kPa, the quantity of oxygen diffusing per min from the total lung space into the blood is 375–700 ml. The diffusion factor for carbon dioxide is about 20 times greater, so that the carbon dioxide diffuses much more rapidly. c) The contact time, during which the circulating blood is accessible to the alveolar air. Usually the blood flow rate in lung capillaries is such that equalization of partial pressures is possible; for example, the partial pressure difference for oxygen between alveolar air and arterial blood is only 0.7–1.2 kPa.

2) G.e. in the tissues. This also depends entirely on diffusion, governed by the same physical laws that apply in the lungs. The direction of diffusion of the two respiratory gases is the reverse of that in the lungs. In tissue cells, oxygen partial pressure varies between 0 and 4kPa, that of carbon dioxide between 6.7 and 8 kPa. The rate of G.e. depends mainly on two factors. Metabolic activity alters the oxygen and carbon dioxide contents of the tissue, which has the effect of altering the pressure gradients of these gases between the tissue and the blood. On the other hand, dilation of blood vessels causes an increase in the rate of blood flow through the tissue and increases the rate of G.e. by increasing the diffusion factor.

In ***plants,*** G.e. occurs in 1) Photosynthesis (see) (carbon dioxide uptake and oxygen release), in 2) Respiration (see) (oxygen uptake and carbon dioxide release), and 3) in the loss of water vapor in Transpiration (see). Uptake of carbon dioxide and release of water vapor occurs mainly through the stomata, and is therefore largely dependent on the degree of stomatal opening. Oxygen exchange is less dependent on stomatal opening, because it diffuses more readily through the cuticle. G.e. in plants is also influenced considerably by the composition of the plant surface and the system of intercellular air spaces.

Gas-plant dittany: see *Rutaceae.*

Gaster: see Digestive system.

Gasteropelecidae: see *Characiformes.*

Gasterosteidae, *sticklebacks:* a family of small fishes (order *Gasterosteiformes),* characterized by their spiny fins and free-standing spines anterior to the dorsal fin. About 7 species and 5 genera are clearly recognized. As a result of great genetic diversity, there are, however, considerable taxonomic difficulties, and some authorites claim the existence of about 200 species. In many species the sides are covered with bony plates. Males build nests out of vegetation, which they bind with a secretion from their kidneys. Later, the males display brood care.

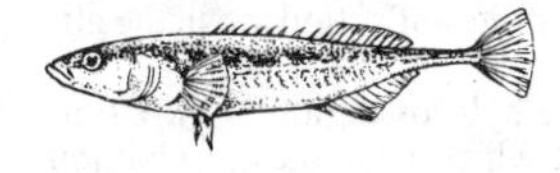

Gasterosteidae. Nine-spined stickleback *(Pungitius pungitius).*

The ***Three-spined stickleback*** *(Gasterosteus aculeatus)* and the ***Nine-spined stickleback*** *(Pungitius pungitius)* occur in marine, brackish and freshwater in the Northern Hemisphere. The Three-spined stickleback has a grayish body with silver sides, but the male becomes brightly colored at spawning time with an orange-red underside with gold opercula. Both the Three- and Nine-spined sticklebacks are popular experimental animals for ethological and physiological studies.

With a length of 18 cm, the ***Fifteen-spined stickleback*** *(Spinachia spinachia)* is the longest member of the family. It is a marine species of the shallows and intertidal zones of the North Sea and Baltic, and it is sometimes kept in cold water aquaria.

Gasterosteus: see *Gasterosteidae,* Pilot-fish.

Gastral filaments: finger-like folds in the gastral cavity of scyphomedusae, which secrete enzymes for external digestion. G. correspond to the mesenteric filaments of anthozoans.

Gastric brooding frog: see *Myobatrachidae.*

Gastric cavity: see Digestive system.

Gastric inhibitory peptide, *GIP:* a polypeptide tissue hormone purified from crude preparations of cholecystokinin. It consists of 43 amino acid residues of known sequence. Produced by the small intestine, GIP exhibits potent enterogastrone activity, i.e. it inhibits secretion of acid and pepsin by the stomach, and inhibits gastric motility.

[J.C. Brown & J.R. Dryburgh, *Canad J. Biochem.* ***49*** (1971) 867–872]

Gastricsin: a proteinase present in human gastric juice, which is responsible for milk coagulation. Its mode of action is the same as that of Rennin (see).

Gastrin: a hexadecapeptide hormone from the gastric antrum. The C-terminal sequence is identical with that of pancreozymin. Human G.I is Pyr-Gly-Pro-Trp-Leu-$[Glu]_5$-Ala-Gly-Tyr-Trp-Met-Asp-Phe-NH_2. In human G.II there is a sulfate group on Tyr 12. In porcine G., Leu 5 is replaced by Met. In sheep G., Leu 5 is replaced by Val. In bovine G., Leu 5 is replaced by Val, and Glu 10 by Ala.

Produced in the G-cells of the gastric antrum, G. enters the circulation and stimulates the gastric mucosa to synthesize and secrete hydrochloric acid. G. production is regulated by the pH of the gastric contents, and by the hormones, secretin and gastric inhibitory peptide. Overproduction of G. leads to excess acid in the stomach, which in turn causes stomach ulcers.

Gastrioceras (Greek *gaster* stomach, *keras* horn): an index fossil goniatite *(Cephalopoda)* genus of the Upper Carboniferous of Europe (Westfalian), Asia, North Africa and North America. The shell is planispiral, and laterally compressed to rotund in cross section, with an open umbilicus of variable width. The umbilical shoulder or edge has well developed tubercles. The suture is simple, with a characteristic outer saddle, and similar to that of other goniatites (see Ammonites).

Gastrocnemius muscle: see Musculature.

Gastrocoel: see Digestive system.

Gastrointestinal hormones: tissue hormones whose sites of secretion and sites of action are in the alimentary canal (Fig.).

In mammals, cells in the pylorus region of the stomach secrete *Gastrin* (see), which promotes production of hydrochloric acid in the cardiac region. Acid production and stomach motility are inhibited by *enterogastrone,* which is released into the circulation by the wall of the duodenum. The following G.h. are produced in the intestinal mucosa. *Pancreozymin (coleocystokinin)* (see) is produced in response to the acidic mixture of partially digested food (chyme) emerging from the stomach; it promotes contraction of the gall bladder and secretion of pancreatic juice. *Secretin* (see) stimulates hydrogencarbonate production in the exocrine pancreas, bile production in the liver, and the secretion of large quantities of pancreatic juice of low enzyme content. *Duocrinin* stimulates Brunner's glands to produce enterokinase, amylase, lipase and a proteinase. *Enterocrinin* stimulates the secretion of mucus and digestive enzymes by the ileum. *Villikinin* increases the movement of the intestinal villi.

G.h. are secreted in response to specific stimuli. Stretching of the stomach wall by food intake initiates gastrin secretion. Gastrin promotes the production of hydrochloric acid, and the acid in turn promotes the secretion of secretin and duocrinin. Fat in the intestine inhibits the secretion of enterogastrone, whereas the degradation products of protein accelerate the release of enterocrinin.

The properties described for enterogastrone are largely those of *Gastric inhibitory peptide* (see).

Some other G.h. produced by the intestine include: 1) *motilin,* a polypeptide of known structure (22 amino acid residues), which promotes movement of the stomach and intestine; 2) *bulbogastrone,* a hypothetical hormone whose production is stimulated by gastric secretion, and which inhibits gastrin-stimulated acid secretion in the stomach, but has no effect on pancreatric secretion; 3) *Vasoactive intestinal polypeptide (VIP),* which relaxes the smooth muscle of the stomach and gall bladder, as well as causing vasodilation and increasing cardiac output; it is produced in large quantities by tumors of the pancreas (possible also by other tumors) and is probably the mediator of the syndrome of chronic watery diarrhea (Verner-Morrison syndrome); 4) *Enteroglucagon,* which is formed in the mucosa of the stomach and intestine, and is thought to stimulate insulin secretion after glucose intake; 5) *serotonin,* which in addition to playing a role in the central nervous system, is also formed in the intestinal mucosa and stimulates intestinal peristalsis.

Gastrointestinal hormones. Sites of synthesis and functions of gastrointestinal hormones

Gastroliths: hard structures, consisting mainly of calcium carbonate, in the stomach wall of lobsters and crayfish.

Gastroma: see Digestive system.

Gastromycetales: see *Basidiomycetes.*

Gastropoda, ***gastropods:*** a class of the *Mollusca* (subphylum *Conchifera),* containing about 110 000 species, representing a wide variety of different forms. *G.* have evolved different modes of locomotion and feeding, and they have invaded almost every type of habitat, displaying a diversity and evolutionary adaptability similar to that of insects and vertebrates. Much detailed information relevant to the *G.* is presented under the entry, *Mullusca,* which should also be consulted.

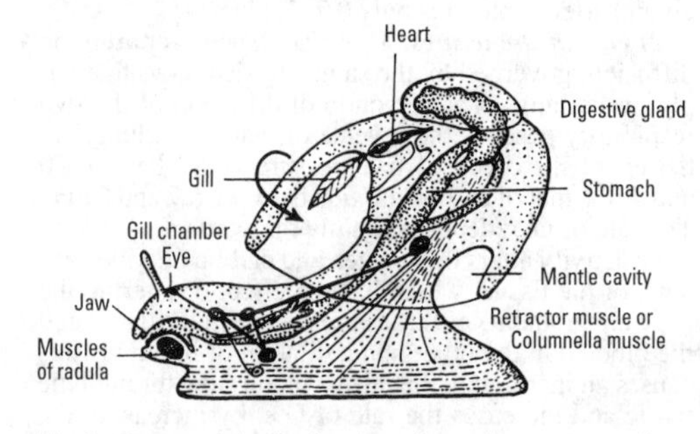

Fig.1. *Gastropoda.* Body plan of a pulmonate snail.

Morphology. There is a distinct head with tentacles, eyes and a mouth, an elongated or disk-shaped, flattened foot, and a visceral hump; the latter, which is enveloped by the mantle, is often coiled, and it exhibits various degrees of torsion. In most species, a multilayered, usually spiral, calcareous shell (consisting of conchiolin and aragonite, and secreted by the mantle) covers the mantle and visceral hump. This shell may also be flattened, reduced or absent (see Slugs). The internal organs within the visceral sac are spirally twisted, following the inner contors of the spiral shell. The most variable character of the *G.* is the often splendidly colored and sculptured shell, sometimes ornamented with spikes, ridges or ribs. In the *Streptoneura,* the shell orifice is usually closed with a stiff operculum.

Biology. *G.* are unisexual *(Streptoneura)* or hermaphrodite *(Opisthobranchiata* and *Pulmonata).* In hermaphrodite forms, eggs and sperms are produced in the same gland. Fertilization is unilateral or reciprocal, or self-fertilization may occur. With few exceptions, the *G.* lay eggs. In marine forms, development occurs via a free-swimming larva (see Veliger larva), whereas development is direct in terrestrial and freshwater forms. There are various sites of respiratory gas exchange, including gills, secondary gills, a vascular network in the mantle cavity ("lung"), and the skin. The majority of species feed on plants, while some are predatory and a few are parasitic. Food is taken by means of the radula, which can be used for many different sources and types of food material. The number of radula teeth is very variable, and their arrangement and development are taxonomically important (see *Mollusca). G.* glide over their substrate by wave-like movements of the foot, usually aided by the secretion of mucilage from an anterior foot gland. Some aquatic forms can swim (e.g. the ***Sea hare,*** *Aplysia),* while others lead a pelagic existence (e.g. ***Sea butterflies,*** *Pteropoda).*

Freshwater and marine forms of *G.* are found from the banks or shore to great depths. Many species are adapted to a terrestrial existence, not only in damp habitats, but also in dry areas.

Economic importance. *G.* have been used variously as currency (cowrie shells), articles of daily use such as shell containers, as sources of dyes (see Tyrian purple), and as food. Shells are still used for ornamentation, but *G.* are now important mainly as a food source, various snails being eaten in different parts of the world (e.g. the ***Giant African land snail,*** *Achatina achatina,* eaten from Guinea to Nigeria; the shell may be 20 cm in length and almost 10 cm wide). Many *G.* are important pests of crops and stored food products (see Slugs). Others act as intermediate hosts of various helminth infections of vertebrates (e.g. Bilharzia).

Classification. Three subclasses are recognized.

1\. Subclass: *Streptoneura (Prosobranchia).* About 60 000 unisexual species. The pallial complex exhibits torsion, so that the ctenidia lie anterior to the heart, and the visceral connectives are crossed. This phenomenon, due to 180 ° torsion, in which the mantle complex with anus, gills and genitalia is moved forward, the visceral connectives become crossed, and the original right-hand organs and nerve centers come to lie on the left hand side, is known as *streptoneury* or *chiastoneury*. As a rule, the shells of the *Streptoneura* are very robust and and often strikingly colored, with bizarre shapes ranging from basins to spirals, and the orifice is usually closed by an operculum. With the exception of a few freshwater families, all the *Streptoneura* are marine, and many species are predators or carrion eaters.

1\. Order: *Diotocardia (Archeogastropoda, Aspidobranchia).* The order contains about 6 000 species, with shells varying from 0.1 to 25 cm in length. The earliest known members belong to the extinct superfamily, *Bellerophontacea* (see *Bellerophon).* They have 2 auricles, 2 nephridia and 2 ctenidia. The ctenidia are usually bipectinate (i.e. with 2 rows of leaflets). The inner layer of the shell is often covered with mother of pearl. Further division of the order is based on the characters of the radula. Thus the radula of the suborder, *Rhipidoglossa,* consists of rows of numerous narrow teeth, which diverge like the ribs of a fan, e.g. *Haliotis, Fissurella* ***(Keyhole limpet).*** The radula of the suborder, *Docoglossa,* consists of rows of a few very long, strong teeth, used for browsing algae on stones, e.g. *Acmaea, Patella* ***(Limpet).***

2\. Order: *Monotocardia (Pectinibranchia).* The order contains about 53 000 species, with shells varying in length from 0.1 to 60 cm. There is only a single auricle, ctenidium and kidney. The ctenidium is fused to the roof of the mantle cavity, and has only one row of leaflets. Mother-of-pearl is always absent from the inner surface of the shell. The order contains parasitic, free-swimming and pelagic sea snails, as well as terrestrial species, some with delicate shells, others with thick, strong shells. The *Monotocardia* are divided into 2 suborders:

1\. Suborder: *Mesogastropoda* (radula is taenioglossate, i.e. a transverse row of 7 teeth). Some members of this suborder are as follows:

Heteropoda (Pterotrachea) are naked or very thinly shelled sea snails, in which the foot (propodium) is modified as a verticle fin. They inhabit the open sea, swimming with the ventral surface uppermost, and some attain a considerable size (e.g. genus *Carinaria,* up to 53 cm in length). They are often voracious predators, feeding on worms, jellyfish, crustaceans and small fish.

Cypraea **(Cowrie snails, Porcelain snails)** are sea snails with glossy, splendidly colored shells, with a slit-shaped, toothed orifice. The shells of *C. moneta* and *C. annulus* serve as currency in parts of Africa and the Pacific islands. As amulets or necklaces, or worked into various ornaments, cowrie shells have been used since antiquity. ***Panther cowries*** *(C. pantherina)* were once carried as amulets against infertility and venereal disease. Other well known decorative cowries are the ***Tiger cowrie*** *(C. tigris)* and the ***Golden cowrie*** *(C. aurantium).*

Crepidula fornicata **(Slipper shell),** native to the east coast of North America, and now introduced into Europe, is often found among oyster catches. It has become a significant pest of oyster beds, where it competes with the oysters for food. They are protandric hermaphrodites, changing their sex from male to hermaphrodite to female during their lifetime. Motile, sexually mature animals are always male, but these soon become sessile by attaching themselves to the top of a growing chain or stack, in which each member is firmly attached to the next: males (destined to become hermaphrodites) at the top , females (no further sex change) at the bottom and hermaphrodites (destined to become females) inbetween. This form of protandry is known as *consecutive hermaphroditism.* The right shell margins are adjacent, so that the penis of the upper individual can reach the female gonophore of the animal below, and females may also be fertilized by wandering males.

Tun shells (genus *Tonna)* are large (up to 40 cm in height), barrel-shaped organisms found at moderate depths in the mud flats of the Mediterranean. They possess a long, mobile proboscis with a terminal suction cup, and 2 large buccopharyngeal glands, in which they produce a secretion containing 2–4% sulfuric acid. Two grooved hooks (modified jaw plates) are used to inject the acid secretion into the prey, which is then ripped apart with the radula.

Species of the genus, *Littorina* ***(Periwinkle),*** have thick, chubby, rounded shells, with a smooth or spirally striped outer surface, and a porcelain-like inner surface. The genus exhibits a tendency toward a terrestrial habit; thus *L. rudis* lives almost at the high water mark and spends more time in air than in water. In some species the filaments of the ctenidium are extended over the roof of the mantle cavity to form a vascular network. The ***Rough periwinkle*** *(L. saxatilis)* is ovoviparous, and displays brood care. *Littorina* displays a unique form of locomotion, in which the 2 halves of the foot (which is divided by a deep longitudinal groove) are alternately raised and pushed forward. Numerous fossil *Littorina* are found from the Eocene (Tertiary) to the present, in the shore zones of all past and present seas. Thus, *L. littorea* is typical of the deposits of the Littorina sea, which existed about 6 000 years ago as a post-Ice Age developmental stage of the Baltic. Since the last Ice Age, the Baltic basin has undergone the following changes: ***Baltic ice lake*** [containing a succession of small, ice-dammed lakes along the southern part of the present baltic basin] → ***Yoldia sea*** [lasted about 1000 years; characterized by the deposits of *Yoldia arctica,* later renamed *Portlandia arctica,* class *Bivalvia,* order *Nuculoida]* → ***Ancyclus lake*** [crustal uplift closed the connection with the sea across southern Sweden, forming a lake, characterized by deposits of *Ancyclus fluviatilis,* class *Gastropoda,* subclass *Pulmonata]* → ***Littorina sea*** [existed about 6000 years ago, characterized by deposits of *Littorina littorea,* class *Gastropoda,* subclass *Prosobranchia,* order *Mesogastropoda]* → ***Lymnaea sea*** [the water became brackish, and the dominant gastropod genus was *Lymnaea,* subclass *Pulmonata].*

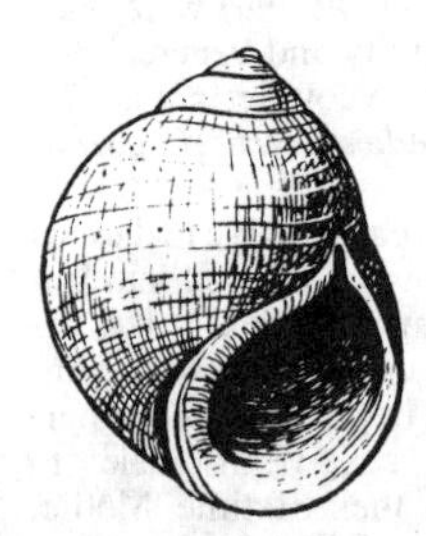

Fig.2. *Gastropoda. Littorina littorea L.* from the postglacial period (natural size).

Some species of ***Conchs*** (family *Melongenidae)* are the largest shelled gastropods. *Syrinx aruanus* from northern Australia measures up to 60 cm from the siphonal canal to the apex, and is exceeded in size only by *Aplysia* (Sea hare).

2. Suborder: *Neogastropoda* (radula is rachiglossate, i.e. a transverse row of 3 teeth). A familiar member of this suborder is the carnivorous *Buccinium* ***(Whelk),*** which feeds on living and dead animals. Its mouth is situated at the end of a proboscis, which can be retracted within a proboscis sheath. The radula is used to rasp flesh, but it can also bore holes in the carapace of crustacea. It lives between the low water mark and about 180 m, and its egg capsules form a familiar sponge-like mass, often washed ashore by the tide.

Also included in the *Neogastropoda* are the ***Drills*** *(Murex, Murosalpinx, Eupleura, Purpura* and *Thais),* the ***Olives*** *(Oliva and Olivella),* the ***Miter shells*** *(Vexillum and Mitra)* and the ***Harp shells*** *(Harpa).* The predatory ***Cone shells*** and ***Turrids*** (family *Conidae)* and the high-spired ***Auger shells*** *(Terebridae)* possess a poison gland, and the radula is modified into grooved or hollow, stiletto-like teeth, which serve for the injection of venom; as a group or superfamily they are therefore known as the *Toxoglossa.*

2. Subclass: *Opisthobranchia.* This subclass contains about 13 000 hermaphrodite species. Opisthobranchs have undergone a reversal of torsion (detorsion), so that the mantle cavity (if present) tends to occupy a position posterior to the pericardium. There are close affinities to the *Streptoneura* (e.g. the family, *Acteonidae,* has a streptoneural nervous system, and the mantle cavity occupies an anterior position), and to the *Pulmonata* (euthyneural nervous system and hermaphroditism). There is a wide variety of forms, ranging from delicate, often bizarre planktonic species, to large plump animals like the Sea hare. The majority have a slug-like habit, i.e. the shell is very thin, internal or totally absent. The ctenidium is usually regressed and replaced by accessory respiratory organs, located around the anus and on the ventral surface, and often spendidly colored.

Apart from a few species that live in brackish and in fresh water, all opisthobranchs are marine. They feed on algae and sponges (often matching these in shape and color), hydroids and sea anemones, other small animals, and on detritus. Some species can swim, while other live pelagically. Massive swarms of planktonic forms are an important food source for fish and whales. Eight orders are recognized, although the taxonomy still contains some uncertainties: *Cephalaspidea, Thecosomata* (shell pteropods), *Anaspidea* (sea hares), *Gymnosomata* (naked pteropods), *Sacoglossa, Acochlidiacea, Notaspidea* and *Nudibranchia* (naked sea slugs) .

The ***Sea butterflies*** have members in both the *Thecosomata* and *Gymnosomata,* and are often grouped together as the *Pteropoda.* They are pelagic gastropods of the open seas, occurring down to depths of 2 000 m. Only a few centimeters in size, they are practically transparent, possessing a transparent, usually uncoiled, vase-shaped shell. The foot is represented by 2 ciliated fins (epipodia or parapodia), which project from the aperture, serving to sift small food organisms and propelling the animal through the water by slow flapping movements like the wings of a butterfly (see Plankton, Fig.). Examples are *Cavolinia* (uncoiled shell) and *Limacina* (coiled shell). Fossils of *Pteropoda* are known from the Cretaceous to the present. Earlier finds from the Paleozoic and the Lower Mesozoic are doubtful. Large accumulations of the shells of dead pteropods often occur at depths between 1 000 and 27 000 m on the sea floor, where they form a *pteropod mud.*

The *Juliidae* are an interesting family of the order, *Sacoglossa,* whose adult members have 2 shell valves, superficially resembling members of the *Bivalvia;* the muscles that close the valves are also similar to those of the *Bivalvia.* The larvae of these ***Bivalved gastropods*** possess a normal coiled shell with an operculum. A valve then begins to develop on the right hand side, the original shell becoming less symmetrical and forming the left valve. Occasionally, the coiled embryonal shell can still be seen in the spire of the left valve. Until 1959 only the valves had been described. The first living specimen *(Berthelinia limax)* was discovered by the Japanese scientists, S. Kawaguti and D. Baba. Only about 12 species are known from Madagascar, Japan, Australia, the Caribbean, and the Pacific coast of North America from California to Peru.

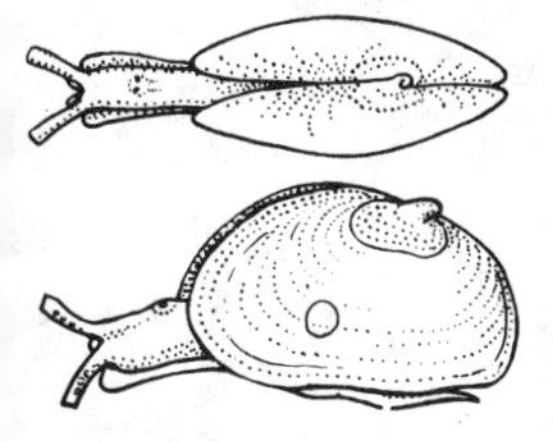

Fig.3. *Gastropoda. Berthelinia limax.*

The olive-green-colored adult **Sea hare** *(Aplysia)* is found between the tides, crawling over green seaweeds, which serve as its food. Young animals are found in deeper water, and they are colored red, matching the red algae on which they feed. On the head are 2 pairs of tentacles, the anterior large and ear-like (hence the name "hare"), while the second pair is olfactory. There are 2 simple eyes, one at the base of each olfactory tentacle. Two flaps (parapodia) arise from the sides of the foot in the posterior region, and these are used for swimming. Glands in the mantle cavity secrete a purple pigment, which the animal ejects when it is molested.

Sea slugs (nudibranchs) are the largest (with respect to the number of species) order of the opisthobranchs. They lack a shell and mantle cavity, and the body is secondarily bilaterally symmetrical. Ctenidia are reduced or replaced by secondary gills. Many species are brightly and beautifully colored as an adaptation to their surroundings, and the body surface frequently has bizarre projections (cerata). They progress like other snails, and some can swim or drift with the aid of lateral parapodia. Many species possess skin glands that secrete sulfuric acid. Their diet consists of coelenterates, bryozoans and tunicates, and in some species the nematocysts from ingested prey are transported to cnidosacs at the distal tips of the cerata, where they serve for defense.

3. Subclass: *Pulmonata.* This subclass comprises about 35 000 recent species. All are hermapohrodite, fertilization being reciprocal or unilateral. As a rule, development from the egg is direct without a free larval stage. Gills are absent and are functionally replaced by a vascular network ("lung") in the roof of the mantle cavity. Except for the respiratory orifice (pneumostome), the mantle cavity is closed to the exterior. Only a few freshwater species have secondary gills. A single auricle and a single nephridium are present. Most forms have a shell [Slugs (see) are an exception], which is spirally twisted, and an operculum is nearly always absent *(Amphibola,* a marine member of the *Basomattophora,* is the only operculate pulmonate). Most species are terrestrial, while others live in fresh, brackish, or salt water. They feed predominantly on plants. Three orders are recognized:

1. Order: *Systellommatophora.* These are slugs with an anus at the posterior end of the body. The intertidal *Onchidiidae* possess a posterior pulmonary sac, which is absent in the tropical *Veronicellidae.*

2. Order: *Basomattophora.* Members of this order are mainly freshwater aquatics, with a few marine forms. They have one pair of nonretractable tentacles, and an eye near each tentacle base. The *Lymnaeidae* are a worldwide distributed freshwater family with thin, pointed shells. In some species, the orifice is greatly extended by water movement, particularly in the surge at the water's edge. Lymnaeids are fairly numerous in most inland waters, where they are important intermediate hosts of parasitic trematodes. Members of the genus, *Lymnaea,* have thin, turret-like shells, which spiral to the right, the final turn of the spiral (housing the aperture) being particularly large. *Lymnaea stagnalis* ***(Great pond snail)*** has a relatively hard shell, which varies in shape depending on the habitat, e.g. short and strong in fast water currents, or indented around the aperture in still, aquatic vegetation. Fossil *Lymnaea* spp. are found from the Malm to the present, and were very numerous in the Tertiary. In the Quaternary, the genus gave rise to the deposits of the Lymnaea sea (see above). The lymnaeid ***Dwarf pond snail*** *(Galba truncatula),* found in ditches, meadow ponds and wet grass in central Europe, is the intermediate host of the liver fluke of sheep *(Fasciola hepatica).* The freshwater *Planorbidae* have posthorn-shaped shells (spirally rolled in one plane), and the orifice of the shell is rounded and very wide. Planorbids are important intermediate hosts for developmental stages of the helminths. The most familiar genus is *Planorbis* ***(Posthorn snail).*** Fossil planorbids are found from the Malm to the present, and they are stratographically important in the Miocene.

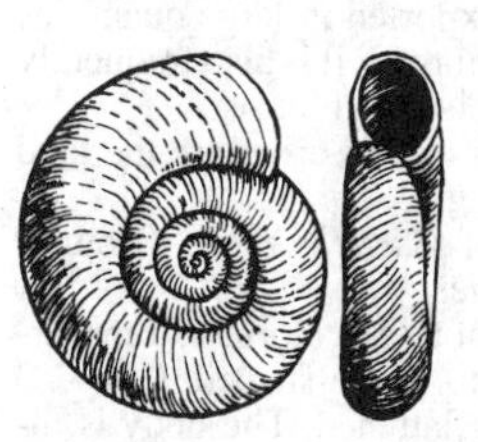

Fig.4. *Gastropoda.* Shell of *Planorbis* (family *Planorbidae)* from the Upper Tertiary (x 3.5).

3. Order: *Stylommatophora* ***(Land snails).*** There are 2 pairs of retractable tentacles, the upper pair with eyes at their tips. Almost all species are terrestrial. However, they have a moist skin and produce a copious mucilaginous secretion, and are only fully active in wet or damp environments. By withdrawing into their shells they can survive long periods of drought, and this ability has enabled them to colonize areas of low rainfall. Familiar European representatives of the family, *Succineidae,* are the amber-colored amber snails, e.g. the ***Amber snail,*** *Succinia putris.* Living mostly on plants in marshes and at the edges of ponds and stream, they can survive long periods of submersion, but they are not amphibious. Amber snails are the intermediate hosts of *Urogonimus macrostomus (Distomum macrostomum),* a trematode parasite of birds. The tubular, green and

red-ringed sporocyte generation of this trematode (known as *Leucochloridium)* migrates into one of the snail's tentacles during the day (preventing the snail from withdrawing this tentacle), where it pulsates (40–70 times per min) and attracts worm-eating birds.

Members of the family *Viviparidae* ***(River snails)*** have a short wide foot, and the head tapers to a club-like snout. The tentacles are relatively long; in males, the thicker right tentacle is used as an organ of copulation. Females retain the fertilized eggs and give birth to young. *Viviparus contectus (= V. fasciatus)* is widely distributed, especially in the lowlands of Central Europe, where it lives in the mud at the bottom of stagnant pools and ditches containing abundant vegetation. The 6–6.5 tumid whorls of the brownish green, semitransparent shell have a very deep suture between them. *V. contectus* can withstand dry conditions and long periods of freezing in ice, during which the aperture is sealed with the stiff, ringed operculum. Fossil species of *Viviparus (= Paludina)* occur from the Lower Carboniferous onward, in brackish and fresh water, and are particularly abundant in the Tertiary, e.g. in the Paludine Pliocene strata of the Balkans and in the Pleistocene. *Paludina diluviana* is an index fossil of interglacial deposits.

The *Helicidae* is a large pulmonate family whose members are widely encountered in undergrowth, open woodland, ditches, edges of pasture, gardens and agricultural land. The shell is variously patterned with dark spiral bands. They include the European ***Grove snail*** *(Cepaea nemoralis),* which has now been introduced into eastern USA, the ***Garden snail*** *(Cepaea hortensis),* also found in Europe and New England, and the ***Common snail*** *(Helix aspersa),* which has been introduced into many countries worldwide. The largest European land snail, *Helix pomatia* ***(Roman snail, Edible snail,*** or ***"Escargot")*** is also a member of this family. It was formerly a vineyard pest in southern Europe, but has become more widely distributed by humans. In certain European countries, especially France, it is an important food, and is therefore exported in large quantities from Central and Southeast Europe. It is also frequently grown commercially on special snail farms.

Other important members of this order are the land slugs (see separate entry: Slugs).

Gastrotricha: a class of *Aschelminthes* (Roundworms) with about 200 species.

Morphology. The bodies of these colorless, transparent animals are cylindrical to bottle-shaped, 0.06–2.0 mm in length and ventrally flattened. The body is divided into head, neck and trunk, and the posterior end carries two prong-like projections or tail forks (caudal furcae). Most *G.* have adhesive glands (cement glands, adhesive tubes, adhesive papillae), which secrete an adhesive fluid and enable the animal to become attached to the substratum. In some species these adhesive glands are numerous (up to 250) and organized in groups or rows over the whole body surface; or they may be present only on the tail forks. The anterior or head end, as well as the whole ventral surface, is covered with numerous cilia. The anterior end also carries tactile hairs. There is no true body cavity (coelom); the weakly developed pseudocoelom lacks an epithelium, and is simply a space between the ectodermis and endodermis.

Biology. *G.* are aquatic organisms living on the bottom, especially in vegetation zones. The feed on unicellular organisms, such as algae, flagellates, ciliates, bacteria, etc, and on decaying organic matter (detritus), which are taken into the mouth by pharyngeal suction. Bands of cilia on the ventral surface enable the animal to glide over the substratum. Some species progress by leech-like movements, alternately becoming attached to the substratum by their anterior and posterior adhesive glands. The extremely long anterior cilia of some species enable them to swim for short distances.

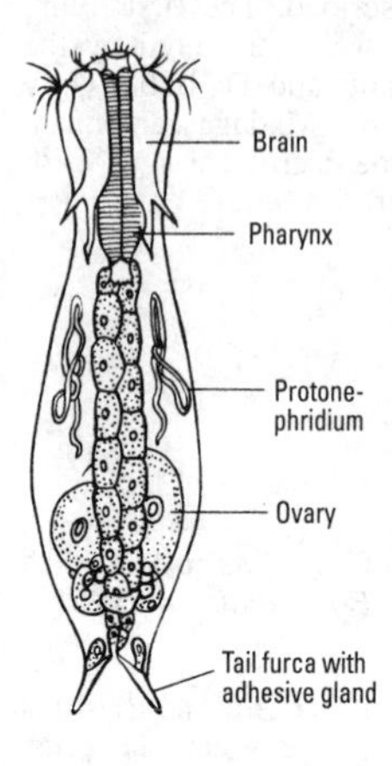

Gastrotricha. Chaetonotus, order *Chaetonotoidea.*

Classification.

1. Order: *Macrodasyoidea.* These are all hermaphrodite with one or two testes or ovaries. They are found only in coastal sea water, mainly in the bottom sand, and more rarely on submerged plants. Dissolved metabolic waste is voided through a "ventral gland", which appears to be homologous with the excretory organ of nematodes.

2. Order: *Chaetonotoidea.* With the exceptions of the genera, *Neodasys* and *Xenotrichula,* members of this order lack male organs and they reproduce parthenogenetically. Paired protonephridia serve as excretory organs. Both marine and freshwater forms are known, many species being found worldwide. Adhesive glands are present only on the tail forks. In the families *Neogosseidae* and *Dasydytidae,* adhesive glands are completely absent. These latter two families also have long, movable spines, which enable them to float and become part of the freshwater plankton. Some freshwater species are also found in wet, mossy areas.

Gastrovascular system: see Digestive system.

Gastroverms: see *Monoplacophora.*

Gastrozooid: see Polyp.

Gastrulation: the first phase of germ layer formation of the embryo. G. starts with the appearance of new and variously localized areas of vigorous cell division in the blastula (see Cleavage). By migration and relocation of cells, the blastula is then converted into a gastrula, consisting of 2 primary germ layers, ectoderm and endoderm. The course of G. varies considerably, depending on the type of blastula and its yolk content. The following types of G. are recognized.

1. *Invagination* or *emboly.* The blastoderm of the coeloblastula invaginates at the vegetal pole, so that it protrudes into the blastocoel. The latter either persists and becomes part of the primary coelom (e.g. in the *Echinodermata),* or it becomes totally obliterated by the invagination process (e.g. in the *Branchiostoma).* The invaginated cell layer becomes the inner germ layer or

Endoderm (see), while the cell material not participating in this process remains on the exterior, forming the outer germ layer or Ectoderm (see). Endoderm and ectoderm meet around the rim of the blastopore or prostoma, and the rim itself forms the blastopore lips. In the *Prostomia,* the blastopore persists and becomes a definitive mouth, whereas in the remaining *Bilateria* the blastopore closes, and a new secondary mouth is formed later as an independent perforation of the ectoderm (e.g. in the *Deuterostomia).*

2. *Epiboly.* This is a process of overgrowth, in which vigorously dividing micromeres spread over the more slowly dividing macromeres, and finally cover them completely (e.g. in many mollusks and annelids, Fig.1). Archenteron formation is prevented by the relatively large macromeres. A gut cavity is formed later by separation of the endoderm cells, which by then have consumed their yolk and consequently have decreased to normal size.

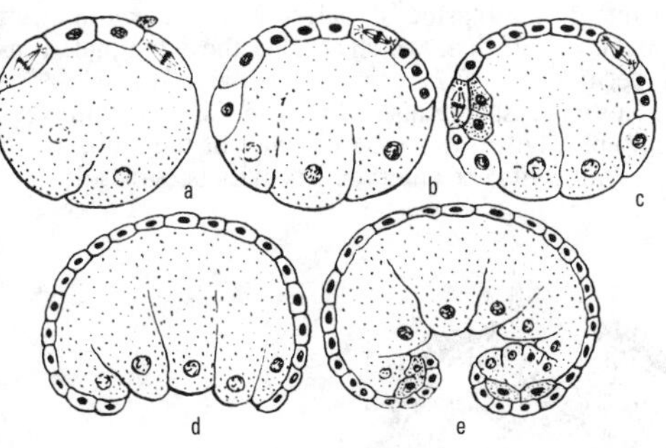

Fig.1. *Gastrulation. a* to *e* Endoderm formation by epiboly in the marine gastropod, *Crepidula* (slipper shell); the epibolic gastrula is formed by proliferation of micromeres around the macromeres.

3. *Immigration.* In this mode of G., the blastoderm cells migrate into the blastocoel, either exclusively at the vegetal pole (unipolar immigration), or at multiple sites in the ciliated coeloblastula (multiple immigration) (e.g. in the *Hydrozoa,* Fig.2). The irregularly shaped mass of blastoderm cells eventually fills the blastocoel, but a central split soon develops, followed by consolidation of the blastoderm cells into an epithelial endoderm.

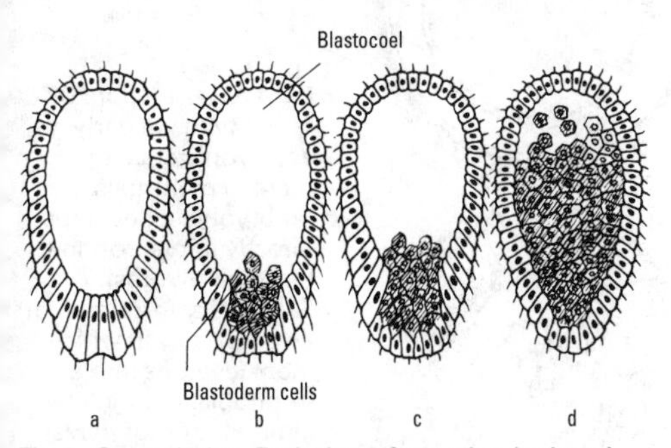

Fig.2. *Gastrulation.* Endoderm formation by immigration in the larva of the hydromedusa, *Aequora. a* Coeloblastula. *b* to *d* Migration of blastoderm cells into the blastocoel.

4. *Delamination.* This term is applied to the formation of endoderm by tangential divisions of the blastoderm cells of the coeloblastula, e.g. in some *Hydrozoa* (Fig.3.) and in the *Scyphozoa.*

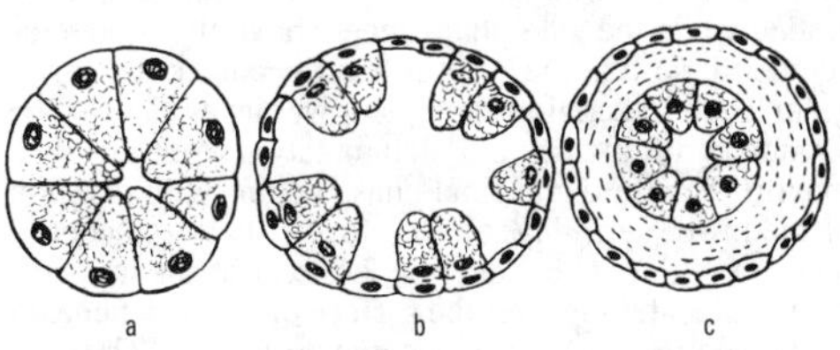

Fig.3. *Gastrulation. a* to *c* Endoderm formation by delamination in the hydromedusa, *Geryonia.*

The posterior rim of the embryonic disk of sharks and rays rolls over and tucks under to form a small pouch, creating a structure that is closely related to the invagination gastrual of a holoblastic egg. The pouch is homologous with the archenteron; its cell lining is called primary endoderm or protoendoderm, but it makes only a small contribution to subsequent gut formation. The main mass of gut cells arises from the secondary endoderm or deuteroendoderm, which is formed by proliferation from the rim of the embryonic disk into the subgerminal cavity, at first irregularly, later becoming organized into an epithelium. G. in reptiles closely resembles the process in fishes. In reptiles, a primitive plate develops from the posterior rim of the 2–3-layered blastodisk by cell movements, and this primitive plate invaginates along a transverse cleft to form a pouch-shaped structure. The pouch corresponds to the archenteron, the cleft represents the blastopore, and the cellular wall of the structure is primary endoderm (protoendoderm). Here again, most of the later gut is formed from secondary endoderm, which arises from the lowest cell layer of the discoblastula.

The type of invagination described above for fish and reptiles G. does not occur in the extreme telolecithal eggs of birds. In birds, a blastodisk is present at the animal pole on the surface of the yolk, where it forms the area vitellina. Cleavage of the blastodisk produces a cap of cells (blastoderm), which separates from the yolk, forming a subgerminal cavity. From above, by transmitted light, the blastodisk then shows 2 distinct areas. A central translucent area (area pellucida) lies above the subgerminal cavity, while the darker zone surrounding it (area opaca) lies above the opaque yolk. The blastodisk is now a 2-layered structure, and this stage is reached at the start of incubation after the egg has been laid. In the region of the area pellucida, the endoderm plate is formed by delamination from the lowest blastoderm layer, and probably complemented by cells from the posterior rim of the germ plate.

Endoderm formation in insects is rather complicated, and it is part of a multiphasic G. process. In the first phase, the blastoderm of primitive hemimetabolous insects (e.g. grasshoppers and crickets) separates very early into large-celled, extraembryonic endoderm and paired bands of presumptive embryonic ectoderm (embryonic primordia), consisting of numerous small cells. These primordia soon unite ventrally into a shield- or heart-shaped germ plate lying on the yolk, with 2 head lobes (cephalic ectoderm) at the anterior end. The second G. phase commences when the germ plate becomes differentiated longitudinally into 3 bands, the central one along the junction of the 2 earlier embryonic primordia. This central band then forms a primitive groove by emboly or epiboly, and the second phase is concluded when the side plates meet above the groove as ectoderm, so that the central band becomes the lower germ layer. In certain cases, the lower layer can also be formed by immigration or delamination from the blastoderm. The third and final phase begins with division of the lower germ layer into 3 further bands: a central band and 2 lateral bands. The 2 lateral bands immediately glide laterally over the surface of the yolk beneath the ectoderm, where they become mesoderm. The central internal band remains in position, and at each end it possesses a group of cells that are evolutionary rudiments of a former endoderm. These 2 endodermal regions grow toward each other from each end, and enclose the yolk in a sac of definitive or secondary endoderm. Later, this sac opens at each end and becomes continuous with the stomodeum (foregut) and protodeum (hindgut), forming the gut epithelium.

All placental mammals (with the exception of humans) show similar types of G. (Fig.4) up to blastocyst formation, with a unicellular trophoblast wall and an embryonic knot (see Cleavage). Subsequent endoderm delamination from the basal embryonic knot cells, and progressive lining of the inner trophoblast wall with endoderm to form a 2-layered blastocyst wall, are also practically indentical in different placental mammals. Subsequent stages differ according to the animal, but many common features are shared.

In carnivores and rabbits, the outer layer of trophoblast cells lying immediately above the embryonal knob (Rauber's cells) disintegrates, and the cells are incorporated as ectoderm into the epithelial layer of the now opened blastocyst; a layer of endoderm also forms beneath the embryonal knob. The latter (having flattened by spreading, it is more appropriately called the embryonal disk) is now 2-layered, and exposed at the surface of the blastocyst; it is the sole origin for subsequent embryonic development.

G. in ungulates and lemurs differs from that in carnivores and rabbits, only in that the embryonal knob first develops a cavity, thus becoming an embryocyst. The latter splits open at its tip and is incorporated into the disintegrating blastocyst wall to form the ectoderm of the embryonal disk, which overlies the delaminated endoderm.

Embryocysts are also formed in the G. of monkeys and apes, but in this case the wall of the embryocyst does not split. The embryocyst cavity becomes the am-

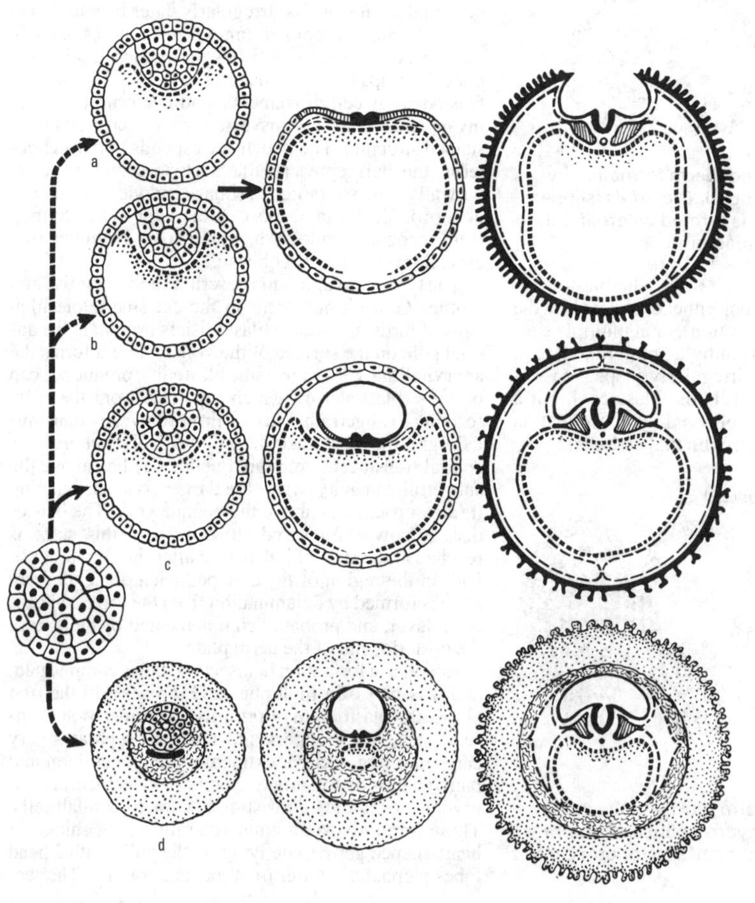

Fig.4. *Gastrulation.* Different types of early embryonic development in mammals. *a* Embryoblast develops directly into a trophoblast (carnivores). *b* Temporary formation of an embryocyst, which then develops into a trophoblast (ungulates, lemurs). *c* Embryocyst and primary amniotic cavity (insectivores, primates). *d* Like *c,* but with primary yolk sac.

niotic cavity (see Extraembryonic membranes) and the floor of the blastocyst becomes the ectoderm of the embryonal disk, which overlies the endoderm.

Human development (see Yolk sac) shows further differences. Organization of the embryo into trophoblast and embryoblast is complete at time of implantation in the uterus (6 days after egg release). The outermost trophoblast functions by transferring nutrients. The embryoblast (embryonal knob or inner cell mass) eventually develops into the embryo. In contrast to rabbits and carnivores, the layer of trophoblast cells above the embryonal cell mass remains intact and does not disintegrate. A netlike mesoderm of trophoblast cells fixes the trophoblast in the uterus wall, and produces enzymes that degrade the uterine mucosa to provide nutrients for the embryo (embryotrophism), until the fetal placenta becomes functional (see Placenta). On the side of the embryoblast that protrudes into the blastocyst cavity, a thickened layer of endoderm cells (hypoblast)

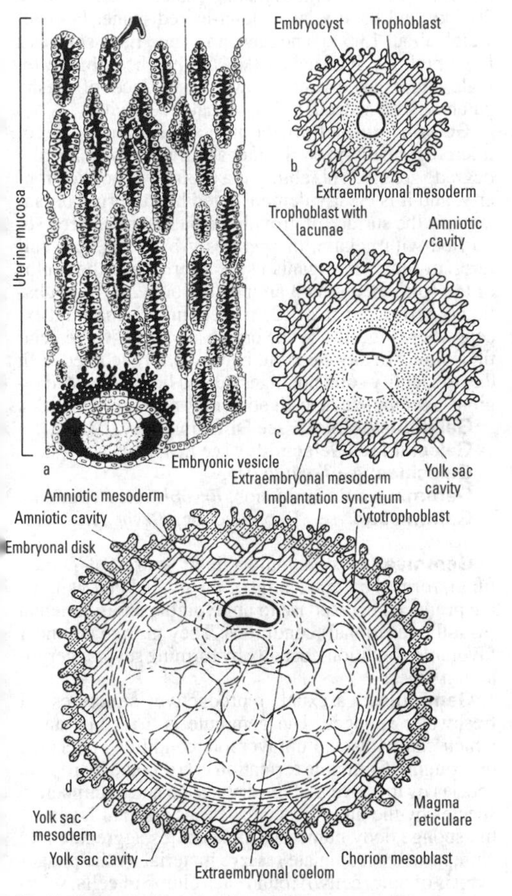

Fig.5. *Gastrulation.* Sections of human embryos between the ages of 6 and 18 days. *a* Implantation of the blastocyst in the uterine mucosa. *b* At 12-13 days: further growth of the trophoblast; loosening of the extraembryonic mesoderm; amniotic cavity formed; embryonal disk (blastodisk) separates into ectoderm and endoderm. *c* 15 days. *d* 17–18 days.

is formed. The amniotic cavity arises by cavitation of the embryoblast. Its roof (amniotic endoderm) is a fairly thin epithelium, whereas its floor is a thick plate of cells arranged as a layer of columnar ectodermal epithelium. This plate and its underlying layer of hypoblast (endoderm) constitutes the bilaminar blastodisk (embryonal disk), from which the embryo proper subsequently develops. Extraembryonic mesenchyme rapidly increases until it fills the blastocyst cavity. Beneath the embryonal disk, a primary yolk sac is formed by a mesodermal membrane (Heuser's membrane). This is subsequently replaced by endodermal cells from the hypoblast, forming a secondary yolk sac which displaces the primary yolk sac as a small vesicle. Meanwhile, one or more new cavities arise in the extraembryonic mesenchyme to the right and left of the rudimentary embryo, then fuse to form the extraembryonic coelom, lined with a mesothelium of cells of the extraembryonic mesenchyme. The resulting extraembryonic coelom (exocoelom) is the largest cavity in the blastocyst, occupying most of the space within the trophoblast.

Gate: adjustable entrance to an ion channel. At rest, the channel is closed by a *gating protein.* Within the membrane, the strength of the electrical field is monitored by a sensor, which controls the extent to which the gating protein extends into and blocks the ion channel. See Sodium channel.

Gated channel: see Gate, Sodium channel.

Gatherers: see Modes of nutrition.

Gattendorfia (Gattendorf is a village in Bavaria): a goniatite genus of the Lower Carboniferous (Gattendorfia stage). The shell is a planispiral coil, usually a thick disk, and smooth. The wavy suture line lacks the characteristic outer saddle that is present in most of the Carboniferous ammonites.

Gaultheria: see *Ericaceae.*

Gaur: see *Bovinae.*

Gause's hypothesis: see Competition-exclusion principle.

Gaussian distribution: see Biostatistics, Normal distribution.

Gavial: see *Gavialidae.*

Gavialidae, *gharials, gavials:* a family of the *Crocodylia* (see) containing a single, exclusively fish-eating species. The snout is slender and greatly elongated, and specially adapted for pulling fish from rock fissures and for sweeping the water for fish. In males the tip of the snout is shaped like a pot. All teeth are the same size and shape, and no teeth are visible when the mouth is closed. The dorsal and lateral plates of the body have dermal ossifications. The male ***Gharial*** or ***Gavial*** *(Gavialis gangeticus)* attains a length of 5–5.5 m (maximum 6.6 m), while the female is smaller (3–4 m). An inhabitant of large rivers in northern and eastern India, Nepal, Pakistan and Bangladesh, it is threatened with extinction, and is a protected species.

Gavialis gangeticus: see *Gavialidae.*

Gaviiformes, *divers:* an order of birds (see *Avies)* containing only 4 living species in a single genus *(Gavia),* with no close affinities to any other avian order. Seven other species are are known as fossils, the earliest in the Eocene. Restricted to the Northern Hemisphere, they breed on inland waters in the northern parts of Eurasia and America. They are , however, essentially marine birds, migrating south in the winter to the ice-free inshore waters of the temperate region. Also known as ***Loons,*** they are relatively large (66–95 cm), torpedo-shaped. long-bodied, thick-necked birds with straight, sharply pointed bills, laterally compressed tarsi, webbed feet, relatively short, pointed wings (but they are strong fliers), and a short but well developed tail. The legs are set far back on the body so that divers progress clumsily on land. Divers dive from the surface, and they swim powerfully and quickly under water, sometimes at considerable depths, in pursuit of fish. They are the only birds whose legs are encased within the body down to the ankle joint, and they are among the very few birds whose bones are solid rather than pneumatic. Adult birds are mostly patterned with sharply contrasting areas of black and white. They feed mainly on fish, but aquatic insects, frogs, crustaceans, mollusks, etc. are also taken.

Gavia: see *Gaviiformes.*

Gayal: see *Bovinae.*

Gazella: see *Antilopinae.*

Gazelles: a varied group of small, dainty antelopes, herds of which inhabit the steppes and deserts of Africa and Asia. Their horns are usually long and pointed, twisted or lyre-shaped. In the strict sense, G. are members of the tribe *Gazellini* of the subfamily *Antilopinae,* in the family *Bovidae* (see). Examples are ***Grant's gazelle*** *(Gazella granti),* ***Thomson's gazelle*** *(Gazella thomsoni),* ***Mongolian goitered gazelle*** *(Gazella subgutturosa hilleriana),* ***Tibetan gazelle*** *(Procapra picticaudata).*

Gazellini: see *Antilopinae.*

Gean: see *Rosaceae.*

Geastrum: see *Basidiomycetes (Gasteromycetales).*

Geckos: see *Gekkonidae.*

Gedinnian stage: see Devonian.

Geese: see *Anseriformes.*

Geitogamy: see Pollination.

Gekko: see *Gekkonidae,* Tokay gecko.

Gekkonidae, *geckos:* a phylogenetically old family of the *Lacertilia* (see), with representatives in all tropical and subtropical regions of the world. They are insectivorous, have a broad, flat head, and their minute, tubercular scales are sparsely distributed on a soft skin. With the exception of *Phelsuma* (see), they are inconspicuously colored, nocturnal animals with a vertical, slit-shaped pupil. There are no movable eyelids, and the original eyelids are fused and transparent, forming a spectacle. With few exceptions, geckos have adherent pads, consisting of transverse lamellae, on the undersides of their toes, which enable them to climb on the vertical surfaces of walls and to walk on ceilings. It is now known that these pads do not adhere by suction or by means of adhesive secretions, but by millions of minute hooks on the edges of the lamellae. The tail is often flat and wide, and serves as a storage organ; it is also capable of autotomy, and in some species it is prehensile. In contrast to other lizards, geckos are capable of specific vocalizations. With few exceptions, they lay parchment-shelled eggs (usually 2), which in many cases are attached to the corners of walls or ceilings, or to tree bark by their sticky surface. About 700 species are known (in about 80 genera), and many of these have become associated with humans, being especially common in human habitations in the tropics. By travelling in ships, some of these have been dispersed worldwide in the tropics. Six species are native to southern, southwestern and southeastern Europe. Some examples are the Common gecko (see), *Phyllodactylus* (see), *Hemidactylus* (see), Tokay gecko (see), *Phelsuma* (see).

Gelada, *Theropithecus gelada:* the only species of this genus of Old World monkeys (see). It inhabits grasslands on the high Amhara plateau in northern Ethiopia, and is thought to be a relict of a population that was once distributed throughout Africa. The G. is totally terrestrial, and its habitat is practically devoid of trees. It sleeps on the vertical sides of canyons, clinging to rocks and bushes, and ascends to the plateau to feed during the day. The social group may contain less than 50 individuals or several hundred. While feeding, the G. progresses by shuffling on its bottom, pulling and digging grassses and their roots, which it chews while grubbing out the next handful. Grasses constitute 90–95% of the diet, which is exclusively vegetarian and may also include a small proportion of seeds, fruits and flowers. Gripping and pulling of grasses is facilitated by highly apposable thumbs. The G. is baboon-like and displays sexual dimorphism. Males weigh about 20.5 kg and have a large, long-haired mane. Females weigh about 13.5 kg and have no mane. Each sex has a large pink patch of naked skin on the chest. In the female, this naked area shows a necklace of fleshy patches, which swell during sexual receptivity.

Gelatin liquifaction: the ability of certain species of microorganisms, in particular bacteria, to enzymatically degrade gelatin. This ability is a useful taxonomic property, and it is clearly demonstrated by culturing organisms on the surface of growth media that have been solidified with gelatin, or preferbaly by using stab cultures, in which the organisms are inserted into the depths of the solid medium on an inoculation wire. If incubation is performed at 20 °C, the gelatin remains solid except where digested by the organism. At 37 °C the gelatin is liquid in any case, but may be briefly cooled each day to identify solid nondigested and liquid digested regions. A complete investigation may take up to 2 weeks.

Gel diffusion plate: see Diffusion plate test.

Gelechiidae (Gelechiids): see *Lepidoptera.*

Gelocidae: see *Traguloidea.*

Gemma: see Reproduction, *Bryophytes, Hepaticae.*

Gemma-cup: see Reproduction, *Bryophytes, Hepaticae.*

Gemmae: asexual multicellular spores of certain fungi, formed by thickening of the hyphal cell wall. G. are produced in response to unfavorable environmental (usually nutritional) conditions. They germinate under favorable conditions, usually by forming several germination tubes.

Gemmules: asexual reproductive structures of freshwater sponges. The gemmule is a resting stage, which can withstand unfavorable conditions (freezing or drought). Gemmule formation is promoted in tropical species by the advent of the dry season and in temperate species by the onset of winter. In the internal layers of the sponge body numerous archeocytes aggregate into clumps, and accumulate reserve material by the phagocytosis of other cells. Around each clump of cells, scleroblasts secrete two layers of protective spongin, between which are found amphidiscs or other sclerites. In the spring or with the onset of the rainy season, archeocytes leave the G. and develop into a sponge, either by invading the skeleton of the old sponge or by generating an entirely new sponge. All the differentiated cells of the new sponge develop from archeocytes. See *Porifera.*

Gempylidae: see *Scombroidei.*

Gemsbock: see *Hippotraginae.*

Gene, *cistron:* an individual element of the genome, which regulates a specific metabolic process or process of differentiation, and determines a typical characteristic of the organism. It is a segment of nucleic acid (normally DNA, but in retroviruses it is RNA), consisting of a specific sequence and number of nucleotide residues. It represents a functional unit and carries genetic information for the regulation of a biochemical structure or process. By the processes of Transcription (see) and Translation (see), the information encoded in structural Gs. leads to the synthesis of protein molecules; or transcription alone results in the synthesis of specific RNA (e.g. transfer and ribosomal RNA). Other Gs. are regulatory, serving for the control of transcription and translation. In contrast to the Gs. of prokaryotes, most of the Gs. of eukaryotes are discontinuous, i.e. the linear sequence of information-carrying sections of a eukaryotic gene *(exons)* are interrupted by noncoordinated sections known as *introns.* Transcription of eukarytoic DNA produces pre-mRNA containing complementary sequences of all the introns and exons. The pre-mRNA then undergo *posttranscriptional modification* or *processing,* in which the intron sequences are excised and the exons are spliced to form "mature" RNA.

Gs. encoding the molecular fine structure and specificity of an enzyme were formerly known as structural Gs., as distinct from regulatory Gs. which control the timing of the transcription of the structural genes (see Operon). It is now known that some structural Gs. also encode regulatory products. Experimental demonstration of the existence of a particular G. depends on the occurrence of the G. in at least two alternative forms, known as alleles (shortened form of allelomorph). These alleles are the result of molecular modifications within the G., known as Gene mutations (see). The different alleles of a G. control the same biochemical reaction and therefore the same phenotypic character.

All the Gs. of an organism (see Plasmid) are arranged linearly and associated in linkage groups; thus, all the Gs. of a single chromosome are linked. Nonhomologous chromosomes possess different linkage groups, whereas homologous chromosomes have identical Gs. in identical sequences. In homologous chromosomes, the respective alleles may be the same (homozygous) or different (heterozygous). The distance between two Gs. of the same linkage group is the sum of the distances between all the intervening genes. A schematic representation of the linear arrangement of linked Gs. with their relative distances of separation is known as a *genetic map.* In the construction of such maps the frequency of recombination during crossing over is used as an index of the distance between linked Gs.

Each functional G. or cistron can be subdivided by Recombination (see), and its molecular structure can be modified at numerous sites by mutation. The smallest genetic unit within the G. that can be recognized by recombination is a single nucleotide; similarly, the smallest mutational unit is also a single nucleotide residue (see Recon, Muton).

Since eukaryotic individuals possess two copies of each gene, the genotype of an individual is expressed by a pair of letters or symbols, one for each allele. Dominant and recessive alleles are denoted by the use of higher and lower case letters, respectively (see the entries, Mendel's laws and Mendelian segregation which quote several examples of this usage). To indicate that a gene is functionally normal it is assigned a plus (+) sign, whereas a damaged or mutated, nonfuncional gene is indicated by a minus sign. This usage is particularly common for prokaryotic organisms, e.g. wild-type or standard *Escherichia coli* is capable of synthesizing its own arginine, and is therefore symbolized as arg^+ . Strains that have lost the ability to synthesize arginine by mutation of one of the Gs. involved in arginine synthesis are symbolized as arg^-. Such strains, known as arginine auxotrophs, will not grow without a supplement of arginine. At this level of definition the plus or minus sign refers to an operon rather than a single G. Finer genetic analysis can be used to reveal the mutated gene, or it may show that the entire operon is deleted.

Before the chemical and physical nature of genetic material was discovered, there was strong evidence that Gs. were located on chromosomes. Otherwise the G. was a unit of heredity, defined on the basis of its phenotypic expression, and used as an algebraic symbol to record its distribution and segregation.

Gene activation: the controlled activation of individual genes at specific times during the development of an organism, which is necessary for the orderly and regulated progress of development. See Operon.

Gene activity pattern: the totality of primary and secondary processes that can be ascribed to a single gene, and which are responsible for the development of a pattern of phenotypic characters. The G.a.p. of a gene can be established by comparing the wild phenotype with the phenotype of mutants, in which an allele of the normal gene has been altered, inactivated or eliminated. It is assumed that all differences between the two phenotypes are the direct or indirect result of the mutation. Certain parts of the G.a.p. may not be lost by mutation, i.e. the normal allele and mutated gene may behave similarly with respect to certain aspects of the phenotypic pattern; the extent to which this occurs depends on the mutation in question. The G.a.p. of lethal factors results in harmful phenotypic characters, which decrease the ability of the organisms to survive.

Gene amplification: production of multiple copies of genes at certain sites on the chromosome. G.a. occurs in the chromosomes of different organs of certain insects, and during oogenesis in many organisms, e.g. the selective replication of genes encoding ribosomal RNA in amphibian oocytes. For example, during oocyte maturation in *Xenopus* a set of 500 ribosomal RNA genes is replicated about 4 000 times, producing about 2 million copies.

Gene bank:

1) A gene library, consisting of collection of vectors (see Gene technology) carrying defined and identifiable genes in the form of sequences of cloned DNA.

2) A collection of seeds of plant species and varieties, made with the purpose of preserving their genotypes, and if necessary of making them reavailable for culture in the future.

Gene center: see Center of diversity.

Gene conversion: nonreciprocal behavior in the crossing over between closely adjacent genetic sites; alleles become segregated in a 3:1 and not a 2:2 ratio in meiosis, i.e. one allele is lost and an extra copy of the other appears. In the region of the crossover site, the segregation of genes can be affected by DNA repair enzymes, which may switch templates and copy from the other homologous sequence, or they may randomly remove and replace one of a mismatched base pair.

Gene death, ***gene elimination:*** elimination of a gene from a population, due to its generally unfavorable effects on carrier organisms. The gene may cause death before sexual maturity, or it may reduce fertility in its carriers.

Gene elimination: see Gene death.

Gene expression: the presence of a genetically determined phenotypic character, resulting from transcription and translation of the information in a gene; the temporal and spatial realization of the information encoded in the DNA of a gene. G.e. starts with the transcription of the gene, i.e. synthesis of messenger RNA (mRNA) which has a nucleotide sequence complementary to that of the gene. Translation of this mRNA leads to production of a structurally important or catalytically active protein (i.e. an enzyme), whose presence or activity is the basis of the phenotypic character in question. Expression of a single gene is regulated by other genes in the genome and is influenced by the environment.

Gene flow: movement of genes between populations of a species. It occurs by interbreeding with immigrant species, by the dispersal of pollen, and by the dispersal of seeds, etc. G.f. may occur in only one direction, or it may be an exchange. It decreases the differences between the Gene pools (see) of separated populations. Differentiation of populations by Genetic drift (see) is prevented by the interaction of a few individuals in each new generation. Differences between incompletely isolated populations can therefore only arise from the existence of different selection pressures.

Gene frequency: the frequency of occurrence of an allele in a population, represented by the number of times it is present at a particular locus or loci divided by the number of such loci in the population, and expressed as a proportion or percentage.

Gene fusion: the joining together of molecules of DNA carrying genetic information, using the methods of Gene technology (see). By G.f. it is possible to join prokaryotic to eukaryotic genes, or to join the regulatory units of a gene or operon to a different gene. These artifical combinations of different genes or gene parts may encode the structure of a hybrid protein molecule, i.e. a new gene product.

Gene mutation: a spontaneous or experimentally induced alteration in the molecular structure of a gene, which modifies the encoded genetic information, thereby producing a new allele (see Mutagen, Gene, Genetic code). G.m. therefore differs from Genome mutation (see) and Chromosomal mutation (see). G.ms. are often *point mutations,* arising from the change of a single nucleotide pair of the DNA of the gene. In a point mutation, only a single base pair is affected: in a *transition,* the purine base is replaced by the other purine base (guanine by adenine or vice versa) and the pyrimidine base replaced by the other pyrimidine base (cytosine by thymine or vice versa); in a *transversion,* the purine is replaced by a pyrimidine and the pyrimidine by a purine. In both cases, the number of nucleotide residues remains unchanged, but part of the sequence (affecting a single triplet) is altered. Deletion or insertion of nucleotides within a gene results in a *frameshift mutation.*

In a frameshift mutation, transciption still commences at the same site on the DNA, and proceeds unchanged to the point where nucleotides have been inserted or deleted. Beyond this point, the number and sequence of nucleotides is changed. These changes are preserved during transcription and translation so that the final translation product is altered. For example, the mRNA sequence UUU, CUC, CCA, GAG would by translated into the peptide sequence Phe-Leu-Pro-Glu; deletion of a single nucleotide, e.g. the first U, produces the sequence ...UUC UCC CAG AG..., which would be translated as -Ser-Ser-Gln- (see Genetic code).

A point mutation that changes a codon for one amino acid into a codon for a different amino acid is called a *missense mutation.* If the new codon is a termination codon the change is known as a *nonsense mutation.* Since the genetic code contains only three termination codons, most single base replacements result in missense rather than nonsense. Many codons are synonymous (redundant), so that a point mutation may have no effect on the encoded amino acid; such mutations (e.g. a change from CUU to CUG both of which code for leucine) are known as *silent mutations.* A silent mutation is different from a *neutral mutation,* which does cause an alteration of the translation product, but the alteration does not prevent the product from performing its normal biological function. A protein (e.g. enzyme) arising from a missense mutation contains a single amino acid replacement. This replacement may be crucial (e.g. it may affect the active site of the enzyme), resulting in a complete loss of activity. A completely inactive mutant enzyme is obviously undetectable by its catalytic activity, but it may still cross react with antibodies to the native enzyme. It is then known as a CRIM protein (CRIM is the acronym of cross reacting material). A missense mutation may also lead to the production of a mutant protein, which behaves abnormally, only partially fulfils its biological function, but under certain conditions permits growth and survival of the carrier organism. Many abnormal hemoglobins are in this category. Alternatively, it may affect the temperature sensitivity of the protein, which functions at certain growth temperatures but becomes inactive at higher temperatures. Such *temperature sensitive mutations* are often encountered among bacteria.

A mutation that restores the original nucleotide sequence is known as a Reverse mutation (see). A mutation that restores the function of the gene product by compensating for the effect of the first mutation is known as a Suppressor mutation (see).

The smallest unit of mutation is a single nucleotide residue. Each gene therefore consists of numerous mutational units, and a mutational change in any one of these can lead to a new phenotype (a mutant organism). Like the allele that it replaced, the new mutant allele is stable, and it is replicated and transmitted from cell to cell. It can only be changed by further mutation. G.m. can occur in somatic tissues or gametes. Most G.m. are recessive to the wild-type allele.

All genes and alleles do not display the same frequency of spontaneous mutation or the same susceptibility to mutagenic agents. Some are extremely stable, others display a great readiness to mutate, while other show intermediate behavior. The numer of different alleles arising by mutation from a single gene in one generation is known as the Mutation rate (see). Complementation (see) is used to determine whether two mutations are present in a single gene or in two different genes.

Gene pool: totality of genes in a population of sexually reproducing organisms. Since new gene combinations arise in each generation, all the genes of the population are assumed to belong to a common pool, from which the genetic constitution of each subsequent generation is derived. This concept is a useful but greatly simplified abstraction. The genes on one chromosome are more often inherited together than those on different chromosomes, so that they cannot be considered as entirely independent components of the pool. Furthermore, the tendency for certain alleles to be inherited together is reinforced by the numerous linkage equilibria (see Genetic equilibrium) that are maintained by selection in each population.

Generation:

1) Those individuals of a population originating at the same time from a common parent or parents.

2) In organisms that display an alternation of generations, the developmental phase between one reproductive process and the next. By definition, the organism has a structurally different form in each G.

Generation time: in microbiology, the time between one cell division and the next of a unicellular organism, or the time taken for a culture of unicellular organisms to double its numbers. G.t. depends on the species and the growth conditions. Some bacteria have a very short G.t., e.g. under optimal conditions, the G.t. of certain strains of *Escherichia coli* is about 15 min. Unless special techniques are adopted, the millions of cells in a culture of microorganisms do not grow and divide synchronously, and the plot of cell numbers against time describes a smooth curve (i.e. it does not occur in steps corresponding to each division) (see Figs. in the entries: Population growth and Population ecology). Moreover, the actual G.t. of a microorganism culture is the average of a more or less narrow range of G.ts. displayed by different cells. The G.t. derived from such a curve is therefore also known as the *mean generation time* (MGT).

Generator process for acetic acid production: see Acetic acid fermentation.

Generic hybrids: the offspring from the crossing of individuals from different genera. Examples of G.h. in the plant kingdom are wheat x couch grass, and raddish x cabbage. G.h. are especially numerous among orchids. In the animal kingdom, many G.h. are found among ducks [e.g. *Anas platyrhynchos* (mallard) x *Cairina moschata* (muscovy duck)] and among pheasants [e.g. *Phasianus colchicus* (ring-necked pheasant) x *Chrysolophus pictus* (golden pheasant)].

Genet cats: see Genets.

Gene technology, *recombinant DNA technology, genetic engineering:* combination of methods enabling the isolation of DNA sequences (genes) from prokaryotic and eukaryotic organisms and their multiplication by *cloning,* G.t. uses a sequence of procedures: 1) Preparation of DNA fragments from the isolated DNA of a donor organism using Restriction enzymes (see), or the synthesis of lengths of DNA from nucleotides. 2) Attachment (ligation) of the resulting fragments to the plasmid or viral DNA of a suitable vector (preparation of recombinant DNA molecules). 3) Cloning of the DNA fragment by multiplication of its vector in a host organism. This host organism may be phylogenetically very different from the organism that originally carried the gene, e.g. eukaryotic genes can be cloned in bacteria. 4) Selection of suitable DNA clones.

One important goal of the genetic engineer is the expression of purified genes in commercially exploitable, rapidly growing species. Human insulin, ovalbumin, fibroblast interferon, somatostatin, human growth hormone and many other eukaryotic proteins can be produced by expression of their cloned genes in *Escherichia coli.* Stages in the isolation and cloning of a gene are outlined in Fig. 1, and the strategy for cloning human growth hormone is shown in Fig. 2.

A more recent goal of G.t. is the transfer of cloned genes to restore absent or defective genes (see Gene therapy).

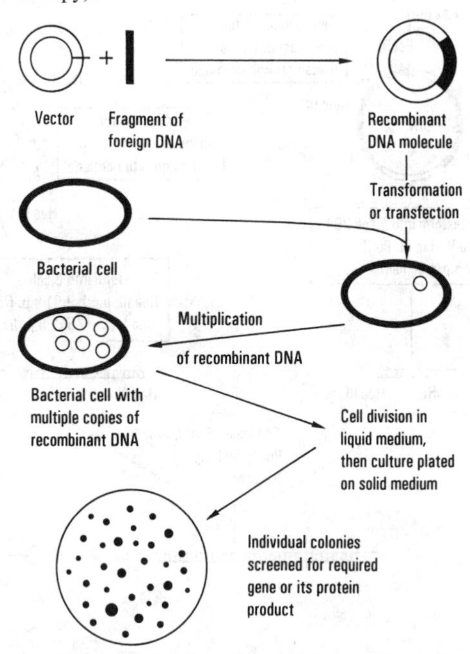

Fig. 1. *Gene technology.* Essential stages in gene cloning using a bacterial plasmid vector.

Gene therapy: see Inborn illnesses.

Genetic analysis: determination of the number and mode of action of all the genes involved in the inheritance of a particular character, by crossing different genotypes.

Genetic balance: the balanced cooperative expression of the individual genes of a Genome (see). Genes do not act independently of other genes in the genome Within the environment of the genome, they display only relative activity, i.e. their intrinsic activity is not expressed fully, and their expression is not necessarily constant or continuous, but determined by interaction with other genes. The composition and maintenance of balanced genotypes are controlled by natural selection.

Genetic block: interruption, by mutation, of a genetically controlled biochemical pathway (see Gene expression). This may be characterized by a failure to produce a metabolite (e.g. the absence of melanin in albinism) or the accumulation of metabolites before the site of the G.b. (e.g. excretion of homogentisic acid in alcap-

Genetic block

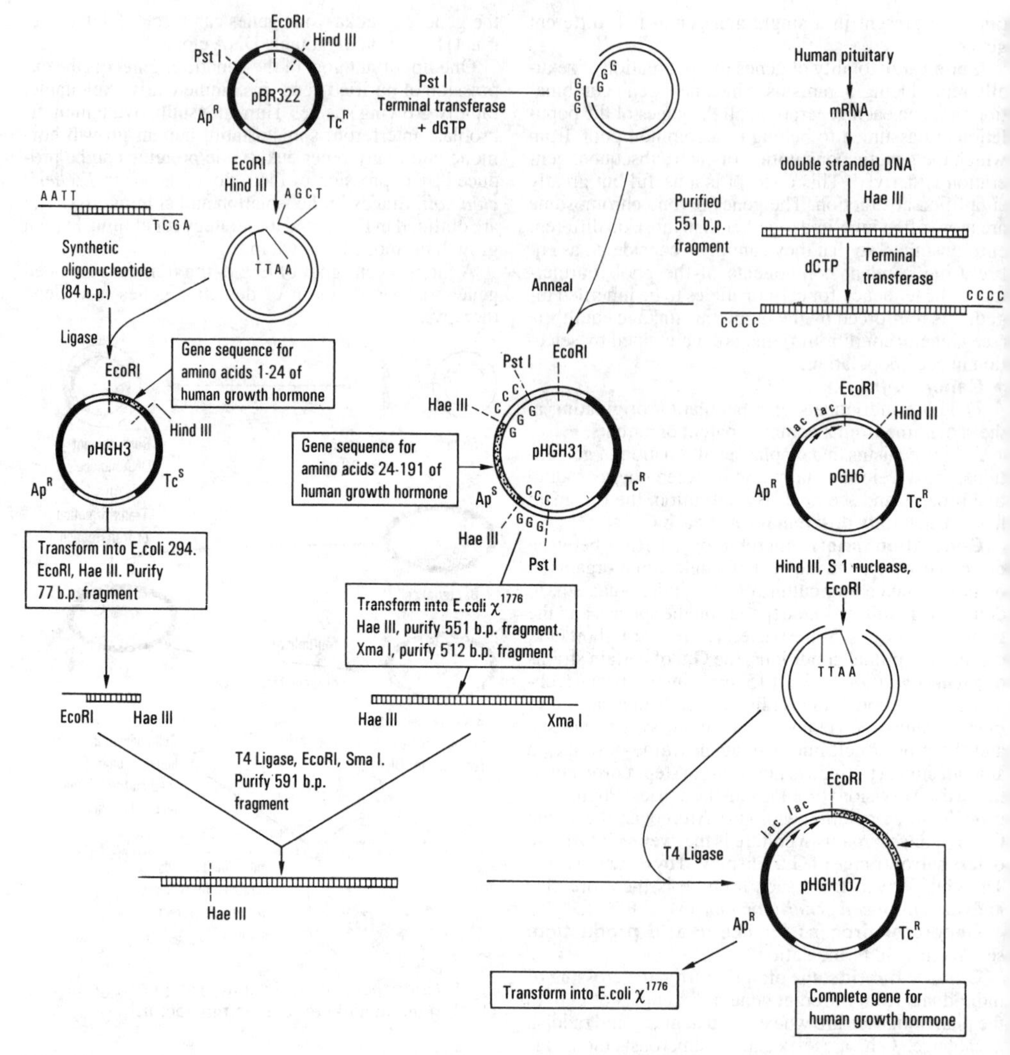

Fig.2. *Gene technology.* Cloning of human growth hormone. Growth hormones are species-specific, and human cadavers were previously the only source of human growth hormone for clinical use, e.g. correction of dwarfism. One liter of a culture of Escherichia coli χ^{1776} carrying pHGH107 plasmids produces about 2.5 mg of human growth hormone; previously about 200 human cadavers would have been needed to prepare this quantity. The protein is not secreted, and it must be isolated and purified from the harvested cell paste. Rigorous purification is necessary to avoid the presence of antigenic bacterial proteins, if the product is to be injected into humans.

Hae III restriction sites are present on the 3' noncoding region of the gene, and in the sequence coding for amino acids 23 and 24. Hae III restriction therefore cleaves the structural gene into 2 fragments, which are cloned separately and later ligated. The smaller cleavage fragment was discarded and replaced by a chemically synthesized oligonucleotide containing an ATG initiation codon. In ligation, restriction enzymes are also used to cleave unwanted dimerization products, e.g in the formation of the 591 b.p. fragment, Eco I cleaves dimers formed from the Eco I sites, and Sma I cleaves Xma I dimers. *Escherichia coli* χ^{1776} is a "safe" host, specially constructed to meet the safety requirements of genetic engineering; it can only survive under specially controlled laboratory conditions, and it cannot colonize, or survive in, the intestinal tract of warm blooded animals. Ap^R and Tc^R represent the genes for resistance to ampicillin and tetracyclin, respectively. If resistance is lost by inactivation of the gene, the same region is denoted by Ap^S or Tc^S (S = sensitive). b.p. = base pair. [D.V. Goeddel et al. *Nature 281* (1979) 544-548] tonuria). The block may be complete or partial, depending on whether the normal function of the gene is completely prevented, decreased or altered.

Genetic code: the code which enables the information encoded in DNA to be transferred to Messenger RNA (mRNA) (see), and then used to establish the correct sequence of amino acids in a polypeptide. The elements of the encoding system are the nucleotide bases, thymine (T), adenine (A), cytosine (C) and guanine (G) in DNA, and the bases A, G, C and uracil (U) in RNA. Thus, RNA contains U in place of C. The nucleotide sequence of DNA acts as a template for synthesis (transcription) of a complementary sequence in RNA. The term, G.c., refers specifically to the sequence of nucleotides in mRNA. Information encoded in these sequences is read in units of 3 nucleotides (triplets), i.e. groups of 3 neighboring nucleotides represent a single letter of the code. Each triplet of the mRNA is called a *codon,* and the corresponding triplet in the DNA is called a *gene triplet.* Each codon corresponds to an amino acid. Since there are 4 bases, there is a theoretical total of 64 (4^3) different triplets. As there are only 20 amino acids, the code is redundant. The G.c. is also degenerate, i.e. several synonymous codons may encode a single amino acid. Translation always starts with the codon AUG and proceeds to a stop (termination) codon. Three different stop codons act as signals for termination of the polypeptide chain. The code has no commas, but spacer sequences exist between cistrons. Normally, the G.c. does not overlap, but in the small bacteriophage, ΦX174, two genes have been found which are entirely contained within other genes. These are translated in different frames, so that the amino acid sequences of the encoded proteins are different. The "reading frame" represents the manner in which a long sequence of bases is divided into codons, e.g. the sequence CATCATCATCAT might be read as CAT/CAT/CAT/CAT, as C/ATC/ATC/ATC/AT, or as CA/TCA/TCA/TCA/T, representing three possible reading frames. The G.c. shown in the table was established for cytoplasmic and bacterial mRNA. It appears to be generally universal, i.e. the same G.c. normally operates in all living organisms, but exceptions to this universality are known. For example, in mitochondrial mRNA, some codons are different from those in the G.c. described above. In human mitochondrial mRNA, AGA and AGG are stop codons. Other differences are as follows.

Codons in mitochondrial mRNA

mRNA	5' CAA	GUC	AUA	CUA	UGA 3'
Yeast mitochondria	Gly	Val	Ile	Thr	Trp
Human mitochondria	Gly	Val	Met	Leu	Trp
Neurospora mitochondria	Gly	Val	Ile	Leu	Trp

The G.c. in ciliates is similar to that in other eukaryotes, except that UAA and UAG code for glutamine, and do not act as stop signals. In ciliates, the only stop codon appears to be UGA.

Genetic counselling: see Human genetic counselling.

Genetic disorders: see Inborn illnesses.

Genetic drift: random fluctuations of gene frequencies in a population. G.d. occurs because each new generation arises from only a random sample of parental gametes, which does not contain a perfectly representative sample of the parental genes. The relative gene frequencies of this sample are therefore different from those of the parent generation. Moreover, certain factors are not transmitted to the new generation. Continuation of this process over many generations can lead to impoverishment of the Gene pool (see).

Second nucleotide of the codon

First nucleotide	Uracil	Cytosine	Adenine	Guanine	Third nucleotide
U	UUU Phenylalanine	UCU Serine	UAU Tyrosine	UGU Cysteine	U
	UUC Phenylalanine	UCC Serine	UAC Tyrosine	UGC Cysteine	C
	UUA Leucine	UCA Serine	UAA Stop	UGA Stop → Trp	A
	UUA Leucine	UCA Serine	UAA Stop	UGG Tryptophan	G
C	CUU Leucine	CCU Proline	CAU Histidine	CGU Arginine	U
	CUC Leucine	CCC Proline	CAC Histidine	CGC Arginine	C
	CUA Leu → Thr	CCA Proline	CAA Glutamine	CGA Arginine	A
	CUG Leucine	CCG Proline	CAG Glutamine	CGG Arginine	G
A	AUU Isoleucine	ACU Threonine	AAU Asparagine	AGU Serine	U
	AUC Isoleucine	ACC Threonine	AAC Asparagine	AGC Serine	C
	AUA Ile → Met	ACA Threonine	AAA Lysine	AGA Arginine	A
	AuG Methionine	ACG Threonine	AAG Lysine	AGG Arginine	G
G	GUU Valine	GCU Alanine	GAU Aspartic acid	GGU Glycine	U
	GUC Valine	GCC Alanine	GAC Aspartic acid	GGC Glycine	C
	GUA Valine	GCA Alanine	GAA Glutamic acid	GGA Clycine	A
	GUG Valine	GCG Alanine	GAG Glutamic acid	GGG Glycine	G

Genetic code. Alternative codons that function in mitochondria are indicated by boxes.

G.d. is most pronounced in small, isolated populations. In numerically large reproductive societies, G.d. is one of the factors that determines whether rare genes are lost, or whether they are retained and become more common in the population. G.d. influences not only the frequencies of selectively neutral genes, but also the frequencies of those that have selective advantages or disadvantages. In extremely small populations, G.d. often overrides strong selection pressures, thereby decreasing population adaptability.

Theoretical distributions of gene frequencies resulting from G.d. have been calculated by Kimura for different starting frequencies and population sizes (Fig.1).

The influence of G.d. on evolution is controversial. According to the hypothesis of Non-Darwinian evolution (see), most evolutionary changes at the molecular level (substitution of nucleotides and amino acids) occur by G.d. and mutation pressure. According to Mayr, G.d. plays an important part in genetic revolution, by allowing the attainment of new fitness peaks (see Adaptive landscape). In large populations, G.d. may cause changes in the gene pool, accompanied by temporary marked decreases in population size, a phenomenon known as the *bottleneck effect.*

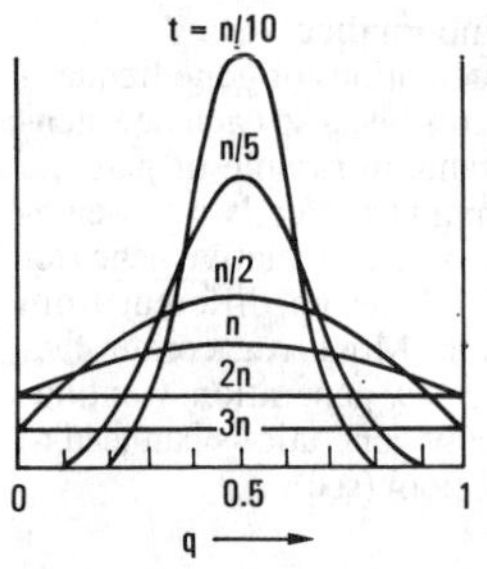

Fig.1. *Genetic drift.* Changes of gene frequency by drift. The starting frequency of allele *A* is q = 0.5. The number of generations is shown as a ratio to population size. With the increasing relative number of generations, the range of possible frequencies displayed by the gene widens, until all frequncies have the same probability. The increasing number of populations that are monomorphic for *A* or for the alternative allele *a* is not shown.

G.d. can be demonstrated in experimental populations (Fig.2). The narrow genetic variability of various natural populations shows that they have at some time passed through a bottleneck generated by G.d. In the 19th century, the population of the sea elephant, *Mirounga angustirostris,* was reduced to only 20 individuals. Electrophoretic investigation of the present population shows that all the animals are genetically impoverished, in that all of the 24 studied protein loci are monomorphic (see Polymorphism).

Genetic engineering: see Gene technology.

Genetic equilibrium, ***linkage eqilibrium:*** random frequencies of possible gametotypes and genotypes in a population, which is polymorphic at two or more gene loci. In G.e. all possible combinations of linked genes are present in the population with equal frequency, i.e. only the frequency of alleles at individual gene loci determines the frequency of occurrence of different gametotypes. Consider 2 gene loci, each with 2 alleles: a_1a_2 and b_1b_2. Gametes a_1b_1, a_2b_2, a_1b_2 and a_2b_1 occur with frequencies of f_{11}, f_{22}, f_{12} and f_{21}. In G.e., $f_{11} \times f_{22} = f_{12} \times f_{21}$. If the gametes and their resulting genotypes display a different frequency, then G.e. has a *relative* value of D, where $D = (f_{11} \times f_{22}) - (f_{12} \times f_{21})$. If certain genotypes are disadvantaged, then G.e. is affected by selection. In addition, genotypes may pair with unequal frequencies, e.g. if a_1b_1 individuals pair preferably with each other, then the population will contain relatively large numbers of a_1b_1 chromosomes. Also, certain genotypes may tend to emigrate or immigrate.

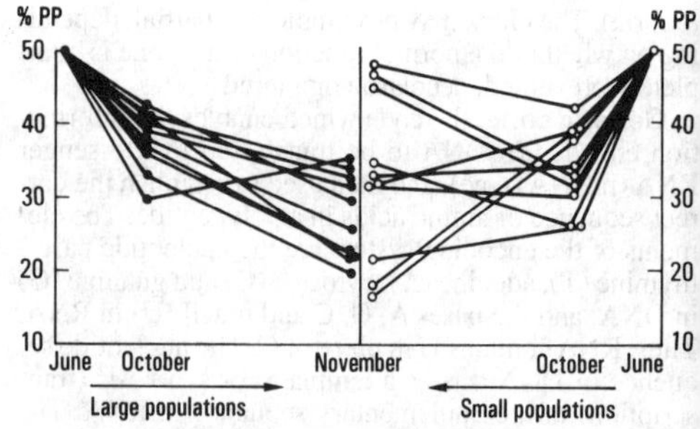

Fig.2. *Genetic drift.* In experimental populations of *Drosophila pseudoobscura,* selection decreases the starting frequency of the PP- chromosome. In the 10 large populations (each starting with 4 000 flies), relatively similar frequencies were attained, whereas in the 10 small populations (each starting with 20 flies), the frequencies show a considerable scatter as a result of drift.

Whereas the Hardy-Weinberg equilibrium is established after only one generation, independently of the starting frequency of genotypes, G.e. is attained much more slowly, especially if the genes in question are close to one another on the chromosome, i.e. if they display linkage. Closely linked alleles only rarely become separated by crossing over during meiosis. The rate of attainment of G.e. between two gene loci on different chromsomes is halved in each successive generation (i.e. G.e. is approached asymptotically), provided it is not disturbed by any of the above-mentioned factors. The rate of attainment of G.e. decreases with the number gene loci involved.

Genetic flexibility: the ability of a genotype or a Mendelian population (see Population) to adapt to changing environmental conditions by realizing part of its potential genetic Variability (see).

Genetic fragment: an incomplete piece or fragment of genetic material or Genome (see). In organisms with true chromosomes, a G.f. may be centric or acentric, according to whether or not it contains the centromere. On account of their unrestricted movement and migration, acentric G.fs. are generally rapidly lost during nuclear division.

Genetic information: information encoded in the nucleotide sequence of the DNA of plants, animals, bacteria and many viruses, and in the RNA of retroviruses, i.e. information encoded in the genome. Utilization of this information to initiate and control the synthesis of a specific molecule of RNA or protein is equivalent to gene expression. A sequence of nucleotides in DNA (a gene or cistron) is transcribed into a complementary sequence of nucleotides in a molecule of messenger RNA. The RNA leaves the nucleus and enters the cytoplasm, where it serves as a template for the synthesis of a specific polypeptide chain (an enzyme or structural protein). See Gene, Genetic code, Transcription, Translation.

Genetic linkage: see Linkage.

Genetic load: the totality of disadvantageous genetic factors in a population. Mathematically, G.l. is expressed as the relative decrease in the average fitness of a population, compared with the fitness of the population if all members possessed the optimal genotype:

$L = (W_{max} - W)/W_{max}$, where L is the G.l., W_{max} the maximal fitness, and W the average fitness of the population. Any population containing individuals of different fitnesses has a G.l., which is made up of factors for genetic diseases and lethal factors, as well as the weakly disadvantageous factors that contribute to normal variability.

In populations of diploid organisms with random pairing, a small proportion of the *total load* is expressed as the *manifested load.* In 1933–1935, Dubinin investigated 2175 wild individuals of *Drosophila melanogaster,* and found 614 abnormal animals, i.e. 2.8%. In 1 million new-born children, 11212 had genetic diseases, i.e. 1.1%.

The *concealed load* can be revealed in animals and plants by inbreeding. Although such experiments cannot be performed on humans, the G.l. of human populations is apparent from the greater frequency of defects in children from consanguineous marriages, compared with the entire population. In *Drosophila melanogaster,* however, the presence of concealed disadvantageous genes can be investigated easily and efficiently. Using Muller's method (see), it has been shown that of 1063 chromosomes in a population of *Drosophila,* 278 (i.e. more than 25%) are lethal or semilethal in the homozygous state. On average, the diploid human genome contains about 4 recessive lethal genes. Since these genes are distributed over all chromosomes, 2 lethal genes are only rarely brought together in a single individual, and they manifest their activity in the population only under exceptional circumstances. With the aid of electrophoretic methods, it can be shown that most natural populations of organisms display high genetic variability in certain proteins, but the significance of these proteins for fitness is unclear.

Some researchers believe that the G.l. is maintained at a constant value primarily by the operation of Mutation-selection balance (see). If this were the case, then the G.l. would be a *mutational load* (i.e. the relative decrease in population fitness would be due to the accumulation of deleterious genes resulting from recurrent mutation). On the other hand, in balanced polymorphism, recessive deleterious genes are not lost from the population, because they are only deleterious in the homozygous state. In contrast, they are beneficial in the heterozygous state, i.e. the heterozygote has an advantage over both types of homozygote, so that selection favors heterozygotic individuals. By segregation (according to Mendel's second law), the offspring from these heterozygotes contain individuals that are homozygous for the deleterious factor. Thus, one can also speak of a *segregation load* (accumulated loss of fitness in a population due to segregation of genes from advantaged heterozygotes to relatively unfit homozygotes). An example of balanced polymorphism is the maintenance of the genes for Sickle cell anemia (see).

Attempts to use the B/A quotient (see) to determine whether the G.l. of natural populations is mainly mutational or segregational have produced no clear answers.

Each developing population carries a *substitutional* or *transitional load* (the cost to the population of replacing one allele by another as part of evolutionary change), because selection occurs by the genetic death of disadvantaged individuals. According to a calculation by Haldane, in order for an advantageous allele with a very low initial frequency to completely displace the disadvantageous allele, a certain number of individuals must be excluded from reproduction, and this number is 30 times the number of individuals in the population. Because of this high substitution load, different disadvantageous genes cannot simply be selected out simultaneously, because the demands placed on the reproductive potential of the population would be too great. The substitution load therefore limits the rate of evolution. Since cases of unusually rapid evolution are known, which according to this calculation are not possible, one speaks of *Haldane's dilemma.*

The concept of load, as well as the calculations of load lead to contradictory results and apparent problems. For example, when an advantageous gene arises for the first time in a population, it generates by definition a G.l., because all individuals of the population not possessing that gene are relatively disadvantaged. According to Wallace, calculations of load are meaningful only when the chances of survival are determined exclusively by the genotype. Usually, however, weak selection dominates, in which fitness is influenced by the genetic background of the particular genotype, and by environmental factors. In weak selection, the calculated load is not an index of the competitve nature of a population.

Genetic manipulation: the specific and deliberate alteration of genetic information in viruses, microorganisms, plants and animals, as a means of achieving desired properties in these organisms. In humans the future use of G.m. is conceivable in the form of Gene therapy. In the long term, Genetic counselling (see), which can be considered a type of G.m., is the most effective means of avoiding hereditary diseases in future generations.

Genetic mosaic: see Chimera.

Genetic locus: the position of a gene on a chromosome. The genetic loci of genes within the same linkage group are normally arranged in the same linear order and are separated by the same relative distances. The term may also be applied to the position of a gene in a linkage map.

Genetic prognosis, ***prognosis of genetic defects:*** determination, for the purposes of Genetic counselling (see), of the probability that genetically determined illnesses and malformations will be manifested in the next generation. Prognosis of monogenic defects can be performed by family anamnesis (analysis of the family tree), and by the detection of microsymptoms of incompletely expressed dominant genes and of recessive genes in heterozygotes (see Heterozygote test). Defects of chromosomal number and morphology can be detected by the cytogenetic analysis of the prospective parents. Prognosis of polygenic defects is performed empirically. Using the largest possible number of affected probands from different families, an estimate made of the frequency of a genetic disorder for different levels of relationship. The resulting *risk coefficient* is used in genetic counselling. An increasing number of genetic defects can now be determined during pregnancy. Such methods of Prenatal diagnosis (see) are almost 100% accurate in the prognosis of selected genetic diseases in the unborn child (e.g. prenatal diagnosis of Down's syndrome; see Chromosomal aberrations).

Genetic revolution: a hypothesis of E. Meyr concerning the nature of genetic processes at the beginning of Speciation (see). According to this hypothesis, if a numerically small population becomes geographically isolated from the main areal of the species, its gene pool undergoes a fundamental change, with a resulting shift toward a new genetic balance.

The isolated population carries only part of the gene complement of the larger main population. After isolation, further alleles are lost at random by Genetic drift (see); in addition, many gene loci become homozygous. In the large main population, alleles are selectively favored which, together with a wide variety of other genes, contribute to a high population fitness. In contrast, in the small isolated population, alleles are selected for their beneficial interaction with a qualitatively and quantitatively different gene complement. These processes, together with new gene combinations and possible changes in environmental conditions, generate pressure for the attainment of a new genetic balance. G.r. places the population at the beginning of a new fitness peak (see Adaptive landscape).

Genetics: the science of the laws and material bases of heredity and organism variability (Luers, 1963). G. deals with the hereditary basis of structural and functional phenotypic characters. It investigates the cellular factors responsible for, and contributing to, the expression and transmission of these characters, and examines the function and behavior of these cellular factors in the hereditary process under different environmental conditions. For these purposes, G. makes use of the analysis of variation within single species, crossing experiments, cell research, genetic analysis, Molecular biology (see) and Gene technology (see). Particularly important contributions are made by Molecular genetics (see). *Analysis of variation* identifies those Modifications (see) that are not genetically based, but caused by environmental influences; and with the aid of statistical methods (see Biometry) it determines the degree of Variability (see) of characters and properties. *Crossing experiments* provide information on the distribution and segregation of genes. *Cell research* investigates the relation between the genetic complement of cells and their structure and function, in particular their chromosomes. By crossing experiments and the construction of hereditary trees, *Genetic analysis* determines the number, mode of action and mode of inheritance of genes involved in the expression of a character or property. G. can be divided into theoretical and applied G. *Theoretical G.* includes Cytogenetics (see), Molecular G. (see), Gene technology (see) and Population G. (see), whereas *applied G.* embraces plant G., animal G., Microbial G. (see), Radiation G. (see) and Human G. (see). Applied G. also includes plant and animal breeding. Gene technology (see) represents the most recent and explosive development of applied G.

Historical. G. did not become an independent discipline until the turn of the 20th century. On the other hand, hypotheses on the nature of heredity had existed since antiquity. According to Stubbe (1953), G. has passed through two periods of development. 1) *The classical period.* This includes the first decade of the 20th century, and it is characterized by the rediscovery of Mendel's laws, the demonstration of chromosomal individuality, gene linkage, crossover and the linear arrangement of genes in chromosomes. In contrast to speculative pre-Mendel hereditary theories, this period established that the material basis of heredity (already known as genes) lay in the chromosomes, i.e. culminated in the *chromosomal theory of heredity.* 2) *The dynamic period.* This period continues to the present and has many areas of overlap with the classical period. It includes studies on extranuclear (cytoplasmic) inheritance, research into the nature of mutation, as well as problems of developmental physiology and evolutionary G. Microbial and molecular G. became important specialized areas of experimental genetic research. But most of this would have been impossible without the proof, in the 1940s, that the genetic material consists of nucleic acids, and the discovery, in the 1950s, of the double helix of DNA. Recognition of the double helix is not only important to G., but is probably the greatest biological advance since Darwin's theory of evolution. Further development of mathematical methods in G. has created the basis for a new specialized area known as population G. With the ever increasing involvement of G. in virus research, biochemistry, biophysics, molecular biology and other areas of biology, it sometimes difficult to clearly separate any of these disciplines from each other. A case might be made for the existence of a third period of development, perhaps the *explosive period,* beginning with the development of Gene technology (see). By definition, gene technology must be considered as part of applied G., but it also clearly illustrates how ill defined the boundaries between disciplines have become, since gene technology is clearly also the province of molecular biology, biochemistry, biophysics, microbiology, etc.

Genetic syndrome: a genetically controlled complex of several to many phenotypic characters, resulting from pleiotropy or aneuploidy.

Genets, ***genet cats,*** *Genetta:* a genus of viverrids (see *Viverridae)* represented by 8–10 species, found mainly in Africa, but with one species in Southwest Europe and the Middle East. They are primarily ground dwellers and do not burrow, and they have sharp, arched, semi-retractile claws on each of their 5 fingers and toes. The coat is darkly spotted.

Genetta: see Genets.

Geneva Convention: see Biological weapons.

Geniculate antennae: see Antennae of insects.

Genin: see Glycosides.

Genital papilla: a conical outgrowth of many crustaceans, which houses the male genital orifice.

Genoelement: see Floral element.

Genome: totality of genes in the haploid chromosome complement of a cell nucleus, or in the DNA of a bacterium.

Genome allopolyploidy: a polyploid condition arising from the hybridization of different species. Since the contributing genomes are different, chromosomes of one genome cannot pair with those of the other uring meiosis I. After polyploidization, the infertile iploid state increases to the fertile tetraploid state by a oubling of chromosome numbers. Each chromosome ien finds a homologous partner, so that Bivalents (see) re formed in the normal way in meiosis I (Fig.). G.a. is ommon in plants but not in animals.

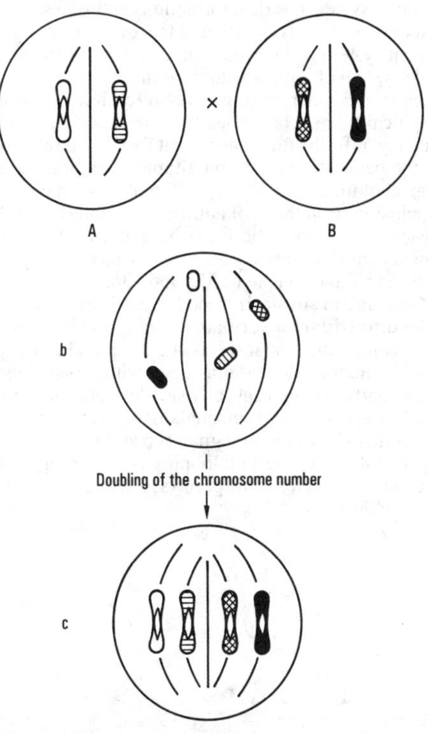

ienome allopolyploidy. *a* Crossing of species A and . *b* A sterile, diploid species hybrid. *c* Spontaneous r induced doubling of the chromosome complenent, and generation of a fertile allopolyploid (alloetraploid) form.

Genome mutation: change in the chromosome umber of a cell or individual, leading to Polyploidy see) or Aneuploidy (see). In polyploidy the number of omplete sets of chromosomes is changed, whereas in neuploidy the normal chromosome number is altered y the loss or addition of individual chromosomes. Two ther types of mutation affecting the cell chromosomes re Gene mutation (see) and Chromosomal mutation see).

Genome separation: a process leading to reduction f the maximal possible number of chromosomes durng mitosis. Depending on the level of ploidy, two or nore entire sets of chromosomes become separated. he frequently observed somatic reversion of polyploid orms to the diploid state is due to G.s.

Genopathy: genetically determined functional or tructural defect, in contrast to an environmentally determined Phenopathy (see). See Genetic diease, Inborn rrors of metabolism.

Genophores: nucleic acid molecules of bacteria prokaryotes) and viruses, which are the functional equivalents of eukaryotic chromosomes. Structurally, they are considerably less complicated than eukaryotic chromosomes. They do not undergo typical meiosis or mitosis, and they are not enclosed in a nuclear membrane. Depending the stage of their growth cycle, bacteria may contain between one and four G. Viruses often contain one, but may sometimes contain up to ten G. Many viral G. and all bacterial G. are circular. The term has not been generally adopted. It is more usual to speak of the bacterial or viral genome, bacterial or viral "chromosome", or simply bacterial or viral DNA.

Genotype: totality of genetic information in the genome of an organism, as opposed to the expression of this information in the Phenotype (see). The G. of well adapted individuals is a combination of genes formed by natural selection, which enables the organism to exploit, tolerate or compensate its environment. See Plasmotype, Plastidotype.

Genotypic milieu: conditions within the cell resulting from the interplay and interaction of all the genes of the genotype. Each gene functions within a G.m. generated by all the remaining genes.

Gentiamarin: see Gentiopicrin.

Gentianaceae, ***gentian*** or ***enzian family:*** a cosmopolitan, mainly temperate family of herbaceous dicotyledonous plants, containing about 1 100 species in 80 genera. Leaves are opposite, entire and exstipulate. Flowers are actinomorphic, 4- to 5-merous, hermaphrodite and insect-pollinated. The ovary is superior and unilocular, with 2 parietal placentas. Each placenta has numerous anatropous ovules, which are sometimes 2-celled due to placental intrusion. The fruit is a capsule, which dehisces along the septa of the ovary. Many species contain bitter glycosides, such as gentiopicrin and amarogentin.

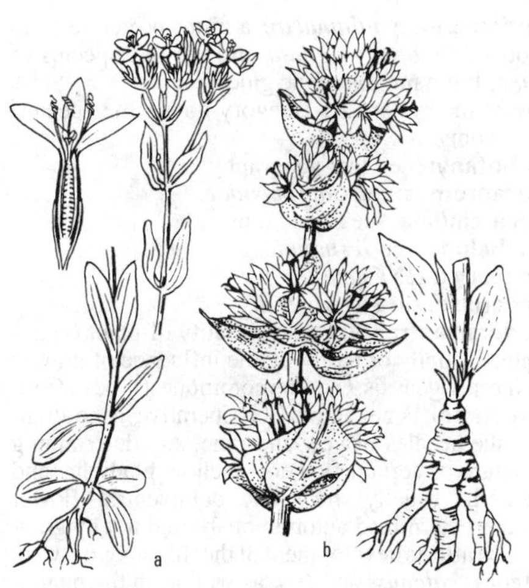

Gentianaceae. a Common centaury *(Centaurium umbellatum* Gilib.). *b* Yellow gentian *(Gentiana lutea).*

The largest genus, *Gentiana,* contains numerous montane species of the northern temperate and cool regions. Many species have beautiful deep blue flowers, others being violet, purple, white or yellow. All European species are protected plants. Enzian liqueur is preparaed from the roots of the ***Yellow gentian (Yellow en-***

zian, Gentiana lutea), which grows in the Alps and Lower Alps of Europe. The blue-flowered *Gentiana septemfida,* which originates from the Far East, is an easily grown ornamental garden plant, suitable for all soils. *Exacum affine* (native to the Old World tropics) has fragrant blue-lilac flowers, and is a popular house plant. The red-flowered *Erythraea centaurium* Pers. *(Centaurium umbellatum* Gilib., ***Common centaury)*** and the Himalayan *Swertia chirata* **(*Chiretta*)** are both official pharmaceutical plants.

Gentianose, ***O-β-D-glucopyranosyl-(1→6)-O-α-D-glucopyranosyl-(1→2)-β-D-fructofuranoside:*** a storage carbohydrate in the roots of various species of gentian. It is a nonreducing trisaccharide consisting of two molecules of D-glucose and one molecule of D-fructose. The fructose is linked to one of the glucose molecules as in sucrose, while the second glucose is linked β-1,6 to the first. It is hydrolysed to sucrose and glucose by the action of gentianase or emulsin, and to gentiobiose and fructose by the action of invertase.

Gentiobiose, ***6-β-glucosidoglucose:*** a reducing disaccharide consisting of two molecules of D-glucopyranose in β-1,6 linkage. It is hydrolysed into Gentianose (see) and fructose by the action of invertase. G. occurs naturally as a component of many glycosides, e.g. amygdalin, and as the ester component of α-crocin (see Crocetin).

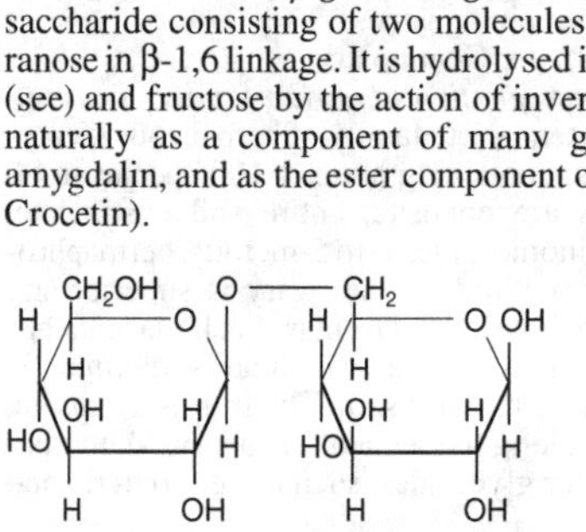

β-Gentiobiose

Gentiopicrin, ***gentiamarin:*** a bitter principle from the roots of *Chlora perfoliata* and various species of *Gentiana.* It is a monoterpene glucoside.

Genus: the systematic category above the species. See Taxonomy.

Geobotany: see Plant geography.

Geocapromys: see *Capromyidae.*

Geocarcinidae: see Land crabs.

Geochelone: see *Testudinidae.*

Geococcyx: see *Cuculiformes.*

Geocorisae: see *Heteroptera.*

Geodorsiventrality: dorsiventrality of plant organs determined wholly or partly by the influence of gravity during morphogenesis. G. is the commonest type of Geomorphosis (see). For example G. is operative in the alignment of the needles of yew and pine, and determining the structure of certain flowers, such as gladiolus and willow-herb (usually, however, dorsiventral flower structure is determined autonomously, and not by gravity). G. also affects development of the rhizomes of sweet flag *(Acorus calanus)* and *Iris,* as well as in the plagiotropic side branches of trees.

Geoduck: see *Bivalvia.*

Geoelement: see Floral element.

Geographical speciation: see Speciation.

Geohelminth: a parasitic worm, whose eggs or larvae pass directly from the earth or similar substrate to the host, without participation of an intermediate host. *Biohelmonths* require an intermediate host, which also mediates transfer to the final host.

Geological time scale, ***geochronology:*** the se quence of events in geological history, which als serves as a calender for dating fossils and evolutiona processes. Geological time is divided into several maj eras, which, in turn, are subdivided into periods, the into epochs. The rocks themselves are properly class fied into systems, series, formations and zones, and th name of a system is the same as that of a period, e.g. D vonian system and period. There is no uniform, worl wide system of nomenclature for the epochs of the Cr taceous and older periods; they often have local ge graphical names, or names that are descriptive of th rock layer. Both animal and plant fossils are referred a time calibration based on animal evolution. Eras plant evolution with slightly different times of onset an conclusion from those of animal evolution (Eo-, Pale Meso- and Cenophytic eras) have been defined (se separate entries) but they are rarely used by Paleontol gists. See Tables on pages 535 and 536.

Geometric similarity: see Isometric growth.

Geometridae: a very large family of the *Lepido tera,* containing about 12 000 species. The imagoe (moths) are mostly medium-sized, with broad, delica wings, and the body is usually slender. Like the pyrali ids, they also possess tympanal organs on the abdome Caterpillars lack feet on segments 6 and 7 (also often o segment 8), resulting in a looping type of progressio The family includes some pests (see Winter moths, Bo dered white).

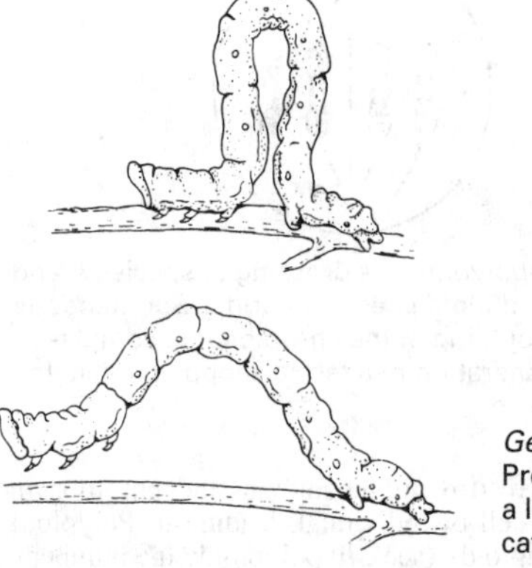

Geometridae. Progress of a looper caterpillar.

Geomorphosis: in plants, the determination of particular shape or structure, or alteration thereof, b the earth's gravity. G. is most commonly seen in the in duction of dorsiventrality (see Geodorsiventrality) an of polarity. The causal mechanism of G. is unknow Under natural conditions, gravity and light usually a in unison in determining the form of plant organs.

Geophilomorpha: see *Chilopoda.*

Geophilus: see *Chilopoda.*

Geophylellidae: see *Symphyla.*

Geosaurus giganteus: see *Thalattosuchia.*

Geospiza: see Galapagos finches, Adaptive radia tion.

Geospizinae: see Galapagos finches.

Geotaxis: taxis in response to gravity. It is dis played, inter alia, by certain bacteria, flagellates an green algae. In *Volvox,* a rhythmic negative G. comper

Eras of zoological evolution	Millions of years since the present	Periods	Epochs
Cenozoic		Quaternary	Recent or Holocene Pleistocene
	1.8–2		
		Tertiary	Pliocene Miocene Oligocene Eocene Paleocene
	65		
Mesozoic		Cretaceous	Upper Cretaceous Lower Cretaceous
	140		
		Jurassic	Upper Jurassic, White Jurassic, Malm. Middle Jurassic, Brown Jurassic, Dogger. Lower Jurassic, Black Jurassic, Lias, Liassic
	195		
		Triassic	Upper Triassic, Keuper. Middle Triassic, Shell limestone, Muschelkalk. Lower Triassic, Variegated sandstone, Bunter, Bundsandstein.
	225		
Paleozoic		Permian	Upper Permian, Zechstein. Middle Permian } Rotliegendes Lower Permian }
	285		
		Carbonifeerous	Upper Carboniferous, Pennsylvanian. Lower Carboniferous, Mississipian.
	350		
		Devonian	Upper Devonian Middle Devonian Lower Devonian
	405		
		Silurian	
	430		
		Ordovician	Upper Ordovician Middle Ordovician Lower Ordovician
	500		
		Cambrian	Upper Cambrian Middle Cambrian Lower Cambrian
	570–600		
Proterozoic		Precambrian	Upper Precambrian
	2000		
	2500		Middle Precambrian
Archean	3000		
			Lower Precambrian
	3700		
	4500		Formation of the Earth

sates the tendency to sink. G. is also known in animals, e.g. honey bees and bumble bees, which possess special gravity organs.

Geotorsion: the twisting of petioles or pedicels under the influence of gravity (see Geotropism), which enables dorsiventral leaves and flowers to return to their normal alignment after being displaced. Spiralling of the ovaries of many orchids is also due to G.

Geotropism: the curvature growth of plant organs in response to gravity. Together with phototropism, G. is one of the most important responses that cause the plant to orientate intself in space. For example, it enables tree trunks and cereal halms to grow vertically on a steep slope. Shoot axes always reorientate themselves vertically when displaced from a vertical position. G. is classified according to the type of response of the plant organ. 1) In *negative G.*, the plant organ grows vertically away from the center of the earth, e.g. most shoot axes. 2) *Positive G.* is displayed by the primary root, which grows directly toward the center of the earth. Positive and negative G. are both types of *orthogeotropism.* 3) *Transverse G.* or *plagiogeotropism* involves more or less horizontal growth, e.g. lateral first order shoots, procumbent and prostrate stems and shoots, many side branches and leaves. Transverse geotropic organs grow at an angle to the vertical that is smaller or greater than 90 °. If growth occurs at exactly 90 ° to the vertical, the response is known as *diageotropism.* 4) *Lateral G.* oc-

Important events during the Precambrian	Millions of years since the present
Appearance of multicellular life; the earliest forms of megascopic algae and invertebrate animals, which characterize the beginning of the Cambrian.	700
Appearance of cropping protists, i. e. unicellular organisms feeding on other organisms. "Cropping" is thought to increase evolutionary diverstíty at the lower end of the food chain, and "croppers" are the evolutionary precursors of herbivores and carnivores.	800
Appearance of meiosis, sexual reproduction and eukaryotic cells. The earliest unequivocal eukaryotic organisms found in the bedded black cherts of the Bitter Springs formation of central Australia, containing putative fungi and several types of unicellular algae.	900
Appearance of mitotic, nucleated cells. Several sites have yielded fossils with eukaryotic affinities, i. e. intracellular structures suggestive of organelles, the oldest being the Bungle Bungle dolomite of northern Australia.	1500
The start of the Upper Precambrian. Atmospheric concentration of oxygen probably greater than 0.1%, and increasing rapidly. Oxygen-producing photosynthetic organisms becoming very widely distributed.	2000
Widespread occurrence of stromatolites. Blue-green beacteria and other bacteria much in evidence. Significant biological production of oxygen, but much of this consumed in oxidation of minerals and volcanic gases. Rate of increase of atmospheric oxygen extremely slow.	2300
Start of the Proterozoic. Fossil record becomes more extensive. Microfossils known from several different sites throughout the world.	2500
Start of the Middle Precambrian. Appearance of photosynthetic prokaryotic organisms.	3000
Occurrence of the oldest konwn stromatolites in the Bulawayan limestone of Zimbabwe. These stromatolites have virtually the same organization as modern stromatolites, which are known to be formed by communities of photosynthetic organisms, and which consist of stacked layers of organic and calcareous material.	3100
Earliest detectable microscopic life, in the form of abundant microfossils in the Swaziland carbonaceous sediments of the eastern Transvaal. These structures closely resemble organic spheroids, and their biotic nature is not universally accepted. The isotopic ($^{13}C/^{12}C$) of the carbon from these sediments indicates the biological fixation of CO_2.	3300
Start of slow organic evolution.	3500
Start of the Archean, marked by the occurrence of the oldest known sedimentary rocks (Isua iron formation of western Greenland).	3700

curs in many twining plants (see Twining): the outward facing side of the shoot (rather than the downward facing surface) is stimulated to grow by the horizontal or oblique position of the shoot axis, thereby causing the shoot to twine around a perpendicular support.

In certain plant organs, the geotropic response may be modulated by internal and/or external factors. Thus, in the poppy plant the tip of the flower stalk is at first positively geotropic, i.e. the young bud hangs downward; later, during flowering, it becomes negatively geotropic, so that the open flower faces upward. In some species the geotropic behavior of the flower stalk changes after fertilization. For example, the flower stalks of the peanut *(Arachis hypogaea)* are at first negatively geotropic, but become positively geotropic after fertilization. As a result, the fruit is pushed into the earth, where it matures. The upper lateral branches of spruce and fir are transversely geotropic, but they become negatively geotropic if the apical shoot is removed (see Apical dominance). Temperature and light are the most important external factors in the modification of G.

Proof that the above orientation movements are induced by the earth's gravity is provided by Pfeffer's *clinostat.* With the aid of this machine, a plant can be revolved (2–20 revolutions per hour) around its long axis. This neutralizes the unilateral effect of gravity without neutralizing gravity itself, and the plant displays no curvature growth. If revolved more rapidly so that centrifugal forces exceed the effects of gravity, the roots bend outward (centrifugally) and the shoots bend inward (centripetally). If the centrifuge speed is such that the centrifugal force is equal to gravity, the roots grow at 45° outward and downward, while the shoots grow at 45° inward and upward, i.e. they grow along the line of the resultant force. This clearly shows that G. occurs in response to the acceleration due to gravity.

Geotropic curvature is caused by different rates of growth on opposite sides of the plant organ (see Nutations). The most reactive zones are therefore the main growth zones immediately behind the root or shoot tip. The nodes of grass halms consist of tissue that remains capable of growth for a relatively long time, so that these can also bend in response to a geotropic stimulus. Despite their woodiness, the trunks and branches of trees can also occasionally display geotropic responses, albeit very slow, due to increased cambial growth. Frequently a geotropic response results in overcurvature, followed by gradual pendulum-like adjustment to the vertical (see Autotropism).

Plants generally exhibit extremely high geotropic sensitivity, as expressed by the usually very short Presentation time (see) of the stimulus. In most cases, the presentation time is about 2 min when the acceleration due to gravity is 9.81 $m.s^{-2}$. The Reaction time (see) is 10–90 min, but it may be several hours if growth processes must first be reactivated (e.g. in the response of a grass halm node). In coleoptiles, susception (the stage of purely physical reception) of the stimulus is the property of the first 3 mm of the coleoptile tip, while in shoots, all zones that exhibit stretching growth are apparently capable of susception. In roots, susception of gravity occurs in the root cap; removal of the root cap renders a root nonresponsive to gravity, until a new root cap is regenerated. According to the *statolith theory* of susception, stimulus-sensitive cells contain relatively heavy particles known as statoliths, e.g. large starch granules (statolith starch) in the root cap, or in the starch sheath of the shoot. According to the orientation of the plant organ, these statoliths exert pressure on different parts of the cell protoplasm, thereby initiating the excitatory stage of the stimulation. When a stem or root is placed in a horizontal position, the starch granules accumulate quickly on the lower sides of the cells. If the starch content is reduced (by keeping the plant in the dark, by low temperature, or by some other means), various plants lose the ability to respond to geotropic stimuli, and this ability is not restored until starch synthesis is renewed. On the other hand, fungi that respond to gravity do not possess starch granules, and a number of other plants contain cells that are capable of susception, but which possess no statolith starch. In such cases, it is possible that a displacement and pressure effect is caused by other (possibly submicroscopic) particles. The unicellular rhizoids of the stonewort *(Chara)* contain highly refractive bodies of unknown composition. If these are displaced to one side of the rhizoid, the Golgi bodies and vesicles on that side are obstructed and are no longer able to migrate to the apical growing region; wall material is therefore delivered mainly to the other side of the tip, and growth is one-sided, i.e. curvature results. This mechanism does not, however, explain the geotropic curvature of multicellular organs.

An alternative theory of susception, known as the *geoelectric effect* or *geoelectric phenomenon,* states that the earth's gravity causes a differential diffusion of cations and anions, so that the underside of a plant organ becomes more positively charged (by a few millivolts) than the upper side. It now appears that the geoelectric effect is in fact an effect of the liquid electrodes used to measure the potential. When vibration electrodes are used, which do not disturb the plant, the geoelectric effect cannot be detected. There are therefore still many uncertainties regarding the mechanism of gravity susception.

Gravity susception is followed directly by an energy-dependent change in cell polarity, known as *transverse polarization,* because the accompanying change in action potential between opposite sides of the cell results in an increased transport of material across the cell. Thus, opposite sides of the cell display differences in sugar concentration, catalase activity and acidity, and in particular in their auxin concentration. It has been proved that transverse transport of auxin in coleoptiles occurs not only at the tip (i.e. at the site of susception), but also over a distance of at least 2 cm extending from the tip. Transverse transport intensifies the basipetal flow of auxin on the underside of the shoot. The underside therefore grows more rapidly, and the shoot becomes directed upward. In addition, the inhibition of auxin transport on the upper side of the shoot increases the auxin gradient across the cell, results in a deficiency of auxin on the upper side, and leads to even more rapid curvature growth. In some cases, e.g. in grass nodes, localized differences in auxin biosynthesis contribute to the formation of the trans-cellular auxin gradient. In roots, the root cap, which acts as the susception site, is unimportant as a site of auxin synthesis. Other mechanisms must therefore operate for the induction and conduction of the gravitational stimulus in roots. On the one hand, it is proposed that the excitation is conducted as an action potential. On the other hand, there is experimental support for the hypothesis that an inhibitor is released or caused to accumulate on the lower side of the root, so that curvature growth is caused by the more rapid growth of the upper, uninhibited side. This inhibition is possibly due to the unequal distribution of a true growth substance, which becomes inhibitory at high concentrations (roots are particularly sensitive to growth substances, and root growth is readily inhibited by concentrations in excess of those required for optimal stimulation). The transverse geotropic orientation of side branches and leaves is the result of the interaction of negative G. with an Epinasty (see).

Gephyrostegus: a genus of the *Gephyrostegoidea,* which serves as a mosaic-type morphological model of an ancestral reptile. It had an amphibian mode of reproduction and a lizard-like appearance. Fossils are found from the Upper Carboniferous to the Permian of America and Central Europe.

Geraniaceae, *geranium family:* a family of dicotyledonous plants containing about 800 species in 11 genera distributed in all parts of the world, but mainly in temperate and subtropical regions. They are herbs or undershrubs (rarely trees). Leaves are usually alternate and usually stipulate. The 5-merous (rarely 4- or 8-merous) flowers are hermaphrodite (very rarely dioecious), actinomorphic (or slightly zygomorphic), and pollinated by insects. The 2-5- (rarely 2- or 8-) celled ovary develops into a lobed capsule, or separates when ripe into 5 single, one-seeded portions. In members of the genera *Geranium* ***(Cranesbill)*** and *Erodium* ***(Storksbill),*** these one-seeded fruits carry a beaklike extension at their upper end, and are also usually equipped with a hygroscopic awn which aids soil penetration by the fruit. *Geranium* species have palmate or palmately lobed leaves, whereas the leaves of *Erodium* species are characteristically pinnate or pinnately lobed. Species and hybrids of the southern and central African genus, *Pelargonium,* are widely planted ornamentals, popularly known as "geraniums"; the flowers are zygomorphic (dorsiventral) with a spur adnate to the pedicel along its entire length. Some *Pelargonium* species yield essential oils which are used in perfumery.

Geraniol, *lemonol, 3,7-dimethyl-2,6-octadien-1-ol,* or ***2,6,-dimethyl-2,6-octadien-8-ol:*** a noncyclic, doubly unsaturated, monoterpene alcohol (an isomer of linalool) with a pleasant smell of roses. It is present free or esterified in many essential oils. Rose oil contains up to 75%, geranium oil up to 50%, and lemon oil up to 40%. Geranylpyrophosphate is the biosynthetic precursor of the tetraterpenes. G. is used in perfumery and food indistries. The butyrate of G. is used for formulating artificial attar of roses, and the formate is a constituent of artifical orange blossom oil.

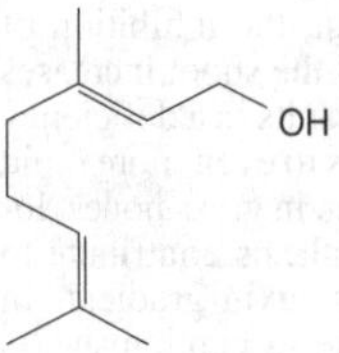

Geraniol

Geranium: see *Geranaceae.*

Gerbera: see *Compositae.*

Gerbillinae, ***gerbils*** and ***jirds:*** a subfamily of the *Cricetidae* (see), containing over 100 species in 16 genera. The smallest species ***(Henley's pigmy gerbil)*** weighs 8–10 g with a head plus body length of 6–7 cm, while the largest species ***(Great gerbil)*** weighs 120–200 g with a length of 15–20 cm. Most species are rat-sized, and they inhabit mainly deserts and steppes in Africa, Asia and the Middle East. They rarely drink, and water balance is maintained by eating food with a high water content, and by concentration of the urine. Most are active diggers and they live in excavated tunnel systems. With the exception of the ***Great gerbil*** (monospecific genus *Rhombomys;* native to Turkestan, China and Mongolia), all are nocturnal. The hindlimbs are large, and these rodents often progress by jumping. In many species the tail is long and serves as a balancing organ. Hindfeet are characteristically elongated, and the middle 3 digits of the hindfeet are longer than the 2 lateral digits.

The genus *Gerbillus* ***(Small gerbils)*** comprises 34 species found in dry or coastal areas in North Africa, the Middle East, Iran, Afghanistan and western India, e.g. the ***Hairy footed gerbil*** *(G, latastei).* Unlike other *Gerbillinae,* ***Wagner's gerbil*** *(G. dasyurus)* has an extraordinary preference for snails. Most jirds belong to the genus *Meriones,* e.g. the ***Clawed jird or Mongolian gerbil*** *(Meriones unguiculatus),* and the ***Indian desert jird*** *(M. hurrianae).* Most of the 10 species of ***Large gerbils*** *(Tatera)* are found in southern and eastern Africa, while the ***Large Indian gerbil*** *(T. indica)* occurs from Sri-Lanka and India to Arabia. The ***Fat-tailed mouse*** or ***Fat-tailed gerbil*** *(Pachyuromis duprasi;* North African countries bordering the Sahara) is the only species of its genus, and the only member of the *Gerbellinae* that stores fat in its tail against periods of food scarcity. Some other members of the subfamily are the ***Bushy-tailed gerbil*** *(Sekeetamys calurus;* eastern Egypt, southern Israel, Jordan, Saudi Arabia), ***Przewalski's gerbil*** or ***jird*** *(Brachiones przewalski;* northern China, Mongolia), ***Namaqua gerbil*** or ***Cape short-toed gerbil*** *(Desmodillus auricularis;* South Africa).

Gerbils: see *Gerbillinae.*

Gerenuk: see *Antilopinae.*

GERL region of the Golgi apparatus: see Golgi apparatus.

Germ:

1) See Embryo.

2) Popular term for a live microorganism, especially a pathogen.

German tinder fungus: see *Basidiomycetes (Poriales).*

Germarium: see Ovary.

Germ band: see Mesoblast.

Germ cells: see Reproduction.

Germ-free animals: see Gnotobiosis.

Germicidal activity: the ability to kill the vegetative stages of microorganisms, displayed by certain chemical compounds and some types of irradiation (e.g. ultraviolet light). Resistent spores may survive.

Germination: the earliest developmental process of plants. In spermatophytes, it represents the resumption of embryo development after its interruption by seed formation. See Seed germination.

Germ line, ***germ cells:*** cells of multicellular organisms from which the next generation is ultimately derived, i.e. gametes and the generative cells and their descendants that give rise to gametes. The G.l. extends from the fertilized egg to functional gametes, and it becomes differentiated from somatic cells in early ontogenesis. Somatic cells form the rest of the body and ultimately leave no progeny.

Germ plasm theory: the theory that there are two types of cells, germ cells and somatic cells, and that germ cells contain germ plasm which is potentially immortal by transmission from generation to generation. This was put forward as a hypothesis in 1885 by A. Weismann, on the basis of his studies of the sex cells of hydromedusae. Thus, the germ plasm was supposed to carry the entire primordial complex for development of a new individual. Part of this remained unchanged and formed the basis of the next and all subsequent generations. Germ plasm was transferred from cell to cell via the "germ line", until it separated again from the body or somatic cells.

Germ warfare: see Biological weapons.

Geronimo legless lizard: see *Anniellidae.*

Geronticus: see *Ciconiiformes.*

Gerontology: the study of aging processes, in particular the relation of disease and illness to age.

Gerrhosaurinae: see *Cordylidae.*

Gerrhosaurus: see *Cordylidae.*

Gerstmann-Straussler syndrome: see Prion.

Gervaisiidae: see *Diplopoda.*

Gestagens, ***progestins, gestins:*** female steroid gonadal hormones, including Progesterone (see) and other natural and synthetic steroids with progesterone-like effects. The synthetic G., norgestrel and chlormadione acetate, are administered orally to correct irregularities in the menstrual cycle, against repeated abortion, and as components of ovulation inhibitors (see Hormonal contraception). They are used increasingly for the regulation of animal reproduction (synchronization of estrus, and initiation of mating).

Gestation, ***gravidity:*** in female mammals, a condition that begins with implantation of the fertilized egg and ends with parturition. In humans it is known as pregnancy. The gestation period varies widely between species, and may also display considerable variation (within certain limits) for a single species (Table).

Gestation time (days)

Species	Average	Range
Horse	336	310–360
Cow	285	270–296
Sheep and Goat	150	156–151
Pig	115	104–133
Dog	63	59–65
Cat	60	56–60
Rabbit	28	27–33
Rhinoceros	520	510–540

Species	Average	Range
Elephant	630	610–650
Chimpanzee	240	210–270
Human	280	265–295

Gestins: see Gestagens.
Gharial: see *Gavialidae.*
Ghorkar: see Asiatic wild ass.
Ghost frogs: see *Heliophrynidae.*
Ghost moths: moths of the family *Hepialidae.* See *Lepidoptera.*
Ghost pipefishes: see *Syngnathoidei.*
Ghost shrimp: see *Amphipoda.*
Giant African land snail: see *Gastropoda.*
Giant African otter shrew: see Water shrews.
Giant Asiatic tree squirrels: see *Sciuridae.*
Giant chromosomes, *polytene chromosomes:* chromosomes that are up to 0.5 mm in length and 25 mm in diameter, consisting of hundreds to more than 30 000 homologous chromatids. They occur in highly differentiated cells of flies, mosquitoes, lepidotera and snails, e.g. the cells of salivary glands, silk glands, etc. They have also been observed in some plant tissues. G.c. arise by multiple endomitotic duplication of the chromosomes, without separation of the chromatids (also known in this case as endochromosomes), thereby producing cable-like bundles of many chromatids (see Polyteny). In accordance with the greatly increased quantity of DNA, the cell nucleus and the entire cell become larger.

The chromatids are largely in an unwound state, i.e. the supercoiled DNA molecules are much more extended than in ordinary chromosomes, and the duplicated molecules remain in register as they lie side by side. This gives rise to a banded structure (observable under the light microscope), consisting of dark bands that are rich in DNA, with DNA-poor bands inbetween (see Chromosomes). The dark, DNA-rich bands contain 95% of the total DNA. When the DNA is genetically inactive (i.e. it is not actively engaged in transcription of RNA), it is relatively tightly coiled and repressed by specific proteins. During transcriptional activity, however, the DNA becomes more relaxed and is derepressed. This relaxation is manifested as transverse swellings that encircle the axis of the G.c., and which are known as *puffs* or *Balbiani rings.* The terms, puff and Balbiani ring, are more or less equivalent, but a Balbiani ring is generally considered to be a more pronounced than a puff. A puff or Balbiani ring therefore represents a region of gene activation. Specific incorporation of RNA precursors in the puffs can be demonstrated by Autoradiography (see).

About 5 000 dark bands (which periodically swell) have been identified in the G.c. of *Drospophila melanogaster,* roughly corresponding to the number of active genes in this species.

G.c. possess no functional spindle attachment sites, and are therefore unable to replicate.

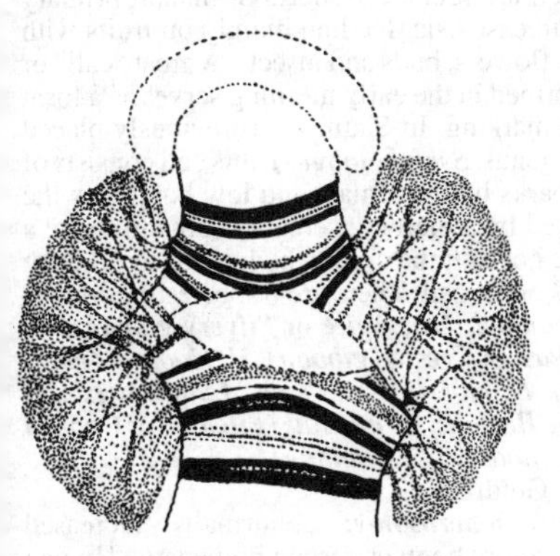

Giant chromosome. A puff in a giant chromosome (left) and a diagram of the course of the individual chromatids in the region of the puff (right).

Giant clams: see *Bivalvia.*
Giant frog: see *Raninae.*
Giant girdled lizard: see *Cordylidae.*
Giant gourami: see *Osphronemidae.*
Giant panda, *Ailuropoda melanoleuca* (plate 38): a very round-bodied bear with black and yellow-white markings. It does not share all the affinities of the bear family, and it is considered by some authorities to be more closely allied to the racoons *(Procyonidae).* An interesting forepaw modification is thought to aid the grasping of bamboo stems: an accessory lobe with a supporting bone is present on the sole pad of each forepaw; the pad of the first digit, and to some extent that of the second digit, can be flexed onto the summit of this accessory lobe. The G.p. lives in mountain forests between eastern Tibet and Szechuan, where it feeds mainly on bamboo shoots, which are essential for its survival. Other plants like gentians, irises, crocuses and tufted grasses may be eated occasionally, and on the Tibetan plateau it possibly hunts fish and small rodents. It is also known to scavenge large dead animals, such as deer. It rarely adjusts well to captivity, but mating and breeding by zoos have been successful. The G.p. is the symbol of The World Wildlife fund, renamed in 1988 The Worldwide Fund for Nature.
Giant salamanders: see *Cryptobranchidae.*
Giant sea scorpions: see *Gigantostraca.*
Giant silkworm moths: moths of the family *Saturniidae.* See *Lepidoptera, Bombycidae.*
Giant toad: see *Bufonidae.*

Giant water beetle: see Respiratory organs.

Giant water bugs: see *Heteroptera.*

Gibberellic acid: see Gibberellins.

Gibberellin antagonists: inhibitors of the effects of gibberellins in plants. Their action can be at least partly overcome by gibberellins. Competitive inhibitors of gibberellin activity, i.e. substances that compete with gibberellins for the same binding site(s), are called Antigibberellins (see) rather than G.a. Natural G.a. include the phytohormone, abscisic acid, as well as tannins. Certain compounds, including some with quaternary structure, have application as synthetic G.a. Most important of these are: CCC (chlorocholine chloride), AMO 1618 (2-isopropyl-5-methyl-4-trimethylammonium chloride), Phosgen D (2,4-dichlorobenzyl-tri-n-butylphosphonium chloride), EL-531 [α-cyclopropyl-α-(4-methoxy-phenyl)-5-pyrimide methanol] and B-995 (succinic mono-*N*-dimethylhydrazide). These substances primarily cause a marked shortening of the shoot, which is desirable in some ornamental plants (e.g. poinsettias, chrysanthemums, etc.), as well as in cereals, as a means of obtaining more compact growth and of avoiding tall plants that are subject to wind damage. The shortening caused by G.a. may be accompanied by other effects. Thus, B-995 increases the number of blossoms on fruit trees and prevents loss of immature fruit. Many G.a. are known to act by blocking gibberellin synthesis, and other G.a. probably act in the same way, although unequivocal proof of this is lacking.

Gibberellins: a class of widely occurring plant hormones which stimulate extension growth. The first member to be discovered was isolated in 1938 from the fungus *Gibberella fujikuroi (Fusarium moniliforme),* the causative agent of of the rice disease, Bakanae. It was subsequently found that G. are widespread natural plant growth regulators (they occur in higher plants in extraordinarily low concentrations, i.e. μg per kg plant material). G. are diterpenes containing the ent-gibberellane ring system, and they are named in order of their discovery as GA_1, GA_2, GA_3, etc. (more than 60 G. have been described and structurally elucidated). On account of its ubiquitous occurrence and high biological activity, GA_3 (gibberellic acid) is the most important representative. G. are divided into two main groups: those with 20 carbon atoms and those with 19 carbon atoms. G. differ from one another essentially in the position, number and stereochemistry of hydroxyl groups at positions 1, 2, 3, 11, 12 and 13, and in the level of oxidation of C20 in the C_{20}-G.

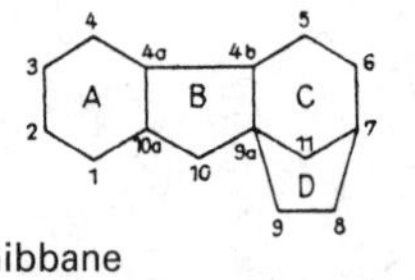

Gibbane

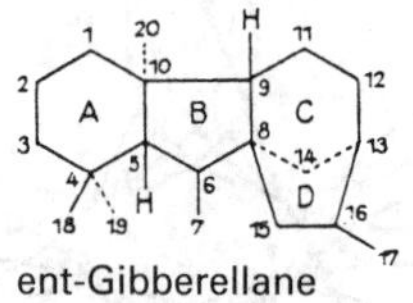

ent-Gibberellane

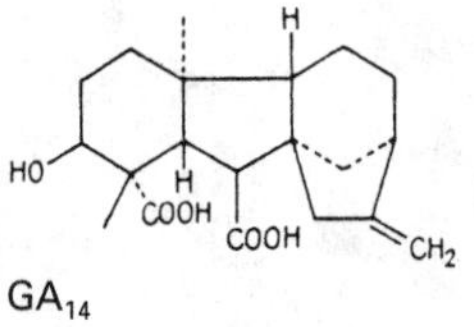

GA_{14}

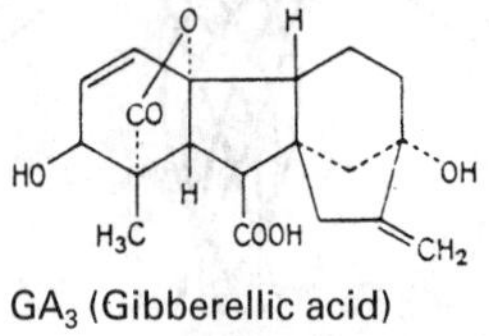

GA_3 (Gibberellic acid)

In addition to free G., plant tissues also contain water-soluble conjugates, such as gibberellin-*O*-glucosides and gibberellin glucose esters, which are possibly transport and storage forms, or products of reversible deactivation. G. act in close cooperation with other plant hormones in the regulation of plant growth and development. They stimulate extension growth, as well as influencing germination, flower induction and fruit ripening. The activities of different G. vary widely according to the biotest used for their assay. Dwarf mutants of maize, rice and peas with genetically blocked G. synthesis are used in assays which have a sensitivity of 1 ng. The lettuce and cucumber hypocotyl tests are also based on the stimulation of shoot growth. Acceleration of germination by G. is the basis of the seed germination test. G. are also assayed by physical chemical methods (e.g. combined gas chromatography and mass spectroscopy of the methyl esters or trimethylsilyl ethers), or by immunological methods.

G. are biosynthesized from geranylgeranylpyrophosphate, which cyclizes to kaurene. Stepwise oxidation and ring reduction of kaurene leads to the pivotal common intermediate in the biosynthesis of all G., i.e. GA_{12}-aldehyde. Thereafter the biosynthetic pathways diverge considerably. The chief sites of synthesis are the meristematic tissue of the shoot and root tips. G. are transported in the xylem and phloem.

Gibberellic acid is prepared technically from cultures of *Gibberella fujikuroi.* It is used in the brewing industry to promote uniform and rapid germination of barley in the malting process. It also promotes blossom formation and fruit growth and both properties are exploited in commercial agriculture. Its ability to induce parthenocarpy in, e.g. tomatoes, grapes and cucumbers, is also used commercially in the production of seedless fruits.

Gibbons, *Hylobatidae* (plate 41): a family of Old World monkeys (see). G. are small, tailless, slender, extremely long-armed, medium sized apes, with long, fluffy, dense fur, which varies from black to gray, brown or gold, depending on the species. They are almost totally arboreal and display considerable acrobatic ability. The thorax is not deep as in monkeys, but broad from side to side. A fully extensible elbow and a forearm capable of full pronation-supination means that G. can hang by one arm and rotate through almost 360 °. The legs, however, are relatively short. G. inhabit primary forest in Southeast Asia, feeding mainly on fruits with some leaves, flowers, buds and insects. A great "call" or "song" performed in the early morning serves as a form of territorial marking. In Siamangs (previously placed in a separate genus, *Symphalangus),* this call consists of booms and barks by the female and low booms by the male, followed by a piercing scream; siamangs have a naked air sac beneath the chin, which helps to produce the two notes of the call. There are 6 species: *Hylobates lar* ***(White-handed, Lar, Agile*** or ***Silvery gibbon),*** *H. concolor* ***(Black*** or ***Crested gibbon),*** *H. Hoolock* ***(Hoolock gibbon),*** *H. klossii* ***(Kloss's*** or ***Mentawai Island gibbon,*** or the ***Beeloh),*** *H. pileatus* ***(Pileated*** or ***Capped gibbon),*** *H. syndactylus* ***(Siamang).***

Gibel: see Goldfish.

Gigantism, ***macrosomy:*** abnormally increased growth of the entire body or certain body parts. Human body heights greater than 2 m represent G. (see Body height). Giant mice have been produced by injecting fertilized mouse eggs with a plasmid containing the coding sequences of rat growth hormone attached to regulatory sequences of the mouse metallothionine gene. The resulting mice grew faster than their normal litter mates, even without adding heavy metals as inducers. For G. in plants, see Hypertrophy.

Gigantostraca, *giant sea scorpions:* an extinct arthropod order containing the genera *Pterygotus* and *Eurypterus* (see) with a superficial resemblance to modern scorpions. The elongated body was almost 2 m in length, composed of a large cephalothorax and an abdomen of 12 free segments ending in a caudal spine or plate. The cephalothorax carried an anterior pair of chelicerate appendages, three pairs of walking legs and a last pair of appendages partly modified as flattened swimming organs. All were entirely aquatic; the earliest forms were purely marine, whereas later forms also invaded brackish and freshwater habitats. Fossils are found from the Ordovician to the Permian.

Gila monster: see *Helodermatidae.*

Gill arch: see Gills, Gill skeleton, Skull.

Gill chamber, *branchial chamber:* a chamber present in many crustaceans, which is formed by a lateral extension of the carapace, and which houses the gills.

Gill respiration: respiratory gas exchange via gills, the respiratory organs of aquatic animals. Water contains less than 1 vol % of dissolved oxygen, so that the respiratory volume of aquatic animals must be greater than that of terrestrial animals. In fish, the ratio of circulated water to that of the blood circulating at the same time in the gills is 16:1, whereas in humans the corresponding ratio (inspired air volume: circulated blood) is 1:1. Between 50 and 80% of the oxygen in the circulated water diffuses into the blood. In G.r., respiratory water is first drawn into the mouth cavity through the open mouth by expansion of the internal volume of the buccal space. At the same time, water is also sucked over the gills into the gill chamber by an outward movement of the gill operculum (inhalent phase). After closure of the mouth, water is then forced to the exterior through the open gill operculum (exhalent phase) (Fig.).

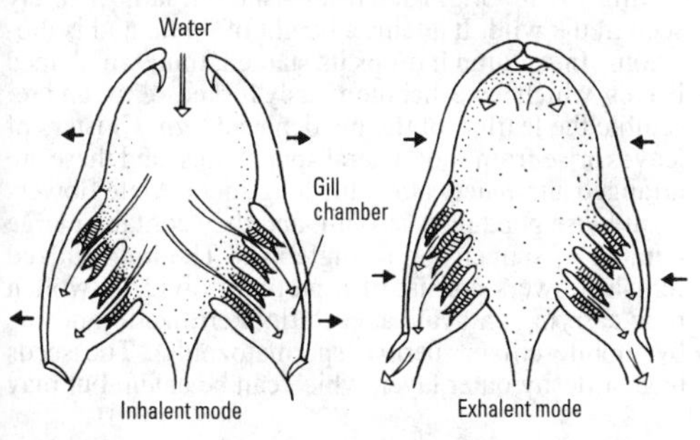

Gill respiration of a teleost fish. The mouth functions as a valve.

Gills, *branchial organs:* organs of respiratory gas exchange (external respiration) present in aquatic animals, e.g. polychaetes, mollusks, crustaceans amd echinoderms, as well as certain chordates such as *Amphioxus,* tunicates, cyclostomes, fish and the larvae of amphibians. G. are usually thin-walled evaginations of the body wall (external G.) or the mucosa of the pharynx (internal G.). Their copious blood supply insures efficient transport of diffused oxygen from the external water to the body tissues.

The surface area of G. is increased by a variety of structural adaptations, e.g. division into numerous adjacent filaments or lamellae results in the filibranch G. of many bivalves and crustaceans, and the G. lamellae or filaments of teleost fishes; multiple branching produces the comb G. of many sea snails and the external G. of frog and salamander larvae; the lamellibranch G. of many bivalves and the branchial chamber of tunicates have large surface areas due to the presence of multiple perforations. To insure efficient gas exchange between the external water and the blood in the G., it is necessary for water to flow continually over the gill surface; this is achieved by appropriate siting of the G., by the generation of water currents by cilia, by the active inhalation and exhalation of water, and by the generation of water flow by locomotion of the animal. In many invertebrates, G. are simply exposed on the body surface, e.g. the external G. of of polychaetes, sea stars and sea urchins. Mollusk G. are usually housed in a more or less closed chamber, through which passes a stream of water, propelled by the cilia of the mantle epithelium and the cilia of the G. In addition to respiratory gases, this water stream also carries food particles. The primitive molluscan gill is the comb gill, which consists of single filaments. As an adaptation to their sedentary mode of existence, however, bivalves possess specially modified filibranch G., or multiply perforated lamellibranch G., which also trap food particles in the water stream.

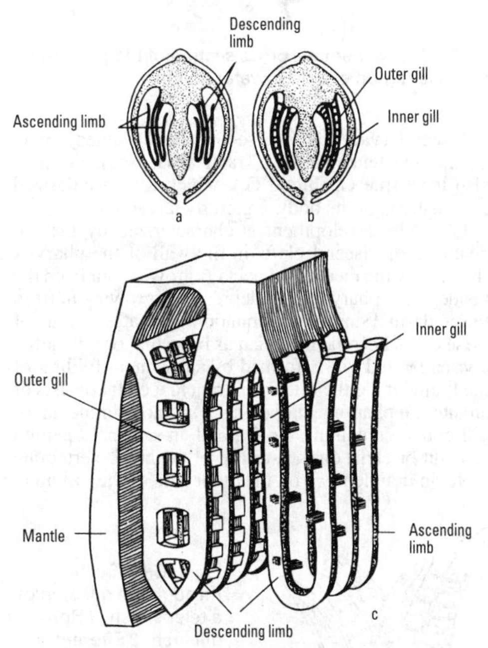

Fig.1. *Gills.* Gill types of the *Bivalvia. a* Filibranch gill. *b* Lamellibranch gill. *c* Schematic structure of a lamellibranch gill (outer) and a filibranch gill (inner).

Evolution of crustacean G. is linked to the evolution of their jointed appendages. Thus, the basal appendages (epipodites) of the walking legs have become modified to lamellar, filamentous or multiply branched G. In more advanced crustaceans, the G. are housed in a gill chamber (branchial chamber), which is formed from a lateral fold (branchiostegite) of the carapace. Specially modified second maxillae (scaphognathites) generate anteriorly directed respiratory currents through the branchial chamber. Crustaceans also often possess secondary G. at various sites, e.g. the tracheal organs of some isopods (see *Isopoda).* In land crabs and hermit crabs, the gill chamber is modified to a "lung"; the G. are reduced and the gill chamber epithelium has taken over the function of gas exchange. Some terrestrial isopods also have Tracheal lungs (see).

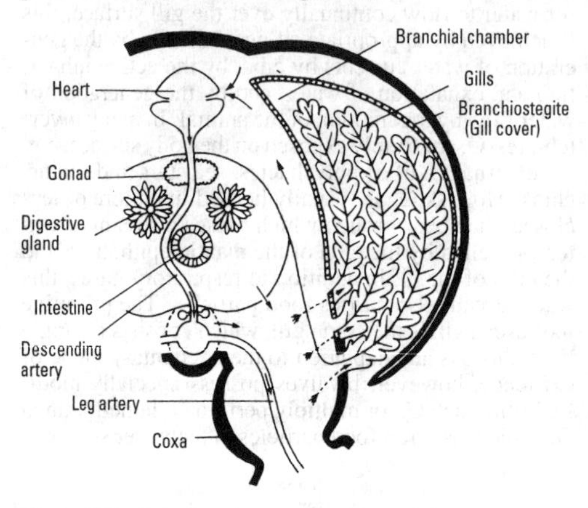

Fig.2. *Gills.* Schematic cross section of the posterior thoracic region of a freshwater crayfish.

Insect larvae that are secondarily adapted to an aquatic existence possess Tracheal G. (see), but they also have true G. (blood G.), which are vascularized evaginations of the body, e.g. in fly larvae.

Chordate development is characterized by the appearance of visceral clefts in the wall of the pharynx (formed by the meeting of pockets growing out from the inside of the pharynx with depressions growing in from the epidermal surface). In aquatic vertebrates, certain of these clefts become modified as branchial or gill clefts; a vascularized gill is formed by subsequent folding of the lining of a gill cleft. Each branchial cleft consists of an internal pharyngeal opening, a spacious branchial or gill pouch containing the G., and an external opening (gill slit or cleft) on the surface of the head. Vertebrate G. lie in double rows on the outer side of the gill arch.

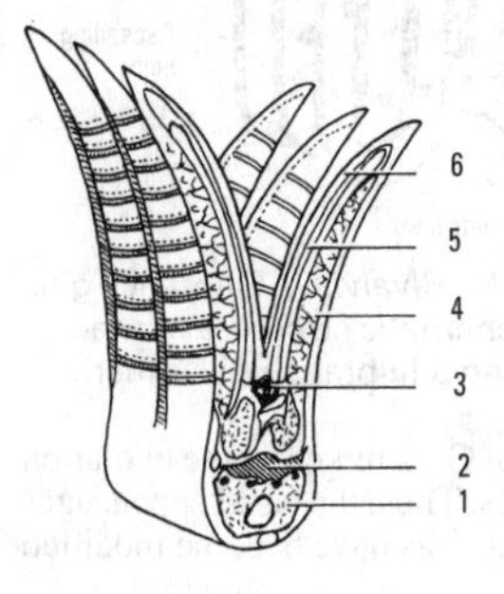

Fig.3. *Gills.* Section through the gill arch of a teleost fish. *1* Bony gill arch. *2* Efferent vessel (vein). *3* Afferent vessel (artery). *4* Gill filament vein (efferent vessel). *5* Cartilaginous gill ray. *6* Gill filament artery (efferent vessel).

In elasmobranchs the G. are reinforced by interbranchial septa, which extend obliquely back from the gill arch to the surface of the body. In teleosts, the septa are reduced, and the G. lie in a common gill chamber, which communicates with the exterior via a protective gill cover or operculum. The gill arch also carries gill rakers, which protect against the entry of large or hard particles that would block the gills, and prevent food from passing out through the gill slits. Cyclostomes possess up to 14 gill pouches, elasmobranchs have between 5 and 7, while teleosts have basically 4.

Gill skeleton, *branchial skeleton:* the skeletal system that forms part of the Visceral skeleton (see), and which lies in the region of the pharynx. The G.s. consists of cartilaginous or bony bars (gill arches or branchial arches), which support the gills. Although the G.s. displays a serial arrangement (branchiometry), this is not part of the metamerism of the rest of the body. Each gill arch consists primarily of 4 parts: the pharyngobranchial, epibranchial, keratobranchial, and hypobranchial, which are linked to one another medioventrally by an unpaired cupula. To protect the gills within the gill chamber, teleost fishes have a gill cover or operculum, which consists of a bony skin flap. The G.s. of cyclostomes contains 8 gill arches, as well as a cardiobranchial which protects the heart. Teleost fishes have 5 gill arches, the last of which is reduced. Following the evolutionary loss of gill respiration, the embryonic G.s. of terrestrial animals gave rise to organs of the anterior visceral region. Thus, the first 2 gill arches develop into the lower jaw, the hyoid bone and the auditory ossicles, while the subsequent arches contribute mainly to the cartilaginous larynx skeleton (see Visceral skeleton).

Gill tails: see *Branchiura.*

Ginger: see *Zingiberaceae.*

Ginglymus-type joint: a hinge joint (see Joints).

Ginkgoaceae, *ginkgo family:* a family of fork- and needle-leaved gymnosperms. The family is largely extinct, and represented by a single living species, the ***Ginkgo tree*** or ***Maidenhair tree*** *(Ginkgo biloba).* This "living fossil" originates from Asia, but is now rarely seen in the wild. It attains a height of 30 m, and is dioecious. In Autumn it drops its stalked, thin, fan-shaped leaves, which have dichotomously forked veins, and resemble the leaflets of the maidenhead fern. Clusters of leaves arise from short lateral spur shoots, and these are arranged alternately along the long shoot. Male flowers form loose pendulous strobili, and they contain a large number of stamens on a single axis. The long-stalked female flowers consist of a partially divided swollen base carrying an ovule at each tip. Fertilization occurs by mobile ciliate sperms (spermatozoids). The seeds have a fleshy outer layer, which can be eaten, but may

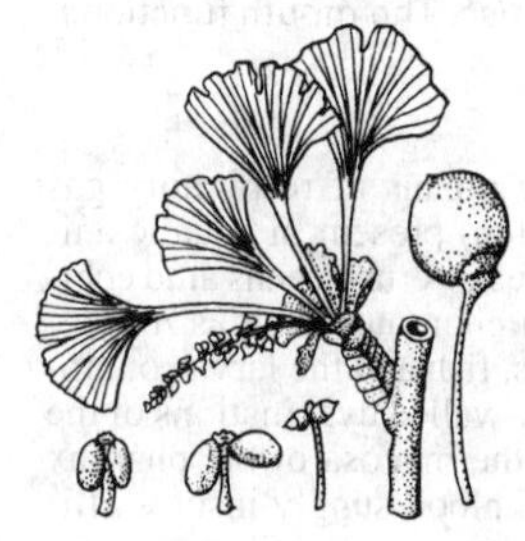

Ginkgoaceae. Maidenhair tree *(Ginkgo biloba);* short spur shoot with male strobilus and young leaves.

cause nausea if smelled. In some individuals, contact with the flesh of the fruit may cause superficial skin lesions. The Ginkgo tree is a popular ornamental, which flourishes in European and North American parks and avenues. Other members of the family are known only as fossils. The evolutionary peak of the G. was in the Mesozoic.

Ginkgo family: see *Ginkgoaceae.*

Ginseng: see *Araliaceae.*

Gipsy moth, *Lymantria dispar* L.: a moth of the family *Lymantriidae* (tussock moths) found from Europe through temperate Asia to Japan and eastern USA and Canada. Its occurrence in USA and Canada is due to accidental introduction in 1869. The subsequent population explosion devastated North American deciduous and evergreen trees. Control measures include the introduction of *Nosema lymantriae,* a microsporidian which specifically attacks the G.m.; the use of synthetic equivalents of the female pheromone to attract males to insecticide-baited traps; and the application of *Bacillus thuringiensis* against the caterpillar.

Giraffa: see *Giraffidae.*

Giraffes: see *Giraffidae.*

Giraffidae (plate 39): an artiodactyl family of large, long-legged, long-necked mammals, with skin-covered horny outgrowths from the frontal bone of the head. The family is represented by only two living genera, found only in sub-Saharan Africa. With their long legs and long necks, ***Giraffes*** are the tallest of all living mammals, attaining a height of almost 6 m to the top of the head and about 3 m at the shoulder. The basic color is light yellowish brown with more or less regularly distributed large dark brown spots. The horns are not shed, and they vary in number (2, 4 or even 5, depending on the subspecies). Only one species is recognized: *Giraffa camelopardalis.* Subspecies include, e.g. *G.c. tippelskirchi* (Masai giraffe), *G.c. camelopardalis* (Nubian giraffe), *G.c. reticulata* (Reticulated giraffe). Giraffes live in small groups on the African savanna, where they feed on leaves and twigs, plucked with their prehensile tongues from the crowns of trees. The ***Okapi*** *(Okapia johnstoni),* first described scientifically in 1901, is only about 150 cm at the shoulder. It has a shorter neck (no longer than that of some antelopes) and shorter legs than the giraffe, a chestnut-brown coat, and light gray markings on the head and legs. Horns are present, but only in the male, and they are pointed and naked at the tips. It lives alone or in pairs in the forests of Zaire, where it feeds on leaves and fruits with the aid of its 40 cm-long, prehensile tongue.

Fossils of heavily built giraffids with branched horns (e.g. *Sivatherium)* have been found in the Pleistocene of Africa and southern Eurasia. Relatively short-legged and short-necked fossil forms resembling the okapi are known from the Eurasian Pliocene, e.g. the genus *Palaeotragus,* which was extremely similar to the modern okapi. The *G.* appear to have diverged from the *Cervidae* during the Miocene.

Girdle-tailed lizards: see *Cordylidae.*

Givetian stage: see Devonian.

Glacial periods: see Ice Ages.

Glacial refuges, ***glacial refugia:*** climatically favorable geographical areas, in which animals and plants were able to survive during the Tertiary and during the last Ice Age. G.r. generally have a rich diversity of species. The concept of G.r. as areas of retreat for migrating species is largely wrong. They are rather survival and maintenance areas for the species already present, but restricted by the surrounding glaciation. Maintenance of a species in several separated G.r. has often led to the differentiation of subspecies or even new species. With the end of the Ice Age, G.r. became Dispersal centers (see) for the recent flora and fauna of the Holarctic region. The Mediterranean region and the Caspian depression were important G.r. for the Palearctic, whereas Alaska, Central America and the extreme southeastern part of North America served as G.r. for the Nearctic. Forest inhabitants and species of steppes and deserts survived in different G.r. Small animals often survived on mountaintops protruding from the ice.

Gladiolus: see *Iridaceae.*

Gladius: see *Cephalopoda.*

Glandotropic hormones: a term sometimes applied to the peptide and protein hormones of the pituitary gland, which regulate the synthesis and secretion of hormones by other glands, e.g. thyroid, adrenal cortex and gonads. Some G.h. are corticotropin, thyreotropin, follicle stimulating hormone, luteinizing hormone and prolactin. G.h. play a pivotal role in the regulation of the entire hormone production in the body. Circulating glandular hormones control the production and secretion of G.h. by feedback. In addition, the production of G.h. is subject to higher levels of control (e.g. by the central nervous system) and it displays biological rhythms. See Releasing hormones.

Glands:

1) In animals, specialized epithelial cells (Fig.1), which serve for the synthesis and release of secretory material. Classically, the definition of a gland infers the separation of materials from blood and their reassembly into new material. Secretion in the sense of separation is derived from the Latin *secretionem* (from the verb *secernere* to separate, which in turn is derived from *se* aside and *cerno* distinguish). G. may be single cells, e.g. cuboid goblet gland cells that produce mucus, and which are present in the skin of fishes and reptiles, and in the intestine of higher animals. In the narrow sense,

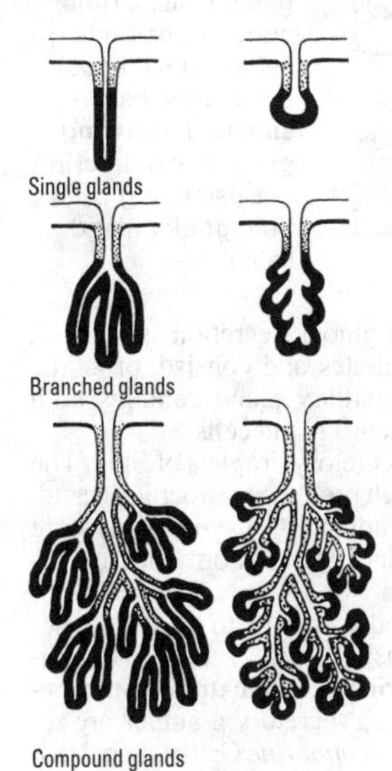

Fig.1. *Glands.* Types of glands. *a* Tubular glands. *b* Alveolar glands.

however, G. are multicellular organs with specific functions. G. that secrete into a body cavity or onto a body surface are known as exocrine G., whereas those that secrete into tissue fluid or blood are known as endocrine, incretory or ductless G.

In exocrine G., the gland cells are grouped around a distal expanded space, known as the acinus or alveolus. Secretory material from the acinus travels via a system of passages to the surface, which may be within the body (e.g. salivary G.) or on the body surface (e.g. sweat G.). According to the nature of the secretion, exocrine G. may be classified as 1) *serous G.,* or 2) *mucous G.*

Serous G. are protein-synthesizing G. (e.g. parotid gland, pancreas, lacrimatory G.), which produce a thin, watery, proteinaceous secretion. The secretory route can be briefly illustrated by describing the rat pancreas cell. Precursors (amino acids and monosaccharides) arrive at the base of the gland cell via the blood capillaries, then pass through the capillary wall and plasmalemma. Using a template of mRNA derived by transcription from the cell nucleus, protein is synthesized on the rough endoplasmic reticulum (described as rough because of the presence of ribosomes). The first protein product is transported in transport vesicles to the Golgi apparatus, where the synthesis of glycoproteins is completed by addition of oligosaccharide chains. The prosecretion, in the form of zymogen granules, accumulates in the apical section of the cell and is then released.

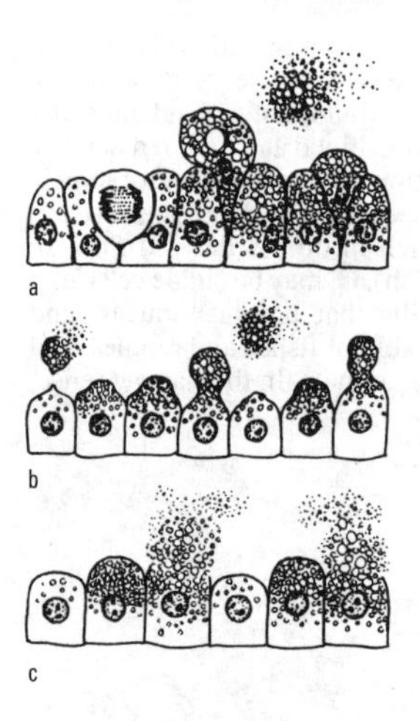

Fig.2. *Glands.* Extrusion mechanism of gland cells. *a* Holocrine extrusion (e.g. sebaceous glands). *b* Apocrine extrusion (e.g. lactating glands). *c* Merocrine extrusion (e.g. cuboid goblet gland cells).

In mucous G. a mucilaginous secretion is formed, which is rich in carbohydrates and consists of acidic glycoproteins. The submaxillary gland contains both serous gland cells and mucous gland cells.

Sebaceous G. contain secretory droplets of lipid. The oxyntic cells of the stomach produce hydrochloric acid.

Endocrine G. (e.g. the adrenal G.) synthesize, store and release peptide hormones, steroid hormones, or catecholamines.

G. may also be classified according to the type of secretory mechanism (extrusion) (Fig.2). In serous, mucous, and in many endocrine G., the extrusion mechanism is *merocrine,* in which secretory granules are released via the cell surface. In *apocrine* G., the apical cell region, which is loaded with secretory material, becomes abstricted (e.g. lactating glands). In *holocrine* extrusion, the entire gland cell disintegrates, e.g. sebaceous G.

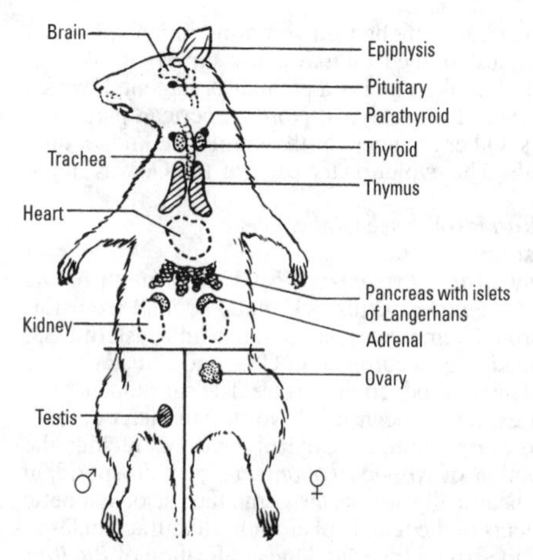

Fig.3. *Glands.* Position of the endocrine glands in a rodent (rat).

Two basic types of internal gland structure are shown in Fig.1.

Endocrine G. are characteristically well vascularized and have no external ducts; their hormones reach their target sites directly in the blood stream. Fig.3 shows the positions of mammalian endrocrine G. In addition, many other organs or parts of organs have an endocrine function, e.g. neurosecretory centers of the brain.

2) In the wider sense, all Excretory tissues (see) of plants have a glandular function.

Glandular cell: see Excretory tissues of plants.

Glandular epthelium: specialized Epithelium (see), whose cells are secretory. G.e. forms the epithelial lining of the acinus region of Glands (see). See Excretory tissues of plants.

Glandular hormones, ***endocrine hormones:*** hormones secreted by endocrine glands. In humans, the most important endocrine glands for normal body function are the hypophysis, thyroid, parathyroid, adrenals, gonads, and the islets of Langerhans in the pancreas. Hyperactivity, hypoactivity or loss of activity of any of these glands leads to severe metabolic disturbances with characteristic clinical symptoms, e.g. *Diabetes mellitus* is caused by decreased production or nonproduction of insulin by the islets of Langerhans.

Glandular tissue: see Excretory tissues of plants.

Glareolidae, ***coursers and pratincoles:*** a small Old World avian family (16 species) in the order *Charadriiformes* (see). *Glareolidae* show their greatest diversity in Africa and southern Asia, with some species in Eurasia and Australia. Both coursers and pratincoles characteristically stretch on tiptoe to look around.

1. Subfamily: *Cursoriinae* ***(Coursers;*** 9 species). These are long-legged, short-winged, short-tailed, extremely fast running birds, which fly only when forced. They have long, pointed, decurved bills, and 3 toes (the hallux is absent). Except for the atypical Egyptian plover, all members have plain and sandy plumage, affording camouflage in their dry sandy habitats. The ***Egyptian plover*** *(Pluvianus aegyptius)* is gray and

white with black and metallic green markings, blue-gray wings, orange underparts, and bold black and white markings on the head. It inhabits the margins of inland lakes and rivers and nearby grasslands, feeding on insects taken on the ground and at the water's edge, and it serves to warn crocodiles of approaching danger.

2. Subfamily: *Glareolinae* ***(Pratincoles;*** 7 species). These dainty birds are brown or olive and lighter below, with short legs, long pointed wings, long forked tails, short bills and a wide gape. They hawk insects on the wing. Efficient locust killers, they are known in South Africa as Locust birds.

Glareolinae: see *Glareolidae.*

Glass frogs: see *Centrolenidae.*

Glasshouse red spider: see *Tetranychidae.*

Glass lizards: see *Anguidae.*

Glass snakes: see *Anguidae.*

Glass sponges: see *Porifera.*

Glasswort: see *Chenopodiaceae.*

Glassy cutworm: see Cutworms.

Glaucidium: see *Strigiformes.*

Glaucium: see *Papaveraceae.*

Gleba: see *Basidiomycetes (Gasteromycetales).*

Glial cells: see Nerve fiber, Neuron, Nervous tissue.

Gliridae, *dormice* (plate 7): a family of Rodents (see) containing 23 species in 7 genera. They are small, arboreal rodents with bushy tails, large eyes and ears, and squirrel-like behavior and appearance. They build small, spherical nests, and are noted for their long hibernation. Dental formula: 1/1, 0/0, 1/1, 3/3 = 20.

The ***Hazel mouse*** *(Muscardinus avellanarius;* head plus body 7.5–9 cm, tail 7–7.5 cm), a protected European species, is found throughout Britain, Europe, Asia Minor and western Russia. Its creamy white throat, chest and toes contrast with its richly colored yellow-brown back and sides. It nests at a height of 1–2 m in thickets and wooded areas.

The squirrel-sized ***Edible dormouse, Common dormouse*** or ***Fat dormouse*** *(Glis glis;* head plus body 15–19 cm, tail 13–16.5 cm), found over wide areas of Europe and Asia Minor, hibernates for 7 months or longer, using nesting boxes or tree hollows for this purpose. Its upper parts are silvery gray to brownish gray, with lighter underparts. It was introduced into Britain by the Romans, who ate it as a delicacy.

The ***Southwest Asian garden dormouse*** *(Eliomys melanurus)* and the ***Garden dormouse*** or ***Orchard dormouse*** *(Eliomys quercinus)* (both species: head plus body 12–17 cm, tail 11–13 cm) are found in a variety of habitats (wetlands, woodlands, cultivated areas, rocky outcrops) in Europe, Russia and North Africa.

The ***Tree dormouse*** or ***Forest dormouse*** *(Dryomys nitedula;* head plus body 8–10 cm, tail 8–9 cm), found in deciduous woodland and orchards in Europe and Asia Minor, is very similar to, but somewhat smaller than, *Eliomys.*

Glischropus: see *Vespertilionidae.*

Gln: standard abbreviation for the amino acid, glutamine.

Globe artichoke: see *Compositae.*

Globe-flower: see *Ranunculaceae.*

Globicephala: see Whales.

Globigerina (Latin *globus* sphere, *gerere* carry): a worldwide distributed genus of planktonic *Foraminiferida* (see), whose spherical or flattened shells attain diameters of up to 2 mm. The chambers are arranged in a rising helicoid spiral, and perforated with long spines. They occur from the Cretaceous to the present, and during geological history have been responsible for the formation of entire calcareous rock strata. Most of the floor of the western Indian ocean, the mid-Atlantic and the equatorial and southern Pacific (i.e. almost 50% of the modern deep ocean floor at depths of 2 000–5 000 m, representing about one quarter of the earth's surface) is covered with a calcium carbonate-rich sediment, known as the *Globigerina ooze,* consisting predominantly of *G.* shells. Although calcareous shells are also deposited at greater depths, they do not survive, because the increased pressure of dissolved carbon dioxide (as carbonic acid) converts the calcium carbonate into soluble calcium bicarbonate.

Globigerina ooze: see *Globigerina.*

Globin: protein component of hemoglobin and other chromoproteins, which contain a high proportion of histidine and lysine residues. Gs. are albumins.

Globodera: see Eelworms.

Globoid: small spherical body of amorphous phytin in the aleurone granules of plant storage tissue. See Storage materials.

Globulins: a group of simple proteins, present in all animal and plant cells and body fluids. Thus, G. are present, e.g. in egg white and egg yolk (livetin), in serum, milk, hemp seeds (edestin) and potatoes (tuberin). The group includes many enzymes and most glycoproteins. Plasma G. have been extensively investigated; their electrophoretic separation (into α_1, α_2, β_1, β_2 and γ-G.) is important in clinical diagnosis, and in animal phylogenetic and genetic studies. G. are practically insoluble in water, but they dissolve easily in dilute salt solutions. They are precipitated by ammonium sulfate: e.g. fibrinogen at 20–25%, the euglobulins at 33% and the pseudoglobulins at 50% ammonium sulfate.

Glochid: see *Cactaceae.*

Glochidium: the parasitic larva produced by members of the freshwater lamellibranch family, *Unionidae* (see *Bivalvia).* The G. is enclosed by two triangular valves, and in some species the edge of each valve carries a movable hook, A larval mantle is present, as well as a rudimentary foot, which carries a long adhesive thread. In the young larva, the alimentary canal is poorly developed, and there is no mouth or anus. Mature glochidia range in size from 0.5 to 5 mm, depending on the species. The hooked G. of *Anodonta* attaches to the fins or other parts of the body surface of the host fish. Hookless glochidia may be picked up by the respiratory currents of the fish, or they may be released in worm-like masses, which are eaten by the fish. In either case, the G. attaches to the gills of the host, whose tissues form a cyst around the parasite. The parasitic period lasts from 10 to 30 days, during which time phagocytic cells of the larval mantle feed on the host tissues, and the larval structures are replaced by adult organs. The immature adult breaks out of the cyst and completes its development in the bottom mud.

Glomerida: see *Diplopoda.*

Glomeridae: see *Diplopoda.*

Glomeris: see *Diplopoda.*

Glomerular filtrate: see Kidney functions.

Glossa: see Tongue.

Glossopharyngeal nerve: the IX cranial nerve; see Cranial nerves.

Glottis: see Larynx.

Glu: standard abbreviation of the amino acid, glutamic acid.

Glucagon: a polypeptide hormone secreted by the A cells of the islets of Langerhans of the pancreas, in response to a decrease in the blood glucose concentration. It promotes the hydrolysis of glycogen and lipids, stimulates gluconeogenesis, raises the blood glucose concentration (increased glucose release from the liver), and is responsible for the mobilization of storage materials. It therefore has the opposite effect to that of insulin. G. acts by stimulation of the adenylate cyclase system (see Hormones). The single chain of 29 amino acid residues (all 29 residues are necessary for receptor binding and biological activity) shows a structural similarity with secretin. Synthesis of G. in the A cells occurs via a much larger prohormone (proglucagon); the latter is converted into G. by limited proteolysis, which is then stored in secretory granules. It is transported in the blood as the free hormone, and is degraded in the liver.

It also increases contractility of cardiac muscle and the rate of the heart beat.

Glucans: straight or branched-chain polysaccharides composed of D-glucose. The glycosidic linkages may be α1,4 (e.g. amylose and bacterial dextran), β1,4 (e.g. cellulose), β1,3 (e.g. leucosin and callose), or 1,6 as in pustulan. Branched G. include amylopectin (α1,4 and α1,6 bonds), dextran, laminarin and lichenin.

Glucocorticoids: a group of Adrenal corticosteroids (see) with pronounced effects on carbohydrate metabolism. They include Cortisol (see), Corticosterone (see) and Cortisone (see).

G. stimulate glycogen synthesis in the liver and inhibit glucose utilization by peripheral tissues. In addition, they inhibit protein synthesis and stimulate protein degradation in muscle, bone and lymphatic organs.

Glucogallin: see Tannins.

Glucogenic amino acid: see Amino acids.

D-Gluconic acid: a carboxylic acid produced in the laboratory by mild oxidation of D-glucose (the aldehyde group is oxidized to COOH), and which also occurs as an intermediate in carbohydrate metabolism. See Aldonic acids.

Gluconobacter: see Acetic acid bacteria.

D-Glucosamine, ***chitosamine, 2-amino-2-deoxyglucose:*** a widely occurring aminosugar, which differs structurally from D-glucose by replacement of the C-2 hydroxyl group by an amino group. As its *N*-acetyl derivative, it is a component of chitin, mucopolysaccharides (e.g. heparin, chondroitin sulfate and mucotin sulfate), blood group substances and other complex polysaccharides.

D-Glucose, ***dextrose, grape sugar, blood sugar:*** an aldohexose monosaccharide. The most stable configuration for the pyranose form is the chair, in which all the hydroxyl groups of the β-form are equatorial. G. has a widespread natural occurrence, free and combined, especially as its stable pyranose chair form. Free G. occurs, e.g. in honey, sweet fruits and nectar, and is present in human blood ("blood sugar") at a concentration of about 0.1%. Blood concentrations greater than 0.1% occur in diabetic hyperglycemia. D-G. is also a component of many oligo- and polysaccharides, e.g. starch, maltose, sucrose, cellulose, lactose and glycogen, as well as numerous glycosides, all of which yield D-G. by acidic or enzymatic hydrolysis. On the industrial scale, it is prepared by the acid or enzymatic hydrolysis of potato or corn starch, or cellulose. Since D-G. is utilized directly by the body, it is administered as a readily available nutrient and energy source. Metabolism of one gram of D-G. releases17.2 kJ of energy. Phosphoric esters of D-G. are extremely important metabolic intermediates, e.g. glucose 1,6-diphosphate, glucose 1-phosphate (Cori ester) and glucose 6-phosphate (Robinson ester). ADPglucose is the substrate of starch synthesis in plants. UDPglucose is the donor of D-G. in the synthesis of glycogen in animals, and in the synthesis of many saccharides in both plants and animals. D-G. is fermented aerobically and anaerobically by many different microorganisms, e.g. to ethanol, acetic acid, citric acid, lactic acid.

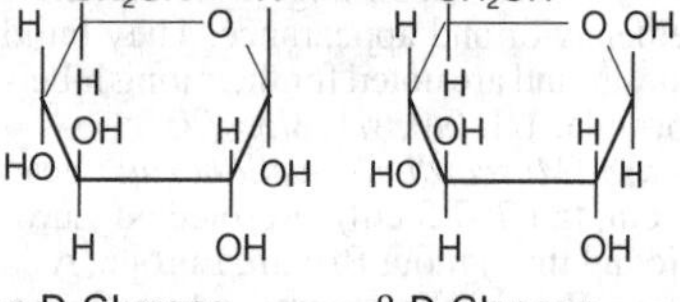

α-D-Glucose β-D-Glucose

Glucosidases: glycosidase enzymes that catalyse the hydrolytic cleavage of α- and β-glucosidic bonds. They are accordingly named ***α-glucosidases*** and ***β-glucosidases.*** α-G. cleave maltose (see Maltase), sucrose and turanose, whereas β-G. cleave cellobiose and gentiobiose.

Glucosolinate, ***mustard oil thioglucoside, mustard oil glucoside:*** a natural plant product with the structure shown (Fig.). Gs. are particularly abundant in members of the *Cruciferae, Capparidaceae* and *Resedaceae.* More than 70 different Gs. are known. Sinalbin was crystallized from white mustard as early as 1831, and sinigrin was isolated from black mustard in 1840. The carbohydrate residue is always a single glucose unit, and all Gs. are 1-β-D-thioglucopyranosides. The term, G., has been in use since 1961. The cation is usually potassium, but in sinalbin it is the basic organic molecule, sinapine. An alternative and earlier nomenclature uses the prefix "gluco-" attached to the appropriate part of the Latin binomial of the plant of origin, e.g. glucotropaeolin from *Tropaeolum majus.*

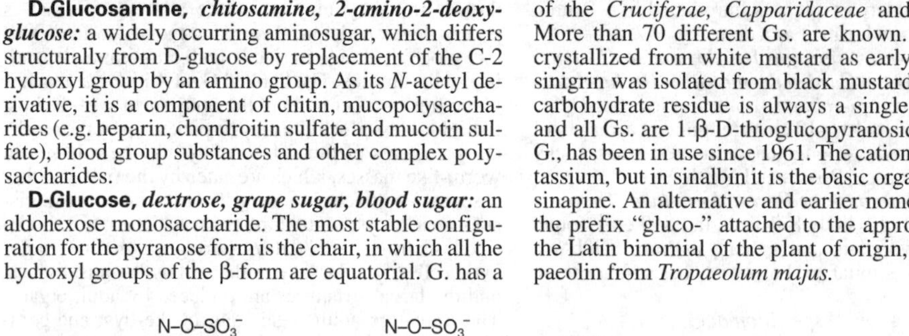

Glucosinolate (Mustard oil glucoside) $\xrightarrow[\text{Thioglucoside glucohydrolase}]{+OH^-}$ $^-S-C(=N-O-SO_3^-)-R$ + glucose $\xrightarrow{H^+ \; \to \; HSO_4^-}$ S=C=N–R Isothiocyanate or Mustard oil (main product

Minor products: N≡C–S–R (Thiocyanate); S + N≡C–R (Nitrile and elemental sulfur)

Damage to the plant results in hydrolysis of Gs. and release of volatile, pungent, lacrimatory isothiocyanates, also known as *mustard oils.* This ability to form mustard oils has led to the adoption of several members of the *Cruciferae* (e.g. horseradish, mustard) as condiments. Mustard oils are normally absent from the plant, and are first formed when tissue damage permits interaction of G. and thioglucoside glucohydrolase (myrosinase). Enzymatic hydrolysis of the *S*-glucoside bond is followed by molecular rearrangement of the aglycon with concomitant production of sulfate and isothiocyanate. The nitrile may also be produced.

Sinigrin (present in black mustard, horseradish and other members of the *Cruciferae),* yields the mustard oil, allylisothiocyanate, which is highly toxic to the pathogen *Peronospora parasitica* (downy mildew). Varieties of cabbage bred for milder flavor (and hence lower sinigrin content) lack resistance to the pathogen. Other Gs. probably also protect their plants against pathogens. On the other hand, sinigrin is a feeding attractant for the cabbage white butterfly larva *(Pieris brassicae)* and an oviposition stimulant for the adult female. It is also a feeding attractant to the cabbage aphis, which prefers mature leaves of medium sinigrin content.

Glucuronic acid: see Uronic acids.

Glume: see *Graminae.*

Glutamic acid, ***Glu:*** a proteogenic, glucogenic amino acid, present in almost all animal and plant proteins, and particularly abundant in wheat and maize gluten, casein, soybeans, plasma albumin (see Albumins) and egg white. Glu occupies a key position in amino acid metabolism. It is an early product of ammonia assimilation in plants and prokaryotic organisms. It functions as an amino group donor in transamination reactions, and is the biosynthetic precursor of the amino acids of the 2-oxoglutarate (α-ketoglutarate) family. Glu is involved in the transport of potassium ions in the brain, and also detoxifies ammonia in the brain by forming glutamine, which can cross the blood-brain barrier and be transported away. In the central nervous system, Glu is decarboxylated to 4-aminobutyric acid by glutamate decarboxylase. Glu is a component of γ-glutamyl peptides. A few of these are found in the central nervous system, but they occur in considerable quantities as storage compounds in plants. Glu is also a component of glutathione. Neutral salts of L-G. are called glutamates. Monosodium glutamate is used as a flavor enhancer in sauces, soups, etc.

Glutaminase: a widely distributed enzyme that catalyses the hydrolysis of L-Glutamine (see) to L-glutamic acid and ammonia.

L-Glutamine, ***Gln:*** H_2N—CO—CH_2—CH_2—$CH(NH_2)$—COOH, the 5-amide of L-glutamic acid. Gln is a proteogenic amino acid, which also plays an important role in other areas of nitrogen metabolism. It serves as a nitrogen storage compound in plants, and as a nitrogen donor in the synthesis of purines, carbamoyl phosphate and other nitrogenous compounds, such as hexosamines. Enzymatic hydrolysis of the 5-amido group (catalysed by glutaminase) converts Gln into L-glutamic acid and free ammonia; this reaction occurs, inter alia, in the kidney, where the free ammonia aids the resorption of potassium and sodium ions.

Gln is biosynthesized from glutamic acid and ammonia, in an ATP-dependent reaction catalysed by glutamine synthetase: $ATP + Glu + NH_4^+ \rightarrow ADP + P_i + Gln$. This reaction is particularly important in the brain, where it serves for the detoxification of ammonia; the resulting Gln can cross the blood-brain barrier and be transported away.

Glutathione, ***L-γ-glutamyl-L-cysteinyl-L-glycine, GSH:*** a naturally occurring tripeptide found in animals, most plants and bacteria. It serves as a biological redox agent, as a coenzyme and cofactor, and as a substrate in certain coupling reactions catalysed by glutathione S-transferase. The intracellular concentration of GSH is 0.1–10 mM. As a redox agent, GSH scavenges free radicals and reduces peroxides, and thus protects membrane lipids from these reactive substances. It is especially important in the lens and in certain parasites which have no catalase to remove H_2O_2.

Glutei muscles: see Musculature.

Glutelins: a group of simple proteins from cereal grains. They are generally insoluble in water, salt solutions and dilute ethanol, but at extreme pH values they do become soluble. They may contain upto 45% glutamic acid and also have a high content of proline. Most familiar are ***glutenin*** from wheat, ***orycenin*** from rice and ***hordenin*** from barley.

Gluten: a mixture of about equal parts of the simple proteins, glutelins and prolamines. On account of its colloidal nature, G. makes flour (in particular wheat and barley flour) capable of rising when used in bread making. Since prolamines are absent from oat and rice grains, their flour is unsuitable for baking.

Glutenin: see Glutelins.

Gluton: see Wolverine.

Glyceraldehyde: the simplest aldotriose. It can be prepared in the laboratory by the careful, mild oxidation of glycerol. In the living cell, free G. has practically no metabolic importance, but glyceraldehyde 3-phosphate is an important intermediate in alcoholic fermentation, glycolysis and photosynthesis. D-(+)-glyceraldehyde is the reference compound for the configuration of asymmetrical (optically active) carbon atoms.

H—C(=O) / H—C—OH / CH_2OH *D-(+)-Glyceraldehyde*

Glyceric acid: $HOCH_2$—CHOH—COOH, a hydroxy acid which occurs naturally to some extent in the free form, but is particularly important in the form of its phosphate esters: 1-phospho-, 2-phospho-, 3-phospho- and 1,3-diphosphoglyceric acid are key intermediates in alcoholic fermentation, glycolysis and photosynthesis. G.a. can be formed metabolically from glyoxylate (2 x glyoxylate $\rightarrow$ tartronate semialdehyde $\rightarrow$ D-glycerate), or from serine via hydroxypyruvate.

Glycerol, ***propan-1,2,3-triol:*** HOH_2C—CHOH—CH_2OH, a syrupy, colorless, odorless, sweet-tasting, hygroscopic liquid, and the simplest trihydric alcohol. It is formed as a byproduct of alcoholic fermentations (up to 3% of the fermentation products) by reduction of dihydroxyacetone phosphate or glyceraldehyde 3-phosphate, followed by hydrolytic removal of the phosphate group. Technically, it is prepared by adding sodium hydrogen sulfite to an alcoholic fermentation; the sodium sulfite forms an addition compound with acetaldehyde, which cannot then be reduced to ethanol, and the reducing power is employed instead in the reduction of dihydroxyacetone phosphate. Commercially, G. is employed in the manufacture of nitroglycerin (explosive constituent of dynamite), dyes, soap and cosmetics. It is also occasionally used in the biology laboratory for storing biological specimens and for temporarily mounting material for microscopic study. All fats, fatty oils and phospholipids are esters of G., and G. can be prepared in the laboratory by the saponification of fats. Free G., however, is not a metabolic intermediate in the biosynthesis or biological degradaion of these compounds.

Glycine:

1) Gly, aminoacetic acid: NH_2—CH_2—COOH, a proteogenic, glucogenic, nonessential amino acid. Structurally, it is the simplest of the amino acids, and its α-C is not asymmetric (i.e. Gly is not optically active). Since its amino nitrogen is easily exchanged, Gly is added to diets as a nitrogen source for general amino acid synthesis. Gly is converted by transamination or oxidative deamination to glyoxylate, which is further metabolized to formate. It is synthesized from glyoxylate by transamination, or from serine by removal of the hydroxymethyl group. The α-C of Gly can be used to synthesize active one carbon units, either directly or via glyoxylic acid. Direct cleavage depends on tetrahydrofolic acid (THF) and yields active formaldehyde: Gly + THF → $N^{5,10}$-methylene-THF + CO_2 + NH_3. The pathway from glyoxylate leads to active formate. The intact Gly molecule is incorporated during purine ring biosynthesis (C atoms 4,5 and 7 are derived from glycine). The α-C and amino nitrogen of Gly are utilized in the biosynthesis of porphyrins and therefore of heme. Methylation of Gly produces sarcosine and glycine betaine.

2) See *Papilionaceae.*

Glyciphagus: see *Acaridae.*

Glycocholic acids: see Bile acids.

Glycogen: an animal polysaccharide which, like amylopectin, consists of D-glucose units. It contains about twice as much branching as amylopectin. Most of the glycosidic bonds are α-1,4, but at branching points there are also α-1,6 links. The side chains consist of 6-12 glucose units. The M_r ranges from 1 to 16 million, which means a maximum of about 10^5 glucose units. G. reacts with iodine to give a brown-violet color. It is most abundant in the liver (up to 10%) and muscles (up to 1%), where it serves as a short-term energy storage material, and is subject to continuous synthesis and degradation.

G. is hydrolysed to D-glucose by dilute acid, whereas amylase hydrolyses G. to maltose.

Glycolic acid, ***hydroxyacetic acid:*** $HOCH_2$—COOH, a hydroxycarboxylic acid found in young plant tissue and unripe green fruits, such as gooseberries, grapes and apples, where it contributes to the sour taste. It has also been found in sugar cane, sugar beet and the leaves of the wild grape vine. Glycolate is formed during in photosynthesis. It arises from active glycolaldehyde, which is formed in transketolation reactions. In some systems it may serve as a precursor of glycine, since it can be oxidized by glycolate oxidase to glyoxylic acid, which in turn can be reduced back to G.a. or converted into glycine.

Glycolipids: compounds in which one or more monosaccharide residues are linked glycosidically to a lipid. The latter is either a mono- or diacylglycerol, a long-chain base like sphingosine, or a ceramide *(sphingosine* is 2-amino-4-octadecene-1,3-diol, i.e. an unsaturated, dibasic, C_{18}, amino alcohol; *ceramides* consists of a fatty acid in amide linkage with sphingosine). In common usage, the term G. refers to sphingosine derivatives that do not contain phosphorus.

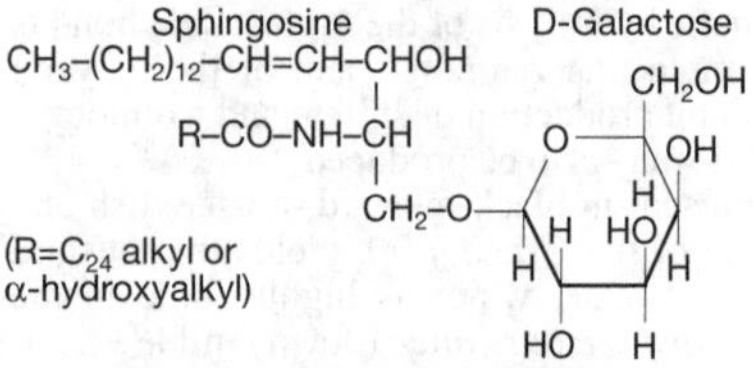

Sphingosine — D-Galactose

CH_3–$(CH_2)_{12}$–CH=CH–CHOH

R–CO–NH–CH

CH_2–O–

CH_2OH, OH, H, HO, H, H, HO, H

(R=C_{24} alkyl or α-hydroxyalkyl)

General structure of a cerebroside

Cerebrosides (Fig.) consist of sphingosine with a fatty acid or hydroxy fatty acid in amide linkage with the 2-amino group, and a single monosaccharide unit (glucose or galactose) linked glycosidically with the primary alcohol group. Examples of naturally occurring cerebrosides are kerasin (containing lignoceric acid), phrenosine [containing cerebronic acid (2-hydroxylignoceric acid)], nervone (containing nervonic acid) and hydroxynervone (containing 2-hydroxynervonic acid). Cerebrosides account for 11% of the dry matter of brain, and are found in smaller amounts in liver, thymus, kidney, adrenals, lungs and egg yolk.

Sulfatides, which also occur in brain, are cerebrosides in which C6 of the sugar is esterified with sulfate. Glycosphingolipids are more complex derivatives of ceramide carrying an oligosaccharide unit on the primary alcohol group of the sphingosine. This oligosaccharide is usually some combination of *N*-acetylglucosamine, *N*-acetylgalactosamine, galactose, glucose and fucose. If the oligosaccharide chain also contains sialic acid *(N*-acetyl or *N*-glycoloylneuraminic acid), the compound is called a ***ganglioside.*** Gangliosides are particularly abundant in the gray matter of the brain and in the thymus, and are also found in erythrocytes, leukocytes, serum, kidneys, adrenals and other organs. They are characteristic components of some neuronal membranes of the central nervous system, and are probably involved in the transmission of impulses along neurons.

Glycolysis, ***Embden-Meyerhof-Parnas pathway:*** the main pathway for anaerobic degradation of carbohydrates in all groups of organisms. Under aerobic conditions, it may be followed by the oxidative stage of respiration, or under sustained anaerobiosis it may be modified as a fermentation pathway. In 11 enzyme-catalysed steps, G. converts glucose into pyruvate (Fig.). All the reactions of G. and of any additional fermentation steps occur in the cell cytoplasm. In preparation for glycolysis, storage carbohydrates (starch in most plants, glycogen in animals) are degraded by phosphorolysis to glucose 1-phosphate. Alternatively, the starting substrate may be free glucose. Free glucose is produced during digestion and absorbed into the circulation; it is produced by the liver to maintain the circulating level of blood glucose. Plants, bacteria and yeasts utilize glucose produced by hydrolysis of small saccharides such as sucrose or maltose.

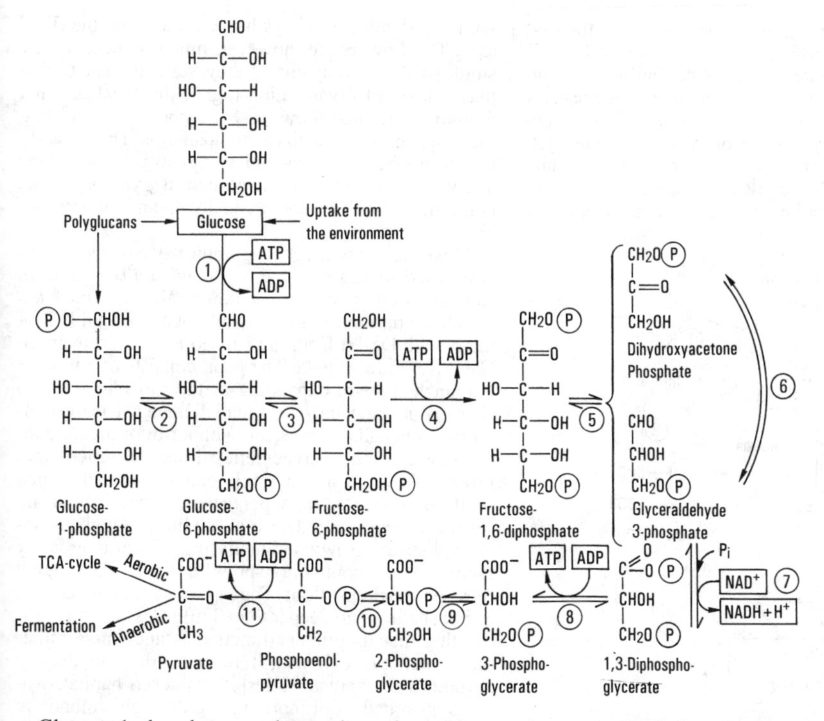

Glycolysis. Enzymes: *1* Hexokinase. *2* Phosphoglucomutase. *3* Hexose phosphate isomerase. *4* Phosphofructokinae. *5* Aldolase. *6* Triose phosphate isomerase. *7* Glyceraldehyde phosphate dehydrogenase. *8* Phosphoglycerate kinase. *9* Phosphoglycerate mutase. *10* Enolase. *11* Pyruvate kinase.

Glucose 1-phosphate or glucose is converted into glucose 6-phosphate by phosphoglucomutase or hexokinase, respectively. After isomerization to fructose 6-phosphate, a further molecule of ATP is consumed in the phosphorylation of glucose 6-phosphate to fructose 1,6-bisphosphate. Two molecules of triose phosphate (glyceraldehyde 3-phosphate and dihydroxyacetone phosphate) are then formed by the action of aldolase. Up to this point, G. has yielded no energy, in fact two molecules of ATP have been consumed (starting from free glucose). In the conversion of the triose phosphates into 1,3-diphosphoglycerate, two molecules of NAD^+ are reduced to NADH, and two molecules of inorganic phosphate are raised to the status of high energy phosphate (each molecule of glucose gives rise to two molecules of triose phosphate). The newly generated high energy phosphate is then transferred to ADP to form two molecules of ATP. Removal of water from 2-phosphoglycerate generates a high energy enol phosphate (phospho*enol*pyruvate), which transfers its phosphate to ADP, forming two molecules of ATP and two molecules of pyruvate.

If G. is linked to aerobic respiration (i.e. if the pyruvate is oxidatively decarboxylated to acetyl-CoA, which enters the tricarboxylic acid cycle), then the NADH formed during G. can be oxidized (albeit indirectly) by the process of Oxidative phosphorylation (see) to yield 3ATP per molecule of NADH. Under anaerobic conditions, this NADH is employed for the reduction of pyruvate to lactate (in muscle), or acetaldehyde (formed by decarboxylation of pyruvate) to ethanol (in yeast). Thus, in the absence of oxygen, G. produces a net yield of 4–2 = 2 molecules of ATP per molecule of glucose. In the presence of oxygen, the yield is increased to a maximum of 8 molecules of ATP, due to the oxidative phosphorylation associated with the NADH.

Glycosidases: group-specific hydrolases which catalyse the hydrolytic cleavage of the glycosidic bonds of carbohydrates, glycoproteins, glycolipids and other glycosides. The hydrolysis products of carbohydrates are mono- or oligosaccharides, whereas glycosides are cleaved into the carbohydrate moiety and the aglycon. G. are more or less specific for the type of sugar (glucose, galactose, etc.) and the type of bond (*O*- or *N*-glycoside, α- or β-linkage). The following G. are listed separately: Amylases, Cellulases, Chitinases, Glucosidases, Maltase, β-Galactosidase, Emulsin, Invertase, Pectinase.

Glycosides: an extensive group of natural products, in which mono- or oligosaccharide units are linked by acetal bonds with hydroxyl groups of alcohols or phenols (*O*-glycosides) or with amino groups (*N*-glycosides). Acid hydrolysis cleaves these compounds into a sugar and a noncarbohydrate portion, the *aglycon* or *genin.* Examples of *O*-glycosides are Saponins (see) and Cardiac glycosides (see); the Nucleic acids (see) and Nucleosides (see) are *N*-glycosides.

Glycosphingolipids: see Glycolipids.

Glycyrrhiza: see *Papilionaceae.*

Glyoxylate cycle: a pathway in microorganisms and plants, but not in animals, which permits the conversion of fats into carbohydrates, enables plant seedling to exploit their fat reserves, and enables microorganisms to grow at the expense of fatty acids, including acetate. It exists in addition to the Tricarboxylic acid cycle (see). As in the tricarboxylic acid cycle, citrate is synthesized from oxaloacetate and acetyl-CoA (from the β-oxidation of fatty acids), then converted into isocitrate. The latter, however, is not oxidized, but is split into succinate and glyoxylate by the action of isocitrate lyase, the first key enzyme of the G.c. (3 in Fig.). Succinate can be converted into oxaloacetate via fumarate and malate, and subsequently serve as a substrate for gluconeogenesis. A second molecule of acetyl-CoA is introduced by the action of the second key enzyme of the G.c., malate synthase (4 in Fig.), which catalyses the formation of malate from glyoxylate and acetyl-CoA. The cycle is completed by dehydrogenation of malate to oxaloacetate.

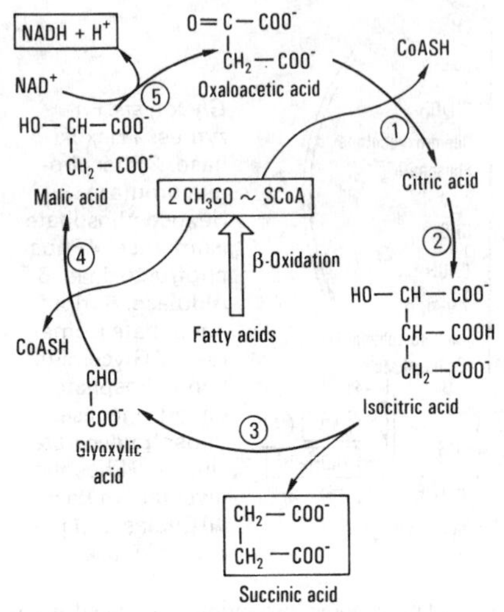

Glyoxylate cycle. 1 Citrate synthase. *2* Aconitase. *3* and *4* (see text). *5* Malate synthase.

In the cotyledons of fat-storing seeds, the G.c., as well as degradation of fatty acids to acetyl-CoA, occurs in the *glyoxysomes.*

Glyoxylic acid, ***oxoacetic acid, glyoxalic acid:*** OHC—COOH, a carboxylic acid found in green fruits (e.g. gooseberries), rhubarb, seedlings and young leaves of certain plants. G.a. is formed metabolically by transamination or oxidative deamination of glycine, or from serine or sarcosine (*N*-methylglycine). It is an intermediate in the Glyoxylate cycle (see) of microorganisms and plants.

Glyoxysome: see Glyoxylate cycle, Peroxysomes.

Glyptostrobus: see *Taxodiaceae.*

Gnat bugs: see *Heteroptera.*

Gnatcatchers: see *Muscicapidae, Sylviinae.*

Gnateaters: see *Conopophagidae.*

Gnathiidea: a suborder of the *Isopoda* (see).

Gnathocephalon: see Mesoblast.

Gnathochilarium: a flattened plate that forms the floor of the preoral chamber of the *Diplopoda* (see).

Gnathosoma: see *Acari.*

Gnat-wrens: see *Sylviinae.*

Gnawing reflex: see Chewing reflex.

Gnetatae, *Chlamydospermae:* a class of gymnosperms, comprising 3 genera, whose phylogenetic relationships to each other and to other gymnosperms is unclear. Members of the *G.* appear to be fragmentary relicts of an ancient and formerly more extensive group. Some authorities consider them to be derived from the extinct gymnosperm order, *Bennettitales,* which they resemble in sometimes having hermaphrodite flowers, with a perianth and ovules borne directly on the floral axis. The flowers are, however, much reduced with a single ovule and only one or a few stamens. The *G.* display a level of organization higher than that of other gymnosperms, and show close morphological and anatomical similarities with the angiosperms. Thus in addition to tracheids, the secondary xylem includes fibers and vessels; in contrast, the phloem is gymnospermic, consisting of sieve cells, parenchyma and albuminous cells.

Most curious of the 3 geographically isolated genera is *Welwitschia* (a single species, *W. mirabilis = bainesii*; subclass *Welwitschiidae,* or family *Welwitschiaceae),* which is confined to the narrow, foggy coastal belt of the Namib Desert from the Kuiseb River northward to Cabo Negro in Angola. The plant consists of a woody, unbranched stem, expanding at the top into a turnip shape which may be up to 1 m in diameter. The top contains a crater-like depression with a rim of green, photosynthetic and meristematic tissue. Inserted into grooves in the rim is a single pair of parallel-veined (with anastomoses) and persistent leaves, which are several meters long. These exceedingly tough, strap-shaped leaves grow continually from basal intercalary meristem, and commonly sprawl in a torn and tangled mess on the desert floor. The tap root is elongated, but rather shallow (no deeper than 3 m) in its substrate, indicating that its primary function is anchorage rather than water uptake. In fact, *Welwitschia* does not display xeromorphic characters, despite its desert habitat. Water is absorbed from morning fog through stomata on the upper surface of the leaves; although the Namib Desert has little rainfall, heavy fog covers the coast for about 300 days of the year. Flowers occur as spike-like florescences on the crown of the stem. The pseudohermaphrodite male flowers (borne on small salmon-colored cones) contain 6 stamens (each with 3 fused pollen sacs) united below into a tube, surrounding a rudimentary ovule. Female flowers consist of an envelope of 2 united perianth segments, and a single terminal ovule with 2 integuments; they are borne on large green to yellowish cones, resembling those of pines. Pollination is by wind (contrary to the former belief that pollination is effected by the large, sucking, nonflying hemipteran, *Probergrothius sexpunctatis,* which feeds on the female cones).

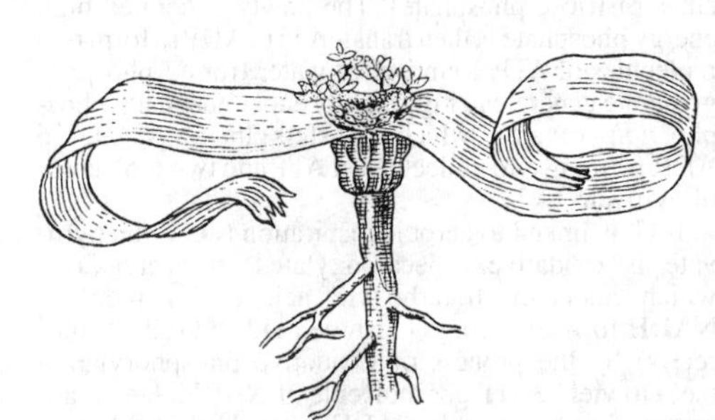

Gnetatae. Welwitschia bainesii. Habit of a young plant with female inflorescences.

Representatives of the genus *Ephedra* (about 35 species; subclass *Ephedridae,* or family *Ephedraceae)* are shrubby with markedly branched shoots and minute foliage leaves, and they display a variety of xeromorphic characters. Species occur in the Mediterranean region, Asia and America. Two species from China yield the medically important alkaloid, ephedrine. Male flowers possess a 2-lobed perianth and a sometimes forked and stalk-like stamen, whose apex carries united groups of pollen sacs. Female flowers have a 2-lobed perianth and a single ovule enclosed by a tubular micropyle.

Gnetum (about 34 species; subclass *Gnetidae*, or family *Gnetaceae)* is a genus of tropical rainforests. Most members are climbers with net-veined leaves, more rarely trees or shrubs. The unisexual flowers are monoecious or dioecious, borne in the axils of ring-shaped, paired scale leaves on spike-like inflorescences. Male flowers have a single stamen and a perianth, while the female flower is a single ovule with 2 integuments (the inner prolonged as a tube) and a perianth.

The *Gnetaceae* and *Ephedraceae* are sometimes called the *joint fir* families.

[C.H. Bornman et al. Welwitschia mirabilis: Observations on general habit, seed, seedling and leaf characteristics. *Madoqua Ser.* II, I, ***53*** (1972)].

Gnetum: see *Gnetatae.*

Gnotobiosis: existence of animals or plants in the absence of all other demonstrable organisms or in the presence of known species. By implication, other organisms or species are microorganisms, and G. is more or less synonymous with germ-free or sterile culture of animals. This is achieved by operative removal of animals from the mother, or from eggs under sterile conditions. They are then maintained in a sterile environment, fed sterile food, and supplied with sterile air. Studies on gnotobiotic (germ-free) animals or gonotobiotes show that they do not suffer from tooth decay, and that wound healing is slower. The entire development of germ-free rats is impeded, compared with that of nonsterile animals. Germ-free animals are particularly useful for studying the metabolism of fed materials by isotopic tracing; distribution of the administered label is due only to the animal's own metabolism, uncomplicated by metabolic reactions of the intestinal flora. Studies with germ-free animals have shown that intestinal flora are normally an important source of certain vitamins; in particular, germ-free rats require a supplement of biotin, in order to maintain growth and to avoid dermatitis and skin ulcers.

Gnotobiotes: see Gnotobiosis.

Gnus: see *Alcelaphinae.*

Goal orientation: orientation behavior that leads an animal to a predetermined destination or goal. In the distance field, the destination cannot be perceived directly by the senses; orientation is then indirect and dependent on the response of the animal to spatial stimuli not directly associated with the final goal (e.g. a homing bird may fly in a particular direction, and may recognize a particular type of terrain). When the animal enters the near field, the destination or goal is in range of the sensory organs (e.g. can be seen, heard, smelled, etc.). See Telotaxis.

Goat-antelope: see *Caprinae.*

Goatfishes: see *Mullidae.*

Goat moths: moths of the family *Cossidae;* see *Lepidoptera.*

Goats, *Capra:* medium-sized, compact, short-tailed ungulates of the tribe *Caprini,* subfamily *Caprinae* (see). The horns of the male are large, slightly curved (saber-like) or spiral, whereas the horns of the female are small and straight and bent slightly backward. Males possess a chin beard. There are no lachrymal grooves or interdigital glands, which are present in sheep. Parts of the skull behind the horns are rather elongated, and the apex of the skull is arched. The underside of the tail is naked, and in the male it houses glands that are responsible for the typical male goat scent. Several different species live in the mountains of Europe, Asia and North Africa. Wild forms include: 1) the ***Markhor*** *(Capra falconeri),* which has corkscrew-like horns, and inhabits the high mountains of Asia; 2) the ***Wild goat*** *(C. aegagrus),* a large, light brown-gray animal, which is found from the Mediterranean region to northern India, and which is the ancestor of the domestic goat *(C. aegagrus* var. *hircus)* (see plate 37); and 3) the ***Ibex*** *(C. ibex),* a magnificent animal, whose saber-like horns are strengthened by transverse ribbing, and which inhabits the Alps and mountains of Central Asia. Since wild goats (like wild sheep) are high mountain animals, populations are widely separated, with specific forms in each mountain area. The exact number of distinct species and subspecies is therefore debatable.

Goatsbeard: see *Rosaceae.*

Goatsuckers: see *Caprimulgiformes.*

Gobies: see *Gobiidae.*

Gobiesocidae, *clingfishes:* a family of fishes in the order *Gobiesociformes (Xenopterygii).* These small, soft-rayed, scaleless, shore fishes have a large flattened head and slender body. They are widely distributed in the tropics, and a few species are found in temperate waters, e.g. the ***Northern clingfish*** *(Gobiesox maeandrium,* syn. *Caularchus maeandricus),* which may attain a length of 10 cm, is found from Califormia to Alaska; and *Lepadogaster bimaculatus* (body length 8 cm) which lives in the North Sea. *Bryssetaeres penniger* inhabits rock pools on the Californian coast, sharing its biotope with sea urchins. Among the larger members of the family is *Chorisochismus* of South African waters. Clingfishes often inhabit the intertidal zone of the benthic environment, living on moist surfaces and in small pools, among stones and loose shells, to which they adhere by means of a ventral holdfast organ (suction disk). This large organ consists mainly of pads of thickened skin, and contributions to its structure are also made by modifications of the widely spaced pectoral fins and the bones of the pectoral girdle. Using specialized movements of the suction disk, clingfishes are able to glide over smooth rocks. When stationary, the suction is extremely powerful, so that heavy stones remain attached to the fish when it is lifted. Members of the related family, *Alabetidae,* are found in Australian seas; they are small, eel-like fishes with a smaller suction disk.

Gobiesox: see *Gobiesocidae.*

Gobiidae, *gobies:* a family of fishes of the order *Perciformes,* containing about 200 genera and at least 1 500 species. They are distributed worldwide, mostly in the tropics and subtropics, in marine or brackish waters, with some species occurring in freshwater. They are small to medium-sized bottom fish, with a rounded, scaly head that is wider than the body, and a large mouth. The pelvic fins are united and modified into an adhesive disk, which lies between the pectoral fins and enables the fish to cling to rocks. Maximal length is 50 cm, and most species are less than 10 cm. The family contains the smallest known fishes; the shortest species is *Trimmatom nanus* (a scaleless fish of the Chagos Archipelago in the Indian Ocean), whose mature females attain only 8–10 mm. *Pomatoschistus microps (= Gobius leopardinus;* ***Common goby****)* occurs along the European shores of the Atlantic, in the Baltic, the Mediterranean, the Black Sea and the Sea of Azov, and it is the most commonly encountered goby in esturaries and intertidal pools of all European coasts. Many gobies live on wet beaches and may spend several days out of water. Of particular interest are the ***Mudskippers*** *(Periophthalmus* and *Periophthalmodon)* of tropical sea coasts of East Africa and Southern Asia, which are sometimes placed in a separate family, the *Periophthalmidae.* They live in the pools and trenches of the intertidal zone, usually holding up their heads, so that their highly placed eyes can see above the water level. With the aid of their muscular pectoral fins they can move overland at considerable speed, and when faced with danger they use their tails to flee across the mud in great leaps (plate 4).

Gobio: see Gudgeon.

Gobius: see *Gobiidae.*

Godetia: see *Onagraceae.*

Godwits: see *Scolopacidae.*

Goettesche larva: see Müller's larva.

Goldcrest: see *Sylviinae.*

Golden algae: see *Chrysophyceae.*

Golden dojo: see *Cobitididae.*

Golden eagle: see *Falconiformes.*

Golden eye: see *Neuroptera.*

Golden-eyed babbler: see *Timaliinae.*

Goldeneyes: see *Anseriformes.*

Golden hamster: see *Cricetinae.*

Golden mountain thrush: see *Turdinae.*

Golden mouse: see *Cricetinae.*

Golden orfe: see Orfe.

Golden oriole: see *Oriolidae.*

Golden plover: see *Charadriidae.*

Golden rain: see *Papilionaceae.*

Goldenrod: see *Compositae.*

Golden snakes, ***flying snakes,*** *Chrysopelea:* a genus of slender, brightly patterned, tree-dwelling snakes of the family *Colubridae* (see), subfamily *Boiginae* (see), found in Southeast Asia and the Indo-Australian islands. By flattening their bodies, they are able to glide for short distances from the higher branches of trees. They prey mainly on lizards and small birds, which are paralysed by the weak venom of their opisthoglyphic venom teeth.

Golden thistle of Peru: see *Papaveraceae.*

Golden tree frog: see *Hylidae.*

Golden whistler: see *Pachycephalinae.*

Goldfinches: see *Fringillidae.*

Goldfish, ***johnny carp,*** *Carassius auratus:* one of the 2 species of carp in the genus *Carassius* (the other is the crucian carp). It is not as high set as the crucian carp, and it lacks the black tail spot of the latter. In the G., the serrations on the third hard ray of the dorsal fin become larger toward the tip of the ray, whereas in the crucian carp, the serrations remain about the same size. Originating in East Asia, the G. has been bred and cultured in China for more than 1 000 years, with the production of many varieties and ornamental forms, which are now usually regarded as a subspecies, *Carassius auratus auratus.* Examples of specially bred varieties are "harlequins" (double tails like small peacocks), "telescope fishes" (protruding telescoped eyes) and "egg fishes" (greatly rounded bodies). The decorative forms are gold, brown or silver-white with a variety of black or white markings. It is widely kept in small ponds and aquaria.

A very close relative is the ***Gibel*** *(Carassius auratus gibelio),* which may also be called a G. Found in the Aral Sea basin and in European waters, it differs from the East Asian G. only in size, and is perhaps descended from the latter.

Gold frog: see *Mantellinae, Brachycephalidae.*

Goldman equation: an equation relating the permeabilities of the main participating ions to the resting membrane potential.

$E = RT/F \times \ln \{(P_{Na}[Na]_O + P_K[K]_O + P_{Cl}[Cl]_i)/(P_{Na}[Na]_i + P_K[K]_i + P_{Cl}[Cl]_O)\}$,

where [ ] denote the extracellular (O) and intracellular (i) concentrations of the respective ions.

Gold-sinny: see Wrasses.

Golgi apparatus, ***Golgi complex, Golgi body:*** a series of flattened, membrane-bounded cisternae, arranged in a stack like a pile of plates (see plate 21, Fig.5; also Fig.2 of the entry, Endoplasmic reticulum). One or more G.as. are present in the cytoplasm of practically all eukaryotic cells. Under the electron microscope, the G.a. appears as a stack of 4–10 parallel cisternae, each cisterna typically about 1 µm in diameter. The G.a. of mammalian cells typically consists of a stack of 5 or 6 cisternae. In plant cells, each G.a. may consist of as many as 20 stacked cisternae. When the G.a. consists of stacked cisternae, it may generally be referred to as a *Dictyosome,* but this term is usually reserved for the stacked Golgi cisternae of plants. In some fungi, there are no stacks, and each G.a. is represented by a single Golgi cisterna. The two faces of the stack are functionally and structurally different, i.e. the G.a. displays a distinct polarity. The *trans* face of the stack is typically concave, whereas the other surface is usually convex and known as the *cis* face. Between the *cis* and the *trans* compartments is the *medial* compartment ("compartment" is a functional term, and each compartment may consist of more than one cisterna). Since the G.a. can be isolated as a unit from cell homogenates, the individual stacked cisternae must be attached to each other. Filaments observed between the cisternae are possibly involved in this attachment. There is, however, no direct continuity between the lumens of neighboring cisternae.

In animal cells, the G.a. is typically seen between the cell nucleus and the centriole, but in plants the dictyosomes are dispersed throughout the cell. Associated with the G.a. are swarms of small vesicles (about 50 nm in diameter), which transport proteins and lipids to and from the Golgi and between the Golgi cisternae. Some of these vesicles are coated with clathrin or other types of coat protein, and such *coated vesicles* can be seen in the process of abstricting from the dilated rims of the Golgi cisternae and on the *trans* side of the G.a. Proteins destined for the G.a. are synthesized in the rough endoplasmic reticulum (ER), while lipids arise in the smooth ER. Such proteins and lipids are the precursors of substances subsequently secreted from the cell or used in the synthesis of membranes. They are transported from the ER to the G.a. in small vesicles, more rarely by direct fusion of the ER with the *cis* face (entry or regeneration side) of the G.a. Proteins received from the ER are processed as they pass from one cisterna to the next. The mechanisms of protein and lipid transfer between neighboring cisternae is uncertain, but coated vesicles seen abstricting from cisternal rims are thought to be responsible for this "traffic". Passage of material through the stack is unidirectional *(cis → medial → trans* compartment) and intervening compartments are never bypassed. Each vesicle must therefore be programmed (i.e. it contains appropriate receptors) for fusion with the next compartment in the sequence. During the processing of proteins, each cisterna functions as a distinct compartment with its own complement of processing enzymes, so that the stack serves as a multistage processing unit. Glycoprotein synthesis occurs in the Golgi cisternae by attachment of oligosaccharide chains to proteins (glycosylation of *N*-glycan-type glycoproteins starts with the transfer of an oligosaccharide chain in the lumen of the ER, and continues in the G.a.). Thus, the cisterna(e) on the *cis* or protein-receiving side of the stack are characterized by the presence of mannosidase I; those in the middle *(medial* compartment) by mannosidase II and *N*-acetylglucosamine transferases; and those on the *trans* or protein-releasing side by galactosyl and sialyl transferases. Synthesis of secretory or cellular oligo- and polysaccharides from glucose also occurs within the Golgi cisternae, and the enzymes for this synthesis (glycosyl transferases) are transported to the G.a. from the rough ER. Further typical enzymes of the G.a. are thiamine pyrophosphatase and acid phosphatase. Also synthesized in the G.a. are: 1) the acidic carbohydrates of the outer layer of the plasmalemma, 2) the glycocalyx of animal cells, 3) the ground substance of the plant cell wall (protopectin and hemicelluloses), and 4) the mucilaginous polysaccharides secreted by root cap cells. Certain highly specialized structures are also known to be synthesized by the G.a., e.g. the complicated fibrillar cell wall scales of planktonic flagellates of the genus *Chrysochromulina.*

The membrane on the *cis* side (synthesis or regeneration side or pole) of the Golgi stack still displays the typical thickness (5 nm) of the ER membrane; it is largely derived from vesicles abstricted from the ER. The *trans* face is extended as a tubular reticulum called the *trans Golgi network* (TGN), in which the proteins for export are sorted before enclosure in specific transport vesicles, according to their final destination. Secretory material is released from the cell by exocytosis, i.e. the transport vesicle membrane fuses with the plasmalemma, and the contents of the Golgi vesicle are emptied to the exterior. The membranes of these transport vesicles have the same thickness as the plasmalemma (9–10 nm), and they contain the transmembrane proteins and lipids required for renewal of the plasmalemma. Alternatively, transport vesicles may contain substances for use by the cell itself (e.g. they may contain newly synthesized lysosomal hydrolases, which they transport to endolysosomes). Thus, from the *cis* to the *trans* side of the Golgi stack, the membrane becomes modified from the type found in the ER to that present in the plasmalemma. By the continual stream of vesicles from the ER to the *cis* side of the Golgi stack, and the continual release of vesicles from the *trans* side, the Golgi cisternae and the entire Golgi stack are in a steady state of constant removal and renewal, i.e. there is a "membrane flow", during which the membranes change both their composition and their structure. In cells that display intense secretory activity, Golgi cisternae can be newly formed in a few minutes. Thus, proteins synthesized within a few minutes in the rough ER of the exocrine pancreas can be detected about 10 minutes later in the G.a., and about 30 minutes after that in transport vesicles (in this case also known as transport granules). Through the accumulation of secretory material, Golgi cisternae may become expanded into vacuole-like structures.

Primary Lysosomes (see) arise by abstriction from the ER or from Golgi cisternae. Direct fusion between the ER and Golgi cisternae containing acid phosphatases may occur on the *trans* side of the Golgi stack (in the GERL region: G.a. bound to ER, serving the formation of Lysosomes), especially in gland cells and other secretory cells.

In the division of plant cells, Golgi vesicles fuse to form the cell plate, which is the first cell wall primordium. The vesicle contents provide the cell plate material (acidic polysaccharides), while the vesicle membranes become the new plasmalemmae. Blood plasma lipoproteins are elaborated in the G.a. of the liver, the protein part being formed in the rough ER, and the lipid moiety in the smooth ER.

A G.a. was first described in nerve cells by C. Golgi in 1898 as a reticulate structure made visible under the light microscope by staining with OsO_4 or $AgNO_3$, but this is not identical with the G.a. seen under the electron microscope, or by phase contrast light microscopy in unstained living cells.

Golgi body: see Golgi apparatus.

Golgi complex: see Golgi apparatus.

Goliath frog: see *Raninae.*

Gommelin: see Dextrins.

Gomphidae: see *Odonatoptera.*

Gomphosis articulation: see Joints.

Gonadotropins, ***gonadotropic hormones:*** glycoprotein hormones synthesized in the anterior lobe of the pituitary and in the placenta. They stimulate the growth of the gonads, and they stimulate the gonads to produce sex-specific hormones (estrogens and progesterone in females; androgens in males). The G. include Follitropin or Follicle-stimulating hormone (FSH), Lutropin or Luteinizing hormone (LH), Human menopausal gonadotropin (hMG), Prolactin and Choriogonadotropin (hCG).

Gonads, ***sex glands:*** animal organs which produce gametes (eggs or spermatozoa). Female G. are called ovaries (see Ovary); male G. are called Testes (see Testis). Many hermaphrodite animals possess separate ovaries and testes, but in pulmonate snails the hermaphrodite gland or ovotestis functions as both ovary and testis.

During embryonic development the G. differentiate later than all other organs. In contrast, the gonocytes (which eventually develop into gametes) arise from the embryonic endoderm very early in development. In vertebrates the G. arise from a pair of genital ridges projecting from the roof of the splanchnocoel median to the mesonephroi. The genital ridges are then invaded by gonocytes which migrate there via blood vessels. Sexual cords (consisting mainly of coelomic epithelial cells with some gonocytes) then grow from the epithelium of the genital ridge down to the connective tissue below. Subsequent development differs for male and female G.

In the male, the sexual cords join with tubules at the anterior end of the mesonephros, and these tubules then become the vasa efferentia. The sexual cords become seminiferous tubules, and the gonocytes develop into spermatogonia. In the female, however, the primary sexual cords do not give rise to eggs. A second downgrowth of cells (secondary sexual cords) become arranged in small groups, known as primary follicles.

See Oogenesis, Spermatogenesis.

Gonane: see Steroids.

Gonatus: see *Cephalpoda.*

Gondwana, *Gondwanaland:* a former extensive southern continent, which consisted of parts of the present land masses of South America, Africa, Madagascar, India, Australia and the Antarctic., and which had its own characteristic flora and fauna. At the end of the Mesozoic and during the Cenozoic, it was split by movements of the ocean floor into the present land masses of the Southern Hemisphere. The *Glossopteris* flora or Antarcto-carboniferous flora *(Glossopteris, Gangomonopteris, Noeggerathiopsis)* is typical of the Upper Carboniferous period of G. Further evidence for the existence of a uniform southern continent during the Permian and Triassic is provided by the distribution of the reptile, *Mesosaurus,* and the occurrence of terrestrial, herbivorous dicynodont reptiles (e.g. *Lystrosaurus* and *Kannemeyeria)* in South America, India and the Antarctic during the Lower Triassic. The existence of marsupials and the monkey puzzle tree *(Araucaria)* in both South America and Australia are examples of modern biological indicators for the previous existence of G. Traces of several Permocarboniferous glaciations (tillites and glacial scouring), found in South America, Africa, India and Australia, are characteristic of G.

The name means "Land of the Gond". The Gond were formerly a tribe in India, and the name is now given to the landscape of Madhya Pradesh.

Goniatites, *Goniatitina* (Greek *gonia* angle): an ammonite order with planispiral, smooth or simply ribbed shells. The suture line is relatively simple (goniatitic type; see Ammonites), and in primitive forms it consists of three undivided lobes. The number of lobes may, however, be increased by subdivision of a saddle on the external or internal surface of the shell.

Fossils are found from the Middle Devonian to the Upper Permian. Individual goniatite genera and species are important index fossils of the later Paleozoic.

Goniatites. Stone cast of a rounded swollen species of *Goniatites,* showing suture lines; from the upper layers of the Lower Carboniferous (natural size).

Goniatitic ammonites: see Ammonites.

Gonidia: a botanical term with two different meanings. *1)* Spores that are not produced by reduction division (see Meiosis). *2)* An old term for the algal symbiont in the thallus of a lichen.

Gonioclymenia (Greek *gonia* angle, *Klymene* daughter of the sea god *Oceanos):* an ancient genus of ammonites, which are index fossils of the Upper Devonian. The planispiral, subdiscoidal shell has a wide umbilicus with a groove on the venter. Faint simple ribbing is apparent on the surface. The simple suture has deep, undivided ventral lobes, and sometimes two pairs of lateral saddles; the latter may be partly or entirely divided by marginals.

Gonococcus: see *Neisseria.*

Gonoduct: a duct that serves for the conduction of gametes. In the female it is known as the oviduct (see), in the male as the vas deferens (see Testis).

Gonomery: see Fertilization.

Gonopodium: copulatory organ of teleost fishes and the viviparous *Poeciliidae* (guppies). In the *Poeciliidae,* the G. has evolved by modification of the anal fin rays, and it has shifted its position to the ventral pectoral region. In male teleost fishes, the posterior ends of the two pelvic fins are fused, and each bears a rod-shaped, grooved process, also known as the clasper, projecting backward from behind the cloaca. The claspers (i.e. the gonopodium) are thrust into the female cloaca, and the seminal fluid (containing spermatozoa) passes along the grooves.

Gonopods: in crustaceans and millipedes, extremities that are modified as copulatory organs.

Gonosomes: see Sex chromosomes.

Gonostomatidae: see *Stomiiformes.*

Good King Henry: see *Chenopodiaceae.*

Goosander: see *Anseriformes.*

Goose: see *Anseriformes.*

Goose barnacles: see *Cirripedia.*

Gooseberry: see *Grossulariaceae.*

Goosefishes, *Lophiidae:* a fish family of the order *Lophiiformes* (see), containing 25 species in 4 genera, found in the Arctic, Atlantic, Indian and Pacific Oceans. Their mobile fishing pole device has a flap of flesh at the tip, which appears to attract prey to within easy reach of the large mouth. The up to 2 m-long ***Angler*** or ***Frogfish*** *(Lophius piscatorius)* of the eastern Atlantic was once thought to be synonymous with *Lophius americanus* ***(Monkfish),*** and is still known to European fisherman as the Monkfish. It has a short, tapering body, a large head, which is broader than it is long, a very wide mouth with a protruding lower jaw, and 2 rows of sharp, movable, backward-directed teeth. Although unattractive in appearance, the head and skin are removed, and the flesh is sold on the fish market.

Goosefoot family: see *Chenopodiaceae.*

Goose grass: see *Rubiaceae.*

Gopher: see *Sciuridae.*

Goral, *Nemorhaedus goral:* a chamois-like horned ungulate (see *Bovidae)* of the subfamily *Caprinae* (see), found in the high mountains (1000–4000 m) of Asia. Two subspecies are recognized: ***Cashmere goral (N.g. goral), North Chinese goral*** *(N.g. caudatus).*

Gordioidea: see *Nematomorpha.*

Gorgonacea: see *Gorgonaria.*

Gorgonaria, *Gorgonacea, horny corals, gorgonian corals:* an order of the *Octocorallia* (see). In addition to individual sclerites, the upright branching colonies also have a noncalcareous, horny internal supporting axis [the axis of *Corallia* (see) has a different composition]. The organic horny material of the axis, known as gorgonin, is composed of proteins and mucopolysaccharides.

The axis is secreted by an invagination of the basal epithelium, and is therefore strictly an external structure, but it is surrounded by a cylinder of coenenchyme with polyps on the outside, and is therefore functionally an internal skeleton. In the coenenchyme are embedded various shapes and colors of calcareous spicules, which are responsible for the yellow and pale blue color of certain species; yellow-brown colors, however, are due to the presence of zooxanthellae. The axis is sometimes used for decorative purposes. The order includes, e.g. *Leptogorgia* ***(Whip corals),*** *Gorgonia* ***(Sea fans)*** and *Corallium* (see) ***(Precious corals).***

Gorgonia: see *Gorgonaria.*

Gorgonian corals: see *Gorgonaria.*

Gorgonocephalus: see *Ophiuroidea.*

Gorilla, *Gorilla gorilla* (plate 41): the largest of the Great apes (see). The 3 subspecies are geographically isolated: *G.g. gorilla* ***(Western lowland G.)*** is found in southeastern Nigeria, Rio Muni, Gabon and Brazzaville. *G.g. beringei* ***(Mountain G.)*** is found at 3 000 m on the slopes of the Virunga volcanoes on the borders of Zaire, Rwanda and Uganda. *G.g. grauerei* ***(Eastern lowland G.)*** lives at lower altitudes between Zaire and Lakes Tanganyika and Edward. The G. is predominantly terrestrial. Adult males rarely climb trees, and they build ground nests. Females and young build tree nests near the ground, and may climb to obtain food. The diet is mainly vegetarian, consisting of leaves, bark, pith, stems, roots, bamboo, various whole ground plants, such as nettles, thistles, wild celery. Large quantities must be consumed to sustain such a large animal. Invertebrates like insects and snails are also sometimes taken. The adult male of each race develops a saddle of gray hairs across its back, and is then known as a "silverback". All 3 races are under pressure from poaching and destruction of their habitat. There is particular concern for the mountain G., of which only a few hundred individuals remain. Because of the civil war in Rwanda, the program for establishment of a fully protected region for mountain G. was suspended in 1994.

Goshawk: see *Falconiformes.*

Gossypium: see *Malvaceae.*

Goura: see *Columbiformes.*

Gouramies: members of the families *Anabantidae* ***(Climbing gouramies*** or ***Labyrinth fishes;*** 3 genera, about 40 species), *Belontiidae* (11 genera, about 28 species), *Helostomatidae* ***(Kissing gourami;*** single species) and *Osphronemidae* ***(Giant gourami;*** single species). The *Belontiidae* are divided into 3 subfamilies: *Belontiinae* ***(Combtail gouramies),*** *Macropodinae* ***(Siamese fighting fishes, Paradisefishes, etc.),*** *Trichogastrinae* ***(Gouramies, etc.).*** The 4 families constitute the suborder *Anabantoidei* of the *Perciformes.* In most species of the *Anabantoidei* the male builds a nest of floating bubbles, eggs are laid among the bubbles, and the male displays brood care. Some species of the *Macropodinae* are mouth brooders. Some members of the *Anabantidae* have a labyrinth organ in a cavity above the gills, in which they store air. This air store must be replenished at intervals to avoid drowning, and these fish can spend up to a day out of water among damp vegetation. The ventral fins are placed forward and often modified into sensory fibers. Some members of the *Belontiidae* are popular aquarium species, e.g. ***Blue gourami*** *(Trichogaster trichopterus sumatranus),* a light bluish gray, freshwater species from Indonesia and the Malay Peninsula; ***Paradise fish*** *(Macropodus opercularis),* which is common in the rice fields of China, where it successfully endures a wide range of temperatures; and the ***Veiltail betta*** or ***Siamese fighting fish*** *(Betta spendens)* of Southeast Asia, originally introduced as an aquarium fish early in the 20th century.

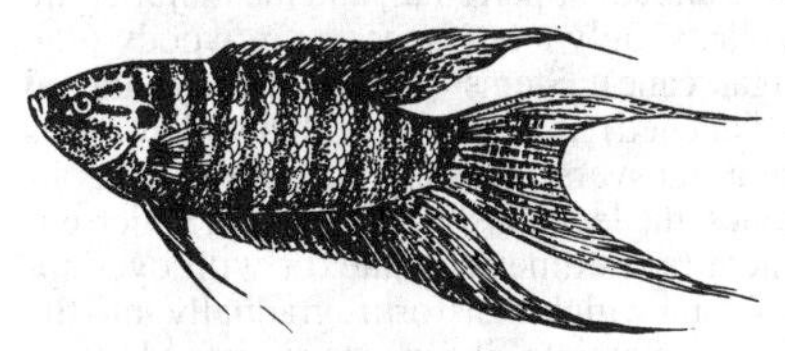

Gouramies. Paradise fish *(Macropodus opercularis).*

Gourd family: see *Cucurbitaceae.*

Gourinae: see *Columbiformes.*

G_0-phase: see Cell cycle.

G_1-phase: see Cell cycle.

G_2-phase: see Cell cycle.

G proteins: membrane-bound, regulatory proteins that bind guanosine triphosphate (GTP). They are part of a signal transducing system, which consists of receptor, G protein and effector. G protein is activated by binding GTP. Activation is then abolished by slow hydrolysis of the GTP to GDP. A G protein called transducin occurs in the membranes of retinal rods, and is involved in the visual process. A G protein (G_k) functions in the activation of K^+ channels in heart cells. Other examples are G_s and G_i, which function in systems that regulates adenyl cyclase.

Graafian follicle: see Oogenesis.

Grackles: see *Icteridae.*

Gracula: see *Sturnidae.*

Gradation: see Population ecology.

Grain moth: see European grain moth.

Grain thrips: see *Thysanoptera.*

Gramicidins: cyclic or linear peptide antibiotics produced by various strains of *Bacillus brevis.* The first to be isolated (1942), was gramicidin S (Fig.). Gramicidin D is now known to be a complex of four components (A, B, C and D). Gramicidin A is a linear peptide of alternating D and L-amino acids: HCO-L-Val-Gly-L-Ala-D-Leu-L-Ala-D-Val-L-Val-D-Val (L-Trp-D-Leu)$_3$-L-Trp-NHCH$_2$CH$_2$OH. A second series, known as the isoleucine G., has an isoleucine instead of valine in position 1. Gramicidin B has Phe in position 11, while gramicidin C has Tyr at this position.

The linear G. act as ionophores, forming channels through biological membranes which allow the passage of monovalent cations, and result in the equilibration of intra- and extracellular N^+ and K^+. The gramicidin A channel has been shown by physical methods to be a single-stranded helix of two gramicidin A molecules.

Gramicidin S is active against Gram-positive but not Gram-negative bacteria, and is therefore used for the teatment of wound infections.

Pro —Val —Orn—Leu—D-Phe
| |
D-Phe—Leu—Orn—Val —Pro   *Gramicidin S*

Graminae, *Poaceae,* ***grass family:*** a family of monocotyledonous plants containing about 8 000 species distributed throughout the world. In many areas they form a predominant component of the vegetation, and they may determine the entire biome, e.g. in the steppes and pastures of Europe and Asia, the prairies of North America, the savannas of Africa and the pampas of South America.

They are are annual or perrenial, and the majority are herbaceous plants; only a few species are woody (e.g. bamboo, sugar cane). Stems (culms) are cylindrical (sometimes flattened) and hollow, with swollen nodes which house a transverse diaphragm. Leaves are solitary at the nodes; the leaf consists of a sheath, which encircles the stem (sometimes connate, or with overlapping margins), and which transforms gradually into the blade [rarely with a petiole-like constriction, and sometimes with thickened projections (auricles) at each side of the base]. At the junction of the sheath and blade is a small flap of tissue known as the ligule, sometimes replaced by a ring of hairs. The blade is usually long and narrow, rarely broad.

Flowers are grouped in partial inflorescences (spikelets) containing one or many flowers. Spikelets are in turn aggregated into various types of inflorescences. Thus grasses can be subdivided into 3 groups according to the structure of the inflorescence: *eared* or *spicate* grasses (spike, in which the spikelets have no stalks and are seated directly on the main axis); *paniculate* grasses (panicle, in which the spikelets are borne on stalks on branches from the main axis); *racemose* grasses (spike or raceme, in which the spikelets have short stalks to the main axis). Each spikelet usually has 2 bracts *(glumes)* at the base. Each individual flower is also usually enveloped by 2 bracts, the lower being membranous to coriaceous and known as the *lemma,* the upper being generally thin and delicate and known as the *palea.* Between the flower and the palea are 2 delicate scales (occasionally absent) known as *lodicules.* The wind-pollinated flowers are hermaphrodite (rarely unisexual) with a superior ovary, usually 2 styles and usually 3 stamens. The ovary wall is nearly always united with the seed coat, forming a type of fruit known as a *caryopsis.*

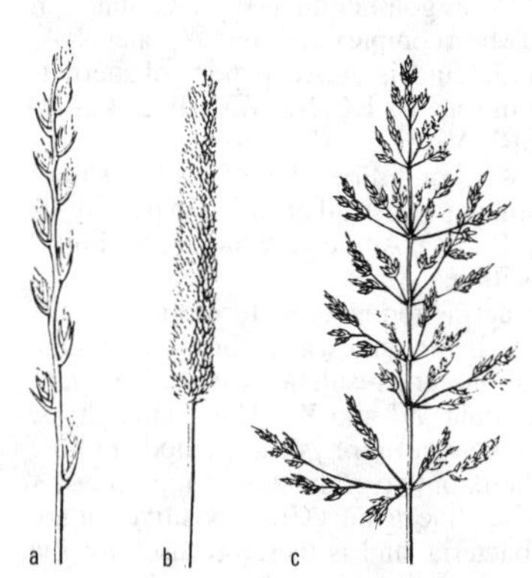

Fig.1. *Graminae.* Types of inflorescence. *a* Eared or spicate. *b* Racemose. *c* Paniculate.

Grasses characteristically contain proteins known as prolamines and glutelins. A combination of these two proteins forms the gluten of wheat and rye grains, which makes the flour capable of rising when made into bread. Oat and rice flours contain no prolamines and are therefore unsuitable for bread making.

Many grasses are economically important, especially the cereal crops, whose grains contain large quantities of carbohydrate, as well as smaller quantities of protein. Important crops of temperate regions are wheat, rye, barley and oats, as well as maize which is originally a crop of warmer climates, where it is grown widely.

Wheat *(Triticum),* an eared grass, is an ancient cultivated plant and one of the most important cereals in the world economy, occupying a greater growth area than any other agricultural plant. Its grains are processed, inter alia, to flour, semolina and starch. Of the 21 wild and cultivated species of wheat, there are 3 evolutionary series, which are distinguished morphologically, but in particular according to their chromosome number: the diploid *one-grained wheats,* the tetraploid *emmer wheats,* and the hexaploid *spelt wheats.* The most widely distributed and economically important species is the spelt wheat, *Triticum aestivum,* commonly known simply as "wheat".

Rye *(Secale cereale),* an eared grass, is an important bread cereal, and its grains are also malted, fermented, and used to prepare spirits. It is also used to some extent as a green fodder.

Barley *(Hordeum;* the most important species is *H. vulgare)* is a racemose grass. The 1-flowered spikelets occur in groups of 3, alternating on opposite sides of the spike axis, the middle one bisexual and stalkless, the 2 lateral ones male or barren and often much reduced in size, on very short stalks. In the polystichous forms, which are planted largely as forage crops, all 3 spikelets are fertile. In the distichous forms, which are widely used for beer brewing, only the middle spikelet is fertile. Also prepared from barley are pearl barley, coffee substitute, etc.

Winter wheat, rye and barley are planted in the autumn. They remain in the ground during the winter, and ripen 8–14 days before the summer forms of these cereals, which are planted in the spring.

Economically the most important paniculate grass of temperate climates is the ***Oat*** *(Avena sativa),* whose grains are excellent as cattle and horse fodder. Oats are also cultivated as forage, as well as being used for the manufacture of rolled oats, oatmeal, etc.

Maize or ***Sweet corn*** *(Zea mais),* which probably originated from Central America, is used for many purposes. The male inflorescences of this large, sturdy grass are situated terminally on the halm, whereas the female inflorescences are enveloped by leaf sheaths and lie in the axes of the leaves. Its grains are used to prepare flour, semolina and starch. They are also used as animal feed, or eaten unripe as a vegetable. Maize is also planted as green fodder.

Rice *(Oryza sativa),* a paniculate grass, is the most important cereal crop of the tropics and subtropics. Dry or mountain rice is grown with constant sprinkling of the soil with water, whereas swamp or water rice, which is considered to be the better quality product, is grown in standing water. Rice grains are eaten steamed or boiled. In East Asia, alcoholic beverages are prepared by fermentation from rice (rice beer, rice wine, arrack). Rice starch is used in the food and textile industries. Rice straw is used for basket making, and as a raw material for paper manufacture. The best quality cigarette paper is also made from rice straw.

Fig.2. *Graminae.* The rice plant. *a* Leaves, stems and inflorescences. *b* Inflorescence (panicle). *c* Flower. *d* ear. *e* Seed (caryopsis).

Common millet *(Panicum miliaceum)* is not as widely planted as formerly. Its grains are eaten as a porridge or bread cakes, and they are also fermented to prepare beer and spirits. In Africa, the most important bread cereal is *Sorghum bicolor* ***(Durra, Brown durra, Sorghum,*** or ***Guinea corn).*** In Africa and Asia, *Eleusine coracana* ***(Ragi millet, Raggee,*** or ***Tocusso)*** is planted as a food and for brewing beer. In Ethiopia, ***Teff*** *(Eragrostis tef = E. abyssinica)* is cultivated as a cereal.

Sugar cane *(Saccharum officinarum)* is also an economically important member of the grass family. Originating from tropical Asia, this sucrose-yielding plant is now cultivated all over the tropics.

Vetiver grass *(Vetiveria zizanioides)* and ***Lemon grass*** *(Cymbopogon confertiflorus)* are sources of essential oils used mainly in the manufacture of cosmetics and toiletries.

Bamboos have various technical uses, and are particularly important as a source of building material. These gigantic grasses, which can attain a height of 30 m, and some of which have strongly lignified stems, occur practically exclusively in the tropics and subtropics. The approximately 250 species of bamboo belong to various genera, the most important being *Bambus, Dendrocalamus, Phyllostachys, Arundinaria* and *Sasa.*

Important fodder grasses of European pastures are as follows. Eared grasses: ***Rye grass*** *(Lolium perenne)* and ***Italian rye grass*** *(Lolium multiflorum).* Paniculate grasses: ***Timothy*** *(Phleum pratense)* and ***Meadow foxtail*** *(Alopecurus pratensis).* Racemose grasses: ***Common bent-grass*** *(Agrostis tenuis),* ***Fiorin*** or ***Creeping bent*** *(Agrostis stolonifera),* ***Tall*** or ***False oat grass*** *(Arrhenathenum elatius),* ***Yellow or Golden oat-grass*** *(Trisetum flavescens),* ***Yorkshire fog*** *(Holcus lanatus),* ***Cock's foot*** *(Dactylis glomerata),* ***Meadow grass*** *(Poa pratensis),* ***Upright brome*** *(Bromus erectus),* ***Meadow fescue*** *(Festuca pratensis),* ***Creeping fescue*** *(Festuca rubra)* and ***Sheep's fescue*** *(Festuca ovina).*

Graminae-type stomata: see Stomata.

Graminicides: herbicides for the elimination of unwanted grasses.

Grampus: see Whales.

Gram stain: staining method, developed by the Danish bacteriologist H.C.J. Gram, for the differentiation of bacteria into two groups, based largely on differences in cell wall structure. In 1884, Gram found that anthrax bacilli stained with Ehrlich's methyl violet, then treated with iodine, could not be destained even with ethanol. Not all bacteria, however, could be permanently stained by this procedure, i.e. the stain was readily removed by ethanol, and the resulting transparent preparation could be counterstained with a different dye. The usual laboratory procedure consists of treatment of a heat-dried bacteriological smear (on a microscope slide) with a dye (usually methyl violet), followed by treatment with iodine-potassium iodide solution, ethanol, and finally a counterstain of fuchsin. In *Gram-positive* organisms, the iodine-methyl violet complex is retained by the cell wall after washing with ethanol, so that the cells are dark violet under the microscope, e.g. streptococci, bacilli, actinomycetes. In *Gram-negative* organisms, the iodine-methyl violet complex is completely removed by the ethanol, and the cells acquire a red color from the counterstain of fuchsin, e.g. coliform bacteria and salmonella. A few species of bacteria give an indefinite or mixed result, and are known as *Gram-variable.*

Granatellus: see *Parulidae.*

Grand postsynaptic potential: see Nerve impulse frequency, Neuron.

Grandry corpuscles: see Tactile stimulus reception.

Granite night lizard: see Night lizard.

Grantia: see *Porifera.*

Granulose: see Insect viruses.

Granulosis virus: see Insect viruses.

Grape: see *Vitaceae.*

Grape erineum mite: see *Tetrapodili.*

Grapefruit: see *Rutaceae.*

Grape phylloxerid: see *Homoptera.*

Grape vine: see *Vitaceae.*

Graptemys: see *Emydidae.*

Graptolites, *Graptolithina, Graptozoa* (Greek *graptos* written, *lithos* stone, *zoon* animal): an extinct class of invertebrates, whose classification has long been in doubt. Most modern authorites would now place them with the *Hemichordata,* living representatives of which include *Rhabdopleura* and *Cephalodiscus.* G. were marine colonial organisms, with small dendritic or saw-blade-like colonial exoskeletons (periderm), which consisted of two scleroproteinaceous layers. The inner layer *(fusellar tissue)* consisted of regular half-ring segments *(fuselles),* interlocking with one another on both sides to form a zig-zag-shaped junction. This inner layer was overlain by an outer *cortex* of thin, parallel laminae. Exoskeletons consisted of a variable number of cups *(thecae)* along one or more branches *(stipes).* The colony *(rhabdosome),* comprising a variable number of stipes, developed from a conical embryonic *sicula* by systematic budding. Some species formed clusters of rhabdosomes radiating from a common center, and known as *synrhabdosomes.* Each theca housed a single zooid (polyp) of the colony. The thecae were set obliquely and were more or less overlapping, and they displayed a variety of forms: elongated, cylindrical, rectangular or conical sacs, with a distal external aperture; this aperture was circular or quadrate and directed upward, or contracted and directed backward (so that the theca resembled a bird's head). Some fossil G., especially the *Dendroidea* and *Tuboidea,* possessed three types of thecae, known as *autothecae, bithecae* and *stolothecae.* Members of the *Dendroidea* also possessed *nematothecae,* minute tubular thecae scattered on the surface of the autothecae, and thought to house defensive zooids or nematocysts. The rhabdosome was long and narrow, or leaf-shaped; elongate rhabdosomes were straight, bent, or spirally enrolled, sometimes with side branches, some species also displaying arborescent to reticulate branching. Thecae were arranged in one, two, or four series along the stipe like the teeth of a single- or double-edged saw blade. In some species a sclerotized stolon passed through the length of the rhabdosome; in those species lacking an obvious stolon, this may have been present but nonsclerotized. Six orders are recognized, but the majority of known G. belong to the *Dendroidea,* or the *Graptoloidea.*

Order: *Dendroidea.* These were sessile forms, mostly benthic, occasionally planktonic. Three types of thecae were present: *autothecae, bithecae* and *stolothecae,* produced by regular triad budding of the stolon. The colony developed from a conical sicula that was typically erect and appears to have been attached by a stalk *(nema* or *virgula)* to a substratum. Adjacent stipes were anastomosed or joined together at intervals by small tranverse bars (dissepiments).

Order: *Graptoloidea.* These were planktonic or epiplanktonic forms, possessing fewer stipes than the *Dendroidea,* and only one type of theca (autotheca), which often had a modified aperture. There was no sclerotized stolon, and the stipes were uniserial, biserial, or quadiserial, suspended from a float to which they were attached by a nema or virgula.

G. are very useful index fossils, since they were abundant, widespread, the result of rapid evolution, and easily identified. Especially planktonic forms (e.g. *Graptoloidea)* became distributed worldwide and are therefore reliable index fossils for intercontinental correlation, and for the delineation of ancient seaways.

G. are found from the Middle Cambrian to the Lower Carboniferous, with their evolutionary peak in the Ordovician. They are commonest and most abundant in black shales, usually preserved as flattened carbonaceous films, resembling scroll saw blades or written hieroglyphics (hence the name); but they are also present in other shales, sandstone and limestone, and in chert nodules.

See *Climacograptus, Cyrtograptus, Dictyonema, Phyllograptus.*

Graptoloidea: see Graptolites.

Grasses: see *Graminae.*

Grasshopper mice: see *Cricetinae.*

Grasshoppers: see *Saltatoptera.*

Grasshopper warblers: see *Sylviinae.*

Grass pea vine: see *Papilionaceae.*

Grassquits: see *Fringillidae.*

Grass snake: see Common European grass snake.

Grass sponge: see Bath sponges.

Grass warblers: see *Sylviinae.*

Gravidity: see Gestation.

Graviportal locomotion: see Movement.

Gray: see Dose.

Gray alder: see *Betulaceae.*

Gray lag goose: see *Anseriformes.*

Grayling, *Thymallus thymallus:* a fish in the order *Salmoniformes.* It inhabits fast flowing rivers and streams in mountainous and hilly regions in the Northern Hemisphere, where it feeds on insect larvae, especially caddis larvae and the larvae and pupae of midges, but also taking flying insects, salmon eggs and small fish. Spawning occurs from March to April. The G. is characterized by a high, long dorsal fin, which is larger in the male. In the mating season, the male dorsal fin is radiant reddish purple with dark spots. At the age of 3–4 years, the body length of a G. is about 30 cm with a weight of 250 g. Up 14 years, individuals rarely exceed 50 cm (1 kg), and the maximal size is about 60 cm (2–3 kg). The G. region of a river lies downstream of the Char region. G. are found only in clear, cool, well oxygenated water with strong currents, preferably with deep holes and large stones on the river bottom. Rivers flowing down a gradient of 20 meters per km appear to be suitable, i.e. the same type of river that favors trout, chars and sculpins. The G. has local importance as a game fish. Although an esteemed table fish, its special flavor is lost a few hours after death. In Scandanavia, the G. is also found in lakes. The Latinized name is derived from the thyme-like smell of the fish.

Grayling region: see Running water.

Gray seal: see *Phocidae.*

Gray squirrel: see *Sciuridae.*

Gray tree frog: *see Rhacophoridae.*

Grazers: see Modes of nutrition.

Great apes, *Pongidae:* a family of Old World monkeys (see), containing the Gorilla (see), Chimpanzees

Great apes (heads). *a* Gorilla. *b* Chimpanzee. *c* Bonobo (Pygmy chimpanzee). *d* Orang-Utan (male).

(see) and Orang-utan (see). Recent G.a. are large, tailless mammals with relatively short legs and long arms, prognathous lower face, and powerful teeth. On the basis of their anatomy, morphology and mental capabilities, they are the nearest animal relatives of humans. They are predominantly vegetarian, feeding on leaves and fruits, but chimpanzees in particular are known to also eat insects and small monkeys, as well as small antelopes. G.a. build nests for sleeping. For the fossil record, see Pimates.

Great basin spadefoot: see *Pelobatidae.*

Greater celandine: see *Papaveraceae.*

Greater weever: see *Trachinidae.*

Great gerbil: see *Gerbillinae.*

Great house gecko: see Tokay gecko.

Great plains toad: see *Bufonidae.*

Great pond snail *(Lymnaea stagnalis):* see *Gastropoda.*

Great tit: see *Paridae.*

Great yellow pond lily: see *Nymphaeaceae.*

Grebes: see *Podicipediformes.*

Greek tortoise: see *Testudinidae.*

Green algae: see *Chlorophyceae.*

Green bacteria: see *Chlorobacteriaceae.*

Green boniella, *Bonellia viridis:* a green-colored species of the phylum *Echiurida,* found on shallow gravelly bottoms and in submarine rock clefts in the Mediterranean. The female (Fig.) has a terminally bifurcated prostomial proboscis, which is very mobile and can be greatly elongated. The minute ciliated male resides in the oviducts of the female. *Bonellia* displays the most extreme form of sexual dimorphism known in the animal kingdom; the ratio of the length of the male to that of the female is about 1:800. It also exhibits a curious and characteristic type of Sex determination (see). Larvae have the potential to develop either sex. Larvae that become attached to the proboscis of a female develop into males, otherwise they become females.

The skin is toxic, probably serving as a protection against predators, since *Bonellia* is avoided by cuttlefish and predatory crustaceans. The green color of the skin is due to a dihydroporphyrin.

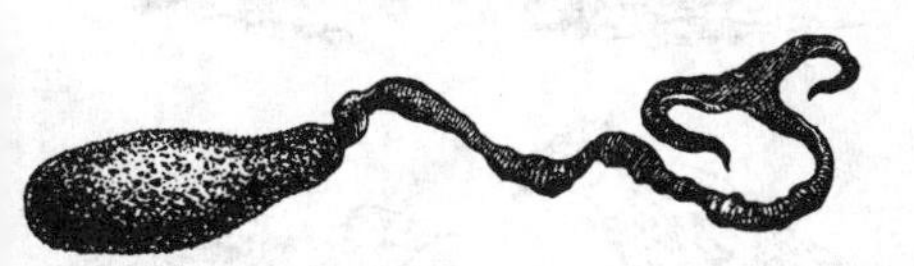

Green boniella (Bonellia viridis)

Greenfinch: see *Fringillidae.*

Green frogs: see *Raninae.*

Green honeycreepers: see *Drepanididae.*

Greenhouse effect: a natural phenomenon (first recognized in the middle of the 19th century), in which the earth's surface is heated by solar radiation, and the resulting heat radiation from the earth is absorbed and re-radiated back to earth by water vapor and carbon dioxide in the troposphere (lower atmosphere).

Solar energy arriving at the surface of the atmosphere consists mainly of radiation in the wavelength range 0.4–0.7 µm (400–700 nm) [it also contains a small proportion (about 7%) of short-wave ultraviolet radiation, which is responsible for ionization reactions in the stratosphere, leading to the formation of ozone; this *ozone layer* is important because it absorbs the dangerous (to life) short-wave UV-radiation that caused its formation, and prevents it from reaching the earth].

This 0.4–0.7 µm radiation band passes through the atmosphere without significant absorption (some is reflected back into space by clouds), and it warms the surfaces of the land and sea. The warm surface of the earth radiates heat mainly at wavelengths above 4 µm. Atmospheric water vapor absorbs radiation between 4 and 7 µm, and atmospheric carbon dioxide absorbs radiation between 13 and 19 µm, leaving a window between 7 and 13 µm, through which more than 70% of the radiation from the earth's surface escapes into space. Carbon dioxide and water vapor (known as greenhouse gases) not only absorb, but also re-emit infrared radiation. Each layer of atmosphere absorbs energy from the layer below, and re-radiates it at a shorter wavelength (i.e. lower temperature) until the radiation finally escaping into space is at a much lower temperature than that leaving the earth. Heat is therefore retained.

Provided the concentrations of water vapor and carbon dioxide in the atmosphere remain constant (and provided the solar radiation arriving at the surface of the atmosphere also remains constant), an equilibrium is maintained. By making comparisons with the temperatures of the moon's surface, which has no atmosphere, it is calculated that the G.e. increases the average temperature of the earth's surface by about 33 °C.

There is current concern that human activity is increasing the atmospheric concentrations of greenhouse gases, thereby increasing the G.e. and leading to an increase in average temperatures throughout the world. Since the beginning of the industrial revolution, large quantities of carbon dioxide have been released into the atmosphere from the burning of fossil fuels. Analyses of bubbles of atmospheric gases trapped in polar ice show that for the10 000 years leading up to the industrial revolution, the carbon dioxide concentration of the atmosphere remained constant at about 270 ppm. Records since 1957 show a steady increase in atmospheric carbon dioxide, and the average concentration is now (1994) greater than 350 ppm (0.035%).

In addition, it is now known that other gases released by human activity absorb radiation in the 7–13 µm window, which normally provides a free escape route for 70% of the earth's heat radiation. Examples of these "anthropogenic gases" are ozone, methane, nitrous oxide (N_2O) and chlorofluorocarbons (CFCs). Although notorious as the chemicals responsible for holes in the ozone layers over the Arctic and Antarctic, CFCs are also very efficient greenhouse gases; thus, one molecule of a CFC has the same geenhouse warming effect as 10 000 molecules of carbon dioxide. Methane is certainly released from gas and oil fields, and relatively large quantities are produced by bacteria in rice paddy fields, and by ruminating cattle. Nitrous oxide arises from the action of denitrifying bacteria on nitrogen fertilizers.

Ozone is beneficial in the upper atmosphere where it shields the earth from short-wave ultraviolet radiation, but close to the earth it behaves as an undesirable greenhouse gas, as well as damaging crops and many materials (e.g. textiles), and accelerating reactions in the atmosphere that lead to the production of Acid rain (see). Ozone is formed near to the earth by photochemical reactions (promoted by sunlight) between nitrogen oxides (nitrogen dioxide, NO_2, and nitric oxide, NO) and hydrocarbons, both produced by automobile exhausts. Nitrogen oxides are also produced by power stations, while hydrocarbons arise from a variety of sources, such as industrial solvents; sources of the hydrocarbon, methane, are mentioned above.

Greenhouse frog: see *Leptodactylidae.*

Greenhouse thrips: see *Thysanoptera.*

Green iguana, *common iguana,* *Iguana iguana:* the largest iguana (see *Iguanidae),* attaining a total length of up to 2.2 m, although only 45 cm is accounted for by the head and trunk. Males have an especially well developed dorsal crest (up to 8 cm high), and a large gular pouch. This insectivorous iguana is found along river banks of tropical Central and South America, where it inhabits trees and undergrowth up to a height of 20 m. It is hunted by man for its flesh and its eggs. If disturbed or threatened, it can jump to the ground from a height of 6 m, or directly into the water where it can swim with the aid of its powerful tail. A clutch of about 30 eggs is buried in the ground.

Greenland collared lemming: see *Arvicolinae.*

Green lizard, *emerald lizard,* *Lacerta viridis:* a handsome, powerfully built member of the *Lacertidae* (see) of central, southern and southeastern Europe. Up to 40 cm in length, it is found in a few isolated areas of Europe north of the Alps, but has a continuous range south of the Alps from Asturias in Spain across France and Italy to the Balkan Peninsula and Asia Minor. It is the largest lizard of Central Europe, and is especially common in the Balkans, where it inhabits sunny, bushy areas that are not too dry. The upper surface is green with fine dark spotting, and the female also has longitudinal stripes. During the mating season (April to June) the throat of the male becomes sky blue. It eats all large insects, and occasionally also fruit (grapes). Eggs are buried in the ground.

Green monkey: see Guenons.

Green peach aphid: see *Homoptera.*

Green pigeons: see *Columbiformes.*

Greenshank: see *Scolopacidae.*

Green toad: see *Bufonidae.*

Green turtle: see *Chelonidae.*

Green two-spotted mite: see *Tetranychidae.*

Gregarinidea: see *Telosporidia.*

Grenadiers: see *Macrouridae.*

Gressoria: see *Cheleutoptera.*

Gribble: see *Isopoda.*

Grinnell's axiom: see Competition-exclusion principle.

Gripping frogs: see *Hylidae.*

Grisein: an iron-containing antibiotic synthesized by *Streptomyces griseus.* It is a cyclic polypeptide which also contains cytosine. The iron is strongly bound as hydroxamate-iron(III) complexes. It is particularly effective against Gram-negative bacteria and rickettsias, but not against fungi. See Siderochromes.

Griseoflavin: an antibiotic produced by *Streptomyces griseoflavus,* which is active against Gram-positive bacteria.

Griseofulvin, *curling factor:* an antifungal agent produced by *Penicillium* species *(P. griseofulvum, P. patulum, P. nigricans).* It is highly active against numerous phyto- and human pathogenic dermatophytes. It was used widely against fungal dermatoses by oral administration, but its clinical applications in human and veterinary medicine are now very limited, since it is known to have harmful effects on the liver. It is a polyketide, biosynthesized from one molecule of acetyl-CoA and 6 molecules of malonyl-CoA.

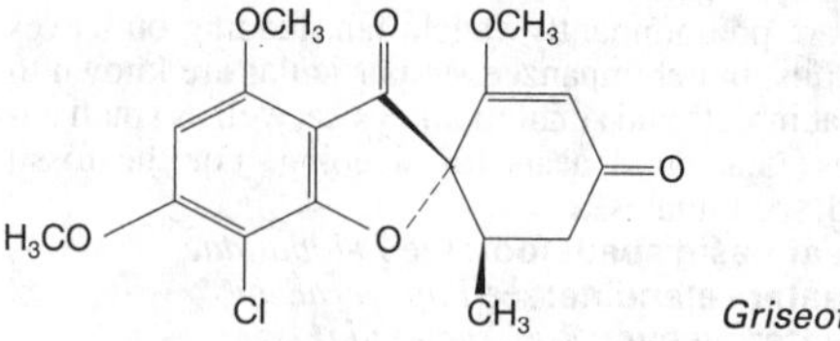

Griseofulvin

Grivet: see Guenons.

Grizzly bear: see Brown bear.

Grooming, *comfort behavior:* behavior directed to body care. Three levels of G. are recognized. 1) G. in which an organism uses organs of its own body. In this context, the organs in question are called cleaning or-

Grooming. Different types of self-grooming displayed by the skylark *(Alauda arvensis). a Sun bathing. b* Shaking and ruffling. *c* Wing cleaning. *d* Wiping beak and head on the ground. *e* Rain bathing. *f* Wing and leg stretching (stretching syndrome). *g* Dust bathing. *h* Head scratching. *i* Stretching both wings (stretching syndrome).

gans, and the parts of the body that are cleaned are known as cleaning areas. Often cleaning occurs according to a fixed pattern or sequence, determined by the order of attention to different cleaning areas and/or the use of different cleaning organs; or it may be characterized by special patterns of movement, e.g. head scratching or head wiping with the forefeet. 2) G. in which environmental facilities are used, e.g. mud wallowing, sand bathing, dust bathing, water bathing, rubbing up against vertical structures, and the use of cleaning tools. 3) Biosocial G. carried out by a partner, which may be unilateral or mutual. An example is the social fur grooming performed by baboons and other primates. Between different species, it is known as *allopreening,* e.g. the removal of gill parasites from large reef fishes by smaller cleaner fishes, and the removal of ticks, etc., from the hides of large game animals and domestic cattle by African oxpeckers *(Buphagus).*

Grosbeaks: see *Fringillidae.*

Grossulariaceae, ***gooseberry family:*** a family of dicotyledonous plants, containing about 150 species, which are distributed mainly in the temperate zone of the Northern Hemisphere. They are woody shrubs or bushes, with simple, often lobed, alternate leaves. The insect-pollinated flowers are usually small, 4- or 5-merous, and arranged in racemes. The unilocular, inferior ovary develops into a berry, and the berries of several species are eaten as soft fruit.

Examples are the native European and Asian ***Gooseberry*** *(Ribes uva-crispa = R. grossularia),* which has characteristic simple or divided thorns; and the western European ***Red currant*** *(Ribes rubrum).* The ***Black currant*** *(Ribes nigrum)* grows wild in damp habitats in Europe and Asia, and it is widely cultivated for its vitamin C-rich fruits. Some members of the genus *Ribes* are also popular ornamental plants, in particular the yellow-flowered ***Buffalo currant*** *(Ribes aureum),* which is also used as a stock for grafting long-stemmed gooseberry and currant varieties, and the curiously scented, red-flowered ***Flowering currant*** *(Ribes sanguineum),* both originating from North America.

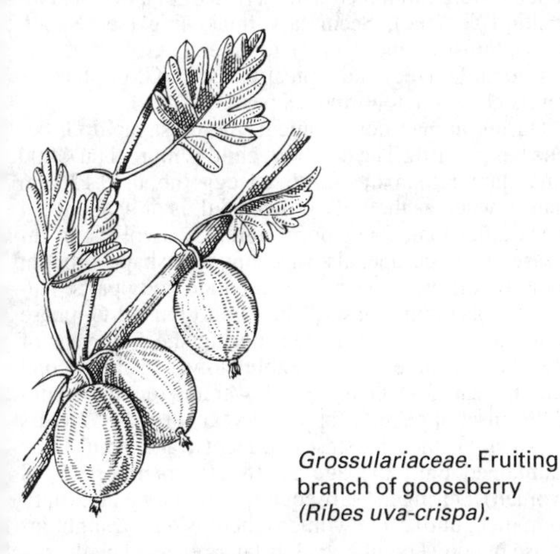

Grossulariaceae. Fruiting branch of gooseberry *(Ribes uva-crispa).*

Ground beetles, *Carabidae:* a large family of *Coleoptera* (see) containing 25 000 species worldwide. They are mostly black or black-brown, sometimes with a green, blue or reddish metallic lustre. Most of the daytime is spent beneath stones, in ground litter or in the soil. The great majority of species is predacious, preying at night on other insects and their larvae. The ***Searcher*** or ***Caterpillar hunter*** *(Calosoma sycophanta* L.) is a well known destroyer of forestry pests; its mobile larva is also predacious, living on pupae of forest lepidoptera. Widely distributed in Europe, the Balkans and elsewhere in the Palearctic region, *Calosoma sycophanta* has been introduced into North America for the biological control of lepidopterous pests. Only a few species feed on plants, e.g. the ***Cereal ground beetle*** *(Zabrus tenebrioides* Goeze).

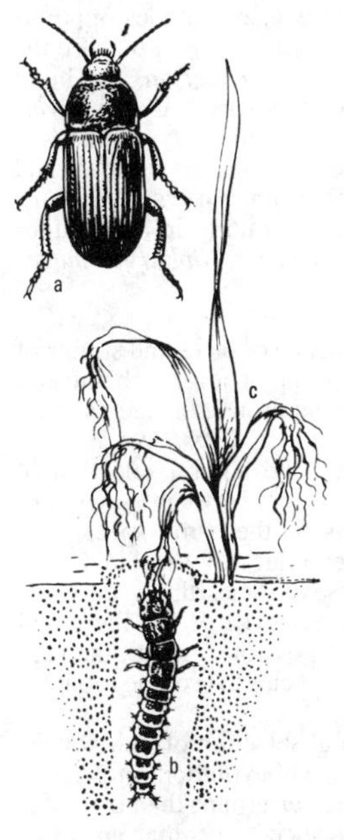

Ground beetles. Cereal ground beetle *(Zabrus tenebroides* Goeze). *a* Beetle (enlarged). *b* Larva in earth burrow. *c* Damaged winter cereal.

Ground cuckoos: see *Cuculiformes.*

Ground finches: see *Fringillidae.*

Ground nut: see *Papilionaceae.*

Ground rollers: see *Coraciiformes.*

Ground squirrel: see *Sciuridae.*

Ground water: water that occurs below ground level, where it is contained in clefts, fissures and pore spaces in regolith and solid bedrock, and is free to move under the influence of gravity.

Ground water fauna: animal inhabitants of Ground water (see). At the bottom of lakes, rivers, etc. and in springs and wells, ground water is in direct contact with surface waters. Ground waters are colonized via these points of contact, and all ground water organisms (stygobionts) are derived from forms found in surface waters. Their distribution is restricted by factors such as darkness, nutrient deficiency, and a constant low temperature, which roughly corresponds to the yearly average of the upper environment. Darkness precludes the presence of autotrophic plants. The nutrients of ground water originate from the earth's surface. Ground water animals feed on detritus or fungi, or they live by predation. Light-sensitive organs are regressed, and the majority of ground water animals are blind. Tactile organs and chemical senses are, however, well developed. A further result of life in complete darkness is the absence of pigmentation in true stygobionts; white or flesh-colored forms predominate. In species long adapted to an existence in ground water, loss of pigmentation is irreversible. Evolutionarily more recent stygobionts, however, are still capable of synthesizing their original pigments when they are exposed to daylight, e.g. the olm or cave salamander *(Proteus anguinus)*. Other characteristics of G.w.f. are their longevity and low reproduction rate, compared with their relatives above gound level; since the nutrient supply is restricted, each species is maintained at a constant low population density. Most ground water species occur in rough limestone country, where the abundance of caves also provides opportunites for their direct study. Well known G.w.f. are the turbellarian worm *Planaria montenegrina,* the blind well-shrimp *Niphargus,* the cave isopod *Asellus cavaticus,* the snail *Lartetia,* the amphopod *Stygodytes balcanicus,* the blind cave fishes *Chologaster agassizii* and *Amblyopsis spelaeus,* and the olm *Proteus anguinus.* Some stygobionts also occur regularly in deep oligotrophic lakes, wells and springs, e.g. *Niphargus puteanus* and *Asellus cavaticus.*

Groupers: see Sea basses.

Group resistance: resistance of races and strains of certain pest organisms to a group of several chemically closely related pesticides, plant protection agents, etc.

Group selection:

1) In population genetics, the natural selection of groups of individuals. G.s. occurs between geographically separated populations of the same species, or between species that are reproductively isolated from each other. During the course of the earth's history, the evolution of living forms as revealed by the fossil record would not have been possible without the suppression and loss of inferior species to provide room for new, superior forms.

Compared with the natural selection of individuals, the effect of G.s. on the course of evolution is insignificant, because there are far fewer groups than individuals. Since a group also lives much longer than any of its individuals, the relative intensity of G.s. is small. G.s. permits the evolution of characters that are useful to the group, but are unimportant or even disadvantageous to the individual.

Selectively important differences between individuals arise by mutation and recombination. There are no analogous mechanisms for conferring selective advantages on a population or relatively large group. Differences between groups are essentially the result of different individual selection processes within these groups. No group can acquire selective superiority by genetic drift alone, which depends on the presence of a large number of genetic factors.

The decline in the reproductive rate of a population that has passed its optimal density may be due to G.s. G.s. is not the same as Kin selection (see).

2) In behavioral research, the selective discrimination between biosocial groups. It is proposed that selection occurs at the level of the group, and is independent of the degree of relationship of individuals within the group, i.e. it works to the advantage of the group and not the individuals within it. Selection operates by influencing the contribution of individual groups to the gene pool of the subsequent generations that make up the population.

Group specificity: see Enzymes.

Group sterility: see Intersterility.

Grouse: see *Tetraonidae.*

Grove snail: see *Gastropoda.*

Growth: an irreversible modification associated with physiological activity of the protoplasm, and often accompanied by an irreversible increase in the volume or mass of the cell. It is an essential attribute of living material. Fundamental to the development of all organisms is G. of cells, irrespective of whether the organism is unicellular or multicellular. There are 3 types of cell G.: cell division, embryonal or Plasmatic G. (see), and Extension G. (see). There are certain fundamental differences between G. in plants and animals.

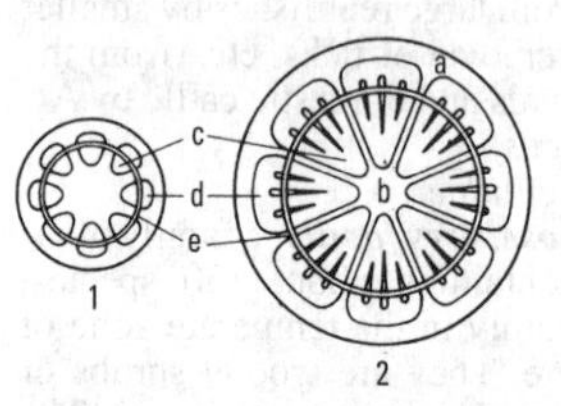

Growth. Transverse section of the stem of a dicotyledon, showing primary permanent tissue (1) and secondary thickening (2). *a* Primary cortex. *b Pith. c* Xylem (woody part of vascular bundle). *d* Phloem (sieve tubes). *e* Cambium.

In plants, cell division and plasmatic G. occur mainly in meristematic zones. Since meristems are normally present throughout the life of the plant, G. can occur continuously up to the point at which the plant dies. In plasmatic G., it is principally the protoplasm that grows, with little if any enlargment of the length or diameter of the cell. In plant cells, plasmatic G. and extension G. generally occur separately, whereas in animal cells both occur simultaneously. Also in plants cells, Cell differentiation (see) is often linked to G., occurring immediately after any increase in cell volume; this phenomenon is called differentiation G. Other types of plant G. are Apposition G. (see), Secondary thickening (see Shoot, Root), Intussusception (see), Intercalary G. (see), Interpositional G. (see) and Apical G. (see). G. in plants is regulated by Phytohormones (see).

During human development, the most rapid G. occurs before birth. The newborn child is more than 4 000 times larger than the fertilized egg (about 0.12 mm diam.), whereas the fully grown adult is only about 3.4 times taller than a newborn child. G. involves an increase in mass, and also a change in shape brought about by different G. rates of separate body parts. Human G. occurs in phases. The rapid prenatal G. phase continues, somewhat abated, during the first year of life, then becomes considerable slower in the second year. In years 5–6, G. is even slower and remains so until the onset of puberty (about 11 years in girls, 12 years in boys). G. then increases suddenly, and rapid G. is maintained until the age of 18–20 (men) or 16–18 (women). During the subsequent stationary phase (until 60 in men, until 50 in women) there is only a slight increase in body height, which is later reversed by the degenerative changes of increasing age. By flattening of the angle of the neck of the femurs, flattening of the foot arches, and irreversible compression of joint cartilages, including intervertebral disks, the body height decreases by about 3% in advanced age.

G. is regulated hormonally and depends on numerous genetic and dietary factors. G. of growing children is inhibited during starvation. The significance of genetic factors is most evident from the study of twins. Differences in the sizes of newborn identical twins (from a single egg and therefore genetically identical) are imposed by their positions in the uterus, but these differences are compensated by G. in later life. Thus, differences between the heights of identical twins are much less than those between nonidentical twins or between separately born siblings.

G. is altered by endocrine disturbances. Removal of the gonads leads to excessive limb G. Enlargment of the pituitary leads to gigantism. Removal of the thyroid or the pituitary in early development leads to dwarfism.

In most animals, G. continues until shortly before the attainment of sexual maturity. Fishes, however, grow continually until they are very old. In incompletel;y grown mammals, premature sexual activity causes G. inhibition, whereas postponement of sexual activity can extend the G. period. Short-lived animals generally grow rapidly and start to reproduce at an early stage of their existence.

Growth curve:

1) In microbiology, the graphical presentation of the change in the number of cells of a unicellular organism in a nutrient medium plotted against time. G.cs. are used in particular to characterize the growth course of bacterial and yeast cultures. The G.c. displays characteristic phases. The *lag phase* commences when the organisms are first inoculated into the growth medium; the cells may increase in size and they gradually start to divide. The duration of the lag phase depends on the age and physiological state of the inoculum and on the composition of the nutrient medium. During the lag phase, the cells synthesize RNA and enzymes, and generally adjust their metabolism for maximal growth. At the end of the lag phase, the maximal rate of division is attained, and the cells enter the *logarithmic* or *exponential growth phase (log phase).* The latter is characterized by a constant, maximal cell division rate, in which the logarithm of cell numbers increases linearly with time. In the large scale culture of microorganisms, e.g. in the production of baker's yeast, the product is normally harvested in the log phase, i.e. while it still displays its maximal metabolic activity.

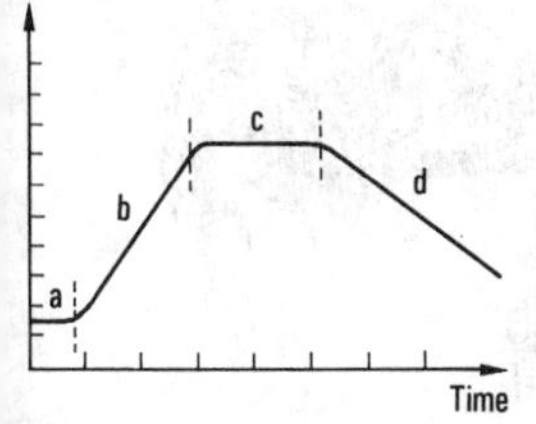

Growth curve. Growth of a culture of microorganisms. *a* Lag phase. *b* Logarithmic phase or phase of exponential growth. *c* Stationary phase. *d* Death phase.

As the concentration of nutrients decreases and the toxic metabolic products accumulate, the cells pass into the *stationary phase,* in which there is no further increase in the number of living cells. Finally, the number of living cells decreases in the *death phase.* Each phase is connected to the next by a brief transition phase.

2) See Population growth.

Growth form: see Life form.

Growth intensity: the rate and extent of growth that is characteristic of each individual organism, and which is expressed as an increase of body mass. Potential G.i. is genetically determined, and it is fully expressed only under favorable environmental conditions, in particular when growth is not limited by the food supply. G.i. also varies with the developmental phase of the organism.

Growth media, ***culture media:*** artificial media for the growth of living organisms, e.g. bacteriological growth media, tissue culture media, hydroponic media for the growth of plants. G.m. may be solid or liquid. They may be used for growing relatively large quantities of biological material for experimental or commercial purposes, for growing relatively small quantities (e.g. bacteria on the surface of nutrient agar) for diagnostic purposes, or for preserving and storing organisms. G.m. contain water, an inorganic or organic nitrogen source, and mineral salts. For the culture of heterotrophic microorganisms, an organic source of carbon and energy is also required, e.g. sugars. Sodium chloride is added to many G.m. to insure an appropriate osmolarity. The pH is also important and must be adjusted appropriately, e.g. fungi usually prefer a slightly acidic environment, whereas most bacteria require neutral to slightly alkaline media. Solid G.m. are prepared by adding agar (more rarely gelatin) to liquid G.m. Depending on the purpose of the medium, further components may also be necessary, e.g. pH indicators, buffer solutions, trace elements, inhibitors of potential contaminating organisms (see Selective culture medium). After preparation, all G.m. must be sterilized.

Natural G.m. consist of naturally occurring substrates, which usually require no further additions, e.g. slices of potato, pieces of carrot, milk. *Artifical G.m.* are prepared from various components according to standard recipes (see e.g. Nutrient agar). They often contain complex additions, e.g. meat infusion, malt, yeast extract, blood. Many of these G.m. are commercially available as dry powders. *Synthetic G.m.* do not contain complex additions, and their composition is known exactly, i.e. they contain known quantities of defined mineral salts, inorganic nitrogen sources (e.g. nitrates or ammonium salts) and simple organic sources of energy and carbon (e.g. glucose). Simple media of this kind can be used for growing organisms such as *Escherichia coli.* Some plant tissue culture media are now entirely synthetic. Synthetic media are also useful for determining the growth requirements of microorganisms.

The term *nutrient medium* is usually applied to media containing material like yeast extract, meat infusions, etc. (e.g. nutrient agar, nutrient broth).

Growth rate:

1) the rate of increase in size or mass of a tissue or organ, but particularly the entire organism, as a result of the interplay of the genetically determined growth intensity, nutrient availability and other environmental conditions. It can be presented as the absolute G.r. at different ages (increase per time unit), or as proportional growth related to total body mass. The ratio of the daily increase in mass to the body mass is called the *growth quotient.*

2) The rate of growth of a Population (see). It can be

expressed as the relative increase per time unit (exponential growth), or as the change in population size over a certain time period (geometric growth).

Growth regulators: organic compounds which promote, inhibit or otherwise influence plant growth in very low concentrations. Natural G.g. include Phytohormones (see) and Inhibitors (see). There are numerous synthetic G.r., such as Herbicides (see) and Growth retardants (see).

Growth retardants: synthetic regulatory growth inhibitors, which cause the shortening of internodes. They have practical importance in the shortening and strengthening of halms in cereal agriculture. Compounds such as CCC, Amo 1618 and phosphon D function by inhibiting gibberellin synthesis. B-995 (Alar) has similar physiological effects, but does not inhibit giberellin synthesis. Camposan, which contains chloroethylphosphonic acid, shortens and strengthens cereal halms by releasing ethylene. Other important G.r. are Maleic hydrazide (see) and succinic acid mono-*N*-dimethylhydrazide.

Gruidae: see *Gruiformes.*

Gruiformes: an order of birds (see *Avies),* most of which feed on the ground, and nest on or very near the ground. Many species fly only rarely, and a few never fly. Some groups such as the *Heliornithidae* and some of the *Rallidae* (e.g. coots) are adapted to an aquatic existence, while many others are inhabitants of marshes and water margins. Except for the polar regions, the order is represented worldwide. The 12 constituent families are rather diverse, but classified in the one order on account of similarities in internal anatomy. Numerically, the *G.* are declining probably more rapidly than any other avian order; many members have become extinct in recent times, and many species are endangered and rare.

1. Family: *Mesitornithidae* or *Mesoenatidae **(Mesites*** or ***Roatelos).*** This strictly Malagasy family is represented by 3 ground living species: *Mesites unicolor **(Brown mesite;*** straight bill; rainforests of eastern Madagascar), *Mesites variegata **(White-breasted mesite;*** straight bill; dry forest of northwestern Madagascar) and *Monias beschi **(Bensch's monia;*** decurved bill; dry brushland of southwestern Madagascar), all 25–30 cm in length. They are unique in possessing 5 powder down patches, more than any other bird. Although the wings are quite well developed, the clavicles are greatly reduced. They prefer to run or walk, and their frail platform nest (2–3 meters above the ground) is reached by climbing. They feed on insects, seeds and small fruit.

2. Family: *Turnicidae **(Hemipodes, Buttonquails,*** or ***Bustard quails).*** These are very short-tailed, slender-billed, round-winged birds, 10–20 cm in length, and bearing a superficial resemblance to true quails. There is no crop, and only 3 toes are present. They are running birds and weak fliers, inhabiting open grassland and brush country, and feeding on small seeds and insects. About 14 species of *Turnix* are distributed from southern Spain through Africa (including Madagascar) and Asia to the islands of the Far East and Australia. A single migratory species is found in China and Korea. Females generally have brighter plumage than males, and they are generally polyandrous, laying several clutches for incubation by her mates. The ***Quail plover*** or ***Lark quail*** *(Ortyxelos meiffrenii;* 12.5 cm; tropical East Africa) is monotypic.

3. Family: *Pedionomidae.* A monotypic family consisting of the ***Collared hemipode*** or ***Plains wanderer*** *(Pedionomus torquatus),* which inhabits the central deserts of Australia. Plumage is cryptic buff, brown and black with a collar of black spots, the female being more brightly colored. It has well developed hindtoes, and differs from other members of the order in having paired instead of single carotid arteries, and strongly pointed rather than oval eggs. It is sometimes classified as a member of the *Turnicidae.*

4. Family: *Gruidae **(Cranes).*** Most of the 14 species are endangered, The family is represented in North America, Africa, Eurasia and Australia. They have long legs, long necks, long bills and striking plumage. The head is often partly bare and colored red, or it may carry a highly decorative crest. Secondary wing feathers are long and decorative, and the long, normally overhanging tail is ruffled, curled and raised in display. Cranes are well known for their elaborate courtship display, which involves dancing , leaping and strutting. Both sexes, but especially the male, have a loud and penetrating call, made possible by the convoluted trachea, which enters a space in the keel of the sternum behind the collar bone. In the breeding season thay inhabit shallow wetlands, but are found on grasslands and agricultural fields in the nonbreeding season. They feed on small fish and other animals, as well as tubers and seeds. Ten species are placed in the genus *Grus.* The ***Brolga*** or ***Native companion*** *(Grus rubicunda)* occurs in Australia. Two species are found in North America, e.g. the ***Whooping crane*** *(Grus americana;* 127 cm), which is uniformly white, except for black wing primaries and coverts, bare red crown and cheeks, black legs and pale yellow eyes; one of the world's rarest species, it now breeds only in Wood Buffalo Park by the Great Slave Lake, overwintering in Arkansas. The ***Common crane*** *(Grus grus)* (Fig.) breeds from the Baltic across Eurasia to eastern Siberia and eastern China; its body plumage

Gruiformes. Common crane *(Grus grus).*

s gray, while the primaries and most of the head and neck are blackish, with some white extending up the sides of the head; there is a red patch on the crown of the head, the bill is greenish with a reddish base, the legs are black and the eyes red. Most of the others species of *Grus* are Asian. Monotypic African genera are *Tetrapteryx, Bugeranus* and *Balearica.* The ***Demoiselle*** *(Anthropoides virgo)* is found in eastern Europe and southern and Central Asia.

5. Family: *Aramidae.* This monotypic family consists of the ***Limpkin*** *(Aramus guarauna),* a strictly New World species, with a range that extends from South Georgia and Florida to Cuba and southern Mexico, continuing south to Argentina. Anatomically rather crane-like, the limpkin is dark olive brown with a greenish iridescence, and broadly streaked with white. It feeds almost exclusively on snails, with some insects and seeds.

6. Family: *Psophiidae* **(*Trumpeters*).** This is a small family of 3 species of the genus *Psophia,* confined to South American tropical rainforests. In habits and structure they are intermediate between cranes and rails. They have purplish black plumage, and are about the size of a domestic hen (but longer in the leg), with a hump-backed appearance and a fowl-like bill. The courtship display involves crane-like dancing, leaping and strutting. They are named for the strident call of the male. South American indians often keep them with poultry as "watchdogs".

7. Family: *Rallidae* **(*Rails, Crakes, Coots*** and ***Moorhens*** or ***Gallinules;*** 124 species in 41 genera). Family characteristics are soft plumage and short tails and weak, leg-trailing flight or flightlessness. Many have evolved in isolation on remote islands to the point where they are monotypic species, and often flightless owing to the absence of predators; especially these isolated monotypic species are endangered and becoming extinct. The family is represented worldwide except in the polar regions, usually in swampy habitats. ***Rails*** (long-billed) and ***Crakes*** (short-billed) are small to medium-sized, generally with drab plumage and unobtrusive habits, but with loud voices. In particular they have thin bodies and long legs and toes, probably an adaptation to walking among closely packed reeds and grasses. In contrast, the ***Coots*** and ***Moorhens (Gallinules),*** have more striking, blackish or purple plumage, and are less inclined to skulk among reeds. Moorhens have narrow flanges of skin along their toes, while coots have lobes between the joints like finfoots, grebes and phalaropes. The ***Moorhen*** or ***Common gallinule*** *(Gallinula chloropus;* 33–38 cm) (plate 6) is found in all the temperate and tropical wetlands of the world, except in Australasia where it is replaced by the closely related *G. tenebrosa.* In Europe, it often occurs in parks and gardens. Upper parts are brownish black, underparts dark gray, with white stripes on the flanks and white under-tail coverts. The legs are green with a red garter stripe. Base and frontal shield of the bill are red, and the bill tip is yellow. The ***Coot*** *(Fulica atra;* 38 cm) is found right across the Palaearctic temperate zone, as well as in India, Southeast Asia and Australasia; it is almost uniformly black with a white wing bar, green legs, white bill and frontal shield, and red eyes. The larger ***American coot*** *(Fulica americana;* 41 cm) is similarly colored, but with some red on the frontal shield; it is distributed from central Canada to Colombia and Ecuador, and is also found in Hawaii.

Notornis mantelli is a small turkey-sized New Zealand gallinule, with brilliant green-blue plumage and a striking red frontal shield. Known to the maoris as the ***Takahe,*** it was last collected in 1898, and thereafter believed to be extinct. At the time only 4 specimens had ever been reported. In 1948, however, about 100 individuals were found in some narrow valleys 2 000 feet above sea level in the Murchison range of South Island. The ***New Zealand wood rail*** or ***Weka*** *(Gallirallus australis)* is one of the few flightless rails that has not been decimated by introduced predators; in fact, this omnivorous bird feeds on introduced rats and mice, as well as raiding garbage bins and chicken runs, and feeding on tide-wrack near the sea.

8. Family: *Heliornithidae* **(*Finfoots*** or ***Sungrebes*).** The 3 species constituting this family are confined to the tropics and subtropics: *Heliornis fulica* (29 cm; southern Mexico to northeastern Argentina); *Heliopais personata* (58 cm; Southeast Asia from Assam to peninsular Malaysia); and *Podica senegalensis* (63 cm; West Africa to eastern Cape Province). They have a long thin head and neck, very broadly lobed toes, and a strongly graduated tail. Mainly aquatic in habit, they have a peculiar forward leaning gait on land. Their pale brown plumage contains varying admixtures of green, gray, black and white, and contrasts with the brightly colored bill and feet. They inhabit the margins of rivers and other inland waters, where thick cover is afforded by overhanging forest or thick vegetation. They are fast swimmers, and they feed mainly on insects, also taking frogs and mollusks.

9. Family: *Rhynochetidae.* This monotypic family consists of the ***Kagu*** *(Rhynochetos jubatus),* a largely nocturnal bird inhabiting the forest floor, and found only on the island of New Caledonia. It is not much larger than a domestic fowl, but with longer legs. The loose plumage is slaty gray with darker bars, but the spread wings reveal a pattern of white, reddish and black markings. The head carries a long loose crest, which can be erected, and the bill and legs are bright red. Worms, insects, etc. form the major part of the diet.

10. Family: *Eurypygidae.* This monotypic family consists of the ***Sunbittern*** *(Eurypyga helias;* 46 cm), the three races of which inhabit the forested regions of America from southern Mexico to Bolivia and central Brazil. It is a graceful long-necked, long-legged bird with barred, spotted and mottled plumage. The head is almost black with white stripes above the eyes and along the lower edge of the cheeks. Each wing is patterned with large round shield of striped white, deep orange chestnut and black set in an area of pale buff, which is revealed when the wings are spread. In addition, the upper mandible is black, while the lower mandible is orange. The diet consists mainly of insects, small crustaceans and small fish, which are captured by foraging along water courses.

11. Family: *Cariamidae* **(*Seriemas*).** A small family of only 2 species inhabiting the drier areas of south central South America. With a long neck, long slender legs, and bushy head crest, they resemble small cranes. They eat small snakes, but are not immune to snake venom. Most familiar is the ***Crested Seriema*** *(Cariama cristata;* tablelands of Brazil, Paraguay and northern Argentina), which stands about 76 cm tall. Young birds are easily domesticated and they are often reared among

farmyard fowls, where they warn of predators with their loud alarm call. The somewhat larger ***Burmeister's Seriema*** *(Chunga burmeisteri)* is found only in northwestern Argentina and western Paraguay.

12. Family: *Otidae, Otididae* ***(Bustards).*** These Old World medium to large birds (37–132 cm in length) are terrestrial in habit, living in open treeless country. They have quite strong legs, only 3 toes (the hallux is absent), and a rather long thick neck. Wary and keen-sighted, they usually run to escape danger, but they can also fly powerfully and fast. They are omnivorous, feeding on leaves, buds, seeds and insects, large species also taking small mammals such as voles. Sixteen species occur in Africa, 6 in central and southern Asia, and one in Australia. Several African species are no larger than a domestic hen, standing only 30 cm high, but most are larger. The ***Kori*** or ***Giant bustard*** *(Choriotis kori;* 137 cm) of the South African Veldt is arguably the heaviest of all flying birds, males weighing up to 23 kg. The ***Great bustard*** *(Otis tarda;* 101 cm) (plate 37) is a Palaearctic species with a discontinuous distribution, occurring in scattered areas of south-central Europe, with its main center in western and central Asia.

Grus: see *Gruiformes.*

Grylloblattoidea (Grylloblattids): see *Notoptera.*

Gryphaea: a bivalve of the subclass *Pteriomorphia* (see *Bivalvia).* The strongly convex left valve is incurved at its anterior end and usually firmly attached to the sea floor, while the right valve forms a flat lid. G. is found from the Lower Jurassic to the present, and many species, especially those of the Cretaceous, are index fossils.

Gryphaea (Liogryphaea) arcuata Lam. from the Lower Triassic (x 0.5).

Grysbock: see *Neotraginae.*

Guaiacol, ***catechol monomethyl ether:*** an oily liquid with a strong, spicy odor. It is present in crude creosote of beech tar, and was first prepared by distillation of guaiacum resin. It is the starting material for the commercial synthesis of vanillin.

OH
OCH_3

Guaiacol

Guaiacum resin, ***resini guaiaci:*** a resin from the trunk of the West Indian tree *Guaiacum officinale* L. The heartwood contains 20–25% resin. The resin of commerce usually occurs as large masses of dark color, often more or less covered with a greenish powder. It breaks with a clean, glassy fracture, the resulting fragments being transparent, and varying in color from yellowish green to reddish brown. It is used medicinally against chronic gout and rheumatism, and against skin complaints.

Guanaco, *Lama guanicoë, Lama glama:* a non-humped member of the *Camelidae* with a red-brown coat, which inhabits the dry plains and high mountain regions of South America. The G. is the ancestor of the domestic llama and alpaca.

Guanine, ***2-amino-6-hydroxypurine:*** one of the two purine bases present in nucleic acids (the other is adenine). It is also a component of certain nucleotide coenzymes, and it serves as a biosynthetic precursor of many natural products, including pteridins, folic acid and riboflavin. Free G. is rare in nature. It was first discovered in 1844 in Peruvian guano. It is the nitrogen excretion compound of spiders. Deamination of G. to xanthine is the first step in purine degradation.

OH
N N
H_2N N N
H

Guanine

Guans: see *Cracidae.*

Guard cell: see Stomata.

Guayule: see *Compositae.*

Gudgeon, *Gobio gobio:* a fish of the family *Cyprinidae,* order *Cypriniformes.* Its maximal length is 22 cm (usual length 10–12 cm), and it has a single barbel at each corner of the mouth. A bottom fish, it lives in shoals in rivers and lakes of Europe and northern Asia. It is preyed on by other fishes, and is useful primarily as a bait fish.

Guelder rose: see *Caprifoliaceae.*

Guenons, *Cercopithecus* (plate 41): relatively small, long-tailed Old World monkeys (see), often with strikingly colored and patterned faces and coats, living in large troops in the forests, jungles, savannas and scrublands of Africa. *C. aethiops* ***(Savanna monkey, Green monkey, Grivet, Vervet, Tantalus monkey)*** shows adaptation to a terrestrial existence, but all other G. are arboreal. Twenty species have been described, and they are the commonest monkeys of Africa.

Guerezas, *Colobus:* a genus of the *Colobinae* (see Colobine monkeys), inhabiting the African equatorial rainforest from Senegal to Ethiopia and southern Angola. The fur is long and silky, and has been widely used for ornamentation by both Europeans and Africans. As in all colobines, the thumb is absent or reduced to a small protuberance. About 20 races are known and 2 main species are recognized: *C. polykomos* ***(Black and white colobus)*** and *C. badius* ***(Red and black colobus).*** Some authorities include the Olive colobus (see) in this genus.

Guillemots: see *Alcidae.*

Guinea corn: see *Graminae.*

Guineafowl: see Pheasant, *Numididae.*

Guinea pig: see *Caviidae.*

Guinea worm, ***Medina worm,*** *Dracunculus medinensis, Filaria medinensis:* a parasitic nematode (order *Dracunculida;* see *Nematoda),* more than 1 m in length, which infects humans and domestic animals in the Old World tropics. Guinea worm infection is also known as dracontiasis. Cocepods are host to the larvae, and the infection is transferred by the ingestion of infected cocepods in impure drinking water. The adult, female worm usually lodges in the subcutaneous connective tissue of the lower leg, more rarely in the forearm, with its vagina situated distally. When the female is ready to release embyonic larvae (microfilaria), the tissues of the host over the site of the worm's vagina are damaged enzymatically, leading to irritation and painful ulceration, and exposure of the vagina at the skin surface. Thousands of microfilaria are released to the exterior when the skin of the host is in contact with water. Treatment includes the gradual extraction of the worm by winding the exposed end onto a stick.

Guira cuckoo: see *Cuculiformes.*

Guizotia: see *Compositae.*

Gulls: see *Laridae.*

Gulo gulo: see Wolverine.

Gulpers: see *Saccopharyngoidei.*

Gum arabic, ***acacia gum, acaciae gummi:*** a hard, friable, colorless to pale yellow substance prepared by allowing the mucilaginous exudate from certain Acacia trees to dry and "ripen" in the sun. It can be obtained from the stem and branches of various species of *Acacia* native to East, Central and West Africa, but most of the gum of commerce originates from orchards of *Acacia senegal* in the Sudan. The exudate is produced naturally through cracks in the bark that appear during growth, and its production is increased by removing strips of bark (0.5–1 m x 2–8 cm) from the trunk and branches. G.a. was traded in antiquity in Arabia. High quality material occurs as rounded or ovoid tears, varying in size from a pea to a hazelnut or larger, which easily break into transparent, glistening, vitreous fragments. G.a. has practically no smell and has a bland, mucilaginous taste. It is structurally similar to pectin, and it yields galactose, arabinose, rhamnose and glucuronic acid on hydrolysis. Medicinally, it was used as a demulcent, but is now used mainly as a binding agent, as a means suspending oils, resins, etc. in aqueous media, and as a glazing and finishing agent. See Plant gums.

Gundi: see *Ctenodactylidae.*

Gunnels: see Butterfish.

Guppy, *Poecilia reticulata:* a small, multicolored, live-bearing, freshwater fish of the subfamily *Poeciliinae,* family *Poeciliidae,* order *Cyprinodontiformes.* It occurs in Central and South America, and feeds primarily on mosquito larvae. *Gambusia affinis* (Mosquitofish) of the same subfamily is used for mosquito control. The guppy is a popular aquarium fish, which has been bred to give numerous color and forms varieties.

Gurnards: see *Triglidae.*

Gustatory organ: see Taste buds.

Gut: see Digestive system.

Gutta: a rubber-like polyterpene of about 100 isoprene units in which the double bonds are in the *trans* configuration. G. is produced on the Malayan peninsula and the Indonesian islands from the latex of *Palaquium gutta.* It is less elastic than rubber, but is more resistant to chemical and environmental influences. Depending on its source, G. occurs in mixtures with other terpenes. The mixture with resins is called guttapercha, which when cold is inelastic and hard, and has been used as a tooth filling material, for the manufacture of surgical splints, as an electrical insulator, and for the manufacture of containers for hydrofluoric acid which does not attack it. The mixture with triterpene alcohols is chicle (starting material for manufacture of chewing gum).

Guttation: extrusion of liquid water from specialized hydathodes (see Secretory tissue). When transpiration is prevented or very slow, due to high atmospheric humidity (e.g. on warm, humid nights, or in the tropical rainforest), water is lost by G. To a certain extent, G. also takes over the function of transpiration, by insuring water movement within the plant for the transport of nutrient salts to the leaves. Large droplets of G. water, resembling dew, can be observed on the leaf edges and tips of Lady's mantle *(Alchemilla), Fuchsia,* nasturtium *(Tropaeolum),* grasses, etc. Submerged aquatic plants can only lose water from their leaves as a liquid. The G. water of saxifrages contains calcium carbonate, which remains as a fine incrustation on the leaves when the water evaporates. Especially high rates of G. occur in some tropical rainforest plants; certain species can lose up to 100 ml per leaf in one night. The driving force for G. via *passive hydathodes* (e.g. in grass leaves) is the root pressure, which forces the xylem contents to the exterior via pore systems. *Active hydathodes,* however, are *water secretion glands,* which actively secrete water and operate independently of root pressure (e.g. in nasturtium, saxifrage, *Phaseolus).*

Guttera: see *Numididae.*

Gymnallabes: see *Clariidae.*

Gymnamebas: see *Rhizopodea.*

Gymnamoeba: see *Rhizopodea.*

Gymnocarp: fruiting body of an ascomycete, in which the spore-forming fertile layer (ascohymenium or thecium) is naked and exposed throughout its development.

Gymnocephalus: see Perches, Ruffe.

Gymnogyps: see *Falconiformes.*

Gymnolaemata: see *Bryozoa.*

Gymnomera: see *Branchiopoda.*

Gymnophiona: see *Apoda.*

Gymnosomata (naked pteropods): see *Gastropoda.*

Gymnosperms, *Gymnospermae:* spermatophytes (seed plants), whose seeds are borne naked and exposed, i.e. they are not enclosed in a seed case. The 2 equally ranked subdivisions of the G., the *Coniferophytina* and the *Cycadophytina,* represent a very ancient level of evolutionary organization within the *Spermatophyta,* but they do not constitute a natural grouping. They were already independent in the Middle Devonian, arising separately from the spore-bearing *Progymnospermae,* then evolving in parallel.

Members of the 2 subdivisions share the following characteristics. They are exclusively woody plants, and nearly all are trees. Their shoots are either unbranched, or they display very regular, usually monopodial branching. Gymnospermous wood consists of annular rings of open-textured, collateral vascular bundles. The water-conducting cells are tracheids covered with large, circular bordered pits. The phloem consists mainly of sieve tubes, which lack companion cells, although a certain amount of phloem parenchyma may be present. Secondary thickening occurs by periodic growth of cambium, which gives rise to annual rings. Resin or mucilage canals are often present. For example, in *Pinus, Picea* and *Larix,* the secondary xylem, which consists almost exclusively of tracheids, also contains a small quantity of parenchyma, which is found around resin canals. These resin canals are schizogenous, i.e. they are formed by the enlargement of intercellular spaces, and they penetrate the wood both longitudinally and radially, linking up to form a network. The vascular bundles of the leaf are unbranched or forked. The ground plant of the gymnospermous leaf is dichotomous *(Coniferophytina)* or pinnate *(Cycadophytina),* but this is often reduced to a thick. leathery, needle-shaped or scale-like structure. Leaves nearly always survive several vegetative periods.

With the exception of a few fossil species, the flowers are always unisexual and usually monoecious, more rarely dioecious, and usually very simple in structure. Microsporophylls ("stamens") or megasporophylls ("carpels") arise from the floral axis and there is no true perianth. Several microsporophylls or megasporophylls may be assembled on one axis to form a cone, which arises from the axil of a bract (e.g. Spruce, Larch). Alternatively, the microsporophyll or megasporophyll may sit singly in the axil of a bract (e.g. Yew). In male flowers the microsporophylls are arranged in a whorl or spiral on the axis. The ventral side of each microsporophyll nearly always carries several pollen sacs, which are opened to release the pollen by a cohesion mechanism of the epidermis. Pollination is almost exclusively by wind. Formation of the male gametophyte commences before pollen release. Within the initially uninucleate pollen grains, nuclear division and cell wall synthesis result in the formation of several cells (prothallial cells) at a particular site against the internal surface of the pollen grain wall. The remaining cell divides into a large vegetative cell (pollen tube cell) and a smaller generative cell, which divides further into a stalk cell and a spermatogenous cell. The latter produces 2 sperm cells, which may develop further into multi-ciliated spermatozoids, or they may be transported to the ovule via the pollen tube without undergoing further development.

The female flowers consist of one or several megasporophylls, which bear the ovules. There is no stigma. The ovules consist of a well developed mass of tissue, the nucellus, surrounded by a single integument, with a micropyle toward the base. Embedded in the nucellus is the embryo sac. The nucellus of each ovule exudes a glutinous material, to which the pollen grains adhere. As this material dries and shrinks, it pulls the pollen grains throught the micropyle, so that they come into contact with the nucellus. Formation of the female gametophyte occurs within the ovule (megasporangium) inside the embryo sac (megaspore). The membrane of the embryo sac is differentiated into 2 layers (exospore and endospore layers), and the embryo sac is completely embedded in nucellus tissue. A multicellular megaprothallus (also called primary endosperm, because it stores nutrients for the embryo) develops inside the embryo sac by free nuclear division and cell wall synthesis. Varying numbers of archegonia also develop, each consisting of a large egg cell, neck wall cells (true neck canal cells are lacking) and sometimes a transitory ventral canal cell or a ventral canal nucleus. The fertilized egg cell develops into an embryo with at least 2 (and in some species 50) cotyledons. After fertilization, the female flower or inflorescence becomes a woody cone with seeds between its scales.

G. attained their greatest evolutionary diversity during the Mesozoic. Most of the approximately 800 species living today are conifers.

Classification.

1. Subdivision: *Coniferophytina (Pinicae),* ***Fork- and Needle-leaved gymnosperms.***

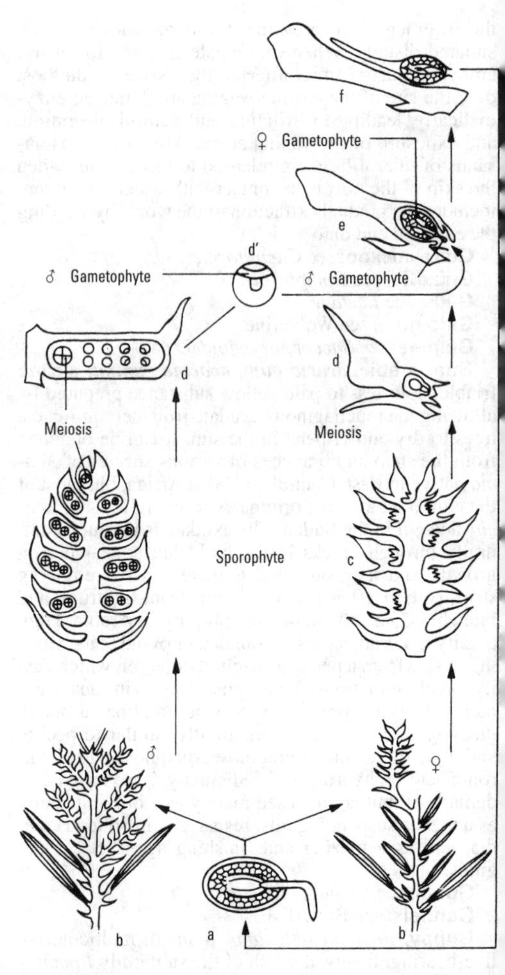

Gymnosperms. Life cycle of a gymnosperm *(Pinus)* with alternation of generations. *a* Germinating seed. *b* Shoot with axes, leaves and inflorescences. *c* Male flower (left), female inflorescence with young cones (right). *d* Left: male cone scale with pollen mother cell and 1- and 2-celled pollen grains (air bladders omitted); Right: cone scale of female flower below the fused "carpels" (ovuliferous scales), showing a naked ovule with its embryo sac cells. *e* Fertilization of female flower, showing ovule, germinating pollen grain and gametophyte. *f* Ripe seed with cone scale.

Class: *Ginkgoatae* (see *Ginkgoaceae)*

Class: *Pinatae*

Subclasses: *Cordaitidae* (see) (extinct), *Pinidae* (see *Pinaceae), Taxidae* (see *Taxaceae)*

2. Subdivision: *Cycadophytina (Cycadicae),* ***Pinnate-leaved gymnosperms.***

Classes: ***Seed ferns*** (extinct, see *Lyginopteridatae),* ***Palm ferns*** (extinct, see *Cycadatae), Bennettitatae* (see), *Gnetatae* (see)

Gymnostomata: see *Ciliophora.*

Gymnotiformes, *knife fish:* an order of Central and South American freshwater fishes considered by some authorities to be a suborder of the *Cypriniformes.* The elongated body is eel-like, or compressed so that it resembles a knife blade. Knife fish are noctural in habit, and they are capable of aerial respiration if they become short of oxygen. Most species have electric organs. The caudal fin is absent or greatly reduced. There is no pelvic girdle, and pelvic fins are absent. An extremely long anal fin, containing more than 140 rays, extends nearly to the tip of the body from near to the origin of the pectoral fin. The anus is under the head or pectorals. The approximately 55 species are placed in 23 genera, 6 familes and 2 suborders.

1. Suborder: *Sternopygoidei.* Body compressed; 12–26 precaudal vertebrae.

2. Suborder: *Gymnotoidei.* Body partially or completely rounded; 33–43 precaudal vertebrae. Families: *Gymnotidae* ***(naked-back knife fishes)*** and *Electrophoridae.* The latter family is represented by a single species, the 2 m-long *Electrophorus electricus* ***(electric knife fish),*** of the Orinoco and Amazon. It can produce individual discharges of up to 600 volts, which are used defensively or for killing prey.

Gymnotoidei: see *Gymnotiformes.*

Gymnure: see Hedgehogs.

Gynandromorphism: the existence of an organism as a mosaic of male and female parts, in which male parts have a male sex chromosome complement and female parts have a female sex chromosome complement. Such an organism is called a *gynandromorph.* True G. is only possible in organisms that lack sex hormones, and whose sex characters are determined by the genetic constitution of their constituent cells. In organisms with sex hormones, the gynandromorphic condition does not result in a clear sexual mosaic, because the presence of male and female hormones leads to intersexual development. See Sex determination, Intersex.

Gynogamete: see Reproduction.

Gynogenesis: development of an egg cell without participation of the sperm nucleus. During fertilization, the male gamete penetrates and activates the egg cell, but its nucleus is then spontaneously or experimentally inactivated and takes no further part in development. The converse is Androgenesis (see).

Gynostemium: see *Orchidaceae.*

Gypaetus: see *Falconiformes.*

Gyphohierax: see *Falconiformes.*

Gyps: see *Falconiformes.*

Gyrocotylidea: see *Cestoda.*

Gyrfalcon: see *Falconiformes.*

Gyttja: see Humus.

H

Habenulae: see Brain.

Habit, *habitus:* the entire outward appearance and posture of animals and plants.

Habitat: the place, area, site or climatic region inhabited by an organism or community, which can be specifically characterized by its physical or biotic properties.

Habitat alternation: a change of habitat, usually linked to a change of climate. Thus, some species are found typically in open meadows near the coast, whereas they occur only in forests in the interior of land masses where the climate is continental. Species that are strongly dependent on specific microclimatic conditions are known as bioclimatic character species.

Habituation: a gradual decrease in the response to repeated stimuli, especially when such stimuli have proved to be meaningless or unworthy of response. At the level of nervous control, H. is due to an increase in the stimulation threshold, i.e. *afferent attenuation.* Restoration of responsiveness is called *dishabituation.* H. may be considered as one of the simplest forms of learning. See Adaptation.

Habronema: see *Nematoda.*

Habu: see Lance-head snakes.

Hackberries: see *Ulmaceae.*

Haddock, *Melanogrammus aeglefinus, Gadus aeglefinus:* a marine fish of the family *Gadidae* (order *Gadiformes),* found in the North Atlantic and adjoining seas. It grows to more than 1 m in length, and has a short chin barbel. The dorsal surface is gray-black with a violet tinge; sides and belly are paler, with a conspicuous dark lateral line, and a black body spot above each pectoral fin. The diet consists of young fish, polychaete worms and small crustaceans. It is commercially important, and the flesh is more highly prized than that of cod.

Hadrome: the xylem of a vascular bundle, excluding any accompanying mechanical tissue.

Haemanthus: see *Amaryllidaceae.*

Haematopodidae, *oystercatchers:* a small cosmopolitan avian family in the order *Charadriiformes* (see). They are mainly birds of the seashore. Some of the geographical races may deserve species status, so that the true number of species (usually placed between 4 and 7) is uncertain. Plumage is black and white, or totally black. The long, laterally compressed, powerful bill (used for striking and prising open mollusks) is bright orange-red. Each foot has 3 slightly webbed toes, and the legs are reddish in color. The tail is short, and the wings are long and pointed. In addition to mollusks, the diet also consists of crustaceans, annelid worms and insects. *Haematopus ostralegus* (43 cm) (Fig.) is the most widely distributed, with several races occurring from Iceland through Europe to the Caspian, across Asia to Australia and New Zealand, as well as in the Canary islands and South Africa.

Haematopodidae. Oystercatcher *(Haematopus ostralegus).*

Haematopus: see *Haematopodidae.*

Haemophilus: a genus of small, nonmotile, slender, aerobic, Gram-negative bacteria, displaying marked pleomorphism in culture. Some strains (e.g. those isolated from cases of meningitis) display elongated, threadlike forms. Species of *H.* occur as parasites in humans and various vertebrates, sometimes causing severe disease, sometimes without apparent effect on their hosts. *H. influenzae* (Pfeiffer's influenza bacillus) was originally though to be the causative agent of influenza. It is now clear that influenza is caused primarily by a virus, and that *H. influenzae* occurs as a secondary, opportunistic pathogen. It is commonly found in the thoats of healthy individuals, and often becomes a secondary pathogen in respiratory infections and other afflictions such as sinusitis, etc. Smooth strains of *H. influenzae* can cause pyogenic meningitis and acute laryngo-tracheitis. Other species of *H.* are responsible for diseases in pigs and poultry.

Haemosporidia: see *Telosporidia.*

Hagfishes, *Myxinidae:* a family of the *Cyclostomata* (see). The body is cylindrical and eel-like, and paired fins are absent. The respiratory organs are essentially similar in structure to those of the Lampreys (see), but the 7 external branchial apertures are absent; on each side the gill pouches are connected by a common tube which opens on the ventral surface of the body at a considerable distance behind the head. There is single nasal aperture, as opposed to the two occurring in higher vertebrates. This leads to a sac which communicates with the pharynx. In contrast, the nasal sac of lampreys forms a cul-de-sac, and does not communicate with the pharynx. Also, in all true fishes, (except *Lepidosiren* and *Ceratodus)* the nasal chambers do not communicate with the pharynx.

The single representative of this family, *Myxine glutinosa,* is found in the soft mud bottoms of northern European waters. It is about 50 cm in length and rose-pink in color, and its external surface is very slimy. It feeds on carrion and invertebrates, as well as often living parasitically inside large fishes (e.g. sturgeon, haddock), which it penetrates by means of a single, serrated, recurved fang arising from the center of the palate (the median palatine tooth). If the hagfish perceives a threat, large quantities of viscous mucus are secreted by glands on the sides of the body.

Haircap moss: see *Musci.*

Hair color: coloration of hair, due to different concentrations of pigments (usually melanin) in the cortex and medulla of the hair shaft. Human hair colors can be grouped into two series: 1) all shades of black and brown (due to eumelanins); 2) all gradations from palest yellow to titian red. The pigments of red hair (trichochromes and phaeomelanins) appear to be formed as a result of an inherited blockage in the synthesis of normal eumelanins. The white hairs of albinos contain colorless pigment granules. Graying and whitening of hair in old age is due to the occurrence of air pockets, the decreased deposition of pigment, or both. Hormonal disturbances can cause premature graying and loss of hair, or conversely they may lead to excessive growth of body hair. For anthropological purposes, human hair is matched against a table of 30 standard hair colors.

Hair moss: see *Musci.*

Hairpin loop countercurrent principle: the concentration of dissolved material by the flow of a solution in a tubular structure that folds back upon itself like a haipin. Within this looped structure, two streams of the solution pass very close to each other in opposite directions, permitting transfer from one stream or current to the other.

1) Kidney (Fig.) (see Kidney functions). The loop of Henle describes a haipin bend, so that two currents of urine flow close to each other in opposite directions; in the proximal tubule urine flows toward the medulla, whereas in the distal tubule it flows toward the cortex. The adrenal cortical hormone, aldosterone, activates ion pumps in the distal tubule, which transfer Na^+ ions from the distal tubule into the interstitial space. The positively charged Na^+ ions are accompanied by Cl^- ions, but the resulting osmotic effect does not cause water molecules to migrate with the NaCl, because the ascending part of the distal tubule is impermeable to water. Since the cells of the proximal tubule are especially permeable to NaCl, the NaCl that enters the interstitial fluid diffuses into the descending limb of the loop of Henle. This process occurs simultaneously and with equal intensity at all levels of the loop of Henle. Integration of the many transport events along the loop, combined with the urine flow, results in an increasing concentration of NaCl toward the hairpin bend of the loop; the concentration decreases again up the limbs of the loop of Henle in the direction of the cortex. This graded NaCl concentration is also reflected within the interstitial tissue. As a result of the increase of concentration toward the point of the hairpin, water can be withdrawn from the collecting tube, and this water can then be taken up by the renal vessels *(vasa recta)* that lie in close approximation to the loop of Henle. A further countercurrent mechanism operates in the hairpin loops of the vasa recta, which prevents transport of osmotically active substances from the kidney medulla, as follows: Na^+ ions concentrated at the hairpin bend diffuse back from the ascending limb of the blood vessels into the descending limb; water diffuses in the opposite direction from the descending into the ascending vessels; this creates a situation in which the osmolarity of the blood leaving the vasa recta is almost constant, and the dissolved substances (in particular NaCl) remain in the medulla. With an increase of blood pressure in the body, the blood flow through the vasa recta increases, back diffusion of water into the vasa recta is reduced, urine is concentrated to a lesser extent in the loops of Henle, and a larger volume of urine is voided. This is known as *pressure diuresis.*

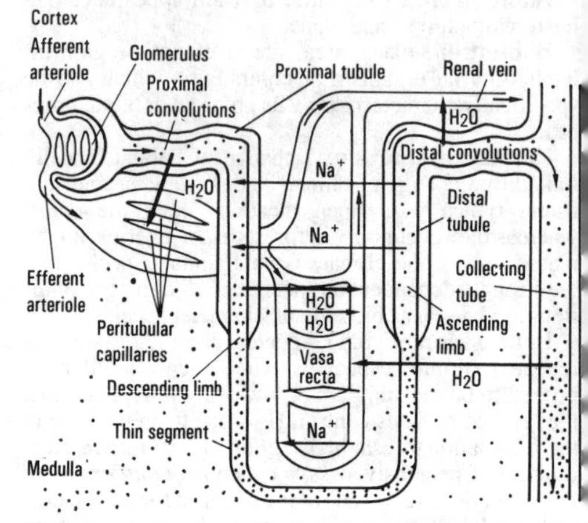

Hairpin countercurrent principle in the kidney

2) Swim bladder (see). The anterior chamber of the swim bladder is the site of the gas gland, which is supplied with abundant capillaries in a structure known as the rete mirabile or "wonder net". This is, however, not a network but a system of capillaries arranged in long hairpin structures, which form an efficient countercurrent system. The capillary walls are unidirectionally permeable to blood gases, which diffuse from the leaving to the entering capillaries, i.e. in the opposite direction to the blood flow, so that the gas concentration continually increases toward the bend of the hairpin loop. The gas gland actively releases gas from the capillary hairpin loops into the swim bladder. This is supported by an increase in glycolysis, the resulting lactic and carbonic acid causing the release of gaseous oxygen, due to the Bohr effect (see). At the same time, this acid effect is reinforced by a small "salting out" effect, in which the solubility of blood gases is decreased by secreted salts. The rete mirabile of an eel contains 500 000 capillaries/cm², and is able to secrete up to 130 ml of gas per minute per bar.

Receptors for the sense of balance respond to the orientation of the fish in space and stimulate the diencephalon. A branch of the vagus nerve from the diencephalon activates gas secretion by increasing glycolysis in the gas gland, and by regulating the blood flow through the two chambers of the swim bladder. In the secretory phase, the vessels of the anterior chamber are opened, while those of the posterior chamber are closed. When gas is removed from the swim bladder, this relationship is reversed. The two chambers are furthermore separated by a movable ring muscle, which migrates posteri-

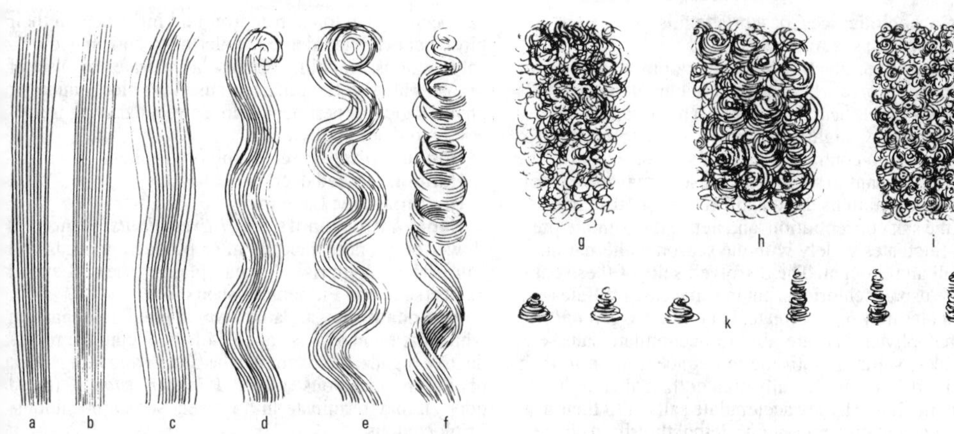

Hairs. Human hair forms (after Martin). *a* Straight. *b* Smooth. *c* Shallow waves. *d* Wide waves. *e* Tight waves. *f* Curly. *g* Frizzy. *h* Loosely frizzy. *i* Densely frizzy. *k* Fil-fil. *l* Spiral. *a* and *b* are lisotrichous; *c* to *f* are kymatotrichous; *g* to *l* are ulotrichous.

orly in the secretion phase and anteriorly in the resorption phase.

Hairs:

1) Epidermal structures of mammals (see Skin, Fig.). A single hair is a slender, unbranched, cylindrical keratinous thread, contained in an invagination of the epidermis called a *hair follicle,* at the base of which the hair shaft and follicle wall are continuous. At the expanded base of each follicle/ hair shaft (the *hair root),* a conical concentration of vascular and nervous tissue (the *hair papilla)* projects into the base of the hair and provides nutrients for hair growth. The hair shaft consists of an outer cornified layer, the *cortex,* and the central part consists of the *medulla* or *pith.* In addition, the outer surface of the cortex may be modified to form a thin *cuticle.* In soft hair (e.g. sheep wool), the cortex is relatively well developed, whereas stiff hairs (e.g. beard hairs of deer) show a typically strong development of the medulla. In very thin hair (e.g. soft human hair, fleecy wool of certain sheep varieties, seal hair), the medulla is absent. Hairs have evolved into defensive *spines* independently in different mammalian groups (spiny anteater, tenrec, hedgehog, etc.).

Glandular evaginations of the follicle wall, called *sebaceous glands,* secrete an oily product *(sebum),* which keeps the hairs supple and waterproof. At rest, hairs normally lie almost parallel to the skin surface, but they can be moved (i.e. erected) by the contraction of smooth muscles associated with the follicle. The follicles of tactile hairs are richly supplied with blood sinuses and nerve fibers.

Hairs are usually aligned in *tracts* and the pattern of these tracts is characteristic of each species. Hair color (see) is due to the presence of pigments. Many animals have a winter pelage and a summer pelage, and therefore change their hair at least twice a year; this may also be associated with a conspicuous color change (e.g. in the stoat). In other animals (e.g. higher apes and humans), the hair is more or less permanent.

Animal hair can be broadly classified into: guard hairs (the longest type of hair, with a pointed tip and a pattern of scales on the basal one third), fur hairs, wool hairs and tactile hairs (vibrissae). Human head hair is generally differentiated into 11 different types (Fig.). Straight,smooth (lisotrichous) types occur mostly on mongoloid races, wavy (kymatotrichous) hair on Europeans, and curly (ulotrichous) hair on negroid races. The follicle of curly hair is markedly curved, that of wavy and lank hair only slightly curved, while the follicle of straight hair is also quite straight. Follicles are inserted more or less at an angle in the scalp; the emergent angle of slightly wavy or lank hair varies between 20° and 70°, whereas straight (oriental-type) hair can emerge at an angle of 90°.

2) See Plant hairs.

Hair seal: see *Phocidae.*

Hairstreaks: butterflies of the family *Lycaenidae;* see *Lepidoptera.*

Hairworms: see *Nematomorpha.*

Hairy footed gerbil: see *Gerbillinae.*

Hairy footed hamsters: see *Cricitinae.*

Hairy frog: see *Astylosterninae.*

Haldane's dilemma: see Genetic load.

Halfbeaks: see *Hemiramphidae.*

Half-collared flycatcher: see *Muscicapinae.*

Haliaeetus: see *Falconiformes.*

Halibut: see *Pleuronectiformes.*

Halichoerus: see *Phocidae.*

Halictophagidae: see *Strepsiptera.*

Haliplankton: see Plankton.

Haller's organ: see *Ixodides,* Olfactory organ.

Halloysite: see Sorption.

Hallstrom dog: see Dingo.

Halmophagy: see Chylophagy.

Halobionts: organisms found only in saline waters or on a substratum consisting largely of salt. In contrast, halophilic organisms display a high preference for, but can exist without, saline conditions. Salt dependency is due to the osmotic state of the organism. H. are iso-osmotic in relation to their substrate, i.e. the osmolarity of their body fluids is the same as that of the environment. *Poikilosmotic* organisms can adapt to changes in the osmolarity of the environment by changing the osmolarity of their body fluids, whereas *homeosmotic* organisms maintain a constant (within certain limits) internal osmolarity.

Halomachilis: see Coast bristletails.
Halophiles: see Archaebacteria.
Halophytes: plants whose development and distribution are favored when the salt content of the available water is greater than about 0.5%. The ocean has an average and only slightly variable salt concentration of about 3.5%. In contrast, the H. of sea coasts and at the edges of salt pans in steppes and deserts may have to tolerate concentrations up to and exceeding 10%. In addition, the salt concentration, and hence the osmotic pressure, fluctuates widely with the seasonal alternation of rainfall and drought. The dissolved salts of these habitats are usually chlorides, but in some cases sulfates and carbonates may predominate. In contrast to *glycophytes* (nonhalophytes), H. are able to accumulate salts, e.g. chlorides, sulfates, until the ion concentration in their cells is equal to, and usually greater than, that of the external medium. H. may accumulate salts until their usually succulent organs contain a lethal threshold concentration [salt accumulators, e.g. glasswort or marsh samphire *(Salicornia)],* wherupon the organ dies or is shed. Alternatively, the salt content of the organs may be regulated by salt excretion via salt glands or salt hairs [salt regulators, e.g. tamarisk *(Tamarix),* orache *(Atriplex)]. Stenohaline* species can grow only on weakly saline soils, whereas *euryhaline* species thrive on strongly salinified soils (up to and greater than 10% salt).
Halorites (Greek *halos* sea, salt): characteristic ammonite genus of the upper alpine Triassic. The shell is planospiral and very narrowly umbilicate, with weakly developed simple radial ribs covered with rows of delicate tuberosities. The shell widens markedly with the last turn of the spiral. The suture line has completely digitated saddles and lobes.
Haloxylon: see *Chenopodiaceae.*
Haltere: see *Diptera.*
Halticinae: see Leaf beetles.
Halysites: see *Tabulata.*
Halys viper, *Agkristrodon halys:* a pit viper (see *Crotalidae)* with several subspecies distributed over a wide area from Asia Minor to Japan. It is red-brown to graybrown in color, with dark-edged bands and spots, especially on the flanks. The subspecies, *A.h. caraganus* ***(Siberian moccasin*** or ***Shistomordnik),*** is the only pit viper on European soil, since its range extends to the Volga area. A more easterly subspecies is *A.h. blomhoffi* ***(Mamush).*** The diet consists of small vertebrates, and insects are also taken. Females give birth to 3–12 live young.
Hamadryad: see King cobra.
Hamamelididae: see *Dicotyledoneae.*
Hamlets: see Sea basses.
Hammerhead: see *Ciconiiformes.*
Hammerkop: see *Ciconiiformes.*
Hammer of the middle ear: see Auditory organ.
Hammond's spadefoot: see *Pelobatidae.*
Hammond's spea: see *Pelobatidae.*
Hamster: see *Cricitinae.*
Hancornia: see *Apocyanaceae.*
Hand animal: see *Chirotherium.*
Handedness: preference for the use of the right or left hand as the working hand. The majority of humans are right-handed; only between 3 and 5% of adults and about 10% of children are left-handed. H. is probably largely inherited, but it can be reversed by training.
Handling: the appropriate handling of laboratory animals, especially during their early development, which has been shown to exert a profound influence on their biosocial behavior, degree of alertness, and physiological functions. Handling can also accelerate the onset of sexual maturity. If animals are used for biochemical or physiological experimentation, correct handling is necessary to reduce stress.
Hanging pitcher: see Carnivorous plants.
Hangul: see Red deer.
Hanuman: see Langurs.
Hapaxanthic plants, ***short-lived plants:*** plants that flower only once, then die after producung seeds and fruit. They comprise Annual plants (see), Biennial plants (see), and Pluriennial plants (see).
Haplocaulous: an adjective applied to plants in which the main axis is capable of producing the reproductive organs. For example, the first germinating shoot of the hypocaulous plant, *Papaver rhoeas* (Field poppy), may terminate in a flower. See Diplocaulous, Triplocaulous.
Haplodioecious: see Heterothallic.
Haplogastra: see *Coleoptera.*
Haplogastrinae: see Stone crabs.
Haplomitrium: see *Hepaticae.*
Haplont: see Alternation of generations, Meiosis.
Hapten: a low molecular mass component of an antigen. The hapten cannot stimulate an immune response on its own, but becomes an antigenic determinant when combined with a carrier molecule. The resulting antibodies react with the uncombined hapten. For example, antibodies to 2,4-dinitrophenol (DNP) are produced in response to a conjugate of DNP with bovine serum albumin, but not to DNP alone. The resulting antibodies are then capable of reacting with free, unconjugated DNP. See Contact hypersensitivity.
Haptera: spiral bands around the spores of *Equisetatae.* Exterior to the endospore and exospore (which form the true wall of the green meiospore) are several layers of perispore (laid down by the periplasmodium). The outermost of these layers consists of two narrow, parallel bands with spathulate ends (the H.). In the moist condition, these are spirally coiled around the spore, but when dry they unroll and extend, but they remain attached at their mid point to each other and to the exospore. These hygroscopic movements serve for dissemination of the spores, but by linking the spores together, the H. may also aid fertilization when the prothalli that arise from them are of separate sexes.
Haptic stimulus: see Contact stimulus, Haptonasty, Haptotropism.
Haptomorphosis: see Mechanomorphosis.
Haptonasty, ***thigmonasty:*** nastic growth movements in response to touch (see Nasty). As in haptotropism, touching or irritation by rigid solids elicit a response, but touching or rubbing with damp gelatin or disturbance by a water stream have no effect.

H. is exhibited by the tendrils of *Bryonia,* cucumber and other *Cucurbitaceae.* In contrast to the tropic response of other tendrils (see Haptotropism), these always bend toward their underside, even when stimulated on their upper surface. The onset of the haptonastic response is due to an alteration of cell turgor, leading to contraction of one flank. This is then followed by growth movements.

Striking haptonastic movements are displayed by certain carnivorous plants, in particular by the tentacles of the sundew *(Drosera).* Only the glandular heads, which produce the sticky secretion, are touch-sensitive. The stimulus is conducted relatively quickly from the head to the base of the tentacle (about 8 mm/min); the base then undergoes marked unilateral growth, causing the tentacle to bend toward the middle of the leaf. The stimulus is then conducted to other tentacles, so that the insect causing the initial response and until now held only by the sticky secretion, is finally entrapped by tentacles on several sides. Some of the subsequent responses are tropic as well as nastic, i.e. they are directional. The reaction of sundew tentacles is also influenced by various chemicals (see Chemonasty).

Haptoreaction, ***haptic reaction, thigmoreaction:*** response to contact with a solid objects. Touch sensitivity is displayed widely by plants, and it leads to different responses: 1) Morphogenic effects, e.g. formation of attachment disks (wild vines), root hairs, haustoria (by fungi). 2) Effects on tissue differentiation, e.g. increased production of mechanical tissue. 3) Growth movements (see Haptotropism, Haptonasty).

Haptoreception: see Tactile stimulus reception.

Haptors: see Adhesion organs.

Haptotropism, ***thigmotropism:*** growth movements of plant organs induced by contact stimuli. The induced movement is usually toward the source of the stimulus (positive H.), e.g. movement of a climbing plant toward a support, and more rarely it may be directed away from the stimulus (negative H.). Like a haptonasty, H. is induced only by the touching or rubbing stimulus of a solid body, and not by constant static pressure or by vibration or shock stimuli. H. is displayed by various organs in many plant species, especially by tendrils. These react to localized and repeated touching by bending markedly toward the stimulated side, a movement that is brought about by a sudden increase in growth on the nonstimulated side. The stimulated site shows no growth. Excitation is transmitted through the plant organ from the touched to the growing site at a rate of up to 4 mm/min. A single, brief stimulation is followed by curvature growth, but this is then compensated by growth on the opposite side as a result of autotropism. Further examples of H. are provided by the leaf stalks of *Tropaeolum* and *Clematis* species, the elongated leaf tip of *Gloriosa,* and the aerial roots of certain orchids. The haustoria-forming parts of Dodder *(Cuscuta)* shoots become adpressed to their host by positive H.

Harbor seal: see *Phocidae.*

Hard amadou: see *Basidiomycetes (Poriales).*

Hardening: see Drought resistance.

Hard ticks: see *Ixodides.*

Hardun: see *Rough-tailed agama.*

Hardy-Weinberg law, ***Hardy-Weinberg distribution:*** a law specifying the expected frequencies of alleles and genotypes in a population of diploid organisms, at various relative allele frequencies. If 2 alleles, ***A*** and ***a,*** are present at relative frequencies p and q (where $p + q = 1$), then the frequencies of the 3 genotypes are given by $(p + q)^2 = p^2{:}2pq{:}q^2$. For example, at $p = 0.6$ and $q = 0.4$, the ratio of the genotypes is 0.36***AA***:0.48***Aa***:0.16***aa***. This is illustrated by the following table:

		Male germ cells	
		0.6*A*	0.4*a*
Female germ cells	0.6*A*	0.36*AA*	0.24*Aa*
	0.4*a*	0.24*Aa*	0.16*aa*

The H-W. law requires that both germ cell types in both sexes are also produced in the ratio 0.6: 0.4, that mating is random, the population is large (i.e. sampling errors are small), the population is isolated, and there is no immigration. In this form, the rule is valid for autosomal gene loci with 2 alleles. Where the polymorphism involves a series of multiple alleles, the genotype frequency is calculated in a similar way, and the H-W. law takes the form of a polynomial square, i.e. $(p + q + r \ldots + z)^2$.

If the gene locus is subject to strong selective pressure, or mating is nonrandom, or the population is not completely isolated, then the relative proportions of genotypes deviates from those predicted by the H-W. law.

Hares: see *Leporidae.*

Harmless snakes: see *Colubridae.*

Harpacticoida: see *Copepoda.*

Harpacticus: see *Copepoda.*

Harpactes: see *Trogoniformes.*

Harpia: see *Falconiformes.*

Harp shells: see *Gastropoda.*

Harpy eagle: see *Falconiformes.*

Harriers: see *Falconiformes.*

Hartebeests: see *Alcelaphinae.*

Harvester termites: see *Isoptera.*

Harvestmen: see *Opiliones.*

Harvest mouse: see *Muridae.*

Hashish: the dried resin from the microscopic glandular hairs on the upper foliage leaves of the female hemp plant *(Cannabis sativa* L.); the variety *indica* produces particularly large quantities of the resin. H. contains 2–8% of the psychotropic Δ^3-tetrahydrocannabinol. Marihuana (often used synonymously with hashish) is the dried and chopped tips of the shoots of the female hemp plant, with a content of 0.5–2% Δ^3-tetrahydrocannabinol. Both hemp drugs have been used for millenia in folk medicine. Today they are, along with alcohol, the most widely used narcotic drugs. The number of consumers worldwide is estimated at 300 million. The drugs are usually smoked alone or mixed with tobacco in cigarettes ("joints") or pipes. Other nonpsychotropic compounds present in H. are cannabinol, cannabidiol and cannabidiolic acid.

Hatchet fishes: see *Characiformes.*

Hatch-Slack-Kortschak cycle: the CO_2-trapping cycle in C4-plants (see) and in Crassulacaen acid metabolism (see).

Haustoria (sing. **Haustorium):**

1) Absorptive organs of parasitic or semiparasitic angiosperms. They are papillary outgrowths of the parasite, which penetrate more or less deeply into the host plant. H. that make contact with host vascular tissue also develop sieve tubes and water conducting tissue, which conduct water and nutrients into the parasite.

2) Hyphal outgrowths of parasitic fungi, which penetrate and withdraw nutrients from host plant cells.

3) Absorptive, organ-like parts of cotyledons (e.g. in the caryopses of grasses).

4) Cells of the embryo sac or endosperm, which are modified for absorption.

5) The foot of the sporangium stalk of mosses, which is embedded in the gametophyte.

Haversian canal: see Bones.

Hawaiian honeycreepers: see *Drepanididae.*

Hawfinch: see *Fringillidae.*

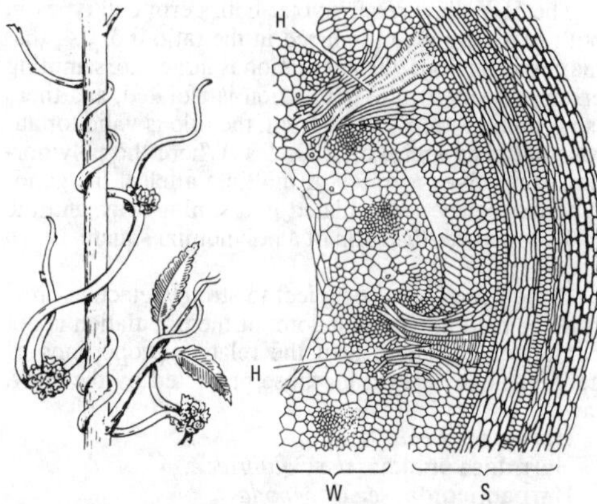

Haustoria. Left: Willow branch entwined by large dodder *(Cuscuta europaea).* Right: transverse section through the stem of the host *(W)* with a short longitudinal section of the parasite stem *(S). H* Haustorium.

Hawks: see *Falconiformes.*
Hawksbill turtle: see *Chelonidae.*
Hawthorn: see *Rosaceae.*
Hay bacilli: see *Bacillus.*
Hazel mouse: see *Gliridae.*
Head lice: see *Phthiraptera*
Headstanders: see *Characiformes.*

Heart, ***cor, cardia:*** central organ of the blood vascular system of different animals. It is a hollow muscular organ, which by rhythmic contractions of its muscle fibers propels the blood in one direction (with the exception of the tunicates, in which the direction of blood flow is continually reversed).

In *polychaete worms,* the H. consists of an elongate, contractile vessel, which lies dorsal to the gut and extends for almost the entire body length. Some of the commissural vessels to the longitudinal ventral vessel are also contractile, and are known as hearts or pseudohearts. In *crustaceans,* the heart is also situated dorsally, and is saclike (branchiopods), elongate and tubular (decapods) or spherical (e.g. *Astacus)* in shape. The H. of advanced *arachnids* and of *mollusks* lies in a pericardium. In the latter, it is divided into a main chamber and one or two atria. *Cephalopods* have auxilliary Hs. (auricles or gill Hs.) which pump venous blood into the gill vessels. The *insect* H. is usually a blind, contractile, hindsection of the dorsal vessel; the blood supply to antennae, legs and wings is insured, inter alia, by additional pulsating organs, also known as accessory hearts.

The vertebrate heart is a hollow muscular organ, which functions as both a suction and a pressure pump, and which plays an important part in regulating the distribution and pressure of the circulating blood (see Blood circulation). In all vertebrates, the H. occupies a characteristic ventral position in the body.

The H. of fishes is divided into four sequential chambers. The caudal *sinus venosus* receives venous blood from the body, and passes it to the *auricle* or *atrium.* From the atrium, blood is pumped into the *ventricle,* and from there into the *conus arteriosus,* which then pumps it into the gill arteries. In most fishes, the H. contains exclusively venous blood.

In *amphibians,* the occurrence of a pulmonary circulation marks the onset of modifications to the H. With increasing evolutionary advancement of the vertebrates, these changes eventually lead to complete separation of atria and ventricles. The amphibian H. has two separate atria, the right one receiving oxygen-depleted blood from the body, the left one receiving oxygen-rich blood from the lungs. Blood is mixed in the single ventricle, but since mixing is incomplete due to the presence of folds in the ventricle wall, blood pumped from the ventricle to the lungs has a fairly high venous content, while that pumped to the head and body is relatively enriched with oxygen.

The *reptilian* H. has completely separated atria, and the ventricle contains an incomplete dividing wall. The aortic arch (which curves to the right) arises from the left part of the ventricle. Since the carotid arteries branch from the aortic arch, the head always receives fresh blood. The left aortic arch contains mixed blood, while the pulmonary artery, which arises on the extreme right, carries oxygen-depleted blood to the lungs.

In *birds* and *mammals* (Fig.), the two atria and two ventricles, are all completely separate chambers. Thus, only in birds and mammals possess two truly separate circulations, the *pulmonary* and the *systemic circulations.* Oxygen-depleted systemic blood passes through the upper and lower vena cavae into the right atrium, and from there into the right ventricle. Contraction of the right ventricle then pumps venous blood through the pulmonary arteries to the lungs. Oxygen-enriched blood then passes via the pulmonary veins into the left atrium, from there into the left ventricle, from where it is pumped into the aorta. Blood flow through the H. and its associated vessels is regulated by Heart valves (see).

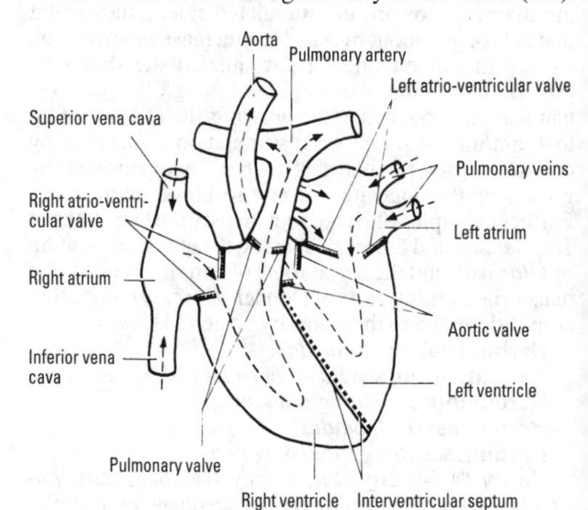

Heart. Schematic representation of the human heart. The direction of blood flow is shown by the arrows.

Considering the avian and mammalian Hs. as evolutionary derivatives of the reptilian H., it is interesting to note that the single carotico-systemic arch of the avian H. is the right one, whereas that of the mammalian H. is the left one, thus reinforcing the view that mammals and birds arose from quite different reptilian stock.

In some mammals (e.g. horse, pig, dog) the regions where vessels leave the H. are reinforced by a cartilaginous cardiac skeleton; ruminants and pachyderms have a similar but bony structure.

The avian and mammalian H. has a specialized sensory system of modified muscle fibers (see Cardiac excitation).

Heart beat: see Cardiac cycle.

Heart contraction: the systole of a heart. It involves two types of muscle contraction: 1) an increase in tension without a change in length *(isometric contraction),* which occurs when both the atrioventricular valves, the pulmonary valve and the aortic valve are closed; and 2) a shortening of muscle fibers without an increase in tension *(isotonic contraction),* which occurs as blood flows into the arteries. Both types of contraction occur during the transition phases.

H.c. is initiated by Cardiac stimulation (see), and consists of the following stages of electromechanical coupling. In the resting state the thin (main component actin) and thick (main component myosin) filaments of the sarcomeres are not connected by cross-bridges and they are separated by tropomyosin. Excitation permits calcium ions to enter the sarcomere from the extracellular space, accompanied by release of calcium ions from intracellular reserves. Troponin molecules (in the actin filaments) bind calcium ions and displace the tropomyosin filaments, so that the actin filaments become attached to the myosin filaments via the knob-like globular heads of the latter. Energy from the cleavage of ATP supports an oar-like movement of the myosin heads, which moves the actin filaments past the myosin filaments; the I-band (actin filaments) then shifts into the A-band (interdigitated myosin and actin filaments). H.c. is ended when pumps remove the calcium ions from the muscle cytoplasm. Uncleaved ATP releases the binding between the actin and myosin filaments. Blood flowing into the heart stretches the relaxed heart muscle (i.e. I-bands reappear) to its original resting state. See Muscle tissue.

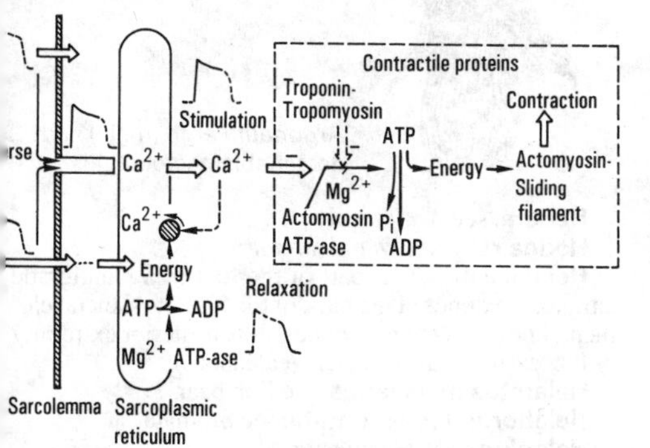

Heart contraction. Electromechanical coupling of the heart.

Heart-lung preparation: see Starling's law of the heart.

Heart minute volume: see Cardiac output.

Heart regulation: see Cardiac regulation.

Heart rot: see Deficiency diseases.

Heart stimulation: see Cardiac stimulation.

Heart urchins: see *Echinoidea.*

Heart valves: projections of the internal wall of the heart or arteries, which, by acting as valves, determine the direction of blood flow in the heart. The heart of higher vertebrates, including humans, contains 4 valves.

The *tricuspid valve* or right atrioventricular valve guards the opening between the right atrium and right ventricle. It consists of three irregular-shaped flaps or cusps, formed mainly of fibrous tissue covered by endocardium. They are continuous at their bases, forming a ring-shaped membrane around the margin of the atrioventricular orifice. Their free ends are directed into the ventricle and are attached by the chordae tendineae to papillary muscles arising from the interior surface of the right ventricle.

The opening between the left atrium and left ventricle is guarded by the *bicuspid* or *mitral valve.* It consists of only two flaps and is stronger and thicker than the tricuspid valve. Cordae tendineae attach the cusps to papillary muscles in the left ventricle.

Blood flow from the atria into the ventricles is permitted by the tricuspid and bicuspid valves, whose flaps point in the same direction as the blood flow. Attempted flow in the opposite direction pushes the flaps backward (upward) and together to form a partition; this is prevented from opening into the atria by the tension of the chordae tendineae, increased by contraction of the papillary muscles.

Semilunar valves guard the orifice between the right ventricle and pulmonary artery *(pulmonary valve)* and between the left ventricle and the aorta *(aortic valve).* Semilunar valves consist of three half-moon-shaped pockets, the opening (i.e. free border) of each pocket pointing away from the heart and into the artery. Semilunar valves permit free passage of blood into the arteries, but any attempted back-flow fills each pocket with blood, causing the free borders to distend and form a barrier to further flow.

Heart wood: see Wood.

Heart work: see Cardiac work.

Heat collapse: circulatory failure due to over-heating. Since H.c. is a circulatory disturbance rather than a failure of temperature regulation, it frequently occurs in the standing position. Excessive dilation of blood vessels leads to a drastic fall in blood pressure and loss of consciousness.

Heat damage: damage to plants caused by exposure to heat, frequently as a result of over-exposure to solar radiation. In general, temperatures higher the 40 °C are detrimental to plants. This also applies to local overheating, e.g. when strong sunlight is concentrated by the lens effects of water droplets on wet plants. Chlorophyll can also be destroyed by excessive illumination. Many terrestrial plants of dry, hot habitats are resistant to both heat and desiccation (see Desiccation damage). In cereal crops, H.d. is manifested as premature grain hardening. Heat-damaged leaves often change color, wilt, and/or become scorched. Heat damage to fruits, e.g. gooseberries, apples and tomatoes, causes cork formation, discoloration and shrinkage.

Different underlying mechanisms give rise to H.d. For example, the temperature optimum for respiration is generally higher than that for photosynthesis or growth, so that at high temperatures a plant usually respires more actively than it assimilates. In the temperature range between optimum and maximum (maximal tolerable temperature), inhibitory processes predominate; e.g. the growth rate gradually decreases, until it becomes zero at the maximal temperature. Temperatures above the maximal value finally lead to cell death. Irreversible damage is usually preceded by Heat rigidity (see). Heat death depends on several interrelated variable factors, e.g. the killing temperature for a particular plant or plant organ is not an absolute quantity, and it can only be expressed as a function of the duration of exposure. For example, the germination rate of wheat grains is decreased from 90% to 50% by exposure to 55 °C for 9 minutes, or 50 °C for 2 hours, or 45° for 15 hours.

Heat death is often due to coagulation and denaturation of cytoplasmic proteins. When heat death occurs below the coagulation temperature of the protoplasm, it is often due to the destruction of cytoplasmic structures, e.g. alterations of protein-lipid interactions or micellar structures. Furthermore, a gradual and continual increase in temperature causes protein degradation, accompanied by increased carbohydrate degradation, and the uncoupling of respiration and phosphorylation, so that the plant itself generates more heat instead of ATP.

Heat death: see Heat damage.

Heath: a term first used in the narrow sense for shrubby vegetation dominated by the genus *Calluna* (heather or ling). It is generally applied to treeless or sparsely treed plant communities, consisting of low, evergreen shrubs, dwarf shrubs and tough-leaved grasses. The soil type varies widely, from the nutrient-poor, acidic, podzolized soils of grass heaths and lichen heaths to the dry, sandy soils of sparse woodland heaths.

Heather: see *Ericaceae.*

Heath family: see *Ericaceae.*

Heat-losing center of the hypothalamus: see Temperature regulation of the body.

Heat of combustion of dietary components: see Calorific value of dietary components.

Heat-promoting center of the hypthalamus: see Temperature regulation of the body.

Heat resistance: the ability of certain plants to withstand high temperatures. Depending on their natural habitat and accompanying adaptations, different plant species display widely varying degrees of sensitivity or resistance to high temperatures. The separate organs of a plant react differently to temperature, and even the cells of a single tissue differ physiologically with respect to H.r. Water economy is crucially linked to H.r. Water-rich plants are particularly susceptible to heat damage, whereas seeds and other water-poor resting organs withstand relatively high temperatures. The protoplasm of extremely heat-resistant plant species is relatively resistant to heat denaturation. Stimulation of protein synthesis by kinins also often contributes to H.r. Other plant species avoid heat damage by reflecting part of the solar irradiation. Excessive absorption of irradiation is also prevented by vertical or profile alignment of leaves, e.g. in compass plants such as *Lactuca serriola* (Prickly lettuce). Alignment of the leaves of *Acacia* and *Eucalyptus* so as to decrease the incident irradiation results in "shadeless" forests. Alternatively, the plant may be shaded by its own dead leaves (leaf insulation, e.g. in *Mesembryanthemum* species), or by layers of bark (e.g. "fireproof" savanna trees). Many plants of wet habitats are prevented from overheating by the cooling effect of transpiration.

Heat rigidity: cessation of metabolism in plants, caused by temperatures above the tolerable maximum. The plants no longer respond to stimuli and are incapable of autonomic movements. Cells no longer display cytoplasmic streaming, and the condition rapidly becomes one of irreversible Heat damage (see).

Heat stroke: a form of Hyperthermia (see) in humans. H.s. occurs when the body temperature is raised to 40–41 °C for a long period. The resulting brain edema destroys nerve cells, leading to a cessation of perspiration, disorientation, hallucinations and cramps. H.s. and sun stroke are often used synonymously, although the vomiting experienced in sunstroke may be caused, at least partly, by overexposure to UV irradiation, i.e. a form a radiation sickness.

Heat torpor: see Cold torpor, Temperature factor.

Hectocotylus: an arm of a male cephalopod, which is modified as an intromittent organ (right or left 4th arm in squids and cuttlefish; right 3rd arm in octopods) for the transfer of spermatophores through the restricted mantle opening of the female. The degree of hectocotyly varies according to the genus. In *Sepia* and *Loligo* the H. possesses several rows of reduced suckers, which form an adhesive area for the transport of spermatophores. In *Octopus* the tip of the arm carries a spoonlike depression, while in *Argonauta* the H. actually possesses a sperm storage chamber. The H. of octopods may detach from the male and remain in the mantle of the female, e.g. in *Argonauta* and *Philonexis.* Detached hectocotyli were originally thought to be parasitic worms, which were assigned to the genus "Hectocotylus".

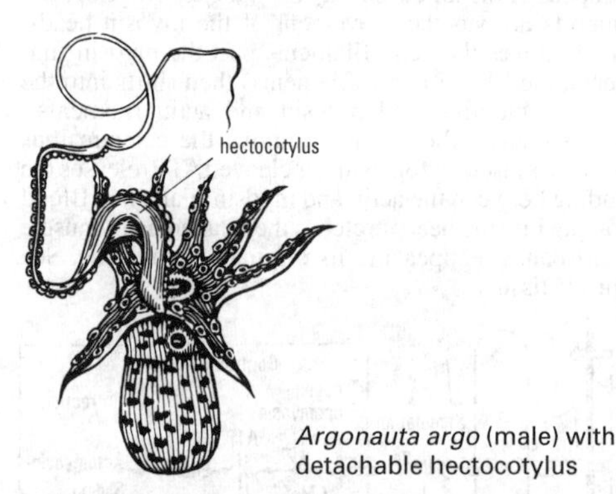

Argonauta argo (male) with detachable hectocotylus

Hedera: see *Araliaceae.*

Hedge rats: see *Octodontidae.*

Heel: the muscular part of the foot of a plantigrade animal. It extends at the back of the foot. Its skeletal element, known as the heel bone, is the posterior extremity or tuberosity of the os calcis (calcaneus).

Helarctos malayanus: see Sun bear.

Heleborus-type stomata: see Stomata.

Helenium: see *Compositae.*

Heleophrynidae, *heliophrynids, ghost frogs:* a family of the *Anura* (see), containing 3 species in a single genus, which inhabit rocks in cascading mountain streams in South Africa. They vary in size from 3.5 to 6.5 cm; their large-mouthed tadpoles lack beaks and have a large oral sucker for clinging to rocks. The belly skin of the adult is thin and almost transparent. The skin also carries hooks or spines, which probably act as an aid to climbing on stones in fast flowing water. Eggs are laid on damp ground or in puddles. Eight ectochordal presacral vertebrae are present with cartilaginous intervertebral joints and imbricate (overlapping) neural arches. The notochord persists in the adult. Presacrals I and II are fused, and the atlantal cotyles of presacral I are closely juxtaposed. Ribs are absent and the pupils are horizontal. The coccyx, which has proximal transverse processes, has a bicondylar articulation with the sacrum. A cartilaginous omosternum and sternum are present on an arciferal pectoral girdle. Maxillary and premaxillary teeth are present.

Heleophryne rosei ***(Table Montain ghost frog)*** is mottled green and brown with a reddish tinge, and is found on Table Mountain near Capetown.

Heleoplankton: see Plankton.

Heliamphora: see Carnivorous plants.

Helianthus: see *Compositae.*

Helicidae: see *Gastropoda.*

Helicoid cyme: see Flower, section 5) Inflorescence.

Helicotylenchus: see Eelworms.

Heliolites: see *Helioporida.*

Heliopais: see *Gruiformes.*

Heliophytes: see Sun plants.

Heliopora: see *Helioporida.*

Helioporida, *Coenothecalia,* ***blue coral:*** an order of the *Octocorallia* (see), with a massive, tubular, external, calcareous skeleton. The order is represented by the single genus, *Heliopora,* which occurs in the Indian and Pacific Oceans. The skeleton is secreted by a layer of ectodermal cells, and it is not composed of sclerites. Relatively large pits (thecae) on the surface of the colony are occupied by polyps, while smaller pits house the tubular processes of a network of solenia (hollow strands of endoderm), which form a continuum between neighboring polyps. The fossil *Heliolites,* a dominant form of Paleozoic coral reefs, displayed similar skeletal characteristics.

Heliornis: see *Gruiformes.*

Heliornithidae: see *Gruiformes.*

Heliozoans, *Heliozoida,* ***sun animalcules:*** a group of protozoa of the phylum (or superclass) *Sarcodina,* class *Actinopodea* (which contains the subclasses, *Radiolaria, Acantharia, Heliozoida* and *Proteomyxida).* H. occur primarily in fresh water, and they may be floating or benthic. Most forms are free-living, but some benthic forms are stalked and attached to the substratum. H. are comparable to the exclusively marine radiolarians, and some forms have siliceous skeletons resembling those of the latter. Their radiating pseudopodia (known as axopodia), however, are even stiffer than those of radiolarians. Each axopodium (axopod) consists of a central axial rod covered with granular, adhesive cytoplasm. The axial rod consists of packed microtubules, which may decrease in size or even be absorbed totally, so that the axopodia become temporarily less stiff and are able to wrap around large prey. Several axopodia work in unison to draw prey into the cytoplasm, where it is enclosed in a food vacuole. H. are spherical or ovoid, and on account of their radial axopodia they have a sun-like appearance. The body consists of: a) an outer ectoplasmic sphere (cortex), which is often highly vacuolated; and b) an inner medulla composed of dense endoplasm, which contains one to several nuclei, together with the bases of the axial rods. The axial rods may all be attached to the membrane of a single, central nucleus, e.g. in *Actinophrys,* or they may be attached to the membranes of several nuclei, e.g. in *Actinosphaerium* and *Camptonemia.* Alternatively, the nucleus may be eccentric with no connection to the axopodia, which then arise from a central centroplast. During mitosis the centroplast behaves as a centriole, and it is involved in the formation of the mitotic spindle. A skeleton is often present, embedded in an outer gelatinous covering. This skeleton may consist of foreign particle such as sand or living diatoms; or it may consist of plates, spheres, tubes or needles of silica secreted by the animal, loosely matted in one or more layers lying tangentially to the body. The sun-like appearance of many species is emphasized by the radial arrangement of skeletal siliceous needles. In a few species the skeleton is composed of tectin in the form of a decorative lattice.

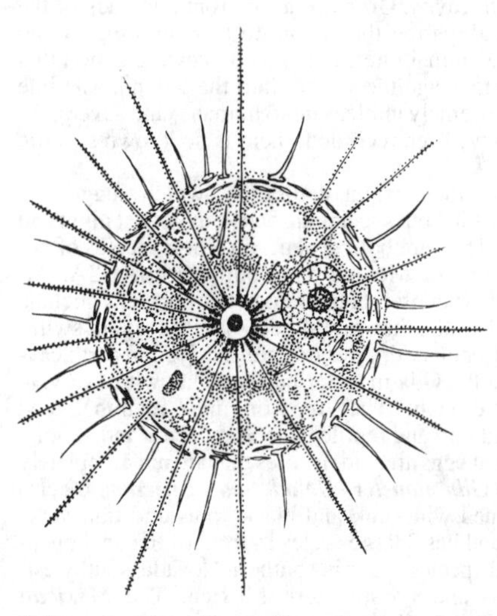

Heliozoans. Acanthocystis aculeata. Axopodia radiate from the central centroplast. The nucleus is situated on the right.

Helix:

1) see *Gastropoda.*

2) The spiral arrangement of certain biopolymeric compounds, e.g. starch, proteins and DNA. In many Proteins (see), a high proportion of the polypeptide chain has the configuration of an α-helix. DNA (see Deoxyribonucleic acid) exists as a double helix of two intertwined helical polynucleotide strands, with a diameter of about 2 nm, an internucletide distance of 0.34 nm (along the length of the DNA fiber), and a helical repeat distance of 3.64 nm; each turn of the helix therefore contains 10 nucleotides.

Hellbender: see *Cryptobranchidae.*

Helleborine: see *Orchidaceae.*

Helmet bird: see *Vangidae.*

Helminthology: the study of parasitic worms, a branch of Parasitology (see).

Helminths: a collective term for various types of worms (trematodes, tape worms, round worms, acanthocephalans, leeches), which have in common the fact that they live parasitically in or on humans and animals (and plants).

Helobiae: see *Monocotyledoneae.*

Heloderma: see *Helodermatidae.*

Helodermatidae, ***beaded lizards, venomous lizards:*** a geologically old family of the *Lacertilia* (see) containing only 2 living species in a single genus. They are 50–80 cm in length, with a plump cylindrical body, broad head, short powerful legs (5 toes on each foot), and a thick rounded tail which stores fat. The back is covered with relatively large, bony, bead-like scales, whereas the flat ventral scales are arranged regularly and are hardly ossified. The 2 species of this family are the only living venomous lizards. The paired venom glands are situated on the rear edge of the lower jaws (cf. snakes, in which the venom glands are found in the upper jaw). There is no direct connection between venom glands and teeth, and the venom is emptied into the mouth cavity. Grooves at the front and back of the lower teeth permit the venom to flow into the wound when the animal bites, aided by a chewing action that helps to massage the venom into the wound. The bite can be extremely unpleasant to humans, and several fatalities have been recorded; there is no known specific antidote.

Despite their bright coloration, which suggests a diurnal habit, both species are active mainly at night and twilight. They are both ground-dwelling species of hot arid regions of Mexico and southwestern USA. Although their chosen habitat does not normally include bodies of water, they are, if necessary, excellent swimmers. Particulary dry periods are survived in earth cavities, and the Gila monster has been known to survive years of drought. They feed on small mammals and birds, and bird and reptile eggs. Between 3 and 13 long cylindrical eggs are laid in holes that are not completely dry. The ***Gila monster*** *(Heloderma suspectum,* 60 cm) is patterned with pink and black spots and transverse stripes, and has 2 large scales in front of the anal opening; 2 subspecies occur in southern Nevada, southwestern Utah and northwestern Mexico. The ***Mexican beaded lizard*** *(H. horridum,* 80 cm), which has no scales before the anal opening, is patterned with yellow and black spots and transverse stripes, the pattern varying in the 3 subspecies, which range from northwestern to southwestern Mexico.

Helodermatidae. Gila monster *(Heloderma suspectum).*

Helophytes: plants of marshy habitats. Their roots and lower shoot parts, which are at least temporarily submerged, are structured externally and internally like those of aquatic plants. The upper shoot, however, resembles that of a terrestrial plant. H. therefore represent a transitional form between aquatic and terrestrial plants.

Helostomatidae: see Gouramies.

Helotiales: see *Ascomycetes.*

Helper cells: see Deficient virus.

Helper virus: see Defective virus, Deficient virus.

Hemagglutination: clumping of red blood cells, e.g. by antibodies or viruses.

Hemal arch: see Axial skeleton.

Hematopoiesis: production of red and white blood cells. Erythrocytes and granulocytes are produced in the red bone marrow. Other leukocytes arise in the lymph nodes and spleen.

Human erythrocytes survive for about 120 days, most leukocytes about 2 days. They are eventually degraded in the bone marrow, liver and spleen. The short life of blood cells means that new cells must be produced at a relatively high rate. Normal rates of production in the human are 15×10^9 leukocytes and 120×10^9 erythrocytes per day. After extensive blood loss, the rate of erythropoiesis may increase up to seven-fold. Erythrocyte precursors (erythroblasts) arise in red bone marrow and develop into reticulocytes in rather less than one week. Reticulocytes are released into the circulation, where they mature into erythrocytes in 1–2 days.

H. is humorally regulated. The stimulus for erythropoiesis is an oxygen deficiency in the blood, which is usually due to erythrocyte degradation, and is intensified by blood loss or low atmospheric oxygen tension (e.g. at high altitudes). Oxygen deficiency stimulates centers in the hypothalamus, which in turn stimulate production of the hormone, Erythropoietin (see), predominantly by the kidneys. Erythropoietin is transported in the blood to the red bone marrow, where it stimulates cell division. Less than 1% of the cells produced in the red bone marrow are normally released into the circulation each day. Loss of blood markedly stimulates the sympathetic nervous system, which causes the reserve of developing cells in the bone marrow to be released into the circulation. In addition, erythropoietin production is increased, so that the loss of blood is completely compensated within 2–3 weeks by an increased rate of synthesis.

Hemelytra: see *Heteroptera.*

Hemerophilic species, ***hemerophilous species:*** animal and plant species which favor the environmental conditions created by humans and human agencies. By closer association with humans, H.s. may find protection from predators, a more constant and improved food supply, and improved conditions for reproduction. If anthropogenic dependence is very strong, the condition is called Synanthropy (see). An Anthropocoenosis (see) consists of hemerophilic animal species in association with humans and their domestic animals. Establishment of animal species in urban settlements is called *urbanization.*

Typical H.s. are the house sparrow, redstart, common swallow, house martin, house mouse, black rat, brown rat, parasites and pests of humans and domestic animals (cockroaches, bedbugs, fleas, flies). H.s. of field and meadow are the gray partridge, skylark, field shrew and hamster. The blackbird and ring-dove have long been associated with towns and cites, and have been joined more recently by the jay.

Plant H.s. are primarily ruderal plants and weeds.

Hemerophobic species: animal and plant species that cannot tolerate anthropogenic alteration or disturbance of their environment, and withdraw to less affected areas, provided this is possible. H.s. are therefore particularly threatened with extinction, e.g. the black stork, white-tailed eagle, lynx.

Species that are neither disadvantaged nor favored by human influence are said to be *culture-indifferent,* e.g. stinging nettle.

Hemerophytes: plants that preferentially colonize cultivated land.

Hemerythrin: see Blood.

Hemicelluloses: high molecular mass polysaccharides of plants, consisting of β-1,4-glycosidically linked pentose and hexose residues, and often uronic acids. They serve as structural compounds, contributing to cell wall structure and woody mechanical tissues, and sometimes as reserve substances. Animals cannot digest them. Common H. are mannans, galactans, xylans, fructans, glucans and arabans.

Hemichordata, *hemichordates,* *Branchiotremata:* a small phylum of the *Deuterostomia.* Hemichordates are exclusively marine, wormlike animals, which are solitary *(Enteropneusta,* ***Acorn worms)*** or live in communities or colonies *(Pterobranchia).*

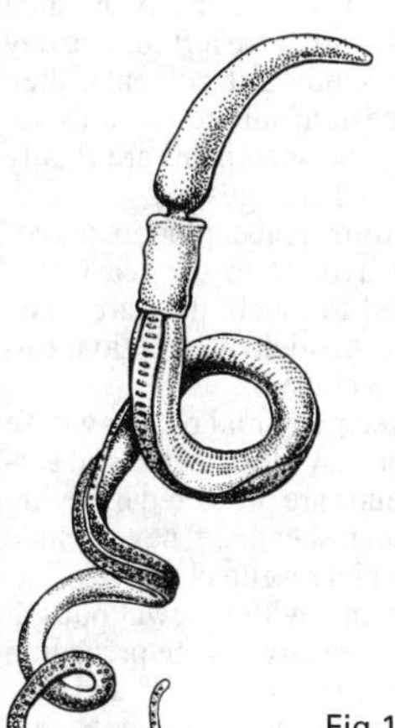

Fig.1. *Hemichordata.* The acorn worm. *Saccoglossus.*

The hemichordate body consists of 3 regions (protosome, mesosome, metasome), of which the hindmost (metasome) is by far the largest. Each region contains a coelomic cavity, formed during embryonal development from evaginations of the archenteron. The foregut is modified to a pharynx with pharyngeal gill clefts. From the foregut, a hollow, blind Stomochord (see) projects into the protosome. A hollow nerve tube lies on the dorsal side of the mesosome. There is a fairly well developed circulatory system, usually with a contractile heart. In the *Pterobranchia,* the mesosome caries tentacles. Acorn worms possess a free-swimming larva (Tornaria, Fig.2), which is very similar to the larvae of some echinoderms. The hemichordates show close affinities to both the echinoderms and the chordates. They are divided into 2 classes.

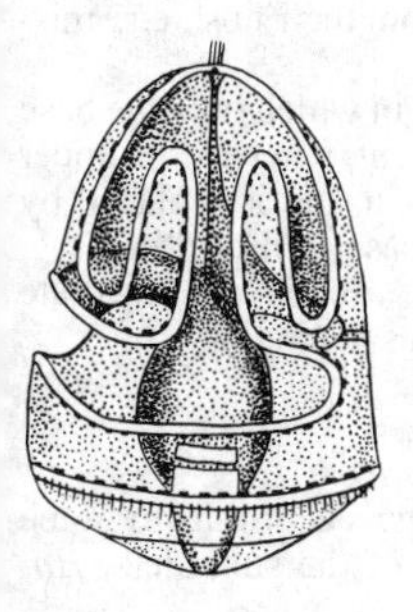

Fig.2. *Hemichordata.* Tornaria larva of an acorn worm.

1. Class: *Enteropneusta* ***(Acorn worms).*** About 60 species of these worm-like animals are known, most of them living in the shallow waters of the tidal zone, where they burrow into the mud and sand, forming burrows lined with mucus and adherent sand grains. The burrows of some genera (e.g. *Balanoglossus, Saccoglossus)* are U-shaped, with an anterior and a posterior opening to the surface. The elongated body is usually between 10 and 15 cm in length. *Balanoglossus gigas* (a Brazilian species) sometimes exceeds 1.5 m in length and lengths up to 2.5 m have been claimed. It consists of a short, conical (hence "acorn") proboscis (protosome), a short collar (mesosome) and a long trunk (metasome). The short, cylindrical collar overlaps the proboscis stalk, and the mouth opens ventrally within this overlapping region. A nerve tube on the ventral side of the collar corresponds embryologically to the spinal cord of the chordates. Sense organs are lacking. On each side of the anterior part of the trunk is a dorsal row of small, external gill openings, leading into deep pouches, each communicating with the pharynx by a tall cleft. Each cleft is almost divided into two by a tongue bar. The anterior gut is perforated by 40–200 pairs (maximally 700 pairs) of these gill clefts, which are the sites of respiratory gas exchange.

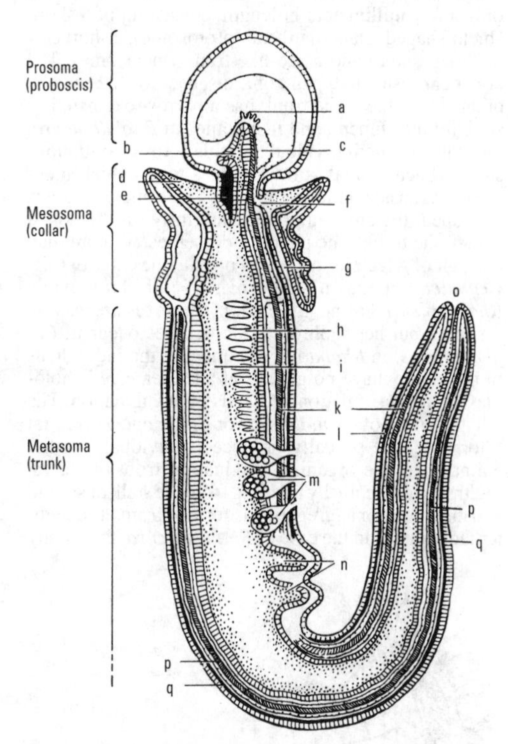

Fig.3. *Hemichordata.* Structure of an acorn worm. *a* Glomerulus. *b* Cecum of notochord. *c* Pericardium. *d* Mouth. *e* Notochord skeleton. *f* Proboscis pore. *g* Tubular collar pore. *h* Pharynx with gill slits. *i* Intestine. *k* Dorsal blood vessel. *l* Dorsal nerve. *m* Gonads. *n* Intestinal ceca. *o* Anus. *p* Ventral blood vessel. *q* Ventral nerve.

Sexes are separate. The gonads, situated in the trunk alongside the gut, consist of numerous saccules opening directly to the exterior. Fertilization occurs in the water. Cleavage is equal, holoblastic and usually radial. The blastopore of the invagination gastrula soon closes, but marks the posterior of the embryo where the anus later breaks through. Most species possess a pelagic tornaria larva with an apical thickening, a simple gut and 3 coelomic cavities (protocoel, mesocoel and metacoel), the latter being formed by evagination of the archenteron. These coelomic cavities are largely filled with connective tissue and muscle fibers, and they have no peritoneal lining; those of the protosome and mesosome open to the exterior via dorsal pores. The tornaria shows close affinities to the auricularia larva of the echinoderms. Burrowing forms consume large quantities of sand and mud, and digest the constituent organic material, producing casts of processed sand at the posterior opening of the burrow. In other forms, suspension feeding is important, in which detritus and plankton come into contact with the proboscis, become trapped in mucus and are carried posteriorly to the mouth by ciliary currents. Three families are recognized: *Harrimanidae, Spengelidae, Ptychoderidae.*

2. Class: *Pterobranchia.* Members of this class nearly always form communities or colonies. Individuals are only a few millimeters in length, consisting of a short, shield-shaped, glandular head (protosome), a short collar (mesosome) and a sac-like trunk (metasome). The collar carries up to 9 pairs of arms, each with 2 rows of ciliated tentacles. The trunk has a narrow extension or stalk for attachment, and in colonies of *Rhabdopleura,* the stalks of the individual animals form a continuous stolon. Each animal is housed in a tube, which it secretes. As an adaptation to its sessile existence, the gut is U-shaped, the anus emptying anteriorly on the dorsal side of the trunk. The pharynx possesses no more than one pair of gill clefts, and these open to the exterior (e.g. *Cephalodiscus* has a single pair of gill clefts, while *Rhabdopleura* has no gill clefts). The sexes are usually separate, but hermaphrodites sometimes occur in *Cephalodiscus;* in *Rhabdopleura,* most of the individuals in the colony have no gonads at all. Cleavage is holoblastic, and gastrulation proceeds by invagination. The ciliated larva (of *Cephaldiscus* or *Rhabdopleura)* is not a tornaria. The sexually produced individual (mother animal), i.e. the organism originating from the larva, multiplies asexually by budding from the stalk or stolon, and the colony or aggregation is formed from the daughter individuals. In the case of *Rhabdopleura,* the colony consists of daughter zooids all permanently attached via the stolon. Barely 20 species are known, and most of these are found at great depths, their colonies or aggregations attached to solid substrata. They feed on plankton, which are caught with the tentacles. Individuals of *Atubaria* do not aggregate and do not secrete a tube, and they live on communities of coelenterates. The class consists of 3 rarely encountered genera: *Atubaria, Cephalodiscus* and *Rhabdopleura.* The majority of species is found in the Southern hemisphere, but *Rhabdopleura* has been found off European coasts and Bermuda.

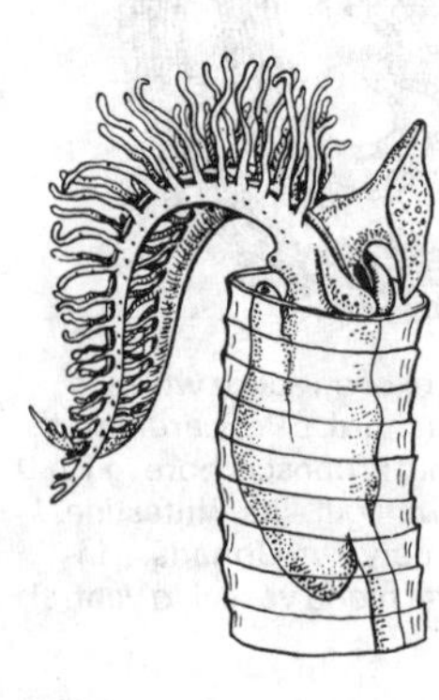
Fig. 4. *Hemichordata.* The pterobranch, *Rhabdopleura normani.*

Collection and preservation. Acorn worms reveal their presence by producing characteristic coiled casts on the surface of tidal flats. Sand and mud in the vicinity of these casts are dug out and carefully washed through a sieve. The exposed worms are left for one day in sea water to empty their sandy gut contents, then slowly narcotized with magnesium sulfate. Bouin's solution is suitable for fixation, and specimens are finally preserved in 75% ethanol with 3 changes.

Pterobranchs brought up from great depths are narcotized by adding 5% chloral hydrate to the sea water. When they no longer respond to touch, they are fixed for 10 min in concentrated formaldehyde solution, followed by preservation in 70% ethanol.

Hemicryptophytes: rooted perennial plants, whose resting buds lie directly on, or very close to, the soil surface. During the winter, the buds are protected by living or dead scales, leaves, or leaf sheaths. They include many of the *Graminae,* perennial rosette plants (e.g. *Taraxacum, Plantago, Beta),* plants with renewal buds at the bases of dead stems that were alive in the preceding summer (e.g. *Artemesia, Urtica, Lysimachia)* and perennial herbs with stolons (e.g. *Fragaria vesca, Potentilla reptans, Ranunculus repens).* See Life form.

Hemidactylus, ***leaf-toed geckos:*** a large genus of the *Gekkonidae* (see), distributed in the tropics. Its species have two transverse rows of attachment lamellae on the underside of each toe. In tropical Asia they are the commonest "house geckos". Some species have been distributed to other parts of the world in ships, e.g. the southeastern European and African *H. turcicus* ***(Turkish gecko)*** was introduced into Central America in this way.

Hemidesmosome: see Epithelium.

Hemiedaphic organisms: Soil organisms (see) that live in the well aerated surface layer of the soil.

Hemignathus: see *Drepanididae.*

Hemilateral hermaphrodite: an individual organism in which the sex chromosome complement and phenotype of one half of the body are male, while those of the other half are female. See Gynandromorphism.

Hemiparasites: plant parasites that still perform photosynthesis, and withdraw only certain materials (e.g. water and mineral salts) from their host, e.g. mistletoe.

Hemiphanerophyte: a plant in which only the base of the stem is woody and persistent, whereas the upper herbaceous part dies off each year and is replaced by young shoots from the woody base in the spring, e.g. Sage *(Salvia officinalis).* The H. is intermediate between a shrub and a herbaceous perennial.

Hemipodes: see *Gruiformes.*

Hemiprocne: see *Apodiformes.*

Hemiprocnidae: see *Apodiformes.*

Hemiptera, *Rhynchota,* ***bugs:*** according to some authorities, an order consisting of the suborders, *Ho-*

moptera (see) and *Heteroptera* (see). Others accord separate ordinal status to these 2 closely related taxa. Common to both is the structure of the piercing-sucking mouthparts. Mandibles and maxillae are modified to slender, bristle-like stylets resting in a grooved labium (stylet sheath or proboscis). Both pairs of stylets are hollow, seta-like structures. The outer 2 stylets are derived from the mandibles, and they are usually serrated at the apex. The inner pair is derived from the maxillae. Each of the inner (maxillary) stylets is finely pointed and carries 2 parallel grooves along its inner surface. As the inner surfaces of the 2 stylets lie together, these paired grooves form 2 fine tubes; the dorsal of these functions as a suction duct, while the ventral tube is an ejection canal for saliva.

H. are the most advanced of those insects that display incomplete metamorphosis *[Parametabola,* see Insects (evolutionary tree)]. The tarsae of the imago never have more than 3 joints. Cerci are absent.

Hemiramphidae, ***halfbeaks:*** a family of teleost fishes of the order *Cyprinodontiformes,* comprising 12 genera and up to 80 species. Most species are marine (Atlantic, Indian and Pacific Oceans), and about 16 species are found in freshwater in the Indo-Australian region. They have a slender body, and in most species the lower jaw is markedly elongate. *Dermogenys pusillus,* a live-bearing freshwater species, is a popular aquarium fish. The *Exocoetidae* (Flying fishes) and the *H.* comprise the superfamily *Exocoetoidea.*

Hemisaprophytes: see Saprophytes.

Hemisinae, ***hemisins, shovel-nosed frogs:*** a subfamily of the *Ranidae* (see), containing 8 species in a single genus *(Hemisus),* found in Natal and possibly further north. They inhabit dryish regions and lay eggs in subterranean chambers not far from water. After hatching, the tadpoles make their way to the water via a tunnel dug by the adult female. The bodies of adults appear to be inflated, with the front of the small head protruding well beyond the mouth; this part of the head is hardened and used for digging.

Hemitragus: see Tahrs.

Hemixis: a process of macronuclear fragmentation and division observed in *Paramecium.* H. is not accompanied by any particular activity of the micronuclei, and it does not appear to be part of the normal life cycle of *Paramecium.* It is probably an aberrant or degenerative process, and in most cases leads to cell death.

Hemizygosity: the occurrence, in somatic cells, of one or several genes, or entire linkage groups, not as allele pairs but as individual genes. All the genes on the X-chromosome of XY-type male organisms are hemizygous. Due to Lyonization (see), genes on the X- chromosome of female cells may also be hemizygous.

Hemlock: see *Umbelliferae.*

Hemochorial placenta: see Placenta.

Hemocoel: see Body cavity.

Hemocyanin: see Blood.

Hemocytes, ***blood cells:*** free cells which are the formed components of blood and hemolymph.

Body fluids that occupy the tissue spaces of coelenterate animals and platyhelminths contain no H., and only migratory, ameboid cells (amebocytes) are present. In the remaining invertebrates, as well as the tunicates and *Amphioxus,* the blood or hemolymph contains freely suspended, colorless cells (amebocytes and/or leukocytes), which are produced in the mesoderm. They function mainly as Phagocytes (see) or for the storage and transport of waste products (excretophores). Insect hemolymph also contains specialized cells (see Oenocytes), which are of ectodermal origin. In only a few invertebrates, e.g. *Nemertini, Echinoidea* and serpulid worms, the blood pigment is contained in H. (see Blood).

Vertebrates possess three main types of H.: red blood corpuscles or erythrocytes, white blood corpuscles or leukocytes, and blood platelets or thrombocytes. In adult mammals, erythrocytes are produced in the red bone marrow and degraded in the liver and spleen. They contain the respiratory pigment, hemoglobin, which accounts for 95% of their dry mass. In addition, they contain a number of enzymes, e.g. carboanhydrase, which functions in the release of carbon dioxide in the lungs. Erythrocytes are oval or round disks, and they are sufficiently elastic to pass through the finest capillaries. Mammalian erythrocytes lack a nucleus (secondary loss), and as a rule they are round and biconcave. All nonmammals possess oval, nucleated erythrocytes. Erythrocyte size varies between 78 x 45 µm in the *Apoda* and 2.5 x 2.5 µm in the musk deer. The number of erythrocytes per mm^3 blood depends on their size and on the evolutionary level of the animal (Table). Whereas cold-blooded animals generally have fewer, larger cells, warm-blooded animals have much greater numbers of smaller cells. The total surface area for gas exchange is thereby markedly increased, in accordance with the higher metabolic rate of warm-blooded species. Erythrocyte numbers per unit blood volume are higher in neonatal mammals than in adults, reflecting the lower availability of oxygen in the maternal body. Life at high altitudes also leads to an increase in the number of erythrocytes. When a human moves from sea level to 1 800 m, the number of erythrocytes eventually increases by 2 million per mm^3. On average, erythrocytes represent 40–50% of the total blood mass.

Size and number of red blood corpuscles

	Size (µm)	Number (millions per mm^3)
Ray	27 x 20	0.23
Lamprey	15 x15	0.13
Eel	15 x 12	1.10
Flounder	12 x 9	2.00
Olm	58 x 34	0.036
Fire salamander	43 x 25	0.095
Centrolenid frog	23 x 16	0.40
Common toad	22 x 16	0.39
European grass snake	17.6 x 11.1	0.97
Wall lizard	15.4 x 10.3	0.96
Sand lizard	15.9 x 9.9	1.42
Ostrich	14.3 x 9.1	1.62
Pigeon	13.7 x 6.8	2.4
Chaffinch	12.4 x 7.5	3.66
Elephant	9.4	2.02
Human	6.6–9.2	4.05–5.5
Cow	6.0	6.28
Cat	4.5–7.1	8.22
Sheep	3.9–5.9	9.13
Llama	7.6–4.4	13.19
Goat	3.2–5.4	18.5

The leukocytes of vertebrates fall into 3 main groups: 1) *granulocytes* (67–80% of total leukocytes; small granules in the cytoplasm; produced in the red bone marrow); 2) *lymphocytes* (20–30% of total leukocytes; transport of fat and carbohydrate; clear, nongranular cytoplasm; produced in lymphatic organs); and 3) much less numerous *monocytes* (6–8%). On the basis of their different staining properties, granulocytes (which have a mainly defensive function) are subdivided into neutrophils (65–75%), eosinophils (2–4%) and basophils (0.5%). All leukocytes exhibit ameboid movement and are rounded only when at rest. They display a wide range of sizes, e.g. lymphocytes 7 µm, monocytes 20 µm. Human blood contains 5 000 to 10 000 leukocytes per mm^3, and this number is considerably increased in many diseases, together with marked changes in the ratios of the different leukocyte types. The blood picture (numbers and ratios of different cell types) is used in the diagnosis of disease. Leukocytes function mainly by destroying disease agents (toxins, bacteria, etc.), and removing cell and tissue debris, which they render harmless by engulfing and digesting (see Phagocytes). Their ameboid motility, allows them to pass through blood and lymph vessel walls, and to accumulate in large numbers at infection sites and wounds; puss consists of massive accumulations of dead leukocytes and other tissue degradation products.

Thrombocytes are the smallest nucleated H. They display ameboid movement, and possess many flagella-like appendages which enable them to attach to the edges of wounds or damaged vessel walls. They contain thrombokinase, an important enzyme of blood coagulation. Thrombocytes are produced in specialized cells of the bone marrow; they remain in the blood for about one week and are then degraded in the spleen.

Hemoglobin, ***Hb:*** the red pigment of blood (present as a 34% solution in human erythrocytes). It is both a chromoprotein and a metalloprotein, consisting of an apoprotein (globin) and a prosthetic group (heme). Vertebrate Hb is a tetramer (M_r 64 500) composed of two pairs of polypeptide chains, each of these 4 chains carrying a heme group. Heme is a porphyrin derivative (protoporphyrin) whose 4 nitrogen atoms bind (partly by coordinate bonds) a central divalent iron atom. The iron (II) is also bonded to two histidine residues of the globin. By amino acid analysis and X-ray analysis, the three-dimensional structure of Hb has been investigated in detail. With a Fe(II) content of 0.334%, the total of 950 g Hb in the human body represents 3.5 g or 80% of the total body iron.

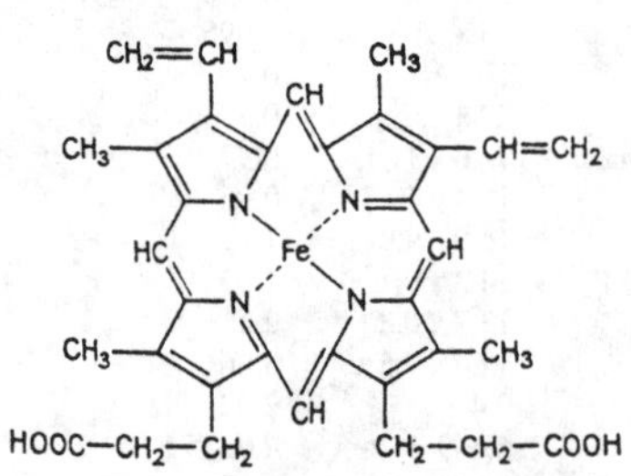

Heme (Ferroprotoporphyrin, Protoheme IX)

Adult human Hb consists of 96.5–98.5% HbA_1 ($\alpha_2\beta_2$) and 1.5–3.5% HbA_2 ($\alpha_2\delta_2$). At least 7 different structural genes specify globin; these are clustered on chromosomes 11 and 16. Three forms of Hb are found very early in embryonic development: $\zeta_2\varepsilon_2$ (Hb Gower), $\zeta_2\gamma_2$ (Hb Portland) and $\alpha_2\varepsilon_2$ (Hb Gower 2). After 3 months gestation, ζ and ε polypeptides are no longer synthesized, and all three embryonic Hbs disappear to be replaced by fetal Hb (HbF, $\alpha_2\gamma_2$)

The prosthetic group is easily removed from Hb by acid treatment. If this cleavage is performed in the presence of chloride ions (e.g. blood is heated with sodium chloride and glacial acetic acid), the product is easily crystallizable *chlorhemin* or *Teichmann's crystals,* which contain Fe(III), and which have long been used for the forensic, microscopic identification of blood.

By forming a sixth coordinate bond, the iron of each heme of Hb can reversibly bind one molecule of O_2 to form bright red oxyhemoglobin (the iron remains in the divalent state). Each Hb molecule can therefore bind 4 molecules of O_2. The physical solubility of O_2 is very low, so that the transport of O_2 in simple solution would be inadequate for the respiratory needs of the tissues. Whether O_2 is bound or released depends on the O_2-partial pressure: at the high O_2-partial pressure in the lungs or gills of the organism, the Hb becomes loaded with O_2. In the tissues, the O_2-partial pressure is low, and the O_2 is released for cellular respiration (see Respiratory chain); see Bohr effect, Oxygen binding by the blood. Hb also plays an important role in transporting CO_2 to the lungs or gills. Carbon monoxide is a very potent inhibitor of O_2 transport; its affinity for Hb is about 320 times greater than that of O_2. The most effective therapy for CO-poisoning is to breath a mixture 98% O_2 + 2% CO_2. By increasing the O_2-partial pressure, the equilibrium is shifted in favor of the O_2-Hb complex. CO-poisoning is therefore due to an inhibition of O_2 transport and not to a direct inhibition of cellular respiration. In contrast, cyanide poisoning is due to inhibition of cytochrome oxidase. By oxidation of the Fe(II) to Fe(III), Hb is converted into brown *methemoglobin,* which no longer transports O_2.

In vertebrates and some invertebrates (e.g. *Nemertini* and *Echinoidea),* Hb is contained in red blood cells (erythrocytes), while in the other invertebrates it is dissolved in the blood plasma. The structure of the globin moiety differs according to the species, but the heme always has the same structure. Hb is the same in all human races, and human Hb is even identical with that of the chimpanzee. Unusual forms of Hb do exist in some individuals, but these Hbs are pathological, they arise by mutation, and they are inherited as blood disorders (hemoglobinopathies). The commonest hemopathy is Sickle cell anemia (see). Hemoglobin-like respiratory pigments of some invertebrates (single polypeptide chain, M_r 16 000) may be evolutionary precursors of Hb.

Hemogram, ***blood picture:*** a written record or graphic representation of the blood count, compiled by combining the results of the separate determinations of erythrocytes, leukocytes, thrombocytes and reticulocytes. The H. gives the absolute number, as well as the relative proportions of each blood cell type. Each animal species has a characteristic H. A *blood count* is performed by counting the different kinds of cells in a sample of blood at known dilution in a special counting chamber, under the microscope, or by placing the blood sample in an automated counting machine that recognizes the differences in size and optical properties of different cells (Coulter counter). The *differential blood* *count,* or *differential leukocyte count* gives the proportion of different kinds of leukocytes. It is usually determined by counting and classifying 100 leukocytes under the microscope, so that the results can be expressed immediately as percentages. Both the H. and the differential blood picture display specific changes in certain diseases, and are therefore useful in diagnosis.

The *total blood picture* includes the concentration of hemoglobin.

Normal total blood picture of the human (average values)

	Hemoglobin per 100 g whole blood	Blood corpuscles per mm³ red	white	Thrombocytes per mm³
Male	16 g	5.0×10^6	6 000–8 000	300 000
Female	14 g	4.5×10^6	6 000–8 000	300 000

Hemolymph: in invertebrates with open vascular systems (mollusks, arthropods), the body fluid that bathes all the cells, tissues and organs, and is made to circulate by body movements or by the heart. It is the functional counterpart of veretebrate blood. H. contains free, usually colorless, ameboid cells (amebocytes, leukocytes), which function as blood cells (hemocytes). H. is generally slightly acidic, and has a relatively high osmotic pressure. It contains inorganic salts and numerous organic substances, in particular proteins and sugars. It may be yellow, red or green, due to the presence of different pigments. Hemoglobin is present in only a few organisms (earthworms, some primitive crustaceans and chironomid larvae). H. transports nutrients, hormones and waste metabolites, and in animals lacking tracheae (mollusks, crustaceans) it is particularly important for the transport of respiratory gases. In certain animals, especially mollusks, H. also serves as a hydraulic fluid, transferring the internal pressure of the body cavity from one region to another during body locomotion. In insects, the internal pressure of the H. is necessary for respiratory activity, molting, and the unfolding of wings after emergence of the imago from the pupa.

Hemolysis: destruction of the membrane of red blood cells and the dissolution of their contents. In blood samples it can be caused by osmotic shock or organic solvents. In the body it can be caused by antibodies (e.g. hemolytic disease due to rhesus antibodies), by bacterial hemolysins, animal toxins (e.g. phospholipases of snake venoms) by parasites (e.g. the hemolytic crisis of malaria), etc.

Hemophage: see Modes of nutrition.

Hemophilia: an inherited blood clotting deficiency. Hemophilia A is caused by the absence from the plasma of antihemophilic globulin A (factor VIII), while hemophilia B is due to the absence of factor IX. In the absence of either factor, thrombin, and therefore fibrin, cannot be formed (see Blood coagulation). Hemophilia is a classic example of X-chromosomal recessive inheritance in humans. Males normally possess only a single X-chromosome, so that all genes on this chromosome are hemizygotic. All males carrying the defective gene therefore manifest the disease. In females, however, such genes are compensated by normal alleles on the second X-chromosome, i.e. a woman may be a heterozygotic carrier of the defective gene, but she does not manifest the disease. On average, female carriers transmit the defective gene to half of their daughters, who are therefore also nonaffected carriers. Similarly, half of her sons also have the defective gene, but in contrast to their sisters they manifest the disease, i.e. they are hemophiliacs. Since the X-chromosome of the male is transferred only to daughters, a male hemophiliac always has phenotypically healthy offspring, although all his daughters are carriers, whereas none of his sons or their descendants possess the gene. Hemophilia is very rare in females, because such individuals must be homozygous with respect to the defective gene, which presupposes a father who is a hemophiliac and a carrier mother.

All sons of female hemophiliacs suffer from hemophilia, and all the daughters are carriers. Female carriers often have an increased clotting time with a tendency to bleeding; in hemophilia A they show a distinct decrease of factor VIII and in hemophilia B a distinct decrease of factor IX. This can be used to determine which women in a family are carriers, and which women do not run the risk of bearing diseased or carrier children. On average, 1 in every 5 000 males born has hemophilia A, and 1 in 25 000 has hemophilia B. (See Fig.)

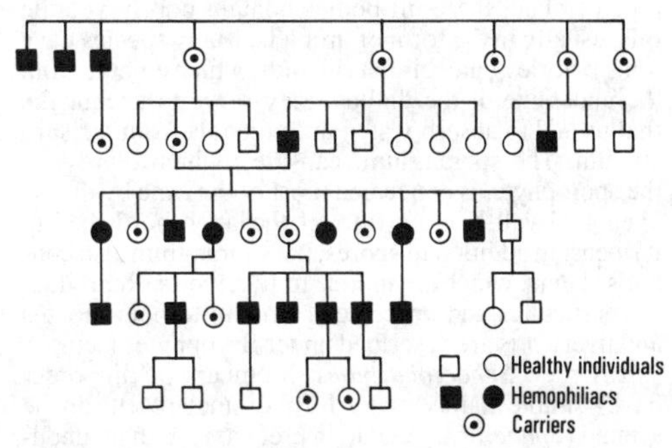

Family tree of a family with hemophilia (after Gilchrist)

Hemoproteins: chromoproteins in which the prosthetic group is heme (iron porphyrin IX). They are responsible for oxygen transport (hemoglobin, hemerythrin), oxygen storage (myoglobin), electron transport (cytochromes) and peroxidation (catalase, peroxidase).

Hemorheology (plate 28): a branch of biomechanics which studies the flow of blood in blood vessels. Whereas the physiology of blood circulation is concerned with values such as blood pressure, flow rates and their regulation, etc., H. also considers the mechanical properties of erythrocytes. The larger part of the pressure decrease that occurs across capillaries is due to the properties of the blood cells. At the low rates of shear that occur in the capillaries, the non-Newtonian behavior of the blood leads to a considerable increase in viscosity. In critical cases, this can lead to a cessation of flow. As erythrocytes pass through narrow capillaries, they become very deformed, a process facilitated by their favorable surface to volume ratio. In addition, a spectrin-actin complex on the internal surface of the erythrocyte membrane actively contributes to the deformation process. During erythrocyte aging, material is increasingly lost from the membrane. Old cells are therefore less elastic and not readily deformed, and would clog the capillaries. By a mechnanism that is unclear, old cells are selectively removed from the circulation and destroyed. In certain diseases, e.g. sickle cell anemia, the viscosity of the erythrocyte interior is increased, so that the cell is less easily deformed. Other diseases affect the stability of the erythrocyte membrane.

Hemorrhage, arrest of: see Staunching of blood flow.

Hemorrhagic jaundice: see Spirochetes.

Hemp: see *Cannabaceae.*

Henbane: see *Solanaceae.*

Henicopidae: see *Chilopoda.*

Henley's pygmy gerbil: see *Gerbillinae.*

Henophidia: see Snakes.

Hensen's node: see Mesoblast.

Hepatica: see *Ranunculaceae.*

Hepaticae, ***liverworts:*** a class of the Bryophytes, containing about 10 000 species in 5 orders. Most members possess either a flat, lobed thallus or a vegetative body consisting of a dorsiventrally winged stalk. In some species, the thallus has a thickening of tissue known as a midrib. Characteristic oil bodies are present in the thalli of most liverworts; unknown in any other type of plant, these oil bodies contain poorly volatile oils, usually in the form of droplets. Many species have long, colorless, unicellular rhizoids, which emerge from the underside of the thallus; they serve to anchor the thallus and to absorb water and minerals from the substratum. The sporangium (capsule), which represents the sporophyte, is concealed most of the time by the archegonial wall, bursting through the latter shortly before it ripens. In addition to spores, the sporangium also contains elaters, which are unique to liverworts. Reproductive structures and processes common to both mosses and liverworts are described under Bryophytes (see)

1. Order: *Sphaerocarpales.* Members of this order have a simple thallus, which forms a small rosette on the ground *(Sphaerocarpus)* or an erect axis with an undulate wing (submerged, aquatic *Riella).* The archegonia and antheridia are enclosed in pear-shaped perianths, which have an upper apical opening. Plant sex chromosomes were first discovered in species of the genus *Sphaerocarpus.*

2. Order: *Marchantiales.* Members of this order have a flat thallus, which is clearly differentiated histologically into assimilatory and storage tissue. Air pores and oil bodies are also characteristically present. *Marchantia polymorpha* is a well known cosmopolitan species of moist habitats. It is easily recognized by the presence of cup-shaped outgrowths with a toothed margin, known as gemma cups. These contain several flat gemmae, which arise by the outgrowth and division of single epidermal cells. The stalk attachment of each gemma abscisses, and the released gemmae develop into new thalli (i.e. a form of vegetative reproduction). The gametangia (archegonia and antheridia) are carried on special, erect, umbrella-like stalks (gametangiophores), which emerge from the thallus. *Marchantia* and *Conocephalum* (common on stream banks and wet stones) were once used as remedies for liver complaints. The thallus of *Conocephalum* closely resembles that of *Marchantia,* but it lacks gemma cups. *Lunulasia cruciata* often occurs on the moist soil surface of plant pots in greenhouses; it originates from the Mediterranean region, and it is recognizable by its crescent-shaped gemma cups. Most species of *Riccia* are found on moist soil, and are especially common on arable soil. The thallus forks repeatedly, so that it has the appearance of small rosettes. *Riccia natans* is a surface-floating aquatic species, while *Riccia fluitans* is a submerged aquatic species with a ribbon-like, forked thallus. The smallest thallose liverwort *(Monocarpus sphaerocarpus* from Australia) belongs to this suborder; its greatly reduced thallus carries a single sporangium within a sheath which contains air chambers.

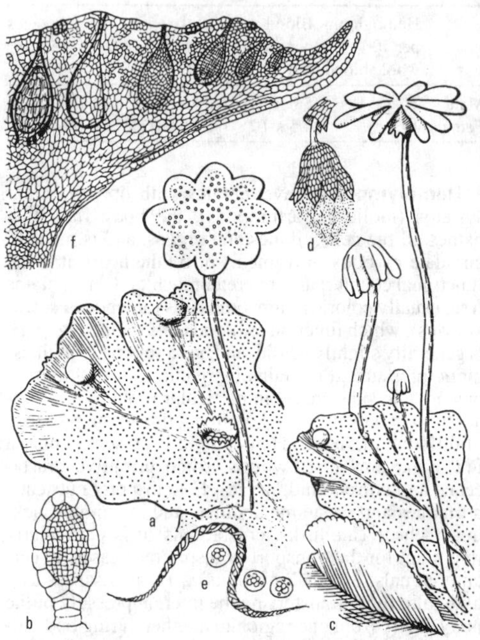

Hepaticae. Marchanta polymorpha. a Male plant with gemma cups and antheridiophore; the small dots visible on the upper surface of the thallus are air pores (x 1.7). *b* Longitudinal section of a nearly mature antheridium (x 125). *c* Female plant with archegoniophores (x 1.7). *d* Dehisced sporangium liberating spores and elaters; the remains of the archegonium can be seen at the base of the seta (x 9). *e* Spores and elater (x 180). *f* Longitudinal section of an antheridiophore (x 20).

3. Order: *Calobryales.* Members of this order have erect stems with three rows of identical leaves; rhizoids are absent. Examples are *Takakia* (an Asian genus with cylindrical leaves) and *Haplomitrium* (an Asian and European genus with flat leaves).

4. Order: *Jungermanniales.* Most of these are small species living on moist surfaces, such as soil, tree trunks and leaves. Most of the approximately 9000 species (which account for about 90% of all the *Hepaticae)* are restricted to tropical habitats. One of the simplest members of the *Jungermanniales* is *Pellia epiphylla,* the thallus of which superficially resembles that of *Marchantia,* but lacks internal differentiation. *Pellia* is widely distributed and common on moist soil, and often

used as a type species for class study. Most species show a clear differentiation into stem and leaves, the latter with no midrib and one cell thick, the former with no vascular tissue. Often the lateral leaves are differentiated into upper (antical) and lower (postical) lobes. In species subject to periodic drought, the postical lobe has the shape of a container which retains water by capillary attraction (e.g. in *Frullania tamarisci,* a branched, brown liverwort found on rock and tree trunks). Most genera also possess an additional row of smaller central leaves (underleaves or amphigrastria). *Jungermanniales* differ from all other liverworts in lacking stomata and air pores. As in other liverworts, the protonema normally has few cells, but in *Metzgeriopsis pusilla* the protonema is the main vegetative organ, producing small plantlets which carry the sex organs.

5. Order: *Anthocerotales* (type genus *Anthoceros).* This order only contains a few calcifuge species. The gametophyte has a very simple structure, being disk-shaped and lobed, a few centimeters in diameter, and attached to its substratum by rhizoids. Numerous antheridia are present inside cavities in the upper surface of the thallus. The archegonia are also beneath the surface of the thallus. Unlike all other bryophytes, each cell contains a single, large, basin-shaped chloroplast with a pyrenoid. This resemblance to the *Chlorophyceae* suggests an ancient evolutionary relationship. The intercellular space beneath the stomatal guard cells (stomata are present on the lower surface of the thallus) is filled with mucilage and normally colonized by *Nostoc.* In contrast to other liverworts, the wall of the sporangium possesses stomata. Moreover, the pod-shaped capsule continually grows from a basal meristem, and all its parts do not mature at the same time.

Hepatic ceca: see Digestive system.

Hepatic duct: see Gall bladder.

Hepatic portal system: see Blood circulation.

Hepatidae: see *Acanthuridae.*

Hepialidae: see *Lepidoptera.*

Heptaene antibiotics: a group of Polyene antiobiotics (see) produced by actinomycetes, and possessing 7 double bounds. They inhibit the growth of fungi.

Heptoses: monosaccharides containing seven carbon atoms, and possessing the empirical formula $C_7H_{14}O_7$. The 7-phosphates of D-mannoheptulose and D-sedoheptulose are intermediates of carbohydrate metabolism.

Herbaceous perrenials: perennial herbaceous plants, whose aerial foliage and inflorescences die back every year to their basal parts, and are renewed by growth from buds directly at soil level (see Hemicryptophytes), or from buds below ground level (see Geophytes).

Herbicides: chemicals used to kill weeds. *Total H.* destroy the total plant population of the treated area. They are used in particular against the weeds and vegetation of roads, pathways, squares, waste land and waterways. Agricultural land treated with total H. must be left for a rather long period before being replanted. *Selective H.* attack only certain plants, and are therefore suitable for attacking weeds among agricultural and horticultural crops. The selectivity may be due to genuine physiological differences between the affected and nonaffected plants, or it may depend on the developmental stage of the plants in question. In the latter case, the time of application is important: before seed sowing; after seed sowing, but before germination; and after seed germination. H. are classified as *root H.* or *leaf H,* according to their site of uptake. *Contact H.* destroy only those parts of the plant with which they make contact. *Systemic H.* often behave as synthetic growth hormones (see, e.g. 2,4-Dichlorophenoxyacetic acid); they are taken up and translocated thoughout the plant by the vascular tissue, causing abnormally excessive and irregular growth, leading to the death of the entire plant. Other systemic H. inhibit or interfere with cellular respiration, seed germination, photosynthesis, chlorophyll synthesis, root formation or mitosis. For example, electron flow from water in photosystem II of photosynthesis is blocked by monuron [3-(4-chlorophenyl)-1,1-dimethylurea or CMU] and by diuron (dichlorophenyldimethylurea or DCMU). Other H. include chlorates (total H.), phenols, carbamates and triazines.

Herbivore: see Modes of nutrition.

Herbst corpuscles: see Tactile stimulus response.

Hercogamous flower: see Pollination.

Hercogamy: see Pollination.

Herding: see Swarming.

Heritability: the degree to which a phenotype is genetically influenced and therefore capable of modification by selection. $h^2 = V_a/V_p$, where h^2 is heritability, V_a is additive genetic variance (variance due to genes with additive effects), and V_p is the phenotypic variance. Heritability in the broad sense *(H^2)* is the phenotypic variance remaining after exclusion of variance due to environmental influences.

Hermaphrodite gland, ***ovotestis:*** unpaired sex gland of hermaphrodite mollusks, formed by the fusion of testis and ovary. The H.g. produces both eggs and sperm, either equally in all parts of the gland, or in separate regions. Eggs and sperm are produced at different times and pass down the hermaphrodite duct. Sperm is then passed to the exterior via the sperm duct and penis, whereas eggs pass to the seminal receptacle where they are fertilized by stored foreign sperm.

Hermaphroditism, ***androgyny, bisexualism:***

1) In animals, the occurrence of male and female sex organs in a single individual, known as a hermaphrodite. In *true* or *primary H.,* hermaphrodite offspring are produced by hermaphrodite parents, and H. is a genetically determined condition typical of the species. In *secondary H.,* hermaphrodite offspring are produced by parents of separate sexes. A hermaphrodite individual possesses both male and female gonads, or they possess a single hermaphrodite gland which produces both sperms and eggs. Various strategies have evolved for the prevention of self fertilization. Often the male gametes mature before the female (proterandry), e.g. in tape worms. Protogyny (female gonads mature first) occurs, inter alia, in ascidians. In other cases, hermaphrodite animals exchange sex products during pairing, and perform a mutual cross-fertilization, e.g. earthworms and many snails. In a few cases, self fertilization is prevented by self sterility.

Lepidoptera and certain other insects occasionally display a special type of H. (hemi-hermaphroditism), in which one half of the animal has only male sex glands, the other only female glands. In external appearance, these animals are also half male and half female.

H. is particularly common in parasites and sessile animals, but also occurs in free-living forms. Sometimes, it is incomplete, in that both types of sex gland are present but only one is functional, e.g. in some insects.

2) In plants, see Flower.

Hermit crabs: crustaceans of the family *Paguridae* (containing about 550 species), infraorder *Anomura,* order *Decapoda* (see). They possess a soft-skinned, asymmetrical abdomen, which offers a surface for respiratory gas exchange in addition to the gills. The abdomen is housed in an empty gastropod shell, which serves as a permanent residence, into which the animal withdraws when disturbed or threatened. An area of wart-covered skin on the abdomen, as well as the last pair of walking legs hold the body tight within the shell. Only when it outgrows the shell does the crab leave it for a larger one. One cheliped is larger than the other, and it is used to close the entrance to the shell. Sea anemones, sponges and *Polyzoa* may grow on the surface of the shell. The stinging capability (nematocysts) of the sea anemones afford protection to the hermit crab, and when changing shells the crab sometimes actually transfers sea anemones from the old to the new shell. All species are marine. The family contains the genera *Diogenes, Pylocheles, Petrochirus, Clibanarius, Pagurus, Pylopagurus.* Land or terrestrial H.c. comprise the genera *Coenobita* and *Birgus* (see Coconut crab).

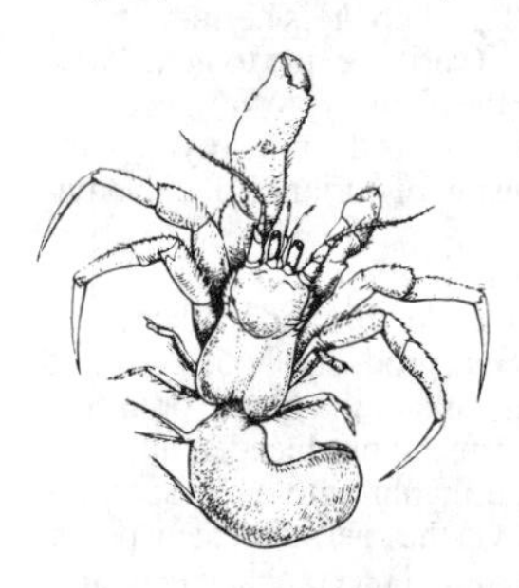

The hermit crab, *Diogenes pugilator,* removed from its snail shell

Hermit spadefoot: see *Pelobatidae.*

Hermit thrush: see *Turdinae.*

Heroin: see Morphine.

Herons: see *Ciconiiformes.*

Herpestes: see Mongooses.

Herpestinae: see Mongooses.

Herpetotheres: see *Falconiformes.*

Herpetotherinae: see *Falconiformes.*

Herring, *Clupea harengus:* a green to blue, silvery marine fish of the family *Clupeidae,* order *Clupeiformes* (see), which lives in large shoals, feeds on plankton, and attains a length of 45 cm (occasionally up to 50 cm). Capable of living up to 20 years, it is one of the most important food fishes of the world. It is caught with ring nets, drift nets, dragnets and trawls. For culinary purposes, a distinction is drawn between the young, sexually immature fish (the German "Matjeshering") and the adult, sexually mature form (the German "Vollhering"). It occurs in the North Atlantic as far south as the Bay of Biscay and in some adjoining Arctic waters. Local forms include *Clupea harengus pallasi* of the northeastern Pacific.

Hesperiidae: see *Lepidoptera.*

Hesperiphona: see *Fringillidae.*

Hesperomyini: see *Cricitinae.*

Hesperoptenus: see *Vespertilionidae.*

Hesperornis (Greek *hesperos* evening or western, *ornis* bird): an extinct genus of pigeon-sized water birds, with rudimentary wings. The hindlimbs were powerfully developed for swimming. Conical teeth were present, housed in a groove in each jaw, and the lower jaws were not fused. Fossils are found in the Upper Cretaceous of North America (Kansas).

Hessian fly: see Gall midges.

Heteralocha: see *Callaeidae.*

Heterauxesis: see Allometry.

Heteroalleles: the alleles of a functional gene (see Cistron), which arise by gene mutation at different sites within the gene, and which are therefore capable of Recombination (see). H. are distinct from from *homoalleles,* which arise by mutation at identical sites of the gene, and which therefore cannot undergo recombination with each other.

Heterobathmy of characters: existence of mosaics of relatively primitive and relatively derived characters in related species and species groups.

Heteroblastic development: in plants, a type of development in which there are distinct differences between juvenile and subsequent (older) forms. The term is generally used only for vegetative growth, but in the strict sense flower formation is also an aspect of H.d. Both primitive and advanced plants display H.d. In spermatophytes, differences between juveniles and later forms are found in leaf shape, general plant anatomy, and certain physiological characters. Examples are found among certain aquatic plants, e.g. Water crowfoot *(Ranunculus aquatilis),* in which the submerged leaves (young leaves) and floating leaves (older leaves) are morphologically different; and in the heterophyllic acacias, in which pinnate (juvenile) primary leaves are followed by simple, flattened, leaf-like petioles (phyllodes). The duration of the juvenile phase and the transition to later forms depend critically on environmental factors; most important are the availability of nutrients, light and temperature. Physiologically, H.d. is closely related to problems of aging. It has been shown that phytohormones may be causally involved in H.d. Thus, after treatment with gibberellins, some plants revert to the juvenile phase. The converse of H.d. is Homoblastic development (see).

Heterocarpy, ***heterocarpous condition:*** the occurrence of different types of fruits on the same plant, usually with different modes of dispersal. Thus, the capitula of various *Taraxacum* (dandelion) species contain achenes with and without pappus.

Heterocera: see Moths, *Lepidoptera.*

Heterocercal fin: see Fins.

Heterocercous fin: see Fins.

Heterochlamydeous: a term applied to a flower whose perianth segments occur in two distinct series, which differ from one another. Usually this takes the form of a relatively inconspicuous, green calyx (sepals) and a conspicuous, colored corolla (petals). The flowers of most dicotyledonous plants are H. The converse is Homeochlamydeous, in which the perianth segments are similar. See Flower.

Heterochromatin: Chromatin (see) that is condensed and highly coiled. *Constitutive H.* is always condensed. *Facultative H.* exists only in certain phases of the cell cycle; at other times the constituent DNA is relaxed and it is then known as euchromatin. H. is stained intensely by basic dyes. Certain regions of chromosomes or entire chromosomes may be heterochromatic, e.g. during interphase, one of the X-chromosomes of female mammals is demonstrable as a structure of intensely staining H. (see Sex chromatin). In contrast to euchromatin, H. is condensed during interphase, e.g. in Satellite-chromosomes (see) and regions in the neighborhood of the centromere. H. may also form complexes known as *chromocenters.* On account of its tight coiling, H. cannot support transcription. During interphase, it is replicated only after the replication of euchromatin. Constitutive H. contains the repetitive sequences that are generally abundant in eukaryotic DNA.

Heterochromosomes: see Sex chromosomes.

Heterochrony:

1) the appearance of a particular flora or fauna in two different regions at different times.

2) see Metaboly.

Heterochthonius: see Oribatei.

Heterodera: see Eelworms.

Heterodon: see Hog-nosed snakes.

Heterodont: a term applied to vertebrate dentition (see Dentition), and a hinge type of the *Bivalvia* (see).

Heterodonta: see *Bivalvia.*

Heterodontidae: see *Selachimorpha.*

Heterodontiformes: see *Selachimorpha.*

Heterogamy:

1) Anisogamy, diptogamy: fusion of gametes of different shape or size. See Reproduction, Fertilization.

2) See Heterogony.

3) In plants, the production of male and female gametes by one kind of flower.

Heterogenote: a partially diploid bacterial cell (see Syngenote), which contains a complete bacterial genome, together with a genetic fragment. The genome and the fragment contain different alleles of one or more genes, so that the cell is heterozygous for these particular genes. If the genes of the genome and fragment are all homozygous, i.e. they contain the same alleles, the cell is said to be a *homogenote.* Homo- and heterogenotes, which are jointly known as merozygotes, can arise by bacterial conjugation, by Transduction (see) and by F-duction (see F-plasmid).

Heterogony, *heterogamy:* a form of homophasic Alternation of generations (see) in the animal kingdom. There is an alternation between bisexual and unisexual (parthenogenetic) reproduction. Often there are several sequential parthenogenetic generations, all of which are parthenogenetic females. These may closely resemble the females of the sexual generation, but they often display a different morphology.

Thus, late in the year, the fertilized, sexually mature females of aphids lay eggs on an appropriate food plant. *Aphis rumicis* lays eggs on *Euonymus* (Spindle tree). In the spring, the eggs hatch into wingless parthenogenetic females, which produce more wingless parthenogenetic females (in parthenogenesis in *Aphis,* the young are produced viviparously, but this is not a necessary feature of H.). After a number of such parthenogenetic generations during the summer, a generation of migratory parthenogenetic winged females is produced, which move to a different food plant (e.g. bean). These produce more wingless parthenogenetic females, whose offspring eventually produce winged parthenogenetic females, which return to *Euonymus.* This generation produces sexual (oviparous) females which are fertilized by winged males (produced by parthenogenesis on the secondary host plant). See Parthenogenesis.

Other examples of animals exhibiting H. are *Rotifera,* Water fleas *(Cladocera)* and *Cynipidae.* See Allometry

Heterokaryocyte: a cell produced by Cell Fusion (see), in which the fusion partners were of different genotypes.

Heterokaryon: see Cell fusion.

Heterokaryosis: the coexistence of genetically different nuclei within one cell. The cell or organism is known as a *heterokaryon.* If the two nuclei are genetically identical, the condition is called *homokaryosis,* and the cell or organisms is a *homokaryon.* See Cell fusion.

Heteromery: see Polygeny.

Heteromorphosis: see Regeneration.

Heteromyarian condition: see *Bivalvia.*

Heteromyota: see *Echiurida.*

Heteromyridae: see Moray eels.

Heteronemertea: see *Nemertini.*

Heteronetta: see *Anseriformes.*

Heteronomy: see Metamerism.

Heteroplasmonic: a term applied to cells in which cytoplasmic inheritance is due to genetically different plasmons. See Cytoplasmom, Homoplasmonic.

Heteroploidy: a condition in which the chromosome number of cells, tissues or organisms differs from the normal chromosome number (homoploidy) in the diploid or haploid phase. H. occurs as Aneuploidy (see), Polyploidy (see), or as abnormal haploidy in the diploid phase.

Heteropneustes: see *Heteropneustidae.*

Heteropneustidae, *Saccobranchidae,* ***airsac catfishes:*** a family of catfishes (see *Siluriformes),* largely similar is structure to the *Clariidae,* but with a different type of air-breathing apparatus. The family is represented by only 2 species *(Heteropneustes fossilis* and *Heteropneustes microps),* which are found in fresh water from Pakistan to Thailand. The pectoral spines have an associated venom gland, and the sting of *Heteropneustes fossilis* is especially painful and potentially dangerous. Extending posteriorly from the gill chamber is a long air sac, which is unbranched and has the form of a simple blind tube (cf. the labyrinthine air-breathing structure of the *Clariidae).* Some authorites do not recognize the *Heteropneustidae* as a separate family, and include *Heteropneustes* with the *Clariidae.*

Heteropoda: see *Gastropoda.*

Heteroptera, *true bugs:* an insect order containing about 40 000 species. The earliest fossil *H.* are found in the Permian.

Imagoes. Lengths vary from 0.1 cm to 10 cm; some temperate species attain 4 cm. Mouthparts are adapted for piercing and sucking (see *Hemiptera),* and the stylet sheath is folded back against the underside of the body when at rest. Antennae are very short in aquatic bugs, and longer in terrestrial forms. Depending on the mode of existence, legs are adapted for walking, seizing prey or swimming; tarsae are usually 2- or 3-segmented. The wings usually lie close against the flattened body, and in some species they are reduced or absent. Fore- and hindwings are dissimilar; the latter are uniformly membranous, whereas the inner two-thirds of the forewings ("hemelytra" or hemielytra) are chitinized (corium) while the outer third is membranous. Many species possess odoriferous (repugnatorial or stink) glands (on the dorsal side of the larval abdomen, and in the metathorax of the imago), which produce an odoriferous substance repellent to enemies (e.g. ants), as well as toxic, corrosive substances, which can be sprayed in defense or attack. Most species feed on plant sap, but some are predators (aquatic bugs and assassin bugs) of insect larvae and other small animals. Many species stridulate. Metamorphosis is incomplete (paurometabolous development).

Eggs. These are elongated, rounded or pear-shaped, laid singly or in clusters. The female bed bug lays 150–250 eggs in its lifetime. In several heteropteran families, the larva possess a toothlike prolongation or egg burster. Only a few species are viviparous.

Larvae. All stages (usually 5) resemble the adult. Wing primordia appear after the second molt. Development from the newly hatched larva to the adult lasts between 8 weeks and 5 years.

Economic importance. A number of plant bugs are agricultural pests, e.g. beet bugs, brassica bugs and cereal bugs. The flattened, wingless bed bug is a very troublesome pest that sucks human blood. Some species of the *Reduviidae* carry the trypanosomes of Chaga's disease (see below).

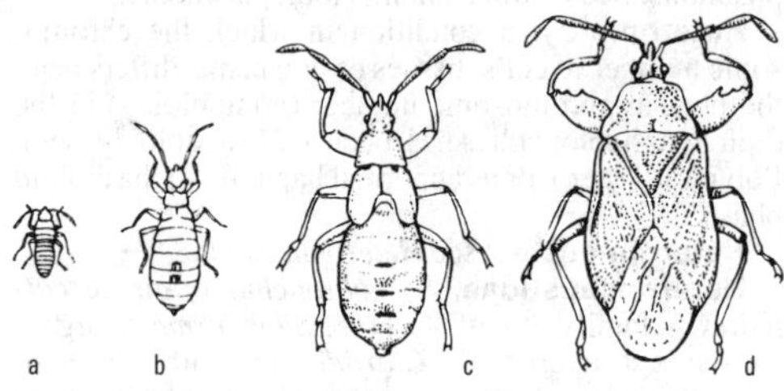

Heteroptera. Metamorphosis of a capsid bug (*Gastrodes* sp.). *a* Young larva. *b* Third larval stage. *c* Fifth larval srtage. *d* Imago.

Classification.

1. Suborder: *Hydrocorisae* ***(Aquatic bugs).*** All members live in water. Antennae are very short and hardly visible.

1. Family: *Corixidae* ***(Water boatmen).*** Generally these insects live under water, holding to various objects by the middle legs, occasionally ascending to the surface, propelled by the hindlegs. They feed on diatoms and the contents of algal cells.

2. Family: *Notonectidae* ***(Backswimmers).*** Members are predatory, even attacking small fish, etc. The dorsal surface is boat-shaped, and the insect swims on its back. They can dive, carrying an air supply trapped beneath the wings, and they can spring into the air and fly.

3. Family: *Peidae.* A family of small backsimmers closely allied with *Notonectidae.* The back is not boat-shaped but highly arched. They can fly, although the hindwings are regressed, and the forewings consist only of the corium.

4. Family: *Naucoridae* ***(Creeping water bugs).*** Members are covered with extremely fine hairs, which retain a layer of air around the insect. The gases of this air layer exchange with those of the surrounding water, so that the insect can respire continually below water (plastron respiration). All 3 pairs of legs are ambulatory. The insect walks, ventral surface upward, just below the water surface, or it may "swim" with the dorsal surface upward. The prehensile forelegs are used for holding prey. Tibia and tarsus fold into a groove on the femur like a clasp knife.

5. Family: *Belostomidae* ***(Giant water bugs).*** This family includes some of the largest insects, adult body length varying from 8 to 11 cm. They feed mainly on frogs and other amphibians, frequently attack fish, and often cause havoc in fish nurseries. *Lethocerus americanus* is found widely in ponds and streams in North America. Other species live in North America, South Africa and India. All species are active fliers, as well as swimmers. The end of the abdomen carries 2 retractable, semicylindrical appendages, which together form a breathing tube when the insect is under water.

6. Family: *Nepidae* ***(Water scorpions).*** These are bottom dwellers, lying in wait for prey, which is seized in the strongly prehensile forelegs. The family also includes the water stick insects *(Ranatra).* All members possess a long, apical breathing tube.

7. Family: *Helotrephidae.* A small, but widely distributed family, occurring in still water with plentiful vegetation.

2. Suborder: *Geocorisae* ***(Terrestrial bugs).*** Some families are water surface dwellers, which do not submerge. Most familes are truly terrestrial.

1. Family: *Dipsocoridae (Cryptostemmatidae, Ceratocombidae).* Small, predatory insects, living among moss, leaf litter, etc.

2. Family: *Enicocephalidae* ***(Gnat bugs).*** A small, but widely distributed family. In some tropical regions, adults (3–4 mm in length) have been observed to swarm like midges

3. Family: *Miridae (Capsidae)* ***(Plant bugs).*** All members are phytophagous, and some species are also facultative blood suckers. The family includes pest species, such as *Lygus lineolaris* ***(Tarnished plant bug),*** feeding on cotton, alfalfa, vegetables, fruit; and *Poecilocapsus lineatus* ***(Four-lined plant bug),*** feeding on gooseberry, currant, rose and various flowers.

4. Family: *Nabidae* ***(Damsel bugs).*** Predatory bugs living on smaller phytophagous insects, and bearing a superficial resemblance to ants.

5. Family: *Cimicidae* ***(Bed bugs).*** Blood-sucking ectoparasites of birds and mammals, including the cosmopolitan human bed bug *(Climax lectularius* L.). All are oval and flattened with very short hemelytra.

6. Family: *Anthocoridae* ***(Minute pirate bugs).*** Small, elongate-oval, flattened insects, feeding on the blood or eggs of other arthropods. Some are myrmecophilous.

7. Family: *Reduviidae* ***(Assassin bugs).*** Predatory species, especially on other arthropods. Some attack vertebrates, including humans. *Triatoma* spp. and *Rhodnius prolixus* carry *Trypanosoma cruzi,* the causative agent of Chaga's disease, a fatal form of South American sleeping sickness in humans.

8. Family: *Phymatidae (Macrocephalidae)* ***(Ambush bugs).*** A mainly tropical family of predatory species, allied with the *Reduviidae.* The prehensile forelegs are adapted for grasping.

9. Family: *Tingidae* ***(Lacebugs).*** A family of phytophagous species with lacelike patterns on the dorsal surface of the head and body and on the wings.

10. Family: *Aradidae* ***(Flat bugs).*** Members have extremely long mouthparts, which are coiled within the head, and used for feeding on fungal sap beneath tree bark and in rotting wood.

11. Family: *Termitaphididae.* Members are closely allied with the *Aradidae,* and found only in termite nests.

12. Family: *Pentatomidae* ***(Shield bugs).*** Most members are phytophagous, but some (subfamily *Asopinae)* are predators on other insects, especially lepidopterous larvae. All are characteristically brightly colored or brightly patterned. They are noted for the emission of foul-smelling defensive secretions from their odoriferous glands. The ***Harlequin cabbage bug*** *(Murgantia histrionica)* of USA and Central America feeds on cabbage and other *Cruciferae.* In Europe the ***Brassica bug*** *(Eurydema oleraceum)* has similar feeding habits.

13. Family: *Cydnidae* ***(Negro bugs).*** These are similar in shape to the *Pentatomidae,* but darkly colored, living under stones and leaf litter. Some are myrmecophilous.

14. Family: *Scuterellidae.* The mesoscutellum is greatly enlarged, extending backward to cover the abdomen and wings.

15. Family: *Lygaeidae* ***(Seed bugs).*** Most members are ground-dwellers, feeding on fallen seeds. A few are predatory. Several are pests, e.g. the ***Cinch bug*** *(Blissus leucopterus),* which attacks cereal crops in USA.

16. Family: *Pyrrhocoridae* ***(Red bugs).*** There are relatively few species, and these occur mostly in the Old World. A notable exception is *Dysdercus,* several species of which are cotton pests in the USA. Their attack introduces a fungus into the cotton bolls, which discolors the cotton fibers, hence the name "cotton stainers" for these bugs. Members are commonly red and black in color.

17. Family: *Piesmidae* ***(Ash-gray leaf bugs).*** All are phytophagous, especially on *Chenopodiaceae,* e.g. *Piesma quadrata* Fieb. ***(Beet bug).***

18. Family: *Berytidae (Neididae)* ***(Stilt bugs).*** Sluggish, secretive insects, found in dense undergrowth. They are small, with a narrow, elongate body and long, slender legs.

19. Family: *Coreidae* ***(Leaf-footed bugs).*** Members are closely allied with the *Lygaeidae* and *Pyrrhocoridae.* All are phytophagous, generally dark in color, and distributed chiefly in Asia, Africa and South America. Like the pentatomids, many species produce obnoxious odors. Some are pests, e.g. *Anasa tristis* ***(Squash bug).***

20. Family: *Gerridae* ***(Pond skaters, Water striders).*** Intermediate-sized, usually slender insects. The ***Common pond skater*** *(Gerris)* is cosmopolitan. The body is covered, especially on the underside, with a silvery, water-repellent coat of hairs. The forelegs are used for catching prey (other insects) on the water surface, and the middle legs are used for rowing. Members of the genus, *Halobates,* are wingless and are found on tropical and subtropical oceans often hundreds of miles from land. These "ocean striders" lay eggs on drifting seaweed and on the violet snail, and they feed on the bodies of dead marine animals.

21. Family: *Veliidae* ***(Ripple bugs).*** A mainly Neotropical and Oriental family. Members are similar to the gerrids, but with a wider mesothorax and metathroax. The short legs are held bent as the insect runs on the water surface.

22. Family: *Hydrometridae* ***(Water measurers).*** A small, mainly tropical family. The genus, *Hydrometra,* which crawls on the surface of stagnant water, is cosmopolitan. Members are elongate with long legs, and bear a superficial resemblance to stick insects.

23. Family: *Mesoveliidae* ***(Water treaders).*** Body lengths are 5 mm or less. Members are found in vegetation at the margins of water.

24. Family: *Hebridae* ***(Velvet water bugs).*** These extremely small bugs are found on floating vegetation, aerial parts of water plants, or at the edges of water.

25. Family: *Leptopodidae.* A mainly Oriental family.

26. Family: *Leotichiidae.* A family of only 2 species from Southeast Asia.

27. Family: *Saldidae* ***(Shore bugs).*** A predatory Holarctic group found on the margins of fresh and salt water. Members are egg-shaped, and are able to fly and run quickly. They possess large, conspicuous, protruding eyes, and they are probably the most primitive of the modern *H.*

Heteropyknosis: a condition in which certain chromosomes or chromosome segments differ from other chromosomes in their degree of coiling and compactness. They may be more compact (positive H.) or largely unwound (negative H.). In particular, positive H. is characteristic of Heterochromatin (see), which occurs, inter alia, as dense structures in the interphase nucleus.

Heterosis, *hybrid vigor:* the vigor displayed by the first generation (F_1-generation) from the crossing of two inbred lines, races or varieties, which exceeds that of either parent. F_1-plants are not only more vigorous, but also produce more fruits and seeds, as well as being more resisitant to disease and adverse environmental conditions. F_1-animals become sexually mature at an earlier age, attain a larger size, produce more eggs, etc. Such *heterotic characters* are expressed maximally only on the F_1-generation, and they are not transmitted with equal intensity to the F_2-generation (from the interbreeding of F_1 males and females). Possible causes of H. are the bringing together of dominant performance-increasing genes, gene interactions, etc.

H. is important in plant and animal breeding, e.g. it is used widely in the breeding of maize, sugar beet, sunflowers and other agricultural plants, and in animal breeding it is particularly important in the breeding of pigs for fattening, and poultry for meat and egg production. Breeding with the intention of exploiting H. is known as *heterotic breeding,* and each parent is usually the result of several generations of Inbreeding (see).

Heterosomes: see Sex chromosomes.

Heterostyly: see Pollination.

Heterotardigrada: see *Tardigrada.*

Heterothallic, *haplodioecious:* a term applied to ferns with prothalli of different sexes (micro- and macroprothalli). See Homothallic.

Heterothripidae: see *Thysanoptera.*

Heterotricha: see *Ciliophora.*

Heterotrophy, *heterotrophic nutrition:* nutritional dependency on organic compounds, displayed by animals and most nongreen plants, bacteria and fungi. In contrast to autotrophic plants (see Autotrophy), heterotrophic organisms cannot perform photosynthesis or chemosynthesis. They often possess devices for the acquisition and ingestion of organic nutrients, as well as enzymes for their conversion (if necessary) into compounds of basal metabolism. *Exoenzymes* are particularly common, i.e. enzymes secreted by the organism, which degrade high molecular mass compounds, e.g. proteins, cellulose, lignin or humus compounds, to compounds which can then be assimilated.

Complete heterotrophs require organic sources of both carbon and nitrogen, e.g. parasites that withdraw only organic nutrients from their host, as well as certain saprophytic microorganisms. *Carbon heterotrophs* are able to assimilate inorganic nitrogen sources but require an organic source of carbon, e.g. green plants, yeasts, molds and certain bacteria. In addition to general organic sources of carbon and nitrogen, *auxotrophic* organisms also require specific organic compounds, such as amino acids, fatty acids and various other organic compounds, which in animals are known as vitamins, and in other organisms as organic growth factors. *Auxotrophic mutants* are organisms that have lost the ability to synthesize an essential metabolite, which must then be supplied in the diet, e.g. *Escherichia coli* may, by mutation, lose the ability to synthesize arginine, which then becomes an essential dietary constituent. The wild type of such organisms, with original, unchanged dietary requirements, is known as a *prototroph* or *prototrophic organism.*

With regard to the type of organic substrates utilized for heterotrophic nutrition, H. displays varying degrees of specialization. Parasitism (see Parasites), saprophytism (see Saprophytes) and Symbiosis (see) are specialized forms of heterotrophic feeding. With respect to their nitrogen and phosphorus nutrition, Carnivorous plants (see) are, to a certain extent, *facultatively heterotrophic.*

Heterotypic interaction: see Relationships between living organisms.

Heterotypic organism collective or population: see Organism collective.

Heteroxeny: see Host.

Heterozetes: see *Oribatei.*

Heterozooid: see *Bryozoa.*

Heterozygosity: a condition arising in a fertilized egg (zygote) or individual from the fusion of gametes that differ in the type, quantity or arrangement of their genes, i.e. the resulting heterozygotes are genic, numerical or structural hybrids. In the narrow sense, H. refers to the presence, in a fertilized egg, an individual or a genotype of different alleles at a particular gene locus, which display Mendelian segregation.

Heterozygosity, degree of: see Heterozygosity index.

Heterozygosity index, ***degree of heterozygosity:*** a little used measure of the number of heterozygous chromosomal structures in a population. The H.i. is the average number of paired chromosomes per individual that differ from each other by the presence of one or more inversions.

Heterozygote test: a method for detecting heterozygous carriers of recessive alleles for genetic diseases. In the case of genetically determined metabolic defects, an affected enzyme may display decreased activity, or a metabolite may accumulate before a partial genetic block (conversely, there may be a deficiency of metabolites after the genetic block). In some cases, heterozygotes can also be recognized by slight, clinically unimportant alterations of morphological or physiological characters, cytological structures, etc. The H.t. is important in Genetic counselling (see).

Hevea: see *Euphorbiaceae.*

Hexacorallia: a subclass of the *Anthozoa.* In the primitive condition, the gastrovascular space is divided into 6 radial compartments. Thus, there are always 6 septa (mesenteries) or a multiple thereof. The mode of formation of new septa is different in the various groups. About 4000 species are known, all of them marine. They are either solitary or colonial, and many species secrete a skeleton. The subclass contains the following orders: *Actiniaria* (see) (sea anemones), *Madreporaria* (see) or *Scleractinia* (stony corals), *Antipatharia* (see) (black or horny corals), *Ceriantharia* (see), and *Zoantharia* or *Zoanthidea* (small anemone-like anthozoans with a single siphonoglyph and no skeleton)

Hexactinellida: see *Porifera.*

Hexanchidae: see *Selachimorpha.*

Hexanchiformes: see *Selachimorpha.*

Hexaploidy: a form of Polyploidy (see), in which cells, tissues or individuals possess 6 sets of chromosomes.

Hexapoda: see Insects.

Hexitols: sugar alcohols with six carbon atoms. Of the 10 possible isomers, D-sorbitol, dulcitol, D-mannitol, iditol and allitol occur naturally.

Hexokinase: an enzyme catalysing the transfer of a phosphoryl group from the magnesium complex of ATP (MgATP) to the C-6 oxygen of glucose or another hexose, such as mannose or D-glucosamine. Phosphorylation of glucose is the first step in Glycolysis (see) and the Alcoholic fermentation (see) of glucose.

Hexoses: monosaccharides with 6 carbon atoms, which occur widely in nature, in both free and bound forms. The most important aldohexoses are D-glucose, D-mannose and D-galactose. The most important ketohexoses (hexuloses) are D-fructose and L-sorbose. The two naturally occurring 6-deoxy-sugars, L-rhamnose and L-fucose, are also H.

Heydemann's rule: see Ecological principles, rules and laws.

Hfr cell: see Conjugation, F-plasmid.

Hibernation: see Overwintering.

Hibiscus: see *Malvaceae.*

Hickory: see *Juglandaceae.*

High energy system: a pigment system of plants that is activated by exposure to light of wavelengths below 500 nm (blue) or above 700 nm (deep red), and which influences the stature and structure of the plant, and chemical reactions within the plant. In contrast to the phytochrome system, the H.e.s. does not display morphogenic activity until the plant has been exposed to large quantities of light of the appropriate wavelengths for several hours. This high energy effect is possibly mediated by an unknown pigment, the main candidate being xanthophyll. On the other hand, at least the deep red peak of the high energy effect could be due to absorption by the phytochrome system. The light response of some plants is mediated entirely by the H.e.s., e.g. inhibition of hypocotyl growth in *Lactuca sativa* (lettuce). In other cases, the light red-deep red system is also active (see Phytochrome). For example, rank growth in *Sinapis alba* (white mustard) is inhibited by illumination with light red, or with blue/deep red light, depending on the duration of exposure.

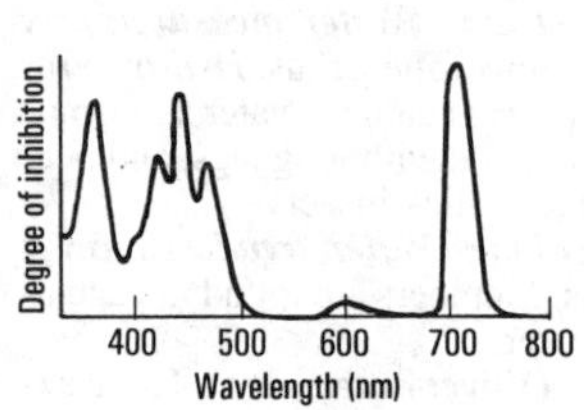

High energy system. Action spectra of the inhibition of extension growth of the hypocotyl of lettuce *(Lactuca sativa);* the inhibition is regulated by the high energy system.

High-moor bog: a region of peat-forming vegetation, usually dominated by *Sphagnum* species. See Bog.

Hilar appendix: see Ballistospores.

Hilum:

1) The site on a fungal spore at which it was attached to its sporophore (sterigma). On the detached spore, the H. is usually recognizable as a distinct scar or projection.

2) The scar on a seed (see Seeds) marking the point at which it was attached to the funicle.

Himantopus: see *Recurvirostridae.*

Hindbrain: see Brain.

Hinge, ***cardo:*** see *Bivalvia.*

Hinge-backed tortoise: see *Testudinidae.*

Hinny: see *Equidae.*

Hipparion (Greek *hippos* horse): extinct genus of pony to zebra-sized *Perissodactyla.* Three toes are developed, but only the central one touches the ground. The enamel lamellae of the high crowned teeth displayed complex folding. H. is an important stage in the evolution of modern horses. Fossils occur from the Pliocene to the Pleistocene, and are especially abundant in Eurasia.

Hippeastrum: see *Amaryllidaceae.*

Hippocampus: see *Syngnathoidei.*

Hippocastanaceae, ***horse-chestnut family:*** a small family of dicotyledonous plants, all trees, and containing 2 genera: *Aesculus* (13 deciduous species of horse-chestnuts and buckeyes, widespread in northern temperate regions) and *Billia* (2 evergreen species, restricted to southern Mexico and tropical South America). Leaves are opposite, palmate and exstipulate. Flowers are zygomorphic, andromonoecious and insect-pollinated. Upper male flowers open before the lower hermaphrodite flowers. Large winter buds are present, sometimes covered with resinous scale leaves, containing a new shoot and inflorescence in an advanced stage of development. The superior, 3-celled ovary develops into a large, leathery, smooth or spiny, usually 1-seeded loculicidal capsule, opening by 3 valves. The large, glossy, nutlike seed is called a chestnut. *Aesculus hippocastanum* ***(Common horse chestnut)*** grows wild in the mountains of the Balkans, and is widely planted as an ornamental in Europe. It grows up to 25 m in height, and its white flowers (with yellow-red flecks) are arranged in long, erect panicles. The bark, leaves and especially the seeds contain the hemolytic saponin, aescin. ***Red buckeye*** *(Aesculus pavia)* originates from North America. ***Red horse-chestnut*** *(Aesculus* x *carnea),* probably a hybrid of *Aesculus pavia* and *Aesculus hippocastanum,* is a popular pink-flowered ornamental.

Hippoglossus: see *Pleuronectiformes.*

Hippoidea: sand or mole crabs, a superfamily of the *Decapoda* (see).

Hippophaî: see *Elaeagnaceae.*

Hippopotaminidae, ***hippopotami:*** a family of the *Artiodactyla* (see). They are plump vegetarian mammals, with naked skin, an extremely large mouth, small eyes set high in the skull, and nostrils which can be closed. Canines and incisors have open roots and grow continuously. The constantly growing canines are very long and modified to large tusks, especially in the lower jaw. Dental formula: $2–3/1–3_i$, $1/1_c$, $4/4_{pm}$, $3/3_m$ = 38–44; the number of upper incisors varies between 2 and 3 pairs; *Choeropsis* usually has only a single pair of lower incisors, whereas *Hippopotamus* has 2 or 3 pairs. Each limb carries four digits, all of them functional. A gall bladder and appendix are lacking, and the large stomach has three sections. There are two monotypic genera. Large herds of the ***Common hippopotamus*** *(Hippopotamus amphibius)* inhabit rivers, lakes and swamps in East Africa. This large animal (almost 4 m in length, weighing between 2 and 3 tons) leaves the water only at night to graze on land. Territory is marked by spraying dung. The young are born and suckled under water. The ***Pygmy hippopotamus*** *(Choeropsis liberiensis)* of West Africa (fragmented distribution from Liberia to Nigeria) is a smaller (length up to 154 cm, maximal weight 250 kg or 550 pounds) solitary forest dweller, first described scientifically in 1841. It lives largely on dry land, feeding on shoots and other forest vegetation, but readily retreats to water when it perceives danger.

Hippopotamus: see *Hippopotamidae.*

Hipposeridae, ***Old World leaf-nosed bats, horse-shoe bats:*** a family of bats (see *Chiroptera),* comprising about 40 species in 9 genera, inhabiting tropical and subtropical regions of Africa and southern Asia, as far east as Australia and the Philippines. They are closely related to the *Rhinolophidae* (see), but differ in the design of the nose leaf, by the presence of only 2 bones in each toe, by the absence of a small, lower premolar, and in certain features of the pelvic and pectoral girdles. The nose-leaf consists of an anterior horseshoe-shaped part, and an erect transverse leaf corresponding to the lancet of the *Rhinolophidae.* This latter structure is usually divided into 3 parts, whose apexes may be prolonged into points. A sella (the central projection of the rhinolophidian nose-leaf) is lacking. Dental formula: 1/2,1/1, 1–2/2, 3/3 = 28 or 30. *Hipposideros gigas* (forearm length 11 cm) is one of the largest bats in the group *Microchiroptera.*

Hipposideros: see *Hipposeridae.*

Hippotraginae, ***roan*** and ***sable antelopes:*** a subfamily of the *Bovidae* (see). The H. are subdivided into 2 tribes.

1. Tribe: *Hippotragini; genus: Hippotragus* (3 species, of which one recently extinct).

2. Tribe: *Orygini;* genera: *Oryx* (5 species) and *Addax* (1 species: *A. nasomaculatus,* ***Addax).***

Hippotragus leucophaeus ***(Blue buck),*** formerly of southwestern South Africa, became extinct in modern times, the last known individual being killed in about 1800. Both *Hippotragus equinus* ***(Roan antelope;*** savanna from Senegal to western Ethiopia, and south to South Africa) and *H. niger* ***(Sable antelope;*** savanna from southeastern Kenya to Angola, and east to South Africa) are powerfully built animals with erect, backward curving horns.

Oryx dammah ***(White*** or ***Scimitar oryx)*** originally inhabited semidesert regions on the southern fringe of the Sahara, but is now extremely depleted and restricted in its range. *Oryx leucoryx* ***(Arabian oryx)*** was originally widely distributed in the Middle East, but is now extinct in the wild and known only in zoos. *Oryx gazella* ***(Gemsbock)*** is still fairly common in arid country from Ethiopia and Somalia, south to Namibia and eastern South Africa. All species of *Oryx* have relatively long, thin horns; in the Gemsbock these are fairly straight, while those of the Scimitar oryx curve backward.

Hippotragini: see *Hippotraginae.*

Hippotragus: see *Hippotraginae.*

Hippuric acid, ***benzoylglycine:*** C_6H_5—CO—NH—CH_2—COOH, a compound synthesized in the kidneys of mammalian herbivores and excreted in the urine. Synthesis of H.a. serves for the detoxication of benzoic acid, which is ingested in the diet (e.g. in plums, pears) or produced metabolically from other sources. H.a. is especially abundant in horse urine, from which it was first isolated in 1829 by Liebig.

Hippurites (Greek *hippos* horse, *ura* tail): a fossil genus of bivalves of the extinct order, *Hippuritoida* (see *Bivalvia).* The right conical valve was longitudinally ribbed or smooth, with three characteristic grooves. The upper (left), caplike valve and the lower (right) valve had interlocking serrated edges. Some species attained a length of 1 m. Only the upper part of the shell was inhabited by the animal, the deeper parts being separated by transverse partitions. These and related rudists (large hornlike or conical bivalves) tended to form reef-like aggregates. Fossils of *Hippurites* occur in the Upper Cretaceous, being especially common in the Alpine zone, where they form the hippurite reefs of the Gosau.

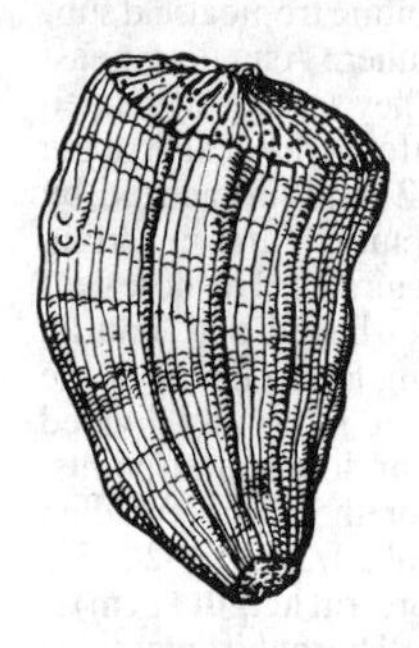

Hippurites gosaviensis from the Upper Alpine Chalk (x 0.3)

Hippuritoida: see *Bivalvia.*

Hirudin: high molecular mass protein secreted by the medicinal leech. In inhibits blood coagulation, and is used clinically against thrombsis.

Hirudinea, ***leeches:*** a class (nearly 300 species) of mainly aquatic predators or blood-suckers in the phylum *Annelida.* With one exception *(Acanthobdella),* they all lack setae.

Morphology. Most species are 1–20 cm in length, but specimens of the American species, *Americobdella (= Cardea) valdivania,* an unusual burrowing form, have been reported up to 76 cm. The shape is generally flattened, being linear, ovate, subcylindrical or foliaceous, but form and length vary with contraction and movement. With one exception *(Acanthobdella),* the number of segments is constant at 33. Coloration is plain brown, green or gray, sometimes vivid with characteristic patterns. Thus, *Hirudo medicinalis* (medicinal leech) is dorsally longitudinally banded with alternate green-gray and rust-red stripes, and its ventral surface is greenish yellow with black markings. Leeches possess a sucker at both the posterior and anterior ends. The anterior sucker contains the mouth opening, and the anus is usually dorsal, immediately in front of the posterior sucker. In the jawed leeches the mouth has 3 pairs of toothed jaws. In the order, *Pharyngobdellida* (freshwater or amphibious carnivorous leeches, e.g. *Erpobdella, Dina, Trocheta),* the mouth opening leads to a nonprotrusible pharynx without jaws or teeth. In the proboscis leeches (order *Rhyncobdellida),* the pharynx is eversible to form a sucking proboscis. The midgut often has paired distensible ceca, in which ingested food can be stored for long periods. Adult leeches lack the internal segmentation that is typical of the annelids. Coelom is more or less obliterated by mesenchyme and powerful muscle fibers, and the residue of the coelom is represented by sinuses, which sometimes carry blood. The excretory system consists of metanephridia, which are always arranged segmentally. The nervous system consists of a suprapharyngeal ganglion joined by short peripharyngeal connectives to a ganglionated ventral nerve cord. Several pairs of light-sensitive organs (pigment spots or eyes) are present at the anterior end, and often also at the posterior end. All species are hermaphrodite. Reproductive organs are always arranged segmentally, consisting of a single pair of ovaries (segment 9, 10 or 11), lying anterior to numerous testes (4 to 10 pairs, the first pair in segment 12 or 13). During the reproductive process, these segments form an annular glandular thickening, the clitellum, whose secretions form the egg capsule or cocoon.

Biology. Most species are aquatic, some living parasitically on marine fish, but the majority occurring in fresh water. Only a few species have adapted to moist terrestrial habitats. Leeches prey on small animals or suck blood from vertebrates, piercing the skin with their toothed jaws. Macrophagous species eat most kinds of protein food, including carrion. Using both suckers, leeches move by creeping: the free end of the body is drawn up to, or extended away from, the attached end, which in turn becomes free, while the previously advancing end becomes attached. Aquatic species also swim by dorsoventral undulations of the flattened body.

Development. All species develop directly, and there is no metamorphosis. Young animals remain in the cocoon, which is filled with a protein-rich nutritive fluid, until they are capable of independent existence. Cocoons are usually attached to stones or submerged vegetation, and may sometimes be buried in the mud of ponds or streams. Terrestrial species bury their cocoons in damp soil beneath stones or other objects, and many members of the *Piscicolidae* attach them to the host. Many members of the *Glossiphoniidae* display brood care, in that cocoons are attached to the adult ventral body surface, and the embryos after hatching are carried on the ventral surface for several weeks by the parent.

Economic importance. Some blood-sucking leeches are pests, e.g. fish leeches in commercial fish ponds, and the terrestrial blood leeches in the jungles of tropical Asia (e.g. *Haemadipsia* and *Phytobdella* of the family *Haemadipsidae),* which are a feared nuisance of animals and humans. Some blood leeches have been used medicinally (see Medicinal leech).

Classification. Taxonomy is based mainly on the presence or absence of setae, and the structure of the pharynx.

1. Order: *Acanthobdellida,* represented by the single family, *Acanthobdellidae,* containing the single genus, *Acanthobdella,* a Russian species parasitic on salmon; unlike other leeches it possesses setae (2 pairs on each side of the 5 anterior segments), it does not have a fixed number of segments, and its coelom is not entirely obliterated.

2. Order: *Rhyncobdellida.* Members are jawless with an eversible pharynx. Marine and freshwater species are both represented. There is a separate blood system with colorless blood.

1. Family: *Glossiphoniidae.* Exclusively freshwater forms, e.g. *Glossiphonia* (parasitic on aquatic snails), *Hemiclepsis* (fish parasite).

2. Family: *Piscicolidae.* Mostly marine species, e.g. *Piscicola, Pontobdella, Branchellion, Pterobdella, Ozobranchus.*

3. Order: *Gnathobdellida.* Aquatic and terrestrial forms, with 3 pairs of jaws and a noneversible pharynx.

1. Family: *Hirudidae.* Members include the common leech *(Hirudo),* the American leech *(Macrobdella),* the free-living carnivorous *Haemopsis,* and the Australian *Limnobdella.*

2. Family: *Haemadipsidae.* Tropical Asiatic forms (see text).

4. Order: *Pharyngobdellida* (5 familes). See text.

Hirudinidae, ***swallows*** or ***martins:*** an avian family containing 80 species in the order *Passeriformes* (see). Most of them hunt flying insects, and they also drink and bathe on the wing. They breed in loose colonies, and mud is used liberally in nest construction. Some species nest in earth burrows, some in tree hollows, while others build on rock faces or house walls, even on the inside of buildings on balconies and ledges. They are 13–23 cm in length, have forked tails, pointed wings with 9 primaries, and small bills with a wide, bristle-edged gape for catching insects in flight. Their most characteristic feature is the possession of closed (complete) bronchial rings; all other members of the *Oscines* (see *Passeriformes)* have half rings with a membrane across the inner face. The family is distributed throughout the world, except for New Zealand, Antarctica and certain larger islands. Many are migratory summer visitors to the temperate zones. The ***Swallow*** or ***Barn swallow*** *(Hirundo rustica)* breeds throughout the Palearctic between 30°N and 70°N; Old World populations overwinter in Africa, India and Australia, while New World populations migrate to Central and South America. It almost invariably nests on or in human artifacts. Other examples are the ***African rock martin*** *(Hirundo fuligula;* sub-Saharan Africa) and the ***Mosque swallow*** *(Hirundo senagalensis;* 3 races in West, Northeast and East Africa). The ***House martin*** *(Delichon urbica)* (plate 6) breeds across Eurasia, and is also found in North Africa, Spain, Asia Minor, and from the Caspian to the northern Himalayas, overwintering mainly in Southeast Asia and East Africa. The large, monotypic ***African river martin*** *(Pseudochelidon eurystomina;* limited area of Zaire) is often placed in a separate subfamily.

Hirudo: see *Hirudinidae.*

Hirudo medicinalis: see Medicinal leech.

Histamine, ***β-imidazol-4(5)ethylamine:*** a biogenic amine formed by decarboxylation of L-histidine. It occurs widely in plants and animals, and is found, e.g. in stinging nettles, *Claviceps purpurea* (ergot), bee venom and in the salivary secretions of blood-sucking insects. As a tissue hormone, it occurs in liver, lungs, spleen, skeletal muscle, gastric mucosa and intestinal mucosa, and it is stored with heparin in mast cells. H. stimulates the glandular cells of the fundus to produce gastric juice, causes dilation of blood capillaries, increases tissue permeability (urtication and reddening after local application of histamine), and causes contraction of the smooth muscles of the alimentary canal, uterus and respiratory passages. H. is degraded to imidazolylacetic acid by the action of diamine oxidases and aldehyde oxidases.

L-Histidine, ***His, imidazolylalanine:*** a proteogenic, basic, heterocyclic, weakly glucogenic amino acid, present in relatively high proportion in globins, and widely distributed in most other proteins. It is a component of the catalytic centers of many enzymes. Since mammals cannot synthesize the imidazole ring, His is an essential dietary constituent. In the putrifaction of protein, it is converted into histamine.

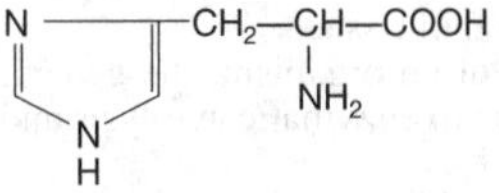

N —CH$_2$—CH—COOH
NH$_2$
N
H

Histidine

Histiophoridae: see *Scombroidei.*

Histioteuthis: see *Cephalopoda.*

Histochemistry, ***cytochemistry:*** detection and identification of components of living cells by studying chemical reactions (usually staining or precipitation reactions) in tissues or microscopic sections, using the light or electron microscope. Identification of the intracellular locations of certain enzymes, ions and metabolic products reveals the relationship between morphology and metabolism, and provides information on the composition of subcellular structures.

Histocompatibility, ***histoincompatibility:*** the compatibilty or incompatibility of tissues, reflected in the extent to which a transplanted tissue stimulates formation of antibodies to itself. The genes responsible for the production of antibodies to foreign tissues are called *histocompatibilioty genes.* H. determines whether transplanted or grafted tissues are accepted or rejected (see Transplatation), and it is also important in blood transfusion (see Blood groups).

Histocompatibility genes: see Histocompatibility.

Histogen concept: the assumption that the fate of individual cells is already determined as they are formed at the shoot apex and the root apex. According to the H.c., the histogens of the shoot apex are (from the innermost to the outermost): 1) the *plerome,* which gives rise to the central cylinder; 2) the *periblem,* which forms the primary cortex, the endodermis arising from the inner periblem layer; 3) the *dermatogen (protoderm),* which gives rise to the epi- and rhizodermis. With respect to the shoot, the H.c. is largely invalid, because individual cells are not predetermined at such an early stage. The H.c. is, however, applicable to the outermost dermatogen, which eventually develops into the primary boundary layer.

Developing roots contain layers of initial cells that can be regarded as true histogens. There is an outermost meristematic layer known as the *calyptrogen,* which gives rise to the root cap. In most dicotyledons, this is derived by periclinal divisions from the primary meristematic cells that give rise to the protoderm, so that this parent meristematic layer (histogen) is usually referred to as the *dermato-calyptrogen.* Below the dermato-calyptrogen is a layer of initial cells (periblem) that produce the cortex and the endodermis, and below this a layer (plerome) that forms the central cylinder and the precambium.

Histoid gall: see Galls.

Histoincompatibility: see Histocompatibility.

Histology: a discipline of biology and medicine, which in addition to the study of tissues in the wider sense, includes the cytology and microscopic anatomy of tissues. At the practical level H. is concerned with the microscopic structure of plant and animals. Up to 1900, H. was predominantly a form of descriptive morphology. Modern H. has a more functional approach, considering the organism and its parts as a structural and functional unity. In medicine, H. form a basis for the diagnosis of healthy and pathologically affected organs, so that a distinction is drawn between *normal H.* and *pathological H.*

Histolysis: disintegration of tissue.

1) Following the death of an organism, the general disintegration of tissues due to enzymatic autolysis and bacterial decomposition.

2) Digestion of organs and body parts during metamorphosis of insects and amphibians.

3) Local destruction of tissues in living organisms, due to disease processes and other damaging influences.

Histones: simple basic proteins that form reversible complexes with DNA, called nucleohistones. They are divided into the following main classes: H1 is lysine-rich; H2a and H2b are moderately lysine-rich; and H3 and H4 are arginine-rich. Octomers containing 2 molecules each of H2a, H2b, H3 and H4 have been isolated. DNA is wound around the octomers, giving chromatin the appearance of a chain of beads. H1 is present in smaller quantities than the other H., and is thought to act as a link between the "beads" (nu bodies) on the chromatin chain. H. function as nonspecific gene repressors.

HIV: acronym of Human Immunodeficiency Virus. See AIDS.

Hoaglanis: see *Clariidae.*

Hoatzin: see *Opisthocomiformes.*

Hobbies: see *Falconiformes.*

Hobby: see *Falconiformes.*

Hofmeister series: see Lyotropic series.

Hog-nosed snakes, *Heterodon:* a genus of North American colubrid snakes (see *Colubridae)* with enlarged rear teeth in the upper jaw, and a flattened, turned up snout. They lay eggs, burrow in the soil or sand, and feed on toads and frogs. Dentition is opisthoglyphic, and prey is bitten only after it has been seized When excited they can blow up the anterior part of the body to twice its normal diameter. The brightly colored ***Eastern hog-nosed snake*** *(H. platyrhinos)* occurs in Florida and regions to the west as far as Texas.

Hog slater: see *Isopoda.*

Holandrous genes, ***holandric genes:*** genes localized on the Y-chromosome (see Sex chromosomes), and transmitted from the male parent only to male offspring, i.e. they display absolute Sex linkage (see).

Only very few cases of holandrous inheritance are known. Among animals it has been observed in the fish species, *Lebistes reticulatus* (order *Cyprinodontoidei).* No unequivocal case of H.g. is known in humans. See Hologynous genes.

Holarctic:

1) A biogeographical area consisting of the continents of Europe, Asia and North America which encircle the north pole.

2) Holarctic kingdom, Holarctic realm: the largest floristic kingdom, which embraces the entire temperate and Arctic zones of the Northern Hemisphere. Characteristic families are *Betulaceae, Salicaceae, Ranunculaceae, Cruciferae, Primulaceae, Campanulaceae.* The Holarctic kingdom is divisible into Arctic, Boreal, Temperate, Submeridional and Meridional floristic zones, corresponding to the changes in floristic composition that accompany the climatic changes from the Arctic north to the warm south. The floristic zones are subdivided into floristic regions, which in turn consist of floristic provinces. The Holarctic kingdom is adjoined on its southern borders by the Paleotropical kingdom (see) and the Neotropical kingdom (see); see also Floristic kingdom. The Arctic floristic zone comprises the circumpolar region beyond the northern tree limit, where vegetation growth is possible for only 2 to 3 months in the year; during these summer months, daylight is practically continuous ("midnight sun"). In the southerly part of this zone, known as the sub-Arctic, the vegetation consists of stunted birch and willow scrub, representing a transition from the more northerly tundra to the forested boreal zone further south. The boreal floristic zone (or circumboreal region) includes the entire northern temperate zone of Eurasia and North America. In oceanic areas, the boreal zone extends as a closed pine forest region, from the northern forest limit southward as far as the deciduous forest zone. On the continents, it extends southward to where pine forests meet steppe vegetation, the former on the northern slopes, the latter on the southern slopes. The predominant vegetation is conifer forest (taiga) of spruce, pine, fir, stone pine and larch. Floristic elements of this boreal floristic zone also occur in southern latitudes, e.g. in mountainous areas of central Europe (see Altitudinal zonation). The temperate floristic zone comprises the temperate regions of Eurasia and North America, and it is subdivided into several floristic regions, e.g. the central European region from the Atlantic coast to the Urals, and the central Siberian region. In the central European region, neither the summers nor the winters are extreme, and the rainfall is relatively high. The dominant vegetation is deciduous forest with beech, hornbeam, oak and small-leaved lime. In the oceanic western part (the Atlantic province) of the central European region, beech woods are accompanied by evergreen trees (e.g. holly, *Ilex aquifolium)* and evergreen *Erica-* and *Ulex*-rich dwarf scrub heaths and marshy areas. Beech, hornbeam and sessile oak are absent from the subcontinental eastern part (the Sarmatic province) of the central European region, which is characterized by forests of pendunculate oak and small-leaved lime. The submeridional floristic zone is a warm, temperate transitional zone of the meridional floristic zone; it extends from the nortern part of the Iberian peninsula, across the southern Alps, northern Italy, the northern Balkans, the Caucasus, and via the Ukraine toward southern Siberia. The deciduous forests of the submeridional zone, interspersed with evergreen floristic elements, show a transition from the more oceanic type vegetation in the west to more continental type vegetation in the east, finally changing to the forest steppes, steppes and desert steppes of the Pontic-southern Siberian region of the inner continent. Plant communities and floristic elements typical of the Pontic-southern Siberian region are found as extrazonal vegetation (see Zonal vegetation) in appropriate habitats in central Europe, where they form Xerothermic vegetation (see). The meridional floristic zone is the southernmost of the Holarctic zones. Its western European sector includes the Macronesian-Mediterranean region, i.e. the Atlantic islands and the Mediterranean area with its low frost climate, wet winters and dry summers (Mediterranean or etesian climate), characterized by evergreen sclerophyllous vegetation. Warm, wet areas, such as the Canary Islands, contain relict evergreen laurel communities, which are characteristic of the ex-

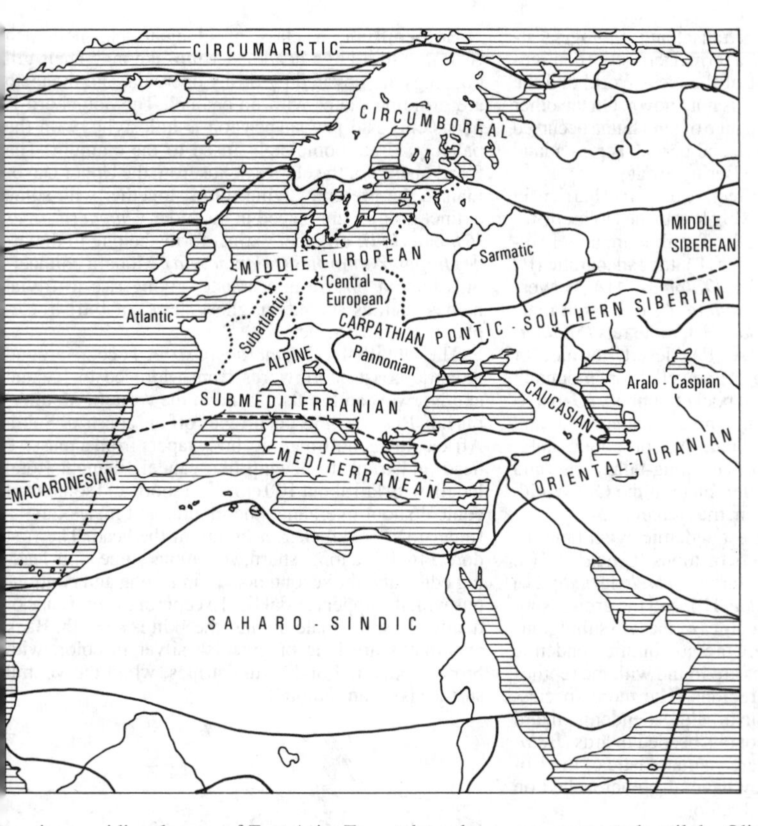

Holarctic kingdom. The floral regions of western Eurasia and North Africa (modified from Meusel, Jäger and Weinert, 1965). Floral regions are printed in large case, floral provinces in small case letters. ——— Region. ------------ Subregion. Floral provinces of the central European region: Atlantic, Subatlantic, Central European and Sarmatic.

tensive meridional areas of East Asia. Forest degradation is followed by development of bush and scrub, e.g. the maqui (see Sclerophyllous vegetation). East of the Mediterranean and sub-Mediterranean regions, the drier continental areas of the meridional floristic zone become the Oriental-Turanian region. The latter shows a transition from deciduous forest in the oriental mountains, via steppes and desert steppes, to the wormwood and saxaul deserts of the wide plains. In the East Asian part of the meridional floristic zone, deserts, steppes and deciduous forests give way to the evergreen forest vegetation of the Sino-Japanese region, with a wet monsoon climate. The Holarctic kingdom of the North American continent displays a succession of floristic zones and an oceanic-continental vegetation transition analogous to those of Eurasia.

3) ***Holarctic region, Holarctic realm:*** the largest zoogeographical region, embracing the Arctic, North America, Europe and a large part of Asia. Its southern boundary is not sharply defined; transitions occur in the Caribbean, in the desert belt of the Old World, and in China (see Transitional region). Various systems for the subdivision of the Holarctic region are favored by different authorities. Usually it is divided into 2 subregions, the Paleoarctic (see) and the Neoarctic (see), considered by some authorities to be independent regions. In view of the uniformity of fauna, the northernmost part of the Holarctic region, i.e. the Arctic (see), is also sometimes treated as a third subregion.

Many changes have occurred during the history of the Holarctic region. For example, Europe and Asia were separated until the Oligocene by water east of the Urals (the Turanian corridor). More than any other zoogeographical region, the Holarctic region was especially affected by the Ice Age. For climatic reasons, the number of species in the Holarctic region was and is always below average, but species were further impoverished by the effects of glaciation. In Europe, the decrease in forest cover, and in tree and bush species due to glaciation was, inter alia, a decisive factor in the decrease of fauna; in addition, the European mountains, stretching eastward and westward, hindered the return and redispersal of species in the postglacial period. On the other hand, the resulting confinement of many species in two or more separated Glacial refuges (see) initiated evolutionary differentiation, which in certain areas resulted in new species. Recognition of these refugia is essential for an understanding of the fauna of the Holarctic region. Many present areal expansions may represent recent phases of postglacial recolonization.

In addition to the biological uniformity of the Arctic, the taigas of the Neoarctic and Paleoarctic are also ecologically very similar, with numerous common species. Similarities in the vertebrate populations of these two subregions are well represented by the native birds and mammals. For example, the brown bear, wolf, elk and red deer are all represented in both the Neoarctic and the Paleoarctic. The total number of common species, including the Arctic is about 5 000. Similar biotic communities also exist in these two subregions (and further south), although they consist of vacarious species; this is also true of invertebrate communities, especially insects.

The extent to which ancient transatlantic biological affinities owe their existence to a formal terrestrial continuity between the eastern part of the New World and the western part of the Old World is not known. On the other hand, it is certain that an exchange of land fauna occurred repeatedly, even in the postglacial period, across a land bridge formerly spanning the Bering Strait.

Owing to the absence of effective dispersal barriers in the south, endemism is relatively low in the Holarctic region. The following mammalian families are largely restricted to the Holarctic region (H.), the Paleoarctic (P.) or the Neoarctic (N.): moles *(Talpidae,* H.), beevers *(Castoridae,* H.), pikas *(Ochotonidae,* H.), birch mice and jumping mice *(Zapodidae,* H.), mole rats *(Spalacidae,* P.), jerboas *(Dipodidae,* P.), desert dormice or Selevin's mice *(Selevinidae,* P.), sewellels or "mountain beavers" *(Aplodontidae,* N.), pocket gophers *(Geomyidae,* N.), pocket mice, kangaroo mice, kangaroo rats and spiny pocket mice *(Heteromyidae,* N.), pronghorn antelopes *(Antilocapridae,* N.). The long-tailed rats and mice *(Muridae),* which predominate in the Old World, are conspicuously absent from the Neoarctic.

The only endemic or almost endemic avian families are auks *(Alcidae,* H.). divers or loons *(Gaviidae,* H.), accentors *(Prunellidae,* P.), turkeys *(Meleagridae,* N.) and phalaropes *(Phalaropidae,* H.), and the grouse subfamily *(Tetraoninae,* H.). Reptiles of the two subregions show little affinity, and there are no common endemic families. Greater similarities are found with the reptiles of the adjoining southern regions. The more species-rich Neoarctic contains 3 small, almost endemic lizard families, including the poisonous beaded lizards *(Helodermatidae).* Salamanders are characteristic of the Holarctic region, although they have also encroached on the Oriental and Neotropical regions.

Freshwater fish species are more numerous in the Neoarctic then in the Paleoarctic. The following families are endemic in the Holarctic region: pikes *(Esocidae),* mudminnows *(Umbridae),* true perches *(Percidae),* salmon *(Salmonidae)* and sturgeons *(Acipenseridae).* In addition, a few small families are endemic in the Neoarctic. On the basis of a comprehensive analysis of 16 000 genera, Holdaus has shown that the proportions of endemics (see Endemic rate) are lower in the Neoarctic and Paleoarctic than in any other regions.

Holbrook's eastern spadefoot: see *Pelobatidae.*

Holcus: see *Graminae.*

Holdfast: see *Cestoda.*

Hollyhock: see *Malvaceae.*

Holobasidiomycetidae: see *Basidiomycetes.*

Holobasidium, ***autobasidium, homobasidium:*** a nonseptate (nonchambered), single-celled basidium of the *Basidiomycetes,* which carries the basidiospores on its upper surface or apex. Holobasidia are typical of the *Holobasidiomycetidae.*

Holocene (Greek *holos* total, *kainos* new or unusual), ***Flandrian, Postglacial:*** a late division of the Quaternary (see Table in Geological time scale), which includes the geological present. In North America, and in central and northern Europe, the H. is characterized by the reestablishment of animal species driven out by the climatic conditions of the Ice Age, by recolonization of the ice-free or tundra-covered regions by plant communities displaced during the Ice Age, and by the gradual development of the present-day patterns of vegetation. See Pollen analysis.

Holocephali: a class of elongate elasmobranch fishes, with a large head and whip-like tail. Their gill openings are covered by fleshy opercula, and the teeth are grinding plates with no enamel. The notochord is persistent. The jaw suspension is holostylic, with the palatoquadrate completely fused to the cranium. The fossil history of the class extends from the Upper Devonian to the present. Living forms, together with some extinct groups, are placed in the order *Chimaeriformes (Chimaerida);* all other extinct forms belong to the order *Iniopterygiformes (Iniopterygia).* Modern chimaeroids have a bizarre appearance, giving rise to trivial names such as spookfish, ghost shark, rabbitfish, elephantfish, ratfish, etc.

The rabbitfish *(Chimaera monstrosa)* occurs around Atlantic coasts northwards from Morocco to Iceland and Norway, and is also found in the western Mediterranean. Reports of its occurrence off the coasts of South Africa are unconfirmed. Its body tapers from a massive head and trunk to a whip-like caudal filament (total length with tail about 100 cm). The short, and somewhat conical snout overhangs the mouth, and males have a finger-like frontal clasper on top of the head. The first dorsal fin has a long, sharp, venomous spine at its leading edge, and the second dorsal fin is long and continuous with the upper caudal fin. Except for the presence of denticles on the male organs, the skin is smooth. Back and flanks are blue or greenish silver in color, with brown spots and undulating stripes, while the ventral surface is cream-colored.

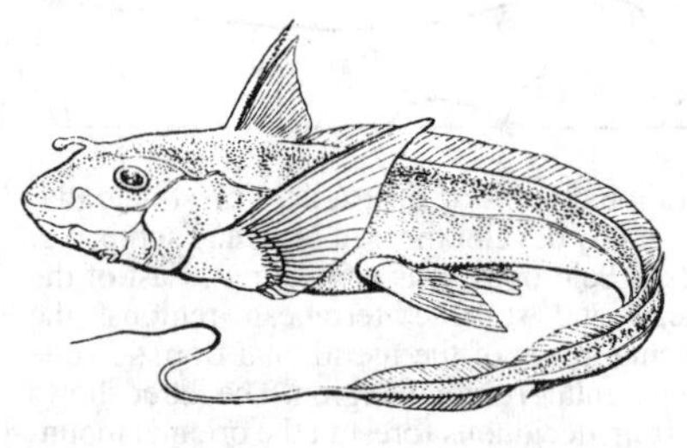

Holocephali. Rabbitfish *(Chimaera monstrosa)*

Holocyclic parthenogenesis: see Anholocyclic parthenogenesis.

Holoenzyme: see Enzymes.

Hologamy: see Reproduction, Fertilization.

Hologastra: see *Coleoptera.*

Hologynous genes: genes localized in the X-chromosome, and transmitted from the female parent only to female offspring. See Holandrous genes.

Holometabolous insects: Insects (see) with a complete metamorphosis.

Holosaprophytes: see Saprophyes.

Holosteans, *Holostei:* a diverse group of fish, differing from the *Chondrostei* in that each supporting element of the fins bears a single fin ray. Most H. (e.g. gars) have rhomboid-shaped ganoid scales, which consist of 3 strata, the outer constituted by a substance resembling tooth enamel, called ganoin. More advanced forms (e.g. the bowfin) have cycloid scales without ganoin. The skeleton is more or less ossified. Extinct orders (mainly marine) are known from the Devonian to the present. There are only two living orders.

1. *Amiiformes.* The bowfin *(Amia calva)* of the fam-

ily *Amiidae* is the only living representative of this order. It prefers warm shallow water, and is found in eastern North America from the Great Lakes southwards. Its gas bladder is septate, and it can function in aerial respiration, enabling the fish to inhabit stagnant water with low oxygen concentration.

2. *Lepisosteiformes*. This order contains a single family: *Lepisosteidae* (gars), found from the Great Lakes of eastern North America to Costa Rica, with one species in Cuba. Gars are elongate, and the body is covered with heavy ganoid scales. Both elongate jaws are armed with several rows of strong sharp teeth. Gars differ from all other living fishes in possessing opisthocoelus vertebrae, i.e. vertebrae that are posteriorly concave and anteriorly convex. They can utilize atmospheric oxygen, as well as dissolved oxygen taken through the gills. Most are freshwater fishes, but the alligator gar *(Lepososteus spatula)* and close relatives may enter salt water.

Holothuria: see *Holothuroidea.*

Holothuroidea, *holothuroids, sea cucumbers:* a class of the *Echinodermata* (see) with elongated cylindrical bodies, averaging 10–20 cm in length. In the extended state, the largest representatives attain a length of 2 m and a diameter of 5 cm. Mouth and anus are situated at opposite ends of the cylinder. The skin is leathery and pliable, since the calcareous skeleton is reduced to minute, very characteristic sclerites or spicules. The mouth is surrounded by retractable tentacles. The axial organ (protocoel) is degenerate, and there is only one gonad. Respiratory trees (see) are often present on the hindgut. Holothuroids are found in all the seas of the world, preferring quiet water. Practically all species live on the sea bottom, and only a few are able to swim or suspend themselves at different levels. A few forms are pelagic. They feed on small animals, which are ingested together with bottom material; some species trap plankton with their tentacles. Development occurs via an *auricularia* larva, which metamorphoses into the doliolaria stage, and finally into the pentactula stage. About 1 100 living species are known, and some of these are commercially important (see Trepang). The most familiar genera are *Cucumaria, Holothuria* and *Stichopus.*

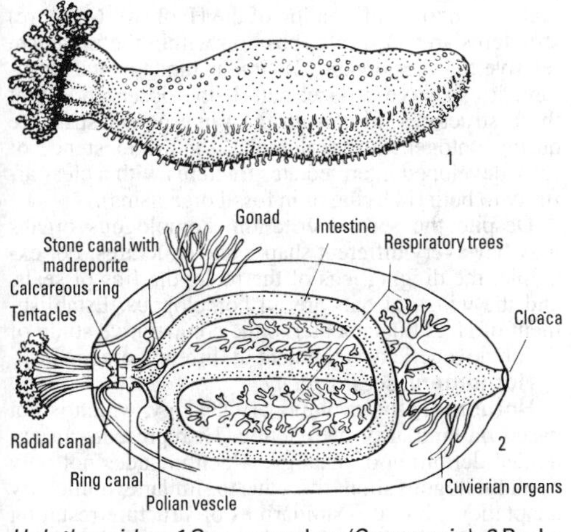

Holothuroidea. 1 Sea cucumber *(Cucumaria). 2* Body plan.

Fossil record. Fossils are found from the Cambrian to the present. Since holothuroids do not have a continuous skeleton, they have little paleontological importance. Impressions of complete animals are very rare. From the Lower Carboniferous onward, however, the microscopically small skeletal elements of holothuroids are relatively common, and these are used for the identification of the strata of Mesozoic sediments.

Collection and preservation. They are best caught with a dredge or bottom grab. Specimens must be thoroughly narcotized, to prevent them from ejecting their digestive tract through the hindgut. Menthol crystals are scattered carefully on the surface of the sea water, or urethane crystals are placed in front of the cloaca. They should then be killed in 30% ethanol and placed in 75% ethanol for short-term preservation. After the first few days, the 75% ethanol must be replaced with 85% ethanol for long-term storage.

Holothuroids: see *Holothuroidea.*

Holotricha: see *Ciliophora.*

Holozygote: a zygote in which two complete genomes are united, as distinct from a merozygote.

Homalorhagida: see *Kinorhyncha.*

Homaridae: see Lobsters.

Homarus: see Lobsters.

Homeochlamydeous: see Heterochlamydeous.

Homeogenetic induction: see Determination.

Homeohydric: see Poikilohydric.

Homeology: a special case of Homology (see). The term was coined in 1922 by L. Plate for structural similarities that arise in homologous organs by parallel and phylogenetically independent development, e.g. the wings (flight membranes) of bats and pterodactyls.

Homeothermic animals, *warm-blooded animals:* animals that maintain their own constant body temperature, which may differ considerably from that of the surroundings. Mammals and birds are H.a.; all other animals are Poikilothermic animals (see). See Temperature factor.

Home range: part of the habitat of a species, defined in time and space, in which the individual spends a relatively long period of time, at least within a certain period or stage of its life. The individual is also capable of responding appropiately to all the environmental demands of the H.r. The H.r. is therefore a characteristic of each individual. Analogously, each group or biosocial unit also has a characteristic H.r.

Hominidae, *hominids:* the primate family containing all fossil and recent human forms and their antecedants. See Hominoids, Anthropogenesis.

Homininae: true humans; a subfamily of the *Hominidae,* containing the single genus, *Homo* (see). See Anthropogenesis.

Hominoids, *Hominoidea,* ***anthropoid apes*** and ***humans:*** a superfamily of the infraorder *Catarrhini* (see Primates), consisting of 4 families embracing all fossil and recent anthropoids and humans.

1. Hylobatids *(Hylobatidae,* gibbons) are represented by the fossil genera, *Aelopithecus* from the Oligocene, *Limnopithecus* (Miocene), *Pliopithecus* and *Epipliopithecus* (Pliocene) and the recent genus, *Hylobates* (gibbons and siamang). Fossils are known from Europe and Africa; the recent *Hylobates* inhabit Southeast Asia.

2. Pongids *(Pongidae,* great apes in the narrow sense) are divided into 2 subfamilies:

a) Dryopithecines, with the fossil genera, *Proconsul*

(Miocene), *Dryopithecus, Sivapithecus* and *Ramapithecus* (Miocene-Pliocene) and *Gigantopithecus* (Pliocene); b) Pongines, with the recent great apes, genera *Pongo* (orang-utan), *Pan* (common and pygmy chimpanzee) and *Gorilla* (3 subspecies of G. gorilla). Fossil pongids are found in Europe, Africa and East Asia, while recent forms are restricted to Africa and East Asia.

3. Oreopithecids *(Oreopithecidae)* are represented by the fossil species, *Oreopithecus bamboli* (Miocene-Pliocene) from the brown coal strata of Tuscany (Italy).

4. Hominids *(Hominidae)* are represented by the subfamilies, Ramapithecines and Australopithecines, as well as the Hominines with the single genus, *Homo* (see).

See Anthropogenesis.

Homo: the only genus of the subfamily *Homininae,* or the family *Hominidae* (see Primates). *Homo* is divided into the species *Homo sapiens, Homo erectus* and *Homo habilis* (?) During the Pleistocene, as at present, species of *Homo* do not appear to greatly overlap in time, i.e. only single species are found in any given geological stratum. Separation of fossil "species" of *Homo,* based on customary procedures is therefore questionable.

All humans belong to the species, *Homo sapiens,* which has inhabited the earth since the Late Pleistocene. Recent humans belong to the subspecies, *Homo sapiens sapiens,* while their immediate ancestors are designated *Homo sapiens praesapiens* (e.g. fossil remains at Steinhein in Germany, Swanscombe in southern England, and Fontechevade in central France). These 2 groups form the neanthropines (late or modern man). *Homo sapiens* also includes the paleanthropines or Neanderthalers (early man) *(Homo sapiens neanderthalensis* and *Homo sapiens praeneanderthalensis),* whose remains occur widely, and are named after a famous skeleton found in the Neander valley near Düsseldorf.

Fossil remains of *Homo erectus,* also known as primitive man or pithecanthropines (= archenthropines), have been found from the Middle to the Late Pleistocene. These include *Pithecanthropus (Homo erectus erectus,* Java man) from Java, *Sinanthropus (Homo erectus pekinenesis,* Peking man) from China, and various finds in Europe *(Homo erectus heidelbergensis, Homo erectus bilzingslebenensis, Homo erectus palaeohungaricus)* and Africa *(Homo erectus mauritanicus, Homo erectus leakyi,* etc.).

Some authorities place *Australopithecus habilis* in the genus *Homo,* as *Homo habilis.*

See Anthropogenesis.

Homoalleles: see Heteroalleles.

Homobasidium: see Holobasidium.

Homoblastic development: in plants, a type of development, in which there is no essential difference between juvenile and subsequent stages, i.e. they merge continuously into one another. The converse is Heteroblastic development (see).

Homocercal fin: see Fins.

Homocercous fin: see Fins.

Homochromy: see Protective adaptations.

Homocysteine, ***Hcy:*** $HS—CH_2—CH_2—CH(NH_2)—COOH$, a nonproteogenic amino acid and a higher homolog of cysteine. Hcy occurs as an intermediate in the metabolic conversion of L-methione into L-cysteine.

Homocytotropic antibodies: antibodies that bind to certain cells of the organism that produces them. They are responsible for atopical allergies.

Homodont: see Dentition.

Homoeologous chromosomes, ***homeoelogs:*** chromosomes that are only partially homologous. It is thought that their ancestral chromosomes were truly homologous, and that their synaptic attraction has been decreased by evolutionary divergence.

Homogamy:

1) In plants, the occurrence of different types of flowers, containing only anthers or ovaries.

2) In a flower or hermaphrodite organism, the simultaneous maturation of male and female reproductive organs.

3) Positive assortative mating (see Assortative mating).

Homogentisic acid: a phenolcarboxylic acid and a *p*-quinone, which is an intermediate in the oxidative metabolism of L-tyrosine. It is normally oxidized further (catalysed by the liver enzyme homogentisate oxidase) to fumaric acid and acetoacetic acid. In the inborn error of metabolism known as *alcaptonuria,* homogentisate oxidase is absent, and H.a. is excreted in the urine, where it undergoes atmospheric oxidation and polymerization to a black pigment.

OH

$CH_2—COOH$

HO

Homogentisic acid

Homoio-: a prefix meaning the same as homeo-, e.g. homeothermic, homeogenetic, etc.

Homokaryocyte: a cell produced by Cell fusion (see) of cells of the same genotype.

Homokaryon: see Heterokaryosis, Cell fusion.

Homokaryosis: see Heterokaryosis.

Homology: the relationship between structures or organs in different organisms, determined by their common phylogenetic origin. The main criteria for the establishment or confirmation of the H. of two (or more) structures are: 1) similar position within the organism relative to other organs; 2) similar fundamental structure, irrespective of size; and 3) the observation that these structures arise from the same parent structure during ontogenetic development, or the existence of fully developed intermediate structures with a clear affinity to both (in living or in fossil organisms).

Despite the second criterion, homologous organs may have very different shapes and structures. For example, the digging legs of the mole, the fins of seals, and the wings of bats are all homologous. Establishment of H. is fundamental to the comparative study of the phylogenetic development of different organisms.

Homomery: see Polygeny.

Homomorphy: evolutionary analogy, which is not based on similarity of function. H. represents an extended definition of analogy, which includes not only nonhomologous similarities due to similar evolutionary adaptations, but also similarities of structure resulting from other unknown or hypothetical factors of phylogenetic development.

Homomyarian condition: synonymous with dimyarian or isomyarian condition, terms applied to members of the *Bivalvia* (see) that possess 2 adductor muscles of about the same size.

Homonomy: similarity of structures in the same individual, e.g. the various leaves of a plant, or the hairs of an animal, e.g. in pine trees, the female cones of the long vegetative shoots and the male cones of the short shoots are homonomous.

Homoplasmonic: a term applied to cells with genetically identical plasmons. The converse is Heteroplasmonic (see).

Homoploidy: occurrence of the normal chromosomal complement (diploidy in the diploid phase, monoploidy in the haploid phase) in cells, tissues or individuals. The converse is Heteroploidy (see).

Homoptera: an insect order containing about 38 200 species worldwide. The earliest fossil *H.* are found in the Permian.

Imagoes. Body length of the adult varies from 0.5 to 80 mm, with a wingspan up to 8.5 cm in temperate species and 18 cm in tropical species. Mouthparts are adapted for piercing and sucking (see *Hemiptera),* and in plant lice they are opisthognathous (opisthorhyncous), i.e. posterioventral in position, lying between the coxae of the forelegs. Forewings are undivided and larger than the hindwings, but like the hindwings they are uniformly membranous. In this regard, the *H.* contrast with the *Heteroptera* (see), whose forewings are divided into strong and fragile sections. In some groups, wings are absent or reduced. The legs, which are reduced in some species, are either normal walking legs or modified for jumping. In the latter case, the musculature is not located in the femur, as in other jumping insects, but in the posterior coxae; tarsae 1–3 are segmented. All species are terrestrial, and all feed on plant sap; some species are gall-formers. When taking plant sap, the insect relies mainly on fluid pressure within the pierced sieve tube, which forces sap into the stylet mouthparts; the term "sucking mouthparts" is therefore inappropriate. In fact, the insect may sometimes be in danger of bursting, and must in some way restrict the influx of sap. With such a copious flow, the insect receives a surplus of plant carbohydrates, which is disposed of in various ways. Many *H.* produce a liquid sugary excrement (honeydew), which in some species is milked from the anus by ants. The honeydew of *Buchneria pectinatae* (a conifer-feeding aphid, family *Aphididae)* is taken by honeybees and is the basis of the highly prized fir honey. Other *H.* convert the carbohydrate into wax, which is secreted as threads or body covering. Wax may also be stored in blood cells, and aphids use their waxy blood as a means of defense. Probably all *H.* harbor symbiotic, nitrogen-fixing bacteria in their alimentary canal in organs called mycetomes. Plant lice are usually gregarious and they often occur in extremely large numbers. Reproduction is either sexual or by parthenogenesis. Metamorphosis is incomplete and varies greatly between different groups.

Eggs. Cicadas, psyllids and whiteflies lay eggs, whereas many aphids and scale insects are viviparous.

Larvae. In cicadas, psyllids and aphids, the wings appear at an early stage of larval development (paurometabolous development), whereas in other groups they do not appear until the last or penultimate larval instar (neometabolous development). In scale insects (in which male and female larval development proceed differently), the larva passes through resting phases. Aphid development is characterized by an alternation of generations.

Economic importance. Some scale insects produce wax, lac or pigments, which are used commercially (pela wax, shellac, cochineal, carminic acid). Some cicadas and psyllids, but especially many aphids and scale insects, are plant pests. Various aphids are vectors of plant disease-causing viruses.

Classification. The *H.* are usually divided into 2 suborders, which have recently been assigned separate ordinal status by some authorities. A third, small, primitive suborder is also recognized.

1. Suborder: *Auchenorrhynchi.* The stylet sheath (labium) of the piercing mouthparts lies on the underside of the head, and appears to arise from the throat. Antennae are very short, and the tarsae are 3-segmented. Two pairs of richly veined wings are always present, and they are held rooflike over the body when at rest. Males, but not females, possess sound-producing tympanic organs (tymbals). Both sexes have auditory organs (tympana).

1. Superfamily: *Fulgeroidea* ***(Plant hoppers)***

Family: *Delphacidae.* Members have a large, sawtoothed spur at the tip of the hind tibia, e.g. *Perkinsiella saccharicida* ***(Sugarcane leafhopper),*** originally from Queensland, which became a major pest in Hawaii at the beginning of the 20th century; it was controlled biologically by importing appropriate parasites from Queensland.

Family: *Fulgoridae.* Numerous brightly colored tropical species with bulbous heads, known as ***Lantern flies,*** although they are not luminous.

Family: *Tettigometridae.* Their heads are flattened in front, and they are usually found among ants.

Other families: *Cixiidae, Derbidae, Dictyopharidae, Achilidae, Issidae.*

2. Superfamily: *Cercopoidea.* Many larvae of the families *Cercopidae* and *Aphrophoridae* live in a mass of froth, and are known as "cuckoo spit insects" or "spittlebugs". Larvae of the *Machaerotidae* live in a calcareous tube. Adults are hopping insects, commonly called "froghoppers".

3. Superfamily: *Cicadoidea* ***(Cicadas).*** These are common in all warm regions of the world. They are well known for the sound they produce and their lengthy period of larval development, which may take several years. The familiar "periodical cicada" or seventeen-year locust *(Magicicada septendecim)* requires 17 years for development in the eastern USA and 13 years in the southern states. These larvae live entirely underground, feeding on roots, and they first emerge above the soil surface for the final molt to the imago.

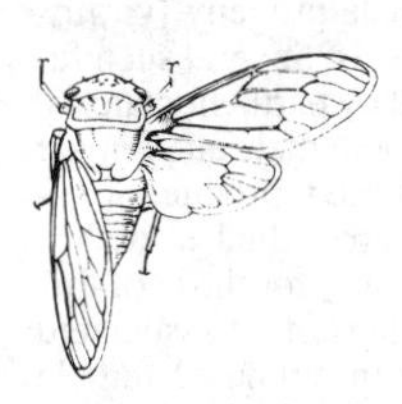

Fig.I. *Homoptera.* Cicada (*Lyristes plebejus* Scop.).

4. Superfamily: *Cicadelloidea.* Most members belong to 1 of 2 large families: *Cicadellidae (Jassidae)* ***(Leafhoppers)*** and *Membracidae* ***(Treehoppers).*** Im-

portant pests are the ***Beet leafhopper*** *(Circulifer tenellus)* and the Potato leaf hopper *(Empoasca fabae).*

2. Suborder: *Sternorrhynchi* ***(Plant lice).*** The stylet sheath (labium) of the piercing mouthparts appears to project from the thorax between the forelegs. Tarsae are 1- or 2-segmented. Wings are not always developed, but when present they may be held rooflike over the abdomen or flat against the body (whiteflies and scale insects). Wing venation is markedly reduced in some groups. Tympanic and auditory organs are absent.

5. Superfamily: *Psylloidea* ***(Jumping plant lice, Psyllids).*** All members belong to a single family *(Psyllidae).* Two pairs of relatively well veined wings are present in both sexes. They can jump with the aid of powerful hindlegs. Two pest species introduced into North America from Europe are *Psylla pyricola* ***(Pear leaf psyllid)*** and Psylla mali (Apple leaf psyllid).

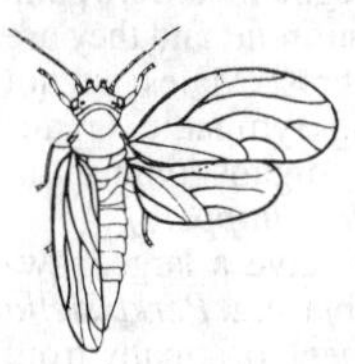

Fig.2. *Homoptera.* Jumping plant louse or psyllid *(Psylla* sp.).

6. Superfamily: *Aleurodoidea* ***(Whiteflies).*** All members belong to a single family *(Aleurodidae* or *Aleyrodidae).* Forewings have 3 veins, the hindwings only a single vein. Adults are small, never exceeding 3 mm in length, and rather mothlike in appearance. The body is covered with a waxy dust. An example is the ***Greenhouse whitefly*** *(Trialeurodes vaporarium* Westw.), which is established in North America, Europe and Australia, but is thought to have originated in Central America; it attacks all types of ornamental and horticultural greenhouse plants.

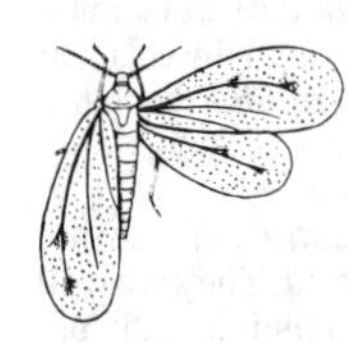

Fig.3. *Homoptera.* Whitefly *(Aleurodes* sp.).

7. Superfamily: *Aphidoidea* ***(Aphids).*** These are nonjumping insects, up to 5 mm in length. Males are usually winged, and females are usually wingless, in particular females that reproduce parthenogenetically. Two secretory tubes are often present on the abdomen. The fertilized aphid egg always develops into a fundatrix or stem mother, which produces female progeny (virgins or fundatrigeniae) by parthenogenesis. Several such female generations may be produced by parthenogenesis, but finally there is a generation of both male and female offspring (known as sexuparae); these reproduce sexually, and the resulting offspring are called sexuales. Some authors use the term female only for those of this generation. This development cycle is often associated with a change of plant host. The inseminated female lays winter eggs on the primary host, which is often a tree. As the tree sap pressure decreases in summer, the virgins transfer to herbaceous plants. The last virginal generation then transfers back to the tree in autumn. This complete cycle of development (holocycle) may be interrupted or shortened, depending on the species, availability of host plants and the climate.

Family: *Aphididae* ***(Plant lice, Greenfly, Blackfly, True aphids).*** Examples are the ***Black bean aphid*** *(Dorlais fabae* Scop.); and the ***Green peach aphid*** *(Myzus persicae),* which attacks a single primary host (the peach tree), but has a wide range of secondary plant hosts; it is known to carry more than 100 virus diseases between different plants.

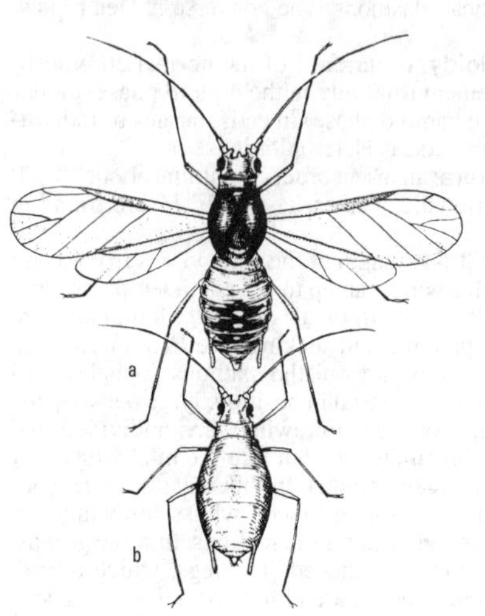

Fig.4. *Homoptera.* Green peach aphid *(Myzus persicae* Sulzer). *a* Winged fundatrix. *b* Wingless fundatrix.

Family: *Pemphigidae (Eriosomatidae).* This family includes many gall-forming species and many ***Woolly aphids,*** so named because they produce woolly waxy filaments, e.g. ***Woolly apple aphid*** *(Eriosoma lanigerum).*

Family: *Adelgidae (Chermidae).* Members are confined to conifers, sometimes forming galls. Most species alternate hosts, the primary host being spruce.

Family: *Phylloxeridae.* This small family includes the important grape vine pest, the ***Vine phylloxera*** or ***Grape phylloxerid*** *(Phylloxera vitifoliae* Fitch).

8. Superfamily: *Coccoidea* ***(Coccoids, Scale insects, Mealy bugs).*** Adult insects are generally 1 to 6 mm in length, but some tropical species attain 3 cm. Sexual dimorphism is strongly marked. Females are wingless, eyeless and legless. These degenerate females may be scale-like, gall-like, or covered with a waxy or powdery layer. Males are either wingless or possess only forewings with a single forked vein. Tarsae have only one segment and a single claw. The largest family is the *Diaspididae* ***(Armored scales)*** whose females are covered with a hard waxy layer. e.g. the ***San José scale*** *(Quadraspidiotus perniciosus* Comst.) and the ***Oystershell scale*** *(Lepidosaphes ulmi).* The *Coccidae* ***(Soft scales, Wax scales)*** is also a large family, containing several pests, e.g. ***Cotton maple scale*** *(Pulvinaria innumerabilis),* a pest of fruit trees in North America, introduced from Europe. The extremely degenerate females of the *Lacciferidae* live in a resinous cell, e.g. the ***Indian lac insect*** *(Laccifer lacca)* produces a secretion used for

preparing shellac. Females of the *Pseudococcidae* ***(Mealy bugs)*** are coverd with a mealy or filamentaous waxy secretion.

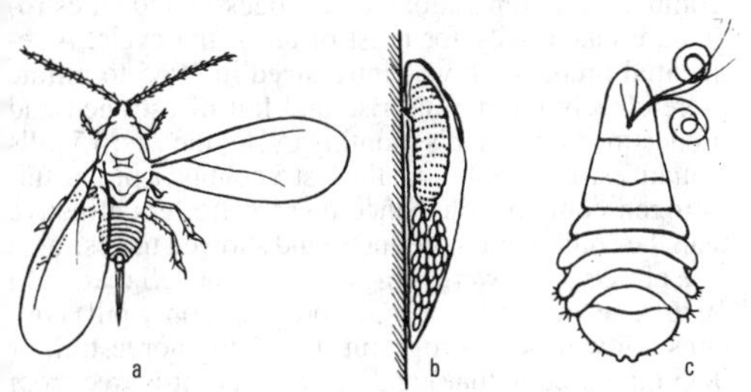

Fig.5. *Homoptera.* Oystershell scale *(Lepidosaphes ulmi* L.). *a* Male. *b* Longitudinal section through a scale with female and eggs. *c* Magnified ventral view of female.

3. Suborder: *Coleorrhynchi.* These are closely related to the fossil *Paleorrhynchi,* which are known from the Permian. There is a single family, the *Peloridiidae,* with 12 species in Australia, New Zealand and South America. They are the only modern *H.* that feed on mosses.

Homothallic, ***haplomonoecious:*** a term applied to ferns whose prothalli are hermaphrodite. The term is also applied to yeasts, in which the mating-type genes of the haploid cells show a rapid rate of interconversion (almost every other cell division).

Homotypic interaction: see Relationships between living organisms.

Homotypic organism collective or population: see Organism collective.

Homoxeny: see Host.

Honey agaric: see *Basidiomycetes (Agaricales).*

Honey ants: see Ants.

Honeybee, *Apis mellifica* L.: one of the most useful insects to humans, serving as a flower pollinator, and producer of honey and wax. Wild H.b. build their nests in hollow trees or other natural cavities. In some parts of the world (e.g. Southeast Asia), honey and wax are still collected from wild bees. H.b., however, have been domesticated since antiquity, by providing hollow tree stumps, baskets and other structures, all generally referred to as hives. Nowdays, the bee keeper uses much larger wooden hives which can be dismantled. A bee colony contains between 10 000 and 70 000 individuals. Most of these are females with regressed sex organs, and known as *workers.* Workers are specially equipped for the collection and transport of pollen and nectar. Their main tasks include the provision of food and building materials, construction of the wax honeycomb of hexagonal cells, and care and feeding of the developing larvae. The single fertile female is known as the *queen.* The queen embarks on a single nuptual flight, during which she is followed by several drones (males) until one succeeds in mating with her. Spermatozoa from this single act of mating are stored by the queen, then used to fertilize between 500 to 2 000 eggs per day during the summer months. The fertilized eggs normally develop into workers. At a certain time in the spring, larvae in queen cells (recognizable from their conical shape) receive a special diet and develop into young queens. Some eggs are laid without fertilization, and these develop into *drones.* After queen has performed her nuptual flight, drones are driven from the colony and die off. If the population becomes particularly large, the colony "swarms", i.e. shortly before emergence of the young queens, the old queen leads a proportion of the workers away, in order to establish a new colony.

Honey bear: see Kinkajou.

Honey buzzard: see *Falconiformes.*

Honeycomb: see Honeybee, Colony formation.

Honeycomb moths: see Wax moths.

Honeycreepers: see *Parulidae.*

Honeydew:

1) A sugary fluid secreted by plants. There are various causes and sites of H. secretion, e.g. it may be produced from the flower heads of cereal plants due to infection with *Claviceps purpurea.* Insects attracted by this H. then contribute to the dispersal of the fungal spores. Under certain conditions, in particular during pathological metabolic disturbances, liquid containing sugars and other organic material is secreted from leaves by guttation.

2) Sugary excrements of leaf aphids and other insects that feed on plant sap. This H. serves as a food source for many insects, e.g. ants, bees and flies. It also provides a growth substrate for powdery mildews.

Honeyeaters: see *Meliphagidae.*

Honeyguides: see *Indicatoridae.*

Honey locust: see *Caesalpiniaceae.*

Honey mesquite: see *Mimosaceae.*

Honey moth: see Wax moths.

Honey pollen analysis: see Melitopalynology.

Honeysuckle: see *Caprifoliaceae.*

Hooded seal: see *Phocidae.*

Hoof: see Claws, hoofs and nails.

Hook-tip moths: moths of the family *Drepanidae.* See *Lepidoptera.*

Hook worm, *Ancyclostoma duodenale:* a parasitic

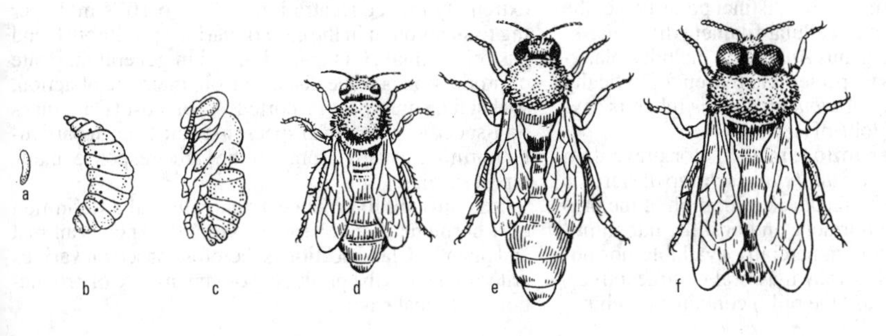

Honeybee (Apis mellifica). a Egg. *b* Larva. *c* Pupa. *d* Worker. *e* Queen. *f* Drone.

nematode (see *Nematoda)* responsible for one of the most common worm infections of humans in the tropics and sub-tropics. Eggs are voided with the host's feces, and they hatch in the soil, where the juvenile worms feed on bacteria and organic detritus. Afer 2 molts, young worms migrate to the soil surface and penetrate the skin of a new host. The worm then migrates from the venous system, via the heart and pulmonary artery, to the lung capillaries, where it may be responsible for temporary inflammation of the lungs. After entering the lung cavity, the adult descends the trachea to the intestine, where it attaches to the intestinal epithelium with the aid of its large, toothed mouth cavity. The worms feed on the intestinal mucosa and blood, causing anemia, although some authors consider that they feed primarily on the intestinal mucosa, accidentally rupturing blood vessels in the process. In severe infections, the large number of worms attached to the intestinal epithelium resembles the pile of a carpet. Females are fertilized in situ within the host intestine.

Hoopoes: see *Coraciiformes.*

Hoop pine: see *Araucariaceae.*

Hopf bifurcation: see Far off equilibrium.

Hoplonemertea: see *Nemertini.*

Hop red spider mite: see *Tetranychidae.*

Hops: see *Cannabaceae.*

Hordeiviruses: see Virus groups.

Hordenin: see Glutelins.

Hordenine, *anhaline, N,N-dimethyltyramine:* a biogenic amine of wide natural occurrence, which is found e.g. in barley seedlings. Like adrenalin and ephedrine, it increases blood pressure but has a low physiological activity.

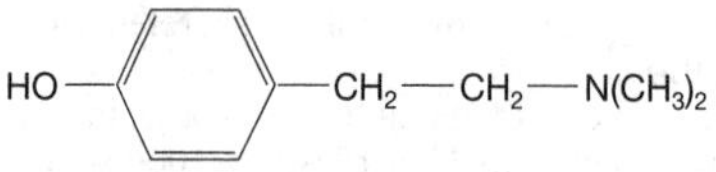

$HO—C_6H_4—CH_2—CH_2—N(CH_3)_2$

Hordenine

Hordeum: see *Graminae.*

Horizontal resistance: see Field resistance.

Hormogonales, *Oscillatoriales:* an order of the *Cyanophyceae.* The cells are associated into filaments, often enclosed in mucilaginous sheaths. The generally unbranched filaments may contain heterocysts (thick-walled, refractory cells, which fix nitrogen), and they may display gliding movements on their substratum. Species of *Oscillatoria* are abundant in all moist habitats, occurring on mud and in water; their filaments consist of similar disk-shaped cells. On the other hand, *Nostoc* forms spherical or irregular masses in water or on damp soil, consisting of necklace-like threads with heterocysts embedded in mucilage. The genus *Rivularia* consists of sessile forms with a distinct polarity, i.e. the base and apex of the multicellular filament differ markedly in structure. The genus *Anabaena* includes planktonic species, as well as species that live endophytically inside other plants, e.g. *Anabaena azollae* inhabits cavities in the upper leaf lobe of *Azolla.*

Hormonal contraception: methods for preventing pregnancy with *ovulation inhibitors,* a group of steroids that inhibit ovulation by feedback inhibition of the production of luteinizing hormone and/or follicular stimulating hormone. Most commercially available hormonal contraceptives (also known as oral contraceptives, and otherwise known as "the pill") contain a combination of a progestin and an estrogen. The progestin is the actual ovulation inhibitor, whereas the estrogen is included to prevent breakthrough (midcycle) bleeding. Combination preparations of a progestin and an estrogen are taken daily for most or all of the cycle. A sequential preparation was introduced in 1965 to mimic more closely the normal rise and fall of estrogen and progestin in a woman's monthly cycle; the first 15 pills contain estrogen only, and the last 5 contain a progestin-estrogen combination. Since these were less effective than the combination products, and showed undesirable side effects, they were removed from the US market in 1976. The "minipill" or POP (progestin only pill) contains a low dose of progestin (0.075 mg norgestrel, or 0.35 mg norethindone) and no estrogen; it is free from many side effects, but users sometimes have irregular bleeding. Many commercially available preparations now contain 0.5–5 mg progestin and 0.03–0.1 mg estrogen per tablet, and one tablet is taken daily for about three quarters of the duration of the cycle. To take just one example of these many preparations: each tablet contains 0.5 mg norethisterone and 0.035 mg ethinylestradiol; one tablet is taken daily for 21 days, followed by 7 tablet-free days. Trials have also been conducted on once-a-month and once-a-week oral contraceptives, and on injectable depot hormonal contraceptives.

Since the natural progestin and estrogens show poor gut absorption, all the steroids used in oral contracptive preparations are synthetic products with efficient gut absorption. Examples of the synthetic progestins used in H.c. are chlormadinone acetate, norethindone, norgestrel, norethynodrel, ethynodiol diacetate, dimethisterone, desogestrel, norethisterone. Synthetic estrogens used in H.c. are ethinylestradiol, mestranol, ethinylestradiol 3-cyclopentyl ether (quingestanol).

For postcoital contraception, high estrogen doses are used, which lead to abortion of the fertilized egg.

Hormones: organic compounds, produced and interpreted within the organism as intercellular signals, which are essential for the normal life processes of nearly all animals and plants. H. are responsible for the regulation and coordination of metabolic processes, and thereby for morphological changes. They induce the biosynthesis of certain enzymes and influence membrane permeability. By acting as chemical signals, they promote or trigger metabolic reactions and physiological processes, but they are not net reactants, i.e. they can be regarded as biocatalysts.

Chemically, H. represent a heterogeneous group of substances. They also display characteristic differences, as well as fundamental similarities, with respect to biosynthesis, storage, secretion, transport, action mechanism and degradation. All H. are encountered in extremely low concentrations (10^{-12} to 10^{-15} mole per mg tissue protein in the case of various pituitary H. and gastrointestinal H. in animals), and in general their site of production is some distance from their site of action, so that they must be transported. Also, most H. are function-specific, rather than species-specific, i.e. a particular hormone usually promotes the same response in different species.

It is difficult to produce a generally valid definition of a hormone that covers all presently known animal and plant H. Classification is therefore based on various criteria, such as the producing organism, site of production, chemical nature.

Thus, a distinction is drawn between Phytohormones (see) and animal H. Of the animal hormones, those of invertebrates are still poorly understood, and only the molting hormones of insects, e.g. Ecdysterone (see) and Juvenile hormone (see), have been thoroughly investigated. Vertebrate H. (those of the human are particularly well investigated) are chemically either steroids (see Steroid H.), peptides or proteins (see Peptide H., Proteohormones), or low molecular mass, amino acid derivatives such as Thyroxin (see) and Adrenalin (see).

H. produced in endocrine glands are called Glandular H. (see), and their synthesis and secretion is generally regulated at a higher level by Glandotropic H. (see). Aglandular H. are produced in paracrine cells (see Tissue H.). Neurosecretory H., like Oxytocin (see), Vasotocin (see) and Releasing H. (see) are synthesized in neurosecretory nerve cells.

Biosynthesis. H. are biosynthesized in specialized secretory cells or glands (Fig.1 of Hormone system).

Secretion and transport. H. are secreted and transported in the free form (peptide H.) or bound to carrier protein (steroid H.), usually in the blood, but also in the lymph or intercellular fluid. After arriving at the target organ, they become bound to a hormone-specific receptor protein.

Action mechanisms. Two different mechanisms of action are known, based on the different cellular location of the hormone receptors.

Specific peptide hormone receptors (e.g. insulin receptor, see Insulin) and adrenalin receptors are membrane-bound with a receptor site on the surface of the target cell. The hormone forms a complex with its receptor, and, as the first stage in signal transmission, causes a change of conformation in the latter. This change in receptor conformation promotes the allosteric activation of an enzyme (e.g. adenylate kinase) on the cytoplasmic side of the membrane. When adenylate kinase is the activated enzyme, it catalyses the conversion of adenosine triphosphate (ATP) into cyclic adenosine 3',5'-monophosphate (cAMP) in the cell interior. cAMP then serves as an intercellular *second messenger* of hormone action, by exerting a specific effect on cell metabolism, such as the activation of a protein kinase. For the role of Ca^{2+} as a second messenger of hormone action, see Calcium.

The specific receptors of steroid hormones and of thyroid hormones are not localized in the cell membrane, but within the target cell. After crossing the cell membrane of the target cell, the thyroid H. form a hormone-receptor complex with the cytoplasmic receptor; this complex is transported into the cell nucleus. Within the nucleus the complex associates with and activates certain gene loci, leading to stimulation of the biosynthesis of specific enzymes, which in turn influence certain metabolic reactions. It was long thought that steroid receptors are present in the cytosol of target tissues, but it is now known that they are located in the nucleus, and that the steroid hormone must cross both the cell membrane and the cytosol before becoming bound in the nucleus. The "cytosolic" receptors are a reproducible artifeact, due to release of the nuclear receptor by homogenization and dilution of the cell contents with homogenizing buffer [J. Gorski et al. *Mol Cell. Endocrinol.* ***36*** (1984), 11–15].

Degradation and inactivation. H. are removed from the circulation and inactivated within a relatively short time, thereby avoiding over-exposure of the organism to hormonal activity. Peptide and protein H. are inactivated by proteolysis or reduction of disulfide bridges. Steroid hormones are generally inactivated by hydroxylation (often at position 3 of ring A), and the inactive metabolites are excreted in the urine, either in the free form or as their glucuronides.

Regulation of hormone action. Hormone action is subject to hierarchical control by positive and negative feedback. In most cases the synthesis and secretion of the first hormone in a system are inhibited (negative feedback) by the hormone whose production it stimulates. A classic example of the coordination of various regulatory circuits is the system composed of the hypothalamus (nervous control of secretion of releasing H.), anterior lobe of the hypophysis (releasing H. promote the release of specific glandotropic hypophyseal H.) and target organ (hypophyseal hormone stimulates release of specific hormone), in which there is feedback at every level. For example, thyrotropin releasing hormone from the hypothalamus stimulates synthesis and secretion of thyrotropin by the anterior hypophysis; thyrotropin stimulates synthesis and secretion of thyroid hormones by the thyroid gland; thyroxin inhibits the secretion of both thyrotropin and thyrotropin releasing hormone.

In animals, the over- or under-production of hormones results in severe disturbances of metabolism, and is usually accompanied by characteristic clinical symptoms.

Hormone system: a communication system, whose signals (hormones) are released into a distribution system (i.e. body fluids such as blood, lymph, hemolymph, coelomic fluid, etc.) by transmitters (hormone-producing cells), then transported to distant target cells where they are bound by receptors as a prelude to promoting cellular reactions. Signals of the H.s. are therefore addressed to cell populations, whereas the signals of the nervous system (transmitters) are addressed to individual cells through synaptic contact between transmitter and receptor. Tissue hormones (paramones) do not depend on transport, because they are produced by, and act on, cells in the same immediate neighborhood. There is no single, generally valid definition of a H.s. In the definition presented here, the same chemical signal may be a true hormone, a tissue hormone, or a transmitter, depending on the sequence of events between signalling and response. For example, the peptide, somatostatin, behaves as a hormone when it is released by neurosecretory cells into the hypophyseal portal system, reaches receptor cells in the adenohypophysis, and inhibits the release of somatotropin. Somatostatin functions as a transmitter, however, when it is released presynaptically in the nervous system and triggers postsynaptic reactions. In the intestine, somatostatin is released from the D-cells of the islets of Langerhans and acts upon neighboring cells; in this case, it behaves as a tissue hormone and inhibits the secretory activity of other endocrine cells.

Hormones can be classified chemically into three types: peptides, low molecular mass amino acid derivatives, and steroids. The prostaglandins, which are derivatives of fatty acids, might represent a fourth type, but these are probably more appropriately considered as tissue hormones.

Chemical signalling is phylogenetically very ancient. At least four of the peptide signals characterized

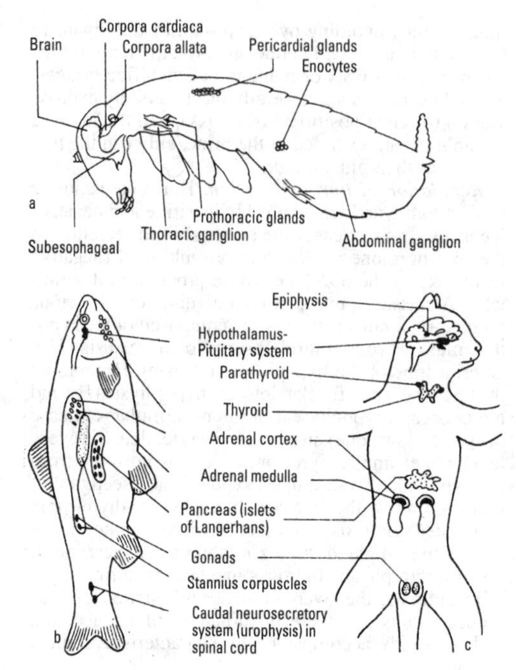

Fig.1. *Hormone system.* Endocrine glands. *a* Insect. *b* Teleost fish. *c* Male cat. (After Penzlinm, 1980)

in the ciliate *Tetrahymena* (insulin, somatostatin, ACTH, β-endorphin) also occur in the H.s. of vertebrates. Furthermore, it is thought that all eukaryotes, as well as many prokaryotes are able to synthesize steroids.

The H.s. is though to have its phylogenetic origin in neurosecretory cells, e.g. of the type found in coelenterates, where nerve cells release peptides which presumably act upon receptor cells in their immediate neighborhood. In the entire animal kingdom, more than 50 hormones have been chemically characterized, which are released by nerve cells or ontogenetic derivatives of nerve cells. The latter include cell populations of the adenohypophysis and glandular cells in the intestine (e.g. islets of Langerhans). In view of the occurrence of peptide signals in protozoa, and for other reasons, the neurosecretory origin of hormones remains, however, a hypothesis. At least most of the hormones of nerve cells and of their ontogenetic derivatives are peptides. Among different vertebrates, each peptide hormone either has an identical structure, or it exists in very similar forms with only minimal differences of amino acid sequence. Numerous peptide hormones first described in vertebrates, have also been demonstrated in invertebrates, sometimes with only small structural differences. Certain other nerve cell hormones are low molecular mass amino acid derivatives, such as indolamines (e.g. 5-hydroxytryptamine) and catecholamines (e.g. adrenalin), which occur in both vertebrates and invertebrates.

Insect steroid hormones are synthesized in the prothoracic gland, e.g. ecdysone. In vertebrates, steroid hormones are synthesised in the gonads (sex steroids) and adrenal cortex (corticosteroids). Specific receptors for a particular hormone may be present in more than one type of cell, and the response is accordingly different. For example, vasopressin causes resorption of water when it acts upon cells of the distal kidney tubule, but also causes constriction of peripheral blood vessels.

Many hormones influence the secretion of other hormones, so that the H.s. can be arranged as a hierarchy (Fig.2). The nervous system occupies a central position in the regulation of the H.s. Neurons of the central nervous system stimulate other neurons (classical neurosecretory cells), which release hormones into the circulation from their nerve endings. These signals bind to the receptors of endocrine cells, and cause the release of glandular hormones. This principle may be extended over several stages, in which case the system is called an "axis", e.g. the hypothalamic-hypophyseal-adrenocortical axis. Such systems are usually subject to internal feedback control.

From this hierarchical system, it is evident that the H.s. and the nervous system are functionally intimately associated. The complex function of organs or individual organisms is therefore only possible if these two systems are efficiently coordinated.

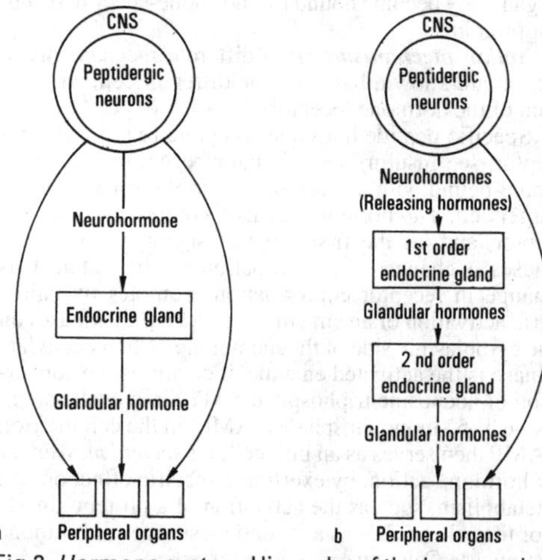

Fig.2. *Hormone system.* Hierarchy of the hormonal system. *a* Hormonal interrelationships in invertebrates that possess endocrine glands (mollusks, arthropods). *b* The multistage system of vertebrates. (Modified from Gersch and Richter, 1981)

Horned frogs: see *Leptodactylidae.*

Horned lark: see *Alaudidae.*

Horned toads:

1) See *Pelobatidae, Leptodactylidae.*

2) The name given to members of the genus, *Phrynosoma,* which are not toads but iguanas (see *Iguanidae),* found in desert and steppe regions of Mexico and western USA. The 6 to 12 cm-long body is flattened and spiny, and the short, broad head carries large horns. They feed mainly on ants, and bury themselves in the sand by lateral shovelling movements. Egg-laying species include the ***Texas horned toad,*** *Phrynosoma coronatum,* while the most northerly species (as far as southern Canada) is the viviparous ***Short-horned toad,*** *Phrynosoma douglassi.*

Horned vipers, *Cerastes:* a genus of sturdy, venomous snakes of the family *Viperidae* (see). They inhabit the North African desert, attain lengths of 60-75 cm. As an adaptation to life in sand, they progress by "sidewinding", i.e. they roll obliquely like a wire spiral by throwing loops of the body sidways. Using the same movements, they can also bury themselves very rapidly in the sand. The ***Horned viper*** *(C. cerastes),* which gives the group its name, has a pointed horn of scales above each eye, which is sometimes lacking in southern Tunisian forms. These "horns" are also absent from the ***Common sand viper*** *(C. vipera).* Both species are able to rattle their scales.

Horns: two sheaths of horny fibers, developed from the epidermis, on the heads of ungulates. Below and extending into each horn is a bony process or horn core which grows from the fontal bone. In contrast to the antlers of cervids, H. are not shed annually, but retained throughout life. Growth occurs by the development of horn rings from the base. H. are usually present in both sexes, but they are often smaller and occasionally absent in females.

Horntails: see *Hymenoptera.*

Horny coral: see *Antipatharia, Gorgonia.*

Horny orb shell: see Freshwater mussels.

Horotely: see Evolutionary rate.

Horse: see *Equidae.*

Horse bean: see *Papilionaceae.*

Horse chestnut: see *Hippocastanaceae.*

Horsehair snakes: see *Nematomorpha.*

Horse mackerel: see Scad.

Horse mushroom: see *Basidiomycetes (Agaricales).*

Horseshoe bats: bats of the 2 families, *Hipposideridae* (see) and *Rhinolophidae* (see).

Horseshoe crab: see *Xiphosura.*

Horseshoe worms: see *Phoronida.*

Horsetails: see *Equisetatae.*

Host: the ecologically necessary partner of a Parasite (see). The parasite lives permanently, or for part of its life cycle, on or in the host, where it feeds and reproduces. In ***brood parasitism*** (e.g. the hatching and rearing of cuckoos by other birds), the host actively provides for the parasite. In all cases of parasitism, the plant or animal host is placed at a competitive disadvantage, as well as possibly being injured physically and/or physiologically. If the host is changed during the life cycle of a parasite, the host in which the parasite reproduces is called the ***final host,*** while the preceding host is known as an ***intermediate host.*** Some parasites have several intermediate hosts, e.g. the Broad fish tapeworm (see), which parasitizes a small crustacean, a small fish, a large fish and finally a human or other mammal. Parasites with intermediate hosts are known as *heteroxenic,* those with a single host as *homoxenic* parasites. *Heteroxeny* is thought to be phylogenetically more advanced than *homoxeny.* The host that plays the main part in the parasite life cycle, i.e. where it is most often encountered, where it resides for the longest period, and where it is most numerous, is called the ***primary host;*** this may be the final or an intermediate host. The term, ***secondary host,*** is not the converse of primary host; it is rather an ***alternative host*** that is not often attacked by the parasite, but which can serve as a host for development of the parasite, if necessary. Secondary or alternative hosts serve to maintain the parasite when the primary host is unavailable. In the absence of the normal host, an alternative or secondary host (also known as a ***reserve host)*** constitutes a reservoir of parasites, enabling renewed spread of the parasite when the normal host becomes available. The existence of alternative hosts therefore complicates the introduction of control measures for parasites. A ***casual*** or ***accidental host*** does not belong to the normally recognized group of hosts, and development of the parasite within such a host is usually restricted and incomplete. A ***transport host*** is usually an intermediate host, and it serves the spread and distribution of the parasite, rather than its nutrition and development. Parasites display varying degrees of ***host specificity.*** Strictly specific parasites have no secondary or alternative hosts, and development is possible only in one type of homoxenic host, or one type of each heteroxenic host. Thus, strict host specificity is exhibited by coccidia of the genus *Eimeria,* by many trematodes, and by certain phytonematodes.

Hottentot apron: a hereditary condition of hottentots and bushmen, in which the labia pudendi minora of the female are often considerably extended, in extreme cases to a length of 20 cm.

Hottentot fig: see *Aizoaceae.*

House cricket: see *Saltatoptera.*

Houseleek: see *Crassulaceae.*

House longhorn: see Long-horned beetles.

House martin: see *Hirudinidae.* See plate 6.

House mite: see *Acaridae.*

House mouse: see *Muridae.*

House rat: see *Muridae.*

House wren: see *Troglodytidae.*

Houting: see *Coregonus.*

Howler monkeys, *Alouatta:* a genus of the subfamily, *Alouatthinae,* family *Cebidae.* H.m. are large (head plus body 570 mm), robust, totally arboreal Cebid monkeys (see) of the infraorder, *Platyrrhini* (see New World monkeys), which are widely distributed throughout the forested areas of Central and South America. The tail is prehensile (600 mm) and naked on the undersurface of its tip. Coloring varies with species, and is indicated in most of the English names: *A. belzebul* ***(Black and red howler),*** *A. caraya* ***(Black howler),*** *A. fusca* ***(Brown howler),*** *A. palliata* ***(Mantled howler;*** black with gold-fringed flanks), *A. seniculus* ***(Red howler)*** and *A. villosa* ***(Guatemalan howler;*** totally black). H.m. are named for their deep roar, which is produced with the aid of a modified hyloid bone in a large swelling beneath the chin. Howling represents a form of territory claim and defense, and occurs at dawn. The vegetarian diet contains varying proportions of leaves and fruits, depending on the species.

Hoya: see *Asclepiadaceae.*

Huchen: see Danube salmon.

Hucho: see Danube salmon.

Huernia: see *Asclepiadaceae.*

Huia: see *Callaeidae.*

Human, *Anthropos, Homo* (plates 8 and 9): the most highly evolved life form on the planet. Zoologically, humans are mammals, but they differ in certain respects from the rest of this class, especially in their upright gait, the use of their hands, and in their possession of a highly differentiated brain. Their social organization, based on their ability to work, speak and think, sets humans apart from all other life forms. They are the only living organisms that consciously modify their natural

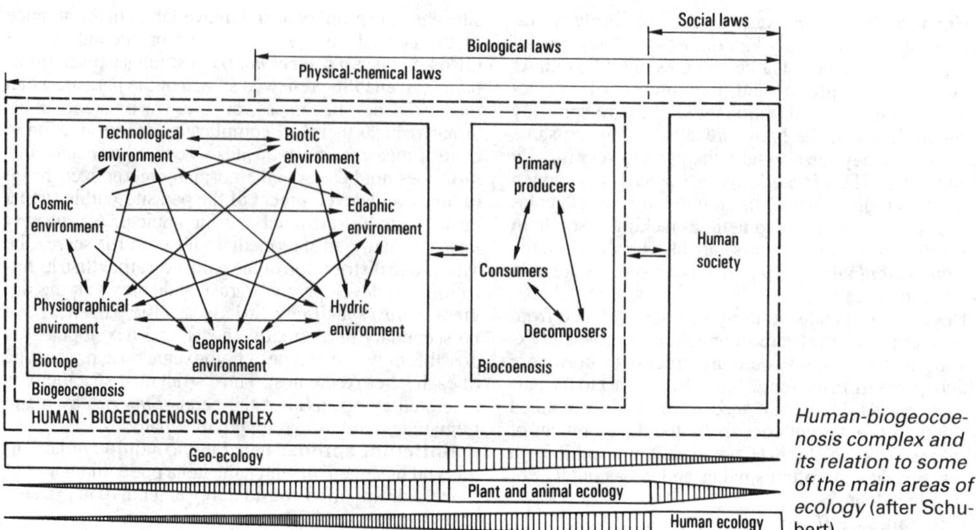

Human-biogeocoenosis complex and its relation to some of the main areas of ecology (after Schubert)

environment, and produce the means of satisfying their own material and spiritual needs. The nearest mammalian relatives of *Homo* are the anthropoid apes, both having evolved from the same phylogenetic root (see Anthropogenesis). The evolutionary and biological study of *Homo* is known as Anthropology (see), while the health and illnesses of *Homo* are the subject of medical science. Human behavior is the subject of psychology. The fundamentally biological aspects of human behavior, like those of other organisms, are the subject of Ethology (see). Finally, the social dimension of human existence is the subject of sociology.

Human-biogeocoenosis complex: a system of natural interrelationships, in which human society exists separately from animal and plant communities, but interacts closely with them. In the course of their phylogenetic development, humans have consciously assumed the role of creators and modifiers of ecosystems, and have thereby generated a new level of organization of living material (Fig.).

Human ecology: science of the interrelationships between humans and the structure and function of their abiotic and biotic environments, which are ever increasingly becoming modified by humans themselves. H.e. studies the system properties of those elements of the biosphere that are relevant to humans, their interactions with each other and with man. The framework of such studies assumes that humans are members of a structured society within an environmental complex, which consists of working environment, a living environment, a relaxation environment, formerly natural environment that is now exploited and modified, and residues of a truly natural environment. Accordingly, H.e. is an interdisciplinary subject, which addresses biological, physiological, human genetic, ethological, demographic, anthropological and sociological questions, as well as problems of environmental protection and town planning. The aim of H.e. is to increase human health, performance and comfort through the physical and psychological affects of an optimal environment. Scientifically based, planned efforts to achieve a human ecological equilibrium are becoming an increasing existential necessity.

Human genetic counselling: a branch of human genetics, which seeks to advise individuals who suffer from an inherited disorder or belong to a family in which the disorder occurs, on the risk of the same disorder occurring in their offspring. Individuals who have been exposed to mutagenic agents, those that have married or contemplate marrying a relative, and older married couples also neeed H.g.c., because in all such cases, there is an increased risk that any offspring will suffer adverse genetic effects.

The aim of H.g.c. is to increase the likelihood that genetically affected parents will not have a child that suffers from a genetic disease or deformity. It depends extensively on a careful analysis of family medical history, and, depending on the perceived degree of risk, uses further measures, such as the Heterozygote test (see), Chromosome analysis (see), investigation of microsymptoms, Prenatal diagnosis (see), etc. H.g.c. is therefore able to offer, on a voluntary basis, a genuine aid to decision making to any couple planning a family.

Human genetics, *anthropogenetics* (plate 27): the study of inheritance in humans. It is concerned with recognition and analysis of the molecular and cytogenetic basis of hereditary information, its expression during ontogenesis, its transmission from generation to generation, and the genetic composition of populations. It attempts to determine the extent to which the physical and psychological variability of individuals in health and illness is genetically determined.

H.g. is subdivided into numerous overlapping disciplines and areas of research, such as biochemical genetics, cytogenetics, serogenetics, immunogenetics, pharmacogenetics, developmental genetics, mutation research, twin research, behavioral genetics, population genetics, medical genetics (inherited disorders). H.g. has made important contributions to the understanding of the basis of numerous diseases, as well as making significant contributions to Genetic prognosis (see), Human genetic counselling (see), establishment of paternity in cases of doubtful or unknown male parentage (see Forensic anthropology), and Twin diagnosis (see).

Human immunodeficiency virus: see AIDS.

Human races: groups within the species *Homo sapiens sapiens* that differ with respect to average body height and proportions, skin pigmentation, facial structure (see e.g. Prognathy), blood groupings, etc. The earliest American fossil humans, dated around 20 000 years ago, are distinctly Amerindian, and on other continents there is evidence for racial differences as early as 35 000 years ago. Possibly racial divergence within *Homo sapiens sapiens* can be dated to about 150 000 years ago, i.e. at the end of the penultimate major glacial advance. During this time, some unknown races may have become extinct, but by 8 000 years ago it is generally agreed that the human species was represented by seven major races: White, Early mongoloid, Late mongoloid, Negro, Bushman, Australian and Pygmy. Further subdivisions of these major groups are contentious and not universally agreed by all authorities.

[Human Variations and Origins, W.H. Freeman and Company, San Francisco and London 1967]

Humic acids: see Humus.

Humid: the adjective applied to the climate of areas where the annual precipitation exceeds the annual evaporation. The soils of these areas are heavily leached, and in colder zones they are characterized by the formation of mor humus (see Humus). The converse is Arid (see).

Humidification: see Misting.

Humification: see Humus.

Humins: see Humus.

Hummingbirds: see *Apodiformes.*

Humming frog: see *Myobatrachidae.*

Hummock: a small domed elevation on the surface of a bog, consisting of bog mosses such as *Sphagnum,* and often colonized by ericaceous plants. Old hummocks cease to grow and become wet hollows, as younger neighboring hummocks grow and rise above them. Thus hummocks and hollows alternate on the same spot as the entire surface of the bog is gradually raised.

Humoral immune reaction: an immune response, involving the formation of antibodies.

Humpback whale: see Whales.

Humus: the totality of dead organic material in or on the soil. In the broad sense, *humification* includes all transformations of dead animals, plants and microorganisms in the soil. In the narrow sense, it is the biochemical synthesis of polymeric, structurally heterogeneous, dark-colored *humins,* which are relatively stable, and which are retained in the soil by binding to clay minerals.

H. may be formed in drained soils (terrestrial H.) or waterlogged soils (semiterrestrial H.), or it may be formed under completely submerged conditions (aquatic or submerged H.). The most well defined terrestrial H. profile is displayed by *mor.* Found mainly in cold and temperate climates, mor arises by slow decay, mainly through the action of fungi (eumycetic or fungal H.), and it forms a deep, dense layer on the mineral soil base (Fig.). Beneath the freshly deposited leaf litter is a layer of slightly altered vegetable material (the L-layer). This is followed by a more strongly decomposed layer, in which plant structures are still recognizable (the F-layer, fermentation layer, or leaf mold layer). This nutrient-rich F-layer contains the highest density of small arthropods (e.g. mites, springtails), which feed on the fungal hyphae. The final layer, next to the mineral soil surface, is the structureless H-layer (Humus layer). Mor is characterized by 1) the sharp division of the H. layer from the mineral soil, due to the absence of appropriate soil fauna (e.g. earthworms) for intermixing the soil and H., 2) the slow rate of humification, resulting in a deep L-layer and a shallow H-layer, 3) acidic conditions, and 4) its formation from components that are relatively resistant to decay, e.g. the mosses and leaf litter of coniferous forests, remains of ericaceous plants on heaths and moors, and to a certain extent the leaf litter of beech and oak.

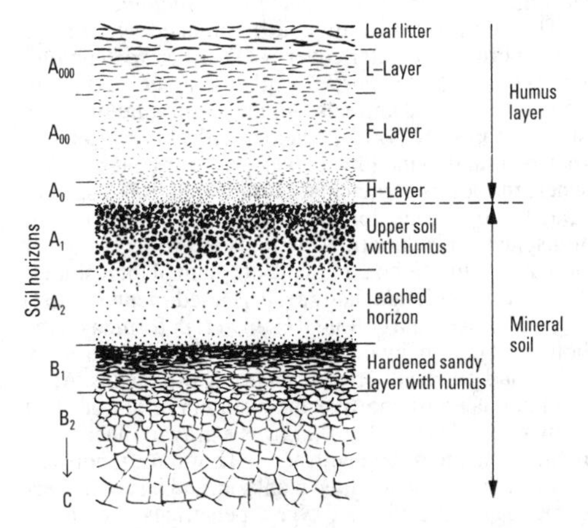

Humus and soil profile of a forest floor with mor

Under less acidic conditions, and with greater participation of fauna (especially arthropods), a different type of H. is formed, known as *moder.* Decomposition is more rapid than in mor formation, the F- and H-layers are deeper, and there is more intermixing with the mineral soil.

Even more rapid breakdown of leaf litter, mainly by bacteria and actinomycetes, with some contribution from earthworms, leads to the formation of *mull* (lumbricid H. or worm mull). This is practically neutral, forms only a temporary H. layer, and is rapidly incorporated into the soil, to form a loose friable mass with a crumbly structure. The resulting soil contains about 10% H., is inhabited by large earthworms, and supports an abundant vegetation of geophytes.

H. formed under very wet conditions (near a permanent ground water table, or in the succession from open water) (semiterrestrial H.) is represented by various forms of *peat.* Peat is formed where oxygen is largely excluded; decay is therefore arrested, and plant remains retain their morphology more or less indefinitely.

Low moor peat is usually formed in nutrient-rich water that contains plentiful calcium, e.g. water that is draining slowly from higher regions. Dominant constituents are the remains of various sedges and reeds, together with certain trees and shrubs, and *Sphagnum* is usually absent. According to the predominant plant, it is classified as *Carex* peat, *Phragmites* peat, *Cladium* peat, etc. Chemically, it has high ash and high nitrogen contents, but a low cellulose content and a low acidity.

High moor peat is characteristically formed in cold or temperate climates with a high rainfall. If the water is not largely provided by atmospheric precipitation, then it is provided by drainage from nutrient-poor soils with low concentrations of calcium. The dominant plants are *Sphagnum, Calluna, Ledum, Andromea* and *Eriophorum.* High moor peat has a high acidity and a high cellulose content, and low contents of ash and nitrogen.

The chemical composition of *forest peat* is intermediate between that of low moor and high moor peat. The water is provided partly by ground waters and partly by atmospheric precipitation. Various trees (beech, oak, alder, pine) are the dominant vegetation, together with plants such as *Andromeda, Calluna,* etc., and the floors of some forests may be carpeted with *Sphagnum.*

The primitive and little differentiated H. formed under permanent water cover is known as *sedimentary* or *lake peat.* It consists mainly of algae and other aquatic plants and animals, with an admixture of insect shells, spores and pollen, together with clay particles. It may be found in the lower layers of other peat profiles, where the peat is permanently submerged. It is also generated independently on the bottoms of lakes and ponds. It displays wide variations with respect to its composition of calcium carbonate, sand, clay and plant residues, as well as its colloidal nature. *Dy* is a sediment, consisting of H. flakes and colloidal H. material, formed on the bottom of extremely nutrient-poor waters. It is strongly acidic and practically free from living organisms. After drying, it is nonsupportive and even inhibitory of plant growth. *Gyttja* (Swedish *Jüttja)* is a gray to black sediment formed in waters with aerated bottoms. It consists mainly of dead, sedimented plankton, and it is rich in organic materials. Since oxygen penetrates the upper layers of the sediment, it has a richly developed fauna. *Sapropel* is a malodorous, slimy sediment, colored deep black by iron sulfide. It is formed on the oxygen-poor bottoms of nutrient-rich waters, primarily in stably layered meromictic lakes. Organic material is degraded anaerobically by bacteria to intermediate products like methane, etc. In contrast to Gyttja, it contains little or no fauna.

A distinction is drawn between nonhumin materials (plant remains and their immediate decay products) and humins, which are formed secondarily from the latter in the soil. Humins represent a series of compounds of ever increasing depth of color, degree of polymerization, and particle mass, with decreasing solubility and tannin character. The series starts with fulvic acids (crenic and apocrenic acids) and finishes with true humic acids (hymatomelanic acids, brown humic acids). These, however, are not definite substances, but fractions defined by solubility properties. Humic acids may be synthesized from lignin (which is converted into lignohumic acids), but the starting material for humic acid synthesis is more commonly the metabolic products of microorganisms, etc. in the soil. Humin synthesis is usually a biological process. Dead material is disintegrated by passage through the alimentary canals of different organisms, admixed with inorganic salts, then subjected to microbial activity. The products of this microbial activity become autoxidized and/or polymerized to form complex humins. Nonbiological processes make a major contribution to humin production in e.g. peat and mor formation. Acidic humins, which are not bound to minerals, are easily leached from the upper soil layers and they form a deep horizon (horizon B) in the lower layers of the soil. The most stable and beneficial (for plant growth) soil structures are formed by the binding of humins to clay minerals (clay-H. compex) to give a friable, crumbly structure.

Nutritive H. (usually nonhumin material) can be rapidly converted into soluble plant nutrients, while *reserve H.* decomposes only slowly and helps to maintain the soil crumb structure. In addition, reserve H. has a high binding affinity for minerals and is important as a nutrient trapping agent and carrier.

The H. content of the soil is approximately equal to its content of organic materials, which is determined from the loss in mass of dry soil when it is ignited. Many colorimetric tests have been used to determine the humin content of soils, in which soil samples are extracted with solutions of sodium carbonate, sodium hydroxide or ammonia, and the color intensity of the extract is measured. Alternatively, an alkali extract can be filtered then acidified, and the precipitate ("humins" or "humic acid") collected and weighed. Acetyl bromide dissolves practically all the organic constituents of soil, but not the dark-colored complexes ("humins"). Thus, soil is extracted briefly with diethyl ether and alcohol, dried at 105 °C, then extracted in a soxhlet with acetyl bromide for 3 days. The dry residue is weighed, ignited and weighed again, the difference between the last 2 weights representing the humin content.

Hun: see *Phasianidae.*

Hungarian: see *Phasianidae.*

Hunger: perception of the need for food; it gives rise to a complex sensation in humans, and instinctive feeding behavior in animals. H. has a nervous and a hormonal basis. An empty stomach undergoes rhythmic contractions, which stimulate fibers of the autonomic nervous system and generate a gnawing feeling of H. Filling of the stomach signals the satisfaction of H. via stretch receptors of the vagus nerve. Section of the vagus leads to an excessive food intake. Injection of insulin or thyroid hormones causes an intense craving for food, whereas injection of adrenalin generates a feeling of satiety; the increased blood sugar concentration caused by adrenalin stimulates hypothalamic glucose receptors which signal satiety.

The hypothalamus contains regulatory centers for food intake. Destruction of the hunger center (nucleus ventralis lateralis) results in refusal of food and emaciation, whereas its stimulation generates a lust for eating. Blockage of the satiation center (nucleus ventralis medialis) leads to increased food intake and a rapid increase of body mass, whereas its stimulation causes food refusal.

Hunters: see Modes of nutrition.

Hurter's spadefoot: see *Pelobatidae.*

Huso: see *Acipenseridae.*

Hutias: see *Capromyidae.*

Hutschinsoniella: see *Cephalocarida.*

Hyaena: see *Hyaenidae.*

Hyaenidae: a family of terrestrial carnivores (see *Fissipedia).* They are large, long-legged mammals with a posteriorly sloping back, which usually carries an erect mane. The African ***Spotted hyena*** *(Crocuta crocuta;* head plus body 95–166 cm, tail 25–36 cm) has a relatively large braincase, greatly reduced upper molars, relatively short ears and a pale, spotted coat; its mane may be only slight or absent. With its powerful jaws the spotted hyena can bite through and crush the strongest bones. In addition, it has a specialized digestive system, which enables it to utilize much of the bone, as well as the flesh of its prey. Spotted hyenas hunt cooperatively in packs, preying on medium-sized ungulates, mainly young, sick or old wildebeest *(Connochaetes),* and sometimes attacking humans. It is practically impossible to distinguish between male and female in the field, because even the external genitalia are very similar in the two sexes.

The ***Striped hyena*** *(Hyaena hyaena;* head plus body 103–119 cm, tail 26–47 cm) inhabits open country in Africa, Arabia, Central Asia and India. Mainly nocturnal or crepescular in habit, it has a long, erectile mane, prefers to be near to fresh water, and avoids deserts. Although its diet is mainly mammalian carrion, it occasionally preys on domestic sheep and goats.

The ***Brown hyena*** *(Hyaena brunnea;* head plus body 110–136 cm, tail 19–26.5 cm) is an endangered species. Small scattered populations exist in the southern half of Africa, and it is most numerous in Botswana. It has a long, coarse, shaggy brown coat, with a gray head and tawny neck and shoulders. The lower legs and the feet have dark brown bars. Mainly nocturnal and crepescular in habit, it is found in open scrub, grassland and semidesert, where it feeds by scavenging the prey of larger predators. Rodents, insects, eggs and fruit may also be taken.

See Aardwolf.

Hyaenodontidae: see *Creodonta.*

Hyaloplasm: see Cytoplasm.

Hyalospongiae: see *Porifera.*

Hybodontiformes: see *Selachimorpha.*

Hybrid: see Hybridization.

Hybrid atavism: see Atavism.

Hybrid cells: see Cell hybridization.

Hybrid complexes: groups of species in which the identifying morphological characters of the diploid starting types have been largely lost during hybridizatrion. H.c. have made an important contribution to the evolution of higher plants, and they are frequently characterized by marked taxonomic complexity.

Hybridization, *crossing:* the fusion of genetically different gametes, leading to the development of heterozygous zygotes and heterozygous organisms. The latter are commonly known as *hybrids,* and are also referred to as *bastards.* (Hybrids may also arise from mutations in a homozygous zygote). H. may occur naturally *(spontaneous H.),* or it may be promoted artificially. Depending on whether the genetic differences between the parents are determined by 1, 2 or more allele pairs, the resulting hybrid is called a mono-, di-, or polyhybrid. Heterozygosity may also be due to differences in chromosomal structure between homologous chromosomes (see Structural hybrids). The heterozygous allele pairs of the hybrid display Mendelian segregation (see) when they separate in the next generation. In an *introgressive H.,* new genetic material is incorporated stepwise by species H. and subsequent backcrossing. Hybrids may possess characters that are lacking in both parents, or they may surpass their parents in certain characters (see Transgression, Hybrid vigor). H. is almost always possible between members of a single species, e.g. different races of a species. H. of different species is more difficult, e.g. lion x tiger. *Species hybrids* are usually fertile, whereas *Generic hybrids (see)* are seldom fertile. Thus, the mule, which is a hybrid from a male horse and a female donkey, is infertile. In some plants, e.g. maize, rye, millet, the endosperm as well as the seed may also have hybrid characters; one nucleus of the pollen tube fuses with the ovum to form the zygote, while the other fuses with the secondary embryo sac nucleus to form the endosperm nucleus (see Xenies). The discovery that H. can occur among prokaryotes and their viruses (bacteriophages) made an essential contribution to the development of Microbial genetics (see) and Molecular genetics (see).

Hybridogenesis: reproductive system of hybrid fishes of the cyprodont genus, *Poeciliopsis.* Three hybrid forms are found in Mexico, each with one set of chromosomes from *P. monacha,* and a further set from either *P. lucida, P. occidentalis* or *P. latidens.* All hybrids are females. During meiosis, the eggs receive only the *monacha* genome. The eggs are fertilized by bisexual male species *(P. lucida, P. occidentalis, P. latidens)* that live with the hybrid population. Since the paternal and maternal genes do not undergo recombination in the hybrid female, the maternal *monacha* chromosonal complement is always transmitted unchanged.

H. also appears to operate in the more complicated and locally variable reproduction system of the frog genus, *Rana,* of which there are 3 forms: 2 true species, *R. lessonae* and *R. ridibunda,* and their hybrid, *R. "esculenta".* Some populations contain only female hybrid *R. "esculenta"* with males and females of *R. lessonae.* Female *R. "esculenta"* is perpetuated by mating between the female hybrid and male *R. lessonae* without mixing the phenotypes.

Hybridoma: a hybrid cell, in which one of the parent cells is a tumor cell. In practice, for experimental purposes and monoclonal antibody production, Hs. are produced between T- or B-lymphocytes and appropriate myeloma cell lines.

An example of the technique for preparing Hs. is as follows. A mouse is immunized to stimulate proliferation of the desired B-lymphocytes. The latter are isolated from the spleen and fused with mouse myeloma cells (lymphocarcinoma cells maintained in tissue culture), using polyethylene glycol to promote fusion. The chosen myeloma cell line no longer produces antibodies; it is also a mutant lacking the enzyme hypoxanthine-guanine phosphoribosyl transferase (HGPRT). The resulting Hs. are able to grow on a medium containing hypoxanthine, aminopterin and thymidine (HAT medium), which supports the growth of neither the parent B-lymphocyte nor the parent myeloma cell. Aminopterin blocks one pathway of nucleotide synthesis, so that the Hs. grow on a salvage pathway dependent on HGPRT, which was contributed by the B-lymphocyte. All the myeloma cells therefore die in the HAT medium because they cannot synthesize nucleic acid precursors, and the B-lymphocytes die because, unlike myeoloma cells, they are not immortal; only true Hs. survive and grow. H. clones are then screened and selected for the production of a particular monoclonal antibody.

The technique can also be applied to human cells. Hybridomas combine the properties of both parent cells, i.e. rapid growth and immortality, and antibody production. Monoclonal antibodies produced from hybridomas are an extremely important tool in clinical diagnosis and in biological and medical research, and may ultimately prove useful in therapy. Hybridomas have also made a considerable contribution to the elucidation of the structure and function of immunoglobulins, and the elucidation of DNA and RNA processing.

Hybrid sterility: the sterility of an organism resulting from its hybrid nature. H.s. may be genetic and/or chromosomal in origin. See Hybridization.

Hybrid vigor: see Heterosis.

Hydatid disease (Hydatidosis): see Dog tapeworm.

Hydnaceae: see *Basidiomycetes (Poriales).*

Hydnum family: see *Basidiomycetes (Poriales).*

Hydothode: see Excretory tissues of plants, Guttation.

Hydra: a freshwater polyp of the order, *Hydroida.* There is no medusoid stage, and colonies are not formed. There is a mouth cone (hypostome) and a circle of about 6 tentacles around the base of the cone. Body (stalk) lengths vary from a few mm to 1 cm and more, but the diameter is rarely greater then 1 mm. *H.* is found on aquatic plants in stagnant or slowly flowing water. It is attached by its basal disk and is essentially sessile, but it can move its location by looping like an inch worm, i.e. the tentacles bend over until they contact the substratum, and the basal disk is released, followed by a somersault. The diet consists mainly of water fleas and other small crustaceans. Nematocysts (see) in the tentacles hold and paralyse the prey, which is then brought to the mouth by the tentacles. *H.* actively reproduces asexually by budding, but sexual reproduction occurs especially with the approach of winter. Most species are dioecious. The germ cells arise mainly from interstitial cells, which aggregate in the epidermis to form testes in the upper half of the stalk, ovaries in the lower half. Eggs are fertilized in situ by water-borne spermatozoa. The encapsulated embryo detaches from the parent, remains dormant through the winter, and hatches into a young *H.* in the spring. [N.B. This type of embryological development is not typical of the majority of cnidarians, which develop via a planula larva].

Hydracarina: see *Hydrachnella.*

Hydrachnella, *Hydracarina, **water mites:*** a group of about 2 800 species of mites (see *Acari),* which have become secondarily adapted to an aquatic existence in both marine and freshwater habitats. The group consists of several families, which have in common their aquatic existence, but are otherwise not necessarily closely related. Almost all are predatory, feeding on small crustaceans by piercing them with their chelicera and sucking out the body contents. Some species can swim, but most remain on the surfaces of submerged structures.

Hydranth:

1) The head of a hydropolyp, which carries one or several rings of tentacles around an oral cone.

2) A single polyp head of a hydroid polyp colony.

Hydration:

*1) **Imbibition, swelling:*** reversible, noncovalent association of high molecular mass substances (e.g. plasma colloids) with water with a resulting increase in volume. H. is a purely physical chemical process, displayed by both living and nonliving material, e.g. gelatine can be made to repeatedly and reversibly swell and contract by alternate H. and dehydration. H. consists of the penetration of water between the macromolecules. Since water molecules have a nonuniform distribution of charge, they have the property of a dipole, and are attracted by both the negative and the positive groups of the macromolecule. By association with the hydrophilic groups of the macromolecule, the water molecules lose part of their kinetic energy, and this is converted into heat (heat of hydration). This heat production is easily measured, e.g. in seeds that are imbibing water. In protoplasm, most of the water is held by hydration. In the cell wall, in addition to H. (primarily association of water with the charged groups of protopectins and hemicelluloses), water is also held by capillary action between the microfibrils and in the intermicellar spaces.

The water potential in the hydrated material is decreased both by electrostatic attraction of the water dipole by the charged groups of the macromolecule and by capillary forces, so that a gradient of water potential exists between the hydrated material and its surroundings. The forces of H., especially in the early stages of the process, are extraordinarily large, amounting to several hundred bar. Thus, hydratable materials, e.g. dry seeds, are used to carefully force open bone sutures in the preparation of skulls. Swelling wood can even split rocks. The entry of water into a hydratable material can be prevented by the application of sufficient force to prevent swelling.

Although macromolecules or molecular aggregates (micelles) are forced apart by imbibed water, they may remain bound by intermolecular forces, forming a network with water-filled internal spaces (e.g. cellulose and starch). Alternatively, the individual particles may be totally separated by water to form a colloidal sol (e.g. gum arabic, egg white, cytoplasmic proteins). Water molecules are then bound (water of hydration) to the particles of the disperse phase. These *hydration* or *solvation shells* maintain the particles in colloidal suspension. By removal of their charge or by withdrawal of water, disperse, hydrophilic colloids can be converted into semisolid gels.

H. is strongly affected by inorganic ions, which can actually bind and withdraw water from the hydrated macromolecule. Thus, proteins can be salted out of solution, e.g. by ammonium sulfate. In addition, inorganic ions may weaken the charge of the hydrated macromolecule, or (after adsorption to the macromolecule) they may reinforce the charge and considerably alter the water-binding properties. Crucial factors in this effect of inorganic ions are the strength and sign of the charge on the ion, the diameter of the ion and the thickness of its solvation shell. Cytoplasmic proteins carry a predominantly negative charge, and positively charged cations generally act by decreasing this charge, i.e. in the presence of positively charged cations, the H. of cytoplasmic proteins is less than in pure water. In contrast, anions favor the H. of cytoplasmic proteins. Divalent ions have a stronger dehydration effect than monovalent ions. H. is inhibited more strongly by Ca^{2+} than by K^+. If both ions are present together, Ca^{2+} exerts a strong dehydration effect, whereas K^+ exerts a certain H. effect, due in particular to the large solvation shell of the K^+ ion. A series of ions arranged in order of their effect on H. is called a Lyotropic series (see).

2) A chemical term used for the addition of the elements of water without cleavage, e.g. the addition of water across the double bond of an alkene to form an alcohol is H.

Hydroactive stomatal movements: see Stomatal movements.

Hydrobates: see *Procellariiformes.*

Hydrobatidae: see *Procellariiformes.*

Hydrobiology: the study of aquatic organisms and their environmental relationships. *Hydrozoology* is concerned with aquatic animals, *hydrobotany* with aquatic plants. *Applied H.* is a specialized branch with relevance to the management of water supplies, waste water disposal, fish farming, etc.

Hydrocaryaceae: see *Trapaceae.*

Hydrochasy: opening movements of fruits and fructescences, which occur when they become damp. See Hygroscopic movements.

Hydrochoeridae, *capybaras, water hogs:* a family of Rodents (see), represented by a single species, the ***Capybara*** or ***Water hog*** *(Hydrochoerus),* which inhabits the banks of marshes and streams in Panama and northern South America east of the Andes. It is the largest living rodent, weighing up to 50 kg, with a plump body, which is about 1 m in length. The legs are relatively short, and the head disproportionately large with a deep rostrum and truncate snout. A vestigial tail is present, and the ears are short. Capybaras are herbivorous, browsing the lush vegetation of their semi-aquatic, tropical habitats. The upper and lower third molars are much larger than the other cheek teeth in their respective rows, and they consist of transverse lamellae with interlamellar cement. Dental formula: 1/1, 0/0, 1/1, 3/3 = 20. Capybaras are excellent swimmers and divers, and their strongly built digits are partly webbed. They hide from predators by immersing themselves among aquatic vegetation with only their nostrils above the water.

Hydrochoerus: see *Hydrochoeridae.*

Hydrochory: see Seed dispersal.

Hydrocoel: the middle of the three coelom sections of echinoderms. It develops from the central section of the paired primordium of the secondary body cavity. The left H. develops into the water vascular system, whereas the right H. degenerates and is totally absent. See Mesoblast.

Hydrocorallina: see *Millepora.*

Hydrocorisae: see *Heteroptera.*

Hydrodamalis: see *Sirenia.*

Hydrodynamics of flow: see Flow hydrodynamics.

Hydrodynamics of blood circulation: the nature of blood flow through the vascular system and its dependence on physical quantities and physiological factors. Within the circulation, flowing blood is layered. The innermost layer has the highest velocity, the flow rate decreases continuously toward the vessel walls, and the velocity profile is a parabola. This laminar flow is a prerequisite for optimal blood transport. A flattening of the flow profile, or the presence of turbulence cause flow resistance and poor circulation.

A Newtonian (homogenous) liquid (e.g. water) under a constant pressure gradient in a rigid tube obeys Poiseuille's law: $V/t = (\Delta P \times \pi \times r^4)/(8 \times L \times \eta)$, where V/t is the volume flowing per unit time, ΔP the driving presure, r the vessel radius, L the vessel length, η the viscosity, and 8 an empirical factor. Essentially, this equation shows that the quantity of liquid flowing per unit time inceases linearly with pressure, but is proportional to the fourth power of the radius. For example, doubling the pressure doubles the flow rate, but doubling the radius increases the flow rate 16-fold. This physical law underlies the physiological regulation of blood flow, which is largely achieved by changes of vessel diameter.

In the animal vascular system, Poiseuille's law is usually not applicable, because the vessel walls are elastic, blood is not a homogeneous liquid (it consists of plasma and blood cells), and the driving pressure is applied rhythmically. In the normal pressure range there is an exponential relationship between blood pressure and blood flow (Fig.). Depending on the structure of a vessel (elastic, contractile, plastic), it opens at different pressures and displays different flow characteristics. When maximally expanded, most vessels show a linear relationship between blood pressure and flow volume, i.e. only then does the flow obey Poiseuille's law. Vessels in certain vital organs, like the kidney cortex and brain, are capable of autoregulation, i.e. they maintain a constant rate of blood flow at different pressures; stretching of the vessel walls by increased blood pressure causes depolarization and therefore contraction of the smooth muscle in the vessel wall.

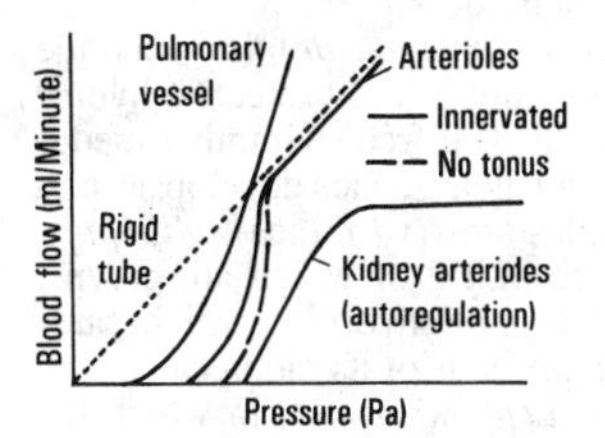

Hydrodynamics of blood circulation. Pressure-volume diagram for blood flow in different vessels.

Hydrogamy: see Pollination.

Hydrogenation: catalytic (including enzymatic) addition of hydrogen to unsaturated carbon-carbon bonds or other unsaturated systems, e.g. nitriles, nitroso compounds and nitro compounds. Catalytical hydrogenation of liquid plant fats is performed on a commercial scale for the manufacture of margarine. In this process, hydrogenation of the double bonds of the unsaturated fatty acid residues raises the melting point of the fat, so that it becomes more solid at room temperature.

Hydrogen, bacterial oxidation of: see Bacterial oxidation of molecular hydrogen.

Hydrogen ion concentration: the concentration of free hydrogen ions, H^+ (more exactly the concentration of hydronium ions, H_3O^+) in an aqueous solution, expressed as the gram equivalents of hydrogen ions per liter. The H.i.c. determines the acidity or basicity of that solution. Following a usage introduced by the Danish biochemist, S.P.L. Sörensen, hydrogen ion concentrations are now usually expressed as pH-values, i.e. the negative logarithm of the gram-equivalents of hydrogen ions per liter: $pH = -\log_{10}[H^+]$. The pH of pure (neutral) water is 7.0, i.e. the dissociation of pure water into H^+ and OH^- ions produces a H^+-ion concentration of 10^{-7} mol/l. The pH of acidic solutions is less than 7, whereas the pH of alkaline solutions lies between 7 and 14. This method has the advantage that all degrees of acidity or alkalinity, from that of a solution containing 1 gram-equivalent of hydrogen ion per liter to that of a solution containing 1 gram equivalent of hydroxide ion per liter, can be expressed by a series of positive numbers from 0 to 14.

pH is an extremely important factor in biochemical reactions. For example, each enzyme has an optimal pH for activity, and relatively small deviations from this optimal value result in sometimes dramatic decreases in activity.

pH can be measured by observing the color changes of indicators, and by electrometric methods using various electrodes. The ultimate reference standard for all pH measurements is the hydrogen gas electrode. Other electrodes include the quinhydrone electrode and the glass electrode.

Hydrogen peroxide, ***H_2O_2*:** a powerful cell poison produced in small quantities in plant and animal tissues by biological redox reactions. It does not accumulate, because it is destroyed immediately by the action of the enzyme catalase and by peroxidases.

Hydroida: an order of the class, *Hydrozoa.* Most members display a marked alternation of generations. The polyps usually form colonies, from which medusae are produced by budding. The medusae, which often remain attached to the polyp colony, reproduce sexually. In a few forms (e.g. *Hydra),* there is no medusoid generation. Most forms are marine, with only a few freshwater species.

H. are divided into 2 suborders:

1) *Athecate hydroids (Anthomedusae),* in which the periderm of the polyp does not form a protective hydrotheca around the hydranth. The free swimming medusae have no statocysts, and their gonads develop around the manubrium. Examples are *Hydra* (see), *Millepora* (see), *Cordylophora* (see), all 3 of which are polyps; *Sarsia* is a medusoid thecate hydroid, which buds daughter medusae from the ends of its radial canals.

2) *Thecate hydroids (Leptomedusae),* in which the polyp periderm forms a protective hydrotheca around the hydranths. The free swimming medusae usually possess statocysts, and their gonads develop on the radial canals. Examples are the polyps, *Campanularia, Laomedea* and *Sertularia* (see), and the medusa, *Obelia.*

Hydrolabile species: see Water economy.

Hydrolases: enzymes that catalyse the Hydrolysis (see) of ester bonds, glycosidic bonds, peptide bonds, amide bonds or acid anhydride bonds, as well as catalysing the hydrolytic cleavage of carbon-carbon, carbon-nitrogen and phosphorus-nitrogen bonds. See Esterases, Glycosidases, Proteases.

Hydrolysis: a reaction in which a chemical bond (e.g. in an ester, peptide, amide, glycoside) is cleaved by addition of the elements of water: $AB + H_2O \rightarrow AOH + HB$. Hydrolytic cleavage is very important in the metabolism of living organisms, where it is catalysed by enzymes known as Hydrolases (see). H. of neutral fats, proteins and carbohydrates is particularly important (see Digestion).

Hydromedusa: see *Chelidae.*

Hydronastic response of stomata: see Stomatal movements.

Hydropassive stomatal movements: see Stomatal movements.

Hydroperoxidases: a collective name for the oxidoreductases, Peroxidase (see) and Catalase (see). The prosthetic group is hemin.

Hydropetes: see Chinese water deer.

Hydrophasianus: see *Jacanidae.*

Hydrophiidae, ***sea snakes:*** a family of entirely aquatic Snakes (see) containing about 50 species. All those that have been investigated posses glands in the head for the elimination of salt. Scales are practically uniform over the entire body (less so in the primitive subfamily, *Laticaudinae),* and there are no broad ventral scutes of the type seen in terrestrial snakes. The laterally compressed tail serves as an oar for swimming. The trunk, particularly the rear portion, is also often laterally compressed. Up to 2.75 m in length, the entire body is usually conspicuously patterned with tranverse colored bands. Situated on the upper surface of the head, the external nasal openings can be closed by folds of skin. The flesh around the teeth contains a dense network of blood vessels, which serves a gill-like function, enabling the *H.* to remain submerged for long periods. *H.* have proteroglyphic dentition and dreaded venomous teeth, as well as stout gripping teeth in the lower jaw. The venom has neurotoxic activity like that of cobra venom, and it is used to paralyse the prey, which consists of fish, especially members of the *Anguilliformes* (eels and allies). Humans have been known to die of the bite, and most at risk are pearl divers. In Southeast Asia, *H.* are caught as food and for their skin. All reports of larger, giant *H.* are untrue, and probably due to confusion with dolphins and large fish. Also deceptive are the often kilometer-long aggregations of sunbathing *H.* on the sea surface. Most species prefer shallow coastal waters.

Of the 2 subfamilies, the more primitive is the *Laticaudinae* ***(Banded sea kraits),*** which have broad scales on the body and tile-like scales on the belly. Members of this subfamily have markedly overlapping scales, and they still seek out the land (islands and coral beaches) to lay their eggs. The common *Laticauda semifasciata* of coastal waters in the Pacific attains lengths of up to 1.9 m; it has a laterally compressed tail, but the trunk is still cylindrical. In contrast, the more advanced *Hydrophiinae* have smooth, uniform, nonoverlapping scaling; they never go onto the land, and they give birth to live young in the water. The rear part of the body is oar-like and usually broader than the front part, and the nasal openings are not as far back as in the *Laticaudinae. Hydrophis cyanocinctus* ***(Blue-banded sea snake)*** inhabits the Asiatic and Indo-Australian coasts. *Pelamis platurus* ***(Yellow-bellied*** or ***Pelagic sea snake),*** which lives on the high seas, is distributed from the American west coast through the Pacific and Indian Oceans to East Africa. The entire length of its body is laterally compressed, and colored black on the upper side and yellow on the underside. Since its range is limited by the colder waters of the southern oceans, it cannot pass Cape Horn or the African Cape, and is therefore not found in the Atlantic. Sections of the Panama canal are controlled by freshwater locks, thereby preventing its colonization of the Atlantic by that route. Entry via the Suez canal has not occurred, because *Pelamis* apparently does not venture into the Red Sea. In view of global warming, and the possibility that sea water may be employed in the enlargement of the Panama canal, it is though to be only a matter of time before *Pelamis* enters the Atlantic.

Hydrophiinae: see *Hydrophiidae.*

Hydrophis: see *Hydrophiidae.*

Hydrophytes, ***water plants, aquatic plants:*** all nonplanktonic plants that liver permanently in water, and whose resting and propagation organs are submerged during the nonvegetative season. Since H. obtain CO_2, O_2 and nutrient salts directly from the water, their submerged shoots have thin epidermal walls and a weakly developed cuticle. The epidermis contains chlorophyll, and its stomata are usually small. In most H., the leaf parenchyma is not differentiated into spongy and pallisade layers, but consists of large parenchyma cells with an abundant system of intercellular spaces. Known as *aerenchyma,* this ventilated tissue serves for gas exchange within the plant, as well as conferring buoyancy. Water conducting vessels are often absent, and mechanical tissue is also largely superfluous. The relatively low concentration of nutrient salts and the low rate of gas diffusion in water are compensated by an increase in the surface area of the submerged leaf laminas, which are usually very delicate, thin and rich in sap, and often perforated by holes and slits, e.g. *Myriophyllum* (water milfoil), *Ceratophyllum* (hornwort), *Utricularia* (bladderwort), *Ranunculus aquatilis* (water crowfoot). In fast flowing water, the leaf lamina is often ribbonlike, e.g. *Zostera* (eel grass), *Ranunculus fluitans* (water crowfoot), *Vallisneria.*

Water plants can be subdivided into 1) free-floating plants, which may be rootless and submerged (hornwort, bladderwort, water milfoil) or with nonattached roots and floating on the surface like *Lemna* (duckweed) and *Hydrocharis* (frogbit); 2) plants well rooted in the bottom substratum, which may be submerged, like eel grass and certain species of *Potamogeton* (pondweed), or with floating leves, like *Nymphaea* (water lilies), *Nuphar* (water lilies) and *Ranunculus aquitilis,* as well as amphibious plants with aquatic and terrestrial forms, like *Polygonum amphibium* (amphibious bistort).

Hydroponic culture (Hydroponics): see Water culture.

Hydrosaurus: see Sail fin lizard.

Hydroschendyla: see *Diplopoda.*

Hydrostable species: see Water economy.

Hydrostatic pressure: the pressure in a resting liquid which acts at right angles on the vessel walls and on the surface of submerged bodies. It is measured in bar (formerly in atmospheres). Osmotic pressure (see Osmosis) is a H.p. and obeys the same rules.

Hydrotaxis: directional movement of freely motile organisms is response to moisture and humidity gradients, e.g. the plasmodia of myxomycetes exhibit H. See Taxis.

Hydrotheca: Protective envelope around the hydranths of thecophore polyps. See Theca.

Hydrothermal regime: specific conditions generated by the joint effects of temperature and humidity at the soil surface and in the soil. The H.r. is crucially important for the existence and distribution of soil fauna.

Hydrotropism: a special type of chemotropism, in which the direction of growth of a plant organ is determined by a gradient of water vapor pressure. H. occurs only in growing plant organs. *Positive H.* is particularly well developed in the roots of higher plants and in the rhizoids of liverworts and fern prothalli, causing these structures to curve and grow towards regions of adequate water availablilty. Stimulus susception occurs only at the extreme tip of the root. Negative H. is relatively uncommon, but is observed in the sporangiophores of certain molds, e.g. *Phycomyces,* and in the thalli of liverworts.

Hydroxyproline, *Hyp:* a naturally occurring cyclic amino acid, formed by the hydroxylation of proline residues in collagen, i.e. by post-translational modification of the protein, and not by hydroxylation of free proline. Hydroxylation occurs in position 4 or 3, and 4Hyp is the more common derivative in collagen. 3Hyp is also a constituent of *Aminata* toxins.

HO, COOH, N, H

Hydroxyproline

Hydrozetes: see *Oribatei.*

Hydrozoa: a class of the *Cnidaria.* H. display either polypoid or medusoid structure, some species passing through both forms during their life cycle. The *hydropolyp* is differentiated into a stalk-like body or *hydrocaulus* and the oral end or *hydranth,* which bears the mouth and one or more circles of tentacles. The gastrovascular cavity is uniform and not divided by septa, and in rare cases it also extends into the tentacles. The polyps reproduce asexually by budding, in which all layers of the body wall form a bulbous evagination, the interior of which is continuous with the gastrovascular cavity; the resulting bud finally develops into a daughter polyp. In a few species the daughter polyp detaches and becomes free living, but in most species it remains attached to the mother polyp, becoming part of a polyp colony, in which all individuals are connected by tubular extensions of the gastrovascular space. Many thousand individuals are often united in a single colony. Polyps can also produce medusae by budding, which regularly detach themselves and become the free living sexual generation. The outer wall of the colony secretes a solid, but elastic periderm, which in many species also encloses the hydranth. Many polyp colonies exhibit a division of labor, which is reflected in the shape and size of the individual polyps (i.e. polymorphism). The feeding polyps, which are responsible for the nutrition of the entire colony, retain their original form, whereas the mouth and tentacles of the specialized polyps regress. The specialized forms include defense and feeding polyps with numerous nematocysts, and blastozoids which produce the medusa generation. The polyp colonies of the order, *Hydroida,* are attached to the substratum. Members of the order, *Siphonophora,* move freely in the water as large pelagic colonies containing both polypoid and medusoid individuals, displaying marked polymorphism. In many siphonophoran colonies, one of the medusoid individuals is modified to a gas-filled sac (pneumatophore), which serves as a float for the colony, e.g. the Portuguese man-of-war *(Physalia physalis)* (see), which occurs in warm seas. Some species (not *Physalia)* can regulate the buoyancy of the float so that it sinks below the surface in stormy weather. Others (e.g. *Nanomia)* live below the surface, but they migrate vertically (by as much as 300 m in less than an hour). Buoyancy of the float is controlled by the secretion or release of its gas content; interestingly, the gas in question consists of more than 90% carbon monoxide.

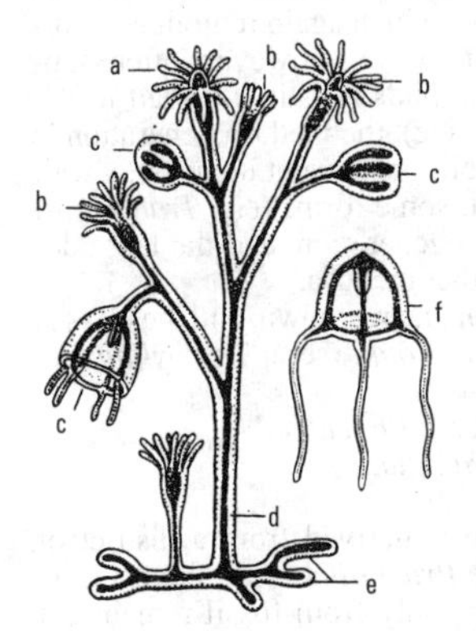

Fig. 1. *Hydrozoa.* Polyp colony. *a* Oldest polyp. *b* Daughter polyp. *c* Medusa bud. *d* Horny perisarc covering the coenosarc. *e* Stolons. *f* Detached, free-living medusa.

Siphonophorans can also travel with the aid of swimming bells or nectophores, which are modified pulsating medusae. In the order, *Trachylina,* there is no polyp generation. Colony-forming hydropolyps are usually no longer than 1 mm, but the largest polyp attains a length of 2 m.

The *hydromedusa* is characterized by a cell-free supporting lamella between the ectoderm and endoderm, and an ectodermal lamella (the velum) which projects inward from the edge of the bell. The velum is muscular, and it can behave as a diaphragm by pulling in the sides of the bell and constricting the hollow space beneath. This action increases the pressure of the ejected water and propels the medusa, topside forward, through the water. The gastrovascular space consists of the central space of the manubrium and, in the primitive condition, 4 radial canals linked by a ring canal in the rim of the bell. During growth, the number of radial canals may be considerably increased. The number of tentacles on the rim of the bell can also increase from 4 (in the primitive condition) to several hundred. Many sensory cells are also present in the rim of the bell, which are responsive to gravity (statocysts) or to light. Hydromedusae are usually unisexual, and the sex cells always arise in the ectoderm, in the manubrium or beneath the radial canals. Many free swimming hydromedusae have a diameter of 2–6 mm, but the largest form attains 40 cm.

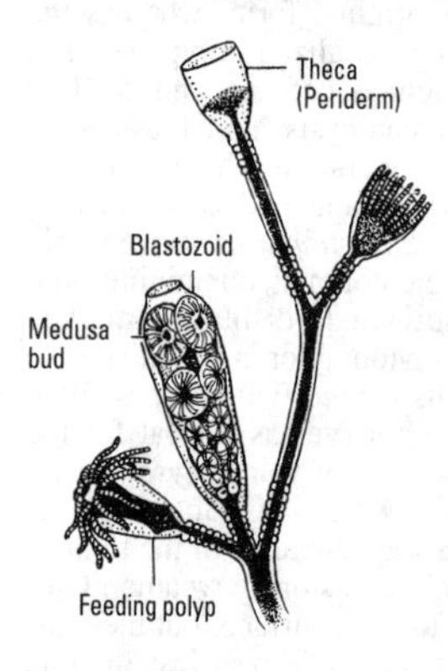

Fig.2. *Hydrozoa.* Part of a branched polyp colony, showing three feeding polyps and a blastozoid.

As a rule, reproduction occurs with an alternation of generations. The fertilized egg develops into a planula larva, which in turn develops into a polyp. The latter reproduces asexually by budding, forming not only more polyps, but also medusae, which again reproduce sexually. The medusa generation is, however, often suppressed, and the medusa buds remain attached to the polyp colony. In *Hydra* (see), the medusa generation is totally absent, and sex cells are present in the body wall of the polyp. In contrast, some forms (e.g. *Trachylina)* exist only as the medusa generation, and the larva develops directly into another medusa.

About 2 700 species of H. are known, and the class is divided into 3 orders: *Hydroida* (see), *Trachylina* and *Siphonophora.* (see).

Hydrurga leptonyx: see *Phocidae.*

Hyemoschus: see *Tragulidae.*

Hyena: see *Hyaenidae.*

Hyeniales (the name is derived from a district of Norway): an order of the *Primofilices* (primitive ancestors of the ferns) known only from fossil remains. It

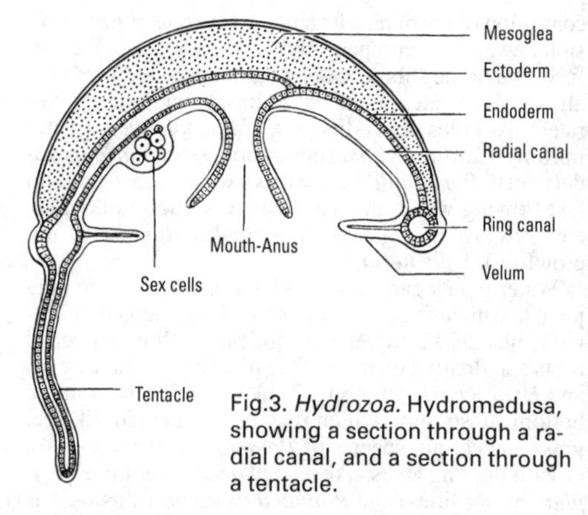

Fig.3. *Hydrozoa.* Hydromedusa, showing a section through a radial canal, and a section through a tentacle.

contains the most primitive forms that still display distinct affinities to the psilophytes, from which the *H.* evolved. The oldest form, *Protohyenia,* which was discovered in the Lower Devonian of western Siberia, also shows the greatest affinity to the psilophytes. It is very reminiscent of *Rhynia,* and appears to lie at the point of separation of the psilophytes and sphenopsids. Members of the genus, *Hyenia,* from the Middle Devonian, displayed a similar habit to that of *Rhynia.* The dichotomous "leaflets" of *Hyenia elegans* were opposite and arranged in more or less distinct whorls. The upper end of the shoot carried sporophylls and showed a tendency to form loose racemose spikes. Internal transverse differentiation of shoot tissues is not evident in any of the different forms of *Hyenia;* this is first observed in the genus, *Calamophyton,* also of the Middle Devonian. The *H.* belong partly to the *Protopteridiales.*

Hygroactive stomatal movements: see Stomatal movements.

Hygronastic response of stomata: see Stomatal movements.

Hygropassive stomatal movements: see Stomatal movements.

Hygromorphosis: modification of plant shape and structure due to an excess or deficiency of water. Growth in moist air leads to extension of the internode distance, whereas dry conditions may promote dwarf growth, as well as anatomical changes such as thickening of the cuticle, increase in the density of surface hairs, and increased formation of mechanical tissues, etc. In their normal dry habitat, certain xerophytes produce shoot thorns in their leaf axes, but in a moist atmosphere these thorns are replaced by leafy shoots. These morphological changes are, however, not due solely to the increased availability of water, but are also influenced by light, increased supply of nutrient salts and other environmental factors. The striking morphological and anatomical differences between the aquatic and terrestrial forms of Amphibious plants (see) are also due to a complex of factors, rather than water alone.

Distinct hygromorphoses are observed in various lower plants, e.g. liverworts, algae and fungi.

Hygronasty: a Nasty (see) induced by a water deficiency in the leaves of a plant, or by variations in the humidity of the surrounding air. Hygronastic reactions make a large contribution to stomatal movements.

Hygrophytes: plants which grow in permanently damp soils and a very humid atmosphere. As an adaptation to damp habitats, H. have thin cuticles and show increased cuticular transpiration. Their wide, thin, delicate leaf blades are richly supplied with sap, and they are either naked or covered with live hairs or papillae, which considerably increase the effective surface area. The chlorophyll-containing epidermis is thin-walled and covered by a delicate cuticle. Stomata are not sunken, and in some H. they are raised above the epidermis. Many H. possess active water-excluding pores (hydathodes). Hygrophytic habitats tend to be weakly illuminated, and H. have large leaf blades which maximise photosynthesis in weak light. Stomatal transpiration is relatively low, and the roots and vascular systems are usually poorly developed.

H. grow mainly in the lower vegetation layer of tropical rainforests, e.g. ferns, arum lilies, begonias. In temperate zones, H. are found as shade plants in damp woods, e.g. *Impatients noli tangere* ***(Touch-me-not).***

Hygroscopic movements: in plants, passive movements due to swelling, shrinkage, or drying out of usually dead cells or membranous structures. They are often reversible and the same structure may repeatedly display H.m. In particular, H.m. serve for the dispersal of seeds, spores and pollen. H.m. are brought about by the differences in the submicroscopic structure of individual layers of cell walls, which lead to anisotropic hydration, so that water imbibition (accompanied by swelling) occurs to different extents and in different directions. Swelling occurs mainly in a direction perpendicular to the line of the cellulose microfibrils. If cell walls or cell wall layers with different planes of swelling are closely adpressed to one another, then bending, rolling or spiral twisting must ensue when the degree of hydration is altered.

H.m. are the basis of opening and closing movements of the fruit capsules of many higher plants, the hapteres of *Equisetatae,* and the spore capsules of *Bryophyta.* For example, the fruit capsules of *Saponaria* open by the outward bending of the membranous capsule teeth when they dry out. The fruit capsules of some species close again when they become damp. In other plants, e.g. *Veronica, Sedum,* etc., opening occurs when the membranes imbibe water from rain or dew (hygrochasy). After flowering and setting fruit, the native North African cruciferous plant known as the Rose of Jericho *(Anastatica hierochuntica)* becomes a dry, skeletonized hollow ball, in which the branches curve in and over and enclose the fruits. The stem breaks at ground level and the plant is blown over the soil surface until it encounters moisture; it then rapidly becomes rehydrated, the branches spread out, and the seeds are released. The fruits of storksbill *(Erodium)* have a long beak or awn, which becomes spirally twisted when dry; it extends again when damp, and providing it acquires a purchase, this extension process aids penetration of the seed into the soil. H.m. are also responsible for the spiral twisting of the pods of many species of the *Leguminosae.*

Hylea: see Biocoenosis.

Hylidae, *hylids, true tree frogs:* a family of the *Anura* (see) containing 630 living species in 37 genera in 4 subfamiles, distributed throughout the world, with the exception of sub-Saharan Africa.

Most taxa have 8 holochordal, procoelous presacral vertebrae, with nonimbricate (nonoverlapping) neural arches. Presacrals I and II are not fused, and the atlantal cotyles of presacral I are widely separated. The coccyx lacks transverse processes, and displays bicondylar articulation with the sacrum. A cartilaginous omosternum (absent in all *Allophryne)* and a sternum are present, and the pectoral girdle is arciferal. Maxillary and premaxillary teeth are present (absent in all Allophryne). As an adaptation to an existence in trees and vegetation, a short, cartilaginous intercalary element is present between the penultimate and terminal phalanges, and adhesive disks are present on the fingers and toes. Individual species usually have characteristic voices, and a marked ability to change color. Most species lay spawn in clumps in water, and the tadpoles are aquatic. Other species display various highly specialized forms of brood care.

The ***European tree frog*** (plate 5, *Hyla arborea,* subfamily Hylinae) is found in central and southern Europe, where it inhabits wet, dense vegetation in tree-lined and bushy zones bordering rivers and lakes, or on the edges of wet forests. Subspecies occur in Asia and North Africa. It is 4–5 cm in length, with a smooth, grass-green back, which can change color to gray-brown, depending on the background colors and the mood of the animal. The fingers are slender and grasping, the pupils horizontal, and the eardrums prominent. Fingers and toes have terminal adhesive disks. Males have an internal vocal sac. Insects are the main food source, and these are caught in the mouth while the frog is jumping, or by a rapid flick of the tongue while the frog is stationary. They enter the water only for mating and spawning. Spawn consists of several walnut-sized clumps, each containing up to 1 000 eggs. Tadpoles hatch after 14 days, and the metamorphosed froglets leave the water for the land from July to August. *Hyla arborea* has a generally attractive and cute appearance, and is frequently sold in European pet shops.

The voice of the South American ***Blacksmith frog*** *(Hyla faber)* resembles the ring of a hammer on an anvil. It builds mud nests 30 cm in diameter and 10 cm high, with a central, water-filled, crater-like depression, where the eggs are laid.

Another form of brood care is displayed by the female of the South American species, *Hyla goeldii,* which carries its few eggs in a groove on its back. Considerable development occurs within the eggs, so that the emerging tadpoles already have fully developed hindlimbs; they are then placed by the mother in the water-filled cistern of an epiphytic bromeliad.

One of the commonest Australian hylids is *Litoria aurea* **(Gold(en) tree frog,** subfamily *Pelodryanidae);* it is predominantly aquatic, the spawn is laid in clumps, and the fingers and toes carry poorly developed adhesive disks.

Hyla regilla ***(Pacific tree frog)*** occurs throughout North America, and is notable for its marked ability to change color between green and brown. The female (about 5 cm in length) lays her eggs in free clumps in the water.

The genus *Acris* ***(Cricket frogs,*** subfamily *Hylinae),* is represented by a number of small North American species, some aquatic and others mainly terrestrial. They lack adhesive disks on the fingers and toes, are noted for their powerful jump, and are especially well known for their shrill, insect-like voices, which are often combined in a chorus. They live in low bushes along rivers and lakes.

Species of *Aparasphenodon* ***(Armor-headed frogs,*** subfamily *Hylinae)* occur in Central and South America. They are tree-dwelling and habitually hide in the cisterns of epiphytic bromeliads. The top of the head carries a bony plate that is fused with the skin.

Marsupial frogs *(Gastrotheca,* subfamily *Hemiphractinae)* occur in South America. The female has 2 lateral folds of skin, which close together over the back to form a brood pouch for the fertilized eggs. *G. marsupiata* ***(Small marsupial frog)*** of Ecuador lays up to 200 eggs. During mating, the female raises her hindparts, so that the eggs roll into the brood pouch. In an action not unlike molting behavior, the female uses the longest (4th) toe of both feet as a hook to open her dorsal brood pouch and release the tadpoles.

Species of *Phyllomedusa* (subfamily *Phyllomedusinae)* are distributed from Mexico to South America. They are nocturnal, tree-climbing frogs with very large eyes with vertical pupils. Eggs are laid on the underside of leaves above water, and the emerging tadpoles simply fall into the water below. *P. lemur* of Central America remains in amplexus for days until a suitable egg-laying site is found. *Phyllomedusa* species are exceptional in that their first 2 fingers and toes are apposable to form gripping hands and feet; they are sometimes called "gripping frogs".

Hylobatidae: see Gibbons.

Hylocichla: see *Turdinae.*

Hylotrupes: see Long-horned beetles.

Hymatomelanic acid: see Humus.

Hymenium: a spore-producing layer in the fruiting bodies of many acsomycetes and basidiomycetes. It has a regular structure of paraphyses (sterile hyphae) and asci *(Ascomycetes),* or paraphyses and basidia *(Basidiomycetes).*

Hymenophore: the hyphal layer of fungi that carries the hymenium.

Hymenoptera: an insect order containing more than 100 000 species, including the bees, wasps, ants, sawflies and ichneumons. The earliest fossil *H.* occur in the Jurassic.

Imagoes. Adult body length varies from 0.2 to 50 mm. The hypognathous, usually very mobile head carries large compound eyes and 3 ocelli. Antennae are usually filiform, with 3 to 36 segments and an elongate stalk; in some species the antennae are bent or carry a clublike terminal thickening. Mouthparts are mandibulate (adapted for chewing), but usually also adapted for sucking; in the *Sphecidae* and *Apidae,* the mouthparts form a sucking proboscis. The 2 pairs of transparent wings, which give the order its name, lie against the body at rest. The hindwings are smaller than the forewings, and interlocked with the latter by hooklets. In most groups the wings have a highly specialized pattern of veins, but in the *Chalcididae* and allied families venation is much reduced, the forewings of certain species possessing a single main vein. Some forms are only temporarily winged, e.g. ant queens, or they are completely wingless, e.g. ant workers, female *Mutillidae* and certain parasitic *Terebrantes.* The legs are normal walking legs, with a 5-segmented foot.

Two suborders are recognized, differentiated on the basis of the type of junction between thorax and abdomen. In the *Symphyta,* the abdomen is broadly attached to the thorax, and there is no marked constriction between the first and second abdominal segments. In the *Apocrita,* the first abdominal segment (propodeum) is nearly always intimately fused with the thorax, and there is a constriction between the first and second abdominal segments. The first abdominal segment is included in the structure of the thorax, and the ninth and subsequent abdominal segments are strongly reduced or regressed. The freely movable abdomen is well armored, the intersegmental membranes being protected by the sclerotized rings of the abdominal segments.

Females of the *Symphyta* and *Terebrantes* have an ovipositor, modified for sawing or piercing, and similar in basic structure to that of the *Orthoptera;* in the *Aculeata,* it is modified to a sting. All *H.* are thermophilous and diurnal, feeding on pollen or nectar, or preying on other insects. Many lead a largely solitary existence, but ants, social bees and social wasps display highly specialized colonial behavior.

Metamorphosis is complete (holometabolous development).

Eggs. All *H.* lay eggs, varying in number from a very few to the several millions laid by the queens of certain ants and bees. They are laid on or in plant tissues *(Symphyta),* in a host (parasitic *H.: Ichneumonidae, Braconidae* and many *Chalcididae),* or in specially constructed brood cells or chambers (colonial species and those displaying brood care). In certain parasitic species, the developing embryo (within an egg of about 1 mm diameter) divides to form several larvae (polyembryony).

Larvae. Larvae of the *Symphyta* resemble lepidopterous caterpillars, but with a greater number of legs (6 to 8 pairs). Larvae of other *H.* are generally apodous and grub-like. Larvae of the *Symphyta* feed independently on plant material. In most families of the *Apocrita,* the larvae live as parasites on the tissues or body fluids of their host; some, however, feed on material collected earlier and stored for this purpose by the egg-laying female (such massed provisions consist of pollen, honey or paralysed or dead insect larvae and spiders) and others are fed in the nest by the mother insect or by other members of the social hierarchy (ants, social bees, social wasps). Some species lay their eggs in the nests of other *H.* (brood parasites), e.g. species of the *Chrysididae* and *Mutillidae.*

Pupae. The final larval instar usually spins a cocoon, in which it pupates. Larvae of parasites pupate either inside or outside the host. All pupae are of the pupa libera type, being adecticous and usually exarate. Most are colorless and soft.

Economic importance. Apart from the obvious value of the honeybee, *H.* are extremely important as flower pollinators, and as controllers of other insect pests. Those attacking insect pests include predatory ants and digger wasps, as well as species with parasitic larvae *(Ichneumonidae, Braconidae, Chalcididae).* Among the *H.,* the only significant pests are certain *Symphyta,* e.g. *Tenthredinidae, Diprionidae, Cephidae.*

Classification. About 80 families are recognized, and a selection of these is given below.

1. Suborder: *Symphyta* (plant wasps without a "wasp waist").

1. Family: *Tenthredinidae,* e.g. the ***Currant worm*** *(Nematus ribesii)* and the ***Larch sawfly*** *(Pristiphora erichsonii).* See Sawflies.

2. Family: *Diprionidae* ***(Conifer sawflies,*** e.g. the North American forest pests, *Neodiprion* and *Diprion).* See Sawflies.

3. Family: *Cimbicidae* (e.g. *Cimbex americana*, which attacks deciduous trees). See Sawflies.

4. Family: *Siricidae* ***(Horntails).*** Larvae burrow in the heartwood of deciduous and evergreen trees, some living symbiotically with wood-rot fungi.

5. Family: *Cephidae* ***[Stem sawflies*** (see)].

2. Suborder: *Apocrita* (members have a "wasp waist").

Section: *Terebrantes*

1. Family: *Ichneumonidae* ***[Ichneumon flies*** (see)]. Larvae are parasitic on other insects, especially larvae or pupae of *Lepidoptera,* but also those of *Diptera Coleoptera, Neuroptera* and other *H.,* as well as spiders.

2. Family: *Braconidae.* Larvae are parsitic and with similar hosts to those of *Ichneumonidae.*

3. Family: *Cynipidae.* The family includes Gall wasps (see) (subfamily *Cynipinae)* which are vegetarian, forming plant galls or living in galls not formed by themselves (inquilines). Other cynipoids are parasitic.

4. Family: *Chalcididae.* A mainly South American family, but with representatives in all regions. Members are parasitic on larvae or pupae of Lepidoptera, Diptera and Coleoptera.

5. Family: *Agaontidae* ***(Fig wasps).*** These are the sole pollinating agents of the internal flowers of figs *(Ficus).* Different species or even varieties of fig are pollinated by different specific species of fig wasp. Seeking an egg-laying site, the female enters the synconium (fig) through the ostiole, using her modified legs, antennae and mandibles to force apart the bracts that fill the ostiole. Her ovipositor is then inserted into the stigma through the style and into the ovary, where an egg is deposited. The ovule endosperm develops parthenocarpically into a gall, which serves as the larval food supply. Emergent males escape by biting through the thin ovary wall. They seek out a gall containing a female, chew a hole through the ovary wall, insert the curved metasoma through the hole and copulate with the female. When the anthers of her home flower are ripe, the female escapes by enlarging the hole made by the male. Pollen collects in pollen pockets on her mesosoma, in corbicula on each foreleg, or in folds of the intersegmental membrane of the metasoma. Pollination then occurs while the female is seeking suitable flowers for oviposition.

Section: *Aculeata* (aculeatus = "equipped with a sting")

6. Family: *Chrysididae* ***(Cuckoo wasps).*** Eggs are laid in the cells of solitary wasps or bees.

7. Family: *Mutillidae* ***(Velvet ants).*** Parasites of social and solitary bees and wasps, and of some *Diptera.* Females are wingless.

8. Family: *Formicidae* ***[Ants*** (see)].

9. Family: *Apidae* ***[Bees*** (see)].

10. Family: *Sphecidae* ***[Thread-waisted wasps, Mud daubers, Sand wasps*** or ***Digger wasps*** (see)]. Females of most species provide a store of mass provisions (paralysed or killed insect larvae and spiders) for the subsequent larvae.

11. Family: *Pompilidae* ***(Spider wasps).*** Large, predatory digging wasps. Spiders are typically paralysed or killed as food for the developing larvae. Some species construct a nest, others use the spider's own burrow.

12. Family: *Vespidae* ***(Social wasps,*** i.e. ***Yellow jackets*** and ***Hornets).*** A nest is constructed of papery material formed from chewed wood mixed with saliva. Larvae are fed by continuous provisioning, and not by the laying in of mass provisions as in solitary forms. See Wasps.

Hymenostomata: see *Ciliophora.*

Hynobiidae, ***Asiatic salamanders:*** a phylogenetically old family of the *Urodela* (see), comprising 33 species in 9 genera. The palatine teeth form a characteristic V-shaped series. Each foot carries 4 toes, and all species are wholly aquatic. The female lays 2 egg sacs directly in the water (usually attached to stones), and they are then fertilized by the male. The family is more or less restricted to Asia, ranging from the east coast of the former USSR throughout Siberia as far as the Urals; it is also represented in Korea and Japan, and Afghanhistan and Iran. In addition to lowland species, there are also many mountainous species (up to 4 000 m), living in well aerated, fast flowing mountain streams. Metamorphosis is complete; adults possess eyelids and have no gill slits.

Hynobius keyserlingii ***(Siberian salamander)*** is found from Kamchatka across Siberia and Mongolia as far as the Urals, and it is the only species that also encroaches on Europe (Gorki region); it is also the only urodele to cross the 66 ° northern limit into the Arctic circle. Its 13 cm-long, slender, brown body carries a glossy, bronze, longitudinal, dorsal band.

Onychodactylus japonicus (up to 16 cm in length) is found in fast-flowing, cold mountain streams in Korea and Japan. Lungs are absent, and external respiration occurs through the skin. The toes of the larvae carry sharp, horny, recurved claws, which are sometimes retained in the adult.

Ranodon sibiricus (up to 25 cm in length), found in eastern Semiryechensk and western Chinese Turkestan, is olive-brown with black spots. The lungs are degenerate.

Hyoid: see Skull.

Hyomandibular: see Skull.

Hyopsodontidae: see *Condylarthra.*

Hyoscyamine: see Atropine.

Hyoscyamus: see *Solanaceae.*

Hyp: abbreviation of Hypoxanthine (see).

Hyperdiploidy: see Aneuploidy.

Hyperhaline: see Brackish water, Hypersaline lake.

Hyperiidea: see *Amphipoda.*

Hyperimmunization: immunization repeated many times, leading to the production of large quanities of antibodies. H. is employed in particular for the preparation of high-activity antisera.

Hypermasty: the presence of supernumerary mammary glands in humans.

Hypermorph: a mutant allele that exerts the same phenotypic effect as the normal allele, but in an intensified form. If the phenotypic effect of a mutant allele is the same in character but less in intensity than that of the normal allele, the mutant allele is described as *hypomorph.* See Amorph, Antimorph, Neomorph.

Hyperoliidae, ***hyperoliids:*** a family of the *Anura* (see) containing 206 living species in 14 genera, distributed mainly in sub-Saharan Africa, with 1 genus (*Heterixalus*, 8 species) in Madagascar and a monotypic genus (*Tachycnemis seychellensis*) in the Seychelles.

Eight, holochordal, procoelous presacral vertebrae are present. In most species and genera the neural arches are nonimbricate (nonoverlapping). In all but 2 genera *(Acanthixalus, Callixalus)* presacral VIII is biconcave and the sacrum biconvex. Presacrals I and II are not fused, and the atlantal condyles of presacral I are widely separated. The coccyx lacks transverse processes and has a bicondylar articulation with the sacrum. Omosternum and sternum may be cartilaginous or ossified, and the pectoral girdle is firmisternal. Maxillary and premaxillary teeth are present. Like the rhacophorids, the hyperoliids also possess short, cartilaginous intercalary elements between the penultimate and terminal phalanges. In *Acanthixalus* (1 species, Nigeria, Cameroon, Zaire), *Chrysobatrachus* (1 species, Zaire) and *Hyperolius* (109 species, sub-Saharan Africa) the pupils are horizontal or round; in all other genera the pupils are vertically elliptical. Hyperoliids are mostly small or moderately sized (1.5–8 cm) tree frogs with adhesive pads. *Chrysobatrachus* and *Tornierella* (2 species, Ethiopia) and some species of *Kassina* (12 species, sub-Saharan Africa) are terrestrial, but all others are arboreal.

Hyperolius ***(Sedge frogs)*** are widespread and familiar frogs in sub-Saharan Africa. They are small (2–4 cm) and very colorful, inhabiting the lower vegetation near to water in tropical forests. They do not build a foam nest, and eggs are laid in small heaps on the stems of water plants, above or below the water level. In some classifications, the hyperoliids are subsumed as a subfamily of the closely related *Rhacophorida.*

Tadpoles of *Kassina* ***(Kassinas, Kassina frogs)*** have unusually high caudal crests. Adults are ground-dwelling and they run instead of hopping. The adult ***Senegal kassina*** *(K. senegalensis*; 2.5–4 cm) has a dark longitudinal stripe on a silver-gray background.

The 23 arboreal species of *Afrixalus*, which occur in sub-Saharan Africa, are exclusively nocturnal with vertical rhomboid pupils. Eggs are laid above water in small balls covered with leaves, and the hatching tadpoles fall into the water below.

Hyperoodon: see Whales.

Hyperparasites, ***2nd degree parasites, secondary parasites:*** parasites that live in or on other parasites, e.g. small ichneumon flies that parasitize the eggs of other ichneumon flies.

Hyperplasia: increase in the size or quantity of a tissue or organ, due to an abnormal increase in cell division.

Hyperploidy: the occurrence of additional individual chromosomes or chromosome segments in the chromosomal complement of an originally haploid (hyperhaploidy), diploid (hyperdiploidy) or polyploid (hyperpolyploidy) cell, tissue or individual. In *hypoploidy,* there is a deficit of individual chromosomes. H. and hypoploidy are types of Aneuploidy (see).

Hyperpolarization: see Excitation, Resting potential.

Hypersaline lake, ***hyperhaline lake:*** a lake with a salt concentration greater than 40%. Between 80 and 90% of the primary product of such waters is due to photoautotrophic bacteria.

Hypersaprophytic conditions: see Saprophytic system.

Hypersensitivity: exaggerated reaction of a plant tissue to a penetrating parasite. H. is an active Resistance (see) mechanism, in which the host cells at the infection site die rapidly, resulting in tissue necrosis. This prevents spread of the infection, and in the absence of healthy host tissue the pathogen usually also dies.

Hypertension: see Blood pressure.

Hyperthely: the presence of supernumerary mammary nipples in humans.

Hyperthermia: an increase in the body temperature of warm-blooded animals to above the normal value. H. arises when heat loss (see Temperature regulation of the body) no longer keeps pace with the heat input, so that the constant desired body temperature cannot be maintained. Body temperature is also increased during Fever (see), but in this case the physiological target value is increased. Thus in H., the processes of heat loss are overloaded, whereas in fever they are not. In humans, in both H. or fever, death occurs when the body temperature exceeds 42 °C.

Hypertragulidae: see *Traguloidea.*

Hypertrophy: an increase of growth and/or development. In humans and other vertebrates, H. is usually the result of disease of the hypophysis. In plants, polyploid forms often display H., in which case the increased growth is genetically based. H. can also be caused by an excessive intake of nutrients, or attack by parasites (see Galls). H. of an entire organism leads to *gigantism* or *macrosomy.* Hypertrophic changes in organs or tissues lead to the appearance of structured or structureless proliferative swellings and excrescences. Hypertrophied cells show increased growth without division.

Hyphalmyroplankton: see Plankton.

Hyphal noose: see Carnivorous plants.

Hyphomicrobium: a bacterial genus found in the soils of all continents. The cells are rod-shaped, oval or bean-shaped, 0.5–1.0 μm x 1–3 μm. In an unusual form of cell multiplication, each cell produces mono- or bipolar filamentous, nonseptate outgrowths, known as hyphae. The ends of these outgrowths are swollen, and they form buds, which break off as flagellated and freely motile daughter cells. These eventually attach themselves to surfaces or to other cells, form clumps and lose their motility.

Hyphosphere: the immediate environment of a hypha, in particular in the earth.

Hypnobryales: see *Musci.*

Hypnum: see *Musci.*

Hypobranchial groove: see Endostyle.

Hypochile: see *Orchidaceae.*

Hypocoliinae: see *Bombycillidae.*

Hypocolius: see *Bombycillidae.*

Hypodiploidy: see Aneuploidy.

Hypoglossal nerve: the XII cranial nerve; see Cranial nerves.

Hypognathy: in insects, the alignment of the mouth opening and mouthparts at right angles or at an acute angle to the long axis of the body, so that they face downward or backward.

Hypogynous (flower or ovary): see Flower.

Hypolimnetic zone: see Lake.

Hypolimnion: see Lake.

Hypomorph: see Hypermorph.

Hyponasty: an unfolding or opening movement caused by increased growth on the underside of a plant organ (e.g. leaf), leading to erection or straightening of the plant part in question.

Hyponeuston: see Neuston.

Hypopenna: see Feathers.

Hypophysis:

1) In plants: 1) a more or less marked swelling at the upper end of the stalk of a moss capsule, which is distinctly displaced by the capsule; 2) the terminal cell of the embryo suspensor; it abuts on the embryo, and after further divisions it may participate in the formation of the root cap and root tip.

2) In vertebrates, synonymous with ***pituitary gland*** or ***pituitary body:*** an endocrine organ, consisting of two lobes, and lying at the base of the vertebrate skull. The smaller lobe *(neurohypophysis* or *posterior pituitary)* develops from the floor of the diencephalon and remains connected to the latter via the infundibulum. The larger lobe *(adenohypophysis* or *anterior pituitary)* arises from the roof of the mouth cavity. There is also an intermediate lobe (pars intermedia), which forms part of the neurohypophysis and lies between the latter and the adenohypophysis; the intermediate lobe is absent from some species, e.g. birds and whales. Different cell types with specific functions are differentiated within the anterior pituitary. There is an intimate interaction between the H. and the other endocrine organs of the body. Thus, the active products of the H., known as *pituitary* or *hypophyseal hormones,* stimulate hormone production in other glands (e.g. the thyroid gland, adrenal glands and gonads); by a process of feedback inhibition, hormones from other glands may in turn inhibit the synthesis and release of hypophyseal hormones.

The following six hormones are formed in the adenohypophysis (see separate entries for each hormone): somatotropin (growth hormone), adrenocorticotropic hormone (ACTH), thyreotropin (TSH), follicle stimulating hormone (FSH), luteinizing hormone (LH) and prolactin. The neurohypophysis releases two neurohormones: Vasopressin (see) and Oxytocin (see). The intermediate lobe produces melanotropin.

Hypoplastic fetal chondrodystrophy: see Achondroplasia.

Hypoplectrus: see Sea basses.

Hypoploidy: see Hyperploidy, Aneuploidy.

Hyporheon: cavities beneath a river bed, which form a habitat for species ill adapted to high rates of water flow, but which contain sufficient oxygen and organic material to sustain life.

Hypositta: see *Sittidae.*

Hypostasis: interference with, or prevention of, the expression of a gene (the *hypostatic gene)* by another nonallelic gene (the *epistatic gene)* in the same genome. See Epistasis.

Hypostome: see Polyp.

Hypotension: see Blood pressure.

Hypothalamus: see Brain.

Hypothecium: the hyphal layer in certain fruiting bodies (in particular in apothecia) of memebers of the *Ascomycetes.* The H. lies directly beneath the hymenium, and it gives rise to the spore tubes (asci) and sterile hyphae (paraphyses).

Hypothermia: a decrease in the body temperature of a warm-blooded animal to a value below normal. In response to the onset of H. there is a decrease of heat loss and an increase of heat production (see Temperature regulation of the body). These countermeasures show their maximal intensity at body temperatures between 26 and 28 °C. Below 26 °C death may occur from ventricular fibrillation of the heart. In surgery, a combination of narcosis and specific inhibition of temperature regulation is sometimes used to bring about H. Metabolic activity decreases with body temperature, thereby increasing the resuscitation time.

In humans, with increasing age, there is a tendency for body temperature to be controlled at a lower value. Very old people often have a body tempertaure of 35 °C. Thus, H. is the thermoregulatory converse of fever.

Hypotony:

1) ***Hypotrophy:*** increased growth on the underside of dorsiventral plant organs, e.g. the branches of conifers.

2) see Blood pressure.

Hypotricha: see *Ciliophora.*

Hypotriplopidy: see Aneuploidy.

Hypotrophy: see Hypotony

Hypoxanthine, *Hyp:* 6-hydroxypurine, a purine derivative formed by hydrolysis of its nucleoside (see Inosine) or by deamination of adenine. It is widely distributed in low concentrations in the plant and animal kingdoms, and is present as a rare base in certain transfer RNAs.

Hypoxanthine

Hypselodont: see Teeth.

Hypsicephaly: see Acrocephaly.

Hypsipetes: see *Pycnonotidae.*

Hypthymis: see *Muscicapinae.*

Hyracoidea, *hyraxes:* an order of mammals about the size of rabbits with rudimentary tails. The order is represented by only 11 living species in 3 genera. With the exception of one hind digit, the claws are hoof-like. The undersurfaces of the feet have wrinkled, rubbery pads containing numerous sweat glands, which enable the animal to cling to, and climb on, perpendicular surfaces. Although the hyraxes resemble tailless rodents, they are placed in a separate order, on account of the hooflike structure of their feet and certain peculiarities of dentition. Dental formula: 2/2, 1/1, 4/4, 0/0 = 28 (deciduous teeth); 1/2, 1/1, 4/4, 3/3 = 38 (permanent teeth). They share many features of primitive ungulates, and are considered to be phylogenetically close to elephants and sirenians, and to the aardvark. The gut is complex with three separate regions of microbial digestion. They are therefore able to efficiently digest fiber, but they do not ruminate. All hyraxes characteristically possess a "dorsal spot", consisting of a dorsal gland, surrounded by a light-colored circle of hairs, which stiffen when the animal is excited; this is most conspicuous in the Western tree hyrax and least noticeable in the Cape hyrax.

Hyrocoidea. Cape hyrax *(Procavia capensis).*

The 5 light to dark brown species of the genus *Procavia* ***(Rock hyraxes*** or ***Dassies;*** head plus body length 44–54 cm) inhabit semi-arid and rocky terrains: *P. habessinica* (= *P. syriacus,* ***Abyssinian hyrax;*** Egypt to Kenya, Sinai to Lebanon, and southestern part of the Arabian peninsula), *P. capensis* ***(Cape hyrax;*** dark brown dorsal spot; southern Africa), *P. johnstoni* ***(Johnston's hyrax;*** Northeast Zaire and Central Africa to Malawi), *P. welwitschii* ***(Kaokoveld hyrax;*** Namibia and Southwest Angola), *P. ruficeps* ***(Western hyrax, Sudan hyrax*** or ***Sahara hyrax;*** yellowish orange dorsal spot; southern Algeria and Senegal to Central African Republic). Some authorities recognize only a single species, *P. capensis,* with various regional subspecies.

Heterohyrax ***(Bush hyraxes, Gray hyraxes*** or ***Yellow-spotted hyraxes;*** head plus body length 32–47 cm) is represented by 3 light gray species with white underparts and a yellow dorsal spot. They occur in semi-arid, rocky areas and scrub in Southwest Africa and from Southeast to Northeast Africa: *H. antinae* ***(Ahaggar hyrax),*** *H. brucei* ***(Bruce's yellow-spotted hyrax),*** *H. chapini* ***(Matadi hyrax).***

The 3 species of *Dendrohyrax* ***(Tree hyraxes*** or ***Bush hyraxes;*** head plus body length 32–60 cm; dorsal spot light to dark yellow) inhabit forests in Africa up to 4 500 m: *D. validus* ***(Eastern tree hyrax;*** eastern Tanzania, Kenya coast, islands of Zanzibar, Pemba and Tumbatu), *D. arboreus* ***(Southern tree hyrax;*** Southeast and East Africa), *D. dorsalis* ***(Western tree hyrax;*** West and Central Africa).

Hyssop *(Hyssopus):* see *Lamiaceae.*

Hysterosoma: see *Acari.*

Hystricidae, ***Old World porcupines:*** a family of Rodents (see) consisting of 11 species in 4 genera, divided into 2 subfamilies. They are large mammals with long quills and spines. Their large, 5-toed, plantigrade feet have smooth soles. The occipital region of the skull is strongly built, and provides attachment for powerful neck muscles. Dental formula: 1/1, 0/0, 1/1, 3/3 = 20.

1. Subfamily: *Atherurinae* ***(Brush-tailed porcupines).*** Head plus body length 37–47 cm. Of the 6 species belonging to this subfamily, 5 are found in forests of South and Southeast Asia, and a single species is native to Africa. Their long, slender tails have a terminal tuft of long, white, stiff hairs, and their bodies are covered with short, flat, sharp bristles. Only a few quills are present, and these are on the back. Members of this subfamily are host to the malaria parasite, *Plasmodium atheruri.* The 6 species are: *Atherurus africanus* ***(African brush-tailed porcupine),*** *Atherurus macrourus* ***(Asiatic brush-tailed porcupine),*** *Thecurus crassispinus* ***(Bornean porcupine),*** *Thecurus pumilis* ***(Philippine porcupine),*** *Thecurus sumatrae* ***(Sumatran porcupine),*** *Trichys lipura* ***(Bornean long-tailed porcupine).***

2. Subfamily: *Hystricinae* ***(Crested porcupines).*** Head plus body 60–83 cm. The short tail of the crested porcupines is encircled by stout, sharp, cylindrical quills, while the tip of the tail is armed with hollow, open-ended quills, which have a stalked base. The latter produce a characteristic warning rattle when the tail is shaken. The posterior two thirds of the back and flanks are covered with black and white spines, up to 50 cm in length, as well as cylindrical quills. These sharp quills are an effective weapon when crested porcupines run backward or sideways into adversaries or intruders. Having penetrated, the quills remain embedded and detach; later the wound often turns septic. The remainder of the body is covered with flat, coarse, black bristles. Most species also possess an erectile crest of long coarse hair from the top of the head to shoulder level. Five species are recognized in a single genus: *Hystrix cristata* ***(African porcupine),*** *H. africaeaustralis* ***(Cape porcupine),*** *H. hodgsoni* ***(Himalayan porcupine),*** *H. indica* ***(Italian porcupine),*** *H. brachyura* ***(Malayan porcupine).***

African porcupines carry fleas responsible for bubonic plague, as well as ticks that carry babesiasis, rickettsias and theilerioses. Their numbers are probably increasing, due to the decline of their natural predators (hyenas, leopards, etc.) and increased production of food crops which they readily eat (groundnuts, melons, maize, etc.).

Hystricinae: see *Hystricidae.*

Hystrix: see *Hystricidae.*

I

I-band: see Muscle tissue.
Iberian water vole: see *Arvicolinae.*
Ibex: see Goats.
Ibidorhyncha: see *Recurvirostridae.*
Ibisbill: see *Recurvirostridae.*
Ibises: see *Ciconiiformes.*
ICBP: acronym of International Council for Bird Preservation.

Ice ages, *glacial periods:* those periods of the earth's history when climatic changes led to persistent polar glaciation and ice-free nonpolar regions, e.g. the glaciation of the Southern Hemisphere in the Paleozoic (see Geological time scale, Table), before the beginning of the Cambrian, in the Ordovician, and at the junction of the Carboniferous and Permian, as well as the glaciation of the Northern Hemisphere in the Pleistocene. Each I.a. is divided into several glacial and interglacial episodes. The I.a. of the Pleistocene is zoogeographically important, because it caused animal migration into climatically more favorable areas (see Glacial refuges), and was often accompanied by the splitting of species into races. After the melting of the glaciers, some of the displaced species returned to their former distribution areas, but even today they still display race differences that developed during their separation. The European flora and fauna of that time were greatly depleted, because the glaciation of the Alps first restricted the inhabitable areas, then prevented escape from the glaciers advancing from the north. In contrast, flora and fauna retreating from the glaciers of North America were not trapped, so that North America now has a richer diversity of species than Europe.

Icelandic moss: see Lichens.
I-cells of hydrazoans: see Regeneration.
Ice plant: see *Aizoaceae.*

Ichneumon flies, *Ichneumonidae* (plate 1): a family (order *Hymenoptera)* containing over 30 000 species, mostly in the northern temperate zone, with over 3 000 species described in central Europe. Tropical species are beautifully colored. Most adult I.f. are slender and agile, with fully developed wings. I.f. are closely allied with the *Braconidae*, which are more sluggish than I.f., and possess a single recurrent vein on the forewing. In contrast, the forewings of I.f. have 2 recurrent veins. A noticeable characteristic of I.f. is the dithering or vibration of the antennae. In some species, the ovipositor of the female, which is also used as a stabbing defensive weapon, may be longer than the body. It is used to pierce and deposit eggs inside living insect larvae; members of the genus *Rhyssa* (among the largest European I.f.) detect the presence of horntail (*Siricidae*) larvae up to 4 cm below the surface of the wood, and use their ovipositor to bore through the wood with unerring accuracy and inject eggs into the host larva. Host animals live until the parasites are ready to pupate, then die. Since I.f. parasitize many insect pests, they render considerable service, especially by preventing or keeping in check pest population explosions.

Ichneumonidae: see Ichneumon flies.
Ichthyophiidae: see *Apoda.*

Ichthyornis (Greek *ichthys* fish, *ornis* bird), ***fish-bird:*** an extinct genus of pigeon-sized birds, whose fossils are found in the Upper Cretaceous of North America (Kansas). The jaws possessed conical teeth, set into cavities. The vertebrae were primitive, with concave articulating surfaces. Both the wings and the large, keeled sternum displayed more advanced evolutionary development.

Ichthyosauria, ***ichthyosaurs*** (Greek *ichthys* fish, *sauros* lizard), ***"fish lizards":*** extinct order of fish-like reptiles, secondarily adapted to a marine existence. These torpedo-shaped animals were up to 12 m in length, with a large head, long pointed snout, large eye sockets with a bony ring, pointed conical teeth in a common groove, and a long, bilobed, propeller-like, heterocercal tail fin. The limbs were modified as fins, and there was a single, large, triangular dorsal fin. *I.* gave birth to live young (ovoviviparous). Fossils are found from the Triassic to the Upper Cretaceous, and are particularly common in the Jurassic.

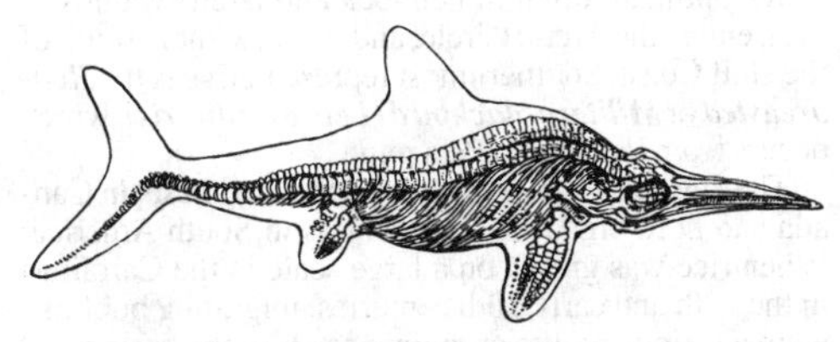

Ichthyosauria. Ichthyosaurus from the Lias e of Holzmaden (Württemberg) (x 0.03).

Ichthyostega (Greek *ichthys* fish, *stegos* roof or cover): an extinct genus of labyrinthodont amphibians from the Upper Devonian of eastern Greenland. They were about 1 m in length with a stegocephalic skull

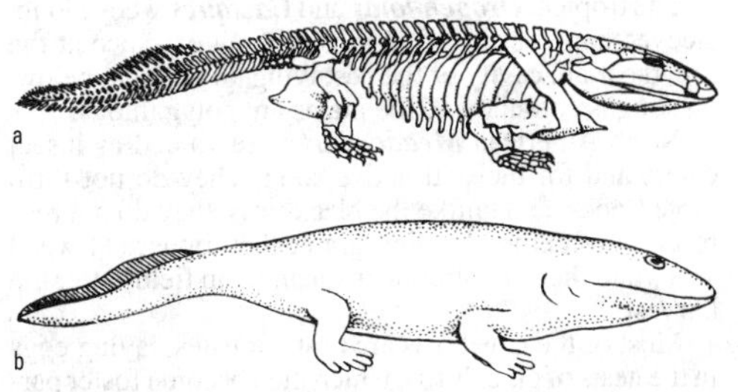

Ichthyostega (after Jarvic). *a* Skeleton. *b* Reconstructed animal.

(roof of strongly ossified plates, and no temporal fossae). As the oldest known quadrupeds, they represent a Connecting link (see) between fishes and amphibians. *I.* possessed four 5-toed walking legs, a fish tail with a dorsal, fin-like border, and conical teeth.

ICRP: acronym of International Commission for Radiation Protection. See Radiation protection.

ICSH: acronym of interstitial cell stimulating hormone.

Icteridae, ***icterids:*** an avian family containing 94 species, and belonging to the order, *Passeriformes* (see). The family contains such diverse groups as ***Oropendolas, Caciques, Grackles, American blackbirds, True American orioles*** or ***Troupials, Cowbirds, Meadowlarks*** and the ***Bobolink,*** representing sizes from 16.5 to 53 cm. All have conical bills, which are unnotched and usually pointed. In orioles and grackles the bill is slender and decurved. Meadowlarks and most blackbirds have straight, sharp bills, while the bills of the cowbirds and the bobolink are short and finch-like. Rictal bristles are usually lacking, and all members possess 9 primaries. In the larger forms (oropendolas and caciques), the bill is heavier and the culmen (dorsal ridge of upper mandible) is expanded or swollen basally to form a conspicuous frontal shield or casque. The family name implies the color yellow (Latin "icter"); although this is fitting for the bright yellow or orange orioles, it is not appropriate for all members of the family, many of which have much black plumage, often relieved by brown, red or yellow. Although this exclusively New World family *(Leistes militaris* has been introduced to Easter Island) is mainly tropical, it is also represented in colder regions. Its northernmost representative is the ***Rusty blackbird*** *(Euphagus carolinus),* which occurs across the North American Continent, often in hemlock and larch swamps; it even enters the Arctic Circle, and it overwinters south of the Gulf Coast. Southernmost representative is the ***Red-breasted*** or ***Military blackbird*** *(Leistes militaris),* which occurs from Panama to Patagonia.

The ***Bobolink*** *(Dolichonyx oryzivorus)* nests in Canada and northern USA, and winters in South America. When rice was grown on a large scale in the Carolinas in the 19th and early 20th centuries, migrating boblinks were a serious pest to sprouting seeds in the spring and to the mature grain in the autumn.

Most brightly colored of the family are the ***Orioles*** *(Icterus),* which are also noted for their singing and their nest architecture. Most of these are tropical and several species migrate north to nest in temperate woodlands during the spring, e.g. the ***Baltimore oriole*** (male, orange and black; female, yellow-olive and brown) is seen in the eastern woodlands of USA in the spring.

The tropical ***Oropendolas*** and ***Caciques*** weeve long, sleeve-shaped nests of grasses with an entrance at the top, the main cavity of the nest lying 90–180 cm below. Nesting is colonial and the males are polygamous.

North American ***Meadowlarks*** are valued as insect eaters and for their attractive song. They do not form large flocks, and unlike the blackbirds, they do not represent a threat to standing grain; they rather eat weed seeds and they are often seen gleaning in fields after the harvest.

Most of the Cowbirds are nest parasites, laying eggs in the nests of other birds, which then become foster parents. Some of them display host specificity. Thus, the ***Screaming cowbird*** *(Molothrus rufoaxillaris)* of Brazil and Argentina victimizes the ***Bay-winged cowbird*** *(Molothrus badius),* the only cowbird that does not practise nest parasitism. *Molothrus badius* pairs for the breeding season and cares for its own brood, and although it may sometimes build its own nest, it prefers to take over the already occupied or abandoned nest of another species, The ***Giant cowbird*** *(Scaphidura oryzivorus)* lays its eggs in the nests of oropendolas and caciques.

Ictidiosauria: fossil reptiles, particularly abundant in the Karroo Formation of South Africa, and considered to be the immediate ancestors of the mammals. They had a secondary palate and paired occipital condyles, with no distinguishable pineal foramen. Prefrontal, postfrontal or postorbital bones were not separate. The small quadrate, together with the squamosal, formed an articulation for the mandible, in which the articular, angular and surangular bones were only rudimentary. A median vomer separated the pterygoids in the palate.

Ictinaetus: see *Falconiformes.*

ICTV: acronym of International Committee on the Taxonomy of Viruses. See Virus families.

Ide: see Orfe.

Identification key: a tabular synopsis of well differentiated and, as far as possible, easily observable characteristics for the identification of plant and animal taxa. I.ks. are also used in minerology, crystallography and soil science. The totality of properties leading to an identification are not necessarily related to evolutionary affinities. I.ks. are therefore referred to as artificial keys to identification.

Various "floras" for plant identification are the most familiar type of biological I.k. The most popular I.k. is the dichotomous key, in which each stage of the identification represents a choice between opposing characteristics. By using appropriate properties, the key may also express systematic relationships, i.e. closely related taxa remain associated throughout the identification system. The following examples illustrate the identification of species of the genus *Melilotus* (sweet clover).

A) Stipules of the middle leaves of the stem distinctly dentate.	*M. dentata* (W.& K.) Pers.
B) Stipules of the middle leaves of the stem smooth-edged or indistinctly dentate	
1) Stipules dentate at base; flowers 2 to 3 mm	*M. indica* (L.) All.
2) Stipules smooth-edged; flowers 5 to 8 mm	
a) Ovaries pilose	*M. altissima* Thuill.
b) Ovaries glabrous	
aa) Corolla yellow	*M. officinalis* (L.) Pall.
bb) Corolla white	*M. alba* Medic.

This key, however, is very space-consuming, and it is tortuous for the identification of rather large taxa, because opposing decisions on a single characteristic may be separated by several pages. For this reason, a numerical key is often used, in which an opposing decision is found under the text number, or under the number given in brackets after the key number.

1. (8) Corolla yellow	
2. (7) Leaflets with fewer than 16 pairs of lateral nerves	
3. (6) Flowers 5 to 8 mm long	
4. Ovaries glabrous	*M. officinalis*

5. Ovaries pilose — *M. altissima*
6. (3) Flowers 2 to 3 mm long — *M. indica*
7. (2) Leaflets with more than 18 pairs of lateral nerves — *M. dentata*
8. (1) Corolla white — *M. alba*

A third type of I.k. is favored for simplified identification systems for general use. Opposing properties are listed one after the other, with directions to the next decision stage at the end of each line. Related species, however, are not always adjacent in the key.

1. Corolla white — *M. alba*
   Corolla yellow — 2
2. Leaflets with more than 18 pairs of lateral nerves — *M. dentata*
   Leaflets with fewer than 16 pairs of lateral nerves — 3
3. Flowers 2 to 3 mm long — *M. indica*
   Flowers 5 to 8 mm long — 4
4. Ovaries pilose — *M. altissima*
   Ovaries glabrous — *M. officinalis*

Idiacanthidae: see *Stomiiformes.*

Idioblasts: individual cells with specialized functions within a plant tissue, which otherwise consists of cells different from the I., e.g. the hairs of the aerenchyma of aquatic plants, or cells with crystalline inclusions.

Idiobotany: the study of individual plants, as distinct from the study of plant communities or plant groups. See Synbotany.

Idiogram: see Karyotype.

Idiosepius: see *Cephalopoda.*

Idiothetic orientation, ***kinesthetic orientation:*** orientation of an animal in its environment without the aid of external clues, and by reference to internal information. Proprioceptors register muscle movements, and this information can be used later. Earlier movement patterns can be repeated by using the stored information on the sequences of muscle activity. In this way, lizards register the route from their hiding place to their sunning site. By using the same information, a lizard can return to its hiding place in fractions of a second without the benefit of external orientation.

Idiotype: the totalily of genes in a cell, comprising the genes of the cell nucleus (genome or genotype) and the extrachromosomal genetic complement (plasmon or plasmotype). The latter consists of the mitochondrial genes (chondriome), and in green plants it includes the genes of the plastids (plastome).

Idoteidae: a family of the *Isopoda* (see).

Idus: see Orfe.

Igapo forest: see Rainforest.

Iguanidae, ***iguanids:*** the most numerous family of Lizards (see). *I.* are related to the *Agamidae* (see), although the geographic regions of these two families are different. They are the most common of the New World lizards, and are found from southern Canada to the extreme south of South America. In the Old World, *I.* are found only on the island of Madagascar (2 genera: *Hoplurus* and *Chalarodon*) and on the Fijian and Tongan islands (1 genus: *Brachylophus*), where there are no *Agamidae*. Possibly the *Agamidae* and *I.* competed for the same area at an earlier time, and the *I.* were supplanted. Most *I.* are 20 to 30 cm long with long tails and a normal lizard-like shape and appearance. There are also 1–2 m long giant species. Like the *Agamidae*, the *I.* have invaded practically every type of habitat in their geographic area, and have evolved a range of appropriate adaptations. In addition to terrestrial species, there are also arboreal species, which are often green and frequently possess adhesive toes. There are also flattened, spiny desert species, and jungle-dwelling, water-loving giant species. The evolutionary convergences of form and habit of the *Agamidae* and *I.* (they possibly also evolved from the same ancestral stock) are so pronounced that family membership is difficult to determine from external appearances alone, without knowledge of their place of origin. The main anatomical difference between the two families is found in their dentition. In the *Agamidae*, the teeth lie on the borders of the jaw bones (acrodont) and do not show replacement growth. The teeth of *I.* are found on the inside of the jaw (pleurodont) and they show replacement growth. Many *I.* possess spiny scales, head, neck and tail crests, and gular pouches; prehensile tails are less common. Erection of crests, dilation of gular pouches and characteristic head nodding often form part of the threat and courtship displays. Certain species of *I.*, just like certain *Agamidae*, can bury themselves in sand. Many *I.* show marked color changes. Most *I.* are insectivorous, while the largest representatives, like those of the *Agamidae*, tend to be herbivorous. They are mostly oviparous, and only a few mountain and desert forms are viviparous. More than 700 species are known in more than 50 genera; the most numerous genus, *Anolis*, contains about 300 species. The following examples of *I.* are listed separately: Green iguana, Crested keeled lizards, Basiliscs, Anoles, West Indian rock iguanas, Spiny-tailed lizards, Horned toads, Marine iguana, Land iguana.

Iguanodon: best known genus of the *Iguanodontidae,* a family of Jurassic-Cretaceous herbivorous, bipedal dinosaurs. Species of *I.* were up to 10 m in length, and 6 m in height, and they ran or walked in their hindlegs in a semi-upright posture. The much reduced forelimbs carried a dagger-like bony spike, which presumably served as a defensive weapon. The skull was relatively small and extended anteriorly as a rather elongated snout. Remains have been found in Europe, Asia and Africa, e.g. in the European Wealdon strata (brackish-limnic Lower Cretaceous deposits). Three-toed, 20 cm-long footprints are relatively common. A particularly important find was that of 29 complete skeletons in the carboniferous limestone of Bernissart, Belgium.

Iguanodontidae: see *Iguanodon.*

Ikeda taenioides: see *Echiurida.*

Ilarviruses: see Virus groups.

Illaenus: a trilobite genus with approximately equal-sized, semicircular cephalon and pygidium, and a body of 10 clearly delineated segments. The cephalon carries an hour glass-shaped glabella with an indistinct anterior border. The two eyes are of medium size and located at the corners of the cephalon. The pygidium has a hardly discernible unsegmented axis.

Illex: see *Cephalopoda.*

Illicium: see *Lophiiformes.*

Illite: see Sorption.

Illness: any deviation from the normal course and pattern of living processes in plants, animals and humans, which is associated with a weakening of the organism or its parts, and which under certain conditions may lead to death.

Imago (plural **Imagoes** or **Imagines**)**:** the adult stage of an insect. See Metamorphosis.

Imbibition: see Hydration.

Imhoff tank: see Biological waste water treatment.

Imitating behavior: copying of the behavior, posture, movements or sounds of another animal by an individual. In *current I.b.*, the individual may temporally coordinate its behavior with that of another (e.g. coordination of type of gait or length of stride), and the two actions may also become synchronized (see Synchronization). In *latent I.b.*, the imitated behavior continues in the absence of the example or prototype, representing a form of motor learning, e.g. the playful tendency of young animals to copy the movements and actions of conspecifics.

Imitation: acquisition of patterns of behavior by copying the behavior of others. *Obligatory I.* is the aquisition of behavior or vocalizations, which constitute an essential part of species-specific behavior, and which are necessary for the maintenance of life or reproduction. *Facultative I.* is not essential, and may even involve the copying of the behavior of other species. Thus, the chaffinch can only acquire its species-specific song by copying other chaffinches, whereas the I. of human vocalizations by a budgerigar is not an essential life process, and it is not a necessary environmental adaptation. Young humans also acquire patterns of behavior by I., although human I. may also be conscious and deliberate. I. is a precondition for Tradition (see).

Immune diseases: diseases caused by immune reactions, e.g. Allergy (see) and Autoimmune diseases (see).

Immunity:

1) Acquired specific resistance of an organism to a disease agent or toxin. I. is the result of immune reactions triggered by the antigen. Thus, antibodies may be secreted into the circulation (humoral immune reaction), or activated lymph cells may be responsible for I. (cell-mediated immune reaction).

Acquisition of I. can be passive or active. *Passive immunization* occurs, e.g. when antibodies are transferred from mother to fetus, leading to I. of the new-born animal. Injection of therapeutic antisera (e.g. against toxins) also confers passive I. *Active immunization* against disease-causing bacteria or viruses can be brought about by infection with a causative organism (followed by recovery !), or by protective inoculation with attenuated or killed disease organisms or their components (see Inoculation). After exposure to appropriate antigenic material, active I. develops over a period of about 1 week, whereas passive I. is operative immediately after injection of an antiserum. On the other hand, active I. persists, in some cases for many years, whereas passive I. is soon lost. Passive I. by injection also carries the risk of Serum sickness (see).

2) In phytopathology, the absolute resistance of a host plant to attack by a pathogen. See also Preimmunity.

Immunoassay: see Serology.

Immunochemistry: a branch of Immunology (see) concerned with the biochemical bases of immune reactions. It analyses the processes of antibody formation, as well as antibody structure, the structure and function of lymphocyte receptors, and the nature of antibody-antigen interactions. For substances with antigenic properties, immunochemical techniques (immunoassays) are among the most sensitive and specific assay and detection methods. Immunoassays are used widely in biology and medicine, e.g. for the assay of hormones in clinical diagnosis, and for the identification of blood proteins and tissue types in forensic medicine.

Immunocompetent cells: mature lymphocytes, characterized by their ability to produce antibodies. They possess antigen-specific receptors, and they produce an immune response after binding an antigen.

Immunocytochemistry, ***immunohistochemistry*** (plate 18): methods for detecting biological compounds in cells and tissue preparations by immune reactions. Antigens of interest are purified and used to raise antibodies in rabbits or goats, etc. For light microscopic investigations, the resulting purified antibodies are coupled to a fluorescent agent or enzyme, then incubated with the tissue or cell section under investigation. The site of antibody binding corresponds to the location of the antigen, as revealed by the fluorescence of the antigen-antibody complex (fluorescence-labeled antibody) or by color production after incubation with an appropriate substrate (enzyme-coupled antibody). For electron microscopic investigations, the antibody is coupled with heavy metals like gold or iron (in the form of ferritin).

Immunofluorescence technique: a sensitive immunological method for the detection of antibodies or antigens. A fluorescent dye is coupled to an antibody, which can then be detected in very small quantities under the fluorescence microscope. A piece of tissue containing the appropriate antigen is incubated with the labeled antibody. The subcellular or intercellular localization of the antigen is then revealed by the fluorescence of its bound antibody. The I.t. can be used, e.g. to identify the site of synthesis of hormones. It is also important in clinical diagnosis.

Immunogenetics: a branch of Immunology (see) concerned with the genetic basis of immune reactions, and with the inheritance of immunologically important substances, e.g. blood group substances.

Immunoglobulins, ***Ig, antibodies:*** specific defense proteins found in blood, plasma, lymph and many body secretions of all vertebrates. The phylogenetic precursors of the Ig, the cell-bound hemagglutinins, are also found in invertebrates including annelids, crustaceans, spiders and mollusks. The most salient features of Ig production are the ability to respond to the presence of any foreign antigen (primary response), to respond more quickly to a previously encounterted foreign antigen (secondary response), and under normal circumstances not to respond to components of the animal's own body (self-nonself discrimination).

Lymphocytes derived from bone marrow in mammals or the bursa fabricus in birds display Ig on their surfaces. Upon exposure to antigen, those cells whose surface Ig can bind the antigen are stimulated to proliferate into plasma cells which secrete antibodies of the same specificity as the ancestor of the clone. The process of stimulation is complex and involves cooperation with other lymphocytes (T-cells) and macrophages; but B-cells can also be stimulated to proliferate by polyclonal activators such as concanavalin A or bacterial lipopolysaccharide. When bound to antigen, some Ig can fix the C1q component of the Complement system (see), thus facilitating lysis of the foreign cell to which they are bound. Because they are multivalent, Ig also cross-link soluble antigens and facilitate their clearance from the blood or lymph by macrophages.

Structure. Ig are tetrameric, consisting of two identical light (M_r 22 000 to 24 000) and two identical heavy, carbohydrate-containing chains (M_r 50 000 to 73 000).

There are two types of light chain, κ and λ, each of which can be associated with any of five types of heavy chain, α, δ, ε, γ or μ. The Greek letter designating the type of heavy chain corresponds to the class of Ig (A, D, E, G or M). These can be distinguished electrophoretically or serologically. Each type of chain consists of a variable and a constant region. The constant region always has the same amino acid composition, whereas the organism synthesizes a wide variety of different structures for the variable region. The variable regions of the heavy and light chains together form the paratope (antigen-binding site) on the antibody.

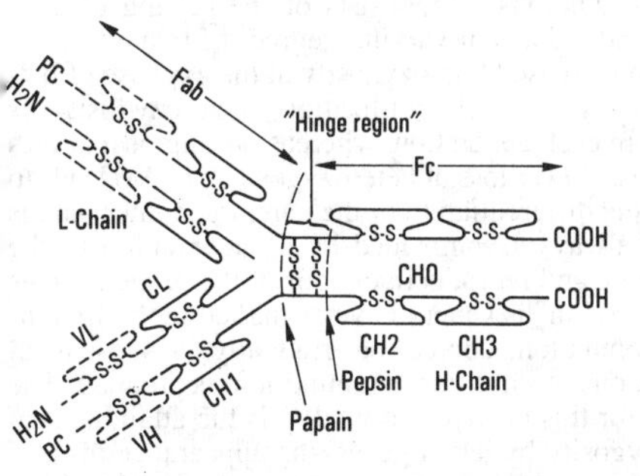

Immunoglobulins. Structure of an IgG molecule. VL and VH are variable domains of the light and heavy chains, respectively. CL and CH are constant domains of the light and heavy chains, respectively. The molecule shown represents an IgG or IgA. Other classes have different numbers of heavy chain domains (2 to 4), and the hinge region may not be present (IgM or IgE). CHO = carbohydrate chain. VL and VH form the antigen binding site. Papain cleavage produces two Fab and one Fc fragment. Pepsin cleavage forms a $(Fab')_2$ fragment and a smaller Fc′ fragment. Pc = pyrrolidone carboxylic acid.

Papain treatmentof Ig releases monovalent antigen-binding fragments (Fab; M_r 50 000) and a complement-binding fragment (Fc; M_r 60 000). Pepsin cleavage, on the other hand, releases a bivalent, antigen-binding $F(ab')_2$ fragment and a somewhat smaller complement-binding fragment Fc′. The $F(ab')_2$ can be dissociated by thiol reagents into monovalent fragments, indicating that the two H chains in the intact molecule are held together by one or more disulfide bonds between the sites of pepsin and papain cleavage (Fig.)

The Y shape of the Ig molecule has been confirmed by electron microscopy and X-ray diffraction. The former shows that the molecule is flexible at the "hinge" region, as was postulated from the susceptibility of this region to proteolytic attack.

Historically, on the basis of their electrophoretic behavior, human Ig were divided into five classes, IgA, IgD, IgE, IgG and IgM. The most abundant of these is IgG. Antibacterial antibodies are of the IgG type, which therefore fulfils the important protective role of Antitoxin (see). Antibodies against viruses consist of the high molecular mass IgM, which also contributes to the destruction of foreign cells by binding and activation of the complement system. IgA-type antibodies are responsible for important immunological defense outside the blood and lymph. Secretory IgA are present in practically all body secretions, and they form a barrier to disease organisms in the mucosae of the body. IgD and IgM are antigen-specific receptors of lymphocytes. IgD presumably does not function as a free antibody. IgE-type antibodies are formed in response to multicellular parasites, particularly intestinal parasites; they are also involved in allergic reactions (see Anaphylaxis).

Immunohistochemistry: see Immunocytochemistry.

Immunological memory: see Anamnestic reaction.

Immunology: science of the immune response. It includes Immunochemistry (see), Serology(see), Immunogenetics (see), Tumor immunology (see), and *immunopathology,* which is concerned with dangerous and deleterious immune reactions such as allergies and autoimmune diseases.

Immunopathology: see Immunology.

Immunoprecipitation, immune precipitation: precipitation of a soluble antigen by a specific antibody. I. reactions are used widely for detection of antigens with antisera, e.g. in the identification of human proteins in blood, saliva, etc. for forensic purposes. I. also occurs when antigen and antibody are brought together in a gel, and routine methods for observing I. in agar gels have been developed. With the aid of such immunodiffusion or gel-diffusion techniques, several specific I. reactions can be studied simultaneously in the same gel. A liquid test sample is placed in a cavity in a layer of agar gel, and different specific antisera are placed around it. Antigens and antibodies diffuse in all directions into the gel from their sites of application. A typical precipitation zone is formed wherever a diffusing antigen encounters a diffusing specific antibody. In immunoelectrophoresis, the antigen mixture is first separated by electrophoresis, and the resulting fractions are tested against specific antisera.

Impala: see *Antilopinae.*

Imperial pigeons: see *Columbiformes.*

Implantation: see Placenta.

Imprinting: a form of obligatory learning, which occurs only at a certain time in an animal's life, and which cannot later be reversed, or only with great difficulty. Often, I. occurs very early in life during a brief critical period or Critical phase (see), which may last only a few hours. For example, I. is strongly developed in precocial birds, which at an early age learn to follow a moving object. This first example of the phenomenon of I. was described by Konrad Lorenz in 1937. Thus, in goslings, the critical phase for this particular learning process or I. lasts only about 24 hours. I. influences subsequent behavior, in particular sexual behavior. If I. does not occur at the appropriate time (i.e. during the critical phase), the animal thereafter displays behavioral abnormalities. Three levels of I. are recognized. a) A programmed motor response is imprinted by a stimulus e.g. precocial birds learn to follow a moving object. b) A programmed motor response is fixed in association with an object, then adjusted, e.g. this type of I. operates in some species for copulatory behavior, or for the capture and killing of prey. c) Patterns of stimuli, experienced during sensitive phases, determine the response of the animal to different patterns of environmental stimuli, i.e. they provide the animal with spatial and temporal identity and orientation.

Inborn behavioral response: see Innate releasing mechanism.

Inborn illnesses, ***genetic disorders:*** illnesses or malformations arising from specific alterations in the genome. They may arise from altered DNA structure within a single gene (monogenic disorders) or within several genes (polygenic disorders), from realtively gross structural changes in individual chromosomes (see Chromosome mutations), or from abnormal chromosome numbers (see Genome mutations).

About 3 000 human I.i. are known. The mode of inheritance of most of these conditions has been elucidated, and rapid progress is being made in the identification of the corresponding specific genetic defects, accompanied by development of diagnostic methods, appropriate therapies and effective prophylactic measures. About 5% of all neonates display an inherited disorder. The proportion of genetically damaged embryos that die prenatally is several fold higher.

An important task of Human genetics (see) is the elucidation of factors causing I.i., as well as the development of suitable means of control. Of particular importance are measures to decrease the frequency of new mutations by consistent prophylaxis against the mutational effects of environmental mutagens, etc., and the prevention of the transmission of defective genetic material from generation to generation by effective Genetic counselling (see).

The first experimental attempts to cure an inborn human disorder by normalization of defective genetic material, i.e. by *gene therapy,* were started in 1990 in the USA. The disorder in question was a rare immune deficiency disease caused by the absence of adenosine deaminase (ADA). A disabled mouse retrovirus was used to carry healthy copies of the ADA gene into white blood cells taken from the patient; the treated white cells were than returned to the blood stream. British and Dutch workers later announced their plans for a similar experimental approach, using stem cells from bone marrow instead of white cells. Such experiments require clearance from medical ethics committees. According to the British Clothier Committee, set up in 1989 to consider proposals for experimental gene therapy, the ethical issues are similar to those associated with organ transplantation, and such experiments should be restricted to alleviating disease for which there is no other effective treatment. In 1990, the National Institutes of Health (USA) approved three proposals to use genetically engineered cold viruses to carry healthy genes into the respiratory passages of cystic fibrosis patients. Cystic fibrosis is the commonest lethal inherited disease of Caucasians, affecting 1/2 000 to 1/2500 babies, whose life expectancy is about 30 years. It is characterized by the accumulation of dense, sticky mucus in the lungs and airway epithelia, leading to bacterial infection of respiratory passages. The airway epithelia of affected patients display defective cyclic AMP-regulated chloride conductance. Similar experimental therapies are planned for muscular dystrophy and severe melanoma.

In 1993, it was shown for the first time, that gene therapy can restore normal function to an organism suffering from an inherited lethal mutation (S. C. Hyde et al., *Nature,* ***362,*** 250–255). These authors enclosed healthy copies of the human cystic fibrosis membrane conductance regulator (CFTR) gene in liposomes, which were then introduced into the lungs of mice with artificially induced cystic fibrosis (the liposomes fuse with cell membranes so that the healthy DNA passes into the cells). Normal ion transport was restored and the accumulation of mucus was prevented.

See Genetic prognosis.

Inborn memory: see Innate behavior.

Inborn reflex: see Conditioned reflex.

Inbred line: a largely homozygous line, strain or variety, produced by repeated inbreeding of an allogamic individual and its offspring. See Inbreeding.

Inbreeding, ***endogamy:*** pairing and reproduction between individuals that are more closely related than average pairs selected at random from the population. The most restrictive form of I. results from self-fertilization. Otherwise, I. consists of the pairing of allogamic individuals of varying degrees of relatedness. I. leads to increased homozygosity at the expense of heterozygosity (in self-fertilization, heterozygosity is halved in each generation, whereas pairing of relatives leads to a slower loss of heterozygosity); it also leads to genotypic differentiation of the parent genomes, and is almost always accompanied by an attenuation of the vegetative and reproductive vigor of the progeny. This "I. depression" is generally very marked in the first inbred population, decreasing gradually in subsequent generations, until a stable I. minimum is attained. The reason for this decrease in vitality is the attainment of homozygosity by lethal genes, the appearance of nonadapted or poorly adapted genotypes, and the breaking up of balanced polygenic systems. Offspring from the crossing of lines that have attained their I. minimum often display hybrid vigor (heterosis effect).

The extent of I. is expressed by the Inbreeding coefficient (see). The frequency of a genotype in a population with I. deviates from the Hardy-Weinberg distribution (see), but conforms to the Wright distribution (see).

Inbreeding coefficient, f**:** the probability that two alleles at the same locus in an individual are copies of the same gene, i.e. identical by descent. In children from the pairing of cousins, $f = 1/16$. In children from the pairing of a brother and sister, $f = 1/4$. The I.c. is inversely related to the panmictic index or panmixis index (P), so that $f = 1 - P$.

See Wright's inbreeding coefficient.

Inca dove: see *Columbiformes.*

Incarbonization: see Carbonification.

Incasement theory: see Preformation theory.

Inceptisols: mineral soils in the early stages of formation of visible horizons, and with an incompletely developed profile. One or more horizons may be present, in which minerals have been weathered or removed. See Tundra.

Incertae sedis: a term meaning "of uncertain taxonomic position".

Incisor: see Dentition.

Inclusion bodies: see Insect viruses.

Incoalation: see Carbonification.

Incompatibility: a condition resulting in the failure to produce offspring, despite the presence of functional gametes. I. between members of the same species or race is called *intraspecific I.* Between different species, it is called *interspecific I.* It often has a genetic basis, and genetically determined I. is usually due to multiple alleles; the latter are also known as *I. factors.* Incompatibility factors in the protoplasm can give rise to *protoplasmic I.* I. is the cause of the self-sterility of many seed plants, and the frequently observed failure for genetically distant plant species to cross with one another.

In the latter case, the pollen tube is inhibited from penetrating the stigma. In physiologically heterothallic fungi, the inability to reproduce in single or paired cultures is also due to I. In *homogenic I.* of fungi, a zygote is not formed because the partners possess the same incompatibility genes. Alternatively, in *heterogenic I.,* two sexually incompatible individuals carry different alleles at their incompatibility loci. In the culture of fruit trees, the inability of certain scions to successfully graft onto a stock is also known as I.

Incrustation: a type of petrification; the encasement of fossils and other objects, on or in the earth, in a mineral layer of SiO_2, $CaCO_3$, FeS_2, or brown or red ironstone. The final form may be a nodule, concretion or geode, containing the petrified object which originally acted as the precipitation focus.

Incubation:

1) In microbiology, the maintenance of microorganisms under conditions of temperature, light, humidity, etc. (in particular temperature) that favor their multiplication, e.g. in an incubation cabinet.

2) The process that occurs between exposure of an animal to an infectious agent and the clinical manifestation of disease. The time that elapses between infection and disease manifestation is the incubation period (or time) of the disease.

3) Maintenance of eggs at a temperature at which embryo development proceeds, culminating eventually in hatching.

Incubation cabinet: a laboratory cabinet, whose internal temperature can be adjusted, and kept constant by a heater and an automatic thermostat. It usually has double metal walls, enclosing a water jacket or insulating material, and it is used mostly for the culture of microorganisms.

Incubation period: see Incubation.

Incurvariidae (Incurvariids): see *Lepidoptera.*

Incus of the middle ear: see Auditory organ.

Indehiscent fruits: see Fruit.

Independent assortment: see Mendel's laws.

Index plant:

1) See Indicator plants.

2) A plant species that characterizes a certain region or plant community. When applied to a particular region (e.g. Brandenburg I.ps.), the term is generally a Floral element (see). As applied to plant communities, the term usually means a diagnostically important species, but it is used more loosely, and is often not based on a statistical vegetation survey.

Index species: species with a high Constancy (see) in the communities of a biotope. Together with Character species (see), I.s. serve for the characterization of biocoenoses. I.s. are sometimes considered as synonymous with differential species.

Indian desert jird: see *Gerbillinae.*

Indian fig: see *Cactaceae.*

Indian laburnum: see *Caesalpiniaceae.*

Indian robin: see *Turdinae.*

Indican: see Indigo.

Indicator: see *Indicatoridae.*

Indicatoridae, ***honeyguides:*** an avian family in the order *Piciformes,* containing 11 species in 4 genera, ranging in size from 11.5 to 20.5 cm. Some species show sexual dimorphism, but all members generally have drab brown and olivaceous plumage, with lighter underparts and spotting on the wings and tail. The claws of the zygodactylous feet are strong and hooked. The bill is short, stout and blunt, with raised rims around the nostrils. It is predominantly an African family, 9 of the known species being found south of the Sahara. Another species inhabits the foothills of the western Himalayas, while a further rare species is found in the jungles of peninsular Malaya and the islands of Sumatra and Borneo. They are nest parasites, often laying their eggs in the nests of related species, such as barbets, woodpeckers, etc. They feed on a wide variety of insects, but show a marked preference for bees and wasps. It was long thought that honeyguides raided hives and nests for honey, larvae and eggs, but it is now known that they feed on the wax of the comb, i.e. they display cerophagy. Digestion of the wax is aided by their specialized intestinal flora. The name, honeyguide, is due to the habit of the ***Black-throated*** or ***Greater honeyguide*** *(Indicator indicator)* of enlisting the help of other animals in robbing the nests of wild honey bees, a type of behavior known as "foraging symbiosis". This is achieved by generally displaying and making itself conspicuous (e.g. by fanning its tail and making repeated churring noises) in front of a ratel (honey-badger) or a human, then leading the way to the nest, keeping 4–5 m ahead. After the nest has been robbed, it feeds on the remains of the broken comb.

Indicator media: see pH-indicator media.

Indicator organisms: organisms whose presence is indicative of certain environmental conditions. See Biological water analysis, Indicator plants.

Indicator plants: plants whose abundance is indicative of environmental conditions and factors. Indicator species have a narrow ecological amplitude with respect to one or more environmental factors. Their presence may indicate the existence of a complex of interactive factors, or the presence of just a few dominant factors. For example, I.p. may indicate the nature of the local climate, soil nitrogen content, soil or water alkalinity or acidity, presence of limestone, degree of salinity, etc. Many of these factors can be quantitatively calibrated by chemical and physical soil and water analysis, then related to the particular species or ecotype of I.p. Species with heavy metal tolerance have been exploited in geobotanical surveys as indicators of the presence of metallic ores. See Calcicoles, Calcifuges, Halophytes, Bioindicators, Biological water analysis.

Indicaxanthin: see Betalains.

Indifference temperature: environmental temperature range in which Temperature regulation (see) is unnecessary, i.e. the Body temperature (see) remains constant without the intervention of physiological mechanisms for the extra generation or loss of heat. For the unclothed human, the I.t. is about 30 °C.

Indigenae: inhabitants of a biotope that live, feed and reproduce in that biotope. They are therefore permanent, native inhabitants, and their allegance to the biotope is absolute. Other inhabitants are classified (in decreasing order of allegance) as: visitors *(hospites),* neighbors *(vicini), accidental guests* and aliens *(alieni).*

Indigo: a dark blue vat dye known in ancient times and particularly valued in the Middle Ages. It was formerly extracted from a few *Indigofera* species (family *Papilionaceae)* like *Indigofera tinctoria,* and from dyer's woad *(Isatis tinctoria;* family *Cruciferae).* The latter contains the colorless ester, indoxyl-5-oxofuranogluconate, whereas *Indigofera* contains the colorless indoxyl-β-D-glucoside (indican). When plants are damaged (I. is prepared by the water extraction of chopped leaves) both of these compounds are enzymatically hydrolysed to the corresponding sugars and soluble indoxyl (3-hydroxyindole). Insoluble I. is then formed by the atmospheric oxidation of indoxyl. Natural I. is a mixture containing up to 95% I. and the accompanying pigments, indirubin and indigo brown. In the original dyeing process, insoluble I. was converted into soluble indigo white under reducing alkaline conditions. Material soaked in this liquid was then hung in the air to oxidize the colorless product into I. Today, I. is produced synthetically. It is fast to light, acid and bases, and is particularly suitable for dyeing cotton and wool. It was formerly a most important organic dye, which was used and traded throughout the world, but it has now been largely replaced by other synthetic dyes.

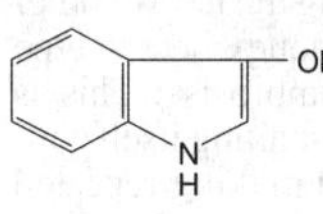

Indican: R = glucose
Indosyl R = H

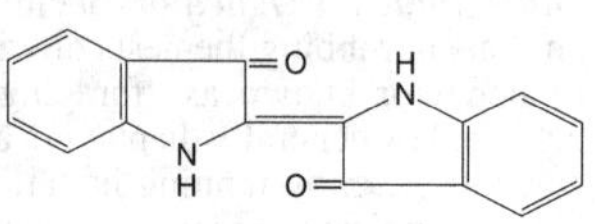

Indigo

Indigofera: see *Papilionaceae.*

Indispensable amino acid: see Amino acids.

Individual: a single living organism, defined by its independent existence in time and space, and the uniqueness of its phenotype. The I. displays all the essential characters of its species, such as the structural plan, substance metabolism, energy metabolism, stimulatory responses, developmental processes, adaptation and reproduction. The uniqueness of the I. is determined by the particular combination of its genetic material, and by the environmental influences operating at the various stages of its ontogenic development.

Assignment of individual identities is not possible within all species (see Organism collective).

Individual distance: the minimal distance between individuals in populations and groups of freely motile animals. I.d. is governed by various types of interaction, but with respect to I.d. there are basically two types of animal: Distance types (see) and Contact types (see).

Individual selection: in behavioral research, the selective discrimination between individuals. In contrast to Group selection (see), the selection process is applied at the level of the individual. Individual Fitness (see) is therefore operative in maintaining the individual genotype in the gene pool of the population.

Indoaustralian transition zone: see Wallacea.

Indole, ***benzopyrrole:*** a heterocyclic, aromatic base, which occurs naturally in the free form and in many derivatives. In very small quantities, pure I. has a pleasant, jasmine-like odor, but large concentrations have a fecal smell, especially in the presence of impurities. Free I. is an important odiferous constituent of numerous essential oils, e.g. jasmine, orange and wall-flower oils. I. produced from L-tryptophan during the putrifaction of proteins, is partly responsible for the unpleasant smell of feces. The bacterial enzyme responsible for the formation of I. from L-tryptophan (tryptophanase) catalyses the reaction: L-tryptophan → indole + pyruvic acid + ammonia. Pyridoxal 5'-phosphate (a vitamin B_6 derivative) is the cofactor of the enzyme. Production of I. from tryptophan is diagnostic of *Escherichia coli.* The microbiological test for indole production is based on the formation of a red compound between indole and Ehrlich's reagent (*p*-dimethylaminobenzaldehyde) under acid conditions. Important plant products derived from L-Tryptophan (see) and containing the indole ring system are: Ergot alkaloids (see), Skatole (see) and Indigo (see).

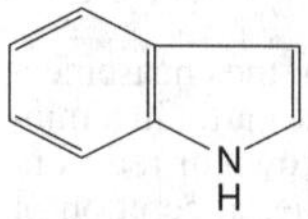

Indole

Indri: see Lemurs.

Indriidae (indriids): see Lemurs.

Induced resistance: in phytopathology, resistance to infection by a parasite, arising from a previous infection or treatment. Resistance can be induced by Inoculation (see) or treatment with other pathogens, with less virulent races of the same disease organism, saprophytes, heat-inactivated disease organisms, cell extracts, substances isolated from the pathogen, or pathogen metabolites from infected plants. After the appropriate treatment, a certain time must elapse before I.r. is demonstrable. Application of substances that act directly on the pathogen (e.g. chemical fungicides) does not qualify as the induction of resistance.

Inducer: see Induction, Enzyme induction.

Induction:

1) "Actuation of a developmental process in one part of an organism by a different part of that organism" (A. Kühn). Different parts of an embryo do not develop independently of each other, but display many types of interdependence and interaction. A combination of inhibition and I. insures the harmonious and balanced development of the entire embryo.

A well known example is that discovered by Spemann in the amphibian embryo, in which an "Organizer" (see) in the region of the blastopore lip induces the overlying ectoderm to differentiate into the spinal cord and brain, while the I. itself develops into notochord and somites.

General principles governing I. are exemplified by lens development. In most amphibians, abstriction of the lens from the ectoderm is preceded by a stimulus from the optic vesicle as it advances from the diencephalon toward the epidermis. If the optic vesicle (e.g. of *Rana fusca* or *Bufo bufo)* is removed before it reaches the epidermis, the lens does not develop. Also in these two species, a piece of epidermis from the abdomen or head will develop into a lens, if it is transplanted into the region of lens formation. Apparently, epidermis from a different site in the embryo also has the ability (competence) to respond faithfully to the inducer (inductor). In other amphibian species (e.g. *Bombinator pachypus),* the entire epidermis does not have lens competence, and only the epidermis of the head region is capable of developing into a lens. In some species, lens competence is restricted to the site of presumptive lens formation.

The lens blastema is competent only for a certain period of time. Competence first appears at the stage of the neurala, after the epidermis has differentiated from the ectoderm. In embryos with widely dispersed lens competence, competence disappears earliest from the epidermal regions most distant from the site of presumptive lens formation, and is finally (at the stage of the tail bud) lost from the presumptive site itself.

I. represents the response of the blastema to chemical substance(s) from the inducer. I. is merely a process of initiation, and the particular structure that eventually develops depends on the nature of the blastema. For example, axolotl epidermis, transplanted into the presumptive lens site of *Triturus vulgaris,* responds to the I. stimulus of the different genus and develops into a lens, but the lens is much too large, i.e. it has the dimensions of an axolotl lens.

I. originates in the notochord mesoderm of the head region (primary I.), followed by lens I. by the optic vesicle (secondary I.). The optic cup and lens are then responsible for I. of the development of the cornea from the overlying epidermis. I. therefore consists of the following chain of processes:

Notchord mesoderm of the head region —**Primary induction**→ Brain → Diencephalon → Optic vesicle —**Secondary induction**→ Lens —**Tertiary induction**→ Cornea

A similar hierarchy of inducers can be constructed for the embryonic development of the ear:

Notchord mesoderm of the head region —**Primary induction**→ Brain → Hindbrain —**Secondary induction**→ Auditory vesicle → Membranous labyrinth —**Tertiary induction**→ Cartilaginous labyrinth —**Quaternary induction**→ Ear drum

In synergistic I., two different tissues influence the development of each other. For example, synergistic I. occurs in the embryonic development of the mammalian kidney from two different components. Nephrons are derived from the caudal section of the nephrogenic tissue, which consists of a strand of loosely packed mesodermal cells. The ureter, however, develops from the mesonephric (Wolffian) duct, thickens at its end and develops into the kidney pelvis and the numerous collection channels, which later unite with the tubules of the nephrons. If either of these two components is removed, further differentiation of the other is stopped. Similar behavior is observed in tissue culture; development proceeds normally only if the two tissues are cultured in contact with each other.

I. is a fundamental process in the development of living structures. It is operative not only in the embryology of higher animals, but also in the development of lower animals and plants. For example, in the development of the insect embryo *(Chrysopa),* differentiation of musculature, fat body and and heart cells from the mesoderm is induced by the overlying ectoderm. This is therefore the reverse of the relationship between notochord mesoderm and ectoderm, in which the former is the inducer and the latter is the blastema.

2) In the sensory physiology of plants, the initial reaction of a chain of stimuli, consisting of *stimulus susception,* e.g. absorption of light by phytochrome, and *stimulus perception,* which is the first physiological reaction resulting from the susception. See Perception

3) In animals, see Receptor, Sensory organ.

4) See Enzyme induction.

Inductor: see Induction *(1).*

Indusium:

1) The delicate envelope that completely surrounds the sori of Ferns (see).

2) The delicate membrane covering the underside of the young cap of various fungi. It is later torn, and forms a collar (see Anulus) around the stipe.

Industrial melanism: increase in the frequency of occurrence of black or dark moth species in areas affected by industrialization. Since 1850, more than 100 of the 780 species of large *Lepidoptera* in the UK have produced an increasing number of melanistic varieties. In an environment darkened by smoke and exhaust gases, formerly rare dark mutants are not easily recognized by predatory birds, and they therefore acquire a selective advantage. Displacement of the light variety of the Peppered moth *(Biston betularia)* by the dark variety is one of the best documented selection processes in any natural population.

Industrial microbiology: a branch of applied microbiology which uses microorganisms for production processes (production microbiology) or other technological processes, as well as developing methods for preventing the harmful effects of microorganisms on raw materials, food and other products. I.m. plays an important part in the fermentation industry for the production of beer, wine and vinegar, and in the food industry for the production of bread, risen pastries, sauerkraut, quark, cheese, yoghurt and numerous other milk products. Many processes of production microbiology are performed with microorganisms on a large industrial scale, e.g. the preparation of antibiotics, vitamins, enzymes, growth substances, alkaloids, alcohols, amino acids, carboxylic acids, polysaccharides, etc. A special case is the synthesis of steroid hormones, in which microorganisms are used to convert inactive steroids into biologically active compounds. Microbiological processes for obtaining fiber from plants (see Flax retting), and microbiological mining (leaching of compounds of valuable metals from ores by bacterial activity) also belong to I.m. Yeasts are grown on a large scale as animal feed and for baking. Mass production of bacteria and algae for the preparation of feedstuffs, using cheap raw materials, is carried out in some countries. A further important field of I.m. is biological water purification.

Infans 1 & II: see Age diagnosis.

Infant releasing mechanism: see Infant schema.

Infant schema, ***infant releasing mechanism:*** a releasing stimulus for human care behavior. In the original sense, this consisted of visual signals, such as rounded contors, relatively small facial skeleton, chubby cheeks and relatively large eyes. Nonvisual signals, however, may also act as releasers, e.g. certain frequencies of the infant voice. These characters are often exaggerated in the construction of dolls. They stimulate hand contact (stroking), the action of pressing the doll (or infant) close to the holder's own body, as well as a characteristic manner of "talking" to the doll (or infant).

Infection:

1) The process of transmission of disease-causing microorganisms to, and their establishment on or in, a host organism. The presence of disease-causing organ-

Infant schema. a Head and facial proportions that are active in the infant releasing mechanism: child, desert jerboa, pekinese, robin. *b* Non-active head and facial types: man, hare, hunting dog, oriole. (After Lorenz, 1943)

isms in a host is also called an I. Subsequent multiplication of such microorganisms usually leads to an outbreak of an infectious disease. Sources of infection include infected organisms (animals, humans, plants), plant and animal products (e.g. food). Some infections may be water-borne or air-borne, and others may be contracted from soil. Transmission of a disease organism may be indirect (e.g. by insects), or direct by bodily contact between infected and noninfected individuals.

2) In microbiology, contaminatioin of a pure culture with other unwanted microorganisms.

Infection chain: in phytopathology, the continued transmission of pathogens or their infection organs from one host generation to the next, and from one vegetative period to the next. In regions of permanent vegetative cover, I.cs. are *continuous,* whereas in regions with a seasonal loss of vegetation I.cs. are discontinuous. In a *homogeneous I.c.,* only one host species is infected, whereas in *heterogeneous I.c.* there is an alternation of hosts (e.g. rust disease of cereals).

Infection potential: in phytopathology, the ability of a pathogen to cause an infection, depending on the density (concentration) of the pathogen and environmental conditions.

Infection threshold: the minimal number of individual infectious particles (e.g. spores) of a plant parasite, necessary to establish an infection when conditions favoring Infection (see) are optimal.

Infectious diseases: diseases of humans, animals and plants, caused by specific disease-causing organisms (usually microorganisms). I.d. are acquired and spread by Infection (see).

Infertility: see Sterility.

Infilling of lakes: see Aging of lakes.

Inflorescence: see Flower.

Influenza virus: see Multipartite viruses, Virus families.

Influx: entry of substances into cells. It is usually an active process, and the term is appled in particular to the active uptake of ions.

Information:

1) A structural characteristic of ecosystems (see Diversity). A biological system at Climax (see) has a high I. content, because it displays a high degree of structural organization, and maximal entropy with respect to the number of species and individuals.

2) See Genetic information.

Information biotransformation: see Biotransformation of information.

Information theory: a branch of probability theory concerned with the likelihood of the accurate transmission of information. The unit of information is the *bit,* one bit representing information with a probabilty content of 1/2. The information content I_i is calculated from the probabiltiy of the occurrence of certain information, e.g. the frequency of the symbol x_i, where $I_i = -\mathrm{ld}(x_i)$. The average information content of an information source containing *N* symbols of information is known as entropy, since it formally corresponds to thermodynamic entropy. This entropy decreases with increasing order in the distribution of the information. If, in a sequence of information, the occurrence of particular information depends on that preceding it, then long series of linked pieces of information must be considered in the calculation of entropy (e.g. the probability of the appearance of a particular letter in a printed text depends on the preceding letters).

In contrast to communication theory, I.t. is concerned with the biased or one-sided transfer of information. The receiver does not influence the source. An important task of I.t. is to investigate the influence of random disturbance or irrelevance on the content of the transferred information. Shannon showed that it is possible, with appropriate coding, to transfer a given quantity of information with an error probability approaching zero. An important quantity of I.t. is *channel capacity,* which represents the maximal possible information flow. The channel capacity of technical information transfer methods is adapted to our own sensory organs. Thus, a channel capacity of a television set of 7×10^7 bits/s is suitable for the receiver channel capacity of the eye of about 3×10^6 bits/s. The telephone channel capacity of about 5×10^4 bits/s can be compared to the receiving channel capacity of the ear (4×10^4 bits/s). Although humans can receive information at a high rate, they are capable of processing it at only a relatively slow rate, e.g. about 40 bits/s when reading. There is therefore a necessary reduction in the amount of information accepted, and this occurs unconsciously in the sensory organs and in the central nervous system.

Communication theory is concerned with the coupling of information transmission and reception, and this coupling is affected by interference and coding. A distinction is drawn between *one-sided* or *biased coupling* (the transmitter is influenced only by its own previous information output), *determined coupling* (information from the transmitter is controlled entirely by the receiver), and the general case of *interactive* or *re-*

ciprocal coupling. In the human context, these types of coupling are equivalent to *monolog, suggestion* and *dialog,* respectively. Communication theory provides a means of quantifying communication in the study of behavior.

Informofers: see Nucleus.

Informosomes: see Nucleus.

Infrared spectroscopy: a spectroscopic method for investigating the absorption of light in the infrared region of the spectrum, i.e. the region where absorption is due to inter- and intra-molecular vibrations. Intramolecular vibrations are classified as follows. a) Stretching vibrations, which occur along the axis of a chemical bond. c) Bending or deformation vibrations, in which atoms move perpendicular to the chemical bond. Atoms may vibrate within or out of the plane of the group, out-of-plane vibrations also being known as wagging vibrations. c) Skeleton vibrations, in which the entire skeleton of a group, e.g. the benzene ring, vibrates. These include skeleton-stretching and skeleton-bending vibrations. d) Relatively low intensity absorptions resulting from the sums or differences of other vibrations. I.s. is particularly useful for investigating hydrogen bonds, and it provides useful information on the conformation of proteins and polynucleotides. Analyses can be performed on extremely small quantities of material.

Infusoria: an old term for the populations of microorganisms that appear after a certain time in hay infusions and the water surrounding other submerged plant material. I. consist largely of protozoa, most of which are ciliates, with some flagellates. Their relatively sudden appearance is due to their emergence from resting cysts, followed by rapid multiplation.

Ingolfiellidea: see *Amphipoda.*

Inheritance of acquired characters: transfer of adaptations, which develop during the life of an individual, to the offspring of that individual. An I.o.a.c. has never been demonstrated experimentally. In view of the unidirectionality of gene expression ("central dogma of molecular biology"), it is also highly unlikely on theoretical grounds. See molecular genetics.

Inherited behavioral response: see Innate releasing mechanism.

Inherited developmental response, ***innate morphogenic response:*** genetically determined specific response(s) of the developing organism to environmental stimuli, where the environmental stimuli may be internal (e.g. degree of cell to cell contact within a growing mass of cells) or external (e.g. temperature, light, ion concentrations in the surrounding medium). An I.d.r. involves the generation of spatial patterns of cells, determination of the site of formation of certain structures, determination of the direction in which cells and organs will grow, and the plane in which cells will divide, etc., all leading eventually to the final morphology of the organism. For example, the zygotes of brown algae germinate on the shaded side when illuminated unilaterally. When placed in various concentration gradients, they germinate toward the side of highest K^+, Ca^{2+} or H^+ concentration. Much evidence suggests that cell membranes are the sites for the perception and transduction of environmental information, and that endogenous electric currents are involved in the control of differentiation and growth.

Inherited releaser mechanism: see Innate releasing mechanism.

Inhibition:

1) In neurophysiology, decreased excitability, due to e.g. an increased Refractory period (see), hyperpolarization, presence of inhibitory substances that block receptors, presence of Transmitters (see) that elicit an inhibitory postsynaptic potential, action of Neuron circuits (see). I. is fundamentally important for the operation of the nervous system; for example sudden, apparently spontaneous excitations can often be explained by the removal of I. from certain neuronal centers. Inhibitory processes reinforce the degree of contrast with which the external environment is perceived (see Perception). Reflexes, such as the patellar reflex, involve an interplay of excitation and I.; thus, the sharp tap on the knee causes a reflex contraction of the extensor muscle with simultaneous inhibition of the flexor.

2) In behavioral research, blockage of a pattern of behavior is called I.

3) In enzymology, a well developed area of study concerned with the inhibition of enzyme activity (see Inhibition).

4) A general term for a decrease in the extent or rate, or the prevention of any process or reaction.

Inhibitor: a substance that retards or prevents a chemical or biochemical reaction, as distinct from a catalyst which accelerates a reaction. In particular, there are many Is. that specifically block the activity of certain enzymes. A distinction is drawn between *reversible* and *irreversible* inhibition. Reversible inhibition is further subdivided into competitive inhibition, noncompetitive and uncompetitive inhibition. In *competitive inhibition,* the I. competes with the enzyme substrate for binding at a particular site (the active center) on the enzyme molecule. According to the law of mass action, the I. displaces the substrate from the enzyme, and substrate conversion is slowed down or stopped. The rate of the inhibited reaction then depends on the substrate/I. concentration ratio. In *noncompetitive inhibition* the I. binds at a site different from the active center, and decreases the affinity of the enzyme for its substrate; it may bind only to the free enzyme or to both the free enzyme and the enzyme-substrate complex. In *uncompetitive inhibition,* the I. binds only to the enzyme-substrate complex. Noncompetitive and uncompetitive Is. are special cases of *allosteric effectors,* which by definition change the activity of an enzyme by binding at a site different from the active center (see Metabolic regulation). Antibiotics (see) are specialized I.

In plants, more or less specific Is. oppose the actions of Phytohormones (see) in the regulation of growth and development. Interactions of phytohormones and Is. are very complex, the more so because growth regulators may be inhibitory or stimulatory, depending on their concentrations. Natural Is. have been demonstrated biologically in practically all plant organs, although many have not been chemically identified. In addition, there are many synthetic Is., belonging to different classes of chemical compounds. On the basis of the applied biological tests, most of the Is. isolated from leaves, stems, tubers, buds, roots, pollen, seeds and fruits are auxin antagonists or antiauxins. Within the plant, however, it is thought that they not only affect extension growth, but also have specific functions in the regulation of developmental processes, e.g. seed germination, bud dormancy, root formation, etc. Conversely, *germination Is.* (blastocolins) can also influence growth processes.

Known biogenic Is. include various types of chemical compounds. Several contain an unsaturated lactone ring. Examples are coumarin, scopoletin, etc. Cinnamic acid can also function as an I. Correlative Is. are responsible for correlative inhibition (see Correlation).

Important synthetic Is. are certain gibberellin antagonists, and especially the various types of chemical compounds used as Herbicides (see).

Inhibitory postsynaptic potential: see Synaptic potential.

Inia: see Whales.

Iniopterygiformes: see *Holocephali.*

Initial cells: see Meristems.

Ink sac: see *Cephalopoda.*

Innate behavior: behavior based on and regulated by a genetic store of information, which is passed from generation to generation ("innate or inborn memory"). I.b. is functional, and it is performed when an animal of an appropriate age and motivational state encounters the appropriate environmetal stimulus or cue. Thus, the usually successful quest for a source of milk by newborn mammals, using specific environmental markers, is an example of I.b. The establishment or alteration of the genetic basis of I.b. is only possible over generations by a process of evolution. In contrast, "learning" can be aquired or modified in a single lifetime.

Innate memory: see Innate behavior.

Innate releasing mechanism, ***IRM, inherited behavioral response, inborn behavioral response, inherited releaser mechanism:*** a genetically determined behavioral system, which reacts to signals (e.g. it reacts to recognition stimuli) without training or conditioning, and which is phylogenetically selected. An I.r.m. is usually species-specific. The converse is an acquired releaser mechanism (acquired behavioral response), which is a behavioral reaction acquired by training or conditioning.

Inner ear: see Auditory organ.

Innidation: see Niche.

Inoceramus (Greek *is* or *inos* muscle or strength, *keramis* brick or tile), *Catillus,* Brong., *Haploscapha,* Conr., *Neocatillus,* Fish.: a fossil bivalve genus of the family *Inoceramidae.* The valves are unequal, although in several species the difference is only slight. Internally, the shell carries an impression of the musculature. The hinge line is long, straight and without teeth, but with numerous small, transverse ligament pits. Each valve is sculptured externally with coarse, concentric undulations. The shell material consists of an outer thick prismatic layer (ostracum) of calcite and a thin internal layer of mother of pearl.

I. is found from the Jurassic to the Cretaceous, and is one of the most important genera of noncephalopod index mollusks of the Upper Cretaceous. It has a more or less homogeneous cosmopolitan distribution, probably because its members spent a comparatively long period of their life history as pelagic larvae. Many Cretaceous pearls have been reported, all produced by species of *I.*

Inoculation:

1) In microbiology, the transfer of living microorganisms onto or into a growth medium for the purposes of cultivation.

2) A procedure for generating potective immunity by the injection or oral administration of antigenic material prepared from disease-causing organisms. Immunization may be achieved by the injection of toxins prepared from a disease organism. Thus, protection against diphtheria is achieved by immunization with exotoxins prepared from the filtrate of cultures of *Corynebacterium diphtheriae.* Diphtheria toxin heated for 1 hour at 70 °C is no longer toxic, but still stimulates antibody production. Such detoxified but antigenic exotoxin is called *toxoid,* and it is used, e.g. for active immunization against diphtheria and tetanus. Alternatively, killed bacteria can be used for immunization. Suspensions of killed bacteria are called *bacterins.* Bacterins are very effective, e.g. in preventing typhoid and paratyphoid. If toxoid or bacterins prove ineffective, immunization is

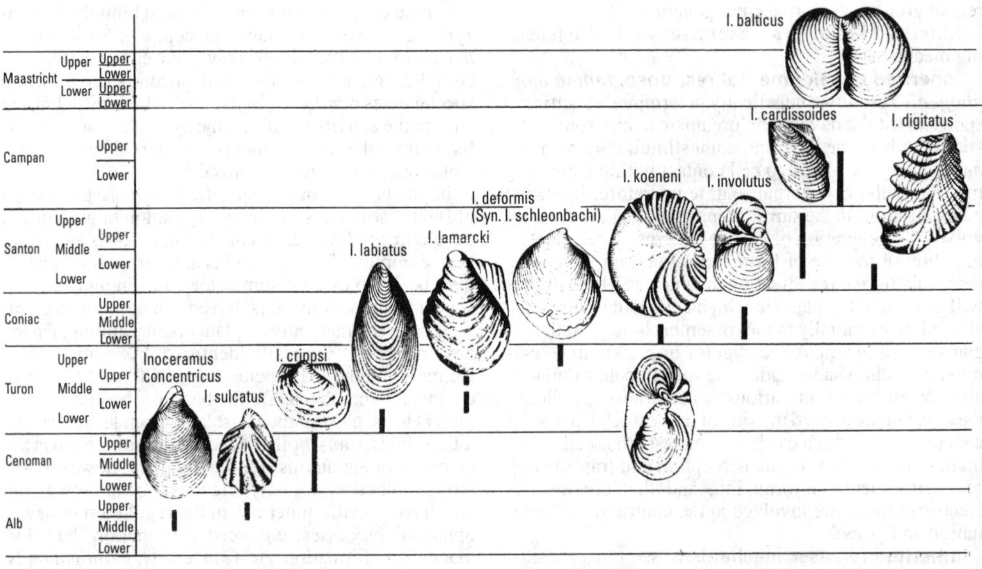

Inoceramus. Index forms of *Inoceramus* from the uppermost Lower Cretaceous to the Upper Cretaceous.

performed with *attenuated, living, infectious agents* that are no longer capable of causing the active disease, e.g. the poliomyelitis vaccine which is administered orally; the modified smallpox virus which is use to produce a localized infection (this disease is now extinct) (see Vaccine); and BCG vaccine, a nonvirulent culture of the tubercle bacillus, which is used to protect against tuberculosis.

Any preparation administered for the purpose of conferring protective immunity is generally known as a Vaccine (see), although the term was originally coined for the immunizing agent against smallpox.

Inoculation wire: an instrument used for the transfer of microorganisms in the microbiological laboratory. It consists of a wire (of platimun or nichrome) set in a metal holder or in a glass rod. It may serve as a straight inoculation needle, or it may be bent into an inoculation loop. Before and after use, it is sterilized by briefly making red-hot in a Bunsen or spirit flame.

Inoculum: a certain quantity of living cells of a microorganism, which are used to inoculate a culture medium (see Inoculation). The appropriate size of an I. depends on the volume or surface area of the culture medium.

Inocybe: see *Basidiomycetes (Agaricales).*

Inosine, *Ino, hypoxanthinosine, hypoxanthine riboside, 9β-D-ribofuranosylhypoxanthine:* a β-glycosidic nucleoside in which D-ribose is linked to hypoxanthine. It occurs free, especially in meat and yeast, and is formed by dephosphorylation of inosine phosphates. It fulfils a special function as a component of the anticodon of certain species of tRNA.

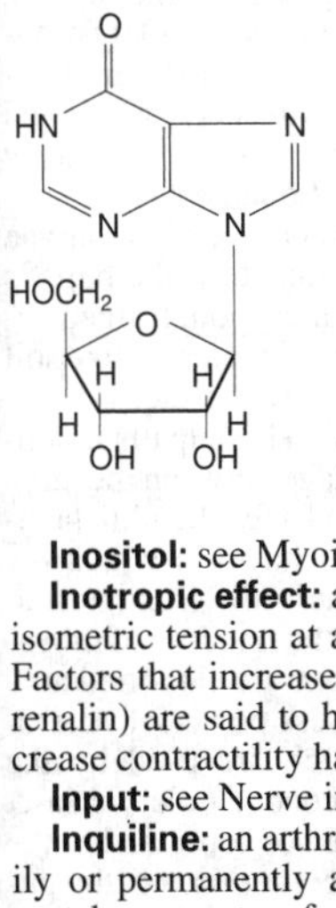

O HN N N N HOCH$_2$ O H H H H OH OH

Inosine

Inositol: see Myoinositol.

Inotropic effect: an effect on the contractility (peak isometric tension at a fixed length) of cardiac muscle. Factors that increase cardiac contractility (e.g. noradrenalin) are said to have a positive I.e.; those that decrease contractility have a negative I.e.

Input: see Nerve impulse frequency.

Inquiline: an arthropod (usually an insect) temporarily or permanently associated with or inhabiting the nest, burrow, etc. of a different species of insect. Examples are: 1) certain species of gall wasp which undergo their larval development in the galls of other gall wasps; 2) various ant guests or Myrmecophiles (see); 3) termitophiles (insects and other arthropods), including synechthrans, syno(e)ketes and symphiles, etc., whose relationship with termites is analogous to that of myrmecophiles with ants.

Insecta: see Insects

Insect antennae: see Antennae of insects.

Insect hormones: a group of hormones that regulate the postembryonic development of insects from the egg to the sexually mature imago. In particular, they are responsible for metamorphosis. Important I.h. are the steroid hormone Ecdysone (see), which is produced in the prothoracic gland, its C-20 hydroxy-derivative Ecdysterone (see), and Juvenile hormone (see). Synthesis of these hormones is in turn regulated by specific hormones of the neurosecretory brain cells.

Insecticidal bait, *insecticidal lure:* plant protection agent containing both a poison and an attractant.

Insecticides: plant protection agents for controlling insects. I. act as feeding, respiratory or contact poisons, or as systemic agents. Some I. are of plant origin (e.g. pyrethrins), and certain inorganic formulations are also used. The most widely used and commercially important I. are synthetic organic compounds, such as chlorinated hydrocarbons (e.g. DDT), organophosphorus compounds and carbamates. They are applied by sprinkling, spraying, dusting or misting, or they are scattered as granules. In addition to serving as plant protection agents, they are also used to control insect disease vectors (e.g. mosquito and tsetse fly) and insect parasites of humans and domestic animals.

Insectivora, *insectivores:* a primitive order of mammals, representing the true ancestors of the higher mammals. Insectivores are small to extremely small animals with various morphologies. Their numerous, very pointed teeth (in some species as many as 48) are used to seize insects, worms and small vertebrates, which form the main components of the diet. The dental formula shows many variations (see individual familes). Canines may be shaped like incisors or premolars. The lower molars often have 5 pointed cusps, while the upper molars characteristically have 3–4 tubercles. Radius and ulna are separate, while the tibia and fibula frequently show distal fusion. Members of the order are found in all parts of the world, except the polar regions and Australia. The following families are listed separately: *Chrysochloridae* ***(Golden moles),*** *Erinaceidae* ***(Hedgehogs*** and ***Gymnures),*** *Macroscelidae* ***(Elephant shrews),*** *Potamogalidae* ***(Water shrews*** or ***Otter shrews),*** *Solenodontidae* ***(Solenodons),*** *Soricidae* ***(Shrews),*** *Talpidae* ***(Moles, Shrew-moles,*** and ***Desmans),*** *Tenrecidae* ***(Tenrecs*** and ***Madagascar "hedgehogs"),*** *Tupaiidae* ***(Tree shrews).***

Insectivore: see Carnivorous plants, *Insectivora,* Modes of nutrition.

Insectivorous plants: see Carnivorous plants.

Insects, *Insecta, Hexapoda* (plates 1 & 2): a class of invertebrate animals, phylum *Arthropoda.* About 850 000 species have been described, representing almost three quarters of all known animal species.

Adults (singular *imago,* plural *imagoes* or *imagines)* all characteristically possess a chitinous exoskeleton, and the body is divided by indentations (Latin *insectum* means "notched"; Greek *entomon* also means "notched", and gives rise to the term, entomology). At these points of indentation or jointing, the outer covering of chitin is thin and pliable, thus permitting movement. The harder and nonflexible parts of the exoskeleton contain sclerotin (cross-linked or tanned protein) in addition to chitin, and are said to be sclerotized. In contrast to vertebrates (which have an internal skeleton), the musculature of I. lies inside the exoskeleton. In most species, the body of the imago is distinctly divided into head, thorax and abdomen. The largest species attain a body length of 330 mm, whereas the smallest are only

0.2 mm. Important sense organs (compound eyes, simple eyes or ocelli, and antennae) are carried on the head, which houses the brain (more appropriately called the cerebral ganglion or supraesophageal ganglion). The head also carries the mouthparts, which are adapted for chewing (original primitive condition), licking, sucking or piercing. The thorax consists of 3 segments, each carrying a pair of jointed legs. In addition, the middle and posterior thoracic segments each carry a pair of more or less veined wings (except in wingless I.; see Apterygota). In most I., the abdomen has 10 or 11 segments. The primitive number of abdominal segments appears to be 12, but the most posterior of these are usually no longer recognizable as segments, because they are modified to form the external genitalia and cerci. In addition to the central nervous system (cerebral ganglion, subesophageal ganglion and ventral nerve cord), the body contains a dorsal heart, alimentary canal, tracheal system and the internal sex organs (Fig.). I. feed on living, dead, paralysed or decaying material of plant or animal origin. Reproduction is usually sexual, but parthenogenesis occurs in some groups. Depending on the species, I. also display brood protection, brood care, parasitism and various levels of social organization. Metamorphosis (see) is either incomplete (egg, larva, imago) or complete (egg, larva, pupa, imago).

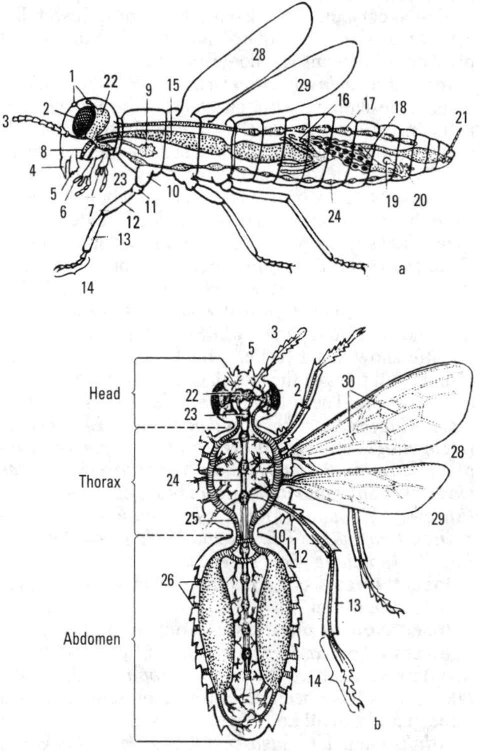

Anatomy of an idealized winged insect. Each diagram shows only one half of the body. *a* Lateral view. *b* Dorsal view. *1* Simple eyes (ocelli). *2* Compound eye. *3* Antenna. *4* Labrum. *5* Mandible. *6* First maxilla. *7* Second maxilla. *8* Mouth opening. *9* Salivary gland. *10* Coxa. *11* Trochanter. *12* Femur. *13* Tibia. *14* Tarsus with five tarsal subsegments. *15* Alimentary canal. *16* Malpighian tubules or vessels. *17* Heart. *18* Ovary. *19* Spermatheca. *20* Colleterial glands or accessory glands. *21* Cercus. *22* Brain (cerebral ganglion, supraesophageal ganglion). *23* Subesophageal ganglion. *24* Ventral nerve cord. *25* main longitudinal trachea. *26* Spiracles. *27* Air sac. *28* Forewings. *29* Hindwings. *30* Wing veins.

Eggs exhibit a wide variety of shape, surface structures and color, and they vary in size from a fraction of a millimeter to 8 mm. Depending on the species, a female may lay a single egg or several million. These are laid individually or in groups (rows, heaps, plates), glued to a substratum, inserted into plant tissue, feces, carrion or a host animal, or placed in earth or water. Some species give birth to larvae, i.e. the young larvae emerge from the egg inside the female.

Larvae. The larval phase commences with emergence of the larva from the egg. It ends either with the last molt which produces the imago (incomplete metamorphosis; hemimetabolous I.), or with pupation (complete metamorphosis; holometabolous I.). Between the newly hatched larval stage and the imago or pupa, the larva molts several times as it grows, each molt giving rise to a distinct larval stage. Larvae of hemimetabolous I. generally resemble the imago in structure and habits, or gradually come to resemble the imago with development of the wing sheaths. In contrast, the larvae of holometabolous I. differ markedly from the imago, e.g. caterpillar/butterfly, maggot/fly, cockchafer larva/cockchafer imago. From these 3 examples and other known larval forms, it is clear that larvae vary very widely in appearance, not only between, but also within, different orders of the *Insecta.* Usually, all larval stages move and feed independently. Imago and larva do not necessarily share the same mode of feeding or the same diet. For pupation, a protected site is usually chosen. Many larvae produce a silken support or attachment, or the pupa is housed in a more or less substantial cocoon, formed entirely of silk or containing extraneous material, like soil particles, fragments of vegetation or wood, etc.

Pupae (chrysalids or chrysalis). The pupa is a nonfeeding, nongrowing, quiescent stage between the larva and imago of holometabolous I. and a few hemimetabolous I. It represents a transitional stage, in which the body and internal organs become redifferentiated to those of the imago. Depending on the freedom of movement of the body appendages, pupae are classified as:

1. *Decticous pupae.* These possess powerful, sclerotized, articulated mandibles, which are used by the pharate imago to escape from the cocoon or larval case. The encased adult appendages are not adpressed to the body, and they can be used for locomotion, i.e. the pupae are exarate (free). Decticous pupae are found in the *Neuroptera, Mecoptera, Trichoptera* and in the families, Micropterygidae and Eriocraniidae.

2. *Adecticous pupae.* In this type the mandibles are not articulated, are not used for escape by the pharate imago, and are also often reduced.

2a. *Exarate adecticous pupae (pupa libera).* The encased adult appendages are not adpressed to the body; found in the *Siphonaptera* and *Strepsiptera,* most *Coleoptera* and *Hymenoptera,* and in the cyclorraphan and most brachyceran *Diptera.*

2b. *Obtect adecticous pupae (pupa obtecta).* The encased adult appendages are adpressed to the body, to which they adhere by a secretion produced during the final larval molt; found in all higher *Lepidoptera,* some *Coleoptera,* the nematoceran *Diptera* and many Chalcidoidea.

2c. *Coarctate pupae (pupa coarctata).* Adecticous, exarate pupae enclosed by a puparium formed from the preceeding larval cuticle; found in cyclorraphan *Diptera.*

A fourth type, the *pseudochrysalid (larva coarctata),* is really a resting larva in the hardened cuticle of the preceeding larval stage, e.g. in the *Meloidae* (oil beetles).

Economic importance. Damage caused by I. is estimated in millions of dollars per year. This is due mainly to I. feeding on agricultural plants, stored foods, textiles, animal hide products, structural timbers and other materials, and by transmission of diseases to plants, humans and domestic animals. On the other hand, the beneficial effects of I. are often underestimated. Although only a few species are exploited directly by man (e.g. honeybee, silkworm), numerous I. are equally if less directly useful. Flower-pollinating I. (bees, bumblebees, flies, butterflies) are necessary for fruit production, and I. in the soil (especially primitive I.) contribute to mechanical breakdown of the soil and increased fertility. Furthermore, I. are a food source for useful animals (fish, birds), some I. are parasitic upon insect pests, while others produce valuable raw materials (wax, shellac, silk, dyes) and pharmaceuticals (cantharidine, bee toxin). In addition, the role of I. in the turnover of dead material in the biosphere must not be underestimated.

Classification. Separation and derivation of the insect orders is still partially disputed. The evolutionary tree shown here is based on the phylogenetic system of Hennig. In line with the standardization of the endings of the Latinized names of tribes *(-ini),* subfamilies *(inae)* and families *(-idae),* recent attempts have been made to attach uniform endings to the names of higher systematic categories, e.g. the names of orders of winged insects *(Pterygota)* should end in *-ptera,* while superorders should end in *-pteria.* Hitherto, the ending *-oidea* has usually been used for superorders, and this ending should now be retained for superfamilies.

Geological record. About 12 000 fossil species are known. The earliest fossil I. are wingless, and they are found in the Middle Devonian. Winged I. first appear in the Carboniferous. Prominent among these are primitive, predatory dragonflies with chewing mouthparts and a wingspan of 75 cm; they include the largest known I. The main period of evolutionary diversification started in the Cretaceous and continued through the Tertiary to the present. This development is probably correlated with the appearance of flowering plants. Tertiarty I. (butterflies, beetles, ants, etc.) are abundant in the Eocene brown coal of the Geisel valley near Halle (Germany) and the Oligocene amber of the Baltic area.

Collection and preservation. I. are usually caught with a net in air or water. They can also be lured with various baits, or attracted by light. There are various screening apparatuses and traps for determining the number of I. in a habitat. Specimens are killed in a wide-necked jar with a cork stopper, using potassium cyanide (admixed with calcium sulfate to form a firm layer on the bottom of the killing jar) or ethyl acetate (preferably soaked into cotton wool in a glass container attached to the cork of the killing jar). Dragonflies are killed with acetone. Some specimens may be preserved in ethanol, but otherwise they are prepared then pinned to a display board or glued to a cardboard base.

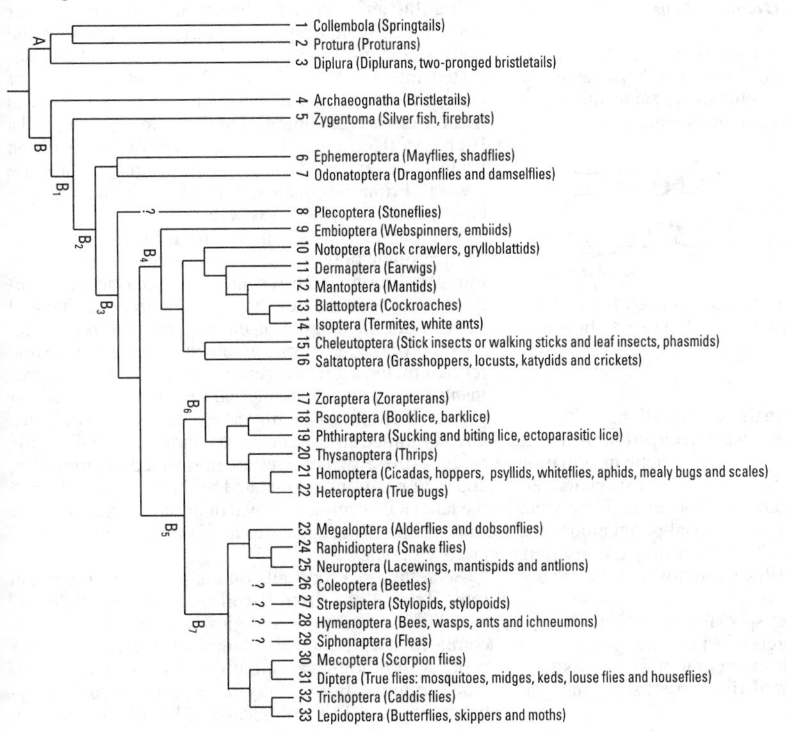

Evolutionary tree of the Insecta, showing all the orders and some higher phylogenetic divisions. *A* Subclass *Entognatha* (orders 1–3). *B* Subclass *Ectognatha* (orders 4–33). *B1 Dicondylia* (orders 5–33). *B2 Pterygota* (orders 6–33). *B3 Neoptera* (orders 8–33). *B4 Paurometabola* (orders 9–16). *B5 Eumetabola* (orders 17–33). *B6 Parametabola* (orders 7–22). *B7 Holometabola* (orders 23–33). The orders may also be grouped as follows: *Apterygota I* (orders 1–5); *Paleoptera* (orders 6 and 7); *Orthopteroid I* (orders 8–17); *Hemipteroid I* (orders 18–22); *Endopterygote I* (orders 23–28); *Panorpoid I* (orders 29–33).

Insect viruses: viruses that cause disease in insects. Mostly affected are insects with complete metamorphosis, especially *Lepidoptera, Hymenoptera, Diptera* and *Coleoptera.* Larval and pupal stages are affected, although adult insects may carry the virus without manifesting a disease state. In certain infections, the cells of afflicted insects contain light microscopically visible, polygonal to elipsoid, paracrystalline, often strongly refractive *inclusion bodies,* containing virus particles embedded in their matrices; in such structures the virus particles may remain infective for many years, even after the death and disintegration of the host. Inclusion bodies may be nuclear or cytoplasmic. When inclusion bodies are pesent, the virus is classified as a *polyhedrose* or *polyhedrosis virus.* Best known of the cytoplasmic polyhedroses is that responsible for the "yellowing disease" of the silkworm. In a second group, known as *granuloses* or *granulosis viruses,* geometrically shaped inclusion bodies are absent, and the infection is characterized by an accumulation of small granular inclusions called capsules in the infected cells, each capsule containing a single rod of viral DNA [e.g. infections of the Cabbage white butterfly *(Pieris brassicae)* and the Turnip dart moth *(Agrotis segetum)].* Sacbrood, which kills honey bee larvae, and is feared by bee keepers, is caused by a noninclusion virus, i.e. neither capsules nor inclusion bodies are present in infected cells.

Some I.v., in particular polyhedroses, are used for biological pest control, e.g. against the Pine sawfly *(Diprion pini)* and the Fir sawfly *(Pritisphora abietina),* two of the most important European forestry pests. The American Environmental Protection Agency has approved the use of viruses for the control of the Cotton bollworm *(Heliothis zea),* Gypsy moth *(Lymantria dispar)* and Tussock moths *(Hemerocampa).*

Insertion:

1) A chromosome mutuation (Fig.), in which a segment (e.g. a copy of a transposon) is incorportated into a nonhomologous chromosome of the same karyotype. I. is also known as *insertional mutagenesis.*

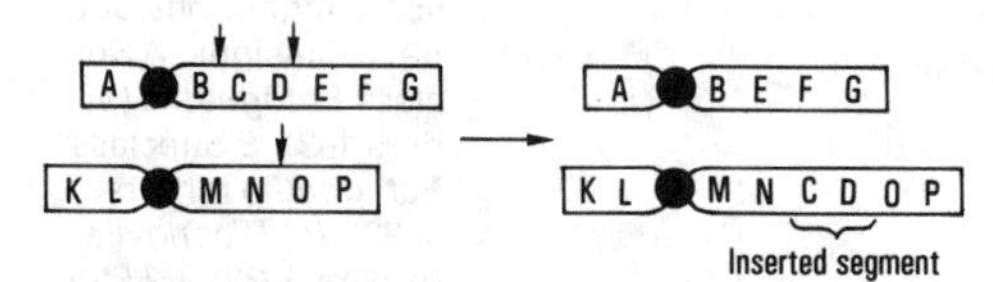

Insertion. Insertion of a chromosome segment into a nonhomologous chromosome of the same chromosome set.

2) see Leaf.

Insertional mutagenesis: see Insertion.

Insertional sequence, ***simple transposon:*** a mobile nucleotide sequence that carries no genetic information except that necessary for transposition. I.ss. characteristically have repeat sequences at both ends. They occur naturally in the genomes of bacterial populations, and they inactivate any gene into which they are inserted. The inactivation is reversible by removal of the I.s. See Transposon

Insertion sequences: special DNA molecules, responsible for the transfer of "jumping genes" (see Transposon) and for their incorporation at a new site.

Inspiratory capacity of the lungs: see Lung volume.

Inspiratory reserve volume of the lungs: see Lung volume.

Instant dried yeast: see Baker's yeast.

Instinct: a controversial term in ethology, which is probably best avoided. Usage varies according to authority. It is usually understood as an innate behavior mechanism, which is manifested as an ordered sequence of movements (fixed action patterns). The behavior mechanism is released by environmental factors, and judgement or understanding on the part of the individual are not involved. "Purposeful drive without awareness of the purpose" (K.E. von Baer).

Instinctive behavior: behavior resulting from Instinct (see). It consists of species-specific, complex patterns of behavior. However, like the noun, "instinct", the term I.b. is best avoided in ethology.

Instinctive movements: Elements of movement that translate Instinct (see) into action. The term is used more or less in the same sense as *reflex coordination* or *normal behavior.* I.m. constitute the species-specific behavior that does not depend on previous experience, and they are released by specific stimuli. However, like the noun "instinct", the adjective "instinctive" must also be considered contentious in ethology, and it is best avoided.

Instruction hypothesis: a theory of antibody formation current for 30 years until the end of the 1950s. The antigen was thought to act on the newly formed antibody, leaving an "imprint" which subsequently served as the antigen binding site. The I.h. is now disproved and replaced by the Clone selection theory (see).

Instrumental conditioning: see Operant conditioning.

Instrumental learning: see Operant conditioning.

Insulin: an important polypeptide hormone of higher vertebrates, which regulates nutrient flow within the body following food intake. It is synthesized and secreted into the blood by the B-cells of the islets of Langerhans in the pancreas. The molecule of I. consists of an A-chain (containing 21 amino acid residues) and a B-chain (30 residues), which are linked by two disulfide bridges. Seventeen different amino acids residues are present. Primary sequence 8–10 of the A-chain is species-specific. Bovine I. has M_r 5 780.

The mature mRNA transcript for I. encodes preproinsulin, a single polypeptide chain carrying an amino-terminal, hydrophilic, 16-residue sequence (the signal sequence), which enables the molecule to penetrate and traverse the membrane of the endoplasmic reticulum. As preproinsulin enters the lumen of the endoplasmic reticulum, the signal sequence is removed, leaving proinsulin. Removal of an internal sequence ("connecting peptide") from proinsulin leaves the A- and B-chains, already united in the correct position by two disulfide bridges (these two bridges are preformed in proinsulin, and their positions are dictated by the tertiary folding of the latter). Proteolyic removal of the connecting peptide starts in the Golgi apparatus and is completed in the secretory vesicles.

Secretion of I. depends on the presence of calcium ions. Release of I. from B-cells is regulated by the extracellular concentration of glucose (e.g. the hyperglycemia following a meal promotes I. secretion). Other primary physiological stimuli include mannose, leucine, arginine, lysine, fatty acids, acetoacetate, β-hydroxybutyrate. The biological half-life of circulating I.

is about 5 min. It is inactivated mainly in the liver by enzymatic cleavage of the disufide bridges. I. affects the entire intermediary metabolism, especially of the liver, adipose tissue and muscle. It increases the permeability of cells to monosaccharides, amino acids and fatty acids, and it accelerates glycolysis and the pentose phosphate cycle. It promotes the synthesis of glycogen, fatty acid and protein in the liver, adipose tissue and muscle, respectively. Of primary importance is the ability of I. to reduce the concentration of blood glucose. It has the opposite affects to those of glucagon, and the flow of nutrients within the organism is largely determined by the opposing actions of these two hormones.

I. becomes bound to the exterior of its target cells by a membrane-bound I. receptor, which has been cloned. The receptor traverses the membrane with an I.-binding site on the cell membrane exterior, as well as an intracellular domain (possessing latent tyrosine kinase activity) on the interior surface of the membrane. Binding of I. on the outside of the membrane activates the tyrosine kinase activity of the intracellular domain. This receptor-tyrosine kinase activity is essential for insulin signal transduction. The major physiological substrate of receptor-tyrosine kinase is a multifunctional "docking" protein (known as IRS-1) containing multiple tyrosine residues that become phosphorylated by the receptor-tyrosine kinase. IRS-1 couples the insulin receptor to multiple signalling pathways via different proteins, all of which contain a domain (SH2 domain) that is recognized by phosphorylated IRS-1. One important insulin signalling pathway results in the phosphorylation of phosphoinositides at position 3, while a second signalling pathway leads to activation of a small GTP-binding protein known as Ras. The activated Ras in turn initiates a serine/threonine kinase cascade, leading to dephosphorylation (activation) of glycogen synthase. Other signalling pathways remain to be elucidated.

The metabolic disease known as Diabetes mellitus can be classified into two main types: 1) Insulin-dependent Diabetes mellitus affects less than 20% of diabetic patients. The appearance of symptoms (high blood glucose, high urinary glucose, ketosis) is rapid and may be life-threatening. The age at onset is usually less than 30 years, following a prolonged period of apparently autoimmune destruction of B-cells of the islets of Langerhans. The condition is therefore due mainly to a lack of I., and it is treated by I. administration. 2) Non-insulin-dependent Diabetes mellitus affects more than 80 % of diabetics. Age of onset is usually greater than 30 years (also known as maturity onset diabetes). Blood glucose is abnormally high and the urine contains glucose, but ketosis is rare. It is often associated with peripheral I. resistance, i.e. the pancreas still secretes I., but the body does not respond. The defect appears to lie in the altered activity of the I. receptor or of associated intracellular proteins, such as sugar transporters. The condition can often be managed by diet control and the use of oral hypoglycemic agents such as sulfonylureas.

The primary sequence of I. was determined in the early 1950s by the English biochemist, Fred Sanger; this was the first protein primary sequence ever to be elucidated. The total synthesis was achieved in 1963.

Integrated plant protection: maintenance of the number of plant pest organisms below the economically important threshold, by using a combination of different measures, including crop management procedures, ecological methods and the application of pesticides. Most important is the exploitation of natural factors. A sequence of measures is required, the last (and preferably avoidable) being the application of chemical agents. I.p.p. has not been fully achieved, and is the subject of applied field research.

Integripalliate: a term applied to members of the *Bivalvia*, in which the pallial line (line of attachment of the shell to the mantle) is entire, i.e. not interrupted by an indentation toward the rear of the shell. See Sinupalliate.

Intention movement: movement(s) exhibited prior to a complete behavioral response. The term is normally applied only to those forms of behavioral movement that have become independent and have acquired signal character. For example, a dog goes down on its front paws as a prelude to springing, but this particular action has also become a signal of readiness to play. The use of I.m. as signals has also been called "mimic exaggeration".

Intention movement. Ritualized intention movements of the house sparrow (after Daanje, 1951).

Interaction of living organisms: Relationships between living organisms.

Intercalary growth: in plants, the growth of a persistent zone of undifferentiated tissue that is interpolated between permanent, differentiated tissue. I.g. occurs, e.g. in the internodes of grass halms directly above each node. Basal I.g. zones are also present in leaves, particularly in conifers and monocots, and also in some dicots. I.g. is displayed, e.g. by the petiole, which represents an I.g. zone between the lamina and leaf base.

Intercellular bridges: see Epithelial tissue.

Intracellular substance: see Extracellular matrix.

Intercellular system: the aeration system of plants, consisting of a network of fine canals, which arise by separation of the walls of adjacent cells when meristematic tissue differentiates into permanent tissue. The I.s. is continuous with the stomata and with the lenticels of the primary or secondary boundary tissue, and it funcions in the diffusion and exchange of gases.

Interference:

1) The effect of crossing-over at one locus of a chromosome on the probability of a crossing-over occurring at another locus. The effect may be negative (crossing over at the second locus inhibited) or positive (crossing over promoted). Alternatively, I. is the phenomenon in which the chromatids in a double crossing-over do not show random behavior. In both cases, I. causes the recombination value to differ from that based on random behavior. The *differential distance* or *interference distance* is the distance between the centromere of a chromosome and the nearest point at which the first crossing over and associated chiasma formation can occur in that arm of the chromosome. The magnitude of the interference distance is determined by the centromere, and the phenomenon is known as *centromere interference.* The *interference range* characterizes that length of the chromosome over which two crossing-over events in the same chromatid pair can interfere with each other. See Coincidence index.

2) The mutual negative influence of two organisms (e.g. psychological incompatibility) in the absence of demonstrable Competition (see). I. is an important factor in the determination of population densities.

Interference distance: see Interference *(1).*

Interference microscopy: a microscopic method for revealing phase differences, which can also be used for quantitative purposes. In modern apparatus, an interferometer is used, in which the light beam is split by a mirror or prism; one part of the split beam is passed through the object, the other through a split-image focusing wedge, after which the two are recombined. The difference in path-length caused by the object can be measured by compensating the reference beam with the focusing wedge. If white light is used, the background appears colored. Any desired color can be selected by moving the focusing wedge, and the object appears in a color different from the background. In addition to the visualization of contrast-poor objects and its use for quantitative studies (e.g. measurement of the thickness of thin layers), I.m. can also be used for the microrefractometric determination of the dry mass and protein concentration of cells and cell organelles.

Interference range: see Interference *(1).*

Interferons: species-specific proteins which induce antiviral and antiproliferative responses in animal cells. They are a major defense against viral infections and neoplasms. I. are produced in response to the penetration of animal cells by viral (or synthetic) nucleic acid, then leave the infected cell to confer resistance on other cells of the organism. In contrast to antibodies, I. are not virus-specific but host-specific. Thus, virus infections of human cells are inhibited only by human I. The human genome contains 14 nonallelic and 9 allelic genes for a family of α-I. (macrophage I.), as well as a single gene for β-interferon (fibroblast interferon), α- and β-I. are structurally related glycoproteins of 166 to 169 amino acid residues. In contrast, γ-interferon (also known as immune interferon) is not closely related to the other two and is not induced by virus infection. It is produced by T-cells after stimulation with interleukin-2. It enhances the cytotoxic activity of T-cells, macrophages and natural killer cells, and thus has antiproliferative effects. It also increases the production of antibodies in response to antigens administered simultaneously with α-I., possibly by enhancing the antigen-presenting function of macrophages.

I. bind to specific receptors on the cell surface, and induce a signal in the cell interior. Two induction mechanisms have been elucidated. 1) I. induce a protein kinase, which, in the presence of double-stranded RNA, phosphorylates the a-subunit of an initiation factor of protein synthesis (eIF-2B), causing the factor to be inactivated by sequestration in a complex. 2) I. induce 2′,5′-oligoadenylate synthetase (2′,5′-oligo A synthetase), which, in the presence of double-stranded RNA, catalyses the polymerization of ATP into oligomers of 2 to 15 adenosine monophosphate residues linked by phosphodiester bonds between position 2′ of one ribose and 5′ of the next. 2′,5′-oligoadenylate activates an interferon-specific RNase (an endonuclease known as RNase L; L for latent, because the enzyme is always present but normally inactive), which cleaves both viral and cellular single-stranded mRNA. I. therefore do not directly protect cells against viral infection, but rather render cells less suitable as an environment for viral replication, a condition known as the *antiviral state.*

Genes for I. have been cloned and expressed in microorganisms and cultured animal cells, and it is hoped that I. may eventually become available as therapeutic agents.

Interkinesis: see Meiosis.

Interlobular veins: see Kidney functions.

Intermediate-day plants: plants that are sensitive to the photoperiodicity of their environment, producing flowers only when exposed to intermediate day lengths, and not under long-day or short-day conditions (see Long-day plants, Short-day plants). Their existence is disputed by some authorities.

Intermediate filaments: tough and durable protein fibers, forming part of the Cytoskeleton (see) in the cytoplasm of most eukaryotic cells. They were first described in muscle cells as fibers with a diameter (8–10 nm) intermediate between thin and thick filaments. Their diameter is also intermediate between actin filaments and microtubules. In most animal cells, I.f. form a network around the nucleus, and extend outward to the cell periphery. They are especially common in cells that are subject to mechanical stress, e.g. epithelial cells; moreover, the desmosomal junctions between adjacent epithelial cells are linked by I.f. They are important structural constituents of nerve axons and the cytoplasm of smooth muscle cells.

I.f. consist of overlapping arrays of dimers of elongated, fibrous polypeptide molecules. Four different tissue-specific types are recognized, which differ in the nature of their constituent polypeptides. *Type I* is found in epithelial cells and structures such as hair and nails; the component polypeptides have M_r 40 000–70 000, and they may be acidic keratins, or they may be mixed neutral and basic keratins. *Type II* comprises 3 subtypes. Many cells of mesenchymal origin (fibroblasts, blood vessel endothelial cells, leukocytes) contain *type IIa,* in which the polypeptide is vimentin *(M_r* 53 000). Muscle cells contain *type IIb,* in which the polypeptide is desmin *(M_r* 52 000). Glial cells (see Neuron) contain *type IIc,* which is a polymer of glial acidic fibrillary protein *(M_r* 45 000). The I.f. of neurons *(type III)* are known as neurofilaments, and their constituent polypeptides are known as neurofilament proteins *(M_r* 60 000, 130 000 and 100 000). *Type IV* is found in the nuclear lamina of all cells; the constituent polypeptides are nuclear lamins A, B and C *(M_r* 56 000–75 000).

Despite their diversity of size, all cytoplasmic I.f. polypeptides are encoded by genes of the same multigene family; they all contain a central homologous α-helix domain (about 310 amino acid residues), which forms a rigid coiled structure when the protein dimerizes. The main function of I.f. seems to be the provision of mechanical support to the cell and the cell nucleus.

Intermediate forms, ***collective types:*** fossil animals possessing a combination of characters, which do not occur together in modern animals, but are found in two or more distinctly separate systematic groups, e.g. *Ichthyostega, Gephyrostegus, Archaeopteryx* (see).

Intermediate montane forest zone: see Altitudinal zonation.

International Commission on Enzyme Nomenclature: see Enzymes.

International Convention on Water Fowl Sanctuaries: see Nature conservancy.

International Union for Conservation of Nature and Natural Resources: see IUCN.

Interneuron: see Neuron, Neuronal circuit.

Interoceptor: see Proprioceptors.

Interparietal bone, ***os incae:*** a bone of the human skull, formed by separation of the upper part of the occipital squama by the transverse occipital suture (sutura occipitalis transversa). It may be a single bone, or it may be divided by two vertical sutures into two symmetrical halves; very rarely it may be divided into three parts. It occurs with relatively high frequency in Peruvian skulls (hence *os incae),* but is otherwise a rare condition.

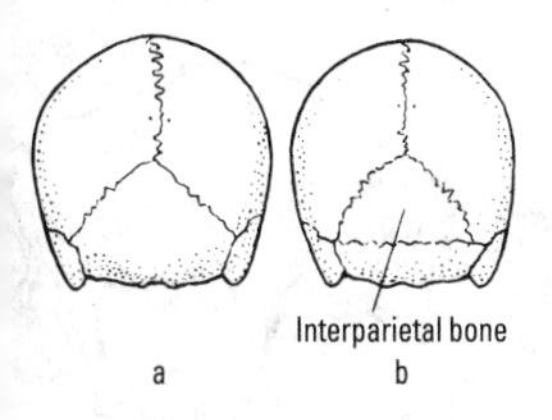

Interparietal bone. Rear view of human skull without *(a)* and with *(b)* an interparietal bone (after H. Bach).

Interphase: see Meiosis, Nucleus, Cell cycle.

Interpositional growth: in plant tissue, a type of secondary extension growth. By dissolution of the middle lamellae and formation of new wall material, the tips of tracheids, wood fibers and vessels grow and insinuate themselves between adjacent cells.

Interrenal bodies: in fishes, patches of cells lying between the kidneys and along the anterior section of the cardinal vein. Embryologically they arise from coelomic epithelial cells near the mesonephros. In more advanced vertebrates, the I.b. have evolved into the cortex of the adrenal glands (the medulla arises from the suprarenal bodies).

Intersex: a condition of abnormal, intermediate sexual development. The term includes all forms of intermediate sex, but excludes organisms in which sexually different cells form a mosaic (see Gynandromorphism, Sex determination).

By combining two mutations, Drews produced two types of I. in mice. a) Individuals with "testicular feminization", i.e. female external phenotype, but with testes and XY-chromosomes. b) XX-males with "sex-reversed" factor on the autosomes (possibly due to attachment of a Y-fragment on the autosomes), which had a male external phenotype. In addition, there were many transitional forms with intersexual genitalia, e.g. vagina and prostate present in the same animal.

Interspecific interaction: see Relationships between living organisms.

Intersterility: the extension of Self-sterilty (see) to different varieties of the same species. Thus, many apple varieties of the same I. group will not cross fertilize each other. Apple orchards of such varietes should therefore contain at least one other variety from a different I. group as a pollinator.

Interstitial cells of hydrazoans: see Regeneration.

Interstitial water: pore water in the bottom sediment of lakes, rivers, etc.

Intestinal breathing: see *Cobitididae.*

Intestinal flora: the totality of microorganisms (bacteria, yeasts and protozoa) inhabiting the intestine. In an adult human, the intestinal tract and its appendages harbor about 10^{14} (one hundred thousand billion) microorganisms. A large proportion of the fecal mass consists of intestinal microorganisms. The neonatal intestine is sterile, and becomes infected with the intake of food.

The normal I.f. of humans and animals consists predominantly of bacteria, one of the most common and numerous being *Escherichia coli.* Especially in herbivorous animals, the normal I.f. contributes to the digestion of food material, e.g. degradation of proteins and cellulose. Intestinal bacteria also synthesize vitamins that are useful to the host (e.g. vitamin K). Furthermore, acids produced by the normal I.f. (e.g. lactic and acetic acid) prevent the development of disease organisms. To a certain extent, intestinal microorganisms are also digested by the host.

Some pathological intestinal bacteria, which cause severe disturbances of intestinal function, are *Salmonella, Shigella* and *Vibrio cholerae (= V. comma).*

Many members of the family *Enterobacteriaceae* (see) usually inhabit, or at times occur in, the intestinal tract.

Intestinal juice, ***sucus enteric us:*** a mucous, protein-rich, alkaline liquid secreted by the duodenum. It contains several enzymes, the most important being erepsin (a mixture of proteinases), lipase, amylase and maltase, and lactase in suckling infants. Secretion of I.j. is primarily due to local mechanical and chemical stimuli. A small influence is also exerted by the autonomic nervous system. Gastrointestinal hormones (see) are critically important in regulating the secretion of I.j.

Intestinal motor activity: Movements of the musculature of the intestinal wall (Fig.). In addition to the site-specific contraction and relaxation of sphincters, 3 types of intestinal movement have been described: peristalsis (see Intestinal peristalsis), segmentation, and pendular movements. In *segmentation* or *rhythmic segmentation,* a length of intestine becomes divided into approximately equal segments by constriction of the muscular wall, followed a few seconds later by formation of a new constriction in the middle of each distended segment, and relaxation of the former constrictions. Segmentation does not transport the intestinal contents, but mixes the food with digestive juices and brings the mixture into intimate contact with the villi, as well as promoting the local circulation of blood and lymph. It is an intrinsic function of the musculature (i.e. it is not under nervous control), and it is initiated by stretching of the intestinal wall; it is therefore called a *myogenic* movement. *Pendular movements,* as seen in the opened abdomen, consist of swaying of loops of the intestine. In the unopened abdomen, these movements probably consist of the backward and forward migration of constrictions over short distances, also failing to transport the intestinal contents, but serving to mix food with digestive juices, improve contact with the villi, and promote the local circulation of blood and lymph.

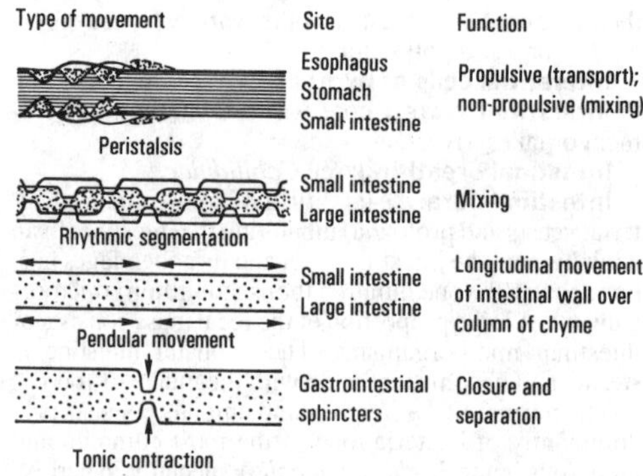

Type of movement
Site
Function
Esophagus
Stomach
Small intestine
Propulsive (transport);
non-propulsive (mixing)
Peristalsis
Small intestine
Large intestine
Mixing
Rhythmic segmentation
Small intestine
Large intestine
Longitudinal movement
of intestinal wall over
column of chyme
Pendular movement
Gastrointestinal
sphincters
Closure and
separation
Tonic contraction

Intestinal motor activity. Types of movement in the gastrointestinal tract and their functions.

Intestinal peristalsis: a type of intestinal movement resulting from nervous stimulation by an axon reflex via Meissner's plexus. It may occur as slow waves of contraction over a short section of intestine, or as rapid waves along the entire length of the intestine. I.p. serves to transport the intestinal contents, and it is controlled by the autonomic nervous system. By alterations in the resting membrane potential (Fig.), the parasympathetic nerves stimulate I.p., whereas the sympathetic nerves inhibit it. I.p. is also subject to thermal and mechanical stimuli (see Digestive system). See Intestinal motor activity.

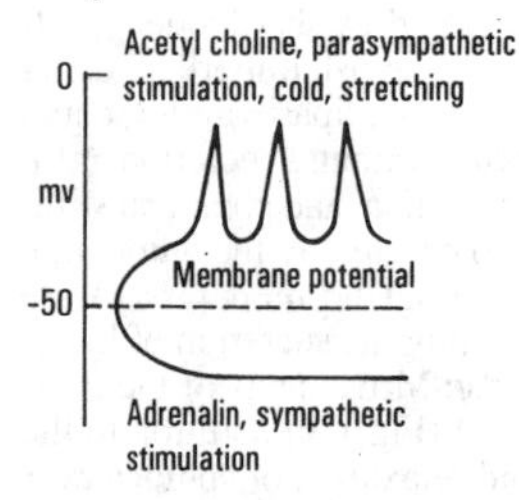

Intestinal peristalsis. Changes in the resting membrane potential of intestinal smooth muscle.

Intestinal villi: small columnar or laminate evaginations of the duodenal mucosa in vertebrates.

I.v. occur in very high density, e.g. 2 000–4 000 per cm^2 in humans, thereby greatly increasing the absorptive area of the duodenum. The absorptive capacity is considerably increased still further by the presence of microvilli on the brush border cells (see Epithelium); digestive enzymes are localized within the membranes of the microvilli. I.v. contract 3–6 times per minute, bringing them into contact with different parts of the duodenal contents, and rhythmically transporting absorbed material out of the villi and into the circulating blood or lymph. This stirring and pumping action also increases the rate of absorption.

Intestinum: see Digestive system.

Intimidation: a form of Agonistic behavior (see) in a near field. It is directed toward a competitor or rival, and is a form of self-imposition, with the function of generating awe or respect, which may make aggressive contact unneccesary. It may also serve as a display of superior status to a sexual partner. Different types of intimidatory behavior involve an increase in apparent body size, change of posture or stance, color changes and other features. It is oriented toward the competitor so as to optimize the effect of the visual (and sometimes acoustic) signals, e.g. it may take the form of *broadside I.* or *frontal I.*

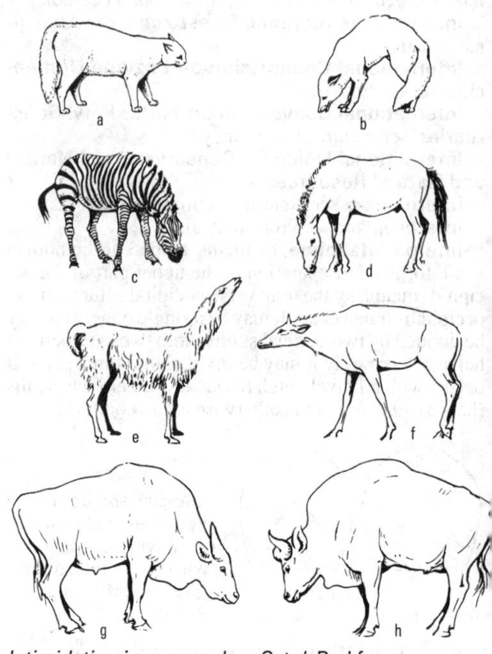

Intimidation in mammals. a Cat. *b* Red fox. *c* Hartmann's zebra. *d* Horse. *e* Llama. *f* Nilgau *(Boselaphus tragocamelus). g* Gayal *(Bos gaurus* v. *frontalis). h* Gaur *(Bos gaurus).*

Intine: the inner, delicate layer of the cell wall of bacterial spores. In spore germination, the I. forms the cell wall of the germ tube. I. is also the inner layer of the spore wall of mosses and ferns, and of pollen grains (see Flower). See Exine.

Intracellular digestion: incorporation of food particles into the cell interior by phagocytosis, where they are enzymatically degraded (digested) in food vacuoles.

Intraspecific evolution: see Microevolution.

Intraspecific interaction: see Relationships between living organisms.

Intrinsic factor: a neuraminic acid-containing glycoprotein, M_r 60 000, in the human gastric mucosa. It forms a pepsin-resistant complex with vitamin B_{12}, and this complex is absorbed in the lower part of the intestinal tract. I.f. is lacking in patients with pernicious anemia.

Introgression: spread of genes of one species into the gene pool of another by hybridization and backcrossing.

Intumescence:

1) In plants, a small pustular swelling on stems, leaves, flowers or fruits, consisting of hypertrophied, thin-walled cells. It arises from the ground tissue, and its formation is especially favored by excessive humidity.

2) See Spinal cord.

Intussusception: the incorporation of new components into the framework of an existing plant cell wall. I. occurs mainly during the growth of young plant cells.

Under pressure from the expanding cell contents, the primary plant cell wall undergoes irreversible (plastic) expansion. Its structure therefore becomes more open, allowing the incorporation of cell wall matrix material. I. of material into existing wall lamellae is of secondary importance. See Stretching growth, Apposition.

Inulin: a high molecular mass vegetable reserve carbohydrate. It is a fructan, consisting of 20 to 30, β-1,2-glycosidically linked fructofuranose units. It occurs as a reserve carbohydrate in the tubers and roots of many members of the *Compositae,* like dahlia and Jerusalem artichoke tubers. It is cleaved to fructose by hydrolysis with acid or enzymes. Since it does not contain glucose, it is used in food for diabetics.

Invasion:

1) Attack of a host by a parasite, not accompanied by multiplication of the parasite, in which the degree of damage is due solely to the number and virulence of the invading organisms. The I. may be temporary, so that the host subsequently becomes free of the parasite, e.g. invasions by blood-sucking and biting insects (gadflies, clegs, horseflies).

2) Mass movement of organisms into new territory, usually as a result of short-term alterations of their environmental conditions (food shortage, over-population, climatic change). Well known examples are the irregular mass migrations of lemmings, and the irregular I. of western Europe by Pallas' sandgrouse *(Syrrhaptes paradoxus)* from Central Asia during or after hard winters. In 1863, 1888 and 1908, these Is. occurred on a large scale, involving thousands of birds, some of which even reached the Outer Hebrides of the British Isles. Is. of locusts may have devestating economic effects. A swarm of *Schistocerca paranensis* in South America was recorded as 120 km long and 20 km wide, containing an estimated ten thousand million (10^{10}) locusts. Locusts are capable of travelling enormous distances; swarms of migratory locusts have been observed over the Atlantic, 2 000 km from the nearest coast.

Inverse pigment cup ocellus: see Light sensory organs.

Inversion:

1) A chromosome mutation involving a single chromosome (i.e. an intrachromosomal mutation), in which a chromosome segment is excised, completely reversed (turned through 180°), then reinserted at the same site. An I. may be heterozygous or homozygous. If the inverted segment includes the centromere of the chromosome, the I. is symmetrical (pericentric or eucentric); if it does not include the centromere the I. is asymmetrical (paracentric or dyscentric). A single chromosome may have several Is., in which case a distinction is drawn between independent, inclusive and overlapping Is.

Chromosomes containing a heterozygous I. display a number of genetic and cytological peculiarities, resulting mainly from unusual pairing and crossing over processes in the inverted region. In the prophase of meiosis, pairing between the inverted and noninverted segments of the respective homologous chromosomes results in the formation of a characteristic loop structure. Crossing over in the inverted region leads to the formation of meiosis products containing various duplications and deletions, and often results in sterility.

2) See Learning sets.

Invertase, *β-D-fructofuranosidase:* a glycosidase from yeast, fungi and higher plants, which catalyses the hydrolytic cleavage (inversion) of sucrose. The resulting mixture of equal parts of D-fructose and D-glucose (see *Invert sugar)* is levorotatory, in contrast to the dextrorotatory sucrose. Bees use I. to promote the inversion of sucrose, so that honey contains up to 80% invert sugar.

Invertebrates: a collective term for all animals which are not members of the *Vertebrata.* Ninety percent of all animal species are I. They include the unicellular animals and 23 phyla of the *Metazoa.*

Invert sugar: A mixture of equal parts D-glucose and D-fructose, which in contrast to dextrorotatory sucrose, is levorotatory. I.s. is generated by acid or enzymatic hydrolysis of sucrose. Bees possess an invertase, so that honey consists of up to 80% I.s. The sweet taste of honey is due essentially to the fructose. See Invertase.

Involute paxillus: see *Basidiomycetes (Agaricales).*

Involution forms: abnormal cell forms of microorganisms, in particular bacteria, which arise under abnormal environmental conditions, e.g. in the presence of certain chemical compounds. The Bacteroids (see) in the nitrogen-fixing root nodules of members of the *Papilionaceae* are I.f. of *Rhizobium* species.

Iodine, *I:* a halogen, widely distributed in plants and animals as iodide and in various organically bound forms. The iodine content of terrestrial plants is low, between 0.07 and 1.2 mg/kg dry mass. In plants, I increases the activity of some degradative enzyme systems, as well as the respiration rate, but large quantities of I are injurious, leading to chloroses and necroses. I was first discovered by heating extracts of seaweed ash with sulfuric acid. Seaweed ash (in particular the ash of kelp; the ash may also be called "kelp") contains 0.1–0.3% iodine, and seaweeds from deep seas appear to contain more than those from shallow waters.

I is an essential element in animal nutrition, and is usually present in the diet as iodide. In particular, it is required for the biosynthesis of thyroid hormones. The normal human thyroid contains 0.007–0.18% iodine, accounting for about 20% of the body's I. The average daily human intake is 150 mg, and the requirement increases during adolescence and pregnancy. I deficiency impairs the ability of the thyroid to synthesize thyroglobulin, the precursor of the hormones, thyroxin and triiodothyronine, leading to a condition known as hypothyroidism.

Ion: an electrically charged atom or molecule. Positively charged cations (so-called because they are attracted to the cathode in an electric field) are produced by the loss of electrons, whereas negatively charged anions are formed by the acquisition of electrons. In aqueous solution, salts, acids and bases are more or less completely dissociated into Is. This dissociation may even occur within the solid, crystalline salt, e.g. sodium chloride is dissociated into sodium cations (Na^+) and chloride anions (Cl^-).

Ionizing radiation: radiation with sufficient quantum energy to release electrons from an atomic or molecular structure. The radiation energy is generally given in electron volts (eV). For conversion to SI units, 1 eV = 1.602 x 10^{-19} J. A distinction is drawn between *directly I.r.* which consists of charged particles (α- and β-particles), and *indirectly I.r.* (e.g. neutron beams, X-rays and γ-rays). Artificially generated radiation often has a very high energy, so that it causes several different ionization processes.

X-rays and γ-rays are energetic forms of electromagnetic radiation, and their absorption by matter is due to three processes: the photoelectric effect, the Compton effect, and pair formation (formation of an electron-positron pair in the field of the nucleus). Which of these mechanisms predominates depends on the energy of the radiation and the atomic number of the absorbing atom. Neutrons dissipate their energy in matter by three processes: elastic scattering, inelastic scattering and neutron capture. Neutron capture produces an isotope (usually radioactive) with release of a γ-quantum.

Ionone: a sesquiterpene ketone with an intense smell of violets. It occurs in two isomeric forms, depending on the position of the double bond in the ring. β-Ionone has been found, e.g. in raspberries. The natural smell of violets is not due to I., but to the structurally similar compound, α-irone. Synthetic I. is an important starting material in the manufacture of perfumes. Cyclic carotenoids and vitamins A_1 and A_2 contain I. ring structures, but I. is not a biosynthetic precursor of the carotenoids.

O

β-Ionone

Ion pumps: metabolic cycles within cell membranes which transport ions against prevailing concentration gradients. Thus, the bioelectric membrane potential of nerves is based on a differential distribution of Na^+ and K^+ ions. The high concentration of K^+ on the inside, and the predominance of Na^+ on the outside of the nerve cell is maintained by transport of Na^+ out of the cell and transport of K^+ into the cell, the entire process being catalysed by Na^+/K^+-ATPase. For each molecule of ATP cleaved, 3 Na^+ are pumped from the inside and 2 K^+ from the outside of the cell. The enzyme is an integral membrane protein, with functional sites on both the interior and exterior membrane surfaces (Fig.). Binding of ions causes a change in the conformation of the enzyme, so that Na^+ and K^+ are transported to opposite sides of the membrane (see Excitation). ATP-dependent I.p. are also known for calcium and magnesium ions (Ca^{2+}/Mg^{2+}-ATPase), as well as sodium-dependent systems in the membranes of erythrocytes and intestinal cells that transport amino acids and sugars. See Adenosine triphosphatase.

Ioras: see *Chloropseidae.*

Ipo arrow poison: see *Moraceae.*

Ipomoea: see *Convolvulaceae.*

IPSP: acronym of inhibitory postsynaptic potential. See Synaptic potential.

Ips typographus: see Engraver beetles.

Irediparra: see *Jacanidae.*

Irenidae: see *Chloropseidae.*

Ireninae: see *Chloropseidae.*

Iridaceae, ***iris family:*** a family of dicotyledonous plants, containing about 1500 species in 70 genera, distributed mainly in the tropics and subtropics, but also represented in temperate regions. All members of the family are perennial herbs with sword-like leaves and subterranean corms, bulbs or rhizomes. The 6-merous flowers have 3 free or partially connate stamens. Flowers are usually actinomorphic, but the family shows a progressive tendency toward zygomorphy. A spathe of

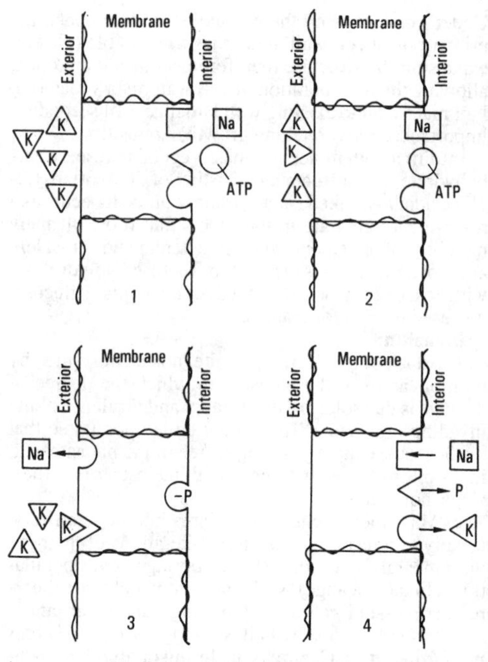

Ion pumps. Model for the action of Na^+/K^+-adenosine 5′-triphosphatase. *1* Na^+ and ATP approach their specific binding sites on the enzyme. *2* Na^+ and ATP bind to the enzyme. *3* Phosphorylation of the enzyme by ATP causes a change in conformation of the binding sites and movement of Na^+ through the membrane to the exterior; K^+ binds to the enzyme on the outside of the membrane. *4* Dephosphorylation of the enzyme, promoted by K^+ binding, is accompanied by movement of K^+ to the cell interior and the return of the enzyme to its original conformation.

1 or 2 bracts is usually present at the base of the flower. The ovary is inferior, 3-locular with axile placentas, or unilocular with 3 parietal placentas, and the style (often petaloid) has 3 branches. Ovules are anatropous and usually numerous The fruit is a loculicidal capsule, dehiscing by 3 valves.

Many members of the family are grown as ornamentals, e.g. the genus *Crocus* ***(Crocus)***, which has actinomorphic flowers and a homeochlamydeous perianth (75 species in Europe and the Mediterranean region to Central Asia). Popular spring-flowering garden plants are the white- or violet-flowered *Crocus vernus* and the yellow-flowered *Crocus flavus (= C. aureus).* The ***Autumn-flowering crocus,*** *Crocus neopolitanus (= C. speciosus),* originiates from Asia Minor and Iran. *Crocus sativus* ***(Saffron crocus)*** whose stigmas yield the yellow coloring agent, *saffron,* has been cultivated for so long that its origin is unknown. The Greeks and Romans used it as a medicine, dye and flavoring agent. It was cultivated in Spain in the tenth century, and Spain still produces most of the European saffron, but the price of the product and the advent of synthetic coloring agents have largely killed the industry. Saffron contains carotenoids, a trace of volatile oil, and the bitter principle, picrocrocin (a colorless glycoside).

Iridaceae. 1 Crocus sativus. 2 Iris sibirica.

Members of the genus *Iris* ***(Irises*** or ***Flags)*** are characterized by their unifacial, sword-like, equitant leaves and their large, showy, actinomorphic flowers in a monochasium (rhipidium); the outer perianth segments ("lips" or "falls") bend outward and backward and hang down, and they are larger than the (usually erect) inner segments ("standards"). Countless hybrids and cultivars are used for display in parks and gardens, most prominent being the group known as "bearded irises". Cultivars of *Iris germanica* are grown widely. The rhizome of *Iris germanica* var. *florentina,* known as *oris root,* is used in perfumery. Some wild species have been adopted directly as garden plants, e.g. *Iris reticulata* (blue flowers) and *Iris danfordiae* (yellow flowers). *Iris sibirica,* found on wet moorlands in Central Europe, is a protected plant. *Iris pseudocrocus* ***(Yellow flag)*** grows in marshes and on the edges of rivers and ponds in western Asia, North Africa and most of Europe.

The genus *Gladiolus* has also provided popular cultivated varieties. Its weakly zygomorphic flowers form a long, unifacial spike, each flower with a (usually herbaceous) spathe. Most ***Gladioli*** grown and sold as cut flowers in temperate regions are hybrids of South African species, and their corms cannot withstand a European or North American winter.

Freesia *(Freesia refracta)* is another popular garden and cut flower originating from South Africa. Like the South African *Gladiolus,* its corm must be dried off and protected during the northern winter. About 5 species of *Crocosmia* occur in South Africa. The hybrid of *C. Pottsii* (female) and *C. aurea* (male), known as ***Montbretia*** *[C.* x *crocosmiiflora* (Lemoine)] is a hardy cultivated plant in temperate regions, which is especially tolerant of shade, and has become widely naturalized. Its inflorescences are long and spike-like with deep orange flowers suffused with red.

Iridoids: natural monoterpenes chemically related to methylcyclopentane. They are widespread in higher plants, and many are bitter principles, e.g. loganin of buckbean *(Menyanthes trifoliata),* gentiopiocrin in the roots of gentian *(Gentiana* species). The name is derived from iridodial, an iridoid present in the defense secretions of ants.

Iris:

1) see Light sensory organs.

2) see *Iridaceae.*

Irish moss: see *Rhodophyceae.*

Iron, ***Fe:*** a bioelement, which is an essential constituent of all living cells. Green parts of plants contain 100–200 mg/kg dry mass. An adult human contains between 4 and 5 g Fe, of which 75% is combined in hemoglobin. In addition to hemoglobin, Fe is found in many other proteins, e.g. cytochromes, peroxidase, catalase and ferredoxin. It is stored in the spleen and liver, where it is bound to ferritin and hemosiderin; these proteins form Fe complexes containing up to 23% and 35% Fe, respectively. Fe is transported in the blood as a complex with transferrin, and it has a variety of physiological activities. It is an essential component of many redox proteins, where it forms a catalytic component of the redox reaction by alternating between the Fe(II) and Fe(III) oxidation states; examples are the respiratory chain cytochromes (where Fe is located within a porphyrin ring) and nonheme iron proteins (e.g. ferredoxin, in which Fe is attached to the protein surface in a complex with inorganic sulfur). It has a regulatory role in many microorganisms, for example as an inhibitor of citrate synthesis in *Aspergillus niger,* and as a promoter of antibiotic synthesis in *Streptomyces species.* It is a coenzyme of aconitase in the tricarboxylic acid cycle

Fe deficiency in plants results in a marked inhibition of chlorophyll synthesis. This is manifested as *iron chlorosis,* in which, starting with the youngest leaves, the leaf areas between the veins become lighter green, and finally yellowish white; fruit trees and grape vines are particularly susceptible.

Iron bacteria: chemoautotrophic bacteria that obtain most or all of their energy from the oxidation of inorganic Fe(II) compounds to Fe(III) compounds. This energy is then used to convert carbon dioxide into carbohydrates. I.b. occur in iron-containing water and soil, and their cells are often thickly incrusted with iron compounds. *Thiobacillus ferrooxydans* occurs in seepage water from coalmines, and it is responsible for the occurrence of thick, conspicuous deposits of ferric hydroxide (see *Thiobacillus).* According to the strict definition of I.b., it is debatable whether organisms such as *Gallionella* (see) or *Leptothrix* (see *Chlamydobacteriaceae)* should be called I.b., since it is not certain that they are truly chemoautotrophic.

Irone, ***6-methyl-ionone:*** a terpene ketone. It exists as 3 isomers (α,β,γ) which differ with respect to the position of the double bond in the 6-membered ring (Fig.). β-I. occurs in orris root and gives rise to the pleasant perfume of the violet. Mixtures of the α-, β- and γ-isomers are found in the snowdrop *(Galanthus),* stock *(Matthiola),* mezereon *(Daphne),* wallflower *(Cheiranthus)* and iris *(Iris).* Synthetic I. is used in perfume manufacture.

O

α-Irone

Irregularia: see *Echinoidea.*

Irreversible thermodynamics: see Nonequilibrium thermodynamics.

Irritability, ***sensitivity:*** the ability of living organisms to react to even the simplest external effects. The environmental and conditional factors that elicit a reaction (e.g. light, gravity, heat, chemicals, electricity, mechanical factors) are called stimuli (see Stimulus). I. is essential for the existence of any living organism, because it enables the continual perception of, and reaction to, environmental changes.

The degree of I., expressed by the Stimulus threshold (see), is an important index of the functional state of any living cell, nerve or muscle fiber.

Phylogenetically, I. is at least as old as the eukarytotic cell, but it is not inseparable from the definition of living material. Where I. has arisen during phylogenesis, it has become an important genetically fixed property for the processing and output of information.

Isauxesis: see Isometric growth.

Ischnochitonida: see *Polyplacophora.*

Island bridges: island chains, found almost exclusively in shallow seas, which enable the stepwise colonization of new terrirory by animals and plants. Before Continental drift (see) was recognized, "island hopping" was often presented as a counter-argument to the existence of hypothetical Land bridges (see).

Island hopping: see Island bridges.

Island night lizard: see Night lizard.

Island theory: see MacArthur-Wilson theory of island biogeography.

Islets of Langerhans: see Pancreas.

Isoalleles: alleles that are so similar in their phenotypic expression that they can only be distinguished by special techniques, e.g. intensification of the phenotypic effects by incorporation of the alleles into an unusual genotype, or by the action of extreme environmental conditions. They can, however, be exactly characterized by methods of Molecular genetics (see). *Mutation I.* are also very similar in their expression, but have different rates of mutation.

Isobryales: see *Musci.*

Isochromosomes: chromosomes with two structurally and genetically identical arms. I. may arise during mitosis by the splitting of the centromere along an abnormal plane; this produces two I., one containing 2 copies of the left arm, the other containing 2 copies of the right arm of the original chromosome. Therefore, after mitosis, one arm of the chromosome is duplicated in one daughter cell and deleted in the other. In *Drosophila,* I. are usually referred to as *compound chromosomes* or *attached chromosomes.* I. can also be formed by Robertsonian translocation, involving homologous acrocentric chromosomes.

Isocitric acid: a tricarboxylic acid isomeric with citric acid. It is found in many plants, and it is an intermediate in the Tricarboxylic acid cycle (see).

```
HOOC—CH—CH—CH2—COOH
     |   |
     OH  COOH
```

Isocitric acid

Isocoenoses: communities of animals and plants, which exist under similar environmental conditions and therefore have a largely similar life forms, although the various ecological niches are occupied by different species, e.g. the prairies of North America and the steppes of Central Asia. See Ecological site equivalence equivalence, Nutritional interrelationships.

Isodont: a hinge type of the *Bivalvia* (see).

Isodynamic equivalencies of foods: quantities of different food materials that are energetically equivalent. Thus, 1 g protein has the same energy value as 1 g carbohydrate, while 1 g fat is energetically equivalent to 2.27 g carbohydrate and 2.27 g protein. These equivalencies apply only to the combustion values of the different food components, in no way reflect their relative qualities, such as their contents of Essential nutrients (see) or the Biological value of proteins (see).

Isoenzymes, ***isozymes:*** enzymes that catalyse the same reaction and have the same substrate specificity, but possess different primary structures. Biologically, they may be immunologically different, and they may be differently affected by inhibitors. The differences in primary structure give rise to differences in physical-chemical properties, which can be exploited for their laboratory separation, e.g. by electrophoresis. Five I. of human lactate dehydrogenase are known.

Isoetales: see *Lycopodiatae.*

Isogamy: see Reproduction, Fertilization.

Isogeny: the genetic identity of all individuals in a group or line. Isogenic lines are most commonly derived from inbreeding or vegetative reproduction.

Isognathous dentition: see Dentition.

Isolates: populations or groups of populations whose members mate preferentially with one another, rather than with members of other populations (i.e. other I.). Genetically, I. are Mendelian populations, which differ from one another in the relative frequency of certain genes.

Isolation:

1) Prevention of gene exchange between populations by geographical barriers. This *geographical I.* is an important evolutionary factor, and is a necessary precondition for allopatric speciation.

2) Prevention of gene exchange between populations of different species by genetic barriers. This *reproductive I.,* which is genetically based, usually arises during speciation as a result of geographical I. The following reproductive isolation mechanisms prevent gene exchange between different species.

a) *Ecological I.,* in which different species prefer different habitats, so that potential pairing partners do not meet.

b) *Ethological I.,* in which potential partners fail to mate because of their different, species-specific behavior.

c) *Mechanical I.,* in which successful pairing is prevented by structural incompatibility of the copulation organs.

d) *Gamete mortality,* which prevents egg fertilization, despite successful insemination.

e) *Zygote mortality,* which terminates development, despite successful fertilization.

f) *Hybrid inferiority,* in which hybrids die young or fail to find a copulation partner, so that gene flow between the species is prevented.

g) *Hybrid sterility,* which inhibits gene flow in the F_1 or F_2 generation of a species hybridization.

a)-d) represent prezygote I. mechanisms, whereas e)-g) are postzygote I. mechanisms. When different species live in the same territory, postzygote I. mechanisms generate a tendency for the selective development of prezygote I. mechanisms a)-c), i.e. selection works against the continuation of unfruitful pairing.

Isoleucine, ***Ile:*** CH_3—CH_2—$CH(CH_3)$—$CH(NH_2)$—COOH, a proteogenic, essential, glucogenic and ketogenic amino acid. Ile is present in many proteins, e.g. serum proteins, hemoglobin, edestin and casein. In plants and bacteria, it is biosynthesized from 2-oxoglutarate (a-ketoglutarate) and pyruvate.

Isomerases: a class of enzymes that isomerize their substrates, and catalyse rearrangements within substrate molecules. They are assigned to enzyme group 5 by the Nomenclature Committee of the International Union of Biochemistry. They include the ***racemases*** and ***epimerases,*** e.g. amino acid racemases, which catalyse the interconversion of the L- and D-forms of amino acids; and pentose phosphate epimerase, which catalyses the epimerization of xylulose 5-phosphate to ribulose 5-phosphate. I. are also responsible for ***cis-trans-isomerization,*** while ***phosphohexose isomerase*** catalyses the conversion of glucose 6-phosphate into fructose 6-phosphate.

Isometric contraction: see Cardiac cycle.

Isometric growth, ***isauxesis, auxesis, geometric similarity:*** growth during which the relative proportions of body parts remain constant with the increase in body size.

Isometric phase: see Cardiac cycle.

Isometric relaxation: see Cardiac cycle.

Isomorph: see Amorph.

Isomyarian condition: synonymous with homomyarian or dimyarian condition, terms applied to members of the *Bivalvia* (see) that possess 2 adductor muscles of about the same size.

Isopoda: one of the largest orders of the *Crustacea* (subclass *Malacostraca).* Most of the described 4 000 (approx.) species are bottom dwelling marine organisms. There is no carapace. The first thoracic segment is fused to the cephalon, and in rare instances the second thoracic segment may also be fused to the first. Antennules are characteristically uniramous. The antennae are usually uniramous, but sometimes carry a small exopodite. The body is generally depressed or dorsoventrally flattened; some members of the *Anthuridea* as well as some *Microcerberidae* (a minor family of the *Asellota)* have cylindrical, in some cases almost worm-like bodies. A pair of sessile, compound eyes is present. In addition to the antennae and antennules, the appendages of the cephalon include 1 pair of mandibles, 2 pairs of maxillae and 1 pair of maxillipeds. The thorax carries 7 pairs of walking legs, all of them similar, although the first pair are often modified into subchelate gnathopods. In females the bases of several pairs of legs carry thin plates (oöstegites) which overlap beneath the sternites to form a brood pouch (marsupium). Ventrally on the abdomen are 5 pairs of biramous lamellar pleopods and a single pair of uropods. The pleopods are branchial, sometimes also natatory (adapted for swimming); in the *Oniscoidea* they are adapted for aerial respiration ("white bodies", see below).

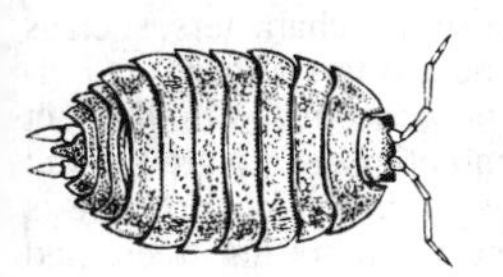

Fig.1. *Isopoda.* Wood-louse *(Porcellio scaber),* female.

Only one order is terrestrial, i.e. the *Oniscoidea,* and these are also the only truly terrestrial crustaceans. Several members of the *Oniscoidea* are common opportunist garden and greenhouse pests, e.g. species of *Porcello* (Fig.1) and *Oniscus,* commonly known a ***Wood lice*** or ***Sow bugs.*** *Porcello scaber* is the ***Common wood louse.*** They conceal themselves during the day in damp habitats (under stones, in cellars, etc.), and they feed on soft, moist, rotting or live plant material, leaving irregularly shaped holes in any leaves that they attack. Also garden pests, but more tolerant of dry habitats, are the ***Pill bugs,*** *Cylisticus* and *Armadillidium,* which can roll themselves into a spherical shape with the antennae and all other appendages completely hidden. *Ligia exotica* is one of many similar terrestrial species common on marine beaches throughout the world. Wood lice still use gills for respiratory gas exchange, but the more advanced families also possess ramifying pseudotracheae, or tracheal organs, also known as "white bodies", in the exopodites of 2 or more pairs of pleopods. The freshwater isopod known as the ***Hog slater*** or ***Water slater****(Asellus aquaticus,* Fig.2) is a widely distributed, common European species, found in standing water among plants and bottom litter. Economically important are the wood-boring isopods of the family, *Limnoriidae* (suborder *Flabellifera),* especially the ***Gribble*** *(Limnoria lignorum),* found in the coastal waters of the North Sea, where it burrows into and eventually destroys submerged wood, such as jetty pilings and other wooden harbor structures.

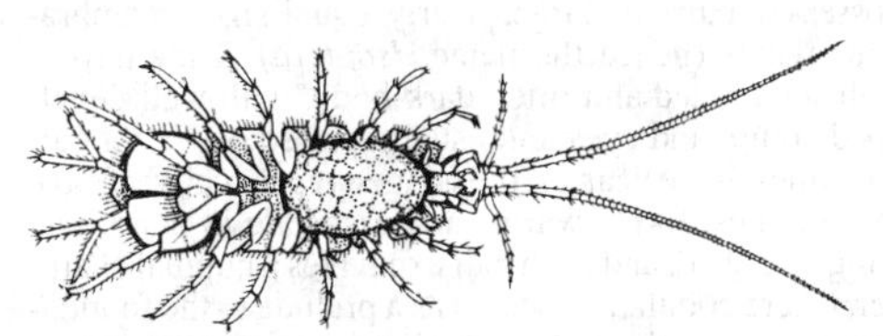

Fig.2. *Isopoda.* Hog slater *(Asellus aquaticus);* ventral surface of female with brood pouch.

Parasitic isopods. Larval (manca) stages of the *Gnathiidea* and adult members of the *Cymothoidae* are ectoparasites on the skin of fish. All members of the suborder, *Epicaridea,* are blood-sucking parasites exclusively of other crustaceans. Epicarids are often highly modified, the female more so than the male, with little resemblance to free living crustaceans, so that their true phylogenetic affinities must be determined from the larval forms; they are either protandrous hermaphrodites, or the sexes are separate with phenotypic sex determination. Most extreme modification is seen in the family, *Entoniscidae,* which live in the gill chambers of crabs.

Classification (only the most important families are listed).

1. Suborder: *Gnathiidea.*
2. Suborder: *Anthuridea.*
3. Suborder: *Flabellifera.*
   Families: *Limnoriidae, Sphaeromatidae, Cymothoidae.*
4. Suborder: *Valvifera.*
   Family: *Idoteidae.*
5. Suborder: *Asellota*
   Family: *Asellidae.*
6. Suborder: *Phreatoicoidea.*

7. Suborder: *Oniscoidea.*

Families: *Ligiidae, Trichoniscidae, Oniscidae, Porcellionidae, Armadillidiidae.*

8. Suborder: *Epicaridea.*

Isoprenoids: see Terpenes.

Isoptera, *Termitina, Termitida, Sociala,* ***termites, "white ants":*** an insect order restricted almost entirely to the tropics, containing 6 families and about 1 860 known species. Two species are native in Mediterranean Europe, and a few species occur in temperate regions of North America. Occasionally termites enter temperate countries in imported wood, and some species are established outside their normal geographical range in heated buildings in temperate and cool regions. The chief food source is wood. In the 5 most primitive families, intestinal protozoa (flagellates) decompose the ingested cellulose, whereas in more advanced termites this function is performed by symbiotic gut bacteria.

Imagoes. Depending on the species, adult body length varies from 0.2 to 2 cm (the gravid queen is longer, see below). The white or pale yellow color of many termites, especially of the workers, has given rise to the misnomer, "white ants". Termites are polymorphic and social, displaying a highly developed division of labor. In most species there are 3 castes: fertile reproductives, sterile workers and sterile soldiers. The 5 less advanced families lack one or both sterile castes, and larger juvenile stages (pseudergates) serve as workers.

Primary reproductive caste. Primary reproductives possess 2 pairs of large, nearly equal-sized membranous wings (hence the name, *Isoptera).* They have a well sclerotized and often dark body, with well developed compound eyes and usually paired ocelli. At certain times of the year, all the winged reproductives leave the colonies, form swarms and associate in pairs. The wings are shed, and each pair excavates a nuptual chamber, where copulation occurs as a prelude to the foundation of a new colony. All the other subsequent members of the colony are the offspring of this fertile king and queen. In many species, the gravid female (queen) is physogastric, i.e. her abdomen becomes swollen, mainly by enlargement of ovaries and fat body, and it may attain a length of 10 cm.

If one of the primary reproductives is lost, other members of the colony of the appropriate sex develop as replacements; one of these survives and becomes dominant. In more highly evolved *I.,* these are known as replacement reproductives, while in the more primitive species they are called supplementary or secondary reproductives. They may be wingless or have wing pads, but they never have fully developed wings. They have a less sclerotized and lighter colored body than the primary reproductives, and their compound eyes are often reduced. Queens that originiate from secondary or replacement reproductives are not physogastric.

Workers are wingless and sterile, usually with pale and weakly sclerotized bodies. Anatomically, they may be of either sex. They are responsible for nest building, provision of food and care of the brood.

Soldiers are wingless and sterile, and of either sex. Responsible for guarding the colony, they possess a very large, well sclerotized head, often with powerful mandibles (mandibulate soldiers) for seizing intruders. In many advanced *I.,* the soldiers have a medial frontal rostrum, and the mandibles are reduced (nasute soldiers); a frontal gland opens at the tip of the rostrum and produces a noxious secretion which is ejected to repel intruders.

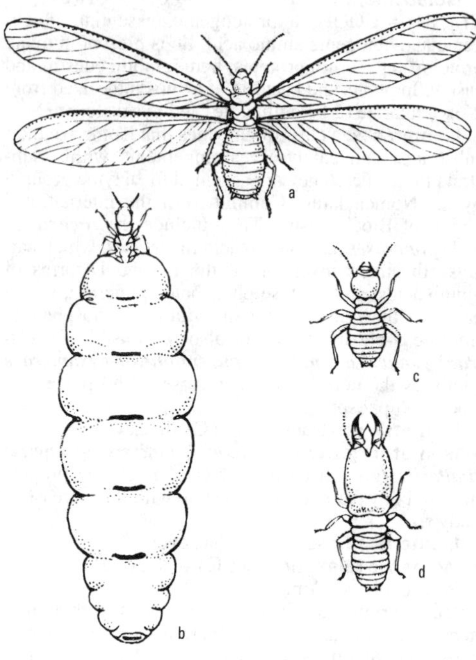

Isoptera. Termite castes *(Macrotermes bellisocus* Smeathman). *a* Winged primary reproductive. *b* Queen. *c* Worker. *d* Soldier.

Eggs. Since each colony contains only one pair of reproductive insects, the female must produce a very large number of eggs, in some cases more than 1 000 per day.

Larvae. The larvae are similar to the adult, and they attain their final form after 4 to 10 molts. At the second molt all larvae are identical, but from the third molt onward they begin to display the characteristics of their subsequent caste

Economic importance. In the tropics, termites cause widespread destruction of living and stored wood and structural timber. Agricultural plants, e.g. cereals, coffee and cocoa, are also attacked. By their burrowing and mining activities, termites are also detrimental to agricultural soils. In some regions of Africa and South America, the swarming reproductive forms, which contain high-value protein and fat, are an important food source for local people.

Classification.

1. Family: *Mastotermitidae.* This family is represented by a single species, *Mastotermes darwiniensis* ***(Darwin termite),*** native to tropical northern Australia. In the Eocene, mastotermitids were distributed worldwide. The family displays primitive characters, such as lobed hindwings and 5-segmented tarsi. There are evident affinities with the *Blattoptera,* including the fact that both flagellates and symbiotic bacteria are present in the intestine, and the eggs are laid in packets. Nests are constructed underground or in rotting trees, and

structural timbers are attacked directly from earth passages beneath them. Eggs are laid by several sexually mature nymphs, and the colony is not founded by a true royal pair.

2. Family: *Kalotermitidae.* The kalotermitids or ***Dry wood termites*** form small colonies of up to 3 000 members, in an irregularly chambered nest hollowed out from sound wood. Older nymphs or pseudergates perform the functions of workers, and a true worker caste does not exist. Many species are native on coasts and islands in southern Asia, causing damage to tea bushes, teak trees, cocoa trees, etc. In the Mediterranean region, *Kalotermes flavicollis* ***(Yellow-necked termite)*** attacks grape vines. In Central America and western USA, species of *Incisitermes* cause serious damage to structural timber. Particularly well known are species of *Cryptotermes* ***(Wood termite*** or ***Powder post termite),*** which are frequently present and concealed in furniture and isolated pieces of commercial timber. They have therefore been widely transported, and are now established in most tropical port cities.

3. Family: *Termopsidae.* The termopsids or ***Damp wood termites*** form colonies of up to 10 000 individuals in decaying trees. As in the *Kalotermitidae,* there is no true worker caste. The 5 species are divided among 3 subfamilies.
1. Subfamily: *Termopsinae. Archotermopsis wroughtoni* lives in dead conifers at high altitudes in Kashmir and the western Himalayas. Species of *Zootermopsis* have a wide range in North America, forming colonies in sequoias on the west coast and cottonwood in the south.
2. Subfamily: *Stolotermitinae.* The single genus, *Stolotermes,* is represented by a single species in South Africa, and 5 species in Australia, Tasmania and New Zealand. All live in rotting wood, and the reproductives and the soldiers have characteristically flattened bodies.
3. Subfamily: *Porotermitinae.* The single genus, *Porotermes,* is represented by 3 species, on the coasts of Chile, South Africa and Southeast Australia/ Tasmania, respectively.

4. Family: *Hodotermitidae.* A family of 16 species in 3 genera, including ***Primitive damp wood termites*** of temperate climates, and the more advanced ***Harvester termites*** of hotter, drier regions. The latter form underground colonies from which the workers forage above ground for plant material. There is no distinct soldier class.

5. Family: *Rhinotermitidae.* A family of subterranean forms, usually living in buried rotten wood. A few species invade wood above ground, some build nests directly in the soil, and some construct mounds on the soil surface. Many species have both mandibulate and nasute soldiers. The 180 species (approx.) are distributed in 14 genera and 6 subfamilies.

6. Family: *Serritermitidae.* The serritermitids or ***Sawtooth termites*** are represented by a single, rare Brazilian species, *Serritermes serrifer.*

7. Family: *Termitidae* ***(Higher termites).*** This family contains about 75% of all extant termite species. The nearly 1 500 species in 150 genera show the greatest variety of social organization and specialization. They lack intestinal protozoa, and digestion is aided by symbiotic bacteria. There are 4 subfamilies.
1. Subfamily: *Amitermitinae.* The most primitive termitids, closely allied with the *Rhinotermitidae.*
2. Subfamily: *Termitinae* ***(True termites).*** The soldiers have snapping mandibles.
3. Subfamily: *Macrotermitinae* ***(Macrotermites).*** This subfamily contains the ***Old World fungus-growing termites,*** which differ from all other termites by their cultivation of fungi. Wood and plant material are taken into the nest, where they are formed ninto spongy, cellulose-rich combs, which may fill the nest chambers. Fungi of the genus, *Termitomyces,* grow on the combs and render the material digestible by the termites. Older parts of the fungus are also eaten. Most macrotermites build subterranean nests. Some species, however, form enormous colonies and build earthern nest hills (termite hills or mounds), also known as *termataria,* which are prominent landscape features, e.g. *Bellicositermes bellicosus* (East Africa).
4. Subfamily: *Nasutitermitinae.* The largest subfamily with 500 species. The soldiers are nasute. *Nasutitermes triodiae* builds massive termataria in northern Australia.

Isoquinoline: a strong organic base present in coal tar. Its heterocyclic ring system is present in the structure of the I. alkaloids.

N

Isoquinoline

Isosaprophytic conditions: see Saprophytic system.

Isosmotic: see Halobiont, Isotonic, Osmosis.

Isothrix: see *Echimyidae.*

Isotonic, ***isosmotic:*** an adjective applied to solutions that have the same osmotic value. See Osmosis.

Isotope exchange measurement: a method for studying systems in dynamic equilibrium, with the aid of radioactive isotopes. The reaction rates of such systems remain unchanged with time. By adding a radioactive label to one compartment of the system, the reaction rates can be determined by measuring the rate at which the isotope appears in the other compartments. The method presupposes that there is no isotope effect, i.e. that the radioactive label behaves in exactly the same way as the material under investigation and is converted at the same rate as its nonradioactive counterpart. The method is used for the study of transport processes and biochemical reactions.

Isotrehalose: see Trehalose.

Isotropic: having the same shape on all sides; having the same properties irrespective of direction. The converse is Anisotropic (see).

Isovaleric acid: $(CH_3)_2CH—CH_2—COOH$, a branched chain fatty acid. In the pure form, it is a colorless liquid with an odor of valerian. It occurs naturally, both free and esterified, e.g. in the roots of valerian *(Valeriana officinalis),* angelica *(Angelica archangelica)* and in the berries of the guelder rose *(Virburnum opulus).*

Isozymes: see Isoenzymes.

Ispaghul: see *Plantaginaceae.*

Istiophoridae: see *Scombroidei.*

Istle: see *Agavaceae.*

Itaconic acid: an unsaturated dibasic carboxylic acid, which is a metabolic product of certain molds, e.g. *Aspergillus itaconicus.* It is formed by decarboxylation of aconitic acid, an intermediate in the tricarboxylic acid cycle. It serves as a reserve material, which is easily remobilized as a carbon source in mold metabolism.

$$HOOC-CH_2-\underset{\underset{CH_2}{\|}}{C}-COOH$$ *Itaconic acid*

Itch mites: see Scabies mite.

Iter: see Ventricle.

IUCN, ***International Union for the Conservation of Nature and Natural Resources:*** an international Nature conservancy (see) organization, founded in 1956 and based in Gland (Switzerland). It is also known under the acronym of its French name, UICN. Its many activities include the publication and continual updating of the Red book (see) or Red list of plants and animals threatened with extinction.

History and organization. In 1934, on a private initiative, the OIPN (l'Office International pour la Protection de la Nature) was founded in Brussels. Following an international conference at Fontainbleau sponsored by Unesco and the Frech government, OIPN became the IUPN (International Union for the Protection of Nature). By 1956, its activities and membership had become so wide that it was renamed the IUCN. Membership consists of states, government departments, private societies and institutions, and international organizations. It is not a United Nations organization, but it is supported by UN agencies, e.g. Food and Agricultural Organization (FAO), Educational, Scientific and Cultural Organization (UNESCO), Economic and Social Council (ECOSOC), the Council of Euroipe and other intergovernmental bodies. It also works with the International Council for Bird Preservation (ICBP). The IUCN works through six commissions: Ecology, Education, Landscape planning, Legislation, National Parks, Survival service.

Ivy: see *Araliaceae.*

Ixobrychus: see *Ciconiiformes.* See plate 6.

Ixocomus: see *Basidiomycetes (Agaricales).*

Ixodidae: see *Ixodides.*

Ixodides, ***ticks:*** a suborder of the *Acari* (see), containing 3 families. They have considerable medical and economic importance. Especially in hot climates, ticks are the vectors of dangerous human and animal diseases. They are relatively large, and they all live as ectoparasites, penetrating the skin of their host with their mouthparts, in order to suck the blood. Morphological characters common to all ticks are are the piercing hypostome with recurved teeth, chelicerae with lateral teeth on the movable digits, a pit containing specialized sensory setae on the tarsus of each first leg *(Haller's organ),* and lateral stigmata without sinuous peritremes (a *peritreme* is a chitinous tube associated with the stigmata and trachea). After several days, when greatly distended with blood, they become detached and fall from the host. Hosts include mammals, birds and reptiles.

The hard ticks (family *Ixodidae)* and the very small family *Nuttalliellidae* have a hard shield or propodosomal plate (scutum) on their backs, although in the *Nuttalliellidae* the scutum is not heavily sclerotized and is similar in appearance to the leathery integument of the rest of the body. The soft ticks (family *Argasidae)* lack this dorsal plate. In the *Nuttalliellidae* the palpal tarsus is terminal, whereas in the *Ixodidae* it is embedded in the ventral apex of the tibia.

Best known of the European hard ticks is the ***Sheep tick*** or ***Castor bean tick*** *(Ixodes ricinus),* which climbs onto the tops of grasses and bushes, from where it can fall onto a host. Its most important host is the sheep, but it also feeds on humans, deer, rabbits, stoats, squirrels, lizards and at least thirty nine different bird species. During feeding, the body length of the female increases from about 4 mm to 11 mm. It then has the appearance of a shiny blue pea, with its mouthparts deeply embedded in the host flesh and held in place by spines that make it difficult to remove. The male does not suck blood, but uses the host animal in its hunt for a female. The larva and nymphs of *Ixodes ricinus* also suck blood. The sheep tick is also the vector for a virus disease of sheep, known as *louping ill;* mild attacks by the same virus have been recorded in shepherds, in contrast to more severe infections accidentally contracted by laboratory research workers.

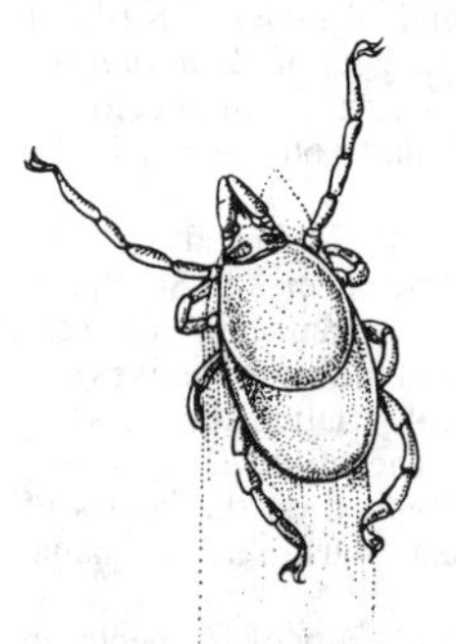

Ixodides. The sheep tick or castor bean tick *(Ixodes ricinus)* in a questing posture on a grass halm.

Viruses carried by ticks belong to the arboviruses (arthropod-borne viruses). Thus, *Ixodes persulcatus* is responsible for transmitting a virus that causes encephalitis among lumber workers in the Taiga forest of northern Asia. Examples of tick-borne Rickettsial diseases of humans are Rocky Mountain spotted fever, South African tick bite fever, Mediterranean fever (fièvre boutonneuse). More than 20 species of the *Ixodidae* and *Argasidae* are known to be vectors of *Coxiella burnettii,* the rickettsia-like diplobacillus that causes the enzootic infection of domestic animals known as Q fever.

The suborder *Ixodides* is usually combined with the suborder *Mesostigmata* to form the *Parasitiformes.*

J

Jacamaralcyon: see *Galbulidae.*

Jacamars: see *Galbulidae.*

Jacana: see *Jacanidae.*

Jacanidae, ***jaçanás, lily trotters:*** an avian family in the order *Charadriiformes* (see). The name, Lily trotter, is applied especially to the African forms. The *Jacanidae* are exceptionally long-legged, medium-sized waterfowl (25–33 cm in length, excluding the tail plumes in the breeding plumage of *Hydrophasianus)* with conspicuously colored plumage. The feet are designed for walking easily over floating vegetation: the toes are excessively elongated with straight claws, the claw of the hindtoe being much longer than the toe and tapering to a fine point. They are agile swimmers and divers, and submerge to conceal themselves. Except in *Hydrophasianus,* the tail is short. The wings are long, and they carry a small spur or blunt knob (sometimes developed into a long spine) on the carpal joint. Most species possess a red, yellow or blue frontal facial shield. The diet consists of all types of animal and vegetable aquatic life. There are 7 species in 6 genera: *Actophilornis africana* ***(African jacana*** or ***lily trotter;*** sub-Saharan Africa); *Actophilornis albinucha* ***(Madagascar jacana;*** Madagascar); *Microparra capensis* ***(Lesser lily trotter;*** East Africa); *Hydrophasianus chirurgus* ***(Pheasant-tailed jacana;*** India); *Metopidius indicus* ***(Bronze-winged jacana;*** Southeast Asia); *Irediparra gallinacea* ***(Lotus bird*** or ***Comb-crested jacana;*** Indonesia and Australia); *Jacana spinosa* ***(American jacana;*** tropical America). With the exception of the Lesser jacana, the female is larger than the male; polyandry (females mating with more than one male) is common, and in the Pheasant-tailed and Bronze-winged jacanas it appears to be the rule.

Jaccard's index, ***q:*** an index of bioc(o)enotic association, expressing the relationship between the number of species common to 2 ecological units and the sum of species occurring in only one of each of the two units. Developed in 1912 for phytosociological comparisons, J.i. has also proved an effective index in animal ecology. It is usually expressed as a percentage: $q\ (\%) = (c/a + b - c) \times 100$, where *a* represents the number of species in sample A; *b* the number of species in sample B; *c* the number of species in both A and B.

Jackals: medium sized (head plus body 56–74 cm) members of the *Canidae* (see). Three species are recognized: *Canis aureus* ***(Golden*** or ***Asiatic jackal;*** Central Africa, Middle East and southern Asia as far as Russian Turkestan); *Canis mesomelas* ***(Black-backed jackal;*** eastern and southern Africa); *Canis adustus* ***(Side-striped jackal;*** eastern and southern Africa). They feed on any small animals that they can catch, including insects, and they scavenge the kills of larger predators. Larger animals, especially those that are ill or old, are also occasionally taken.

Jackass: see *Equidae.*

Jacob's ladder: see *Polemoniaceae.*

Jacobson's organ: see Vomeronasal organ.

Jaegers: see *Stercoraridae.*

Jaguar: see *Felidae.*

Jak fruit: see *Moraceae.*

Jalap: see *Convolvulaceae.*

Japanese deer: see Sika deer.

Japanese fox, *Nyctereutes procyonoides:* a predatory member of the dog damily, resembling a racoon. Native to East Asia, it has been introduced sporadically into eastern Europe for its fur, and is already spreading westward. It feeds on small animals (including toads), birds' eggs and fruits, and sometimes causes damage in bird colonies. In cold climates, it shows winter dormancy (semi-hibernation).

Japanese fox (Nyctereutes procyonoides)

Japanese galls: see Tannins.

Japanese giant salamander: see *Cryptobranchidae.*

Jarales: see Sclerophylous vegetation.

Jararaka: see Lance-head snakes.

Jasmine: see *Oleaceae.*

Jasminum: see *Oleaceae.*

Jasus: see Spiny lobsters.

Jaundice: see Bile pigments.

Javelina: see *Tayassuidae.*

Jays: see *Corvidae.*

Jay thrushes: see *Timaliinae.*

Jellyfish: see *Scyphozoa, Aurelia.*

Jelly fungi: see *Basidiomycetes.*

Jenny: see *Equidae.*

Jerboas: see *Dipodidae.*

Jerdon's babbler: see *Timaliinae.*

Jerusalem artichoke: see *Compositae.*

Jerusalem cherry: see *Solanaceae.*

Jery: see *Timaliinae.*

Jesuits' bark: see Quinine.

Jet ants: see Ants.

Jewel beetles: see *Coleoptera.*

Jew-fishes: see Sea basses.

Jew's ear fungus: see *Basidiomycetes (Auriculariales).*

Jigger: see *Siphonaptera.*

Jingle shells: see *Bivalvia.*

Jirds: see *Gerbillinae.*

John Dory, *St. Peter's fish,* *Zeus faber:* a marine teleost fish, up to 70 cm in length, with a high, laterally compressed body, large head and prominent jaws; family *Zeidae* (Dories), order *Zeiformes.* Head and back are dark to yellowish brown, the sides bear light yellow stripes, and the belly is silvery; there is also a distinctive round dark, yellow-edged spot on each side. The anterior dorsal fin is supported by 9–10 spines, which are elongated beyond the fin into threads. There are rows of spines on the ventral surface, and spines are also found on the opercula, and in front of and behind the eyes. It feeds on small fish, such as pilchards and anchovies, and it is frequently caught (mainly with dragnets) off Atlantic coasts from southern Norway to the Cape of Good Hope, and in the Mediterranean, and it is occasionally encountered in the Black Sea. The flesh has an excellent flavor.

Johne's disease: see *Actinomycetales.*

Johnny carp : see Goldfish.

Johnston's organ: see Chordotonal organ.

Jointed antenna, articulated antenna: a common type of antennae in entognathous insects and other arthropoda. It consists of numerous similar (almost identical) segments, each provided with internal muscles. Only the terminal segment is different. See Antenna.

Joint fir: see *Gnetatae.*

Joints, *arthroses, articulations:* close contacts between the bones of the skeleton. They are classified, according to the amount of movement they display, as *synarthroses* (immovable), *amphiarthroses* (slightly movable) and *diarthroses* (freely movable). In all cases, the opposing surfaces are clothed with cartilage or fibrous tissue.

Four types of synarthrosis are recognized: 1) *suture* (interlocking processes and indentations, or simple apposition of rough surfaces, e.g. the lamboidal, sagittal and coronal sutures of the skull consist of interlocking indentations); 2) *schindylesis* (insertion of a thin bone in a fissure of another bone, e.g. the vomer is inserted between the maxillae and the palatine bones); 3) *gomphosis* (insertion of a conical process into a socket, e.g. insertion of tooth roots in the maxillae and mandible; 4) *synchrondrosis* (a temporary articulation, in which the intervening cartilage ossifies later in life, e.g. between the occipital and sphenoid).

There are two types of amphiarthrosis: 1) *symphysis* (bones separated by a thick layer of fibrous cartilage, e.g. the pubic symphysis and the sacroiliac articulation); 2) *syndesmosis* (opposing surfaces united by an interosseus ligament, e.g. lower tibiofibular articulation).

Diarthroses include most of the articulations of the body that are commonly referred to as joints. Opposing surfaces of the bones are covered with hyaline cartilage and surrounded by a fibrous articular capsule, strengthened by ligaments and lined with synovial membrane. Six types of diathrosis are recognized. 1) *Gliding joints* or *arthrodia* permit only gliding movment, e.g. between carpal bones of the wrist, and the articular processes of the vertebrae. The articulated surfaces are almost flat. 2) *Hinge* or *ginglymus-type joints* permit angular movement in a single plane, and the articulated surfaces are shaped to allow only backward (flexion) and forward (extension) movement, e.g. the knee joint (between the humerus and ulna). 3) *Condyloid joints* or *condylarthroses,* formed by insertion of an oval head or condyle into an elliptical cavity, permit angular movement in two planes, e.g. the wrist (radiocarpal) joint. 4) *Saddle* or *reciprocal reception joints* also permit angular movement in two planes, but the articulated surface of each bone is concave in one direction and convex in another, the two directions being at right angles to each other, e.g. the articulation of the metacarpal bone of the thumb with the greater multangular bone of the carpus. 5) *Pivot* or *trochoides joints* perform a rotary movement in one axis; thus, a process rotates within a ring, or a ring rotates around a pivot, e.g. in the articulation of the axis and atlas, the front of the ring is formed by the anterior arch of the atlas and the back of the transverse ligament, while the odontoid process of the axis forms a pivot; in the articulation of the radius and ulna, the head of the radius rotates within the ring formed by the radial notch of the ulna and the annular ligament, turning the palm of the hand upward (supination), or downward (pronation). 6) *Ball-and-socket joints* or *enarthroses* have angular movement in all directions, as well as pivot movement; a more or less rounded head is inserted in a cup-like cavity, e.g. head of the femur in the acetabulum, and head of the humerus in the glenoid cavity of the scapula.

Joule: the SI unit of work, energy and heat. It is the unit used for the Heat of combusion (see) of food materials and for the Basal metabolic rate (see). It replaces the calorie (1 J = 0.24 cal).

Juglandaceae, *walnut family:* a family of dicotyledonous plants, comprising 64 species in 7 genera, distributed in southeastern Europe, northern temperate and northern subtropical regions of Asia and America, and in the Andes. All members are monoecious trees with alternate (opposite in *Alfaroa),* exstipulate, pinnate leaves. The perianth of the unisexual flowers in inconspicuous, consisting of 3–5 fused segments with a bract and 2 bracteoles. Male flowers occur in catkins, which

Juglandaceae. Flowering branch and fruit of the walnut *(Juglans regia).*

arise from axillary buds on shoots of the previous year. Between 3 and 40 stamens are present, flowers on the upper part of the tree containing more than those lower down. The wind-pollinated female flowers occur at the tip of the current year's growth. The 4-toothed, 4-merous calyx is free above and adnate to the ovary below. The unilocular, inferior ovary consists of 2 united carpels, with a single erect, orthotropous ovule, and a short style with 2 stigmas. The fruit is a drupe or nut. Bracts, which usually envelop the flower, subsequently form a fleshy casing of the fruit, or they become modified to form wings.

The ***Walnut tree*** *(Juglans regia)* is cultivated for its fruit and its high quality hard wood, which is used for furniture and veneers. Its fruit is an aromatic drupe, consisting of a glabrous, green, glandular epicarp, a fleshy mesocarp and an endocarp that finally becomes lignified to form the characteristic walnut. The deeply contored, edible seed contains about 50% oil. Ranging naturally from southeastern Europe to the Himalayas and Burma, the walnut tree is also planted in parklands throughout the world. ***Black walnut*** *(Juglans nigra),* widely cultivated in North America, also yields edible seeds and useful wood. The genus, *Juglans,* is also represented in the Caribbean. Various species of the North American genus, *Carya* **(Hickory),** yield valuable wood and edible winged fruits, e.g. Pecan nuts *(Carya illinoinensis).* Other edible fruits are Caucasian wing nut *(Pterocarya fraxinifolia;* the genus occurs in temperate Asia from the Caucasus to Japan); butternut *(Juglans cinerea,* North America); and mockernuts *(Carya tomentosa).*

Remaining genera are: *Engelhardtia* (India to southern China and Malaysia; fruits with small to medium wings); *Oreomunnea* (Central America; fruits with large (10 cm) rigid wings); *Platycarya* (northern China; spikes of both sexes erect and covered with imbricate bracts; fruit winged); *Alfaroa* (Costa Rica; ellipsoid, softly hispid, nonwinged fruits; leaves opposite).

Juglans: see *Juglandaceae.*

Juglone, ***5-hydroxy-1,4-naphthoquinone:*** a highly toxic, red-brown quinoid plant pigment, which occurs as the glucoside of its dihydro-derivative in the leaves of members of the *Juglandaceae,* in particular in the shells of unripe fruits of the walnut tree *(Juglans regia),* as well as in the inflorescences of members of the *Betulaceae* and *Salicaceae.* Walnut juice was formerly used as a standard treatment for the human fungal parasite, ringworm; the search for the active principle resulted in the discovery and isolation of J. During processing of the plant material, free J. is produced by hydrolysis and oxidation. It was formerly used for coloring walnut oil and as a component of hair dye.

HO O

O *Juglone*

Jugum: see Moths.

Julidae: see *Diplopoda.*

Juliidae: see *Gastropoda.*

Jumping mice: see *Zapodidae.*

Jumping gene: a gene carried by a Transposon (see).

Jumping spiders: see *Salticidae.*

Juncaceae, ***rush family:*** a family of about 300 species of monocotyledonous plants, mostly in the temperate zones of the Northern and Southern Hemispheres. They are herbaceous, grass-like or rhizomatous plants, but lacking the stem nodes and the sharply angular stems of the grasses. The inconspicuous, wind-pollinated flowers are usually in crowded cymes, and sometimes condensed into heads. There are 6 or 3 stamens, 1 style with 3 stigmas, and the fruit is a capsule.

Juncaceae. a Jointed rush *(Juncus articulatus). b* Grooved wood rush *(Luzula nemorosa).*

The largest genus is *Juncus* **(Rushes);** its members grow predominantly in wet habitats, and possess stem-like, glabrous leaf blades. The fibrous stems of larger species are used for plaiting and weaving of mats and baskets, etc. Most members of the genus, *Luzula* ***(Wood rushes),*** have flat leaves fringed with long colorless hairs, and they grow in moderately damp habitats.

Juncaginaceae, ***arrow grass family:*** a family of monocotyledonous plants, containing about 18 species in 4 genera, cosmopolitan in distribution, but most strongly represented in the temperate and cold regions of both hemispheres. They are annual and perennial marsh and aquatic plants. All the foliage leaves of the flowering stem are slender and radical (scapigerous). The small (usually), wind-pollinated, hermaphrodite flowers are 2–3-merous with no bracts, actinomorphic or slightly oblique, and carried in spikes or racemes.

Perianth segments are green or reddish and arranged in 2 series. Flowers are protogynous, i.e. the ovary matures before the stamens. The ovary is superior and the style is short or absent. There are 6 or 4 carpels. When 6 carpels are present, 3 of these may be sterile. There are 6 or 4 stamens with very short filaments.

Triglochin palustris (***Marsh arrow grass***) is found in swampland and in marshy estuaries, even near brackish water, from Europe to the Arctic, in North Africa, northern Asia, and North America. The 3 carpels of its fruit are arranged like the legs of a tripod.

Triglochin maritima (***Sea arrow grass***) is found in salt marsh turf and grassy places near the coast, from Europe to the Arctic, in North Africa, western and northern Asia, and North America. The oblong-ovoid fruit has 6 carpels.

Jungemanniales: see *Hepaticae.*

Jungle babblers: see *Timaliinae.*

Jungle cat: see *Felidae.*

Junglefowl: species of the genus *Gallus* (see *Phasianidae).* Species of the genera *Megacephalon, Eulipoa* and *Megapodius* are also sometimes referred to as junglefowl (see *Megapodiidae).*

Jungle runners, *Ameiva:* a genus of the lizard family *Teiidae* (see). They have a normal lizard-like appearance, and are often brightly colored with striking patterns of spots and stripes.

Juniper (Juniperus): see *Cupressaceae.*

Jurassic (named after the Swiss Jura mountains): the middle system of the Mesozoic (see Geological time scale). It is divided into three epochs which can be recognized as a sequence of blue-gray shales and muddy limestones *(Black J.* or *Liassic),* iron-containing sandstones *(Brown J.* or *Dogger)* and a light-colored calcareous layer *(White J.* or *Malm).* Finer subdivisions give a total of 11 stages. The J. lasted about 55 million years. The most important index fossils are the ammonites, which display a peak of evolutionary diversity in the J., and provide a basis for the biostratigraphic division of the J. into zones and stages, as well as the paleobiostratigraphic separation of provinces. The first smooth ammonite forms appear in the Liassic, accompanied in particular by simple-ribbed and crescent-ribbed forms. The Dogger is characterized by fork-ribbed ammonites, whereas those in the Malm display increasingly complex rib branching and subdivision, and already show evidence of transition to the degenerate ribbing and aberrant forms of the Cretaceous. Foraminifera, belemnites and ostracods also serve for the fine classification of strata. Certain individual brachiopods are very abundant, but there is a decrease in the number of brachiopod genera. Bivalves show great evolutionary diversity and abundance, and some are used for biostratification. Reptiles show an enormous phylogenetic expansion, and a second evolutionary peak. By the end of the J., reptile evolution had produced the most massive terrestrial vertebrates of all time, the dinosaurs. The *Ichthyopterygia* and *Sauropterygia* attained the greatest diversity. In addition to purely marine and terrestrial forms, the Liassic also contains the remains of the first flying dinosaurs, which possessed flight membranes between their digits. Of great phyogenetic importance are the fossil remains of *Archaeopteryx* (see) in the Malm of Solnhofen; this primitive bird possessed both avian features (feathers) and reptilian features (toothed jaws, tail with vertebrae), and is therefore thought to represent a link between reptiles and birds. Different groups of mammals existed at the beginning of the J. as shown by the presence of tooth and jaw fossils; it is thought that these mouse- to rat-sized mammals were mostly crepuscular and nocturnal.

The Jurassic flora is notably uniform throughout the world, consisting primarily of gymnosperms (conifers, cycads, bennettites and ginkgos) and pteridophytes. See Geological time scale, Mesophytic.

Jute: see *Tiliaceae.*

Juvabione, ***paper factor:*** a monocyclic sesquiterpene ester from the wood of the North American balsam fir *(Abies balsamea).* J. was the first Juvenile hormone (see) to be isolated from a plant, and it is specific for the insect *Pyrrhocoris apterus.* Its discovery arose from the observation that filter paper made from the wood of the balsam fir contains a factor which produces developmental anomalies in the larvae of *Pyrrhocoris.*

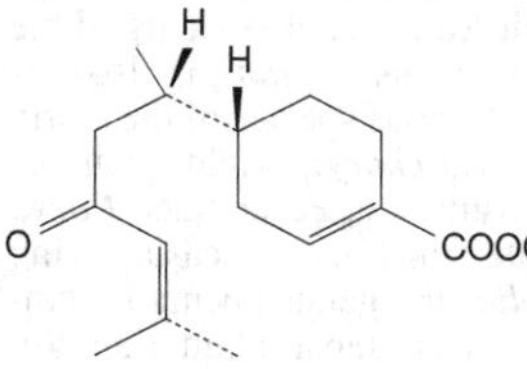

H H O $COOCH_3$

Juvabione

Juvenile age range: in humans, the age range from 14 to 20 years. See Age diagnosis.

Juvenile hormone, ***larval hormone, status-quo hormone:*** a hormone responsible for larval molting and the regulation of metamorphosis in insects (see Insect hormones). The first J.h. to be isolated and structurally elucidated was obtained from the abdomens of male silkworm moths *(Hyalophora cecropia).* Two active homologs were subsequently described, so that three J.hs. are now known (I, II and III), all of them sesquiterpenes. J.hs. are formed in the *Corpora allata* (accessory glands of the insect brain). Together with the steroid hormone, ecdysterone, they are responsible for molting from one larval stage to the next. In addition, they are responsible for the expression of larval characters and for maintenance of the larval stage during molting.

Juvabione (see) and farnesol derivatives are among the natural compounds with J.h. activity Antijuvenile hormones have been isolated from certain plants, e.g. precocene, which blocks the activity of the insect's own J.h. and causes faulty development.

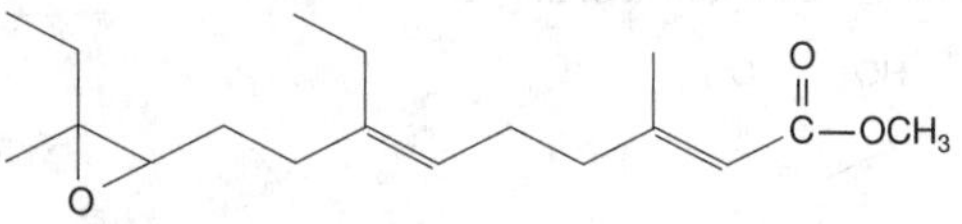

O C—OCH_3 O

Juvenile hormone I

Juvenis: see Age diagnosis.

Jynginae: see *Picidae.*

Jynx: see *Picidae.*

K

Kachuga: see *Emydidae.*
Kagu: see *Gruiformes.*
Kainophytic: see Cenophytic.
Kainozoic: see Cenozoic.
Kaka: see *Psittaciformes.*
Kakapo: see *Psittaciformes.*
Kakatoeinae: see *Psittaciformes.*
Kalanchoe: see *Crassulaceae.*
Kaldana seeds: see *Convolvulaceae.*
Kallidin: see Bradykinin.
Kamchatka brown bear: see Brown bear.
Kamptozoa: see *Entoprocta.*

Kanamycin: an aminoglucoside antibiotic produced by *Streptomyces kanamyceticus.* Like streptomycin, it inhibits protein synthesis on 70S (bacterial) ribosomes. It is used against otherwise intractable *Proteus* infections.

Kangaroos, *Macropodidae:* a family of Marsupials (see), found in Australia and on neighboring islands, consisting of more or less large, long-tailed, agile mammals, which progress by hopping on their hindlegs. A few species can also climb trees. The diet consists of grass, foliage and fruits. Some examples are listed below.

Unlike all other members of this family, ***Tree K.*** *(Dendrolagus)* are adapted to an arboreal existence. About 5 species inhabit forested areas of Papua New Guinea, and 2 species are found in northeastern Queensland. They have powerful hands and long claws, and relatively short hindlegs.

The 4 species of ***Pandemelon wallabies*** *(Thylogale)* are small, sometimes beautifully patterned K., which live mainly among bushes and tall grass, where they construct tunnel-like passages: *T. thetis* (New South Wales to the borders of Queensland), *T. stigmatica* (Papua New Guinea, New South Wales and Queensland), *T. bruijni* (Papua New Guinea and Bismarck Archipelago), *T. billardieri* (southern Australia and Tasmania).

The 11 species of ***Scrub*** or ***Bush wallabies*** *(Wallabia)* include the ***Red-necked wallaby*** *(W. rufogrisea),* which has been introduced outside its native Australia into New Zealand and parts of Europe.

The largest members of the family are the 3 species of ***Macropus.*** Powerful males attain the height of a man, and when fleeing can cover 10 m in a single hop at a height of 3 m. *M. giganteus* ***(Great gray kangaroo)*** inhabits open forested areas and dense bush in Australia and Tasmania. *M. rufus* ***(Red kangaroo)*** is found on inland plains throughout most of Australia. *M. robustus* ***(Wallaroo*** or ***Euro)*** inhabits mainly mountainous areas on Australia.

Kannabateomys: see *Echimyidae.*
Kaolinite: see Sorption.
Kapok: see *Bombacaceae.*
Kaposi's sarcoma: see Tumor viruses.
Karyogamy: see Fertilization.
Karyogram: see Karyotype.
Karyolymph: see Nucleus.

Karyotype: a description of the chromosomal complement of a eukaryotic cell nucleus, with respect to the size, number and shape of all the chromosomes, as seen

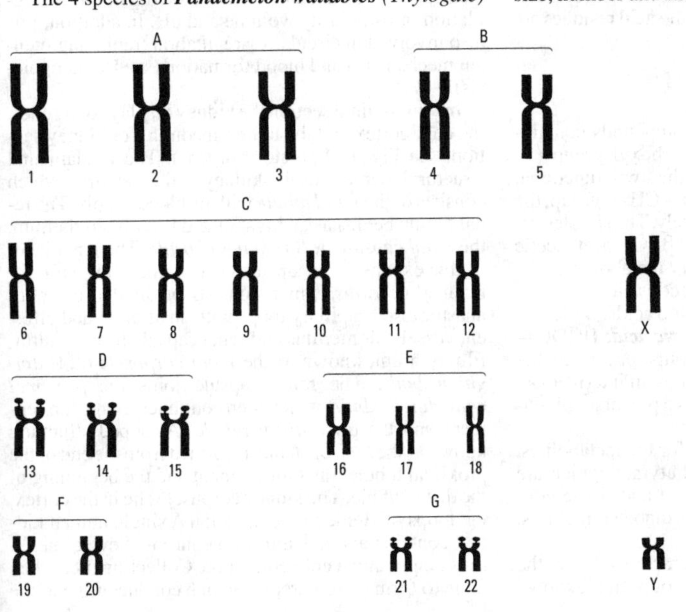

Idiogram of the human male. Only one copy of each homologous autosome is shown (normally these would be displayed as pairs). There 22 autosomes, as well as a single X- and a single Y-chromosome. Secondary constrictions can be seen on the autosomes of groups D and G.

during the metaphase of mitosis of a diploid cell. In all the somatic cells of a single species, the K. is constant, e.g. in humans there are always 46 chromosomes (44 autosomes and 2 sex chromosomes). The K. is usually displayed by cutting out the chromosomes from a photomicrograph and arranging them in homologous pairs; such a display is also called a *karyogram* or *idiogram* (Fig.).

Diploid cells contain two copies of each chromosome (homologous chromosomes). Sex chromosomes are, however, an exception. Female organisms of most plant and animal species possess two X-chromosomes, whereas males have one X- and one Y-chromosome. In birds, butterflies and certain other organisms the male is homogametic (male: XX), and the female heterogametic (female: XY).

In Polyploidy (see) the nucleus contains more than one copy of the K.

Kassinas: see *Hyperoliidae.*

Katal: see Enzymes.

Katydids: see *Saltatoptera.*

Kauri: see *Araucariaceae.*

Kava: see *Piperaceae.*

Kea: see *Psittaciformes.*

Keds: see *Diptera.*

Keel petals: see *Papilionaceae.*

Kelp: see *Phaeophyceae.*

Kelp flies: see Anthomyiids.

Kendall's substances: see Cortisol, Corticosterone, Cortisone.

Kenozooid: see *Bryozoa.*

Keratins: insoluble, cystine-rich, fibrous structural proteins, which are responsible for the mechanical durability of skin and other epithelial structures. They are found in hair, scales, wool, feathers, nails, claws, hoofs, horns, etc. K. typically contain numerous hydrogen bonds and disulfide bridges, which greatly contribute to their mechanical properties. α-K. (in the intermediate filaments of epithelial tissue) are relatively elastic. β-K. are very resistant and flexible, but less extensible, e.g. silk fibroin, in which 60% of the amino acid residues are glycine, alanine and tyrosine.

Keratto fiber: see *Agavaceae.*

Kerodon: see *Caviidae.*

Kestrels: see *Falconiformes.*

Ketoacids, *oxoacids:* organic compounds containing a keto group (>C=O) and a carboxyl group (—COOH). According the whether the two functional groups are adjacent or separated by a—CH_2—group, the acid is called an α- or β-K., respectively. The simplest α-K. is pyruvic acid, while the simplest β-K. is acetoacetic acid. Both are important intermediary metabolites.

β-Ketobutyric acid: see Acetoacetic acid.

Ketogenic amino acid: see Amino acids.

α-Ketoglutaric acid, *2-oxoglutaric acid:* HOOC—CO—CH_2—CH_2—COOH, a naturally occurring ketoacid. It is metabolically important as an intermediate in the tricarboxylic acid cycle, and as a precursor of glutamic acid by transamination.

Ketone bodies: collective term for the metabolites, acetoacetate, acetone and hydroxybutyrate, which are produced in ketogenesis. Their blood and tissue concentrations are particularly high in diabetes mellitus, when they are excreted in the urine.

Ketoses: monosaccharides characterized by the presence of a nonterminal —C=O group. In all known naturally occurring K. this carbonyl group is in position 2. Depending on the number of carbon atoms in the chain, they are classified as tetruloses, pentuloses, etc. Some have trivial names, e.g. D-fructose.

Keuper deposits (Franconian *Kipper, Keiper* or *Köper,* colored cotton material): the upper layer of the Triassic in the Germanic basin, named for the colored marl sandstone deposits in the area of Coburg. See Muschelkalk, Geological time scale.

Kewda attar: see *Pandanaceae.*

Keyhole limpet: see *Gastropoda.*

Key stimulus: see Sign stimulus.

Key West spadefoot: see *Pelobatidae.*

Khellin, *5,8-dimethoxy-2-methylfuranochromone:* a yellow-green, bitter-tasting, crystallizable compound (Fig.) from *Ammi visnaga* (family *Umbelliferae),* which is found in the Mediterranean region. K. dilates the coronary arteries, and is used clinically for the treatment of angina and asthma. The compound, *visnadin,* which occurs in the same plant, and also possesses coronary activity, is a pyranocoumarin.

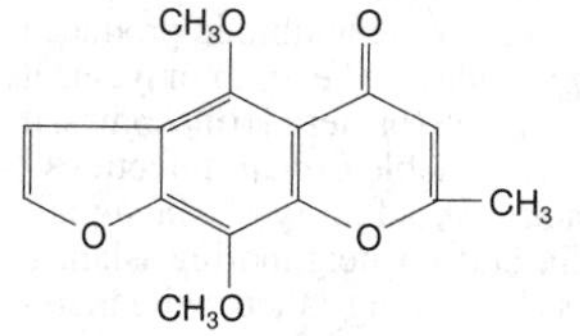

CH_3O O

CH_3

CH_3O

Khellin

Khur: see Asiatic wild ass.

Kiang: see Asiatic wild ass.

Kidney functions: the functions of the kidney, consisting primarily of excretion and homeostasis. Excretion by the kidney is not a simply a process whereby material is removed from the body. Unlike the lungs, skin and intestine, the kidney is also intimately involved in maintaining the constancy of the internal environment (milieu) of the body, i.e. it contributes to *homeostasis.* It is the main homeostatic organ for the regulation of osmolarity, volume and pH. In addition, it is also involved in circulatory regulation (renin-angiotensin mechanism) and blood formation (see Hematopoiesis).

In longitudinal section the kidney displays two zones, the dark cortex and the lighter medulla (for kidney section, see Fig. in Excretory organs). The fundamental structural element of the kidney is the *nephron,* which consists of the *renal tubule* with its blood supply. The renal tubule begins as an invaginated layer of epithelium, the *renal capsule* or *Bowman's capsule.* The inner layer of this expansion encapsulates a capillary tuft called a renal *glomerulus,* which consists of divided but nonanastomosed capillary loops with an afferent and efferent vessel. Glomerulus and renal capsule form an ultrafiltration unit known as the *renal corpuscle* or *Malpighian body.* The renal capsule joins the *proximal convoluted tubule,* which then continues as the thin segment and the thick *distal tubule.* A looped structure, known as the *loop of Henle,* is formed from the end of the proximal tubule, the thin segment and the beginning of the distal tubule. The kidney corpuscles lie in the cortex, the loops of Henle in the medulla. A single human kidney contains about 1 million nephrons. Several nephrons empty into a collecting tube. Collecting tubes then unite to form paired ureters, which conduct urine to the

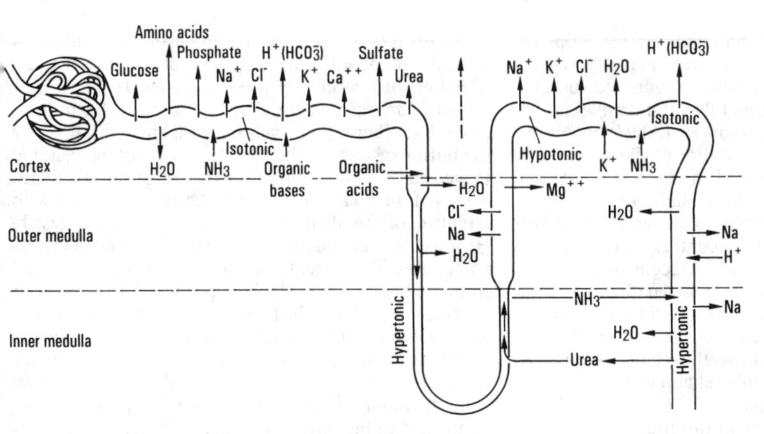

Fig.1. *Kidney functions.* Production of urine in the kidney. The figure in the entry, "Hairpin countercurrent principle", is also relevant.

urinary bladder. The exit of the bladder, which is provided with two ring muscles, continues as the urethra.

The blood vessels of the kidney are arranged in clearly defined patterns. An afferent arteriole from the *renal artery* enters the glomerulus, emerging again as the efferent arteriole, which then divides into a branching network of capillaries *(second capillary system, capillary plexus,* or *peritubular capillaries)* in close approximation with both the convoluted tubule in the cortex and the loop of Henle in the medulla. It then extends into the medulla as a system of vessels that fold back on themselves like hairpins (the *vasa recta).* Peritubular capillaries and vasa recta unite to form the *interlobular (cortical)* and *medullary veins,* which empty their contents into the *arcuate veins* lying at the boundary of cortex and medulla. Arcuate veins converge to form *interlobar veins;* these merge into the *renal vein,* which emerges from the kidney at the hilum.

Compared with the kidneys of other organisms, the mammalian kidney receives a very copious blood supply. Despite its small proportion of body mass (0.5% in humans) it receives a large proportion of the cardiac output (20% in humans). Within the kidney, 90% of the blood supply flows through the cortex, and 10% through the medulla. Cortical vessels are autoregulated, whereas medullary vessels show passive, pressure-dependent behavior. In the intermediate range of 12–27 kPa, *autoregulated* vessels maintain a constant blood flow, despite fluctuations of blood pressure; increased blood pressure stimulates stretch receptors, leading to contraction of the vessel wall. Since this property is absent from medullary vessels, the kidney medulla experiences an increased blood flow with increasing blood pressure.

Urine is produced by ultrafiltration, resorption and secretion. *Primary urine* or *glomerular filtrate,* produced in the glomerulus by ultrafiltration, passes into the lumen of the renal capsule. This primary urine contains all the constituents of blood plasma, excluding fats and high molecular mass proteins. As the primary urine passes through the tubule system, many substances are removed by resorption, while others are added by secretion, so that the composition of final urine is very different from that of primary urine or plasma. The efferent arterioles are always narrower than the afferent arterioles, so that the blood in the capillaries of the glomerulus is at a rather high pressure of about 10 kPa. The ultrafilter is the basal membrane of the capillaries, which has a pore diameter of almost 10 nm. This allows the passage of only comparatively small molecules, with a relative molecular mass less than 70 000, and it holds back the high molecular mass plasma albumins and globulins. The colloidal osmotic pressure of the plasma proteins (3 kPa) and the hydrostatic pressure of the urine column in the tubule (0.7 kPa) act against the capillary pressure of 10 kPa, giving an *effective filtration pressure* of 6.3 kPa. This effective filtration pressure determines the volume of the ultrafiltrate. If the blood pressure within the glomerulus falls by more than 50%, urine formation stops, due to the cessation of ultrafiltration.

The volume of primary urine is regulated by an internal mechanism. At high capillary pressure, ultrafiltration is increased and the Na^+ concentration in the distal tubule is increased. In some nephrons the close proximity of afferent arteriole and distal tubule forms an optically dense region of tissue, the macula densa. When the Na^+ concentration increases in the distal tubule, renin (a kidney enzyme) is formed in the afferent arteriole. Renin partially hydroyses angiotensin (produced in the liver) to produce angiotensin II, which causes the afferent arterioles to contract, thereby decreasing the blood pressure in the glomerulus and reducing the rate of ultrafiltration.

The combined activity of both human kidneys produces 120 ml of primary urine per minute. This is equivalent to 170 liters per day, i.e. more than twice the body mass. Only about 2 liters of urine are voided per day, i.e. 168 liters or 99% of the primary urine are resorbed and only 1% of the volume is excreted. Water resorption in the kidney occurs in two stages. The proximal tubule absorbs 80% of the water of the primary urine. Since this is always a constant quantity, it is known as the *obligatory water resorption.* Depending on the regulatory requirements of the body, a variable quantity of water is removed from the remaining 20% of primary urine (facultative water resorption). The *obligatory isotonic water resorption* is driven by the high colloidal osmotic pressure within the peritubular capillaries, and it is accompanied by the aldosterone-controlled intracellular Na^+ secretion mechanism. Charge neutrality is maintained, because Na^+ export is accompanied by Cl^- export; the osmotic activity of the NaCl leads to the simultaneous transport of 300 molecules of water per molecule of NaCl.

Facultative water resorption occurs in the loops of Henle by the *Hairpin countercurrent principle* (see), which in the human kidney leads to a four-fold concentration of the urine. Collecting tubes run parallel to the loops of Henle. Antidiuretic hormone (ADH) or adiuretin increases the water permeability of the membrane between the collecting tube and the loop of Henle, so that water is resorbed back from the collecting tube. Thus, by controlling ADH secretion, a large volume of dilute or a small volume of concentrated urine is produced according ot the homeostatic needs of the body. Some nephrons have a short loop of Henle, and these have a relatively limited ability to concentrate the urine. In others the loop is long and the concentration effect is high. The kidneys of desert-dwelling animals possess extremely long loops of Henle, enabling considerable water retention (conservation).

Changes occurring in the ultrafiltrate as it passes through the tubule, due to active resorption and secretion, can be studied by micropuncture techniques. Fine glass capillaries are inserted into the tubule lumen, fluid is withdrawn, and its composition is determined by microanalysis. Amphibians are convenient experimental animals, because the renal capsules lie directly beneath the kidney surface. Rodents are also suitable, because their renal tubules extend to the surface of the cortex. Glucose and amino acids are completely resorbed in the proximal tubule. Water and inorganic ions, with the exception of potassium, show a progressive decrease up to the loop of Henle. Inulin remains unchanged. Creatinine and *p*-aminohippuric acid increase by secretion in the proximal tubule. Mammalian final urine contains only small quantities of relatively few secreted substances. In teleost fishes, which lack renal corpsucles, tubular secretion is the only means of urine formation.

Active resorption of glucose in the proximal tubule is very important. Like every active transport process, it can be blocked by metabolic poisons. A specific poison of glucose transport is Phlorhizin (see), which is used to induce experimental diabetes, i.e. excretion of glucose in urine. The glucose transport system also has a saturation limit. If the diet contains too much carbohydrate, or insulin is deficient (diabetes mellitus), glucose is excreted in the urine.

By excreting acidic or alkaline urine, the kidney is crucially involved in the regulation of body pH (Fig.2). Metabolism continually produces considerable quantities of carbonic, sulfuric, phosphoric and lactic acid, as well as other organic acids. These are neutralized by the buffer systems of the blood, but cannot be excreted in neutralized form without exhausting the alkali reserves of the body. Excretion of the acidic cations, with retention of the alkali anions, occurs in three ways. 1) H^+ ions are secreted in exchange for Na^+ ions in the renal tubules. The H^+ required for this exchange is derived from carbonic acid, which is produced from carbon dioxide and water in the tubule cells, a reaction accelerated by the enzyme carbonic anhydrase. This can lead to a 1000-fold concentration of H^+ ions between primary urine and final urine. 2) Hydrogen carbonate is resorbed from the tubule. Exchange of H^+ for Na^+ produces carbonic acid in the tubule. This carbonic acid dissociates into carbon dioxide and water, and the carbon dioxide diffuses into the tubule cells, where it is reconverted into hydrogen carbonate. In this way 99% of the filtered hydrogen carbonate is retained by the organism. 3) Ammonium ions are excreted. Ammonia arises in the tubule cells from the deamination of amino acids, in particular glutamic acid. It is released passively into the urine, where it forms ammonium ions with H^+, which are excreted mainly as ammonium chloride. This withdrawal of H^+ ions as $NH_4^+Cl^-$ allows further secretion of H^+ by the tubule, thereby preventing over-acidification of the blood.

Release of urine (micturition) is a reflex process.

Kidney bean: see *Papilionaceae.*

Kidneyworm: see *Nematoda.*

Kieselguhr: see Diatoms.

Killer whales: see Whales.

Kinases: transferase enzymes which catalyse the transfer of a phosphate residue from adenosine triphosphate to an appropriate substrate, e.g. to the 6-OH group of glucose to form glucose 6-phosphate. Mechanistically, a phosphoryl group is transferred rather than a phosphate group.

Kinesis:

1) Change in the rate of random movement of a freely motile organism in response to a stimulus. Although the

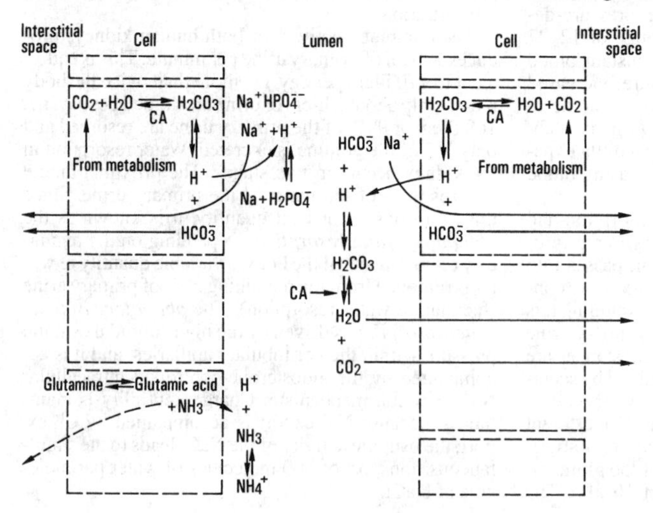

Fig.2. *Kidney functions.* pH regulation in the tubule. CA = carbonic anhydrase.

linear or angular velocity changes according to the intensity of the stimulus, the organism does not become orientated with respect to the direction of the stimulus. In *orthokinesis,* the linear velocity of random movement changes with the intensity of the stimulus, so that the organism attains an optimal position with respect to stimulus intensity. In *klinokinesis,* the frequency of change of direction varies with stimulus intensity.

2) See Mitosis.

Kinesthesia: susception of body movements. Movements of the body and its parts produce excitation patterns in proprioceptors. Integration of these patterns by the central nervous system provides information on the position and movement of limbs, e.g. in cycling. In some animals (e.g. rodents), kinesthetic patterns appear to be efficiently stored and readily retrieved, so that when a sought objective has been found once, the animal can immediately find it again by using its "kinesthetic sense". Similarly, after travelling some distance, the animal easily retraces the same route on the return journey. See Idiothetic orientation.

Kinesthetic orientation: see Idiothetic orientation.

Kinetin: see Cytokinins.

Kinetochore: see Centromere.

Kinetochore breakage: see Centromere misdivision.

Kinetonucleus: see Parabasal body.

Kinetoplast: a greatly modified mitochondrion near the basal body of the trypanosome flagellum. It contains an extensive network of highly catenated circular DNA molecules.

Kinetosome: see Basal body, Centriole, Cilia and Flagella.

King cobra, ***hamadryad,*** *Ophiophagus hannah:* a snake of the family *Elapidae* (see), and the longest venomous snake in the world. The longest recorded length is 5.58 m. Its range extends from India (it is absent from Sri Lanka) to southern China, and southward into the Sunda islands, the Malay Archipelago and the Philippines. Body color varies from olive-brown through gray to glossy black, with 40–50 narrow, pale, tranverse dorsal bands, and the eyes have a characteristic bronze hue. The extended hood is more elongated than the disk-shaped hoods of other Cobras (see). It feed almost exclusively on other snakes. Exceptionally among the *Elapidae,* the K.c. displays brood care. The eggs are laid in a nest of dry plant material which is constructed by both male and female; the female coils her body around the nest, and the male usually also remains in attendance nearby. The 2 fangs in the upper jaw are up to 10 cm in length, usually with 3 small teeth behind them. It is the most aggressive of all cobras, and its bite can kill an elephant. Unlike other cobras, it can advance with its body raised and hood extended in the threat posture. The preferred habitat is dense jungle near to water, but it also occurs in areas of human habitation. In the Nilgiri Mountains in India, it occurs up to 2 000 m. When pursued it often retreats into the water.

King crab: see Stone crabs.

Kingcup: see *Ranunculaceae.*

Kingfish, ***opah,*** *Lampris regius, Lampris guttatus, Lampris luna:* the only known member of the family *Lampridae,* order *Lampridiformes.* It attains a maximal length of 150–180 cm (usually 80–100 cm), and weighs up to 270 kg (usually 50–100 kg). This relatively rare fish occurs predominantly in warm seas throughout the world, mostly in open water at depths of 100–400 m, and it is usually caught accidentally. It bears a number of distinctive features, and it cannot possibly be confused with any other fish. Thus, in side view it is oval in shape, the lateral line arches high over the insertion of the pectoral fin, the back is steely blue-black or violet, fading into a shimmering bluish green on the sides, while the belly is rusty or purplish silver; all the fins are red, and numerous milky-white spots are scattered over the body surface. It feeds primarily on squids, octopuses and crustaceans. The tasty pink flesh of the K. is highly prized.

Kingfishers: see *Coraciiformes.*

Kinglets: see *Sylviinae.*

King lobster: see Lobsters.

King-of-the-herring: see Bandfishes.

King snakes, *Lampropeltis:* a genus of North and Central American, egg-laying, ground-dwelling snakes of the family *Colubridae* (see). They have smooth scales, the head is not strongly demarcated from the body, and they are often strikingly colored and patterned. Several subspecies of the ***Common king snake*** *(L. getulus),* which have a prominent diamond or chain pattern, are found in Mexico and the southern USA. Other species are ringed with red, yellow and black, and are also known as ***False coral snakes.*** This is assumed to represent mimicry of the true coral snakes, thereby affording protection. Small rodents are taken, but many species [e.g. the vividly colored ***Mountain king snake*** *(L. zonata)* of the western USA] also eat other snakes, which are killed by constriction.

Kinixys: see *Testudinidae.*

Kinkajou, ***American potto,*** *Potos flavus:* a monotypic genus of the *Procyonidae* (see). It inhabits tropical forests in southern Mexico and Central America, ranging at least as far as the Matto Grosso in Brazil. It has a rounded head and short face, and is the only procyonid with a prehensile tail. Kinkajous are easily tamed and those kept as pets are often called "honey bears".

Kinkimavo: see *Oriolidae.*

Kinorhyncha (Greek *kinema* motion, *rhynchos* snout), *Echinodera:* a class of *Aschelminthes* consisting of about 100 benthic and mostly marine species.

Morphology. *K.* are 0.2–1 mm in length, bilaterally symmetrical, vermiform but short, and somewhat flattened on the ventral surface. The body wall has both circular and longitudinal muscles. Externally, the body is divided into 13 or 14 segments or zonites, the epidermis of each zonite comprising a single dorsal, and a pair of ventral, cuticular plates. The first zonite carries 10–20 large hooks arranged around the mouth opening, and it can be retracted so that the cuticular plates of the neighboring zonites form a closure. Dorsal and lateral surfaces are each provided with a row of spines, and zonite 4 also carries a pair of attachment structures, each containing the exit pore of a cement gland. The posterior end carries 1 or 2 long spines. The body cavity is a pseudocoel derived from a persisitent blastocoel. There is a

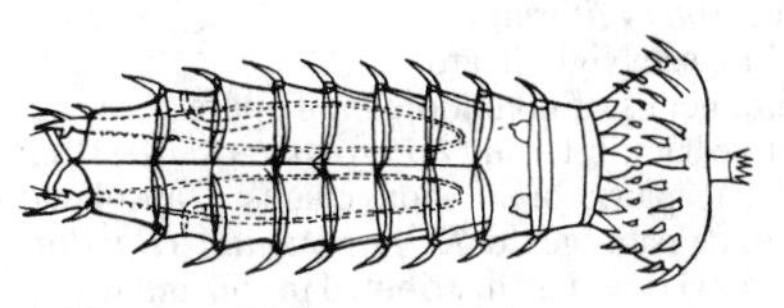

A kinorhynchid

posterior anus, a muscular pharynx, and a protrusible mouth cone. Protonephridia are present in the 11th zonite. The nervous system consists of an anterior circumenteric ring and a ventral nerve cord with clusters of ganglion cells. External cilia are totally absent. Sexes are separate, with paired testes or ovaries, which exit on the last zonite.

Biology. Almost all species are members of the marine meiofauna, living on the sea bed at depths up to 400 m. They feed on detritus or unicellular algae, especially diatoms. Movement is by creeping, in which the body alternately contracts and extends. Simultaneously, the anterior end is alternately everted and retracted, so that its hooks provide a temporary purchase on the substratum. Development proceeds via a barrel-shaped larval stage with five molts.

Classification. The *K.* are regarded as a phylum by some authorities. Two orders are recognized: *Cyclorhagida* (only the first zonite can be retracted) and *Homalorhagida* (both the first and second zonites retractable).

Kinosternidae, ***American mud*** and ***musk turtles:*** a family of the order *Chelonia* (see), containing 20 species in 2 genera, distributed from eastern Canada to Argentina. They display the primitive character of a row of inframarginal plates separating the carapace from the plastron. There are 22 marginal scutes on the carapace, and 1 cervical scute. Lengths vary from 11 cm ***(Flattened musk turtle,*** *Sternotherus depressus)* to 20 cm ***(Scorpion mud turtle,*** *Kinosternon scorpioides).* They inhabit lakes and oxbow lakes, where they feed on small aquatic animals, dead fish and some vegetation. ***Musk turtles*** (genus *Sternotherus),* which are found in southeastern USA, have a single plastral hinge at the rear. The plastron is rather degenerate, with soft uncornified areas of scute around its center. Scent glands near the bridge eject a powerful malodorous secretion when the animal is excited. ***Mud turtles*** (genus *Kinosternon)* are distributed in the inland waters of North, Central and South America. The plastral scutes are completely cornified, and both the anterior and posterior parts of the plastron are hinged with connective tissue, so that they can be folded up to enclose all the soft parts of the animal. See *Staurotypidae.*

Kinosternon: see *Kinosternidae.*

Kin selection: selection of genes within a group of relatives, so that survival and reproduction of certain individuals is increased or decreased. A group of related individuals thus becomes a unit of selection, in which the selective power depends on total fitness. Characters can therefore evolve which have a negative effect on individual fitness, but actually increase total fitness, e.g. the existence of "helpers" in brood care, which support relatives in the rearing of young, but themselves have no offspring. The phenomenon is particularly marked in eubiosocial insects.

Kinta weed: see *Orchidaceae.*

Kites: see *Falconiformes.*

Kittiwake: see *Laridae.*

Kiwis: see *Apterygiformes.*

Klauberina: see Night lizard.

Klebsiella: genus of nonmotile, non-spore-forming, Gram-negative bacilli (family *Enterobacteriaceae)* that occur singly, in pairs, or in short chains. Capsulated forms tend to be pathogenic. *Kl. pneumoniae (Bacillus lactis aerogenes)* is widely distributed in soil and water, as well as in the intestinal tract of humans and animals; it is responsible for a small proportion (less than 1%) of bacterial inflammatory lung infections.

Klinotaxis: movement (taxis) of a motile organism, which orients itself by alternately comparing stimulus intensity on each side of the body. This is achieved by monitoring the surroundings by sweeping sensory organs from side to side. For example, fly maggots move the light-sensitive end of their body from side to side to determine the direction of incident light, from which they then retreat, i.e. they exhibit negative photoklinotaxis. See Telotaxis, Tropotaxis

Klipspringer: see *Neotraginae.*

Knife fishes: see *Gymnotiformes.*

Knallgasbakterien: see Bacterial oxidation of molecular hydrogen.

Knot: see *Scolopacidae.*

Knotweed: see *Polygonaceae.*

Knysna touraco: see *Cuculiformes.*

Koala: see *Phalangeridae.*

Kob: see *Reduncinae.*

Kobus: see *Reduncinae.*

Kochia: see *Chenopodiaceae.*

Koch's liquid plating method: see Plating.

Kodiak bear: see Brown bear.

Kogia: see Whales.

Kojic acid: a metabolite produced by different species of *Aspergillus* and *Penicillium,* and by acetic acid bacteria. It has weak antibiotic activity.

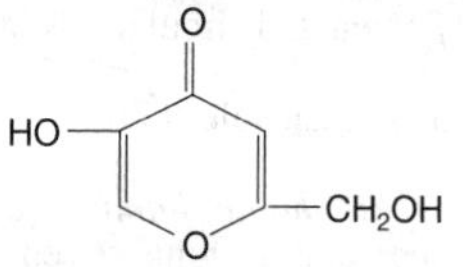

O
HO
CH_2OH
O
Kojic acid

Kokako: see *Callaeidae.*

Kokoi poison-arrow frog: see *Dendrobatidae.*

Komodo dragon, ***ora,*** *Varanus komodoensis* (plate 5): the largest living lizard, and a member of the family *Varanidae* (monitor lizards) (see). It attains a length of 3 m (half of which is accounted for by the tail) and a mass of 135 kg. It is a strictly protected species, found only on the island of Komodo and a few neighboring small islands of the Sunda Archipelago. It preys on animals up to the size of young deer and young wild pigs. The 12 cm eggs are buried in the ground. This aggressive reptile will attack humans. It is not venomous, but the bite is infected with pathogenic bacteria that resist modern antibiotics and result in the death of the victim.

Kookaburra: see *Coraciiformes.*

Korber method: see Estrogens.

Korean gray rat: see *Cricetinae.*

Kori: see *Gruiformes.*

Kouprey: see *Bovinae.*

Kraits, *Bungarus:* a genus of the *Elapidae* (see), 1–2 m in length, distributed from India to southern China. Twelve species are recognized, all with distinct, colorful patterns of transverse black and white, or black and yellow rings. Nocturnal in habit, they feed mainly on other snakes. They lay 5–12 eggs. The deadly ***Banded krait*** *(B. fasciatus)* of Southeast Asia and the Malay Archipelago is banded black and yellow, and sometimes exceeds 2 m, but is more usually about 1.5 m.

Krause's end bulbs: see Tactile stimulus reception.

Krebs cycle: see Tricarboxylic acid cycle

Krebs-Henseleit cycle: see Urea cycle.

Krill: planktonic crustaceans which serve as food of fishes and baleen whales. In the northern oceans, K. consists primarily of copepods (genus *Calanus),* whereas most Antarctic K. belong to the order *Euphausiacea* (genus *Euphausia).*

K-selected species, *K-strategist:* a species whose evolutionary strategy is directed to the greatest possible exploitation of the environment. This strategy is characteristic of organisms that inhabit a relatively constant or predictable environment, and are poorly equipped to withstand changes in that environment. They are strong competitors with a low specific reproduction rate, and they maintain a stable population size, which is close to the carrying capacity of the environment. In contrast, ***r-selected species*** (r-strategists, opportunistic species, reproduction strategists) have a high reproduction rate, are poor competitors, and display a tendency for mass population increases and wide fluctuations in population size; if a stable population size is actually achieved, this lies far below the carrying capacity of the environment. r-Selected species are usually small, simple life forms, whose environment is variable and unpredictable.

Both strategies are extremes of the r-K-continuum. Viruses, bacteria and protozoa are r-strategists, in contrast to vascular plants and mammals, which are largely K-strategists Certain small mammals (e.g. mice, voles), however, show a tendency to be r-strategists, whereas large mammals like the elephant, brown bear and blue whale are typical K-strategists. Predators lie further on the K-side of the spectrum than their prey. r-Strategists are able to rapidly invade new environments, and establish large populations in a relatively short time, after which they are slowly superseded by K-strategists.

A population of a single species can also display varying degrees of r- or K-selection.

K-Selection: the natural selection of species for their ability to fully exploit the environment, i.e. the carrying capacity of the environment for such species is higher than that for less strongly K-selected species. If a population of animals or plants lives permanently under optimal conditions, then their individual species density is high and close to the carrier capacity of the environment. Under such conditions a high reproduction rate is disadvantageous, because many offspring would perish. The selective advantage is held by those genotypes that utilize the available resources most sparingly and most economically, i.e. those that survive by minimal exploitation of the available food supply. K-S. leads to the establishment of the largest possible population density.

K-strategist: see K-selected species.

Kudu: see *Tragelaphinae.*

Kuehneotheriidae: see Triassic.

Kulan: see Asiatic wild ass.

Kulezinski coefficient: an index of similarity between two biological communities, which is relatively little influenced by dissimilar sample sizes. $C_k = c/(a + b)$, where c = number of species common to both communities, and a and b are the numbers of different species occurring in communities A and B, respectively.

Kumquat: see *Rutaceae.*

Kupferschiefer: a stratum of bituminous shales containing copper pyrites, and famous for the abundance of fossil fishes, especially *Palaeoniscus* and *Platysomus.* It is part of the Permian Zechstein deposits, which are particularly well developed in Thüringia, and it rather resembles the English marl slate strata.

Kuru: see Prion.

Kwashiorkor: a chronic form of malnutrition, occurring chiefly in the second year of life. The name is from the Ga language, and was introduced into modern medicine in 1933 by Cicely Williams *[Archs. Dis. Childh.* ***8*** (1933) 423]. The word means "deposed child", i.e. deposed from the breast by the advent of another pregnancy. The earlier literature emphasizes the relationship between K. and various tribal customs and taboos, which result in the administration of a low protein, high carbohydrate diet to children. It has now been shown, however, that there are no essential differences in the dietary protein/energy ratio of marasmus and K. victims [C. Gopolan, in R.A. McChance and R.M. Widdowson, eds. *Calorie deficiencies and Protein Deficiencies,* Churchill Livingstone (1968) pp. 49–58]. Like all other forms of protein-energy malnutrition, K. is often precipitated and exacerbated by microbial infections and intestinal worms. It is more common in rural than in urbanized areas of the developing world. K. victims show retarded growth and anemia. Plasma albumin concentrations fall below 20 or even 10 g/l (in marasmus, plasma albumin is usually about 25 g/l). The resulting decrease in the osmolarity of the blood is thought to be partly responsible for the accumulation of fluid in the body, and the watery, bloated state of the tissues; edema is a common feature of K. The hair becomes sparse, and especially in negro children, it may show patches or streaks ("flag sign") of red, blonde or gray. These hair lesions are probably due to a specific tyrosine deficiency, since the periodic administration of tyrosine to phenylketonurics causes very similar alternating bands of deeply pigmented hair. The skin shows a very characteristic dermatosis, with areas of pigmentation, depigmentation and peeling, the most severely affected areas being the lower limbs and buttocks. Muscles are always wasted, so that walking or crawling may not be possible. Fatty liver is also very characteristic of K.; an average lipid content of 390 g/kg liver has been reported (cf. 35 g/kg in marasmus); this increase is due to triacylglycerols, the level of phospholipids being relatively unaffected. In K., but not in marasmus, plasma levels of triacylglycerols and cholesterol are low. Subcutaneous fat is retained in K., in contrast to marasmus, in which it is severely depleted. Changes in plasma amino acids are similar in most types of protein-energy malnutrition. In more than 50% of cases, a mixture of all the essential amino acids effects a complete cure of K. In the remaining cases the response is partial or negative.

K. is common in hot, humid areas, but not in dry, hot areas. It is especially prevalent in the wet season. Furthermore, it is never found in temperate regions. These observations suggest the contribution of other factors, in addition to protein deficiency. There is now strong evidence that K. is caused by a combination of malnutrition and aflatoxin poisoning [R.G. Hendrickse, *British Medical Journal* ***285*** (1982) 843–846; S.M. Lamplugh & R.G. Hendrickse, *Anmals of Tropical Paediatrics* ***2*** (1982) 101–104]. In hot, humid countries of the developing world, many market foods contain aflatoxins, especially the groundnut oil used for cooking. Well nourished infants can degrade and excrete these relatively small quantities of toxins, but this ability is weakened by protein deficiency. In the resulting vicious cycle, accumulation of aflatoxins causes further liver damage and a marked decrease in the protein synthesizing ability of the liver.

Kynurenic acid: a carboxylic acid containing the quinoline ring system, which is formed metabolically from the amino acid, L-tryptophan, via Kynurenine (see): transamination of the side chain amino group of kynurenine produces an oxogroup, which then condenses with the aromatic amino group.

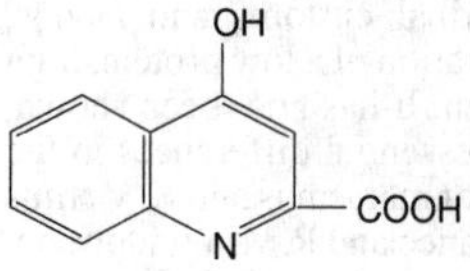

Kynurenic acid

Kynurenine: a metabolic product of the amino acid, L-tryptophan. K. is further metabolized to 3-hydroxyanthranilic acid, which in turn is converted into nicotinic acid and alanine. L-Tryptophan therefore has a sparing effect on the dietary nicotinic acid requirement of animals, but its conversion to nicotinic acid via K. is not rapid enough to satisfy the total requirement. By transamination and cyclization, K. also gives rise to Kynurenic acid (see). 3-Hydroxyanthranilic acid is also the precursor of the arthropod Ommochromes (see).

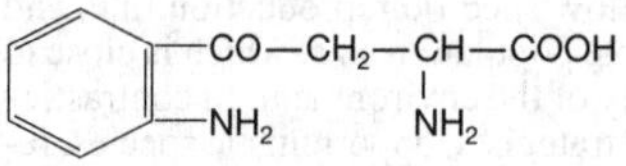

Kynurenine

L

Label: a spectroscopically detectable substance which has a minimal influence on its environment, and whose behavior provides information on the molecular properties of its environment. Important examples are fluorescence and spin Ls. For example, spin Ls. (e.g. compounds containing a stable nitroxide radical) have been applied to synthetic membranes to study lipid phase transitions, and to determine the relaxation times for the rotation of lecithin molecules about their long axes and relaxation times for the conformational oscillations in the carbon chains of the fatty acids. In biological membranes, spin labeling can be used to study the effect of hormones or drugs (e.g. anesthetics) on the fluidity and average order parameter, etc. Fluorescent Ls. have also been used for the study of membrane behavior, and in particular for the investigation of antibody-antigen reactions with the aid of fluorescent antibodies. Thus, an antibody covalently linked to a fluorescent dye (fluorescein or rhodamine) retains its immunochemical activity, and can be used to locate its specific antigen in a microscopic preparation of cells or tissues.

For radioactive L., see Radioactive isotope.

Labellum: see *Orchidaceae.*

Labial nephridia: see Maxillary glands.

Labiatae: see *Lamiacaeae.*

Laboulbeniales: see *Ascomycetes.*

Labrisomidae: see *Blennioidei.*

Laburnum: see *Papilionaceae.*

Labyrinth: see Respiratory organs.

Labyrinthine catfishes: see *Clariidae.*

Labyrinthine reflexes, ***equilibrium reflexes:*** reflex maintainance of body equilibrium and posture by varying the degree of tone of different muscles, initiated mainly by the vestibular system or labyrinth (utriculus, sacculus, semicircular canals and their ampullae; see Auditory organ). The maculae (sensory structures in the utriculus and sacculus) and the cristae ampullaris (sensory structures within the ampullae) transmit signals to the central nervous system (vestibular nuclei, bulboreticular formation, cerebellum and flocculonodular lobes of the cerebellum) via the vestibulocochlear nerve (cranial nerve VIII). The cerebellum is particularly well developed in vertebrates that are highly mobile around all body axes, e.g. various elasmobranch and teleost fishes, birds and mammals. The effectors of the reflex are muscles of the limbs, trunk and eyes, which are responsible for mobility and for changes of position and posture. Stimulation of the vestibular system gives rise to alterations in the positions of the eyes (see Nystagmus). The vestibular system of humans can be stimulated by rotating a person at about 1 revolution per second. This causes the gaze to become fixed on one point, so that as the subject is rotated, e.g. to the right, the eyes move slowly to the left (slow phase of nystagmus). When this fixation point passes out of sight, another point is chosen by rapid movement of the eyes to the right (rapid phase of nystagmus). If rotation is stopped suddenly, sensory impulses from the vestibular system are altered as if in response to rotation to the left, and the nystagmus occurs in the opposite direction. Nystagmus is also exhibited by closed eyes.

Analogous movements of the eye stalks of crustaceans in response to rotation have been observed. Crustacean statocysts, which monitor body movement and alignment, are situated in the basal segment of each first antenna, and they are entirely different from the vertebrate vestibular system.

Vertebrate L.r. are not explainable on the basis of a simple reflex arc, but as part of a complicated system for maintenance of the normal or chosen alignment of the body, which involves other reflexes. For example, the vestibular system monitors only the position of the head, and not the body. Proprioceptors in the neck function in concert with the vestibular system, e.g. if the head is tilted backward this change of position and angle is monitored by the vestibular system, and at the same time the neck proprioceptors signal that the head has changed its angle in relation to the rest of the body. The brain processes both of these signal inputs and concludes that the position of the body has not changed.

Excessive stimulation of the vestibular system by rotational movement about a horizontal axis, may produce "sea sickness" (vertigo, nausea, salivation and vomitting). The feeling of giddiness or loss of equilibrium after being rotated is probably due to confusion between visual signals and signals from the vestibular system. Removal of the vestibular system at first produces an unsteady posture with loss of muscle tone and ataxia, but (at least in higher vertebrates) the animal learns to compensate with the aid of its visual input and proprioceptor signals from the limbs and body surface. Similarly, in humans, damage to the vestibular nuclei and vestibular system by the antibiotic, streptomycin, also produces unsteadiness and lack of muscular tone.

Labyrinthfishes: see Respiratory organs.

Labyrinthodontia (Greek *labyrinthos* maze, *odontes* teeth): extinct subclass of salamander-like amphibians. The tooth dentine was folded into labyrinthine grooves and ridges (hence the name). The skull had no temporal fossae, a character linking the *L.* with their immediate ancestors, the Devonian freshwater *Crossopterygii* (see Coelacanths). All the larger amphibians of the Paleozoic and Triassic were *L.,* including the most ancient, as well as the later transitional forms that link the amphibians to the reptiles. Fossils occur from the Upper Devonian to the Upper Triassic with an evolutionary peak in the Permian and Triassic.

Labyrinth test: see Maze test.

Laccases: oxidase enzymes, which oxidize *p*-hydroquinones to *p*-quinones. They are widely distributed in plants and animals, and are particularly important in the degradation of lignin.

Lacciferidae: see *Homoptera.*

Laccifer lacca (Indian lac insect): see *Homoptera.*

Lacebugs: see *Heteroptera.*

Lacerta: a genus of the family *Lacertidae* (see), comprising 32 species and numerous subspecies, distributed in Europe, western Asia and North Africa. A great variety of forms occurs in the Mediterranean region, and their systematic relationships are not always clear. Similar difficulties are encountered in the taxonomy of various Caucasian forms and subspecies. There is a well defined scaly collar. Male animals are often more strikingly colored than females, but the ability to change color is only weakly developed and hardly noticeable. With the exception of the Viviparous lizard (see) (the most northerly species), all species lay eggs. The following selected species are listed separately: Wall lizard, Sand lizard, Common lizard, Green lizard, Ocellated green lizard.

Lacertidae, ***true lizards, wall and sand lizards:*** a family of the *Lacertilia* (see) containing almost 200 species distributed in Europe, Africa, Asia and the Indoaustralasian Archipelago. Members of this family are the Old World counterparts of the *Teiidae* (see). They are generally small and nimble, with well-developed, typically 5-toed limbs, a long (usually longer than the body) autotomous tail, a conical head distinct from the body, a pointed snout, and usually a skin fold across the throat. There are separate movable eyelids, and some species have a transparent window in the lower lid. Head scales are large and symmetrically arranged, and fused to the skull by their thin bony underlays. The scales on the back are sometimes small and irregular, whereas the belly scales are usually large, thin and rectangular, and arranged in well defined rows. Tail scales are arranged in whorls, and in some species they are elongated into spines. The tongue is flat, long, narrow and deeply forked, and covered on its upper surface with small projections or folds. In contrast to other groups (e.g. geckos, agamids and iguanas), the lacertids possess no specialized structures, such as adhesive foot pads, dorsal crests or gular pouches. Limb regression, leading to snake-like and burrowing forms, is not displayed by any lacertid species. Teeth are pleurodont (attached to the sides of the jawbone), and they are also often present on the palate. With few exceptions they are ground-dwelling species of sunny, dry habitats. They feed mainly on insects, also taking worms and snails. With the exception of the Viviparous lizard (see) and some species of *Eremias* (see), they lay eggs. It is the only family of the *Lacertilia* with strong representation in Europe. The following selected genera are listed separately: *Lacerta, Psammodromas, Acanthodactylus, Takydromas, Eremias.* Some Caucasian species of *Lacerta* reproduce parthenogenetically.

Lacertilia, *Sauria,* ***lizards and allies:*** the most primitive of the 3 suborders of the *Squamata* (see). They usually have 4 well developed limbs, but in some families *(Anguidae, Scincidae, Pygopodidae)* there are species whose limbs have degenerated. All limbless species, however, still possess the vestiges of limb girdles. Conspicuous ear openings are generally present, with a visible ear drum. Most species have separate, movable eyelids, but in some the upper and lower lids are fused and transparent. In contrast to the snakes, the rami of the lower jaw are united by a suture. The lower border of the temporal vacuities is absent, but the upper is usually retained. Absent only from the polar regions, lacertilians are represented throughout the world, their greatest species diversity occurring in the tropics.

Numerous structural modifications are found, representing adaptations to a terrestrial, arboreal, or aquatic existence, e.g. adhesive toe pads, zygodactylous feet, prehensile tails and laterally compressed tails for swimming. Protective adaptations include tails with spines and tails that can be used for striking, as well as specialized threat behavior, the ability to change color, and in some families the ability to jetison the tail by beakage at a predetermined site (autotomy) (the tail is later regenerated). Specialized body appendages include crests and gular pouches. They are mostly predators, small species feeding mainly on insects. The largest members of some families, e.g. *Iguanidae,* feed on vegetation and fruit. The largest living lacertilian is the Komodo dragon (see).

The 18 families (containing about 3 000 species) form 5 phylogenetically related groups: ***geckos and allies*** *(Gekkonidae, Pygopodidae, Dibamidae),* ***iguanas and allies*** *(Iguanidae, Agamidae, Chamaeleonidae),* ***skinks and allies*** *(Scincidae, Feylinidae, Anelytropsidae, Cordylidae, Xantusiidae, Teiidae, Lacertidae),* ***slow worms and allies*** *(Anguidae, Anniellidae),* and ***monitors and allies*** *(Helodermatidae, Varanidae, Lanthanotidae).* Each family is listed separately.

Lacewings: see *Neuroptera.*

Lachesis mutus: see Bushmaster.

Lackey, *Malacosoma neustria* L.: an ocher-yellow to red-brown moth, occurring from Europe, through Asia to Japan. It belongs to the family, *Lasiocampidae* (eggars, lappet moths), whose caterpillars damage orchard and deciduous woodland trees by eating the buds and leaves. Eggs are laid round a small twig in a characteristic broad band, and they overwinter. The multicolored (white, black, blue, red and yellow) caterpillars hatch in the spring and spin a communal web or "tent" from which they devour the young shoots.

Lactarius: see *Basidiomycetes (Agaricales).*

Lacteal vessels: see Lymph.

Lactic acid: CH_3—CHOH—COOH, a naturally occurring, optically active, aliphatic hydroxy-carboxylic acid. The optically inactive DL-form is a clear, syrupy liquid, which crystallizes (m.p. 18 °C) when very pure. The dextrorotatory (L-form) lactate anion is the end product of anaerobic glycolysis in muscle and it is a substrate of gluconeogenesis. In fermenations by L.a. bacteria, carbohydrate (e.g. lactose, glucose, starch) is converted into the racemic DL-form of L.a. It occurs in many acidic fermentation products, e.g. silage, sour milk, sauerkraut, pickles. Gastric juice also contains L.a. It is used in numerous commercial processes, such as drug preparation, acidification of wine for vinegar production, preparation of foods, plastics, laquers, etc.

Lactic acid bacteria: bacteria that produce lactic acid by the fermentation of lactose and other carbohydrates (see Fermentation). L.a.b. have a wide natural distribution, and they occur in particular on plants and dead plant material. Some species commonly occur on the intestinal mucosae of animals and humans, while other species favor food products. They belong to different systematic groups, and they occur as rods or cocci. All L.a.b. are Gram-positive, and no sporulating forms are known. They are divided into two physiological groups: *homofermentative L.a.b.* [e.g. Streptococci (see) and Lactobacilli (see)] produce predominantly lactic acid, whereas *heterofermentative L.a.b.* [e.g. *Leuconostoc* and Bifidus bacterium (see)] produce relatively large quantities of other organic compounds in addition to lactic acid.

Different L.a.b. are commercially important in the production of lactic acid and the preparation of milk products (see Streptococci, Lactobacilli, Lactic acid).

Lactobacilli: rod-shaped, mostly chain-forming, nonmotile, non-spore-forming, anaerobic, Gram-positive lactic acid bacteria. They occur in milk and milk products, in fruit juices, decaying plant material, and in the mouth and gut of homeothermic animals. They are employed in the production of lactic acid, and in the preparation of foods such as sauerkraut, pickles, cheese and fermented milk products. In certain countries L. have been used for centuries in combination with certain yeasts and streptococci to produce fermented milk beverages (e.g. busa of Turkestan, kefir of the Cossacks, koumiss of Central Asia, leben of Egypt). Yoghurt is prepared with the aid of *Lactobacillus bulgaricus.* L. are also present in sour dough. In agriculture, L. perform the fermentation that produces Silage (see).

Lactoferrin: see Siderophilins.

Lactose, ***milk sugar:*** a reducing disaccharide, consisting of galactose β-1,4 glycosidically linked to glucose. It is converted to lactic acid by lactic acid bacteria, and it is fermented by special yeasts (e.g. kefir), but not by ordinary yeasts. It is qualitatively and quantitatively the most important carbohydrate in all mammalian milk. Human milk contains 6–8%, and cow's milk contains 4–5%. It is also found in fruits and pollen. Industrially, it is used as a carbon source in the culture of microorganisms, e.g. in penicillin production.

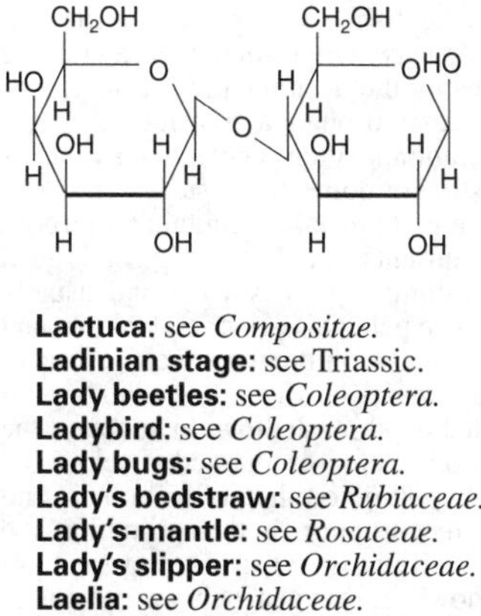

β-*Lactose*

Lactuca: see *Compositae.*

Ladinian stage: see Triassic.

Lady beetles: see *Coleoptera.*

Ladybird: see *Coleoptera.*

Lady bugs: see *Coleoptera.*

Lady's bedstraw: see *Rubiaceae.*

Lady's-mantle: see *Rosaceae.*

Lady's slipper: see *Orchidaceae.*

Laelia: see *Orchidaceae.*

Laemodipodea: see *Amphipoda.*

Lagena:

1) See *Foraminiferida.*

2) See Auditory organ.

Lagenaria: see *Cucurbitaceae.*

Lagenodelphis: see Whales.

Lagenorhynchus: see Whales.

Lagidium: see *Chinchillidae.*

Lagomorpha, *Duplicidentata:* an order of mammals. Formerly included with the rodents, it is now thought that similarities between these two orders are very superficial; for example, the serology of lagomorphs and rodents shows no relationship between the two. Ancestral lagomorphs probably diverged from a primitive eutherian stock at about the same time as the rodents. Lagomorphs appear to have some slight affinities with certain groups of hoofed animals. Unlike rodents, the lagomorphs are *duplicidentate,* a condition in which two sets of incisors are present; those at the center of the front of each jaw are large and have persistent pulps, while the upper jaw also contains two small peglike incisors, one to either side of the large incisors. There is a wide diastema between the incisors and the cheek teeth. The order is represented by two living families: *Leporidae* (see) (hares, rabbits and cottontails) and *Ochotonidae* (see) (picas), and by the extinct *Eurymylidae.* Members of both families produce two types of fecal material: moist pellets that are eventually eaten, and dry pellets that are not eaten. By recycling fecal material they benefit nutritionally from the vitamins synthesized by their intestinal bacteria (see Cecotrophy).

Lagopus: see *Tetraonidae.*

Lagostomus: see *Chinchillidae.*

Lagothrix: see Woolly monkeys.

Laguncularia: see Mangrove swamp.

Lake: a landlocked mass of water, lying in a depression, and not in direct contact with the sea. Once a pan of water has established itself, various forces begin to operate, which alter the ground relief. Ground at the edges become eroded by the wash and surge of the water, and the eroded material is deposited as a submerged, flat, slender strip that broadens into a shelf. This shelf, which forms the shallow-water periphery of the lake, is soon colonized by higher plants. Biologically, this shelf is known as the Littoral (see). As the depth of water increases, there comes a point at which light can no longer penetrate to the bottom, so that photosynthesizing plants are absent. This marks the beginning of the deep zone or Prufundal (see) of the lake. Between the littoral and prufundal lies the *sublittoral* or *littoriprofundal* zone, which extends from the lower limit of plant growth to the upper limit of sedimentation. The body of free water above the littoral and profundal is called the Pelagial (see). The totality of inhabitants of the lake floor is known as the Benthos (see), and the lake floor as a biological habitat is called the Benthal (see). Inhabitants of the free water are called Plankton (see) or Necton (see).

Lake water originates from aerial precipitation, which falls on the lake surface, or enters the lake from subterranean or surface ground water. The mineral economy of the lake is largely determined by the inflowing rivers and streams. Concentrations of plant nutrients (phosphate, nitrate, etc.) and the dynamics of the entire body of water are the most important factors in the maintenance of living organisms in the lake. Trophogenic processes are possble only in the upper, illuminated water layer. Phytoplankton and littoral plants produce organic materials by the assimilation of inorganic substances. Dead plants sink into deeper water, where the organic materials are released by tropholytic processes. Assimilation does not occur in the deep, nonilluminated (aphotic or dysphotic) regions. For the maintenance of the cycles of nutrients, part of the nutrient-rich deep water must therefore be transported upward to the trophogenic surface layer. This occurs by vertical streaming, which is largely a result of the fact that water attains its greatest density at 4 °C. In winter, the lake water consists of stable layers, and this gives rise to a state of *winter stagnation;* the deep water has a temperature of about 4 °C, while the surface temperature is about 0 °C, i.e. the water with the higher density lies below that with the lower density; this is also known as *inverse layering*. As the surface water becomes warmer in the spring, it becomes denser and sinks. During this *spring circulation,* first the upper and then the lower layers become mixed, until the total column of water is at 4 °C and the entire body of water is circulating. Nutrients then re-enter the upper layer and are utilized by the phytoplankton. The same process delivers oxygen-rich water to the deep regions. If the surface water becomes warmer than 4 °C, its density decreases. Stable layering then occurs once more, with the lighter surface water layered above the denser deep water, giving rise to *summer stagnation.* Day and night temperature differences, heat loss by irradiation and wind disturbance cause vertical streaming in the upper layers, resulting in an equalization of temperature. The upper part of the pelagial with a uniform temperature is called the *epilimnion* or *epilimnetic zone.* The level at which this vertical mixing ceases marks the lower boundary of the epilimnion and the beginning of the *metalimnion* or *metalimnetic zone.* Here, the temperature suddenly falls by 1 to 3 °C or more per meter. Below the metalimnion, the temperature then remains more or less uniform to the lake bottom. This usually tropholytic region is called the *hypolimnion* or *hypolimnetic zone.*

In the autumn, the surface waters cool, thereby initiating the *partial autumn circulation,* which finally develops into the *total autumn circulation,* which in turn is terminated by winter stagnation.

In very deep lakes, the temperature of the deep water remains constant throughout the year and does not participate in vertical mixing. This is also true, if the bottom water is saline and of high density. The deep water of such *meromictic* lakes is usually oxygen-deficient and has a high content of nutrients. The metalimnion is characterized by marked vertical changes in the concentrations of dissolved materials, e.g. salts and hydrogen sulfide, and is therefore also known as the *chemocline.* Hydrogen sulfide is almost always present, often in high concentrations (300 mg/l). The lake bottom is covered with black, putrid mud. The entire deep stagnant layer, the *monimolimnion* or *monimolimnetic zone* is uninhabited. Lakes in cold climatic zones, which are covered with ice throughout the year, and whose waters therefore cannot undergo vertical mixing, are called *amictic* lakes. Waters of *monomictic lakes* (e.g. in high mountains) circulate only in the summer. Waters of tropical lakes, where the environmental temperature, and therefore the water temperature are relatively constant all the year round, circulate only when the surface is physically disturbed, usually by strong wind; these are known as *oligomictic* lakes. *Polymictic* lakes occur in the mountainous regions of the tropics; their waters undergo frequent mixing, due to wind action and especially due to the marked day and night temperature differences.

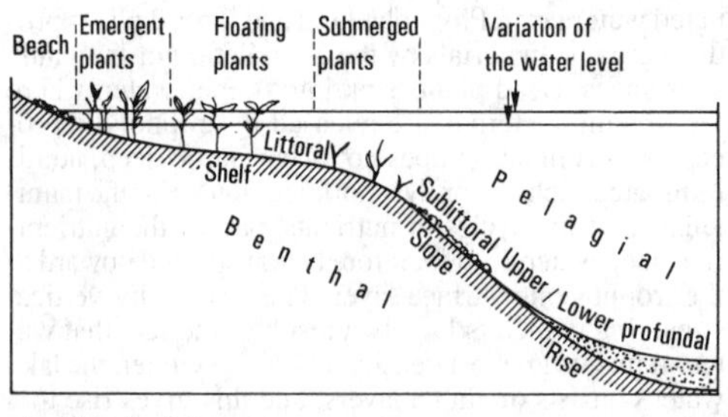

Profile of a lake

Lakes, aging of: see Aging of lakes.

Lakes, types of: units of fresh water classified according to the quantity and type of plant nutrients and humus materials in the water and on the bottom, and according to their biological productivity. Lakes with a low content of humus material are called *clear waters,* as distinct from *brown waters,* which are colored yellow to brown by humus substances.

Clear waters include oligotrophic, mesotrophic and eutrophic lakes, which differ with respect to their contents of plant nutrients.

Oligotrophic lakes (e.g. subalpine lakes) are deep and deficient in nutrients, and they have low biological productivity. The shelf is narrow and there is little development of littoral plant life. Due to nutrient deficiency, there is no mass development of phytoplankton, and water blooms are rare. The water appears blue or green, and there is a large visual depth. There is a low sedimentation rate, and the sediment contains only small quantities of organic constituents. Mud containing decaying or putrifying material is absent. Throughout the entire year, the oxygen concentration decreases uniformly from the surface downwards, and may still be as high as 60% saturation at the bottom. The plankton consists mainly of green algae. Although the bottom fauna contains a wide variety of different species, their individual numbers are small (200–2 000 animals/ m^2). Index forms are chironomid larvae of the genus, *Tanytarsus,* which, like most bottom organisms of oligotrophic lakes, have a high oxygen requirement (stenoxybionts). Characteristic fishes are species of the salmonid genus *Coregonus.* Oligotrophic lakes mature into eutrophic lakes by the accumulation of nutrients, passing in the process through a transitional *mesotrophic* form. Maturation is often accelerated by the introduction of industrial or domestic waste water.

Eutrophic lakes (e.g. Baltic lakes) contain a high concentration of plant nutrients, and display high biological productivity. The littoral region is wide and usually densely vegetated. The pelagial is rich in plankton, and subject to frequent mass outbreaks of algae (water blooms). Water color varies between green and brown-green, and the visual depth is shallow. In summer, the oxygen concentration decreases markedly below the metalimnion. The hypolimnion then often contains no oxygen. Above the metalimnion, in the epilimnion, the water is supersaturated with oxygen. During total autumn circulation, however, oxygen-rich water sinks to the bottom, so that the oxygen distribution is similar to that of an oligotrophic lake. The sediment consists primarily of dead plankton, and it is rich in organic material (see Gyttja). Only a few species are represented in the bottom fauna, but the population is large (up to 10 000 individuals/m^2). On account of the poor oxygen supply, the bottom organisms are euroxybionts. Characteristic forms are chironomid larvae of the genus *Chironomus.* As a eutrophic lake ages, it becomes a pond and eventually a bog (see Aging of lakes).

The only type of brown water lake is the *dystrophic lake.* Its water is deficient in calcium and nutrients, but especially rich in humus materials, the latter being washed into the lake from nearby bogs. Large numbers of dystrophic lakes occur in the boglands of Scandanavia. The littoral contains only a few aquatic plants, and there is little phytoplankton. On the other hand, zooplankton is often abundant. Depending on the content of humus material, the water is yellow to brown, and the visual depth is shallow. In the deeper layers, the oxygen concentration is constantly low. The sediment consists of peat mud (Dy). Only low densities of bottom fauna are present (0 to 20 animals/m^2), consisting mainly of chironomid larvae of the genus *Chironomus.* Aging of a dystrophic lake produces an upland moor or bog.

Lama: a genus of the camel family *(Camelidae),* comprising the Guanaco (see), Llama (see), Alpacca (see), and Vicuña, all of which do not have humps.

Lamarckism: evolutionary theory of Jean Baptiste de Lamarck (1744–1829). Lamarck proposed that environmental changes generated new needs and habits in animals, which acted via "nerve fluid" to cause appropriate modification the organism. Thus, according to Lamarck, the flow of nerve fluid into the head was responsible for the development of deer antlers. Such alterations, including the stengthening of organs by continual use, or their degeneration through lack of use, were supposedly inherited, and these mechanisms led to the gradual evolution of higher forms from lower forms.

L. is often used as a synonym for the Inheritance of acquired characters (see).

Lamellate antennae: see Antennae of insects.

Lamellibranchia(ta): see *Bivalvia.*

Lamellicorn beetles, *Scarabaeoidea, Lamellicornia:* a superfamily or group of the *Coleoptera* (see), in which the antennae have an "elbow" bend (geniculate antennae), and in which the terminal antennal segments are enlarged into a leaflike or fanlike structure. The larvae of L.b. are scarabaeiform (Fig.1). The group contains 3 families:

1. Family: *Scarabaeidae.*

1. Subfamily: *Scarabaeinae (= Coprinae)* ***(Dung beetles,*** including the sacred ***Scarab beetle*** of ancient Egypt).

2. Subfamily: *Melolonthinae* ***(Cockchafers).*** The ***June bug*** *(Melolontha,* see Fig. 2) is represented by 3 species in central Europe; it was formerly a common forestry and orchard pest, but its numbers have decreased considerably.

3. Subfamily: *Rutelinae* ***(Shining leaf chafers).***

4. Subfamily: *Dynastinae* ***(Rhinoceros*** or ***Elephant beetles).***

5. Subfamily: *Cetoniinae* ***(Flower beetles),*** e.g. the ***Rose chafer*** *(Cetonia aurata)* and the ***Flower eater*** *(Cetonia nitidia).*

6. Subfamily: *Aphodiinae* ***(Dung beetles).***

2. Family: *Lucanidae* ***(Stag beetles).*** The Stag beetle, *Lucanus cervus* L., is the largest European beetle.

3. Family: *Passalidae* ***(Betsy beetles).***

Fig.1. *Lamellicorn beetles.* Scarabaeiform larva.

Fig.2. *Lamellicorn beetles.* June bug *(Melolontha melolontha* L.).

Lamiaceae, *Labiatae,* ***mint family:*** a family of dicotyledonous plants containing about 3 500 species in 180 genera, cosmopolitan in distribution, and especially numerous in the Mediterranean region. They are mostly terrestrial herbs or shrubs, which are easily recognized by their quandrangular stems, their opposite, exstipulate, simple leaves, and their usually strong aromatic odor, which arises from glands containing volatile oils. Bitter principles, polyphenols and tannins are also widely distributed and abundant in the *L.* Flowers display bilateral symmetry, and are pollinated by insects. The calyx is usually 5-toothed, and often 2-lipped. The corolla is 5-lobed, the upper 2 lobes usually united to a single lip; the lower 3 lobes often partially fused into a lower lip, the middle lobe usually the largest of the 3. Four stamens are generally present, and these are usually didynamous (one pair longer than the other pair); anthers are introrse. When young, the superior ovary consists of 2 united carpels, but each carpel soon divides into 2 loculi, resulting in a mature 4-locular ovary; consequently, the fruit is usually a group of 4 achenes (nutlets), each containing a seed. The stigma is 2-lobed. Nectar-secreting tissue is present at the base of the ovary. Inflorescences are racemose or cymose, the cymes at the nodes condensed into a false whorl or verticillaster. At the stem apex these false whorls are often close together, forming a spike-like or head-like inflorescence. In some species the flowers are solitary in the axils of each bract.

Systematic division of this species-rich family is difficult, due to the uniformity of many characters.

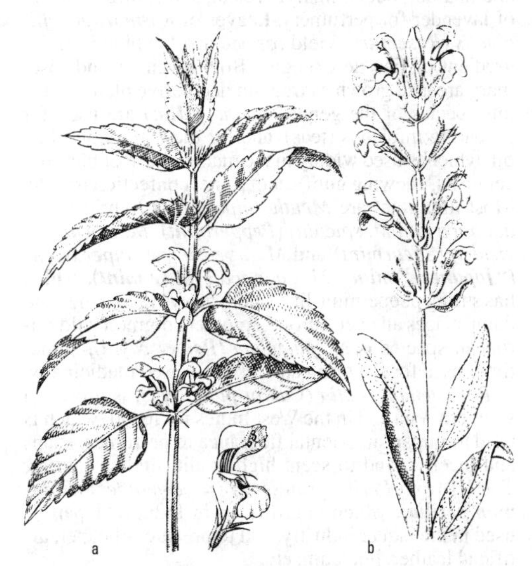

Lamiaceae. a Flowering shoot and single flower of the white dead- nettle *(Lamium album). b* Flowering shoot of sage *(Salvia officinalis).*

Mainly on account of their volatile oils, many L. are medicinal and culinary plants. Some of the oils are extracted commercially for use in the pharmaceutical industry, in food technology, and in the manufacture of cosmetics. Important representatives indigenous to the Mediterranean region are: *Salvia officinalis* ***(Garden sage)***, *Satureja hortensis* ***(Savory)***, *Majorana hortensis* ***(Sweet marjoram)***, *Thymus vulgaris* ***(Common thyme)***, *Ocimum basilicum* ***(Sweet*** or ***Hoary basil)***, *Hyssopus officinalis* ***(Hyssop)***; they are also cultivated to a greater or lesser extent in Europe and North America. *Melissa officinalis* ***(Common*** or ***Lemon balm)***, widely distributed in the Orient, is cultivated in Europe and North America; its leaves are used to prepare a tea, and to prepare "spirit of melissa", a herbal remedy and tonic manufactured in South Germany. The distribution center of the 28 species of *Lavandula* ***(Lavender)*** is in the Mediterranean region, but the genus extends westward to the Atlantic islands, including Cape Verde, and eastward to Somalia and as far as India. ***Common*** or ***True lavender*** *(Lavandula angustifolia = officinalis = vera)* is a small under-shrub indigenous to southern France, Italy and Spain, and cultivated extensively in southern France and elsewhere (e.g. Norfolk, UK) for the commercial extraction of lavender oil, which is used in perfumery and soap manufacture. The oil is prepared by steam distillation of the inflorescences, which take the form of a terminal spike, in which the flowers are arranged in small verticillasters, each arising from the axil of a bract. Most of the oil is present in oil glands, which are visible with a lens among the hairs of the ribbed, blue-violet calyx. Chief constituents of oil of lavender are linalool, linalyl acetate and cineole, with lesser concentrations of geraniol, borneol and nerol. *Lavandula spica (=latifolia)* is distinguished from True lavender by its linear bracts (rhomboidal bracts in *L. angustifolia*), its spathulate leaves and more compressed inflorescence; it yields *oil of spike*, which contains high concentrations of cineole and linalool, but no linalyl acetate, and is inferior to oil of lavender for perfumery. Leaves of *Rosmarinus officinalis* ***(Rosemary)*** yield *rosemary oil*, which is an ingredient of Eau de Cologne. Both lavender and rosemary are also grown as fragrant decorative plants. Various species of the genus *Mentha* ***(Mint)*** are used for preparing infusions (teas), and for the extraction of an oil, which is used widely in the manufacture of pharmaceuticals, chewing gum, tooth paste, confectionary, etc. Most important are *Mentha piperata* (a hybrid of *M. aquatica and M. spicata)* ***(Peppermint)***, *M. spicata (= viridis)* ***(Spearmint)*** and *M. arvensis* var. *piperascens* ***("Japanese" mint)***. *M. aquatica* ***(Water mint)***, which has strong peppermint-like smell, grows in marshes and damp places all over Europe. Another common wild European species is *M. pulegium* ***(Penny royal)***, sometimes used for flavoring and formerly used medicinally.

Pogostemon cablin ***(Cablin patchouli)***, is grown in southern Asia and in the West Indies for its oil, which is used to impart an oriental fragrance to perfumes, soaps and incense, and to scent high quality Indian fabrics. The oil from *Perilla frutescens (= ocymoides)* ***(Common perilla)***, which is grown from India to Japan, is used in the laquer industry, and to prepare oil paper, artificial leather, linoleum, etc.

Laminaria: see *Phaeophyceae.*

Laminariales: see *Phaeophyceae.*

Lämmergeier: see *Falconiformes.*

Lammergeyer: see *Falconiformes.*

Lamna: see *Selachimorpha.*

Lamnidae: see *Selachimorpha.*

Lamniformes: see *Selachimorpha.*

Lamninae: see *Selachimorpha.*

Lamp-brush chromosomes: very large chromosomes, up to 1 mm long, which are seen during the meiotic prophase in the oocytes of many birds, amphibians, reptiles, elasmobranch fishes, as well as in the oocytes and spermatocytes of some invertebrates, and in some plants (e.g. the green alga, *Acetabularia*). They are unwound, extended paired chromosomes (bivalents), with loops along the side, which give them a brushlike appearance. The loops (unwound individual chromomeres, consisting of DNA, protein and RNA) represent highly active transciptional sites, like the Balbiani rings and puffs of giant chromosomes. They are loaded with gene products (RNA) and ribonucleoproteins, and are therefore visible under the light microscope. At the end of the oocyte growth phase, the loops recede and disappear, and the chromosomes take up the form of typical metaphase chromosomes.

Lampetra: see Lampreys.

Lampreys, *Petromyzonidae:* a family of the *Cyclostomata* (see). The body is cylindrical and eel-like, paired fins are absent, and the mouth is provided with horny teeth. Parasitic species, e.g. the River lamprey *(Lampetra fluviatilis)* and Sea lamprey *(Petromyzon marinus)* attach themselves to fishes by means of their large suctorial mouth, then use their file-like tongue to rasp a hole in the body wall. Thereafter, they live parasitically on muscle brei and blood.

There is a single nasal aperture (as opposed to the two occurring in higher vertebrates) which leads to a blind sac, and does not communicate with the pharynx.

When viewed externally, instead of the gill slits seen in true fishes, each side of the neck presents 7 round holes placed far back in a line. These are the external apertures of the gills, which in these fishes have the form of sacs or pouches (hence the older name of *Marsipobranchii* for the *Cyclostomata*). The lining membrane of the sac-like gills is thrown into multiple folds over which the branchial vessels ramify. Internally, the sacs open into a common respiratory tube which communicates with the pharynx. Thus, water can be admitted to the gills without passing through the mouth, an obvious advantage when the mouth is attached to a host fish. Water circulation is assisted by an elastic cartilaginous framework that supports the entire respiratory apparatus and acts somewhat like the ribs of higher vertebrates.

The blind larvae live 3–5 years in the bottom mud of flowing fresh waters, where they feed on microplankton and detritus. The larval mouth has a lobed border, and the horny teeth are not formed until metamorphosis. After metamorphosis, the young animals migrate to the sea, where they live parasitically. Later, they return to rivers and migrate upstream to spawn in that part of the river where they lived as larvae. Some nonparasitic species live permanently in freshwater, e.g. the Stream lamprey *(Lampetra planeri)*.

Lampris: see Kingfish.

Lampropeltis: see King snakes.

Lance-head snakes: pit vipers (see *Crotalidae*) belonging to the 2 genera, *Bothrops* (from Mexico over the whole of South America) and *Trimeredurus* (wide areas of South Asia). A third monotypic genus, *Lachesis mutus* (see Bushmaster), is sometimes also included, but is treated separately here. The venom of the American species is extraordinarily powerful, and it still claims many human lives each year. Although still potent, the venom of the Asiatic species is less dangerous. Most are jungle-dwelling species. Terrestrial forms are patterned with dark dorsal spots on a brownish background, while arboreal species are green. The female ***Fer-de-Lance*** *(Bothrops atrox)* gives birth to up to 70 live young. The vividly patterned (brown, reddish and yellow) ***Jararaka*** *(Bothrops jajaraca)* grows up to 1.6 m in length. Some species, like the very variably patterned, yellow, or brown-spotted *Bothrops schlegeli* ***(Eyelash viper*** *or* ***Schlegel's pit viper),*** hold onto the branches of trees with their rolled tail. The largest Asian member is the ***Okinawa habu*** or ***Yellow-spotted lance-head snake*** *(Trimeresurus flavoviridis).* Only 70 cm in length, the ***White-lipped tree viper*** *(Trimeresurus albolabris)* is a blue-green tree snake of the Indian plains. The ***Chinese mountain viper*** *(Trimeresurus monticolor)* is distributed from the Himalayas to Malaysia, and is the only lance-head snake that lays eggs.

Lancelet: see *Amphioxus lanceolatus.*

Land bridges: physical links which are thought to have existed between continents that possess largely similar fauna. L.b. are a crucial paleobiogeographical factor in the interpretation of the evolution of Tertiary mammals. Paleogeographical L.b., e.g. between North and South America *(Central American land bridge),* and between North America and Eastern Europe/Asia *(Bering land bridge)* were used as migration routes by mammals. Isolated land masses, unconnected to Asia or America by L.b., developed unique fauna, e.g. Australia and Tasmania (egg-laying monotremes and marsupials). See Continental drift.

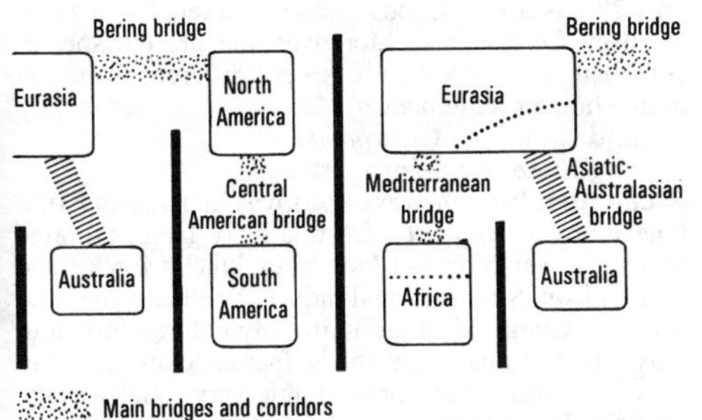

Land bridges. Barriers and land bridges that influenced the dispersal of organisms during the Cenozoic (time of the mammals).

Land crabs, *Geocarcinidae:* a family of true crabs (see *Brachyura),* inhabiting terrestrial habitats in tropical America, West Africa, and the Indo-pacific. An example is the large Caribbean land crab, *Cardisoma guanhumi.* L.c. return to the sea only for reproduction. Their excavated tunnels can be an agricultural nuisance.

Land iguana, *Conolophus subcristatus:* a relatively large (over 1 m long) iguana (see *Iguanidae)* found in the interior of the Galapagos Islands. It has a powerfully developed neck crest, and its ritualized display does not include the head butting which is observed in the Marine iguana (see). It feeds mainly on land plants, including cacti which it consumes complete with spines. Formerly very common, the L.a. is now threatened with extinction, and like all the fauna of the Galapagos Archipelago, it is strictly protected. Some authorities consider the L.a. of Santa Fe Island to be a separate species, *Conolophus pallida,* but it is probably only justified as a subspecies.

Landolphia: see *Apocyanaceae.*

Landolphia rubber: see *Apocyanaceae.*

Landscape (plate 47): a specific part of the earth's surface characterized by a uniform structure and the complementary interaction of its natural and anthropogenic components. The *geofactors* responsible for the establishmnet of a L. can be categorized as inorganic, organic and cultural. Inorganic factors include surface topography, type of rock, soil, atmosphere and hydrosphere. Flora and fauna constitute the organic factors, while the cultural influence arises from human activity and the presence of buildings and other structures.

Every L. is a mosaic or typically patterned complex of specific, interrelated *ecotopes* (more or less uniform units, also known as *L. cells).* The ecotopes of an adjacent L. may be different or similar but differently interrelated. The nature of each ecotope is determined by a combination of all factors: rock type, alignment of strata, surface topography, climate, surface and ground water economy, and the type of soil, as well as human economic activity and its consequences, such as the depletion of ground water or the addition of extra water.

Each ecotope contains very complex *L. elements,* i.e. components of the L. that fulfil special functions, and which are often determined by human agency, e.g. fields, meadows, woodlands, stretches of water and copses, as well as technological structures such as roadways, power transmission lines, buildings, etc.

Ls. can be classified in various ways. Thus, they may be tropical, subtropical or temperate. Alternatively, they can be categorized according to dominant environmental and topological features, which may be abiotic, biotic or anthropogenic, e.g. mountain, morain, coastal, forest, steppe, desert, city and industrial Ls.

It is very common to classifiy Ls. according to the extent to which human activity has influenced their development. A *natural L.* is determined only by inorganic and organic factors. Such Ls., with no evidence of human influence, are rare; they can still be found in nonaccessible parts of the world, e.g. tropical rainforests, central regions of big deserts, rock and glacier regions of high mountains, and the polar regions. In *near natural Ls.,* nonintensive agriculture is practised, and the influence of geofactors is largely unaltered and unrestrained by human activity, e.g. in central Europe during the Middle Ages and up to the beginning of the modern era.

Further development of industrial production and increases in population gave rise to *cultural Ls.* The basis of any L. consists of those geofactors that are practically unchangeable, like rock formations, surface topography and macroclimate. In the most intensively managed, cultivated or populated cultural L., the less stable natural geofactors of plant cover, animal population, microclimate and soil and water economy are altered to the greatest possible extent.

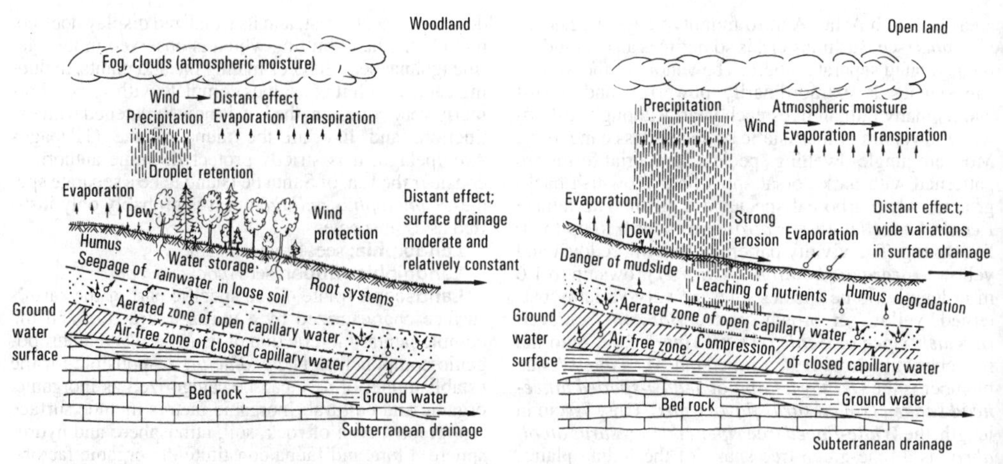

Landscape. The water economy of a region before and after deforestation.

Most cultural Ls. of central Europe would be better described as *civilization Ls.,* because the changes so far brought about by human agency are largely spontaneous and determined by local considerations. The modifications of the L. are usually designed for rapid increases in productivity, without consideration of possible long-term effects.

L. planning is intended to counteract the negative effect of industrialization and human settlement on the L., and to conserve high quality and/or traditional Ls. Assessment of L. as a scenic resource is subjective, but it is necessary as a prelude to L. planning.

One important organic factor particularly affected by human economic activity is woodland and forest. These are not only valuable for their scenic and recreational appeal, but more importantly for their role in maintaining the water economy and for consolidating the soil surface. In cool temperate regions, the precipitation rate is only slightly increased by woodland. Water consumption by woodland is, in most cases, greater than that of other natural or cultivated plant formations, so that the run-off is less. Extensive clearance of woodlands and forests in the Middle Ages therefore increased the quantity of freely available water. However, the total amount of available water is less important than its distribution (allocation), and the increased run-off is a potential cause of flooding. Since tree cover moderates the extremes of soil temperatures, the soil is frozen for a shorter period, seepage is greater, the ability to take up water is greater, and the rate of drainage is more uniform throughout the year than from a treeless surface. The entire L. is affected by the general water flow. The main hydrological functions of woodland must be seen as the evening out of water flow (uniform flow throughout the year), the decreased transport of suspended material and detritus, and improvement of water quality. In contrast, any influence of woodland on the macroclimate is hardly detectable. Locally, the wind in sparsely treed L. is better intercepted if the trees are appropiately spaced and distributed. This decreases wasteful evaporation and to a certain extent promotes assimilation. Depending on their structure, however, woodlands have a marked effect on the microclimate, which benefits adjoining agriculture and human recreation. They also improve air purity by filtration of suspended matter and removal of certain pollutant gases, and by oxygen release from the trees. Woodlands and forests also act as a refuge for animals and plants.

All the functions of woodland depend on its vertical structure. A layered, mixed woodland is the most beneficial, although in special cases pure conifer forest also has many advantages. Within a cultural L., the beneficial effects of woodland are very dependent on its density and distribution. Hedges and residues of previously larger woodlands are still useful, in particular as wind breaks.

Before human setlement, more than 90% of central Europe was covered with forest or woodland. Today, only 28% is covered, and between different Ls. this varies from 5% to 75%. Moreover, the present species composition of woodlands is very different from that before human settlement.

Land snails: see *Gastropoda.*

Langouste: see Spiny lobsters.

Langurs, ***leaf monkeys, Presbytis:*** a genus of Colobine monkeys (see). L. are relatively large, arboreal monkeys, weighing 7–18 kg. They inhabit forests and jungles from Sri-Lanka and India to Southeast Asia and southern China. Most are uniformly colored brown to gray. The hanumans have black faces and are sacred to Hindus. Thumbs are present but very small. Body, hands and tail are long and slender. Sixteen species are classified in 4 groups: ***(l.m.*** = leaf monkey)

1. Group *(P. melalophus* group): ***Banded l.m.*** *(P. melalophus),* ***Javan l.m.*** *(P. comata),* ***White-fronted l.m.*** *(P. frontata),* ***Hose's l.m.*** or ***Everett's l.m.*** *(P. hosei),* ***Mentawai island l.m.*** *(P. potenziana),* ***Maroon l.m.*** *(P. rubicunda),* ***Thomas's l.m.*** *(P. thomasi).*

2. Group *(P. cristata* group): ***Silvered l.m.*** *(P. cristata).* ***Francois' l.m.*** *(P. francoisi),* ***Golden l.m.*** or ***Gee's l.m.*** *(P. geei),* ***Dusky l.m.*** *(P. obscura),* ***Phayre's l.m.*** *(P. phayrei).* ***Capped l.m.*** *(P. pileata).*

3. Group *(P. vetulus* group): ***Purple faced l.m.*** *(P. vetulus),* ***John's l.m.*** *(P. johnii).*

4. Group *(P. entellus* group): ***Hanuman langur*** or ***Entellus langur*** *(P. entellus).*

Laniidae, ***shrikes:*** an avian family of more than 70 small to medium-sized (18–25 cm) species in the order

Passeriformes (see), suborder *Oscines.* Apart from 2 species that occur in North America, the family belongs to the Old World. No representatives are found in Central or South America, Australia or the Pacific islands. Although essentially insectivorous, they also feed on small vertebrates, such as frogs, lizards, rodents and even other birds, and they are the most predatory of all the passerines. Unlike the nonpasserine birds of prey, they kill with their sharply hooked bills rather than with their talons (their legs and feet are, however, strong and their claws are sharp). Popularly known as "butcher-birds", they are known for their habit of impaling their prey on thorns. Three subfamilies are recognized.

True shrikes *(Laniinae;* 25 species) are found in Africa and Eurasia. One member, the ***Northern*** or ***Great gray shrike*** *(Lanius excubitor;* 24 cm) has a circumpolar range, and is found widely in the spruce forests across northern North America and Eurasia. The ***Loggerhead shrike*** *(Lanius ludovicianus)* occurs from the Nearctic to southern Mexico. The well known ***Red-backed shrike*** *(Lanius collurio)* (plate 6), which favors open scrubland, is widely distributed throughout Europe and the Mediterranean region, extending into Scandanavia and across Asia to China, migrating to Africa, India and Arabia in the winter. ***Bush shrikes*** *(Malaconotinae;* 39 species) are restricted to Africa. ***Helmet shrikes*** *(Prionopinae;* 9 species), also restricted to Africa, have black and white plumage, caruncles of skin around the eyes, and prominent crests.

Laniinae: see *Laniidae.*

Lanius: see *Laniidae.*

Lanner: see *Falconiformes.*

Lanosterol, ***kryptosterol:*** 5α-lanosta-8(9),24-dien-3β-ol, a tetracyclic triterpene alcohol, containing the 5α-lanostane ring system. L. is a zoosterol which occurs in large amounts in the fat of sheep wool. It is biosynthesized from squalene, via 2,3-epoxysqualene, and is the biosynthetic precursor of all further tetracyclic triterpenes of the lanostane type, and of the steroids.

H HO H

Lanosterol

Lanternfishes: see *Myctophidae.*

Lantern flies: see *Homoptera.*

Lanthanotidae, ***earless monitor:*** a monospecific family of the *Lacertilia* (see), represented by *Lanthanotus borneensis* of Northwest Borneo. This rare, nocturnal lizard attains a length of 43 cm, lacks external ear openings, has transparent lower eyelids, and a relatively long tail that does not exhibit autotomy. Its anatomy is intermediate between that of the *Vanaridae* and the *Helodermatidae,* and it may have evolved from a lizard group ancestral to the snakes.

Lanthanotus: see *Lanthanoditae.*

Laomedea: see *Hydroida.*

Laphygma: see Cutworms.

Laportea: see *Urticaceae.*

Lappet moths: moths of the family *Lasiocampidae* (see).

Lapwings: see *Charadriidae.*

Larch: see *Pinaceae.*

Larch shoot borer: see *Yponomeutidae.*

Larch shoot moth: see *Yponomeutidae.*

Large gerbils: see *Gerbillinae.*

Large-leaved lime: see *Tiliaceae.*

Largemouth bass, *Micropterus salmoides:* a member of the family *Centrarchidae* (sunfishes), order *Perciformes.* It attains a maximal length of 60 cm, but most specimens are 30–35 cm. This predatory fish is indigenous to the Mississippi basin and the Great Lakes of North America. It was introduced into Europe at the end of the 19th century, and is now found in the Danube basin and most countries of western Europe, where it is a regionally important angling fish. The mouth is large with numerous teeth, and the low front of the dorsal fin is separated from the higher hind part by a deep cleft. It has golden or silvery green sides and a yellowish white belly.

Laridae, ***gulls*** and ***terns:*** an avian family in the order *Charadriiformes* (see), containing 82 species, which are clearly separated into 2 subfamilies. They lack the fleshy cere that characterizes their close relatives, the skuas. Although they are obviously marine birds, adapted to life in coastal waters, some species have moved inland and established stable populations far from the sea.

1. Subfamily: *Larinae* ***(Gulls;*** 43 species; 28–81 cm). These are some of the most familiar birds around harbors and beaches, and ships leaving port (but only as far as the continental shelf). Only the Kittiwake is normally found on the high seas, all other species remaining in sight of land. Gulls are heavy-bodied birds. The long powerful bill is curved downward at the tip into a sharp hook, and carries a projection at the distal end of the lower mandible, called the gonys. The ***Kittiwake*** *(Rissa tridactyla;* 40.5 cm) breeds around the Gulf of St. Lawrence and the coasts of Labrador, Arctic Canada, Greenland, and Arctic Europe and northern France, becoming truly pelagic outside the breeding season, when it inhabits the open ocean between the tropic of Cancer and 60° north. The ***Herring gull*** *(Larus argentatus;* 56–61 cm; circumpolar in Northern Hemisphere) (see Fig. Also see entry: "Species") is migratory, overwintering in the Mediterranean region, Persian Gulf, Arabian Sea, Red Sea, and the coasts and inland lakes of tropical Africa. Mantle, back and scapulars are pale blue-gray, and the rest of the body is white, with gray wings tipped with black and white. The legs are pinkish or yellow, and the bill is deep yellow with a red spot at the end of the lower mandible. A yellow bill with a red spot is characteristic of certain other gulls, e.g. ***California gull*** *(Larus californicus;* 55.5 cm; western North America), ***Great black-backed gull*** *(Larus marinus;* 73.5 cm; North Atlantic coasts), and it acts as a sign stimulus for the feeding response of the young (see Releaser). Gulls do not normally catch live fish, but they are expert scavengers and great opportunists, taking practically any form of garbage, offal and dead animal remains. Shellfish are much favored, and the hardest shells are broken by dropping them from a height onto rocks. Insects, worms and grubs are obtained from agricultural areas near the coast, and flocking gulls are a fa-

miliar sight during ploughing. Swarming locusts commonly attract gulls from far and wide.

Most gulls are colonial ground-nesters, preferring coastal islands and inaccessible cliffs, sometimes forming mixed colonies with other species.

Laridae. Herring gull *(Larus argentatus).*

2. Subfamily: *Sterninae* ***(Terns;*** 20.5–58 cm; 39 species). Terns are generally smaller and more slender than gulls, with narrower, more pointed wings, and thin, sharp-pointed bills. They fly with steady, syncopated wing beats, rather than soaring and gliding like gulls. They catch live fish by snatching them from the surface, or by plunging under the surface from the air. Inland, they capture insects on the wing. Plumage is generally white, with gray back and wings, most species also having a black crown, and many having a nuchal crest. Some exceptions to this general coloration are, e.g. the ***Brown noddy*** *(Anous stolidus),* which is uniformly brown with a pale gray crown; and the ***Black Tern*** *(Chlidonias nigra),* an entirely black inland species, which nests in marshes and reed-bordered lakes in North America and Eurasia. A more familiar species is the ***Common tern*** *(Sterna hirundo;* 35.5 cm) (plate 6), which breeds on the coast or by lakes in Europe, around the Mediterranean, and inland to Iran and Tibet, as well as coastal and inland areas from Labrador and Central Ontario to the West Indies, overwintering along the coasts of Africa and Arabia, and the coastline from Florida to the Megellan Strait. *Sterna paradisaea* ***(Arctic tern;*** 38 cm) breeds in salt marshes and low-lying tundra on the islands and entire circumpolar coastline of the Arctic ocean, from British Columbia to Hudson Bay, Greenland, northern Europe, the Baltic, Scandanavia and northern Siberia, migrating in the winter to the oceans of the Southern Hemisphere, where it becomes truly pelagic.

Larinae (gulls): see *Laridae.*

Lark quail: see *Gruiformes.*

Larks: see *Alaudidae.*

Larkspur: see *Ranunculaceae.*

Larus: see *Laridae.*

Larva: early stage in the life cycle of animals that do not develop directly from egg to sexually mature adult. Transition from L. to adult involves a reorganization of body structure, a process known as metamorphosis. In addition to specific larval organs (e.g. bands of cilia, adhesive glands, external gills), the morphology and mode of existence of the L. are usually very different from those of the adult. In the condition known as Neoteny (see), larval organs are retained by the adult. The larval types of different animal groups are listed below. More information can usually be found under the entry for the particular animal group.

Mesozoa	Ciliated L.
Porifera	Amphiblastula, Parenchymula.
Coelenterata	Planula (see), Actinula (see), Ephyra (see).
Turbellaria	Müller's L. (see).
Trematodes	Miracidium, Sporocyst, Redia, Cercaria (see Liver fluke).
Cestoda	Coracidium, Procercoid, Plerocercoid, Oncosphere, Cysticercus and Cysticercoid, Coenurus.
Entoprocta	Free swimming trochosphere type with alimentary canal.
Nemertini	Pilidium, Desor's L., Schmidt's L.
Nematomorpha	Worm-like with anterior boring organ.
Acanthocephala	Acanthor, Acanthella and Cystacantha.
Priapulida	Primary L. (nonciliated), Secondary L.
Mollusca	Trochophore (see), Veliger (see), Glochidium (see).
Echiurida	Trochophore-like swimming L.
Annelida	Trochophore (see), Mitraria.
Pentastomida	Primary L. with 2 large pairs of hooks and boring apparatus.
Crustacea	Nauplius (see), Metanauplius (see Nauplius), Zoea, Cypris L. (see), Mysis stage.
Insecta	Caterpillar (grub), nymph and numerous other larval forms (see Metamorphosis).
Phoronida	Actinotrocha.
Bryozoa	Cyphonautes L.
Brachiopoda	Free swimming, ciliated L.
Enteropneusta	Tornaria.
Pterobranchia	Free swimming, ciliated L.
Holothuroidea	Auricularia, Doliolaria, Pentactula.
Asteroidea	Bipinnaria, Brachiolaria.
Echinoidea	Pluteus (Echonopluteus).
Ophiuroidea	Pluteus (Ophiopluteus), Crinoid L.
Ascidiaceae	Free swimming L. with a tail.
Fishes	Ammocoete L (lamprey), Leptocephalus (eels), Larvae with external gills (coelacanths and lungfishes), and others.
Amphibians	Tadpole (see).

Higher vertebrates (reptiles, birds and mammals) do not develop via a free living larval stage.

Larva coarctata: see Insects.

Larval hormone: see Juvenile hormone.

Larvicides: plant protection agents for controlling the larvae of insects and mites.

Larynx: the organ of voice, lying at the entry of the trachea. It is protected from the pharynx by a lid of fibrocartilage called the *epiglottis.* The skeleton of the mammalian larynx consists of 2 main single cartilages and a paired cartilage. The *thyroid* cartilage incompletely surrounds the L., while the *cricoid* cartilage forms a complete ring around the base of the L. The paired *arytenoid* cartilages project inward from the back of the L. The paired *corniculate* (cornicula laryngis or Cartilages of

Santorini) and *cuneiform* cartilages are functionally and structurally less significant, and the single epiglottis (epiglottideal cartilage) may also be considered as one of the skeletal cartilages of the L. In humans, the thyroid cartilage is recognizable externally as the Adam's apple (pomum Adami). The individual cartilages are connected by elastic ligaments, and can be moved against each other by a complex laryngeal musculature. The L. is constricted by two pairs of folds; the lower inferior or true vocal folds function in voice production, whereas the ventricular folds function in holding the breath, in protecting the larynx during swallowing, as well as providing moisture for lubrication of the vocal folds. The free edges of these folds, known as the *true vocal cords* and *false vocal cords,* respectively, stretch from each arytenoid cartilage to the thyroid cartilage. The elongated fissure created by the folds is called the *glottis.* Voice is generated by vibrations of the vocal cords, caused by the passage of air during expiration. The shape of the glottis and the tension of the vocal folds is altered by the action of muscles on the walls of the L. Thus, during deep inspiration the glottis is quite rounded, whereas during the voicing of a high note the edges of the vocal folds are drawn tight and the glottis has the shape of a fine slit. Sounds produced by vibration of the vocal cords may be amplified, e.g. in the adult male orang-utan, laryngeal air sacs (outpouchings of the larynx wall) extend into a large throat pouch and serve as resonance chambers. Cricoid and arytenoid cartilages are present in the anuran L., but the thyroid cartilage is a later evolutionary acquisition present only in mammals. In birds, the L. is reduced, but at the bifurcation of the trachea they possess a Syrinx (see), which has evolved by modification of the tracheal cartilages, and development of vibratory lips and resonance chambers.

Lasiocampidae: see *Lepidoptera* for classification. See Lackey.

Late interglacial: a period following a major glaciation, in which the ice is receding and is already absent in many previously glaciated regions. It occurs between the last temperature minimum of the glaciation and the onset of the rapid temperature rise at the beginning of the subsequent post-glacial. It is characterized by the absence of forests and the presence of tundra-like vegetation. The term is usually specifically applied to the L.i. between the Würm (see) glaciation and the Holocene (see) post glacial. See Pollen analysis.

Latency: see Resting state.

Latent attack: an attack on a plant by a disease or pest, without the appearance of typical symptoms.

Latent learning: formation of associations that bring no reward or punishment. L.l. is related to reconnaissance or inquistive behavior, in which the novelty of a new experience may resemble a reward. L.l. is also favored by play behavior. There is also a connection between L.l. and insight learning. See Learning.

Latent period, ***latent time:***

1) In physiology, the time that elapses between the application of a stimulus and the observation of a response. The term is justified for the Excitation (see) of a single structure. If several processes occur between stimulation and reaction, e.g. excitation, conduction of excitation, processing of excitation, then the term *reaction time* is more appropriate.

2) In virology, the time that elapses between infection of a cell with a virus and the first appearance of virus progeny in the cell. The L.p. of bacteriophage T2 in *Escherichia coli* is 25 minutes.

Latent time: see Latent period.

Lateral fold lizards: see *Anguidae.*

Lateralization: in neurophysiology, the excercise of different functions by left and right paired brain regions, in particular the left and right cerebral hemispheres of the human brain. L. was first demonstrated by the clinically necessary severence of the corpus callosum cerebri, which links the two halves of the brain. In humans, L. is manifested in the dominance of one side of the brain for particular functions, e.g. handedness. In right-handed persons, the left cerebral hemisphere controls movement of the right hand, and it dominates speech, abstract thought, linear reasoning and analysis. In contrast, the right hemisphere is more intuitive, and is associated with contemplative, artistic and musical ability and the appreciation and recognition of patterns. In "split brain" experimental animals (the two halves of the brain are separated surgically), conditioned reflexes can be established in each half independently of the other half.

Lateral line system: see Tactile stimulus reception.

Latex duct: see Excretory tissues of plants.

Lathyrus: see *Papilionaceae.*

Laticauda: see *Hydrophiidae.*

Laticaudinae: see *Hydrophiidae.*

Latimera: see Coelacanths.

Lattice fungus: see *Basidiomycetes (Gasteromycetales).*

Laudanosine: an opium alkaloid, which in relatively large doses causes muscle tetanus. Opium contains less than 1% L., which is a close structural relative of Papaverine (see).

Laughing jackass: see *Coraciiformes.*

Laughing thrushes: see *Timaliinae.*

Laura: see *Cirripedia.*

Lauraceae, ***laurel family:*** a family of dicotyledonous plants containing about 2 500 species in 32 genera, distributed in tropical and subtropical regions, often forming forests. They are trees or shrubs with entire, leathery leaves, and actinomorphic, hermaphrodite or unisexual, usually 3-merous, insect-pollinated flowers. The perigynous perianth usually consists of 6 segments, in 2 whorls of 3, united at the base. Stamens (usually 12) are arranged in 4 whorls of 3, some reduced to stami-

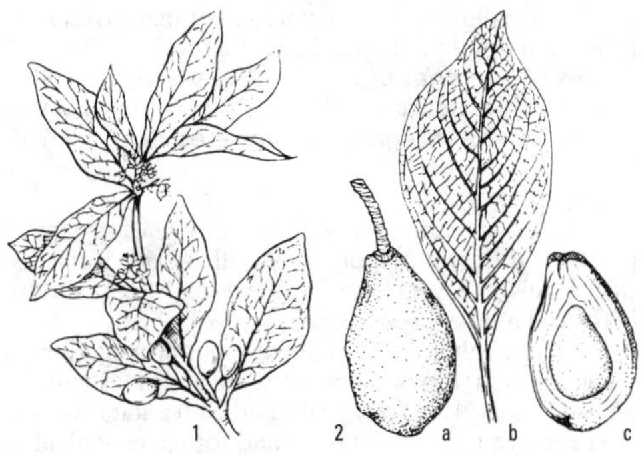

Lauraceae. 1 Sweet bay *(Laurus nobilis). 2* Avacado *(Persea americana): à* leaf, *b* fruit, *c* longitudinal section of a fruit.

nodes. Anthers are 2- or 4-celled, usually introrse (often extrorse in the 3rd whorl), and the pollen escapes through valves. The usually superior ovary consists of a single carpel (which can be interpreted as a unilocular structure consisting of 3 united carpels), which develops into a berry or drupe.

The ***Camphor tree*** *(Cinnamomum camphora),* native to East Asia, attains a height of 40 m. It is cultivated in Sri-Lanka, Malaya, India, Burma, Java, East and West Africa, California, etc. Numerous cells, containing a volatile oil, are present in the leaves, petiole, bark, wood and pith. The volatile oil contains the crystalline substance, Camphor (see), which is also found in the solid state in cavities in the trunk.

Cinnamon (a food spice) and *cinnamon oil* (used pharmaceutically and in cosmetics) are derived from the young bark of 2 species of *Cinnamomum*, i.e. *C. zeylanicum (= ceylanicum)* ***(Cinnamon tree),*** which is indigenous to Sri-Lanka, and widely cultivated in other tropical countries; and *C. aromaticum*, which grows in southern China.

Laurus nobilis ***(True laurel, Bay*** *or* ***Bay laurel),*** native to the Mediterranean region, is grown as a small, ornamental, evergreen tree, and its leaves (bay leaves) are used for flavoring.

Persea americana ***(Avacado pear)*** is widely cultivated for its edible fruit, which has an oily, fleshy mesocarp.

Laurel: see *Lauraceae.*

Laurus: see *Lauraceae.*

Lavandula: see *Lamiaceae.*

Lavender: see *Lamiaceae.*

Lavender oil: a fragrant essential oil obtained from the flowers of *Lavandula angustifolia.* The main components are linalool, linaloyl acetate, geraniol, borneol, other terpenes and coumarin. It is used in toiletries and medicinal baths; in sufficient quantities it is also a narcotic and antiseptic. See *Lamiaceae.*

Law of diminishing yield: see Mitscherlich's law of plant growth factors.

Law of independent assortment: see Mendel's laws.

Law of independent combination: see Mendel's laws.

Law of minima: an ecological law describing the combined effect of different environmental factors. The competitiveness and efficiency of an organism (e.g. abundance, distribution, activity, etc.) are always limited by the single environmental factor that exists at its pessimum (minimum) (see Ecological valency).

Law of segregation: see Mendel's laws.

Law of the heart: see Starling's law of the heart.

Law of the integrity of gametes: see Mendel's laws.

Layering:

1) In a plant community, the existence of groups of plants of different heights, which therefore form distinct levels or layers (see Synusia), e.g. trees, shrubs, herbs and mosses. See Relevé.

2) A horticultural technique of plant propagation, in which a stolon (e.g. of blackberry, red and black currant, and gooseberry) are fixed to the soil with a staple or peg, to encourage it to put out roots and form a new plant.

3) See Lake.

Layer silicates: see Sorption.

Leaf: a lateral organ on the Stem (see) of higher plants, which undergoes a relatively short period of growth before becoming a permanent part of the plant structure. Ls. are primarily organs of photosynthesis and transpiration.

1) *L. development.* L. primordia arise on the apical meristem (see Stem) in the form of small meristematic protuberances. These rapidly increase in size, showing an early separation into an upper and lower part, which are separated by a constriction near the base. The L. blade or lamina [generally flat with an upper surface and a lower (or under-) surface] develops from the upper part of the primordium. The *L. base,* which attaches the L. to the shoot, develops from the lower part of the primordium. One or two *stipules* often develop laterally from the L. base. In many plants these are brown, membranous structures, which at first protect the shoot bud, but usually fall off at an early stage (e.g. beech and lime). In some cases, e.g. members of the *Polygonaceae,* the stipules take the form of a protective organ, the *ochrea,* which at first totally covers the shoot bud, and then remains behind as a dry collar. In other species (e.g. cherry and pea) the stipules are similar to the actual Ls., but smaller. The stipules of certain plants (e.g. the grass pea vine, *Lathyrus sativus)* are the sites of photosynthesis and transpiration, while the actual Ls. are modified to form Tendrils (see). In some plants (e.g. grasses) the L. stalk or petiole may develop into a L. sheath, which envelops the stem and protects the intercalary meristems, and at the same time increases the rigidity of the halm. The apical or terminal growth of Ls., in contrast to that of the shoot, is usually of only short duration; exceptions to this generalization are, e.g. ferns and palm ferns. Normal L. development is completed with the formation, between the L. base and the lamina, of the *petiole* (L. stalk), which arises from the upper part of the primordium. The petiole transports water and mineral substances to, and assimilated material from, the L. lamina. In addition, the petiole supports the L. and extends the lamina into space and toward the light. The site of attachment of the petiole is called the *insertion.* Ls. without a petiole are referred to as *sessile.* Sessile Ls. are usually attached to the stem by a broad L. base. If the latter envelops the stem, the L. is said to be *amplexicaul.* At the site of insertion, the angle between the stem and the attached L. is called the *axil.*

2) *L. size and shape.* Ls. may be undivided (simple) or divided (compound). Simple Ls. may be ovate, elliptic, lanceolate, linear, reniform, cordate, sagittate, or spear-shaped. Margins of simple Ls. may be entire, crenated, dentate, serrate, or lobed. Deeply lobed forms represent the transition from simple to compound Ls. Compound Ls. may be *digitate* or *pinnate,* and a distinction is drawn between *odd pinnate with a single terminal leaflet* and *abruptly pinnate with two terminal leaflets.* In *bipinnate* Ls., each pinna is also pinnate. L. lamina are penetrated by L. *ribs* or *veins,* often in relief on the underside of the L., and usually lighter in color than the remainder of the lamina. A L. rib or vein consists of vascular bundles enclosed in a sclerenchyma sheath, which serve for material transport and lend rigidity to the L. The totality of vascular bundles, which branch off from the shoot axis into the L., is known as the *L. trace.* Ls. of ferns and dicotyledons as a rule have a *network of veins.* Running down the center of the L. lamina is a well developed midrib, which gives rise to several interconnected lateral veins. Ls. of monocotyledons are of-

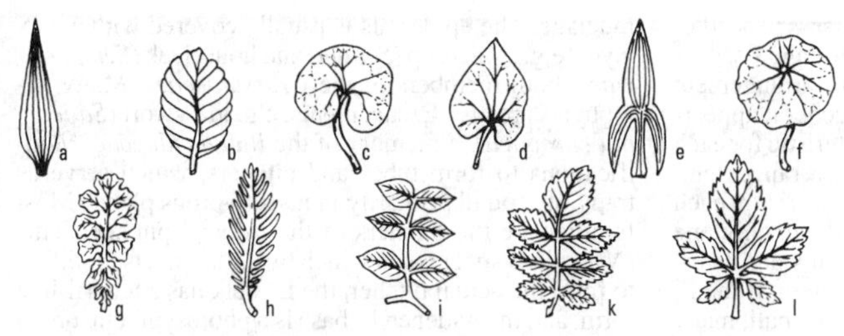

Fig.1. *Leaf.* Leaf shapes. *a* Lanceolate. *b* Elliptic. *c* Reniform. *d* Cordate. *e* Sagittate. *f* Scutiform or peltate. *g* Pinnate lobed. *h* Pinnatifid. *i* Abruptly pinnate or paripinnate. *k* Odd-pinnate or imparipinnate. *l* Palmate.

ten *parallel veined.* The needle-like Ls. of many conifers have only one or two veins.

3) *L. structure.* The upper surface of the L. is covered by an epidermis. The epidermis itself is covered by a cuticle, which protects against water evaporation. Immediately below the epidermis is the actual assimilation tissue; this is a chloroplast-rich *palisade parenchyma,* consisting of one or more layers of elongated, chloroplast-rich cells arranged perpendicular to the L. surface. Adjoining the palisade tissue is the *spongy parenchyma,* consisting of irregular cells with large intercellular spaces. Palisade and spongy parenchyma together form the *mesophyll.* The lower surface of the L. is also bounded by an epidermis, usually strewn with numerous Stomata (see). Directly beneath each stomatum is a large substomatal space, which connects with the intercellular spaces of the mesophyll. Loss of water vapor and the loss and uptake of oxygen and carbon dioxide are regulated by the stomata. Ls. with stomata only on their lower surfaces (as in the majority of plants), are known as *hypostomous* Ls. *Amphistomous* Ls. possess stomata on both upper and lower surfaces. The *epistomous* condition (stomata only on the upper surface) is found, e.g. in the floating Ls. of aquatic plants. Deviations from the L. morphology described above often arise as a result of adaptation to special habitats. Thus, the Ls. of plants of dry habitats may have a multilayered epidermis. Plants of extremely sunny habitats often have palisade parenchyma on both the upper and underside of their Ls. In contrast, immersed Ls. of aquatic plants usually have only spongy parenchyma with well developed intercellular spaces (see Parenchyma) and no palisade parenchyma. Palisade parenchyma is also often lacking in conifer needles; the two surfaces of such Ls. have a similar external appearance as well as a similar internal structure. They are therefore called *equifacial,* whereas Ls. whose mesophyll is separated into dorsal palisade parenchyma and ventral spongy parenchyma are called *bifacial.* If the entire surface of the L. lamina develops only from the lower part of the L. primordium, which grows around the suppressed upper part, the L. is *unifacial.* Unifacial Ls. are as a rule cylindrical in shape (unifacial cylindrical Ls., e.g. chives, onion); they can, however, be secondarily flattened (unifacial flattened Ls., e.g. iris, leek).

4) *L. sequence.* Differently shaped Ls. are formed during the development of a single shoot. The various shapes arise in a definite sequence, known as the *L. sequence.* This includes the juvenile Ls., foliage Ls. and bracts, as well as the sepals, petals, stamens and carpels (see Flower). As a rule, only the foliage Ls. possess the fully developed internal structure and external form described above. All other L. forms are the result of retarded or inhibited development, sometimes leading to considerable changes in shape. The lowest Ls. of the main shoot are the cotyledons (one in moncotyledons, two in dicotyledons, and two or more in most gymnosperms). As a rule, the cotyledons are simpler in shape and structure than the foliage Ls., and they usually persist for only a relatively short time. During seed germination the cotyledons may remain buried in the ground and enclosed in the seed coat (testa) (hypogeal germination, see Seed germination). Cotyledons of hypogeal seeds are either fleshy structures containing depots of reserve material, or they release enzymes which convert the reserve materials of the endosperm into soluble compounds, e.g. starch into sugars, or proteins into amino acids, which are then assimilated by the developing embryo. Alternatively, the extending hypocotyl pushes the cotyledons above the soil surface (epigeal germination, see Seed germination). The cotyledons then turn green, attain the shape of simple, often ovoid foliage L., and serve as photosynthetic organs. They are, however, nonpermanent, and they persist for only a short time. As a rule, the cotyledons are followed by reduced foliage Ls., which are often scale-like and then known as *scale Ls.* (only the L. base is fully developed and the petiole is absent). Intermediate forms between scale and foliage Ls. are quite common, e.g. in the annual growth of perennial herbaceous plants. Subterranean shoots, like rhizomes, runners and tubers possess only small to medium-sized, colorless scale Ls. Finally, the bud scales of all woody plants are modified scale Ls.; in accordance with their protective function they are fairly tough and robust in structure, sometimes colored green but usually brown. The apical or terminal Ls., which follow the foliage Ls., are formed in association with the flowers; they represent regressed Ls., are usually formed from the lower part of the L. primordia, and often functionally modified, becoming the *bracts* of the flower or inflorescence. L. forms intermediate between bracts and foliage Ls. may occur, or the bracts may show transitions into tepals. Strikingly colored bracts occur, e.g. in the *Araceae,* and in the tropical Poinsettia *(Euphorbia pulcherrima).* The bract (bracteole) of the lime tree *(Tilia grandifolia)* is fused (adnate) with the inflorescence. Even after the juvenile phase, different forms of Ls. may occur side by side. In dorsiventral shoots, adjacent Ls. on the upper and lower sides are frequently different in size, even at the same node *(anisophylly),* e.g. the Ls. of the silver fir *(Abies alba)* are arranged in two lateral sets, the lower spreading horizontally, the upper much shorter, pointing upwards and outwards and notched at the apex. In *heterophylly,* the Ls. display a difference in form, e.g. the submerged Ls. of the water crowfoot *(Ranunculus aquatilis)* are finely dissected, while the higher, floating Ls. are laminate; ar-

borvitae or the "tree of life" *(Thuja)* possesses needle-like juvenile Ls. and scale-like mature foliage Ls.

5) *Positioning of Ls. (Phyllotaxis).* The arrangement of Ls. on the shoot axis determines the general appearance or habit of the plant, and it is characteristic for each species. There are two basic patterns of L. arrangement: 1) *opposite* or *whorled,* i.e. two or more Ls. arise at each node, and 2) *alternate, spiral* and *helical,* i.e. only one L. arises at each node. These arrangements are determined to a large extent in the apical meristem. If the L. primordia formed in the meristem are very small, many Ls. can be formed at a single node, and a whorl of many Ls. is produced, e.g. in the horsetails *(Equisetum),* hornworts or coontails *(Ceratophyllum)* and mare's tails *(Hippuris).* The angles between the Ls. of a whorl are usually all the same, so that the Ls. show a uniform distribution around the shoot (equidistance rule). The Ls. of a whorl usually lie opposite the spaces between the Ls. of the next whorl (alternancy rule). In accordance with these rules, the insertions of the Ls. of a two-L. whorl lie exactly opposite one another, and successive pairs are oriented at 90 ° to each other. This arrangement, which is found, e.g. in all members of the *Lamiaceae,* is called alternate-opposite or decussate. If the L. primordia are very large and arise from a broad base on the apical meristem, only a single L. can be formed at each node. The L. arrangement is then disperse. Successive Ls. may then lie on opposite sides of the shoot axis, i.e. a *two-row* or *distichous* arrangement, as displayed by many monocotyledons, e.g. all grasses. If successive Ls. are not exactly opposite each other and the angle between them is less than 180 ° the arrangement is known as *helicoid.* The shortest imaginary line joining the successive Ls. represents a spiral or helix, known as the *basic spiral.* The angle (which is constant in accordance with the equidistance rule) is called the *angle of divergence.* Divergence is usually quoted as a fraction, in which the numerator is the number of revolutions of the stem required to reach the next L. lying directly above, and the denominator is the number of Ls. passed on the way. In general, the L. arrangement of the plant insures optimal illumination of all L. surfaces by minimizing the mutual shadowing caused by overlap.

6) *L. metamorphosis.* Modified Ls. often serve to store reserve material, to trap insects, as defensive organs, or for climbing. Storage Ls. are usually fleshy and thickened. Ls. of the bulbs of many monocotyledons store reserve materials. The bulb consists of a broad, cone-shaped, strongly compressed axis with densely crowded fleshy storage Ls., which overlap each other or form a closed cylinder. Bud scales are either entire scale Ls. or the lower thickened part of foliage Ls. Fleshy thickening of Ls. serves for water storage in L. succulents, which are adapted to dry habitats. Their mesophyll, which is formed as a water tissue, often contains muclage. The epidermis is usually covered with a wax layer [e.g. stonecrop *(Sedum)* and houseleek *(Sempervivum),* both members of the *Crassulaceae].* Many halophytes are also L. succulents, e.g. glasswort *(Salicornia europaea),* a member of the *Ranunculaceae.* Modifications to form tubes and pitchers, which serve as traps, are found primarily in insectivorous plants. Most familiar are the pitchers of the tropical pitcher plants *(Nepenthes* species), in which the L. lamina is modified to form the actual pitcher, the L. stalk has a tendril-like form, and the widened L. base is a photosynthetic organ (see Carnivorous plants). L. thorns frequently arise by modification of the midrib of foliage Ls. or stipules, and they are hard and durable due to lignification or the deposition of collenchyma. L. thorns have a protective function (e.g. against animal predation), and are characteristic of many plants of dry habitats (e.g. cacti), as well as occurring in several other plants (e.g. thistles). In *Berberis,* the midribs of the Ls. are modified to thorns, and they are hardened by an accumulation of sclerenchymatous elements, while the formation of parenchymatous tissue is suppressed. The thorns of locust or false acacia *(Robinia)* are modified stipules. In many climbing plants, all the main L. veins, or only part of the L. lamina are modified to contact-sensitive L. tendrils, which are able to wind around trees, walls and other supports, and raise the plant into the sunlight. The Ls. of some members of the *Cucurbitaceae* have degenerated so that only the midrib remains. In the *Papilionaceae,* the terminal leaflets of the pinnate L. are modified to tendrils. In the Yellow vetchling (*Lathyrus aphaca),* the whole L. is modified to a tendril, and the stipules serve as photosynthetic organs. Certain plants adapted to dry habitats, possess modified L. stalks which have become laminate, photosynthetic organs. In such cases, the true L. lamina is usually reduced. This type of *phyllode* is especially characteristic of species of *Acacia* native to the Australian floral region. The various flower organs (sepals, petals, stamens, carpels) are also modified Ls.

7) *Lifespan of Ls.* Many Ls., e.g. those of most trees and bushes, survive only one vegetation period; the plant is then said to be *deciduous.* In contrast, in *evergreen* plants, the foliage Ls. survive for several years. Before the Ls. of deciduous plants are lost, many L. constituents are degraded and/or exported, followed by formation of a corky separation layer between the L. stalk and the shoot axis. After separation of the L., this corky layer is seen as a *L. scar.*

Leaf beetles, *Chrysomelidae:* a family of the *Coleoptera* (see) containing 25 000 species worldwide. They are mostly less than 10 mm in length (some as small as 1 mm), but larger species measure about 20 mm. Their elytra are often brightly colored, or with a metallic sheen. The dorsal surface is often strongly vaulted, and the head is more or less hidden beneath the neck shield. Both the

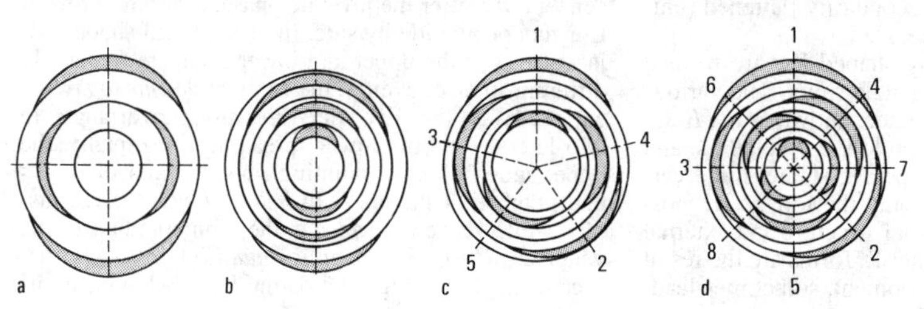

Fig.2. *Leaf.* Types of leaf insertion. *a* Decussate. *b* Distichous. *c* Spiral (2/5) arrangement. *d* Spiral (3/8) arrangement.

imagos and larvae feed on plants. The majority are diurnal. They are found in parks, gardens and fields, all types of wooded and nonwooded grassy areas, as well as marshes and water edges. Most notorious is the North American ***Colorado beetle*** *(Leptinotarsa decemlineata),* which attacks potato plants. Other pests are found in particular in the subfamilies *Halticinae* and *Cassidinae.*

Leaf beetles. Colorado beetle *(Leptinotarsa decemlineata* Say) with larva.

Leafbirds: see *Chloropseidae.*

Leaf cutter ants: see Ants.

Leaf-fingered geckos, *Phyllodactylus:* a large genus of Geckos (see) distributed in tropical America, Africa and Asia as far as Australia. Each of the terminal phalanges of the spread toes carries two large, leaflike, adhesive pads. The European leaf-fingered gecko *(P. europaeus)* is found on Corsica, Sardinia and other Mediterranean islands. Males are velvet black with silver-gray dots.

Leaf-footed bugs: see *Heteroptera.*

Leaf hopper: see *Homoptera.*

Leaf insects: see *Cheleutoptera.*

Leaf litter: see Necromass.

Leaf miners: see Miners.

Leaf mold: a humus of temperate, deciduous, hardwood forests and temperate grasslands. It is characterized by a slow rate of decomposition, usually a weakly acidic reaction, and only minimal mixing with the lower layer of mineral soil. The main agents of decomposition are fungi, accompanied by relatively large arthropods.

Leaf monkeys: see Langurs.

Leaf movements: movements displayed by the foliage leaves of plants. They are triggered by various stimuli (autonomic or induced), and operated by different mechanisms (see Nutation movements, Turgor movements). The unfolding of buds is, for example, an autonomic nutation movement. Autonomic turgor movements are particularly striking, e.g. in wood sorrel, red clover and other clover species. Movements known as Nyctinasties (see) are the result of both autonomic and external factors. Nasties (see) and Tropisms (see) are induced movements. Optimal illumination of leaves is often insured by phototropic movements of the petiole (see Phototropism).

Leaf primordium: see Meristem, Leaf, Stem.

Leaf roller moths: see *Tortricidae.*

Leaf roll viruses: see Virus groups.

Leaf scald of sugar cane: see *Xanthomonas.*

Leaf scorch: see Deficiency diseases.

Leaf scrapers: see *Furnariidae.*

Leaf spot disease of strawberries: see *Xanthomonas.*

Leaf-toed geckos: see *Hemidactylus.*

Leaf warblers: see *Sylviinae.*

Leander: see Shrimps.

Learned avoidance behavior: see Avoidance conditioning.

Learning: a long-lasting modification of behavior brought about by practice and experience (i.e. by the reception, processing and storage of information and signals from the environment), as well as the improvement and perfection of this behavioral change by experience. L. generally represents a relatively prolonged change of behavior on the basis of individual experience. Although rarely stated in definitions of L., it is usually assumed that L. involves changes in the central nervous system. In the *sensory* stage of L., stimuli or patterns of stimuli are perceived before or during the satisfaction of a drive. In the *motor* stage of L., the stimulus is linked to specific appetitive behavior. The L. process is completed by the *conditioning* stage, in which the association of the new releasing and directing stimulus with appetitive behavior is more or less permanently established in the nervous system of the organism. Thus L. can also be defined more simply as "appetitive behavior based on experience", or as "conditioned appetence". Various forms of L. are widespread in all organisms capable of behavior. The ability to learn is genetically determined, and different organisms are genetically disposed to different degrees and types of L. The different types of L. may be classified as follows.

1) *Habituation* (see).

2) *Associative L.* is the sensory linkage of information (a stimulus) with certain patterns of movement; this is also known as *linear L.,* because the crucial determining factor is the time-space proximity of sensory input and movement, and "feedback" is not involved. This type of L. has also been called "classical conditioning", but this is not necessarily equivalent to classical conditioning as defined by Pavlov (see Conditioned reflex). The latter is considered by some authors to be an artificially isolated part of the L. process. Expressed in another way, the conditioned response is often not identical to the unconditioned response, and is often only part of it.

3) *Trial and error L.,* in which the organism tries different, self-conceived methods for achieving a goal, until it meets with success. According to Thorpe's definition, trial and error L. is "the development of an association, as the result of reinforcement during appetitive behavior between a stimulus or a situation and an independent motor action as an item in that behavior when both stimulus and motor action precede the reinforcement and the motor action is not the inevitable inherited response to the reinforcement".

4) *Latent L.,* which is the perception and association of neutral stimuli without reward (or punishment). For example, previous experience of a maze that offers no reward improves the ability of rats to learn their way through the same maze when a reward (food) is offered.

5) *Insight L.,* in which behavior is suddenly modified in response to experience and empirical knowledge. This includes, inter alia, a) *imitation,* i.e. the copying by one animal of otherwise improbable behavior when in

the immediate presence of another animal, and b) *observational L.,* in which one animal observes another animal undergoing a process of L.

Alternatively, L. can simply be divided into 2 types: classical conditioning (see Conditional reflex) and Operant conditioning (see); elements of both are often present in natural L. processes.

As in all forms of conditioning, L. can be *reinforced,* i.e. it can be made more complete and more stable by repetition of the L. process. The obverse of reinforcement is Habituation (see), which may also be considered as a type of L. (above).

Learning sets: experimental protocols designed to test the ability to learn. An animal is first required to solve a problem, and it is then determined whether and in what way this initial experience has affected the ability to solve subsequent problems. For example, using food as a reward, an animal learns that a circle has positive associations, and it learns to distinguish it from a square. In a subsequent experiments, the circle becomes a negative signal while a star is the positive signal; then a rod shape is made positive while the star is negative, etc.; each new discriminatory stage of the experimental procedure is known as an *inversion.* Experiments of this type measure the ability of an animal to "learn how to learn", which is a function of the performance of the central nervous system. If an animal is able to acquire a learning strategy (i.e. retain information from the solution of a problem and apply it to subsequent problems), then its score of mistaken responses decreases as the experiment progresses. Such studies reveal differences in the learning ability of animal taxa and species, i.e. L.s. can be applied to comparative phylogenetic studies.

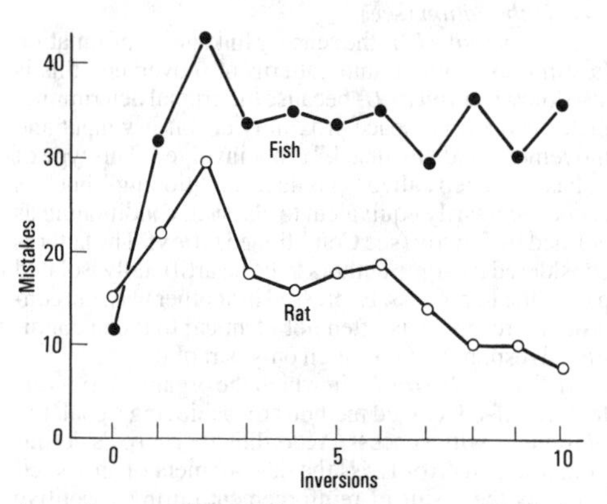

Learning sets. The results of a sequence of exercises in discrimination between two symbols. In each subsequent exercise, one symbol was changed and the sign of the remaining symbol was reversed.

Leatherback sea turtle: see *Dermochelyidae.*

Leatherjackets: see *Balistidae.*

Leccinum: see *Basidiomycetes (Agaricales).*

Lechwe: see *Reduncinae.*

Lecithoepitheliata: see *Turbellaria.*

Lecithus: see Yolk.

Leconora: see Lichens, Lichen pigments.

Lectins, *phytohemagglutinins:* as defined by the Nomenclature Committee of the International Union of Biochemistry, a lectin is a "sugar-binding protein or glycoprotein of nonimmune origin which agglutinates cells and/or precipitates glycoconjugates". This definition may be broadened to include similar proteins which, although they specifically bind complex saccharides, do not precipitate or agglutinate them. Such proteins can be called "monovalent L.". L. are found in almost every major taxon of flowering plants, and in some nonflowering plants as well. Vertebrate and microbial L. have also been identified. L. bind to erythrocytes, leukemia cells, yeasts and several types of bacteria. As the binding is saccharide-specific, L. will not agglutinate cells which do not carry appropriate surface saccharides. They very selectively agglutinate the erythrocytes of different animals, and are therefore useful for the determination of blood groups. It may be expected that as more kinds of surface oligosaccharides are used in screening assays, more L. will be found. Indeed, they may be ubiquitous. L. may account for as much as 10% of the soluble protein in extracts from mature seeds, and they are present in lower concentrations in other plant tissues. L. present in vegetative tissues are often different from the seed L. of the same plant. The physiological function of plant L. is unknown, but they may provide protection against attack by insects, e.g. by interfering with chitin synthesis, or protection against bacterial infections by immobilizing pathogenic bacteria. The ability of L. to stimulate division and differentiation of certain lymphocytes into lymphoblasts (i.e. they display mitogenic activity) is exploited in the *lymphocyte transformation test,* which serves to distinguish between T-lymphocytes (stimulated by P.) and B-lymphocytes (not stimulated by P.). L. are also used in the diagnosis of leukemia cells.

The best known plant L. are concanavalin A (Con A) from jack beans *(Canavalia ensiformis),* which was crystallized in 1919 by Sumner, and the agglutinins from wheat germ (WGA), lima beans, green beans *(Phaseolus),* castor beans *(Ricinus)* and potatoes. WGA is very well characterized, and L. with similar specificites and structure are present in rye and barley. Seeds from 90 other members of the *Triticeae* tribe of the grass family have L. which are immunochemically identical to WGA, although their specificities are not all the same. Within the grass seed, the L. are present only in the embryo. Legume seeds are very rich in L., and complete amino acid sequences are known for a number of them, including Con A. There are extensive homologies among L. from related legumes.

Con A (consisting of 4 identical subunits, each of M_r 27 500) shows maximal sequence similarity with the L. of lentil, soybean and fava bean when its amino terminus is positioned near the middle of the sequences of the other L. It is now known that the primary translation product undergoes transpeptidation, i.e. a new type of protein maturation.

[D.J. Bowles et al. *J. Cell Biol.* ***102*** (1986) 1284–1297; M.E. Etzler, *Ann Rev. Plant Physiol.* ***36*** (1985) 209–234; *Lectins: Biology, Biochemistry, Clinical Biochemistry,* vol ***5,*** de Gruyter, Berlin, 1986]

Ledum: see *Ericaceae.*

Leech: see *Hirudinea.*

Leghemoglobin: see Root nodules, Nitrogen fixation.

Legoglobin: see Root nodules, Nitrogen fixation.

Leguminoseae: see Leguminous plants

Leguminous plants: a group of dicotyledonous

plants formerly regarded as a family *(Leguminoseae)*, with the subfamilies *Mimosaceae* (see), *Caesalpiniaceae* (see) and *Papilionaceae* (see). These former subfamiles are now accorded separate falilial status, and they constitute the order, *Fabales.* All members are characterized by the legume or pod, which is derived from a single superior carpel, originally many seeded and opening both ventricidally and dorsicidally.

Leiocephalus: see Crested keeled lizards.

Leiopelmatidae, *leiopelmatids:* a primitive family of the *Anura* (see), represented by 3 living, terrestrial species in a single genus, *Leiopelma*, found in wet mountain woodlands of New Zealand, where they are the only native anurans. The structure of the vertebrae and the relict tail musculature *(pyriformis* and *caudaliopuboisciotibialis)* indicate great evolutionary age. Vocal sacs are absent. They live independently of free standing water, and the young emerge fully developed from the few large-yolked eggs, which are laid on moist terrestrial sites.

Leiopelmatids are small (2.8–4.7 cm) with 9 ectochordal, presacral vertebrae with cartilaginous invertebral joints and nonimbricate (nonoverlapping) neural arches. Presacrals I and II are not fused, and the atlantal cotyles of presacral I are closely juxtaposed. Free ribs occur on presacrals II–IV and sometimes on presacral V in adults. The diapophysis of the sacrum is narrowly dilated, and is connected to the coccyx by cartilage. The coccyx possesses proximal transverse processes. Both the omosternum and sternum are cartilaginous, and the pectoral girdle is arciferal. Maxillary and premaxillary teeth are present. The pupils are vertically elliptical.

In some classification systems, the closely related *Ascaphus* (see *Ascaphidae)* is subsumed in the *L.*

Leipoa: see *Megapodiidae.*

Leishmanial stage of trypanosomes: see Trypanosomes.

Lek: see *Tetraonidae.*

Lemma: see *Graminae.*

Lemmings: see *Arvicolinae.*

Lemmini: see *Arvicolinae.*

Lemnaceae, *duckweed family:* a family of monocotyledonous plants containing about 25 species, many of them cosmopolitan. They are all very small, free floating or submerged aquatic plants with a strongly reduced, lens-shaped or stalked-lanceolate vegetative body, which is generally called the thallus, and which is variously interpreted as a modified stem or leaf. The unisexual flowers lack a perianth and consist either of a single stamen or a single carpel. Flowers are not always produced regularly and are sometimes rare. Pollination is by water insects. ***Duckweed*** or ***Duck's-meat*** *(Lemna minor)* is cosmopolitan, except for the polar regions and tropics, and it is the commonest European species; each floating thallus is 1.5–4 mm in diameter, usually asymmetrical and rounded at the base, and carries a single root (up to 15 cm in length). The minute inflorescence, consisting of 1 female and 2 male flowers, is borne in a sheath in a pocket in the margin of the thallus. *Lemna minor* often forms a green carpet on the surface of stagnant or slowly flowing, nutrient-rich waters, and it represents an important part of the diet of water birds. ***Great duckweed*** *(Spirodela polyrrhiza = Lemna polyrrhiza)* has a similar distribution to that of *Lemna minor;* each ovate to almost orbicular thallus (5–10 mm in diameter) is often reddish to purplish below and carries a tuft of roots up to 3 cm in length. The smallest flowering plant in the plant kingdom is the ***Rootless*** or ***Dwarf duckweed*** *(Wolffia arrhiza = Lemna arrhiza);* the ovoid to ellipsoid thallus has no roots and is 0.5–1.0 mm in diameter; daughter thalli are commonly produced by budding.

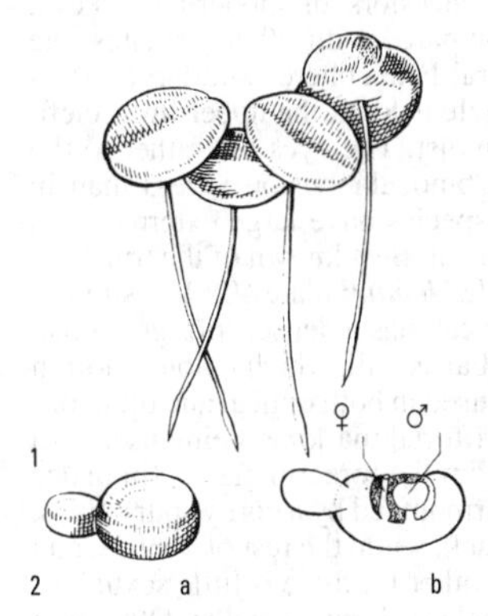

Lemnaceae. 1 Duckweed *(Lemna minor).* 2 Rootless or dwarf duckweed *(Wolffia arrhiza), a* plant, *b* flowering shoot in longitudinal section.

Lemnisci: see *Acanthocephala.*

Lemon: see *Rutaceae.*

Lemon balm: see *Lamiaceae.*

Lemon grass oil: an essential oil with a high citral content. It is obtained from the indigenous Indian grasses, *Cymbopogon flescuosus* (East Indian type, 70–85% citral) and *Cymbopogon citratus* (West Indian type, 53–83% citral). It is used in the manufacture of perfumes and soaps, and as an aroma agent in confectionary and candy.

Lemurs: Prosimians (see) in the superfamily, *Lemuroidea* (see Primates). They are long-tailed, predominantly herbivorous mammals found only in Madagascar (some species are also found on islands adjacent to Madagascar). There are 4 families.

1. Family: *Cheirogaleidae* ***(Mouse L.*** and ***Dwarf L.).*** These have large eyes and a short muzzle. They are entirely nocturnal, and they estivate during the dry season, living on fat reserves accumulated mainly at the base of the tail. Mouse L. (genus *Microcebus)* are found in forested areas, and are represented by 3 species. *M. murinus* ***(Gray lesser mouse lemur)*** is the smallest (head plus body 133 mm, tail 140 mm). It feeds mainly on insects, but other live prey (e.g. chameleons) are also taken. *M. coquereli* ***(Coquerel's mouse lemur)*** from the drier forests of western Madagascar is the largest (head plus body 230 mm, tail 320 mm); it is mainly herbivorous, but also feeds on a sweet exudate from a bug, which it obtains by "milking" the insect. *M. rufus* ***(Brown lesser mouse lemur)*** is of intermediate size.

Dwarf L. are herbivorous. The fat deposit is especially prominent in *Cheirogaleus medius* ***(Fat-tailed dwarf lemur),*** the smaller of the 2 species (head plus body 190 mm, tail 170 mm), which is found in the drier forests of northern, western and southwestern Madagascar. *C. major* ***(Greater dwarf lemur)*** from the tropical rainforest of eastern Madagascar is larger (head plus body 270 mm, tail 250 mm). The ***Hairy-eared dwarf lemur*** *(Allocebus trichotis),* the only species of the genus, is about the same size as *Macrocebus murinus;* it has only been recorded in the eastern forest of Madagascar, and is extremely rare. The ***Fork-marked dwarf lemur***

(Phaner furcifer), the only species of the genus, from northern and western Madagascar, is about the same size as *Cheirogaleus major.* A dark stripe along the spine divides on the head to join the dark eye rings.

2. Family: *Lemuridae* ***(True L.;*** genera *Lemur, Varecia, Hapalemur, Patterus).* This family is thought to closely resemble the ancestors of modern monkeys, apes and humans. Compared with other primates, the brain has small cerebral hemispheres and large olfactory regions. The muzzle is long, the upper lip is cleft, and the rhinarium is moist. The eyes are rather to the side of the head, and binocular vision is less than in other primates. Most species have large external ears. All are arboreal and social. Best known of the true L. is *Lemur catta* ***(Ring-tailed lemur;*** plate 40). It is slightly larger than a domestic cat, has a dense, soft, gray coat, and a black and white barred tail, which it holds aloft in an S-shape. The tail is used in both optical and olfactory recognition, and in territorial marking, being used as a censer to waft scent from the forearm glands (in males the forearm gland is surmounted by a horny spur). Eyes, mouth and nose are black, while the rest of the face and ears are white. Unlike other L., there is little sexual dimorphism, but the female is slightly smaller. Other true L. are *Lemur coronatus* ***(Crowned lemur),*** *L. fulvus* ***(Brown lemur),*** *L. macaco* ***(Black lemur),*** *L. mongoz* ***(Mongoose lemur),*** *L. rubriventer* ***(Red-bellied lemur),*** *Varecia variegata* ***(Ruffed lemur,*** largest of the true L, head plus body 612 mm, tail 612 mm), *Hapalemur griseus* ***(Gray gentle lemur),*** *Hapalemur simus* ***(Broad-nosed gentle lemur).*** *Patterus,* with about 10 species, was recognized recently. *Hapalemur aureus* was first discovered in 1985 in the Hanomafana region of Madagascar. Chromosome analysis confirms its status as a separate species. Only about 500 individuals were thought to exist in 1988, and there are plans to turn their forest habitat into a wildlife reserve.

3. Family: *Lepilemuridae, Megaladapidae.* The ***Sportive lemur*** *(Lepilemur mustelinus)* is the sole extant representative of this family. It inhabits all forested parts of Madagascar, as well as Nosey Be Island, and it is entirely herbivorous. On the basis of chromosome differences, some authorites assign seven species to this genus, but they are barely distinguishable on the basis of anatomy, color, etc. It was formerly grouped with the true L. Extinct members of this family were very large in size, and they survived until a few centuries ago, when they were eliminated by human activity.

4. Family: *Indriidae* (genera *Propithecus, Indri, Avahi).* This family includes the largest of the living L. They are entirely arboreal and exclusively vegetarian. ***Sifakas*** (genus *Propithecus)* are fairly large (head plus body 500 mm, tail 520 mm) with soft, fluffy, predominantly white coats, black or dark brown faces and ears, and dark patch on the head. The muzzle is shorter than that of the true L. They possess a typical prosimian tooth comb, but this consists of only 4 teeth. Their dentition differs fom that of other prosimians, consisting of a total of only 30 teeth instead of 35 (the lower canine is absent, and there is only one upper, and one lower premolar). *Propithecus verreauxi* ***(Verreaux' sifaka)*** inhabits the western deciduous forest of Madagascar, whereas *Propithecus diadema* ***(Diademed sifaka)*** is found in the eastern rainforest. The ***Indri*** *(Indri indri,* a single species) is found in northeastern Madagascar and is the largest of the *Indriidae* (head plus body 700 mm, extremely short tail). Its fur is long, soft and fluffy, with striking black and cream patches. The ***Woolly lemur*** *(Avahi laniger,* a single species) of eastern Madagascar is the smallest indriid (head plus body 300 mm, tail 320 mm); it is nocturnal, whereas the other indriids are diurnal.

Lens:

1) See Light sensory organs.

2) See *Papilionaceae.*

Lens aberrations: defects of lenses resulting in imperfect images.

1. ***Spherical aberration*** occurs when peripheral rays converge at a different point from those passing through the center of the lens. The image of a single point therefore appears as a watery or blurred spot. This defect is corrected by using a combination of a convex lens and a concave lens, but the correction is effective only for monochromatic light.

2. ***Chromatic aberration*** occurs because the blue light in the spectrum is refracted more strongly than the red, leading to images blurred by colored edges. In achromatic lenses, the defect is corrected by using combinations of elements of different types of glass with different refractive indices.

3. ***Curvature of field*** is a defect in which the sharp image produced by the lens lies on a spherical surface. The edge of the image of a thin object is therefore out of focus when the center is in sharp focus, and vice versa. This defect is corrected by combining several elements to product an almost perfectly flat plane of focus.

4. ***Astigmatism*** is caused by radial symmetry in a lens, so that the focal length in one axial plane differs from that in a perpendicular axial plane. It arises in the electron microscope from imperfections in grinding the lens bore, and from inhomogeneities in the polepiece material. Correction is achieved by a weak cylindrical lens of variable azimuth (direction) and amplitude (strength) called a "stigmator", which applies an equal and opposite asymmetry to the electron beam emerging from an astigmatic lens.

Astigmatism of the eye is due to differences in refractive index in the crystalline lens, or to different radii of curvature of the refractive surfaces of the lens; it is corrected with a cylindral glass lens.

Lentibulariaceae: see Carnivorous plants.

Lentil: see *Papilionaceae.*

Leontopodium: see *Compositae.*

Leopard: see *Felidae.*

Leopard moths: moths of the family *Cossidae.* See *Lepidoptera.*

Leopard seal: see *Phocidae.*

Leopardus: see *Felidae.*

Lepadogaster: see *Gobiesocidae.*

Lepas: see *Cirripedia.*

Lepidodendrales: see *Lycopodiae.*

Lepidopleurida: see *Polyplacophora.*

Lepidoptera, ***butterflies and moths:*** a highly evolved insect order with more than 150 000 species, represented in all regions of the world where plants grow as a food source. The earliest fossil *L.* occur in the Baltic amber (Early Tertiary), but their evolutionary line was initiated in the Mesozoic.

Imagoes. The body, which is covered with scales, varies in length from 1 to 60 mm, and the wingspan varies from 3 mm to 30 cm. Adult insects have relatively large compound eyes. Two ocelli may also be present, usually concealed by scales. Antennae are very variable in size and structure. Some very primitive forms *(Micropterygidae)* have chewing mouthparts. Most adult *L.*, however, lack mandibles, and the maxillae are modified as a suctorial proboscis, which is coiled beneath the thorax when not in use. In most species, the prothorax is reduced. All three thoracic segments are joined immovably to each other to form a rigid thorax, which is strongly chitinized. The second and third thoracic segments (meso- and metathorax) each carry a pair of wings, which are covered on their upper- and under-surfaces with overlapping scales. The wings are often decoratively colored and patterned due to the presence of pigmented scales. Alternatively, the scales generate iridescent colors by optical interference, e.g. in the purple emperor *(Apatura iris)*. In flight, the fore- and hindwings are held together by various coupling mechanisms. In the females of some species, the wings are partly or completely regressed, e.g. some geometrid moths (e.g. Winter moth, Mottled umber) and bagworm moths. The abdomen, which is almost always covered with scales and hairs, consists of 10 segments, but only 8 are externally visible in the male, and only 7 in the female; the eighth, ninth and tenth posterior segments are modified in relation to the genitalia. Adults are largely short-lived, feeding on nectar or sometimes not feeding at all. Many species have several generations in one year. Metamorphosis is complete (holometabolous development).

Eggs. The eggs display an extraordinary variety of shape, size, surface structure and color. Up to several thousand eggs are laid on the food substrate of the caterpillar or in its direct vicinity, individually in rows, small clusters or large masses, glued to the substrate with a glandular secretion, and often covered with fine scales. A few species (*Hepialidae* and some *Satyridae*) scatter their eggs randomly in flight.

Larvae (caterpillars). These are mostly cylindrical in shape, and segmented into a strongly chitinized head, three thoracic segments, and eleven abdominal segments, the last three being fused. The integument is smooth or granular; it may be naked (e.g. subterranean or endophagous caterpillars), or it may carry hairs (e.g. processionary caterpillars, tussock moth caterpillars, tiger moth caterpillars), papillae (e.g. giant silkworm) or spurs (e.g. caterpillars of the *Nymphalidae*). The thoracic segments carry three pairs of jointed true legs. The abdominal segments carry unjointed prolegs, varying in number from 8 pairs (*Micropterigidae*) to 1 pair (*Geometridae*); in some bagworms (*Psychidae*), prolegs are totally absent. Most lepidopteran caterpillars, however, possess four pairs of abdominal prolegs on the third to sixth abdominal segments. The penultimate segment also carries a pair of prolegs, which may be modified as claspers. Leaf mining caterpillars often lack legs of any kind. Most caterpillars are plant eaters, and only a few live on animal products (e.g. caterpillars of the *Tineidae* eat wool, keratin etc.). Caterpillars live singly or in groups, free or in spun shelters ("nests"), usually eating the exterior parts of plants (ectophagous); they are more rarely endophagous, i.e. eating into the interior of stems, shoots, roots, fruits, leaves (leaf miners) or forming galls. Depending on the species the larval stage may last between a few weeks and several years. Usually, the caterpillar molts 4 or 5 times.

Pupae. Depending on the freedom of movement of the appendages and body segments, pupae are classified in three types: 1) pupa libera (e.g. *Micropterigidae* and *Eriocraniidae*); 2) pupa incompleta (e.g. *Hepialidae, Cossidae, Aegeriidae, Zygaenidae, Tineidae, Psychidae*); 3) pupa obtecta (all the more highly evolved lepidopteran families). Most pupae possess a hooked caudal extremity (cremaster). For pupation, many caterpillars prepare a loosely woven support of silk or a stout cocoon. Caterpillars of butterflies either attach themselves only by the tip of the abdomen so that the pupa hangs head downward, or they spin an additional silk girdle, which holds the pupa upright. Many caterpillars pupate in the ground.

Economic importance. Caterpillars of several species are agricultural pests (e.g. European corn-borer, white-lined dart, army cutworm, cabbage white), horticultural and garden pests (e.g. yponomeutids, tortricids, winter moths, lackey), forestry pests (e.g. bordered white or pine looper, pine beauty, black arches), pests of stored food products (e.g. European grain moth, Mediterranean flour moth), or of various other commodities (e.g. clothes moths, wax moths). On the other hand, many adult *L.* play an important role in flower polllination; and the silkworm has been exploited economically for centuries.

Classification. Several nonscientific and unnatural divisions are used by collectors, e.g. *Macrolepidoptera* (see), *Microlepidoptera* (see), micro- and macro-moths (see Moths). Division into Butterflies (see) and Moths (see) is widely used but trivial, since these two divisions do not have equal rank. There is still no universally accepted system. According to the most recent research, the *L.* are divided into 3 types based on the number and position of the femal genital apertures: 1) *Exporia:* separate openings for fertilization and egg laying in fused abdominal segments 9 and 10; 2) *Monotrysia:* intestine and genitalia have a common opening (cloaca) on fused abdominal segments 9 and 10; 3) *Ditrysia:* opening for sperm reception on the 8th abdominal segment, and separate opening for oviposition on fused abdominal segments 9 and 10. The ranking of the *Exporia* is not certain, but most authorities agree that it should be placed earlier than the large group of the *Ditrysia.*

Family groups	Families (a selection)
Exporia	*Hepalidae* (Ghost or Swift moths)
Monotrysia	*Micropterigidae* (Micropterigids) *Eriocraniidae* (Eriocraniids) *Nepticulidae* (Nepticulids) *Incurvariidae* (Incurvariids) *Adelidae* (Long-horned moths)
Ditrysia	
Tineformes	*Cossidae* (Goat, Carpenter or Leopard moths) *Zygaenidae* (Burnets) *Aegeriidae* (Clearwing moths) *Tineidae* (True millers) *Psychidae* (Bagworm moths) *Yponomeutidae* (see) *Gelechiidae* (Gelechiids) *Coleophoridae* (Casebearers)
Tortriciformes	*Tortricidae* (see) *Phaloniidae* (see) *Pterophoridae* (Plume moths)

Family groups	Families (a selection)
Pyralidiformes	*Pyralididae* (see) *Geometridae* (see) *Drepanidae* (Hook-tip moths)
Noctuiformes	*Arctiidae* (see) *Lymantriidae* (see) *Noctuidae* (see) *Notodontidae* (Prominents and Puss moths) *Thaumetopoeidae* (Processionary moths)
Bombyciformes	*Lasiocampidae* (Tent caterpillar moths, Eggars and Lappet moths) *Saturniidae* (Giant silkworm or Emperor moths) *Bombycidae* (see)
Sphingiformes	*Sphingidae* (see)
Diurna (butterflies)	*Hesperidae* (Skippers) *Lycaenidae* (Blues, Hairstreaks and Coppers) *Nymphalidae* (see) *Satyridae* (Browns) *Pieridae* (see) *Papilionidae* (Opollo, Birdwing, Swallowtail and Swordtail butterflies)

Lepidorhombus: see *Pleuronectiformes.*

Lepidosaphes ulmi (oystershell scale): see *Homoptera.*

Lepidosiren: see Lungfishes.

Lepidote species: see *Ericaceae.*

Lepidotrichia: see Fins.

Lepidurus: see *Branchiopoda.*

Lepiota: see *Basidiomycetes (Agaricales).*

Lepisma saccharina: see *Zygentoma.*

Lepismatidae: see *Zygentoma.*

Lepisosteiformes: an order of fishes (class *Osteichthyes,* infraclass *Neopterygii),* represented by the single family, *Lepisosteidae* (gars), which contains only one genus, *Lepisosteus.* They are predatory, pike-like fishes with long pointed jaws and needle-like teeth, heavy ganoid scales, and an abbreviated heterocercal tail. The largest species, *Lepisosteus spatula,* attains a length of 3 m. They are distributed in eastern North America, in Central America and in the West Indies south to Cuba and Costa Rica. The northernmost limit of the order (genus) is southern Quebec, which marks the limit of *L. osseus.* They are popular sporting fishes and are locally important as food. The vertebrae are opisthocoelous (convex anteriorly, concave posteriorly) as in some reptiles, and unlike any other fish except the blenny *(Andamia).* Most are freshwater species, inhabiting shallow, weedy areas. Some are occasionally found in brackish water, while *L. tristoechus* is known to enter the sea around Cuba.

Lepomis: see Sunfishes.

Leporidae, *hares, rabbits* and *cottontails:* a family (11 genera) of the *Lagomorpha* (see). Leporids are long-eared, short-tailed mammals with somewhat elongated hindlegs. They are usually accomplished runners, and their diet is vegetarian. The genus *Lepus* ***(True hares)*** contains 26 species with a combined natural original range that includes most of Eurasia as far as Japan, most of Africa, except for the rainforest and the Gulf of Guinea, and most of North America, except for the eastern half. The ***European hare (Lepus europaeus)*** is distributed naturally throughout Europe, extending into the Near East and southward as far as East Africa. It has now been introduced, and is common in, South America, Australia, New Zealand and parts of northeastern USA. The ***Blue, Variable, Mountain*** or ***Alpine hare (Lepus timidus)*** of northern Europe, Asia and North America, as well as the Central European Alps, is usually gray-brown in summer, with a white coat and black ear tips in winter. ***Snowshow rabbits*** or ***Arctic hares (Lepus americanus)*** inhabit the evergreen forests and forested swamplands of North America. The slender-bodied, big-eared ***Jack rabbits*** of western North America are actually hares of the genus *Lepus.* The ***European wild rabbit (Oryctolagus cuniculus),*** originally indigenous only to Spain and Northwest Africa, has been introduced by humans into many parts of Europe, North America, Chile and various islands, and into Australia where it has become a major pest. The various types of domestic rabbit were derived from the wild rabbit by selection and breeding. In trivial names, some true hares are mistakedly called rabbits (e.g. Jack rabbits). There are, however, fundamental differences between *Oryctolagus* and *Lepus.* Thus, *Lepus* has a short palate, the postorbital process is broad and triangular, and the sutures of the interparietal bone are fused in the adult; these characteristics are opposite in *Oryctolagus.* Moreover, the ears, limbs and tail of the rabbit are shorter than those of true hares, and unlike true hares, the rabbit excavates a subterranean burrow. Young rabbits are born naked and blind in a burrow, whereas hares are born above ground, fully furred, and with their eyes open.

The genus, *Sylvilagus,* contains 13 species of ***Cottontails*** (e.g. the ***Eastern cottontail,*** *Sylvilagus floridarus)* with a combined range from southern Canada to Argentina, as well as the ***Swamp rabbit*** *(Sylvilagus aquaticus)* found in the Mississippi basin. Cottontails do not excavate burrows, but may use burrows abandoned by other animals.

The remaining genera are: *Brachylagus* (monotypic, *B. idahoensis,* the ***Little Idaho rabbit*** or ***Pigmy rabbit;*** restricted to the basin between Cascades, the Sierra Nevada and the Rocky Mountains); *Bunolagus* (monotypic, *B. monticularis,* ***Rock hare;*** South Africa; resembles *Pronolagus* externally, but the skull has rabbit-like characteristics); *Poelagus* (monotypic, *P. marjorita,* ***Scrub*** or ***Uganda hare;*** southern Sudan, northeastern Zaire, northwest Uganda); *Nesolagus* (monotypic, *N. netscheri,* ***Sumatran hare;*** forested mountains of Sumatra); *Pentalagus* (monotypic, *P. furnessi,* ***Ryukyu rabbit;*** Ryukyu islands south of Japan); *Pronolagus* (3 species; ***Red hares;*** southern Africa); *Romerolagus* (monotypic, *R. diazi,* ***Volcano rabbit;*** a relict form occurring only at 300–3 600 m on the slopes of volcanos 50 km southeast of Mexico city); *Caprolagus* (monotypic, *C. hispidus,* ***Bristly*** or ***Assam rabbit;*** foothills of the Himalayas).

Leprosy: see *Actinomycetales.*

Leptailurus: see *Felidae.*

Leptinotarsa: see Leaf beetles.

Leptodactylidae, *leptodactylids:* a family of the *Anura* (see) containing 710 living species in 53 genera and 4 subfamilies (*Leptodactylinae*, *Ceratophryinae*,

Telmatobiinae, *Hylodinae*), distributed throughout South America, southern North America and the West Indies.

Leptodactylids occupy a wide range of different habitats and they display a correspondingly diverse range of size, structure and superficial appearance. All possess 8 presacral vertebrae with procoelous centra, which are holochordal in most species, but stegochordal in some talmatobiins. In ceratophryins and some telmatobiins and leptodactylins the neural arches are imbricate (overlapping), but nonimbricate in hylodins, some telmatobiins and most leptodactylins. Presacrals I and II are separate in all genera, except in *Telmatobufo* where they are fused. In ceratophryins and some telmatobiins the atlantal cotyles of presacral I are closely juxtaposed, but widely separated in hylodins, leptodactylins and other telmatobiins. Ribs are absent. The coccyx lacks proximal transverse processes (except in some telmatobiins), and it has a bicondylar articulation with the sacrum, which has rounded diapophyses. In most species the pectoral girdle is arciferal, but it is pseudofirmisternal in *Insuetophrynus*, *Sminthillus* and *Phrynopus peruvianus*. When present, the omosternum is cartilaginous. An omosternum is lacking in *Lepidobatrachus*, *Macrogenioglottus*, *Odontophrynus*, *Proceratophrys* and some species of *Eleutherodactylus*. A cartilaginous sternum is present in all species. In most genera the pupils are horizontal, but they are vertically elliptical in *Caudiverbera*, *Hydrolaetare*, *Hyloria*, *Lepidobatrachus*, *Limnomedusa* and *Telmatobufo*.

The 50 living species of *Leptodactylus* (***Nest-building frogs***; subfamily *Leptodactylinae*) are widely distributed from southern North America through Central America and far into South America. Ecologically, they are the counterparts of the green frogs or aquatic frogs (see *Raninae*) of the Old World. Leptodactylins, however, build foam nests on or near to water. The emerging tadpoles then wriggle their way to the water, except those of *Leptodactylus marmoratus*, which complete their development independently of free standing water inside the terrestrial foam nest. The largest species, *Leptodactylus pentadactylus* (***South American bullfrog***), which is eaten in many parts of South America, is darkly mottled green and red brown, and attains lengths greater than 20 cm. When breeding, the limbs of the male become bright red-orange. The forelimbs of the male are hypertrophied and the humerus carries broad flanges for attachment of the extra muscle mass. Breeding males also have a thorn on each prepollex, which fits into a double ridged horny structure on the female. The tadpoles of *L. pentadactylus* are predators on other species of tadpoles.

The 398 living species of *Eleutherodactylus* (***Robber frogs***; subfamily *Telmatobiinae*) bear a certain resemblance to tree frogs. They are small and mostly inconspicuously colored. Inhabiting moist forest floors in Central and South America, species of *Eleutherodactylus* are remarkably independent of free standing water. Only a few eggs are laid at a time on the ground, and development is direct. At least one species, *E. jasperi*, is viviparous, and at least 2 (*E. coqui* and *E. jasperi*) have internal fertilization. *E. ricordii* (***Greenhouse frog***), originally from Cuba, has been introduced into other Caribbean islands, Florida and South America, where it is common in agricultural areas. At the completion of development, the young adult releases itself from the egg casing with the aid of an egg tooth.

Ceratophrys (***Wide-mouthed toads***, ***Horned toads***; subfamily *Ceratophryinae*) is represented by 6 species, which are distributed discontinuously in tropical and subtropical South America. They are usually vividly colored with large heads and horn-like protuberances above the prominent eyes. Water is used only for mating and egg laying. They habitually bury themselves up to the head in the earth. Some species have very large mouths and several rows of teeth, which are used as weapons, as well as being an adaptation to a cannibalistic diet; *Ceratophrys dorsata* can inflict a serious wound on the hand of a human. The 12 cm-long *Ceratophrys ornata* (***Ornate***, or ***Painted horned frog***) is bright green above and yellowish below, with extensive black and red mottling; its tadpoles feed on the tadpoles of other frogs. The adult has enlarged dagger-like teeth in the upper jaw. *Ceratophrys cornuta* (***Brazilian horned toad***) derives its name from horn-like projections of skin on its upper eyelids.

Leptodeira: a genus of opisthoglyptic, only weakly venomous, egg-laying colubrid snakes, subfamily *Boiginae,* found in northern South America and as far north as the southern USA. They live in trees and bushes, and they are nocturnal in habit. The dorsal surface is red-brown with dark rings and spots. Certain species, especially the ***Cat-eyed snake*** (*L. annulata*), sometimes reach Europe in consignments of bananas, and are therefore known as "banana snakes".

Leptogorgia: see *Gorgonaria.*

Leptolepiformes (Greek: *leptos* fine or thin; *lepis* scale): an extinct order of the oldest known primitive teleosts (see *Osteichthyes*), whose fossils are found from the Upper Triassic to the Middle Cretaceous. They closely resembled the Holosteans in their possession of a cephalic lateral line, and in the presence of a ganoid-like (enamel-like) coating on some dermal bones and scales; the latter probably evolved from more primitively structured precursors. Fossils of the slender, sprat- to herring-sized, swarming members of the genus *Leptolepis* occur from the Liassic to the Lower Cretaceous. *Anaethalion sprattiformis* is common in the Solnhofen limestone (Württemberg).

Leptolepis: see *Leptolepiformes.*

Leptome: the phloem of a vascular bundle, excludung any accompanying mechanical tissue.

Leptomedusae: see *Hydroida.*

Leptomicrurus: see Coral snakes.

Leptomonad stage of trypanosomes: see Trypanosomes.

Leptopterus: see *Vangidae.*

Leptoptilos: see *Ciconiiformes.*

Leptosomatidae: see *Coraciiformes.*

Leptosomus: see *Coraciiformes.*

Leptospira: see Spirochetes.

Leptospiroses: see Spirochetes.

Leptostraca: an order of the *Crustacea,* subclass *Malacostraca,* represented by about 25 small, exclusively marine species, varying in length from 0.6 to 4 cm. The body is elongated, and the anterior is enclosed in a large, 2-valved carapace, with a small anterior hinged rostral plate covering the head. They differ from other malacostracans in having 8 abdominal segments (7 segments plus the telson), instead of 7 (6 plus the telson). The 8 pairs of similar, foliaceous, biramous thoracic appendages produce a water current, from which they filter food particles and detritus; filtered material is transferred to a ventral groove and moved forward to the mouth. Almost all species live on the sea bottom, where they disturb the mud then filter the suspension. Eggs are carried for a relatively long time on and between the thoracic appendages of the female; development is direct, each egg hatching as a post-larva called a manca stage, which has an incompletely developed carapace. The most familiar genus is *Nebalia.* Leptostracans appear to be the most primitive of the malacotracans.

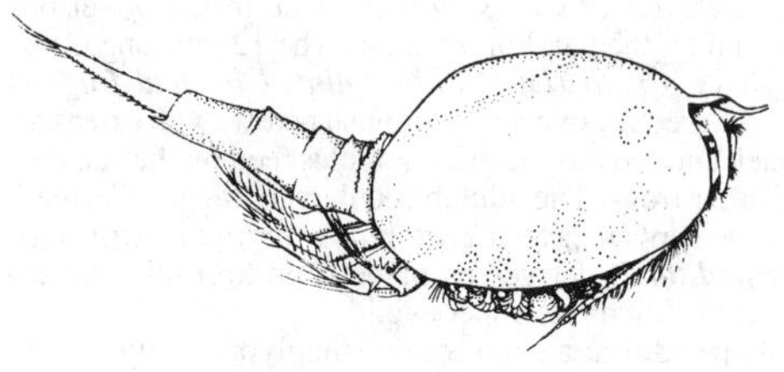

Leptostraca. Female of *Nebalia extrema* carrying eggs between its phyllopodia.

Leptotene: see Meiosis.

Leptothrix: see *Chlamydobacteriaceae.*

Leptotyphlopidae, *slender blind snakes:* A family of ground-burrowing snakes (15–30 cm in length), closely related to the *Typhlopidae* (see). They are found mainly in the tropics of Africa and Arabia, with just a few species in subtropical and tropical America. The family is represented mostly by one genus, *Leptotyphlos,* which contains about 40 species. Remnants of the pelvic girdle are present, consisting of 2 or 3 small bones. In contrast to the *Typhlopidae,* remnants of the femur are also still present. The upper jaw is toothless, while the lower jaw has powerful teeth. They feed mainly on termites.

Lepus: see *Leporidae.*

Lernaeopoda: see *Copepoda.*

Lessepian migration: migration between the Red Sea and the Eastern Mediterranean though the Suez Canal. It is named for Ferdinand de Lesseps, who designed the Suez Canal and who was responsible for pioneering work on the Panama Canal. A potential for L.m. exists at the isthmus of Panama. The locks of the Panama Canal are served by fresh water, which forms a barrier to the migration of marine life. The use of sea water (which is planned) could lead to L.m. between the Pacific and the Caribbean/Atlantic.

Lesser panda: see Red panda.

Lesser weever: see *Trachinidae.*

Lestidae: see *Odonatoptera.*

Lethal concentration, *LC:* the concentration of an active substance that causes death within a specific time interval. LC50 is the concentration that causes the death of 50% of experimental animals in a given time interval.

Lethal dose, *LD:* the dose of a pathogen or toxic substance that causes death in a test organism (animal or plant). It can be measured as the 50% lethal dose (LD50) or as the minimal lethal dose (MLD).

1) ***50% lethal dose (LD50).*** This is the dose that kills 50% of test organisms within a specified time. LD50 is usually determined from the linear mid-section of the dose-response curve (% deaths between 25% and 75% in the test organism population plotted against the logarithms of the doses).

2) ***Minimal lethal dose (MLD).*** The MLD is the dose that just kills the test organism or all of several test organisms within a specified time after administration. Individual test organisms may show differences in susceptibility, so that the accuracy of the determination depends on the number of organisms used. For this reason, LD50 is a more significant parameter.

Lethal factors: dominant or recessive genes which cause death of the carrier individual at a specific stage of development (effective lethal phase) and before the attainment of sexual maturity. According to their degree of Penetrance (see), L.f. can be classified into 1) L.f. in the narrow sense, which always lead to the death of their carriers, 2) semilethal factors which cause the death of more the 50% of their carriers, and 3) subvital factors which cause the death of no more than 50% of their carriers. According to the developmental phase in which they act, L.f. are gametic, gonadic or haplophasic and zygotic. They are also classified according to their chromosomal location, i.e. autosomal or sex chromosome-linked L.f.

Lettuce: see *Compositae.*

Leucaspius: see Moderlieschen.

Leucifer: see Shrimps.

Leuciscus: see Chub, Dace.

Leucemia: alternative spelling of leukemia (see Cancer cell).

Leuciana: see Cutworms.

L-Leucine, *Leu:* $(CH_3)_2CH—CH_2—CH(NH_2)—COOH$, a proteogenic, essential, ketogenic, aliphatic, neutral amino acid. It is widely distributed in proteins, and is particularly abundant in serum albumins and globulins, in keratin and in milk proteins.

Leucobryum: see *Musci.*

Leucojum: see *Amaryllidaceae*

Leucon: see *Porifera,* Digestive system.

Leuconostoc: a genus of nonmotile Cocci (see), 2 mm in diameter, which occur in short chains. They belong to the Streptococci (see), but differ from the genus *Streptococcus* by forming carbon dioxide from glucose. In carbohydrate-rich media they synthesize masses of gummy polysaccharide (dextran). *Leuconostoc mesenteroides,* a common species, is sometimes responsible for spoilage in the sugar industry; sugar solutions infected with this organism contain masses of gummy dextran, which renders them unfit for sugar production.

Leucosticte: see *Fringillidae.*

Leucovorin, *citrovorum factor:* a derivative of 5,6,7,8-tetrahydrofolic acid with a formyl group on N5. It is a growth factor for the bacterium, *Leuconostoc citrovorum.*

Leukemia: see Cancer cell.

Leukotrienes: Lipid hormones derived from arachidonic acid. Structurally, they are closely related to the Prostaglandins (see). All L. contain 3 conjugated double bonds; many also contain a fourth double bond, but this is not conjugated. L. mediate immune hypersensitivity (e.g. anaphylaxis, branchial asthma), and they are identical to the "slow reacting substances of anaphylaxis". All L. cause powerful contraction of the lungs (hundreds to thousands of times more potent than histamines), and they stimulate the release of thromboxanes and prostaglandins. Leukocytes are a major source of L.

Levantine sponge: see Bath sponges.

L-form: an atypical growth form of bacteria, characterized by unregulated longitudinal and transverse growth. L-fs. were named after the Lister Institute in London, where they were first studied. In L-fs. the cell wall is absent or greatly reduced. On semisolid growth media their colonies are shaped like "fried eggs", i.e. deeper in the centre than around the periphery. After extensive subculturing in the laboratory, some bacteria spontaneously form L-fs., and L-fs. can be induced with certain factors, e.g. penicillin. According to whether or not they are capable of reverting to the normal form with a cell wall, they are referred to as stable or unstable L-fs.

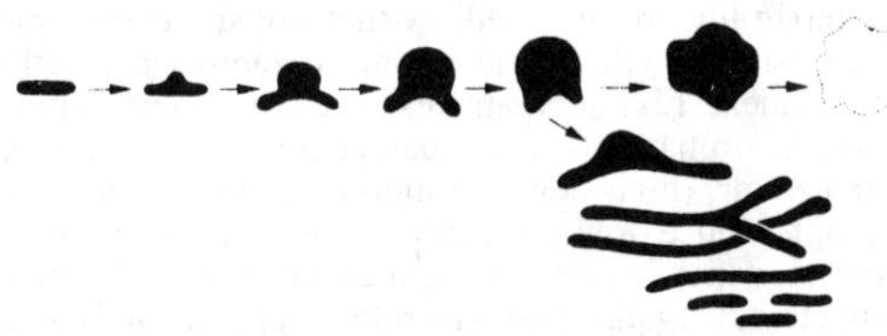

L-Form. Stages in the formation of L-forms and their reversion to the rod form.

Lialis: see *Pygopodidae.*

Lianes: see Climbing plants.

Lias, *Liassic:* the lowest stage of the Jurassic (see). In Britain and France it contains extensive outcrops of fossil-containing blue-gray shales and muddy limestones.

Libellulidae: see *Odonatoptera.*

Liberins: see Releasing hormones.

Licareol: see Linalool.

Lice: see *Phthiraptera.*

Lichen acids: phenolic carboxylic acids present in many lichens, e.g. orsellinic acid (4,6-dihydroxy-*o*-toluic acid). They often occur as their depsides. F.a. are chelating agents, and their ability to bind e.g. trace elements enables the lichen to survive on barren rock surfaces.

Lichenes: see Lichens.

Lichenin, *moss starch:* a polysaccharide (M_r 25 000–30 000) composed of 150–200 D-glucose units linked mainly by β1,4 linkages, with about 25% β1,3 linkages distributed at random in the molecule. It serves as both a structural and reserve material in many lichens, and has been shown to possess antineoplastic properties.

Lichen pigments: pigments or dyes prepared from certain lichens. Many L.p. (e.g. litmus and orcein) are derivatives of orcinol, which occurs free and combined in many lichens of the genera *Roccella* and *Leconora.* Orcein, formed by exposure of orcinol to air in ammoniacal solution, is a reddish brown compound which forms the chief constituent of the dye orseille (French purple). See Lichens.

Lichens, *Lichenes* (plate 33): a division of the plant kingdom containing about 20 000 species. A lichen is a morphological and physiological unit formed by the symbiotic association of a fungus with a member of the *Chlorophyceae* (eukaryotic green algae), or a member of the *Cyanophyceae* (prokarytoic cyanobacteria). The fungus is usually an ascomycete, more rarely a basidiomycete. The form of a lichen is generally determined by the fungus, less often by the nonfungal component. The nonfungal component is either uniformly distributed throughout the thallus *(homeomerous structure),* or it is restricted to a layer (formerly called the *gonidial layer)* between the upper cortical layer and the internal medullary region of the lichen (heteromerous structure). Development of the typical lichen shape and habit requires the presence of both partners, although artificially cultured lichen fungi do occasionally form lichen-like structures. Lichen *Cyanophyceae* or *Chlorophyceae* can also be grown in pure culture, so that in principle, L. can be generated synthetically. Some L. contain a second nonfungal component which is quite unrelated to the first. A second fungus may also be present, and this is generally thought to be parasitic. L. can withstand prolonged periods of drought, and when water again becomes available it is rapidly taken up by specialized hyphae of the fungus. Nutrient uptake from the environment is predominantly the function of the fungal partner, which also receives carbohydrates and other organic compounds from the photosynthesizing nonfungal partner. Usually the nonfungal partner is surrounded by hyphal fungi, or it is actually penetrated by fungal haustoria.

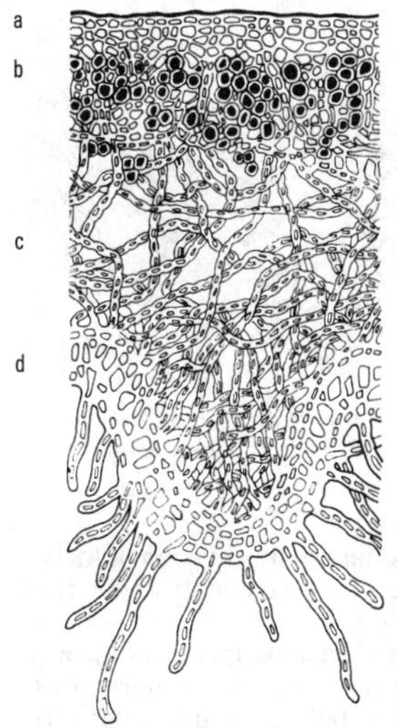

Fig.1. *Lichens.* Transverse section of a lichen thallus with heteromerous structure. *a* Upper cortical layer. *b* Algal layer. *c* Medullary region. *d* Lower cortical layer with rhizoids.

There are several morphological types of L. The slow growing *crustose L.* live as incrustations on rocks, earth, or bark, firmly attached to, or even penetrating the substratum. *Fruticose L.* display shrub-like branching and are attached to the substratum by a narrow base. *Foliose L.* are generally flat and lobed and attached to the substratum by hyphal strands (rhizinae). The habit of fruticose and foliose L. is reminiscent of a higher plant. In *gelatinous* and in *filamentous L.* the form of the colony is dominated by the nonfungal partner.

F. characteristically produce certain metabolites, such as Lichen pigments (see) and Lichen acids (see), which are formed only when the two symbionts are in partnership, and not produced by either partner in isolation. The pigments usually occur as crystals on the outer surface of the fungal hyphae, and they impart a characteristic color to many L.

Lichen *Cyanophyceae* and *Chlorophyceae* reproduce only vegetatively, whereas the fungus develops fruiting bodies, the hymenium usually being free from the nonfungal partner. Therefore, a new lichen can develop only if a fungal spore germinates near to a cell of an appropriate nonfungal partner. In a few cases, nonfungal cells are carried with the fungal spores when the latter are discharged. Most foliose and fruticose L. display purely vegetative reproduction, which is achieved by means of wind-dispersed soredia, i.e. groups of nonfungal cells entwined by fungal hyphae. The thalli of some species produce cylindrical or coralloid vegetative propagules, known as isidia. In addition, a fragment of any lichen can grow into a larger thallus.

L. often colonize sites that are unavailable to other plants, e.g. tree bark and rock surfaces, although some are also found on the soil surface. F. are often extensively involved in the degradation of rock surfaces, and therefore with the beginning of soil formation. Foliose and fruticose L. are found predominantly in the colder zones (Arctic tundra and high mountains), while crustose L. also occur in the tropics. L. are almost totally absent from the stone surfaces of cities, where their growth is inhibited by low humidity and atmospheric pollution.

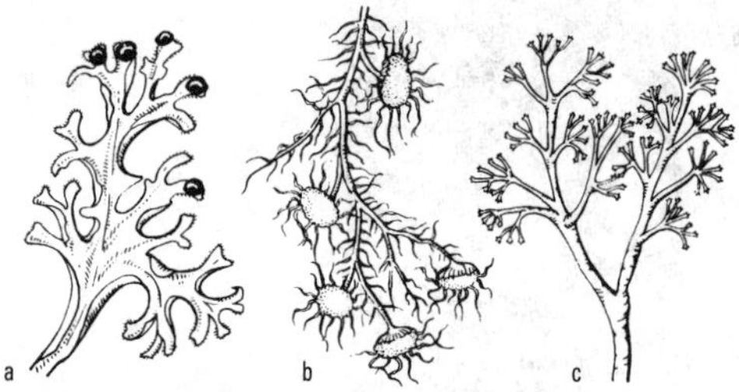

Fig.2. *Lichens. a* Iceland moss *(Cetraria islandica). b* Bearded lichen *(Usnea). c* Reindeer moss *(Cladonia rangiferina).*

Classification is uncertain. The division into two classses and several orders is based mainly on the fructification of the fungal partner.

1. Class: *Ascolichenes* (fungal partner is an ascomycete). Most L. belong to this class. The majority of L. containing *Chlorophyceae* belong to the order *Cyanophilales* of this class. Different species of *Roccella* occur primarily on the Mediterranean coast and are the sources of the dyes litmus and orseille (French purple) (see Lichen pigments). *Lobaria pulmonaria* was formerly used as a remedy for lung complaints. *Rhizocarpon geographicum* forms large yellow-green incrustations on rocks in mountainous regions. Most *Cladonia* species, e.g. *Cladonia rangifera* ("reindeer moss") grow in the Arctic and sub-Arctic tundra, where they form a predominant and characteristic component of the vegetation and serve as the staple food of reindeer. *Lecanora esculenta* (manna lichen), native to the North African and Oriental steppes, is eaten by humans. *Cetraria islandica* (Iceland moss), a fruticose lichen of northern regions, is official and was long used medicinally as a demulcent. *Usnea* (bearded L.) is represented worldwide by about 500 species, which produce the antibiotic, usnic acid.

2. Class: *Basidiolichenes* (fungal partner is a basidiomycete). This class contains only a few genera and species, found mostly in the tropics. Known associations are *Thelephoraceae/Cyanophyceae, Clavariaceae/Chlorophyceae* and *Agaricaceae/Chlorophyceae.*

Lichen woodlands: sparse conifer woodland with ground cover dominated by lichens (e.g. *Cladonia).* L.w. are found on the northern fringes of the European and Siberian Boreal forest.

Licking and lapping feeders: see Modes of nutrition.

Licking sickness: see Deficiency diseases.

Lidless skinks: see *Ablepharus.*

Life: a specific type of organization and function displayed by certain combinations of substance; the mode of existence of living organisms. Phenomena that characterize life are intermediary metabolism, energy metabolism, growth, reproduction, excitability and active movement. Living organisms differ from other natural objects primarily by their high degree of structural organization (high negative entropy), comprizing both simple and extremely complex molecules (biomolecules), which in turn is associated with a high degree of functional organization. The rules of physics and chemistry apply to living organisms, although their interpretation is often difficult, owing to the great variety and interrelated nature of living phenomena. All living organisms not only possess many different proteins and nucleic acids [different molecules of ribonucleic acid (RNA) and deoxyribonucleic acid (DNA)], but are also able to synthesize all of them autonomously. One outstanding property that sets L. apart from all other natural phenomena is the ability of living organisms to continually reproduce themselves, made possible by the self-replication of genetic material (DNA and viral RNA). Since a living cell possesses all the characteristics that also define a living organism, the cell can be considered as the fundamental structural and functional unit of L. Some organisms consist of a single cell (unicellular organisms), whereas others consist of numerous cells, specifically organized into organs and tissues (i.e. the cells display a division of labor). Living phenomena can only be understood and interpreted in terms of the organization of the whole organism and the structural and functional interrelationships of its parts. However, the phenomenon of L. embraces not only the individual organism, but the interaction of organisms with each other and with their nonliving environment. Living organisms are part of a hierarchical system (organism, population, biocoenosis, biosphere) with interactions between the different levels, all levels being in dynamic equilibrium, as are the components of an individual organism.

Although much knowledge has been gained by the reductionist approach to the study of isolated living phenomena, the true aim of Biology (see) is to understand the organism as a whole and as an integrated part of its environment. This holistic understanding is becoming increasingly important, both practically and theoretically.

Closely associated with the problem of the nature of L. is the question of its origin (see Abiogenesis). The materal basis of this process or event, i.e. the evolution of L. from inanimate material, is not in doubt, but the mechanism of the process is a subject of different and conflicting theories.

See Vitalism.

Life form, ***growth form, functional phenotype:***

1) PLANTS: the structure, shape and growth habit of a plant, resulting from adaptation to environmental conditions. Since survival depends on adaptation to the environment, phylogenetically unrelated plants occupying similar ecological niches may have similar L.fs. In the L.f. classification of Rauniaer, plants are grouped according to the length of life of the shoots, together with the position of resting buds and the way they are protected during the unfavorable season (winter dormancy period, summer dry period).

1) Plants rooted in the soil (German *Radikante,* "radicants").

a) *Macrophanerophytes.* Large woody plants with resting buds more than 2 m above the soil surface, e.g. evergreen and deciduous trees, bamboo, palms, tree ferns, etc.

b) *Nanophanerophytes.* Bush-like plants whose resting buds are more than 25–30 cm and less than 2m above the soil surface, e.g. many large, upright, herbaceous plants, small shrubs, etc.

c) *Chamaephytes.* Plants with resting buds very near to, and not further than, 50 cm from the soil surface, e.g. ericaceous plants of oceanic heaths *(Calluna, Erica),* cushion plants and xeromorphic colonizers of desert regions. In northern tundra and Alpine regions, the prostrate and creeping chamaephytes are often protected from frost damage by an insulating covering of snow. In high Alpine and polar regions, cushion-form chamaephytes are the predominant type of vegetation.

d) *Hemicryptophytes.* Perennial and biannual plants with aerial shoots, whose resting buds lie directly on the soil surface, protected during the infavorable season by other organs such as living or dead leaves, e.g. tufted plants such as grasses, rosette plants *(Taraxacum, Plantago, Beta),* plants whose renewal buds lie at the bases of dead stems *(Artemisia, Urtica, Lysimachia vulgaris),* and perennial stoloniferous herbs such as *Fragaria vesca, Potentilla reptans* and *Ranunculus repens.* More than one half of the species in the temperate zone are hemicryptophytes, rising to 60–70% in the northern tundra and on high mountains.

e) *Cryptophytes* (or *Geophytes).* Plants whose resting buds lie beneath the soil surface, on or in storage organs, e.g. root tubers, stem tubers, bulbs, rhizomes.

f) *Therophytes.* Annual, short-lived plants, whose resistant seeds survive the unfavorable season. With the onset of the unfavorable season, the entire plant body dies, and no resting buds or subterranean storage organs are formed. Many field crops, e.g. summer cereals and

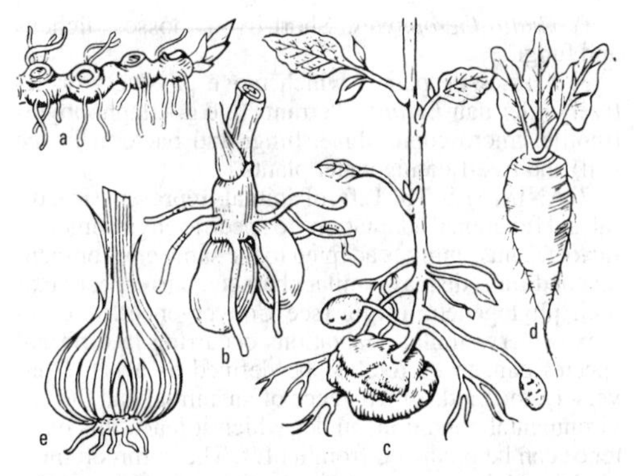

Fig.1. *Life forms.* Subterranean overwintering organs of plants. *a* Rhizome (Solomon's seal, *Polygonatum). b* Root tuber (orchid, *Orchis). c* Stem tuber (potato, *Solanum tuberosum). d* Swollen tap root (sugar beet, *Beta saccharifera). e* Bulb (garden onion, *Allium cepa).*

their seeds, are therophytes. Winter annuals or overwintering annuals, e.g. autumn-sown cereals, which germinate in the autumn and overwinter as seedlings, may be regarded as therophytes or hemicryptophytes. Certain therophytic rosette plants may also occasionally survive the winter as overwintering annuals.

2) Plants which do not root in the soil (German *Adnate,* "adnates" or "adherants"). These plants are attached to stones, bark, etc., or they may be non-root-forming plants in the soil.

a) *Cormo-epiphytes.* Vascular epiphytes; vascular plants attached to other plants above ground level.

b) *Thallo-epiphytes.* Lower plants, e.g. mosses, lichens, fungi, algae.

c) *Thallo-chamaephytes.* Lower plants with resting buds near to the substrate, e.g. cushion mosses, tussock mosses, carpet mosses, fruticose lichens.

d) *Thallo-hemicryptophytes.* Plants very close to the substrate such as mosses, foliate lichens and encrusting lichens.

e) *Thallo-cryptophytes (Thallo-geophytes).* Soil-dwelling fungi, as well as fungi, lichens and algae found among stones.

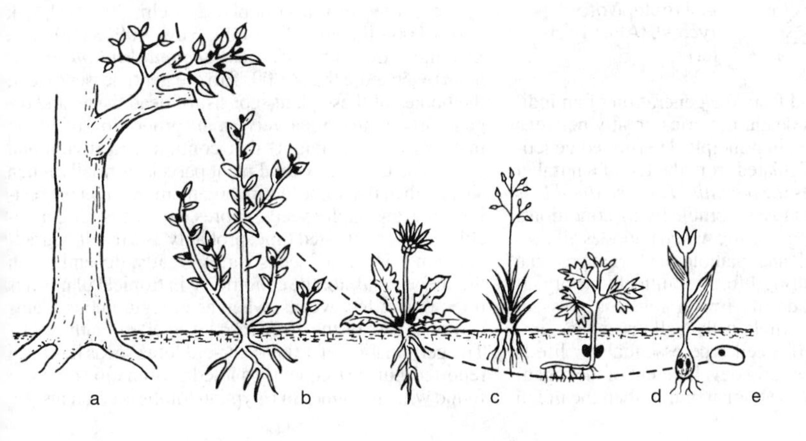

Fig.2. *Life forms.* The dotted line shows the position of the resting buds or overwintering structures. *a* Phanerophyte. *b* Chamaephyte. *c* Hemicryptophytes. *d* Cryptophytes (or Geophytes). *e* Therophyte.

f) *Thallo-therophytes.* Short-lived mosses, lichens and fungi.

3) Nonsessile plants, which move passively or actively (German *Errante,* "errants"), e.g. edaphophytes (motile, microscopic algae, fungi and bacteria in the soil) and free-floating water plants.

2) ANIMALS. The L.fs. of animals represent structural and functional adaptations to specific environmental factors. Thus, animals adapted to the same environmental conditions display similar characters, irrespective of their phylogenetic origin (see Convergence). A complex of very similar adaptations occurring in different species can, in its totality, be defined as a L.f. Conversely, the mode of existence of an animal and the environmental conditions under which it leads this existence can be predicted from its L.f. The common morphological features of a L.f. are often referred to as the *structural type.*

L.f. are usually classified according to the method and medium of locomotion (e.g. aquatic, terrestrial, fossorial, saltatorial, arboreal), the method of obtaining food (see Mode of nutrition), the main organs of respiratory gas exchange (e.g. skin, gills or lungs), or the relationship to the substrate (e.g. planktonic, subterranean, surface dwelling). An ecological classification of animals is possible on the basis of L.fs. The occurrence of different L.fs. in one habitat allows maximal exploitation of the life-supporting potential of that habitiat, with minimal competition.

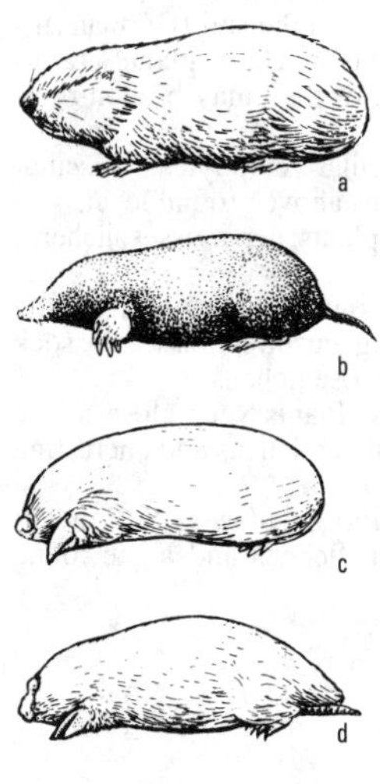

Fig.3. *Life forms.* Body shapes of small fossorial mammals. *a* Rodent (mole rat, *Spalax). b* Insectivore (European mole, *Talpa). c* Insectivore (golden mole, *Chrysochloris). d* Marsupial (marsupial mole, *Notoryctes).* (After Tischler)

Lifespan: the period from the generation of an individual organism to its death. L. varies greatly between different species, and is in principle determined genetically. *Average L.* is calculated from the Ls. of a number of individuals, whereas the *potential L.* or *maximal L.* is that attained under the most favorable living conditions.

All organisms undergo *aging,* which includes all irreversible physiological and pathological changes that arise endogenously during life. In multicellular organisms, these changes lead to the irreversible loss of essential components, such whole cells, cell organelles, important proteins, etc. If a cell type essential for life is formed only once in the early development of the organism (e.g. nerve cells of the human brain), then the life of the entire organism is dependent on the maximal L. of these cells. Cell *senescence* is characterized in particular by a decreased synthesis of RNA and protein. Aging cells also display morphological changes of subcellular structure, e.g. of the endoplasmic reticulum, mitochondria or cell nucleus. The cytoplasm of some cell types is altered by the deposition of aging pigments, which are oxidation products of lipoproteins, probably derived from degradation of cell membranes. For certain plant cells, aging and death are part of the normal process of differentiation, e.g. cork cells, stone cells, sclerenchyma fibers and vessels. In other cases, gradual accumulation or incomplete removal of metabolic end products, e.g. tannins, leads to aging. Aging processes are different in plants and animals. In plants, aging does not occur uniformly throughout the entire organism, but only in the bases of the shoots and roots, while the apical parts often retain an ability for new growth. Continually dividing and growing cells in the vegetative tips of plants are practically immortal, but even these age over a long period. Senescence of plants, progressing from the base to the apex, leads to a decrease in auxin biosynthesis and to the formation of transportable senescence factors, e.g. Ascisic acid (see), which are transported to, and cause aging in, other parts of the plant. Senescence is delayed by all factors that promote RNA and protein biosynthesis, in particular Cytokinins (see).

Senscence factors are notably synthesized by reproductive organs. If flowers and fruit are removed from an annual or biennual plant, the L. of the leaves is increased. If flower formation is prevented (e.g. by growing vernalization-dependent plants under tropical conditions), the plants may live for many years. The L. of agaves is extended from 8–10 years to 100 years if they are kept in a vegetative state.

In many plants, transition to the *aging phase* is associated with striking morphological changes. Often young and old plants differ externally, in particular in leaf shape (e.g. ivy, acacias). They also exhibit considerable physiological differences, e.g. flower formation is generally not possible in young plants, whereas older plants show decreased ability for vegetative reproduction, e.g. poor rooting of cuttings, etc.

In addition to wide interpspecies differences in average L., factors capable of influencing L. also differ from species to species. Flowering plants are classified on the basis of their L. as annuals (entire developmental cycle completed in a single vegetative period), as biennuals or as perrenials. Some trees can grow to an extremely old age, e.g. (age in years) poplar and elm 300–600, oak 500–1 000, linden 800–1 000, yew 900–3 000, and mammoth trees or redwoods *(Sequoiadendron giganteum)* even more than 4 000. To a greater or lesser extent, the bodies of these plants consist of dead tissue and organ parts, with a relatively small proportion of living material (e.g. cambium) that continues to divide and grow. The L. of individual plant parts is normally much shorter than that of the entire organism. Even in the resting state, e.g. in dry seeds, spores, etc., cells remain viable only for a limited time, probably as a result of gradual aging. The L. of seeds varies greatly, depending on the species and storage conditions; in tropical plants it is often only a few weeks, whereas in leguminous plants and members of the *Malvaceae* it is often many years. The germination of 1 000-year-old lotus seeds has been reported, but the frequently quoted germination of wheat found with mummies in Egyptian tombs is erroneous.

Aging in animals and humans is associated with far-reaching physiological changes. In the wild, this process usually lasts only a short time, because aging individuals perish when they can no longer acquire food and are unable to flee from, or resist, predators. In contrast, wild animals in captivity often live to a considerable age. The physiology of aging processes in animals (at least in vertebrates) is similar to that in humans. Detailed information on the physiology of aging in invertebrates is lacking, but the symptoms are similar to those seen in vertebrates. In beetles, the protective chitin exoskeleton becomes more brittle with age; mouthparts, antennae and legs frequently break off (especially the peripheral parts), and the thorax musculature becomes weaker. Movement becomes increasingly labored, until the animal becomes immobile and dies because it cannot feed. Aging changes have also been observed in the brains of invertebrates, e.g. *Cyclopidae.* In the *Cyclopidae,* aging is also associated with marked changes in the intestinal epithelium. Signs of the onset of aging are a decrease in bodily and mental ability, and decreased resistance to damaging environmental influences. Bones become more brittle, the skin loses its elasticity and becomes creased and folded, centers of calcification and fatty infiltration appear, metabolism is decreased, and individual tissues and organs are no longer adequately nourished. Time of onset differs according to the organ, and once aging has started it does not proceed at the same rate in all organs. Existing side by side in the same organism are organs in an early developmental stage and those already in an advanced state of degeneration. Also, some human individuals age prematurely, whereas others show no signs of the aging process even in advanced years. Aging is governed by genetic and environmental factors, and in humans the process starts immediately after fertilization. It is most evident in the nervous system and associated sense organs. Of all the somatic cells of the body, the nerve cells are the first to lose their capacity for regeneration. Other organs and organ systems (e.g. heart, skeleton, muscles, endocrine glands) also eventually display signs of aging, and in combination they sooner or later contribute to the death of the organism. Unicellular organisms which reproduce by binary fission are potentially immortal. After division, both daughter cells grow to maturity, then redivide. Thus unicellular plants and animals have no body and theoretically there is no death, provided the cells are not killed by external causes or as a result of self-poisoning with metabolic products. All multicellular animals pass through a *progressive phase,* a *mature phase* and a *regressive phase.* A number of animals, e.g. insects and aschelminths, die in the mature phase, i.e. directly after performing reproduction. During the development of higher plants and animals, some body cells become superfluous to requirements and die, as a result of tissue and organ differentiation. The germ cells of plants and animals, which must always exist for production of the next generation, are potentially immortal.

Average lifespan of the human and probable maximal age of some animals (years)

Human	60
Elephant	more than 100
Horse, Camel, Bear	50
Hippopotamus, Rhinoceros	40–50
Cow, Red deer	30
Lion, Chamois, Beaver	25
Sheep, Goat, Roe deer, Wolf, Dog	15
Fox, Cat, Squirrel	10
Eagle, Goose	80
Stork, Eagle owl, Raven	70
Crane, Pigeon, Parrot	50
Peacock, Cuckoo	40
Blackbird and small song birds	up to 25
Turtle	more than 100
Crocodile	more than 40
Toad	40
Brook salamander	20
Carp, Wels, Pike	more than 100
Pearl mussel	more than 100
Crayfish, Earthworm, Leech	20
Ant	10–15
Queen bee	5

Among animals, true death from old age (physiological death) occurs practically only in house pets. On the other hand, the potential L. of these animals is probably decreased by their changed living conditions (e.g. diet, restricted movement), compared with that in the wild.

Ligament: see Connective tissues.

Ligases: a class of enzymes, which catalyse the linkage of two different substrate molecules or two ends of the same molecule, by forming a C–O, C–N, C–C or C–S bond, in a reaction coupled to the cleavage of a high-energy pyrophosphate bond of ATP or other high energy compound. An example is glutamine synthetase, which catalyses the synthesis of glutamine from glutamate and ammonium ions, coupled with ATP cleavage.

Light as an environmental factor: The totality of radiation of all wavelengths that is available to plants and animals at their growth site or throughout their geographical range.

1) Plants. Light provides the energy needed for photosynthesis by plants, as well as exerting a wide variety of different influences as a regulator of plant growth and development. Thus, light affects the respiration and germination of Seeds (see) and the development of rhizomes and tubers. Flowering and bud dormancy often depend on the length of daylight (or length of darkness) (see Photoperiodicity). Furthermore, light is a photomorphogenic factor (see Photomorphogenesis), which is evident, inter alia, in the fact that in very weak light plants develop an extended shoot with small, often rudimentary leaves. Orientation of plants and plant organs in space is also greatly influenced by light (see Phototropism, Photonasy, Phototaxis). The effect of light on morphogenesis is largely determined by the genetic constitution and regulation of the cell in question. For example, in the mustard hypocotyl, certain epidermal cells (trichoplasts) develop into hair cells under the influence of light, whereas all cells in the subepidermal layer produce anthocyanins in response to illumination. Light normally influences growth and development via a chain of reactions, which are only partly elucidated. The process often starts with the absorption of long-wave light by Phytochrome (see). The intensity of illumination is relatively unimportant, since the maximal response of the plant is illicited by small quantities of light, and increased light intensity or duration of illumination have no further effect. In some cases of photoregulation, light is absorbed by a pigment system with absorption maxima in the blue and far red regions of the spectrum. This system, which e.g. inhibits growth of the *Lactuca* hypocotyl, requires a higher radiation energy than the very sensitive phytochrome system, and is therefore known as the High energy system (see).

The mechanism of action of ultraviolet light is still unclear. UV irradiation often causes ionization. In higher plants, the *UV-effect* is very dependent on temperature. At suboptimal temperatures, UV generally causes growth inhibition. At supraoptimal temperatures growth may be promoted, whereas UV usually has little action at optimal growth tempertaures. Studies on the dwarf habit of many high mountain plants have shown that, contrary to earlier beliefs, growth inhibition is never due to UV-irradiation alone.

Phylogenetic adaptations to habitat illumination are shown by light and shade plants (see Strong light plants). In *chromatic adaptation,* the plant is adapted to the predominant color of the habitat by a shift in the relative quantities of assimilatory pigments, and partly by the synthesis of accessory pigments. It is often difficult and sometimes impossible to separate the effects of light and temperature. Thus, the water economy and temperature relationships of plants are influenced by long wavelengths, in particular infrared irradiation,

2) Animals. All animals are indirectly influenced by light, in that they ultimately depend on plant photosynthesis for their food supply. Most animals are also directly influenced by light, by both its intensity and wavelength. *Euryphotic animals* tolerate a wide range of light intensities, whereas those tolerating a narrower range are said to be *stenophotic.* Most animals display a preference for an optimal range of light intensity. *Light tolerance* is related to both intensity and wavelength, and it is increased by various protective adaptations (pigments, thickened body surface layers, behavior). Unprotected animals (protozoa, earthworms, some insects and their larvae) are killed by light when the intensity of irradiation exceeds their narrow limit of tolerance. The influence of light on the evolution of body structure is evident, e.g. in eye development. For example, crepescular animals have large eyes for the perception of dim light, while Troglobionts (see), endoparasites and deep sea fishes have degenerate or regressed eyes (Fig.1).

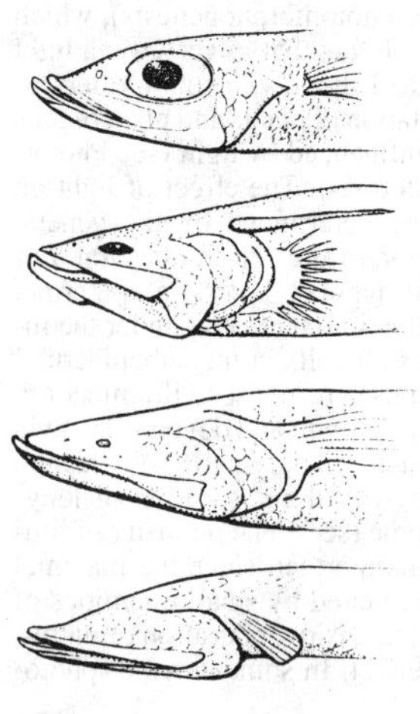

Fig.1. *Light as an environmental factor.* Regression of the eyes of scopelide fish species from different ocean depths. Depths in order from top to bottom: 575 m; 800–1 000 m; 3 000 m; 5 000 m (after Hess-Doflein, 1943).

Shape and color of morphological structures, body size, and physiological and ethological characters are influenced by photoperiodicity (relative proportions of light and dark in the 24-h cycle). In the seasonal dimorphism of the European cicada, *Euscelis plebejus,* the short-day form has a smaller penis, while the long-day form is less pigmented, and has longer wings and a larger penis. In aphids, the development of parthenogenic or bisexual generations is regulated by the photoperiod. The photoperiod also governs the deposition of fat stores and the readiness for migration in migratory birds. Many animals change color in direct response to light; this may occur by chemical alteration of pigments (bleaching), de novo pigment synthesis (browning of human skin), or migration and expansion of chromatophores (prawns, cuttlefish).

The influence of light on activity is expressed in the terms diurnal, crepescular and nocturnal. Diurnal birds first awake, then start calling (singing) at particular daylight intensities (Fig.2). *Photokinesis* is the influence of light on movement, e.g. on the motility of insects. Since environmental light intensity undergoes regular cyclic changes, animals whose activity is influenced by light also display a corresponding photoperiodicity. This periodicity may be daily (monophasic with an active and a quiescent phase,or polyphasic with repeated phase change), monthly (according to the phases of the moon, e.g. in some marine animals), or yearly (e.g. reproductive cycles). Gonadal development in birds and mammals is influenced by the duration of the light phase. In many invertebrates, the rate of development, diapause and molting are promoted, inhibited or synchronized by light (see Periodicity).

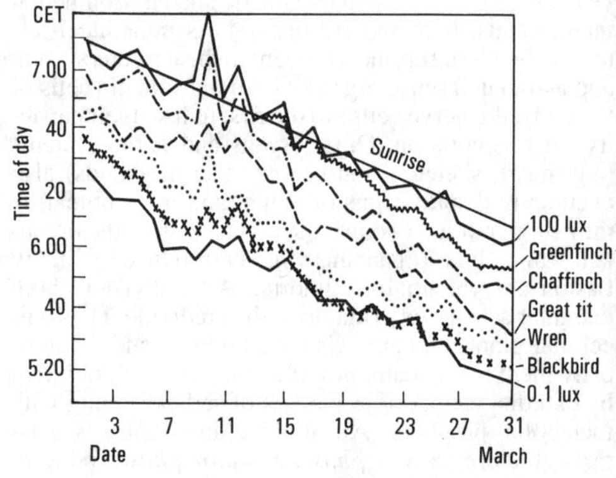

Fig.2. *Light as an environmental factor.* Relationship between light intensity in lux and the onset of singing by some bird species in March (after Scheer. 1952).

Light is particularly important as an orientation factor. Physiological responses are phototropism (orientation movements of sessile animals), phototaxis (linear movements toward or away from a light source), light compass movements (maintenance of a certain angle to a light source during linear progression; Fig. 3). In the last case, simultaneous allowance for movement of the sun with time results in absolute orientation, as demonstrated for bees, crustaceans and birds. The most extreme exploitation of light sources for orientation is observed in the navigation of migratory birds. Finally, orientation may depend on wavelength (color) (e.g. flower-visiting insects), or may depend on the recognition of contrasts and shapes. Thus, the cockchafer flies toward dark silhouettes of forests. Bees are guided by both color and pattern. Pattern recognition reaches a peak of perfection in the star and landscape orientation of birds and mammals.

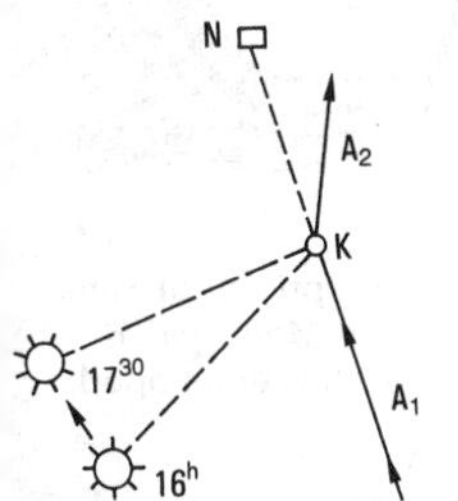

Fig.3. *Light as an environmental factor.* Route taken by an ant *(Lasius niger)* to its nest (N). After proceeding in direction A_1, the ant was kept at K for 1.5 hours in the dark. The angle of deviation of the new direction (A_2) from A_1 is equal to the angle described by the sun during the 1.5 hours of darkness (after Scherdtfeger, 1978).

Light compensation point: the illumination intensity at which photosynthesis and respiration are exactly balanced, i.e. the rates of CO_2 uptake and release are the same, and the rates of O_2 uptake and release are the same, so that there is no net change in the concentration of either gas. Strong light plants and light leaves respire more actively than weak light plants or shade leaves, so that their L.c.p. occurs at a higher light intensity.

Light-dark adaptation: the ability of the eye to adapt to changes in light intensity. The working range of the human eye covers 12 orders of magnitude of light intensity. In *dark adaptation,* the photopic system first ceases to function, followed much later by the scotopic system. *Light adaptation* consists of a rapid neurophysiological alpha-adaptation (see Receptive field), and a slow beta-adaptation, involving biochemical conversions of the visual pigments. Both components are involved in the opposite sense in dark adaptation.

Lightfishes: see *Stomiiformes.*

Lightiella: see *Cephalocarida.*

Light scattering: deflection of light from its original path by small particles. Certain properties of biological molecules (e.g. molecular mass, molecular size, diffusion constant, association-dissociation behavior, molecular vibration) can be determined by measuring the intensity and frequency of the scattered light. In *elastic scattering,* the scattered and incident light have the same wavelength. In *inelastic scattering,* the scattered light shows a shift of frequency, i.e. the molecule converts part of the light energy into translational and vibrational energy.

Elastic scattering. If the molecule is much smaller than the wavelength of the incident light, and the wavelength of the incident light is far removed from the absorption band of the particle, then the entire molecule becomes an oscillating dipole, and no phase differences occur. This oscillating electric dipole emits electromagnetic radiation which propagates in all directions perpendicular to the vector of the electric field stength of the incident light. The intensity of the emitted radiation is proportional to the 4th power of the wavelength of the incident light (Rayleigh scattering). In dilute solution, the intensity of the scattered light is proportional to the square of the relative molecular mass.

Larger particles, however, no longer behave like a point dipole. The various parts of a larger molecule are excited in different phases. The coherent wave trains of the scattered light then interfere with one another, and the overall effect is a reduction of the scattered intensity, in particular a reduction of backscattering intensity. The dependence of scattered intensity on the angle of scatter provides an index of molecular size.

Quasi-elastic scattering. Translational and rotational motion of molecules in solution cause a Doppler shift, so that the spectral width of the scattered light is broader than that of the incident light. These effects can be measured with instruments with very high spectral resolution, and can be used to determine diffusion and rotational diffusion coefficients.

Inelastic scattering. Part of the incident light energy is absorbed by molecular vibration, so that some of the scattered light is shifted in frequency relative to the incident light. These lines of shifted frequency constitute the *Raman spectrum* of the molecule. The Raman spectrum of a molecule in its native state in aqueous solution provides information on structure and conformation.

Light sensory organs, ***light-sensing organs, visual organs, eyes:*** light-sensitive organs of animals, displaying a range of evolutionary complexity. Primitive L.s.o. enable animals to perceive the direction and intensity of light, whereas more advanced L.s.o. generate an image of the environment. The light-sensitive cells of L.s.o. are always primary sensory cells, i.e. nerve cells.

In protozoa, sponges and cnidarian polyps (but not in cnidarian medusae), the entire protoplasm of the body surface is light-sensitive, although specialized L.s.o. are absent. Light-sensitive organelles occur sporadically in protozoa, e.g. in flagellates, where the eyespot (stigma) and flagellum act as a phototactic unit directing the organism toward light.

In each metazoan group, the L.s.o. display an evolutionary sequence. Between metazoan groups, the L.s.o. vary widely in their structure and origin, but with respect to their functional adaptation they display similar developmental stages. According to their degree of perfection as light-detecting and/or image-forming systems, L.s.o. can be generally classified as 1) diffuse sensory cells, 2) eyespots, 3) pigment cup ocelli, 4) optic pits (epithelial cups or grooves), 5) open vesicles, 6)

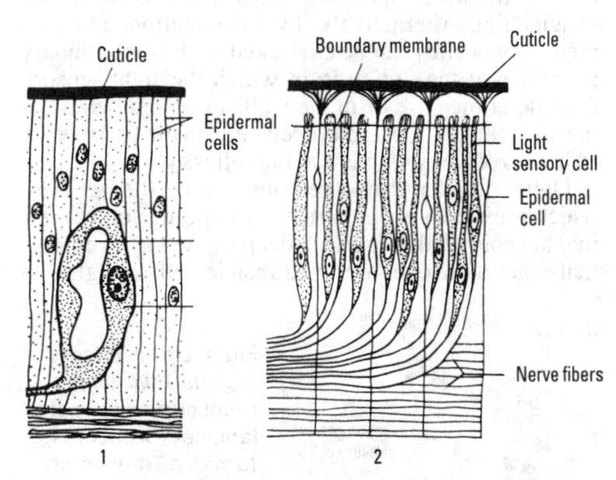

Figs.1 & 2. *Light sensory organs.* *1* Single light-sensitive cell in the epidermis of an earthworm (diffuse light sensory organ). *2* Eyespot of a starfish.

closed vesicles with lens, and 7) compound eyes. Simple metazoan L.s.o. are generally known as *ocelli* (singular *ocellus),* i.e. most metazoan eyes, excluding the camera-type eyes of vertebrates and cephalopods, and arthropod ommatidia, are ocelli.

Diffuse L.s.o. occur in, e.g. annelids, bivalves, sea urchins and ascidians, where they take the form of light-sensitive cells scattered in the epidermis (Fig.1). They merely permit the animal to distinguish between light and dark. An aggregation of light-sensitive cells at a certain site (sometimes screened on one side by a pigment layer) is known as an *eyespot* (Fig.2). Eyespots allow the animal to establish the direction of a light source; they occur in cnidarian medusae, turbellaria, annelids and starfishes.

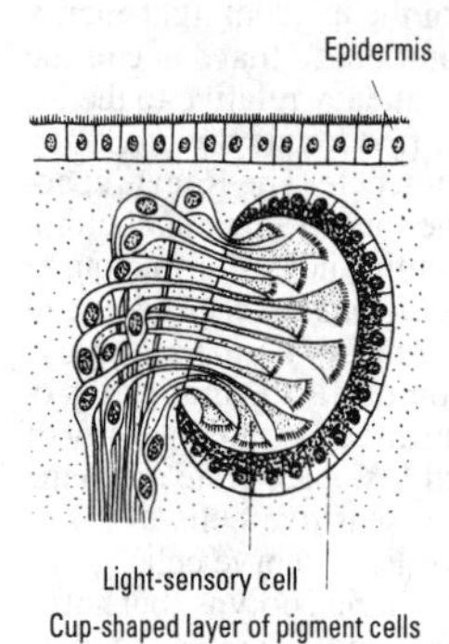

Fig.3. *Light sensory organs.* Inverse pigment cup ocellus of a planarian worm.

In many animal groups, the *inverse pigment cup ocellus* or *pigment cup ocellus* is the most advanced light sensory organ. In principle, one or several light-sensory cells (retinal cells) are surrounded by a hemispherical envelope of one or several pigment cells (Figs. 3 and 4). Since the pigment cells shade the sensory cells from light in all directions but one, the animal is able to perceive the direction of light and orientate itself accordingly. Thus, turbellarians possess pigment cup ocelli, and most turbellarians are negatively phototactic. A few members of the order *Ocoela* (class *Turbellaria)* have patches of photoreceptor cells and pigment cells in the epidermis, known as *pigment spot ocelli,* from which pigment cups have evolved by evagination. Alternatively, by invagination, the pigment spot ocellus can be envisaged as the evolutionary precursor of the *optic pit,* in which the light-sensory cells lie at the base or in the walls of an epithelial depression (Fig.5). Optic pits occur, e.g. in the bell margin of hydrozoan medusoids and in mollusks.

Optic pits have evolved into *camera-type eyes,* which project an image of the surroundings onto a retina: the epithelial depression deepens, while its epidermal edges close over to form a vesicle with a single visual aperture. Thus, the optic pit of the prosobranch gastropod shown in Fig.5 can be interpreted as the evolutionary precursor of the eye of *Nautilus* (Fig.6). The latter behaves as a pinhole camera, producing a reversed and weakly illuminated image of the light source on the internal surface (retina) at the back of the eye. The interior of the *Nautilus* eye communicates with the exterior via the central aperture, so that the retina is bathed in sea water. Most camera-type eyes, however, possess a lens, and the interior of the eye does not communicate directly with the exterior. Eyes with lenses have evolved independently in the more advanced representatives of different animal groups, e.g. polychaetes, mollusks and vertebrates, while the ommatidium of the insect compound eye is a special case of lens-aided light perception. In some polychaetes, the optic pit is provided with a dioptric apparatus in the form of a biconvex lens, while the vertebrate eye has both a lens and a vitreous body. Light rays are focused by the lens onto the sensory epithelium (retina), which usually consists of many sensory cells. Since many light rays from one point on the object are focused at a single point on the retina, the images perceived by a camera-type eye with a lens are considerably brighter than those of the optic pit. On the other hand, the images produced by small camera-type eyes, like the ocelli or ommatidia of insects, are very diffuse. Simple lens-eyes therefore serve to determine the direction of light and to perceive movement.

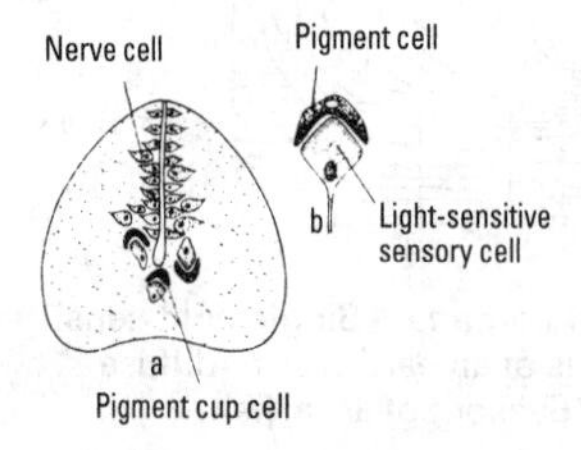

Fig.4. *Light sensory organs.* Inverse pigment cup ocellus of a lancelet *(Branchiostoma). a* Transverse section of neural tube. b Individual pigment cup ocellus.

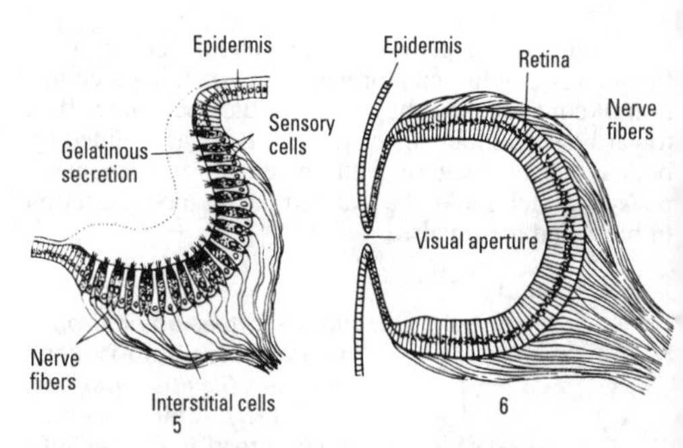

Fig.5 & 6. *Light sensory organs. 5* Optic pit of *Patella* (a prosobranch gastropod; phylum *Mollusca). 6* Open vesicle, camera-type eye of the cephalopod, *Nautilus* (phylum *Mollusca).*

Separate evolutionary series from very simple L.s.o. to camera-type lens-eyes are found in mollusks, annelids and arthropods (Figs. 5,6,7 and 8).

Functionally and structurally, the compound eye of arthropods (Fig.9) differs considerably from other types of L.s.o. Whereas the efficiency of the arthropod ocellus is determined by the number of light-sensory cells, the performance of the arthropod compound eye depends on the often large number of individual ommatidia. The distal end of each ommatidium is covered by a translucent, chitinous *cornea* (derived from the skeletal cuticle), which functions as a lens. The external face of the corneal lens, known as a facet, is usually square or hexagonal, so that all the lenses on the surface of the compound eye form a 2-dimensional repeating pattern, giving the eye a faceted appearance. Beneath the cornea is a cylindrical or tapered element called the *crystalline cone,* which serves as a cylindical lens (its refractive in-

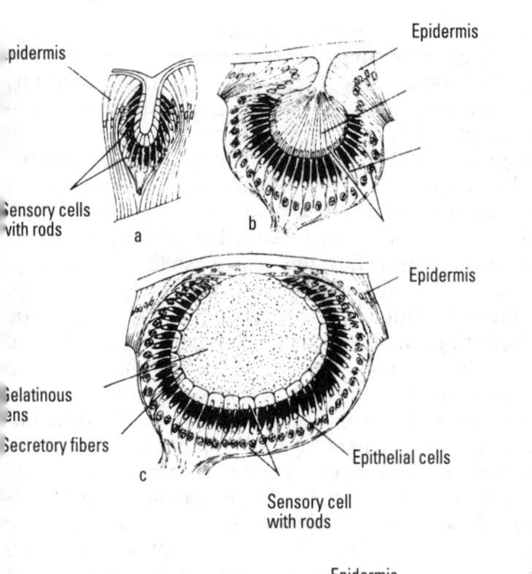

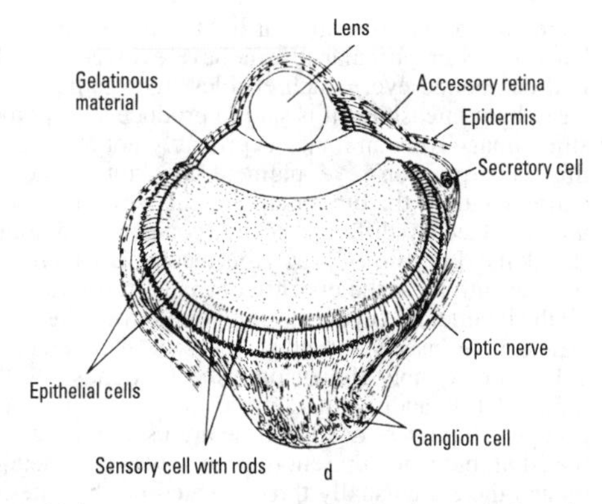

Fig.7. *Light sensory organs.* Stages in the evolution of polychaete and annelid eyes. *a Ranzania. b Syllis. c Nereis. d Alciope.*

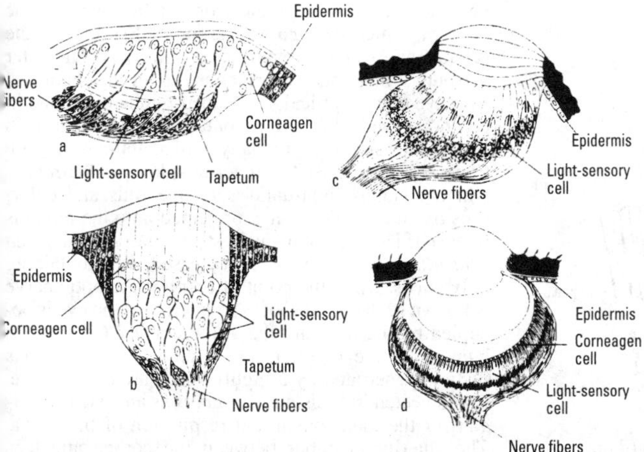

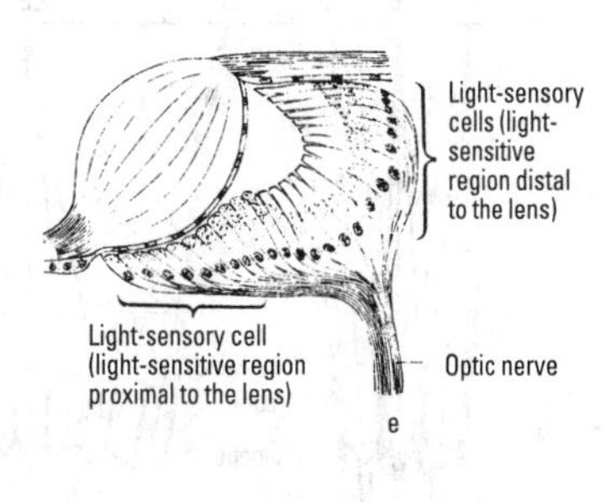

Fig.8. *Light sensory organs.* Insect eyes. *a* Lateral ocellus of a springtail (order *Collembola). b* Ocellus of a grasshopper (family *Acrididae),* order *Orthoptera). c* Dorsal ocellus of a plant aphid (family *Aphididae),* order *Hemiptera). d* Ocellus of a robber fly (family *Asilidae,* order *Diptera). e* Ocellus of a hoverfly (family *Syrphidae,* order *Diptera).*

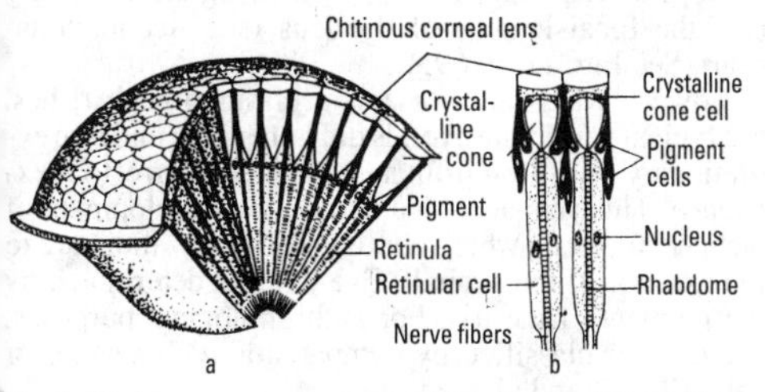

Fig.9. *Light sensory organs. a* Structure of the arthropod compound eye. *b* Two ommatidia.

dex increases from the outside toward the central axis). The crystalline cone is usually derived from 4 cells, and its 4-part structure is often still apparent upon microscopic examination. The basal (proximal) end of the ommatidium is formed by the light receptor element or *retinula.* This consists of elongated cells (retinular cells) lying parallel to the long axis of the ommatidium. There are usually 7 similar retinular cells and a single eccentric cell, arranged in the form of a cylinder with a central core. Many of the retinular cells have tubular folds or microvilli on their inside surface, which lie across the long axis of the ommatidium. The total complement of microvilli on one retinular cell is called the *rhabdomere.* All the rhabdomeres project centrally. They may meet and completely fill the central core, or they may project without meeting, and enclose a fluid-filled central cavity. In either case, the retinula contains a highly refractive central core, known as the *rhabdome.* Individual ommatidia are separated from one another by a sheath of pigment cells. White-eyed mutants lack this pigment sheath. If the pigment sheath extends the entire length of the ommatidium, then the pigment, together with the spherical curvature of the eye surface, ensures that only light from a narrow part of the visual field can enter any single ommatidium (Fig.10a). The resulting image, known as an "apposition image", therefore consists of a grid of light dots. On the other hand, many crepuscular insects have evolved a modified compound eye, which enables light gathering in weak light intensities; it is said to produce a "superposition image", but an image is probably not formed. In this "superposition" eye, pigment does not extend the entire length of the ommatidia, so that several images are projected by different crystal cones onto a single rhabdome (Fig. 10b). Clearly, this mechanism increases light sensitivity at the expense of image resolution. Of all the invertebrate eyes, the compound eye is the most efficient. In insects and crustaceans, it also perceives color. For example, the vision of bees extends into the ultraviolet. In addition to compound eyes, some insect groups still possess ocelli. These are usually absent in the adult, but when present (e.g. in ants and hymenopterans) there are usually three, located on the anterior dorsal surface of the head, between the compound eyes.

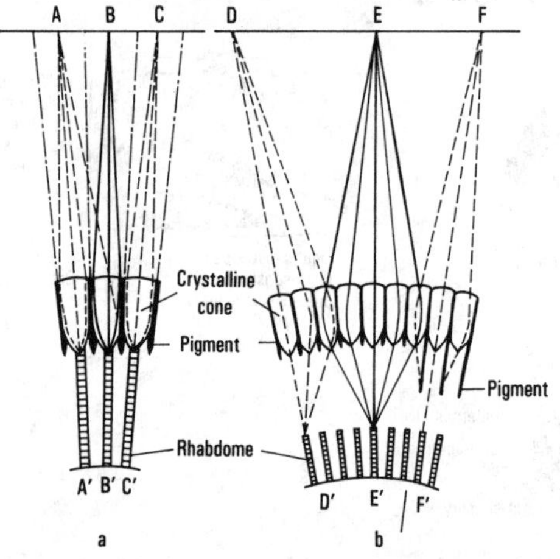

Fig.10. *Light sensory organs.* Formation of apposition and superposition images in the arthropod compound eye (after Kühn). *a* Compound eye adapted for apposition image formation. Each rhabdome receives light from only one crystalline cone. *b* Compound eye adapted for superposition image formation. Pigment is retracted. Light rays from D and E enter a number of crystalline cones and are concentrated on the rhabdome of a single ommatidium. Light from F is received as an apposition image, because the extended pigment sheath prevents light from crossing between neighboring ommatidia.

Among the mollusks, the L.s.o. of cephalopods are the most highly evolved. With respect to its functional parts, the cephalopod lens-eye is completely analogous to the vertebrate lens-eye, although it has an entirely different phylogenetic origin. Like the vertebrate eye, it is provided with a variety of auxiliary structures. Thus, a muscular fold of skin functions as an iris by widening or constricting the visual aperture, in the latter case protecting against intense light and sharpening the image. Eyelids are also present, and there is an outer cornea which prevents the intrusion of foreign bodies.

The paired eyes of vertebrates arise from the ventrolateral region of the diencephalon, and are first recognizable as outpushings or evaginations on the wall of the embryonic brain. These evaginations develop further to form prominent bulges, known as *primary optic vesicles,* which continue to grow outward toward the epidermis. A constriction then appears between the brain and the outer part of each optic vesicle, so that the latter remain attached by a narrow *optic stalk.* The ventrolateral wall of each stalked primary optic vesicle invaginates, transforming the vesicle into a cup, known as the *secondary optic vesicle* or *optic cup.* Meanwhile, embryonic epidermis opposite each optic vesicle thickens to form the rudimentary, ectodermal lens, which becomes abstricted when the optic cup makes contact with the epidermis. The outer, noninvaginated layer of the optic cup eventually differentiates into the pigment layer of the retina. The nervous and light-sensory layers of the retina are derived from the cells lining the optic cup, and are therefore derivatives of the brain. A typical vertebrate eye contains the following three layers. 1) The external *sclera,* which becomes the *cornea* at the front of the eye. 2) The *choroid,* which lies against the internal surface of the sclera, and extends forward to form the *iris.* The latter surrounds the pupil and is reponsible for the characteristic eye color. Vertical slit-shaped pupils are found in crocodiles and geckos, horizontal slit-shaped pupils in many toads and frogs; in many hoofed animals they are transversely oval. 3) The innermost layer is the *retina,* which contains light-dark- sensitive rods and color-sensitive cones. The image projected onto the retina is reversed. The region of sharpest perception is the *fovea centralis* or yellow spot, a depressed area containing only cones, while the point of entry of the optic nerve is known as the blind spot. The interior of the eye is occupied by the lens and the vitreous body (or vitreous humor). In the reptilian and avian eye, the vitreous body is penetrated by a highly vascularized structure, called pecten in birds, which serves as an auxilliary organ for the metabolism and respiration of the retina. The anterior chamber between the cornea and lens contains aqueous humor. Primitive aquatic vertebrates have practically spherical lenses, which are moved forward or backward by ciliary muscles. In contrast, the lenses of higher vertebrates are biconvex, and the ciliary muscles focus the eye by altering the convexity (i.e. the focal length) of the lens (see Accommodation). See Fig. 11, p. 695.

Lower eyelids are present in elasmobranch fishes, but typical vertebrate eyelids did not evolve until the evolutionary transition from an aquatic to a terrestrial existence. This was accompanied by the development of lachrymal glands which continually supply moisture to the exterior of the eyeball. Eye color is determined by the pigment of the iris. For anthropological purposes, eye color is classified by comparision with a series of naturally colored glass eyes.

Ligia: see *Isopoda.*

Ligiidae: a family of the *Isopoda* (see).

Lignans: plant products formally equivalent to two n-propylbenzene (phenylpropane) residues linked at the central carbons of their side chains. The two benzene rings are usually identically substituted, and the type of substitution is similar to that present in ring B of the C_6C_3 residues of flavonoids. It is therefore generally accepted that L. are biosynthesized by dimerization of a

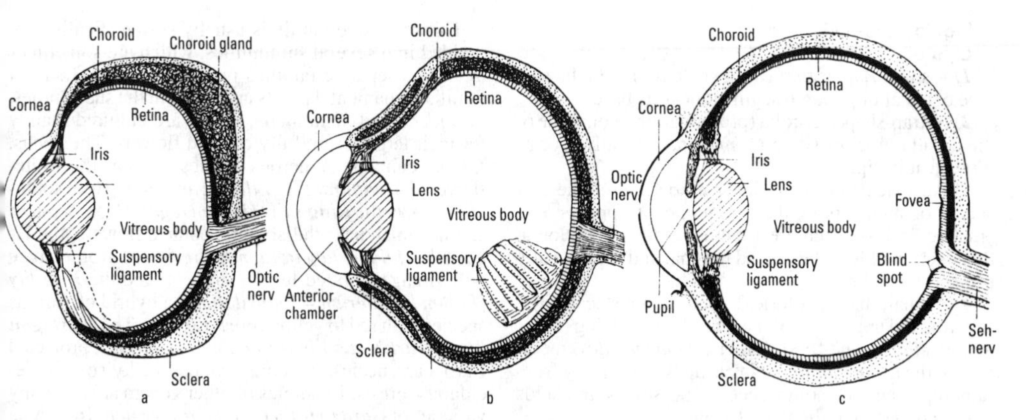

Fig.11. *Light sensory organs.* Vertical sections through different vertebrate eyes. *a* Fish eye. *b* Bird eye. *c* Human eye.

C_6C_3 precursor, but direct experimental evidence for this is lacking. According to Freudenberg, L. represent stages in the biosynthesis of Lignin (see).

Glucuronides of two L. (enterolactone and enterodiol) have been found in primate urine, with maximal production during the luteinization phase of the ovulation cycle and the early stages of pregnancy [S.R. Stitch, *Nature* **287** (1980) 738–740; D.R. Setchell et al., *Biochem J.* **197** (1981) 447–458].

Lignin: a highly polymerized aromatic plant constituent, which next to cellulose is the main component of wood. It is responsible for the thickening and strengthening of plant cell walls. The properties associated with wood are due to the incrustations of plant cell walls with L. Chemically, L. cannot be exactly defined. According to Freudenberg, L. is a highly cross-linked, macromolecular, branched polymer, formed irreversibly by dehydrogenation and condensation. According to Adler and Gierer, L. is an essentially acid-resistent, polymorphic, amorphous incrustation material found in wood, consisting of methoxylated phenylpropane units linked by ether linkages and C–C bonds. It has also been decribed as a "random polymer of hydroxyphenylpropane units".

The chemical composition of L. differs according to the plant species. Beech L. is the most extensively studied. Primary precursors of L. are coniferyl, sinapyl and *p*-coumaryl alcohol, which are derived from L-hydroxycinnamic acid. L. from conifers (i.e. from softwoods) is derived chiefly from coniferyl alcohol with variable but small proportions of sinapyl and *p*-counaryl alcohols. L. from dicotyledonous angiosperms (i.e. from hardwoods) is formed chiefly from sinapyl (~ 44%) and coniferyl (~ 48%) alcohols, with about 8% *p*-coumaryl alcohol. L. in grasses is formed from *p*-coumaryl (~ 30%), coniferyl (~ 50%) and sinapyl (~ 20%) alcohols. These primary L. precursors are formed from the aromatic amino acis, L-phenylalanine and L-tyrosine.

D-Coniferin (glucoside of coniferyl alcohol), D-glucocoumaryl alcohol (glucoside of *p*-coumaryl alcohol) and D-syringin (glucoside of sinapyl alcohol) are storage forms of L. precursors. β-Glucosidase in cells between the cambium and the mature wood hydrolyse the glucosides, releasing the alcohols for L. synthesis.

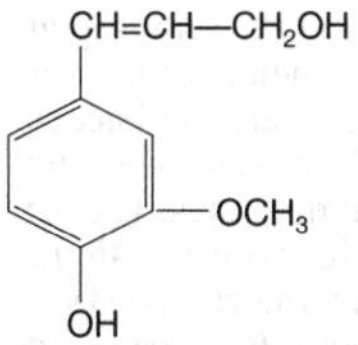

Fig.1. *Lignin.* Coniferyl alcohol (a biosynthetic precursor of lignin).

Total biosynthesis of L. represents the interplay of enzymatic phenol dehydrogenation and nonenzymatic coupling of radicals generated by the loss of an electron from a phenolate ion; it is referred to as reductive polymerization.

L. is insoluble in hot 70% sulfuric acid and hot concentrated hydrochloric acid. After the acid hydrolysis of wood with either of these acids, pure L. remains as a yellowish, amorphous powder with M_r greater than 10 000. Conversely, L. can be solubilized and extracted from wood by heating with a solution of sodium hydroxide or potassium hydrogen sulfite, leaving a residue of cellulose. This solubilized L. (L.-sulfonic acid) is the main component of sulfite liquors, a byproduct from the industrial production of cellulose (paper manufacture). In the dry distillation of wood, the methoxy groups of L. give rise to methyl alcohol. Degradation of deciduous hardwoods produces vanillin. The latter is also extractable from the woods used in barrel making, and it contributes to the aroma of, e.g. old cognac. L. is also an important component of Humus (see).

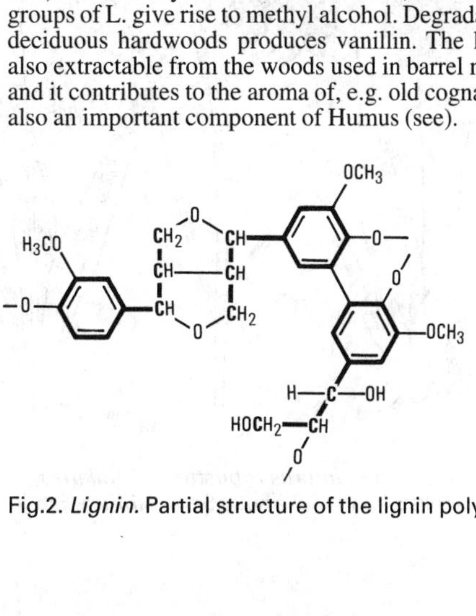

Fig.2. *Lignin.* Partial structure of the lignin polymer.

Ligula: see Bootlace worm.

Ligule:

1) A small flap of tissue or a scale borne on the surface of a leaf or perianth segment near its base.

2) A strap-shaped corolla (prolongation of one side of the corolla tube) of some *Compositae,* usually 3-or 5-toothed at its tip.

3) An extension of the inner epidermis of the leaf sheath of members of the *Graminae;* it appears as a small, colorless, often erect membranous projection at the junction of leaf sheath and lamina, and it physically protects the halm.

4) A small, basal, colorless, membranous scale arising from the upper side of the leaf of some living (order *Selaginellales)* and fossil (order *Lepodendrales)* members of the *Lycopodiatae.* It takes up water rapidly from raindrops, and in some species possesses tracheids which connect it to the vascular bundle.

Liguliflorae: see *Compositae.*

Ligustrum: see *Oleaceae.*

Lilac: see *Oleaceae.*

Liliaceae, ***lily family:*** a monocotyledonous plant family of about 3 500 species, with members in every geographical vegetation zone. They are herbaceous plants, mostly with linear leaves and subterranean storage organs, such as bulbs, corms and rhizomes. The perianth is regular and usually petaloid, with six similar free or fused segments in two similar whorls. The flowers are usually hermaphrodite, contain 6 stamens, and usually a 3-celled superior ovary with axile placentation, rarely 1-celled with parietal placentation. Pollination is mostly by insects, and the fruits are capsules or berries. Chemically, most members of the family are characterized by the presence of steroid saponins and chelidonic acid.

This extensive family is usually systematically subdivided into several subfamilies, which are sometimes treated as separate families in the literature. The most familiar ornamental plants are found in the subfamilies, *Lilioideae* and *Scilloideae*, which are cultured mainly for their large, beautifully colored flowers. The genus, *Lilium*, contains numerous species, such as the white-flowered ***Madonna lily*** *(Lilium candidum)* from southern Europe; the ***King's lily*** *(Lilium regale)* from western China, which has whitish pink, conical flowers; the ***Tiger lily*** *(Lilium lancifolium = tigrinum)* from Japan, with orange, spotted turban flowers; and the ***Fire lily*** *(Lilium bulbiferum = umbellatum),* a hybrid of various species with red to yellow, erect flowers. The ***Martagon lily*** or ***Turk's cap lily*** (*Lilium martagon*) is a protected European species, which grows on chalky soils in deciduous forests. Examples of other genera are the many varieties of ***Garden tulip*** *(Tulipa gesneriana)* from Asia Minor; ***Crown imperial*** *(Fritillaria imperialis)* from the western Himalayas; European ***Snake's-head fritillary*** *(Fritillaria meleagris);* European and Asian ***Day lilies*** *(Hemerocallis);* ***Hostas*** *(Hosta)* from East Asia; various species of *Eremurus* (e.g. *Eremurus robustus* from Turkestan; Siberian ***Squill*** *(Scilla sibirica)* from southeastern Europe, ***Hyacinth*** *(Hyacinthus orientalis)* from the eastern Mediterranean; and ***Grape hyacinths*** *(Muscari),* also from the eastern Mediterranean. The ***Sea onion*** or ***Maritime squill*** *(Urginea maritima),* a member of the *Scilloideae*, occurs mainly on the European and African coasts of the Mediterranean; it contains the official cardiac glycosides, scillarin-A and -B.

Cardiac glycosides also occur in the European ***Lily of the valley*** *(Convallaria majalis),* a protected species belonging to the subfamily *Asparagoideae*. The most numerous genus of this subfamily is *Asparagus*. The ***Veg-***

Fig.1. *Liliaceae. a Eremurus robustus. b Fritillaria meleagris* (snake's head fritillary). *c Scilla sibirica* (Siberian squill). *d Fritillaria imperialis* (crown imperial). *e Aloe ferox. f Haworthia planifolia.*

etable asparagus *(Asparagus officinalis)* from Europe, North Africa and Western Asia has been a popular cultivated plant since ancient times; the young shoots are eaten white (formed below ground after earthing up) or green. Some species are used for decorative cut foliage, e.g. *Asparagus setaceus* and *Asparagus densiflorus* from South Africa. An interesting plant native to the Mediterranean region and Western Europe is ***Butcher's broom*** *(Ruscus aculeatus),* whose flowers and fruits are borne directly on the dorsal surface of the ovoid cladodes. ***Herb Paris*** *(Paris quadrifolia)* grows in deciduous forests of Europe and Asia; it is very poisonous, due to the presence of the glycoside, Paris-typhnine, in its leaves and rhizomes.

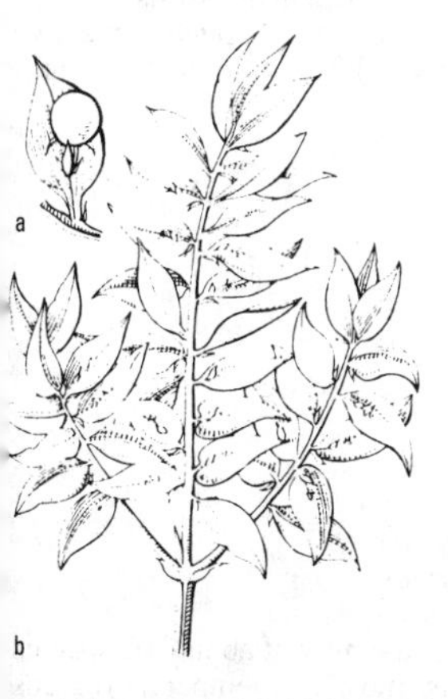

Fig.2. *Liliaceae. Ruscus aculeatus* (butcher's broom). *a* Flowering branch. *b* Cladode.

Allium, with about 300 species, is the largest and most important genus of the subfamily *Allioideae.* Most species of *Allium* contain antibiotic alkyl sulfides, which are responsible for the typical smell and taste of onions, leeks, garlic and other *Allium* species. Some important vegetable, seasoning and medicinal plants are: ***Onion*** *(Allium cepa),* which originates from western Asia; ***Garlic*** *(Allium sativum);* ***Leek*** *(Allium porrum),* the mildest of all *Allium* species, whose natural origin is uncertain; ***Chives*** *(Allium schoenoprasum),* which also occurs wild in Europe, North and Central Asia and North America; ***Shallot*** or ***Eschalot*** *(Allium ascalonicum),* which is known only in the cultivated state; ***Welsh onion*** *(Allium fistulosum),* also unknown in the wild, and cultivated mainly in China, Japan and tropical countries.

The subfamily, *Wurmbaeoideae,* contains the genus *Colchicum,* whose members contain the alkaloid colchicine; this alkaloid inhibits nuclear division, and it is used to generate gene mutations in plant breeding. ***Meadow saffron*** or ***Autumn crocus*** *(Colichicum autmnale)* is native in damp European meadows, and it is used medicinally. The *Asphodeloideae* is an extensive subfamily, which contains, e.g. the yellow-flowered garden plant, ***Yellow asphodel*** *(Asphodeline lutea)* from the Mediterranean region; and the genera *Aloe, Haworthia* and *Gasteria,* which are native mainly to South Africa, and are used as ornamentals. The coagulated sap of some *Aloe* species (aloe resin) contains the bitter principle, aloin, and is used as a purgative.

A popular house plant, also from South Africa, is *Chlorophytum comosum,* which has white-variegated, linear-lanceolate leaves.

Liliatae: see *Monocotyledoneae.*

Liliidae: see *Monocotyledoneae.*

Lily trotters: see *Jacanidae.*

Limanda: see Dab, *Pleuronectes.*

Limax amebas: see Amebas.

Limbs, *extremities:* jointed appendages of animals and humans, which serve as legs, wings or fins for locomotion and propulsion, as prehensile arms or jaws for holding or breaking up food material, as sensory feelers, and even as organs of respiratory gas exchange or reproduction.

Invertebrate L. show various modifications, a wide variety of differently structured appendages being found among the *Arthropoda,* especially in the *Arachnida, Crustacea* and *Insecta.*

In the vertebrates, the forelimbs are distinct from the hindlimbs. The pectoral fins of fish, the forelegs of amphibians, reptiles and mammals, the wings of birds and the arms of primates are all forelimbs, whereas the pelvic fins of fish, the hindlegs of amphibians, reptiles and mammals, and the legs of birds and primates are all hindlimbs.

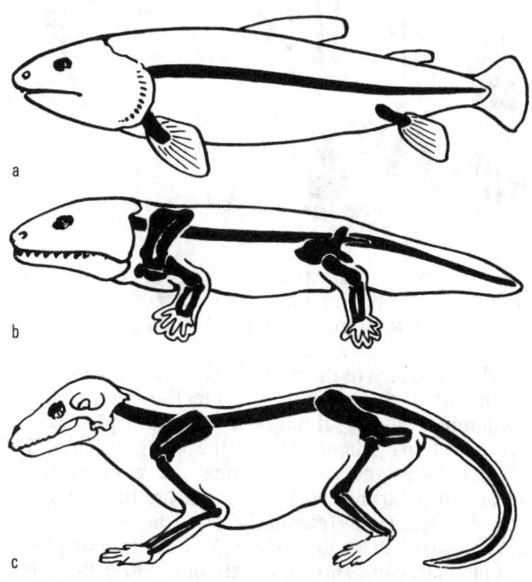

Fig.1. *Limbs.* Evolution of the limb skeleton of vertebrates. *a* Fish, showing absence of linkage between the limb skeleton and the axial skeleton. *b* Fossil amphibian, showing articulation of the pelvic girdle with the axis. *c* Mammal, showing crossover of the radius and ulna of the forelimb.

The *free limb skeleton* (also known as the *pentadactyle limb skeleton)* is associated with the Axial skeleton (see) via two limb girdles, i.e. the Pectoral girdle (see) and the Pelvic girdle (see). The combination of the limb girdles and the skeletons of the free limbs is called the *appendicular skeleton.* In fish, the limb skeleton is represented by the radiating supporting elements of the pectoral and pelvic fins. In quadrupeds, the limb skeleton is divided into 3 main sections: *stylopodium* (humerus of the forelimb, femur of the hindlimb); *zeugopodium* (ulna and radius of the forelimb, tibia and fibula of the hindlimb); and the *autopodium* (hand and foot); the hand consists of the carpals (wrist bones), metacarpals (middle bones of the hand) and phalanges (finger bones), while the foot comprises the tarsals (ankle bones), metatarsals (middle bones of the foot) and phalanges (toe bones).

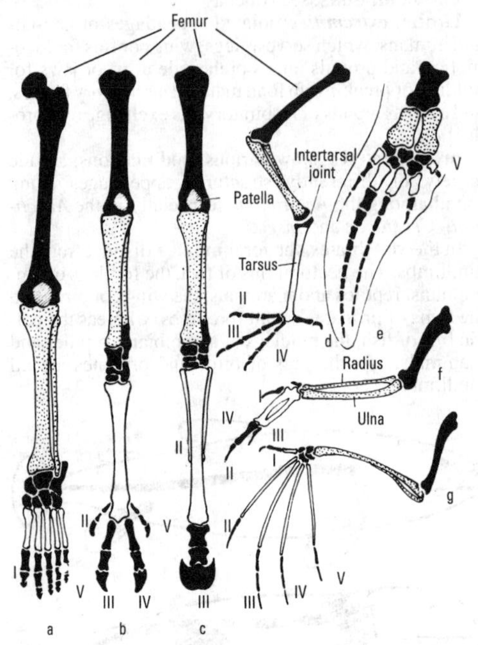

Fig.2. *Limbs.* Pentadactyle limb structure of different vertebrates. Left hind limb: *a* human, *b* deer, *c* horse, *d* bird. Right forelimb: *e* whale, *f* bird, *g* bat.

The basic plan of the pentadactyle fore and hindlimb is modified according to its function (ambulatory, swimming, digging, flying). Bird limbs display extensive functional adaptations. In the wing skeleton, the carpals, metacarpals and phalanges are reduced. In the leg, the metatarsals and some tarsals are fused to form the tarso-metatarsus (popularly called the "tarsus"). According to the disposition of the foot during locomotion, mammalian limbs show a variety of modifications, especially in the autopodium (Fig.2; a,b,c). See also Digitigrade, Plantigrade, Unguligrade.

Lime: see *Rutaceae, Tiliaceae.*

Limnadia: see *Branchiopoda.*

Limnic: pertaining to fresh water; inhabiting or existing in (e.g. sediments) fresh water. Converse terms are Terrestrial (see) and Marine (see).

Limnobacteriology, ***limnomicrobiology:*** bacteriology of freshwater environments, studied in close association with the biochemistry of the microbial metabolism and the microbial degradation of organic materials.

Limnokinetics: totality of all water movements in stagnant lakes.

Limnology: study of freshwater ecosystems, especially lakes, their chemical and physical properties and their constituent organisms. L. developed from hydrobiology. Applied L. includes fishery biology, the biology of water used for industrial and domestic purposes, and the biology of effluents. Paleolimnology studies the history of lakes, based largely on the analysis of bottom sediments.

Limnoplankton: see Plankton.

Limnoria: see *Isopoda.*

Limnoriidae: a family of the *Isopoda* (see).

Limnosaprophytic conditions: see Saprophytic system.

Limonene, ***cinene, cajeputene, kautschin, 1-methyl-4-(1-methylethenyl)cyclohexene,*** or **p-*mentha-1,8-diene:*** an unsaturated, cyclic, monoterpene hydrocarbon, with a pleasent lemon-like odor. In the dextrorotatory, levorotatory or racemized form, it is an important constituent of many essential oils. (+)–L. constitutes up to 90% of Seville orange oil and up to 40% of caraway oil. (–)–L. is present in the oils from silver fir *(Abies alba)* and peppermint. Optically inactive L. (also known as ***dipentene)*** occurs in nutmeg, camphor and turpentine oils.

CH_3 / H_3C / CH_2

Limonene

Limpet: see *Gastropoda.*

Limpkin: see *Gruiformes.*

Limulus: see *Xiphosura.*

Linaceae, ***flax family:*** a family of about 500 species of dicotyledonous plants, mostly in temperate regions of the Northern Hemisphere. They are mostly herbaceous plants, rarely bushes or trees. Leaves are simple, entire or weakly toothed, and usually alternate. The perianth is regular, usually with 5 petals and 5 sepals (rarely 4). Stamens are equal in number to the petals, with which they alternate. The ovary is 5-celled, and the fruit is a capsule, rarely a drupe. Self-pollination is rare and pollination is mostly by insects. One of the oldest known cultivated plants is the pale blue-flowered ***Flax (Linum usitatissimum).*** In the northern ranges of the

Linaceae. Flax *(Linum usitatissimum). a* Flowering plant. *b* Transverse section of a fruit.

temperate zone of Europe, it is grown mostly as the small-seeded, weakly branched variety, Fiber Flax, whose stem fibers are used for the manufacture of linen. In the Mediterranean region, India and Argentina, the Oil Flax variety, which possesses relatively large seeds, is grown for the preparation of linseed oil. Linseed oil is used for cooking, and for the preparation of oil paints, varnishes and linoleum.

Linaloe oil: see Linalool.

Linalool, ***linalol, 3,7-dimethyl-1,6-octadien-3-ol,*** or ***2,6-dimethyl-2,7-octadien-6-ol:*** a noncyclic, doubly unsaturated, optically active monoterpene alcohol (an isomer of geraniol), which smells strongly of lily of the valley *(Convallaria).* It is the main constituent of linaloe oil (a volatile oil from the Mexican wood, *Bursera delpechiana,* family *Burseraceae),* and it is used as a replacement for French lavender oil in perfumery. The dextrorotatory form is also known as *coriandrol,* and the levorotatory form as *licareol.*

```
CH3                    OH
   \                   |
    C=CH—CH2—CH2—C—CH=CH2
   /                   |
CH3                    CH3          Linalool
```

Lincoln index, ***recapture method, capture-recapture method:*** a method for calculating the population density of a species. A number of live individuals are trapped, marked, then released. After a certain period, trapping is repeated, and the ratio of marked to unmarked individuals in the trapped sample is determined. $N/m = n/r$, so that $N = mn/r$, where N is the total size of the population, m is the number of marked animals released into the population, n is the total number of animals in the second trapping, and r is the number of marked animals in the second trapping. It is assumed that the marked animals after release become randomly distributed in the population, and that all animals, marked or unmarked, have an equal probability of being trapped. In the first instance, the method does not take account of population dynamics, but continued sampling at measured intervals can provide useful information on factors such as immigration, emigration, birth rate, and death rate. For types of traps, see Activity density.

Linden: see *Tiliaceae.*

Lingua: see Tongue.

Linguatulida: see *Pentastomida.*

Linkage, ***genetic linkage:*** tendency for genes localized on the same chromosome not to display free and random recombination during meiosis, but to occur together in the gametes. Such genes are inherited together and are said to belong to the same L. group. In contrast, genes localized on different chromosomes display free recombination. During unimpaired recombination of two heterozygous allele pairs (a^+b and ab^+), meiosis results in 50% recombination, producing gametes with the genetic constitution ab and a^+b^+. If L. is present, the proportion of recombination gametes is always less than 50% (partial L.); in special cases, recombinants are totally absent (total L.). Recombination of linked genes occurs by crossing over. The percentage of recombination gametes produced by the crossing over of two linked allele pairs is an index of the degree of L. and the distance separating the allele pairs on same the chromosome (see Recombination value), and it can be used to construct gene maps for a particular L. group.

In genetic formulas, L. is represented by placing the linked genes in the order of their occurrence and underlining them. In hybrids, the alleles of linked genes are normally symbolized as a fraction, with the genetic formula of the mother as the numerator and that of the father as the denominator, e.g. +b+d / a+c+ for four linked allele pairs in one homologous chromosome pair.

Linkage equilibrium: see Genetic equilibrium.

Linkage group: a group of genes localized together on one chromosome. During meiosis, the distributed (segregated) units are L.gs. rather than individual genes. L.gs. are not closed units. Thus, the homologous L.gs. contributed by both parents during fertilization or parasexual processes can perform reciprocal exchanges (see Crossing over), leading to Recombination (see) of the genes of a L.g. Similar reciprocal exchanges occur between the chomosomes of different viruses in cells with mixed virus infections. With the aid of recombinant analysis, the genes of each L.g. can be unequivocally arranged in the order of their occurrence within the L.g. Every genome contains L.gs., and the number of linkage groups that can be identified by breeding experiments is always equal to, or less than, the number of chromosome pairs of that organism.

Linkage value: the percentage of individuals in which linked genes (see Linkage) displayed no recombination by crossing over during meiosis. The ratio of the L.v. to the Crossing over value (see) is called the *linkage number,* and it represents an index of the frequency of recombination between the genes under consideration and the strength of their linkage, i.e. the strength of linkage between two pairs of genes varies inversely with the amount of recombination or crossing over between them.

Linnaea: see *Caprifoliaceae.*

Linnet: see *Fringillidae.*

Linoleic acid: CH_3—$(CH_2)_3$—$(CH_2$—$CH{=}CH)_2$—$(CH_2)_7$—COOH, a doubly unsaturated higher fatty acid, which occurs esterified in the acyltriglycerols of many plant oils (see Fats), e.g. linseed oil (30–40% of the total fatty acid content), poppy seed oil (66%), soybean oil (up to 60%). Like linolenic and arachidonic acid, L.a. is an essential dietary component for some mammals, and is therefore a member of the group of unsaturated acids collectively known as vitamin F.

Linolenic acid: CH_3—$(CH_2$—$CH{=}CH)_3$—$(CH_2)_7$—COOH, a triply unsaturated higher fatty acid, which occurs in esterified form in the acyltriglycerols of many plant oils (see Fats), e.g. it constitutes up to 50% of the total fatty acid content of linseed oil, cannabis oil, poppy seed oil and lentil oil. Like linoleic and arachidonic acid. L.a. is an essential dietary component for some mammals, and is therefore a member of the group of unsaturated acids collectively known as vitamin F.

Lion: see *Felidae.*

Lion tamarin: see Marmosets.

Liparis: see *Cyclopteridae.*

Lipases: hydrolases that function as carboxylic acid esterases, and catalyse the stepwise hydrolytic cleavage of triacylglycerols (neutral fats) to fatty acids and glycerol. According to their substrate specificity, they are classified as ***triacyl-, diacyl-*** and ***monoacylglycerol L.*** Their pH optimum is 7, and they are particularly active and widespread in animals. Especially rich sources are the pancreas, intestinal wall and the liver. Dietary fats are cleaved in the small intestine by pancreatic lipase

(M_r 35 000), which is activated by bile acids (e.g. taurocholic acid) and acts upon emulsified fats. It cleaves the ester bonds at C-1 and C-3. The resulting 2-monoacylglycerol is either absorbed directly, or hydrolysed further to glycerol and fatty acid.

Lipid bilayer: the lamellar arrangement of lipids in biological and artificial membranes (Fig.). The thickness of the L.b. is about 5 nm. The constituent lipids display a high degree of lateral mobility. For physical-chemical reasons, however, transverse movement in pure lipid phases is practically nonexistent. For experimental purposes, artificial L.bs. are prepared as either Bilayer lipid membranes (see) or liposomes (see).

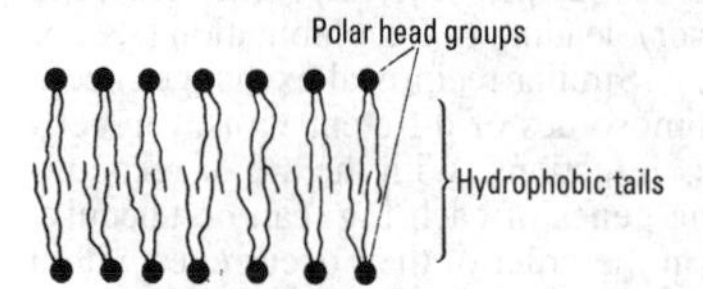

Lipid bilayer. Lamellar arrangement of phospholipids in a lipid bilayer. The lipid "tails" display a high degree of thermal mobility.

Lipid filter theory of permeability: in plant physiology, a theory that the efficient and sometimes selective uptake of relatively large, lipid-soluble molecules occurs because they are able to dissolve in, and therefore diffuse through, the lipid phase of the cell membrane.

Lipids: water-insoluble natural substances. Many L. are esters of long-chain fatty acids with alcohols or alcohol derivatives. Thus, neutral fats (triacylglycerols) and waxes are L. Glycolipids and phospholipids are *complex L.* Substances such as sterols and carotenoids are known as *nonsaponifiable L.,* since they are not esters and cannot therefore by hydrolysed (saponified) by aqueous sodium or potassium hydroxide.

Lipochromes: see Carotenoids.

Lipoic acid, ***thioctic acid:*** 6,8-dithiooctanoic acid (Fig.), a cyclic disulfide with a carboxylic acid side chain. L.a. occurs in all living organisms. It is a coenzyme component of the 2-oxoacid dehydrogenase complexes, where it is bound by formation of an amide linkage between its carboxylic acid group and the ε-amino group of a lysyl residue of the protein. These multienzyme complexes catalyse the decarboxylation of 2-oxoacids, like pyruvic acid and 2-oxoglutaric acid, with transfer of hydrogen and acyl residues. During this process, the sulfur-containing ring of L.a. is opened with the formation of two thiol groups. The resulting dihydrolipoic acid (Fig.) is reoxidized to L.a. by the action of dihydrolipoate dehydrogenase. Interconversion of L.a. and dihydrolipoic acid represents an active redox system.

L.a. was discovered in 1950 as a growth factor of various microorganisms.

COOH S—S COOH SH SH

Lipoic acid   Dihydrolipoic acid

Lipolysis: chemical or enzymatic hydrolysis of fats and fatty oils into fatty acids and glycerol. See Lipases.

Lipoproteins: proteins in combination with lipids, e.g. lecithins, cephalins or other phospholipids. They are found mainly in the blood and cell cytoplasm, in the membranes of cells and cell organelles, and in egg yolk. They function in the transport and distribution of lipids, such as cholesterol, fatty acids and fat-soluble vitamins. The lipid moiety of a L. is not covalently bound.

Liposome: a vesicle-like particle consisting of one or several lipid bilayers. They are generated from lipid suspensions by ultrasonic irradiation, and by other techniques. Ls. are appropriate models for biological vesicles, e.g. synaptic vesicles, pinocytosis vesicles and lysosomes. They are also exploited clinically for the target-directed application of pharmaceutical agents. Receptors for target cells can be incorporated into their structure, and the pharmaceutical agent is not released until the Ls. fuse (see Fusion) with the target cells.

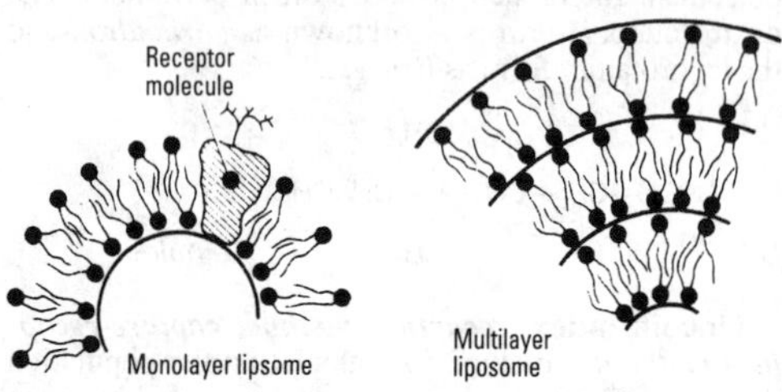

Liposome. Monolayer and multilayer liposomes. A receptor molecule is shown incorporated into the monolayer liposome.

Lipotes: see Whales.

Lipotropins: polypeptide hormones secreted by the pituitary, which promote the release of free fatty acids from adipose tissue. They activate the adenylate cyclase system of their target tissue. An α- and a β-form are known. The β-form consists of 91 amino acid residues, and sequence 41–58 is identical with the sequence of β-melanotropin. L. exhibit no hormonal activity in humans, and are therefore considered to be prohormones of β-endorphin.

Liquid junction potential: see Diffusion potential.

Liquid plating method: see Plating.

Liquor amnii: see Extraembryonic membranes, Amnion.

Liquor folliculi: see Oogenesis.

Liquorice: see *Papilionaceae.*

Liriodendron: see *Magnoliaceae.*

Lissodelphininae: see Whales.

Lissodelphis: see Whales.

Liszt monkey: see Marmosets.

Lithobiomorpha: see *Chilopoda.*

Lithobios: see *Chilopoda.*

Lithobiidae: see *Chilopoda.*

Lithops: see *Aizoaceae.*

Lithocholic acid: see Bile acids.

Lithode crabs: see Stone crabs.

Lithodes: see Stone crabs.

Lithodidae: see Stone crabs.

Lithodinae: see Stone crabs.

Lithodomus: a mytiloid bivalve with equal-sized valves, which together form an elongate cylinder, rounded on both sides. The hinge is toothless. It is burrowing species found mainly in relatively hard, calcareous substrata, which it erodes by the action of carbon dioxide released by respiration from the mantle surface. Fossils are found from the Carboniferous to the present, mainly in deposits near sea coasts.

Litmus: a pigment obtained from the lichens, *Rocella fuciformis* and *Rocella tinctoria,* which grow on the coasts of Scandanavia, England, France and the Mediterranean. It is a derivative of orcinol, but the commercial product is not chemically pure. L. was one of the earliest chemical indicators (red in acid, blue in alkali), and it is still used for this purpose, either in solution or as L. paper.

Litmus milk: a bacteriological test medium prepared by adding 2.5% by volume of an alcoholic litmus solution to freshly steamed milk, from which the cream has been removed. The preparation is then sterilized by steaming for 20 min on three successive days. [The alcoholic litmus solution is prepared by extracting 80g litmus granules with two 150 ml quantities of boiling 40% ethanol for about 1 min; just sufficient 1M HCl is then added to make the solution purple.] L.m. is used for testing the fermentation of lactose and the clotting of milk. The medium is inoculated with the bacteria under test, and placed in an incubation cabinet. A red coloration indicates fermentation of lactose with acid production. Decolorization is indicative of reducing activity. Coagulation without acid formation shows the production of rennin, and degradation of milk proteins (peptonization) is due to the production of proteinases.

Litocraniini: see *Antilopinae.*

Litocranius: see *Antilopinae.*

Litorea: see Biocoenosis.

Litter: see Necromass.

Little brown bats: see *Vespertilionidae.*

Littoral: that region of the Benthal (see) extending from the shore to a depth to which sunlight can still penetrate (about 200 m). Through photosynthesis, the L. contributes to the net primary production of the aquatic system. In the transition from land to water, the L. can be subdivided into: *epilittoral* (the land zone, which is not submerged or splashed by the water); *supralittoral* (the bank zone, which is affected by spray and splashing); *eulittoral* (the intertidal zone; in feshwater the growth zone for higher aquatic plants); *sublittoral* (the continental shelf). See Sea.

Littorella: see *Plantaginaceae.*

Littorina: see *Gastropoda.*

Littorina sea: see *Gastropoda.*

Lituites (Latin *lituus* curved rod, crozier): an extinct nautiloid genus of the Ordovician. In the early forms the shell is spirally rolled in one plane and has an overall disk-like shape. In successive forms the shell becomes less rolled, until the final form, which is elongated and straight. See Orthogenesis (Fig.). The cylindrical siphuncle lies close to the inside surface of the shell. Fossilized L. constitute the *Lituites limestone* of the Swedish island of Øland.

Live-bearing toads: see *Bufonidae.*

Livelong: see *Crassulaceae.*

Liver, ***hepar:*** an accessory gland or organ of the midgut, and the largest metabolic organ of vertebrates. It is a lobed structure, of soft consistency, red-brown in color, and weighing about 1 500 g in the adult human. It lies in the right subphrenic space of the abdominal cavity, where its upper, convex surface (parietal surface) fits closely against the undersurface of the diaphragm, while its lower, concave surface (visceral surface) faces downward and rests upon the abdominal viscera. A wide cleft (the portal or transverse fissure) on the visceral surface houses the portal vein and hepatic artery as they enter the L., and the hepatic (bile) duct as it leaves the L. The hepatic vein exits on the upper surface of the organ. In the adult human, 1 to 2 l blood per minute is supplied by the portal vein and liver arteries.

Together with the muscles, the L. is the most important site of carbohydrate metabolism. The L. cell (hepatocyte) is freely permeable to glucose, in contrast to the cells of other organs whose permeability to glucose is regulated by insulin. The L. stores glycogen and releases it as needed to maintain a constant blood sugar level. Glugagon stimulates glycogenolysis in the L., while glucocorticoids promote gluconeogenesis, or synthesis of glucose from precursors such as lactate (produced in muscle) and glucogenic amino acids. Fatty acids from food or fat depots are degraded by β-oxidation in the L. When necessary, the L. converts carbohydrates into fatty acids, which are transported in the form of phosphatides to adipose tissue, where they are converted into neutral fats (acyltriglycerols). The L. also synthesizes cholesterol, which it exports to other tissues or converts into bile acids.

The L. is crucial in the nitrogen metabolism of the organism. Thus, amino acids are degraded by transamination or deamination, or synthesized from α-keto acids; waste nitrogen from the metabolism of amino acids is converted into urea (see Urea cycle). The L. is also the site of synthesis of blood plasma proteins (albumin, some glubulins and fibrinogen) and the hepatogenic blood-clotting factors (prothrombin, accelerator globulin, proconvertin, Christmas factor). Some hemoglobin is degraded to bile pigments, which are excreted in the bile.

A large number of substances, both natural (e.g. steroid hormones) and foreign (e.g. drugs) are transformed or degraded in the L.

Given the number and variety of metabolic processes occurring in the L. , it follows that any disorder will result in more or less severe metabolic disturbances.

Liver flukes (plate 42): trematodes of the order *Digenea,* which parasitize the liver (in particular the bile ducts and gall bladders) of herbivorous animals.

The ***Common liver fluke*** *(Fasciola hepatica)* is a flattened, leaf-shaped trematode, growing up to 4 cm in length. Almost cosmopolitan in its distribution, it lives in the bile canals of sheep and cattle and less commonly in other animals, including man. Numerous small spines project from the body wall. An attachment sucker is situated on the ventral surface, a short distance behind the mouth. The anteriorly situated mouth (surrounded by a muscular sucker) leads into a muscular pharynx and from there, via very short esophagus, into a bifurcated intestine. Numerous branches of the intestine extend into all parts of the animal.

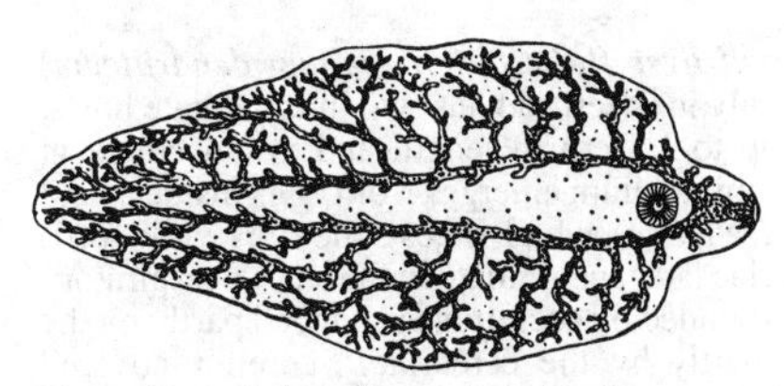

Fig. 1. *Liver flukes. Fasciola hepatica.*

There is an alternation of generations, linked to a change of host. If an egg encounters moist conditions, it hatches into a *miracidium,* a tiny larva covered with long cilia. The anterior papilla of the miracidium houses a terminal mouth opening into a simple sac-like gut, and shortly behind the mouth are two eye-spots. The body also contains a group of germ cells. If the miracidium encounters the freshwater snail, *Lymnaea truncatula,* it bores through the skin, attaches to the wall of the respiratory chamber, and transforms into a *sporocyst.* The sporocyst, the first stage of the asexual generation, is little more than a sac-shaped structure containing mostly germ cells (with a few delicate muscles and protonephridia). Without fertilization, each germ cell develops into a worm-like *redia,* representing the second asexual generation. Rediae have a simple sac-like gut and a terminal mouth, and they also contain germ cells. On reaching maturity, they break through the sporocyst wall, then migrate through the snail's tissues to the digestive gland. In the winter, each redial germ cell then develops, without fertilization, into a *cercaria,* which is the larva of the sexual generation. In the summer, however, the redial germ cells give rise to a second generation of rediae, whose germ cells eventually develop into cercariae. The cercariae escape through an aperture in the body wall of the redia, migrate from the snail's digestive gland to the respiratory cavity, and from there to the exterior. Spherical to ovoid in shape, the cercaria has an oar-like swimming tail. Internally, it closely resembles to the adult liver fluke, but lacks sex organs. After swimming for some time, cercariae encyst on plants at the water's edge, and development continues if a cyst is ingested by a suitable grazing animal. The cyst wall is digested in the alimentary canal of the host, and a tailless young liver fluke emerges. This penetrates the wall of the alimentary canal, enters the peritoneal cavity, bores through the connective tissue covering the liver and enters the bile canal. About 3 months later it is fully developed with a hermaphrodite reproductive system. Eggs leave the fluke through the genital pore (just in front of the ventral sucker) and pass to the outside world in the host's feces.

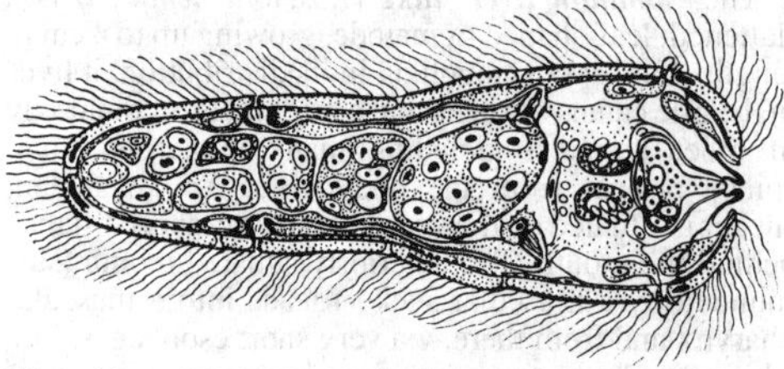

Fig.2. *Liver flukes.* Longitudinal section of the miracidium of *Fasciola hepatica.*

The ***Small liver fluke*** *(Dicrocoelium dendriticum)* occurs mainly in sheep and has two intermediate hosts. It grows up to 1.5 cm in length and is lanceolate in shape. The miracidium emerges from eggs inside a land snail (e.g. *Helicella),* which ingests the eggs while feeding. Cercariae later accumulate in the snail's respiratory cavity, surrounded by secretions produced partly by the snail and partly by the cercariae. Secretion-covered balls of cercariae are ejected by the snail; they adhere to plant surfaces and are ingested by ants. The cycle of infection is completed when a ruminant ingests an infected ant while feeding.

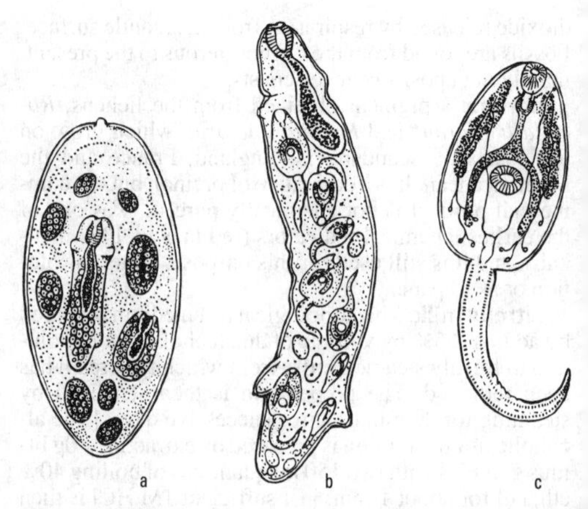

Fig.3. *Liver flukes.* Developmental stages of *Fasciola hepatica. a* Sporocyst containing rediae. *b* Redia. *c* Cercaria.

Clonorchis sinensis infects the bile ducts of humans, dogs, cats, pigs, rats and other animals in the Far East. The life cycle differs from that of *Fasciola hepatica* mainly in the involvement of a third host (any of a number of freshwater fishes), which is attacked by the cercariae. The infection is transferred by eating raw fish.

Clonorchis sinensis infects the bile ducts of humans, dogs, cats, pigs, rats and other animals in the Far East. The life cycle differs from that of *Fasciola hepatica* mainly in the involvement of a third host (any of a number of freshwater fishes), which is attacked by the cercariae. The infection is transferred by eating raw fish.

Liver rot: see Fascioloiasis.

Liverworts: see *Hepaticae.*

Living stones: see *Aizoaceae.*

Liza: see *Mugilidae.*

Lizards: see *Lacertilia.*

Llama, *Lama guanacoë* var. *glama:* a domestic memeber of the *Camelidae,* which is derived from the Guanaco (see). It is used in South America as a beast of burden, as well as a source of meat and wool. It is somewhat larger and has a longer coat than the Guanaco, and it is bred in various color varieties.

Llanos (Spanish *plain):* South American plains, particularly those of the Orinoco basin, formed from Tertiary lakes by infilling with alluvial river sediments. They are mainly savanna grasslands with very few trees.

Lloyd-Ghelardi equitability component: see Diversity index.

Loaches: see *Cobitididae.*

Loa loa: see Filariasis.

Lobaria: see Lichens.

Lobeline: the main alkaloid of *Lobelia inflata* L., a herbaceous member of the *Campanulaceae,* native to North America. L. is a piperidine alkaloid (Fig.), which stimulates the respiratory center, and is therefore used therapeutically in cases of central paralysis of respiration. It has certain importance as an aid to curing addiction to tobacco smoking.

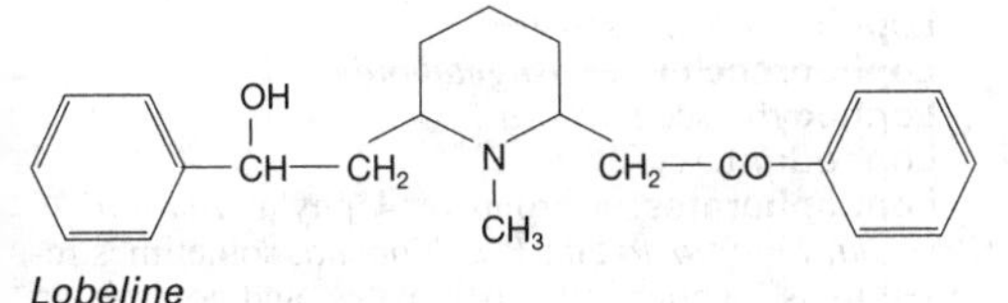

Lobeline

Lobipes: see *Phalaropodidae.*

Lobopodium: see *Rhizopodea.*

Lobsters: large crustaceans of the family *Nephropidae (= Homaridae),* infraorder *Astacidea,* order *Decapoda* (see). All are marine, and unlike the Spiny lobsters (see) they have large claws (chelipeds), the gripping edge of the larger and heavier chela bearing serrations and tuberosities. The ***European lobster*** *(Homarus gammarus)* attains a a considerable size and has large, heavy chelipeds. The ***American lobster*** *(Homarus americanus)* is similar but somewhat larger, attaining lengths of 60 cm and weighing up to 22 kg. They live in rock fissures and under stones, remaining hidden during the day and when they are molting, and feeding mainly on bivalves and to some extent on carrion. The ***King, Norwegian*** or ***Norway lobster*** *(Nephrops norvegicus)* is caught off the coasts of Brittany and Norway and sometimes in the Mediterranean; it prefers soft bottoms in deeper waters, has characteristically slim chelipeds, and may attain a length of 22 cm. Lobsters are caught commercially for their meat, and the enormous chelipeds of some species may contain as much meat as the abdomen. See Crayfish, Spiny lobsters.

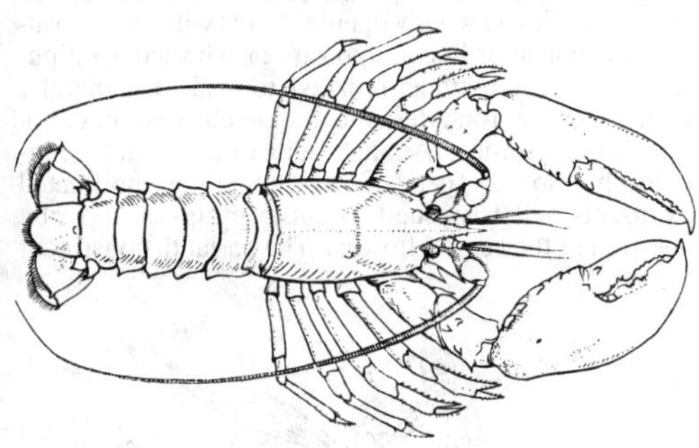
European lobster (Homarus gammarus)

Local potential: see Excitation.

Lockjaw: see Tetanus bacterium.

Locomotion: see Movement.

Locomotory apparatus: the totality of organs involved in the locomotion of an animal. In vertebrates, a distinction is drawn between the *passive L.a.* (skeleton) and the *active L.a.* (musculature).

Loculicidal capsule: see Fruit.

Loculomycetidae: see *Ascomycetes.*

Locust birds: see *Glareolidae.*

Locust borer: see Long-horned beetles.

Locust finch: see *Ploceidae.*

Locust lobsters: see Spiny lobsters.

Locusts: see *Saltatoptera.*

Locust tree: see *Caesalpiniaceae, Papilionaceae.*

Lodicule: see *Graminae.*

Loess: a substratum or sediment of wind-deposited fine quartz (0.015–0.05 mm diameter) found widely in central USA, northern Europe, former USSR, China and Argentina. It gives rise to high quality soils, e.g. the black earths of the European steppes.

Loggerhead: a name applied to the North American Loggerhead shrike *(Lanius ludovicianus)* (see *Laniidae)* and to Steamer ducks *(Tachyeres)* in the Falkland Islands (see *Anseriformes).*

Loggerhead sponges: see *Porifera.*

Loggerhead turtle: see *Chelonidae.*

Lognormal distribution, *logarithmic-normal distribution:* a theoretical asymmetrical distribution, which is steeper on the left when the random values are plotted on the abscissa. Many biological properties that do not decrease below a certain limiting value (e.g. pulse rate, blood pressure) display a distribution that approximates closely to a L.d. A L.d. distribution can be converted into a normal distribution by plotting the random values (x) on a logarithmic abscissa (or by plotting log x on a linear abscissa scale).

Loiasis: see Filariasis.

Loiseleuria: see *Ericaceae.*

Loligo: see *Cephalopoda.*

Lolium: see *Graminae.*

Lolliguncula: see *Cephalopoda.*

Lomentum: see Fruit.

Lonchura: see *Ploceidae.*

London forces: see Van der Waals forces.

London pride: see *Saxifragaceae.*

Long-billed waders: see *Scolopacidae.*

Longclaws: see *Motacillidae.*

Long-day plants: plants that are sensitive to the photoperiodicity of their environment, and which under natural conditions produce flowers only when the period of darkness does not exceed a certain critical maximum. Although they are called long-day plants, they are really "short-night"plants, the crucial factor in the initiation of flowering being the period of uninterrupted darkness. The critical length of the dark period varies according to the species. Many temperate cereal species are long-day plants, e.g. rye, wheat, barley and oats, as well as sugarbeet, dill, spinach, mustard, poppy, henbane *(Hyoscyamus niger),* etc. In addition, flowering in a number of plants is favored by, but is not necessarily dependent on, long days, e.g. onions, lettuce, flax. Many long-day plants grow with a rosette habit until induction, when extension of the shoot is coupled with formation of the floral meristem. See Photoperiodicity, Flower development.

Long-fingered frogs: see *Arthroleptinae.*

Long-glanded coral snakes, *Maticora:* brightly patterned (longitudinally striped) yellow, red and black front-fanged snakes (see *Elapidae)* of eastern India and Southeast Asia. The large venom glands extend from the upper jaw through the entire anterior third of the body, causing the heart to be displaced further to the rear than in other species. The venom appears to be less toxic than that of other cobra species.

Long-horned beetles, *Cerambycidae:* a family of the *Coleoptera* (see), whose numerous, mostly slender species can be recognized from their antennae (usually with 11 segments), which are often longer than the body, especially in the male. Larvae live in the wood of diseased or dead trees, rarely in the earth. Some members occur as pests in living trees and in building timber. European pest species include *Hylotrupes bajulus* L. ***(Old house borer*** or ***House longhorn),*** a pest of dry coniferous timber, which often destroys telegraph poles and house timbers; *Acanthocinus aedilis* L. ***(Timberman beetle),*** and *Ergates faber* L., particularly common in eastern Europe, which lives in old pine stumps and also attacks building timber. American pest species include *Saperda candida* ***(Roundheaded apple tree borer),*** *Saperdia calcarata* ***(Poplar borer)*** and *Megacyllene robiniae* ***(Locust borer).***

Long-horned moths: moths of the family *Adelidae.* See *Lepidoptera.*

Long-nosed squirrel: see *Sciuridae.*

Long-nosed tree snake, *Ahaetulla nasuta:* an extremely slender, brilliant green, totally arboreal, viviparous Southeast Asian snake of the subfamily *Boiginae* (see). It has a long tail, a pointed, turned up nose, and feeds mainly on tree lizards and tree frogs.

Long-nosed viper: see Sand viper.

Long-short-day plants: plants that are sensitive to the photoperiodicity of their environment, and which do not flower until they have experienced first long-day conditions and later short-day conditions (see Short-day plants, Long-day plants, Short-long-day plants). Examples are *Aloe bulbifera,* various species of *Bryophyllum (Crassulaceae),* and Jessamine *(Cestrum nocturnum).* They do not flower during the short days of spring, and under natural conditions do not produce flowers until late summer or autumn. See Photoperiodicity, Flower development.

Long-snouted squirrel: see *Sciuridae.*

Long-tailed field mouse: see *Muridae.*

Long-tailed lizards: see *Takydromas.*

Long-tailed tits: see *Paridae.*

Lonicera: see *Caprifoliaceae.*

Loofah: see *Cucurbitaceae.*

Loons: see *Gaviiformes.*

Loop of Henle: see Kidney functions.

Loosejaws: see *Stomiiformes.*

Loose smut: see *Basidiomycetes (Ustilaginales).*

Loosestrife: see *Primulaceae.*

Lophiidae: see Goosefishes.

Lophiiformes, *Pediculati,* ***anglerfishes:*** an order of plump to monstrous fishes with disproportionately large heads and large mouths. The pelvic fins are set far back and often have a muscular stump or root. In many species the first ray of the spinous dorsal fin is present on the head, and it is modified to an illicium, i.e. a structure resembling a line and bait (esca), which attracts prey to the mouth. Familiar families are *Antennariidae* ***(Frogfishes),*** *Lophiidae* ***(Goosefishes),*** *Ogcocephalidae* ***(Batfishes).***

Lophiodon (Greek *lophos* hill, *odontes* teeth): an extinct, tapir-like genus of the order *Perissodactyla.* The forefeet had 4, the hindfeet 3 toes. Fossil remains are found in the European and American Eocene, and they are characteristic fossils of the brown coal measures of the Geisel valley near Halle in Germany.

Lophiomyinae, ***maned rats:*** a subfamily of the *Cricetidae* (see), represented by a single species *(Lophiomys imhausi)* with several subspecies, found in dense forests at altitudes above 1 200 m in East Africa from Kenya across the Ethiopian mountains to the Sudanese border. As an adaptation to climbing, the thumb is apposable and the fingers are mobile. Pairs live together and build nests in tree or rock cavities. The tail is bushy and the body carries a series of black and white markings.

Lophius: see Goosefishes.

Lophobranchii: see *Syngnathoidei.*

Lophocyte: see *Porifera.*

Lophodont: see Teeth.

Lophophorates: a group of 4 phyla: *Phoronida, Bryozoa, Entoprocta* and *Brachiopoda,* sometimes referred to as lophophorate coelomates, and comprising about 4 300 aquatic species. Encircling the mouths of all L. is circular or horseshoe-shaped extension of the body wall, known as a lophophore. The lophophore carries ciliated tentacles, which are hollow outgrowths of the body wall, i.e. each tentacle contains an extension of the coelom. Ciliary action generates a flow of water through the lophophore, enabling the trapping of plankton. There is evidence of tripartite body organization, although the protocoel is not well developed. A transverse septum divides the coelom into mesocoel (anterior) and metacoel (posterior). Nearly all members are sessile with a reduced head and protective covering. With the exception of a few brachiopods, a U-shaped digestive tract is present. For further details of structure and organization, see the entries for each phylum.

Lophophore: see Lophophorates.

Lophopoda: see *Bryozoa.*

Lophortyx: see *Phasianidae.*

Lophotrichous flagella: see Bacterial flagella.

Lophozosterops: see *Zosteropidae.*

Loquat: see *Rosaceae.*

Loranthaceae, ***mistletoe family:*** a family of dicotyledonous plants containing about 1400 species in 36 genera, found predominantly in the tropics, with some species in temperate regions. They are mostly small, semiparasitic (on woody plants) shrubs with entire, usually lanceolate or linear, opposite or whorled, exstipulate leaves (sometimes reduced to scales), generally with a low chlorophyll content and yellowish in color. The 4- to 6-merous, hermaphrodite or unisexual, actinomorphic flowers (petals free or united) are pollinated by insects, and they usually occur in groups of 3 (or 2 by abortion of the central flower). The perianth consists of

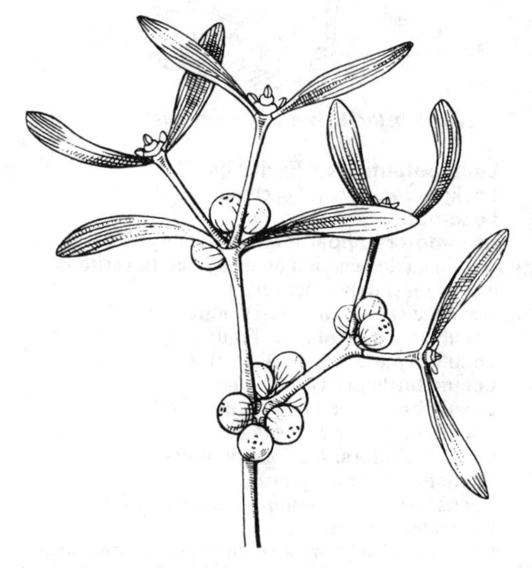

Loranthaceae. Mistletoe *(Viscum album)*

2 whorls, the inner often brightly colored, the outer sometimes suppressed and the inner then sepaloid. Stamens are epipetalous. Male flowers often have a rudimentary ovary, and female flowers often possess staminodes. The ovary is inferior (style simple or absent) and the ovules are usually not differentiated from the placenta. The fruit is a 1- to 3-seeded berry or drupe, clothed in a sticky, slimy layer of viscin. Each fruit contains a single seed with no testa; endosperm is usually abundant, and the embryo is large (sometimes 3 embryos per seed). Attachment to the host is by a sucker-like outgrowth called a "haustorium". *Nuytsia,* a monospecific genus of Western Australia, is exceptional in that it forms a tree, up to 9 m in height, which roots in the ground and parasitizes neighboring herbage with the aid of parasitic roots

Viscum album ***(Mistletoe)*** is dioecious with leathery, evergreen leaves, parasitizing mostly a wide variety of deciduous trees in Europe and southern Scandanavia, southward to North Africa and eastward to Central Asia and Japan. In northern Europe, it is most commonly found on apple. Three host-specific races are found in Central Europe, parasitizing silver fir, pine and deciduous trees, respectively. The European *Viscum* found on fir and pine has also been decribed as the single species, *Viscum laxum.* The southern European *Loranthus europaeus* is found only on oak *(Quercus)* and sweet chestnut *(Castanea sativa).*

Exracts of *Viscum album* have been shown to inhibit cancerous growths in animal experiments, an activity probably ascribable to basic glycoproteins and the peptide, viscotoxin. *Viscum album* and species of *Phoradendron* ***(American mistletoe)*** are used for Christmas decoration. The family is otherwise of no economic or ornamental importance.

Lord Derby's zonure: see *Cordylidae.*

Lords and ladies: see *Araceae.*

Loricata: see *Polyplacophora.*

Lories: see *Psittaciformes.*

Loriinae: see *Psittaciformes.*

Lorikeets: see *Psittaciformes.*

Loris:

1) Prosimians (see) of the subfamily *Lorisinae.* The index finger is much reduced and the thumb has migrated so that it diverges 180 ° from the other digits, forming the hand into a powerful clasping organ. L. comprise the following genera and species.

Loris tardigradus ***(Slender L.),*** the only species of the genus, lives in small dispersed groups in forest and woodland in India and Sri-Lanka, feeding mainly on insects. The animal is small (head plus body 225 mm) and lightly built, with a short, woolly, light gray or fawn coat, short pointed muzzle, large mobile ears and no tail. Movement in the tree canopy is smooth and agile, but slow and stealthy.

Nycticebus coucang ***(Slow L.***; plate 40) is found throughout the mainland of South-East Asia and in Java, Sumatra and Borneo. Head plus body measure 325 mm, and the tail is short and stumpy. The diet includes leaves, fruit, insects, birds and birds' eggs. Birds are caught by a lightning strike, which is atypical of the animal's otherwise slow movements.

Nycticebus pygmaeus ***(Pygmy slow L.)*** is restricted to Vietnam and Laos. Head plus body measure 190 mm.

Perodicticus potto ***(Potto),*** the only species of the genus, is the largest L. (head plus body 350 mm). It inhabits tropical rainforest from West Africa to Kenya, feeds on leaves, fungi and gums, but also includes insects and slugs in its diet.

Arctocebus calabarensis ***(Angwantibo*** or ***Golden potto),*** the only species of the genus, is a golden to yellowish brown African L., inhabiting the area between the Niger and Zaire rivers. Its range overlaps that of the Potto, but the Angwantibo prefers the scrub layer and feeds mostly on insects, whereas the Potto prefers the canopy. Head plus body measure 250 mm, with no tail.

2) see Parrots.

Lorius: see *Psittaciformes.*

Lotka's laws: see Volterra's laws.

Lotus bird: see *Jacanidae.*

Louvar: see *Scombroidei.*

Lovage: see *Umbelliferae.*

Lovebirds: see *Psittaciformes.*

Lower fungi: see *Phycomycetes.*

Loxia: see *Fringillidae.*

Loxocemus: see *Xenopeltidae.*

Loxops: see *Drepanididae.*

Loxosomatidae: see *Entoprocta.*

LSD: acronym of Lysergic acid diethylamide (see).

L-tubules: see Muscle tissue.

Lucanidae: see Lamellicorn beetles.

Lucerne: see *Papilionaceae.*

Lucifer: see Shrimps.

Luciferase: a low molecular mass oxidoreductase. See Luciferin.

Luciferin: a collective term for the substrates of luciferases. L. is the reduced form of a compound, which is converted into an electron-excited, oxidized state by luciferase in the presence of oxygen, ATP and (usually) Mg-ions. Energy released from the excited state is given out almost entirely (96%) as visible light.

$$\text{Luciferin} \xrightarrow[\text{Luciferase}]{O_2} \text{oxy-luciferin*} \rightarrow \text{oxy-luciferin} + h\nu$$

Ls. are responsible for the Bioluminescence (see) of many lower organisms (bacteria, fungi, coelenterates, worms, insects, marine animals). The L. of the firefly, *Photinus pyralis,* is a benzothiazine derivative.

Lucioperca: see Perches, Pike-perch.

Lucky clover: see *Oxalidaceae.*

Ludlow: a stage of the Upper Silurian, which takes its name from an area of the same name on the Wales/Shropshire border (UK).

Ludwig effect: see Annidation.

Ludwigia: see *Onagraceae.*

Luffa: see *Cucurbitaceae.*

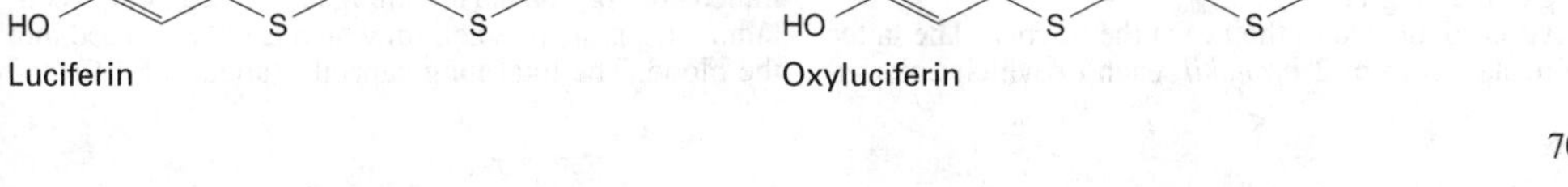

Luciferin Oxyluciferin

Lugworm: see *Sedentaria*

Lullula: see *Alaudidae.*

Lumbar: of, pertaining to, or near the loins. Thus, the human vertebral column contains 5 lumbar vertebrae (Nos. 20–25) in the region of the loins.

Lumbricidae: see *Oligochaeta.*

Lumbricus: see *Oligochaeta.*

Luminescence: collective term for all forms of light generation in solid, liquid or gaseous material at normal temperatures, i.e. cold light, as distinct from light generated by heating materials to high temperatures. See Bioluminescence.

Luminous moss: see *Musci.*

Lumisterol: see Vitamins (vitamin D).

Lumpenus: see Bandfishes.

Lumpsucker: see *Cyclopteridae.*

Lunar rhythms: Biorhythms (see) related to the phases of the moon.

Lundegårdh's law of biological relativity: see Biological law of relativity.

Lung: see Lungs.

Lung compliance: see Pressure-volume curve of the respiratory apparatus.

Lungfishes, *mudfishes,* *Dipteriformes, Dipnoi:* an order of elongated to eel-shaped teleost fishes belonging to the subclass *Sarcopterygia,* class *Osteichthyes* (see). They occur in South America, Africa and Australia. *Lepidosiren* has paired pectoral and pelvic appendages, which are moved alternately. These appendages are long and awl-shaped, each supported by a single, jointed cartilaginous rod. *Lepidosiren paradoxa* is found in the Amazon. There are 4 species of African lungfishes *(Protopterus)* variously distributed throughout all the freshwaters of tropical Africa, e.g. *Protopterus dolloi* of the Zaire basin (see plate 4), and the rare *Protopterus annectus* of the Gambia. The ***Barramunda*** *(Ceratodus forsteri)* of Queensland, Australia, attains a length of 1.8 m and more. Its fins are not awl-like, but consist of a fringe of fin rays surrounding a central muscular lobe. The swim bladder, which serves as an efficient lung, is connected ventrally to the pharynx by an air passage. South American and African L. survive the dry season by burying themselves in the ground in a cocoon of mucous, and may remain there for half of the year. At such times, the air bladder (auxiliary lung) is supplied with air through a small tunnel in the dry mud. *Ceratodus* occasionally comes to land and uses its lungs, but it is essentially an aquatic animal. L. lay large eggs, and the hatched larvae have external gills.

Fossil remains of L. are found from the Middle Devonian to the present. The L. are represented by relatively few living and extinct genera. The order was most numerous in the Upper Devonian, when both freshwater and marine forms existed. All living L. are freshwater forms.

Lungless salamanders: see *Plethodontidae.*

Lungs, *pulmo:* the paired organs of respiratory gas exchange in air-breathing vertebrates. In fine structure, L. are compound alveolar glands, derived embryonically and phylogenetically from a ventral evagination of the foregut (see Respiratory organs). They are homologous with the swim bladder of fishes. Phylogenetically, L. make their earliest appearance as the unpaired lung of the lungfishes.

Air is conducted to the L. via the trachea. The latter bifurcates to form 2 *bronchii,* each of which enters a lung, then divides further into *bronchioles.* The walls of the bronchii and the trachea are reinforced with cartilaginous bars. Dust in the inspired air is largely trapped by the nasal mucus and continually carried back toward the buccal cavity by the action of cilia in the bronchial epithelium.

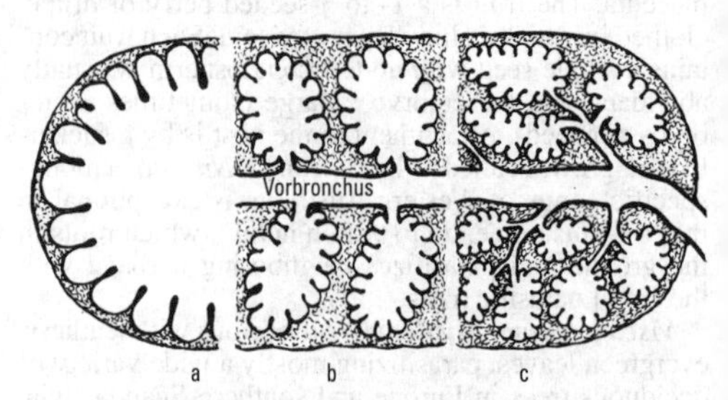

Lungs. Enlargement of the area for respiratory gas exchange with the evolution of the quadruped lung. *a* Wall ridges (amphibians). *b* Septa with alveoli and bronchus (reptiles). *c* Bronchial tree (mammals).

The simplest L. are found in amphibians, which possess sac-like, smooth-walled or slightly ridged (weakly chambered) structures, whose respiratory surface is lined with respiratory epithelium. Reptile L. consist of a more highly developed system of separate chambers. In snakes, only the right lung is developed, and the left lung is degenerate. Only lizards and turtles are able to rhythmically compress and expand their L. by muscular action. Externally, the L. of birds appear very small, but they have a very complex internal structure. Two bronchii arise from the lower end of the syrinx, then give rise to further dorsal and ventral bronchii. The latter are interconnected by a system of narrow tubes, known as *parabronchii,* which serve as the actual sites of respiratory gas exchange in the avian lung. The bronchii terminate in air sacs, which extend throughout the body, even into the bones (pneumatized bones). During flight, each intake of air is circulated twice through the parabranchii, i.e. when the air sacs are filled by inspiration, and when the air sacs are emptied by expiration. Mammalian and reptilian L. are similar in structure. In humans, they consist of 2 distinctly separate L., which are surrounded by 2 delicate membranes, the internal *pleura pulmonaris* (pulmonary pleura) and the external *pleura parietalis* (parietal pleura), which enclose the *cavum pleurae* (pleural cavity) (the pleural cavity is a potential rather than an actual cavity, because the surfaces of the pulmonary and parietal pleurae adhere to one another). The human right lung consists of 3 lobes, and the left lung of 2 lobes. The lung tissue is spongy, elastic and rose-red in color. The bronchii entering each lung display multiple tree-like branching *(bronchial tree)* into bronchioles, which terminate in blind alveoli. Alveoli are supplied with capillaries from the pulmonary artery, and they serve as the actual site of respiratory gas exchange.

For further details of the operation of the human lung, see Respiratory mechanics.

Lung volume, *lung capacity:* air or gas capacity of the lungs. *The total lung capacity* can only by determined after prolonged breathing in a closed system containing a gas such as helium which is not absorbed into the blood. The total lung capacity (about 5.5 l for the adult human) is then calculated from the dilution of the test gas. Part of the total lung capacity (called the *residual volume;* about 1.5 l) remains in the lungs even during the deepest breathing, and even in a corpse; when the thoracic cavity of a corpse is opened, the resulting collapse expells about 1.2 l of air, leaving 0.3 l still in the interstices of the tissues.

Quiet or resting inspiration and expiration via a spirometer, shows that the *resting tidal volume* for an adult is about 0.5 l When breathing to the limit of his or her capacity, however, an individual can inhale a further 2 l (approx.), known as the *inspiratory reserve volume.* Similarly, there is an *expiratory reserve volume* of about 1.5 l The sum of these three volumes is known as the *vital capacity* (Fig.). It depends on body size and physical fitness, and it is larger in men than in women. The sum of the expiratory reserve volume and the residual volume is called the *functional residual capacity.*

As a rough approximation, the vital capacity (in ml) of a man is equal to body mass in kg x 65, whereas the value for women is body mass in kg x 55. Since the respiratory rate of adults at rest is 10 to 20 per minute, the *respiratory volume per minute* for normal respiration is 5–10 liters (assuming a resting tidal volume of 0.5 l). During severe physical exertion , this value can excede 100 liters per min.

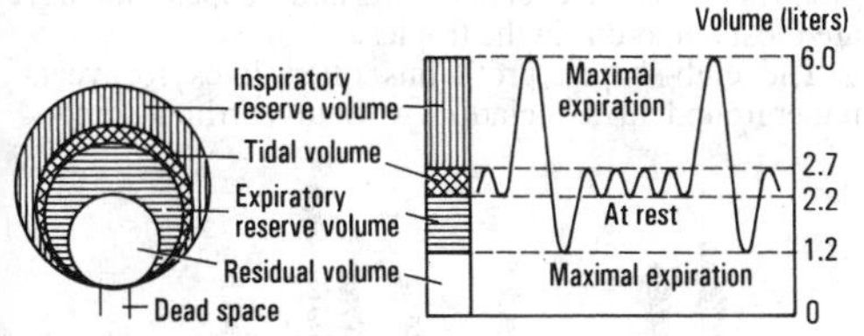

Lung volume. Subcompartments of the lung volume of an adult human.

Lunula: see Claws, hoofs and nails.

Lunulasia: see *Hepaticae.*

Lupeol: a pentacyclic triterpene alcohol, which occurs free and esterified as the aglycon of triterpene saponins in many plants, e.g. in the latex of *Ficus* spp., in the seed coat of yellow lupin *(Lupinus luteus)* and in the leaves of mistletoe. It has also been detected in the cocoons of the silkworm moth, *Bombyx mori.*

Lupin: see *Papilionaceae.*

Lupinus: see *Papilionaceae.*

Lupulone: a bitter principle in hop flowers *(Humulus lupulus).* L. and other compounds of this type (e.g. humulone) have sedative, antibiotic and estrogenic properties.

Luscinia: see *Turdinae.*

Lutein, ***xanthophyll, 3,3′-dihydroxy-α-carotene:*** a xanthophyll, and after the carotenes the most widely distributed, naturally occurring Carotenoid (see). It is present as a yellow pigment, together with carotene ands chlorophyll, in all green parts of plants, and free or esterified in many yellow and red flowers and fruits, e.g. dandelion *(Taraxacum),* marsh marigold *(Caltha palustris),* arnica *(Arnica montana),* daffodil *(Narcissus)* and sunflower *(Helianthus annuus).* Together with zeaxanthin, L. is the main pigment of the yellow yolk of birds' eggs. It is also present in bird feathers, and in the mammalian corpus luteum. Largely in esterified form, it also contributes to the color of autumn leaves. A number of other plant pigments are derived from L., e.g. eloxanthin and flavoxanthin.

Luteinizing hormone, ***LH, lutropin, gonadotropin II, corpus luteum ripening hormone, interstitial cell stimulating hormone, ICSH:*** an important gonadotropin. Together with follicle stimulating hormone, LH stimulates growth of the gonads and the synthesis of male and female sex hormones. It is a glycoprotein, containing about 20% carbohydrate. The polypeptide moiety consists of an α-chain (96 amino acids, M_r 10 791) and a β-chain (119 amino acids, M_r 12 715). The α-chain is identical with the that of follicle-stimulating hormone, thyrotropin and chorionic gonadotropin. The β-chain is hormone-specific. LH is synthesized in the basophilic cells of the anterior pituitary. Its secretion is regulated by, and dependent upon, the regulatory hormone, gonadoliberin, and it is also controlled by the circulating concentrations of estrogen and progesterone. LH is not sex-specific, and it functions via the adenylate cyclase system (see Hormones). In the human male it regulates testosterone synthesis in the testes, and in the female it causes ovulation of the mature follicle and formation of the corpus luteum, as well as stimulating progesterone synthesis. The female cycle is regulated hormonally by the close concerted action of LH, follicle stimulating hormone, estradiol and progesterone.

Oral steroid contraceptives (estrogen plus a gestagen) act essentially by suppressing the synthesis of LH, which leads to the suppression of ovulation.

Luteolin: a flavone and yellow flower pigment. L. is present in high concentration in dyer's rocket or dyer's greenweed *(Reseda luteola),* which up to the middle of the 19th century was cultivated in Europe and America for dying wool. L. is also present in the flowers of dandelion *(Taraxacum),* snapdragon *(Antirrhinum majus),* and as its 7-glucoside in the yellow foxglove, *Digitalis lutea.*

HO O OH OH HO O

Luteolin

Luteoviruses: see Virus groups.

Lutra: see River otters.

Lutreola: see Mink.

Luvaridae: see *Scombroidei.*

Luvarus: see *Scombroidei.*

Lyases: a major group of enzymes, which catalyse cleavage of their substrates into two products, with the formation of a double bond. Alternatively, they may act as synthases by catalysing the reverse process, i.e. combination of two substrates to form a single product by an addition reaction at a double bond. Coenzyme A, biotin or thiamine pyrophosphate (see Vitamins) often function as coenzymes of synthases.

According to the type of bond that is cleaved or formed, L. are classified as:

4.1 C-C lyases: these include the decarboxylases, which are widely distributed in animals and plants, and which catalyse the irreversible cleavage of carbon dioxide from the carboxyl group of carboxylic acids. Enzymes catalysing decarboxylation of amino acids to the corresponding amines employ pyridoxal phosphate (see Vitamins) as their prosthetic group. The coenzyme for the decarboxylation of pyruvate to acetaldehyde is thiamine pyrophosphate (see Vitamins). The C-C lyases also include aldolase, an enzyme of glycolysis and alcoholic fermentation, which catalyses the reversible cleavage of fructose 1,6-bisphosphate to dihydroxyacetone phosphate and glyceraldehyde 3-phosphate.

4.2 C-O lyases: the most important representatives of this group are the dehydratases, e.g. Aconitase (see) and Enolase (see), which catalyse removal of the elements of water from their organic substrates. Enzymes catalysing the addition of water are called hydratases, e.g. Fumarase (see).

4.3 C-N lyases: e.g. aspartase, which catalyses the formation of fumarate and ammonia from aspartate (-plants and bacteria); and histidase, which catalyses the formation of urocanic acid and ammonia from histidine (animals and certain bacteria).

4.4 C-S lyases: a relatively small group, including e.g. homoserine desulfhydrase (L-homocysteine + H_2O $\rightarrow$ 2-oxobutyrate + NH_3 + H_2S).

4.5 C-Halogen lyases: only one enzyme of this type has been well characterised, i.e. DDT-dehydrochlorinase. This enzyme is present in DDT-resistant insects, and the degree of resistance is proportional to the quantity of active enzyme present. It catalyses the conversion of 1,1,1-trichloro-2,2-bis(4-chlorophenylethane) to 1,1-dichloro-2,2-bis(4-chlorophenyl)-ethylene and HCl $(p\text{-}ClC_6H_4)_2CHCCl_3 \rightarrow (p\text{-}ClC_6H_4)_2C{=}CCl_2 + H^+ + Cl^-$).

4.6 P-O lyases: e.g. adenylate cyclase (ATP $\rightarrow$ 3′:5′-cyclic AMP + pyrophosphate), and 3-dehydroquinate synthase (7-phospho-3-deoxy-D-arabino-heptulosonate $\rightarrow$ 3-dehydroquinate + orthophosphate), a reaction in the biosynthesis of the aromatic amino acids in bacteria and plants, and of lignin in plants.

4.99 Miscellaneous lyases: a very small group, including ferrochelatase (protoporphyrin + Fe^{2+} $\rightarrow$ protoheme + $2H^+$), and alkyl mercury lyase (CH_3Hg^+ + HS-cysteine $\rightarrow$ CH_4 + $^+$HgS-cysteine), an important enzyme in connection with environmental mercury poisoning.

Lycaenidae: see *Lepidoptera.*

Lycaon: see African hunting dog.

Lycomarasmine, *aspergillomarasmine B, tomato wilting agent:* a Wilt toxin (see) produced by the fungus, *Fusarium lycopersici,* the causative organism of tomato wilt.

```
            OH
            |
CH2—NH—C—COOH
|           |
|           CH3
|
CO—NH—CH—CH2—CO—NH2
        |
        COOH
```

Lycomarasmine

Lycopene: a Carotenoid (see). An unsaturated, aliphatic, yellow-red tetraterpene hydrocarbon, L. is the pigment of ripe tomatoes and of hips and haws. It is also present in carrots, yellow peaches, cranberries and many other fruits.

Lycoperdon: see *Basidiomycetes (Gasteromycetales).*

Lycopersicum: see *Solanaceae.*

Lycophore: see *Cestoda.*

Lycopodiales: see *Lycopodiatae.*

Lycopodiatae, *Lycopsida,* ***clubmosses:*** a class of the *Pteridophyta,* whose members possess dichotomously branched roots and shoots. The shoot has small, spirally arranged scale-leaves (microphylls). The upper surface of the microphyll usually bears a small, basal, membranous scale (ligule). Sporangia always occur singly on the upper surface of certain leaves (sporophylls), and these are combined into spike-like aggreggations at the shoot terminus. Evolutionary development shows a transition from isosporous to heterosporous forms.

Clubmosses probably evolved from *Asteroxylon*-like forms of the *Psilophytatae;* there are various similarities between the two classes, especially in the structure of the microphylls. They display their greatest evolutionary diversity in the Carboniferous, when tree-like species were common, especially in the coal-forming forests. Modern living clubmosses are herbaceous plants. European species are protected.

1. Order: *Lycopodiales* ***(Clubmosses).*** These are herbaceous, evergreen, isosporous plants, without secondary thickening. Ligules are absent. They inhabit all regions of the world, except deserts and steppes, and they are most numerous in the tropics.

The club-shaped prothallus often lives for years underground in association with mycorrhizal (endo-

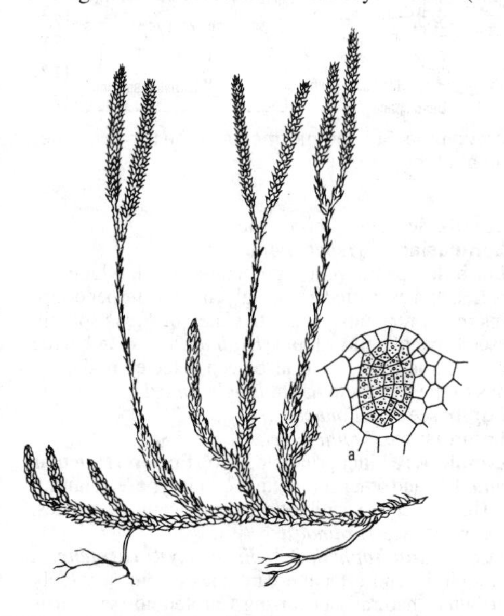

Fig. 1. *Lycopodiatae.* Clubmoss *(Lycopodium clavatum)* with terminal aggregations of sporophylls (cones). *a* Section of unopened antheridium.

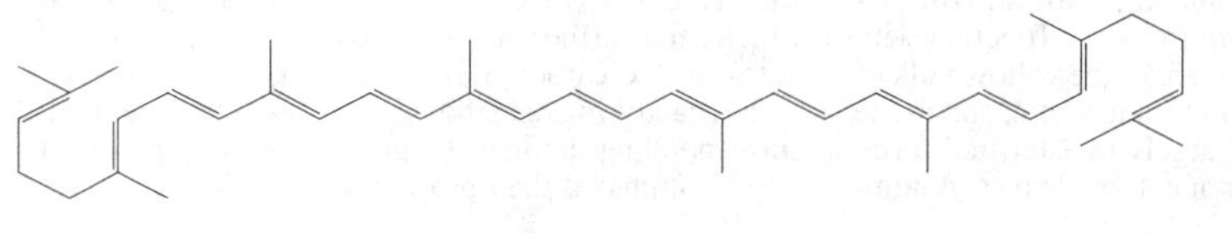

Lycopene

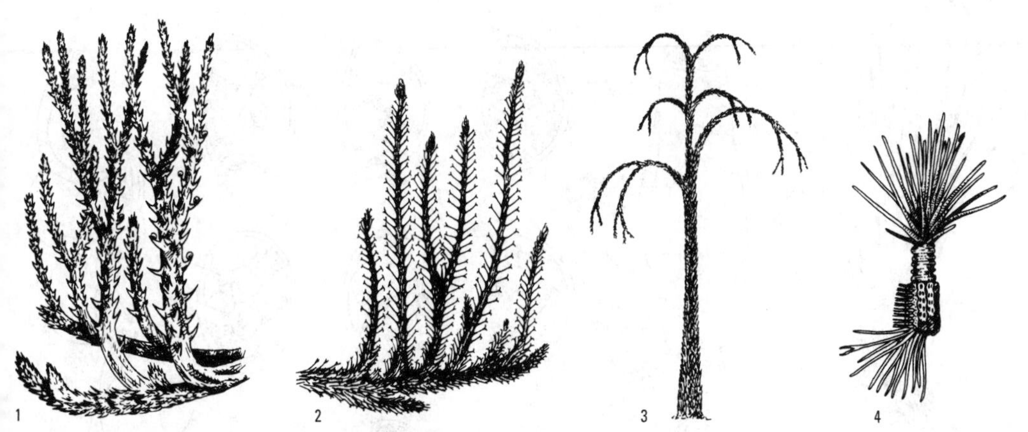

Fig.2. *Lycopodiatae.* Reconstruction of fossil forms. *1 Drepenophycus spinaeformis* (Lower Devonian). *2 Protolepidodendron scharyanum* (Middle Devonian). *3 Sigillaria* (Upper Devonian). *4 Nathorstiana arborea* (Lower Cretaceous).

phytic) fungi. During this time, the prothallus becomes sexually mature, and antheridia and archegonia are formed. Development of the true plant commences after fertilization of the archegonium.

The largest genus is *Lycopodium* ***(Clubmoss),*** with about 400 species distributed worldwide. One of the commonest European species is *Lycopodium clavatum* ***(Stag's horn moss),*** especially in open pine forests.

Fossil members of the *Lycopodiales* are known from the Lower and Middle Devonian, and they display certain close affinities with fossil *Psilophytatae,* such as *Drepanophycus* and *Protolepidodendron.* On the other hand, modern clubmosses have hardly changed their general shape and appearance for over 300 million years, their growth habit closely resembling that of *Lycopodites* from the Upper Devonian.

2. Order: *Selaginellales* ***("Moss ferns").*** These are herbaceous, mainly tropical plants. Their sporophytes are similar to those of the *Lycopodiales,* but the microphylls (leaves) posses a ligule. The small, scale-like microphylls are sometimes spirally arranged, but more commonly decussate and dorsiventral in 4 rows. Sporophylls, each bearing a single axillary mega- or microsporangium, are arranged in a terminal cone. Each cone contains both types of sporangium, thereby representing the evolution of the heterosporous condition. The sexually differentiated gametophytes are short-lived and strongly reduced; they develop within the megaspores (female) or microspores (male) in the respective sporangia.

The only recent genus of the *Selaginellales* is *Selaginella,* with about 700 species, mostly in the tropics and especially in tropical rainforests. The only two European *Selaginella* species are rare mountain plants. Fossil *Selaginellales* occur from the Carboniferous onwards *(Selaginellites).* They show very close affinities with recent forms. Like the *Lycopodiales,* the order *Selaginellales* is therefore also a very ancient group, which has hardly changed since the Paleozoic.

3. Order: *Lepidodendrales* ***("Clubmoss trees").*** Known only from its fossils, this is the most important group of the extinct *Lycopodiatae.* It contains the *Sigillariaceae,* whose stems were covered with longitudinal rows of more or less hexagonal leaf cushions, and the

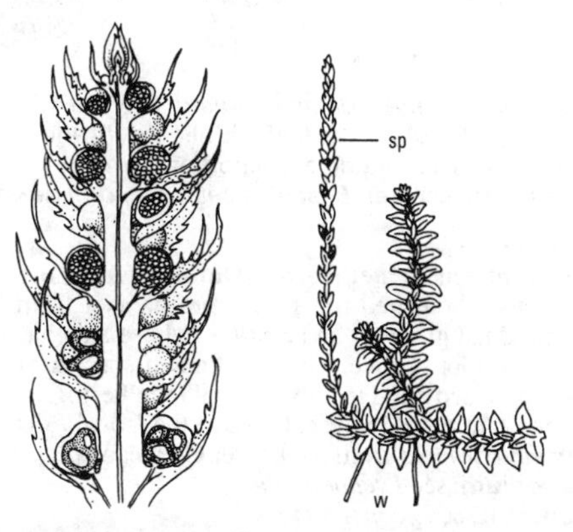

Fig.3. *Lycopodiatae.* Left: longitudinal section of a cone of *Selaginella selaginoides* ("moss fern") with megasporangia below and microsporangia above. Right: plant of *Selaginella helvetica* with cone; *sp* cone, *w* rhizophore).

Lepidodendraceae with rhombic leaf cushions. Together with the *Equisetatae, Lyginopteridatae* and *Cordaitatae,* the *Lepidodendrales* are among the most important plants of the Carboniferous, and they are major contributors to coal formation. They were all trees, 30–40 m in height and 5 m or more in diameter, anchored in the ground by numerous, dichotomously branched roots. The highest level of evolution is displayed by the *Lepidospermae* ***("Seed clubmosses"),*** which show a greater gametophyte reduction than any other member of the *Lycopodiatae,* and whose megasporangia remained on the plant until formation of the embryo.

4. Order: *Isoetales.* These are perrenial herbs, which live submerged in water or on the surface of swamps. They display anomalous secondary growth, which leads to the formation of a tuberous, compressed shoot axis. The long leaves have a ligule and are arranged in a rosette on the axis. Members of the *Isoetales* are heterosporous, and the strongly reduced gametophytes are formed within the spores.

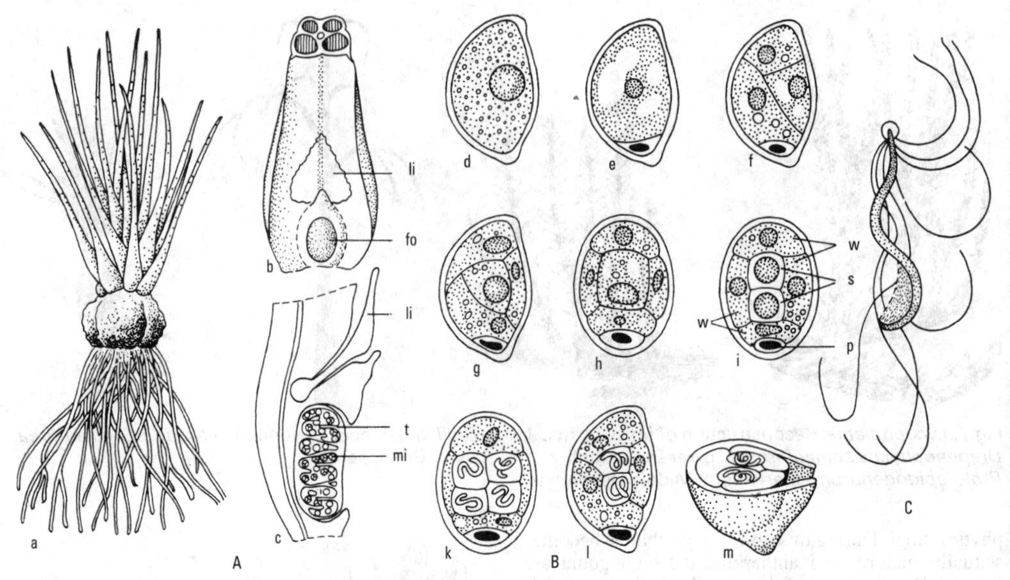

Fig.4. *Lycopodiatae. A Isoetales lacustri, a* entire plant (x 0.5), *b* Basal part of leaf with ligule *(li)* and fovea *(fo)* (x 2), *c* longitudinal section of *b (mi* microspores, *t* trabeculae). *B Isoetes setaceae, d-m* development of the male prothallus and spermatozoids *(p* prothallial cell, *s* spermatogenous cell, *w* wall cells) (x 500). *C* Spermatozoid of *Isoetes malinverniana* (x 1 100).

Two recent genera are: *Isoetes* ***(Quillworts),*** whose dichotomously branched roots arise on the axis from 2–3 longitudinal grooves; and *Stylites,* whose axis has only a single root groove. The two known species of *Stylites* were discovered in 1954 and 1956 in Peru.

An important fossil representative of the *Isoetales* is *Nathorstiana arborea* from the Lower Cretaceous.

Lycopodium: see *Lycopodiatae.*

Lycopsida: see *Lycopodiatae.*

Lycosa: see Wolf spiders.

Lycosidae: see Wolf spiders.

Lydekker's line: see Wallacea.

Lyginopteridales: see *Lyginopteridatae.*

Lyginopteridatae, *Pteridospermae,* ***seed ferns:*** an extinct class of fern-like Gymnosperms (see), belonging to the subdivision *Cycadophytina.* Flowers were absent, and groups of pollen sacs or ovules were borne at individual sites on the copiously branched fronds (sporotrophophylls), and they were never aggregated on short shoots of limited growth. Intermediate forms and evolutionary trends of individual plant parts can be recognized among the diversity of forms of fossil seed ferns. Thus, leaves evolve from frond-like, copiously branched forms, via less divided forms, to entire leaves. Similarly, the pollen sacs display a transition in which the number of pollen sacs increases within a three-dimensional grouping, finally evolving into flat, leaf-like structures. The ovules are formed from megasporangia, which have only a single fertile megaspore, and they are enclosed by an integument. Sometimes one or several ovules are surrounded by a cupule. Seed ferns displayed their greatest evolutionary diversity in the Carboniferous and in the Middle and Lower Permian (Rotliegendes), and they became extinct in the Cretaceous.

Two orders are recognized. In the *Lyginopteridales (Cycadofilicales),* the pollen sacs were borne on three-dimensional branches, which formed part of a leaf. The *Caytoniales* possessed branched stamens, which carried groups of 4 fused pollen sacs (synangia).

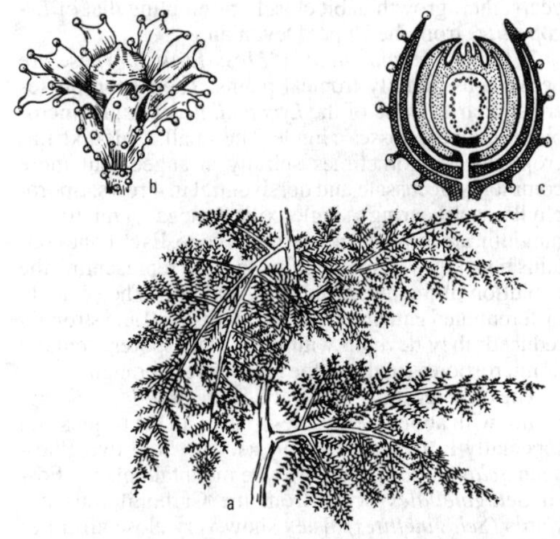

Lyginopteridatae. Lyginopteris oldhamia from the Upper Carboniferous. *a* Reconstruction of stem and leaves. *b* Seed with cupula. *c* Longitudinal section of a seed.

Lymantriidae, ***tussock moths:*** a family of the *Lepidoptera* containing about 2 000 species, closely related to the *Noctuidae* and *Arctiidae* (some *Arctiidae* are also called tussock moths). The adult lacks a functional proboscis and is therefore unable to feed. Males have exceptionally long antennal pectinations. Caterpillars are always hairy and often brightly colored, and many have tufts or tussocks of hairs (usually a dorsal row of four tussocks). In many species, irritant hairs serve as a general means of protection; these are present on the caterpillar, and they are also woven into the cocoon. Irritant hair scales are also present at the end of the female abdomen, and eggs are normally covered with these as they are laid. Some caterpillars are serious pests, e.g. Gypsy moth *(Lymantria dispar)* (see), Black arches *(Lymantria monacha)* (see), Brown-tail *(Euproctis chrysorroea)* (see).

Lymnaea sea: see *Gastropoda.*

Lymnaea stagnalis (Great pond snail): see *Gastropoda.*

Lymnaeidae: see *Gastropoda.*

Lymph: in animals with a closed vascular system, a yellowish to colorless body fluid which drains from intercellular spaces into a network of L. vessels, called the lymphatic system. L. coagulates outside the body. Its composition is similar to that of blood plasma, consisting of varying numbers (300–10 000 per mm^3) of colorless cells (88–97% lymphocytes).

At the sites of contact between L. capillaries and blood capillaries there is a rapid and extensive exchange of material, whereby nutrients are passed from the blood to the tissues via the L., and metabolic products pass in the opposite direction.

Three types of L. are present in vertebrates. 1. *Capillary exudate* or *blood L.* is exuded from blood capillaries. 2. *Tissue fluid* is contained in the intercellular spaces of tissues. 3. L. that flows in lymph vessels *(intestinal L.,* see Chyle). After food intake, the intestinal L. has a milky appearance due to the presence of suspended fat. Fat (absorbed from the intestine during digestion) is transported via the thoracic duct, which empties into the left innominate vein close to the heart. Lymphatics that have their origin in the villi of the small intestine (e.g. the thoracic duct) are also called *lacteals* or *lacteal vessels.*

In addition to transport, L. also has a protective function. Foreign materials, bacteria, etc. that gain access to the body are taken up by the L. and removed by filtration through L. nodes (see).

In contrast to blood, which is restricted to the closed vascular system, L. is also present in the intercellular spaces, where it supplies every cell with nutrients. It collects in certain lymph vessels, where it is kept in circulation by the continual movement and compression effect of the body musculature. Back-flow is prevented by valves in the L. vessels. Lower vertebrates possess "lymph hearts" which ensure L. circulation. Frogs possess extensive L. sacs beneath the skin. The inner spaces of the brain and spinal cord are filled with lymph-like cerebrospinal fluid, and synovial capsules contain a lymph-like synovial fluid. Lymph-like fluids are also found in coelomic cavities, skin blisters, edemas and the cysticerci of tapeworms.

Lymph glands: see Lymph vascular system.

Lymph nodes: see Lymph vascular system.

Lymphocytes: mobile, ameboid, nonphagocytic white blood cells that produce antibodies and carry specific receptors for antigens. They are encountered throughout the entire organism, being present in blood, lymph and intercellular spaces, and they are the most important cells in the immune reaction of the body.

Lymphoid tissue: see Connective tissues.

Lymph vascular system: a system starting with lymph capillaries which arise from intercellular spaces. These capillaries terminate in lymphatic vessels (lymphatics), which unite to form larger vessels and ultimately two terminal trunks, i.e. the thoracic duct and the right lymphatic duct. The latter open into the venous system at the commencement of the left and right innominate veins, respectively. The larger of these is the thoracic duct (ductus thoracicus) (see Lymph). Lymphatics carry tissue fluid from the tissues to the veins. Since they collect part of the lymph that exudes through the thin capillary walls and return it to the innominate veins, lymphatics may be considered to supplement the function of the capillaries and veins.

Along the length of mammalian lymphatics are small oval or bean-shaped bodies, varying in size from that of a pinhead to that of an almond. These *lymph nodes* (sometimes erroneously called *lymph glands)* are sites of lymphocyte formation and therefore play a defensive role. Nonmammals no not possess lymph nodes.

Lynceus: see *Branchiopoda.*

Lynx: see *Felidae.*

Lyomeri: see *Saccopharyngoidei.*

Lyonization: reversible, random inactivation of one of the two X-chromosomes in the cells of female (XX)-mammals; it occurs during embryonal development, and was first described by M. Lyon. At the time of nuclear division and during interphase, the inactivated X-chromosome remains condensed and displays abnormal staining, i.e. it consists of facultative heterochromatin. Inactivation is apparently random, i.e. in each cell either of the two X-chromosomes may be inactivated. A lyonized female organism therefore develops as a mosaic of equal numbers of two types of cells, in which one or the other X-chromosome is inactivated. L. is important in medical genetics; since it results in functional Hemizygosity (see), it enables heterozygote testing for the detection of X-chromosome-localized recessive mutant alleles.

Lyotropic series, *Hofmeister series, hydration series:* a series of anions or cations arranged in order of magnitude of their effect on reactions in colloidal solution, in particular their effect on the hydration or dehydration of macromolecules, e.g. the alkali metal ions can be arranged in the series $Li^+ > Na^+ > K^+ > Rb^+ > Cs^+$. These properties are determined by charge (or charge density), and by the diameter and degree of hydration of the ions (which depend on the charge density). Thus, the large but weakly hydrated Cs^+ cation strongly inhibits or reverses the hydration of (uptake of water by) negatively charged substances (e.g. proteins), whereas the small and highly hydrated Li^+ cation promotes the hydration of negatively charged proteins.

Lyrurus: see *Tetraonidae.*

Lysergic acid: see Ergot alkaloids.

Lysergic acid diethylamide, *LSD:* a synthetic derivative of D-lysergic acid, first prepared in 1938. It is the most potent psychotomimetic substance known. Related hallucinogenic derivatives of lysergic acid (lysergic acid amide and lysergic acid hydroxyethylamide) occur in the Mexican ritual drug, ololiuhqui (seeds of *Rivea corymbosa).* LSD is also structurally closely related to the ergot alkaloids, from which it can be prepared semisynthetically. The hallucinogenic action of LSD, which is characterized by a state resembling schizophrenia, was discovered accidentally in 1943 by A. Hofmann during the recrystallization of a sample of LSD tartrate.

For the structure of Lysergic acid, see Ergot alkaloids.

L-Lysine, ***Lys:*** $H_2N—(CH_2)_4—CH(NH_2)—COOH$, a basic, proteogenic, essential amino acid. It is present in most proteins, in particular albumin, fibrin and globulins, and is especially important for bone formation in growing children and young animals. Cereal proteins (wheat, barley, rice) are rather poor in Lys. In molds and yeasts, it is biosynthesized via α-aminoadipic acid. In bacteria, brown and green algae, higher plants and higher fungi it is biosynthesized via diaminopimelic acid.

Lysogenic conversion, ***virus-induced conversion:*** expression of a typical character of a cell, due to the influence of the genome of a bacteriophage or other virus. Thus, diphtheria toxin is produced only under the control of a bacteriophage gene.

Lysogenic secretory reservoir: see Excretory tissues of plants.

Lysosomes: organelles, 0.2–2 nm diameter, found in the cytoplasm of eukaryotic cells, and discovered in 1959 by De Duve (Nobel prize 1974). L. are bounded by a single lipoprotein membrane, but otherwise show no fine structure. Under the light or electron microscope, L. are markedly polymorphic. They can be characterized biochemically or histochemically, but not morphologically. L. are sites of intracellular digestion, particularly of biological macromolecules, such as proteins, polynucleotides, polysaccharides, lipids, glycoproteins, glycolipids, etc. Approximately 40 different lysosomal hydrolases are responsible for this degradative activity; they all show optimal activity at acidic pH values. The marker enzyme for the characterization of L. is acid phosphatase. Prolonged anaerobiosis of aerobic cells leads to destruction of the L. and release of the lysosomal enzymes into the cytoplasm; subsequent digestion of the cell contents by the lysosomal enzymes is known as autolysis. Autolysis is a characteristic post mortem process.

Primary L. are formed in the Golgi apparatus. Fusion of L. with phagocytosing and pinocytosing vacuoles (phagosomes) produces digestive vacuoles known as secondary L. Excess or old cell parts, including mitochondria, may be digested by L. (phagolysosomes). In ameba, L. apparently provide the digestive enzymes.

Lysozyme, ***endolysin, muramidase, N-acetylmuramide glycanohydrolase:*** a widely occurring hydrolase found in phages, bacteria, plants, invertebrates and vertebrates. In the latter, it is found particularly in egg white, saliva, tears and mucosas. L. acts as a bacteriolytic enzyme by hydrolysing the β-1,4 linkage between *N*-acetylglucosamine and *N*-acetylmuramic acid in the proteoglycan of the bacterial cell wall, thereby affording protection against bacterial invasion. It provides a very important defense against infectious bacterial diseases. All animal Ls. consist of a single chain of 129 amino acid residues with 4 disulfide bridges (M_r 14 200–14 600), with homologous amino acid sequences and similar tertiary structures. The tertiary structure is known in detail, consisting of 42% α-helix with hydrophobic tryptophan residues on the outer surface, and prominent hydrogen bonding between the side chains of Ser, Thr, Asn, and Gln. Like other hydrolases (e.g. papain, ribonuclease), the molecule of L. has a cleft which houses the active center of the enzyme and serves for the attachment of the substrate, which in this case is a hexasaccharide unit of the proteoglycan molecule. There is a remarkable correspondence between the primary and tertiary structures of L. and α-lactalbumin (123 residues). It is thought that both proteins arose from a common precursor with lysozyme activity, an example of divergent evolution by gene duplication. On the other hand, there is no structural relationship between animal L. and bacteriophage L. The latter contains 157 amino acid residues (λ phage endolysin, M_r 17873) or 164 residues (T4 and T2-phage L., M_r 18720), and they are either endoacetylmuramidases (T4, T2) or endoacetylglucosaminidases (streptococcal L.)

M

Mabuya: the most widely distributed genus of the *Scincidae* (see), with species in Indonesia, Southeast Asia, Africa, Madagascar, Central and South America, and the Caribbean Islands, but not in Australia, Papua New Guinea or Polynesia. They are ground-dwelling, mostly ovoviviparous reptiles, 15–25 cm in length, with well developed limbs and a normal lizard-like external appearance. Many species are splendidly colored or striped.

Macaques, *Macaca:* medium sized Old World monkeys (see) with a moderately prognathous muzzle and strong teeth. With the exception of humans, M. are the most widely distributed primate genus, being found throughout the islands and mainland of Southeast Asia, China, Japan, India to Afghanistan and the Himalayas, North Africa (Morocco, Algeria) and even Gibraltar. The 16 species have been divided into 4 groups on the basis of the anatomy of the genitalia [J. Fooden, *Folia primatol., 25* (1976) 225–236].

Group 1 *(Macaca sylvanus* group): *M. sylvanus* ***(Barbary macaque, Barbary "ape"),*** *M. silenus* ***(Lion-tailed macaque),*** *M. maurus* ***(Moor macaque, Celebes macaque),*** *M. nemestrina* ***(Pig-tailed macaque),*** *M. ochreata* ***(Sulawesi booted macaque, Sulawesi crested macaque, Celebes black "ape"),*** *M. nigra* ***(Sulawesi black "ape"),*** *M. tonkeana* ***(Tonkean macaque).***

Group 2 *(Macaca fascicularis* group): *M. fascicularis* ***(Long-tailed macaque, Crab-eating macaque, Philippine macaque, Cynomolgus monkey),*** *M. fuscata* ***(Japanese macaque),*** *M. mulatta* ***(Rhesus macaque, Rhesus monkey),*** *M. cyclopis* ***(Taiwan macaque).***

Group 3 *(Macaca sinica* group): *M. sinica* ***(Toque macaque),*** *M. assamensis* ***(Assamese macaque),*** *M. radiata* ***(Bonnet macaque),*** *M. thibetana* ***(Tibetan stump-tailed macaque).***

Group 4 *(Macaca arctoides* group): *M. arctoides* ***(Red-faced stump-tailed macaque, Bear macaque).***

M. sylvanus is a tailless species with a rather dense coat, inhabiting the temperate cedar and oak forests of Morocco and Algeria. With human protection and food provision, it also thrives on the rock of Gibraltar, where it was introduced 2–300 years ago. This geographically isolated species is the only African primate found north of the Sahara and the only African macaque.

M. mulatta (the well known rhesus monkey; plate 40) is a robust, medium sized macaque, inhabiting both lowland and mountainous areas of India, Southeast Asia and China. Captured from the wild, is used extensively in medical research. Pressure on the wild population has now been partially relieved by the introduction of *M. fascicularis* for experimental purposes. *M. fascicularis,* common throughout Southeast Asia, was used extensively in the development of the polio vaccine.

M. silenus, the most arboreal of the macaques, inhabits the Western Ghats of southern India. Its body hair is black, and the head is framed by a ruff of long gray hair.

M. fuscata is protected in its natural habitat in Japan. Some groups of this species occur further north than any other nonhuman primate (40 °N at the tip of Honshu Island). It displays a remarkable ability to discover and learn advantageous behavior. For example, on Honshu Island, a group, living in the mountains has learned to luxuriate in hot springs during the winter cold, a form of behavior first discovered in 1963 by a 2-year-old female, known as Mukubili to her human observers. Similarly, in 1953, on the southern Japanese island of Koshima, a young 16-month-old female (Imo to her human observers) discovered how to wash sand from sweet potatoes in a stream running into the sea. By 1962, the monkey's mother, one other adult female, and all Imo's contemporaries were washing sweet potatoes before eating them.

MacArthur model of diversity: see Diversity index.

MacArthur-Wilson theory of island biogeography, ***island theory:*** statistical treatment of the colonization of islands, allowing the calculation of the equilibrium established between the immigration and death of animal and plant species. The theory is based on the premise that island populations are relatively short-lived (small individual numbers, limited genetic reserves, absence of refugia in extreme climatic conditions, lack of space, etc.), an effect that increases as the size of the island decreases. Immigration, which is important both for the introduction of new species and the maintenance of existing populations, decreases with the distance from the mainland. There is less immigration into small than into large islands, and it is also restricted by the extent to which appropriate ecological niches for incoming species exist or are already occupied by other species. Conversely, the extinction rate increases with the number of species present (Fig.). In numerous studies, the calculated statistical probabilities agree well with existing island population structures.

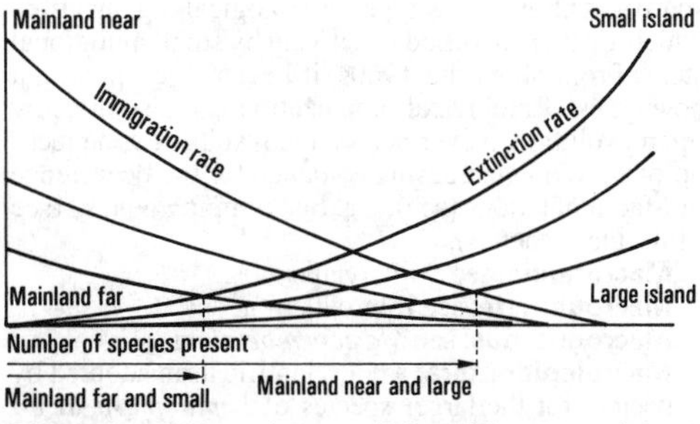

MacArthur-Wilson theory of island biogeography. Equilibrium model for the flora and fauna of islands of different sizes and different distances from the mainland (modified from MacArthur and Wilson).

Originally developed for specific islands, the theory also explains and predicts the species composition of isolated areas of mainland (high mountains, swamps, etc.). In particular, it has been usefully applied to problems of species protection. For example, it is possible to predict the decrease in the number of species in a reserve, in which there is no recognizable direct threat to the protected animals and plants. The theory can also explain the otherwise puzzling disappearance of a species from an unchanged and unchanging habitat X; this may be due to loss of a habitat Y, which hitherto was responsible for replenishing losses from population X, or even for reestablishing population X in extreme situations.

The theory is valid for oceanic islands that have never been in contact with the mainland. If an island was formerly connected to a continent, it starts with a complement of species similar to that of the mainland. An oceanic island can never posses such a number and variety of species, because many biota simply cannot cross the ocean. The number of species on a continental island then gradually decreases to "island level". The same happens in mainland biotopes that become isolated by environmental changes.

Macaws: see *Psittaciformes.*

Macchia: see Sclerophylous vegetation.

Mace: see *Myristicaceae*

Machaerhamphinae: see *Falconiformes.*

Machaerhamphus: see *Falconiformes.*

Machaeroides: see *Creodonta.*

Mackerel, *Scomber scombrus:* a 30–60 cm-long predatory, shoaling fish [family *Scombridae,* suborder *Scombroidei* (see), order *Perciformes],* distributed from European and Scandanavian coasts of the Atlantic to the coast of North America It favors coastal waters and rarely leaves the continental shelf. It is one of the ten most commercially valuable fish species of the world, and its flesh has an excellent flavor. The back is bluish green in color and decorated with numerous dark transverse bands that hardly reach to the lateral line.

Macleaya: see *Papaveraceae.*

Maclura: see *Moraceae.*

Macracanthorhynchus: see *Acanthocephala.*

Macrodasyoidea: see *Gastrotricha.*

Macrodichopatric: see Dichopatric.

Macroevolution, ***trans-specific evolution:*** the formation of relatively large systematic units (e.g. at and above the rank of genus) of the plant and animal kingdoms. It has often been assumed that M. involves a sharp transition to new types of biological organization, which are then modified adaptively by small mutational steps. From about the 1960s, it became accepted that new types of biological organization can also arise by microevolution. Nevertheless, the results of gene technology provide increasing evidence for the occurrence of Macromutations (see), e.g. by "jumping genes" (see Molecular genetics).

Macrofauna: see Soil organisms.

Macrogamete: see Reproduction.

Macrognathus: see *Mastacembeloidei.*

Macrolepidoptera: a nonscientific term adopted by collectors for the larger species of *Lepidoptera.* It includes the *Hepialidae, Cossidae, Zygaenidae, Aegeriidae* and all families from the *Geometridae* through *Papilionidae* in the classification list under *Lepidoptera* (see). See Microlepidoptera.

Macrolepiota: see *Basidiomycetes (Agaricales).*

Macromere: see Mesoblast.

Macro-moths: see Moths.

Macromutation, ***system mutation:*** a type of mutation postulated around 1940 by the paleontologist, O.H. Schindelwolf, and by the geneticist, R. Goldschmidt, as a means of understanding and explaining macroevolution. M. "from small inversions or transfers within a chromosome to a complete reorganization of the pattern of all chromosomes can have a very large overall effect, if the resulting life forms are viable" (Goldschmidt 1955). The existence of M. was disputed until very recently, because Molecular genetics (see) had produced a wealth of detailed information on micromutations, which were latterly thought to be responsible for macroevolutionary processes. More recently, however, it has been shown that M. does occur, and can be actuated, inter alia, by "jumping genes".

Macronyx: see *Motacillidae.*

Macrophages: large, ameboid, phagocytic white blood cells. They consume and destroy microorganisms that enter the body, and therefore play an important part in defense. They also reinforce the immune response by taking up antigens and presenting them to the lymphocytes. In addition, phagocytosis is promoted by the presence of specific antibodies in the M.

Macrophanerophyte: see Life form.

Macroplankton: see Plankton.

Macropodidae: see Kangaroos.

Macropodinae: see Gouramies.

Macropodus: see Gouramies.

Macropus: see Kangaroos.

Macroscelidae, ***elephant shrews:*** a family of the order *Insectivora* (see) comprising 15 species in 4 recent genera. Dental formula: 1–3/3, 1/1, 4/4, 2/2–3 = 36–42. Elephant shrews are distinctive mammals found in dry areas of Africa. They possess a trunk-like extended snout, and long hindlegs specialized for bounding.

Macroskelia: in humans, the possesssion of long extremities in relation to body length. The converse is Brachyskelia.

Macrosporogenesis: see Megasporogenesis.

Macrostomida: see *Turbellaria.*

Macrotrema: see *Synbranchiformes.*

Macrouridae, *Coryphaenoididae,* ***grenadiers, rattails:*** a family of fishes of the order *Gadiformes,* containing about 30 genera and about 260 species, which attain lengths of up to 1 m, but usually less. They are deep sea (200–2 000 m), marine fishes, found from the Arctic to the Antarctic, although most species occur in tropical and subtropical latitudes. The anal fin and the second dorsal fin are continuous with the tail, which tapers to a sharp point, and there is no caudal fin. Some species possess a light organ along the midline of the abdomen.

Macula: see Auditory organ.

Maculae adhaerentes: see Epithelium.

Madagascar geckos: see *Phelsuma.*

Madagascar hedgehog: see Tenrecs.

Madder: see *Rubiaceae*

Madia: see *Compositae.*

Madoqua: see *Neotraginae.*

Madoquini: see *Neotraginae.*

Madreporaria, ***stony corals, scleractinian corals, madreporian corals:*** an order of corals of the *Hexacorallia.* A heavy external skeleton of calcium carbonate is secreted by the epidermis of the lower half of the column and by the basal disk. The basal disk also secretes sclerosepta, which are hexameric radial plates, extending upward from the basal disk into the polyp, and forming ribs between the pairs of mesenteries. The skeleton grows continually upward, but only the upper section is inhabited by the animal, the lower zones being separated by transverse platforms. Most members are colonial, with very small polyps (1–3 mm diam.), but an entire colony can be very large. Massive colonies are formed by budding, sometimes containing many millions of individuals. In colonial forms, all the polyps are interconnected by a horizontal sheet of tissue. A few species are solitary, e.g. *Fungia* of the Indo-Pacific reefs. M. are the main formative organisms of Coral reefs (see). About 2 500 recent species are known, from the littoral zone to depths of 6000 m, and twice that number of fossil forms have been described. The M. are closely related to the sea anemones *(Actiniaria),* but they lack siphonoglyphs, and the mesenterial filaments consist of a single glandular lobe. Examples are *Fungia, Acropora* ***(Stag's horn coral)*** *Porites, Astrangia, Oculina* ***(Eyed coral),*** and *Diploria* ***(Caribbean brain coral).***

Madreporian corals: see *Madreporaria.*

Mad cow disease (bovine spongiform encephalopathy): see Prion.

Maggot: a trivial name for an insect larva which lacks visible legs, e.g. the larvae of flies are Ms.

Magnesium, ***Mg:*** an alkali earth metal, distributed widely as its cation (Mg^{2+}) in biological systems, with many different biological functions. It is the fourth most abundant cation in vertebrtates, and it has great biological significance as a constituent of the porphyrin ring system of chlorophyll. Mg is metabolically important as a cofactor of many different enzyme systems. In particular, it is involved in reactions of energy metabolism, including the enzymatic cleavage of phosphate esters and the transfer of phosphate groups. It is an activator of phosphatases and a cofactor of practically all ATP-dependent phosphorylation reactions, e.g. hexokinase, phosphofructokinase and adenylate kinase; in every case one of the substrates is an ATP-Mg complex (1:1), rather than free ATP. Mg^{2+} is also required for the function of the ribosome in protein synthesis, in particular for the correct interaction of the ribosomal subunits, and for the release of the newly synthesized polypeptide from the ribosome. Intracellularly, Mg is present as Mg^{2+} and $MgOH^+$. About 0.05% of the animal body is Mg, 60% in the skeleton and 1% in extracellular fluids.

Mg is also an essential nutrient for plants, which require it in relatively large quantities. It is present in the soil in numerous minerals, e.g. biotite, augite, serpentine and olivine, and as its carbonate in magnesite and dolomite. Although it is released only slowly from minerals by weathering, soils normally contain adequate amounts of Mg^{2+} for plant growth. However, the Mg^{2+} ion is easily lost by drainage, so that plants may suffer Mg-deficiency in regions of high precipitation, especially when the soils are light and acidic. Mg^{2+} uptake by plant roots is decreased by low pH and by high concentrations of other cations (e.g. K^+). Mg transport in the plant occurs mainly from the root to the tip of the shoot, with little movement of Mg in the reverse direction. Plant tissue normally contains 0.1–0.5% Mg on a dry mass basis. Leaves of potato, tobacco and sugar beet contain relatively high quantities of Mg, but even here it hardly exceeds 1% of the dry mass. Mg-deficient plants show light patches with necrotic centers between the main veins of the leaves. These first appear in the most mature leaves, followed by rolling and drying of the leaf margins. The entire plant has a wilted appearance, but with some erect, nonwilted leaves. In cereals and other graminaceous plants, the light leaf patches contain dark green islands of chlorophyll-containing tissue, often arranged in a pattern like a string of beads. Mg-deficient maize leaves show a striped or "tiger" pattern.

Magnolia: see *Magnoliaceae.*

Magnoliaceae, ***magnolia family:*** a family of dicotyledonous plants containing about 200 species, distributed mainly in Asia and America. They are woody plants with entire, simple or lobed leaves, and large, solitary, usually brightly colored flowers. *M.* are thought to be the least advanced of the Angiosperms, and many of their characteristics are considered to be primitive, e.g. the flower structure, in which numerous stamens and carpels are arranged spirally on an elongated, conical floral axis; and the production of a cone-like compound fruit.

About 75 species of *Magnolia* are native to Southeast Asia and North America, and several of these are familiar ornamental garden plants. A popular parkland tree is the North American ***Tulip tree*** *(Liriodendron tulipitera),* which has characteristic curiously lobed leaves.

Magnoliaceae. a Longitudinal section of a flower *(Magnolia). b* Tulip tree *(Liriodendron tulipifera).*

Magnoliatae: see *Dicotyledoneae.*

Magnoliidae: see *Dicotyledoneae.*

Magnoliophytina: see Angiosperms.

Magpie goose: see *Anseriformes.*

Magpie robins: see *Turdinae.*

Magpies: see *Corvidae.* See plate 6.

Mahonia: see *Berberidaceae.*

Ma Huang: see Ephedrine.

Maidenhair tree: see *Ginkgoaceae.*

Mail-cheeked fishes: see *Scorpaeniformes.*

Maintenance activity, ***maintenance behavior:*** activity and behavior that serve for the regulation and maintenance the body, avoidance of hazards, etc. Such activity has no signal content. It represents an interaction of the individual with its ecosystem or with its own body, and serves to exploit the environment for the requirements of the individual. Examples are preening, bathing, feeding, self-grooming, thermoregulation (e.g. moving to a cooler place, sweating, or both). In contrast, signal behavior (see Biocommunication) serves the interaction of individuals in an animal society or group.

With the evolution of animal society and increased group activity, some M.a. has become ritualized into signal behavior. Thus, feeding is M.a., whereas the feeding call of the farmyard cock is signal behavior.

Maize: see *Graminae.*

Maize smut: see *Basidiomycetes (Ustilaginales).*

Majidae: see Spider crabs.

Majorana: see *Lamiaceae.*

Malaclemys: see *Emydidae.*

Malacology: the study of mollusks. Conchology is similar but tends to be devoted to the shells alone.

Malaconotinae: see *Laniidae.*

Malacosoma neustria: see Lackey.

Malacosteidae: see *Stomiiformes.*

Malacostraca, ***higher crustaceans:*** the largest subclass (about 17 000 species) of the *Crustacea.* All malacostracans have 8 thoracic segments (thoracomeres), and in the primitive condition 7 abdominal segments (pleomeres), the seventh and last being modified as the telson. The eyes are often carried on movable stalks.

Classification (each order is listed separately).

1. Superorder: *Phyllocarida.*
   Order: *Leptostraca.*
2. Superorder: *Haplocarida.*
   Order: *Stomatopoda.*
3. Superorder: *Syncarida* (55 freshwater species)
   Orders: *Anaspidacea, Bathylnellacea.*
4. Superorder: *Eucarida.*
   Orders: *Euphausiacea, Decapoda.*
5. Superorder: *Pancarida.*
   Order: *Thermosbaenacea.*
6. Superorder: *Peracarida.*
   Orders: *Mysidacea, Amphipoda, Spelaeogriphacea, Cumacea, Tanaidacea, Isopoda.*

Malagash: see *Pelicaniformes.*

Malagasy rats: see *Nesomyinae.*

Malapterurus: see *Siluriformes.*

Malaria: see *Telosporidia.*

Malayan black-striped squirrel: see *Sciuridae.*

Malayan gharial: see False gharial.

Malayan sun bear: see Sun bear.

Malcohas: see *Cuculiformes.*

Maleic hydrazide, ***MH, 1,2-dihydro-3,6-pyridazinedione:*** a synthetic plant growth retardant, which is used as a herbicide against grasses. Applied once or twice, shortly before the appearanc of new shoots (end of April to the beginning of May), it decreases the number of necessary annual mowings of ornamental lawns from 12 to 3. Treated areas also often acquire a deeper green color. It prevents the formation of suckers in tobacco and tomatoes, and prevents the sprouting of onions, beets and potatoes, thereby decreasing losses during storage. The effects are limited to green plants, and there is little or no effect on any other organism. There is nevertheless some evidence for carcinogenic activity, and MH is listed as a suspected carcinogen.

Malformations, ***deformities:*** in plants and animals, genetically or environmentally determined (e.g. by pathogens) disturbances of development, leading to abnormalities in organs, organ parts or organ systems.

In plants, M. are manifested as the distortion of organs (due to depressed or accelerated growth and development) and as the multiplication of organs. M. are known in nonflowering plants (cryptogams), but they occur more frequently in flowering plants (phanerogams). Most commonly affected are the root, stem, leaves and inflorescences, but M. of cotyledons, flower organs, fruits and seeds also occur. Familiar M. of plants are fasciation (coalescent development of the branches of a shoot system, often forming a single, strap-shaped axis), phyllody (conversion of floral parts into leaf-like structures), proliferation, and curling or twisting.

In animals, a distinction is drawn between single, double and multiple M. *Single M.* include supernumary M., defect M. and nonconformist M. *Supernumary M.* are the result e.g. of doubled organ primordia. *Defect M.* are exemplified by the absence of limbs or organs, and by *teratomas* which are due to the developmental failure of individual organ parts. *Nonconformist M.* are represented by changes in the normal positions of organs, e.g. location of the heart on the right side of the body.

Double M. are divided into independent and linked M., and each of these is subdivided into symmetrical (equal) and asymmetrical (unequal) M. Linked symmetrical double M. can then by further divided into complete and incomplete M. An example of an independent symmetrical double M. is homozygotic twins; Siamese twins represent a linked symmetrical complete double M. Examples of linked symmetrical incomplete M. are animals with two heads and only one torso. *Multiple M.* are relatively rare in animals.

Malic acid, ***monohydroxysuccinic acid:*** a dicarboxylic acid isolated in 1785 from apple juice by Scheele. Its structure was elucidated in 1832 by Liebig. The L-form of M.a. occurs in many plant juices, e.g. unripe apples, *Berberis* berries, quinces, gooseberries, blackcurrants, cherries. Since M.a. is an intermediate in the Tricarboxylic acid cycle (see for formula), it occurs in all aerobic organisms, at least in small quantities.

Malimbus: see *Ploceidae.*

Malkohas: see *Cuculiformes.*

Mallards: see *Anseriformes.*

Mallee: see Sclerophylous vegetation.

Mallee fowl: see *Megapodiidae.*

Malleus of the middle ear: see Auditory organ.

Mallomonas: see *Chrysophyceae.*

Mallow: see *Malvaceae.*

Malm: a general term for a calcium and phosphorus-rich loam soil. More specifically the calcareous sedimentary rocks of the Late Jurassic (see Jurassic).

Malpighian body: see Excretory organs, Kidney functions.

Malpighian layer: see Epithelium.

Malpighian tubules: see Excretory organs.

Malpighian vessels: see Excretory organs.

Malpighiella: see Amebas.

Malpolon monspessulanus: a very quick, temperamental and shy snake of the family *Colubridae* (see) subfamily *Boiginae* (see), found in the dry regions of southern Europe, North Africa and Asia Minor. It attains a length of 2 m or more. The preferred diet consists of lizards, but small rodents are also taken, and these are paralysed with a poisoned bite before swallowing. Dentition is opisthoglyphic. Humans are rarely attacked, but the consequences of its bite are very unpleasant.

Maltase, ***α-1,4-glucosidase:*** a glycosidase enzyme that catalyses the hydrolysis of maltose into 2 molecules of D-glucose. M. often occurs together with the starch-cleaving enzyme, amylase, e.g. in barley malt, yeast, intestinal juice and pancreas.

Malthusian parameter: the intrinsic rate of natural increase of a population, as well as a relative index of the fitness of a genotype. $m = \ln N_0 - \ln(N_t/t)$, where N_0 is the number of individuals at time 0, and N_t is the number of individuals after time t. The value of t is arbitrary (e.g. a week, month or year).

In contrast to the measure of fitness (W), which is used for populations with discrete generations, m is used for populations in which birth and death occur continuously.

Maltol: a heterocyclic compound found in pine needles and larch bark. It is also produced by dry distillation of wood, cellulose and starch. It contains a γ-pyrone ring (Fig.).

Maltol

Maltose, *malt sugar:* a naturally occurring reducing disaccharide consisting of 2 molecules of glucose in α1,4-glycosidic linkage. Starch and glycogen can be considered as repeating units of M. Hydrolysis of M. with dilute acids or α-glucosidase produces D-glucose. M. serves as a fermentable substrate in brewing, as a component of prepared bee food, as a substrate in microbiological growth media, and generally in food and pharmaceuticals as a nutrient and sweetner (one third as sweet as glucose).

α-Maltose

Malurinae: see *Muscicapidae.*

Malurus: see *Sylviinae.*

Malva: see *Malvaceae.*

Malvaceae, *mallow family:* a family of dicotyledonous plants containing about 1 200 species. The majority occur in the tropics, and only a few are cosmopolitan. They are trees, bushes, or herbaceous plants with simple or lobed, palmately veined, alternate leaves. Flowers are actinomorphic, hermaphrodite, usually 5-merous, and often large and brightly colored. Below the calyx there is often an epicalyx of 3 to several segments resembling an outer set of sepals. The numerous stamens are almost always fused into a tube, which is commonly adherent to the petals. The superior, 3- to multicellular ovary develops into a capsule or schizocarp. Many species have hairy seeds.

The best known and most commonly grown (predominantly in subtropical regions) commercial plant of this family is the ***Cotton plant*** *(Gossypium herbaceum).* Its unicellular seed hairs are several centimeters in length, and they yield cotton, which is the most important commercial plant fiber worldwide. In addition, an oil is obtained from the dehaired seeds, which is used technically and for the manufacture of margarine. ***Hollyhock*** *(Althaea rosea),* a garden plant from the Mediterranean region, has been known since the 16th century. On account of their mucilage content, various members of the *M.* are medicinally important, e.g. ***Marshmallow*** *(Althaea officinalis).*

Fig.1. *Malvaceae. Cotton (Gossypium herbaceum). a* Shoot with flowers and fruits. *b* Seed with seed hairs. *c* Seed hairs magnified.

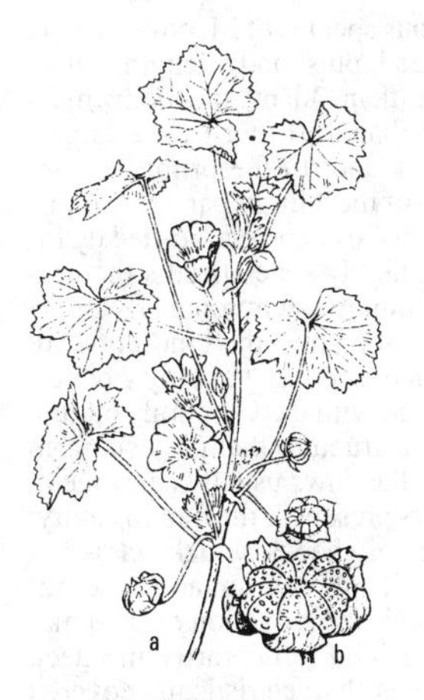

Fig.2. *Malvaceae. a* Hollyhock *(Althaea rosea). b* Fruit of common mallow *(Malva sylvestris).*

Hibiscus cannnabinus (***Gambo hemp, Deccan hemp*** or ***Ambari hemp)*** of tropical Africa and India yields a textile fiber, which is used as a substitute for hemp and jute in the manufacture of cordage and coarse canvas. *Hibiscus esculentus* (***Okra***) is widely cultivated in the tropics, where its unripe fruits are cooked and eaten as a vegetable. The fleshy calyx of ***Rama*** or ***Roselle*** *(Hibiscus sabdariffa)* is used to prepare a refreshing drink; this plant also yields a strong, silky, light brown fiber, which makes a good substitute for jute in textile and paper making. *Hibiscus rosa sinensis,* a native of China, is grown as an ornamental throughout the tropics for its beautiful large red flowers; also known as "shoe flowers", its mucilaginous blooms are used in the West Indies for polishing shoes. The ***Cotton rose*** or ***Blushing hibiscus*** *(Hibiscus mutabilis)*, also grown in the tropics, produces white morning flowers which change to deep pink by evening. They may be picked early, placed in a refrigerator, then brought to the evening dinner table, where they entertain by turning pink during the meal. Various species of *Malva* grow in temperate regions as wayside and ruderal plants, e.g. *M. moschata* ***(Musk mallow)*** and *M. sylvestris* ***(Common mallow)***.

Mambas, *Dendoaspis:* a genus of the snake family *Elapidae* (see). M. are slender, tree-living snakes of tropical Africa, which attain lengths of up to 3.5 m. The eyes are large with round pupils, the head is long and narrow, and the body scales are arranged in diagonal rows. They bear a superficial resemblance to the harmless *Colubridae* (see), but they are dangerous on account of their speed and powerful venom. In addition to the venom fangs of the upper jaw, the front teeth of the lower jaw are also long and powerful, serving to hold birds and tree lizards which form the main part of the diet. They lay 10–15 oval white eggs. The ***Green mamba*** *(D. viridis)* is the most dangerous tree snake of West Africa. Even more feared is the widely distributed ***Black mamba*** *(D. polylepis)*, the largest poisonous snake of Africa.

Mammals, ***Mammalia*** (plates 7, 38, 39, 40, 41): a class of vertebrates. They are warm-blooded (homeothermic), and with the exception of the *Monotremata* they are viviparous (give birth to live young). Characteristic of the class is that the young are fed by the mother with secretions from milk glands (mammary glands). About 6 000 different living species of M. are recognized.

Morphology. The various species of M. cover a wide range of sizes. Thus, head plus body length varies between 6 cm and more than 30 m, the body mass between 1.5 g and more than 130 tonnes, the largest mammal being the Blue whale. In the primitive (unmodified) condition, each of the 4 limbs carries 5 digits, but in many species, these have become modified by fusion or reduction to form highly specialized structures for running, jumping, climbing, swimming, flying or excavating. In whales and sea cows the hindlimbs are markedly regressed so that they consist of only 2 bones, which are loosely associated with the vertebral column, but are no longer part of a structure that projects from the body surface. Mammalian jaws usually carry teeth, and a very mobile tongue is present in the mouth cavity; together with the presence of cheeks, which act as lateral walls to the mouth cavity, these adaptations permit the efficient mastication and size reduction of food material. Some species store food temporarily in cheek pouches. The body surface is characteristically covered with hair, although this may be very sparse or almost absent, especially in certain aquatic species, e.g. whales. Mammalian hair can be classified into 3 main types: 1) long, stiff, coarse hair, 2) softer, ruffled or woolly hair. and 3) individual long, erect hairs or vibrissae. The latter are found particularly in the region of the head and face, and they have a sensory function. In some species the hairs are modified to bristles, spines or scales. Further dermal structures are the horn sheaths of the bovids, as well as nails, claws and hooves. The skin generally contains numerous sebaceous and perspiration glands. Milk glands, scent glands and odor glands that become active in rut are all modified dermal glands. Thoracic and abdominal cavities are separated by a diaphragm. Complete separation of the heart into a right venous and a left arterial side enables M. (like birds) to maintain a high metabolic rate; this in turn is associated with the maintenance of body temperture within certain limits. Efficient sense organs and a large, highly developed brain have led to the evolution of many different and complex patterns of behavior.

Reproduction. After copulation, the fertilized egg develops into an embryo within the uterus of the female, where it is supplied with nutrients and oxygen via the placenta and umbilical cord (except in monotremes and marsupials). At birth, the physical connection between mother and young is broken, and the young are at first fed with the secretion from the milk glands, which are usually located on the ventral surface of the female. Gestation time and the stage of development of the young at birth differ according to the species. For example, the young of terrestrial carnivores are born naked and blind, whereas those of ungulates and whales are born in an advanced stage of development, and can often follow the mother immediately after birth.

Mode of existence. As a result of their ability to maintain a constant body temperature, M. are found in all climatic zones. With appropriate adaptations, M. are also found in all types of habitat, i.e. in water (aquatic species), on land (terrestrial), beneath the earth (fossorial) or in trees (arboreal). Bats have even conquered the air with true flight, while other species are able to glide. Food consists of various types of plant material, and of other animals, the latter being overpowered or occasionally eaten as carrion. Omnivores and highly specialized feeders, as well as all intermediate forms between these two extremes are found among M. Some M. are solitary and form pairs only when mating. Others live in packs or herds, which are led by one dominant animal. In addition to odors and visual signals (body movements, postures, etc.), vocalizations are also often used for communication. Many species are active in daylight, others in the dark. For seclusion, and for raising their young, M. often seek out caves or hollows, or they construct various sites of their own (dens, nests, etc.). In temperate and cold regions, reproduction occurs in a yearly cycle. Food availability and climatic conditions may lead to large migrations (see Animal migration), e.g. in some ungulates, in seals, whales, bats and lemmings. In the coldest season of the year, most M. of temperate and cold regions (e.g. bats, hedgehogs, marmots, dormice, hamsters and other M.) conceal themselves, decrease their metabolism, and enter into a fairly long period of hibernation.

Classification. The following 18 orders are recognized: *Monotremata, Marsupialia, Insectivora, Dermoptera, Chiroptera, Primates, Edentata, Pholidota, Lagomorpha, Rodentia, Cetacea, Carnivora, Tubilidentata, Proboscidia, Hyracoidea, Sirenia, Perissodactyla, Artiodactyla.*

Economic importance. The larger M. are the most important of all exploited animals. Working animals (beasts of burden, draught animals, herding dogs, etc.) are all M. Domestic and hunted M. provide meat and raw materials. In addition, some M. are indispensable experimental animals in medical research. Some M. transmit disease, e.g. rabies, while others are food pests, attacking crops and taking all forms of stored foodstuffs.

Fossil record. Fossils of M. are found from the Upper Triassic to the present. The oldest remains are the teeth of multituberculates; these marsupials have their origin in a special group of the *Sauria,* the *Ictidosauria.* Most Mesozoic M. are primitive forms, and more highly evolved representatives do not appear until the Upper Cretaceous. M. underwent an explosive development in the Tertiary, which was most marked in the Eocene and Pliocene. Important fossil orders are marsupials, insectivores, carnivores, ungulates and primates, and some of these are useful index fossils of the Tertiary and Quaternary.

Collection and preservation. Small M. like mice and rats are widely caught in spring traps. Larger M. (e.g. martens, beavers, foxes) are preferably caught in snares or box traps. M. are killed with chloroform, ether, coal gas, or by injection. For anatomical studies, they are preserved by injection of formaldehyde solution into the body cavity.

Mammillaria: see *Cactaceae.*

Mammoth, *mammoth elephant, cold steppe mammoth,* *Mammuthus primigenius:* an elephant-like, extinct species of the *Proboscidea.* The species is well known from the discovery of entire corpses deep frozen in the Siberian and Alaskan permafrost. Although reputed to be very large, it was not quite as big as the African elephant. The upper jaw carried 2 tusks, which were up to 5 m in length, and curved upward and outward. The large, highly specialized molars, aligned one after the other in the chewing area, had numerous transverse lamellar ridges, usually reaching a maximum of 27 on the third molar. As a protection against the cold, the M. had a thick, woolly coat. Fossils are found in the Pleistocene. In Siberia the M. survived until the last Ice Age.

Mammoth trees: see *Taxodiaceae.*

Mammuthus: see Mammoth.

Mamo: see *Drepanididae.*

Mamush: see Halys viper.

Manatee: see *Sirenia.*

Mandarin duck: see *Anseriformes.*

Mandarin orange: see *Rutaceae.*

Mandibular arch: see Skull.

Mandibulata: a group of arthropods consisting of the 2 subphyla, *Branchiata* (*Diantennata)* and the *Tracheata.* In contrast to the *Amandibulata,* the first appendages behind the mouth are developed as mandibles. The next 2 appendages are also often modified to maxillae. The group may also be defined as arthropods that possess antennae. It is now considered to be an artificial group.

Mandragora: see *Solanaceae.*

Mandrake: see *Solanaceae.*

Mandrill, *Mandrillus:* a genus of Old World monkeys (see). There are two species: *M. sphinx* ***(Mandrill)*** and *M. leucophaeus* ***(Drill).*** They are large, powerfully built animals with short, almost naked tails. Males have very large muzzles with bony swellings on either side of the nose. The male drill has a black face with a scarlet lower lip. The male mandrill has a red nose with bright blue grooved nasal swellings, and yellow-orange beard and cheek whiskers. Males of both species have bright blue, red and violet perineum and genitals. Females and young lack these spectacular colors, and the coats of both sexes of both species are brown to gray. Mandrills and drills inhabit forested areas of western central Africa, the drills occupying a range north of the Sanaga river in Cameroon, the mandrills being found south of this limit. *Mandrillus* is closely related to *Papio* (Baboons); ecologically, the 2 genera can be seen as mutually exclusive, *Papio* preferring a nonforested habitat. The diet consists of fruits, leaves, ants and termites, and feeding forays are often conducted in cassava and oil palm plantations.

Maned rats: see *Lophomyinae.*

Maned wolf, *Chrysocyon brachyurus* (plate 36): A large (head plus body about 125 cm), particularly long-legged predator of the family, *Canidae* (see). The M.w. is a solitary, nocturnal animal found in the pampas regions of South America, where it hunts guinea pigs and other small mammals, and also takes various fruits. It is fast running and ranges over a wide territory. The coat is yellow-red to red-brown in color, and the lower parts of the legs are almost black. *Chrysocyon* is a monotypic genus.

Mangabeira: see *Apocyanaceae.*

Mangabeys, *Cercocebus:* a genus of large, long-legged, long-tailed, slender Old World monkeys (see) inhabiting forested areas of Africa. The thumbs are fully apposable and the big toes are divergent. The diet is mainly fruits, with some leaves, fungi and insects. Capacious cheek pouches are used to transport food to be eaten elsewhere. The 4 species are divided into 2 groups.

Group 1: *C. torquatus* ***(White collared*** or ***Cherry-capped sooty*** or ***White-crowned mangabey)*** and *C. galeritus* ***(Tana river, Agile*** or ***Golden bellied mangabey).*** These are uncrested and mainly brown or gray, inhabiting West and Central Africa, with isolated populations in Kenya and Tanzania.

Group 2: *C. albigena* ***(White-cheeked*** or ***Mantled mangabey)*** and *C. aterrimus* ***(Black*** or ***Black crested mangabey).*** These are mainly black, inhabiting western parts of Central Africa and Zaire, extending to Uganda and Rwanda.

Manganese, ***Mn:*** a bioelement present in all living cells, usually at a concentration of less than 1 ppm on a dry mass basis, or less than 0.01 mM in fresh tissue. Bone contains 3.5 ppm. Bacterial spores contain high levels of Mn(II) (0.3% dry mass). Mn(II) is necessary for sporulation in *Bacillus subtilis,* which can maintain an intracellular Mn concentration of 0.2 mM against an external concentration of 1 μM. Mn is an essential micronutrient for animals and plants. In animals, Mn deficiency leads to degeneration of the gonads and to skeletal abnormalities; the characteristic skeletal abnormality in chickens is called slipped tendon disease, or perosis.

Many glycosyl transferases (in particular galactosyl and *N*-acetylgalactosaminyl transferases) require Mn for activity, which explains the impairment of mucopolysaccharide metabolism associated with the symptoms of Mn deficiency. Mn is required by the enzyme farnesyl pyrophosphate synthetase, which catalyses a stage in cholesterol biosynthesis. Mn is also required at an earlier stage of cholesterol biosynthesis, probably at some stage in the conversion of acetate into mevalonate. Lactose synthetase requires Mn. Pyruvate carboxylase contains 4 atoms of Mn (i.e. one for each biotin molecule); Mn(II) is essential for the transcarboxylation, and the initial ATP-dependent carboxylation of biotin requires either Mn(II) or Mg(II). Superoxide dismutase from *Escherichia coli* contains 2 atoms of Mn(III). Yeast mitochondrial superoxide dismutase is a tetramer containing 1 atom of Mn per subunit. Similar enzymes are present in chicken liver mitochondria. Presumably the Mn of superoxide dismutase alternates between the III and II states during catalysis. Avimanganin, a protein of unknown function from avian liver, contains one atom of Mn(III) per molecule.

Mn is taken up by plants roots as the divalent cation, Mn(II), which is present in the soil as adsorption complexes or in free solution. Mn(III) and Mn(IV) compounds are also present in the soil, but must be reduced to Mn(II) before they can be ultilized by plants. In humus-rich, calcareous, neutral soils, with large populations of microorganisms, Mn is present mainly in its high valency forms; Mn(II) is therefore deficient, and Mn-uptake is decreased. Conversely, on wet, poorly aerated, acidic soils, the concentration of Mn(II) can attain toxic levels. Calcium, magnesium, iron and other heavy metal ions compete with Mn in the uptake process.

Most soils contain adequate Mn in a form available to plants. Mn deficiency can, however, occur on marshy and on very sandy soils. It results in chlorosis and mottling, manifested by the presence of light yellow spots between the leaf veins. The symptoms often appear first in the youngest leaves. In oats, Mn deficiency is manifested as gray to brown-gray dry spots with dark borders, a condition known as dry spot disease. The condition is abolished by fertilization of the soil with Mn sulfate (50–150 kg/ha), or by direct application to the plants (7–8% Mn sulfate solution, 800 l/ha). It can be prevented by the application of modern, Mn-containing mineral fertilizers. Treatment of alkaline soils with acidic fertilizers has the effect of releasing bound Mn, so that it becomes available to plants.

The reaction center of photosystem II of the chloroplast contains 2–4 Mn atoms. The effect of Mn deficiency on photosynthesis resembles poisoning by dichlorophenyl dimethyl urea (DCMU), i.e. evolution of O_2 by photosystem II is inhibited, while photosystem I is unaffected. The vitamin C-content of various plants is increased by Mn. Through its activation of the peroxidase that oxidizes β-indolylacetic acid, Mn plays a part in the control of growth (see Phytohormones). In plants it is also involved in the respiratory chain, in the decarboxylation of oxalosuccinic acid in the tricarboxylic acid cycle, and in nitrate reduction. Concanavalin from jack bean contains one atom of Mn(II) per molecule. Manganin, a protein present in peanuts, contains one atom of Mn per molecule.

Animals require much less Mn than do plants.

Mange: see Scabies mite.

Mangifera: see *Anacardiaceae.*

Mango: see *Anacardiaceae.*

Mangrove snake: see Tree snakes.

Mangrove swamp: a tropical coastal woodland, consisting of halophytic trees and shrubs growing in sheltered tidal mud flats (saline silts or sandy muds) in bays, lagoons or estuaries. The bottom is covered at high tide and exposed at low tide, revealing numerous breathing roots (pneumatophores) projecting upward from the mud. Under the specialized conditions of this habitat, only certain plant species, known as mangroves, are able to thrive. They belong to different families, but they all display similar secondary adaptations, such as the possession of stilt roots, aerial roots and breathing roots; also their seedlings germinate on, and hang down from, the parent plant (viviparous seedlings), eventually falling like an arrow and lodging in the mud below. Important mangroves are: *Rhizophora* (America, Africa, Asia and Australia), *Bruguiera* (India and Africa), *Ceriops* (Africa, India and Australia) [family *Rhizophoraceae]; Laguncularia* (tropical America and West Africa) [family *Combretaceae]; Avicennia* (Tropical America, Africa, Asia and Australia) [family *Verbenaceae;* sometimes placed in the monotypic family *Avicenniaceae]; Sonneratia* (Africa and Asia) [family *Sonneratiaceae* or *Lythraceae].*

Manidae: see *Pholidota.*

Manihot: see *Euphorbiaceae.*

Manioc: see *Euphorbiaceae.*

Manis: see *Pholidota.*

Manna: see *Oleaceae.*

Manna ash: see *Oleaceae.*

Manna lichen: see Lichens.

Manna sugar: see D-Mannitol.

Mannans: polysaccharides that are widely distributed in plants as reserve material and as hemicelluloses. M. of plants consist of D-mannose predominantly in α1,4-glycosidic linkage. M. occur in the seed shells of the ivory nut *(Phytelephas),* the mucilage of orchid tubers, seeds of the carob tree and seeds of lucerne, as well as in grasses and conifers.

The M. of yeast cells consist of a main chain of α1,6-linked mannose with short branches (1–3 mannose units) attached by α1,2- and α1,3-glycosidic linkages.

Mannikin: see *Ploceidae.*

Mannite: see D-Mannitol.

D-Mannitol, ***mannite, manna sugar:*** HOH_2C—CHOH—CHOH—CHOH—CHOH—CH_2OH, a hexitol related to D-mannose, found widely in plants and plant exudates, and in fungi and seaweeds. The exudate of the manna ash *(Fraxinus ornus)* contains 75% D-M. D-M. is used as a purgative and as a sugar substitute for diabetics.

D-Mannose: an aldohexose monosaccharide, which is seldom found in the free state, and occurs mostly as a component of high molecular mass polysaccharides. It is metabolized via glucose 6-phosphate.

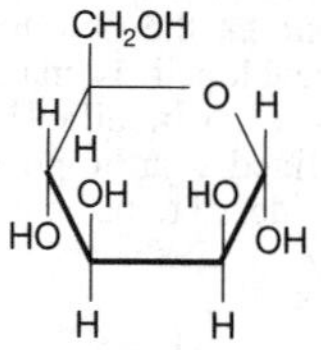

D-Mannose

Mannuronic acid: see Uronic acids.

Man-of-war birds: see *Pelicaniformes.*

Mantellinae, ***mantellins:*** a subfamily of the *Ranidae* (see), containing 60 species in 3 genera, all restricted to the island of Madagascar: *Laurentomantis (= Microphryne = Trachymentis)* 3 species; *Mantella* 4 species; *Mantidactylus (= Gephyromantis = Hylobatrachus)* 53 species. The brilliant orange-red "Gold frog" *(Mantella aurantiaca;* body length 2–3 cm) has been described as one of the most beautiful anurans in the world. Internal fertilization, which is very rare in the *Anura,* is known to occur in *Mantella aurantiaca ;* it also occurs in *Nectophrynoides* (Live bearing toad; family *Bufonidae)* and *Ascaphus* (Tailed frog; family *Ascaphidae).*

Mantids: see *Mantoptera.*

Mantispids: see *Neuroptera.*

Mantoptera, ***Mantodea, mantids:*** an insect order containing about 1 700 species, restricted almost entirely to the tropics. Only one species, the praying mantis *(Mantis religiosa),* is also found in warm habitats in southern central Europe and temperate parts of Asia; it has been introduced into northeastern North America, where it is now common. The earliest fossil *M.* are found in the Lower Tertiary amber.

Imagoes. These are slim, usually green or brown, 1 to 16 cm in length. The head carries multisegmented filiform antennae, powerful mandibulate biting and chewing mouthparts, 2 large compund eyes and 3 ocelli. For detection and trapping of prey, the insect relies on acute vision. Two pairs of wings are present, which are folded closely along the body when at rest. The prehensile forelegs are characteristically modified for seizing and holding prey; the femur is folded closely against the long, free-hanging coxa, and the tibia is folded back against the femur. Femur and tibia both carry 2 rows of spines which are apposed to one another when the leg is bent. In addition, the end of the tibia is equipped with a long, sharp, curved hook. When waiting in ambush for prey, both legs are held close together outward and upward, resembling praying hands. When an insect alights or walks within range, the coxae are jerked forward, as the femora and tibiae extend outward, and the prey is trapped by the hooks and held by the spines between tibia and femur, the entire movement taking less than one twentieth of a second. Small vertebrates, frogs and birds may also be caught and eaten. The female eats the male even before copulation is finished, and this is thought to provide the female with extra protein to aid rapid egg production.

Eggs. A female produces up to 1 000 eggs. These are laid in separate lots (maximally 400) in a frothy secretion which hardens on exposure to the air to form a foam-like, papery capsule (ootheca), attached to plants or stones, etc.

Larvae. In their morphology and mode of existence, the larvae are similar to adults. They molt 5 to 9 times before attaining maturity. Metamorphosis is therefore incomplete (paurometabolous development).

Classification. All species are placed in the single family, *Mantidae.*

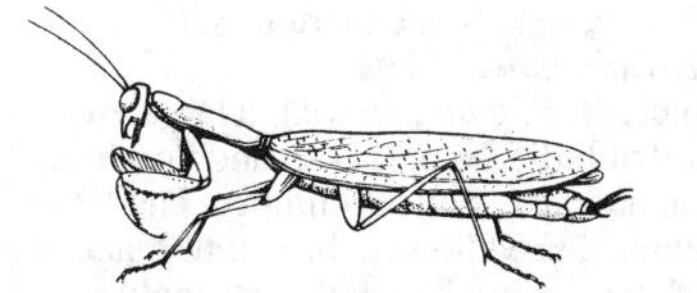

Mantoptera. Praying mantis *(Mantis religiosa* L.).

Manual: see *Felidae.*

Manubrium: the basal part of the caudal furcula of springtails *(Collembola).*

Manx shearwater: see *Procellariiformes.*

MAP: acronyn of microtubule associated protein; see Microtubules.

Maple: see *Aceraceae.*

Map turtles: see *Emydidae.*

Maquis: see Sclerophylous vegetation.

Mara: see *Caviidae.*

Marabou stork: see *Ciconiiformes.*

Maral: see Red deer.

Marasmines: see Wilt toxins.

Marasmus: a nutritional deficiency disease, mostly in children under 1 year, due to lack of both dietary energy and protein. The condition is usually compounded by mineral and vitamin deficiencies, and by gastrointestinal infections. Body mass is less than 60% of the standard for the age. Dehydration is common, there is little or no subcutaneous fat, and plasma albumin levels are about 25 g/l (normal levels are above 40 g/l). The characteristic hair and skin lesions of kwashiorkor are absent. M. victims often have a history of abrupt and early weaning, followed by dilute (and often unhygenic) artificial feeds that are low in both energy and protein. Subsequent gasteroenteritis may be treated by withholding food, thus precipitating and/or exacerbating the disease. A distinction is drawn between nutritional M. (resulting from inadequate food alone) and M. produced by infection (although inevitably against a background of poor nutrition). M. is more common in urbanized than in rural areas of the developing world. Unlike kwashiorkor, M. shows no dependence on climate, and it was once well known in industrial areas of Europe and North America. M. in children is equivalent to starvation in adults.

Marbled salamander: see *Ambystomatidae.*

Marchantia: see *Hepaticae.*

Marchantiales: see *Hepaticae.*

Marey's reflex: see Cardiac reflexes.

Margaritana: see Freshwater mussels, Pearls.

Margaritifera: see Freshwater mussels, Pearls.

Margarornithinae: a subfamily of the *Furnariidae* (see).

Marguerite: see *Compositae.*

Marigold: see *Compositae.*

Marihuana: see Hashish.

Marine hatchetfishes: see *Stomiiformes.*

Marine iguana, *Amblyrhynchus cristatus:* a large (up to 1.7 m long) iguana (see *Iguanidae),* inhabiting the rocky coasts of the Galapagos Islands. There are many subspecies, associated with different islands. The neck carries a high crest, and the blunt-ended head is surmounted by several cone-shaped excrescences. It is an excellent swimmer and diver, feeding almost exclusively on seaweed. Males show marked territorial behavior, and at breeding time they collect several females and employ a ritualized display to expel competitors from their territory. During this display, the legs are stiff, the body raised, crest erect, gular region expanded, and the mouth is agape, revealing a red spot on the tongue. Head bobbing also occurs as two males place themselves laterally to each other. They then face each other and perform head butting. The M.i. is a strictly protected species.

Marine turtles: see *Chelonidae.*

Marjorum: see *Lamiaceae.*

Marker gene: usually a gene, whose action and site on the chromosome are known. It is used as a reference gene for mapping the position and distribution of other genes, or for detecting the presence of certain chromosomes.

Markhor: see Goats.

Marlins: see *Scombroidei.*

Marmosets, *Callitrichidae:* a family of fully arboreal, very small, squirrel-like New World monkeys (see), with thick fur, a nonprehensile tail, and a quadrupedal gait. Unlike other monkeys, which normally have nails on their digits, M. possess claws on all fingers and toes, except the big toes. They live in family groups of 3 to 8 individuals, sometimes more, and litters usually consist of 2 or 3 young. The subfamily, *Callitrichinae,* comprises the ***True marmosets,*** while ***Goeldi's marmoset*** *(Callimico goeldii)* is placed in a separate subfamily *(Callimiconinae)* of only one species. The constituent genera of the *Callitrichinae* are *Callithrix, Cebuella, Saguinus* and *Leontopithecus.* M. have 32 teeth, including 2 molars in each half jaw. Dental formula: 2/2, 1/1, 3/3, 2/2 = 32. *Callimico* is an exception, having 3 molars in each half jaw, a total of 36 teeth, and a similar dental formula to that of Cebid monkeys (see).

Callithrix is represented by 3 species: *C. jacchus* ***(Common*** or ***White fronted marmoset;*** plate 41), *C. argentata* ***(Silvery marmoset)*** and *C. humeralifer* ***(Buffy-tufted-ear*** or ***Santarem marmoset***). They inhabit the tropical rainforest, savanna or dry scrub of eastern Brazil. Coat colors vary from white to dark brown, *C. jacchus* being marbled black and gray with a white face and black and gray-ringed tail. The ears are ornamented with tufts of hair, which arise around the ears *(C. jacchus)* or from the edge of the ears *(C. humeralifer). C. argentata* lacks ear tufts, and some races of this white species possess a contrasting black tail. The diet consists mostly of insects with some plant gums.

The single species, *Cebuella pygmaea* ***(Pygmy marmoset),*** inhabits the tropical rainforest of the eastern Amazon basin. It is the smallest New World monkey (head plus body 133 mm, tail 200 mm).

The ***Tamarins*** (genus *Saguinus,* with 11 species) are distributed throughout tropical America. They are conspicuously colored, black, red or brown, and some species have white eyebrows, white moustaches and white head crests. Head plus body length is about 245 mm (tail 360 mm) and they are somewhat larger than true M. The diet includes both insects and fruit. The territories of tamarins and true M. tend to be geographically separate, with the exception that *Saguinus midas* and *Callithrix argentata* are sympatric between the Xingu and Gurupi rivers.

The ***Lion-tamarin*** or ***Liszt monkey*** *(Leontopithecus rosalis;* plate 40) is the only species of its genus. It inhabits the tropical rainforest of the coastal mountains of southeastern Brazil, a habitat that is becoming rapidly eroded by tree felling. There are 3 subspecies: *L.r. rosalia* (completely golden), *L.r. chrysomelas* (black with golden mane), *L.r. chrysopygus* (black with golden rump and legs). It is generally the same shape and size as *Saguinus* spp., but is distinguished by a mane of long hair, its striking golden color, and its slender elongated hands with long webbed digits. It is listed as an endangered species.

Marmots: see *Sciuridae.*

Marquesas flycatchers: see *Muscicapinae.*

Marsdenia: see *Asclepiadaceae.*

Marsh frog: see *Raninae.*

Marsh hawks: see *Falconiformes.*

Marsh helleborine: see *Orchidaceae.*

Marsh marigold: see *Ranunculaceae.*

Marsh samphire: see *Chenopodiaceae.*

Marsh spot: see Deficiency diseases.

Marsh tit: see *Paridae.*

Marsh wren: see *Troglodytidae.*

Marsipobranchii: see *Cyclostomata.*

Marsupial anteater: see Numbat.

Marsupial carnivores: see *Dasyuridae.*

Marsupial cats: see *Dasyuridae.*

Marsupial frogs: see *Hylidae.*

Marsupialia: see Marsupials.

Marsupial mole, *Notoryctes typhlops:* the only species of the marsupial family, *Notoryctidae,* which inhabits the sandridge deserts of central Australia. The M.m. is a mouse-sized mammal with light colored fur, which has a golden or silver sheen. It leads a subterranean existence similar to that of the placental moles of other continents, and feeds largely on invertebrate larvae. The nose is equipped with a broad cornified shield, and the short powerful limbs are adapted for burrowing. Its eyes are vestigial, covered by skin, without lenses, and nonfunctional.

Marsupials, *Marsupialia* (plate 38): an order of mammals, whose young are born at a relatively early stage of development, and continue growing in a brood pouch or marsupium. Without assistance from the mother, the newly born young find their own way to the pouch, where they become firmly atached to the mammary teats. The teat expands inside the mouth of the young animal, which for a period is therefore inseparable from the mother. In many M. the pouch is merely a fold of skin, or it may even be absent. M. are predominantly native to Australia and some neighboring islands, where they have evolved to fill the ecological niches occupied by placental mammals in America and Eurasia. As a result of this parallel evolution, they display striking similarities of form and behavior with the mammals of other continents. The existence of a few species of M. in the New World reflects the ancient continuity of the continental land masses (see Gondwanaland). The following M. are listed separately: Opossums, *Dasyuridae,* Numbat, Marsupial mole, Bandicoots, Wombats, Kangaroos, and *Phalangeridae.*

Marsupial wolf: see Tasmanian wolf.

Marsupium:

1) In marsupials and echidnas, a pouch or fold of skin supported by the epipubic bones. It serves to carry the eggs of echidnas and the young of marsupials. See Marsupials, *Monotremata.*

2) A structure on the ventral surface of some crustaceans (e.g. *Amphipoda* and some *Isopoda).* Overlapping oostegites (see) on the 2nd, 3rd and 4th thoracic limbs form a M. or egg-carrying pouch.

Martes: see Pine marten, Stone marten, Sable.

Martins: see *Hirudinidae.*

Masked palm civet, *Paguma larvata:* a large viverrid (see *Viverridae)* of India, Nepal, Tibet and Southeast Asia. It is arboreal and an excellent climber. The body coat is uniform grayish or yellowish brown to black in color, and the dark face is marked with a strongly contrasting light frontal streak, or spots under the eyes and in front of the ears.

Masking: see Protective adaptations.

Massoutiera: see *Ctenodactylidae.*

Mastacembelidae: see *Mastacembeloidei.*

Mastacembeloidei, *Opisthomi:* a suborder of freshwater fishes of the order *Perciformes* (see). They have an eel-like or ribbon-shaped, elongate body. Pelvic fins are absent, and anal fins are continuous to or continuous with the small caudal fin. Between 70 and 95 vertebrae are present, and there is no temporal bone. The pectoral arch (supracleithrum) is attached to the vertebral column by a ligament, and the swim bladder is physoclistic, i.e. it has no air duct. Two families are recognized.

1. Family: *Mastacembelidae* ***(Spiny eels).*** Members (maximal length 0.9 m) occur in tropical Africa and through the Middle East to the Malay Archipelago and China, and are notable for their fleshy, proboscis-like rostral appendage. There is no basisphenoid. The dorsal fin is preceded by a series of spines, while the anal fin usually has 2–3 spines and 30–130 soft rays. They are valued locally as food fishes. The body is covered with small scales. There are two genera: *Macrognathus* (3 species in southern Asia) and *Mastacembelus* (about 60 species).

2. Family: *Chaudhuriidae.* The body is naked, there is no rostral appendage, no dorsal spines, and the basisphenoid is present. The three species (1 species of *Chaudhuria,* 2 species of *Pillaia)* are found in northeastern India, Burma and Thailand, and are sometimes placed in a separate order.

Mastacembelus: see *Mastacembeloidei.*

Mast cell: a connective tissue cell found widely in different body tissues. It contains large quantities of histamine and other pharmacologically active substances, which are released at sites of tissue damage. Some of these substances cause constriction of lymph vessels, resulting in localized swelling. Histamine increases capillary permeability, so that leukocytes are able to migrate from blood vessels and aggregate at the site of injury. M.cs. also possess receptors for immunoglobulin E (IgE) and play an essential role in anaphylaxis. See Allergy.

Mastic: see *Anacardiaceae.*

Mastication, muscles of: see Muscles of mastication.

Mastication reflex: see Chewing reflex.

Mastigont: in flagellates (see *Polymastigina),* the totality of organelles arising from a group of basal bodies, consisting of flagella, undulating membrane, basal fibrils, axostyle and parabasal body.

Mastigophora: see *Mastigophorea.*

Mastigophorea, *Mastigophora, Flagellata:* a class of protozoa characterized by the possession of one or more flagella, which are used for locomotion and for generating a feeding current. Their external shape and internal organization differ among the various orders.

Many *M.* have chlorophyll-containing chromatophores (chloroplasts), and therefore display physiological affinities with algae. Since these chromatophores contain varying proportions of chlorophyll, xanthophylls and carotenes, etc., they may be green or yellow (xanthoplasts). To a large extent, however, photoassimilation is supplemented by the uptake of organic material. Under appropriate conditions, the assimilatory pigments are lost, and the resulting colorless organism displays purely animal-type, heterotrophic nutrition. Chlorophyll-containing species are grouped together as the *Phytomastigophorea* (see), and the colorless species as the *Zoomastigophorea (Zoomastigina, Zooflagellata).* More advanced forms have an oral aperture for the intake of organic food material.

Every order of the *Phytomastigophorea* contains species that are able to pass into a rounded resting stage without flagella, which is encased in a cyst or layer of gelatinous material. In this condition, the organism may divide to form a pseudocolony known as a *palmella.* The palmella may then develop into a branched structure resembling similar structures of primitive algae. This is highly suggestive of a common evolutionary origin of the algae and the *Phytomastigophorea* (separation of modern *Phytomastigophorea* from the algae is difficult).

Some *M.* can assume an ameboid form by developing pseudopodia, sometimes with loss of their flagella. Thus, the *M.* display marked affinities to the *Rhizopodea* (see), as well as close relations to the algae.

M. multiply by longitudinal cleavage. Sexual reproduction has been described sporadically in different groups, and it appears to be a general property only in the *Phytomonadina.* Some orders also contain colonial forms.

M. are generally marine and freshwater planktonic organisms, a few are sessile (some with attachment stalks), and there are also parasitic forms (see Trypanosomes).

The class is divided into the following orders: *Chrysomonadina* (see *Phytomastigophorea), Cryptomonadina* (see), *Dinoflagellida* (see *Phytomastigophorea), Euglenida* (see *Phytomastigophorea), Phytomonadina* (small spherical forms), *Chromomonadina, Protomonadina* (including the trypanosomes), *Polymastigina* (see), and *Opalinata* (see).

Mastodon (Greek *mastos* nipple, *odontes* teeth): extinct genus of the *Poboscidea,* which diverged from the evolutionary line leading to modern elephants. They were much smaller than modern elephants, with different skull and dental structures, but they were the first true proboscideans. The skull was short and high, and the jaw was longer than that of elephants; both the upper and lower jaws usually carried a pair of tusks, which were usually strikingly large and curved upward and outward. Two main evolutionary lines are recognized: 1) species with high-ridged (bunodont) molars, and 2) species with transversely ridged (zygodont) molars. *M.* never developed the high crowns and lamellar teeth of modern elephants. The following sequence of bunodont forms is recognized in the Upper European Tertiary: *Mastodon (Bunolophodon) angustidens* → *Mastodon (Bunolophodon) longirostris* → *Mastodon (Anancus) arvernensis.* The zygodont line finished in Europe in the Pliocene with *Mastodon (Zygolophodon) borsoni,* and in North America toward the end of the last Ice Age with *Mastodon (Zygolophodon) americanus.* Fossils are found from the Miocene to the Pleistocene. In South America *M.* survived after the last Ice Age.

Mastodonsaurus giganteus: see Triassic.

Matamata: see *Chelidae.*

Maticora: see Long-glanded coral snakes.

Matilija poppy: see *Papaveraceae.*

Matjeshering: see Herring.

Matorral: see Sclerophylous vegetation.

Matricaria: see *Compositae.*

Maturus: see Age diagnosis.

Mauritanian toad: see *Bufonidae.*

Mauthner's cells: a pair of neurons in the medulla oblongata of teleost fishes and urodeles. Each consists of a cell body with a lateral and a ventral dendrite (about 20 μm diameter and 0.5 cm long), and an axon (40 μm diameter) that passes down the spinal cord. The lateral dendrite receives input from the region of the acoustic nucleus, while the ventral dendrite receives input from the cerebellum, the roof of the mesencephalon, and the trigeminus. In fishes, excitation of M.c. leads to rapid movements. Owing to their greater than average size, the geometry of their synaptic input and the variety of synaptic contacts, M.c. are a popular system for neurophysiological research. In addition to excitatory and inhibitory postsynaptic parts, electrical synapses are also present. M.c. may therefore be considered as "miniature nervous systems".

Maxillary glands, *labial nephridia:* a pair of glands opening to the exterior at the bases of the second maxillae (labia) of some lower arthropods (e.g. *Diplopoda, Archeaognatha,* nonmalacostracan crustaceans). They are derived from, and function as, excretory nephridia. In all adult, lower (nonmalacostracan) crustaceans, the M.g. are the only excretory organs. They are also present in certain higher crustaceans, e.g. *Cumacea* and *Isopoda.* M.g. are particularly important in freshwater species, where they also serve for the regulation of the water economy and mineral salt metabolism (osmoregulation).

Maximal concentration at the workplace: see MCW.

May blobs: see *Ranunculaceae.*

Mayflies: see *Ephemeroptera.*

Maypole builders: see *Ptilinorhynchidae.*

Maze test, *labyrinth test:* a test in which animals have to find their way through a maze containing systematically arranged blind pathways. At the exit there is a reward. The learning process is affected by the structure of the maze, and it is expressed quantitatively by the number of "mistakes" (false choices of alternative paths). According to the learning problem under investigation, the design of the maze can be modified, e.g. from a simple T-maze to very complicated labyrinths, and it can even be made 3-dimensional. It is used to study learning processes, the nature of information reception and processing, motivation structures, as well as genetic and environmental influences on learning.

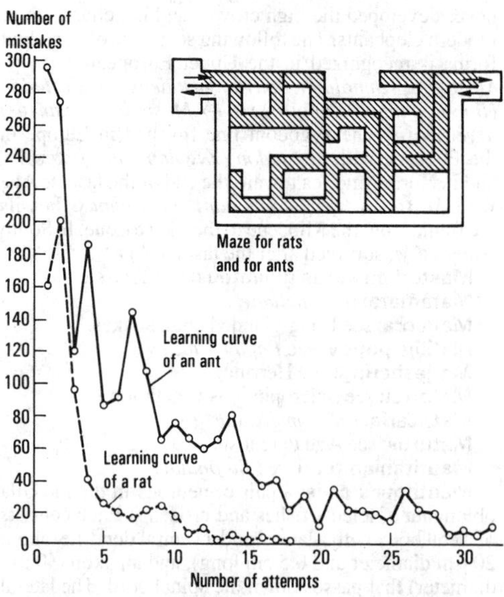

Maze test. Learning graph for rats and ants as determined by the maze test.

MCW, *maximal concentration at the workplace:* the maximal permissible concentration of a harmful substance at a place of work, i.e. the maximal concentration that has no detectable effect on health. The value is determined as the average over a period of 8 hours or 1 working day, and it is expressed as the quantity in 1 cubic meter of air at 20 °C and 101 325 Pascal.

Meadow: a treeless plant formation, typically consisting of a continuous turf of perennial *Cyperaceae* and *Poaceae* and herbs, which is used agriculturally for grazing and for making hay.

Meadowlarks: see *Icteridae.*

Meagre: see *Sciaenidae.*

Meal mite: see *Acaridae.*

Mealy bug: see *Homoptera.*

Mean generation time: see Generation time.

Mecanopsis: see *Papaveraceae.*

Mechanical tissues, *support tissues:* plant tissues responsible for the strength and elasticity of plant organs. In small herbaceous plants, the turgor pressure and resulting tension of the cell walls are sufficient to maintain the shape and rigidity of organs and of the entire plant. Larger plants possess special tissues with thickened cell walls, which impart strength and elasticity. In the absence of such M.t., larger plants would collapse whenever they suffered water stress. There are 2 types of specialized mechanical tissue.

1) ***Collenchyma.*** This is a living tissue, whose elongated cells are reinforced by thickening, either at the angles of the cells (angular collenchyma), or on the tangential walls (plate collenchyma). The thickening substance contains a small proportion of cellulose and a very high proportion of strongly swollen propectin, whose presence imparts a considerable resistance to tearing. Since some of the walls remain unthickened, the cellular exchange of metabolites can continue unimpeded. Collenchyma is quite elastic and still capable of growth by cell division It occurs primarily in young, growing parts of the plant.

2) ***Sclerenchyma.*** This is a dead tissue, which functions mainly as a support in plant parts that have finished growing. Where sclerenchyma is required to withstand pressure or crushing forces, it consists of thick-walled, isodiametric *stone cells (sclereids),* whose walls are strongly lignified. The wall thickening is laid down in layers. Numerous tubular, branched or unbranched pits remain open, and these enable the young, live stone cells to exchange metabolites. As the tissue develops, the cell walls become increasingly lignified. The cell lumen therefore becomes increasingly smaller and metabolite exchange more difficult, until the protoplasm finally dies. Stone cells sometimes occur in isolated groups in the mesocarp of pear. The hardness of the rigid pericarp (shell) of nuts and the stones of drupes is due to stone cell sclerenchyma.

Where the plant requires resistance to tensile forces or bending forces, the sclerenchyma consists of *sclerenchyma fibers.* These are usually fusiform, elongated fiber cells, whose cell walls are either elastic and unlignified, or nonelastic and strongly lignified. In cross section, the fiber is polygonal, and again the cell lumen is small, usually appearing as merely a dot. The simple pits are slit-like, few in number or absent; their oblique alignment testifies to the spiral arrangement of the layers of cell wall thickening, which confers increased resistance to tension. Sclerenchyma fibers are found in most shoots and roots of terrestrial plants. For their small diameter, they are unusually long (normally 1–2 mm), compared with other plant cells. In some plants they are exceptionally long, e.g. in flax *(Linum usitatissimum)* they attain lengths up to 6.5 cm, in the stinging nettle *(Urtica dioica)* up to 7.5 cm, and in ramie *(Boehmeria nivea)* about 30 cm. Such long and mechanically strong fibers are suitable for spinning into textile yarn, and Fiber plants (see) are cultivated for this purpose.

Mechanical resistance to bending is especially well developed in stems, in which bundles of longitudinally arranged sclerenchyma fibers are laid down as strands or straps at the periphery of the organ (i.e. as near as possible to the external surface), the interior consisting of parenchyma. This peripheral arrangment of reinforcing material is also a principle used in engineering to achieve *columnar rigidity* (see Biostatics). It is well exemplified by grass halms, and by the stems of many *Lamiaceae,* in which the reinforcing material actually protrudes from the periphery as longitudinal ridges. Conversely, in plant parts subject to tension (e.g. roots), the sclerenchyma fibers are arranged in the form of a central cable, which is surrounded by parenchyma.

Mechanomorphosis: in plants, the modification of shape and size in response mechanical stimuli. Frequently, mechanical stimulation resembles stimulation by light, in that it inhibits longitudinal growth. Mechanical stimuli may also influence plant anatomy, e.g. promotion of the growth of tendrils. Wind may act as a mechanical stimulus, causing increased thickening of tree trunks. Alterations of form and shape induced by touching or contact are often grouped together as *thigmomorphoses* or *haptomorphoses.* Thus certain algae form rhizoids in response to contact with the substratum. Various tendrils (e.g. those of *Parthenocissus tricuspidata,* Virginia creeper) develop attachment disks in response to contact with a solid surface. In common dodder *(Cuscuta epithymum),* contact stimulates shoot formation.

Mechanoreception: generation of a primary electrical impulse in response to the deformation of a sensory cell. Mechanical sensitivity is a fundamental property of living organisms. The latent period between the beginning of the stimulus and the change in membrane conductivity is in the range 15–100 ms. By analogy with the regulation of Na^+ channels, this short latent period suggests that gated channels are also involved in the response to mechanical stimuli (see Gate), although specific membrane molecules have not been demonstrated for any mechanoreceptor. It is therefore assumed that M. is triggered by unspecific stretching of the sensory cell membrane. On the other hand, the energy required to trigger an electrical response is very small, so that a gated system would involve very few molecules. Examples of M. are sound reception, M. in muscles, and M. by sensory hairs of insects and gravity-sensitive organs.

Meconic acid, ***3-hydroxychelidonic acid, 3-hydroxy-4-oxo-4H-pyran-2,6-dicarboxylic acid:*** a heterocyclic dicarboxylic acid containing a γ-pyrone ring. It occurs in many plants and is particularly abundant in the milky exudate of *Papaver somniferum* (Opium poppy). Raw Opium (see) contains 4–6% M.a. In the plant, the opium alkaloids are associated with organic acids, such as tartaric, citric, malic, etc., but especially with M.a. It decomposes with evolution of CO_2 when boiled in water or heated to 120 °C.

O, OH, HOOC, O, COOH

Meconic acid

Mecoptera, *Panorpatae, Panorpina,* ***scorpion flies:*** a small insect order of about 400 described species, which are particularly common in the Northern Hemisphere. Most species belong to the genera *Panorpa* and *Bittacus.* In males of the family, *Panorpidae,* the terminal abdominal segment is curved upward, with a superficial resemblance to a scorpion's tail (Fig.). The earliest fossil *M.* occur in the Permian.

Imagoes. Adult body lengths vary between 3 and 30 mm. The head capsule is prolonged ventrally into a broad, beak-like rostrum, which incorporates the prolonged mouthparts (clypeus, labrum and maxillae). Head and thorax are joined by a membranous neck. All species have well developed compound eyes, but ocelli may be present or absent. There are usually 2 pairs of identical, delicate, transparent, primitively veined (netlike) wings, which are usually carried horizontally and flat against the body when at rest. Some species lack wings (family *Bittacidae),* or the wings are greatly reduced (family *Boreidae).* The long slender legs are generally similar, but the hindlegs of the snow scorpion flies are longer than the others, and serve as jumping legs. Metamorphosis is complete (holometabolous development).

Eggs. All species lay eggs, which the female inserts into the soil.

Larvae. The larva is eruciform (caterpillar-like) with biting mouthparts. There are 3 pairs of thoracic legs, and abdominal feet may be present or absent. Pupation occurs after 3 to 6 molts. Larve live in the soil, feeding on plant and animal debris.

Pupae. The pupa is decticous and exarate (pupa libera), and it rests in the ground without a cocoon.

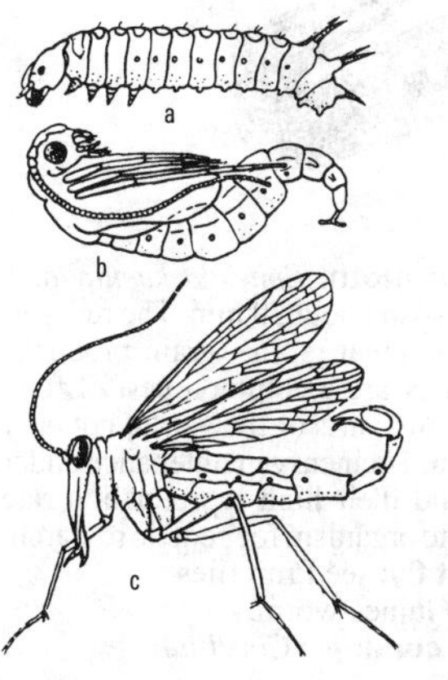

Mecoptera (scorpionflies). *a* Larva. *b* Pupa *(Chorista). c* Imago (male, *Panorpa).*

Classification. Scorpion flies were formerly placed in the *Neuropteria,* but they are clearly more closely allied with the *Trichoptera, Lepidoptera* and *Diptera.*

1. Suborder: *Eumecoptera.* Wings are elongate and narrow, and male genitalia are usually swollen.

Family: *Parnorpidae* ***(Common*** or ***True scorpion***

flies). A cosmopolitan family feeding on animal and vegetable debris. Ocelli are present.

Family: *Bittacidae* ***(Hanging scorpion flies).*** A cosmopolitan family, whose members hunt other insects. The very long, prehensile legs possess a terminal, tarsal segment, which can be snapped back like a trap to hold other insects. Wings are absent, and ocelli are present.

Family: *Boreidae* ***(Snow scorpion flies).*** A family of Holarctic species, which are active in the winter, feeding on dead and dying insects. In place of wings, the male has 2 pairs of bristle-like vestiges, and the female has a single pair of scale-like lobes. Ocelli are present.

Family: *Nannochoristidae.* A small, primitive family in the Southern Hemisphere. Adults live near water and larvae are aquatic.

Family: *Choristidae.* Four species only, restricted to Australia.

2. Suborder: *Protomecoptera.* Wings are shorter and broader than in *Eumecoptera.* Male genitalia are not swollen.

Family: *Notiothaumidae.* A single species, which occurs in Chile. Ocelli are present.

Family: *Meropidae:* Two genera, each with a single species, 1 in western Australia, and 1 in eastern USA.

Mediators: see Tissue hormones.

Medicago: see *Papilionaceae.*

Medicinal leech, *Hirudo medicinalis:* a leech of the family *Hirudidae,* order *Gnathobdellida* (see *Hirudinea),* 10–15 cm in length, and brown to olive-green in color, with 6 red-brown, black-bordered longitudinal stripes on the dorsal surface. It is found in swamps and in ponds that are rich in vegetation. In antiquity in India and Greece it was used medicinally for bleeding patients, a practice that continued in Europe and elsewhere well into the 19th century, even into the 20th century. A single leech is capable of sucking 3–6 g of blood. When it bites, the leech injects the anticoagulant, *hirudin,* which insures a continuous flow of blood.

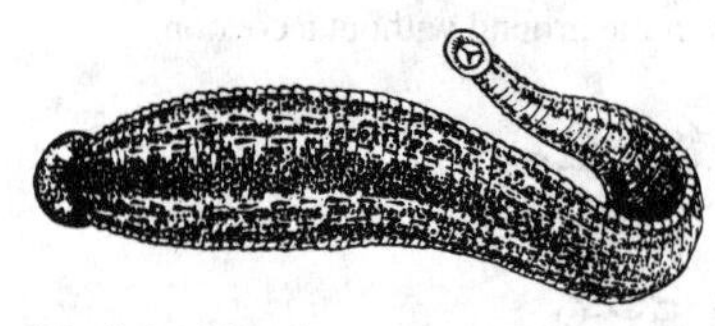

Medicinal leech

Mediterranean flour moth, *Ephestia kueniella:* a gray pyralid moth, wingspan about 20 mm. The red-yellow caterpillars feed on cereal grains, grain products, seeds and pulses, and they are a common pest of food stores and bakeries in many parts of the world, not only the Mediterranean region. Their easy maintenance under laboratory conditions and their high reproduction rate have made them a favorite organism for genetic research.

Mediterranean fruit fly: see Fruit flies.

Medina worm: see Guinea worm.

Mediterranean red coral: see *Corallium.*

Medlar: see *Rosaceae.*

Medulla oblongata: see Brain.

Medulla spinalis: see Spinal cord.

Medusa: one of the 2 morphological forms of the *Cnidaria* (see). Only a few species attach themselves to a substratum, the majority being free swimming. The body is shaped like a bell or umbrella, with an upper aspect (exumbrella) and a lower aspect (subumbrella). Tentacles with dense concentrations of nematocysts arise from the margin of the bell. An oral tube, containing most of the gastrovascular space, hangs from the center of the subumbrella, and the rest of the gastrovascular space consists of several radial canals and a ring canal inside the bell. Medusae propel themselves by pulsating contractions of the bell, in which the subumbrella constricts and generates a pressurized, backward-directed water stream. Medusae are solitary, and they are usually formed asexually by budding or cleavage from a Polyp (see). The medusa produces polyps by sexual reproduction (see Alternation of generations). In rare cases, the polypoid form is absent, and the medusa forms more medusae by sexual reproduction. Conversely, in *Hydra* (see), the medusoid form is absent.

Most medusae are predatory, catching relatively large prey with their tentacles and pushing it into the oral tube. A few species feed on small suspended organic particles. The medusoid body consists of 3–4% solid material and 97–96% water. They therefore have approximately the same density as the surrounding medium, so that they can float, move with the current and remain suspended at various depths with practically no effort.

Collection and preservation. They are allowed to swim in a container of sea water and fixed by the slow addition of 10% formaldehyde. If possible, the animals should first be anesthetized by passing CO_2 through the water.

Meeowing frog: see *Myobatrachidae.*

Megacephalon: see *Megapodiidae.*

Megaceryle: see *Coraciiformes.*

Megacyllene: see Long-horned beetles.

Megafauna: see Soil organisms.

Megagaea: see Arctogea.

Megaladapidae (Sportive lemurs): see Lemurs.

Megalaima: see *Capitonidae.*

Megalodon (Greek *megas* large, *odontes* teeth): an extinct genus of very large bivalves whose equal-sized valves are strongly vaulted, smooth or concentrically ringed, and very thick. The umbo is characteristically very prominent, with a marked backward curve. The hinge has a very broad border. Fossils are found in the Triassic (Alpine).

Megalopidae, ***tarpons:*** a family of streamlined marine fishes of the order *Elopiformes.* Only 2 living species are recognized: *Megalops cyprinoides* (East Africa to the Society islands) and *Megalops atlanticus (= Tarpon atlanticus;* up to 2 m in length and a popular sporting fish) of the western Atlantic and tropical West African coasts. The last ray of the dorsal fin is elongated. Tarpons are the only elopiform fishes in which the swim bladder lies against the skull, and there is no close association between the swim bladder and the perilymphatic cavity, as in the clupeoids and notopteroids.

Megalops: see *Megalopidae.*

Megaloptera, *Sialoidea,* ***alderflies*** and ***dobsonflies:*** a small insect order represented in all regions of the world, but containing only about 100 species. About 50 species are known in North America, and a similar number in Europe. The earliest fossil M. are found in the Permian. See *Neuroptera.*

Imagoes. The adult body length of European species is about 1.5 cm. *Acanthocorydalis* from China and Japan is 7 cm long with a wingspan of 16 cm. The ***American dobsonfly*** *(Corydalis cornutus)* attains a wingspan of 13 cm. In many tropical species, the mandibles of the male are elongated and may equal half the body length; they are used as clasping organs during copulation. The head is prognathous (i.e. the chewing mouthparts extend forward), with long antennae and large compound eyes. There are two pairs of brownish wings with primitive, net-like venation. *M.* are found near to water, usually at rest on plants. Metamorphosis is complete (holometabolous development) and lasts from 1 to 3 years.

Eggs. The more than 1 cm-wide, brown, velvety cluster of eggs (several hundred to 2 000 eggs) is found on partly submerged plants, stones or sticks above the water level. The emerging larvae fall into the water.

Larvae. These are aquatic, freshwater predators, which usually live on the bottom. They breathe with the aid of tracheal gills, which are found on the first 7 abdominal segments. The larva may molt up to 10 times before migrating to the land to pupate.

Pupae. These are decticous and exarate (pupa libera). They lie in damp earth, in moss, or under stones without a cocoon or any form of woven structure.

Classification. There are only 2 recent families: *Sialidae* (alderflies) (3 ocelli present) and *Corydalidae* (dobsonflies) (ocelli absent).

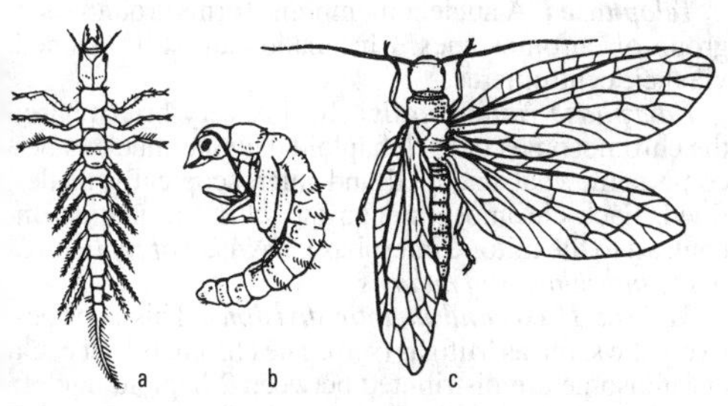

Megaloptera. Alderfly *(Sialis* sp.). *a* Larva. *b* Pupa. *c* Imago.

Megalurus: see *Sylviinae.*

Meganeura (Greek *megas* large, *neura* muscle ligament): an extinct, primitive, predatory dragonfly, whose fossils are found in the Upper Carboniferous. One of the largest known insects, it possessed biting mouthparts and attained a wingspan of 75 cm.

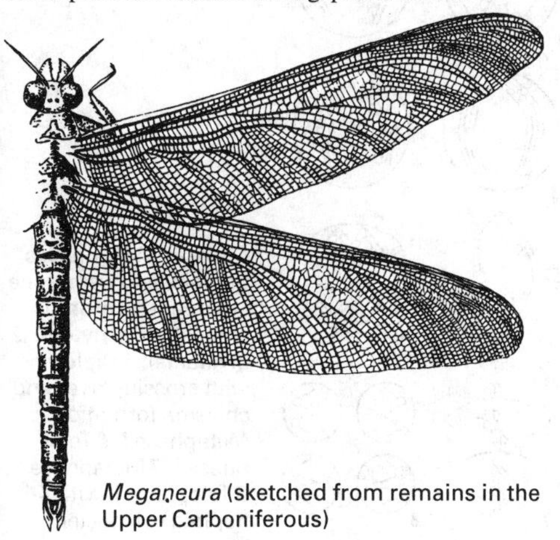

Meganeura (sketched from remains in the Upper Carboniferous)

Megaphanerophyte: same as macrophanerophyte, see Life form.

Megaplankton: see Plankton.

Megapodiidae, ***megapodes*** or ***mound builders:*** a small avian family (12 species in 7 genera) in the order *Galliformes* (see). Megapodes are confined to Australia and islands to the northeast. All have very large and strong feet, legs and claws, as well as short, strong, slightly curved bills. Like the *Cracidae,* they differ from other *Galliformes,* in having large hindtoes on the same level as the 3 front toes. Eggs are laid in warm soil or in specially built mounds of rotting vegetation, which are tended by the male to control the temperature. The young emerge in the absence of the parents, and they are totally independent and able to fly within a day. ***Scrub fowl*** (sometimes called ***Junglefowl)*** belong to the genera: *Megacephalon* (1 species), *Eulipoa* (1 species) and *Megapodius* (3 species). *Megapodius,* which is represented in the Philippines, Marianas and Nicobar Islands, Samoa, New Guinea and the New Hebrides, has the widest distribution of all the megapodes. ***Brush turkeys*** belong to the genera: *Alectura* (1 species), *Aepypodius* (2 species) and *Talegalla* (3 species). The family is completed by the monotypic ***Mallee fowl*** ***(Leipoa ocellata).***

Megapodius: see *Megapodiidae.*

Megascleric shells: see *Foraminiferida.*

Megasporangium, ***macrosporangium:*** a sporangium in which megaspores (macrospores) are produced. In spermatophytes, it corresponds to the nucellus of the ovule.

Megasporogenesis, ***macrosporogenesis:*** formation of the female gametophyte (embryo sac) in angiosperms, involving meiosis and a series of postmeiotic cell divisions (Fig.). Four haploid cells (usually ordered in a row, one behind the other) are formed by meiosis of a cell of the ovule nucellus (usually a subepidermal cell, and known as the *embryo sac mother cell).* Three of these cells die and disintegrate, and only one is retained, usually the lowest of the four which faces the chalaza. This remaining cell enlarges, becoming equivalent to the macro(meio)spore of the pteridophytes. It develops into the female gametophyte or embryo sac; its nucleus (the primary embryo sac nucleus) undergoes 3 phases of division to form 8 nuclei. After the first of these divisions, the 2 daughter nuclei migrate to the cell poles,

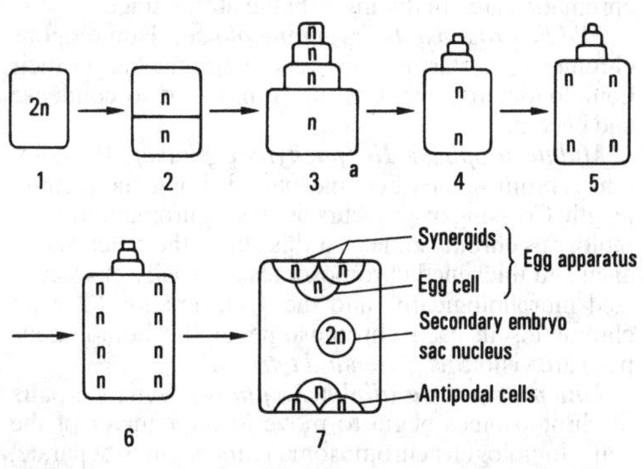

Megasporogenesis. 1 Embryo sac mother cell. *2* First meiotic division. *3* Second meiotic division *(a* macrospore with the primary embryo sac nucleus). *4-6* Postmeiotic divisions. *7* Embryo sac.

where they undergo 2 further divisions. At each pole, 3 of the resulting nuclei become enclosed by cell walls. At the pole facing the micropyle of the ovule, the *egg apparatus* is formed, i.e. the actual egg cell and 2 *synergids*. At the other pole, facing the chalaza, a group of 3 cells is formed, which are called *antipodal cells* (in accordance with their position in relation to the egg apparatus). The remaining 2 pole nuclei that are not enclosed by cell walls migrate to the middle of the embryo sac, fuse with one another, and form the *diploid secondary embryo sac nucleus*.

Megasporophyll: see Flower.

Megatera: see Whales.

Megellanic flightless steamer duck: see *Anseriformes*.

Megrim: see *Pleuronectiformes*.

Mehari: see Dromedary.

Meiofauna: see Soil organisms.

Meiogamete: see Reproduction.

Meiosis: a form of nuclear division, accompanied by cell division, which results in the production of four haploid cells (known as gametes) from a single diploid cell (a pregametic cell). Meiosis involves two separate stages, known as meiosis I (first meiotic division) and meiosis II (second meiotic division). Before meiosis can begin, the nucleus of the pregametic cell passes through the *interphase*, during which genetic material is replicated, i.e. each of the two haploid sets of genetic material (paternal and maternal) is replicated so that the chromosomes (when they subsequently become microscopically visible) exist as paired sister chromatids. *No further DNA replication occurs during meiosis*. As prophase I commences, the 2 copies of each chromosome condense into microscopically visible structures. These homologous chromosomes then become paired, i.e. each maternal chromosome of a haploid set pairs with the corresponding paternal chromosome of the other haploid set. Homologous pairing can be quite prolonged, lasting hours, months or even years, so that prophase I of meiosis is much longer than the prophase of mitosis. The following phases or stages of meiosis can be distinguished.

Early prophase I (leptotene phase). DNA replication has already occurred. Chromosomes become microscopically visible as long, thin threads, each containing the DNA for two chromatids, although separate chromatids are not distinguishable at this stage.

Middle prophase Ia (zygotene phase). Homologous chromosomes start to pair, becoming attached in their homologous regions. Chromosomes start to condense and become shorter.

Middle prophase Ib (pachytene phase). Homologous chromosomes become paired along their entire length. Crossing over occurs between chromatids of homologous chromosomes. At this stage, the much shortened and thickened chromosomes are easily characterized morphologically, and the 4 chromatids (2 sister chromatids in each chromosome of the homologous pair) are visible as *chromatid tetrads*.

Late prophase Ia (diplotene phase). Synaptic pairs of chromosomes begin to move to the equator of the cell. Homologous chromosome pairs begin to separate, remaining attached only by the chiasma (crossing over site) between two chromatids (one from each chromosome). [A chiasma is formed by breakage of two nonsister chromatids at homologous sites, followed by rejoining of the broken ends to new partners, i.e. a *cross over* is performed. The position and number of chiasmata determine the shape of the tetrad (e.g. cross or circle). See Crossing over.]

Late prophase Ib (diakinesis). The chromosome pairs separate further and crossed over strands are pulled away from one another, so that the chiasmata move to the ends of the chromatids (terminalization). The attachment between homologous chromosomes is finally broken during diakinesis, or a little later in metaphase I. The spindle is nearly complete and the nuclear membrane begins to disappear.

Metaphase I. The nuclear membrane has totally disintegrated, and the spindle poles become fully formed. Synaptic pairs become aligned in the equatorial plane of the cell. In contrast to the metaphase of mitosis, the centromeres do not lie on the equatorial plate; rather the centromere of each synaptic pair is orientated toward one of the spindle poles. This polar orientation is random, i.e. there is a 50% chance that the "paternal" or "maternal" chromosome will eventually migrate to a particular pole of the cell.

Anaphase I. Synaptic pairs separate and chromosomes move to opposite poles.

Telophase I. A nuclear membrane forms around each group of chromosomes. Chromatids are still attached via their centromeres.

Interphase (interkinesis). In this very brief phase, the chromosomes of each haploid daughter nucleus become somewhat uncoiled and microscopically ill defined. Each chromosome consists of 2 chromatids. In contrast to the mitotic interphase, *DNA is not replicated in the interkinesis of meiosis*.

Meiosis II (second meiotic division). This is effectively the same as Mitosis (see). The chromatids of each chromosome are distributed between 2 haploid nuclei, resulting in a total of 4 haploid nuclei from the 2 nuclei present at the end of telophase I. By cytokinesis, 4 cells (gametes) are formed, each containing one of the haploid nuclei. If crossing over occurred in prophase I, then the assortment of genetic information in each gamete is different from that in any one of the original halpoid sets.

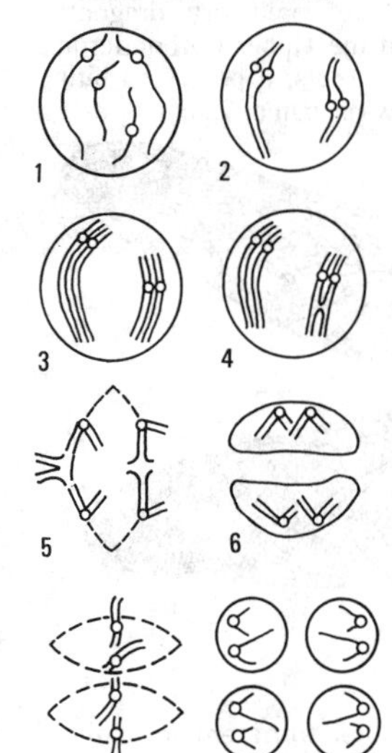

Meiosis. Main phases of meiosis. *1* Leptotene (4 unpaired chromosomes). *2* Pachytene (2 bivalents). *3* Diplotene with crossing over and chiasma formation. *5* Metaphase I. *6* Telophase I. *7* Metaphase II. *8* Telophase II with 4 haploid cell nuclei.

In some unicellular organisms (e.g. *Telosporidia*, certain flagellates, algae and fungi), meiosis occurs in the zygote, so that all cells, apart from those of the zygote, are haploid; such organisms are called *haplonts*. In contrast, in *diplonts* (all metazoa and ciliates, some heliozoans, flagellates, algae and fungi), meiosis precedes fertilization, and only the mature gametes are haploid.

See Alternation of generations.

Meissner's corpuscles: see Tactile stimulus reception.

Melanerpes: see *Picidae.*

Melanins: High molecular mass, nitrogenous, amorphous pigments, which occur naturally in the animal kingdom in vertebrates and insects, and occasionally in microorganisms, fungi and higher plants. Mammalian colors are determined chiefly by two types of M.: black or brown insoluble *eumelanins,* and the lighter colored, sulfur-containing, alkali-soluble *pheomelanins.* The *trichochromes,* a group of low molecular mass, yellow, red and violet pigments, are biosynthetically related to the M. In mammals, M. occur in a cell layer below the epidermis, and they are responsible for the coloration of skin, hair and eyes. Moles and freckles represent areas of especially high M. concentration. The pigments of many bird feathers, the skins of reptiles and fishes, the exoskeletons of insects, and the colored component from the ink sac of the cuttlefish are all M., sometimes diffusely distributed and sometimes present in granules. In animals, the deposition of large quanitites of M. results in the condition known as melanism. M. serve a protective function by absorbing the UV-irradiation of sunlight, which in turn causes increased pigmentation (sun tan). All M. are biosynthesized from the amino acid, L-tyrosine, by the action of a copper-containing polyphenol oxidase called tyrosinase. Eumelanins are formed from tyrosine via dopaquinone and indole 5,6-quinone. Pheomelanins and trichochromes are synthesized from tyrosine via dopaquinone, with the participation of cysteine. Synthesis occurs in melanosomes, which are present in cells called melanocytes. M. are also synthesized in the retina of the eye. In the native state, M. are often bound to proteins, forming melanoproteins with a protein content of 10–15%.

Absence of tyrosinase (usually an autosomal inherited defect in the ability to synthesize the enzyme) results in Albinism (see). Most groups of mammals occasionally produce albino individuals, which completely lack any pigmentation of eyes, skin, hair, feathers, etc. Albinism may also result from 1. deficient melanin polymerization, 2. failure to synthesize the protein matrix of the melanin granule, 3. lack of tyrosine, and 4. presence of tyrosinase inhibitors.

Melanocyte stimulating hormone: see Melanotropin.

Melanogrammus: see Haddock.

Melanoptila: see *Mimidae.*

Melanostomiidae: see *Stomiiformes.*

Melanosuchus: see *Alligatoridae.*

Melanotropin, ***melanocyte stimulating hormone, MSH:*** a peptide hormone or neuropeptide produced by the opiomelanotropinergic cells of the pars intermedia of the hypophysis and by neurons of the central nervous system. In species lacking the pars intermedia (e.g. birds and whales), MSH is produced by the neurohypophysis. Production of MSH is regulated by melanotropin releasing factor (MRF) and melanotropin release inhibiting hormone (MIH) of the hypothalamus. MSH belongs to a family of peptide hormones including ACTH, β-lipoprotein, endorphins, enkephalin and others, all derived from a common precursor protein, pro-opiomelanocortin. MSH was first recognized and named for its ability to cause dispersion of melanin granules in the melanophores of the skin of cold-blooded vertebrates (see Color change). It also causes increased deposition of melanin in mammals, but its role in human pigmentation is doubtful. MSH-like activity of blood and urine increases during pregnancy in humans (the frog skin test described below was originally designed for detection of human pregnancy). This MSH-like activity may be of fetal origin, but its true source and identity have not been identified. Adult humans do not appear to possess circulating MSH, but it is extensively present throughout the central nervous system of humans and other animals, where the deacetylated form is more abundant than the acetylated form. It is synthesized and secreted by an opiomelanotropinergic multineuronal transmitter system, the specific neurons of which have been located in the brain by immunological methods. Behavioral effects of MSH in mammals include arousal, increased motivation, longer attention span, memory retention and increased learning ability. It promotes somatotropin secretion and inhibits secretion of prolactin and luteinizing hormone. The classical assay for MSH is the in vitro frog *(Rana pipiens)* skin bioassay, which is based on the centrifugal dispersion of melanin granules in the dendritic processes of dermal melanophores, leading to skin darkening. It is capable of detecting minimal MSH concentrations of 10^{-11} M. A later method measures the increase in the conversion of [α-^{32}P]ATP into [^{32}P]cAMP by plasma membranes from mouse S-19 (Cloudman) melanoma cells.

Mammalian ***α-MSH*** is an acetyltridecapeptidamide, identical with amino acid sequence 1–13 of ACTH (corticotropin). The material formerly known as ***β-MSH*** contains 22 amino acid residues, and it corresponds to a different region of the pro-opiomelanocortin precursor. In the fog skin test, it displays biological activity similar to that of α-MSH. It may exist naturally, but it is now known that β-MSH isolated from pituitaries is a fragment of β-lipoprotein formed during the extraction procedure.

Melatonin: *N*-acetyl-5-methoxytryptamine, a hormone of the pineal gland and retina, formed from serotonin by *N*-acetylation and *O*-methylation. It influences pigmentation in amphibians, where it opposes the action of Melanotropin (see) (see also Color change). In mammals it inhibits the development of gonadal function in young animals and the action of gonadotropins in mature animals. Synthesis of M. is suppressed by light, acting via the eyes and nervous system, and it is maximal in the dark. A corresponding circadian rhythm has been observed in the activity of serotonin *N*-acetyltransferase in rodent pineal glands and the retinas of frogs and chickens. It seems likely that M. is an effector of mammalian and avian behavioral rhythms. Rats kept in continuous light are in continuous estrus, due to the lack of M. which would otherwise exert antigonadotrophic activity. In amphibians, M. mediates the response of skin pigmentation to light; it reverses the darkening effect of melanotropin by causing the melanin granules within the melanocytes to aggregate.

M. is inactivated in the liver by hydroxylation, followed by conjugation with sulfate or glucuronate.

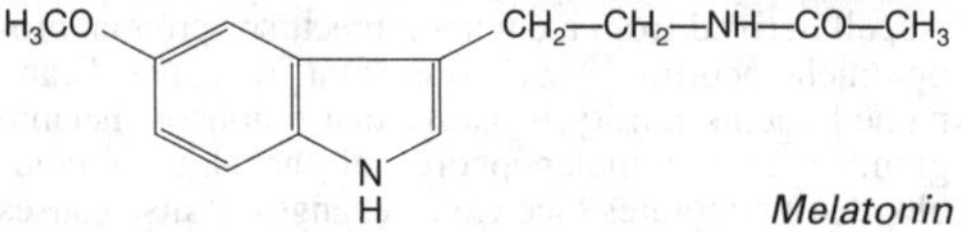

$$H_3CO-C_8H_4N(H)-CH_2-CH_2-NH-CO-CH_3$$

Melatonin

Melba finch: see *Ploceidae.*

Meleagrididae, *turkeys:* a small avian family (only 2 species) of very large woodland game birds, in the order *Galliformes* (see). In many respects they resemble pheasants and guineafowl, but the 2 monotypic species constituting the family are sufficiently different to justify their separate familial status. Thus, the wishbone is straight and slender, they have fanned, square-ended tail feathers, and the tail differs from that of the pheasants in that it is not V-shaped in cross section. In addition, the head is bare and warty, and male turkeys have spurs and rudimentary webs between the toes. They are indigenous to the New World, and they were brought to Europe by the Spanish, who found the ***Wild*** or ***Common turkey*** *(Meleagris gallapavo;* 122 cm; southeastern USA and Mexico) already domesticated by the Mexican Indians. This species is the ancestor of the domestic turkey, which is bred widely throughout the world for its meat, and until recently had hardly been changed by selective breeding. A smaller, white strain has, however, been developed in recent decades. The ***Ocellated turkey*** *(Agriocharis ocellata;* 92 cm; Yucatan, Belize, Guatemala) is smaller than the Common turkey, and lacks the chest tuft. The head is blue rather than red, and its wattles and protuberances are different in pattern, color and shape from those of *Meleagris.* It is named for the presence of an eyespot at the end of each tail feather.

Meleagris: see *Meleagrididae.*

Meliphagidae, *honeyeaters* and ***sugarbirds:*** an avian family in the order *Passeriformes* (see). The family contains 160 species in 38 genera, many of them monotypic. Plumage is generally dull, except in the genus *Myzomela,* which has extensive, brilliant areas of scarlet. Often parts of the head are bare of feathers, revealing strikingly colored, red, yellow, or blue areas of skin; these bare areas may also develop lobes, wattles or other projections. All species have a long, protrusible tongue, which is adapted to nectar feeding: the proximal end is curled on each side to form 2 channels, and the distal end is deeply cleft and tipped with bristles. Nectar can pass directly from the esophagus into the intestine, due to the close juxtaposition of the respective openings. As nectar feeders, they are important flower pollinators of many Australian trees and bushes, particularly Eucalyptus. Sometimes, pollination is species-specific, e.g. pollination of the Western Australian plant *Anigozanthos manglesii* ***(Kangaroo paw)*** depends on a honey feeding visit from the ***Western spinebill*** *(Acanthorhynchus superciliosus).* No species feeds exclusively on nectar, and insects are also taken. The ***Painted honey eater*** *(Conopophila picta)* lives mainly on mistletoe berries and does not take nectar. The majority of species belong to the subfamily *Meliphaginae* ***(Honey eaters),*** which are exclusively Australasian in distribution, forming a very characteristic part of the Australasian fauna, e.g. the New Zealand ***Tui*** or ***Parson bird*** *(Prosthemadera novaeseelandiae)* and the New Zealand ***Bellbird*** *(Anthornis melonura).* The remaining subfamily, *Promeropinae* ***(Sugarbirds),*** contains only 2 species in a single genus, both confined to southern Africa: ***Cape sugarbird*** *(Promerops cafer)* and ***Natal sugarbird*** *(P. gurneyi).* Both have tongue and intestinal structures similar to those of the honeyeaters, and *P. cafer* fertilizes the tree flowers of *Protea.*

Melissa: see *Lamiaceae.*

Melitopalynology, *honey pollen analysis:* a method for establishing the geographical origin of honey by identification of its constituent pollen.

Melitose: see Raffinose.

Meloidogyne: see Eelworms.

Melolonthinae: see Lamellicorn beetles.

Melon: see *Cucurbitaceae.*

Melonechinoidea: see *Palaeechinoidea.*

Melopsittacus: see *Psittaciformes.*

Membrana granulosa: see Oogensis.

Membrana vitellina: see Ovum.

Membrane, *biological membrane, biomembrane, unit membrane*: a structural and functional component of all cells and many cell organelles. It consists predominantly of lipids and proteins, and its detailed structure is apparent only under the electron microscope. Biological membranes are complex, dynamic structures with many different functions. They act as permeability barriers, and at the same time as regulators and mediators of the specific exchange of materials between the cell and its environment, and between different compartments of the cell. Membrane structure and function is currently one of the more active areas of cell biological research.

Structure and function. The chemical and ionic composition of the cell interior differs fundamentally from that of the external environment. For example, the cell maintains a high intracellular K^+ ion concentration (90–100 µmol/ml) and a low Na^+ ion concentration (10–20 µmol/ml), a relationship which is reversed in the extracellular fluid. This differential distribution of ions between the inner and outer surfaces of the membrane gives rise to a potential difference across the membrane, known as the *membrane potential (resting potential).* Membrane potentials ranging from a few mV to 100 mV occur in all living cells. The pH within the cell is kept relatively constant (about 7.5) by the cell membrane. Biological membranes selectively regulate the passage of materials across them, and only those molecules and ions required for metabolic processes within the cell are permitted entry by the plasma membrane. Many prokaryotic cells possess only a single membrane, i.e. the *plasma membrane*, which encloses the cell contents. In addition to the plasma membrane, eukaryotic cells also possess intracellular membranes (cytomembranes), which divide the cell into compartments. With the aid of its surrounding membrane, each of these compartments maintains its own internal physical-chemical environment, which enables it to serve as the specific site of certain metabolic processes. With such compartmentation, metabolic processes operating counter to one another (e.g. degradation and synthesis) can occur simultaneously within the same cell.

All biological membranes are 5–10 nm thick, and they have a relatively uniform structure; hence the term unit membrane. As early as 1925, on the basis of the structural properties of membrane lipids, Gortner and Grendel proposed that biologial membranes consist of a lipid bilayer. Developing this hypothesis, Davson and Danielli proposed their own model of membrane structure in the 1930s, and refined it further in subsequent years. The modern interpretation of membrane structure is still based on the lipid bilayer. Under the electron microscope, all membranes appear to consist of 3 layers. The middle layer is osmiophobic and hydrophobic, and therefore takes up practically no osmium from the aqueous fixation medium. Consequently it is electron-transparent and appears light. This light layer is sandwiched between two dark layers (i.e. hydrophilic, osmiophilic layers that are electron-opaque). At very high electron-optic magnification (x 50 000–80 000) it is possible to recognize proteins within the membrane. These appear as rather large, roundish particles of non-uniform size and distribution. A more detailed investigation of membrane structure is possible by X-ray analysis. For example, irregularities in the lipid layer cause diffraction, and variations in structure are accompanied by an alteration in electron density.

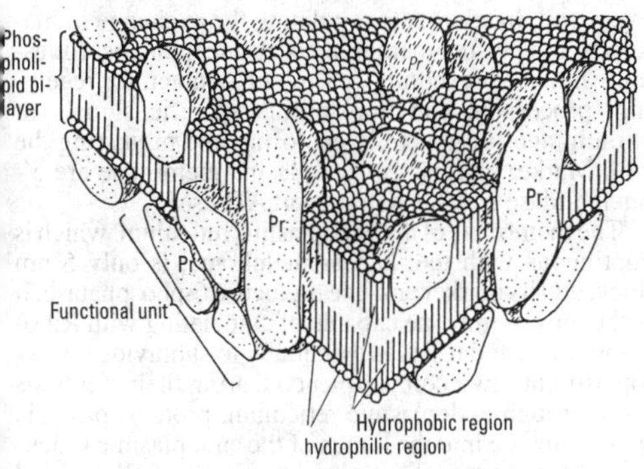

Fig.1. *Membrane.* Structure of a biomembrane (fluid mosaic model after Singer and Nicholson, 1972).

Membrane lipids are predominantly phospholipids, e.g. lecithin, which consists of a hydrophobic hydrocarbon moiety (esterified fatty acid residues) and a hydrophilic moiety (e.g. a phosphate residue linked to a base such as choline). The hydrocarbon chains can undergo conformational changes. At low temperatures, their arrangement resembles that of a crystal lattice, and the membrane then has a gel-like consistency. At physiological temperatures, membranes are in a liquid crystalline state, which is determined by the intense molecular vibrations of the hydrocarbon residues. Within a lipid layer, each lipid molecule acquires new neighbors 10^6 times (average) per second. As the content of unsaturated fatty acid residues increases, the melting point is lowered and the membrane becomes "more fluid". The *fluid mosaic* model of membrane structure, proposed in 1972 by Singer and Nicholson, is based on the results of modern membrane research (Fig.1). According to this model, at physiological temperatures, the lipid bilayer forms a viscous medium, in which the proteins are anchored and more or less immersed. These *integral membrane proteins* may span the membrane or penetrate the membrane without actually spanning it, and some proteins may be aggregated within the membrane. Integral membrane proteins are bound to the lipids by hydrophobic interaction, and they are therefore difficult to separate from their lipid environment. Many integral membrane proteins form channels or pores in the membrane, which mediate active transport and catalyse permeation (such proteins may be regarded as enzymes, and they have been called "permeases"; the term "transport protein" is, however, preferred). In addition, certain easily detachable *peripheral membrane proteins* are associated with the surface of the membrane. The fluid mosaic model allows that lipids and proteins are capable of lateral movement within the plane of the membrane. Cap formation in lymphocytes and the uniform distribution of the surface antigens of a heterokaryon after cell hybridization are evidence for the lateral movement of membrane components.

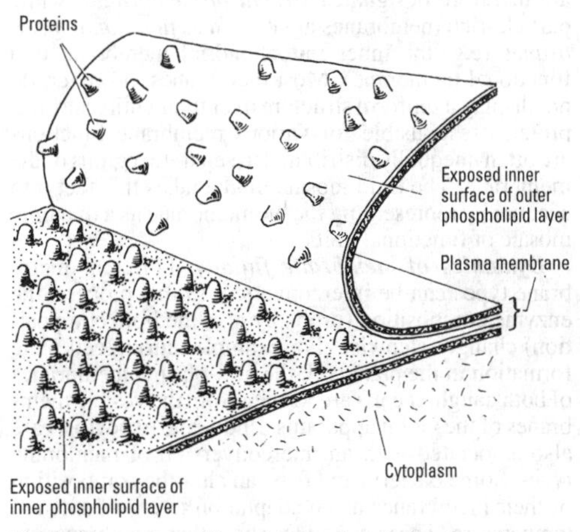

Fig.2. *Membrane.* Plasma membrane fractured and folded back (schematic representation of the freeze-fracture technique). The "particles" in each layer represent membrane proteins.

The mass ratio, proteins:lipids, lies between 1:4 and 4:1. Although membrane structure is apparently homogeneous or monotonous, different membranes exhibit a variety of specific functions, which are largely due to the component proteins. Plasma membranes and the membranes of the endoplasmic reticulum, Golgi apparatus, lysosomes, mitochondria, and other structures contain different proteins, especially enzymes. Examination of freeze-fractured (see Freeze fracture) preparations under the electron microscope not only allows a 3-dimensional portrayal of the inner and outer membrane surfaces, it also reveals the structure of the internal surfaces of each layer of the bilayer. This is because cleavage occurs preferentially between the two layers of the bilayer (Fig.2). On the other hand, if an image of the membrane surface is required, the process of Freeze etching (see) must be applied. Membrane cleavage results in two complementary surfaces, e.g. in the case of the plasma membrane, one of these surfaces is the internal surface of the layer associated with the cytoplasm (inner fracture face), while the other is the internal surface of the layer next to the extracellular space (outer fracture face). These two fracture surfaces, which consist chiefly of structural lipids, are generally smooth in appearance, but they also reveal varying quantities of small particles (diameter 6–18 nm), which are very probably integral membrane proteins, especially enzymes. The inner fracture face usually displays more particles than the outer fracture face, and the total number of particles varies for different types of membranes. Depending on the metabolic function of the membrane, the number of particles varies between a few hundred and several thousand per mm^2. Membranes whose function is mainly that of insulation (e.g. myelin sheath) contain only about 20% protein, and they reveal a correspondingly small number of particles on the two faces of freeze-fractured preparations. In contrast, the microvillus membrane of the cells of the proximal kidney tubule are extremely rich in particles, about 3500/mm^2 lying on the inner fracture face. Particle-poor membranes are therefore designated *structural membranes*, while particle-rich membranes are known as *functional membranes* (e.g. the inner mitochondrial membrane is a functional membrane). Most membranes, however, do not display a uniform structure over their entire surface; proteins responsible for various membrane functions are often unequally distributed in separate regions of the membrane. The fluid mosaic model takes this fact into account by representing the biomembrane as a dynamic mosaic of functional units.

Dynamics of membrane fluidity. Different membrane types can be interconverted. In this process, the enzyme composition (probably also the lipid composition) changes stepwise. For example, during cell plate formation in the plant cell, the new plasma membranes of both daughter cells are formed from the vesicle membranes of the Golgi apparatus. Endo- and exocytosis are also associated with an interconversion of membrane types. Some bacteria and fish can alter the composition of their membranes as an adaptation to environmental temperature. These examples show that membranes do not have fixed structures, but rather undergo constant formation and change.

Transport across membranes. Materials can pass through biological membranes by simple diffusion, catalysed (facilitated) diffusion, or by active transport. Permeation by diffusion requires the presence of channels in the membrane, together with a concentration gradient or electrochemical potential difference. Important factors in this process are particle size, hydrophilicity (and its converse, lipophilicity), as well as the ability for hydrogen bond formation. Since lipids are qualitiatively and quantitatively fundamental to membrane structure, penetration of the membrane increases with the lipophilicity of the permeant material. Charged particles, especially ions, penetrate membranes with difficulty owing to their hydrophilicity. For an organic molecule to penetrate a membrane by simple diffusion, its relative molecular mass must be no greater than 70. Many molecules pass through membranes more rapidly than can be accounted for by their molecular size, degree of hydrophilicity and their potential for hydrogen bond formation. For example, the rate of glucose transport across the erythrocyte plasma membrane is about 10 000 times higher than that calculated for simple diffusion. This phenomemon is termed *facilitated diffusion,* in which specific transport proteins (carriers) facilitate the passage of water-soluble molecules without expenditure of energy. On the other hand, certain ions and molecules are transported ("pumped") against electrochemical and concentration gradients at the expense of ATP cleavage. See Active transport.

Artificial membranes. When phospholipid solutions are placed in water, the forces of surface tension cause a lipid bilayer to form on the water surface. Such artificial membranes can be used as model systems for the study of membrane transport, and proteins can also be incorporated during their preparation. Lipid vesicles, 20–50 nm in diameter, known as *liposomes,* are formed when an aqueous suspension of phospholipids is treated with ultrasound. These vesicles, which are bounded by a lipid bilayer, serve as models for biomembranes and transport vesicles. Enzymes, antibodies, drugs, hormones, etc. can be trapped inside liposomes, then targetted at specific sites within the organism.

Types of membranes. According to their structure and function, biomembranes can be classified into plasma membrane, membrane of the endoplasmic reticulum and nucleus, membrane of the Golgi apparatus, membrane of the mitochondria and plastids, membrane of the myelin sheath, excitable membranes, bacterial membranes and virus membranes.

The *plasma membrane* (width 9–10 nm) envelops every cell. In additon to regulating the exchange of materials, the plasma membrane is also responsible for establishing or preventing cell-to-cell contact, for receiving, processing and transmitting specific signals or stimuli, and for generally enclosing and protecting the cell. Marker enzymes of the plasma membrane are 5′-nucleotidase and Na-K-transport ATPase.

The membrane of the endoplasmic reticulum, which is continuous with the nuclear membrane, is only 5 nm thick, and its marker enzyme is glucose 6-phosphatase. It is the only membrane capable of associating with Ribosomes (see) and it serves essentially to subdivide the cytoplasm into two compartments. During their synthesis on the rough endoplasmic reticulum, proteins pass via the membrane into the lumen of the endoplasmic reticulum. According to the *signal hypothesis* of Blobel and Doberstein (1975), which is well supported by experimental evidence, a membrane recognition sequence is first synthesized during protein translation on free ribosomes. This sequence then binds the large ribosomal subunit to the membrane, where it causes certain membrane proteins to aggregate and form a channel. Finally, the growing polypeptide chain passes through the channel into the lumen, and the signal sequence is removed by proteases. The membrane of the smooth endoplasmic reticulum contains enzymes of phospholipid and steroid metabolism, as well as hydrolases and cytochrome P450.

The nuclear membrane regulates the exchange of material between nucleus and cytoplasm. Microsome fractions consist mainly of fragments of endoplasmic reticulum membrane, which have become sealed into vesicles.

Marker enzymes for the approximately 7 nm-thick membrane of the Golgi apparatus (see) are glycosyl transferases for polysaccharide synthesis, thiamin pyrophosphatase and acid phosphatase. Small vesicles, which pinch off from the endoplasmic reticulum, fuse with the membrane on the proximal side of the Golgi apparatus. Like all membranes of the endoplasmic reticulum, the membranes of these vesicles are only 5 nm thick. Within the Golgi, the vesicle contents as well as the lipid and enzyme composition of the vesicle membrane are altered. The vesicle membrane on the distal side of the Golgi apparatus resembles the plasma membrane in thickness (9–10 nm) and density (relative contrast under the electron microscope after osmium staining). Vesicles formed from the distal side of the Golgi apparatus perform the function of exocytosis. After release of their contents to the cell exterior, they fuse with and become incorporated into the plasma membrane.

The *outer membrane of plastids and mitochondria* contains approximately equal proportions of lipids and proteins, and it is quite permeable to ions, dissolved compounds, and even protein molecules. On the other hand, the *inner mitochondrial membrane,* which contains about 20% lipid (including cardiolipin, a complex phospholipid otherwise found only in bacteria) and about 80% protein, is practically impermeable to all materials including small ions. Structurally and functionally, the inner mitochondrial membrane is the most complex of all membranes. It is the site of oxidative phosphorylation, and about 60 different proteins have been separated from it experimentally, some 25 of these being enzymes of the respiratory chain. Most enzymes of the respiratory chain are located in the outer layer of the membrane, whereas cytochrome *a* spans the entire thickness of the phospholipid bilayer. In addition, the inner mitochondrial membrane contains the regulatory proteins of energy metabolism, as well as specific transport proteins for the movement of substances between the cytoplasm and the mitochondrial matrix.

In its fully differentiated state, the *inner membrane system of chloroplasts* (the *thylakoids*) has no contact with the outer membrane. Thylakoids are the site of the light-dependent reactions of photosynthesis, and functionally they may be regarded as homologous with the inner mitochondrial membrane. Several chlorophyll-binding proteins have been isolated from thylakoids, e.g. P700-chlorophyll *a* -protein and the light-collecting chlorophyll *a/b* -protein. Both are strongly hydrophobic, integral membrane proteins.

In the *membrane of the myelin sheath,* several lipid layers lie one upon the other with regular spacing. Isolated myelin sheath consists mainly of the plasma membrane of Schwann cells or oligodendrocytes, and it is formed by the repeated growth of these cells around a nerve fiber. The high cholesterol content (about 25% of the total lipid) of the myelin sheath membrane insures that this particular membrane exists in a liquid crystalline state.

Excitable membranes are those of nerve, sensory and muscle cells. The membrane of nerve fibers, the axolemma (see Neuron) is specialized for receiving, conducting and transmitting stimuli. When the plasma membrane receives a stimulus (e.g. a neurotransmitter substance) it becomes permeable to Na^+ ions for a few milliseconds (see Sodium channel). The subsequent flux of Na^+ ions into the cell alters the membrane potential, producing an action potential or nerve impulse (a change of potential of about 0.1V, lasting a few milliseconds).

The *plasma membrane of bacteria* contains a high concentration of densely packed proteins, and its complex structure is comparable to that of the inner mitochondrial membrane and thylakoids. Since bacteria lack intracellular membranes, the plasma membrane is responsible for a great variety of functions. It contains the enzymes of the respiratory chain and oxidative phosphorylation, as well as the various proteins and enzymes involved in active transport. The bacterial genome encodes a greater number of membrane proteins than can be accommodated simultaneously within the plasma membrane. Different membrane proteins are synthesized in response to changes in the environment, e.g. proteins involved in the transport of new nutrients. Bacteria also possess mechanisms for altering the lipid composition of the plasma membrane in response to changes in environmental conditions. The outer, lipoprotein-rich membrane of Gram-negative bacteria contains fewer phospholipids and fewer different proteins than the plasma membrane. The proteins form a regular pattern, and in additon to various enzymes and receptors, the membrane contains pore-forming proteins (porins), which allow the passage of substances of low molecular mass. These pores open or close, depending on the distribution of charge on the membrane surface. It has been shown experimentally that one of the functions of this outer membrane is the maintainance of the shape of the bacterial cell.

Virus membranes. The nucleocapsid of many viruses is surrounded by a membrane, whose structure is essentially similar to that of eukaryotic membranes. The lipid components are derived from the host cell, whereas the proteins are encoded by the virus. Oligosaccharide residues, which are attached to both the proteins and the lipids of the membrane, are synthesized predominantly with the aid of host-specific enzymes.

See Plasma membrane.

Membrane filter: a filter manufactured from cellulose acetate or similar material with a standardized pore size, which is impermeable to bacteria. M.fs. are used for cold-sterilization of liquids and for the removal of bacteria from gases (see Filtration, Sterilization), as well as for the determination of the bacterial count in liquids such as drinking water. In the latter case, a measured volume of water is sucked through a M.f., which is then laid directly on a plate of nutrient medium and incubated until colonies develop sufficiently to be counted.

Membrane fusion: see Fusion.

Membrane length constant: see Cable theory.

Membranelles, *membranellae*: small, paddle-like organs consisting of united cilia, found in some of the more advanced forms of the *Ciliophora.* Large numbers of M., usually in a double row, form the adoral wreath along the outer edge of the peristome (see *Ciliophora,* Fig.2).

Membrane models: concepts of the structure of biological membranes. Numerous data show that all plasma membranes and organelle membranes have a common fundamental structure, irrespective of whether they are animal or plant in origin. Models explain the arrangement and dynamics of the structural components of membranes, and serve as a basis for understanding their functional properties.

Historical. In 1855, K. Nägeli coined the idea of a plasma membrane, in order to explain the osmotic sensitivity of plant cells. In 1881, W. Pfeffer conducted numerous experiments in osmosis, which confirm the existence of a functional boundary layer (the plasma membrane), which prevents the entry of water and dissolved materials into the cell. The work on permeability by Ch. Overton in 1899, showing that the plasma membrane displays apolar behavior, can be seen as the beginning of membrane research. The bilayer structure of membranes was first proposed in 1925 by Gorter and Grendel. In 1935, Danielli and Davson recognized the presence of proteins as membrane components.

Close visual inspection of membranes became possible in 1950 with the introduction of electron microscopy. Simultaneously, it became evident that living cells contain many different membranes. Mitochondria, chloroplasts, endoplasmic reticulum, Golgi apparatus, lysosomes, etc. are all delimited by membranes. In the light of these discoveries, the functional aspects of existing membrane models were improved.

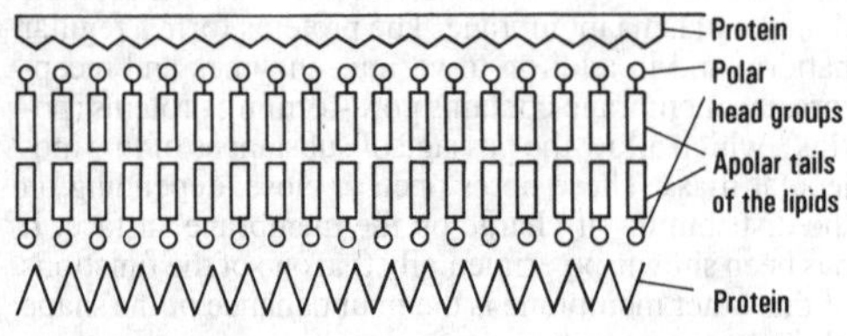

Fig.1. *Membrane model.* The Danielli-Davson-Robertson model.

The Danielli-Davson-Robertson model (Fig.1) is based on a structure consisting of lipids and proteins. This first model of a complete membrane was supported by physical-chemical data and by scientific reasoning. Membrane proteins are attached to each surface of a lipid bilayer. The apolar portion of the lipids ("tails") are orienated inwards, while the polar head groups interact with the polar groups of the attached proteins. Realizing that all membranes display a common 3-layered structure under the electron microscope, Robertson introduced the concept of the ***unit membrane***.

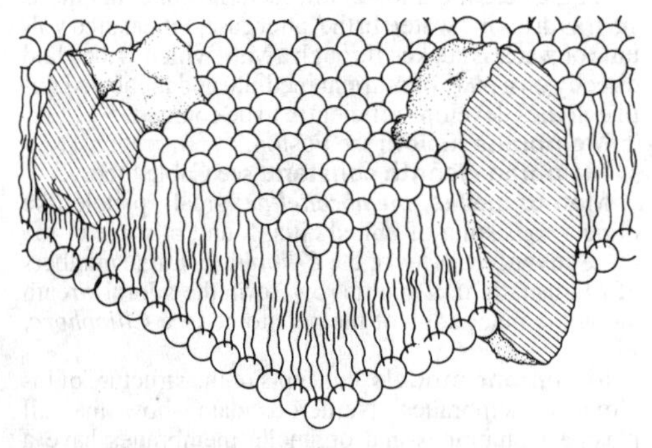

Fig.2. *Membrane model.* The fluid mosaic model.

Fluid mosaic model (Fig.2). Singer and Nicholson combined the experimental findings on the dynamics of membrane components with those on membrane transport. In this model the proteins are either inside the membrane or swimming or floating on the membrane like "ice bergs", i.e. the proteins either span the lipid bilayer or are embedded in one of the two lipid layers.

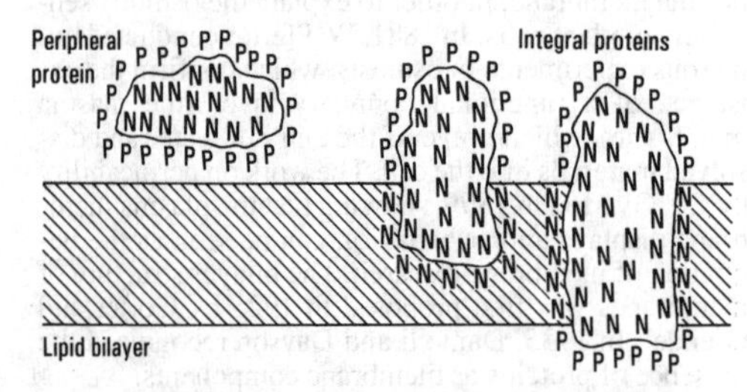

Fig.3. *Membrane model.* Model of peripheral and integral proteins.

In the model of peripheral and integral proteins (Fig.3), apolar amino acids on the surface of proteins interact with lipids in the bilayer, while polar amino acids interact with lipids on the outer surface of the bilayer.

In the functional model of the human erythrocyte membrane (Fig.4.), K^+-Na^+-ATPase (6) and the anion transport protein (3) are integral proteins. Glyceraldehyde dehydrogenase (5), acetylcholinesterase (4) and proteins of the cytoskeleton are peripheral.

Most recent findings indicate that a nonlamellar arrangement of lipids can occur in special circumstances. See Plasma membrane.

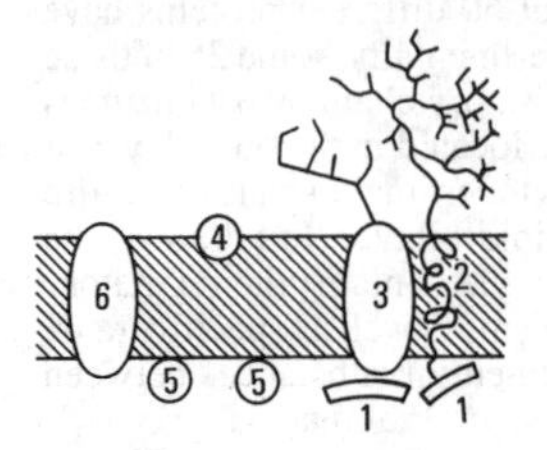

Fig.4. *Membrane model.* Functional model of the human erythrocyte membrane (see text for identity of numbered structures).

Membrane potential: the difference in electrical potential between the cytoplasm of the cell and the external milieu. All biological membranes posses a M.p., and its value can be determined with microelectrodes (see), or by measuring the distribution of permeant ions. The M.p. is due to the asymmetric distribution of ions, which arises from the active transport of ions and/or the impermeability of the membrane to polyions. Accordingly, a distinction is made between diffusion potential and Donnan potential. Existence of the M.p. is an essential factor in the excitability of muscle and nerve membranes, and for energy conservation in mitochondria and chloroplasts. See Resting potential.

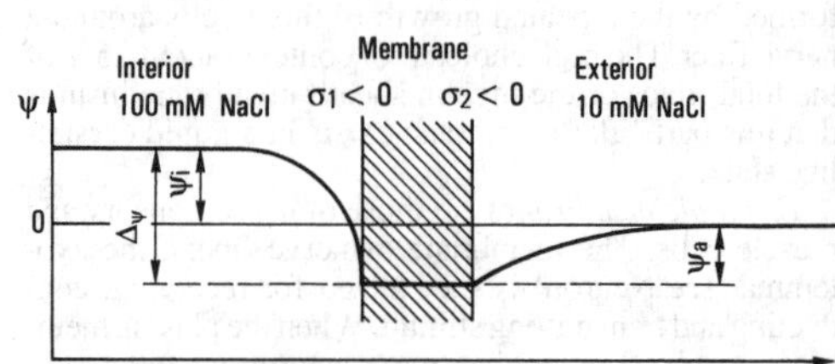

Membrane potential. The membrane potential, Dy, and the surface potentials, y_i and y_a, of a cell membrane with different charge densities (s_1/>/s_2) on its two sides. The membrane separates different aqueous concentrations of sodium chloride, and it is permeable to chloride ions.

Membranous labyrinth: see Auditory organ.

Memory: the ability of living organisms to store learned information, and to recall the same information later. A certain type and level of M. probably exists in protozoa, and the existence of M. is well established in all other animal phyla. Extensive systematic investigations have been performed, in particular on planarians, various mollusks (e.g. *Aplysia, Octopus),* numerous insects and many vertebrates, especially mammals, including humans. The general conclusion regarding the

Number of independent allele pairs	Number of different gamete types produced by F_1 hybrid	Number of gamete combinations in F_2	Number of homozygotes in F_2	Number of hetero zygotes in F_2	Number of different genotypes in F_2	Number of different phenotypes in F_2	Frequency of the different F_2 phenotypes
1	2	4	2	2	3	2	$(3+1)^1 = 3+1$
2	4	16	4	12	9	4	$(3+1)^2 = 9+3+3+1$
3	8	64	8	56	27	8	$(3+1)^3 = 27+9+9+9+3+3+3+1$
n	2^n	4^n	2^n	$4^n - 2^n$	3^n	2^n	$(3+1)^n = 3^n + 3^{n-1} + 3^{n-2} + 3^{n-(n-1)} + 1$

Mendelian segregation. Mathematical analysis of the genetic and phenotypic constitution of the F_2 generation from mono-, di-, tri- and polyhybrid crosses.

M. process are based on mammalian studies. Several types of information storage are differentiated: *immediate memory* or *association,* in which parts of a current input are processed, e.g. in humans the sound of bells may be immediately associated with the clock tower; *short-term memory,* which has little storage content and lasts for only a few hours or less; and *long-term memory,* in which information may be stored for a lifetime. The different types of information storage appear to occur by different storage mechanisms. Immediate and short-term storage probably occurs as oscillation of impulses in circularly connected neurons, i.e. in reverberating circuits (see Neuron circuit), because information stored in this way is lost if electrophysiological processes in the brain are largely suppressed, e.g. by narcosis. It is thought that long-term storage occurs by the improvement of synaptic efficiency in certain neuronal circuits, e.g. by increasing the number or size of synaptic contacts, possible promoted by increased neuronal metabolism (increased synthesis of proteins and glycoproteins). M. storage in molecules, known as "memory molecules", now seems highly improbable. On the other hand, various endogenous substances appear to improve M. and learning processes when administered in extra quantities, e.g. peptides like vasopressin and ACTH.

Different types of learning processes can lead to information storage, e.g . Imprinting (see), Conditioned reflex (see), Operant conditioning (see), observation and imitation, and insight learning (see Learning).

Memory capacity: the storage capacity of nervous systems. M.c. can be quantified by the number of different patterns that an organism can learn to distinguish. Teleost fishes can be trained to distinguish between two visual patterns (a pattern pair), and the memory of up to 6 such pairs may be retained. Mice can memorize 6–7 pairs, rats 8, a horse up to 20. Dogs can memorize up to to 53 acoustic commands. Chimpanzees can memorize a deaf and dumb language of more than 130 signs or gestures. In analogy with computer information storage, the M.c. of humans is often quoted in bits (binary digits), the smallest unit of information. The basis for the determination of such values is so much in dispute, that the data can only be viewed as "upper limits, assuming that certain preconditions are met". Thus, quoted values for the capacity of human long-term memory are as low as 10^8 bits and as high as 10^{14} bits.

Memory cells: see Anamnestic reaction.

Menarche: the first menstruation. The age of onset of M. is influenced by genetic factors, climate, lifestyle, nutrition and health. In Central Europe, the average age of onset is 12 years 10 months, which is 1–3 years younger than in earlier generations (see Acceleration). Early onset of M. is not associated with an early menopause.

Mendelian population: see Population.

Mendelian segregation: the segregation of heterozygous allele pairs according to Mendel's second law (see Mendel's laws). Mathematically, it can be predicted that a single allele pair will always segregate to give the genotype ratio 1AA: 2Aa: 1aa, and the phenotype ratio 3A: 1aa (where A represents both AA and Aa, since these two cannot be distinguished phenotypically if dominance is complete). A cross involving a single heterozygous allele pair (e.g. Aa x Aa) is called a monohybrid cross. If 2,3 or more allele pairs are involved, it is called a di-, tri- or polyhybrid cross (a dihybrid cross is represented by e.g. AaBb x AaBb). In the case of di-, tri- or polyhybrid crosses, the F2 generation contains new combinations as predicted by Mendel's third law.

The phenotype segregation ratios of di-, tri- or polyhybrids can generally be derived from the consideration that they must arise by free combination, and that each allele pair undergoes monohybrid segregation to give a phenotype ratio of 3:1. The mathematical predictions for different hybrid crosses are summarized in the table.

These normal segregation ratios may be modified by different degrees of dominance, and by the influence of other nonallelic genes. In the following, the most common modifying influences are described and illustrated by reference to the normal separation of phenotypes in a dihybrid cross: 9A?B?: 3A?bb: 3aaB?: 1aabb, in which A is dominant over a, and B is dominant over b (I in Fig.1).

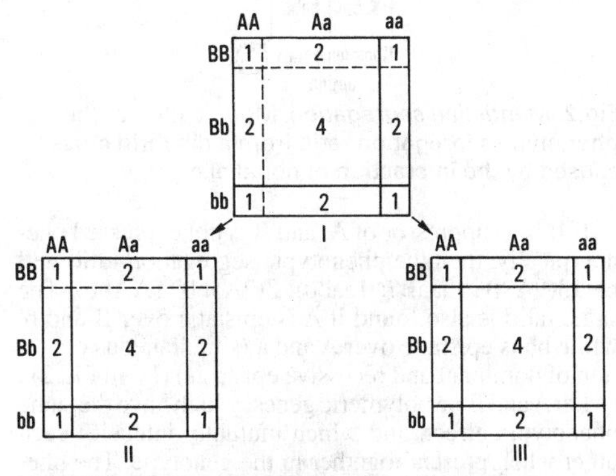

Fig.1. *Mendelian segregation.* Modification of the phenotype segregation ratio from a dihybrid cross. *I* represents the normal segregation. *II* represents the altered segregation when one allele pair (Aa) shows incomplete dominance. *III* represents the altered segregation when both allele pairs show incomplete dominance.

If one or both allele pairs display incomplete dominance (intermediate behavior), there is a shift in the ratio of phenotypes (II and III in Fig.1) (see Mendel's laws, table).

Epistasis (see) leads to a phenotype ratio of 9A?B?: 3A?bb: 4aa?? (where 4aa?? comprises 1aaBb: 2aaBB: 1aabb), if aa is epistatic over B and b (recessive epistasis, I in Fig.2). On the other hand, if A is epistatic over B and b, and genotype A?B? is phenotypically the same as A (dominant epistasis, II in Fig.2), then the phenotype ratio will be 12A (9A?B?: 3A?bb): 3aaB?: 1aabb.

Double recessive epistasis, in which aa is epistatic over B and b, and bb is epistatic over A and a, results in a segregation ratio of 9A?B?: 7(33A?bb: 3aaB?: 1aabb). The same ratio results from the complementary action of A and B (III in Fig.2) (see Complementary genes).

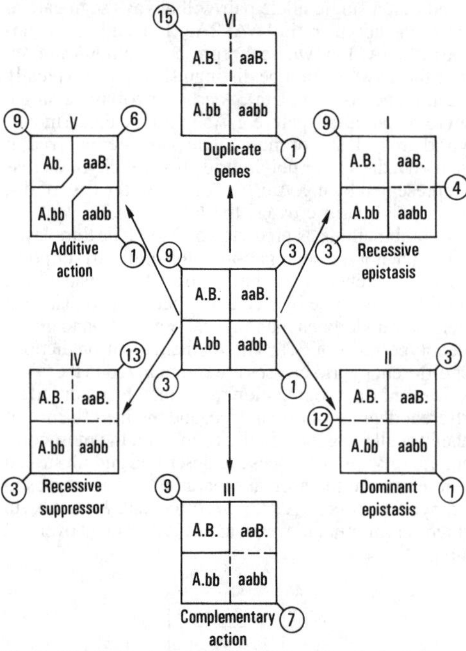

Fig.2. *Mendelian segregation.* Modification of the phenotype segregation ratio from a dihybrid cross, caused by the interaction of nonallelic genes.

If B is a suppressor of A, and B is not expressed phenotypically, then the phenotypic segregation ratio will be 13(9A?B?: 3aaB?: 1aabb): 3(2Aabb: 1AAbb). The same ratio is also found if A is epistatic over B and b, while bb is epistatic over A and a (simultaneous operation of dominant and recessive epistasis) (IV in Fig.2).

Let A and B be polymeric genes, which have the same phenotypic effect, and which mutually intensify each other when present together in the genotype. The phenotype segregation ratio will then be 9A?B?: 6(3aaB?: 3A?bb): 1aabb (V in Fig.2). This can also be interpreted as incomplete, double epstasis, in which A?bb and aaB? are phenotypically the same, whereas A?B? and aabb are phenotypically different.

If A and B have the same phenotypic effect, but do not mutually reinforce each other (duplicate genes), the phenotypic segregation will be 15(9A?B?: 3A?bb: 3aaB?): 1aabb (VI in Fig.2). The same segregation would be observed for double dominant epistasis, in which A is epistatic over B and b, while B is epistatic over A.

Segregation ratios will also differ from those predicted, if allele pairs are linked, and therefore do not undergo independent assortment (see Linkage, Linkage group).

Mendelize: a verb applied to genes as phenotypic determinants, meaning to undergo Mendelian segregation.

Mendel's laws: laws pertaining to the distribution of nuclear chromosomal genes (see Allele). These laws were first discovered by Gregor Mendel (1865), then rediscovered by Correns, Tschermak and de Vries (1900).

Mendel's first law (the law of segregation). The first law states that in sexual reproduction, during gamete formation, the two members of an allele pair or paired homologous chromosomes become separated and each gamete receives only one member of the pair. This law is derived from a consideration of the first filial generation (F1 generation) from the crossing of two genetically pure (homozygous) parents, which differ in one or more allele pairs and therefore display different phenotypic characters. According to the first law, the progeny from such a cross will be uniform, i.e. they will all have the same phenotype, irrespective of which phenotypes served as mother or father. Thus, the reciprocal crosses, A (female) x B (male) or B (female) x A (male), both result in the same uniform offspring, i.e. they possess one allele from each parent. If only one allele of the pair is expressed, this allele is said to be *dominant,* and the offspring will display the phenotype of the parent that supplied it; the unexpressed allele is said to be *recessive.* In the absence of this dominant-recessive relationship, the relevant phenotypic character will have an intermediate quality or intensity (see Dominance).

Mendel's second law (the law of independent assortment, or ***the law of the integrity of gametes).*** The second law refers to the independent segregation of heterozygous allele pairs that are not linked (see Linkage). It states that alleles are distributed randomly and equally into gametes, and that this is the result of the random orientation of the chromosome during meiosis. The law also requires that during fertilization the genetically different gametes undergo random fusion. Derivation of the law is based on a consideration of the second filial generation (F2 generation) from the crossing of two genotypically similar, heterozygous F1 siblings,

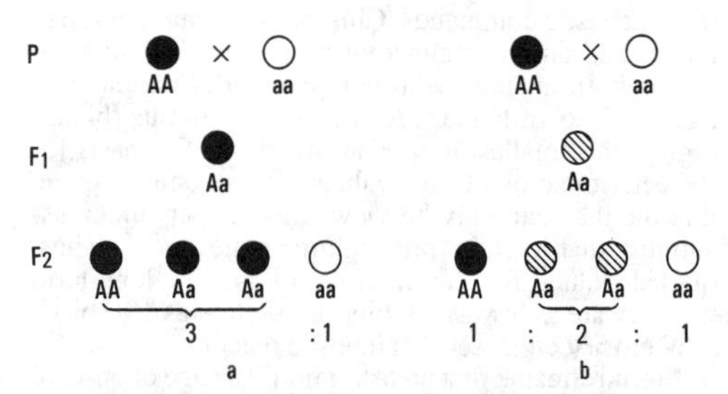

Fig.1. *Mendel's laws.* Genotypes and phenotypes according to Mendel's 1st and 2nd laws. *a:* A is dominant over a. *b:* Dominance is incomplete (intermediate phenotypic expression). *p* Parents. F_1 First filial generation. F_2 Second filial generation.

or from the self-fertilization of a hermaphrodite, heterozygous F1 individual. The progeny from such a cross will contain predictable ratios of genotypically and phenotypically different individuals (see Mendelian segregation). If a particular phenotype is determined by dominant allele A (allele a is recessive), then self-fertilization of the heterozygous F1 organism of genotype Aa will produce a genotype ratio 1AA: 2Aa: 1aa in the F2 generation. The phenotype ratio is therefore 3:1 (Fig.1a). If dominance is incomplete, so that intermediate expression of the phenotype occurs, then the phenotype ratio will be 1:2:1 (Fig.1b).

Mendel's third law (law of the independent combination or new combination of hereditary factors). The third law states that if the heterozygous nature of the F1 progeny involves several unlinked allele pairs (see Linkage), each allele pair is subject separately and independently to the law of independent assortment. Thus, new genotypes and phenotypes may arise that are not present in the parents. This may be exemplified by a cross between a black-spotted and a uniformly red-brown race of cattle. In cattle, the allele for black coat (A) is dominant over that for a red-brown coat (a), while that for uniform coloration (B) is dominant over that for spotting (b). The predicted result of such a cross is shown in Fig.2. Four phenotypes are expected: uniform black, black-spotted, uniform red-brown and red-brown-spotted, in the ratio 9:3:3:1, where the genotypes AABB (box 1) and aabb (box 16) represent genetically pure (homozygous) new combinations.

If both allele pairs show incomplete dominance, then 9 genotype and phenotype groups would be expected. If only one of the allele pairs displayed incomplete dominance, 6 phenotypic groups would arise, as shown in the table.

Table. ***Mendel's laws.*** Genotypes and phenotypes from the cross, AaBb x AaBb, where A is dominant over a, but the allele pair Bb shows incomplete dominance.

Genotype	AA BB	Aa BB	aa BB	AA Bb	Aa Bb	aa Bb	AA bb	Aa bb	aa bb
Ratios of genotypes in F2 progeny (see Fig.2)	1:	2:	1:	2:	4:	2:	1:	2:	1
Ratios of phenotypes	3		1	6		2	3		1

Mengaidae: see *Strepsiptera.*

Meningococcus: see *Neisseria.*

Menotaxis: a tactic response of a motile organism to a directional stimulus, in which the body of the organism adopts a constant angle to the source of the stimulus. Examples are the orientation of organisms in relation to the sun (see Orientation) and the adjustment of the long axis of the body of insects (e.g. dung beetles) according to the wind direction *(anemo-menotaxis).*

Mentha: see *Lamiaceae.*

Menthene, ***Δ³-menthene:*** an unsaturated, cyclic,

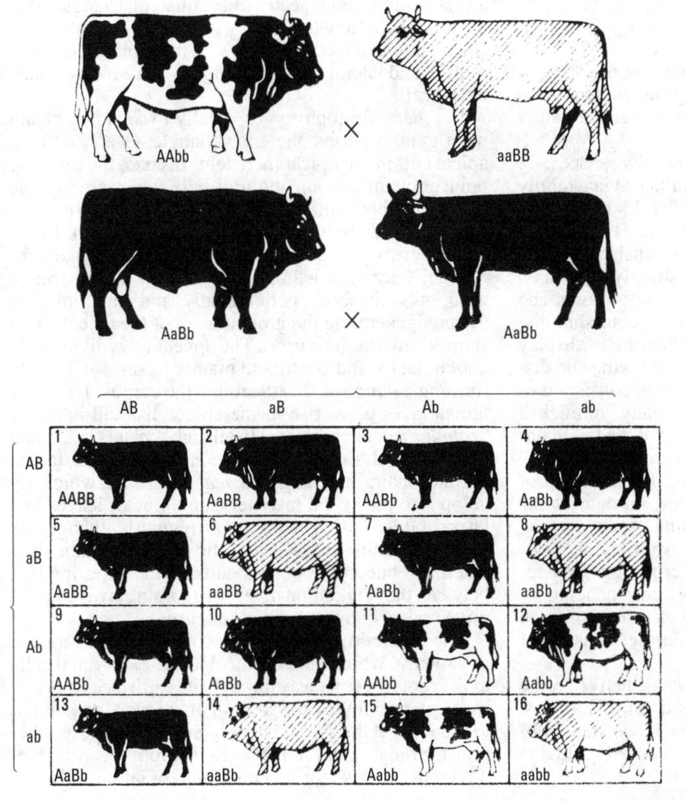

Fig.2. *Mendel's laws.* Results of a dihybrid cross between a black and white spotted variety of cattle and a uniformly colored red-brown variety.

monoterpene hydrocarbon, and biologically the most abundant of the six menthene isomers. It is present in certain essential oils, e.g. peppermint and thyme oils.

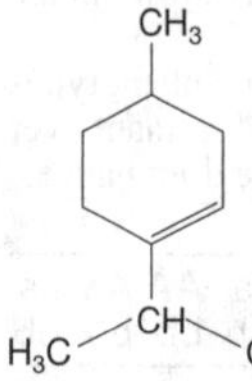

CH_3

H_3C CH CH_3 *Menthene*

Menthol: a cyclic monoterpene alcohol. M. crystallizes as prisms, and possesses a strong, peppermint-like smell. The levorotatory form is the main component of oil of peppermint *(Mentha piperita)* and it occurs in even greater proportion (up to 90%) in the Japanese *Mentha arvensis*. M. is antiseptic, and has antispasmodic activity towards the stomach, intestines and gall bladder. It soothes itching and causes a pleasant cold sensation on the skin. M. is the active principle of peppermint tea, and it is used widely in pharmaceuticals and cosmetics, e.g. in creams, tablets and toothpaste.

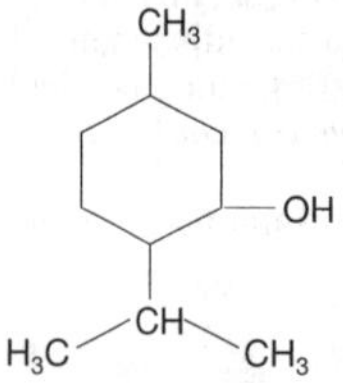

CH_3

OH

H_3C CH CH_3 *Menthol*

Menthone: a cyclic monoterpene ketone, structurally related to menthol (the alcohol group of menthol is replaced by a carbonyl group). It is present, together with menthol, in peppermint oil.

Menurae: a suborder of the *Passeriformes* (see).

Menuridae, ***lyrebirds:*** an Australian avian family (order *Passeriformes),* containing only 2 species. The ***Superp lyrebird*** *(Menura novaehollandiae)* is found in forests and scrubland from South Queensland to Victoria. Its upper parts are dark brown, slightly reddish on the wings, and the underparts are light gray-brown. The eyes are brown with pale orbits. For his courtship display the male constructs several earthern display mounds over a distance of more than 0.8 km. The display is performed at each mound in turn, consisting of wide variety of calls (including many mimicked sounds) and culminating in the fanning of the tail feathers with the 2 lyre feathers forward over the head, forming a shape resembling a lyre, and also reminiscent of the frond of a tree fern. Various sites, on or near the ground, are chosen for nest building, such as rock ledges, tree stumps, clumps of fern. After laying a single, large egg (the largest of the passerine eggs), the female leaves it for about a week before commencing incubation. The polygamous male makes no contribution to egg incubation or brood care. The diet consists of worms, insects and snails.

Prince Albert's lyrebird *(M. alberti),* native to the rainforests of Southeast Queensland and Northeast New South Wales, closely resembles *M. novaehollandiae* in appearance and habits, but it does not build display mounds.

Mephitis: see Skunks.

Mercurialis: see *Euphorbiaceae.*

Mergansers: see *Anseriformes.*

Mergini: see *Anseriformes.*

Mergus: see *Anseriformes.*

Mericarp: see Fruit.

Meristemoids: see Meristems.

Meristems, ***meristematic tissues:*** plant tissues, whose cells are capable of division [cf. Permanent tissues (see) whose cells no longer divide]. Primary M. of higher plants, also known as *promeristems,* are derived directly from embryonal tissue. When promeristem cells become located at the termini of shoot or root primordia, they behave in one of two ways: they become cells of the apical meristem, which retain the ability to divide; or they differentiate into various kinds of Permanent tissue (see), which have permanently or temporarily lost the ability to divide. *Apical meristem* consists of small embryonic cells of uniform diameter, with dense protoplasm and a relatively large nucleus. The thin cell walls contain very little cellulose and consist mainly of propectins. All the cells are closely apposed to each other, and there are no intercellular spaces. Apical M. vary in shape, often having the form of a conical or flattened dome, hence the occasional use of the term, *apical cone.*

In the more advanced thallophytes and in most pteridophytes, the shoot apical meristem originates from a special *apical cell* or *initial cell* lying at the growing point. This cell often has the shape of a tetrahedron, with its convex base facing outward. Thus, from above, a single initial cell is seen as an equilateral triangle. This apical cell continually divides, producing daughter cells basipetally in a spiral sequence. In these plants, leaf primordia and lateral buds also originiate from single new initial cells.

In other pteridophyes (especially *Lycopodiales)* and most gymnosperms, there is no single, clearly defined apical cell in the apical mersitem. Instead, the growing point contains a group of *initial cells,* which are capable of dividing both anticlinically (perpendicular to the surface) and periclinically (parallel to the surface). In the angiosperms and same advanced gymnosperms, the initial cells are grouped into layers. Only the innermost group divides both periclinically and anticlinically, thereby generating the ground mass of the apical meristem, known as the *corpus.* The upper layers divide only anticlinically and contribute mainly to surface growth, forming a *tunica* or sheath around the corpus. The outer tunica layer gives rise to the epidermis, and is called *protoderm* or *dermatogen.* In all higher plants, localized periclinical divisions in the second layer of the tunica produce lateral swellings, the *leaf primordia,* which develop into the apical meristematic tissue of leaves and lateral buds. The growing point regularly changes its form and position; this is a cyclic change, and the time taken for one cycle corresponds to the time interval between the origination of leaf primordia at two consecutive nodes, known as the *plastochrone.*

The root meristem is usually covered by the root cap or calyptra, which protects the delicate embryonal cells as the root forces its way through the earth (see Root). In many pteridophytes, the summit of the growing point of the root (as of the shoot) is occupied by a single, tetrahedral, terminal apical cell, whereas the roots of spermatophytes possess two or several groups of initial cells.

During the differentiation of promeristem cells into specific types of permanent tissue (behind the vegetation points of root and shoot), certain cell layers and groups retain their embryonal form and properties, as well as a latent capacity for division. These are known as *residual M.* Residual M. are important, e.g. for intercalary growth in the basal segments of the internodes of monocotyledons, as fascicular cambium in the vascular bundles of dicotyledons, and as the pericycle in roots.

Secondary M. consist of new meristematic tissue, derived from certain groups of cells in permanent tissues, which regain the ability to divide. They are therefore not descended directly from promeristem. Examples are cork cambium (phellogen) and the interfascicular cambium that occurs in the parenchyma between vascular bundles.

Small secondary M. consisting of only one or a few cells are called *meristemoids.* They usually arise among nondividing cells, they display especially high rates of division, and they are destined finally to become differentiated into permanent tissue. Dividing meristemoids are often surrounded by a zone of inhibition, which prevents the division of other cells in the immediate vicinity, so that the differentiated structures of meristemoids, e.g. stomata or hairs, are arranged in regular patterns on the plant surface.

Merkel's tactile disks: see Tactile stimulus reception.

Merlangius: see Whiting.

Merlin: see *Falconiformes.*

Mermis: see *Nematoda.*

Mermithida: see *Nematoda.*

Merogamy: see Reproduction.

Merogony: development of an animal organism from a nonnucleated egg fragment that has been "fertilized" by a sperm. In a certain sense, M. is the counterpart of Parthenogenesis (see). The latter process shows that the egg cell alone contains all the developmental factors that are necessary for the development of a normal organism. M. shows that the same is true of the male germ cell. Since the sperm consists largely of nuclear material, however, a certain amount of cytoplasm must be contributed from elsewhere for embryo development. The first successful development of an animal by M. was reported in 1889 by the Würzburg zoologist, Theodor Boveri. He prepared fragments of unfertilized sea urchin eggs by shaking, some of the resulting fragments containing no nucleus. "Fertilization" of a nonnucleated egg fragment with a sperm resulted in the normal development of a sea urchin. An embryo from the fusion of egg cytoplasm and a sperm is called a *merogon,* but strictly speaking it should be called an *andromerogon,* since development is determined by the male chromosomal complement. The converse is *gynomerogon* (embryo developed from a nucleated egg fragment with no contribution from a sperm).

Andromerogons were prepared by F. Baltzer (Bern) from newt eggs *(Triton taeniatus).* Shortly after fertilization, he tied an egg with a fine hair, restricting the egg nucleus to one half of the egg. The newt egg is generally penetrated by several sperms (polyspermy), so that both of the abstricted halves contained sperms. The half without the egg nucleus developed into a merogon, which successfully completed metamorphosis.

Clearly, merogons are always haploid. This is possibly why most newt merogons showed pathological defects, and died sooner or later during larval development. Deleterious genes would be fully expressed, because they are not counteracted by "healthy" partners. Normally, such sub-vital or even lethal factors are expressed only in homozygotes. Merogons of mollusks, annelids and frogs have also been generated. Of particular interest are *hybrid merogons,* in which the egg cytoplasm and sperm nucleus are derived from different species. The viability of hybrid merogons increases with the phylogenetic relatedness of the two partners (Table).

These experiments show that the developmental factors in the nuclear chromosomes must be "matched" with the cytoplasm in which they operate. The greater the phylogenetic separation of nucleus and cytoplasm, the earlier the appearance of defects leading to premature death of the hybrid merogon.

Combination

Egg cytoplasm	Sperm nucleus	Stage of development attained
Newt	Toad or Salamander	First developmental step only
Triturus taeniatus or *palmatus*	*Triturus cristatus*	Embryo with neural tube primordium and optic vesicles
Triturus taeniatus	*Triturus apestris*	Embryo with optic cups, labyrinth vesicles and somites
Triturus taeniatus	*Triturus palmatus*	Embryo with limbs and functional circulation, etc.

Meromictic lake: see Lake.

Meromixis: in bacteria, the formation of a merozygote by transfer of only part of the genetic material from the donor cell to the recipient cell during transformation, transduction, or conjugation.

Meropidae: see *Coraciiformes, Mecoptera.*

Meroplankton: see Plankton.

Merops: see *Coraciiformes.*

Merostathmokinesis, ***partial C-mitosis:*** mitosis in which the spindle mechanism is affected by weak doses of C-mitotic agents, but not completely blocked. A multipolar anaphase is therefore possible.

Merostomata: a class of arthropods represented by only 5 known living species, all of them marine. The majority are extinct. The dorsal prosoma is not segmented, and the opisthosoma carries a large pointed telson. Two orders are recognized: *Xiphosura* (see) and *Gigantostraca* (see). Fossils are found from the Cambrian to the present, mostly in the Paleozoic.

Merothripidae: see *Thysanoptera.*

Merozoite: see *Telosporidia.*

Merozygote: a Zygote (see), in which part of the genetic material is diploid and part is haploid. Ms. are produced as a result of Meromixis (see) during the transformation, transduction, or conjugation of bacteria.

Merulius: see *Basidiomycetes (Poriales).*

Mescal: see *Agavaceae.*

Mescaline, ***3,4,5-trimethoxy-phenylethylamine:*** a biogenic amine, and the principal component of the drug, Peyotl or Peyote, used as a ceremonial intoxicant by Indians in northern Mexico. Peyote is the cut and dried parts of the cactus, *Lophophora williamsii.* It causes hallucinations, colored visions and euphoria, and it is classed as a narcotic. The synthetic derivative, 2,5-dimethyl-4-methylamphetamine (DOM), is a more potent hallucinogen than M.

OCH_3, CH_3O, $CH_2—CH_2—NH_2$, OCH_3

Mescaline

Mesembryanthemum: see *Aizoaceae.*

Mesencephalon: see Brain.

Mesenchyme: see Mesoblast, Connective tissues.

Mesenterial filaments: thick, convoluted, epithelial, trilobed filaments present in the *Anthozoa.* They are located in the gastrovascular space on the free internal edges of the gastral septa. The lateral lobes consist of ciliated cells, which aid in water circulation, while the middle lobe contains numerous unicellular digestive glands and nematocysts. They correspond to the gastral filaments of the scyphomedusae (see *Scyphozoa),* and they completely enfold the dietary material during digestion. In some sea anemones, the middle lobe extends as a thread (acontium) from the base of the mesentery into the gastrovascular cavity.

Mesentery: a fold in the dorsal part of the membrane lining the abdominal cavity. It supports the intestine and associated organs, and contains the blood vessels and nerves that supply the intestine.

Mesites: see *Gruiformes.*

Mesitornithidae: see *Gruiformes.*

Mesobilirubin: see Bile pigments.

Mesobilirubinogen: see Bile pigments.

Mesoblast: the median germ layer of the triploblastic animal embryo. It is a cell mass consisting of a compact band of epithelial cells (mesoderm), or of loose parenchymatous tissue *(mesenchyme).* As a result of highly integrated movements during gastrulation, the M. comes to rest between the ectoderm and endoderm. Before the onset of gastrulation, the origin and timing of M. formation are variable. M. is a true germ layer, i.e. it is predetermined to ultimately differentiate into particular tissues. The presence of M. is one of the determining characteristics of the *Bilateria,* and it must be distinguished from the non-organ-forming mesoglea of the *Coelenterata.* The most primitive *Bilateria,* e.g. *Platyhelminthes* (flatworms) and *Nemertini* (proboscis worms) develop exclusively mesenchyme, and they are called acoelomates *(Acoelomata;* the region between the digestive tract and epidermis is filled with mesenchyme and muscle fibers, and a body cavity is lacking). Other phyla *(Acanthocephala, Aschelminthes, Entoprocta)* belong to the *Pseudocoelomata* (both mesoderm and mesenchyme are formed, and the body cavity is a pseudocoelom). Remaining phyla (e.g. *Annelida, Arthropoda, Mollusca,* etc.) are coelomates *(Coelomata;* exclusively mesoderm is formed, which is the origin of the secondary body cavity or coelom).

A) ***Mesenchyme formation.*** Blastomeres predestined to give rise to mesenchyme are determined at an early stage of cleavage. Thus, in flatworms of the orders *Polycladida* and *Acoela* (subclass *Archoophora),* and in the *Annelida* and *Mollusca,* Cleavage (see) is determinate and spiral. In contrast, cleavage in the turbellarian subclass, *Neoophora,* is irregular (see Gastrulation).

In spiral cleavage, two division phases produce 4 *macromeres,* A, B, C and D (Fig. 1). Four subsequent division phases then convert each macromere into 4 further macromeres and 4 *micromeres.* At each of these division phases, the resulting 4 micromeres form a micromere quartet (1a to 1d, 2a to 2d, 3a to 3d, 4a to 4d); these micromeres divide again during each successive phase of cleavage. Thus, with formation of the third quartet, the embryo contains 32 cells. Cleavage (blastula formation) stops at the fourth micromere generation (64 cell stage). By further division, quartets 1 to 3 give rise to ectoderm (leading eventually to epidermis, nervous tissue, sense organs and stomodeum) and the larval mesenchyme of Müller's larva (protrochula). Cells 4a to 4c, together with macromeres A to C, become histolysed and provide nutrient material. Cell 4d divides to produce definitive mesenchyme (parenchyma) and endoderm. The mesenchyme cells migrate inward, while the fore- and midgut are being formed. Later, the 4d daughter cells form the parenchyma (which acts as a support and packing tissue, and as glycogen storage tissue), as well as the layers of circular, longitudinal, diagonal and dorsoventral integumentary muscles.

Embryos of probiscis worms possess a recognizable protomesenchyme cell at the 32-cell stage, or they produce endomesenchyme cells during gastrulation. The delicate mesodermal epithelia of the rhynchocoel and the blood system are formed directly in and from the mesenchyme.

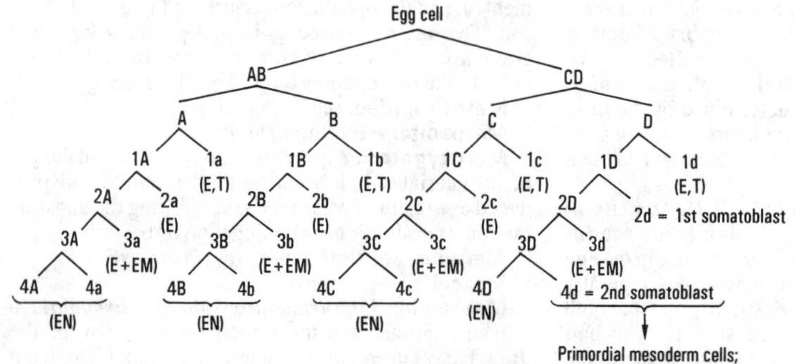

Fig. 1. *Mesoblast.* Cell lineage in spiral cleavage (platyhelminths, annelids and mollusks).

Members of the classes *Gastrotricha* and *Kinorhyncha* (phylum *Aschelminthes)* do not develop mesenchyme. In other aschelminths *(Rotifera* and *Acanthocephala),* the origin of the small amount of mesenchyme is unknown or disputed. As in *Ascaris* and other nematodes, determination occurs early in cleavage.

The weakly developed mesenchyme is derived partly from the EMSt blastomere (endoderm/ mesenchyme/stomodeum), which is already present at the 4-cell stage after the second cleavage. Two further cleavages produce the MSt blastomere, and one further cleavage (i.e. the 32-cell stage is attained) produces the first independent cell (M-cell) destined to become mesenchyme only. The M-cell migrates inward, together with the endoderm and primitive germ cells, by invagination of the stomodeum. Within the embryo, the M-cell divides into two smaller cells; each of these produces a small band of mesenchyme cells by budding, and each band then increases by cell division.

Mesenchyme formation in the *Nematomorpha* (hairworms) differs from the other types described above. It may not occur until shortly before invagination (e.g. by ectomesenchymal migration from the blastoderm in *Gordius),* or it may not become detectable until after invagination (e.g. formed as entomesenchyme from the archenteron in *Paragordius).*

In some echinoderms *[Echinoidea* (sea urchins and sand dollars) and *Ophiuroidea* (brittlestars)], mesenchyme is formed in two stages. Primary larval ectomesenchyme is formed from the blastoderm near the invagination pole. Somewhat later, secondary mesenchyme (secondary mesenchyme is also characteristic of the other classes of echinoderms) is formed from the tip of the archenteron; this eventually gives rise to connective tissue and the deeper layers of the integument, comprising muscular and skeletal tissue in the adult.

B. ***Mesoderm formation.*** Whereas mesenchyme formation is a relatively uniform process in different organisms, the origin and development of mesoderm are very variable.

1. *Mesoderm determination at early cleavage stages from primordial mesoderm cells.* In the spiral cleavage of mollusks and annelids, mesoderm is derived from the second somatoblast of the fourth micromere quartet. This primary mesoderm cell (4d in Fig.1) rapidly divides into two ($4d^L$ and $4d^R$), which are transferred inward during invagination. The annelid gastrula develops into a *trochophore* (Fig.2, a and b), which has a fluid-filled coelom interlaced with muscle and mesenchyme fibers, as well as a pair of protonephridia (derived from the first 3 micromere quartets). The two primary mesoderm cells come to lie one on each side of the posterior end of the gut. At the onset of metamorphosis to the adult (after a planktonic existence of a few days to several weeks), the two primary mesodermal cells proliferate by budding to form two bands of primary mesoderm, which develop ventrally toward the mouth opening, but do not progress beyond the prototroch (pre-oral ciliated zone) into the episphere of the larva. These two bands then divide almost synchronously into a number of segments *(metameres* or *somites,* formed by *metamerism* or *segmentation;* Fig.3, a and b). Within these segments, small cavities appear by cell separation (i.e. the mesoderm begins to split into two layers), which later enlarge into *coelomic cavities* bounded by a cellular monolayer of mesodermal epithelium *(coelomic epithelium* or *mesothelium),* and separated from each other by 2-layered walls *(dissepiments).* The mesothelium, which is next to, and later combines with, the ectoderm, is called the parietal layer of the coelomic epithelium or the *somatopleure,* while the mesothelium apposed to the gut is known as the splanchnic or visceral layer of the coelomic epithelium or the *splanchnopleure.* The coelomic cavities of each segment become extended above and below the gut, so that their epithelia meet dorsally and ventrally, forming dorsal and ventral 2-layered mesenteries. These coelomic sacs, together with those arising subsequently (see below), form the segmented *coelom.* At this stage, the somites have become deuterometamers. The two primary mesoderm cells (M-cells) are still present, lying between

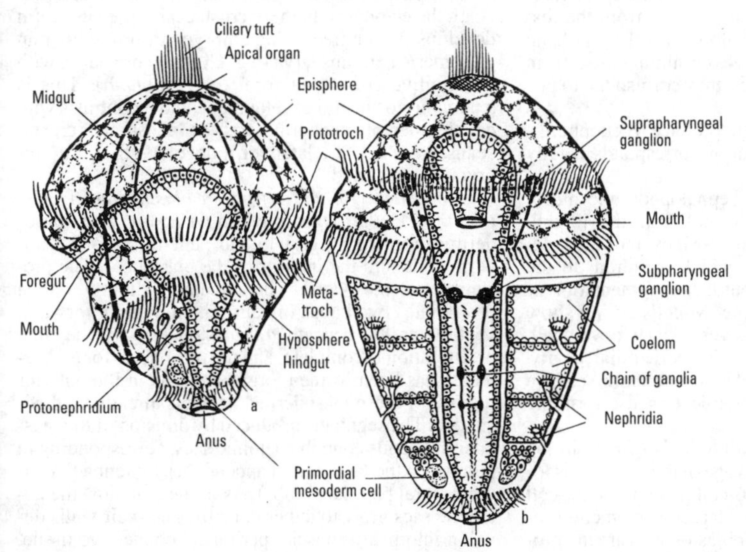

Fig.2. *Mesoblast.* Polychaete trochophore *(a,* viewed from the left) and metatrochophore *(b,* ventral view).

the deutometamers and the anus. After a pause in segmentation, the M-cells proliferate to produce more somites, which also form pairs of coelomic sacs. This growth process is characteristic of the *Articulata* (annelids and arthropods), and it is known as teloblastic somite formation or *teloblasty;* the resulting somites are called *tritometamers.* Occasionally, the descendants of the 4d cell lose their capacity for cell division. Cell material from the ectoderm of the growth zone *(ectomesoderm)* then migrates inward and forms further teloblastic coelomic sacs, which are identical to those formed earlier. The segmented body therefore always arises between the preoral episphere (which becomes the prostomium) and the anal zone (which becomes the pygidium). The extreme anterior and posterior regions of polychaetes remain mesenchymatous and unsegmented. Muscles and connective tissue are of ectomesodermal origin, i.e. they are derived from the first three micromere quartets. Formation of coelomic cavities is not necessarily accompanied by segmentation, but where segmentation occurs, the coelomic sacs are involved in the development of ectodermal structures, such as integument, bristle structures, nervous system and appendages.

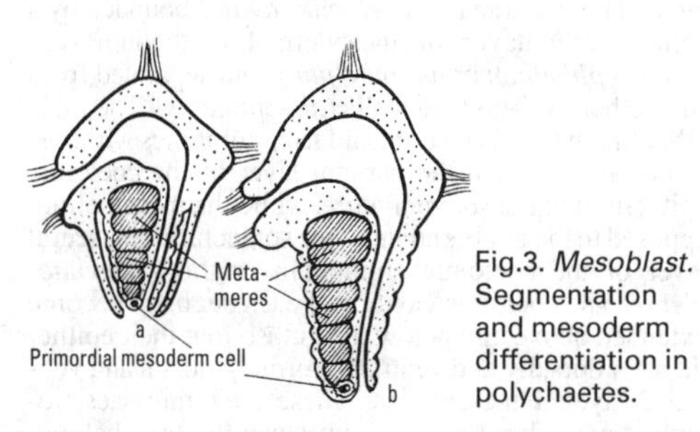

Fig.3. *Mesoblast.* Segmentation and mesoderm differentiation in polychaetes.

In the rapidly ensuing organogenesis, certain organs arise from the mesothelium: the longitudinal musculature of the integument is derived from the somatopleure; gut muscle and peritoneum from the splanchnopleure; certain muscles and ring vessels from the dissepiments; dorsal and ventral blood vessel walls from the mesenteries. Since blood vessel lumina arise from gaps between the coelomic sacs, they correspond to primary coelom.

Although oligochaetes do not produce trochophores, their mode of mesoderm formation corresponds exactly to that of the polychaetes.

With the exception of the cephalopods, the mesoderm of the *Mollusca* also develops from the 4d-cell, which forms mesodermal bands. Many mollusks develop a veliger larva (Fig.4), which is similar to the trochophore of the annelids, but lacks segmentation or teloblastic somite formation (i.e. M-cells do not show teloblastic activity). Coelom is very poorly developed, and it is represented only by the pericardial cavity, which is formed by fusion of two coelomic sacs; in some classes of mollusks, coelomic sacs also form the coelom of the gonads.

Early mesoderm determination is also found in the *Bryozoa* (moss animals). Cleavage in the *Bryozoa* is total and equal, and the lower 16-cell plate of the 32-cell blastula contains four somewhat larger blastomeres (endomesoderm cells). These four blastomeres are the pro-

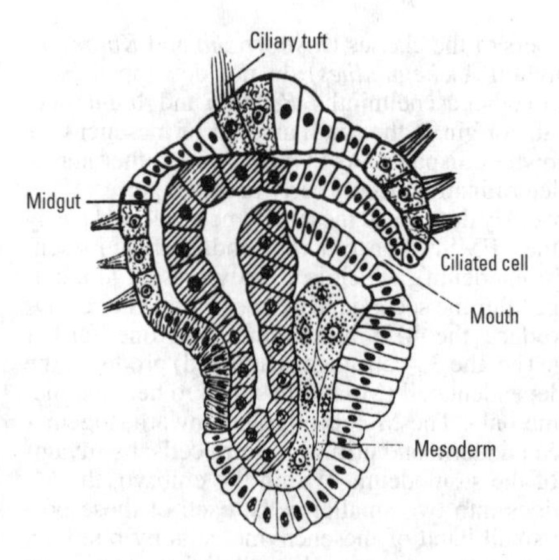

Fig.4. *Mesoblast.* Trochophore larva of *Dentalium* (class *Scaphopoda,* phylum *Mollusca)* with mesodermal band.

genitors of cell material that later separates by an unknown route into endoderm and mesoderm.

Blasomere determination similar to that of annelids is also found in some crustaceans, e.g. *Cladocera* and *Ostracoda* show holoblastic but nonspiral cleavage, while the cleavage of *Daphnia* is holoblastic at first and superficial in its final stages. Otherwise, cleavage of arthropod eggs is typically superficial. Total and equal cleavage of the eggs of *Polyphemus pediculus* produces 6 small ectomesoderm cells at the 16-cell stage, which then divide into 6 ectoderm and 6 mesoderm cells. When cleavage has finished, 12 mesoderm cells, together with endoderm and primordial germ cells, migrate into the interior of the hollow blastula without forming a surface depression. Similar patterns of development have been observed in eggs of the crustacean genera *Lepas, Balanus* and others, where mesodermal immigration occurs by invagination or by epiboly; the mesodermal cells then proliferate to form bands beneath the ectoderm. In these crustaceans, the later form develops from a free-living embryonic form, either an oligomeric nauplius larva (e.g. *Cyclops)* or a larva with a definitive crustacean form (e.g. *Daphnia).* This is analogous to annelid development, and it contrasts with development in other arthropods and most other crustaceans, in which the later form develops from an embryonic germ band.

2. *Mesoderm formation from the blastula during gastrulation.* In *Peripatus* (phylum *Onychophora),* mesoderm is formed by growth and penetration of blastoderm cells into the interior of the blastocoel. The presumptive mesoderm invaginates slightly, then proliferates two bands of cells which migrate forward, ventrolaterally, beneath the blastoderm. By the time proliferation is complete, and the mesoderm for all segments has been formed, organogenesis in the anterior regions of these mesodermal bands is already well advanced. The segments produced by division of the mesodermal bands contain coelomic sacs, corresponding in number to the later paired appendages (antennae, mandibles, oral papillae, limbs). As in the annelids, the coelomic sacs grow together dorsally, and their walls develop into heart, muscle, peritonea, connective tissue,

etc. Coelom, cut off from the original coelomic cavities, persists as the sacculi of the nephridia and the gonadial vesicles; these are lined with a true mesothelium, which also forms the peritoneum of the organs, e.g. the gut. In sea spiders (class *Pycnogonida,* subphylum *Chelicerata),* invagination involves the migration of 8 to 10 mesodermal cells and a single primordial endoderm cell from the blastoderm into a very small cleavage cavity, without forming an archenteron (Fig.5, a and b). The mesoderm cells produce an irregular mesoderm band, which does not form coelomic cavities, but gives rise to the musculature of the extremities. Priapulids develop two strings of cells, which proliferate from the region of the blastopore of the invaginating gastrula into the blastocoel. Individual cells separate from these strings to become mesoderm, which fills the primary coelomic space. The true origin of the adult coelom is uncertain.

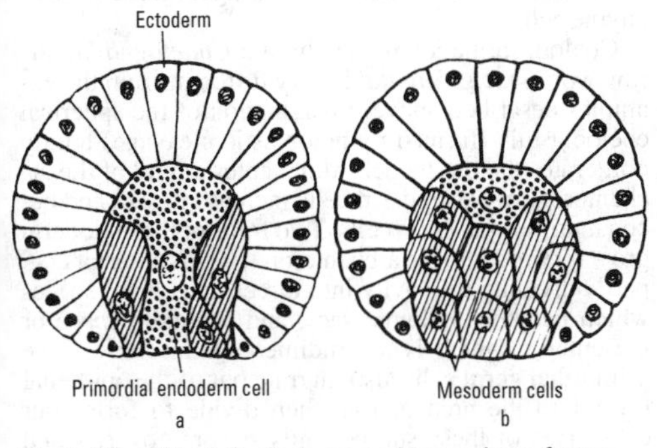

Fig.5. *Mesoblast.* Endoderm and mesoderm formation in the embryonic development of sea spiders *(Pycnogonida).*

3) *Formation of the endomesodermal band or ventral plate.* In insects (insect ontogeny is well documented from extensive studies), blastoderm on the side of the egg that later becomes the ventral part of the insect forms a thickened band or ventral plate. The thickening process progresses forward to the predetermined anterior, and backward to the posterior of the insect, from a midpoint known as the differentiation center. By invagination along its length, the *ventral plate* or *germ band* is turned into a 2-layered structure. This occurs by a median longitudinal strip of cells sinking into the interior, leaving a groove. The groove is closed by inward growth of the two outer remaining strips of cells (presumptive ectoderm), which come together along the ventral midline. The inner band of cells enclosed in this way becomes the mesodermal layer. It continues to divide, becomes thicker and multilayered, and it divides almost completely down the midline, producing two mesodermal bands which become segmented into metamers or somites. Longitudinal clefts are formed in each mesodermal band by a process called schizocoely. These clefts become the paired, mesothelium-lined coelomic cavities of each segment. At the onset of segmentation, the primary head region is composed only of the following three segments. 1. The preantennary or protocerebral segment *(protocephalon),* which includes the acron (homologous with the annelid prostomium). 2. The deuterocerebral segment *(gnathocephalon* or *deutocephalon),* bearing a ventral lobe which later becomes the labrum, and a pair of appendages which will become antennae. The protocephalon and deutocephalon are therefore homologous with the deutometameres of the annelids. 3. The tritocerebral segment, which is devoid of appendages. Later, the first three body segments are added to the head, their appendages forming the mandibles, maxillae and labium. The head is therefore a six-segment structure, but the head of the adult insect shows little trace of this original segmentation. Segments usually appear in anterioposterior succession, and their origin varies according to the site of the differentiation center. In orders whose embryonic primordium is very short *(Orthoptera, Cheleutoptera, Isoptera),* the differentiation center lies close to the posterior pole of the egg, so that the body is formed teloblastically by budding from a segmentation zone of one of the deutometameres, i.e. posterior proliferation is the source of all the mesoderm behind the head lobes, a process similar to that in the annelid trochophore. In contrast, posterior proliferation of the mesoderm after gastrulation does not occur in the exceptionally long primordia of the higher *Diptera* and *Hymenoptera.* The latter lack a segmentation zone, i.e. the differentiation center is situated anteriorly, and the region destined to become the whole larva is included in the mesodermal bands, which differentiate into segments under the influence of the differentiation center. In intermediate forms (e.g. *Forficula),* the abdomen, or parts of it, are proliferated from a segmentation zone. The coelomic cavities mentioned above do not normally persist beyond the stage of the ventral germ band. After the limb buds and embryonal paired ganglia of each segment have been formed from the ectoderm, and during the upward spread of mesoderm that accompanies dorsal closure, the walls of the secondary coelom start to differentiate into body tissues and organs; the outer layer of mesoderm (the parietal or somatic layer) thickens and differentiates into skeletal muscle, while the inner layer (the visceral or splanchnic layer) forms the muscles of the viscera and most of the fat body; the mesodermal cells at the lateral junction of these two germ layers become cardioblasts of the heart, and the rest of the fat body proliferates from the medial end part of the parietal layer.

Germ band formation with mesoderm differentiation also occurs in the *Chelicerata.* For example, the ventral side of the blastoderm of spiders becomes richer in cells and compressed to form a hemispherical germ disk or ventral plate. At a site known as the blastopore, some of the cells sink inward to form primary endomesoderm. Mesoderm and endoderm soon separate. Two mesodermal bands are formed, which undergo regular division into segments, even though the final animal may not show segmentation. Four deutometameres are proliferated, while the remaining segments are produced from a growth zone between the pygidium and the last deutometamere. At dorsal closure of the germ band, the coelomic cavities (in contrast to those of insects) are as extensive as those in the annelids; they extend dorsally, but they are laterally considerably compressed by the large yolk mass. The walls of the somatic coelomic cavities and the dissepiments differentiate into dorsal and ventral longitudinal muscles, the dorsoventral muscles, the heart and the aorta and lateral blood vessels. The splanchnic layer gives rise to visceral musculature and peritoneum. Other parts become the gonadal ducts and the fat body.

Certain crustaceans do not form a ventral germ band, and they resemble annelids in that development commences directly with formation of an embryonal structure. Germ band formation is, however, the norm for most crustaceans. Crustacean mesoderm may arise in a variety of ways. The mesoderm of the first three segments *(larval* or *nauplius mesoderm)* is hardly segmented, and it is equivalent to the deutometameres. The imaginal mesoderm is formed teloblastically from a preanal proliferation zone behind the mandibular segment, which corresponds to the tritometameres. Crustacean somites possess only small coelomic cavities, which never participate in dorsal closure.

4) *Formation of enterocoelomic mesoderm.* Enterocoelomic mesoderm is typical of all *Deuterostomia,* including tunicates and vertebrates, and it also occurs sporadically in the *Protostomia,* e.g. in brachiopods and tardigrades

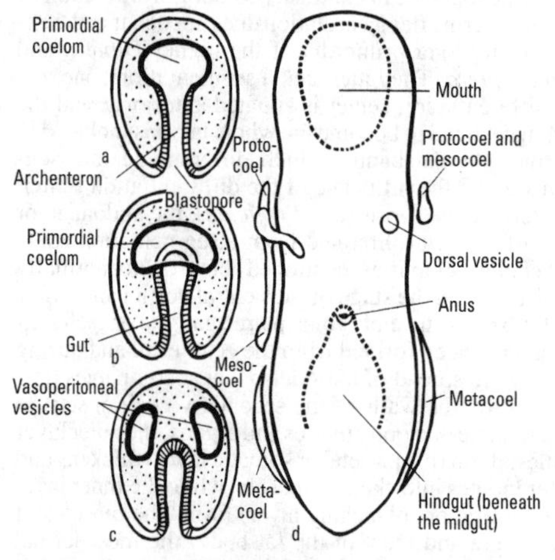

Fig.6. *Mesoblast. a-c* Enterocoelom formation in the sea urchin. *d* Formation of coelomic vesicles in a late pluteus (epidermis not shown).

The echinoderms will be used here as the first example. Their free-swimming, ciliated blastula becomes a 2-layered gastrula by infolding of the cells at one pole; in some cases, further folding of the ectoderm occurs. The inner surface of the infolding blastoderm buds off cells into the blastocoel, which later form the larval ectomesenchyme. The end of the archenteron then swells and pinches off mesoderm to left and right in the form of two vasoperitoneal vesicles (Fig.6, a to c), i.e. two coelomic sacs enveloped by mesoderm. These coelomic sacs divide rapidly into three sections, one behind the other, a process in which a transverse fold on both sides produces a posterior section (the *metacoel* or *somatocoel)* and anterior section. The left anterior section divides further into *protocoel* (or *axocoel)* and a *mesocoel* (or *hydrocoel),* which remain connected by a canal. The right anterior section does not divide further, but atrophies and becomes the pulsating dorsal vesicle (madreporic vesicle), which is considered to be a fragment of the protocoel. Thus, all the coelomic cavities arising in this way originate as vesicles budded from the archenteron, which are also called *archenteron diverticula.* The left protocoel produces the unpaired axial organ, consisting of the axial gland and the protocoel ampulla, which opens into the madreporic plate. The left mesocoel gives rise to the ambulacral system, of which the ring canal arises by a complicated growth process. The ambulacral system is linked to the exterior via the protocoel ampulla and the stone canal, which is derived from the canal linking the protocoel and mesocoel. Radial and side canals are simply derived from outgrowths of the mesocoel. Finally, definitive coelom, the hyponeural or perihemal canal, and the genital canal bearing five gonadal sacs are all formed from the paired somatocoel.

In many enteropneusts and some echinoderms, the coelom does not arise as enterocoel, but from clefts in a solid mesoblast. The latter is derived either by proliferation from the posterior and lateral walls of the archenteron (e.g. acorn worms) or from aggregated mesenchyme cells.

Coelom formation in the phylum *Chaetognatha* (arrow worms), e.g. *Sagitta,* is very different from the examples described above. Invagination of the spherical coeloblastula (formed by holoblastic cleavage) totally obliterates the blastocoel. At the anterior end of the archenteron, two endoderm cells increase in size and become primordial germ cells. Two folds of the endoderm grow forward into the archenteron from its anterior end, partly dividing the cavity into three (two lateral spaces which become coelomic sacs, and a middle space or mesenteron which is the rudimentary intestine). The primordial germ cells also migrate out of the epithelial layer into the archenteron, then divide to form four cells. Two of these subsequently become ovaries and two the testes. During these processes, the blastopore gradually closes. Head and body coelom become separated by a circular transverse fold. Genuine trimeric segmentation therefore does not occur.

In the *Acrania,* endodermal material (in the form of an endodermal plate) moves inward by invagination of the spacious coeloblastula. A mesodermal crescent, consisting of small cells bounding the endodermal plate on its lateral and posterior borders, is also carried inward at a later stage of the gastrulation process. The arms of the mesodermal crescent form areas of rapidly dividing cells *(mesodermal grooves)* along the lateral walls of the archenteron beneath the ectoderm of the dorsal side of the later embryo. The middle of the mesodermal crescent remains on the ventral lip of the blastopore. Mesodermal cells invaginated at the dorsal lip of the blastopore represent the primordial notochord (Fig.7, a to d). These changes are accompanied by elongation of the embryo. In the formation of the neural tube (Fig.7, e and f), a strip of ectodermal cells, located middorsally along the roof of the archenteron, enlarges to form the neural plate, which rolls up into the neural tube. The mesodermal grooves become segmented into a series of metameres (mesodermal pouches) containing coelomic cavities. Remaining endoderm of the archenteron forms the gut wall. The overall process is therefore one of enterocoel formation, which is temporarily halted after the appearance of coelomic pouches, then later continued; well defined trimeric segmentation is no longer present. The large number of metameres (polymeric segmentation) is rather reminiscent of the annelids. In addition to possession of a notochord, other characteristic features should also be noted. For

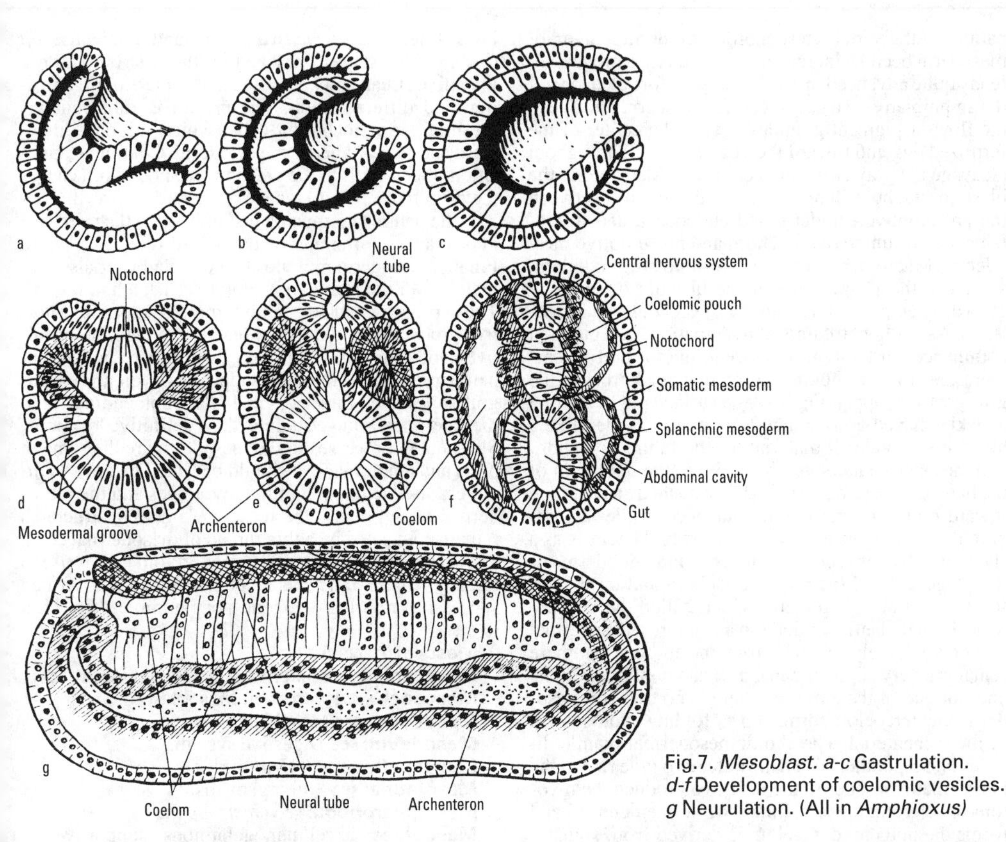

Fig.7. *Mesoblast. a-c* Gastrulation. *d-f* Development of coelomic vesicles. *g* Neurulation. (All in *Amphioxus)*

example, the primordial segments *(archimetameres)* become very rapidly partitioned, by transverse mesodermal folding, into the dorsal myotome containing the myocoels, and the ventral splanchnotome or lateral plates containing the splanchnocoels (Fig.7, g). A uniform splanchnocoel is now formed by breakdown of the ventral mesenteries of the dissipements between the successive smaller splanchnocoels. Metamerism therefore persists only in the myotomes, which contain 4 mesodermal layers: the dermatome or corium layer, the myotome or muscle layer, the fascial layer and the sclerotome or skeletogenic layer (all named for the tissues they eventually form) (Fig.8, a and b). These layers arise from a fold which progresses mediodorsally from the ventral region of the myotomes. Differentiation of the cuticle, the powerful musculature, the fascia and the sheath of the neural tube is accompanied by the complete disappearance of the myocoel by infilling (Fig.8, c). Mesoderm formation in the family *Petromyzonidae* (order *Cyclostomata)* is almost identical with that of the *Acrania.* In the former, however, a single massive mesodermal plate is first formed on all sides from the archenteron, and this plate gives rise much later to separate myotome and splanchnotome, Coelom, with *somato-* and *splanchnopleures,* originates from clefts in the myotome and splanchnotome.

5) *Formation of chordomesoderm.* The most advanced process of mesoderm formation leads to chordomesoderm. It is considered to be an evolutionary derivative of enterocoelomic mesoderm formation in the

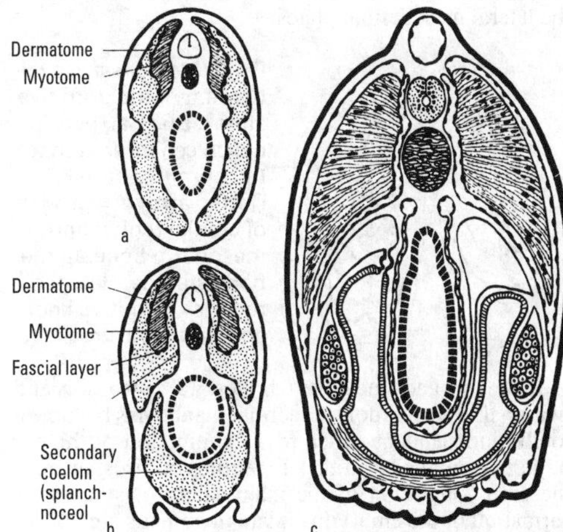

Fig.8. *Mesoblast.* Mesoderm formation in the *Acrania. a* and *b* Segmentation of the coelomic vesicles. *c* Transverse section of the pharyngeal region of a fully differentiated animal.

Branchiostoma, in which mesoderm originates from diverticula of the archenteron. In lancelets, primordial chordal mesoderm, still closely joined to the later endomesoderm, enters the interior of the embryo by invagination. In the vertebrates, chordomesoderm and endomesoderm become increasingly separated, a process already quite advanced in amphibians. Before cleavage of the amphibian egg, a gray crescent is discernable, due to the flow of pigment granules toward the center of the fertilized egg and toward the vegetal pole. Throughout cleavage the gray crescent becomes divided among the blastomeres, but still retains its integrity; it lies between the presumptive ectoderm and endoderm, and it contains the presumptive notochord and presumptive mesoderm. Macromeres of the presumptive endoderm spread to the ventral side of the blastula by epibolic growth, then proliferate into the blastocoel as a shapeless mass, without folding or invagination. Invagination commences in the region between micro- and macromeres, i.e. the blastopore is formed on the border of the gray crescent, appearing first as a surface pit, becoming a sickle-shaped groove, and later more rounded like a horseshoe. Total cell material for the future notochord and mesoderm enters via this groove (the dorsal lip of the blastopore). Endoderm is at first pushed ahead of the inward migrating mesoderm, but once inside the embryo the mesoderm spreads anteriorly and laterally as a sheet of cells between the ectoderm and endoderm. At this stage, cells of the future notochord and mesoderm form a continuous epithelial sheet, called the chordomesodermal mantle. Endoderm and chordomesodermal mantle are largely separate from one another, but there remains a very slight and insignificant continuity at the anterior end of the embryo. There is no trace of mesoderm or entercoelom formation by folding or outgrowth of the archenteron. The chordomesodermal mantle divides into 5 parallel strips; the outer two give rise to the lateral plate mesoderm, the next two produce the myotomes, sclerotomes or somites, while the central band forms the notochord. Coelom is derived from splits in the lateral mesodermal plates.

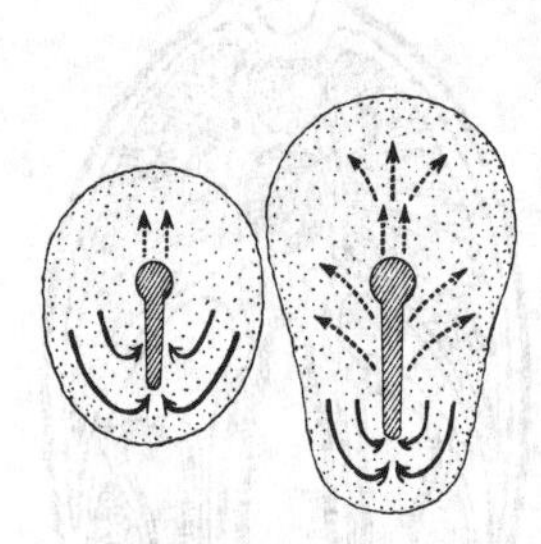

Fig.9. *Mesoblast.* Development of the primitive streak. Black arrows indicate cell migration to the primitive streak. Dotted arrows = growth of cranial region and mesoderm beneath the body surface. Hatched region = primitive knot and primitive streak.

In the most advanced vertebrates, i.e. sauropsids and mammals, the chordomesodermal mantle has become a totally independent structure. Delamination of the endoderm from the blastodisk in reptiles or birds, or from the embryonal knot of the mammalian blastocyst; or formation of the embryonal plate in humans are all followed by active migration of ectoderm cells as shown in Fig.9 (a and b). Lateral cells start to migrate toward the posterior end of the embryo. They move in toward the center, then reverse their direction and move anteriorly along the midline of the embryo. Cells therefore converge and become crowded toward the midline. The resulting long heap of cells forms the *primitive streak,* and a distinct swelling of the anterior end of the streak is termed the *primitive knot* or Hensen's node. The raised central ridge of the primitive streak and knot finally sink inward, leaving a *primitive groove* and a *primitive pit* (in birds this stage is reached on the first brood day). Primitive streak and knot represent centers of organization and differentiation. The primitive knot contains presumptive notochordal cells, which migrate downward and forward under the ectoderm. Continued apposition of new cells to the presumptive notochord causes the primitive streak to recede backward, and as this occurs the neural system area just in front of the primitive knot induces formation of the neural tube (Fig.10, a). Finally, mesoderm migrates laterally and later also cranially from the primitive streak (Fig.10, a and c to h). These processes justify the interpretation of the primitive groove as an elongated blastopore, and the primitive knot as its dorsal lip. They do not represent permanent structures, and they disappear as soon as cell migrations have finished, and the notochordal and mesoderm cells have attained their definitive locations. Subsequent processes in the mesoderm are the same in birds and amphibians. It should be added that in all amniotes, after separation of the myotomes and the mesodermal lateral plates, these two mesodermal structures remain connected by a thin thread of mesodermal cells, known as the *nephrotome.* The nephrotome gives rise to the pro-, meso- and metanephros, which are formed successively during ontogenesis in amniotes (Fig.11, a to d).

Mesocarp: see Fruit.

Mesocephaly: an intermediate Cranial index (see).

Mesocoel: see Hydrocoel, Mesoblast.

Mesoderm: see Mesoblast.

Mesodeum: see Digestive system.

Mesoenatidae: see *Gruiformes.*

Mesofauna: see Soil organisms.

Mesogastropoda: see *Gastropoda.*

Mesogloea: a cellular, gelatinous, supportive lamella between the ectoderm and endoderm of many coelenterates. It arises by cell migration (in particular from the ectoderm) into the gelatinous intermediate layer, M. is not a 3rd germ layer; it does not originate like a germ layer, and it does not contribute to organ development. Structurally and functionally, it resembles vertebrate connective tissue.

Mesohaline: see Brackish water.

Mesohyl: see *Porifera.*

Mesome: see Telome theory.

Mesonephric duct: see Excretory organs.

Mesonephros: see Excretory organs.

Mesophilic: a term applied to organisms which grow optimally at intermediate temperatures (e.g. 20–45 °C) and are adapted to environments that are neither very wet nor very dry.

Mesophytes: plants which grow preferentially in moderately wet habitats with well drained soils. They occupy an intermediate position between hygrophytes and xerophytes. Leaves of M. are dorsiventral, usually relatively large and lacking hairs, wax and water storage tissue. Most cultured plants are M., e.g. bean, pea, tomato, sunflower.

Mesophytic, ***gymnosperm age:*** an epoch of plant evolution, following the Paleophytic. The M. extends from the Upper Permian (Zechstein) to the Lower Early Cretaceous (see Geological time scale, Table). The term, gymnosperm age, refers primarily to the Northern Hemisphere, where the boundary between the Paleophytic and the M. is extremely sharp. With few excep-

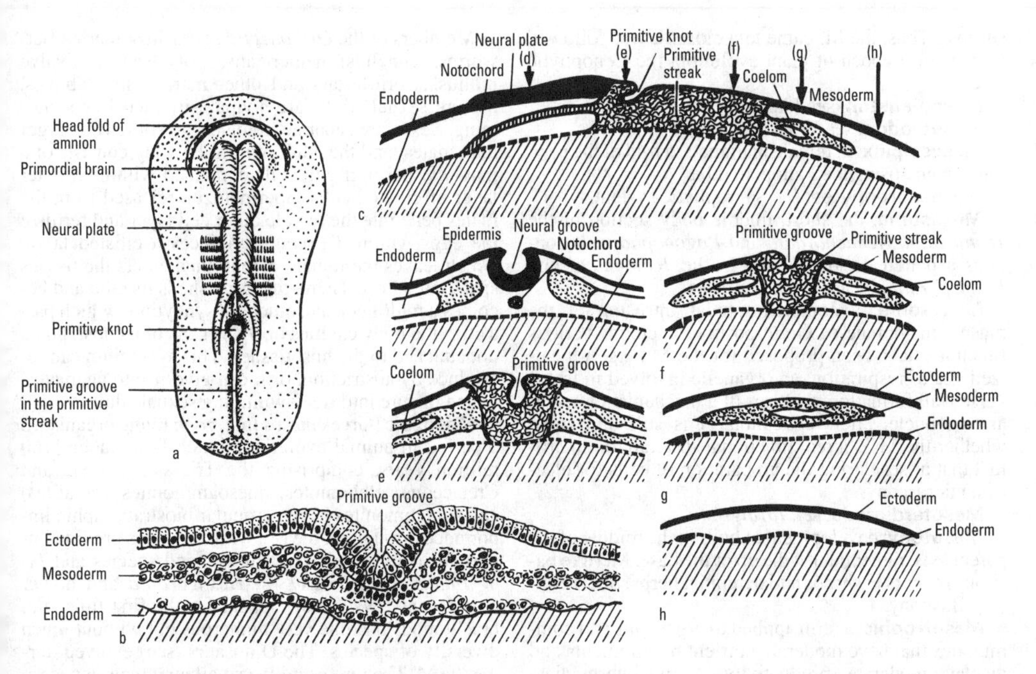

Fig.10. *Mesoblast.* Development of the primitive streak, mesoderm formation and neurulation in birds (semischematic).

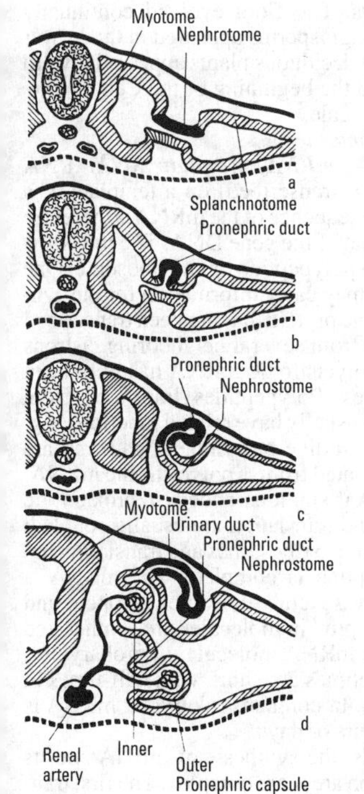

Fig.11. *Mesoblast.* Nephrotome formation in birds.

tions, the tree-like pteriodphytes and the older gymnosperm groups die out with the advent of the M. in the Northern Hemisphere. They are replaced by an abundance of gymnosperms such as *Pinidae*, *Ginkgoatae*, *Cycadatae* and *Bennettitatae*. Only a few of the primitive gymnosperms of the Carboniferous period survive into the M. (e.g. some *Cordaitidae* and *Lyginopteridae).*

Fossils are far less numerous in the Lower Triassic, owing to the extremely dry climate of that period, and only certain forms of *Pleuromeia* are found. With the wetter climate of the Upper Triassic, the vegetation becomes richer, displaying a rapid spread of *Equisetites* species, calamites and ferns of the order *Marattiales*, as well as various gymnosperms (e.g. *Bennettitatae*, see).

In the Southern Hemisphere (Gondwana) the change from the Lower Permian to the Upper Permian is not marked by contrasting plant types. The *Glossopteris* flora continues unchanged; its gradual decline then starts during the Triassic, and it finally disappears in the Late Upper Triassic.

In the Jurassic, the vegetation of the whole earth once again becomes uniform, i.e. the fossil evidence reveals no apparent geographical differentiation. The reasons for this observation are unknown; even continental drift does not provide a satisfactory explanation. *Cycadatae* and *Ginkgoatae* attain their climax in the Jurassic, and the *Bennettitatae* are also very numerous in this period. Angiosperms probably originated during the Jurassic. The Jurassic flora still predominate in the lower Cretaceous, and the fossils of this period also display a striking geographical uniformity. In the Late Lower Cretaceous (Apt or Gault) the total character of the vegetation undergoes a considerable change. Most of the gymnosperms almost or totally disappear, especially the *Bennettitate*, *Ginkgoatae* and *Cycadatae*, and they are suddenly replaced by angiosperms. Only certain conifers survive. Thus, the M. came to a close. It was followed by the latest epoch of plant evolution, the Cenophytic (see).

Mesoplankton: see Plankton.

Mesoplodon: see Whales.

Mesosalpinx: in female mammals, a fold of the peritoneum enclosing the Fallopian tube.

Mesosaprophytes: see Saprophytic system.

Mesosoma: the short, middle body section of the *Tentaculata, Hemichordata* and *Pogonophora.* It contains a paired mesocoel, and in the *Tentaculata* and *Hemichordata* it carries the tentacles.

Mesosome: a complex tubular intrusion of the plasma membrane of Gram-positive bacteria. Various functions have been proposed for the M., e.g. a localized site of respiration, an organelle involved in transverse wall formation, and part of a mechanism for separating nuclei after replication. It is still undecided whether the M. is a real functional structure or an artifact that arises during the preparation of cells for electron microscopy.

Mesotardigrada: see *Tardigrada.*

Mesotony: development of buds in the middle of the parent axis, rather than at the tip or the base. M. (like basitony) occurs in shrubs and perennial herbs. See Acrotony, Basitony.

Mesotrophic: a term applied to organisms and communities that have moderate nutrient requirements and display moderate productivity. It is intermediate between oligotrophic and eutrophic.

Mesozoa (Greek *mesos* middle, *zoon* animal): a small group (phylum) of metazoans comprising about 50 known species. They were originally thought to be intermediate between the *Metazoa* and *Protozoa.* The phylum contains two orders, the *Dicyemida* and the *Orthonectida,* and it is possible that these are largely unrelated. Some authorities consider that the *Orthonectida* are degenerate flatworms, whereas the *Dicyemida* are possibly an offshoot of early metazoan stock.

Members of the *Dicyemida* live parasitically in the nephridia of marine cephalopods (squids, cuttlefish and octopods). The adult body (0.5–7 mm long) consists of only 20–30 cells, i.e. a long, central axial cell surrounded by a single layer of ciliated cells, the anterior of these ciliated cells being used for attachment to the host (Fig.). In young cephalopods, the parasite reproduces asexually by division of axoblasts (from the axial cell) to form vermiform larvae. In mature cephalopods, however, the axoblast becomes modified to a "hermaphrodite gonad" or *infusorigen,* which produces both eggs and sperm. Eggs fertilized by sperm from the same infusorigen develop into *infusoriform larvae.* The asexual form of the parasite (the *nematogen)* is identical in appearance to the sexually reproducing form (the *rhombogen)* and they are distinguishable only by the ultimate behavior of the axoblast.

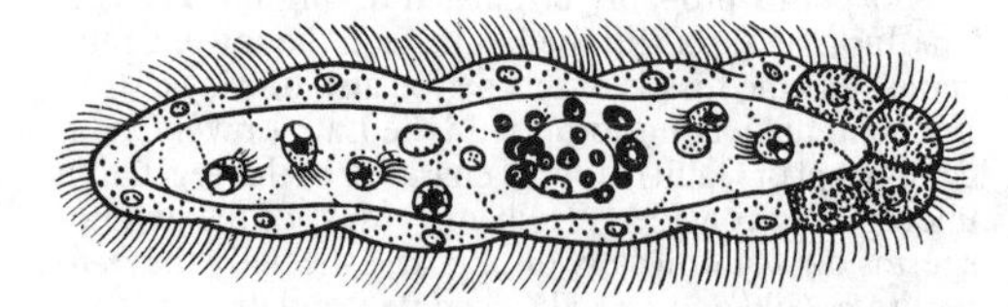

Mesozoa. A dicyemid mesozoan.

Members of the *Orthonectida* parasitize marine flatworms, annelids, nemerteans, polychaetes, bivalve mollusks, brittlestars and other marine invertebrates. Mature sexual adults are nonparasitic and free-swimming. Sexes are separate, females are somewhat larger than males, and the adult, wormlike body consists of a mass of sperm or egg cells contained within an outer layer of ciliated cells. Spermatozoa released from the males penetrate the female outer cell layer and fertilize the eggs within. Cleavage produces a ciliated larva, which leaves the female parent, then enters the tissues spaces of a host. There, the larva sheds its cilia and becomes a multinucleate, ameboid syncytium, which parasitize the body cavities or sex organs of its host, but is not attached to the host tissues. This syncytium can reproduce by abstriction, or it can give rise to agametes, which mature into free-swimming sexual adults.

Mesozoic (Greek *mesos* mid, *zoon* living organism): a period in animal evolution, which lasted about 160 million years, comprising the Triassic, Jurassic and Cretaceous. Belemnites, mesoammonites (ceratites) and neoammonites have particular biostratigraphic importance throughout the M., while bivalves provide important index fossils (e.g. rudists and species and *Inoceramus)* in the early M. *Hexacorallia* and higher crustaceans *(Decapoda)* appear for the first time, and brachiopods become very numerous but without much diversity of species. The Dinosaurs (see) evolved during the M. The first primitive small mammals appear in the M. As precursors of the later mammals, these are of great phylogenetic importance, but on acount of their infrequent occurrence they have little stratigraphic significance. The end of the M. marks the most notable turning point in evolution, with the extinction of the belemnites, ammonites, some bivalve families, and in particular the dinosaurs. The flora evolved continually throughout the M. Angiosperms appeared in the Lower Cretaceous. The first deciduous plants appeared toward the end of the M and the beginning of the Cenophytic. See Geological time scale.

Mesquite: see *Mimosaceae.*

Messenger RNA, *mRNA, template RNA:* RNA which is synthesized (transcribed) on a template of a gene. The nucleotide sequence of the mRNA is therefore complementary to that of the gene DNA. The mRNA is then translated into a polypeptide on the ribosome. One molecule of mRNA may carry information for the synthesis of more than one protein, having been transcribed without interruption from several neighboring cistrons of the DNA. This polycistronic mRNA has only been found in prokaryotes. Polypeptides translated from polycistronic mRNA usually have related functions, e.g. 10 enzymes in the histidine biosynthesis pathway are encoded in, and translated from, a polycistronic mRNA.

Functional mRNA is single-stranded. In prokaryotic cells, transcription and translation are usually coupled: mRNA becomes bound to ribosomes and translation begins before transcription is complete. The mRNA is usually translated by several ribosomes at once, and thus several to many protein molecules are synthesized on a template of one mRNA molecule. In prokaryotes, however, mRNA lifetimes are short, with half-times in the order of minutes. In contrast, eukaryotic mRNA is usually stable for hours or days.

In eukaryotic cells, the synthesis of mRNA and its subsequent translation are more complex. The first transcriptional product is a heterogeneous, very long RNA *(giant messenger-like RNA, mlRNA, heterogeneous nuclear RNA, Hn RNA),* which is synthesized in the nucleoplasm (in contrast to ribosomal mRNA which is synthesized in the nucleolus). HnRNA contains both coding and noncoding sequences (exons and introns, respectively). During posttranscriptional modification or "processing", the introns are excised and the exons are spliced together. The resulting mature mRNA (carrying the necessary genetic information for translation of the gene product) enters the cytoplasm, where it is translated on ribosomes. Processing is so extensive that as little as 10% of the HnRNA enters the cytoplasm as mRNA. Eukaryotic mRNA differs from prokaryotic mRNA in that it usually carries a "cap" of 7-methylguanosine triphosphate linked 5′ to 5′ to the first "normal" nucleotide. The 3′ end of eukaryotic (but not prokaryotic) mRNA also usually carries a nontranslated poly(adenine) sequence (the 3′-poly(A) tail).

Mestranol: a synthetic estrogen used in oral contaceptives; see Hormonal contraception.

Metabasidium: see Promycelium.

Metabolic control: see Metabolic regulation.

Metabolic rate, ***energy turnover:*** the rate of conversion of the physiologically utilizable chemical energy of food material into heat and work, and its utilization for the synthesis of body material.

Living organisms are, of course, subject to the first law of thermodynamics (the law of energy conservation), so that any decrease in the stored energy of an organism is equal to the sum of the energy lost as heat and work.

At the lowest M.r. consistent with the maintenance of life, the organism cannot respond (perform work) immediately to the provision of further energy. This is only possible for cells with a higher M.r., known as the Basal metabolic rate (see).

Direct measurement of the M.r. is very difficult, because the energy content of an organism is extremely high in comparison with its daily energy turnover. All types of activity, and especially physical work, increase the M.r. of the organism. Strenuous physical exertion can increase the human metabolic rate to 10–15 times its basal value. Mental activity increases the value by only 5–10%.

Metabolic regulation, ***metabolic control:*** a system of control mechanisms that underlies the high degree of adaptation displayed by living organisms, and guarantees the efficient and economic operation of living processes. It avoids wastage of energy by compartmentalizing and regulating the operation of all enzyme-catalysed reactions.

The main functions of M.r. can be listed as follows. 1) Each metabolic reaction must be coordinated with the preceding and the following reactions. Concentrations of intermediates in all metabolic pathways are normally maintained at a steady value, which should not greatly increase or decrease, i.e. intermediates must not accumulate. 2) Energy-yielding metabolic reactions (exergonic metabolism) should be balanced by energy-consuming reactions (endergonic metabolism), i.e. the rate of ATP production must be more or less balanced by the rate of ATP utilization. Accordingly, metabolism must adjust to changes in the rate of energy utilization, e.g. in the transition from a resting state to a state of motor activity, and vice versa. 3) According to the metabolic needs of the organism, which may vary, e.g. with environmental temperature, degree of activity, quantity and type of available nutrients, certain metabolic pathways must be accelerated or decelerated.

There are four broad types of control.

1) *Neural (nervous) control.* A type of intercellular regulation. The nerve impulse is an electrical signal, and the regulatory response may also be electrical (e.g. further impulses to a muscle) or chemical (e.g. production of a hormone).

2) *Hormonal (humoral) control.* A type of intercellular regulation. Hormones act as chemical signals in a regulation system that is superimposed on the more basic levels of M.r. Cyclic AMP acts as a second messenger for many hormones.

3) *Differential gene expression.* The signals (or triggers) of differential gene expression may be chemical (hormones) or environmental (e.g. light). Differential gene expression is responsible for the regulation of differentiation and growth at a molecular level.

4) *Feedback and feedforward.* Metabolites themselves act directly as signals in the control of their own breakdown or synthesis. Feedback is negative or positive. *Negative feedback* results in the inhibition of the activity or synthesis of an enzyme or several enzymes in a reaction chain by the endproduct. Inhibition of the synthesis of enzymes is called Enzyme repression (see). Inhibition of the activity of an enzyme by an endproduct is an allosteric effect; this type of feedback control is well known for amino acid biosynthesis in prokaryotic organisms, and is variously known as *endproduct inhibition, feedback inhibition* and *retroinhibition.* In *positive feedback* or *feedback activation,* and endproduct activates an enzyme responsible for its production, e.g. thrombin activates factors VIII and V during blood clotting, thus contributing to the speed of the cascade system and the rapid formation of a clot. An example of *feedforward enzyme activation* is found in the activation of glycogen synthetase by glucose 6-phosphate, i.e. a metabolite activates an enzyme concerned in its utilization. *Enzyme induction* (see) is a positive feedforward process.

The mechanisms discussed under 4) are all intracellular, and they have been studied mainly in prokaryotic organisms.

4a) *Chemical (covalent) modification of enzymes.* This occurs by forming or breaking covalent bonmds. Two mechanisms have been studied in detail: phosphorylation-dephosphorylation by protein kinases and protein phosphatases (e.g. glycogen synthase), and adenylylation-deadenylylation (e.g. glutamine synthetase).

4b) *Physical modification of enzymes* is the basis of the allosteric effect.

M.r. is also achieved by competition of enzymes for common substrates and cosubstrates, cofactor stimulation, regulation of coenzyme synthesis, product inhibition and stoichiometric inhibition by metabolites. An important principle of M.r. in eukaryotes appears to be the regulation of active enzyme concentrations by a change in the turnover number of enzyme proteins. This type of control therefore depends on the control of enzyme proteolysis, e.g. the active concentration of tryptophan synthetase in yeast is controlled by group-specific proteases, called inactivases; further regulation is achieved by an inactivase inhibitor, which is also a protein.

Defects in M.r. may have severe, and sometimes lethal consequences, e.g. diabetes mellitus and cancer.

Metabolism: the totality of chemical reactions within a living organism, which result in the continual turnover (degradation and renewal) of all cell components, and which generate the energy consumed by the organism. In a process of *assimilation,* materials are taken from the environment and converted into organic cell substances. The totality of these assimilatory metabolic processes is known as *anabolism.* Most anabolic (biosynthetic) reactions are endergonic (i.e. they require energy). In green plants, the necessary energy is derived from sunlight (see Photosynthesis), and in other autotrophic organisms it arises from the conversion of inorganic compounds (see Chemosynthesis). In heterotrophic organisms, energy is generated by exergonic reactions during the degradation of organic nutrients or the cell's own organic materials. This part of metabolism is known as *dissimilation,* and the totality of dissimilatory processes is known as catabolism. *Intermediary metabolism* links the energy-generating processes of catabolism to the energy-consuming reactions of anabolism. Since living organisms are open systems, M. consists of numerous linked reactions, which continually strive toward a fixed position of dynamic equilibrium without attaining it, i.e. it is a steady state, rather than an equilibrium.

Metaboly: the metamorphic and/or embryonal phase of development of an organism. Evolutionary changes arise through alterations in M. Thus, certain developmental stages may be omitted, a phenomenon known as *Abbreviation* (see); or extra stages may be added, a phenomenon known as *prolongation. Anaboly* (see below) is a special case of prolongation. An increase in the rate of ontogenetic development, leading to the appearance of characters earlier in the ontogeny of a descendant than in that of an ancestor, is called *acceleration.* The converse, i.e. a decreased rate of ontogenetic development, so that a character appears later in the ontogeny of a descendant than in that of an ancestor, is called *retardation.* Any evolutionary change in the onset or timing of the development of a character, relative to the appearance or rate of development of the same character in the ontogeny of an ancestor, is called *heterochrony.*

According to the stage of ontogenic development affected by the change, the process is referred to as *archallaxis, deviation* or *anaboly.* Archallaxis is the addition of new characters early in embryonic morphogenesis, usually resulting in a major alteration of subsequent development, e.g. the closely related annelid worms, *Tubifex rivulorum* and *Pachydrilus lineatus* display very early developmental differences, which are reflected in their egg structure and pattern of cleavage. The term, deviation, is applied to alterations in the intermediate stages of embryonic morphogenesis, e.g. the banded snails, *Cepaea nemoralis* and *Cepaea hortensis,* differ with respect to the relative sizes of their dart sacs and mucus glands, but the species differences in these organisms do not appear until the intermediate stage of embryonic development. Introduction of an extra developmental stage at the end of the embyonic period of morphogenesis is called *anaboly.* Anaboly (as defined by A.N. Sewertzoff in 1931) is the most important and most frequently effective phylogenetic evolutionary process. Thus, immediately after hatching, young plaice *(Pleuronectes)* resemble other young fish; the typical flat-fish characteristics (both eyes on one side of the body, dorsal fin extended to the head) do not develop until later. As a further example, both baleen whales and toothed whales pass through an embryonic toothed stage, but the baleen whale embryo has evolved an extra toothless stage. New evolutionary characters that first arose in an ancestral juvenile stage may be displaced forward by neoteny until they appear in the adult stage of a descendant, a process known as *proterogenesis.*

Metacestode: see *Cestoda,* Dog tapeworm.

Metacoel: see Somatocoel, Mesoblast.

Metagenesis: a form of homophasic Alternation of generations (see) in multicellular animals. At least two morphologically different generations alternate with one another, one of which reproduces only asexually, and the other usually only sexually. The asexual generation reproduces itself many times by cleavage or budding, whereas the sexual generation forms male and female gametes. Both generations are diploid, and only the gametes of the sexual generation are haploid. For example in many coelenterates, medusae are formed asexually from polyps (see Strobilation), and polyps are then formed by sexual reproduction by the medusae. Thaliaceans (salps) also display M.

Sometimes there are 2 sequential asexual generations, before the recurrence of a sexually reproducing generation.

Metalimnetic zone: see Lake.

Metalimnion: see Lake.

Metallic woodborers: see *Coleoptera.*

Metalloflavin enzymes: see Flavoproteins.

Metalloflavoproteins: see Flavoproteins.

Metalloproteins: proteins containing complexed metals, in particular iron, zinc, manganese or copper. In metalloenzymes, the metal is a functional component, whereas in metal-transporting M. (e.g. blood proteins, transferrin, ceruloplasmin) and metal-storage depot proteins (e.g. ferritin), metal binding is reversible and the metal is a temporary component. Important iron-containing M. are the cytochromes, respiratory pigments (e.g. hemoglobin) and enzymes (e.g. catalase). Zinc is present in insulin and in carbonic anhydrase. Some glycolytic enzymes and proteases contain manganese. Copper is present in certain oxidoreductases (e.g. tyrosinase, cytochrome a_3). Catalytic M. also include enzymes that require specific cations for activity, e.g. sodium-dependent membrane ATPases. Molybdenum (Mo) is an essential component of animal and bacterial xanthine oxidase and animal aldehyde oxidase, and it is particularly associated with enzymes of inorganic nitrogen metabolism. Thus Mo functions as a redox component of the nitrate reductase of green plants and the nitrogenase of nitrogen-fixing root nodule bacteria.

Metamerism: division of the animal body into a linear series of segments or metameres. In *homonomeric M.* (the primitive condition) the metameres are the same, whereas in *heteronomeric M.* (a derived condition) the metameres differ in size and/or internal structure. Annelids are an example of homonomeric M., while all insects display heteronomeric M. In insects and other arthropods, the head, thorax and abdomen represent groups of metameres with different sizes, shapes and functions. Such groups of metameres are known as *tagmata* (sing. *tagma).* The existence of functionally different metameres is called *heteronomy.*

M. is primarily a division of muscles and coelom (i.e.

division of the mesoderm). This in turn imposes segmentation on the nervous system, vascular system and excretory organs. M. may be externally visible (insects, annelids, etc.), but in more advanced forms (e.g. vertebrates) M. is recognizable only during embryonic development. M., with each metamere carrying appendages, probably evolved as an adaptation to efficient locomotion.

Metamorphosis (Greek *meta* after, *morphe* form): any abrupt change of form or shape. In this sense, the term is applied widely, e.g. a phylogenetic change in the form of a plant organ as an adaptation to environental conditions. It is, however, more strictly applied to the development of an animal via one or more larval stages, involving changes of size and major alterations of form, e.g. in insects and amphibians. See Metaboly.

Metanephridia: see Nephridia.

Metanephros: see Excretory organs.

Metaphase: see Meiosis, Mitosis.

Metasaprophytic conditions: see Saprophytic system.

Metasequoia: see *Taxodiaceae.*

Metasoma: the hindmost and by far the longest body section of the *Tentaculata, Hemichordata* and *Pogonophora.* It contains a paired metacoel.

Metatetranychus: see *Tetranychidae.*

Metatheria, *Didelphia:* an infraclass of the mammals (subclass *Theria),* represented by the single living order *Marsupialia* (marsupials), together with their extinct relatives. D. have a paired vagina and, with the exception of the *Peramelidae,* they do not develop a placenta. The young are born in an incompletely developed state and continue their development in the marsupium. See *Eutheria, Prototheria, Theria.*

Metazoa, ***metazoans, multicellular animals:*** animals consisting of numerous cells, which are differentiated at least into somatic and reproductive cells. M. always start life as a single, fertilized egg cell, and development proceeds via an embryo (unless reproduction is asexual). M. are often divided into three groups (divisions): *Mesozoa* (all multicellular animals not included in the other two groups), *Parazoa* (sponges), and *Eumetazoa* (hydroids, jelly-fishes, anemones, corals, etc). It is more satisfactory, however, to arrange all metazoan phyla according to their levels of organization.

Methacyclin: see Tetracyclins.

Methamphetamine, ***Pervitine®:*** an Antidepressant (see) with strong excitatory action on the central nervous system. Since it causes a state of euphoria and is addictive, it is also classed as a narcotic.

Methane bacteria: a small group of strictly anaerobic, rod-shaped or round, non-spore-forming, Gram-positive or -negative bacteria, which obtain energy by using CO_2 as an electron acceptor for anaerobic respiration. The CO_2 is thereby reduced to methane (CH_4). All M.b. can obtain sufficient energy for growth by coupling the reduction of CO_2 with the simultaneous oxidation of molecular hydrogen. Some can also utilize one-carbon organic compounds such as methanol or formic acid, and some can utilize acetic acid. In mixed cultures with other microorganisms, M.b. can be used for the preparation of biogas. M.b. occur in the bottom mud of stagnant waters and in sewage. *Methanobacterium ruminantium* occurs in the rumen of cattle, where it is responsible for the large quantities of methane generated by these animals.

Methanol, ***methyl alcohol:*** CH_3OH, the simplest primary alcohol. It occurs as the free alcohol in some plants and grasses, but it is very widely distributed in combined form. Thus, methyl esters are present in most essential oils, and methyl ethers occur in numerous plant materials. Dry distillation of wood produces wood acetic acid containing up to 3% M., which arises from the methoxy groups of lignin. M. is a poisonous, colorless liquid, and its consumption leads to blindness and death.

Methemoglobin: see Hemoglobin.

L-Methionine, ***Met:*** $H_3C—S—CH_2—CH_2—CH(NH_2)—COOH$, a sulfur-containing, proteogenic, essential amino acid, which is biosynthesized from L-homoserine, L-cysteine and 5-methyltetrahydrofolic acid. Met serves as a methyl donor in transmethylation reactions. The biological value of many plant proteins is limited by their Met content.

Methyl alcohol: see Methanol.

Methyl red test: a qualitative test for the production of acid from sugars by certain bacteria. Thus, when grown in phosphate-buffered glucose peptone, *Escherichia coli* produces a pH of about 5, so that the medium has a red color after addition of methyl red. In contrast, growth of *Enterobacter aerogenes* does not cause the pH to fall so low, and the medium is yellow after addition of methyl red.

Metopidius: see *Jacanidae.*

Metridium: see *Actiniaria.*

Metzgeriopsis: see *Hepaticae.*

Mevalonic acid: an important biosynthetic intermediate, formed from 3 molecules of acetyl-CoA. By subsequent phosphorylation, decarboxylation and dehydration, it is converted into isopentenyl pyrophosphate ("active isoprene"), the biosynthetic precursor of all terpenes, e.g. camphor, abscisic acid, gibberellins, carotenoids, steroids, caoutchouc, polyprenols, etc.

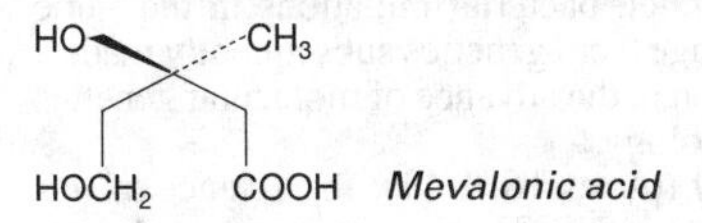

Mevalonic acid

Mexican blind lizard, *Anelytropsis papillosus:* a legless. worm-like, blind, burrowing lizard, reported from Mexico. It has been observed only a few times, and its mode of life is imperfectly understood. Specimens up to 25 cm in length have been reported. The sole living representative of the family *Anelytropsidae,* it appears to be phylogenetically related to the skinks *(Scincidae),* and is thought by some authorities to have evolved from a skink-like ancestor.

Mexican burrowing toad: see *Rinophrynidae.*

Mexican musk turtles: see *Staurotypidae.*

Mezcal: see *Agavaceae.*

Mezereon: see *Thymelaeaceae.*

Mice: see *Muridae.*

Michaelmas daisy: see *Compositae.*

Micraster: see *Echinoidea.*

Micrastur: see *Falconiformes.*

Micrathene: see *Strigiformes.*

Microaerophilic microorganisms: microorganisms that grow optimally in an oxygen concentration lower than that of the atmosphere (often in the presence of only a trace of free oxygen) and often prefer an increased concentration of carbon dioxide. Thus, if deep tubes of nutrient agar are inoculated with bacteria of various oxygen relationships, the microaerophilic bacteria (e.g. *Brucella abortus)* grow in a narrow band of tolerance some distance below the surface, between the upper, well oxygenated zone and the lower anaerobic zone.

Microarthopods: a term used in soil biology for small arthropods that inhabit the cavities between soil particles. In particular, they include mites, springtails and other apterygote insects, and small myriapods.

Microbates: see *Sylviinae.*

Microbes: see Microorganisms.

Microbial genetics: a branch of genetics concerned with the transmission of hereditary characters in microorganisms, in particular bacteria, viruses (notably bacteriophages) and fungi. Hereditary processes in microorganisms are analogous to those in higher organisms. Their genetic material is DNA (with the exception of RNA viruses), which undergoes self-replication, and their genes are arranged linearly. They can be crossed and they display recombination. Mutation occurs spontaneously, and the rate of mutation can be increased with mutagenic agents. For genetic studies, microorganisms offer the advantages that they have a short mean generation time, they are easily cultured in a small space under easily controlled conditions, and they have a relatively uncomplicated structure.

The first attempts to use lower life forms for genetic studies were made in the USA shortly before the second world war, when Beadle and Tatum employed the fungus, *Neurospora,* to investigate the genetics of tryptophan metabolism and nicotinic acid synthesis. This work led to the development of the "one gene one enzyme" hypothesis. Bacteria were not considered suitable material for genetic analysis, until 1947, when Lederberg and Tatum demonstrated the exchange of genetic factors in *Escherichia coli* K12, and Braun and Kaplan demonstrated bacterial mutations in the same year. Bacteriophage (see) genetics subsequently made a major contribution to the advance of molecular genetics and molecular biology.

Microbiology (plates 16 & 17): the science of microorganisms and viruses. These organisms are characterized by their small size, their structure, their unmeasurably large number of individuals, their general distribution in all ecosystems and environments, and their special and indispensable role in the natural cycling of nutrients. On account of their microscopic size, special laboratory and field methods have been developed for the isolation, differentiation and investigation of microorganisms and viruses, e.g. methods for the establishment and maintenance of cultures in liquid or on semisolid nutrient media, staining methods, microscopic and electron-microscopic visualization methods, serological typing, etc. Since microorganisms occur practically everywhere, microbiological studies must always be conducted under sterile conditions, so that methods for achieving sterile conditions are crucial to M.

In accordance with the group of organisms under investigation, M. is subdivided into virology (viruses), Bacteriology (bacteria) (see), Mycology (fungi) (see), algology (algae) and protozoology (protozoa). Algology is also a branch of botany, and protozoology a branch of zoology.

General M. investigates morphology, cytology, physiology, genetics, biochemistry, taxonomy, classification, phylogeny and ecology. *Applied M.* is divided into the following areas. 1) *Medical M.* is concerned with disease-causing micoorganisms and viruses, their distinguishing properties, their action on the human organism, methods for their diagnosis, and methods for combatting or controlling them as agents of disease. 2) *Veterinary medical M.* has much in common with medical M. It is concerned with disease agents in animals, especially domestic animals. 3) *Technological* M. (see). 4) *Agricultural M.* studies soil microorganisms (see Soil microbiology) and plant disease agents (see Phytopathology). It also includes the study of microorganisms used in the preparation of silage.

The beginnings of M. are to be found in the invention of the microscope in the 17th century. Its promotion to an independent discipline was due largely to the work of L. Pasteur, R. Koch and others from the middle of the 19th to the beginning of the 20th century. In this period, the essential methodology of the subject was developed (plating on semi-solid media, preparation of pure cultures, staining for microscopic investigation, sterilization procedures), important aspects of microbial physiology were studied (fermentation, anaerobiosis, symbiosis, nitrogen fixation), and the causative agents of various infectious diseases were discovered (leprosy, anthrax, malaria, tuberculosis, cholera, and others). In addition, the first successful measures against infection were reported (active and passive immunization, chemotherapy). Later important stages in the development of M. were, e.g. the discovery of viruses and antibiotics, the introduction of electron microscopy, the rapid growth of biochemistry with its methods for studying metabolism, and in particular the rapid growth of molecular genetics which was at first based largely on microbial genetics.

Microbiotheriidae: see Opossums.

Microbodies: see Peroxisomes.

Microcavia: see *Caviidae.*

Microcerberidae: see *Isopoda.*

Microcirculation: flow of blood through capillaries. In skeletal muscle (Fig.) arterioles branch to become metarterioles, and the arterioles are also in direct communication with venules via transverse anastomoses. Some metarterioles become venules, while others divide and become capillaries (internal diamter 5–9 μ). Walls of arterioles, metarterioles and venules are muscular. Capillary walls lack muscle tissue, but the site at which a capillary branches from an arteriole or metarteriole is provided with a circular muscle (precapillary sphincter). Venules receive blood from both metarterioles and capillaries.

Blood normally flows via arterioles and capillaries into venules. If the precapillary sphincters contract, blood flows via metarterioles into venules. If the metarterioles are also closed, blood passes via anastomoses directly from arterioles into venules.

The regulatory mechanisms of the M. are clearly defined. The internal diameter of arterioles is controlled by their sympathetic nerve supply. Precapillary sphincters are controlled by metabolic products. Thus, in contracting muscle, the low oxygen tension causes precapillary sphincters to relax, and the blood flow through the capillary bed is increased 20–50-fold, compared with that in resting muscle.

The capillary wall is about 0.5 μ thick, consisting of a unicellular layer of endothelial cells with a thin outer basement membrane of collagen and proteoglycan. The effective pore diameter of the capillary wall is about 8 nm, i.e. sufficient to allow the free passage of all low molecular mass components of blood and extracellular fluid without generating an osmotic pressure. Proteins of blood and extracellular fluid are retained by a pore diameter of 8 nm, and are therefore the only components capable of generating an osmotic pressure; this pressure is generally referred to as colloidal osmotic pressure or oncotic pressure.

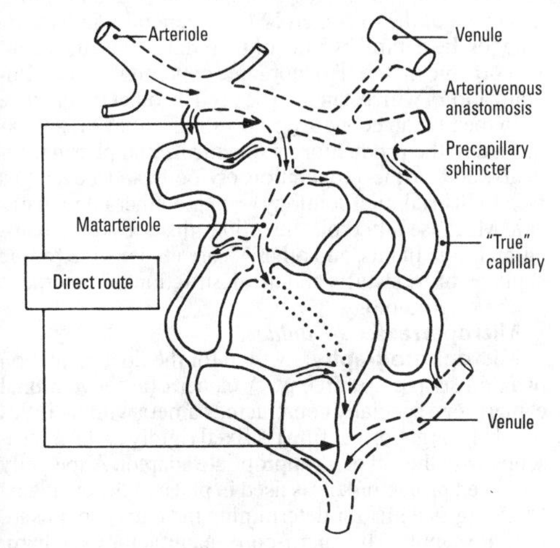

Microcirculation. Diagrammatic representation of a capillary bed with associated arterioles and venules.

Blood pressure tends to move fluid out of the capillary (into the interstitial fluid), whereas the colloidal osmotic pressure tends to move fluid (from the interstitial fluid) into the capillary. The resulting balance between the two forces insures that the blood volume remains constant. Where blood pressure is high at the arterial commencement of the capillary, the blood pressure drives the filtration of blood plasma into the interstitial fluid, but as pressure falls along the length of the capillary, the colloidal osmotic pressure becomes dominant and fluid is resorbed. In this way, dissolved oxygen and nutrients are transported to the cells of the organ, and waste metabolic products are transferred from the interstitial fluid to the blood. Blood pressure at the beginning of a capillary is about 4 kPa, and it approaches 1.3 kPa at the venous end, i.e. an average pressure of about 2.3 kPa. It has been shown that interstitial fluid is under a slight negative pressure of about –0.8 kPa, so that the effective average blood pressure in the capillary is about 3 kPa. The colloidal osmotic pressure of blood plasma is about 3.7 kPa, while that of the interstitial fluid is about 0.7 kPa, leaving an effective balancing osmotic pressure of 3 kPa.

Micrococci, *Micrococcaceae:* in the wider sense a family of spherical, nonmotile, Gram-positive Cocci (see) that form irregular or packet-shaped arrangements of cells. They include the Staphylococci (see) and the genus *Micrococcus* (M. in the restricted sense). Members of the genus *Micrococcus* are facultative aerobes, up to 3.5 μm in diameter, and widely distributed in milk, soil, air, dust and fresh water. They are especially common in putrifaction processes, and they are common contaminants of laboratory cultures.

Microcoryphia: see *Archaeognatha.*

Microelectrode: a glass capillary filled with salt solution, used for measuring the electrical potential difference across the membrane of a single cell. The tip of a M. is drawn out to a diameter of as little as 0.1 μm. and can therefore be inserted into a cell with the aid of a micromanipulator without causing significant damage. The membrane potential can then be measured with the aid of a reference electrode and a voltmeter with a high internal resistance.

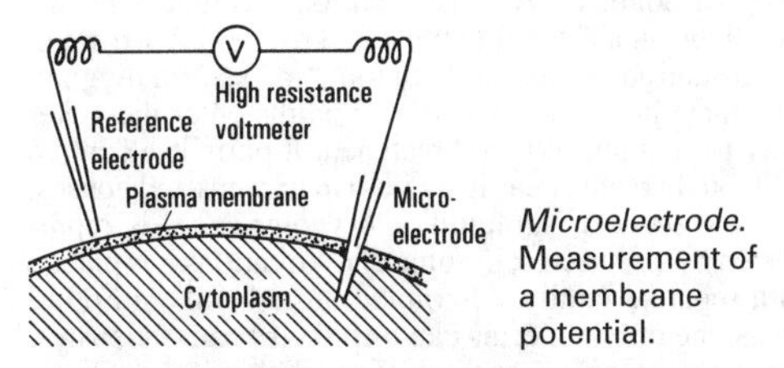

Microelectrode. Measurement of a membrane potential.

Microevolution, ***intraspecific evolution:*** evolutionary differentiation and divergence within species and up to the status of species. To a certain extent M. can be observed, and it is accessible to genetic experimentation. It also forms the basis for the production of higher systematic units.

Microfauna: see Soil organisms.

Microgamete: see Reproduction.

Microglial cells: see Nervous tissue.

Microhylidae, ***microhylids, narrow-mouthed frogs***: a highly specialized family of the *Anura* (see) containing 279 living species in 61 genera and 9 subfamilies, distributed in North, South and Central Ameria, sub-Saharan Africa, Madagascar, southern and eastern Asia, and the northern tip of Queensland. Body shape varies widely among the microhylids and all intermediate forms occur from a conventional, slender frog-like shape to practically spherical frogs with very short limbs. The snout is noticeably small and often pointed. Some species are adapted to an arboreal existence, while many burrow in the ground. In aquatic tadpoles the spiracle is posterior and midventral (type II tadpoles), except in *Otophryne* (subfamily *Microhylinae)* where the spiracle forms a sinistral elongated tube. Some species show direct development, i.e. they lack an aquatic larval stage.

Eight holochordal, procoelous presacral vertebrae are present with imbricate (overlapping) or nonimbricate neural arches. In the cophylins and genyophrynins, all vertebrae are procoelous, but in all other subfamilies presacral VIII is biconcave and the sacrum is biconvex. Presacrals I and II are not fused, and the atlantal cotyles of presacral I are widely separated. The coccyx lacks transverse processes, and has a bicondylar articulation with the sacrum. Most taxa lack an omosternum. A cartilaginous sternum is present, and the pectoral girdle is firmisternal. Ribs are absent. Maxillary and premaxillary teeth are usually absent, but they are present in dyscophins and some cophylins.

Members of the genus *Microhyla* (subfamily *Microhylinae*) are known as ***Answering frogs.*** *Microhyla carolinensis* ***(Carolina narrow-mouthed frog)*** is a 3 cm-long species native in southern North America. During the day it burrows into the earth. The spawn is deposited in water, and the tadpoles have suction pads on the underside of the head.

Breviceps (subfamily *Breviceptinae)* are known as ***Rain frogs.*** They are probably more independent of free water than any other anurans. The South African species, *Breviceps gibbosus,* has a round body and very short snout, and it leaves its burrow only after heavy rain. Egg laying and larval development occur in earth burrows in the absence of water.

Kaloula (subfamily *Microhylinae)* are also sometimes called rain frogs. *Kaloula pulchra* ***(Asiatic painted frog)*** of Southeast Asia is 8 cm long with dark brown markings on a distinct ochre background. It burrows in the ground and possesses hard foot "spades" for digging. The male has a loud voice, and adults enter the water only for mating. When threatened, it puffs itself into a balloon. In many areas it is found in gardens and houses.

Phrynomerus (subfamily *Phrynomerinae)* is represented by six species; in some classifications it is placed in a separate family, *Phrynomeridae*. Unlike all other frogs, the phrynomerins can turn their heads sideways, hence their German name of Wendehalsfrösche ("Wryneck frogs"). They possess adhesive pads, but are not often found in trees, except the crowns of banana trees, more commonly inhabiting termite hills, cavities in rocks and rotting tree stumps. They dig in the ground for termites and ants, and only enter water for breeding. *Phrynomerus bifasciatus* ("Two-striped Wryneck frog") has 2 bright red longitudinal stripes on the back, with further red blotches on limbs and rump on an otherwise dark background. It digs itself into the ground to escape danger.

Microlepidoptera: a nonscientific term adopted by collectors for the smaller and smallest species of *Lepidoptera.* In the classification list under the entry *Lepidoptera* (see), the M. consist of the families *Micropterigidae* through *Pyralidae,* but excluding the *Cossidae, Zygaenidae* and *Aegeriidae.* These latter three families, as well as the *Hepialidae,* are included with the Macrolepidoptera (see). See also *Bombyces.*

Micromanipulator: a device for the surgery of individual cells with microscopic monitoring. The M. consists of a high performance light microscope with adjustable holders on each side, which carry tools such as needles, hooks, injection syringes and electrodes. The tools are moved with the aid of fine levers and screws, similar to those used for the movement of microscopic specimens. With the aid of Ms., cells can be divided, nuclei removed and transplanted, and potentials applied with microelectrodes.

Micromere: see Mesoblast.

Micro-moths: see Moths.

Microorganisms, ***microbes:*** the smallest, mostly unicellular life forms, usually observable only with a microscope. According to their cell structure, M. are divided into two main groups: *eukaryotic* and *prokaryotic* M. The cell structure of eukaryotic M. (algae, fungi and protozoa) is essentially similar to that of higher plants and animals. Prokaryotic M. (see Prokaryotes) are represented by bacteria and cyanobacteria. Since viruses are not cells, they are not classified as M.

M. have a wide natural distribution, occurring in soil, water, air, and in or on other organisms. A wide variety of metabolic conversions and physiological processes are carried out by M. Most M. are *heterotrophic,* living saprophytically at the expense of dead organic material. Other M. (algae, cyanobacteria and photosynthetic bacteria) are *autotrophic,* exploiting light energy to assimilate CO_2 (see Photosynthesis), while other autotrophic bacteria derive energy from the oxidation of inorganic compounds (see Chemosynthesis). Various M. live *anaerobically* (i.e. in the absence of oxygen) by performing fermentation or anaerobic respiration (see Anaerobe, Fermentation). Certain M. fix molecular nitrogen, thereby increasing the availability of nitrogen for plant growth, and playing an important role in the nitrogen cycle of the biosphere. M. are generally indispensable for the natural cycling of materials, since they convert organic material to inorganic compounds (see Mineralization), which can then be used by other organisms.

In the human economy, many different of M. are exploited in the production of chemicals and pharmaceuticals, and in the preparation of foods and beverages (see Industrial microbiology). On the other hand, parasitic M. cause numerous infectious diseases of humans, animals and plants, and other M. are responsible for the spoilage of foodstuffs and the destruction of materials.

See Microbiology.

Microparra: see *Jacanidae.*

Microphotography: method for the documentation of microscopic specimens. A camera (a conventional camera, or a specially constructed camera with facilities for using large format film) is fixed rigidly to the microscope with the aid of an appropriate adaptor. A specially corrected projector lens is used in place of the eyepiece. Metering systems for determining the correct exposure time are supplied by microscope manufacturers. A hard, fine grain film is usually used. To increase the contrast of certain colors, filters are placed in the light path of the illumination system, e.g. red can be accentuated by using a green filter. Color film must be chosen according to the color temperature of the light source. See Microscope.

Microplankton: see Plankton.

Microprobe analysis, ***X-ray microprobe analysis:*** a method for the quantitative determination of elements in biological material with the aid of the scanning or transmission electron microscope. Electrons excite the elements in the biological material, causing the emission of X-rays, whose wavelength and energy are diagnostic of the element that produces them. In *wavelength dispersive M.a.,* the X-rays are separated into different wavelengths by special crystals, and each peak is assigned to an element. In *energy dispersive M.a.,* the X-rays are separated according to their energies with the aid of semiconductors, then stored in a multichannel analyser. The data are processed and corrected in a computer and can be displayed on a screen or printed out. Both methods provide information on the concentrations of elements in tissues, cells and organelles.

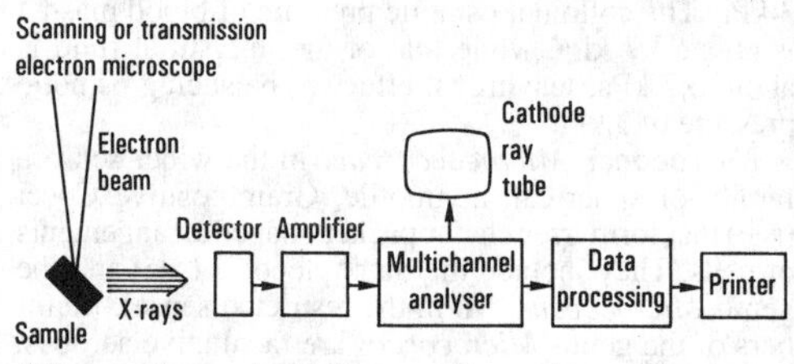

Microprobe analysis. Energy dispersive method (schematic).

Micropsittinae: see *Psittaciformes.*

Micropterigidae (Micropterigids): see *Lepidoptera.*

Micropterus: see Sunfishes.

Micropyle: see Flower (section on ovule), Fertilization, Egg, Seed.

Microscleric shells: see *Foraminiferida.*

Microscope: an optical instrument for enlarging the image of small objects. In principle, the image is generated by enlarging the visual angle, which increases the Resolving power (see). In the light (optical) microscope, the resolving power is limited by the wavelength of the light used for illuminating the specimen, and is about 0.2 µm.

The objective and eyepiece (ocular) are housed and kept in alignment within a metal tube. This entire lens system is mounted on a stand, which also carries an illuminator, a stage for the specimen, and a mechanical device for moving and focusing the specimen.

The illuminator consists of an adjustable light source, iris diaphragm (field stop) and condenser. In the absence of an integral light source, a movable mirror is used to direct light from an external source into the condenser. Various light sources are used, such as low voltage lamps, halogen lamps, and for special applications mercury lamps and xenon lamps. In critical illumination, the light source (e.g. sun, wick of an oil lamp, filament of a bulb) is focused to appear in the field of view, with the major disadvantage that illumination is uneven. This is overcome by a system introduced by Köhler (Köhler illumination), which focuses the light source sharply on the condenser diaphragm, so that the specimen is transilluminated by a pencil beam of parallel light; the light source therefore lies at virtual infinity with respect to the specimen, and inhomogeneities in the light source have no noticeable effect.

Eyepiece (ocular)
Intermediate real image
Objective
Specimen
Condenser
Iris diaphragm (field stop)
Iris diaphragm
Collector lens
Light source

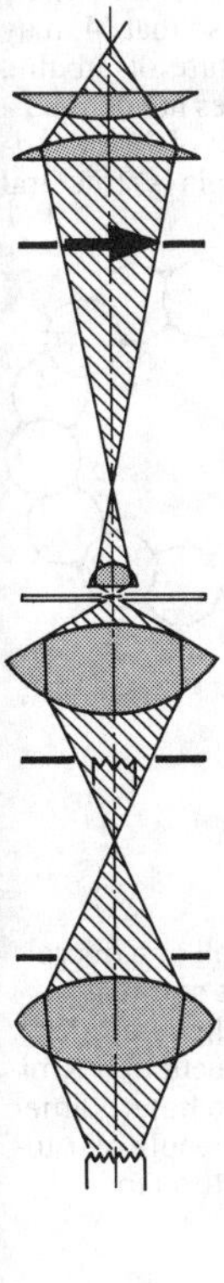

Microscope. Light path in a light microscope (schematic).

The stage carries the specimen, and its distance from the objective is adjustable. For routine microscopy, the alignment of the specimen is adjustable with the aid of vernier screws, so that the observer can return to any chosen point on the focused specimen. In modern Ms. the prepared specimen is housed in a movable casette, which overcomes the inconvenience of refocusing for different thicknesses of glass slides.

The objective generates a real, enlarged, reversed image in the lower focal plane of the eyepiece. Any fault in image generation by the objective affects the quality of the final image. Different correction objectives are manufactured for different applications. For visual observation, achromatic objectives are used, which are corrected for the yellow-green region of the spectrum. Apochromatic objectives are corrected for blue, green and red, and are suitable for microphotography with panchromatic film and color film. For the highest quality microphotography, planar objectives are used, which correct for the curvature of the field of the image.

The eyepiece serves as a magnification system for visual obervation of the intermediate real image generated by the objective. In microphotography the eyepiece functions as a projector lens, projecting an enlarged, reversed, real image onto the film.

Microsome fraction: see Endoplasmic reticulum, Membrane.

Microspectrophotometry: performance of spectrophotometric measurements on microscopic samples. The method is used for the quantitative determination of nucleic acids, proteins, chlorophyll, hemoglobin, etc. in individual cells or organelles. Substances that possess no specific absorption maxima are converted chemically into (usually) colored derivatives prior to M. A microscope is used in place of the conventional absorption cell, and the monochromatized sample and reference beams pass through separate microscopes. The region of the specimen to be measured is isolated as a magnified microscopic image. M. is also used for scanning a microscopic sample at any given wavelength, by moving the magnified area of measurement backward and forward over the specimen. Scattered light is a major source of error, which is obviated by computer processing of the signal.

Microsporangium: a sporangium containing microspores. In flowering plants the M. corresponds to the pollen sac.

Microspore: the smaller of the two types of spore formed by heterosporous ferns, or the pollen grain of seed-forming plants. An M. germinates to form a male gametophyte.

Microsporidia: see *Cnidosporidia.*

Microsporogenesis: pollen formation in the pollen sacs (microsporangia) of angiosperms. Within the two pollen sacs of an anther, repeated division of the archesporium cells leads to the formation of a sporogenic tissue, whose cells separate, become rounded and are transformed into numerous *pollen mother cells* (Fig.). By meiosis, followed by cell wall formation, each pollen mother cell gives rise to 4 uninucleate, haploid pollen grains (meiospores). These may remain attached to each other, and are then known as *pollen tetrads.* As a rule, however, they separate to form individual pollen grains. Each pollen grain then undergoes mitosis to

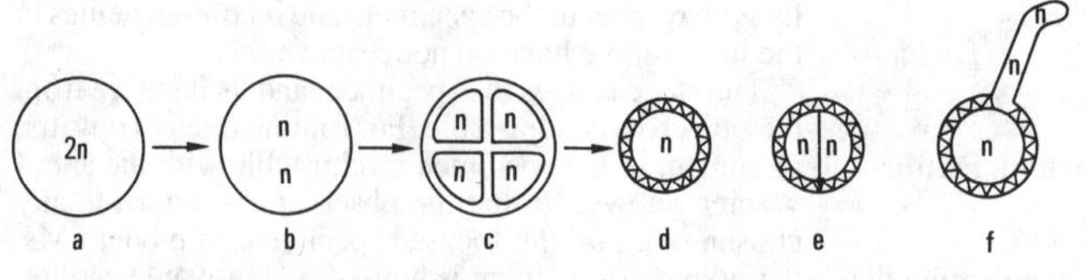

Microsporogenesis. a Diploid pollen mother cell. *b* First meiotic division. *c* Second meiotic division. *d* Pollen grain. *e* First pollen grain mitosis. *f* Germinating pollen grain with the vegetative nucleus and both generative nuclei.

form a small haploid *generative cell (antheridial cell),* and a large haploid *vegetative cell (pollen tube cell).* The generative cell divides mitotically into two male gametes, one of which normally fertilizes the egg cell, while the other combines with the diploid secondary embryo sac nucleus, to form the triploid endosperm tissue. See Macrosporogenesis.

Microsporophyll: see Flower.

Microstomus: see *Pleuronectiformes.*

Microtine rodents: see *Arvicolinae.*

Microtinae: see *Arvicolinae.*

Microtini: see *Arvicolinae.*

Microtome: apparatus for preparing thin sections of embedded material for light and electron microscopy. In the early *rotary M.,* the specimen was mounted on the periphery of a wheel, and sections were cut by passing the specimen across a steel knife, the knife being advanced a chosen distance before the next section. In a later development of the instrument (C.S. Minot, 1887), a fine screw was used to advance the specimen and control the section thickness, and the specimen was moved vertically. Further refinements produced the modern rotary M., which combines a fine feed screw drive and an inclined plane for the control of section thickness. The difficulty of performing a slicing cut with the rotary M. led to development of the *sliding M.* The latter allows the knife to be set for the best slicing stroke, and it is suitable for delicate and less easily sectioned material. The angle of the knife edge is chosed according to the type of embedding material and hardness of the specimen. Sections between 5 and 20 µm thick can be obtained from tissues embedded in paraffin wax. Much thinner sections (0.5 to 2 µm) can be prepared from specimens embedded in resin. Most Ms. are provided with a freezing platform and an arrangement for cooling the knife, so that sections can be made from frozen specimens. Much thinner sections (50–70 nm) are required for electron microscopy. These are prepared from material embedded in epoxy or polyester resin, using an *ultramicrotome* with a glass or diamond knife. In this case, the embedded specimen is passed across the rigidly mounted knife. The specimen is advanced by the thermal expansion of a cooled specimen holder, or by controlled heating. Other methods for controlling section thickness in ultramicrotomes are based on Piezoelectric, magnetoconstrictive and torsional expansion, as well as hydraulic systems. Ultramicrotomes are also adaptable for the sectioning of frozen specimens

Microtubule associated protein: see Microtubules.

Microtubule organizing center: see Microtubules, Spindle.

Microtubules: tubular structures present in all eukaryotic cells, consisting mainly of a protein called tubulin. Individual M. are too small for resolution by the light microscope, but they are observable under the electron microscope. M. are important components of the cytoskeleton, regulating cell shape and providing a system of fibers along which membrane-bound organelles can travel. As components of the mitotic spindle they are important in nuclear division. By forming cross bridges with the plasma membrane they stiffen protoplasmic processes and invaginations. They are not themselves contractile and do not have ATPase activity, but in cilia and flagella they are moved by *dynein,* an associated protein that does possess ATPase activity. Delayed GTP hydrolysis is, however, associated with the assembly of M. (see below).

Structure. The structure of M. was elucidated in 1963 by Slautterback, Ledbeter and Porter, by the electron microscopic obervation of glutaraldehyde-fixed cell preparations. Since that time, M. have been intensively studied as structural and functional elements of the Spindle apparatus (see), Cilia (see) and flagella, the Cytoskeleton (see) and other cell structures. Each microtubule is a polymer of tubulin, which is itself a heterodimer of two tightly linked (by hydrogen bonding), closely related globular subunits, known as α-tubulin and β-tubulin. Both α- and β-tubulin have M_r values of about 55 000, and the M_r of the dimer is about 110 000. The dimers are arranged linearly in protofilaments, 13 parallel protofilaments forming a tube of diameter 15–25 nm. Each protofilament is out of register with its neighbor, so that neighboring dimers are arranged in a spiral or helix (Fig.). By virtue of their specific surface structure and charge, the dimers combine to form M. by a process of cooperative self assembly, so that M. may also be regarded as a quaternary structure of tubulin. This polymer or quaternary structure does not contain a fixed number of subunits, i.e. the length of M. is variable. By definition, any heterodimer displays structural

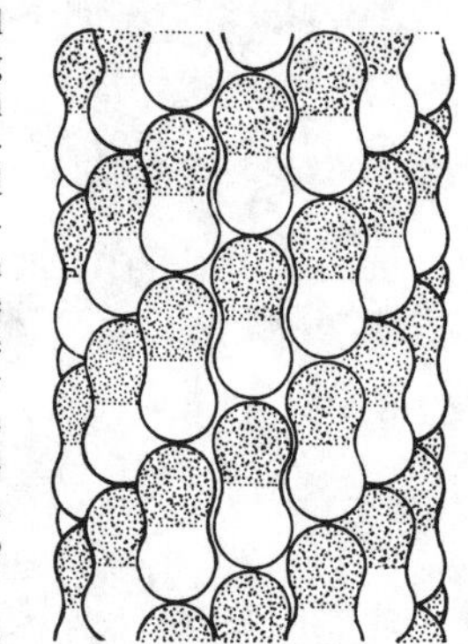

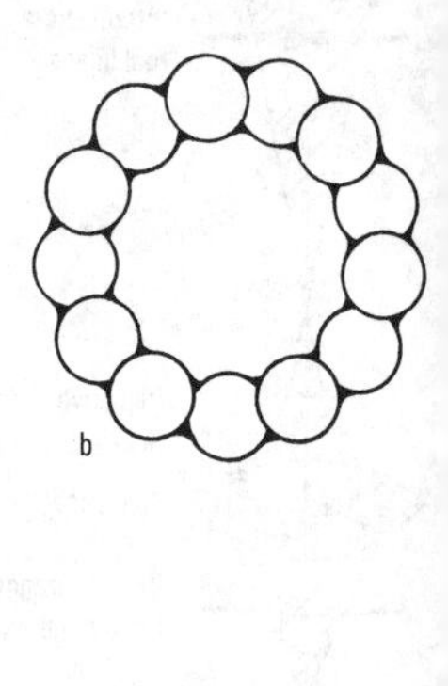

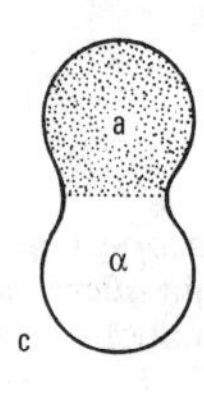

Microtubules. a Wall of a microtubule showing the spiral arrangement of tubule heterodimers. *b* Transverse section of a microtubule. *c* Tubulin heterodimer containing one molecule of α-tubulin and one of β-tubulin.

polarity. Since all the tubulin subunits of M. have the same orientation, M. also have an inherent polarity. By allowing purified tubulin molecules to polymerize onto fragments of M. in vitro, it has been shown that one end (the *plus* end) grows at about 3 times the rate of the other (the *minus* end). Under special in vitro conditions, when free tubulin is added to M. it does not associate with the ends of the M., but forms short, curved protofilaments on the sides of the M. In cross section, these protofilaments are curved in a clockwise or counterclockwise direction, depending on the polarity of the M.

The cell contains a pool of tubulin heterodimers which is in equilibrium with polymerized tubulin (i.e. with M.). In cilia and flagella, however, tubulin is present only in the polymerized form. In the presence of GTP or D_2O, the equilibrium is shifted toward polymerization, whereas colchicine, low temperature, Ca^{2+}, or increased hydrostatic pressure favor depolymerization. Since the M. of the mitotic spindle are in a state of rapid polymerization and depolymerization (assembly and disassembly), they are particularly sensitive to drugs (*antimitotic drugs)* that bind to tubulin. Most familiar of these is *colchicine,* which inhibits the addition of tubulin molecules to M., especially in the mitotic spindle, causing M. to dissociate into their heterodimers; the mitotic spindle disappears, mitosis is blocked, and the chromatids cannot be distributed between the daughter cells. *Colcemid* and *nocadazole* act similarly to colchicine. Other antimitotic drugs are *vinblastine* and *vincristine,* which induce the formation of paracrystalline aggregates of tubulin, also causing M. to dissociate. Since disruption of the mitotic spindle kills rapidly dividing cells, vinblastine and vincristine drugs are used in cancer therapy. In contrast, the antimitotic drug, *taxol,* binds tightly to microtubules and stabilizes them, as well as causing unpolymerized tubulin to form M. In dividing cells, however, the effect is similar, i.e. mitosis is arrested. Since the length, number and distribution of M. change during the cell cycle, mechanisms must operate for the initiation of M. formation, assembly and disassembly. In vitro studies, using biochemical techniques and electron microscopy, have revealed the existence of intermediate structures, e.g. rings of tubulin molecules. Various proteins associate with M., stabilize them against disassembly and mediate their interaction with other cell components. These proteins are collectively known as MAPs (microtubule associated proteins). Each MAP has two domains, one binding to M., the other available for binding to other cell components. Since the tubulin-binding domain associates with several tubulin molecules simultaneously, MAPs also serve as signal molecules for the initiation of polymerization. For the interaction of M. with the proteins, *dynein* and *nexin,* see Cilia and Flagella. Immunofluorescence microscopic studies have shown that during interphase the M. in animal cells exist as single filaments, concentrated around the nucleus, and radiating outward through the cytoplasm. At the beginning of mitosis they become concentrated between the spindle poles and form the mitotic spindle, and they are then absent from the rest of the cytoplasm.

Each tubulin molecule binds a molecule of GTP, which undergoes delayed hydrolysis after the tubulin has been incorporated into the growing M. It should be stressed that the energy of GTP hydrolysis is not required for polymerization. Rather, the plus ends of growing M. are "capped" with GTP-bound tubulin molecules, which depolymerize about 100 times more slowly than tubulin molecules lacking GTP. As long as the rate of elongation is sufficiently rapid, and since the GTP hydrolysis is delayed, the growing end of the polymer is protected against disassembly. If polymerization slows down, i.e. GTP is hydrolysed before addition of subsequent tubulin dimers, then depolymerization may become more rapid than polymerization. For this reason, M. are said to be dynamically unstable.

Since M. are dynamically unstable, they can only persist as effective elements of cell reinforcement or fibers for intracellular transport, etc. if both ends are prevented from disassociating. The minus ends are normally protected within the organizing centers from which the M. grow. It is thought that the plus ends are stabilized by capping by specific proteins. However, M. are also stabilized by a maturation process, so that they become relatively permanent structures of established tissue. Maturation occurs through interaction of M. with MAPs, and by modification of tubulin molecules after their incorporation into M., e.g. tubulin acetyl transferase catalyses the acetylation of a specific lysine residue of α-tubulin, and a detyrosinating enzyme removes the carboxyl-terminal tyrosine residue of α-tubulin. Acetylated or detyrosinated M. do not appear to be inherently more stable than unmodified M., but these modifications are thought to act as signals for the binding of specific stabilizing proteins.

Function. M. confer a certain degree of rigidity on parts of the cytoplasm, e.g. densely packed, parallel arrangements of M. are present in the axopodia of helozoans. In nonspherical animal cells, e.g. thrombocytes, erythrocytes, fibrocytes, M. form part of the cytoskeleton and are therefore involved in the maintenance of cell shape. After treatment with colchicine, low temperatures, high pressure, or high Ca^{2+} concentrations, each of which leads to M. depolymerization, these cells assume a spherical shape, and the axopodia of helozoans are withdrawn. This process is reversible, so that some time after the re-establishment of normal conditions, tubulin dimers repolymerize, the cells regain their original shape and helozoan axopodia are extended. M. do not appear to be directly involved in the generation of cell shape, but they become aligned according to the shape, which they then support. Using immunofluorescence microscopy with monspecific antibodies to tubulin, the formation of the cytoplasmic network of M. has been followed in fibroblasts. It was found that during attachment of the cells to a substratum, the M. accumulate around the nucleus, then begin to spread out toward the cell periphery. Later they become concentrated in strands, which are aligned perpendicular to the plasma membrane. By the time the cell has attained its final shape, a network of M. exists throughout the entire cytoplasm. Some M. are curved in accordance with cell shape, and they appear to serve as support structures in most cell processes. The origin of M. can be demonstrated by depolymerizing with colchicine, then allowing them to regenerate. Under these conditions, M. first appear as the radiating arms of a starlike structure near the nucleus, called an *aster.* The arms then elongate toward the cell periphery. The center of the aster is known as the *microtubule organizing center* (MTOC).

The shape of plant cells is maintained by the cell wall, and the cytoplasm is usually a sol. During interphase, the M. of plant cells are concentrated on the internal suface of the plasma membrane. At the onset of the prophase of mitosis, the M. collect as a "pre-prophase band" at the equator of the cell, finally forming the mitotic spindle. In late telophase, M. are involved in the formation of the cell plate in the region of the phragmosplasts (see Mitosis).

The M. of neurons are known as *neurotubules.* They appear to be responsible for the rapid transport (0.1–1.0 cm/day) of substances in the axon and dendrites, whereas *microfilaments* are responsible for slow transport (a few mm/day).

Microvilli (sing. **Microvillus**) (plate 21): slender, fingerlike, evaginated extensions on the surface of many animal cells. With a length of 1 μ and and diameter of 0.08–1 μ (80–100 nm), they can only be clearly visualized with the aid of the electron microscope. Under the light microscope, intestinal M. have the appearance of a *brush border.* M. considerably increase the cell surface area, and are found particularly on the absorptive surfaces of epithelial cells of absorptive organs, e.g. in the small intestine, gall bladder and urinary tubules. They are particularly abundant on the epithelial cells of the mucosa of the small intestine (50 million M. per mm^2), where they increase the absorptive area some 20–50-fold compared with that of cells lacking M.

Within the intestinal microvillus is a longitudinal, rigid bundle of 20–30 parallel actin filaments (each separated from its neighbors by about 10 nm) extending from the tip of the microvillus to the cell cortex. All the actin filaments are arranged with their plus ends toward the tip of the microvillus, and they are prevented from moving in relation to one another by small cross-linking proteins, known as *actin bundling proteins* (e.g. fimbrin, fascin; small polypeptides with two distinct actin-binding sites). The lower part of the bundle is anchored in a specialized region of the cell cortex called the *terminal web,* which contains a network of spectrin molecules overlying a layer of intermediate filaments; this anchorage probably serves to maintain the projection angle of the bundle and therefore of the microvillus at 90 ° to the cell surface. The plus ends of the actin filaments are connected to the plasma membrane at the tip of the microvillus (via an amorphous cap of densely staining material). At regular intervals along its length, the bundle is connected by cross filaments to the plasma membrane.

Enzymes, e.g. lipase and esterase, are found within intestinal M., while other enzymes, e.g. maltase and lactase, are associated with the microvillus membrane. The plasma membrane of M. has a particularly thick glycocalyx, which contains, inter alia, digestive enzymes. Micropinocytosis vesicles are usually observed at the base of M.

M. also occur, e.g. on hepatocytes, ependymal cells, egg cells and sensory cells. They play a part in cell-cell interaction and in the establishment of cell contact. In cell culture, the contact regions of adjacent hepatocytes develop M. which become enmeshed; once attachment is fully established, the M. are resorbed. Cell-cell agglutination is promoted by the presence of M., leading to the intimate association of adjacent cells. M. also play an essential role in cell fusion. For example, virus-treated cells display numerous M.; when two or more virus-treated cells make contact, the tips of the M. fuse, and the fusion process then spreads throughout the membranes of the fusion partners. M. also increase the surface area of vertebrate photoreceptor cells, increasing the available area for the deposition of membrane-bound visual pigments, and thereby increasing the sensitivity of the receptors. M. in the photoreceptor cells of *Octopus* are oriented with their axes vertical to one another, forming a system for detecting the plane of rotation of polarized light. Many malignant cells possesss more M. than their normal counterparts.

Microwhipscorpions: see Palpigradi.

Micruroides: see Coral snakes.

Micrurus: see Coral snakes.

Micturition: emptying of the urinary bladder. Urine produced in the nephrons of the kidney passes through collecting ducts into the kidney pelvis, and is then carried by peristaltic contractions of the ureter to the bladder. In each ureter, peristalsis is controlled by a pacemaker zone near the exit from the kidney. Extension of the pacemaker cells by urinary pressure from the kidney results in an increase of peristalsis in the ureter, so that urine is squirted into the bladder.

M. is a reflex process (Fig.). Excitation of stretch receptors in the bladder wall causes stimulation of fibers of the pelvic nerve, which transmit sensory signals to the sacral region of the spinal cord. In the gray matter of the sacral region, the excitation passes via interneurons to motor fibers of the same nerve, then exits via efferent neurones to the vesical plexus. The vesical plexus sends fibers to the inner sphincter of the bladder (causing it to relax), and to the bladder wall (causing increased tonus). Nerve fibers from the lumbar region of the spinal cord pass to the inferior mesenteric ganglion and from there, via the hypogastric nerve, to the bladder wall and to the inner sphincter. The hypogastric nerve decreases the tonus of the bladder wall and increases the contraction of the inner sphincter, thereby opposing the action of the pelvic nerve. The outer sphincter of the bladder consists of voluntary muscle, and it permits the voluntary retention or release of urine; it is innervated by the pudic nerve (nervus pudendus) and is under control of the central nervous system.

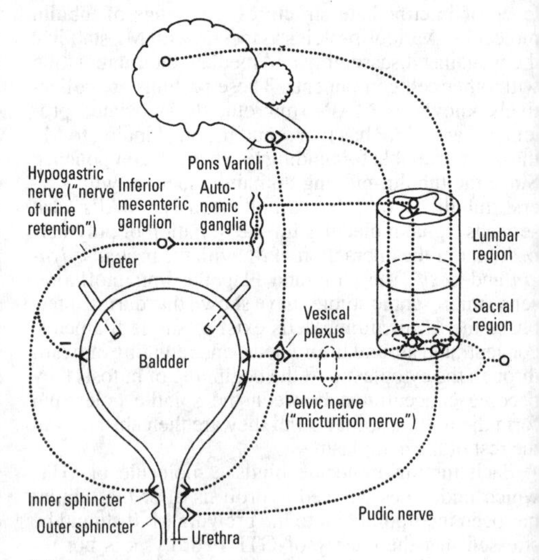

Micturition reflex

Midbrain: see Brain.
Middle ear: see Auditory organ.
Midwife toad: see *Discoglossidae.*
Miesher's tubes: see *Toxoplasmida.*
Migration: a unidirectional, purposeful movement from one area of habitation to another over more or less large distances. Direction and purpose are important in the definition of M. Random movement from one location to another is termed *nomadism.* M. represents an inbuilt response to a complex of external and internal factors. In *passsive M.* the organism exploits water or air as a transport medium, provided the direction and force of flow are appropriate. In *active M.,* the animal must move under its own power, by walking, flying or swimming. *Seasonal migrants* leave their original habitat at a fixed time of the year. Later, again at a predetermined time, they or their offspring (e.g. many species of insects and fishes) return to the first habitat, e.g. bird Ms. and some lepidopteran Ms. (see Migratory lepidoptera) are typically seasonal, and River eels (see) provide an interesting example of M. between generations. Seasonal migration is also known as "annual remigration". Remigration may also be daily, e.g. birds may roost in one area and feed in another. Some aperiodic Ms. occur in only one direction, e.g. Ms. of lemmings and locusts, and are often the result of population pressure. *First order intraterritorial migrants* migrate within their normal area of distribution. *Second order intraterritorial migrants* may cross the boundaries of their normal range and settle in new areas, irrespective of whether their offspring can survive there. Many interesting aspects of migratory behavior can be studied separately, e.g. geographical orientation and direction finding; means of travel and transport; group behavior in preparation for, and during M.; the search for, and selection of, resting places; behavior at migratory resting places; feeding behavior; changes in biorhythm, etc. In additon to these general aspects, there are many species-specific migratory phenomena that are only partially understood. See Animal migration.

Migration theory: part of population theory, concerned with the mathematical analysis of the dispersal of a population, in particular the statistical evaluation of the random dispersal of individuals. It is based on the observation of unicellular organisms and other life forms, whose random distribution can be treated mathematically in the same way as diffusion processes. The data are difficult to handle without a computer, or at least an electronic calculator.

Migratory behavior of birds: Some animals, e.g. certain birds, fishes and *Lepidoptera,* display the ability to migrate over very great distances to specific destinations with almost unerring accuracy. This behavior has been most intensively studied in bird populations. Bird migrations involve seasonal flights between two regions. At a specific time of the year, migratory species leave their breeding areas for their wintering grounds, returning to the breeding areas later in the year, i.e. the migration is seasonal and correlated with the breeding cycle. The journey may occur in stages, with predetermined resting and/or feeding sites on the way, or it may be uninterrupted. Special migrations may occur, e.g. to molting sites. It is estimated that 4–5 x 10^9 (American 4–5 billion; English 4–5 thousand million) birds migrate annually from breeding grounds in the Palearctic to overwintering habitats in Africa, returning north in the summer to breed. Probably the longest journey is undertaken by the Arctic tern, whose breeding range extends to within 700 miles (1126 km) of the North Pole; it winters off the coast of South Africa and even roves the Antarctic to the edge of the pack ice. Similar migrations occur in the Americas. Thus, the Alaskan population of Golden plovers flies over the Pacific Ocean from Alaska to the Hawaiian Islands and other islands further south without the aid of landmarks. The Canadian population makes a long flight over the Atlantic (again largely without landmarks) to its winter range in southern South America, but returns in the spring by a different and mainly overland route. Several migratory species cover distances in one direction that exceed 9 000 km. Some species migrate along a wide front, whereas others follow a narrow corridor. Critical factors that determine the success of these migrations are, e.g. the ability of a bird to determine the correct direction for the start of the migration, its ability to navigate the route, and its ability to recognize the destination and to know when to stop. The entire process is sometimes called "homing", although this term is often applied to the later stages of a migration. Obviously, any accurate, long-distance migration is the culmination of a series of complex interactions between internal (neurophysiological) and external (environmental) factors.

Many questions regarding direction finding and navigation are still unanswered. The ability to navigate to the species' overwintering area and back again is instinctive and inherited. Young birds less than a year old, and therefore with no migratory experience, may migrate before their parents; they then follow the same route to the same overwintering site used by generations of the same species before them. Some birds are guided by the sun, and others by the stars. If the sun or stars are obscured by bad weather, such birds tend to scatter and become confused. Navigation by the sun or stars also requires an internal clock, in order to make allowance for the apparent movement of heavenly bodies across the sky. Thus, when flying due south, the sun must be ahead at noon, but by 15.00, the sun will have moved 45 ° to the west. The process of relating the position of the sun or specific star constellations to the direction of flight is instinctive and based on the neurophysiological integration of signals from a biological clock and from the sensory output of the eyes. Failure to make allowance for the apparent movement of the sun probably explains the broad migration paths taken by some birds. The superp eyesight of birds is probably also important, at least in the later stages or final "homing" stages of a migration. Thus, ordinary landmarks may be used as a guide to the last few kilometers of a migration. Swallows return each year to the nests that they occupied the previous year, probably by visual recognition of the nesting site.

Depending on the species, birds show varying abilities to navigate correctly once they have been diverted from their fixed course, e.g. by high winds, or removal by humans to a different part of the world. In this situation, the bird must determine where it is in relation to its required destination, then navigate a new route. A Manx shearwater, taken to Boston, Massachusetts, returned to its nest on Skokholm Island (Pembrokeshire, UK), covering the 3 050 miles (4 908 km) in 12 days. Similar feats of original navigation have been reported for other birds, especially ocean birds.

The first unequivocal evidence for the use of the magnetic vector for direction finding was reported in 1965 for the European robin. Orientation with respect to the geomagnetic field has now been demonstrated in many nocturnal migrating birds: whitethroat *(Sylvia communis),* garden warbler *(S. borin),* subalpine warbler *(S. cantillans),* blackcap *(S. atricapilla),* pied flycatcher *(Ficedula hypoleuca),* dunnock *(Prunella modularis),* American indigo bunting *(Passerina cyanea),* savanna sparrow *(Passerculus sandwichensis),* bobolink *(Dolichonyx oryzivorus).* There is also considerable but less firm evidence for the same ability in some diurnal migrants. Experimentally, it is shown that the directional tendency of caged birds in a closed room, in the absence of sun or stars, is in excellent agreement with the migratory direction of free-flying birds outside. Reduction of the geomagnetic field by conducting the experiment in a steel vault markedly reduces the orientation ability. The compass is an inclination compass, i.e. it is based on the axial course of the field lines and their inclination in space. It does not distinguish between north and south, but between equator and pole, and it is therefore not affected by reversal of the field. The situation becomes ambiguous at the equator, where the field lines are horizontal; it is not understood how long distance migrants overcome this problem. The nonmigrant homing pigeon uses a magnetic compass when the sky is totally overcast, but otherwise relies on the sun. Experiments have been performed by attaching a pair of small battery-operated Helmholtz coils to the heads of homing pigeons; these are designed to invert the vertical component of the field around the head of the bird. The birds then fly in an opposite direction from the controls, but only if the sun is totally overcast.

Although the existence of an internal geomagnetic compass has been clearly demonstrated in some birds, its mechanism is less certain. Magnetite (which has a ferromagnetic dipole) has been discovered between the skull and the brain, in the Harderian gland, in the sphenoid sinus, and in the olfactory bulbs, depending on the species; this possibly forms part of a mechanism for transducing geomagnetic information in the brain. There is also evidence that the visual pigment, rhodopsin, may be capable of transducing magnetic information when excited by light.

[Current Ornithology, volume 5, edit. Richard F. Johnson, Plenum Press, New York, London, 1988]

Migratory Lepidoptera: butterflies and moths that leave the area in which they hatch, undertaking directional migrations, either singly or in large numbers (see Animal migration). The factors that induce this migratory behavior are unknown. Classification of M.L. into seasonal migrants and first and second order intraterritorial migrants is not accepted by all authorities (see Migration). The study of lepidopteran migration is a relatively new branch of entomology, pursued especially by zoologists in North America and Central Europe. About 40 species are known, which migrate occasionally or regularly, in more or less large numbers, from North Africa and Mediterranean countries into Central Europe [e.g. Painted lady *(Vanessa cardui* L.), Red admiral *(Vanessa atalanta* L.), Death's-head hawk moth *(Acherontia atropos* L.), Humming-bird hawk moth *(Macroglossum stellatarum),* Silver-Y moth *(Autographa gamma* L.)], or take part in migrations within Europe [e.g. Large or Cabbage white *(Pieris brassicae* L.), Small tortoiseshell *(Aglais urticae* L.)]. The migratory urge of the Painted lady is so great that this species occurs all over the world, except in South America. The former group includes some southern species that cannot survive a Central European winter, and these return south in late summer or autumn. Internal migratory species are native to Central Europe, and they and/or their pupae survive the winter by hibernation. In addition, the populations of some native European species are supplemented by migration from further south, e.g. Camberwell beauty *(Vanessa antiopa* L.), Pale clouded yellow *(Colias hyalae),* Queen of Spain fritillary *(Issoria lathonia).*

One of the most impressive, and indeed most carefully studied lepidopteran migrations is that of the Monarch butterfly *(Danaus plexippus).* Three subspecies are recognized: *Danaus plexippus erippus,* a migratory species occurring south of the Amazon drainage system; *Danaus plexippus megalippe,* a rather variable, nonmigratory species occurring north of the Amazon drainage; and *Danaus plexippus plexippus,* a species that migrates between Mexico or the Gulf States of North America and the woodlands around the Great Lakes of North America. Generations of *D. plexippus plexippus* are produced throughout the summer around the Great Lakes, but about two thirds of those arising in the early autumn show a marked tendency to move purposefully southward. The remaining one third of the population stays in the north, hibernating during the winter, often beneath the bark of dead trees. The southward migrating Monarchs follow a direct course, steering by the sun, and compensating for its movement. Each night during the migration, the butterflies settle before the air tempertaure sinks too low, and they continue their flight on the next day when the sun has warmed them sufficiently. Often the same overnight roosting sites are used year after year, and during years of maximal abundance these roosting trees may be covered with many thousands of Monarch butterflies. The true direction of migration is southwest, rather than due south, so that depending on their point of departure the butterflies eventually arrive in Florida, northern Mexico or California, the longest migration (to northern Mexico) being about 3000 km. On arrival they roost in large numbers on certain coniferous trees that have been used for this purpose for generations. When the roosting sites in northern Mexico were eventually discovered, the migratory Monarchs were found clustered in their millions on certain trees in just one or two specific valleys. During this roosting period the butterflies are quite inactive, but in the spring they mate, then start to migrate northward, feeding and laying eggs on the way. Only a few accomplish the return journey to southern Canada, most dying on the way. A new population arises mainly from those Monarchs that stayed in the north.

Migratory unrest: typical spontaneous behavior in migratory birds as a prelude to the actual migration. It is also observed in caged birds. In birds that migrate at night, M.u. occurs in the dark.

Migroelement: see Floral element.

Milfoil: see *Compositae.*

Miliola: see *Foraminiferida.*

Milk proteins: soluble proteins present in milk, consisting of the Caseins (see) and whey proteins. Of particular importance are α-, β- and κ-casein, lactoglobulin, lactoferrin and α-lactalbumin.

Milk teeth: see Dentition.
Milkweed: see *Asclepiadaceae.*
Millipedes: see *Diplopoda.*
Millepora: an athecate, colonial polyp of the order, *Hydroida.* The polyp colonies are reinforced by an internal skeleton of calcium carbonate, which is covered by only a thin layer of epidermis. Defensive polyps arise from separate pores encircling a central gastrozooid. The nematocysts of the defensive polyps are very active, and they cause a sharp burning sensation when touched [hence the German name "Feuerkorallen" (fire corals) for this genus]. The colony has an incrusting growth form and it is deceptively similar in appearance to a stony coral.

In the related *Stylaster* and *Allopora,* the skeleton is overlaid by a thick layer of tissues, and defensive and feeding polyps are located in a common pit. The latter 2 genera are placed by some authorities in the suborder, *Stylasterina,* while *Millepora* is placed in the *Milleporina,* both suborders comprising the order, *Hydrocorallina.*

Milleporina: see *Millepora.*
Miller's thumb: see *Cottidae.*
Millet: see *Graminae.*
Millipedes: see *Diplopoda.*
Milu: see Père David's deer.
Milvinae: see *Falconiformes.*
Milvus: see *Falconiformes.*
Mimesis: see Protective adaptations.
Mimicry: see Protective adaptations.
Mimic thrushes: see *Mimidae.*
Mimidae, *mockingbirds* or ***mimic thrushes, thrashers*** and ***tremblers:*** an all-American avian family (order *Passeriformes)* containing 34 species. The mockingbirds are named for their ability to imitate many different sounds, including the songs of other birds. Phylogenetically, the family occupies a position between the *Troglodytidae* (wrens) and the *Turdinae* (true thrushes). Plumage is usually shades of brown and gray, but the ***Blue*** and the ***Blue-and-White mockingbird*** *(Melanotis caerulescens* and *M. hypoleucus)* are more colorful. The type genus is *Mimus* (9 species, Canada to Argentina), e.g. *M. polyglottos* ***(Northern mockingbird).*** A single representative, *Nesomimus trifasciatus,* is found on the Galapagos Islands. A strongly isolated monotypic genus, *Mimodes graysoni* ***(Socorro thrasher),*** inhabits the Island of Socorro in the Revillagigedo Archipelago off the west coast of Mexico. Ten species of typical ***Thrashers*** *(Toxostoma)* are found in North America. Some members are called ***Catbirds,*** e.g. the ***Black catbird*** *(Melanoptila glabrirostris;* Yucatan), which has iridescent black plumage. The common gray ***Catbird*** *(Dumetella carolinensis)* breeds across most of North America from southern Canada to the northern parts of the Gulf states, and winters in the southern states, West Indies and Central America.; it possesses an unmistakable black cap and russet undertail coverts. The ground-dwelling forest ***Tremblers*** *(Cinclocerthia ruficauda* and *Ramphocinclus brachyurus)* are confined to the West Indies.

Mimodes: see *Mimidae.*
Mimosa: see *Mimosaceae.*
Mimosaceae, *mimosa family:* a family of about 2 000 dicotyledonous, herbaceous and woody plants, distributed mainly in the tropics and subtropics. Leaves are stipulate and usually bipinnate and paripinnate. Flowers are small, actinomorphic and usually 4-merous, combined into globular or spike-like inflorescences. The inflorescences are conspicuous, because the stamens (often secondarily increased and therefore numerous) possess long, colored filaments. Pollination is by insects or birds. The superior ovary develops into a pod. Together with the *Caesalpiniaceae* and *Papilionaceae,* the M. formerly constituted the family *Leguminosae.*

Mimosaceae. Sensitive plant *(Mimosa pudica).*

The largest genus is *Acacia* ***(Acacia*** or ***Wattle),*** the species of which are especially numerous in Australia and South Africa as savanna trees. Many Acacias, especially those of the dry forests in Australia, have leaf-like phyllodes (see Leaf). Many are also armed with hollow thorns, which are colonized by ants. Some species, in particular the North African *Acacia senegal,* yield gum arabic. Hot aqueous extraction of the heartwood of *Acacia catechu* yields a dying and tanning agent, known as black catechu or cutch. Common in India and Burma, *Acacia catechu* also provides valuable timber, and its bark is also used in tanning.

Mainly indigenous to tropical and subtropical America, the genus *Mimosa* contains 450–500 species. Well known is the ***Mimosa*** or ***Sensitive plant*** *(Mimosa pudica),* the leaves of which rapidly collapse when touched (Fig.). Today, *Mimosa pudica* is common throughout the tropics as a ruderal plant. ***Mesquite*** or ***Algarrobe*** *(Prosopis juliflorae),* also known as ***Common mesquite*** and ***Honey mesquite,*** provides a very hard and useful timber, and its edible fruits are especially favored by American Indians.

Mimus: see *Mimidae.*
Mineralization:

1) Complete degradation of dead organic material, leaving a mineral residues (see Decomposition).

2) Conversion, by soil organisms, of organic, nitrogenous compounds to low molecular mass inorganic compounds such as ammonia (or ammonium salts), nitrite and nitrate (see Nitrogen).

Mineral metabolism:

1) In plants, the uptake, conduction and processing of mineral macro- and micronutrients. Most nutrients are taken up from the soil as ions by plant roots, some heavy metals possibly as chelates. *Ion uptake* displays the following important characteristics. 1) Most ions are present in the cell sap in higher concentration than in the surrounding medium, and this accumulation or concentration of ions is an energy-dependent. 2) The uptake of ions by plants is selective, i.e. not all ions are absorbed equally, and the plant even discriminates between chemically closely related ions, such as K^+ and Na^+. However, selectivity is not absolute, i.e. there is also some uptake of nonnutrient ions. 3) The diurnal rhythm of ion uptake (the process is particularly active in daylight) and other physiological phenomena indicates a close relationship with the metabolic activity of the plant.

The actual uptake process can be divided into 2 phases.

1) The first phase consists of the reversible, energy-independent, nonselective penetration and accumulation of ions in the "free space" of the root. This free space consists primarily of the interfibrillar and intermicellar spaces of the cell walls of the root hairs, rhizodermis and root cortex. Ions move within these spaces, due to diffusion, due to the electrochemical potential of the cell wall, and particularly due to the flow of the transpiration stream through the cell walls. Deposition of lignin and lipid substances in the endodermal cells (see Casparian strip) means that ions can only enter the pericycle by first passing through the cytoplasm of endodermal cells. The cell wall surfaces of the "free space", as well as the plasmalemma that is apposed to the cell wall, are occupied by nondiffusible anions, e.g. by the carboxyl groups of polygalacturonic acids (pectinic acids) and proteins, and by the phosphate groups of phospholipids. At individual sites, these nondiffusible ions generate an electrical force field; cations are drawn into the field, whereas anions are repelled. This results in a *Donnan distribution.* Since anionic charges predominate on the surfaces bounding the "free space", mainly cations are adsorbed. Since the negative charges are occupied by adsorbed ions, *exchange adsorption* plays an important role, i.e. ions in excess displace other ions from the adsorption sites. Thus, a root that has been kept in a Ca^{2+} solution loses its adsorbed Ca^{2+} when it is placed in a solution containing K^+. The root therefore has the properties of an ion exchanger. On the other hand, Donnan distribution and ion exchange adsorption play practically no part in the uptake of ions into the cytoplasm. Exchange adsorption is, however, very important in the uptake of cations from the soil: cations adsorbed on soil colloids are exchanged for cations secreted by the root (in particular H^+) and thereby made available to the plant.

2) The second, irreversible phase of ion uptake is often called *metabolic absorption* or *accumulation.* This involves the passage of ions through the plasmalemma into the cytoplasm, where the nutrients exist in much higher concentration than in the soil or in the "free space". The process is energy dependent, and the energy is derived from respiration. Ion uptake is accordingly accompanied by an increase of respiration. The energy is required for the operation of carrier systems that "pump" ions through the relatively impermeable plasmalemma. This carrier transport is selective, and it is responsible for the observed competition for uptake between chemically similar ions, e.g. K^+ and Rb^+. Since carrier systems only exist for nutrients, these or chemically similar ions are taken up preferentially. All other ions in the soil enter the root passively in accordance with their concentration gradient. Uptake selectivity is therefore not absolute, and the plant is unable to totally exclude any ion that is present in solution in the soil in high concentration.

There are no specific sites for the passage of ions from the free space into the cytoplasm (from apoplasts into symplasts). Nutrient ions are absorbed metabolically, with the aid of carrier systems, across the plasmalemmas of all cells of the root cortex. The tonoplast membrane also contains abundant carriers, which transport ions into the vacuoles. Transport into vacuoles is also energy-dependent; it represents a cul-de-sac, but serves for the temporary storage of nutrients.

Following their entry into the cytoplasm, the different ions are more or less rapidly metabolized. For example, NH_4^+, NO_3^- and SO_4^{2+} are utilized for the synthesis of amino acids, PO_4^{3-} is incorporated into ATP, Ca^{2+} is required for the synthesis of phytin, and Fe^{2+} is incorporated into the porphyrin ring system. Removal of ions by metabolism maintains their concentration gradient, which in turn stimulates carrier transport.

Transport of nutrient salts or their metabolic products from cell to cell (symplastic transport) occurs primarily via the plasmodesmata and pits at a rate of about 1–6 cm/h, i.e. it is faster than diffusion. The mechanisms of symplastic transport are unclear. It is not prevented by the cessation of protoplasmic streaming. In addition to diffusion and convection, it seems likely that the endoplasmic reticulum and possibly microtubules are involved. The symplastic transport of inorganic ions in particular does not appear to be energy-dependent. Secretion of ions from the symplast into the vessels is also poorly understood. It is assumed that the same carrier systems are involved that transport ions from the apoplast into the symplast, but that the direction of transport is reversed.

See also Transport.

2) In animals, the mechanisms of uptake, transport, action and excretion of mineral substances. Dissolved minerals pass from the intestine into the blood. Minerals are required by the body for a variety of functions. They are 1) dissolved components of the blood and other body fluids, and 2) structural components of bones and teeth. They are required 3) for the maintenance of a certain osmotic pressure in the cells, and 4) as charged entities for the expression of the electrical potential of each cell. They are intimately involved in 5) the water economy and acid-base balance of the body, and 6) anabolism and the turnover of protein and glycogen. Mineral metabolism is regulated by hormones, in particular parathormone and corticosteroids. Mineral salts are secreted by the kidneys and large intestine. In contrast to organic materials, they enter and leave the animal organism in the same chemical combination, and must be continually supplied in appropriate quantities and in the correct proportions. Thus, the ratios of calcium to phosphorus and potassium to sodium, as well as the acid-base ratio are particularly important. Prolonged mineral deficiency is very harmful, and total absence of dietary minerals soon leads to death.

The individual nutrient ions have specific functions, and they can be divided into 2 groups: 1) structural components, which are always present in the organism in large quantities, and 2) those that serve a catalytic function, and are present only in trace amounts. The first group includes sodium, potassium, calcium, magnesium, phosphorus, sulfur, chlorine. The second group comprises iodine, fluorine, bromine, copper, manganese, iron, cobalt, zinc, silica.

Sodium and *potassium* are present in the blood and body fluids as chlorides, sulfates, phosphates and carbonates. They are required for the generation and propagation of excitatory impulses in muscles and nerves, and for the regulation of osmotic pressure. Their rate of absorption depends on their chemical combination; chlorides are taken up rapidly, whereas sulfates diffuse slowly. Sodium is found mainly in body fluids, whereas potassium occurs mainly in the cells. *Calcium* and *magnesium,* in particular their phosphates, are essential for bone formation. Brain and muscle have very high contents of magnesium. On the other hand, bones and blood contain more calcium, which is required for the "coupling" of contraction and excitation in cardiac, skeletal and smooth muscle. The major part of the body's calcium is contained in the skeleton. *Phosphorus* is present as calcium and magnesium phosphate in the bones, and as the phosphates of alkali metals in the blood, body fluids and muscles. Organically bound phosphorus occurs in nucleic acids, phosphoproteins and phosphatides. *Sulfur* is present in almost all proteins (in the sulfur amino acids, methionine, cystine and cysteine). Hair and all keratinous substances are rich in sulfur. *Chlorine* is a component of the hydrochloric acid of the gastric juice, and the skin also has a high chlorine content; sodium chloride and potassium chloride are present in lymph, urine and perspiration. *Iodine* is present in the thyroid hormones, in the blood and in all body fluids. *Fluorine* is organically bound in muscles and nerves, and it forms part of the inorganic matrices of bones and teeth.

Among the *trace elements,* copper, iron, cobalt and zinc are particularly important. *Copper* occurs in relatively high concentration in the liver, and it is also present in blood, various body secretions and the brain. Together with iron and manganese, copper is also involved in the synthesis of hemoglobin. *Manganese* also occurs in blood, bile and urine, and in hair and bone. As a structural component of hemoglobin and the cytochromes, *iron* is particularly important in metabolism; it is stored in the liver and spleen. *Cobalt* is required by rumen bacteria. *Zinc* is present in large quantities in the pancreas. In the form of silicic acid, *silica* occurs in the skin and its associated structures, as well as in blood, pancreas, spleen and eggs; it is also an important constituent of the skeletons of rotifers and sponges.

Mineralocorticoids: a group of Adrenal corticosteroids (see), which regulate the mineral metabolism of the organism by causing retention of sodium, chloride and water, with simultaneous potassium diuresis. The highest activity is displayed by Aldosterone (see); other M. are Cortexone (see) and cortexolone.

Miners: phytophagous insect larvae, which feed by burrowing inside the tissue of their food plant. *Leaf miners* feed upon succulent interior leaf tissue while tunnelling between the upper and lower cuticles. The resulting leaf mines may be serpentine (Fig.) or blotch-like. The apple leaf miner, *Lyonetia clerkella* L., is often found on cultivated apples, as well as birch, crabapple, hawthorn, cherry, etc. The apple small ermine moth, *Yponomeuta malinellus* Zeller, an easily controlled minor pest of apple orchards, produces reddish brown blotches. The cabbage leaf miner (larva of the cabbage web moth or diamond-back moth, *Plutella xylostella* L.) is a serious pest in continental Europe. *Stem miners* bore cylindrical tunnels in grass stalks, or in the pith of shrubs, occasionally in woody twigs. Many larvae of the *Cephidae* (stem sawflies) are stem borers, e.g. the serious agricultural pest, *Cephus pygmaeus* (European wheat-stem sawfly) and *Cephus cinctus* (American wheat-stem sawfly), a severe pest in grain growing areas of Canada. An example of a *fruit miner* is the apple sawfly *(Hoplocampa testudinea* Klug), a widespread and often serious apple pest; the young larvae bore through the receptacle and down to the ovary, where they feed on one or more seeds. *Seed miners* are the larvae of seed beetles, e.g. *Callosobruchus chinensis (Bruchidae),* which is one of the most destructive pests of stored legume seeds, causing serious damage to chickpea, lentil, cowpea and pigeon pea. In some cases, M. also cause gall formation.

Miners. Cherry leaf with serpentine leaf mines of the apple leaf miner *(Lyontia clerkella L.).*

Minima, law of: see Law of minima.

Minimal area: the smallest area in which a particular biological community is able to develop fully, so that it contains the number and combination of species that characterize that community. M.a. is particularly important in the study of island biotas. The M.a. is determined with the aid of the species-area curve, which is a graph of the relationship between the number of species found and the size of the area surveyed (the area surveyed is equivalent to the number of equal-sized samples taken in parallel). In a sufficiently large, homogeneous community, the species-area curve asymptotically approaches a limiting value (Fig.). The *M.a. of a species* is the smallest area or space necessary for that species to become stabilized *(= Minimal area of constancy).*

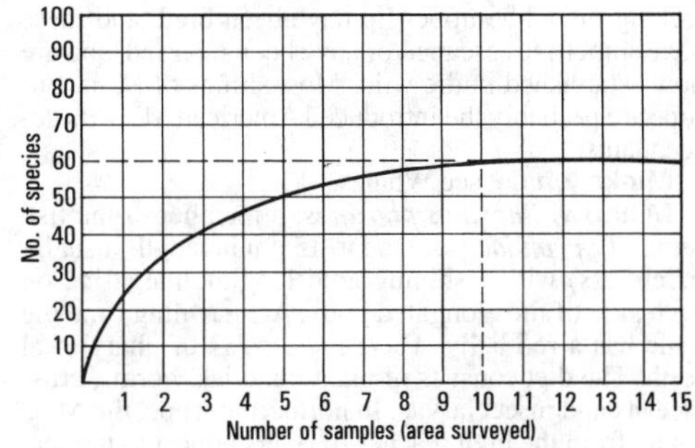

Minimal area. A species-area curve. In this example, the number of species (60) in the studied region is obtained by counting no fewer than 10 samples.

Minimal protein (or nitrogen) requirement: the amount of dietary protein required daily to compensate the nitrogen lost by excretion. Adult humans require 25 to 35 g complete (containing optimal amounts of essential amino acids) protein per day. Growing organisms have a proportionally higher M.p.r., because some nitrogen is retained as new body protein. Lactation also increases the M.p.r. Several of the amino acids present in protein cannot be synthesized by animals; these are known as essential or indispensable amino acids, and they must be supplied preformed in the diet (see Amino acids). Animals are therefore incapable of synthesizing proteins from fats or carbohydrates. On the other hand, proteins are converted into fats and carbohydrates, especially under starvation conditions. Thus, starvation also increases the M.p.r. Conversely, the provision of carbohydrate and fat has a sparing effect on the protein requirement.

Three types of Mp.r. are recognized in human nutrition: 1) the *absolute* M.p.r. accounts for the loss of protein in the form of dead cells and hair, and in the synthesis of hormones and enzymes; it is equivalent to the amount of nitrogen excreted on a protein-free but calorically adequate diet, and is equal to about 2.4 g N = 15 g animal protein per day for adults; 2) the *physiological* M.p.r. represents the protein required to maintain nitrogen balance (i.e. the daily nitrogen intake in the form of protein must equal the quantity of nitrogen excreted), and is equal to 25 g animal protein per day for adults; 3) the *hygienic* M.p.r. or *recommended daily allowance of protein* is the quantity of protein in a mixed diet required for a general state of well being. The United States Dept. of Agriculture's *recommended daily allowance* for adults has been revised downward in recent years from 70 g to 40 g protein for adults.

Mink, *Mustela, Lutreola:* Two species of mustelid *(see Mustelidae)* with a uniformly deep brown coat. The ***European M.*** *(Mustela lutreola)* was once distributed throughout most of Europe, but has now disappeared from many parts, and is threatened with extinction especially in Eurasia and Siberia; it has white upper and lower lips. The ***American*** or ***Eastern M.*** *(Mustela vison)* is found in northern North America and from the Gulf of Mexico to the Pacific coast of southern USA; most individuals lack the white upper lip. M. live near to water and lead an amphibious existence. They are excellent swimmers and divers, feeding mainly on frogs and fish. On account of its highly valued fur, the American M. is farmed commercially. Many color variations lacking the white upper lip have been bred, and these have sometimes escaped or have been released, and are now established in the wild. Most sitings of M. in Europe are probably the introduced American M. or its descendants.

Minke whale: see Whales.

Minnow, *Phoxinus phoxinus:* a member of the fish family *Cyprinidae* (see), up to 15 cm in length (usually much less) with a shining gold longitudinal stripe on each side of the elongated body. At spawning time the male has a red belly. There are 2 rows of pharyngeal teeth. The diet consists of plant material, worms, crustaceans and insect larvae. In northern Europe, the M. is found from the high reaches of rivers almost to the sea. Elsewhere in Europe, it more typically inhabits mountain waters of streams and rivers, and the shores of lakes, as well as relatively small ponds *Phoxinus phoxinus* was the "minnow" used by Karl von Frisch for his classic studies on the auditory sense of fishes. It is often used as angling bait.

Mint: see *Lamiaceae.*

Miracidium: see Liver flukes.

Mirafra: see *Alaudidae.*

Marine: pertaining to the sea; inhabiting or existing in (e.g. sediments) the sea. Converse terms are Terrestrial (see) and Limnic (see).

Mirounga: see *Phocidae.*

Misgurnis: see Weatherfish.

Misgurnus: see Respiratory organs.

Misophria: see *Copepoda.*

Misophrioida: see *Copepoda.*

Missense mutation: see Gene mutation.

Missing link: see Connecting link.

Mississippian subperiod: see Carboniferous.

Mistle thrush: see *Turdinae.*

Mistletoe bird: see *Dicaeidae.*

Miter shells: see *Gastropoda.*

Mites: see *Acari.*

Mitochondria, ***chondriosomes*** (obsolete): organelles in the cytoplasm of all eukaryotic cells. They possess two very different membranes (the inner and the outer membrane), and they contain their own autonomous DNA. M. contain the enzyme systems responsible for the oxidation of carbohydrates, lipids and amino acids to CO_2 and water in the presence of molecular O_2, some of the available free energy appearing in the form of ATP. They are thus the site of of aerobic cellular respiration, and they are responsible for generating most of the energy (as ATP) needed by the aerobic cell. M. of muscle cells are known as *sarcosomes.*

Occurrence. With a diameter of 0.5–1 μm and a length of 1–5 μm, M. are on the threshold of visibility under the light microscope. The average mitochondrion is about the same size as a cell of *Escherichia coli.* In living cells M. are made more readily observable by vital staining with Janus green B, or by phase contrast. Depending on the type of cell and its physiological state, M. are spherical, rod-shaped or thread-like; branched forms also occur. Cinematographic records of living cells show that M. continually alter their shape and are very mobile. During cell fractionation procedures, M. are sedimented by centrifugation at 10 000 g for 10–15 min, i.e. their sedimentation properties are intermediate between those of the nuclear and microsomal fractions.

A mammalian liver cell contains about 1000–2000 M., occupying about 20% of the cell volume, whereas an amphibian oocyte contains about 300 000, and the protozoan, *Chaos chaos,* contains about 500 000. M. display functional cooperation by accumulating in cells or regions of cells with an especially large ATP requirement, e.g. they are especially numerous in heart muscle cells, being packed between adjacent myofibrils, where they provide ATP for contraction; if a resting striated muscle is repeatedly stimulated to contract, the number of M. increases 5–10-fold; M. accumulate near the basal lamina in the epithelial cells of kidney tubules, where they provide ATP for active ion transport and resorption; M. are wrapped tightly around the flagellum in the central body of the mammalian sperm, where they provide ATP for flagellar movement. Yeast cells growing in a glucose-containing medium in the absence of O_2 contain only small *promitochondria,* consisting of a vesicle bounded by a double membrane and with no internal structure. In the absence of O_2, aerobic cellular respiration is not possible, and all the necessary energy is obtained by anaerobic respiration (i.e. fermentation). Promitochondria lack the enzymes that are characteristic of M., including the cytochromes of the respiratory chain. If the yeast culture is exposed to oxygen, the promitochondria develop into typical M., and aerobic cellular respiration commences. Mature mammalian erythrocytes contain no M. Cancer cells usually contain fewer M. than the corresponding normal cells, and their aerobic cellular respiration is often impaired. Prokaryotic cells contain no M. and their respiratory enzymes are located in the plasma membrane or in folded structures derived from it (see Mesosomes).

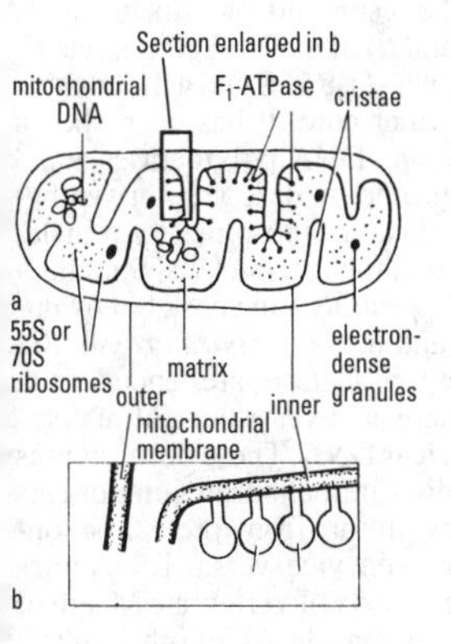

Fig. 1. *Mitochondria.* Schematic representation of a mitochondrion with cristae.

Structure and composition. Electron microscopy of M. shows 2 concentric membranes, each 5–7 nm thick, the inner membrane usually folded or invaginated (Fig.). These two membranes partition the mitochondrion into two compartments. Between the two membranes lies the outer *intermembrane space* (also called the *external matrix* or *outer mitochondrial space).* The inner membrane encloses the *matrix* or *stroma,* which is a contractile network of structural proteins embedded in an aqueous phase. The matrix contains many of the functional components of M., and it is not in contact with the cellular cytoplasm, facts which lend support to the Endosymbiont hypothesis (see). According to this hypothesis, M. and Plastids(see) have evolved from intracellular symbionts. The theory is also supported by the fact that the two membranes have different functions and compositions. The density of the outer membrane is about 1.1 g/cm³, and it is permeable to most substances of M_r 10 000 or less. It contains a high proportion of phospholipid (phospholipid to protein ratio is about 0.82 by weight); after extraction of 90% of mitochondrial phospholipid with acetone the inner membrane remains intact, whereas the outer membrane is destroyed. The lipid fraction of the outer membrane contains a low concentration of cardiolipin, a high concentration of phosphoinositol and cholesterol, and no ubiquinone. The density of the inner membrane is about 1.2g/cm³, with a low content of phospholipids (phospholipid to protein ratio is about 0.27 by weight) and about 20% cardiolipin. Neutral substances, e.g. sugars of M_r less than 150, seem to cross the inner membrane freely, but the passage of all other substances is subject to tight control. Specific transport systems are present for certain materials (e.g. pyruvate, malate, succinate, citrate, H^+ ions, phosphate ions, ATP, ADP, etc.).

The surface area of the inner mitochondrial membrane is greatly increased by folding or invagination. These invaginations may be flat and sheet-like *(cristae)* or tubular *(tubuli).* In cells with high metabolic activity, the cristae or tubuli penetrate deep into the matrix and they are numerous and densely packed, whereas in metabolically less active cells they are smaller and less numerous. Tubuli-type M. are found in many protozoa, as well as in certain metazoan cells (e.g. cells of the adrenal cortex), but the M. of most metazoan cells are of the cristae-type.

The *inner membrane* contains all the components of the electron transport chain (respiratory chain) and oxidative phosphorylation, succinate dehydrogenase, NADH dehydrogenase, ATPase, 3-hydroxybutyrate dehydrogenase, ferocheletase, 5-aminolevulinate synthetase, carnitine palmitoyltransferase, fatty acid elongation system (C_{10}), $NAD(P)^+$ transhydrogenase, and choline dehydrogenase. The *intermembrane space* contains adenylate kinase, nucleoside diphosphate kinase and creatine kinase (heart muscle M.). Marker enzymes for the *outer membrane* are flavin-containing amine oxidase (monoamine oxidase) and NADH dehydrogenase (also known as cytochrome *c*-reductase, or NADH-cytochrome b_5-reductase; this enzyme is insensitive to rotenone, amytal or antimycin, and it is similar to the NADH-cytochrome b_5-reductase of microsomes). In additon, the outer membrane contains kynurenine 3-monooxygenase, ATP-dependent fatty acyl-CoA synthetase, glycerophosphate acyltransferase, acylglycerophosphate acyltransferase, lysolecithin acyltransferase, cholinephosphotransferase, phosphatide phosphatase, phospholipase A_2, nucleosidediphosphate kinase, and a fatty acid elongation system (C_{14}, C_{16}). It is noteworthy that the outer membrane contains many enzymes of phospholipid metabolism. The *matrix* contains all the enzymes of the tricarboxylic acid cycle, with the exception of succinate dehydrogenase. Since succinate dehydrogenase is present only in the inner mitochondrial membrane and nowhere else in the eukaryotic cell, it is used as a marker for the histochemical demonstration of M. and for monitoring the presence of M. during cell fractionation and separation of organelles. Also present in the matrix are glutamate dehydrogenase, pyruvate carboxylase, aspartate aminotransferase, carbamoyl phosphate synthetase (utilizing ammonia and concerned in urea synthesis; unlike a similar enzyme in the cytosol which utilizes glutamine and is concerned in the synthesis of pyrimidines), ornithine carbamoyl transferase, GTP- and ATP-dependent fatty acyl-CoA synthetases, and enzymes for the oxidation of fatty acids.

On the matrix side of the inner membrane are attached small particles (diameter about 9 nm), visible under the electron microscope with negative staining (Fig.); these were earlier known as oxysomes, but are now called coupling factor 1 (F_1 or F_1-ATPase). The knob-like F_1 structure is attached to a stem, which is associated with special proteins in the membrane. This total enzyme complex (F_1 + associated membrane proteins) constitutes the ATP-synthesizing system of oxidative phosphorylation. Electron-dense granules (intramitochondrial granules) in the matrix (Fig.) consist of calcium and magnesium phosphates, the respective ions being actively transported across the inner mitochondrial membrane.

Occasionally the outer mitochondrial membrane is observed in combination with cisternae of the endoplasmic reticulum. M. can also act as storage sites, e.g. for yolk lipoproteins in egg cells. Certain hydroxylation reactions in the synthesis of adrenocorticoids are catalysed by mitochondrial cytochrome P450 monooxygenases in the adrenal cortex.

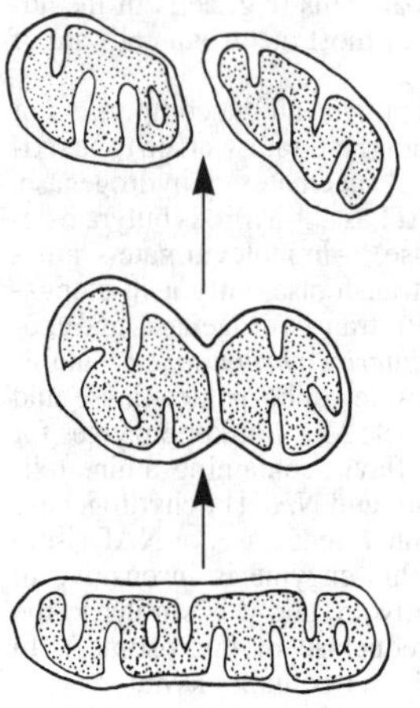

Fig.2. *Mitochondria.* Division of a mitochondrion.

Multiplication, Genetic system, Mitochondriome. M. cannot be formed de novo from other cell structures. They are formed by division (Fig.) of other M., i.e. they are autoreplicative. In most cells, division of M. occurs at random throughout the cell cycle (i.e. it is not restricted to the S-phase of the cell cycle), and it is not synchronized with cell division. There is thus an apparent lack of regulation, and some M. may replicate more than others. There is, however, an overall regulation, which insures that the total number of M. and the total quantity of mitochondrial DNA (mtDNA) are doubled during the cell cycle. M. possess genetic information (DNA) and their own specific systems for RNA transcription and protein translation. This genetic information, however, encodes only a relatively small number of mitochondrial proteins, most mitochondrial proteins being encoded by genes of the cell nucleus, and synthesized on cytoplasmic ribosomes outside the M. Thus, M. are semiautonomous cell organelles. The totality of the genes encoding the M. of a cell are known collectively as the *mitochondriome (chondriome).* The extrachromosomal (intramitochondrial) genes act in unison with the chromosomal (extramitochondrial) genes. mtDNA, which lies in less dense regions of the gelatinous matrix, is circular, histone-free and double stranded; its base composition differs from that of the nuclear DNA of the cell; like other DNA, it replicates semiconservatively. The length of the mtDNA molecule varies from 5 mm (mammals) to 35 mm (plants and fungi), and several copies are usually present in a single mitochondrion. In most cells the total quantity of mtDNA represents less than 1% of the DNA of the cell nucleus. In cells containing extremely large numbers of M. or relatively small quantities of nuclear DNA, the proportion of mtDNA may be considerably higher (greater than 20% in trypanosomes; about 30% in *Physarum* ; greater than 50% in certain oocytes). Although plant mtDNA may be much larger (e.g. 2.5 x 10^6 base pairs) than animal mtDNA (e.g. 15 000 base pairs), it does not contain many more transcribable genes; many of the genes of plant mtDNA contain introns, and much of the remaining plant mtDNA appears to be nontranscribable, redundant or "junk" DNA. The plant family *Cucurbitaceae* is exceptional, in that mtDNAs from its various members vary in size by as much as 7-fold. The mtDNA from *Chlamydomonas* is linear and contains only 16 000 base pairs, i.e. it is similar to that of animals. Yeast mtDNA, mtDNA from other lower eukaryotes, and mammalian mtDNA encode essentially the same gene products, i.e. 4 of the 7 subunits of cytochrome *c* oxidase, 2 of the 3 subunits of cytochrome *b,* one of the subunits (subunit 6) of the F_1-ATPase, one ribosomal protein (termed *var-1),* a 15S and a 21S rRNA, and multiple tRNAs. Furthermore, it has been shown that mitochondrial RNA- and DNA-polymerases and 3 of the subunits of cytochrome oxidase are not synthesized in M., but on cytoplasmic ribosomes by translation of mRNA transcribed in the cell nucleus. Subunit 9 of F_1-ATPase is encoded by mtDNA in yeast, but by nuclear DNA in mammals and in *Neurospora crassa.* Six subunits of NADH-CoQ reductase are encoded in mammalian mtDNA, whereas in yeast this entire complex is encoded in the nuclear DNA. These observations suggest that genes encoding mitochondrial components have migrated during evolution from the mitochondrion to the nucleus, or even vice versa. Ribosomes (see) in the mitochondrial matrix of vertebrate M. are of the small, 50–55S-type, whereas the M. of other eukaryotic organisms contain 70S ribosomes (which also occur prokaryotic cells); in contrast, cytoplasmic ribosomes of eukaryotic cells are of the 80S-type.

A random evolutionary drift appears to have occurred in the genetic code of M., so that it not only differs somewhat from the "universal code", but also between different organisms (see Table).

Comparison of some codons in the "universal" genetic code and in mitochondrial genetic codes.

Codon	„Universal code"	Mammals	*Drosophila*	Yeast	Plants
UGA	Stop	Trp	Trp	Trp	Stop
AUA	Ile	Met	Met	Met	Ile
CUA	Leu	Leu	Leu	Thr	Leu
AGA AGG	Arg	Stop	Ser	Arg	Arg

Mitochondriome: see Mitochondria.

Mitosis: the most common form of nuclear division in eukaryotic organisms. Mitosis results in the formation of two daughter cells with the same chromosome number and the same genetic information as the mother cell. Before mitosis can begin the chromosome must be replicated, a process which occurs in the interphase, during the S-phase of the Cell cycle (see). Under the light microscope, during interphase, the only visible subnuclear structure is the nucleolus; chromosomes are not visible as distinct structures. Shortly before the onset of the prophase, a daughter Centriole (see) begins to appear at right angles to each of the two centrioles. Most higher plant cells do not possess centrioles; an analogous region of the cell acts as a *microtubule organizing center,* from which the spindle tubules subsequently radiate.

The subsequent process of mitosis (lasting from 30 to 180 minutes) can then be divided into 4 phases: prophase, metaphase (2–20 minutes), anaphase (2–20 minutes) and telophase (duration very variable).

1. ***Early prophase.*** Specific cell functions are suspended, and animal cells often become more rounded in shape (since plant cells have a rigid cell wall, they show only slight changes of shape in mitosis). The Golgi apparatus collapses into small vesicles, and the endoplasmic reticulum and other organelles in the cytoplasm undergo partial degeneration. By increased twisting and folding of the nucleohistone strands, the chromatin condenses into light-microscopically visible chromosomes, which first appear as a tangled mass of long thin threads. The nucleolus begins to disperse.

2. ***Middle prophase.*** Chromosome condensation is complete, and each chromosome can be seen to consist of two chromatids attached at their centromeres. At this stage, each chromatid contains one of the two daughter DNA duplexes that were synthesized during the S-phase of the cell cycle. The spindle of microtubules begins to radiate from regions adjacent to the centrioles, which are still migrating to the poles of the cell.

3. ***Late prophase***. The centrioles attain their polar positions. Some spindle fibers extend from the poles to the equatorial region, while others extend from the poles to the kinetochores of the chromatids. The nuclear membrane disintegrates into vesicles, which aggregate in the polar regions.

4. ***Metaphase.*** Contact between spindle fibers and kinetochores is completed, and the centromere of each chromosome lies in the equatorial plane of the spindle. Chromatids have not yet separated, but the individual chromatids are clearly visible under the light microscope, still attached in the region of the centromere. The chromosomes have now attained their shortest length and greatest diameter, and at this stage they are most easily counted and morphologically characterized.

5. ***Early anaphase.*** The two daughter chromatids of each chromosome separate and migrate toward opposite poles, a process that occurs simultaneously throughout the entire set of chromosomes. During this migration the chromatid forms a V-shape as the centromere moves ahead, apparently dragging the rest of the chromatid behind it. The former equatorial arrangement of the chromosomes ensures that the chromatids of each chromosome move to different poles. At the same time, the cell elongates, as do the polar spindles, whereas the kinetochore spindles become shorter in accordance with the polar migration of the chromatids.

The forces responsible for separation of the chromatids and their movement to opposite poles are not fully understood. Spindle fibers consist of microtubules. Possibly the force is generated by shortening of the microtubules attached to the kinetochores. There is evidence for depolymerization of kinetochore microtubules at their point of attachment to the kinetochore, so that the kinetochore may be envisaged as "eating" its way along the spindle fiber. Alternatively, a force could be generated by a sliding action between the kinetochore microtubules and the polar microtubules.

6. ***Late anaphase***. Daughter chromatids are usually referred to as chromosomes. Polar migration of chromosomes is almost complete and a cleavage furrow starts to form.

7. ***Telophase.*** The chromosomes form a dense knot at each pole, and the microtubules of the spindle are degraded. Simultaneously, cell division (cytokinesis) begins. The nuclei of the daughter cells take on the structure of interphase nuclei, i.e. the chromosomes become longer by unwinding and unfolding of the nucleohistone strands, finally forming more or less homogeneous Chromatin (see) with no light-microscopically visible chromosomes. Parts of the endoplasmic reticulum combine to form a new nuclear membrane, and the nucleolus is reformed at the nucleolus organizer.

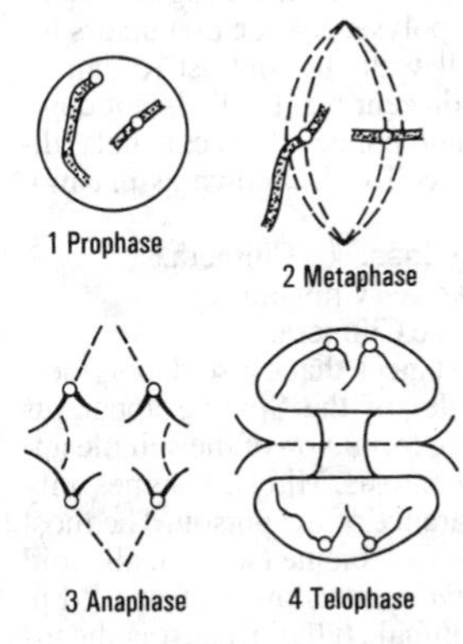

Fig.1. *Mitosis.* Major phases of mitosis in a cell containing 2 chromosomes.

Daughter cells are finally separated by ***cytokinesis.*** In animal cells, the cleavage furrow (in a plane perpendicular to the long axis of the mitotic spindle) continues to deepen until opposing edges make contact. The membranes then fuse, forming two daughter cells. Contractile forces responsible for forming the furrow are probably generated by actin. The narrow, dense region between daughter nuclei formed by contraction of

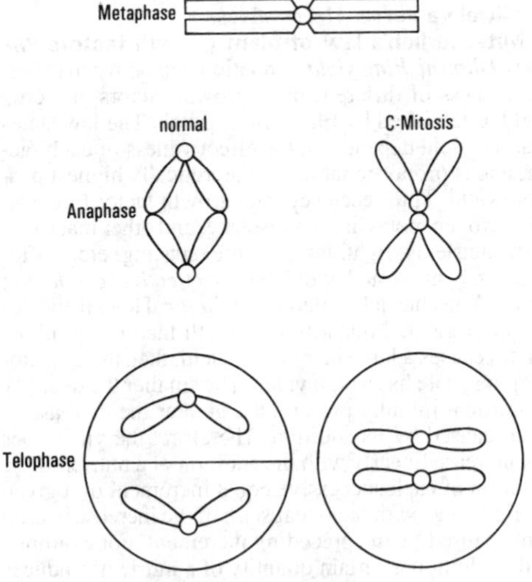

Fig.2. *Mitosis.* Schematic representation of normal mitosis and C- mitosis.

the cleavage furrow is sometimes called a *Flemming body*. Since plant cells possess an inelastic cell wall, cytokinesis cannot occur by the contraction of a cleavage furrow. In the mitosed plant cell, a new cell membrane and cell wall are formed from membrane vesicles (arising from the endoplasmic reticulum or Golgi apparatus), which aggregate in the equator of the cell. These vesicles first appear during metaphase, when they have a more scattered distribution, extending into the mitotic apparatus. This cytoplasmic mass of vesicles and microtubules between the daughter nuclei of plant cells has been called the *phragmoplast* by some authors. In late anaphase, the vesicles of the phragmoplast fuse into sheets at the equator of the spindle, forming the plasma membrane of the daughter cells. These vesicles also carry polysaccharide precursors for synthesis of the new cell wall. In contrast to animal cells, the cytoplasm of adjacent plant cells is not completely separate, being interconnected via channels (diameter 20–60 nm) in the cell wall, known as plasmodesmata.

Mitotic chromosome loss: see Chimera.

Mitotic crossing over: see Chimera.

Mitotic disjunction: see Chimera.

Mitotic poisons: substances that cause depolymerization of the microtubules of the Spindle apparatus (see), as well as preventing formation of the spindle apparatus, thereby blocking mitosis. Mitosis resumes only after removal or disappearance of the poison. The most familiar of these agents is Colchicine (see), an alkaloid from *Colchicum autumnale* (meadow saffron). M.p. bind specifically to microtubule tubulin, causing the tubulin to dissociate into its subunits. Only the microtubules of cilia and flagella are resistant to colchicine. It does not prevent separation of chromatids during Mitosis (see), but since it does destroy the spindle apparatus, all movements of chromatids during anaphase are restricted to one cell nucleus, and all the chromatids remain in that nucleus. Colchicine is used to induce polyploidy. The alkaloid, vinblastine (from *Vinca*, periwinkle) also causes disintegration of tubulin. Arsenicals also act as M.p.

Mitral valve: see Heart valves.

Mitscherlich's law of plant growth factors, ***the law of diminishing yield:*** a relationship between the effectiveness of different plant growth factors and crop yield, established by E.A. Mitscherlich. The law states that the yield depends on the effectiveness of each factor, and is proportional to the theoretically highest possible yield. Thus, each separate growth factor (e.g. water, nitrogen, potassium, phosphate, and other macro- or micronutrients, light, temperature, spacing, etc.) influences the harvested yield by an *effectiveness factor,* which Mitscherlich called the *c-factor.* Thus, if the addition of a small quantity of growth factor (e.g. phosphate) causes a large increase in yield, then the *c*-factor for phosphate has a high value. The smaller the quantity of nutrient initially present, the greater the increase in yield caused by its addition. Therefore, the yield does not increase linearly with the addition of a nutrient. The addition of each successive equal increment of a given nutrient or growth factor causes half the increase in crop yield caused by the preceding increment. For example, if the addition a certain quantity of a nutrient produces a yield equivalent to 50% of the maximal theoretical yield, then twice this quantity of nutrient will produce a yield that is 75% of the maximal theoretical yield, and three times the quantity will produce a yield that is 87.5% of the maximal theoretical yield, etc. [The quantity of any growth factor which will produce one half of the theoretically highest yield is known as a *Baule unit].* The relationship can be expressed as: $\log (A - y) = \log A - c\,(x + b)$, where A is the maximal theoretical yield, y is the actual yield, x is the quantity of nutrient applied, b is the quantity of the given nutrient in the soil in an available form, and c is the effectiveness factor. In view of this relative decrease in yield with the addition of nutrients (i.e. fertilizer), the curves obtained by plotting yield against fertilizer input have considerable economic importance, in addition to their obvious scientific interest. At the same time it should be noted that the highest theoretically possible yield (A) will vary with the supply of other nutrients, climatic conditions, system of cultivation, etc., so that a growth factor exerts its optimal effect only if other growth factors are present in sufficient quantity or intensity.

Mitscherlich was professor of Agriculture at the University of Königsberg in East Prussia. His law of diminishing yield resulted from his efforts to find a method for determining the quantity of plant nutrients farmers should apply to increase crop yields. The reliability of his results is supported by numerous pot tests and about 30 000 field trials (E.A. Mitscherlich: *Die Bestimmung des Düngungsbedürfnisses des Bodens.* P. Parey, Berlin, 1930). The most important aspect of his work is the determination of specific c-values for numerous growth factors.

Mitten crab: see Chinese mitten crab.

Mixoeuhaline: see Brackish water.

Mixohaline: see Brackish water.

Mixomesohaline: see Brackish water.

Mixonephridia: see Nephridia.

Mixo-oligohaline: see Brackish water.

Mixophyceae: see *Cyanophyceae.*

Mixoploidy: variations in the number of chromosomes among the cells of tissues, organs or organisms. See Endomitosis, Polyploidy.

Mixopolyhaline: see Brackish water.

Mnium: see *Musci.*

Mnium-type stomata: see Stomata.

Moas: see *Dinornithidae.*

Mockernuts: see *Juglandaceae.*

Mockingbirds: see *Mimidae.*

Moco: see *Caviidae.*

Model system: a system with physical and/or biological parameters resembling those of a usually larger system. For various reasons (legal, financial, ease of handling, etc.), the larger system may be less amenable to study, so that the M.s. is investigated instead. The classical example is the use of scale models to analyse the aerodynamic and hydrodynamic flow charcteristics of aeroplanes, birds, ships, fish, etc. in the laboratory, using M.ss. that have the same Reynolds number (see). The use of M.ss. is also called *simulation analysis.* The correspondence of the behavior of geometrically, kinematically and dynamically similar systems is called *similitude.* In pharmaceutical research, e.g. in drug testing, laboratory animals are used as models of the human system for the investigation of physiological parameters such as cardiac frequency, respiratory volume and heat production.

Moder: see Humus.

Moderlieschen, *Leucaspius delineatus:* a small, slender, freshwater fish of the family *Cyprinidae,* order *Cypriniformes,* found in large shoals in European lakes and slow-flowing rivers that contain plenty of vegetation. It has a brownish green back and silvery sides, with a pale silver-blue band from each operculum to the tail. The bases of the paired fins are usually tinged red. It is not commercially important, but is sometimes eaten in central Russia, and is also used as a bait fish in Europe.

Modes of nutrition (plate 44): different strategies for acquiring and taking in food by plants and animals. For the M.o.n. of plants, see Assimilation, Autotrophy, Heterotrophy, Symbiosis. A simplified classification of M.o.n. in animals is based on the type of diet and on the means of acquiring it, both of which display a relationship with body structure and Life form (see).

1) ***Type of diet.*** Animals that take up living organic material are said to be *biophagous* and are called *biophages.* Of these, the plant eaters are known as *phytophages,* and the animal eaters as *zoophages.* Animals living on dead organic material are called *necrophages. Omnivores* take both dead and living, plant and animal material. Phytophagous animals can be further subdivided into *phytoepisites* (e.g. grazing hoofed animals) and *phytoparasites* (e.g. *ectoparasitic* leaf aphids and shield bugs, or *endoparasitic* apple codlin moth larvae, root gall nematodes and leaf miners). According to the type of dietary plant material, phytophages can be subdivided into *herbivores* (eaters of green plant parts), *fructivores* (fruit eaters), *mycetophages* (fungus eaters), *xylophages* (wood eaters), and others.

Zoophagous animals may be *episites* (predators) or *parasites* (see Relationships between living organisms). According to the type of dietary animal material, they may also be classified as *carnivores* (meat eaters), *insectivores* or *entomophages* (insect eaters), and *hemophages* (blood suckers).

Necrophagous animals display numerous transitions to biophagy, e.g. bark beetles start their development in a living tree and finish it in the dead trunk. The majority of necrophagous animals feed on material that is already in a state of decomposition, so that they also consume the living saprophages that are present. Others feed on excrement (see *Coprophage).*

2) ***Mode of food capture and intake*** (plate 44). *Suspension feeding* is widespread among aquatic animals. Small, suspended dietary particles are removed by adhesion to secreted mucus, usually from a water current which is generated along or into the body by the action of cilia, e.g. *Paramecium, Rotifera,* many marine polychaetes, bivalves and *Amphioxus;* the same water stream often also serves for the supply and removal of dissolved respiratory gases. Aquatic *filter feeders* retain suspended dietary particles with various filtration structures (pharyngeal basket-shaped filters, sheets of perforated gill tissue, gill bars in the form of a straining grid, horny sieves), e.g. large particles are filtered from water entering the test of the *Appendiculariae* (see) and finer dietary particles are retained by the filtering apparatus within the test. Other examples are the gill filters of the fish, *Coregonus,* the sieving apparatus in the bills of ducks, geese and flamingos, and the baleen of baleen whales. *Tactile feeders* are mostly marine, and they search their environment with tentacles and siphons, e.g. sessile *Cnidaria. Substrate feeders* live in water, soil or decomposing matter, and they feed by taking in the unfiltered and unsorted surrounding substrate, from which they then extract nutrients. Thus, tellins differ from all other clams in that they are not filter feeders, but use a long siphon to suck up sediment from the bottom (e.g. the peppery furrow shell, *Scrobicularia plana);* coarse material is then crushed in the stomach before it reaches the digestive glands. Other examples of substratum feeders are earthworms, marine polychaetes, sea cucumbers, crabs and certain fishes (e.g. mullets). *Grazers* have well developed mouthparts, with which they remove their food material by tearing, biting or rasping; before digestion, the food is usually crushed and broken mechanically by the same mouthparts, e.g. terrestrial snails, numerous phytophagous insects, algae-grazing sea urchins and certain fishes (e.g. parrot fish), while grazing mammals include many rodents and ungulates. *Gatherers* include birds which take smaller plants or animals from their surroundings without hunting. Their mouthparts are often specialized (conical beaks, probing beaks); often there is a crop and/or storage stomach; mechanical reduction of the dietary material does not occur, or only to a small extent, e.g. galliform birds, sandpipers, tits. *Hunters* must first detect their prey then pursue or stalk it. They have well developed sense organs, are capable of rapid movement, and they usually possess efficient grasping organs or biting mouthparts, e.g. predatory mammals (including bats), birds of prey, snakes, wolf spiders, dragonflies, predatory bugs, ground beetles, centipedes and some nematodes. *Ambush hunters* do not search for their prey, but wait for it at a stationary site, e.g. the praying mantis. *Trap layers* prepare traps with substances from their own body or with foreign material, e.g. webs of web-spinning spiders, mucous net of worm snails *(Vermetidae),* capture nets of caddis fly larvae, sand funnels of ant lions, extensible tongues of frogs and chameleons). *Licking and lapping feeders* take in liquids or lick up small animal prey with a proboscis or tongue, e.g. bees, butterflies, hummingbirds, honeyeaters, woodpeckers, flying foxes, anteaters and armadillos. *Piercing and sucking feeders* pierce animal or plant tissue and withdraw liquids such as blood, lymph, plant juices or liquified tissue, e.g. many parasites such as trematodes, aschelminths, leeches, numerous insects (leaf aphids, cicadas, bugs, midges, mosquitoes, fleas), ticks and lamprey.

A wide variety of different mouthparts has evolved, each appropriate to a different type of food acquisition.

Modification:

1) Changes in a Phenotype (see) of a living organism caused by its conditions of development (see Environmental factors). These changes are not inherited by the offspring. During development of an organism, sensitive phases of physiological processes are effected by environmental conditions. The different morphological and physiological characters of the fully developed organism therefore display varying degrees of dependence on the environment, being environmentally stable (not altered by the environment) or environmentally unstable (altered by the environment). If a species differentiates into characteristically distinct forms under different living conditions (e.g. wet meadows, dry rock outcrops, high altitudes, saline soils), the condition is known as *polyphenism.* Sometimes, modifications persist for several generations in the absence of the appropriate environmental conditions. Such modifications, known *dauer-modifications,* or *semipermanent modifications* gradually die out, but at first give the false impression that they are inherited. It appears that semipermanent modifications are due to alterations in self-reproducing components of the cytoplasm, so that they persist in vegetative reproduction. If the cytoplasm of the egg is affected, the modification may persist for several generations; otherwise, modifications disappear abruptly in sexual reproduction. A modification due to poor nutrition may also persist in a well fed subsequent generation, e.g. poorly nourished and poorly irrigated radishes fail to produce a swollen hypocotyl and produce seeds with deficient food reserves; the seedlings are therefore malnourished and also fail to produce a "radish". Similarly, environmentally determined modifications in the morphology of an arthropod cannot be "corrected" until the next molt.

2) See Restriction, Restriction enzymes.

Modification enzyme: an enzyme that catalyses the introduction of minor bases into RNA or DNA. The normal base is modified after its insertion into the polynucleotide.

Modification gene: a gene whose product is a Modification enzyme (see).

Modifiers, ***modifying genes:*** nonallelic genes that decrease or increase the degree of phenotypic expression of other main genes, but are not generally expressed phenotypically in their own right.

Nonspecific M. affect the relevant phenotypic character in the same way in the normal genotype and in mutants. The action of *specific M.* is restricted to a certain genotype. They act either on the normal genotype or on mutant genotypes; they may also act on both, but in different ways.

Modifying genes: see Modifiers.

Mola mola: see Sunfish.

Molar absorption coefficient: see Absorbance.

Molar absorptivity: see Absorbance.

Molar extinction coefficient: see Absorbance.

Molar teeth: see Dentition.

Molas: see *Tetraodontiformes.*

Molds: a collective term for surface-living fungal mycelia. It is commonly used for saprophytic members of different groups of fungi, which cover their solid (-rarely liquid) substrate with a "ghostly" fur of hyphae. On paper and leather, etc., M. are also known as mildew. Among the commonest M. fungi are *Aspergillus* and *Penicillium* (class *Ascomycetes,* order *Plectascales)* and *Mucor* (class *Phycomycetes,* order *Zygomycetales). Capnodium,* one of the "sooty molds" found on leaves, is an ascomycete.

Mole cricket: see *Saltoptera.*

Molecular biology: a branch of biology devoted to the molecular basis of living processes. Since the 1950s, M.b. has developed extremely rapidly and has provided theoretical and practical results of such significance that it was described by M.W. Keldysch as the "science of the century". The term. M.b., was used in the 19th century by J.E. Purkinje. In 1938, it was applied by W. Weaver to "those border areas in which physics and chemistry come into contact with biology". Weaver remarked that "a new branch of science, M.b., is gradually taking shape, which is beginning to decipher many of the final secrets of the living cell".

The modern development of M.b. started in the mid 1940's in parallel with the development of Molecular genetics (see). There were essentially three sources of impetus for this modern development. First, physicists, chemists and other nonbiologists began to take an interest in the basis of living processes, in particular in genetics. They were largely motivated by the work of N.W. Timofeeff-Ressovsky, M. Delbrück and E. Schrödinger in their search for "new physical laws in biology". Some of these physical scientists also rejected the political use of atomic energy, and wished to work in a field with no relevance to the arms race. Second, the founders of M.b. adopted a new and justifiable reductionist approach by introducing experimental biological systems that were much simpler than the mouse, evening primrose, *Drosophila* or the flour moth. By using fungi, bacteria and bacteriophages, they simultaneously founded Microbial genetics (see). Third, new methodology played a significant part in the development of modern M.b. Thus, the development of the ultracentrifuge, electron microscope, X-ray diffraction analysis, stable isotopes and radioisotopes, and separation techniques (chromatography and electrophoresis), etc. enabled detailed investigations of macromolecules. The truly triumphal march of M.b. began in the 1950s, and was marked in 1959 by the founding of the very successful Journal of Molecular Biology.

It is difficult to define the research material of M.b. Like Gene technology (see) and Molecular genetics (see), M.b. is characterized by its methodology rather than its experimental material. Furthermore, during the development of M.b., attention has been focused on different biological systems at different times, and this evolution of emphasis can be clearly seen in the yearly contents of the Journal of Molecular Biology since 1959.

At first, M.b. was largely identical with molecular genetics, and it was mainly concerned with the structure and function of nucleic acids and proteins. Modern M.b. includes, or overlaps with, parts of immunology, membrane biology, virology, cell biology and other biological disciplines. In this sense, M.b. embraces everything concerned with the molecular structure and function of living organisms.

As pointed out by W.A. Engelhardt in 1971, M.b. is characterized by "its three-dimensional character. It is precisely this peculiarity that distinguishes M.b. from its predecessor, biochemistry. Three-dimensional thought is totally alien to the latter".

Furthermore, according to Engelhardt, the concept of information is essential in the development and characterization of M.b. This is manifested in a variety of different ways and in different areas of the subject ..."The starting point is the linear primary structure of macromolecules, which carries information, in one case for the generation of secondary and tertiary structure (i.e. the spatial configuration of protein molecules), in the other case for the specification of the genetic code. New levels of complexity, as manifested in the conformational changes of protein molecules, are all ultimately derived from this linear information, and they have far-reaching consequences for the properties and biological functions of protein molecules. ... Special information for biological integration, i.e. for the transition from lower to higher levels of structural and functional order, is likewise always encoded by structure".

In summary, W.A. Engelhardt believes that the determining features of M.b. are its three-dimensional character, the dependence of function on molecular structure, and the principles of information storage.

In general, M.b. is concerned primarily with the realization of implicit genetic information, with the epigenetic process, and with the molecular structures that determine this process. In contrast, molecular genetics is concerned with explicit genetic information, its carriers and its expression.

Molecular biophysics: a branch of biophysics which studies the physics of the molecular components of living material. Starting from the physical and chemical properties of low molecular mass components, M.p. attempts to elucidate the structure and function of biological macromolecules. This provides a basis for understanding the transmission and processing of genetic information, enzymatic catalysis, transduction of information, material and energy by membranes, cooperative self assembly of molecules, etc The methodology of M.b. includes high discrimination quantum physical and spectroscopic techniques, e.g. ESR-spectroscopy (see), NMR-spectroscopy (see), Fluorescence spectroscopy (see), X-ray diffraction methods, etc. Analysis of the experimental data nearly always requires computer assistance.

In recent years M.b. has made extensive contributions to the study of the structures of proteins and nucleic acids, and is making an increasing contribution to the study of the dynamics of macromolecules.

Molecular clock: the hypothesis that the amino acid composition of homologous proteins changes at practically the same rate during evolution. If the number of amino acid replacements in a given protein (e.g. cytochrome *c*) in a certain period of evolutionary time is known, then the differences between the cytochromes of two organisms is an index of the time that has elapsed (in millions of years) since the respective evolutionary lines diverged from their last common ancestor.

Protein analyses, and more recently DNA sequence analysis, show that the evolutionary rate of amino acid substitution differs between proteins, i.e. the "M.c. runs at different speeds", depending on how strongly the function of a protein depends on its amino acid sequence (which in turn depends on the base sequence of its structural gene). Thus, the M.c. of histones is extremely slow, whereas that of fibrinopeptides is relatively fast. It is nevertheless possible that the mutation rate per nucleotide pair is very similar for all genes, but that mutations in some genes (e.g. histone genes) are more often lethal and lead to elimination of the mutants.

Molecular genetics: a branch of genetics concerned with the structure, function and modification of genes. Founded in the 1940s as Microbial genetics (see), it developed in parallel with Molecular biology (see), and eventually led in the 1970s to the growth of Gene technology (see).

Growth and development of M.g. were made possible by the introduction of bacteria (in particular *Escherichia coli),* bacteriophages (in particular T-series coliphage and lambda phage) and certain fungi as experimental organisms by Delbrück, Demerec, Hershey, Jacob, Lederberg, Luria, Lwoff, Monod and others. This enabled the formulation of the *one gene one enzyme hypothesis* (Lederberg and Tatum, 1945), the demonstration that DNA is the carrier of genetic information (Avery et al., 1944, Hershey and Chase, 1952), the elucidation of the structure of DNA (Watson and Crick; Franklin and Wilkins, 1953), the demonstration that viral RNA also carries genetic information (Fraenkel-Conrat; Gierer and Schramm, 1956), the discovery of messenger RNA (Brenner, Jacob and Meselson, 1961) [which in accordance with the "central dogma of molecular biology" (Crick, 1957) serves as a template for gene-regulated protein synthesis], elucidation of the genetic code (1961–1966 by Crick, Khorana, Matthei, Nirenberg, Ochoa, Wittmann and others), as well as the development of Gene technology at the beginning of the 1970s (by Arber, Berg, Boyer, Nathans, Smith and others).

The essential contributions and findings of M.g. are as follows.

1. In order to serve as the *genetic material,* a biological molecule must both store information, and be capable of self replication (Muller, 1922). DNA and the RNA genome of RNA viruses are the only molecules that satisfy these requirements. Normal cellular RNA (messenger, transfer and ribosomal RNA) does not qualify as genetic material since it is not capable of self replication, and is synthesized on a template of DNA. Information is encoded in the variable linear sequence of four different DNA or RNA components (nucleotides), which are designated A, C, G and T (or U in RNA). Replication is enabled by specific pairing of complementary nucleotides [A with T (or U); C with G] via hydrogen bonding. Replication is catalysed by polymerases in a complex with several other enzymes. During replication, a complementary daughter strand is produced on the template of the parent polynucleotide strand, a process known as semiconservative replication.

Bacterial (prokaryotic) genetic material is more or less naked, whereas the DNA of viruses that attack eukaryotes forms DNA-histone complexes, known as "minichromosomes". Since eukaryotic and bacterial DNA are functionally homologous, the bacterial nuclear equivalent or nucleoid is also referred to as the bacterial chromosome. In addition, bacteria also possess smaller, extrachromosomal DNA molecules, called plasmids. Plasmids that can be reversibly incorporated into the bacterial chromosome are called episomes. Each of these DNA molecules is an autonomous replication unit or replicon.

In eukaryotes (characterized, inter alia, by the possession of a nuclear membrane), numerous replicons are combined into extremely long DNA molecules. Together with histones and acidic structural proteins, these DNA molecules constitute the chromatin of the cell nucleus. At certain stages of the cell cycle, chromatin condenses into microscopically visible chromosomes (see Cell cycle, Meiosis, Mitosis).

In addition to the genetic material of the cell nucleus, eukaryotes also possess extranuclear genetic material. In animals this is found in mitochondria, and in plants in both mitochondria and plastids. As in prokaryotes, the extranuclear genetic material of eukaryotes consists of naked DNA molecules.

2. The genetic material contains explicit *genetic information.* This takes the form of a code that specifies the specific amino acid sequences of unique polypeptide gene products and the nucleotide sequences of transfer and ribosomal RNA. Explicit genetic information is directly accessible from the nucleotide sequence of the codogenic DNA (or viral RNA). In contrast, implicit genetic information, which determines the func-

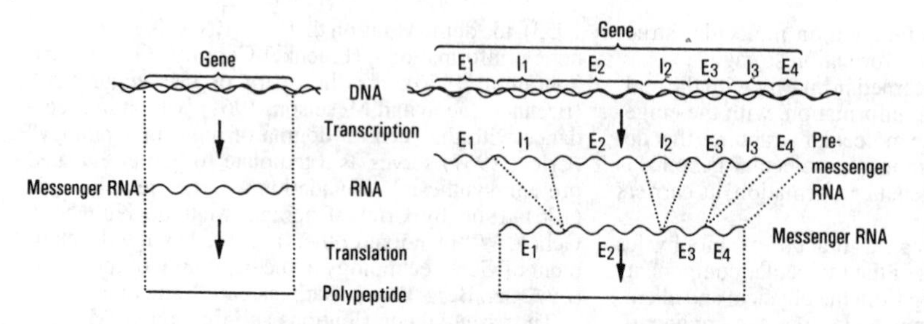

Fig.1. *Molecular genetics.* Prokaryotic and eukaryotic genes and their expression. Most eukaryotic genes are split and consist of codogenic exons (E) and noncodogenic introns (I).

tion and interaction of gene products, is not directly accessible from the structure of the genetic material.

3. *Genes* are sections of genetic material, each encoding unique portions of explicit genetic information. Their existence was first revealed in 1986 by Mendel in hybridization experiments, and they were named genes in 1909 by Johannsen. In prokaryotes and their viruses, genes consist of uninterrupted sequences of about 1 000 nucleotides, separated from one another by start and stop codons (Fig.1). Changes of nucleotide sequence produce different forms of the gene with a greater or lesser information content, which are known as alleles. These occupy a specific site on the chromosome, known as the gene locus.

Genes of eukaryotes and their viruses are usually discontinuous and may consist of tens of thousands of nucleotide pairs (Fig.1). Sections containing coded genetic information are called exons, and these are separated by introns.

Viral genes may overlap, so that a single section of genetic material may encode several different pieces of explicit genetic information.

Viruses possess a few to several hundred genes. Bacteria have several thousand genes, *Drosophila* about 5 000 genes, and the human genome some 50 000 genes. Evolutionary advancement is not correlated directly with genome size, but on the genetic level it is correlated with increasing integration of genetic information, i.e. with an increase in the content of implicit genetic information.

4. In accordance with the "central dogma of molecular biology", the transmission of *explicit genetic information* always proceeds unidirectionally from DNA (or viral RNA) to protein, and never in the reverse direction (Fig.2). This largely guarantees the stability of genetic information and excludes the "inheritance of acquired characters". Elucidation of the individual stages of genetic information transfer and their regulation has been a central research theme of molecular biology. The process proceeds by transcription of DNA to produce messenger RNA, followed by translation of the Genetic code (see) of the messenger RNA. The existence of a universal genetic code indicates that life has a monophyletic origin, i.e. it arose successfully only once, thus lending significant support to the theory of the unity of living material, and to the theory of evolution. This does not exclude the possibility that other unsuccessful episodes of molecular self-organization occurred during the prebiotic phase of evolution.

Replication
DNA
Transcription
Reverse transcription of tumor virus RNA
In the presence of certain antibiotics
RNA
Translation
Protein
Replication of viral RNA

Fig.2. *Molecular genetics.* Central dogma of molecular biology.

5. An essential property of genetic material is its ability to change by mutation and/or recombination. *Genetic variability* is a crucial precondition for biotic evolution.

Mutations are more or less drastic changes in genetic material, evoked by temperature changes, radiation, chemical agents, some viruses, etc., and which also occur spontaneously, e.g. as a result of errors during replication. Many mutagens act indirectly by causing damage to DNA or viral RNA; errors in the subsequent repare process may then introduce mutations.

Point mutations are due to substitution of nucleotide pairs, or the introduction or deletion of nucleotide pairs. Loss or gain of nucleotide pairs leads to a shift in the reading frame ("frame shift mutation"). In both cases, mutant alleles are produced. Owing to the degenerative nature of the genetic code, mutations do not always cause a detectable change in the genetic information, e.g. in a "same sense" mutation a codon is substituted by a synonymous codon that encodes the same amino acid. A "permitted error" mutation produces a mutant codon that encodes a chemically similar amino acid. Most mutant alleles are moreover recessive, i.e. they are not expressed in the heterozygous condition in the presence of the corresponding wild type or normal allele; they are only expressed in the homozygous state or in hemizygotism (i.e. in a male organism, when the mutant gene is located on the X-chromosome).

Chromosome mutations are due to structural changes in chromosomes. At least in prokaryotes, these are caused, inter alia, by "jumping genes" (mobile elements of DNA). In both prokaryotes and eukaryotes these are DNA sequences with shared structural characteristics, which are preferentially incorporated at specific sites in chromosomes, thereby splitting genes and (at least in prokaryotes) causing a wide variety of different chromosome mutations: deletions (loss of entire sections), duplications and further amplifications (multiple gene copies), translocations or transpositions (shift of a gene to a different locus), as well as inversions (turned through 180 °). The effects of chromosome mutations are very variable, and depend on the type and extent of the structural change. Amplification and translocation have considerable evolutionary significance. Chromosome mutations in somatic cells are important in ontogenesis; for example, the genetics of the immune system is based essentially on chromosome restructuring, in which hundreds of thousands of different immunoglobulin (antibody) genes arise from a few hundred precursor genes.

Genome mutations (numerical chromosome mutations) have so far been little studied by the techniques of M.g. Since they are responsible for numerous genetically determined illnesses, they have largely been a subject of cytogenetics and medical genetics. The causative molecular processes are still mostly unknown, but faulty regulation mechanisms appear to be primarily responsible.

6. In addition to mutation, *recombination* also contributes to genetic variability. The independent assortment of allele pairs in dihybrid crosses, as observed by Mendel, is based on *interchromosomal recombination.* It also occurs in viruses whose genome consists of more than one nucleic acid molecule. For example, in a mixed infection of cells with two different strains of influenza virus, completely new types of viruses may arise by interchromosomal recombination. Uncoupling of "linked genes" that are localized on the same chromosome is due to *intrachromosomal recombination.* Discovery of this phenomenon in 1910 permitted the preparation of chromosome maps, in which the distance separating genes is proportional to their recombination frequency.

The developed methodology of M.g. enabled the investigation of the molecular basis of recombination. This led to the discovery of numerous enzymes that catalyse recombination ("recombinases"). These catalyse either homologous recombination (in which homologous chromosome sections are paired and exchanged against each other), or nonhomologous ("illegitimate") recombinations between nonhomologous chromosomes. Jumping genes induce chromosome mutations by their nonhomologous incorporation into the genome.

7. M.g. led ultimately to the development of *Gene technology* (see).

Mole lemmings: see *Arvicolinae.*

Mole rats: see *Cricetinae, Spalacidae.*

Mole salamanders: see *Ambystomatidae.*

Molidae: see *Tetraodontiformes.*

Mollisols: an order of mineral soils found mainly in grasslands with a seasonal water deficiency (e.g. central North American plains and South American pampas). They have a deep, dark surface horizon containing well decomposed organic material (at least 1%). M. are some of the most fertile soils in the world, and they support most of the world's cereal crops.

Mollusca, *mollusks*: a phylum of nonsegmented invertebrates, containing a wide variety of forms and structural types, ranging from millimeter-sized snails and mussels to the giant cephalopods, which can attain a length of 18 meters with their tentacles. With respect to the number of species (about 130 000 known species), the *M.* constitute the largest invertebrate phylum after the arthropods.

***Morphology*.** The ancestral molluscan form was bilaterally symmetrical, but in some modern groups this archeotype condition has been altered (e.g. the *Gastropoda* have undergone twisting or *torsion,* i.e. when viewed dorsally most of the body behind the head, including the visceral mass, mantle and mantle cavity, is rotated anticlockwise through 180 degrees). The skin contains mucus glands, which are especially abundant on the foot, and which represent an adaptation to an aquatic existence or to a life in wet terrestrial habitats. The head is more or less differentiated from the body. In some members of the *Gastropoda* and *Aculifera,* the head is reduced, and it has disappeared entirely in the *Bivalvia.* The head contains the mouth opening, the central control ganglia of the nervous system, and the most important sense organs. Within the arched, dorsal part of the body (the *visceral hump)* are found the internal organs (sometimes referred to as the *visceral mass),* while the ventral part of the body is developed into a flattened and usually very muscular, unpaired creeping sole or *foot.* The latter varies in shape and size, depending on the type of locomotion. The dorsal surface is covered by an external calcareous shell (in the *Conchifera)* or by spicules (in the *Aculifera);* it is assumed that the shell of the ancestral mollusk was oval, convex and shield-like. The shell is secreted by the underlying epidermis, which is known as the *mantle* (or *pallium).* Secretion is most active around the edge of the mantle, but new material is also added to the older, more central parts of the shell, which therefore grows in both thickness and diameter. A fold of the mantle overhangs the body; in some species it extends sufficiently to also overhang the head. Posteriorly, the fold or overhang creates a chamber called the *mantle cavity.* Only the cephalopods possess an internal skeleton, consisting of a cartilaginous capsule around the central nervous system.

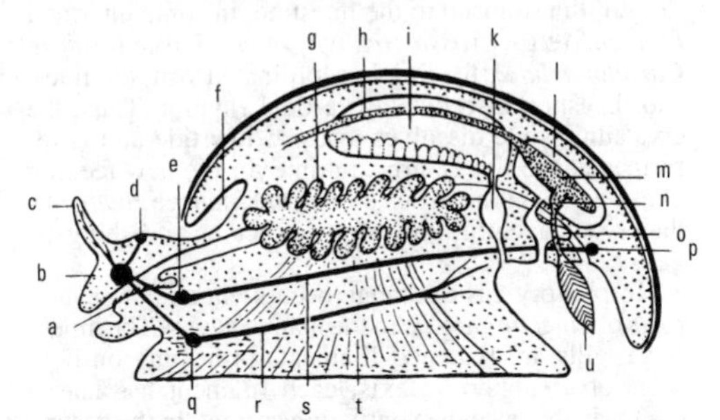

Mollusca. Generalized body plan of a mollusk. *a* Radula. *b* Head. *c* Tentacle. *d* Eye. *e* Pleural ganglion. *f* Mantle cavity. *g* Digestive gland (hepatopancreas). *h* Aorta. *i* Genital coelom with gonads. *k* Gonoduct. *l* Pericardium. *m* Ventricle. *n* Auricle. *o* Coelomoduct ("kidney"). *p* Visceral ganglion. *q* Pedal ganglion. *r* Pedal cord. *s* Pleurovisceral cord. *t* Foot. *u* Gill (ctenidium).

Anatomy. The alimentary canal consists of the foregut (comprising the buccal cavity and esophagus), the midgut (including the intestine and stomach; the latter usually has a protostyle, as well as a very extensive digestive gland or hepatopancreas, where absorption occurs), and the hindgut (rectum), which empties posteriorly into the mantle cavity, or (in species with an asymmetrical visceral sac) into the mantle cavity at one side of the body.

The buccal cavity contains a toothed Radula (see), except in the *Bivalvia,* which are filter feeders. Mucus, secreted into the buccal cavity by salivary glands, lubricates the radula and traps food particles. Strings of mucus with adherent food particles pass from the buccal cavity, along the esophagus and into the stomach, drawn by the winding action of the *protostyle* or *fecal rod.* The protostyle, found in the primitive *Bivalvia,* is a rod of consolidated, amylase-containing mucus, secreted by gland cells in the style sac (posterior, conical region of the stomach) and rotated by the action of cilia in the style sac wall. The acidity of the gastric fluid (pH 5–6), as well as the action of enzymes, decreases the viscosity of the mucus, so that the food particles are released. In the noncarnivorous *Mesogastropoda* and the higher *Bivalvia,* the protostyle is a highly specialized structure known as a *crystalline style;* this flexible rod of hyaline glubuloprotein contains carbohydrate- and lipid-cleaving enzymes (secreted by the style sac during style formation), and it may attain a length of 3 cm or even longer (5 cm in *Tagelus).* As the crystalline style is rotated by the action of the style sac cilia, it acts as a windlass for the food-carrying mucus strand, and at the same time stirs the stomach contents. Close to the head of the crystalline style the stomach wall is protected by a cuticulogastric shield, which abrades the style; presumably the enzymes released during this abrasion contribute to digestion. Enzymes are also released from the stomach wall. Except in the ridged, ciliated sorting region, the anterior part of the stomach is lined with chitin. In the sorting region, fine particles and digestion products are swept by cilia to the duct openings of the digestive diverticula, where digestion is completed, while larger, heavier particles are carried in a deep groove along the floor of the stomach to the intestine. In some intertidal *Bivalvia* (e.g. *Crassostrea virginica, Lasaea rubra, Cardium edule),* feeding is coordinated with the tides, and digestion also displays a tidal rhythm. Thus, the crystalline style dissolves away at low tide and is reformed as the tide returns, in time for the new feeding phase. In the carnivorous *Septibranchia* (see *Bivalvia),* the muscular, chitin-lined stomach acts as a crushing organ.

Respiratory gas exchange occurs via one or several pairs of *dipectinate gills* in the mantle cavity. The dipectinate gills are flattened filaments projecting on both sides of a supporting axis; each filament has lateral cilia, which create the ventilating current. In the gastropods, torsion has reduced the number of gills, so that only a single gill is present. The single gill may be monopectinate rather than bipectinate, with the gill axis attached directly to the mantle wall. A network of respiratory vessels is often present on the inner, upper mantle surface, or in the vicinity of the anus, or on the upper side of the body. Respiratory gas exchange through the skin in also important.

There is an open blood circulatory system with an arterial heart, consisting of a single anterior ventricle and one or two posterior auricles; an arterial canal system and a venous canal system expand into a more or less extensive hemocoel; the respiratory pigment is hemocyanin. The circulation of bivalves and cephalopods may appear to be closed, because the blood lacunae often develop into a form resembling vessels. The coelom displays varying degrees of development, but it is always represented by the *pericardium,* by the nephridial cavity (which communicates with the pericardium and contains the sac-like, paired or unpaired, tubular metanephridia), and by the *gonadocoel,* which contains the sex organs.

The ground plan of the nervous system consists of a nerve ring or collar around the esophagus *(circumesophageal ring* or *commissure),* the underside of which gives rise to two pairs of nerve cords. The ventral pair (the *pedal cords)* innervates the foot, while the dorsal pair (the *visceral, pleural, pleurovisceral* or *palliovisceral cords)* innervates the mantle and visceral organs. In the *Aculifera,* the nervous system lacks definite ganglia, and ganglion cells are evenly distributed along the lengths of the nerve cords; longitudinal pedal and pleural cords are present, and each pair is united by a posterior commissure dorsal to the rectum. In the *Conchifera,* ganglion cells are condensed into well defined ganglia, which form part of the circumesophageal commissure; the cerebral ganglion sends nerves to the head and foregut, while the pleural and pedal ganglia give rise to the visceral loop (innervation of the viscera and mantle) and the pedal cords; peripheral ganglia are also present. Sense organs include tactile organs (especially on tentacles), taste organs at various sites, and osphradia. An *osphradium* (or *Spengel's organ)* is a patch of thickened, ciliated, sensory epithelium on the posterior margin of each afferent gill membrane. Osphradia function as chemoreceptors, and are therefore equivalent to olfactory organs; they also respond to the quantity of sediment in the inhalent current. Light-sensitive organs include light-sensitive cells, the simple ocellus, the highly evolved cephalopod eye with lens and retina, as well as all intermediate forms between these extremes. Statocysts are generally found associated with the cerebral or pedal ganglia.

Biology. The *M.* are monosexual or hermaphrodite. Reproduction is always sexual, and self-fertilization can occur in hermaphrodite species. The larval stage possesses a shell gland, but otherwise resembles a trocophore larva. In marine *M.,* this larval stage often develops into a Veliga larva (see). *Pulmonata* often develop directly without a larval stage. Due to their evolutionary adaptability, the *M.* occupy a wide range of different habitats, including the sea, fresh water and land, from cold regions to the tropics; desert-dwelling *M.* estivate during dry periods, sealing their shells with an operculum, or by the production of a mucous or calcareous epiphragm. With the exception of the cephalopods, which can react very quickly and cover long distances with speed, the *M.* are generally sessile or move only slowly. The majority of the *M.* are omnivorous, preferring small animals and plants or plant parts, or products of decay. Many gastropods and cephalopods are carnivorous predators, and the aglossate gastropods are commonly parasitic. Nearly all the lamellibranchs are filter feeders.

Economic importance. Many species, such as oysters, squids and various snails are important food sources, while others provide materials for jewellery and decoration (see Pearls, Mother of pearl). Frequently, *M.* are pests of plants and stored foodstuffs (e.g. see Slugs), and some serve as the intermediate hosts of parasites (e.g. of the schistosomiasis parasite, see Bilharzia). *M.* were exploited more widely in former times, e.g. as a source of the dye, Tyrian purple (see), as currency (cowrie shells), as various articles in daily use (shell containers, etc.), and as a valuable source of calcium carbonate for industrial purposes.

Classification. (See separate entries for each each subphylum and class)

1. Subphylum: *Aculifera* ("spiny mollusks") with 2 classes: *Aplacophora* (solenogasters); *Polyplacophora* (chitons).

2. Subphylum: *Conchifera* ("shelled mollusks") with 5 classes: *Monoplacophora; Gastropoda* (snails, whelks, etc.); *Scaphopoda* (tusk or tooth shells); *Bivalvia* (bivalves: clams, oysters, mussels, etc.); *Cephalopoda* (octopus, squid, etc.).

In an alternative classification, subphyla are not recognized, and the phylum is divided into 6 classes; here, the *Aplacophora* and *Polyplacophora* are groups within the class *Amphineura* (which is therefore equivalent to the subphylum *Aculifera).*

Phylogeny. The affinities of the *M.* are still largely unclear.

Geological distribution. Fossils are found from the Cambrian to the present. The cephalopods and the mussels provide important index fossils, while the snails are less important.

Collection and preservation. Specimens may be collected by hand and by raking and netting on the sea shore and in shallow waters, or by dredging in deeper waters. Chitons are found in the tidal zones of rocky coasts. Suitable preservatives are 80% ethanol or 2% formaldehyde. Histological fixation is performed in Bouin's mixture. In order to preserve specimens of *Conchifera* in an extended state, they must be narcotized before fixing. Freshwater forms can be narcotized by sprinkling menthol crystals in the water, while the dropwise addition of urethane to the water is effective for marine forms. If necessary, specimens can be suffocated by placing in boiled out (i.e. air-free) water. For the preservation of entire specimens, 80% ethanol is superior to formaldehyde solution, since the latter attacks the shells. If only the shell is to be collected, the specimen is placed in boiling water for 1–2 minutes. The body is then pulled out as completely as possible, using a hook of bent wire. If an operculum is present, this should be saved together with the shell. Shells should never be dried in the sun. Cephalopods are killed by injection of chloroform or diethyl ether into the mantle cavity. Large cephalopods should be cut open, the arms extended, and the animal mounted on a suitable base, followed by fixing by treatment with 4% formaldehyde. For preservation, the formaldehyde is then replaced with 70% ethanol.

Moloch: see Thorny devil.

Molossidae, ***free-tailed bats, bulldog bats, mastiff bats:*** a family of bats (see *Chiroptera),* comprising about 80 species in 12 genera, distributed in southern Europe, southern Asia, Africa, Malaysia and east to the Solomon Islands, and from the middle of the USA southward through the West Indies and Central America into the southern half of South America. The tail projects far beyond the free edge of the tail membrane. The head is thick, with a broad, truncate muzzle, while the lips are large and the upper lip is often vertically wrinkled, giving the impression of a bulldog. The long, thick, leathery wings are relatively narrow, enabling very skilful and agile flight. A nose leaf is lacking. Spoon-shaped bristles on the outer toes of each foot are used for cleaning and grooming. Accumulated feces in caves inhabited by the guano bat *(Tadaria brasiliensis)* are a valuable fertilizer.

Molothrus: see *Icteridae.*

Molting, ***ecdysis, exuviation:*** growth-associated, periodic shedding and renewal of the outer covering of the body of crustaceans, scorpions, insects, spiders, snakes, lizards, frogs and salamanders. In practically all insects and in many crustaceans, M. occurs only up to the stage of sexual maturity, but in springtails and in some crustaceans (e.g. isopods), several molts occur after the attainment of sexual maturity. M. is controlled endogenously by hormones, e.g. in reptiles by thyroxin from the thyroid gland, and in insects by molting hormones (see Ecdysone, Ecdysterone). In arthropods, a few days before the beginning of M., the chitin exoskeleton (including the chitin lining of the tracheae, foregut and hindgut) becomes separated from the epidermis by a M. fluid, secreted by the epidermis or by specialized Molting glands (see). At the same time, the epidermis forms a new cuticle, which at first is soft and pliable, becoming rigid after few days. The old cuticle is shed quickly in its entirety, or it may be lost over a longer period (2 days) in two parts. Cast-off skins (also shells and other covering of animals, as well as fossil remains of animals) are called *exuviae.* Snakes cast their skin in its entirety, producing a ghost-like replica of the animal, whereas the old skin of lizards is lost in flakes and shreds.

In insects, M. is classified according to the growth stage attained by the molt, i.e. the larval, nymphal, pupal or imaginal molt.

Molting glands, ***exuvial glands, Verson's glands:*** small, one to three-celled dermal glands, which are present in many insect larvae and in springtails. They may be irregularly distibuted over the entire body surface, or they may display a symmetrical and segmental organization. It is thought that during Molting (see) the M.g. secrete a liquid between the cuticle and the epidermis, enabling separation and shedding of the cuticle. Since, however, all epidermal cells can produce molting liquid, and M.g. are absent from many insects, the true function of M.g. is in doubt. Nowdays, the term M.g. rather refers to the Prothoracic glands (see) and ventral glands of insects (which produce molting hormones) or to the Y-organ (see) of crustaceans.

Molybdenum, ***Mo:*** An essential trace element for plants and bacteria, and presumably for animals. It is taken up by plant roots as the molybdate anion. Like phosphate, molybdate is present in the soil in adsorption complexes, and its adsorption is increased in acidic soils. Uptake of molybdate is inhibited by sulfate and promoted by phosphate.

In plant and microbial metabolism, Mo is a constituent of certain metalloflavoproteins, which function as oxidoreductases. Thus, *nitrate reductase,* which catalyses the reduction of nitrate to nitrite, contains Mo, which functions as an electron transfer system during catalysis by alternating between two valency states (Mo^{5+} and Mo^{6+}). Nitrate reduction is particularly important, since this is the first stage in the utilization of nitrate by plants (see Nitrogen). Molybdenum is also an essential component of *nitrogenase,* a multienzyme complex present in prokaryotic, nitrogen-fixing organisms, and responsible for catalysing the reduction of atmospheric nitrogen to ammonia. Members of the *Leguminosae* (or their symbionts) therefore have a relatively high requirement for Mo, which is an essential trace element for symbiotic nitrogen fixation. Mo is also thought to be active in other plant enzyme systems, in particular in certain oxidoreductases. For example, in Mo-deficient plants both photosynthesis and ascorbic acid synthesis are decreased, whereas respiration is increased. Partial antagonism has been observed between Mo and copper: Mo inhibits cytochrome oxidase and acid phosphatase, whereas copper increases the activity of these enzymes.

Most soils contain sufficient Mo in a form available to plants. Under certain circumstances, Mo deficiency may occur on acidic soils which bind Mo very strongly. The effect is particularly noticeable in cultivated plants with a high Mo requirement, e.g. legumes (see above) and members of the *Cruciferae,* especially cauliflower and Brussels sprouts. A characteristic sign of Mo deficiency in cauliflower and other crucifers is a condition known as whiptail, in which the leaf laminas are deformed and poorly developed. They remain thin and display yellow spots between the veins, and around the edges they are gray-green and flaccid. In the middle region of the stem the lamina dies, leaving mainly bare midribs. In extreme cases the petiole stumps are abnormally twisted. These effects can be largely abolished by liming the soil, but occasionally it is necessary to use a molybdenum fertilizer (containing sodium molybdate) which is applied to the soil or directly to the leaves. Care must be taken, however, to insure that the Mo level is not too high, since ruminants are adversely affected by high concentrations of Mo in their food.

Mo has not been demonstrated to be essential for animals, but it is known to be a constituent of at least three animal enzymes. Xanthine oxidase, an enzyme of purine degradation, oxidises hypoxanthine and xanthine to uric acid. Particularly high activities of xanthine oxidase occur in milk, where its activity serves as an indicator of whether or not milk has been heated. Sulfite oxidase (isolated from mammalian and avian liver) is a Mo-containing enzyme, and aldehyde oxidase is also a molybdoenzyme.

Momordica: see *Cucurbitaceae.*

Momotidae: see *Coraciiformes.*

Monacanthinae: see *Balistidae.*

Monasa: see *Bucconidae*

Monarch flycatchers: see *Muscicapinae.*

Monarchinae: see *Muscicapinae.*

Monascidiae, *Ascidiae simplices:* solitary (non colonial) tunicates, which, although they form buds, never occupy a common tunic or test. The name is not systematic, and closely related forms may be solitary or colonial. See *Synascidiae.*

Monellin: an intensely sweet-tasting dimeric protein, isolated from the fresh fruits of an African berry, *Dioscoreophyllum cumminsii* ("wild red berry") (family *Menispermiaceae).* The A-chain contains 44, the B-chain 50 amino acid residues. It is 3 000 times (mass basis) or 90 000 times sweeter (molar basis) sweeter than sucrose. See Thaumatin.

Moneywort: see *Primulaceae.*

Mongolian gerbil: see *Gerbillinae.*

Mongoloid eye: see Epicanthic fold.

Mongoloid spot: see Sacral pigment spot.

Mongooses: members of 2 subfamilies of the *Viverridae* (see). The *Herpestinae **(African** and **Asian M.;*** 27 species in 13 genera) do not have an ear bursa (a pocket on the flap of each ear) or a perineal scent gland, and the female has 2 or 3 pairs of teats. Members of the *Galidiinae **(Madagascar M.;*** 4 monospecific genera) have an ear bursa and a perineal scent gland, and the females of 2 species have a single pair of teats.

*Mungos mungo **(Banded mongoose)*** is found in sub-Saharan Africa except Zaire and Southwest Africa. It has dark brown bands across a brownish gray back; the feet are dark brown to black and the tail has a black tip. *Herpestes edwardsi **(Indian gray mongoose)*** is found from East and Central Arabia to Nepal, India and Sri lanka. In habit it is diurnal and solitary, and in color it is gray to light brown, finely speckled with black. It was made famous in Kipling's "Jungle Book", where its description is based on a keen and accurate observations. This does, however, give the misleading impression that M. feed predominantly on snakes, or that they agressively hunt snakes. Although they are not immune to snake venom, they are not intimidated by snakes. They can overcome even a cobra by virtue of their rapid reactions. Although versatile predators, they normally seek easier prey than snakes, as well as eating plant material. Many M. have been introduced (e.g. Caribbean islands, Hawaii) to control rats, usually with devastating consequences for the native fauna. It is no longer clear whether this introduced species is *Herpestes edwaredsi* or *Herpestes javanicus* (India to Java).

Monia: see *Gruiformes.*

Moniezia: see *Oribatei.*

Moniliform antennae: see Antennae of insects.

Monimolimnetic zone: see Lake.

Monimolimnion: see Lake.

Monitor lizards: see *Varanidae.*

Monkey bread tree: see *Bombacaceae.*

Monkey puzzle: see *Araucariaceae.*

Monkeys: see Simians.

Monkey tamarind: see *Bombacaceae.*

Monkfish: see Goosefishes.

Monkshood: see *Ranunculaceae.*

Monoblepharidales: see *Phycomycetes.*

Monocarpus: see *Hepaticae.*

Monochasium: see Stem, Flower.

Monochromasia: see Color blindness.

Monoclonal antibody: an antibody produced by a single clone of Hybridoma (see) cells, which therefore consists of a single species of antibody molecule. A M.a. binds to a single site (epitope) on its antigen. In contrast, the antibodies produced by an animal in response to an antigen normally consist of a polyclonal mixture, i.e. a mixture of several different types of antibody that bind to different epitopes of the antigen.

Monocots: see *Monocotyledoneae.*

Monocotyledoneae, *Liliatae, **monocotyledonous plants, monocotyledons, monocots*** (plate 14):a class of angiosperms, whose seedlings possess only a single cotyledon. This cotyledon develops at the apex of the embryo, and its sheath surrounds the laterally displaced growing point of the stem. When the stem is viewed in transverse section, the closed vascular bundles appear irregularly scattered. The vascular bundles travel the length of the stem and leaves as parallel strands. Stem and leaves generally display little branching. Since the vascular bundles contain no cambium, there is also no secondary thickening of the type found in dicotyledonous plants. A rare type of anomalous secondary thickening does occur, however, in, e.g. *Dracaena, Cordyline, Yucca, Aloe.* With few exceptions, monocot leaves have parallel veins and they lack a petiole; they are generally simple, and their shape is linear, ensiform, lanceolate, or eliptical. In the great majority of species, the flowers consist of trimerous whorls. The primary root is usually short-lived, and is replaced by a permanent system of adventitious roots (a condition known as secondary homorhizy). Chemically, monocots are characterized by the absence of ellagic acid and ellagitannins, as well as the relatively rare occurrence of tannins, essential oils and alkaloids. Almost all monocot saponins are steroid saponins, and calcium oxalate occurs as raphides. Monocots probably share the same evolutionary ancestors as the dicots of the subclass *Magnoliidae,* since certain characteristics are common to both groups, e.g. trimerous flowers, scattered vascular bundles, secondary homorhizy. Conversely, a few monocots display typically dicot characters, such as branched and reticulate leaf veins, cylindrical arrangement of vascular bundles, etc.

The approximately 54 000 species of monocots are normally subdivided into 3 (sometimes 4) subclasses: *Alismatidae (Helobiae), Liliidae* and *Arecidae (Spadiciflorae).* Members of the *Alismatidae* (e.g. *Alisma, Sagittaria, Elodea)* exhibit the most primitive characters, e.g. the gynaecium is choricarpous (a large number of free carpels). The *Liliidae* are a little more advanced than the *Alismatidae,* in that they usually have coenocarpous gynaecia in the form of 3 more or less united carpels. None of their characteristics is particularly primitive or advanced, and they frequently display special adaptations for insect pollination (e.g. orchids) or wind pollination (e.g. grasses). The *Arecidae* (e.g. palms and arum lilies) appear to be a phylogenetically very old group; some species have reticulate leaf venation, and nearly all have numerous inconspicuous flowers borne on a club-shaped inflorescence axis (spadix) subtended by a conspicuously colored bract (spathe). The superior ovaries are occasionally choricarpous but more usually coenocarpous.

Monocotyledonous plants: see *Monocotyledoneae.*

Monocotyledons: see *Monocotyledoneae.*

Monocyclic parthenogenesis: see Anholocyclic parthgenogenesis.

Monodelphia: see *Eutheria.*

Monodon: see Whales.

Monodontidae: see Whales.

Monogenea: see *Trematoda.*

Monognathidae: see *Saccopharyngoidei.*

Monognathus: see *Saccopharyngoidei.*

Monogony: see Reproduction.

Monograptus (Greek *monos* single, *graptos* written): a graptolite genus with straight to arched, unbranched rhabdosomes, furnished on one side with thecae. The thecae are tightly packed, aligned at a more or less sharp angle to the axis, and shaped like a proboscis. Fossils are found from the Silurian to the Lower Devonian, and those of the Lower Devonian are important index fossils.

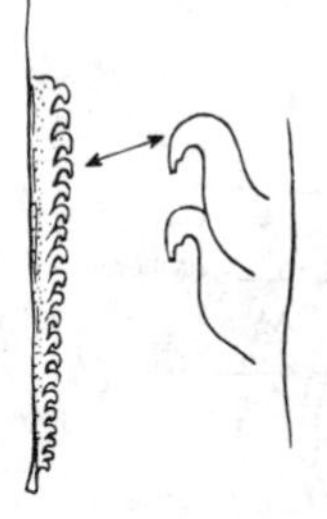

Monograptus. Rhabdosome of *Monograptus priodon* Bronn., natural size (left) and enlarged section (right).

Monohaploidy: a type of haploidy which arises from a pure diploid form, and in which only a single set of chromosomes is present. It therefore differs from polyhaploidy which arises from a polyploid condition by parthenogenesis, and in which there is more than one set of chromosomes. During meiosis, homologous paired chromosomes cannot be formed by monohaploids, but they can be formed by polyhaploids.

Monohybrid: offspring of a cross, in which the parents differ in one allele pair or one pair of phenotypic characters (AA x aa), and which is heterozygous for this allele pair (Aa). The parents of a di-, tri- or polyhybrid differ with respect to 2 , 3 or more allele pairs. See Mendelian segregation.

Monomictic lake: see Lake.

Monophyodont: see Dentition.

Monoplacophora (Greek *monos* one, *plax* plate, *phoreus* carrier), ***gastroverms:*** a class of mollusks, which until recently was known only from the fossils of 50 primitive species found in the Cambrian and the Devonian. In 1952, the first living species was discovered at 3 570 m off the Pacific coast of Costa Rica. Seven different species are now known, all in the genus, *Neopilina.*

Morphology. In the recent *M.,* the mantle covers the entire animal and secretes a single, uniform, flattened, symmetrical, shield-like shell, which varies in length from 2 mm to more than 35 mm. The foot is flat, broad and round with tentacle-like appendages. The mouth is located at the front of the foot. Around the foot are 5 or 6 pairs of unipectinate ctenidia arranged in a broad pallial groove, which at its anterior end also carries the anus on a papilla. In addition to paired ctenidia, the presence of paired muscles, nerves, sex organs and heart chambers is indicative of a segmented condition. The heart is completely paired, in that the 2 pairs of auricles open into 2 ventricles, one on each side of the rectum. Two pairs of gonads lie in the middle of the body (the sexes are separate). On each side of the body are 6 pairs of nephridia.

Classification. The fossil record indicates the existence of 2 evolutionary lines. In the subclass, *Cycloma,* the shell became planospiral, accompanied by an increase in the dorsoventral axis of the body, and these forms may have been ancestors of the gastropods. In the subclass, *Tergomya,* the shell remained flattened, and this form has survived in the recent genus, *Neopilina.* The position of the *M.* within the *Mollusca* is debatable. They exhibit morphological and anatomical affinities with all mollusks, but the presence of segmentation excludes them from any of the recent classes.

The very existence of *M.* indicates that mollusks evolved from segmented animals, and that they probably shared a common ancestor with the *Articulata.*

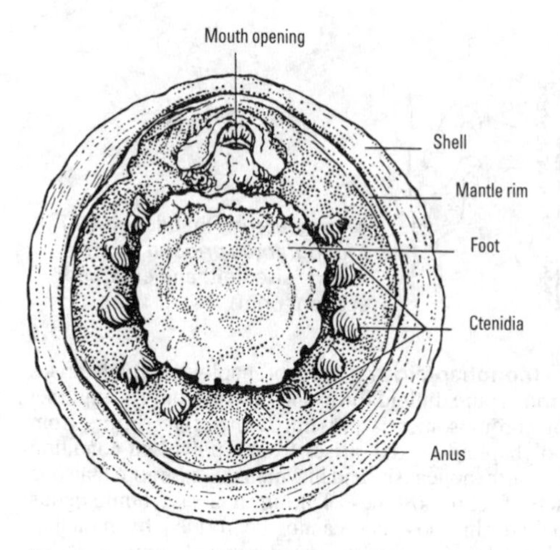

Monoplacophora. Ventral view of *Neopilina.*

Monoploidy: a condition in which the chromosomal complement consists of the lowest recorded or theoretically predictable haploid chromosome number. This number is also known as the basic number and is symbolized by x. See Monohaploidy, Diploidy.

Monopolar flagella: see Bacterial flagella.

Monopterus: see *Synbranchiformes.*

Monosomy: a form of aneuploidy, in which one chromosome of one homologous pair is absent from the diploid chromosome complement (2n – 1). See Aneuploidy, Disomy, Nullisomy, Polysomy, Tetrasomy, Trisomy.

Monothalamia: see *Foraminiferida.*

Monotocardia: see *Gastropoda.*

Monotremata, ***monotremes, egg-laying mammals, cloacal mammals:*** an order of highly specialized, primitive mammals, which do not give birth to live young, but lay and incubate eggs with parchment-like shells. Like reptiles, their eyes possess a nictating membrane, and the exits of the urinary tract, alimentary canal and sex ducts are all contained in a common chamber known as the cloaca. In the male the penis is attached to the ventral wall of the cloaca, and divided at the tip into paired channels used only for the delivery of sperm; the testes are abdominal. In the female separate uteri enter the cloaca via a common urogenital passage. The pectoral girdle is reptilian, i.e. it possesses distinct coracoid bones and an interclavicle. Epipubic or marsupium bones are associated with the pelvis. This skeletal feature is also found in the marsupials, where the marsupium bones help to support the pouch; in the monotremes, however, the marsupium bones are equally developed in both sexes; even when present, the rudimentary female monotreme pouch is not homologous with the marsupial pouch. Possibly these bones are an evolutionary relic, which once served for the attachment of the hindquarter muscles of a reptilian ancestor. Despite the presence of reptilian characters, the monotremes also possess mammary glands, and are therefore by definition mammals. The mammary glands, however, have no teats; milk oozes from glandular areolar patches (milk patches), trickles down the hair of the coat, and is sucked up by the young. Also, the heart is 4-chambered, although the right atrioventricular valve is incomplete, and there is a single aortic arch on the left. Average body temperature is lower than that of other mammals, but monotremes are nevertheless warm-blooded. In monotremes and other mammals, each side of the lower jaw is represented by a single bone, whereas reptiles typically possess several bones in this position. There are 3 middle ear bones (1 in reptiles). As in other mammals, the most strongly developed part of the cerebral hemisphere is the pallial region, whereas the striatal region is developed in reptiles and birds. The corpus callosum (a bridge of nervous tissue connecting the two hemispheres), which is found in most mammals, is absent in reptiles, marsupials and monotremes. Fossil remains of monotremes are known from the Miocene to Recent in Australia, and from Recent deposits in New Guinea. The order contains 2 families:

1. *Tachyglossidae* ***(Spiny anteaters*** or ***Echidnas).*** This family contains 2 genera and 5 living species. Teeth are never present, and the actual grinding surfaces are not even part of the jaw. Food is ground between a pad of horny spines on the back of the tongue and similar spines on the palate. The broad, powerful feet are modified for digging. Males of the various species of echidna possess rear ankle spurs and venom gland systems (see platypus, below), but these are nonfunctional. All species possess spines and a vestigial tail, and the nostrils are situated toward the end of a characteristically long, bare snout. During the mating season, the female develops a crescent-shaped skin fold on her abdomen, which contains the openings of the mammary glands, and into which she lays a single egg while lying on her back.

The 2 species of *Tachyglossus* ***(Short-beaked echidnas, Common spiny anteaters, Common echidnas)*** have a head plus body length of 35–53 cm, a stump-like tail, and the long snout is used for feeding on ants and termites. The 2nd claw of each hindfoot is specialized for grooming. The coat is black to light brown with spines on the back and sides. A shallow burrow may occasionally be dug, and a burrow is used by the incubating or suckling female. Otherwise, no fixed shelter appears to be used, and the animals conceal themselves in hollow logs, thick vegetation and rubble, etc. *Tachyglossus aculeatus* ***(Australian short-nosed spiny anteater)*** is found in Australia and new Guinea. *Tachyglossus setosus* ***(Tasmanian short-nosed spiny anteater)*** is native to Tasmania and the Bass Strait isalands, and considered by some authorities to be a subspecies of *T. aculeatus.*

Some authorities recognize 3 species of *Zaglossus* ***(Long-beaked echidnas, Long-nosed echidnas, Long-nosed spiny anteaters;*** head plus body length 45–90 cm) while others recognize these as subspecies of *Z. bruijni.* The 4th and 5th claws of the fore and hindfeet display various degrees of reduction. The snout is much longer than that of *Tachyglossus,* and it turns downwards. Main dietary components are earthworms and noncolonial insects. The tongue carries a longitudinal spiny groove, which enables the animal to align an earthworm lengthways for entry into the long mouth. Compared with *Tachyglossus,* the spines on the body are shorter and less numerous. *Zaglossus bartoni* ***(Barton long-nosed spiny anteater;*** northeastern and southern New Guinea) has 5 claws on each foot, and its very short, white spines are hidden by long, black hair. *Z.*

bruijni ***(Bruijn long-nosed spiny anteater;*** western New Guinea and the Salawati islands) has 3 or 4 claws on each foot. *Z. bubuensis* ***(Bubu long-nosed spiny anteater;*** Bubu river region of new Guinea) has 5 claws on each foot, and its short white spines protrude from the brown dorsal hair.

Monotremata. Platypus *(Ornithorhynchus anatinus).*

2. *Ornithorhynchidae.* This family is represented by a single living species, *Ornithorhynchus anatinus* ***(Duck-billed platypus),*** whose fossil remains are known from the Pleistocene of mainland Australia and recent deposits in Tasmania. It is a semi-aquatic mammal, whose elongated snout is modified into a broad, flat, pliable bill. The body is streamlined with dense, waterproof fur (consisting of large hair bristles and a very fine, woolly undercoat), a broad, flat tail (which stores fat), partially webbed hindfeet, and fully webbed forefeet with large webs. The dimensions of the male are: head plus body 45–60 cm, bill 5.8 cm (average), tail 10.5–15.2 cm; the female is proportionally smaller with a head plus body length of 39–55 cm. The platypus is found in eastern Australia and in Tasmania, where it burrows into the banks of rivers and lakes. Using its touch-sensitive bill, it scours the mud of the river or lake bottom for insect larvae, worms, snails and mussels. The rear ankles of the male carry a hollow, horny spur, connected by a duct to a venom gland behind the knee. Injected by a frightened or aggressive male, the venom is sufficient to kill a dog or cause agonizing pain in a human. Together with certain shrews, the platypus is one of the very few venomous mammals. Normally 2 eggs are laid, and these are incubated by the female in a breeding burrow. The female does not possess a brood pouch. Newly born platypuses have 34 tooth buds, corresponding to the dental formula: 0/5, 1/1, 2/2, 3/3 = 34, but only 12 of these erupt, according to the formula: 0/0, 0/0, 1/0, 2/3 = 12. The teeth do not persist, and are lost shortly after emergence from the breeding burrow (3–4 months after birth). The animal then develops horny, wrinkled and ridged plates, which are located in internal cheek pouches behind the bill and near the front of the mouth, where food is stored and chewed.

The extinct *Obdurodon insignis* from the Miocene of South Australia is also placed in *Ornithorhynchidae.*

Monotrichous flagella: see Bacterial flagella.

Monotypic: an adjective applied to a taxon that contains only a single subordinate taxon. For example, a monotypic genus is represented by only one species; a monotypic family contains a single genus, etc.

Monsoon forest: tropical forest that is heavily foliated only during the wet monsoon period, whereas most of the trees lose their leaves during the dry season. On the other hand, many of the leafless trees produce flowers during the dry season. M.fs. are found particularly in the Indo-Malaysian region, and they are characterized primarily by teak *(Tectona grandis)* (e.g. in Burma) and species of the hardwood families, *Burseraceae, Sapindaceae, Anacardaceae* and others; they also contain lianes and herbaceous epiphytes.

Monstera deliciosa: see *Araceae.*

Monstrilla: see *Copepoda.*

Monstrilloida: see *Copepoda.*

Montbretia: see *Iridaceae.*

Monte: see Sclerophylous vegetation.

Monticola: see *Turdinae.*

Montifringilla: see *Ploceidae.*

Montmorillonite: see Sorption.

Monuron: see Herbicides.

Moon rat: see Hedgehogs.

Moor frog: see *Raninae.*

Moorhens: see *Gruiformes.*

Moorish idol: see *Acanthuridae.*

Moose, *elk, Alces alces:* a monotypic genus, and the largest deer (head plus body 240–310 cm, shoulder height 140–235 cm), with massive, widely spread spatulate antlers and a broad, overhanging upper lip. The name "moose" is used in North America, while "elk" is strictly a European term for the same animal. The body is black-brown (grayer in winter) with lighter colored legs. *Alces* inhabits marshy woodlands and bogs in northern and eastern Europe, Asia and Canada. Occasionally individuals move westward into Central Europe. A characteristic growth of skin and hair, up to 50 cm in length and known as the "bell", hangs beneath the throat. See Deer.

Mor: see Humus.

Moraceae, *mulberry family:* a family of dicotyledonous plants, containing about 1 500 species in 53 genera, distributed mainly in warm regions of the world. They are latex-containing, woody plants, with simple or lobed, stipulate leaves. The small unisexual flowers are carried in infloresences, which may be cup-shaped, or take the form of a cyme or capitulum. The usually 4-merous flowers are wind- or insect-pollinated. Fruits may be single-seeded achenes or drupes, but are often compound infructescences, formed by the union of the fruits of several flowers with the addition of fleshy receptacles. Many species are economically important on account of their constituents, such as latex, resins, waxes or even toxins (cardenolides). Others are exploited for their edible leaves or fruits, or for their strong timber.

Economically, the most important genus is *Morus* ***(Mulberries).*** The leaves of the Chinese ***White mulberry*** *(Morus alba)* have been used for almost 5 000 years to feed silkworms, while the ***Black mulberry*** *(Morus nigra)* is cultivated mainly for its edible, blackberry-like, red-black fruits.

The family also includes the genus *Ficus* ***(Figs),*** represented by about 700 species, and one of the largest genera of higher plants. Most familiar of these species is the ***Edible fig*** *(Ficus carica),* which is native to the Mediterranean region, and whose fleshy sweet fruits are an important food in the Orient. In its native tropical Asia, the ***Rubber tree*** *(Ficus elastica)* grows to a height of 20–25 m; its latex is the source of caoutchouc, which is converted into rubber by vulcanization. In temperate climates, the rubber tree is a popular decorative indoor plant. Some *Ficus* species are lianas, known as "strangling figs", which strangle and kill their support trees. The East Indian ***Banyan*** *(Ficus bengalensis)* starts life

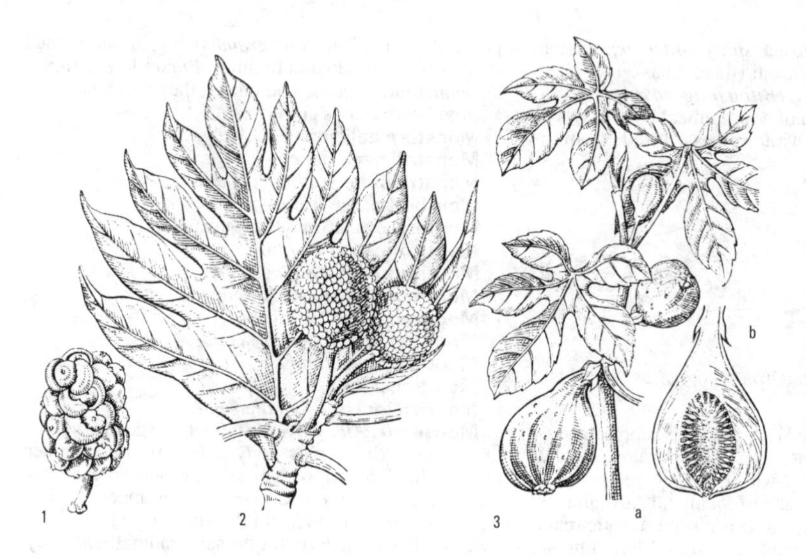

Moraceae. 1 Infructescence of mulberry *(Morus nigra). 2* Breadfruit tree *(Artocarpus),* branch with fruits. *3* Edible fig *(Ficus carica), a* branch with fruits, *b* fruit (longitudinal section).

as an epiphyte which sends roots down to the ground. These develop into columnar trunks, and the supporting tree is eventually strangled and killed; the single banyan tree then forms an entire grove consisting of horizontal branches with trunks to the ground.

Caoutchouc for the preparation of *Panama rubber* is obtained from the latex of the Mexican *Castilloa elastica.* On account of their edible fruits (collective infructescences), the ***Breadfruit tree*** *(Artocarpus incisa)* and the ***Jakfruit*** *(Artocarpus heterophyllus)* are cultivated in the tropics. Breadfruits weigh up to 20 kg; they are eaten raw or cooked, and the seeds taste good when roasted. Jakfruits may attain 80 kg; they are also eaten by humans, but more often used for animal fodder.

The inner bark of the East Asian ***Paper mulberry*** *(Broussonetia papyrifera)* provides a fiber that is used in paper making. The milky latex of the Southeast Asian ***Upas tree*** *(Antiaris toxicara)* is used directly as an arrow and dart poison (Ipo arrow poison); contains two highly toxic cardioactive glycosides (antiarin x and z). ***Osage orange*** *(Maclura pomifera)* is used as a hedging plant in the USA, but its toxic latex is suspected of causing livestock losses in Arkansas and Texas; in Europe it is grown as an ornamental plant for its orange-like but inedible fruits. Species of the genus, *Dorstenia,* which are native in tropical South America and in Africa, have plate-like infloresences. Roots of *Dorstenia brasiliensis* are used as a folk remedy against wound infections and snake bite.

Moray eels, *Muraenidae, Heteromyridae:* a family of eel-like teleost fishes (order *Anguilliformes)* inhabiting inshore waters of rocky coasts, in the tropics and temperate regions. Scales are absent, and the deeply cleft gape is well armed with fang-like teeth. They are nocturnal in habit, and conceal themselves in tunnels and fissures. Some species have venom glands in the roof of the mouth. Some make good eating. There are 2 branchial pores and the fourth branchial arch is strengthened and supports the pharyngeal jaws. Lateral line pores are present on the head but not on the body.

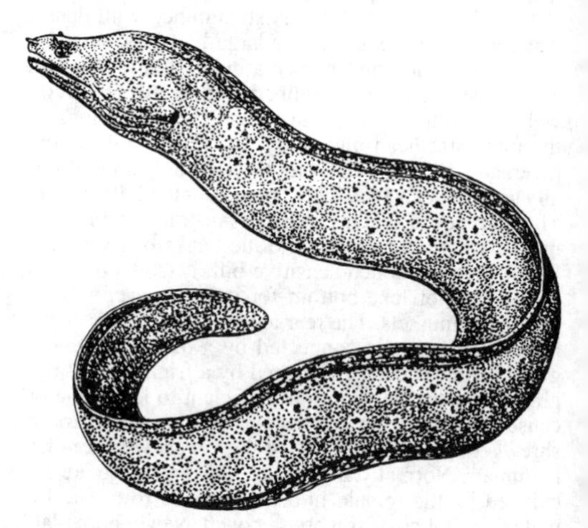

Moray eel (Muraena helena)

Morel: see *Ascomycetes.*

Moreton Bay pine: see *Araucariaceae.*

Morgan unit: the distance between two genes on homologous chromosomes, in which one crossover (on average) occurs per 100 gametes, i.e. the distance is equal to a crossover value of 100%.

Mormyridae, ***elephantfishes:*** a family (about 190 species in about 16 genera) of telost fishes of the order *Osteoglossiformes (Mormyriformes).* They are found in fresh water in Tropical Africa, and in the Nile. Eyes and mouth are small, and the tail fin is forked. They possess a very large cerebellum and an electric organ. In some species the snout is elongate and proboscis-like with a terminal mouth, whereas others have a rounded snout with a ventral mouth; a few species have an elongate lower jaw. Lengths vary from 9 to 50 cm, and the larger species are fished commercially as food.

Morning glory: see *Convolvulaceae.*

Morphactins: a group of highly potent synthetic plant growth regulators. M. are derivatives of fluorenol

(9′-hydroxy-fluorene-[9]-carboxylic acid) (Fig.), primarily by esterification of the carboxyl group and/or substitution in the ring system. Over a wide concentration range, M. inhibit and modifiy plant growth by acting as gibberellin antagonists. Since they act synergistically with certain herbicides (e.g. 2,4-dichlorophenoxyacetic acid), they can also be used as broad spectrum herbicides. Among the many physiological effects of M. are: shortening of shoot internodes, abolition of apical dominance (causing compact bushy growth habit), reversible inhibition of mitosis, effects on root growth, geotropism and phototropism, delayed induction of flowering, production of seedless fruit, and an effect on the sexual development of bisexual flowers.

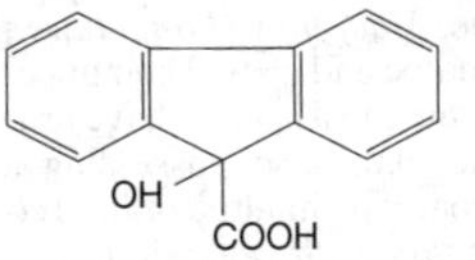

OH COOH

Fluorenol (ring system of the morphactins)

Morphallaxis: see Regeneration.

Morphine: the most abundant of the opium alkaloids (see Opium) and the first ever alkaloid to be isolated and crystallized (Sertürner, 1806). It is a levorotatory tertiary amine, crystallizing as white, bitter-tasting crystals. The chemical synthesis, first achieved in 1952, is not economically viable, so that all commercial morphine is isolated (over 1000 tons/year) from opium; most of this is converted into Codeine (see). M. is a narcotic poison, but in small doses it is an analgesic, in fact the most powerful analgesic known. In large doses it eventually causes death by paralysis of the respiratory center. Medically, it is used to relieve pain in conditions where other analgesics are no longer effective, but owing to its properties as a dangerous hypnotic and narcotic, its use is restricted. Repeated use quickly leads to dependence and addiction. *Heroin,* the synthetic diacetyl derivative of M., is an even more dangerous narcotic than M.

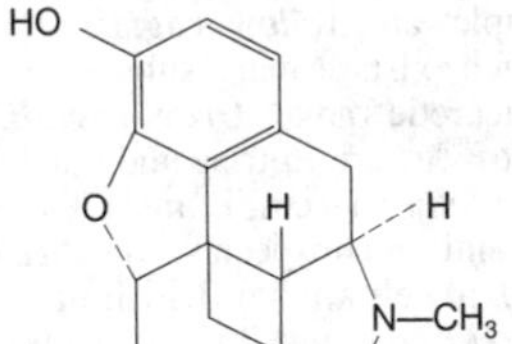

HO O H H N—CH_3 HO

Morphine

Morphology: study of the external form and structure of organisms and the spatial relationships of their organs and parts.

Comparative M. is concerned with external body form, and the structure and relative positions of organs within a systematic unit (e.g. the comparative M. of vertebrates), and their phylogenetic or evolutionary relationship to the basic structural plan of that systematic unit. This evolutionary interpretation of M. provides information on the relationship between the structure and function of organs.

Morphoses, *morphogenetic reactions:* modifications of the shape and form of plants, caused by external, mostly physical influences. Most external morphogenetic signals (e.g. light, temperature, gravity, air humidity) affect the entire plant or many plant parts simultaneously, so that practically every cell reacts directly to the signal. Involvement of transportable morphogenetic substances is therefore unlikely, and has been demonstrated in only a few exceptional cases. The causal basis of most M. is unknown. In complex M., numerous external factors are involved to a more or less equal extent. Morphogenetic reactions are also known, which are determined largely by a single factor, e.g. thermomorphoses, chemomorphoses, photomorphoses. See Photomorphogenesis, Geomorphosis, Hygromorphosis, Mechanomorphosis.

Mortality: the proportion of a population lost by death in a given time. The *death rate* is expressed as the number of deaths in a particular area in a year divided by the average or midyear population of the area. M. and natality are crucial factors in population dynamics. M. is determined by endogenous factors (natural aging, loss of functional efficiency) and external influences (intraspecies competition, antibiosis, food shortage, abiotic factors). See Population dynamics.

For the purposes of human medical and sociological statistics, the *mortality rate* is the number of deaths under one year of age occurring in one year divided by the total live births in that year. The *maternal mortality rate* is the number of deaths attributed to puerperal causes in one year divided by the total live and still births in that year. The *perinatal mortality rate* is the number of stillbirths and deaths in the first week after birth divided by the total live births in that year. The *neonatal mortality rate* is the number of deaths in the first 28 days in one year divided by the total live births in that year. Alternatively, each of these mortality rates may be expressed as the number of deaths occurring per 1 000 live births.

Morus: see *Moraceae.*

Mosaic-cycle concept of ecosystems: see Ecosystem.

Mosaic egg: see Regulation.

Mosaic evolution: a term expressing the fact that different characters evolve at different rates *(Watson's rule).* As a result of M.e., certain organisms possess combinations of characters from different taxa. For example, the primitive bird, *Archaeopteryx,* combines typical reptilian characters (teeth, free metacarpals, etc.) with typical avian characters (feathers, fused metatarsals, etc.).

Mosaic forms: individuals with sectors of tissue that differ genetically from the norm. Such differences are caused by differences in chromosome number (chromosomal mosaicism), by mutations in genes, chromosomes or plastids (see Positional effect), or by somatic crossing over. A phenotypic sexual mosaic is produced by changes in sexual development during ontogeny, i.e. development proceeds in the direction of one sex, then switches to development of the other sex.

Mosaic viruses: see Virus groups.

Mosasaurus (Latin *mosa* → Dutch *meuse* river, Greek *sauros* lizard): an extinct genus of carnivorous marine reptiles, belonging to the *Lepidosauria.* Some members attained a length of 12 m, with a 1 m-long skull and a short neck. The paddle-like limbs served as oars, while the very long, serpentine tail was the main locomotary organ. Fossils are found in the Upper Cretaceous of all continents except Antarctica and New Zealand.

Moschus: see Musk, Musk deer.

Mosque swallow: see *Hirudinidae.*

Mosquitoes, *Culicidae:* a family of the *Diptera* with about 2 500 known species worldwide. Imagoes have scaly wings and a body length not exceeding 1 cm. The blood-sucking females possess a long piercing proboscis, and are potential disease vectors for warm blooded animals and humans, e.g. avian malaria, human malaria, yellow fever and elephantitis are all transmitted by M. Eggs are laid on water, and the larvae and pupae are aquatic in standing water and swamps.

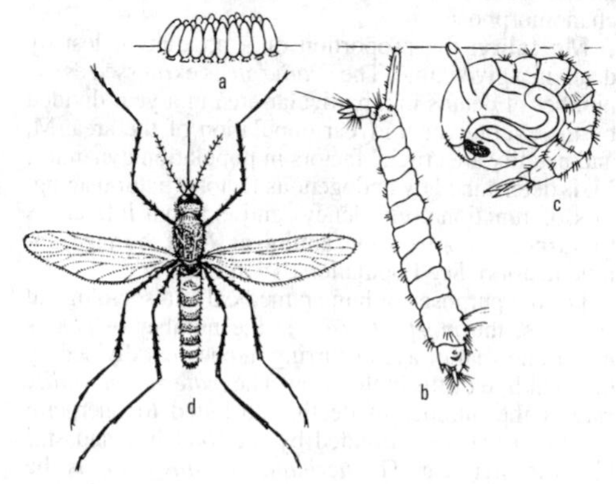

Mosquitoes. Culex sp. *a* Egg raft on water surface. *b* Larva with respiratory tube. *c* Pupa (with two respiratory tubes). *d* Imago.

Mosquitofish: see Guppy.

Moss animals: see *Bryozoa.*

Mössbauer spectroscopy: measurement of the energy absorption of nuclear gamma radiation by ^{57}Fe (the abundance of ^{57}Fe in naturally occurring iron is 2.2%). The nucleus of the isotope ^{57}Fe is excited by gamma radiation emitted during the decay of a ^{57}Co source. This excited state lies about 14.4 keV above the ground state. The source is moved, thereby shifting the frequency of the monochromatic gamma radiation, and the absorption spectrum is determined from the resulting Doppler effect. Such studies, based on the hyperfine splitting of the absorption spectrum, give an insight into the electronic structure of iron in hemoproteins, and provide direct information on the structure of the catalytic center of the protein.

Mosses: see *Musci.*

Moss ferns: see *Lycopodiatae.*

Motacilla: see *Motacillidae.*

Motacillidae, ***pipits*** and ***wagtails:*** an avian family [order Passeriformes (see)] of 54, mostly sparrow-sized species. The family has a worldwide distribution, but breeding members are not found in New Guinea or Oceana. All species possess 9 primaries, and they are more or less adapted to open habitats, where they run, walk or fly after their insect prey. They nest on or near the ground. There 2 related groups: ***Pipits*** (3 genera) and ***Wagtails*** (2 genera).

The 34 very similar pipits of the genus, *Anthus,* constitute the largest genus of the family. These have generally brown, streaked or dappled plumage, and their tails (shorter than those of wagtails) show white outer feathers. The bill is long and straight, and the hindclaw is particularly long. The ***European Meadow pipit*** *(Anthus pratensis;* 15 cm) is found on wet mountain meadows and sea coasts, often sharing the same habitat as the ***Water*** or ***Rock pipit*** *(A. spinoletta;* 17 cm), which nests in the Arctic tundra and winters southward. *A. cervinus* ***(Red-throated pipit)*** also lives in the tundra, and is a passage migrant in central Europe. The ***Tree pipit*** *(A. trivialis;* 16 cm) favors forest edges and parkland, while the sandy-colored ***Tawny pipit*** *(A. campestris;* 17.5 cm) inhabits wasteland and dand dunes. There are at least 27 subspecies of ***New Zealand pipit*** *(A. novaeseelandiae;* 18 cm) ranging from East Asia and Sibera to New Zealand, Australia and Africa, occasionaly occurring in central Europe as a vagrant from Siberia. The 8 pipit species of the African genus, *Macronyx* ***(Longclaws),*** have an exceedingly long hindtoe and claw. Their upper parts are cryptically colored, but most species have brilliant yellow underparts, the male also possessing a darker border around the throat. The hindtoe of the ***Yellow-throated longclaw*** *(Macronyx croceus;* 20 cm) is about 5 cm-long, and the foot spans almost half the length of the bird. *Tmetothylacus tenellus* ***(Golden pipit;*** 15 cm) is found in dry bush country in southern Sudan, eastern Ethiopia, Somalia and south to Tanzania. The upper parts of the male are mottled olive-green, while the wings, tail and underparts are canary yellow with a black breast band. The underparts of the female are buff-brown with a yellow center belly and underwing coverts.

Tree Wagtails *(Dendronanthus)* are represented by a single species *(D. indicus;* 14 cm), a forest-dwelling East African bird, which moves its tail from side to side rather than up and down; it appears to be phylogenetically intermediate between pipits and wagtails. The 10 species of ***True wagtails*** *(Motacilla)* have a long tail, which they characteristically pump up and down. Most of them prefer to live near water, in swampy meadows, by lake shores and by mountain streams. Like pipits, they have white outer tail feathers, which show conspicuously in flight. They are more boldly patterned than pipits, and their sexual dimorphism is more pronounced than that of *Anthus.* Examples are: ***Yellow wagtail*** *(M. flava;* 17 cm) (plate 6), which exists as many subspecies and has a continuouis Palearctic range; ***Gray wagtail*** *(M. cinerea;* 18 cm; most of western Europe and much of Asia); ***White wagtail*** *(M. alba;* 19 cm; Eurasia, Orient, marginally Nearctic) and its subspecies, the ***Pied wagtail*** *(M. alba yarrellii),* a well known British bird. The ***African pied wagtail*** *(M. aguimp;* 20 cm) is distributed throughout the Ethiopian region.

Mother-of-pearl: the inner layer (hypostracum) of the shells of different gastropods and bivalves. In consists of fine calcareous lamellae aligned parallel to the surface of the shell, resulting in iridescent interference colors and a characteristic luster. It is used for decoration and for fashion jewellery. See Pearls.

Mother of vinegar: see Acetic acid bacteria.

Moth owl: see *Caprimulgiformes.*

Moths, *Heterocerca:* a traditional and nonscientific term for *Lepidoptera* which fold their wings against the body and often fly at night. The antennae of M. are usually not club-shaped (cf. butterflies), although the antennae of some day-flying M. have swollen tips. More often, the antennae of M. show lateral processes, so that they appear feathery (pectinate). Most M. possess a bristle (frenulum) on each hindwing, which locks into a hook (retinaculum) on the underside of each forewing. More primitive M. have a wing-coupling lobe (jugum) at the base of the forewing, which overlaps the hindwing. Both structures enable the fore and hindwings to beat together. In the classification list under *Lepidoptera* (see), the M. are represented by the families *Hepialidae* through *Sphingidae,* i.e. those families which are not Butterflies (see). Large M. may also be called macro-moths or "macros", e.g. *Noctuidae, Geometridae* and *Sphingidae,* which are popular with collectors. The micro-moths or "micros" include the more primitive and usually smaller M.

Motilin: see Gastrointestinal hormones.

Motivated behavior: translation of a readiness for species-typical behavior (see Motivation) into a behavioral pattern. In higher animals, this normally occurs in three phases, starting with orienting appetence behavior, i.e. a searching phase in the *distance field.* If stimuli are perceived, which identify the sought object and its position, then the *near field* is entered, and the second phase of oriented appetence behavior is established. This ends with establishment of the *contact field,* when appropriate stimuli activate the final stage of M.b., e.g. food intake. Evolution of M.b. is analogous to that of ontogenesis, i.e. it evolves "from behind" (see Fig.).

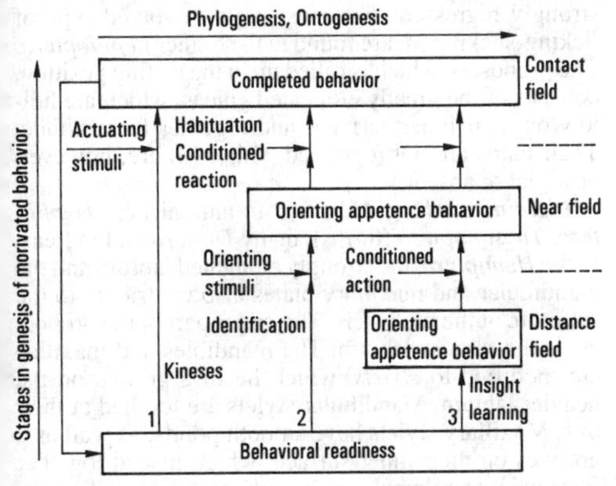

Motivated behavior. Stages in the initiation and completion of motivated behavior, showing possible involvement of learning processes and the three event fields of the behavioral pattern.

Motivation: readiness of an animal for species-typical behavior, which has an essential or vital function. For example, the readiness to feed is a M., which declines when the act of feeding has been performed. In humans, the readiness to perform learned behavior (e.g. riding a bicycle) is known as secondary M., even though it may represent the expression of a primary M. (e.g. the readiness to change location, possibly with a functional goal).

M. can also be variously interpreted as mood or drive, i.e. the internal processes that arouse and direct behavior. The M. for any behavioral pattern declines when the behavior is enacted, then increases again. Stimuli for M. are both external and internal (e.g. hormones).

Motmots: see *Coraciiformes.*

Motoneuron: see Motor neuron.

Motor cells of the anterior horn: large ganglion cells *(somatomotor cells)* of the anterior gray column of the spinal cord, whose axons pass via the anterior roots to the skeletal musculature. Numerous nerve fibers of the voluntary motor system originate from ganglion cells in higher levels of the central nervous system and converge on the somatomotor cells. The anterior horn motor cells and their axons therefore represent the final stage for all the pathways of the voluntary motor system.

Motor endplate: see Neuromuscular synapse.

Motor neuron, ***motoneuron:*** a nerve cell, whose cell body lies in the spinal cord or medulla of the brain, while its single axon extends from the soma into the peripheral nerve. The axon terminates in the presynapse of a neuromuscular endplate. Thin projections of the soma, known as dendrites, extend up to 1 mm into surrounding areas of the cord. M.ns. are particularly numerous in the anterior horn of the spinal cord.

Mottled umber: see Winter moths.

Mouflon: see Sheep.

Mould: British spelling of mold.

Mound builders: see *Megapodiidae.*

Mountain ash: see *Rosaceae.*

Mountain beaver: see *Aplodontidae.*

Mourning dove: see *Columbiformes.*

Mousebirds: see *Coliiformes.*

Mouse-eared bats: see *Vespertilionidae.*

Mouse hares: see *Ochotonidae.*

Mouse-like hamster: see *Cricetinae.*

Mouse-tail: see *Ranunculaceae.*

Moutan: see *Paeoniaceae.*

Mouth brooding, ***buccal incubation*** (plate 3): in some fish species, the incubation of eggs in the mouth of a parent, usually the female. In a very few species, both parents perform M.b. The fertilized eggs are carried in the mouth until the young fishes hatch. In some species, the young are at first allowed out of the mouth only from time to time, and are finally released completely when they attain a certain size. The incubating parent does not feed. M.b. has evolved independently in different fish groups. As a rule, M.b. species produce fewer but larger eggs than related, non-M.b. species. The anal fin of some male M.b. cichlids displays a pattern resembling eggs (egg spots).

Mouth-brooding frogs: see *Rinodermatidae.*

Mouthparts: arthropod appendages grouped around the entrance to the alimentary tract. They usually display considerable modification, and they serve for the ingestion of food. Between 2 and 6 head segments may carry M.

Crustacean M., which are the most primitive, often retain a biramous structure. Usually, however, the protopodite and its processes (endites) are strongly developed at the expense of the endo- and expodite. In the nauplius stage and in some copepods, the paired *upper jaws* or *mandibles* retain 2 branches, but in other crustaceans the mandibles are short and heavy with opposing grinding and biting surfaces, with or without palps. Close behind the mandibles are 2 pairs of *lower jaws,* i.e. the 1st and 2nd *maxillae,* and these usually possess well developed chewing surfaces. In the more advanced crustaceans, M. are also formed by modification of the 1st, 2nd and 3rd thoracic appendages to form *gnathopods* or *maxillipeds.*

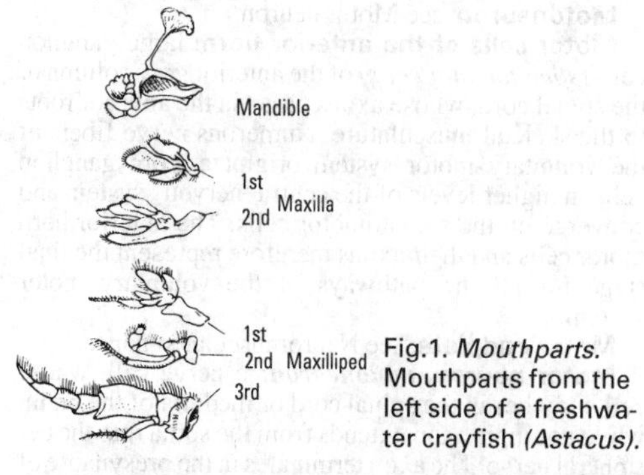

Fig.1. *Mouthparts.* Mouthparts from the left side of a freshwater crayfish *(Astacus).*

In addition to a pair of mandibles, a pair of 1st maxillae (whose palps are rudimentary or absent) and a pair of leglike 2nd maxillae, the predatory chilopods (centipedes) also have a pair of modified thoracic appendages, known as the 1st maxillipeds. The central fused part of the 1st maxillipeds covers the underside of the head, and each maxilliped ends distally in a venom claw with a venom gland at its tip.

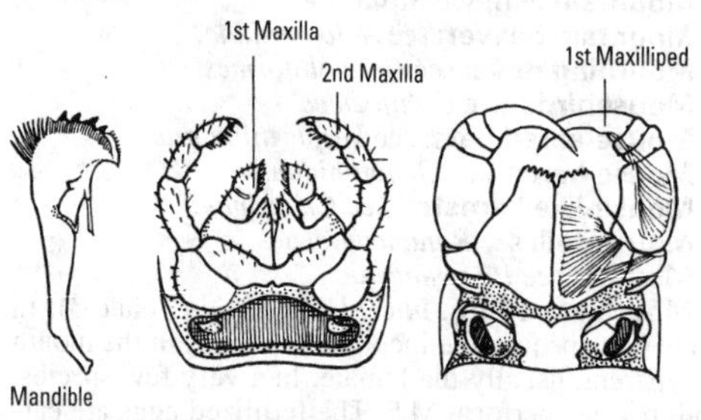

Fig.2. *Mouthparts.* Mouthparts of a centipede *(Chilopoda).*

The plant-eating *Diplopoda* (millipedes) have jointed mandibles, but the 1st maxillae are fused medially to form a plate, called the *gnathochilarium,* which includes the atrophied 2nd maxillae. The gnathochilarium functions as a lower lip, but it is not homologous with the lower lip (labium) of insects.

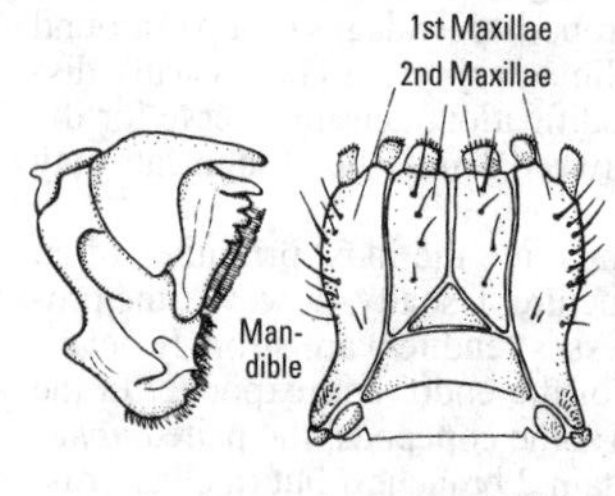

Fig.3. *Mouthparts.* Mouthparts of a millipede *(Diplopoda).*

Insects have various types of M. Most primitive are the ***biting-chewing M.*** (orthopteroid type), which consist of the paired, unjointed *upper jaws* or *mandibles,* followed by the paired, jointed *lower jaws* or *maxillae* (with palps), and the jointed *lower lip (2nd maxilla* or *labium);* the latter is fused at its base and forms the floor of the oral cavity. Although single, the labium represents a pair of fused maxillae. Anteriorly, the mandible is covered by a shelflike extension of the head, forming an unpaired *upper lip* or *labrum.* A median lobelike *hypopharynx* arises behind the maxillae near the base of the labium, and projects from the floor of the prebuccal cavity. The hypopharynx is not a modified appendage in the true sense, but part of the ventral surface of the head. Each maxilla consists of a basal member, an elongated main member with maxillary palps, and two chewing surfaces (the outer galea and the inner lacinia). The labium consists of a basal flat central part, which is subdivided by the labial suture into a postlabium or postmentum (submentum + mentum) and a prelabium or prementum. The prementum has lateral lips or labial palps, as well as distal tongues (glossae) and secondary tongues (paraglossae). Glossae and paraglosaae may be fused to a ligula. All other types of insect M. can be interpreted as derivatives of the biting-chewing M., which are present in the *Orthoptera, Coleoptera, Neuropteria, Odonatoptera, Plecoptera,* termites and many insect larvae.

The ***licking-sucking M.*** of the more advanced *Hymenoptera* consist of a complicated extendable sucking proboscis, which is formed from the markedly elongated galeas of the maxillae and the glossae of the labium, the latter being modified to a long hairy tongue. Whereas the labrum, like the mandibles, shows normal development, the lacinias and the paraglossae are strongly regressed. The most highly evolved type of licking-sucking M. are found in the higher *Lepidoptera.* The proboscis, which is rolled up in the resting position, consists of the greatly elongated galeas, which are hollow on their inner surfaces and together form a tube. Their palps are not regressed. Other M. are, however, vestigial or absent.

Piercing-sucking M. occur in animal lice, *Hemiptera, Thysanoptera* (thrips), many *Diptera* and in fleas. In the *Hemiptera,* the strongly elongated labrum and the mandibular and maxillary plates also contribute to the structure of the proboscis. The major part of the proboscis is the tubular labium. The mandibles and maxillae are modified to *stylets,* which lie in a groove on the heavier labium. Mandibular stylets are toothed at their tips. Maxillary stylets have smooth points, as well as 2 grooves on their inner surface, which placed together form the larger dorsal *sucking tube* and the smaller ventral *salivary duct.* There are no palps. In the piercing-sucking M. of midges and mosquitoes, the stylets also lie in the groove of the labium. The sucking tube (the *labrum-epipharynx),* however, is formed from the elongated, ventrally grooved labrum, while the salivary duct lies inside the greatly elongated *hypopharynx,* which is a tongue-like eversion of the soft-skinned region of the mouth. The maxillary palps are well developed. Mandibles and 1st maxillae are absent from the more advanced flies, e.g. the common housefly *(Musca domestica).* The proboscis is formed by a strongly developed labium, which carries the sucking tube (formed from the labrum) in an upper groove. The salivary duct lies in the hypopharynx, which widens at its tip into a powerfully developed, soft-skinned *haustellum.* Maxillary palps are well developed.

The M. of bees and wasps are adapted for both chewing and sucking. In bees, nectar is gathered by the elongated maxillae and the labium, while the labrum and mandibles (which have retained a chewing function) handle wax and pollen.

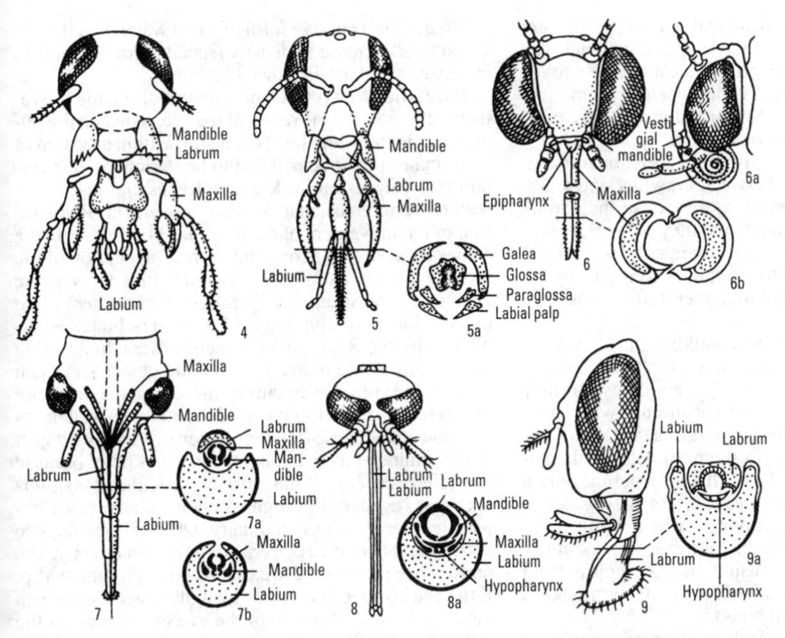

Fig.4-9. *Mouthparts. 4* Biting-chewing mouthparts of a cockroach *(Blattoptera). 5* Chewing-sucking mouthparts of the honeybee *(Apis). 6* Licking-sucking mouthparts of a butterfly or moth *(Lepidoptera) (6a* side view). *7* Piercing-sucking mouthparts of a bug *(Heteroptera). 8* Piercing-sucking mouthparts of a midge or mosquito *(Culicidae). 9* Lapping and sucking apparatus of a fly (e.g. *Musca domestica). 5a, 6b, 7a, 7b, 8a* and *9a* are transverse sections of the respective proboscis at the site indicated.

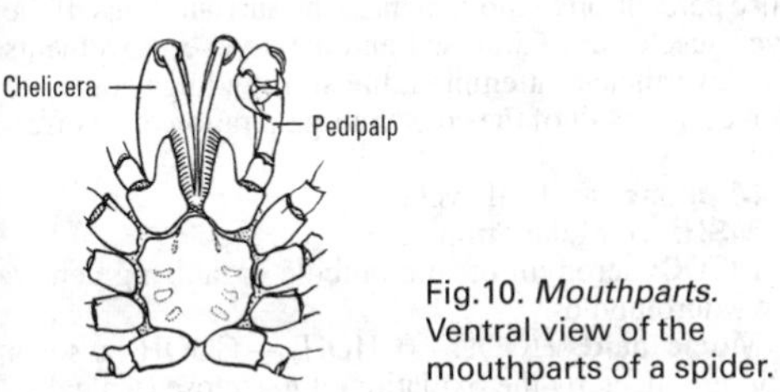

Fig.10. *Mouthparts.* Ventral view of the mouthparts of a spider.

Spiders have very simple M. The first pair, the *chelicera,* consist of a basal member and a claw. The claw, which carries the opening of the venom gland at its tip, can be closed against the basal member. The second pair, the *pedipalps,* have a primitive leglike structure in some spiders. In the more advanced spiders, the pedipalps are small and resemble palps, and in males they function as part of a complicated copulation apparatus.

Movement: a change in the position of an organism or its parts. In ***plants,*** locomotory M. is found only in less advanced forms, e.g. unicellular algae and fungi. In numerous plant groups, the gametes are also capable of locomotion, mainly by the action of cilia or flagella, or by ameboid movement. Plants rooted in the soil display M. of organs. Reversible bending often occurs by asymmetrical changes in turgor pressure on opposite sides of a cylindrical organ, usually at a node or other jointed or hinged region, such as a leaf cushion (see Turgor movements). Many plant bending Ms. are due to unequal growth on opposite sides of an organ, e.g. a lateral shoot (see Nutation Ms.). M. can be *autonomic* or *endogenous,* i.e. not requiring an external stimulus, or it may be *induced* by an external stimulus, e.g. light, temperature change, certain chemicals, etc. Locomotory M. in response to a stimulus is called *Taxis* (see). Directional movement of part of a rooted plant in response to a stimulus is called a *Tropism* (see), whereas a movement that is not governed by the direction of the stimulus is called a *Nasty* (see). Ms. of plant organs due to swelling, shrinkage, or desiccation (see Hygroscopic Ms.) involve only dead tissues, and often occur very suddenly. Such Explosive mechanisms (see) can be classified as Catapulting mechanisms (see), Squirting mechanisms (see), or Cohesion mechanisms (see). They serve especially for the distribution of spores, pollen and seeds. See also: Chloroplast Ms., Flower Ms., Leaf Ms., Plastid Ms., Tendril Ms., Bending Ms., Cyclic Ms., Nodding Ms., Orientation Ms., Pendular Ms., Unfolding Ms., and Torsion.

In animals, different types of locomotory M. have evolved as a result of the close interaction of body structure, behavior and habitat (see Life form). Sessile animals are restricted mainly to aquatic habitats, with the exception of certain parasites, such as scale insects and nematodes. Terrestrial animals can be classified according to their types of locomotion, as follows.

1. ***Fossorial*** animals (Fossores) lead a wholly or partly subterranean existence. Those that bore through the earth have slender bodies with a pointed and often reinforced anterior end (e.g. earthworms, wireworms). The shovellers or excavators have evolved especially efficient forelimbs (e.g. mole, mole-cricket, larvae of cicadas, dung beetles). Certain fossorial animals may not live permanently in the earth, but build burrows by scraping or scratching with the aid of sharp claws (e.g. fox, many rodents, bank swallow, natterjack toad, digger wasps). The mouth excavators use their mouthparts (e.g. ants, kingfisher).

2. ***Crawling animals*** (Reptantia) progress either by *gliding* (e.g. snails) on a broad sole that adheres to the substratum, or by *undulation* (e.g. nematodes, snakes), which requires the exertion of a thrust against the substratum. *Stretching crawlers* attach the front end and progress by drawing the hindpart of the body up to the front (many insect larvae, especially looper caterpillars).

3. ***Cursorial animals*** (Currentia) have paired extremities and they progress by running or walking. The body is lifted from the substratum, and a rigid exo- or endoskeleton is necessary for this type of locomotion. The most important type of cursorial progress is *striding* (spiders, insects, birds, mammals).

Locomotion by lizards and certain amphibians (salamanders) is intermediate between crawling and walking. As these animals progress, they raise their trunk only slightly from the ground, and they move their outwardly extended legs forward alternately so that the hindfoot is set down near to the spot just vacated by the forefoot; the vertebral column therefore bends in a snake-like manner.

Many beetle larvae combine walking and stretching.

4. ***Climbing animals*** (Nitentia) may have *suction* equipment on their extremities (geckos, flies), or they may *anchor* themselves to the substratum with sharp claws, and use their tail for support (e.g. woodpecker). Other climbers actually grip their supports with specially adapted hands and feet. This may permit only a more or less fixed position (e.g. lice, sloths), or suspensory hand over hand progression may be possible, as in many arboreal mammals (e.g. squirrels, lemurs, monkeys), or even arm suspension and swinging from branch to branch, i.e. brachiation (e.g. many monkeys and apes and especially gibbons).

5. ***Saltatorial animals*** (Andantia) temporarily leave the ground during locomotion. They may have long extremities with an efficient lever movement (e.g. grasshoppers, jerboas, kangaroos), or they may have a spring-loaded apparatus, such as a furcula, spike or peg (e.g. springtails, clickbeetles). Some authors differentiate between *bipedal saltation* (also called *ricochetal locomotion)* and jumping on all fours. Kangaroo rats usually move by jumping on all fours, but they can increase their speed by bipedal saltation. Kangaroos, however, always move by bipedal saltation. Certain gliding animals start their aerial passage by jumping from a high point, and might appropriately be called *gliding jumpers;* they possess a membranous structure which serves as a parachute, and they are incapable of true flight (e.g. flying frog, flying lemur).

6. ***Flying animals*** (Volantia) have evolved wings for active flight. In birds, powerful wing muscles, stiffening of the leading edge and appropriate adjustment angle of the wings permit high performance flying. Relatively inefficient fliers have short broad wings that "flutter" (e.g. butterflies, bats), or slender wings with a high beat frequency that "buzz" (e.g. hoverflies, hummingbirds).

7. ***Graviportal*** locomotion refers to progress on pillars carrying a great mass, e.g. the locomotion of elephants.

In water, swimming is the main form of active M. Undulatory movements have evolved in nematodes, some annelids, and in sea snakes. Swimming in which the extremities are used as oars or paddles is found widely among swimming mammals, as well as many crustaceans, frogs and toads, and water birds. In sharks and a great many teleost fishes, the vertebral column acts as a rigid support, and rhythmic contractions of the lateral body muscles are transmitted to the tail region, where they generate a forward thrust. *Recoil* propulsion (ejection of a jetstream of water from contracting body cavities) is employed by jellyfishes and cephalopods.

Floating is a passive form of M. in water, and it is favored by a decrease in density (specific gravity) and an increase in surface area (see Plankton).

Movement perception: registration of the movement of images on the retina of the eye, in which the images originate from objects that shift within a system of coordinates that it considered to be stationary. It is certain that various mollusks, crustaceans and insects, as well as vertebrates, can perceive movement. Registration of the movement of an image is not necessarily the same as the simultaneous and correct perception of the moving object itself. For example, insects perceive changes in the speed and direction of movement of an object, not its morphology. M.p. in vertebrates is explained by the Reafference principle (see). In a resting eye, a shift of the image of a moving object acts as an exafference; the movements of images on the retina are registered in higher centers of the nervous system as movements of the object. Perception of a moving object with simultaneous eye movement is possible, because the efference copy (and possibly additional exafferences) is registered in higher centers. Conversely, objects are perceived as stationary, when the reafference of the actively moved eye erases the efference copy. The reafference principle also explains misrepresented M.p. If the eye is moved passively by light finger pressure on the eyeball, the afference of the eye muscle causes the false perception of movement in the surroundings. If the eye muscles are paralysed and active eye movements are nevertheless attempted, the surroundings appear to move as a result of the conscious generation of an efference copy.

M-phase: see Cell cycle.

MSH: see Melanotropin.

MTOC: acronym of microtubule organizing center; see Microtubules.

Mucic acid: HOOC—$(CHOH)_4$—COOH, a sugar acid produced by the oxidation of galactose (via galactonic acid).

Mucopolysaccharides: polysaccharides consisting of glycosidically linked amino sugars and uronic acids, and containing additional *N*- acetyl or *O*-sulfate groups. In animals M. serve as support, protective and lubricating substances. Most important examples are mucoitin sulfate, chondroitin sulfate and hyaluronic acid.

Mucoproteins: compounds of polysaccharides and proteins. M. that contain neutral polysaccharides are known as *glycoproteins.* Those containing polysaccharides with a high content of amino sugars are called *mucoids.*

Mucor: see *Phycomycetes.*

Mucorales: see *Phycomycetes.*

Mucro: the terminal section of the caudal furcula of springtails. The M. has one or more barbs, and is often hooked.

Mudfishes: see Lungfishes.

Mudminnows: see *Umbridae.*

Mudpuppy: see *Proteidae.*

Mudskippers: see *Gobiidae.*

Mud turtles: see *Kinosternidae.*

Mugger: see *Crocodylia.*

Mugil: see *Mugilidae.*

Mugilidae, *mullets:* a family of medium-sized, shoaling fishes of the order *Perciformes,* represented by about 95 species in about 13 genera. They favor the coastal regions of warm seas, and also ascend rivers.

Many are good to eat, e.g. the ***Common mullet*** *(Mugil cephalus)* found in tropical and warm seas worldwide; and the ***Golden gray mullet*** *(Liza aurata)* found in the Black Sea, Mediterranean, and the west coast of Europe and Africa. The dorsal fins are widely separated, the anterior with 4 spines, the posterior being soft-rayed. [Members of the *Mullidae* (see) are also known as mullets].

Mugwort: see *Compositae.*

Muhlenberg's turtle: see *Emydidae.*

Mulberry: see *Moraceae.*

Mule: see *Equidae,* Hybridization.

Mule deer, *Odocoileus hemionus:* a medium-sized deer, (head plus body 85–210 cm, shoulder height 55–110 cm) distributed from southern Yukon and Manitboa to Baja Califormia and southern Mexico. Unlike the closely related White-tailed deer (see), the antlers of *O. hemionus* have 2 main branches, forming 2 nearly equal parts. It also has a characteristic black-tipped tail. See Deer.

Mull: a type of Humus (see) in biologically very active soils. M. is produced in practically neutral soils from biologically easily and rapidly degradable straw and plant litter. At the microbiological level, this degradation is due mainly to the action of bacteria and actinomycetes, and to earthworms at the zoological level. The top layer of humus becomes rapidly incorporated into the mineral soil, favoring the formation of structurally important humins.

Müllerian mimicry: see Protective adaptations.

Müller's larva, ***protrochula:*** a planktonic, free-swimming larval stage of many *Turbellaria* in the order *Polycladida.* The spindle-shaped body characteristically has 8 projecting processes or lobes, whose outer edges carry a continuous band of cilia. A special form of M.l. with only 4 appendages is called the Goettesche larva.

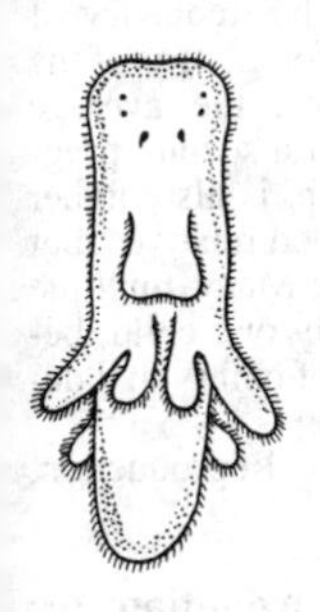

Müller's larva

Muller's method: a scheme of genetic crossing which enables the recognition of deleterious recessive genes on a chosen chromosome. In particular, it was used by Muller to identify the presence of lethal mutations on chromosomes of *Drosophila melanogaster.* The method does not register individual factors, but the total effect of the chromosomes that carry them. In principal, crossing experiments are performed to increase the frequency of occurrence of the chromosome in question. Two copies of this chromosome are then combined in a single individual, so that its deleterious genes become homozygous. The decreased fitness of flies that are homozygous for the chromosome is determined by comparison with heterozygous carriers.

A marker strain is also required. For example, Muller constructed a marker chromosome of chromosome II (also known as a "balancer" chromosome; see Balanced lethal stock) containing the dominant mutation "curly" which is lethal in the homozygous condition, while its homolog carried the dominant, homozygously lethal mutation "plum". All flies with the curly mutation have rolled wings, and all flies with the plum mutation have an abnormal eye color. Furthermore, the balancer chromosomes also carry long, overlapping inversions which serve as crossover suppressors, i.e. they prevent recombination of genetic markers. The crossing scheme is shown in the Fig. A wild fly is crossed with an individual of the marker strain (top row). Each fly of the resulting progeny (1st generation) carries one of the two marker chromosomes. The frequency of occurrence of the wild chromosome (designated X in Fig.) is then increased by crossing a fly of this first generation with an individual of the marker strain. In the progeny from this crossing (2nd generation; 3rd row in Fig.), all viable curly flies that are not also plum carry copies of the same wild chromosome. If two of these curly animals are crossed, Mendel's 3rd law predicts that 25% of the resulting progeny (bottom row in Fig.) will be homozygous curly, 50% will be heterozygous for the wild chromosome, and 25% will be homozygous for the wild chromosome. Since all homozygous curly flies die, the two surviving genotypes must be present in the ratio 2:1. Usually, however, there are too few wild flies, which indicates the presence of deleterious recessive factors on chromosome II. If the chromosome carries one or more lethal factors, then no wild type flies appear at all.

With the aid of this method, H.J. Muller first showed that X-rays cause mutations. See Fig. p. 788.

Mullets: see *Mugilidae, Mullidae.*

Mullidae, ***goatfishes:*** a family of fishes of the order *Perciformes,* containing 55 species in 6 genera. They are marine, bottom-dwelling fish, with 2 long chin barbels. The dorsal fins are widely separated, the anterior with 6–8 spines, the posterior being soft-rayed. The anal fin has 1 or 2 small spines. Many species are brightly colored, and many are important as food, e.g. the brownish red to purplish ***Red mullet*** *(Mullus barbatus)* from the Mediterranean, Black Sea and Sea of Azov.

Mullus: see *Mullidae.*

Multienzyme complex: an aggregate of several enzymes, each catalysing a different reaction and possessing a different structure. Each catalysed reaction is one of a sequence of reactions in a pathway. Examples are the fatty acid sythetase complex (which consists of 7 different enzymes), the α-ketoacid dehydrogenase complex, and tryptophan synthase.

Multimanimate rats: see *Muridae.*

Multipartite viruses, ***coviruses:*** plant viruses whose genome is divided and packed into more than one virion, often in unequal sized portions. Divided viral genomes packaged in two or three different virions, all required for infectivity and production of new virus progeny, are called dipartite and tripartite, respectively. Thus, the dipartite genome of *comoviruses* (e.g. cowpea mosaic virus) is divided between two similarly sized icosahedral virions, containing RNAs of M_r 2.3 and 1.4 x 10^6, respectively. *Almoviruses* (e.g. alfalfa mosaic virus) consist of 4 nonhelical, bacilliform rods of different lengths, which according to their positions in density gradient centrifugation are classified as bot-

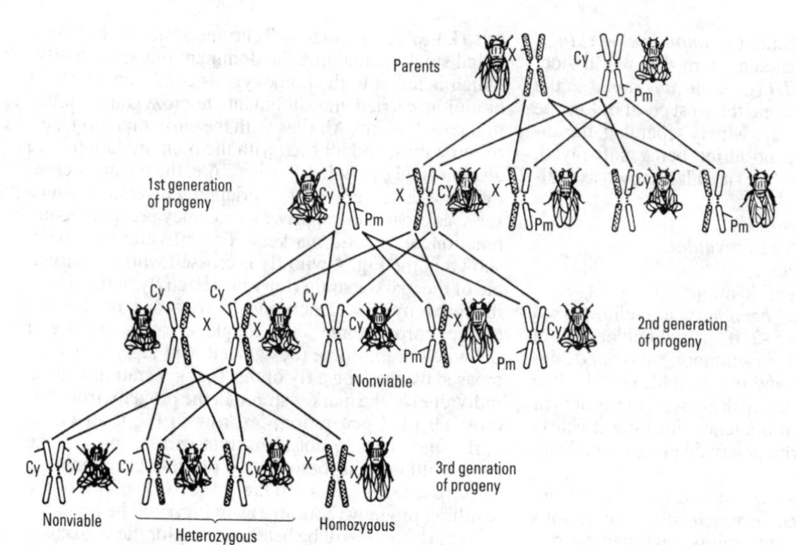

Muller's method. Scheme of genetic crosses designed to reveal the presence of deleterious recessive genes.

tom (B), middle (M) and top (T) components. B particles are the largest (length 58 nm) and they consist of two types: those containing RNA component 1 (M_r 1.1 x 10^6) and those containing RNA component 2 (M_r 1.0 x 10^6). The RNA of the M particles has M_r 0.8 x 10^6. Some T particles (Tb) contain no RNA, while the Ta particles contain RNA of M_r 0.3 x 10^6. Only the 3 largest RNA components are required for infection and production of progeny, so the genome is tripartite. The genomes of *bromoviruses* (e.g. brome mosaic virus) and *cucumoviruses* (e.g. cucumber mosaic virus) consist of 4 RNA components (2 molecules of M_r about 10^6 and one each of M_r about 0.7 and 0.3 x 10^6) separately packaged in icosahedral virions. Viral infection and production of progeny require the presence of only the three largest RNAs, so these genomes are also tripartite.

In other ***divided genome viruses,*** portions of the genome are not packaged in freely separable virions, but within the same virion. For example, the genome of influenza viruses consists of 8 different RNA molecules. Within the influenza virion, each RNA molecule is weakly encapsulated in a delicate helical, filamentous nucelocapsid, about 10 nm in diameter and of varying length. The frequency of recombination of this multiplicity of genome components accounts for the frequent changes of the influenza virus, leading to new strains, some with increased virulence and pathenogenicity, and different serological properties. The only other animal viruses displaying such a high degree of physical segmentation are the *Orthomyxoandreoviridae.* The reoviruses and several plant viruses (e.g. wound tumor mosaic virus) are not a multipartite, but their genomes consist of several elements of double-stranded RNA, all packaged in a single virion.

Multiple births: birth of several neonatal animals which have developed simultaneously in the uterus. In many animals, M.b. occur regularly, but they are relatively rare in humans. About one in eighty human pregnancies results in the birth of twins. Triplets, quadruplets, etc (up to septuplets) are much less frequent. *Monozygotic* M.b. arise from a single fertilized egg, which splits in the course of early development into 2 or more parts, each developing into an individual animal. In *polyzygotic* M.b., each individual arises from a separate fertilized egg. Thus, polyzygotic twins resemble each other no more than other siblings, whereas monozygotic twins ("identical twins") are genetically identical and resemble one another so closely that it is difficult to distinguish between them. Incomplete splitting of a fertilized egg results in "*Siamese twins*", whose bodies are joined to a greater or lesser extent. The frequency of polyzygotic twins is 3–4 times greater in women between 35 and 40 years than in women less than 20 years, whereas the frequency of monozygotic twins is about the same for all ages of the mother. If a mother produces twins, the average chance that she will produce twins in a second pregnancy is 3–6%. The frequency of M.b. is also higher than average when the parents are blood relatives, but no genes for M.b. have been found. Many multiple pregnancies end with the birth of only one child, because the others die at an early stage of embryonal development.

Multiple fission: see Schizogony, Reproduction, Cell multiplication.

Multiplication cyst: see Cyst.

Muliregional model of human evolution: see Anthropogenesis.

Multisubstrate enzymes: see Enzymes.

Multituberculata, *multituberculates* (Latin *multi* many, *tuberculatus* bumpy): an extinct order of primitive, mouse to marmot-sized mammals. Their multituberculate cheek teeth suggest that they were vegetarians. Fossils are found from the Upper Triassic to the Tertiary (Eocene).

Multivesicular body: a secondary lysosome, of variable shape, diameter 0.15–0.5 µm, containing cytoplasmic vesicles (about 50 nm diam.). In the formation of an M.b., an invagination of the lysosome surface abstricts to form a closed vesicle of cytoplasm, which moves into the lysosomal matrix (Fig.). The content of the lysosomal marker enzyme, acid phosphatase, is variable. M.bs. may fuse with primary lysosomes.

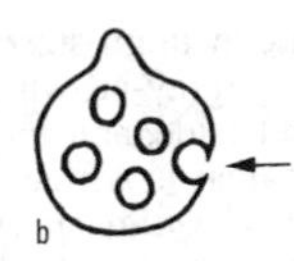

Multivesicular bodies. There is more catalytically active acid phosphatase in *a* than in *b*. The arrow indicates the uptake of cytoplasm and the abstriction of a cytoplasmic vesicle.

Multivesicular hydatid disease: see Dog tapeworm.

Münch's pressure flow theory: see Pressure flow theory.

Mungos: see Mongooses.

Municipal irrigation fields: see Biological waste water treatment.

Muntiacus: see Muntjacs.

Muntjacs, ***barking deer, Muntiacus:*** six species of small deer (head plus body 80–113 cm): *Muntiacus muntjak* (India, Sri Lanka, Nepal to southern China, Malay Peninsula, Sumatra, Java), *M. rooseveltonum* (Laos), *M. reevesi* (southern China, Taiwan), *M. crinifrons* (east-central China), *M. feae* (Tenasserin and adjoining parts of Thailand). Only males have antlers, whereas the females have small bony knobs and tufts of hair. The antlers are relatively short (12.5–15 cm) and they emerge from long, bony, hair-covered pedicels. The upper canines on the male are elongated into tusks, which curve outward from the lips, and serve as effective weapons.

The sixth species was reported by the joint Vietnamese/WWF team in 1994 in the Vu Quang Nature Reserve in Vietnam. This new species, weighing 40–50 kg, is about one and a half times larger than the other M. Its long antlers are curved forward rather than backward, and it possesses massive canines. It lacks the tuft of hair normally found at the base of the antlers of other species.

M. are found from sea level to moderate elevations in hilly country, and they prefer forest cover or dense vegetation. See Deer.

Muntjac (male)

Muraenidae: see Moray eels.

Muraena helena: see Moray eels.

Muridae, ***murids, Old World rats and mice*** (plate 7): a family of Rodents (see) containing about 457 species in 98 genera. Dental formula: 1/1, 0/0, 0/0, 3/3 = 16. The cusps of the molars are arranged in 3 rows (cf. 2 rows of cups in the *Cricetidae*). Murids are usually relatively small with pointed snouts and long tails. Although originating in the Old World, the family is cosmopolitan in distribution, certain members having become established in or near to human habitations throughout the world.

The ***Black rat*** or ***House rat*** *(Rattus rattus)* has probably been associated with humans ever since the present human form evolved. Its center of distribution is Southeast Asia. It was present in Europe during the last Ice Age, and it has been an inhabitant of the Mediterranean region for many thousands of years. In the completely wild state, it appears to be arboreal in habit. Certainly in its association with man it prefers to climb, favoring the upper, drier parts of wooden buildings (in contrast to the Brown rat, which prefers the lower, damp parts of buildings). Ships, especially early wooden ones, are also greatly favored, hence the spread of the Black rat to practically all other parts of the world. It is thought to have arrived in South America in 1540. It appeared in Europe and colonized other parts of the world before the Brown rat (see below), but in most areas it is nowdays less numerous than the Brown rat. In the Mediterranean region, the Black rat can still be found far away from human habitations, in trees and bushes, especially on fruit plantations. It is mainly vegetarian, and is notorious for its destruction of stored grain and other vegetable food supplies. Through its fleas, urine and feces, it is a carrier of disease, made especially efficient by its extremely close association with man. The European bubonic plague during the Middle Ages was carried mainly by the Black rat. Measuring 16–22 cm in length, it is somewhat smaller than the Brown rat. Its tail (17–24 cm) is always longer than its body, whereas that of the Brown rat is always shorter. Most individuals are very dark, usually black, with white markings on the underside. There are 3 main subspecies: *Rattus rattus rattus* ***(House rat,*** almost entirely in human habitations), *Rattus rattus alexandrinus* ***(Roof rat)*** and *Rattus rattus frugivorus* ***(Corn rat),*** the latter 2 occurring in the wild as well as in buildings in the tropics and subtropics.

The ***Brown rat*** or ***Norway rat*** *(Rattus norvegicus),* while originating from the Asian steppes, now also has a cosmopolitan distribution. It may have become closely associated with humans a few thousand years ago, or less than a thousand years ago, but certainly the length of this association does not match that of the Black rat. It is larger (body length 22–26 cm) and more powerfully built than the Black rat, and its tail (18–22 cm) is always shorter than its body. Its ears are also shorter than those of the Black rat. Sometimes black individuals are found, often with lighter underparts and forefeet, but most specimens are brown to gray with lighter underparts. The albino white rats widely used in laboratory research (e.g. Wistar strain) are varieties of the Brown rat. By the middle of the 18th century, the Brown rat had invaded European sea ports and was being carried in ships. Consequently, it is now established in sea ports worldwide. In contrast to the Black rat, it is not arboreal in habit, and in its original Central Asian home it still burrows in the ground. However, its ability to adapt to and exploit other habitiats is phenomenal. In human habitations, it prefers the damp, lower parts, such as cellars and sewers. It is also common in refuse tips, it is even an excellent swimmer and diver, and it catches fish. As evidenced by the fact that laboratory rats are maintained on a pelleted vegetable chow, the Brown rat can clearly exist satisfactorily as a vegetarian. In the wild, however, animal food is preferred. It eats mice, chickens, etc, even biting pieces from larger animals like pigs. Large Brown rat colonies are also a threat to the eggs of ground-nesting birds. Like the Black rat, it is also a major disease carrier, contributing to the spread of bubonic plague in the Middle Ages, and especially to the later Indian plague (1892–1918) which killed 11 million people in the Indian subcontinent.

The ***House mouse*** *(Mus musculus)* occurs throughout the world in human habitations, as well as in fields and forest fringes. Various color varieties are kept as pets, and it is also used as a laboratory animal. Individuals are 6–12 cm in length, with a long tail (6–11 cm). The first molar is very large and the third molar is small. A sharp-edged ridge on the back of the upper incisors aids the identification of this species. Several subspecies are recognized. The ***Short-tailed*** or ***Pale-bellied Eastern house mouse*** *(Mus musculus spicilegus)* is occasionally found in human habitations, but normally lives outside all year. It is the ancestor of the ***Northern house mouse*** *(Mus musculus musculus),* a full or part time inhabitant of human dwellings, found east of the Elbe in eastern, southeastern and northeastern Europe. The ***Bactrian house mouse*** *(Mus musculus bactrianus)* from Central Asia is the ancestor of the ***Western house mouse*** *(Mus musculus domesticus),* which occurs west of the Elbe in western and northwestern Europe.

Similar in size to the House mouse, the ***Striped field mouse*** *(Apodemus agrarius)* has brown-red upper parts with black stripes along the back. It excavates subterranean burrows, and from Siberia to the Rhine it is a pest of cereal crops, invading barns and stables in the winter.

Also resembling the House mouse and somewhat smaller than the Striped field mouse, the ***Long-tailed field mouse*** *(Apodemus sylvaticus)* frequently invades buildings and food stores. Most specimens have a dark yellow area on the underside. In addition to human habitations, the preferred habitats are fields and bushy areas

The ***Harvest mouse*** *(Micromys minutus)* (plate 7) is a very small (body length 5.5–7.5 cm, tail 5–7 cm), agile climber with a prehensile tail. It prefers a habitat of young grasses or rushes, and commonly invades cereal fields. It is not a particular pest of European cereal crops, but causes widespread damage in East Asian rice fields.

Spiny mice *(Acomys)* are medium-sized mice (body length 7–12 cm) with spiny, gray-brown to yellowish hair on their backs. The underside, nonspiny fur is gray to white. They inhabit dry areas in Africa and Asia Minor, e.g. *Acomys cahirinus* **(*Egyptian spiny mouse*)**, *A. minous* **(*Cretan spiny mouse*)**, *A. dimidiatus* **(*Sinai spiny mouse*).** Most other mice have a gestation time of about 22–24 days, giving birth to naked, blind, helpless young. In contrast, the gestation time of Spiny mice is 36–38 days, and they give birth to well developed young, already able to maintain their own body temperature, with open or nearly open eyes and with a considerable covering of body hair.

Female ***Multimammate rats*** *(Mastomys)* possess 12–24 nipples. These medium-sized (body length 9.5–16 cm) murids have a high reproduction rate, and they are found in dry areas throughout sub-Saharan Africa. They often enter human habitations, and they are responsible for transmitting plague, as well as destroying food stores. Although they are not easy to handle, species of *Mastomys* (e.g. *M. coucha)* are used in laboratory research for the study of bubonic plague, Bilharzia and cancer; malignant growths are common in the glandular stomach (proventriculus) of these animals.

Murre: see *Alcidae.*

Musaceae, *banana family:* a family of monocotyledonous plants, containing about 220 species, native in the tropical regions of Africa and Asia. They are handsome herbaceous plants with tuberous thickened rhizomes, and long leaf sheaths which give the appearance of a stem. The pinnately veined (pinninervate) leaves are stalked; they possess a large surface area, and they often become split by the mechanical stress of wind action. The flowers are irregular, and grouped into large overhanging inflorescences. Five fertile stamens are present, and pollination is by birds or bats. The fruits are berries, but most cultivated forms lack seeds and are propagated vegetatively.

Precise taxonomy of some of the 60 species (approx.) of ***Banana*** *(Musa)* is difficult. Cultivated bananas are derived mainly from *Musa acuminata* and *Musa balbisiana*; the 200 different varieties are of two types, i.e.

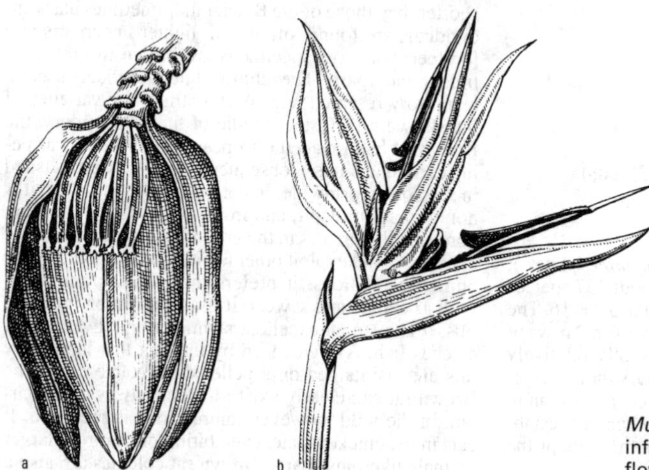

Musaceae. a Terminus of the pendulous inflorescence of banana, showing male flowers. *b* Flower of *Strelitzia.*

the sweet banana which is eaten as a fruit, and the plaintain which has a more bland taste and is normally cooked and seasoned. The leaf fibers of ***Abaca*** or ***Abaca banana*** *(Musa textilis)* yield Manila hemp. Other members of the *M.* are ***Bird of paradise*** *(Strelitzia),* with splendidly colored flowers; and ***Traveler's tree*** *(Ravenula),* whose leaf sheaths act as water reservoirs.

Muscaaurins: see Betalains.

Muscarin, ***2S,4R,5S-(4-hydroxy-5-methyltetrahydrofurfuryl)-trimethylammonium:*** a quaternary ammonium base, which is concentrated in the red skin of the cap of the fruiting body of the fly agaric *(Amanita muscaria)* and present in high concentrations in species of the fungal genera *Inocybe* and *Clitocybe.* M. causes slowing of the heart beat, increased secretion of saliva, perspiration, mucus and tears, dilation of the pupils, and gastrointestinal cramps, which are not the symptoms of fly agaric poisoning. Toxicity of fly agaric is primarily due to other, centrally acting water-soluble compounds, namely *ibotenic acid, muscimol* and *muscazone.* Atropine is an antagonist of M. and abolishes the physiological effects of the latter.

[Structural formula: HO, H, H_3C, H, O ring; H, CH_2—$N(CH_3)_3$ ⊕; OH^-]

Muscarin

Muschelkalk, ***mussel limestone:*** the middle division of the Triassic in the Germanic basin of north-central Europe. M. deposits include limestones, evaporites, and marginal sandstones. Marine fauna consist of only a few species, but these were present in large numbers, suggesting that the M. sea was characterized by unusual salinities, varying between brackish and hypersaline. The M. is sandwiched between two predominantly nonmarine sequences, the Bunter deposits (below) and the Keuper deposits (above), hence Triassic (3 divisions). See Geological time scale.

Musci, ***mosses:*** a class of Bryophytes. The actual moss plant, the gametophyte, is always differentiated into stem and leaves. The leaves are arranged spirally (rarely in 2 rows) and possess a midrib. Multicellular, branching rhizoids, with oblique cross walls, attach the plant to its substratum. The young sporophyte is initially enclosed by an embryotheca, which is derived from the archegonial tissue and is therefore haploid. As the sporophyte extends and matures to form the sporogonium, the embryotheca breaks transversely, leaving the lower part as the vaginula, the upper part becoming the calyptra. The sporogonium consists of the capsule (containing spores) on a short stalk (seta) with a foot (haustorium), which is always embedded in the end of the pseudopodium, i.e. it is embedded in the gametophyte, since the pseudopodium is formed by extension of the gametophyte stem after fertilization. Inside the capsule, the spore sac surrounds a central columella, which stores water and conducts nutrients to the developing spores. Many other features of the *Musci* are described under Bryophytes (see).

1. Subclass: *Sphagnidae* ***(Bog mosses, Peat moss).*** This subclass contains a single family *(Sphagnaceae),* which contains a single genus *(Sphagnum),* represented by about 300 species. Most are native to the cold and

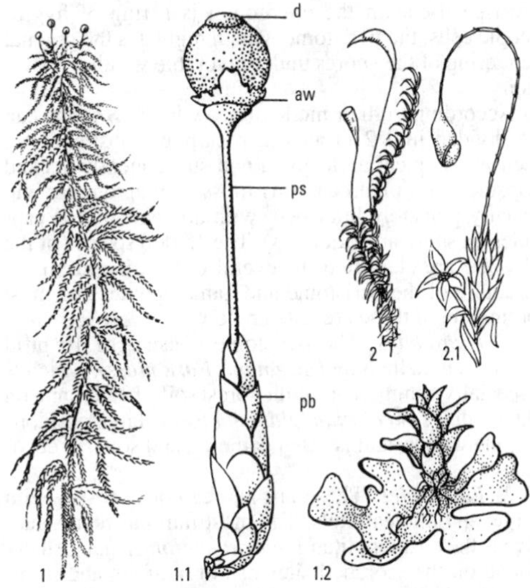

Fig.1. *Musci* (mosses). *1 Sphagnum acutifolium,* plant with sporogonia (x 2/3) (subclass *Sphagnidae). 1.1 Sphagnum squarrosum,* mature sporogonium at the apex of a branch (x 10), *d* operculum, *aw* vaginula (bottom part of broken embryotheca), *ps* pseudopodium, *pb* perichaetial leaves. *1.2 Sphagnum acutifolium,* protonema with young plant (x 100). *2 Dicranium scoparium,* 3-year-old plant, natural size (subclass *Bryidae). 2.1 Funaria hygrometrica* (x 2) (subclass *Bryidae).*

temperate zones, and their taxonomic differentiation is difficult. Spore germination requires the presence of a mycorrhizal fungus, and the resulting protonema is lobed, thallose and one cell thick, producing a single leafy gametophyte with rhizoids. The stem of sphagnum mosses carries clustered branches, which form a dense tuft at the extremity of the plant. They grow in boggy areas, forming hummocks and blankets, growing upward each year, the lower parts dying and eventually forming peat. As calcifuge plants, they are good indicators of acidic soils. The stem cortex consists of layers of dead, empty cells, which absorb water by capillary attraction. Leaves consist of a network of elongate, live, chloroplast-rich cells, the meshes between this network being occupied by empty, dead cells, which also serve to absorb water. Thus, by capillarity, sphagnum mosses are able to store large quantities of water. They are the only mosses with a direct economic value; dried sphagnum has been used to dress wounds, while peat is used widely in horticulture.

2. Subclass: *Andreaeidae.* This small subclass contains a single family *(Andreaeceae)* with 3 genera. *Andreaea* is represented by 120 species, which are native to the European Alpine region, the Arctic and the Antarctic, where they grow as small, dense, dark brown cushions on noncalcareous rocks. Their protonema is branched and ribbon-like.

3. Subclass: *Bryidae.* Most of the "typical" mosses belong to this subclass. Structure and shape of the capsule are characteristic, the upper part forming the operculum, which opens as a circular lid when the spores

are ripe. Beneath the operculum is a ring of hygroscopic cells, the peristome, which regulates the gradual scattering of the spores under favorable weather conditions.

According to their mode of growth, the *Bryidae* can be divided into 2 groups: acrocarpous mosses (erect stem with apical archegonia and subsequently stalked capsule), and pleurocarpous mosses (plagiotropic and usually pinnately branched, with archegonia and capsules on short side branches). The 15 000 species of the *Bryidae* are classified in several orders, based on the structure of the peristome and gametophyte. The most prominent of these are listed below.

1) *Dicranales.* The peristome consists of 16 bifid teeth. Genera include *Dicranum* ***(Fork moss),*** which is especially common on acidic forest soils, heaths, and on the tundra. *Leucobryum glaucum* forms large cushions in pine woods, and is often gathered and sold for decoration.

2) *Funariales.* These are acrocarpous mosses with large, smooth leaf cells. The most familiar representative is the cosmopolitan *Funaria hygrometrica,* usually found on the scorched sites of old bonfires and in ruderal habitats.

3) *Schistostegales.* This order is represented by a single species, *Schistostega pennata* ***(Luminous moss),*** which is characterized by the absence of a peristome, and by the persistence of the protonema, which grows as a weft of fine threads. It can grow in low light intensity, e.g. in cave entrances, because the cell wall is structured like a lens that concentrates light on the chloroplasts.

4) *Eubryales.* Acrocarpous mosses with a double peristome. *Bryum* ***(True mosses),*** with over 800 species, is one of the most numerous moss genera. Most familiar is the cosmopolitan *Bryum argentum,* which thrives between paving stones and on roofs, and has even been found on sea cliffs and by the roots of cacti in the Arizona desert. Species of the genus *Mnium,* which has parenchymatous leaf cells, are common on damp forest floors.

5) *Isobryales.* Pleurocarpous mosses with a double peristome, the inner peristome being strongly reduced. Many genera are common in woods, on the ground and on tree trunks, e.g. the ***Tree mosses,*** *Climacium* and *Thamnobryum.* ***Willow moss*** *(Fontinalis antipyretica),* however, grows attached to stones and wood in lakes and slow flowing rivers; the name is derived from its former use as a nonflammable insulating material in the walls of houses in Lapland. Its dark green, keeled leaves are unmistakable.

6) *Hypnobryales.* Mosses with a hanging, long-stalked capsule and a double peristome. They are widely distributed and often form the characteristic vegetation of their habitats, e.g. *Thuidium tamariscinum, Brachythecium, Scleropodium purum, Pleurocium schreberi, Plagiothecium* and *Hypnum.*

7) *Polytrichales.* On account of their peristome structure (the peristome teeth consist of horseshoe-shaped cells) members of this order are sometimes placed in a separate subclass. They also differ from other mosses in having a hairy calyptra. Examples are *Polytrichum* ***(Hair mosses*** or ***Haircap mosses),*** which are especially common in wet habitats, and *Atrichum undulatum* ***(Catherine's moss),*** which grows in forests on nutrient-rich soils.

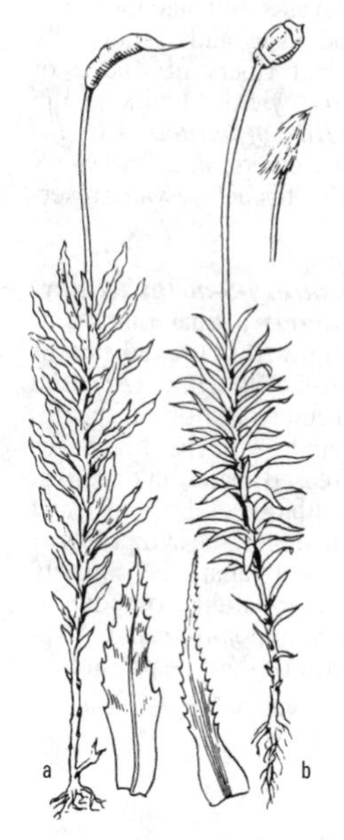

Fig.2. *Musci* (mosses). *a* "Catherine's moss" *(Atrichum undulatum). b* Hair moss *(Polytrichum commune).*

Muscicapa: see *Muscicapinae.*

Muscicapidae: an avian family of the order *Passeriformes* (see). In some taxonomies the *Muscicapidae* is the family of the Old World flycatchers. In the present account, the name embraces a much larger grouping of several subfamilies: *Cinclosomatinae* ***(Rail babblers), Malurinae*** ***(Australian wren warblers),*** *Muscicapinae* ***(Old World flycatchers)*** (see), *Pachycephalinae* ***(Thickheads*** or ***Whistlers)*** (see), *Paradoxornithinae* ***(Parrotbills),*** *Polioptilinae* ***(Gnatcatchers),*** *Sylviinae* ***(Old World warblers)*** (see), *Timaliinae* ***(Babblers, Wren tit)*** (see), *Turdinae* ***(Thrushes, Chats, Robbins)*** (see).

Muscicapinae, ***Old World flycatchers:*** an avian subfamily of the family *Muscicapidae* (see), order *Passeriformes* (see). The Old World flycatchers are widely distributed over the Old World, with representatives even in New Zealand *(Rhipidura)* and the Hawaiian islands *(Chasiempis).* They hunt insects on the wing, and in the New World their ecological niche is occupied by the tyrant flycatchers *(Tyrranidae).* The plumage of young flycatchers, like that of young thrushes *(Turninae)* has light colored spots. The ***Spotted flycatcher*** *(Muscicapa striata)* is a well known European bird, with several subspecies in Northwest Africa, Eurasia and as far as the western Himalayas. It leaves Central Europe in the autumn for its overwintering area, which extends from northwest India through the Arabian Peninsula to southern Africa. The sub-Saharan African flycatcher, *Muscicapa gambayae,* is possibly the ancestor of *M. striata.* Also well known in Central Europe as a forest and park bird is *Ficedula hypoleuca* ***(Pied flycatcher);*** its breeding range extends from Northwest Africa, through Europe to central Siberia. Very closely related to *Ficedula hypoleuca* are the ***Collared flycatcher*** *(F. albicollis;* Central, southern and eastern Europe) and the ***Half-collared flycatcher*** *(F. semitorquata;* Balkans and Caucasus into central Asia), and hybrids occur where the ranges overlap; these 3 species are therefore known as a triplet species. *Muscicapa* and *Ficedula* have characteristic broad beaks and rictal bristles; they nest in holes and cavities, and they are migratory.

The long-tailed ***Paradise flycatchers*** *(Terpsiphone)* are larger and far more strikingly colored than *Ficedula* or *Muscicapa.* The ***Asiatic paradise flycatcher,*** which is almost pure white with a black head, measures 53 cm, of which 42 cm is tail. ***Monarch flycatchers*** *(Monarcha;* Bismarck Archipelago, islands from Celebes to Carolinas, Tasmania) have velvet-textured plumage on part or all of the head, and the various plumage colors (red-brown, blue-gray, black and white) are sharply demarcated. Like the paradise flycatchers, the monarchs hunt for insects in foliage, rather than chasing it on the wing. Paradise flycatchers and monarchs are sometimes grouped with the ***Marquesas flycatchers*** *(Pomarea)* and the ***Black-naped flycatchers*** *(Hypthymis)* in a separate subfamily known as the *Monarchinae.*

Other genera are *Batis* (***Puffbacks,*** e.g. ***Chin-spot puffback*** or ***White-flanked flycatcher,*** *Batis molitor,* with a wide range in East Africa), the Asian *Platysteira* and *Diaphorophyia* ***(Wattle eyes),*** *Arses* (New Guinea) and *Zeocephus* (Philippines).

Members of the *Rhipidura* ***(Fantails)*** are sometimes placed in a separate subfamily, called the *Rhipidurinae.* The ***Gray fantail,*** *Rhipidura fuliginosa,* is found in Australia, New Zealand, Tasmania, New Caledonia and the New Hebrides. *Rhipidura leucorphrys* ***("Willie wagtail"),*** an inhabitant of open woodlands, is very common around Australian settlements.

Muscle: a contractile organ or tissue of animals, responsible for movement of the body and its parts and the constriction of hollow organs. See Muscle tissue, Musculature.

Muscle cell: structural element of smooth muscle and cardiac muscle. Skeletal muscle, on the other hand, consists of multinucleated muscle fibers (see Muscle tissue). Smooth M.cs. occur widely in animals, but are particularly abundant in the walls of viscera, glands and blood vessels, and in the skin. The elongated, spindle-shaped cells have a centrally situated, oval cell nucleus, and their cytoplasm (sarcoplasm) contains myofibrils without transverse striations (in contrast to skeletal and cardiac muscle which are transversely striated). Both thick and thin filaments are present, but they are not arranged in the orderly pattern found in skeletal and cardiac muscle. Distinct myofibrils are not observed, but rather a loosely arranged contractile apparatus aligned approximately with the long axis of the cell, and attached obliquely to the plasma membrane at disklike junctions between adjacent cells. Actin and myosin filaments are present in the ratio of 10:1. Contraction and relaxation of smooth M.cs. are much slower than those of striated muscle fibers.

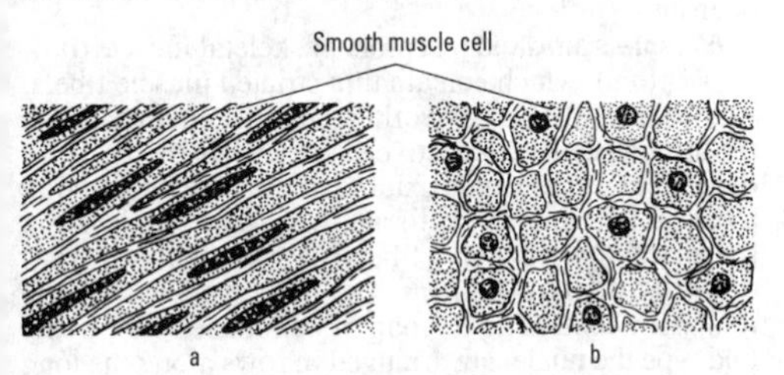

Fig.1. *Muscle cell.* Smooth muscle: *a* longitudinal section, *b* transverse section.

Cardiac M.cs. are branched, anastomosing with each other to form a network structure. The centrally situated, single nucleus is surrounded by a fibril-free sarcoplasmic region. Cardiac M.cs. are joined end to end by special contact sites known as intercalated disks. In stained preparations (especially after iron-hematoxylin staining) the intercalated disks appear as densely stained, transverse bands, whereas under the electron microscope, they are recognizable as cell boundaries with gap junctions. In cardiac M.cs., the arrangement of actin and myosin filaments within the myofibrils is similar to that observed in the cells of striated muscle, i.e. cardiac muscle is also striated. Myofibrils of cardiac muscle actually continue across the tissue, ignoring cell boundaries by penetrating plasma membranes in the region of the intercalated disks.

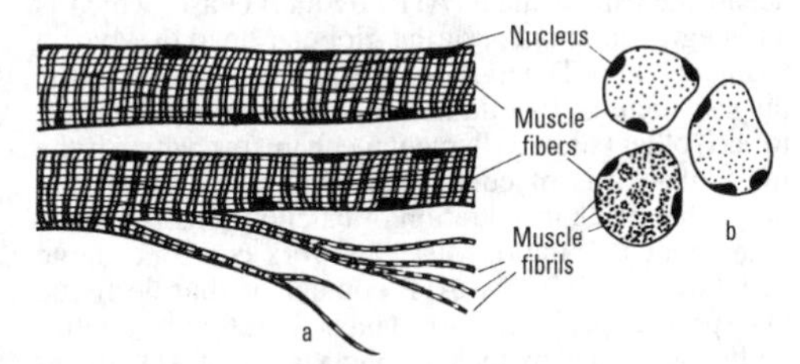

Fig.2. *Muscle cell.* Striated muscle: *a* longitudinal section, *b* transverse section.

Muscle contraction: reversible change in the mechanical state of a muscle, caused by nerve impulses or experimental stimulation. In *isotonic* M.c., the length of the muscle decreases and the force acting on the muscle remains unchanged. In *isometric* M.c., the length of the muscle remains unchanged, but its tension increases. Most muscular movements involve both isotonic and isometric contraction. Some muscle fibers contract phasically (rapid or twitch fibers), whereas others contract tonically (slow fibers). The membrane of phasic fibers generates action potentials, whereas in tonic fibers contraction is promoted by a spread of local depolarization. Muscles of the extremities of mammals consist almost exclusively of rapid fibers, while others contain a higher proportion of slow fibers, whose contraction occurs in the order of minutes. Invertebrate muscles contain a high proportion of slow fibers. The single contraction or *twitch* of a muscle is divided into the *latent period, contraction period* and *relaxation period.* These phases or periods differ according to the muscle and the animal. In skeletal muscle the latent period is the shortest and the relaxation period in the longest. Total twitch time for many mammalian skeletal muscles is only a few milliseconds, whereas the contraction period alone of smooth muscle lasts for several to many seconds (see Twitch time). Under natural conditions, individual muscle twitches can be observed in certain reflexes, such as blinking. Usually, however, muscle fibers are stimulated again in the relaxation period, causing the onset of a new contraction phase. If the time between stimuli is decreased, the relaxation phase no longer occurs, and the muscle remains in a permanent state of contraction, a condition known as *tetanus.* Tetanic M.c. leads to a relatively marked shortening of the muscle, eventually to a minimal length, because each contraction is superimposed on the one before it (superposition). The stimulation frequency necessary to cause complete tetanus is called the fusion frequency. In smooth muscle, e.g. adductor muscles of bivalves, the fusion frequency is only a few stimuli per second. In contrast, the eye muscles of mammals, with their very short twitch time, have a fusion time of more than 300 stimuli per second. Brief refractory times are a prerequisite for tetanus. In vertebrate cardiac muscle, the refractory time extends into the final third of the relaxation period, so that tetanus is impossible. Many muscles continually generate a basal tension, known as Muscle tonus (see). Muscle cells transduce chemical energy into mechanical work. Under optimal conditions, this conversion of chemical into mechanical energy procedes with an efficiency of one third or more, two thirds or less being given out as heat. Energy is released directly by the cleavage of adenosine triphosphate (ATP) by an ATPase, which is an integral component of the globular head of Myosin filaments (see). The transduction process consists of the displacement of F-actin and myosin filaments relative to each other (sliding filament mechanism), which is the molecular basis of contraction. ATP cleavage occurs only when the concentration of calcium ions is greater than about 10^{-6} mol/l. Muscular work consumes large quantities of ATP. The ATP content of muscle tissue (1–5 µmoles per gram tissue) must therefore be continually regenerated by the phosphorylation of ADP by reaction with creatine phosphate (vertebrates) or arginine phosphate (invertebrates), and by anaerobic and aerobic respiration. ATP is produced aerobically by oxidative phosphorylation during the respiration of fats and carbohydrates (glucose, glycogen). Under anaerobic conditions, ATP is produced by glycolysis, with the formation of lactate. During the first minutes of muscle contraction, the increased ATP requirement is met by synthesis from ADP and a phosphagen (creatine phosphate or arginine phosphate). This period allows time for the readjustment of muscle metabolism and the mobilization of glycogen, followed by increased ATP production by glycolysis. Aerobic ATP production soon slows down in a powerfully contracting muscle, due partly to metabolic oxygen depletion (oxygen debt), but also due to the fact that the contracting muscle squeezes blood out of the muscle so that freshly oxygenated blood is poorly available. Emphasis is therefore placed on anaerobic glycolysis, leading to an accumulation of lactate. Excess of lactate may eventually cause overacidification, muscle tiredness and cramps. It is removed by the circulation from resting muscle and metabolized mostly in the liver.

Muscle contracture: persistent but reversible muscle contraction or spasm, which is not due to repeated natural stimulation. It can be caused, e.g. by toxins such as caffeine, by acetylcholine, by acids and bases, as well as by mechanical damage. It may develop into irreversible muscle rigor or paralysis, in which the muscles no longer contain ATP.

Muscle elasticity curve: a plot of muscle length against a stretching force applied to the muscle, e.g. the increase in length can be measured when the muscle is loaded with different weights. The resulting plot shows that the relative stretch of muscle is greater when it is contracted than when it is relaxed, which lessens the danger of tearing muscles or tendons during muscle contraction. In contrast, the stretching of a piece of rubber is linear in response to the force applied to it.

Muscle noise: audible sounds produced during tetanic muscle contraction, due to mechanical oscillation of the repeatedly stimulated contractile filaments.

Muscle proteins: proteins constituting up to 20% of of muscle tissue. The *insoluble* or *contractile proteins* of the myofibrils are all involved in contraction. They account for 60% of M.p., and they include Myosin (see) which forms the thick filaments, as well as Actin (see) which is the main component of the thin filamemnts. See Actomyosin. The *soluble proteins* include the enzymes of glycolysis, the tricarboxylic acid cycle and the respiratory chain, and the muscle pigment, myoglobin.

Muscle relaxants: natural or synthetic substances, which relax muscles by interrupting the transmission of impulses from nerves to muscles. A well known M.r. is curare, isolated from the roots and bark of various *Strychnos* species, and used as an arrow poison by South American indians. M.r. are used widely in clinical medicine. See Endplate potential.

Muscle relaxation: a phase or period during the overall process of Muscle contraction (see). It is usually somewhat longer than the contraction period, while in heart muscle relaxation and contraction are of equal duration. M.r. occurs when the concentration of calcium ions in the sarcoplasm falls below 10^{-8} mol/l. Cross bridges between actin and myosin are then broken, and ATPase activity ceases (see Myosin filaments, Muscle contraction). During the contraction phase, the sarcoplasm contains 10^{-5} to10^{-6} mol/l calcium ions. At the beginning of M.r. calcium ions are pumped out of the sarcoplasm into the tubules by a calcium pump in the membranes of the longitudinal canals of the sarcoplasmic reticulum. As the calcium concentration falls, troponin separates from the tropomyosin filaments, which then shift their position and block the attachment of myosin heads to the actin filament, and the muscle passes into a relaxed, resting state. Since myosin ATPase activity depends on activation by calcium ions, ATPase activity ceases.

Muscles of mastication: muscles operating the jaw and enabling biting and chewing movements. In elasmobranch fishes, the adductor mandibularis is a uniform muscle, whereas in teleosts it is divided into several individual muscles concealed by the lateral bones of the skull. The superficial masticatory muscles of quadrupeds have evolved from the muscles of the visceral arches (branchiomeric muscles), which in fishes serve as levators, depressors and constrictors of the gills. Quadrupeds possess an outer group consisting of the temporalis and the masseter (the masticatory muscle proper of mammals), and an inner muscle layer is formed by the pterygoid. Temporalis, masseter and inner pterygoid close the mouth, while the outer pterygoid opens it.

Muscle spindles: receptors of skeletal muscle (proprioceptors), which contain thin striated muscle fibers. They run parallel to the working muscle, but they end at the intramuscular connective tissue and are therefore independent of the latter. Externally they are enclosed in a capsule of connective tissue. Internally they contain sensory and motor nerve fibers. M.s. contain two types of *intrafusal muscle fibers*. The first type has a central dilatation containing a group of cell nuclei. In the second type the nuclei are arranged in rows along the long axis of the interior of the fiber. M.s. are primarily responsible for the perception of movement.

Muscle system: see Musculature.

Muscle tissue: contractile tissue of multicellular organisms, characterized morphologically by its possession of myofibrils, functionally by its ability to contract, and chemically by its relatively high content of actin and myosin. M.t. serves for movement of individual body parts and for locomotion. Morphologically and functionally, a distinction is drawn between *smooth (involuntary), striated (skeletal* or *voluntary)* and *cardiac muscle.* Most M.t. arises during embryonal development from the middle germ layer, i.e. it is of mesenchymal or mesodermal origin. An exception is found in the coelenterates, whose epithelial cells contain contractile fibers (see Epithelial muscle cells). Contractile elements may be unicellular Muscle cells (see) or multicellular muscle fibers, the latter being either smooth or transversely striated.

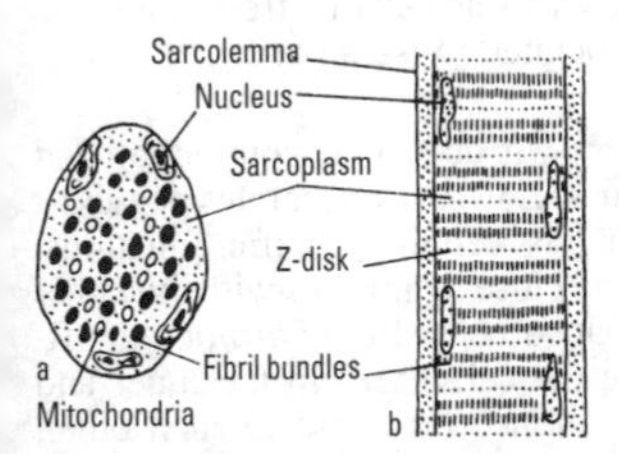

Fig.1. *Muscle tissue.* Striated muscle fiber consisting of packed fibrils and peripheral nuclei. *a* Transverse section. *b* Longitudinal section.

In primitive unicellular organisms, M.t. is represented by filamentous cytoplasmic structures known as myonemes, which can contract rapidly along their longitudinal axis. Sponges possess primitive M.t. around their pores, in the form of fusiform cells with smooth surface fibers. The specialized M.t. of the *Cnidaria* consists of modified epithelial cells of the ectoderm and endoderm (see Epithelial muscle cells). Long fibrils lie at the base of the cell, transverse to the longitudinal cell axis; in the ectoderm, the fibrils run parallel to the longitudinal body axis of the animal, whereas in the endoderm they run transverse to the body axis. In most species, these epithelial muscle cells are smooth, but in medusae the edge of the bell contains some striated elements. Comb jellies, on the other hand, have branched, often multinucleate muscle cells, which arise from myoblasts in the middle germ layer, and are similar to the muscle cells of multicellular organisms. The M.t. of worms consists of elongated cells with contractile fibrils on one side (contractile filament cells). In some cases, the fibrils form a tube around the outside of the muscle cell, e.g. in leeches. The pharyngeal muscles of annelids are striated. Arthropods have predominantly striated M.t., only the muscles of the gut wall and oviducts being smooth. Striated muscle of arthropods is structurally very similar to that of vertebrates. The fibers consist of granular *sarcoplasm (myoplasm),* with a structureless thin envelope known as the *sarcolemma* or *myolemma.* Sometimes there is also a thin envelope of connective tissue (see Perimysium). The myofibrils are either distributed throughout the entire transverse section of the fiber (with cell nuclei at the edge of the fiber), or they are arranged in longitudinal bands known as septa (with centrally situated cell nuclei). Attachment of the M.t. to the chitin exoskeleton is mediated by specialized Tonofibrils (see), or by inclusion of processes of the epidermis or even the cuticle. In mollusks, striated muscle is found only in certain organs, e.g. in the heart, sometimes in the pharynx. In fast swimming cephalopods, certain body muscles are also striated. Most of the M.t. of mollusks, however, consists of smooth muscle cells branching outward to the skin. Similarly, the M.t. of echinoderms consists of smooth fibers with laterally situated nuclei. In vertebrates the visceral musculature consists of elongated, fusiform smooth muscle cells, while the musculature of the skeleton and heart consists of striated muscle cells (see Muscle cell). Smooth muscle cells may occur separately in the connective tissue, or they may be combined in bundles. Some bundles of smooth muscle cells have terminal tendons. The number of fibrils in striated fibers determines the different colors of skeletal muscle. Fibril-deficient, water-rich fibers appear white. Fibril-rich, red-colored fibers contract more rapidly, but tire sooner. In the heart, the delicate fibers are arranged in a spatial lattice and joined to one another by anastomoses. Under the microscope, the myofibrils of striated muscle display light and dark bands, which are usually in register in neighboring fibrils (hence the striations). In polarization microscopy, the dark bands are seen to be anisotropic, and are therefore called *A-bands.* The light bands are isotropic, and are known as *I-bands.* Within each I-band is a more strongly refractive, fine, anisotropic line, the *Z-band* or *Z-disk.* The section of myofibril between two Z-bands is known as a *sarcomere.*

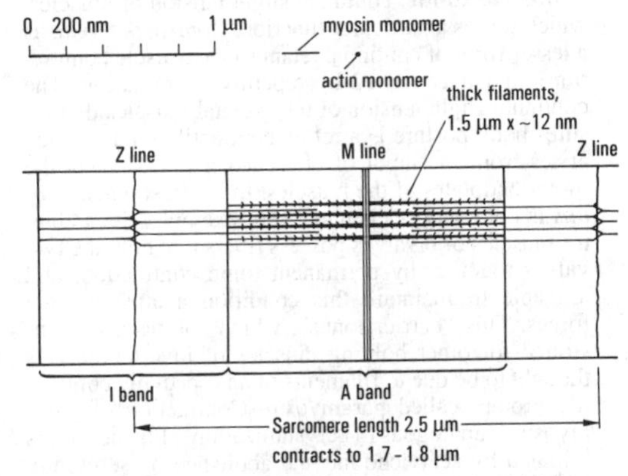

Fig.2. *Muscle tissue.* Interdigitation of thick and thin filaments in striated muscle. (Used with permission from D.E. Metzler, *Biochemistry. The Chemical Reactions of Living Cells,* Academic Press, 1977)

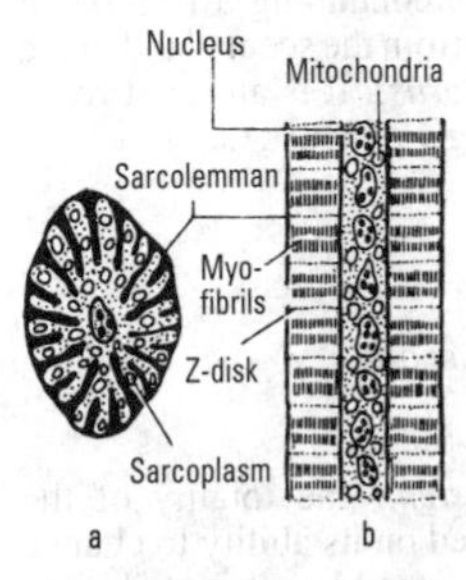

Fig.3. *Muscle tissue.* Striated muscle fiber with fibrils organized in septa and with centrally located nuclei. *a* Transverse section. *b* Longitudinal section.

In striated muscle the endoplasmic reticulum forms a reticulate system of tubes, which is distributed around the myofibrils. In each sarcomere, the longitudinal meshes of the reticulum *(L-tubules)* form two parallel cisternae arranged transverse to the axis of the myofibril. A third tubular system *(T-tubules)* arises from invaginations of the sarcolemma and extends into the interior of the muscle fiber. Points of contact between L- and T-tubules are known as triads. In amphibian striated muscle, the triads are in register with the Z-band; in mammalian striated muscle, they lie at the border between the A- and I-bands. T-tubules serve to conduct excitation from the surface to the interior of the muscle fiber, whereas the channels of the sarcoplasmic reticulum serve for the uptake, storage, transport and release of calcium ions during the relaxation phase.

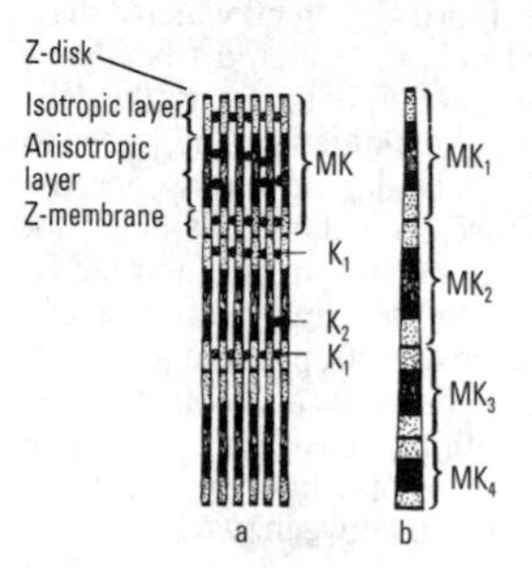

Fig.4. *Muscle tissue.* Layers of striated muscle fibrils (schematic). *a* Uncontracted. *b* Increasing degrees of contraction from top to bottom. *MK* = sarcomere.

Muscle tonus: continual slight tension of muscles, which serves a particular function. *Contractile tonus* is a lesser form of continual tetanus (see Muscle contraction), and it is caused by repetitive stimulation. The continual slight tension of the skeletal muscles that insures body posture is a reflex contractile tonus, which arises from the input of efferent nerve fibers via the motor endplates of the muscles. In contrast, *plastic tonus* is not energy-dependent. For example, the adductor muscles of bivalves possess fibers that hold the two valves together by permanent tonic contraction, and are able to maintain this condition against outside forces. This "barrier tonus", which has been demonstrated for other holding muscles of invertebrates, is thought to be due to filaments of an especially contractile protein called paramyosin. Contraction of paramyosin is analogous to "crystallization". Plastic M.t. is initiated by acetylcholine and abolished by serotonin. Molecular mechanisms of the tonus of vertebrate smooth muscle, e.g. in the gastrointestinal system and blood vessel walls, are not as well understood as those of striated muscle.

Muscone: a cyclic ketone containing a ring of 15 carbon atoms (Fig.), released from the scent gland of the male ibex *(Moschus moschiferus).* It is an oil at room temperature and smells strongly of Musk (see).

CH_2
$H_3C—CH\ \ C=O$
$(CH_2)_{12}$

Muscone

Muscovy duck: see *Anseriformes.*

Musculature, *muscle system:* the totality of the muscles of an organism. Based on its ability to change shape, i.e. to contract and relax, the M. is responsible for movement of the body and its parts, for blood circulation (cardiac muscle), and for the generation of body heat.

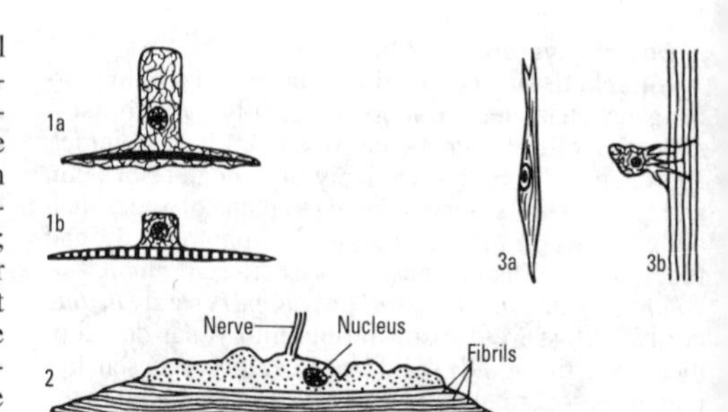

Fig.1. *Musculature. 1a* Epithelial muscle cell of hydra. *1b* Epithelial muscle cell of a medusa, showing transverse striations of the fibrils. *2* Mononucleated muscle cell of an ascarid nematode. *3a* Smooth muscle cell of a trematode. *3b* Muscle cell of a trematode separate from its associated muscle fibrils.

Movement in *protozoa* is due to the contraction and relaxation of myonemes. Lower invertebrates possess a primitive M., which has certain specific functions. Thus, *sponges* open and close their pores with the aid of contractile spindle-shaped cells. In *cnidarians,* the fibers of Epithelial muscle cells (see) in the inner and outer germ layer are aligned tranverse to each other, enabling the polyp to contract or elongate. *Platyhelminths* and *aschelminths* are the most primitive organisms in which M. occurs as a distinct structure clearly differentiated from other tissues. These organisms are enclosed by a muscular body wall, consisting of an outer circular muscle layer and an inner longitudinal muscle layer, as well as possessing variously arranged transverse muscle fibers within the body. The muscles act against the fluid-filled column of the body cavity. *Annelids* also have a muscular body wall, but in addition their alimentary tract also consists of circular and longitudinal muscle. In *arthropods,* the body wall consists entirely of longitudinal muscle, with pronounced divisions corresponding to the separate sections of the exoskeleton. The body M. often forms 2 dorsal and 2 ventral muscle bundles. Mouthparts, legs, wings, heart and other organs have their own M., which usually originates from the transverse M. The *insect* intestine has specific circular and longitudinal M., whereas the *crustacean* intestine has only a circular muscle layer. The M. of *mollusks* is very variable and nonuniform. Primitive forms like the *Placophora* have a muscular body wall. In *snails,* only the retractor muscle remains, its individ-

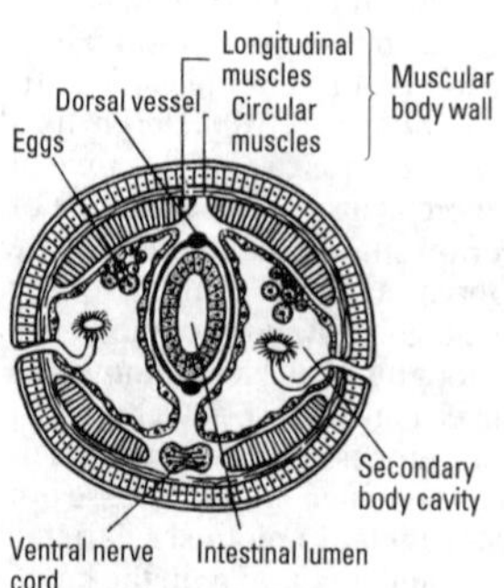

Fig.2. *Musculature.* Transverse section of an annelid showing the muscular body wall.

ual strands being responsible for retraction of the foot, head and tentacles into the shell. In *bivalves*, the muscles of the foot and mantle edge have evolved from the primitive body wall M. The M. of *echinoderms* consists predominantly of longitudinal M. In *starfishes*, this longitudinal M. is responsible for movement of the arms and ambulacral feet, while in *sea urchins* it supplies the spines, pedicellariae and other body parts.

Individual *vertebrate* muscles are named, inter alia, according to their sites of attachment (insertion) on the skeleton (e.g. coraco-brachialis), their position (e.g. pectoral muscle), their shape (e.g. trapezoid muscle), or their function (e.g. abductor hallucis). Since the Ms. of the various vertebrate groups have sometimes evolved to a stage where they are very different, a comparative presentation of M. is difficult. Despite these differences, however, the nerve supply has remained constant, and can be used for establishing the identity and homology of muscles.

According to its embryonic origin, vertebrate M. is subdivided into *somatic* or *skeletal M.* (e.g. body, tail, limb and eye M.) and *splanchnic* or *visceral M.* (e.g. intestinal M.). In all triploblastic vertebrate embryos, the M. is derived from the mesoderm, which differentiates during growth into an upper epimere, a middle mesomere and a ventral hypomere. The hypomere divides tangentially into an outer somatic layer (which forms the lining of the body cavity) and an inner splanchnic layer (which forms the smooth muscle and connective tissue of the archenteron). The epimere is divided into vertical segments, separated by myosepta. The epimere differentiates further to form the cells of the vertebral column (scerotome), the deep layers of the skin (dermatome), and the somatic muscles of the trunk and appendages (myotome). The myotomes grow and extend ventrally between the integument and somatic mesoderm.

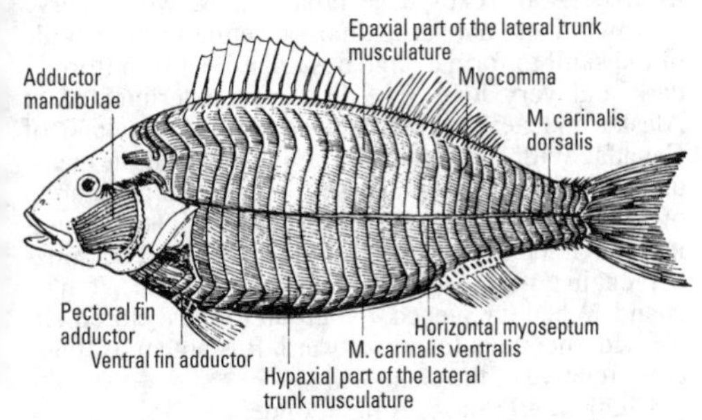

Fig.3. *Musculature.* Superficial musculature of a perch.

In *fishes*, the trunk plays a dominant role in locomotion, and the associated M. is consequently powerfully developed (Fig.). The sheet of muscles formed by the downward growth of the myotomes is, however, not continuous, but divided into dorsal (epaxial) and ventrolateral (hypaxial) sections by a horizontal partition (horizontal myoseptum or horizontal skeletogenous septum). Muscles have also evolved for movement of the fins. In some fishes (e.g. electric eel and electric catfish), part of the body M. is modified to an electric organ. In *Amphioxus* and cyclostomes, the trunk muscles are generally arranged like those of fishes. In fishes, muscles arising from the mesoderm of the visceral arches (the *branchiomeric* M.) are differentiated into levators, depressors and constrictors of the gills.

In *amphibians*, the transition from an aquatic to a terrestrial existence is accompanied by changes in somatic M. Epaxial muscles are less prominent than in fishes (the trunk is no longer the primary agent of locomotion), and the vertebral column is now a true articulating structure; some deep fibers of the epaxial muscles (now called the dorsalis trunci) become attached to the articular processes and spines of the vertebrae. The anterior end of the dorsalis trunci divides into two extensor muscles which are attached to the head; amphibians represent the earliest evolutionary stage at which the head is movable on the axial skeleton. As in fishes, the branchiomeric muscles of larval (aquatic) amphibians form gill muscles, but in metamorphosed amphibians, the loss of gills and reduction and modification of visceral arches have given rise to marked changes in the branchiomeric M., which includes, inter alia, pharyngeal constrictors and levators.

In *reptiles and birds*, the epaxial trunk M. is divided into three columns: iliocostalis, longissimus and transversospinalis. The most lateral of these (iliocostalis) extends from the ilium to the head. In creeping reptiles, this is highly developed, since it moves the trunk column from side to side. The most medial (transversospinalis) covers the laminae of the vertebrae; its deeper parts retain a segmented character, consisting of short muscles between adjacent vertebrae.

In terrestrial vertebrates, the branchiomeric muscles have generally evolved into head and neck M., e.g. jaw muscles, the trapezius which connects the axial skeleton to the pectoral girdle, the sternocleidomastoid (prominent muscle projecting on the side of the neck and largely concerned with turning the head) which extends from the occiput to the sternum, as well as constrictor muscles of the pharynx. In quadrupeds, the lateral body M. is differentiated into *dorsal, pectoral* and *abdominal* muscles. The *limb* M. is derived from the ventral part of the lateral M., and it displays marked specialization according to the animal group, e.g. the flight M. of birds.

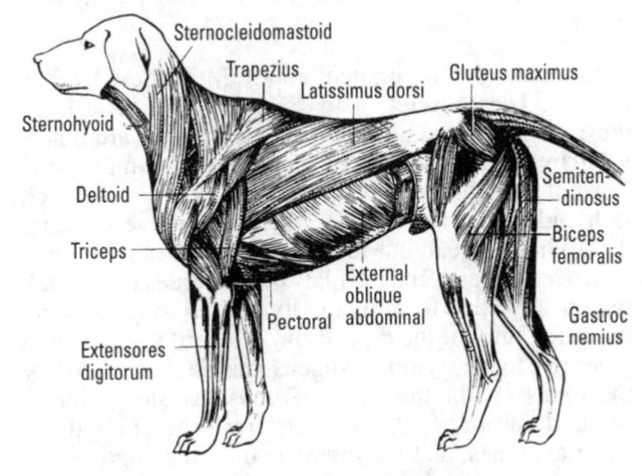

Fig.4. *Musculature.* Superficial musculature of a dog.

The most important superficial somatic muscles of the *mammalian* body are shown in Fig.4. Several muscles are involved in the movement of the lower arm against the upper arm, the most important being the two flexor muscles known as the *Musculus biceps* and the *Musculus brachioradialis,* and the extensor known as the *Musculus triceps.* In the hindlimb, the *Musculus quadriceps,* which is inserted in the patella, serves as an extensor of the lower leg. The upper leg is moved backward by the muscles of the buttocks *(glutei* muscles). The heel is lifted by the calf muscle *(Musculus gastrocnemius),* which is connected to the calcaneum (heelbone) by the Achilles' tendon. Only mammals possess a Diaphragm (see).

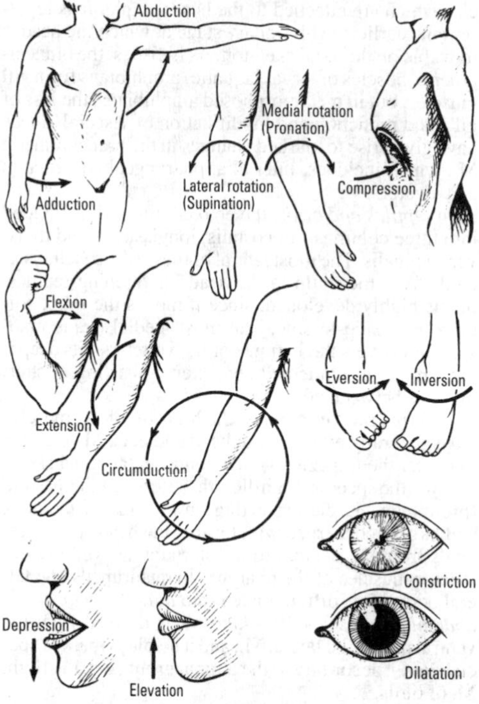

Fig.5. *Musculature.* Types of joint and muscle movement.

In mammals, the branchiomeric M. has evolved as follows. The Muscles of mastication (see) arise from the 1st visceral arch (mandibular arch), and are innervated by the mandibular branch of the 5th cranial nerve. Most of the muscles arising from the 2nd visceral arch or hyoid (innervated by the 7th cranial nerve or facial nerve) have spread out to form the muscles of expression over the face, and the platysma myoides in the neck region, as well as forming the stylohyoid, stapedius and posterior belly of the digastrium. The 3rd visceral arch gives rise to the stylopharyngeus muscle (innervated by the vagus), while the laryngeal muscles, sternocleidomastoid and trapezius originate from the 4th and 5th visceral arches, and are innervated by the vago-accessory (spinal accessory) nerve.

Mushroom: see *Basidiomycetes (Agaricales).*

Mushroom bodies: see Corpora penduncalata.

Musk, ***moschus:*** the odiferous, dried secretion from the preputial follicles of the male musk deer, *Moschus moschiferus* (a deer species found mainly in China and adjoining regions of Asia), and found in smaller amounts in other animals, e.g. musk-ox and musk-rat. The secretion was formerly used widely as a constituent of perfumes, but has now been largely replaced by synthetic materials with a similar odor. The odiferous components of M. are cyclic ketones and alcohols with 14–18-membered rings, in particular Muscone (see).

Musk deer, *Moschus:* a small deer (head plus body 80–100 cm, shoulder height 51–61 cm) of forest and brushland in the mountainous regions of central and eastern Asia. The rump is slightly higher than the shoulder. Three species are recognized: *M. moschiferus* (Siberia, northern Mongolia, Manchuria, Sakhalin), *M. chrysogaster* (Himalayas, central and southern China, northern Vietnam), *M. sifanicus* (Himalayas, central China, Burma). Antlers are absent from both sexes. The upper canines are elongated into tusks, which are about 7.5 cm-long in the male and somewhat shorter in the female. *Moschus* differs from all other deer in its possession of a gall bladder. Males of 3 years of age and older have an abdominal musk gland (musk sac or musk pod), which is situated behind the navel and just in front of the preputial orifice. This gland contains a treacly or waxy, brownish material, known as musk, which has a remarkably intense, penetrating and persistent odor. The outer surface of the gland is covered with hairs and is provided with a small canal for discharge of the secretion. About 28 g may be obtained from a single animal. Musk is used mainly in the manufacture of perfumes. See Deer.

Musk mallow: see Ambrettolide.

Muskrats: see *Arvicolinae.*

Musk ox, *Ovibos moschatus:* a monotypic bovid genus of the tribe *Ovibovini,* subfamily *Caprinae* (see), family *Bovidae* (see). Three subspecies are recognized: ***Alaskan musk ox*** *(O.m. moschatus),* ***Wager musk ox*** *(O.m. niphoecus), and* ***Greenland musk ox*** *(O.m. wardi)* . Both sexes have broad horns, which curve downward and outward, almost meeting in the middle of the skull to form a large boss. The coat is uniformly dark and very long. The M.o. was exterminated in Alaska and nearly exterminated on the mainland of Canada. With the introduction of conservation measures, it has recovered to an estimated world population of 25 000. It inhabits the Arctic regions of North America and Greenland, and has also been released on Spitzbergen, in northern Quebec, in Norway and on Wrangel island. When threatened, the group forms a closed circle with members facing outward. It is not an ox, but a close relative of the chamois.

Musk seed oil: see Ambrettolide.

Musk turtles: see *Kinosternidae, Staurotypidae.*

Musophagidae: see *Cuculiformes.*

Mussel limestone: see Muschelkalk.

Mussels: see *Bivalvia.*

Mussurano, *Clelia clelia:* one of the most familiar members of the South American colubrid snakes, subfamily *Boiginae* (see). It is a ground-dwelling species, glossy blue-black in color, and 2 m or more in length. Since it feeds on other snakes, even large poisonous ones, it is protected as a useful animal in some native areas.

Mustard oils: see Glucosolinate.

Mustela: see Polecat, Ermine, Weasel, Mink.

Mustelidae, ***mustelids:*** a large family (about 63 species in 23 genera) of terrestrial carnivores (see *Fissipe-*

dia). They are small to medium-sized mammals, usually with long, slender bodies and long tails. They have a nearly cosmopolitan distribution, but the family is not represented in Madagascar, Australia and the Oceanic islands. All members possess anal scent glands, and these are usually well developed. The 5-toed feet are either plantigrade or digitigrade, and the claws are never completely retractile. Dentition is variable, but is generally 3/3, 1/1, 3/3, 1/2 = 34. In the arboreal mustelids *(Martes),* the dentition is 3/3, 1/1, 4/4, 1/2 = 38. The skull has a long brain case and a short rostrum. In many species the carnassial teeth are trenchant, but in others they are modified to crushing teeth. The following representatives are listed separately: Pine marten, Stone marten, Sable, Polecat, Ermine, Weasel, Mink, Wolverine, Badger, American badger, Skunks, River otters, Sea otter.

Mutagens, *mutagenic agents:* chemical and physical agents that increase the rate of gene and chromosome mutation. Chemical M. include various types of compounds, e.g. nitrites, urethane, formaldehyde, peroxides, hydroxylamine, base analogs, acridine dyes (e.g. proflavin), methyl- and ethylmethane sulfonic acid, nitrosomethylguanidine. The most important physical M. are different types of ionizing radiation (X-rays, β-rays, γ-rays, neutrons), UV-light and temperature shock. See Mutator genes.

Mutant: an individual organism, which as a result of Mutation (see), shows characteristic visible or measurable differences from the parent or wild-type organism. The minimal change necessary for the generation of a M. is the alteration of a single nucleotide in the genome.

Mutation: a spontaneous or experimentally induced (see Mutagens) qualitative or quantitative alteration of genetic materal. The following types of M. are listed separately: Gene mutation, Chromosomal mutation, Genome mutation, Plasmon mutation, Plastid mutation.

Mutational load: see Genetic load.

Mutation isoalleles: see Isoalleles.

Mutation rate: the proportion of gametes with new gene mutations (gene M.r.) or new chromosome mutations (aberration rate) in the gametes of one generation, i.e. the probability of a mutation occurring in one generation. Studies on *Drosophila* show that the natural M.r. is very low. According to Timofeeff-Ressovsky, each generation of *Drosophila* contains 2–3% of animals with one or more newly mutated genes. According to Muller, 10–40% of sex cells in each human generation carry new mutations.

The M.r. varies considerably for different genes. Genes with relatively low M.rs. are known as *stable genes,* whereas those with relatively high M.rs. are said to be *labile* or *unstable.* M.r. increases with temperature, and may be increased several-fold by Mutagens (see). The low natural M.r. has often been used as an argument against the mutation-selection theory. Statistical calculations have shown, however, that the natural M.r., in combination with long periods of geological time and other evolutionary factors, forms an adequate basis for the known rate of evolution. In addition to the total M.r. of an organism, it is also possible to test the M.r. of a particular chromosome, the M.r. of genes at a chosen locus, or the M.r. of the genes of a single allele. Among different organisms, genotypes, genes and alleles, the M.r. may vary over a wide range. The average M.r. of a single gene is calculated to be 10^{-5} to 10^{-7}. In germ cells of *Drosophila,* the average M.r. of a normal allele is estimated to be 0.0005%. Rates as high as 10^{-2} are also sometimes observed.

The M.r. in the absence of any known mutagenic influence, is known as the *spontaneous M.r.* If the M.r. is influenced experimentally, it is known as the *experimental M.r.*

Frequency of spontaneous gene mutations in maize (after Stadler)

Gene	Seed phenotype	Number of tested gametes	Observed mutations	Mutations per million gametes
R	(background pigmentation factor)	554 768	273	492
J	(color inhibition factor)	265 391	28	106
Pr	red aleurone layer	647 102	7	11
su	sugary endosperm	1 678 736	4	2.4
Y	yellow endosperm	1 745 280	4	2.2
Sh	shrunken grain	2 469 285	3	1.2
Wx	waxy endosperm	1 503 744	0	0

Mutation-selection balance: the constant frequency of deleterious hereditory factors in a population, resulting from recurrent mutation and continual selecting out. The frequency of a recessive deleterious gene in the M-s.b. is $q = \sqrt{u/s}$, where u is the mutation rate and s is the selection coefficient (see Fitness). The *elimination rate* (proportion of deleterious genes removed in each generation compared with the totality of all genes at that locus) is $E = u = sq^2$. The selection coefficient of the lethal factor is 1. For a lethal mutation rate of $u = 10^{-5}$ (i.e. one gene in 100 000 mutates to a lethal factor in each generation), $q = 10^{-5} \approx 0.00316$. In this case, the frequency of recessive lethal factors in the M-s.b. is therefore about 0.3%. The frequency of a dominant deleterious factor in the M-s.b. is $q = u/hs$. The value of h (degree of dominance) lies between 0 (complete recessivity) and 1 (complete dominance). For a completely dominant lethal allele, both h and s are equal to 1, so that the frequency of such a factor in the M-s.b. is equal to u.

The rate of elimination of dominant deleterious factors from the M-s.b. is $E = u = qhs$. On the other hand, the proportion of individuals eliminated by this factor from the total population in each generation is $2u$.

Mutation-selection theory: a simplified term for the Synthetic theory of evolution (see).

Mutator genes: genes that increase (up to 10-fold) the mutation rate of other genes in the same genome.

Mutative atavism: see Atavism.

Mutillidae: see *Hymenoptera.*

Mutton bird: see *Procellariiformes.*

Mutualism: see Relationships between living organisms.

Mya (Soft-shelled clams): see *Bivalvia.*

Mycelium: vegetative body of a fungus, consisting either of hyphae (a ramified system of unbranched or branched, septate or nonseptate filaments), or of chains of spherical to cylindrical, often sharply delineated cells with little or nor branching (pseudomycelium). In some yeast species, the pseudomycelium may be reduced to single cells, which therefore also function as fruiting bodies. In the formation of a fruiting body, fungal hyphae grow together to form a plectenchymatous pseudoparenchyma. The cord-like rhizomorphs responsible for the spread of parasitic fungi (e.g. the honey fungus, *Armillaria mellea)* consist of interwoven hyphae. Scerotia also consist of tuberous masses of hyphae.

See Sclerotium, Aerial mycelium, Promycelium, Substrate mycelium.

Mycerobas: see *Fringillidae.*

Mycetome: in some insects, an intestinal organ which harbors symbiotic bacteria or fungi (generally yeasts). Several Ms. are usually present. The symbiotic organisms provide enzymes for the degradation of lignin and/or cellulose. Very many insects harbor symbiotic microorganisms, and these are frequently confined to specialized cells (mycetocytes), which may be scattered in the gut wall or fat body. Ms. represent groupings of these cells into definite and sometimes very conspicuous organs. In many female insects the symbionts migrate to the maturing eggs and are transmitted to the offspring.

Mycetophage: see Modes of nutrition.

Mycobacteriaceae: see *Actinomycetales.*

Mycobacterium: see *Actinomycetales.*

Mycocastoridae, ***mycocastors, coypus*** or ***nutrias***: a family of Rodents (see) represented by a single species, *Mycocastor coypus* ***(Coypu*** or ***Nutria).*** The Coypu is indigenous to the temperate zones of South America, where it occupies the same ecological niche as the Muskrat *(Ondatra zibethica)* in North America. The body length of the adult ranges from 43 to 63 cm, and the weight is usually 7–9 kg, but weights up to 17 kg can be attained. Males are larger than females. The skull is triangular with short postorbital processes and long paraoccipital processes, and the head is relatively large and sturdy. Webbed membranes are present between the first, second and third digits of the hindfeet. The tail, which is round in cross section, is covered with scales and sparse hair. The incisors are large and orange-colored. Burrows are dug in the sloping banks of rivers and lakes. It is a powerful swimmer, and much time is spent in the water. The name, "nutria" is given to both the animal and its fur. On account of the commercial value of its fur, the coypu is bred on fur farms. Escape from these farms or the intentional introduction of the animal for clearing aquatic vegetation have led to the establishment of the coypu in the southern parts of North America and in Eurasia.

Mycoins: a group of antibiotics that inhibit the growth of *Brucella.* They are produced by different fungi of the genera *Penicillium, Aspergillus, Fusarium, Cephalosporium, Microsporum* and others.

Mycophenolic acid: an antibiotic produced by different species of the fungal genus *Penicillium.* It has a broad spectrum of activity, and is particularly active against fungi.

Mycoplasmas, ***pleuropneumonia-like organisms, PPLO:*** prokaryotic, bacteria-like organisms, similar in size to the larger viruses (ranging downward from 0.2 μm), lacking a true cell wall, and with an irregular, variable round or elongated shape (discoid and ameboid structures, rings, clubs, stellate forms and pleomorphic filaments up to 100 μm long). Colonies have a characteristic "fried egg" morphology on agar media. They are parasitic or saprophytic, various parasitic species causing disease in humans, animals and plants. The best known M. have been known for many years as the causative agents of severe respiratory diseases in cattle, sheep and goats. Others cause chronic respiratory infections in poultry and a wide range of other animals. Primary atypical pneumonia in humans is caused by *Mycoplasma pneumoniae. Mycoplasma (= Astercoccus) mycoides* causes infectious pleuropneumonia in cattle. M. are insensitive to sulfonamides and penicillin, but are sensitive to chloramphenicol, tetracyclin, neomycin and kanamycin.

Mycorrhiza: symbiotic or mutual parasitic association between the roots of advanced terrestrial plants and fungi. In the wild, the roots of practically all terrestrial plants may be associated with fungi, with the exception of the *Cyperaceae, Brassicaceae* and many *Centrospermae.* For many tree species (e.g. Scots pine) a M. is obligatory for survival. There is an exchange of material between the mycorrhizal partners, in which the fungus receives primarily carbohydrate from the plant, while the plant is provided mainly with water and mineral salts (e.g. nitrogenous and phosphorus-containing compounds) by the fungus.

Based on morphological-anatomical differences, the three following types of M. are recognized. 1) *Ectotrophic (epiphytic)* Ms. are typically found between forest trees and basidiomycetes, e.g. between birch and the birch fungus *(Boletus scaber).* The short, thick, absorptive rootlets are enveloped in a dense network of fungal hyphae. Root hairs are suppressed, and they are replaced functionally by the fungal hyphae, which connect the fungal sheath of the rootlets to the mycelium in the soil. The hyphae of an ectotrophic M. penetrate the intecellular spaces of the first cortical layer of the root. 2) *Endotrophic* Ms. are found, inter alia, in members of the *Ericaceae* and in orchids. In this case, the hyphae penetrate into the interior of the cortical cells. Transitions exist between the two types, and these are known as 3) *Ectendotrophic* M. See Fig. p. 801.

Mycostatin: see Nystatin.

Mycosterols: sterols that occur in fungi, e.g. Ergosterol (see) and Zymosterol (see), both from yeasts.

Mycotoxins: metabolic products of certain fungi, which are harmful to other organisms, especially vertebrates, including humans. The same M. may be produced by more than one fungal species. Of about 100 000 described species of fungi, 50 are known to produce M. These may damage the host directly (e.g. plant pathogenic fungi), or indirectly by causing illnesses in animals and humans when ingested in food. Ergotism is a classical example of a mycotoxicosis (see Ergot alkaloids). M.-producing organisms frequently develop on improperly stored foodstuffs (see e.g. Aflatoxins). The nephritic toxin, citrinin, is produced by *Pencillium notatum,* and sporodesmin is synthesized by *Pithomyces chartarum.* Other important M. producers are *Penicillium islandicum, Pencillium rubrum, Paecilomyces varioti, Fusarium sporotrichoides* and *Stachybotrys atra.* Bacterial toxins are also often included with the M. The division between antibiotics and M. is ill defined.

Mycoviruses: viruses that attack fungi. M. have been found in members of the *Deuteromycetes, Phycomycetes, Ascomycetes* and Basidiomycetes. Virus infection of cultured mushrooms *(Agaricus)* leads to considerable economic losses, due to weak mycelial growth, watery, elongated stipes, abnormal small caps and mummification of the fruiting bodies.

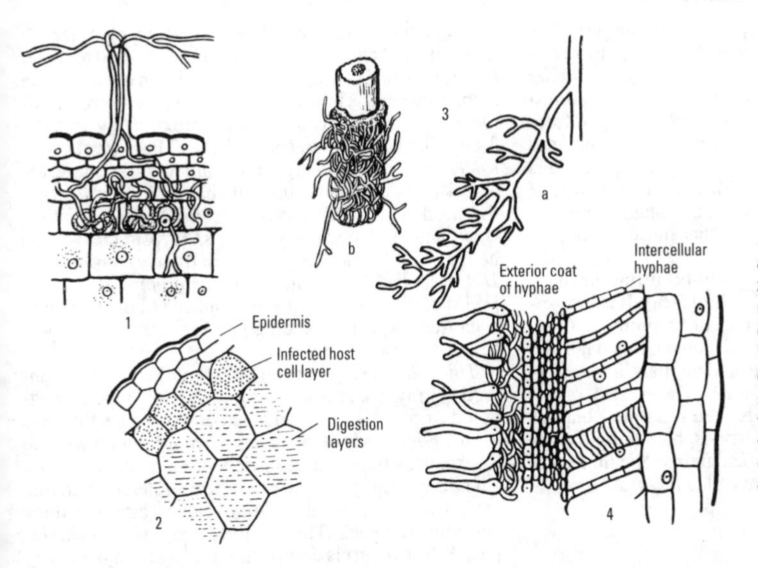

Mycorrhiza. Endotrophic and ectotrophic mycorrhizas. *1* Section of an orchid root *(Platanthera)* with endotrophic mycorrhiza. *2* Root of the bird's nest orchid *(Neottia)* in transverse section. *3* Short, coral-like, fungal root of the beech tree *(a)* and root tip covered with a fungal mat *(b)*. *4* Ectotrophic mycorrhiza of the oak; longitudinal section of a fungus-covered root.

Myctophidae, ***lanternfishes:*** a family of marine teleost fishes of the order *Myctophiformes,* containing about 235 species in about 32 genera, distributed in all the oceans of the world. Except for one species, all have luminescent organs, consisting of small photophores arranged in groups and rows on the head and body. Some are monstrous deep sea species.

Myelin sheath: see Nerve fiber, Neuron.

Myeloencephalon: see Brain.

Myeloma proteins: immunoglobulins in the blood of certain cancer patients. They are produced by transformed plasma cells, known as *plasmocytomas.* Since they are structurally homogeneous and relatively easy to isolate from blood, they have been widely used for investigations of immunoglobulin structure.

Myiophoneus: see *Turdinae.*

Myiopositta: see *Psittaciformes.*

Mynahs: see *Sturnidae.*

Myobatrachidae, ***myobatrachids:*** a family of the *Anura* (see) containing 99 living species in 20 genera and 2 subfamilies *(Limnodynastinae, Myobatrachinae).* Eight presacral vertebrae are present. In subadults the notochord is persistent and the intervertebral disks are free (these are absent in *Lechriodus* and *Mixophys*), but are usually fused to the posterior end of the centrum in adults to form procoelous vertebrae. In most limnodynastins the neural arches are imbricate (overlapping) and nonimbricate in most myobatrachins. Presacrals I and II are fused in limnodynastins (except *Lechriodus* and *Mixophys*) and they are separate in myobatrachins. In limnodynastins the atlantal cotyles of presacral I are juxtaposed, and in myobatrachins they are widely separated. Ribs are absent. The sacrum has a bicondylar articulation with the coccyx, which may or may not have proximal transverse processes. A cartilaginous omosternum is normally present, but absent in *Myobatrachus* and *Notaden.* The sternum is also cartilaginous and the pectoral girdle is arciferal. Pupils are usually horizontal, but vertically elliptical in *Heleioporus, Megistolotis, Mixophys* and *Neobatrachus.*

The genus *Crinia* (subfamily *Myobatrachinae)* is represented by 13 species found in Australia, Tasmania and New Guinea. Some are water-spawing species with gilled tadpoles, while there are also terrestrial species showing direct development; various intermediate forms also exist. The call of these 2 cm-long frogs resembles the chirping of insects.

Twelve predominantly aquatic species of *Limnodynastes* (subfamily *Limnodynastinae)* are distributed in Australia, Tasmania and New Guinea. *Limnodynastes tasmaniensis* ***(Spotted grass frog)*** of Tasmania builds a foam nest at the water's edge. The voice of *Limnodynastes dorsalis* of Australia and Tasmania resembles the strumming of a stringed instrument, hence its name of ***Banjo frog***.

Frogs of the genus *Neobatrachus* (7 stout-bodied, burrowing species; South and West Australia; subfamily *Limnodynastinae)* are notable for their unusual voices, hence the names ***Trilling frog*** *(N. centralis),* ***Meeowing frog*** *(N. pictus),* ***Humming frog*** *(N. pelobatoides)* and ***Shoemaker frog*** *(N. sutor).*

The genus *Notaden* (Australian ***Shovelfoots***; subfamily *Limnodynastinae)* is represented by 3 species in North and Southeast Australia. *Notaden bennetti* is a popular source of edible frog legs. The ***Desert shovelfoot*** *(N. nichollsi)* burrows deeply during the dry season and spawns in puddles in the Australian rainy season (January to February).

The 2 species of *Rheobatrachus* (subfamily *Myobatrachinae)* are found in the mountains of Queensland, Australia. *R. silus* ***(Gastric brooding frog)*** displays an extraordinary form of brood care, in which the female ingests as many as 20 fertilized eggs. Hormones secreted in the oral mucus of the tadpoles inhibits HCl secretion in the stomach and peristalsis in the gut and thereby prevent feeding by the parent. Development proceeds in the stomach of the female, which eventually "vomits" froglets together with tadpoles at an advanced stage of metamorphosis.

Myoglobin: a red-colored, single chain (153 amino acid residues; M_r 17 200), heme protein of skeletal muscle. It can therefore be classified as a chromoprotein or a metalloprotein. It is structurally similar to a single hemoglobin subunit. Like hemoglobin, M. combines reversibly with oxygen, but its affinity for oxygen is higher than that of hemoglobin. M. therefore functions in muscle by storing oxygen and by transferring oxygen from circulating oxyhemoglobin to the respiratory enzymes (cytochromes) of the muscle cells. Very active muscle is particularly rich in M. The cardiac muscle of diving mammals, such as whales and seals, contains up to 8% M., compared with 0.5% in the cardiac muscle of dog.

Myoida: see *Bivalvia.*

Myoinositol, *inositol*: a widely occurring hexahydric alcohol, which is also a Cyclitol (see). It is biosynthesized from D-glucose, and the configuration of C-atoms 1–5 is preserved during the conversion. The free form is found widely in plants and animals, and is notably present in muscle, brain and liver. M. is also a component of many phospholipids. The hexaphosphate of M. *(phytic acid)* is an important phosphate storage compound in plant tissues, e.g. cereal grains. The mixed Ca and Mg salts of phytic acid are called *phytin.*

Myoinositol

Myometrium: peripheral muscle layer of the mammalian uterine wall. It consists of closely interwoven bundles of smooth muscle cells, with vascularized connective tissue. It also contains elastic fibers and very fine nerve plexuses.

Myophoria (Greek *myo* muscle , *phoros* carrier), *Costatoria:* a fossil genus of bivalves of the subclass *Palaeoheterodonta* (see *Bivalvia).* The shell is obliquely oval or triangular, often with one or more ridges, and its surface is smooth or ribbed. There is a simple or bifid triangular hinge tooth on the left valve, sometimes with crenulate lateral teeth. The genus is found in the Triassic, while the family as a whole *(Myophoridae)* is found from the Devonian to the Triassic, with numerous species serving as index fossils in the Germanic and Alpine Triassic.

Myophoria from the upper strata of the Middle Triassic (x 0.5)

Myoprocta: see *Dasyproctidae.*

Myopsida: see *Cephalopoda.*

Myosin: a protein present in muscle and accounting for 1/3 of total muscle proteins and 2/3 of the contractile muscle proteins. It consists of a long, slender stem or tail (about 135 nm long, 2 nm diam.) and a globular head region (designated S-1, about 10 nm long). The tail is formed from 2 polypeptide chains (about 2 000 amino acid residues each), which terminate in globular regions; these 2 chains *(heavy chains)* are twisted together in a tight α-helix, with the globular region of each chain at the same end. Thus, the globular head region consists two globular S-1 "heads", each approximately 15 x 4 x 3 nm. The head region contains an ATPase (part of the system producing energy for contraction). There are also 2 pairs of smaller protein chains *(light chains)* in the head region. M. therefore consists of 2 identical heavy and 4 identical light chains. Trypsin cleaves the heavy chain into a heavy fragment *(heavy meromyosin, HMM)* and a light fragment *(light meromyosin, LMM).* HMM corresponds to the globular head of the M. chain; it carries a specific binding site for actin, and it possesses ATPase activity

Thick filaments (diam. about 10 nm) consist of hundreds of myosin tails packed together in a regular staggered array, with a regular pattern of projecting S-1 heads. Each thick filament has a central region with no S-1 heads, where two oppositely orientated sets of myosin tails come together. The S-1 heads interact with thin actin filaments, forming cross bridges between thick and thin filaments. The resulting *actomyosin complex* (see Actomyosin) is contractile. Contraction occurs by the *sliding filament mechanism,* in which the cross bridges between the S-1 heads and actin are temporarily broken, then reformed further along the actin filament, energy for the process being derived from operation of the S-1 ATPase. Contraction is regulated by Ca^{2+} and by the troponin-tropomyosin complex. Nonmuscle cells contain no thick filaments, but they do contain oligomeric M.

M. is easily prepared from homogenized muscle by extraction with concentrated potassium chloride solution, which causes the thick filaments to depolymerize into their constituent M. molecules.

Myosin filaments: see Myosin.

Myospalacini: see *Cricetinae.*

Myosurus: see *Ranunculaceae.*

Myotis: see *Vespertilionidae.*

Myoxocephalus: see *Cottidae.*

Myriangiales: see *Ascomycetes.*

Myriapoda, *myriapods:* formerly a class of uniramian arthropods (arthropods with unbranched appendages), comprising the subclasses, *Chilopoda* (centipedes), *Diplopoda* (millipedes), *Pauropoda* and *Symphyla,* all of which share the characteristics of a long trunk (often consisting of a very large number of legbearing segments) and separate muscles in each antennal segment. The *Diplopoda, Pauropoda* and *Symphyla* are sometimes grouped together as the *Progoneata,* because their genital opening (gonopore) (paired or unpaired) occurs on the third or fourth body segment. In contrast, the gonads of the *Chilopoda* open to the exterior in front of the anus on the last segment. These 4 groups are now considered to be separate classes, and each one is listed separately.

Myricaceae, *bayberry family:* an almost cosmopolitan family of dicotyledonous plants, containing a single genus and 56 species, distributed in the subtropics and temperate regions of both Hemispheres. Members are exclusively trees or bushes, with greater or lesser densities of resinous glands, which contain aromatic compounds. Leaves are usually simple, alternate or exstipulate. Flowers are monoecious or dioecious, without a perianth, forming a spike or catkin. The fruit is a drupe or nut. The family is characterized chemically by an abundance of polyphenols and triterpenes. ***Bog myrtle*** or ***Sweet gale*** *(Myrica gale)* grows in bogs and wet heaths across Europe to Scandinavia and North West Russia, and North America from Labrador and Alaska to North Carolina and Oregon; its bark was formerly used for tanning, while its leaves and fruits were used to prepare a stomachic liqueur. A subspecies is found in East Asia. ***Bayberry*** *(Myrica carolinensis* Mill., *M. pensylvanica* Lois.) is indigenous in eastern North America from Newfoundland to North Carolina; it is occasionally cultivated, and has been naturalized in the New Forest (Hampshire) in the UK.

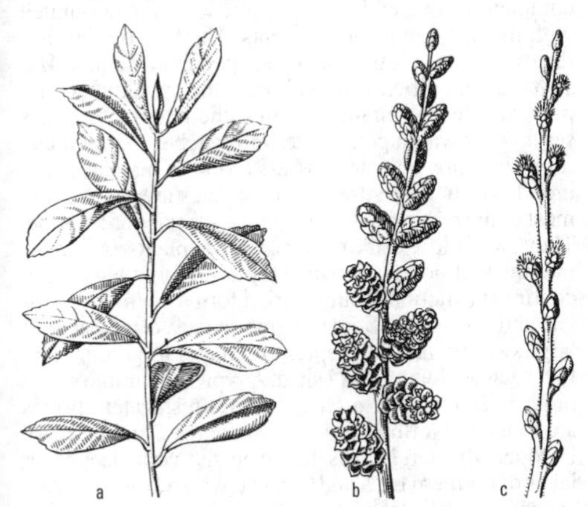

Myricaceae. Bog myrtle or sweet gale *(Myrica gale* L.). *a* Shoot with leaves. *b* Twig with male inflorescences. *c* Twig with female inflorescences.

Myrientomata: see *Protura.*

Myristica: see *Myristicaceae.*

Myristicaceae, *nutmeg family:* a family of dicotyledonous plants containing about 380 species in 16 genera, restricted to the tropics, mainly in rainforests in New Guinea, Africa and tropical America, and especially the Amazonian rainforest, where species of *Virola* are important forest trees. All species are trees with simple, exstipulate leaves, which often carry small glands containing an aromatic oil. Petals are lacking, and the inconspicuous, unisexual, dioecious, flowers have a calyx of 3 partially fused sepals. Inflorescences are capitate, fascicled or corymbose. Female flowers have a unilocular ovary containing a single basal ovule. Male flowers have 2–30 stamens fused in a column, and the pollen is shed through 2 longitudinal slits. The fruit is a fleshy, dehiscent berry, which splits into 2 or 4 valves. In most species the seed has a red aril, and the large oil- and starch-rich endosperm of the seed is deeply fissured by ingrowth of the perisperm.

Commercial *nutmegs,* used a flavoring and spice, are the dried, ripe seeds of *Myristica fragrans* ***(Nutmeg tree),*** native to the Moluccas and now cultivated widely in the tropics. The spice known as *mace* is the aril of the nutmeg. Nutmegs contain 25–40% of a fixed oil and 5–15% of a volatile oil. The hallucinogenic and toxic properties of nutmegs are thought to be largely due to the terpene, myristicin, which is a major constituent of the volatile oil.

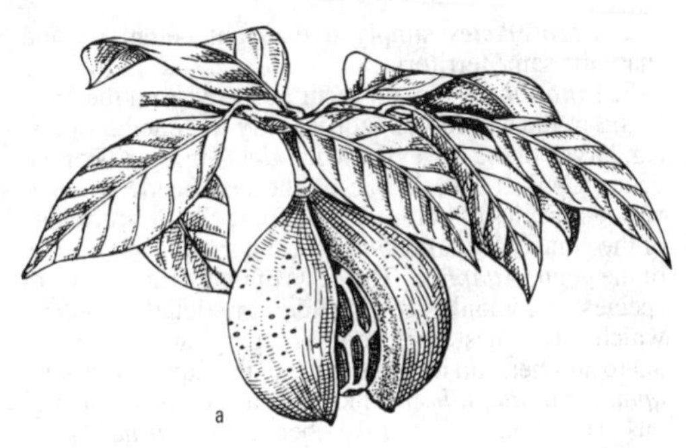

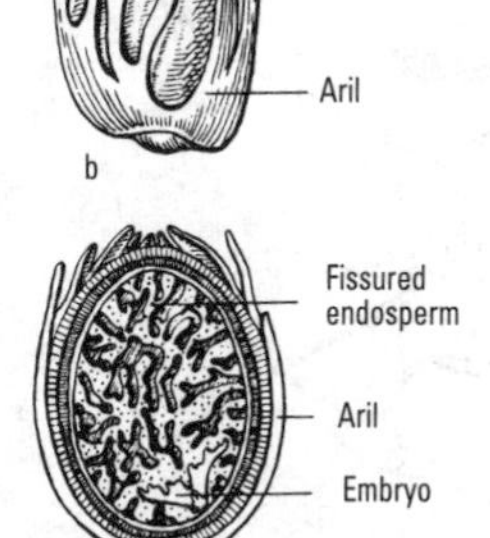

Myristicaceae. Nutmeg tree *(Myristica fragrans). a* Branch with mature fruit. *b* Seed. *c* Longitudinal section of seed.

Myrmicinae: see Ants.

Myrmecobiidae: see Numbat, *Dasyuridae.*

Myrmecobius: see Numbat.

Myrmecochory: see Seed dispersal.

Myrmecocochla: see *Turdinae.*

Myrmecolacidae: see *Strepsiptera.*

Myrmecophagidae, *anteaters* (plate 3): a family of the *Edentata* (see), containing 3 small genera, found in Central and South America. M. are toothless mammals, with a greatly elongated head and a very extensible, wormlike, sticky tongue. They feed on ants and termites, whose nests they tear apart with powerful claws on their forepaws. The genus *Myrmecophaga* contains a single species, *M. tridactyla* ***(Great anteater*** or ***Yurumi),*** which possesses a large, long-haired, brush-like tail. The main representative of the genus *Tamandua* is the arboreal ***Tamandua*** or ***Caguare*** *(T. tetradactyla),* which has a prehensile tail. Even more strongly adapted to an arboreal existence is the squirrel-sized ***Dwarf anteater*** *(Cyclopes didactylus).*

Myrmecophiles, *myrmecophilous species*, *ant guests:* animal species, other than ants, which associate with ants or their nests; more than 2 000 M. have been described. Myrmecophily (myrmecophilism) may be determined by the favorable microclimate within the ant nest, or it may provide a source of food or shelter; relationships between ants and M. are therefore varied.

1. ***Synechthrans*** are carnivorous rove beetles *(Staphylinidae)* which behave as scavengers or actual predators, and are treated with hostility by the ants. These beetles may be large enough to defend themselves, or smaller species may conceal themselves, often lurking near the entrance of a nest. Sickly or solitary ants are often attacked, or eggs or larvae may be seized.

2. ***Paro(e)ketes*** simply live as ant neighbors and share the same territory.

3. ***Syno(e)ketes*** do not incur hostility from the ants, living in the ant nest as indifferently tolerated lodgers, e.g. larvae of the hover fly, *Microdon;* springtails of the genus *Cyphoderus;* larvae of the leaf beetle, *Clytra;* various species of hump-backed flies; wingless crickets of the genus *Myrmecophila* (Fig. 1); small cockroaches of the genus *Attaphila.* The nests of many European ant species commonly contain the bristletail, *Atelura,* which intervenes when food is regurgitated from one ant to another, and imbibes some of the liquid. *Myrmecophila* and *Attaphila* lick the host ants to obtain cutaneous secretions. The rove beetle, *Dinardia,* is a syno(e)kete of some *Formica* species.

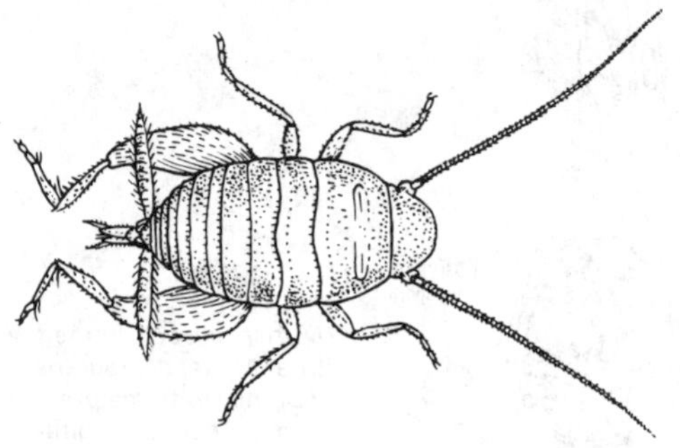

Fig.1. *Myrmecophiles.* The ant cricket *(Myrmecophila acervorum)* (female).

4. ***Symphiles*** (true ant guests) are usually *Coleoptera.* They show various adaptations of color and structure, which make them less conspicuous among the ants, and they produce exudates which ingratiate them with their hosts. Rove beetles of the genera *Lomechusa* and *Atemles* display highly developed forms of symphilic behavior, in which the ants assiduously rear and care for the beetle larvae as if they were ant larvae; in addition, the guest larvae devour many host eggs and larvae. The caterpillar of the Large Blue butterfly *(Maculinea arion* L., subspecies *M. arion arion)* feeds at first on flowers of thyme *(Thymus).* After the third molt, it falls to the ground. If encountered by a passing red ant, the ant caresses the caterpillar with its antennae, and the caterpillar exudes a liquid from the tenth segment, which is attractive to ants. The ant leaves and returns at intervals to "milk" the caterpillar, eventually carrying it to the nest. Here, the caterpillar feeds on ant larvae, overwinters, then continues to feed on ant larvae before pupating. The emergent adult crawls through the passages of the nest to the exterior.

5. ***Parasites.*** Ants are parasitized by the larvae of various species of chalcids and hover flies, and by certain mites. Endoparasites include hymenopteran larvae, stylopids and the nematode, *Mermis.*

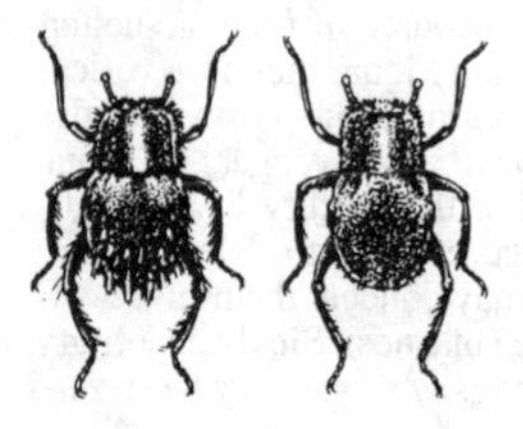

Fig.2. *Myrmecophiles.* The beetle, *Symphilister pilosus,* a guest of army ants. The secretory hairs of the animal on the right have been licked off.

See Inquiline.

Myrmecophilism (Myrmecophily): see Myrmecophiles.

Myrmecophilous plants: see *Rubiaceae.*

Myrmecophilous species: see Myrmecophiles.

Myroxylon: see *Papilionaceae.*

Mysidacea, *opossum shrimps, mysids:* an order of *Crustacea,* subclass *Malacostraca.* They are shrimp-like, and the majority are 1.5 to 3 cm in length, a notable exception being the primitive deep sea species, *Gnathophausia,* which may attain 35 cm. The thorax is almost completely covered by a carapace, which is not united with the last 4 thoracic segments. The first pair of thoracic appendages (also the second pair in the family, *Mysidae)* are modified as maxillipeds; the remaining 6 or 7 pairs are slender biramous limbs, the interior branches serving for walking, the exterior branches (exopodites) for swimming. Females possess a ventral marsupium or brood pouch. A few freshwater species are known, but most of the recognized 500 (approx.) species are marine. They swim close to the bottom, cling to plants, or burrow into the soft bottom; in some the swimming legs are in continuous motion. Many marine forms are found in algae and tidal grass. Some marine species often occur in large swarms, and they represent a major food source for fish such as flounder and shad. A typical example of the order is *Boreomysis arctica* (Fig.). Freshwater species are found in North America, Scandanavia, and northern Europe. *Mysis relicta* is found in the cold deepwater lakes of northern USA and Canada (where large numbers are eaten by lake trout), as well as in Scandanavian lakes and in the Mecklenberg lake region of northern Europe. This species requires cold water, and in the Mecklenberg lakes breeds only in the winter; it appears to have originated in the Arctic and to be a relict of the Ice Age.

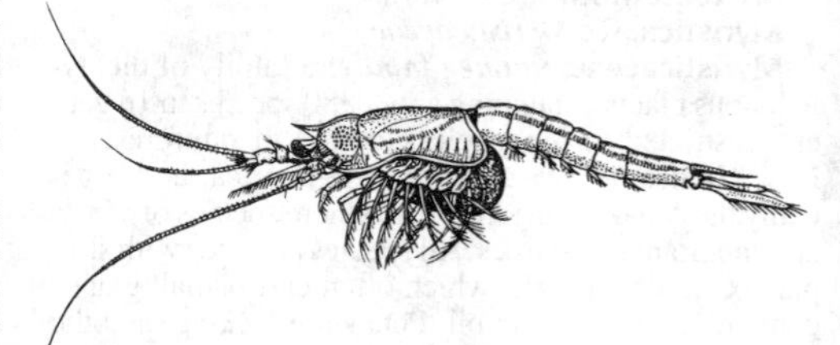

Mysidacea. Boreomysis arctica (female with marsupium).

Mystacocarida: a class or subclass of crustaceans, first described in 1943, and known only from 3 species in a single genus, *Derocheilocharis.* The elongated, cylindrical body is only 0.5 mm long, consisting of the head, a thoracic segment (carrying setose feeding maxillipeds) and 10 practically identical abdominal segments, of which the first 4 carry uniramous appendages. Both pairs of antennae are well developed, and a single nauplius eye is present. Mystacocarids live between the intertidal sand grains of sea coasts. They are allied to the copepods and barnacles.

Mystacoceti: see Whales.

Mysticetes: see Whales.

Mysticeti: see Whales.

Mystromyini: see *Cricetinae.*

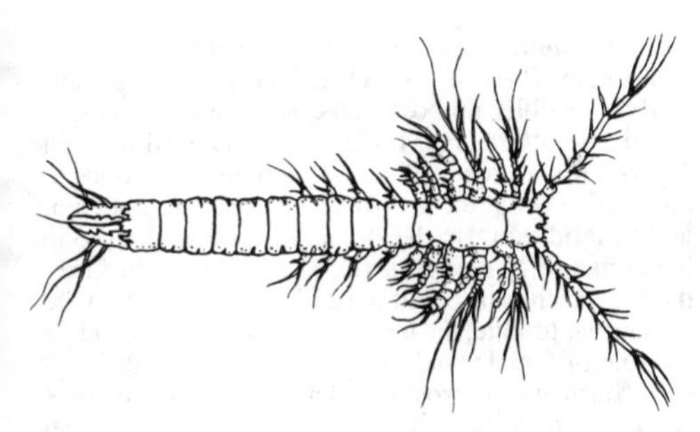

Mystacocarida. Derocheilocaris remanei.

Mytiloida: see *Bivalvia.*

Myxameba: see *Myxomycetes.*

Myxamoeba: see *Myxomycetes.*

Myxinidae: see Hagfishes.

Myxobacteria: rod-shaped, mucilage-forming, aerobic bacteria, characterized by their ability to glide over the substratum, and by their formation of colored fruiting bodies that are large enough to see with the naked eye. They occur in the soil, on animal feces, and on decomposing plant remains, where they feed on insoluble organic material. Insoluble material is utilized by secreting degradative enzymes. When the nutrient supply is plentiful, M. remain in loose colonies or swarms, all members benefitting jointly from the secretion of degradative enzymes. When nutrients are in short supply, the cells aggregate to form a multicellular fruiting body, within which the bacteria differentiate into resistant spores. Under favorable conditions, these spores germinate to produce a new bacterial swarm. M. are classified primarily according to the shape and color of their fruiting bodies.

Myxogastrales: see *Myxomycetes.*

Myxomycetes, *slime molds:* a class of primitive and phylogenetically ancient fungi. The vegetative stage consists of naked cells with ameboid movement, which either fuse into a multinucleate mass of protoplasm (a *plasmodium),* or simply aggregate without fusing (an *aggregation plasmodium).* They multiply by spores, which are formed in sporangia, also called fruiting bodies or fructifications. Four orders are recognized, but their relationship is questionable. It is perhaps not a natural assembly of organisms, but rather a polyphyletic one composed of members of differing ancestry.

1. Order: *Myxogastrales.* This, the largest order, contains about 500 species. In contrast to that of other orders, the course of development of the *Myxogastrales* is well understood. In the presence of moisture, the spores germinate to produce flagellated swarmers (myxoflagellates). These either unite in pairs, multiply by cleavage, or lose their flagella and become *myxamebas (myxamoebae).* The latter can also divide, or they may unite in pairs to form *amebozygotes.* Several diploid amebozygotes then fuse to form multinuclear plasmodia, which live saprophytically or by engulfing small organisms like bacteria. Finally, the plasmodium creeps away from moist substrates and toward a source of light, loses a considerable proportion of its water content and develops numerous sporangia or fructification. Each fructification has a firm, often calcareous investment (the *peridium),* and it contains numerous small, uninucleate, haploid spores, each with a proteinaceous wall.

In some representatives, a reticulate system of fine tubes or fibers, known as the *capillitum,* is formed from the remaining cytoplasm in the fructifications. In some species hygroscopic movements of the capillitum assist liberation of the spores from the fructification. The most familiar species of this order is "flowers of tan", *Fuligo varians (= Aethalium septicum),* whose yellow fructifications, consisting of several united sporangia, are common on damp forest litter and on tan bark.

2. Order: *Ceratiomyxales.* Members of this order are found mostly on rotting wood. They produce stalked spores on the surface of a columnar fructification.

3. Order: *Acrasiales.* These are the cellular slime molds. Flagellated stages are absent. The spores germinate directly to myxamebas, which creep together without fusing to form an aggregation plasmodium. This becomes highly organized to form a fructification with a cylindrical stalk and ovate head. *Dictyostelium* is a popular laboratory organism.

4. Order: *Plasmodiophorales.* Parasitic slime molds, e.g. *Plasmodiophora brassicae,* the causative agent of club root or finger and toe disease of members of the *Cruciferae.* Spores germinate in the soil, liberating biflagellate swarmers. These enter the root hairs, lose their flagella and develop into multinicleate ameboid protoplasts, which in turn produce biflagellate gametes by differentiation. The resulting zygotes penetrate the root cells where they form diploid, multinucleate plasmodia, causing goitrous swellings on the root. Spores (with chitinous walls) are produced by meiosis, and are released when the root tissues rot in the spring.

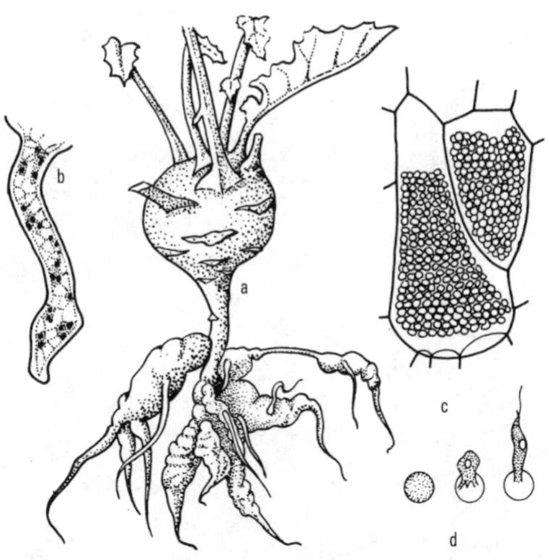

Myxomycetes. Plasmodiophora brassicae. a Club root (finger and toe disease) on the roots of kohlrabi. *b* Plasmodia in a root hair. *c* Spores inside the cells of the root cortex. *d* Spore germination.

Myxophaga: see *Coleoptera.*

Myxosporidia: see *Cnidosporidia.*

Myzomela: see *Meliphagidae.*

Myzostomida, *myzostomes:* a class of parasitic and commensal marine annelids, containing about 130 species, which live only on echinoderms, especially crinoids. They are flat, disk-shaped animals, 0.3–3 cm in length, and their thin edge carries finger-like processes (cirri). On the ventral surface are 5 pairs of unsegmented parapodia, each provided with a strong, curved bristle. These correspond to 5 body segments, which are no longer evident externally or in the structure of the body cavity. Segmentation is, however, recognizable in the nervous sytem. An anterior transverse opening and a posterior anus are present. The anterior part of the body can be everted through the anterior opening. Ontogenic development proceeds by spiral cleavage, via a free-swimming trochophore larva. All M. are hermaphrodite.

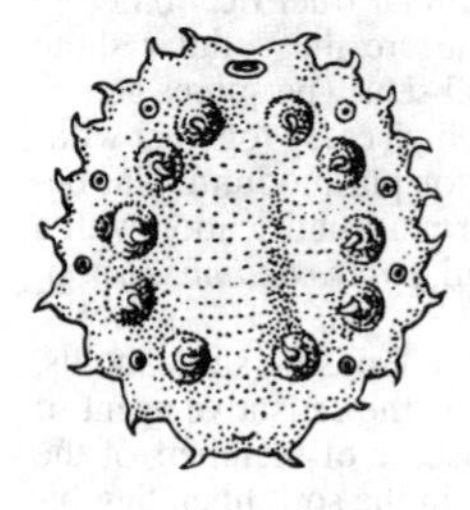

Myzostomida. Ventral surface of *Myzostoma.* The mouth is anterior (top), the anus posterior (bottom). Five parapodia and 4 glandular organs are present on each side. The sexual opening is situated at the level of the third parapodium.

Classification. Two orders are recognized.

1. Order: *Proboscifera.* Over 100 species. The anteriorly eversible, proboscis-like structure contains the mouth, esophagus and brain, and is referred to as the "head". Most proboscifers are commensals, living on the host's skin in the grooves near the mouth opening, and benefitting from the stream of food leading to the host's mouth. Some species cause galls on the surface of the host, each gall containing 2 parasites. Although they are hermaphrodite, the larger member of the pair adopts the role of female, and the smaller serves as the male. Galls formed by myzostomes have been found on fossil sea lilies from the Devonian (over 300 million years ago).

2. Order: *Pharyngidea.* About 25 species. The anteriorly eversible structure contains the mouth and esophagus, but not the brain, and it is called the "proboscis". Pharyngids are mostly true internal parasites in the body cavity or intestine of their host. The gonads of brittle stars may carry 300–400 parasitic pharyngids.

N

NAD: acronym of Nicotinamide-adenine-dinucleotide (see).

NADH: acronym of reduced Nicotinamide-adenine-dinucleotide (see).

NADP: acronym of Nicotinamide-adenine-dinucleotide phosphate (see).

NADPH: acronym of reduced Nicotinamide-adenine-dinucleotide phosphate (see).

Nagana: see Trypanosomes.

Nail: see Claws, hoofs and nails.

Naja: see Cobras.

Na^+/K^+-ATPase: see Ion pumps.

Namaqua gerbil: see *Gerbillinae.*

Namurian stage: see Carboniferous.

Nanism, ***nanosomy, dwarfism:*** occurrence of dwarf forms or poorly developed individuals, usually resulting from dietary deficiency or a genetic disposition to N. In humans, an adult body height less than 120 cm is regarded as N. Most cases of N. have a pathological basis, whereas others are genetically determined. In *primordial* or *chondrodystrophic N.,* the newborn child is abnormally small, whereas in *infantile* or *hypoplastic N.*, growth inhibition is first manifested between the 2nd and 8th year, and subsequent development of primary and secondary characters remains incomplete. As a racial characteristic, N. occurs in the pygmies of central Africa and other small races in southern and Southeast Asia (see Body height). In animals, N. has essentially the same basis as in humans; it may be genetically determined, due to inadequate feeding, or due to other disturbances of development, e.g. malfunction of certain endocrine glands.

Different types of N. occur in plants. Dwarf plants that arise by mutation are selected by plant breeders and developed into new cultivars, e.g. dwarf peas, bush varieties of beans and tomatoes, dwarf asters, etc. The physiological differences between normal and *genetically dwarf plants* are largely unknown and apparently very complicated. In some cases, there may be differences in the absolute and relative quantities of phytohormones. Thus, normal longitudinal growth can be induced in dwarf peas, bush beans and others by the application of gibberellins. In many species, e.g. peas, N. occurs only in the light. Others remain dwarfed even in the dark, e.g. certain dwarf maize mutants.

Nongenetic N. can be caused by water and nutrient deficiency. Seedlings (but not older plants) of some members of the *Rosaceae* (those in which seed germination is promoted by low temperatures, e.g. apple, peach and hawthorn) may display temporary N., depending on their growth temperature. In fruit trees, e.g. apple, N. can be induced by grafting. Certain synthetic gibberellin antagonists cause N. by inhibition of internode growth, and this is abolished by gibberellins.

Nannochoristidae: see *Mecoptera.*

Nanomia: see *Hydrozoa.*

Nanophanerophyte: see Life form.

Nanoplankton: see Plankton.

Nanosomy: see Nanism.

Napaeozapus: see *Zapodidae.*

Narcissus: see *Amaryllidaceae.*

Narcotics, ***narcotic drugs:*** substances which act predominantly on the central nervous system. Depending on the dose, N. show different phases of activity: small doses have a sedative action, somewhat higher doses are stimulatory, while even higher doses cause loss of consciousness (narcosis).

The World Health Organization classifies N. into seven groups: 1. alkaloids (LSD, mescalin, opium, etc.), 2. barbiturates and other sleeping drugs, 3. alcohol, 4. cocaine, 5. hashish and marihuana, 6. hallucinogens, 7. stimulants or antidepressants (e.g. amphetamines).

N. may be usd in psychiatry and psychotherapy, but misuse can lead to acute and chronic physical and mental deterioration, and to addiction. Regular use of N. leads to tolerance, i.e. an increased rate of breakdown of N. by the body, so that 10–20 times the normal dose may be necessary to achieve the desired effect.

Narcotine: an opium alkaloid (see Opium). Its narcotic activity is very low, but it intensifies the activity of morphine.

Narrow-mouthed frogs: see *Microhylidae.*

Narwhal: see Whales.

Nasal worms: see *Pentastomida.*

Nasua: see Coatis.

Natality: see Population dynamics.

Natal plum: see *Apocyanaceae.*

Natal sugarbird: see *Meliphagidae.*

Natantia: see Shrimps.

Nates: see Brain.

National park: see Nature conservancy.

Native companion: see *Gruiformes.*

Natterjack toad: see *Bufonidae.*

Natural engineering: in the broad sense, the study of the biological effects of structural alterations of the landscape and/or the use of biological methods to change or stabilize the landscape.

In essence, N.e. consists of the consolidation of river banks, lake shores, embankments and dunes, and their protection against slippage, erosion and drifting, with the aid of suitable plants, especially, trees and grasses. These methods are used primarily in water management and in the consolidation of the steep sides of spoil heaps from mines. Initially, they are more expensive than the building of physical support structures, but in the long term they provide a stonger and more durable protection, are easier to maintain, and blend more naturally with the landscape.

Naturalization: the successful establishment of animals and plants (see Adventitious plants) in an area in which they had not previously existed. In the restricted sense, N. is used only when the process is fully intentional and planned by human agency. In the broader sense, N. also includes the introduction of animals (e.g. raccoon, mink) that inadvertently escape from human care, and those whose introduction is entirely accidental and whose presence at the time of introduction may be unknown (e.g. seeds brought into areas in the mud of car tyres; insect pests introduced in exported food materials). Important examples of animals that have become naturalized following their planned introduction are game animals for hunting, fishes, carnivores (e.g. foxes and mongooses, introduced for the biological control of other species), predatory insects and parasites. Other European examples are fallow deer, mouflon, chamois (in the Elbe sandstone mountains of Germany), rabbit, pheasant, carp, American freshwater crawfish. The occurrence of the muskrat and Japanese fox in Central Europe is also due to N. and subsequent dispersal.

Countless attempts at N. have failed. For example, the N. of birds is particularly difficult. In the 19th century, there were many attempts by colonialists to introduce species from Europe, leading sometimes to a considerable "Europeanization" of local fauna. An extreme example of this is found in New Zealand, where mammals were formerly almost absent, and 15 different ungulate species were introduced. Nowdays, the agrarian landscape of New Zealand contains predominantly foreign bird species, about 35 of these introduced intentionally. Even the sparrow has been imported into many countries. N. of new species has often led to severe damage to the agricultural and forest economy, as well as the destruction of natural biotopes. Often the success of new species is linked to the decline of indigenous species. By competition for food supplies, even herbivores can contribute to the extinction of other species. Particular hazards are associated with the N. of predators. Occasionally, diseases are also introduced with naturalized animals.

Natural selection: a theory founded independently by Charles Robert Darwin (1809–1882) and Alfred Russel Wallace (1823–1913) to explain the mechanism of evolution. Natural populations produce a large excess of offspring. Since individuals display genetic variation, those that survive the "struggle for existence" will be those that are better adapted to the environment than their competitors. In this way, advantageous characters gradually spread through a population. By N.s. (see Selection), all present living organisms and all earlier but now extinct organisms are supposed to have arisen from one or a few primitive forms. In 1859, Darwin brought this evolutionary theory to the attention of biologists and the world with his publication "The Origin of Species". The arguments for N.s. are compelling, and the theory rapidly gained acceptance after 1859. N.s. is an essential component of the modern Synthetic theory of evolution (see).

Nature conservancy, ***conservation*** (see plates 47 and 48): the totality of legal, administrative, scientific and practical measures for insuring the continued existence of certain "protected" natural environments and living organisms. This "protected" status, which is legally enforceable, is accorded to biosystems and organisms of particular scientific or cultural importance. N.c. serves for the preservation and continued development of natural scenery, natural habitats and individual natural features, as well as the maintenance and preservation of animal and plant species that are endangered or threatened with extinction.

N.c. is necessary, because human economic activities have led to the alteration of many natural Ecosystems (see), with a corresponding decrease in the variety of plant and animal components. The tendency to impoverish and to impose uniformity on Nature was recognized in the nineteenth century by responsibly minded scientists, who promoted social and legal measures for the prevention or restriction of these negative influences. Early in the twentieth century, the first N.c. movement was established. It saw as its primary function the preservation of individual natural systems and organisms of outstanding interest or value, and their protection from wanton destruction and extermination. In 1904, the pioneer work of E. Rudorff in Dresden led to the establishment of the "Deutsche Bund Heimatschutz" (Federal German Homeland Protection Movement). In 1906, Prussia introduced a "Staatliche Stelle fur Naturdenkmalspflege" (State Department for the Care of Sites of Special Natural Interest), under the direction of H. Conwentz. Around this time, similar offices were established in other German States. N.c., however, had practically no legal status.

During the early decades of the twentieth century, the movement for the protection of nature was very successful, and it generated an increased awareness of the natural environment. This success, however, was due primarily to the untiring efforts of numerous enthusiasts. At that time, emphasis was placed on *preservation* and *conservation*. The basic policy was to ban access to certain areas, as well as to enact legislation to protect indigenous plants and animals, and to preserve characteristic scenery, thereby halting the impoverishment of flora and fauna.

A significant shortcoming of these first attempts at N.c. was that the relationship between Nature and society was not, or was insufficiently, appreciated; N.c. was seen as the antithesis of human economic activity; Nature had to be protected from the clutches of humans, and where not yet destroyed or altered, it had to be preserved in its original state.

Modern N.c. embraces wider principles, which may be summarized as follows. The organized protection of Nature should be scientifically based, drawing upon both the natural and the social sciences. Since the exploitation and the protection of Nature are closely interrelated, they can only be effectively pursued as a unity. Regional interests should always take precedence over local or individual interests. Demands and claims of the present should always be subordinate to the interests of future generations. N.c. cannot be restricted to the preservation or establishment of the status quo. It should also actively participate in the modification of the human environment, in order to ensure that this process is planned, carefully controlled, and as far as possible in harmony with the natural environment.

Nature reserves. A nature reserve is an area protected for its outstanding cultural or scientific value, or because it is the habitat of rare or threatened species of plants and animals. They are protected by legislation from encroachment or alteration; building is prohibited; the use of biocides (e.g. herbicides and insecticides) is prohibited; visitors are not allowed to leave footpaths; and it is forbidden to make noise, to camp or cause pollution. Sign boards are placed at entrances to nature reserves to draw attention to the special status of the area.

There are essentially 3 types of nature reserve:

1. *Documentary reserves* serve to illustrate various aspects of a natural environment, e.g. they are useful for comparison with similar areas more profoundly influenced by human activity. They are used for education and training in field biology and ecology. Some may also contain important geological formations.

2. *Refugial reserves* offer a home for plants, animals and biogeocoenoses, which, due to intense exploitation of the area or region, have no other sanctuary. Such reserves therefore serve predominantly to maintain species variety by preserving the natural environment.

3. *Experimental reserves* are areas for observation and research, and are also known as outdoor laboratories. These are used for the study of natural processes and phenomena and the possibility of their regulation or exploitation by man, the reaction of ecosystems to planned intervention, and the response of natural systems to environmental changes.

In practice, most nature reserves fulfil two or all three of these functions. A special function is served by *total reserves,* which totally exclude all direct human influence, thus enabling the observation of natural development over an extended period.

Nature reserves can also be classified into several types, according to their natural composition and the reasons for their reservation status. *Woodland reserves* are intended for the maintenance of natural stands of woodland, e.g. deciduous components of evergreen forests, relicts of former forest in agricultural land, or woodlands that display evidence of earlier methods of forest exploitation and management. *Botanical reserves* serve for the protection of rare plant species from extinction, the protection of scientifically or economically valuable plants, and the maintenance of endangered or especially interesting plant communities. Plants occurring on the borders of the reserve are also protected. The primary purpose of botanical reserves is to monitor and regulate the habitat, in order to guarantee the continued existence of certain plants or plant communities. *Zoological reserves* are habitats of threatened animal species. These include areas that are permanently inhabited by breeeding colonies, as well as the gathering, resting and overwintering sites of migratory birds. *Hydrological reserves* are waters and wet areas of special interest, including bogs and swamps, which are important in the water economy of the landscape. In *geological reserves,* geologically interesting structures, characteristic rock formations and similar natural phenomena are protected. These five types of reserve rarely occur as distinct or pure entities; most are *complex reserves,* i.e. they usually serve more than one purpose in N.c.

Since our whole landscape, including nature reserves, is formed and influenced by human activity, protection is only possible by some degree of intervention and control. This point is well illustrated by mountain meadows or alps, which were originally created from cleared woodland for pasture or haymaking. Without human intervention, these would rapidly revert to woodland, and the charactersitic meadow flora would be lost. For this reason, the measures necessary for protection must be established. In the case of mountain meadows, these measures consist of mowing and the removal of young bushes and saplings.

Protected Landscapes (National Parks). National parks are notable for their natural beauty and for their recreational value. They are model landscapes, in that they are both protected and tended. In these protected areas, activities that change the landscape, especially boring and surface engineering works, quarrying, mining, erection of agricultural buildings, etc. can only be pursued with permission of the legal authority. National parks are intended primarily for recreation and tourism. Prerequisites for the designation of a national park are therefore natural diversity, interesting topography, woodlands alternating with open landscape, and ample lakes and rivers, or at least several of these components. Furthermore, as recreational areas, national parks should be environmentally clean, with minimal air and water pollution and low noise levels, and they should offer a wide choice of activities for tourists and hikers.

Sites of Special Scientific Interest (SSSI). These have special scientific interest with respect to living organisms, natural habitats, local natural history, geological structures and local topography, and they often play a valuable role in education and training. They are protected by law from damage or destruction, and they may be altered only in exceptional circumstances. They receive special maintenance and protection, and they are identified by a plaque or signboard. Examples of SSSI are stands of trees, natural springs, wet areas, small lakes, ponds and streams with important animal and plant communities, caves, erratics and other unique and/or interesting geological formations, small communities of theatened or rare animals and plants, breeding colonies of birds, sites of bat colonies, and even single trees of great age, rarity or exceptional size.

Protected plants. Wild plants of especial value for research and teaching, and those of economic value, which are rare or threatened in their habitat, are designated for legal protection. It is forbidden to dig up or to pick these plants, or to alter their habitat in any way that would endanger their continued existence. The necessity for protection arises from the increased intensity of land utilization, which goes hand in hand with a certain impoverishment of local flora, so that many plants decrease in numbers or are even threatened with extinction. Examples of threatened plants in Mainland Europe, which are protected by legislation in Germany, are the orchids, *Dactylorhiza sambucina* and *Leucorchis albida.* It is fundamentally important to maintain the diversity of species, which provides the genetic potential (gene pool) for plant breeding and other scientific purposes. In addition, various branches of the biological sciences, e.g. ecology, geobotany, taxonomy, etc., depend to great extent on the availability of a complete spectrum of species for research purposes. Not least, many plants are protected for ethical-esthetic reasons, as beautiful and attractive representatives of the local natural scene.

Many countries legislate for the protection of local plant species. For example, protected plants in Germany include all native species of orchids *(Orchidaceae),* Gentian *(Gentiana, Gentianella),* Primrose *(Primula)* and Sundew *(Drosera);* the Globeflower *(Trollius europaeus),* European Columbine *(Aquilegia vulgaris),* Arnica *(Arnica montana),* Daphne *(Daphne mezereum),* Goatsbeard *(Aruncus dioicus),* Liver-leaf Hepatica *(Hepatica nobilis),* Yellow Foxglove *(Digi-*

talis grandiflora), Gas-plant Dittany *(Dictamnus albus)* and Stemless Carline Thistle. *(Carlina acaulis).*

Protected animals. These are animals that need to be protected because of their economic value, their rarity and their scientific value for research and teaching, or because they are threatened with extinction. Game animals, however, are protected by special hunting laws that restrict the numbers that may be taken and impose close seasons. Protected animals may not be killed or disturbed, trapped or held in captivity. Breeding sites and habitats are also protected, as well as the eggs of protected oviparous species and the pupae of insects.

The necessity for protection arises from the fact that many animal species have decreased in numbers, or have even become threatened with extinction. This is caused by changes in their natural environment, and by the unwarranted persecution of certain species that are alleged to cause damage or carry disease. In general, the decrease in the populations of threatened species has been halted, and the numbers of many protected species show evidence of increasing. Animals placed under protection because of their threatened extinction include the beaver population of the River Elbe *(Castor fiber albicus),* Wild cat *(Felis silvestris),* Hen Harrier and Montagu's Harrier *(Circus cyaneus* and *C. pygargus),* Black Stork *(Cicona nigra),* Eagle Owl *(Bubo bubo),* Great Bustard *(Otis tarda),* Common Crane *(Grus grus),* and all species of eagle of the genera *Haliaeetus, Pandion, Aquila* and *Circaetus.* Other protected species are, e.g. the Eurasian Hedgehog *(Erinaceus europaeus),* all bats *(Chiroptera),* Roman Snail *(Helix pomatia),* Freshwater Mussel *(Margaritana margaritifera),* Stag Beetle *(Lucanus cervus),* many species of butterflies, most amphibians and reptiles, and most birds.

When populations of game animals are threatened , they can be protected by a total ban on hunting. Such animals include the Otter *(Lutra lutra),* Badger *(Meles meles),* Sparrow Hawk *(Accipiter nisus),* Goshawk *(Accipiter gentilis),* Buzzard *(Buteo buteo)* and the Capercaillie *(Tetrao urogallus).*

Protective measures are extended to the provision of nesting sites or other aids to breeding, the provision and care of copses, and the designation of protected areas for the maintenance of particular species, e.g. the conservation of small ponds and streams as spawning sites for amphibians. Special protective measures include the designation of areas containing the eyries and nesting sites of certain bird species. To ensure the successful rearing of the young birds, forestry work is forbidden within a radius of 100 m around the nesting site, and may only take place to a limited extent within the subsequent 200 m zone. Protected zones have also been established for e.g. the Elbe beaver and the Great bustard; in these zones, agriculture and forestry are regulated to provide normal living conditions for the animal species, and the populations of these two species are now no longer decreasing. Protected zones, known as wild fowl refuges, are also provided for native and visiting water birds.

Protected parks. Town or country parklands, if not already protected for their scientific value, may be designated as protected parks. Especially suitable for this designation are wooded areas in agrarian landscapes that are poor in woodlands, or any areas of natural greenery in cities and urban districts. Since they often provide a habitat for many plants and animals, and contribute to improving the local environment by counteracting dust and pollution and by decreasing noise, they have cultural and esthetic value.

Protected trees. In addition to woods and forests, protected status may also be given to hedges, copses and avenues of trees, which have environmental value. Such trees and bushes may act as wind breaks and/or decrease water erosion. They preserve soil fertility, protect river banks, decrease the danger of soil desiccation, act as barriers against dust, cut out noise in the proximity of residential areas and factory sites, and generally beautify the landscape, as well as acting as breeding and feeding grounds and as sanctuaries for many wild animal species.

Nature trails. These are not actually protected areas or objects, but they are very important for the dissemination of information and knowledge about N.c. They may be laid down inside or outside protected areas. They serve to transmit information about local natural history, and at the same time familiarize the visitor with protected objects. In protected areas, visitors are guided by signposted trails. which also keep them away from sites that must be left undisturbed for purposes of species protection.

International N.c. Although N.c. applies to national situations and is regulated by national laws and legislation. N.c. has international importance, extending beyond the borders of any single country. International N.c. is required for the protection of natural objects that are unique in the world or in a particular region, and for the protection of animal and plant species that are threatened with extinction worldwide or in a particular area, as well as the protection of migratory birds, which recognize no national boundaries.

The interests of International N.c. are served by the *International Union for the Conservation of Nature and Natural Resources* (IUCN). This organization functions as an international documentation and coordination center for N.c., and it advises UNESCO and other institutions of the United Nations. A further international movement, *Project MAR,* is concerned with the maintenance of open waters and swamp areas in Europe and North Africa, as well as their protection against pollution and obstruction. The former *World Wildlife Fund* (WWF), renamed the *Worldwide Fund for Nature* in 1988, gives financial support in individual countries for the protection of threatened species and the provision of reserves, as well as for education in N.c. and the training of N.c. workers. For the protection of water birds, especially migratory species, there is the *International Convention on Water Fowl Sanctuaries.*

In 1970, UNESCO gave birth to the program *Man and the Biosphere* (MAB). This is concerned with the rational exploitation and mantenance of the resources of the biosphere, and the improvement of the relationship between humans ans their environment. A component of the MAB program is the establishment of biosphere reserves. These differ administratively from national conservation areas, in that they belong to a UNESCO network of conservation areas in all regions of the world. Within biosphere reserves, the maintenance of characteristic ecosystems and plant communities takes priority over individual objects or species. They also contain clearly defined and demarcated specific conservation zones, and they serve for international research and environmental surveillance programs.

Nature reserve: see Nature conservancy.

Nature trail: see Nature conservancy.

Naucrates: see Pilot-fish.

Nauplius larva: a primitive, planktonic larva of lower crustaceans *(Entomostraca).* It has a single, median nauplius eye at the front of the head, and only 3 pairs of appendages: the first antennae (antennulae), second antennae and the mandibles. The appendages serve for both locomotion and feeding. Additional trunk segments and appendages appear with successive molts, and this later stage is called a *metanauplius.* In copepod development, the metanauplius is succeeded by the copepodite stage, while in the *Cirripedia* it is succeeded by the cypris larva.

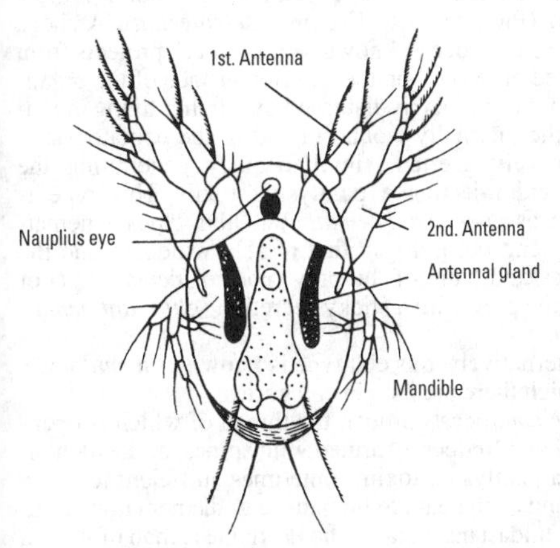

Nauplius larva of *Cyclops*

Nautilus: see *Cephalopoda.*

Nautiloidea: see *Cephalopoda.*

Navigation: self-orientation of an animal toward a goal in unknown surroundings. For this type of Orientation behavior (see), the animal must determine its own position relative to its goal. In *vector N.,* the direction and distance to the goal is determined; the determination and measurement of the time of day may also be involved. In *inertial N.,* it is thought that the organism records the spatial displacement of its body on the way from given site, and is able to reconstruct the neccessary movements for an accurate return to the same site. In *bicoordinate N.,* it is assumed that the animal determines its present geographical position, and is able to derive from this the direction to its destination. This means that the animal must use a network of coordinates, probably two intersecting gradient fields, from which it derives a compass bearing.

Neanderthal: see Anthropogenesis.

Neanthropines, *Neanthropinae:* see Anthropogenesis, *Homo.*

Nearctic, ***Nearctic subregion, Nearctic faunal subregion:*** an animal geographical subregion, formerly recognized as an independent region, but now combined with the Palearctic (see) to form the Holarctic (see). It includes North America south to Mexico, and does not include tropical South and Central America. The most northerly part of the subregion is often considered as separate and combined with the northernmost parts of Eurasia and the Arctic islands into a separate subregion, the Arctic (see). There is an agreed southern boundary (see map, Animal geographical regions), but this is a compromise, and the transition zone is wide. Neotropical animals that are also found in the N. are, e.g. the opossum *(Didelphis),* jaguar *(Panthera),* coati *(Nasua)* and tree porcupine *(Erethizon),* and especially birds, such as humming birds *(Trochilidae),* American vultures *(Cathartidae),* tanagers *(Thraupidae),* icterids *(Icteridae)* and even parrots, as represented by the recently extinct Carolina parakeet *(Conuropsis carolinensis).* Since many Nearctic animals, mammals in particular, have penetrated southward, it is understandable that earlier authors sometimes attached greater importance to the relationship between the N. and more southerly regions than between the N. and the Palearctic.

In the north, however, there are numerous Holarctic genera and species, as well as endemic families (see Holarctic). In addition to taxa that are widely distributed in both subregions (e.g. wolf, brown bear), there are numerous Palearctic insects with relatively small areals in either East Asia or Europe, i.e. there are marked transpacific and transatlantic relationships. The Pacific connection is explained by the earlier land bridge at the Bering strait, but an Atlantic land bridge is still hypothetical.

Compared with the fauna of the Palearctic, the N. fauna were less severely affected by the Ice Age, since there were no mountain ranges to prevent areal movement in the face of the advancing ice. The dominant true mice *(Muridae)* of the Palearctic are notable absent from the N. Despite the immigration of southern animals, the N. has less variety of mammals than the Palearctic; thus, primates, horses, viverrids, hyenas, antelopes and gazelles are all absent. The reptilian taxa of the N. clearly reveal the influence of the Neotropical region, with numerous species of the *Urodela.*

Near field behavior: motivated behavior in relation to an object or place, which has been identified and its existence in time and space has been established. For example, a search for a field mouse by a falcon represents Distance field behavior (see). When the falcon sees the mouse and locates its position, it enters the near field and exhibits N.f.b., i.e. it stoops toward the mouse. As it seizes the mouse with its tallons, the contact field is established, which provides the stimuli for feeding (i.e. contact field behavior).

Necator: see *Nematoda.*

Necromass: the quantity of dead organic material in a biome that represents part of the primary product of that biome. It is usually recorded as quantity per year per m^2 of surface area. It includes all nonliving plant material, such as shed leaves and pine needles, wood and bark, dead roots, secretions such as honeydew and resin, and harvest residues. Relatively recent N. forms an upper, incompletely transformed layer known as *litter,* and in many biomes this consists largely of *leaf litter.* Otherwise defined, the litter layer is the organic layer that lies on top of the mineral soil. The quantity and type of the N. are important criteria in the dynamics and development of the ecosystem. N. is subject to the complicated processes of Decomposition (see); it forms the dietary basis for saprophytic organisms, and it is one of the crucial factors in determining the formation of Humus (see).

Necrophage: see Modes of nutrition, Nutritional interrelationships.

Necrophytes: plants that live on, and at the expense of, dead organisms. N. are also saprophytes.

Necrosis: in plants and animals, local cell or tissue death resulting from the activities of disease organisms, toxins or physical factors (e.g. temperature), as well as inadequate nutrition or starvation (e.g. as a result of an interrupted blood supply). N. is a typical symptom of many plant diseases.

Necrosyrtes: see *Falconiformes*.

Nectar: a sugary liquid secreted by the nectaries of plants. Major components of N. are glucose, fructose and sucrose. Nectaries consisting of glandular epithelia or hairs are situated mainly within flowers (floral Ns.), where their secretions attract pollinating insects. Extrafloral Ns. (e.g. on the petioles of the bird cherry, *Prunus padus,* and on the stipules of of the horse bean, *Vicia faba)* have a different structure, have no significance in pollination, and are usually functional only during the vegetative period.

Nectarinia: see *Nectariniidae*.

Nectariniidae, ***sunbirds:*** an Old World avian family of more than 100 splendidly colored, small to very small (9.5 cm upwards) species, in the order *Passeriformes* (see). They are the Old World counterparts of the humming birds: the bill is usually long, thin and curved, and as an adaptation to feeding on nectar the tongue can be rolled into a tube. Although they can hover, they usually perch to take nectar, either directly or by piercing the corollas of large flowers. The family ranges from Africa via Palestine to India, Southeast Asia, Oceana and Australia, with the majority of species in Africa. Important genera are *Nectarinia, Cinnyris, Aethopyga* and *Anthreptes*. The 9 species of ***Spiderhunters*** *(Arachnothera)* of the Oriental region are larger, have duller plumage, show no sexual dimorphism, and feed on invertebrates.

Nectary: see Excretory tissues of plants.

Necton: pelagial organisms with well developed means of locomotion. In contrast to plankton, N. can resist water currents by active swimming. In freshwater environments, the N. consists solely of fish. In the sea, the N. consists primarily of fish, seals and whales. The boundary between N. and plankton is blurred by the existence of many intermediate forms.

Nectonematoidea: see *Nematomorpha*.

Nectophore: see *Hydrozoa*.

Necturidae: see *Proteidae*.

Needle-leaved gymnosperms: see *Gymnosperms*.

Negative staining, ***negative contrast technique:*** method for the observation of fine structures in the electron microscope. The specimen (viruses, cell particles, protein crystals) is surrounded by a layer of electron-dense material, such as phosphotungstic acid, phosphomolybdic acid or uranyl acetate, which fills the cavities of the specimen and follows its contors. The specimen is therefore observed in negative contrast, i.e. it appears light against a dark background. See plate 18 (Fig.4).

Negro bugs: see *Heteroptera*.

Neiris: see *Polychaeta, Errantia*.

Neisseria: a genus of small (up to 1 μm diameter) Gram-negative cocci, usually arranged in pairs (diplococci), each cell characteristically flattened where it makes contact with its partner. Species of *N.* are aerobic or facultatively anaerobic, and under certain culture conditions they form capsules and pili. They generally parasitize the mucosae of humans and other mammals. The gonococcus, *N. gonorrhoeae,* was discovered as the causative agent of the sexually transmitted disease, gonorrhea, by A Neisser. The meningococcus, *N. meningitidis,* is responsible for epidemic meningitis.

Nema: see Graptolites.

Nematicides: plant protection agents for the control of nematodes. Important nematicidal products are dazomet, metham-sodium, aldicarb and oxamyl.

Nematocyst: complex ectodermal cells present in the *Cnidaria,* which serve for protection against aggressors and for the capture of prey. They are usually distributed over the entire body surface, but are especially numerous on the tentacles. Each nematocyst consists of a double walled vesicle containing a wound hollow thread (Fig., see also Fig. in *Coelenterata).* A short, bristle-like process, known as a *cnidocil,* projects from one end of the cell on the exterior surface of the organism. If the cnidocil is disturbed by another living organism, the internally wound thread of the nematocyst is explosively everted, simultaneously penetrating the prey and injecting a paralysing toxin. This type is known as a *penetrant nematocyst* Other types of nematocyst send out a thread that merely winds around the bristles and hairs of the prey *(volvent nematocyst)* or traps the prey with a sticky secretion *(glutinant nematocyst).*

Alternatively, this cell type is known as a *cnidocyte,* of which there are 3 main categories:

1) *Nematocysts* proper, the thread of which is open-ended and frequently armed with spines, and which injects a paralysing toxin, sometimes sufficient to cause pain and even death to humans, e.g. the nematocysts of some cnidarians, notably the Portuguese man of war. In other nematocysts, known as *desmonemes,* the thread is closed, lacks spines, and serves to entangle the prey.

2) *Spirocysts,* the thread of which carries numerous fine tubules that discharge an adhesive net over the prey. The undischarged cell possesses receptive microvilli rather than a cnidocil.

3) *Ptychocysts,* found only in the *Ceriantharia,* in which the undischarged thread is pleated rather than wound; the everted thread lacks spines and is adhesive.

Nematocysts or cnidocytes discharge only once. They are then released from the epidermis and replaced by the growth and differentiation of neighboring interstitial cells.

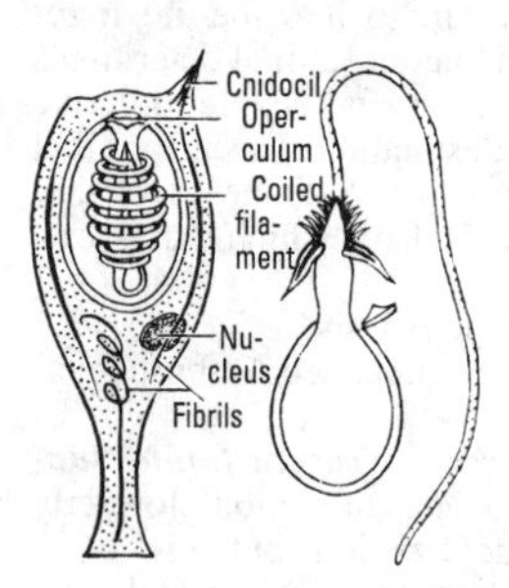

Nematocysts (before and after discharge)

Nematoda, ***nematodes, roundworms:*** a class of unsegmented worms (phylum *Aschelminthes* or *Nemathelminthes)* containing about 10 000 species, including free living and parasitic forms.

Morphology. The spindle or thread-shaped body is

circular in cross section, and it varies in length from a few millimeters to more than 1 meter. The largest member is the female of *Placentonema gigantissima* (length 8.4 meters, diameter 2.5 cm) from the placenta of the sperm whale, *Physeter catodon*. *N.* are transparent or translucent, or various shades of white to yellow. Some animal parasites appear red due to ingested blood. The round body shape appears to be maintained by internal hydrostatic pressure, e.g. the internal pressure in ascarids is 0.3 atmosphere, so that body fluids squirt out when the animal is punctured. The mouth (which opens into a buccal cavity) is situated at the anterior end of the body, and it is often surrounded by sensory bristles, tactile papillae or small lips. The posterior end of the body is rounded or pointed, and in males it is curled into a hook. The body is covered with a tough cuticle secreted by the underlying ectoderm or syncytium (containing many nuclei but lacking cell walls). Beneath the syncytium is a layer of longitudinal muscle cells, divided into 4 bands by projections of the syncytium. Externally, these bands are visible as dorsal and ventral lines, and as the slightly more prominent right and left lateral lines. Locomotion appears to be random, resulting from thrashing movements in a dorsoventral plane. Between the outer muscular layer and the alimentary canal is an extensive fluid-filled space (pseudocoelom), formed by coalescence of syncytial vacuoles. Unlike the true coelom of higher animals, this pseudocoelom has no epithelial lining.

The alimentary canal is divided into a pharynx (which acts like a suction pump), an extended mid-gut, and a short hindgut, exiting shortly before the posterior end of the body. In males, the genital duct opens into the terminal part of the hindgut to form a cloaca, whereas in females the anus and vagina each has its own exit. A dorsal pocket of the male cloaca contains 2 chitinous copulatory spicules, the points of which are held together by a grooved connecting piece. The pocket also contains a muscle which protrudes the spicules from the cloaca. Males possess an auxillary reproductive organ (a copulatory bursa, consisting of lateral extensions or lobes of the epidermis), which acts as a sucker, keeping the body of the male against that of the female until the copulatory spicules are inserted through the female aperture. With a few notable exceptions, *N.* are unisexual, and the tubular sex organs lie in the pseudocoelom. Males possess a single tubular testis, while females usually have 2 tubular ovaries. Most *N.* lay eggs, and only some of the parasitic species are ovoviviparous.

The excretory system consists either of a pair of lateral longitudinal canals lying in the epidermis, or an unpaired excretory gland; both systems exit on the ventral surface toward the anterior end of the body. In primitive marine species, excretion is served by the renette gland, which consists of one or several large cells, lying ventral to the pharynx, and emptying anteriorly to the exterior. Some species possess two unicellular glands or phasmids (thought to have a sensory function), each in a cuticular pouch on either side behind the anus. The nervous system consists of a nerve ring around the pharynx, which contains 2 or more lateral pairs of ganglia. Longitudinal nerves run backward from the pharyngeal nerve ring, the main ones being the dorsal and ventral nerves, which run in dorsal and ventral thickenings of the ectoderm. Only a small number of peripheral nerve fibers are present.

Cell lineage (see) has been shown for various tissues and organs of *N.*, e.g. pharynx, nervous system, etc.

Biology. *N.* have a cosmopolitan distribution in every conceivable type of habitat. They are found in marine and freshwater habitats, and even thermal springs, as well as practically all terrestrial biotopes. They include parasitic species that live on, but mostly in many animals and plants. Free living species and the free living juvenile stages of parasitic species feed on living and dead organic material, and many species are predatory. Parasitic forms feed on plant sap, or on the body fluids and tissues of their host.

Development. *N.* develop in stages by molting, but without metamorphosis. They usually pass through 3–4 juvenile stages, which resemble the adult except for the absence or smaller size of the reproductive organs. In free-living forms, the third juvenile stage may remain encysted in previous larval integument as a resistant cyst or dormant larva. In many parasitic forms, the third larval stage is the infective agent. Animals become infected in various ways: 1) Ingestion of the free-living larvae or eggs; 2) Active penetration of the skin by larvae; 3) Transfer of larvae from an intermediate host by a blood-sucking insect. In a few species, the parasitic generation reproduces parthenogenetically, and the free-living form reproduces sexually.

Economic importance. Parasitic *N.* attack crop plants, as well as humans and domestic animals, e.g. the potato and turnip eelworms, and the parasitic ascarids, hookworms and trichinellas of humans and animals. Host organisms are damaged by toxins, nutrient deprivation and physical effects such as blockage of blood and lymph vessels.

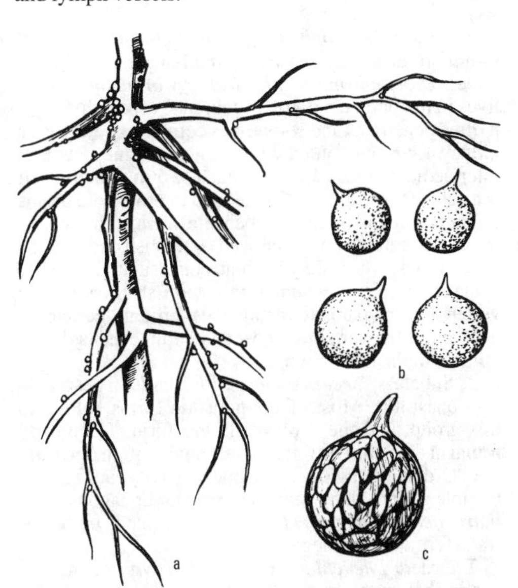

Nematoda. Potato eelworm *(Heterodera rostochiensis). a* Roots with cysts. *b* Cysts highly magnified. *c* Cyst with eggs.

Classification. The phylogenetic affinities of the *N.* are unclear. About 14 orders are recognized, and these can be separated into 2 subclasses:

1. Subclass: *Adenophorea* or *Aphasmida.* These lack phasmids, and the majority are free-living in salt or fresh water, only a few members being parasitic. Dormant larvae have not been demonstrated for any member of this subclass.

1. Order: *Enoplida.* A primitive order consisting of free-living, predatory species, most of which are marine. There are 3 anterior circles of sensory papillae or cirri, a unicellular renette gland, and 3 caudal glands. Three hollow teeth are usually present, which form the ducts of the pharyngeal glands.

2. Order: *Dorylaimida.* These are common in fresh water and soil. Most have a hollow stylet, which is used for sucking plant and animal juices.

3. Order: *Mermithida.* The threadlike adults (up to 50 cm in length) are free-living in soil or fresh water, while the juvenile stages parasitize land and freshwater mollusks and arthropods. *Mermis subnigrescens* lives deep in the soil, but emerges to attach its eggs to grasses, which are eaten by grasshoppers. In the grasshopper intestine, the larvae hatch and penetrate into the body cavity. After feeding on the host tissues, they exit by boring through the body wall, then develop to maturity in the soil.

4. Order: *Trichurida.* The pharynx contains 1 or 2 rows of large cells, known as stichosomes, which are thought to be glandular. *Trichuris trichiura* (the ***Whipworm;*** 3–5 cm) is quite common in the appendix of humans and pigs. Transmission is by the ingestion of the eggs, which are found in water and soil; the infection is rarely troublesome and may even pass unnoticed. In contrast, one of the most dangerous human parasites is the ***Trichina worm,*** *Trichinella spiralis* (see Trichinosis).

5. Order: *Dioctophymida.* Members of this small group are characterized by the bell-shaped tail of the males. *Dioctophyme renale* ***(Kidneyworm;*** up to 1 m) lives in the body cavity of vertebrates, especially dogs; it often penetrates and sometimes completely destroys a kidney. Eggs are voided with the host's urine. The first intermediate host is an oligochaete worm, which itself is a parasite of crayfish. The second intermediate host is a fish that feeds on the oligochaete worm. The worms move through the intestinal wall of the fish and into the mesenteries, where they become encapsulated. The life cycle is completed when an infected fish is eaten by a vertebrate; the parasite then passes through the intestinal wall of the vertebrate host and completes its development in the body cavity.

2. Subclass: *Secernentea* or *Phasmida.* These possess phasmids. Most of the parasitic forms belong to this group, and the few free-living forms are mostly found in the soil. The third larval stage is often dormant.

6. Order: *Rhabditida.* The buccal cavity lacks a protrusible stylet. Many members are saprophagous. *Turbatrix aceti* ***(Vinegar eelworm;*** 2 mm) feeds on bacteria in fermenting vinegar.

7. Order: *Tylenchida.* The buccal cavity contains a protrusible stylet. In addition to soil inhabitants, many are plant parasites (see Eelworms). Others are parasites of insects.

8. Order: *Rhabdiasida.* Members of this order are characterized by complicated life cycles. After feeding on the tissues of their mother, the ovoviviparously produced young worms of *Rhabdias bufonis* infect amphibians by ingestion or skin penetration. In the lungs of the amphibian host, they develop into hermaphrodite worms, the fertilized eggs of which pass up the trachea, down into the intestine, then to the exterior in the feces. In the soil, the eggs hatch into free-living forms with separate sexes.

9. Order: *Oxyurida.* These are obligatory parasites, with simple life cycles, e.g. *Enterobius vermicularis* (see Enterobiasis).

10. Order: *Ascarida.* These are obligatory parasites of vertebrate intestines. The mouth is surrounded by 3 lips. *Ascaris lumbricoides* or ***Maw worm*** (female up to 40 cm x 5 mm; male up to 25 cm x 3 mm) (plate 43) inhabits the small intestine of humans and pigs, and is also found as a parasite of sheep and cattle. It has a cosmopoliotan distribution, and is one of the commonest human worm infections in the tropics. Heavy infestations cause serious disablity by blockage or perforation of the intestine, allergic reactions to toxic metabolic products (especially during lung passage of the larvae), as well as blockage of the bile and pancreatic ducts. Low numbers, however, have only slight pathological effects. Females produce about 20 000 eggs daily, which pass out with the host's feces. Once outside the host and in the presence of oxygen, an embryo larva develops and remains viable within the egg casing for up to 2 years under humid conditions. Infection occurs by ingestion of the eggs, usually on uncooked, unclean vegetables or fruit. Larvae hatch in the intestine, burrow through the intestinal wall, and migrate with the blood stream to the lung, where they lodge for some time and grow. After penetrating into the alveoli of the lung, they ascend the bronchi and the trachea to the mouth, are swallowed, and establish themselves as adults in the intestine. Related species parasitize horses, dogs, cats and birds.

11. Order: *Strongylida.* Lips are absent, but cuticular teeth are present on the edge of the mouth. The order includes many animal parasites, e.g. *Strongylus vulgaris* (horse parasite), *Syngamus tracheae* (chicken parasite), *Necator americanus* (human parasite, known as American hookworm), and *Ancylostoma duodenale* or oriental hookworm (see Hookworm).

12. Order: *Spirurida.* All are obligatory parasites with an intermediate arthropod host. The lips are unlobed or trilobed. Examples are *Thelazia* (infects eyes of birds and mammals), *Spirura* (infects mammalian stomach), and *Habronema* (infects horse stomach).

13. Order: *Dracunculida.* See Guinea worm.

14. Order: *Filarioida.* See Filariasis; Onchoceriasis.

Collection and preservation. *N.* can be collected from practically any habitat: earth, moss cushions, sea water, fresh water, and as parasites of plants and animals. To isolate *N.* from the soil, they are washed out through sieves of ever decreasing mesh size. Small plants or soil samples are placed on cotton wool in a fine sieve, and placed above a water sample in a funnel (see Soil organisms). After a time, the *N.* migrate into the water, where they can be separated and studied. The selected animals are carefully warmed in water on a concave microscope slide until they stop moving. The water is removed with a pipette and replaced with the following fixative: concentrated formaldehyde solution: glacial acetic acid: water (1:1:8, by vol.). *N.* can be kept indefinitely in this mixture. Intestinal parasites are removed from post-mortem material by washing out with 1% sodium chloride solution, immersed in hot 70% ethanol, then stored in 70% ethanol: glycerol (1:1, by vol.).

Nematodes: see *Nematoda.*

Nematognathi: see *Siluriformes.*

Nematomorpha, ***hairworms, "horsehair snakes":*** a class of unsegmented worms (phylum *Aschelminthes),* containing more than 200 known species.

Morphology. Hairworms are threadlike or hairlike, 2–160 cm in length, and rarely exceeding 2.5 mm in width. The yellow or gray to black-brown body is rounded anteriorly. Posteriorly, it may be rounded, or provided with two or three plump tail lobes. The space between the body wall and the more or less degenerate alimentary canal may be filled with parenchyma (which contains primary coelomic cavities), or the parenchyma may be reduced, leaving a typical pseudocoelom. Sexes are separate. The testes or ovaries are laterally situated, paired, tubular structures, and the genital ducts empty into the hindgut to form a cloaca.

Biology. Adults are free-living, usually in fresh water, rarely in moist earth or in the sea. Most adults lack a mouth, and those with a mouth probably do not feed. Most of the lifespan is spent in the process of development from larva to adult, and during this period the animal lives parasitically, in particular in beetles, orthoptera, myriapods, crabs and hermit crabs. Eggs are laid in strings, which are wound about water plants. The hatched larva has a spiny proboscis, with which it bores into a host animal (usually an aquatic insect larva). Development occurs in the body cavity of the host. The sexually mature hairworm does not mate and produce eggs until it has left its host. During the parasitic phase, most hairworms obtain food material from their host by osmosis.

Classification. Two orders are recognized, the *Gordioidea* (fresh water) and the *Nectonematoidea* (marine).

Collection and preservation. Usually dark-colored hairworms are found in large tangled masses in shallow water in most parts of the world (the genus name, *Gordius,* is derived from "Gordian knot"). For demonstration purposes, specimens are fixed and stored in 70% ethanol. For histology, worms must be fixed in hot, saturated mercuric chloride solution containing 5% acetic acid.

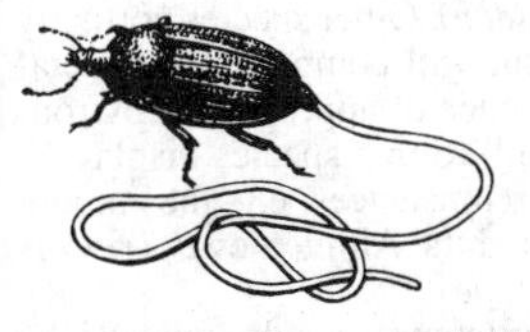

Nematomorpha. Mature hairworm leaving its beetle host.

Nematophora: see *Diplopoda.*

Nemertea: see *Nemertini.*

Nemertinea: see *Nemertini.*

Nemertini, *Rhynchocoela, Nemertea, Nemertinea,* ***proboscis worms, ribbon worms:*** a phylum of the *Protostomia* containing about 800 living species. Most species are truly aquatic. Externally, they resemble flatworms *(Platyhelminthes),* but they are more complex than the latter, because they possess a vascular system, as well as a long, straight alimentary canal with a mouth at one end and an anus at the other. In some species the alimentary canal is provided with lateral diverticula.

Morphology. Most species have a threadlike body, which is elongated and flattened, and often strikingly colored red, orange or green and/or patterned with stripes and bars. Lengths vary from a few millimeters to 30 m (!). The entire body epithelium is ciliated. Situated at the anterior end, above the foregut, is the *proboscis sheath,* consisting of a long fluid-filled cavity *(rhynchocoel),* housing a retractable proboscis; the rhynchocoel is considered to be a true coelom. The head is more or less differentiated from the body, and it carries between 2 and 250 simple eyes (pigment cups), as well as ciliated grooves or slits, and paired dorsal ciliated tubes with external openings above the brain area; these ciliated grooves and tubes are thought to be chemoreceptors. The nervous system resembles that of higher flatworms. The 4-lobed brain consists of dorsal and lateral ganglia on each side of the rhynchocoel, connected by dorsal and ventral commissures, so that a nerve ring is formed around the rhynchocoel. A pair of ganglionated nerves extends posteriorly in the body wall. Lateral blood vessels run the length of the body, and are connected by lacunae. The blood is colorless, yellow, red or green, and often contains oval blood corpuscles. In most species the sexes are separate. Numerous testes or ovaries lie in rows at the sides of the body, each gonad with its own separate opening to the exterior. Eggs are fertilized after they have been released into the surrounding water.

Fig.1. *Nemertini.* General habit of a proboscis worm.

Biology. Only a few species inhabit fresh water or moist earth. The majority are marine, and most of these are found in coastal waters. Progression is usually by creeping on the substratum, but some species can swim and may descend to 3 000 m.

All species feed by predation on aquatic animals. Prey is captured with the long tubular proboscis, which is rapidly extruded by contraction of the fluid-filled rhynchocoel. Depending on the species, the proboscis may or may not be armed with a piercing stylet. The proboscis is attached at the anterior end of the proboscis sheath, and it becomes everted as it is extruded; it is drawn in again by retractor muscles attached to the posterior end of the sheath. The animal prey (e.g. an annelid worm) may be swallowed, or its body contents may be sucked out.

Development is direct (most species), or it may procede via a helmet- or bell-shaped pelagic larva, known as a *pilidium larva.* The latter possesses an apical ecto-

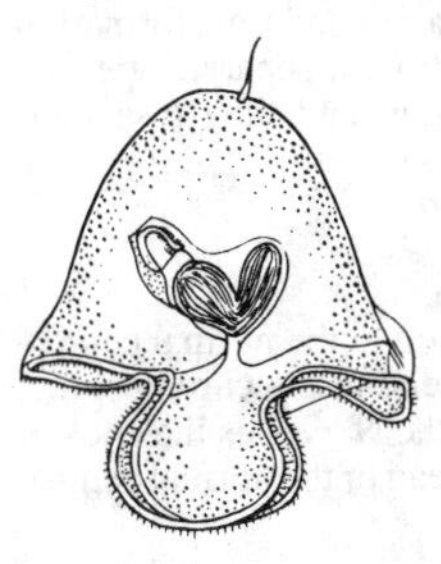

Fig.2. *Nemertini.* Pilidium larva of a proboscis worm.

dermal thickening, known as the apical organ, which is surmounted by a tuft of stiff flagella. From the edge of the base hang two lobes, which resemble the earlaps of a helmet (Fig.2). The larva swims with the aid of a band of cilia around the base (the prototroch).

Classification. Separation into two classes is based on the position of the mouth (posterior or anterior to the brain), and the absence or presence of a stylet on the proboscis.

1. Class: *Anopla.* The mouth is posterior to the brain, and stylets are absent.

Order: *Paleonemertea.* These are the most primitive *N.* Muscles of the body wall occur in either 2 or 3 layers. In 3-layered walls, a longitudinal layer lies between two circular layers. Nerve cords lie outside the muscle layers, or within the central longitudinal layer. Most species lack eyes and other sense organs.

Order: *Heteronemertea.* The body wall consists of a circular muscle layer between two longitudinal layers. Nerve cords lie in the central layer. Eyes, other cerebral sense organs, mid-dorsal blood vessels, and intestinal diverticula are all present.

2. Class: *Enopla.* The mouth is anterior to, or below, the brain. The proboscis displays distinct regions, and a stylet may or may not be present.

Order: *Hoplonemertea.* This order contains all those species that possess a stylet. The intestine has lateral diverticula and a cecum. A dorsal blood vessel is present.

Order: *Bdellonemertea.* This order contains only a single genus *(Malacobdella)* with 4 species. Three species are commensals in the mantle cavities of marine clams; one species is a commensal in the mantle cavity of a freshwater snail. There is no stylet, and this may be due to evolutionary loss. Eyes and other cerebral sense organs are absent.

Collection and preservation. Marine specimens can be collected in quantity only by the so-called environmental deterioration method. Large samples (5–10 liters of sea water) are placed in a darkened glass tub, with one corner exposed to daylight. As the environmental conditions in the water deteriorate, worms accumulate in the light corner. Since proboscis worms are very contractile, they must be narcotized with a calibrated dose of chloral hydrate, urethane or acetone/chloroform, before they are fixed. Bouin's fluid or 80% ethanol is used for killing and storage.

Nemichthyidae, *snipe eels:* a family (3 genera, 9 species) of extremely long, bizarre, deepsea fishes of the Atlantic, Indian and Pacific Oceans. They have have large eyes and extremely long, nonocclusible jaws, the upper jaw longer than the lower. Pectoral fins are present; the dorsal, anal and caudal fins form a continuous, confluent structure. On the attainment of sexual maturity, males undergo such a marked transformation that they were once thought to be a separate species; their jaws become much shorter, and they lose their teeth.

Nemorhaedini: see *Caprinae.*

Nemorhaedus: see Goral.

Neoblast: see Regeneration.

Neocentromeres: in addition to the normal centromere, regions near to or at the end of the chromosome. Contact of spindle fibers with the N. causes the ends of the chromosome to advance ahead of the normal centromere during nuclear division.

Neo-Darwinism:

1) A term for the evolutionary concepts of August Weismann (1834–1914).

2) See Synthetic theory of evolution.

Neodrepanis: see *Philepittidae.*

Neofelis: see *Felidae.*

Neogastropoda: see *Gastropoda.*

Neogea, *neogaea, dendrogaea:* a zoological area originally comprising both the Nearctic and the Neotropical regions, but now applied only to the Neotropical region (see). See Animal geographical regions.

Neo-Lamarckism: a discredited theory of evolution, based on the views of Lamarck (1744–1829). N. opposes partly or completely the theory of natural selection, and claims that evolution is the result of the inheritance of acquired characters or of Psycho-Lamarckism (see). N. was formerly particularly favored by paleontologists. Prominent proponents of N. were E.D. Cope (1840–1897) and O. Abel (1875–1946).

Neomeniida: see *Aplacophora.*

Neomorph: a mutant allele which exerts a qualitatively different phenotypic effect from that of the normal allele. See Amorph, Antimorph, Hypermorph.

Neomorphinae: see *Cuculiformes.*

Neomycins: collective term for a group of antibiotics synthesized by *Streptomyces fradiae.* They inhibit the development of certain aerobic Gram-positive and Gram-negative bacteria and streptomycetes. Owing to undesirable side-effects, they have limited clinical use.

Neoophora: see *Turbellaria.*

Neopallium: see Brain.

Neophema: see *Psittaciformes.*

Neophoca: see *Otariidae.*

Neophocaena: see *Whales.*

Neophron: see *Falconiformes.*

Neophytic: see Cenophytic.

Neopilina: see *Monoplacophora.*

Neorhabdocoela: see *Turbellaria.*

Neositta: see *Sittidae.*

Neoteny: shortening of the ontogenic development of an animal by the development of sexually mature larval or infantile stages. Some species of the *Urodela* display permanent N., i.e. they never metamorphose to the terrestrial form (see *Proteidae).* Other species normally reproduce neotenically, but will complete their metamorphosis under the influence of appropriate environmental factors (see Axolotl). Some species display N. only under exceptional circumstances, e.g. the smooth newt *(Triturus vulgaris)* and the Alpine newt *(Triturus alpestris)* (see *Salamandridae).*

Neotraginae, *dwarf antelopes:* a subfamily of the *Bovidae* (see). The *N.* are divided into 5 tribes.

1. Tribe: *Neotragini* ***(Dwarf antelopes);*** genera: *Neotragus* (1 species: *N. pygmaeus,* ***Royal antelope)*** and *Nesotragus* (2 species: *N. bateri,* ***Bate's dwarf antelope;*** *N. moschatus,* ***Suni).***

2. Tribe: *Madoquini* ***(Dik-diks);*** genera: *Madoqua* (3 species: *M. saltiana,* ***Salt's dik-dik;*** *M. phillipsi,* ***Redbellied dik-dik;*** *M. swaynei,* ***Swayne's dik-dik)*** and *Rhynchotragus* (2 species: *R. guentheri,* ***Guenther's dik-dik;*** *R. kirki,* ***Kirk's dik-dik).***

3. Tribe: *Dorcatragini* ***(Beira antelopes);*** genus: *Dorcatragus* (1 species: *D. megalotis,* ***Beira antelope).***

4. Tribe: *Oreotragini;* genus: *Oreotragus* (1 species: *O. oreotragus,* ***Klipspringer).***

5. Tribe: *Raphicerini* ***(Steinbocks);*** genera: *Nototragus* (1 species: *N. melanotis,* ***Grysbok),*** *Raphicerus* (1 species; *R. campestris,* ***Steinbock, Steenbock)*** and *Ourebia* (1 species: *O. ourebi,* ***Oribi).***

The 5 species of ***Dik-diks*** (head plus body 52–67 cm) have a soft pelage, which is yellow-gray to red-brown above, and gray to white below. Only the males have horns, which are ringed, stout at the base and somewhat longitudinally grooved. Dik-diks have a characteristically elongated snout, which is more pronounced in *Rhynchotragus.* They are distributed from northeastern to southeastern Sudan, through Ethiopia and Somalia, into Uganda, Kenya, Tanzania, Angola and Namibia.

The ***Klipspringer*** is a small, very agile climbing antelope, inhabiting many rocky and mountainous areas of sub-Saharan Africa. Horns are usually present only on the male. Its habitat in northern Nigeria is possibly disjunct with the rest of its range in the whole of eastern and southern sub-Saharan Africa, excluding the western forest regions.

Neotragini: see *Neotraginae.*

Neotragus: see *Neotraginae.*

Neotrehalose: see Trehalose.

Neotropical kingdom, ***Neotropis:*** the floristic kingdom which embraces the Tropical, Subtropical and Australian floristic zones of the New World (southwest North America, Central America and most of South America). Some characteristic families are *Cactaceae, Bromeliaceae, Cannaceae* and *Tropaeolaceae.* The main vegetation types are tropical rainforest, dry tropical forest and savanna.

Neotropical region, ***neotropical realm, Neotropis:*** a zoogeographical region which comprises South America, Central America, the southernmost part of North America, the Caribbean and the Galapagos Islands (see map under Zoogeographical regions). The northern limit of the N.r. merges into a transitional zone. With the exception of Chile, the mainland fauna of the N.r. is very species-rich.

Some animals of the N.r., especially invertebrates, are derived from the original fauna of Gondwanaland. The N.r. was originally linked to Australia and New Zealand via the Antarctic; the connexion with Africa was broken much earlier. Repeatedly during geological history, land connexions negotiable by terrestrial animals have existed between the Nearctic and the N.r. The fossil mammals of the Cretaceous reveal two migratory waves, one in the Eocene and the other much later. The latter occurred via the Central American land bridge at the end of the Pliocene about 3–4 million years ago. This was exploited by mammals mainly for north to south migration, while it enabled bird movements mostly from south to north.

Only two endemic marsupial families are found in the N.r., and the population numbers of these relatively small, rat- or mouse-like marsupials are relatively small. Insectivores are almost completely absent. Bat species (all are *Microchiroptera)* are numerous, including many fruit- and flower-feeding species, and a few species that feed on blood; the larger bats or flying foxes *(Megachiroptera)* are absent. New World monkeys are exclusively arboreal (often with prehensile tails), and certain *Callithricidae (= Callitrichinae)* (marmosets and tamarins) resemble squirrels. Notwithstanding their slight encroachment into the Nearctic, the endemic *Edentata* or Xenarthrans (tree sloths, *Bradypodidae*; anteaters, *Myrmecophagidae*; armadillos, *Dasypodidae)* are characteristic of the N.r. Rodent species are especially numerous, e.g. *Cavioidae* (true guinea pigs, *Cavia*; Patagonian cavies, maras or pampas "hares", *Dolichotis patagona*; mocos or rock cavies, *Kerodon rupestris), Erethizontidae* (New World porcupines) and *Hydrochoeridae* (capybaras). Carnivores (e.g. maned wolf, jaguar, puma, spectacled bear, otter) are much less specific to the N.r.; viverrids and hyenas are absent. There are no Neotropical horses, cattle, sheep, goats, antelopes, giraffes, rhinoceroses or indigenous pigs, the only representatives of the ungulates being tapirs, llamas and some deer species.

With some 2 500 species of birds, the N.r. has the richest avian fauna of any region. About 90% of the species and about half the families are endemic. In addition, widely distributed families of other regions are also well represented in the N.r., e.g. parrots. Some families found only in the N.r. are: tinamous *(Tinamidae),* rheas *(Rheidae),* curassows *(Cracidae),* screamers *(Anhimidae),* toucans *(Ramphastidae),* ovenbirds *(Furnariidae)* and cotingas *(Cotingidae).* Some families that also occur in the Nearctic are also well represented in the N.r., e.g. humming birds *(Trochilidae),* flycatchers *(Tyrranidae)* and tanagers *(Thraupinae).*

There are only a few species of tortoises and turtles, whereas the crocodilians are represented by an above average number of 9 species. Dominant lizards are the whiptails and race runners *(Teiidae)* and the iguanas *(Iguanidae),* which are also present in the Nearctic. The snakes are also species-rich, including about 20 constrictors *(Boinae),* numerous harmless snakes *(Colubridae),* front-fanged snakes *(Elapidae)* and pit vipers *(Crotalinae).* Notable among the amphibians are about 40 caecilians *(Gymnphonia).* Salamanders extend southwards only as far as the Amazon basin. A great variety of frogs is found in the N.r., in particular poison arrow frogs *(Dendrobatidae),* leptodactylid frogs *(Leptodactylidae)* and true tree frogs *(Hylidae).* The primitive clawed and surinam toads *(Pipidae)* are present on both sides of the Atlantic.

With 2 400 to 2 700 fish species, the N.r. contains more different fish than any other region; some 28 families are endemic. The ancient order of lung fish *(Dipteriformes),* which is represented in the Ethiopian region (family *Protopteridae)* and the Australian region (family *Ceratodontidae),* is represented by a single species in the N.r. (South American lung fish, *Lepidosiren paradoxa*, family *Lepidosirenidae).* Dominant fish are catfish *(Siluriformes)* and characins *(Characoidae,* which are otherwise only found in Africa). There are also numerous species of tooth carps *(Cyprinodontiformes)* and cichlids *(Cichlidae,* which are otherwise only found in Africa). Although the Andes have presented an almost impassable barrier to dispersal since the Lower Tertiary, the areas on both sides possess many common species.

The many orders of insects of the N.r. are species-rich, and often represented by large, magnificently colored forms. The large number of ant species includes leaf cutters and driver ants. Neotropical bird spiders are probably the largest spiders in the world. Finally, the number of different snails and slugs is greater than in any other region. The earlier trans-Antarctic continuity of the N.r. with the Australian region (see) and New

Zealand is evidenced by the existence of marsupials, numerous related insects and crustaceans, the *Onychophora*, and certain mollusk groups. Many animals are threatened with extinction by the clearing of forests in the N.r.

Neotropical toad: see *Bufonidae.*

Neottia: see *Orchidaceae.*

Neozoic: see Cenozoic.

Nepenthaceae: see Carnivorous plants.

Nepenthes: see Carnivorous plants.

Nephelodes: see Cutworms.

Nephridia, ***metanephridia, segmental organs:*** paired, segmentally arranged excretory organs in annelids, entoprocts, horseshoe worms, mollusks and in modified form in crustaceans. They arise ontogenetically from the ectoderm. N. begin in the coelom with a slit-shaped, open, ciliated funnel *(nephrostome)* which continues as a tubule, lined internally with cilia and consisting of a monolayer of epithelial cells. The tubule penetrates the dividing wall (dissepiment) between adjacent segments, passes through the coelom of the next segment, then empties to the exterior via the *nephropore (uropore).* In addition to N., many annelids possess other excretory ciliated tubules *(coelomoducts),* which open into the coelom via expanded rather than slit-shaped funnels, and are derived ontogenetically from the mesoderm. In many polychaete annelids, N. and coelomoducts are associated in a variety of ways. If coelomoducts and N. are not completely united and are well differentiated from each other, the condition is called *nephromixy.* If they are totally united, they are called *mixonephridia.* In addition to excretion, the funnel and tubule also serve for the conduct of eggs and sperm to the exterior. The most primitive N. are closed Protonephridia (see).

Nephrocyte, ***pericardial cell:*** a large, rounded, bi- or multinucleate phagocytic cell found in insects and mollusks. Ns. take up and store excretory products.

Nephromixy: see Nephridia.

Nephron: see Excretory organs.

Nephropidae: see Lobsters.

Nephrops: see Lobsters.

Nephrostome: see Excretory organs.

Nepoviruses: see Virus groups.

Nepticulidae (Nepticulids): see *Lepidoptera.*

Nereocystis: see *Phaeophyceae.*

Nereum: see *Apocyanaceae.*

Neritic plankton: see Plankton.

Nernst equation: a simplified case of the Goldman equation (see), in which only one membrane-permeable ion is present.

$E = RT/zF \times ln\ (c_o/c_i)$, where z is the valency of the permeant ion ($z < 0$ for anions), and c_o and c_i are the bulk concentrations of the ion in the extracellular (o) and intracellular (i) solutions.

Nernst-Planck equation: see Electrochemical potential.

Nerol, ***2,6-dimethylocta-2,6-dien-8-ol:*** a doubly unsaturated, acyclic, monoterpene alcohol, present in nerolis oil and bergamot oil. The double bond at position 2 has a *cis* configuration (Fig.). N. can be prepared by *cis-trans* isomerism of its stereoisomer, geraniol, by heating in an alcoholic solution of sodium ethoxide. N. is the most valuable of the acyclic monoterpene perfume constituents.

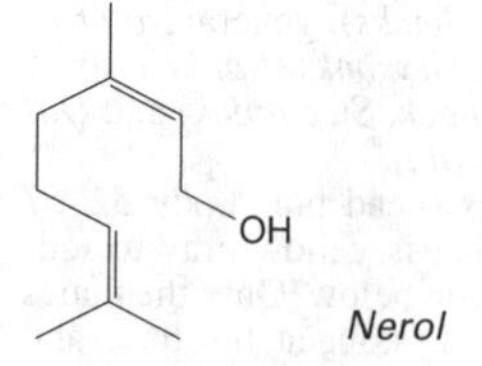

Nerol

Nerophis: see *Syngnathoidei.*

Nerve: see Nerves.

Nerve degeneration: see Degeneration.

Nerve fiber: a long process of a nerve cell, surrounded by an envelope, and responsible for the propagation of nervous impulses. In humans, N.fs. can attain lengths up to 1 m (from the spinal cord to the foot). The central bundle of fibrils (the *axis cylinder, axon,* or *neurite)* is surrounded by a sheath. In *white* or *medullated fibers,* this sheath consists of an outer nucleated *neurilemma* (more correctly *neurolemma)* and an inner lipid-containing *myelin sheath.* In peripheral nerves, the myelin sheath is formed by specialized *glial* or *neuroglial cells,* known as *Schwann cells,* whereas in the central nervous system it is formed by *oligodendrocytes.* During embryonic development, these cells wrap layers of their own plasma membrane in a tight spiral around each axon. A myelinating Schwann cell associates with a single axon, forming a segment of sheath about 1 mm in length, with an apparent lamellar structure of up to 300 apparently concentric layers (see Neuron: Fig.4). Careful microsopic inspection, however, reveals that the layers form a tight spiral rather than concentric layers. Oligodendrocytes function similarly, but are associated with a few axons simultaneously. At regular intervals of about 1 mm, the myelin sheath in interrupted by circular constrictions known as *nodes of Ranvier,* where the axon is covered only by the neurilemma. In the white matter of the central nervous system (spinal cord and brain), the lamellar structure of the myelin sheath does not arise by a wrapping process, but by secretion of membranes by specialized glial cells. For-

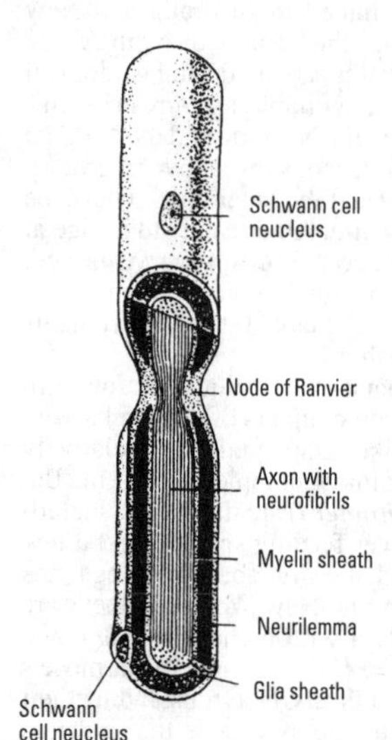

Nerve fiber. Structure of a myelinated nerve fiber (as seen under the light microscope).

merly, a distinction was drawn between white fibers and gray or nonmedullated fibers. It has now been shown, however, that nonmedullated fibers possess a thin myelin sheath, which can be visualized by polarization microscopy and electron microscopy, although not under the ordinary light microscope. In addition to these myelin-poor N.fs., the autonomic nervous system also contains true myelin-free N.fs., in which axon bundles are embedded in the cytoplasm of Schwann cells, i.e. they are not enclosed by a sheath derived from Schwann cell membrane. Naked N.fs., in which the axon lacks any kind of envelope, are rare. N.fs. can be classified according to their functional, physical, or chemical properties.

1) Functionally, N.fs. can be classified according to a) the direction in which they propagate nervous impulses, and b) whether they are part of the *autonomic* (vegetative or visceral), or *somatic* (cerebrospinal) nervous system. *Afferent* or *sensory* N.fs. conduct nervous impulses from the periphery to the central nervous system, whereas *efferent* N.fs. carry impulses in the opposite direction, i.e. from the central nervous system to a target organ. *Motor* N.fs. are efferent, and they belong to the somatic nervous system; they carry impulses from the central nervous system to skeletal muscle. *Autonomic efferent* N.fs. innervate and regulate the activity of internal organs, such as glands, smooth muscle, blood vessels, the gastrointestinal tract, heart, etc.

2) Physically, N.fs. differ according to the rate of propagation of impulses, the size and duration of different phases of the action potential, the duration of the refractory period, and the effect of oxygen deficiency. Vertebrate N.fs. are divided into 3 categories or groups, designated A, B and C, each with further subgroups. The highest rates of impulse propagation are found in the motor N.fs. that innervate skeletal muscle, and these are placed in group A. Group B is occupied almost exclusively by N.fs. of the autonomic nervous system. N.fs. in group C are characterized by very low rates of impulse propagation, and they consist largely of the nonmedullated postganglionic fibers of the sympathetic and parasympathetic systems (see Autonomic nervous system).

3) N.fs. can also be classified according to the chemical nature of the Neurotransmitters (see) at the nerve synapses (see Synapse). Thus, *cholinergic* N.fs. (neurotransmitter: acetylcholine) include motor N.fs., all the preganglionic N.fs. of the autonomic nervous system, almost all the postganglionic N.fs. of the parasympathetic system, a few postganglionic N.fs. of the sympathetic sytem (e.g. the fibers that innervate sweat glands), and possibly sensory N.fs. *Adrenergic* N.fs. (neurotransmitter: noradrenalin and to a lesser extent adrenalin) include most of the postganglionic fibers of the sympathetic system. "*Histaminergic*" N.fs. are responsible for dilating blood vessels in the skin; the neurotransmitter is thought to be histamine or a related substance. Many peptides probably also function synaptically as neurotransmitters (e.g. substance P, enkephalin, etc.), and the corresponding N.fs. may be termed *peptidergic*. There is now increasing evidence that bioactive peptides also play a part in impulse propagation in cholinergic and adrenergic N.fs. The chemical status of sensory N.fs. is uncertain, and some authorities consider that substance P (a polypeptide extracted from the dorsal roots of the spinal cord) is the actual neurotransmitter, rather than acetylcholine. The central nervous system contains additional neurotransmitters, which may be excitatory (e.g. glutamic acid), or inhibitory (γ-amino butyric acid, glycine), or able to show both effects (e.g. dopamine, serotonin). See also Nitric oxide.

The model for the spread and conduction of an action potential in a N.f. or muscle cell is based on the Cable theory (see).

For more useful information relevant to this entry, see Neuron.

Nerve growth factor, *NFG:* a highly (noncovalently) associated dimer of identical polypeptides, each of M_r 13 259, which stimulates division and differentiation of sympathetic and embryonic sensory neurons in vertebrates. The mitogenic effect of NGF is restricted to early embryonic life; it has no effect after 9 days post partum in mice. A sensitive assay for NGF measures the stimulation of the outgrowths of neurites from cultured ganglia.

NGF occurs in high concentration in adult mouse submaxillary gland, and in some snake venoms, but it is unlikely that these high local concentrations play any part in the stimulation of nerve growth. The first reported source of NGF was mouse sarcoma. In cannot be detected in innervated tissues, but denervation leads to the appearance of measurable NGF in the target tissue; when nerve regeneration is complete, NGF again becomes undetectable. It is therefore thought to carry information from innervated end organs or tissues to the cell bodies of neurons by retrograde axonal transport.

Nerve impulse frequency, ***rate of firing of nerves:*** the frequency with which one action potential follows another in a nerve fiber. Although the Action potentials (see) in an excited nerve fiber have a constant amplitude in accordance with the All-or-nothing-rule (see), their frequency is variable. The number of impulses per second generated and propagated by a nerve fiber depends on the type of nerve fiber, the species of animal possessing the nerve fiber, and the *input.* The length of the Refractory period (see) is determined by the first two factors. Nerve fibers possess a spontaneous impulse frequency, i.e. even in the absence of an input they produce a small number of impulses per second. This frequency increases with the input, which is equivalent to the *grand postsynaptic potential,* i.e. the summation of all the stimuli received. If excitatory inputs predominate, the result is depolarization, whereas a predominance of inhibitory inputs usually leads to hyperpolarization. The temporal summation of incoming signals (effects of signals received at different times) and spatial summation of incoming signals (effects of signals received at different sites on the membrane) jointly determine the input received by a single postsynaptic cell. The postsynaptic cell then generates action potentials and relays them to other cells. Since the amplitude of each action potential is the same, the magnitude of the grand postsynaptic potential can only be encoded by the frequency of firing of action potentials. This encoding is achieved by a special set of voltage-gated ion channels at the base of the axon, adjacent to the cell body, in a region known as the *axon hillock.* Maximal frequencies of up to 1 000 impulses per second have been recorded. At nerve fiber terminals, the frequency of the action potentials is transmitted as the frequency of quantized release of neurotransmitter molecules.

Nerves:

1) Leaf veins or ribs of plants.

2) In animals, the whitish colored strands of the nervous system. They consist of varying numbers of nerve fibers inside a layered envelope of connective tissue, known as the perineurium. Separate bundles of nerve fibers, each within its own perineurium, lie alongside one another and are enclosed by the epineurium which covers the surface of the nerve and extends between the fiber bundles. The epineurium contains fat cells, as well as blood and lymph cells.

Nervous system: a coordination system consisting of nerve sells or sensitive cells of certain organs, plus specific support cells (neuroglia and connective tissue cells). It serves to conduct stimuli from receptor organelles or organs (see Sensory organs) of the body interior or the external environment to target organs (e.g. muscles and glands), thereby evoking an appropriate body response. Invertebrates display a wide variety of N.ss., which differ in their structure and in their location and distribution within the organism. Although the *Protozoa* do not possess a N.s. in the usual sense, the entire body (cell) surface is usually capable of receiving stimuli. In addition, sensory organelles may also be present (e.g. sensory cilia, flagella and eyespots), and the protoplasm may be differentiated to form an *excitomotor (neuromotor) apparatus.* Even in ameboid forms, the ectoplasm is more sensitive than the rest of the protoplasm, and is capable of propagating an excitation wave. *Paramecium* possesses a primitive coordination center and conducting system, consisting of a central neuromotor mass connected by fibrils to the base of each cilium. An even higher level of development is found in *Diplodinium ecaudatum,* in which the strands of a neuromuscular network form connections between a central neuromuscular mass and retractor strands, which in turn are associated with adoral and oral cilia and membranelles.

The *Porifera* do not possess a N.s. Stimuli are passed directly from cell to cell, and each cell functions as both receptor and effector. In some sponges, however, the osculum is surrounded by a ring of primitive muscular tissue, suggesting an evolutionarily early neuromuscular mechanism.

The simplest and evolutionarily most ancient true N.s. is found in the *Coelenterata,* where it consists of two uniform networks of true nerve cells, one in the endoderm and the other in the ectoderm. Both *nerve nets* extend over the entire body, merging into one another in the oral region. It is a *diffuse N.s.,* i.e. stimuli are conducted equally in all directions from any point, each receptor influencing the entire body. In addition to this nerve net, epithelial cells are differentiated into neurosensory elements, so that the cell surface functions as a receptor, while nerve fibers are prolonged from the cell base to make contact with underlying muscle cells.

In corals, the ecto- and endodermal nerve nets are united not only in the oral region, but also throughout the mesoglea. In medusas, the nerve net becomes concentrated to form a nerve ring in the rim of the bell; nerve strands radiate from this nerve ring to the sensory organs, often with a ganglion at the base of each sensory organ.

In the *Turbellaria* and practically all subsequent animal groups, the N.s. is of purely ectodermal origin. In the primitive condition, the turbellarian N.s. probably consists of a subepithelial nerve net with three to four pairs of longitudinal nerve cords, interconnected along their length by commissures, and an anterior brain (cerebral ganglion) containing a statocyst. More advanced turbellarians have lost the statocyst, and they display increased centralization of the N.s., with a reduction in the number of pairs of nerve cords, and increasing prominence of the ventral pair. This may represent the evolutionary origin of the (embryonically paired) ventral nerve cord of the annelid-arthropod line.

As centralization of the N.s. progresses, the connectives and commissures decrease in length and the *dermal nerve net* (the *orthogon)* sinks into the body interior. At the same time, however, a morphological and therefore physiological connection is established between the brain and the dermal nerve net via connectives that extend from the brain *(endon)* (1st integration phase of the central nervous system). The N.s. of the *Trematoda* consists of a brain (a pair of anterior cerebral ganglia), which gives rise to three paired, posteriorly extending, longitudinal nerve cords. In contrast, in most members of the *Cestoda,* the brain (an anterior nerve mass in the scolex) gives rise to only two lateral longitudinal nerve cords, which travel the entire length of the strobila. Some cestodes, however, also possess a dorsal and a ventral pair of nerve cords, often with lateral nerves; the longitudinal nerve cords in each proglottid are connected by ring commissures.

The N.s. of the *Nemertini* is similar to that of the more advanced platyhelminths, consisting of a brain (4-lobed) with several radiating branches. Dorsal and lateral ganglia of the brain are connected by commissures, forming a ring around the rhynchocoel. A major pair of longitudinal, ganglionated nerve cords extends posteriorly in the body wall, but other, minor longitudinal nerves are also often present, e.g. a mid-dorsal nerve (extending from the dorsal commissure) and a pair of esophageal nerves (arising from the ventral ganglia). In the primitive condition, these longitudinal nerves are located in the epidermis or dermis, but in most species they are found in deeper layers within muscle tissue.

In the various classes of the *Aschelminthes,* the brain (usually a nerve ring around the pharynx with associated ganglia) gives rise to an unpaired, ganglionated dorsal nerve cord and an unpaired, ganglionated ventral nerve cord, the two cords being linked by ring commissures. In the *Nematoda,* the large ventral cord arises as a double cord from the ventral side of the nerve ring, but these two cords subsequently fuse to form a single chain of ganglia. In the *Kinorhyncha,* a single midventral ganglionated nerve cord is present, but dorsal and lateral clusters of ganglion cells in each segment are often not interconnected to form a distinct nerve cord. Free-living aschelminths generally possess 5–6 longitudinal nerve cords.

Among the *Mollusca,* the simplest N.s. is found in the *Placophora,* where pedal and pleurovisceral nerve cords (interconnected by numerous commissures) extend from a weakly developed brain. In addition to the central nervous system, all mollusks have a peripheral dermal nerve network, which is composed of many different types of nerve cells, and is capable of independent reflexes. In most of the advanced mollusks, all the cell bodies of the central nervous system are aggregated into ganglia, and the interconnecting nerve cords consist of parallel nerve fibers. The most important ganglia of mollusks are the cerebral (or supraesophageal), pleural, pedal, buccal, visceral and parietal ganglia. Prosobranchs and opisthobranchs characteristically display chiastoneury (see *Gastropoda* for explanation of chiastoneury) and possess long connectives. With evolutionary advancement, the connectives of the opisthobranchs and pulmonates become shorter and chiastoneury is abolished by the reversal of torsion. Bivalves do not display chiastoneury and they lack a buccal ganglion.

In the *Cephalopoda,* the connectives aand commissures are much reduced. The cerebral, pedal and visceral ganglia are combined into a single mass, and additional ganglia (brachial, labial and stellate ganglia), which are absent from other mollusks, are formed from the peripheral nerve network.

The N.s. of the *Annelida* consists of a brain (cerebral ganglia or suprapharyngeal ganglia), a subpharyngeal ganglion, and a pair of ventral nerve cords. In addition, many polychaetes, in particular those with an eversible pharynx, have a system of *stomatogastral nerves* involved in the control of the proboscis or pharynx. In most annelids, the two ventral cords are combined as a single ventral gangion chain, consisting of two fused ganglia in each segment linked by two longitudinal connectives; in most cases these two connectives are fused and invested in a common connective tissue sheath, so that the dual nature of the ventral nerve cord is not immediately apparent. The presence of two separate ventral cords (e.g. in the *Hirudinea)* represents the primitive condition. Annelids and arthropods do not possess a diffuse dermal nerve net.

In the *Arthropoda* and *Onychophora,* the central nervous system consists of the brain, the pharyngeal connectives and a ventral ladder of two ganglionated cords linked transversely by commissures. All arthropods display an evolutionary tendency toward the concentration and fusion of ganglia. Thus, the 1st pair of ventral ganglia are fused to form a part of the brain known as the tritocerebrum. Whereas the ventral cords of onychophorans have no ganglia, the ventral cords of arthropods are almost always furnished with ganglia and conspicuous commissures; double commissures are found only in the *Branchiopoda.* In many crustaceans, insects and all terrestrial *Chelicerata,* the ganglia and connectives fuse to form an unpaired chain. Centralization of the ventral ganglion chain can, however, also occur by shortening of the connectives, e.g. the subesophageal ganglia (mandibular and maxillary ganglia I and II) of many crustaceans, insects and other arthropods are formed in this way, as are the abdominal ganglia of all insects (in the 8th to the 11th segments of insects), all chilopods and some crustaceans. As a further development, all the thoracic ganglia, or all the abdominal ganglia (in crabs, some copepods, cirripedes, many insects, arachnids, pycnogonids) may become fused.

Lower deuterostomes *(Enteropneusta, Pterobranchia, Pogonophora)* have a dense subepithelial nerve net which extends to the mouth cavity, as well as a weakly developed unpaired nerve cord.

The primitive N.s. of the *Echinodermata,* as exhibited by starfishes, has a similar level of organization to that of the enteropneusts (ectodermal, endodermal and mesodermal nerve net). At certain sites (in the ambulacral groove, around spines and pedicellariae), the nerve net forms cords and nodes.

In the *Tunicata,* the N.s. consists of a dorsal ganglion that gives rise to five nerves: 1 pair of oral and pharyngeal nerves, 1 pair of egestion nerves, and 1 unpaired visceral nerve.

In vertebrates, the central nervous organs have migrated into the body interior, and many complex groups of neurons are associated into a central organ. Both of these features confer considerable selective advantage, i.e. improved coordination and greater integration of the activities of different parts of the nervous system and a hierarchical structure of the control and regulation systems, all of which result in an increased response rate of the organism, and an increased ability to respond appropriately to a range of different environmental stimuli. The vertebrate N.s. is divided into the *central nervous system (CNS)* which consists of the Brain (see) and Spinal cord (see), and the *peripheral N.s.* which includes the totality of nerves from the brain and spinal cord, together with their ramifications, as well as the Autonomic nervous system (see).

Nervous tissue: a highly evolved tissue, from which the nervous system is formed in coelenterates and in all animals that are evolutionarily more advanced than coelenterates. It is derived from ectodermal epithelial tissue. In vertebrates, it arises during embryonal development from the dorsal ectodermal neural plate, which develops into the neural tube. Both peripheral and central N.t. is closely associated with a support tissue called the *neuroglia,* containing relatively large numbers of *glial cells,* which surround the neurons and occupy the spaces between them (mammalian brain contains about 10 glial cells to every neuron). Glial cells known as *Schwann cells* (in vertebrate peripheral nerves) and *oligodendrocytes* [in the vertebrate central nervous system (CNS)] wrap around axons to form the electrically insulating myelin sheath. The CNS also contains *microglia,* which are functionally similar to macrophages. Whereas neurons and all other types of glial cells share a common embryonic ectodermal origin, the microglial cells are exceptional in being derived from hemopoietic tissue. In the CNS, glial cells called *ependymal cells* line the internal cavities of the brain and spinal cord; also known as *epithelial glial cells,* they recall the phylogenetic origin of the CNS from a tube of ectoderm. The most abundant CNS glial cells are the *astrocytes* (named for their possession of radiating cell processes), which are important in the maintenance of Nt. structure, as well as apparently playing a role in regulating the chemical and ionic environment of the neurons. *Radial glial cells* of the CNS appear to be cells that have persisted from the original columnar epithelium of the neural tube, and which serve as scaffolding during embryonic brain development; each cell stretches from the inner to the outer surface of the neural tube, a distance which may be as much as 2 cm in the cerebral cortex of the developing primate brain. Migrating neurons crawl along the radial glial cells. Toward the end of embryonic brain development, radial glial cells disappear from most parts of the CNS, and it is thought that many transform into astrocytes. The main cellular components of N.t. are Neurons (see). These are morphologically and functionally independent units, linked within the N.t. into chains that propagate nervous impulses. They are not connected directly with one another or with target organs, but via chemical synapses. Certain neurons (neurosecretory cells) secrete neurohormones.

Nervus abducens: the VI cranial nerve. See Cranial nerves.

Nervus accessorius: the XI cranial nerve. See Cranial nerves.

Nervus facialis: the VII cranial nerve. See Cranial nerves.

Nervus glossopharyngicus: the IX cranial nerve. See Cranial nerves.

Nervus hypoglossus: the XII cranial nerve. See Cranial nerves.

Nervus oculomotorius: the II cranial nerve. See Cranial nerves.

Nervus olfactorius: the I cranial nerve. See Cranial nerves.

Nervus opticus: the II cranial nerve. See Cranial nerves.

Nervus statoacusticus: the VIII cranial nerve. See Cranial nerves.

Nervus trigeminus: the V cranial nerve. See Cranial nerves.

Nervus trochlearis: the IV cranial nerve. See Cranial nerves.

Nervus vagus: the X cranial nerve. See Cranial nerves, Autonomic nervous system.

Nesoctites: see *Picidae.*

Nesolagus: see *Leporidae.*

Nesomimus: see *Mimidae.*

Nesomyinae, ***Malagasy rats:*** a subfamily of the *Cricetidae* (see), containing about 15 species in 7 genera, found only in Madagascar and some neighboring islands. Most members are rat-sized, with few exceptions, e.g. the ***Votsotsa*** *(Hypogeomys antimena)* has a body length of 30–35 cm. This animal lives in dense, virgin forest, where it burrows extensively and occupies a similar ecological niche to that of the rabbit in other parts of the world. ***Lamberton's Malagasy rat*** *(Nesomys lambertoni)* sometimes grows to the size of a muskrat, while a slightly smaller species *(Brachytarsomys albicauda)* is found on a neighboring island. The smallest species belong to the genus *Macrotarsomys,* members of which possesses long ears and a tail longer than the body.

Nesotragus: see *Neotraginae.*

Nest-building frogs: see *Leptodactylidae.*

Nest frog: see *Raninae.*

Nestor: see *Psittaciformes.*

Nestorinae: see *Psittaciformes.*

Net photosynthesis, ***apparent photosynthesis, primary photosynthetic product:*** the quantity of photosynthetic assimilate remaining after subtraction of that lost by respiration and especially by Photorespiration (see). If losses due to discarded plant material (leaf fall, fruit dispersal, etc.) are also subtracted, the result is the *net photosynthetic product.*

Net photosynthetic product: see Net photosynthesis.

Nettapus: see *Anseriformes.*

Nettle: see *Urticaceae.*

Nettle trees: see *Ulmaceae.*

Neural canal: see Axial skeleton.

Neural spine: see Axial skeleton.

Neuraminic acid: 5-amino-3,5-dideoxy-D-glycero-D-galactononulosonic acid, a polyhydroxyaminocarboxylic acid (Fig.), widely distributed in the animal kingdom in mucolipids, mucopolysaccharides, glycoproteins, and the oligosaccharides of milk. It readily cyclizes in vitro. The *N*- and *O*-acetyl derivatives are called sialic acids.

$$HOH_2C-CH(OH)-CH(OH)-C(OH)H-C(NH_2)H-CH(OH)-CH_2-CO-COOH$$

Neuraminic acid

Neuraminidase: a hydrolase that removes *N*-acetylneuraminic acid from glycoproteins and gangliosides. It occurs in mixoviruses, various bacteria, blood plasma, and the lysozomes of many animal tissues. The richest source is the culture filtrate of the cholera organism, *Vibrio cholerae.*

Neurine: a very toxic organic base (Fig.), formed by removal of the elements of water from choline during putrifaction. See Ptomaines.

$$\left[(H_3C)_3\overset{\oplus}{N}-CH=CH_2\right]\ OH^-$$

Neurine

Neurite: see Nerve fiber, Neuron.

Neuroactive peptides: a pharmacological term for peptides that cause a response in the nervous system or in nerve cells. See Neuropeptides.

Neuroblastoma: see Cancer cell.

Neurocranium: see Skull.

Neurocrine: an adjective applied to substances (in particular hormones), which are synthesized in neurons and released into the blood, e.g. Neurosecretory hormones (see).

Neurodynamics: a collective term for the study of the trophic components of nerve cell function. Primary areas of investigation are Degeneration (see), Regeneration (see) and Neuronal transport (see).

Neuroendocrine cells: see Brain.

Neurofibrils, ***neurofilaments:*** see Neuron.

Neuroglial cells: see Nerve fiber, Neuron, Nervous tissue.

Neurohemal gland, ***neurohemal organ:*** a neurosecretory storage organ, usually adjoining blood lacunae or blood vessels. Different forms of N.o. are found in different animal phyla. The simplest structure occurs in the trematodes, where it consists of cavities in the inner wall of the gastrointestinal tract. In the more highly evolved N.o., secretory nerve fibers form branches within an envelope of connective tissue, and secretory material is stored in the more or less expanded nerve endings. This type includes the Sinus gland (see) and the pars-distalis-X-organ (see X-organ) of crustaceans, as well as the N.o. of *Nemertini, Polychaeta, Oligochaeta, Hirudinea, Gastropoda, Cephalopoda* and some other groups. Insects possess highly differentiated N.os. (see Corpora allata, Corpora cardiaca).

Neurohormones: see Neurosecretory hormones.

Neurohypophysis: see Hypophysis.

Neuromast: see Tactile stimulus reception.

Neuromuscular endplate: see Neuromuscular synapse.

Neuromuscular junction: see Neuromuscular synapse.

Neuromuscular synapse, ***neuromuscular junction, neuromuscular endplate, motor endplate:*** a chemical Synapse (see), in which the presynapse is the ending of a motor nerve fiber, the postsynapse is the sarcolemma of a striated muscle fiber, the neurotransmitter is Acetylcholine (see), and the synaptic cleft is about 20 nm wide. The presynaptic vesicles have a diameter of 30–50 nm, and are limited by a single membrane. In the active zone of the presynapse, large numbers of presynaptic vesicles can be seen attached to the presynaptic membrane in ordered bands. Directly opposite this active zone, the postsynaptic sarcolemma displays synaptic folds, whose outer borders carry large numbers of receptors (appearing as small particles under the electron microscope). Since a motor nerve and its muscle can be dissected free from surrounding tissue and maintained in a medium of controlled composition, the N.s. has been the subject of extensive investigations, and has provided much information about synapses in general. Muscle relaxants act by blocking the function of the N.s. The electric organs of certain teleost fishes (e.g. electric eel, *Electrophorus electricus)* and elasmobranch fishes (e.g. electric ray, *Torpedo nobiliana)* have evolved from N.ss.

Neuron (plates 21 & 25): the highly differentiated animal nerve cell, including all its extensions and processes, which is specialized for the reception, propagation and transfer of informational (signal) stimuli. Even protozoa are able to perceive and respond to stimuli, and they generate and propagate excitations. Metazoans have evolved special cells for these functions, which are known as Ns., and which arise from the ectoderm during embryonal development. Recognition of these relatively large cells as structurally and functionally independent units of the nervous system is due largely to the studies of the Spanish neurohistologist, R. Cajal (1852–1934), the most renowned proponent and founder of the neuronal theory. The cell processes of the N. make contact with other N. and with muscle, glandular and sensory cells, at specifically differentiated zones or sites known as *synapses.* The synapse contains an electron-microscopically visible gap of 20–30 nm, known as the *synaptic cleft.* Most vertebrate Ns. consist of the actual cell body *(perikaryon* or *soma)* and its associated processes (Fig.1). Cell processes develop during embryonal development of the organism; they are mainly of two types: *dendrites,* which display treelike branching, and a single rather long and usually relatively thin *axon (neurite).* The axon may form lateral branches, known as collaterals, which form lateral contacts with the processes of other Ns. A *nerve fiber* is an axon with its associated *glial (neuroglial) cells.* At a greater or lesser distance from the soma, the axon divides into a number of short branches with knoblike thickened termini. Most vertebrate Ns. are multipolar, possessing several dendrites and a single axon. Unipolar Ns. possess only a single axon and no dendrites, while bipolar Ns. have one axon and one dendrite. Most invertebrate Ns. are unipolar, with no dendrites; in this case, the single, branched, treelike axon serves the functions of both axon and dendrites. In advanced metazoans, Ns. are always associated with glial (neuroglial) cells. Different types of glial cells (e.g. Schwann cells, oligodendrocytes, astrocytes) provide an envelope for the N. and its processes, and they are presumably important in the metabolism of the N., and in the regulation of the ion concentration of the extracellular space.

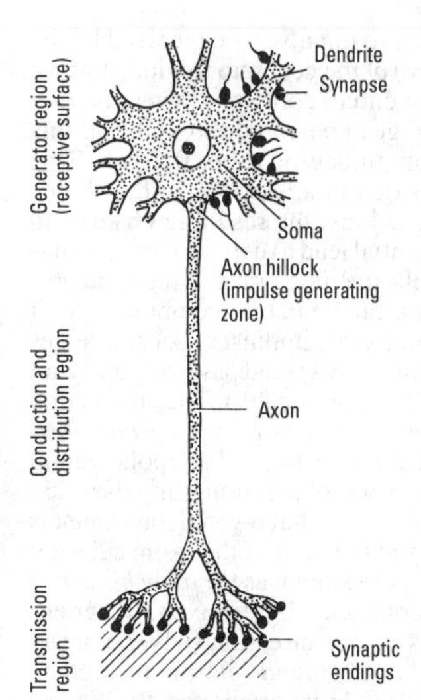

Fig.1. *Neuron.* The different functional parts of a neuron.

Function and fine structure. Each N. collects numerous inputs from other Ns. or from sensory cells. It integrates these inputs into a pattern of impulses, which are conducted along the axon and transferred to many other Ns. or to target organs. This *convergence-divergence* principle is the basis for the formation of neuronal networks. Ns. of higher vertebrates, including humans, are so highly differentiated that they are permanently in the G_0-phase of the Cell cycle (see), and have lost their ability to undergo cell division. However, they remain viable for several decades.

Ns. are functionally bipolar: stimuli are received at the *receptor* pole, and excitations are transferred to neighboring Ns. at the *effector* pole. This polarity is also expressed morphologically, in particular in bipolar, pseudo-unipolar and multipolar Ns. (Fig.2). Dendrites receive excitatory information (stimuli) via synapses from other Ns. or from sensory cells. In vertebrate Ns. this receptive surface of the dendrites, together with the receptive surface of a greater or larger part of the soma, forms the *generator region* of the N. (Fig.1). In vertebrates, including humans, the total receptive surface of dendrites and soma is more or less fully occupied by hundreds to thousands of synapses. In invertebrates, synapses are absent from the surface of the soma, and they are present only on branched processes of the N.

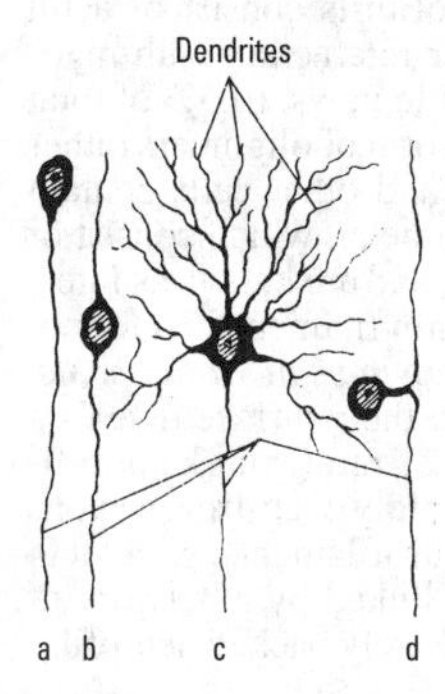

Fig.2. *Neuron.* Types of nerve cells. *a* Unipolar. *b* Bipolar. *c* Multipolar. *d* Pseudounipolar.

In vertebrate Ns., nerve impulses are received by the Plasma membrane (see) of the generator region. A nerve impulse (see Action potential) consists of a very brief (a few milliseconds) change of potential across the plasma membrane, with an amplitude of about 0.1 volt. This change of potential is due to alterations in membrane permeability to specific ions; the resulting changes in membrane resting potential lead to depolarization or hyperpolarization. The plasma membrane of the generator region is, however, not an excitable membrane, i.e. it cannot generate and propagate impulses. Local changes of potential can therefore only spread passively and with decreasing intensity (electrotonically). The summation of all the stimuli received is known as the *grand postsynaptic potential* of the cell body. If depolarization predomionates over hyperpolarization, impulses are generated by a special set of voltage-gated ion channels present in high density at the base of the axon, adjacent to the cell body, in a region known as the *axon hillock*.

Impulses are then conducted by the axon. In vertebtate Ns. conduction is in one direction only, i.e. away from the cell body. The *axolemma* (plasma membrane of the axon) is conductile. In invertebrates, the plasma membrane of the branched processes of the N. is also conductile, and impulses are conducted in both directions, i.e. toward and away from the soma. Nevertheless, it is usually not possible for an impulse to be transferred from a thinner to a thicker branch.

The soma and dendrites contain rough Endoplasmic reticulum (see) (see also plate 21), which supports intense protein synthesis. These regions of rough endoplasmic reticulum stain strongly with basic dyes, appearing as granules in stained sections under the light microscope, and known as *Nissl's granules* after their discoverer. Rough endoplasmic reticulum is absent from the axon and from the axon hillock. The nucleus of the soma contains a relatively large nucleolus, and is surrounded by several dictyosomes of the Golgi apparatus. Numerous mitochondria and lysosomes are also present in the soma. Ns. are rich in Microtubules (see), which in this situation are known as *neurotubules*. Axons and dendrites contain particularly large numbers of neurotubules, as well as various other fibrils. In growing axons, tubulin accounts for 10–20% of the total cell protein. *Neurofibrils* can be seen under the light microscope, whereas microfibrils (5–10 nm diam.) require an electron microscope. Neurotubules are probably involved in the rapid, energy-dependent transport (0.2–5.8 µm per second) of material in the axon (see Neuropeptides), whereas microfibrils are involved in the slow transport (2.3–46 nm per second) of specific axon components. Some microfibrils consist of actin (shown experimentally by their interaction with myosin). Myosin is also demonstrable in Ns. (0.5% of total protein), but it is present in the form of oligomers rather than fibrils. Furthermore, Ns. and other cells contain microfibrils of about 10 nm diameter, which consist of a keratin-like structural protein, and are known as Intermediate filaments (see) or "10 nm filaments". Intermediate filaments from Ns. are known as *neurofilaments*. Under the electron microscope, their surface reveals a structural periodicity of about 21 nm (as in keratin fibrils). Freeze etching (see) of rapidly deep-frozen axons shows bundles of individual neurofilaments, parallel to the long axis of the axon, and linked by a network of finer transverse fibers. These densely packed neurofilaments and/or neurotubuli are the "neurofibrils" observed under the light microscope. The tubulin content of Ns. is sufficiently high to be demonstrated by indirect immunofluoresence using anti-tubulin antibodies.

The surfaces of dendrites may also carry small thorn-like processes, about 0.2 mm in length and known as *spinae* or spines. Numerous axons synapse with these spines or directly with the dendrites. The terminus of the axon, which delivers stimuli to the next cell, is known as the *presynaptic cell region, terminal knob, bouton, endfoot, presynaptic terminal* or *presynapse.* In accordance with their various specialized functions, these presynapses have different anatomical forms, but most resemble round or oval knobs. The cell receiving the excitatory information is called the *postsynaptic cell.* Cell membranes in the pre- and postsynaptic regions show a characteristic differentiation, e.g. membrane thickening by deposition of material from inside the cell. In a small minority of synapses, the postsynaptic cell can be depolarized electrically through close contact of the pre- and postsynaptic membranes, assisted by other specialized structures. Such electrical synapses are found, e.g. in the ventral nerve cord of annelids. Direct electrical communication between cells is also possible via gap junctions (see Plasma membrane). In most synapses, however, the pre- and postsynaptic membranes are separated by a gap of 20–30 nm (the synaptic cleft), and stimuli are carried across this cleft by chemical neurotransmitters. The neurotransmitters (e.g. acetylcholine, noradrenalin, dopamine) of these chemical synapses (Fig.3) are synthesized in the soma and stored in small synaptic vesicles (average diameter 30 nm) in the presynapse. Depolarization of the presynaptic membrane is accompanied by release of the neurotranmitter from the vesicles. The neurotranmitter then enters the synaptic cleft and becomes bound to receptors on the postsynaptic membrane. Binding of the neurotransmitter causes a conformational change in the receptor, leading to an opening of ion channels in the postsynaptic membrane. In general, each neuron synthesizes and releases only one type of neurotransmitter molecule. Neurotransmitters are not the only substances synthesized and secreted by Ns. Practically all animals, including humans, possess Ns. that function as neurosecretory cells. For example, peptidergic Ns., which are functionally equivalent to endocrine glands, synthesize neurosecretory compounds in the soma and release them from the ends of the axon. Since these compounds enter the blood, they are by definition *Neurosecretory hormones* (see) (e.g. oxytocin and vasopressin in vertebrates). In addition to elements of the axonal cytoskeleton (neurotubules and microfibrils), the cytoplasm of the axon (axoplasm) also contains smooth endoplasmic reticulum (axoplasmic reticulum),

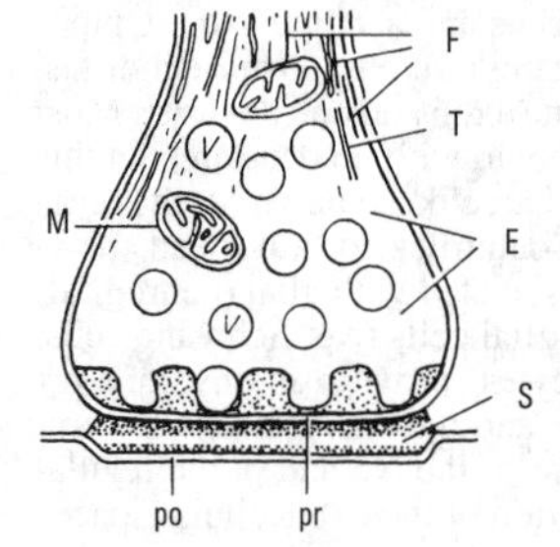

Fig.3. *Neuron.* The chemical synapse (schematic). *E* Nerve ending (presynaptic cell region). *F* Microfilaments. *M* Mitochondrion. *po* Postsynaptic membrane. *pr* Presynaptic membrane. *S* Synaptic cleft. *V* Synaptic vesicle. *T* Microtubules.

mitochondria, lysosomes, multivesicular bodies, vesicles and lipid granules.

Axons of vertebrate neurons are covered for most of their length by a myelin sheath. Myelinated nerve fibers are also found in some crustaceans. In peripheral nerves, the myelin sheath is formed by specialized *glial* or *neuroglial cells,* known as *Schwann cells,* whereas in the central nervous system it is formed by *oligodendrocytes.* During embryonic development, these cells wrap layers of their own plasma membrane in a tight spiral around each axon (Fig.4). This system of duplicated cell membranes, known as the *mesaxon,* matures into true myelin sheath by the transfer of Schwann cell cytoplasm to the exterior. A myelinating Schwann cell associates with a single axon, forming a segment of sheath about 1 mm in length, with an apparent lamellar structure of up to 300 apparently concentric layers. Careful microsopic inspection, however, reveals that the layers form a tight spiral rather than concentric layers. Several to many (a few hundred) Schwann cells may be engaged in the formation of the myelin sheath of a single axon. The myelin sheath functions primarily as an insulator, providing a high electrical resistance between the interior of the axon and the extracellular space. Oligodendrocytes function similarly to Schwann cells, but are associated with a few axons simultaneously. In unmyelinated nerve fibers the mesaxon is short, and one or more axons are embedded in a single Schwann cell. At regular intervals of 0.2–1.5 mm, the myelin sheath is interrupted by circular constrictions known as *nodes of Ranvier,* where the axon is covered only by the neurilemma. The section of nerve between two nodes is known as the *internode.* Myelinated nerve fibers conduct nerve impulses much more rapidly (up to 130 m/sec) than unmyelinated fibers (0.5–2.5 m/sec), because the excitation jumps from node to node *(saltatory conduction),* rather than continuously along the entire fiber. Saltatory conduction is made possible by the fact that ions cannot flow to any significant extent through the thick myelin sheath, whereas they can flow easily through the nodes of Ranvier, where the membrane is 500 times more permeable than the membranes of some unmyelinated fibers. Thus, electrical current flows through the surrounding extracellular fluid and through the axoplasm, forming a circuit between each node, and exciting successive nodes. The rate of impulse conduction increases with the diameter of the axon, the thickness of the myelin sheath and the length of the internode.

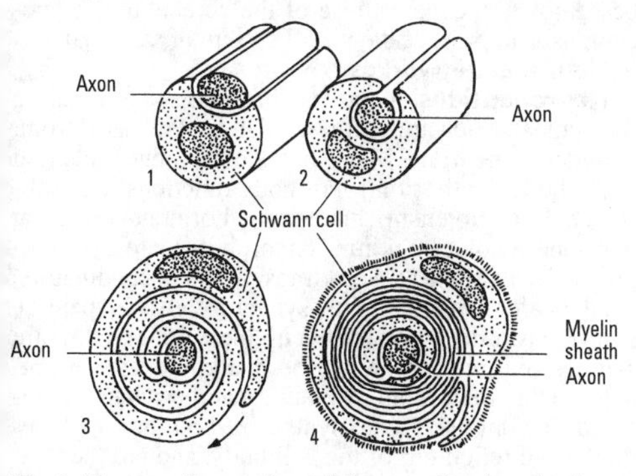

Fig.4. *Neuron.* Formation of the myelin sheath of peripheral nerve fibers (schematic).

Neuronal networks. Ns. never occur singly. They are found in neuronal networks, where the approximate number of Ns. and the pattern of their synaptic connections are largely genetically determined. These networks process and store excitatory information. Many such networks have stereotyped structures and functions, which are relatively easy to interpret. In contrast, learning processes are explained by the activation of existing synapses and the creation of new synaptic connections induced by sensory inputs. Neuronal networks appear to have acquired a certain plasticity in the course of evolution, and long-term memory in particular is thought to result from permanent functional or structural changes. Relatively simple neuronal circuits have been elucidated in some primitive invertebrates, e.g. in the nematode *Caenorhabditis elegans,* which has a total of 300 Ns. In more advanced invertebrates, and especially in vertebrates, the analysis of complete neuronal networks is extremely difficult owing to the large number of Ns. and the complexity of their interconnections. For example, the nervous system of *Drosophila* contains about 10^5 Ns., the human nervous system about 10^{12} Ns. Nevertheless, some progress has been made in the elucidation of these more complex systems.

Some Ns. are specialized to *primary sensory cells,* or to *sensory nerve cells.* The former transduce incoming stimuli to excitations and conduct the resulting nerve impulses directly to the central nervous system via their axon. The soma of sensory nerve cells is located deep in the tissue or in the central nervous system. Their processes, which serve for the reception of stimuli, branch out on the surface of the tissue (e.g. skin) into free nerve endings. *Secondary sensory cells* are not Ns., but specialized epithelial cells; they transfer excitatory information to Ns., with whose axons they form synapses.

Motor neurons are Ns., whose axons innervate striated muscle. Interneurons form connections between Ns., and usually have a short axon.

Neuronal circuit: a system of nerve cells interconnected via their synapses (see Synapsis) to form a circuit. N.cs. may be diverging or converging, and they may give rise to presynaptic inhibition, recurrent inhibition, amplification of excitation, or an inhibition of excitation (Fig.). *Convergence* occurs when a single neuron is controlled by the converging signals from two or more separate input fibers; the input signals may arise from the same source or from several different sources. Convergence is typical of many sensory cells and neurons. Thus, the human eye contains 3–6 million cones and about 125 million rods. Excitations from these sensory cells converge on about 1 million neurons of the optic nerve (see Receptive field). Convergence can also result in Summation (see).

Neurons of sensory organs also display *divergence* (stimulation of multiple output fibers by a single input fiber), but this phenomenon is particularly important in the central nervous system. For example, collateral branches from a single moss fiber in the mammalian cerebellum diverge to about 400 other nerve cells. In *amplifying* divergence an input signal spreads to an increasing number of neurons in a single tract. For example, stimulation of a single pyrimidal cell in the motor cortex transmits a single impulse to the spinal cord, but this impulse can stimulate several hundred interneurons, each of which stimulates an anterior motor neuron, which in turn stimulates between 100 and 300 mus-

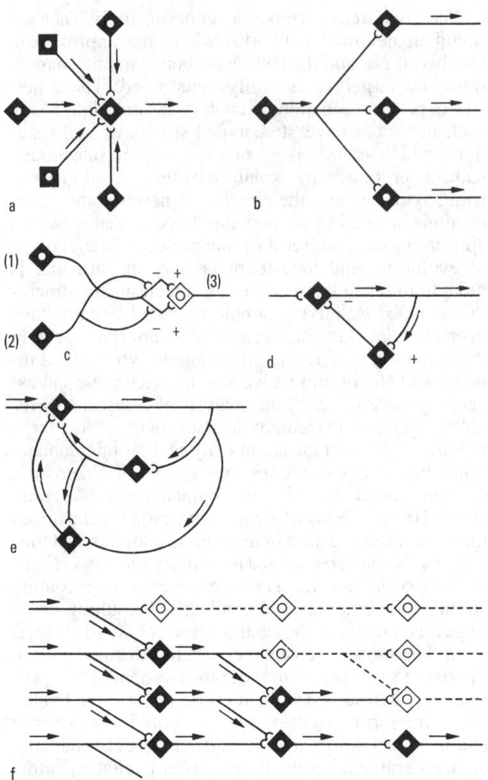

Examples of neuron circuits. a Convergence. *b* Divergence. *c* Presynaptic inhibition. *d* Recurrent inhibition. *e* Reverberating circuit. *f* Lateral inhibition. "+" signifies the generation of postsynaptic excitation, "–" the generation of postsynaptic inhibition. Excited neurons are shown black with solid lines, non-excited neurons open with dashed lines. In *c* the activity of neuron (2) prevents excitation of neuron (3).

cle fibers; the total divergence or amplification is therefore in the order of 10 000-fold. Alternatively, divergence may occur into multiple tracts, i.e. the signal is transmitted in different directions to different parts of the nervous system.

The excitation frequency (see Frequency modulation of excitation) of a particular N.c. can be considerably influenced by the nature of its Neurotransmitters (see), with consequent effects on subsequent connected circuits. *Presynaptic inhibition* occurs when a neurotransmitter is released by a presynapse, whose postsynaptic element is the presynaptic region of a different nerve cell (Fig. *c*). The presynaptic inhibitory knobs secrete a neurotransmitter that partially depolarizes the terminal fibrils and excitatory synaptic knobs, i.e. the neurotransmitter released by the first nerve cell inhibits the second nerve cell, so that the latter does not release its own neurotransmitter. Thus, the excitatory state of any subsequently connected cell or circuit remains unchanged (Fig. *c*). Presynaptic inhibition occurs in many circuits of the central nervous system.

Recurrent inhibition occurs in Motor neurons (see) of the spinal cord, when a collateral branch of a motor neuron axon stimulates a *Renshaw cell.* A Renshaw cell is an inhibitory spinal interneuron, which receives direct excitation from collateral branches of spinal motor neurons, and whose axon synapses back to many motor neurons, including the one that gave rise to its input. Neurotransmitter released by Renshaw cells generates inhibitory postsynaptic potentials. There are thought to be two types of Renshaw cell, one producing glycine as its neurotransmitter, the other producing γ-aminobutyric acid. Some N.cs. produce a continuous output, even in the absence of an input signal. This may be due to the repetitive discharge of neurons, whose membrane potentials are constantly above certain threshold levels; this *intrinsic neuronal excitability* is observed in some cells of the cerebellum and in some interneurons of the spinal cord. Alternatively, continuous signal output can result from the *reverberation* of excitatory currents within a closed loop, i.e. neural activity is maintained by excitatory feedback connections between neurons. It has been suggested that short-term learning (information storage) may be encoded by reverberating circuits, but this is disputed (see Memory). Reverberating circuits are formed not only by individual neurons (Fig. *e*), but also by bundles of interconnected neurons and fibers, some of which may or may not be connected to other circuits, e.g. Papez' circuit (see) of the limbic system of mammals.

N.cs. form the functional basis of the nervous system, and many are still unrecognized and/or unelucidated. The primitive diffuse nervous sytem of the *Cnidaria,* consisting of a reticulum of multipolar nerve cells, appears to be a phylogenetically early form of N.c. Excitations spread out in this network with decreasing amplitude. Similar *nerve nets,* known as *nerve plexi,* are also found in other invertebrates. In these accumulations of nerve cells, N.c. are concentrated in the smallest possible space. Nerve nets also occur in mammals, e.g. the plexus myentericus between the muscle layers of the intestinal wall. The diverse N.cs. of any part of the central nervous system, especially when these are unknown and have not yet been elucidated, are often loosely referred to as nerve nets. This is the case for many nerve nuclei and strata, including those of the central nervous system of vertebrates and mammals. Considering that the human brain alone contains about 10^{11} microscopic, interconnected nerve cells, the existence of unelucidated N.cs. is hardly surprising. The best known N.cs. are those of the cerebellum, hippocampus, and parts of the visual system of mammals.

Neuronal network: see Neuron.

Neuropeptides: peptides that serve as signaling molecules, which are synthesized in, and released from, peptidergic neurons. Many peptide hormones released in the body for the control of body functions (e.g. substance P, angiotensin, luteinizing hormone releasing hormone, cholecystokinin, endorphins), are also employed by neurons as neurotransmitters or modulators, i.e. they are secreted into the synaptic cleft. Nonpeptide neurotransmitters (e.g. acetylcholine, noradrenalin) are synthesized in both the cell body and the axon terminals, so that their rapid availability at the synapse is insured. In contrast, N. are synthesized on the rough endoplasmic reticulum of the cell body, and are then exported to the axon terminals by fast axonal transport. In many synapses employing a N., a nonpeptide neurotransmitter is also released, and the two transmitters act jointly but differently. Thus, presynaptic axon terminals in some autonomic ganglia of the bullfrog secrete both acetylcholione and a N. closely related to luteinizing hormone releasing hormone. At least 3 different types of receptor have been identified on the postsynaptic cell membrane: a fast responding acetylcholine receptor (channel-linked), a slow responding acetylcholoine receptor (G-protein-linked), and receptor for the N. that mediates an extremely slow response. The N. also diffuses widely and evokes a postsynaptic potential at the synapses of other neighboring neurons.

Neurophysiology, ***nerve physiology:*** a branch of Physiology (see) concerned with the laws governing nervous action in the human and animal body, and all causally related processes involved in the release, propagation and transfer of nervous excitation. N. is closely connected with sensory physiology, the physiology of muscle action and biocybernetics.

Neuroptera, *Planipennia,* ***lacewings, mantispids*** and ***antlions:*** an insect order, containing 4 500 species (approx.), with representatives in all regions of the world, but especially common in warmer climates. The earliest fossil N. occur in the Permian. See *Neuropteria.*

Imagoes. Adult body length varies between 2 and 70 mm. The head is orthognathous, i.e. the chewing mouthparts are directed downward. There are 2 pairs of approximately equal-sized wings with primitive, netlike venation; the wings are greenish and transparent (green lacewings), cloudy and brown (brown lacewings), glassy (antlions) or strikingly patterened and colored *(Ascalaphidae).* All *N.* are predatory on other insects. Metamorphosis is complete (holometabolous development).

Eggs. In many species the eggs are stalked (green lacewings, brown lacewings). Eggs are laid singly or in groups (green lacewings lay up to 30) on plant parts, with the exception of the antlion which spreads its eggs over dry sand.

Larvae. Almost without exception, larvae are predatory and terrestrial. The prey is seized with long sickle-shaped mandibles, which are grooved along the inside surface, forming a tube for sucking out the body contents. Antlion larvae bury themselves at the bottom of a funnel-shaped depression in loose sandy soil, where they lie in wait for prey (especially ants). Larvae of green lacewings ("aphid lions") and brown lacewings live mainly on aphids. The midgut of neuropteran larvae is blind, and feces are not produced; residues from the digestion process are partly regurgitated and otherwise eventually excreted by the imago.

Pupae. All larvae pupate within a silk cocoon. The silk thread is manufactured by the Malpighian tubules, the tips of which are secondarily attached to the rectum from which the silk emerges. The pupa is decticous and exarate (pupa libera).

Economic importance. Whereas antlion larvae consume useful ants, the larvae of green and brown lacewings are voracious predators on plant aphids.

Classification.

1. Family: *Ithonidae.* Fewer than 10 species, mostly from Australia. Members display a generalized structure, and represent an evolutionary link with the *Megaloptera.*

2. Family: *Coniopterygidae.* About 100 species. Adults are 1–2 mm in length and resemble aphids. Their bodies are covered with white powdery exudate.

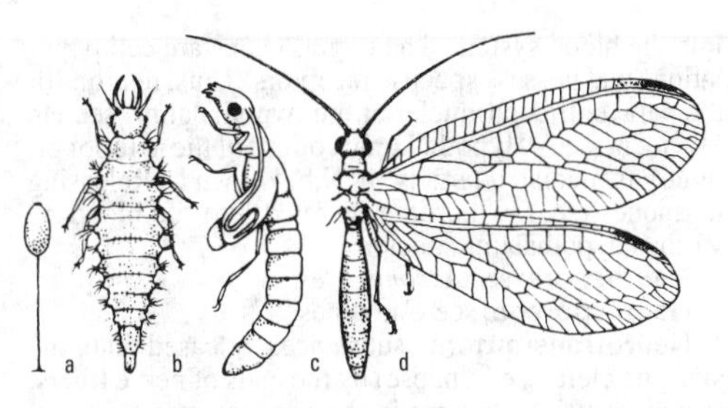

Neuroptera. Green lacewing *(Chrysopa* sp.). *a* Egg. *b* Larva. *c* Pupa. *d* Imago.

3. Family: *Osmylidae.* Large insects found near water. Larvae are terrestrial or semiaquatic. Members occur mostly in the Southern Hemisphere.

4. Family: *Berothidae.* A small but widely distributed family. Wings are hairy, especially along the posterior margins. Female wings also carry seed-like scales on the posterior edge or on certain veins.

5. Family: *Sisyridae.* Larvae aquatic, feeding on freshwater sponges, and carrying 7 pairs of abdominal gills.

6. Family: *Mantispidae* (mantispids). Adults resemble the praying mantis, with large, raptorial forelegs. Members are found only in warmer regions.

7. Family: *Hemerobiidae* (brown lacewings). Widely distributed nocturnal insects of wooded areas, mainly in temperate climates.

8. Family: *Chrysopidae* (green lacewings). Also called "stink flies" or "golden eyes". Certain species emit an unpleasant smell from a pair of prothoracic glands when caught. Many have bright green bodies and wing veins.

9. Family: *Psychopsidae.* A small family, similar to the *Hemerobiidae,* with members in Australia, South Africa and Asia.

10. Family: *Myrmeleontidae.* The larvae are called "antlions". The larvae of relatively few species construct the familiar funnel-shaped pits. Most lie in wait under stones or debris for passing insects.

11. Family: *Ascalaphidae.* Larvae are similar in habit to antlions, and in some species the adult flies actively and catches its prey on the wing.

12. Family: *Nymphidae.* A small relict group of Australian insects, whose ancestors probably also gave rise to the *Myrmeleontidae.*

13. Family: *Stilbopterygidae.* Members found in Australia and South America.

14. Family: *Nemopteridae.* Highly specialized with elongate, ribbon-like hindwings, and head usually prolonged into a rostrum. Members are found in the Old World and Australia.

Neuropteria, *Neuropteroidea:* a superorder of insects, comprising the orders *Megaloptera, Raphidioptera* and *Neuroptera.* Characteristic of all members is the presence of two pairs of approximately equal-sized, transparent wings with net-like venation, which are held together rooflike over the abdomen when at rest. They were formerly classified with the *Mecoptera.*

Neurosecretory cells: see Brain.

Neurosecretory hormones, ***neurohormones:*** a group of peptide hormones containing 3–12 amino acid residues. They are synthesized in neurons and secreted into the blood system. The target organs are cell populations that possess specific receptors. Thus, neurons of the paraventricular nuclei of the hypothalamus secrete Oxytocin (see). Hypothalamic control of the anterior pituitary is mediated largely by N.h. known as Releasing hormones (see), which are secreted into the blood by various hypothalamic nuclei.

Neurospora: see *Ascomycetes.*

Neurosporene: see Carotenes.

Neurotransmitters: substances released into the synaptic cleft (see Synapse) by the ends of nerve fibers, thereby causing changes in the postsynaptic electrical potential. T. are therefore responsible for the chemical tranmission of nervous impulses. Some known T. are acetylcholine, noradrenalin, serotonin (5-hydroxytryptamine), dopamine, γ-aminobutyric acid, glutamate, and glycine. Numerous peptides (see Neuropeptides) have also been implicated as N., e.g. substance P, vasopressin, oxytocin, endorphins, releasing hormones, angiotensin, proctolin (in insects). Some N. are common to both invertebrates and vertebrates, while others are restricted to one group. N. are responsible for postsynaptic excitation or inhibition. They may therefore be referred to as excitatory or inhibitory, but this can cause confusion, since excitation or inhibition may depend on the type of postsynaptic cell. Changes of electrical potential caused by many peptide N. persist longer than those caused by other N. Peptide N. should therefore possibly be considered as modulators of synaptic activity. See also Nitric oxide.

Neurotrophic factors, *trophic factors:* essential nutrients or other factors supplied to nerve cells by other cells. If motor neurons are removed from an embryonic spinal cord and grown in a defined cell culture medium, they die in 2–3 days, but they survive for weeks or months if the medium is supplemented with muscle extract. This suggests that target muscle cells support their own motor neurons by supplying N.f. Each of the following known N.f. supports a distinct group of neurons: nerve growth factor; brain-derived N.f.; neurotrophin 3; ciliary N.f.

Nerve cells themselves also perform a *trophic function* by providing specific substances for their own maintenance and for export to other cells. See Neurodynamics.

Neurotubules: see Microtubulues, Neuron.

Neuston: the biotic community living in the water surface, at the air-water interface. It generally consists of unicellular algae. Microorganisms of the epineuston, e.g. *Chromophyton rosanoffi, Botrydiopsis arhiza* and *Nautococcus emersus,* lie on the water surface on a disk of hydrophobic material, and they are adapted to an aerobic existence. The hyponeuston occupies the underside of the water surface, consisting of organisms such as *Codonosiga botrytis, Navicula* and *Arcella,* which project into the water (Fig.). On small areas of still water, the N. may form a continuous layer with striking colors (see Water bloom).

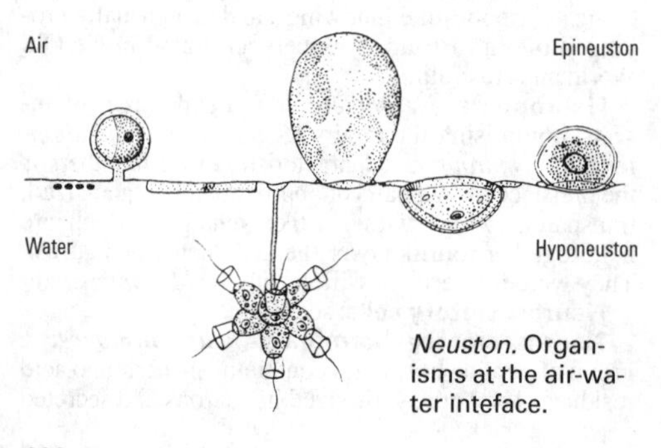

Neuston. Organisms at the air-water inteface.

Neutral fats: see Fats.

Neutral mutation: see Gene muttaion.

Neutral stimulus: see Conditioned reflex.

New Guinea plateless turtle: see *Carettochelyidae.*

Newt: a member of the family, *Salamandridae* (see). It has a laterally compressed tail, and it is adapted to an aquatic existence. Ns. spend their time in the water, at least for breeding purposes, and usually longer.

New World monkeys, *platyrrines,* *Platyrrhini:* an infraorder of the Simians (see). All members are restricted to the Americas. They possess a broad, cartilaginous nasal septum, with laterally directed nostrils. The ear drum (tympanic membrane) is supported by a simple bony ring. The tail is usually prehensile and functions for clasping and hanging. N.W.m. include the Cebid monkeys (see) and the Marmosets (see), which differ in their dental formulas. All N.W.m., however, possess a total of 12 premolars (i.e. 3 in the dental formula), whereas Old World monkeys have 8 (2 in the dental formula).

New Zealand bellbird: see *Meliphagidae.*

New Zealand wrens: see *Xenicidae.*

Nexin: see Cilia and Flagella.

NGF: acronym of Nerve growth factor (see).

Niche, *ecological niche:* the role or functional position of a species within the community of an ecosystem. It is sometimes loosely used to mean the microhabitat in the sense of the physical space occupied by a species, but this usage is condemmed and should be avoided. The true definition includes the geographical range as plotted on a map, the type or range of environments occupied by a species within a habitat, the period of time for which it is present, the way in which a species relates to, and interacts with, other species in the same community, and the resources that it obtains. The total potential for different modes of existence and different life forms offered by an environment is fulfilled by the different Ns. occupied by the species of the occupying community. During the course of its development, a species can occupy different Ns., i.e. the role of a species in an ecosystem may vary with time. The evolutionary process whereby a species attains its N. is known as *innidation;* the critical force in this process is Competition (see), which leads to optimization of the way in which a species exploits the available physical environment and its resources, and its interaction with other species. The greatest competition occurs between the members of a single species, because their Ns. are largely identical in all dimensions. A divergence of structure or behavior within a species (character displacement), perhaps resulting in new species, leads to *N. diversification* and a decrease of competition. The greater the *N. breadth,* and the smaller the specific separation of the resource optima of two species, then the greater is their *N. overlap.* This statement, however, is always applicable only to one or a few *N. dimensions* (Fig.). If the N. is considered as a multidimensional space, then *N. breadth* is the range of any given *dimension* in that space, within which the species can function. *N. overlap* results in direct competition for a given resource by 2 or more species; considering the N. as a multidimensional space, N. overlap means the joint occurrence of 2 or more Ns. along all or part of the same resource axis. For any dimension or resource within a N., innidation results in a spectrum of life forms that perform different functions with respect to that resource or dimension. Thus, plant sap-feeding cicadas form different groups that are adapted to different plant communities, different seasons and different climates.

Close similarities in the Ns. of 2 species leads to ecological Vicariance (see) and the operation of Gause's exclusion principle (see).

[Niche: Theory and Application (Benchmark Papers in Ecology 3), edit. R.H. Whittaker and S.A. Levin, Dowden Hutchinson & Ross, Inc. Distributed by Halstead Press, a division of John Wiley & Sons, Inc., 1975]

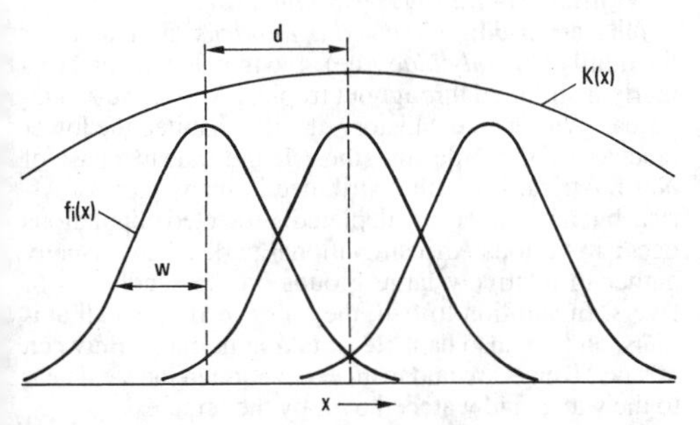

Niche. Utilization of a resource continuum by 3 species, represented by the curve *K(x)* [e.g. the quantity of food *(K)* as a function of the number of feeding species *(x)]*. The function $f_i(x)$ describes the way in which species *i* uses this resource. The ratio of the separation of the utilization optima *(d)* to half the average width of the utilization curve *(w)* is an index of the niche separation of the species concerned.

Nickel, *Ni:* a heavy metal that occurs only in traces in living organisms, mostly in association with nucleic acid. In plants, its physiological action is similar to that of cobalt. Like the latter, it tends to form chelates, and is able to displace other heavy metals from their functional sites. In plant culture media, high Ni concentrations cause root damage, which interferes with the uptake of nutrients. In aerial plant parts, Ni damage is often manifested as chlorosis. Most soils contain extremely low concentrations of Ni, so that Ni toxicity is usually not a problem. Serptentine-rich soils, however, contain relatively large quantities of Ni. Among agricultural plants, high Ni levels are tolerated well by cereal species, with the exception of oats. Ni toxicity in agricultural soils can be counteracted by the extra application of potassium and calcium fertilizers. In animals, Ni is found in particular in liver, pancreas and egg yolk.

Nicobar breadfruit: see *Pandanaceae.*

Nicobar pigeon: see *Columbiformes.*

Nicoletiidae: see *Zygentoma.*

Nicotiana: see *Solanaceae.*

Nicotianamine: see Siderophores.

Nicotinamide-adenine-dinucleotide, NAD^+*:* a pyridine nucleotide, discovered by O. Warburg, which serves as the hydrogen transferring coenzyme of numerous dehydrogenases and reductases. It is important in many biological oxidation and reduction processes. Chemically, NAD^+ consists of nicotinamide riboside linked to adenosine via pyrophosphate. Nicotinamide riboside consists of the pyridinium cation of nicotinamide (see Vitamins) linked N-glycosidically to D-ribose.

NAD^+ can accept hydrogen from a suitable substrate and transfer it to a hydrogen acceptor. Specificity for the hydrogen donor and acceptor is determined by the enzyme protein. The accepted hydrogen is attached stereospecifically to position 4 of the pyridine ring. As a result of this reduction, the pyridine ring loses both its aromatic character and the positive charge on the heterocyclic nitrogen. The reduced form is represented as NADH. Since 2 hydrogens are transferred, the positive charge can be envisaged as becoming associated with the second hydrogen, which therefore becomes a proton. Subsequent transfer of the hydrogen results in the reformation of the pyridinium cation. The interconversion $NAD^+ \rightleftharpoons NADH + H^+$ can be monitored spectroscopically, because the reduced form (and not the pyridium cation) is characterized by absorption at 340 nm. In aerobic organisms, much of the NADH formed metabolically is oxidized via the respiratory chain. It is also involved in fermentation and glycolysis, especially in anaerobic metabolism. Reducing power for biosynthesis is more commonly derived from Nicotinamide adenine dinucleotide phosphate (see).

NAD^+ and NADH are present in all living cells, partly in a free form and partly bound to enzyme proteins.

NAD^+ was formerly known as ***diphosphopyridine nucleotide (DPN), codehydrogenase I, coenzyme I,*** and ***cozymase.***

NAD^+ : R = H
$NADP^+$: R = (P)

Nicotinamide adenine dinucleotide

Nicotinamide-adenine-dinucleotide phosphate, $NADP^+$*:* a pyridine nucleotide, which, like Nicotinamide-adenine-dinucleotide (see), is a coenzyme of numerous dehydrogenases and reductases, and is important in many reductive processes of metabolism. Reducing power for biosynthesis is more commonly derived from $NADPH^+$ than from $NADH^+$. $NADP^+$ is structurally very similar to NAD^+, differing from the latter merely in the possession of an additional phosphate residue at position 2′ of the adenosine moiety. The mechanism of hydrogen transfer is the same in NAD^+ and $NADP^+$, involving reduction of the pyridinium cation to NADH (+ H^+), which absorbs at 340 nm. $NADP^+$ is biosynthesized from NAD^+ by phosphorylation.

$NADP^+$ was formerly known as ***triphosphopyridine nucleotide (TPN), codehydrogenase II,*** and ***coenzyme II.***

Nicotine: a pyridine alkaloid, and the main alkaloid of *Nicotiana tabacum,* where it is accompanied by other tobacco alkaloids, e.g. Nornicotine (see). It forms a colorless levorotatory liquid, which turns brown on exposure to the air. Depending on the plant variety, fresh tobacco leaves contain 0.06–8% N. It is present in all parts of the tobacco plant, being synthesized in the roots, then transported to the aerial parts. It is also found in several plants of other families.

N. is a powerful plant poison. Through tobacco smoking it is the most widely consumed addictive alkaloid in the world. Between 50 and 100 mg (the content of half a cigar) is fatal in man, causing paralysis of the vasomotor and respiratory centers in the medulla oblongata. When tobacco is smoked, most of the N. is destroyed by burning. The concentration in smokers' blood plasma is 5 to 50 ng/ml. In small doses it stimulates the autonomic nerve endings, causing nausea, perspiration, salivary flow, and constriction of the pupils. N. is also used as an insecticide for pest control.

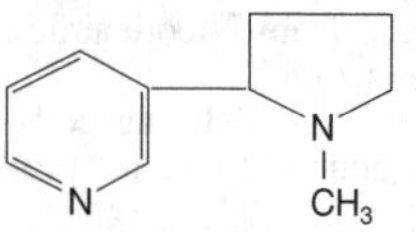

N N CH_3 *Nicotine*

Nidamental glands: glands of the female genital tract, which secrete the horny egg capsule of cephalopods.

Nidicolous:

1) A term applied to young birds that stay in the nest until they are able to fly. Newly hatched, N. young are usually naked.

2) A term applied to animals (usually arthropods) that inhabit the nests of warm-blooded animals. Such animals are beneficiaries, profitting from the constant temperature conditions of the nest, as well as finding food and protection. They include many ectoparasites. The rove beetle, *Quedius ochripennis,* occurs in both mammalian and birds' nests, as well as in the nests of social insects. The nests of hamsters are home to the clavicorn beetle, *Cryptophagus schmidti,* while the rove beetle, *Microglossa nidicola,* is found in the burrows of the bank swallow (sand martin). Mites also occur widely as nest inhabitants. See Myrmecophiles.

Nidifugous: a term applied to young birds that leave the nest soon after hatching. Newly hatched N. young usually possess down feathers.

Nidation: see Placenta.

Niger seed: see *Compositae.*

Nighthawks: see *Caprimulgiformes.*

Nightingale-like thrushes: see *Turdinae.*

Nightingales: see *Turdinae.*

Nightjars: see *Caprimulgiformes.*

Night lizards, *Xantusiidae:* a family of Lizards (see) comprising 4 genera and 12 species, distributed in North and Central America and the Antilles. They are 10–15 cm in length, inconspicuously colored, and typically lizard-like in appearance. There are 4 normally developed limbs, each with 5 toes. The dorsal scales are small, whereas the ventral shields are large and rectangular. Movable eyelids are absent. Instead, the eyelids are fused and transparent, forming a "spectacle", which covers a vertical pupil. They feed on insects and some species also consume plants; e.g. the ***Common*** or ***Island night lizard*** *(Klauberina riversiana),* restricted to certain rocky islands off the coast of California, is a plant eater. The ***Granite night lizard*** *(Xantusia henshawi),* native to California, hides during the day in narrow rock clefts. Other species conceal themselves in tree hollows. The ***Cuban night lizard*** *(Cricosaura typica)* is one of the world's rarest lizards, found only in a 200 km^2, wooded, rocky area inland from Cabo Cruz in the Oriente Province of Cuba.

N.l. are truly viviparous, i.e. the embryo is nourished by a placenta.

Nightshade family: see *Solanaceae.*

Nile crocodile, *Crocodylus niloticus:* a member of the family *Crocodylidae* (see), 4–6 m in length, and formerly distributed throughout tropical Africa and Madagascar. Within recent history, it still inhabited the lower reaches of the Nile and the Mediterranean coast of North Africa. It is still distributed in many parts of Africa, but its numbers are depleted; protected populations occur in various African National parks. They usually gather in relatively large groups on the sandbanks of rivers. In addition to fish, they also feed on small animals, and are also capable of taking humans. Between 40 and 50 eggs are laid in an excavation in the sand near to the water, and watched over by the female.

Nile monitor, *Varanus niloticus:* a 2 m-long monitor (see *Varanidae)* distributed over large areas of Africa, in arid areas, as well as in the vicinity of water, but now extinct in some parts of North Africa. It is a good swimmer and climber, and lays its eggs in termite mounds.

Nilgai: see *Tragelaphinae.*

Nisin: a polypeptide antibiotic produced by lactic acid bacteria, e.g. *Streptococcus lactis.* It inhibits the growth of various bacterial species, and it is used in the food industry as a preservative.

Nissl's granules: see Neuron.

Nitentia: see Movement.

Nitrate respiration: respiration by certain bacteria, in which nitrate is reduced instead of atmospheric oxygen. Thus, the type of metabolism normally associated with aerobic conditions (e.g. the tricarboxylic acid cycle) may operate, but the hydrogen removed from organic substrates serves for the reduction of nitrate rather than free oxygen, i.e. the terminal electron acceptor of the respiratory chain is not oxygen but nitrate. The product of nitrate reduction varies according to the species. Thus denitrifying bacteria reduce nitrate to molecular nitrogen (see Denitrification), whereas other species reduce it to ammonia (see Ammonification). Other species may reduce nitrate to nitrite. Thus, *Escherichia coli,* in the absence of oxygen and in the presence of nitrate, will maintain its "aerobic" metabolism by reducing nitrate to nitrite. In wet and poorly aerated (i.e. oxygen-deficient) agricultural soils, the loss of nitrate by N.r. can lead to a decrease of fertility.

Nitric oxide, *NO:* a highly reactive, naturally occurring gas, produced and utilized by a wide range of animals. Synthesis of NO has been demonstrated in arthropods (e.g. king crabs) and in several mammals, including humans. It is speculated that the biological production of NO arose in response to increased levels of atmospheric oxygen early in the earth's history. Animals have subsequently exploited NO in two ways: 1. as an assassin molecule (macrophages produce NO when stimulated by bacterial lipopolysaccharide or γ-interferon from immune cells), and 2. as a signalling or messenger molecule in both the nervous system and the vascular system.

NO carries an unpaired electron and is therefore highly reactive. After it has been released by a cell, it rapidly combines with oxygen to form nitrogen dioxide (NO_2), then nitrite and nitrate ions, or it binds to hemoglobin. It therefore has short half life, and can act only over a relatively short range. There appear to be no specific membrane receptors for NO. The gas gains access to the cell interior by rapid diffusion across the membrane. Inside the cell it stimulates guanylate cyclase, which catalyses the production of cyclic GMP from GTP. Cyclic GMP is a messenger that activates many processes, such as muscle relaxation and changes in brain chemistry.

NO is now known to be the "endothelium-derived relaxing factor" discovered by R. Furchgott et al. (State University, New York) in the early 1980s. It was identified as NO in 1987 by S. Moncada et al. at the Wellcome Research Laboratories, UK. These latter authors also showed that NO is generated biologically from arginine. Synthesis of NO in blood vessel epithelia occurs in response to the distortion of blood vessels by blood flow. The gas then rapidly diffuses into the surrounding muscle layers, causing them to relax. One possible cause of high blood pressure may be the defective production of NO by blood vessel epithelia. NO therefore represents a new system for blood pressure control, in addition to nervous control from brain centers and the effects of vasoconstrictor and vasodilator hormones.

As a neurotransmitter, NO occurs in certain nerve networks, e.g. it is active in the dilation of arteries supplying the penis, and in the relaxation of muscles of the corpora cavernosa. NO released from stomach nerves causes the stomach muscles to relax in order to accommodate food. Intestinal nerves also cause relaxation of the intestinal muscles by releasing NO. In addition, nervous activity in the cerebellum is increased by NO, which acts by promoting the release of cyclic GMP. There is evidence that NO is an important neurotransmitter in nervous processes concerned with memory.

NO synthase of macrophages is inducible, enabling macrophages to kill parasites and tumor cells by the poisonous action of NO. An inducible as well as a constitutive NO synthase is also present in blood vessel epithelium. Septic shock is thought to be due to the induction of NO synthase, e.g. by bacterial lipopolysaccharide, resulting in overproduction of NO and excessive dilation of blood vessels.

Nitrifying bacteria, *Nitrobacteriaceae:* bacteria capable of nitrification, i.e. capable of oxidizing ammonia to nitrite *(Nitrosomonas, Nitrosococcus, Nitrosospira, Nitrosocystis and Nitrosogloea),* and nitrite to nitrate *(Nitrobacter* and *Nitrocystis).* This is an important phase of the nitrogen cycle, because nitrate is often the nutrient in least supply in the soil, and nitrates are one of the most expensive forms of fertilizer in agriculture. See Nitrogen, *Nitrobacter, Nitrosomonas.*

Nitrobacter: a genus of nitrate-forming, rod-shaped, nonmotile, Gram-negative, aerobic bacteria, which live autotrophically, obtaining energy by the oxidation of nitrite to nitrate. The genus has a worldwide distribution in soil (especially alkaline soils, pH 7–8) and water. See Nitrifying bacteria.

Nitrobacteriaceae: see Nitrifying bacteria.

Nitrogen, *N:* a nonmetallic, fundamentally important bioelement, present in all living organisms as a component of nucleic acids, proteins and numerous other biomolecules. Compounds of N are essential nutrients for all living organisms.

In the N-cycle of the biosphere, the high N-content (80%) of the atmosphere represents a permanent reservoir of the element, which, however, cannot be utilized directly by plants or animals. Only certain prokaryotic microorganisms are capable of assimilating atmospheric N_2 (dinitrogen), e.g. free living bacteria such as *Azotobacter chroococcum* and *Clostridium pasteurianum,* free-living cyanobacteria such *Anabaena,* and the symbiotic bacteria, *Rhizobium* and *Frankia* (see Root nodules). In this process of biological N-fixation, N_2 is reduced to ammonia, which is then assimilated as the amino N of amino acids, which in turn serve as a source of N for the synthesis of a variety of other nitrogenous compounds. An insignificant proportion of atmospheric N_2 is oxidized to nitrate by electrical atmospheric discharge and washed into the soil by rain. Industrially, large quantities of atmospheric N_2 are reduced to ammonia, which is further converted into N-containing fertilizers, e.g. ammonium sulphate, urea, sodium nitrate.

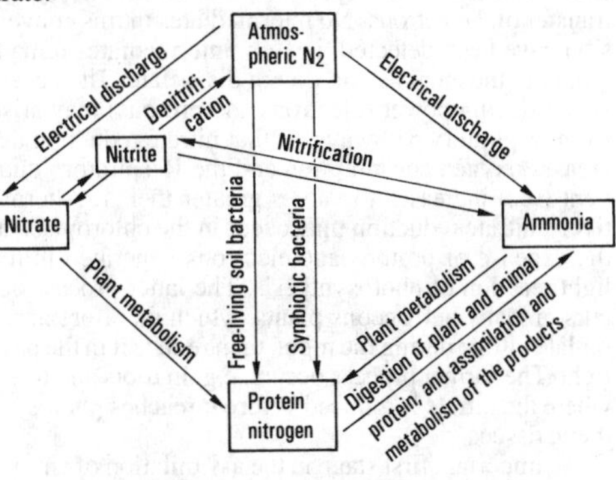

Fig.1. *Nitrogen.* The nitrogen cycle.

In addition to biological N-fixation, soil microorganisms are responsible for other conversions of nitrogenous compounds. Organic nitrogenous compounds in the dead remains of organisms (plant residues, dead microorganisms, animal remains), organic fertilizers and animal excreta are attacked by numerous bacteria and converted into inorganic N-compounds. The primary product of this process of *mineralization* is ammonia (NH_3). In well aerated, weakly acidic or neutral soils, this ammonia is rapidly oxidized to nitrite (NO_2^-), then nitrate (NO_3^-) (see Chemosynthesis), a process known as *nitrification.* Nitrate and nitrite-forming bacteria (nitrifying bacteria) always occur together in the soil. Under unfavorable conditions, microbial denitrification may occur, i.e. reduction of nitrate to ammonia, gaseous N_2, nitric oxide and nitrous oxide. Available N is also lost from the soil by adsorption of ammonium ions (NH_4^+) by soil colloids, and by the washing out of soluble nitrogenous compounds into ground waters. Mineral degradation in the soil does not produce nitrogenous compounds, since minerals are free from N. Although denitrification represents a loss of N from the soil, the activity of denitrifying organisms in large lakes and particularly in the sea would appear to be important in the natural circulation of N. Thus, the N-compounds leached from the soil into ground waters eventually appear in streams, rivers, lakes and the sea, and would accumulate there, but for the activity of denitrifying organisms; the resulting N_2 returns to the atmosphere and is available for N-fixation.

The most important nitrogenous plant nutrient is nitrate. Some plants also take up ammonium salts, and organic compounds (e.g. urea) may also be utilized. These are normally absorbed by the root, but it is possible to fertilize plants with both organic and inorganic N-compounds by application to the leaves. Nitrate ions taken up by the root are first reduced to ammonia (or ammonium ions): $HNO_3 + 8e^- + 8H^+ \rightarrow NH_3 + 3H_2O$. This nitrate reduction occurs in 2 stages :

Nitrate + $2e^- \rightarrow$ Nitrite (catalysed by Nitrate reductase), and

Nitrite + $6e^- \rightarrow$ Ammonia (catalysed by Nitrite reductase).

Nitrate reductase catalyses the reduction of nitrate to nitrite by the transfer of 2 electrons, and nitrite reductase catalyses the reduction of nitrite to ammonia by the transfer of 6 electrons, i.e. while nitrite is bound to nitrite reductase, it is reduced in 3 stages, each by the transfer of 2 electrons. No intermediates in this conversion have been detected, i.e. any intermediates remain bound to the enzyme and do not dissociate. The necessary reducing power (electrons and protons) may arise from respiratory pathways, so that nitrate reduction decreases oxygen consumption and the Respiratory quotient (see) increases to values greater then 1. Alternatively, nitrate reduction may occur in the chloroplast, at the expense of protons and electrons generated in the light reaction of photosynthesis. The latter process occurs in many herbaceous plants, which therefore accumulate nitrate during the night, then reduce it in the daylight. The former process occurs, e.g. in roots and trees, where the nitrate is reduced before it reaches photosynthetic tissues.

An important first stage in the assimilation of ammonia into organic compounds is the ATP-dependent amidation of glutamic acid to form glutamine (Fig.2). Glutamine then reacts with 2-oxoglutarate (α-ketoglutarate) to produce 2 molecules of glutamic acid; this reaction requires reducing power (provided by respiratory pathways or by the light reaction of photosynthesis) and it is catalysed by the enzyme, glutamate synthase. The amino group of one of these molecules of glutamic acid is therefore derived from the ammonia of the first glutamine synthetase reaction. Glutamic acid is a partner in numerous transamination reactions, so that the N can then be transferred for the synthesis of other amino acids, as well as serving as a N-source for other syntheses. The reactions can be envisaged as a cycle, from which the amino N of one of the molecules of glutamic acid is continually removed for syntheses, and the resulting α-ketoglutarate is reconverted to glutamic acid in the glutamate synthase reaction.

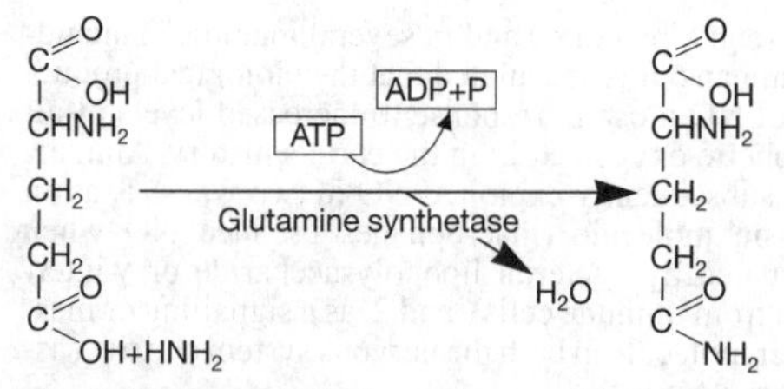

Fig.2. *Nitrogen.* ATP-dependent amidation of glutamic acid to glutamine.

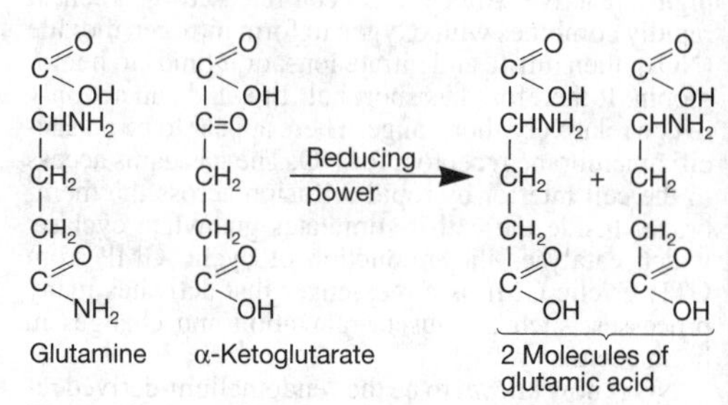

Fig.3. *Nitrogen.* Synthesis of glutamic acid catalysed by glutamate synthase. The reducing equivalents are derived from NADPH (e.g. in bacteria) or from reduced ferredoxin (in plant chloroplasts).

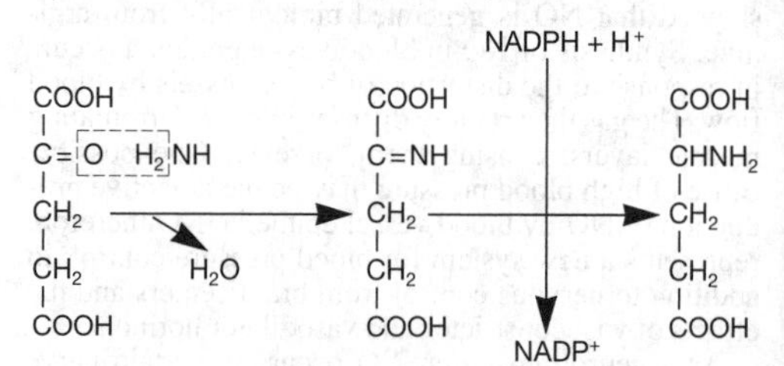

Fig.4. *Nitrogen.* Reductive amination of α-ketoglutaric acid to glutamic acid

Another assimilation pathway is the reductive amination of α-ketoglutarate to glutamic acid (Fig.4), which operates in some organisms.

Nitrogenous compounds are easily transported in plants. The transport form is usually organic, although the transpiration stream of herbaceous plants may contain high concentrations of nitrate. Transport forms of N also serve for N-storage and ammonia detoxification, and they are species-, genus- and family-specific, e.g. glutamic acid and glutamine *(Ricinus,* carrot, gherkin), aspartic acid and asparagine (asparagus, members of the *Cruciferae,* legumes), citrulline (members of the *Betulaceae),* allantoin (members of the *Boraginaceae)* and allantoic acid (members of the *Aceraceae).*

The relative values of nitrate and ammonium salts as plant N-sources have been the subject of numerous investigations. Although nitrate must be reduced to ammonia after entering the plant, it is often the physiologically the more effective N-source. In the presence of high nitrate concentrations, the uptake of other anions is suppressed, while high concentrations of ammonium suppress the uptake of other cations, and favor the uptake of anions, especially phosphate. Ammonium uptake is strongly dependent on the soil pH, whereas this has little effect on the uptake of nitrate. In the pH-range 5.5–6.5, ammonium salts are very good fertilizers, whereas nitrates support optimal growth over a much wider pH-range (4.5–7). Ammonium-containing mineral fertilizers cause soil acidification, because ammonium ions are taken up by the plant root in exchange for protons. Individual plants respond differently to these two forms of N. Ammonium salts are especially suitable for plants that prefer or tolerate acidic soils, e.g. potatoes, rye, oats, maize and rice. For fertilization in late autumn, ammonium salts are more suitable than nitrates, because in late autumn and winter nitrification and therefore loss of N (as nitrate) by leaching is minimal. On the other hand, nitrate fertilizers (sodium and potassium nitrate) act very quickly and are therefore more effective for fertilization of crops in the spring. Pure ammonium fertilizers are ammonium chloride and ammonium sufate. Mixed fertilizers containing both ammonium and nitrate are calcium and ammonium nitrate and ammonium sulfate/ ammonium nitrate.

N-deficiency affects the entire metabolism, so that N-deficient plants are small and poorly developed. They have drab yellow or reddish leaves, which are often shed prematurely. In cereals, tillering is restricted, flowers are formed and seeds are set earlier than usual, and the ears remain short and contain incompletely formed grains. In the presence of excess N, plants are dark green, with abundant sap and large leaves. The tissues, however, are soft and spongy, due to the inadequate formation of mechanical tissues, and the plant is consequently very susceptible to attack by diseases and pests; cereals have very weak stems.

Heterotrophic organisms (see Heterotrophy) and Carnivorous plants (see) represent special cases of N-nutrition.

In contrast to plants, the N-nutrition of animals is heterotrophic, and animals are generally incapable of utilizing inorganic N-sources.

Nitrogenase: see Nitrogen fixation.

Nitrogen balance: see Biological value of proteins.

Nitrogen fixation: the conversion of atmospheric nitrogen into ammonia, as a prelude to the assimilation of the ammonia into various nitrogenous compounds of the cell. The activation and reduction of molecular nitrogen is catalysed by the enzyme nitrogenase.

Nitrogenase consists of 2 proteins, both of which are required for activity. One of these proteins contains iron, molybdenum and acid-labile sulfur (Mo-Fe-protein; component I; molybdoferredoxin) and the other contains iron and labile sulfur (Fe-protein; component II). The 2 components are present in the molecular ratio, 1 Mo-Fe-protein: 2 Fe-protein. Both have been isolated and characterized from various nitrogen-fixing organisms. Nitrogenases from different sources are very similar. Fe-protein from one organism (e.g. *Azotobacter)* will often combine with Fe-Mo-protein from another (e.g. *Klebsiella)* to give a functional enzyme. On the other hand, heterologous nitrogenase components may generate catalytically inactive complexes. Thus, the nitrogenase of *Clostridium* is inhibited by the Mo-Fe-protein of *Azotobacter,* because *Clostridium* Fe-protein forms a tight, inactive complex with *Azotobacter* Mo-Fe-protein, even in the presence of excess *Clostridium* Mo-Fe-protein. Nitrogenase is inactivated by oxygen, so that anaerobic conditions must be employed in the purification and study of the enzyme. Nitrogenase requires relatively large quantities of ATP in order to activate molecular nitrogen prior to reduction. The generation of this ATP in aerobic organisms requires a high respiration rate, and therefore a plentiful oxygen supply. In aerobic, nitrogen-fixing organisms, the respiratory enzymes are located in the cell membrane, and the nitrogenase is protected from oxygen by its location in the relatively anaerobic interior of the cell (the respiratory enzymes rapidly consume oxygen and thereby form a "cordon sanitaire" that protects the nitrogenase). Nitrogenase in the intact cell is possibly less sensitive to oxygen than is the purified enzyme.

N.f. is fundamentally important for the nitrogen economy of soils and waters, and it is an essential stage in the nitrogen cycle of the biosphere. Certain free-living soil microorgansms, especially of the genera *Clostridium* and *Azotobacter,* are capable of N.f. Other microorganisms fix nitrogen in symbiosis with higher plants (see Root nodules). In water, especially in the ocean, the most important nitrogen fixers are cyanobacteria. N.f. by cyanobacteria is also economically important for the cultivation of rice in the tropics. Lichens (symbiotic associations between a cyanobacterium and a fungus) and the symbiotic system *Nostoc-Gunnera* (i.e. symbiosis between a cyanobacterium and an angiosperm) are ecologically very important, because they are able to colonize habitats that have extreme climates or are poor in nutrients. The carbon and nitrogen requirements of the lichen are met by photosynthesis and N.f., so that these systems are sustained largely by the atmosphere, and their nutritional demands on the remaining environment are relatively small. Lichens pioneer the exploitation of barren environments and pave the way for later colonization by plants with more exacting nutritional requirements. In poor soils, the *Nostoc-Gunnera* system fixes about 70 g atmospheric nitrogen per m^2 per year.

The ability to perform N.f. (equivalent to the ability to produce the enzyme, nitrogenase) is a special characteristic of relatively few prokaryotic organisms; it has never been detected in a eukaryote. See Root nodules, Rhizobium, Bacteroids, Nitrogen.

Nitrosomonas: a genus of nitrite-forming, oval to rod-shaped, Gram-negative, aerobic bacteria, which occur singly or in chains. The cells contain numerous vesicular membrane structures. They live autotrophically, obtaining energy by oxidizing ammonia to nitrite, and are found in soil and water. See Nitrifying bacteria.

Nival zone: see Altitudinal zonation.

Noble decay of grapes: see *Ascomycetes.*

Nocadazole: see Microtubules.

Nocardia: a genus of the family *Actinomycetaceae.* They are aerobic, acid-fast bacteria. Filament formation may be present or absent, and mechanically disturbed filaments break down into separate cells. Most species live saprophytically in the soil; some species are parasitic on man or animals, causing tuberculosis-like diseases or ulcerative lesions.

Nociceptors: see Pain.

Noctiluca: see *Phytomastigophorea.*

Noctuidae, *noctuids, owlet moths:* a highly evolved family of the *Lepidoptera* (see), containing more than 20 000 species, found in all habitats from the tropics to the polar regions. Of these, 1 800 species occur in the Palaearctic, 3 500 in North America and 500 in Central Europe. Almost all species are nocturnal and consequently dull and inconspicuous in color, although a few species are brightly colored. The forewings carry characteristic design elements (Fig.), while the hindwings are mostly uniformly gray or brownish. The smallest species has a wingspan of 5 mm, while *Thysania agrippina* (South America) has the largest wingspan (about 32 cm) of all *Lepidoptera.* Large European species have wingspans up to 10 cm. The distinguishing character of the noctuids is the presence of paired auditory organs (tympana or tympanal organs) which lie at the posterior edge of the thorax, in contrast to the abdominal tympana of other lepidopteran families. With the aid of these tympana, noctuids can detect the echolocation signals of bats, and are able to take evasive action to avoid capture. Some species even confuse bats by producing bat-like sound frequencies. The noctuid thorax is heavy and stout. Almost all species have two ocelli. A frenulum is always present, maxillary palps are vestigial and the proboscis is usually well developed, only rarely atrophied. The proboscis is used for sucking nectar from flowers. Adults of the genus *Calpe* use their proboscis to pierce fruit, and in one species it is even used to suck blood, usually from cattle. Caterpillars are mostly naked, and pupation usually occurs in a cavity (cell) below ground. Many noctuid caterpillars are agricultural and forestry pests, known as cutworms and armyworms (see Cutworms). See also Silver-Y moth, Pine beauty, Underwing moths.

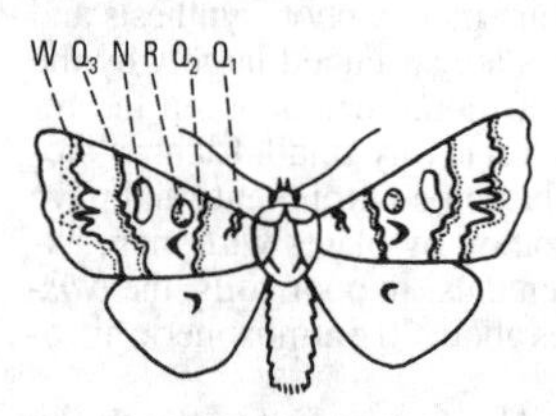

Noctuidae. A stylized noctuid moth, showing wing markings. Q_1 Basal or subbasal band. Q_2 Transverse anterior band, also called the t.a. line or first line. Q_3 Transverse posterior band, also called the t.p. line or second line. *R* Orbicular or round spot (below this is the claviform spot). *N* Reniform spot. *W* Subterminal band. When present, the terminal band (not shown) lies along the periphery of the wing tip. A median transverse band is sometimes present running between the reniform spot and the orbicular spot.

Noctuid moths: see *Noctuidae.*

Noctule bat: see *Vespertilionidae.*

Nocturnal species: see Biorhythm.

Nodding movements: movements of flower stalks that result in the raising and lowering of flowers or infloresecences. Rhythmic diurnal N.m. occur, e.g. in water lilies and other plant species. See Nyctinasties.

Noddy: see *Laridae.*

Node of Ranvier: see Nerve fiber, Neuron.

Nodosaria: a genus of perforated calcareous *Foraminiferida* (see). The elongated shell consists of a straight row of spherical to cylindrical chambers, and has the appearance of a string of beads. The shell surface may be decorated with ribs or spines, which may attain a length of 16 mm. Questionable fossils have been found in the Late Paleozoic. Otherwise, fossil N. occur from the Triassic to the present.

NOEL: acronym of no-observed-effect level (see Acceptable daily intake).

Noise makers: a trivial name for the suborder, *Tyranni,* of the *Passeriformes* (see).

Nomenclature: an international system for the scientific naming of taxa. Unlike Common names (see), scientific names are internationally recognized, and form a basis of communication and understanding in all areas of biology. The *N. rules* and their updating are established and agreed at international congresses. Since they differ in certain essential aspects, the rules of botanical and zoological N. are agreed separately. In both botany and zoology, however, every plant and animal species is defined by a binomial, consisting of the name of the genus (spelled with a capital letter), followed by the species name (spelled with a small letter), e.g. *Pinus nigra* (Austrian pine) or *Gonepterix rhamni* (brimstone butterfly). These Latinized binomials for species were introduced by the Swedish botanist, Linneus (1707–1778). In botanical N. the genus and species name must be different, whereas tautonyms are allowed on zoology, i.e. genus and species name may be the same, e.g. *Bufo bufo* (common European toad). A species is not completely named, unless the species name is followed by the name of the author (authority) who introduced the name for that species, e.g. *Carlina acaulis* Linneus (carline thistle). In zoology, the year of publication is also given, e.g. *Canis lupus* Linneus 1758 (wolf). The author's name is often abbreviated, e.g. *Carlina acaulis* L.

If, on the basis of new information, a species is placed in a different genus, the name of the first author is placed in brackets. In botanical, but not in zoological N., this is followed by the name of the revising author, e.g. *Melilotus officinalis* (L.) Pallas (common melilot); *Dama dama* (Linneus 1758) (fallow deer).

The names of all taxa ranked higher than species consist of only a single Latin or Latinized word. The endings of the names for different taxa are internationally agreed, e.g. plant families end in *-aceae* (e.g. *Liliaceae),* and animal families end in *-idae* (e.g. *Canidae).*

In zoological N., subspecies, are designated by the simple addition of a third name, and referred to as races (e.g. *Mus musculus musculus,* the common race of house mouse). In botanical N., the term subspecies (abbreviated to subsp.) is placed between the species and subspecies names (e.g. *Trifolium pratense* subsp. *sativum,* cultivated red clover). Similarly, in the botanical N. of varieties, the term variety (abbreviated to var. or v.) is placed between the species and variety names.

A scientific name is not considered to be established or current, until it has been published in the scientific literature, accompanied by a description (diagnosis) of the taxon in question, and a written presentation of a type organism on which the description is based. For botanical N., the diagnosis must be written in Latin. Samples of the type organism should be kept in scientific institutions, and be available for further scientific investigation.

If a taxon is named more than once, then according to the *priority rule,* the valid name is the one that was published first. Thus, in the N. of higher plants, many names date back to the *Species plantarum* published by Linneus in 1753, and in zoology to the *Systema naturae* (10th edition) published by Linneus in 1758.

A later, different name for the same taxon is known

as a *synonym,* whereas a *homonym* is the same name for a different taxon.

Non-Darwinian evolution: the hypothesis that most evolutionary changes occur on a molecular level by mutation pressure and genetic drift, rather than by natural selection. Accordingly, most nucleotide and amino acid replacements during evolution are selectively neutral, i.e. nucleotides and amino acids are usually replaced by their functional equivalents. An analogous concept (but without a molecular basis) was also expressed by Charles Darwin, so that the term, non-Darwinian, is inappropriate. Furthermore, the concept fits readily into the framework of the Synthetic theory of evolution (see). With some small changes of emphasis, N-D.e. corresponds to already widely held views on the relative roles of different factors in evolution. Its adherents do not dispute that functionally important properties are subject to natural selection.

Arguments in support of the hypothesis run as follows. If it were not selectively important, the tremendous interspecies genetic variation that exists on the molecular level would represent an intolerable genetic burden. Similarly, the enormous intraspecies differences can hardly have arisen by selection, because of the accompanying burden of substitution. In the course of phylogenesis, DNA changes more rapidly than proteins. In the context of the hypothesis, this can be explained by the fact that many mutations are same-sense or permitted error mutations, resulting in synonymous codons or codons for chemically similar amino acids. Since these mutations do not affect protein structure, they are selectively neutral. Different proteins appear to alter at different but constant rates, i.e. they are subject to constant (unchaging) selection influences (see Molecular clock). Proteins whose functions are sensitive to small changes in structure, e.g. cytochrome *c,* evolve (change) more slowly than those possessing several sites at which amino acids can be exchanged without greatly affecting protein function, e.g. fibrinopeptide A. If the evolution of proteins were due mainly to selection, then the converse would apply.

The debate as to the validity of N-D.e. is known as the *selectionist-neutralist controversy.*

Predictions of the genetic composition of populations, made on the assumption of N-D.e., are found to be invalid, e.g. the percentage heterozygosity in large natural populations is much less than predicted from the basic assumptions of N-D.e. On the other hand, sequence analysis of DNA confirms the prediction of the hypothesis that the third nucleotide of a codon (or gene triplet) is exchanged more rapidly than nucleotides in positions 1 and 2 of the triplet; the Genetic code (see) shows that the nucleotides in positions 1 and 2 of the triplet cannot be changed without changing the encoded amino acid, whereas an exchange of the third nucleotide has either a 50% chance (e.g. histidine, glutamine, tyrosine, cystine) of changing the encoded amino acid, or it can be changed to any of the other 3 nucleotides without affecting the encoded amino (e.g. arginine, glycine, alanine, valine).

The selectionist-neutralist controversy is still open, but it seems increasingly likely that the two views are not mutually exclusive, and that selection, mutation pressure and genetic drift all make important contributions to the evolutionary process.

Nondeciduate placenta: see Placenta.

Non-disjunction: the failure of homologous chromosomes or sister chromatids to separate during meiosis or mitosis, respectively. Homologous pairs therefore migrate to the same cell pole, resulting in the formation of hyperploid and hypoploid daughter cells, which in turn can result in aneuploid organisms or tissues (Fig.) (see Aneuploidy). B-chromosomes often display N-d., a phenomenon referred to as directed or ordered N-d.

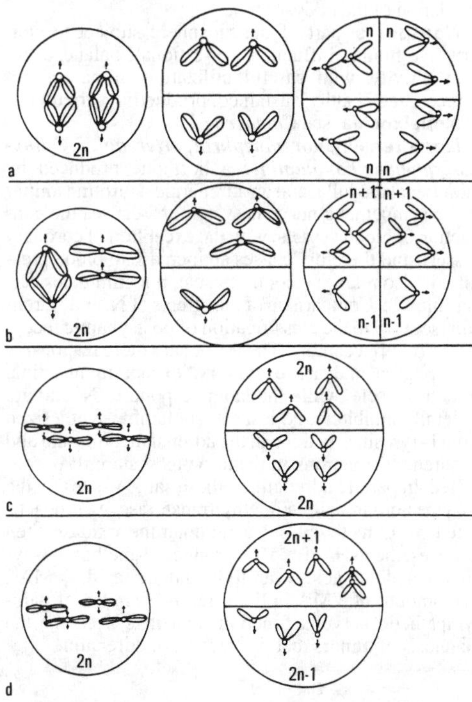

Non-disjunction. a Normal meiosis with formation of 4 haploid daughter cell nuclei. *b* Non-disjunction in meiosis I, leading to production of daughter cell nuclei with different chromosome numbers. *c* Normal mitosis. *d* Non-disjunction in mitosis, leading to production of daughter cell nuclei with different chromosome numbers.

Nonequilibrium thermodynamics, ***irreversible thermodynamics:*** study of processes outside thermodynamic equilibrium. N-e.t. considers changes in systems in terms of generalized flows (fluxes) and forces. Flows are caused by forces, which arise from the nonequilibrium distribution of mass and energy. Examples of flow in biological systems are the mass flow of dissolved substances and solvents, the flow of heat and the flow of electric charge; the respective forces being concentration gradients, pressure gradients, temperature gradients and electric fields. Chemical reactions are also considered in terms of flow, in which the generalized force, the difference between chemical potentials, is a scalar quantity.

Linear N-e.t. can be successfully applied to membrane phenomena, such as osmosis, diffusion, diffusion potential, and electrokinetic transport. The great majority of biological processes, however, proceed under conditions that are "Far off from equilibrium" (see), and do not fulfil the requirements for linear N-e.t.

Entropy (see) is an important quantity in N-e.t.

Nonessential amino acid: see Amino acids.

Nonpersistent viruses: see Virus vectors.

Nonproteogenic amino acid: see Amino acids.

Nonsense mutation: see Gene mutation.

Non-shivering thermogenesis: see Temperature regulation of the body.

Noosphere: part of the biosphere subject to constructive human influence, i.e. rational, balanced human activity, with careful utilization of natural resources, and healthy, sustained, productive agriculture.

Nopalxochia: see *Cactaceae.*

Noradrenalin, *norepinephrin, arterenol, 3,4-dihydroxyphenylethanolamine:* a hormone produced by the adrenal medulla, and an adrenergic neurotransmitter in the sympathetic nervous system. N. causes the contraction of blood vessels, with the exception of coronary vessels, and therefore causes an increase in blood pressure. It also relaxes smooth muscle, but stimulates cardiac muscle. Comparison of the effects of N. and Adrenalin (see) led to the classification of postsynaptic receptors as α or β receptors. The former are more responsive to adrenalin and are excitatory, except in intestinal smooth muscle, while the latter respond to N. and are generally inhibitory. N. is a catecholamine synthesized from L-tyrosine via dopa in the adrenal medulla and and in adrenergic neurons in the nervous system. It is converted, in part, to adrenalin in these same tissues by the enzyme noradrenalin *N*-methyltransferase. N. is deactivated by *O*-methylation; a monoamine oxidase then removes the NH_2-group to produce 3-methoxy-4-hydroxymandelic acid (vanillylmandelic acid, VMA). The amount of VMA in the urine is an index of parasympathetic nervous function, and it is used for the diagnosis of tumors that produce N. or adrenalin.

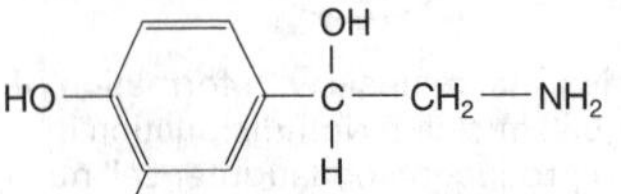

HO, OH, HO, C, CH_2, NH_2, H

Noradrenalin

Nordkapper: see Whales.

Norepinephrin: see Noradrenalin.

Norethindone: a synthetic progestin used in oral contraceptives. See Hormonal contraception.

Norethisterone: a synthetic progestin used in oral contraceptives. See Hormonal contraception.

Norethyndrol: a synthetic progestin used in oral contraceptives. See Hormonal contraception.

Norfolk Island pine: see *Araucariaceae.*

Norgestrel: a synthetic progestin used in oral contraceptives. See Hormonal contraception.

Norian stage: see Triassic.

Norma basalis: see Skull planes.

Norma facialis: see Skull planes.

Norma lateralis: see Skull planes.

Normal distribution, *Gaussian distribution:* theoretical continuous frequency distribution of the values of completely random samples or results of random experiments. The N.d. is influenced by a large number of random factors. It is particularly important in biostatistics, since the distribution of many biologically relevant Random values (see) shows a close approximation to a N.d. For the curve and equation of the N.d. function and distribution density, see Biostatistics.

Norma occipitalis: see Skull planes.

Norma verticalis: see Skull planes.

Nornicotine: one of the tobacco alkaloids. It is a minor alkaloid of *Nicotiana tabacum,* and the major alkaloid of *Nicotiana glutinosa.* It is similar in structure to Nicotine (see), but lacks the *N*-methyl group.

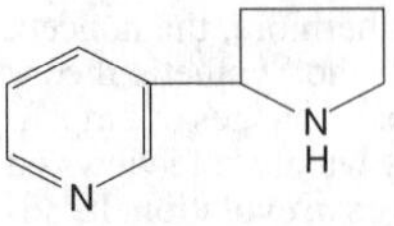

N, N, H

Nornicotine

North American beaver: see *Castoridae.*

Northern house mouse: see *Muridae.*

Northern winter moth: see Winter moths.

North sea shrimp: see Shrimps.

Norway haddock: see Redfish.

Norway lobster: see Lobsters.

Norway rat: see *Muridae.*

Norwegian lobster: see Lobsters.

Nosema: see *Cnidosporidia.*

Nostoc: see *Hormogonales.*

Notaspidea: see *Gastropoda.*

Notharchus: see *Bucconidae.*

Nothofagus: see *Fagaceae.*

Nothosauria (Greek *nothos* not genuine, *sauros* lizard): a suborder of the *Sauropterygia.* About 3 m in length, these animals possessed a long, slender skull with large temporal fossae and pointed, fluted fangs, a long neck containing 22 vertebrae, a relatively slender, flexible trunk, and only slightly modified limbs with webbed digits. Although largely adapted to a terrestrial existence, the N. were adept swimmers, and possibly inhabited beaches or other areas close to water. Fossils are found in the European, Southeast Asia and North American Triassic. Teeth and ribs of *N.* are well known from the Muschelkalk of the German Triassic. The youngest known genus of the line was *Nothosaurus.*

Notiothaumidae: see *Mecoptera.*

Notochord: see Axial skeleton.

Notodontidae: see *Lepidoptera.*

Notoedres: see Scabies mite.

Notogea, *Notogaea:* the name of a zoogeographical region, generally equivalent to the Australian region (see) in the widest sense, but sometimes taken to also include Antarctica and the Hawaian islands. Huxley (1868) formerly included the Neotropical region (see) in the N., which was therefore the southern counterpart of the northern Arctogea (see). According to Schmidt (1954), N. comprises the Australian region and the Oceanic region (see). See Animal geographical regions.

Notophthalmus: see *Salamandridae.*

Notoptera, *Grylloblattoidea,* ***rock crawlers, gryllobatids:*** an insect order discovered in 1914 (E.M. Walker, *Can. Entomol.,* ***46*** (1914) 93–99). At present, 16 species are known: 9 in Canada, 6 in Japan and 1 in Siberia. Imagoes are up to 30 mm in length and resemble earwigs. Wings are totally absent. There are no ocelli, and the compound eyes are reduced or absent. Antennae are long and filiform. Both sexes carry two long, 8-segmented, caudal cerci. They live at relatively high altitudes, between rocks and moss, or in cavities, favoring low temperatures (optimally about 4 °C). They are commonly predaceous, but may also feed on plant and animal debris. Metamorphosis is incomplete (paurometabolous development) and still largely obscure. The adult female of *Grylloblatta campodeiformis* (a Canadian species) lays black eggs when she is about one year old; these hatch after about one year, followed by a further 5-year period for development of 8 nymphal instars. [J.W. Kamp, *Can. Entomol.*, ***111*** (1979) 27–38; J.W. Kamp, *Can. Entomol.*, ***105*** (1973) 1235–1249]

Notornis: see *Gruiformes.*

Notoryctidae: see Marsupial mole.

Notoryctes typhlops: see Marsupial mole.

Notostigmophora: see *Chilopoda.*

Notostraca: see *Branchiopoda.*

Nototragus: see *Neotraginae.*

Novumbra: see *Umbridae.*

N/P/K ratio: the ratio of the nutrients, nitrogen (N), phosphorus (P, expressed as P_2O_5) and potassium (K, expressed as K_2O) in a fertilizer. The ratio is calculated to give an optimal crop yield. See Nutrient ratio, Fertilizers.

Nucellar embryony: see Adventitious embryony.

Nucifraga: see *Corvidae.*

Nuclear fragmentation: see Amitosis.

Nuclear phases: the haploid and diploid phases of the nucleus, which occur in the life cycle of most eukaryotes. The haploid phase is generated by Meiosis (see), while the diploid phase arises by fertilization. Meiosis must be preceded by fertilization, and fertilization cannot occur without prior meiosis, so that these two cellular processes occur alternately. In different organisms, meiosis and fertilization occur at different stages, so that the haploid and diploid states occupy different phases of the developmental cycle (Fig.). In diplonts (all *Metazoa* and *Ciliophora,* as well as some fungi, algae, *Heliozoa* and *Flagellata)* meiosis occurs immediately prior to fertilization, so that only mature gametes are haploid. In haplonts *(Gregarinida* and *Coccidia,* as well as some algae, fungi and *Flagellata)* meiosis occurs immediately after fertilization, so that only the zygote is diploid. Most plants and *Foraminifera* are diplohaplonts, i.e. there are separate haploid and diploid generations, which are responsible for fertilization and meiosis, respectively.

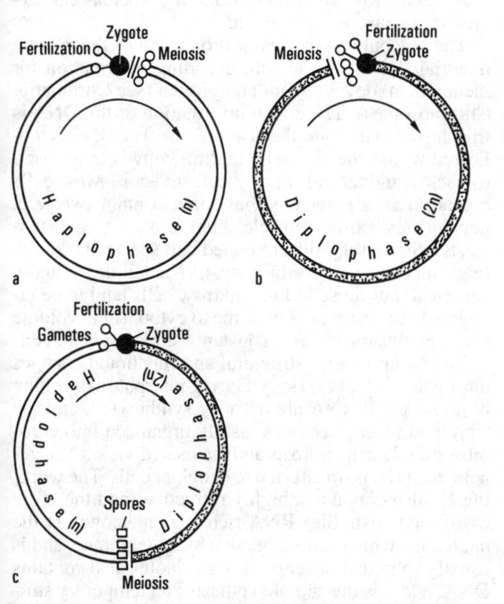

Nuclear phases. a Haplont (purely haploid organism, diploid zygote). *b* Diplont (purely diploid organism, haploid gametes). *c* Diplohaplont (intermediate change of nuclear phase; alternation of haploid and diploid developmental phases; heterophasic alternation of generations).

Nuclear sap: see Nucleus.

Nucleases: a group of hydrolytic enzymes that cleave nucleic acids. Exonucleases attack the nucleic acid molecule at its terminus, whereas endonucleases are able to catalyse a hydrolytic cleavage within the polynucleotide chain. Deoxyribonucleases (DNAases) are specific for DNA, ribonucleases (RNAases) for RNA. All N. are phosphodiesterases, catalysing hydrolysis of either the 3′ or 5′ bond of the 3′,5′-phosphodiester linkage.

Nucleic acids: polynucleotides with a relative molecular mass of 20 000 to several millions. The monomeric unit is a nucleotide consisting of one molecule of a purine base (adenine or guanine) or pyrimidine base (thymine, cytosine or uracil) linked to one molecule of a monophosphorylated monosaccharide (D-ribose or 2-deoxy-D-ribose). According to their carbohydrate component, N. are classified as Deoxyribonucleic acids (see) or Ribonucleic acids (see), which are among the most important components of all living cells, viruses and bacteriophages. DNA is the carrier of genetic information, and it undergoes self replication during cell division, so that each daughter cell receives an identical copy. RNA is responsible for the transfer and realization of the genetic information encoded in the DNA. N. were first prepared in 1869 by Miescher from pus and fish sperm.

Nucleocapsid: see Viruses.

Nucleoid, ***nuclear equivalent:***

1) An aggregate of DNA not bounded by a nuclear envelope, and analogous to the nucleus of a eukaryotic cell. Ns. occur in prokarytes, plastids, mitochondria and viruses. See Cell, Nucleus.

2) The RNA core of a RNA virus, which is surrounded by a protein capsule.

Nucleolus: see Nucleus.

Nucleolus organizer: see Nucleus.

Nucleoprotein: heteropolar complexes of nucleic acids (in particular nuclear DNA) with basic, acid-soluble proteins (histones or protamines), and with acidic base- or detergent-soluble nonhistone chromatin proteins. N. occur mainly in the chromatin of the cell nucleus in its quiescent state, and in the chromosomes when the nucleus is active, i.e. dividing. Many viruses consist entirely of N., but N. are absent from bacteria. N. are concerned in DNA replication, and in the control of gene function during protein biosynthesis. The protein-RNA complexes of the ribosome are also N.

Nucleoside: a compound in which a heterocyclic, nitrogenous base is linked N-glycosidically with a sugar. Of particular biological importance are those Ns. in which a purine or pyrimidine is linked to ribose or 2-deoxyribose, e.g. adenosine, deoxyadenosine, guanosine, deoxyguanosine, cytidine, deoxycytidine, uridine, deoxyuridine, ribothymidine and thymidine, all of which are present in the structures of RNA or DNA. Ns. and deoxyribonucleosides can be converted into the corresponding nucleotides by the action of specific kinases. They are produced by the alkaline hydrolysis of nucleic acids and nucleotides.

Nucleosomes: spherical chromatin components containing DNA and histones. Electron microscopy of isolated chromatin (prepared by spreading) reveals smooth strands (20–30 nm diam.), a lesser proportion of thinner filaments (3 nm diam.), and stretches of N. resembling strings of beads. The N. are discrete particles, about 10 nm in diameter, consisting of a closely packed octamer of histones $(H2A, H2B, H3, H4)_2$, encircled by about 140 nucleotide pairs of DNA wrapped around the octamer in 1¾ shallow turns (Fig.). Histone octamers without DNA are also observed. The DNA content of N. from different organisms ranges from 160 to 240 nucleotide pairs, but digestion with nuclease releases a nucleosome core particle containing 140 nucleotide pairs, irrespective of the origin and DNA content of the N. preparation. The remaining DNA (linker DNA) constitutes the connecting regions between the N. Histone 1 (H1) binds to the outside of the core particle and to the linker DNA. In this way, H1 promotes association between the different N., and may serve to make chromatin more compact. It is thought that N. inhibit transcription, but not replication. N. are also known as ν bodies.

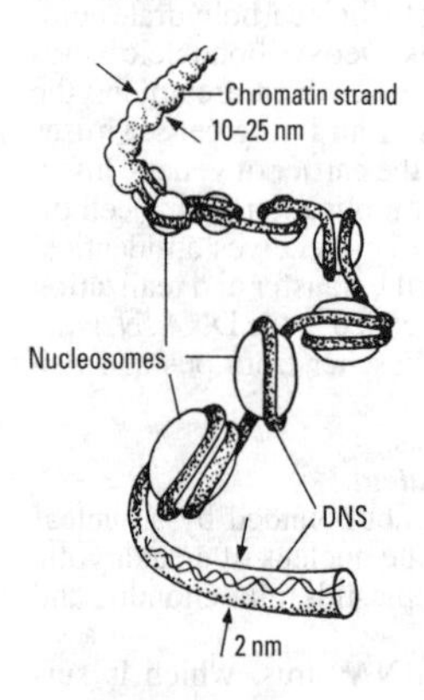

Nucleosomes. Structure of a chromatin strand (nucleofilament). Spherical histone complexes (nucleosomes) are held within loops of double stranded DNA.

Nucleotide: a phosphate ester of a nucleoside. *o*-Phosphoric acid is esterified with the free OH-group of the sugar. If the sugar is D-ribose, the N. is called a ribonucleotide or ribotide. If the sugar is 2-D-deoxyribose, the N. is a deoxyribonucleotide, deoxynucleotide or deoxyribotide. The phosphate may be present on position 2′, 3′ or 5′ (in deoxyribonucleotides, only the 3′ and 5′ phosphates are possible). The 5′-nucleoside phosphates are metabolically very important; they may be mono-, di-or triphosphorylated, e.g. guanosine 5′-monophosphate, cytidine 5′-diphosphate and adenosine 5′-triphosphate.

The cyclic N. (cyclic 3′,5′-monophosphates) have important regulatory properties (see cyclic 3′,5′-adenosine monophosphate).

Nucleoside monophosphates are synthesized de novo in the course of purine biosynthesis and pyrimidine biosynthesis. They are then phosphorylated stepwise by the action of kinases, to produce nucleoside di- and triphosphates. The 2-deoxyribose moiety is formed by reduction of the ribose in ribonucleotides.

Nucleoside triphosphates are precursors in the biosynthesis of DNA and RNA, a process in which a pyrophosphate group is lost from each nucleoside triphosphate when it is incorporated into the growing nucleic acid polymer. The monomeric components of oligonucleotides, polynucleotides and nucleic acids are therefore nucleoside monophosphates.

Enzymatic degradation of oligo- and polyribonucleotides (but not the corresponding deoxyribonucleotides) produces cyclic 2′,3′-nucleoside phosphates. Ns. are cleaved hydrolytically to nucleosides by the action of 5′- or 3′-nucleotidases, which function as phosphomonoesterases. Ns. are cleaved to free bases and phosphoribosylpyrophosphate in a pyrophosphate-dependent reaction catalysed by N. pyrophosphorylases. Certain coenzymes contain nucleotide structures, e.g. NAD contains the structure of adenosine 2′,5′-diphosphate, and coenzyme A the structure of adenosine 3′,5′-diphosphate. In all living cells, Ns. (especially adenosine 5′-triphosphate) act as high energy compounds in the storage and transfer of chemical energy.

Nucleus, ***cell nucleus*** (plates 21 & 22): a structural unit which serves as the information center of the eukaryotic cell. It is bounded by a nuclear membrane (nuclear envelope), so that its contents are separated from the cytoplasm. Prokaryotes (bacteria and cyanobacteria) do not possess a nuclear membrane, and the genetic material of these organisms (usually located in the cell center) is known as the *nuclear equivalent* or *nucleoid.* Since they lack a visible, membrane-bounded N., prokaryotes are known as *anucleobionts,* whereas eukaryotes are known as *nucleobionts.*

The N. contains the major proportion of the genetic material (DNA) of the cell, encoding information for metabolism, development and growth (see Chromatin, Chromosomes). The coded information of the DNA is transferred to the cytoplasm as mRNA. The DNA is replicated within the N., and a faithful copy is transferred to each daughter cell during cell division. Most cells contain a single nucleus, but some contain two (e.g. hepatocytes, cardiac muscle cells) or more (e.g. osteoclasts). Some fully differentiated cell species lack a N. (e.g. mammalian erythrocytes). Nuclei are usually spherical, but those in long, narrow cells tend to be elongated. The ratio of N. volume to cytoplasmic volume (nucleus-plasma ratio) is constant for each type of cell.

The N. undergoes structural and functional changes during the Cell cycle (see). During interphase (working N.), DNA and RNA are actively synthesized, and the Chromatin (see) becomes visibly organized into chromosomes. During mitosis and meiosis (division N.), genetic material is transferred to daughter cells. The working N. also contains a highly refractive (and therefore easily demonstrable) RNA-rich region known as the *nucleolus,* which is not bounded by a membrane, and is usually spherical in shape. The nucleolus also contains DNA, and it is the site of synthesis and temporary storage of ribosome precursors. Nucleoli may occur singly, or there may be two or more, but they are rarely absent (e.g. in spermatocytes and yeast cells). Nuclei of cells that synthesize large quantities of protein (e.g. gland and nerve cells) usually have several nucleoli. In the division N., ribosomal RNA (rRNA) synthesis is interrupted and the nucleolus disappears. After nuclear division, the nucleolus is reformed at a site known as the *nucleolus organizer,* a loosely packed part of a Satellite chromosome (see): a prenucleolar body is formed from ribonucleoprotein (RNP) fibrils, followed by the deposition of RNP granules, the entire structure finally developing into the nucleolus. Nucleoli contain more than 80% protein (histones are absent), about 15% RNA, as well as the DNA of the nucleolus organizer which contains genes for the synthesis of 45S-rRNA (precursor-rRNA). These genes exist in multiple copies arranged in tandem (hundreds or thousands of repeated copies, known as redundant genes). Some amphibian oocytes contain more than 10^6 rRNA genes per nucleolus.

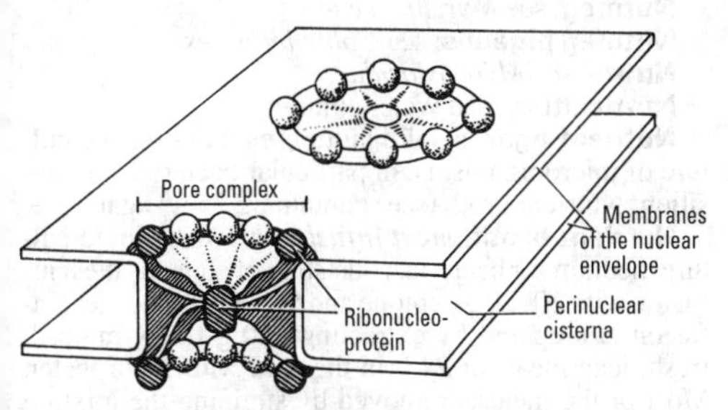

Nucleus. Pore complexes of the nuclear envelope.

In the electron microscope, the nucleolus is seen to consist of a poorly delineated central region of fine fibrils (pars fibrosa) surrounded by a granular region (pars granulosa). The pars fibrosa contains newly formed RNP fibrils (diam. 5–10 nm) and the loosely packed chromatin of the nucleolus organizer (intranucleolar chromatin). The entire nucleolus may also be surrounded by loosely packed chromatin (perinucleolar chromatin). In the cells of higher animals, the nucleolus often carries a cap of Heterochromatin (see) (nucleolus-associated chromatin). The pars granulosa contains RNP granules (diam. 15–20 nm). Under the light microscope, some nucleoli are seen to contain a network (nucleolonema), which probably consists of chromatin surrounded by RNP fibrils.

The RNP fibrils of the pars fibrosa consist of 45S-rRNA coated with proteins. The sequences of 25–28S rRNA, 16–18S rRNA and 5.8–7S rRNA are contained within this 45S-preRNA, and they are excised from it in a multistage process. As each rRNA is formed it associates with ribosomal proteins from the cytoplasm. The RNP granules of the pars granulosa are thought to represent the various intermediates of this process. The final 40S and 60S ribosomal subunits (direct precursors of the small and large ribosomal subunits) enter the cytoplasm separately through the pore complex of the nuclear membrane. Some ribosomal subunits remain in the N., forming ribosomes for protein synthesis within the N.

The nucleoplasm differs from the cell cytoplasm in its high density and strong refractive properties. It contains, inter alia, fibrillar and granular RNP structures and ribosomes. Precursors of mRNA (pre-mRNA) are stabilized immediately after their synthesis by binding to protein, and the resulting RNP complexes (M_r about 800 000) account for most of the RNP structures in the nucleoplasm. These RNP complexes of pre-mRNA (heterogeneous nuclear RNA, HnRNA) are known as *informofers.* After extensive processing, 80–90% of the pre-mRNA is lost, the remaining 10–20% being mRNA. This mRNA associates with specific, nonribosomal proteins, forming smaller *informosomes,* which protect the mRNA against nuclease activity, and serve as a vehicle for conveying it to the cytoplasm. The *nuclear protein matrix* of the nucleoplasm is physically and chemically similar to the cytosol of the cytoplasm. It contains many enzymes, e.g. enzymes for the synthesis of NAD (which is synthesized only in the nucleus) and enzymes of glycolysis. Also known as *nuclear sap* or *karyolymph,* this nuclear protein matrix is a relatively fluid phase of the N., serving as a medium for the transport of material within the N. and between the N. and the cytoplasm.

The nuclear envelope is a locally differentiated part of the endoplasmic reticulum, consisting of a double membrane, enclosing a compressed space or perinuclear cisterna (15–75 nm wide). In certain regions of the nuclear surface, there are visible transitions of the outer membrane layer into the membrane system of the endoplasmic reticulum (plate 21). The nuclear envelope serves as a boundary between the nuclear material and the cytoplasm, and it regulates the exchange of material between these two compartments. In most cells it contains *pore compexes* (diam. 60–100 nm), which are visible under the electron microscope (Fig.), and which function as gates and pumps. These admit proteins synthesized in the cytoplasm (ribosomal proteins, histones, nonhistone nuclear proteins, poymerases) into the N., and allow RNA and ribonuclear proteins to leave the N. and enter the cytoplasm. These transfer processes are at least partly ATP-dependent, and ATPase activity is associated with the pore complexes. At high magnification, the circular, thickened pore rim (annulus) is seen to consist of usually 8 granules. A further single granule occupies the narrow central opening (diam. 15 nm) of the pore. The granules consist of ribonuceloprotein, while the remainder of the pore complex consists of amorphous material, which prevents the passage of any material with a molecular radius greater than 4.5 nm. According to cell type and metabolic activity, nuclear pore complexes show different distribution densities. Particularly high densities (>70 per μm^2) are found in the large nuclei of insect salivary gland cells and in amphibian oocytes. In some cells, the surface area of the N., and therefore the number of pore complexes, is increased by modification of nuclear shape by folding or corrugation, or by the development of lobes (e.g. in peptidergic neurons of vertebrates).

Heterochromatin accumulates on the inner face of the nuclear envelope, while in many cells ribosomes or polysomes are aggregated on the outer nuclear surface. In the prophase of mitosis, the nuclear envelope collapses into small vesicles. During telophase it is formed anew from parts of the endoplasmic reticulum.

In ciliated protozoa, the nuclear functions are shared between a macronucleus and a micronucleus. The polyploid macronucleus regulates metabolism, while the micronucleus contains the genetic information for cell multiplation.

Nuculoida: see *Bivalvia.*

Nudibranchia (naked sea slugs): see *Gastropoda.*

Null hypothesis: see Biostatistics.

Nulliplex: in autopolyploidy, a locus containing only recessive alleles (aaa in triploidy, aaaa in tetraploidy, etc.). If recessive alleles are replaced by dominant alleles, the corresponding genotype is known as a simplex

(Aaa or Aaaa), duplex (AAa or AAaa), triplex (AAA or AAAa), or a quadruplex (AAAA) for triploid and tetraploid genotypes, respectively.

Nullisomy: a form of aneuploidy, in which the chromosomal complement of a diploid cell or organism lacks one chromosome pair (2n – 2). Nullisomic forms in which the missing chromosome pair can be identified are designated with the number of the missing pair, e.g. nulli-I, nulli-II, etc. See Monosomy.

Numbat, ***banded anteater, marsupial anteater,*** *Myrmecobius fasciatus:* a squirrel-sized Australian marsupial with an elongated snout. It feeds mainly on termites and ants, which it catches with the aid of a sticky, very extensible tongue. The dorsal fur is patterned with white, transverse stripes. Some authorites regard it as a divergent dasyurid, and classifiy it as the sole member of a dasyurid subfamily. Others place it in a separate family, the *Myrmecobiidae*.

Numenius: see *Scolopacidae*.

Numerical aperture: see Resolving power.

Numididae, ***guineafowl:*** a small avian family in the order *Galliformes* (see), containing 7 species in 5 genera. Truly wild guineafowl are found only in sub-Saharan Africa, although escaped domesticated birds have reverted to the wild in various parts of the West Indies.

The ***Crested guineafowl*** *(Guttera edouardi;* 50 cm) is found in primary forest and thick bush. It is gray, dotted with bluish white, with a black crest and a bare neck. The bare skin of the face is blue, and the throat is crimson. It feeds on insects (especially termites) and snails, as well as seeds, roots and shoots. The largest species is the ***Vulturine guineafowl*** *(Acryllium vulturinum;* 56–66 cm), so named because its head resembles that of an Old World vulture. The gray-blue facial skin sets off the red eyes, and the bill is bluish white. The plumage is generally black, streaked and spotted with white. There is a long cape of black, white and blue feathers, and a chestnut tuft at the nape. The underparts are blue, and the secondaries are edged with pink. As well as the usual diet of seeds and other plant parts, it also takes snails and insects, and even small reptiles and amphibians.

Nummulites (Latin *nummulus* small cap): an extinct genus of *Foraminiferida* (see) with large lens-shaped or discoid shells, which attained diameters of up to 12 cm. They had a complicated internal skeleton with numerous spirally arranged chambers and canals. As important index fossils of the Tertiary (Paleocene to Oligocene), they existed primarily in subtropical seas, and in particular they flourished in the Tethys sea (see). They were responsible for the formation of the Nummulitic limestones of the Eocene, which are well known for their use in the construction of the Egyptian pyramids.

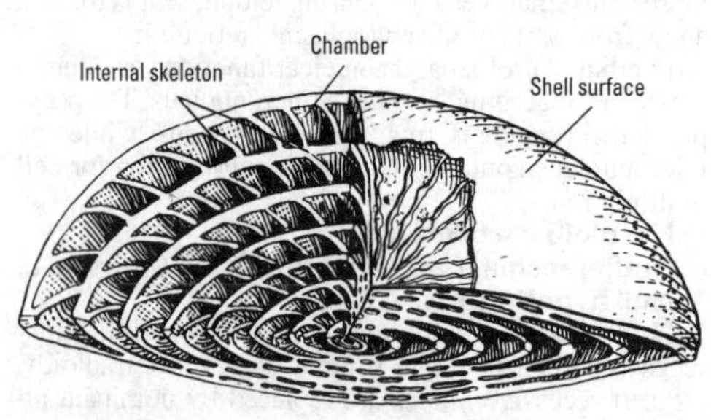

Nummulites. Structure of the test or shell (x 3).

Nunbirds: see *Bucconidae*.

Nuphar: see *Nymphaeaceae*.

Nurse cell: see Trophocyte, Ovary.

Nurse frog: see *Dendrobatidae*.

Nut: see Fruit.

Nutations: movements of the organs of sessile plants (e.g. of tendrils or the shoot terminus of twining plants), due to unequal growth rates on different sides of the organ (see Circumnutation). N. may be induced by external stimuli, or they may be autonomous. In contrast, Variational movements (see) are due to differential changes in cell turgor. See Bending movements.

Nutcracker: see *Corvidae*.

Nuthatches: see *Sittidae*.

Nutmeg: see *Myristicaceae*.

Nutmeg pigeons: see *Columbiformes*.

Nutria: see *Myocastoridae*.

Nutriculture: see Water culture.

Nutrient agar: a solid culture medium for the culture of microorganisms, in particular bacteria. Composition: Nutrient broth (see) containing 25 g/l agar.

Nutrient broth, ***meat infusion broth:*** a liquid culture medium for bacteria, consisting of aqueous meat infusion with 10–20 g peptone and 5% NaCl. The meat infusion is prepared by extracting 500 g finely minced, fresh, lean meat for 24 h in the cold with 1 l of water. Most of the meat is removed by straining the mixture through muslin, and any fat is skimmed from the surface. The mixture is steamed at 100 °C or boiled for 15 min, and coagulated proteins are removed by filtration. The peptone and NaCl are added and dissolved by heating. Since the solution is acidic, due to lactic acid from the meat, the pH is adjusted (usually to 7.5) at this stage. After dispensing into the final growth containers, the medium is sterilized by autoclaving at 120 °C for 15 min.

Nutrient cycles: natural cycles, in which nutrients assimilated by plants are released into the biosphere, then reassimilated. Release occurs by respiration (CO_2) and secretion, and by mineralization of dead plant material. All animal food chains ultimately lead to plant sources, and waste animal products and dead animals are also recycled by decay and mineralization. Important nutrient cycles are the carbon cycle (see Carbon), nitrogen cycle (see Nitrogen) and sulfur cycle (see Sulfur).

Nutrient ratio: the ratio of the available quantities of the plant nutrients, nitrogen (N), phosphorus (P), potassium (K) and calcium (Ca), which is critical for a maximal crop yield. Crop yield is very dependent on the ratio of available nutrients, rather than their absolute quantities. A large excess of a single nutrient is more likely to cause crop damage than to have a beneficial effect. Agricultural plants generally require a N.r. in the following range: N:P:K:Ca = 1:0.2:1.0–1.8:0.2–0.8. However, the optimal N.r. varies according to the type of plant, each species displaying a different absolute requirement for each nutrient. Thus, root and tuber crops have a high potassium requirement, brassicas a high nitrogen requirement, and legumes a high calcium requirement. Potatoes in addition need rather high levels of phosphate. Besides species differences, the optimal N.r. is also influenced by the variety or cultivar, the stage of plant development, environmental conditions, and total quantity of available nutrients. It should also be noted that the quality as well as the quantity of a crop is important in assessing the appropriate N.r.

For practical purposes, the N.r. of fertilizers is usually based only on nitrogen, phosphorus and potassium (see N/P/K ratio).

Since nutrients are already present in the soil, the true optimal N.r. for a plant is not the same as the N.r. that should be applied as fertilizer. See Fertilizers

Nutrimentary oocyte nutrition: see Oogenesis.

Nutrition: the uptake of nutrients for the synthesis of body material, maintenance of the body's physiological functions, and the performance of work.

Animals require preformed organic nutrients; unlike plants, they cannot convert inorganic into organic material. Apart from the major nutrients (protein, neutral fat and carbohydrate), animals also require a number of minerals, vitamins and trace elements. In animals, the structure of the digestive organs is determined by the nature of the diet (see Digestive system).

Nutritional interrelationships: a system of interrelationships in which one type of organism serves as a food source for another, thus resulting in a flow of materials and energy through the relevant ecosystem. Organisms that produce their body materials from inorganic sources, using the energy of sunlight or energy from the oxidation of inorganic compounds are known as *autotrophic organisms* or *autotrophs* (e.g. green plants and chemoautotrophic microorganisms; see Autotrophism, Chemosynthesis). In contrast, *heterotrophic organisms* or *heterotrophs* (fungi, animals, most microorganisms) cannot exist without an external supply of organic compounds.

Autotrophs (e.g. green plants) are *primary producers,* and these serve as a food source for heterotrophs (e.g. animals), which are the *consumers.* The corpses and waste products of the consumers are then degraded (mineralized) by microorganisms *(destruents or decomposers),* the resulting inorganic compounds being reutilized by the primary producers.

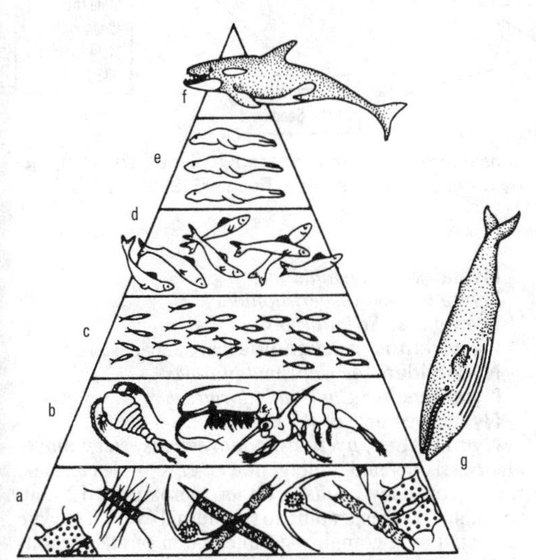

Fig.1. *Nutritional interrelationships.* Food pyramid of marine organisms. *a* Primary producers (phytoplankton). *b* Zooplankton. *c* Plankton eaters. *d* Predatory fish. *e* Seals. *f* Toothed whales (e.g. killer whales). *g* Baleen whales (e.g. blue whales and fin whales).

Due to the interaction of consumers at several different trophic levels, natural N.i. may be very complicated. In the simplest case, N.i. are represented as a *food chain,* e.g. green plant → phytophagous insect (1st order consumer) → predator insect (predator I; 2nd order consumer) → insectivore (predator II; 3rd order consumer) → medium sized predatory mammal (predator III; 4th order consumer) → large predatory animal (predator IV; 5th order consumer). In this hierarchy of trophic levels, the total number of individuals at each level characteristically decreases from the base of the chain to the final consumer, whereas the average size of individual predators increases (this progressive increase in size, however, is observed only for predators; terrestrial autotrophic producers and certain phytophagous organisms do not display this correlation). Food chains rarely consist of more than 4 or 5 trophic levels. The processes by which organisms assimilate their food are not 100% efficient, so that a net loss of nutrients occurs at each trophic level. Within a single ecosystem, the total biomass of producers is therefore greater than that of subsequent consumers. Trophic levels can be represented by a food pyramid (Fig.1). Alternatively, the different trophic levels, as well as the individuals within a level, can be linked to form a intricate *food web* (Fig.2), which represents graphically the degree of *connectedness* or *connectance* between the organisms of an ecosystem. Within this web, plants, phytophagous organisms, predators, parasites and hyperparasites interact at every trophic level; carrion eaters (necrophages), feces eaters (coprophages), and primary and secondary decomposers ensure that organic substances are consumed on every trophic level, and are themselves consumed by parasites and predators, so that the degree of connectance within the food web is very high. While key groups at the base of the food chain can easily replenish their losses, the population size of final consumers (e.g. large predators) is very sensitive to disturbances within the chain.

Organisms which at some stage of their existence feed only on certain parts of a specific food source or host species are said to be *monophagous.* Organisms preferring a relatively restricted group of different food sources (e.g. spurge hawk moth caterpillars feed on various *Euphorbia* species) are said to be *oligophagous.* The close relationships between food plants and monophagous feeders, or between host and parasite lead to coevolution. Thus, the evolution of genetically distinct populations from an original host may be accompanied by corresponding evolutionary changes in the parasite species, so that the monophagous condition is retained. On the other hand, if the parasite does not form new species, but is nevertheless still able to feed on all species of the host, then by definition the parasite becomes oligophagous. *Polyphagous* organisms feed on a broad spectrum of species, but usually prefer certain parts of, or sites within, their food source (e.g. leaf eaters, wood miners, intestinal parasites). *Panthophagous* organisms or *omnivores* (e.g. pig, human, bear) feed on a wide range of different animal and vegetable organisms. Extreme omnivores like the pig are able to feed not only on living plants, fungi and animals, but also on their dead remains (e.g. kitchen waste and carrion).

The special dependence of some herbivores on certain food plants is the result of complicated adaptive processes. Thus, the mustard oil glycosides of the *Cruciferae* are poisonous to many animals, but they act as

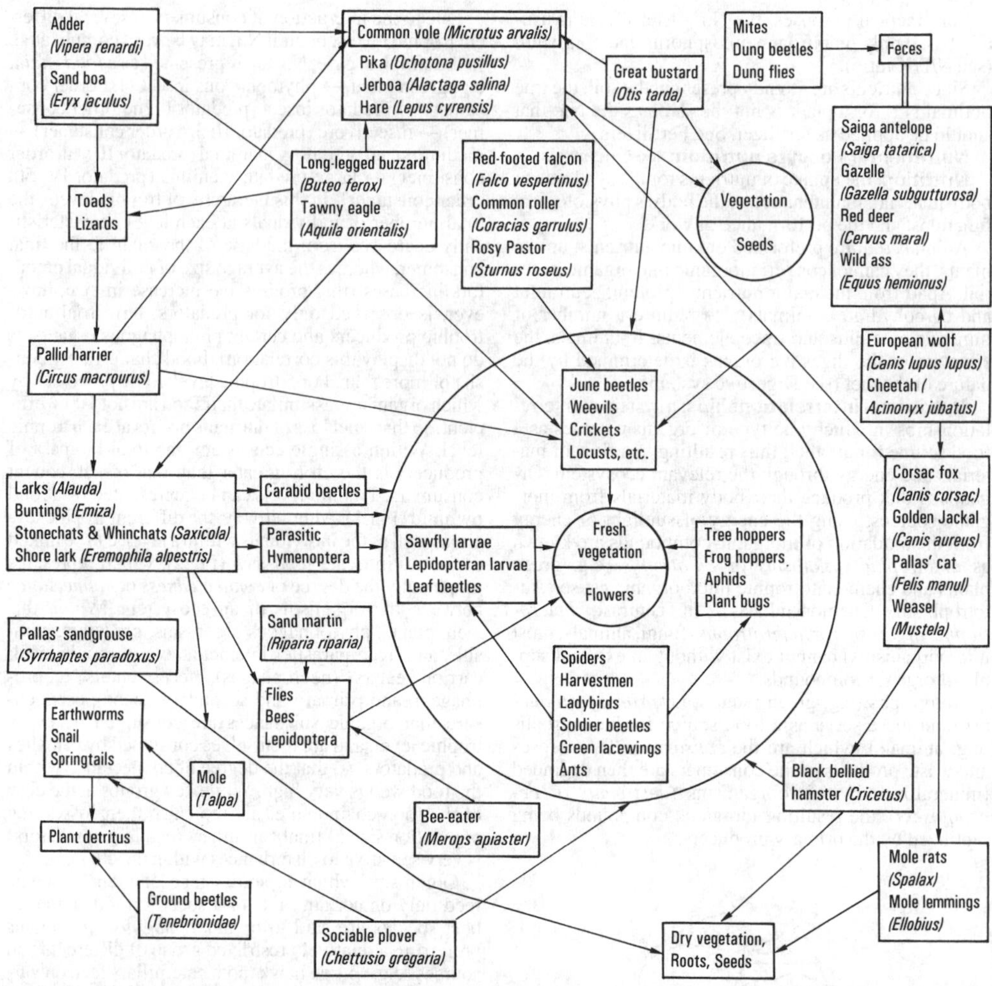

Fig.2. *Nutritional interrelationships.* Comparison of the food webs of the Eurasian steppes (A) and the North American prairies (B). Both represent similar ecosystems, in which analogous niches are occupied by different species (after Tischler, 1955).

attractants and egg laying stimuli to the cabbage white butterfly. Some species of plant bugs of the family *Lygaenidae* are able to store toxic cardiac glycosides and hydrolyse the hemolytic glycosides which they consume when feeding on the foxglove. Any plant that produces poisonous compounds gains a selective advantage, because it is then less likely to be damaged or destroyed by herbivores. Phytophagous feeders, which can nevertheless utilize such a chemically protected natural food source, have an advantage over competitors. Thus, the adaptive strategies of plants and phytophagous species are mutually opposed.

Nutritional physiology: a branch a physiology concerned with the uptake, conversion and utilization of nutrients by plants and animals. In the wider sense, it is the study of the nutrition of all living organisms.

Nutritive cell: see Trophocyte, Ovary.

Nutritive mycelium: see Substrate mycelium.

Nyala: see *Tragelaphinae.*

Nyctalus: see *Vespertilionidae.*

Nyctea: see *Strigiformes.*

Nyctereutes procyonoides: see Japanese fox.

Nyctibiidae: see *Caprimulgiformes.*

Nyctibius: see *Caprimulgiformes.*

Nycticryphes: see *Rostratulidae.*

Nyctinasties, ***nyctinastic movements, sleep movements:*** rising and falling movements of leaves and flowers of many plants, with an approximate 12 hour periodicity corresponding to the diurnal cycle. They are regulated by a complicated interaction of endogenous rhythms and external stimuli, in particular light and temperature. N. do not represent recovery or fatigue.

Most nyctinastic leaf movements occur by the leaf lamina or petiole sinking downward at a joint in the evening, then reversal of this joint movement in the morning, so that the leaf is raised. Many plants, e.g. *Phaseo-*

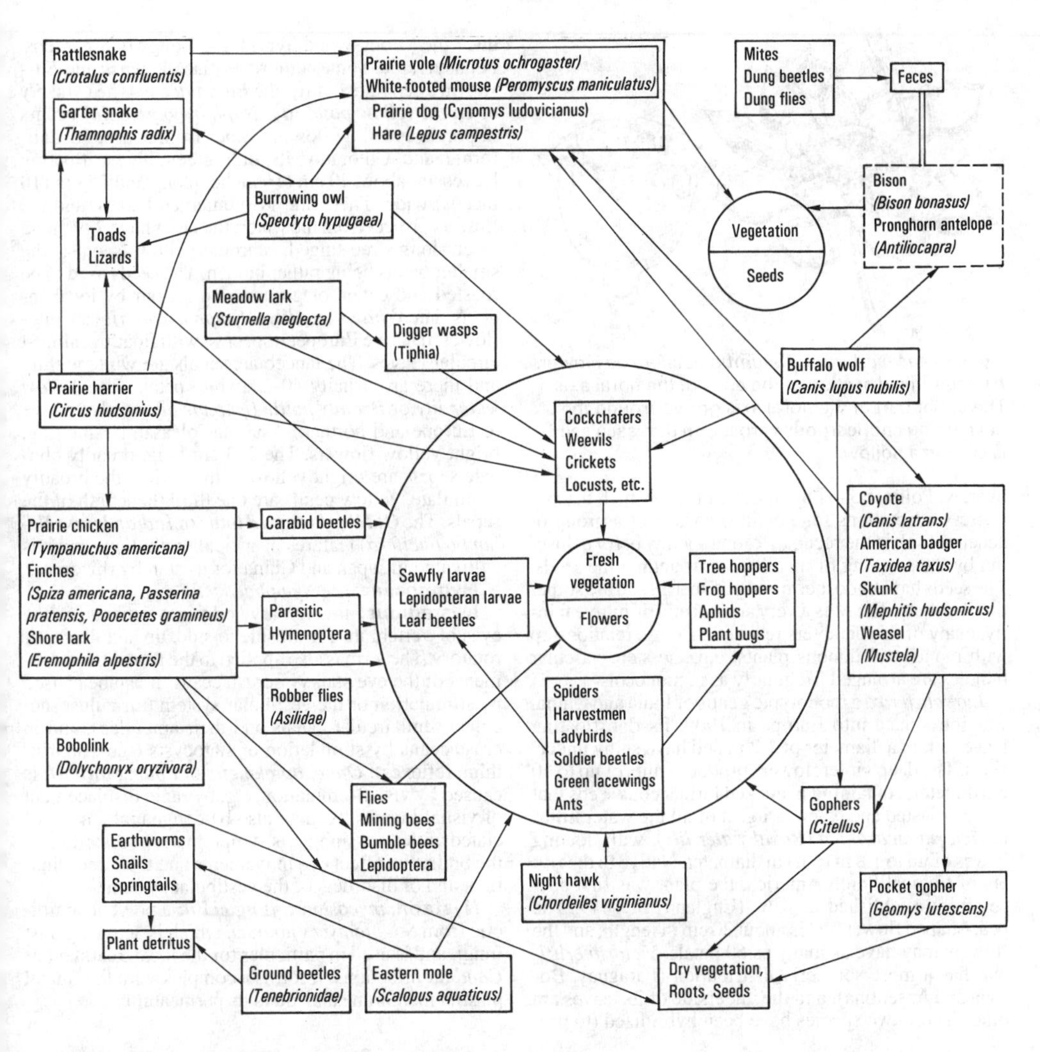

lus, Robinia pseudo-acacia and *Oxalis acetosella* can, however, adapt to a shorter cycle of alternating light and dark. In *Trifolium pratense,* the terminal leaflets remain stiffly erect in the evening and collapse early in the day. *Mimosa* displays nyctinastic movements, so that in the evening the leaves appear as if they have responded to a mechanical shock (see Seismonasty), but in this collapsed condition they remain capable of displaying a seismonastic response. Nyctinastic leaf movements are due mainly to periodic changes of turgor (see Variational movements). Similar movements of flowers and inflorescences are known (see Nodding movements), which are due to differential growth.

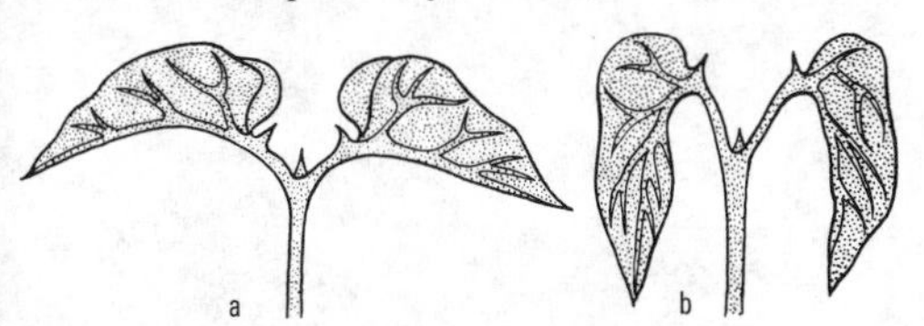

Nyctinasties. Primary leaves of a bean plant showing, *a* the day position, and *b* the night position. In this species, the night species is characterized by drooping of the leaf laminae and raising of the petioles.

In continuous light or constant darkness, N. are at first maintained autonomically, then gradually cease.

Nymphaeaceae, ***water lily family:*** a family of perennial dicotyledonous plants, containing about 80, mostly tropical species; a few species are cosmopolitan. They are exclusively aquatic plants, with stout, creeping rhizomes and large floating leaves, and often also with submerged or with aerial leaves. Flowers are hermaphrodite, actinomorphic, hypogynous to epigynous, solitary, terminal, and generally floating. The perianth usually consists of both sepals and petals, which occur either in trimerous whorls (e.g. *Cabomba),* or in spirals (e.g. *Nuphar).* In some cases (e.g. *Nymphaea)* there is an obvious gradation between the petals and the numerous spirally inserted stamens, and the green calyx was originally the perianth. Each flower contains 3 to many (usually many) stamens and the same number of free or fused

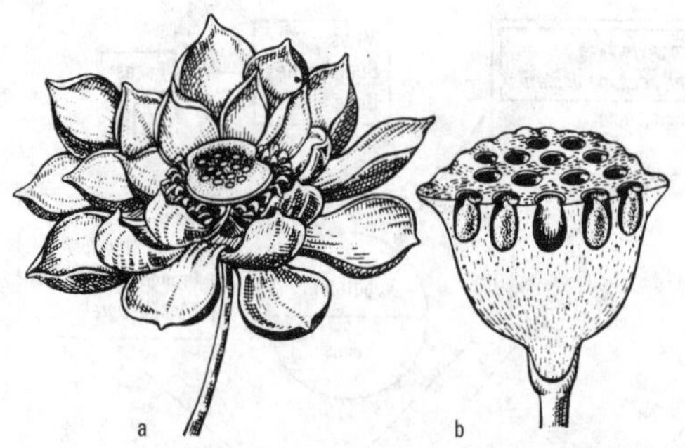

Nymphaeaceae. Lotus *(Nelumbo nucifera). a* Flower. *b* Longitudinal section of the apex of the floral axis. The upper part of the floral axis grows around the developing choricarpous carpels, so that each seed is sunk in a hollow.

ovaries. Pollination is by insects. Fruits (which usually possess adaptations for floating) consist of a group of achenes sunk in the receptacle, or a spongy berry dehiscing by the swelling of mucilage surrounding the seeds. The seeds have endosperm and perisperm, and are sometimes arillate. The *N.* is a very ancient and primitive family, many of its characters revealing a close relationship with monocotyledonous plants, e.g. the stem vascular bundles are arranged irregularly as in monocots.

Euryale ferox, a monotypic genus of India and China, was introduced into Europe in 1909. Its flat, floating leaves attain a diameter of 1.2 m and have spiny undersides. The deep violet flowers produce fruits of up to 10 cm dimeter, containing pea-sized farinaceous seeds that can be roasted and eaten. Largest of all the water lillies is *Victoria amazonica* ***(Royal water lily),*** with floating leaves of up to 1.8 m (6 ft) in diameter. Native to the rivers of tropical South America, the plant was first successfully established at Kew (England) in 1849. The pear-shaped flower bud is about 20 cm in length, and the flowers may have as many as 60 petals. *Victoria cruziana* has a more southerly distribution (Paraguay, Bolivia and Argentina), and the underside of its leaves are bluer. These two species have been hybridized (to produce the "Long wood hybrid") at Longwood gardens, Pennsylvania. Some authorities place *Victoria* and *Euryale* in a separate family, the *Euryalaceae.* The ***Cow lily*** or ***Great yellow pond lily*** *(Nuphar polysepala)* occurs in ponds and slow flowing water from Alaska to California and Colorado. Its dull green, thick, floating leaves are about 40 cm (16 inches) long and 25 cm (10 inches) wide. There are no submerged leaves, and the flowers project from the water; the latter have 7–9 thick, 5 cm-long, red-tinged, chrome-yellow sepals, the smaller petals being rather hidden. The seeds used to be roasted and eaten, or ground into a flour by local Indians. The ***White water lily*** *(Nymphaea alba)* is a white-flowered native European species, with floating, almost circular leaves. The lanceolate sepals are white within, and there are usually 20–25 white petals. The ***Yellow water lily*** or ***Brandy bottle*** *(Nuphar lutea),* also native to Europe and northern Asia, has pleasantly smelling, bright yellow flowers. The 2–3 cm-long, broadly obovate sepals are bright yellow within, while the broadly spathulate, yellow petals are one third the length of the sepals. The ***Old World sacred lotus*** or ***Indian lotus*** *(Nelumbo nucifera)* features in ancient mythology, and it is cultivated in Japan and China for its starchy rhizomes.

Nymphaeae: see *Nymphaeaceae.*

Nystagmus: involuntary jerky movements of the eyes of vertebrates, from side to side, up and down, or rotatory. The term is also applied to the to and fro movements of the eye stalks of crustaceans. It can be caused by stimulation of the vestibular system (utriculus, sacculus, semicircular canals and their ampullae), and in crustaceans by stimulation of statocysts (see Labyrinthine reflexes). *Optic, optokinetic* or *optomotoric N.* is caused by visual simulation, e.g. by rapid displacement of visual images. N. may also be congenital and associated with poor sight; it is symptomatic of disorders of the brain that affect eye movements and their coordination, and of disorders of the vestibular system.

Nystatin, *mycostatin, fungicidin:* a polyene antibiotic from *Streptomyces noursei,* which is active against fungi, and is used in particular for the local treatment of *Candida* infections. It forms a complex with the fungal plasma membrane and alters its permeability.

O

Oak: see *Fagaceae.*

Oak apple: see Galls, Gall wasps.

Oarweed: see *Phaeophyceae.*

Oats: see *Graminae.*

Obelia: see *Hydroida.*

Obligate aerobe: see Aerobe.

Obligate anaerobe: see Anaerobe.

Obligatory learning: the learning of species-specific behavior. O.l. is determined by orienting or releasing stimuli, fixed in space and time, and adapted to a predetermined behavioral framework or setting. Since there is an inner readiness for the learning of species-specific behavior, the time required for O.l. is relatively short. Moreover, the animal is receptive to O.l. only at certain times in its development, known as *critical phases.* A familiar example of this type of learning is Imprinting (see). In humans, the learning of the native language is an example of O.l., whereas the acquisition of a foreign language is Facultative learning (see).

Obolus (Latin from Greek *obolus* an Attic coin): an extinct, primitive genus of the *Brachiopoda* (see), whose fossils are found from the Cambrian to the Ordovician (obolus sandstone). In outline, the nonhinged, chito-phosphate shell is rounded, oval or lens-shaped. A calcareous support for the lophophore is lacking. The valves were held together by six pairs of muscles, and the pedicle extended through a furrow at the posterior of the shell.

Obtect adecticous pupa: see Insects.

Oca: see *Oxalidaceae.*

Occipital condyle: see Skull.

Occiput: see Skull.

Oceanic region, ***oceana:*** a recently defined, climatically very heterogeneous animal geographical region, which forms part of the Notogea (see) and is subdivided into 3 subregions: New Zealand, the Oceanic subregion (Poly and Micronesian islands and Hawaii) and Antarctica. Most of the O.r. was formerly considered to be part of the Australian region (see). It has close connections with both the Australian region and the Oriental region (see). Due to their common origin in the former continent of Gondwana, there are similarities in the invertebrate fauna of New Zealand and the Neotropical region, and the fauna of the Hawaiian Islands shows a strong influence from the Nearctic subregion.

With the exception of some bat species, terrestrial mammals are absent from the O.r. As a result of its insularity or for climatic reasons, the fauna is otherwise impoverished, and the rate of Endemism (see) is relatively high. Endemic bird families are the kiwis *(Apterygidae)* and the extinct moas *(Dinornithidae)* of New Zealand, and the Hawaiian honeycreepers *(Drepanididae)* which show considerable evolutionary variety because they have adapted to numerous free ecological niches. Also notable are the kaka *(Nestor meridionalis)* and kakapo *(Strigops habroptilus)* (both of the family *Psittacidae)* of New Zealand. Despite its size and favorable climatic conditions, New Zealand has only 65 species of native land birds, while Hawaii had 41 or 42 (fewer now that some honey creepers have become extinct).

The relatively poor reptilian and amphibian faunas display certain peculiarities. The tuatara *(Sphenodon punctatus)* of New Zealand is an endemic "living fossil", which is the only representative of its order. The islands of Fiji and Tonga are home to a single representative of the iguanas *(Iguanidae),* which are otherwise found only in America and Madagcascar. Several species of Candoia boas *(Boinae)* (e.g. *Candoia bibroni, Candoia aspera),* which are representatives of a neotropical group, occur on Fiji, Tonga and other islands stretching to New Guinea and into the Indoaustralian transition zone (see Wallacea). New Zealand has a few species of geckos and skinks, but no snakes. A family of very primitive frogs *(Leiopelmatidae)* inhabits the uplands of the North Island of New Zealand. The Hawaiian Islands were invaded by neither reptiles nor amphibians. They do, however, offer a notable example of adaptive radiation by certain snails, with no fewer than 300 species of the endemic family, *Achatindellidae.*

On many islands, especially New Zealand and the Hawaiian Islands, the endemic faunas have been both enriched and harmed by the introduction of foreign species. Numerous native species have given way to newly introduced species and are now threatened with extinction or are already extinct.

The Antarctic subregion has an extraordinarily species-poor fauna, consisting of a few marine mammals and birds, and a small number of invertebrates, namely insects (including parasites of vertebrates), mites, nematodes, rotifers, tardigrades and small freshwater crustaceans.

Oceanic rule: see Ecological principles, rules and laws.

Oceanites: see *Procellariiformes.*

Oceanodroma: see *Procellariiformes.*

Ocean perch: see Redfish.

Ocellated green lizard, ***eyed lizard,*** *Lacerta (Gallotia) lepida* (plate 5): the largest of the lacertid lizards (see *Lacertidae),* with a maximal total length of 80 cm, found in southern France, Spain and Northwest Africa. Its back is mottled yellow-green, and the flanks are decorated with rows of blue spots. In addition to large insects, it feeds mainly on small lizards and young mice.

Ocellus: see Light sensory organs.

Ocelot: see *Felidae.*

Ochotonidae, ***picas, pikas, whistling hares, calling hares, mouse hares, rock conies:*** a family of the *Lagomorpha* (see). Ochotonids are small, short-haired mammals, with no externally visible tail, and with a characteristic shrill, piping voice. They inhabit the steppes and mountainous areas of Asia, North America and southeastern Europe, sometimes at extremely high altitudes where many mammals cannot survive. They do not hibernate, and they feed during the winter on plant material gathered during the summer and autumn. The single recent genus, *Ochotona,* is represented by 12 species in Asia, and 2 species in North America. Dental formula: 2/1, 0/0, 3/2, 2/3 = 26.

Ochotonidae. Daurian pika *(Ochotona daurica).*

Ocimene: a triply unsaturated, open chain monoterpene hydrocarbon. It is a pleasantly smelling oily liquid, and a component of many essential oils.

Octobrachia: see *Cephalopoda.*

Octocorallia, *Alcyonaria:* a subclass of the *Anthozoa.* The gastrovascular space is always divided into 8 compartments by 8 mesenteries. In addition, there are always 8 tentacles, which are invariably pinnate (i.e. with feather-like side branches). The colonial polyps are interconnected by a thick mass of mesoglea, known as coenenchyme. This is perforated by gastrodermal tubes, which are continuous with the gastrovascular cavities of the polyps. A layer of epidermis covers the coenenchyme and the polyp column, so that only the upper part of the polyp projects. Only the polypoid stage exists, and practically all species are colonial. A skeleton is secreted, composed of individual small sclerites. About 2 000 species are known, all of them marine.

The subclass contains the following orders: *Alcyonaria* (see) ***(Soft corals),*** *Gorgonaria* (see) ***(Horny*** or ***Gorgonian corals),*** *Helioporida* (see) ***(Blue coral),*** and *Pennatularia* (see) ***(Sea pens).***

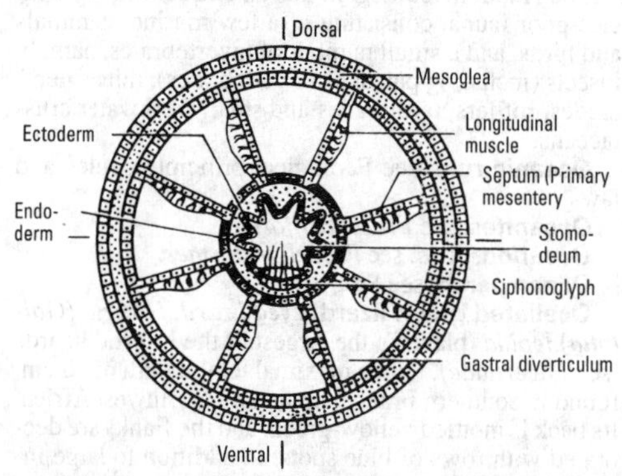

Octocorallia. Schematic transverse section through *Octocorallia* in the region of the stomodeum.

Octodon: see *Octodontidae*

Octodontidae, ***octodont rodents, hedge rats:*** a family of Rodents (see) comprising 6 genera and 8 species, found from moderate heights up to 3000 m in South America. The head is relatively large with a blunt snout and long vibrissae. Each limb has 5 digits, and the thumb has a nail instead of a claw. The toes of the hindfeet have stiff bristles extending beyond the claws. A pattern of grooves on the surface of the molars describes a figure of 8, hence the name of the family. The ***Degu*** *(Octodon degus,* body length 15 cm) is common in central Chile. Large colonies live in earth burrows, and spend much time climbing in hedges and bushes. The ***Cururo*** *(Spalacopus cyanus,* body length 14–16 cm) has large, protruding incisors, and the animal is generally adapted for a fossorial existence. Also an active burrower is the ***South American rock rat*** *(Aconaemys fuscus,* body length 15–18 cm). The ***Viscacha rat*** *(Octomys mimax),* like the ***Degu,*** is not specially adapated for digging.

Octomys: see *Octodontidae.*

Octopoda: see *Cephalopoda.*

Octopus: see *Cephalopoda.*

Oculina: see *Madreporaria.*

Oculomotor nerve: the II cranial nerve. See Cranial nerves.

Ocypodidae: see Fiddler crabs.

Odiferous gland: see *Heteroptera.*

Odiferous principles, ***scents:*** volatile substances produced by living organisms, which are released into the air or into water, often in extremely small quantities, and which are perceived by individuals of the same or different species by olfaction. O.p. are represented by many different types of chemical structure, and they are often difficult to identify. O.p. produced by animals serve for defense and deterrence (e.g. the O.p. of skunks), territorial marking, route marking, and as sexual attractants. Some plants also produce O.p. that attract animals, often as an aid to pollination.

Odobenidae: a family of the *Pinnipedia* (see) represented by a single species, the ***Walrus*** *(Odobenus rosmarus).* The walrus is found on shorelines in the Arctic waters of the Pacific and Atlantic, sometimes straying further south. This robust pinniped weighs up to nearly 1 300 kg, and its skin is practically hairless. Progress on land is ponderous and slow, but aided by the hindflippers, which can be brought beneath the body. The upper canines are modified into long tusks, which in the adult are entirely lacking in dentine, i.e. they consist of ivory. Incisors are absent, and there are usually 12 cheek teeth; dental formula: 1–2/0, 1/1, 3–4/3–4, 0/0 = 18–24. The tusks serve as weapons and for raking mollusks from the sea floor. Two subspecies are recognized: *O.r. divergens* ***(Pacific walrus;*** nostrils not visible when viewed from front) and *O.r. rosmarus* ***(Atlantic walrus;*** nostrils visible when viewed from front). Walrusses are gregarious and polygynous, often gathering in groups of about 1000 individuals.

Odobenus rosmarus: see *Odobenidae.*

Odocoileus: see White-tailed deer, Mule deer.

Odonata: see *Odonatoptera.*

Odonatoptera, *Odonata:* an insect order containing about 5 000 known species from different areas of the world, divided into 2 suborders: *Zygoptera* ***(Damselflies)*** and *Anisoptera* ***(Dragonflies).*** The earliest fossil *O.* are found in the Upper Carboniferous.

Imagoes. In some species, the elongated body is up to 13 cm in length. The head carries large, prominent, compound eyes, chewing mouthparts, and short, setaceous, inconspicuous antennae. The long, usually, slim and often brightly colored abdomen is normally distinctly differentiated from the somewhat broader thorax. There are 2 pairs of large, approximately equalsized, richly veined, mostly transparent wings. Dragonflies are particularly powerful fliers, catching their prey (other insects) in flight, with the aid of their long legs. In pairing, the male grasps the female's prothorax (damselflies) or head (most dragonflies) with his terminal abdominal claspers, so that the two insects form a chain or tandem. The female then loops her abdomen forward to bring its tip into contact with the accessory genitalia on the male's second and third abdominal segments (Fig.1). *O.* sometimes form migratory swarms, but the reason for this is unclear. Metamorphosis is incomplete (hemimetabolous development).

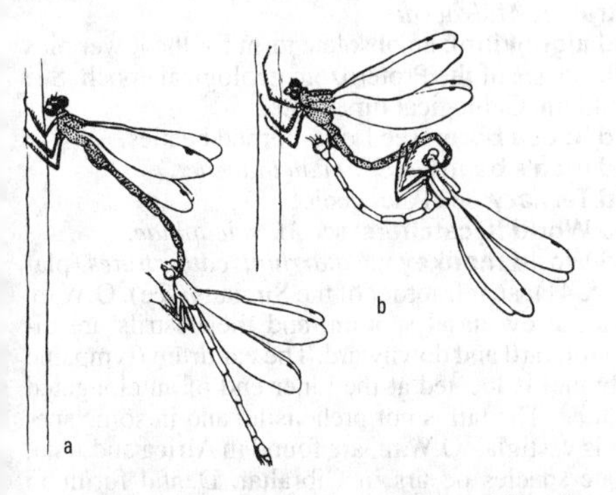

Fig.1. *Odonatoptera.* Pairing of damselflies *(Lestes dryas). a* "Chain" or "tandem" position; male grasping female's prothorax with his terminal abdominal claspers. *b* Sperm transfer in the "wheel" position; female loops her abdomen forward, bringing tip into contact with male's accessory genitalia. The male is shaded gray.

Eggs. Oviposition commences directly after copulation. 200 to 1 600 eggs, measuring 0.5 to 2 mm, are simply dropped into the water, or they are laid on or in water plants, or on moist earth at the water's edge.

Larvae. In temperate climates, eggs laid late in the year remain dormant through the winter. Otherwise, the prelarva hatches 2 to 5 weeks after oviposition. In seconds or minutes, the prelarva molts to produce the first true larva. The latter has large eyes and a rapidly extendable mask, carrying hooklike labial palps, which are used for seizing prey (small water animals). The otherwise slow moving larvae live in both standing and flowing water. The last of the 10 to 15 molts, which releases the imago, occurs out of the water. In most species, larval development lasts 1 to 3 years.

Economic importance. Dragonflies and damselflies are neither beneficial nor harmful, except that the larvae of some species are an intermediate host of a parasitic trematode *(Prosthogonimus pellucidus* v. Linst) which attacks the oviducts of poltry

Classification.

1. Suborder: *Zygoptera* ***(Damselflies).*** These are predominantly small species, with widely separated eyes. Fore and hindwings are identical in shape and venation, and folded together above the insect at rest. Larvae have 3 well developed terminal processes which form caudal gills. Rectal tracheal gills are absent.

Family: *Coenagriidae (Coenagrionidae).* This is a cosmopolitan family, and some genera (e.g. *Coenagrion)* occur worldwide. Females are usually uniformly dark, while males have blue and black patterning on their dorsal surface. Nymphs are found on the vegetation of still or slow moving water.

Family: *Agrionidae (Agridiidae, Calopterygidae).* These have brilliant metallic colored bodies; males also have colored wings. Nymphs are found at the edges of fast flowing water.

Family: *Lestidae.* A large cosmopolitan family. Imagoes are metallic colored. Still water is preferred, and eggs are laid on emergent vegetation. The elongate nymphs clamber actively in submerged vegetation.

2. Suborder: *Anisoptera* ***(Dragonflies).*** Predominantly medium to large species. Eyes are contiguous or nearly so. Fore and hindwings have similar venation, the hindwings are usually wider than the forewings, and the wings are held outward, horizontally or depressed when at rest. Nymphs have concealed rectal tracheal gills.

Family: *Aeschnidae.* A primitive, large, cosmopolitan family, containing the largest and strongest flying of the recent dragonflies. Females have a zygopteran-type ovipositor. The very large eyes meet dorsally in the midline of the head. Nymphs are stout and elongate, preferring vegetation is still or slow moving water.

Family: *Libellulidae.* The largest and most heterogeneous dragonfly family, distributed worldwide. Still water is preferred. Nymphs are found in rotting vegetation on the bottom of lakes or ponds; some are adapted for living among moist forest floor debris.

Family: *Gomphidae.* A primitive family with widely separated eyes. Imagoes are black and yellow. The ovipositor is rudimentary. Eggs are laid directly into fast flowing water, but they have adhesive properties which prevent them from being washed away. Nymphs burrow or sprawl, and burrowing species have an elongated 10th abdominal segment, which protrudes from the substratum and permits respiratory gas exchange with the water.

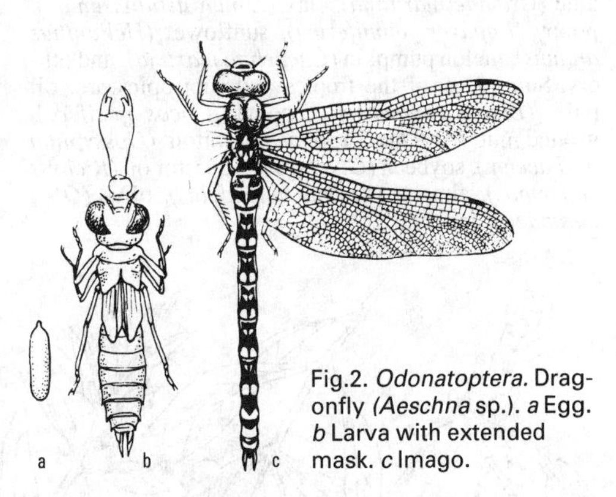

Fig.2. *Odonatoptera.* Dragonfly *(Aeschna* sp.). *a* Egg. *b* Larva with extended mask. *c* Imago.

Odontaspididae: see *Selachimorpha.*
Odontoceti (Odontocetes): see Whales.
Odontophorinae: see *Phasianidae.*
Odontophorus: see *Phasianidae.*
Oecium: see *Bryozoa.*
Oegopsida: see *Cephlopoda.*
Oenanthe: see *Turdinae.*
Oenocyte: see Enocytes.

Oenothera: see *Onagraceae.*

Oenotheraceae: see *Onagraceae.*

Oesophagus: British spelling of esophagus. See Digestive system.

Off center field: in neurophysiology, a basic type of Receptive field (see), which possesses Off elements (see) in the central region and On elements (see) at the periphery. See On center field.

Off elements: sensory or nerve cells which produce no or few impulses when exposed to a stimulus, but produce numerous impulses when the stimulus is removed.

Ogcocephalidae: see Batfish.

Oidia (sing. **Oidium):** thin-walled spores resembling mycelial fragments (see Arthrospores), produced by lower fungi, rarely by higher fungi, by the disarticulation of chambered hyphae into individual cells, e.g. in *Geotrichum candidum.* In a few higher fungi, O. are abstricted from the tip of a short hyphal branch known as the oidiophore. They may be produced by monokaryotic or dikaryotic hyphae, thus forming uninucleate or binucleate O. The resulting O. may germinate to produce a uninucleate, primary mycelium. A uninucleate oidium may also behave as a spermatium and unite with a uninucleate hypha, thereby producing a new dikaryotic hypha.

Oidium: the singular form of Oidia (see), or the former name of the fungal genus, *Acrosporium,* of the powdery mildews *(Erysiphales).*

Oilbird: see *Caprimulgiformes.*

Oil bodies: see *Hepaticae.*

Oil of Theobroma: see Cocoa butter.

Oil palm: see *Arecaceae.*

Oil palm squirrels: see *Sciuridae.*

Oil plants: plants whose fruits or seeds yield technically useful oil. O.p. cultivated for their oil in temperate climates include: rape *(Brassica napus* var. *napus),* colza or cole-seed *(Brassica napus* var. *oleifera),* black and white mustard *(Brassica nigra, B. alba),* oil-rocket *(Eruca sativa),* gold of pleasure *(Camelina sativa),* seakale *(Crambe maritima),* flax *(Linum usitatissimum),* poppy *(Papaver somniferum),* sunflower *(Helianthus annuus),* melon pumpkin *(Cucurbita maxima),* and others. Some O.p. of the tropics and subtropics are: oil palm *(Elaeis guineensis)* coconut *(Cocos nucifera),* ground nut *(Arachis hypogaea),* cotton *(Gossypium herbaceum),* soybean *(Glycine max),* castor oil *(Ricinus communis),* almond *(Prunus amygdalus),* olive *(Olea europaea),* and others.

Many O.p. also yield fat, e.g. coconut fat. On account of their content of unsaturated fatty acids, plant fats and oils represent a valuable dietary supplement to animal fats.

Oils: water-insoluble, liquid organic compounds. They are combustible, lighter than water, and soluble in ether, benzene and other organic solvents. Most naturally occurring O. originate from plants (fish liver oil is a notable exception). They may be saponifiable acylglycerols (see Fats), or nonsaponifiable essential oils.

Okapi: see *Giraffidae.*

Okapia: see *Giraffidae.*

Okasaki fragment: see DNA replication.

Okinawa habu: see Lance-head snakes.

Okra: see *Malvaceae.*

Old algonkium: an obsolete term for the lower plus middle stages of the Proterozoic geological epoch. See Algonkium, Geological time scale.

Old house borer: see Long-horned beetles.

Old man's beard: see *Ranunculaceae.*

Old Tertiary: see Paleogenic.

Old World flycatchers: see *Muscicapinae.*

Old World monkeys, *Catarrhini,* ***catarrhines*** (plates 40 & 41): an infraorder of the Simians (see). O.W.m. have a narrow nasal septum, and the nostrils are directed forward and downward. The ear drum (tympanic membrane) is located at the inner end of an elongated bony tube. The tail is not prehensile, and in some species it is vestigial. O.W.m. are found in Africa and Asia, and one species occurs in Gibraltar. Dental formula: 2/2, 1/1, 3/2, 3/2 = 32. The following O.W.m. are listed separately: Macaques, Baboons, Patas monkey, Mangabeys, Allen's swamp monkey, Talopins, Gelada, Mandrill, Guenons, Colobine monkeys, Gibbons, Great apes. See Primates.

Old World porcupines: see *Hystricidae.*

Old World rats and mice: see *Muridae.*

Old World robins: see *Turdinae.*

Old World seed eaters: see *Ploceidae.*

Old World sparrows: see *Ploceidae.*

Old World tree frogs: see *Rhacophoridae.*

Old World vultures: see *Falconiformes.*

Old World warblers: see *Sylviinae.*

Oleaceae, ***olive family:*** a family of dicotyledonous plants containing about 600 species, cosmopolitan in distribution, but especially well represented in warm regions. They are trees or shrubs with simple or pinnate, opposite, exstipulate leaves, and actinomorphic, usu-

Oleaceae. 1a Ash fruits *(Fraxinus excelsior),* commonly called ash keys. *1b* Olive fruits *(Olea europea). 1c* Lilac fruits *(Syringa vulgaris). 2* Flowering shoot *(d)* and flower *(e)* of olive.

ally 4-merous flowers, which are insect- or wind-pollinated. The superior, 2-celled ovary develops into a variety of different fruits.

The fruit of the ***Common ash*** *(Fraxinus excelsior)* is a winged, one-seeded nut, known as an ash key. The common ash has both hermaphrodite and unisexual flowers; these have no sepals or petals, are wind-pollinated, and open in the spring before the leaves. Ash wood is relatively hard, but bendable (by steaming), and it is used in furniture making and for tool handles. The strongly scented, insect-pollinated flowers of the southern European ***Manna ash*** *(Fraxinus ornus)* have a white corolla and are grouped in conspicuous pannicles. The manna ash, which attains a height of 8 m, is planted as an ornamental. It is also exploited as a source of manna (prepared from the exudate of the wounded bark, and consisting mainly of mannitol), which is official as a gentle laxative. The fruit of the Mediterranean ***Olive tree*** *(Olea europaea)* is an edible drupe, the mesocarp of which contains 22% of an edible oil, used in cooking, for technical purposes, and in pharmacy. The southern European ***Lilac*** *(Syringa vulgaris)* and the East Asian ***Forsythia*** *(Forsythia suspensa)* both produce dehiscent capsules, and they have become widely distributed as ornamentals. ***Privet*** (the native European *Ligustrum vulgare,* and the Japanese *L. ovalifolium)* and ***Jasmine*** *(Jasminum;* various species from China, the Himalayas, western Asia and the eastern Mediterranean) are also widely planted as ornamentals, but have berry fruits. Jasmine flowers are used in the perfumery industry for extraction of their abundant and pleasantly-scented essential oils.

Olea europaea: see *Oleaceae.*

Oleander: see *Apocyanaceae.*

Oleaster: see *Elaeagnaceae.*

Oleic acid: Δ^9-octadecanoic acid (see Formula), the most widely distributed naturally occurring unsaturated fatty acid. It is a colorless, odorless liquid, which gradually autoxidizes and turns brown when left exposed to the air. It is a constituent acid of triacylglycerols in many plant and animal oils, e.g. olive oil, almond oil and fish liver oil. It is also present in practically all the acylglycerols of depot and milk fats, as well as occurring in phospholipids. The double bond has the *cis*-configuration. The *trans*-isomer is called elaidic acid.

Commercially, liquid fats (triacylglycerols) containing unsaturated fatty acids are converted into solid fats by catalytic, high-pressure hydrogenation ("fat hardening"). In this way, triolein (the triester of oleic acid with glycerol) is converted into tristearin (m.p. 71 °C), which is a component of margarine. Triolein and tristearin are among several fat components used in the manufacture of soap.

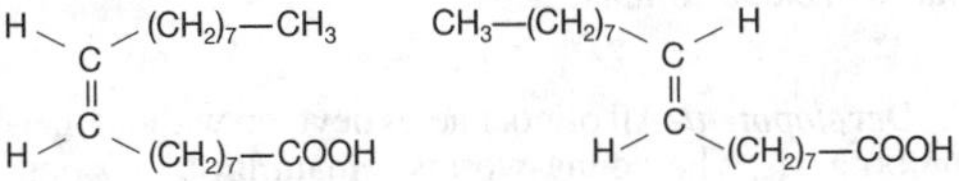

$$\begin{array}{c} H\diagdown\quad\diagup(CH_2)_7{-}CH_3 \\ C \\ \| \\ C \\ H\diagup\quad\diagdown(CH_2)_7{-}COOH \end{array} \qquad \begin{array}{c} CH_3{-}(CH_2)_7\diagdown\quad\diagup H \\ C \\ \| \\ C \\ H\diagup\quad\diagdown(CH_2)_7{-}COOH \end{array}$$

Oleic acid   Elaidic acid

Olenellus stage: Lower Cambrian (see Cambrian).

Olenus stage: Upper Cambrian (see Cambrian).

Oleoresins: see Balsams.

Oleosomes, ***spherosomes:*** lipid-rich, spherical plant-cell organelles. They are bounded by a membrane, and have a diameter of 0.8–1 mm. O. are particualrly abundant in the cotyledons of fat-storing seeds, where they are found in close juxtaposition to the glyoxysomes (see Peroxysomes), which contain enzymes for the degradation of triacylglycerol and metabolism of the resulting fatty acids. Membranes of both O. and glyoxysomes contain lipase. Acid phosphatase, which is a marker enzyme for lysosomes, has also been demonstrated in O. It is thought that O. are formed by abstriction from the endoplasmic reticulum. During their formation they probably contain enzymes for oil and fat synthesis. Older O. may degenerate into oil droplets.

Oleum betulae empyreumaticum: see Birch tar.

Oleum ricini: see Castor oil.

Oleum rusci: see Birch tar.

Olfactory nerve: the I cranial nerve; see Cranial nerves.

Olfactory organ: animal sensory organ for the perception ("smelling") of substances in the gaseous state or dissolved in water. The two chemical senses are taste and smell; O.os. are responsible for the sense of smell, while gustatory organs are responsible for taste. Olfactory sensory cells are situated in various parts of the body, usually in the region of the mouth, but in some animals they distributed over the entire body. Olfactory cells are often aggregated into sensory buds.

In lower animals, there is often only a single type of chemical sensory cell, which serves for both taste and smell, e.g. *Turbellaria, Polychaeta* and *Hirudinea.* In arachnids and crustaceans, the O.os. are often located in the legs, e.g. in ticks *(Ixodidae),* Haller's organ (a chemoreceptor located in a sensory pit) is found on the tip of each of the first walking leg tarsi.

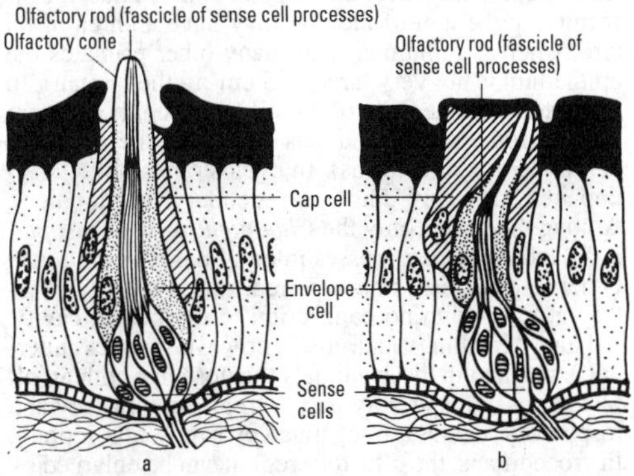

Olfactory organ. Insect chemoreceptors. *a* Basiconic sensillum. *b* Placoid sensillum or pore plate.

Insect chemoreceptors consist of sensilla formed by the parallel aggregation of processes (hairs) from primary neurons. 1) *Sensilla trichoid olfactoris* (olfactory trichoid sensilla) are thin-walled hairs observed on the tarsi and labella of *Tabanus,* the tarsi of *Pyrameis* and *Musca,* and on the antennae of many insect species. *Sensilla chaetina* are very similar but shorter and stouter. 2) *Sensilla basiconica* and *styloconica* (basiconic and styloconic sensilla) (Fig.1, ***a)*** are peg- or cone-like organs. They closely resemble olfactory trichoid sensilla, but the projecting part is more heavily built. Particularly common on insect antennae, they also occur on the palps of *Hydrophilus piceus* and *Agriotes* larvae, and on the antennae and palps of *Peripla-*

neta and larvae of the *Lepidoptera.* Very similar sensilla occur on the antennae of muscids and other dipterans and on the palps of butterflies, where they are aggregated in groups and sunk into a pit. 3) *Sensilla coeloconica* lie below the level of the integument, but are otherwise similar in structure to basiconic sensilla. In some species, the sensillum is deeply sunk at the bottom of a tube, known as a *sensillum ampullacea.* Coeloconic sensilla are found on the antennae of various *Coleoptera* and *Hymenoptera.* 4) *Sensilla placodea* (pore plates) (Fig.1, ***b)*** consist of a thin cuticular plate covering a cap cell and an envelope cell, which envelop the aggregated distal processes from the bipolar neurons. The terminal strand ends in the pore plate. Pore plates are particularly abundant on the terminal 8 segments of the antennae of *Apis mellifera,* and they also occur in the antennae of many other insect species. Perception of certain olfactory substances is highly developed in some insects; the stimulatory threshold varies greatly for different materials, and is particularly low for essential stimuli like those of food and sex pheromones.

In mollusks, O.g. are aggregated on the large tentacles. In addition, there is strong evidence that the Osphradium (see) and less certainly the Rhinophore (see) have an olfactory function.

The olfactory sensory cells of vertebrates are concentrated in the olfactory mucosa of the nose. Each sensory cell carries a sensory hair, which is embedded in the mucus of the olfactory mucosa. In many species, the surface area of the mucosa is increased by folding. For example, the internal surface of the olfactory chamber may be highly complicated due to spreading of the olfactory epithelium over the turbinal bones. Thus, the olfactory epithelium of the dog may have an area of 85 cm^2, whereas in humans and many other primates the epithelium is not very large (2.5 cm^2 in the human). In all vertebrates the cells of the olfactory epithelium are also neurons, and their axons constitute the olfactory nerve (nervus olfactorius). In cyclostomes, the O.o. is unpaired, probably due to a secondary development. Among the fishes, only the *Choanichthyes (Crossopterygii* and *Dipneusti)* posses internal nares or *choanae.* Otherwise, all fishes have paired, round to oval olfactory grooves with no connection to the buccal cavity. All lung-breathing quadrupeds, however, possess internal nares, and in these animals the nose has both an olfactory and a respiratory part. In reptiles, olfactory and respiratory regions are separated by a muscular septum. In crocodilians, the olfactory region can be enlarged by protrusion of the mucosa, in birds and mammals by the action of a varying number of nasal muscles, which may also be branched (dog, deer).

Oligochaeta, ***oligochaetes, earthworms and allies:*** a subclass of the *Annelida,* containing about 3 100 species. More than 80% of these are burrowing earthworms, most of the remainder being small or very small terrestrial or aquatic freshwater worms (a few marine species are also known). The chaetae are less numerous and less strongly developed than those of *Polychaeta.*

Morphology. Chaetae are usually present in bundles, embedded in ectodermal pits (chaetigarous sacs) in the body wall . Parapodia are absent. Body lengths vary from 1 mm to more than 2 m. The spindle-shaped to cylindrical body is circular in transverse section, and it consists of between 10 and more than 600 segments. At the anterior end, a flap or prostomium projects over the mouth in the first segment. Body segments display the typical internal structure of other annelids. Two (more rarely 1 or 3) bundles of chaetae are present on each side of each segment. Chaetae may be modified for creeping, attachment or swimming, or they may serve a sensory function. The last segment is perforated by the anus, and is always without chaetae. In some aquatic species, the rear portion of the body carries gill appendages, and in some parasitic species it is modified to a posterior sucker. Small species are often transparent. Larger forms possess skin pigments, and are often red, brown or violet, sometimes vividly multicolored. The vascular system is closed. The nervous system consists of cerebral ganglia ("brain") connected by a nerve collar (circum-pharyngeal connectives) to the ventral nerve cord. Sensory organs include sensory hairs, epidermal sense organs and photosensitive organs scattered over the body. Some species have anterior ciliated olfactory grooves and one or two pairs of cup-shaped pigment cells. All oligochaetes are hermaphrodite and the reproductive organs (1 or 2 pairs of testes and ovaries) are restricted to a few adjacent anterior segments, the testes anterior to the ovaries. These segments swell to form a glandular, saddle-like structure (clitellum), which secretes the egg cocoon. In some some species the clitellum is always present, while in others it only appears during reproduction.

Biology. Oligochaetes are found on the bottom and in the vegetation zone of all freshwater habitats (a few marine species are also known), or in damp earth, sometimes in burrows at depths of several meters. The majority of species move by creeping, and a few are able to swim. Oligochaetes feed chiefly on dead plant and animal remains, but live plant material or small aquatic animals may also be taken. Some species live parasitically on freshwater crustaceans and fishes. During copulation, two worms come to lie with their anterior ends together, facing in opposite directions. Sperm is released and stored in the sperm pockets (spermathecae) of the partner. After copulation, a mucous tube (the future cocoon) is secreted around the clitellum. Eggs are released into this cocoon, and as the worm withdraws backward from it, sperm is released over the eggs.

Oligochaeta. Two earthworms (co-copulants) lying "head to tail" in the pairing position with ventral surfaces in close contact.

Development. All oligochaetes develop without metamorphosis. The young worms remain in the cocoon, which is filled with a protein-rich nutritive fluid, until they are capable of an independent existence. They grow by development of new segments at the posterior end. Some species can reproduce asexually by transverse cleavage, with subsequent regeneration of both body parts to form 2 new individuals. Certain species display a regular alternation of asexual and sexual reproduction.

Classification. Three (sometimes 4) orders are recognized, based on the position of the male sexual opening behind the testes segment:

1. Order: *Plesiopora* or *Tubificida* . A single pair of testes is followed by a single pair of ovaries in adjacent segments. The male gonopores are located in the segment before or behind the testicular segment. Setae may be hairlike or otherwise modified; they occur in bundles of two or more, and are rarely absent. There are 5 families, including the *Tubificidae* and *Enchytraeidae.*

2. Order: *Prosopora* or *Lumbriculida.* At least one pair of male funnels is present in the same segment as the male gonopores. There are 4 pairs of setae per segment. The order is represented by a single freshwater family, *Lumbriculidae.*

3. *Opisthopora* or *Haplotaxida.* In the original condition, 2 segments with testes are followed by 2 segments with ovaries. Often, however, one pair of testes or ovaries, or one pair of each has been lost. The male pores are one or more segments behind their funnels. *Lumbricidae* is one of the 18 families.

Earthworms (e.g. *Lumbricus,* family *Lumbricidae)* vary in length from a few centimeters to 3 m. They burrow tubular passages in the earth or in decaying organic material, and they improve soil quality by generating a looser soil structure, increasing aeration, improving drainage, carrying humus to lower layers, and fertilizing the soil with their excrement. It is calculated that agricultural yields are increased by more than 50% by the activity of earthworms.

Members of the family *Tubificidae* live in slime tubes in the mud at the bottom of fresh waters. Between 2 and 20 cm in length, some smaller species, e.g. the small, red *Tubifex tubifex,* are used as food for aquarium fishes. Members of the family *Enchytraeidae* (see) are also used to feed aquarium fishes.

Oligogene: a gene which alone is responsible for the significant expression of a qualitative phenotypic character. Oligogenes are easily demonstrable by crossing experiments, and they display Mendelian segregation. They differ from *polygenes,* which alone are hardly expressed phenotypically, or not at all. Polygene combinations, however, exert a cumulative effect, which determines the quantitative expression of a phenotypic character, e.g. pigmentation, size, etc.

Oligohaline: see Brackish water.

Oligomictic lake: see Lake.

Oligopause: see Dormancy.

Oligosaprophytes: see Saprophytic system.

Oligotricha: see *Ciliophora.*

Oligotrophication: re-establishment of the original composition and quality of water, mainly by elimination of excess nutrients, e.g. by chemical precipitation, diversion of sewage, and deep water drainage.

Olive: see *Oleaceae.*

Olive colobus, *Van Beneden's colobus,* *Procolobus verus:* a member of the *Colobinae* (see Colobine monkeys), inhabiting rainforests from Sierra Leone and southeastern Guinea to Benin. The mother carries the infant in her mouth for several weeks after birth, a form of behavior unique to the *Anthropoidea.* It is the only species of the genus *Procolobus,* but some authorities place it in the genus *Colobus* (see Guerezas).

Olive oil: a fatty oil expressed from the pericarp of the ripe fruits of the olive tree *(Olea europaea).* The whole fruit contains up to 14 % of the oil, while the pericarp alone contains up to 22%. O.o. is used as a edible and cooking oil, in soap manufacture, and for technical and medical purposes.

Olive shells: see *Gastropoda.*

Olm: see *Proteidae.*

Olophoridae: see Shrimps.

Olympic salamander: see *Dicamptodontidae.*

Omasum: see Digestive system.

Ommastrephes: see *Cephalopoda.*

Ommatidium: see Light sensory organs.

Ommatin: see Ommochromes.

Ommin: see Ommochromes.

Ommochromes: a class of natural pigments containing the phenoxazine ring system, which is biosynthesized from 3-hydroxykynurenine (a metabolic derivative of tryptophan). Colors range from yellow through red to violet. O. are especially common in arthropods, such as insects crustaceans, and spiders, and have also been found in the eyes and skin of *Sepia officianalis (Cephalopoda),* and in the eggs of the marine worm *Urechis caupo (Echiuridae).* They were named for their occurrence in the ommatidia of the insect eye. In the organism they are often bound to proteins as chromoprotein granules. Low molecular mass, dialysable, alkali-labile O. are known as *ommatins,* whereas the high molecular mass, alkali-stable O. are called *ommins.* Crystalline dihydroxanthommatin was first prepared in 1954 from the meconium (post-pupal secretion) of the tortoiseshell butterfly *(Vanessa urticae).* It is found universally in insects as an eye pigment, accompanied in most orders by a greater ammount of ommin.

Our knowledge of the chemical structure of O., and the biochemistry and genetic control of their synthesis is due largely to the work of Butenandt et al.

Omosternum: see *Anura.*

Onager: see Asiatic wild ass.

Onagraceae, *Oenotheraceae, willow-herb family:* a family of dicotyledonous plants containing 21 genera and about 650 species, cosmopolitan in distribution. Most genera are herbaceous, but species of *Fuchsia* are woody. Leaves are opposite and exstipulate; flowers are usually hermaphrodite, actinomorphic and nearly always 4-merous (sometimes 2-merous) with a brightly colored corolla. In most species the axis is greatly elongated beyond the inferior ovary to form a hypanthium or "calyx tube". Pollination is by insects, especially night-flying moths, and self-pollination occurs in some species. Many tropical species of *Fuchsia* are pollinated by humming birds, while others are wind-pollinated. The fruit is usually a loculicidal capsule, exceptions being *Fuchsia* (berry), *Trapa* (single-seeded drupe, which soon looses its fleshy layer) and *Circaea* ***(Enchanter's nightshade;*** indehiscent fruit with barbs). Seeds are nonendospermic, and some (e.g. those of *Epilobium)* carry a basal plume of hairs.

Fuchsia (Fig.b) is a disjunct genus of about 100 species. Its major areas of distribution are Central and South America, but 4 species occur in New Zealand and a single species in Tahiti. As ornamental plants, Fuchsias are valued for their beautifully colored pendulous flowers. At the height of their popularity in the Victorian era, over 1 500 horticultural varieties were grown, derived mainly from *Fuchsia magellanica.*

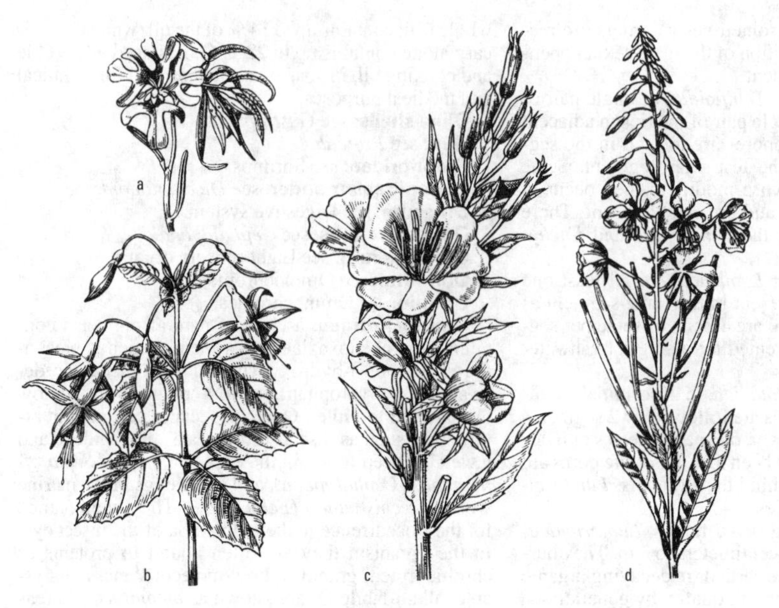

Onagraceae. a Clarkia pulchella. b Fuchsia. c Common evening primrose *(Oenothera biennis). d* Rosebay willow-herb or fireweed *(Epilobium angustifolium).*

The genus *Oenothera* ***(Evening primroses,*** Fig.c) contains over 120 species, although some authorities subdivide the genus into several smaller genera, e.g. *Megapterium, Pachylophus, Camissonia,* mostly distributed in North America. Evening primroses are especially common in ruderal habitats. Their large, long, tubular, usually yellow flowers open only at night.

Clarkia (Fig.a) is represented by about 40 species, mostly in western North America, but also as far south as Chile. Some varieties of *Clarkia* are cultivated as ornamental annuals. *Clarkia amoena* ***(Farewell to spring)*** is sometimes placed in the genus *Godetia.* All horticultural varieties known by the English name, ***Godetias***, are derived from *Clarkia amoena.*

Rosebay willow-herb or ***Fireweed*** *(Chamaenerion angustifolium = Epilobium angustifolium,* Fig.d) is a common plant in Europe, Asia and North America, sometimes cultivated, and common in ruderal habitats. Several other species of *Epilobium* ***(Willow-herbs)*** are widely distributed in Europe, Asia and North America, e.g. *E. hirsutum* ***(Great hairy willow-herb, Codlins and cream),*** *E. parviflorum* ***(Small-flowered hairy willow-herb),*** *E. adnatum* ***(Square-stemmed willow-herb),*** *E. palustre* ***(Marsh willow-herb).***

The genus *Ludwigia (= Jussiaea)* contains some 20 species distributed in temperate and warm regions, especially North America. They are all water or swamp plants, sometimes free-floating, and some tropical species are weeds of rice paddies.

On center field: a complex of sensory or nerve cells with On elements (see) at the center and Off elements (see) at the periphery. For example, in the retina the existence of on and off center fields and the juxtaposition of such fields, together with certain other special properties and the subsequent level of information processing of the visual system, account for the ability to perceive a high contrast scene or picture, including fine details and perspective.

Onchocera: see Onchoceriasis.

Onchoceriasis, *river blindness:* a disease of the tropics caused by the parasitic nematode , *Onchocerca volvulus,* of the order *Filarioida* (see *Nematoda;* Filiariasis), which is transmitted by blood-sucking blackflies of the genus *Simulium.* The host reacts by actively encapsulating adult worms in subcutaneous tissue, forming nodules 1–6 mm in diameter in the subcutaneous connective tissue and in the musculature. Within these nodules, the adult worms live up to 16 years. The migratory route from the infective bite to subcutaneous encapsulation is unknown. Microfilaria released by the females pass out of the host capsule and circulate in the host skin. The time from the first infective insect bite to the release of microfilaria by the encapsulated females is 15–18 months; the circulating microfilaria remain viable for up to 2 years. The life cycle is complete when these microfilaria are ingested by the insect vector. Within the insect vector, the microfilaria molt twice to become infective third stage larvae. Symptoms include irritation and chronic dermatitis, depigmentation of the skin, and a tendency for the skin to hang in slack folds, due to destruction of the elastic fibers. Most seriously, however, the eyes are affected, and heavy and frequent infections lead to the tropical disease of river blindness.

Onc(h)osphere: see *Cestoda.*

Oncogenic viruses: see Tumor viruses.

Oncorhyncus: see Salmon.

On elements: sensory or nerve cells that produce impulses when stimulated. See Off elements.

One-toed amphiuma: see *Amphiumidae.*

Onion maggot: see Anthomyiids.

Onion thrips: see *Thysanoptera.*

Oniscidae: a family of the *Isopoda* (see).

Oniscoidea: a suborder of the *Isopoda* (see).

Oniscomorpha: see *Diplopoda.*

Oniscus: see *Isopoda.*

Onobrychis: see *Papilionaceae.*

Onototragus: see *Reduncinae.*

Ontogeny: the totality of all processes of structural

development that occur in the life history of an individual, starting with the fertilized egg (zygote) (or unfertilized egg in the case of parthenogenesis), leading to sexual maturity, and ending with the gradual regression of organs and natural death. O. forms part of Biogenesis (see). *Formal O.* describes the course of development in space and time, while *causal O.* investigates the mechanics and physiology of development. The early stages of O. are sometimes referred to as Progenesis (see), which includes meiosis and fertilization, as well as the segregation of the egg or zygote cytoplasm observed in certain cases. O. is generally divided into 4 phases.

1) *Embryogenesis* or embryo development comprises all processes starting from the egg and culminating in the hatching or birth of a young animal capable of an independent existence. Embryogenesis includes: a) cleavage or *blastogenesis,* b) the two phases of germ layer formation, i.e. gastrulation and mesenchyme or mesoderm formation, c) appearance of organ primordia, followed by organ formation and differentiation *(morphogenesis* and *organogenesis),* d) histological differentiation or *histogenesis* of organs, e) the totality of all growth processes of the embryo or fetus, f) the formation of embryonal organs, in particular the extraembryonal membranes, and g) the process of hatching or birth.

2) *Postembryonal* or *juvenile development* extends from the juvenile to the adult phase. If development is *direct,* this phase consists of the progressive growth and development of organs (especially the reproductive organs and secondary sex characters) that are essentially similar in the juvenile and adult stages. *Indirect development* involves the complicated processes of metamorphosis via one or more larvae.

3) In the *adult phase* all bodily functions are developed maximally, sexual maturity or *adolescence* is attained, and reproduction occurs. This phases merges gradually into:

4) Aging and senescence.

Onychophora: a protostomian animal phylum, containing about 90 species, all of them terrestrial, distributed discontinuously in permanently damp habitats in the tropics and subtropics. The approximately cylindrical and sluglike body has closely packed transverse bands of tubercles covered by scales. Many are brightly colored blue, green or orange, while others are black. The head is not differentiated from the body, and there is not true external segmentation. Each pair of legs and each pair of metanephridia, however, belong to original segments. The anterior head lobe (prostomium, acron) carries a pair of small, simple vesicular eyes. This is followed by the first true segment, which carries two ringed antennae (first pair of appendages). The second pair of appendages consists of pointed, clawlike mandibles surounded by a ring of papillae. These are followed by a pair of oral papillae, which house the openings of defensive slime glands. All other appendages are ambulatory legs, varying in number from 14 to 43 pairs. The legs are hollow, fleshy evaginations of the body, ending in a pair of claws; they differ from both the parapodia of annelids and the jointed appendages of arthropods. Almost every segment contains a pair of metanephridia; in the third segment (which carries the oral papillae), the metanephridia are modified to long salivary glands. Respiratory gas exchange occurs via tracheae. In all species the sexes are separate, the genital opening lying between the last or the penultimate pair of legs. The male attaches a spermatophore to the body of the female, and the bodies of females may carry numerous adherent sperm capsules from different males. Spermatozoa penetrate the skin of the female, then travel in the body cavity fluid to the ovaries, where they fertilize the eggs. Reproduction may be oviparous, ovoviviparous, or viviparous; most species are viviparous. The lifespan is at least 6–7 years, and during this time the animals molt at regular intervals of 13–18 days.

O. are markedly nocturnal in habit, living under leaf litter, stones, etc., usually near to water. They are carnivorous, but observations on their natural prey are lacking. In captivity they take insects, snails, isopods and similar organisms. For defense, a sticky mucus is discharged by a pair of slime glands on the oral papillae. Large quantities of the mucus can be ejected over distances of 15–30 cm, entangling an adversary and cementing it to the ground.

On the one hand, the *O.* appear to be related to the *Annelida* (uniform segmentation of the body, metanephridia in nearly all segments, and vesicular eyes). On the other hand, they display affinities to the *Arthropoda* (mandibles, leg claws, tracheae, perivisceral hemocoel). *Aysheaia pedunculata,* a fossil marine invertebrate from the Middle Cambrian, is thought to be an ancestor of the modern *O.* Its segmented, wormlike body had 10–43 pairs of legs, a terminal mouth and frontal papillae.

The 90 species (approx.) are divided into 2 families. The most familiar genus is *Peripatus.*

Collection and preservation. Specimens are found under leaf litter, rotting tree stumps or stones, and they leave these sites only if the relative humidity of the outside atmosphere approaches 100%. They can be killed with ethyl acetate or chloroform. The body cavity should then be injected with a mixture of formaldehyde and ethanol, and the animal preserved in the same mixture.

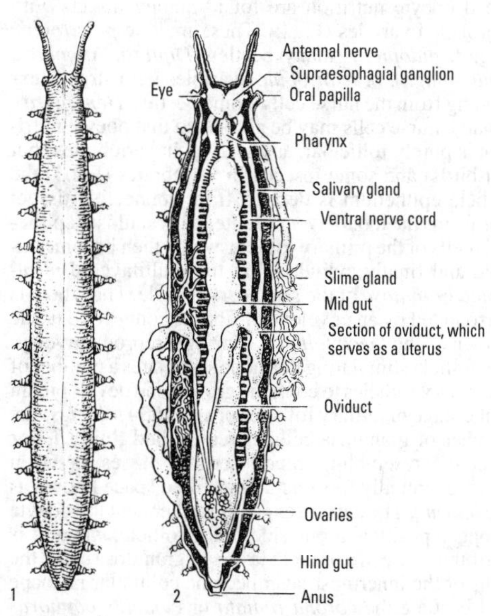

Onychophora. 1 Dorsal view of *Peripatopsis. 2* Internal anatomy.

Onychoteuthis: see *Cephalopoda.*

Onychura: see *Branchiopoda.*

Oocyst: see Cyst.

Oocytes: see Gametogenesis, Oogenesis.

Oogamy: see Reproduction, Fertilization.

Oogenesis, ***egg cell formation:*** development of female sex cells from primordial germ cells via oogonia (the entry, Gamete, and its accompanying Fig. should also be consulted). The process occurs in several phases. In the *multiplication phase,* which in most mammals is complete at the time of birth, *oogonia* are produced by more or less numerous mitotic divisions of the primordial germ cells. In the *growth phase,* cell division ceases, yolk accumulates, and there is often a considerable increase in cell volume (with reference to the laying down of yolk, the growth phase is sometimes called *vitellogenesis).* In only a few species, the oocytes are not associated with accessory cells, but float freely in body fluids or tissues, from which they withdraw the nutrients necessary for growth (a process known as *solitary O.,* observed in some *Turbellaria,* coelenterates, mollusks and echinoderms). In the great majority of animals, the oocytes display *alimentary, auxiliary* or *follicular O.,* in which the oocytes are intimately associated with an investment of somatic cells, which are directly involved in the provision of nutrients. In *nutrimentary O.,* oocytes are associated with sibling cells derived from the same primordium. Incomplete cytokinesis often results in development of a syncytial complex of oocytes and nurse cells connected by cytoplasmic bridges or ring canals (e.g. in coelenterates, insects, snails). The number of nurse cells varies widely [1 in *Ophryotrocha* (Fig.1), *Forficula* and *Chironomus;* 3 in lower crustaceans; up to 50 in *Piscicola* and *Apis mellifica].* In invertebrates, the supply of nutrients by nurse cells is often supplemented by provision of material from the monolayer of epithelial somatic cells (follicle cells) that surround the primordial germ cell. Typical examples of this type of combined oocyte nutrition are found among insects with meroistic ovarioles (Fig.2). These may be *polytrophic* (e.g. *Lepidoptera,* many beetles, *Diptera, Anoplura, Mallophaga),* or *telotrophic* ovarioles with strands extending from the nurse cells (some beetles, *Hemiptera).* Finally, nurse cells may be absent, so that oocyte nutrition is purely follicular, as observed in vertebrates (but not birds) and some insects. In vertebrates (Fig.3) the follicle epithelium is derived from connective tissue; this forms the monolayer of flattened granulosa (epithelial) cells of the primary follicle, which then become cuboid and finally cylindrical in the multilayered *membrana granulosa* of the secondary follicle. The whole is surrounded by an envelope of fibrous connective tissue known as the *theca folliculi.* During a reproductive cycle, follicle stimulating hormone stimulates a number of secondary follicles to continue growth and development to the stage of tertiary follicles or *Graafian follicles;* the number of granulosa cells increases, and fluid *(liquor folliculi;* resembling blood serum) passes between them, eventually forming a fluid-filled space known as the *antrum.* The antrum expands further, and the oocyte adopts a position at one side of the follicle, where it is surrounded by two or more layers of granulosa cells; the cells of the innermost layer become columnar in shape and become the *corona radiata* or *cumulus oophorus* which remains attached to the egg for a period after ovulation. At the same time, blood capillaries invade the inner layer of the theca folliculi to form a vascular layer, the *theca interna,* which is surrounded by fibroblasts of the *theca externa,* i.e. the theca folliculi becomes a bilayer.

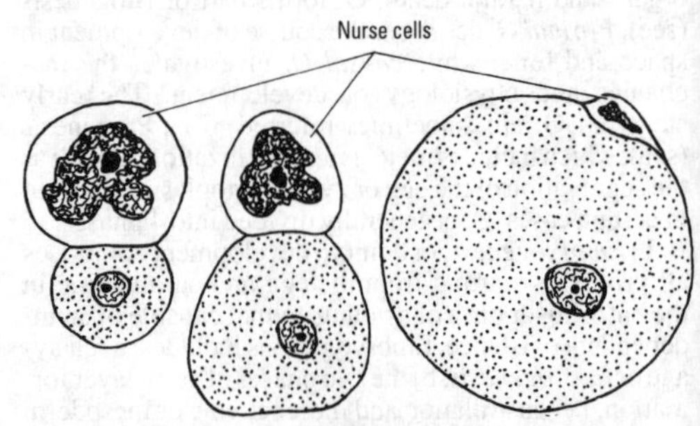

Fig.1. *Oogenesis.* Three stages in the growth of the egg of the annelid, *Ophryotrocha,* showing the nurse cell which eventually becomes completely depleted.

Cells at the end of the growth phase are known as primary oocytes. These enter the third phase of oogenesis, known as *maturation* (see Meiosis), in which the mature haploid egg is produced. In vertebrates, maturation commences in the Graafian follicle, but is not completed until after ovulation, in which the egg is released still enveloped by its *zona pellucida* and *corona radiata.* The process of *follicle maturation* is similar in most mammals, but differs with respect to the number of Graafian follicles [1–2 in humans, more (corresponding to litter size) in rodents, dogs, pigs, etc.]. The first ovulation marks the onset of sexual maturity in the female.

Follicle maturation and ovulation are hormonally regulated. In all vertebrates, the relevant hormones are secreted by the usually very small (0.5 g in humans) ectodermally derived *hypophysis (pituitary).* Three hormones of the adenohypophysis (anterior pituitary) play a crucial role in follicle maturation, ovulation and subsequent modification of the follicle. These are collectively known as *gonadotropic hormones* or *gonadotropins:* 1) Follicle stimulating hormone (see) is responsible for the onset and continuation of follicle maturation, as well as the growth of granulosa cells and theca tissue, but it does not cause follicle rupture or estrogen formation by the follicle. 2) Luteinizing hormone (see) together with follicle stimulating hormone helps to maintain follicle maturation until completion. It stimulates estrogen formation by the follicle, and together with prolactin it stimulates progesterone formation by the corpus luteum. It stimulates rupture of the follicles (ovulation) and is responsible for formation of the corpus luteum, i.e. for the transformation of granulosa cells into lutein cells immediately after ovulation, a process marked by the accumulation of fat and the yellow pigment *lutein.* 3) Prolactin (see), in addition to its well known mammotropic and lactogenic activity, has a luteotropic action similar to that of follicle stimulating hormone and luteinizing hormone, e.g. it controls the secretion of progesterone by the corpus luteum and maintains the corpus luteum in an active state. In humans, formation of the corpus luteum is completed in 3–4 days after ovulation, and in the absence of fertilization it survives for about 12 days.

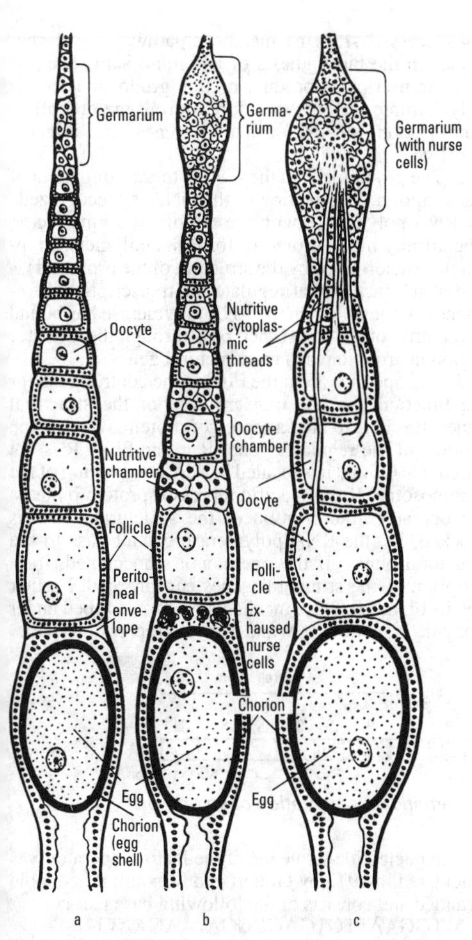

Fig.2. *Oogenesis.* The three main types of insect ovarioles in longitudinal section.
a: Panoistic. This most primitive of the three types is found, e.g. in *Japyx, Orthoptera, Dictyoptera, Isoptera, Odonata, Siphonaptera* and others. Distinct nurse cells are absent.
b: Polytrophic. Nurse cells are grouped together in chambers (nutritive chambers), which alternate with oocyte-containing chambers along the ovarian tube. In the *Neuroptera, Adephaga* and *Hymenoptera,* neighboring chambers are separated by a constriction, whereas in the *Lepidoptera* and *Diptera* this constriction is lacking.
c: Acrotrophic. The nurse cells remain in the germarium, which therefore also becomes a nutritive chamber *(Polyphaga, Hemiptera).* Some *Heteroptera* have *telotrophic* ovarioles, i.e. acrotrophic ovarioles in which the nurse cells in the germarium are connected to the developing oocytes by protoplasmic (nutritive) threads (illustrated). Types *b* and *c* (those possessing distinct nurse cells) are often grouped together as *meroistic* ovarioles.

None of these gonadotropic hormones is sex- or species-specific. Follicle stimulating hormone is also produced in males, where it stimulates spermiohistogenesis and growth of seminal tubules. Luteinizing hormone is also a male hormone which stimulates synthesis and secretion of testosterone and growth of accessory sex organs. Although prolactin is present in the male circulation, its role is not clear.

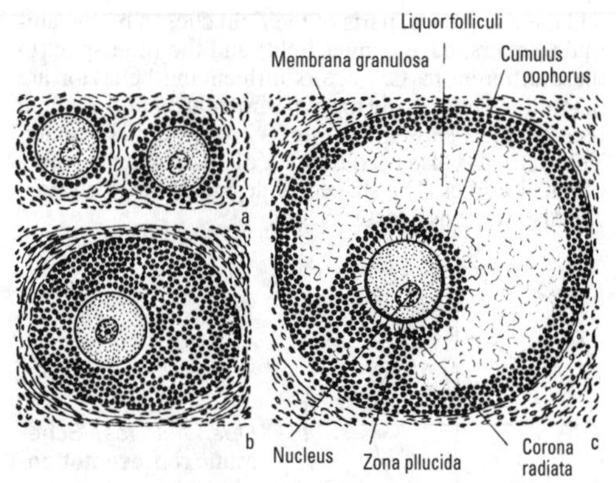

Fig.3. *Oogenesis.* Human primary *(a),* secondary *(b)* and tertiary or Graafian *(c)* follicles.

Follicle and corpus luteum are themselves endocrine glands, whose specific hormonal function is discussed under the entry Placenta (see). It should be mentioned here that follicular hormone (estradiol) and corpus luteum hormone (progesterone) have a regulatory function in the action and production of the gonadotropic hormones.

Oogonium: see Reproduction, Gametogenesis, Oogenesis.

Oomycetes: see *Phycomycetes.*

Oophagy: a specialized type of cannabalism, in which eggs are eaten by the mother animal or other members of the same species. O. may result from overpopulation (see Fluctuation). In the establishment of a new insect colony, the queen may at first satisfy her food requirements by consuming some of her own eggs (e.g. in the ant genera *Myrmica* and *Atta);* until the colony has grown, she has no other way of acquiring food.

Oostegite: an epipodite on the basal segment of each of the biramous abdominal limbs (coxopodites) of certain crustaceans (e.g. *Isopoda, Amphipoda).* The Os. extend inward to form a ventral brood chamber (see Marsupium).

Opah: see Kingfish.

Opalinata: a superclass of flagellated protozoa, which live parasitically in the intestines of amphibians. The flagella are arranged in uniform longitudinal or transverse series, thereby resembling the cilia of the *Ciliophora.* The *O.* possess two to many nuclei, all of the same type (cf. *Ciliophora,* which possess macronuclei and micronuclei). Reproduction and development resemble those of the *Mastigophora (Flagellata).* Organic nutrients are absorbed through the body surface.

Open field test: a standardized behavioral test for the qualitative and quantitative determination of reactions to certain stimuli and other behavioral interactions with the environment. The experimental animals are placed in a walled room *(enforced open field behavior),* or they can choose to enter the field from a box with an open sliding door *(voluntary open field behavior).* The following are recorded: particular forms of behavior and their sequence, parts of the field chosen by the animal (corners, edges, inner field) and the time spent in these different parts. Factors influencing behavior are determined by varying the experimental conditions, and recording any subsequent effects in open field behavior. The method is also used for toxicological and pharmacological studies. The shape and design of the open field can also serve as a test variable.

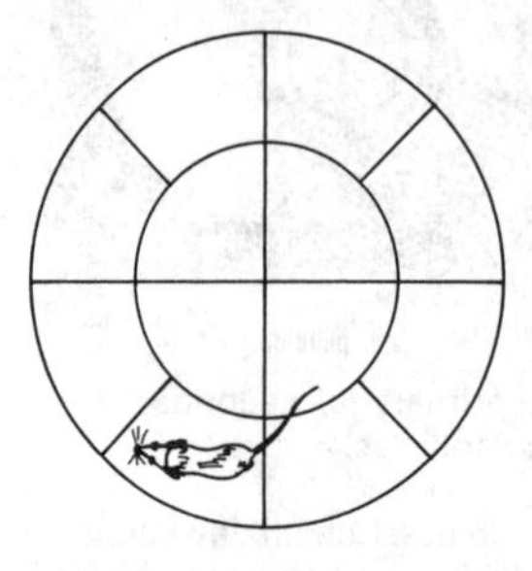

Open field test. Schematic representation of an elementary open field containing 12 fields of equal area.

Opening movements: in flowers, movements mostly determined by thermo or photonastic growth of the underside of petals (see Flower movements, Unfolding movements). Opening of seed capsules and sporangia is due to Hygroscopic movements (see) and Cohesion mechanisms (see).

Operant box: see Skinner box.

Operant conditioning, ***operant learning, instrumental conditioning, instrumental learning:*** a type of Conditioned reflex (see), in which the animal's own behavior serves as the neutral stimulus that eventually becomes the conditioned stimulus. For example, an animal can learn that pressing a lever will deliver food. Also, if a hungry pigeon is given food when it just happens to be spreading its left wing, it will eventually associate this with the reward of food. In O.c., the reinforcer is presented only after the subject has given the appropriate response, and the reinforcer may be a reward (positive conditioning) or a punishment (negative conditioning). In contrast, in classicial conditioning (see Conditioned reflex), the unconditioned stimulus is presented after the conditioned stimulus.

Operant learning: see Operant conditioning.

Operational area: see Activity range.

Operator: see Operon.

Operculum:

1) A small lid that closes the spore tube (ascus) of fungi of the order *Pezizales* (class *Ascomycetes).* When the spores are ripe, it separates along a preformed line and opens laterally like a hinge, accompanied by ejection of the spores.

2) The dehiscent lid of moss capsules.

3) A rounded horny or calcareous plate on the foot of most snails of the subclass *Streptoneura,* which closes the aperture when animal withdraws into its shell.

4) A hard but flexible cover forming the outer wall of the gill chamber of teleosts. It protects the gill clefts and helps to regulate the water flow over them. The gills of certain amphibia are also protected by an O.

Operon: a group of neighboring genes representing a functional system. An O. contains the following interacting units.

1. Structural genes (S_1 to S_4 in the Fig.), which encode the primary structures of enzyme proteins catalysing successive steps in a metabolic pathway, e.g. the enzymes in the biosynthesis of an amino acid. The primary transcription product of this group of genes is polycistronic mRNA; therefore, during the control of transcription all the structural genes are affected equally.

2. The promotor (P in the Fig.) is the starting point of transcription. This section of the DNA is "recognized" by RNA polymerase with the aid of the sigma factor. The affinity of the promotor for RNA polymerase (apparently determined by the structure of the promotor) is one of the factors that regulate the transcriptional frequency of the O. When the RNA polymerase has bound to the promotor, it must then pass through the operator region in order to reach the structural genes.

3. The operator (O in the Fig.) is the control gene for the function (i.e. the transcription) of the structural genes. It is able to bind a repressor protein, which is the product of the regulator gene (R in the Fig.). R is not part of the O., and it is located in a different region of the chromosome. If the specific repressor protein binds to the operator, transcription of the structural genes is blocked, i.e. the RNA polymerase cannot pass to the structural genes. If the operator is unoccupied, transcription of the structural genes can proceed. Further details of these control mechanisms are described under Enzyme induction (see) and Enzyme repression (see).

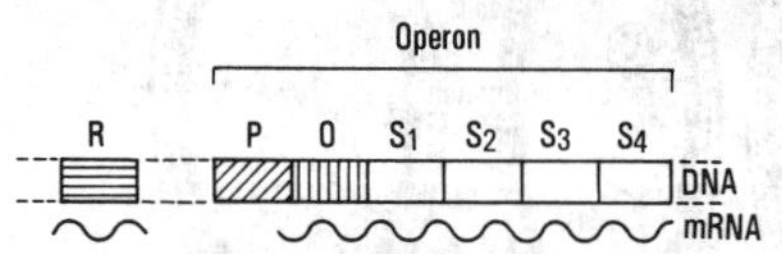

Schematic representation of an operon

The nucleotide sequence of the lactose operator was elucidated in 1973 by Gilbert and Maxam; it is double stranded and consists of the following base pairs:

5' TGGAATTGTGAGCGGATAACAATT
3' ACCTTAACACTCGCCTATTGTTAA

The O. model was developed by Jacob and Monod, and it has only been demonstrated in prokaryotic organisms.

Operophthera: see Winter moths.

Ophidia: see Snakes.

Ophiognomon: see Snake teiids.

Ophiophagus: see King cobra.

Ophiothrix: see *Ophiuroidea.*

Ophiotoxins: see Snake venoms.

Ophisaurus: see Sheltopusik, *Anguidae.*

Ophisternon: see *Synbranchiformes.*

Ophiura: see *Ophiuroidea.*

Ophiuroidea, ***ophiuroids, brittle stars, serpent stars:*** a class (or subclass) of the *Echinodermata* (see), whose bodies consist of a compressed central disk, with five sharply demarcated, slender and usually long arms. There is no hindgut or anus, and the blind alimentary canal is nonprotrusible. Progress is made by pushing and pulling on neighboring objects, or they crawl over the ocean floor by agile serpentine movements of the arms (hence the name, Serpent star). The flexible arms often break off, hence the name, Brittle star. The arms are provided with upper, lateral and lower series of skeletal plates. Spines on the lateral plates provide grip, while the plates on the undersides of the arms cover the ambulacral groove, providing protection for the nerve cord. Pairs of these ambulacral plates fuse to form "vertebrae", which articulate along the length of the arm, and can be moved against each other in various directions by four muscles. These relatively large "vertebrae" reduce the perivisceral cavity of the arm to a narrow canal. The arms no not contain ceca of the alimentary canal.

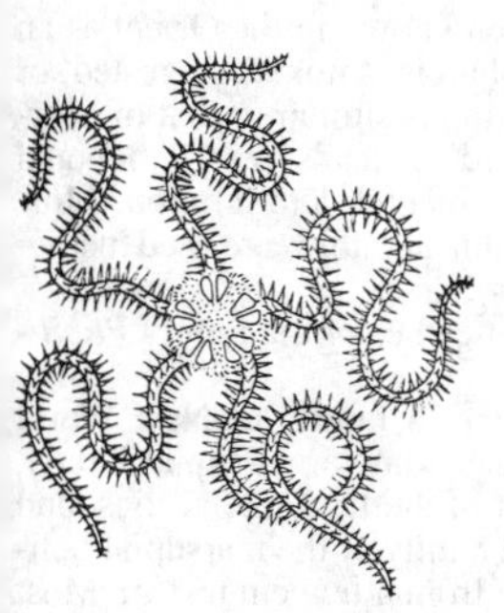

Ophiuroidea. The brittle star, *Ophiocomina.*

Brittle stars live on the sea bottom in all regions of the world, but only a few species are found in the deep oceans or in brackish water. Many species bury themselves in the sea bottom. Smaller species often occur in large numbers. The largest species with unbranched arms attains a disk diameter of 5 cm with an arm length of 23 cm. Among those species with branched arms, the largest representative attains a disk diameter of 14 cm and an arm length of 70 cm. They feed mainly on small animals. Development to the juvenile stage may occur inside the mother (brooding ophiuroids), in which case the larvae are reduced ophiopluteus larvae, produced from a yolk-rich eggs. In nonbrooding ophiuroids, yolk-poor eggs hatch to planktonic ophiopluteus larvae, and subsequent development closely resembles that of asteroid larvae. *Ophiopluteus* larvae have 4 pairs of elongated arms supported by calcareous rods and bearing bands of cilia. About 1 900 living species of ophiuroids have been described. The most familiar genera are *Amphiura, Ophiothrix, Ophiura* and *Gorgonocephalus.*

Basket stars are ophiuroids with branching arms. They can roll into a ball or spread out their arms over a large area to capture planktonic organisms and organic particles. The most familiar basket stars are members of the genus *Gorgonocephalus,* whose arms fork at the periphery of the disk and continue to branch repeatedly. Toward the ends of the arms, the spines are modified to hooks, and the smaller branches of the arms have girdles of hooked granules. There are up to several thousand gonads. *G. arcticum* (at depths down to 1 500 m) and *G. eucnemis* (1 850 m) are found in Boreal and Arctic waters. *G. caput-medusae* (150–1 200 m) is a boreal species.

Fossil record. Fossils are found from the Ordovician to the present, but their occurrence is erratic, only about 180 fossil species are known, and they have no value as index fossils. The greatest variety of forms is found in the Old Paleozoic, while those of the Mesozoic already display a close similarity to modern living species.

Collection and preservation. They are collected by hand on stony sea bottoms, or by dredging from deeper bottoms. Specimens are very successfully caught at night by shining a light into the water near the shore. Since britle stars immediately shed any arm that is grasped, the animals should be picked up by their central disk. They are narcotized by placing in fresh water, followed by the slow, dropwise addition of ethanol. After spreading out on a board, they are preserved in 85% ethanol.

Ophiuroids: see *Ophiuroidea.*

Ophthalmoscope: an apparatus invented in 1851 by von Helmholtz for examining the inside of the eye, especially its rear wall. In principle, light is shone through the pupil to illuminate the optic nerves, blood vessels and fovea. This permits the observation of abnormalities, such as hemorrhages and anomalies in the refractive index of the dioptric apparatus in relation to the length of the eyeball. The simple O. consists of a small concave mirror with a central hole.

Opiliones, *harvestmen:* an order of the class *Arachnida* (see). The almost egg-shaped body is divided into a prosoma and a broad, segmented opisthosoma. The prosoma usually consists of a proterosoma and two free segments. Ambulatory legs are often extremely long, and the tarsus is divided into many individual segments, enabling it to obtain a firm grip by curving around stalks and twigs. Other appendages are the powerful chelate chelicerae, and the pedipalps which resemble walking legs. Both the ovipositor of the female and the penis of the male are often very long. Eggs are laid in heaps in earth fissures, under bark, or under stones. About 3 200 species have been described. They live in ground litter, under stones, in mosses, in bushes, even on trees, and some are cavity dwellers. Living and dead animal material is the major component of the diet, especially small insects, mites and spiders, and rarely small snails. Plant material is also taken. Prey is pulled to pieces with the chelicerae. They have no specific defenses, but protect themselves by feigning death, or by shedding a leg if it is seized by a predator (autotomy).

The largest members attain a body length of 0.6 cm, and a posterior leg length of 16 cm, whereas the bodies of the smallest, mite-like representatives are only 2 mm in length.

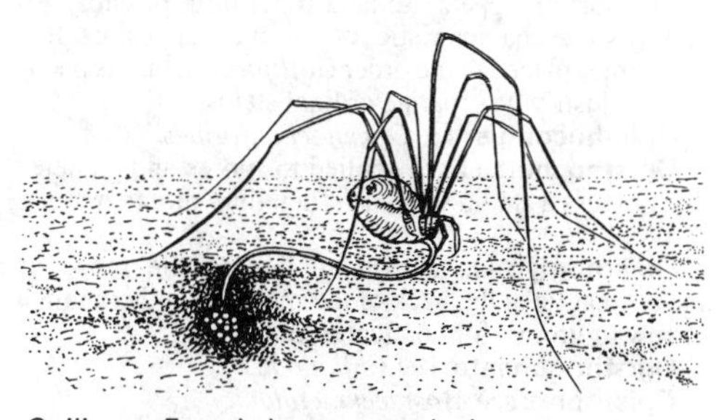

Opiliones. Female harvestman laying eggs.

Opioid pepides: see Endorphins.

Opisthobranchia: see *Gastropoda.*

Opisthocoela: see *Anura.*

Opisthocoelous vertebra: *see Anura,* Axial skeleton.

Opisthocomiformes: an order of birds (see *Aves)* represented by a single species, the ***Hoatzin*** *(Opisthocomus hoatzin),* which inhabits the riverine, marshy areas of the Amazon forest. It is a slender bird, 56–66 cm in length and superficially similar to a pheasant or a chachalaca, but weighing little more than 810 g. Its upper parts are predominantly brown, flecked white on the neck, nape and mantle. The tail has a yellow tip, the outer primaries carry a reddish patch, the wing coverts have white margins, and the underparts are orange to red. The remarkably small head carries a prominent red-brown crest. It has eyelashes (a rare avian feature found only in a few birds such as ostriches, hornbills and a few cuckoos and hawks), red eyes, and bare, blue facial skin. The sexes are almost identical. On account of its strong musky smell, it is sometimes called the "stink bird". The neck is very long, and the huge crop occupies one third of the forward part of the body. This muscular, 2-chambered crop takes over the function served by the gizzard in other birds; it serves for both the storage and the start of digestion of the hard leaves that form the main part of the very selective diet. Survival is only possible where certain marsh plant foods are available, especially *Montrichardia* (a large member of the *Araceae)* and the arboreal White mangrove *(Avicennia).* Fish and small river crustaceans are, however, also sometimes taken, and these are captured in the river mud or shallow water. Two or three eggs are laid in a nest built in the branches above the water. It flies extremely reluctantly and poorly, and can achieve little more than a glide from higher to lower levels within the forest, usually used as a means of crossing rivers. It climbs among the branches using both feet and wings. It cannot grip tightly with its feet, and equilibrium is maintained by flapping of the wings and tail. If necessary, adult birds can swim to save themselves from a fall. On account of its restricted ability to travel, the hoatzin exists in the Amazon jungle in separate, isolated areas, some nesting and feeding sites serving for generations. The chicks are far more agile than the adults. They are born with large claws on the 1st and 2nd digits of each wing, which are moved by special muscles. These unique wing claws, which disappear after 2–3 weeks, are used to scramble among the branches. If they fall into the water, the chicks can swim well, scramble out and make their way up the tree to their original position.

The hoatzin appears to be a relict bird, possibly retaining some characteristics of pre-avian ancestors. It is sometimes placed in the order *Galliformes,* but its peculiarities justify its separate ordinal status.

Opisthocomus: see *Opisthocomiformes.*

Opisthocont: a term applied to spores and gametes that possess a single whiplash flagellum directed to the rear.

Opisthoglyptic: a term applied to the fangs of snakes that lie at the back of the maxillae and contain a furrow or groove for conduct of the venom.

Opisthogoneata: see *Chilopoda.*

Opisthomi: see *Mastacembeloidei.*

Opisthopora: see *Oligochaeta.*

Opisthosoma: the hindbody section or abdomen of the *Chelicerata.* In the *Xiphosura* the O. carries true appendages. The O. of the *Arachnida* never carries true appendages, but may possess modified limb rudiments in the form of lung books and spinnerets. See *Xiphosura* and *Arachnida* for further details.

Opisthospermophora: see *Diplopoda.*

Opisthoteuthis: see *Cephalopoda.*

Opium, ***gum opium:*** the dried, milky exudate from the incised, unripe capsules of the Opium poppy *(Papaver somniferum; Papaveraceae).* It is a brown, acid-reacting powder, containing about 25 opium alkaloids, most of which are isoquinoline compounds. The main alkaloids are morphine (10–16%), codeine (0.8–2.5%), thebaine (0.5–2%), papaverine (0.5–2.5%) and narcotine (4–8%). Also present are water (12–25%), mineral substances, meconic acid (4–6%), resins and waxes. The alkaloids are associated with organic acids, e.g. tartaric, citric and malic acid, and especially Meconic acid (see). Opium has long been known in the Orient as an hallucinogenic drug, which is smoked, ingested or drunk as a tincture. Its action is similar to that of Morphine (see), and its repeated use leads to addiction and ultimately to physical and mental deterioration. Morphine and certain other opium alkaloids are used medicinally (see separate entries).

Opollo butterflies: butterflies of the family *Papilionidae;* see *Lepidoptera.*

Opossums, *Didelphidae:* a family of New World marsupials. O. are the oldest known marsupial family, dating from the early part of the Late Cretaceous, and considered to be the basal family of the marsupial radiation. They vary in length from a few cm to 1 m. Most species have a prehensile tail, and are agile climbers, leading an arboreal and nocturnal existence. One species, *Chironectes minimus,* has webbed hindfeet and is adapted to an aquatic existence. *Dromiciops australis* inhabits the southern Andean forests, and is regarded by some authorities as the only species of the subfamily, *Microbiotheriidae.* O. occur from southeastern Canada to southern Argentina, represented by 70 recent species in 14 genera, e.g. the ***American opossum*** *(Didelphis virginiana)* and the ***Patagonian opossum*** *(Lestodelphis halli).*

Opportunistic species: see K-selected species.

Opsonin: the name given by Wright and Douglas in 1903 to the thermostable material present in serum, which stimulates phagocytosis of bacteria. It was shown to act directly on bacteria and not on the phagocytes. It is probably identical to C3b of the complement system (see Opsonization), although other complement components may be active to a lesser extent. In addition to its role in the opsonization of immune complexes, C3b binds to various structures, such as foreign erythrocytes and bacterial cells, and renders them more readily phagocytosed by immune adherence to phagocytes.

Opsonization: the ability of serum to render an immune complex more readily phagocytosed. O. is a property of the complement system. Although immune complexes are subject to phagocytosis, interaction with complement greatly increases the rate of this process. Component C3b (activated C3 of the complement system) has a labile binding site(s), which permits it to bind tightly to antigen-antibody complexes (opsonic adherence). Other stable binding sites on C3b enable the opsonized immune complex to bind to polymorphonuclear leukocytes, monocytes and macrophages (immune adherence); this binding results in enhanced phagocytosis. The phagocytosis can be abolished by metabolic inhibitors, whereas the immune and opsonic adherence (due to physical-chemical interaction of receptor and binding sites) are unaffected.

Optical density: see Absorbance.

Optical rotation: rotation of the plane of polarized light by an optically active material. O.r. is due to the different refractive indices for right and left circularly polarized light. Measurement of O.r. is used in the analysis of molecular conformation.

Optical specificity: see Enzymes.

Optic cup: see Light sensory organs.

Optic lobes: see Brain.

Optic nerve, *nervus opticus*: the II cranial nerve; the sensory nerve in the stalk of the eyeball. Where the two O.ns. come together after leaving the orbits, they join the form the optic chiasma ventral to the hypothalamus. See Cranial nerves.

Optic pit: see Light sensory organs.

Optic vesicle: see Light sensory organs.

Optimal period: an essential phase of learning in young animals in response to received stimuli from the environment. For example, in the O.p., young animals learn to recognize the characters of their own species and their parents, and they learn to differentiate between different qualities of food objects.

Optoacoustic spectroscopy: see Photoacoustic spectroscopy.

Optocoel: see Ventricle.

Opuntia: see *Cactaceae.*

Ora: see Komodo dragon.

Orache: see *Chenopodiaceae.*

Oral contraception: see Hormonal contraception.

Orange: see *Rutaceae.*

Orange-brown lactarius: see *Basidiomycetes (Agaricales).*

Orang-utan, *Pongo pygmaeus* (plate 41): one of the Great apes (see). Two subspecies are recognized: *P.p. pygmaeus* (Borneo) and *P.p. abelii* (Sumatra), both inhabitants of lowland tropical rainforest with complex canopy structure. The reddish brown coat (darker in the Bornean subspecies) is coarse and shaggy. Adult males of both subspecies possess a large throat pouch, which contains laryngeal air sacs, extending under the arms and over the shoulders. Old males also have greatly enlarged cheeks. It is a peaceful animal, almost entirely aboreal in habit, building sleeping nests in the forest canopy. The Bornean subspecies is observed more regularly on the ground, possibly because the canopy structure is less dense than in Sumatra, and there are no large ground predators, whereas the Sumatran subspecies still has to contend with the Sumatran tiger. Individual adult females occupy territory with one or two young, whereas the male ranges alone over the territories of several females. Being large, heavy animals (males 77 kg, females 37 kg; Sumatran subspecies somewhat smaller), locomotion in trees is sometimes cautious, although very skilful and acrobatic. The diet is mainly fruits with some ants, termites and honeybees, while leaves, bark and pith are eaten in times of food scarcity. They have occasionally been observed to take meat, such as young monkeys.

In January 1993 there were an estimated 20 000 individuals in the two islands, compared with 30 000 ten years earlier. About 900 are known to be in zoos throughout the world, and about 600 are thought to be captive in Taiwan as pets or nightclub attractions, or in private zoos. The species is threatened by the pet trade and deforestation, as well as being hunted for food by local natives. See Hominoids.

Orb-web spiders: see *Araneidae.*

Orca: see Whales.

Orcaella: see Whales.

Orcein: a reddish brown amorphous mixture of pigments, obtained from the almost colorless lichens of the genera *Rocella, Lecanora* and *Variolaria* by treatment with ammonia. Orcinol (see) in the lichens in converted into O. by the action of ammonia and atmospheric oxygen. O. is the chief constituent of the dye *orseille* (French purple). It is closely related to litmus, which is also prepared from lichens of the above genera.

Orchard dormouse: see *Gliridae.*

Orchidaceae, *orchid family:* a monocotyledonous plant family of over 20 000 species in more than 450 genera, and therefore one of the largest *Phanerogam (Spermatophyte)* families. The family has a worldwide distribution, the number of species being greatest in the humid tropics and decreasing toward the poles; a few are found within the Arctic Circle in Alaska, Greenland and Siberia. Other species occur above the tree line in mountainous regions. They are all perennial herbs, sometimes taking years to reach maturity. The majority live as epiphytes on trees in tropical rainforests, especially in moist tropical cloud forests on mountain slopes. A small number of tropical species is also terrestrial. Temperate species occupy a variety of habitats (grassland, woodland, marsh, bog, mountainous rocky outcrops), and all of them are terrestrial; a few temperate woodland species are saprophytic. Tropical species are usually larger, more elaborate and more colorful than those native to temperate zones. Orchids are almost always mycorrhizic (see Mycorrhiza). The fungal mycorrhiza are indispensable for seed germination, and this fungal association continues during the young stages of plant growth; saprophytic species remain dependent on mycorrhiza throughout their existence.

Storage organs are frequently present; in terrestrial orchids, these are usually root tubers or rhizomes, whereas epiphytic species have fleshy leaves or shoot tubers. Many orchids form *pseudobulbs,* which are normal vegetative stuctures formed each year by the thickening of one or more stem internodes, and containing reserves of water and nutrients. In dry periods the plant may shed all its leaves and hibernate as a pseudobulb. Flower stalks arise at the top, side or base of the pseudobulb, thereby forming a succession of stems of limited growth, i.e. the plant displays sympodial growth. In contrast, some other orchids (e.g. the *Vanda* tribe of Southeast Asia) display monopodial growth, i.e. the tip of every stem grows indefinitely, producing roots at intervals, so that older parts may die while the tip continues to grow.

The fleshy, green, hanging aerial roots of epiphytic orchids, which have much the same role as leaves, are covered with spongy dead tissue *(velamen),* which serves to absorb water, but protects from desiccation when conditions are dry. Anchor roots attach the plant to its support (usually a tree branch), while other roots form an entanglement between the plant and its supporting branch, creating a reservoir of nutrient humus by trapping dead leaves and debris.

Flowers are zygomorphic, epigynous and usually hermaphrodite, and often arranged in a spike, raceme, or racemose panicle. In epiphytic species, the inflorescence is commonly pendulous. There are 6 perianth segments in 2 whorls, the outer whorl representing sepals, the inner whorl petals. The median posterior petal is larger and different in shape from the others, and forms a lip *(labellum),* which serves as a landing site for pollinating insects. The basal or proximal part of the labellum is called the *hypochile,* while the distal part is known as the *epichile.* Due to rotation of the ovary through 180 ° (resupination), the individual flower parts are generally directed in the opposite direction from their sites of insertion, and the labellum usually points downward in the center of the flower. A spur containing nutrient material or nectar is often present at the base (hypochile) of the labellum. Many orchids display complicated pollination mechanisms, dependent on a single species of pollinating insect. The pollen is not released as a dust, but remains in packages known as *pollinia,* which adhere to the insect. Usually there is only a single fertile stamen (rarely 2), and this is fused with the styles or stigmas to form the *column* or *gynostemium,* situated in the center of the flower. In the subfamily, *Diandrae,* there are 3 fertile stigmas and 2 fertile stamens, and the pollen is not united in pollinia. The majority of orchids, however, belong to the subfamily, *Monandrae,* which possess pollinia, and in which there is a single fertile stamen and 2 fertile stigmas; the third, sterile stigma forms a beak-like process *(rostellum)* between the anthers and fertile stigmas. In some types, the rostellum produces 1 or 2 viscid bodies *(viscidia)* to which the pollinia are attached, or it may be entirely absent, or represented only by viscidia. The fruit (a capsule formed by the fusion of 3 carpels) contains a very large number of minute seeds, which are remarkable for their lack of endosperm and undifferentiated embryo. The seeds are widely distributed by wind and by light air currents: 3 million seeds weigh only 1 g, and almost 4 million seeds are produced, e.g. by a single fruit capsule of the orchids, *Cynoches* and *Anguloa.*

Cypripedium calceolus ***(Lady's slipper),*** which grows on calcareous soils in deciduous woods, is the sole European representative of the *Diandrae* . The brown-red perianth segments are 6–9 cm in length, while the slipper-shaped (obovoid with a rounded tip) labellum is rather shorter, and yellow in color with faint darker veins. The proximal end of the labellum and the column bear reddish spots. Also belonging to this subfamily is the tropical Asian genus, *Paphiopedilum* ***(Slipper orchids),*** which includes some of the most spectacular flower species known. About 70 species of slipper orchids have been described, some of them extremely rare. About half of these are threatened in the wild, due to the destruction of their habitats by the spread of agriculture and logging, and due to illegal and indiscriminate collecting. All other known orchids belong to the *Monandrae.* A small selection of *Monandrae* genera is given below.

1) *Orchis* (about 80 species in Europe, temperate Asia, North Africa and Canary islands), e.g. *O. purpurea* ***(Lady orchis)*** occurs in copses and open woods, on chalk or limestone, in Europe (rare in Britain) from Denmark southwards, in the Caucasus, and in Asia Minor. *O. masculata* ***(Early purple orchis*** or ***Blue butcher)*** is found in woods and particularly in intensively farmed pastures, mainly on basic soils, in Europe (common in Britain), North Africa and western Asia.

2) *Ophrys* (about 30 species in Europe, western Asia and North Africa). The flowers of this distinctive genus (often bearing a striking resemblance to an insect, and named accordingly) have a large, spurless labellum and 2 rostellar pouches. E.g. *O. apifera* ***(Bee orchid),*** *O. insectiflora* ***(Fly orchid),*** *O. sphegodes* ***(Early spider orchid),*** *O. fuciflora* ***(Late spider orchid;*** more appropriately known in the German as Hummel-Ragwurz or "Bumble-bee orchid").

3) *Platanthera* (about 80 species in northern temperate and tropical zones), e.g. *P. chlorantha* ***(Greater butterfly orchid),*** found on calcareous soils in woods and on grassy slopes in Europe (including Britain), Caucasus and Siberia.

4) *Epipactis* (about 20 species in the northern temperate zone), e.g. *E. palustris* ***(Marsh helleborine),*** found in marshes and fens in Europe (including Britain), temperate Asia and North Africa.

5) *Neottia* (3 species in temperate Europe and Asia).

Fig.1. *Orchidaceae. a* Lady's slipper *(Cypripedium calceolus). b* Lady orchis *(Orchis purpurea). c* Red helleborine *(Cephalanthera rubra). d* Bird's nest orchid *(Neottia nidus-avis).*

All species of *Neottia* are saprophytic with short creeping rhizomes concealed in a mass of short, fleshy, blunt roots ("bird's nest"). *N. nidus-avis* ***(Bird's-nest orchid),*** has uniformly yellow-brown leaves, stem and flowers, and is found in shady beech woods, especially on calcareous soils, in Europe (including Britain), Scandanavia, the Caucasus and Siberia.

6) *Brassia* (about 50 species confined to tropical America). The name commemorates the plant collector, William Brass (died 1783). Several species have great ornamental merit, e.g. the native Costa Rican *B. longissima,* which has extremely long, tail-like sepals, and is colored yellow with brown spots.

7) *Laelia* (about 30 species in Central America and eastern South America). The purple-flowered *L. anceps* ***(Flor de San Miguel)*** is native to Mexico, where it gows on rocks and trees on the fringes of dense forest. It is one of the most popular greenhouse orchids in temperate climates.

8) *Catasetum* (Amazonia) is one of the minority of orchids that have unisexual flowers. Taxonomic studies of herbarium material have sometimes been confused, because the male and female flowers differ considerably in appearance. Thus, the green female flowers of *C. saccatum* are borne singly and erect, whereas the numerous maroon male flowers are borne in a pendulous inflorescence.

9) *Caleana* (5 species in Australia and New Zealand). The name commemorates the plant collector, George Caley (died 1829). *C. nigrita* ***(Flying ducks),*** native to Western Australia, is terrestrial (like many Australasian orchids) and inconspicuously colored. The hinged lip is sensitive, and when at rest it is poised above the mouth of the pouch-like column. When irritated by an insect, the lip snaps down, and pollen becomes attached to the trapped creature before it can escape.

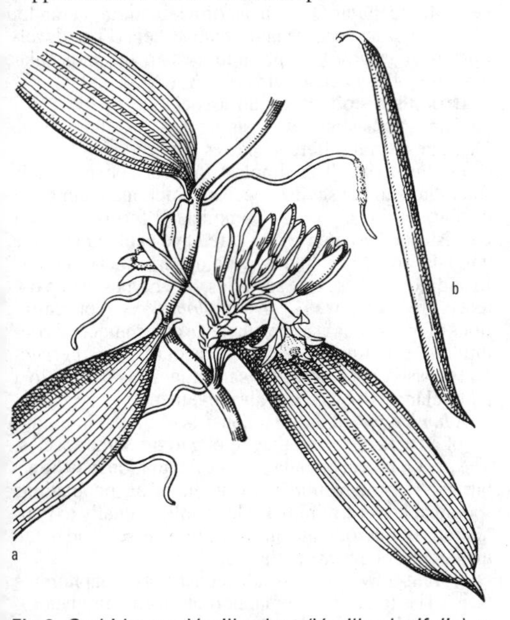

Fig.2. *Orchidaceae.* Vanilla plant *(Vanilla planifolia). a* Part of shoot with young inflorescence. *b* Fruit capsule.

10) *Arachnis* (15 species in Burma, Malaya and Indo-China). *A. flos-aeris* ***(Common spider orchid)*** of Indonesia and Malaya is found mainly in limestone areas, where it climbs over other vegetation.

11) *Vanda* (Southeast Asia). The deep pink flowered *V. hookerana* ***(Kinta weed)*** is found in swampy coastal areas, often climbing on small shrubs. Many species of *Vanda* hybridize easily, and *V. hookerana* is the parent of several garden hybrids, some of which are popular in the cut flower trade, e.g. "Miss Joaquim" *(V. hookerana* x *V. teres).*

The world's largest orchid is the ***Malaysian tiger orchid*** *(Grammatophyllum speciosum),* with 2 m-long pseudobulbs (occasionally twice that size) and flowers up to 12 cm across.

About 50 different genera are commonly cultivated commercially for their decorative value, e.g. the tropical American *Cattleya,* and the Southeast Asian *Cymbidium, Dendrobium* and *Vanda,* and especially many species and hybrids of *Paphiopedilum.* Otherwise, only one species, *Vanilla planifolia* ***(Vanilla plant),*** is exploited economically. This climbing orchid is indigenous to the east coast of Mexico, and it is now cultivated on the islands of Réunion, the Seychelles, Mauritius, Tahiti, Java, etc., where the climate is similar to that of eastern Mexico, being warm and humid, but with a dry season. Like so many other orchids, fertilization of wild *Vanilla* depends on specific insects. In other countries, where these insects are absent, fertilization is performed by inserting a pointed stick into one flower after another. Pods are picked before they are ripe, and cured by a process of slow drying and fermentation to develop the full flavor and fragrance of vanilla, which are due to crystals of Vanillin (see). The cured fruits (official *Fructus Vanillae)* are dark brown, nearly black, slender, flexible, stick-like pods. Vanillin is also present in some other orchids.

On account of their conspicuously beautiful (even bizarre) flowers, orchids are highly prized. Their fascination is increased by the fact that some species are extremely rare, as well as being difficult to cultivate. All European species are protected, and the export of rare tropical species from the wild is banned under the Convention for the Illegal Trade of Endangered Species (CITES).

[Phylogeny and Classification of the Orchid family, R.A.Dressler, 1993, Cambridge University Press, Cambridge, UK., Melbourne, Australia]

Orchinol, ***9,10-dihydro-2,4-dimethoxy-7-phenanthrol:*** a Phytoalexin (see) formed by the orchid *Orchis militaris* as a defense against infection by the fungus *Rhizoctonia repens.* The fungus penetrates the plant but inhibition of mycelial growth by O. prevents deeper penetration and further infection.

H_3CO — — OH

OCH_3

Orchinol

Orchis: see *Orchidaceae.*

Orcininae: see Whales.

Orcinol: a phenol present in some lichens *(Rocella,*

Lecanora, Variolaria). It is converted into the lichen pigments, litmus and orcein, by treatment with alkali in the presence of atmospheric oxygen.

CH_3 HO OH *Orcinol*

Orcinus: see Whales.

Order:

1) A systematic unit in Taxonomy (see).

2) A high ranking phytosociological unit, consisting of a characteristic combination of species growing in close proximity to one another. See Character species.

Ordovician (named after a Celtic tribe of North Wales, the Ordovices): a system of the Paleozoic lying between the Cambrian and the Silurian, and covering a period of 70 million years. Its plant life belonged to the Eophytic and was characterized by the conspicuous development of calcicolous algae. Floristically, it belongs to the thallophyte age, and the first vascular plants were already living in its shallow seas. The fauna of the O. was more diverse than that of the Cambrian. Graptolites are important index fossils of the central parts of Ordovician seas. In contrast, trilobites and brachiopods predominated in the shallow shore waters. Typical cephalopods are the calcareous-shelled nautiloids. Gastropods were abundant, and their fossils show a high species diversity. A great variety of bivalves was also present. During the O., brachipod evolution can be traced from nonhinged, horny-shelled species to forms with hinged, calcareous shells. The first unequivocal tabulate and rugose coralls also appear in the O. Cystoids are characteristic of certain parts of the O., e.g. in the echinosphere chalk, where they form the so-called "crystal apples" of the Baltic. Of particular interest is the appearance of the first known fossil vertebrates; these were primitive jaw-less and finless "fish-like" forms, known as the *Agnatha.*

Oreamnos: see Rocky Mountain goat.

Orectolobidae: see *Selachimorpha.*

Oreoica: see *Pachycephalinae.*

Oreomunnea: see *Juglandaceae.*

Oreopithecids, *Oreopithecidae:* an extinct family of the *Hominoidea* from the Upper Miocene to the Lower Pliocene in Europe and Africa. The family apparently represents a blind evolutionary branch. Possible ancestors were primitive primates of the *Apidium* type from the Lower Oligocene. According to absolute dating methods, the O. lived 12–14 million years ago. See Anthropogenesis, Hominoids.

Oreortyx: see *Phasianidae.*

Oreotragini: see *Neotraginae.*

Oreotragus: see *Neotraginae.*

Orfe, ***silver orfe, ide,*** *Idus idus:* a cypriniform fish, up to 75 cm in length (usually much smaller), widely distributed in fresh water in central and northern Europe, in western Siberia, and in the Baltic region (where it also occurs in brackish water). The mouth is small and directed forward. There are no barbels, and the lateral line is complete. The dorsal side is olive-brown to blackish, and the belly is silvery. Posteriorly, the pale flanks become silver with a blue-green sheen. Dorsal and caudal fins are opaque gray, and all other fins are pale red, becoming deeper red at spawning time. They live in the upper layers of flowing waters and lakes, feeding on crustaceans, water snails, worms, etc. Breeding occurs from April to July. A golden or yellow-red variety, known as the ***Golden orfe,*** occurs naturally in various parts of South Germany, notably in the Dinkelsbühl district of Bavaria. It is bred in large numbers for ornamental ponds and is hardier than the true goldfish *(Carassius auratus).*

Organ: in multicellular organisms, a body structure consisting of numerous cells, usually in association with various tissues. Each O. has a characteristic structure and a specific function (e.g. eye, kidney). An Organism (see) is the totality of its interrelated and interacting Os. Os. can be classified as *vegetative* (nutritional, excretory, reproductive) or *animalic* (sensory and locomotory Os., nervous system). *Transitory* Os. are functional only during a certain developmental stage (e.g. uterus, gills of tadpoles). *Rudimentary* Os. are present but undeveloped (e.g. human appendix, ostrich wings). If the functions of a number of Os. are integrated, they form an *apparatus* (e.g. locomotary apparatus). Os. that have a common function, but are distributed throughout the body, form an O. *system* (e.g. muscular, nervous and vascular systems).

Organ genus: temporary, artificial, pseudosystematic unit of fossil plant parts, which can be assigned to a family. See Form species.

Organic selection: see Baldwin effect.

Organism:

1) A totality of interacting organs (see Organ).

2) In the most general sense, a living O. is an individual system, which is capable of self replication, and displays the properties of metabolism, regeneration, growth, development, multiplication and response to external stimuli. In the simplest case, an O. is a single cell. Multicellular Os. consist of a structural hierarchy of cells, tissues, organs and organ systems. The development of each O. is programmed in the molecular structure of its nucleic acids (DNA, RNA).

Organism collective: an association of living organisms, consisting of a single species (homotypic O.c.), or several different species (heterotypic O.c.).

1) Homotypic. The simplest form of coexistence is found in colonial sessile species, which maintain body contact, e.g. via stolons. Complicated forms of social coexistence based on brood care have evolved in some animals, and this is highly developed in social insects. In addition to these true societies, other types of coexistence are recognized, e.g. *conglobations* (concentrations of organisms at a site where an environmental condition(s) is optimal, e.g. at a food source), or *aggregations* (passive accumulations of organisms carried to a particular site by water or air movements).

A *homotypic population* consists of all the individuals of an area or region that belong to the same species. The geographical boundaries of a population may be arbitrary, but interaction with neighboring populations should at least be minimal. This applies equally to populations of microorganisms in culture vessels and to the megafauna of entire continents.

2) Heterotypic O.cs. usually consist of Semaphoronts (see). The totality of semaphoronts forms the heterotypic population, i.e. the Biocoenosis (see) of an area. A phytocoenosis is formed by the plants, and a zoocoenosis by the animals.

In plant sociology, the floristically defined Association (see) is used to classify vegetation types. Groups of species with a similar mode of existence and similar position in the ecosystem are known as *synusia* (e.g. semiedaphic *Coleoptera).* Examples of larger units are the epedaphon, atmobios, etc.

Organization center: see Organizer.

Organizer: any part of an animal embryo that regulates the development of another part of the embryo. By Transplantation (see) of embryonic tissue to different sites of the same or a different embryo, it is possible to determine the time at which different tissues become determined. Special techniques of embryo tissue transplantation were developed by Hans Spemann and coworkers, in the 1920s and 1930s, at the University of Freiburg, and the work of this group led to some of the most fundamental discoveries in developmental physiology.

When presumptive neural plate tissue of the early gastrula was transplanted to the presumptive abdominal region, it developed into abdominal epidermis in accordance with its position. When the same experiment was perfomed at the end of gastrulation, the result was entirely different, i.e. the transplant developed in accordance with its origin. Thus, if the transplanted tissue originated from the presumptive diencephalon region from which the optic vesicle would later evaginate, then an optic vesicle developed in the abdominal region. Determination (programing) of the neural plate to develop into neural tissue therefore occurs during gastrulation.

A surprising result was obtained by Spemann and his pupil Mangold in 1921, when they transplanted the dorsal lip of the blastopore (region of presumptive notochord and somites) of the amphibian gastrula to the presumptive abdominal region of another embryo of the same age. The transplant developed, according to its origin, into notochord and somites. In addition, a second embryonic axis developed at the site of the transplant, leading to the autonomous invagination of a secondary blastopore which spread out beneath the ectoderm. This particular layer of ectoderm then developed into a secondary neural plate, which later folded and closed to form a ventral neural tube. In rare cases, the transplant developed into a complete secondary embryo, with head, gills, eyes, limbs, trunk, tail, muscle segments, notochord and alimentary lumen. Usually, however, development progressed no further than the production of incomplete secondary embryonal primordia.

The term "organizer" was originally coined by Spemann for that part of the dorsal lip of the blastopore of the amphibian gastrula, which is capable of "organizing" the development of complete secondary primordia. The area displaying this ability is called the *organization center.* This organization center corresponds essentially to the gray crescent of the frog embryo, which marks the position of the dorsal lip of the blastopore when it is first formed; it is the region of presumptive notochord mesoderm which forms the roof of the archenteron after gastrulation, and later gives rise to the notochord and somites.

By transplantation between embryos of unpigmented *Triturus taeniatus* and pigmented *Triturus cristatus,* it was shown that tissue transplanted from the dorsal lip of the blastopore not only folds and differentiates independently into notochord and somites *(autonomous differentiation)* and induces formation of a secondary neural plate in the underlying ectoderm *(constituent induction),* but also incorporates undifferentiated mesodermal host tissue when it develops into notochord and somites *(assimilatory induction).* These experiments also show that the O. is not species-specific. The embryo of *Triturus taeniatus* responds in the same way, irrespective of whether the transplanted part of the organization center is derived from its own species, a different species of salamander, or even a frog. It appears that the fundamental processes of O. action are the same in all vertebrates, including humans. For example, it is possible to induce formation of a secondary embryonal primordium by implantation of an amphibian O. in an avian embryo.

Experiments of this kind have largely elucidated the causal relationships during normal embryonal development. During gastrulation, the dorsal lip of the blastopore rolls over and grows forward beneath the ectoderm to become a lining layer of the archenteron roof. The neural plate differentiates from that part the ectoderm in close contact with the roof of the archenteron, and the remainder of the ectoderm becomes epidermis. In the absence of the underlying archenteron, neural tissue is not formed. Therefore, explants from the presumptive neural plate region of the early gastrula (i.e. archenteron formation has only just started) only develop into atypical epithelium, and never into neural tissue.

If an axolotl blastula is removed from its covering and placed in hypertonic salt solution, gastrulation movements occur, but they do not result in the folding of the endoderm and the notochord mesoderm into the interior of the embryo; these two tissues remain on the exterior of the embryo, and the process is known as exogastrulation. The ectoderm therefore does not become underlain by the archenteron roof, and it remains as an empty, folded sac displaying little differentiation. Mucoid cells appear later inside this exogastrula, but the neural plate does not form, and a typical epidermis does not differentiate. In the rest of the exogastrula, however, a notochord is formed, and somites are differentiated, together with coelom, pronephritic ducts and blood vessels. The surface endoderm differentiates histologically into the epithelia of the buccal cavity, esophagus, stomach, small intestine and large intestine. Gill pockets, as well as the liver and pancreas primordia, are folded inward rather than outward. Endoderm and notochord mesoderm therefore differentiate typically, albeit under abnormal circumstances.

A half embryo develops into a complete embryo, only if it contains sufficiently large part of the organization center (see Regulation).

The action of the O. is due to the effect of chemical substance(s) on the responding tissue, and it does not depend on the simultaneous presence of living tissue in the organization center. The transplant from the dorsal lip of the blastopore can be boiled, dried or saturated with substances such as alcohol, xylene or liquid paraffin, without loss of its inducer activity. On the other hand, insertion of an impermeable barrier between the ectoderm and roof of the archenteron prevents induction of the neural plate. The chemical inducer(s) has not been identified. A variety of tissues (kidney, adrenal, thymus, brain) have yielded preparations that exhibit O. activity, when they are incorporated into small agar blocks and applied to an amphibian embryo. It has also been shown that presumptive neural plate tissue can be stimulated to develop into nerve tissue by controlled injury in the absence of an O. This effect was first demonstrated by Holtfreter in explants, by briefly raising or lowering the pH-value of the culture fluid. Exposure of the explant to calcium-free or hypertonic culture medium has the same effect. It might therefore be hypothesized that inducers act unspecifically by injuring the responding tissue.

Finally, it must be pointed out that the roof of the archenteron does not display uniform induction properties, but is differentiated longitudinally. In addition to inducing differentiation of parts of the brain, the anterior archenteron roof also induces head organs such eyes and auditory vesicles, whereas the posterior part induces differentiation of the spinal cord, trunk and tail. It is therefore possible to speak of an anterior O. and a posterior O. Subdivision of the neural plate into brain and spinal cord is therefore not an inherent function of the responding tissue, but is determined by the inducer.

The insect embryo also contains an organization center, which controls the important stages in the differentiation of the germ band. It lies in the region of the later boundary between head and thorax, and is known as the differentiation center (see Mesoblast).

Organ of Corti: see Auditory organ.

Organoid gall: see Galls.

Organ species: a pseudosystematic artificial unit for the temporary classification of fossil plant remains, which can be assigned to a particular natural family. See Form species.

Oribatei, ***beetle mites, oribatids:*** a cosmopolitan suborder of the *Acari* (see), order *Sarcoptiformes.*

These small mites (0.1–1.5 mm) are very common and numerous consumers of dead vegetable material, and are therefore important in soil ecology. As a protection against desiccation and predators the mature form is heavily sclerotized (except in the *Palaeacaridae,* which have a soft integument). Strong pigmentation is also common. The integument of young stages is soft or leathery. Many members of the *Apterogasterina* can become spherical in shape by pressing the legs into depressions in the body. Several families are grouped together as the *Pterogasterina.* Most of these are heavily armored and dark brown in color, and they are all characterized by the possession of movable, hinged, lateral winglike chitinous extensions of the integument, known as *pteromorphs,* which cover the legs and give the animal a roughly spherical shape (e.g. genus *Galumna;* Fig.1). The *O.* are easily divided into 2 distinct groups: the *Aptyctima* (propodosoma not movably hinged with the hysterosoma), and the *Ptyctima* (propodosoma movably hinged with the hysterosoma). In the *Ptyctima,* a spherical shape is acquired by withdrawal of the anterior part of the body into a cavity in the hysterosoma, the opening being covered by the propodosomal plate (aspis) (Fig.2).

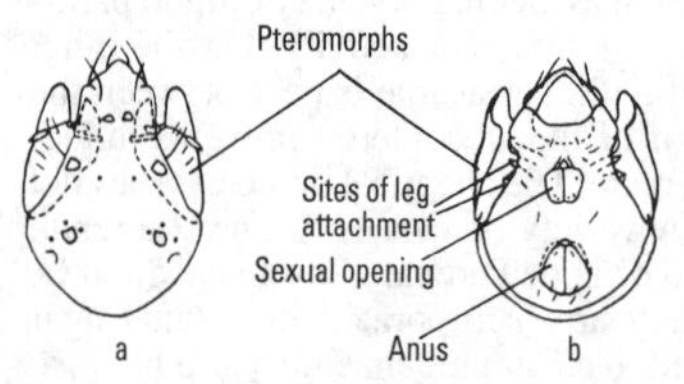

Fig.1. *Oribatei. Galumna virginiensis* Jacot. *a* Female from above. *b* Female from below. (After Jacot, 1934)

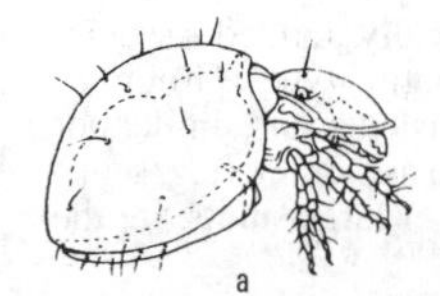

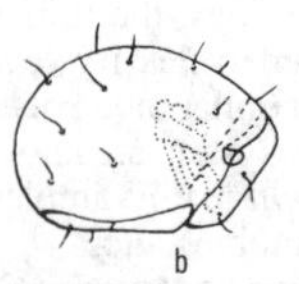

Fig.2. *Oribatei. a Pseudotritia ardua* Koch, lateral view of walking female (after Jacot, 1930). *b Phthiracarus setosellum* Jacot, propodosoma and legs (dotted) withdrawn into cavity in upper part of hysteroma, with aspis covering the opening (after Jacot, 1930).

Soft-skinned young and mature stages often lack tracheae. In armored forms, the stigmata of the tracheae are usually located at the bases of the legs; since these are withdrawn into the body the stigmata are hardly visible and they are known as cryptostigmata. Oribatids characteristically possess a pair of setiform *pseudostigmatic sense organs,* which usually arise dorsally on both sides from depressions in the propodosomal plate, or in this region when the plate is lacking; they serve as a diagnostic feature of both armored and unarmored forms. They probably serve for the perception of air vibrations, and they are absent from aquatic forms. The great majority of beetle mites lack any kind of eye, but sensitivity to light is common. In a very few species simple eyes are present on the propodosoma, e.g. in *Heterochthonius gibbus.*

Soft skinned forms, as well as most armored forms, require air that is almost totally saturated with moisture. Nymphs often feed inside conifer needles or leaf stalks, which they leave only after the final molt to maturity. The heaviest concentrations of beetle mites occur in the moss-covered leaf litter of damp coniferous forests, while dry soils are the least populated. The upper 5 cm of the soil are preferred. Members of long-legged groups *(Belbidae* and others) frequently carry earth-incrusted larval and nymphal molt skin on their backs as a protection. Some species (family *Zetorchestidae)* can jump with the aid of their elongated fourth pair of legs. Most beetle mites inhabit terrestrial soils. A few species live in the intertidal zone. *Hydrozetes* is found in fresh water on aquatic plants; *Heterozetes* walks on the water surface.

Beetle mites are oviparous, ovoviviparous, or rarely viviparous. The hatched larva has three pairs of legs, and attains sexual maturity via three four-legged nymphal stages. One or two generations are produced each year, development lasts from one to six months, and the life span is at least one year.

They feed on fungi, algae, mosses, deciduous and coniferous leaf litter and decomposing wood, rarely on the dead remains of soft-bodied animals or insect eggs and pupae. Under favorable conditions, one square meter of soil surface may contain 200 000 to 300 000 individuals. They make a considerable contribution to humus formation.

Some beetle mites are, however, harmful. Thus, *Monieza, Anoplocephala* and other genera are the intermediate hosts of tapeworms that parasitize sheep, horses, cattle, rodents, etc.

Oribatids: see *Oribatei.*

Oribi: see *Neotraginae.*

Oriental grass lizards: see *Takydromas.*

Oriental plague: see *Yersinia pestis.*

Oriental region, ***Oriental faunal region, Orientalis:*** a relatively small but sharply divided and species-rich animal geographical region, which includes India, Indo-China, western and southern China, Taiwan and some of the islands of Indo-Malaya, i.e. roughly the area encompassed by India and Asia south of the Himalayan-Tibetan mountain barrier (see Animal geographical regions for map). In northwestern India and in China, the O.r. is not sharply separated from the Palearctic region (see). The southern slopes of the Himalaya, which are covered in tropical vegetation, belong to the O.r. In the valleys, the border between the O.r. and Palearctic is sometimes sharply defined by altitude and land profile. The O.r. is separated from the Australasian region by an intermediate or transitional region, consisting of islands and known as Wallacea (see).

At least part of the O.r. (India) belonged to the continent of Gondwana. The O.r. contains fewer primitive taxa than the Ethiopian region (see), but like the Ethiopian region the O.r. was inhabited by numerous animals that became extinct in the Pleistocene. Since there are marked similarities in the families, genera and even species of the Ethiopian and Oriental regions, some authorities combine these two regions as the Paleotropical region (see).

The climate of the O.r. is predominantly tropical. Most of the region was originally forested. The high density of human settlement, however, has resulted in the clearance of large areas of forest, accompanied by the extinction of many animal species, and severe restriction of the distribution of others, e.g. Siamese crocodile, Ganges gavial, Orang-utan, wolf, lion, tiger, cheetah, leopard, one-horned Indian rhinoceros, Javan rhinoceros, Sumatran rhinoceros, and the wild Indian buffalo.

Since the O.r. is the region of origin of numerous taxa, and it was a land corridor before the folding of the Himalaya, it does not display a high degree of Endemism (see). Endemic mammalian families are the *Cynocephalidae* (flying lemurs), *Tupaiidae* (tree shrews) and *Ailuropodidae* (red panda). The *Tarsiidae* (tarsiers) are endemic to O.r. plus Wallacea. There is a rich variety of monkeys, and the *Hylobatinae* (gibbons) are characteristic of the O.r. The family, *Manidae* (pangolins and scaly anteaters), is represented in both the Ethiopian and the Oriental regions. Rodents are very numerous (1 endemic family), as are predators (numerous viverrids and 3 bear species). Most of the ungulates are forest inhabitants, with the notable occurrence of 3 rhinoceros species, the Asian tapir *(Tapirus indicus)* and the chevrotain *(Tragulidae).* The O.r. contains a greater variety of *Bovidae* than any other region. Most of the 66 avian families (excluding sea birds) are widely distributed, and many bird genera and species occur in both the Ethiopian and Oriental regions. Well represented families are the *Phasianidae, Cuculidae, Pycnonotidae, Timaliidae* and *Sturnidae.* Only the leafbirds and fairy bluebird (placed by different authors in the *Chloropseidae* or in the *Irenidae)* are endemic.

There are numerous reptilian species, including. e.g. 6 crocodiles, more than 30 freshwater turtles *(Emydidae)* and about 12 soft-shelled turtles *(Trionychidae).* The O.r. is the distribution center of the agamid lizards *(Agamidae).* Two lizard families and the shield-tailed snakes *(Uropeltidae)* are endemic. As in the Ethiopian region, the *Colubridae* are very prominent among the snakes. The *Elapidae* of the O.r. are well known (cobra, king cobra, krait). Caecilians *(Apoda)* are represented.

The newts and salamanders *(Urodela)* of the O.r. are largely restricted to Indo-China. Throughout the O.r. there is a considerable variety of amphibians of the families *Rhacophoridae* and *Microhylidae* (narrow-mouthed frogs). There are no less then 12 endemic families of primitive freshwater fishes; members of the orders *Cypriniformes* and *Siluriformes* are the most numerous.

As a result of their great age, the invertebrates of the O.r. are even less specific to the region than the vertebrates. The distribution of certain taxa of the *Diplopoda* and *Lepidoptera* suggests that they originated on the Gondwana Continent.

The fauna of the large Sunda Islands displays evidence of a previous long-standing association with the mainland fauna. Java is notable for its relative paucity of mammalian species.

Orientation: a response to direction in space and/or time. *Spatial O.* is achieved by bodily movements, which may be Tropisms (see) or Taxies (see). *Temporal O.* may be determined by exogenous "Zeitgeber" (see Biorhythms) or/and Physiological clocks (see), as well as the memory-based recognition of timed events. Several sensory systems are almost always involved in animal O., which is jointly influenced by both spatial and temporal influences, e.g. positional O. in response to both light and gravity. Animals usually locate their prey with the aid of all available senses. Highly specialized forms of O. depend on a particularly low threshold response of certain sensory systems, or on sensory systems that are absent from humans. For example, *echo-O.* (echolocation) is used by whales, bats and some birds (Fig.1), and *electro-O.* (electrolocation) is

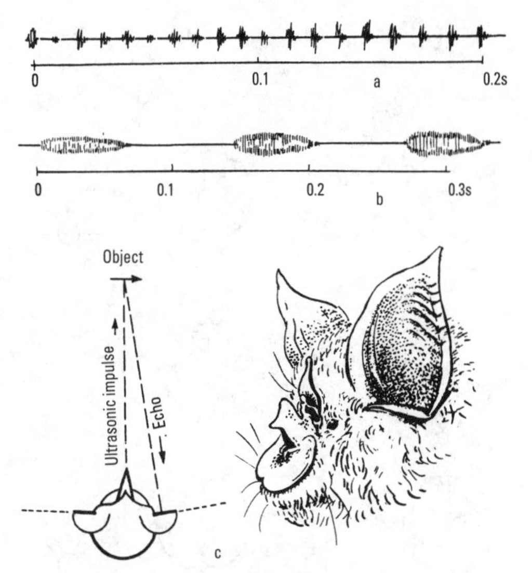

Fig.1. *Orientation.* Orientation in bats. *a* Oscillograph record of a series of ultrasonic orientation impulses from the little brown bat *(Myotis). b* Oscillograph record of three ultrasonic directional impulses from a horseshoe bat. *c* Plan of impulse generation and echo reception by a horseshoe bat, and head of horseshoe bat (modified from Kulzer, 1957).

found in some fishes. With the aid of low-voltage impulses certain fishes generate an electric field between the head and tail region, e.g. modified muscles in the tail region of *Gymnarchus niloticus* generate 300 impulses/s at 3–7 V per impulse; with the aid of electrical sensory organs they can detect distortions of this field caused by nearby objects. Some complex forms of O. depend on a highly programed interaction of sensory inputs from spatial and temporal factors, as well as memory, e.g. the *sun-compass-O.* of bees, and the homing ability of migratory fishes and birds.

In bee colonies, the direction and distance of nectar and pollen sources is communicated to other bees by the "dance" of returning workers. In the honey bee colony, a "round" dance conveys the presence and quality of a food source near to the hive, but provides no directional information; the bee runs in small circle, and the frequency with which it changes direction is proportional to the quality of the food source. In contrast, the "waggle" dance, performed on the vertical face of the comb, provides information on the direction, distance and quality of more distant food sources, and serves to recruit other workers to find and exploit the foraging site. The waggle dance can be viewed as a compressed figure of eight pattern, the waggle action being performed in the straight run between the two half rounds of the figure. The inclination of the waggling progression indicates the angle between the line from the foraging site to the hive and the line from the hive to the sun. If the foraging site lies on a line between hive and sun, i.e. in the same direction as the sun, then the bee dances perpendicularly upward. If the site lies on a line with the sun but behind the hive, the bee dances perpendicularly downward. The vigor of the waggling progressions, the loudness and duration of associated wing buzzing, as well as the number of waggling progressions per unit time are indicative of the distance of the site from the hive (Fig.2). This ritualized language of communication varies somewhat between geographical races of bees.

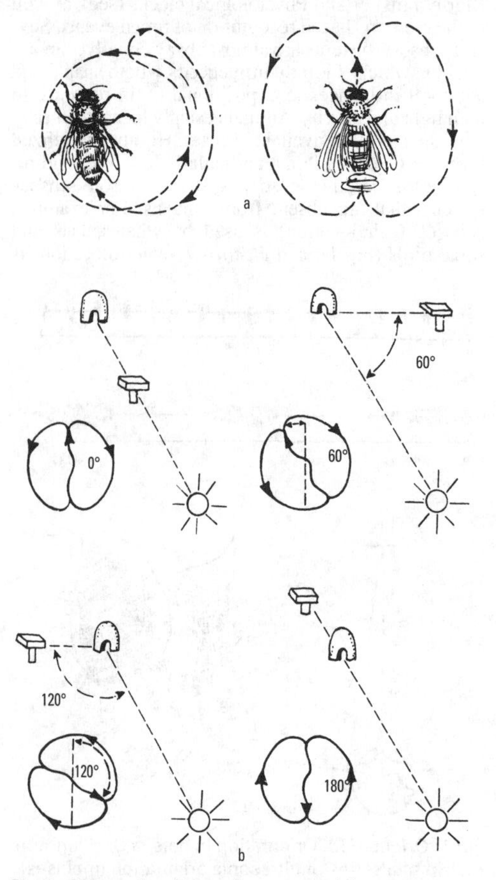

Fig.2. *Orientation.* Dance language of the worker honeybee (after von Frisch, 1950). *a* Round dance and waggle dance. *b* Indication of the direction of a foraging source by the waggle dance on the verticle comb in the hive.

The homing ability of migratory fishes, e.g. salmon, is due essentially to the imprinting of the young animal, during a sensitive period of its development, with the olfactory stimulus of the water where it hatched. In addition, it is thought that migratory marine fishes use the sun to establish their direction of migration, and are able to compensate for the apparent movement of the sun. Having entered an appropriate river, O. is then governed by the mechanical stimulus of the direction of water flow, as well as the olfactory stimulus of the "home water" from upstream.

Orientation behavior: behavior which serves for the orientation of an organism in space. In the wider sense, O.b. also includes behavior related to time orientation. *Spatial orientation* involves the determination of direction (taxis component) and distance (elasis component). *Time orientation* involves the determination of phase (e.g. time of day based on the alternation of light and dark) and measurement of the duration of time. In spatial orientation, a distinction is drawn between: *contact orientation* (in response to physical contact with objects in the surroundings, partners, etc); *near field orientation* (based on sensory location of objects, partners, etc. in the immediate vicinity); *distance orientation* (sometimes also called *goal orientation:* based on spatial factors, objects or partners not in direct sensory range). *Directional orientation* may be achieved by responding to the direction of a stimulus, or it may represent a response to a stimulus gradient. *Multimodal orientation systems* involve more than one sensory modality. Some types of distance orientation depend on an ability to monitor the position of the stars or the direction and strength of the earth's magnetic field (see Navigation). Echo-location has evolved in some animal groups (bats, whales, some birds) as a means of near field orientation.

Orientation movements: spatial orientation of a plants or its parts by movement of the entire plant or its parts away from (negative) or toward (positive) an external stimulus. In motile plants, O.m. occur by Taxis (see), while in sessile plants O.m. consist of the Tropism (see) of organs.

Origin of life: see Abiogenesis.

Orioles (Old World): see *Oriolidae.*

Oriolidae, Old World orioles: an avian family (order *Passeriformes)* of about 30, mainly tropical species. Males have luminous black and gold plumage, but are nevertheless difficult to see in the evergreen forest trees that form their natural habitat. They feed on insects and fruit, and they build hammock nests in forked branches. The family is widely distributed in the Old World, with representatives in Africa, Europe, Asia and Australia. Some isolated species have evolved on islands, such as Sao Tomé. A Malagassy bird known as the ***Kinkimavo*** *(Tylas)* is possibly an aberrant oriole. All orioles, but especially the ***Golden oriole*** *(Oriolus oriolus),* are noted for their fluting song. The Golden oriole is a tropical African species, which makes irregular summer visits to the Palearctic. Similarly, the tropical Southeast Asian ***Black-naped oriole*** *(Oriolus chinensis)* visits the Palearctic in summer. The 4 Australasian ***Fig birds*** *(Sphecotheres)* have bare skin around the eyes.

Oriolus: see *Oriolidae.*

Oris root: see *Iridaceae.*

Orn: accepted abbreviation of L-ornithine.

Ornate frog: see *Leptodactylidae.*

Ornithaptera: see *Raphidae.*

L-Ornithine, ***Orn, α,δ-diaminovaleric acid, 2,5-diaminopentanoic acid:*** $H_2N—CH_2—(CH_2)_2—CH(NH_2)—COOH$, a nonproteogenic amino acid. Orn is biosynthesized from L-glutamate, and in turn is a biosynthetic precursor of arginine. In mammals, this pathway is exploited as the Urea cycle (see), in which arginine is converted back to ornithine (L-glutamate → 2-*N*-acetyl-L-ornithine → L-ornithine → L-citrulline → argininosuccinate → L-arginine → L-ornithine). Most nonmammalian animals, as well as plants and bacteria perform the reactions from L-glutamate to L-ornithine, but lack the enzyme that catalyses the conversion Orn into L-arginine. Orn is also a component of various antibiotics such as gramicidin S. The biogenic amine, putrescine, is formed by decarboxylation of Orn.

Ornithine cycle: see Urea cycle.

Ornithischia: see Dinosaurs.

Ornithogamous flower: see Pollination.

Ornithogamy: see Pollination.

Ornithology: a branch of zoology devoted to the study of all aspects of the biology of birds.

Ornithophily: see Pollination.

Ornithopods: see Dinosaurs.

Ornithopus: see *Papilionaceae.*

Ornithorhynchidae: see *Monotremata.*

Ornithosis: a disease of birds caused by the bacterium *Chlamydia psittaci.* It is a zoonosis, i.e. humans can also be infected. In humans, the disease is known as psittacosis.

Orobanchaceae, ***broomrape family:*** a family of dicotyledonous plants, closely related to the *Scrophulariaceae,* and containing about 150 species in about 11 genera, distributed mainly in the warmer parts of the northern temperate zone of the Old World. They are all annual to perennial herbaceous root parasites, devoid of chlorophyll, with alternate scale-like leaves. The zygomorphic, hermaphrodite and hypogynous. flowers are usually in dense bracteate terminal racemes or spikes. The corolla has 2 lips and a curved tube, while the calyx is tubular below, with 2–5 teeth, or 2 upper lateral lips. There are 4 stamens in 2 pairs, and the single style has a 2-lobed stigma. The superior, single-celled ovary contains numerous ovules on 2–6 parietal placentae. Fruits are dehiscent capsules containing many microscopically small seeds, which apparently do not germinate unless they are in the vicinity of an appropriate host root. Some species of the large genus, *Orobanche* ***(Broomrapes),*** are specific for only a few host species, while others can parasitize a wide range of species, and cause considerable damage to commercially important plants, e.g. *Orobanche ramosa* ***(Branched broomrape),*** which can be a serious pest of hemp, tobacco, potatoes and tomatoes.

Orobanchaceae. Branched broomrape *(Orobanche ramosa)* parasitizing the root of a hemp plant *(Cannabis).*

Orobanche: see *Orobanchaceae.*

Oropendolas: see *Icteridae.*

Orpine: see *Crassulaceae.*

Orseille: see Lichens, Lichen pigments.

Orsellinic acid: see Lichen acids.

Ortalis: see *Cracidae.*

Orthis (Greek *orthos* straight): an extinct articulate genus of the *Brachiopoda.* The brachial skeleton (a calcified structure, present in some genera, which arises from the dorsal valve and supports the lophophore) is absent. Both valves are convex, approximately semicircular in outline, with prominent radial ribbing. Fossils are found worldwide from the Cambrian to the Permian, with particularly numerous species in the Ordovician.

Orthoceras (Greek *keras* horn), *Michelinoceras:* an extinct genus of nautiloids with smooth, straight, elongate shells, sometimes up to 3 m in length. The shell is circular in cross section with a thin, narrow, central siphuncle. Fossils are found from Ordovician to the Triassic.

Orthodont: see Dentition.

Orthogenesis, ***orthoevolution, rectilinear evolution, uniform evolution:*** continuous, unidirectional, long-term phylogenetic evolution, proposed by early paleontologists to account for the apparent linear progression of certain fossil characters. Well documented examples of O. are the continuous changes of shell shapes during the Cretaceous, and the evolution of the horse during the Tertiary. The Eocene fossil record reveals a small, forest-dwelling animal, *Eohippus* or *Hyratherium,* with a shoulder height of 28 cm, 4 toes on the forefeet, 3 functional toes on the hindfeet (1st and 5th toes represented by tiny splints) and a set of 44 low-crowned and rooted teeth lacking cement. Gradual modifications are found in *Miohippus* or *Mesohippus* (Oligocene; 61 cm shoulder height; 3 functional toes on each foot; taller, rooted molars), *Merychippus* (Miocene; 102 cm shoulder height) and *Pliohippus* or *Hipparian* (Pliocene; 109 cm shoulder height), leading finally to the modern plains-dwelling horse, *Equus* (recent Pleistocene; 152 cm shoulder height; high-crowned molars with cement; single functional toes or hoofs on each foot). The progression appears to follow known changes in Tertiary landscape from moist forest to relatively dry grasslands. The observed structural changes are representative of changes throught which the ancestors of the modern horse must have passed, but the above fossil forms are not necessarily on a direct line to *Equus.*

The original concept implied that O. was the result of internal directional factors. In its modern usage the term is purely descriptive. There is no evidence for the existence of an internal phylogenetic program, and there is not even an acceptable hypothetical mechanism for such a system. Like other evolutionary processes, O. can be attributed to natural selection.

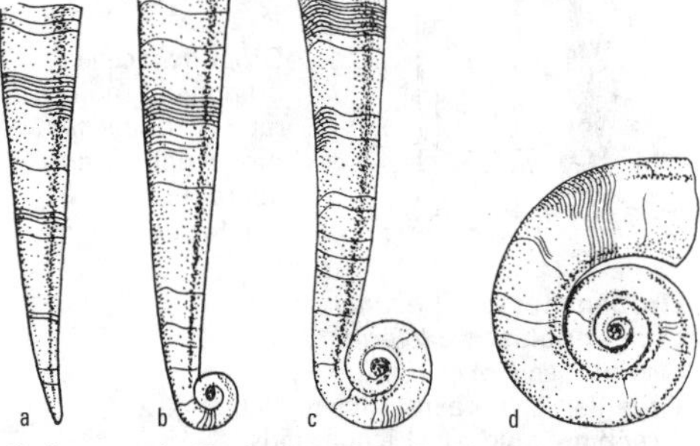

Orthogenesis. Evolution of shell spiralling in the lituites from the Scandanavian Lower Ordovician. *a Rhynchorthoceras. b Ancistroceras. c Lituites. d Cyclolituites.*

Orthogeotropism: see Geotropism.

Orthognathic profile: see Facial profiles.

Orthomyxoandreoviridae: see Multipartite viruses.

Orthotomus: see *Sylviinae.*

Orthotropism: see Tropism.

Ortyxelos: see *Gruiformes.*

Orycenin: see Glutelins.

Oryctolagus: see *Leporidae.*

Orygini: see *Hippotraginae.*

Oryx: see *Hippotraginae.*

Oryza: see *Graminae.*

Osage orange: see *Moraceae.*

Oscarella: see *Porifera.*

Oscillations: periodic alterations of physical, chemical and biological quantities. In biological systems, O. occur with periodicities varying from fractions of a second to years. For example, the rhythmic alterations of nerve potential represent rapid O., whereas very slow O. are described by changes in the sizes of entire animal populations. Often several O. with different periodicities are observed in a single system. In such cases, it appears to be the rule that the periods of the different O. are widely different. For example, in the human, neuronal O. occur in the millisecond range, cycles of cardiac activity occur in the order of seconds, physiological functions display a periodicity of hours, and female hormonal regulation shows a periodicity of one month. Chronobiology (see) is based on the application of O. to Systems theory (see). O. are described with the aid of nonlinear differential equations, which also include the effects of Feedback (see). Systems with high amplification normally also display oscillatory behavior, e.g. regulation of the water economy of plants, alteration of pupil diameter. The periodicity of each oscillating system is probably regulated by coupling between all or many of the oscillating systems present.

[I.R. Cowan, Oscillations in stomatal conductance and plant functioning associated with stomatal conductance: observations and a model, *Planta* ***106,*** 185–219, 1972; D.J. McFarland, *Feedback mechanisms in animal behaviour,* Academic Press, New York and London, 1971; Th. Pavlidis, *Biological oscillators: their mathematical analysis,* Academic Press, New York and London, 1973]

Oscillatoriales: see *Hormogonales.*

Oscines, *Passeres, "song birds":* a suborder of the *Passeriformes* (see). Oscines differ from other passerine birds in having 5–9 pairs of song muscles, and the syrinx is of the trachobronchial type.

Osculum: see Digestive system, *Porifera.*

Osiers: see *Salicaceae.*

Osmerus: see Smelt.

Osmophilic microorganisms: microorganisms that are able to grow in media of high osmolarity, e.g. in highly concentrated sugar or salt solutions. O.m. can cause food spoilage, e.g. fermentation of honey by osmophilic yeasts.

Osmoreceptors: see Osmoregulation.

Osmoregulation: in animals, the maintenance of a constant water content and salt concentration. Evolution has produced two O. mechanisms.

1) In *aquatic animals,* the problem consists largely of counteracting the loss of water from marine forms, and the entry of water into freshwater forms. Body fluids of marine invertebrates and elasmobranch fishes are isotonic with the surrounding sea water. In elasmobranchs this is due to the urea content of the blood, whereas isotonicity of the body fluids of marine invertebrates is due to their salt content. Such organisms are unable to live in fresh or brackish water. On the other hand, body fluids of both marine and freshwater teleost fishes have a similar osmolarity, which is lower (hypotonic) than that of sea water, and higher (hypertonic) than that of fresh water. Marine teleosts meet their water requirement by continually drinking sea water and excreting excess salt via the kidneys and specialized cells in the gills. Freshwater telosts absorb water through their gills, and satisfy their salt requirement by retaining appropriate ions from their diet; excess water is removed by excreting large quantities of very dilute urine throught the kidneys (up to on third of the body weight daily). The skin of all fishes is practically impermeable to water and dissolved materal. Freshwater invertebrates have well developed excretory organs, which reexport the water that enters through the body surface.

2) In *terrestrial animals,* water and salts are not exchanged directly with the environment. Here, emphasis lies with the prevention of water loss, and O. often con-

sists largely of volume regulation (see Blood volume regulation). In addition to arterial receptors, osmotic pressure is continually monitored by *osmoreceptors* in the hypothalamus. The stimulus for O. is an increase in *serum osmolarity* which leads to loss of water from the osmoreceptors. This results in the sensation of Thirst (see), and in the increased production of antidiuretic hormone (ADH) by both the paraventricular nucleus and the supraoptic nucleus of the hypothalamus. An ADH-containing neurosecretion from the hypothalamus enters the posterior lobe of the pituitary, where it may be stored or released into the circulation. Via the circulation, ADH reaches the kidneys, where it increases facultative water resorption, with a resulting decrease in serum osmolarity. Fine control of O. involves additional mechanisms, such as excitation of the sympathetic nerve supply to the kidneys, and the production and action of the kidney enzyme, renin. Finally, ADH secretion may be increased by emotional stimuli (e.g. pain), or by agents such as nicotine, acetylcholine, histamine or insulin, as well as by many narcotics. Estrogens reinforce the action of ADH on the nephron. The above control mechanisms operate in reverse when serum osmolarity is decreased (Fig.).

In mammals, the osmoregulatory functions of the kidneys are supplemented to a small extent by sweat secretion and insensible perspiration from the skin. Marine birds and reptiles that drink sea water possess additional regulatory mechanisms, e.g. nasal glands that secrete a concentrated salt solution.

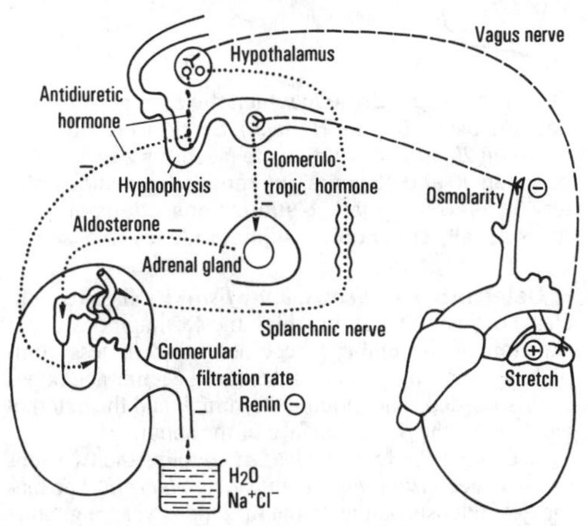

Osmoregulation. Main mechanisms for the regulation of osmolarity and volume.

Osmotaxis: a special case of chemotaxis or hydrotaxis, in which taxis is determined by a change in osmolarity. For example, withdrawal of water from one side of a plant cell leads to negative O. or positive hydrotaxis of chloroplasts.

Osmosis: equalization of concentration between two solutions of unequal concentrations, or between a solution and the pure solvent, by diffusion of the solvent through a semipermeable membrane. A semipermeable membrane is permeable only to the solvent, or it may also permit the very limited diffusion of dissolved substances. Under these conditions, the solvent diffuses through the membrane from the weaker (hypotonic) solution into the stronger (hypertonic) solution. In *endosmosis,* a limited volume of a concentrated aqueous solution is enclosed by a semipermeable membrane and immersed in water, so that water migrates into the solution. In *exosmosis,* these conditions are reversed, i.e. the stronger solution lies outside the membrane, and water is withdrawn from the enclosed solution. The plasmalemma and the tonoplast of plant cells are natural semipermeable membranes. For demonstration and experimental purposes, suitable semipermeable membranes are pig's bladder, collodium membranes or precipitated layers of calcium phosphate, Prussian blue, etc. Ideal semipermeable membranes of copper hexacyanoferrate (II), $Cu_2[Fe(CN)_6)]$, were first described in 1866 by Moritz Traube. In order to obtain a membrane sufficiently strong to withstand a pressure of several atmospheres, the plant physiologist Wilhelm Pfeffer in 1877 produced a precipitate of copper hexacyanoferrate (II) in the walls of a porous battery pot by filling the pot with cupric sulfate solution and immersing it in a solution of potassium hexacyanoferrate. Where the two salts met in the walls of the pot the membrane of copper hexacyanoferrate (II) was precipatated. Such a pot, containing a membrane of copper hexacyanoferrate (II) and fitted with a monometer, is known as Pfeffer's apparatus. If it is filled with sucrose solution and placed in a large vessel of water, then water diffuses in until the hydrostatic pressure of the liquid column balances the osmotic pressure within the aparatus, at which point net water diffusion ceases. Pfeffer's apparatus can therefore be used as an osmometer.

For a nondissociating solute, $\pi^* = c \times R \times T$, where π^* is the osmotic pressure in bar (1 atmosphere = bar x 0.987), c the concentration of dissoved substance in mol/l, T the absolute temperature and R the gas constant. A molar solution at 0 °C has an osmotic pressure of 22.7 bar (= 22.4 atmospheres). At the same molar concentration, nondissociating substances have the same osmotic pressure, i.e. they are *isosmotic* or *isotonic.* Osmotic pressure is a *colligative property,* i.e. it depends on the number of dissolved particles and not on their type or size, so that a solution of a dissociated salt has a higher osmotic pressure than an equimolar solution of a nondissociating compound. Thus, the osmotic pressure of a dilute, totally ionized potassium nitrate solution is 1.69 times greater than that of an equimolar nondissociated solution of sucrose; accordingly, potassium nitrate has an osmotic coefficient of 1.69. It does not attain a value of 2, because the free mobility of the ions is decreased by ionic interaction.

Pfeffer's apparatus was constructed as a model of the plant cell, in which the porous pot represents the easily permeable plant cell wall, while the semipermeable membrane of copper hexacyanoferrate (II) represents the layer of protoplasm with its two semipermeable membranes, the plasmalemma and the tonoplast. The liquid inside Pfeffer's apparatus corresponds to the cell sap with its dissolved material. Similar laws govern the behavior of the two systems.

Within the plant cell, the vacuole contains dissolved material, which causes water to enter by O., thereby generating a hydrostatic pressure known as *turgor pressure (P).* This contributes to the rigidity of the plant cells and plant organs, because it causes the plasmalemma to press against the cell wall. To a limited extent, the latter stretches elastically, until finally the reactive pressure of the stretched wall prevents the further entry of water, i.e. an equilibrium is attained between water entry and water exit. The ability of the plant cell to take up water by osmosis depends therefore not only on the osmotic pressure, but also on the turgor pressure: $S = \pi^* - P$, where S is the suction pressure of the cell.

If the cell is not isolated but part of a tissue, cell stretching is opposed by *tissue tension,* which arises because the outer layer or epidermis prevents the inner layers from attaining their maximal extension. This further decreases the suction pressure, so that $S = \pi^* - (P + A)$, where A is the pressure on the cell from surrounding tissue.

If living plant cells (e.g. pieces of epidermis) are placed in a hypertonic solution (e.g. a strong solution of sucrose), water is withdrawn from the cells into the hypertonic solution. The cell vacuole shrinks and the protoplasm surrounding it detaches from the cell wall. This phenomenon, known as *Plasmolysis* (see), is easily observed under the light microscope. When a plasmolysed but still living cell is transferred to water, plasmolysis is reversed (i.e. *deplasmolysis* occurs). The osmotic pressure of individual cells can be found by determining the strength of the exterior solution that barely causes plasmolysis, and which must therefore be isotonic with the cell interior. Osmotic pressure of whole tissues or organs can also be found by measuring the concentration of dissolved material in carefully expressed cell sap by determining the depression of freezing point (cryoscopic method).

Osmotic values differ between plant species, and between the tissues and organs of the same plant. Furthermore, individual cells are able to adjust their S-values within certain limits, in response to changes in environmental conditions. Generally the S-values of parenchyma cells in the root cortex lie between 5 and 15 bar. S-values in the stem increase with distance from the root, and maximal values of 30 to 40 bar are found in leaves. In the roots of halophytes and some desert plants, S-values can exceed 100 bar.

For O. in animals, see Osmoregulation.

Osmotic pressure: hydrostatic pressure due to dissolved molecules, and proportional to the number of dissolved molecules per unit volume. The O.p. of very dilute solutions obeys the ideal gas laws. O.p. is proportional to concentration, decreases proportionally with temperature, and can be determined experimentally with an osmometer. Since osmosis is a colligative property (i.e. it depends on the number of dissolved molecules, irrespective of size or type), measurement of the O.p. of a solution of known mass concentration can be used to calculate the relative molecular mass of the dissolved material. See Osmosis.

Osphronemidae, *giant gouramis:* a monospecific family of freshwater teleost fishes of the order *Perciformes,* suborder *Anabantoidei.* The single representative, *Osphronemus goramy* ***(Giant gourami)***, is a heavy, powerful fish, up to 60 cm in length, with a head that appears short in proportion to the body. The ventral fins are elongated and threadlike, the lower jaw appears lopsided, and the body is humped. Both parents guard the eggs, which are laid in a large bubble nest of plant material. In Southeast Asia the Giant gourami is an important food fish, with few bones and very tasty meat.

Osphronemus: see *Osphronemidae.*

Osprey: see *Falconiformes.*

Os Sepioe: see Cuttlefish bone.

Osseus labyrinth: see Auditory organ.

Ossicles of the middle ear: see Auditory organ.

Ossification: see Bone formation.

Ossification age: determination of age from the regular succession of bone nuclei, the ossification of cartilaginous primordia (in particular epiphyseal disks), and the closure of cranial sutures. O.a. is used to estimate the biological age of children and young adults, and for the age diagnosis of human skeletal remains.

Particularly important indices are: 1) the ossification of limb bones, which is not complete until the body has grown to its maximal height; and 2) closure of the cranial fontanelles; the smaller of these are already closed at and shortly after birth, whereas the larger fontanelles on the crown of the cranium are not completely ossified until 9 to 16 months of age. Toward the end of the 3rd year of life, closure of the cranial sutures begins and continues into old age. Both the order of occurrence and the rate of cranial and skeletal ossifications are subject to marked individual variability.

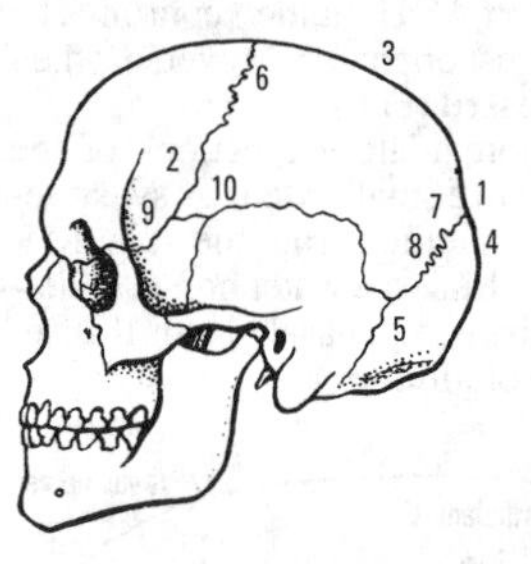

Ossification age. Order in which the cranial sutures become ossified (after H. Bach). Section 1 ossifies between 20 and 30 years of life, sections 2 to 4 between 30 and 40 years, the remaining sections after 40 years. Non-marked sutures or sections of sutures usually show only a slight tendency to ossify.

Osteichthyes, *teleosts, bony fishes:* a class (or subclass) of Fishes (see), in which the cartilaginous endoskeleton of the embryo becomes more or less completely replaced by bone in the adult. Membrane bones are also added. The mouth is terminal, and the external nares lie on the dorsal surface of the snout.

Accepting the O. as a class, it contains the two subclasses, the *Actinopterygia* and *Sarcopterygia.* The majority of teleosts belong to the *Actinopterygia,* e.g. Sturgeons *(Chondrostei)* (see), Holosteans *(Holostei)* (see), and the *Teleostei,* the latter containing the vast majority of modern bony fishes. The *Sarcopterygia* are divided into Lungfishes (see), Coelacanths (see), and members of the extinct *Osteolepidoti* (Paleozoic sarcopterygians related to the earliest known amphibians).

Fossil remains of teleosts are found from the Lower Devonian to the present. The earliest representatives were freshwater inhabitants, and marine environments were not invaded until the end of the Paleozoic. The *Teleostei* evolved in the Triassic.

Osteoblast: see Bone formation.

Osteoclast: see Bone formation.

Osteocyte: see Bones.

Osteogenesis: see Bone formation.

Osteoglossidae: a relatively primitive family of

freshwater teleost fishes, which possess strikingly large scales, and are found in inland tropical waters. The family belongs to the suborder *Osteoglossoidei,* order *Osteoglossiformes* (bonytongues). The ***Arapaima*** or ***Pirarucu*** ***(Arapaima gigas)*** is one of the largest freshwater fishes, attaining a length of 3 m and a weight of 200 kg. It inhabits the Amazon and Orinoco and other South American rivers, where it is prized as an edible fish. Osteoglossids are also distributed in Africa and Southeast Asia, while a single species of *Sclerophages* is the only strictly freshwater teleost of Australia.

Osteolaemus: see African dwarf crocodile.

Osteolepis (Greek *osteon* bone, *lepis* scale): a genus of the extinct family *Osteolepidae* of the extinct order *Osteolepiformes,* subclass *Crossopterygii* (Fringe-finned or Tassel-finned fishes). It possessed thick, rounded to rhombic scales, paired and unpaired fins, and an asymmetrical (heterocercal) tail fin. Fossils occur in the Devonian.

Osteometry: measurement of animal or human postcranial skeleton. See Anthropometry.

Osteosclerosis congenita: see Achondroplasia.

Osteostraci (Greek *osteon* bones, *ostrakon* shell): an extinct order of jawless, fishlike vertebrates *(Agnatha),* in which the flattened head and forebody were covered with an armor of bony plates or scales, with a series of lateral gill openings and perforations for the eyes and olfactory organs. Lateral fins were present. The mouth was situated on the ventral surface, and the tail was asymmetrical (heterocercal). Fossils are found from the Upper Silurian to the Upper Devonian.

Ostia (singular **Ostium):**

1) Slits or openings in the heart wall of arthropods, through which blood enters the heart directly from blood vessels (closed vascular systems), or from the body cavity (via the pericardial sinus) in open systems (see Blood circulation). Pairs of O. are arranged segmentally, and their edges extend into the heart lumen to form valves that permit blood flow in only one direction. Depending on the length of the tubular heart, the number of O. pairs varies widely between different arthropods. The greatest number is found in insects (13 pairs, 2 of which lie in thoracic segments).

In vertebrates, the openings between atria and ventricles (atrioventricular O.) and the openings of the ventricles into the aorta and pulmonary artery are also called O.

2) In the wider sense, any opening that provides communication between organs is an ostium.

3) In *Porifera,* one of the openings through which water is drawn into the interior cavity.

Ostraciidae: see *Ostraciontidae.*

Ostraciinae: see *Ostraciontidae.*

Ostraciontidae, *Ostraciidae,* ***boxfishes:*** a family of fishes (about 30 species in 13 genera) of the order *Tetraodontiformes,* which inhabit coral reefs. The body is encased in a rigid bony carapace, and ventral fins are absent.

Subfamily *Aracaninae* (about 10 species, 6 genera): carapace open behind dorsal and anal fins; caudal fin has 11 principal rays; very abundant around Australia.

Subfamily *Ostraciinae* (about 20 species, 7 genera): carapace closed behind anal fin; caudal fin has 10 principal rays.

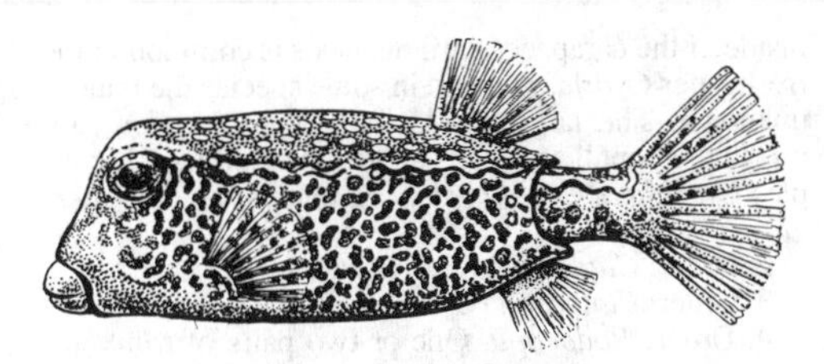

Ostraciontidae. Ostracion lentiginosus.

Ostracoda, ***mussel shrimps, seed shrimps:*** a sublass of the *Crustacea* (see), mostly 1–2 mm in length, and fully enclosed by a bivalve carapace, which possesses an adductor muscle; the eliptical valves are impregnated with calcium carbonate. The cephalic limbs (head appendages) are well developed, and there are never more than two recognizable pairs of trunk limbs; these are uniramous and they are not phyllopodia. One or both pairs of trunk limbs may be absent. All appendages are highly specialized, varying in structure according to the mode of life and the feeding mechanism. The 2000 (approx.) known living species occur in salt and fresh water. They live on or within the bottom mud or sand of their aquatic habitiat, or on aquatic plants, swimming short distances with the aid of their antennulae and antennae, or their antennae alone. They include predators, carrion eaters, plant eaters and detritus feeders, while some species are exclusively filter feeders.

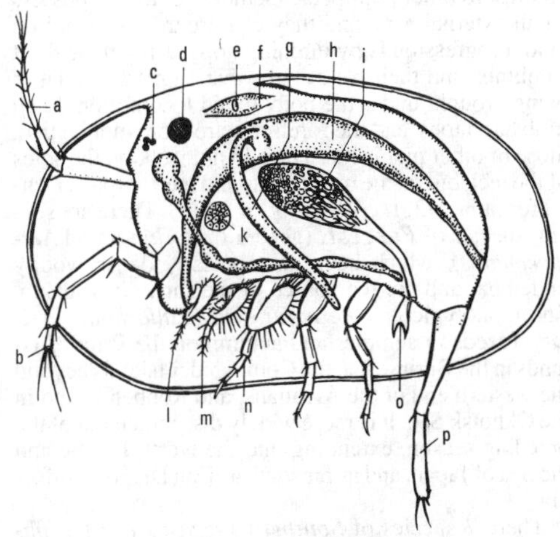

Ostracoda. Lateral view of the organization of an ostracod (left half of shell removed). *a* Antennule. *b* Antenna. *c* Nauplius eye. *d* Compound eye. *e* Ostium. *f* Heart. *g* Hepatic cecum. *h* Alimentary canal. *i* Testis with large spermatozoa. *k* Adductor muscle. *l* Mandible. *m* First maxilla. *n* Second maxilla. *o* Penis. *p* Furca.

Classification.

1. Order: *Myodocopa.* The only order possessing sessile compound eyes. The second antennae are usually adapted for swimming, and the expodite is larger than the endopite. Two pairs of trunk appendages are present, the second pair being adapted for cleaning the

inside of the carapace. Parthogenesis is common in the freshwater *Cyprididae,* and in some species the male is unknown. The large, predacious *Gigantocypris* captures other small crustacea and even small fish. Examples:*Cypridina, Conchoecia, Gigantocypris, Skogsbergia.*

2. Order: *Cladocopa (Polycope,* etc.)

3. Order: *Platycopa (Cytherella)*

4. Order: *Podocopa.* One or two pairs of trunk appendages are present. On the second antennae, the endopite is equal to or larger than the expodite. The order includes both marine and freshwater species. Examples: *Candona, Notodromas, Cypris, Cythere, Pontocypris, Cypridopsis, Mesocypris, Darwinula.*

Geological distribution. Some 10 000 described fossil species (based entirely on shell size and structure) provide an abundant, widespread and continuous fossil record from the Cambrian to the present. Together with the *Foraminifera,* they are extremely useful index fossils. Even in the early Paleozoic they occur in a great variety of forms, attaining their peak of evolutionary development in the Middle Devonian. From the Jurassic onward they occur with a more or less constant frequency in all subsequent formations.

Ostrea: see *Bivalvia.*

Ostreasterol: see Chalinasterol.

Ostrich: see *Struthioniformes.*

Otaria: see *Otariidae.*

Otariidae, ***fur seals*** and ***sealions, eared seals:*** a family of the *Pinnipedia* (see), distributed over the seas of the Southern Hemisphere and the North Pacific. In contrast to other pinnipeds, members of the *O.* possess small external ears, and they can progress forward on land. Progression is by lurching hops, using their short forelimbs and their rudder-like rear flippers, the latter being brought under the body to aid locomotion. Each limb has 5 toes, and the forelimbs are more mobile than those of other pinnipeds. They do not walk on the soles of the feet, but on the metatarsals and metacarpals. Dental formula: 3/2, 1/1, 4/4, 1–3/1 = 34–38. There are several species of ***Fur seals*** (genera *Callorhinus* and *Arctocephalus),* which have an especially dense woolly undercoat and are the source of the furrier's sealskin. The monotypic ***Northern fur seal*** *(Callorhinus ursinus)* breeds in summer and autumn on the Pribilof Islands in the Bering Sea, the Commander Islands beyond the western end of the Aleutians, and Robben Island in the Okhotsk Sea. It is more widely distributed out of the breeding season, extending into the North Pacific and the Sea of Japan, and as far south as Dan Diego, California.

There 7 species of ***Southern fur seals*** *(Arctocephalus),* all found in the Southern Hemisphere, where they remain in more or less defined areas and are largely nonmigratory. Largest and most numerous of the genus is the ***South African fur seal*** *(A. pusillus),* which is found on the coasts of Southwest Africa around the Cape to Algoa Bay. Only a small remnant herd remains of the ***Australian*** or ***New Zealand fur seal*** *(A. forsteri),* while the ***Juan Fernandaz fur seal*** *(A. philli),* found on Juan Fernandez Island off Chile, is nearly extinct. Only a few individuals remain of the ***Galapagos fur seal*** *(A. galapogoensis).* The smallest species is the ***Kerguelen fur seal,*** small herds of which still survive on Kerguelen island and other sub-Antarctic neighboring islands. The ***South American fur seal*** *(A. australis)* is found around the coasts of southern South America and offshore islands. Formerly very abundant on the North American Pacific coast, the ***Guadalupe fur seal*** *(A. townsendi)* is now found only around the small islands of Guadalupe, Baja and Santa Barbara.

In contrast to the fur seals, the ***Sealions*** (6 species) have shorter, less dense hair. Two monotypic sealion genera, *Phocarctos hookeri* ***(New Zealand sealion)*** and *Neophoca cinerea* ***(Australian sealion),*** have a skull structure intermediate between those of fur seals and the sealions, and the young of *Phocarctos* have a thick, fur seal-like coat. Largest of all the *Otariidae* is the monotypic ***Northern*** or ***Stellar sealion*** *(Eumetopias jubata).* The monotypic ***South American sealion*** *(Otaria byronia)* is found around the coasts of South America from the mouth of La Plata around the Cape and north to Peru. Most familiar of all sealions is the *Californian sealion* ***(Zalophus californianus)*** from the Pacific coast of America. A close relative ***(Galapagos sealion,*** *Z. wollebaeki)* is found on the Galapagos Islands. *Z. californianus* is generally more agile on land than the other sealions; it has a more tapered and dog-like head, is seen in almost every zoo, and is the performing sealion of the circus.

Otidae: see *Gruiformes.*

Otididae: see *Gruiformes.*

Otis: see *Gruiformes.*

Otocolobus: see *Felidae.*

Otocyon: see *Canidae.*

Otolemur: see Galagos.

Otters: see River otters, Sea otter.

Otter shrews: see Water shrews.

Otus: see *Strigiformes.*

Ouabain: see Strophanthins.

Ounce: see *Felidae.*

Ourebia: see *Neotraginae.*

Out-of-Africa model of human evolution: see Anthropogenesis.

Oval of the swim bladder: see Swim bladder.

Ovariole: see Ovary.

Ovary (plate 23): a more or less site-specific gonad of female animals, in which eggs are elaborated.

Sponges lack a definitive O., and their eggs are produced by the maturation of germ cells distributed in the dermal layers. In the simplest case, in coelenterates, the O. is an accumulation of female germ cells in the exoderm or mesoderm. When a body cavity is present, the primitive germ cells migrate either into the parenchyma of the primary coelom (e.g. in *Platyhelminthes)* or into the peritoneal epithelium of the secondary coelom (e.g. *Annelida).* In bilaterlly symmetrical animals the Os. are paired. In radially symmetrical animals (e.g. medusae and sea stars) the Os. are arranged radially.

Flatworms *(Platyhelminthes)* possess paired or unpaired Os., which are associated with a complex sexual apparatus for export of the eggs. In most species, a region of cells has become separated from the O., lost the ability to develop into eggs, and become specialized for the production of yolk and shell material (see Vitellarium).

The Os. of mollusks are paired or secondarily unpaired. They are hollow structures (e.g. in snails and cephalopods), which are connected with the gonocoel (the residue of the secondary coelom). In hermaphrodite snails, the Os. and testes are united to an Ovariotestis (see).

In insects, the Os. are paired. Each O. consists of numerous ovarioles or ovarian tubules, in which the eggs are elaborated. These ovarioles lie side by side, and are connected anteriorly by a terminal filament, which serves to attach the O. to the inner surface of the body wall. Each ovariole contains a single row of oogonia, recognizable externally as swellings, which become progressively larger and the ovariole correspondingly wider from the anterior to the posterior end, i.e. toward the oviduct. Three types of ovariole can be recognized, depending the mode of egg cell nutrition. 1) In panoistic ovarioles, the egg cells are surrounded only by the follicular epithelium, while nutritive or nurse cells are lacking (e.g. *Orthoptera).* 2) In polytropic ovarioles, chambers containing nurse cells alternate with oogonia

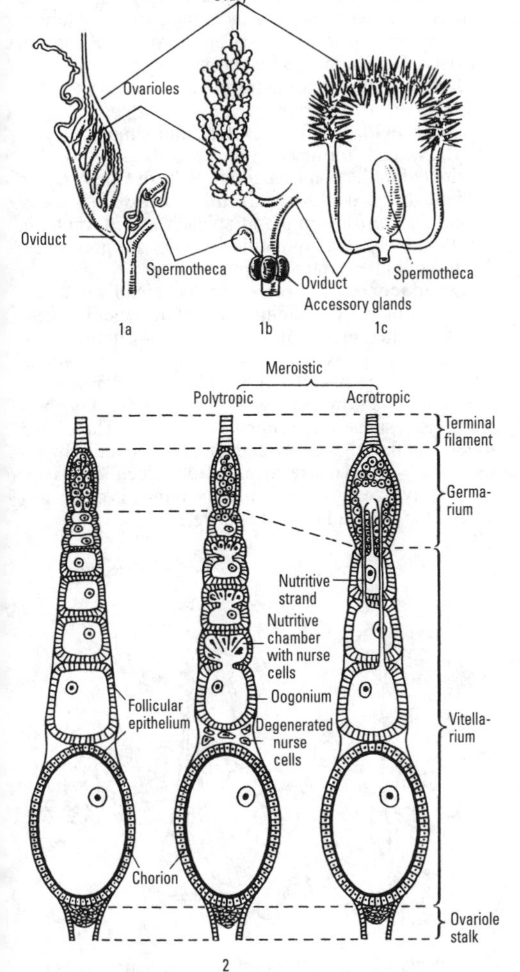

Ovary. Insect ovaries. *1* Different types of insect ovaries. *1a* Comb- shaped ovary of the meadow grasshopper *(Dissosteira carolina* L.). *1b* Clustered ovary of a shield bug *(Dactylopius* sp.). *1c* Horseshoe-shaped ovary of the stonefly *(Perla maxima* Scop.). *2* The three types of ovarioles found in insects.

(e.g. *Neuroptera, Hymenoptera).* 3) In telotropic or acrotropic ovarioles, nutritive cells are present in the germarium and connected to the egg cell in the yolk sac by nutritive strands (e.g. *Hemiptera).* The anterior and narrowest part of the ovariole is known as the germarium, in which the egg cells are initially formed. This is attached to the vitellarium, in which the egg cells grow and develop and acquire their shell. On each side, the stalks of the vitellaria unite and empty into a common right or left Oviduct (see), and the two oviducts unite to form a median channel or vagina, which in turn opens into the genital pouch.

In *Amphioxus* and in *Cyclostomata,* the Os. are arranged segmentally, and the eggs enter the peribranchial space by rupture of the body wall.

The Os. of vertebrates are paired, although in birds and the platypus the right O. is regressed. Elasmobranch fishes also usually have only one functional O.

Ovenbird: see *Furnariidae.*

Overtopping: see Telome theory.

Overwintering: different strategies adopted by animals in temperate and cold regions for surviving the low temperature and food scarcity of winter.

Many poikilothermic animals seek out suitable places of concealment even before the onset of the cold season, a form of behavior often triggered by the photoperiod (shortened daylength). With decreasing temperature, they then sink into a state of Cold torpor (see), which is reversible by warmer environmental conditions. Such inactivity in response to environmental factors is generally known as *quiescence.* In contrast, *diapause* (see Dormancy) is linked to a specific stage of development or metamorphosis, and in many insects it represents a characteristic overwintering strategy.

Some homeothermic animals migrate before the onset of winter to a milder climatic region (see Animal migration). Certain bats decrease their metabolism and their body temperature as the environmental temperature decreases, and they fall into a state of *cold lethargy,* i.e. they are heterothermic. Similar torpid states with a temporary decrease in body temperature are displayed by some birds (nightjars, common swift, humming birds, swallows). This enables the birds to survive food shortages during periods of bad weather, or very low night temperatures, and is therefore not specifically a form of overwintering.

Some mammals (badger, brown bear, raccoon) extend and deepen their normal sleep in the cold season, without a decrease in body temperature. This *winter rest* is interrupted by occasional phases of hunting and feeding activity.

Winter sleep or *true hibernation* is preconditioned by the daylength and it is hormonally regulated. The temperature of onset of hibernation depends on the species [common hamster *(Cricetus cricetus)* about 9 °C, common dormouse *(Muscardinus avellanarius)* about 15 °C, European hedgehog *(Erinaceus europaeus)* 17 °C, fat dormouse *(Glis glis)* 18–20 °C]. Body temperature falls to a fixed minimal value (barely above 0 °C), with an accompanying decrease in blood flow, heart beat, respiratory frequency and response to sensory stimuli. If the body temperature begins to fall below its critical value, there is an increase in metabolism, the body temperature rises again, and the animal often then wakes and feeds. Animals prepare for hibernation by constructing specially protective winter dens and nests (marmots, susliks. dormice), or by laying in food stores (common hamster), but primarily by increasing their own fat reserves.

Homeothermic animals that remain active in the winter tend to grow a denser coat (hairs or feathers), often with a change of color (e.g. white as camouflage in snow). They also lay down fat reserves and may change to a different diet.

Invertebrates overwinter during a particular stage of development (egg or pupa of insects). Some live and overwinter as larvae in water (dragonflies, stoneflies), or survive as cysts (protozoa), gemmules (freshwater sponges), or winter eggs *(Daphnia,* aphids). Others remain active in the winter without special auxillary aids (some spiders, mites, springtails).

The converse of O. is Estivation (see).

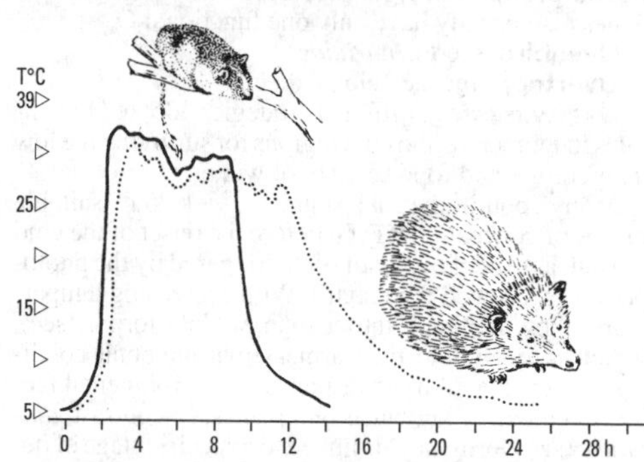

Overwintering. Interruption of winter sleep by a brief waking period. Body temperature rises very rapidly to above 30 °C. Continuous line: garden dormouse *(Eliomys quercinus).* Dotted line: European hedgehog *(Erinaceus europaeus).*

Ovibovini: see *Caprinae.*

Ovibos: see Musk ox.

Ovicell: see *Bryozoa.*

Ovicides: plant protection agents that kill the eggs of insects and mites.

Oviduct, ***tuba uterina:*** a duct for the conduct of mature eggs from the ovary or ovaries. In teleost fishes the O. is joined directly to the ovary. In quadrupeds the eggs pass first into the body cavity, and from there via a funnel-like expanded opening of the O. (ostium tubae) (which in mammals is usually very close to the ovary) into the O. Embryonically, the O. is derived from the Müllerian duct, from which it displays varying degrees of differentiation in different classes of vertebrates. Frequently, special glands are present (albumin, shell, calcium carbonate-secreting and gelatinous glands), which secrete the secondary and tertiary egg membranes or shells (see Egg).

Ovipary: the laying of eggs, which at the time of exit from the body still consist of a single cell. Fertilization occurs either later outside the female parent (fish and amphibians), or during the laying process (insects and spiders). Many echinoderms, mollusks, insects, spiders, fishes and amphibians are oviparous. See Ovovivipary, Vivipary.

Ovis: see Sheep.

Ovotestis: see Hermaphrodite gland.

Ovotransferrin: see Siderophilins.

Ovovivipary: the laying of eggs in the which the embryo is already more or less well developed. O. is displayed by many reptiles, some insects and worms, and in a certain sense by birds, whose eggs are laid when the embryonal disk is at an early stage of cleavage. In some cases of O., the egg membranes are ruptured very shortly after birth.

Ovulation inhibitors: see Hormonal contraception.

Ovum: see Egg.

Owlet frogmouths: see *Caprimulgiformes.*

Owlet moths: see *Noctuidae.*

Owl monkey, ***night monkey, douroucouli,*** *Aotes trivirgatus:* the only nocturnal monkey, and the only species of the subfamily, *Aotinae* (see Cebid monkeys, New World monkeys). The eyes are very large, without retinal cones, but with a fovea. Family groups of one male, one female and young inhabit tropical forests from the Amazon to southern Central America. The diet consists of fruits, insects and birds' eggs.

Owl parrot: see *Psittaciformes.*

Owls: see *Strigiformes.*

Oxalic acid: HOOC—COOH, the simplest dicarboxylic acid. It forms colorless crystals and is widely distributed in the plant kingdom. Salts of O.a. occur in particular in wood sorrel *(Oxalis acetosella),* sorrel *(Rumex acetosa),* turnip and spinach leaves, rhubarb, etc. Relatively large quantities of O.a. are poisonous, due to its ability to bind calcium.

Oxalidaceae, ***wood sorrel family:*** a family of dicotyledonous plants, containing about 950 species in 8 genera, distributed mainly in the tropics and subtropics of the Southern Hemisphere. They are herbs, rarely woody plants, with alternate, compound leaves and 5-merous, actinomorphic flowers. The 10 (rarely 15) weakly obdiplostemous stamens are connate at the base. There are 5 styles, and the stigmas are capitate. Pollination is by insects; the 5-celled overy develops into a loculicidal capsule, rarely a berry. Chemically, the family is characterized by the accumulation of oxalate.

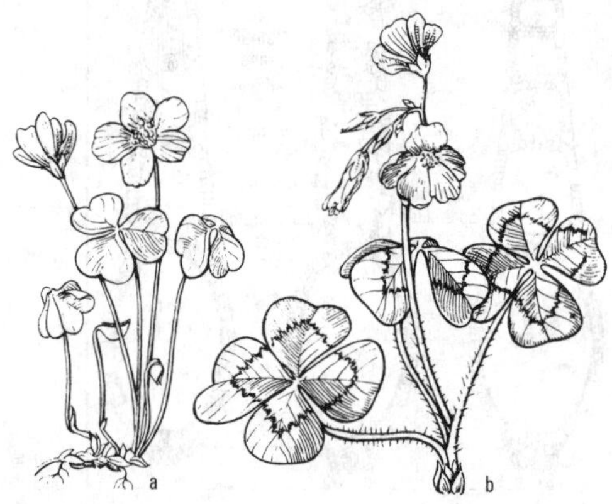

Oxalidaceae. a Oxalis acetosella (wood sorrel). *b Oxalis deppel* (sold as "lucky clover").

The most numerous genus is *Oxalis* (about 800 species); it is cosmopolitan, but represented mainly in Central and South America and South Africa. The perennial white-flowered ***Wood sorrel*** *(Oxalis acetosella)* is widely distributed in damp, shady deciduous and conifer forests throughout most of Europe and North and Central Asia. It has a slender, creeping rhizome, and long-stalked leaves with 3 obcordate leaflets, indented at the apex, which provide a fine ground cover for shady areas in gardens and parks. In eastern North America, the same ecological niche is occupied by the very similar *Oxalis montana* (see Species pairs). Various other *Oxalis* species, mostly originating from America, whose leaves consist of 4 leaflets, are sold in the garden trade as "lucky clover", e.g. *Oxalis deppel,* which has red flowers. *Oxalis tuberosa,* distributed in the Andes from Colombia to Chile, has edible tubers, known as ***oca.***

The genus *Averrhoa* is represented by only 2 species of evergreen trees, which probably originated in Brazil, but are now grown widely in the tropics for their fruits: the ***carambola*** or ***starfruit*** *(Averrhoa carambola),* and the ***bilimbi*** *(Averrhoa bilimbi).*

Oxalis: see *Oxalidaceae.*

Oxaloacetic acid: HOOC—CO—CH_2—COOH, a ketocarboxylic acid (oxocarboxylic acid) first found in higher plants (e.g. red clover, peas). As an intermediate in the tricarboxylic acid cycle and as the transamination product of aspartic acid, O.a. is an important metabolic intermediate. It is present, at least in small quantities, in all living organisms.

Oxen: see *Bovinae.*

Oxidases: oxidoreductases that catalyse the transfer hydrogen and electrons to oxygen, and are able to transfer 2 or 4 electrons. The protein-bound cofactor is FMN, FAD, or a heme group. The 2-electron-transferring O., which oxidize their substrate with the formation of hydrogen peroxide, include Amino acid oxidases (see) and Xanthine oxidase (see). Examples of 4-electron-transferring O. are cytochrome oxidase (see Cytochromes), Laccases (see) and ascorbic acid oxidase.

Oxidation: removal of electrons, e.g. $Fe^{2+} \rightarrow Fe^{3+} + e^-$, by an oxidizing agent, thereby increasing the electropositive value of the oxidized material. Dehyrogenation is also a form of O. Every O. must be coupled with a simultaneous Reduction (see). Enzymes called oxidoreductases, which couple oxidations and reductions (redox reactions), are important in cell metabolism (see Respiratory chain).

Oxidative deamination: conversion of α-amino acids into 2-oxoacids (α-ketoacids) in two stages: R—CH(—NH_2)—COOH → R—C(=NH)—COOH → R—CO—COOH + NH_3. In D-amino acid oxidases (dehydrogenases), the hydrogen removed in the first stage is transferred to the coenzyme FAD, while in L-amino acid oxidases this role is performed by FMN. Glutamate dehydrogenase is strictly specific for the L-form of glutamate, and it utilizes NAD as coenzyme for the transfer of hydrogen. In the second stage, the intermediate imino acid is hydrolysed spontaneously (second stage) to produce the corresponding oxoacid and ammonia (mostly in the form of NH_4^+ at physiological pH). The first (enzymatic) stage is rate-limiting. The ammonia (in particular that produced by the O.d. of glutamate) is incorporated into the urea cycle where it contributes to urea synthesis.

Oxidative decarboxylation: an important metabolic process in which a 2-oxoacid (α-ketoacid) is decarboxylated and oxidized with formation of the CoA-derivative of the product fatty acid. O.d. of pyruvate to acetyl-CoA, and of 2-oxoglutarate to succinyl-CoA are reactions of major metabolic importance, and each is catalysed by a multienzyme complex (known as a 2-oxoacid dehydrogenase). Bound, active groups of both multienzyme complexes are thiamin pyrophosphate, lipoic acid and FAD. Other substrates are CoA and NAD^+. Pyruvate dehydrogenase converts the pyruvate produced by glycolysis (or by transamination of alanine) into acetyl-CoA, which is an essential substrate of the tricarboxylic acid cycle (see equation). Conversion of 2-oxoglutarate to succinyl-CoA by 2-oxoglutarate dehydrogenase is an essential stage of the tricarboxylic acid cycle.

CH_3—CO—COOH + HS—CoA + NAD^+
Pyruvate

→ CH_3—CO ~ SCoA + CO_2 + NADH + H^+
Acetyl-CoA

Oxidative phosphorylation: synthesis of adenosine triphosphate (ATP) from adenosine diphosphate (ADP) and inorganic phosphate, coupled with the operation of the cytochrome chain in mitochondria and bacterial cell membranes. The yield of O.p. is expressed as the P/O ratio, i.e. the ratio of the number of molecules of inorganic phosphate esterified (or the number of molecules of ATP synthesized) to the number of atoms of oxygen consumed by the respiratory chain. This is same as the number of molecules of ATP synthesized each time the respiratory chain goes into action and transfers 2 electrons to oxygen. For substrates that are dehydrogenated by NAD-dependent dehydrogenases (e.g. malate; see Tricarboxylic acid cycle), and for NADH itself, the P/O ratio is 3. Substrates dehydrogenated by flavoenzymes (e.g. succinate) yield a P/O ratio of 2. Energy transduction is by *chemiosmosis,* a process in which electron transport in the respiratory is coupled to the transfer of protons across the inner mitochondrial membrane. The resulting proton gradient drives the synthesis of ATP, as protons re-enter the inner mitochondrial space via specific ATP-synthesis sites.

Oxidoreductases: a very large and important class of hydrogen- and electron-transferring enzymes, which catalyse biological oxidations and reductions. Many O. that transfer hydrogen from organic substrates are necessary for the generation of energy in heterotrophic cells, whereas those that transfer hydrogen to organic substrates are important in biosynthesis. Coenzymes and prosthetic groups of O. are NAD^+, $NADP^+$, FMN, FAD and heme. All reductions must be accompanied by a corresponding oxidation, and vice versa. For example, dehydrogenation of 2-oxoglutarate in the tricarboxylic acid cycle is coupled with the hydrogenation of NAD^+, while hydrogenation of β-ketoacids to β-hydroxyacids in fatty acid biosynthesis is coupled with the dehydrogenation of NADPH + H^+. The cytochromes of the respiratory chain are also O.; by transfering electrons from cytochrome c_1 to cytochrome *a,* cytochrome *c* catalyses the oxidation of cytochrome c_1 and the reduction of cytochrome *a.* Some O. (oxygenases) transfer oxygen directly to their organic substrates, whereby the product represents a more oxidized form of the organic substate and a more reduced form of oxygen. Cytochrome a_3 of the respiratory chain is a particularly important O., which transfers electrons from (i.e. oxidizes) cytochrome *a* to molecular oxygen; the oxygen is therefore reduced and ultimately combines with protons to form water. See Oxidases (see), Dehydrogenases (see), Peroxidases (see), Hydroperoxidases (see), Oxygenases (see) and Alcohol dehydrogenase (see).

Oxlip: see *Primulaceae.*

Oxoacids: see Ketoacids.

2-Oxobutyric acid: see Acetoacetic acid.

2-Oxoglutaric acid: see α-Ketoglutaric acid.

Oxpecker: see *Sturnidae.*

Oxyaena: see *Creodonta.*

Oxyaenidae: see *Creodonta.*

Oxycephaly: see Acrocephaly.

Oxycoccus: see *Ericaceae.*

Oxygenases: Oxidoreductases (see) that transfer molecular oxygen directly to their organic substrates. Bonds attacked by O. include C—C, C=C, C—O, C—N, N—H and particularly C—H. Reaction products include compounds with C—OH or N—OH groups, N-oxides, epoxides and compounds containing keto groups. ***Dioxygenases*** transfer both atoms of the oxygen molecule to their substrate, usually with cleavage of a carbon-carbon double bond: $-C{=}C- + O_2 \rightarrow -C{=}O + O{=}C-$. Examples are the oxidative ring cleavage of 3-hydroxyanthranilic acid and of tryptophan. ***Monooxygenases*** transfer only one atom of the oxygen molecule to their substrate (S), while the second atom is reduced to water by a hydrogen donor (DH_2): $S + O_2 + DH_2 \rightarrow SO + D + H_2O$.

Monooxygenases are responsible for many important metabolic hydroxylations and are often called ***hydroxylases*** or ***mixed function oxidases.*** For example, the cytochrome P450 system functions as a mixed function oxidase which hydroxylates xenobiotics such as pharmaceutical agents, environmental, pollutants, etc., making them more hydrophilic, and therefore more easily excreted by the kidney; the introduced hydroxyl group can also be conjugated with glucuronic acid or sulfate to make a more water-soluble and excretable derivative. Endogenous substrates are also hydroxylated by O., e.g. proline residues of precollagen are hydroxylated to the hydroxyproline residues of collagen, phenylalanine is hydroxylated to tyrosine, estrogens are hydroxylated to catechol estrogens. Phenol oxidases (see) are also O.

Oxygen binding by blood: the uptake and release of blood oxygen. During passage of blood through mammalian lungs, only 3 ml of oxygen become dissolved in 1 liter of blood plasma, which is much too little for the oxygen requirement of the tissues. The amount of oxygen carried by the blood is, however, increased to an adequate level by the high oxygen binding capacity of hemoglobin: 1 g hemoglobin is capable of binding 1.34 ml oxygen. At *maximal oxygen binding capacity* or *oxygen saturation,* 1 liter of male human blood contains about 160 g hemoglobin and about 210 ml oxygen.

The oxygen binding capacity of blood is influenced by four factors (Fig.): 1) oxygen itself, 2) protons, 3) diphosphoglycerate, and 4) blood temperature.

1) The sigmoid oxygen binding curve of hemoglobin is an expression of heme interaction. Binding of one oxygen molecule to one of the four subunits of hemoglobin alters the conformation of the remaining three unoccupied subunits and increases their affinity for oxygen. The increase of oxygen binding with increasing partial pressure is called *oxygenation.* At 13 kPa (the oxygen partial pressure in the lung alveoli), the hemoglobin is practically 100% saturated. When blood is transported to the tissues, the partial pressure falls to about 4 kPa, and the hemoglobin loses part of its oxygen, which is then available for tissue respiration. 2) If the pH value falls, oxygen is released more readily. Conversely, when hemoglobin is oxygenated, protons are released. In particular, carbon dioxide is effective in promoting the release of oxygen from hemoglobin (see Bohr effect). 3) Fetal blood always removes oxygen from maternal blood, not because it binds oxygen with higher affinity, but because fetal erythrocytes contain a lower concentration of diphosphoglycerate than that present in adult erythrocytes. 4) Increased temperature favors oxygen release, whereas cooling increases the uptake of oxygen.

In actively contracting skeletal muscle, three of the above factors play an important part in the release of oxygen from the blood to the tissues: muscle work results in 1) increased carbon dioxide production and 2) lactic acid production, and 3) there is a local increase in temperature.

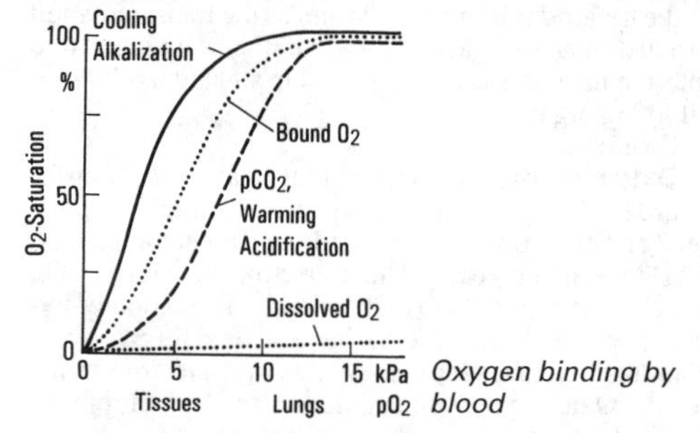

Oxygen binding by blood

Oxygen binding capacity of blood: see Oxygen binding by blood.

Oxygen deficiency: a condition in which the supply of oxygen to the tissues is less than normal, or less than that required for normal tissue function. Under normal conditions and at rest, vertebrate tissues have an oxygen reserve which compensates any small or temporary decrease in the oxygen supply. A severe oxygen shortage causes temporary or permanent tissue damage, if it cannot be abolished by the stimulated mechanisms of respiration and circulatory regulation.

In humans, the organ most sensitive to O.d. is the brain. In the early stages of O.d. there is a decrease in awareness, attentiveness and the ability to concentrate. Severe O.d. leads to loss of consciousness. The function of other internal organs is also markedly decreased. Increasing O.d. is generally associated with four phases of damage: disturbance of organ function, abolition of organ function, irreversible changes in the organ as a result of cell death, and finally tissue disintegration.

Oxynotinae: see *Selachimorpha.*

Oxypetalum: see *Asclepiadaceae.*

Oxyrhyncha: decorator and Spider crabs (see), a section of the infraorder *Brachyura* (true crabs) of the *Decapoda* (see).

Oxystomata: Box crabs (see), a section of the infraorder, *Brachyura* (true crabs) of the *Decapoda* (see).

Oxytetracyclin: see Tetracyclins.

Oxytocin: a neurohypophysial, peptide hormone

(Cys-Tyr-Ile-Gln-Asn-Cys-Pro-Leu-Gly-NH_2; the two Cys residues are linked by a disulfide bond) which acts via the adenylate cyclase system of its target organs, causing contraction of the smooth muscle of the uterus and of the mammary gland (milk ejection). O. is structurally related to the other neurohypophysial hormone, Vasopressin (see); these two hormones have a common ancestry and there is an overlap in their physiological activities. In lower vertebrates, four other neurohypophysial hormones have been identified, which are variants of O. and vasopressin, with different amino acid residues in positions 4 or 8, or both. [Arg^8]O. is a hybrid analog, in which the tripeptide tail of vasopressin is attached to the ring of O.; it occurs naturally in chicken pituitary, and it is thought to be the common ancestor of all the other related neurohypohysial hormones. [Ser^4Ile^8]O. (also called isotocin) occurs in teleost fishes; [Ser^4Glu^8]O. (also called glumitocin) occurs in elasmobranchs. [Ile^8]O. (also called mesotocin) is present in amphibians and reptiles. O. is synthesized in the Nucleus paraventricularis of the hypothalamus, and transported to the posterior lobe of the pituitary via the tractus paraventriculo-hypophyseus, in combination with the transport protein, neurophysin I. O. is released into the circulation in response to psychological and tactile stimulation of the genitalia, or sucking stimulation of the mammary gland. More than 300 analogs of O. have been synthesized, some of which have higher biological activity than the natural product. The role of O. in parturition is unclear. Although it is probably involved in uterine contraction, it is not generally accepted that its release determines the onset of parturition.

Oxyura: see *Anseriformes.*

Oxyuranus: see Taipan.

Oxyurida: see *Nematoda.*

Oxyurini: see *Anseriformes.*

Oxyuris: see Enterobiasis; *Nematoda.*

Oystercatchers: see *Haematopodidae.*

Oysters: see *Bivalvia.*

Oystershell scale: see *Homoptera.*

Ozone and Ozone layer: see Greenhouse effect.

P

Paca: see *Agoutidae.*

Pacemaker: that part of an autorhythmic biological system which provides the impulse. For example, the *sinus nodes* of the vertebrate heart generate the rhythm of the heart beat. The *suprachiasmatic nucleus,* a bilaterally paired structure in the mammalian anterior hypothalamus, contains a master circadian clock that regulates most, if not all, endogenously generated circadian rhythms. See Biorhythm.

Pachycephala: see *Pachycephalinae.*

Pachycephalinae: an avian subfamily of the family *Muscicapidae* (see), order *Passeriformes* (see). The approximately 50 species (in 11 genera) are found in Australia and the islands of Southeast Asia from New Guinea to Polynesia. The largest genus is *Pachycephala* ***(Thickheads*** or ***Whistlers).*** They are robust birds with yellow and green plumage often with black and brownred, but the dominant color tends to be yellow. They rarely have wattles, but some species have crests. The ***Golden whistler*** *(Pachycephala pectoralis;* Australia eastern Java, Lord Howe Island, Santa Cruz Islands, New Caledonia, Fiji) is green above and yellow below, with a white throat and black gorget.

The ***Australian Crested bellbird*** *(Oreoica gutturalis)* is notable for decorating the rim of its nest with semi-paralysed insect larvae.

Pachydont: a hinge type of the *Bivalvia* (see).

Pachytene: see Meiosis.

Pacific giant salamander: see *Dicamptodontidae.*

Pacific mole salamanders: see *Dicamptodontidae.*

Pacific newts: see *Salamandridae.*

Pacific tree frog: see *Hylidae.*

Pacinian corpuscles: see Tactile stimulus reception.

Pacus: see *Characiformes.*

Paddlefishes: see *Chondrostei.*

Paedogamy: see Pedogamy.

Paeonia: see *Paeoniaceae.*

Paeoniaceae, ***peony family:*** a small family of dicotyledonous plants, containing the single genus, *Paenia* ***(Peonies).*** The 33 species (approx.) occur mainly in the northern temperate zone. They are large herbs or shrubs, with alternate, divided, exstipulate leaves. Flowers are hermaphrodite, actinomorphic, hypogynous, and usually solitary, large, terminal and brightly colored. The corolla consists of 5–10 large free petals, with a calyx of 5 free sepals. After insect pollination, the superior ovaries develop into a group of 2–5 large follicles, each with several seeds.

Many species, often with double flowers, are grown as ornamentals, e.g. the usually red-flowered *Paeonia officinalis* ***(Common peony)*** from southern Europe; *P. lactiflora* ***(Shaoyao)*** from China and Siberia, with white or pink flowers; and *P. suffruticosa (= arborea)* ***(Tree peony,*** or ***Moutan)*** from China, with pink to white petals with dark purple bases, and with woody stems.

Paeoniaceae. Flowering stem of Shaoyao *(Paeonia lactiflora).*

Paguma: see Masked palm civet.

Paguridae: see Hermit crabs.

Paguroidea: Hermit crabs (see) and allies, a superfamily of the *Decapoda* (see). See Coconut crab, King crab.

Pagurus: see Hermit crabs.

Pain: a reflex response to the stimulation of receptors by tissue injury or stress, as well as a sensation interpreted via the cortex. Almost all body tissues contain free nerve endings connected to small afferent fibers. Many of these free nerve endings are P. receptors (nociceptors), differing from other receptors in that they can be excited by any of several stimuli: mechanical, chemical and thermal, as well as pathological changes (e.g. inflammation). The free nerve endings are stimulated chemically by e.g. ions, kinins and prostaglandins released from stressed or destroyed cells. There is much evidence that substance P serves as a neurotransmitter in the primary afferent fibers associated with P. receptors. P. impulses are transmitted to the central nervous system by two fiber systems: unmyelinated (nonmedullated), slow-conducting (0.5–2 m/s) C-fibers, and myelinated (medullated), faster conducting (12–30 m/s) Aδ-fibers. Both types of nerves enter the dorsal horn of the spinal cord. Aδ-fibers terminate mainly on neurons in laminas I and V, whereas the dorsal root C-fibers terminate on neurons in laminas I and II. The axons of the relevant dorsal horn neurons have various routes and destinations. Many end in the reticular system which projects to the thalamus and from there to different parts of the cortex. Others project to the hypothalamus, some ending in an area concerned with P., known as the periaqueductal gray. A few ascend in the posteriolateral portion of the cord, and some of these project to the specific sensory relay nuclei of the thalamus, and from there to the post-central gyrus. Some of the dorsal horn neurons enter the anterolateral system, including the lateral spinothalamic tract.

The existence of fast- and slow-conducting neurons is correlated with two types of P., known as "fast and slow P." or "first and second P." These correspond to the sharp, localized sensation of P. due to the activity of the fast neurons, followed by a dull, more diffuse, but intensely unpleasant sensation, due to the operation of the slow neurons. Accordingly, the as the distance between the site of P. and the brain increases, the sensations of fast and slow P. become separated by a greater time interval.

The cortex is required for the interpretation of the sensation of P., but not for the perception of P. Thus, if a stimulus painful to an intact animal is applied to the foot of an animal with its spinal cord severed in the neck region, a flexion reflex is initiated, resulting in withdrawal of the foot.

Substances that decrease the sense of, and the reponse to, P. are called analgesics, e.g. morphine. They act via various types of molecular receptors in different regions of the central nervous system. Natural ligands of the opiate (morphine) receptors are known as Endorphins (see).

It can be assumed that all vertebrates can experience P., but the existence of a sense of P. in nonvertebrates is uncertain.

Painted horned frog: see *Leptodactylidae.*

Painted pigeons: see *Columbiformes.*

Painted redstart: see *Parulidae.*

Painted snipe: see *Rostratulidae.*

Painted turtle: see *Emydidae.*

Painter's mussel: see Freshwater mussels.

Pair bonding: see Sexual partnerships.

Palaeechinoidea (Greek *palaios* old, *chinos* hedgehog, *eidos* similar), *Melonechinoidea:* an extinct order of sea urchins with a spherical to melon-shaped test. The interambulacral fields are formed from one or more rows of plates, while the ambulacral fields consist of two or more rows of plates. The spines are all similar, each arising from a small boss. Fossils of this relatively isolated group of paleozoic *Echinoidea* are found from the Silurian to the Permian.

Palaeacaridae: see *Oribatei.*

Palaemon: see Shrimps.

Palaemonidae: see Shrimps.

Palaeoheterodonta: see *Bivalvia.*

Palaeoniscum (Greek *palaios* old, *oniskos* little donkey): a genus of the extinct holostean order, *Palaeonisciformes.* The body was torpedo-shaped with a large head and a heterocercal tail fin. Teeth were small, conical and somewhat nonuniform. The entire body was covered with rhombic and round ganoid scales. Fossils occur in the Permian, and serve as mineralized index fossils of the Germanic Kupferschiefer.

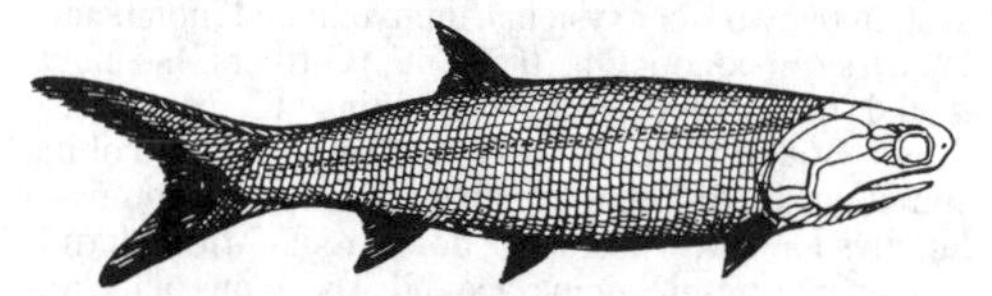

Palaeoniscum. Reconstruction of the ganoid fish, *Palaeoniscum,* from the Lower Zechstein (x 0.25).

Palaeotaxodonta: see *Bivalvia.*

Palaeotragus: see *Giraffidae.*

Palatoquadrate: see Skull.

Palea: see *Graminae.*

Palearctic, ***Palearctic subregion:*** a faunal subregion, formerly treated as a separate region, but now considered as part of the Holarctic (see). It consists of nontropical Eurasia, North Africa and much of Arabia.

The P. is the original evolutionary center of many animal groups, and it has also received numerous species and higher taxa from other regions, in particular the Oriental region. Animals originating in the P. have penetrated as far as the Neotropical and Australian regions, but no animals originating in these regions are found in the P. During the postglacial period, northern areas were repopulated from Glacial refugia (see) which acted as Dispersal centers (see). A special feature of the P. is the Arctic-alpine distribution (see) of many taxa.

The P. possesses far fewer characteristic faunal taxa than the Nearctic (see). Among the vertebrates, the mammals display the greatest diversity, with numerous endemic genera. The following orders and families occur in the P., but are absent from the Nearctic: primates *(Primates),* viverrids *(Viverridae),* hyenas *(Hyaenidae),* true rats and mice *(Muridae),* horses *(Equidae),* camels *(Camelidae)* and *Bovidae* (represented by antelopes and gazelles). Of the endemic groups of the Holarctic, the *Glirinae* (true dormice) and *Rupicaprini* (goat-antelopes) are restricted to the P.

Almost all the bird families of the P. are also widely distributed in other faunal regions. The immense area of the P. contains only about 1 100 bird species, somewhat fewer than the number of species found in Columbia (> 1 500), which has the greatest number of bird species of any country. In view of the ability of birds to disperse, and to overwinter in neighboring regions, there is an obvious close relationship between the bird populations of the P. and the Ethiopian and Oriental regions. On the other hand, the eastern and western parts of the P., which contain approximately equal numbers of bird species, display considerable differences in species composition.

There is a poor evolutionary variety of reptiles and amphibians. Since the Tertiary many invertebrates have been lost from the P., as evidenced by the Baltic amber, which contains many groups that are now only found further south.

Palechinoidea: see *Echinoidea.*

Paleoanthropology: a branch of anthropology which studies the origin and biological evolution of hominids up to the present, the evolution of human culture, and the role of abiotic and biotic factors in these evolutionary processes (see also Anthropogenesis).

P. is based on the analysis of skeletal remains of recent humans and their ancestors as far back as the phylogenetic root of the hominids. It includes the study of fossil remains of related hominids, comparative studies of genetically close, nonhominid, living primates, and it is supplemented by the results of numerous allied disciplines, in particular the analysis and reconstruction of early societies and their cultural-technological development.

For periods that yield relatively representative series of skeletons (in Europe, the last 6 000 to 8 000 years), it is becoming increasingly possible to derive an accurate picture of population-specific changes in human morphology and size, and ethnogenetic processes, the dynamics of human demographic quantities (see Paleodemography) such as mortality, sex ratios, population size, etc, as well as the type of medicine practised by early cultures, and the epidemiology of diseases like dental caries, periodontopathies, degenerative skeletal changes, etc. The body of knowledge of P. represents a significant addition to our understanding of human history, as well as making an important contribution to our understanding of the biological status of modern humans.

Paleobotany, ***paleophytology, phytopaleontology*** (plate 30): a branch of paleontology devoted to the study of plant life of earlier periods of the earth's history. It is commonly based on the investigation of macroscopic plant fossils, i.e. large residues, although these may be studied by microscopic methods. The preparation and study of fossil pollen and spores is discussed under Pollen analysis (see).

P. has considerably inceased our understanding of plant evolution. It also forms an indispensable basis for the practical geology of certain geological periods, in particular the Carboniferous.

The oldest records of fossil plants date from ancient Greece. No further progress was made in our knowledge of plant fossils until after 1630, but until 1820 the finds were sporadic and accidental, so that the period before 1820 is known as the prescientific period. The scientific period was introduced by the work of Schlotheim (1765–1832), soon followed by the investigations of other authors, in particular Brongniart (1801–1876), Sternberg (1761–1838) and Göppert (1800–1884).

Paleoclimatology: the study of climate and climatic changes during the earth's history. Since climate is crucial factor in the distribution of animal and plants, fossil remains serve as indicators of the position of former climatic boundaries.

Paleodemography: a specialized area of Paleoanthropology (see). It investigates the population biology of primitive and early humans, and is based primarily on the study of human skeletal remains and the residues of cremations. From the diagnosis of age and sex, and the study of pathological changes in skeletons, it is possible to analyse the age composition, mortality, sex ratio, size and rate of increase of populations, and to identify the type and frequency of occurrence of diseases in populations for which there are no written records. Information from skeletal remains is supplemented by the number, density and type of cultural artifacts, which to a certain extent are also indicative of population size, as well as providing clues to the nature of the environment and the economic system.

Paleoecology: a specialized branch of paleontology, which attempts to determine the ecological relationships between fossil organisms and the relationships of these organisms to their environment. It is based on the reconstruction of the functional activities of organisms and of their biocoenoses.

Paleogea, ***Palaeogaea:*** a zoogeographical region originally comprising the Palearctic, Ethiopian, Oriental and Australian regions.

Paleogenic, ***Old Tertiary:*** lower part of the Tertiary, consisting (in order of decreasing age) of the Paleocene, Eocene and Oligocene. See Geological time scale.

Paleogeography: the science of geographical-geomorphological relationships during the earth's history. P. investigates the contors of former continents, relying heavily on the evidence of marine and terrestrial sedimentary rocks and particularly on the fossil record.

Paleolimnology: see Limnology.

Paleonemertea: see *Nemertini.*

Paleontology: the study of fossil flora and fauna. It is divided into Paleobotany (see) and Paleozoology (see), both subdivided into further independent research areas [e.g. Pollen analysis (see) is a branch of Paleobotany]. Other independent branches of P. are Paleoanthropology (see), Paleogeography (see), Paleoclimatology (see) and Paleopathology (see).

Through its research material and historical perspective, P. is closely linked to both the geosciences and the biosciences.

As a biological discipline, P. investigates 1) the ontogeny of individual organisms, 2) the mode of existence and environmental conditions of organisms (Paleoecology), 3) the distribution of organisms on former continents and in the oceans (Paleobiogeography), and 4) the evolution of organisms during the earth's history (Phylogeny).

Micropaleontology is of practical importance for the stratigraphic interpretation of many geological sediments. By using specially developed methods for the identification of microscopic animal and plant residues (generally smaller than 2 mm), strata can be rapidly and reliably dated. Applied to bore samples, this technique of *applied P.* is used in mineral prospecting, particularly in the oil industry.

Paleopallium: see Brain.

Paleopathology: a branch of paleontology and anthropology, which studies pathological alterations in the fossils and subfossils of animal and human skeletons, as evidence of early inherited and infectious deseases and adverse environmental effects.

Paleophytic: the era of plant evolution that followed the Eophytic. It began in the Upper Silurian and ended in the Rotliegendes (Middle and Lower Permian) (see Geological time scale). The P. is usually divided into the Early P. or *psilophyte period* (Upper Silurian to Middle Devonian) and the Late P. or *pteridophyte period* (Upper Devonian to the Rotliegendes). The beginning of the psilophyte period is characterized by the appearance of widespread terrestrial vegetation. This first land flora was dominated by psilophytes and the earliest primitive representatives of different classes of pteridophytes, such as *Archaeoleptidophytales, Hyeniales* and *Primopteridae.* Early P. vegetation was strikingly uniform throughout the world. By the Upper Devonian the psilophytes had nearly disappeared, and the individual classes of pteridophytes were already highly differentiated, sometimes into treelike forms. This marked the beginning of the pteridophyte period in the narrow sense, when this particular group of plants was dominant. In the Lower Carboniferous, the following groups are strongly represented: lepidodendra, asterocalamites, sphenophylls, calamites, horsetails, and numerous ferns and lyginopterids. There was still no geographical differentiation in the Upper Devonian and Lower Carboniferous. In the Upper Carboniferous, there was a marked increase in evolutionary diversity, and the plants of the P. became separated into two geographical kingdoms: the *Northern Hemisphere* or *Arctocarboniferous kingdom* and the *Southern Hemisphere* or *Antarctocarboniferous kingdom.* The boundary between the two did not follow the present equator. The southern part of South America, central and southern Africa and Madagascar, India, Australia and Antarctica formed the Antarctocarboniferous kingdom, whereas all other regions formed the Arctocarboniferous kingdom. The latter was made up of the following three floral regions. 1) *Eurameric flora* of North Africa, characterized by tall pteridophyte trees (lepidodendra, sigillaria, calamites), *Cordaitales* and the large-leaved, richly pinnate pteridophylls. 2) *Angara flora* of northwestern and central Asia from the Urals to Petschora, typified by a species-poor, mixed flora. 3) *Cathaysia* or *Gigantopteris flora* of southeast Asia, typified by the occurrence of *Gigantopteris.*

In contrast, the flora of the Antarctocarboniferous kingdom was geographically uniform, characterized in particular by *Glossopteris, Gangamopteris, Schizoneura* and *Phyllotheca.* This flora appears to have contained an extraordinarily small number of different species, and is generally called the *Glossopteris* or *Gondwana flora.*

The only satisfactory explanation for the geographical relationships of the P. is provided by Continental drift (see). As the P. came to an end in the Rotliegendes, the flora of the Southern Hemisphere had hardly changed from the Upper Carboniferous. In the Northern Hemisphere, however, the Carboniferous flora had become increasingly depleted, which was certainly due to a climatic change. At the same time, the pteridophytes were receding and gymnosperms were becoming more prominent. The latter then became dominant in the Mesophytic (see). In this sense, the Rotliegendes was a time of transition.

Paleophytology: see Paleobotany.

Paleotropical kingdom, ***paleotropical realm, Paleotropis:*** a floristic kingdom embracing the tropical and subtropical floristic zones of the Old World (Africa, India, Indochina, Malay archipelago and the Polynesian islands). The families, *Dipterocarpaceae, Pandanaceae* and *Nepenthaceae* are particularly characteristic. Many members of the *Moraceae* (especially *Ficus* species) are found in the P.k., as well as many stem succulents belonging to the *Euphorbiaceae.* The main types of vegetation are tropical rainforest, monsoon- forests, dry tropical forests, savannas and deserts.

Paleotropical region, ***paleotropical realm, Paleotropis:*** a zoogeographical region comprising tropical Asia and Africa south of the Sahara. These two areas are usually considered as two separate regions, the Oriental region (see) and the Ethiopian region (see).

Paleosuchus: see *Alligatoridae.*

Paleozoic: the era of animal evolution that lasted about 350 million years, embracing the Cambrian, Ordovician, Silurian, Devonian, Carboniferous and Permian systems. It started with a rich fauna of 11 animal phyla, which underwent explosive evolutionary development. Toward the end of the P. a number of these groups had become extinct, e.g. rugose corals *(Tetracorallia), Tabulata,* goniatites, trilobites, *Cystoidea, Blastoidea,* graptolites and placoderm fishes, as well as primitive amphibians and reptiles. Other animal groups, e.g. brachiopods, nautiloids, giant sea scorpions *(Gigantostraca)* and coelacanths *(Crossopterygii),* attained a degree of evolutionary diversity, then underwent a permanent decline. Ammonites, sea urchins and advanced fishes, birds and mammals did not appear until the end of the P.

Paleozoology (plate 31): study of animals, usually extinct, that lived at earlier times in the earth's history; a branch of Paleontology (see).

Pale western cutworm: see Cutworms.

Palila: see *Drepanididae.*

Palindrome: a nucleic acid sequence that is identical to its complementary strand, when both are read in the 5'-3' direction. In the region of a P. there is therefore a twofold rotational symmetry. Perfect Ps., e.g. GAATTC, are often recognition sites for restriction enzymes. Imperfect Ps., e.g. TACCTCTGGCGGTGATA, often act as binding sites for proteins such as repressors. Interrupted Ps., e.g. GGTTXXXXXAACC, make possible the formation of a stem with a loop (hairpin structure) as in tRNA.

Palingenesis: the occurrence of embryonal or juvenile stages of an animal, which display certain characters of the adult evolutionary forms of their ancestors, i.e. the recapitulation of ancestral phylogeny in the ontogeny of descendants. P. is therefore used for the determination of phylogenetic relationships. For example, the juvenile forms of many eyeless coelenterates possess functional eyes, indicating that their adult evolutionary ancestors also had similar functional light-sensitive organs.

Palinura: Spiny lobsters (see) and shovel-nosed lobsters, an infraorder of the *Decapoda* (see).

Palinuridae: see Spiny lobsters.

Palinurus: see Spiny lobsters.

Pallas's cat: see *Felidae.*

Pallial complex: the various organs and structures of the mantle cavity of mollusks. The P.c. consists of: 1. the anus, which lies posteriorly on the mid line; 2. the osphradia, which lie either side of the anus; 3. the gills or ctenidia, which hang from the mantle skirt anterior to the osphradia; and 4. the excretory openings (nephridiophores) and genital openings, which discharge into the mantle cavity medial to the ctenidia.

Palm (tree): see *Arecaceae.*

Palmae: see *Arecaceae.*

Palmate newt: see *Salamandridae.*

Palm chat: see *Bombycillidae.*

Palm civets, *Paradoxurus:* a genus of the *Viverridae* (see) containing 3 species. The semiarboreal ***Common palm civet*** or ***Toddy cat*** *(Paradoxurus hermaphroditus)* occurs from India and Sri Lanka eastward throughout Southeast Asia. It is brown in color, usually with dark stripes on the back, and spots on the sides; the face mask is spotted and the forehead is streaked. The arboreal ***Golden palm civet*** *(P. zeylonensis)* occurs in Sri Lanka. Red-brown to golden brown in color, it has barely visible spots and stripes. The nap of the hair is directed forward on the neck and throat. ***Jerdon's palm civet*** *(P. jerdoni)* of southern India is also arboreal. It is deep brown to black with gray speckling, and the nap of its hair is also directed forward on the neck and throat.

In addition to animal food and fruit, P.c. are also particularly fond of coffee berries, and the undigested excreted seeds are reputed to yield an especially delicious coffee.

Palmella: see *Mastigophorea.*

Palmetto: see *Arecaceae.*

Palmitic acid, ***n-hexadecanoic acid:*** CH_3—$(CH_2)_{14}$—COOH, a colorless, water-insoluble, higher fatty acid. P.a. is one of the most widely distributed natural fatty acids, and is present in esterified form in practically all natural fats and fatty oils, e.g. 36% in palm oil, 29% in bovine carcass fat, 15% in olive oil; it is also present in phospholipids and waxes. In beeswax, P.a. is esterified with myricyl alcohol, and in spermaceti with cetyl alcohol (palmityl acohol).

Palmityl alcohol, ***cetyl alcohol, n-hexadecanol:*** $C_{16}H_{33}OH$, a straight-chain alcohol which occurs esterified with higher fatty acids in many natural waxes. Spermaceti consists mostly of palmityl palmitate.

Palm lilies: see *Agavaceae.*

Palms: see *Arecaceae.*

Palolo worm: see *Errantia.*

Palpigradi, ***microwhipscorpions:*** a worldwide-distributed order of arachnids (the majority, however in southern Europe, South America and southern USA), containing about 50 known, minute species (0.6–2.8 mm). The abdomen is attached by a stalk, and has a large mesosoma and short metasoma, with a long multisegmented flagellum. The pedipalps are leglike, and the first pair of walking legs are modified as tactile organs. These thin-skinned, delicate animals live in humid environments beneath stones or in soil fissures.

Paludina: see *Gastropoda* (subclass *Streptoneura,* order *Stylommatophora).*

Palynology (Greek *palynein* to scatter dust): inclusive term for all areas of study concerned with pollen or spores. Special branches of P. are Pollen morphology (see), Pollen analysis (see), Spore paleontology (see), Melitopalynology (see), etc.

Pampas: level, temperate grassland of South America. In the moister regions it consists of bunch grasses with bare earth between the tussocks, while in the drier areas (largely in the south and west of Argentinia) it consists of short grasses and xerophytic shrubs. See Steppe.

Pampas hare: see *Caviidae.*

Panax: see *Araliaceae.*

Pancreas: a gland containing both exocrine and endocrine cells. The exocrine part of the P. consists of purely serous-type, pancreatic acinar cells that produce enzyme-rich secretions (hydrolases catalysing the degradation of protein, triacylglycerol and carbohydrate). The fine structure of pancreatic acinar cells is generally similar to that of other serozymogenic cells, e.g. serous cells of the salivary galands, or the pepsinogenic cells of the gastric glands. All share the common characteristic of a great abundance of rough endoplasmic reticulum and the presence of zymogen granules. In the pancreas, clusters of acini and their duct systems are separated by areolar connective tissue, which is continuous with that surrounding the pancreatic lobules. The acinus consists of a layer of pyrimidal cells with their apical ends bordering the lumen and their bases resting on a basement membrane and reticular connective tissue. The termini of the pancreatic duct system extend into the acini so that the flattened duct cells are interposed between some of the acinar cells and the lumen. Characteristic of the exocrine pancreas are the duct cells within the acinus, which are known as the *centroacinar cells.* The apices of the acinar cells contain acidophilic zymogen granules, representing storage forms of the various digestive enzymes that are destined for secretion via the pancreatic duct (ductus pancreaticus) into the duodenum. The basal portions of the acinar cells display intense basophilic staining properties, due to the presence of high concentrations of ribonucleoprotein (formerly known in these cells as "ergastoplasm").

The endocrine part of the P. is represented by groups of cells known as the *islets of Langerhans,* which occur as scattered groups throughout the organ, each group consisting of a few to hundreds of cells embedded in the acinar tissue. Their pattern and density of distribution depends on the species, but in humans the islets are more numerous in the distal part of the P. (the human P. contains an estimated 10^6 islet cells). The endocrine and exocrine parts of the P. are separate (i.e. not interspersed with each other) only in the *Petromyzonidae.* In histological preparations, the islets are seen as groups of lightly stained cells surrounded by a thin layer of reticular fibers. The islet cells are clearly smaller than the acinar cells of the exocrine tissue, and they have a much more abundant capillary supply than the latter. With appropriate fixation and staining, the following 4 cell types can be differentiated: large, flame-shaped, glucacon-producing alpha or A-cells (about 20% of total); smaller, insulin-forming beta or B-cells (about 75% of total); delta or D-cells (about 5% of total), which are most numerous in primates, but are also present in other species; and clear or C-cells, which contain apparently empty vesicles. By immunohistochemical methods, D-cells have been shown to contain the polypeptide hormone, somatostatin, which is a release-inhibiting factor with activity in the hypothalamus. Glucagon stimulates glycogen degradation, and its action is antagonistic to that of insulin. Somatostatin appears to regulate the activity of the islets of Langerhans.

Pancreozymin, ***cholecystokinin:*** a tissue hormone consisting of 33 amino acids (porcine), M_r 3838. It is formed in the E-cells of the small intestine, and released into the circulation in response to the acidic pH of the chyme, and in particular in response to the fatty acids, amino acids and peptides that it contains (chyme is an acidic mixture of partially digested food). It can also be released by nervous stimuli. It promotes contraction of the gall bladder and secretion of pancreatic juice.

Panda: see Giant panda, Red panda.

Pandalidae: see Shrimps.

Pandanaceae, ***screw-pine family:*** a family of monocotyledonous plants containing about 880 species, favoring damp habitats, and found exclusively in tropical and subtropical regions of the Old World. The family includes trees and climbing shrubs (lianes), which often have stilt roots and dichotomous stems. Stems and trunks characteristically bear the annual scars of the leaf bases. The name, ***Screw pine,*** is derived from the spirally arranged tuft of long, narrow leaves at the end of the main axis; more accurately, there are 3 ranks of terminal leaves, and they appear to be spirally arranged because the stem is twisted. The flowers are unisexual and dioecious. Both male and female flowers lack a perianth. In the genus *Sararanga* (2 species) the flowers are stalked and arranged in a panicoid inflorescence. In other genera [e.g. *Pandanus* ***(Screw pines,*** about 600 species)and *Freycinetia (about 100 species)]* the flowers are arranged in a spherical or club-shaped racemose spadix subtended by a spathe, which is sometimes brightly colored. Male flowers have numerous stamens. Female flowers have a superior ovary consisting of a ring of many carpels, which is sometimes reduced to a single carpel or a row of carpels; female inflorescences sometimes contain regressed stamens (staminoids). The fruit is a berry or multilocular drupe. Some species have large, cone-like, edible starchy fruits, which superficially resemble a pineapple, e.g. *Pandamus leram* ***(Nicobar breadfruit),*** *P. utilis* and *P. andamanensium.* Leaves of the very common *Pandamus odoratissimus* are used for thatching and basket work, and fibers from its aerial roots are used for cordage and brushes. Ornamentals include the hothouse plant *Freycinetia banksii* from New Zealand, and *Pandamus veitchii,* which has dark, glossy leaves with silvery white borders. Kewda attar, a well known Indian perfume, is prepared by boiling the fragrant white male flowers of *Pandamus odorus* in water, and absorbing the vapor in sandalwood oil.

Pandion: see *Falconiformes.*

Pandionidae: see *Falconiformes.*

Panicle: see Flower, section 5) Inflorescence.

Panicum: see *Graminae.*

Pangolins: see *Pholidota.*

Panmixis: random mating between all sexually members of a population.

Panmixis index, ***panmictic index:*** the inverse of the Inbreeding coefficient (see).

Panonychus: see *Tetranychidae.*

Panorpatae: see *Mecoptera.*

Panorpidae: see *Mecoptera.*

Panorpina: see *Mecoptera.*

Pansporoblast: see *Cnidosporidia.*

Pansy: see *Violaceae.*

Panthera: see *Felidae.*

Panther cap: see *Basidiomycetes (Agaricales).*

Panther toad: see *Bufonidae.*

Pantholops: see *Saiginae.*

Pantopoda: see *Pycnogonida*

Panurus: see *Timaliinae.*

Papain: a plant protease from the latex and unripe fruit of *Carica papaya* (papaw) It is a carbohydrate-free, basic, single chain protein, M_r 23 500, containing 212 amino acid residues. It is unusually stable to high temperatures and has a pH optimum of 5–5.5. It catalyses the hydrolysis of proteins to amino acids via polypeptides. In addition to endopeptidase activity, P. also displays amidase and esterase activity. It is a thiol protease; a single free SH group is essential for activity, and SH blocking agents (e.g. iodoacetate) are powerful P. inhibitors.

Papaver: see *Papaverceae.*

Papaveraceae, ***poppy family:*** a family of dicotyledonous plants containing about 200 species, cosmopolitan in distribution, but concentrated mainly in the temperate and subtropical regions of the Northern Hemisphere. They are mostly herbs, rarely shrubs or trees, and all plant parts produce a milky or colored latex when cut. Leaves are usually alternate, and often pinnate or deeply divided. Flowers are hermaphrodite and actinomorphic. The 2 sepals usually fall before the flower is fully open. There are 4 (2 + 2) petals, numerous stamens, and a superior ovary composed of 2 or more united carpels. The fruit is a capsule, opening by pores or valves. Seeds possess a minute embryo in an oily endosperm, and they often have membranous or fleshy appendages that attract the ants that aid their dispersal.

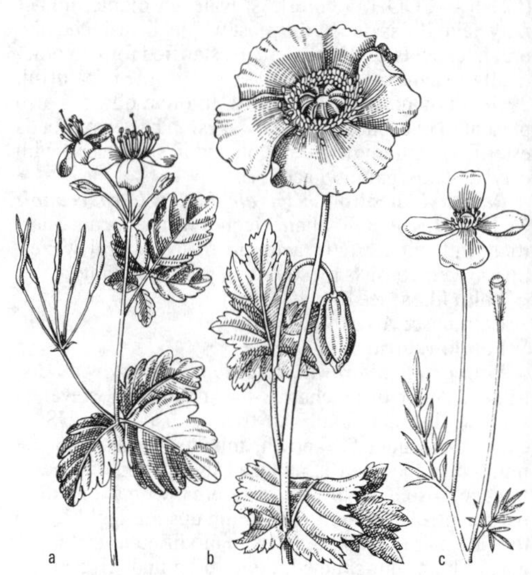

Papaveraceae. a Flowering and fruiting shoot of *Chelidonium majus* (greater celendine). *b* Flowering shoot of *Papaver somniferum* (opium poppy). *c Papaver argemone* (long prickly-headed poppy).

Economically, the most important member is the ***Opium poppy*** *(Papaver somniferans),* an ancient oil and medicinal plant; its seeds contain about 50% oil, which is used as an edible oil, as well as medicinally (official Oleum papaveris) and as a drying oil by artists. The latex contains several sedative and analgesic alkaloids (see Opium alkaloids), e.g. morphine, codeine, narcotine, thebaine. The opium poppy grows to a height of 1.2 m, and its flowers range in color from white through pale lilac, purple, and white tinged with red. Having been exploited and spread by man for centuries, the origins of the opium poppy are now uncertain, but its original wild habitat was probably in the eastern Mediterranean or Middle east.

Latex from other members of the family also contains alkaloids, e.g. the orange-colored latex of the ***Greater celandine*** *(Chelidonium majus),* a common European ruderal plant. Many species of the family have been adopted as ornamental and garden plants, e.g. ***Californian poppy*** *(Eschscholzia californica),* a native American plant found from the Columbia river to New Mexico; its bright yellow flowers open only in direct sunlight. Another ornamental is *Macleaya cordata* ***(Plume poppy),*** which can attain a height of 3 m; its flowers have no petals, but 24–30 pure white feathery stamens on a large panicle. The ***Oriental poppy*** *(Papaver orientale)* has unusually dark, scarlet or blood/brick-red flowers. It is closely related to the somewhat larger (up to 1 m in height) and sturdier *Papaver bracteatum,* which has broader petals and persistent bracts. The 2 separate species have a similar wild distribution in eastern Turkey, Caucasus, Armenia and northern Iran, but many intermediate forms occur in cultivation.

Well known and conspicuous agrarian weeds are the ***Field poppy*** *(Papaver rhoeas)* with smooth capsules, and the ***Long prickly-headed poppy*** *(Papaver argemone)* with bristly capsules. The ***Devil's fig*** or ***Golden thistle of Peru*** *(Argemone mexicana)* is neither a thistle nor a native of Peru. The genus, *Argemone,* is found from southeastern to western USA, as well as in the Caribbean, with 1 species in Hawaii. *A. mexicana,* which is spiny and contains a bitter-tasting principle, originates from Mexico, but is now a widely distributed tropical weed, especially in coastal regions. Other genera are: *Platystemon* (restricted to western North America), e.g. *P. californicus* ***(Cream cups);*** *Romneya* (California and northwestern Mexico), e.g. *R. coulteri* ***(Matilija poppy);*** *Stylomecon heterophylla* (monotypic genus from California); *Glaucium* (western Asia, Europe, Mediterranean, Scandanavia), e.g. *G. flavum* ***(Yellow-horned poppy);*** *Mecanopsis* (southern-central temperate Asia, especially Himalayas, with one species in western Europe), e.g. *M. cambrica* ***(Welsh poppy)*** of southwest England and Wales; and *Roemeria* (southern and central Europe), e.g. *R. hybrida* ***(Violet horned poppy).***

Papaverine: an opium alkaloid (see Opium). It is a smooth muscle relaxant and cerebral vasodilator, used medicinally as a soporific and anticonvulsive agent. Unlike morphine, it is not addictive.

Paper factor: Juvabione.

Paper nautilus: see *Cephalopoda.*

Papez' circuit: a complex reverberating neuronal circuit that interconnects the association areas of the cerebral cortex with the hypothalamus. The association areas project to the cingulate gyrus, from which information is relayed via the cingulum bundle to the parahippocampal gyrus, and from there to the hippocampus. The hippocampus then projects through the fornix to the mammillary bodies of the hypothalamus, and from there to the anterior nuclei of the thalamus. Information is finally relayed from the thalamus back to the cingulate gyrus. See Neuronal circuit.

Paphia: see *Ericaceae.*

Paphiopedilum: see *Orchidaceae.*

Papilionaceae, ***pea family:*** a family of dicotyledonous plants containing 9 000–12 000 species in about 300 genera. Together with the *Caesalpiniaceae* and the *Mimosaceae,* the *P.* formerly constituted the family *Leguminosae.* Members of the *P.* always possess root nodules, which harbor symbiotic, nitrogen-fixing bacteria. They are therefore used agriculturally for soil improvement, and their value for this purpose was discovered empirically by prehistoric agriculturalists. Trees and bushes belonging to the family are found mainly in the tropics, but the majority of species are herbaceous and occur in the temperate zones. Leaves are usually alternate, stipulate, sometimes simple, but more usually compound (pinnate or 3-foliate), and sometimes with a terminal tendril. Flowers are zygomorphic and papilionate, possessing 5 petals: a single large, often erect, ***standard*** (adaxial) petal, 2 ***wings*** (lateral petals), and 2 lower petals that form a ***keel*** by varying degrees of fusion of their lower margins. The calyx usually consists of 5 more or less united sepals. In the bud, the standard encloses the other petals. There are generally 10 stamens, which are usually fused (monadelphous condition) into a tube surrounding the ovary. Often the tube is formed by 9 fused stamens, leaving one stamen free-standing (diadelphous condition). Pollination is by insects or birds. The ovary is superior, consisting of 1 carpel containing 2 rows of ovules, which alternate to form a single vertical rank. The fruit is usually a typical multiseeded, dehiscent pod, and the seeds possess a hard coat that does not readily imbibe water. Members of the *P.* contain a wide range of secondary metabolites, e.g. alkaloids, toxic proteins, saponins and isoflavones, some of which have taxonomic significance. Nonprotein amino acids are also found widely in this family.

Many seeds of the *P.* are exploited as food, since the large seed cotyledons contain rich stores of fat, starch and protein. In fact, some members of the *P.* were eaten by Neolithic man in Southwest Asia, and the same or genetically related crops are still important in modern agriculture, e.g. the ***Garden*** or ***Field pea*** *(Pisum sat-*

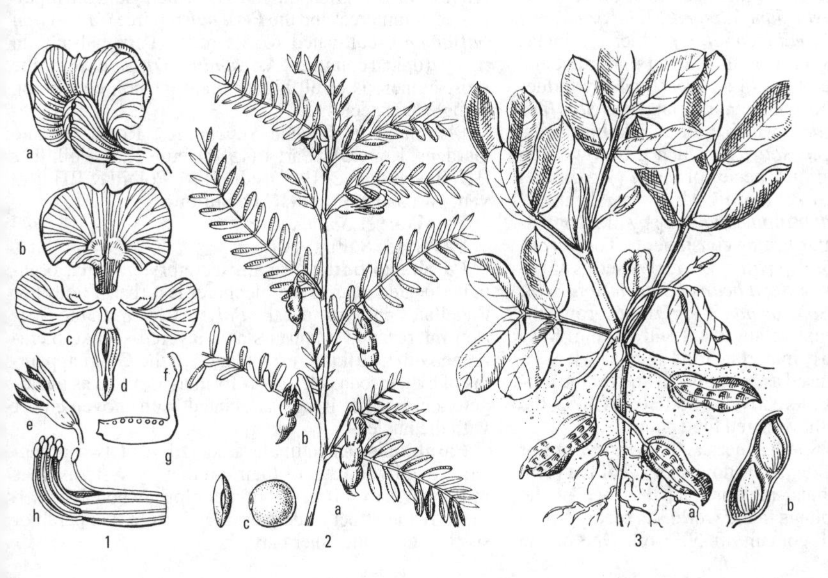

Fig. 1. *Papilionaceae. 1* Garden pea *(Pisum sativum), a* flower, *b* standard (adaxial) petal, *c* wings (lateral petals), *d* keel, *e* calyx, *f* stigma, *g* ovary, *h* stamens. *2* Lentil *(Lens culinaris), a* flowering and fruiting plant, *b* pod, *c* seeds. *3* Groundnut *(Arachis hypogaea), a* plant with flowers and fruits in various developmental stages, *b* pod in longitudinal section.

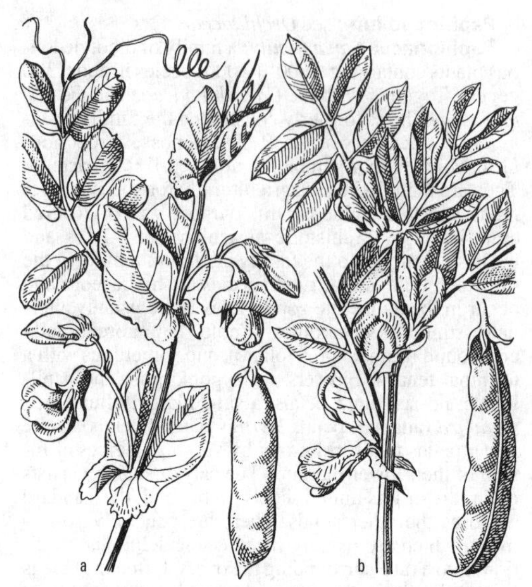

Fig.2. *Papilionaceae. a* Flowering stem of the garden pea *(Pisum sativum)* with pod on the right. *b* Flowering stem of the broad bean *(Vicia faba)* with pod on the right.

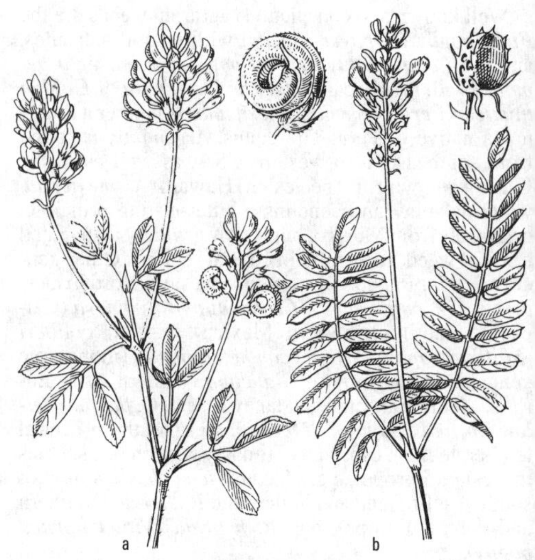

Fig.3. *Papilionaceae. a* Flowering plant and seeds of lucerne or alfalfa *(Medicago sativa). b* Flowering plant and seeds of sainfoin *(Onobrychis viciifolia).*

ivum), ***Lentil*** *(Lens culinaris),* ***Broad*** or ***Horse bean*** *(Vicia faba),* and the ***Chick pea*** *(Cicer arietinum).*

The ***Grass pea vine*** *(Lathyrus sativus)* and the ***Pigeon pea*** *(Cajanus cajan = C. indica)* are used especially by the indigenous peoples of tropical countries as food, fodder and green manure. Important vegetable food plants like the ***French*** or ***Kidney bean*** *(Phaseolus vulgaris)* and the ***Scarlet runner*** *(Phaseolus coccineus)*, originally native to South America, are now cultivated widely throughout the world. Other important fodder and green manure plants, which can also be grown on nitrogen-poor soils, are the various species of ***Clover*** *(Trifolium pratense, T. hybridum, T. repens, T. incarnatum);* ***Lucerne*** or ***Alfalfa*** *(Medicago sativa),* which originates from Central Asia and is one of the most important and at the same time the oldest of all green and dry fodder plants of nontropical continental countries; ***Sainfoin*** *(Onobrychis viciifolia),* which prefers alkaline soils; ***Seradella*** *(Ornithopus sativus)*, which grows well on poor sandy soils; various species of ***Vetch*** *(Vicia);* and various strains of "***Sweet lupin***", which have been developed from the ***Bitter lupin*** *(Lupinus)* by plant breeding to remove the bitter-tasting constituents. The ***Perennial lupin*** *(Lupinus polyphyllus)* is widely used as an ornamental. ***Soybeans*** or ***Soya beans*** *(Glycine soja = G. max = G. hispida = Soja hispida = S. max)* are grown for fodder, and for the extraction of oil and protein. Soybeans are a particularly important food item in East Asia; oil from the seeds is used as an edible oil and for the production of margarine. As early as 2 800 BC in China, the soybean was one of the 5 sacred food plants.

Originating from South America, the ***Ground nut*** or ***Peanut*** *(Arachis hypogaea)* is now widely cultivated in tropical and subtropical regions, and it is one of the most important oil plants in the world. Its seeds, which ripen below ground, contain 40–50% oil. One of the oldest dye plants is ***Indigo*** *(Indigofera tinctoria).* ***False acacia, Lucust,*** *or* ***Black locust*** *(Robinia pseud(o)acacia),* from North America, is used in various parts of the world as a pioneer plant for the reclamation of slag heaps, waste tips and other barren land. Introduced into Europe in 1600 from North America, *Robinia* also yields a useful, durable timber.

Ornamental plants include the poisonous, southern European ***Laburnum*** or ***Golden rain*** *(Laburnum anagyroides);* ***Wistaria*** or ***Wisteria*** *(Wistaria = Wisteria sinensis),* which is native to East Asia; ***Sweet pea*** *(Lathyrus odoratus);* and ***Broom*** *(Sarothamnus = Cytisus scoparius),* the tops of which are official as a diuretic. Many species of *P.* are medicinal plants, e.g. ***Liquorice*** *(Glycyrrhiza glabra),* the roots of which yield the liquorice of commerce; and the ***Peru balsam*** tree *(Myroxylon balsamum),* cultivated for its resin (Peru balsam) in many tropical countries. *Astragalus* ***(Tragacanth)*** is the most numerous of all higher plant genera, containing some 2000 species.

[*Advances in Legume Systematics,* Royal Botanic Gardens, Kew, UK. Part 1 (1981), edit. R.M. Polhill & P.H. Raven; Part 2 (1982), edit. R.M. Polhill & P.H. Raven; Part 3 (1987), edit. C.H. Stirton]

Paprika: see *Solanaceae.*

PAPS: see Sulfur.

Parabasal body: a small, secretory, granular organelle located near to the blepharoplast (basal body) of flagellates, in particular the *Polymastigina.* In the electron microscope it appears as a layered saclike membrane with a striking resemblance to the Golgi apparatus of a metazoan cell. It was formerly known as the kinetonucleus, but is not associated with movement or with the nucleus.

Parabiosis: an intimate association of two organisms, e.g. by linkage of their circulatory systems. P. is important in experimental immunology, where it is used to study the effect of defense reactions in one parabiosis partner on the other partner.

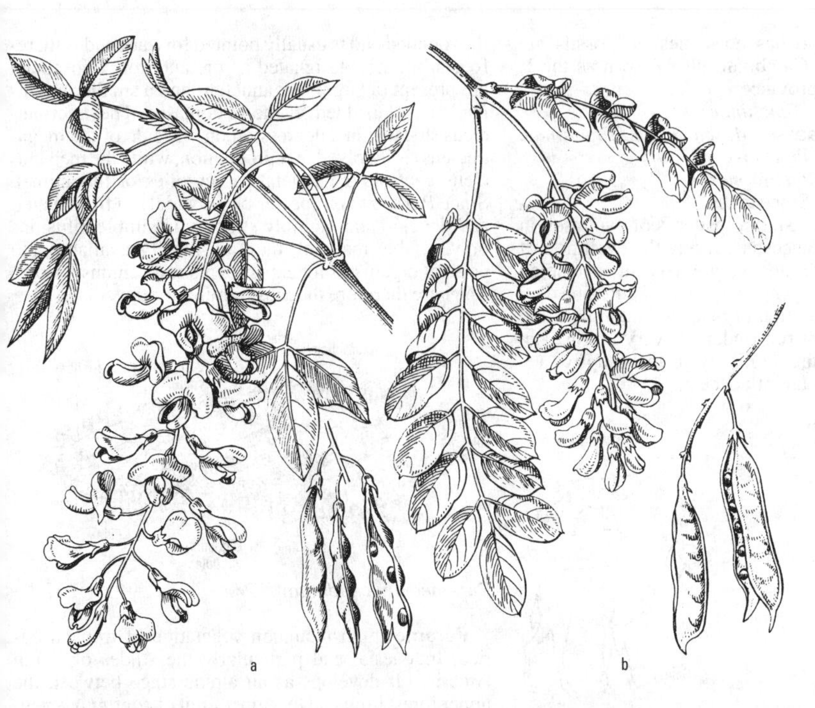

Fig.4. *Papilionaceae. a* Flowering branch and fruits of laburnum or golden rain *(Laburnum anagyroides). b* Flowering branch and fruits of the locust tree or false acacia *(Robinia pseudoacacia).*

Paraboloid condenser: see Darkfield illumination.

Paradisaea: see *Paradisaeidae.*

Paradisaeidae, ***birds of paradise:*** an avian family of the order *Passeriformes* (see). The approximately 40 species are found mainly in New Guinea and neighboring islands, with only a few representatives in Northeastern Australia and the Moluccas. Close relatives of the *Corvidae,* they are very active, robust birds with strong beaks and strong feet, and often loud, harsh voices. They are essentially inhabitants of tropical forest trees, often found at considerable altitudes. All species feed on insects, small tree frogs and lizards, and some species also take fruit. There are a few black species that lack decoration and display no sexual dimorphism. These are in stark contrast to the many species, in which the females are relatively plain in comparison with the males, which are arguably the most colorful, decorative and bizarre of all birds. Examples are the ***Greater bird of paradise*** *(Paradisaea apoda;* Aru Islands and New Guinea; introduced into Little Tobago in the West Indies in 1909): wire-like central tail feathers up to 76 cm in length; head and neck orange-yellow; forehead and chin black with a green gloss; throat iridescent green; upper breast brown to dark maroon, depending on the race; underparts and upper parts, including wings and tail, bright yellow shading to mauve; large lacy plumes on each side of the breast, up to 58 cm in length, yellow to orange at their bases, becoming pale cinnamon; eyes lemon yellow; legs pale brown; beak light gray-blue, paler at the tip. The ***Magnificent bird of paradise*** *(Diphylodes magnificus)* has a red back, glossy golden ruff, iridescent green breast shield and 2 long, thin, curled tail feathers. The ***King of Saxony bird of paradise*** *(Pteridophora alberti)* is notable for its head plumes, which emerge from the sides of the head and are more than twice the length of the bird.

The splendid head and tail plumes, back capes, breast shields and flank plumes of the male are erected and spread in the courtship display, which also involves ritualized dances and poses. New Guinea tribesmen used and still use the feathers for decoration and ornamentation, and at one time many were also used in the European millinery trade.

Paradisefishes: see Gouramies.

Paradise flycatchers: see *Muscicapinae.*

Paradoxical frog: see *Pseudidae.*

Paradoxides (Greek *paradoxus* eccentric, conspicuous): a trilobite genus with a large cephalon and long, sickle-like genal spines. The glabella extends forward to a broadly rounded front, and the 3–4 glabellar furrows are usually well defined. The facial sutures run from the posterior margin of the cephalon, over the crescent-shaped, large eyes, to the anterior margin of the cephalon. There is no ocular ridge. The body consists of 16–21 segments. The small, lobed pygidium is

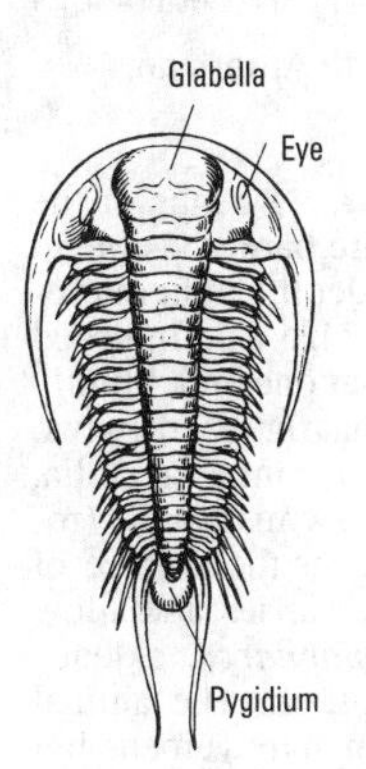

Paradoxides gracilis Boeck from the Middle Cambrian of Czechoslovakia (x 0.3)

hardly segmented and has no spines. Its fossils are found in the Middle Cambrian (also known as the P. stage) of the Alantic province.

Paradoxornis: see *Timaliinae.*

Paradoxornithinae: see *Muscicapidae, Timaliinae.*

Paradoxurus: see Palm civets.

Parakeets: see *Psittaciformes.*

Paralithodes: see Stone crabs.

Parallel evolution: similar patterns of evolution in lineages of common ancestry. During the evolution of the ammonoids, separate evolutionary lines display similar changes in the complexity of the suture line. Despite the separate evolution of placental mammals and marsupials, their forms remain largely very similar. P.e. is generally due to the similarity in the adaptive response to environmental influences.

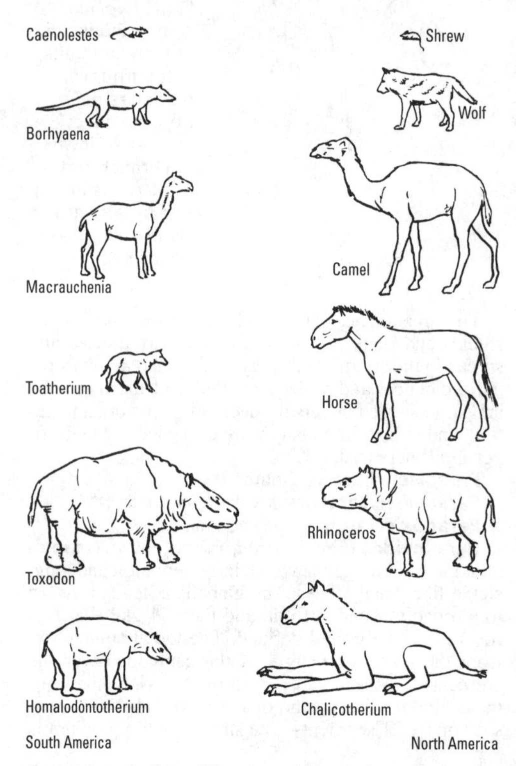

Parallel evolution of South and North American mammals

Paramecium, ***slipper animalcules:*** a genus of the *Ciliophora* (see), whose members are barely visible to the naked eye, but in sharp focus under the light microscope resemble the sole of a slipper. *P.* has a definite and constant shape (elongated, rounded at one end, bluntly pointed at the other end, and somewhat flattened on two sides). By the coordinated beating of numerous cilia, which cover the entire cell surface, *P.* swims rapidly (rotating at the same time) or glides over the surface of solid objects. One flattened surface carries a shallow *oral groove,* from which the blind *cytopharynx* extends obliquely backward into the endoplasm. The animal may occasionally reverse its direction of movement, but the rounded end is usually pointed forward and is therefore arbitrarily designated as the anterior. Two nuclei are present (a larger meganucleus and a smaller micronucleus) embedded in the endoplasm. The micronucleus (located in a depression on one side of the meganucleus) is involved in reproduction, while the meganucleus controls the vegetative activities of the animal. Since *Paramecium* species occur widely in fresh water, and develop and multiply strongly in simple infusions of vegetable material, they are readily available and popular organisms for experiments and demonstrations. They are therefore the most familiar of the ciliates.

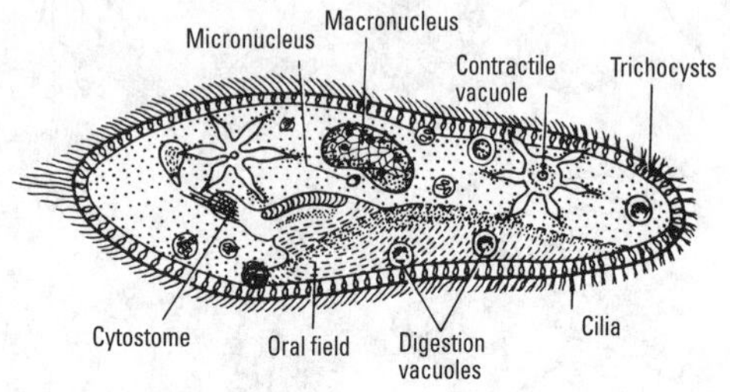

Paramecium caudatum

Paramo: high mountain vegetation of tropical Africa, Indonesia, and particularly the Andes of South America. It develops as an alpine stage between the upper forest limit and the upper limit of continuous vegetation (or the snow line). Even during the dry season the soil remains wet, since moisture is derived from clouds and mist. Nights are frosty and the days warm. The vegetation contains grasses, dwarf bushes and prostrate and creeping woody plants, and is characterized in particular by scattered tall plants, tree forms with foliage concentrated at the top [the "cabbage trees" *(Espeletia)* of the Andes; East African tree species of *Senecio* with an upper rosette of cabbage-like leaves (also called cabbage trees); and the bushy *Anaphalis* of Indonesia], and giant perennial herbs *(Lobelia* in East Africa, *Puya* in the Andes).

Parapatric: used of populations whose geographical ranges are contiguous but not overlapping, so that gene flow between them is possible.

Parapatric speciation: speciation occurring is disjunct populations, which later become contiguous but remain separate.

Parapause: see Dormancy.

Paraphyletic: see Paraphyly, Taxonomy.

Paraphyly: a term from phylogenetic or cladistic systematics. In contrast to monophyletic taxa (see Monophyly), a paraphyletic group does not contain all the recent descendants of a single ancestor species. Descendants of a younger species of a paraphyletic group are excluded from the group. For example, the terrestrial carnivores *(Fissipedia)* represent a paraphyletic group, because they are separated from the *Pinnipedia* which evolved from fully adapted terrestrial carnivores.

Paraphyses:

1) Sterile, haploid, branched or unbranched hyphae, which occur in the fructification of the *Ascomycetes* between the asci; together with the asci, they may form the spore-bearing layer (hymenium); they are also present, but less numerous, in the *Basidiomycetes.*

2) Sterile hairs in the Conceptacles (see) of the genus *Fucus (Phaeophyceae).*

3) Multicellular vascular hairs, often with spherical, terminal cells, found between the gametangia of mosses.

4) Hair structures in ferns, which may arise from the stalk of the sporangium, or from the placenta (receptacle).

Paraphysis: see Parietal organs.

Paraphysoclistic fishes: see Swim bladder.

Parapineal organ: see Parietal organs.

Paraplasmatic inclusions: various types and quantities of materials which are stored in the cytoplasm of various cells, and are permanently or temporarily excluded from metabolism. Most P.i. are products of the cells that store them, e.g. secretions, incretions, reserve materials (glycogen, lipid droplets, etc.), pigments (including melanin and lipofuscin).

Parapodia:

1) Paired, biramous (usually), more or less laterally compressed, setaceous, muscular appendages on the body segments of polychaetes. In free-swimming forms they serve for locomotion, and in sessile forms they are used to generate water currents.

2) In opisthobranch mollusks, lateral projections of the foot, which serve as fins for swimming.

Parapophysis: see Axial skeleton.

Paraproteins: abnormal immunoglobulins, or normal proteins present in increased quantities in blood plasma in various hematological disturbances, known as paraproteinemias.

Known P. are Bence-Jones proteins, amyloid proteins, Waldenström-type immunoglobulin M and cryoglobulin. Owing to their homogeneous character and relative ease of isolation on a preparative scale, P. are among the best studied immunoglobulins. P. are often produced by cancer cells (see Myeloma proteins).

Parasite (plates 42 & 43): an organism that lives at the expense of another (usually larger) organism, the Host (see). The host provides food, a dwelling site, and a site for reproduction. A P. is therefore physiologically dependent on its host, and the host itself becomes the ecological habitat of the P. Specific host-P. relationships are the result of co-evolution. The host is put at a competitive disadvantage, but is normally not killed by the P.

Facultative Ps. can develop normally and complete their life cycle as free-living organisms, or as Ps. (e.g. certain bacteria and fungi). ***Obligate Ps.*** are dependent on the host, and must always live parasitically. Plant Ps. are either ***holoparasites*** or ***hemiparasites.*** Holoparasites are totally dependent on the host for all nutrients (e.g. broomrape), whereas hemiparasites obtain only part of their nutrient requirement from the host (e.g. mistletoe).

Ectoparasites are associated with the exterior of the host (e.g. lice, mosquitoes, bed bugs). Ecologically, the host represents only part of the environment of an ectoparasite. ***Endoparasites*** live inside their hosts. They may actively penetrate the host from the exterior (some trematodes and nematodes), may be taken in with food (tapeworms), or transmitted by an intermediate host (e.g. malaria Ps.). Ecologically, the host represents the total environment of an endoparasite. Endoparasites of animals live in the alimentary canal, organs, tissues, or blood of their hosts. Blood Ps. are unicellular organisms or worms, which live in the circulatory system, either free or inside blood cells, and which are usually transmitted by blood-sucking arthropods. Thus, the migratory larvae of parasitic nematodes live free in the blood fluid, whereas the sporozoa of the malaria P. invade the red blood cells.

If all developmental stages of a P. occur during its parasitic existence, it is known as a ***permanent P.*** If only certain stages of the life cycle occur while living parasitically, it is called a ***periodic P.*** (e.g. imaginal or larval Ps.). ***Stationary Ps.*** remain permanently, or for very extended periods, on the host (e.g. lice). If the host is visited only briefly, usually to obtain food, the P. is called a ***temporary P.*** (e.g. midges, mosquitoes, bed bugs). Classified on the basis of their hosts, Ps. may be human Ps., zooparasites (Ps. of animals), or phytoparasites (Ps. of plants). According to the part of the plant attacked, plant Ps. are known as root, leaf, or flower Ps. Diseases caused by Ps. are called ***parasitoses.***

Not all host-P. relationships can be defined as true parasitism. There are many intermediate forms, representing smooth transitions to commensalism, mutualism and predation. See Relationships between living organisms.

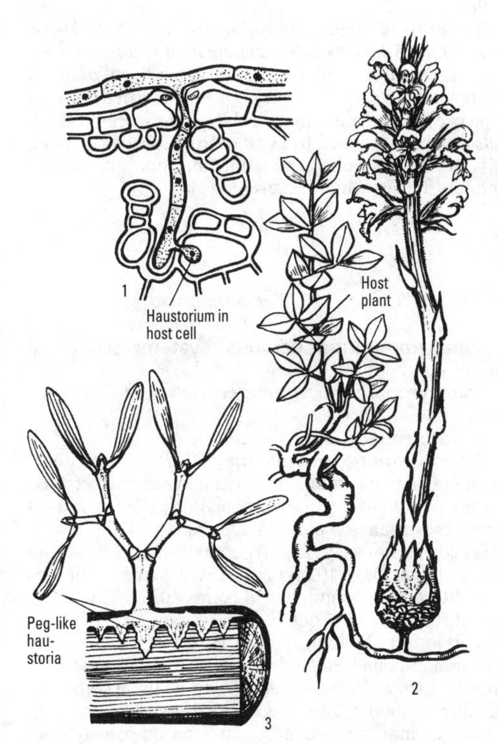

Parasites. Parasitic plants. *1* Mildew; a lateral hypha has penetrated the host plant via a stomatum. *2* Broomrape *(Orobanche)* parasitizing the root of a host plant. *3* Young mistletoe *(Viscum)* plant on a host branch.

Parasitic castration: weakening or destruction of the function of a host's gonads by a parasite, e.g. snails may be sterilized when they are strongly infected by trematode larvae. Crabs are castrated by the parasitic crustacean, *Sacculina carcini.*

Parasitiformes: see *Ixodides.*

Parasitism: interaction and coexistence of different organisms, to the unilateral advantage of one of the partners and at the expense of the other. See Parasites.

Parasitoid: an endoparasitic insect (usually an ichneumon wasp), whose development within the host (usually another insect or an archnid) leads to the death of the latter.

Parasitology: the study of parasites, their mode of life and measures for their control. P. has close links with ecology, contributes to the understanding of phylogenesis, and makes an important practical contribution to medicine, veterinary medicine and agriculture.

Parasitophyletic laws: laws for interpreting the level of evolution and phylogenetic position of parasite hosts from the systematics and variety of their parasites. P.l. are based on the assumption that hosts and parasites have evolved largely in parallel, so that phylogenetically ancient hosts have predominantly primitive parasites.

Parasitosis: see Parasite.

Parasol mushroom: see *Basidiomycetes (Agaricales).*

Parasorbic acid, ***S-5,6-dihydro-6-methyl-2H-pyran-2-one, 5-hydroxy-2-hexenoic acid lactone:*** an unsaturated lactone, which is the sole constituent of an oil obtained by steam distillation of the acidified juice of ripe berries of the mountain ash *(Sorbus aucuparia).* It has also been isolated from other fruits, seeds and plant oils. It inhibits seed germination and root growth, and inhibits the growth of bacteria.

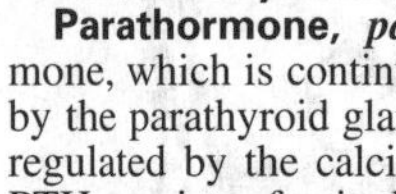

H_3C O O *Parasorbic acid*

Parasympathetic nervous system: see Autonomic nervous system.

Parasympathetic tone: see Autonomic nervous system.

Paratetranychus: see *Tetranychidae.*

Parathormone, ***parathyrin, PTH:*** a proteohormone, which is continuously synthesized and secreted by the parathyroid gland without storage. Synthesis is regulated by the calcium concentration of the blood. PTH consists of a single polypeptide chain of 84 amino acid residues, M_r 9 402 (porcine), with some small species differences in amino acid composition. The *N*-terminal sequence of about 30 amino acids is particularly important for biological activity. PTH activates the membrane-bound adenylate cyclase of osteoclasts, thereby increasing the uptake of Ca^{2+} by these cells. The resulting mobilization of Ca^{2+} causes an increase in blood calcium, which of necessity is accompanied by an increase of free phosphate. The phosphate is excreted by the kidneys. Thus, PTH promotes phosphate secretion in the distal part of the kidney tubule, and inhibits phosphate resorption in the proximal tubule. PTH also promotes calcium absorption by the intestine. Interaction of PTH with its antagonists, Calcitonin (see) and 1,25-dihydroxycholecalciferol, is primarily responsible for maintaining constant Ca^{2+} and phosphate concentrations in blood and tissues. Secreted PTH has a half life in the body of about 20 min. It is degraded in the liver to smaller fragments, which are excreted in the urine.

Parathyroid glands, ***glandula parathyreoidea, Sandstroem's body:*** endocrine organs, which develop from the endodermal lining of the third and fourth pharyngeal pouches of the embryo. P.g. are present in all vertebrates, excluding fishes. As a rule there are four P.g., but their location and number may vary considerably. Parathyroid tissue is sometimes found in the mediastinum. In humans the P.g. are small, oval bodies about 5 mm long and 4 mm wide, embedded in the posterior surface of the lobes of the thyroid gland (2 in the superior, and 2 in the inferior lobes). In nonmammals, the P.g. are not embedded in the thyroid. Although the P.g. are usually physically associated with the thyroid, there is no functional relationship between the two glands. P.g. secrete parathormone, which raises the concentration of blood calcium and causes resorption of bone calcium.

Paratope: see Antibodies.

Paratylenchus: see Eelworms.

Paratyphoid fever: see *Salmonella.*

Paratypic atavism: see Atavism.

Parazoa: see *Porifera.*

Pardalotus: see *Dicaeidae.*

Pareinae: see Snail-eating snakes.

Parencephalon: see Parietal organs.

Parenchyma:

1) Ground tissue of plants. It usually consists of regularly-shaped, living cells with only slightly thickened, elastic, cellulose cell walls. The cytoplasm may contain chloro-, leuco-, or chromoplasts. The cytoplasm usually forms a thin layer around a large vacuole, which (especially in storage parenchyma) is rich in stored food materials.

According to its function, P. may be classified as follows. a) ***Assimilatory P. (Assimilation tissue),*** which is the chlorophyll-containing tissue of leaves and shoots. It serves for the photoassimilation of carbon dioxide, and as an aid to gas exchange it usually contains large intercellular spaces and thin (gas-permeable) cell walls. Assimilatory P. includes palisade P., green cortical P., and spongy P. Palidase P. and spongy P. together form the *mesophyll.* b) ***Storage P.*** occurs mainly in storage organs (swollen tap roots, tubers, seed endosperm), but also in the pith and cortex of shoots and roots. The cells contain large quantities of food material, often starch. Wood parenchyma (an interconnecting network of living cells among the nonliving xylem) forms an important food-storage tissue of woody plants. Parenchymatous rays and bast P. serve for both food storage and conduction. In the broad sense, water-storage tissue (e.g. of succulents and epiphytes) is also storage P. c) ***Conduction P.*** or ***Transfusion P.*** This consists of P. cells which are elongated along the main directional axis of food transport, and which aid the conduction of organic food material and water. d) ***Ventilation tissue*** or ***Aerenchyma.*** This type of P. is specialized for ventilation. It occurs in most aquatic and marsh plants, where its large intercellular spaces facilitate gas exchange in submerged organs; it is connected to the external atmosphere via stomata in leaves and shoots above the water level.

2) The mesenchymal packing tissue between and around the organs of platyhelminths.

3) The actual organ tissue of the compact organs of vertebrates, e.g. liver, kidney and pancreas parenchyma, as opposed to the interstitial connective and epidermal tissue of these organs.

Parenchymula larva: see *Porifera.*

Parental care: see Brood care.

Parental provision: see Brood provision.

Paridae, ***titmice:*** an avian family containing 51 species in the order *Passeriformes* (see). In the present classification, the 11 species of Penduline tits, sometimes considered to be a subfamily of the *Paridae,* are assigned separate familial status (see *Remizidae).* The Bearded tit *(Panurus biarmicus)* belongs to the family *Timaliidae.*

Titmice feed on spiders and insects, as well as on seeds (especially seeds with a high fat content such as nuts) which they hold in their feet and hammer open with their powerful, somewhat conical bills. The strong bill is also used to penetrate bark in search of insects. Food is often stored in bark crevices for future consumption. They nest in holes and cavities, and therefore readily adopt nesting boxes. In winter they form large, loose flocks, which are often mixed, including other birds such as nuthatches and woodpeckers.

Paridae are mainly distributed in the Palearctic, but with some representatives in Africa and North America the family has an extremely wide range. Characteristic plumage colors are yellow, blue and gray-green, with black head caps and white cheeks. They are tree-living rather than ground-dwelling birds, with rounded bodies, 8–20 cm in length. The subfamily *Parinae* ***(Parus tits)*** contains 43 species, e.g. the Eurasian species of the genus *Parus:* ***Great tit*** *(P. major),* ***Blue tit*** *(P. caeruleus),* ***Crested tit*** *(P. cristatus),* ***Coal tit*** *(P. ater),* ***Marsh tit*** *(P. palustris),* ***Willow tit*** *(P. montanus).* The subfamily, *Aegithalinae,* comprises 8 species of ***Long-tailed tits.***

Parietal eye: see Parietal organs.

Parietal organs: four unpaired evaginations which arise medially, one behind the other, from the roof of the vertebrate diencephalon. The most anterior is known as the *paraphysis,* followed by the *dorsal sac,* the *parietal organ,* and finally the *pineal organ (pineal body, epiphysis).*

The paraphysis is evident in the embryos of all classes of vertebrates, as a branch or fold of the telencephalic roof. In birds and in the majority of quadrupeds, however, it does not persist in the adult. It is well developed in fishes and amphibians, where its epithelial cells store glycogen and display secretory properties. The dorsal sac ("parencephalon") remains as a vesicular structure, and it is often overlain by the epiphysis ("epiphysial cushion").

At their highest level of development, the parietal and pineal organs have the character of eyes, and in fish and lizards they are still sensitive to light, despite their rudimentary nature. In contrast to the lateral eyes, they are vesicular with an everted retina. In fish, amphibians and reptiles the early primordia of the parietal and pineal organs lie next to each other in the same transverse plane. It is therefore assumed that early vertebrates possessed a pair of light-sensitive organs in the forehead, which then evolved unequally. The left anterior organ moved forward and remains at the developmental level of an eye only in *Sphenodon* and some lizards, where it is also known as the *parietal eye.* It disappeared completely in other vertebrates, with the exception of *Petromyzon* (lamprey) and *Amia* (bowfin), in which it is still present but very degenerate. In *Petromyzon* it is also known as the *parapineal organ.* The right posterior of these two organs (the pineal organ) still has the character of an eye only in *Petromyzon,* whereas in other vertebrates its diencephalic root is modified to a well vascularized Epiphysis (see) which functions as an endocrine gland (pineal gland). The hindmost pineal organ is frequently shifted forward over the parietal eye.

The pineal organ has been shown to be involved in the regulation of endogenous biological rhythms.

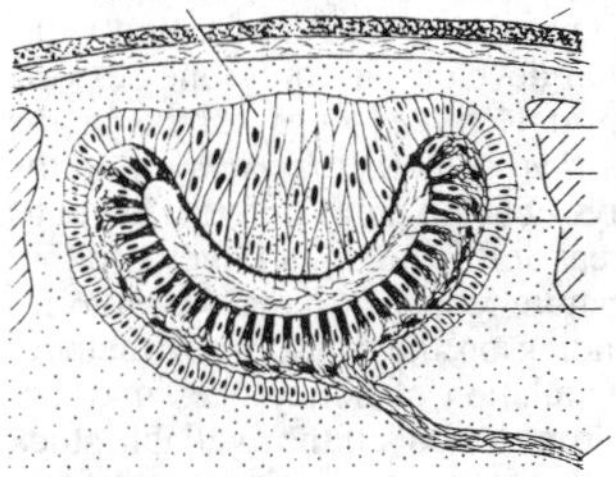

Parietal organs. Parietal eye of a lizard in longitudinal section.

Parinae: see *Paridae.*

Paroecia: see Relationships between living organisms.

Paro(e)kete: see Myrmecophiles.

Paroophoron: a collection of rudimentary tubules in mammals enclosed by layers of the mesosalpinx, but lying nearer to the uterus than the Epoophoron (see). In the male, these rudimentary tubules represent the paradidymis or Giraldés' organ. They are derived embryonically from the part of the mesonephros lying near the caudal end of the embryo.

Parovarium: see Epoophoron.

Parrotbills: see *Muscicapidae, Tamaliinae.*

Parrotfishes: see *Scaridae.*

Parrots: see *Psittaciformes.*

Parrot's disease: see Achondroplasia.

Pars intermedia: see Hypophysis.

Parsley: see *Umbelliferae.*

Parsley frogs: see *Pelodytidae.*

Parsnip: see *Umbelliferae.*

Parson bird: see *Meliphagidae.*

Parthenium: see *Compositae.*

Parthenocarpy: fruit formation without fertilization. Natural P. occurs in bananas, pineapples and various citrus plants, resulting in the production of seedless fruits or fruits with degenerate seeds. Natural P. depends on a high phytohormone concentration in the nucellus tissue. In cases of incompatibility or gamete sterility, P. may be actuated by auxin from the pollen grains. *Artifical P.* can be brought about by treatment of plants with phytohormones, e.g. auxin causes P. in tomato, tobacco, squash or fig, whereas gibberellin is effective when sprayed on rose, cherry, apple or grape.

Parthenocissus: see *Vitaceae.*

Parthenogenesis: asexual reproduction, in which progeny arise from unfertilized eggs. P. occurs naturally in some invertebrates: *Trematoda* (flukes), *Rotifera* (rotifers), *Nematomorpha* (hairworms), *Acarina* (mites), some of the *Araneae* (spiders), and several insect groups such as *Cheleutoptera* (phasmids), *Aphidoidea* (aphids), *Coccoidea* (shield bugs), *Cynipidae* (gall wasps), *Cecidomyiidae* (gall midges), *Apidae* (bees), *Vespidae* (wasps), *Bombinae* (bumble bees), *Formicidae* (ants), some *Curculionidae* (weevils) and *Lepidoptera* (butterflies and moths), etc.

If unfertilized eggs usually perish and only occasionally develop further (e.g. some *Araneae* and *Sphingidae),* the condition is known as *exceptional P.* In those species where unfertilized eggs regularly develop to produce progeny, the condition is known as *normal* or *physiological P.* In *facultative P.,* the eggs develop whether they are fertilized or unfertilized; most cases of facultative P. are also physiological P. In facultative P., unfertilized eggs often develop into only one sex, the other sex arising from fertilized eggs. Thus, in the honey bee *(Apis mellifica)* and other *Hymenoptera,* the males (drones) always develop from unfertilized eggs *(arrhenotoky),* whereas fertilized eggs develop either into sexually fully mature females (queens) or into female workers, depending on the nutrition of their respective larvae. A population of honey bees may become a degenerate "drone brood", if the available stock of sperm becomes exhausted, or a queen is not fertilized. The population suffers a similar fate, if a worker lays eggs in the absence of a queen. In the shield bug, *Lecanium hesperidum,* only females hatch from unfertilized eggs *(thelytoky),* while in some *Lepidoptera* unfertilized eggs develop into either males or females *(amphitoky* or *deuterotoky).* There are many different versions of obligatory P., in which the eggs are never fertilized, or there is at least always one generation in which parthenogenetic development is obligatory: parthenogenetic females may occur over many generations with the occasional and rare appearance of males (e.g. some rotifers and ostracods, and the phasmid, *Carausius morosus),* or there may no males at all (i.e. pure thelotoky) (e.g. some members of the *Plecoptera, Thysanoptera, Homoptera, Coleoptera,* and the family *Ichneumonidae* of the *Hymenoptera). Cyclic P.* or Heterogony (see) represents a special case of an Alternation of generations (see), in which several successive parthenogenetic generations are followed by sexual reproduction. Occasionally in heterogony the respective eggs exhibit dimorphism. Thus, in rotifers and water fleas, certain eggs are capable of fertilization, and if fertilized they are destined to lie dormant through the winter; these eggs are larger, richly endowed with yolk, and develop more slowly. In contrast, the eggs that are laid parthenogenetically in large numbers during the summer are small, have little yolk, and develop rapidly. Especially complicated examples of heterogony have been discovered in the aphids and gall wasps, in which the different generations of parthenogenetically produced females exhibit a pronounced polymorphism and also have different modes of existence (see Alternation of generations for a description of heterogony in *Aphis rumicis).* If sexual reproduction is omitted from the cycle, e.g. because the climate is favorable, as may happen in the case of the vineyard pest, *Pylloxera vitifoliae,* then there is a continual succession of generations of parthenogenetic females.

Finally, obligatory P. may occur in the larval stage, i.e. larvae produce more larvae by P. This was first observed in larvae of the gall midge, *Miastor,* then later in other species, e.g. larvae of *Oligarces,* pupae of *Chironomidae* (midges), larvae of the beetle, *Micromalthus debilis,* and in the juvenile stages of some aphids. The successive larvae of the sheep liver fluke, *Fasciola hepatica,* probably also belong in this category, i.e. development of the redia stage from unfertilized eggs of the sporocysts, and cercaria development from the unfertilized eggs of the rediae. This development is so rapid that it is difficult to determine whether it corresponds to P. or to asexual vegetative reproduction. The deciding factor, however, is the presence of meiosis. Larval P. is part of the phenomenon known as Paedogenesis (see).

P. can be further classified with regard to chromosome complement. In *hemizygoid, generative* or *haploid P.,* the progeny arise from haploid eggs. Thus, honey bee drones are haploid, while the workers and queens are diploid; a similar situation exists in wasps, bumble bees and ants. Haploid parthenogenetic progeny also occur in the shield bugs, *Icerya* and *Trialeurodes,* and in some mites and rotifers. Haploid P. has also been observed in a number of flowering plants, e.g. thorn-apple *(Datura stramonium),* tobacco, rice, maize and wheat. Diploid parthenogenetic progeny (produced by *zygoid, diploid* or *somatic P.)* can arise by various mechanisms: the haploid egg nucleus may fuse with the second polar body; 2 haploid cleavage nuclei may fuse; or the egg may be diploid (i.e. meiosis does not occur during egg formation). Diploid P. is known in the gall wasp, *Neuroterus lenticularis,* in aphids, gall midges, some beetles, roundworms and flukes, as well as in various flowering plants, e.g. dandelion, *Cucurbita,* tobacco, and hops. Diploid P. is much more common than haploid P.

P. also occurs naturally in the *Protozoa,* e.g. during sporogeny in the parasitic *Sporozoa* (see), and in *Paramecium,* if the micronucleus is absent from one conjugant, and the nuclear complement of both new animals is derived from the haploid products of the micronucleus of the other conjugant. In the ciliates, however, formation of a new nucleus from the micronucleus of a would-be conjugant is now considered to be a case of automictic fertilization, rather than P. (with endomixis).

In addition to natural P., it is possible to induce *artificial P.* This has been achieved with representatives of practically all animal phyla, including amphibia and mammals. Nonfertilized eggs can be induced to develop by a variety of means: by exposure to hyper- or hypotonic solutions of sodium chloride, by exposure to various salts (potassium chloride, magnesium chloride, calcium chloride), by treatment with hydrochloric acid, carbonic acid, fatty acids, alkalis, toxins, narcotics, sucrose, diethyl ether, acetone, or citrate, by brief immersion of the eggs in concentrated sulfuric acid, by ultraviolet or radium irradiation, or by brief electrical stimulation, by puncture with a glass or platinum needle, by mechanical agitation, and by temperature change of the medium. In *pseudogamy, gynogenesis,* or *merospermy,* the sperm enters and activates the egg, but it degenerates without its nucleus fusing with that of the egg. Gynogenesis is known to occur naturally in roundworms. It can be performed experimentally in sea urchin or amphibian eggs, which are allowed to be penetrated by sperm of the same species, followed by destruction of the sperm nucleus by irradiation; alternatively, egg development is induced by penetration (without fertilization) by sperm of a different species.

The counterpart of gynogenesis, i.e. *male P.* or *androgenesis,* in which progeny develop from a male gamete, is known to occur naturally only in certain isogamous algae *(Chlamydomonas, Ectocarpus).* Artificial androgenesis has been performed with sea urchin and amphibian sperm, by allowing it to penetrate denucleated egg fragments of the same species, and also by damaging the egg nucleus of a normal sperm-penetrated and activated egg by radium irradiation. Artificial androgenesis has also been performed with mouse sperm, by allowing egg penetration by the sperm nucleus, then removing the female nucleus by suction. Androgenetic development involving egg protoplasm is known as Merogony (see).

Partial pressure: the pressure contributed by a gas in a mixture of gases. The relation between the total pressure of a mixture of gases and the pressures of the individual gases was expressed by J. Dalton (1801) in his *law of P.ps.* The P.p. of each gas in a mixture is defined as the pressure the gas would exert if it alone occupied the entire volume of the mixture at the same temperature. The total pressure of a mixture of gases is equal to the sum of the P.ps. of the constituent gases. Under normal conditions, atmospheric pressure is about 100 kPa. Oxygen accounts for about 21% of the volume of air, and has a P.p of 21.3 kPa. In lung alveoli, the P.p. of oxygen is less than in the external air, because alveolar air contains water vapor with a P.p. of 6.2 kPa at 37 °C. The Oxygen content is then 14% by volume, with a P.p of 13.3 kPa.

Percentage volumes and P.ps. of oxygen and carbon dioxide in human respiratory air are shown in the table.

		Inspired air	Expired air	Alveolar air
O_2	Vol%	21	16	14
	kPa	21.3	15.2	13.3
CO_2	Vol%	0.03	4	6
	kPa	0.03	3.8	5.7

Partner exchange: a change of chromosomal partner by one member of a bivalent. This is a necessary event for the formation of a multivalent, because only two chromosomes can pair at a single site.

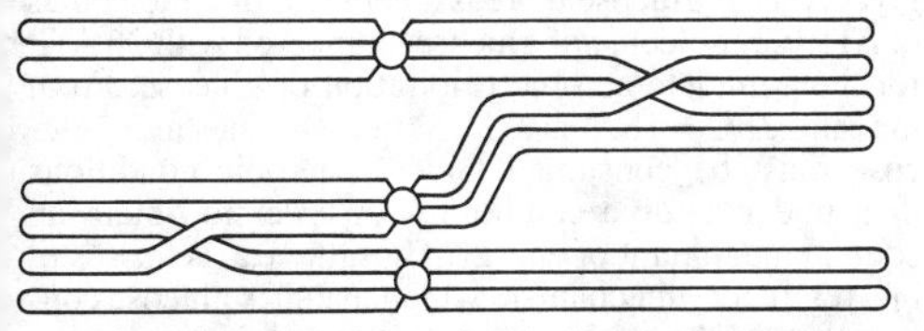

Partner exchange. Change of partner in the formation of a multivalent during the prophase of meiosis.

Partridges: see *Phasianidae.*

Parturition, ***birth:*** expulsion of developed progeny from the mother's body. P. normally occurs when the progeny becomes capable of independent existence outside the mother's body, or when it exceeds the available growth space. The term *premature birth* is applied to P. of progeny that have not attained the usual neonatal level of development, but are nevertheless viable. In contrast, a *miscarriage* produces a nonviable, early fetus. A product of a *still birth* may be quite developed at P., but is dead. The normal level of development at P. varies according to the species. Some animals are still naked and blind, e.g. mouse and rabbit, whereas others have a hairy coat and are able to see and walk, e.g. guinea pig, cattle, horses. Animals that normally give birth to single progeny are called *uniparous* (horse, cattle), whereas those that have several are multiparous (pig, dog, cat). Most animals give birth while lying or sitting. P. is hormonally regulated, in particular by oxytocin from the posterior pituitary. Oxytocin causes contraction of the uterus, which together with abdominal contractions results in expulsion of the progeny. P. can be considered in three phases: 1) relaxation and expansion of the birth canal; 2) contraction phase, with opening of the mouth of the uterus and passage of progeny along the birth canal (delivery); and 3) expulsion of the afterbirth (placenta and embryonal membranes). The duration of P. varies widely according to the species. In horses, the opening of the birth canal takes about 10 hours, and delivery lasts 1/2 hour. In the cow, sheep and goat, the first stage lasts 4–6 hours, delivery 3–4 hours. The actual delivery (starting with movement of the progeny into the birth canal and finishing with its expulsion from the mother) is usually relatively brief: horse (5–30 min), cow (15–60 min), sheep and pig (15–30 min).

Parulidae, ***American*** or ***Wood warblers:*** an avian family of 119 species in the order *Passeriformes* (see). In general appearance, they resemble the Old World warblers *(Sylvidae),* Old World flycatchers *(Muscicapinae),* or small finches, but they are phylogenetically most closely related to the tanagers *(Thraupidae).* Two subfamilies are recognized. The 109 insectivorous species of the *Parulinae* are widespread throughout the Americas. Common plumage colors are yellow, orange, black and white, and several tropical species have some red coloration. Examples are: ***Rose-breasted chat*** *(Granatellus pelzelni;* 14 cm; Venezuela to Bolivia); ***Black-throated blue warbler*** *(Dendroica caerulescens;* 13 cm; eastern North America); ***Prothonotary warbler*** *(protonotaria citrea;* 14 cm; eastern USA); ***Painted redstart*** *(Setophaga picta;* 13 cm; Arizona to Nicaragua).

The 10 species of the *Coerebinae* ***(Bananaquits*** and ***Conebills)*** were formerly given familial status as the *Coerebidae* or ***Honeycreepers;*** at one time they were also included with the buntings and tanagers as a subfamily of the *Emberizidae.* They are essentially insectivorous, but like the Old World sunbirds *(Nectariniidae),* they also take nectar from flowers. They inhabit the lowland forests of the West Indies and Central and South America. The ***Bananaquit*** *(Coereba flaveola;* 11.5 cm) is found throughout the West Indies (except Cuba), and from Southeast Mexico to Paraguay and Argentina.

Parulinae: see *Parulidae.*

Parus tits: see *Paridae.*

Pasque flower: see *Ranunculaceae.*

Passage: transfer of microorganisms to a new culture medium or animal host, etc., for a given time before further transfer or harvest. The term is also applied to cell and tissue culture. P. may be accompanied by a change in environmental conditions. For example, pathogenic bacteria may be transferred from an artificial medium to an experimental animal, then later reisolated from the animal. In this case, the single P. in an experimental animal may be necessary to maintain viability. Alternatively, P. in artificial media can be used to attenuate disease-causing organisms.

Passalidae: see Lamellicorn beetles.

Passenger pigeon: see *Columbiformes.*

Passer: see *Ploceidae.*

Passeres, *Oscines, "song birds":* a suborder of the *Passeriformes* (see). They differ from other passerine birds in having 5–9 pairs of song muscles, and the syrinx is of the trachobronchial type.

Passeriformes, ***passerines, "perching birds":*** a widely distributed order of birds, including more than half of all bird species. All possess 3 forward toes and 1 well developed, nonreversible hindtoe, which enable them to grip a perch. All the toes originate from the same level on the tarsus, and they are generally free at the base. Passerines are generally small to medium sized in comparison with nonpasserines. All members possess 14 cervical vertebrae, except the broadbills, which have 15. A distinct sternal keel is always present, and the appendix is vestigial. The bony palate of the passerines is designed differently from that of most other orders. Nest building is often elaborate, involving weaving or sewing, or the use of mud; many species nest in holes and hollow structures. The young are altricial.

The order is divided into 4 suborders, based on the muscles of the Syrinx (see), and on the flexor tendons of the toes. On account of their joined tendons, the *Eurylaimi* are also referred to as the *Desmodactylidae;* the remaining 3 suborders with free tendons may then be referred to collectively as the *Eleuterodactylae.* Based on the insertion of the song muscles, suborders 1 and 2 may be grouped together as the *Acromyodae,* while suborders 3 and 4 form the *Diacromyodae.* The number of song muscles, however, sets suborder 4 quite apart from the others, earning it the name of *Oscines,* while suborders 1,2 and 3 are called the *Suboscines.*

1. Suborder: *Eurylaimi,* represented by a single family, the *Eurylaimidae* ***(Broadbills)*** (see). The front toes are fused at their bases. The flexor tendons of the third toe (middle front toe) are joined with those of the rear.

2. Suborder: *Tyranni, Clamatores* or ***"noisemakers".*** The flexor tendons of the toes are separate. In the lower syrinx, 1 or 2 pairs of tensor muscles are inserted into the half rings of the trachea, either in the middle, through the entirety, or only at its end. The *Tyranni* are divided into 2 superfamilies.

Superfamily: *Funarioidea,* containing the familes, *Furnariidae* ***(Ovenbirds)*** *(see), Formicariidae* ***(Antbirds)*** *(see), Conopophagidae* ***(Antpipits*** or ***Gnateaters)*** *(see), Dendrocolaptidae* ***(Woodcreepers*** or ***Wood hewers)*** *(see), Rhinocryptidae* ***(Tapaculos)*** *(see).*

Superfamily: *Tyrranoidea,* containing the families, *Pittidae* ***(Pittas)*** (see), *Philepittidae* ***(Asities)*** (see), *Xenicidae* or *Acanthisittidaeae* ***(New Zealand wrens)*** (see), *Tyrranidae* ***(Tyrant flycatchers)*** (see), *Muscicapidae* ***(Flycatchers)*** (see), *Pipridae* ***(Manakins)*** (see), *Cotingidae* ***(Cotingas, Becards, Tityras*** and ***Bellbirds)*** (see).

3. Suborder: *Menurae,* in which the flexor tendons of the toes are separate, and the lower syrinx has 1 or 2 pairs of tensor muscles inserted at both ends of the tracheal half rings. The suborder is represented by only 4 species in the 2 families, *Menuridae* ***(Lyrebirds)*** (see) and *Atrichornithidae* ***(Scrub birds)*** (see).

4. Suborder: *Passeres, Oscines* or ***"true song birds",*** in which the flexor tendons of the toes are separate, and the lower syrinx has 4–9 pairs of tensor muscles inserted at both ends of the tracheal half rings. The suborder *Passeres* contains about 4000 species in about 50 families. All members of the *Passeres* show close affinities. Many distinguishing characters may be functional and secondary rather than phyletic, so that taxonomy is difficult and many disagreements exist. Each of the following families is listed separately: *Alaudidae* ***(Larks),*** *Bombycillidae* ***(Waxwings, Silky flycatchers, Hypocolius),*** *Callaeidae* ***(Wattlebirds),*** *Certhiidae* ***(Treecreepers),*** *Chloropseidae* or *Irenidae* ***(Leafbirds),*** *Cinclidae* ***(Dippers),*** *Corvidae* ***(Crows, Magpies, Jays),*** *Dicaeidae* ***(Flowerpeckers),*** *Dicruridae* ***(Drongos),*** *Drepanididae* ***(Hawaiian honeycreepers),*** *Fringillidae* ***(Finches),*** *Hirundinidae* ***(Swallows, Martins),*** *Icteridae* ***(Icterids),*** *Laniidae* ***(Shrikes),*** *Meliphagidae* ***(Honeyeaters),*** *Mimidae* ***(Mockingbirds*** or ***Mimic thrushes),*** *Motacillidae* ***(Pipits, Wagtails),*** *Muscicapidae* ***(Thrushes, Chats, Robbins, Babblers, Wren tit, Rail babblers, Parrotbills,Gnatcatchers, Old World Warblers, Australian wren warblers, Old World Flycatchers, Thickheads*** or ***Whistlers),*** *Nectariniidae* ***(Sunbirds),*** *Oriolidae* ***(Old World Orioles),*** *Paradisaeidae* ***(Birds of paradise),*** *Paridae* ***(Titmice),*** *Parulidae* ***(American*** or ***Wood warblers),*** *Ploceidae* ***(Weavers, Sparrows),*** *Ptilonorhynchidae* ***(Bower birds),*** *Pycnonotidae* ***(Bulbuls),*** *Remizidae* ***(Penduline tits),*** *Sittidae* ***(Nuthatches),*** *Sturnidae* ***(Starlings, Mynahs, Oxpeckers),*** *Troglodytidae* ***(Wrens),*** *Vangidae* ***(Vanga shrikes),*** *Zosteropidae* ***(White eyes).***

Passerinae: see *Ploceidae.*

Passive resistance: in phytopathology, a type of resistance due to defense mechanisms already present before the host has experience of the pathogen. Axeny (see) is a form of P.r.

Pastern: part of the foot of an ungulate, extending from the fetlock to the hoof.

Pasteur effect: inverse relationship between the rate of glucose utilization and the availability of oxygen. This was first observed in 1860 by L. Pasteur, who noted that yeasts decompose more sugar under anaerobic conditions than they do aerobically. In the anaerobic glycolysis of glucose there is a net yield of 2 molecules of ATP per molecule of glucose, compared with 36 ATP for the complete aerobic respiration of glucose. From the ratio 36/2 = 18, it is evident that 18 times more glucose must be consumed under anaerobic conditions than under aerobic conditions, in order to obtain an equivalent amount of energy. Thus, the P.e. is the result of a regulatory mechanism which matches glucose consumption to the energy needs of the cell. Naturally, the effect is only observed in facultative cells, i.e. cells that can adjust their metabolism to either aerobic or anaerobic conditions, e.g. yeast cells, muscle cells.

In the absence of oxygen, animal cells perform anaerobic glycolysis and produce lactate, whereas yeast cells ferment glucose to various fermentation products, e.g. glycerol and notably ethanol. When oxygen is admitted, the production of lactate quickly ceases and lactate rapidly disappears; at the same time the rate of glucose uptake is markedly decreased. The P.e. can largely be explained by the allosteric properties of phosphofructokinase, the rate-limiting ("pacemaker") enzyme of glycolysis. This enzyme is inhibited by ATP and activated by ADP and AMP. The ATP inhibition is overcome by AMP, cAMP, inorganic phosphate, fructose 1,6-bisphosphate and fructose 6-phosphate, but is increased by citrate. The citrate effect is significant in yeast, and in aerobic muscle which also uses fatty acids and ketone bodies as major energy sources. The effect of cAMP may be inportant in adipose tisue, where it overcomes the citrate inhibition.

The P.e. is especially important in skeletal muscle; the quantity of ATP in muscle (5–7 μmoles per g fresh weight) is sufficient to support contraction for 1–5 seconds. ATP synthesized from creatine phosphate (ADP + creatine phosphate ⇌ ATP + creatine) supports a further short period of contraction. During this early period the muscle becomes relatively anaerobic as the oxygen stored in the myoglobin is consumed and contraction restricts the blood supply by squeezing the blood vessels. Rephosphorylation of ADP to ATP then depends on the anaerobic glycolysis of glycogen to lactate; the activity of phosphofructokinase is markedly increased, with a corresponding increase in the rate of glycolysis. However, this effect cannot simply be explained by the change in level of ATP; in fact the steady state concentration of ATP in contracting muscle is only slightly lower than in resting muscle; for example, in insect flight muscle a 10% decrease in the ATP level results in a 100-fold increase in the rate of glycolysis. Two complementary mechanisms have been proposed for this amplification: 1. Adenylate kinase catalyses the interconversion, 2ADP ⇌ ATP + AMP, with an equilibrium constant of approximately 0.5. Under the steady state conditions in the cell, the ATP concentration is much higher than those of ADP and AMP. Thus a small change in ATP concentration causes a much larger percentage change in AMP concentration. 2. Simultaneous operation of phosphofructokinase and fructose 1,6-bisphosphatase would constitute a futile cycle. In resting muscle phosphofructokinase is largely inhibited by the ATP/AMP ratio. Furthermore, much of the fructose 1,6-bisphosphate that is formed is converted back to frucose 6-phosphate by fructose 1,6-bisphosphatase, i.e. the futile cycle does operate at a low rate, and the net flow rate of glycolysis is equal to the difference between the activities of phosphofructokinase and fructose 1,6-bisphosphatase. When muscle contracts, the small decrease in ATP causes a large relative rise in AMP; the activity of phosphofructokinase increases and that of fructose 1,6-bisphosphatase decreases, and there is a marked increase in the rate of production of fructose 1,6-bisphosphate.

The first reaction of glucose utilization is catalysed by hexokinase, which is inhibited by high concentrations of glucose 6-phosphate. In cells containing a reserve carbohydrate (e.g. glycogen in muscle cells), glucose 6-phosphate is also formed via glucose 1-phosphate. Glucose 6-phosphate accumulates during inhibition of phosphofructokinase in sufficient quantity to inhibit hexokinase, thus preventing even the first stage of glucose utilization.

A quantitative index for the effect of oxygen on glycolysis and fermentation is given by the Meyerhof oxidation quotient:

(Rate of anaerobic fermentation) – (Rate of aerobic respiration) / (Rate of O_2 uptake), which is usually about 2. The consumption of one molecule of oxygen generally inhibits the production of two molecules of lactate or ethanol.

The P.e. is not observed, or only slightly, in malignant tumors, intestinal mucosa, red blood cells (mammalian) and retina (mammalian and especially avian). These tissues obtain their energy by anaerobic degradation of glucose, even in the presence of oxygen.

Uncoupling agents (e.g. dinitrophenol; see Oxidative phosphorylation) abolish the P.e., so that the full rate of anaerobic carbohydrate utilization in maintained in the presence of oxygen.

Pasteurella pestis: see *Yersinia pestis.*

Pasteurization: a method developed by Pasteur for the protective heat treatment of heat-sensitive foodstuffs, especially beverages, which kills the majority of microorganisms present. P. increases the shelf life of the food, as well as killing disease-causing organisms, without greatly affecting the taste or nutritive value (e.g. vitamin content). The method finds its greatest use in the treatment of milk, but it is also applied to fruit juices, wine and beer. The temperature is raised to between 64 and 85 °C for no longer than 30 min. For example, in the P. of milk, the temperature is held at 71–74 °C for 15–40 seconds, or 85 °C for 3 seconds. Since spores and particularly resistant microorganisms are not killed by P., pasteurized foods are not sterile, but contain a very low concentration of microorganisms. See Sterilization.

Patagium: see Flight membranes; see *Sciuridae.*

Patagonian cavy: see *Caviidae.*

Patagonian hare: see *Caviidae.*

Patas monkey, *nisnas monkey, red monkey, hussar monkey, Erythrocebus patas:* the only species of this genus of Old World monkeys (see), and closely allied to the Guenons. It is found in savanna woodland, semidesert, rocky outcrops and the Ennedi Plateau in Chad, feeding mainly on grasses, fruits and seeds, with some insects, lizards and birds' eggs. It has a coarse, shaggy, red-brown coat, with orange underparts and a bright blue scrotum, The face has a white moustache and face color is rather variable. Sexual dimorphism is marked (head plus body: males 625 mm; females 490 mm). Much time is spent on the ground, but trees are used for sleeping.

Patchouli: see *Lamiaceae.*

Patch reef: see Coral reef.

Paternity, determination of: see Forensic anthropology.

Pathogen: the causative agent of a disease; any pathogenic virus, mycoplasma, bacterium, or fungus.

Pathogenesis: the course of a disease from the onset of infection to the end of the diseased condition (return to health or death of the affected organism or its parts). The course may be monocyclic or polycyclic.

Pathogenicity: a general term for the ability of a pathogen to produce the manifestations of disease in its host. The term, Virulence (see), is almost synonymous.

Pathosystem: a parasitic relationship of organisms, which represents a subsystem of an ecosystem.

Pathotoxins, *phytopathogenic toxins:* products (excluding enzymes) of phytopathogenic microorganisms, which in relatively low concentrations damage the cells of higher plants or interfere in their metabolism. They may be host-specific or nonspecific.

Pathotype: in phytopathology, members of a pathogenic species which differ in their genetically determined virulence. The term is frequently used as a synonym of Physiological race (see).

Patriofelis: see *Creodonta.*

Patristic distance: see Divergence index.

Pattern of gene activity: see Gene activity pattern.

Pauropoda, ***pauropods:*** a class of myriapodous arthropods (see *Myriapoda).* They are small (maximal length 1.5–2.0 mm), thin-skinned, soft-bodied animals, with 9 pairs of legs and usually 11 segments. The genital opening is in the third trunk segment. Respiratory gas exchange occurs through the skin. Eyes are absent, and most species lack pigmentation. Dorsal plates are fused to form syntergites, each covering two segments. The head carries two forked (biramous) antennae; one division terminates in a flagellum, the other terminates in two flagella and carries spherical sensory organ. A disk-like sensory organ is present on each side of the head, and this is perhaps homologous with the organ of Tömösvary of other myriapods (see *Symphyla).* There are both slender, long-legged species (e.g. *Pauropus sylvaticus)* and sturdy, short-legged species (e.g. *Eurypauropus ornatus).* They occur widely in temperate and tropical regions, living in leaf litter, under bark, and in the upper layers of the soil. They feed saprophytically on decaying material, or by sucking out the contents of fungal hyphae. Development proceeds through 4 larval stages, which possess, successively 3, 5, 6 and 8 pairs of legs. About 360 species are known, divided among 8 families. Damp forest floors may contain 300–500 individuals per square meter.

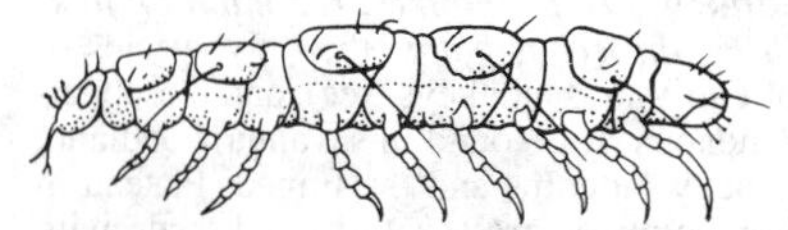

Pauropoda. Pauropus sylvaticus.

Pauropus: see *Pauropoda.*

Pavement epithelium: see Epithelium.

Paxillus: see *Basidiomycetes (Agaricales).*

Peach: see *Rosaceae.*

Peacock: see *Phasianidae.*

Peafowl: see *Phasianidae.*

Pea moth, *Cydia nigricana* F.: a moth of the family *Tortricidae* (see), and a pest of cultivated peas *(Pisum sativum).* Eggs are usually laid on the sepals around young pods. Usually no more than two newly hatched caterpillars bore into and eat away the peas in each pod. Crop damage is often considerable. It is a serious pest in parts of England and Europe, and has been introduced into the USA and Canada.

Peanut: see *Papilionaceae.*

Peanut worms: see *Sipunculida.*

Pear: see *Rosaceae.*

Pearl fish, *Rutilus meidingeri:* an elongated chub-like fish, up to 70 cm in length, with extremely small scales. It belongs to the family *Cyprinidae,* order *Cypriniformes,* and is found in the deep lakes of the Lower Alps. At spawning time, the male produces a pearly milt and has a red belly. It is a popular sporting fish.

Pearlfishes: see *Carapidae.*

Pearl mussel: see Freshwater mussels, Pearls.

Pearls: calcareous bodies secreted by the mantle epithelium of bivalves, e.g. the marine pearl oyster, *Pteria margaritifera* and the freshwater pearl mussel, *Margaritifera (= Margaritana) margaritifera.* Several other species also occasionally produce P., e.g. the pen shell, edible mussels and edible oysters. They are rarely formed by gastropods (e.g. the ormer, *Haliotis tuberculata* L.) or cephalopods (e.g. *Nautilis).* P. are formed as a protections against foreign bodies such as sand grains, or parasites, which are encapsulated by successive layers of pearly aragonite (calcium carbonate) similar to the internal mother-of-pearl layer of the bivalve shell. P. also contain a small proportion of conchiolin, the main horny organic constituent of the outer layer of the mollusk shell. In Japan and China, pearls are cultured by the introduction of foreign bodies between the shell and mantle of oysters. Natural marine P. ("Oriental" P.) originate mainly from the Indopacific ocean, whereas natural freshwater P. are collected in central Europe (southern Germany) and from certain freshwater locations in the USA. P. may be attached to the inside surface of the shell, or they may develop unattached between the mantle and the shell. P. are prized as articles of jewellery, and the most valuable consist of concentric layers around a single nucleus. Otherwise their value depends on size, evenness, shape, color and luster.

Pearly nautilus: see *Cephalopoda.*

Pear midge: see Gall midges.

Peat: residues of bog plants, mainly *Sphagnum,* which are only partially decomposed due to the absence of oxygen. See Humus.

Peat formation: a process which occurs when the decay of plant materials is strongly inhibited by low temperatures and water immersion. The accompanying oxygen deficiency and acid production inhibit organisms that would otherwise cause rapid decomposition. Dead vegetation remains on the surface and is not drawn into the soil by burrowing fauna (e.g. earthworms). The decomposition that does occur is due mainly to the activity of fungi, and it is only partial. It stops with the formation of a dark, compact mass of humic materials, known as peat.

Peat moss: see *Musci.*

Pebrine: see *Cnidosporidia.*

Pecan: see *Juglandaceae.*

Peckhammian mimicry: see Protective adaptations.

Peck order: see Rank order.

Pectase, ***pectin esterase:*** an esterase that catalyses the hydrolysis of pectins into pectinic acids and methanol. It is widely distributed in plant tissues.

Pectinase: a glycosidase that catalyses the hydrolysis of pectic acids into galacturonic acid. It occurs in seeds and in various molds. It is active during the retting of flax and hemp, and in the fermentation of tobacco and tea.

Pectinate antennae: see Antennae of insects.

Pectinate hyphae: fungal hyphae which send out branches mostly on only one side. They are common in certain dermatophyes.

Pectinator: see *Ctenodactylidae.*

Pectinibranchia: see *Gastropoda.*

Pectinic acids: see Pectins.

Pectins: high molecular mass polyuronides, consisting of α1,4 glycosidically linked D-galacturonic acid residues. Some of the carboxyl groups are present as their methyl esters. The free acids are called *pectinic acids.* Varying extents of esterification and polymerization give rise to a wide variety of P. They are distributed widely in plants, and are found in association with cellulose; they are structurally important as cement and support substances, particularly in the middle lamellae (where they occur as calcium pectinate) and in primary walls of plant cells (where they occur as insoluble protopectin). P.are especially abundant in fleshy fruits, roots, leaves and green stems. They are prepared from sliced sugar beet and apple and citrus residues (following juice extraction) by gentle extraction with hydrochloric, lactic or citric acids.

Due to the presence of hydrophilic groups, P. have a high capacity for binding water. They are therefore powerful gelling agents, a property which is widely exploited in the food industry, where they are used as setting agents in jams, etc., and in the preparation of pharmaceuticals and cosmetics.

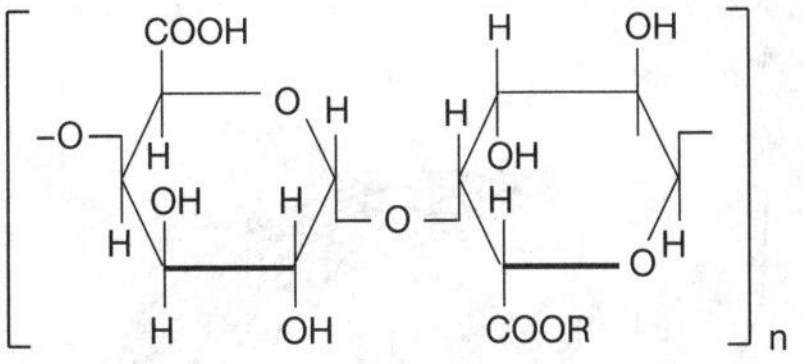

Pectin (R = H or CH_3)

Pectoral girdle, ***shoulder girdle:*** the skeletal structure of vertebrates which serves for attachment of the forelimbs. In contrast to the pelvic girdle, the P.g. is connected only indirectly to the axial skeleton, via the thorax. In its fully evolved form, it consists of 3 pairs of bones: *scapula* (shoulder blade), *clavicle* (collar bone) and *coracoid bone.*

In birds the clavicles are fused at their lower ends to form a V-shaped *furcula* (wishbone), the coracoid bones are joined to the *sternum,* and the scapulas are long and slender and lie on the thorax parallel to the vertebral column.

With the exception of the *Monotremata,* mammals lack separate coracoid bones, but the coracoid process on each scapula is interpreted as a vestige of a coracoid bone. The clavicles, inserted between the forelimb and sternum (in humans each is bent into a flat S-shape), may also be degenerate (e.g. in the *Carnivora),* or absent (e.g. in whales and ungulates). The scapulas are always present, and situated in the dorsal musculature. Each mammalian scapula is triangular in shape, with a ridge on the outer surface, and it carries an articulatory surface for the forelimb.

In fishes, the P.g. is joined to the head. Its various elements consist of more or less broad plates. In addition to the three basic components mentioned above, there is always a *cleithrum* (a covering bone, which is not present in amniotes) and a variable number of other epidermal bones, which are the remnants of the protective armor of the primitive fossil *Placodermi.*

The amphibian P.g. is described in detail under *Anura* (see).

Pedal gland, ***mucous gland:*** a gland present in snails, which runs dorsal to the foot and opens ventral to the mouth. It produces copious slime, which spreads out as a smooth layer and serves as a lubricant for the progress of the animal.

Pedetidae, ***spring hare:*** a family of Rodents (see) represented by a single species *(Pedetes capensis, Spring hare),* found in semi-arid areas of eastern and southern Africa. The size of a large rabbit, the spring hare has short and robust forelimbs with long claws, which are used for digging elaborate burrows. The tail is long with thick fur, and the eyes are exceptionally large. The long, powerful hindlimbs enable the animal to leap more than 6 m, although it moves slowly on all four feet when feeding. The fibula is reduced and fused distally with the tibia, and only 4 toes are present on each foot. As in several saltatorial rodents, the cervical vertebrae are partially fused. An anterior division of the medial masseter passes through the infraorbital foramen, which is extremely large. Dental formula: 1/1, 0/0, 1/1, 3/3 = 20.

Pedetes: see *Peditidae.*

Pedicel:

1) In spiders, a short, narrow part of the body connecting the prosoma and opisthosoma.

2) In insects, the second segment of the antenna (see Antennae of insects).

3) In bees and wasps, the short, narrow part of the body or "wasp waist" between the thorax and abdomen, where it is also known as the petiole. In the *Hymenoptera,* the first abdominal segment forms part of the thorax, and it is separated from the gaster (abdomen without the first abdominal segment) by the petiole.

4) A scale-like or nodular segment on the abdominal stalk of an adult ant.

Pedicellaria: see *Entoprocta.*

Pedicellariae: small stalked appendages attached to turbercles on the calcareous skeletons of the *Echinoidea* and *Asteroidea.* Each consists of a stem with a movable head built of three or two pincer-like valves. They serve for grasping, defense and scavenging. In many species they are provided with venom glands.

Pedicellinidae: see *Entoprocta.*

Pediculati: see *Lophiiformes.*

Pedionomidae: see *Gruiformes.*

Pedionomus: see *Gruiformes.*

Pedipalpi: an order of arachnids, of which the *Uropygi* and *Amblypygi* are considered to be suborders. The body is either elongated (scorpion-like uropygids) or compact (spider-like amblypygids), varying in length between 0.7 and 7.5 cm. The pedipalps are developed either as raptorial organs or they form a spiny capture basket. Each of the first pair of walking legs has a multisegmented, whiplike tarsus which serves as a tactile organ. There are about 190 species, mostly in the tropics and subtropics, a few being found in northerly parts of Asia.

Pedogamy: formation of a zygote by fusion of 2 gametes from the same individual, but not necessarily from the same meiosis. See Automixis, Autogamy, Fertilization, Reproduction.

Peewit: see *Charadriidae.*

Pegusa: see *Solea.*

Peireskia: see *Cactaceae.*

Pelagia: see *Semaeostomae.*

Pelagial (Greek *pelagos* the high seas): the open water of the ocean and inland freshwater lakes. It is inhabited by plankton, nekton, neuston and pleuston. The upper, illuminated (penetrated by sunlight) part of the P. is called the *epipelagial,* and the lower, dark (too deep for sunlight to penetrate) part is known as the *bathypelagial.* The region between 3 000 and 6 000 m below the sea surface is called the *abyssopelagial.* See Lake, Sea.

Pelamis: see *Hydrophiidae.*

Pelargonidin: 3,5,7,4′-tetrahydroxyflavylium cation, an aglycon of many anthocyanins (see). Its various glycosides are widely distributed in higher plants, where they are responsible for the rose, orange-red and scarlet colors of petals and fruits. About 25 natural glycosides are known, e.g. callistephin (3-β-glucoside) of the red aster and *Dianthus* (pinks and sweet william); pelargonin (3,5-di-β-glucoside) of various *Pelargonium* species and red garden Dahlias; fragarin (3-β-galactoside) of strawberry.

Pelargonium: see *Geranaceae.*

Pelea: see *Reduncinae.*

Pelecanoididae: see *Procellariiformes.*

Pelecanoides: see *Procellariiformes.*

Pelecanus: see *Pelicaniformes.*

Pelecus: see Ziege.

Pelecypoda: see *Bivalvia.*

Peleini: see *Reduncinae.*

Pelican eels: see *Saccopharyngoidei.*

Pelicanidae: see *Pelicaniformes.*

Pelecaniformes, *Steganopodes:* an order of birds, comprising about 55 species in 6 families.

Classification:

1. Suborder: *Phaethones,* containing the single family *Phaethonidae* **(*Tropic birds*).**

2. Suborder: *Pelecani,* divided into two superfamiles: *Pelecanoidea* [containing the single family *Pelecanidae* **(*Pelicans*)]** and *Suloidea* [comprising the three families *Sulidae* **(*Gannets and Boobies*),** *Phalacrocoracidae* **(*Cormorants*),** and *Anhingidae* **(*Darter*)**].

3. Suborder: *Fregatae,* containing the single family *Fregatidae* **(*Frigate birds*).**

The order is cosmopolitan in distribution, although some familes are restricted to southern latitudes. All members are rather large, fish-eating birds, closely associated with water. All 4 toes are webbed, the hallux being turned forward and connected with the second digit. Except in the *Phaethonidae* and *Fregatidae,* the 8th and 9th of the 17–20 cervical vertebrae are shaped so that they articulate at an angle with both neighbors. Nesting usually occurs in colonies, sometimes of immense size.

The ***Tropic birds*** are represented by a single genus *(Phaethon)* and 3 species, found over tropical and subtropical waters. They nest mainly on oceanic and offshore islands, flying far from land between breeding seasons. Their long central tail feathers may be longer than the body, giving a characteristic profile in flight. They are usually white with a black bar across the eyes and black markings on all or some of the flight feathers. A subspecies of *P. lepturus (P. lepturus fulvus)* has orange-pink plumage. The stout bill is slightly decurved and usually brightly colored. The webbed feet serve for paddling and for scraping a shallow nest, but movement over ground is achieved by shuffling, because the short legs cannot support the weight of the body on land. Nests are usually made in cavities and on ledges on the sides of high, steep cliffs. A single egg is incubated in turns by both parents. They feed on fish and squid, which are captured by diving into the water from flight. The ***Red-tailed tropic bird*** *(P. rubracauda),* of the southern tropical Indian Ocean and the tropical Pacific, is the largest of the 3 species, measuring 46 cm in length (excluding the central tail feathers), with a wingspan of 89–92 cm, and an orange to red bill. The smaller (34 cm in length) ***Red-billed tropic bird*** *(P. aethereus)* has a coral-red bill and barred plumage. It ranges from the Galapagos through the Caribbean and tropical Atlantic into the Red Sea and northern Indian Ocean. The slightly smaller members of the populations in the Indian Ocean and the Gulf of Aden are sometimes regarded as a separate species *(P. indicus).* The ***Yellow-billed*** or ***White-tailed tropic bird*** *(P.lepturus)* (about 30.5 cm in length) is found in the Caribbean, Atlantic, Indian Ocean and Southwest Pacific. The adult plumage is white or tinted, but with no barring.

Fig.1. *Pelecaniformes.* Dalmatian pelican *(Pelecanus crispus).*

Depending on taxonomic opinion, there are between 6 and 8 species of ***Pelican,*** all of the genus *Pelecanus.* The genus is widely distributed in the tropics, also occurring in some temperate parts. Pelicans are some of the largest birds (125–183 cm in length), and they are all adapted to an aquatic existence. The upper mandible, which is long, straight and flattened with a median ridge, ends in a hook; in certain species it carries a small "horn" in the breeding season. From the lower mandible is suspended a large distensible pouch, which serves as "dredge" for fish. The pelican's large bill and pouch are its most immediately recognizable features, making it the subject of legend and fairy tales, and instantly distinguishable from all other birds. The feet and the short, thick legs are powerful. Only a short tail is present, but the wings are large and pelicans are strong fliers. Adult plumage is mostly white, sometimes suffused with pink in the breeding season, but dark markings also occur. Crests are present in some species, and in some the bill, pouch and naked part of the face are brightly colored yellow or red. There is little or no sexual dimorphism. Immature birds are brownish in color. Pelicans are very gregarious birds. They display cooperative fishing behavior, in which they remain in a line while advancing and sweeping shallow water with their pouched bills, or they may swim together, plunging their heads beneath the surface in unison (see Allelomimetic behavior). Breeding is colonial, and nests are built in a variety of sites, depending on the immediate environment, i.e. in swamps, on the sides of cliffs, on small islands, in trees. Both parents care for the young, which feed by thrusting their bills into the adult's gullet (the pouch is not used to transport or store food).

The 5 species of Old World species of pelican are found mostly on inland waters, but they also occur on estuaries and coastal lagoons. 1. *P. crispus* ***(Dalmatian pelican)*** ranges from extreme southeastern Europe to Central Asia, migrating to Egypt and northern India in the Winter. 2. *P. onocrotalus* ***(White, Rosy*** or ***Spotted-billed pelican)*** exists as 2 subspecies: *P.o. onocrotalus* (parts of Africa, northern India, southeastern Europe) and *P.o. roseus* [East Africa (migrating to Southeast Asia in winter) and South Africa]. 3. *P. rufescens* ***(Pink-backed pelican)*** is a smaller species, with darker wings, and a pink back in the breeding season. The commonest African species, it is widely distributed southward from 16 ° north of the equator; it is also found in southern India and Madagascar. Nests are usually made in trees, often baobobs, side by side with nests of the Maribou stalk *(Leptoptilos crumeniferus)*. 4. *P. philippensis* ***(Gray pelican)*** of Southeast Asia is rather similar to *P. rufescens*. 5. Two races of *P. conspicillatus* ***(Australian pelican)*** are found in Australia, New Guinea and neighboring islands.

Only 2 species of pelican are found in the New World. *P. occidentalis* ***(Brown pelican)*** is the one species of pelican fully adapted to a maritime existence. It has an extremely wide distribution, with at least one subspecies breeding in southern USA and the West Indies, and another ranging from British Columbia to the Galapagos and Chile. It has dark plumage and is the smallest species of the genus. Unlike any other pelican, it is a sea bird, and it catches fish by falling into the sea from a height of several meters, rather in the fashion of a gannet. *P. erythrorhynchos* ***(White*** or ***Rough-billed pelican)*** breeds from as far north as the Great Slave Lake to as far south as southern Texas; migratory routes extend as far as Central America.

Gannets and ***Boobies*** are large (66–102 cm in length), heavily built sea birds, with a wedge-shaped tail and long, pointed wings. The stout bill is of moderate length, tapering toward a slightly decurved tip. Adult plumage is commonly white, sometimes with some dark feathers in the wings and tail. They lead an exclusively maritime existence, but they are offshore rather than oceanic. Food consists mainly of fish and squid, which they capture by dropping into the sea with partly closed wings from as high as 30 meters, or by diving from the surface. Nesting colonies are often extremely large and densely populated, situated on the ledges of cliffs, or on the flat tops of inaccessible islands; some nest in trees or bushes. Most species lay only a single egg, and although some may lay 2–3 eggs, only one chick is reared. Both parents incubate and care for the young.

Gannets and boobies belong to a single genus *(Sula)* with 9 species. *S. (= Morus) bassama* ***(North Atlantic gannet)*** breeds on both sides of the Atlantic in the Northern Hemisphere. *S. capensis* ***(Cape gannet*** or ***Malagash)*** breeds off South Africa and migrates northward up both coasts of Africa. *S. serrator* ***(Australian gannet)*** breeds in the Bass Strait and off Tasmania, and on the coast of the North Island of New Zealand. *S. leucogaster* ***(Brown booby)*** is found in the Caribbean and tropical Atlantic, and westward through the Pacific to northern Australia. Its head and upper parts are mainly dark, and the naked facial skin is dark blue in the male and bright yellow in the female. *S. sula* ***(Red-footed booby)*** and *S. dactylata* ***(Masked, Blue-faced*** or ***White booby)*** are both found in the tropical regions of the Atlantic, Pacific and Indian Oceans; *S. sula, dactylata* and *leucogaster* all breed on Ascension Island. *S. nebouxii* ***(Blue-footed booby)*** ranges from Mexico to Peru and the Galapagos. The lower back plumage of *S. variegata* ***(Peruvian booby;*** off Peru and Chile) and *S. abbotti* ***(Abbott's booby;*** a true nester, found in tropical regions of the Indian ocean) is chequered black and white.

Fig.2. *Pelecaniformes.* Flightless cormorant *(Nannopterus harrisi)* of the Galapagos Islands (after Beebe).

Cormorants or ***Shags*** have a practically worldwide distribution, being found on coasts from Greenland to sub-Antarctic islands (but not in the central Pacific) and on the inland waters of most land masses (but absent from northern Asia). These medium-sized birds (47–102 cm in length) have a long neck and body, a slender, cylindrical and strongly hooked bill, short legs, large feet, relatively short wings, and a rather long, stiff tail. The plumage is mainly dark, often black with a greenish or bluish metallic sheen. White underparts and throat markings are present in some species, and some species are predominantly gray in color. The iris, bill and naked facial skin are also sometimes brightly colored (red, orange, yellow, green, blue), and in a few species the legs are yellow or red. There is little or no sexual dimorphism. Fish are caught by underwater pursuit after diving from the surface. Most species are colonial breeders, nesting on rocky islets, cliff ledges or trees. Thirty species have been described in 2–3 genera. The largest and most widely distributed species is *Phalacrocorax carbo* ***(Common, Great*** or ***Black cormorant),*** whose breeding range extends from eastern Canada and Greenland through Europe, Africa and much of Asia to Australia and New Zealand. The adult plumage is mainly black with some bronze and dark gray, with small white patches on the throat; during breeding there is also a short head crest. Both *P. carbo* and *P. capillatus* ***(Japanese cormorant)*** have been domesticated for fishing in the east; a leather collar prevents them from swallowing the fish, which are collected by the fishermen. *P. neglectus* ***(Bank cormorant)*** is found around the coast of South Africa. *P. aristotelis* ***(Green cormorant*** or ***Shag),*** found in Europe (including Britain) and Northwest Africa, is smaller, with a green sheen to its plumage. In North America, the most widespread species is *P. auritus* ***(Double crested cormorant),*** while *P.*

penicillatus ***(Brandt's cormorant)*** is the most common species on the Pacific coast of the USA. *P. olivaceus* ***(Olivaceous cormorant)*** ranges from Mexico to Tierra del Fuego. *P. urile* ***(Red-faced cormorant)*** breeds in the Bering Sea and migrates to Taiwan in the winter. *P. varius* ***(Pied*** or ***Yellow-faced cormorant)*** inhabits swamps, rivers, and coastal areas of Australia. Four smaller species, possessing relatively longer tails, and inhabiting mainly rivers and lakes in the tropics and subtropics, are often placed in the separate genus, *Halietor: H. pygmaeus* ***(Pigmy cormorant;*** Hungary to Algeria to Afghanistan); *H. africanus* ***(Long-tailed shag*** or ***Reed duiker;*** large rivers and lakes of Africa and Madagascar); *H. niger* ***(Little cormorant;*** India to Borneo); *H. melanoleucos* ***(Little pied cormorant;*** New Guinea, Australia and New Zealand). The ***Galapagos flightless cormorant*** *(Nannopterum harrisi)* of the Galapagos Islands, one of the largest species, is assigned to a separate genus. It has mainly dark plumage, and its irises are dark blue. The wings are much reduced and the sternum has no keel. Like all other cormorants, it is fast and agile in the water, but ungainly and vulnerable on land.

Fig.3. *Pelecaniformes.* Common cormorant *(Phalacrocorax carbo).*

Darters belong to a single genus *(Anhinger),* with about 6 different geographical forms, which have been variously classified as a single or 4 different species. Some authorities include the darters with the cormorants, but they display certain fundamental differences that justify their separate classification. Thus, darters possess a single carotid artery (cormorants have 2), and their stomach has a pyloric lobe lined with hair-like processes. In addition, the unusual arrangement of the cervical vertebrae of the darter enables the bill (pointed, not hooked) to be projected forward like a triggered spear. The head is relatively small, and the neck is very long and slender. Darters are found only on inland or tidal waters in the tropics and subtropics, where they inhabit wooded shores. They nest in colonies in trees or bushes near to the water. Both parents incubate the eggs and care for the young. Fish and other water animals are caught by underwater pursuit. They resemble cormorants very closely in their habits. Various races of *Anhinger anhinger* ***(Anhinger*** or ***Snake bird)*** are found from southern USA to northern Argentina. *A. rufa* ***(African darter)*** is found throughout sub-Saharan Africa. *A. melanogaster* ***(Indian darter)*** is distributed in southern Asia from India to Celebes. *A. novaehollandiae* ***(Australian darter)*** is found in New Guinea and Australia.

Frigate birds or ***Man-of-war birds*** are found in tropical and subtropical oceanic areas, and they remain rather close to their breeding ground throughout the year, although individual birds have been sighted 800 km (500 miles) from land. The wings are very large (in the larger species, the wingspan is greater than 2 meters), and the bill is long and hooked. They are powerful and agile fliers, adept at catching flying fish as they leave the water, and often taking material from the sea surface. Their plumage is not well oiled and easily becomes saturated. They therefore rarely land on the water. The defenseless chicks of terns and other small sea birds also serve as food, and frigate birds are famous for their habit of chasing home-coming sea birds, especially boobies, and forcing them to disgorge their catch, which the frigate bird then catches in mid air. There is marked sexual dimorphism, the female being considerably larger. In the breeding season, the male develops a crimson, inflatable gular sac. Both parents incubate the single, large, white egg and care for the chick. With the existence of many isolated populations, the taxonomy is uncertain, but 5 species are tentatively listed. The largest is *Fregata magnificens* ***(Magnificent frigate bird;*** Galapagos and eastern Pacific to the Caribbean, Cape Verde Islands and West African coast). *F. andrewsi* ***(Christmas Island frigate bird)*** is found on Christmas Island and elsewhere on islands in the eastern Indian ocean. The population of *F. aquila* ***(Ascension Island frigate bird),*** found only on Ascension Island, is 2–3000. Also recognized are *F. ariel* ***(Lesser frigate bird)*** and *F. minor* ***(Great frigate bird),*** both found in the Indian and Pacific Oceans and in the southern Atlantic.

Pelicans: see *Pelecaniformes.*

Pellia: see *Hepaticae.*

Pellicle: a proteinaceous membrane of fixed shape immediately below the cell membrane and surrounding the cytoplasm of many protozoa. It is permeable to metabolites. Electron microscopy usually reveals 3 layers.

Pelobatidae, ***pelobatids, spadefoot toads*** and ***horned toads:*** a family of the *Anura* (see) containing 83 living species in 9 genera and 2 subfamilies *(Pelobatinae*, distributed in Europe, West Asia, North Africa, southern North America to southern Mexico; *Megophryinae*, distributed in Indo-China, northwards into China and southwards into the East Indies, including Sumatra, Borneo and the Philippines). A third, extinct subfamily (*Eopelobatinae*) is also recognised.

Pelobatids are toad-like in appearance, but have webbed feet. Free ribs are absent and the pupils are vertical and elliptical. As a modification for digging, each foot carries a broad, sharp-edged tubercle, the bony core of which is formed by the greatly enlarged prehallux. Vocal sacs are absent, and the spawn is deposited in short, thick strings. There are 8 stegochordal presacral vertebrae with imbricate (overlapping) neural arches. Ossified intervertebral disks are present, and in adult pelobatids these fuse with the centrum to form procoelous vertebrae; the disks remain free in megophyrins. Presacrals I and II are not fused, and the atlantal cotyles of presacral I are closely juxtaposed. The sacrum is fused with the coccyx. In most taxa the sternum is ossified, the omosternum is always cartilaginous, and the pectoral girdle is arciferal. Maxillary and premaxillary teeth are present.

Pelobates fuscus ***(European common spadefoot;*** subfamily *Pelobatina)* grows up to 8 cm in length, and is colored gray-brown with darker patches and small brick-red spots. It is nocturnal and hides during the day in self-excavated holes, usually in sandy soil. Spawn is laid in 2 short, thick strings, and the large tadpoles (up to 17 cm in length) sometimes overwinter before the completion of metamorphosis.

Members of the North American genus, *Scaphiopus* (subfamily *Pelobatinae),* are also nocturnal. They bury themselves by backing into the ground and digging with the hindfeet, using their extremely well developed "spades". *Scaphiopus hammondii hammondii* ***(Western spadefoot toad, Hammond's spadefoot, Hammond's spea;*** California to Lower California, eastwards through Arizona, Southwest Colorado, New Mexico to Texas and Mexico) ranges in size from 3.7 to 6 cm, and is noted for its call, which has been described variously as "rolling or bubbling, the loud purr of a cat with the metallic sound of grinding gears, weird, plaintive, ventriloqual". Its relatively smooth skin is dotted with tubercles. Forward underparts are white or buffy, while the rear underparts are purplish. The back is greenish with green spots and the sides are yellowish, light gray or yellow-green. Breeding from mid February to August is dependent on heavy rainfall. Eggs are laid in cylindrical masses attached to grass or plant stems. The tadpoles sometimes prey on their own kind, and they are also effective in the control of mosquito larvae.

Scaphiopus couchii ***(Southern spadefoot, Couch's spadefoot, Sonoran spadefoot, Cape St. Lucas spadefoot;*** Texas to Arizona and Utah, Northern Mexico and Lower California) breeds from April to August in temporary pools after heavy rainfall, and its black tadpoles are dotted with old gold or fawn. The adults are nocturnal, living during hte day in self-excavated subterranean burrows, or under logs and stones, emerging only after heavy rain. The call is often heard in chorus and has been described as "harsh and noisy, a great caterwauling, like someon in pain".

Further similar species and subspecies of North American *Scaphiopus* are, e.g. *S. hammondii bombifrom* ***(Central plains spadefoot, Cope's spea, Western spadefoot;*** range extending southwards from the Canadian border in the north and slightly overlapping the ranges of *S. holbrookii hurterii* and *S. hammondii hammondii* at its southern limit); *S. hammondii intermontanus* ***(Great Basin spadefoot, Western spadefoot;*** Utah, Nevada, Northern Arizona, Northern Colorado, Southwest Wyoming, Idaho, Southwest Washington, East Oregon, and extending northwards into British Columbia); *S. holbrookii holbrookii* ***(Holbrook's eastern spadefoot, Hermit spadefoot;*** West Virginia and Ohio along the Ohio River to Southeast Missouri and Northeast Arkansas to extreme Southeast of Texas, along the coast to Massachusetts); *S. holbrookii albus* ***(Key West spadefoot);*** Florida Keys and extreme south of Florida); *S. holbrookii hurterii* ***(Hurter's spadefoot, Hurter's solitary spadefoot;*** eastern half of Texas).

More than 40 species of spadefoot toads occur in Southeast Asia and the Indo-Australian archipelago. They differ systematically from the European and North American spadefoots in that they possess free intervertebral elements, and are therefore assigned to a separate subfamily, the *Megophryinae.* In *Megaphrys monticola* ***(Malaysian horned toad),*** the maxillary teeth are well developed, but vomerine teeth may be reduced or absent. Two subspecies are recognized: *M. monticola monticola* and *M. monticola nasuta,* the latter being the "large-nosed form". They are found in Thailand, Malaysia, Indonesia, Borneo and the Philippines. In *M. monticola nasuta,* sensitive folds of skin protrude beyond the snout and above the eyes. Coloration is generally light and dark brown, providing camouflage in the jungle habitat.

Pelodytidae, ***pelodytids, parsley frogs:*** a Eurasian family of the *Anura* (see) containing only 2 small (50 mm), living species in a single genus.

There are 8 stegochordal presacral vertebrae with procoelous centra and imbricate (overlapping) neural arches. The arches of presacrals I and II are fused and the atlantal cotyles of presacral I are closely juxtaposed. Proximal transverse processes are present on the coccyx, which has a bicondylar articulation with the sacrum. The sternum is ossified, the omosternum is cartilaginous, and the pectoral girdle is arciferal. Ribs are absent and the pupils are vertically elliptical. Maxillary and premaxillary teeth are present. The hindlegs provide a powerful jump, and the hindfeet lack digging structures.

The slender, 4 cm-long ***Western parsley frog*** *(Pelodytes punctatus)* occurs from Spain to Belgium and Northwest Italy; its name derives from the speckled green of its back, which resembles chopped parsley. *Pelodytes caucasicus* from the northern Caucasus and western Transcaucasus is a little known species.

Pelomedusidae, ***Afro-American side-necked*** or ***snake-necked turtles:*** a family of the order *Chelonia* (see), containing 24 species in 5 genera, distributed in South America , Africa and Madagascar. The neck is withdrawn sideways beneath the carapace, and there is no cervical scute on the anterior edge of the carapace. In some species the plastron has a transverse joint. The meat and eggs of several species are economically very important as a food source, especially in South America.

The genus, *Pelusios,* is represented in Africa and Madagascar. Members are up to 40 cm in length, with a deep brown or black carapace and a transverse joint behind the anterior third of the plastron. Members are almost exclusively aquatic, feeding on fish and small aquatic animals. *Pelusios niger* ***(African black terrapin)*** is one of the commonest species of turtle in West Africa.

The genus, *Podocnemis,* is represented in South America, with a single species in Madagascar. Members are almost exclusively aquatic, feeding on fish and small aquatic animals, as well as vegetation. *Podocnemis expansa* ***(Arrau;*** length 90 cm) is a threatened species of the Amazon and Orinoco basin. The eggs, which are are laid on small sandy islands, used to be collected in large numbers for their oil (over 40 million per year).

Pelomyxa palustris: see Amebas.

Peloric condition: see Flower, section 4) Floral symmetry.

Peltogaster: see *Cirripedia.*

Pelusios: see *Pelomedusidae.*

Pelvic girdle, ***pelvis:*** a skeletal support structure, present in vertebrates, which serves for the attachment of the hindlimbs to the axial skeleton. In fishes, the P.g. is incompletely developed, and exists only as a rod of bone or cartilage (the ischiopubic bar, embedded in the ventral body muscles), which serves for the attachment of the pelvic fin musculature, and in elasmobranchs is articulated with the cartilaginous skeleton of the fins. In all vertebrates, the P.g. is primarily associated with the cloacal region, or the region of exit of the separate urinary and genital ducts. In quadrupeds, the P.g. (unlike the pectoral girdle) is linked crosswise to the axial skeleton. In birds and mammals, a *sacrum* (the largest bony structure of the axial skeleton) is formed by fusion of sacral vertebrae. Linkage of the P.g. to the axial skeleton occurs via the *ilium,* which is attached to the sacrum. The number of fused sacral vertebrae carrying the P.g. (i.e. united with the ilium) varies in different vertebrates: 1 in amphibians, 2 in reptiles, up to 23 in birds, and 2–6 in mammals. Each half of the P.g. consists of three elements: the dorsal *ilium,* the ventral-anterior *pubic bone* or *pubis* (plur. *pubes),* and the posterior *ischium.* Right and left halves of the P.g. are articulated ventrally at the *pelvic symphysis,* which is absent from birds (except the ostrich and rhea) and the limbless vertebrates (e.g. snakes, whales and sea cows). The pelvic symphysis of mammals is generally formed only by the pubic bones, and is known as the *pubic symphysis.* In mammals, the ilium, ischium and pubis of each half girdle are fused to form the large *innominate bone* or *hip bone;* all three elements meet and alkylose in a socket called the *acetabulum* (cotyloid cavity) into which the head of the femur fits. For practical purposes in human anatomy the *coccyx* is also considered as part of the P.g.

Pemphigus: see Epithelium.

Penaeidae: see Shrimps.

Penaeidea: primitive shrimps, an infraorder of the *Decapoda* (see). See Shrimps.

Penaeus: see Shrimps.

Pendular movements of the intestine: see Intestinal motor activity.

Penduline tits see *Remizidae.*

Pendunculate bodies: see Corpora pendunculata.

Penellus: see *Copepoda.*

Penelope: see *Cracidae.*

Peneroplis: see *Foraminiferida.*

Penicillata: see *Diplopoda.*

Penicillinases: enzymes, present in many microorganisms, especially bacteria, which catalyse hydrolysis of the β-lactam ring of penicillins. The resulting penicilloic acids have no antibiotic activity. Production of P. is adaptive, and P.-producing (i.e. penicillin-resistant) strains can be isolated by treating bacterial populations with penicillin. They are single chain enzymes, and both membrane-bound and soluble forms are known.

Penicillins: sulfur-containing antibiotics produced by fungi of the genera *Penicillium, Aspergillus, Trichophyton* and *Epidermophyton.* All P. contain a condensed β-lactam-thiazolidine ring system, whereas the acyl group (R) is variable.

The growth inhibitory properties of P. were first observed in 1928 in a *Staphylococcus* culture by Alexander Fleming, and its use against bacterial infections in humans was first tested in 1941. Against sensitive bacteria, penicillin G is active at a dilution of 1:50 000 000; 1 mg of pencillin G corresponds to 1 670 international units. P. kill cocci and Gram-positive bacteria, including many pathogens. They are unstable to acids, and they are readily hydrolysed to inactive penicilloic acids by the action of Penicillinases (see). Commercially, P. are prepared by large scale fermentation in steel vats (up to 100 m^3), using high yielding strains of *Penicillium notatum* and *P. chrysogenum.* These commercial strains produce about 20 000 times more P. than the original isolates, and the yield is further increased by addition of precursors, e.g. phenylacetic acid for pencillin G.

6-Aminopenicillanic acid is an important precursor for the organic synthesis of new P. The compound itself has no antibiotic activity; it is isolated as a fermentation product from cultures of *Penicillium chrysogenum,* or prepared by the enzymatic hydrolysis of benzylpenicillin. Thousands of new P. have been prepared by the acylation of 6-aminopenicillanic acid, but only a few are therapeutically useful, e.g. pencillin V is relatively stable to acid and is not hydrolysed in the stomach, so it may be administered in tablet form. Ampicillin (the aminophenylacetyl derivative of 6-aminopenicillanic acid) has a wider spectrum of activity than most other P., including activity against various Gram-negative bacteria *(Typhus, Escherichia coli,* etc.). Pencillin G (the original "penicillin") is still used widely.

Penicillin V serves as the starting material for the semisynthesis of a group of antibiotics related to cephalosporin, which are active against penicillin-resistant strains.

P. are biosynthesized from the amino acids α-aminoadipic acid, cysteine and valine, which become linked by pepide bonds. The residue of α-aminoadipic acid is usually subsequently lost and replaced by a different acyl residue, but it is retained in penicillin N.

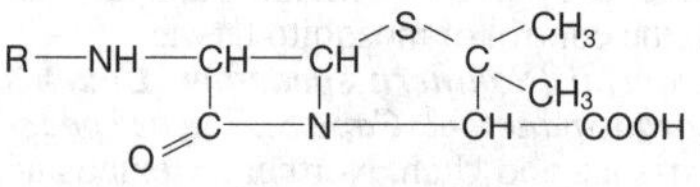

R—NH—CH—CH—S—C(CH_3)$_2$—CH—COOH (β-lactam C=O—N ring fused to thiazolidine)

Name	R
Benzylpenicillin (penicillin G)	C_6H_5—CH_2—CO—
Pentenylpenicillin (penicillin F)	CH_3CH_2—CH=CH—CH_2—CO—
n-Heptylpenicillin (penicillin K)	$CH_3(CH_2)_6$—CO—
Penicillin N	HOOC—CH(NH_2)(CH_2)$_3$—CO—
6-Aminopenicillanic acid ("6 APS")	H—
Penicillin X	HO—C_6H_4—CH_2—CO—
Penicillin V	C_6H_5O—CH_2—CO—
Ampicillin	C_6H_5—CH(NH_2)—CO—

P. act by inhibiting cross linkage of the muropeptide in the murein (peptidoglycan) layer of the cell wall. The cell wall is therefore weakened and cannot withstand the high internal pressure of the bacterial cell (about 30 atmospheres). P. are the most widely and intensively used used antibiotics; they are well tolerated by the animal organism and have a relatively broad spectrum of activity. Bacteria may, however, become resistant to penicillin by the development of Pencillinases (see). Care must therefore by taken to avoid suboptimal dosages of P. and to complete all courses of treatment to the point of total anihilation of the disease organism. Development of resistant strains can also be prevented by using two different antibiotics.

Pennaceous: see Feathers.

Pennales: see Diatoms.
Pennatae: see Diatoms.
Pennatula: see *Pennatularia.*
Pennatulacea: see *Pennatularia.*
Pennatularia, *Pennatulacea, **sea pens:*** an order of the *Octocorallia* (see). The axis of the colony is a relatively long (sometimes 3–4 m), single primary polyp with a skeleton of calcareous sclerites. Two types of secondary polyps are budded from the primary individual: autozooids, which feed the colony; and siphonozooids with reduced mesenteries and an enlarged siphonoglyph, which aid in maintaining the water circulation in the colony. The lateral branches formed by these secondary polyps give the colony a feather-like appearance. Colonies are free but relatively sessile, with the bottom of the main axis buried in the soft sea bottom. About 300 species are known, some of them capable of producing light. Examples are *Umbellula, Pennatula* and *Renilla* ***(Sea pansy).***
Penicillata: see *Diplopoda.*
Penicillium: see *Ascomycetes.*
Pennsylvanian subperiod: see Carboniferous.
Penny royal: see *Lamiaceae.*
Penshells: see *Bivalvia.*
Pentalagus: see *Leporidae.*
Pentameral symmetry: see *Echinodermata.*
Pentanoic acid: see *n*-Valeric acid.
Pentaploidy: a form of polyploidy, in which cells, tissues or individuals possess 5 sets of chromosomes, i.e. they are pentaploid.
Pentastomida, *Linguatulida:* an animal phylum of the *Protostomia,* containing about 80 exclusively parasitic species. The body is up to 14 cm in length, and displays external annulation. The external rings may be secondary, or they may correspond to true internal segmentation, although coelomic cavities have not been observed, even in the embryo. Almost all species are elongate and vermiform, but others are tongue-shaped. The short forebody carries two pairs of claws, one pair on each side of the mouth. The remainder of the body is a relatively long trunk. Sexes are separate, and the female stores sperm in a seminal receptacle.

Adult organisms parasitize the respiratory organs of terrestrial vertebrates: the lungs of reptiles, air sacs of birds, nasal passages of carnivorous mammals. They suck blood or mucus, the foregut serving as a suction pump.

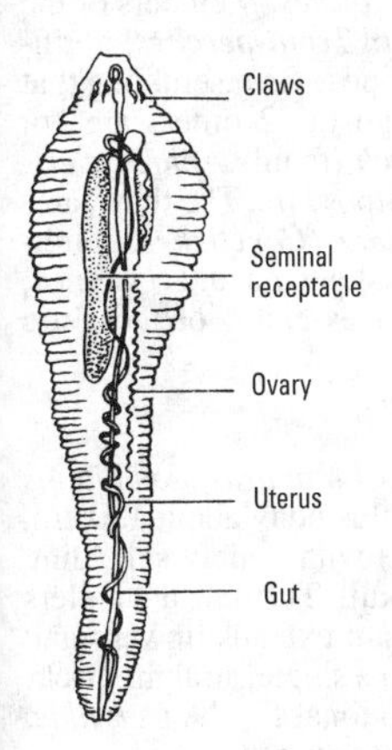

Pentastomida. Female pentastomid worm.

Females produce enormous quantities of eggs. Larval development occurs within the egg in the uterus of the female. When the egg is released, the larva within it already has with pairs of claws, but its body trunk is still short. In some species, the larva first infects an intermediate host, e.g. a fish, amphibian, or terrestrial vertebrate, which serves as the prey of the final host. Once inside the alimentary tract of the final host, the larva migrates forward to the respiratory organs. Thus, the intermediate host of the nasal worm, *Linguatula serrata,* is a herbivore (e.g. a rodent or an ungulate). The adult worm establishes itself in the nasal passages of the final host, e.g. a fox, wolf, or dog. The eggs, containing well developed larvae, are sneezed out by the final host, and they fall onto food plants of the intermediate host.

Classification. The phylum is divided into 2 orders: *Cephalobaenida* and *Porocephalida.* Since the larvae resemble certain parasitic mites (e.g. *Demodex folliculorum,* the human hair follicle mite), the *P.* were formerly considered as a class of the *Arachnida.*

Collection and preservation. *P.* are rarely found in temperate regions. The parasite is removed from the host at autopsy and washed in physiological saline. It is relaxed in 10% ethanol, and preserved in a mixture of 40% formaldehyde and alcohol.

Pentifylline: see Theobromine.
Pentitols: C_5 sugar alcohols. Naturally occurring P. are D- and L-arabitol, ribitol and xylitol.
Pentoses: aldoses containing 5 carbon atoms. P. are an important group of naturally occurring monosaccharides, including, e.g. D- and L-arabinose, L-lyxose, D-xylose, D-ribose and 2-deoxy-D-ribose, and the ketopentoses (pentuloses) D-xylulose and D-ribulose. P. occur chiefly in the furanose form. They are not fermented by the usual yeasts. By distillation with dilute acids they are converted into furfural, a reaction which serves for the detection of P. and their differentiation from hexoses.
Peony: see *Paeoniaceae.*
Peperomia: see *Piperaceae.*
Peplomer: see Spike.
Pepper family: see *Piperaceae.*
Peppermint: see *Lamiaceae.*
Peppery furrow shell: see *Bivalvia,* Modes of nutrition.
Pepsin: an important digestive protease in the stomach of all vertebrates with the exception of stomachless fishes (e.g. carp). P. preferentially catalyses hydrolysis of peptide bonds between two hydrophobic amino acids (Phe-Leu, Phe-Phe, Phe-Tyr). Purified P. shows maximal activity at pH 1–2, but in the stomach the optimal pH is 2–4. The end product of P. digestion of dietary proteins is a mixture of polypeptides of M_r 600–3 000, formerly known as *peptone.*

P. is a highly acidic (IP 1), single chain phosphoprotein (327 amino acid residues of known primary sequence, M_r 34 500; single phosphate group attached to a serine residue), which is secreted in the stomach as its inactive zymogen, pepsinogen *(M_r 42 500).* It is derived from the latter by autocatalysis in the presence of hydrochloric acid. Ps. from different animal species are extremely similar in structure and activity. P. is sometimes used clinically to supplement deficient P. production.

Peptide hormones: a group of hormones with the chemical structure and physical properties of peptides. They differ from each other in their primary structure (number, sequence and identity of their constituent amino acids). The variety of biological actions displayed by different P.h. are determined by specific sequence regions and by peptide conformation.

The P.h. of vertebrates can be classified according to the number of constituent amino acids. Examples of *oligopeptide hormones* (up to 10 amino acid residues) are Oxytocin (see), Vasopressin (see), Bradykinin (see), Releasing hormones (see). *Polypeptide hormones* (containing more than 10 amino acid residues; M_r of monomers up to 5 000) are represented by Melanotropin (see), Gastrin (see), Secretin (see), Gastric inhibitory peptide (see), Glucagon (see), Calcitonin (see), Pancreozymin (see), Corticotropin (see), Lipotropins (see). Some *proteohormones* (M_r of monomers between 5 000 and 30 000) are Insulin (see), Parathyrin (see), Relaxin (see), Somatotropin (see), Prolactin (see), Luteinizing hormone (see), Choriogonadotropin (see), Thyreotropin (see) and Follicle stimulating hormone (see).

P.h. are synthesized in endocrine glands of vertebrates (parathyroid, thyroid, pancreas, hypothalamus, anterior pituitary). Some P.h. are also tissue hormones, e.g. gastrin, secretin, pancreozymin and bradykinin. P.h. are biosynthesized either directly from the appropriate amino acids, or by partial proteolysis of a larger precursor peptide or protein (prohormone). Most P.h. act via the adenylate cyclase system of their target organs (see Hormones). They are inactivated by proteolysis, and they have biological half-lives of 30 min or less.

Peptides: organic compounds consisting of two or more amino acids joined covalently by peptide bonds (amide bonds between the carboxyl group of one amino acid and the α-amino group of the neighboring amino acid). The number of amino acid residues in a peptide is indicated by a Latin prefix, e.g. dipeptide, tripeptide. *Oligopeptides* contain 10 or fewer residues. More than 10 residues constitute a polypeptide. P. with M_r 10 000 (about 100 amino acid residues) lie on the borderline between polypeptides and proteins. P. dialyse through natural membranes, whereas proteins do not. Homodetic P. contain only peptide bonds, whereas heterodetic P. contain at least one other type of bond, e.g. ester, disufide, thioether.

In the naming of P., an amino acid contributing its carboxyl group to the peptide bond is given the ending *-yl,* and only the amino acid retaining its free carboxyl group retains its usual name, e.g. glycyl-histidyl-lysyl-alanine. Using the accepted abbreviations of the amino acids, the name of this tetrapeptide becomes Gly-His-Lys-Ala.

Naturally occurring P. are e.g. glutathione, phalloidin, peptide hormones.

Peptolides: see Depsipeptides.

Peptonization:

1) Enzymatic degradation of protein to peptones.

2) In microbiology, the solubilization of coagulated milk proteins by bactera to produce a clear yellow solution. The ability to peptonize milk protein is a taxonomic feature of certain bacteria.

Peramelidae: see Bandicoots.

Perca: see Perches.

Perception:

1) In the sensory physiology of plants, the process of Induction (see) that follows Susception (see). P. is also used to mean the total process of induction and susception.

2) In the sensory physiology of animals and humans, the process whereby signals from receptors are processed in the brain.

Perch, *Perca flaviatilis:* a widely distributed, Eurasian freshwater predatory fish (family *Percidae,* order *Perciformes).* Its green-gold backgound color is broken by dark transverse bands. Pectoral, ventral, anal and caudal fins are yellowish orange to red. It is a popular angling fish. The white flesh is tasty but bony.

Perches, *Percidae:* a family of predatory fishes with an elongated body (sometimes slightly laterally compressed) and a large mouth with pointed teeth. They belong to the order *Perciformes,* and they occur primarily in fresh water in the Northern Hemisphere. Members include, e.g. ***Perch*** *(Perca fluviatilis),* ***Pike-perch*** *(Lucioperca lucioperca),* a ***Pope*** *(Acerina = Gymnocephalus schraetzer),* which is found in the Danube and its tributaries (length about 20 cm; lemon-colored sides and large, pointed head) and a related ***Pope*** *(Acerina = Gymnocephalus cernua),* which is found throughout the whole of northern Asia, parts of Scandanavia and in eastern Europe (length about 25 cm; olive to brown-green in color with dark spots on fin membranes; sides and belly yellowish to white; thick head with prominent mucous pits; blunt snout with thick, fleshy lips; gill cover with strong spine). Pike-perch and perch are economically important as food sources, as well as being popular sporting fishes.

Perching birds: see *Passeriformes.*

Perching ducks: see *Anseriformes.*

Percidae: see Perches

Perciformes: the largest and most diverse order of fishes, in fact the largest order of vertebrates. The scales are typically ctenoid, although some species lack scales, while in others the scales are mainly cycloid as in lower teleosts. Perciform fins are usually supported by spines, whereas spines are absent from the fins of lower teleosts. The order contains both marine and freshwater forms, as well as species that prefer brackish water. Important families are: Perches (see) *(Percidae),* Butterflyfishes (see) *(Chaetodontidae),* Cichlids (see) *(Cichlidae),* Wrasses (see) *(Labridae),* Damselfishes (see) *(Pomacentridae),* Sunfishes (see) *(Centrarchidae),* Sea basses (see) *(Serranidae).*

Percopsidae: see *Percopsiformes.*

Percopsiformes: a small North American order (3 families, 9 species) of relatively small, freshwater, teleost fishes, related to the *Perciformes.* Members of the family *Percopsidae* (2 species of ***Trout-perches;*** northern North America) have an adipose fin resembling that of the *Salmoniformes,* although the 2 orders are not closely related. The ***Pirate perch*** (family *Aphredoderidae;* eastern USA) lacks the adipose fin. The third family of the order, the *Amblyopsidae* ***(Cavefishes;*** southern and eastern USA) contains 4 genera and 6 species; 5 species inhabit limestone caves and 4 of these are blind.

Perdicinae: see *Phasianidae.*

Perdix: see *Phasianidae.*

Père David's deer, ***milu,*** *Elaphurus davidianus* (plate 39): a large deer (head plus body about 150 cm, shoulder height about 115 cm) with widely spreading hooves and a relatively long skull. The unusual antlers lack a brow tine; the main beam extends upward and usually forks only once to form a single, straight, backward-pointing tine. Also exceptional for the *Cervidae,* it has a long tail. The summer pelage is reddish mixed with gray, becoming grayish buff in the winter. A mane is present on the neck and throat. Originally it inhabited marshy lowlands in northeastern China. When discovered in 1865 by the French missionary, Abbé Armand David, it was already extinct in the wild, and existed as a single herd in the Emperor's walled hunting park in Peking. It is now kept in several zoological gardens. See Deer.

Peregrine falcon: see *Falconiformes.*

Pereiopods: see Pereopods.

Perennating spores: durable, thick-walled spores, which remain quiescent but viable, for more or less extended periods, when environmental conditions are unfavorable, e.g. drought and/or low seasonal temperatures ("winter spores"). Germination occurs when conditions are favorable, e.g. winter spores germinate in the spring. Examples of P.s. are Chlamydospores (see), Teleutospores (see) and Zygospores (see).

Perennial plants: plants which normally live for more than 2 vegetative periods (2 growing seasons), and produce flowers annually. In contrast to annuals and biennials, the lifespan of a perennial is not determined by flowering. Two requirements for the perennial habit are the storage of food reserves and the formation of winter buds (see Bud). See Life form, Lifespan, Annual plants, Biennial plants.

Pereopods, ***pereiopods:*** locomotary limbs (walking and swimming limbs) of crustaceans, as distinct from those appendages that are adapted for feeding.

Pereon: in crustaceans, the totality of free thoracic segments (thoracomeres), which are not fused with the head. The individual segments of the P. are called *pereomeres.*

Perianth: see Flower.

Periblem: see Histogen concept.

Peribranchial space: a cavity surrounding the pharynx of lancelets and the pharyngeal basket of sea squirts. It receives respiratory water from the pharyngeal clefts and conducts it to the exterior. In lancelets, the P. arises by enlargement and ventral fusion of two lateral body folds *(metapleural folds),* which form a protective shield around the sensitive pharynx. At its posterior end the P. communicates with the exterior via a ventral respiratory pore (branchial pore). In sea squirts, the P. lies within the mantle and the outer cylindrical muscle layer, and its final section opens into a cloaca shared by the intestine and sex organs.

Pericardial cell: see Nephrocyte.

Pericardial space: a cavity arising from part of the coelom, and enclosing the heart.

Pericardium: a sac consisting of an outer fibrous layer (the fibrous P.) and an inner serous layer (the serous P.), which surrounds the heart. The serous P. forms attachments to the surface of the heart. Within the P. the heart is able to beat smoothly without friction. In mammals, the P. is fused with the diaphragm. The narrow cavity around the heart is known as the pericardial cavity, but it is sometimes also called the P.

Pericarp: see Fruit.

Perichondrium: see Cartilage.

Periclimenes: see Shrimps.

Periclinal cell walls: cell walls that are oriented parallel to the surface of the vegetative point. See Stem.

Periderm:

1) ***Perisarc:*** a rigid, chitinous (rarely calcareous) tube produced by the ectoderm of colony-forming *Hydrozoa,* and which supports and protects the heads and stalks of the polyps.

2) See Boundary tissue, Stem.

Peridium: see *Myxomycetes.*

Peridroma: see Cutworms.

Perigon: see Flower.

Perigynous (flower and ovary): see Flower.

Perikaryon: see Neuron.

Perilla: see *Lamiaceae.*

Perimetrium: the serous membrane of the uterus.

Perimysium: delicate connective tissue associated with muscle. The perimysium internum is an envelope of connective tissue which holds muscle fibers together in bundles (fasciculi); it is connected to the cell wall of the muscle cell (sarcolemma) and to the coarser perimysium externum. The latter continues externally with the aponeurosis enclosing the muscle.

Periodicity:

1) In ***plants,*** changes in the intensity of metabolism, growth, development and movement, which are often but not always linked to daily or annual climatic and environmental variations.

Diurnal P. (see Circadian rhythm) is seen in many plant movements (e.g. sleep movements, and flower and stomatal movements), as well as numerous other processes, like photosynthesis, material transport, and transpiration. Some of these processes are regulated exclusively by external factors, especially light and temperature, whereas others are also subject to an endogenous rhythm.

Many perennial plants in climatic zones of variable rainfall or variable temperature (e.g. Eurasian and North American deciduous trees) show an annual cycle of quiescence and growth. After the autumnal leaf fall, the buds remain dormant and do not sprout until the following spring, The laying down of woody tissue is also subject to an annual cycle (annual ring formation from early and late wood). Seed development and germination also display a marked seasonal rhythm. The repeated cycle of flower formation is particularly crucial to the developmental cycle of all plants.

The numerous periodic annual alterations in the type and intensity of plant growth and development are certainly regulated and synchronized by climatic factors. Nevertheless, in the absence of external influences, plants still display a tendency for an autonomic P. of behavior, often with a more or less marked endogenous P. Thus, in the tropics, trees may annually shed their leaves, but this is not strictly synchronized with any season; neighboring trees of the same species may shed their leaves at different times, and even separate branches of the same tree may behave differently in this respect. The general aspect and composition of a plant community may also exhibit P., i.e. a yearly cycle of changes.

2) Animals. see Biorythms.

Periophthalmidae: see *Gobiidae.*

Periophthalmodon: see *Gobiidae.*

Periophthalmus: see *Gobiidae.*

Peripatus: see *Onychophora.*

Periphyses: sterile hyphae situated at the openings of ascomycete perithecia.

Periplocoideae: see *Asclepiadaceae*

Perisperm: nutritive tissue in the Seeds (see) of the *Caryophyllaceae* and certain other plants. Unlike the endosperm, the P. is derived from the nucellus and is therefore situated outside the embryo sac.

Perispore: the external cell layer of the oospores of some fungi, e.g. *Peronosporaceae,* or the meiospores of pteridophytes. In the latter case, the P. consists of several cell layers, the outer layer taking the form of two slender, parallel bands that are wound helically around the spore (see Haptera).

Perissodactyla, *odd-toed ungulates:* an order of large, hooved, herbivorous mammals with an odd number of toes. The third toe of each foot is strongly developed, whereas the others are regressed or absent. Hooves are well developed and the carpals and tarsals of the feet are not fused (see Unguligrade). The hindlimbs cannot be flexed sufficiently to allow the animal to raise its body on its hindfeet from a lying position. Antlers and horns are absent (dermal horns are present in the rhinoceroses) and the skull is elongated. The order comprises the families: *Equidae* (see), *Tapiridae* (see) and *Rhinocerotidae* (see).

Peristalsis: progressive waves of contraction, e.g. in the wall musculature of digestive systems (see Intestinal peristalsis). P. can also occur in single cells, e.g. peristaltic waves of nerve fibers, which transport neuronal substances.

Peristedion: see *Triglidae.*

Persitome:

1) Oral field, oral disk, the region of the mouth in many animals, e.g. in spirotrichous ciliates, polyps and sea urchins. *See Ciliophora.*

2) See *Musci.*

Perithecium: flask-shaped or spherical fruiting body of certain ascomycetes and lichens, with an internal hymenium of asci and paraphyses (sterile hyphae). Spores are discharged via an apical pore, and they may also be released by disintegration of the P.

Peritreme: see *Ixodides.*

Peritricha: see *Ciliophora.*

Peritrichous flagella: see Bacterial flagella.

Peritrophic membrane: see Digestive system.

Perivisceral space: see Body cavity.

Periwinkle (a dicotyledonous plant): see *Apocyanaceae.*

Periwinkle (a mollusk): see *Gastropoda.*

Perlaria: see *Plecoptera.*

Permanent community: a plant community that hardly changes over a long period, and which is in equilibrium with the environmental conditions of its habitat. See Succession.

Permanent modification, *semipermanent modification:* an alteration in the form of a living orgnaism, usually as a result of extreme environmental influences, which even after removal of these influences is retained for a long period by the haploid or diploid generations of the offspring, then gradually becomes less pronounced and finally disappears. As shown by back crossing, the transmission of P.m. is entirely cytoplasmic. The mechanism of its development and subsequent decrease in transmission to succeeding generations is largely unclear.

Permanent optimal region: that part of the distribution area of a species, in which the optimal conditions for reproduction and development are permanently present. Thus, the centers of mass outbreaks of pest organisms are P.o.rs. See Fluctuation.

Permanent teeth: see Dentition.

Permanent tissues: fully differentiated and more or less highly specialized plant tissues, which in contrast to meristematic tissues have lost the ability for further cell division. The cells of P.t. are larger than the embryonal cells of meristematic tissue, and they often possess large vacuoles. Some P.t. die early in the life of the plant, e.g. tracheids or sclerenchyma cells.

Primary P.t. arise from promeristems, while seconary P.t. are derived from secondary meristems. Histologically, however, primary and secondary P.t. often consist of similar elements, and they usually cannot be distinguished. P.t. belong to one of the following categories (each is listed separately): Parenchyma, Boundary tissue, Absorptive tissue, Conducting tissue, Mechanical tissue and Excretory (secretory) tissue.

Permease: see Carrier.

Permian (from the town of Perm in eastern Russia, where the formation exists): the youngest system of the Paleozoic. In the Germanic basin (northern central Europe), the P. divides into a lower, predominantly terrestrial formation (see Rotliegendes) and an upper marine formation (see Zechstein). The P. embraces a period of 60 million years. Delineation of the P. from the Lower Carboniferous and the Upper Triassic is often difficult in terrestrial rocks. Marine rocks and their substrata are more clearly defined, due to the presence of foraminifera and ammonites. The boundary between the P. and the Triassic marks an important stage in animal evolution. At this time, the rugose corals, trilobites, goniatites, blastoid echinoderms, and the primitive amphibians and reptiles became extinct. Mesoammonites appear for the first time, and they serve as important, easily recognized index fossils, because the lobes of their suture lines are divided into numerous foliaceous elements. Bryozoans are particularly important as reef-forming organisms. Brachiopod evolution also passed through an important stage at the junction of the P. and the Triassic, and a number of Paleozoic groups, e.g. *Horridonia,* became extinct.

Bivalves began to show similarities with those of the subsequent Mesozoic, whereas gastropods retained their Paleozoic features. Gastropods contributed to the formation of Alpine sedimentatry rocks. Echinoderms displayed a further evolutionary peak. Labyrinthodont amphibians evolved into primitive reptiles.

For the plant kingdom, the P. was a transitional period between the Paleophytic (see) and the Mesophytic (see). For the first time in the earth's history there was a clear differentiation of floral provinces. Pteridophytes and pteridosperms displayed great diversity in the Rotliegendes, and were then superseded by gymnosperms *(Ginkgoaceae* and *Pinidae).*

Pernis: see *Falconiformes.*

Peronopsis: see *Agnostus.*

Peronospora: see *Phycomycetes.*

Peronosporales: see *Phycomycetes.*

Peroxidase: a widely distributed oligomeric oxidase in plants and animals, which utilizes hydrogen peroxide as an oxidant in the dehydrogenation of various substrates (e.g. ascorbic acid, quinones, cytochrome *c).* The hydrogen peroxide is reduced to water: $AH_2 + H_2O_2 \rightarrow A + 2H_2O$. In the cell, P. is often accompanied by catalase and localized in the peroxisomes. The prosthetic group of plant Ps. is ferriprotoporphyrin. Animal Ps. contain an unidentified green hemin. The most widely studied P. is the crystallizable horseradish P., which can be separated into apoenzyme and hemin by treatment with acetone-hydrochloric acid. The nonenzymatic, heat-stable P. activity of hemoproteins is called pseudoperoxidase, e.g. as shown by hemoglobin and its degradation products. The P. reaction is diagnostically important, because myelotic granulocytes have high P. activity, whereas lymphocytes, reticular cells, carcinoma, sarcoma and myeloma cells have no P. activity.

Peroxisomes, *microbodies:* structurally similar, but functionally heterogeneous cytoplasmic organelles, which are present in all eukarytoic cells, and which are the sites of certain oxidative metabolic reactions. Oxidative reactions in the P. give rise to hydrogen peroxide, which is converted into water and oxygen by the action of catalase. Catalase therefore serves as a detoxifying agent by preventing the accumulation of toxic hydrogen peroxide. This catalase constitutes up to 40% of total peroxisomal protein, and serves as a marker enzyme for P., i.e. for the histochemical demonstration of P. The various oxidative reactions of P. are not accompanied by ATP synthesis, since P. possess no phosphorylation mechanism; the resulting energy is dissipated as heat.

P. were first identified in 1954 in kidney tubules. They are approximately spherical in shape (diameter 0.5–2 μm), bounded by a single membrane, often contain a crystal of urate oxidase, and do not contain DNA or ribosomes. P. are thought to be formed in the same way as chloroplasts and mitochondria, i.e. by selective protein and lipid import from the cytosol. They were formerly thought to arise by abstriction from the endoplasmic reticulum, which they resemble in being a self-replicating organelle without a genome. Subsequent studies, however, indicate that P. always arise by growth and fission of existing P.

Two types of P. have been studied in plants. The type present in leaves catalyses the oxidation of glycolate to glyoxylate (the glycolate is derived from phosphoglycolate, a byproduct of CO_2 fixation in C3 plants; see Photorespiration). The other type is present in germinating seeds, where it converts fatty acids (from stored triacylglycerols) into carbohydrates. This latter type is called a *glyoxysome,* because the conversion of fatty acids into sugars involves the glyoxylate cycle.

P. are particularly abundant in liver and kidney cells (plate 21), e.g. a single rat liver cell contains 350–400 P. The half-life of P. is 1.5–2 days. Other cell types usually contain smaller P. with homogeneous contents *(microperoxisomes).* In addition to catalase, P. contain, inter alia, D-amino acid oxidase, α-hydroxyacid oxidase and urate oxidase. The latter enzyme is often present as a large crystal in the otherwise homogeneous matrix. These enzymes are particularly important in the oxidative degradation of metabolic intermediates (e.g. purine bases) and in the formation of carbohydrates from amino acids and other materials.

Persea: see *Lauraceae.*

Persicaria: see *Polygonaceae.*

Persistent pulp: see Dentition.

Persistent viruses: see Virus vectors.

Perspiratio insensibilis and sensibilis: see Perspiration.

Perspiration: a term used as a noun synonymous with Sweat (see), or a general term for the process of water loss from the surface of the skin, which serves for Temperature regulation of the body (see). 1) Insensible P. *(perspiratio insensibilis)* occurs through the upper layers of the skin. 2) Sensible P. *(perspiratio sensibilis)* or sweating occurs in response to high environmental temperatures, and is observed only in ungulates, primates and humans. Sweat (see) is secreted by sweat glands in response to excitation by cholinergic fibers of the sympathetic nervous system. Sweat secretion is increased by acetylcholine and parasympathicomimetic agents (e.g. pilocarpine), and it is inhibited by atropine. Evaporation of 1 liter of water extracts 2 400 kJ from the organism, while the evaporation of 1 liter of sweat extracts 2 240 kJ. Birds and carnivores (e.g. the dog) exploit the cooling effect of evaporation by panting. By panting, the dog increases its ventilation rate 25–40-fold.

Perthophytes: see Saprophytes.

Peru balsam: an aromatic smelling, dark brown, viscous balsam exuded from the trunk of *Myroxylon balsamum (= M. pereiroe)* after the bark has been beaten and scorched. P.b. consists of about 60% cinnamein, a mixture of benzyl benzoate and benzyl cinnamate. P.b. is applied externally in certain skin diseases, and was formerly employed as a stimulant and disinfectant expectorant in bronchitis. It is also used as a mounting agent in microscopy, and in the manufacture of perfumes. See Balsams.

Peruvian bark: see Qunine.

Pervitine®: see Methamphetamine.

Peryphyton: plant growth (mostly algae and bacteria) on stones, larger plants and other substrata in aquatic environments.

Pessulus: see *Alaudidae,* Syrinx.

Pest control: measures to decrease or exterminate pests of plants, animals and humans, their living places and their stored goods, using physical, chemical and biological methods. The integrated approach to P.c. avoids unneccessary disturbance or burdening of the ecosystem, by employing all possible measures, including optimization of the living conditions of the organism to be protected.

See Biological pest control.

Pest control activity index: a value for assessing plant protection agents and other measures taken for the control of plant pests.

Activity index (AI) according to Abbott: AI% = $[(C - T)/C]$ x 100, where C is the number of living pests in an untreated control, and T is the number of living pests remaining in a matched, treated sample.

Activity index according to Henderson and Tilton:

AI% = 100 x $[1 - (T_a \times C_b / T_b \times C_a)]$, where T is the number of living pests in the test sample, *(a)* after treatment, *(b)* before treatment, C is the number of living pests in the matched control, *(a)* after treatment of the test sample, *(b)* before treatment of the test sample.

Activity index according to Scheider and Orelli:

AI% = $(b - k / 100 - k)$ x 100, where b is the % of dead individuals in the treated test sample, k is the % dead individuals in the untreated matched control.

Pests (plates 45 & 46): organisms whose presence and activities are contrary to the interests and well being of humans. They influence the development, performance and yield of domestic and agricultural animals and plants, destroy, or decrease the value of, economically important goods, or are simply an unpleasant nuisance.

The attacked object usually suffers loss or damage, e.g. a pest may feed on plant leaves, roots or sap, or on animal blood. Marked changes in the normal life processes of the attacked organism are known as illness or disease, e.g. in animals: inflammations, defense reactions and other manifestations of impaired health; in plants: deformities, yellowing, stunted growth, etc.

Pest animals include aschelminths, mites, wood lice, snails, birds, rodents and game animals, but insects are the most significant and prominent. ***Pest plants*** include parasitic plants and weeds, as well as numerous fungi. The latter represent a transition from pest plants to ***pest microbes,*** which embrace viruses, bacteria and mycoplasmas. P. are responsible for the yearly loss of about 30% of agricultural crops

According to their occurrence, P. are classified as field, garden, orchard or forest P., or as P. of stored foodstuffs. P. of humans and domestic animals may simply cause inconvenience, or may be disease organisms.

Although many organisms can be classified as P., only those that occur in large numbers and regularly cause losses to human health and economic interests are considered to be important.

In a natural ecosystem, no organism can be classified as a pest or as a beneficial organism. The definition of a P. rests on human economic considerations, and an organism cannot be classified as a pest until its activities have exceeded the economic damage threshold, and its control is therefore justified. The opposite of P. is Beneficial organisms (see).

Pesticide efficiency: see Pest control activity index.

Petaurus: see *Phalangeridae.*

Petiole:

1) see Leaf.

2) see Pedicel.

Petiole climbers: plants in which the leaf stalk serves as a tendril-like climbing organ, displaying haptotropic bending movements, e.g. nasturtium *(Tropaeolum).*

Petrels: see *Procellariiformes.*

Petri dish: a round, flat dish with vertical walls, and a similarly shaped but slightly larger diameter lid, whose walls overlap those of the bottom dish. The commonest type has a diameter of 10 cm, and is made of glass or transparent plastic. P.ds. are used for the preparation and maintenance of plate cultures of microorganisms, e.g. on nutrient media solidified with agar. See Culture of microorganisms.

Petrochirus: see Hermit crabs.

Petromyzon: see Lampreys.

Petromyzonidae: see Lampreys.

Petronia: see *Ploceidae.*

Petropedetinae, ***petropedetins:*** a subfamily of the *Ranidae* (see), containing 10 African species in 2 genera: *Arthroleptides* (2 species; mountains of Kenya and Tanzania) and *Petropedetes* (8 species; Sierra Leone to Cameroon, and Fernando Po island). The terminal phalanges of the fingers are broadened and T-shaped, and the tops of the finger tips carry a pair of dermal scutes resembling those of the South American *Dendrobatidae* (see). *Arthroleptides* is much longer than *Petropedetes*. The skeleton of the *Petropedetinae* is very similar to that of the *Raninae*, with a bony omosternum that is entire or slightly forked posteriorly.

In an alternative classification, the *Petropedetinae* contain 86 species in 11 genera, all of them restricted entirely to Africa. The most numerous genus of this classification is *Phrynobatrachus* (63 species; sub-Saharan Africa).

Petunia: see *Solanaceae.*

Peyer's glands: see Peyer's patches.

Peyer's patches, ***Peyer's glands*** (Peyer, a Swiss anatomist 1653–1712): circular or oval aggregations (2.5–10 cm diam.) of lymphoid tissue interspersed at intervals among the intestinal villi. They are largest and most numerous in the ileum, small and few in number in the lower part of the jejunum, and occasionally present in the duodenum. As the sites of immune reactions, they play an important part in defense against infections. They may become inflamed and ulcerated in typhoid fever and intestinal infections.

Peyote: see *Cactaceae,* Mescaline.

Peyotl: see *Cactaceae,* Mescaline.

Pezizales: see *Ascomycetes.*

Pezophaps: see *Raphidae.*

Pezoporus: see *Psittaciformes.*

Pfeiffer's influenza bacillus: *Haemophilus influenzae.* See *Haemophilus.*

pH: see Hydrogen ion concentration.

Phacidales: see *Ascomycetes.*

Phacochoerus aethiopicus: see Wart hog.

Phacops (Greek *phakos* lens, *ops* eye): a trilobite genus found from the Silurian to the Devonian. The cephalon has a club-shaped, anteriorly expanded glabella without furrows. There are 2 large compound eyes, and the thorax consists of 8–19 segments. The semicircular pygidium is somewhat smaller than the cephalon, and is clearly divided into 3 regions by a pair of furrows. Phacopids were presumably able to roll themselves hedgehog-like to protect their nonarmored underside.

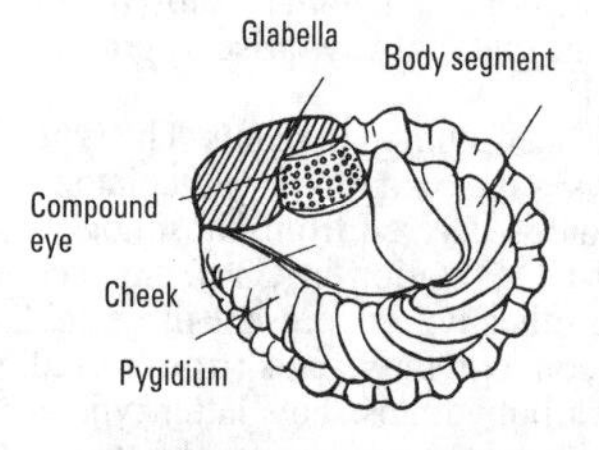

Phacops. Rolled up specimen of *Phacops schlotheimi* Bronn from the Middle Devonian (x 2).

Phacus: see *Phytomastigophorea.*

Phaenicophaeinae: see *Cuculiformes.*

Phaeococcus: see *Phytomastigophorea.*

Phaeophyceae, ***brown algae:*** an exceptionally diverse class of algae, containing about 2 000, almost exclusively marine species, found in temperate and cooler regions. Some species are small and filamentous, while others have thalli many meters in length, which are differentiated into rhizoids, cauloids and phylloids. Most species are firmly attached to the substratum. All species are generally known as seaweed, and the larger robust forms are also called kelp, oarweed or wrack.

Cells of the *P.* are uninucleate, and their complement of pigments includes chlorophyll *a,* β-carotene and brown fucoxanthin, which masks the other pigments. Starch is never present. Photosynthetic assimilation products and storage material include laminarin, oil and mannitol. In addition to cellulose and other materials (e.g. pectic substances), the cell wall contains the polysaccharide algin (the calcium salt of alginic acid) which is economically important (see Alginic acid). The asexual zoospores closely resemble the gametes; both are motile with two unequal flagella, the longer flagellum possessing flimmer hairs and directed forward in motion, while the shorter flagellum is directed backward. Near the point of insertion of the flagella is a red-brown eyespot, and there is a brown chromatophore at the opposite end of the cell. Asexual reproduction can also occur by nonmotile spores, which are formed in sporangia The latter may be undivided (unilocular), or separated into chambers by transverse and longitudinal walls (multilocular). In sexual reproduction, the gametes are always produced in multilocular gametangia, and the gametes may exhibit isogamy, anisogamy, or oogamy.

Most *P.* display an alternation of generations. In primitive species, the sporophyte and gametophyte are morphologically similar. More advanced species show a progressive reduction of the gametophyte, until finally only the sporophyte has a separate existence. Thallus growth is either intercalary, or by means of a large apical cell. There is also true tissue differentiation into outer photosynthetic tissue and inner conducting elements and storage cells.

The *P.* belong to a very ancient plant group; fossil species are known from the Silurian and the Devonian. It is thought that they evolved from the ancestors of the *Chrysophyceae,* as evidenced by the similar pigment composition, and the similar type of flagellation in the two groups.

Six orders of *P.* are usually recognized.

1. Order: *Ectocarpales.* Mostly filamentous forms with similar gametophytes and sporophytes. Growth is intercalary, and reproduction is by isogamy, e.g. the genera *Ectocarpus* and *Pylaiella.*

2. Order: *Sphacelariales.* Represented by only a few species, all differing from the *Ectocarpales* by possessing an apical cell, and therefore by displaying apical growth.

3. Order: *Cutlariales.* Reproducion is by anisogamy, e.g. *Cutleria multifida,* which occurs in warmer European seas.

4. Order: *Dictyotales.* Oogamy is the predominant form of sexual reproduction. Thalli grow with the aid of an apical cell. The widespread marine alga, *Dictyota dichotoma,* has a hand-sized, regularly dichotomous thallus.

5. Order: *Laminariales.* The thallus is generally differentiated into rootlike, shootlike and leaflike sections, and there is a true differentiation of tissues. Sexual reproduction is by oogamy, and female and male gametes are morphologically different. In the colder seas of the Southern Hemisphere, the thallus of *Macrocystis pyrifera* can exceed 100 m in length. Various species of *Laminaria,* which occur in the North Sea and the Atlantic, can attain a length of 5 m. The perrenial cauloid or stipe of *Laminaria* (attached to the substratum by its holdfast), carries a phylloid or lamina that is many cells thick, and which is renewed each year at the base by an intercalary mersitem. The ***Sea belt*** *(Laminaria saccharina),* as well as other *Laminaria* species and other genera of the order, are used as food, fodder and fertilizer. Iodine is prepared from the ash of some *Laminariales,* and members of this order also yield alginic acid, mannitol, etc. Other examples are *Laminaria digitata* ***(Oarweed, Tangle)*** and *Chorda filum* ***(Sea laces, Dead man's ropes).*** *Nereocystis* of the North American Pacific coast has a ropelike cauloid, up to 100 m in length, carrying a large air bladder with attached phylloids.

6. Order: *Fucales.* There is no obvious alternation of generations. The gametophyte is reduced to the oogonia (containing the ova) and antheridia (containing the spermatozoids) in depressions or conceptacles in the sporophyte. The differentiated thallus often contains gas-filled bladders, e.g. *Fucus vesiculosus* ***(Bladder wrack),*** which is common in northern seas, and whose ash is also used for the preparation of iodine. The leathery, dichotomous thallus of *Fucus* species has a reinforcing midrib, and is attached to rocks by a disk-shaped holdfast. About 250 species of *Sargassum* occur

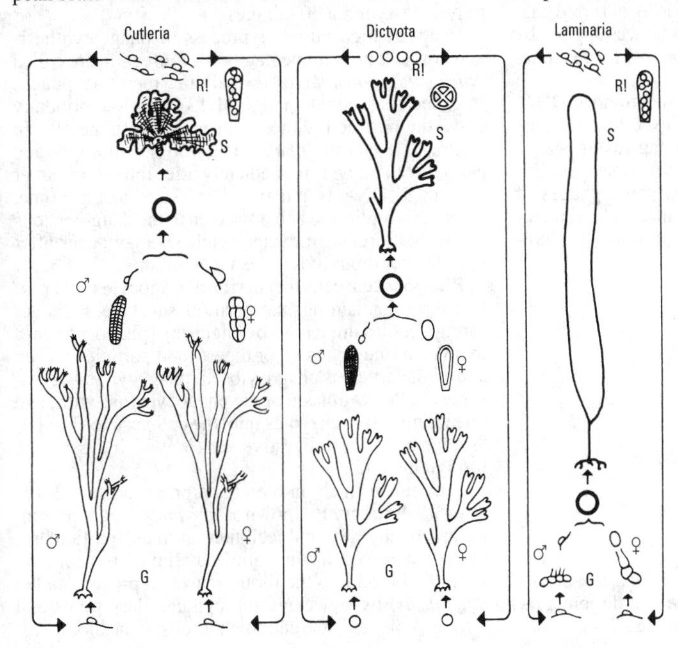

Fig.1. *Phaeophyceae.* Alternation of generations and nuclear phases in some brown algae. *G* Gametophyte. *S* Sporophyte. *O* Zygote. *R!* Meiosis. Haploid phase: thin lines. Diploid phase: thick lines. (After Harder)

as free-floating (by means of air bladders) thalli in the mid-Atlantic Sargasso sea, where they reproduce entirely by fragmentation.

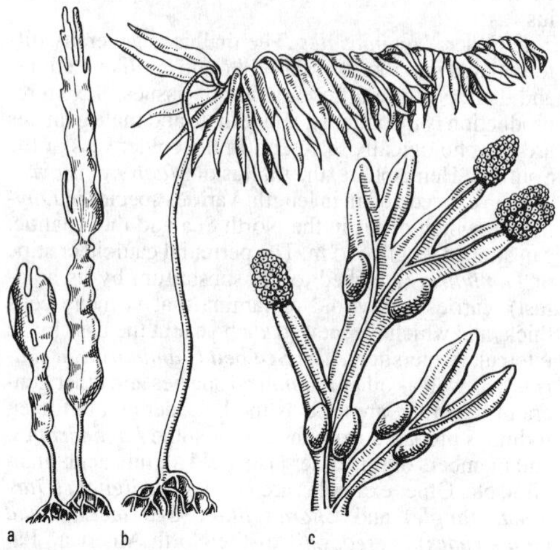

Fig.2. *Phaeophyceae.* Seaweeds. *a Laminaria saccharina* (x 0.025). *b Macrocystis pyrifera* (x 0.04). *c Fucus vesiculosus* (bladder wrack) (x 0.04).

Phaethon: see *Pelicaniformes.*
Phaethonidae: see *Pelicaniformes.*
Phages (Greek *phagein* eat) (plate 19): viruses that attack prokaryotic organisms. A distinction may be drawn between viruses of bacteria *(bacteriophages)* and of cyanobacteria *(cyanophages).* Viruses of actinomycetes are called *actinophages.* Before the role of viruses as bacterial parasites was fully understood, Twort and later D'Hérelle showed that certain bacterium-free filtrates cause the rapid clearing of turbid bacterial cultures. The term, bacteriophage, was applied by D'Hérelle (1914) to the agent responsible for this phenomenon.

Most phage particles contain double-stranded DNA, while a few contain single-stranded DNA. In others, the genetic material is RNA. The nucleocapsids of phages (see Viruses) display various geometrical shapes. Extensive studies on coliphages, in particular phages of the T series in *Escherichia coli,* have made a fundamental contribution to the development of molecular biology and modern virology.

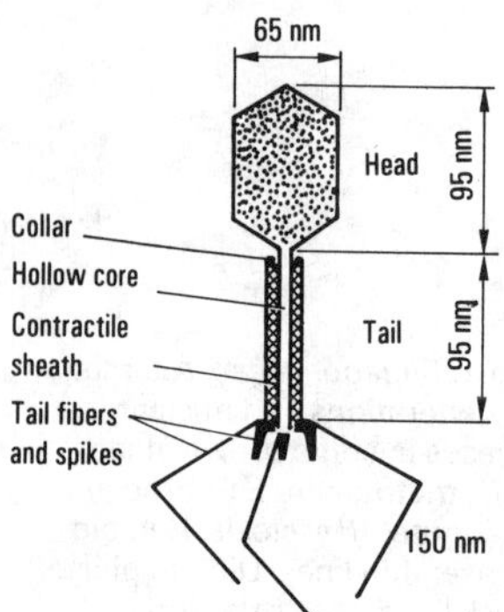

Fig.1. *Phages.* Structure and dimensions of phage T2.

Multiplication of T-even-phages, e.g. T_2 or T_4 (plate 19), starts with adsorption of the phage tail to specific receptors on the bacterial wall. The tail sheath then contracts, and the tail core is pushed into the outer soft layers of the bacterial cell wall. Following the onset of this *penetration phase,* the inner, more resistant cell wall layers are dissolved by phage lysozyme. Finally, the DNA strand (length 50 000 nm) leaves the phage head and passes through the tail core (internal diam. 2.5–3.5 nm) into the interior of the bacterial cell. The entire process of adsorption and penetration lasts about 1 minute. This is followed by about 8 minutes of active transcription and translation, during which *early proteins* are synthesized. Early proteins include phage-specific enzymes required for the synthesis of phage-specific DNA. They also include phage-specific DNA polymerase, as well as enzymes for the synthesis phage-specific precursors, such as deoxy-5-hydroxymethylcytidine phosphate, an unusual nucleotide found in phage DNA. Early proteins also include *regulatory proteins,* which prevent the synthesis of host nucleic acids and proteins, e.g. a thermosensitive factor which associates with host ribosomes and allows only the translation of phage mRNA. When early proteins have been synthesized in sufficient quantity, the phase of virus DNA replication commences. Thus, about 10 minutes after the initial infection, the bacterial host cell contains a relatively large quantity of viral DNA, the synthesis of several early proteins comes to a halt, and the synthesis of the remaining early proteins is greatly decreased. The synthesis of *capsid proteins* then commences, followed later by the synthesis of lysozyme. In the later phase of *morphogensis,* entire virus particles are assembled; DNA is packaged into phage heads and tail parts are attached. About 45 genes are involved in the synthesis of the capsid of phage T-4D (Fig.2). The host cell wall is attacked by lysozyme, the cell lyses, and the newly synthesized phage particles are released. From adsorption to lysis lasts about 30 minutes.

Phages which initiate a process of phage synthesis and host cell lysis immediately after infection are called *lytic* or *virulent phages* In addition, there are phages (e.g. the filamentous phage M 13) which continually leave the host cell without causing lysis (plate 19). In contrast, *Temperate phages* (see) do not multiply and cause host cell lysis immediately after infection; rather the phage DNA is integrated into the host genome, where it is replicated and passed on to the daughter cells of the host organism; phage synthesis may occur after hundreds or thousands of host generations.

Phagocytes: cells that perform phagocytosis. P. protect the organism against harmful substances, render foreign bodies innocuous by engulfing (phagocytosing) them (see Endocytosis), or digest food particles. When a foreign particle is phagocytosed (phagocytized), lysosomes in the cytoplasm of the phagocyte fuse with, and release digestive enzymes into, the cell vesicle. The latter thus becomes a digestive vesicle, which digests the foreign material.

P. circulate freely in blood and/or are present in the tissues. Migratory P., known as *amebocytes,* are present in practically all multicellular animals, particularly those that possess a lymphatic body fluid. They migrate from the blood into the tissue spaces (a process called *diapedesis)* by squeezing through capillary pores and through holes in some endothelial cells of the blood ves-

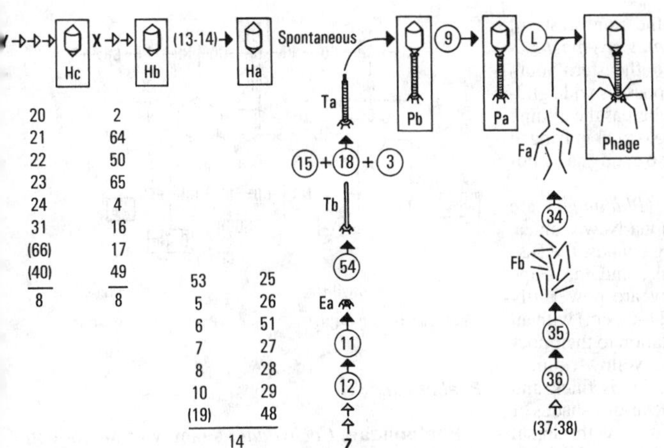

Fig.2. *Phages.* Synthesis and assembly of bacteriophage T-4D. Numbers refer to the genes involved in phage synthesis. Two series of genes *(Z)* are necessary to produce the precursors of the endplate. *Hc* Head precursor. *Hb* Complete head. *Ha* Head prepared for attachment to tail. *Ea* Endplate. *Tb* Endplate joined to core. *Ta* Sheath protein added. *Fb* Incomplete tail fibers. *Fa* Complete tail fibers. *L* Also known as gene 63. *Pb* Immature head-tail combination. *Pa* Mature head-tail combination.

sels. The phagocytic *monocytes* of vertebrate blood are formed in the bone marrow; they consume and digest foreign bodies, bacteria and the body's own disintegration products. After moving from the blood into the tissues, monocytes become much larger and display an extremely high potential for phagocytic activity; they are then called *macrophages.* These tissue macrophages are normally homogeneously distributed within the intercellular spaces, some attached within the tissue (fixed tissue macrophages) while others remain migratory (mobile macrophages). In response to tissue degradation products, macrophages and monocytes migrate to sites of tissue damage and infection and clean the affected area by phagocytosis. The combined phagocytic system of monocytes, mobile macrophages and fixed tissue macrophages is generally known as the *reticuloendothelial system.* Tissue macrophages differ according to their tissue of residence. They are known as *Kupffer cells* in the liver, *alveolar macrophages* in the lung alveoli, *tissue histiocytes* or *fixed macrophages* in subcutaneous tissues, and *micoglia* in the brain. Entrapped macrophages *(reticulum cells)* in lymph nodes, spleen and bone marrow phagocytose foreign particles or organisms that enter the circulation. Connective tissue cells *(fibrocytes)* may also become modified to macrophages. Opsonins (see) and complement (see Complement system) increase the phagocytic activity of P. Independently of complement, antigens also bind to the cell surface of P. via cytophilic receptors, where they induce phagocyte-specific immune reactions.

P. also play a very important part in normal metamorphic processes of tissue and organ modification, degradation and reformation. For example, during metamorphosis of insects and amphibians, larval organs are degraded by ameboid cells, and their components are made available as reserve or precursor materials for the growth and development of new tissues and organs. In vertebrates, specialized P., *osteoclasts* and *chondroclasts,* digest old, unwanted tissue during the development of bone and cartilage. In many animals (with the exception of insects and vertebrates), P. are responsible for digesting food material. In sponges, coelenterates and flatworms a large proportion of dietary material is consumed (phagocytosed) and digested intracellularly by ameboid cells in the alimentary system. In bivalves and echinoderms, ameboid cells migrate into the lumen of the stomach or intestine, where they endocytose food material. Intracellular digestion then occurs after the cells have moved into the intercellular spaces of the surrounding tissue.

Phainopepla: see *Bombycillidae.*

Phainoptila: see *Bombycillidae.*

Phalacrocoracidae: see *Pelicaniformes.*

Phalacrocorax: see *Pelicaniformes.*

Phalangers: see *Phalangeridae.*

Phalangeridae, ***possums, cucuses, phalangers*** and ***koala:*** a family of mouse-sized to fox-sized Australian Marsupials (see) containing 46 species in 16 or 17 genera. They all possess a long, prehensile, and often naked tail, which is used while climbing among rocks and branches. The diet consists of insects, leaves, fruits and nectar. As an adaptation to climbing, some of the finger and toe bones (phalanges) are modified, hence the name "phalanger". The name "possum" is preferred, in order to distinguish the *P.* from the American opossums.

The two tree-dwelling, fox-sized ***Brush-tailed possums*** or ***Vulpine phalangers*** *(Trichosurus)* have a furry tail, which is naked on the lower side. *T. caninus* inhabits the wet, mountainous forests of eastern Australia. Most Australian cities and parks have a resident population of *T. vulpecula,* which is found throughout Australia, on small neighboring islands and on Tasmania. It has also been introduced into New Zealand. It also lives in treeless areas, using old rabbit burrows for shelter.

Gliding possums have membranes between their fore and hindlimbs, which enable the animal to extend the distance of a leap by gliding up to 50 m. They are found in Australia and neighboring islands, where they occupy the same ecological niche as the "flying squirrels" *(Glaucomys)* of the Holarctic region; unlike *Glaucomys,* the gliding possums have furry rather than flattened tails. There are 3 genera: *Petaurus* ***(Lesser gliding possums*** or ***Sugar gliders),*** *Schoinbates* ***(Greater*** or ***Dusky gliding possums),*** and *Acrobates* ***(Pygmy gliding possum*** or ***Feather-tail glider).*** They are active at night and at dusk. *Petaurus breviceps* occurs in Papua New Guinea and neighboring small islands, eastern Australia and Tasmania. *P. australis* and *P. norfolcensis* occur in the higher coastal regions of eastern Australia, *Schoinbates volans* in the wooded mountainous areas of eastern Australia, and *S. minor* in the rainforests of Queensland. The monotypic *Acrobates pygmaeus* is found from Cape York Peninsula to southeastern South Australia. *Petaurus* and *Schoinbates* possess wide gliding membranes from wrist to ankle, whereas the gliding membrane of *Acrobates* is a narrow fold of skin fringed with long hairs, its hair-fringed tail also acting as a gliding plane.

The seven species of ***Cuscuses*** *(Phalanger)* are found in forest and thick scrub in Papua New Guinea, west to Celebes, on small neighboring islands, the Bismarck Archipelago, Solomon Islands, and the Cape York Peninsula of Queensland. They are powerfully built (head plus body 32–65 cm, tail 24–61 cm) with an extremely prehensile tail. As an adaptation to their nocturnal habit, the eyes are notably large, yellow-rimmed and protruding. The fur of most species is thick and woolly, ranging in color from white to various shades of gray-brown and red-brown, sometimes with dorsal stripes. *P. maculatus* ***(Spotted cuscus)*** is cat-sized and patterned with large dark spots.

Phascolarctos is a monotypic genus *(P. cinereus,* ***Koala*** or ***Koala bear;*** body length 50–80 cm) with a vestigial tail. The posteriorly opening marsupium extends upward and forward. During weaning the young animal feeds on partially digested feces, which it takes from its mother's anus. Native to eastern Australia, it is an accomplished but slow climber, moving slowly in the tree branches, and feeding almost exclusively on the foliage of certain *Eucalyptus* species. Digestion of the bulky, fibrous leaves is aided by cheek pouches and by the presence of a large cecum (1.8– 2.5 m in length).

Phalaropes: see *Phalaropodidae.*

Phalaropodidae, ***phalaropes:*** a small avian family (order *Charadriiformes),* containing only 3 species of dainty little specialized wading birds. They all have colorful, patterned plumage. Females are larger and more brightly colored than the males, and the males incubate the eggs and rear the young. Each toe carries lobate webs. They are the most aquatic of all the wading birds, spending much time swimming rather than wading. *Phalaropus (= Lobipes) lobatus* ***(Red-necked*** or ***Northern phalarope;*** 20.5 cm) has a circumpolar distribution, breeding in the Arctic tundra and overwintering at sea in temperate latitudes. *Phalaropous fulicarius* ***(Gray*** or ***Red phalarope;*** 23 cm) is also circumpolar, breeding in the high north and overwintering at sea as far as the Southern Hemisphere. *Phalaropus (= Steganopus) tricolor* ***(Wilson's phalarope;*** 25.5 cm) breeds on the North American prairies and overwinters on the Argentine pampas.

Phalaropus: see *Phalaropodidae.*

Phallotoxins: heterodetic, cyclic heptapeptides present in *Amanita phalloides.* Together with the amatoxins, P. are the main toxic components of this fungus. The chief P. are phalloidin, phalloin and phallacidin (Fig.). Phallacidin is similar to phalloidin, but the D-threonine-alanine grouping is replaced by valyl-D-erythro-β-hydroxyaspartic acid. P. are not as toxic as the amatoxins, but they act more quickly. The structural requirements for toxicity are the cycloheptapeptide structure, and the characteristic thioether bridge between the indole residue of the tryptophan and the mercapto group of the cysteine. LD_{50}-values in white mice: phalloin 1.35 mg, phalloidin 1.85 mg, phallacidin 2.5 mg per kg body mass.

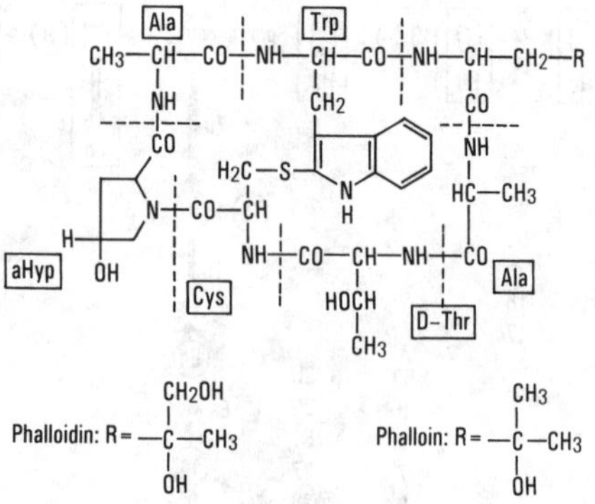

Phalloidin: $R = -C(CH_2OH)(OH)-CH_3$ Phalloin: $R = -C(CH_3)(OH)-CH_3$

Phallotoxins

Phaloniidae, *Cochylidae:* a family of micro-moths abundant in North America and Europe, with species also in Japan, India and Indonesia. A single species is known in Australia. The family is closely allied to the *Tortricidae* (see) and there is a close superficial visual resemblance between the members of the two families. Phaloniid larvae are internal feeders on flowers, seed heads and stems. See Vine moths.

Phalloidin: see Phallotoxins.

Phalloin: see Phallotoxins.

Phallus: see *Basidiomycetes (Gasteromycetales).*

Phanerophyte: see Life form.

Pharbis seeds: see *Convolvulaceae.*

Pharoah's chicken: see *Falconiformes.*

Pharomachrus: see *Trogoniformes.*

Pharyngeal basket: the specialized pharynx of ascidians and acraniates, which functions not only in the exchange of respiratory gases, but also as a filtration apparatus for the retention of food particles. In cyclostomes, the P.b. is part of the visceral skeleton surrounding the gill pouches. See Pharynx, Gill skeleton, Gills.

Pharynx:

1) In many invertebrates, the anterior section of the foregut between the mouth opening and the esophagus. It is often developed as an organ for the uptake and disintegration of food, being supplied with special musculature and additional hard tooth-like structures or abrasive plates.

2) In vertebrates, the common connection between the buccal cavity and the esophagus, and between the nasal cavity and the larynx, so that the P. is the site where the alimentary and respiratory tracts cross.

3) In acorn worms and chordates, it is the anterior part of the alimentary canal which is perforated by stigmata or gill slits, and serves as the respiratory tract.

The P. of the *Tunicata* and *Amphioxus* not only serves for respiratory gas exchange, but is also a highly specialized food trap, and in these organisms is also known as the *pharyngeal basket* or *branchial chamber.* Water is driven by cilia through the mouth opening into the P., then passes through the gill slits into the peribranchial space, which communicates with the exterior via a respiratory pore. Food particles are retained by the gill slits *(Amphioxus)* or stigmata (tunicates). Trapped in a viscous mucilage which is secreted in a ventral groove (see Endostyle), the food particles are carried dorsally by the action of cilia of the pharyngeal epithelium; here they are carried by a stream of mucilage in the epibranchial

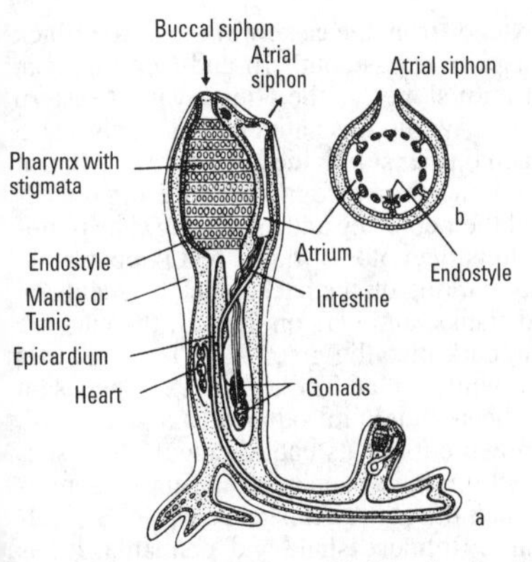

Fig.1. *Pharynx. a* Schematic longitudinal section through a colony- forming sea squirt. *b* Cross section of the pharyngeal region; the walls of the pharynx contain horizontal rows of slits, and each slit is divided into adjacent stigmata by verticle bars, producing a grid-like structure; the stigmata are bordered by cilia, which beat outwardly toward the atrium, generating a water current through the pharynx.

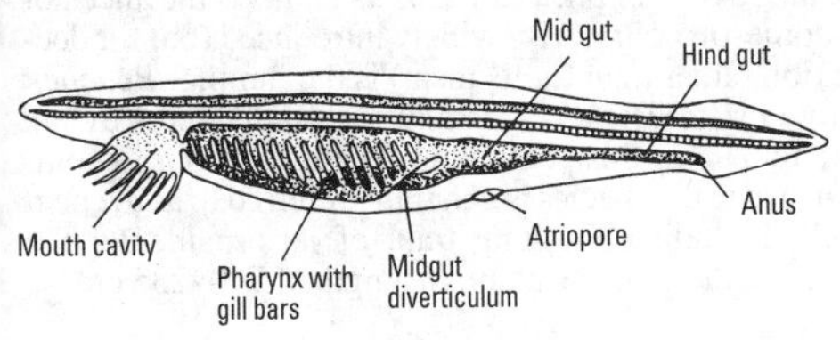

Fig.2. *Pharynx.* The alimentary canal of *Amphioxus.*

groove back into the midgut. In *Amphioxus* there are 50 or more primary gill slits, each of which is divided into anterior and posterior halves ("secondary" gill slits) by downgrowths from the dorsal wall of each slit; each gill bar contains a blood vessel.

In primarily aquatic vertebrates (fish, amphibian larvae), the P. is developed essentially as a respiratory organ, with Gills (see) on the walls of the gill slits. Elasmobranch fishes generally have 6 gill slits, the first being developed as a siphon or spiracle. The gill slits of amphibian larvae close during metamorphosis, whereas in amniotes their primordia regress during embryonal development.

The embryonal P. of terrestrial vertebrates give rise to ductless glands (e.g. thymus, parathyrois, thyroid). The lung primordia also develop in the region of the embryonal P.

Phascogalinae: see *Dasyuridae.*

Phascolarctos: see *Phalangeridae.*

Phascolomidae: see Wombats.

Phase contrast microscopy: method for observing unstained, contrast-poor, living or fixed biological specimens under the light microscope. Unstained living cells are particularly transparent and lacking in contrast. When light passes through an unstained specimen, only a little light is absorbed. On the other hand, the phase of the light (but not its amplitude) may be changed. P.c.m. is a technique for exploiting such phase changes by converting them into visible changes of intensity. This is achieved by replacing the substage iris diaphragm with

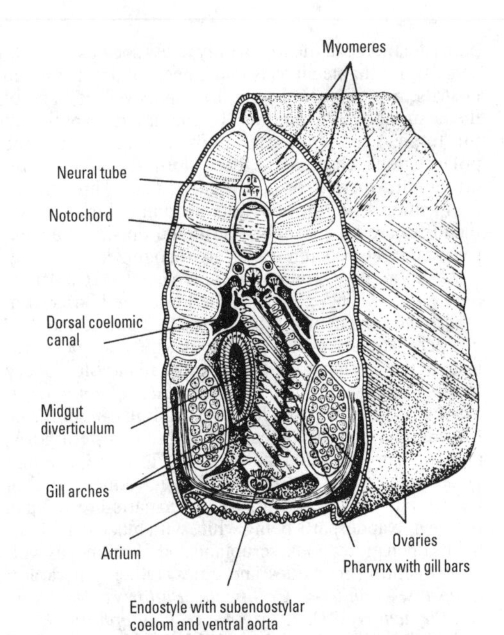

Fig.3. *Pharynx.* Block section of the pharyngeal region of *Amphioxus.*

an annular diaphragm, and by placing a phase plate in the objective. The phase plate is a transparent disk containing an annular groove, which coincides with the image of the substage annulus. Direct light therefore passes through the groove, while diffracted light (from the specimen) passes through the entire phase plate. The resulting phase difference depends on the thickness of the groove, which is usually one quarter of a wavelength thicker or thinner than the rest of the plate. Since the amplitude of the incident light is usually much greater than that of the diffracted light, their mutual interference does not result in total blackness. A layer of absorbing material (e.g. a thin metal film) is therefore usually incorporated in the groove of the phase plate to decrease the amplitude of the incident light.

In addition to facilitating the microscopic observation of living unicellular organisms and unstained tissue, P.c.m. is also used to determine the refractive index of living cells, and (since refractive index varies linearly with concentration) the concentration of dissolved material (roughly equivalent to the protein concentration) within the cell. It is a null point method in which the specimen is observed in media of different refractive index. When the refractive indices of medium and cell are equal, the phase change becomes zero and the cell becomes invisible.

Phaseolin: 3-hydroxy-9,10-dimethylchromenylpterocarpan, a phytoalexin (see) produced by various bean species in response to infection by phytopathogenic microorganisms (e.g. *Phytophthora, Monilinia)* and other forms of stress, such as treatment of the plant with heavy metal salts.

Phaseolus: see *Papilionaceae.*

Phasianidae: the largest family (178 species) of the order *Galliformes* (see). Members are characterized by their naked feet (the tarsus is also usually naked, and is partly feathered at the top in only a few species), the absence of inflatable air sacs in the neck, and unfeathered nostrils. many species also carry spurs on the back of the tarsus above the hindtoe. The natural distribution of this tropical and temperate family does not include the polar regions, Oceana, northern North America, southern South America, or northeastern Asia. Through domestication and introduction as game birds, many members are now widely distributed outside their native range. Three subfamilies are recognized.

1. Subfamily: *Odontophorinae* ***(New World*** or ***American Quails).*** These small to medium-sized birds form a fairly homogeneous group. They lack spurs, and they characteristically possess a "tooth", i.e. a single serration on the cutting edge of the upper mandible. A very familiar representative is the ***Bobwhite*** *(Colinus virginianus;* 22–27 cm), which occurs naturally in southeast Canada, throughout the eastern and midwestern USA, and south to Mexico, the Gulf of Florida, Cuba and Guatemala. It has also been widely introduced, e.g. in Britain and New Zealand. Its upper parts are reddish brown, the underparts being white with black bars. It inhabits open pinewoods, scrubland and farmland, as well as frequenting roadsides and cities. Other genera are: *Oreortyx, Lophortyx, Callipepla, Philortyx, Dactylortyx, Rhynchortyx, Dendrortyx* and *Odontophorus.*

2. Subfamily: *Perdicinae* ***(Old World quails, Partridges*** and ***Francolins).*** These rather plain colored, small to medium-sized birds form a very varied group. There is no "tooth" on the cutting edge of the upper mandible, and most species carry spurs on the back of the tarsus. Represented by about 95 species in 30 genera, the subfamily is widespread over most of Eurasia, Africa and the Australomalayan region. The smallest members belong to the genus *Excalfactoria* ***(Painted quails*** or ***Blue quails)*** which are hardly larger than sparrows. This subfamily also includes the only truly migratory members of the *Galliformes,* i.e. the 4 species of the genus *Coturnix.* Most familiar of these is the mottled brown ***Quail*** *(Coturnix coturnix;* 20 cm) of Eurasia. It formerly migrated in phenomenal numbers from Europe and western Asia to its wintering ranges in Africa, but it has been severely depleted by exploitation for food, especially on its migratory path in the Middle East. ***Partridges*** are slightly larger than quails. The ***Common*** or ***Gray partridge*** *(Perdix perdix;* 30.5 cm) is native to Scandanavia, Ireland and Spain across Europe and Asia Minor to Central Asia, where it frequents farmland, rough grassland, heaths and steppes. A very popular game bird, it has been introduced to the USA, where it is known as the ***Hungarian*** or ***Hun.*** The male has a striking horseshoe-shaped, chocolate-brown breast shield. The largest representatives of this subfamily are the ***Francolins*** and ***Spurfowl,*** some attaining lengths of 46 cm. They are mostly drab-colored inhabitants of brush and open grassland, e.g. ***Swainson's francolin*** or ***spurfowl*** *(Francolinus swainsoni;* 40.5 cm) of sub-Saharan Africa.

3. Subfamily: *Phasianinae* ***(Pheasants*** and ***Peafowl).*** The cocks usually have bright colored plumage, and certain parts like collars, tail fans and eye spots are used for display, especially during courtship; the female is always drab. Spurs are usually present. The long, ornate tail is an inverted V-shaped in cross section. Most widespread of the pheasants is the ***Ring-necked, English*** or ***Mongolian pheasant*** *(Phasianus colchicus).* Its natural distribution extends from the eastern shore of the Black Sea to the Caspian across southern and Central Asia, along the northern slopes of the Himalayas to eastern China and Korea. Within this native range it exists as 5 subspecies, each further subdivided into races, so that at least 40 different races are recognized, differing mainly in the details of their coloring and patterning. The plumage of the male is the color of burnished copper, with pale U-shaped marking on the back, black crescents on the breast and flanks, and bars on the tail; the ears are tufted, and the dark metallic green head is sometimes separated by a white collar. The ring-necked pheasant has, however, been widely introduced as a game bird, and is well known outside its native range. Thus, successful introductions have been made in Europe, America, New Zealand, the Hawaian islands, Chile, St. Helena, King Island, Rottnest Island and Tasmania. It has not been successfully introduced into Australia. Since such care is taken in the preservation of game stocks (e.g. populations often supplemented with hand-reared birds), it is sometimes regarded as a rather artificial member of the local avifauna. There are also 2 native subspecies of the ***Japanese green pheasant*** *(Phasianus versicolor),* in which the plumage of the male is predominantly green with a purplish tinge. ***Junglefowl*** *(Gallus)* are also a type of pheasant. *Gallus gallus* is the ancestor of domestic poltry. Also widely introduced (but for decoration rather than for its meat) is the familiar ***Peacock*** *(Pavo cristatus),* which is native to India and Ceylon. The female of this peafowl is the ***Peahen,*** but in common usage the species is generally referred to as the peacock. Including the long train of its ornamental tail feathers, the peacock attains a length of 200–235 cm.

Phasianidae. Game pheasant *(Phasianus colchicus).*

Phasianus: see *Phasianidae.*

Phasmatodea: see *Cheleutoptera.*

Phasmida: see *Nematoda;* see *Cheleutoptera.*

Phasmatidae: see *Cheleutoptera.*

Phasmodea: see *Cheleutoptera.*

Pheasants: see *Phasianidae.*

Pheasant's eye: see *Ranunculaceae.*

Phegornis: see *Charadriidae.*

Phelsuma, ***day geckos, Madagascar geckos:*** a genus of the *Gekkonidae* (see), found predominantly on Madagascar and neighboring islands. They have round pupils, and are often strikingly colored. In contrast to other geckos, they are active during the day. In addition to insects, many also eat fruit. The most familiar species is *P. madagascariensis,* which is colored bright green with red markings.

Phenocopy: a phenotypic modification, which has no genetic component and arises in response to environmental influences, and which closely resembles a phenotypic modification that is genetically based and arises by mutation. The term was coined in 1935 by Goldschmidt for *Drosophila* variants obtained by heat shock of the larval stages, which resembled certain mutants.

Phenocritical phase: a specific period during phenogenesis, when the developmental processes of different genotypes (e.g. wild type and mutants) diverge, and the expression of different characters becomes recognizable.

Phenogenesis: expression of inherited phenotypic characters during ontogenesis by the interaction of genotype and environment.

Phenogenetics: study of the role of genetic factors in the development of phenotypic characters during ontogenesis.

Phenol oxidase, ***tyrosinase, phenolase:*** an enzyme complex with the activity of a mixed function oxygenase. It catalyses the oxidation of monophenols to the corresponding diphenols, e.g. L-tyrosine to 3,4-dihydroxyphenylalanine (dopa). Further oxidation by P.o. leads to the formation of quinones, in the case of dopa to the formation of dopaquinone and dopachrome, which is then metabolized to brown-black melanins. P.o. is widely distributed in plants, where it is responsible for the darkening of the cut surfaces of plant parts and fruits, e.g. potatoes and bananas. In insects it is responsible for melanin synthesis during sclerotization of the cuticle.

Phenopathy: structural or functional abnormality induced by environmental factors, in contrast to a genetically determined Genopathy (see)

Phenotype: the totality of characters of an individual, which in the widest sense includes all external and internal structures and functions of the organism, which are investigated and described by morphological, anatomical and physiological methods. Resulting from the interaction of genotype and environment, the P. is not a constant property, but can be changed within genetically determined limits by the action of external and internal environmental factors during the development of an individual. Environmental factors include extrachromosomal effects of the cytoplasm on the zygote, as well as the influence of the maternal environment on egg and embryo development. They also include effects on the zygote of substances from outside the organism, and all other internal and external factors that influence the formation of genotypically determined characters during the life of an individual.

L-Phenylalanine, ***L-α-amino-β-phenylpropionic acid, Phe:*** an aromatic, proteogenic amino acid, which is essential in the animal diet, and is both ketogenic and glucogenic. Phe is degraded metabolically via L-tyrosine to fumarate and acetoacetate. It is the precursor of melanin, the neurotransmitter dopamine, the hormones adrenalin, noradrenalin and thyroxin, and other compounds. See Phenylketonuria.

In plants and bacteria, Phe is synthesized by the shikimic acid pathway of aromatic biosynthesis.

$C_6H_5—CH_2—CH(NH_2)—COOH$

L-Phenylalanine

Phenylketonuria, ***phenylpyruvic oligophrenia, Fölling's syndrome:*** a hereditary (autosomal recessive) defect in the synthesis of phenylalanine hydroxylase (the enzyme may be absent or inactive), which affects about 1 infant in 10 000. These individuals are unable to convert phenylalanine into tyrosine, so that the major route of phenylalanine metabolism is blocked. Phenylpyruvate and phenylacetate are excreted in the urine. The condition is characterized by defective pigmentation, seizures, skin eczema and severe mental retardation (hence the name, phenylpyruvic oligophrenia) and decreased life expectancy. The urine of newborns is now routinely tested (Guthrie test) for the presence of phenylketones; the condition can be compensated and mental retardation avoided by a diet low in phenylalanine. A normal diet may be resumed later in life, i.e. in the mid teens when childhood development is complete.

Phenylpyruvic oligophrenia: see Phenylketonuria.

Pheromone: a usually low molecular mass substance produced by animals, especially insects, and secreted outside the body for the purpose of communication with members of the same species. The main structural types are lower terpenes and higher unbranched fatty acids, and there are also ali- and heterocyclic representatives. Depending on their mode of perception, Ps. are classified as oral or olfactory. An animal may show a direct reaction to a P., which ceases when the P. disappears (i.e. a releasing effect) or the response may constitute a long term physiological change (i.e. a priming effect). In most cases a mixture of Ps. is necessary, together with appropriate biotic and abiotic environmental factors.

For the interaction of individual members of a species the most important Ps. are *sexual attractants.* For social insects (e.g. honeybees, ants, termites), *aggregation Ps., alarm Ps.* and *trail Ps.* are also important.

Sex Ps. (sexual attractants) are usually mixtures of various components (activity is, however, usually dominated by a single compound) secreted by one sex of the species to attract the other. Activity of sex Ps. can be studied by electrophysiological measurement of the nerve impulse of isolated receptor organs such as antennae. Known quantities of a P. (10^{-6}–10^{-2} µg) are blown over the isolated antennae in a stream of air, and the resulting cell potential is fed to a recorder with the aid of microelectrodes. The olfactory cells of the silkworm moth or the cockroach produce a measurable nerve impulse in response to a single molecule of the appropriate P. Generally the quantity of a P. produced by one animal is less than 1 µg. The quantity of bombykol (the sexual attractant of the silkworm moth) produced by one female moth is sufficient to attract all male members of the species in existence, if they were within range.

Aggregation Ps. serve e.g. as signals that attract members of the same species to a host plant, or they may repel members of the same species if the population density is too high; they do not usually discriminate between sexes.

Alarm Ps. are secreted by many social insects to communicate danger to other members of the species, e.g. they are used by ants and termites to signal to the rest of the colony that the nest is endangered or an intruder has been detected.

Trail Ps. are used to mark trails which can be recognized and followed by members of the same species.

Ps. are used for the investigation of biological information transfer (signalling), receptor theories and structure-activity relationships. Some P. are used for the control of insect pests (see Biological pest control).

Pheomelanins: see Melanins.

Phialide: a conidia-forming cell of ascomycete fungi. It is fusiform truncate, fusiform beaked, or acuminate in shape, and located at the terminus of a hypha Often entire chains of conidia (phialospores) are abstricted from the apex of the P., e.g. in the mold *Penicillium.* Cross-walls separating the spore from the phialide are not formed until the spore is fully grown. In other cases, e.g. in the family *Sordariaceae,* the spores are formed within the P.

Phialospores: spores formed on a Phialide (see).

Philepitta: see *Philepittidae.*

Philepittidae, ***asites:*** a relict avian family (4 species in 2 genera) in the order *Passeriformes* (see), found only in the forests of Madagascar. The ***Velvet asity*** *(Philepitta castanea)* of eastern Madagascar is 15 cm in length and resembles a tree-living, fruit-eating pitta. The little-known ***Schlegel's asity*** *(P. schlegeli)* is found in the west of the island. The 10 cm-long ***False sunbird*** *(Neodrepanis coruscans)* has brilliant plumage and resembles the true sunbirds in its habits. *N. hypoxantha* of western Madagascar is little known. Both species of *Neodrepanis* are solitary and keep to dense cover.

Philodendron: see *Araceae.*

Philomachus: see *Scolopacidae.*

Philortyx: see *Phasianidae.*

Philydorinae: subfamily of the *Furnariidae* (see).

pH-indicator media: bacteriological growth media containing a pH indicator, which reveals the production of acid or alkali during bacterial growth. Their importance lies in the wide use of fermentative ability as a marker in genetics and taxonomy. Such media are used particularly in medical bacteriology for the differentiation of enterobacteria. A few selected examples are as follows. MacConkey medium, which contains bromocresol purple, is initially blue in color but turns yellow due to acid production by *Escherichia coli.* Phenol red and different sugars (e.g. maltose, sucrose, lactose, mannitol, inositol, etc) may be incorporated into broth media; if the tested bacterium is capable of fermenting the particular sugar, the medium turns from red to yellow due to acid production. Agar media containing eosin and methylene blue, as well as a selected sugar, are initially blue-black in color; nonfermenting colonies are colorless on the dark background, whereas fermentors grow as black colonies with a metallic sheen. Agar media containing sodium deoxycholate, neutral red and an appropriate sugar are initially transparent and reddish brown in color; acid-producing colonies become bright red and opaque, whereas nonfermenters are colorless and transparent.

Depending on the medium and the bacterial species, fermentation does not always result in a decrease of pH, because the alkali produced by deamination of amino acids may exceed the acid produced by the fermentation of carbohydrate. Thus, *Alcaligenes* gives an alkaline reaction in the litmus milk test.

Phlaeothripidae: see *Thysanoptera.*

Phlebitis: see Thrombus.

Phlebobranchiata: see *Ascidiacea.*

Phleum: see *Graminae.*

Phlobophenes: see Catechins.

Phloem: see Conducting tissue, Shoot.

Phlorizin: a crystallizable, sweet-tasting, poisonous glucoside, which is present in the roots and bark of apple, pear, plum and other members of the *Rosaceae.* Hydrolysis of P. produces glucose and the aglycon, *phloretin.* In animals, P. causes a condition of glucosuria (excretion of glucose in the urine), which is used as an experimental model of diabetes. It specifically blocks resorption of glucose by kidney tubules, possibly by inhibition of mutarotase.

Phlox: see *Polemoniaceae.*

Phoca: see *Phocidae.*

Phocarctos: see *Otariidae.*

Phocidae, ***earless seals:*** the most abundant of the aquatic carnivores (see *Pinnipedia).* The family contains 18 species in 13 genera. On land, the hindflippers are useless, and progress is made by humping and writhing the body. External ears are absent. Representatives are found along most northern and southern coastlines, sometimes ascending large rivers. The ***Hair seal*** or ***Harbor seal*** *(Phoca vitulina)* inhabits the ocean shores of the Northern Hemisphere, often ascending large rivers in pursuit of fish. It is preyed on by the Killer whale *(Orca)* and the Polar bear *(Thalarctos).* The large, long and sinuous ***Leopard seal*** *(Hydrurga leptonyx)* is the only seal that preys regularly on warm-blooded animals. It feeds on penguins, also taking other birds, and even other seals, as well as cephalopods. *Hydrurga* has a scattered distribution in the Antarctic, where it frequents the pack ice in summer and migrates to more northerly, ice-free islands in the winter. It has 2 main types of pelage: dark gray to black above and light gray below, or light gray above and below with conspicuous black spots. Largest of the pinnipeds are the 2 species of *Mirounga* ***(Elephant seals*** or ***Sea Elephants):*** *M. angustrirostris* (yellowish or grayish brown, lighter underneath; Californian coast and Galapagos Islands) and *M. leonina* (bluish gray, lighter underneath; sub-Antarctic waters). Elephant seals are gregarious and polygamous. Bulls attain a length of about 6.5 m, and can weigh up to 3.5 tonnes. Females are smaller. The male has short, trunk-like proboscis. which he inflates when excited. They feed on fish and cephalopods, and are able to dive to depths of several hundred meters. The ***Hooded seal*** *(Cystophora cristata)* (Fig.) is a large species (males 2.1–3.3 m, females up to 3.1 m) associated with the ice pack in the deep waters of the Atlantic and Arctic Oceans. Both sexes have a nasal hood or pouch, which is larger in the male. When excited or annoyed, males inflate one side of the thin nasal septum and the hood, to form a brilliant red bladder. After losing their white birth coat, young hooded seals (known as "bluebacks") acquire an attractive gray coat, which is sought by sealskin hunters. The ***Ringed seal*** *(Pusa hispida)* is probably the smallest of the pinnipeds (140 cm in length, weighing up to 90 kg). Although its coat color and pattern are highly variable, white spots with dark centers are often present. The underparts are usually whitish. It is distributed along, and follows the seasonal movements of, the edge of the ice of the North Pole. One of the most common and widely distributed seals, it is also found in certain land-locked lakes such as Lake Baikal and the Caspian Sea. The ***Gray seal*** *(Halichoerus grypus)* inhabits the temperate waters of the North Atlantic, and the young are born on the ice in late winter and spring. Separate population groups in the Northwest Atlantic, Northeast Atlantic, and the Baltic Sea have rather different breeding times.

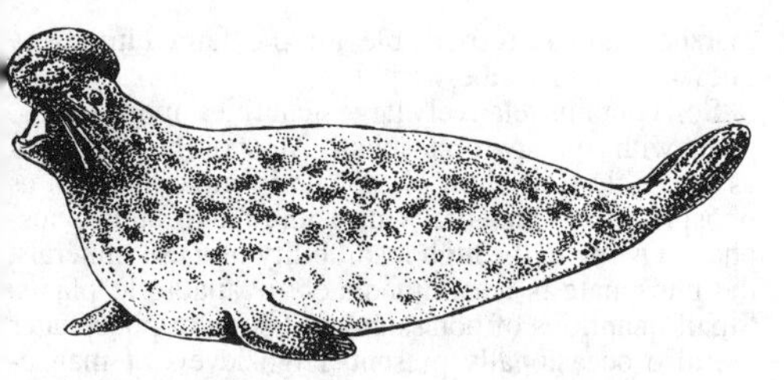

Phocidae. Hooded seal *(Cystophora cristata).*

Phocoena: see Whales.
Phocoenidae: see Whales.
Phocoenoides: see Whales.
Phodopus (dwarf hamsters): see *Cricetinae.*
Phoebetria: see *Procellariiformes.*
Phoeniconaias: see *Ciconiiformes.*
Phoenicoparrus: see *Ciconiiformes.*
Phoenicopteridae: see *Ciconiiformes.*
Phoenicopterus: see *Ciconiiformes.*
Phoenicurus: see *Turdinae.*
Pholadomyoida: see *Bivalvia.*
Pholididae: see Butterfish.

Pholidota, ***pangolins*** or ***scaly anteaters:*** an order of mammals containing only 7 species in a single genus *(Manis;* family *Manidae).* The body surface is covered with large, sharp-edged, pointed, movable horny plates, which overlap like roof tiles, and which are used for attack as well as defense. The underside of the body and the inner surfaces of the limbs lack scales, but by rolling itself into a ball the animal presents an effective shield to an aggressor. The powerful digging claws of the forefeet are used to tear open ant and termite hills, whose inhabitants are collected with the aid of the long, vermiform, sticky tongue. There are no teeth. The 7 species are: ***Common*** or ***Chinese pangolin*** *(Manis pentadactyla;* the smallest species: length about 1 m; Nepal, Burma, northern Thailand, southern China); ***Indian pangolin*** *(M. crassicaudata;* India and Sri Lanka); ***Malayan pangolin*** *(M. javanica;* Burma, Thailand, Indo-China, Peninsular Malaya, Borneo and other islands); ***Giant pangolin*** *(M. gigantea;* the largest species: length about 2 m; scrub and grassland in East Africa from Ethiopia to South Africa); ***Cape pangolin*** *(M. temmincki;* scrub and grassland in East Africa, especially in the south); ***Small-scaled tree pangolin*** or ***White-bellied pangolin*** *(M. tricuspis;* an arboreal species in the lower levels of the rainforest from West Africa to Rift Valley); ***Long-tailed pangolin*** *(M. tetradactyla;* same range as *M. tricuspis,* but in the tree canopy). All 3 Asian species are considered to be endangered, and appear in appendix 2 of CITES. The Cape pangolin is in appendix 1 of CITES.

Pholidota. Cape pangolin *(Manis temmincki).*

Phonotaxis: positioning or movement of the body in relation to the direction of sound.

Phoresy: see Relationships between living organisms.

Phormium: see *Agavaceae.*

Phoronida, *Phoronidea,* ***horseshoe worms:*** a uniform phylum (or class of the *Tentaculata)* containing only a few species, all of them marine.

Morphology. The mouth opening is partly covered by an epistome and surrounded by a horseshoe-shaped filter feeding organ (lophophore), which is spirally coiled at its ends, and carries ciliated tentacles. The body is long and wormlike, but the intestine is U-shaped, so that the anus opens to the exterior on the dorsal surface near to the lophophore. Adult organisms are sedentary, and live inside a secreted chitinous tube.

Biology. Some P. are hermaphrodite. Gametes leave the organism via the nephridial canals. Eggs remain among the lophophore tentacles, where they develop into a free-living actinotrocha larva, which has an anterior ("cerebral") ganglion, and two rings of cilia (posterior ring and a circumanal ring). The definitive tentacles of the adult organism develop during metamorphosis.

All *P.* are suspension feeders. Microorganisms and detritus are trapped on a mucilage film and transported into the mouth. The secreted chitinous tubes of the sessile adults range in length from 1 mm to 45 cm, and often contain other extraneous material. They are found individually or in colonies, embedded vertically in the mud or sand substratum. The 18 known living species are all placed in the single family, *Phoronidae.*

For collection and preservation see *Tentaculata.*

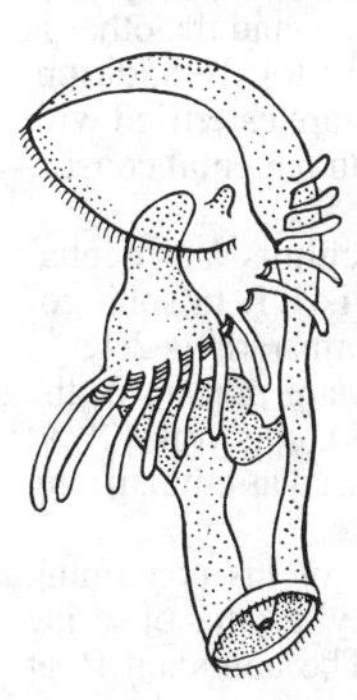

Phoronida. Actinotrocha larva of a horseshoe worm.

Phoronidae: see *Phoronida.*
Phoronidea: see *Phoronida.*
Phorozooids: see *Doliolida.*

Phosphagen: an energy-rich guanidinium or amidine phosphate, which functions as a storage compound for high energy phosphate in muscle. The phosphoryl group is transferred from excess ATP to the P., from which ATP can be regenerated when required. In invertebrates, arginine phosphate in the commonest P. (ATP + arginine $\rightleftharpoons$ arginine phosphate + ADP; catalysed by arginine kinase). Other invertebrate Ps. include lombricine phosphate and taurocyamine phosphate. The P. of vertebrate muscle is creatine phosphate (ATP + creatine $\rightleftharpoons$ creatine phosphate + ADP; catalysed by creatine kinase).

Phosphatases: hydrolases which catalyse the hydrolytic cleavage of phosphate monoesters, with the release of inorganic phosphate. They have a wide natural distribution and are important in many areas of metabolism. Common substrates are the phosphate esters of carbohydrates, in particular hexoses. P. are also involved in the synthesis of inorganic bone material. According to their pH-optimum, P. are classified as acid P. (pH optimum about 5) or alkaline P. (pH optimum 7–8).

Phosphatides: see Phospholipids.

Phosphatidic acids: nitrogen-free phospholipids, consisting of *sn*-glycerol-3-phosphate esterified with a saturated fatty acid at C1 and an unsaturated fatty acid at C2. Small quantities of P. occur in mammals, but are more common in plants, e.g. in cabbage.

Phosphoadenosine phosphosulfate: see Sulfur (section on Assimilation).

Phosphodiesterases: esterases which catalyse hydrolytic cleavage of phosphodiesters. Important examples are Deoxyribonucleases (see) and Ribonucleases (see).

Phospholipases: a collective term for the carboxylic acid esterases, A_1, A_2 and B, and the phosphodiesterases, C and D, which are specific for lecithins. A_1 catalyses the release of the fatty acid at C1 of the glycerol, producing a lysophosphatide (a 2-acylglycerophosphocholine) which hemolyses erythrocytes. A_2 removes the unsaturated fatty acid at C2. B also removes the unsaturated fatty acid, but only from lysophosphatides. C releases the base in its phosphorylated form. D cleaves on the other side of the phosphoryl group and releases the nonphosphorylated base. P. activity is particularly high in liver and pancreas (A_1), bee and snake venom (A_1 and A_2), bacteria (C) and plants (D).

Phospholipids, ***phosphatides:*** phosphate diesters, in which one esterification partner of the phosphoric acid residue is glycerol or sphingosine, and the other is choline, ethanolamine, serine or inositol. In addition, saturated and unsaturated fatty acids are esterified with the glycerol or sphingosine. P. are fundamental constituents of cell membranes.

In glycerophospholipids (e.g. lecithins and cephalins), the nitrogen/phosphorus ratio is 1:1. In sphingophospholipids, the nitrogen/phosphorus ratio is 2:1.

Particularly large quantities of P. are present in the brain (30%), liver (10%), heart (7%), egg yolk (20%) and soybeans (2.5%). Lecithins are used as emulsifying agents.

Phosphoproteins: conjugated proteins containing phosphate esterified with the hydroxyl groups of serine or (less often) threonine residues. The abundant P. of milk and egg yolk are nutritionally important because they contain essential amino acids. See Caseins, Vitellin.

Phosphorescence: see Absorption.

Phosphorolysis: a process catalysed by phosphorylases, in which the glycosidic linkages of starch or glycogen are cleaved by addition of phosphoric acid. The degradation occurs stepwise from the end of the polysaccharide molecule. At each stage, one molecule of glucose is removed as glucose 1-phosphate, and the polysaccharide substrate decreases in length by one glucose unit.

Phosphorus, ***P:*** a nonmetal element, and an essential constituent of all living organisms, where it is present as inorganic or organically bound (esterified) phosphate. Its uptake, assimilation and utilization all occur at the oxidation level of phosphate, and more reduced forms have no biological relevance [with the exception of phosphine (PH_3), which is formed under the reducing conditions in the bottom mud of bogs and marshes, and is responsible for the faint luminosity known as "will-o'-the-wisp"].

Soil contains relatively large quantities of orthophosphate with some pyrophosphate. Most of this is present as insoluble tricalcium phosphate in minerals such as phosphorite and apatite, with some magnesium phosphate. Owing to the high insolubility of these minerals, the phosphate is not immediately available to plants. Small quantities of potassium and sodium phosphates are also occasionally present. High levels of magnesium and iron phosphates occur in acidic soils. Variable quantities of organically bound phosphate are also present in the residues of dead plants and microorganisms, but this phosphate is also not immediately available to plants. Weathering of phosphate minerals by CO_2-containing soil water, as well as the rapid degradation of organic material (mineralization) produce small quantities of soluble phosphate, which can be utilized by plants. Most agricultural soils are potentially deficient in available phosphate, and it is necessary to add phosphate fertilizers. These must contain phosphates that are water-soluble or soluble in weak acid (e.g. citric acid). On very basic soils, particularly chalk soils or freshly limed soils, part of the available phosphate is converted rapidly into insoluble, unavailable calcium phosphate. In strongly acidic soils containing free aluminium and iron ions, insoluble aluminium and iron phosphates are formed. Phosphate fertilizers are most efficient on soils with pH-values between 6 and 7.

Plant roots take up the phosphate ions, $H_2PO_4^-$ and HPO_4^{2-}, which are formed from insoluble phosphates by the action of protons and organic acids (e.g. citric and malic acids) secreted by plant roots. Specific energy-dependent carriers are responsible for the uptake. Once inside the plant, phosphate is rapidly incorporated into organic cell constituents. Free phosphate ions are also always present in living cells and tissue fluids. The subsequent distribution and concentration of phosphate depend strongly on the developmental stage of the plant. The average content generally corresponds to 0.1–0.4% phosphorus on a dry mass basis. Reproductive organs contain more phosphate than vegetative parts. Young leaves contain relatively large amounts of organically bound phosphate, whereas old leaves contain relatively more inorganic phosphate.

Phosphate is a component of nucleic acids (see Ribonucleic acid, Deoxyribonucleic acid), Phospholipids (see) of cell membranes, and phosphoproteins (in which it is esterified with serine and more rarely with threonine residues). Many coenzymes contain one or more residues of phosphate, e.g. flavin nucleotides, pyridine nucleotide coenzymes, thiamin pyrophosphate, and coenzyme A. Adenosine triphosphate (ATP) and other phosphorylated nucleosides occupy a central position in energy metabolism (see Oxidative phosphorylation, Respiratory chain). Transfer of phosphate from ATP to glucose to form glucose 6-phosphate serves to introduce glucose into the pathway of glycolysis. By phosphorolysis, stored glycogen and starch are mobilized for glycolysis. Phosphate esters are intermediates in numerous metabolic processes, in fact phosphate is involved in practically all the central pathways of metabolism. In animals, phosphate is a constituent of bones and teeth (see Mineral metabolism). Many seeds and fruits contain a phosphate reserve material called phytin, which is a calcium-magnesium salt of phytic acid (myoinositol hexaphosphate); e.g. globoids of amorphous phytin are commonly present in the aleurone layer of cereal grains

P. deficiency causes widespread impairment of metabolism. Phosphate-deficient plants ars small and rigidly upright with a thin stem and stiff, dirty green leaves, often with tones of red and bronze. Older leaves are often shed prematurely. In cereals, tillering is decreased. Formatioin of seeds and fruits is particularly affected. Deficiency can be alleviated by adjusting the soil pH to 6–7, but the main practical measure consists of applying phosphate fertilizers. Potatoes have an especially high requirement for phosphate, and they respond markedly to phosphate fertilizers. High phosphate requirements are also shown by sugar beet, red clover and lucerne.

Photichthyidae: see *Stomiiformes.*

Photoacoustic spectroscopy, ***optoacoustic spectroscopy:*** a method for the determination of absorption spectra in gases, suspensions, gels and solids. The sample under investigation (e.g. a leaf, a whole cell) is irradiated with modulated light of tunable wavelength. Absorption of the light leads to modulation frequency-dependent changes of temperature and density in the sample (i.e. sound waves). The intensity of the resulting acoustic signal (detected and recorded with a microphone and amplifier) is a measure of the wavelength-dependent light absorption.

Photochemical reaction: a chemical reaction that is initiated by the absorption of light by one of the reactants. The excited state of the molecule differs from its ground state with respect to electronic configuration, symmetry or some similar property. Generally, the pathway of a chemical reaction promoted by thermal agitation of the ground state is different from the reaction pathway of a photo-excited molecule.

P.rs. have considerable biological significance, e.g. in photosynthesis and in the formation of atmospheric pollutants. An important unimolecular P.r. is the *cis-trans* isomerization of retinal in the primary reaction of the visual process. Catalysts of P.rs. are called *sensitizers.* Incompletely degraded products of hemoglobin metabolism may be involved in the development of skin cancer, by acting as sensitizers of the photochemical excitation of molecular oxygen. The use of light-induced oxidation via stable sensitizers (i.e. the *photodynamic effect)* is now used for cancer therapy (see Photodynamic therapy).

Photodinesis: see Dinesis.

Photodynamic therapy: selective incorporation of photosensitive materials into malignant tumors, followed by irradiation with laser light of an appropriate wavelength. The resulting photochemical reaction generates toxic radicals that kill the cancer, or slow its growth so that other therapies are more effective. Several different photosensitizers have now been used with varying success on hundreds of patients. The most commonly used photosensitizers are hematoporphyrin derivatives, e.g. 5,10,15,20-tetra(hydroxyphenyl)porphyrins, and in particular a compound known as Photofrin, which was the first photosensitizer to be approved for clinical use in any country (Canada). More recently, a promising new compound called benzoporphyrin derivative monoacid (BPDMA) has been announced, which clears much faster from the body than photofrin, and does not induce skin photosensitivity more than one day after injection. *[Photodynamic Therapy and Biomedical Lasers,* edit. P.Spinelli, M. Dal Fante, R. Marchesini, Excerpta Medica, Amsterdam/London/New York, 1992; *Photodynamics Newsletter,* Issue 7, 1993, edit S.B. Brown, ISSN 0960-6459]

Photoinduction: see Light as an environmental factor.

Photomorphogenesis: regulation of plant morphology by the influence of light during differentiation and development, without involvement of chlorophyll or photosynthesis. The importance of light in morphogenesis is well illustrated by the deviations from normal growth and differentiation that occur when light is deficient (see Etiolation). In dicotyledons, light affects primarily the longitudinal growth of the shoot (inhibited by light), the number and size of leaves (strongly promoted by light), chloroplast differentiation as well as the biosynthesis of chlorophyll and anthocyanins (these occur only in light), differentiation of vessels and support elements (strongly promoted by light) and many other phenomena. Most of these light-dependent photomorphoses are positive, i.e. they are promoted by light. Onset of illumination and the beginning of the response are often separated by a time lapse of several hours. In such cases, it is conceivable that light exerts its effect through gene activation. In contrast, the inhibition of extension growth by light (a negative photomorphosis) occurs within a few minutes of illumination, indicating that gene repression is not responsible for the inhibition of extension growth.

The light signal is received and transduced by different *photoreaction systems,* in particular by the phytochrome system (red-far red pigment system). One particular pigment system, which displays absorption maxima in the blue and in the red region of the visible spectrum and responds to only relatively high light energy (i.e. prolonged or intense illumination), is thought to be responsible for high energy responses. Every photomorphosis is not necessarily directly induced via a photoreceptor or pigment system; some may be initiated by preceding photomorphoses. Feedback control and correlation may also operate in P.

In addition to the morphogenetic processes of normal (nonetiolated) growth and development, light also influences the polarity of the egg cells and gonospores of *Fucus,* mosses, ferns and horsetails. Light also induces dorsiventrality in the gemmae of the liverwort, *Marchantia,* and in fern prothalli. The branching pattern of many trees is determined by light, because well illuminated buds extend before those on the shaded side. Also, well illuminated leaf buds on the southern sunny side of a tree (e.g. beech) develop into leaves with a deeper palisade tissue than the shade leaves in the interior of the crown and on the northern side of the tree. Strong light is also responsible for the modification of the radial shoots of *Opuntia* into flattened shoots. Numerous photomorphoses have also been described in fungi, where blue and ultraviolet light are the most effective. Formation of fungal sporangia, sporangiophores and conidia are induced by light, and the sex of gametangia is determined by light. Etiolation of fungi (development of a long stipe and small cap) is prevented by blue or ultraviolet light.

Photonastic response of stomata: see Stomata.

Photonasty: nastic movement in response to a change of light intensity, the direction of the light being immaterial. Photonastic opening and closing movements of flowers and leaves are exhibited by many plants. Thus, the flowers of gentian, water lily and evening primrose, as well as the capitula of *Compositae* (e.g. dandelion) open when illuminated and close in the dark or in dull weather. Night-flowering species, e.g. Nottingham catchfly *(Silene nutans)* and white campion *(Melandrium album)* show the inverse behavior. Also some leaf laminas (e.g. of water balsam, *Impatiens glandulifera;* and wood sorrel, *Oxalis acetosella)* are raised in the light and lowered in the dark. These photonastic movements are very often due to differential growth; in day-flowering plants, growth on the upper surface of the petal base is promoted by light and inhibited in the dark. In some cases, e.g. in the fully grown leaves of wood sorrel, photonastic raising and lowering of the leaf is due to turgor changes in a pulvinus at the junction of the petiole and axis.

Photoperiodic flowering stimulus: see Flower development.

Photoperiodism: regulation of growth and developmental processes in plants by the duration of the light period *(day length* or *photoperiod)* in the 24-hour cycle.

In many plants, the following phenomena of growth and development are influenced by the photoperiod: growth rate, internode length, cambium activity, leaf shape, formation of tubers, bulbs and other storage organs, the shedding of leaves, onset of bud dormancy, frost resistance, seed germination, in violets the development of normal flowers (short day) or cleistogamous flowers (long day), and other phenomena. Best studied is the photoperiodic regulation of the induction of flowering. The entry, Flower development (see), should be read in conjuction with the present text.

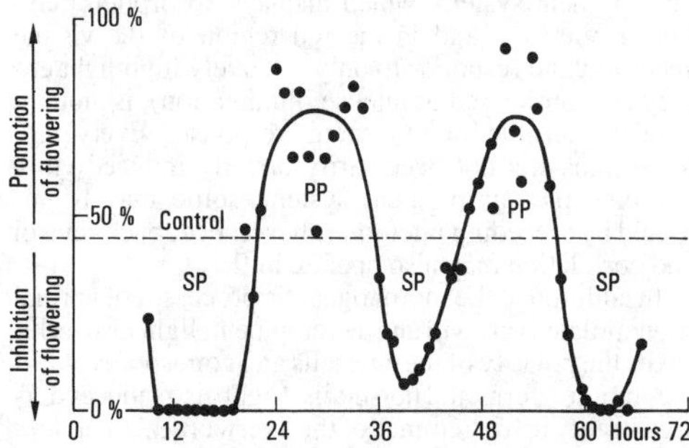

Fig. 1. *Photoperiodism.* Flower development in soybean (a short day plant) is sensitive to interfering illumination during a 64-hour dark phase, and this sensitivity displays a circadian rhythm. The plants received 7 treatment cycles, each consisting of 8 hours light and 64 hours dark. The controls, which received no interfering illumination, developed an intermediate number of flowers. The other plants received 4 hours of interfering light at different times during the 64-dark period. Absissa; onset of interfering light (hours after the beginning of the illumination period). Ordinate: flower development (number of flowers as a percentage of the maximum). *SP* = scotophilic phase. *PP* = photophilic phase.

A photoperiodic response is induced by low illumination intenstities (0.1 to 1 lux). The number of inductive cycles required to induce flower development differs widely according to the species, varying from a single cycle (short day plant, *Xanthium,* 1 short day; long day plant, *Lolium temulentum,* 1 long day) to many cycles (short day plant, *Salvia occidentalis,* 17 short days; long day plant, *Plantago lanceolata,* 25 long days). The photoperiodic signal is perceived by the foliage leaves, usually by Phytochrome (see), sometimes (e.g. in *Hyoscyamus)* by the High energy system (see). It is transduced into a chemical signal, which is transported to the target organ. As discussed under Flower development (see), the length of the dark period, rather than the light period, is critical for the photoperiodic effect. If the period of darkness of a short day plant is briefly interrupted by illumination (in extreme cases only one minute of illumination) the plant fails to flower. Long day plants will flower under short day conditions, if the dark period is interrupted by a relatively long period of illumination. The timing of this *interfering illumination* is, however, important. Thus, interfering illumination in the dark phase of a short day plant (e.g. soybean; light phase 10 h) is most effective when applied 16, 40 (= 16 + 24) or 64 (= 16 + 24 + 24) hours after the start of the light phase (Fig.1). At other times (and other times plus 24 hours) the effect is less. At certain times of interfering illumination, e.g. 28 and 52 hours after the start of the last light phase, flower development is even promoted. Thus, the circadian (24–hour) rhythm contains a *scotophilic phase* (in which light inhibits flower development) and a *photophilic phase* (in which light promotes or initiates flower development). The latter displays peaks at 4, 28 and 52 hours after the start of the light period. Long day plants often display only one photophilic phase, which recurs every 24 hours, and which is shifted by 12 hours in comparison with the photophilic phases of short day plants. These findings show that the peak of the photophilic phase occurs 4 hours after the start of the light phase, whereas the peak of the scotophilic phase occurs 12 hours later. Plants in which the peak of the photophilic phase is about 12 hours after the start of the light phase are in fact long day plants. In long day plants, the first few hours of illumination are ineffective in the initiation of flower development, which does not occur unless illumination is continued. The start of illumination in photoperiod-sensitive plants therefore acts as an environmental cue (Zeitgeber) for measuring the duration of illumination (Fig.2), and this measurement is performed by an endogenous physiological clock (see Biorhythm).

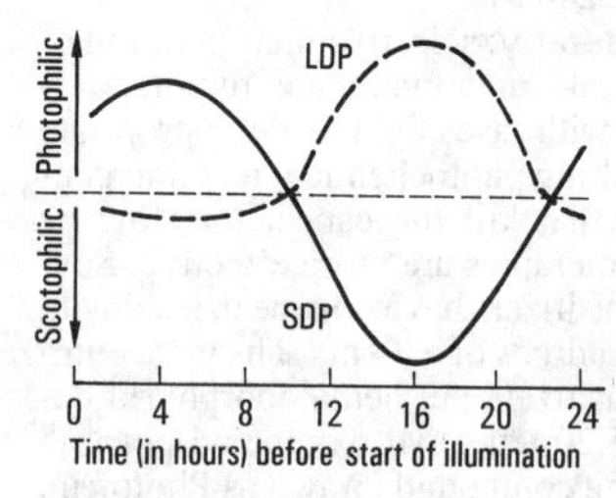

Fig. 2. *Photoperiodism.* Relative positions of the scotophilic and photophilic phases in the circadian rhythms of short day plants (SDP) and long day plants (LDP).

Photophores: light-producing organs (modified glands) in the skin of deep sea fishes, sometimes equipped with auxilary structures, such as lens, reflector and pigment. The light is either produced by resident bacteria, or it is generated by oxidative processes of the secretory cells (see Bioluminescence). P. serve to deter predators, to attract prey, and in the recognition of sexual partners.

Photoreception: the generation of a primary electrical response in light-sensory cells by the absorption of light quanta. A light sensory cell can be stimulated by a single light quantum, which promotes a flow of energy through the membrane stacks of the outer segments, resulting in the release of stored metabolic energy in the form of an electrical potential gradient. The resulting amplification factor can attain a value of 10^6.

After absorbing a light quantum, rhodopsin (a membrane-bound chromoprotein) undergoes several conformational changes, leading to release of the pigment, retinal. The actual light-sensitive photochemical reaction is a cis-trans isomerization of the retinal in the rhodopsin.

Photorespiration: light enhanced oxygen consumption and carbon dioxide production in photosynthesizing plant cells. P. differs fundamentally from normal respiration, with respect to the metabolic reactions involved and the organelles in which the process occurs. Furthermore, P. occurs only in illuminated photosynthesizing cells, whereas normal respiration occurs in all aerobic cells in both the presence and absence of light. P. results in a loss of photosynthetic yield of C3 plants.

The Fig. shows the reactions of P. and their compartmentalization in different organelles of the cell. The substrate of P. is ribulose diphosphate (RuDP), which is produced by Photosynthesis (see) in the Calvin cycle. At high partial pressures of CO_2, RuDP is carboxylated to form 2 molecules of phosphoglyceric acid, which are subsequently converted to glyceraldehyde 3-phosphate and finally to glucose. In contrast, at high partial pressures of O_2 (which are rapidly produced in C3 plants during vigorous activity of photosystem II), RuDP is converted to phosphoglycolate and phosphoglyceric acid (PGS) with the uptake of oxygen and water, a reaction catalysed by ribulose bisphosphate carboxylase which also has oxygenase activity (enzyme 1 in Fig.). If the natural O_2 partial pressure of the atmosphere decreases, the P. of C3 plants decreases and Net photosynthesis (see) increases.

The phosphoglycolate is dephosphorylated by phosphoglycolate phosphatase (enzyme 2), and the resulting glycolate leaves the chloroplasts and enters the peroxisomes, which in leaves are usually closely associated with chloroplasts. In the peroxisomes, the glycolate is oxidized to glyoxylate by the action of glycolate oxidase (enzyme 3); this reaction is coupled to the production of hydogen peroxide, which is rapidly cleaved to water and oxygen by catalase (enzyme 4, the marker enzyme of peroxisomes). The glyoxylate can be degraded, via oxalate, to formate and CO_2 (reaction 8). Alternatively, it may return to the chloroplasts, where it is reduced back to glycolate by the action of glyoxylate reductase (enzyme 7) (the NADPH + H^+ consumed in this reaction is produced in the light reaction of photosynthesis). Circulation of glyoxylate in this way, between chloroplasts and peroxisomes, leads to the futile oxidation of considerable quantities of the NADPH + H^+ produced in the light reaction. However, glyoxylate can be transaminated to glycine (reaction 5), which leaves the peroxisomes. By the action of serine synthase (enzyme 6), 2 molecules of glycine give rise to one molecule of serine and one molecule of CO_2. It is not clear whether this last reaction occurs in the chloroplasts, mitochondria or both. One possible function of P. is the synthesis of glycine and serine. In C4 plants, P. is either absent or extremely low, primarily because the O_2 partial pressure in C4 plants is always below that required to promote P.

Photosynthesis: the assimilation of carbon dioxide by green plants and by pigmented, photosynthetically active prokaryotes, and its conversion into carbohydrate, using energy from sunlight. The conversion of carbon dioxide into carbohydrate can be described by the equation:

$$6CO_2 + 12H_2O \xrightarrow{+\text{ energy}} C_6H_{12}O_6 + 6CO_2 + 6H_2O.$$

P. is the most important biochemical process for the maintenance of life on this planet. It supports the existence of autotrophic green plants and prokaryotes, and these organisms in turn (especially green plants) are the ultimate source of assimilated, high-energy food material for all other forms of life, i.e. nearly all food chains, however long or tortuous, eventually lead to a photosynthetic food source. It is estimated that photosyn-

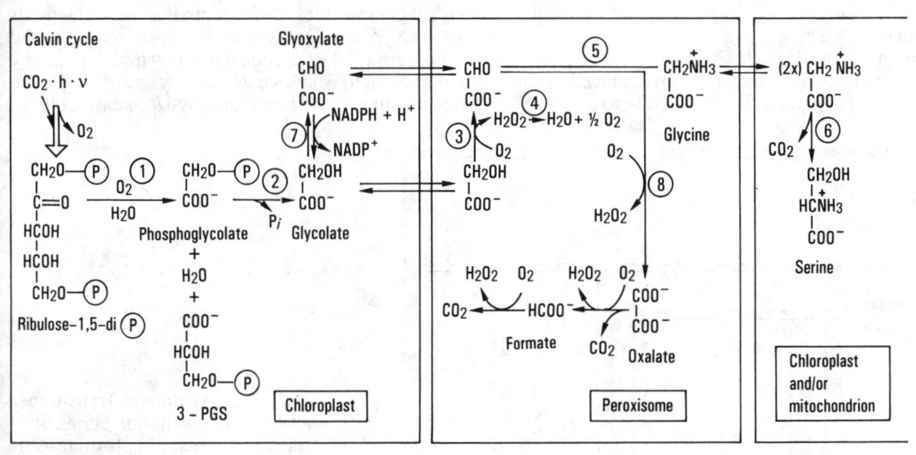

Photorespiration. Reactions of photorespiration in different cell compartments.

thetic plants, including marine algae, remove 150 billion (150 x 10^9) tonnes of carbon in the form of CO_2 from the atmosphere each year. Of this, 20 billion tonnes are accounted for by land plants. These quantities represent about 3% of the carbon dioxide of the atmosphere and 0.3% of that in the oceans (see Net photosynthesis).

Reactions of P. P. consists of two functionally and spatially separated processes: 1) the *light reaction* (generation of energy), which occurs in the thylakoids of the chloroplasts, and 2) the *dark reaction,* which occurs in the matrix of the chloroplasts.

1) The *light reaction,* which is initiated by the absorption of light quanta (photons) by the photosynthetic pigments, results in the reduction of $NADP^+$ to NADPH + H^+, and the phosphorylation of adenosine 5'-diphosphate (ADP) to adenosine 5'-triphosphate (ATP). The overall process can be formulated as:

$$12H_2O + 12(NADP^+) + n(ADP + P_i) \xrightarrow{\textbf{light } (h\nu)} 6O_2 + 12(NADPH + H^+) + n(ATP).$$

Light quanta are absorbed by chlorophylls and carotenoids (carotenes and xanthophylls). Chlorophylls absorb in the blue and red regions of the spectrum, while carotenes absorb in the blue and blue-green regions. Red algae and cyanobacteria also possess phycobiliproteins, which absorb in the green and yellow regions. In all other photosynthetic organisms, however, the green and yellow regions of the spectrum are not absorbed and make no contribution to P. Quantum absorption raises the pigment molecule from its ground state to a short-lived, high-energy state, i.e. an electron or electron pair is raised to a higher energy level. As the molecule reverts to the ground state, the absorbed energy may be released as light quanta (i.e. fluorescence) or as heat; under these circumstances no further work is performed. However, in order to contribute to P., the energy is transferred to other molecules such as chlorophyll $a_{I\,700}$ (pigment 700, see below) and chlorophyll $a_{II\,680}$ in photocenters 1 and 2 (see below). The transitions between the excited states of chlorophyll, following the absorption of blue or red light, are shown in Fig. 1. Accessory pigments (see) perform no photochemical work, but serve to absorb light and transfer the energy more or less without loss to the photosynthetically active pigments.

Transfer of electronic excitation energy to the photochemical center occurs essentially by induced resonance, i.e. as an excited electron falls back to a low energy state, it excites an electron in a neighboring pigment molecule (i.e. an electron of this molecule is raised to a higher energy level) by a dipole-dipole interaction, which does not involve radiation. When this molecule in turn falls to a low energy state, the excitation is passed on similarly to another neighboring chlorophyll molecule. For this energy transfer to occur, the fluorescence emission spectrum of the first excited molecule must overlap the absorption spectrum of the second molecule. The greater the overlap, the more efficient is the energy transfer. Thus, with minimal loss, the electronic excitation energy can be transferred from one molecule of chlorophyll *a* or *b* to another. On the other hand, the transfer from carotene to chlorophyll occurs with greater losses of energy. This type of energy transfer is only possible, however, if the energy required for excitation of the second molecule is equal to or less than the energy being transferred. Thus, although the direction of energy flow is in itself random, it naturally proceeds from pigments of high excitation energy to those whose excited state are less energetic. For example, in photosystem 1 (see below) the following electron transport pathway can be envisaged: Carotene → chlorophyll *b* → chlorophyll a_{660} → chlorophyll a_{670} → chlorophyll a_{678} → chlorophyll a_{685} → chlorophyll a_{690} → chlorophyll a_{700}. In this scheme, chlorophyll a_{700} has the longest absorption wavelength, i.e. it has the lowest excitation energy. Chlorophyll a_{700} (= pigment 700 = P_{700}) therefore acts as the final receptor for all the electrons passing through the system, i.e. it is an *energy sink.* In photosystem 2, the ultimate energy sink is chlorophyll a_{680} (= pigment 680 = P680). These respective chlorophyll *a* molecules are the components of a quasi-crystalline structure, which behaves as a semiconductor, so that the excited electrons of any component molecule are rapidly transferred to P700 or P680. Of course, P_{700} and P_{680} can also be excited directly by absorption of light at 700 and 680 nm, respectively, so that each photosystem can utilize light energy from a wide range of wavelengths.

Although all the pigments of a photosystem can absorb light energy, only a few molecules (perhaps only a single molecule) of chlorophyll a_I (P100) or chlorophyll a_{II} (P680) are able to perform photochemical work. These molecules are anchored in the thylakoids of the chloroplasts and they receive absorbed energy from many other molecules of chlorophyll *a* and carotenoids (also from phycobiliprotein molecules in red algae and cyanobacteria). The functional unit that transfers energy to chlorophyll a_I is known as photosystem 1, while the corresponding system for chlorophyll a_{II} is called photo-

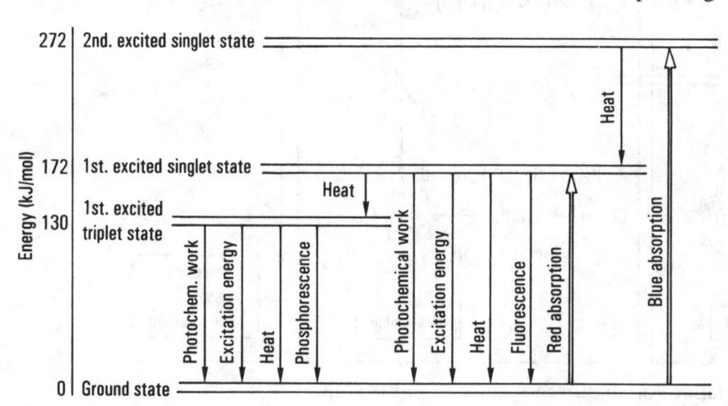

Fig. 1. *Photosynthesis.* Transitions between the excitation states of chlorophyll, resulting from the absorption of blue or red light.

system 2. A functional photosynthetic unit, made up of a single photosystem 1 and a single photosystem 2, contains about 200 molecules of chlorophyll *a,* 70 molecules of chlorophyll *b,* and 50 carotenoid molecules, as well as many accessory molecules, such as proteins. Each photosynthetic unit contains a small photosystem 1 particle and a large photosystem 2 particle (consisting of 3–4 subunits). The photosystem 2 particle is often called a *quantasome*. In order for the light reaction to occur, the whole, undissociated photosynthetic unit must be present. The P_{700} and P_{680} molecules transfer excited (high energy) electrons to an electron acceptor, i.e. a compound with a strongly negative redox potential, e.g. ferredoxin or plastocyanin. The electron acceptor therefore becomes reduced, and a pigment cation (Chl^+) is formed, which in turn receives an electron from an electron donor with a more positive redox potential.

In green plants, two light reactions are coupled in tandem, whereas photosynthetic bacteria possess only one light reaction. The first light reaction, which occurs in both green plants and photosynthetic bacteria, depends on photosytem 1 (so called, because it presumably evolved before photosystem 2). In the ground state, the electron sink (P_{700}) of photosystem 1 has a redox potential of about +0.45 V, which is changed to about – 0.60 V in the excited state, representing an energy difference of 96 J. The extremely high energy electrons of excited P_{700} are transferred to an as yet unknown acceptor (substance X), and from there to ferredoxin. Ferredoxin is a red-brown iron-sulfur protein, which is the most powerful biological reductant known, with the lowest recorded redox potential for a biological agent of – 0.43 V (Fig.2). From ferredoxin, the electrons can then flow to cytochrome b_{564} (cytochrome b_6) and back to photosystem 1 via plastoquinone and cytochrome *f*, i.e. the electrons return to their origin (photosystem 1), where they once again reduce P^+_{700}. This cyclic electron flow is coupled with the esterification of inorganic phosphate and ADP to form ATP, a process known as *cyclic photophosphorylation.* Most of the electrons received by ferredoxin from photosystem 1 do not, however, participate in cyclic photophosphorylation. They are normally transferred via ferredoxin-$NADP^+$-oxidoreductase (a flavoprotein) to $NADP^+$, which is thereby reduced to $NADPH + H^+$. This reduction requires 2 electrons and 2 protons. The latter are derived from the water, which is dissociated to a certain extent (10^{-7}) into H^+ and OH^-.

The flow of electrons to $NADP^+$ leaves pigment system 1 with an electron deficiency, and this deficiency is made up by electrons that are ultimately derived from water. Electrons are taken up from water by chlorophyll a^+_{II} (P_{680}) of photosystem 2. The details of this process are not clear, but there is evidence that 4 molecules of P^+_{680} withdraw 4 electrons from a redox system (called Z), and that the latter is reduced again by the simultaneous oxidation of water. The process can be represented as:

$$4H_2O \rightarrow 4OH^- + 4H^+$$
$$4OH^- \rightarrow 2H_2O + O_2 + 4e^-$$

Manganese is involved in these reactions, possibly as a component of redox system 2. Chloride is also thought to be involved. The resulting molecular oxygen represents a waste product of P. The protons produced in this reaction formally compensate for the protons consumed in the reduction of $NADP^+$. The pH of the aqueous medium is therefore unaffected by the cleavage (photolysis) of the water.

Photoexcitation of chlorophyll *a* II (P680) provides sufficient energy to transfer electrons at redox potential 0.81 V (from the photolysis of water) to an as yet unidentified acceptor (Q). The redox potential of Q is more negative than 0.00 V. Acceptor Q is possibly a bound form of plastoquinone. The electron deficit of P_{700} is replaced by electrons from Q, which pass to P_{700} via a series of coupled redox systems, including cytochrome *f* and plastocyanin (Fig.2). The redox potentials of plastoquinone and cytochrome *f* differ by 0.32 V, so that the energy released in the transfer of an electron pair between

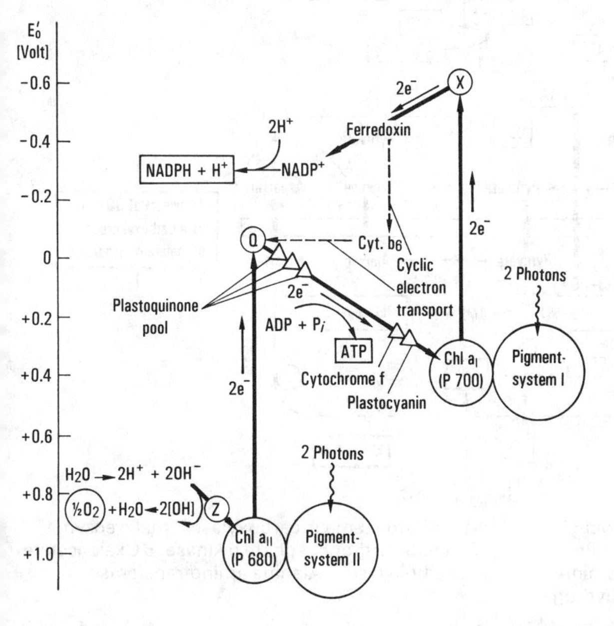

Fig.2. *Photosynthesis.* The two stages in the light reaction of photosynthesis.

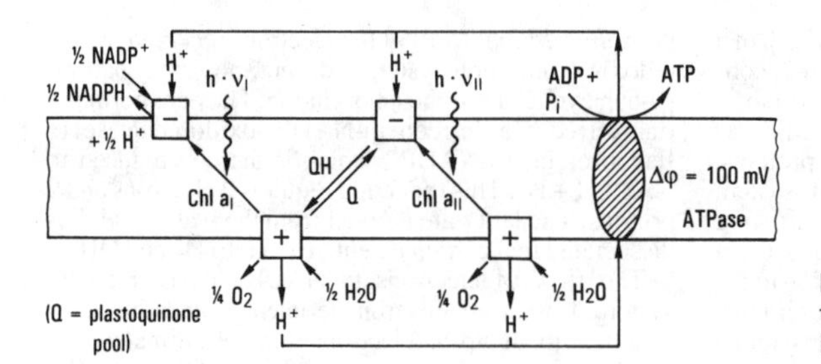

Fig.3. *Photosynthesis.* Arrangement of the two photosystems in the thylakoid membrane (after Witt, 1978).

Fig.4. *Photosynthesis.* Carboxylation of phospho*enol*pyruvate (PEP) and subsequent reactions in C4-plants. *1* Phospho*enol*pyruvate carboxylase. *2* Aspartate aminotransferase. *3* NADP⁺-dependent malate dehydrogenase. *4* Pyruvic-malic decarboxylase ("malic enzyme"). *5* Pyruvate, orthophosphate dikinase.

these two carriers is sufficient for the synthesis of one or 2 molecules of ATP. In green plants, most of the ATP synthesized in the light reaction of P. arises from this process of *noncyclic photophosphorylation.*

Fig.2 is an energy diagram of the components of the light reactions of P. It does not necessarily represent the spatial juxtaposition of these components in the thylakoid membrane. Information on the actual location of these components in the thylakoid membrane has been obtained by biophysical methods, including the use of pulsed laser light in the nano- and pico-second range. The results of these studies up to 1978 are summarized in Fig.3.

Products of the light reaction, i.e. NADPH + H^+ and ATP, leave the thylakoids and enter the matrix of the chloroplast, which is the site of the dark reaction of P.

2) The *dark reaction*, in which CO_2 is reduced to carbohydrate, can be represented by the equation: $6CO_2 + 12(NADP + H^+) + 18ATP \rightarrow C_6H_{12}O_6 + 12NADP^+ + 6H_2O + 18(ADP + P_i)$. It is, however, a cyclic process, which consumes CO_2 and produces carbohydrate. The intermediate stages were elucidated by Calvin and coworkers, using radioactive CO_2, so that the dark reaction is also called the *Calvin cycle.* It commences with

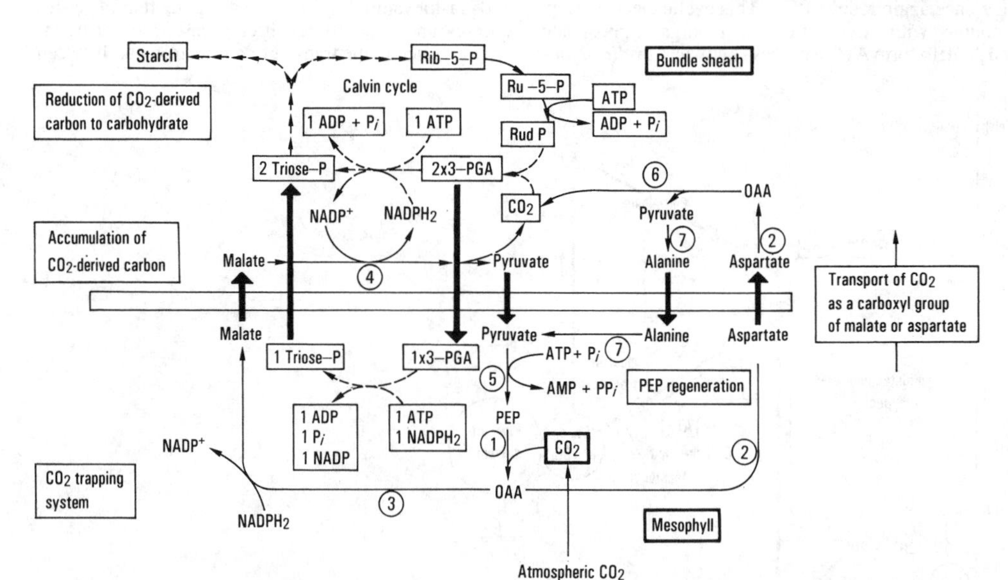

Fig.5. *Photosynthesis.* Reactions in the mesophyll and bundle sheath of the leaf of a C4-plant. *1* Phospho*enol*pyruvate carboxylase. *2* Aspartate aminotransferase. *3* NADP⁺-dependent malate dehydrogenase. *4* Pyruvic-malic carboxylase ("malic enzyme"). *5* Pyruvate, orthophosphate dikinase. *6* Oxaloacetate β-decarboxylase. *7* Alanine aminotransferase.

the carboxylation of ribulose 1,5-diphosphate to form an unstable intermediate, which dissociates into two molecules of 3-phosphoglyceric acid. Since this first detectable product is a 3-carbon compound, the Calvin cycle is also referred to a the *C_3-pathway.* Plants that possess only this pathway of CO_2 incorporation are called C_3-plants, in contrast to C_4-plants (see) and CAM plants, which have other pathways in addition to the C_3-pathway (see Figs.4 & 5).

The 3-phosphoglyceric acid is then reduced to glyceraldehyde 3-phosphate in the reduction phase of the Calvin cycle, which utilizes ATP and NADPH + H^+ from the light reaction. Two molecules of glyceraldehyde 3-phosphate ultimately give rise to fructose, which can be converted into glucose. Phosphates of sedoheptulose, tetroses and pentoses are involved in a series of interactions, leading to the regeneration of ribulose 1,5-diphosphate, which accepts another molecule of CO_2. The individual reactions of the Calvin cycle are shown in Fig.6, which also gives the names of the most important enzymes.

In contrast to higher plants and cyanobacteria, the photosynthetic bacteria, e.g. purple sulfur bacteria and green photosynthetic bacteria, are unable to photolyse water. Instead, they use hydrogen sulfide or simple organic compounds as a source of protons and electrons for the reduction of CO_2. In the case of purple sulfur bacteria, P. can be represented by the equation:

$$6CO_2 + 12H_2S \xrightarrow{\textbf{light}} C_6H_{12}O_6 + 6H_2O + 12S.$$

The chlorophylls of photosynthetic bacteria, known as bacteriochlorophylls, differ structurally from those of higher plants and cyanobacteria.

Phototaxis: directional movements of freely motile organisms or cell organelles (see Taxis) in response to unilateral light intensity. Among the motile lower plants, P. is displayed primarily by those that are capable of photosynthesis. Generally, the green flagellates and algae are positively phototaxic, but they become negatively phototaxic if illumination becomes too intense. Photosynthetic purple bacteria display positive P., especially to long-wave red light and infrared light; these organisms also show a strong phobic reaction to weakly illuminated areas, so that it is possible to restrict them to a small illuminated area, as if in a trap. Plasmodia of *Myxomycetes* are negatively phototaxic, until they form fruiting bodies, when they become positively phototaxic. The chloroplasts of higher plants are markedly phototaxic (see Chloroplast movements). Animals exhibit photophobotaxis, photomentotaxis, phototelotaxis and phototropotaxis (see Orientation).

Phototropism, ***heliotropism:*** bending movements that serve to orientate plant parts in relation to a source of unilateral illumination (see Tropism). This response to light is strongly developed in plants, and together with the response to gravity (geotropism) it is one of the most important bases of plant orientation in space. Different plant organs respond to light in different ways. 1) In *positive P.,* growth is directed toward the light source, e.g. the aerial shoots of many higher plants, the fruiting bodies of higher fungi and the sporangiophores of molds. 2) In *negative P.,* growth is directed away from the light source; this relatively uncommon form of P. is found in e.g. adhesive rootlets, various aerial roots, the hypocotyl of mistletoe, tendrils of the wild grape which end in adhesive disks, the primary roots of *Sinapis* and *Helianthus,* and the rhizoids of fern prothalli and of liverworts. The subterranean roots of most plants are in-

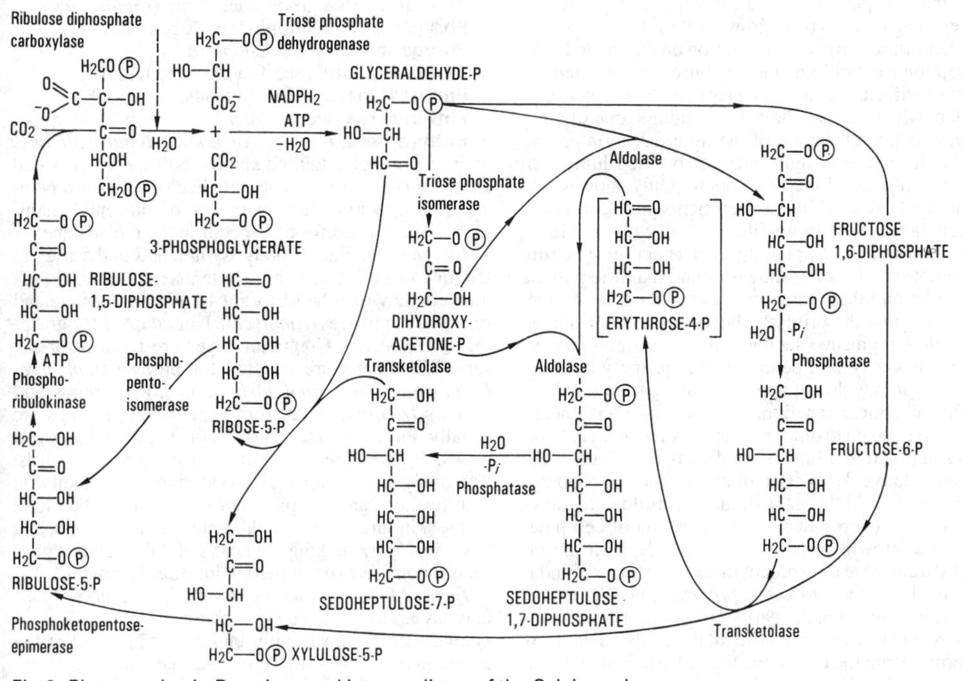

Fig.6. *Photosynthesis.* Reactions and intermediates of the Calvin cycle.

sensitive to light. 3) *Transverse P.* or *diaphototropism* results in orientation across the light path, and it is displayed mostly by dorsiventral organs. Usually, the upper suface of laminate foliage leaves is turned toward the light, and the lower surface from it, the entire organ lying transverse to the incident light, but not necessarily at right angles to it. In some plants, the upper surface of the leaves is held perpendicular to the incident light only in the weaker light of morning and evening; as the light intensity increases, the leaves rotate so that by midday only the edges of the leaves are presented to the light source, thereby avoiding direct illumination of leaf surfaces at the time of most intense irradiation (see Compass plants). Aberrant, panphotometric leaf orientations are also exhibited by certain *Eucalyptus* species that form open "shadeless" forests. Thallose liverworts also usually display transverse P. In certain plant organs, the type of phototropic response may be modulated by internal and/or external factors. Flower stalks of several plant species, e.g. ivy-leaved toadflax *(Linaria cymbalaria),* are initially positively phototropic, but become negatively phototropic after fertilization. Oat coleoptiles are positively or negatively phototropic, depending on the quantity of incident light, i.e. the product of the illumination intensity (lx) and time (s). In the range of 10–10 000 lx.s, the response is positive, and it becomes positive again at 100 000 lx.s. Between these two values, oat coleoptiles show a negative response, while other germinating plants are indifferent. In many plants, a more prolonged exposure to light causes stunted growth. Extreme temperatures can decrease or abolish P. (heat torpor and cold torpor).

Plants are generally extremely sensitive to light. For example, a single, brief flash of light is sufficient to cause positive phototropic bending in oat coleoptiles. With the exception of certain rare phototropic turgor movements, P. arises from growth inhibition on the illuminated side and growth stimulation on the shaded side. Susception (see) of light and response to light often occur in specific and different zones of a plant organ. Susception, which occurs above the bending zone of a coleoptile, is not a function of the light direction *per se,* but the difference in light intensity on the illuminated and shaded sides of the plant organ. Only short wavelength light causes P., and experiments on P. can safely be conducted in a room illuminated with red light, which has no phototropic effects. The action spectrum for phototropic bending shows a small maximum in the ultraviolet and three large maxima in the blue region, suggesting that the primary photoreceptors are orange and yellow pigments, such as carotenoids and flavins. Carotenoids may well be involved in primary photoreception, but they do not appear to take part in any kind of photochemical reaction, i.e. the absorbed energy must be passed on to other reactants by resonance. It has been suggested that flavins are directly involved in the photo-oxidative destruction of auxin, but this is doubtful. Susception of the light stimulus is followed immediately by an energy-dependent alteration of cell polarity in the growth response zone; this leads to an increased transverse transport of material in the cell, and is therefore known as *transverse polarization.* When a coleoptile is growing in the dark or is growing directly towards a light source, growth-regulating substances flow uniformly from the coleoptile tip and affect all sides of the coleoptile equally. Under unilateral illumination, however, the sugar concentration, catalase activity, acidity, and especially the concentration of auxin all decrease on the illuminated side in the growth response zone below the coleoptile tip, while increasing on the shaded side. At the same time, basipetal auxin transport from the tip is inhibited on the illuminated side. It is possible that this unilateral inhibition of transport on the illuminated side leads to an accumulation of material, which therefore becomes distributed on the shaded side. The shaded side therefore grows more rapidly and becomes convex.

In oat coleoptiles, sensitivity is especially high in the first 0.25 mm of the tip, thereafter decreasing rapidly (e.g. at 2 mm behind the tip, it is less than 0.003% of the value at the tip). Shading of the coleoptile tip, e.g. with a small tinfoil cap, prevents the phototropic response. Alternatively, if almost the entire coleoptile is shaded, leaving only the first 0.25 mm of the tip exposed, unilateral illumination of the tip causes curvature in the lower part of the coleoptile toward the light.

Phoxinus: see Minnow.

Phragmobasidia, ***heterobasidia, protobasidia:*** multicellular basidia, produced by *Basiomycetes,* and typical of the *Phragmobasidiomycetidae.* In the order, *Tremellales,* the P. generally have 4 longitudinal, axial septa, and the basidospores are abstricted at the apex. In the order, *Auriculariales,* the P. are divided by cross walls into 4 superposed cells, each carrying a lateral sterigma with a single spore.

Phragmobasidiomycetidae: see *Basidiomycetes.*

Phragmocone: the chambered part of the cephalopod shell, in particular the belemnite shell. The septa of the P. are perforated and traversed by the siphuncle, a slender, tubular elongation of the visceral hump.

Phragmoplast: see Mitosis.

Phreatoicoidea: a suborder of the *Isopoda* (see).

Phrygana: see Sclerophylous vegetation.

Phryganoidea: see *Trichoptera.*

Phrynocephalus: see Toad-headed lizards.

Phrynosoma: see Horned toads.

Phthiracarus: see *Oribatei.*

Phthiraptera, *Pseudorhynchota,* ***ectoparasitic lice:*** an insect order containing about 3 600 species, divided about equally between chewing lice (mostly bird parasites) and sucking lice (parasites of placental mammals). The earliest fossil *P.* occur in the Pleistocene.

Imagoes. The adult body is flattened and varies in length from 0.8 to 14 mm. Antennae are 3- to 5-segmented; they may be filiform *(Anoplura),* capitate and recessed in grooves *(Amblycera),* or adapted for grasping *(Ischnocera).* Compound eyes are reduced or absent. Mouthparts are adapted for biting *(Amblycera, Ischnocera, Rhynchophthirina),* or for piercing and sucking *(Anoplura).* All forms are wingless. Legs are usually short, and they carry 1 or 2 retractable tarsal claws. *P.* live as ectoparasites on mammals or birds. The biting lice do not suck blood; they feed mainly on hair, feathers, cast-off pieces of skin, and clotted blood on the skin surface. The sucking lice pierce the skin, and live entirely by sucking the host's blood. Metamorphosis is incomplete (paurometabolous development).

Eggs. All species lay eggs; the female body louse may lay up to 300. The eggs (those of sucking lice are known as nits) are usually elongate with a lid at the upper pole, and they are cemented to the hairs or feathers of the host. Biting lice often lay their eggs in clumps.

Larvae. There are 3 larval stages, each very similar in appearance and mode of life to the adult. Development from egg to imago is usually complete after 2 to 5 weeks.

Economic importance. Infestation of domestic animals with biting lice is usually harmless; this occasionally causes irritation, which in rare cases leads to skin damage by scratching and rubbing, and emaciation of the host. In humans and animals, skin puncture by sucking lice causes irritant swellings, eczema or abcesses. Some lice are dangerous disease vectors (see below: *Pediculidae* and *Trichodectidae*).

Classification. Some authorities combine the *Amblycera* and the *Ischnocera* in a single family *(Mallophaga),* which contains all the biting lice, except the *Rhynchophthirina.*

1. Suborder: *Amblycera* (biting lice, in part). Maxillary palps 4-segmented. Mandibles horizontal. Antennae capitate, 4-segmented, lying in grooves. Mesothorax and metathorax usually separate.

1. Family: *Menoponidae.* The largest family; cosmopolitan distribution. All species infest birds, and several are poltry pests, e.g. *Menacanthus stramineus* (chicken body louse).

2. Family: *Laemobothriidae.* Represented by a single genus, *Laemobothrion,* whose members parasitize water birds and hawks.

3. Family: *Ricinidae.* A family of only 2 genera, found on humming birds and several passerines.

4. Family: *Boopiidae.* Formerly found only on Australian marsupials, mainly kangaroos, but one species *(Heteradoxus spiniger)* has transferred to the dog, which is not indigenous to Australia.

5. Family: *Trimenoponidae.* Found only on South American histricomorph rodents.

6. Family: *Gyropidae.* Found mostly on histrocomorph rodents in South and Central America. Guinea pig lice *(Gyropus ovalis* and *Gliricola porcelli)* have been introduced to other parts of the world by commerce.

2. Suborder: *Ischnocera* (biting lice, in part). Maxillary palps absent. Mandibles vertical. Antennae 3- to 5-segmented and filiform, not in grooves and exposed. Mesothorax and metathorax usually fused.

1. Family: *Philopteridae.* Worldwide distribution. Parasites of birds, including some of poltry, e.g. *Cuclutogaster heterographus* (chicken head louse) and *Chelopistes meleagridis* (large turkey louse).

2. Family: *Trichodectidae.* Restricted to placental mammals, including domestic animals, e.g. *Damalinia (= Bovicola) bovis* (cattle), *D. ovis* (sheep), *D. equi* (horses), *Felicola subrostratus* (cats), and *Trichodectes canis* (dogs), which is an intermediate host of the dog tapeworm *(Dipylidium caninum).*

3. Suborder: *Anoplura* (true or sucking lice). Mandibles absent. Mouthparts styliform.

1. Family: *Echinophthiridae.* Parasites of pinnipede carnivores (seals, sealions, walruses). The body is covered with spines or scales, which trap a film of air when the host submerges. The sea elephant louse *(Lepidophthirus macrorhini)* lives in deep passages, which it burrows in the skin of the host.

2. Family: *Linognathidae.* Parasites of dogs *(Linognathus setosus)* and ruminants *(L. ovillus* and *L. pedalis),* cattle *(L. vituli)* and goats *(L. stenopsis).*

3. Family: *Neolinognathidae.* Parasites of African elephant shrews.

4. Family: *Haematopinidae.* Parasites of ungulates, including serious pests of domestic animals, e.g. *Haematopinus eurysternus* (cattle), *H. suis* (hogs), *H. asini* (horses).

5. Family: *Hoplopleuridae.* Parasites mainly of rodents and rabbits, and some insectivores and primates.

6. Family: *Pediculidae.* Parasites of primates, e.g. human lice: *Peduculus capitus* Deg. (head louse), *Pediculus corporis* Deg. (body louse), *Phthirus pubis* L. (crab louse). According to some authorities, head and body lice are subspecies: *Pediculus humanus capitis* and *Pediculus humanus corporis (= vestimenti).* Head and body lice carry typhus and other epidemic fevers caused by blood-borne rickettsias, and relapsing fever caused by a spirochaete.

Subfamilies: *Pedicininae* (Parasites of Old World monkeys) and *Pediculinae* (Parasites of lemurs, anthropoid apes and humans).

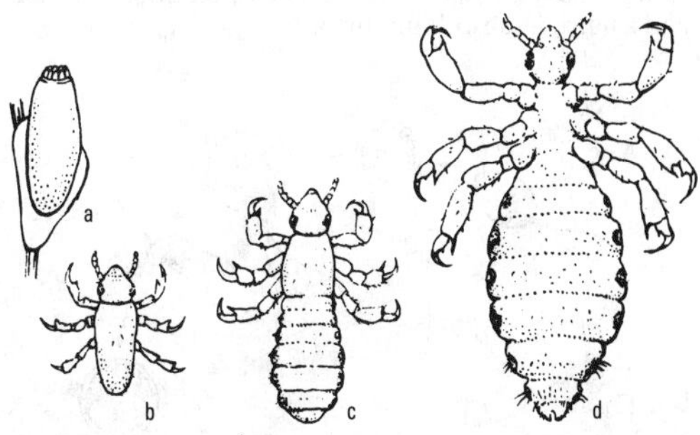

Phthiraptera. Human body louse *(Pediculus corporus* Deg.). *a* Egg. *b* Young larva. *c* Older larva. *d* Imago (female).

4. Suborder: *Rhynchophthirina.* Mandibles are located at the apex of a rostrum; they do not function as pincers, but operate outward. Only 2 species are known: *Haematomyzus elephantis* Piaget, which parasitizes both Indian and African elephants; and *H. hopkinski,* which infests warthogs.

pH-value: the negative common logarithm of the Hydrogen ion concentration (see).

Phycobiliproteins: see Bile pigments.

Phycoerythrins: see Accessory pigments.

Phycomycetes, *algal fungi, lower fungi:* formerly a class of fungi, but now a collective term for the three fungal classes: *Chytridiomycetes, Oomycetes* and *Zygomycetes,* which differ sufficiently in their characters to justify separate status. The term, algal fungi, was used by early investigators, because the few forms known at that time resembled certain known algae. Hyphal walls are always of chitin or cellulose plus glucans. The hyphae are as a rule tubular (siphoneous), branched and multinucleate. With a few exceptions among the advanced forms of the *Zygomycetes,* the hyphae are aseptate (lacking transverse walls). As a group, the *P.* display an evolutionary transition from an aquatic to a terrestrial existence, which finds expression especially in the different types of spores. Thus, primitive forms produce free-swimming spores, whereas advanced forms produce sporangiospores or conidia. Sexual reproduction shows a transition from iso-, aniso- and oogamy to gametangiogamy. *P.* also exhibit the evolutionary beginnings of fruiting body formation. The vegetative body is usually haploid; with few exceptions, the diploid phase is restricted to the zygote.

1. Class: *Chytridiomycetes.* These are either unicellular organisms, or they have a multinucleate, aseptate thallus with a chitin wall. Gametes and zoospores are opisthocont (i.e. they have a single whiplash flagellum directed to the rear).

1. Order: *Chytridiales.* These are parasites in or on aquatic plants and animals. Some are also found in the soil, where they parasitize terrestrial plants, while other representatives trap and feed upon unicellular animals with the aid of fine hyphae (rhizopodial processes). The latter may therefore be regarded as carnivorous plants. The order also contains, e.g. the causative agent of the wart disease of potatoes, *Synchytrium endobioticum.* In the higher *Chytridiales,* e.g. *Rhizophydium* (a parasite of planktonic algae and pollen grains), there is a division of labor between a reproductive chitinous vesicle and a nutrient-absorbing rhizoid.

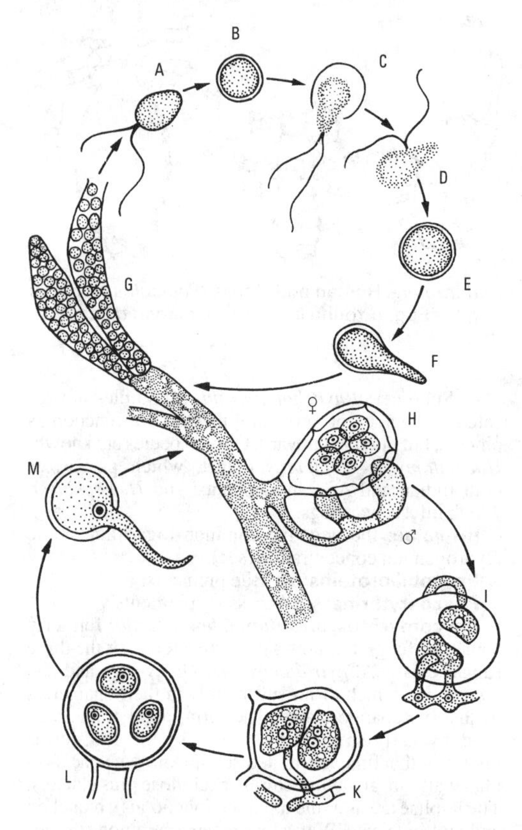

Fig.1. *Phycomycetes.* Life cycle of *Saprolegnia* sp. *A-G* Asexual. *H-M* Sexual. *A* Primary biflagellate zoospore. *B* Encapsulated zoospore. *C* Germination. *D* Secondary zoospore. *E* Encapsulated zoospore. *F* Germination. *G* Thallus with zoosporangia; one zoosporangium is liberating biflagellate zoospores. *H* Oogonium and antheridium. *I* Penetration of the fertilization tube. *K* Plasmogamy. *L* Oospores. *M* Germination.

2. Order: *Blastocladiales.* Members of this order form richly branched, multinucleate hyphae with chitin walls. Sexual reproduction is by uniflagellate, opisthocont iso- or anisogamy, sometimes with an alternation of generations. Most species are soil organisms.

3. Order: *Monoblepharidales.* Saprophytes of dead aquatic vegetation. Sexual reproduction is oogamous; uniflagellate opisthocont zoospores are produced in large numbers in club-shaped sporangia.

2. Class: *Oomycetes.* All species have a tubular mycelium, the walls of which give a reaction for cellulose and glucans. Sexual reproduction is always oogamous.

1. Order: *Saprolegniales.* Most species are aquatic saprophytes of decaying plant remains and dead insects, and a few are fish parasites. Vegetative reproduction is by club-shaped sporangia, which produce uninucleate zoospores. The zoospores have two flagella, one whiplash, the other tinsel (with flimmer hairs).

2. Order: *Peronosporales.* Parasites of terrestrial plants. The mycelium penetrates the plant tissue, and the zoosporangia are carried on characteristic sporangiophores. The zoosporangia are shed intact, and they do not release their zoospores (swarmers) unless they land on a new host plant and encounter suitable moist conditions (e.g. *Plasmopara).* Zoospores have two unequal lateral flagella, the shorter with flimmer hairs. In some species, as an adaptation to a terrestrial existence, the formation of zoospores is suppressed and the entire wind-borne sporangium develops an infective germ tube, so that the sporangium has become a conidium (e.g. *Peronospora).* The order contains many causative agents of economically important, sometimes cosmopolitan plant diseases. For example, *Plasmopara viticola* is an intercellular parasite of the leaves and fruits of the grape vine, causing a disease known as downy or

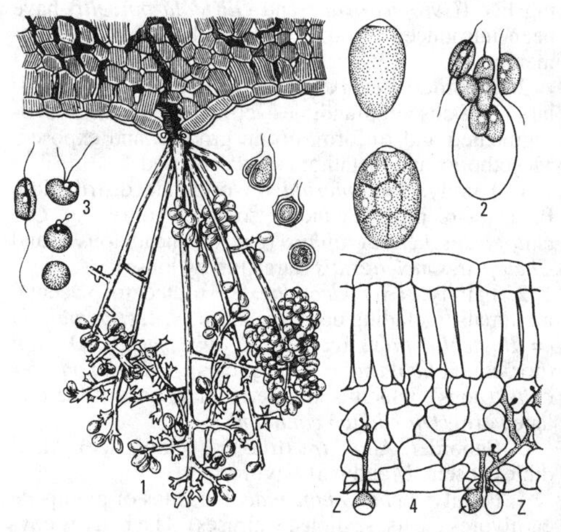

Fig.2. *Phycomycetes.* Downy or false mildew *(Plasmopara viticola)* which attacks the grape vine. *1* Sporangiophore emerging from a stomatum; on the right hand side are shown oogonia (with antheridium) and zygotes (x 100). *2* Formation and liberation of zoospores (x 600). *3* Withdrawal of flagella (x 600). *4* Penetration of stomata and intercellular spaces by germinating zoospores *(z)* (x 250). *(1* after Millardet, *2-4* after Arens)

false mildew. Species of *Peronospora* also cause downy or false mildews of various agriculatural plants. *Phytophthera infestans,* the causative agent of potato blight, displays both methods of infection; in a humid atmosphere the sporangium releases swarmers, whereas under other conditions it produces a germ tube. The most primitive genus is *Pythium,* whose sporangia release zoospores while they are still attached to the mycelium. *Pythium debaryanum* is commonly present in soil, where it is responsible for the lethal damping off and foot rot of seedlings.

3. Class: *Zygomycetes.* Most members are saprophytes. There is a well developed mycelium with chitinous walls. Members of the *Zygomycetes* are more adapted to a terrestrial existence than the *Oomycetes.* Zoospores are not produced, and asexual reproduction is by endospores, which do not require moisture for their dispersal. Sexual reproduction is by gametangiogamy (see Reproduction).

1. Order: *Mucorales.* Widely distributed molds, commonly found on plant and animals substrates, e.g. *Mucor mucedo (= Rhizopus stolonifera)* ***(Pin mould),*** which forms white cottony growths on dung, bread, food residues, etc. Dung also supports the growth of species of *Pilobolus,* which can eject its sporangia up to distances of 2 m.

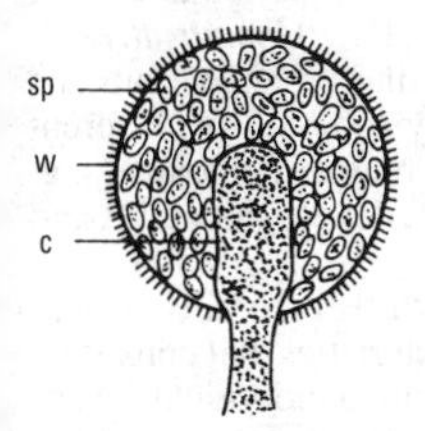

Fig.3. *Phycomycetes.* Pin mold *(Mucor mucedo).* Section of a sporangium. *c* Columella. *w* Wall. *sp* Spores.

2. Order: *Entomophthorales.* Parasites of insects and various green algae. Some species have hyphae with transverse walls. Asexual reproduction is by sporangia that have evolved into conidia. Most familiar is the causative agent of an epidemic disease of flies, *Empusa (Entomophthora) muscae.*

Phylactolaemata: see *Bryozoa.*

Phyletic speciation: see Speciation.

Phyllastrephus: see *Pycnonotidae.*

Phyllidae: see *Chelentoptera.*

Phylloceratidus (Greek *phyllon* leaf, *kera* horn): an order of ammonites whose fossils occur from the Lower Liassic to the Upper Cretaceous. The planispiral shell is flat and disk-shaped with a narrow umbilicus. The surface is either smooth or patterned with fine raised veins. Leaf-like division of the suture line is typical of the genus.

Phyllograptus (Greek *phyllon* leaf, *graptos* written): an index graptolite genus of the Lower Ordovician (see Graptolites). The rhabdosome consisted of two, leaf-shaped parts, crossing one another at 90 ° and fused for their entire length. There were four uniserial rows of prismatic thecae, which were fused along the entire length of their dorsal margins; the apertures of the thecae often had two spines.

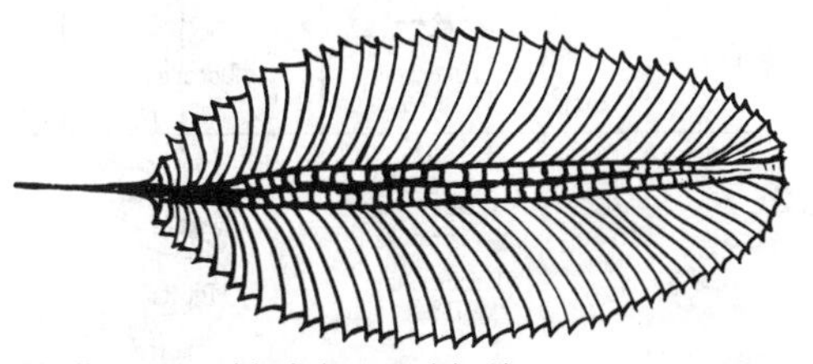
Phyllograptus (rhabdosome) (x 2)

Phyllopodia (sing. **Phyllopodium**): paired trunk appendages of some crustaceans (anostracans, branchiopods, leptostracans). P. have a thin, soft cuticle, without obvious joints, and their shape is maintained largely by their internal blood pressure. By complicated cooperative movements, rows of P. serve simultaneously for trapping food material and for respiratory gas exchange, and sometimes also for locomotion.

Phylloscopus: see *Sylviinae.*

Phyllostachys: see *Graminae.*

Phylogenetic trend: the continued occurrence of

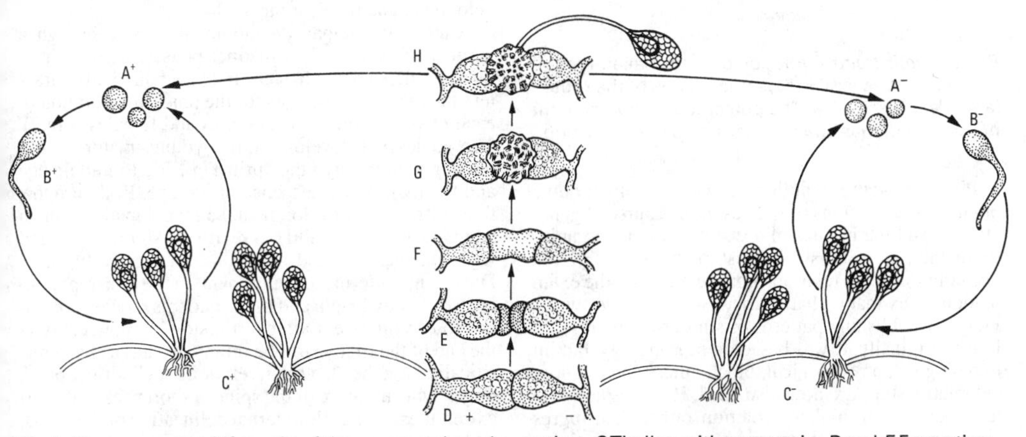

Fig.4. *Phycomycetes.* Life cycle of the common bread mold *(Rhizopus stolonifer). A* to *C* Asexual reproduction. *D* to *H* Sexual reproduction. *A* Spores. *B* Germination. *C* Thallus with sporangia. *D* and *E* Formation of gametangia. *F* and *G* Zygospore formation. *H* Germinating zygospore with sporangium.

evolutionary changes over a relatively long geological period, always in the same direction. Related organisms often display similar P.ts. For example, certain mammalian characters, such as heterodont dentition and the typical articulation of the lower jaw with the skull, also developed more or less completely in various evolutionary lines of the reptilian order, *Therapsida* (Fig.). See Orthogenesis.

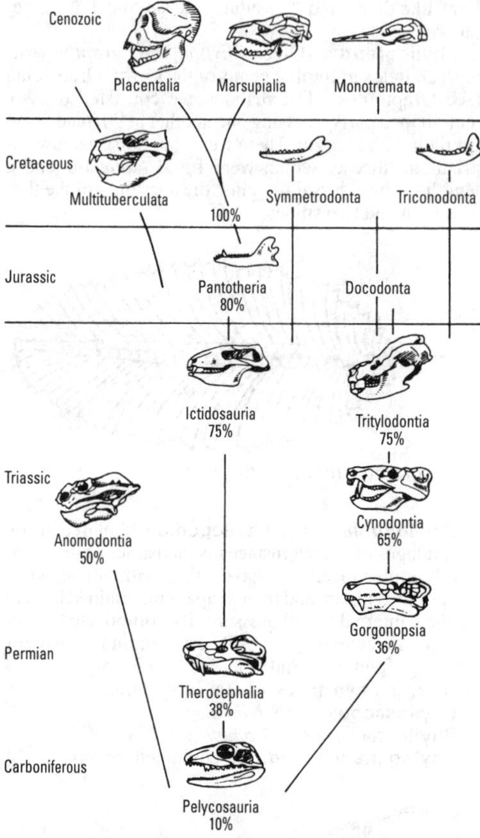

Phylogenetic trend. Evolution of mammalian-type characters in several independent lines of the reptilian order, *Therapsida.* The percentages represent the proportion of mammalian characters in each group.

Phylogeny: the evolutionary history of any group of organisms; alterations of species in the course of generations, resulting in lines of descent from one organism to another, as well as relationships between different taxa. Phylogenetic relationships are based on the examination of as many characteristics as possible, with a view to establishing patterns of similarity and dissimilarity. Originally, P. was based on morphology, but immunological, physiological, biochemical and cytological relationships are now also used. P. shows a general tendency toward higher evolution rather than regression. Some groups of organisms evolve more slowly than others, so that even today there are organisms with similarities to Precambrian forms.

The course of P. can be reconstructed from fossils and their historical sequence. This information is supplemented by analysis of the morphological, biochemical, physiological, ethological and biogeographical characters of modern species. The driving forces of P. are a subject of the Synthetic theory of evolution (see).

Physalia: see Portuguese man-of-war.

Physalin: see Zeaxanthin.

Physalis: see *Solanaceae.*

Physeter: see Whales.

Physeteridae: see Whales.

Physignathus: see Water dragons.

Physiological clock: see Biorhythm.

Physiological race, ***physiological form, biotype:*** a term used in microbiology and botany for a race differing from other races in its physiological, biochemical, pathological or cultural properties, rather than morphologically.

Physiology: a branch of biology concerned with the functions and performance of plants and animals and their cells, tissues and organs, with the aim of elucidating the causal relationships of living processes. Methods of physiological research are taken largely from physics, chemistry and other sciences. *General P.* deals with those principles and relationships that are valid for all organisms (plants and animals, including humans). *Specialized P.* deals with the P. of one particular type of organism (animal P., plant P., etc.). *Comparative P.* compares the functions of particular organs within one group of organisms (e.g. kidney function of different vertebrates). Other specialized areas of P. include, e.g. neurophysiology, reproductive P., developmental P. and physiological chemistry.

Physique, ***constitution, somatotype:*** totality of anatomical and physiological characteristics that constitute the external appearance of the human individual. These characteristics are established by the interaction of hereditary and environmental factors; they do not include transitory alterations, and they are not synonymous with racial characteristics. Classification of the numerous types of P. is based primarily on morphological, physiological, functional, psychological and pathological properties. Of particular importance in human biology are the patterns of male and female ontogenic development and the P. of the adult.

After birth, human development passes through a number of more or less distinct phases, each with its own age-dependent physical and psychological characteristics. The average ages for the onset and duration of each phase are different in males and females (Fig.2). The adolescent develops into a young, mature adult, passing then through the "prime" of life to middle age and finally to the senescence of old age. Body proportions differ with developmental age; compared with an adult, the newborn child has relatively shorter arms and legs, whereas its head and trunk are relatively large. During intrauterine development there is a rapid increase of body length and mass, and this process continues to certain extent after birth, slowing noticeably at the end of the first year. The infant also has a characteristically large head, a relatively long, cylindrical body, only slight curvature of the spine, a short neck and short extremities, with little external delineation of the joints, and with conspicuous subcutaneous fat (Fig.1). A distinct change of form occurs between the 5th and 7th years, culminating in the preadolescent child. In the preadolescent, the head is much smaller in relation to the rest of the body; the face acquires a stronger profile, especially the lower face which is affected by development of the dentition; the thorax becomes flatter; the curvature of the spine becomes more pronounced; and the extremities are relatively longer and more slender (Fig.1).

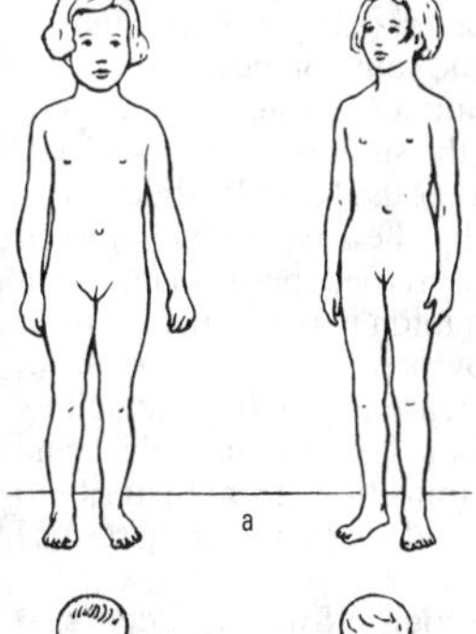

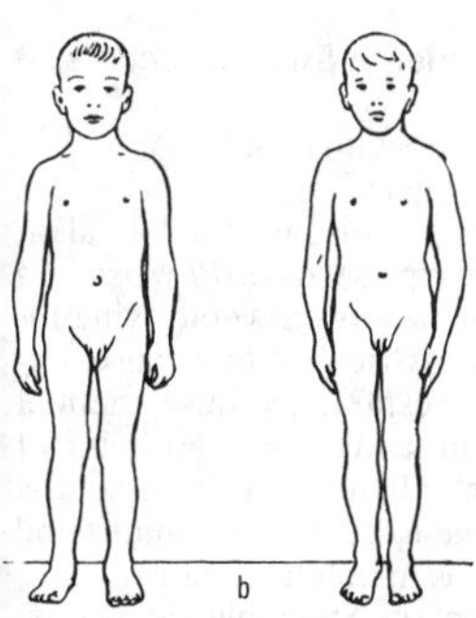

Fig.1. *Physique. a* Preschool type (left) and school type (right) girls of the same age (5.75 years). *b* Boys aged 3 years (left) and 7 years (right). The scale has been adjusted so that all four children have the same height.

Adolescence commences between the 11th and 12th years in girls, and between the 12th and 13th years in boys. This phase is characterized by a striking surge of growth, mainly of the extremities, and by secondary sexual development. The alterations of P. leading to puberty usually occur in a definite sequence. In the female, these are: broadening of the pelvis, development of the breasts, growth of pubic hair, growth of axillary hair, and menarche; in the male: enlargement of the testicles and penis, growth of hair in the pubic region and on the upper lip, increased prominence of the thyroid cartilage and completion of the voice change, growth of axillary hair, first production of semen. This pubescent phase is followed by the postpubescent phase, accompanied by further alterations of body proportions. Growth then ceases, and the individual is a mature, young adult. Women usually attain maturity in their 18th year, men in their 20th year (Fig.2). These developmental changes are regulated primarily by a complex interplay of the hormones of the thyroid, thymus and the anterior lobe of the pituitary, as well as the activity of the gonads and the diencephalon. Age-related psychological changes also occur in association with the alterations in P. and the development of sexual maturity.

The age span of optimal performance and efficiency, sometimes referred to as the "prime" of life, corresponds very approximately to the 20–25 years following the onset of maturity. It is marked by full development of sexual differences and large individual variations of the psycho-visual image.

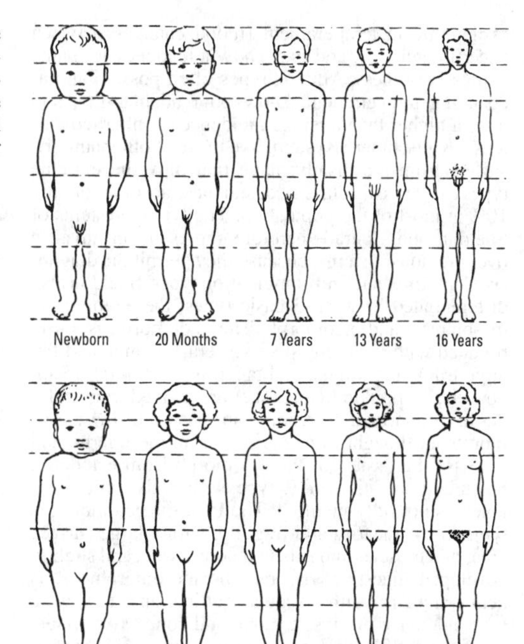

Fig.2. *Physique.* Changes in body proportions during growth (after Conrad, 1963).

Different classifications of P. have been proposed. The system of Kretschmer is well known, not least because it demonstrates a close association between P. and temperament, as well as relationships with certain mental illnesses.

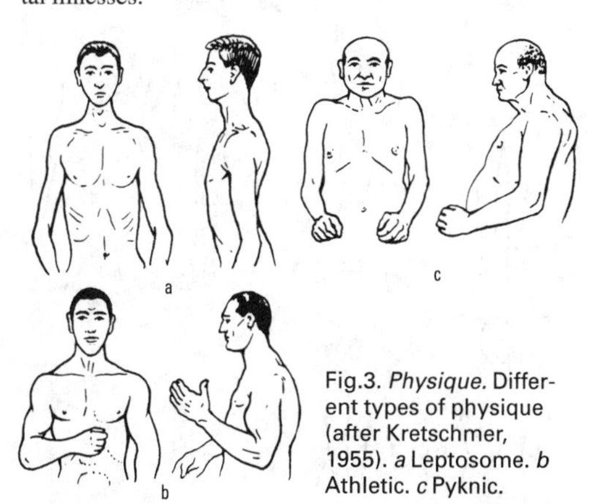

Fig.3. *Physique.* Different types of physique (after Kretschmer, 1955). *a* Leptosome. *b* Athletic. *c* Pyknic.

Kretschmer defined 3 extreme types: leptosome, pyknic and athletic (Fig.3). Leptosomatic types show a greater than average frequency of schizothymic temperament (irritability and fine sensitivity) and a disposition to schizophrenia. Pyknic types are predominantly cyclotomic in temperament (mood changes between cheerful and sad), and they show a tendency to manic depressive illness. Athletic types often possess an unchanging temperament, show poor adaptability, and have a higher than average incidence of inherited epilepsy. Kretschmer's system, like those of other authors, suffers from the disadvantage that only the extreme types can be easily classified, and these account for only 10% at most of the population. The typing systems of Sheldon and Conrad therefore represent an advance over previous systems, because they permit the diagnosis of intermediate and combination types. Sheldon also differentiated between 3 basic types; the endomorph, mesomorph and ectomorph. The endomorph is softly rounded with large digestive viscera, accumulated fat, large trunk and thighs, and tapering extremities. Supposedly, the predominant tissues are derived from endoderm. Similarly, the predominant tissues of the mesomorph are thought to be derived from mesoderm, and likewise the tissues of the ectomorph from ectoderm, but this association of P. type with embryonic germ layers is not fully justified. The P. of the ectomorph is typified by linearity and fragility, with a large surface area, thin muscles and subcutaneous tissue, and slightly developed digestive viscera. The mesomorphic P. is heavy and firm with a regular outline, and it shows a preponderance of muscle, bone and connective tissue. According to Sheldon, endomorphic, mesomorphic and ectomorphic components all contribute in varying degree to the P. of an individual. Each component is subdivided into 7 grades of intensity, thus permitting a more precise characterization of any P. According to Conrad, P. arises from the interaction of two independent variants. The primary variant is derived from a range of types between the two extremes of leptosomatic and pyknic; the secondary variant represents a form on the scale from hypoplastic to hyperplastic. Supposedly, there exists a close relationship between the primary variant and ontogenic development, i.e. with respect to morphological proportions, pyknomorphic and leptomorphic Ps. are related in the same way that an ontogenetically early stage of development is related to a later stage. A similar relationship can be demonstrated between the physiology of an individual P. type and ontogenesis. The classification systems of Conrad and Sheldon both permit a quantitative diagnosis, in which any individual can be characterized by a simple numerical expression, known as the somatotype.

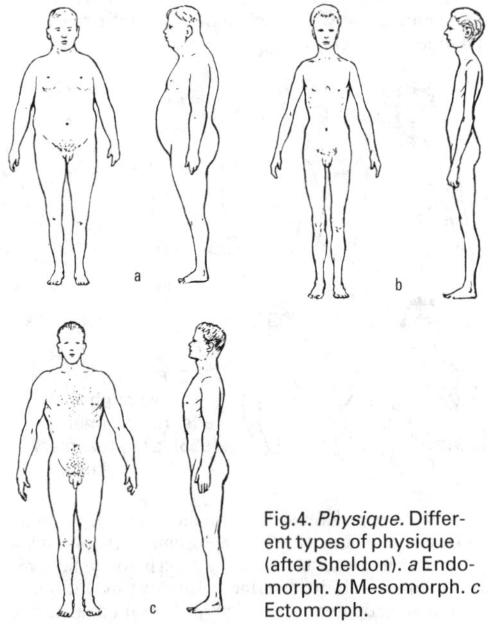

The P. of the mature individual usually undergoes only slight changes during the subsequent years of life, and these changes do not alter the basic P. Middle age, however, is accompanied by the first visible signs of physical regression and decreased performance, but there is essentially no diminution in the ability to work. Old age is characterized by predominant physical regression and a marked decrease in working capability. Depending on the demands made on the individual, and his/her predisposition, the transitions from maturity to middle age to old age show wide variations with respect to duration and time of onset.

a, b, c

Fig.4. *Physique.* Different types of physique (after Sheldon). *a* Endomorph. *b* Mesomorph. *c* Ectomorph.

Physoclistic swim bladder: a Swim bladder (see) with no air duct.

Physoclistous fishes: see Swim bladder.

Physopoda: see *Thysanoptera.*

Physostigmine, ***eserine:*** a poisonous indole alkaloid from calabar beans, the ripe seeds of *Physostigma venenosum* (family: *Papilionaceae;* a woody vine indigenous to the west coast of Africa). P. first causes excitation, then paralysis of the central nervous system; it increases glandular secretions and causes narrowing of the pupils. The alkaloid inhibits cholinestrase, and is used clinically as an antidote to curare, strychnine and atropine poisoning, as well as in ophthalmology.

Physostomatous fishes: see Swim bladder.

Phytic acid: see Myoinositol.

Phytin: see Myoinositol.

Phytoalexins, ***stress compounds:*** substances with antibiotic activity, produced in greater quantities or synthesized de novo by plants, in response to injury or stress, e.g. infection with fungi, bacteria or viruses, mechanical wounding, UV-irradiation, cold and treatment with phytotoxic chemicals (e.g. heavy metals) or Elicitors (see). They are absent from healthy plants, or present only in very small concentrations. [P. have also been called Phytoncides (see), but this term should be reserved for compounds that are always present in the plant in appreciable quanitity.] Most known P. have been discovered in the *Papilionaceae* followed by the *Solanaceae,* while many examples are known from other families. P. do not belong structurally or biosynthetically to any one class of compounds. Many are isoflavonoids and several of these are pterocarpans. There are also terpenes, acetylenic compounds, α-hydroxydihydrochalcones, stilbenes and polyketides. They function as growth inhibitors of pathogens, chiefly fungi. They have also been defined as *novel post-infectional metabolites produced by plants in response to fungal infection, and which, because of their antifungal activity, protect the plant from attack by fungi.* Under certain conditions, P. may also be antibacterial. P. do not include antifungal compounds already present in the uninfected plant. The original concept of P. included compounds formed from immediate precursors, e.g. the antifungal dihydroxyme-

Main characteristics of physiques according to Kretschmer

	Leptosome	Pyknic	Athletic
Trunk	Flat, long chest; sharp rib angle: relatively wide pelvis	Short, deeply vaulted chest; wide rib angle	Broad, strong shoulders trapezoid trunk with relatively narrow pelvis
Surface topography	Thin or sinewy with little subcutaneous adipose tissue	Softly rounded shape, resulting from well formed adipose tissue	Powerful, prominent muscles on a sturdy bone structure
Extremities	Long, thin extremities with long, slender hands and feet	Soft, smooth, relatively short extremities; delicately boned, short, broad hands and feet	Powerful, sturdy arms and legs; large hands and feet; possible acrocyanosis
Head and neck	Relatively small head; long, thin neck	Relatively large, rounded head with flattened crown; wide, heavy neck	Powerfully built, high head; freely mobile, strong neck with sloping, taught trapezius muscle
Face	Pallid, delicate face, truncated oval in shape; pointed, slender nose; sometimes an angular profile	Broad, flushed face with soft, supple features; profile lacking	Strong, bony face with emphasized prominent angles; steeply oval in shape
Hair	Strong head hair, sometimes tightly curled; sparse terminal hair	Soft head hair; tendency to baldness; moderate to strong terminal hair	Strong head hair; terminal hair unexceptional

thoxybenzoxazine released from its glucoside, which is present in cells of wheat and maize. The modern definition of P. excludes such compounds, and is restricted to compounds synthesized by remote pathways, which are first activated by infection or stress. Since there are no immediate precurors, the appearance of a P. is delayed until the biosythetic pathway has been activated. See Orchinol, Phaseolin, Pisatin.

Phytobenthos: plant organisms of the Benthos (see).

Phytocecidium: see Galls.

Phytochrome: a ubiquitous plant pigment, present in algae, mosses, liverworts, ferns and spermatophytes. It is a water-soluble, blue chromoprotein, with a tetrapyrrole prosthetic group that is related to the c-phycocyanin of cyanobacteria. It is an important photoreceptor, which mediates the light sensitivity of many growth and developmental processes. P. exists in two forms, a physiologically inactive form with an absorption maximum around 660 nm (P_r), and the physiologically active form which absorbs maximally around 730 nm (P_{fr}). When P_r absorbs red light it is converted into P_{fr}. Irradiation with red light therefore represents susception of the light stimulus and initiates the first stage of the physiological response. Subsequent irradiation with dark red light is inhibitory, because it converts P_{fr} back into P_r. The P. system is therefore often called the *reversible red-far red pigment system,* and the associated mechanism is the *reversible red-far red reaction system.* Since the P. system responds to very low energy (weak light intensity), the resulting reaction is referred to as the low energy response, as distinct from the high energy response, which is attributed to a different pigment system. The reactions linking the formation of P_{fr} and the final response, e.g. increased longitudinal growth, or seed germination, are largely unelucidated. P. is essential for the normal growth of green plants, and in many cases it is responsible for the regulation of important developmental and metabolic processes, as well as nastic and tropic movements, e.g. for seed germination and induction of flower development. P. was discovered in the course of investigations into the type of light required to promote the germination of lettuce seeds. When the seeds were kept in the dark, they germinated only after exposure to red light for at least 1 minute, the most effective wavelength being 660 nm. If the seeds were irradiated shortly afterward with dark red light of wavelength 730 nm, the inductive effect of the red light was abolished. Repeated alternation of the irradiation wavelength showed that germination or nongermination was always determined by the wavelength of the last exposure.

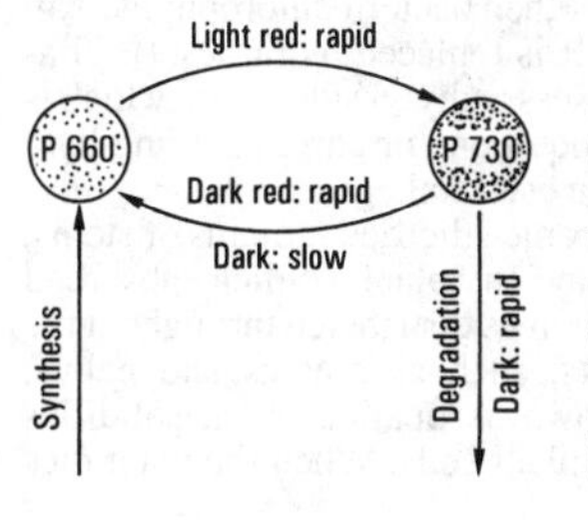

Phytochrome. The arrows indicate the rapid and slow conversions of phytochrome 660 (P 660) and phytochrome 730 (P 730) in light red and dark red light, and in the dark.

Phytoc(o)enosis: see Plant community.

Phytoene: see Carotenes.

Phytoepisite: see Modes of nutrition.

Phytofluene: see Carotenes.

Phytogeography: Plant geography.

Phytohemagglutinin: see Lectins.

Phytohormones, ***plant hormones:*** endogenous plant growth regulators which control and coordinate growth and development of higher plants. They have a ubiquitous occurrence in all higher plants, but the level of each P. depends on the plant organ and its stage of development. P. are synthesized in plant tissue in very small quantities, then transported locally from cell to cell, or from organ to organ within the plant. By enzymatic reactions, they are either irreversibly inactivated or temporarily deactivated by conjugation with monosaccharides or amino acids.

Chemically, P. do not belong structurally or biosynthetically to any single class of compound. Known P. are the growth-promoting Auxins (see), Cytokinins (see) and Gibberellins (see), and the more inhibitory Abscisic acid (see) and Ethylene (see). P. are active at very low concentrations. Regulation of growth and development often depends on an intimate interaction of the effects of two or more Ps. Frequently, therefore, it is not the absolute concentration of a single P. but the relative concentrations of two or more Ps. that crucially determine whether a hormonal effect will be inhibitory or stimulatory. They regulate the growth of root, stem and leaves, the development of seeds and fruits, senscence and abscission, apical dominance, dormancy of plants and their parts, geotropism, phototropism, and other processes.

As distinct from animal hormones, P. have a very broad activity spectrum and low action specificity; furthermore, sites of synthesis and target sites are not always clearly separated.

P. are detected and determined by various sensitive biotests, by modern physical-chemical methods (e.g. high performance liquid chromatography linked to mass spectroscopy), and more recently by immunological methods.

P. and related synthetic growth regulators are used widely in agriculture and forestry, and in horticulture.

In addition to the ubiquitous P. of higher plants, there are numerous other plant constituents which display growth regulating activity when applied exogenously, e.g. some phenolic compounds and steroids, which, however, do not meet the criteria for classification as P.

Phytol: an unsaturated, C_{20} diterpene primary alcohol (Fig.), discovered in 1907 by Willstätter. In chlorophyll *a* and and bacteriochlorophyll *a,* the propionic acid residue on C7 (ring D) of the porphyrin ring system is esterified with P., forming a long-chain lipophilic residue that anchors the chlorophyll molecule in the thylakoid membrane. (In other bacteriochlorophylls, e.g. from *Chlorobium,* the P. is replaced by farnesol). Vitamins K_1 and E also possess a phytyl side chain, which is attached to the naphthoquinone or chromane ring by a C-C bond and not by an ester linkage.

Phytoliths: microscopic silicified remains of stems, leaves and fruits of ancient plants. Silica, absorbed through plant roots, becomes distributed throughout the plant. In certain plants, such as grasses and palms, amorphous silica (known as opaline) even solidifies within and around the plant cells. When the plant dies and decays, a durable silica cast or phytolith is left in the soil. The term, phytolith, is also applied to sedimentary rocks derived from plant remains; see Bioliths.

Phytomastigina: see *Phytomastigophorea.*

Phytomastigophorea, *Phytomastigina, Phytoflagellata:* a subclass of the *Mastigophorea* (see). They are the only organisms that photosynthesize like plants, but are classified as animals. Some authorities regard them as algae, in which case these unicellular organisms represent the most primitive level of organisational development of the thallus. They possess flagella, which are used for locomotion, and they are found in all marine and freshwaters throughout the world. Their cells usually have a wall of cellulose or pectin, and may be solitary or combined in colonies. Chromatophores are present, which range in color from green to brown, depending on their relative contents of chlorophyll, xanthophylls and carotenes; photoassimilation products include various carbohydrates and fatty oils. Reproduction is usually asexual by longitudinal cell cleavage. Sexual reproduction, which occurs by iso- and anisogamy, is rarely observed, and only in a few species. Under unfavorable environmental conditions (usually lack of water), resting cysts are formed, surrounded by a membrane of gelatinous or other material, which protects the organism from desiccation.

1. Order: *Euglenida* (euglenoids) [If classified with the algae, this order becomes the class *Euglenophyceae].* These are relatively large organisms, with 2 flagella (a relatively long one and a very short one). The flagella arise within an interior invagination of the cell which consists of a tubular canal and a pyriform chamber, the so-called reservoir. An orange-red eyespot (stigma), containing β-carotene and other carotenoids, is present against the neck of the reservoir. The stigma functions as a light sensor. Chromatophores are green, and they contain chlorophylls *a* and *b* (as in green algae), but they also contain a xanthophyll that is otherwise unknown in the plant kingdom. Reserve materials are fatty oils and the carbohydrate, paramylon. The cell may be rigid, or it may be capable of changing its shape. Important genera are: *Euglena,* whose species often undergo mass population explosions in nutrient-rich waters; *Phacus,* which is found only in unpolluted waters; and *Colacium,* whose species adhere to other organisms by means of a gelatinous stalk.

2. Order: *Cryptomonadina.* Members of this order have 2 anteriorly directed flagella. They may be green, yellow, brown or colorless. Their chromatophores contain chlorophylls *a* and *c,* with various carotenoids, xanthophylls and phycobilins. A gullet or longitudinal groove is present. Ameboid forms are only rarely encountered. Species of *Cryptomonas* have a gullet and 2 green chromatophores; marine and freshwater species are known, and they are notably common in mesotrophic waters. *Chilomonas* lives saprophytically in freshwater; it has a very deep and narrow gullet, and no chromatophores. Both marine and freshwater forms of *Phaeococcus* are known, and they are normally found in the palmella phase (see *Mastigophorea).*

3. Order: *Dinoflagellata* (dinoflagellates) [If classified with the algae, this order becomes the class *Pyrrophyceae* or *Dinoflagellatae].* This order contains many marine organisms, which make a major contribution to the marine plankton. Some species are ameboid, but most are covered with an armor of numerous indi-

vidual cellulose plates, which are perforated by pores. There is a complex vacuole system and two flagella. One flagellum is directed backward, usually in a longitudinal groove (sulcus), while the other is directed transversely, usually in a transverse or spiral groove (annulus). Many species have processes or wing-like structures that help the cells to remain suspended in the water. Some species, in particular *Noctiluca miliaris,* are responsible for ocean phosphoresence. Others, especially marine species, contain or excrete toxins that kill fish.

4. Order: *Chrysomonadina.* (If classified with the algae, this order becomes the class *Prymnesiophyceae* or *Haptophyceae).* These are predominantly marine forms, with the same pigment composition as the *Chrysophyceae* (see). See Coccolithophorids.

Phytomimesis: see Protective adaptations.

Phytomonadina: see *Mastigophorea.*

Phytoncides: defense substances of higher plants that are active against microorganisms. Mutually exclusive definitions of P. and Antibiotics (see) are not possible. There is also a conceptual overlap between P. and the allelopathically active colins (see Allelopathy). See also Phytoalexins.

P. are generally distributed in plants. They are particularly conspicuous and therefore most extensively studied in herbs and spice plants. P. are secreted as gaseous or liquid substances by aerial plant organs or roots *(excretory P.),* or they are active within the living cell, and/or are released only when cells are injured *(nonexcretory P.).* They generally have a broad activity spectrum. Antiprotozoal, antibacterial and antifungal P. are known. Chemically, P. belong to various classes of compounds (alkaloids, tannins, alkanes, alcohols, etc.). They are present partly as glycosides, from which they are released by enzymatic cleavage.

Phytopaleontology: see Paleobotany.

Phytoparasite: see Modes of nutrition., Parasite.

Phytopathology, ***plant pathology:*** the study of plant diseases and plant damage by pests, and the prophylactic and therapeutic measures for their control. P. was established as an independent discipline in the mid 19th century by J. Kühn.

Phytophaga (Hessian fly): see Gall midges.

Phytophage: see Modes of nutrition, Nutritional interrelationships.

Phytophage rule: see Ecological principles, rules and laws.

Phytophthora: see *Phycomycetes.*

Phytoplankton: see Plankton.

Phytosaurs: see Triassic.

Phytosociology: see Sociology of plants.

Phytosterols: sterols synthesized by plants. The following important P. are listed separately: Stigmasterol, Brassicasterol, Campesterol, Fucosterol, Sitosterol.

Pica: see *Corvidae.*

Picas: see *Ochotonidae.*

Picathartes: see *Timaliinae.*

Picidae, ***woodpeckers:*** the largest family in the avian order *Piciformes,* represented in every zoogeographical region except Australia and Madagascar. As in all other *Piciformes,* the toes are zygodactylous. The 210 species are divided into 3 subfamilies:

1. Subfamily:*Picinae* ***(True woodpeckers,*** 179 species). The bill is typically straight, strong and pointed, and used for chiselling into bark and wood in search of insects, larvae and pupae. This type of food source does not vary greatly with the seasons, and in the winter of northern regions hibernating insects and pupae are readily available under bark; northerly *Picinae* therefore do not need to migrate, although some show a tendency to move a little farther south in the winter. The long tongue projects far beyond the bill, and its tip is armed with bristles or barbs. True woodpeckers usually perch along a branch or upright on a tree trunk, in contrast to most other birds which perch across a branch. The stiff tail

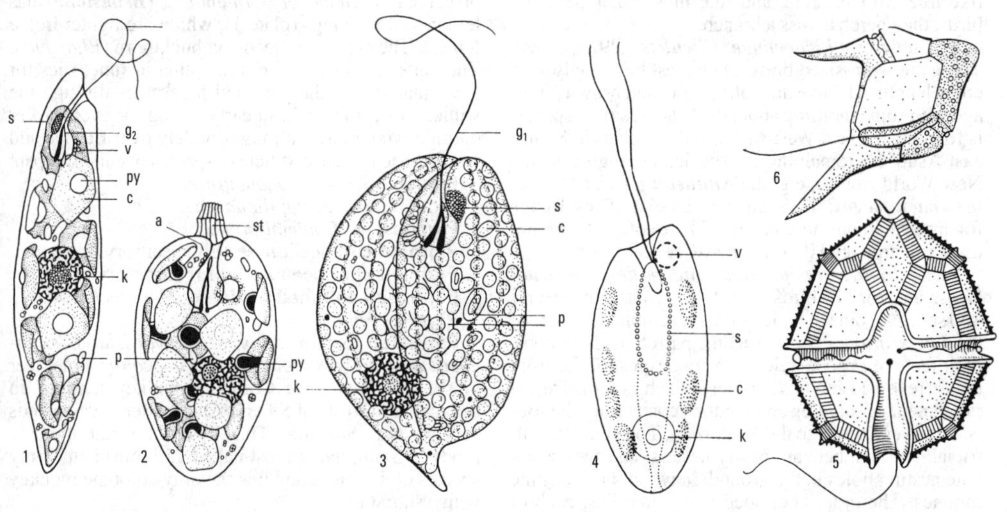

Phytomastigophorea. 1 Euglena gracilis (x 600). *2 Colacium mucronatum* (x 500). *3 Phacus triqueter* (x 600), *c* chloroplast, g_1 motile flagellum, g_2 second flagellum, *k* nucleus, *p* free paramylon, *py* pyrenoid with paramylon sheath, *s* eyespot, *st* mucilage stalk, *v* contractile vacuole. *4 Cryptomonas* (x 600), *c* chromatophore, *k* nucleus, *s* gullet, *v* contractile vacuole. Dinoflagellates: *5 Peridinium tabulatum* (x 600). *6 Ceratium hirundella* (x 350).

feather have strong shafts, and are used as a support when the bird holds itself erect against the surface of a tree. Plumage colors vary, but are mainly mixtures of black, white, green and brown, the male often having additional red coloration on the head and nape. Most familiar are the 30 species of small to medium-sized members of the genus *Dendrocopos* ***(Pied woodpeckers;*** black or grayish brown, barred or mottled with white), which are widely distributed across North America and Eurasia, e.g. the Eurasian ***Lesser spotted woodpecker*** *(Dendrocopos minor;* 14.5 cm), and the Eurasian ***Greater spotted woodpecker*** *(Dendrocopos major;* 23 cm). Of more restricted distribution is the ***Green woodpecker*** *(Picus viridis;* 32 cm) (plate 6), which is found in woods, large gardens, farmland and parks in Britain (not Ireland) and Europe, extending marginally into western Asia and southern Scandinavia. Other allied *Picinae* include the 2 species of ***Three-toed woodpeckers*** *(Picoides),* which lack one of the 2 hindtoes: ***Ladder-backed*** or ***Northern three-toed woodpecker*** (23 cm; circumpolar in evergreen forests of North America and the Eurasian taiga); and the ***Black-backed*** or ***American three-toed woodpecker*** (25.5 cm; spruce-larch forests from Alaska to Labrador and south to Nevada). Other American genera are, e.g. *Melanerpes* (e.g. the widely distributed ***Red-headed woodpecker)*** and *Colaptes* ***(Flickers)*** (e.g. the ***Yellow-shafted flicker*** of eastern North America)

2. Subfamily: *Jynginae* ***(Wrynecks,*** 2 species) The 2 wrynecks are the most primitive members of the family, so called for their ability to twist their heads. One wryneck is found in Africa, the other ***(Eurasian wryneck,*** *Jynx torquilla;* 18 cm) breeds across northern Eurasia from Britain to Japan, migrating south in the winter. Unlike true woodpeckers, their bills are not adapted as chisels. They do not peck or bore for food, but pick it up from the ground or tree surface by flicking their long tongues; they also catch flying insects on the wing. Unlike true woodpeckers, and like most other perching birds, they perch across a branch.

3. Subfamily:*Picumninae* ***(Piculets,*** 29 species). These are wren-sized birds, the largest being only 12.5 cm in length. All have drab olive, gray and brown plumage with some mottling about the head. A single species is found in tropical West Africa and 3 species in Southeast Asia, the remaining 25 species belonging to the New World tropics, e.g. the ***Antillean piculet*** *(Nesoctites micromegas;* 10.5 cm; Hispaniola). They forage for insects among low bushes. Their soft tails are not used as a prop, and they sit crosswise on a perch.

Piciformes, ***woodpeckers and allies, picarian birds:*** an order of birds (see *Avies)* containing nearly 400 species, some of which rather resemble members of the *Passeriformes.* Like trogons, parrots and cuckoos, their feet are zygodactylous (the outermost of the front toes is reversed and lies parallel to the hindtoe). The arrangement of muscles and tendons controlling the toes is, however, unique to the *Piciformes.* The young are altricial. All members are cavity nesters, and they excavate nesting holes in the ground, in wood, or in termite mounds. The order is divided into 6 families, each of which is listed separately: *Galbulidae* ***(Jacamars),*** *Bucconidae* ***(Puffbirds),*** *Capitonidae* ***(Barbets),*** *Indicatoridae* ***(Honeyguides),*** *Ramphastidae* ***(Toucans)*** and *Picidae* ***(Woodpeckers).***

Picinae: see *Picidae.*

Pickleblennies: see *Blennioidei.*

Picoides: see *Picidae.*

Picrocrocin: a bitter principle in crocus species, e.g. *Crocus sativus* and *Safran aquila (= S. electus).* It is a terpene glucoside, and it is also thought to be a sex determining substance (gynotermone) in *Chlamydomonas.* Hydrolysis of P. produces glucose and safranal (2,6,6-trimethyl-1,3-cyclohexadien-1-al). The aglycone is 2,6,6-trimethyl-4-hydroxy-1-cyclohexaen-1-al, which is converted into safranal by loss of water when it is released by hydrolysis.

Picrotoxin, ***cocculin:*** a molecular compound of one molecule of picrotoxinin and one molecule of picrotin. P. is a neurotoxin, which occurs in the seeds of the Indian plant *Anamirta cocculus,* and in *Tinomiscium philippinense.* As a specific antagonist of 4-aminobutyric acid, it acts as a central and respiratory stimulant, and is an intidote to barbiturates. It is extremely toxic to fish. The active component of the molecular compound is picrotoxinin, the picrotin being physiologically inactive.

Piculets: see *Picidae.*

Picumninae: see *Picidae.*

Picus: see *Picidae.*

Piddocks: see *Bivalvia.*

Pied flycatcher: see *Muscicapinae.*

Pied wagtail: see *Motacillidae.*

Piercing and sucking feeders: see Modes of nutrition.

Pieridae, ***whites and sulfurs:*** large, cosmopolitan family of the *Lepidoptera* (see), containing many common and well known species. Most are white, yellow or orange in color. *Pieris brassicae L.* ***(Large cabbage white)*** (see Cabbage white), a well known migrant, is common throughout its range, which includes Europe, North Africa and all Mediterranean countries. The caterpillars (plate 45) are yellowish green with dark green bands and black spots on either side; they feed on several members of the *Cruciferae* and are a particular pest of cabbages. *Gonepteryx rhamni L.* ***(Brimstone)*** has lemon-yellow wings (plate 1), which are lighter in the female. The caterpillar feeds on buckthorn *(Rhamnus).* The butterfly emerges from the pupa in June, flies for the remainder of the year and hibernates through the winter, emerging again in early spring, or even earlier on mild winter days. It ranges widely over Europe and North Africa, and a distinct subspecies occurs in Japan.

Pigeon milk: see *Columbiformes.*

Pigeon pea: see *Papilionaceae.*

Pigeons: see *Columbiformes.*

Pigment cup ocellus: see Light sensory organs.

Pigment spot ocellus: see Light sensory organs.

Pig-nosed softshell turtle: see *Carettochelyidae.*

Pigs: see *Suidae.*

Pig-tailed leaf monkey, ***Mentawai islands snub-nosed monkey, simakobu,*** *Simias concolor:* a member of the *Colobinae* (see Colobine monkeys), found in tropical rainforests of Siberut island, Mentawai islands and western Sumatra. This medium sized monkey (body 515 mm; tail almost naked 155 mm) is the only species of its genus, and it is the only colobine monkey with a short tail.

Pigweed: see *Chenopodiaceae.*

Pikas: see *Ochotonidae.*

Pike cichlids: see Cichlids.

Pike-perch, *Lucioperca lucioperca, Stizostedion lucioperca:* the largest freshwater European member of the family *Percidae,* attaining a maximal length of 130 cm. A predatory species, it inhabits fresh and brackish waters of north, central and eastern Europe. In many areas it is a valuable food fish, with high quality, tasty flesh, as well as a popular sporting fish. It has been introduced into western Europe and the USA.

Pilchard, *Sardina pilchardus:* a small, shoal-forming marine fish (family *Clupeidae,* order *Clupeiformes),* 20–25 cm in length (maximum 30 cm), with a bluish green back, gleaming gold sides and a silver-white belly. There is a line of dark spots along the upper flank and a single dark spot on each operculum. It is a monotypic genus, characterized by loosely attached, large body scales that continue onto the caudal fin. The dorsal fin lies in a groove formed by the scales. The finger-like young Ps. are known as ***sardines.*** The P. is one of the most important commercial fish of the Mediterranean and coastal regions of the Atlantic as far north as Norway.

Pilea: see *Urticaceae.*

Pili: filamentous appendages of certain Gram-negative bacteria. P. are only 25 nm in diameter, consist of protein, and sometimes arise in large numbers from the cell suface. In contrast to flagella, P. are not responsible for motility. They attach the bacteria to other cells or surfaces. The F-pili or sex pili are hollow and serve as conjugation tubes for the transfer of genetic material between bacterial cells.

Pilidium larva: see *Nemertini.*

Pillaia: see *Mastacembeloidei.*

Pill bugs: see *Isopoda.*

Pill millipedes: see *Diplopoda.*

Pilobolus: see *Phycomycetes.*

Pilot-fish, *Naucrates ductor, Gasterosteus ductor:* a teleost fish (family *Carangidae,* order *Perciformes),* found almost exclusively in warm seas. Up to 70 cm in length (usually 30–50 cm), the torpedo-shaped body has a pale bluish background with dark transverse rings. Small shoals of P.fs. migrate long distances by accompanying larger fish (mainly sharks and whales), as well as ships, behavior that was familiar to ancient mariners.

Pinaceae, ***pine family:*** a family of conifers containing about 250 species, found almost exclusively in the Northern Hemisphere. They grow as forests, which are often a dominant feature of the landscape, they form the tree limit of high mountains, and they mark the boundary of the Arctic tundra. With few exceptions, they are evergreen trees with spirally inserted needles and woody cones. The male flowers (strobili) consist of numerous stamens with two pollen sacs on the underside. In most species (larch and the common Douglas fir are exceptions), the pollen grains have two air sacs or vesicles, formed between the intine and exine (Fig. 5), which increase the ability of the pollen grain to float in the air. There may be a long delay between pollination and fertilization, e.g. one year in the Scotch pine. Before fertilization, the female strobili are always erect, and this position usually alters when the seeds are ripe. The unilaterally winged seeds are wind-distributed. The timber from pine trees has many varied uses. In addition, most pines contain essential oils and balsams. Turpentine is obtained by distillation from pine balsam (an oil/resin mixture occurring primarily in the wood), leaving a residue known as colophony.

Pines are divided into 3 groups, depending on the mode of growth of their needles on short or elongated shoots.

Fig.1. *Pinaceae.* Fir shoot with a mature cone and a cone which has already lost some seeds.

1. Needles arising separately along the length of an elongated shoot (Fig. 1). This type is represented by the genera *Abies, Picea, Tsuga* and *Pseudotsuga.* The ***Silver fir*** *(Abies alba)* is cultivated for forestry in Central and Southern Europe. On the underside of each needle are two light-colored lines of wax. The wood is durable and easily worked. As in all species of *Abies,* the ripe cones remain upright. The ***Balsam fir*** *(Abies balsamea)* is native to northern North America; its wood is used for paper making, and its bark is the source of Canada balsam used in microscopy.

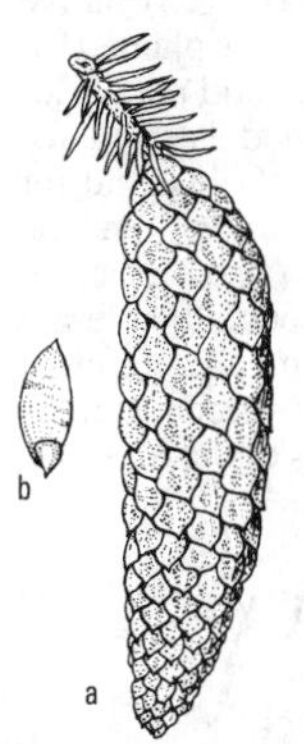

Fig.2. *Pinaceae.* Norway spruce. *a* Shoot with cone. *b* Seed.

Norway spruce *(Picea abies = Picea excelsa)* (Fig. 2) is the most widely distributed conifer of northern and central Europe. Its wood is rather soft and easy to work and it is used for planks, pit props, telegraph poles, crates and boxes, and for the sounding boards of pianos and organs.

Sitka spruce *(Picea sitchensis)* is planted for forestry in the USA and various European countries, and is indigenous to the west coast of North America. Its light wood has many uses. A popular ornamental tree is the ***Colorado spruce*** or ***Blue spruce*** *(Picea pungens);* it has gray-green to silver-white, hard, prickly needles. The common ***Douglas fir*** *(Pseudotsuga menziesii),* a native North American tree, is also cultivated in Europe for its valuable wood. Also indigenous to North America is the ***Canada hemlock***, ***Canadian hemlock*** or ***Eastern hemlock*** *(Tsuga canadensis),* which is also used for forestry in Europe; it is a source of pitch and its bark has been used for leather tanning.

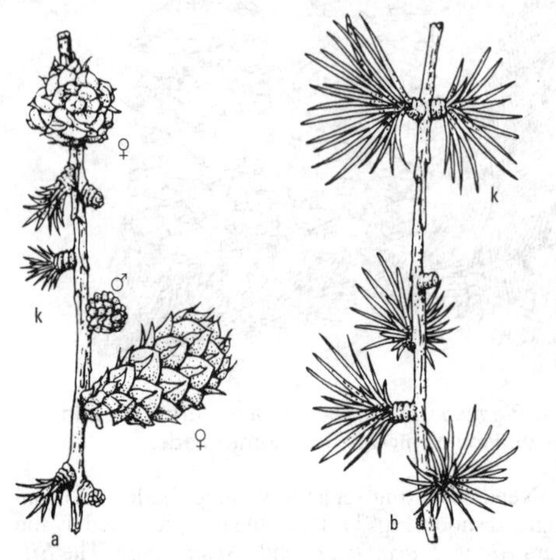

Fig.3. *Pinaceae.* European larch. *a* Elongated shoot in spring, showing male and female strobili and emerging short shoots *(k)*. *b* Elongated shoot in summer, with short shoots carrying needle clusters *(k)*.

2. Needles in clusters on short and elongated shoots. This type is represented by the genera *Larix* and *Cedrus.* The ***Larches*** *(Larix* spp.) shed their needles in autumn, e.g. the ***European larch*** *(Larix decidual)* (Fig. 3), native to northern and central Europe, which is now planted for forestry in North America, New Zealand and many European countries. Its soft, durable wood is used for planks and pit props, for steps and thresholds, and for furniture. The ***Japanese larch*** *(Larix kaempferi)* and the ***Siberian larch*** *(Larix sibirica)* are also used for commercial forestry. The latter species is especially tolerant of air pollution. ***Cedar*** trees *(Cedrus* spp.) are similar in habit to the larches, except they are evergreen. Cedars are indigenous in the mountainous areas of the Mediterranean region and in the western Himalayas. Most familiar of these is the ***Cedar of Lebanon*** *(Cedrus libani)* (Fig. 4), which yields a very valuable wood; its numbers have decreased markedly in the Lebanese mountains.

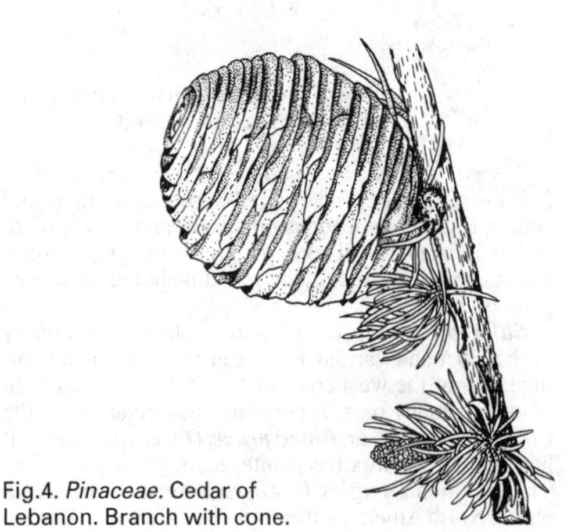

Fig.4. *Pinaceae.* Cedar of Lebanon. Branch with cone.

3. Needles in groups of 2 to 5, always on short shoots. This group is represented by the single genus, *Pinus* (pines in the narrow sense), which has 105 known species. The natural range of the ***Scotch pine*** *(Pinus sylvestris)* (Fig. 5) is the northern temperate belt of Eurasia, but it is found more widely than this as a commercial forestry tree. It yields turpentine, tar and pitch, and its soft, durable wood is used for many purposes. The ***Aleppo pine*** *(Pinus halepensis)* and the ***Italian stone pine*** *(Pinus pinea)* are native to the Mediterranean region. Other frequently cultivated *Pinus* species are the North American ***Eastern white pine*** or ***Weymouth pine*** *(Pinus strobus),* which was introduced into Europe in the 18th century, and the ***Austrian pine*** *(Pinus nigra),* which is native to the eroded limestone soils of southern Europe and Asia minor. The ***Swiss stone pine*** *(Pinus cembra),* which is found primarily in the Alps and the Carpathians, produces an edible seed called a pine nut. ***Bristle cone pines*** *(Pinus aristata)* are some of the oldest living trees; one specimen in California has been dated at about 4 900 years.

Fig.5. *Pinaceae.* Scotch pine. *a* Flowering and fruiting shoot. *b* Pollen grain with two air sacs. *c* Seed.

Pinacocyte: see *Porifera.*
Pinacoderm: see *Porifera.*
Pine: see *Pinaceae*
Pineal body: see Epiphysis.
Pineal gland: see Epiphysis.

Pine beauty, *Panolis flammea* D. & S.: a noctuid moth, wingspan 35 to 38 mm, occurring in Europe and western Asia. The wings are patterned with wavy lines and white spots, the predominant effect being red-brown. About 200 eggs are laid in rows on pine needles, which are the sole food source. The caterpillars are longitudinally striped green and white, which effectively camouflages them against the pine needles. Occasional mass infestations cause considerable damage, especially to young pine trees and new shoots.

Pine forest community: see Character species.
Pine grosbeaks: see *Fringillidae.*
Pine leaf-cast: see *Ascomycetes.*
Pine looper: see Bordered white.
Pine marten, *Martes martes:* a yellow-brown arbo-

real mustelid (see *Mustelidae)* found in Eurasia. It conceals itself in tree-holes, feeding mainly on squirrels. It also robs birds' nests for their eggs and chicks, and even attacks larger mammals. Insects and fruit are also taken.

Pinene, ***2,6,6-trimethylbicyclo[3.1.1]hept-2-ene:*** an unsaturated, bicyclic monoterpene hydrocarbon; a colorless, optically active liquid with a characteristic odor. P. is the main constituent of turpentine oil (see Turpentine) and a component of many other essential oils, e.g. juniper, eucalyptus and sage oils. α- and β-P. differ with respect to the position of the double bond. P. is the starting material for the preparation of synthetic camphor, and it is used as a perfume base.

CH_3 H_3C CH_3

α-Pinene

Pin-eyed flower: see Pollination.

Pinfeather: see Feathers.

Pinguicula: see Carnivorous plants.

Pinguinus: see *Alcidae.*

Pinicae: see Gymnosperms.

Pinicola: see *Fringillidae.*

Pinidae, *Coniferae,* ***conifers*** (plate 35): the most numerous subclass of the *Coniferophytina* (see Gymnosperms). Conifers are distributed worldwide. They are profusely branched trees, rarely shrubs. The stem is usually monopodial with lateral branches, which are arranged in regular tiers, and which often show distinct differentiation into long and short shoots. The xylem consists almost entirely of tracheids, and it is usually strongly developed. Resin canals are common in the bark and leaves, as well as in the wood. Bark is often well developed. The thick, tough-textured, xeromorphic, evergreen leaves are reduced in size and simple in structure, being needle-shaped or scale-like. In the most primitive conifers the leaves are arranged spirally, but in more recent species they are whorled or decussate. Flowers are always unisexual, monoecious or dioecious, and usually arranged in typical strobili or cones. The male strobilus consists of a dense spiral of numerous scale-like stamens with 2–20 pollen sacs on the underside. Spermatozoids are not formed by the male gametophyte (see Gymnosperms). Female flowers form a cone-like inflorescence or strobilus, containing numerous spiral, whorled or decussate bracts with ovuliferous scales in their axils. Ovuliferous scales carry a few ovules, rarely a single or several ovules. This female cone represents a modified inflorescence, in which each ovuliferous scale is a reduced flower. At maturity, the cone becomes woody; this may occur by lignification of the ovuliferous scale alone, or of both the ovuliferous scale and its associated bract. The female strobilus rarely remains unlignified after seed formation. Embryos possess at least 2 cotyledons.

According to the fossil record, the evolutionary development of the conifers started over 250 million years ago in the Upper Carboniferous. Recent conifers are subdivided into several families, according to the structure of their cones and leaves, and other characteristics. The following important conifer families are described under separate entries: *Araucariaceae, Pinaceae, Taxodiaceae, Cupressaceae, Podocarpaceae, Cephalotaxaceae* .

Pinitol, ***abietol:*** the 5-O-monomethyl ether of D-inositol. It occurs, inter alia, in the needles of the silver fir *(Abies abies).* See Cyclitols.

Pin mold: see *Phycomycetes.*

Pinna of the ear: see Auditory organ.

Pinnate-leaved gymnosperms: see Gymnosperms.

Pinnipedia, ***pinnipeds, aquatic carnivores:*** a suborder of the *Carnivora. P.* are mammals, which are adapted to an aquatic or amphibious existence. They have fusiform bodies with very short tails, and their short limbs are modified to oar-like flippers. Skilful and agile swimmers and divers, *P.* are ungainly on land. Their nose and ear openings can be closed under water. They are found on all sea coasts, but are more numerous in the colder and polar regions. Some species inhabit large inland waters. They feed on fish, crustaceans, bivalves, cephalopods and water birds. The young are born on land. Three families are recognized: ***Fur seals*** and ***Sealions*** (see *Otariidae),* ***Walrus*** (see *Odobenidae),* and ***Earless seals*** (see *Phocidae).*

Pinta: see Spirochetes.

Pintails: see *Anseriformes.*

Pinworm: see Enterobiasis.

Pionnotes: mucilaginous masses containing the embedded micro- or macroconidia of the genera *Fusarium* and *Cylindrocarpon (Fungi imperfecti,* assigned to the *Ascomycetes,* order *Sphaeriales).*

L-Pipecolic acid: piperidine-2-carboxylic acid (Fig.), a heterocyclic, nonproteogenic amino acid, which is found in many plants. The 4- and 5-hydroxy derivatives of L-P.a. are particularly abundant in mimosas and palms.

N H COOH

Pipecolic acid

Pipefish: see *Syngnathoidei.*

Piper: see *Piperaceae.*

Piperaceae, ***pepper family:*** an exclusively tropical family of dicotyledonous plants containing about 3 100 species in 8 genera. They are herbaceous or woody, with whorled or spirally inserted simple leaves. There is no calyx. Pollination is probably by insects, and the minute, unisexual or hermaphrodite flowers are arranged in dense spikes. The fruits are berries or small drupes. The most numerous genus is *Piper* (about 2 000 species). *Piper nigrum* ***(Pepper vine),*** a perennial climbing plant indigenous to southern India, is widely cultivated in the tropics. The unripe, dried fruits (drupes) are sold as black pepper. White pepper is obtained by removal of the outer part of the pericarp from the nearly ripe, fermented fruits. The sharp taste is due to the alkaloid, piperine, and the odor is due to the essential pepper oil. Cubebs (formerly used pharmaceutically for their stimulant and antiseptic properties) are the fruits of *Piper cubeba,* a woody climber indigenous to Java, Sumatra and Borneo. *Piper methysticum* ***(Kava pepper)*** is found mainly on the Pacific Islands. Its roots contain a resin (kavalin), which is stimulatory, diuretic, sudorific and intoxicating, and they are used for making the Polynesian beverage known as kava. Leaves of the Southeast Asian *Chavica (= Piper) betle* ***(Betel pepper)*** are mixed with the seeds of the betel palm (betel nuts) and lime, and chewed as a stimulant. The second largest genus is *Peperomia* with about 1 000 species. Mainly on account of their attractive leaves (usually somewhat succulent, sometimes corrugated, sometimes with silvery effects), some species of *Peperomia* are popular ornamental plants.

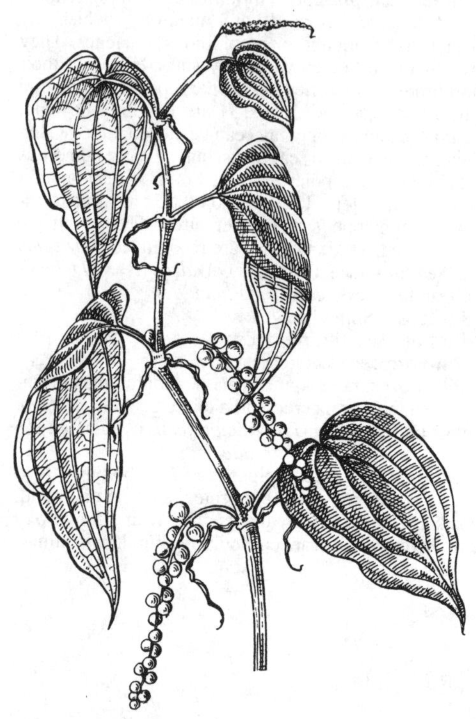

Piperaceae. Pepper vine *(Piper nigrum),* branch with infructescences.

Piperidine: a liquid heterocyclic base, which has a strong ammonia-like smell. The piperidine ring system is present in the structure of numerous alkaloids *(piperidine alkaloids),* e.g. coniine, piperine and the tropane alkaloids (see Tropane).

Piperine: a piperidine alkaloid, consisting of piperic acid and piperidine linked by an amide bond. The sharp taste of black pepper, *Piper nigrum,* is due to its content (5–9%) of P.

Pipe snakes: see *Aniliidae.*

Pip-fruit: see *Rosaceae.*

Pipidae, *pipids, clawed* and ***Surinam toads:*** a family of the *Anura* (see), containing 26 species in 4 genera: *Xenopus* (14 species; sub-Saharan Africa); *Pseudhymenochirus* (1 species; West Africa); *Pipa* (7 species; northern South America east of the Andes, and eastern Panama); *Hymenochirus* (4 species; Equatorial Africa).

Pipids possess 6–8 epichordal, opisthocoelous presacral vertebrae with imbricate (overlapping) neural arches. Presacrals I and III are fused in 3 genera, but separate in *Xenopus.* The atlantal cotyles of presacral I are juxtaposed in 3 genera, but separated in *Pipa.* In some species of *Pipa,* presacral VIII is fused with the sacrum. Free ribs are present in the tadpole, but they fuse with the transverse processes of presacrals II–IV after metamorphosis. The sacrum has expanded diapophyses, and it is fused with the coccyx, which usually lacks proximal transverse processes. An omosternum is lacking, and the pectoral girdle is pseudofirmisternal. The sternum is greatly expanded, and the epicoracoid cartilages are found posterior to the coracoids. Nonpedicellate maxillary and premaxillary teeth are present in *Xenopus* and in some species of *Pipa.*

Adults live permanently in water, and their broad, flattened bodies are inconspicuously colored gray, brown or green on black, usually paler on the under surface. Movable eyelids are absent (except in *Pseudohymenochirus*), the pupils of the eyes are round, and there is no tongue. In some species the adult retains the larval lateral line.

See Surinam toad, African clawed toad.

Pipilo: see *Fringillidae.*

Pipistrelle: see *Vespertilionidae.*

Pipistrellus: see *Vespertilionidae.*

Pipridae, *manakins:* an avian family (59 species in 20 genera) in the order *Passeriformes* (see). Manakins are fruit-eating, Neotropical forest birds, nearly all of which are less than 15 cm in length, although the tail may be elongated. Wings are rounded and short, and the bills are short, straight and slightly hooked. Males usually have brilliantly colored plumage, while the female is dull, but some genera (e.g. *Schiffornis)* show no sexual dimorphism. Fruit is plucked in flight. Males perform unique mating displays accompanied by vocal, and in some species by mechanical noises, which are produced by modified flight feathers.

Piranha: see *Characiformes.*

Piraruc˙: see *Osteoglossidae.*

Pirate bugs: see *Heteroptera.*

Pirate perch: see *Percopsiformes.*

Piroplasmids, *piroplasms,* *Piroplasmida:* very small protozoa, which parasitize mammalian blood cells and various cells and body organs of ticks. They are transmitted by ticks to their mammalian host, and vice versa. No alternation of generations occurs; the parasite retains the same form in both hosts. P. are restricted to warm climates. Despite their wide range of mammalian hosts, they have never been observed in humans. The taxonomy of P. is unclear. Although spores have never been observed, they are tentatively placed in the *Sporozoa.* Two families are recognized.

1. *Babesidae* (babesians) are found in the erythrocytes of their mammalian hosts, and in the salivary glands and other organs of ticks. They infect the tick eggs and thereby subsequent generations. Erythrocytes are destroyed and hemoglobin appears in the urine. They reproduce in the erythrocyte by binary division, and quadruple division may also occur. *Babesia bigemina* is responsible for Texas cattle fever, which is encountered not only in USA, but in Africa and all warm climates.

2. *Theileridae* (theilerians) are found in the white blood cells of their mammalian hosts, and in the salivary glands of ticks. The tick's ovaries are not infected. Reproduction is by binary division. *Theileria parva* (transmitted mainly by *Rhipicephalus appendiculatus*) is responsible for East Coast fever in African cattle.

Pisatin: a fungitoxic substance produced by the endocarp pod tissue of *Pisum sativum* in response to infection by *Monilina fructicola.* See Phytoalexins.

Pisidium: see Freshwater mussels.

Pistachio nuts: see *Anacardiaceae.*

Pistol shrimps: see Shrimps.

Pisum: see *Papilionaceae.*

Pit: see Pits.

Pitcher plants: all plants belonging to the following genera: *Nepenthes, Sarracenia, Heliamphora, Darlingtonia, Cephalotus.* See Carnivorous plants.

Pitfall trap: see Carnivorous plants.

Pithecanthropus, *Homo erectus erectus:* the name given to the fossil human remains discovered in 1891/92 by the Dutchman, Eugen Dubois, at Trinil on the island of Java. The name had already been coined in 1866 by Ernst Haekel for a hypothetical transitional form between ape and human. See Anthropogenesis, *Homo.*

Pits: isolated, unthickened areas in plant cell walls. In cells with heavily thickened cell walls (e.g. sclerenchyma cells), P. appear in cross section as narrow tubular channels. P. of adjacent cells meet one another and are separated by the pit membrane. The latter consists of the middle lamella with primary membranes from the respective cells on either side. Pit membranes contain numerous perforations (plasmodesmata), which are penetrated by fine threads of protoplasm containing extensions of the endoplasmic reticulum. Thus, the cisternae of the endoplasmic reticulum of neighboring cells are in open contact via tubular channels. This facilitates an exchange of material between cells, so that groups of cells are able to act as a physiological unit. In addition to these *simple P.,* some plant cells (in particular the tracheids of conifers) possess *bordered P.,* in which the pit channel broadens like a funnel toward the pit membrane. In surface view, bordered P. appear as two concentric circles. The inner ring represents the boundary of the pore, i.e. the opening of the channel into the cell lumen, whereas the outer ring is the boundary of the channel at its greatest diameter (immediately next to the pit membrane). In bordered P., the middle of the pit membrane may be thickened to form a *torus.* If the torus lies centrally between the two pores, material can move freely through the pit from cell to cell. The pit membrane may, however, arch to one side, so that the torus closes the pore on that side like a valve. Exchange of material through a bordered pit can therefore be regulated. Where dead, water-conducting elements adjoin living cells, pits are usually bordered only on the water-conducting side.

Pittas: see *Pittidae.*

Pittidae, *pittas:* an Old World avian family (23 species in a single genus) in the order *Passeriformes* (see). The pittas are insectivorous, tropical forest species, concentrated in Southeast Asia, with 3 species reaching Australia and 2 restricted to Africa. They have a distinctive long tarsus, a stout body, sometimes a degenerate tail, and rounded wings, the latter permitting short bursts of flight. All species have erectile crests and brilliant plumage. They build large, domed nests near the ground, but roost at higher levels.

Pituitary: see Hypophysis.

Pit vipers: see *Crotalidae.*

Pizonyx: see *Vespertilionidae.*

Placenta: a richly vascularized structure linking the embryo and maternal organism in higher mammals *(Monodelphia).* [The placenta of the plant embryo is described under the entries for Flower and Seed]. Nutrients and oxygen for the maintenance of embryonic growth pass across the P. from the maternal organism, while metabolic waste products from the embryo pass in the opposite direction. Since mammalian eggs have no yolk, the P. is necessary for maintenance of the embryo during its lengthy period of growth and development in the uterus. The P. is known to be an endocrine organ in many species. It also stores vitamins for use by the embryo. Although it displays a great variety of shapes in different species, the P. has a basically uniform structure. The embryonal part of the P. *(P. foetalis)* is derived from the diverticulae of the allanto- or amnio-chorion (see Extraembryonic membranes), while the maternal part *(P. materna* or *uterina)* is derived from the uterine mucosa (uterine epithelium or endometrium).

A *yolk-sac P.* occurs in live-bearing sharks. Certain reptiles also have a primitive P. It is doubtful whether the fish or reptilian Ps. are in any way homologous with the mammalian organ. Most marsupials produce only a weakly developed and briefly temporary P., or none at all. In some marsupials, e.g. the bandicoot *(Perameles;* family *Peramelidae),* the allantois and chorion together form a limited vascular area closely apposed to the uterine wall, which is known as an *allantoic* or *chorion P.* In others, e.g. the Tasmanian devil *(Sarcophilus = Dasyurus;* family *Dasyuridae),* the splanchnopleure of the yolk-sac joins the chorion and forms a P.-like vascular area against the uterine wall, known as a *yolk-sac P.*

Placentation in mammals, including humans, begins with *implantation* or *nidation* of the trophoblast in the uterine mucosa. In most species this occurs immediately after the fertilized egg has entered the lumen of the uterus. In some species, e.g. bears, martens and seals, implantation is delayed; in the roe deer, implantation does not occur until 4 months after fertilization. Implantation is the process whereby the embryo becomes fixed in position and establishes physical contact with the maternal organism. Part of the embryo consists of tissue that is specialized for interaction with the receptive uterine mucosa; this tissue is known as the *trophoblast* (see Cleavage). Trophoblast and uterine mucosa interact with varying degrees of intimacy. Thus, in cow, sheep, horse, pig, whale, dugong and lemurs, the embryo remains in the uterine lumen, and the trophoblast (which later becomes the chorion) simply makes close contact with the uterine mucosa. Such Ps. are called *syndesmochorial, epitheliochorial,* or *nondeciduate.* In carnivores, however, the uterine mucosa is broken down and chorionic villi from the trophoblast invade the deeper uterine tissue until they reach the endothelium of the maternal blood vessels; this is known as an *endotheliochorial* or *deciduate P.* In the *hemochorial P.* of rodents and humans, the chorionic villi break down the maternal blood vessels, so that they become bathed in maternal blood.

The following description of implantation and P. development applies largely to the human organism.

The uterus is prepared for implantation by the action of ovarian hormones, i.e. estrogens produced by the Graafian follicles, and progesterone produced by the Corpus luteum (see Follicle cells). Estrogens produced by the theca cells of the mature follicles initiate estrus and various cyclic phenomena in the uterus and vagina, and they are essential for sensitization of the uterine mucosa, so that it becomes receptive to implantation.

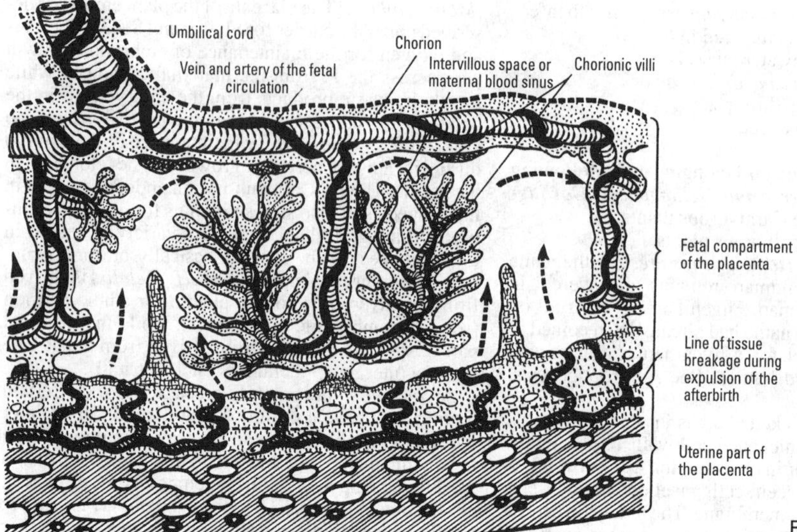

Fig.1. *Placenta.* Structure of the placenta (semischematic).

Implantation occurs during the secretory phase when the proliferated mucosa has acquired a looser and more open texture.

Ovulation is initiated by *Luteinizing hormone* (LH) (see) from the anterior pituitary. Immediately after the follicular cavity in the ovary has become sealed with coagulated blood to form the *Corpus rubrum,* LH causes luteinization of granulosa cells to form the secretory yellow body known as the *Corpus luteum,* which synthesizes progesterone. Secretion of progesterone by the corpus luteum is maintained by the luteotropic complex (LH plus prolactin, both hormones of the anterior pituitary). Progesterone maintains the viability of the egg during its passage to the uterus; it inhibits the formation of Graafian follicles in both ovaries by inhibiting the action of follicle stimulating hormone (FSH) from the anterior pituitary; and it causes dilation of blood vessels in the uterus. An additional important function of progesterone is to to continue the process of pre-implantation sensitization of the uterine mucosa, which was initiated by hormones from the Graafian follicles, i.e. continuation of the transformation of the uterine mucosa as it passes into the 13-day-long secretion phase. At this stage, the mucosal cells contain stored glycogen for nutrition of the implanted embryo, and they are known as *decidual cells;* glandular spaces in the mucosa (uterine glands) become larger and filled with secreted fluid; the tissue becomes more vascularized; and the mucosa becomes the *decidua.*

Progesterone secretion reaches a maximum on the 3rd to 6th day after ovulation, then decreases on the 7th day. The nature of all subsequent changes in the decidua and the corpus luteum depends on whether the released egg becomes fertilized.

If fertilization occurs, the embryo enters the uterus lumen on its 10th day of development, having travelled down the oviduct for 9–10 days during the first part of the secretory phase. By this time, the multilayered trophoblast has developed (see Cleavage). The trophoblast now secretes large quantities of chorionic gonadotropin (CG), which takes over the function of LH (LH secretion by the anterior pituitary is suppressed by estrogens and progesterone). CG has the following functions: 1) prevents breakdown of the corpus luteum and extends its active life (now called the *Corpus luteum graviditatis);* 2) suppresses desquamation of the decidua (now known as the *Decidua graviditatis);* 3) stimulates further secretion of hormones by the Graafian follicles and corpus luteum graviditatis; and 4) stimulates development of more Graafian follicles. Trophoblast secretion of CG continues during the first 3 months of pregnancy, and sometimes longer, and its action is supported by prolactin. Estrogens and progesterone, acting synergistically with other hormones (prolactin, growth hormone, corticosteroids), promote development of the mammary glands and growth of the milk ducts.

After implantation, prolactin and CG production is so active that the excess hormones in the urine can be used for the diagnosis of pregnancy.

In humans, implantation can occur at any site on the internal surface of the uterus. As soon as the trophoblast makes contact with the mucosa, it adheres to it, breaks down the upper surface by histolysis and phagocytosis (using proteolytic enzymes), and sinks through into the underlying stroma (intradecidual implantation). The point of entry becomes sealed with an operculum of clotted blood, below which the mucosa reseals itself. The embryo is then surrounded on all sides by uterine tissue (Fig.2), representing the first stage of the *P. foetalis.* Food reserves in the maternal tissue are taken up by the trophoblast and transferred to the developing embryo. For this purpose, trophoblast cells are transformed into sponge-like tissue with lacunae, known as the *syncytial trophoblast.* During breakdown of the mucosa, maternal blood vessel walls are also degraded. Arterial maternal blood then passes without coagulation into the lacunae of the syncytial trophoblast, and is taken up again by uterine veins. Thus, the early histotropic stage of P. devel-

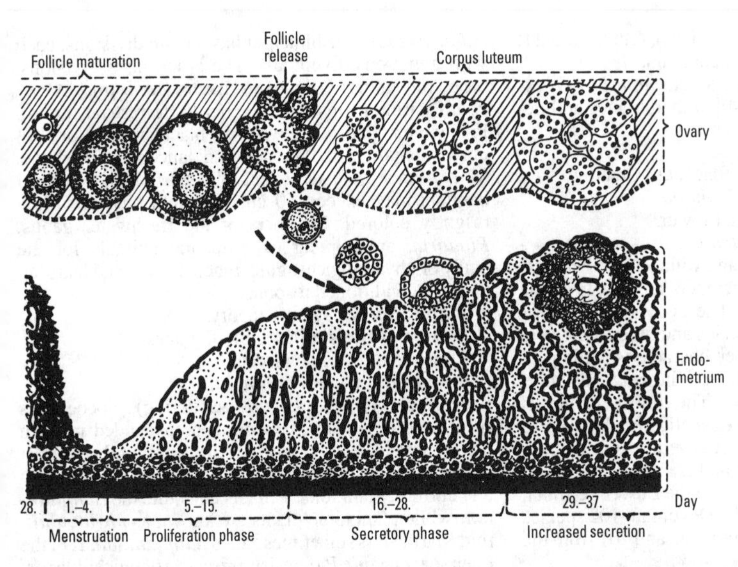

Fig.2. *Placenta. Above:* follicle maturation and formation of the corpus luteum in the ovary. *Below:* implantation of the egg in the uterine mucosa.

opment is replaced by the hemotropic stage, which then remains functional for the remainder of the pregnancy. On the 15th day, the trophoblast develops into a cellular *cytotrophoblast* (see Gastrulation, Fig.5). The Basalis (see Uterus) represents the maternal compartment of the P., whereas the fetal compartment is formed by the chorion (derived from the epithelia of the amnion and trophoblast). Extending from the chorion are finger-like diverticulae, known as *chorionic villi,* which are characteristic of all mammalian Ps. (Fig.1). The chorion is homologous with the serosa of other amniotes (see Extraembryonic membranes). Chorionic villi (placental villi) continue to branch and penetrate the maternal blood sinus (derived from the yolk-sac cavity), which thereby becomes the *intervillous space.* Normally, the chorionic villi are vascularized by vessels from the allantois. The villi become pillars (with vascularised branches), which physically connect the embryonal and maternal parts of the P. In addition, the P. is divided into regions by nonvascualrized septa *(cotyledons)* that arise from the trophoblast. Only certain regions of the chorion, known as the *chorion frondosum,* are supplied with villi. Blood vessels of the villi are in direct connection with the embryo via the vessels of the umbilical cord. The P. therefore insures close physical and physiological contact of mother and fetus; it exchanges materials between the two organisms; filters and actively processes material from the fetus and from the mother; receives waste fetal material, including carbon dioxide; passes nutrients and oxygen from the mother to the fetus; and transfers hormones and vitamins to the fetus. During the first 3 months of pregnancy, the P. develops an endocrine function, and from about the 4th month onward it takes over the production of progesterone and estrogens; the Graafian follicles and corpus luteum then degenerate and disappear. Toward the end of pregnancy, estrogens promote oxytocin production, which initiates uterine contractions. Oxytocin further accelerates and reinforces these contractions, leading to labor.

Ps. are classified into different types, according to the arrangement of the chorion frondosum (Fig.3). If the entire surface of the chorion is supplied with villi (as in

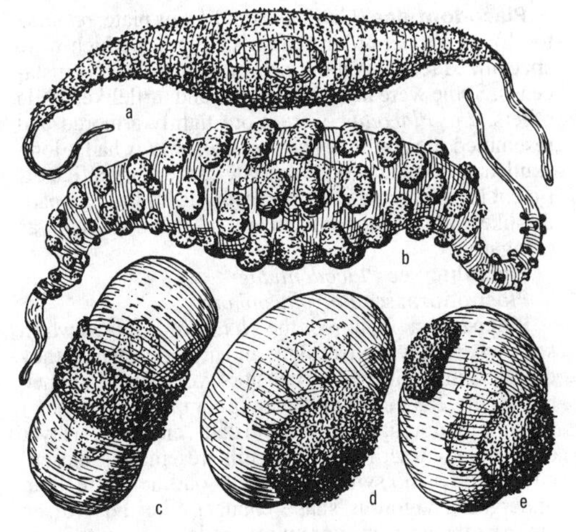

Fig.3. *Placenta.* Types of placentas (outer shape and form). *a* Diffuse. *b* Cotyledonary. *c* Zonary. *d* Discoidal. *e* Bidiscoidal.

the whale, pig, horse and many other ungulates), the P. is called diffuse *(P. diffusa).* In the cotyledonary P. *(P. cotyledonaria)* the villi are localized in numerous patches or clusters of varying sizes (e.g. 5–6 patches in deer, and 60–100 patches in cattle). Both of these first two types form a relatively loose association between the mother and the embryo.

In other types of P. the fetal and uterine compartments form a more intimate association, so that parturition is accompanied by extensive bleeding and removal of part of the uterine mucosa (decidua). This is the case in carnivores, in which the P. forms a broad band or or zone encircling the chorion at a position about midway between head and tail of the fetus, and is called a zonary P. *(P. zonaria).* Also in this category is the discoidal P. *(P. discoidalis),* in which the villi are restricted to a single, relatively large area of the chorion (insectivores, bats, rodents, higher primates, humans). A bidiscoidal P. *(P. bidiscoidalis)* occurs in some lemurs.

At a certain time after parturition, the *afterbirth* is expelled; this consists of embryonal membranes and decidua, and it is eaten by many animals.

Placentalia: see *Eutheria.*

Placental circulation: see Placenta, Allantois.

Placental mammals: see *Eutheria.*

Placentation: see Placenta, Flower.

Placentonema: see *Nematoda.*

Placodermi, ***placoderms:*** an extinct class of jawless fishes (see *Agnatha).* Most members are placed in the order *Arthodira (Coccostei).* The body form was diverse, including shark-like species and others that were flattened like rays. All, however, characteristically possessed an armor of bony plates on the head and forebody, and small, rhombic scales. The endoskeleton was at least partially ossified. Phyogenetically, they represent a heterogeneous group. They were predominantly bottom fishes of both marine and freshwater habitats. Fossil P. occur from the Silurian to the Lower Permian, with an evolutionary peak in the Devonian. 102 species are known from the Middle Devonian, and 201 from the Upper Devonian.

Placodontidae (Greek *plax* or *placos* plate, *odontus* tooth): an extinct order of Triassic reptiles which were specialized feeders on mussels brachiopods and crustaceans. Some were heavily armored and turtlelike, while others, e.g. *Placodus,* were more lightly armored and resembled the *Nothosauria* (see). *Placodus* had a long skull, and its palate and lower jaw carried large, flat, efficient breaking and crushing teeth. Its limbs were paddle-like with webbed fingers and toes, and the tail was elongate.

Placodus: see *Placodontidae.*

Placophora: see *Polyplacophora.*

Placozoa: a phylum of the *Metazoa. P.* live in warm seas, where they feed on algae and protozoa, which they digest extracellularly. The dorsoventrally flattened body, only 2–3 mm in diameter, is composed of an outer layer of monociliated epitheloid cells, enclosing a layer of loose, contractile, stellate cells, resembling mesenchyme. It has no symmetry and its outline is irregular, undergoing vigorous shape changes like those of an ameba. Organs, nerves or muscle cells are absent. There appear to be 2 species, the most familiar being *Trichoplax adhaerens,* which can reproduce sexually as well as asexually. Nielsen (1985) believes that they represent a flattened blastaea, while Grell (1981) believes that they are ancestral to the bilateral gastraea (see Colonial theory).

Plagiodontia: see *Capromyidae.*

Plagiogeotropism: see Geotropism.

Plagiothecium: see *Musci.*

Plagiotropism: see Tropism.

Plague: see *Yersinia pestis.*

Plaice, *Pleuronectes platessa, Platessa platessa:* a commercially important, fine-tasting flatfish (family *Pleuronectidae,* order *Pleuronectiformes),* found on sandy seabeds at depths of 1–125 m on eastern Atlantic coasts from Gibraltar to the Barents Sea, Iceland, southern Greenland, the Mediterranean and the Baltic. The upper surface is greenish brown with red-orange spots.

Plainrunners: see *Furnariidae.*

Plains wanderer: see *Gruiformes.*

Planarians, *Tricladida, Seriata:* an order of the *Turbellaria* (see). The blind gut has 3 main divisions, each with numerous diverticula. The order has 3 divisions, based on habitat: *Paludicola* (freshwater forms), *Maricola* (marine forms), and *Terricola* (terrestrial forms). Freshwater species are white to blackish in color, 1–4 cm in length, and flattened and lanceolate in shape. More rounded, slender forms of wet, tropical, terrestrial habitats may exceed 50 cm in length, and are often brightly colored. Members of the freshwater genus, *Planaria,* are popular experimental animals for the study of physiological aging, regeneration, and learning through conditioned response.

Planation: see Telome theory.

Planes of the skull: see Skull planes.

Plane tree: see Platanaceae.

Planipennia: see *Neuroptera.*

Plankton (Greek *planktos* wandering): biocoenoses of aquatic organisms, floating and suspended in open water, with little or no power of locomotion. In contrast to the necton, P. cannot resist water flow, and it drifts with any current. Planktonic animals consitute the ***zooplankton,*** while planktonic plants are known as ***phytoplankton.*** Marine P. is sometimes called ***haliplankton.*** P. of the open sea ***(oceanic P.)*** consists primarily of unicellular organisms, medusae, siphonophores, small crustaceans, pteropods and tunicates, whereas the ***neritic P.*** of shallow coastal waters consists of larvae of otherwise nonplanktonic organisms (e.g. larvae of coelenterates, advanced crustaceans, bivalves, gastropods, various worms, echinoderms and fishes). Freshwater P. or ***limnoplankton*** contains mostly unicellular organisms, rotifers and small crustaceans. P. of brackish water is called ***hyphalmyroplankton,*** while the P. of saline inland waters is known as ***salinoplankton.*** Limnoplankton is subdivided into eulimno-, heleo-, telmato-, creno- and potamoplankton. The best adapted and most constant of these is ***eulimnoplankton*** which inhabits the open waters of large lakes; it consists of a number of unicellular organisms, small crustaceans such as *Bosmina coregoni* and *Eudiaptomus graciloides,* the rotiferan, *Notholca longispina,* and others. The organisms of ***heleoplankton*** (pond P.) can lead a benthic as well as a planktonic existence, e.g. species of the rotiferan genus, *Brachionus,* and the small crustacean *Eudiaptomus coeruleus.* Organisms of ***telmatoplankton*** are likewise facultatively planktonic; as inhabitants of small ponds and puddles, they must be able to withstand wide daily variations of temperature, and when the water dries out, they bury themselves in the mud or form resistant resting stages. Specific ***crenoplankton*** (spring P.) and ***potamoplankton*** (river P.) do not exist as such. Potomoplankton is washed into rivers from regions of still water. In large rivers, however, it continues to multiply and makes a big contribution to the Biological self-purification (see) of water. ***Perennial P.*** occurs throughout the year, whereas ***periodic P.*** develops and declines according to an annual cycle. The majority of known planktonic organisms belong to the ***euplankton*** and are only found in open water. Planktonic developmental stages of otherwise nonplanktonic organisms are called ***meroplankton.*** Organisms that actually belong to other biocoenoses, but are transferred by water currents to become part of the P. of open waters, are known as ***tychoplankton.*** Very often pond and puddles contain tychoplankton in the form of soil and plant inhabitants, e.g. testate amebas *(Arcellinida),* benthic rotiferans, nematodes and chironomid larvae.

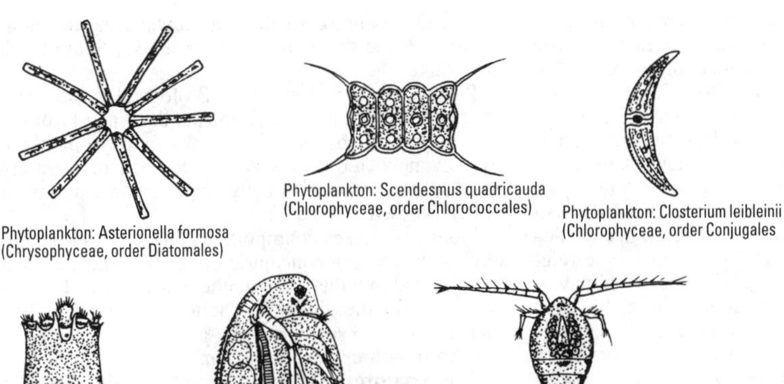

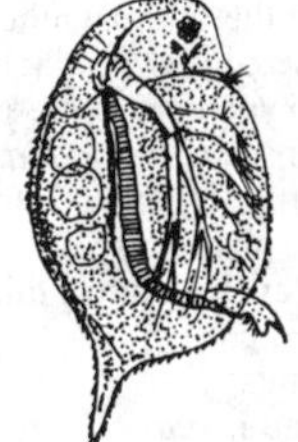

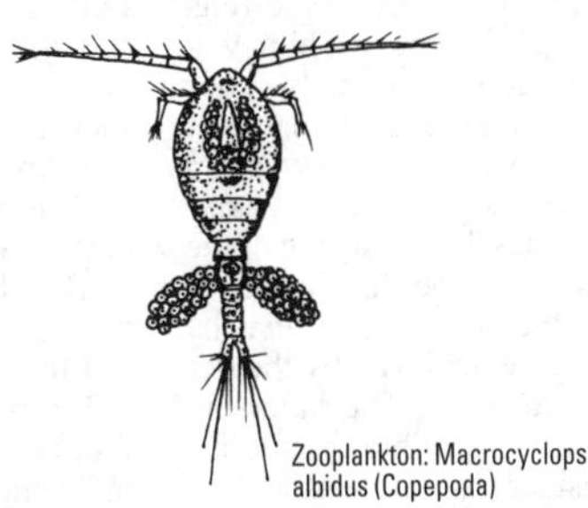

Fig.1. *Plankton.* Some organisms of the freshwater plankton.

Planktonic organisms larger than 5 mm are found only in the sea. Known as ***macroplankton,*** these consists mainly of medusae, many higher crustaceans, and tunicates. Rare forms larger than 1 m are known as ***megaplankton*** (e.g. the medusa *Cyanea arctica,* diameter 2 m, tentacles longer than 30 m). Most small planktonic crustaceans belong to the ***mesoplankton,*** which consists of organisms in the size range 1–5 mm. ***Microplankton,*** practically invisible to the naked eye, includes all planktonic organisms between 0.05 and 1 mm, i.e. most planktonic plants, most unicellular planktonic animals, all rotiferans and some small crustaceans. All the above sizes of P. are retained by a fine P. net and are collectively referred to as "net P." Organisms of the ***nanoplankton*** are smaller than 50 µm, so that they pass through the finest P. net. They must therefore be collected by centrifugation, filtration or by allowing them to slowly settle. Nanoplankton consists largely of small algae and ciliates, especially phytoflagellates (in the sea these are members of the *Silicoflagellatae, Coccolithophoridae* and *Dinoflagellatae).*

For an organism to remain suspended in open water, the rate at which it sinks must be reduced to a minimum. This can be achieved by 1) a decrease in specific grav-

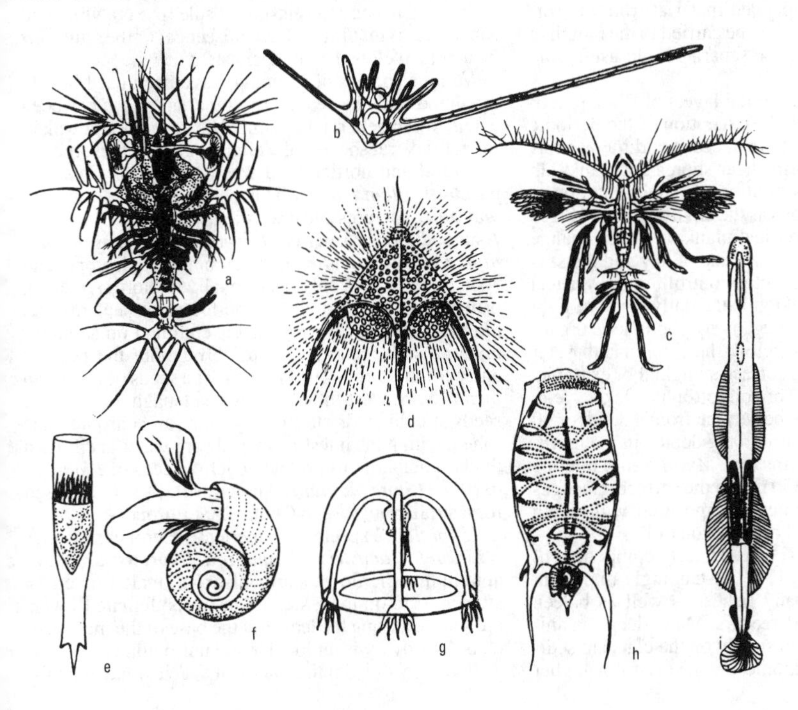

Fig.2. *Plankton.* Some organisms of the marine zooplankton. *a* Prawn larva. *b* Brittlestar larva. *c* Copepod. *d* Radiolarian. *e* Ciliate. *f* Shelled pteropod. *g* Medusa. *h* Tunicate (class *Thaliaceae). i* Arrow worm.

ity by deposition in the organism of low density substances like fats, oils and gases, 2) a decrease in specific gravity by an increased water content of the organism, and 3) a large surface area relative to volume. Fatty inclusions are very common in aquatic organisms; storage of fat instead of heavy starch is widespread among diatoms, cyanobacteria and green algae; and many small crustaceans have a particularly high fat content. Apart from the swim bladder of fishes, gas inclusions are also found in siphonophores *(Physalia, Velella),* in cephalopods *(Nautilus, Argonauta)* and in insect larvae (e.g. *Corethra).* Low body specific gravity due to a high water content is common among aquatic animals. Thus medusae consist of 98% water, with the result that their density is almost equal to that of water, i.e. they have a specific gravity of almost unity. A large surface area relative to volume generally increases the resistance of the organism to sedimentation; this may be the simple result of small size (the smaller the body, the greater the relative surface area), or it may be achieved by the development of external body processes. The latter are found on radioloarians, and are resposible for the bizarre appearance of many crustaceans. The seasonal alteration in the morphology of a planktonic organism is known as Cyclomorphosis (see), and the individual stages of this change are called *temporal variations.* Cyclomorphosis is particularly common in water fleas *(Cladocera)* and rotiferans; it is genetically programmed, and not correlated with variations in the temperature or chemical composition of the water.

Despite all the adaptations for a decreased sedimentation rate, the majority of planktonic organisms cannot remain permanently suspended without the assistance of water turbulence.

Heterotrophic P. feeds on detritus and bacteria, or on animals and plants. Most planktonic organisms indiscriminately filter all suspended material from the surrounding water; this may then be carried to the mouth in its entirety, or it may first be separated into useful and noningestible particles.

The populations of horizontal layers of P. are practically homogeneous, with the exception of the periodic development of shoals of crustaceans and the appearance of euplanktonic forms near shores and banks. In contrast, the vertical distribution of planktonic organisms shows wide variations as the environment changes with increasing depth. Vertical plankton layers change their position according to the time of day and the season. The most important factor controlling the vertical migration of the zooplankton is the daily cycle of light intensity. During the day, negatively phototactic organisms remain in the dark below the illuminated water layers, but they rise to the surface at night. For example, freshwater crustaceans of oligotrophic lakes (e.g. *Daphnia hyalina)* rise to the surface from a depth of 60 m in the evening, and return to the depths in the morning. Similarly, marine forms (e.g. *Mysidaceae)* migrate between depths of 250–350 m and the surface. Changes in temperature and chemical composition with depth also influence the vertical distribution of P.

Directly or indirectly (i.e. as an early component of an extended food chain), P. is outstandingly important in the diet of fishes. Many fishes, as well as baleen whales, are exclusively P. feeders. Many deep sea animals depend for their food supply on the constant sedimentation of dead planktonic organisms from higher levels. Oil pollution of the sea prevents the development of P. and therefore threatens the survival of fish and whales.

P. plays an important role in Biological waste water treatment (see), e.g. in oxidation ponds. Water turbidity may be caused by mass outbreaks of phytoplankton (e.g. cyanobacteria) (see Water blooms), more rarely zooplankton, or occasionally large accumulations of pollen (known as false P.).

Finally, P. makes an important contribution to maintaining the oxygen concentration of the atmosphere; it is estimated that the photosynthetic activity of P. is responsible for the global production of 3.5–4.1 x 10^{10} metric tons of oxygen per year.

Planorbidae: see *Gastropoda.*

Planozygote: a flagellated, freely motile zygote of certain algae and phycomycetes, which later withdraws its flagella and develops into a thick-walled, nonmotile, resting cystozygote.

Plant: see Plants

Plantaginaceae, ***plantain family:*** a cosmopolitan family of dicotyledonous plants, containing 269 species in 3 genera. They are usually annual or perennial herbs, rarely bushes, with entire, alternate, exstipulate leaves, which are usually all basal or nearly so. In several species the leaves are narrow, parallel-veined and without clear differentiation of stalk and leaf blade. The inconspicuous, wind-pollinated (rarely insect-pollinated) 4-merous flowers are usually hermaphrodite, actinomorphic and bracteate, and usually in a head or spike. The 4 petals of the imbricate, membranous corolla are united, as are the 4 sepals of the calyx. The 4 stamens possess very long filaments and versatile anthers with an abundance of fine powdery pollen. The superior ovary consists of 2 united carpels, which are 2-locular with 1 to many semi-anatropous ovules on axile placentas. The fruit is often a circumscissile capsule (pyxis), with a lid (operculum) that lifts off at dehiscence; or the fruit may be a nut enveloped by the persistent calyx.

Various species of *Plantago* have been used in folk medicine and the leaves of some species have been eaten, but in general the family has no economic value; in fact, several species of *Plantago* are troublesome agricultural and horticultural weeds. ***Great*** or ***Common plantain*** *(Plantago major)* is widespread on footpaths, waysides, pastures and grassy areas throughout Europe, Asia and North America. ***Ribwort*** or ***Ribgrass*** *(Plantago lanceolata)* is equally common in Europe and Asia, but is found only occasionally in North America, probably introduced with imported grass seed. ***Saline plantain*** *(Plantago eriopoda)* is common on saline or alkaline soils, river flats, etc. throughout the prairies and parklands of North America. The seeds of *Plantago ovata* (Northwest India), known as Ispaghul or Spogel seeds, are taken dry to treat dysentry and chronic diarrohea; within the intestine they absorb water and form a thick, mucilaginous protective jelly. Seeds of *Plantago psyllium* (***Sand plantain***), known as flea seeds, are used for a similar purpose in Continental Europe.

Littorella (2 species in America; 1 European species = *Littorella lacustris* = *L. uniflora*, ***Shoreweed)*** is notable in that the flowers are unisexual; male flowers (4 stamens) are on long, slender pedicels; female flowers are sessile among the leaves at the base of the male pedicels. The ovary is unilocular and uni-ovulate.

Bougueria (1 South American species) has indehis-

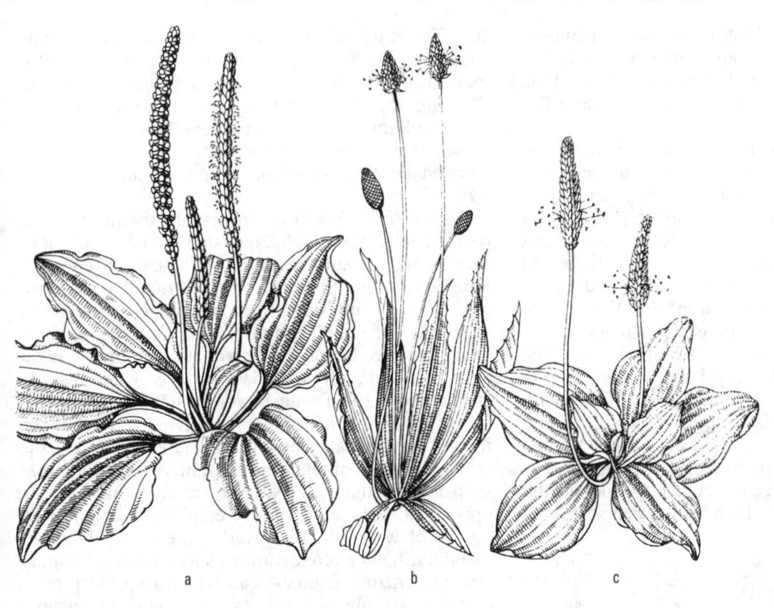

Plantaginaceae. a Great plantain *(Plantago major). b* Ribwort *(Plantago lanceolata). c* Hoary plantain *(Plantago media).*

cent fruits. Its flowers are spicate and capitate, a few at the top being bisexual, the others female. There are 1–2 stamens and the ovary is unilocular and uni-ovulate.

Plantago: see *Plantaginaceae.*

Plantain eaters: see *Cuculiformes.*

Plantain family: see *Plantaginaceae.*

Plant ash: the solid residue after complete burning of a plant or plant part. It consists mainly of nonvolatile oxides and carbonates: water-soluble potassium and sodium carbonate, sulfate and chloride; and the insoluble carbonates, phosphates, silicates and oxides of calcium, magnesium and iron. Also present are small quantities of the corresponding salts of aluminium, manganese, boron, copper and zinc, as well as traces of numerous other inorganic salts and oxides. *Analysis of P.a.* provides no information on the status of the various constituents in plant metabolism. Some components are essential macro- and micronutrients, some are nonessential trace elements, while others simply represent ballast material.

The *ash content* of different plant species and plant parts varies widely. In leaves (e.g. of tobacco) the ash may amount to 10–20% of the dry mass, whereas the ash of seeds and fruits represents only 1–5% of the dry mass. The ash content of plants and plant parts increases with age. It is also influenced by the mineral content of the soil and by fertilizer application. The composition of the P.a. of a single species is also subject to wide variations.

Plant bugs: see *Heteroptera.*

Plant community (Plate 11), ***phytoc(o)enosis, phytoc(o)enose:*** a group of different plant species, sharing the same habitat, which have the same or similar ecological requirements. P.c.s are characterized primarily by their constituent species, which are more or less constant and predictable. P.c.s are not random associations. They arise by selection and competition between numerous species. Their existence in any given habitat is the result of the interaction of a totality of environmental factors. They may therefore be used more or less accurately to identify the habitats in which they are found (see Indicator plants), a property which is used to advantage in phytosociology and ecology.

The term P.c. is used in both a concrete and an abstract sense. The units of classification are abstract, while concrete P.c.s have been referred to as stands. The term, phytoc(o)enose, has been proposed for the concrete P.c., while phytoc(o)enon has been proposed for the abstract P.c. (A phytoc(o)enon is derived from the characterization of a group of phytoc(o)enoses that correspond with each other in all characters considered to be typologically relevant).

Various methods can be used for analysis (see Relevé) and classification (see Phytosociology) of P.c.s.

For the main Central European P.c.s., see Character species.

Plant evolutionary history: major stages of plant evolution, with respect to time and geography. The following stages are recognized: *algal period* (see Eophytic), *pteridophyte period* (see Paleophytic), *gymnosperm period* (see Mesophytic), *angiosperm period* (see Cenophytic). These periods are not named for the first appearance of particular plant groups, but for the time of their greatest evolutionary diversity and distribution. See Geological time scale.

Plant galls: see Galls.

Plant geography, *phytogeography, geobotany:* study of the geographical distribution of plants. It is subdivided into *floristic P.g.* (see; see also Chorology, Areal), *ecological P.g.* (concerned mainly with the ecological basis of plant distribution), *historical* or *genetic P.g.* (identification of earlier vegetations of an area, and the history of their development into the existing vegetation), and *sociological P.g.* (see Sociology of plants).

Plant gums: certain plant polysaccharides, which occur in wood, are exuded to protect wounds, and often occur as a protective layer on the outside of bud scales. Familiar examples are Gum arabic (see), cherry gum (an exudate from cherry, plum and other fruit trees, which consists of a mixture of galactans and arabans), yeast gum (a mixture of D-glucose and D-mannose from certain yeast species), gum ghatti (from an Indian tree, *Anogeissus latifolia)* and gum tragacanth (dried gummy exudate of *Astragalus gummifer).* Many P.g. are official.

Plant hairs, *trichomes:* plant epidermal structures, usually consisting of one (sometimes several) cells, and derived exclusively from epidermal meristem cells. Single-celled, papilla-shaped hairs are responsible for the velvety sheen of many foliage leaves and petals (pansy, lupin). Root hairs (see Root) are unicellular and tubular, as are the hairs on fuchsia leaves and on the seed coat of the cotton plant (cotton). Bristles (e.g. in the *Boraginaceae)* are unicellular, rigid hairs, sometimes with lignified, calcified or silicified walls The unicellular, awl-shaped stinging hairs of the stinging nettle have a thin-walled, distended base (filled with cell sap) enclosed by a raised upgrowth of epidermal cells; the remaining wall is calcified, except for the small, inclined, silicified tip and a small area of unthickened wall below it. Even gentle contact breaks off the tip, leaving a small canula which readily penetrates the

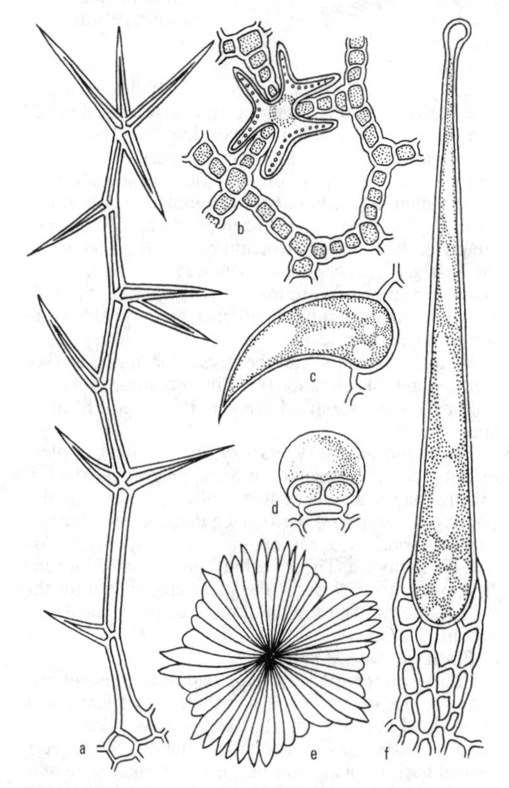

Plant hairs. a Branched hair of mullein *(Verbascum)* which forms a woolly tomentum. *b* Internal hair of yellow water lily *(Nymphaea lutea)* (a stellate calcified cell in the aerenchyma). *c* Climbing hair of goosegrass or cleavers *(Galium aparine). d* Multicellular glandular hair of sage *(Salvia). e* Scale hair of sea buckthorn *(Hippophaë rhamnoides). f* Stinging hair of stinging nettle *(Urtica dioica).* Magnification 50–300 times.

skin. The stinging sensation and local inflammation are caused by the discharged hair contents of sodium formate, acetylcholine, histamine and other substances. Stinging hairs protect against grazing animals. Branched unicellular hairs are found, inter alia, on the leaves of rock-cress *(Arabis)* and in the form of stellate hairs on the leaves of shepherd's purse *(Capsella bursa-pastoris).*

Multicellular hairs may consist of unbranched single rows of cells. The regular, repeated branching of multicellular hairs results in a woolly tomentum (e.g. mullein, *Verbascum).* If branching is restricted to the terminal cell, the result is a stalked or unstalked disk, e.g. the scale hairs of sea buckthorn *(Hippophae rhamnoides).* Dead hairs are largely filled with air, and therefore appear as a white felt on the leaf surface, where they decrease the rate of transpiration and protect against strong sunlight. On the other hand, live hairs on the surface of young leaves effectively increase the leaf surface area and promote transpiration. *Absorption hairs,* such as root hairs and specialized scale hairs on the upper leaf surface of tropical epiphytes, serve for the uptake of water and dissolved nutrients. Club-shaped glandular hairs excrete various substances, which discourage grazing animals, and which trap the prey of Carnivorous plants (see). In some contact-sensitive plants (e.g. venus' fly trap, *Dionaea), sensory hairs* serve for stimulus perception. In climbing plants (e.g. hops, *Humulus),* the grip of the climbing shoot on its support is increased by the presence of hooked climbing hairs. Flight hairs decrease the rate of fall and facilitate the wind dispersal of fruits and seeds.

Hairs may also be present in the walls of internal plant cavities, e.g. stellate, calcified cells in the aerenchyma of water lilies. Some aquatic plants lack hairs entirely.

Hair-like structures incorporating deeper cell layers are called Emergences (see).

Plantigrade: walking on the sole or undersurface of the entire foot, i.e. on the ventral surface of the metacarpus or metatarsus and the digits. The entire sole of the foot is placed on the ground at once. Typical P. animals are human, badger and bear. See Digitigrade, Unguligrade.

Plant nutrients: compounds that contain the elements required by plants, and which are available to, and utilizable by, plants. Thus, nitrate in the soil is the plant nutrient that supplies nitrogen, atmospheric CO_2 supplies carbon, and soil sulfate supplies sulfur, etc. According to the quantities in which they are required, they are classified as macro and micronutrients.

The essential *macroelements* contained in P.n. are carbon (C), hydrogen (H), oxygen (O), nitrogen (N), phosphorus (P), sulfur (S), potassium (K), calcium (Ca), magnesium (Mg) and iron (Fe). Those elements required in much smaller quantities, i.e. *microelements* or *trace elements,* are manganese (Mn), copper (Cu), zinc (Zn), molybdenum (Mo), boron (B), sodium (Na), and under certain circumstances chlorine (Cl) and silica (Si). Certain microorganisms also require cobalt (Co), others require vanadium (V), and some algae require iodine (I). Iron and manganese occupy an intermediate position between micro and macronutrients.

The nonmetallic elements, carbon, hydrogen and oxygen, are fundamental components of practically all organic compounds (hydrocarbons do not contain oxygen). The carbon in all the organic compounds of autotrophic green plants is derived from CO_2, and the hydrogen is derived from water (see Autotrophy, Photosynthesis). C-heterotrophic plants require an organic source of carbon and energy, e.g. carbohydrates (see Heterotrophy).

The remaining nonmetals, nitrogen, phosphorus, sulfur and boron, and the metals, potassium, sodium, calcium, magnesium, iron, manganese, copper, zinc and molybdenum are normally available in the soil as mineral salts. In water and sand culture, all these elements must be present in the nutrient solution. Uptake, transport and biochemical conversions of mineral substances in the plant is known as Mineral metabolism (see).

Plant pigments:

1) Pigments responsible for the coloration of plants and their parts. The green color of plants is due to a combination of green chlorophylls *a* and *b* (see Chlorophyll), yellow Xanthophyll (see) and orange-yellow Carotenoids (see), which occur together in fixed proportions in the chloroplasts. All are crucially involved in the utilization of light for photosynthesis. Certain marine algae contain additional pigments, e.g. Fucoxanthin (see). Various types of pigments are responsible for the sometimes vivid flower colors. Yellow flower colors can be due to carotenoids, which are present in chromoplasts, or to flavonoids (see Flavones) dissolved in the cell sap. Red and blue colors are due to the presence of Anthocyanins (see), which are also dissolved in the cell sap. The color of red leaves (e.g. copper beech, copper hazel) is also due to anthocyanins. White flowers are generally pigment-free, the white color being due to total light reflection by air-filled intercellular spaces.

In the coloration of autumn leaves, which commences before abscission, chlorophyll is degraded so that the yellow pigments predominate. In addition, anthocyanins may be synthesized, and these, especially in combination with carotenoids, produce vivid yellow-red color tones. The later brown coloration is caused by the deposition of Phlobaphenes (see) in the cell walls.

Light has a critical effect on the growth and development of plants, on seed germination and on photonastic movements. Susception of light stimuli depends on the presence of light-absorbing pigments, such as Phytochrome (see). By attracting pollinating insects, P.p. are also important for the existence and maintenance of the plant species. Other P.p. have a protective function, or are active as metabolic regulators. Numerous P.p. also appear to be metabolic endproducts without an obvious function.

2) Pigments extracted from plants for industrial use, especially as fabric dyes, e.g. madder, woad, indigo and safflower, and now almost completely replaced by synthetic dyes.

Plant protection agents, ***crop protection agents:*** substances for counteracting pests and disease organisms, and for preventing the growth of weeds. They include the following, all of which are listed separately: Acaricides, Algicides, Antifeedants, Aphicides, Aboricides, Respiratory poisons, Attractants, Bacteriocides, Seed dressings, Feeding poisons, Fungicides, Graminicides, Herbicides, Insecticides, Larvicides, Molluscicides, Repellants, Rodenticides, Dusting agents, Systemic agents, Food store protection agents, Animal repellants.

Plant protection agents, application of: application of plant protection agents for counteracting pests, for improving the keeping properties of vegetables and fruits, and for preventing loss of quality, by Spraying (see), by Misting (see), or by dusting. Certain preparations are applied as granules.

Plants: organisms which, in contrast to Animals (see), generally live autotrophically, i.e. they synthesize their organic material from carbon dioxide and water by Photosynthesis (see) or Chemosynthesis (see), using the energy of sunlight or exergonic chemical reactions, respectively. Since P. do not hunt or seek food, they are generally nonmotile; vascular P. are rooted in one position. For the purpose of collecting a large number of light quanta, the external surface area of P. is large in relation to body mass. In contrast to animals, P. represent an "open system". A precondition of such an open system is the development of robust, mainly cellulose-containing cell walls. P. also differ from animals, in that growth is restricted to the division of thin-walled meristematic cells, mainly at the terminus of the shoot, but also in intercalary zones.

The open system has led to the development of mechanical tissue, which differs markedly from animal support tissue. Materials are not transported by a humoral circulatory system. There is no need for the rapid coordination of the action of different tissues or parts of the plant. Specific, anatomically defined pathways for the conduction of stimuli and characteristic organs for sensory perception are absent or only weakly developed.

At lower evolutionary levels, in particular in the *Schizophyta* and among the *Protista,* there is often no clear difference between P. and animals, which indicates a common evolutionary origin for both. Also, in the course of their phylogenetic development, many P. have lost the ability to live autotrophically.

Plant sociology: see Sociology of plants.

Planula: an elongate, ovoid, ciliated larva of many coelenterates. After fertilization, the egg segments by equal division to form a single layer of cells (ectoderm) enclosing a central cavity, the resulting structure being known as a *blastula.* The cavity then becomes filled with tissue (endoderm) by the inward migration or delamination of cells, and more rarely by invagination, while the ectoderm becomes ciliated. The resulting larva, with a solid core of endoderm, is known as the *planula.* The planula may be free-swimming and become widely distributed by water currents, or it may crawl on the surface of aquatic vegetation, etc. Eventually, a split appears in the endoderm, representing the start of enteron formation. At this point, the larva attaches itself to the substratum by its aboral end, a mouth and tentacles develop at the opposite end, and the entire structure becomes a *polyp.* In certain forms (e.g. *Tubularia)* the polyp stage is omitted, and the planula develops directly into a tentacled, ciliated, free-swimming *Actinula* (see), which in turn becomes a *medusa.*

Plaque: a transparent area in a lawn of bacteria on the surface of a solid growth medium. A P. is due to lysis of bacteria by bacteriophage. Under controlled conditions, each P. represent a center of infection initiated by a single infective bacteriophage particle. The number of P. produced after evenly spreading a known volume of phage suspension over the surface of the bacterial culture is used as a simple assay of the number of infective phage particles.

The term is also used in a general sense for an area of lysed cells in any type of lawn. In immunology, a P. assay is used to detect immunoglobulin-producing cells on a lawn of red blood cells. When complement is added to the serum, those erythrocytes that have bound to the immunoglobulin molecule are lysed, leaving a transparent P. Since most antigens can be artificially bound to erythrocyte membranes, the technique can be used for practically any antibody capable of fixing complement.

Plasma cell: a highly differentiated cell type with the ability to form antibodies. P. develop from small lymphocytes, after they have been activated by contact with an antigen. Their cytoplasm contains abundant rough endoplasmic reticulum (i.e. numerous membrane-bound ribosomes), which synthesizes antibodies for export. A large fraction of synthesized antibodies is also demonstrable inside the P.c.

Plasmakinins: physiologically highly active oligopeptides with hormone-like properties. P. cause contraction of the smooth muscle of the gastrointestinal tract, uterus and bronchi, dilation of blood vessels and increased vessel permeability. Important representatives are bradykinin and kallidin which have the same qualitative action, but different potencies.

Plasmalemma: see Plasma membrane.

Plasmalogens: glycerophospholipids in which C3 of the glycerol moiety carries a 1-alkenyl ether group, the secondary alcohol group is esterified with a higher fatty acid, and the nitrogenous base is choline or ethanolamine.

Plasma membrane, ***plasmalemma*** (plate 22): the complex surface system (boundary layer) of all cells. It is involved in the exchange of material and energy between the cell and its environment, the promotion and prevention of cell-to-cell contact, the reception, processing and conduction into the cell of specific information from the environment, as well the release of specific signals into the environment. It also serves for physical support and protection of the cell. The ability of the P.m. to regulate entry and exit of material enables the living cell to maintain a specific chemical composition that is qualitatively and quantitatively different from that of ts surroundings. All cells interact with their environment via their P.m., which contains a wide variety of receptor molecules or molecular complexes capable of recognizing other cells or specific chemical ligands. Binding of a ligand causes a signal to be transmitted to the cell interior, and this in turn triggers a specific reaction (e.g. a membrane-bound adenylate cyclase may be activated; the resulting cAMP then promotes the first in a series of coupled reactions in a self-amplifying cascade). Most of these receptor molecules (usually glycoproteins or gangliosides) are integral components of the P.m. (e.g. insulin receptor, acetylcholine receptor). Acetylcholine receptors display a specific arrangement in the postsynaptic membrane (see Neuron). Binding acetylcholine (a neurotransmitter) by these receptors, causes the permeability of the postsynaptic membrane to increase very briefly due to the opening of ion channels. A single adipocyte carries about 10 000 insulin receptors on the outer surface of its P.m., as well as numerous receptors for other hormones, all of which influence adenylate cyclase.

Antibodies (see) may also serve as membrane-bound receptor molecules. In addition to their occurrence in serum and other body fluids, antibodies are also present in the P.m. of lymphocytes and plasma cells. The constant parts of the antibody H-chain are anchored in the P.m. of B-lymphocytes, whereas the specific binding domains project outward from the P.m., where they are free to interact with specific antigens.

Uptake and release of macromolecules and relatively large particles occurs by endocytosis or exocytosis, respectively.

Under the electron microscope, the P.m. of animal cells is seen to have an asymmetric structure. The bimolecular lipid layer generally consists of phospholipids, glycolipids and cholesterol, with various types of protein inclusions, but the inner and outer lamellae have different compositions. Glycosylated regions of glycoproteins project to the cell exterior from the surface of the outer lipid layer. Proteins are generally found in both lamellae, nonpolar regions of their polypeptide chains partly or completely penetrating the lipid layer. Various types of membrane proteins have been demonstrated by cytochemical and immunological methods, e.g. immunoglobulins (antibodies) and ATPase (involved in active, ATP-dependent transport).

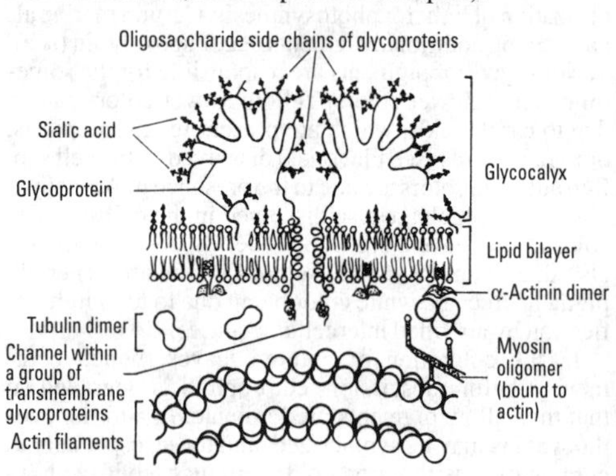

Fig.1. *Plasma membrane.* Schematic representation of plasma membrane structure.

In contrast to other Membranes (see), the animal P.m. consists of 2–10% carbohydrate (as glycoproteins and glycolipids), which is located on its outer surface. The hydrophilic parts of the glycoprotein molecules, together with the sugar chains of the glycolipids form the carbohydrate-rich *glycocalyx* (Fig.1). Many of the oligosaccharide side chains carry terminal residues of sialic acid, which give the outer surface of the P.m. a strong negative charge. Plant protoplasts (see) also carry a strong negative surface charge, but this is due to phosphate groups. The oligosaccharides of the animal cell glycocalyx are important in cell-cell interactions, and they are largely responsible for the antigenic properties of the cell (e.g. the A, B and O-antigenic determinants of erythrocytes are carbohydrate structures of the glycocalyx). Glycocalyx carbohydrates also serve as receptors for steroid hormones and viruses.

The inner surface of the eukarytoic P.m. is underlain by a *cytoskeleton,* consisting of contractile proteins (see Actin, Myosin), intermediate filaments and/or Microtubules (see). The latter appear to have no direct contact with the P.m., and are separated from it by microfilaments. In prokaryotes and in most plants and fungi, the outer surface of the P.m. is covered by a cell wall. Bacterial cell walls contain, inter alia, antigens and phage receptors. Adjacent plant cells are connected via thin cytoplasmic channels in the cell wall *(plasmodesmata,* diameter 60 nm), which are lined with P.m. Plasmodesmata enable the direct transport of material between neighboring cells. Their tubular lumen is almost always associated with cisternae of the endoplasmic reticulum. Thus, all the living cells of a plant are connected via plasmodesmata.

The surface area of some animals cells is considerably increased by the presence of Microvilli (see). Deep folding of the basal P.m. *(basal labyrinth)* of water-transporting cells (e.g. epithelia of the main and intermediate sections of the kidney tubules) causes a marked increase in the surface area available for active ion transport. This transport results in a concentration gradient between the cytoplasm and extracellular space, causing passive water transport into the latter.

In *transfer cells* of plants (e.g. modified xylem parenchyma cells or xylem companion cells) the surface area of the P.m. is greatly increased by invaginations of the cell wall. It has been shown by radioactive tracing that transfer cells can take up dissolved substances from the xylem sap, the necessary energy being provided by the numerous mitochondria present in the transfer cells.

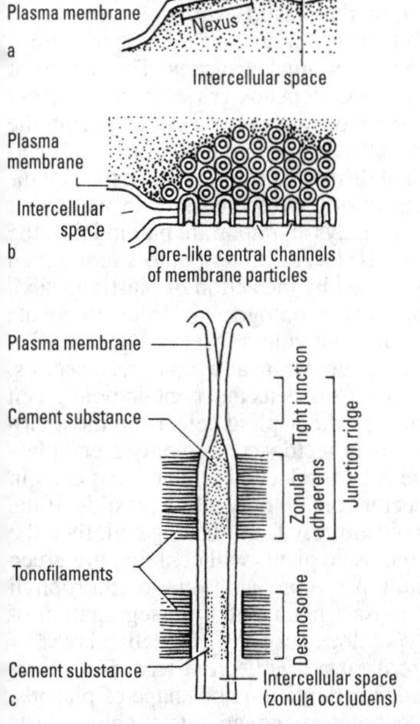

Fig.2. *Plasma membrane.* Cell contacts. *a* Gap junction. *b* Enlarged section of *a,* showing hypothetical structure of the nexus. *c* Tight junction between epithelial cells.

In animal tissues, various differential structures are displayed by the P.m. in the interfacing region of neighboring cells (Fig.2). Thus, a *tight junction* is a variably sized region of the P.m. located immediately below the cell apex, in which the membranes of neighboring cells are fused to form a pentalaminate structure. In epithelia, the P.ms. of neighboring cells may form tight junction ridges, which enclose a continuous layer of intercellular space *(zonula occludens),* and separate it from contact with the rest of the intercellular space. Such an occluded zone is usually associated with the formation of an attachment zone *(zonula adhaerens).* The expanded intercellular space then contains a cement substance that binds the two cells, but does not prevent transport, even of macromolecules. Noncontractile microfilaments *(tonofilaments)* extend deep into the cytoplasm from both sides, indicating the importance of such attachment zones for the maintenance of the cell association. *Maculae adhaerens* or *desmosomes* presumably serve a similar function. Desmosomes are local, patchlike, oval or circular intercellular junctions (cf. zonulae adhaerens that encircle the cell apex); the intercellular space of a desmosome contains a glycoprotein, and in this region the P.m. of each cell contains a plaque of electron-dense material, and behind this a less dense area of microfilaments. An area of particularly close cell contact is formed by the *gap junction.* Here, the intercellular separation (normally 15–20 nm) decreases to about 2 nm, and small molecules and ions (up to M_r 600) pass rapidly between the two cells via fine connecting channels (internal diameter about 1 nm). Gap junctions are a particular feature of nerve and heart muscle cells, i.e. cells that are electrically coupled with one another.

Plasma proteins: a complex mixture of mostly conjugated proteins present in the blood plasma of vertebrates. They function in the regulation of blood pH, and in the transport of ions, hormones (thyroxin- and steroid hormone-binding proteins) and vitamins (retinol-binding protein). Plasma albumin is important in the transport of free fatty acids. P.p. that serve for defense (against foreign proteins and microorganisms) are the components of the complement system and immunoglobulins. With few exceptions (e.g. albumin), P.p. are glycoproteins. P.p. also include the proteins of the blood clotting cascade (prothrombin, fibrinogen, etc.) With the exception of the immunoglobulins, P.p. are synthesized in the liver.

Plasmatic growth, *embryonal growth:* increase in the quantity of protoplasm and vital structural components of the cell. In contrast to Extension growth (see), P.g. results in only an insignificant increase in cell volume. A unicellular organism that multiplies by cell division must double its cell material between each cell division. In the vegetative apex of spermatophytes, the period of P.g. between cell divisions lasts for 15–20 hours. During this time, there is a continuous, albeit nonuniform synthesis of RNA, proteins, lipids, etc. DNA synthesis is, however, restricted to the synthesis phase of mitosis.

Many heterotrophic plants require certain substances for P.g. Thus, yeast cultures require traces of biotin, myo-inositol and pantothenic acid, while other fungi require vitamin B_1, and many bacteria require *p*-aminobenzoic acid. Since these substances are synthesized endogenously by autotrophic plants, it is difficult to determine their significance for the P.g. of such organisms. Some information can be obtained from tissue cultures of autotrophic plants, which often require the addition of certain vitamins or plant hormones for P.g.

Plasmid, *plasmagene:* an extrachromosomal, circular DNA molecule in some bacterial cells, which carries one or several (usually several) genes. Ps. may be free in the cytoplasm or they may be integrated into the chromosomal DNA. Ps. capable of such integration are known as *episomes.* Ps. are autonomous cytoplasmic replicons (see DNA replication), i.e. their replication and segregation during cell division proceeds independently of the bacterial chromosome, provided they are not integrated in the latter. Ps. are important manipulable elements of Gene technology (see). Genes for antibiotic resistance are often carried on Ps.

Plasmin, ***fibrinolysin:*** a trypsin-like enzyme, containing two polypeptide chains. Its inactive precursor or zymogen (plasminogen, profibrinolysin) occurs in blood at a concentration of 50–100 mg/100 ml serum. P. is responsible for fibrinolysis, i.e. dissolution of blood clots by the proteolytic degradation of fibrin to soluble peptides. In addition, the products of fibrinolysis inhibit thrombin and the conversion of fibrinogen into fibrin, thus forming a delicately balanced mechanism for the prevention of clotting. The natural inhibitor of P. is α_2-macroglobulin. Plasminogen is converted into P. by limited proteolysis by specific tissue proteases, which are released under various conditions, e.g. emotional stress, during exercise, following the injection of adrenalin, and from blood vessel walls in response to vascular injury. The best known activator is the kidney enzyme, urokinase. One of the most potent exogenous activators of human (not animal) plasminogen is streptokinase, produced by β-hemolytic steptococci. Streptokinase is not generally considered to be a proteolytic enzyme; it forms a complex with plasminogen, which then releases active P.

Plasmodiophora: see *Myxomycetes.*

Plasmodiophorales: see *Myxomycetes.*

Plasmodium:

1) A multinucleate cell, which arises by nuclear division without cell division. The fusion P. of myxomycetes is a multinucleate mass of protoplasm, formed by the fusion of numerous amebozygotes. It displays ameboid movement and its growth is accompanied by synchronous nuclear division.

2) The causative agent of malaria; see *Telosporidia.*

Plasmogamy: see Fertilization.

Plasmolysis: retraction of the protoplast from the cell wall, which occurs when water is withdrawn osmotically from the cell vacuole. P. can therefore be caused by placing cells in an hypertonic solution. During P., the protoplast contracts around the ever decreasing vacuole. If the protoplast detaches totally from the cell wall, the microscopic picture is one of *convex P.* If parts of the protoplast remain attached to the cell wall, only the detached surfaces contract inward, giving a picture of *concave P.* Extreme convex P. is known as *cap P.,* characterized by a caplike swelling of the contracted cytoplasm. This occurs if the hypertonic plasmolytic medium contains hydration-promoting lithium, sodium or potassium salts, in particular potassium thiocyanate, which can penetrate into the protoplast.

If plasmolysed cells are placed in an hypotonic solution, the vacuole enlarges, and the protoplast becomes pressed again against the cell wall. This reversal of P. is called *deplasmolysis.*

Plasmon: the total extranuclear hereditary complement in the cytoplasm and its organelles (plasmids). See Cytoplasmon, Plastome.

Plasmon mutation: spontaneous or induced alteration of cytoplasmic genes (see Cytoplasmic inheritance). P.m. differs in many respects from genome mutation. One major difference is that mutation of just one of the many copies of the plasmon is not manifested or detectable unless and until it accumulates by selection. For this to occur, the mutation must have a selective advantage, so it is not surprising that P.ms. are rarely encountered.

Plasmon segregation: separation or demixing of genetically different plasma constituents, leading to different cytoplasmic types. Whereas the separation of heterozygous allele pairs of the genome normally occurs during meiosis, P.s. can occur in the course of mitosis. The distribution of different plasmids is then generally nonuniform, because there is no mechanism to insure that each daughter cell receives an identical genetic complement (cf. mitotic nuclear division).

Plasmopara: see *Phycomycetes.*

Plasmoptysis: the bursting of living plant, bacterial or yeast cells due to the osmotic uptake of water. The cell contents are often released explosively. P. is easily demonstrated by placing thin-walled cells with a high osmotic value (see Osmosis) in water.

Plastid inheritance: inheritance of plastid characters. P.i. is part of Cytoplasmic inheritance (see). New plastid characters can arise by Plastid mutations (see). Although the plastome, plastidome or plastidotype (the hereditary system of plastids) is extranuclear and independent (i.e. plastid characters are determined by genetic material within the plastid), situations are known in which plastid characters (e.g. development of chlorophyll) are influenced by nuclear genes. The nature of the inheritance process depends critically on whether the plastids of the zygote are derived entirely from the egg cell, or from both egg cell and pollen. *Plastid demixing* (separation of different plastid types in the cell divisions that occur during ontogenesis or in macro- or microsporogenesis) plays an important part in P.i. If the plastids (chloroplasts) have different colors (e.g. green or white, as determined by the action of plastid genes), plastid demixing during ontogenesis leads to plants with mosaics of different colors (see Variegation), i.e. there are pure green, pure white and light green sectors. The latter are formed from cells that contain both green and white plastids. All the haploid cells from the meiosis of cells in the green sectors contain only green plastids. Some of the cells produced by meiosis of cells in the light green sector contain only white plastids. If the plastids are derived entirely from the egg cell, then the offspring of such mosaic plants will contain pure green types, pure white types (nonviable due to chlorophyll deficiency) and mosaic plants, and the segregation of these different types does not obey Mendelian laws.

Plastid movements: a collective term for alterations of the position, orientation and shape of plastids. The directional locomotory movements of chloroplasts are especially marked and extensively studied (see Chloroplast movements).

Plastid mutations: spontaneous or induced mutations of genes present in plastids, leading to changes in the plastid phenotype. Characters that change as a result of P.m. display Plastid inheritance (see). Nuclear genes have been demonstrated that increase the spontaneous rate of P.m.

Plastids: cell organelles that are bounded by a double membrane and contain their own genetic system.

They are present in the cells of green plants. The structurally and functionally different types of P. include chloroplasts, leukoplasts, amyloplasts, etc.

1) *Chloroplasts* are hemispherical or ovoid, relatively large organelles (average size 3–4 μm x 5 μm). They are the sites of photosynthesis, transducing light energy into reducing power (NADPH) by light-driven electron transport, and into chemical energy (ATP) by photophosphorylation. This stored energy is employed in the synthesis of glucose and other organic materials from CO_2 and water. Oxygen is released during photosynthesis. Chloroplasts contain chlorophyll and carotenoids, and a complex internal structure consisting of dark granules known as *grana.* In surface view under the electron microscope, a granum is seen to consist of a stack of practically round thylakoids (Fig.1) The chloroplast matrix is called the *stroma.* Such *granular* chloroplasts are typically found in foliage leaves and outer stem tissue of higher plants, including most ferns and mosses and liverworts. They also occur in green algae, whereas other algae possess nongranular chloroplasts with an homogeneous internal structure.

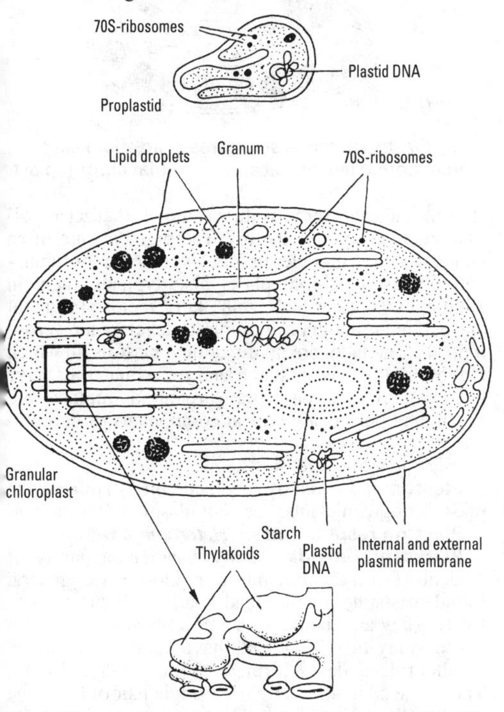

Fig.1. *Plastids.* Internal structures of a proplastid and granular chloroplast.

The outer and inner membranes of P. are structurally and functionally different [i.e. like those of Mitochondria (see)]. The outer plastid membrane consists of about equal parts of lipid and protein, permitting the passage of ions, dissolved substances and even proteins. The surface area of the inner membrane of granular chloroplasts is increased (even more than in mitochondria) by the development of a photosynthetic pigment-containing membrane system that extends into the interior of the organelle. In fully developed chloroplasts, these double membranes or flattened vesicles (thylakoids) are no longer continuous with the inner chloroplast membrane. Thylakoid membranes consist of about 50% lipids and 50% proteins. In addition to phospholipids, the membrane also contains mono- and digalactosyldiacylglycerols, which are characteristic of thylakoid membranes. Embedded in the membrane lipids are large and small complexes of protein-bound pigments (chlorophylls *a* and *b,* and carotenoids). These protein-pigment complexes probably correspond to photosystems I and II, which have different absorption spectra. In both systems, chlorophyll *a* is the ultimate light-absorbing pigment, but various wavelengths are also absorbed by other pigments. Grana consist of different numbers of thylakoid disks, and the different grana are linked to each other by stroma thylakoids. The thylakoid surface facing the matrix (stroma) carries particles of ATP synthase (CF_1-ATPase), which is responsible for the light-driven ATP synthesis or photophosphorylation. Isolated thylakoids are also capable of photophosphorylation. Thylakoids also contain plastoquinone (a redox system). Large quantities of the enzyme, ribulose bisphosphate carboxylase, as well as numerous other enzymes of photosynthetic CO_2 fixation (Calvin cycle) are present in the finely granular matrix. Ribulose bisphosphate carboxylase catalyses the first reaction in the incorporation of CO_2 into organic material. Lipid droplets, starch granules, starch synthesizing enzymes, storage proteins (sometimes as crystals, e.g. phytoferritin) and plastid DNA are observed not only in chloroplasts, but also in the matrix of other plastids.

In contrast to mitochondria, chloroplasts operate relatively independently of the rest of the cell. When isolated carefully, they carry out the entire process of photosynthesis in vitro. In addition to the pigments and enzymes of the light reaction, they also contain the enzymes for carbohydrate synthesis. The light reaction occurs in the thylakoids, the dark reaction in the matrix (stroma). Under appropriate conditions, nonpigmented, simply structured chloroplast precursors, known as *proplastids* develop into chloro- and leukoplasts, and sometimes into chromoplasts. Both proplastids and fully developed chloroplasts are capable of rapid division. In homogeneous chloroplasts, similarly structured thylakoids are distributed homogeneously throughout the entire chloroplast. The chloroplasts of red algae contain single, unassociated thylakoids (containing chlorophyll *a,* carotenoids and phycobilins). In cryptophytes, the thylakoids occur in pairs, and in all other algae (including *Euglena)* they occur in triplets. All homogeneous chloroplasts lack chlorophyll *b,* and the storage polysaccharide (similar to starch) occurs in the cell cytoplasm, not in the chloroplast.

Prokaryotes do not possess typical membrane-bounded P. Photosynthetic bacteria lack chlorophyll *a,* and therefore do not photolyse water, but other inorganic substances, such H_2S. In these organisms, invaginations of the plasma membrane extend into the cell interior, where they contain pigments, such as bacterial chlorophyll and carotene; some species contain bacteriorhodopsin rather than bacterial chlorophyll. These membrane invaginations are also referred to as thylakoids. In cyanobacteria, the plasma membrane also invaginates to form thylakoids; these extend loosely into the peripheral chromatoplasm, and when fully formed they become largely separate from the plasma membrane. In addition to carotenoids and phycobilins, the thylakoids of cyanobacteria also contain chlorophyll *a,* so that these organisms, like green plants, photolyse water and produce oxygen.

2) *Genetic system.* All types of P. are semiautonomous organelles. Like Mitochondria (see), they contain genetic material (DNA). The totality of genes in a P. is known as the *plastome.* Several regions of loosely packed, double-stranded, histone-free, circular DNA occur in the P. matrix (plastid DNA, ptDNA). It consists of several circular DNA molecules (more than 50 in chloroplasts, some almost 50 μm in length), and it differs from the nuclear DNA in the degree of methylation of its cytosine residues. A single plastid contains more DNA than a mitochodrion. ptDNA accounts for between 1% *(Euglena)* and 25% (tobacco) of the total DNA of the cell. DNA-polymerase (for replication of the ptDNA), RNA-polymerase (for transcription of the ptDNA), the small subunit of ribulose bisphosphate carboxylase and certain other plastid-specific proteins are encoded by the nuclear DNA and synthesized in the cytoplasm outside the P. ptDNA is too small to encode all plastid-specific proteins. The ribosomes in the plastid matrix are of the smaller 70S-type (i.e. like those of mitochondria). P. can only arise from other P. by cleavage. The plasma of P. does not communicate directly (i.e. does not mix) with the cell cytoplasm. This and other facts support the Endosymbiont hypothesis (see) for the evolution of P. and mitochondria from intracellular prokaryotic symbionts.

3) *Movement of P.* In addition to being moved within the cell by cytoplasmic streaming, chloroplasts can also change their position in response to a stimulus. This positional movement is usually regulated by receptors for blue light. For example, chloroplasts of the green alga, *Mougeotia,* and the moss, *Funaria hygrometrica,* become orientated at right angles (weak light) or parallel (strong light) to the direction of illumination. Probably, P. are turned around their long axis by actomyosin fibrils (see Actin, Myosin). In *Mougeotia,* this occurs within 30 min of changing the light intensity.

4) *Structural types of P. and their development.* Chloroplasts and leukoplasts arise from *proplastids* (Fig.1). The inner membrane of proplastids is already partly invaginated, but pigments are absent. In angiosperms, especially monocots, light is necessary for the development of chloroplasts from precursor proplastids. In the absence of light, *etioplasts* are formed, whose *prolamellar bodies* contain chlorophyll *a* precursors, i.e. protochlorophyllide *a* and protochlorophyll *a.* Prolamellar bodies are paracrystalline systems of branched tubules which, in response to weak light, are transformed into perforated (primary) thylakoids, accompanied by oxidation of chlorophyll *a* precursors to chlorophyll *a.* If the illumination includes blue (high-energy) light, chlorophyll synthesis continues, and the grana and stroma thylakoids of the typical higher plant chloroplast develop. In root and epidermal cells, proplastids develop into nonpigmented *leukoplasts,* which may store starch *(amyloplasts)* or oil *(elaioplasts).*

Chromoplasts occur mainly in flowers and fruits, and they develop from immature, pale green chloroplasts, proplastids, or leukoplasts. In their fully mature state, chromoplasts contain no chlorophyll and relatively few membranes. They are yellow, orange or red, due to the presence of carotenoids, and the color they impart to

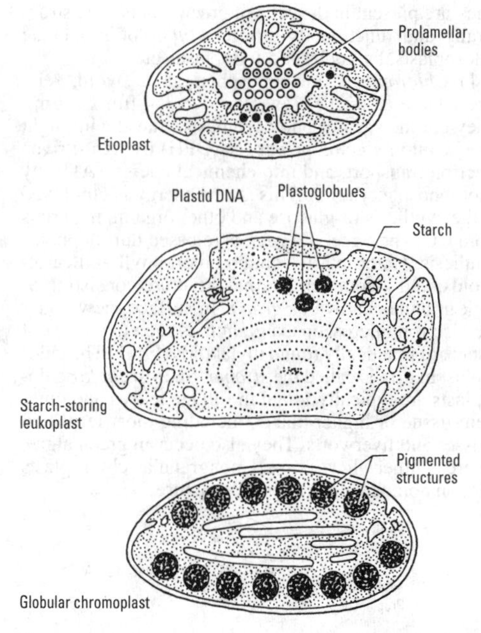

Fig.2. *Plastids.* Internal structures of an etioplast, a starch-storing leukoplast and a globular chromoplast.

flowers and fruits serve to attract animals that act as pollinators and seed distributors. The pigments are often concentrated in globules or tubules (globule- or tubule-chromoplasts). Chromoplasts also typically contain xanthophylls esterified with fatty acids (pigment waxes). In the autumn, chloroplasts develop into *gerontoplasts.* The latter are nonfunctional plastids, in which chlorophyll, proteins, starch and other materials have been degraded. Some of the carotenoids remain, and xanthophyll esters are typical components of gerontoplasts.

Plastochrone: see Meristems.

Plastome: the totality of genetic information in the plastids of green plants. See Cytoplasmon, Plasmon.

Plastron respiration: see *Heteroptera.*

Platacanthomyidae, ***spiny dormice:*** a family of Rodents (see), comprising 2 monospecific genera, found in southern China and southern India, respectively. They are about the size of a house mouse, with dense, spiny fur. Cheek teeth have a series of oblique, parallel ridges; dental formula: 1/1, 0/0, 0/0, 3/3 = 16. The palate contains a series or a single pair of foramina between the rows of teeth. The thumbs are replaced by a pad, and the limbs show modifications for an arboreal existence.

Platanaceae, ***plane tree family:*** a family of dicotyledonous plants containing 7 species, native from southern Europe to India and in North America. All species are large trees with alternate, lobed, stipulate leaves. The unisexual, monoecious flowers are wind-pollinated, and arranged in inconspicuous, spherical inflorescences. The fruits are nutlets. Flakes of bark of varying sizes continually detach from the trunk, which consequently has a flecked or blotchy appearance.

One of the most familiar avenue trees, especially in

large cities, is the ***Plane tree*** *(Platanus hybrida),* a hybrid of *Platanus occidentalis* (North America) and *Platanus orientalis* (eastern Mediterranean region). It flourishes practically anywhere and attains a height of 40 m.

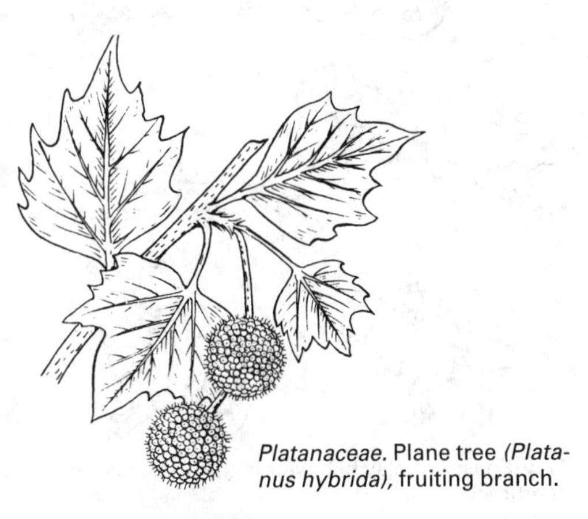

Platanaceae. Plane tree *(Platanus hybrida),* fruiting branch.

Platanistidae: see Whales.

Platanthera: see *Orchidaceae.*

Platax: see Batfish.

Plated lacertids: see *Psammodromas.*

Plated lizards: see *Cordylidae.*

Platelea: see *Ciconiiformes.*

Plateosaurus (Greek *platys* flat, *sauros* lizard): a genus of heavily built, omnivorous or herbivorous prosauropods, up to 8 m long and 3 m high. They were partly bipedal, walking upright on their hindlegs, with a long tail which served as a body support. The muscular neck carried a very small head with powerful teeth. All 4 toes of the powerfully developed hindlimbs carried claws. The forelimbs were much smaller, each provided with 5 clawed fingers. *P. gracilis* grew to 6–7 m in length and weighed 2 tonnes. *P.* probably represents a side branch of the sauropod lineage. Fossils are largely restricted to the Upper Triassic.

Platform builders: see *Ptilinorhynchidae.*

Platichthys: see *Pleuronectiformes.*

Plating: a general term for culturing microorganisms on solidifed (agar or gelatin) nutrient media. "Plating out" normally means the preliminary transfer of a sample of microorganisms (e.g. from a soil specimen, liquid culture, wound swab, etc.) to a solid growth medium, with a view to identification, detection of contaminants, preparation of a pure culture, etc. (see Streaking).

Koch's liquid plating method is used for enumerating and isolating bacteria in material such as soil, water, milk, blood, feces. The material under investigation is thoroughly shaken with a measured volume of sterile water, and a series of dilutions of this water is prepared. 1 ml of each dilution is mixed with about 15 ml of melted (45 °C) nutrient agar in a sterile Petri dish. After incubation of the solidifed medium, those cultures prepared from appropriate dilutions display discrete colonies. Assuming that each colony is derived from a single cell, the number of discrete colonies appearing after incubation can be used to calculate the number of organisms in the original specimen. Moreover, any colony can be isolated and further subcultured to obtain a pure culture. This method has the advantage that only living cells are counted, but it relies on two assumptions that are not always valid: that each colony is derived from a single cell, and that each living cell develops into a colony.

Platycarya: see *Juglandaceae.*

Platyhelminthes, ***flatworms:*** an animal phylum of morphologically diverse, bilaterally symmetrical, acoelomate, triploblastic worms, varying in length from 1 mm to more than 10 m. Metameric segmentation is absent. The phylum includes free-living aquatic species, terrestrial species of moist habitats, and parasitic species living in or on other animals. Most species are markedly dorsoventrally flattened (hence the name), and only a few are oval in cross section. The volume between the gut and the muscular layers of the body wall is filled with parenchymatous tissue or mesenchyme (derived from the mesoderm), which serves as a packing around the viscera and as a transport medium for soluble substances. The gut is always blind, i.e. there is no anus, and in parasitic species it may be vestigial or absent. There is no blood vascular system. Protonephridia serve as excretory and osmoregulatory organs. Most species are hermaphrodite, with complicated sex organs; in particular, the ovaries are often associated with vitellaria and shell glands. Cleavage is spiral, or it is irregular, total and unequal. Development may be direct or indirect. Indirect development often proceeds via several different larval stages. In some parasitic forms, development proceeds via different asexually reproducing generations in a succession of different hosts, and the mature, sexually reproducing form infects the final or primary host.

The phylum is divided into 3 classes: *Turbellaria* (see), *Trematoda* (see) and *Cestoda* (see).

Collection and preservation. Large turbellarians are collected by hand from water plants and submerged stones, and from the bottom surface of ponds and streams, etc. Some species can be attracted by laying a bait of small pieces of meat. Smaller species are best collected from water samples by waiting until they migrate to the surface due to oxygen shortage. Specimens are allowed to extend in a little water, then killed by pouring hot 2% nitric acid over them; they are then stored in 70% ethanol.

Trematodes are found on the skin or gills or fishes, and many are found in the intestine and other internal organs of vertebrates. They are washed out with physiological saline, narcotized in 10% ethanol, or in 1% chloral hydrate or urethane, then fixed in a mixture of 40% formaldehyde and ethanol. For histological purposes, they should be fixed in Bouin's fluid. To prevent specimens from rolling up, they are pressed between two glass strips, and stored in 70% ethanol or 2% formaldehyde containing a few drops of glycerol.

Tapeworms are washed out of the small intestines of vertebrates with physiological saline, killed in 1% aqueous chromic acid, then fixed in a mixture of 40% formaldehyde and ethanol, containing a little acetic acid. They are stored in formaldhyde-ethanol without the acetic acid.

Platymantidae, ***platymantids:*** a family of the *Anura* (see) containing 2 diverse genera, distributed widely in Africa, Asia and the Indo-Malaysian islands as far as Australia, including the Solomon Islands and the Fijis. The entire group is characterized by broadened finger tips, members of the genus *Amlops* possessing well developed finger disks. Members of the genus *Platymantis* are good climbers and fully independent of standing water. A few large-yolked eggs are laid on leaves or stems, or on leaf litter at ground level, and development is direct. In contrast, species of *Amlops* are found in fast-flowing rivers in Borneo and China; their eggs hatch into tadpoles, which attach themselves to rocks by a suction disk, extending from behind the mouth to the belly. In many systems of classification, both *Amlops* and *Platymantis* are subsumed in the *Raninae.*

Platypsaris: see *Cotingidae.*

Platypus: see *Monotremata.*

Platyrrhina: see New World monkeys.

Platysma myoides: in mammals and humans, a thin quadrilateral sheet of muscle extending from the chest to the face over the side of the neck, between the superficial and deep faciae.

Platysomus (Greek *platys* broad or flat, *soma* body): a genus of holostean fishes (order *Palaeonisciformes,* suborder *Platysomoidei,* family *Platysomidae)* with a very deep, compressed, rhombic shape. They were mainly reef inhabitants. Their truncated conical teeth were adapted to a diet of hard-shelled creatures. Fossils are found in the Carboniferous, Permian, and questionably in the Triassic; they are relatively common in the Germanic red shale.

Platyspiza: see Galapagos finches, Adaptive radiation.

Platysteira: see *Muscicapinae.*

Platystemon: see *Papaveraceae.*

Platysternidae: a family of the order *Chelonia* (see), represented by the single living species, *Platysternum megacephalum* ***(Big-headed turtle),*** found in cool, strong flowing mountain rivers in southern China, Indochina, Burma and Thailand. The olive-brown carapace attains a length of 20 cm. The plastron is cream to pale orange, sometimes with dark markings. The muscular tail is the same length as the shell. *Platysternum* has an unusually large and completely roofed skull, and the large head cannot be withdrawn beneath the rather flat carapace. In some specimens the brown head is decorated with orange spots or stripes behind the eyes. It is an agile climber, most active at night, and feeds mainly on water snails, which it breaks open with its powerful hooked jaws.

Platysternum: see *Platysternidae.*

Plautus: see *Alcidae.*

Play: patterns of voluntary behavior with no functional goal, and often not pursued to a conclusion. P. stengthens motor activity and reinforces awareness and perception, especially in young animals. Readiness for P. is species-specific, and it occurs only in species that also exhibit well developed self-motivated exploratory behavior, i. e. in particular in higher mammals. In *solitary P.* the animal plays by moving its own body, or it plays with, or with the aid of, objects. *Biosocial P.* occurs between two or more animals, involving contact P., near field P. (usually chasing games) and distance field P. (hidung and searching games). Solitary and biosocial P. may be combined.

Playa: see Saline lakes.

Pleasure and pain system: see Self stimulation.

Plecoptera, *Perlaria,* ***stoneflies:*** an insect order containing about 1 700 species. The earliest fossil P. are found in the Permian.

Play. Phases of play behavior in the arctic fox (after Tembrock).

Imagoes. Stoneflies are delicate insects ranging in body length from 0.5 to 3 cm. The head carries elongate, setaceous antennae and weak, biting mouthparts. There are two pairs of equally long, membranous and richly veined wings, the hindwings being somewhat wider than the forewings. At rest the wings are laid horizontally, one above the other, flat over the back of the abdomen, which carries two long, multi-articulate cerci. The body is hairless and of a brownish dusky color. Imagoes usually remain concealed and at rest on stones near to the water where they emerge from the final molt; they tend not to fly, but when they do the flight is ungainly, the almost equal-sized fore and hindwings moving independently. Metamorphosis is incomplete (hemimetabolous development).

Eggs. The eggs are dropped into water during flight or at rest by placing the tip of the abdomen into the water. Eggs of the family *Nemouridae* (suborder *Filipalpia*) are held together in a clump of 300 to 1 000 by a mass of sticky substance at the end of the abdomen, and may be carried around by the female in this form for a few hours or days. Immediately on entering the water, the clump dissociates and the individual eggs sink to the bottom or become attached to plants or stones. Usually two to three egg clumps are laid at specific time intervals.

Larvae. Larvae (naiads or nymphs) are found in clear, flowing water. After the last of the approximately 20 aquatic larval stages, the mature naiads possess wing primordia and closely resemble the imago. They closely resemble the larvae of the mayflies, but differ from the latter in possessing only 2 cerci and 2 tarsal claws. Some larvae are vegetarian, others feed by predation on the aquatic nymphs of other insects. They breathe with the aid of tracheal gills, which are variable in position but usually located in the thorax. Total development lasts 1 to 3 years.

Classification. Authorities differ as to the number of suborders. The system proposed by Illies (*Annu. Rev. Entomol.* ***10*** (1965) 117–140) has 3 suborders.

1. Suborder: *Archiperlaria.* Many primitive features are present, e.g. large size; brightly colored (red, yellow

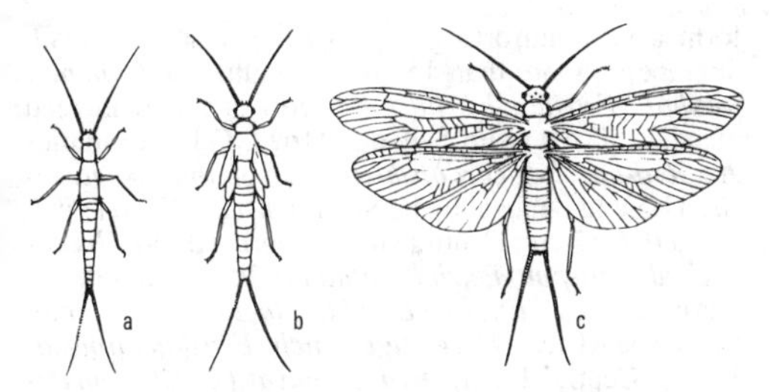

Plecoptera. Metamorphosis of a stonefly *(Perla* sp.). *a* Wingless larval stage. *b* Older larva with wing sheaths. *c* Imago.

or green) body and wings; anal fan in the hindwing with 8 or 9 anal veins; larvae have 4 to 6 pairs of abdominal gills.

2. Suborder: *Filipalpia.* Palps are threadlike and of equal diameter from base to tip; mouthparts are simplified, but mandibles can still handle solid food particles. Larvae are vegetarian.

3. Suborder: *Setipalpia.* Palps are bristle-like, narrower toward the end; mandibles are vestigial. Imago can only drink. Larvae are predators.

Plecotus: see *Vespertilionidae.*

Plectascales: see *Ascomycetes.*

Plectenchyme: a false or pseudo-tissue formed by the intertwining of richly branched filamentous cells, sometimes with postgenital fusion of filaments. P. can form highly organized structures, which superficially resemble those composed of true tissues in higher plants. Well developed P. is found in the *Rhodophyceae* and in the fructifications of higher fungi.

Plectognathi: see *Tetraodontiformes.*

Plectopterus: see *Anseriformes.*

Plectrophenax: see *Fringillidae.*

Plegadis: see *Ciconiiformes.*

Pleiochasium: see Stem, Flower.

Pleocyemata: Shrimps (see), a suborder of the *Decapoda* (see).

Pleomere: see Pleon.

Pleomorphism: a term used mainly in microbiology for the occurrence of different living forms within a single species. During growth, a single bacterial species may vary considerably in size and shape, forming cells which are spherical, pear-shaped, elongated filaments, filaments with localized swellings, etc. P. is readily observed in species of *Streptobacillus* and *Pasteurella,* some important causes being the age of the culture and environmental conditions (e.g. type of culture medium). The pathogenic skin fungus, *Candida albicans,* grows partly as oval or spherical yeast cells, which reproduce by budding, and partly as a pseudomycelium of nonbranching filamentous cells, which divide by constriction. Development of the yeast form is favored by a ready availability of nutrients and good aeration, whereas mycelial growth is favored by poor aeration and a shortage of nutrients.

Pleon: hindmost body section or abdomen of crustaceans of the class *Malacostraca.* The individual segments of the P. are called pleomeres, and its biramous appendages (usually slender and fringed and used for swimming) are called pleopods. See *Crustacea.*

Pleopod: see Pleon.

Plerocercoid: see *Cestoda.*

Plerome: see Histogen concept.

Plesiomorphic: a term applied to "primitive" characters, e.g. reptile scales are P. with respect to bird feathers, which are their evolutionary derivatives. Similarly, the toothed jaws of reptiles are P. with respect to the toothless jaws of birds. In any one organism, a character can be P. in relation to its later derivatives, and apomorphic in relation to its earlier, more primitive form. The opposite of P. is Apomorphic (see).

Plesiopora: see *Oligochaeta.*

Plethodontidae, ***lungless salamanders:*** the most numerous family of the *Urodela* (see), containing 209 species in 24 genera. Two genera occur in South America, and 2 species of *Hydromantes* occur in southern Europe, but the family is distributed primarily in North and Central America, the major center of differentiation being Mexico and Central America. Adults do not have lungs, and respiratory gas exchange occurs via the body skin and the buccal mucosa; in cases of neotenic development, external respiration also occurs via the persistent gills. Movable eyelids are present and there are verticle rib grooves along the sides of the long, slender body. The tail is long and cylindrical. There are 4 fingers and 5 toes. Fertilization is internal. Characteristic of the family is the presence of 2 glandular grooves (nasolabial grooves) from the lip border to each nostril. The first lungless salamanders inhabited mountain streams, but evolutionary diversification has led to the invasion of a variety of different habitats. Most species display courtship behavior with head rubbing and moving around each other. Reproduction shows all intermediate forms, from mating and spawning in the water with the emergence of free-living, gilled larvae, to complete development of the eggs on land with the emergence of fully metamorphosed young animals.

Members of the genus *Aneides* (e.g. *A. aenus,* ***Green salamander)*** are totally adapted to a terrestrial existence, and they are able to climb trees and rocky clefts with the aid of adhesive toe disks.

Typhlomolge rathbuni ***(Texas blind salamander),*** which inhabits the waters of deep wells and caves in the southwest USA, closely resembles the Olm (see *Proteidae).* This 12 cm-long, neotenic species is an unmetamorphosed "permanent larva", with thin legs and pink, external gills which are retained throughout life. The skin is colorless and the eyes vestigial. It is a threatened species.

The North American genus, *Plethodon* ***(Forest salamanders),*** contains many small species. Eggs are laid under moss and stones. The gilled larvae develop inside the egg, but the gills are lost on hatching, so that the young emerge as miniature adults.

Members of the genus, *Batrachoseps* (e.g. *B. aridus,* ***Desert slender salamander)*** of western North America, are slender and elongate with degenerate limbs. They progress by serpentine movements, and they lead a mainly subterranean existence, often in earthworm burrows. Development of the young occurs directly on the land, inside the egg, without a free-living larval stage.

Certain species of the genus *Bolitoglossa* occur in South America well south of the Equator.

Two species of *Hydromantes* are found in Southern Europe: *H. italicus* ***(Italian*** or ***Cave salamander)*** in the wet, rocky regions of the French and Italian Alps; and *H. genei* ***(Sardinian salamander)*** of Sardinia. They are the only lungless salamanders found outside the New

World, and this disjunction indicates that the family was formerly much more widely distributed. Members of the genus *Hydromantes* have slender, 10–12 cm-long bodies. The tongue stalk is long, and the tongue is used for catching prey. They are oviparous, and the young develop and metamorphose inside the egg.

Other cave genera are *Haideotriton* and *Typhlotriton*. *Typhlotriton spelaeus* ***(Grotto salamander),*** from the Ozarks region of USA, is pale pink to white with vestigal eyes, but no external gills. The larva is typical and leads a normal aquatic existence, with functional eyes, external gills, a large tail fin and gray-brown coloration. At metamorphosis, the tail fin is lost, together with the gills and pigmentation, the eyes stop growing and become covered with skin, and the animal retreats to a subterranean habitat.

Pleurocarpous growth: see *Musci.*

Pleurocium: see *Musci.*

Pleurodeles waltl: see *Salamandridae.*

Pleurodira: a suborder of the *Chelonia* (see).

Pleurodont: see Teeth.

Pleuronectes: see Plaice.

Pleuronectiformes, ***flatfish:*** an order of bottom living, marine teleost fishes, which have a flattened asymmetrical body with compressed sides. They lie on one of their sides which is light in color. The other side is pigmented and forms the upper, dorsal surface. The eye of the underside is shifted to the topside, and the mouth is oblique. Larvae are free-swimming and at first symmetrical like those of other fishes, but they metamorphose into asymmetrical forms which then take up life on the sea bottom. Many are able to change their pigmentation to blend with that of the substratum, and most species bury themselves in the soft sand until only their eyes are exposed. About 538 species are recognized in about 117 genera and 6 families. On account of their tasty flesh, many are fished commercially, e.g. ***Scaldfish*** or ***Sole*** *(Arnoglossus laterna),* ***Brill*** *(Scophthalmus rhombus),* ***Turbot*** *(Scophthalmus maximus),* ***Megrim*** *(Lepidorhombus whiffiagonis),* ***Halibut*** *(Hippoglossus hippoglossus),* ***Dab*** *(Limanda limanda),* ***Lemon sole*** *(Microstomus kitt),* ***Flounder*** *(Platichthys flesus),* ***Plaice*** *(Pleuronectes platessa),* ***Dover sole*** *(Solea solea).*

Pleuropneumonia-like organisms: see Mycoplasmas.

Pleurostigmophora: see *Chilopoda.*

Ploceidae, ***Old world seed eaters:*** an avian family of about 260 species in the order *Passeriformes* (see). It shows no obvious close affinity to any other Old World family. Similarities with the New World *Fringillidae,* e.g. stout, conical seed-eating bills, do not imply a close phylogenetic relationship. The goldfinches and their allies *(Carduelinae)* are considered by some authorities to be a subfamily of the *Ploceidae,* but here they are placed in the *Fringillidae.*

The largest members are found in the subfamily, *Bubalornithinae* ***(Buffalo weavers),*** which contains only 2 members, both ground-foraging species, which inhabit the drier parts of Africa, and have nothing in particular to do with buffaloes: *Bubalornis albirostris* ***(Black buffalo weaver;*** 25 cm;) and *Dinemellia dinemelli* ***(White-headed buffalo weaver;*** 23 cm).

The subfamily, ***Estrildinae*** *(Waxbills),* contains 107 species in 15 genera, many of which are well known as cage birds. They display a wide variety of often bright plumage colors, but behaviorly and anatomically they form a very uniform group. All are relatively small, none being more than 15 cm in length. The ***Common waxbill*** *(Estrilda astrild;* 11.5 cm) is found in reed marshes all over sub-Saharan Africa. Other examples are: ***Red-cheeked cordon-bleu*** *(Uraeginthus bengalensis;* 11.5 cm; Senegal to the Sudan); ***Melba finch*** *(Pytelia melba;* 13 cm; Central and southern Africa); ***Green-backed twinspot*** *(Estrilda nitidula;* 7.5 cm; South Africa); ***Locust finch*** *(Estrilda locustella;* 9 cm; Southeast Africa); ***Green tiger finch*** *(Estrilda formosa;* 10 cm; Central India); ***Red avadavat*** *(Estrilda amandava;* 7.5 cm; India to Indochina and East Indies); ***Chestnut Mannikin*** *(Lonchura ferruginosa;* 12 cm; India to Philippines and East Indies); ***Blue-faced parrot finch*** *(Erythrura trichroa;* 11.5 cm; Micronesia). The *Estrildinae* are highly gregarious birds and they build large, loose, dome-shaped nests with a side entrance.

The 30 species of the *Passerinae* ***(Old World sparrows)*** include 8 species of African ***Sparrow weavers*** (e.g. *Plocepasser mahali,* ***Black-billed sparrow weaver),*** 2 African ***Scaly weavers*** *(Sporopipes),* 7 ***Snow finches*** *(Montifringilla),* 4 ***Rock sparrows*** *(Petronia),* and the well known sparrows of the genus, *Passer.* The latter include the ***House sparrow*** *(Passer domesticus),* which has been widely introduced and is now probably the most familiar and widespread small bird in the world.

Most of the approximately 90 species of ***True weavers*** (subfamily *Ploceinae)* are African, and 5 occur in Southeast Asia and India. Plumage of the true weavers characteristically shows red and yellow tinting, especially in the genera *Ploceus* (57 species) and *Malimbus* (10 species). These are arboreal species, building complex nests with entrance tunnels, sited over water and often associated with wasp nests. ***Diochs*** or ***Queleas*** *(Quelea),* ***Fodies*** *(Foudia),* ***Bishops*** *(Euplectes)* and the nonviduine ***Whydahs*** or ***Whidas*** *(Steganura)* are noted for the brilliant plumage of the male. Male bishops and whydahs are are also noted for their long tails, that of the ***Sakabula*** *(Euplectes progne)* being 51 cm in length.

Lastly, the ***Widow weavers*** or ***True widowbirds*** (subfamily *Viduinae),* also called Whydas and referred to as viduine Whydas, are represented by 9 species that inhabit grasslands and savannas in sub-Saharan Africa. Sexual dimorphism is very pronounced, the female being brown and inconspicuous, while the male during the breeding season has a long tail and striking plumage. They are both polygynous and parasitic, and the courtship display of the male is very elaborate. Parasitism is highly selective, each species of widowbird laying its eggs in the nest of only one or a few species of waxbill *(Estrildinae);* their eggs are not marked like those of other weavers and sparrows, but white like those of waxbills; the young also have brightly colored gapes like those of young waxbills.

Ploceinae: see *Ploceidae.*

Plocepasser: see *Ploceidae.*

Ploceus: see *Ploceidae.*

Plover: a member of the avian family, *Charadriidae* (see), order *Charadriiformes.* The name is also applied to certain species in other families of the *Charadriiformes,* e.g. Egyptian plover *(Pluvianus aegyptius; Glareolidae),* Upland plover *(Bartramia longicauda; Scolopacidae),* Crab plover *(Dromas ardeola; Dromadidae),* Stone plover *(Esacus magnirostris; Burhinidae),* Norfolk plover *(Burhinus oedicnemus; Burhinidae).*

The Quail plover *(Ortyxelos meiffrenii)* belongs to the family *Turnicidae* in the order *Gruiformes.*

Plum: see *Rosaceae.*

Plumaceous: see Feathers.

Plumage: see Feathers.

Plume moths: moths of the family *Pterophoridae;* see *Lepidoptera.*

Plumeria: see *Apocyanaceae.*

Plum fruit moth, ***red plum maggot,*** *Cydia (Grapholita) funebrana* Tr.: a moth of the family *Tortricidae* (see), which lays its eggs on young plums and other stone fruits. The caterpillar (fruit "maggot") tunnels into the fruit, which soon falls. Second generation caterpillars are found in partially eaten harvested fruit.

Plumulae: see Feathers.

Pluriennial plants: Hapaxanthic plants (see), whose vegetative period lasts from several to many years. They flower and produce fruit only once, then die. Various species of palm and *Agave* are P.p.

Pluvialis: see *Charadriidae.*

Pluvianus: see *Glareolidae.*

Pneumatic duct of swim bladder: see Swim bladder.

Pneumatophore: a terminal, gas-filled, sterile medusa present in many members of the order, *Siphonophora.* The medusa is modified as a buoyancy aid, which by storing or releasing gas enables the colony to move vertically in the water, or it may project above the water like a sail, so that the colony is transported by the wind. The gas (more than 90% carbon monoxide) is generated in special gas glands. See *Siphonophora,* Portuguese man-of-war, *Velella.*

Pneumogastric nerve, ***vagus:*** the X cranial nerve; see Cranial nerves, Autonomic nervous system.

Pneumonic plague: see *Yersinia pestis.*

Pneumothorax: see Respiratory mechanics.

Poaceae: see *Graminae.*

Pochards: see *Anseriformes.*

Pocket plums: see *Ascomycetes.*

Pod: the fruit of a member of the *Leguminoseae* or *Fabales* (see Leguminous plants). See Fruits.

Podarcis: see Wall lizard.

Podargidae: see *Caprimulgiformes.*

Podargus: see *Caprimulgiformes.*

Podica: see *Gruiformes.*

Podicipediformes, ***grebes*** (plate 6): an order of birds (see *Avies)* containing 18 species in 4 genera. The largest genus, *Podiceps,* is cosmopolitan in distribution, the remaining genera being monotypic and confined to the New World. Each toe is fringed with stiff, horny flaps, a condition known as lobate webbing. The laterally compressed tarsi have serrated hind edges, and the legs are placed at the very rear of the body. The tail is vestigial. They breed on inland waters, where they build floating nests, and they cover the nest with vegetation before leaving it unattended. The sexes are similar, and the mutual courtship display is elaborate. Young are characteristically carried on the backs of the adults. A few species migrate to marine inshore waters in the winter. Grebes display the peculiar habit of eating quantities of their own body feathers (and feeding them to their chicks), possibly as a digestion aid. Their food consists of fish, insect larvae, and other aquatic animals. Examples are: *Podiceps cristatus* ***(Great crested grebe;*** 48 cm; Eurasia, Africa, Australia, New Zealand); *Podiceps grisegena* ***(Red-necked grebe;*** 33 cm; Eurasia, North America); *Podiceps auritus* ***(Slavonian*** or ***Horned grebe;*** 33 cm; Eurasia, North America); *Podiceps ruficollis* ***(Little grebe;*** 25.5 cm; Eurasia, Africa, East Indies); *Podilymbus podiceps* ***(Pied-billed grebe;*** 48 cm; Canada to southern Argentina). Colors vary according to the species, but the breast plumage is always tight and satin-like in texture. The flightless ***Titicaca*** or ***Short-winged grebe*** *(Centropelma micropterum),* found only on Lake Titicaca in Peru and Bolivia, has a ragged crest and white face with a red and yellow bill; the feet are bright yellow above and black below. The ***Great western grebe*** *(Aechmophorus occidentalis;* 56–74 cm; western North America, migrating to the coast and southward in the winter) is black and slaty gray above and white below, with a long and largely white neck, a mainly yellow bill, olive green legs and red eyes.

Podiceps: see *Podicipediformes.*

Podilymbus: see *Podicipediformes.*

Podocarpaceae: a family of the *Pinidae (Coniferae)* containing about 140 species, distributed mainly in mountainous regions of the tropics and subtropics in the Southern Hemisphere. They are trees, rarely shrubs, with spirally inserted scale- or needle-shaped leaves, some species possessing leaflike short shoots and dioecious flowers. The female inflorescences are greatly reduced, and woody cones are not formed. The ovuliferous scale develops into a unilateral fleshy seed covering. Characteristic resins and essential oils have been isolated. The genus, *Podocarpus,* represented by about 100 species, is the largest coniferous genus of the Southern Hemisphere, where it is widely distributed. Another important genus is *Dacrydium,* which is found predominantly in Southeast Asia.

Podocarpus: see *Podocarpaceae.*

Podocnemis: see *Pelomedusidae.*

Podophthalmus: see Swimming crabs.

Poecilia: see Guppy.

Poecilostomatoida: see *Copepoda.*

Poelagus: see *Leporidae.*

Pogoniulus: see *Capitonidae.*

Pogonophora, ***beard worms:*** an exclusively marine phylum of the *Protostomia,* most members occurring at depths below 100 m, where they live in secreted tubes in the bottom ooze. The tube is thought to be orientated vertically. The body is thread-like, and the lengths of known species vary from 0.5 cm to 2 m, with diameters no greater than 2.5 mm. Pogonophorans are found in all the world's seas, especially along continental slopes. They often occur in dense aggregations (up to 200 per square meter). A few species are found in decaying wood. The 2 m-long pogonophorans were discovered in 1977 around warm water vents on the 9000 m floor of the Galapagos rift.

The body consists of a short anterior part, a long trunk, and a short opisthoma. The anterior part consists of a cephalic lobe and a posterior glandular region, the latter secreting material for tube formation. Tentacles arise below the cephalic lobe; these vary in number from 1 to more than 250, lying parallel to one another and arranged in the form of a cylinder or tube. The tentacles may be straight or spirally arranged; in *Siboglinum* the anterior tube is formed by the spiralling of the single tentacle. Cilia drive a water current throught the tentacle tube from the anterior to the posterior end. Small appendages of the tentacles (pinnules) on the inside surface of this cylinder appear to trap suspended food particles brought in with the water current. As free-living animals, the pogonophorans are unique in lacking a mouth or digestive tract. The trapped food particles are possibly digested externally within the tentacle tube, and then absorbed by the tentacular epithelium; at least the epithelium has microvilli, like those of absorptive cells in other animals. It has also been suggested that the animal absorbs organic compounds from the surrounding mud. The tentacles may also be the site of respiratory gas exchange. Below the tentacle-bearing anterior part is an extremely long trunk, which is responsible for the great length of the animal Halfway along the trunk are two belt-like regions studded with minute, hard, toothed structures, called uncini. Numerous papillae are distributed over the hindmost body region. The extreme posterior end (opisthoma) is composed of 5–23 short segments that carry setae. Nephridial structures are present. There is a blood-vascular system, and each tentacle is supplied by an afferent and an efferent vessel. Reports on the existence and the nature of pogonophoran coelom are contradictory. Sexes are separate, but there is no sexual dimorphism. Cleavage is holoblastic and unequal. There is no blastocoel, and gastrulation is by epiboly. A yolk-filled, posterior blastopore is formed briefly. The larva has an anterior and a posterior ciliary ring.

The phylogeny of the pogonophorans is a subject of debate. Before the discovery of the segmented, setiferous, annelid-like opisthoma, they were regarded as deuterostomes. Most authorites now place them in the *Protostomia* and many regard them as close relatives of the *Annelida.* Others consider that they occupy a position intermediate between the deuterostomes and protostomes, or that they represent a separate evolutionary line.

More than 100 species have been described, but new species are still reported from deep ocean surveys. The best known genera are *Siboglinum* (lateral nephridiopores, pericardial sac, one tentacle), *Lamellisabella* (medial nephridiopores, no pericardial sac, several tentacles fused into a cylinder) and *Spirobranchia* (medial nephridiopores, no pericardial sac; some species with more than 200 tentacles, which are fused and spiralled at the base).

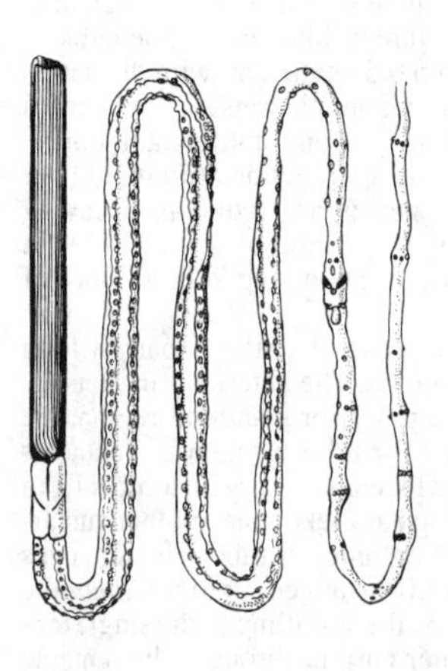

Pogonophora. The beard worm, *Spirobrachia,* removed from its tube. The hind end of the body is omitted.

Pogostemon: see *Lamiaceae.*

Poikilohydric: a term applied to a system whose moisture content and water vapor pressure are very variable and dependent on the moisture content and vapor pressure of the environment. P. plants, like mosses, lichens and aerial algae have the same degree of hydration as their environment, and they withstand desiccation without injury. The converse is *homeohydric.* Homeohydric plants are vascular plants that can regulate their own water economy.

Poikilothermic animals, ***cold-blooded animals:*** animals that do not maintain a constant body temperature, and whose body temperature approximates to that of the surroundings. The body temperature of aquatic poikilotherms approximates very closely to that of the surrounding water. All animals except birds and mammals are poikilothermic. Mammals and birds are Homeothermic animals (see). See Temperature factor.

Poinsettia: see *Euphorbiaceae.*

Point mutation: see Gene mutation.

Poison-arrow frogs: see *Dendrobatidae.*

Poisonous plants: plants whose constituents (plant toxins; usually alkaloids or glycosides) adversely affect humans or animals, and under certain conditions can also cause death. Notable examples are ricin from Castor bean and amanitin from the Death cap fungus. In small doses, certain plant toxins are also used medicinally, e.g. strychnine from *Strychnos nux-vomica.*

Polar bear, *Thalarctos maritimus:* a white bear of Arctic regions. It covers large distances on land and is a proficient swimmer. Seals, birds, fish and carrion are eaten, as well as berries and leaves of tundra plants. The female gives birth in winter in a snow cave, which she forms by allowing herself to become snowed in.

Polar flagella: see Bacterial flagella.

Polecat, *Mustela putorius:* a mustelid (see *Mustelidae)* with a dark brown coat and light face. Distributed over wide areas of Europe, it spends the day sleeping in caves and other hiding places, and feeds mainly on small vertebrates, which it hunts at night. The domesticated form is known as a ***ferret,*** and this is often bred as an albino.

Polemoniaceae, ***phlox family:*** a family of dicotyledonous plants containing 275 species in 20 genera, distributed mainly in Pacific North America, but also in the South American Andes, Central America, Europe and northern Asia. They are perennial or annual herbs, rarely shrubs, climbers or trees. Leaves are exstipulate, simple or compound, alternate or opposite. Flowers are actinomorphic and 5-merous, and the calyx is a fused tube. The superior ovary, which usually consists of 3 fused carpels, develops into a capsule. Most species are pollinated by bees, but some are specifically pollinated by humming birds, bats, or lepidopterous insects such as hawk moths. The only central and northern European species is *Polemoneum caeruleum* ***(Jacob's ladder),*** which is becoming rare in certain areas, but is also widely introduced as a garden escape. Many species are cultivated as garden plants for their showy colored flowers, in particular different species and hybrids of the North American *Phlox.*

Polemoneum: see *Polemoniaceae.*

Poliohieracinae: see *Falconiformes.*

Polioptila: see *Sylviinae.*

Polioptilinae: see *Muscicapidae, Sylviinae.*

Polish mushroom: see *Basidiomycetes (Agaricales).*

Pollachius: see Coley, Pollack.

Pollack, *Pollachius pollachius:* a predatory fish (family *Gadidae,* order *Gadiformes),* which feeds mainly on herring. It is found in small shoals in Atlantic coastal waters from northern Norway to North Africa, off the Icelandic coast, in the western Baltic and the Mediterranean. It attains lengths greater than 1 m and is a good angling fish. Although it has some commercial value, the flesh is not greatly prized. It is generally dark brown with yellow-green on the sides and irregular brown-yellow spots. There is a conspicuous dark lateral line.

Pollen: see Flower; section 2) Stamens; see Pollen analysis.

Pollen analysis: determination of the pollen content of core samples from peat or sediment deposits, in order to establish the nature of earlier vegetation, climate and foods in the source area. P.a. is a branch of Palynology (see).

Systematic P.a. began in 1893, and by 1930 the methodology had been extended to prequaternary strata.

P.a. and spore paleontology are possible because the outer walls of pollen grains and spores (exine or the exospore) are highly resistant to decay and decomposition.

Wind-distributed pollens are especially widespread in geological sediments. Most gymnosperms, but relatively few angiosperms are wind-pollinated (anemophilous). The predominant trees of central Europe (e.g. beech, oak, hornbeam, birch, elm, ash), grasses and members of the *Ericaceae* are, however, also wind-pollinated. Fossil pollens from insect-pollinated (entomophilous) species are encountered only rarely, so that only a small proportion of the plants in an area or stratum can be identified from P.a. Since pollen production varies with species and environmental conditions, the concentration of a particular pollen in a core sample is not indicative of the population density of the species producing it. The percentage of each pollen species in a sample constitutes the pollen spectrum. Pollen spectra can be ordered according to sampling depth, and the result presented diagrammatically as a pollen diagram. The term, pollen spectrum, is also applied to the P.a. of honey, which provides qualitative information about forage plants at the collecting site.

One of the most important achievements of P.a. has been to provide a detailed record of vegetation changes since the last Ice Age. As determined by P.a. at many diferent sites, European postglacial vegetation passed through the following major phases (as defined by Firbas and Overbeck).

1) Cold Arctic conditions of the Lower Dryas with no trees. Before 15 000 BC.

1a) Pollen zone I of the Lower Dryas (see Dryas).

1B) Pollen zone II. Bälling period (see).

1b) Pollen zone II. Upper Dryas or early sub-Arctic period.

2) Pollen zone III. Alleröd period (see) or intermediate sub-Arctic period.

3) Pollen zone IV. Later sub-Arctic period, ending about 8 000 BC.

Periods 1 to 3 are characterized by the absence of forests, and they are classifed as late glacial in P.a. chronology. The subsequent periods 4 to 10 are postglacial.

4) Pollen zone V. Preboreal. Birch and pine period.

5) Pollen zone VI. 1st Boreal (see Boreal period). Pine and hazel.

Pollen zone VII. 2nd Boreal. Mixed oak forest, pine and hazel.

6) Pollen zone VIIIa. Early Atlantic (see Atlantic period). Mixed oak forest.

7) Pollen zone VIIIb. Late Atlantic. Mixed oak forest.

8) Pollen zones IX and X. Subboreal. Mixed oak forest with alder, becoming beech.

9) Pollen zone XI. Late Subatlantic. Beech forest and mixed woodlands containing predominantly beech.

10) Pollen zone XII. Recent Subatlantic. Forest exploited and managed. 1 000 AD.

Pollen cement: see Flower; section 2) Stamens.

Pollination: transfer of pollen to the stigma of angiosperms or the ovule of gymnosperms. P. within the same hermaphrodite flower is called ***self-pollination*** or ***autogamy.*** P. between plants of the same clone is known as ***adelphogamy.*** P. of a flower in the open state is called ***chasmogamy,*** in contrast to P. within an unopened flower (e.g. violet), which is known as ***cleistogamy.*** P. between different flowers on the same plant is ***geitogamy,*** while P. between flowers of different plants is called ***cross-pollination, allogamy*** or ***xenogamy.***

Self-pollination enables the setting of fruit and reproduction by isolated individual plants. It is therefore widespread among pioneer plants and weeds, as well as among island flora, and in areas where pollinating animals are scarce. In order for self-pollination to be successful, the flowers must not be self-sterile. Self-pollinating plants usually possess inconspicuous, scent-free and nectar-free flowers, which produce only small quantities of pollen. Since genetic recombination is restricted by self-pollination, the offspring from this type of P. often show signs of degeneration and decreased vigor. Also, the efficacy of pollination as measured by the number of seeds produced is usually greater for cross-pollination than for self-pollination.

In contrast, cross-pollination promotes genetic recombination, favoring more vigorous offspring. The efficiency of cross-pollination may, however, be restricted by the decreased compatibility of the male and female elements of the two plants; in the extreme case, the two plants in question are genetically totally isolated, and cross-fertilization is impossible, i.e. they are different species. In dioecious plants cross-pollination is the only possible type of P., but even in the unisexual flowers of monoecious plants and in hermaphrodite flowers, cross-pollination is enforced or favored by self-sterility, heterostyly, dichogamy and hercogamy. In Self-sterility (see) (see also Incompatibility), the pollen grains often fail to germinate on the stigma of the same flower, or the pollen tubes grow too slowly, or they atrophy (e.g. many orchard fruits, the olive, most varieties of rye, etc.). ***Heterostyly*** is found in, e.g. primulas and forsythia. Thus, in some *Primula* plants (about 50% of the population), the 5 stamens are attached about half way up the corolla tube, and the style of the ovary is long, so that the stigma is above the stamens (pin-eyed flowers). In other plants the stamens are attached at the top of the corolla tube and the style is short, thus placing the stigma below the stamens (thrum-eyed flowers) (Fig.). In addition, in the pin-eyed flower, the pollen grains are relatively small and the papillae of the stigma are relatively large, a condition that is reversed in the thrum-eyed flower. Insects penetrate to the same depth in both types of flower, transferring pollen from high

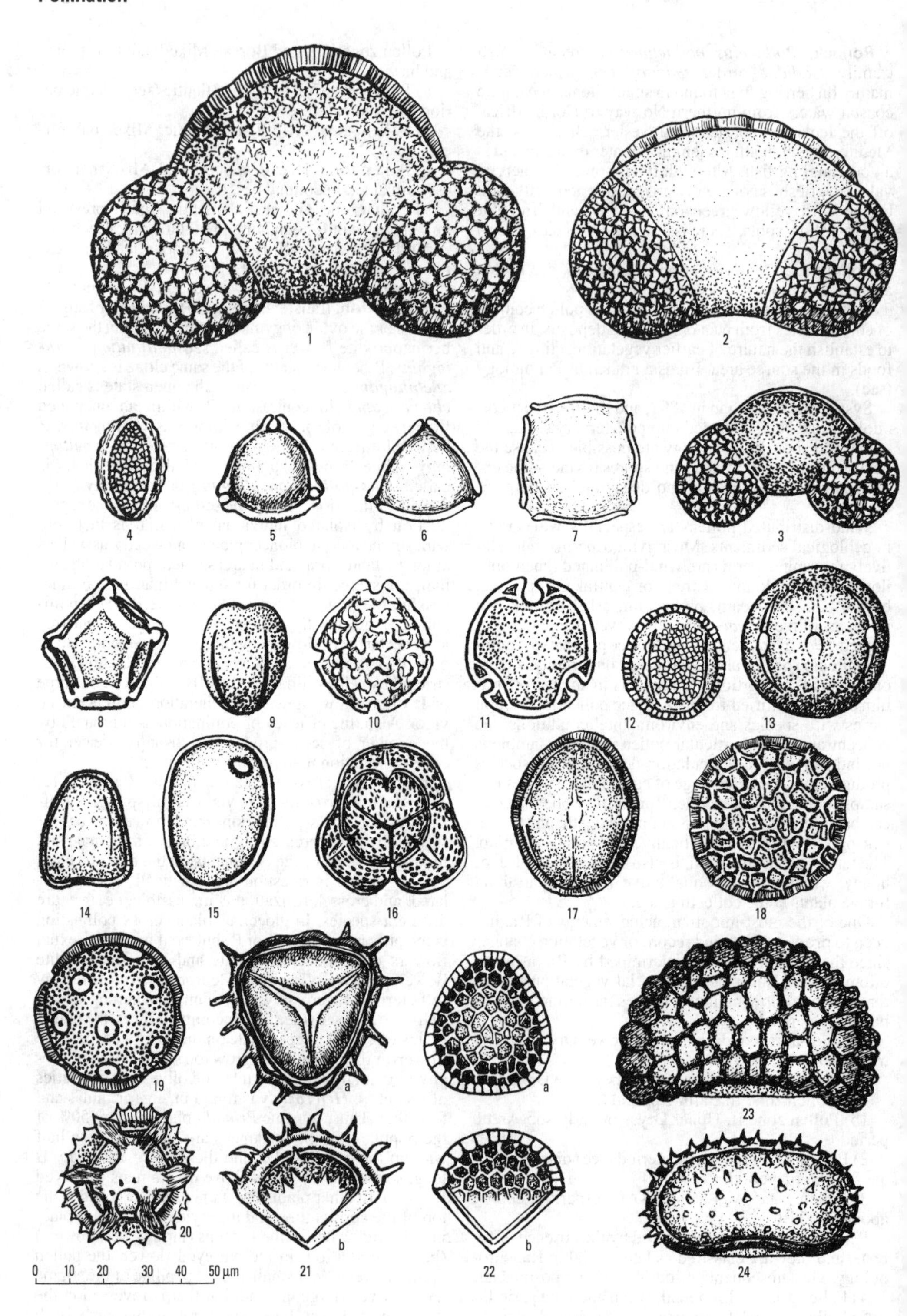
1
2
3
4
5
6
7
8
9
10
11
12
13
14
15
16
17
18
19
20
a
b
21
a
b
22
23
24
0 10 20 30 40 50 µm

stamens to high stigmata, and from low stamens to low stigmata; moreover, the transferred pollen is compatible in size with the stigmatic papillae to which it is transferred.

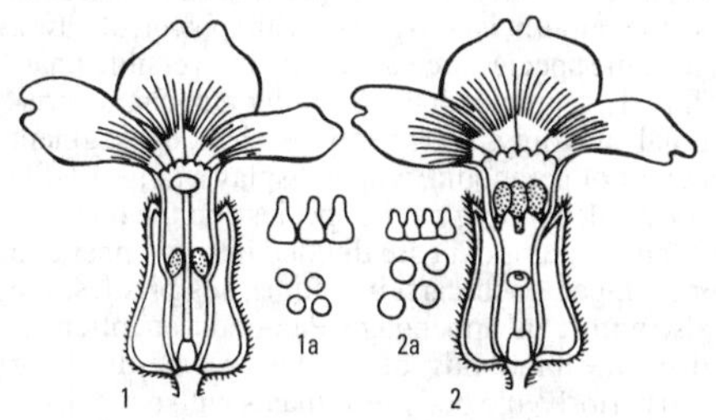

Pollination. Heterostyly in the Chinese primula *(Primula sinensis). 1* Pin-eyed flower, showing the long style and attachment of stamens in the middle of the corolla tube; the stigmatic papillae are large and the pollen grains are small *(1a). 2* Thrum-eyed flower, showing the short style and attachment of stamens toward the top of the corolla tube; the stigmatic papillae are small and the pollen grains are large *(2a).*

In ***dichogamous*** flowers, the stamens and stigmata mature at different times, thus insuring that self-fertilization cannot occur. Dichogamous flowers may exhibit Protandry (see) in which the male organs mature first, or protogyny in which the female organs mature first. The protandrous condition is very common in e.g. *Compositae, Campanulaceae* and *Umbelliferae.* Protogyny is encountered less often, and is found, e.g. in the *Araceae* and *Plantaginaceae.*

In ***hercogamous*** flowers the juxtaposition of stamens and style is so unfavorable for pollen transfer within the same flower that self-pollination is hardly possible. Thus, in *Iris* the 3 large petaloid styles arch over the stamens then turn upward, the stigmatic surface being located on the outside of this turning point. As a pollinating insect enters the flower, it bends the stigmatic surface downward so that it comes into contact with its back or head, which is carrying pollen from another flower. Continuing into the flower, the insect then brushes its head against the dehisced anthers and collects more pollen. The backward retreating insect presses the stigmatic surface against the style, which therefore cannot receive pollen from the same flower.

In self-pollinating hermaphrodite flowers, P. is automatic. In ***cross-pollination,*** pollen is transferred primarily by the wind or by animals, and more rarely by water. ***Wind-pollination (anemophily, anemogamy)*** is a random process. Accordingly, anemophilous flowers characteristically produce large quantities of pollen, and they have free stigmata with a large stigmatic surface area. The pollen of anemophilous flowers is usually floury or dust-like, separating readily into individual grains. Since the pollen of gymnosperms is almost exclusively wind-distributed, anemophily is considered to be the most primitive mechanism for cross-pollination. However, certain genera and species in otherwise animal-pollinated plant families have reverted to anemophily. Examples of such plants, in which anemophily is a secondary adaptation are *Thalicrum* (Meadow rue), *Sanguisorba minor* (Salad burnet), *Fraxinus excelsior* (Ash) and *Artemisia* (Mugwort, Wormwood, etc.). Anemophilous plants often occur in large stands, presumably an adaptation that increases the rate of P. by increasing the number of available stigmata in the immediate vicinity of the source of the wind-borne pollen; examples are grasses and many important forest trees (oak, beech, hornbeam, birch, conifers). Animal P. ***(zoophily, zoogamy)*** is carried out predominantly by insects ***(entomophily, entomogamy).*** In tropical and sub-tropical regions, however, birds are also important pollinators ***(ornithogamy, ornithophily).*** P. by bats ***(chiropterogamy, chiropterophily)*** and small marsupials is quantitatively less important. Chiropterogamy is found in some members of the exclusively tropical families, *Bombacaceae* and *Bignoniaceae.*

Significant pollinating insects are *Hymenoptera,* especially the *Apidae* (honey bees and bumble bees) (47%), *Diptera* (flies, hover-flies) (26%) and *Lepidoptera* (especially butterflies, hawk moths and noctuids) (10%). P. by beetles occurred relatively early in gelogical time (e.g. in the Permian); it is found in the *Polycarpicae (Ranales),* but is regarded as a primitive type of animal P. that is largely accidental. Zoogamy requires that the pollinating animal is induced to make regular visits to the species of flower in question, and then remains in the flower for a sufficient length of time. The flower must also meet certain mechanical requirements, insuring that the pollen and stigma are regularly touched by the animal, and that the pollen is attached sufficiently firmly to an appropriate part of the pollinator. Zoogamous flowers therefore produce attractants, e.g. pollen and nectar which are collected as food. *Rosa, Anemone* and *Papaver* are some of the genera that produce excess pollen rich in protein, fat, carbohydrate and vitamins. The exploitation of pollen as a food is thought to have evolved earlier than feeding on nectar. In rare cases, the animal is lured by the plant into a trap, or the plant may imitate a sexual partner of the animal, so that the animal attempts to mate with the flower, e.g. in species of *Ophrys,* such as the Fly orchid and Bee orchid. The relationship between pollinating insect and flower is often so close and specific that it can be considered as a true form of symbiosis, in which the partners are mutually dependent, e.g. *Yucca* and the Yucca moth.

Adaptations of the flower for entomophily take the

Pollen analysis. Pollen and spores of central European species (after Overbeck). *1* Fir *(Abies alba). 2* Spruce *(Picea abies). 3* Pine *(Pinus sylvestris). 4* Goat willow *(Salix caprea). 5* Silver birch *(Betula verrucosa). 6* Hazel *(Corylus avellana). 7* Hornbeam *(Carpinus betulus). 8* Alder *(Alnus glutinosa). 9* Common or pendunculate oak *(Quercus robor). 10* Fluttering elm *(Ulmus laevis). 11* Small-leaved lime *(Tilia cordata). 12* Ash *(Fraxinum excelsior). 13* Beech *(Fagus sylvatica). 14* Deer-grass *(Trichophorum caespitosum = Scirpus caespitosus). 15* Rye *(Secale cereale). 16* Heath *(Erica* spp.). *17* Rockrose *(Helianthemum* spp.). *18* Water pepper *(Polygonum hydropiper). 19* Red campion *(Melandrium silvestre = M. rubrum = M. dioicum = Silene dioicum). 20* Marsh Hawk's-beard *(Crepis paludosa). 21* The club moss *Selaginella* or "resurrection plant" *(a* ventral view, *b* lateral view). *22* Stag's-horn moss *(Lycopodium clavatum). 23* Common polypody *(Polypodium vulgare). 24* Marsh buckler fern *(Thelypteris palustris = Polystichum thelypteris).*

form of levers, clamps and ejection devices, and the production of a copious oily pollen cement, so that the pollen remains clumped together and adheres efficiently to the pollinator. Spines and ridges on the pollen grain surface also facilitate attachment to the hairs of a pollinating insect. Some flowers or inflorescences form traps into which insects are attracted and are unable to leave until cross-pollination has been effected. Thus, in various species of *Arum* the basal part of the large bract or spathe forms a chamber with a constricted mouth. The unisexual flowers are borne on monoecious inflorescences on the lower part of the spadix inside the chamber: female flowers at the bottom, followed by a ring of elongated, sterile, bristle-like flowers, then a zone of male flowers, followed by an uppermost ring of sterile hairs. On the first morning after opening, the spathe generates heat and emits a carrion-like odor, which attracts beetles and carrion flies, some of them already laden with pollen from other *Arum* inflorescences. Due to the smooth, oily nature of the epidermal surface, insects are unable to cling to the spadix or inner surface of the spathe, and they fall into the chamber. Escape is not possible, because the upper part of the chamber wall is also slippery, and in any case the 2 rings of bristles bar the exit. Pollen brought in by the restless insect is dusted onto the stigmata. When P. has been achieved the stigmata wither, and during the following night the male flowers scatter pollen over the insect(s) in the chamber. The epidermal cells of the bristles and spadix then wither, allowing the pollen-laden insect(s) to depart, and to serve the purpose of P. by falling once again into a similar trap. The spadix of the successfully cross-pollinated inflorescence no longer emits a carrion-like odor.

P. by birds is carried out mainly by Hummingbirds *(Trochilidae)*, Sunbirds *(Nectariniidae)* and Honeyeaters *(Meliphagidae)*. The main requirements on the part of the flower are pure and brightly contrasting, garish colors, including pure blue, yellow and even green, and an especially striking red. In addition, nectar is produced in large quantities, and it is often of thin consistency and mucilaginous, representing an important source of water for the pollinating bird as well as a food. A sticky pollen is also necessary, which adheres to the bill or other parts of the body. Ornithogamous flowers are almost always without scent.

Water fertilization ***(hydrogamy)*** is rare, and it is encountered only in certain submerged water plants, e.g. Eel grass *(Zostera marina)* and in the Horn-wort *(Ceratophyllum)*, which release their pollen into the water, where it is carried to the stigmata. In *Vallisneria* and *Elodea* (Canadian pondweed), the female flowers lie on the surface of the water; the submerged male flowers detach and float to the surface where they reach the stigmata.

Polyanthes: see *Agavaceae.*

Polyanthus: see *Primulaceae.*

Polyborinae: see *Falconiformes.*

Polyborus: see *Falconiformes.*

Polycarpiaceae: see *Dicotyledoneae.*

Polychaeta, ***polychaetes, bristleworms:*** a class of mostly marine annelids with distinct metameric segmentation. Bristles are borne on the parapodia of each body segment.

Morphology. The elongated, somewhat flattened body varies in length from a few millimeters to 3 meters, and it consists of up to 700 segments. The first 2–3 segments are united to form the head, which carries an anterior lobe extending over the mouth, several pairs of antennae, and sometimes a crown of long feathery tentacles (in sessile, filter-feeding species). In predaceous species, the mouth is provided with powerful jaws, which in some species are carried on an eversible pharynx. All body segments have a similar and characteristic internal anatomy. Externally, each body segment carries a pair of parapodia, which display a range of different forms, depending on the species and the body region. Often the parapodia are divided into an upper and a lower member, each carrying numerous bristles, the upper also with a gill appendage. Parapodia are often regressed on the hindbody of sessile or tube-dwelling forms. Posteriorly, the body terminates with a modified segment, the pygidium, which carries the anus and anal feelers, but no parapodia. *P.* are variously colored, sometimes brilliantly. The vascular system is closed. The brain (suprapharyngeal ganglion) is connected by two nerves to a subpharyngeal ganglion. A double ventral nerve cord from the subpharyngeal ganglion travels the length of the body, expanding in each segment into a double segmental ganglion, which sends nerves to all parts of the segment.

Light-sensitive organs are located on the head, but also on the sides of the body. Most species are dioecious. Testes or ovaries are paired, and they are usually present in the posterior body segments, rarely in all body segments. Fertilization is external, after the eggs have been released into the water.

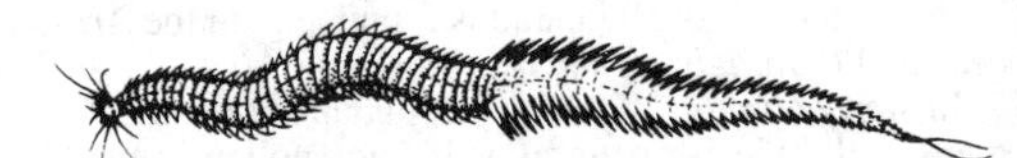

Fig.1. *Polychaeta. Neiris* sp., a common, free-living, marine polychaete.

Biology. Most *P.* are marine, only a few tropical species occurring in fresh water, and a small minority is terrestrial. There are free-swimming forms, as well as species that creep on the bottom with the aid of their parapodia. Other species remain in self-excavated passages in the mud of the sea bottom, and some live inside a secreted tube that is firmly cemented to the substratum. Feeding habits are various. Predatory species feed on small marine animals, others on plant parts, microorganisms and detritus. Many sessile species generate a water current with the aid of their tentacles, from which they sieve suspended food particles. In favorable sites, many tube-dwelling *P.* build their tubes on top of one another, contributing to the formation of marine reefs. In the reproductive period, some species, e.g. the Palolo worm (see *Errantia),* lose the posterior body section containing the mature gonads, and these sections form large swarms on the ocean surface.

Development. Development usually proceeds via a free-swimming larval stage, known as a Trochophore (see). The structure of the trochophore larva displays family differences.

Economic importance. As members of the marine food chain, some P. are eaten by fishes, and are therefore indirectly important in human nutrition. The Palolo worm (see *Errantia)* is a delicacy in the South Pacific.

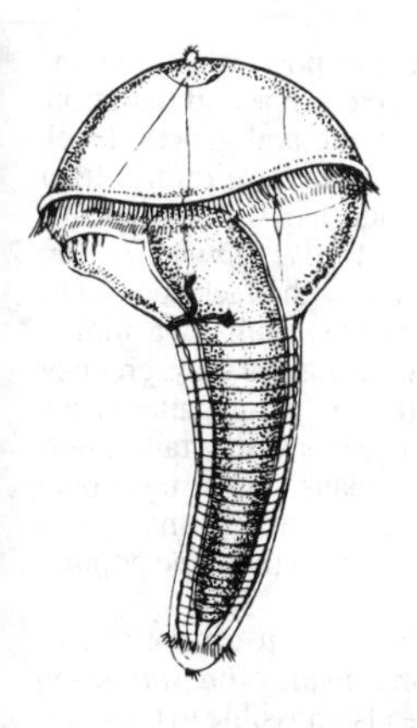

Fig.2. *Polychaeta.* Trochophore larva of a polychaete in the early stages of segmentation.

Some species are used as fishing bait, e.g. the Lugworm (see *Sedentaria).* Some Serpulids (see *Sedentaria)* have nuisance value, on account of their calcareous tubular deposits on ships, sluice gates, marine industrial pipelines, etc.

Classification. The taxonomy of the *P.* is not finalized. About 13 000 species are known. They are the only annelids with a fossil record, making their first appearance in the Cambrian. Recent phylogenetic studies suggest the existence of 17 orders with 79 families. The most familiar system, however, contains 3 orders or subclasses: *Archiannelida* (see), *Errantia* (see), *Sedentaria* (see).

Collection and preservation are as for the *Annelida* (see).

Polychaetes: see *Polychaeta.*

Polycladida: see *Turbellaria.*

Polycystic hydatid disease: see Dog tapeworm.

Polydemic: dwelling in several regions.

Polydesdae: see *Diplopoda.*

Polydesmidae: see *Diplopoda.*

Polydesmus: see *Diplopoda.*

Polyene antibiotics: antibiotics containing conjugated, unsaturated carbon chains. They are classified according to the number of conjugated double bonds. Thus, fungicidin and fumagillin contain 4 conjugated double bonds and are known as *tetraene antibiotics.* Eurocidin, with 5 double bonds, is a *pentaene antibiotic.* Similarly, *hexaene* and *heptaene antibiotics* contain 6 and 7 conjugated double bonds, respectively. About 100 P.a. are known, most of them from actinomycetes. They are particularly active against fungi.

Polyenergid: see Energid.

Polygene: see Oligogene.

Polygeny: regulation of the expression of a phenotypic character by several nonallelic genes. In *complementary P. (cryptomery)* the character is only expressed, if at least one dominant allele is present for each of the participating genes. In *additive P. (polymery)* several allelic pairs influence the same character, and their actions are cumulative. In *homomery,* each allele has the same quantitative effect. In *heteromery,* each has a different qualitative effect.

Polygonaceae, *buckwheat family:* a family of dicotyledonous plants, containing about 800 species, distributed throughout the world, but mainly in the temperate regions of the Northern Hemisphere. Members are usually herbs, rarely bushes, with a jointed stem, alternate leaves and stipules which sheath the stem (ochreae). Flowers are unisexual or hermaphrodite, usually small and aggregated in large numbers in an inflorescence. Wind-, insect- and self-pollination all occur, depending on the species. The superior ovary develops into a solitary nut. Chemically, the family is characterized by the presence of various polyphenols and anthraquinone glycosides, as well as high levels of calcium oxalate.

Polygonaceae. a Sorrel *(Rumex acetosa). b* Bistort or snake root *(Polygonum bistorta). c* Buckwheat *(Fagopyrum asculentum).*

Several species of the extensive genus, *Polygonum* ***(Knotweeds),*** are cultivated as ornamental plants, e.g. ***Japanese knotweed*** *(Polygonum cuspidatum)* which attains a height of 2–3 m. Many species of knotweed are ruderal plants or agrarian weeds, e.g. ***Red shank***, ***Willow weed*** or ***Persicaria*** *(Polygonum persicaria),* and ***Pale persicaria*** *(Polygonum lapathifolium).* ***Snake-root***, ***Easter-ledges*** or ***Bistort*** *(Polygonum bistorta)* is a common meadow plant with a striking pink, cylindrical inflorescence. Many agrarian weeds and field and wayside plants also belong to the genus *Rumex* ***(Docks*** or ***Sorrels).*** The acid-tasting leaves of *Rumex acetosa* ***(Garden sorrel)*** are eaten in salad in some areas. Various species of the native Central Asian genus, *Rheum* ***(Rhubarb),*** are used as vegetable, pharmaceutical or decorative plants. ***Common buckwheat*** *(Facopyrum esculentum),* which originates in East Asia, is now grown in sandy soils in other parts of the world. It is used as a cereal substitute, and is also a useful bee food plant. The whole plant is used as green fodder, but it is unsuitable for light colored animals, because it contains fagopyrin and other compounds, which lead to photosensitization of the skin.

Polyhedrose: see Insect viruses.

Polyhedrosis virus: see Insect viruses.

Polymastigina: an order of medium-sized to large flagellates with two to many (usually 4) flagella, an axostyle and a parabasal body (Fig.) (see Mastigont). Certain large, ciliate-like species possess numerous flagella arranged in distinct sets. Most known P. live symbiotically in the hindgut of termites, where they digest cellulose and enable their hosts to exploit dietary wood. Other P. are preferential intestinal parasites of arthropods and vertebrates, where they feed on bacteria and organic particles (e.g. starch grains).

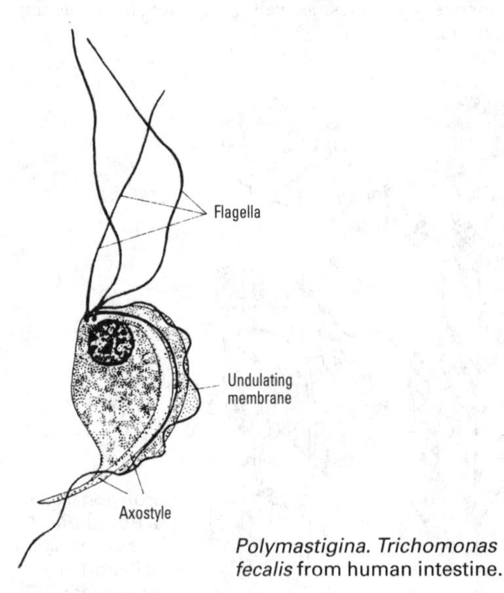

Polymastigina. Trichomonas fecalis from human intestine.

Polymery: see Polygeny.

Polymictic lake: see Lake.

Polymorphism: the simultaneous appearance of two or more genetically determined, discontinuous forms of a single species in the same habitat, in numerical proportions that exclude the possibility that the least frequent forms do not arise by repeated mutation. P. can occur at both the phenotypic and genetic level. The shells of grove snails *(Cepaea nemoralis)* are either yellow, pink or brown, with a pattern of 0–5 dark bands. The wing covers of ladybirds *(Adalia bipunctata)* are either black with red spots, or red with black spots. The male dragonfly, *Enallagma cyathigerum,* is a monomorphic blue, whereas the female may be gray, green or reddish in color. Several different wing patterns are known in the female of the African swallowtail butterfly, *Papilio machaon.* Some humans cannot taste phenylthiourea, whereas others find it bitter. Humans with different blood groups live together in the same population.

By gel electrophoresis of proteins it can be shown that practically all populations display *Enzyme polymorphism* (see). Variations are also possible in the number of nuclear chromosomes in individuals of a population, and there may also be structural differences between homologous chromosomes. Many populations of various species of the fruit fly, *Drosophila,* display *inversion polymorphism* (see Inversion). The relative frequencies of inversions in different populations of the North American species, *Drosophila pseudoobscura,* show a regular pattern of fluctuation during the course of the year. These fluctuations are superimposed on longer term alterations of frequency.

If polymorphic characters influence Fitness (see), then they are responsible for a Genetic burden (see). Selection against suboptimal genotypes (which would cause polymorphic populations to become monomorphic) is prevented mainly the selective advantage of heterozygotes, and occasionally by Frequency-dependent selection (see).

Genuine genetically determined P. must be clearly differentiated from *polyphenism* (see Modification), which is the result of environmental influences (e.g. social polyphenism in colonial insects). In bisexual species, the sexual dimorphism between male and female is a special case of P.

P. enables the individuals of a population to exploit different environmental resources. Polymorphic populations are more adaptable than monomorphic populations; they can meet the demands of a changed environment by selection, leading to a shift in the proportion of environmentally relevant variants; alternatively, inherited characters may always be potentially present, but expressed as a function of environmental variables, such as temperature, light, trophic conditions, etc.

Polymyxins: heteromeric, homodetic, cyclic, branched peptides, produced by *Bacillus polymyxa,* and possessing antibiotic activity against Gram-negative bacteria. They are used to treat infections of the intestinal and urinary tracts, sepsis and endocarditis. Various P., designated A, B_1, B_2, C, D, E and M are known. For example, polymyxin B_1 is a cyclic decapeptide containing a ring of 7 amino acids with a tripeptide side chain.

Polyodontidae: see *Chondrostei.*

Polyp: one of the 2 morphological forms of the *Cnidaria* (see). Ps. are nearly always sessile and attached to a substratum. They have a tubular or cylindrical body with a basal disk, a trunk (syn. column, stalk), and an oral disk (or peristome, which is raised into a mound, known as a hypostome). The internal gastrovascular space may be uniform or divided by sepeta into gastral pouches. Tubular tentacles, armed with batteries of nematocysts, arise between the stalk and the oral disk. The body can be stretched or contracted, and usually bent or arched over in all directions; the tentacles are especially mobile.

Ps. are either solitary, or they form colonies by budding. Within a colony, the Ps. usually exhibit polymorphism, i.e. they are differentiated into gastrozooids (feeding Ps.), dactylozooids (defensive Ps.) and blastozooids (which bud off medusae). Many are predatory, capturing prey with their tentacles and pushing it into the gastrovascular space where it is digested. Others feed on suspended organic particles.

Ps. arise from fertilized eggs via a Planula (see). They may themselves possess gametes (e.g. *Hydra* and members of the *Anthozoa).* Alternatively, by asexual budding or cleavage, they may produce free-swimming sexual medusae. The latter case represents an Alternation of generations (see) (see also Metagenesis).

Collection and preservation. Small Ps. are collected with the material to which they are attached, then picked off under a binocular lens. Larger species are preferably collected by hand with the aid of a mask and snorkel. They are kept in water (fresh or saline depending on their origin), and narcotized by the dropwise addition of 10% magnesium chloride or sulfate. Formaldehyde (4%) or ethanol (80%) are suitable for preservation.

Polyphaga: see *Coleoptera.*

Polyphemus: see *Branchiopoda.*

Polyphenism: see Modification.

Polypide: see *Bryozoa.*

Polyplacophora, *Placophora, Loricata,* ***chitons, coat of mail shells:*** a subclass of the *Amphineura* (see), containing about 1 000 species. Chitons are up to 33 cm in length, with a broad creeping foot which occupies the entire ventral surface of the body. All possess a radula. Dorsally, the head and body are covered by the shell, which consists of 8 (rarely 7) overlapping plates, secreted by the dorsal surface of the mantle. Neighboring plates articulate and can be moved relative to one another. Each plate is calcareous and composed of 2 layers, the upper tegmentum (covered by a thin periostracum) and the lower articulamentum, which is underlain by the innermost hypostracum; the articulamentum serves to articulate neigboring plates via apophyses. The tegmentum is penetrated by parallel canals containing ectodermal tissue, which emerge on the surface as sense organs, many of which are typical esthetes. The mantle with its embedded shell forms a rooflike dome over the visceral mass, then extends beyond the shell forming a fold of tissue known as the *perinotum* or *girdle.* Between the foot and the girdle is the mantle cavity or groove, which extends around the entire body. Within this groove on each side of the body is a row of ctenidia (gills).

Development proceeds by metamorphosis via a trocophore larva. Some species display brood care. One species is known to be viviparous. Chitons move sluggishly while grazing algae on rocks and stones in the intertidal and surf zone of the sea. In the surf zone they remain firmly attached to the substrate. When removed from a rock surface, they roll into a ball like an isopod, using special musculature in the girdle; this presumably serves as a protection against desiccation and the effects of rainwater during the ebb tide.

Three orders are recognized: *Lepidopleurida, Ischnochitonida, Acanthochitonida.*

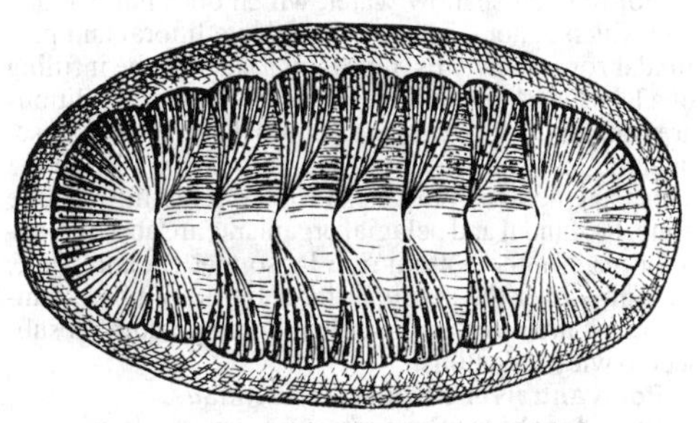

Polyplacophora. Shell of a chiton.

Polypody: the occurrence of several pairs of limbs on one body segment, e.g. in *Notostraca (Branchiopoda)* a hind segment may carry up to 6 pairs of legs.

Polyporus: see *Basidiomycetes (Poriales).*

Polyporaceae: see *Basidiomycetes (Poriales).*

Polyprion: see Sea basses.

Polysaprophytes: see Saprophytic system.

Polysomy: a form of aneuploidy, in which one or more (but not all) chromosomes appear more than twice in otherwise diploid cells, tissues or organisms. If a particular chromosome is present in three or four copies, the condition is known as tri- or tetrasomy. If two different chromosomes are each present in three copies, the condition is *double trisomy.*

The presence of more than two identical chromosomes and the accompanying increase in the number of the corresponding alleles in the cell nucleus results in *polysomal inheritance,* which is characterized by an altered genetic segregation (see Mendelian segregation). See Disomy, Monosomy, Nullisomy, Tetrasomy, Trisomy.

Polyspermy: penetration of an egg by several sperms. P. occasionally occurs under unusual conditions, e.g. when sea urchin eggs are over-ripe, or when the supply of spermatozoa is excessive. P. results in defective larvae. *Physiological P.* has been observed in bryozoans, insects, certain spiders, elasmobranchs, amphibians, reptiles and birds. In all cases, however, only one sperm nucleus fuses with the egg nucleus. The remaining spermatozoa disintegrate or behave as mesocytes in yolk resorption.

Polystomella: see *Foraminiferida.*

Polysymptomatic similarity analysis: see Forensic anthropology, Zygosity analysis.

Polytene chromosomes: see Giant chromosomes.

Polythalamia: see *Foraminiferida.*

Polytrichales: see *Musci.*

Polytrichous flagella: see Bacterial flagella.

Polytrichum: see *Musci.*

Polyxenus: see *Diplopoda.*

Polyzoa: see *Bryozoa.*

Polyzoarium: see *Bryozoa.*

Polyzonium: see *Diplopoda.*

Pomacentridae: see Damselfishes.

Pomarea: see *Muscicapinae.*

Pomatoschistus: see *Gobiidae.*

Pome: see Fruit, *Rosaceae.*
Pomegranate: see *Punicaceae.*
Pomoxis: see Sunfishes.
Pompilidae: see *Hymenoptera.*
Pond: open shallow water, which does not dry out, and which is not deep enough to have littoral and profundal zones. Many Ps. represent a stage in the infilling of a lake (see Aging of lakes). The biological conditions are similar to those in the littoral lake zone. Consequently, most of the plants and animals in a P. are the same stagnicolous organisms found in the littoral lake zone. Profundal and pelagial organisms are absent. Sunlight reaches the bottom of a P., so that aquatic plants can grow over the entire surface. In the summer, temperatures are often high, and daily temperature is subject to wide variations.
Pond and river turtles: see *Emydidae.*
Pond loach: see Respiratory organs.
Pond skaters: see *Heteroptera.*
Ponerinae: see Ants.
Pongidae: see Great apes, Old World monkeys.
Pongid theory: a theory based on the assumption that the evolutionary line leading to humans first became phylogenetically independent after diverging from the pongids.
Pons (varoli): see Brain.
Pontic:
1) A geographical term for the region of the Black Sea.
2) A phytogeographical term for the floral region on the north side of the Black Sea, together with its floral elements, which extend into the xerothermal grasslands of the dry areas of Central Europe. See Holarctic.
Pontoporia: see Whales
POP: acronym of progestin only pill. See Hormonal contraception.
Pope: see Perches, Ruffe.
Pope and flounder region: see Running water.
Poplar borer: see Long-horned beetles.
Poplars: see *Salicaceae.*
Poppies: see *Papaveraceae.*
Population: the totality of individuals of one species in a region that is more or less isolated from other regions containing the same species. For example, a P. is formed by black rats *(Rattus rattus)* on a small island, and by red deer in an isolated forest. Of particular interest are *Mendel Ps.,* whose diploid individuals reproduce sexually, with unrestricted (random) pairing. In each generation the genes in a Mendel P. form new combinations. These genes therefore belong to a common gene pool, more or less irrespective of their distribution among individuals at any one time. In Mendel P., the alleles of many genes are distributed according to the Hardy-Weinberg rule. The different gene loci are either in coupling equilibrium or coupling disequilibrium.

The P. is the evolutionary unit. Evolutionary processes, like selection, recombination and genetic drift, cannot occur in isolated individuals, and they are only possible within Ps. Also, different Ps. of the same species can evolve in different or totally opposite directions.

Gene flow (see) may occur between Ps. that are not totally isolated from each other. Ps. of organisms that multiply exclusively asexually do not behave like Mendel Ps.; their evolutionary potential is presumably restricted.

Population control: see Population planning, Family planning.
Population density: the number of humans per square kilometer, living in a particular region. See Demography.
Population dynamics: variations in the size and composition of a population, due to births, deaths, immigration and emigration. A distinction is drawn between *natural P.d.* (births and deaths) and *spatial P.d.* caused by migrations. The positive effect of the birth rate on population size (or rate of population increase) is known as *natality,* whereas the negative effect of the death rate is called the *mortality. Dispersion* may also be negative (emigration) or positive (immigration).

Natural P.d. pass through different phases during the historical development of a population. During the *first phase,* population numbers become adapted to the food supply, primarily by the influence of the death rate. The death rate among infants and children is very high, and the average life expectancy is very short. Thus, although the birth rate may be high, the population increase is only very small. The *second phase* is characterized by a marked increase in the population size. This is primarily due to a decrease in mortality, especially of infants and children, resulting from advances in medicine and hygiene. The birth rate changes only slightly, if at all; it may even be somewhat increased, because the population contains a greater number of individuals of reproductive age. The result is therefore a widening difference between the birth rate and the death rate, as the former increases slowly, and the latter decreases rapidly. In the *third phase,* family planning leads to a fall in the birth rate, which eventually equals the death rate. The population then grows only slowly, or even shows a decrease. This sequence of 3 phases represents a simplified analysis of population development. There are numerous other factors that modify and influence P.d., and not all populations of the world are in the same stage of development.

Population ecology, ***demecology:*** study of the relationship between homotypic organism collectives (populations in the narrow sense) and their environment. P.e. and population genetics form the basis of general population theory.

Important characteristics of a population are: population density (see Abundance), the distribution of individuals within the populated space (see Dispersion), the sexual index (see Sex ratio), reproductive sexility, fertility, natality, morbidity and mortality.

The growth of a population depends on its specific reproduction rate, and at first it proceeds exponentially (Fig.). Indefinite unrestricted exponential growth is, however, not possible, because the environmental resources for each species are limited. Therefore, as the population size approaches the capacity limit of the environment, there is a large increase in the negative effects of density-dependent factors (competition, interference). The result is a sigmoid growth curve (Fig.).

In its competition with other populations for limited environmental resources, a population may adopt different strategies (see r- and K-strategies).

Relationships between populations of predators and prey (or parasite and host) are governed by Volterra's laws (see) of population kinetics. Extreme changes in

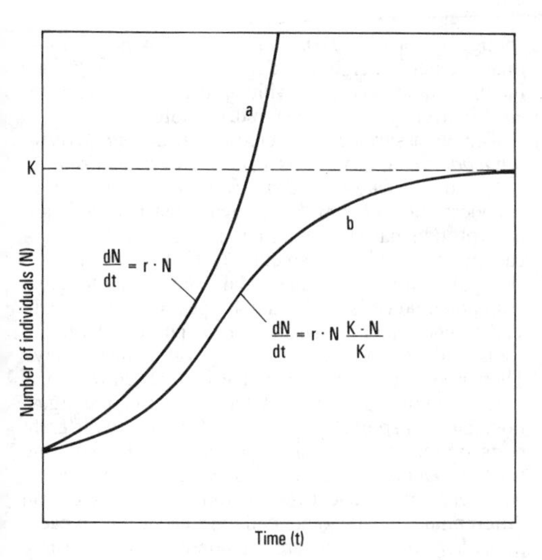

Population ecology. Growth curve of a population. *a* Exponential growth, dependent only on the reproduction rate (r). *b* Growth limited by the environmental capacity (K).

abundance, e.g. a *gradation* (a mass outbreak of a pest organism, including its period of increase, peak of abundance, and decrease) can be explained by the specific relationship between the rate of increase and the environmental capacity.

Study of the biological and social basis of the growth of human populations is becoming increasingly important. The world population is still in the exponential (logarithmic) growth phase. The maximal number of humans that the earth can support is thought to be 15 billion (15 thousand million or 15 x 10^9). At the present growth rate, this limiting capacity would be reached in about 120 years.

Population fitness: the arithmetic mean of the individual fitness of all members of a population (see Fitness). P.f. depends on, and changes with, the relative frequency of coexistent genotypes. Natural selection increases P.f., which therefore increases with the frequency of the optimal genotype within the population. If the optimal genotypes of two populations are different, then the respective P.f. values do not indicate which population is the fitter.

All criteria for the absolute evaluation of P.f. (e.g. population size under different environmental conditions, or the genetically determined rate of population increase) are more or less inadequate.

Population genetics: study of the distribution and changes in frequency of genes in populations. P.g. deals primarily with natural populations that are subject to natural selection. At the same time, some of the findings of P.g. are also important in the artifical breeding and selection of both animals and plants.

Since all evolutionary processes occur in populations, P.g. is central to evolutionary theory. Problems of microevolution are easily analysed by the methods of P.g., but only in exceptional cases have questions of macroevolution proved amenable to this approach.

In 1908, Hardy and Weinberg independently established an important law of P.g. (see Hardy-Weinberg rule). In 1930, P.g. developed into an independent branch of biology. Its founders are S.S. Cetverikov, R.A. Fisher, J.B.S. Haldane and S. Wright. The latter three authors laid the foundations of a developed system of mathematical models for the investigation of problems of P.g. The predictions of these models can often be verified or disproved on natural or experimental populations. Nevertheless, it is always difficult to determine the extent to which these greatly simplified models, often using imprecise quantities, truly correspond to reality (see Genetic burden).

Population growth: the increase in the number of individuals in a population. In the absence of any restraints, the number of individuals in a young growing population shows an initial rapid increase. The actual rate of this exponential increase is genetically determined. Since there is a limit to the number of individuals that can be supported or accommodated by a given habitat, the rate of population increase slows down as the population size approaches this limiting carrying capacity *(K)* (Fig.).

With unrestricted growth, a population of size N_0 would require $t = 1/r \times \ln (N_1/N_0)$ years to attain a size of N_1, where r is the annual rate of increase. If growth is restricted, then $t = 1/r \times \ln [N_1(K-N_0)/N_0(K-N_1)]$. For example, if the genetically determined annual rate of increase is 1% $(r = 0.01)$ and the carrying capacity (K) is 5 000, then the population would require $t = 1/0.01 \times \ln (2\,000 \times 4\,000/ 1\,000 \times 3\,000) = 98$ years, in order to increase from a size of $N_0 = 1000$ to $N_1 = 2000$. If growth is unrestricted, only 69 years are required for the same increase. The number of individuals in the population at any given time is $N_1 = K/(1 + C_0^{-rt})$, where $C_0 = (K-N_0)/N_0$.

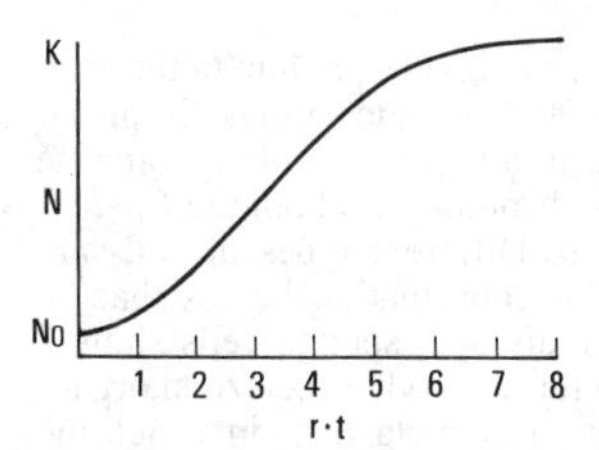

Population growth. Idealized curve of population growth. The growth curves of natural populations deviate to varying extents from this theoretical curve.

Population planning, ***population control:*** birth control in relation to economic development. Certain socioeconomic factors are important in assessing the desirability and practicality of any P.c. program. These include the level of education of the population, the social status of women, the degree of Urbanization (see), the living standard, and the ensuing predicted increase in output per head of the population. See Family planning.

Population structure:

1) The number and type of subdivisions of a species within a population. The existence of a species in only one or a few large panmictic populations imposes a restriction on its evolutionary potential. Advantageous genetic changes spread only slowly through the population, which is in equilibrium. Large populations do not attain new adaptive peaks (see Adaptive landscape).

According to Wright, the greatest evolutionary potential exists among relatively small and partly isolated populations. Favorable gene combinations can arise by drift and selection in the smaller populations. Some of these populations then attain new adaptive peaks, and the superior genotypes spread through the other populations.

Mayr claims that, in order to evolve successfully in a new direction, a population must be completely isolated (see Genetic revolution). According to this concept, the evolutionary potential of a species is increased to a maximum when it is divided among many completely isolated small populations.

2) Subdivision within a population of a single habitat. Whereas an entire habitat may be occupied by a single large panmictic population of a species, other species become divided into more or less independent reproductive units. Panmixy does not even exist among the mice of a barn.

3) See Balance model of population structure.

4) See Classical model of population structure.

Populus: see *Salicaceae.*

P/O ratio: see Respiratory chain, Oxidative phosphorylation.

Porbeagle: see *Selachimorpha.*

Porcelain shell: see *Gastropoda.*

Porcellionidae: a family of the *Isopoda* (see).

Porcupinefishes: see *Tetraodontiformes.*

Porcupine huernia: see *Asclepliadaceae.*

Porcupines: see *Hystricidae, Erethizontidae.*

Pore plates: see Olfactory organ.

Porgies: see *Sparidae.*

Poriales: see *Basidiomycetes.*

Porifasterol, ***(24S)-5α-stigmasta-5,22-dien-3β-ol:*** a marine zoosterol, which is a characteristic sterol of sponges. It has been isolated from, e.g. *Haliclona variabilis, Cliona celata* and *Spongia lacustris.* It differs from Stigmasterol (see) by an altered configuation at C24.

Porifera, ***Spongiaria, sponges:*** a phylum of the *Metazoa,* containing about 5 000 aquatic species. Some authorities place sponges in a separate subkingdom, the *Parazoa.* They display affinities with both the *Coelenterata* and the *Protozoa.* Different types of cells are present, but intercellular coordination is less than in other metazoans, and individual sponge cells show a greater autonomy than those of other metazoan organisms. Also, they are the only metazoans in which the main opening to the exterior (the *osculum)* is an exit, rather than an entrance. The *choanocytes* of sponges are found in no other metazoans (although they closely resemble choanoflagellate protozoa). It is obviously a successful phylum, but since no other higher group of animals is known to have evolved from the sponges, they appear to lie on a lateral branch of the evolutionary tree.

Morphology. All sponges are sessile, and they display a wide variety of internal and external structure. Some forms grow as incrustations or as rounded clumps, especially where water currents are strong. Many species have a branched, reticulate, plantlike appearance, especially in calm water. Due to the effects of water currents, the nature and inclination of the substrate, and the availablility of nutrients (some sponges thrive in sewage-polluted water), the same species may have a different appearance and size under different conditions. Other species are tubular or funnel-shaped. Heights and diameters vary from a few millimeters to 2 m. Certain members of the *Spheciospongia* (loggerhead sponges of the class *Demospongiae)* form masses more than a meter in height and diameter. Many sponges are inconspicuously colored whitish to gray, whereas others are bright yellow, green, red, or violet.

The outer surface is covered by the *dermal layer* or *pinacoderm,* which consists of an external mosaic of flattened *covering cells* or *pinacocytes.* Below the pinacoderm lies the *mesohyl,* which consists of a gelatinous protein matrix, containing a supporting skeleton of calcareous or siliceous spicules (sclerites), or fibers of spongin (a fibrous protein similar to collagen), together with ameboid cells. There are 4 types of ameboid cell: 1) *Archeocytes* or *mesenchyme cells* posses a large nucleus, and they move around freely within the mesohyl. They are phagocytic and take partly digested food particles from the collar cells. After completing the digestion, they transport the products of digestion to different parts of the sponge wall, probably also transporting waste products to sites where they can be removed by water currents. Archeocytes are totipotent, i.e. they can differentiate into any other cell type, according to need. 2) *Collencytes,* which are anchored by cytoplasmic strands, secrete dispersed collagen fibers. 3) *Lophocytes* are present in many sponges; they also secrete collagen fibers, but they are not anchored. 4) *Sclerocytes* or *spongocytes* secrete the spicules of calcium carbonate or silicic acid, or the fibers of spongin. Below the mesohyl is the *gastral layer* or *choanocyte layer,* consisting of a layer of specialized cells known as *choanocytes* or *collar cells,* which remove dietary particles from the water that passes over them. The free end of each collar cell carries a long flagellum, and is encircled by a collar of protoplasm. The water current through the sponge is generated by the beating of the choanocyte flagella. The combination of mesohyl and pinacoderm is also known as the *dermal layer.* The body wall (dermal plus gastral layers) encloses an interior cavity, known as the *atrium* or *spongocoel,* which opens to the exterior via a large opening (the *osculum)* at the top of the sponge.

According to the design of the channels that perforate the sponge wall, sponges are classified into three types (see Fig. in the entry, Digestive system). 1) *Ascon type.* This simple type is tubular in shape, and always small. The gastral layer forms a flat lining to the spongocoel, which is in direct contact with the exterior through simple perforations. 2) *Sycon type.* In this more advanced form, the body wall is folded, so that the choanocyte

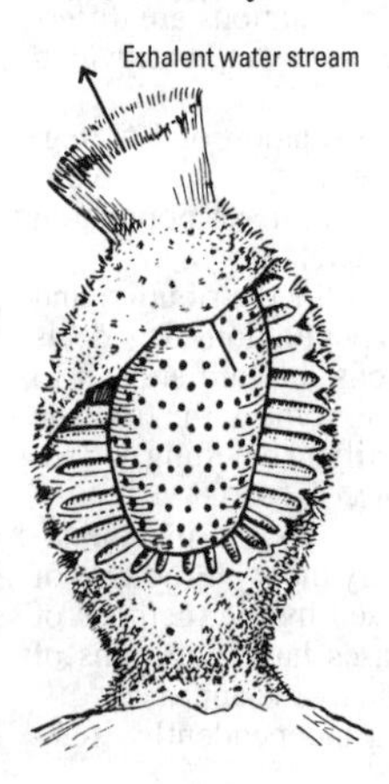

Porifera. A calcareous sponge (genus *Sycon)* with body wall partly removed.

layer forms the lining of several *flagellated chambers* that extend radially from the spongocoel. 3) *Leucon type.* The gastral layer forms the lining of vesicular structures *(flagellated chambers)* within the dermal layer, which open to the exterior via inhalation channels (incurrent pores or dermal pores) and to the spongocoel via exhalation channels (excurrent pores).

Biology. The majority of sponges are marine, inhabiting mainly coastal regions down to depths of 50 m. They also occur at greater depths, and some species have been found below 5 000 m. Some marine species also penetrate brackish estuarine waters, and about 150 species (notably of the family *Spongillidae)* are found in freshwater habitats, even in the high lakes of extinct volcanic craters. Sponges feed on unicellular organisms and detritus, which are filtered from the water current that passes through the dermal pores directly into the spongocoel (ascon type) or via the flagellated chambers (sycon and leucon types). The exhalant water then passes out from the spongocoel through the excurrent opening or osculum. Sponges reproduce both sexually and asexually. In sexual reproduction, spermatozoa are derived by the differentiation of choanocytes, and the eggs are formed from archeocytes. Dioecious species are known, but most species are hermaphrodite, although their eggs and spermatozoa are produced at different times. Most species possess a *parenchymula larva,* in which the posterior pole is often nonflagellated, but the larval surface is otherwise entirely covered with flagellated cells. The parenchymula larva leaves the mesohyl of the parent, then swims for no more than two days (an aid to dispersal of the species) before becoming attached and developing into a new sponge. Other species (e.g. some calcareous sponges, such as *Grantia* and *Sycon,* as well as *Oscarella* of the *Demospongiae)* have a hollow, blastula-like larva, known as an *amphiblastula,* in which one hemisphere consists of small flagellated cells, the other of large nonflagellated macromeres. The amphiblastula attaches immediately to a substratum and develops into a new sponge. Asexual reproduction occurs by budding. Buds become detached and develop into new sponges. The ability to grow and spread indefinitely over the surface of the substratum, and the readiness with which a complete sponge develops from practically any fragment, must also be seen as a form of asexual reproduction. Many sponges also form asexual reproductive structures known as Gemmules (see). Sponges posses remarkable powers of regeneration; even when pieces of sponge are pressed through a fine silk screen, the separated cells can reaggregate and form a new sponge.

Economic importance. Commercial sponges are found in all tropical and subtropical waters, but their main source is the Mediterranean. They are still important for cleaning purposes, since their absorbent properties cannot be matched by any synthetic product. Most familiar is the Bath sponge (see) *(Spongia officinalis).*

Geological distribution and importance. Fossils of siliceous sponges are found from the Cambrian to the present, whereas calcareous sponges appear first in the Devonian. Evolutionary peaks occur in the Malm, Upper Cretaceous and Lower Tertiary. Almost without exception, the only fossil forms are those with calcareous or siliceous skeletons. In certain geological strata, their skeletal elements are the only recognizable remains. However, complete and variously shaped fossil forms with resistant skeletons are also found, as well as fossils of incrusting forms on foreign bodies. In certain gelogical periods, sponges were also responsible for the formation of reefs, e.g. reefs of the south German Jurassic and of Provence in France. As biostratigraphic index fossils, sponges are important only in the Upper Jurassic and the Upper Cretaceous.

Collection and preservation. Specimens are removed from the substratum by hand, or with a suitable tool. Large specimens should be dried in a warm, shaded place. Pieces of sponge can be fixed in 75% ethanol, which is then replaced with fresh 75% ethanol for subsequent storage. Some siliceous sponges can be stored in sea water containing 4% formaldehyde.

Classification. The following four classes are recognized, based mainly on the nature of the skeleton.

1. Class: *Calcarea* or *Calcispongiae* (calcareous sponges). This class includes the orders *Homocoela* and *Heterocoela.* The supporting skeleton consists of spicules of calcium carbonate, which are separate from each other, and which may have single, double, or triple points (monaxons, diaxons, or triaxons). Most species grow to less than 10 cm in height, and ascon, sycon and leucon types are all represented. They are found mostly in shallow coastal waters in all the oceans of the world.

2. Class: *Demospongiae.* This class contains about 95% of all sponge species, many being brilliantly colored due to pigment granules in the amebocytes. All are of the leucon type. The skeleton may consist of siliceous spicules, or spongin, or both. The genus *Oscarella* is unique in lacking either a spongin or a spicule skeleton. Six-pointed spicules (hexaxons) are never observed (cf. the *Hexactinellida).* This class contains the important boring sponges (family *Clionidae),* which bore into calcarious substrata, such as mollusk shells, and make an important ecological contribution to the natural destruction and recycling of shells. The worldwide family, *Spongillidae,* contains most of the freshwater species.

3. Class: *Hexactinellida* or *Hyalospongiae* (glass sponges). The skeleton consists of siliceous, six-pointed spicules (hexaxons), or needle forms derived from them. In many species some spicules are fused into a lattice-like skeleton of long siliceous fibers, hence the name, "glass sponge". Such fused skeletons retain the shape of the living sponge, the most notable example being Venus' flowerbasket *(Euplectella).* The osculum is sometimes covered by a sieve plate of fused spicules. Most structures tend to be of the sycon type. There is no pinacoderm, and the outer surface is covered by a syncytium of interconnected amebocyte pseudopodia, through which project long siliceous spicules. These are mostly deepwater sponges (450–900 m), and some have been dredged from the abyssal zone. They are found in all the oceans of the world, and are the dominant type in the Antarctic. The class is divided into the subclasses *Hexasterophorida* and *Amphidiscophorida.*

4. Class: *Sclerospongiae.* This class comprises a small number of species that inhabit cavities and tunnels in coral reefs. They all have a leucon-type structure, and the internal skeleton consists of siliceous spicules and spongin fibers. In addition, there is an outer casing of calcium carbonate, so that the oscula are seen as raised structures on a calcareous mass.

Porites: see *Madreporaria.*

Pork tapeworm: see *Taenia.*

Porocephalida: see *Pentastomida.*

Porogamy: see Fertilization.

Porpema: see *Velella.*

Porphyrins: biologically important cyclic compounds of plants and animals, containing four pyrrole rings (see Pyrrole), linked via their α-carbons by methine groups (=CH—). The unsubstituted parent system of four rings is called porphyrin in the narrow sense (Fig.). In other P., the 8 β-hydrogens (on C-atoms 2, 3, 7, 8, 12, 13, 17 and 18) are variously substituted with alkyl residues, propionic acid groups, vinyl groups, etc. In all naturally occurring P., the two side chain substituents of each pyrrole ring are dissimilar. When only two types of side chain substituent are present there are only four possible isomers. With three different side chains the number of theoretically possible isomers increases to 15. Protoporphyrin occurs only as its IX isomer (sometimes also called protoporphyrin III because it is derived biologically from coproporphyrinogen III) P. readily form metal complexes (e.g. with iron, magnesium, copper) which often display high catalytic activity. Heme (the prosthetic group of hemoglobin) is the complex of protoporphyrin (Fig.) with Fe(II). The complex with Fe(III) (known as hemin, hematin or ferriprotoporphyrin IX) is the prosthetic group of catalase, plant peroxidases and certain cytochromes (see Enzymes). Chlorophyll (see) is a magnesium complex of a dihydroporphyrin, while Bacteriochlorophyll (see) is a tetrahydroporphyrin. Bile pigments (see) are also porphyrin derivatives.

Porphyrin Protoporphyrin

Porpita: see *Velella.*

Porpoises: see Whales.

Portal systems: see Blood circulation.

Portlandia: see *Gastropoda.*

Portuguese man-of-war, *Physalia physalis:* a colonial polyp of the order *Siphonophora.* Found in warm seas throughout the world, *Physalia* possesses a large pneumatophore, which floats on the surface of the water. This crest-like pneumatophore, which may attain a length of 20 cm (even 30 cm-specimens have been reported), acts as a sail and it is aligned so that the colony sails at 45 ° to the wind direction, some populations sailing to port, others to starboard. Fishing tentacles, sometimes as long as 50 m, hang down like a drift net, combing the water for prey (fish, copepods, etc.). *Physalia* is capable of catching and consuming a fully grown mackerel, and the toxin of its nematocysts is sufficiently virulent to endanger human life.

Portunidae: see Swimming crabs.

Portunus: see Swimming crabs.

Posidonia, *Posidonomya* (Greek *poseidon* the god of the sea): a fossil anisomyarian genus of the *Bivalvia.* The valves are thin and equal, and they have a shallow contor. In shape they are inequilateral, oblique oval with concentric ribs or folds. The hinge line is short and straight and lacks teeth. The genus is found from the Silurian to the Jurassic. *P. becherei* is an index fossil for the Lower Carboniferous (Kulm facies). *Steinmannia bronnii,* an index fossil of the Posidonian shales of the Upper Lias (Lias ε), is very similar to *Posidonia* and was formerly included in this genus.

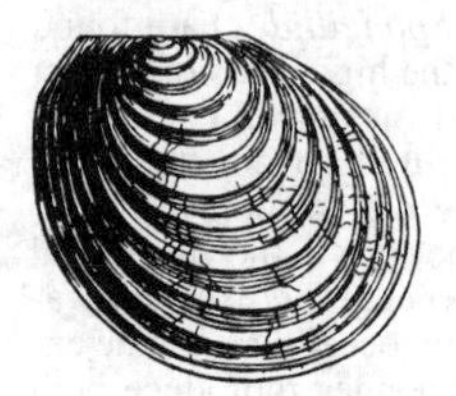

Posidonia. Drawing of *Posidonia becheri* Bronn from the Lower Carboniferous (x 0.5).

Positional effect: alteration in the mode of action of certain genes when they are placed in a new position within their linkage group or with respect to other chromosomes. Such changes of position may be caused by chromosome mutation. When the gene in question is restored to its former position, the change of activity is usually reversed. Separation of two neighboring genes may cause a P.e. in both or only one of them.

Possum: see *Phalangeridae.*

Postantennal organ, ***postantennal organ of Tömösvary, Tömösvary's organ, pseudoculus:*** a sensory organ, varying in shape from a simple ring to a rosette-like, multituberculate sensory area on the heads of springtails and proturans. It appears to function as a humidity detector, temperature sensor or olfactory organ, and is considered to be phylogenetically equivalent to the organ of Tömösvary found in other primitive *Mandibulata* (see, e.g. *Symphyla).* In soils that periodically dry out, the P.os. of the native springtails and proturans are particularly complex.

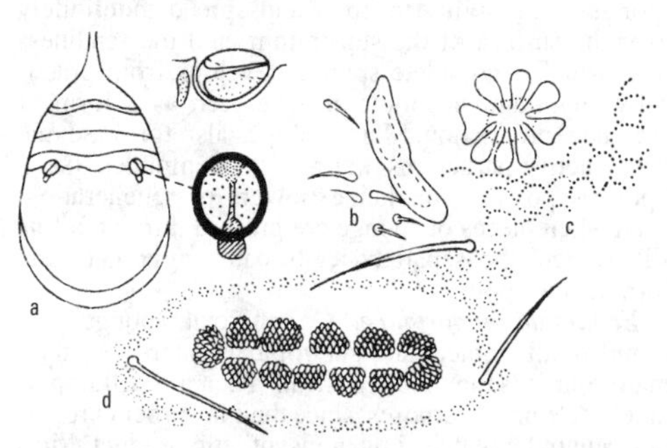

Postantennal organ. a Head of a proturan with postantennal organs; surface view and transverse section of an organ. *b-d* Postantennal organs of springtails *(b Isotomidae, c Neanuridae, d Onychiuridae).*

Postglacial: see Holocene.

Posterior pituitary: see Hypophysis.

Posthorn snail: see *Gastropoda.*

Postical leaves: see *Hepaticae.*

Postsynapse: see Neuron, Synapse, Synaptic potential.

Posture: alignment of the upright human body. P. is determined by inherited variations in skeleton and musculature, as well as lifestyle and profession. Four basic types of P. can be recognized, from the most desirable type A to the least desirable type B (Fig.). In *type A,* head, trunk and leg axes lie in a straight line, the back is slightly curved, the rib cage is high and well vaulted, and the abdomen is held in and flattened. In *type B,* the axes of the head and legs are shifted slightly forward, the axis of the trunk is slightly bent, and the head is not held as far back as in type A. *Type C* displays all the characteristics of type B in a more pronounced form. In *type D,* the leg axis slopes markedly forward and the curve of the back is extreme, so that the trunk axis is markedly bent. The chest is flat and drawn in, and the slack abdomen arches forward.

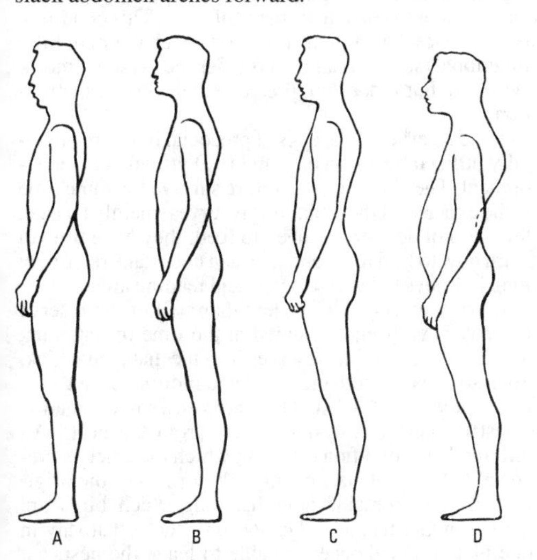

Posture. Different types of human posture, See text.

Potamic: pertaining to rivers, or transport by river currents.

Potamochoerus porcus: see African bush pig.

Potamodromous: the adjective applied to an organism that migrates only within freshwater.

Potamogalidae, ***water shrews*** or ***otter shrews:*** a family of the order *Insectivora* (see). Dental formula: 3/3, 1/1, 3/3, 3/3 = 40. The ***Giant African otter shrew*** *(Potamogale velox),* from Western and Central Africa, is the only species of its genus, and one of the largest of the insectivores (head plus body length 290–35 cm); as an adaptation for swimming, it has a laterally flattened, oar-like tail, and it feeds on crabs, fish and amphibians. The somewhat smaller ***Dwarf*** or ***Small African otter shrews*** *(Micropotamogale)* have a round tail; 2 species are known: *M. lamottei* (Guinea to the Ivory Coast) and *M. ruwenzorii* (Ruwenzori and Zaire rivers).

Potamon: the lower reaches of rivers and streams, i.e. the sand-mud zone of flowing water, which is significantly warmed by the sun.

Potamonidae: see Freshwater crabs.

Potamophyte, ***potomad:*** a river plant.

Potamoplankton: see Plankton.

Potassium, ***K:*** an alkali metal, whose cation (K^+) is an essential nutrient required in relatively large quantities by animals and plants.

K is present in various clay minerals (mica, feldspar, etc.). Black earth soils are also rich in K, whereas laterite soils have a low K content. Like all plant nutrients, the availability of K depends crucially on the nature of its binding. Adsorbed or dissolved K^+ is readily available, whereas K combined in minerals becomes available only after prolonged weathering. Uptake of K^+ by plants is strongly dependent on the moisture content of the soil. Under dry conditions, K is so firmly fixed that the application of K-fertilizers is ineffective, and is only of benefit to subsequent crops grown under wetter conditions.

Plant roots actively take up more K^+ than any other cation (see Carrier, Ion pumps). Active K^+ uptake is inhibited by oxygen deficiency in the roots, e.g. as a result of poor soil structure. After uptake, it is rapidly transported from the roots to the shoot in the transpiration stream, as well as being transported from shoot to root in the phloem. In this way, K^+ circulates in the plant several times each day via the xylem and the phloem, accumulating mainly in metabolically active young leaves and meristematic tissue. Young leaves may contain as much as 60 mg K^+/g dry mass, compared with only 20 mg/g in old leaves.

In the cell, K^+ is present partly as the free cation and partly adsorbed to macromolecules. The degree of hydration of cytoplasmic colloids is increased by K^+ and decreased by Mg^{2+} and Ca^{2+}. Vacuoles contain a high concentration of K^+, which makes a significant contribution to their osmolarity. Accordingly, K^+ strongly influences the water economy of plants, and it has the effect of sparing their water requirement. K-deficient plants tend to be limp, their water loss by transpiration is increased, and their cell turgor pressure is decreased.

K^+ also influences photosynthesis and respiration. It is an indispensable factor for photosynthesis, and there is evidence that it stabilizes chloroplasts by adorption to various sub-chloroplast structures, and that it is responsible for maintaining the optimal activity of enzymatic reactions within chloroplasts. It also influences the swelling and the structure of mitochondria, and is necessary for mitochondrial phosphorylation reactions, as well as activating the cytochrome system. Furthermore, K^+ is required for protein synthesis, and this effect is particularly marked when nitrogen is in short supply. More than 40 enzymes are known that are activated more or less specifically by K^+, including numerous kinases and dehydrogenases. Loose but selective binding of K^+ to the enzyme protein appears to be necessary for the establishment of the active protein conformation.

On intensively worked agricultural land, the uptake of large quantities of K^+ by crops necessitiates the regular use of K^+ fertilizers. In mineral fertilizers, K^+ is present as KCl (potassium chloride) or K_2SO_4 (potassium sulfate). When using KCl, it must be remembered that some crops are sensitive to high concentrations of chloride (see Chlorine). Cabbages, beetroots, celery, squash, tomatoes, carrots, fruit trees and grape vines, as well as many root crops, red clover and lucerne have particularly high requirements for K.

K deficiency symptoms appear more readily if the plant is also water deficient. There is an increased tendency to wilt, and the margins of older leaves become lighter in color, finally displaying brown necrotic areas. K-deficient plants accumulate sugars, rather than converting them into high molecular mass carbohydrates, e.g. cellulose. K-deficient cereal crops are therefore mechanically weak and fiber plants are of poor quality. In addition, only small, sessile leaves are formed. K-deficient grape vines sometimes have blue leaves, and their flowers die back.

In animals, K^+ ions are required for the activity of many enzymes, as well as being indispensable for nervous conduction. The normal daily diet of an adult human contains about 4 g K. K-deficiency leads to muscular weakness and lethargy.

Potassium-argon dating: see Dating methods.

Potato: see *Solanaceae.*

Potato virus X: see Virus groups.

Potato virus Y: see Virus groups.

Potential difference: see Electrical potential differences.

Potential microflora: see Autochthonous microflora, Soil organisms.

Potetometer: see Potometer.

Potexviruses: see Virus groups.

Potomad, ***potamophyte:*** a river plant.

Potometer: a simple apparatus for measuring transpiration in plants. The cut end of a transpiring branch is inserted through an airtight and watertight seal into a vessel, which is filled with water and carries a horizontal side arm consisting of a narrow-bore, graduated tube. The end of this side tube bends downward through 90 °, and is placed in an open vessel of water. The rate at which water is sucked up by transpiration is shown by the movement of an air bubble in the water-filled horizontal tube.

Potoos: see *Caprimulgiformes.*

Potos: see Kinkajou.

Potto: see Kinkajou.

Potyviruses: see Virus groups.

Powder down: see Feathers.

Powdered growth media: Commercially available, dry mixtures of the inorganic and organic constituents of Culture media (see) Before use, the mixture is dissolved in a recommended quantity of water, adjusted to a specific volume, and sterilized. Such mixtures are of guaranteed quality and have a long shelf life.

Powder post termite: see *Isoptera.*

Powdery mildews: see *Ascomycetes.*

Prairie: temperate grassland of continental North America, dominated by more or less xeromorphic grasses with an admixture of herbaceous annual and perennial dicotyledonous plants. It is ecologically equivalent to the steppes that extend from the Pontic region of the Black Sea across Central Asia.

Prairie chicken: see *Tetraonidae.*

Prairie dog: see *Sciuridae.*

Prairie lizard: see Spiny lizards.

Pratincoles: see *Glareolidae.*

Pratylenchus: see Eelworms.

Prawn: see Shrimps.

Praying mantis: see *Mantoptera.*

Preadaptation: a property or character of an organism which proves to be beneficial when the organism enters a new environment. For example, the paired fins of the *Crossopterygii,* which were supported by a bony skeleton (unlike most other fish fins), and originally used for swimming, provided a suitable basis for the evolution of walking legs. They were therefore preadapted for a terrestrial existence.

Preantennary segment: see Mesoblast.

Preboreal: oldest period of the postglacial, in which the final dispersal of forests occurred. It lasted from about 8 300 to 6 800 BC, and was characterized by the predominance of pine-birch or birch-pine forests. See Pollen analysis.

Precambrian: see Geological time scale.

Precious corals: see *Corallium.*

Precious red coral: see *Corallium.*

Precipitins: antibodies which react specifically with an antigen to form a precipitate (see Immunoprecipitation). P. must be at least bivalent, i.e. they must possess at least two antigen binding sites.

Precocial: an adjective applied to young animals which immediately or a short time after birth or hatching are capable of a high degree of independent activity, and are able to leave their place of birth. This condition presupposes that systems involved in behavior are fully functional (sense organs, central nervous system, motor systems), but it does not preclude their further maturation.

Food reserves in the eggs of precocial birds are not totally utilized by embryonic and fetal growth and development. The chicks can therefore survive for some time at the expense of the remaining reserves (mainly fat), i.e. they do not necessarily need to feed, they have time to learn how to find food, and they can undertake relatively long journeys between the nest and feeding areas.

In truly precocial birds, less than half of the reserve fat may have been consumed at the time of hatching, leaving considerable reserves for the hatched chick. Such species are both precocial and nidifugous, e.g. domestic duck and fowl, and they are known in German as "Nestflüchter" (those that escape from the nest). An intermediate condition exists, in which the chick is precocial but nidicolous, i.e. only 40% or less of the original fat store remains after hatching. Such birds are known in German as "Platzhocker" (those that stay in one place). Platzhocker are able to leave the nest, but they remain in a more or less restricted area, e.g. gulls, terns, penguins, which are fed by the parents, but are able to survive for relatively long periods on their remaining fat reserves if the parents are away for a long time gathering food.

Animals born in the converse condition, i.e. in a state of helplessness, are described as Altricial (see).

Precursor:

1) Any compound that is converted metabolically into another compound, e.g. glucose is a precursor of starch.

2) A chemical compound used by an industrial microorganism for the synthesis of an end product. Ps. are therefore added to microbiological fermentation processes to increase yield. For example, in the industrial production of penicillins by *Penicillium* species, phenylacetic acid is added to increase the yield of penicillin G.

3) A structure, organ, form of behavior, etc. that has evolved further, e.g. the endostyle in lower chordates is the evolutionary P. of the thyroid gland in higher vertebrates.

Predation: see Realtionships between living organisms.

Predetermination: direct influence of the maternal organism on the phenotype of the next generation, due to factors operating in the egg before fertilization. For example, a premeiotic oocyte of genotype *Aa* produces a haploid egg cell containing the *a* allele, but before meiosis the dominant *A* allele so modifies the egg cytoplasm that its action persists into the next generation. Therefore, although fertilization with a paternal gamete containing allele *a* produces a genotype with the genetic constitution *aa,* the phenotype is that of the dominant *A* allele. The resulting offspring are *matroclinal hybrids.*

Predisposition: in phytopathology, the nongenetically determined condition of a plant, which determines its susceptibility to a disease. P. is affected by the age of the plant and all environmental conditions, such as nutrient availability, water aconomy, temperature, illumination, etc., as well as treatment with agicultural plant protection agents. Comparable terms are *disease proneness, acquired susceptibility, acquired disposition, disease potential* and *physiological susceptibility.*

[Plant Pathology, vol.1, edit. J.G. Horsfall & A.E. Dimond, Academic Press, 1959; Chapter 14: *Predisposition,* pp. 521–562]

Preference: preference by an organism for a certain valency range of an ecological factor.

Preformation theory, *encasement theory, incasement theory, preformism:* the disproved theory that germ cells contain preformed miniature adults, which simply unfold and increase in size during development. See Biology (Historical section).

Preformism: see Preformation theory.

Pregametic cell: see Meiosis.

Pregnancy: see Gestation.

Pregnane: see Steroids.

Preheterodont: a hinge type of the *Bivalvia* (see).

Preimago: the developmental stage of certain insects which immediately precedes, and is very similar to, the imago. See Subimago

Preimmunity, *infection-linked immunity:* resistance of a diseased organism to attack by a second, related pathogen. Superinfection is therefore prevented as long as the first infection persists. P. is observed in animals, and it also occurs in various fungal and virus diseases of plants (see Virus interference). It is not related to specific immunity, which only occurs in vertebrates.

Prelarva: the first postembryonal developmental stage of some insects (e.g. *Protura),* which differs from subsequent larval stages.

Premolar: see Dentition.

Preneanderthal man, *Homo sapiens praeneanderthalensis:* a group of humans from the Riss-Würm interglacial period of the Pleistocene. They possessed the typical characters of the Würm Ice Age Neanderthal humans, but in less pronounced form. See Anthropogenesis.

Prepatent period: the time between entry of a parasite into a host and the detection of its presence by the appearance of reproduction products (eggs, larvae).

Presapiens, *Homo sapiens praesapiens:* the direct phylogenetic ancestor of *Homo sapiens sapiens.* See Anthropogenesis.

Presbyopia, *"old sight":* a decrease in the youthful elasticity of the crystalline lens of the eye as a result of aging. This hardening of the crystalline lens reduces the amplitude of accommodation, i.e. it decreases the ability of the eye to accommodate (see Accommodation). Consequently, the near point of vision (minimal distance from the eye at which an object is still in sharp focus) increases. P. is corrected with the aid of convex lenses (i.e. reading glasses).

Presbytis: see Langurs.

Presence: see Constancy.

Presentation time: a term from sensory plant physiology, meaning the minimal duration of a stimulus required to cause a response. The P.t. decreases with the increased intensity of the stimulus. The P.t. may be much shorter than the reaction time, e.g. in Geotropism (see).

Pressed yeast: see Baker's yeast.

Pressoreceptors: see Baroceptors.

Presssaft (Preßsaft): see Zymase.

Pressure chamber function: a term borrowed from engineering technology to describe the pressure storage properties of the central arteries. The large arteries of mammals have elastic properties which enable them to function as a pressure storage chamber. During the ejection of blood from the left ventricle, the systolic pressure drives one half of the Stroke volume (see) into the peripheral circulation and simultaneously stretches the artery wall. During diastole, the potential energy of the stretched artery is converted into kinetic energy, which then transports the other half of the stroke volume. In this way the pressure chamber behavior of the artery converts the rhythmic pumping impulses of the heart into a uniform blood flow. Since the artery is relatively long, it is able to act as a pressure storage chamber by increasing its normal (unstretched) diameter by only 2–10%.

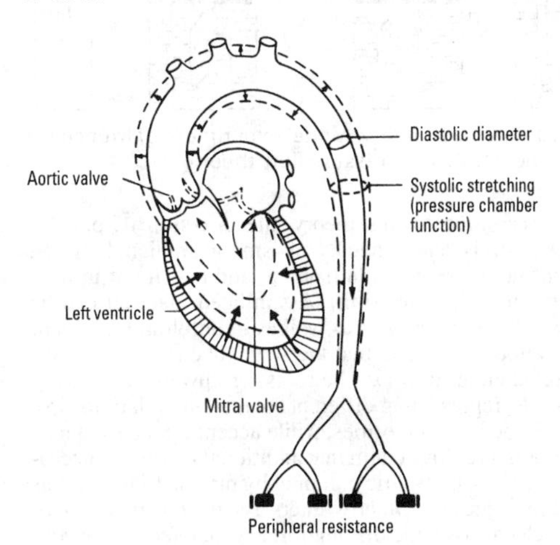

Pressure chamber function of the arterial system

Pressure diuresis: see Hairpin loop countercurrent principle.

Pressure flow theory: a mechanism proposed in 1926 by Münch to explain mass transport by fluid flow in the sieve tubes of the phloem of plants. Münch demonstrated the physical basis of his theory with the illustrated model: compartment A contains a 10% sucrose solution plus a coloring agent (congo red); compartment B contains water. The colored 10% sucrose solution extends only part of the length of connecting tube R, the remaining length as far as compartment a being occupied by water. All other vessels contain water. Compartment A takes up water through its semipermeable membrane, generating a hydrostatic pressure which forces an equivalent amount of water through the semipermeable membrane covering compartment B. Thus, sucrose (plus coloring agent) is transported from A to B, while water flows through the system. In the living plant, A represents sieve tubes in the assimilation tissue (e.g. leaves), into which the assimilate is actively transferred, possibly via companion cells. B is the receptor tissue (e.g. roots), which utilizes or stores the assimilate. R represents the sieve tube transport system, permitting longitudinal movement of fluid (across the sieve tube plates), but with no substantial lateral entry or exit of material. The pressure within this sieve tube system is evident from the exudation produced from an incision. Sap "sucking" insects, such as aphids, do not actually need to generate a negative pressure to withdraw sap from the plant; once they have punctured the phloem, the sap is delivered into their mouthparts by the internal pressure of the sieve tubes. If the inserted mouthparts of an aphid are cut off and left standing from the plant surface, they behave as microcannulae, though which a considerable quantity of exudate passes to the exterior under the internal osmotic and elastic pressure of the phloem.

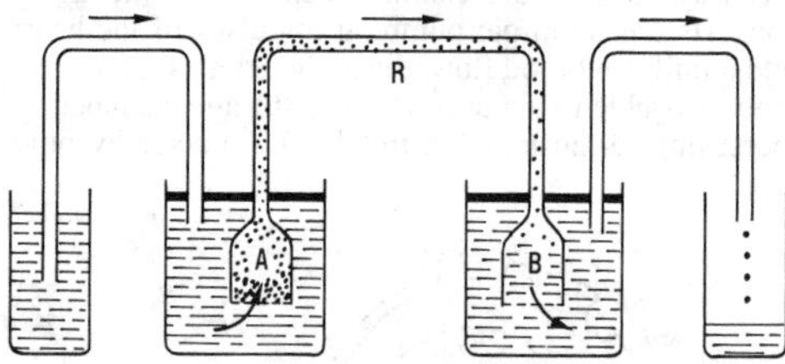

Pressure flow theory. Model experiment of Münch to demonstrate the pressure flow theory.

According to this theory, the direction of phloem transport is determined by an osmotic gradient between the site of assimilation (source) and the site of utilization of assimilate (sink). Assimilated material can be used both in the roots as well as in unfolding leaves or developing fruit, so that the sieve tube flow can be directed either down to the roots, or upward to growing points, representing different source-sink relationships.

Some later authorities, while accepting the existence of pressure flow, claim that additional forces are necessary. Thus, peristaltic pumping by plasma filaments has been suggested. Other workers assume the existence of an electro-osmotic driving force, generated by concentration differences of K^+ across the sieve plate.

Pressure receptors: see Baroceptors.

Pressure-volume curve of the respiratory apparatus, *respiratory hysteresis loop, breathing loop:* a plot of lung volume against the intrapulmonary pressure (pleural surface pressure) during a single cycle of expiration and inspiration. Owing to frictional resistance to air movement and the viscosity of the tissues (i.e. owing to nonelastic resistance), the P-v.c. is not linear. Thus, to draw a certain volume of air into the lungs the intrapulmonary pressure must decrease more than that given by a linear relationship between lung volume and intrapulmonary pressure. When inspiration is complete, there is no movement and only elastic forces are acting; at this point, the linear pressure-volume plot and the true P-v.c. for inspiration become coincident. For the same reasons (nonelasic resistance), expiration also results in a pressure-volume plot that is curved, its curvature being in the opposite direction to that of the plot for inspiration. The P-v.c. therefore has the shape and properties of a hysteresis loop, and its curvature becomes more pronounced with increased depth and rate of respiration. Each arm of the hysteresis loop represents an exponential relationship, and corresponding *compliance* (elasticity of the respiratory apparatus; see below) is the dynamic compliance. To move the respiratory apparatus during inspiration, it is necessary to overcome the elastic forces of the lungs and thorax, airflow resistance, and tissue viscosity. During quiet respiration in a healthy body, the elastic forces that operate during inspiration are greater than the frictional resistance in the respiratory apparatus during expiration (expiration is passive). In labored breathing, if the resistance becomes greater than the elastic forces, or if the elasticity of the apparatus decreases with age or illness, then expiration must be performed actively. To overcome the elastic and nonelastic resistance of the respiratory apparatus, work must be done. From the P-v.c. this work can be visualized as the area of the hysteresis loop. The work is equivalent to (pressure x volume), and it also has the dimensions of (force x distance). In humans at rest, respiratory work is estimated at about 0.5 kg-m/min. Very heavy physical exertion can increase this value 400–fold to 200 kg-m/min.

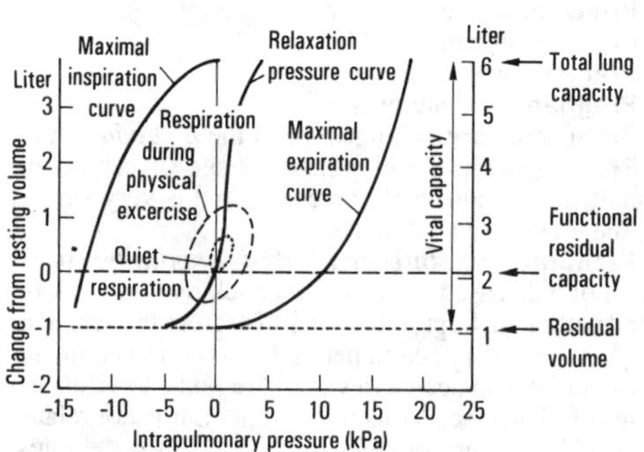

Pressure-volume curve of the respiratory apparatus. The relaxation pressure curve is the plot of the intrapulmonary pressure against the lung volume, measured with the airway closed (i.e. while the pressure is static). The maximal inspiration and expiration curves display the airway pressures that are developed during maximal inspiratory effort and the accompanying expiration.

To construct a *relaxation pressure curve* for the respiratory apparatus of a mammal or human, breathing is briefly stopped at various stages of inspiration and expiration, and the corresponding lung volume and intrapleural pressure are measured. The resulting plot provides an index of the elasticity of the respiratory apparatus (see Fig.). Elasticity is expressed as the ratio of the volume change to pressure, and is known as *compliance.* The static compliance ("static" because the airflow is temporarily arrested at each successive measurement) of an adult human during quiet respiration is 1.3 ml/Pa, i.e. under a pressure difference of 1 cm water, the volume of the lungs increases by 130 ml, or the lungs expire 130 ml of air (1 kPa is approximately equal to a pressure of 10 cm water). This, however, represents the elasticity of the entire respiratory apparatus, i.e. the lungs and thorax combined. To determine the elasticity of the lungs alone, the procedure is repeated, but with measurement of the intrapleural pressure. With a value of 2.5 ml/Pa, the elasticity of the lungs is about twice as great as that of the entire respiratory apparatus. Compliance alters with age, and varies according to the animal species. A knowledge of compliance is a prerequisite for the correct application of artificial respiration. Lung compliance is influenced by the surface tension of the alveoli, which acts against the elastic forces. Healthy lungs, however, are coated internally with a film of lipoprotein, which decreases the surface tension to about one tenth of the uncoated value.

Presumptive: in embryology, an adjective applied to a still undifferentiated part of a developing embryo, which will (under normal conditions) develop into a certain structure. For example, the presumptive notocord is that part of the roof of the archenteron of an early gastrula, which later gives rise to the notocord.

Presynapse: see Neuron, Synapse.

Presynaptic inhibition: see Neuronal circuit.

Priapulids: a small group of marine animals of uncertain affinities, often placed in a separate minor coelomate phylum, the *Priapuloidea.* Only two genera *(Priapulus, Halicryptus)* and 9 species are known, united in a single order, *Priapulida.* They are nonlophophorate protostomes.

Morphology. The cylindrical, 1.5–1.8 cm-long, unsegmented body possesses an anterior retractile proboscis (or introvert); a trunk, which carries spines and is divided superficially into annuli; and a caudal appendage. The mouth, surrounded by several concentric rings of coarse, hooked teeth, is situated at the anterior end of the cylindrical to pear-shaped proboscis, which carries 25 longitudinal rows of spines. The caudal appendage consists of one or two extensions of the body, and it carries a large number of papillae. There is a spacious fluid-filled body cavity, which communicates with the hollow papillae of the caudal appendage. Since the body cavity is lined with a thin, nucleated peritoneun, it appears to be a true coelom. For this reason the P. may be classified as coelomate (some authorities still classify P. as pseudocoelomate), although the embryonic origin of the body cavity is unknown. The sexes are separate. Gonads and excretory protonephridia form a simple, paired urogenital system with ducts opening near the anus.

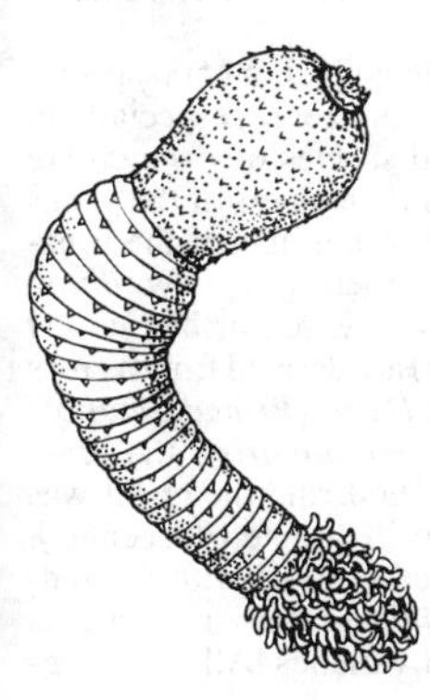

Priapulida. The unicaudate priapulid, *Priapulus caudatus.*

Biology. P. are found in the temperate seas of both Northern and Southern Hemispheres, where, depending on the species, they live in the mud of the intertidal zone and up to depths of 8 000 m. They move forward by contraction and stretching movements of the body. Their diet consists of slow-moving marine animals (e.g. other P. and polychaete worms) and rarely vegetable matter. P. develop from a larval stage (the loricate larva) which resembles the adult, but is encased in a chitinous armor or lorica. Growth occurs through several molts, the adult appearing after about 2 years.

Collection and preservation. P. are captured by bottom scooping or dredging, carefully narcotized by dropwise addition of 70% ethanol or cocaine in saline, then killed and stored in the fully extended state by placing a mixture of alcohol and 40% formaldehyde solution.

Prickly cup: see *Basidiomycetes (Agaricales).*

Primary photosynthetic product: see Net photosynthesis.

Primary sensory cell: see Neuron.

Primary urine: see Kidney functions.

Primase: see DNA replication.

Primates, *Primates* (plates 40 & 41): an order of mammals, which includes humans and contains the suborders, Prosimians (see) and Simans (see). Two slightly different schemes of classification are shown below. In the second scheme, humans are placed in the same family as the gorilla and chimpanzees on the basis of molecular evidence. There is a strong clavicle and rotatable scapula, and the pectoral and pelvic girdles permit extensive, agile limb movements. Radius and ulna are separate and jointed, and the tibia and fibula are separate (except in *Tarsioidea).* Hands and feet possess 5 digits and are mainly developed as organs for gripping. All P. possess an opposable pollex, and the hallux is opposable in many species. Eyes face forward, many forms having binocular vision. Although the brain case is large, the olfactory region is often reduced, and the sense of smell is accordingly less developed (the snout is also commonly reduced). The diet is generally omnivorous and the teeth unspecialized. Simians have a single uterus, and fetal and maternal circulations are separated only by the walls of the fetal vessels (see Placenta). Only small numbers of young are produced, and parental care continues long after birth. There is usually a single pectoral pair of mammae. Behavior is highly differentiated and capable of considerable adaptation, a feature that is correlated with advanced brain development.

Geological record. Fossil remains are found from the Tertiary to the present, but they are scarce and therefore difficult to interpret geologically. The few documented remains are, however, of considerable interest, because the evolution of P. culminates in the appearance of humans. Prosimians are widely distributed in the Eocene of Eurasia and North America, and they are unknown in the Upper Tertiary. They reappear in the Pleistocene and in recent times, but are then restricted to the tropical regions of the Old World. The sole recent representative of the *Tarsioidea,* a suborder or superfamily representing transition from prosimians to simians, is the Tarsier. Tarsioids were relatively numerous in the European and North American Paleocene and Eocene. Simians show an unchanged geographical distribution since their appearance in the Oligocene.

Classification. SCHEME 1.

Order: *Primates.*

1. Suborder: *Prosimii **(Prosimians).***

1. Infraorder: *Loisiformes,* containing the single superfamily, *Lorisoidea,* containing the single family *Lorisidae,* with the subfamilies *Galaginae **(Bushbabies*** and ***Needle-clawed galagos)*** and *Lorisinae **(Loris, Slow loris, Angwantibo*** and ***Potto).***

2. Infraorder: *Lemuriformes.*

1. Superfamily *Lemuroidea.*

1. Family: *Lemuridae,* with the subfamilies, *Lemurinae **(True lemurs)*** and *Cheirogaleinae **(Dwarf lemurs).***

2. Family: *Indriidae **(Indrids).***

2. Superfamily: *Daubentonioidea,* with the single family, *Daubentoniidae* (containing the single species, the ***Aye-aye).***

2. Suborder: *Tarsioidea,* with the single family, *Tarsiidae **(Tarsiers).***

3. Suborder: *Anthropoidea **(Simians: Monkeys, Apes*** and ***Humans).***

1. Infraorder: *Platyrrhini **(New World monkeys).***

Superfamily: *Ceboidea.*

1. Family: *Cebidae **(Cebid monkeys),*** with the subfamilies, *Cebinae **(Capuchins),** Aotinae **(Owl*** or ***Night monkey),** Callicebinae **(Titis),** Saimirinae **(Squirrel monkeys),** Pithecinae **(Uakaris, Sakis and Bearded skis),** Alouattinae **(Howler monkeys),*** and *Atelinae **(Spider monkeys, Woolly spider monkey*** and ***Woolly monkeys).***

2. Family: *Callitrichidae,* with the subfamilies, *Callitrichinae **(Marmosets, Pygmy marmoset*** and ***Tamarins)*** and *Callimiconinae* (containing the single species, ***Goeldi's marmoset).***

2. Infraorder: *Catarrhini **(Old World monkeys).***

1. Superfamily: *Cercopithecoidea.*

Family: *Cercopithecidae.*

1. Subfamily: *Cercopithecine,* with the genera, *Allenopithecus **(Swamp monkey),** Cercocebus **(Mangabeys),** Cercopithecus **(Guenons),** Erythrocebus **(Patas monkey),** Macaca **(Macaques),** Miopithecus **(Talapins),** Papio **(Baboons),** Mandrillus **(Mandrill*** and ***Drill)*** and *Theropithecus **(Gelada).***

2. Subfamily: *Colobinae **(Colobine monkeys),*** with the genera, *Colobus **(Guerezas*** and ***Colobus monkeys),** Procolobus **(Olive colobus),** Presbytis **(Langurs),** Nasalis **(Proboscis monkey),** Simias **(Pig-tailed leaf monkey),** Rhinopithecus **(Snub-nosed monkeys)*** and *Pygathrix **(Douc).***

2. Superfamily: *Hominoidea **(Anthropoid apes),*** with the families, *Hylobatidae **(Gibbons),** Pongidae **(Great apes: Gorilla, Orang-utan, Chimpanzees).***

Classification. SCHEME 2.

Order: *Primates.*

1. Suborder: *Strepsirhini.*

1. Superfamily: *Lemuroidea,* with the families, *Cheirogaleidae **(Dwarf lemurs),** Daubentoniidae **(Aye-aye),** Indriidae **(Sifakas, Indri*** and ***Woolly lemur),** Lemuridae **(True lemurs)*** and *Lepilemuridae **(Gentle lemur*** and ***Sportive lemur).***

2. Superfamily: *Lorisoidea* (further classification as in scheme 1).

2. Suborder: *Haplorhini.*

1. Infraorder: *Tarsii* or *Tarsiiformes.*

1. Superfamily: *Omomyoidea* (known only from fossil remains).

2. Superfamily: *Tarsioidea,* with the single family, *Tarsiidae **(Tarsier).***

2. Infraorder: *Platyrrhini **(New World monkeys).***

Superfamily: *Ceboidea,* with the families, *Callimiconidae **(Goeldi's marmoset),** Callitreichidae **(Marmosets, Pygmy marmoset*** and ***Tamarins),*** and *Cebidae* (subfamilies as in scheme 1).

3. Infraorder: *Catarrhini **(Old World monkeys, Apes*** and ***Humans).***

1. Superfamily: *Parapithecoidea* (known only from fossil remains).

2. Superfamily: *Cercopithecoidea* (further classification as in scheme 1).

3. Superfamily: *Hominoidea **(Anthropoid apes*** and ***Humans).***

1. Family: *Hominidae,* with the subfamilies, *Gorillinae **(Gorilla*** and ***Chimpanzees),** Homininae **(Humans),** Ramapithecinae* (?) (known only from fossil remains) and *Australopithecinae* (known only from fossil remains).

2. Family: *Hylobatidae,* with the subfamilies, *Pliopithecinae* (known only from fossil remains) and *Hylobatinae **(Gibbons).***

3. Family: *Pongidae **(Orang-utan).***

Primer effect: changes in the endocrinal and metabolic status of an organism. arising from the action of stimuli on the central nervous system. The affected organism displays an altered attitude toward the environment. The term was originally used for the effects of chemical stimuli, in particular Phermones (see), but it is now used generally for an analogous response to any stimulus.

Primin: a golden yellow *p*-benzoquinone derivative from *Primula obconica* (top primrose). It is a skin irritant.

Primitive groove: see Mesoblast.

Primitive knot: see Mesoblast.

Primitive pit: see Mesoblast.

Primitive streak: see Mesoblast.

Primordial germ cell: see Gametogenesis, Oogenesis, Spermatogenesis.

Primosome: see DNA replication.

Primrose: see *Primulaceae.*

Primulaceae, *primrose family:* a family of dicotyledonous plants (about 1 000 species in 20 genera), cosmopolitan in distribution, but found mainly in northern nontropical regions. Most members are herbs with simple, variously shaped leaves, and regular, 5-merous hermaphrodite flowers, which are normally insect-pollinated. Some species are heterostylous. The superior ovary develops into a capsule.

The largest genus is *Primula* with about 500 species in the northern cold and temperate zones, especially in mountainous regions. All wild species of *Primula* are protected in Europe, and several members of the genus are popular garden plants. Most familiar of the ornamentals are the multicolored varieties of ***Polyanthus;*** these large-flowered white, yellow, red or blue-violet hybrids *(Primula* x *polyanthus)* are derived from a cross between the native European ***Oxlip** (Primula elatior)* and the wild ***Primrose** (Primula vulgaris);* the cross was first performed in about 1660, and the result was improved by amateur gardeners in the late 19th century. *Primula vulgaris* has also been crossed with the red-flowered Caucasian *Primula juliae* to produce various early spring-flowering garden varieties. All these species, as well as the European ***Cowslip*** *(Primula veris = P. officinalis)* possess characteristically wrinkled and easily perishable leaves, whereas the popular rockery plant, ***Auricula,*** has smooth, fleshy and largely durable leaves. This garden auricula *(Primula* x *hortensis)* is a hybrid of the alpine plants, *Primula auricula* and *Primula hirsuta.* Another common garden species is the pink and white-flowered Himalayan ***Drumstick primrose*** *(Primula denticulata),* which has compact globular inflorescences. Best known of all the house plants of this genus is the ***Top primrose*** *(Primula obconica)* from Southwest China, whose glandular hairs contain the poisonous benzoquinone, primine, a powerful skin irritant responsible for primrose disease. Also from Southwest China is the ***Fairy primrose*** *(Primula malacoides),* which has rather replaced *Primula obconica* in recent decades, since it does not possess irritant hairs. An old Chinese garden plant, the ***Chinese primrose*** *(Primula sinensis),* with a wide range of color, was also for a time very popular in Europe.

Cyclamen *(Cyclamen persicum)* from the eastern Mediterranean is a widely known ornamental. Characteristic of *Cyclamen* are its hypocotylar tuber, the backward inflection of its 5 corolla segments, and its downward facing flowers. About 15 species of *Cyclamen* occur in southern Europe, North Africa and East Asia, mainly in mountainous areas. The most northerly species are *Cyclamen purpurascens (= C. europaeum = C. neapolitanum)* ***(Sowbread)*** and *Cyclamen hederifolium,* which occur in the stony forest and scrub of the central European mountains. Sowbread has been introduced even further north and is found (but very rare) in hedgebanks and woods of England and Wales.

Scarlet pimpernel *(Anagalis arvensis)* is a common annual on cultivated clay soils in many parts of the world except the tropics. Moist habitats are preferred by the mostly yellow-flowered species of *Lysimachia,* e.g. ***Yellow loosestrife*** *(Lysimachia nummularia).* ***European starflower*** *(Trientalis europaea)* has a 7-lobed white corolla, and is found mainly in pine forests of the northern temperate zone. ***Pygmy flowers*** *(Androsace* spp.) and snowbells *(Soldanella* spp.) are mostly typical high mountain plants.

Prinia: see *Sylviinae.*

Prion, *proteinaceous infectious agent:* one of a group of infectious agents with properties that distinguish them from both viruses and viroids. The only macromolecules associated with infectivity are proteins. The infective agent of the sheep and goat disease, scrapie, is a sialoglycoprotein, M_r 30 000. Other diseases caused by Ps. are human Creutzfeldt-Jacob disease, bovine spongiform encephalopathy (BSE, "mad cow disease"), transmissable mink encephalopathy, chronic wasting disease of the mule deer and elk, Kuru (in the Foré tribe of New Guinea; transmitted by eating human brains), and the Gerstmann-Straussler syndrome of humans.

Prionodura: see *Ptilinorhynchidae.*

Prionopinae: see *Laniidae.*

Pristiophoridae: see *Selachimorpha.*

Privet: see *Oleaceae.*

Proband: see Propositus.

Probiosis: see Relationships between living organisms.

Proboscidea: an isolated and formerly very successful order of mammals comprising the Elephants (see) and their extinct relatives. Fossil *P.* are found from the Tertiary to the present, in the Americas, Eurasia and Africa. The oldest known representatives, from the Upper Eocene, were relatively small. Otherwise, all fossil and modern members of the order are exceptionally large animals. Since the Late Miocene, most members have possessed a trunk, developed from the nose and upper lip, which is used to obtain food and water, and for the generation of sounds. The geological succession of fossils shows an evolutionary modification and reduction of dentition, so that adults finally possess three molars in each side of the jaw; these are used one at a time, old worn teeth being shed and replaced by those behind. The upper incisors are enlarged to form tusks. There is

Primulaceae. a Top primrose *(Primula obconica). b* Sowbread *(Cyclamen europaeum). c* Oxlip *(Primula elatior). d* Cowslip *(Primula veris). e* Drumstick primrose *(Primula denticulata).*

a well developed brain and social organization is complex. Some extinct representatives are Dinotherium, Mastodon (see) and Mammoth (see).

Probosciger: see *Psittaciformes.*

Proboscis monkey, ***long-nosed monkey,*** *Nasalis larvatus:* a member of the *Colobinae* (see Colobine monkeys). The genus has only one species, which inhabits the low mangrove swamps and delta regions of the island of Borneo. The male grows to the size of a baboon and possesses a fleshy, 10 cm-long, pendulous nose; the female is smaller with a shorter nose.

Proboscis monkey (male)

Proboscis worms: see *Nemertini.*

Procapra: see *Antilopinae.*

Procaryotes: see Prokaryotes.

Procellariidae: see *Procellariiformes.*

Procellariiformes, ***tube-noses,*** *Tubinates:* an order of birds (see *Avies),* containing 80–100 species in 12–24 genera (the precise numbers depending on the taxonomic authority). All members of the order are totally marine, feeding on zooplankton, fish, crustaceans and cephalopods in the ocean, often coming ashore only to breed. All lay a single white egg. Only the albatrosses build an open nest; all other species nest in an underground burrow. They have thick plumage, webbed feet and a deeply grooved, hooked bill. Very characteristic is the design of their nostrils, which extend onto the bill in short tubes. In addition, all species have a clear yellow stomach oil, which they feed to their newly hatched young. The greater variety of species and genera is found in the Southern Hemisphere. The order is divided into 4 families.

1. Family: *Diomedeidae* **(Albatrosses;** 71–135 cm; 10 species in the genus *Diomedea,* and 2 in the genus *Phoebetria).* These are large, stoutly built ocean birds with long narrow wings and spectacular powers of gliding flight. They dive to take their prey from the ocean, and rarely land on the water In most species the sexes are alike, the plumage being mainly white or brown, often with dark brown or black on the wings, back and tail. They breed in large colonies on remote islands mainly south of the Tropic of Capricorn; a few are found in the North Pacific.

2. Family: *Procellariidae* **(Shearwaters, Fulmars and Petrels).** Of the 53 species comprising this family, 23 belong to the genus *Pterodroma,* while 17 belong to the genus *Puffinus.* The greatest nember of species is found in the Southern Hemisphere, but the family is represented on all the unfrozen seas and oceans of the world. For example, the ***Atlantic Fulmar*** *(Fulmarus glacialis;* 45.5 cm) (Fig.) is found from the temperate to the Arctic regions of the Atlantic and the north Pacific; and the ***Manx shearwater*** *(Puffinus puffinus;* 35.5 cm) is found in the northern Atlantic from the Azores to Iceland and across to Bermuda, overwintering off Brazil and Argentina. The ***Gould*** or ***White-winged petrel*** *(Pterodroma leucoptera;* 30.3 cm) is found in both the north and south Pacific oceans, while the ***Pintado petrel*** or ***Cape pigeon*** *(Daption capensis;* 35.5 cm) is restricted to the Southern oceans. The ***Slender-billed*** or ***Short-tailed shearwater*** *(Puffinus tenuirostris;* 35.5 cm) breeds on islands in the Bass Strait between Tasmania and Australia. Both parents incubate the egg in shifts, and the young chick is fed on stomach oil and partly digested shrimp. Growth is rapid, and the parents leave the fat young bird at about 14 weeks of age. About a week later, after losing some excess fat and developing its flight feathers, the young bird instinctively follows the species-specific migration route, returning 3–4 years later. The migration route passes the northern coasts of New Zealand, proceeds north past the east coast of Japan, further north toward Kamchatka, then east, remaining on the high seas, but more or less in line with the coast of Alaska, Canada and the USA, until crossing the entire width of the Pacific from the latitude of southern California back to the Bass Strait. Tasmanians harvest the young birds after they have been deserted by the parents, both for the oil, which is drained from the stomach, and for their meat. For this reason, *Puffinus tenuirostris* is also known as the ***Mutton bird.***

Procellariiformes. Atlantic fulmar *(Fulmarus glacialis).*

3. Family: *Hydrobatidae* **(Storm-petrels;** 15–25.5 cm). These familiar oceanic birds commonly follow ships on the high seas, snatching marine organisms in flight from the disturbed water of the ship's wake. Many of the 22 species are very dark, almost black above, with a conspicuous white patch on the rump, e.g. the ***Storm petrel*** *(Hydrobates pelagicus;* 15 cm; eastern North Atlantic and Mediterranean); ***Wilson's petrel*** *(Oceanites oceanicus;* 18 cm; Antarctic to the North Atlantic and South Pacific); and ***Leach's petrel*** *(Oceanodroma leucorhoa;* 20 cm; North Atlantic and North Pacific), the latter possessing a distinctive forked tail.

4. Family: *Pelecanoididae* ***(Diving petrels;*** 16.5–25.5 cm). With their compact bodies and short wings, these Southern Hemisphere birds superficially resemble certain auks *(Alcidae)* of the Northern Hemisphere. They do not soar over the ocean taking food from the surface like other *Procellariiformes,* but fly close to the surface with a whirring wing action, diving and swimming underwater. All 5 species are placed in the genus *Pelecanoides.*

Procercoid: see *Cestoda.*

Processionary moths: moths of the family *Thaumetopoeidae.* See *Lepidoptera.*

Prochlorophytes: a group of recently discovered green prokaryotic organisms. They differ from all other prokaryotes by their possession of chlorophyll *b,* which hitherto was known only in eukaryotes. They also lack phycobilins, which are present as red and blue pigments in cyanobacteria. P. live as symbionts on various tunicates.

Procnias: see *Cotingidae.*

Procoela: see *Anura.*

Procoelous (procoelic) vertebra: see Axial skeleton, *Anura.*

Procreation: see Reproduction.

Proctodeum: see Digestive system.

Procyon: see Raccoons.

Procyonidae: a family of terrestrial carnivores (see *Fissipedia),* containing 19 species in 7 genera. They are medium-sized, long-tailed mammals. As an adaptation to their omnivorous feeding habits, their premolars are not reduced, and their carnassial teeth are low-cusped and modifed for crushing rather than shearing. The 5 toes on each foot are usually plantigrade, and the claws are nonretractile or semi-retractile. The following representatives are listed separately: Ringtail, Raccoons, Coatis, Kinkajou, Red panda.

Prodenia: see Cutworms.

Production biology: an area of synecology concerned with the turnover of material and energy in the ecosystem. The totality of organic material in an ecosystem at a given time (related to surface area or volume) is known as the Biomass (see). The change (increase) of biomass is the *production,* which on a time basis becomes the *production rate* or *productivity.* The basis for the existence of all ecosystems is the primary production by autotrophic plants *(primary producers),* which with the aid of sunlight are able to perform a net synthesis of organic material. At the expense of this primary production, a complicated system of heterotrophic consumers is then responsible for *secondary production.* The *net production* is obtained by subtracting the material respired for the maintenance of metabolism from the total production. The net production of long-lived terrestrial plants is relatively easy to determine (harvest method), because the synthesized material accumulates in the plant. In aquatic ecosystems it is not possible to determine primary production by the harvest method, because the lifespan of the producers (phytoplankton) is too short, and the turnover of organic material occurs very quickly. In this case, net production is determined by measuring the *assimilation rate* by C^{14} labeling. The total quantity of organic material on the planet is calculated to be 10^{20} g, of which 99.99% is in primary autotrophic producers. The annual net production of the oceans is equivalent to 58 x 10^9 tonnes of glucose, that of the land 50 x 10^9 tonnes of glucose.

Productivity: in a general physiological sense, the net production of material by plants. From a practical standpoint, the most important indices of P. are the quality and quantity of specific products of exploited plants. The net production of material is the result of plant growth and development, but the increase in dry mass does not occur uniformly throughout the growth and development. This is particularly marked in annual plants, where the plot of dry mass against time is generally sigmoid. During the initial phase of germination and root and shoot formation, the increase of dry mass is relatively slow; at first the root dry mass increases more rapidly than that of the shoot. The most rapid phase of dry mass increase starts with the development of the photosynthetic apparatus and ends after the onset of reproduction. The later stages of growth and development are characterized by the synthesis and storage of reserve materials.

A variety of physiological processes is involved in P., in particular mineral metabolism, photosynthesis and respiration, carbohydrate and protein metabolism, as well as synthesis of numerous secondary plant products. All these processes are closely interrelated, and the extent to which their genetically determined potential is realized is dependent on environmental factors. Thus, the potential maximal P. is genetically determined, but the actual P. is largely determined by climate, irrigation and soil fertility.

Productus (Latin *productus,* extended, stretched out): an extinct hinged genus of brachiopods without a brachial skeleton. Fossils are found from the Upper Devonian to the Permian. The valve carrying the stalk is highly arched with an incurved umbo, and the arm valve is markedly concave. Both valves are covered with hollow spines or processes which serve for attachment. The genus includes stratigraphically important index forms of the Carboniferous limestone and of the Permian *(Productus* chalk).

Proechimys: see *Echimyidae.*

Profundal: the deep sea floor below the limit of light penetration, and therefore below the limit for the growth of photosynthesizing plants (see Sea). The biocoenoses of the P. consist entirely of consumers, i.e. they are dependent on nutrients from the pelagic waters (see Pelagial) above them and from the Littoral (see). Such biocoenoses contain organisms that occur in other regions, but which thrive especially in the P. *(bathyphilic species),* as well as those that have evolved in, and are restricted to, the P. *(bathybiont species).* Bathyphilic species comprise 1) *eurybathic forms,* which are not restricted to a specific depth, and which are also able to colonize shallow waters (the majority of bathyphilic species); 2) animals of ground waters, which have migrated into the P., e.g. the blind cave crustaceans, *Niphargus puteanus* and *Asellus cavaticus,* which occur in large numbers in some oligotrophic Alpine lakes; 3) Ice-Age relicts like the crustaceans, *Mysis relicta, Pontoporeia affinis* and *Pallasea quadrispinosa;* originally part of a cold-loving ice-age fauna, these organisms still survive in the cold water of the P. in some deep lakes of northern Europe and North America.

The only known bathybiont species are certain forms in the deep regions of Lake Baikal and Lake Tanganyika, and probably the deepwater char *(Salmo salvelinus profundus).* This latter subspecies is restricted to certain deep Alpine lakes, where it evolved.

Progenesis:
1) A form of pedomorphosis, in which ontogenesis is abbreviated by the early onset of maturity, so that an adult descendant closely resembles a juvenile stage of its ancestor.
2) The early stages of Ontogeny (see)

Progesterone, *corpus luteum hormone, luteohormone:* pregn-4-ene-3,20-dione; Δ^4-pregnane-3,20-dione (m.p 128 °C). P. is structurally related to the parent hydrocarbon, pregnane. Produced in the corpus luteum, it is the natural progestin, and therefore an important female steroid hormone. P. promotes proliferation of the uterine mucosa, and having thus prepared the uterus, it then promotes implantation and further development of the fertilized ovum. During pregnancy, it prevents further maturation of follicles and stimulates development of the lactatory function of the mammary glands. During pregnancy, some P. is also produced by the placenta, and the production of P. by the corpus luteum is markedly increased. In the circulation, P. is bound to carrier proteins. During the menstrual cycle, the level of circulating P. displays phasic behavior. It increases in the second half of the menstrual cycle under the influence of Luteinizing hormone (see) from the pituitary. The human cycle is controlled hormonally by the interplay of the effects of follicle-stimulating hormone (FSH) and luteinizing hormone (LH), as well as the actions of the steroid hormones, P. and estradiol, in which P. acts as an antagonist of the estrogens. For more details of these interactions and the timing of P. secretion, see Placenta, Uterus. P. is also formed in the adrenal cortex, where it serves as an intermediate in the biosynthesis of adrenocortical hormones, and in the testes, where it serves as a precursor of the androgens. P. itself is synthesized from cholesterol via pregnenolone.

P. was first isolated in 1934 by Butenandt et al. from corpus luteum. It is used in veterinary medicine to correct irregularities of estrus and to control habitual abortion.

H_3C C=O H_3C H_3C HO

Progesterone

Progestins: see Gestagens.

Proglottid: see *Cestoda.*

Proglottis: see *Cestoda.*

Prognathy: see Facial profiles.

Prognosis of genetic defects: see Genetic prognosis.

Prognosis of inborn illnesses: see Genetic prognosis.

Progoneate: the trunk segment of the *Diplopoda* (see) that contains the gonopores; counting the collum as the first trunk segment, the progoneate is the third segment.

Programed cell death: hypothesis that the lifespan of cells is determined genetically. It can be shown, particularly in cell culture, that most cells have a finite lifespan. For example, the number of mitotic divisions of human fibroblasts in cell culture depends on the age of the donor; cultured cells from embryos divide about 50 times, whereas cells from older individuals divide only a few times. Cancer cells often display an unlimited capacity for division. Plant cells in tissue culture (e.g. callus cultures) also appear to be capable of unlimited division.

One possible cause of P.c.d. in animal cells is the accumulation of oxidation products.

Mononucleate unicellular organisms can be regarded as immortal, since their total cell material is transferred to the daughter cells at each division.

Multicellular organisms possess a small number of germ cells, which are potentially immortal and totipotent, since they are responsible for the perpetuation of the species. Most of the cells in a multicellular organism are, however, somatic cells with a limited lifespan; their totipotnecy (and therefore their unlimited capacity for division) is lost in the process of functional specialization (differentiation).

Prokaryotes, *prokaryotic organisms:* a group of organisms comprising the Bacteria (see) and the *Cyanophyceae* (see). Their cell structure differs from that of all other organisms, which are collectively known as eukaryotes. The nucleoplasm of P. is not separated from the cell cytoplasm by a nuclear membrane, there is no nucleolus, and the DNA is not associated with basic proteins. During cell division, chromosomes and a spindle apparatus (typical of eukaryotic cell division) are not formed. P. have no mitochondria or plastids, and the functions perfomed by these organelles in eukaryotes are performed in P. by the plasma membrane, invaginations of the plasma membrane, and simple chromatophores. Vacuoles occur only rarely, and there is no endoplasmic reticulum. Prokaryotic ribosomes are smaller than those in the cytoplasm of eukaryotes, but they are very similar in size to those in the mitochondria and plastids of eukaryotes (see Ribosomes). P. usually have a rigid cell wall, which always contains proteoglycan (murein) in addition to other materials. When present, flagella (which serve for locomotion) consist of a few protein fibrils wound around each other.

P. are unicellular. The single cells sometimes form associations (usually filaments), in which all the cells are identical. Cell differentiation is restricted to the formation of e.g. spores, cysts and adhesion cells. Cell size varies from 0.2 to 10 mm.

There are autotrophic, heterotrophic, aerobic, anaerobic, free-living, parasitic and symbiotic forms, representing a wide variety of different modes of existence and nutrition. Nutrients are always in molecular form, and they enter the cell by crossing the cell wall and membrane.

P. are widely distributed, occurring in water, soil, air, and on or in other organisms. Some tolerate conditions that are extremely unfavorable for eukaryotes (e.g. high temperatures, high salt concentrations, adverse pH, etc). They are primitive life forms, representing a very early stage in the evolution of living organisms.

Prolactin, *lactotropin, lactogenic hormone, mammotropin, luteomammotropic hormone, luteotropic hormone, LTH, luteotropin:* a gonadotropin. Phylogenetically, P. is one of the oldest adenohypophyseal hormones. It acts primarily on the mammary gland by promoting lactation in the postpartal phase, and in rodents it also acts on the ovary. Its activity in males is not clear, although males do produce it. Bovine P. is a single chain polypeptide (198 amino acid residues, M_r 22 500, three disulfide bridges) of known primary structure.

P. synthesis in the α-cells (eosinophils) of the anterior pituitary (adenohypophysis) is under the control of the hypothalamus hormones, prolactin releasing hormone (PRH) and prolactin release inhibiting hormone (PIH). It is produced in increasing amounts during pregnancy and during suckling; it can be detected in the blood (ng/ml range) by radioimmunological assay. P. promotes metabolism and growth, influences osmoregulation, pigment metabolism and parental behavior, and suppresses reestablishment of the menstrual cycle post partum. It acts on its target cells via the adenylate cyclase system.

Prolamines: a group of simple (unconjugated) proteins, which are soluble in 50–90% ethanol. They contain up to 15% proline and 30–45% glutamic acid, and have only low contents of essential amino acids. They occur in cereal grains, and the chief representatives are gliadin (wheat and rye), zein (maize; contains no tryptophan or lysine) and hordein (barley; contains no lysine). Oats and rice do not contain P.

Prolecithophora: see *Turbellaria.*

L-Proline, ***Pro, pyrrolidine-2-carboxylic acid:*** a proteogenic and glucogenic amino acid. Pro is very soluble in both water and ethanol, so it can be separated from other amino acids by ethanol extraction. Being an imino acid, it forms a yellow color with ninhydrin, rather than the purple color which is characteristic of α-amino acids. Pro is particularly abundant in collagen, gliadin and zein. It does not participate in α-helix formation, and therefore has special importance in protein tertiary structure. It is biosynthesized from L-glutamate via glutamic 5-semialdehyde and pyrroline carboxylate. Some may be formed from exogenous ornithine via glutamic 5-semialdehyde.

```
H2C ——— CH2
 |       |
H2C     CH—COOH
   \   /
     N
     H
```

L-Proline

Promastigote: see Trypanosomes.

Promeristems: see Meristems.

Promerops: see *Meliphagidae.*

Promicrops: see Sea basses.

Prominent moths: moths of the family *Notodontidae;* see *Lepidoptera.*

Promitochondria: see Mitochondria.

Promycelium, ***metabasidium:*** a tubular, septate or nonseptate outgrowth found in the smut fungi, which is formed from a brand spore after its paired nuclei have fused. The diploid nucleus then undergoes meiosis and each of the resulting 4 haploid nuclei is incorporated into a sessile primary sporidium (gonospore), which is abstricted from the tip of the P. The 4 primary sporidia conjugate in pairs by means of short conjugation tubes. Each of the resulting H-shaped pairs of primary sporidia abstricts a single, sickle-shaped, binucleate spore on a lateral sterigma; this binucleate spore is sometimes called the secondary sporidium, secondary conidium, or sickle conidium. Sporidia are thus homologous with basidiospores, and the P. is a rudimentary basidium. See *Basidiomycetes* (*Ustilaginales*).

Pronephros: see Excretory organs.

Pronghorn: see *Antilocapridae.*

Pronolagus: see *Leporidae.*

Proostracum: see *Cephalopoda.*

Propagation: see Reproduction.

Propagule: see Reproduction.

Prophase: see Meiosis, Mitosis.

Propionic acid: $CH_3—CH_2—COOH$, a simple fatty acid, which is a colorless, sharply smelling liquid. It occurs as its salts and esters in many plants, e.g. propionic esters occur in apples. See Propionic bacteria.

Propionic bacteria, *Propionibacterium:* a genus of pleomorphic, nonmotile, anaerobic, Gram-positive bacteria, which produce considerable amounts of acid, mainly propionic acid, by the fermentation of glucose and other substrates (see Fermentation). They occur in the gastrointestinal tract of animals and in milk products, and they are exploited in cheese manufacture for the production of the flavors and holes in Emmenthal cheese.

Propodosoma: see *Acari.*

Propositus, ***proband:*** in human genetics, the person whose condition first gives occasion for the genetic analysis of the remaining family. In the family tree, the P. is usually indicated by an arrow. In the general sense, probands are all the human subjects investigated as part of a scientific survey, e.g. in order to determine the reference limits of a blood quantity for clinical chemical purposes.

Proprioceptors, ***interoceptors, visceroceptors:*** in the broad sense, receptors that respond to stimuli within the body. These include receptors for monitoring the circulation and respiration, and for registering the position and movement of body parts relative to each other, thereby signaling the active and passive movements of the body. In a more restricted sense, P. are Mechanoreceptors (see) of the locomotory apparatus and body surface, e.g. the hair sensillae of arthropod joints. In vertebrates, they include muscle and tendon spindles, various types of joint receptors, and pacinian corpuscles.

Prorodon: see *Ciliophora.*

Prosauropoda: see *Plateosaurus.*

Proscylliinae: see *Selachimorpha.*

Prosencephalon: see Brain.

Prosimians (plate 10): a suborder of the Primates (see). P. are nocturnal, arboreal, monkey-like animals, with wet noses that aid smell location. Nose and upper lip are relatively immobile. Their finger and toes have mostly flat nails, and the second toe carries a long claw. The eyes lack a fovea, and a light-reflecting tapetum behind the eye enhances nocturnal vision. P. possess a characteristic dental comb or scraper, formed by forward projection of the front teeth of the lower jaw. A horny structure (sublingula) under the tongue is used for cleaning the comb. Dental formula: 2/2, 1/1, 3/3, 3/3 = 36, with the exception that the Indriids (see Lemurs) have 30 teeth and the Aye-aye has 18. Many genera are found only in Madagascar. See Lemurs, Aye-aye, Loris, Galagos. Tarsiers (see) may also be classified with the P., but they are usually placed in a separate suborder between the simians and the P. Tree shrews (see) are considered to be P. by some authorities.

Prosobranchia(ta): see *Gastropoda.*

Prosoma: see *Chelicerata.*

Prospora: see *Oligochaeta.*

Prostaglandins, ***PG:*** a family of unsaturated, hydroxylated fatty acids, with hormone-like action, which are widely distributed in mammals. Chemically, they are formal derivatives of prostanic acid, individual, PG differing in the number and positions of their double bonds, hydroxyl groups and keto groups. They are classified in the series A to H (e.g. PGA, PGF), and the number of double bonds in the side chain is given by the subscript, e.g. PGE_2.

Prostanic acid

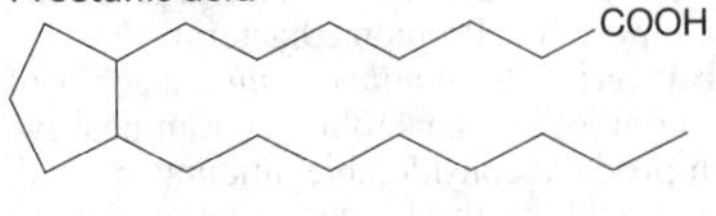

COOH

Prostaglandin E_2

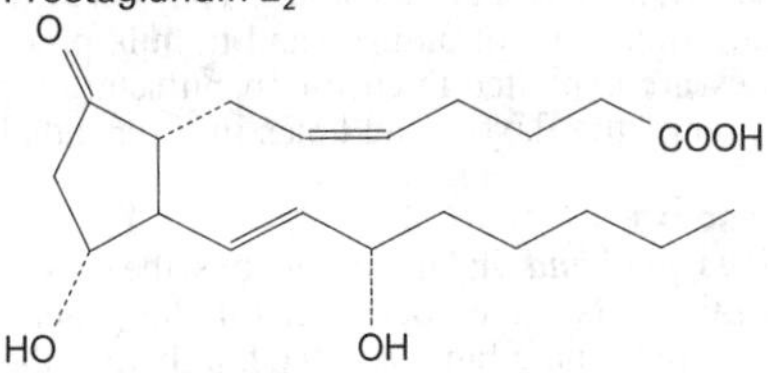

O

COOH

HO OH

Prostaglandin $F_{2\alpha}$

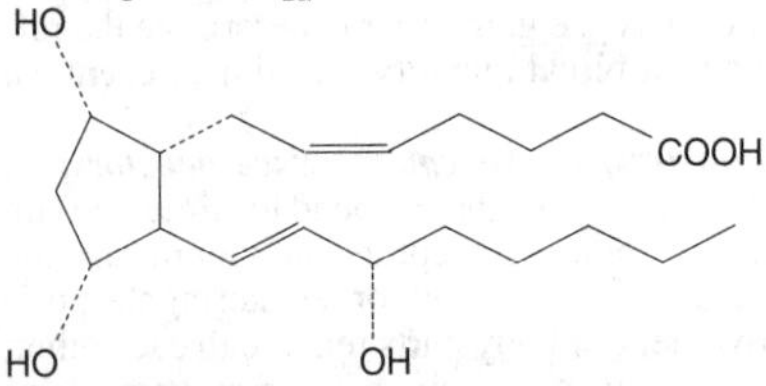

HO

COOH

HO OH

Animal tissues contain only about 1 mg PG per g fresh mass. Seminal fluid is particularly rich in PG (about 300 mg/ml). PG are determined by sensitive methods, such as biotests, radioimmunoassay, or combined gas chromatography/mass spectrometry.

PG are biosynthesized in several stages by the oxidation of long-chain, polyunsaturated fatty acids. Arachidonic acid is a major precursor. Together with PG, arachidonic acid also gives rise to the structurally related *thromboxanes, prostacyclins* and *leukotrienes,* all of which are grouped together as *eicosanoids.* The eicosanoids display a variety of biological actions, and all possess exceptionally high biological activity (potency).

PG are involved in the control of blood pressure and the inhibition of tissue inflammation, and they stimulate uterine contraction. Functionally, they appear to occupy an intermediate position between tissue hormones and transmitters. They have very high biological activity (1 ng PG/ml causes contraction of smooth muscle), but their action is usually short-lived.

Prostate, ***prostate gland:*** a gland at the base of the urethra of male mammals. It secretion makes a considerable contribution to the volume of the male ejaculate. In humans there is tendency for enlargement of the P. with increasing age, with a resulting restriction of urinary flow.

Prosthemadera: see *Meliphagidae.*

Prosthetic group: see Enzymes.

Protactinium-thorium dating: see Dating methods.

Protamines: strongly basic proteins, consisting of 80–85% arginine (e.g. salmine), which occur in association with DNA. They can be conveniently isolated from fish and bird sperm, and they function in gene repression.

Protandrous: the adjective applied to organisms that display Protandry (see).

Protandry:

1) In hermaphrodite organisms, the maturation of the male gametes earlier than the female gametes, e.g. in many parasitic worms and certain *Pulmonata.* Like Protogyny (see), protrandy largely prevents self fertilization.

2) In insects, the appearance of males before females during the course of the year.

3) See Pollination.

Protanopy: see Color blindness.

Proteases: hydrolases that catalyse hydrolytic cleavage of the peptide bonds (R—NH—CO—R) of proteins and peptides, to produce smaller peptides and/or amino acids. They are widely distributed and very important in the metabolic degradation and conversion of proteins in all cells and tissues, as well as playing a crucial role in the digestion of dietary proteins in the alimentary canal of higher animals and in extracellular digestion processes of lower organisms. Proteases of the alimentary canal are among the longest known and most extensively studied P., e.g. Trypsin (see), Chymotrypsin (see), Pepsin (see) and Rennin (see). Blood coagulation is the result of several sequentially acting P., each activating the next member of the cascade by partial proteolysis. Intracellular P., localized in the lysosomes, are responsible for degradation of the cell's own proteins. Some lysosomal P. have no apparent prostheic groups or cofactors, while others are metalloenzymes, in particular, zinc-enzymes. Most proteases are formed from enzymatically inactive, higher molecular mass precursors (see Zymogens), which are converted into the active enzymes by specific limited proteolysis by other P., or by autocatalysis. Some P. exhibit high substrate specificity (e.g. rennin is specific for casein), whereas others (e.g. pepsin, trypsin, chymotrypsin) attack practically all dietary proteins, but may be specific for certain types of peptide bonds. Thus trypsin attacks only arginyl and lysyl peptide bonds, whereas no marked amino acid specificity is displayed by pepsin.

According to the site of attack, P. are classified into 2 groups. 1) ***Endopeptidases (proteinases)*** cleave proteins and large polypeptides in the interior of the polypeptide chain, producing variously sized, relatively large cleavage products. These P. are further subdivided into ***serine proteases*** (the active center contains serine, e.g. trypsin, chymotrypsin), thiol proteases (the active center contains cysteine, e.g. papain), ***acid proteinases*** (activity optimum at an acid pH, e.g. pepsin), and ***metalloproteinases*** (a particular metal ion is specifically bound and necessary for activity). 2) ***Exopeptidases (peptidases)*** cleave one amino acid at a time from the end of the peptide chain. ***Carboxypeptidases*** attack the *C*-terminus, whereas ***aminopeptidases*** attack the *N*-terminus. The end result of this stepwise degradation is a dipeptide, which is hydrolysed into two amino acids by a dipeptidase.

Protective adaptations (plate 2): animal adaptations, which protect their carriers from the unfavorable effects of other organisms, mostly predators. Such adaptations occur in shape, posture, color, patterning, or movement capability. Resistance (see) and Immunity (see) are physiological adaptations against invasion by pathogenic organisms and parasites. Passive P.a. include protective coverings of chitin or calcareous shells, as well as spines. The distinction between active and passive P.a. is less clear in the case of chemical defenses, which may be feeding deterrants, or may be employed actively (see Venomous animals).

Similarly, the ability to change color represents an active adaptation, whereas adaptive coloration (protective coloration) (often also associated with appropriate changes in body shape) is a passive adaptation. In most cases, active or passive color adaptations serve as camouflage, e.g. the white fur or plumage of Arctic animals (polar fox, snow hare, snow owl), the yellow or brown hues of desert and steppe inhabitants (lion, fennec fox, antelope), and the green color of animals that inhabit green plants (grasshoppers, tree frogs, tree snakes) represent an adaptation to the environment, known as *homochromy.* Examples of camouflage are also seen in the differently colored summer and winter coats of certain birds and mammals (willow grouse, stoat), and in the inconspicuous coloration of female animals (mainly birds). Even more effective camouflage is achieved by *mimesis,* i.e. the imitation of other objects through adaptative alterations of both color and shape. Adaptive similarity to another animal is called *zoomimesis.* Similarity to a plant or plant part is called *phytomimesis.* An adaptive resemblance to an inanimate object, e.g. a stone or animal dropping, is known as *allomimesis.* Examples are: leaf insects and stick insects (orthopteroid insects); various larvae and imagoes of the lepidopteran family *Geometridae;* beetles and spiders that inhabit lichens; and the fish known as the Leafy sea dragon *(Phyllopteryx eques),* of Australian coastal waters, which has numerous reddish brown frondlike and spiny processes, and closely resembles a piece of seaweed.

These adaptations are made more effective by body postures that make the animal less conspicuous. This may mean a total absence of movement, known as *akinesis.* Thus, young birds (primarily those hatched in ground nests) and hares press themselves close to the ground. Other animals (e.g. the stick insect) stiffen and appear to be paralysed *(catalepsy),* showing a greatly decreased reponse to stimulation. Bitterns adopt an extended vertical posture like a post among the reeds. As a response to a perceived threat, many arthropods simply fall and become rigid as if dead *(thanatosis),* e.g. click beetles and dung beetles.

Adaptations that render prey indistinguishable from their background are known as *crypsis.* Crypsis may involve homochromy, countershading, or disruptuve coloration. If, as a result of crypsis, the animal to blends totally with its surroundings, the condition is known as *somatolysis.*

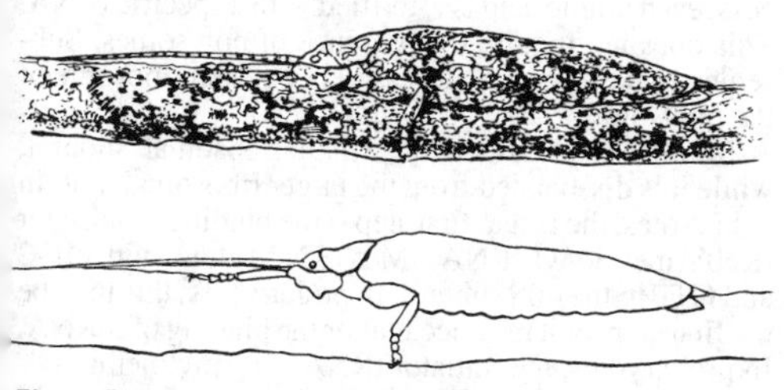

Fig.1. *Protective adaptations.* The grasshopper, *Satrophyllia femorata,* resting on a lichen-covered twig; the same picture is drawn in outline below (after Hesse-Doflein, 1943).

Some animals make themselves less conspicuous by *masking.* Thus, many crabs, but also sea urchins, cover themselves with material from the surroundings, such as seaweed and even sea anemones; larvae of the fly bug *(Reduvius personatus;* an assassin bug, family *Reduviidae)* use dirt and dust; larvae of the cereal leaf beetle cover themselves completely with their own feces.

Threat, fright and *warning* colorations and patterns are adaptations intended to deter an aggressor. They are often accompanied by appropriate noises or vocalizations, and by threatening postures. Aggressors are deterred simply by the unusual and conspicuous nature of such colors, as in the firebug *(Pyrrhocoris),* or by the sudden display of a striking color pattern, like the eyespots of many butterflies, or the red hindwings of the locust *Psophus stridulus.* Warning colors are displayed by dangerous or unpalatable animals, e.g. coral snakes or salamanders. Coloration intended to attract a predator's attention, i.e. to provide a deliberate warning, is known as *aposematic coloration.*

Mimicry is a special type of P.a. In *Batesian mimicry,* a well protected (e.g. venomous or unpalatable) animal with warning coloration is imitated by a different species. This latter species may not be venomous or unpalatable, but the imitation is so accurate that predators avoid it. Thus, the hornet clearwing *(Trochilium crabroniformis;* family *Aegeriidae)* benefits from its close resemblance to a hornet. The strikingly colored and venomous coral snakes are mimicked by various harmless snakes. On the other hand, *aggressive mimicry* or *Peckhammian mimicry* serves to conceal a predator from its prey, or even to attract the prey to the predator, e.g. some tropical mantids resemble flowers. The close resemblance of the saber-toothed blenny *(Aspidontus taeniatus)* to the wrasse *(Labroides dimidiatus)* represents another type of aggressive mimicry. As a cleaner fish, *Labroides* removes ectoparasites from larger fishes, which therefore readily accept it, whereas *Aspidontus* feeds on skin that it tears from larger fishes. *Wasmannian mimicry* is found among inquilines of ant colonies; the close resemblance of these ant guests to their hosts probably serves to deter predators (it is not thought to deceive the ants, which rely largely on chemical recognition signals). In *Müllerian mimicry,* two or more dangerous or unpalatable species adopt the same warning coloration; this increases the rate at which predators gain experience, and therefore decreases the likelihood of being "sampled" by an inexperienced predator.

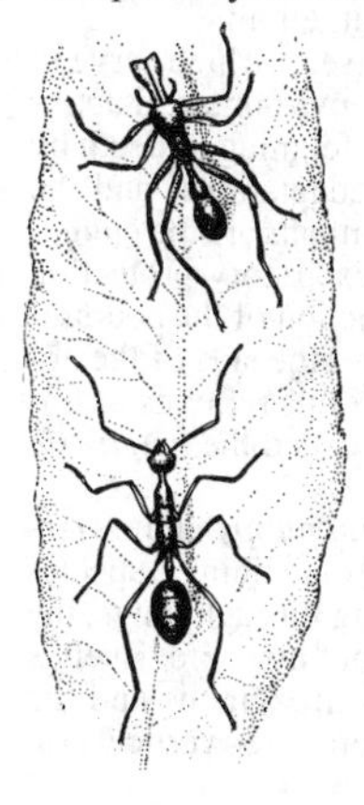

Fig.2. *Protective adaptations.* The spider, *Myrmarachne* (above), acquires protection by imitating a weaver ant worker (below) (after Hesse-Doflein, 1943).

The imitation of an object collectively by several animals is known as *collective mimesis* (e.g. small cicadas may aggregate in imitation of an infloresence). *Ethomimicry* is the imitation of the behavior of other species (e.g. imitators of cleanerfish). *Automimicry* is the existence of polymorphism for palatability to predators. For example, depending on the food plants chosen by the ovipositing female, the feeding larvae of the monarch butterfly may or may not consume materials that are toxic to birds. These toxins do not affect the larvae, and they are passed on to the adult insect. For the automimicry to be effective, the unpalatable insects must be more numerous and therefore more frequently encountered by birds.

Behavior such as burrowing, the construction and seeking of places of concealment, and the general ability to flee must all be looked upon as P.a. This kind of protective behavior can be triggered by acoustic, mechanical or chemical signals between members of the same species, and sometimes by signals from different species (e.g. the warning call of the jay). Termites signal alarm by knocking. Examples of chemical signals are the alarm chemicals of shoaling fishes, and the alarm pheromones of social insects.

Structures that can be used as weapons (horns, antlers, claws, mouthparts), the electric organs of certain fishes [electric ray *(Torpedo),* electric eel *(Electrophorus),* electric catfish *(Malapterurus)],* and the wide variety of chemical defenses (secretions, blood, body cavity fluids, nematocysts) also represent P.a.

All P.a. confer a relative advantage on their carriers, rather than absolute protection.

Protective behavior: behavior that defends or protects against an aggressor. It may be preventative (see Protective adaptations), in that the animal covers itself or withdraws into a place of security; it may involve the active avoidance of danger by flight or akinesis; or it may be an active defense, e.g. by release of repellant substances (see Pheromones). Predators can be confused or exasperated by sudden changes in the prey, e.g. an increase in size by inflation. In order to protect their offspring, birds attract the attention of predators and lead them away from the nest. Swarming animals perform certain monoeuvres that confuse potential attackers. Finally, in response to a perceived threat, certain body parts or appendages may be shed by detachment at a predetermined site (see Autotomy), e.g. shedding of the tail by certain lizards, and the loss of large feathers by some birds (fright molting). See Protective adaptations.

Protective epithelium: see Epithelium.

Protective reflexes: Reflexes (see), which serve to protect or defend against adverse environmental activity. Examples are the "freezing" or feigning of death by insects (see Akinesis, Protective adaptations) and the *wiping reflex* of the frog. P.r. in mammals are the *coughing reflex, sneezing reflex* (see Respiratory protection reflexes), *abdominal reflex* (contraction of the abdominal wall muscles when the skin over the side of the abdomen is stroked; the upper part of this reflex is a contraction at the epigastrium, and is called the *epigastric reflex)* and *blinking reflex.*

Proteidae, *Necturidae,* ***olm, mudpuppy*** and ***waterdogs:*** a family of the *Urodela* (see) containing 6 species in 2 genera. In addition to breathing through lungs, the adults have external gills throughout their life. Eyelids are lacking. The skull is largely cartilaginous and the limbs are very degenerate. There are vertical rib grooves along the sides of the body. All members of this family are "permanent larvae" in a fixed state of Neoteny (see), and they have never been observed to undergo metamorphosis.

Proteus anguinus ***(Olm)*** has a 30 cm-long, milk-white, elongate, slender body, with widely spaced, weak limbs carrying 3 fingers and 2 toes. The tail is laterally compressed, and the animal progresses by eel-like body undulations. The degenerate eyes are covered by skin. On the neck are 3 external, blood-red gills. The latter are never lost, but respiration also occurs via the lungs. It inhabits cold, subterranean cave pools near the Yugoslavian coast; the best known (and strictly protected) habitat is the Karst cave system of Postojna. They can survive briefly out of water. In the wild, the young are probably borne live, but eggs are laid at higher temperatures under laboratory observation. The spawning site is guarded by the female.

Necturus is a native North American genus. *Necturus maculosus* ***(Mudpuppy)*** is 40 cm in length with a dark, spotted, stout body, a large, flat head and a short tail. It is not subterranean, and it lives in mud or stagnant water where there is abundant vegetation. There are 4 fingers and 4 toes. Mating takes place in the autumn, and the eggs are laid singly in the water in the springtime. Larvae become sexually mature after 5 years. The remaining 4 species are called ***Waterdogs*** (e.g. *N. beyeri,* ***Gulf Coast waterdog)*** with a similar appearance and mode of life to those of the mudpuppy.

Proteinaceous infectious agents: see Prion.

Protein biosynthesis: a cyclic, energy-dependent, multistage process, in which free amino acids of the cell are polymerized to polypeptides in a genetically determined sequence. P.b. represents the translation of genetic information carried by mRNA. It requires the presence of mRNA, Ribosomes (see), tRNA, amino acids and a number of enzymes and protein factors, many of which are integral components of the ribosome. Various cations, ATP and GTP (guanosine triphosphate) are also required. The mechanism and the individual steps of P.b. were elucidated mainly by careful preparation of the functional components, by the use of synthetic polynucleotides, and in particular by the use of specific inhibitors in cell-free systems. The process is similar in prokaryotes and eukaryotes, althought it involves more factors in the latter. A standardized nomenclature exists for the factors involved in initiation (IF-1, IF-2, etc.), elongation (EF-1, etc.) and termination (release factor, RF). To distinguish them from prokarytoic factors, eukaryotic factors are designated "eIF-1", etc. There are 4 phases of P.b.

Activation of amino acids. In an ATP-dependent process, each amino acid is esterified with a specific tRNA. This does not require the presence of polysomes. Subsequent phases occur successively on each ribosome of the polysome.

Initiation occurs on the small ribosomal subunit, while it is dissociated from the larger ribosomal unit. In eukaryotes, the actual first step is the binding of initiator tRNA (methionyl-tRNA = Met-tRNA) along with eIF-2 and GTP to the 40S subunit. In prokaryotes, this may be the first step, or it may occur after the binding of mRNA. In prokaryotes, the initiator tRNA is formylmethionyl-tRNA (fMet-RNA). The assembled complex of 30S subunit, IF-1, 2 and 3, GTP and mRNA with the fMET-tRNA is called the 30S-initiation complex. The corresponding initiation complex in eukaryotes involves another factor, eIF-4, and ATP, which are required for binding the mRNA. The 40S initiation complex is therefore 40S:eIF-3;Met-tRNA:eIF-2:GTP:eIF-4:mRNA.

The initiation codon for both pro- and eukaryotes is AUG. The initiation complex is prepared for the addition of the large ribosomal subunit by the release of IF-3. In bacteria, the 50S subunit appears simply to replace IF-3, then IF-1 and IF-2 leave the complex afterward. In eukaryotes, another factor, eIF-5, catalyses the departure of the previous initiation factors, and the attachment of the 60S subunit. In both cases, release of IF-2 involves hydrolysis of the GTP bound to it. Also, the Met-tRNA becomes bound to the P site of the large ribosomal subunit.

Elongation. The ribosome can accommodate two tRNA molecules at once. One of these carries the Met-tRNA (or fMet-tRNA) or the peptide-tRNA complex, and is therefore called the P-site. The other accepts the incoming aminoacyl-tRNA and is therefore called the A-site. On the ribosome, a reaction is catalysed between the carboxyl of the P-site occupant and the (free) amino group of the A-site occupant. The peptidyl transferase activity which catalyses this is intrinsic to the ribosome. The final step of elongation is the motion of the ribosome relative to the mRNA, accompanied by translocation of the peptidyl-tRNA from the A- to the P-site.

Termination. The end of polypeptide synthesis is signaled when, as a result of the movement of the ribosome relative to the mRNA, a termination codon is positioned at the A-site.

Although this account describes the process on a single ribosome, it should be remembered that the functional system is the Polysome (see). At any one time, several ribosomes are positioned along the mRNA; those nearest the 3'-end carry the longest newly synthesized polypeptide chains, whereas those at the 5'-end have translated fewer codons and therefore carry a shorter length of peptide. Thus, initiation, elongation (at various stages) and termination proceed simultaneously on the same length of mRNA.

After the termination of P.b., many newly synthesized polypeptides or proteins are subject to further reactions to convert them into a biologically active form (posttranslational modification), e.g. covalent attachment of certain groups, such as the carbohydrate moieties of glycoproteins, or removal of certain amino acid sequences. Thus, although methionine or formylmethionine is the first amino acid to be incorporated in the biosynthesis of all proteins, many proteins do not have

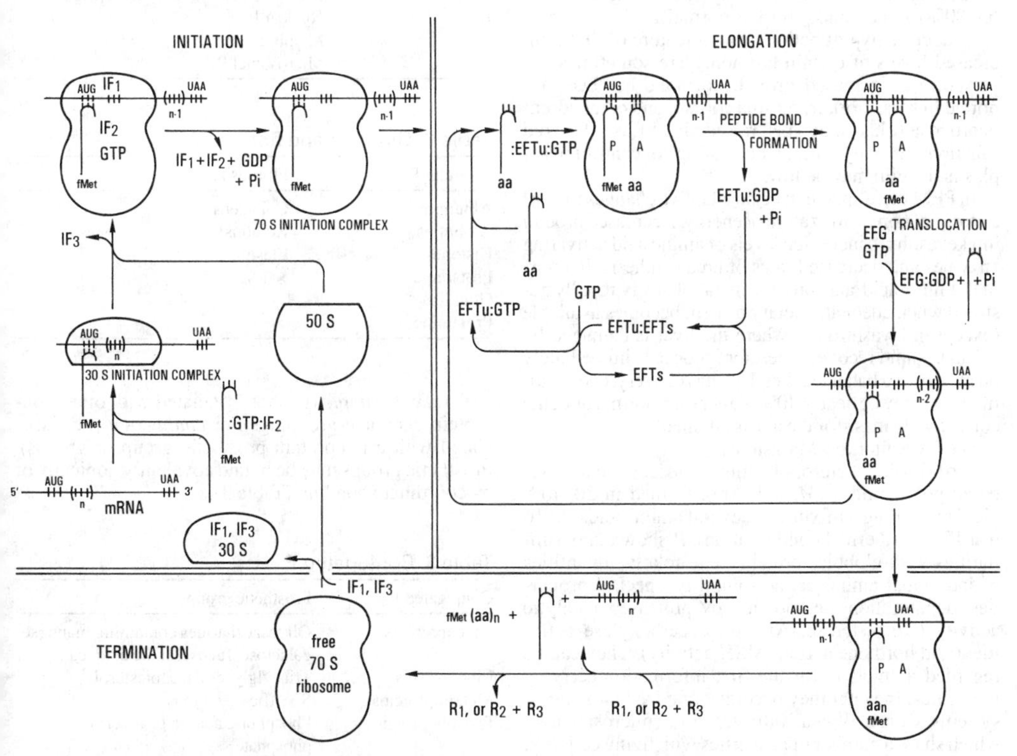

Protein biosynthesis. Diagrammatic representation of translation on prokaryotic ribosomes. The elongation cycle starts by interaction of the 70S initiation complex with fMet-tRNA:EFTu:GTP. In all subsequent rounds of the cycle, fMet-tRNA:EFTu:GTP interacts with the mRNA:ribosome complex carrying the growing polypeptide chain. Termination occurs when n amino acids have been incorporated, where n represents the number of codons between the initiation codon AUG and the termination codon (in this example UAA).

methionine as their N-terminal amino acid, because this is removed during posttranslational processing.

See Genetic code, Ribonucleic acid.

Protein-energy malnutrition, ***PEM, protein-calorie malnutrition, PCM:*** a spectrum of nutritional deficiency states, occurring characteristically in children under 5 years, although no age is immune. Marasmus and kwashiorkor are often regarded as the two extremes of this spectrum, but it is now known that there are no essential differences between the dietary protein: energy ratios of these two conditions (see Kwashiorkor). Plasma albumin concentrations below 35 g/l are characteristic of protein deficiency and of PEM. Gamma globulin, on the other hand, may be raised in response to the presence of infection. There is always a decrease in metabolic rate, but this is roughly proportional to the decrease in cell mass. Plasma concentrations of branched chain amino acids and tyrosine are lower than normal, whereas the concentrations of some nonessential amino acids may be increased; similar abnormal patterns of plasma amino acids appear after only 4 days on a protein-free diet.

Loss by diarrhea is probably largely responsible for the low plama potassium often observed in PEM; values less than 2.5 mmol/l have been recorded. Plasma magnesium may be low for the same reason. Plasma sodium is usually normal. Total body water increases to 65–80% of body mass (60% is normal).

Endocrine hypofunction is not a feature of PEM. Increased levels of certain hormones are sometimes observed, e.g. increased growth hormone has been reported in kwashiorkor; plasma cortisol and other adrenocorticosteroids may be raised in PEM. Thyroid function is usually normal, but fasting concentrations of plasma insulin may be low.

In PEM there appear to be no adaptive changes toward a more economic utilization of energy. Decreased protein intake results in increased levels of amino acid activating enzymes and decreased rates of urea synthesis, but normal amino acid and protein metabolism is rapidly restored when adequate dietary protein becomes available (except in kwashirokor, where the liver is damaged by toxins). Rapid recovery therefore proceeds immediately on receipt of a balanced diet. During recovery, energy utilization may be nearly 40% higher than normal until the correct body mass for the age is attained.

See Kwashiorkor, Marasmus.

Proteinoids: heteropolyamino acids; artifically prepared polypeptides (M_r > 1 000), formed in 20–40% yield by heating a mixture of several amino acids for 16 h at 170 °C (thermal condensation). P. show many similarities with globular proteins, e.g. relative quantities of individual amino acids, solubility, spectral properties, denaturation, degradation by proteases, catalytic activity (e.g. esterase, ATPase, decarboxylase activities) and hormone action (MSH activity). They can be regarded as models for the first information-carrying molecules. In water they become organized into microsystems with a typical ultrastructure (microspheres), which share a number of properties with living cells, i.e. they possess a bilayer membrane, which exhibits a certain degree of semipermeability; they can also multiply in the absence of nucleic acids, by budding.

Proteins: natural substances of high molecular mass, consisting of polymers of amino acids, often with other chemical groups also attached. They are essential components of all living cells, representing up to 50% of organic cell material, and performing a wide variety of different functions. As enzymes they are responsible for the operation and regulation of biochemical reactions. They also serve as proteohormones, structural P. of the cell, storage P., transport P., defense P., blood coagulation P., as well as being responsible for blood group specificity, etc.

Although the chemical structures of many P. are known, classification is not based on structure, but mainly on their function, occurrence and physical properties (Tables 1 & 2). *Globular P.* are soluble in water and dilute salt solutions, and they have a more or less spherical shape. *Fibrous P.* are practically insoluble in water and salt solutions; their molecules have a long axis, and they consist of polypeptide chains arranged in parallel.

Table 1. Classification of P. according to occurrence and function.

Organism	Organ	Function
Plant P	Plasma P.	Enzyme P.
Animal P.	Muscle P.	Structural P.
Bacterial P.	Milk P.	Transport P.
Viral P.	Egg P.	Storage P.
	Ribosomal P.	Receptor P.
	Nuclear P.	
	Membrane P.	
	Microsomal P.	

Table 2. Globular and fibrous P.

Globular P.	Fibrous P.
Albumins	Collagens
Globulins	Keratins
Histones	Elastins
Protamines	
Glutelins	
Prolamines	

P. may be *simple* (i.e. not associated with other nonprotein chemical groupings), or *conjugated* (i.e. associated with a nonprotein prosthetic group or groups). Prosthetic groups may be bound covalently, ionically or by coordinate bonding (Table 3).

Table 3. Conjugated P.

Conjugated P.	Prosthetic group
Glycoproteins	Oligosaccharides containing mannose, galactose, fucose, sialic acid, etc.
Lipoproteins	Triacylglycerols, cholesterol
Metalloproteins	Specific metal ions
Phosphoproteins	Phosphoric acid (e.g. as serine phosphate)
Nucleoproteins	Nucleic acids
Chromoproteins	Pigment molecules (e.g. heme)

The relative molecular mass (M_r) of single chain P. lies between 10 000 and 100 000, whereas that of multichain P. may be several millions. P. can be separated on the basis of their different sedimentation rates in the ultracentrifuge. Other widely used separation methods include molecular sieve chromatography and electrophoresis.

Determination of the identity and sequence of the amino acid residues that constitute any P. is now a relatively routine and automated laboratory procedure, based on the chromatographic identification of amino acids and peptides, following total or complete degradation by chemical, enzymatic and physical methods. The linear sequence of amino acids is known as the *primary sequence* or *primary structure* of a protein. The primary sequences of more 1 000 proteins have been elucidated.

The *secondary structure* describes the configuration of the polypeptide chain, which is stabilized by hydrogen bonds between the carboxyl oxygen and amido nitrogen of different peptide bonds. Well documented secondary structures are the *α-helix,* stabilized by hydrogen bonds within the peptide chain, and the *pleated sheet* or *β-structure,* stabilized by hydrogen bonds between polypeptide chains. *Tertiary structure* refers to the overall three-dimensional arrangement of the polypeptide chain, which is maintained by intramolecular forces. Recognition of the structure and position of active centers (e.g. catalytic centers of enzymes) depends on the elucidation of the tertiary structure. The *quaternary structure* arises from noncovalent intermolecular interactions between two or more polypeptide chains, which associate or aggregate to form a stable oligomeric P. The quaternary structures (number, identity and arrangement pattern of the protein monomers) of nearly 1 000 oligomeric proteins have been elucidated. Secondary, tertiary and quaternary structures are determined largely by X-ray diffraction.

Properties. P. possess both basic and acidic functional groups, and as typical ampholytes they have important buffering properties. Their basic or acidic properties predominate, depending on the pH of the medium in which they are dissolved. At their isoelectric point, P. exist as zwitterions, and their solubility and hydration attain a minimal value. Solubility of P. is also determined by electrolyte concentration, type of solvent, and their amino acid composition. The presence of polar amino acid residues on the molecular surface favors the formation of a hydration layer and prevents precipitation. Nonpolar sovents and high concentrations of neutral salts cause precipitation, a phenomenon exploited for the crude separation of different P. In P. with an asymmetric charge distribution (e.g. serum glubulins) a certain salt concentration suppresses association and aggregation of the P. molecules, and stabilizes the P. in solution (salting in effect).

Heating, irradiation, physical agitation, treatment with acids, bases or detergents, all cause *denaturation* by breaking bonds between peptide chains. The process is usually reversible, provided it is performed under mild conditions, and only noncovalent interactions are affected. Most denaturation reactions, however, are accompanied by other changes, such as disulfide bond exchange, unspecific oxidation reactions, or the formation of new covalent bonds, leading to an irreversible structural alteration and loss of biological activity.

Protein yeasts: see Yeast protein.

Proteles cristatus: see Aardwolf.

Protelidae: see *Hyaenidae.*

Proteocephaloidea: see *Cestoda.*

Proteogenic amino acid: see Amino acids.

Proteolepas: see *Cirripedia.*

Proteolysis: hydrolytic cleavage of the peptide bonds of proteins by enzymes (see Proteases) or acids.

Proteolytic organisms: a general term for microorganisms that produce proteolyic enzymes (proteases, peptidases). They degrade proteins and utilize the resulting amino acids as nutrients. As agents of putrifaction and decomposition P.o. play an important part in the cycling of elements in the biosphere. They include both fungi and bacteria, e.g. *Proteus vulgaris, Clostridium* spp. and streptomycetes.

Proteroglyphic dentition: see Snakes.

Proterozoic: an early period in the evolution of animal life, 2 600 million to 570 million years ago, and an important period for the evolution of the early eukaryotes. It corresponds to the end of the Precambrian, and it is the earliest period with an abundant fossil record.

Proteus: a genus of Enterobacteria (see). Species of *P.* are rod-shaped, Gram-negative, facultatively anaerobic bacilli, 1.5-3 µm x 0.5 µm, but also exhibiting pleomorphism. The cells occur individually, in pairs or in chains, and they are notable (e.g. *P. vulgaris)* for their ability to spread rapidly over the surface of solid media, forming a thin, almost transparent film. Motility is due to the presence of numerous flagella. *P.* is an organism of putrifaction and decay, found in decaying animal and vegetable matter, and in sewage, and often in the human intestine. It is often found as a secondary opportunistic pathogen in human infections, especially cystitis. The ability of *P.* to hydrolyse urea is used diagnostically.

Prothonotary warbler: see *Parulidae.*

Prothoracic glands: paired, independent molting glands of holometabolous insects, which are homologous with the ventral glands of hemimetabolous insects. They have very different shapes, depending on the insect order, but they are usually elongate and often branched, and situated laterally between the head and thorax. Cells of the P.g. synthesize and secrete molting hormone (see Ecdysone), which is necessary for each ecdysis (molt). They therefore display cyclic activity, and are most active immediately before the emergence of the imago from the pupa. They finally degenerate in the imago.

Protists, *Protista:* a kingdom, superphylum or phylum including all unicellular organisms (protozoa and protophytes). Haeckel regarded the P. as a third, neutral kingdom, between the animal and plant kingdoms, since the boundaries between animals and plants are often indistinct in uncellular organisms. Even when P. associate into colonies, each cell remains undifferentiated and identical to its neighbors. Some P. show differentiation into generative and somatic cells, but different tissues are never formed.

Protoascomycetidae: see *Ascomycetes.*

Protobranchia: see *Bivalvia.*

Protoceratidae: see *Traguloidea.*

Protocoel: see Coelom, Mesoblast.

Protocephalon: see Mesoblast.

Protocephalopoda: see *Cephalopoda.*

Protoceratops: see Dinosaurs.

Protocerebral segment: see Mesoblast

Protocerebrum: see Brain.

Protoconch (Embryonic chamber): see Ammonites.

Protoderm: see Histogen concept, Meristems.

Protogynous: the adjective applied to organisms that display Protogyny (see).

Protogyny:

1) In hermaphrodite organisms, the maturation of the female gametes earlier than the male gametes, e.g. in the *Tunicata.* Like Protandry (see), protogyny prevents self fertilization.

2) See Pollination.

Protomecoptera: see *Mecoptera.*

Protomonadina: see *Mastigophorea,* Trypanosomes.

Protonema: see Bryophytes.

Protonephridia: excretory organs of all acoelomates and pseudocoelomates, and particularly characteristic of flatworms and aschelminths. P. are paired, often profusely branched tubules which communicate with the exterior via excretory pores. Each tubule commences in the parenchyma with a blind inner end consisting of a club-shaped cell *(solenocyte)* bearing a tuft of continually moving cilia that direct material into the lumen of the tubule. These ciliated solenocytes are also known as *flame cells.* Secretory fluid is formed by (ultra?)filtration through the wall of the solenocyte. Specialized P. are present in some annelids and lancelets. In this case the tubule extends into the coelom and it is furnished with many terminal tubular cells (solenocytes), each bearing a single flagellum instead of a tuft of cilia, and in which the cell nucleus is located in a terminal thickening. The term, solenocyte, is sometimes reserved for cells possessing a single flagellum, as distinct from flame cells.

Protonephros: see Excretory organs.

Protonotaria: see *Parulidae.*

Protophytes: unicellular autotrophic organisms, as distinct from heterotrophic protozoa. They comprise primarily the phytoflagellates, diatoms, unicellular green algae, and in the wider sense also bacteria and cyanobacteria.

Protoplasm: the complex, colloidal, living substance of the cell, including the cell membrane. It is differentiated into the P. of the nucleus (nucleoplasm) and the P. of the rest of the cell (cytoplasm). See Cytoplasm, Cytosol, Nucleus.

Protoplasts: naked, spherical cells of bacteria or plants, from which the cell wall has been removed experimentally with a suitable enzyme preparation. A protoplast is therefore bounded only by its plasmalemma, and retains its shape only in a medium of the correct osmolarity. P. can be cultured and will divide in appropriate nutrient media. They can also fuse to produce hybrid cells, e.g. P. of *Nicotiana glauca* and *Nicotiana langsdorfii* have been fused, and the resulting cell hybrids cultured and allowed to differentiate into plants. Progeny from the seeds of these plants were identical with hybrids produced by sexual crossing of the two species.

For removal of cell walls of bacteria and plants, see Cell wall.

For more information on cell hybridization, see Fusion, Cell fusion.

Protopterus: see Lungfishes.

Protosoma: the short, most anterior body section of the *Tentaculata, Hemichordata* and *Pogonophora.* In the *Pogonophora* it is reduced to a small epistome, which lacks a coelomic cavity, and which carries the tentacles. In the other phyla it contains an unpaired coelom.

Protostomia, *Gastroneuralia,* ***protostomes:*** a branch of the *Bilateria* (see) of the animal kingdom, in which the blastopore of the gastrula divides to give rise to both mouth and anus (hence *protostome,* "first mouth"). The central strand of the nervous system runs ventrally. Cleavage in P. is determinate, i.e. the fate of embryonic cells is fixed at a very early stage. For example, after only two cleavage divisions of the egg, each of the resulting 4 blastomeres is destined to develop into one fixed quarter of the gastrula and larva. Radial cleavage is rare in *P.,* which normally display spiral cleavage. All the entomesoderm arises from the mesentoblast cell, which is formed at the 6th cleavage. Also known as the 4d cell, this single mesentoblast cell gives rise to 2 teloblasts or primordial mesoderm cells, which are originally situated at the posterior end of the animal, and which proliferate to form two masses of mesoderm cells, one on each side of the body. In segmented *P.,* these mesodermal masses develop into linearly arranged blocks of cells, which correspond to the segments. Within each mesodermal mass, a split appears, which enlarges to form the coelom. This type of coelom formation is known as *schizocoely.* Coelomate *P.* are therefore known as *schizocoelous coelomates* or *schizocoelomates.* Chitin is commonly found in the *P.* The typical phosphagen of *P.* is arginine phosphate. Most animal phyla belong to the *P.,* which comprise the *Platyhelminthes* (platyhelminths or flatworms), *Priapulida* (priapulids), *Mollusca* (mollusks), *Sipuncula* (sipunculans or peanut worms), *Echiura* (echiurans), *Onychophora* (onychophorans), *Tardigrada* (tardigrades or water bears), *Pentastomida* (pentastomids), *Arthropoda* (arthropods) and the *Tentaculata.* The other main branch of the *Bilateria* is represented by the *Deuterostomia* (see), which differ from the *P.* embryologically and in other ways. See Blastogenesis.

Protostyle: see *Mollusca.*

Prototheria: a subclass of the mammals containing certain extinct orders *(Docodonta, Triconodonta, Multituberculata),* and only one surviving order, the *Monotremata* (platypus and echidnas). In the *P.,* the squamosal and alisphenoid bones of the skull wall are characteristically small (cf. the *Theria,* where these bones are large). The dentition of the *P.* also differs from that of the *Theria.* These two subclasses probably diverged in the late Triassic.

Protozoa: a phylum of small (mostly microscopic) unicellular eukaryotic organisms at the evolutionary base of the animal kingdom. All life functions are performed by the single cell (cf. multicellular animals, where these functions are shared between different tissues and organs). The protozoan cell often contains complex intracellular structures or *organelles,* which are reminiscent of the organs of higher animals.

Classification. Four main morphological groups (classes or subphyla) of P. are recognized: *Mastigophorea* (see), *Rhizopodea* (see), *Sporozoa* (see), *Ciliophora* (see).

Fossils. Those forms that secrete hard body parts have left a fossil record from the Cambrian to the present. In particular, a wide variety of *Rhizopodea* have left fossils, sometimes contributing to the formation of sedimentary rocks.

Protozoan theory: an outdated view that soil protozoa supposedly decrease soil fertility by consuming soil bacteria. It is now known that bacterial populations, partially depleted by protozoan feeding, undergo physiological rejuvenation, and become ecologically more effective as agents of decomposition (see).

Protozoology: the study of *Protozoa* (see). Since this phylum contains many species of serious parasites, P. is an important part of parasitology.

Protrochula: see Müller's larva.

Protura, *Myrientomata:* an order of primarily wingless, entognathous *Apterygota* (see), discovered in 1907 by Silvestri. They are small (0.5–2.5 mm) whitish to yellowish, soft- and thin-skinned, elongated insects with 12 abdominal segments (Fig.). Antennae, compound eyes, ocelli and caudal appendages are absent.

Morphology. The head carries two dorsolateral sensory organs (see Postantennal organ), which appear to serve as humidity detectors. The piercing-sucking or tearing-sucking mouthparts are styliform and extensible. When the animal progresses, the first pair of legs, which carry sensory bristles, are held aloft like antennae. Rudimentary unsegmented or 2-segmented appendages with eversible vesicles are present on the ventral side of each of the first three abdominal segments. Members of the family, *Eosentomidae,* have a simple tracheal system, but this is absent from other *P.* True Malpighian tubules are lacking, but six coronoid papillae (reduced Malpighian tubules) at the junction of the midgut and hindgut serve as excretory organs.

Life history. There are five nymphal stages, and the fourth instar is the first to possess the full number of 12 abdominal segments. The newly hatched prelarva and the subsequent two prelarval stages have 9, 9 and 10 abdominal segments, respectively; the fourth and fifth instars (preadults) have 12 abdominal segments (Fig.). Larval development lasts 4 to 7 months, and the generation time is usually about 1 year. Parthenogenesis is observed occasionally.

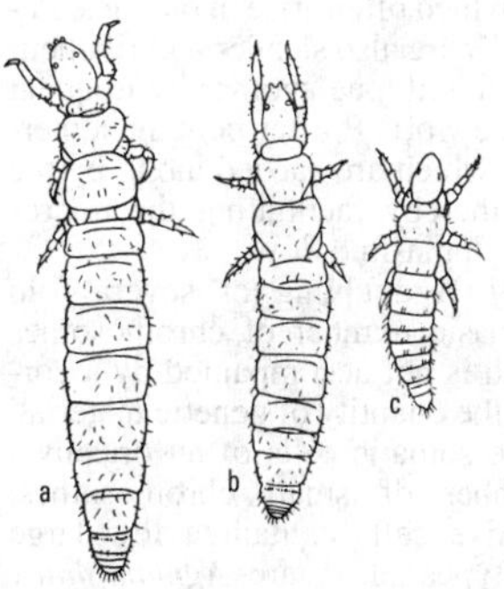

Protura. Acerentulus danicus. a Imago. *b* Preimago. *c* Prelarva.

As far as is known, *P.* feed by sucking out the contents of fungal hyphae, in particular mycorrhizal hyphae. They are usually found 2 to 5 cm below the soil surface, and all stages appear to require a very humid soil atmosphere for activity. Humus-rich soils may contain up to 10 000 individuals per square meter. Their significance as pests or as beneficial organisms is not known, but their presence is a good indication of a healthy, fertile soil.

Distribution. *P.* are found worldwide, including the polar regions. Currently about 300 species are known, divided into 3 suborders *(Eosentomoidea, Sinentomoidea, Acerentomoidea),* 8 families and 35 genera.

Provitamin D_2: see Ergosterol.

Przewalski's jird: see *Gerbellinae.*

Psalidont: see Dentition.

Psammechinus: see *Echinoidea.*

Psammodromas, ***sand lizards, plated lacertids:*** a genus of the family *Lacertidae* (see) found in the western European and North African parts of the Mediterranean region. They have large, sometimes imbricated, keeled scales on the back.

Psammon: flora and fauna of the interstitial spaces of sand grains of marine and freshwater habitats. The *epipsammon* is the relatively sparse fauna of the sand surface. In contrast, the *mesopsammon* of the cavities and spaces within the sand is relatively abundant with a wide variety of species, e.g. cnidarians, ciliophorans, nematodes, polychaetes and copepods *(Harpacticoidea),* as well as bivalves (usually very small, elongated forms) are all represented. The *endopsammon* includes burrowing animals such as the sand worm, *Arenicola marina.*

Psammophytes: plants that are typical of, and adapted to, sandy habitats.

Pselaphognatha: see *Diplopoda.*

Pseudaletia: see Cutworms.

Pseudanthium, ***false flower:*** a flower-like inflorescence, functioning biologically as a single flower, e.g. the heads or capitula of the *Compositae*, the cyathium of *Euphorbia*, and the spadix of the *Araceae*, which is often enclosed in a brightly colored bract or spathe. Formation of pseudanthia appears to be especially characteristic of terminal members of evolutionary lines. See Flower, section 5 Inflorescence.

Pseudanthium theory: a theory of plant evolution, in which the angiosperm flower is regarded as homologous with the entire gymnosperm inflorescence. The single flower is therefore regarded as a pseudanthium, each stamen or carpel corresponding to a single, much reduced male or female flower. An opposing theory is the Euanthium theory (see).

Pseudidae, ***pseudids:*** a family of the *Anura* (see) containing 4 species in 2 genera (*Lysapsus* and *Pseudis*), distributed in tropical South America in the lowlands east of the Andes and in the Magdalena Valley in Colombia. They have yellow and black striped limbs, large protruberant eyes and robust hindlimbs. As an adaptation to swimming the feet are fully webbed, and an extra, ossified, intercalary element (phalanx bone) is present between the terminal and penultimate phalanges of the fingers and toes. Eight, holochordal, procoelous presacral vertebrae are present with nonimbricate (nonoverlapping) neural arches. Presacrals I and II are not fused, and the atlantal cotyles of presacral I are juxtaposed. The coccyx lacks transverse processes, and has a bicondylar articulation with the sacrum. Omosternum and sternum are cartilaginous, and the pectoral girdle is arciferal. Ribs are absent, and the pupils are horizontal. All are aquatic with body lengths up to 8 cm.

Pseudis paradoxus ***(Paradoxical frog),*** found in the Amazon basin, is mottled green, brown, yellow and black, with a pointed snout and prominent nostrils and eyes. The adult frog is 6–8 cm in length, but the tadpole attains the relatively enormous length of 25 cm, and decreases in size during metamorphosis.

Pseudoallele: closely linked genes (cistrons) which have related functions, occupy a specific region of the linkage group, and in rare cases can be recombined by crossing over. P. are generally transferred together as a single unit (known as a complex locus), and in the heterozygous state they show a positional effect, depending on their configuration, i.e. they give rise to different phenotypes according to whether they have a cis or trans configuration. For example, if A and B represent the normal alleles of two neighboring pseudoallelic loci, and the cis configuration, AB/ab, gives the wild phenotype, then the trans configuration, aB/Ab, produces the mutation phenotype, and the mutated a and b alleles behave like the alleles of a single gene. If A and B show incomplete dominance, and the configuration AB/ab produces the mutant phenotype, then the mutant character of aB/Ab is always more strongly expressed.

Pseudoarcifery: see *Anura.*

Pseudobranchi: see Tracheal gills.

Pseudobulb: see *Orchidaceae.*

Pseudocalyptomena: see *Eurylaimidae.*

Pseudocells: rounded skin pores closed by a thin chitin membrane, which are found in the *Onychiuridae* [a family of the *Collembola* (springtails)]. In a process known as reflex bleeding, P. can exude a viscous, repellant-containing hemolymph as a defense against predators (e.g. predatory mites).

Pseudochelidon: see *Hirudinidae.*

Pseucoculus: see Postantennal organ.

Pseudo-extinction: see Extinction.

Pseudofirmisterny: see Anura.

Pseudogene: a DNA segment whose nucleotide sequence resembles that of a functional gene, but does not give rise to mRNA. There is therefore no polypeptide product.

Pseudohaploid: the term applied to haploid forms which arise from autopolyploids (see Polyploidy), and which contain more than one set of chromosomes. See Monohaploidy.

Pseudois: see Blue sheep.

Pseudolamellibranchs: a morphological term for bivalves whose ctenidia have evolved to the stage where they resemble lamellar ctenidia, but in which the individual gill filaments are not yet fused with one another (cf. eulamellibranchs in which the gill filaments are fused). Examples are the genera *Ostrea* (oysters), *Meleagrina* (pearl mussels) and *Pecten* (scallops).

Pseudomonas: a large genus of rod-shaped, motile, Gram-negative bacteria, carrying one or more monopolar flagella. *P.* species live free in the soil and in water, or parasitically in animals, humans or plants. In natural environments they make an important contribution to the mineralization of organic materials. *P. fluorescens* is found in all fertile soils and in surface waters; it produces a greenish yellow fluorescent pigment in culture. *P. aeruginosa (= P. pyocyanea)* is distinguished from other *P.* by the production of a blue pigment, known as pyocyanin. Small numbers of *P. aeruginosa* are present in the normal intestinal flora of humans and animals, and the organism is also found on healthy human skin. It is sometimes found in infected wounds and burns, where it is responsible for the greenish blue color of the pus. Infections with *P. aeruginosa* are usually localized, but may lead to a fatal septicemia in infants or debilitated individuals.

Pseudomonotis (Greek *pseudos* erroneous, *monos* single, *ous* or *otus* ear): an extinct genus of anisomyarian bivalves with an oval, radially ribbed shell with practically equal valves. The long, straight border of the hinge extended anteriorly into a small ear-like shape, and posteriorly into a similar but larger "ear". Fossils occur from the Lower Carboniferous to the Cretaceous, and are particularly abundant in the Alpine Triassic where they serve as index fossils.

Pseudomycorrhiza: a form of Mycorrhiza (see), in which the fungus is parasitic.

Pseudomyrmecinae: see Ants.

Pseudonavicella: see *Telosporidia.*

Pseudoparaphyses: sterile hyphae with degenerate pairs of nuclei. They are found in the hymenium of basidiomycetes together with sterile cystidia.

Pseudoparenchyma: tissue-like cell associations that are not true tissues. For example, P. is formed in red algae by bundles of cell filaments enveloped in a water-insoluble gelatinous material, the latter being formed by swollen cell walls. Alternatively, in the fruiting bodies of many fungi, P. consists of closely intertwined fungal hyphae (see Plectenchyma). In transverse or longitudinal section, the microscopic structure of P. is deceptively similar to that of true plant tissues.

Pseudoperidium: see Aecidium.

Pseudophyllidea: see *Cestoda,* Broad fish tapeworm.

Pseudopodia (sing. **Pseudopodium):** temporary extensions of the cytoplasm of members of the *Rhizopodea* (see) (see also Amebas), which serve for progression on a solid substratum, and for entrapping and engulfing food material. They arise and are extended when parts of the cytoplasm flow in one direction. They are withdrawn by cytoplasmic flow in the opposite direction, or when the rest of the cell flows into and past them during locomotion of the organism. Locomotion usually occurs by the extension of *lobopodia* (broad, blunt extensions consisting of ecto- and endoplasm), whereas some members of the *Testacida* (an order of testate amebas) produce *filopodia* (slender, tapering P. of hyaline ectoplasm), which often arise in bundles. *Reticulopodia* or *rhizopodia* are also slender and filamentous, but they are linked with one another by frequent anastomoses to form a network. P. often contain numerous refractive granules which are carried along by the streaming cytoplasm, thereby faciltiating the microscopic observation of cytoplasmic flow.

Pseudopolyplopidy: a doubling or several-fold multiplication of the basic number of chromosomes (see Polyploidy), which is not accompanied by a corresponding increase in the quantity of genetic material. In *fragmentation P.* the somatic cells of an organism contain a large number of small chromosomes, whereas the reproductive cells contain a few large chromosomes. Other types of P. are *Agmatoploidy* (see) and *fusion P.* In the latter case, small, monocentric chromosomes fuse to form larger polycentric chromosomes, so that within one systematic unit there are forms with many small chromosomes, forms with a few large chromosomes, and intermediate forms containing both sizes of chromosomes; this gives the erroneous impression of a polyploid series, and the complement of small monocentric chromosomes can be mistaken for a state of polyploidy.

Pseudorca: see Whales.

Pseudoresistance: in phytopathology, a type of Resistance (see), in which an effective encounter of host and parasite is prevented (incoincidence). P. may be determined by temporal, spatial or biological factors. A host may also "escape" attack by growing or developing beyond the stage at which it is susceptible.

Pseudorhynchota: see *Phthiraptera.*

Pseudoryx nghetinensis: see Vu Quang ox.

Pseudosphaeriales: see *Ascomycetes.*

Pseudostigmatic organ: see *Oribatei.*

Pseudostrabism: see Epicanthic fold.

Pseudothecium: see *Ascomycetes.*

Pseudotrachea: see *Isopoda.*

Pseudotriakinae: see *Selachimorpha.*

Pseudotritia: see *Oribatei.*

Psila rosae: see Carrot fly.

Psilophytales: see *Psilophytatae.*

Psilophytatae, *Psilopsida:* the most primitive class of pteridophytes. The shoot is dichotomously branched. Leaves are absent or developed only as scales or spines. There are no true roots, and sporangia are situated at the ends of the shoots. The class is known mainly from its extinct fossil forms; only a few living species survive.

1. Order: *Psilophytales.* These are the oldest terrestrial plants with vascular bundles and stomata. Their fossils appear in the Silurian/Devonian transition, and the order was extinct by the beginning of the Upper Devonian. The family, *Rhyniaceae* (named for Rhynie in Scotland, which is near to the fossil discovery site), contains the most primitive of all known pteridophytes, and within this family the genus, *Rhynia,* is considered to be the "original ancestral land plant". Two fossil species are known. Both were 50 cm-high, leafless marsh plants, growing in clumps or beds like reeds, The horizontal rhizomes (with unicellular rhizoids but lacking true roots) bore erect, cylindrical dichotomous shoots with stomata (Fig.1a). Sporangia (containing homosporous tetrads) were terminal with a multilayered wall, but lacked a dehiscence mechanism. The rhizome was probably the gametophyte, since it has been shown to possess archegonia.

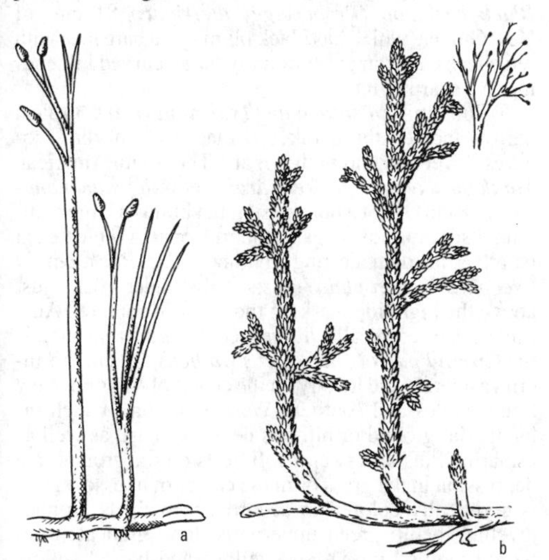

Fig.1. *Psilophytatae.* Reconstruction of: *a Rhynia major, b Asteroxylon mackei* (both from the Middle Devonian).

The *Zosterophyllaceae* grew in shallow water, resembled algae, and possessed slender, ribbon-like shoots. Fossils occur throughout the world from the Silurian to the Middle Devonian. Important genera are *Zosterophyllum* (Fig.2) and *Taeniocrada.*

Fig.2. *Psilophytatae.* Reconstruction of *Zosterophyllum rhenanum* from the Lower Devonian.

The *Asteroxylaceae* bore widely or closely spaced photosynthetic microphylls or needles, which had numerous stomata but no vascular tissue. Several fossil species of *Asteroxylon* are known (Fig.1b).

2. Order: *Psilotales.* This order consists of 2 recent genera *[Psilotum* (**Whisk fern**) and *Tmesipteris],* each represented by only 2 species. They are small, herbaceous, dichotomously branched, rootless plants. The shoot carries scale-like emergences or somewhat larger leaf-like emergences without vascular bundles (Fig.3). Fossils are unknown and the taxonomic position of the order is obscure.

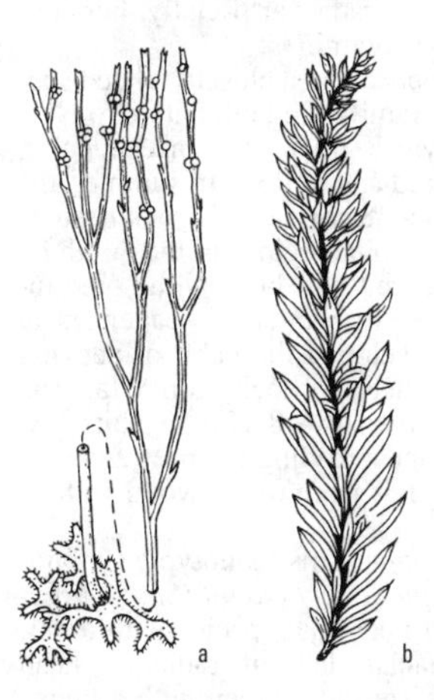

Fig.3. *Psilophytatae.* Reconstruction of: *a Psilotum, b Tmesipteris.*

Psilopsida: see *Psilophytatae.*

Psilotales: see *Psilophytatae.*

Psilotum: see *Psilophytatae.*

Psithyrus: see Bumble bees.

Psittacidae: see *Psittaciformes.*

Psittaciformes, ***parrots and allies:*** an order of

birds (see *Avies)* containing more than 300 species in 82 genera, all in a single family *(Psittacidae).* The main centers of dispersion are the Amazon basin and Australasia. Although some types are extreme, and sizes range from 9 to 100 cm, all resemble one another closely and are immediately recognizable as members of the same family. All possess a large head and short neck. The strongly decurved, hooked bill has at its base a bulging cere, which houses the nostrils and is sometimes feathered. They have powerful, grasping, zygodactylous feet with strong, hooked claws, and they are unique among birds in holding up their food in one foot to eat it. The upper mandible is not rigidly attached to the skull, but articulated via a special hinge structure, so that it is somewhat mobile. In fact, the bill is a highly adaptable structure, used for delicate preening, for crushing hard nuts and seeds, and as a very effective "grappling hook" when climbing and clambering. Parrots are largely vegetarian, fruit and nuts being typical articles of diet. Most are truly sedentary, but a certain degree of nomadism is sometimes necessary in order to follow the seasonal ripening of fruit, especially in dry Australian habitats.

With a few exceptions, parrots are gregarious birds, often forming large, noisy flocks. The majority are also tree-dwellers and tree-nesters, nesting in holes in the trunks or limbs of trees, often at a considerable height above the ground. Nesting holes may be excavated or stolen from other hole-nesting birds like woodpeckers. Parrot eggs are always round and glossy white, varying in number from 1 for some of the larger species to as many as 10 for some smaller species. Incubation is normally performed by the female, and during this time she is fed by the male, which also shares responsibility for feeding the young. The young are altricial, and development to independence may take as long as 3.5 months. As a rule the sexes are similar. Notable exceptions are the Australian ***King parrot*** *(Alisterus scapularis;* males brilliant scarlet, females green), and the ***Eclectus parrot*** (see below). The family is rather artificially, but conveniently divided into 6 subfamilies.

1. Subfamily: *Nestorinae.* The 2 closely related species comprising this subfamily are both native to New Zealand. Both are brownish green or bronze in color, variously marked with red and yellow. In summer, the ***Kea*** *(Nestor notabilis;* 46–48 cm) nests in rock fissures above the tree line in the alpine regions of the South Island, where it feeds on fruit and buds with some insects, grubs and worms. In winter, it scavenges at lower levels, and has developed the habit of pecking into the backs of living sheep for their kidney fat. The ***Kaka*** *(Nestor meridionalis;* 46–48 cm) inhabits low lying forests on both islands, feeding on fruits and nectar, and on grubs that it digs from rotten wood with its bill.

2. Subfamily: *Strigopinae.* This monotypic subfamily consists of the ***Kakapo*** or ***Owl parrot*** *(Strigops habroptilus;* 58–66 cm), a nocturnal species with a yellow-brown, owl-like facial disk of radiating head feathers. Its upper parts are moss green with a bluish tinge, the underparts greenish yellow, and the flight and tail feathers are barred brown and dull yellow. This rare and threatened bird is now confined to Fiordland on the South Island of New Zealand. It is very short-winged and flightless, and maintians its own beaten tracks through the vegetation of the forest floor. It also climbs trees, then glides for about 100 yards (about 90 m).

3. Subfamily: *Loriinae* (***Lories*** and ***Lorikeets;*** Papua, Australia, Polynesia and New Guinea). Members of this distinct subgroup have sleek plumage, and include some of the most brilliantly colored species of parrots. Their bills are rather more elongated, narrower, and less powerful than those of other parrots, and the gizzard is relatively thin-walled and weak. The tongue is longer than in other parrots, and carries a tuft of thread-like papillae at the tip, which are used to mop up nectar and pollen. Flowers are also crushed with the beak, in order to feed on the exuded juices. When at rest, or when the bird is feeding on fruit or seeds, the papillae are enclosed in a protective sheath. Most widespread is the ***Rainbow lorikeet*** *(Trichoglossus haematodus;* 25–28 cm) of Indonesia, New Guinea, eastern Australia and western Pacific islands.

4. Subfamily: *Micropsittinae* (***Pygmy parrots;*** 6 species; jungles of New Guinea and adjacent islands). The smallest of the pygmy parrots is little more than 9 cm in length, the largest no more than 12.5 cm. They are not particularly gregarious, feeding singly or in pairs, not only on fruits, but also on fungi and insects (especially arboreal termites), which they pry from bark crevices like woodpeckers. Bearing a certain resemblance to miniature woodpeckers, they have stiff tails with spiny tips, and long claws which enable them to creep over the trunks of trees in their search for insects. They excavate a nesting burrow in a tree termatarium.

5. Subfamily: *Kakatoeinae* (***Cockatoos;*** Papua, Australia and adjacent islands). Kockatoos differ from other parrots in having a long, erectile crest of long, pointed feathers. Most are fairly large, white birds, often tinged with pink or yellow, most familiar being the ***Sulfur-crested cockatoo*** *(Cacatua galerita;* 46 cm) of Australia and New Guinea, which is white with a yellow crest. Cockatoos are notoriously gregarious birds travelling in small, loose, noisy flocks. The largest member is, however, neither gregarious nor white; this is the solitary ***Black cockatoo*** *(Probosciger aterrimus;* 81 cm) of New Guinea, which has black plumage, a bare face with red cheeks, and an exceptionally large, curved bill ending in a sharp point.

6. Subfamily: *Psittacinae* (***Typical parrots).*** Typical parrots include the parakeets, macaws, amazons, conures, lovebirds and budgerigars. The South American ***Monk parakeet*** or ***Quaker parakeet*** *(Myiopsitta monachus;* 28 cm) is not a hole-nester, but builds a large communal structure of twigs in the treetops, in which each pair has its own nesting chamber. The ***Rock parrot*** *(Neophema petrophila)* nests only under rocks just above the high tide mark on the coast of southern Australia. Also "atypical" with respect to its nesting habit is the ***Ground parrot*** *(Pezoporus wallicus;* 33 cm; northern Tasmania, and locally on the coasts of Victoria, New South Wales and southern West Australia), which inhabits flat ground or hilltops near the coast, as well as estuarine flats and swamps. It nests on the ground in a depression in the grass or in the center of a tussock, and is cryptically colored in accordance with its ground-dwelling habit: green upperparts, long graduated tail barred mottled brown and yellow, and barred yellow underparts with a red frontal patch. All the 15 species of ***Macaws*** are inhabitants of tropical rainforests from Mexico to South America; they are notable for their

gaudy plumage and long tails. Familiar representatives in European and North American zoos are the ***Scarlet macaw*** *(Ara macao;* 90 cm; Mexico to Bolivia) and the ***Gold-and-blue macaw*** *(Ara ararauna;* 78 cm; Panama to Argentina). The New World ***Amazons*** (about 25 species) have mainly green plumage and square or rounded tails. One of the best known of all parrots is the ***African Gray parrot*** *(Psittacus erithacus;* 33 cm; forests from West to East Africa). In the wild it causes damage to grain crops, and travels in screaming, fast-flying flocks. In captivity, this gray-plumaged, red-tailed parrot is an entertaining mimic, and is sold for high prices by bird dealers. Wild ***Lovebirds*** form large flocks, often causing damage to crops. They are heavy-bodied, pointed-tailed parrots of the Old World, ranging from Africa and Madagascar to the Pacific islands. In captivity, they huddle together in pairs. In the gaudy lovebird known as the ***Eclectus parrot*** *[Eclectus (= Lorius) roratus;* 51 cm; New Guinea, extreme northern Queensland, Moluccas, Bismark and Solomon Islands, and other lesser Southeast Asian islands], sexual dimorphism is so pronounced that the 2 sexes were once thought to be different species; the male is bright green with scarlet underwings and flanks, while the female is crimson with violet underparts. ***True parakeets*** are a widely distributed Old World group of small species, mostly with long pointed tails. Many inhabit farmland, and they form large flocks, often feeding on the ground on grain and fruit. Most familiar of the parakeets is the Australian ***Budgerigar*** *(Melopsittacus undulatus;* 18 cm), which, as a popular cage bird, has been subject to controlled breeding and selection to produce a range of patterns and colors not seen in the wild. The hanging parakeets are small green species, occurring from India to the Philippines, which sleep at night hanging upside down.

Psittacinae: see *Psittaciformes.*

Psittacosis: see Ornithosis.

Psittacus: see *Psittaciformes.*

Psittirostra: see *Drepanididae.*

Psittirostrinae: see *Drepanididae.*

Psocida: see *Psocoptera.*

Psocomorpha: see *Psocoptera.*

Psocoptera, *Copeognatha, Corrodentia,* ***psocids, barklice, booklice:*** an insect order containing almost 2 000 species, distributed worldwide. The earliest fossil *P.* occur in the Permian.

Imagoes. Body length varies from 0.1 to 1 cm. The relatively large, bulbous head carries filiform antennae and chewing mouthparts. Some species are wingless. Others possess 2 pairs of fully developed or reduced membranous wings. The whole body is covered in fine hairs or scales. In the wild, P. live beneath tree bark, on lichens or in animal nests, and feed mainly on algae, fungi and lichens. Some species, e.g. the booklice, *Trogium pulsatorium* L. and *Liposcelis divinatorius* Müller, live in houses, on paper, leather and upholstery, where they feed mainly on molds. Metamorphosis is incomplete (paurometabolous development). Reproduction is sexual or by parthenogenesis.

Eggs. The female lays up to 100 eggs (about 0.5 mm in diameter) individually or in small groups. Some species cover their eggs with cement substance, dust, nutrient material or silk threads. The eggs of many species lie dormant through the winter.

Larvae. These are similar to the adult. Wing primordia appear after the first molt, and the mature insect emerges after 15 to 30 days and 5 to 8 molts. Free living species in temperate climates usually produce only one generation per year, but those living in buildings produce several.

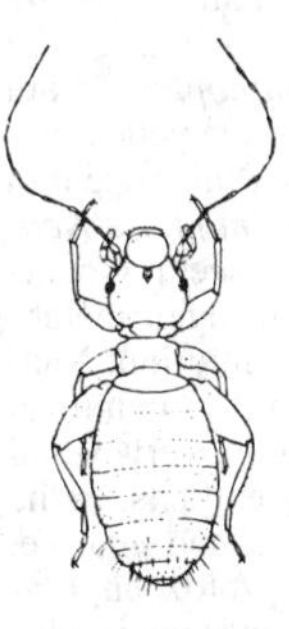

Psocoptera. Book louse *(Liposcelis divinatorius,* Müller).

Economic importance. Booklice have practically no economic importance, the free living species none at all. Large numbers do not occur in buildings unless damp has already set in and fungal growth is extensive. So far, the only damage attributable to P. has occurred in insect collections and herbaria.

Classification. Thirty families are recognized. There are various possible classifications, e.g.

1. Suborder: *Atropida* (prothorax well developed). 2. Suborder: *Psocida* (prothorax markedly reduced).

Alternatively:

1. Suborder: *Trogiomorpha* (antennae with more than 20 segments; tarsi 3-segmented). 2. Suborder: *Troctomorpha* (antennae 12-17-segmented; tarsi 2-3-segmented; labial palps 2-segmented). 3. Suborder: *Psocomorpha* (antennae almost always 13-segmented; tarsi 2-3-segmented; labial palps unsegmented).

Psophia: see *Gruiformes.*

Psophiidae: see *Gruiformes.*

Psychidae: see *Lepidoptera.*

Psycho-Lamarckism: part of Lamarckian evolutionary theory, according to which animals possess an "inner evolutionary drive". The concept is now largely discredited.

Psychological arousal: see Vigilance.

Psychrophyllic organisms, ***psychrophiles:*** microorganisms whose optimal growth temperature is below 20 °C. Some can still multiply at 0 °C. P. are common in the sea, and they occasionally infect refrigerated foods.

Psyllid: see *Homoptera.*

Psylloidea: see *Homoptera.*

Ptarmigan: see *Tetraonidae.*

Pteranodon: see *Pterosauria.*

Pteraspis (Greek *pteron* wing, *aspis* shield): a genus of the *Agnatha.* Head and forebody were enclosed in an armor of body plates. The markedly heterocercal tail and the hindbody carried scales. Fossils are found from the Cambrian to the Lower Devonian.

Pterichthys (Greek *pteron* wing, *ichthys* fish): a fish genus of the extinct class *Placodermi* (order *Antiarchiformes).* Head and body were enclosed by an armor of tightly jointed bony plates, and there were two armored, oar-like forelimbs. A pineal organ was present between the closely placed, dorsally situated eyes. Fossils are found in the Middle Devonian of Europe.

Pteridines (Greek *pteron* wing): a group of compounds containing the pteridine ring system. The majority of naturally occurring P. are chemically related to pterine (4-hydroxy-2-amino-pteridine). A smaller number are derived from lumazine (2,4-dihydroxypteridine). P. have a ubiquitous distribution in biological material, usually in small concentrations. Familiar P. are the white-colored *leucopterin* from the wings of the cabbage white butterfly, and yellow *xanthopterin* from the wings of the brimstone butterfly and the yellow abdominal segments of wasps and hornets. P. also occur in the imagoes of other *Lepidoptera, Hymenoptera, Neuroptera* and *Hemiptera*. P. that serve as insect pigments may be regarded as end products of nitrogen metabolism, which have been exploited for coloration. Xanthopterin regularly occurs in small quantities in human urine. Biopterin is widely distributed in very small quantities; it has been found, e.g. in fly extracts, in the royal jelly that is fed to queen bee larvae, and in urine. Biopterin is also a growth factor for the protozoan, *Crithidia fasciculata*. The pteridine ring system is also present in folic acid and its derivatives (see Vitamins), which accounts for its ubiquitous distribution. The isoalloxazine ring system of Flavins (see) is equivalent to a pteridine ring system fused with a benzene ring system; early stages in the biosynthesis of the P. and flavins are similar.

Elevated excretion of neopterin is correlated with certain malignant diseases, viral infection and graft rejection, apparently because neopterin is secreted by macrophages in the course of T-lymphocyte activation.

Leucopterin Xanthopterin

Pteridophora: see *Paradisaeidae*.

Pteridophytes, *Pteridophyta:* a phylum of the plant kingdom comprising 12 000 species, which occur in all climatic regions of the world, but are most widespread and numerous in the tropics. Like the mosses, the P. show a distinct Alternation of generations (see), but in the P. the gametophyte and sporophyte are organically independent. The haploid gametophyte (prothallus), which develops from a spore, is usually a thallose, liverwort-like structure carrying antheridia and archegonia. The archegonia are similar in structure to those of the mosses. Mosses and P. are therefore also jointly classified as *Archegoniatae*. Under suitably moist conditions, the spermatozoids swim to the egg cells and fertilize them. The resulting diploid zygote then develops into the asexual generation (sporophyte). This is the actual pteridophyte plant, and it is very different in appearance from the gametophyte. It possesses the cormophyte characters of root, shoot and leaves; the roots arise laterally on the shoot and have a root cap. In most respects, the shoot and leaves are similar to those of spermatophytes, so that P. are classified as cormophytes. The vascular bundles consist of sieve tubes and vessels and may contain lignified tracheids, and they extend throughout the whole plant. P. are therefore able to achieve a greater variety of vegetative forms, in some cases even developing into tree-like structures (tree ferns, etc.). Since the epidermis is covered by a protective cuticle, gas exchange occurs through well-developed stomata, a structural feature which especially illustrates the adaptation of P. to a terrestrial existence. The spores are formed in sporangia, which are found predominantly on the leaves (sporophylls). Some P. may possess separately differentiated assimilatory leaves (trophophylls) and sporophylls, and the latter may be aggregated into distinct structures called strobili, which are sometimes referred to as primitive flowers. In the sporangia, spores arise from spore mother cells by meiosis. Spores usually have two membranes, the outer resistant exospore, and an inner, thin cellulose layer, the endospore. Following germination, the spores develop into the gametophyte. In some P. there are two different types of spores, which give rise to male and female prothalli. Large female prothalli develop from megaspores, which contain a large reserve of nutrients; and small male prothalli develop from microspores. Mega- and microsporangia contain mega- and microspores, respectively. Species producing only one type of spore are called isosporous or homosporous, whereas species forming mega- and microspores are called heterosporous. In some cases, the gametophytes are reduced to such an extent that the prothalli and archegonia, and even the embryo after fertilization

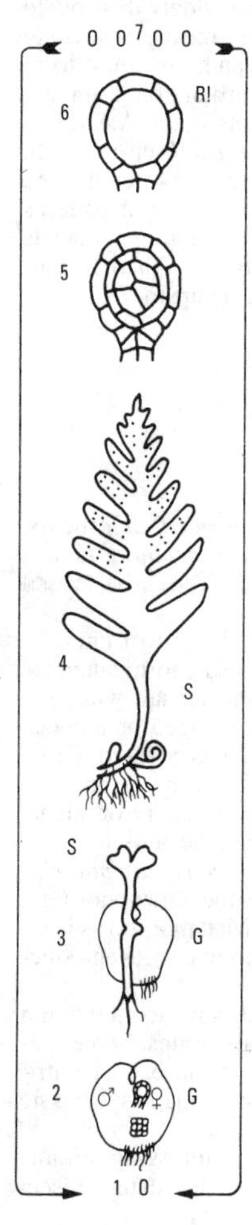

Pteridophytes. Life cycle of a fern. *G* Gametophyte (Haploid). *S* Sporophyte (diploid). *R!* Meiosis. *1* Spore. *2* Prothallus with female and male gametangia. *3* Prothallus with young sporophyte. *4* Sporophyte (greatly reduced). *5* Immature sporangium (greatly enlarged) from a sorus. *6* Mature sporangium with spore tetrads. *7* Spores. (After Harder)

arise within the megaspore. The completed embryo eventually separates and becomes independent. The entire process clearly resembles seed formation in spermatophytes; it seems probable that heterospory in different classes of P. and seed formation are examples of parallel evolution.

P. are a phylogenetically ancient group, whose relationship to primitive plants is uncertain. It is thought that they evolved in parallel with the mosses from unknown ancestral algae. Mosses, P. and green algae share certain characteristics, such as the presence of chlorophylls *a* and *b* and certain types of carotenoids.

P. are systematically divided usually into four classes: *Psilophytales* (see), Clubmosses (see), Horsetails (see) and Ferns (see).

Pteridospermae: see *Lyginopteridatae.*

Pterioidea: see *Bivalvia.*

Pteriomorphia: see *Bivalvia.*

Pterocarya: see *Juglandaceae.*

Pterocles: see *Columbiformes.*

Pteroclidae: see *Columbiformes.*

Pterodactyl: see *Pterosauria.*

Pterodroma: see *Procellariiformes.*

Pterogasterina: see *Oribatei.*

Pteroglossus: see *Ramphastidae.*

Pteromorphs: see *Oribatei.*

Pterophoridae: see *Lepidoptera.*

Pterophyllum: see Angelfishes.

Pteropoda: see *Gastropoda.*

Pterosauria (Greek *pteron* wing, *sauros* lizard): an extinct order of bird-like, flying reptiles, whose fossil remains are found from the Upper Triassic to the Upper Cretaceous. A flight membrane stretched between the body and an elongated 4th finger. The first 3 fingers were free and clawed, whereas the 5th finger was sometimes rudimentary. The bones were hollow, and the long-beaked reptilian skull was aligned at 90 ° to the powerful neck. Teeth were usually conical and pointed. Earlier forms had long tails, which were lacking in the more advanced forms. A well developed sternum was present, but it lacked a keel. The robust pectoral girdle, mechanically bracing the sternum and vertebral column, is suggestive of active flight rather than gliding.

Pterodactylus (pterodactyls) (Greek *pteron* wing, *dactylos* finger) was a genus or family of sparrow-sized to hawk-sized pterosaurs, with long necks and practically no tails. They probably moved erratically by fluttering, or climbed in tree branches with the aid of their claws. The 5th finger of the hindlimbs was undeveloped, and the pointed conical teeth were present only in the front part of the long beak or jaw. Fossils of pterodactyls are found in the Malm strata, and especially well preserved fossils have been retrieved from the Solnhofer chalk layers in Bavaria.

Pteranodon (Greek *pteron* wing, *an* without, *odontes* teeth) was a genus or family of pterosaurs with wingspans ranging from more than 10 m to 21 m. They were the largest flying animals of all time. Their pointed jaws lacked teeth, and the 1 m-long skull carried a long, dorsal bony crest, which probably served for steering in flight. The bones were largely hollow and relatively light. Fossils are found in the Upper Cretaceous of Kansas and Texas.

Pterotrachea: see *Gastropoda.*

Pterygota: winged insects. The larger of the two subclasses of insects whose members have wings, or are wingless by secondary loss. It includes all insects except the *Apterygota.*

Pterygotus: see *Gigantostraca.*

Pterylae: see Feathers.

Pterylography: see Feathers.

Pterylosis: see Feathers.

Ptilinopinae: see *Columbiformes.*

Ptilinopus: see *Columbiformes.*

Ptilinorhynchidae, ***bowerbirds*** and ***catbirds:*** a small avian family in the order, *Passeriformes* (see). The 17 living species (23–28 cm) are native to the forests of northern Australia and New Guinea. They are very closely related to the Birds of paradise, and considered to be more advanced than the latter. The male of most species builds an elaborate structure for his courtship display, which he decorates with flowers and other articles. Although mating may take place within this structure, it does not serve as the nest, and the male contributes nothing more to reproduction. The female builds the nest, incubates the eggs and cares for the young alone. They are divided into 2 subfamilies.

1. Subfamily: *Ailuroedinae* ***(Catbirds;*** *Ailuroedus* and *Scenopoeetes).* This subfamily is considered to be the most primitive of the 4, and it contains 3 living species. The males of the ***Green catbird*** *(Ailuroedus crassirostris;* northern and eastern Australia) and the ***White-throated catbird*** (western New Guinea) do not build a bower of any kind. The male ***Tooth-billed bowerbird, Tooth-Billed catbird,*** or ***Stagemaker*** *(Scenopoeetes dentirostris;* highland forests of Queensland) clears an area of the forest floor 1–1.5 m in diameter, which he decorates each morning with freshly cut leaves; both sexes have notches near the tips of the upper and lower mandibles, and these are used by the male to cut the leaves for his stage or playground.

2. Subfamily: *Ptilinorhynchinae.* The subfamily is divided into 3 groups according to the type of display structure built by the male.

Group 1: ***Platform builders*** or ***yard builders*** *(Archboldia).* This group contains only 2 members. ***Sanford's golden-crested bowerbird or Tomba bowerbird*** *(Archboldia papuensis sanfordi;* high palm forests of New Guinea) builds a display floor covered with fern fronds, and decorates the edges with beetle wings, snail shells, pieces of resin, etc. In addition, he suspends a curtain of bamboo strands and wilted ferns from nearby vines, in which he places pieces of bark, berries etc. The ***Archbold bower bird*** *(Archboldia papuensis)* is also found in the forests of New Guinea.

Group 2: ***Maypole builders*** *(Prionodura* and *Amblyornis).* This group contains 5 species. The small ***Golden bowerbird*** *(Prionodura newtoniana)* of northern Australia builds an extremely high (up to 2.8 m) roofed structure. The other 4 maypole builders are found in the high mountain forests of New Guinea, where they plant mosses in their display or dancing grounds. They are therefore called "gardeners": ***Crested gardener, Mocca-breasted bowerbird*** or ***MacGregor's bowerbird*** *(Amblyornis macgregoriae);* ***Orange-crested gardener, Streaked bowerbird, Striped bowerbird*** or ***Eastern bowerbird*** *(Amblyornis subalaris);* ***Crestless gardener*** or ***Vogelkop bowerbird*** *(Amblyornis inornatus);* ***Golden-fronted gardener*** *(Amblyornis flavifrons).* The latter species is known only from 3 specimens shipped to Europe in the 19th century. The male maypole builder constructs a cone of sticks and twigs horizontally around the base of a sapling until they reach the desired height (about 3 m in the case of the Golden bowerbird). This cone may be surmounted by a conical or hemispherical roof, with or without a stockade at the front. The walls and the ground beneath and around the bower are decorated with ferns, mosses and flowers.

Group 3: ***Avenue builders*** *(Ptilonorhynchus, Chlamydera* and *Sericulus).* The 9 species constituting this group are considered to be the most highly specialized members of the family. Their bowers are smaller than those of the maypole builders and only partly roofed, if at all, but their construction is more intricate. The male prepares a floor of well trodden sticks up to 20 cm deep. In the center of this floor he erects 2 parallel walls of firmly implanted upright sticks, entwined with one another and sometimes arched over. The bower is then decorated with leaves, flowers, stones, shells, etc. The Australian Regent and Satin bowerbirds provide rare examples of the use of tools by birds: they daub the inside of the bower with a blue-green mixture prepared from plant material and saliva, using a wad of leaves or a wad of bark, respectively. Examples are: the ***Great gray bowerbird*** *(Chlamydera nuchalis;* Australia); ***Regent bowerbird*** *(Sericulus chrysocephalus;* eastern Australia); ***Satin bowerbird*** *(Ptilorynchus violaceus;* Australia); ***Black-faced golden bowerbird*** *(Sericulus aureus;* New Guinea and northern Australia), ***Yellow-breasted golden bowerbird*** *(Chlamydera lauterbachi;* New Guinea), and ***Baker's, Fire-maned*** or ***Madang bowerbird*** *(Sericulus bakeri;* New Guinea).

Ptilinorhynchinae: see *Ptilinorhynchidae.*

Ptilogonatinae: see *Bombycillidae.*

Ptilogonys: see *Bombycillidae.*

Ptomaines: toxic products of bacterial putrifaction, e.g. Neurine (see).

Ptychites (Greek *ptyche* fold): an index genus of ammonites from the Middle Triassic. The planispiral shell was thick and disk-shaped, with a narrow umbilicus, and a rounded exterior edge. Indistinct radial folds decorated the shell surface, and the suture line was weakly serrated.

Ptychocyst: see Nematocyst.

Ptyctima: see *Oribatei.*

Ptyophagy: digestion of endotrophic mycorrhiza, primarily in orchids. Fungal hyphae burst inside the host cells. The released protoplasm then becomes surrounded by a membrane (formed by secretions of both the host and the fungal protoplasm), forming free fungal structures known as ptyosomes. These are gradually degraded by host digestive cells.

Pudu: the smallest deer (head plus body 77–93 cm, shoulder height 32–41 cm). Two species are recognized: ***1)*** *Pudu pudu* (southern Chile, southwestern Argentina; preorbital glands present; premaxillaries reach the nasals; hooves of medium width and not pointed; normal sized rhinarium), and ***2)*** *Pudu mephistophiles* (Andes of Colombia, Ecuador and northern Peru; no preorbital glands; premaxillaries of the skull do not extend to the nasals; hooves narrow and pointed; small rhinarium). The coat of long, coarse, brittle hairs is uniformly dark brown to gray *(P. pudu)* or deep rich brown *(P. mephistophiles).* The male has tiny antlers, each consisting of a single spike. There are no canine tusks and practically no visible tail. See Deer.

Puff adders, *Bitis:* a genus of plump, broad-headed vipers (see *Viperidae)* which are found in sub-Saharan Africa. They have vertical pupils and long venom fangs. Undulation is only rarely used for locomotion, and P.a. normally progress with the aid of large ventral scales, which are parted by rib movements. Desert forms are generally light in color, whereas rainforest inhabitants are darkly patterned. When excited, they puff themselves up, and hiss loudly when breathing out. About 10 species are recognized, and all are vivi-oviparous. The ***Puff adder*** *(B. arietans),* which gives the group its name, is one of the most widely distributed poisonous snakes in Africa, preferring open country, and feeding on rats and mice. It has a short tail, and is patterned with crescent-shaped, light markings on a gray-brown background. Although this 1.5.m-long snake appears sluggish, it strikes with lightening speed, and its potent venom is potentially fatal without medical intervention. Largest of the P.a. is the ***Gaboon viper*** *(C. gabonica)* of the West and Central African rainforests. It attains lengths of over 2 m, its venom fangs are up to 3.8 cm in length, and the body is patterned with bright purple, brown, yellow, blue and dark brown like a carpet. The somewhat smaller, and also vividly colored and patterned ***River jack*** or ***Rhinoceros viper*** *(B. nasicornis)* has 2–3 small, pointed horns of scales above each nostril. Several smaller, mostly sandy brown species of *Bitis* are found in the dry areas of South Africa.

Puffbacks: see *Muscicapinae.*

Puff-ball: see *Basidiomycetes (Gasteromycetales).*

Puffbirds: see *Bucconidae.*

Puffers: see *Tetraodontidae.*

Puffins: see *Alcidae.*

Puffinus: see *Procellariiformes.*

Puffs: see Giant chromosomes.

Puku: see *Reduncinae.*

Pulmonata: see *Gastropoda.*

Pulmonate snails: see *Gastropoda.*

Pulque: see *Agavaceae.*

Pulsatilla: see *Ranunculaceae.*

Pulse: the effect of heart contractions on the rate of blood flow (flow P.), vessel wall tension (pressure P.) and vessel diameter (volume P.).

1) *Flow P.* At the beginning of the expulsion phase of blood from the heart, the blood flow rate rapidly increases to a maximal value in the aorta; in humans this maximal value may be as high as 100 cm/s. The rate then decreases rapidly, until diastole, when the blood is hardly moving. 2) *Pressure P.* Heart contraction causes a pressure increase in the arterial system, which rises to a peak (about 17 kPa in humans) during systole. At the end of systole, the arteries near to the heart undergo a brief, steep pressure decrease known as the incisure. The Blood pressure then continues to fall to its diastolic end value of barely 11 kPa. A complete pressure decrease to the value of the left ventricle is prevented by closure of the aortic valve, by the Pressure chamber function (see) of the aorta which is responsible for a continuing pressure increase during diastole, and the resistance of peripheral vessels which causes a build-up of blood and therefore of pressure. Since each heart contraction causes a pressure P., the P. rate corresponds to the heart rate. In young humans, the *P. wave rate* (rate of propagation of the pressure wave along blood vessels) is about 6 m/s. It is higher in the peripheral arteries, and it generally increases with age. The P.wave rate is always greater than the blood flow rate. 3) *Volume P.*

During systole, the increase in blood pressure stretches the aorta wall. Thus, the volume P. is the change in cross-sectional area of the aorta with the blood pressure.

Pulse cytophotometry: a method for the quantitative determination of substances (nucleic acids, proteins) in single cells.

P.c. is represents the further development and automation of cytophotometry. Measurements are made on suspensions of fixed cells, in which the analyte of interest has been labeled with a fluorescent dye. As the cell suspension flows through the measurement chamber, the emitted fluorescence of each cell is converted into an electrical signal and classified in a multichannel analyser. The results are processed electronically to give the frequency distribution of the measured compound, which in turn provides information on the degree of differentiation, growth and cell multiplication. P.c. is used in the early diagnosis of cancer and in the monitoring of cancer therapy.

Puma: see *Felidae.*

Pummelo: see *Rutaceae.*

Pumpkin: see *Cucurbitaceae.*

Punaré: see *Echimyidae.*

Punctuated equilibrium, ***punctuated evolution:*** a hypothesis founded by Eldredge and Gould (1972), in which species are relatively stable and long-lived, and in which the majority of evolutionary changes occur by concentrated bursts of speciation. The sudden appearance of new species is then followed by the establishment of a new equilibrium, i.e. certain species prove to be more successful than others. This hypothesis is supported by the observation that fossil species often remain unchanged over long geological periods, and that when new species make their first appearance they are fully developed with no apparent intermediate stages, i.e. they appear to have evolved very rapidly.

The concept that species usually exist in a stable equilibrium, which is only disturbed by speciation processes in isolated peripheral populations is in harmony with the hypothesis of Genetic revolution (see) within the framework of the Synthetic theory of evolution (see). It contradicts, however, the widely held neodarwinistic view that gradual adaptive processes within the species have made a significant contribution to evolution.

In its extreme form, the hypothesis of P.e. states that selection of individuals is relatively unimportant, that only the selection between species is important, and that these species arise suddenly in a few generations in small, isolated populations, possibly by chromosome mutations and mutations of regulatory genes.

On the basis of the fossil record, it is probably not possible to decide between the P.e. theory and the conventional neodarwinistic theory of phyletic gradualism or Quantum evolution (see). Different interpretations of the same fossil evidence (e.g. fossil human remains) can be made in support of both theories. It is likely that neither theory alone can claim to explain evolution, and that evolution has proceeded (proceeds) both gradually and in bursts.

See also Adaptive radiation, Typostrophism, Virescence period.

Punctuated evolution: see Punctuated equilibrium

Pungitius: see *Gasterosteidae.*

Punica: see *Punicaceae.*

Punicaceae, ***pomegranate family:*** a family of dicotyledonous plants, containing only 2 species, which occur in the tropics and subtropics. They are woody plants with simple leaves clustered on short shoots, and 5-7-merous, actinomorphic flowers. The ovaries, whose carpels are arranged in 2 or 3 superposed tiers, develop into a berry-like pseudocarp. The fruit of the ***Pomegranate*** *(Punica granatum)* has a bright red, leathery rind, and contains numerous seeds with pulpy, edible seed coats. Originally native to the Orient, where its fruit is very popular, the pomegranate is now cultivated widely in the tropics and subtropics. Double-flowered varieties are frequently planted as ornamentals. In addition, the bark of the shoot and root contains alkaloids, and it is used for the treatment of tapeworm infections. The other species of the family, *Punica protopunica,* is known only in the wild.

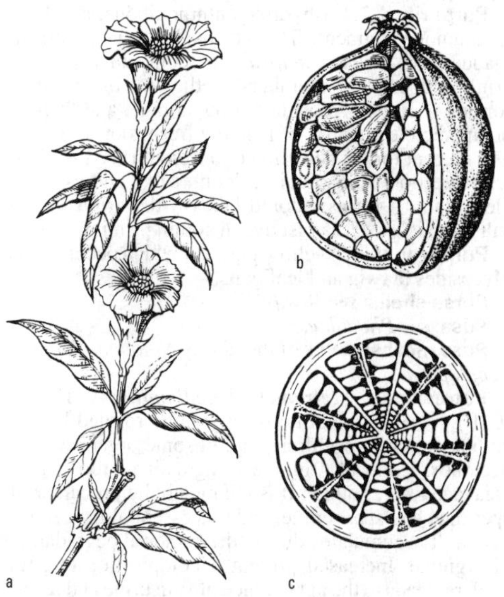

Punicaceae. Pomegranate *(Punica granatum). a* Flowering shoot. *b* Fruit. *c* Transverse section of fruit.

Punk: see *Basidiomycetes (Poriales).*

Pupa: see Insects.

Pupa coarctata: see Insects.

Pupa libera: see Insects.

Pupa obtecta: see Insects.

Pure culture: a culture of microorganisms containing only a single species or strain. P.cs. are indispensable for exact microbiological work, both in research and medical diagnosis. P.cs. of particular strains (high yield or high production strains) are used for the industrial production of many compounds, e.g. citric acid, antibiotics, etc., and P.cs. of special yeast strains are important in brewing. P.cs. are prepared by Plating (see) or Streaking (see).

Purine: a heterocyclic compound with a condensed pyrimidine-imidazole ring system. The free compound is not known to occur naturally, but the otherwise unsubstituted P. ring system is found in combination with ribose in the nucleoside antibiotic, nebularine. Various substituted and oxidized derivatives of P. occur naturally and are of considerable biological importance. The

P. derivatives, adenine and guanine (commonly referred to as P. bases) are present in DNA and RNA. Other metabolites include uric acid, hypoxanthine, kinetins, caffeine, xanthine, etc. P. was first prepared in 1898 from uric acid by E. Fischer.

Purine

Purple bacteria: see Rhodobacteria.

Purple finches: see *Fringillidae.*

Purple sulfur bacteria: see Sulfur bacteria, Rhodobacteria.

Purpura: see *Gastropoda,* Tyrian purple.

Purpurin: 1,2,4-trihydroxyanthraquinone, a red anthraquinone pigment. The glycoside of P. occurs in madder root *(Rubia tinctorum)* (accompanied by alizarin), and in other members of the *Rubiaceae.* P. is formed from its glycoside during storage, and there is no appreciable quantity of P. in the fresh root. It is used as a reagent for the detection of boron, for the histological detection of insoluble calcium salts, and as a nuclear stain. It forms colored lakes with various metal salts, and is used as a fast dye in cotton printing.

Purpurogallin: a yellow pigment which occurs as its glycosides in twig and leaf galls.

Purse shells: see *Bivalvia.*

Pusa: see *Phocidae.*

Puss moths: moths of the family *Notodontidae;* see *Lepidoptera.*

Putrescine, ***tetramethylenediamine:*** $H_2N—(CH_2)_4—NH_2$, a ubiquitous polyamine formed by decarboxylation of ornithine and, in some organisms, by decarboxylation of arginine to agmatine followed by cleavage to urea and P. It is the metabolic precursor of spermine and spermidine, and it is essential for cell division. It accumulates during the bacterial degradation of arginine. Increased protein decomposition (e.g. in cholera) leads to the appearance of P. in urine and feces. It is also a product of the bacterial putrifaction of protein.

Putrifaction: a special type of Decomposition (see), in which nitrogen-rich substances are decomposed in the relative absence of oxygen and light. In consists largely of fermentation reactions performed by bacteria and fungi. Important end products of P. are ammonia and hydrogen sulfide. If small animals are also present, they have little effect on the biochemical course of P. Such animals include bacteriophagous nematodes. Large accumulations of putrifying material may also contain dipteran larvae and carrion beetles.

Pycnidia: see Pycnospores.

Pycnogonida, ***pycnogonids,*** *Pantopoda,* sea spiders: a class of arthropods, containing over 1 000 species, distributed at all depths in oceans throughout the world, but most abundant in coastal waters of the far north and far south. Body length varies from 0.8 to 6 cm and the walking legs are often exceptionally long (spanning up to 50 cm). The rod-shaped body is differentiated into a large anterior cephalothorax or prosoma and a very much reduced, short stumpy abdomen or opisthosoma. There is no well defined head, but the anterior tip of the body carries a protruberance with 4 eyes and a proboscis with a powerful sucking pharynx and a terminal mouth. The proboscis is followed by three pairs of nonambulatory appendages: the chelicera (with claws), the pedipalps, and the ovigerous legs (egg carrying legs). In some species, all three pairs of nonambulatory appendages may be absent in the female, and the first two may also be absent in the male, but the male always possesses ovigerous legs. These are followed by 4 pairs (less commonly 5 or 6 pairs) of walking legs, which contain blind extensions (digestive ceca) of the small stomach. The branched gonads also extend into one or more pairs of legs, where they open on the fourth segment. *P.* live by predation or parasitically on *Hydrozoa, Octocorallia,* sponges, etc. Juices are sucked from the prey by insertion of the proboscis. Excretion and exchange of respiratory gases occur by direct diffusion; specific organs or structures for these functions are absent. The male takes the eggs from the laying female, and carries them until they hatch as spherical masses adhering to his ovigerous legs. The newly hatched protonymphon (a type of free larvel stage found otherwise only in the *Crustacea)* possesses only three pairs of appendages: chelicerae, pedipalps and one pair of walking legs. These appendages are not homologous with those in the crustacean nauplius larva. During a series of molts, the body grows, segments are added to the legs, and further appendages appear. *P.* therefore display a unique combination of characteristics not found in any other arthropod group. Their taxonomic position is controversial, but they are usually placed in the chelicerates, because they have chelicerae. Well known genera are *Nymphon* and *Pycnogonum.*

Collection and preservation. *P.* can be found in drag nets or attached to seaweed, hydroid polyps, sponges, etc. They should be fixed in 96% ethanol, and stored in 70% alcohol containing 5% glycerol.

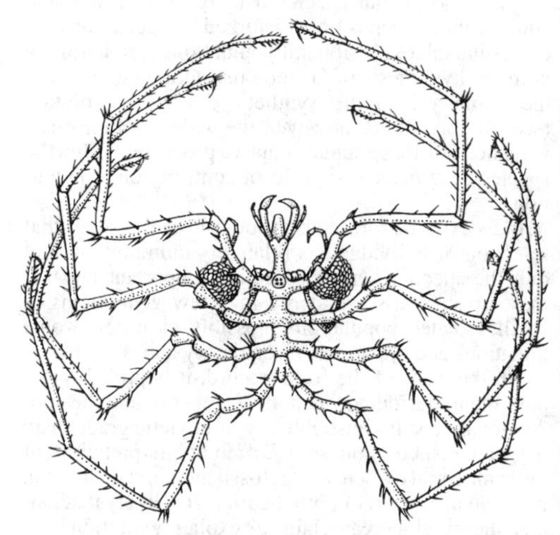

Pycnogonida. *Nymphon rubrum* (male) with egg masses attached to the ovigerous legs.

Pycnonotidae, ***bulbuls:*** an avian family in the order *Passeriformes* (see), containing 120 species in 15 genera, distributed from Africa to the Moluccas. The natural range extends to Wallace's line, but recent introductions have extended it to Australia. The widely distributed type genus, *Pycnonotus,* contains 50 sparrow to thrush-sized species. Plumage is generally drab olive-green, yellow or brown, except for brighter patches on the cheek or vent, hence the naming of species such as the ***Red vented bulbul*** *(Pycnonotus cafer;* India, Burma, Sri-Lanka), ***Yellow vented bulbul*** *(P. goiavier;* southern Asia) and the ***White vented bulbul*** *(P. barbatus;* Africa). Bills are straight and slender, usually with well developed rictal bristles. Many species are crested. The most distinctive character is a patch of often long and striking hair-like feathers on the nape of the neck. Originally forest birds, feeding primarily on berries and other fruit, many species have adapted to man-made habitats and have become common in agricultural areas and gardens. The 2 species of ***Finch-billed bulbuls*** *(Spizixos canifrons, S. semitorquas)* are unusual in having short, thick, finch-like beaks; they are often seen in large flocks near to mountain jungle villages in southern Asia. Also an inhabitant of mountainous jungle in southern Asia is the unusually insectivorous ***Striated bulbul*** *(Pycnonotus striatus),* which is olive with white-striped upper parts. Other genera are, e.g. *Chlorocichla* (5 species of African scrubland), *Pyllastrephus* (tropical forests of Africa and Madagascar), *Criniger* (Africa to Southeast Asia). The ***Black bulbul*** *(Hypsipetes madagascariensis)* is found from Madagascar to East Asia.

Pycnonotus: see *Pycnonotidae.*

Pycnospores, ***spermatia:*** minute uninucleate, elliptical asexual conidia formed by rust fungi (see *Basidiomycetes,* Fig. 3.2), in special fruiting bodies called pycnidia or spermogonia. P. are generally incapable of infecting healthy plants. The flask-shaped pycnidia contain both sterile hairs and a dense, central mass of short hyphae, from which the pycnospores are abstricted. They are usually pustulate and located on the upper surface of the host plant, where they erupt through the epidermis.

Pygmy falcons: see *Falconiformes.*

Pygmy flower: see *Primulaceae.*

Pygmy geese: see *Anseriformes.*

Pygmy mouse: see *Cricetinae.*

Pygmy parrots: see *Psittaciformes.*

Pygopodidae, ***snake lizards:*** a family of the *Lacertilia* (see), consisting of about 15 species in 8 genera, distributed in Australia and Papua New Guinea. Anatomically, they are closely related to the *Gekkonidae.* Like the geckos they lack free eyelids and have a transparent spectacle, and the pupil is vertical and slit-shaped, but they differ from the latter in possessing an elongated, snake-like body without forelimbs. The hindlimbs have degenerated to small flap-like, scaly appendages. The tail is twice the length of the trunk, and it is capable of autotomy. Snake lizards are nocturnal and terrestrial in habit, and they progress by lateral serpentine movements. Some species burrow in the ground. Two cylindrical eggs are laid. They feed mainly on insects, as well as small lizards. The biology of this family is still not known in detail. One of the largest representatives is the 70 cm-long *Pygopus lepidopodus* of Australia. The most widely distributed of the snake lizards is *Lialis burtoni,* which occurs throughout Australia, including Queensland, and in Papua New Guinea.

Pygopus: see *Pygopodidae.*

Pygostyle: see Axial skeleton.

Pylocheles: see Hermit crabs.

Pylopagurus: see Hermit crabs.

Pylorus: see Digestive system.

Pyloric cecum: see Digestive system.

Pyracantha: see *Rosaceae.*

Pyralidae: see *Pyralididae.*

Pyralididae, *Pyralidae,* ***pyralid moths, snout moths:*** a worldwide family of the *Lepidoptera* containing about 10 000 species. The wingspan of most species is 2 to 3 cm, and wing shape is variable; usually both pairs of wings have relatively short hairy fringes. In many species the palps are medium sized or long, but they often project beak-like, hence the name "snout moths". As in the *Geometridae,* the adults possess abdominal tympanal organs, which are often visible as a small groove on the first or second abdominal segment. The *P.* are divided into several subfamilies, considered by some authors as distinct families. The caterpillars display a variety of habits, e.g. aquatic species (genus *Acentropus)* are known, and some are cactus feeders. *Cactoblastis cactorum* was introduced from South America to Australia to control prickly pear cactus. Many are pests, e.g. European corn-borer (see), Mediterranean flour moth (see), Wax moths (see).

Pyralid moths: moths of the family *Pyralididae* (see).

Pyramidal tracts: conducting pathways in the white matter of the spinal cord of mammals and humans. They begin in the cerebral cortex and end in the Motor cells of the anterior horn (see). All fibers of the P.t. conduct impulses of the voluntary nervous system, and they are responsible for the very finely graded movements and responses displayed by the voluntary system.

Pyramonodontinae: see *Carapidae.*

Pyrenoid: a dense area, practically free from thylakoids, within the Plastids (see) of phytoflagellates, algae, mosses and some ferns. Ps. exist singly or in numbers. They represent the sites of synthesis of storage materials, consisting of starch grains or other polysaccharides (e.g. paramylon in *Euglena).*

Pyrenomycetidae: see *Ascomycetes.*

Pyrethrum: see *Compositae.*

Pyridine nucleotide cycle: a salvage pathway in which the nicotinamide produced by degradation of NAD^+ is reutilized to synthesize more NAD^+. The P.n.c. probably operates in all organisms, whether or not they are capable of synthesizing the pyridine ring system, and irrespective of the pathway of biosynthesis (from L-tryptophan in animals, *Neurospora* and *Xanthomonas pruni;* from aspartate and dihydroxyacetone phosphate in plants and most bacteria).

Pyridine nucleotides: nucleotides in which the nitrogenous base is pyridine or a pyridine derivative. The pyridine nucleotide coenzymes are Nicotinamide-adenine-dinucleotide (see) and Nicotinamide-adenine-dinucleotide phosphate (see).

Pyrola: see *Pyrolaceae.*

Pyrolaceae, ***wintergreen family:*** a family of dicotyledonous plants containing about 45 species, found in wooded areas of the northern temperate region, and in cool areas at high altitudes in the tropics. They are evergreen perennial herbs or dwarf shrubs with alternate leaves, and 4-5-merous, hermaphrodite, actinomorphic flowers, usually with free petals. The superior ovary develops into a capsule with numerous small seeds. P. are mycorrhizal plants, i.e. they live in symbiosis with a fungus. In Europe, ***Larger wintergreen*** *(Pyrola rotundiflora)* and ***Common wintergreen*** *(Pyrola minor)* are relatively common in pine woods and in mixed woods with acidic soils.

Pyrolaceae. Common wintergreen *(Pyrola minor),* flowering plant.

Pyrosomida: an order of free-swimming colonial tunicates (see *Tunicata)* of the class *Thaliacea* (see). In shape, the individual organism resembles a hollowed-out pine cone. There is no larval stage, and the solitary oozoid is degenerate. The colony consists of a gelatinous cylindrical tube, in which each blastozooid lies perpendicular to the colony surface, with the mouth openings directed outward and the cloacal openings emptying into a common cloacal space. *P.* reproduce sexually and by budding. Each individual first produces an egg, after which its spermatozoa mature and are released (proterandry). Older individuals in the colony have therefore ceased to produce eggs and produce only spermatozoa. With increasing age, each individual also produces buds, so that the oldest individuals reproduce mainly by budding, which enables the colony to grow to a large size (up to 4 m in length). They are intensely luminescent (phosphorescent), due to the presence of light-emitting bacteria in their light organs. The bacteria are taken up by the follicle cells of the egg and transferred to the embryo. The emitted light is sufficient to illuminate the sails of ships.

The order comprises the single genus, *Pyrosoma.* For collection and preservation, see *Tunicata.*

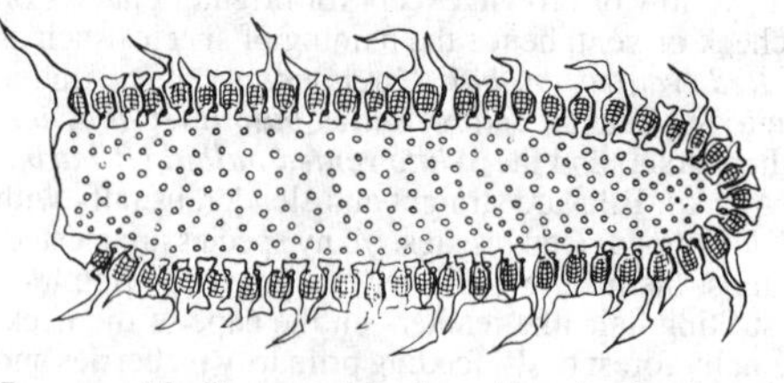

Pyrosomida. Pyrosomid colony (longitudinal section).

Pyrrhocorax: see *Corvidae.*

Pyrrhula: see *Fringillidae.*

Pyrroles, *pyrrole derivatives:* compounds containing the pyrrole ring. They are subdivided into mono-, di-, tri- and tetrapyrroles. Tetrapyrroles may be noncyclic or cyclic. Bile pigments and the chromophores of biliproteins and phytochrome are linear tetrapyrroles, whereas prophyrins and corrinoids are cyclic tetrapyrroles. A reduced pyrrole ring system is also present in many alkaloids (e.g. atropine, nicotine), and in the amino acids, proline and hydroxyproline. Unsubstituted pyrrole was first discovered in coal tar and bone tar, and it is also present in the distillation products of bituminous shale. It has not been found in living organisms. When a pine splint moistened with hydrochloric acid is held in the vapors of pyrrole it develops a fiery red color, hence the name of the compound.

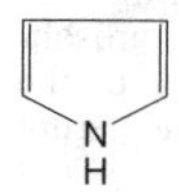

Pyrrole

Pyrrophyceae: see *Phytomastigophorea.*

Pyruvic acid: CH_3—CO—COOH, the simplest of the α-ketoacids (2-oxoacids) that occur in intermediary metabolism. It is an important intermediate in alcoholic fermentation and glycolysis. It can also be formed by the deamination or transamination of excess L-alanine, and it can serve as the precursor of L-alanine by transamination.

Pytelia: see *Ploceidae.*

Pythium: see *Phycomycetes.*

Pyxidium: see Fruit.

Pyxis: see Fruit, *Plantaginaceae.*

Q

Q_{10}: see Temperature coefficient.

Q-fever: see *Rickettsiae.*

QS-value: see Quotient of similarity.

Quadrate: see Skull.

Quadriceps muscle: see Musculature.

Quadrigeminal bodies: see Brain.

Quadrivalent: a group of four homologous chromosomes, formed during the first meiotic division of a polyploid nucleus, and held together by chiasmata from the diplotene until early anaphase I.

Quadrupedy, ***quadrupedalism:*** in the narrow sense, progessing on four feet. It was a taxonomic characteristic of the early animal taxon, *Tetrapoda,* which contained the quadruped (tetrapod) vertebrates, i.e. amphibians, reptiles, birds and mammals. Linneus used the term *Quadrupeda* only for the mammals. Birds are considered to be extremely divergent Q., in which the forelimbs are modified to wings.

Quagga: see Zebras.

Quahog: see *Bivalvia.*

Quail plover: see *Gruiformes.*

Quails: see *Phasianidae.*

Quantum evolution, ***phyletic gradualism:*** a mode of evolution proposed by G.G. Simpson to account for the relatively sudden appearance of higher systematic units. Evolution within an adaptive zone (which Simpson termed phyletic evolution) is relatively continuous, whereas new adaptive zones are attained by small populations in "quantum jumps". The process usually begins with a disturbance of the environmental equilibrium of the organism, leading to an unstable phase, which is survived by only a few populations (Q.e. is therefoe an "all or nothing" event). This is followed by a preadaptive phase with high selection pressure, in which populations are striving for a new equilibrium with the environment. In the final, adaptive phase, this equilibrium is achieved. An example of Q.e. would be the evolution of grazing animals from browsing animals during the phylogenetic evolution of the horse. Simpson was a peleontologist, and viewd Q.e. on a geological time scale. Substantial changes can occur, however, in shorter periods, measured in scores or hundreds of years (e.g. in living plant populations). In Simpson's original view, speciation was not necessarily involved in Q.e., but it seems likely that a very rapid evolutionary quantum jump would be accompanied by a speciation shift.

Quaternary, ***Anthropogene, Pleistogene:*** the youngest geological system, which covers the last 1.8 to 2 million years. It is subdivided into the Pleistocene (see) and the more recent Holocene (see) Epochs, and it is characterized by numerous glacial and interglacial periods in the Northern Hemisphere. By the Pleistocene, the world flora and fauna were largely similar to those of the present. Humans first appeared in the Q. See Geological time scale.

Quebrachine: see Yohimbine.

Quebrachitol: see Cyclitols.

Queen of the night: see *Cactaceae.*

Quelea: see *Ploceidae.*

Quercetin: 3,5,7,3′,4′-pentahydroxyflavone, a naturally occurring flavonol, which has been found as a yellow pigment in 56% of all angiosperms, notably flowers of the cotton plant, wallflowers, hops, horse chestnuts, onion skins and China tea. It occurs free, as in mono and dimethyl ethers, and as various glycosides, About 35 glycosides are known, e.g. rutin (3-rutinoside) from garden rue *(Ruta graveolens),* yellow pansies and forsythia, and quercitrin (3-rhamnoside) from quercitron bark.

Quercus: see *Fagaceae.*

Quezal: see *Trogoniformes.*

Quick process for acetic acid production: see Acetic acid fermentation.

Quiescence: see Dormancy, Resting state.

Quill: see Feathers.

Quillworts: see *Lycopodiatae.*

Quingestanol: a synthetic estrogen used in oral contaceptives; see Hormonal contraception.

Quinic acid: an optically active carboxylic acid containing a cyclohexane ring system. It occurs naturally in the levorotatory form, e.g. in cinchona bark, coffee beans, blueberry and cranberry leaves, and in sugar beets, and is otherwise widely distributed in the plant kingdom, either free, or as a component of chlorogenic acid.

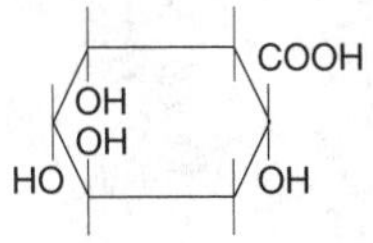

COOH
OH
OH
HO
OH

Quinic acid

Quinidine: see Quinine.

Quinine: the most abundant of a family of about 30 alkaloids, all present in cinchona bark, and characterized chemically by possession of the quinoline ring system. Cinchona bark (also known as Countess bark, Peruvian bark, or Jesuits' bark), containing up to 18% of alkaloids, is the branch and trunk bark from various species of the genus *Cinchona.* This large tree, often exceeding 35 m in height, grows at elevations of 1 500 to 2 500 m on western spurs of the Andes chain in Bolivia and Peru, where the climate is warm and moist. It grows singly, not forming forests, or even small groups. Cinchona alkaloids (as salts of quinic and other acids) are located almost entirely in the bark, the major sites being the epiderm and periderm. Cinchonidine differs chemically from quinine by the absence of the methoxy group (Fig.). Quinine and cinchonidine share the same molecular geometry, and they are the respective stereoisomers of quinidine and cinchonine (Fig.). Quinine has been synthesized chemically, but the route is not economically viable. All commercial quinine is therefore derived from cinchona bark.

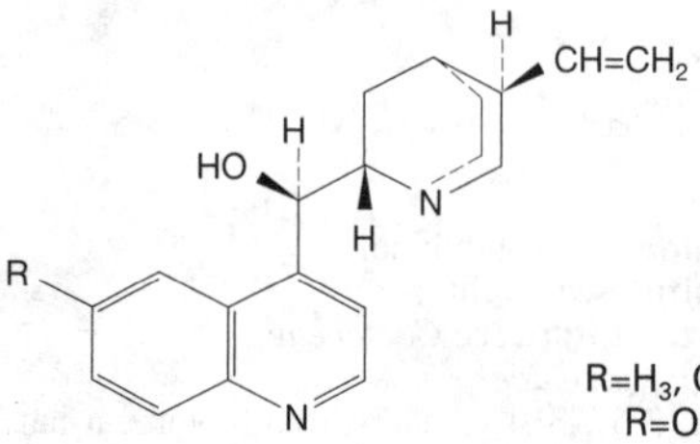

$R=H_3$, Cinchonidine
$R=OCH_3$, Quinine

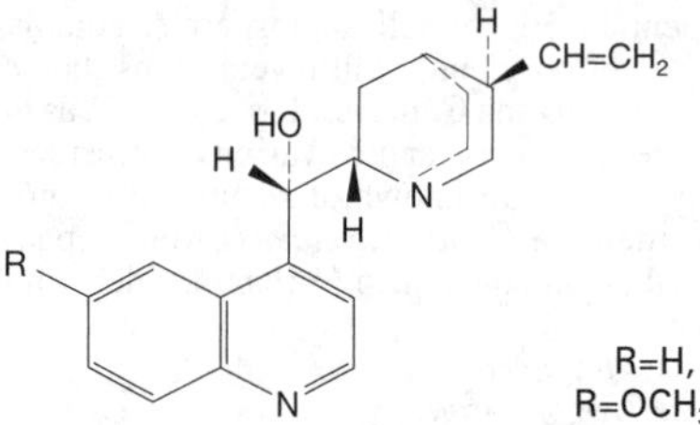

$R=H$, Cinchonine
$R=OCH_3$, Quinidine

Quinine was the first effective therapeutic agent against malaria, and it acts on the asexual blood forms of *Plasmodium.* By reducing the rate of tissue respiration, it has an antipyretic effect. It is also an analgesic and inhibits cardiac excitation, although quinidine surpasses it in this last respect. In higher doses, it is toxic, leading to deafness and blindness, and 10 g of quinine represents a fatal human dose. It is debatable whether the Peruvian indians used cinchona bark as a herbal remedy in pre-Columbian times (the name is derived from the Quechuan word, "quina" or "kina", meaning bark), or whether its therapeutic properties were discovered by the Spaniards when they reached South America. It was subsequently replaced as an antimalarial agent by various synthetic compounds, especially chloroquine. Since the advent of resistance to some synthetic drugs by *Plasmodium falciparum,* the use of quinine has been revived. About 40% of quinine production is used in the food and drink industry as a bittering agent, and about 30–50% is converted into its diastereoisomer, quinidine, which is used in the USA as an arrhythmic agent. The annual demand for cinchona alkaloids is 300–500 tonnes.

See *Rubiaceae.*

Quinoa: see *Chenopodiaceae.*

Quinoline: a heterocyclic, liquid, organic base. The ring system of Q. is present in the structures of the quinoline alkaloids, which are found in both microorganisms (e.g. viridicatine from *Penicillium)* and higher plants (e.g. quinine from *Cinchona).*

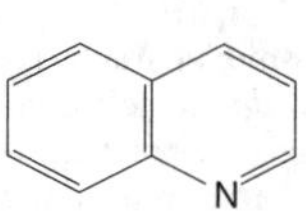

Quinoline

Quotient of similarity, ***QS-value:*** an index, proposed by Sorensen in 1948, which expresses the degree of association for each species in pairs of ecological samples. It is one of the many quantitative techniques used (see, e.g. Jaccard's index, Dominance, Diversity index) when an ecological sample contains no obvious characteristic group of species. QS (%) = $200c/(a+b)$, where a represents the number of species in sample A; b the number of species in sample B; c the number of species common to both A and B.

R

Rabbitfish: see *Holocephali.*

Rabbits: see *Leporidae.*

Raccoons, *Procyon:* a genus of the *Procyonidae* (see), represented by 7 species. *P. lotor* has dense, gray-brown fur, a long, bushy tail, and a pointed muzzle. More nocturnal than diurnal, it is frequently active at dusk, feeding on frogs, fish, small vertebrates, insects, nuts and fruit. It regularly washes its food in water, holding it between its forepaws. Naturally distributed from southern Canada through most of the USA and into Central America, the raccoon has been introduced into Europe for its fur, but it is not well established there. It favors timbered and brushy areas, and it is a good climber and swimmer.

Race: a group of organisms with shared, inherited characteristics, which differs in appearance, physiological characters and/or environmental requirements from other groups of the same species. A good example of biological R. is provided by the cuckoo, in which an inherited dependence on a particular host species is associated with the production of specifically colored and patterned eggs. Ecological Rs. are also known as Ecotypes (see). Geographical Rs. are commonly called subspecies.

Raceme: see Flower, section 5) Inflorescence.

Racemization of amino acids for dating: see Dating methods.

Racemose inflorescence: see Flower, section 5) Inflorescence.

Racer: see Whip snakes.

Racerunners: see *Eremias, Teiidae,* Whiptail lizards.

Rachis: see Feathers.

Racquet organ: see *Solifugae.*

Rad: see Dose.

Radial symmetry: symmetrical arrangement of the body parts of an animal around a central axis. It is typical of sessile animals which are attached posteriorly and have an anterior mouth, e.g. most coelenterates and sessile echinoderms. For the R.s. of plants, see Flower.

Radiated tortoise: see *Testudinidae.*

Radiation biophysics: a branch of biophysics concerned with the action of Ionizing radiation (see) on biological material. R.b. is closely related to radiation chemistry and photobiology. An important aspect of R.p. is that minimal radiation energies can evoke very large effects by triggering chain reactions. A model for the description of radiation effects is provided by the Target theory (see).

R.b. was one of the earliest subjects of biophysics, originating from the clinical need for radiation protection and the medical applications of radiation therapy.

Radiation effect: the response of biological systems to absorbed ionizing radiation. An effect may be direct or indirect. Indirect R.es. consist of the generation of highly reactive chemical compounds, which are normally short-lived and react immediately with other molecules. In this respect, the radiolysis of water is particularly important, resulting after 10^{-7} s in the following reactive products: H·, OH·, e_{aq}^-, H_2, H_2O_2. These reactive species then damage biological macromolecules, mainly by oxidation with the release of molecular oxygen. The resulting organic radicals can be reduced to their original state by the protective action of SH compounds:

$$R\cdot + R'SH \rightarrow RH + R'S\cdot$$
$$R'S\cdot + R'S\cdot \rightarrow R'S\text{-}SR'$$

This protective action is only possible, however, if the SH compound is present at the site of radiation ionization. Otherwise, the bioradical is converted by oxygen into a peroxyradical ($RO\cdot_2$) and the process is no longer reversible by an SH compound.

Biological material varies in its radiation sensitivity. Damaging radiation for proteins in vitro are in the range of 10 Gray and more, and the damage is due mainly to indirect effects. Nucleic acids are damaged mainly by direct effects, because the histones provide protection against indirect effects. Membrane processes are affected by very low doses. Damage to lysosomal membranes is particularly dangerous, due to the release of lysosomal enzymes which themselves can cause severe cell damage. At extremely low radiation doses, cells display an increased mutation rate; the rate of mutation in human cells is doubled by a dose of 0.5 Gray.

Radiation genetics: a branch of genetics which studies the effects of ionizing radiation on genetic material in the cell, and the accompanying cytological and genetic changes. R.g. addresses three main questions. 1) What type of mutations are induced by ionizing radiation ? 2) What is the frequency of different types of mutation in response to a particular radiation dose ? 3) What is the genetically tolerable radiation dose for a human ? R.g. was founded by the American biologist, H.J. Muller. Using the fruit fly, *Drosophila melanogaster,* he provided the first proof that mutations can be caused by X-rays.

Radiation protection: methods for decreasing the radiation burden. Fundamental to R.p. are the acquisition of precise data on the natural background radiation, and the determination and monitoring of artificial radiation. Average natural radiation exposure for an individual is 1 mSv/year (Sv = Sievert). This value may be increased 100-fold by local sources of natural radiation in the soil. Natural background radiation throughout the world has been approximately doubled by atomic tests. Additional environmental radiation exists in the vicinity of atomic power stations, but by law this should not exceed 1/3 of the natural radiation. The medical use of radiation sources also contributes to the total radiation burden; as an average for the total population, this is also equivalent to an increase in the background burden of 1/3.

Internationally binding limits for radiation exposure are recommended by the International Commission for Radiation Protection (ICRP).

Radicant: see Life form.

Radioactive isotope: an isotope that decays with the emission of charged particles and/ or high energy electromagnetic radiation (e.g. γ-rays). It decays to a new element or to a different isotope of the same element. The unit of radiactive decay (radioactivity) is the Becquerel (Bq). 1 Bq corresponds to one radioactive event per second. Different methods are used for the detection and measurement of radioactivity.

The *Geiger-Müller tube* consists of a chamber of ionizable gas with a central, insulated probe. A high potential difference is maintained between the walls of the chamber and the central probe. A charged particle from a radioactive source passes into the chamber through an end window, ionizes the gas and gives rise to a brief discharge, which is recorded as a radioactive event (a single "count"). Since the soft β-rays from ^{14}C have poor penetrating power, a thin mica end window is necessary to allow their passage into the ionizing chamber; even then ^{14}C is usually counted with less than 10% efficiency, much of the radiation being absorbed by the end window or earlier. In contrast, the powerful β-rays emitted by ^{32}P are able to pass through a copper end window.

Scintillation counting exploits the observation that certain substances (e.g. 2,5-diphenyloxazole) emit light when excited by charged particles, i.e. they scintillate. The light pulses (each corresponding to a single radioactive event) are recorded with photomultipliers. Scintillation counters are much more sensitive than Geiger-Müller tubes, and can be used to quantify the very weak β-emissions from ^{3}H (tritium).

Autoradiography detects radioactivity by its ability to blacken a photographic film. Thus, it can be used to locate the position of radioactive compounds by laying a chromatogram or electropherogram on a photographic film, which is then developed. The method is also used to great effect in electron microscopy. Thus, after treatment of the organism, organ or tissue with a radioactive substance, electron microscope sections are prepared and layered with photographic emulsion. The subcellular location of radioactively labelled metabolites is then identified from the position and number of silver grains observed under the electron microscope.

R.is. have become an indispensable tool of biological, medical and industrial research. Extremely small amounts of radioactivity can be detected and quantified, so that methods using R.is. are extremely sensitive They are used in archeological dating, in the study of the transport and distribution of natural compounds and drugs, and for the elucidation of biochemical pathways. However, on account of the associated hazards of working with radioactive material and the problems of radioactive waste disposal, alternative methods are used where possible, e.g. enzyme or fluorescent markers are now frequently used in routine immunoassays, rather than radioactive iodine.

Radiocarbon dating: see Dating methods.

Radiolarians: marine actinopod amebas placed in the subclass *Radiolaria* (class *Actinopodea*, superclass *Sarcodina*), or in the order *Radiolarida* (subclass *Actinopodia*, class *Sarcodinea*). They are usually approximately spherical with a skeleton of silicic acid. The cytoplasm is foam-like with colored inclusions. Most R. possess a lattice-like or fine filigree skeletal structure. In some R. the skeleton is lacking or represented by primitive spines. Both free living unicellular and colonial forms are known. The typical division into ectoplasmic layer and central body with their respective zones and inclusions are as described for Acantharians (see), although the symbiotic organisms (zooxanthellae) of R. have often been described in the ectoplasmic layer rather than the central body. The vacuolated calymma may be so extensive that certain unicellular R. can attain diameters up to 3 cm, and colonies up to 25 cm.

Reproduction is by budding. In species with a solid skeleton, daughter cells emerge then develop their own skeletons. From time to time, gametes (isospores) are produced, but their subsequent development in the ocean depths is incompletely understood.

Some of the oldest recorded fossils are those of R. and acantharians, found in Precambrian rocks. A red clay (radiolarian mud) rich in radiolarian skeletons is found in the very deep parts (4 000–8 000 m) of the Indian and Pacific oceans. In former geological times, in both shallow and deep oceans, solidification of radiolarian mud gave rise to structures known as radiolarites.

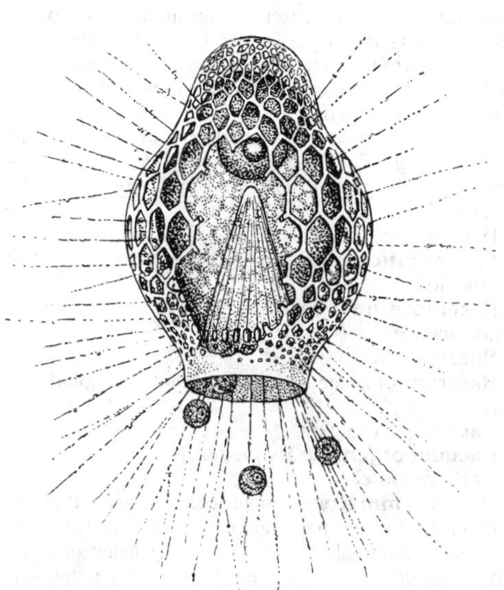

Radiolarians. Cyrtocalpis urceolus. The central capsule and nucleus are seen within the lattice shell. Below are the sieve-like opening and pyramidal filament cone (after Haeckel).

Radiomimetic substances: chemical compounds whose action on cells and tissues is similar to that of mutagenic irradiation, including the causation of gene and chromosome mutations. See Mutagens.

Radiosa amebas: see Amebas.

Radiosaprophytic conditions: see Saprophytic system.

Radula:

1) A uniquely molluscan feeding organ present in the pharynx of most mollusks, but absent from the bivalves and from some gastropods and solenogasters (see *Mol-*

lusca). The R. consists of a belt of recurved chitinous teeth or denticles (arranged in longitudinal and transverse rows), which passes over a cartilaginous base known as the odontophore. Depending on the species, the number of teeth varies from a few (e.g. only 2) up to 750 000. The R. is formed by dorsal and ventral epithelial cells in a blind invagination of the pharynx, known as the radula sac. From its site of synthesis (posterior or caudal end) in the radula sac, the R. grows forward over the odontophore (in most species at a rate of 1–5 rows of teeth per day) to replace the teeth and membrane lost by wear at the anterior end. The newly synthesized teeth are softly pliable, but as they pass forward they acquire a hard, impervious surface. For feeding, the R. is projected forward and pressed against the food source. The change in tension caused by this forward projection stretches the radula membrane over the odontophore tip, so that the teeth become erect. By alternate projection and partial retraction with an upward movement, the R. therefore acts as a scraper or rasp, and as a food collector (the curvature of the teeth is directed posteriorly).

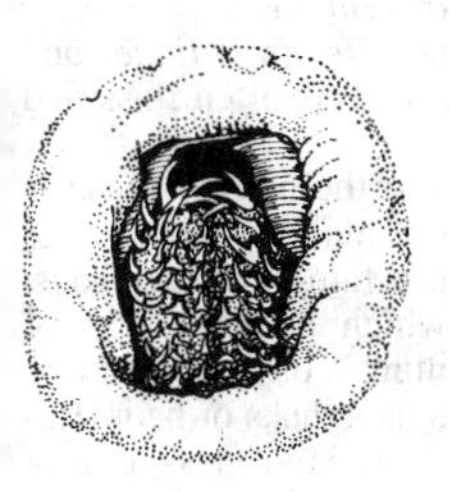

Radula. External opening of the proboscis of a streptoneurous gastropod, showing the radula.

In addition, the radula can also be moved backward and forward to a small extent over the odontophore. In the gastropods, the R. is of taxonomic importance, and taxa can be differentiated according to the number, shape and arrangement of the R. teeth, e.g. *Taenioglossa* (7 robust teeth in a transverse row, for powerful abrasion of the substrate), *Ptenoglossa* (a broad expanse of small, blade-like teeth), *Rhachiglossa* (3 sharp cutting teeth in a transverse row, as an adaptation to a carnivorous diet), *Rhipidoglossa* (a primitive type; the radula is a wide sheet with rows of numerous narrow teeth diverging like the ribs of a fan; it functions by sweeping flexibly over irregular surfaces, sweeping up or abrading fine particles), *Toxoglossa* (two highly specialized, elongated teeth, which are hollow and used for harpooning prey and for injecting a poisonous secretion of the salivary glands), *Docoglossa* (limpets; only a few very stout teeth in each row, and an exceptionally long radula).

2) The hypha of a filamentous fungus carrying a Radulaspore (see).

Radulaspore: a type of Conidium (see) or spore formed by filamentous fungi. Other types of conidia include the blastospore, porospore and arthrospore. Rs. resemble blastospores in most respects [i.e. they arise by budding from a somatic hypha (known as a radula) or a conidiophore, and they are not borne on a stipule], but they are characterized by the fact that rasp-like denticles remain at the site of their abstriction on the conidiophore. Rs. are produced, e.g. by *Cercospora, Botrytis,* and *Aureobasidium pullulans*. [Talbot, P.H.B., *Principles of fungal taxonomy,* London, MacMillan, 1971].

Raffinose, ***melitose:*** a nonreducing disaccharide consisting of one molecule each of D-galactose, D-glucose and D-fructose. The galactose and glucose are linked α-1,6, while the frucose in linked to the glucose by an α,β-1,2-glycosidic bond. Next to sucrose, R. is the most common sugar of flowering plants. Particularly high concentrations are found in sugar beets, molasses and some seeds. It is readily fermented by yeasts.

Raggee: see *Graminae.*

Rail babblers: see *Muscicapidae.*

Rails: see *Gruiformes.*

Rainbow bird: see *Coraciiformes.*

Rainbow boa, *Epicrates cenchria:* a snake of the family *Boidae* (see), 2–4 m long, and native to the Central American mainland. The larger (up to 4.5 m) and closely related species, *Epicrates angulifer* (Cuban boa), is found in Central and South America and the Antilles. Both have colored iridescent scales, live mainly in trees, feed mostly on birds, and are viviparous (10–30 young).

Rainbow lizard, *Agama agama:* a member of the lizard family, *Agamidae* (see Agamids). It is the commonest of all African lizards, existing as several subspecies, distributed widely over the dry grassy plains and bush areas of Africa. It is particularly common on farmland, and it is fond of human habitations, where it is useful as an eater of insects. The gray-brown body is up to 30 cm long with a reddish head and spiny scales. In the breeding season the male has an orange head and indigo body. Each adult male shares territory with one or more duller females and sometimes several young. Courtship and territorial behaviour are highly developed. Eggs are buried in the sand. One of the most detailed published accounts of reptilian social life is that of the R.l. (A. Harris, *The Life of the Rainbow Lizard,* Hutchinson Tropical Monographs, 1964). Other species of *Agama* are found in the Near East and Central Asia, e.g. *A. caucasica* (Caucasian agama).

Rainbow plant: see Carnivorous plants.

Rainforest: a tropical evergreen vegetation formation with rampant and luxuriant growth. It occurs in regions of uniform high temperature (never below +18 °C) and a regularly high rainfall (2 000 to 4 000 mm per year) that is not interrupted by marked periods of drought. Rs. contain a rich variety of species. The trees consist of a large number of mostly evergreen, leathery-leaved species (40 to more than 100 species per hectare). Trunks are generally tall and straight and relatively slender, with buttress roots at the base. The usually open canopy is uneven in height, the tallest trees attaining a height of 55 m. Branches are often festooned with numerous climbing plants and epiphytes.

The largest area of tropical R. is found in the Amazon basin of South America. Lower lying and often flooded areas are covered with a specialized type of R. called *Igapo forest,* whereas higher, nonflooded parts are covered with *Eté forest.* At higher altitudes, the temperature decreases, fog is common, and R. becomes *mountain R.,* characteristically containing many epiphytic ferns and mosses. In its extreme form, mountain R. becomes *cloud forest,* which is shrouded in mist most of the time (e.g. the Alakai swamp of Hawaii). The often small trees of the cloud forest form a dense canopy. Compared with the forest at lower altitudes, there are relatively few species, and the branches and ground are draped with ferns, mosses, orchids and other epiphytes.

Rain frogs: see *Microhylidae.*
Raisins: see *Vitaceae.*
Rallidae: see *Gruiformes.*
Rama: see *Malvaceae.*
Ramapithecines: see Anthropogenesis.

Ramphastidae, ***toucans:*** an avian family in the order *Piciformes,* consisting of 37 species (in 5 genera) of medium-sized to large birds (30.5–61 cm), confined to the tropics of continental America. Their very large bills, which are sometimes almost as long as their bodies, are even larger, more brightly colored and more canoe-shaped than those of the Old World hornbills. Unlike the hornbills, toucans do not possess a casque. The toucan bill is surprisingly light, being constructed of a honeycomb of cells with bony struts, and the tongue is long, thin and flat, notched at the sides and bristly at the tip. They feed mainly on fruit, occasionally taking insects and even fledgling birds. As in other picarian birds, the feet are zygodactylous. The upper end of the tarsus characteristically carries prominent pads; similar heel pads are also found in other hole-nesting species which rear their young on hard, unpadded surfaces (e.g. jacamars and hornbills), but they show their greatest development in the toucans. ***Hill toucans*** *(Andigena)* move seasonally up and down the Andean slopes, while the smaller ***Toucanets*** *(Aulacorhynchus)* live in high mountain forests at about 3 000 m. All toucans are cavity nesters. Other genera are *Pteroglossus* ***(Araçaris),*** which are notably the most gregarious toucans, and *Ramphastos,* which are the largest representatives.

Ramphastos: see *Ramphastidae.*
Ramphocaenus: see *Sylviinae.*
Ramphocinclus: see *Mimidae.*
Ramus: see Feathers.
Random variable: see Variate.
Rangifer: see Caribou.

Ranidae, ***ranids, true frogs:*** a family of the *Anura* (see) containing 666 species in 47 genera. The family is distributed worldwide, except for polar and desert regions, southern Australia and New Zealand.

Members are usually streamlined in shape with a pointed head, and the eyes (horizontal pupil) and eardrums are large. Long, powerful hindlegs enable them to jump considerable distances, and the toes are often webbed. The omosternum is usually ossified and the sternum ossified (the sternum is usually cartililaginous in the petropedetines). Ribs are absent. Eight holochordal, procoelous presacral vertebrae are present, usually with nonimbricate (nonoverlapping) neural arches (imbricate in astylosternines). In most taxa, presacrals I and II are not fused (fused in *Hemisus),* and the atlantal cotyles of presacral I are widely separated. Presacral VIII is usually biconcave and the sacrum biconvex. Sacral diapophyses are cylindrical or slightly dilated, and they display a bicondylar articulation with the coccyx. Transverse processes are not present on the coccyx. The pectoral girdle is usually firmisternal (arciferal in a few species of *Rana).* The digits characteristically lack intercalary cartilages.

Most true frogs live in water or wet terrestrial habitats. Mating usually occurs in water (except in the *Dendrobatinae* and *Sooglossinae).* In many species, brood care is displayed by the male or female. The tadpoles usually have horny jaws (except in the *Sooglossinae)* and their respiratory opening lies to the left. Adult males often possess an expandable vocal sac and a loud croaking voice. The diet consists mainly of insects, while large species also eat water birds and fish. Members of this family, especially in the subfamily *Raninae,* are used widely as experimental animals in medicine, physiology and experimental morphology. In many countries the thigh muscles are eaten. The widely varying members of this family are divided into 6 subfamilies: *Arthroleptinae, Astylosterninae, Raninae, Petropedetinae, Mantellinae, Hemisinae* (see separate entries for each subfamily). Four of these subfamilies *(Arthroleptinae, Astylosterniae, Petropedetinae, Hemisinae)* are found exclusively in Africa, while the *Mantellinae* are found only in Madagascar, so that Africa is thought to be the evolutionary center of radiation of the family.

A further subfamily, the *Phrynopsinae,* is recognized by some authorities. This comprises 2 genera of small African frogs (Cameroon and Mozambique) possessing horizontal pupils, and cartilaginous unforked omosternum and sternum: *Phrynopsis* (large head; elongated, spike-like, single-pointed teeth) and *Leptodactylon* (small head and small disks on finger tips). Many authorities, however, place *Phrynopsis* in the *Raninae,* and *Leptodactylon* in the *Astylosterninae.*

The families, *Sooglossidae, Dendrobatidae* and *Platymantidae* (see separate entries) are often classified as subfamilies of the *R.*

Raninae, ***ranins:*** true frogs in the narrow sense; a subfamily of the *Ranidae* (see), containing 440 species in 24 genera. Ranins are the most familiar of all frogs, especially the genus *Rana,* which contains 258 of known ranin species. The sternum is ossified and ribs are absent. The tapering fingers lack disks or pads. Ranins are represented on every large land mass, except Greenland. They are absent from new Zealand, Central and South Australia, and the southernmost part of South America.

The ***Small European Water frog*** or ***Pool frog*** *(Rana lessonae)* is grass green (rarely brown), changing to yellowish with round dark spots in the breeding season. The dorsal surfaces of the forelimbs are yellow. The underside of each foot carries an inner metatarsal tubercle that is deeply vaulted and semicircular in shape. The ratio, toe length:tubercle length is less than 2.1. Males have a white vocal sac. The Pool frog is found throughout Europe in small stagnant lakes and ponds containing dense aquatic vegetation. Mating occurs from May to June, somewhat later than for the ***Marsh frog*** *(Rana ridibunda).*

The ***Pond frog*** or ***Edible frog*** *(Rana esculenta)* is the result of hybridization between the Marsh frog and the Pool frog, and in most characteristics it is intermediate between these two species. The inner metatarsal tubercle is more strongly vaulted than in the Marsh frog, but it does not form a semicircle as in the Pool frog. The ratio, toe length:tubercle length is 2.0–2.8. Color and markings are similar to those of the Pool frog, although the vocal sac is usually gray and the forelimbs less yellow. In Central Europe, *R. esculenta* occurs up to altitudes of 800 m. It spends the entire year in water. After overwintering on the floor of a pond, it emerges with a brown-black coloration, which soon turns to bright green.

The ***Marsh frog*** *(Rana ridibunda)* is the largest European frog (up to 15 cm in length). Its dorsal surface is olive green or olive brown, with large irregular dark spots, and the vocal sac of the male is smoke gray to blackish in color. The forelimbs, unlike those of *R. esculenta* or *lessonae,* are never yellow. The ratio, toe length:tubercle length is greater than 2.5. *R. ridibunda* is an accomplished swimmer, diver and jumper. In addition to insects, snails and worms, it also eats small frogs, birds and mice. It is found in Central and eastern Europe, where it normally inhabits large lakes and ponds containing dense aquatic vegetation. Mating occurs in April and May, somewhat earlier than for *R. lessonae* and *esculenta.* The spawn masses contain up to 10 000 eggs. *R. lessonae, esculenta* and *ridibunda* are collectively known as "Green frogs" or "Water frogs". They are typically aquatic, and on land they are rarely found more than one jump away from the water.

The ***European Common frog*** *(Rana temporaria)* is the most commonly used European species for edible frog legs. The squat body measures up to 10 cm, with a blunt snout, medium length hindlegs and slightly webbed toes. A distinctive brown-black triangular area is seen around the eardrum. The back is red to dark brown, uniformly colored or darkly spotted, occasionally with a pale longitudinal line. The belly is whitish and usually spotted gray-brown. Males have internal, nonextrusible vocal sacs. *R. temporaria* inhabits wet areas on the plains and in the mountains of the northern and temperate Palearctic region as far as the most northerly point of Norway (North Cape). Mating occurs in February to April, so that *R. temporaria* is the earliest spawning European frog. Spawn is laid in large masses containing up to 4000 eggs. It restricts itself to one site, always returning to the same water to spawn.

The ***Moor frog*** *(Rana arvalis),* which inhabits the lowlands of Central and South Europe, is a slim, brown frog, up to 7 cm in length, with a more tapered head than that of *R. temporaria.* It has a large, dark ear spot, and usually a pale stripe along the center of the back. The belly is unspotted, and there are weakly formed webs between the toes. At mating time the male is often a striking sky blue. The vocal sacs of the male are internal and nonextrusible. *R. arvalis* prefers damp meadows and bogs. Mating occurs from March to April, and spawn is laid in clumps containing 1000–2000 eggs.

The ***Spring frog*** *(Rana dalmatina)* is found in lowland areas of Central and eastern Europe, also occurring sporadically in coastal areas. It prefers deciduous forest, often some distance from water. It is a slim frog, 6–7 cm in length, with a tapering head. The back is light brown, red-brown or brick-red, and the belly is white and unspotted. With the aid of very long hindlegs, which can be extended forward and far beyond the head, *R. dalmatina* can jump up as high as 0.75 m, over distances of up to 2 m. The male does not possess a vocal sac. Mating occurs from March to April, and the spawn is laid in small clumps, each containing about 1000 eggs. *R. temporaria, arvalis* and *dalmatina* are collectively known as "Brown frogs". They typically live in water only during the mating season, and they overwinter on land.

In North America, *Rana* is represented by 10 different species, similar in appearance and preferred habitat to their European relatives. Being resistant to cold, the widely distributed ***Wood frog*** *(Rana sylvatica)* is found as far north as Alaska. One of the largest North American anurans is the dark green ***American Bullfrog*** *(Rana catesbeiana),* which has massive limbs, attains lengths of up to 20 cm, and is notable for the resounding, hollow-toned call of the male. In many parts of America it is farmed and eaten. It has been introduced from the east into the Pacific States of North America, and into some Caribbean islands, e.g. Cuba, where it has become naturalized in the wild. Leather goods are prepared from the skin.

The largest anuran in the world is the ***Goliath frog*** or ***Giant frog*** *(Rana goliath;* 40 cm head plus body length), found in deep rivers of West Africa. It is a poor jumper, retreating into depressions in the river when threatened. It differs from other species of *Rana,* in that the coracoid cartilages ("epicoracoids") anterior to the coracoids are only weakly calcified. Some authorities therefore place it in a separate genus as *Gigantorana goliath.* Also, the extensively webbed toes end in thick dilations.

The ***Chinese Nest frog*** *(Rana adenopleura)* of East Asia exhibits brood care. The male digs a groove at the water's edge, where the female deposits the spawn. The ***Asian Bullfrog*** *(Rana tigrina)* of South and Southeast Asia is a 15 cm-long, powerful aquatic frog, which eats other frogs, and in turn is used extensively as a human food source. Its mottled brown-green skin carries longitudinal folds. The commonest frog of Southeast Asian rice paddies is the ***Southeast Asian Rice frog*** *(Rana limnocharis).* It is eaten in Korea. Like many anurans that breed in shallow, still waters, the spawn consists of a thin film with all the eggs at the surface of the water

Rank order, dominance hierarchy, social hierarchy, status system: an order of rights and duties among biosocial animals, based on dominance. The highest ranking member of a group is called the *alpha-individual,* the lowest ranking member the *omega-individual.* High ranking members are accorded, e.g. priority at food and drinking sites, mating priority, first choice of resting and sleeping sites, etc., but this may be associated with certain duties, such a guarding, defense and settlement of conflicts between other individuals in the group. Within any social group there may be more than one dominance hierarchy, and these may not be linear. Thus, 3-way relationships and loops of dominance may exist, males and females may have separate rank orders, groups may dominate an individual that would dominate each member of the group separately, etc. The concept of dominance was first discussed in 1922 by T. Schjelderup-Ebbe *(Zeit. Psychol.* ***88,*** 225–252), who studied the *peck order* in domestic chickens.

The dominance value of an individual refers to its position in a hierarchy, and it has no units.

Ranunculaceae, *buttercup family:* a family of dicotyledonous plants containing about 2 000 species, most of which are distributed in the nontropical Northern Hemisphere. They are predominantly herbaceous, with some woody plants and some woody climbers. Leaves are usually alternate exstipulate, palmately lobed or dissected (opposite in *Clematis* and stipulate in *Thalictrum* and *Batrachium).* The brightly colored flowers are usually hermaphrodite and hypogynous, and they display a diversity of forms. The original condition is actinomorphic, but some species have dorsiventrally zygomorphic flowers (e.g. *Aconitum, Delphinium).* All flowers have numerous stamens, usually arranged spirally on the conical receptacle and usually several to many free carpels. Perianth members are inserted in whorls or spirally on the convex receptacle. The perianth may be undifferentiated, or it may consist of a distinct calyx and corolla. Stamens are often devel-

Ranunculaceae. a Columbine or aquilegia *(Aquilegia vulgaris). b* Monkshood *(Aconitum anglicum = A. napellus). c* Horticultural hybrid of *Clematis viticella* (southern Europe) with *Clematis lanuginosa* (China) *(C.* x *jackmani* Th. Moore). *d* Lesser celandine or pilewort *(Ranunculus ficaria). e* Christmas rose *(Helleborus niger).*

oped as honey-leaves, which carry nectar in a cavity or outgrowth. Pollination by birds and wind is rare, and the majority of species is pollinated by insects. Fruits are various, but often consist of one or more many-seeded follicles, or a cluster of indehiscent, single-seeded fruits, especially nutlets; berries are less common.

Many species are planted as ornamentals for their attractive flowers, including many horticultural varieties which have much larger flowers than their wild counterparts, e.g. ***Delphinium*** or ***Larkspur*** *(Delphinium),* ***Aquilegia*** or ***Columbine*** *(Aquilegia),* ***Monkshood*** *(Aconitum),* ***Anemone*** *(Anemone)* and ***Clematis, Traveller's joy*** or ***Old man's beard*** *(Clematis).* On account of their content of alkaloids or glycosides, some species are poisonous or medicinal plants, e.g. ***Adonis*** *or* ***Pheasant-eye*** *(Adonis vernalis),* ***Aconite*** or ***Monkshood*** *(Aconitum napellus),* ***Pasque flower*** *(Pulsatilla vulgaris = Anemone pulsatilla),* and ***Christmas rose*** *(Helleborus niger).*

Some species are widely distributed meadow plants and agricultural weeds. Very familiar are the poisonous ***Meadow buttercup*** *(Ranunculus acris)* and the ***Creeping buttercup*** *(Ranunculus repens),* which is found especially in wet cultivated fields and wet meadows. The ***Kingcup, Marsh marigold*** or ***May blobs*** *(Caltha palustris),* which has reniform-deltoid leaves and intensely yellow flowers ("egg-yolk yellow"), is found in marshy habitats. Agricultural weeds of calcareous soils in Europe are ***Pheasant's eye*** *(Adonis aestivalis),* ***Forking larkspur*** *(Delphinium consolida)* and ***Corn crowfoot*** *(Ranunculus arvensis).*

A common species of acidic soils is the ***Mouse-tail*** *(Myosurus minimus),* so-called on account of its conspicuously long fruit hairs.

Some native European species are protected, e.g. the ***Globe-flower*** *(Trollius europaeus)* of mountain meadows, species of *Aconitum* found in wet habitats in mountain woodlands, ***Liver-leaved hepatica*** *(Hepatica nobilis* found in calcareous woodland, all species of *Pulsatilla,* the Spring-flowering ***Pheasant's eye*** *(Adonis vernalis),* and the ***Large wood anemone*** *(Anemone sylvestris).*

Ranunculus: see *Ranunculaceae.*

Raphicerini: see *Neotraginae.*

Raphicerus: see *Neotraginae.*

Raphidae, ***dodo*** and ***solitaires:*** an extinct avian family of the order *Columbiformes.* The family contains 3 well authenticated species: **Dodo** *(Raphus cucullatus;* Mauritius), ***Réunion solitaire*** *(Raphus = Pezophaps solitarius;* Island of Réunion), and ***Rodriguez solitaire*** *(Pezophaps = Ornithaptera solitaria;* Rodriguez Island). The former existence of a fourth species on Réunion ***(White dodo,*** *Victoriornis imperialis)* has been claimed, but the evidence for this is not universally accepted.

Inhabiting only the protected and predator-free Mascarene Islands of the Indian Ocean, the raphids lost the ability to fly and evolved into relatively large (turkey-sized) birds. The dodo and the Réunion solitaire became extinct toward the end of the 17th century, whereas the Rodriguez solitaire survived into the 18th century. All were massively built with strong feet and bills, with external nares far out toward the tip of the upper mandible. The wings were rudimentary and the sternum only slightly keeled. The largely blue-gray dodo had an apparently clumsy shape, with a large head, heavy, hooked bill, short legs and a little tail of a few curly feathers. It was slow moving and trusting, hence its demise, due to hunting by sailors and the ravaging of the bird and its eggs by introduced pigs, rats and dogs. The solitaires were less heavily built with a longer neck and longer legs, but they were no less vulnerable. Solitaires were predominantly brown, except the female Réunion solitaire, which is thought to have been white with yellow legs and bill.

Osteological remains of the raphids are quite abundant, and many almost complete skeletons of the dodo and the Rodriguez solitaire have been recovered. Mounted museum specimens of the dodo also exist. Classification under the *Columbiformes* has been questioned, and an affinity with the *Rallidae* has been suggested.

Raphidioptera, ***snake flies:*** a small insect order, represented on all continents except Australia. About 120 species are known, 20 in western North America and 12 in central Europe. The earliest fossil *R.* are found in the Jurassic. See *Neuroptera.*

Imagoes. Body length is 1–2 cm. The head is elongate and heart-shaped, with prognathous, chewing mouthparts, bulging compound eyes and long antennae. Especially characteristic is the prolongation of the first thoracic segment to form an elongate "neck". There are two pairs of approximately equal-sized, transparent wings, with primitive, net-like venation, and a dark oblong patch on each leading edge. The female possesses a long thin ovipositor at the tip of the abdomen. Cerci are absent. *R.* are predatory and are usually found in shade. Metamorphosis is complete (holometabolous development).

Eggs. All species lay eggs, which are deposited in cracks in tree bark and in rotting wood.

Larvae. The elongate larvae live in cracks in tree bark or in the tunnels and channels formed by other wood insects, whose larvae they waylay. They can move quickly backwards or forwards, and they molt 3–4 times.

Pupa. The pupa is decticous and exarate (pupa libera) and closely resembles the adult. At first it rests in a chewed out hollow (pupal cell) in a substrate of bark or decaying wood; before emergence of the adult, the pupa becomes active and leaves the pupation cell.

Economic importance. Adult *R.* consume aphids, bark beetles and caterpillars, but since they never occur in very large numbers they are not especially important in pest control.

Classification. There is a single family *(Raphidiidae)* embracing the main genera, *Raphidia* and *Inocellia.* Some authorities recognize two families, *Raphidiidae* and *Inocelliidae.*

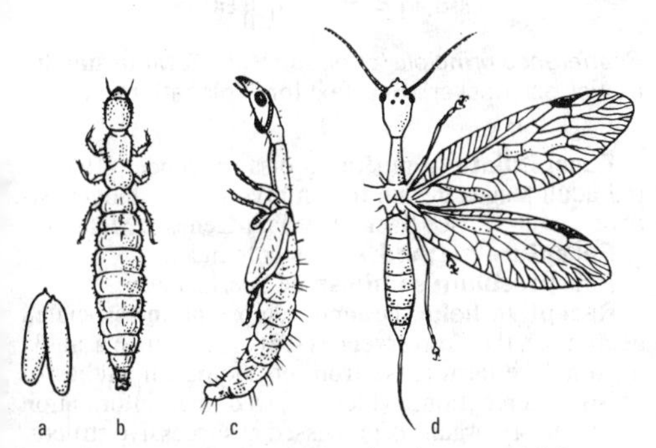

Raphidioptera. Stages in the development of *Raphidia notata. a* Eggs. *b* Larva. *c* Pupa (male). *d* Imago (female).

Raphus: see *Raphidae.*

Raptor: a Bird of prey (see).

Raspberry: see *Rosaceae.*

Raspberry cane maggot: see Anthomyiids.

Rastrites (Latin *rastrum* rake): an index genus of graptoliths of the Lower Silurian. The rhabdosome has the appearance of a rake, with numerous separate, straight thecae.

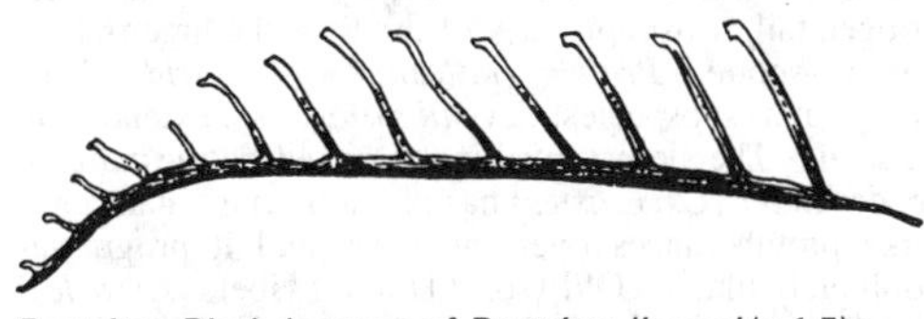

Rastrites. Rhabdosome of *Rastrites linnaei* (x 1.5).

Rat bite fever, ***sodoku:*** a name given to 2 different specific infections. One is caused by *Spirillum minus,* which lives naturally in the blood of wild rats (see *Spirillum),* and may be injected by a rat bite if the animal has a damaged and bleeding mouth. It is a relapsing, febrile illness with a local inflammatory lesion, enlargement of regional lymph glands and a macular skin eruption. The other is caused by *Streptobacillus moniliformis,* which is a normal inhabitant of the rat nasopharynx. This is also a febrile condition, resembling in many ways that caused by *S. minus,* but differing by the occurrence of acute rheumatic symptoms.

Rat chinchilla: see *Abrocomidae.*

Rate of evolution: see Evolutionary rate.

Rat flea: *Xenopsylla cheopis.* See *Yersinia pestis.*

Ratitae (Ratites): see *Struthioniformes, Rheiformes, Casuariiformes, Apterygiformes.*

Ratlike hamsters: see *Cricetinae.*

Rato de Taquara: see *Echimyidae.*

Rats: see *Muridae.*

Rat snakes, *Elaphe:* a genus of large, egg-laying snakes of the family *Colubridae* (see), distributed in America, Europe and Asia. Many species are specially adapted to an arboreal existence, e.g. ventral ridges, formed by the angulate sides of the belly scales, enable the snake to gain a hold when climbing. There are also many ground-dwelling species. Except for *E. dione,* found in the sandy steppes of Asia, most species live primarily in forested areas and in human settlements. The teeth in each jaw are equally long, and the eyes have round pupils. Examples are the Aesculapian snake (see) and the Four-lined rat snake (see).

Rattails: see *Macrouridae.*

Rattlesnakes, *Crotalus:* a genus of the *Crotalidae* (see) containing about 15 species with many subspecies and local races. They possess curved venom fangs, which in some species attain a length of 2.5 cm, and they are characterized by a "rattle" at the end of the tail. The rattle consists of loosely joined keratinized material (mostly tail scales) retained from the previous molt. When the snake is irritated, the rattle is held erect and shaken rapidly, emitting a penetrating whirring sound, as a warning to would-be attackers.

R. are distributed in North and Central America, and the range of one subspecies extends into South America. They are mostly brownish or gray in color, often with a light diamond or angled design on the dorsal surface. The face has a "threatening" expression, with large yellow eyes, vertical pupils, and small eyebrow shields. They live mostly in dry stony areas (but are also found in forests), feed mainly on rodents, give birth to live young, and usually overwinter in groups. On account of their potent venom, they are among the most feared of snakes. The largest species (2.5 m) is the ***Eastern diamond-back rattlesnake*** *(C. adamanteus)* from the southern USA. The ***Western diamond-back rattlesnake*** *(C. atrox)* is characterized by a black and white-ringed tail. Most northerly of the R. is the green-olive, brown-spotted ***Prairie rattlesnake*** *(C. viridis);* the range of its subspecies, *C. viridis oreganus,* extends into Canada. The desert-dwelling ***Horned rattlesnake*** or ***Sidewinder*** *(C. cerastes)* has pointed, triangular, horn-like protuberances over the eyes, and it progresses obliquely like the Old World Horned vipers *(Cerastes)* by throwing loops of its body sideways. A subspecies of the Central American ***Tropical rattlesnake*** or ***Cascaval*** *(C. durissus terrificus)* is found as far south as northern Argentina; its venom contains both hemotoxins and neurotoxins (see Snake venoms); it is therefore especially dangerous, and its bite can be counteracted only with a specially prepared monovalent serum.

In America, R. are eaten and processed into conserves.

Rattlesnakes. Western diamond-back rattlesnake *(Crotalus atrox).*

Rauwolfia: see *Apocyanaceae.*
Raven: see *Corvidae.*
Ray-floret: see *Compositae.*
Razorback: see Whales.
Razorbill: see *Alcidae.*
Razor shells: see *Bivalvia.*

Reaction time: a term from sensory plant physiology, meaning the time that elapses between the onset of stimulation to the appearance of the reaction. In certain types of movement response, e.g. seismonasty, the R.t. is also known as the Latent time (see). The R.t. for the folding movements of the mimosa leaf is 0.08 s, compared with 25–60 min for the phototropic bending of graminaceous coleoptiles.

Reafference principle: a regulatory principle described in 1950 by von Holst and Mittelstaedt and by Sperry for self-initiated orientational changes in direction, involving the active positioning of limbs, an awareness of the interplay and relative positions of body parts, and perception of the environment. A command from a higher center in the nervous system (C_1) causes a change of state in connected centers (C_2, C_n), the last of which (C_n)sends a command to the motor apparatus, resulting, e.g. in a limb movement and change of direction. The response of this effector is registered by a receptor and transmitted to C_n. Normally, afferent impulses signalling a change of posture or direction, etc. cause a reaction that re-establishes the previous orientation. If the changes are initiated by the animal itself (i.e. the central physiological state is altered prior to the event), this does not occur. To explain this phenomenon, it was proposed that an "efference copy" (called "diverted excitation" by Sperry) is made in a lower center (C_n in the Fig.). The feedback afference (the "reafference") from the receptor is compared with, and anulled by, this efference copy. The impulses therefore have opposite signs, the efference (as well as its copy) and reafference being designated plus (+) and minus (–), respectively. Alternatively, these two impulses may be unequal, thereby leaving a residual positive or negative impulse; this is registered by the higher centers (C_1, C_2), which then modify the command (i.e. the efference), so that the reafference is balanced by the efference copy. However, nonsystemic, external afferences (exafferences) can interfere. Centers C_2 and C_1 can respond to exafferences by modifying the efference, provided the additional load remains in the normal range. Independently of this correction, exafferences are also perceived by the entire system, and may be registered as sensory errors (see Movement perception). Thus, R.p. represents an evolved function of the nervous system, based on an expected or target value for reafference. It explains how the central nervous system of an animal transmits a target value for the direction of a body movement, and is able to correct for any discrepancy between the ultimate value and the target value.

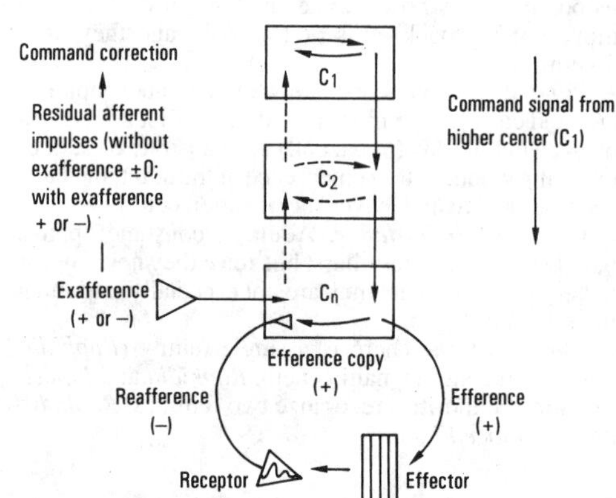

Reafference principle (after von Holst & Mittelstaedt, as well as Trincker). See Text for explanation.

Recapitulation: repetition of a sequence of ancestral adult stages during the embryonic morphogenesis of descendants. See Baer's law, Palingenesis, Metaboly.

Recapture method: see Lincoln index.

Receptaculum seminis: see Spermatheca.

Receptive field: the input region for an associated neuron, i.e. the Convergence (see) of stimuli on a single neuron. The term arose from physiological studies of sensory perception, which showed that information from sensory organs is processed at successive structural levels. Visual information processing in vertebrates starts with neuronal responses in receptors (see), continues through successive layers of neurons in the retina, reaches a switching region in the thalamencephalon, then is completed in stages in the cerebral cortex. Transformation and integration of informatrion at each of these structurally defined levels can be characterized in relationship to the R.f. The R.f. of an individual neuron in the visual system is that region of the retina occupied by receptors, which are connected to that neuron and therefore able to stimulate it. Each R.f. therefore represents a unit or building block for the synthesis and Perception (see) of the complex outside world. The resulting high-contrast, detailed picture of the environment perceived by the organism is based mainly on the perception of contrasting properties. See Off center field, On center field.

Receptor: in the general sense, a sensitive element that initiates a reaction when it is stimulated or receives a signal, i.e. it is part of the receiver in a chain of communication. It may be defined on the molecular level (biochemical definition) or on the cellular level (physiological definition).

A molecular R. is the reaction partner of a second molecule or ion (known as the ligand), and contact or binding between R. and ligand initiates a reaction. R. molecules are specific for their ligands (also known as agonists). Antagonists also bind to the R. molecule, but do not initiate a response, and at the same time they block the binding of the specific ligand (see Blocker). For example, acetylcholine is the ligand of the postganglionic acetylcholine receptor, which is blocked by atropine. Binding of acetylcholine to the receptor increases the ion permeability of the cell membrane (i.e. it generates a nervous impulse), but atropine binds to the receptor, displaces acetylcholine, and thereby blocks nervous transmission. Many receptor molecules (e.g. hormone receptors, including the insulin receptor) have been isolated and chemically characterized, enabling closer study of R.-receptor interactions.

Physiologically, Rs. are cells that respond to internal or external stimuli or signals by generating a receptor potential and subsequently an action potential. R. are classified according to the nature of their specific stimulus, and they can also be grouped into *exoreceptors* (on the periphery of the body) which react to stimuli from the outer environment, and *enteroreceptors* in the body interior, which react to internal conditions (e.g. R. involved in blood pressure regulation, and R. that give rise to the internal sensation of pain). *Propioceptors* (proprioreceptors) are located in the same organ that manifests the response (e.g. the muscle spindle of skeletal muscle).

Receptor cells display functional polarity. One pole (the input region) is a specialized, sensitive terminal structure (e.g. membrane stacks in light-sensitive cells) in which the receptor potential is generated. Stimuli are transduced into receptor potentials by different mechanisms. A strong receptor potential depolarizes relatively large regions of the cell membrane and travels to the beginning of the axon. There it act as an electrical stimulus (generator potential) and initiates an Action potential (see). The generator potential (-40 to -60 mV) is higher than the critical membrane potential (threshold potential), and it generates a rhythmic action potential (see Frequency modulation of excitation). The impulse is then conducted to the presynaptic or output region of the R. cell via the axon membrane. Rs. are capable of Adaptation (see).

Recessivity: complete or nearly complete prevention of the expression of a recessive allele of a gene by dominant allele (see Dominance). Recessive alleles are symbolized in genetic formulae by the use of small case letters, dominant alleles by large case letters (e.g. A^+a, Ert ert).

Recessus lateralis: see *Clupeiformes.*

Reciprocal reception joint: a saddle joint (see Joints).

Recognition stimulus: see Sign stimulus.

Recoil locomotion: see Movement.

Recoil swimming: a jerky swimming movement, usually by the contraction of muscles surrounding a water-filled cavity. Ejection of the water causes propulsion of the animal in the opposite direction. Examples are medusae, cephalopds, scallops *(Pecten).* In some crustaceans, R.s. occurs by rapid folding and extension of the abdomen.

Recombinant: a new genotype resulting from genetic recombination.

Recombinant DNA: see Gene technology.

Recombination: the totality of processes that result in new gene combinations, during meiosis or mitosis. Recombined genes may be localized on different chromosomes (unlinked genes) or they may belong to the same linkage group (linked genes).

1) Meiotic (sexual) R.

a) Unlinked genes. The allele pairs of the genes in question (which belong to different linkage groups) can recombine freely, due to the random orientation of bivalents and the random distribution of chromosomes during meiosis, and the random union of gametes during fertilization. The process obeys Mendelian laws.

b) Linked genes. The alleles of the genes in question belong to the same linkage group and behave as a unit in meiosis, i.e. they cannot freely recombine. R. between paired chromosomes in meiosis may be intergenic (nonallelic) or intragenic (allelic). In *intergenic (nonallelic) R.* there is a reciprocal exchange of segments between the chromatids of the paired chromosomes. The exchange involves functional genes (cistrons), and it is reciprocal and symmetrical, i.e. each crossing over during meiosis results in a cell tetrad containing two cells with exchange chromatids and two cells with nonexchange chromatids. *Intragenic (allelic) R.* occurs between the subunits of a functional gene (cistron), which is present in two different allelic forms in the homologous chromosomes. In this case, in contrast to classical crossing over, R. is nonreciprocal (asymmetric).

Thus in *reciprocal R.,* parent types a+ and +b give rise to a tetrad of cells with equal proportions of genotypes a+, +b (parent combinations), ab and ++ (recombinants), provided crossing over occurs between the two genes. In *nonreciprocal, intragenic R.,* the four genotypes do not occur with equal frequency, i.e. many cells are wild-type recombinants, and the double mutants expected from R. are absent (or vice versa). A further difference is that intergenic R. generally shows positive interference, whereas intragenic R. usually shows negative interference.

2) Somatic or *parasexual R.* This occurs in the mitosis of somatic cells, and not during meiosis. Like meiotic R., it may be intergenic or intragenic. Recombinable linked structures are found in diploid, heterokaryotic and heterogenotic cells. Somatic R. is rare in higher organisms, but it is the normal process of R. in certain fungi and in bacteria and viruses. The heterogenetic state of bacteria, which is a precondition for genetic R., is attained by conjugation, transduction or transformation.

In Gene technology (see), recombination is performed in vitro, in order to prepare DNA containing new gene combinations.

It is unclear whether the mechanisms of inter- and intragenic R. are in principle the same or fundamentally different. The totality of all the factors that result in a specific case of genetic R. is known as the recombination system of the organism.

Recombination value, ***percentage recombination:*** the relative frequency with which two genes on the same chromosome become separated by crossing over and recombination during meiosis. It is expressed as the percentage of gametes in which two genes appear in new combinations. Under standardized conditions the R.v. for two particular genes of a coupling group is constant, and can be used to locate (map) the relative position of these genes on the chromosome, i.e. it provides an index of the relative distance between two genes. Genes located close together on a chromosome display close linkage, i.e. they assort with one another more regularly than genes located far apart on the chromosome. The unit of gene separation is taken as a recombination frequency of 1%, e.g. two genes displaying a recombination frequency of 8% are said to be separated by 8 map units. In view of the high probability of double crossover between widely separated genes, this assignment of map distances is only accurate for closely linked genes. Because of the occurrence of multiple crossovers, even very widely separated genes (e.g. at opposite ends of a long chromosome) display at least 50% apparent linkage (i.e. they assort together in at least 50% of the gametes). If an even number of crossovers occurs between them, they will eventually come to lie on the same chromosome in the gamete, i.e. each separation by crossover is reversed by the next crossover. Separation occurs only after an odd number of crossovers.

Rectal ampulla: see Digestive system.

Rectal diverticulum: see Digestive system.

Rectal glands, ***rectal papillae:*** single- or double-layered thickened patches of epithelium in the rectum of most insects. They occur in varying numbers (usually 4 or 6), and are always richly supplied with tracheae. They are particularly active in water resorption from feces, and therefore important in the regulation of the water economy of the insect. See Digestive system.

Rectal papilla: see Digestive system.

Rectal pouch: see Digestive system.

Rectal sac: see Digestive system.

Rectum: see Digestive system.

Recurrent inhibition of neurons: see Neuronal circuit.

Recurvirostra: see *Recurvirostridae.*

Recurvirostridae, ***stilts*** and ***avocets:*** an avian family in the order *Charadriiformes* (see), containing only 7 species (4 Avocets, 3 Stilts and the Ibisbill), which are all characteristically long-legged. In proportion to body size, stilts have the longest legs of any bird, except possibly the flamingo. All have slender necks and small heads, with very long, thin, straight (Stilts), recurved (Avocets) or decurved (Ibisbill) bills. Stilts and the Ibisbill probe deeply into the mud for food, while Avocets sweep their partly open bill backward and forward in a wide arc on the surface or near the bottom. All species feed on a wide range of aquatic invertebrates and small vertebrates, including small fish, while wading in relatively deep water. They are all good swimmers and strong fliers. The ***Avocet*** *(Recurvirostra ovoretta;* 43 cm, bill about 8.25 cm) (Fig.) is white with patterns of black on the head, nape, scapulars, wing coverts and wing primaries; the legs are blue-gray and the eyes red-brown. It inhabits the shores of inland or coastal waters, preferring shallow salt or brackish areas. Scattered populations occur in western Europe, Tunisia, the Nile Delta, Lake Victoria and South Africa, but the main center of distribution is central Asia , extending south to the Persian Gulf. The ***North American*** and the ***Australian avocet*** have similar plumage, except for a brown wash on the head and neck. The ***Andean avocet*** *(Recurvirostra andina)* is found only on saline lakes above 3 600 m in the Andes of Chile and Bolivia. *Himantopus himantopus* ***(Black-winged*** or ***Red-necked stilt;*** 38 cm, bill about 6.25 cm) is the most widely distributed member of the family, breeding between 50 ° north and 50 ° south in central Europe, the Mediterranean region, Egypt, Southeast Asia, Australasia, sub-Saharan Africa, and from southern USA to northwestern South America, on the vegetated shores of shallow freshwater and brackish lakes, rice fields and coastal lagoons. Its plumage is white (duskier in the summer) with black wings and back. *Ibidorhyncha struthersii* ***(Ibisbill),*** an aberrant member of the family, is uniformly gray with a bright red bill. It inhabits mountain lakes and streams between 1 600 and 4 400 m in the Himalayas.

Recurvirostridae. Avocet *(Recurvirostra avosetta).*

Red admiral: a migratory butterfly (see Migratory lepidoptera), family *Nymphalidae.* Two subspecies are recognized: *Vanessa atalanta atalanta* L. of the Old World (Britain, mainland Europe, North Africa, Turkey, Asia, North India) and *Vanessa atalanta rubria* Fruhstorfer of the New World (North America, Mexico, Guatemala, Bermuda, Cuba, intoduced into Hawaii). The Wings are patterned red and black. Caterpillars are generally black with rows of yellow spots, with white warts along the sides, and with black spines issuing from orange bases. Members of the *Urticaceae* are favored food plants for the caterpillars, especially nettles *(Urtica)* and sometimes hops *(Humulus).* In late summer, adults are often seen sucking the juices of fallen, over-ripe fruit. It is a protected species in Europe.

Red algae: see *Rhodophyceae.*

Red-and-blue poison-arrow frog: see *Dendrobatidae.*

Red book, *red list:* a catalogue of animal and plant species threatened with extinction in a particular area. In 1970, the Survival Service Commission of the IUCN published the *Red data book* which attempts to list all the endangered species in the world, and in which the most threatened species are printed on red paper. It is continually updated. This has given rise to a series of similar lists for individual countries and smaller regions. Some red lists also include invertebrates and lower plants, but the overall coverage is incomplete, mainly due to a shortage of specialist biologists for many groups (e.g. insects).

Red-bug: see *Heteroptera, Trombiculidae.*

Red coral: see *Corallium.*

Red currant: see *Grossulariaceae.*

Red data book: see Red book.

Red deer, *Cervus elaphus:* a deer with magnificent, well developed, branched antlers, which exists as numerous subspecies in Eurasia, Northwest Africa and North America. It has now disappeared from much of Europe. The pelage is red-brown in summer and grayer in winter. The North American subspecies *(C. elaphus canadensis),* considered by some authorites to be a separate species *(C. canadensis),* is known as the ***Wapiti*** or ***American elk;*** it originally ranged from western North America to northern Mexico, but its numbers are now severely depleted. The Siberian subspecies *(C. elaphus sibiricus;* ***Altai Maral*** or ***Altai Wapiti)*** ranges from the Altai mountains to China. It is more yellow in color than the western subspecies, and its very large antlers do not usually form a branched crown. Depending on the subspecies, *C. elaphus* is also variously known as a Hangul, Shou, Bactrian deer or Yarkand deer. See Deer.

Red-eared turtle: see *Emydidae.*

Red fir gall louse: see Galls.

Redfish, *rosefish, Norway haddock, ocean perch,* *Sebastes marinus:* a red-colored, live-bearing marine fish of the family *Scorpaenidae* (order *Scorpaeniformes),* which occurs from the Arctic White Sea and Greenland as far south as Scotland and eastern Norway, and in the cold Labrador current to the latitude of New York. It is an economically important fish with tasty flesh, sold on the European market, and caught with dragnets at depths of 300–600 m in the North Atlantic. It is predatory on other fish and invertebrates, and in turn is eaten by sperm whales.

Redia: see Liver flukes.

Red-legged tortoise: see *Testudinidae.*

Red list: see Red book.

Red mullet: see *Mullidae.*

Red oak: see *Fagaceae.*

Redox potential: an electrical potential difference, E_0 [V], measured against a reference hydrogen elecrode. The redox potential is a measure of the oxidizing or reducing potential of a redox system. The defined standard potential in physical chemistry, or Normal potential, E_0 (pH = 0, pH_2 = 1 atmosphere) is, however, inapproriate for biological purposes. Instead, the Normal potential, E'_0, measured at pH 7 is usually taken as the reference point. The E'_0 value is only valid under standard reaction conditions, i.e. all reactants at unit activity. Such conditions do not obtain in the living cell, and the ratio of concentrations of oxidized and reduced forms of the redox pair is included in the calculation. The actual redox potential, E′, is therefore expressed as: $E' = E'_0 + 0.06/n \times \log([Ox]/[Red])$ (30 °C). Strong reducing agents have high negative E'_0 values, and strong oxidizing agents have high positive values. On the basis of their redox potentials, the redox systems of the respiratory chain can be arranged in the order of their oxidizing or reducing power. Thus, the redox pair, $NAD^+/NADH$, has the most negative R.p. and operates at the beginning of the chain, whereas cytochrome *a* has the most positive R.p. and is the final member of the chain.

Red panda, *lesser panda,* *Ailurus fulgens:* a monotypic genus of the *Procyonidae* (see), found in bamboo forests from Nepal to Szechuan. It has a long, soft-haired coat, a bushy tail and large, pointed ears. The upperparts are rust-red, and the underparts are blackish, in contrast to the face, which is largely white around the muzzle, lips, cheeks and ears, with small, dark eye patches. The tail is inconspicuously ringed. It is easily tamed and tends to lead a nocturnal existence. Bamboo shoots, grasses, roots and fruits constitute the major part of the diet, but it also takes birds' eggs, and according to some observers it may also take birds and mice.

Red plum maggot: see Plum fruit moth.

Redpoll: see *Fringillidae.*

Red-river hog: see African bush pig.

Red scorpion fish, *Scorpaena scrofa:* a large-headed bottom fish of the family *Scorpaenidae* (order *Scorpaeniformes),* found in the Atlantic and Mediterranean. It favors sandy or stony seabeds at depths of 20–100 m, where it conceals itself among seaweeds and stones. The foremost spines of the dorsal and anal fins, as well as some head spines, carry small venom glands. The R.s.f. regularly sheds its outer layer of skin, usually once, sometimes twice a month. Its wide mouth allows it to engulf prey over half its own size, and its diet consists of fish, crabs and other crustaceans.

Redshank: see *Scolopacidae.*

Red shank: see *Polygonaceae.*

Red spider mites: see *Tetranychidae.*

Red squirrel: see *Sciuridae.*

Redstarts: see *Turdinae.* See plate 6.

Reduction:

1) Withdrawal of electrons by a reducing agent, with a resulting decrease in the electropositive value of the material being reduced. Every R. must be coupled with an Oxidation (see). Addition of hydrogen or withdrawal of oxygen are also R. In cell metabolism, R. is catalysed by oxidoreductases.

2) In genetics, the decrease in (halving of) the number of chromosomes during meiosis. In special cases, during mitosis of somatic cells, the chromosomes may become separated into two or more numerically equal or unequal groups, so that cells are formed with decreased numbers of chromosomes; this process is known as *somatic* or *mitotic R.* Autopolyploids (see Polyploidy) can display *double R.,* i.e. meiosis results in the presence, in the same gamete, of two sister chromatid segments and therefore two completely identical alleles of a gene, due to crossing over and appropriate segregation of the chromosomes. Preconditions for a double R. are multivalent formation, heterozygosity and crossing over between the centromere and the genes in question.

Redunca: see *Reduncinae.*

Reduncinae, *reedbucks:* a subfamily of the *Bovidae* (see). The *R.* are subdivided into 2 tribes.

1. Tribe: *Reduncini;* genera: *Kobus* (5 species) and *Redunca* (3 species: *R. rundinum,* ***Reedbuck;*** *R. redunca,* ***Bohor reedbuck;*** *R. fulvorufula,* ***Mountain reedbuck).***

2. Tribe: *Peleini;* genus: *Pelea* (1 species: *P. capreolus,* ***Gray rhebuck).***

Kobus ellipsiprymnus ***(Water buck)*** is a deer-sized African antelope with a straggly, oily coat. It inhabits the sub-Saharan African savannas and prefers to be near to water, often fleeing into the water for safety. Other species are *K. megaceros (= Onototragus megaceros,* ***Nile lechwe*** or ***Mrs. Gray's waterbuck;*** swamps of southern Sudan and western Ethiopia), *K. leche (= Onototragus leche,* ***Lechwe;*** wetlands from Zaire to northern Botswana), *K. kob (= Adenota kob,* ***Kob;*** savanna from Senegal to western Kenya), *K. vardoni (= Adenota vardoni,* ***Puku;*** savannas of Zaire, Angola, Zambia, Tanzania, Malawi, Botswana).

Reduncini: see *Reduncinae.*

Redwing: see *Turdinae.*

Redwoods: see *Taxodiaceae.*

Reedbuck: see *Reduncinae.*

Reed duiker: see *Pelicaniformes.*

Reedling: see *Timaliinae.*

Reed warblers: see *Sylviinae.*

Reeve: see *Scolopacidae.*

Reflectance microscopy: a method for observing the surface of opaque objects. Light is deflected through 90 ° in a vertical illuminator, and shines down onto the object from above. For more detailed investigations of surface structure and topography at higher magnification, the scanning electron microscope is used.

Reflecting bodies: see Amebas.

Reflex: an autonomic, stereotypic reaction to a stimulus, which occurs via a *reflex arc.* The R. arc consists of 1) a Receptor (see), 2) afferent neurons that conduct impulses to the central nervous system, 3) junctional neurons of the central nervous system, 4) efferent neurons, and 5) the effector that receives impulses from the efferent neurons, and which displays the response. In a *monosynaptic R.* the R. arc contains a single synapse, e.g. the patellar (knee jerk) R. This single synapse is situated in the spinal cord, so that the presynaptic pathway consists of the efferent neuron, and the postsynaptic pathway is the efferent motor neuron. In *polysynaptic Rs.,* varying numbers of Interneurons (see) operate between the afferent and efferent pathways, so that the R. arc contains two or more synapses. Cleaning, defense and fleeing Rs. are polysynaptic. The time between reception of the stimulus and the response (reflex time) is short when the R. arc contains only a single synapse, and longer when it contains numerous synapses. Monosynaptic Rs. show no Facilitation (see) or Adaptation (see), whereas these properties are pronounced in polysynaptic Rs.

Genetically, Rs. are classified as unconditioned (i.e. inherited) and conditioned (i.e. aquired after birth by training). Unconditioned Rs. are mainly associated with the removal or counteraction of interference or disturbance, e.g. reflex withdrawal of the hand from a hot stove. Conditioned Rs. (see) are a form of learning.

Refractory period: the period following the generation of an Action potential (see), during which a nerve fiber (or any other excitable system) no longer responds to stimulation. A stimulus above the threshold value cannot generate an action potential in an excitable fiber as long as the membrane is still depolarized from the preceding action potential. The second phase of the action potential involves activation of K^+ channels (i.e. there is an efflux of K^+) and closure of Na^+ channels. The Na^+ channels then remain inactivated for a short period, and during this period no stimulus is able to increase the rate of influx of Na^+ so that it exceeds the rate of efflux of K^+. The R.p. consists of an *absolute R.p.,* during which no depolarization can occur in response to a stimulus, and a *relative R.p.,* in which stronger than normal stimuli are required to excite the nerve fiber. The relative R.f. is usually shorter than the absolute R.f. In a myelinated nerve fiber the absolute R.f. is about 1/2500 second, and the relative R.f. is about one quarter of this value. The sum of the absolute and relative R.fs.represents the rate of repolarization of the nerve membrane.

Refugium: see Glacial refuges.

Regalecus: see Bandfishes.

Regeneration: the ability of a plant or animal to more or less completely replace injured, dead or lost body parts. In the broad sense, R. also includes the development of structures and functions that compensate for such losses.

R. in plants. Plants generally display a very marked ability for R. *Direct R.* is the new development of organs (e.g. propagation by cuttings) and the direct formation of organs from meristematic tissue (e.g. root R. and formation of adventitious roots). Shoot buds can also arise on roots, and plants that display this ability can be propagated by root cuttings, e.g. raspberry, blackberry, cherry and plum. Shoots can also arise on the foliage leaves of *Bryophyllum* species. This marked ability of plants to undergo direct R. is exploited in horticulture for the vegetative propagation of many plants. The annual production of new shoots by perennial plants can, to a certain extent, be looked upon as *spontaneous R.*

Indirect R. is the formation of new organs from wound callus (see Wound healing). Wound callus is a parenchymatous growth, which serves primarily for sealing the wound, but in many cases it develops endogenous shoot and/or root primordia.

Development of tillers on the stumps of deciduous trees may be the result of direct [oak, lime *(Tilia)]* or indirect (poplar, beech) R. Leaf cuttings often exhibit only *partial R.,* i.e. root formation from a more or less extensive callus, e.g. ivy *(Hedera),* orange *(Citrus sinensis),* camelia *(Camellia japonica).* Leaves of other plants are capable of total R., forming both roots and shoot buds, e.g. stonecrop *(Sedum),* mother-in-law's tongue *(Sansevieria),* and especially *Begonia.* The detailed mechanism of R. is unclear. There is no doubt that R. is subject to regulation by the injured organism, and the same factors are involved as in normal plant organ development.

R. in animals. In animals, *reparative R.* may follow the accidental loss of a body part (e.g. by the bite of an adversary), or the active loss of a part due to a stimulus [e.g. spontaneous shedding of organs or Autotomy (see)]. In addition, the periodic or continuous replacement that occurs in the normal life of the animal is also a form of R., known as *physiological R.* Thus, the surface keratinized cells of the skin are continually shed and replaced by new cell layers from below. Another example is the continual degradation and replacment of blood components, e.g. old erythrocytes are continually degraded and replaced by new ones. Physiological R. also occurs at a molecular level; all the constituents of living organisms are in a state of continual degradation and renewal, so that chemically the body is in a dynamic state. In this sense at least, every living organism is capable of R.

Reparative R. is, however, not a uniform property of all animals. The greatest regenerative potential is possessed by hydrozoans *(Coelenterata)* and turbellarian worms *(Platyhelminthes).* A single fragment of the green freshwater polyp, *Chlorohydra viridissima,* representing about 1/200 of the mass of the original animal, can regenerate to form a new small polyp. In *Planaria maculata,* complete R. has been reported from 1/279 of the starting size.

Even close relatives of these animals, however, display very limited or no regenerative ability. Thus, among the hydrozoans, the *Siphonophora* have no regenerative ability. Whereas the various species of freshwater planarians regenerate well, the marine triclades and polyclades have a much lower regenerative capacity. *Lecitophora* (a turbellarian worm) does not regenerate at all. Sometimes species of the same genus, or even races of the same species have very different R. capacities, e.g. there are two species of *Lineus* (phylum *Nemertini),* which can hardly be distinguished morphologically and were therefore described earlier as races of the same species: *Lineus ruber* can replace very extensive losses, whereas *Lineus viridis* has almost no R. ability.

Regenerative ability is very unevenly distributed in the animal kingdom. Some lower animal groups display no R., e.g. *Sporozoa, Rotifera* and *Nematoda,* whereas some groups of highly differentiated animals are capable of extensive replacement R., e.g. adult *Urodela* and larval *Anura.* In addition, R. is not a constant property of the animal; it may be present in the juvenile stages but disappear after metamorphosis, e.g. insects and *Anura.* In starfishes *(Asteroidea)* the reverse is true, i.e. larvae do not regenerate lost body parts, whereas adult forms of the genus, *Linkia,* can regenerate to a complete 5-armed animal if one arm is cut off. In fish and amphibia, matrix zones in the region of the ependymis of the brain ventricle retain mitotic activity, so that these animals can regenerate nerve cells in certain regions of the brain; in contrast, mammals are incapable of nerve R.

R. can be divided into 3 overlapping phases: 1) formation of the *R. blastema,* 2) growth, and 3) differentiation of the regenerating tissue. Before blastema formation is complete, growth begins, and growth continues after the onset of differentiation.

Blastema formation occurs beneath the apical ectodermal cap (a single layer of cells that proliferates to cover the wound surface). Beneath this ectodermal cap the cells become dramatically dedifferentiated, i.e. bone cells, cartilage cells, fibroblasts, muscle cells, neuroglial cells all become morphologically identical and detach from each other. This accumulation of undifferentiated cells is the R. blastema; it may arise at the site of the wound or in its immediate vicinity (e.g. limb R. in amphibians), or it may form from cells that migrate to the wound site from distant parts of the body (e.g. in planarians). The blastema contains high activities of cathepsins and dipeptidase, as well as a relatively high concentration of free SH-groups and a relatively low pH.

Increased mitosis in the blastema leads to the formation of a rapidly growing meristem, followed finally by differentiation. The degree of metaplasia (change in cell type) in the blastema is still disputed and of great theoretical interest; the two extreme alternatives are: 1) the dedifferentiation of tissue cells to mesenchymal blastema cells is so complete that every other type of cell can arise from them (i.e. differentiation is reversible), or 2) dedifferentiation is only a transitory loss of all morphological characters of the tissue cells, followed later by redifferentiation to the starting form. In some cases (e.g. growth of the neural tube in the tail of anurans), new tissue also arises simply by growth from the remainder of the damaged stump.

The omnipotent replacement parts of planarians, which are scattered in the parenchyma, and which accumulate in large numbers at the R. site, are called *neoblasts.* Analogous cells *(interstitial cells* or *I-cells)* are also present in hydrozoans, where they occur in groups between the epithelial muscle cells of the supporting lamella. Neoblasts and I-cells are relatively small, and spindle-shaped to oval in appearance. Like all embryonal cells, they contain a high level of RNA, and all other types of cell in the animal can be derived from them. They can be selectively killed with X-rays, thereby destroying the capacity of the animal for R.

In addition to the type of R. known as *epimorphosis,* in which the lost body part is regenerated directly at its site of loss, some lower animals display *morphallaxis,* i.e. after the loss of large body parts, the remainder of the body undergoes a complete reorganization to form an entire but smaller organism. Some protozoa regenerate by epimorphosis, e.g. *Paramecium,* and some regenerate by morphallaxis, e.g. *Stentor.* Other organisms combine both forms of R., e.g. most hydrozoans, *Turbellaria* and *Annelida.*

Lost parts are not always regenerated exactly. Moreover, R. is often incomplete. Thus, in the R. of the tarsus of various insects, only 4 joints are regularly formed, rather than 5. Fragile lizard tails, which are easily lost by autotomy, never regain their typical axial skeleton, which is replaced by an incompletely ossified cartilaginous tube.

Conversely, more may be regenerated than was lost (super R.). Thus, there are frequent naturally occurring examples of the replacement of a single structure with two, three or even more copies. These Malformations (see) can certainly arise from the splitting of organ primordia, due to damage during embryonal development. But they can also arise as a function of R., as shown by many examples of the experimental generation of malformations. Of particular importance in this context is the formation of supernumerary limbs. If an amphibian limb is deeply incised from the side, without completely severing it from the body, it can then be bent back at an obtuse angle. R. of the distal limb section then occurs from both the proximal and the distal wound surface. Together with the original distal limb section, the two regenerated limbs form a 3-fold structure. The central component of this malformation is structurally the mirror image of the other two components (Bateson's law).

If the structure arising at the amputation site is totally different from that lost, and is a structure normally never found at that site, then the process is called *heteromorphosis,* e.g. formation of an antenna-like structure in place of the amputated eye stalk of a decapod crab, when this is so deeply cut at the base that the ganglion opticum is removed with it. Similarly, in the Indian stick insect *(Carausius morosus),* an amputated antenna may be replaced by a leg-like structure.

R. is affected by many external and internal factors. For example, numerous studies have been reported on the influence of the nervous system. In insects, R. proceeds independently of a nerve supply, whereas it has long been known that blastema formation and the R. of amphibian limbs can be prevented by severing all the nerves to the stump. A certain minimal number of unsevered neurons is necessary at the amputation site, irrespective of whether they are motor, sensory or sympathetic. This number corresponds to one half to one third of the normal number. There is no correlation between impulse conduction and R. It is thought that the neurons release trophic factors, which increase proliferation of the blastema cells, but are not necessarily required for subsequent growth and differentiation of the new tissue. One such substance known to be produced by these neurons is fibroblast growth factor. From the seventh to ninth day of R. onward, severing of the neurons no longer has an effect on R. Growth and differentiation of the regenerating organ then proceed independently of a nerve supply.

R. is also influenced by certain hormones. In adult salamanders (not the larvae), R. depends on the production of growth hormone by the hypophysis (pituitary). Hypophysectomized salamanders are no longer capable of R. Thyroxin from the thyroid gland is also important for normal R. In lower animals (hydrazoans, planarian worms, annelids, crustaceans and insects) the neuroendocrine system also plays a role in R.

Regeneration blastema: see Regeneration.

Regolith: the layer of unconsolidated weathered rock (rock fragments and mineral granules) resting on the solid bedrock. In the humid tropics it may attain a depth of several hundred meters. Soil is simply R. with sufficient nutrients to support plant growth.

Regularia: see *Echinoidea.*

Regulation: the mechanisms whereby an organism totally or partly reestablishes its normal state, following a disturbance or alteration; the phenomenon of the reestablishment of the normal state. For example, if the two blastomeres of the two-cell stage of the sea urchin embryo (called 1/2-blastomeres) are separated by shaking, the cells do not necessarily die or develop into incomplete embryos. At the beginning of the 20th century, Driesch showed that 1/2-bastomeres and even 1/4-bastomeres can develop into a complete pluteus larva, albeit with a corresponding reduction in size, i.e. he produced twins or quadruplets experimentally. The development of a blastomere that would have developed into one half or one quarter of the embryo is apparently totally reoriented when it is isolated from the embryo; it then develops into a complete larva. These experiments with sea urchin embryos stimulated a large number of further investigations, which played an important part in Driesch's theory of neovitalism.

The sea urchin embryo is limited in its regulatory ability. If the animal and vegetal halves of the embryo are separated at the 8- or 16-cell stage, each half develops into a blastula, but only that derived from the vegetal half develops into a gastrula and then into a larva. In most cases, this larva remains incomplete, lacking those structures that arise from the animal half in normogenesis, e.g. larval arms *(vegetative anormogenesis* as defined by F.E. Lehmann).

This deficient regulatory ability of the animal half is even demonstrable in the unfertilized egg. If the animal half of an egg is excised with a fine glass needle, then fertilized, neither a gastrula nor a skeleton will develop. In contrast, a similarly treated vegetal half develops into a more or less complete pluteus larva. Thus, even in the oocyte of the sea urchin there is a nonuniform distribution of developmental factors along the animal-vegetal axis. Only the vegetal half contains all the factors required for development of a complete larva.

Special developmental factors are bound in the dorsal gray crescent of the frog egg. The first cleavage plane is always meridional, but not at a fixed angle to the subsequent dorsoventral axis of the embryo. It may divide the embryo sagitally (medially), in which case both blastomeres receive one half of the gray crescent. The cleavage plane may also be frontal, so that only one (dorsal) blastomere receives the entire material of the gray crescent. All intermediate stages between these two extremes are also possible.

If the half blastomeres are carefully separated with a hair loop, a complete larva develops only from cells that have received an adequate proportion of material from the gray crescent. Otherwise, cells develop into a section of abdomen without differentiated organs. Twins can therefore arise only when the first cleavage plane is more or less sagittal. Any appreciable deviation from a sagittal plane leads to a complete larva and a piece of abdomen. In the amphibian embryo, the second cleavage plane is meridional and at right angles to the first cleavage. 1/4-Blastomeres therefore develop into either two normal larvae and two abdominal pieces, or a normal larva and three abdominal pieces.

R. leading to the development of twins is still possible at the stage of the late blastula, provided this is carefully sectioned in the sagittal plane. R. does not disappear until after gastrulation. Each half then develops into only half an embryo, with only one eye, one forelimb and one hindlimb, etc.

Heider (1900) proposed that embryos displaying an ability for R. be called *regulation eggs,* whereas those eggs or embryos displaying no or hardly any ability for R. are called *mosaic eggs.*

Like the frog egg and the sea urchin egg, the mammalian egg (including the human egg) is of the R.-type, as evidenced by the occurrence of monozygotic twins. In the armadillo, *Dasypus novemcinctus,* the embryo always separates at an early stage of development into four, so that quadruplets are regularly produced.

In contrast to R. eggs, the 1/2-blastomeres of a mosaic egg always develop into incomplete embryos. In the extreme case, it leads to the development of exact half embryos. An example is provided by the comb jelly, *Beroe.* Like all comb jellies, this disymmetric Mediterranean form has eight longitudinal rows of comb plates (ciliated bands). A 1/2-blastomere develops into a larva with eight instead of four ciliated bands. Correspondingly, a 1/4 or an 1/8-blastomere yields a larva with 2 or a single ciliated band, respectively. R. does occur to the extent that the 1/2 and 1/4-larvae develop a complete stomach and a complete static organ at the aboral pole. Typical 1/2-larvae develop from the 1/2-blastomere of the sea squirt, *Styela.* Nematodes and some insects (e.g. flies) also produce mosaic eggs.

Unequivocal demonstration of whether an insect egg is of the R. or mosaic type is problematic. Well defined cases of R. or mosaic behavior are easily classified, but there is a complete spectrum (in particular among insects) of intermediate behavior between the two extremes.

Regulation egg: see Regulation.

Regulatory genes: genes that regulate the expression of structural genes, by affecting their rates of Transcription (see) or Translation (see). They encode repressors or activators, which have an affinity for a specific operator or analogous base sequence, and block the production or translation of messenger RNA. See Operon.

Regulinae: see *Sylviinae.*

Regulus: see *Sylviinae.*

Reichstein's substances: see Cortexone, Corticosterone, Cortisol, Cortisone,

Reindeer: see Caribou.

Reindeer moss: see Lichens.

Reinforcement: an increase in the permanence and/or intensity of the response to a stimulus. In *primary R.,* an inborn release mechanism is reinforced by a positive consummation. The most familiar example is R. by giving a "reward" of food or water for a specific response to a stimulus. In *secondary B.,* the valency of a neutral stimulus is so altered by its combination with a primary stimulus that it acquires the property of acting as a stimulus or reinforcer on its own. The witholding of a negative (aversive) stimulus can also cause reinforcement. The basis of all these processes is associative learning. See Learning, Conditioned reflex.

Reissner's membrane: see Auditory organ.

Relapsing fever: see Spirochetes.

Relationships between living organisms, ***interactions of living organisms:*** A distinction is drawn between 1) relationships within a species *(homotypic* or *intraspecific interactions),* which occur mostly within one group and lead to the formation of communities (see Sociology of animals), and 2) relationships between members of different species *(heterotypic* or *interspecific interactions).*

Heterotypic interactions are classified according to the benefits to the partners: 1) *Probiosis* or *carposis,* which is beneficial to one partner, while the other is not obviously injured or disadvantaged; 2) *Symbiosis,* in which both partners benefit from the interaction; and 3) *Antibiosis,* in which one partner is clearly disadvantaged, while the interaction is of obvious benefit to the other (generally with respect to feeding). Each type of interaction can be further subdivided according to the its intensity and permanence.

The seeking out of other animals and benefitting from their proximity is known as *shared settlement* or *paroecia.* Thus, a less conspicuous species may find protection by forming a neighborly group with a different species, e.g. eider ducks build their nests in tern colonies to protect their eggs and young from predatory gulls. In *attachment settlement* or *epoecia,* an animal uses another species as a substratum, e.g. sessile, freshwater and marine inhabitants, such as barnacles, on the skins of whales and the shells of muscles. *Tenancy* or *etoecia* is practised by animals which occupy the nests and living quarters of other species, e.g. the Inquillines (see) of bird and rodent nests, and the nests of colonial insects; the shelduck *(Tadorna tadorna)* nests in occupied fox or rabbit burrows. Tenants do not harm the host species, but they profit from the favorable physical conditions and/or food residues (inquillines in the nests of colonial insects). The spatial interaction is even closer when a tenant lives in the body cavities of the host, e.g. ants in the hollow organs of tropical plants, and the numerous worms and crustaceans in the cavities of sponges; the oyster crab *(Pinnoteres pisum)* actually lives inside the shells of Mediterranean bivalves. Clearly, small changes in this type of probiosis may produce a situation in which the host is disadvantaged. In *phoresy,* one organism merely uses another species as a means of transport; attachment to the host may be active or passive. Phoresy is practised by many parasites (larvae of the oil beetle on bees, ectoparasitic lice of flies), and by many inhabitants of feces and putrifying material (nematodes on insects, mites on beetles). A well known example is that of the remoras *(Echeneididae),* small fish that use larger, faster fish (sharks, tunafish) as transport vehicles. Even in the absence of specific means of attachment to the host, transport may occur (passive transport) incidentally and accidentally, e.g. small organisms may be carried in mud on the feet of larger species.

A probiotic relationship may confer a feeding advantage on one partner, e.g. jackals follow other predators in order to profit from the remains of their kill; this is known as *Commensalism* (see).

Symbiosis (see) in the broad sense includes all transitions from loose associations to close interactions, the latter often being necessary for mutual survival of the partners. A loose association with mutual benefit is called an *alliance,* e.g. various animal species live with each other for greater protection (e.g. antelopes and ostriches); oxpeckers have access to a plentiful food supply by picking ticks and other organisms from the hides of large mammals, the latter benefitting from the removal of pests and parasites. In *mutualism,* the species interaction is closer, and often necessary for survival, e.g. animals from a variety of systematic groups visit flowers for food and thereby effect cross-pollination. A further example is the distribution of seeds that are enclosed in nutritious mesocarp (e.g. elder, rowan, mistletoe); the fruits are eaten, especially by birds, and the seeds are excreted undigested. *Symphily* is a special case of mutualism, exemplified by the feeding relationship of ants and aphids; the aphids yield honeydew for the ants, which husband and protect the aphids. Similar relationships exist between ants and other guests (symphiles).

The closest relationship for mutual benefit is Symbiosis (see) in the strict sense.

Antibiosis, which includes Parasitism (see), is characterized by a negative effect on one of the partners and its population. If an animal is colonized by microorganisms which impair its life processes, the condition is called *pathogenesis.* Such conditions are directly important to human society, when humans themselves, their domestic animals, or their agricultural plants are the damaged partners in the association. Knowledge of the ecological factors that influence infection with parasites and pathogens (e.g. existence of vectors, pathogenicity and virulence of causative agents, and susceptibility of the host) is necessary for the design of control measures.

A most striking form of antibiosis is *predation* or

episitism. The suitability of various animal species as prey depends on their size, ability to avoid capture, and their abundance. Many insect larvae, small crustaceans, small birds and small mammals serve as prey. Their abundance attracts numerous predators *(episites),* which in turn often react to the ready availability of prey by increasing their rate of population growth (e.g. the increase in the breeding rate of owls in "mouse years").

R.b.l.o. are fundamental to the equilibria of natural ecological systems. An understanding of R.b.l.o. is of practical importance for pest control in forestry and agriculture, for biological pest control, for maintenance of the health of humans and their domestic animals, and for restoring and maintaining natural environments.

Relative biological effectiveness: see Dose.

Relative resistance: see Field resistance.

Relaxation pressure curve: see Pressure-volume curve of the respiratory apparatus.

Relaxin: a female sex hormone produced in mammals during pregnancy. It is a heterodetic, cyclic polypeptide, M_r 12 000; the A-chain contains 22, the B-chain 26 amino acid residues. It is probably produced by the ovary and placenta and/or uterus under the influence of progesterone. It causes relaxation of collagenous connective tissue (sensitized by estrogens) in the symphysis and ileosacral joints, thus enabling enlargement of the pelvic girdle during pregnancy and at birth.

Release inhibiting hormones: see Releasing hormones.

Releaser, ***sign stimulus:*** an external stimulus, which acts via an innate or inherited release mechanism, and releases motivational energy for a specific type of behavior in the recipient. Animals possess a variety of devices for producing sign stimuli, such as brightly colored structures, stereotyped movements and postures, scent-emitting organs and sound-producing structures. Within the same species, individuals possess the specific response mechanism to these signals or sign stimuli. Typical sign stimuli are contained in courtship displays, which by their interaction lead to the functional co-ordination of the two partners. Rs. have also been called triggers.

The use of dummies and models has contributed much to behavioral research and the study of Rs. Such models permit the experimental isolation of individual factors in the sign stimulus, i.e. they permit the identification and quantitation of the stimuli responsible for species-specific behavior. The strength of the response to a sign stimulus can also be determined experimentally.

A well known example is the use of 2-dimensional flat cardboard models of the heads of adult herring gulls to study the elicitation of the pecking response of herring gull chicks in the parent-offspring feeding interaction (Fig.). This experiment appears to show that the red spot on the adult bill releases and directs the pecking response from the chicks, which then feed. Further work has shown that motion of the model is also important, and that older chicks learn to recognize other details of the adult head and require more contextual information in addition to the simple R.

An unlearned or instinctive response to R. in a model is evidence for the existence of inherited release mechanisms.

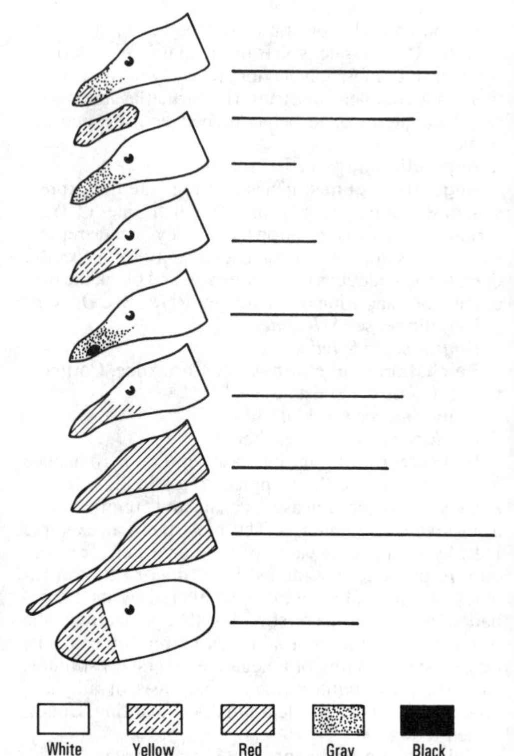

Releaser. Model experiments on the releasers for the pecking response of herring gull chicks in parent-offspring behavior (after Tinbergen). The models of adult heads are made from cardboard. The "standard" model (uppermost in the Fig.), painted in natural colors (white head, yellow bill, red spot on the bill), provided the basis for comparison with the other models. Response was scored as the number of pecks during a 30 s presentation of the model. The lengths of the horizontal lines opposite each model are proportional to the pecking response.

Releaser stimulus: see Releaser, Sign stimulus.

Releasing factors: see Releasing hormones.

Releasing hormones, ***releasing factors, liberins:*** Neurosecretory hormones (see) synthesized in various distinct nuclei of the hypothalamus. They are released into the capillaries of the portal vessels in the median eminence of the hypothalamus, then carried to the anterior pituitary (adenohypophysis) where they stimulate the synthesis and secretion of tropic hormones. All known R.h. are peptides containing between 3 and 10 amino acid residues. They are produced in neurons, and their synthesis represents the conversion of an electrical stimulus into hormonal signals. In the median eminence of the hypothalamus, R.h. are stored in nanogram quantities. In the anterior pituitary, R.h. act via the adenylate cyclase system, and are then degraded by proteolysis.

The following R.h. have been identified and their structures elucidated: 1) Corticotropin releasing factor (CRH, corticoliberin); 2) Follicle stimulating hormone releasing factor (FRH, folliliberin); 3) Luteinizing hormone releasing factor (LRH, luliberin); 4) Thyrotropin releasing hormone (thyrotropic hormone releasing factor, TRF, thyroliberin); 5) Melanotropin releasing factor (MRF, melanotropin releasing hormone, melanoliberin); 6) Prolactin releasing factor (PRF, prolactin releasing hormone, prolactoliberin); 7) Somatotropin releasing factor (SRF, somatotropin releasing hormone, somatoliberin).

Three release-inhibiting hormones (statins) are also known: Prolactin release inhibiting hormone (PIH, prolactostatin); Somatotropin release inhibiting hormone (SIH, somatostatin); and Melanotropin release inhibiting hormone (MIH, melanostatin). Thus, the secretion of at least three tropic hormones is regulated jointly by the relevant releasing hormone and release inhibiting hormone.

Relevé: a record of the analysis of the most significant features of a stand of vegetation. It is usually presented in tabular form with additional notes. Analysis of vegetation starts with reconnaisance for suitable stands of phytoc(o)enoses. Sample plots are taken from these stands, then analysed. The resulting analysis is called the R. of that stand. The chosen stand must satisfy three requirements. 1. It must be sufficiently large to include all the regularly occurring species of the plant community. The minimal size of such a stand depends on the plant community in question. 2. It must be homogeneous, i.e. it must have no gaps, and the dominant species must not change in different parts of the stand. 3. It must be environmentally uniform, e.g. constant slope or level, uniform illumination, water supply, etc.

The R. lists all the plants occurring in the stand, together with their species, number, distribution, vitality, etc. In addition, notes are also made on the growth forms of individual species, and the general form and structure (configuration) of the vegetation. If possible, the geology of the substrate is also investigated. A compete R. provides information on the following.

1. *Vegetation layering or stratification.* Four principal layers are distinguished: tree, shrub, herb, moss. The degree of coverage by each layer is also determined. This is measured as the vertical projection of all aerial parts of a given layer as a percentage of the total stand area, i.e. the proportion of the stand that is shaded in vertical illumination.

2. *Quantitative occurrence of each species.* This includes abundance, i.e. frequency (number of individuals) of a species, and the degree of coverage (sometimes inappropriately referred to as dominance), which define the space requirements of a species within a plant community. Abundance and degree of coverage are usually estimated together, using a 7-point scale of Cover-abundance (see).

3. *Sociability* (see), which defines the degree of clustering and type of distribution of individuals of a species. Some species grow in definite communities, e.g. some always grow in patches, while others form extensive mats, etc.

4. *Vitality.* A scale of relative vitality is used, with four levels indicated by symbols or numbers. ●, 1: Plant fully developed; life cycle complete; if growth is especially luxuriant, this is indicated by “lux”. ⊙, 2: Vegetative propagation occurs, but life cycle not completed. ○, 3: Poor vegetative propagation; feeble growth; life cycle not completed. ○○, 4: No vegetative propagation; germination occurs occasionally.

Also included are the date, topographical locality (preferably with map references), altitude, exposure, inclination, substrate geology, depth and differentiation of root system, soil profile, character of adjoining vegetation, and notes on other influences, such as animals, irrigation and agriculture.

Rem (Roentgen-equivalent-man): see Dose.

Remiges: see Feathers.

Remiz: see *Remizidae.*

Remizidae, ***penduline tits:*** a small avian family (11 species, 3 genera) in the order *Passeriformes* (see). Penduline tits occur in Africa and the Holarctic region. They have finer bills than other tits. Like other tits, they are tree dwellers, but prefer more open country, rather than woods or forest. They are noted for their finely felted, globular or bottle-shaped nests, usually with a tubular side entrance that closes when the bird has passed through it. The range of the ***Penduline tit*** ***(Remiz pendulinus)*** extends from Central Europe across Asia, and also into southern France, Italy and Spain. It inhabits reed marshes and dense bushes by rivers and lakes, and its bottle-shaped nest hangs from the tips of twigs over the water, sometimes at considerable heights (up to almost 20 m). The 6 species of *Anthoscopus* are found in Africa. *Anthoscopus caroli* builds a firm, bag-shaped nest, the entrance of which the bird closes by pressing the sides together when it leaves; its texture is so tough and durable that old nests are used locally as purses. The montypic ***Verdin*** ***(Auriparus flaviceps),*** the only New World species, inhabits semi-desert in western North America and Mexico.

Remmert's rule: see Ecological principles, rules and laws.

Remoras: see *Echeneidae.*

Renal capsule: see Kidney functions.

Renal corpuscle: see Kidney functions.

Renal portal system: see Blood circulation.

Renal tubule: see Kidney functions.

Rendzina: a brown earth soil characteristic of grasslands, in which the parent soil material is calcareous. The term is now obsolete according to the classification system of the United States Department of Agriculture. Rendzinas are now subdivided and assigned to the orders Inceptisols or Mollisols.

Renilla: see *Pennatularia.*

Renin-angiotensin mechanism: see Kidney functions.

Rennet enzyme: see Rennin.

Rennin, ***rennet enzyme, chymosin:*** a pepsin-like proteinase which is present in, and prepared from, the abomasum (fourth stomach) of young ruminants, and is probably present in the stomach of all suckling mammals, where it is responsible for the coagulation of milk. R. is formed as the inactive precursor, *prorennin,* which is converted into R. by autocatalysis or by the action of pepsin. In the presence of calcium ions, R. specifically cleaves a phenylalanine-methionine bond in the soluble casein (caseinogen) of milk, producing insoluble casein (paracasein) and a glycoprotein. In humans, the similar enzyme in gastric juice is known as gastricsin.

Renshaw cell: see Neuronal circuit.

Repantia: see *Decapoda.*

Reparation: a type of restitution, in which replacement of a damaged organ proceeds from the wound surface by renewed meristematic activity the damaged cells. R. is more common in animals than in plants. Certain fungi are able to replace their fruiting bodies by R.

The phenomenon also occurs in the siphonaceous alga, *Acetabularia,* while in higher plants the damaged growing points of roots display true R.

Repellants: chemical compounds which repel potentially dangerous organisms without killing them, e.g. protective chemicals used to repel the clothes moth in textiles, and R. produced by plants, which repel insects or inhibit their feeding. See Antifeedants, Pheromones, Attractants.

Repetitive DNA: in eukaryotic organisms, DNA that contains many copies of different short nucleotide sequences. *Slightly repetitive* DNA contains 1–10 copies per haploid genome, *moderately repetitive* DNA contains 10^2 to 10^4 copies, while *highly repetitive* DNA contains about 10^6 copies. The genes of moderately repetitive DNA are transcribed into rRNA, tRNA and the mRNA of histones. Highly repetitive DNA often serves as a spacer between structural genes.

Replacement plant community: see Substitute plant community.

Replicase: see DNA replication.

Replicon: see DNA replication.

Replisome: see DNA replication.

Repressor: see Operon, Enzyme induction, Enzyme repression.

Reproduction, *propagation, procreation:* the production of new living organisms by existing living organisms (the parent or parents). R. usually, but not necessarily, results in multiplication, i.e. an increase in the total number of individuals. It is a fundamental property of living material, which guarantees the continuity of life, notwithstanding the death of the individuals. R. is closely coupled with active metabolism (energy metabolism and biosynthesis), and it guarantees the relative constancy of the species in the subsequent generation by transfer of specific information from parent to offspring (hereditary transmission, or inheritance). The first product of R. may be a young organism, in which all parts have equal morphological and physiological status and are involved in subsequent growth leading to the new adult. Alternatively, parts of the young organism may have different developmental status, so that only smaller primordia are destined to develop into the adult. Often single cells, which separate from the parent organism, are capable of proliferation to a new organism; these are known as germ cells, and R. via germ cells is called *Zygotony.* If part of the parent develops directly into a new individual, the process is known as asexual R.; if the new organism develops only after the fusion of two specialized germ cells known as gametes, the process is known as sexual R. Asexual and sexual R. often occur in the same organism, and environmental factors may determine which process predominates. In some organisms, asexual and sexual R. alternate (see Alternation of generations).

1. ***Asexual R., vegetative R., monogony:*** a process in which the new organism is derived from part of a single parent. The progeny is therefore genetically identical to the parent (see Clone).

Plants. Asexual R. is widespread in the plant kingdom. The simplest form of asexual R. is the binary fission of unicellular organisms, such as bacteria and cyanobacteria (see Prokaryotes) and unicellular algae; the total material of the parent cell becomes equally divided between two daughter cells, each of which then grows to parent size prior to further division. This type of R. usually leads to a large population increase as each phase of division is followed by another. Thus, under favorable conditions of temperature and nutrient supply, certain bacterial species can attain a population of 2^{48} cells from a single cell in 24 hours.

In the multiple fission or schizogany of various algae and fungi, the cell nucleus divides repeatedly, followed by division of the cell into a corresponding number of daughter cells or spores. In budding, which is found widely in the *Saccharomycetaceae* (yeast fungi), a small outgrowth of the parent cell buds off as a daughter cell, which then grows to the size of the parent cell.

The simplest form of multicellular (polycytogenic) R. in plants occurs by fragmentation, e.g. in the cyanobacteria, *Plectonema* and *Oscillatoria,* as well as in the more highly organized marine algae, like *Caulerpa* and *Fucus,* and lichens, in which the thallus simply becomes fragmented, each fragment growing into a new thallus. Flowering plants can also reproduce by fragmentation or separation of plant parts, where the older dead parts of the plant rot away, and the remaining separated parts develop into new plants. Examples of this are found in lily of the valley *(Convallaria)* and the Canadian pondweed *(Elodea canadensis).* Only the female plant of *Elodea* was introduced into Europe in the mid 19th century, and it has multiplied exclusively by fragmentation to the point where it can clog waterways.

Asexual or vegetative R. in plants often occurs by the separation of propagules, of which there are three types: spores (single cells), gemmae (groups of cells), bulbils (entire organs or groups of organs). Spores (see), which grow directly into a new individual, are commonly produced by multicellular algae and fungi. Many mosses produce gemmae, which are formed inside gemmacups on the thallus surface; after becoming scattered, each gemma develops into a complete moss plant. In some ferns, especially tropical species, the gemmae are complexes of meristematic cells, which have persisted from the juvenile stage, and are located in the angles formed by the serrations on the edge of the frond. These develop into miniature plants complete with roots, which when released will continue growing to form complete fern plants. Some flowering plants form bulbils in their leaf axes, which are shed and develop into new plants, e.g. *Cardamine bulbifera* (coral-wort).

Aerial and subterranean runners and stolons represent a further means of vegetative R. in flowering plants. These structures carry buds at intervals along their length, which develop into new individuals [e.g. the strawberry plant *(Fragaria)* forms aerial runners; couch grass *(Agropyrum)* and bindweed *(Convolvulus)* form subterranean stolons]. Vegetative R. also occurs by Tubers (see) and Bulbs (see). Familiar examples are the subterranean tubers of the potato *(Solanum tuberosum),* and the bulbs of the snowdrop *(Galanthus)* and the tulip *(Tulipa),* which can divide further to form daughter bulbs. A particularly striking form of vegetative R. in some higher plants is Vivipary (see).

Vegetative R. is exploited for the agricultural and horticultural propagation of many plants. Thus, leaf or stem cuttings may be rooted simply by placing them in the ground. Shoots and buds of valuable varieties can be propagated by grafting onto the root-stock of a vigorous less valuable variety, e.g. in fruit, vine and rose propagation. One particular advantage of these methods is that plant species can be propagated, which have lost the ability to reproduce sexually, due to pollen sterility, degenerate ova or chromosome disorders. In addition, vegetative propagation may be the only means of preserving a strain or variety which does not breed true, i.e. the characters of earlier ancestral types become separated in the progeny from sexual R.

Animals. Asexual R. is found in both unicellular *(Protozoa)* and multicellular *(Metazoa)* animals. In the simplest form of asexual R., i.e. cytogony or monocytogenic R. (also called agamogony), which is normally observed in the *Protozoa,* single cells separate from the parent. Thus, in binary fission (monotomy) the parent organism (cell) loses its independent identity by dividing into two daughter organisms (cells); the plane of cleavage may be characteristic (longitudinal in flagellates, transverse in ciliates). On the other hand, simple and multiple budding, e.g. in *Acanthocystis,* leads to unequal products, i.e. small progeny cells plus the parent organism, which retains its identity as a mature cell. Multiple division is characteristic of the *Sporozoa,* some amebas, and flagellates. It is encountered in two forms: 1. schizogony produces merozoites, which then differentiate into gametes and undergo sexual fusion without encystment; 2. sporogony or sporogenesis produces encysted sporozoites, a process which is preceded by sexual fusion (see e.g. the life cycle of the malaria parasite, plate 43). Multinucleate *Protozoa* usually reproduce by division, in which the nuclei are distributed among the daughter cells. Monocytogenic R. is rarely observed in the *Metazoa;* one of the few examples is the formation of young dicyemids (cephalopod bladder parasites, phylum *Mesozoa*, order *Dicyemida)* from axoplasm produced by the axoplasts of the parent animal.

Metazoa multiply asexually by vegetative polycytogenic R. (previously also called somatogenic monogony); in this case, the progeny develops from more or less large cell complexes, which arise from an adult parent or from larval or embryonal stages. The process takes several forms. Longitudinal or transverse cleavage of the whole organism is frequently observed in *Hydrozoa, Actinozoa* (*Anthozoa*) and *Turbellaria;* in this case, the growth of new organs may commence before or after the cleavage, e.g. the new tentacles of *Gonactina prolifera* (one of the few sea anemones reproducing by horizontal cleavage) develop before asexual cleavage. Cleavage of the larval stage is observed in the aciliate planula larva of *Polypodium* (a hydroid), which parasitizes sturgeon eggs; the larva grows to a stolon with more than 30 buds; when the fish spawns, the stolon is released, the buds become detached and develop into sexually mature medusae. *Craspedacusta (Limnocodium)* (a hydroid first known from the *Victoria Regia* tank in the Royal Botanic Gardens at Kew, UK), displays two types of asexual R.: swellings occur in the trunk wall or at the oral pole, then become detached and form new polyps; alternatively, new polyps are formed by lateral budding, and under suitable conditions (e.g. rise of water temperature) medusae may form directly from the buds. Division of embryonal stages results in the phenomenon of polyembryony, e.g. the regular birth of quadruplets to the nine-banded armadillo *(Dasypus novemcinctus),* and the occasional birth of identical offspring (twins, triplets, etc.), which have developed from a single egg, to humans and other mammals. In budding, an elliptical or round bud is formed by protrusion of all layers of the body wall, which becomes separated from the tissues of the parent by a deep dividing cleavage. In the simplest case, the bud consists of an undifferentiated complex of cells, which develops later into a new organism. Normally, however, the young organism differentiates as the bud is formed, so that a separately differentiated organism may be released as an independent individual (e.g. *Hydrozoa*, Fig. 1); colony formation starts with a primary polyp, which forms a system of tubes at its base (the stolon of hydrorhiza) covering the substrate, followed by budding of new polyps; an analogous process of budding and stolon formation is also found in the *Tunicata.* Larval budding occurs in *Cestoda* (tapeworms); for example the hydatid cyst or cysticercus of *Echinococcus granulosa* (= *Taenia echinococcus)* proliferates surface vesicles (brood capsules) on its inner surface; each of these vesicles produces up to 30 scolices by internal budding (Fig. 2). Frustules are planula-like bodies, which constrict from the sides of polyps, or detach from the ends of hydroid stolons; they eventually develop into polyps. Internal budding occurs in some organisms, often in response to unfavorable conditions; it results in the formation of resistant structures (possessing a special outer layer and air chambers) which enable the organism to overwinter. Such structures may, however, also be formed under more favorable conditions. Thus, the gemmules of freshwater sponges (e.g. *Ephydatia)* have the appearance of a small hard ball; they consist of a mass of archeocytes from the old sponge, enclosed by an outer wall (two layers of cuticular membrane with a median layer of amphidiscs) with a pore to the exterior called a micropyle. In the *Phylactolaemata* (freshwater *Polyzoa)* internal budding in the coelom, from a mesodermal cell mass at the funiculus, gives rise to small, round, sessile or swimming statoblasts, with a chitinous ectodermal double outer layer; the polypides die in the winter, and the statoblasts germinate to produce new colonies in the spring.

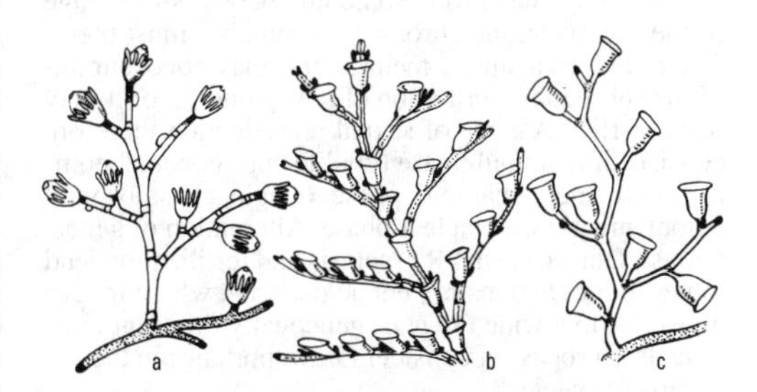

Fig. 1. *Reproduction.* Colonies of hydrozoan polyps. *a* Monopodial branching with terminal polyps. *b* Monopodial branching with terminal vegetative zone and lateral polyps. *c* Sympodial branching.

2. ***Sexual R., amphigony:*** a reproductive process in which the new individual arises from a zygote, which is formed by fusion of the nuclear material of a male gamete and a female gamete. If the entire organism functions as a gamete, e.g. in certain unicellular organisms, such as *Chlamydomonas,* gamete fusion is referred to as hologamy. Normally, however, sexual R. occurs by merogamy or gametogamy, i.e. the two gametes are distinct from the parent organism or organisms that pro-

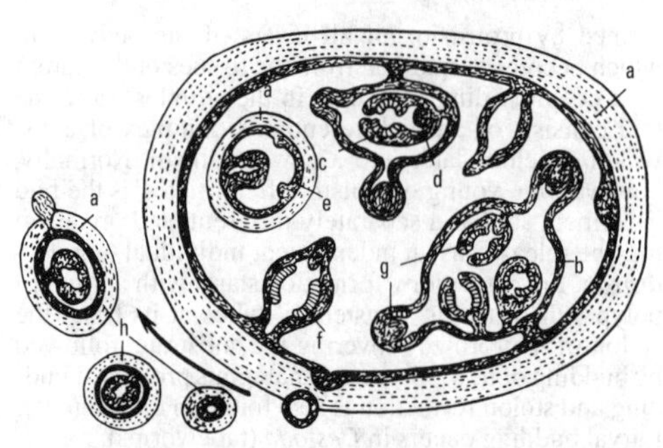

Fig.2. *Reproduction.* Transverse section of the hydatid cyst (cysticercus) of *Echinococcus granulosus,* a minute dog tapeworm. The hydatid cyst occurs in several mammals, including humans, other primates, cattle, cats, dogs, horses, pigs, etc., usually in the liver. Owing to the vesicle-like appearance of the hydatid cyst, it is also called a bladder worm. *a* Lamellated cyst wall (cuticula). *b* Brood capsule. *c* Sucker of the scolex. *d* Rostellum of the scolex with rostellar hooks. *e* Free endogenous daughter cyst. *f* Scolex. *g* Formation of scolices by budding within a brood capsule; left, outward budding; right, inward budding. *h* Stages in the budding and growth of an exogenous daughter cyst.

duce them. Many lower organisms and all higher plants and animals possess specialized cells (germ cells) which produce gametes. Especially in algae and fungi, the cells or organs which produce and contain the gametes are called gametangia. In some lower organisms (e.g. unicellular algae, the filamentous green alga *Spirogyra),* however, any cell may be capable of producing a gamete.

Fusion of gametes is called Fertilization (see). Since the cell nuclei also fuse, fertilization is accompanied by a doubling of the chromosome number. At some stage of the life cycle, the chromosome number must therefore be halved again by meiosis; this may not occur until the subsequent formation of new gametes, or it may occur earlier. A cycle of sexual reproduction therefore consists of a diploid phase (double chromosomal complement) and a haploid phase (single chromosomal complement) (see Nuclear phase, Alternation of generations). During sexual R., meiosis and fertilization lead to new combinations of genetic material, which in turn gives rise to a wide range of genetically different individuals, a process of considerable importance in the evolution of species.

In diatoms, some green algae (e.g. *Acetabularia, Codium,* etc.) and some brown algae (e.g. *Fucus),* gamete formation is accompanied by meiosis, the resulting gametes being known as meiogametes; but gametogenic meiosis is otherwise rare in plants. In contrast, it is common in animals. Typical of plants is the occurrence of a haploid generation between meiosis and gametogenesis; the gametes are then formed by mitosis and they are known as mitogametes.

The simplest form of sexual R. is isogamy, which occurs in many algae and lower fungi. In this case, the gametes are similar in shape and size, but physiologically differentiated with respect to sex (Fig. 3). Such gametes are not referred to as male and female, but as plus (+) and minus (–) gametes. These naked, ciliated, motile isogametes often resemble (and have apparently evolved from) the asexual zoospores produced by the same species. Thus, under certain conditions (especially when they arise from over-ripe sporangia), the zoospores of *Olpidium* behave like isogametes and fuse with each other. Similarly, the ciliated swarmers of *Chlamydomonas eugametos* may reproduce asexually or function as isogametes, depending on the environmental conditions. Such behavior is known as facultative sexuality.

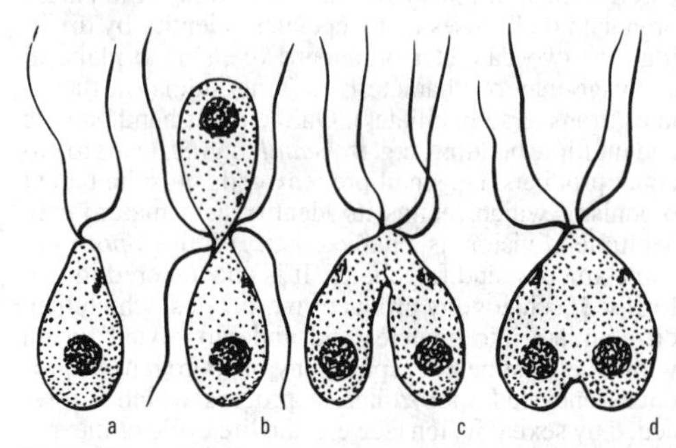

Fig.3. *Reproduction.* Isogamy in *Chlamydomonas steinii. a* Isogamete. *b-d* Fusion of isogametes.

Often in algae and fungi, even in unicellular forms, and regularly in mosses and ferns, the gametes are of unequal size. They are then known as anisogametes, and the corresponding type of R. is called anisogamy. The larger gamete (macrogamete, gynogamete) almost invariably contains reserve nutrients, and is referred to as the female, while the smaller gamete (microgamete, androgamete) is the male (Fig. 4); both are freely motile.

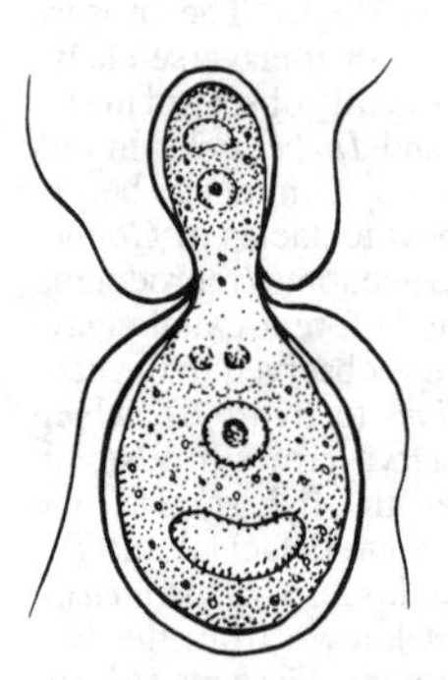

Fig.4. *Reproduction.* Anisogamy in *Chlamydomonas braunii.*

If the female macrogamete remains stationary, it is called an egg cell (oosphere), while the male microgamete is called a spermatozoid. In this condition, known as oogamy, the spermatozoid is attracted chemotactically to the female to effect fertilization. Egg cells and spermatozoids are always formed in different sex organs.

The female sex organs of oogamous algae and fungi are called oogonia. They are usually unicellular, and contain one or more egg cells. In mosses and ferns, only a single egg cell is formed in a multicellular, usually urn-shaped container called the archegonium; the corresponding male organs (antheridia) always contain numerous spermatozoids. Algal and fungal antheridia are unicellular, those of mosses and ferns being multicellular.

Sexual reproductive organs of gymnosperms and angiosperms are restricted to the flowers; the egg cell is contained in an ovary, which subsequently develops into a seed after the egg cell has been fertilized by a sperm cell from a pollen grain.

Two special types of sexual R. are displayed only by fungi: random fusion of simple vegetative cells (somatogamy), and fusion of whole gametangia (gametangiogamy).

Many forms of sexual R. are found in the various phyla of the animal kingdom. *Protozoa* exhibit two main types of sexual R.:

1. *Gametogamy* represents amphigony in the true sense. It occurs in two forms, hologamy and merogamy (or gametogamy). In hologamy, two protozoan cells fuse with each other; each individual cell therefore functions as a gamete, and it usually does not differ in appearance from the asexual form (e.g. in *Rhizopoda, Flagellata* and *Sporozoa).* As a rule, hologamy is followed by vigorous agamous multiplication. Merogamy involves the fusion of gametes (merogametes), which are formed from a protozoan cell (the gamont) by a single or multiple divisions, and which are distinctly different from the agamous cells. If the two gametes are morphologically identical, but physiologically differentiated as male and female, the process is called isogamy, e.g. *Chlamydomonas steinii* (Fig. 3). If the gametes are morphologically differentiated into macro- or gynogametes and micro- or adrogametes, the condition is called anisogamy (heterogamy), e.g. *Chlamydomonas braunii* (Fig. 4). A special case of anisogamy is represented by oogamy, in which the macrogamete is unflagellated and nonmotile and is equivalent to an egg cell, e.g. *Chlamydomonas coccifera, Eimeria stiedae, Plasmodium malariae.* In the great majority of cases of iso-, aniso- and oogamy the copulatory gametes arise from different gamonts. Some *Protozoa* are known, however, in which such amphimixis has not been observed, and an obligatory self-fertilization or Automixis (see) appears to operate. In such cases, gametes or gamete nuclei from the same individual, and possessing identical hereditary material, fuse with each other within a mucus envelope or a tough cyst. Pedogamy is the fusion of two haploid daughter animals from the same parent (Fig. 5). In autogamy, meiosis produces two daughter nuclei, which after maturation immediately reunite within the same cell.

2. *Gamontogamy* includes all cases of sexual R. in which gamonts unite. These processes are often complicated, and sometimes difficult to distinguish from the superficially similar hologamy. In the *Foraminifera, Gregarinidea* and *Coccidae,* two or more gamonts come together and surround themselves with a common gamontocyst; gametes are then formed and they copulate within the cyst. A special case of gamontogamy is found in the conjugation of the *Ciliophora* (see) (except *Opalinidae* and *Chonotricha),* in which the two conjugants retain their individuality, while performing a mutual exchange of nuclear material; the meganucleus of each cell disintegrates and the micronucleus divides to form several nuclei, all but one of them disappearing; the surviving micronucleus divides into a male pronucleus and female pronucleus; the male pronucleus passes to the cell of the partner conjugant and fuses with the resident female pronucleus. *Paramecium caudatum* is illustrative of isogamontous gametogamy, where both partners of the congress are of equal size and physiological status. In contrast, the peritrichous ciliates *(Vorticella, Opercularia, Lagenophrys, Trichodina)* and the *Suctoria (Ephelota)* display anisogamontous gametogamy, i.e. they form nonmotile, normal-sized macrogamonts and small, free-swimming microgamonts (the smaller arises from the ordinary individual as a bud or by repeated fission); after the exchange of nuclei a zygote nucleus is formed only in the macrogamont, which absorbs the microgamont.

In the *Metazoa*, sexual R. is without exception gametogamy in the form of gamogony. Sexually differentiated gametes (egg cells and spermatozoa) are formed in gonads, which are usually distinct organs, and rarely diffuse in the mesenchyme of the body. Egg cells and spermatozoa are usually produced by separate female and male individuals but a minority of metazoans are hermaphrodite, i.e. egg cells and spermatozoa are produced by a single individual. Fertilization of the egg cell produces a zygote, which normally represents the beginning of Ontogeny (see). Parthenogenesis (see) is a special form of gamogeny.

Some lower animals display an alteration of sexual and asexual forms (see Alternation of generations).

Reproduction strategist: see K-selected species.

Reproductive mycelium: see Aerial mycelium.

Reproductive organs:

1) In plants: the organs serving the processes of asexual and sexual reproduction. They are usually structurally well differentiated from the vegetative organs. See Reproduction, Flower, *Bryophyta*, *Pteridophyta*.

2) In animals (Figs. 1 and 2): the sexual organs of multicellular animals, which are directly involved in fertilization, egg and sperm production, storage and transport, fetus development and birth. The internal

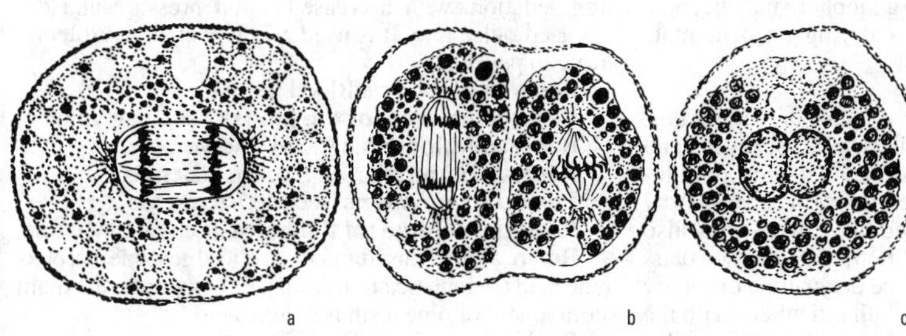

Fig.5. *Reproduction.* Pedogamy in *Actinophrys sol. a* Onset of mitosis with the diploid number of 44 chromosomes. *b* Second maturation division with halving of the chromosome complement to the haploid number of 22. *c* Fusion of the haploid nuclei, following fusion of the two unicellular sister animals.

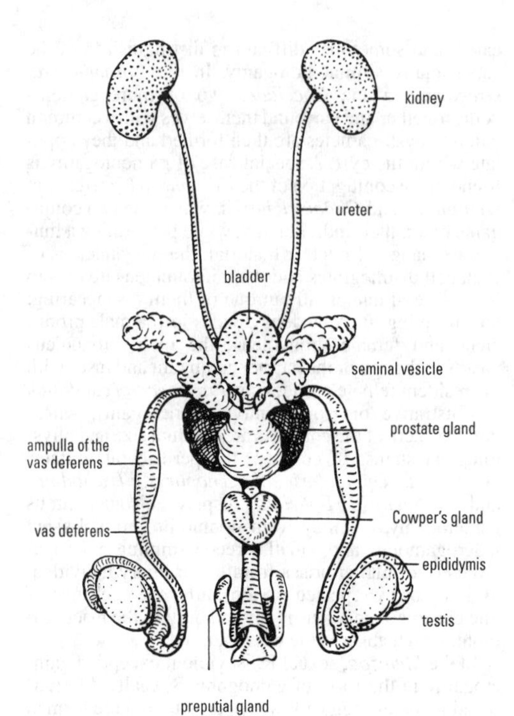

Fig. 1. *Reproductive organs.* The urogenital system of the male hamster.

R.o. include the gonads (see Ovary, Testes), the gamete transporting ducts (see Oviduct, Vas deferens), in some animals the Yolk sac (see), the Uterus (see), the Vagina (see), and in pregnant mammals the Placenta (see). The external R.o. are the Copulatory organs (see) and their accessory glands.

R.o. are largely absent in the *Porifera* and *Coelenterata.* In the *Porifera,* the ova and male gametes, which arise in the dermal layer, simply move to the exterior via the water canals. The gonads of the *Coelenterata* split open and release their contents directly, or via the gastrovascular system, into the surrounding water. In the colonial *Coelenterata (Hydrozoa* and *Siphonophora)* the sexual medusae show varying degrees of differentiation, and they may be free-swimming or remain attached to the colony. Gonophores are formed by budding. Depending on the species, buds may remain attached as sessile gonophores in which the gametes form and ripen, or they may achieve various intermediate stages of development, or each gonophore may become a fully differentiated medusa carrying its own male and/or female gonophores.

Reptantia: see Movement.

Reptilia, ***reptiles*** (plate 5): a class of vertebrates, comprising the lizards, crocodiles, snakes, turtles and their allies, both living and extinct. They are poikilothermic, tailed animals, and their dry skin carries horny epidermal scales and scutes, and bony plates may also be present. In the original condition, there are two pairs of clawed limbs, but these may be degenerate or totally regressed (e.g. in snakes). The skull articulates with the vertebral column by a single condyle. In most reptiles the outer casing of dermal bones over the cranium is incomplete owing to the presence of temporal vacuities, which allow greater space for the housing of jaw muscles. The number of temporal vacuities and their relations to the dermal bones of the skull are taxonomically important. Lungs are the only means of respiratory gas exchange. The heart consists of 2 atria and 2 incompletely separated ventricles (with the exception of crocodiles, in which the ventricles are completely separated). The kidney is a metanephros and associated with the need for water conservation. Uric acid is the main excretory product of nitrogen metabolism. Adaptation to a terrestrial existence is complete, and even embryonic development does not require an aquatic environment. Fertilization is internal; most species lay large-yolked eggs enclosed in a shell, and some give birth to live young. The embryo is enclosed in embryonic membranes (amnion, chorion, allantois) and there is no larval stage. Brood care is found only in the *Crocodylia,* some snakes and a very few lizards, although most reptiles bury their eggs in the ground or lay them in situations with a favorable microclimate.

Strict herbivores, omnivores and exclusively predatory species (e.g. all snakes) are all represented. The main senses are vision and chemical senses (see Vomeronasal organ).

Reptiles are found in all parts of the world except the polar regions, the variety of species increasing toward the equator. They have adapted to practically all environments, and some groups, such as marine turtles and sea snakes, are secondarily adapted to a marine existence.

From the great diversity of forms in the Mesozoic, the class is now represented by only 4 living orders: *Testudines* (see), *Rhynchocephalia* (see), *Crocodylia* (see) and *Squamata* (see). Originating in the Carboniferous, the turtles *(Testudines)* are phylogenetically so ancient that they represent a "sister" group of other recent reptiles, which appeared later from other evolutionary branches.

Repugnatorial gland: see *Heteroptera.*

Requienia: an extinct bivalve genus with very unequal valves. The greatly elongated and spirally twisted umbo of the left, larger valve was attached to the sea floor. The right valve was flat and lid-shaped, and also possessed a spiral umbo. Fossils are found in the Cretaceous, particularly in the Alpine sediments.

Reserpine: an alkaloid of complicated structure, containing both indole and isoquinoline ring systems. It occurs in the roots of *Rauwolfia* species and was isolated in 1952 from *Rauwolfia serpentina.* It is also the main active principle of the Mexican plant, *Rauwolfia heterophylla.* After a latent period, R. causes long-lasting sedation, with decreased blood pressure and decreased pulse rate. It is used as a powerful neuroleptic drug in psychiatry.

Reserve blood: see Blood reserves.

Reserve volume (of the lungs): see Lung volume.

Reservoir blood: see Blood reserves.

Reserve materials: see Storage materials.

Residual meristems: see Meristems.

Residual volume (of the lungs): see Lung volume.

Resin acids: unsaturated, cyclic diterpene carboxylic acids, which occur in resins. For example, the main component of pine resin is abietic acid.

Resini guaiaci: see Guaiacum resin.

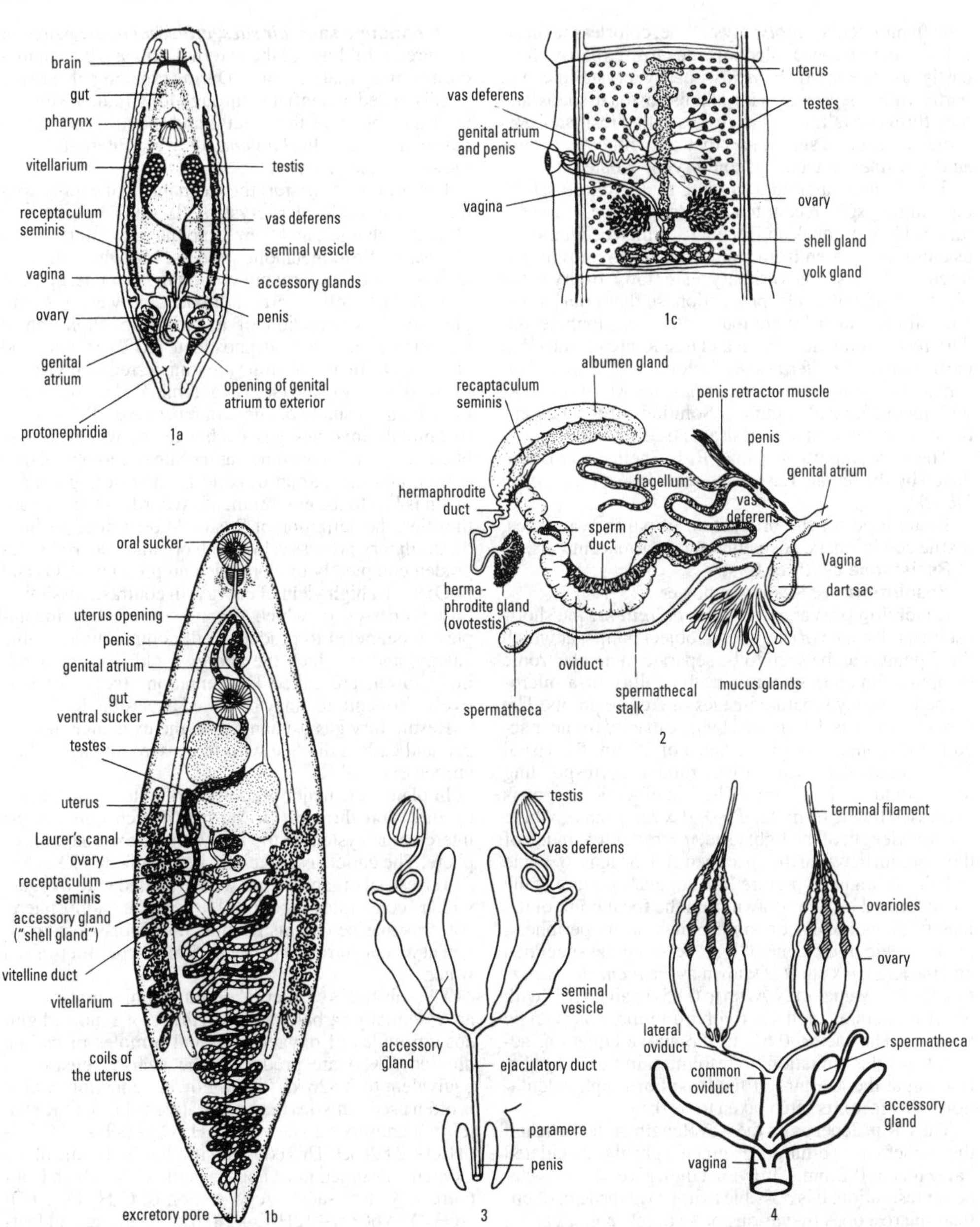

Fig.2. *Reproductive organs* (invertebrate). *1a–1c* Flatworms *(Platyhelminthes)*. *1a Proxenetes* (a free-living flatworm, class *Turbellaria)*. *1b Dicrocoelium dendriticum* (small liver fluke, class *Trematoda)*. *1c* Sexually mature proglottid of *Taenia saginata* (bovine tapeworm, class *Cestoda)*. *2* Hermaphrodite sexual organs of the Roman snail *(Helix pomatia)*. *3* Basic plan of the internal male reproductive organs of insects. *4* Basic plan of the internal female reproductive organs of insects.

Resins: largely amorphous, solid or semi-solid, yellow to brown, transparent odorless and tasteless organic substances, usually of vegetable origin.

R. are supercooled melts, soluble in nonpolar solvents. They are complex mixtures of acids, alcohols, esters, hydrocarbons, phenols, etc., in which di- and triterpenes and aromatics predominate. The structures of resin components are not well clarified, but they can be classified according to their chemical properties into: *Resinotannols:* aromatic phenylpropane resin acids, allied to the tannins, occurring partly free, but more generally esterified with aromatic acids or with umbelliferone (tannol R); *Resinols:* crystalline, colorless resin alcohols, e.g. terpene alcohols occurring partly free, partly as esters; *Resin acids:* partly crystalline and partly free; they react with alkalis to form soaps and they form crystalline salts or esters called resinates; *Resenes:* indifferent substances that are neither esters nor acids; *Resines:* esters, e.g. coniferyl benzoate.

Tree R. are classifed according to age into fossil R. e.g. Amber (see), recent fossil R. (several years to centuries old), e.g. copal, and recent R., which occur mostly as balsams released from resin ducts when trees are injured for the first time (primary resin flow). Injury to the plant also stimulates R. production, so that injury is often followed later by copious, often year-long secondary resin production. Plant and tree R. are produced in particular by *Coniferae* (see Colophony), *Euphorbiaceae, Umbelliferae* and *Anacardiaceae.* Mixtures of R. with mucin are called gum R. Solutions of R. in essential oils are referred to as Balsams (see).

The most important animal R. is Shellac (see), produced by the female East Asian scale insect *(Tachardia lacca).*

R. are used for the production of paints, varnishes, textile conditioners, cosmetics and pharmaceuticals.

Resistance cyst: see Cyst.

Resolution: see Resolving power.

Resolving power, ***resolution, definition:*** the shortest linear distance *(d)* between 2 object points, at which the 2 points can be seen to be separate in an electronic or optical imaging system, i.e. the ability of a microscope to display separate images of 2 close points. The R.p. of the eye is determined by its capacity for near accommodation. At a sight distance of 25 cm, the visual angle is equivalent to about 1 arc minute, corresponding to a resolution of 0.07 mm. The R.p of a microscope is expressed by the formula: $d = 0.61 \times \lambda / n \times \sin\alpha$, where λ = wavelength of the light used, n = refractive index of the medium between the specimen and the lens, α = sine of half the angular aperture (the angular aperture is the angle formed by lines drawn from the focal point of the lens to diametrically opposed points on its periphery, i.e. the angle of the cone of light between the specimen and the lens). $n \times \sin\alpha$ is known as the *numerical aperture* (N.A.). Values of N.A. up to 0.85 (in air), 1.25 (with water immersion) and 1.4 (with oil immersion) can be attained. The factor, 0.61, in the above equation, accounts for the limitation on resolution imposed by diffraction at the aperture of the lens. For rough calculations, this factor is often taken to be 0.5.

Thus, R.p. depends on the wavelength of the light and the numerical aperture. For green light the calculated value for *d* is 0.2 mm. Ultraviolet light gives a somewhat better resolution. It is possible to improve the R.p. of optical microscopes by various procedures, e.g. *d* can be decreased by using oil immersion (i.e. by changing the refractive index of the medium so that it is very similar to that of the glass of the lens), but the possibilities of improving the R.p. of optical systems have now been more or less exhaused. A very large increase of R.p. is possible only with the electron microscope, where the limiting factor is not the wavelength of the electrons, but aberrations of the electron lenses and the thickness of the visualized specimen. For whole specimens with an optimal size, shape and composition, the R.p. of the electron microscope is about 0.3 nm. For ultra-thin sections, as used in biology and medicine, the R.p. is about 1.0 nm.

Resonance sacs, ***air sacs, croaking bags:*** paired or unpaired infoldings of the external skin and buccal mucosa of many male anurans. During croaking, these sacs are distended and inflated through longitudinal slits behind the angles of the mouth, and they serve as resonance chambers. In *Bombina,* they are internal, below the skin of the pharynx.

Resonance transfer: the mechanism of energy conduction in molecular crystals. It does not involve electrical charges, and there is no mass transfer.

Respiration: metabolic processes in which the oxidation of reduced organic compounds to energy-poor end products (often carbon dioxide and water) is coupled to the conversion of adenosine 5'-diphosphate (ADP) to adenosine 5'-triphosphate (ATP). Protons and electrons from the substrate are transferred stepwise via the Respiratory chain (see) to a terminal electron acceptor, which is usually oxygen. In most cases, R. in plants or animals involves gas exchange, in which atmospheric oxygen is consumed at the site of biological oxidation, and the carbon dioxide formed during metabolism is lost to the environment. According to some authorities, the definition of R. should be limited to those dissimilatory processes in which organic material is degraded completely to inorganic end products (CO_2 and H_2O) with a high yield of energy. In contrast, dissimilatory processes in which organic material is incompletely degraded to products still containing available energy, and in which the energy yield is correspondingly lower, are called Fermentations (see). Alternatively, fermentation may be called anaerobic R.

Respiratory gas exchange, i.e. the exchange of oxygen and carbon dioxide with the environment, is often termed external R.

In plants, the major proportion of external R. occurs by diffusion through the stomata which connect the intercellular system of the leaf with the external atmosphere. The cuticle is especially permeable to CO_2.

Terrestrial organisms are well provided with oxygen, which constitutes about 21% of the atmosphere, whereas marine organisms must utilize dissolved oxygen at a concentration of about 5 ml O_2 per liter of sea water.

The substrates of internal or intracellular respirations are essentially carbohydrates. In the R. of 1 mole of glucose, 6 moles of oxygen (O_2) and 6 moles of carbon dioxide (CO_2) are produced; the energy released is equivalent to 2 826 kJ. The following equation, which is often used, satisfies the quantitative balance of R., but it is not entirely satisfactory: $C_6H_{12}O_6 + 6O_2 \rightarrow 6CO_2 + 6H_2O + 2\,826$ kJ. This equation implies that some of the oxygen consumed must appear in the CO_2, which is incorrect. A more satisfactory version is: $C_6H_{12}O_6 + 6O_2 + 6H_2O \rightarrow 6CO_2 + 12H_2O + 2826$ kJ. Many salient facts of the respiratory process are embodied in this second equation, i.e. in the degradation of glucose, CO_2 is formed on six occasions by the decarboxylation of oxoacids (ketoacids); the respiratory chain goes into operation 12 times, each time consuming a half molecule of oxygen and producing a molecule of water; hydration reactions (addition of water) occur 6 times.

The ***respiratory quotient*** is the ratio of CO_2 produced to the O_2 consumed, and it can be seen from the above equation that the respiratory quotient for the total R. of glucose is 1.0. In higher plants, proteins are respired only in extreme conditions of energy deprivation.

Energy production in R. occurs stepwise during the operation of many reaction steps, involving more than 50 enzymes and a variety of intermediates. Important stages of carbohydrate degradation are Glycolysis (see), conversion of the resulting pyruvate to acetyl-CoA, metabolism of the latter in the Tricarboxylic acid cycle (see), and the Respiratory chain (see). Part of the energy released is stored in chemical form, mostly as adenosine 5′-triphosphate (ATP), which in turn provides energy for biosynthetic reactions. A further portion of the released energy appears as heat, which in warm blooded animals contributes to the maintenance of body temperature. Plants also utilize respiratory energy to effect a rise of temperature, e.g. in germinating seeds. In certain flowers and inflorescences, the temperature may rise by 10 to 17 °C above the ambient as a result of intense R. Even higher temperature rises are possible in dense masses of plant material, e.g. in haystacks due to the respiratory activities of certain bacteria and fungi. Plant respiration must be investigated in the dark, in order to exclude reactions of photosynthesis, which are the reverse of those in R.

In plants, the intensity of R. shows wide variations, depending on the tissue and its current energy requirements. Respiratory activity increases markedly at the start of germination, during growth, and after a tissue is wounded or stimulated. In contrast, the R. of dormant plant parts, e.g. air-dried seeds, is hardly detectable. Mature, permanent tissue, e.g. foliage leaves, shows a constant low rate of basal R., which is just sufficient for the maintenance of necessary cell function. In total, a green plant excretes 5 to10 times its volume of respiratory CO_2 in 24 hours; this represents 1/5 to 1/3 of the quantity of CO_2 bound per day by photosynthesis. Of the environmental conditions that influence R., temperature is especially important. In normal temperature ranges, the temperature coefficient (Q_{10}) is generally 2.0 to 2.5. Temperature increases above normal are accompanied by corresponding increases of R., almost to the point at which the plant becomes heat-damaged. The effect of oxygen concentration on R. is much less in plants than in animals. In the presence of excess respiratory substrates, R. may increase beyond the energy requirement. Photorespiration (see) involves an exchange of O_2 and CO_2, but it is otherwise an entirely different process from normal R.

Special types of R. are Nitrate respiration (see) and Sulfate respiration (see Sulfur).

In animals, a variety of structures is involved in physiological gas exchange. In some primitive invertebrates, respiratory gas exchange occurs entirely through the epidermis, e.g. in freshwater polyps and earthworms. In more advanced invertebrates and vertebrates, the epidermis makes a variable contribution to the overall respiratory gas exchange; in humans it amounts to about 1.5%. In Gill respiration (see), which occurs in many invertebrates and in fishes, gases are exchanged through the very thin epithelium of the highly vascularized gills. Gill surface area is very large, amounting to about 60 cm² in a 20 g crayfish, and about 27 cm² in a 10 g crucian carp. Gas exchange in insects occurs via trachea (tracheal respiration).

The most highly evolved organ of respiratory gas exchange is the lung. Respiration via the lung can be divided into 3 phases. 1) A phase of external R., in which the respiratory muscles work against a mechanical resistance (see Respiratory mechanics), so that air is exchanged between the internal lung cavity and the outer atmosphere. Oxygen and carbon dioxide diffuse in opposite directions through the walls of the lung alveoli. The efficiency of the external phase depends on the volume of air involved, the vascularization of the lungs, the composition and distribution of the respiratory gases in the alveoli, and the rate of diffusion between the alveolar and capillary wall. 2) Transport of gases between the lungs and the tissues by the blood. The efficiency of this process depends on the quantity of circulating blood, the blood supply to the active organs, and the binding capacity of the blood for carbon dioxide and oxygen. 3) Internal R., which includes the exchange of gases between the blood and tissues, and cellular oxidation processes. The activity of this phase depends on the blood supply to the tissue, local diffusion processes, and the cellular oxygen requirement. Finally, the utilization of oxygen and the production of carbon dioxide are accompanied by the release or storage of energy. The carbon dioxide is taken up by the buffer system of the blood, transported back to the lungs, released into the lung cavity, then into the atmosphere by operation of the external phase of R.

Respiratory center: a group of autonomic neurons and synapses in the mammalian brain stem, which receives, evaluates and transmits the nervous impulses that control the pattern of breathing.

By experimental elimination, stimulation and bypassing, different R.c.s have been located in the mammalian brain stem; the concept of a single R.c. controlling the total breathing process is no longer tenable. The precise location reported for a R.c. may vary somewhat, depending on the methodology and on the animal species. The lowest section of the medulla oblongata contains the inspiratory and expiratory centers. These lie fairly close together, the inspiratory center being centrally situated and deep in the interior of the medulla, with the expiratory center above it and in a more lateral position. The pneumotaxic center is found in a more anterior section of the brain stem, i.e. in the pons. The apneustic center lies between the other three centers (see Fig.). A fifth center in the uppermost region of the spinal cord of some animals is rarely active, but it operates, for example, in prematurely born animals, or shortly before death. An inspiratory and an expiratory center also exist in the spinal cord, but their activity is normally suppressed by the higher R.c.s.

It is not fully understood how the reflex or autonomic breathing rhythm is established. All models are based on reciprocal inhibition between discharging neurons, i.e. a natural oscillator. Certainly, the pneumotaxic center does not appear to be a rhythmic pacemaker in itself, and it depends on impulses from elsewhere. There is no clear evidence for or against the existence of pacemaker cells in the brain stem. Alternatively, an oscillating output might be maintained by time delays in interneuronal transmission. The following interdependent processes appear to be involved. 1) The resting potential of an autonomic neuron rises and falls spontaneously in a self-regulated cycle of stimulation and inhibition. 2) In a positive feedback process, the neurons of the R.c.s. interact to produce a rhythmic discharge. 3) Autonomic neurons control their own activity by stimulating inhibitory intermediate neurons (feedback inhibition). 4) Inspiratory and expiratory motor discharges are produced alternately. When the inspiratory center is stimulated, the expiratory center is inhibited and vice versa.

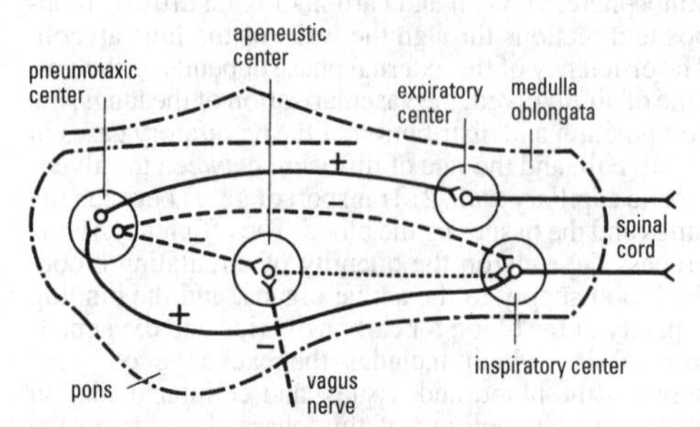

Respiratory center. Functional nerve connections in the respiratory center.

Usually, stimulation commences in the inspiratory center, causing firing of the respiratory nerves which bring about inspiration (phrenic and intercostal nerves). This is followed by stimulation of the pneumotaxic center, which in turn activates the expiratory center. The expiratory center then triggers expiration by acting peripherally, and it also inhibits the activity of the inspiratory center in the medulla oblongata. The timing of the switch from the stimulation of the inspiratory center to stimulation of the expiratory center is thus determined by the pneumotaxic center. Under normal conditions, the apneustic center is so powerfully inhibited by the pneumotaxic center that it remains inactive. Removal of this inhibition results in deep convulsive inspirations with deep, rapid expirations, a type of breathing known as apneusis.

The autonomic rhythm of breathing, established by interaction of the R.c.s, can be altered considerably by physical stimulation of the respiratory apparatus, and by peripheral or central chemical effectors (see Respiratory regulation).

Respiratory chain, ***electron transport chain:*** a multienzyme complex located in the inner mitochondrial membrane of eukaryotes and in the cell membrane of prokaryotes. It provides most of the energy requirement of aerobically respiring cells. The R.c. consists of a series of coupled redox catalysts, which transport electrons stepwise from respiratory substrates to molecular oxygen. The energy of this electron flow is partly converted into chemical energy, i.e. it is coupled to the phosphorylation of ADP to form ATP, a process known as Oxidative phosphorylation (see). Dehydrogenases catalyse the removal of hydrogen from respiratory substrates, and this hydrogen is transferred to a coenzyme or prosthetic group of the dehydrogenase. For example, NAD is the cofactor of the specific dehydrogenases of isocitrate, malate, glutamate, pyruvate, α-ketoglutarate, 3-hydroxyacyl-CoA, etc., while FAD is the prosthetic group of the dehydrogenases that attack succinate, acyl-CoA, and glycerol 3-phosphate, etc. Hydrogen from reduced NAD is transferred to a flavoprotein complex containing nonheme iron and inorganic sulfur (= complex I of the respiratory chain, known as NADH:ubiquinone oxidoreductase), and from there to ubiquinone (coenzyme Q). Hydrogen from reduced FAD (e.g. from the dehydrogenation of succinate) is transferred directly to ubiquinone, without using the NADH:ubiquinone oxidoreductase. The resulting reduced ubiquinone loses 2 H^+ ions, and the electrons released are taken up by a valency change in the heme iron of cytochrome *b* ($Fe^{3+} + e^- \rightarrow Fe^{2+}$). Electrons are then transferred along the cytochrome chain via cytochromes *b*, c_1, *c*, *a* and a_3, to molecular oxygen, each cytochrome transporting electrons by changing the valency of its heme iron. Thus, a single operation of the R.c. results in the dehydrogenation of 1 molecule of respiratory substrate, and the production of 2 H^+ and 1 hyperoxyl ion (O^{2-}), which combine to form water.

The R.c. is functionally closely associated with the Tricarboxylic acid cycle (see), which is responsible for the dehydrogenation of many of the important respiratory substrates. Depending on the tissue and its metabolic state, other pathways, e.g. fatty acid degradation, may be more important than the tricarboxylic acid cycle in providing reduced NAD or FAD for the R.c.

The R.c. illustrates the fact that true chemical equilibria do not exist in living organisms. Operation of the R.c. is only possible with continual dehydrogenation of substrates and continual provision of oxygen either directly from the atmosphere or by a respiratory carrier such as hemoglobin. Thus, the system is in a "steady state", in which the degree of reduction of its components decreases from NAD to cytochrome a_3, and in which the overall degree of oxidation or reduction of the system depends on the respective concentrations of oxygen and respiratory substrate.

Respiratory compliance: see Pressure-volume curve of the respiratory apparatus.

Respiratory effectors: unspecific effectors with a marked action on the rate and depth of external respiration. The most significant R.e. are temperature, pain, hormones and blood pressure (see Baroceptors). Sudden chilling of the skin surface causes deep inspiration. Prolonged temperature stimuli may cause hyperventilation (e.g. fever, moderate chilling) or hypoventilation (e.g. deep chilling). Pain stimuli reinforce respiration. High blood pressure, acting via baroceptors, reduces respiratory activity. Low doses of adrenalin increase respiration, whereas high doses cause an increase of blood pressure, leading to a decrease of respiration. Progesterone, a steroid hormone, reinforces quiet breathing both in the time between ovulation and menstruation, and during pregnancy. Thyroid hormones promote respiration, acting directly by stimulating oxygen consumption, and causing a rise in the overall metabolic rate.

Respiratory gas exchange: see Gas exchange.

Respiratory hysteresis loop: see Pressure-volume curve of the respiratory apparatus.

Respiratory mechanics: the totality of mechanical processes leading to the exchange of gases in the respiratory organs.

1. ***Mammals, including humans.*** Prerequisite for expiration and inspiration is the existence of a pressure difference between the lung space and the external environment of the organism. During the first half of inspiration ("breathing in") in humans, the pressure within the lungs falls to 0.4 kPa below atmospheric pressure, and this pressure difference disappears during the later stage of inspiration. During expiration ("breathing out"), the situation is reversed, i.e. the internal lung pressure rises to 0.4 kPa above atmospheric, then falls away to zero at the end of the intake

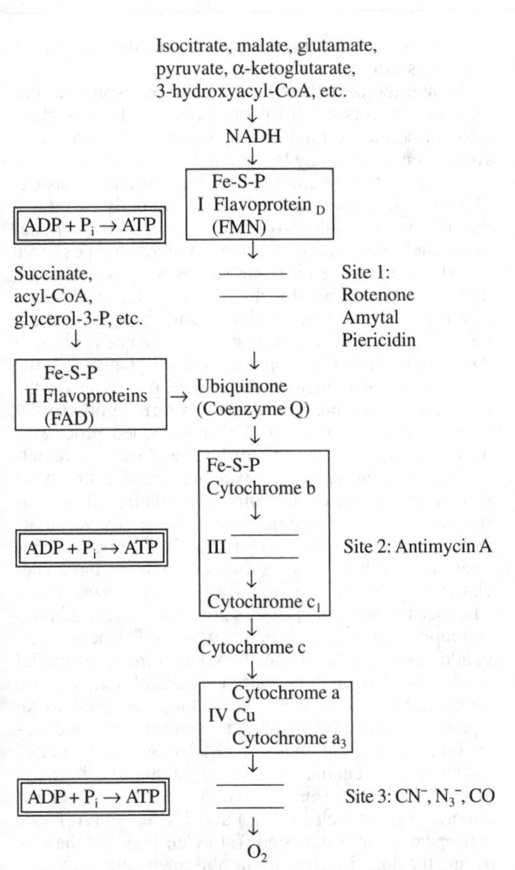

Respiratory chain. Boxes indicate the composition of complexes I to IV, which have been prepared from mitochondria by treatment with detergents. Fe-S-P = iron-sulfur protein, i.e. a redox protein containing centers of nonheme iron and inorganic sulfur (sulfur attached via the iron, and not part of sulfur-containing amino acids). Arrows indicate the direction of electron flow. Sites 1, 2 and 3 (indicated by horizontal bars) are the sites of action of certain inhibitors, and at each of these sites, sufficient energy is available for the synthesis of one molecule of ATP, i.e. the difference in redox potential between the 2 components at the site is equivalent to, or greater than, the free energy of hydrolysis of ATP. This statement must, however, be interpreted in the light of the chemiosmotic mechanism of Oxidative phosphorylation (see).

of breath (Fig.1). During very deep inspiration, the negative intrapulmonary pressure can fall to -10 to -20 kPa, and during maximal expiration (e.g. during coughing) it can rise to +13 to +33 kPa. Since the lungs are nonmuscular, these pressure changes are caused by expansion or contraction of the thoracic volume. Expansion of the thorax occurs by contraction of the diaphragm and the external intercostal muscles. In its relaxed state, the diaphragm is dome-shaped and arched into the thoracic cavity. By contraction of radial muscle fibers, it becomes flattened, extending the thoracic cavity toward the abdominal cavity (Fig.2). At the same time, the viscera are displaced downward and the abdominal wall curves outward. The fibers of the external intercostal muscles lie diagonally from above and behind to below and in front. Contraction of these muscles lifts the ribs, so that the thoracic cavity becomes enlarged anteriorly.

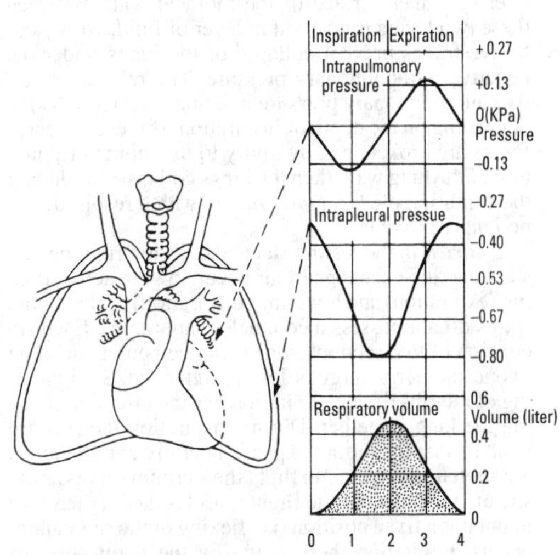

Fig.1. *Respiratory mechanics.* Pressure and volume changes during inspiration and expiration.

Expiration is usually passive, and it is active only during greatly increased respiratory effort. Relaxation of the muscle fibers of the diaphragm and of the external intercostal muscles results in a contraction of the thoracic cavity, because elasticity of the stretched respiratory apparatus (thoracic wall, lungs, diaphragm, abdominal wall) causes the latter to return to its resting position. Active expiration occurs by contraction of the internal intercostal muscles and the musculature of the abdominal wall. The internal intercostal muscles lie diagonally from above and in front to below and behind, so that by contracting, they pull the ribs downward and decrease the thoracic volume; in addition, increasing contraction of the muscles of the abdominal wall causes the diaphragm to arch further into the thoracic space.

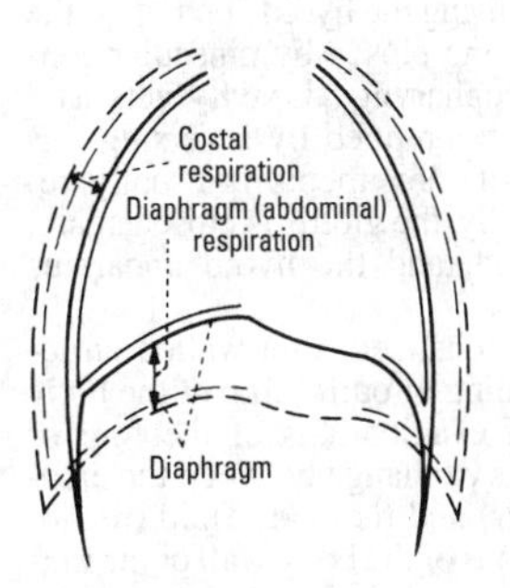

Fig.2. *Respiratory mechanics.* Movements of the thoracic cavity during deep inspiration.

Diaphragm and intercostal muscles usually make an approximately equal contribution to inspiration. If the functional contribution of the diaphragm is increased, the process is known as *abdominal* or *diaphragm respiration.* If the activity of the external intercostal muscles predominates, the process is called *thoracic* or *costal respiration.*

The lungs are not fused to the thoracic wall. During respiration, the outer surface of the lung slides freely over the inner surface of the thoracic wall. Between these two layers is a very thin layer of fluid, whose cohesive forces prevent collapse of the lungs under the negative intrapulmonary pressure. The relevant negative intrapulmonary pressure in humans is 0.3–1.6 kPa, depending on the depth of inspiration. If these cohesive forces are broken, e.g. by injury to the thorax or puncture of the lung wall, then the lungs collapse, producing the condition of *Pneumothorax,* in which respiration is no longer possible.

2. ***Birds.*** In the resting state, and during running and standing, the exchange of air occurs by the active raising (expiration) and lowering (inspiration) of the thorax (birds do not possess a complete diaphragm). Each rib consists of two sections, which can be bent at an angle to one another or more or less straightened. Straightening of the ribs during expiration has the effect of pressing the lungs together. During inspiration the ribs are bent, so that the thorax expands laterally and the lungs are stretched outward. In flight the sternum serves as the site of attachment of the flight muscles, and it therefore maintains a fixed position (i.e. flexing of the ribs cannot occur). Expiration then occurs by the application of pressure to the air sacs (which are connected to the lungs), causing air within the air sacs to flow through the lungs to the exterior. During inspiration, air is forced into the lungs and air sacs by the forward air pressure generated by flight.

3. ***Reptiles and amphibians.*** In reptiles and those amphibians that have lungs, air is gulped (reptiles and amphibians do not have a diaphragm). With the exception of turtles, reptiles (like mammals and birds) also employ rib movements. Turtles possess lung muscles, and these are assisted by certain abdominal muscles.

In terrestrial amphibians there are two phases of respiration. In the first phase, or *buccal respiration,* the glottis is closed and the external nares are open, and air is exchanged between the buccopharyngeal cavity and the exterior by rhythmic movements of the hyoid apparatus in the floor of the buccopharyngeal cavity. In the second phase, the glottis is open and the external nares are closed, and air in the buccopharyngeal cavity is forced into the lungs by raising the hyoid apparatus (the entry to the esophagus being closed by muscular contraction). Thus, the buccopharyngeal cavity acts as a force pump. The lungs are emptied by reversing this procedure, and this set of movements is usually repeated several times. Finally, the glottis is closed again, the external nares opened, and the hyoid apparatus raised to force the air out.

Respiratory organs: localized, thin-walled structures formed by the infolding or outfolding of the body surface, or consisting of evaginations of the fore or hindgut. R.o. facilitate gas exchange between the exterior medium (water or air) and the body fluid (hemolymph, blood). Evaginations of the body wall or the mucosa of the foregut, which are associated directly with vessels of a circulatory system, are known as Gills (see). Invaginations of the body wall that lie between other organs, as well as ventral evaginations of the foregut are known as Lungs (see).

In invertebrates, the simplest form of respiratory gas exchange occurs by diffusion across the body surface (skin or dermal respiration). Specialized R.o. are therefore absent from many lower animals, e.g. sponges, coelenterates, platyhelminths and aschelminths. Gas exchange also occurs exclusively via the body surface in animals with small body volumes, e.g. small crustaceans and many mites. In many annelids that possess a blood system, the entire skin serves for gas exchange, and it is richly supplied with blood vessels; but there are also often gills of various shapes and sizes, all of them basically thin-walled evaginations of the body. Thus, in *Arenicola* (class *Chaetopoda,* order *Polychaeta),* the gills consist of branched, segmented processes on the segments of the median region. In terrebellids (order *Polychaeta),* the gills consist of branched processes, usually 3 pairs, just behind the head and full of circulating blood. In mollusks, gas exchange may occur via the skin or gills. In terrestrial pulmonate snails, gills are absent, and gas exchange occurs via the richly vascularized roof of the mantle cavity ("lung"). Among the echinoderms, the holothuroids (sea cucumbers) have specialized gills known as Respiratory trees (see). Crustaceans normally possess gills. The terrestrial onychophorans and the arthropods possess Tracheae (see). In addition to tracheae, the abdomen of many terrestrial arachnids also contain paired Tracheal lungs (see). Many arthropods that are secondarily adapted to an aquatic existence (water spiders, water beetles and water bugs, as well as many insect larvae) have special structures that enable them to trap a supply of air beneath the water. Thus, *Dytiscus fasciventris* (North American giant water beetle) and *Dytiscus marginalis* (European giant water beetle) store air between the elytra and the dorsal surface of the abdomen (spiracles having migrated dorsally in this family), while water bugs and water spiders trap a layer of air on their hairy bodies. The trapped layer serves not only as a supply of air, but also as a "physical gill"; the oxygen requirement can be satisfied completely by diffusion into this air layer from the surrounding water. Insect larvae, e.g. fly larvae, breathe with the aid of secondarily evolved true gills (vascularized gills) or via Tracheal gills (see).

In vertebrates, R.o. are formed predominantly by differentiation of the foregut (see Pharynx). Lamellar dermal foldings of the walls of the pharyngeal clefts are known as gills. Endodermal buddings of the foregut, which extend into the body interior by branching, are known as Lungs (see). Furthermore, in those lower vertebrates that possess a thin, moist skin, e.g. amphibians, the skin plays an important part in gas exchange, and this property is retained to a slight extent in mammals.

The labyrinthfishes, e.g. the southern Asia climbing perch *(Anabas testudineus),* can utilize atmospheric oxygen directly with the aid of a supplementary respiratory organ, known as the labyrinth. The labyrinth, situated in a protrusion of the gill cavity inside the head, consists of bony lamellae covered with respiratory mucosa. Intestinal respiration occurs in "air-swallowing" fishes, whose intestinal epithelium functions as an organ of gas exchange, e.g. the pond loach *(Misgurnus fossilis).*

Respiratory poisons: substances that specifically inhibit stages in the transport of electrons from substrates to oxygen in the Respiratory chain (see). Certain R.p. are used as plant protection agents against insect pests, e.g. Derris dust contains rotenone, which inhibits on the oxygen side of the nonheme iron of NADH dehydrogenase.

Respiratory protection reflexes: reflexes that protect the respiratory system from injury. Receptors of the R.p.r. are located in the nasopharyngeal cavity, in the upper respiratory passages and in the pulmonary circulation. R.p.r. consist of sneezing, sniffing and apnea. Sneezing is caused by chemical or mechanical stimulation of the nasal mucosa, acting via the trigeminal nerve. Sneezing begins with a deep inspiration, followed by an explosive expiration. Weak chemical stimulation illicits a rapid, superficial respiration, known as sniffing. Sharp-smelling, irritant gases lead to a cessation of breathing (apnea), nervous stimulation being transmitted via the trigeminal nerve and the optic nerve. Mechanical stimulation of the entire area between the throat and bronchi excites the coughing center in the pons, causing deep inspiration, followed by rapid and powerful expiration. During a cough, pressures of 40 kPa and air flow rates up to 300 m/s are developed. Inspired dust and other irritants stimulate receptors in the upper lung passages, and cause apnea. Strong dilation of the pulmonary circulation is usually accompanied by shallow, rapid breathing, with increased functional residual capacity, and apnea of variable duration. See Respiratory regulation.

Respiratory quotient, *RQ*: ratio of the quantity of carbon dioxide produced to the quantity of oxygen consumed. The magnitude of the RQ depends on the nature of the diet. There is therefore a relationship between the RQ and the size of the "calorific equivalent" (heat produced per liter of oxygen consumed), which also depends on the diet (Table).

1 g	Oxygen consumed (ml)	Carbon dioxide produced (ml)	RQ	Heat produced (kJ) per g	per l O_2
Carbohydrate	828	828	1.0	17.5	21
Fat	2019	1427	0.7	38.5	19.6
Protein	966	774	0.8	18	18.8

Respiratory rate: the number of respiratory cycles (inspiration and expiration) per minute. The R.r. is subject to wide individual variations, and it varies greatly between different animal species: horse (10–16), dog (11–38), guinea pig (70–100), mouse (84–230), domestic fowl (12–30), turtle (2–5). It also varies with age. At rest, the adult human breathes 10 to 18 times per minute, whereas the R.r. for a newborn human is 30–80. During the first year of life, this value falls to between 20 and 40.

The R.r. is also influenced by several physiological factors. Respiratory regulation (see) is achieved partly by alteration of the R.r.

Respiratory regulation: the sum of the processes that adjust respiration to internal physiological requirements, in accordance with certain environmental conditions.

In lung respiration, R.r. is achieved by alterations of both the frequency and volume of breathing, and by the optimal coordination of these two factors. Depending on the nature of the regulatory stimulus, R.r. is defined as chemical or physical. The complicated mechanism of R.r., however, is the result of the summation or interaction of various stimuli, and it is never purely chemical or physical. The variety of effectors and their stimulatory or inhibitory actions are shown in Fig. 1.

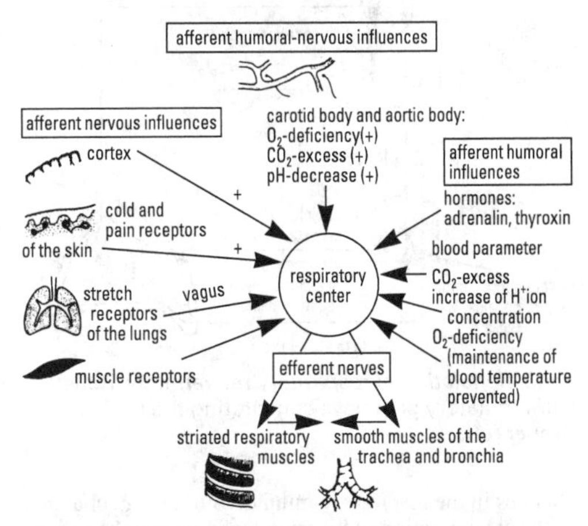

Fig.1. *Regulation of respiratory movements.* Summary of afferent and efferent regulatory mechanisms.

Physical (nervous) R.r. is achieved by alteration of the minute volume. This depends mainly on the operation of the Hering Breuer reflex, which accounts for the self-regulatory nature of each breath. During inspiration, the resulting extension of the lungs stimulates stretch receptors in the lung wall, which are served by the vagus nerve. The vagi (vagus nerve fibers) are rapidly conducting and have a high action potential. During lung extension they adapt only slowly, their discharge frequency increasing linearly with lung volume. As the lung wall is stretched further, additional vagi are activated. Stimulation of the stretch receptors inhibits the inspiratory center and inspiration is rapidly terminated (Fig. 2). In the reverse process, the inspiratory center is stimulated when the lung wall relaxes during expiration. This reflex cycle of inhibition and stimulation of the inspiratory center represents the basic mechanism for the regulation of breathing. The period between the onset of inspiration and its inhibition can be altered. If the blood bicarbonate concentration increases, this period becomes longer and the depth of breathing (i.e. the minute volue) increases, accompanied by an increased expiration of carbon dioxide (see below; Fig. 3).

Other vagus reflexes participate in physical R.r. 1) When the breath volume is too small, inspiration is increased. 2) A vigorous, arbitrary increase in breathing is followed by a fairly long pause. 3) At lengthy intervals (about every 15 min. in humans) an excessive inspiration occurs, which exercises the lung tissue.

Compression or even puncture of the chest cavity, leading to collapse of the lungs (pneumothorax), accelerates breathing and greatly intensifies inspiration. Thus the thorax may be greatly extended, but inspiration and expiration occur in a series of gasps.

In chemical R.r., various chemical factors cause

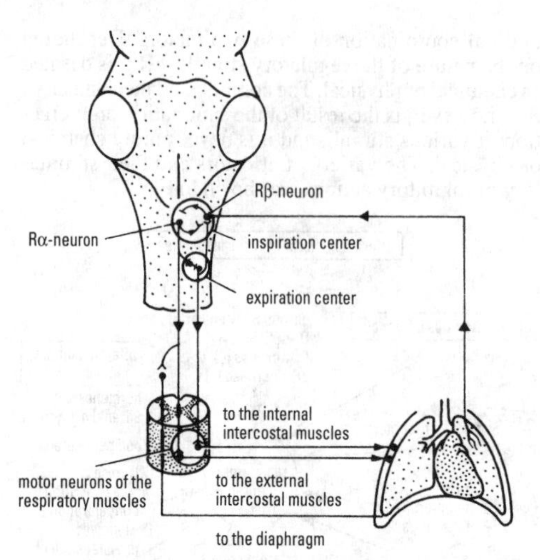

Fig.2. *Regulation of respiratory movements.* Reflex self-regulatory pathways constituting the Hering-Breuer reflex.

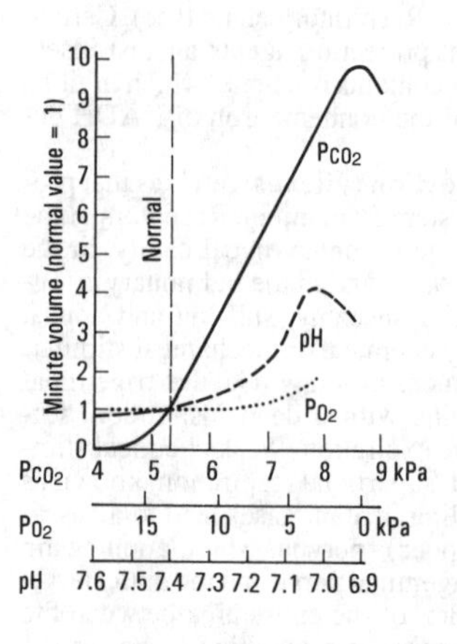

Fig.3. *Regulation of respiratory movements.* Influence of blood oxygen, CO_2 and pH on the minute volume.

changes in the respiratory minute volume, e.g. changes in the concentration of bicarbonate in the cerebrospinal fluid and in the blood, changes in blood oxygen concentration, and changes of hydrogen ion concentration in the brain, blood and muscle tissue. The most potent chemical effector is bicarbonate or carbon dioxide. Even normal blood concentrations of carbon dioxide stimulate Chemoreceptors (see) in the arterial circulation and in the lowest sections of the brain stem. A lowering of the carbon dioxide tension decreases respiration, and an increase can stimulate respiration up to ten times the resting value (Fig. 3). Excessive carbon dioxide concentrations cause respiratory paralysis. The most sensitive and most active receptors are located in the central nervous system, and they are normally responsible for 70 to 90% of the increase in breathing. The arterial chemoreceptors are less sensitive and less active, being responsible for 10 to 30% of the respiratory increase. Breathing is only slightly affected by the concentration of blood oxygen. Although the arterial chemoreceptors are stimulated by the normal blood oxygen tension, and respond exactly to every increase or decrease, their reflex activity is very slight. Even in the presence of a life-threatening oxygen deficiency, the minute volume is increased no more than twofold. Oxygen deficiency in the region of the Respiratory centers (see) decreases their ability to respond. Very small variations of hydrogen ion concentration in the arterial blood cause alterations in breathing. Acidification (decrease of pH; increase of hydrogen ion concentration) results in an increase of respiration up to four times the resting value, whereas an increase of pH above normal leads to a decrease of respiration. Excess hydrogen ions stimulate both central receptors in the brain stem and peripheral chemoreceptors in the arterial circulation and in skeletal muscles, where acidic metabolic products arise from muscle contraction. Since the hydrogen ion concentration and the carbon dioxide tension both contribute to the blood buffer system and influence each other, it is often difficult to establish which of the two factors is acting as an effector.

Respiratory trees, ***water lungs:*** paired, tree-like, multiply branched, thin-walled extensions of the wall of the hindgut in many members of the *Holothuroidea* (Sea cucumbers). Ramifications of the R.t. extend into much of the coelomic cavity. Their numerous branches end blind in thin-walled ampullae, which are bathed in coelomic fluid. Oxygen-rich water is pumped into the R.t. by contraction of the cloaca, and oxygen is transferred directly from the R.t. to the coelomic fluid and the organs. It is pumped out again by contraction of the vessel walls. During this process the mid gut is closed off from the hindgut by muscular contraction.

Resting condition: see Resting state.

Resting cyst: see Cyst.

Resting potential, ***membrane potential:*** the electric potential between the two liquid layers separated by the cell membrane. The membrane resting potential can be measured by inserting into the cell a very thin electrode (a glass capillary with a diameter at its drawn out tip of 0.1 mm or less, containing an electrolyte solution of e.g. 3 molar KCl). A second microelectrode (this does need to have a drawn out tip) is placed in the fluid surrounding the cell.

Rps. have been demonstrated for many different types of cell. In cells capable of transmitting excitatory impulses, the R.p. has a value of 70–100 mV, the interior surface of the membrane being negative in relation to the exterior surface of the cell.

Resting and action potentials of excitable cells (after Penzlin)

	Resting potential (mV)	Action potential (mV)
Loligo (giant axon)	73	112
Carcinus (nonmedullated fiber)	82	134
Periplaneta (50 mm axon)	77	99
Rana (medullated fiber)	71	116
Cat (motor cell in anterior horn of gray matter)	70	90–100
Locusta (leg muscle)	60	60
Rana (skeletal muscle)	85	112
Dog (heart ventricle)	82	102
Electrophorus (electric organ)	84	151

The membrane R.p. arises from the unequal distribution of ions, as described by the ionic theory of Excitation (see). At rest, there is a high concentration of potassium ions and a low concentration of sodium ions in solution on the inside of the cell membrane. The cell interior contains an excess of sulfate and protein ions, whereas chloride ions predominate in the liquid surrounding the cell (Table).

Internal and external concentrations of some ions in the freshly isolated axon of Loligo, and the calculated equilibrium potentials

	Concentration (mmol/l)		
Ion	Axoplasm	Blood	Calculated equilibrium potential
K^+	400	20	$E_K = 75$ mV (internal surface negative)
Na^+	50	440	$E_{Na} = 55$ mV (internal surface positive)
Cl^-	108	560	$E_{Cl} = 41.5$ mV internal surface negative)

The permeability of the cell membrane varies for different materials; K^+ permeates the membrane more easily than Na^+ and Cl^-. Potassium ions are therefore allowed into the cell, where they generate an inequality of charge distribution, because the appropriate counterion (Cl^-) is less permeant. In the solution outside the cell membrane, the positive and negative charges for Na^+, K^+ and Cl^- are approximately balanced, whereas the charges for Na^+ and K^+ in the cell interior are not balanced by the small number of Cl^- ions. The balance is accomplished by organic anions, most of them carboxyl ions, and some of these partially fixed as the carboxyl groups of cytoplasmic proteins. Thus, there are approximately equal quantities of positive and negative charges on the two sides of the membrane, but an unequal distribution of small cations and ions. The resulting equilibrium potential can be calculated from the Nernst equation:

$E_K = (RT/F) \times \ln ([K^+]_e / [K^+]_i) = k \times \log_{10} ([K^+]_e / [K^+]_i)$, where $K \sim 58$ mV at 20 °C); $[K^+]_e$ and $[K^+]_i$ are the exterior and interior potassium concentrations, respectively, R is the gas constant, T the absolute temperature and F the Faraday constant.

Since the R.p. is a mixed potential (other ions are permeant as well as potassium), it is more appropriately calculated with the generalized equation derived by Goldman from the Nernst equation.

The R.p. is formally similar to a *demarcation potential* or *injury potential.* The site of injury of a nerve or muscle is also negative with respect to the undamaged surface. Demarcation potentials are lower than normal R.ps., because the damage leads to some mixing of fluids from the interior and exterior of the cell. See Excitation, Action potential.

Resting stages: see Resting state.

Resting state, ***resting condition:*** in plants, a period of strongly decreased metabolic activity (see Dormancy), during which growth is temporarily suspended. The R.s. is usually associated with particular morphological and anatomical adaptations of the resting organ and tissue, i.e. with the production of special *resting stages.* In higher plants, the R.s. is displayed prominently by meristematic tissue, e.g. the embryonal tissues of the seed (see Seed dormancy) and apical meristems of shoots, as well as secondary meristematic tissue, especially cambium. Resting meristems are protected against the rigors of unfavorable climatic periods by seed coats and bud scales. The intensity of the resting condition is variable. In the *partial resting state* only growth is suspended, whereas metabolism (respiration, turnover of cell material) continues at a decreased rate, e.g. in winter buds. In the *total resting state,* sometimes called *anabiosis* or *latency,* there is considerable dehydration of the cytoplasm, and all life processes are decreased to a minimum, e.g. in resting seeds and spores.

In *compulsory rest* or *quiescence,* growth of the resting organ is prevented by the absence of suitable external growth conditions, by the action of inhibitory external factors, or by certain organ structures, e.g. impenetrable seed coats, or by correlative inhibition (see Correlation). Growth can be re-established at any time by removal of the source of the inhibition. The *true* or *autonomous resting condition,* also called *dormancy,* can only be ended by specific conditions, e.g. a particular temperature range or photoperiodicity. As a rule, this involves gene activation, which overides the influence of other inhibitory factors such as an unchanged level of certain phytohormones. The *relative resting state* represents a transition between quiescence and dormancy; plant organs display decreased growth, but in principle are still able to grow under a very narrow range of external conditions.

In some plants, the R.s. shows a progressive series of changes, e.g. the following sequence can occur: quiescence → relative rest → dormancy → relative rest → quiescence.

The β-inhibitor complex is substantially involved in the physiological regulation of rest and activity in plant tissues. In addition, changes in the levels and activities of certain phytohormones (e.g. auxins, gibberellins and phytokinins) are also thought to be crucially involved. All changes in the rate of plant growth and development are also intimately associated with changes in respiration and metabolic turnover. A part is also played by the water content (dehydration causes a widespread suppression of metabolism and respiration) and mineral metabolism. The R.s. can be artificially shortened or abolished by different types of physical and chemical treatment.

Restitution:

1) Replacement of lost or damaged organs or parts of organs. The ability to undergo or perform R. is more widely distributed and better developed in plants than in animals. Totipotency (see) is an essential basis of R. An extremely important phenomenon of R. is Regeneration (see), and R. may also occur by Reparation (see). Special types of R. occur in grafting and other forms of artificial propagation.

2) Recombination of chromosome fragments with restoration of normal (prefragmentation) chromosome structure. In contrast, a new combination of fragments produces a chromosome mutation.

Restitution nucleus: a diploid or polyploid nucleus, resulting from an error in the first or second meiotic division. Such errors are usually due to a faulty spindle, which prevents migration of chromosomes or chromatids to the cell poles.

For R.s. in animals, see Hibernation.

Restriction: see Restriction enzymes.

Restriction enzymes, ***restriction endonucleases:*** enzymes present in a wide variety of prokaryotic organisms, where they serve to cleave foreign DNA molecules (e.g. phage DNA). R.e. recognize specific palindromic sequences (see Palindrome) in double-stranded DNA. Many R.e. are known, all with different specificities. Host DNA is protected from the activity of the host R.e. by methylation, a process known as *modification.* The phenomenon whereby foreign DNA introduced into a prokaryotic cell becomes ineffective owing to the action of R.e. is known as *restriction.*

When isolated and purified for laboratory use, R.e. represent a powerful set of tools for the analysis of chromosome structure and for genetic engineering. Cleavage of duplex DNA at a palindromic sequence generates two cleavage products, each with a short terminal sequence of single-stranded DNA. With the aid of ligases, the single-stranded ends of such fragments can be spiced with the single-stranded termini of other appropriate fragments of duplex DNA. See Gene technology, Restriction map.

Restriction map: a map constructed with the aid of restriction enzymes, showing the recognition sequences for restriction enzymes (see Palindrome) in a molecule of DNA. R.ms. are used particularly for the differential diagnosis of viruses and for the characterization of eukaryotic genomes.

Retardation: inhibition of development, manifested as a slowing of the growth of individual organs and body parts or of the entire organism. R. is one of the mechanisms of phylogenetic change.

Rete mirabile: a complex system of intertwined small arteries and veins, which permit heat transfer between arterial, and venous blood. The temperatures of warm blood passing to an appendage and cool blood from an appendage become equalized in the R.m., thereby conserving energy that would otherwise be needed to warm the cool blood from a poorly insulated appendage, e.g. in the *Lorisidae* (lorises). See Hairpin loop countercurrent principle, Swim bladder.

Retia mirabilia: see Hairpin loop countercurrent principle, Swim bladder.

Reticular fibers, ***reticulin fibers, argyrophil fibers:*** fine fibers of the intercellular substance of vertebrate support tissue. Typically inelastic, very pliable and of high tensile strength, they form fine, dense networks at the boundaries between epithelial tissue and connective tissue, between connective tissue and muscle, between connective tissue and capillary endothelia, and around several organs. They stain readily with silver salts (hence argyrophil). Reticulin (the main protein constituent of R.f.) is also present in collagen fibers, and like collagen it is not digested by trypsin.

Reticulated python, *Python reticulatus:* a snake of the subfamily *Pythoninae* in the family *Boidae* (see). With an authenticated maximal length of 9 m, it is the second largest boid snake after the anaconda. It is marked with a dark brown reticulate pattern on a light brown background. The R.p. is found throughout southern Asia and in large areas of Wallacea. It does not usually climb, but is a good swimmer, and is also found in human settlements. Rats, other rodents, domestic mammals and poultry are eaten. The female coils herself around a heap of eggs (up to 100 are laid); the temperature between the coils is up to 7 °C higher than ambient, and the process is one of true incubation with thermoregulation. The 50–70 cm-long young hatch after 3 months. The skin is used in the manufacture of small leather goods, and in certain areas the R.p. and other boid snakes have decreased in numbers due to commercial hunting.

Reticuloendothelial system: a term coined by L. Aschoff for a complex of different cells, which in addition to performing metabolic reactions also play a role in the defense of the human body. Cells of the R.s. are found in the liver, bone marrow, lymph, connective tissue, spleen and brain. They attack, destroy and remove foreign bodies (e.g. bacteria) from the body. The R.s. also produces antibodies, and in the liver it is involved in the synthesis of bile pigments and storage of vitamins. Some reactions or iron metabolism also occur in the R.s. See Macrophages.

Reticulum (of ruminant digestive system): see Digestive system.

Retina: see Light sensory organs.

Retinaculum: see *Collembola,* Moths.

Retinula: see Light sensory organs.

Retinular cells: see Light sensory organs.

Retrices: see Feathers.

Retrocerebral complex: in insects, a structure of ectodermal origin, closely associated with the brain as part of the protocephalic pathway, and consisting of 2 parts: the Corpora cardiaca (see) and the Corpora allata (see), with contributions from the cerebral ganglia. The R.c., which forms an open or closed ring of nervous tissue, extends anteriorly above the optic lobes of the brain, surrounding the aorta on all sides.

Retroinhibition: see Metabolic regulation.

Retrotransposons: see see Transposon.

Réunion solitaire: see *Raphidae.*

Reverberation of neuronal signals: see Neuronal circuit, Memory.

Reward and punishment system: see Self stimulation.

Reynolds number (R_e), ***Damköhler number V (DaV):*** a dimensionless parameter of fluid dynamics, used in studying the flow of fluids through pipes, and also important in the design of model flying and swimming systems, in which flow patterns are influenced by the viscosity of the medium. $R_e = v \times l/V$, where v = velocity, l = the geometric size, V = kinematic viscosity.

In swimming and flying, R_e roughly expresses the relationship between resistance and velocity. For values of $R_e < 1$, resistance is due almost entirely to the viscosity of the medium. For very large values of R_e ($> 10^3$), resistance due to the inertia of the medium must be taken into account. Resistance increases with the square of the velocity, and it has two components, i.e. friction and pressure. Friction depends on the total surface area, and is due to the viscosity of the medium. Pressure effects depend mainly on the shape (contors) of the swimming or flying object. The R_e-values involved in swimming cover a range of 13 orders of magnitude. Thus, flagellates swim with an R_e of about 10^{-5}, whereas large whales have an R_e of about 10^8. Flow characteristics therefore change with the size of the organism, and the resulting evolutionary adaptations for swimming are totally different.

Rhabdiasida: see *Nematoda.*

Rhabdite: see *Turbellaria.*

Rhabditida: see *Nematoda.*
Rhabdocoela: see *Turbellaria.*
Rhabdome: see Light sensory organs.
Rhabdomere: see Light sensory organs.
Rhabdornis: see *Sittidae.*

Rhabdosome (Greek *rhabdos* rod, *soma* body): a colony of Graptolites (see), or the chitinous exoskeleton of a colony of graptolites.

Rhacophoridae, ***rhacophorids, Old World tree frogs:*** a family of the *Anura* (see) containing 186 living species in 10 genera, distributed in Africa, Madagascar and especially in South and East Asia. Members are well adapted to an arboreal existence, possessing large adhesive disks on the fingers and toes, whose flexibility is increased by the presence of short, cartilaginous (ossified in some taxa) intercalary elements between the penultimate and terminal phalanges. Color is very variable, the predominant tones being green and brown. Many species build foam nests, usually in the branches above water. The female secretes a liquid, which she beats into a ball of foam with her hindfeet. Eggs are then laid in this foam and fertilized by the male as he sits on the female's back. When the larvae are ready to hatch, the foam liquifies and the tadpoles fall into the water. The genera that make foam nests are *Chiromantis, Buergeria, Chirixalus, Polypedates* and *Rhacophorus*. Some species of *Nyctixalus, Philautus* and *Theloderma* lay eggs in tree holes and develop via an abbreviated, nonfeeding larval stage. *Aglyptodactylus* and *Boophis* have small aquatic eggs and tadpoles.

There are 8 holochordal, procoelous presacral vertebrae with nonimbricate (nonoverlapping) neural arches. Presacrals I and II are not fused, and the atlantal cotyles of presacral I are widely separated. The coccyx lacks transverse processes, and has a bicondylar articulation with the sacrum. In some taxa, presacral VIII is biconcave and the sacrum biconvex. Omosternum and sternum are ossified and the prectoral girdle is firmisternal. Maxillary and premaxillary teeth are present. The pupils are horizontal.

Chiromantis is a genus of tropical African tree frogs. Both the male and the female of *Chiromantis xerampelina* ***(Gray tree frog)*** participate in beating the female exudate into a foam. The foam nest is kept moist by the female, who occasionally adds water from the pool. Tadpoles hatch after 3–4 days and fall into the water.

The African *Leptopelis* ***(Bush frogs)*** have rudimentary adhesive pads, and live mostly on the ground. The female produces large, yolk-rich eggs, which she buries in the earth near the water; the newly hatched tadpoles must wriggle their way to the water.

Rhacophorus reinwardti ***(Flying frog)*** is a jungle dwelling, 6–8 cm-long frog of Southeast Asia, with large webs on hands and feet. These webs act as parachutes and permit the frog to glide downwards over a distance of up to 15 m. There is also a flap of skin on each heel, a fringe of skin on each forearm, a projecting fringe on the inner side of the fifth toe, and a transverse flap above the vent. The adult is grass-green above, and yellow below. Young animals also have large blue patches on the webs of the hands and feet and behind the armpits. Both partners participate in beating the female exudate into a foam. The nest, which is placed between leaves, contains 60–100 eggs. A smaller species of flying frog *(R. pardalis)* inhabits the jungles of Borneo and the Philippines.

Rhacophorus schlegeli is an East Asian species which builds a foam nest like that of *R. reinwardti*, but places it in a hole in the ground near the water's edge. Larvae develop into adults within the nest, without passing through an aquatic stage.

Rhacophorus leucomystax ***(Banana frog, Malayan house frog)*** is a common, 6 cm-long, brownish species, which occurs from the eastern Himalayas to southern China, and from the Malay peninsula to Java and Borneo and the Philippines, often being found near human settlements and even in cities. Like *R. reinwardti,* it builds foam nests over open water, including flooded rice fields, or nests may be placed in the mud around buffalo-wallows.

In some classifications, the closely related *Hyperoliidae* (see) are placed in the *Rhacophoridae.*

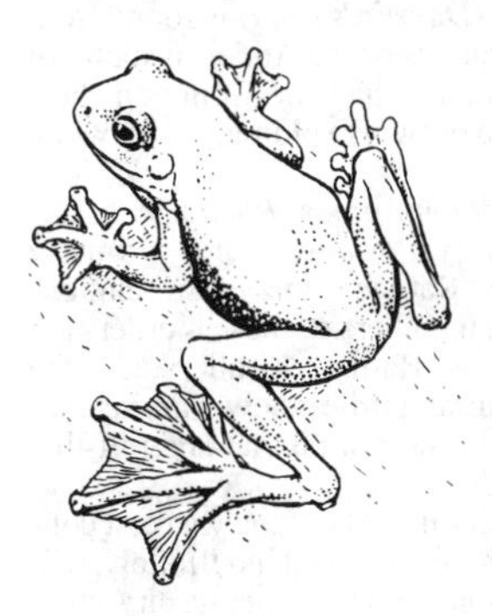

Rhacophoridae. Flying frog *(Rhacophorus reinwardtii).*

Rhaetian stage: see Triassic.

Rhamnose: 6-deoxy-L-mannose, a deoxyhexose present in many glycosides (e.g. anthocyanins) and in plant polysaccharides. It is biosynthesized from glucose.

Rhamphorhynchus (Greek *rhamphos* beak, *rhynchos* snout): a small primitive genus of pterosaurs. The long tail carried a lobe of skin which served for steerage and stabilization. The jaws contained pointed teeth, and the hind limbs were more weakly developed than the forelimbs. The forelimbs supported a stretched flight membrane (maximal span 175 cm), which probably permitted progression by fluttering and scrambling. Fossils are found in the Malm (Upper Jurassic) of Europe, East Africa and India.

Rheas: see *Rheiformes.*
Rhebuck: see *Reduncinae.*

Rheiformes, ***rheas:*** a South American order of large, long-necked, long-legged, flightless, running birds, sometimes called "American ostriches". Two species are known. *Rhea americana* ***(Common rhea),*** about 152 cm to the top of the head and weighing 20–25 kg, is found from Northeast Brazil to Central Argentina. The plumage is gray-brown and inconspicuous; white individuals also occur. Males are slighly taller, and in the male the base of the neck is black, but sexual dimorphism is otherwise not pronounced. Like ostriches, they possess powerful legs, and they are fast and agile runners. Each foot has 3 well developed toes. The wings are relatively large for a flightless ratite, and they cover the upper part of the rump like a cloak when at rest. The diet consists of leaves, roots, seeds, insects, and small vertebrates. Rheas congregate in flocks of 20–30, sometimes mixing with herds of the deer, *Dorcelaphus bezoarticus,* or with grazing cattle. In the breeding season, these large flocks break up into smaller groups consisting of a single male with 3–8 females. The cock prepares a shallow hollow with his bill or chooses an existing depression, which he lines with dry vegetation. He leads the hens to this "nest", and all lay at least one egg (golden when newly laid, rapidly becoming off-white), returning every 2–3 days to lay another. When the nest contains between 10 and 20 eggs, the cock begins to incubate them, a process lasting 35–40 days. The hens take no further part in the breeding and rearing process, but they may return, when the cock is temporarily absent from the nest, to lay more eggs. As many as 30 eggs may therefore be incubated. The newly hatched birds are led from the nest by the cock.

Pterocnemia pennata ***(Darwin's rhea)*** is found from Patagonia to the high plateaus of the Andes in southern Peru. It is somewhat smaller than the Common rhea, lays green eggs, and has a brownish plumage with white spots.

Members of the *Struthioniformes, Rheiformes, Casuariiformes* and *Apterygiformes* were formerly grouped together as the "Ratitae"; this close relationship is now doubted, but members of these 4 orders are still collectively known as Ratites. In all ratites, the wing skeleton and pectoral girdle show varying degress of reduction, and the sternum is flat and raft-like with no keel. See *Avies.*

Rheobase: the intensity in volts of the weakest constant direct current which, if continued indefinitely, will eventually cause excitation, i.e. will generate an Action potential (see).

Rheotaxis: taxic alignment of the body with the direction of water flow.

Rheotropism: bending movements of plant parts with respect to the direction of water flow. R. is displayed in particular by some plant roots. See Tropism.

Rhesus monkey: see Macaques.

Rhexigenous: arising by the tearing of cells. For example, hollow regions are often formed inside plant stems by the tearing apart of parenchymatous pith cells due to unequally distributed growth (rhexigenous formation of hollow pith).

Rhinatrematidae: see *Apoda.*

Rhincodontidae: see *Selachimorpha.*

Rhinencephalon: see Brain.

Rhinoceros: see *Rhinocerotidae.*

Rhinoceros beetle: see *Lamellicorn beetles.*

Rhinoceros iguana: see West Indian rock iguanas.

Rhinoceros viper: see Puff adders.

Rhinocerotidae, ***rhinoceroses:*** a family of odd-toed ungulates (see *Perissodactyla).* They are large, plump, three-toed mammals with one or two horns lying in tandem on the dorsal surface of the nose. These horns consist of compacted hair and they are essentially dermal structures, not associated with the bone beneath. The skin is thick and only sparsely haired. Usually solitary, they feed on twigs and grasses. A single calf is born after a gestation lasting 1.5 years, and suckling lasts for 2 years.

The ***Indian*** or ***One-horned rhinoceros*** *(Rhinoceros unicornis)* is a large native species of Nepal and northeast India. Formerly more widely distributed, the other one-horned species, *Rhinoceros sondaicus,* is now restricted to the tall grass and reed beds in swampy jungle areas of Java, where it is an endangered species.

The rare ***Asiatic two-horned rhinoceros*** *(Didermocerus sumatrensis)* occurs in dense hill forests, usually near water, in Assam, Burma, Thailand, Indochina, the Malay Peninsular, Sumatra and Borneo. Gray to black in color, it is the smallest living rhinoceros (head plus body 2.5–2.8 m; shoulder height 1.1–1.5 m; weight about 1 tonne).

Most numerous of the living species of rhinoceros is the ***African black rhinoceros*** *(Diceros bicornis).* Yellowish brown to dark brown in color, and possessing 2 horns, it is common locally in thorn bush country of eastern and southern Africa, also occurring in dense mountain forests in Kenya. Its upper lip is prehensile and protrudes slightly in the middle.

Largest of the rhinoceroses is the ***Square-lipped*** or ***White rhinoceros*** *(Cerathotherium simum)* (head plus body 3.6–5 m; shoulder height 1.6–2 m; weight 2.3–3.6 tonnes). Yellowish brown to slaty gray in color, its square lip is not prehensile and shows no trace of protrusion. "White" does not refer to color, but is derived from the Afrikaans "weit" meaning "wide" as applied to the lip. In South Africa it is now restricted to the Kruger National Park, and it is also found in southern Sudan and Uganda.

Fossil record. Fossil rhinoceros remains are found from the Upper Eocene to the present. They showed their greatest evolutionary diversity in the Early Tertiary. The European Pleistocene also contains remains of the ***Woolly rhinoceros*** *(Coelodonta antiquitatis),* a woolly-haired, 2-horned species.

Rhinocoel: see Ventricle.

Rhinocryptidae, ***tapaculos:*** an avian family (27 species in 12 genera) of the order *Passeriformes* (see), consisting of small, brown, insectivorous species found in the temperate and montane regions of Central and South America. They are characterized by an operculum or movable nostril cover. Tapaculos and Antpipits are the only passerines possessing 4 notches to the back edge of the sternum. The weak, rounded wings are little used. The birds have long legs and they scratch like farmyard fowl. Some species are called Gallito, from their habit of cocking the tail upward.

Rhinodermatidae, ***rhinodermatids, mouth-brooding frogs:*** a family of the *Anura* (see), containing 2 species in a single genus: *Rhinoderma rufum* and *R. darwinii,* found in humid forests in southern Argentina and Chile.

There are 8 holochordal, procoelous presacral vertebrae with imbricate (overlapping) neural arches. Presacrals I and II are fused, and the atlantal cotyles of presacral I are juxtaposed. The coccyx has no transverse processes, and it has a bicondylar articulation with the sacrum, which has broadly dilated diapophyses. The pectoral girdle is pseudofirmisternal, with a cartilaginous sternum and omosternum. Ribs are absent. The pupils are horizontal.

Both species are small (up to 30 mm in length), and have a trunk-like, fleshy extension of the snout. A relatively small number (up to 40) of large eggs is laid on land. Newly hatched tadpoles are picked up by the male. Tadpoles of *R. darwinii* complete their development in the vocal sac of the male, at the expense of the yolk reserves of the egg. The male of *R. rufum* carries tadpoles in his mouth to water, where they complete their development.

Rhinolophidae, ***horseshoe bats:*** a family of bats (see *Chiroptera)* inhabiting tropical and temperate regions of Africa and Asia and eastward to Australia, Japan, New Guinea and the Philippines. The family comprises 2 genera: *Rhinolophus* (about 50 species) and *Rhinomegalophus* (a single species known only from a type specimen from Tonkin and now in the Paris museum). Surrounding the nostrils is a complex expansion of skin known as the nose-leaf, which consists of 3 parts: a *horseshoe*-shaped lower part covering the upper lip, a pointed erect appendage above the nostrils *(lancet),* and a laterally flattened *sella* between the horseshoe and the lancet. They fly with the mouth closed and emit ultrasonic signals via the nostrils. All toes have 3 bones, except the first which has 2. Unlike most bats, which close their wings alongside the body, the R. wrap their wings around the body. Dental formula: 1/2, 1/1, 2/3, 3/3 = 32. See *Hipposeridae.*

Rhinolophus: see *Rhinolophidae.*

Rhinomegalophus: see *Rhinolophidae.*

Rhinophyrnidae, ***rhinophrynids:*** a family of the *Anura* (see), represented by a single, highly specialized, living species, *Rhinophrynus dorsalis* (the ***Mexican burrowing toad),*** inhabiting lowland areas from the Rio Grande valley of Texas to Costa Rica. In contrast to almost all other anurans, the tongue is free at the front, and protrusible in mammalian fashion. The adult is up to 7.5 cm in length, with a robust body, short limbs, and a minute head. The inner metatarsal tubercle is greatly enlarged and cornified. *Rhinophrynus* burrows rapidly into the soil by lateral movements of the feet. It spends most of the time underground and emerges only after heavy rain. The pools formed after rain serve for mating and egg laying. Although deposited in masses, the eggs subsequently float singly on the surface.

The arciferal pectoral girdle is quite distinctive, with a rudimentary omosternum and no sternum. Teeth are lacking. There are 8 ectochordal, modified opisthocoelous presacral vertebrae (the intervertebral disk adheres to the anterior end of the centrum, but does not fuse with it) with imbricate (overlapping) neural arches. Presacrals I and II are not fused, and the atlantal cotyles of presacral I are closely juxtaposed. Free ribs are absent. The coccyx lacks transverse processes, and it has a bicondylar articulation with the sacrum, which has expanded diapophyses.

Rhipidium: see Flower, section 5) Inflorescence.

Rhipidura: see *Muscicapinae.*

Rhipidurinae: see *Muscicapinae.*

Rhiphophore: a sensory organ of mollusks, which probably responds to chemical stimuli. Examples of Rs. are the hind antennae of opisthobranchs and an organ beneath the eye of *Nautilus.*

Rhithral: the summer-cold, stony/sandy zone of flowing water. Collectively, the organisms in this region are known as the *rhithron.* It corresponds essentially to the salmonid region.

Rhizobium, ***root nodule bacteria:*** a genus of rod-shaped, motile, Gram-negative soil bacteria, which form a symbiotic relationship with leguminous plants. As plant symbionts they fix atmospheric nitrogen, i.e. they reduce nitrogen to ammonium, which can be utilized by the plant for the synthesis of nitrogenous cell material. *R.* cells penetrate the root hairs of the plant, then migrate to deeper parts of the root cortex, where they multiply and become less differentiated (more globular and irregular in shape, and less easily stainable), and are known as Bacteroids (see). The resulting swelling of the root tissue produces a Root nodule (see). There are several species of *A.,* which differ in their specificity for the host leguminous plant, e.g. *R. leguminosum* is specific for pea plants, *R. lupin* for lupins, and *R. trifoli* for red and white clover.

Rhizocalin: a hypothetical root formation factor, which arises in aerial plant organs, and is thought to be involved in root development, together with auxin and probably other factors. It has not been chemically identified.

Rhizocarpon: see Lichens.

Rhizocephala: see *Cirripedia.*

Rhizoids:

1) Root-like thallus structures, usually formed as attachment organs, and also often serving for the uptake of water and nutrients. They occur in green, brown and other algae, mosses, and the independent prothalli of ferns.

2) Root-like branched hyphae which serve for attachment and/or water and nutrient uptake in certain fungi, e.g. *Rhizopus.* They are also called *appressoria* (sing. *appressorium).*

Rhizomorph: a very durable string-like structure formed from intertwined fungal hyphae. Rs. can attain lengths of several meters, and they are heavily cutinized on the outer surface. They serve for the transport of water and nutrients over long distances, and for the spread of the fungus. They are produced, inter alia, by the honey agaric, the dry rot fungus, and other fungi that infect trees and timber.

Rhizomyidae, ***bamboo rats:*** a family of Rodents (see) consisting of 18 species in 3 genera, native to Africa and the southern part of Eurasia. The skull is modified for a fossorial existence, and the ears are hidden in the pelage. Dental formula: 1/1, 0/0, 0/0, 3/3 = 16.

Rhizophora: see Mangrove swamp.

Rhizopoda: see *Rhizopodea.*

Rhizopodea, *Rhizopoda,* ***rhizopods:*** a class of protozoa, which feed and move by pseudopodia. The pseudopodia constantly change their shape, and may be lobopodia (broad and blunt) or filopodia (slender and tapering). *R.* are related to the *Mastigophorea.* The *R.* include the Gymnamebas (subclass *Gymnamoeba:* amebas lacking shells and with lobopods), *Testacea* (see), *Foraminiferida* (see), Heliozoans (see), Radioloarians (see) and Acantharians (see). The last 3 orders are also combined as the *Actinopoda,* which are separated from the *R.* in the narrow sense. Fossil remains are found from the Cambrian to the present. Many important geological formations consist of the remains of *R.,* in particular, the fossil calcareous and siliceous skeletons of foraminiferans and radioloarians.

Rhizopus: see *Phycomycetes.*

Rhizosphere: a zone of high biological activity on the surface of roots and in the immediately surrounding soil. Dead cells, parts of cells and excretory products of the roots serve as nutrients for microorganisms and saprotrophic animals (e.g. nematodes), so that the R. contains a much higher density of organisms than soil that is free from plant roots. As a secondary consequence of their close proximity, plant roots and inhabitants of the R. interact in a variety of ways. Other interactions occur between microorganisms, saprotrophic fauna and predatory fauna. The character of the R. is largely determined by material excreted by the roots. Organisms in the R. may be beneficial or nonbeneficial to root gowth and development. See Mycorrhiza.

Rhizostomae: see *Scyphozoa.*

Rhizothamnia: see Root nodules.

Rhodeus: see Bitterling.

Rhodobacteria, ***purple bacteria:*** photosynthetic bacteria, which are colored purple, red or brown by cellular pigments. Most R. are aquatic, and they assimilate carbon dioxide with the aid of light energy, under anaerobic conditions, without the release of oxygen. Unlike cyanobacteria and green plants, R. do not photolyse water, but obtain reducing equivalents from reduced sulfur compounds, molecular hydrogen, or simple organic compounds.

A distinction is drawn between the *Rhodospirillaceae* (formerly *Athiorhodaceae),* which do not use sulfur compounds, and the *Chromatiaceae* (purple sulfur bacteria; formerly *Thiorhodaceae),* which utilize sulfides and sulfur, and are therefore also classified as Sulfur bacteria (see).

Rhododendron: see *Ericaceae.*

Rhodopechys: see *Fringillidae.*

Rhodophyceae, ***red algae:*** a class of algae containing about 4 000 species, which are predominantly marine (maximal depth 200 m), with only a few freshwater species (e.g. *Batrachospermum).* The thallus is usually red to violet (rarely dark purple or reddish brown) due to the presence of phycoerythrin; some species also contain the blue pigment, phycocyanin, and all species also contain chlorophyll *a* (chlorophyll *b* is absent, but a small amount of chlorophyll *d* may also be present) and various carotenoids. With the exception of a small number of unicellular species in the group *Bangiodeae,* all other *R.* have multicellular thalli, but truly organized tissues are never present. The thallus is always attached to the substrate by filaments or a disk-shaped holdfast. Assimilation products are oily triacylglycerols and a carbohydrate (floridean starch), which is a 1,4-polyglycan structurally related to amylopectin. Asexual reproduction is always by nonmotile spores. Also in sexual reproduction, the gametes are nonmotile, and only oogamy occurs. The female gametangium is known as a carpogonium. Eggs are fertilized by male gametes (spermatia), which are transported passively by water movement. The zygote germinates within the carpogonium, and grows into a diploid carposporophyte. The sporogenic filaments of the latter produce vegetative spores (carpospores), most of which grow into diploid tetrasporophytes. Tetraspores are then formed by meiosis, and these haploid spores develop into haploid gametophytes (Fig. 1). Thus the alternation of generations and mode of reproduction are characterized by the insertion of a third generation, which is never found in other algae. Fossil *R.* are known from the Jurassic to the Cretaceous. The evolutionary origin and phylogenic position of this totally isolated algal group are obscure. In view of their similar pigments, the *R.* may have originated from the *Cyanophyceae;* alternatively, they may be close relatives of the red-colored *Cryptophyceae* (see *Flagellatae).*

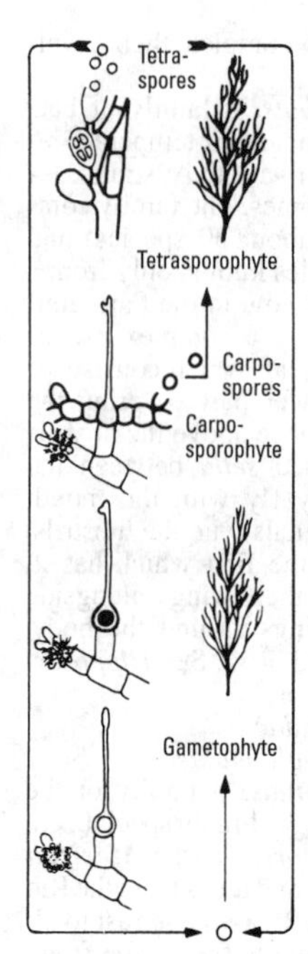

Fig.1. *Rhodophyceae.* Alternation of generations and nuclear phases of a red alga. Haploid phase: thin lines. Diploid phase: thick lines. *R!* Meiosis. (After Harder).

The *R.* are divided into several orders. Only a few species of interest are mentioned here. Various species of *Gelidium* and *Gracilaria* yield the widely used gelling agent, agar-agar. Both *Chondris crispus* and *Gigartina mamillosa,* when dried, yield carrageen or Irish moss, which has demulcent properties and is official in several national pharmacopeas; it is also used for various technical purposes, such as calico dressing, and as a cheap substitute for gum arabic.

The frogspawn alga, *Batrachospermum* (Fig. 2), occurs widely in unpolluted fresh water. Its thallus, which consists of aggregations of branched filaments positioned regularly along a central filamentous stem, is enveloped in mucilage, resembling frogspawn.

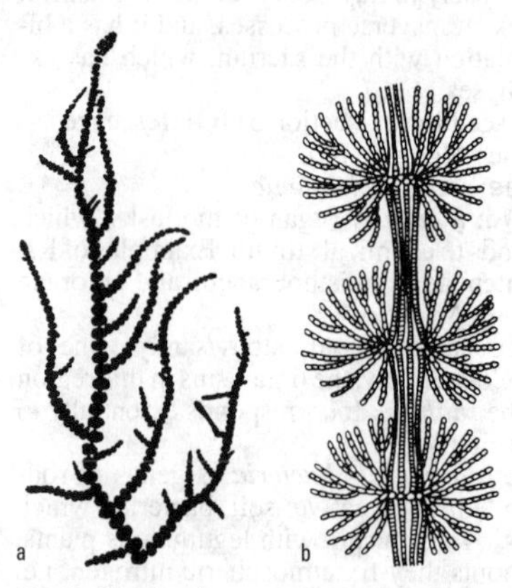

Fig.2. *Rhodophyceae. Batrachospermum moniliforme. a* Habit (natural size). *b* Enlarged section.

Rhodopsin, ***visual purple:*** a red, labile chromoprotein, consisting of the carotenoid pigment, 11-*cis*-retinal, and the protein, opsin. The retinal is bound covalently to the opsin by the formation of a Schiff's base. R. is present in the retina, where it functions in the visual process, and it is particularly important in twilight vision. In the primary event of visual excitation, light isomerizes the 11-*cis*-retinal to all-*trans*-retinal. Since all-*trans*-retinal does not fit the binding site for 11-*cis*-retinal, the R. molecule becomes unstable and undergoes a series of conformational changes, followed by hydrolysis of the Schiff's base linkage, with the release of all-*trans*-retinal. The resulting mixture of all-*trans*-retinal and opsin is yellow in color, and was formerly known as "visual yellow". In the recovery process, all-*trans*-retinal is isomerized to 11-*cis*-retinal (directly or via retinol), which recombines with opsin.

The rods of the retina, which are responsible for black and white vision at low light intensities, contain R. Color vision in the cones is mediated by closely related *iodopsins* (three in the normal human eye), which contain 11-*cis*-retinal but different opsins.

Rhodospirillum: a genus of aquatic, purple nonsulfur bacteria. The cells are spiral in shape, red to brown in color, and equipped with polar flagella. In the light they live autotrophically by photosynthesis. In the dark and in the presence of oxygen they live heterotrophically.

Rhodoxanthin: 3,3′-dioxo-β-carotene, a xanthophyll present in brown-red ("copper") leaves, in yew seeds and in some bird feathers.

Rhombencephalon: see Brain.

Rhombifera: see *Cystoidea.*

Rhopalium: a variously shaped sensory organ in the bell rim of many medusae. It is an organ of balance, and it is frequently associated with light sensory organs.

Rhopalocera: see Butterflies, *Lepidoptera.*

Rhubarb: see *Polygonaceae.*

Rhus: see *Anacardiaceae.*

Rhynchocoel: see *Nemertini.*

Rhynchocoela: see *Nemertini.*

Rhynchotragus: see *Neotraginae.*

Rhynchortyx: see *Phasianidae.*

Rhynchocephalia: next to the turtles, the most ancient of the surviving orders of reptiles. Rhyncocephalians are found as early as the Jurassic, where they are represented by the genus *Homoeosaurus,* and they became widespread and numerous in the Mesozoic. There are 2 complete temporal fossae, so that the skull has a two bony arches, as in the crocodilians. The teeth are fused with one another to form chewing surfaces. The pineal body is sensitive to light, and serves as a third visual organ. They are the only reptiles lacking a male copulatory organ, and the cloaca is transverse to the line of the body. The order is now represented by a single surviving species, the ***Tuatara*** *(Sphenodon punctatus),* the only member in the family *Sphenodontidae.* The tuatara has hardly changed for 200 million years. It somewhat resembles an iguana in appearance, with a high head and a crest of elongated, movable horny plates along the neck and back. The 3 subspecies are found only on some small islands in the Cook Strait and off the northern coast of the North Island of New Zealand. The tuatara died out more than 100 years ago on the North and South Islands of New Zealand. It lives in self-excavated earth cavities, or in the nesting holes of petrels. It is active at night (the eyes have vertical pupils), when it feeds on large insects, worms, snails, birds and birds' eggs. Whereas all other reptiles reach peak activity at a body temperature of 25–38 °C, the optimal temperature for the tuatara is 12 °C. In the spring, the female lays between 5 and 15 parchment-shelled eggs in a hole in the ground; after covering them with earth, she then pays them no further attention. The eggs hatch after 12–15 months, representing the longest known embryonic period in reptiles; embryonic development probably passes through a dormant period in the winter. Despite their numerous primitive anatomical characteristics, the R. were not direct antecedents of the recent *Squamata* (see). The tuatara is a great rarity in museums and zoos, and it is strictly protected in its natural habitat.

Rhynia: see *Psilophytatae.*

Rhyniaceae: see *Psilophytatae.*

Rhynochetidae: see *Gruiformes.*

Rhynochetos: see *Gruiformes.*

Rhythmic activity: see Biorhythm.

Rhythmic behavior: see Biorhythm.

Ribbon community, ***ribbon biocoenosis:*** see Biocoenosis

Ribbon worms: see *Nemertini.*

Ribes: see *Grossulariaceae.*

Riboflavin adenosine diphosphate: see Flavin adenine dinucleotide.

Riboflavin 5′-phosphate: see Flavin mononucleotide.

Ribonuclease: a pancreatic phosphodiesterase specific for the phosphodiester bonds of ribonucleic acid (RNA). The cleavage products are 3′-ribonucleoside monophosphates and oligonucleotides with a terminal pyrimidine nucleoside 3′-phosphate. Although R. is present in most vertebrates, high activities are only found in the pancreas of ruminants, certain rodents and some herbivorous marsupials. *Ribonuclease A* from bovine pancrease is a single chain, basic protein (124 amino acid residues, M_r 13 700) of known primary, secondary and tertiary structure. The molecular mechanism of action of R.A is also known. The pH optimum is 7.0–7.5 and the isoelctric point is 7.8. Other Rs. have been isolated from fungi and bacteria.

Ribonucleic acid, ***RNA:*** a biopolymer of ribonucleotide units, present in all living cells and some viruses. The mononucleotides of RNA consist of ribose phosphorylated at C3, and linked by an N-glycosidic bond to one of four bases: adenine, guanine, cytosine or uracil. Many other bases (chiefly methylated bases) also occur, but are less common. The mononucleotides form a linear chain via 3′,5′-phosphodiester bonds. RNA does not form a double-stranded α-helix (cf. DNA). By hydrogen bonding between complementary bases, the single chains show partial folding into α-helical regions, which are separated by regions of single stranded, unordered RNA. There are three main types of RNA, classified on the basis of their function (but also displaying marked differences in secondary and tertiary structure): messenger RNA (mRNA), Transfer RNA (tRNA) (see) and ribosomal RNA (rRNA) (see Ribosomes). In eukaryotic cells, RNA is present in the nucleus, cytoplasm and cytoplasmic organelles (ribosomes, mitochondria, chloroplasts).

RNA, including cytoplasmic RNA, is synthesized mainly in the cell nucleus. Besides tRNA, the cytoplasm contains rRNA (in the ribosomes) and mRNA (in polysomes). Mitochondria and plastids also contain mRNA, tRNA and rRNA, which are, however, transcribed from the DNA of the organelles.

RNA is crucially important in all living cells: mRNA carries genetic information from DNA to sites of protein synthesis; tRNA carries amino acids to their correct site of incorporation during the translation of mRNA into protein; and rRNA is a structural and functional component of ribosomes. See Protein biosynthesis.

D-Ribose: a monosaccharide pentose. It is a component of RNA, various nucleosides and their phosphates (e.g. ATP), some coenzymes (e.g. NAD), vitamin B_{12}, ribose phosphate and various glycosides. It is not fermented by yeasts.

HOCH$_2$ OH, O, H H, H H, OH OH

β-*D-Ribose*

Ribosomes: cell organelles, consisting of ribonucleic acids and proteins, and visible only in the electron microscope. R. are the sites of Protein biosynthesis (see). The number of R. in a cell depends on the functional state of the cell, the cell type and its growth phase, and it is directly correlated with the capacity of the cell for protein biosynthesis. A rat liver cell under normal conditions contains about 6 x 10^8 R. R. are normally present in the cytoplasm as polysomes (several to many R. attached along the length of a strand of mRNA) (see plate 21). There two main types of R., according to their size and origin (Table): 80S-R. from the cytoplasm of eukaryotic cells; and 70S-R. from prokaryotic cells, plastids and some mitochondria. The mitochondria of vertebrates contain 55S-R.

Comparison of some properties of 70S (prokaryotic) and 80S (eukaryotic) ribosomes.

	70S (*Escherichia coli*)	80S (mammals)
M_r (x 10^6)	2.7	4.0
S-values of subunits	50 + 30	60 + 40
% RNA	65	50
S-values of high M_r rRNA (x 10^6)	23 + 16	28 + 18
M_r of high M_r rRNA	1.1 + 0.56	1.7 + 0.7
GC content of high M_r rRNA (%)	54 +54	67 + 59
Number of ribosomal proteins	34 + 21	about 70
Initiation of protein biosynthesis by	formyl-Met-tRNA$_F$	Met-tRNA$_{Met}$
Inhibition of protein biosynthesis by:		
chloramphenicol	+	–
cycloheximide	–	+

The subunit composition of R. depends essentially on the ionic composition of of the suspension medium, especially the Mg^{2+} concentration. If this less than 0.001 M, the R. dissociate into two morphologically and functionally dissimilar subunits. Thus 70S-R. consist of a large 50S and a small 30S subunit, whereas 80S-R. consist of 60S and 40S subunits. In the complete absence of Mg^{2+} (or in the presence of about 1 M monovalent cations, such as Li, Cs or K), the subunits dissociate into still smaller, discrete ribonucleoprotein particles, known as core particles; at the same time, certain proteins (called split proteins) are removed. Core and split proteins are inactive in protein biosynthesis, but they can still reassociate into functionally competent ribosomal subunits.

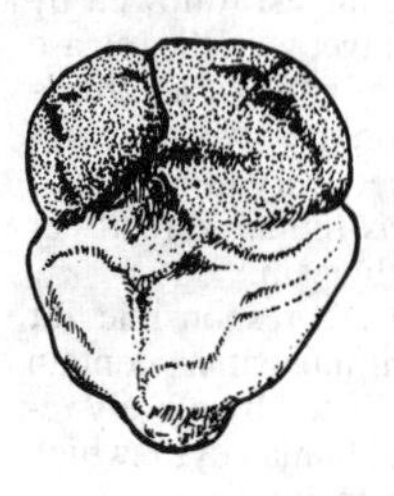

Fig.1. *Ribosomes.* Model of an animal (80S) ribosome.

The three-dimensional structure of R. (Fig.1) has been studied by electron microscopy (whole R., subunits, and subunits cross-linked by antibodies against individual ribosomal proteins); by low angle diffraction of X-rays and neurons; and by Fourier reconstruction of electron micrographs of crystallized subunits.

R. structure has been highly conserved during evolution, and it is similar for the most distantly related organisms. Nevertheless, electron microscopy reveals small but distinct differences in the shapes of R. from eubacteria, archebacteria, eocytes and eukaryotes, which have been used to interpret phylogenetic relationships between these groups.

In the rough endoplasmic reticulum, the large ribosomal subunit is associated with the membrane (Fig.2). Binding to the membrane occurs when the protein synthesized on the R. is for export (secretion) from the cell; such proteins are normally released by exocytosis (e.g. release of hormones from the exocrine pancreas).

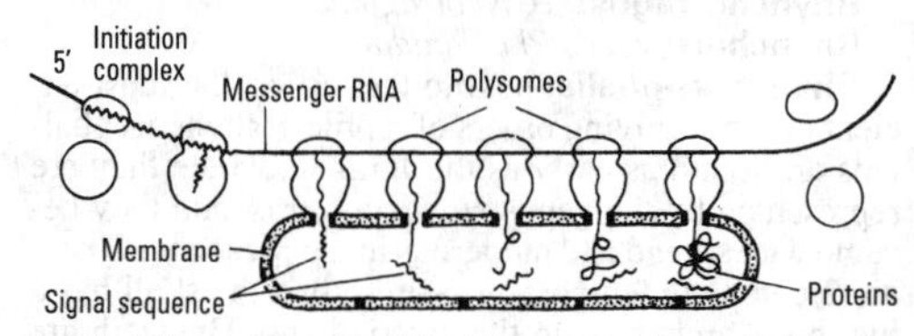

Fig.2. *Ribosomes.* Model of the signal hypothesis. As proteins are synthesized they pass through the membrane and into the lumen of the endoplasmic reticulum.

Translation begins with the formation of a protein signal sequence or leader (consisting of 15–30 amino acid residues); at this point the translation of secretory proteins is arrested until the polypeptide chain binds to a soluble signal recognition particle (SRP). The complex of SRP and nascent peptide subsequently binds to a "docking protein" in the membrane. The SRP is released, translation of the polypeptide resumes, and the leader sequence penetrates and forms a channel in the membrane. As synthesis proceeds, the polypeptide chain passes through a groove in the large ribosomal subunit and into the lumen of the endoplasmic reticulum. The signal sequence is removed hydrolytically by a specific enzyme inside the endoplasmic reticulum. In contrast, most undifferentiated embryonal cells contain a high proportion of free polysomes (i.e. not bound to the endoplasmic reticulum), which synthesize proteins for cell growth and development.

D-Ribulose: a monosaccharide pentulose. Ribulose 1,5-diphosphate is the CO_2 acceptor in the dark reaction of photosynthesis. Ribulose 5-phosphate is an intermediate in the pentose phosphate cycle.

Ribwort: see *Plantaginaceae.*

Riccia: see *Hepaticae.*

Rice: see *Graminae.*

Rice frog: see *Raninae.*

Rice rats: see *Cricetinae.*

Richmondena: see *Fringillidae.*

Richmond River pine: see *Araucariaceae.*

Ricinoleic acid, *12-hydroxyoleic acid:* $CH_3—(CH_2)_5—CHOH—CH_2—CH = CH—(CH_2)_7—COOH$, a fatty acid present in the acylglycerols of castor oil, where it accounts for 80–85% of the total fatty acids. It is also present in maize oil, wheat oil and various other vegetable oils.

Ricinulei: an order of small (5–10 mm long), slow-moving arachnids, represented by 45 species in tropical Africa and America. They have a thick and often sculptured cuticle, and they differ from all other arachnids in their possession of a hood on the anterior rim of the prosoma, which can be lowered over the mouthparts and chelicerae.

Ricinus: see *Euphorbiaceae.*

Ricinus oil: see Castor oil.

Rickettsiae, *Rickettsias* (after H.T. Ricketts, who died of typhus while studying its cause): very small (0.2–2 µm; intermediate in size between viruses and bacteria), nonmotile, Gram-negative organisms, which only divide and multiply in live host cells. They usually live as parasites or symbionts of arthropods, and cause disease in vertebrates after transfer by lice, fleas, ticks, etc. *Rickettsia prowazekii* is the causative agent of European epidemic typhus fever of humans; its intermediate animal host is the rat and it is transmitted by body lice. *Coxiella (= Rickettsia) burneti* causes Q fever in humans in USA and Europe; there is probably a reservoir of the disease in cattle, and it is transmitted by ticks. Other human rickettsial diseases are Endemic typhus *(R. mooseri;* USA, Europe; carried by rats; transmitted by body lice); Rocky mountain spotted fever *(R. rickettsia;* USA, Brazil; carried by sheep, rodents, dogs; transmitted by ticks); Tsutsugamushi *(R. orientalis;* Japan; carried by field mice; transmitted by akamushi mites); Rickettsial pox *(R.akari;* New York City; carried by house mice; transmitted by mites)

R. are variously shaped, like minute bacilli, cocci, or diplococci. They sometimes form relatively long filaments. Spores are not formed. They are difficult to stain with ordinary basic dyes, but are readily colored with Giemsa's stain. With the exception of *Rickettsia burneti,* they are nonfiltrable like bacteria. *R.* are cultured for investigative purposes in the live embryos of hens' eggs and in tissue cultures.

Rickettsial pox: see *Rickettsiae.*

Ricochetal locomotion: see Movement.

Riella: see *Hepaticae.*

Rifamycins: a group of antibiotics (including e.g. rifampicin and rifamycin SV) produced by strains of *Streptomyces mediterranei* and *Nocardia.* They inhibit DNA-dependent RNA synthesis in prokaryotes, chloroplasts and mitochondria, but not in the nuclei of eukaryotes. Inhibition is due to formation of a stable complex between RNA polymerase and the antibiotic; DNA still binds to the enzyme, but incorporation of the first purine nucleotide into RNA is prevented. R. are active against mycobacteria, and are used in tuberculosis therapy.

Rifleman: see *Xenicidae.*

Right whales: see Whales.

Ringed seal: see *Phocidae.*

Ring gland: see Corpora allata, Corpora cardiaca.

Ring ouzel: see *Turdinae.*

Ringspot viruses: see Virus groups.

Ringtail, *ring-tailed cat,* *Bassariscus astutus:* a monotypic genus of the *Procyonidae* (see), which occurs in North and Central America. It is omnivorous and nocturnal, feeding on plants, mammals, birds and invertebrates. The tail is banded black and white for its entire length. Although a good climber, the Ringtail it is not especially arboreal in habit.

Ring-tailed cat: see Ringtail.

Ripening of fruit: see Fruit ripening.

Ripple bugs: see *Heteroptera.*

Rissa: see *Laridae.*

Ritualization: coversion of maintenance activity into Signal behavior (see). For example, in sparrows, the breaking up of food or the preparation of nest cavities involves drumming with the beak, which has evolved into a form of signaling. Similarly, activities such as self-grooming, feeding, defense and protection, nest building, etc. may, in the course of phylogenetic evolution, become ritualized and acquire a signaling function. A ritualized activity chacteristically has a typical intensity. Thus, when a fox is involved in the maintenance activity of earth excavation, the duration and intensity of the activity are more or less governed by the prevailing conditions. If, however, the forepaws are used in a scratching display against a rival, the activity is a form of signaling, and is largely stereotyped. Such behavior patterns may also become "emancipated", i.e. they become separated from the original motivation, achieve independence, and are incorporated into a different motivation system, often into sexual motivation as part of courtship behavior. About 20 different types of behavioral changes are known that are due to R.

Ritualization. Stooping posture with tail fanning in the killdeer *(Charadrius vociferus).* This threat posture is derived by the ritualization of elements of nest building activity.

River blindness: see Onchoceriasis.

River eels, ***freshwater eels,*** *Anguillidae:* a family of the *Anguilliformes* (see Eels). Members of this family live partly in the sea and partly in fresh water, and are widely distributed throughout the world. Females are up to 1 m in length; males are smaller. Unlike other eels, the R.e. possess rudimentary scales, which are arranged irregularly, and are relatively well embedded in the upper layer of the corium below the epidermis.

Two ecotypes of the European River eel are known: the broad headed eel, which eats mainly fish, and the narrow headed eel, which lives predominantly on insects and worms. After living 9–12 years in fresh water, mature R.e. migrate to the sea in the autumn. During this migration, the eyes become larger, the ventral surface takes on a silvery sheen, and the stomach atrophies. These silver or bronze eels travel to the Sargasso sea, where they spawn in deep water, then die. The delicate, transparent, willow leaf-shaped larvae (leptocephali) swim near the surface, feeding and growing. These develop into unpigmented glass eels, and in the course of 3 years they arrive on the coasts of Europe in the Gulf stream. With the onset of pigmentation, they migrate up-river and are known as elvers. The elver develops into a fully pigmented form called a yellow eel. The latter term is rather inappropriate, because pigmentation of the dorsal surface may vary from dark green to black. R.e. are exploited commercially. In many areas the glass eel is caught directly, or cultured and stored live in eel ponds. Eels are caught with eel baskets, stow nets, eel traps, and by bottom fishing with trawls or baited hooks.

River hog: see African bush pig.

River jack: see Puff adders.

River otters, *Lutra:* a genus of about 12 species of mustelids (see *Mustelidae),* inhabiting streams, lakes and coastal waters over most of North and South America, Africa, Europe and Asia. The upperparts are brownish and the underparts are paler, while the lower jaw and throat may be whitish. As an adaptation to an amphibious existence, the trunk is cylindrical with a very flexible, muscular tail, which tapers from a thick base, the toes of both the forefeet and hindfeet are webbed, the head is flattened and rounded, the neck short and wide, and the small ears and nose can be closed under water. They feed on crayfish, frogs, fish and other aquatic animlas, but also take rodents, birds and rabbits on land. *Lutra lutra* is found in Europe, Asia and North Africa, while *Lutra canadensis* ***(Canadian otter)*** is native to northern North America.

River pea shell: see Freshwater mussels.

River snails: see *Gastropoda.*

Rivularia: see *Hormogonales.*

Roadrunners: see *Cuculiformes.*

Roan antelope: see *Hippotraginae.*

Roatelos: see *Gruiformes.*

Robber crab: see Coconut crab.

Robber frog: see *Leptodactylidae.*

Robin-chats: see *Turdinae.*

Robinia: see *Papilionaceae.*

Robins: see *Turdinae.*

Robin snipe: see *Scolopacidae.*

Robin's pincushion: see Galls.

Roborovsky's dwarf hamster: see *Cricetinae.*

Roccella: see Lichens, Lichen pigments.

Rock crab: see Cancer crab.

Rock crawlers: see *Notoptera.*

Rock dove: see *Columbiformes.*

Rockfishes: see *Scorpaenidae.*

Rockfowl: see *Timaliinae.*

Rock pigeon: see *Columbiformes.*

Rock python, *Python sebae:* a snake of the family *Boidae* (see), subfamily *Pythoninae.* Up to 7.5 m in length, the R.p. is the third largest boid snake. It is found in Africa south of the Sahara, mainly in savanna, where it preys on ground birds, pigs and small antelope. Attacks on humans have been recorded. 30–40 eggs are laid.

Rock thrushes: see *Turdinae.*

Rocky Mountain spotted fever: see *Rickettsiae.*

Rodenticide: a plant protection agent for controlling rodents, e.g. chlorphacinon, caphechlor, aluminium phosphide, zinc phosphide. Legal permission must be obtained for the use of the more potent agents.

Rod-shaped bacteria: see Bacillus.

Rock conies: see *Ochotonidae.*

Rock sparrows: see *Ploceidae.*

Rock wren: see *Xenicidae.*

Rocky Mountain goat, *Oreamnos americanus:* a medium-sized North American horned ungulate (see *Bovidae)* of the subfamily *Caprinae* (see). It is not a true goat, but a member of the tribe *Rupicaprini* (Goat-antelopes). A very agile climber and jumper, it inhabits steep slopes and cliffs in alpine and subalpine areas of the North American Rocky Mountains, where the snowfall is heavy and the temperatures persistently low. The pelage is white to yellowish white, with a thick, woolly underfur. Long, soft hair along the midline of the neck and shoulders forms a raised ridge. A beard is present. Although the horns are only about 25 cm in length, they are very sharp, and may be used very effectively; hunting dogs have been killed in self defense, and there is at least one authentic report of a grizzly bear suffering the same fate.

Rodents, *Rodentia* (plate 38): the most numerous order of mammals (more than half of all mammalian species are R.), representing a wide variety of forms and sizes. Canines and premolars are characteristically absent, leaving a large gap (diastema) (see Dentition). Two pairs of chisel-edged incisors are present (1 pair in upper jaw, 1 pair in lower jaw), with enamel only on their front surfaces. The incisors have no roots, and they grow continuously from persistent pulp. The masseter muscle and its skeletal attachments display adaptations associated with gnawing. The muscle is divided into lateral and internal parts, which may be further subdivided. The zygomatic process of the maxilla is often broadened into a plate, which acts as an attachment for the lateral masseter. The horizontal arm of the zygoma may serve as an attachment for the internal masseter, or part or all of this muscle may pass through the enlarged infraorbital aperture and attach to the side of the rostrum.

R. are predominantly herbivores, and they are ecologically important as converters of plant material into animal protein. The following families of R. are listed separately: *Abrocomidae* (Rat chinchillas), *Agoutidae* (Pacas), *Aplodontidae* (Mountain beavers), *Capromyidae* (Hutias), *Castoridae* (Beavers), *Caviidae* (Guinea pigs, Cavies and Patagonian "hares"), *Chinchillidae* (Viscachas and Chincillas), *Cricetidae* (Hamsters, Jerbils, Voles and Lemmings), *Ctenodactylidae* (Gundis),

Ctenomyidae (Tuco-tucos), *Dasyproctidae* (Agoutis), *Dipodidae* (Jerboas), *Echimyidae* (Spiny rats), *Erethizontidae* (New World porcupines), *Gliridae* (Dormice), *Hydrochoeridae* (Capybaras), *Hystricidae* (Old World porcupines), *Muridae* (Mice and Rats), *Myocastoridae* (Coypus or Nutrias), *Octodontidae* (Octodont rodents), *Pedetidae* (Spring hares), *Platacanthomyidae* (Spiny dormice), *Rhizomyidae* (Bamboo rats), *Sciuridae* (Squirrels), *Spalacidae* (Mole rats), *Zapodidae* (Jumping mice and Birchmice).

Rodriguez solitaire: see *Raphidae.*

Roe deer, *Capreolus capreolus:* a monotypic genus of small, graceful deer (head plus body 95–135cm, shoulder height 65–76cm), native to Europe and Asia. In summer the coat is uniformly brown to black-brown, with a gray face and white chin, and a black band from the angle of the mouth to the nostrils. In winter it is gray-brown, with white patches on the throat and a white rump patch. The small (23 cm long) branched antlers of the male usually have only 3 (sometimes 4) points. A visible tail is absent, but in winter the prominent anal tuft that develops on the female gives the appearance of a tail. Either solitary or in small groups, *Capreolus* inhabits open woodland, forest edges and open moorland. See Deer.

Roemeria: see *Papaveraceae.*

Rohrer index, ***corpulence index:*** a sensitive numerical expression for the relationship of human body weight (mass) to height. R.i. = (body mass x 100)/body height. The average value for women is somewhat greater than that for men.

Rollers: see *Coraciiformes.*

Rolling circle mechanism of DNA replication: see DNA replication.

Roman snail: see *Gastropoda.*

Romerolagus: see *Leporidae.*

Romneya: see *Papaveraceae.*

Rondomycin: see *Tetracyclins.*

Roob juniperi: see *Cupressaceae.*

Roofed turtles: see *Emydidae.*

Roof rat: see *Muridae.*

Rook: see *Corvidae.*

Root: the basal organ of pteridophytes and seed plants, which is usually situated in the earth. It anchors the shoot in the ground, takes up water and nutrients and conducts them to the shoot, and often stores reserve material. Most Rs. fulfil two or three of these functions simultaneously. Important characteristics of the R. are the complete absence of leaves, protection of the vegetative point by a R. cap, the radial arrangement of the vascular bundles in the central cylinder, the endogenous development of lateral Rs. (i.e. in the interior of the main R.), and possession of a zone of R. hairs.

The growing *primary R.* is divided into four successive regions: 1) the apical meristem, 2) the zone of extension growth, 3) a R. hair zone, and 4) the zone of R. branching, which lacks R. hairs. Rs. of dicotyledons have a fifth region, which is a zone of secondary growth. The *apical meristem* is blunt and conical; in pteridophytes it is formed from a single apical cell, whereas in seed plants it is derived from several *initials.* In this region, embryonal cells are produced constantly by cell division, so that the apical mersitem continually penetrates deeper into the earth. The delicate embryonic tissue is protected by the *R. cap* or *calyptra,* which consists of a cap of persistant parenchymatous cells. The middle lamellae of the outer cells of the calyptra become mucilaginous, so that they become detached, thereby facilitating movement of the R. tip between the soil particles. As calyptral cells are lost anteriorly, they are replaced by division of deeper cells. The cells of the calyptra usually contain large, freely mobile starch grains, which are thought to function in the geotropic orientation of the R. *(statolith starch).* The *zone of extension growth* starts at the base of the apical meristem, where embryonal cells differentiate into persistent cells by extension growth. It is usually only 5–10 mm in length, and it merges directly into the R. hair zone. R. hairs are thin-walled, tubular evaginations of the epidermis, and they increase many times the area of the R. available for water uptake. Since R. hairs remain viable for only a few days, they are normally present on only a short zone of the newly developed R. Many water and marsh plants lack R. hairs, and in this case the *rhizodermis* (an epidermal monolayer of thin-walled cells) serves for the uptake of water and nutrients. The rhizodermis lacks a cuticle and stomata, and is just as short-lived as R. hairs. In older parts of the R., it is replaced by the *exodermis,* a secondary boundary tissue, whose cells walls are suberized and covered by a cuticle, and which contains irregularly spaced, thin-walled, nonsuberized absorption cells.

All cells become differentiated in the R. hair zone to produce the primary internal root structure. This consists of the *R. cortex* and the *central cylinder.* The cortex consists of parenchymatous storage tissue, the innermost layer of which is differentiated into *endodermis,* which sharply delineates the cortex from the central cylinder. The radial walls of the layer of endodermal cells adjoining the central cylinder are impregnated with a suberin-like material, which greatly decreases their permeability and prevents apoplastic transport in this region (see Casparian strip). In older tissue, the inner walls of the epidermis usually become considerably thickened, and often lignified, leaving only a few passage cells unthickened and unlignified. Immediately beneath the endodermis lies the *pericycle* of the central cylinder; this consists of one or more layers (usually a single layer) of parenchyma cells, which later divide for the initiation of lateral Rs. and the synthesis of periderm (see Boundary tissue).

In the central cylinder, the tract of conducting tissue consists of a radially symmetrical vascular bundle, in which strands of xylem alternate with strands of phloem. According to the number of xylem strands present in the bundle, the R. is said to be monarch, diarch, triarch, tetrarch, etc., or polyarch.

Gymnosperm and angiosperm Rs. can increase their thickness by secondary growth in the region adjoining the R. hair zone. In this process, radially arranged parenchyma plates (separating the xylem and phloem in the vascular bundles) divide to produce strips of cambium which meet the pericycle on both sides of each xylem strand. This results in a continuous belt of secondary cambium, which is wavy or stellate in transverse section. By unequal thickening (particularly by active xylem production on the adaxial side of the phloem) the concavities of this wavy outline are evened out, so that the cambium becomes circular in transverse section, resembling that of the stem. As in the shoot, the R. cortex is ultimately ruptured as a result of dilatation; but before fissures actually form, a tertiary insulating layer, known

as *R. periderm,* is laid down by the growth and differentiation of cells deep in the periderm.
The swollen R. of *Beta* arises from the activity of numerous cambial layers, which form a system of concentric growth zones, each surrounding a ring of phloem and a ring of xylem.

The initials of *lateral Rs.* arise endogenously, i.e. within the R. In seed plants they arise in the periderm, whereas in pteridophytes they arise in the inner layers of the cortex. Although the developing lateral R. is weaker than its parent R., it must penetrate the entire cortex of the parent. Branching of lateral Rs. ultimately produces a richly branched *R. system* in the earth. This system may be centered around a dominant *tap R.,* whose main axis continues vertically downward into the earth, e.g. oak, pine, lucerne and others. Other plants lack a main central R. and their Rs. extend profusely in all directions at a relatively shallow depth, e.g. fig tree and grasses. Rs. formed on the shoot system (on stems or leaves) are called *adventitious Rs.* They arise normally from subterranean and creeping stems, and may also be produced in response to injury or treatment with hormones. In the strict sense, "adventitious" should be reserved for Rs. initiated by injury or hormone treatment, since the term means "arising at an unusual place or time".

Rs. display a variety of anatomical-morphological modifications (metamorphoses), representing adaptations to specialized functions. *R. tubers*, which are modified adventitious Rs., serve as storage organs. If the main R. becomes swollen as a storage organ, it is not called a tuber, but simply a swollen tap R. Other R. modifications are the slender, partition-like *buttress Rs.* of some tropical trees, the *respiratory Rs.* of tropical marsh plants and mangrove plants, the *climbing Rs.* of lianes, the *aerial Rs.* (see Absorption tissue) of orchids and other epiphytes, some of which contain chlorophyll and carry out photosynthesis, and are therefore also *assimilatory Rs.* The Rs. of many semi and totally parasitic plants are modified to Haustoria (see).

Root nematodes: see Eelworms.

Root nodule bacteria: see Bacteroids, Root nodule, *Rhizobium.*

Root nodules: variously shaped swellings on the roots of different plant species, which contain symbiotic bacteria. Bacteria in the nodule convert atmospheric nitrogen into a form that can be assimilated by the plant. In return, the bacteria are provided with a favorable and stable environment, and they receive nutrients from the plant. The R.n. of leguminous plants contain *Rhizobium* (see). Cells of *Rhizobium* aggregate in the rhizosphere of the leguminous plant, and the root hairs undergo marked bending, apparently under the influence of hormones secreted by the rhizobia. It is not clear whether bending of the root hair causes breakage of the cell wall, allowing entry of the rhizobia, or whether the cell wall actively invaginates. An infection thread extends from the point of entry in the root hair into the deeper parts of the root cortex. The wall of an infection thread is lined with cellulose, which is continuous with the cell wall of the root hair, i.e. the lumen of the infection thread is continuous with the exterior of the plant, and rhizobia within the thread are separated from the plant cell cytoplasm by the cellulose wall of the thread. On its way to the interior, the infection thread passes through several plant cell walls, and the cellulose of these becomes continuous with the cellulose of the infection thread. Rhizobia progress to the interior of the cortex by travelling along the infection thread, eventually reaching a point where cellulose is no longer synthesized. They are then released into a plant cortical cell. Even then, they do not lie free in the cell cytoplasm, but are enclosed in a membrane (the membrane envelope), which is derived from the plasmalemma of the cell. At this stage, the rhizobia multiply rapidly and change their form to become Bacteroids (see). A red pigment, known as *leghemoglobin* or *legoglobin* is formed inside the membrane envelope (not in the cell cytoplasm, and not in the bacteroids), i.e. the bacteroids are bathed in a solution of leghemoglobin. As a rough guide, the degree of pigmentation of the nodule interior is an index of nitrogen fixing activity. Legoglobin is structurally related to mammalian myoglobin, and it is totally absent from free-living *Rhizobium* and from uninfected leguminous plants. The structural genes for the globin of leghemoglobin are present in the genome of the plant, whereas the heme moiety of leghemoglobin is synthesized by the symbiotic *Rhizobium.* Leghemoglobin appears to function by facilitating the diffusion of oxygen to the bacteroids (since nitrogen fixation is strongly energy-dependent, the bacteroids must respire rapidly in order to synthesize sufficient ATP). R.n. of perennial legumes can continue to grow and remain functional for several years.

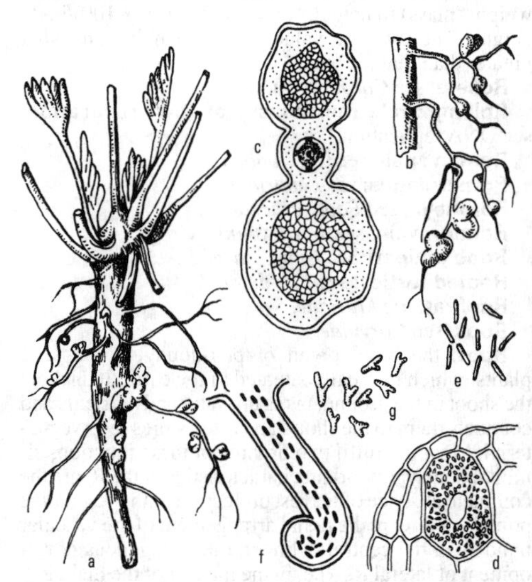

Root nodules. Root nodules of sainfoin *(Onobrychis viciifolia). a* Root with nodules. *b* Nodules somewhat magnified. *c* Transverse section of root and nodules. *d* Nodule tissue showing an enlarged cell filled with groups of bacteroids. *e* Free-living nodule bacteria. *f* Penetration of nodule bacteria into a root hair. *g* Modified bacteria (bacteroids) from a nodule.

With the aid of the symbiotic bacteria in their R.n., leguminous plants considerably increase the available nitrogen content of the soil in which they grow, and they are able to thrive in conditions of low nitrogen availablility. Thus, for centuries, it has been part of empirical agricultural knowledge that a crop of legumes (e.g.

clover) increases the fertility of the soil, and that soil fertility can be maintained by alternating other crops with legumes. Thus, 1 hectare of lupins can fix up to 200 kg of nitrogen during one growth season. If, for agricultural purposes, particular legumes are introduced to an area for the first time, it is advisable to inoculate the soil with the *Rhizobium* species specific for the plants in question.

Nitrogen-fixing R.n. are not only found on leguminous plants. They also occur on the roots of plants such as alder *(Alnus),* sea buckthorn *(Hippophae rhamnoides),* bog myrtle *(Myrica gale)* and others. In these plants, the nodules are often branched structures known as *rhizothamnia,* and the symbiotic bacteria are actinomycetes of the genus *Frankia* (see).

See Nitrogen fixation.

Rorquals: see Whales.

Rosaceae, *rose family:* a family of dicotyledonous plants, containing about 3 000 species in 100 genera, cosmopolitan in distribution, but especially numerous in cold and temperate regions. They are trees, shrubs or herbs, with alternate (except in *Rhodotypos),* simple or compound leaves, which are always stipulate. Flowers are 5-merous, usually hermaphrodite, actinomorphic, and terminal on the stems. The often large number (usually more than 10) number of stamens, and the participation of the flower axis in flower and fruit structure are characteristic. The ovary is usually apocarpous and superior, rarely syncarpous or inferior. Fruits of the more primitive species are multi-seeded follicles, whereas the more highly evolved species have achenes or drupes, which often form a compound or aggregate fruit. This very extensive family is divided into 4 subfamilies (sometimes regarded as separate families), mainly according to the structure of the female flower parts of the fruits.

Fig.1. *Rosaceae. a* Dog rose *(Rosa canina),* flowering and fruiting branches. *b* Blackthorn or sloe *(Prunus spinosa),* flowering and fruiting branches. *c* Bird cherry *(Prunus padus),* inflorescence and infructescence.

1. Subfamily: *Spiraeoideae.* Fruits are characteristically multiseeded follicles. Several members are familiar decorative plants of gardens and parks, e.g. the genera *Spiraea* ***(Spirea)*** and *Sorbaria* ***(False spirea),*** which originate from temperate regions of Asia. ***Goatsbeard*** *(Aruncus dioicus)* is a protected European species, found predominantly in the woods and forests of mountain slopes and ravines.

2. Subfamily: *Rosoideae.* Members of the *Rosoideae* are herbs or bushes, with apocarpous ovaries and single-seeded, indehiscent fruits, usually achenes. The subfamily includes a number of useful and decorative plants. Examples are the edible fruits (aggregate drupes, borne on raised receptacles) of the genus *Rubus,* e.g. ***Raspberry*** *(Rubus idaeus),* and many species of ***Blackberry*** *(Rubus fruticosus* coll.spp.). The ***Strawberry*** *(Fragaria)* forms aggregate achenes and its fleshy receptacle is eaten. The cultured ***Garden strawberry*** *(Fragaria ananassa)* is derived from several crosses between South American wild species; some crosses include the ***European wild strawberry*** *(Fragaria vesca),* whose fruits are also collected in the wild. Rose fruits (hips) are used in the preparation of jams, wine and tea. These consist of an urn-shaped fleshy receptacle enclosing many free carpels or achenes. The roses, however, are primarily ornamental plants, grown for their flowers. Countless varieties have been produced by hybridization and selection from about 150 species. The most important parent species of our garden roses are: the ***Tea rose*** *(Rosa odorata)* from western China; ***Bengal*** or ***China rose*** *(Rosa chinensis)* from China; ***Damask rose*** *(Rosa damascena)* of unknown origin; ***Provence*** or ***Cabbage rose*** *(Rosa centifolia)* from the Caucasus; and the ***French*** or ***Gallica rose*** *(Rose gallica)* from Europe and Western Asia. Yellow flower color is derived from the yellow ***Austrian briar*** *(Rosa foetida).* Floribunda roses are derived from crosses with the ***Polyantha rose*** *(Rosa multiflora)* from eastern Asia. Rambling and climbing roses usually owe their habit to ancestral crossing with the ***Wichuraian*** or ***Memorial rose*** *(Rosa wichuraiana)* from eastern Asia, while most bush roses are derived from the ***Musk rose*** *(Rosa moschata,* from the Mediterranean region and western Asia) and from the Japanese ***Rugosa rose*** *(Rosa rugosa).* In Balkan countries, the petals of some cultivated species, in particular *Rosa centifolia* and *Rosa damascena,* are used to prepare rose oil or attar of roses. The commonest wild European species is the ***Dog rose*** *(Rosa canina),* whose rootstock is used for grafting many cultivated varieties. Thorny members of the *Rosoidea,* especially those with a climbing or rambling habit, are often called briars.

Some species of the above two subfamilies are official, e.g. ***Tormentilla cinquefoil*** *(Potentilla erecta; Rosoideae)* and European ***Meadowsweet*** *(Filipendula ulmaria; Rosoideae),* which grows on damp meadows and river banks.

In some genera the flowers are regressed, e.g. the wind-pollinated ***Burnets*** *(Sanguisorba),* and the large genus of ***Lady's mantle*** *(Alchemilla),* in which the corolla is absent.

3. Subfamily: *Maloideae (Pyroideae, Pomoideae).* These are exclusively woody plants. The flower receptacle is deeply concave, containing the inferior ovary which is partly fused with the flower axis. Some members (hawthorn, medlar) form drupes, in which the fruit flesh is derived from the flower axis, but in most genera the carpels form a parchment-like casing, which forms the core of a pip-fruit or pome. One of the economically most important pome fruits is the ***Apple*** *(Malus sylves-*

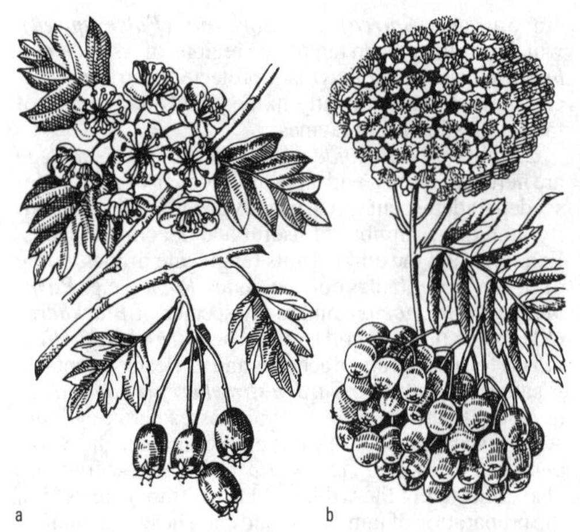

Fig.2. *Rosaceae. a* Hawthorn *(Crataegus monogyna),* flowering and fruiting branches. *b* Rowan or mountain ash *(Sorbus aucuparia),* flowering and fruiting branches.

tris var. *domestica),* a favorite fruit of temperate and subtropical regons. The ancestors of the cultivated apple are probably Transcaucasian species, brought to central Europe via Greece and Italy by the Romans. Even before this, cultivated forms existed in central Europe, together with the spiny, indigenous ***Crab apple,*** which has sour fruits. The ***Cultivated pear*** *(Pyrus domestica)* is probably derived from the indigenous ***European wild*** or ***Common pear*** *(Pyrus communis).* Other cultivated plants are the ***Quince*** *(Cydonia oblongate)* from southeastern Europe and southwestern Asia; ***Medlar*** *(Mespilus germanica)* from Central Asia; the edible ***Mountain ash*** or ***Rowan*** *(Sorbus aucuparia* var. *edulis),* whose wild form is a forest tree of acidic soils, especially in mountainous areas; and ***Loquat*** or ***Japanese medlar*** *(Eriobotrya japonica),* grown in subtropical regions for its early ripening fruits, which taste like apricots.

Many species of this subfamily are popular ornamental bushes and trees of parks and gardens. Examples are the ***Japanese flowering*** or ***Ornamental quince*** *(Chaenomeles speciosa),* with scarlet-crimson, cup-shaped flowers and yellow-green, fragrant fruits; the southern European scarlet ***Firethorn*** *(Pyracantha coccinea)* with abundant splendid scarlet fruits; and various species of the predominantly Asian ***Cotoneaster*** *(Cotoneaster* spp.), also cultivated for its decorative fruits.

The most extensive genus of this subfamily is *Crataegus* **(*Hawthorn*),** with about 1 000 species, which are especially numerous in North America. Some species, e.g. European or English hawthorn *(Crataegus oxyacantha),* are medicinal plants.

4. Subfamily: *Prunoideae.* The main characteristic of this subfamily is the presence of only one carpel, which is free of the hypanthium and forms a drupe. It contains the stone fruits, almost all of which belong to the genus *Prunus.* Most important of these is the ***Garden plum*** *(Prunus domestica),* native to southeastern Europe and western Asia, and now cultivated in many different varieties. Also economically important are the ***Apricot*** *(Prunus armeniaca)* from Central Asia, which is now cultivated in many warmer parts of the world; the ***Peach*** *(Prunus persica)* from China, which has large, fleshy fruits; and the ***Almond*** *(Prunus amygdalus),* which in contrast produces a hard, nonfleshy fruit, and is grown for its seeds (almonds). Peach and almond can be hybridized; the progeny are fertile, producing almond-like fruits. The ***Flowering almond*** *(Prunus triloba),* which produces pink blossom in the early spring, is a widely planted ornamental garden tree, originating from China.

Fig.3. *Rosaceae. a Cotoneaster integerrima,* fruiting branch. *b* Japanese flowering or ornamental quince ("japonica") *(Chaenomeles = Cydonia speciosa),* grown for its flowers. *c* Pear-shaped quince *(Caenomeles = Cydonia oblonga),* grown for its fruits. *d* Goatsbeard *(Aruncus dioicus);* left, male flowering stem and flower; top right, female flowering stem and flower; bottom right, fruiting stem and fruit.

Cherries are also stone fruits of the genus *Prunus*. The ***Sour cherry*** *(Prunus cerasus),* which has sour fruit, grows wild from southeastern Europe to western Asian, and is the progenitor of many cultivated varieties. The ***Wild sweet cherry, Mazzard cherry*** or ***Gean*** *(Prunus avium),* of which there are many cultivated varieties, possesses a large, sweet-tasting fruit; it is common throughout Europe, except in the extreme north, and it is also found in western Asia. *Prunus avium* has a beautifully grained wood, which is often used for furniture. The pleasantly smelling wood of the ***Mahaleb cherry*** or ***St. Lucy cherry*** *(Prunus mahaleb)* is used for walking sticks and tobacco pipes, etc. *Prunus padus* produces small, oval, astringet-tasting black fruits, which attract only birds, hence the name ***Bird cherry;*** it is widely distributed in the wild throughout Europe, Asia Minor, the Caucasus, northern Asia, northern China, Korea and Kamchatka; in the British Isles, it is common in glens and valleys of Scotland and northern England and in central Wales, but otherwise scarce except where cultivated as an ornamental.

Favorite ornamental trees are the pink-flowering ***Japanese flowering cherry*** *(Prunus serrulata),* and the evergreen ***Cherry-laurel*** *(Prunus laurocerasus).*

Blackthorn or ***Sloe*** *(Prunus spinosa)* is a thorny bush, often found on the edges of woodland and in hedges; its small, blue-black, spherical fruit are palatable only after the first frost.

Rose: see *Rosaceae.*

Rose apple: see Galls.

Rosebay willow-herb: see *Onagraceae.*

Rose-breasted chat: see *Parulidae.*

Rose chafer: see Lamellicorn beetles.

Rosefish: see Redfish.

Roselle: see *Malvaceae.*

Rosemary: see *Lamiaceae.*

Rosenmüller's organ: see Epoophoron.

Rosidae: see *Dicotyledoneae.*

Rosiflorae: see *Dicotyledoneae.*

Rosin: see Balsams.

Rosmarinus: see *Lamiaceae.*

Rossia: see *Cephalopoda.*

Rostellum: see *Orchidaceae.*

Rostratula: see *Rostratulidae.*

Rostratulidae, ***painted snipes:*** an avian family in the order *Charadriiformes* (see), consisting of 2 monotypic genera, living in marshy areas of the tropics and subtropics of Africa, Asia, Australia and South America. As in the *Jacanidae* (order *Charadriiformes)* and the *Gruidae* (order *Gruiformes),* the sternum of the painted snipes has 2 notches on its posterior border. They are unobtrusive, skulking birds that run with lowered heads, and fly noiselessly with legs hanging down; they are more active at night than in the day, feeding on insects (especially *Orthoptera),* small mollusks, worms and vegetable material. In the ***Old World painted snipe*** (24 cm) (2 races: *Rostratula benghalensis benghalensis* in Africa and Asia; *R. b. australis* in Australasia) sexual dimorphism is pronounced; females are larger and more brightly colored than males, and incubation is performed by the male. The nostrils of *R. benghalensis* lie in deep grooves on either side of the long, slightly curved upper mandible. The smaller ***American painted snipe*** *(Nycticryphus semicollaris)* is confined to South America. Its bill is more curved and flattened at the tip, there is a slight web between the middle and outer toes, the sexes are alike, and the female incubates.

Rostrhamus: see *Falconiformes.*

Rostrum: a usually pointed appendage on the head or carapace of many crustaceans. Any anterior extension of the head of an animal may also be referred to as a rostrum. See *Cephalopoda.*

Rose finches: see *Fringillidae.*

Rosy finches: see *Fringillidae.*

Rosy pastor: see *Sturnidae.*

Rotalia: see *Foraminiferida.*

Rotary shaker: see Shaking platform.

Rotator: a muscle which rotates a limb about its longitudinal axis.

Rotatoria: see *Rotifera.*

Rotliegendes: the older division of the Permian in the German basin (northern parts of eastern and western Germany and Poland). See Geological time scale (Table).

Rotifera, *Rotatoria,* ***wheel animalcules:*** a class of the *Aschelminthes,* consisting of acoelomate, unsegmented, bilaterally symmetrical, animals, and containing about 1 800 known species.

Morphology. *R.* are 0.4 to 2.5 mm in length, usually fusiform or sac-like (bottom dwellers), but sometimes also spherical (surface floating forms) or with a bizarre flower-like shape (sessile forms). In many species the anterior end is bordered by a double ciliated ring known as the *corona* or *velum,* which serves for locomotion and food collection. The outer, more powerfully ciliated, preoral part of the velum is known as the *cingulum,* while the inner, postoral part is called the *trochus.* Between these two rows of cilia is situated the mouth, leading to a muscular pharynx (mastax) with well developed jaws, which serve for grasping and chewing. A complete alimentary canal is present in most species. When active, the velum has the appearance of a rotating wheel, and is sometimes called the "wheel organ". The entire body is protected by a cuticle, which in many species consists of thick, stiff plates, forming a *lorica* into which the head may be withdrawn. Many species have simple eyes and lateral antennae. A circulatory system is lacking. Excretion is served by flame cell systems,

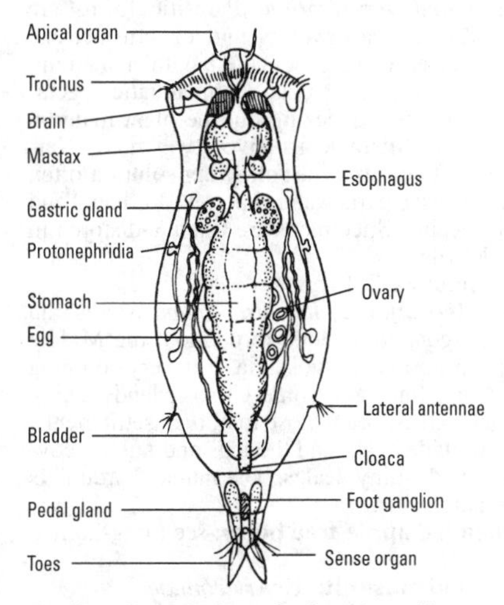

Rotifera. Dorsal view of a female rotifer.

and the terminal part of the excretory tube is enlarged into a pulsating bladder, which ejects its contents into the posterior part of the intestine. The posterior body is divided into several telescoped sections, which can be retracted into the anterior end. In most species the body terminates in a postanal *foot,* which is furnished with 1–4 conical processes or toes; these are provided with an adhesive glandular secretion from cement glands in the foot, and are used by the animal to anchor itself temporarily. The sexes are usually separate, with one or 2 testes or ovaries. Males have never been observed in the *Bdelloidea,* and all members of this order appear to consist exclusively of parthenogenetic females. In other groups, males are rare and often small and degenerate, sometimes lacking excretory and digestive systems. All *R.* display cell constancy, i.e. each species consists of a fixed number of cells at specific locations determined during cleavage.

Biology. *R.* are sessile or free-swimming, on the bottom or among vegetation, in fresh water or in coastal marine waters. They are rarely parasites of other aquatic animals. Predatory species feed on small animals, while other species feed on microoganisms that are swirled into the mouth by the cilia or removed by filtration. Development occurs without metamorphosis. In the *Monogononta,* reproduction occurs for several successive generations by parthenogenesis, followed by a single sexually reproducing generation. Each successive parthenogenetic generation often has a different typical body (lorica) shape (see Cyclomorphosis). Some *R.* are remarkably resistant to drying, and can survive in the dried state for years, either as the dehydrated animal or as its resistant eggs. Feeding recommences as soon as the dehydrated animal encounters water. They can therefore live in temporarily wet habitats (roof gutters, plant surfaces, etc.), and they easily survive wide distribution by wind and animals.

Classification. Some authorities regard the *R.* as a phylum. Three subclasses are recognized, based on the structure of the sex organs, pharynx and velum: *Seisonidea*, *Bdelloidea* and *Monogononta.* Most *R.* belong to the *Monogononta,* which is divided into 3 main orders: *Ploima, Flosculariida* and *Collothecida.*

Collection and preservation. Parasitic forms are found in slugs and earthworms and on crustaceans. Planktonic species can be collected with a medium mesh plankton net. Species that inhabit aquatic vegetation are collected by squeezing out the plant material into a plankton net (many *R.* also live in wet moss). Narcotization is difficult, but the following solution often gives good results: 6 parts water, 1 part 96% ethanol and 3 parts 2% cocaine. Specimens are fixed and stored in 1% formaldehyde.

Rotylenchus: see Eelworms.

Rough-tailed agama, *hardun,* *Agama stellio:* an agamid (see *Agamidae*) found throughout the Middle East and as far north as Yugoslavia. It is very common in Egypt, Asia Minor and some Greek islands, being found in rocky areas, often close to human settlements. It has a short, wide head, and the legs and tail are covered with keeled, spiny scales. The name "hardun" is from the Arabic.

Roundheaded apple tree borer: see Long-horned beetles.

Round-tailed muskrat: see *Arvicolinae.*

Roundworms: see *Nematoda.*

Rous sarcoma: see Tumor viruses.

Rove beetles: see *Coleoptera.*

Rowan: see *Rosaceae.*

Royal water lily: see *Nymphaeaceae.*

RQ: acronym of Respiratory quotient (see).

r-Selected species: see K-selected species.

r-Strategist: see K-selected species.

Rubber tree: see *Euphorbiaceae, Moraceae.*

Rubiaceae, *madder family:* an extensive family of dicotyledonous plants, containing about 6 000 species in 500 genera, distributed mainly in the tropics and subtropics, but with some species in temperate and even Arctic regions. They are herbs or woody plants with opposite and decussate, simple, entire (rarely toothed), stipulate leaves. The stipules are often united to one another and to the petioles; they are sometimes large, resembling foliage leaves, in which case the plant appears to possess whorls of leaves, but unlike true leaves they possess no axillary buds. Flowers are 4- or 5-merous, the corolla consisting of united petals. Sepals are free or united and often greatly reduced; in some genera (e.g. *Mussaenda)* one sepal is larger than the others and brightly colored. Pollination is by insects. The usually inferior ovary develops into a capsule, schizocarp or berry.

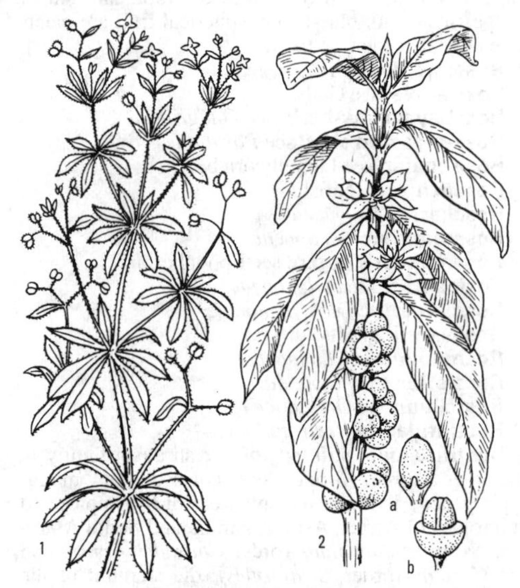

Rubiaceae. 1 Goosegrass or cleavers *(Galium aparine). 2* Coffee *(Coffea arabica),* flowering and fruiting shoot; *a* coffee drupe, *b* drupe with top fleshy part removed, revealing two seeds (coffee beans).

Economically, the most important plant of this family is the ***Coffee bush*** *(Coffea arabica)* (Fig.), whose seeds ("beans") are roasted to produce commercial coffee. To produce the distinctive coffee aroma, the coffee beans must be roasted between 200 and 250 °C. The stimulatory effect of coffee is due to the presence of caffeine (1–2.5%), as well as theobromine and theophylline. Since *Coffea arabica* is very prone to attack by plant diseases and pests, more resistant coffee species are now often planted in many areas, e.g. *Coffea robusta, Coffea liberica,* etc.

Quinine (see), a bitter-tasting alkaloid, is isolated from the bark of the South American ***Cinchona tree (Cinchona succirubra)*** and related species. Quinine was earlier used as an antimalarial drug, as an antifebrile agent in other diseases, and as a cardiac stimulant. Parasite resistance to synthetic antimalarials (e.g. chloroquine) is increasing in many areas of the tropics, so that quinine has now been reintroduced for treatment of the disease. This family also contains the ***Madder (Rubia tinctorum)*** whose roots were earlier used for the preparation of the red dye, alizarin. ***Sweet woodruff*** *(Galium odoratum = Asperula odorata)* is a well-known member, which grows in shaded habitats, especially in beech woods, and possesses whorled, spiky leaves and cymes of pure white flowers. When wilted or dried, sweet woodruff has a high coumarin content, and it is used as a flavoring plant for wines, sweet desserts and confectionery. Another *Galium* species is ***Catchweed, Cleavers*** or ***Goosegrass (Galium aparine)*** (Fig.), which is a weed of fields, deciduous forests and thickets; the stems and fruits are covered with hooked bristles, which readily adhere to animals, birds, fabrics, etc. *Galium verum* ***(Yellow bedstraw*** or ***Lady's bedstraw)*** is a yellow-flowered species, growing predominantly on grassland and dunes; it is coumarin-scented, the flowers are used for coagulating milk, and the stolons yield a red dye. Some genera are myrmecophilous (i.e. the plants form an association with ants). For example, species of *Myrmecodia* and *Hydnophytum* (native to tropical Asia and Australia) are epiphytes, whose roots are used only for attachment; these roots develop large swellings, which contain a network of cavities inhabited by ants.

Rubidium, ***Rb:*** an alkali metal found in all investigated plants at a concentration of 1 to 100 mg/kg dry mass. It is not an essential nutrient. To a limited extent, Rb can replace potassium in the nutrition of fungi, various algae and lower plants.

Ruby throat: see *Turdinae.*

Rudbeckia: see *Compositae.*

Rudd, *Scardinius erythrophthalmus:* a minnow, up to 40 cm in length, and a member of the family *Cyprinidae* (see). There are 2 rows of jagged pharyngeal teeth. The eyes are yellowish to orange, and the fins (except the dorsal fin) are bright red. It is found throughout Europe (but not the Iberian Peninsula), extending well to the east, as well as into northern Turkey and southern Scandanavia. It inhabits almost exclusively standing waters that are rich in vegetation, feeding mainly on aquatic plants, and occasionally invertebrates.

Ruddy duck: see *Anseriformes.*

Ruderal coenosis, ***ruderal community:*** one of several types of community that arise as a result of human influence (see Culture community). Ruderal habitats are usually rich in nutrients, and they may be subject to wide hydrothermal fluctuations. Often, however, they tend to be warm and dry with areas of exposed soil. As a rule, R.cs. have no characteristic horizon, although the majority are treeless. Typical ruderal sites are rubble heaps and rubbish dumps, but they also include cultivated areas and waysides.

Ruderal plants (plate 11): plants of rubbish tips and waysides, especially near to human settlements. Soils of such areas are usually rich in mineral salts and nitrates. R.p. show variable toleration of high salt concentrations, and frequently accumulate high concentrations of nitrate in their leaves. Familiar R.p. are *Urtica dioica* (stinging nettle), *Chenopodium* and *Atriplex.*

Rudimentary organs, ***vestigial organs:*** organs that have partially or totally lost their function, and therefore appear as degenerate or regressed structures. Organisms with R.o. have evolved from forms in which the organ was fully developed and fully functional. Examples are the stumplike hindlimbs of the *Boidae,* and the degenerate eyes of many cave-dwelling animals.

Rudists (Latin *rudis* rough, crude): a suborder of mussels with thick, unequal valves. The righthand, conical or horn-shaped valve is fused with the substratum of the sea bottom. Large, conical teeth on the lefthand, lid-shaped valve are seated in a groove in the lower valve. The R. are important index fossils of the Cretaceous, in particular in the Alpine region (Urgon facies). See *Hippurites, Bivalvia.*

Rue: see *Rutaceae.*

Ruff: see *Scolopacidae.*

Ruffe, ***pope,*** *Gymnocephalus cernua, Acerina cernua:* a fish of the family *Percidae,* order *Perciformes,* found in ponds and rivers of northern Eurasia, but absent from the Iberian Peninsula. It is also found in brackish water. The body is laterally compressed with a slightly arched back. The single dorsal fin is divided into two distict lobes, the anterior lobe with supporting spines. The R. is gray-green to brown in color with a dark spotted back and a yellow-white belly; the opercula have a slight metallic blue sheen. Adults take insect larvae, worms, fish fry and eggs, while larvae feed on zooplankton. The flesh makes good eating.

Ruffini's end-organ: see Tactile stimulus reception.

Rufous crowned babbler: see *Timaliinae.*

Ruminantia, ***ruminants:*** a suborder of the *Artiodactyla* (see). All members possess a complex stomach and perform some form of Rumination (see). Canines are usually reduced and the molars are selenodont (with crescent-shaped ridges).

Rumination: regurgitation of partially digested, cellulose-containing plant food material (cud) from the reticulum of the complex stomach to the buccal cavity, where it is thoroughy chewed before being returned to the stomach. To aid the return of the cud to the mouth, the esophagus is first lubricated by swallowing saliva; the glottis closes, and the regurgitation phase begins with a sudden inspirational movement of the diaphragm; this decreases the pressure within the thoracic cavity, and the esophagus expands. The reflex opening of the cardiac sphincter allows the brei of partially digested food to flow into the esophagus. The cardiac sphincter then loosely closes, the diaphragm relaxes, and a phase of antiperistalsis occurs. While the glottis is closed, the entire thoracic wall performs an expiration movement, and, supported by contraction of the abdominal muscles, the contents of the esophagus are propelled into the mouth cavity under pressure. Here, excess fluid is expressed and swallowed, and R. begins. When the contents of the rumen have been thoroughly chewed and liquified, they pass to the omasum and abomasum where normal digestion occurs. See Digestive system for the structure of the complex stomach of ruminants. In some ruminants, the rumen and reticulum form a single chamber.

Rumen: see Digestive system.

Ruminant reflex: see Chewing reflex.

Runner: see Stolon.

Running water: precipitated water from glaciers, springs, lakes or high swampland, flowing from highland to lowland in artificial or natural channels.

The first section of R.w. is the stream or brook. Its course often changes and the stones on its bed are frequently disturbed, due to the steep gradient and frequently changing water input. The water is very clear with a low and constant temperature, and on account of the high turbulence it has an extremely high oxygen content. Higher plants are absent, and torrenticolous communities of organisms live beneath or on the stones. A characteristric inhabitant is the brown trout *(Salmo trutta fario),* so that this region is also known as the *trout region.* As the gradient and flow rate decrease, the trout region gives way to the *grayling region,* characterized by the presence of the grayling *(Thymallus thymallus).* The water course is more winding and deeper, with deep depressions and soft deposits of silt on the bends. Its bed is covered with coarse stones, which are exposed only when the water level is low. The water temperature is less constant and the oxygen content is high. Torrenticolous forms still predominate, and only the deep areas on the bends provide a habitat for lenitic or stagnicolous forms. Trout and grayling regions are often referred to jointly as the *salmonid region* (see Rhithral).

As the stream flows onto the plain it becomes the *upper reaches of a river,* or a *small river.* Raised water levels and floods occur less frequently, and the shapes of the banks and river bed are more constant. In the upper reaches, river bottoms are sometimes still stony, often covered with gravel or sand, and soft sediments are more common. The water oxygen content is still high, and the annual temperature fluctuations are greater. Water flow rate is still relatively high, but cold water organisms that favor fast currents are no longer found, and they are replaced by eurythermal quiet water organisms (e.g. bivalves). Higher plants are abundant. The region is known as the *barbel region* from the characteristic presence of the barbel *(Barbus barbus).*

On the plain, the river enters its *lower reaches,* i.e. it becomes a *large river.* The water volume is greater, but the flow rate is less. The water is cloudy, much previously suspended material is deposited, and the deep layers near the river bed are oxygen-deficient. Water temperature is variable, depending largely on the environmental temperature. Ox-bow lakes are formed when the water cuts through tight river bends. A rich flora develops near to the banks, but is absent from the central current. The abundant fauna associated with the plant communities, the groynes and the stony bars consist of phytophilic (plant-inhabiting) and lithophilic (stone-inhabiting) organisms. Mud and silt deposits in coves and near the bottom contain numerous individual organisms (up to 12 000 individuals/m^2). A genuine river plankton or potamoplankton does not exist. The plankton found in rivers originate from ox-bow waters or from river lakes. These lower river reaches are also known as the *bream region,* from the characteristic occurrence of the common bream *(Abramis brama).* Finally, a region of *brackish water* serves as the transition zone between the fresh water and the sea. Here the water flow is sluggish, leading to large depositions of sediment, and in summer the water is often very warm. The oxygen concentration is usually very low near the bed of the brackish zone. There are wide fluctuations in salinity, causing the death of many sensitive marine and freshwater organisms, which therefore form a sediment. The bed is usually very silted and inhabited by large numbers of bottom-dwelling and burrowing organisms. Dense vegetation grows on the banks. Common fishes of the brackish zone are the pope *(Acerina = Gymnocephalus cernua),* smelt *(Osmerus eperlanus),* eel *(Anguilla anguilla)* and flounder *(Pleuronectes = Platichthys flesus),* so that this zone is also known as the *pope and flounder region.*

If R.w. are used as an outfall, the biological equilibrium is often destroyed. The natural processes of self-purification are inadequate to deal with the heavy burden of organic and inorganic material introduced by the inflow of industrial and domestic waste water. Thus, in the most heavily polluted rivers, entire stretches may be devoid of all forms of animal life.

Rupicapra: see Chamois.

Rupicaprini: see *Caprinae.*

Rupicola: see *Cotingidae.*

Rusa deer: see Sambar.

Rushes: see *Juncaceae.*

Russell's viper, *daboia,* *Viper russelli:* a deadly Southeast Asian member of the *Viperidae* (see). Together with the cobra, R.v. is one of the most feared venomous snakes in India, responsible for deaths of many people each year. The venom acts as a powerful coagulant. Up to 1.5 m in length, the body is brown with 3 longitudinal rows of red-brown, black-bordered, oval markings (25–30 in each row), which resemble the links of a chain. In addition to rodents, it also feeds on frogs, birds and lizards. It gives birth to up to 60 live young.

Russian desman: see Desmans.

Russian olive: see *Elaeagnaceae.*

Russula: see *Basidiomycetes (Agaricales).*

Rust fungi: see *Basidiomycetes (Uredinales).*

Rusty blackbird: see *Icteridae.*

Rutaceae, *rue family:* a family of about 1 600 species of dicotyledonous plants, occurring mainly in tropical, subtropical and warm temperate regions. They are mostly trees or bushes, rarely herbs. The alternate or opposite, exstipulate leaves frequently carry translucent, schizolysigenous secretory sacs, or cavities filled with essential oils. The insect-pollinated flowers are mostly regular with a 4- or 5-lobed perianth. Fruits may be berries, capsules, drupes or schizocarps.

Many species are important medicinal or fruit-bearing plants. The most familiar and economically important genus is *Citrus,* originally native in southern Asia. Numerous varieties of the different species of *Citrus* are now cultivated in most tropical and subtropical countries. Best known and most widely grown is the ***Orange*** or ***Sweet orange*** *(Citrus sinensis).* Other important species are ***Lime*** *(Citrus aurantium* var. *spinosissima; Citrus limonum);* ***Pummelo*** or ***Shaddock*** *(Citrus maxima);* ***Grapefruit*** *(Citrus paradisi);* ***Lemon*** *(Citrus limon);* ***Citron*** *(Citrus medica* var. *bajoura);* ***Mandarin orange*** or ***Tangerine*** *(Citrus reticulata);* ***Bergamot orange*** *(Citrus aurantium* spp. *bergamia);* and ***Seville orange or Bitter orange*** *(Citrus aurantium* spp. *aurantium).* For the preparation of jams and preserves, the East Asian ***Marumi kumquat*** or ***Round kumquat*** *(Fortunella japonica)* is favored commercially; this genus is closely related to *Citrus,* and the fruits have the appearance of small (2–3 cm) oranges with a very aromatic-tasting skin. The only native European species is the protected, thermophilous calcicole, ***Gas-plant dittany,***

Rutaceae. a Gas-plant ditany *(Dictamnus albus). b* Common rue *(Ruta graveolens).*

Fraxinella or ***"Burning bush"*** *(Dictamnus albus = D. fraxinella).*

Common rue *(Ruta graveolans)* was earlier widely cultivated as a medicinal plant.

Rutelinae: see Lamellicorn beetles.

Rutilis: see European roach.

Rutilus: see Pearl fish.

Rye: see *Graminae.*

Rynchopidae, ***skimmers:*** a small avian family (3 species in 1 genus) in the order *Charadriiformes* (see). They capture food (fish and shrimps) by skimming the water with the lower mandible just cutting the surface. When prey is encountered, the upper mandible clamps down, the prey is flipped from the water, then swallowed in flight. As an adaptation to this mode of fishing, they are the only birds in which the lower mandible is longer than the upper. In additon, as an aid to grasping prey, the outer part of the lower mandible has a sharp edge which fits into a groove on the upper. Extra bony structures attach the skull firmly to the neck vertebrae, and the neck muscles show extra development, both being adaptations that enable the bird to withstand impact with its prey. Skimmers hunt in groups at dusk or at night, spending most of the day at rest on the shore or river bank. The ***Black skimmer*** *(Rynchops nigra;* 51 cm) is found by coastal waters from Massachusetts to northwest Mexico, and down both coasts of South America. It moves inland to the South American grasslands in the winter. All underparts (from the forehead to the tail forks) are white, while the upperparts (from the base of the tail to the crown of the head) are black. The black wings have white trailing edges. The legs are red, and the black bill has a red base. Generally similar to the Black skimmer in appearance, the other 2 species are slightly smaller, with brown rather than black upper parts. The African skimmer has a bright yellow bill, and is found by most African coasts, rivers and lakes. The bill of the Indian skimmer is black at the base and yellow at the tip, and this bird inhabits inland waters from India to Southeast Asia.

Rynchops: see *Rynchopidae.*

S

Sable, *Martes zibellina:* a mustelid (see *Mustelidae)* of northern Asia, with a smoky brown to blue-gray coat, a pointed conical head and relatively large ears. It lives in coniferous and mixed forests, preferably around the upper reaches of small rivers, feeding mainly on small vertebrates, but also taking plant food. The S. is valued for its fur. Strict protection measures have been introduced in Russia and the former USSR to increase its numbers and insure its survival.

Sable antelope: see *Hippotraginae.*

Sacbrood of honeybees: see Insect viruses.

Saccharic acids, ***tetrahydroxyadipic acids:*** hydroxylated dicarboxylic acids with the general structure HOOC—$(CHOH)_n$—COOH, which are produced by the oxidation of aldohexoses with relatively strong oxidizing agents, e.g. nitric acid. D-Glucose is oxidized to D-saccharic acid, D-mannose to D-mannosaccharic acid, and D-galactose to mucic acid.

```
   COOH
    |
H—C—OH
    |
HO—C—H
    |
H—C—OH
    |
H—C—OH
    |
   COOH
```

D-Saccharic acid

Saccharomyces cerevisiae: see *Ascomycetes,* Baker's yeast, Brewer's yeasts.

Saccharomycetales: see *Ascomycetes.*

Saccharum: see *Graminae.*

Sacchiphantes abietis: see Galls.

Saccobranchidae: see *Heteropneustidae.*

Saccocoma (Greek *sakkos* sac, Latin *coma* hair): a genus of free-swimming sea lilies (crinoids), lacking a stalk, and consisting of a small, spherical chalice and five long, bifurcated, spiral arms, which give rise to undivided lateral branches. Fossils are found from the Upper Jurassic (Malm) to the Upper Cretaceous.

Saccoglossa: see *Gastropoda.*

Saccopharyngidae: see *Saccopharyngoidei.*

Saccopharyngoidei, ***Lyomeri, pelican eels:*** a suborder of the *Anguilliformes.* These highly aberrant, eel-like, scaleless, deep sea fishes of the Atlantic, Indian and Pacific Oceans have an enormous mouth and a highly distensible pharynx, which can accommodate large prey. Pelvic fins are absent, and the caudal fin is absent or rudimentary. Dorsal and anal fins are long. Jaws, hyomandibular and quadrate are very elongated. There is no symplectic bone, opercular bones, branchiostegal rays, ribs, pyloric ceca or swim bladder. The small eyes are set far forward. Three familes have been described. The family *Saccopharyngidae* ***(Swallowers)*** is represented by a single genus *(Saccopharynx)* with 4 species; only 16 specimens have been described, all with curved teeth and well developed pectoral fins. The family *Europharyngidae* ***(Gulpers)*** is represented by a single species *(Europharynx pelecanoides)* with minute teeth and greatly reduced pectoral fins; it is the only teleost fish with 5 gill arches and 6 visceral clefts, and its gill openings are closer to the anus than to the end of its snout. Six species of *Monognathus* constitute the family *Monognathidae;* maxilla and premaxilla are absent (i.e. there is no upper jaw); pectoral fins are lacking, and the dorsal and anal fins have no skeletal support.

Saccopharynx: see *Saccopharyngoidei.*

Sacculina: see *Cirripedia.*

Sacculus (saccule) of the inner ear: see Auditory organ.

Saccus vasculosus: see Brain.

Sacral pigment spot, ***Mongoloid spot:*** accumulation of pigment (chromatophores) between the cells of the chorium in the sacral region of humans. Depending on the pigment concentration, the S.p.s. varies in color from gray to blue-gray. It varies widely in size, and has been observed in all races, being especially common in mongoloids (80–90%) and very rare in Europeans. Generally, it is present at birth, then disappears during the early years of life.

Sacrum: see Axial skeleton, Pelvic girdle.

Saddleback: see *Callaeidae.*

Safflower: see *Compositae.*

Saffron: see *Iridaceae.*

Saffron milk cup: see *Basidiomycetes (Agaricales).*

Sagartia: see *Actiniaria.*

Sage: see *Lamiaceae.*

Sagitta: see *Chaetognatha.*

Sagittaridae: see *Falconiformes.*

Sagittarius: see *Falconiformes.*

Sagittocyst: see *Turbellaria.*

Sago palm: see *Arecaceae.*

Saiga antelope, *Saiga tatarica:* a gray-yellow antelope with a large, fleshy, inflatable nose, which has the appearance of a short proboscis lined with hair. Only the male has horns, and these are long, vertical, ringed, pale-colored and translucent. The curious nose structure is thought to be a dust and sand filter, representing an adaptation to the frequent sandstorms in the animal's Central Asian habitat. Two subspecies are recognized: the ***Russian saiga*** *(S.t. tatarica)* and the ***Mongolian saiga*** *(S.t. mongolica).* Once numerous over all the steppes of Europe and Asia as far as Siberia, the S.a. was greatly depleted by large scale hunting and trapping. After the introduction of protective measures, it increased again, and it is now culled as a food source. Some authorities place *Saiga* in the subfamily *Saiginae* (see) of the family *Bovidae* (see), while others place it with sheep, goats, etc. in the subfamily *Caprinae.*

Saiginae: a subfamily of the *Bovidae* (see). It is represented by 2 monotypic genera: *Saiga* (see Saiga antelope) and *Pantholops.* The ***Tibetan antelope*** or ***Chiru*** *(Pantholops hodgsoni)* inhabits plateau steppes in the highlands of Central Asia.

Sail fin lizard, ***soa-soa,*** *Hydrosaurus amboinensis:* an agamid (see *Agamidae),* up to 1 m in length, inhabiting the banks of jungle rivers in Celebes, the Moloccas and New Guinea. The male has a high crest of skin on the base of the tail, supported by bony projections of the tail vertebrae; the dorsal crest is low. Somewhat in appearance and especially in its mode of life, it closely resembles the South Amrican basiliscs. Since the latter are iguanas, this is an example of ecological convergence.

Sailfishes: see *Scombroidei.*

Saimiri: see Squirrel monkeys.

Sainfoin: see *Papilionaceae.*

Saint Peter's fish: see John Dory.

Sakabula: see *Ploceidae.*

Saké: see *Ascomycetes.*

Saker: see *Falconiformes.*

Saki: see *Ascomycetes.*

Sakis, *Pithecia:* a genus of the subfamily, *Pitheciinae.* S. are medium-sized Cebid monkeys (see) of the infraorder, *Platyrrhini* (see New World monkeys). They are widely distributed throughout the Amazon basin in tropical rainforest that does not become seasonally flooded. They have long, coarse coats, in which dark hairs are tipped with white. The head hair grows forward into a hood, and the tail is the same length as the body. Two species are recognized: *P. pithecia* ***(White-faced saki;*** male black with white mask; female dark brown with red underparts; head plus body 420 mm) and *P. monachus* ***(Monk saki;*** both sexes dark brown to black, with white lines from each eye to the chin; head plus body 470 mm). *P. pithecia* feeds on fruit and seeds, whereas *P. monachus* eats leaves and ants. Bearded sakis (see) belong to a separate genus.

Salamander: see *Salamandridae.*

Salamander alkaloids: toxic steroid alkaloids secreted by the skin glands of salamanders (see Salamander toxins). S.a. are modified steroids, in which the A-ring is extended to a seven membered ring by a nitrogen between C2 and C3 (A-azahomosteroids) (Fig.). They excite the central nervous system and cause paralysis. The chief representative, ***samandarin*** (1α,4α-epoxy-3-aza-A-homoandrostane-16β-ol), also causes hemolysis (lethal dose for mice 1.5 mg/kg). ***Samandarone*** possesses a keto group in place of the hydroxyl group at position 16. ***Samandaridine*** contains a lactone bridge (—CH_2—CO—O—) between C16 and C17, and also displays local anesthetic properties.

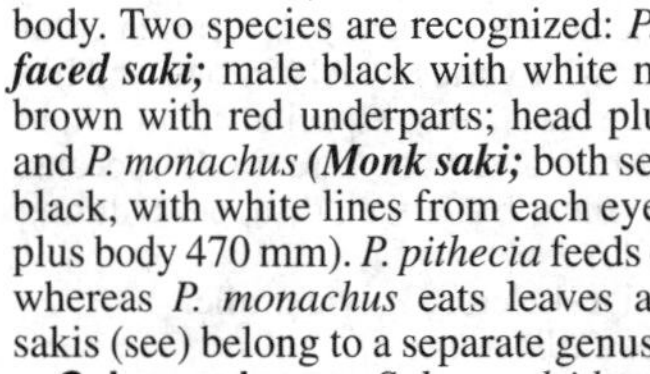

Samandarine

Salamander toxins: toxins secreted by the skin glands of *Salamandra maculosa* (European fire salamander) and *Salamandra atra* (alpine salamander). S.t. include the Salamander alkaloids (see), biogenic amines (tryptamine, 5-hydroxytryptamine), and high molecular mass substances that cause skin irritation and hemolysis.

Salamandra: see *Salamandridae.*

Salamandridae, ***newts*** and ***salamanders:*** a large family of the *Urodela* (see) distributed throughout Europe, central and eastern Asia, and just extending into Northwest Africa; 2 genera are endemic in North America. The palatine teeth are often in long, S-shaped series, which are divergent posteriorly. Adults are aquatic or terrestrial. Except for isolated cases of Neoteny, all adults breathe with lungs. Eggs or live young are usually produced in water. Fertilization is internal. Following courtship display, the male produces a spermatophore, which the female takes up into her cloaca. Eggs are usually laid singly on water plants, and the gilled larvae generally attain an advanced state of metamorphosis before leaving the water. There are usually 3 distinct forms during ontogeny: the aquatic larva (about 3 months), a terrestrial, sexually immature form called an eft (2–3 years), and finally the sexually mature, aquatic adult. Salamanders in the narrow sense are those species that lead a terrestrial existence and are characterized by a cyclindrically round tail (e.g. *Salamandra salamandra* and *Salamandra atra).* Species that are more adapted to an aquatic existence and have a laterally compressed tail are known as newts (e.g. *Triturus, Pleurodeles waltl, Notophthalmus, Taricha, Euproctus).*

Salamandra salamandra ***(Fire salamander)*** (plate 5) is found in damp hilly and mountainous regions over large areas of central and southern Europe. It is up to 20 cm in length, with a cylindrically round tail and a smooth, glossy skin. The upper surface is black, either with irregular, bright yellow patches (eastern subspecies, *Salamandra salamandra salamandra),* or with 2 discontinuous yellow to orange-red longitudinal stripes (western subspecies, *Salamandra salamandra terrestris).* Hybrid populations of the 2 subspecies occur in intermediate areas, e.g. in the Thuringian forest. Courtship, involving body contact, mutual clasping, and rubbing of the cloacae, occurs on land. The female stores the spermatophore, and in the following spring, up to 70 tadpoles are borne in shallow water; their 4 limbs are already formed at birth and they still possess gills. The fire salamander produces an irritant skin secretion, which is toxic to small animals; its main active constituent is the neurotoxin, samandarine. The diet consists chiefly of slugs and worms.

Salamandra atra ***(Alpine salamander)*** is coal black and up to 15 cm in length, with a cylindrical tail. It inhabits mountainous European forests at heights of 700–3 000 m. Of all the salamanders, it shows the greatest adaptation to a terrestrial existence, not even using water for reproduction. Mating occurs in summer on land. The female gives birth, usually after about 2 years, to 2 fully developed, lung-breathing young animals.

Pleurodeles waltl ***(Iberian sharp-ribbed newt)*** is an aquatic species found in small ponds and pools on the Iberian peninsula. When the animal bends, the pointed ends of the free ribs can protrude through preformed holes in the skin.

The genus, *Triturus,* is found in the temperate zone of Eurasia. Species have a laterally compressed tail, usually with a granular skin and a red or spotted belly. They lead an aquatic existence, at least during the reproductive period, and often longer. When seeking a mate, the male becomes highly colored and patterned, and in most species the male also possesses a dorsal crest which extends along the tail. When the animal is not sexually receptive and leading a terrestrial existence, the body color is relatively drab. Deposition and uptake of the spermatophore occurs after courtship, during which the male uses his tail to waft chemical attractants over the female. This contrasts with the courtship behavior of *Notophthalus* and *Taricha,* during which there is intimate body contact. Fertilized eggs are finally laid singly on water plants. The tadpoles leave the water when they have developed lungs.

Triturus cristatus ***(Crested*** or ***Warty newt)*** (Fig.), which occurs from western Europe to the Urals, has a yellow belly with black spots. The male grows up to 15 cm in length, the female 18 cm. During the mating season (March to June), the male has a high, dorsal, toothed crest, which is interrupted at the end of the body, then continues as a high tail fin. Between February and March, it lives in water with abundant aquatic vegetation, such a ponds, lakes and large ditches. For the rest of the year it leads a concealed terrestrial existence.

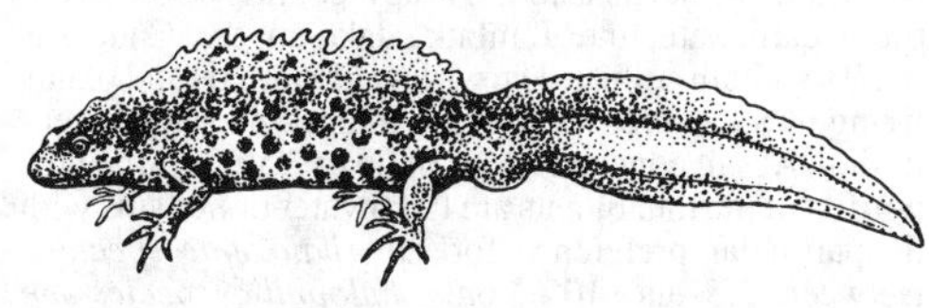

Salamandridae. Crested or warty newt *(Triturus cristatus);* male in the mating season.

Triturus vulgaris ***(Smooth newt)*** grows up to 11 cm in length, and is very common in northern, central and southern Europe. The belly is whitish yellow with round black spots. During the mating season (March to May), the male has a high, wavy, dorsal crest, which continues without interruption as the tail fin. It inhabits all kinds of standing water from January/February to August, the rest of the year being spent concealed on land.

Triturus alpestris ***(Alpine newt),*** found in hilly and mountainous areas of Europe up to 3 000 m, attains a length of 8–11 cm. The belly is orange-red without spots. During the mating season (April to May) the male has a low, ridge-like crest, which continues without interruption as the tail fin. It lives from March to September in water, and in rare cases it also overwinters in water. Some individuals display neoteny.

Triturus helveticus ***(Palmate newt)*** is found in streams and forest ponds in hilly areas of western Europe. It grows up to 9 cm in length and has an angular back. The belly is whitish with a central yellow area. During the mating season, the male has a low dorsal ridge and a 6–8 mm-long, thread-like extension to the tail. Spawning takes place from April to May.

Triturus marmoratus ***(Marbled newt),*** native to southwestern Europe, is olive-green with a black-marbled back. It grows up to 16 cm in length and has a broad head. During the mating season (March to April) the male has a straight crest.

The genera, *Notophthalmus* ***(Eastern American salamanders,*** native to eastern North America) and *Taricha* ***(Western American salamanders*** or ***Pacific newts,*** native to western North America), are closely allied to the Old World newts. During the mating season, males have a dark pad on the inner surface of each hindlimb, and they do not have a dorsal crest. In contrast to the Old World newts, the male clasps the female with his legs during courtship.

Members of the genus, *Euproctus* ***(Brook salamanders),*** inhabit the mountains of southern Europe. They have a flattened head and a laterally compressed, prehensile tail. For reproduction, they seek out ice-cold mountain streams. The male has no dorsal crest. During mating, partners clasp one another, and sperm is transferred directly into the female's cloaca.

Salicaceae, ***willow family:*** a family of dicotyledonous plants containing about 350 species, most of which occur in the northern temperate region. They are deciduous trees or bushes with simple, alternate, stipulate leaves, and naked, inconspicuous, usually dioecious flowers, which lack a perianth. Each flower is solitary in the axil of a scale-like bract, and flowers are grouped in inflorescences known as catkins. Generally the flowers mature before the leaves appear, and pollination is by wind or insects. The superior ovary develops into a 2-valved capsule with numerous seeds that are enveloped by long silky hairs arising from the funicle.

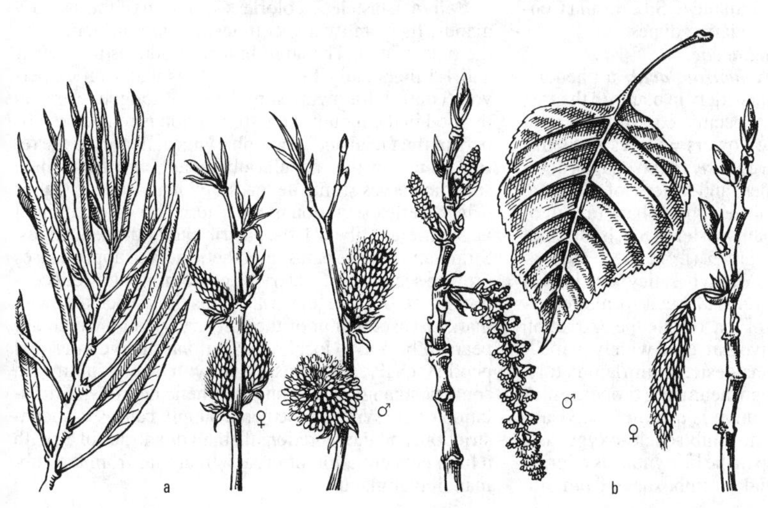

Salicaceae. a Common osier *(Salix viminalis). b* Black poplar *(Populus nigra).*

The family is represented by 2 genera: *Populus* ***(Poplars)*** and *Salix* ***(Willows*** or ***Sallows).*** Willows are also known as ***Osiers,*** especially those that are used for "cane work" and basket weaving. Some native European poplars are the ***Aspen*** *(P. tremulata)* with almost circular leaves that are readily shaken and vibrated by the wind; ***White poplar*** or ***Abele*** *(P. alba),* whose leaves have a white felt of hairs on the underside; ***Black poplar*** *(P. nigra),* with deeply fissured, nearly black bark; its trunk and larger branches usually have swollen bosses, and its spreading branches arch downward to form a wide crown. Native of Central Asia and Afghanistan is a variant of *P. nigra,* known as the ***Lombardy poplar*** *(P. nigra* v. *italica* Duroi, or *P. pyramidalis* Rozier). In contrast to *P. nigra,* the bark of *P. pyramidalis* is gray or brown above, and fissured only on the lower part of the trunk, bosses are absent from the trunk and larger branches, and the branches and twigs are suberect, so that the tree has a narrow, more or less fusiform outline. Poplars are quick-growing trees, but their wood has little value. They are often used as pioneer plants in the reclamation of slag heaps and waste tips.

Many hybrids are found among the willows. Some Central European species are: the ***Almond willow*** *(S. triandra),* which is often planted as an osier; the ***Common osier*** *(S. viminalis),* whose young cane-like branches are woven into baskets; the ***White willow*** *(S. alba),* whose long (5–10 cm) narrow leaves are covered on both surfaces with white silky appressed hairs, so that the tree appears silver-gray when in leaf. *S. alba* is often planted as an ornamental. *S. alba* v. *coerulea* ***(Cricket bat willow)*** is considered to yield the best wood for cricket bats, and it is often planted for that purpose in England. The ***Weeping willow*** *(S. babylonica)* probably originated in China, and it is widely planted for its decorative weeping habit, the long recurved branches nearly reaching the ground. The ***Creeping willow*** *(S. repens),* found on moors and in other wet palces, is a low bush with a subterranean creeping rhizome, and a prostrate to erect stem. *S. daphnoides* Vill. is a large shrub or tree (7–10 m), whose twigs carry a bluish waxy bloom for about 3 years, hence the German name "Schimmelweide". Two of its subspecies, *S. daphnoides daphnoides* (Europe to western Russia) and *S. daphnoides acutifolia* ***(Caspian willow;*** Finland to Siberia and Central Asia), are used for consolidating dunes.

Salicornia: see *Chenopodiaceae.*

Salicylic acid, o-*hydroxybenzoic acid:* a phenolic carboxylic acid which occurs widely in plants in the free or combined form. Free S.a. occurs, e.g. in flowers of *Anthemis nobilis* (camomile flowers = Flores Anthemidis) and in the roots of *Polygala senaga* (senaga root = Radix Senegae). Glycosides and esters of S.a. are present in many essential oils, e.g. birch bark oil, oil of violets, and clove oil. The methyl ester of S.a. is the chief constituent of oil of wintergreen (from the leaves of *Gaultheria procumbens).* Methyl salicylate is not present in the oil cells of *Gaultheria,* but is formed enzymatically from a glucoside of S.a. during the extraction process. S.a. and its derivatives are used widely in medicine, on account of their analgesic, antiinflammatory, antipyretic, and specific antirheumatic properties. A particularly important derivative is aspirin (acetylsalicylic acid). Aspirin is known to inhibit cyclo-oxygenase, which catalyses the first step in the biosynthesis of prostaglandins, prostacyclins and thromboxanes from arachidonate. Daily doses of aspirin are also used for the prevention of coronary thrombosis and strokes. Most of the beneficial effects of S.a. derivatives are due to S.a. itself, which is liberated from its derivatives after ingestion. Free S.a. is not administered directly, because it has a disagreeably irritant effect on the alimentary canal. S.a. is also used to a lesser extent as a preservative.

Saline lakes, *playa:* inland waters containing dissolved salts. In dry regions, these are usually lakes with no outflow, in which the water economy is regulated by inflow and evaporation. This leads to an enrichment of dissolved salts, in particular sodium chloride, sodium carbonate and sulfates. In humid regions, S.l. are often fed by mineral springs. Different S.l. contain different salt concentrations, and the salt concentration may vary widely at one locality, depending on the relative loss and gain of water; concentrations can exceed 300 g/l. The Great Salt Lake of North America contains 15–30 % sodium chloride. At the surface of the Dead Sea, the concentration of sodium chloride is 26.84%, increasing to 32.66% at a depth of 50 m. Lake Gusgundag on Little Ararat in northeastern Turkey contains 38.6% sodium carbonate, the Tambuka lake in the Caucasus 34.7% sodium sulfate. The variety of animals and plants living in S.l. decreases with the increasing salt concentration. At salt concentrations of up to 2.5%, a large proportion of the inhabitants are freshwater organisms with no particular preference for salt *(haloxenic species).* Between 2.5 and 10%, only *halophilic species* are present, together with true saltwater species *(halobionts).* The halophiles (e.g. the sticklebacks *Gasterosteus aculeatus* and *G. pungitius,* the copepod *Cyclops bicuspidatus,* and the midge larva *Chironomus halophilus)* are also widely distributed in freshwater, whereas the halobionts are restricted to (inland) saline waters. At salt concentrations above 10%, only halobionts survive, and they often attain high individual, population densities. They develop optimally at 12%, and can withstand salt concentrations up to 20%, e.g. the anostracan *Artemia salina,* the salt fly *Ephydra riparia,* and others. Halobiont flagellates, like *Dunaliella salina* and *Asteromonas gracilis* have been found in the Elton lake of the Caspian depression at salt concentrations of 28%.

Salinoplankton: see Plankton.

Saliva: a tasteless, colorless secretion of the salivary glands. Its quantity and composition are influenced by the type of diet. The adult human produces between 1 and 1.5 liters daily. Secretion of S. is under reflex nervous control, following stimulation of tongue receptors by food in the mouth. This stimulation is relayed via fibers of the facial and glossopharyngeal nerves to the reflex center in the medulla oblongata, and from there specific nerves stimulate the individual salivary glands (Fig.). Reflex secretion is also caused by stimulation of vagal afferent fibers at the gastric end of the esophagus. Stimulation of the parasympathetic nerve supply causes profuse secretion of watery S., containing a low concentration of organic material. This is associated with marked vasodilation of the salivary glands, which appears to be due to local release of vasoactive intestinal peptide (VIP, a cotransmitter with acetylcholine in some postganglionic parasympathetic neurons). Stimulation of the sympathetic nerve supply causes vasoconstriction, and the secretion of small quantities of S. with a high concentration of organic material from the submandibular glands.

As shown by Pavlov's classical experiments, salivary secretion is easily conditioned, so that secretion can be caused by the sight, smell and even the thought of food.

S. lubricates the food for its passage down the esophagus. It also contains amylase (ptyalin), so that carbohydrate digestion commences in the mouth. As a protective function, copious amounts of dilute S. are produced to wash away or dilute foreign or irritant substances. The buffering power of S. depends largely on its protein content, and this may be increased in response to acidic or basic substances in the mouth.

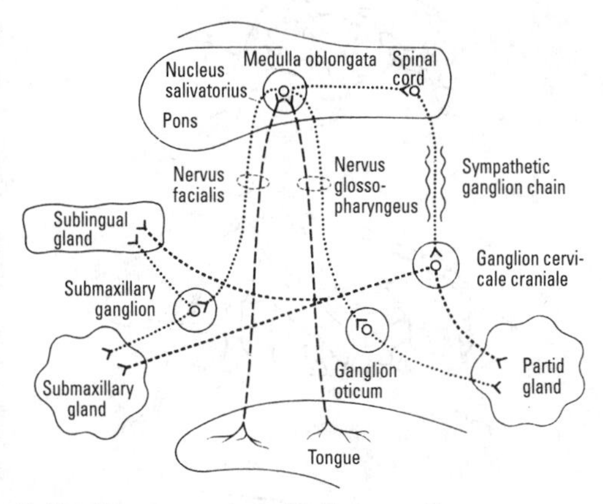

Saliva. Nervous control of saliva secretion.

Salix: see *Salicaceae.*

Sallows: see *Salicaceae.*

Salmo: see Salmon.

Salmon, *Salmo salar:* a commercially important predatory fish of the order *Salmoniformes,* which occurs in the North Atlantic, and is highly prized for its flesh. The largest specimens attain 1.5 m, but lengths of 60–100 cm are more usual. Before spawning the flesh is pink, and after spawning gray-white. In the mating season, males acquire a purple-red ventral surface, and the head becomes blue with red zig-zag lines. For spawning, they ascend rivers to their source, overcoming weirs and other obstacles by leaping long and high. Young S. grow in freshwater, then migrate to the sea.

The genus, *Salmo,* is closely related to *Oncorhyncus* (Pacific salmon), which is also commercially important. Species of this genus occur on the coasts of the Pacific, also ascending rivers for spawning, e.g. *O. keta* (Chum S.), *O. nerka* (Sockeye S.) and *O. gorbuscha* (Humpback or Pink S.).

Salmonella: a genus of rod-shaped, Gram-negative, usually motile Enterobacteria (see). A few are saprophytic and found in soil and water, while many others are parasitic, disease-causing organisms in humans and animals. Toxins produced by *S.* are responsible for the symptoms of febrile gastrointestinal diseases and other general illnesses caused by these organisms. *S. typhi* causes typhoid, *S. paratyphi* causes paratyphoid fever, while *S. typhimurium* is the most commonly encountered cause of food poisoning. More than 1000 different types of S. can be differentiated antigenically. The genus is named after the American bacteriologist D.E. Salmon.

Salmonid region: see Running water.

Salmoniformes: an order of teleost fishes containing 320 species, 90 genera and 15 families, distributed worldwide in oceans and freshwater, but mainly in the Arctic and Boreal regions An adipose fin is often present between the dorsal and caudal fins. Most species are excellent and powerful swimmers, and some species are deep sea inhabitants. Predators include salmon, Danube salmon, smelt, trouts and charrs. A nonpredatory species is, e.g. *Coreganus.*

Salpida, *Hemimyaria, Desmomyaria:* an order of the *Thaliacea* (see) with semicircular muscle bands on the ventral or dorsal side of the body, a single pair of gill slits, and a heart that alternates the direction of its beat. There is no tailed larva. *S.* display an alternation of generations. The egg hatches to the oozoid, which forms a stolon (see Stolo prolifer) on its ventral side. Numerous buds (blastozooids) are formed from the stolon, and these grow into a continuous chain (chain salpid). The chain detaches and further chains are formed. Chain salpids have gonads, and they represent the sexual generation, in which each individual of the chain produces either an egg or spermatozoa. The fertilized egg develops within the blastozooid to form an oozoid. *S.* are planktonic tunicates, occurring in most seas, and they are most abundant between the surface and a depth of 200 m.

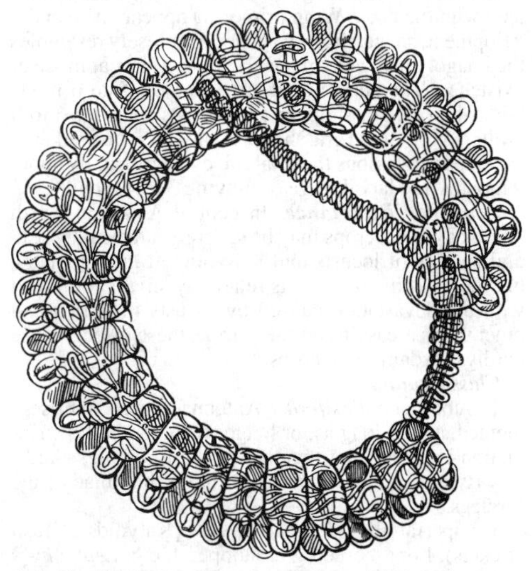

Salpida. Chain-forming generation of *Salpa democratia* Forsk.

Salpornis: see *Sittidae.*

Salsify: see *Compositae.*

Salsolaceae: see *Chenopodiaceae.*

Saltatoptera, *Saltatoria:* an insect order comprising 15 500 species worldwide, including grasshoppers, locusts, katydids and crickets. The earliest fossil *S.* occur in the Upper Carboniferous.

Imagoes. Adults vary in length from 0.3 to 15 cm (central European species do not exceed 6.5 cm). They have downward-directed, biting mouthparts, filiform or bristle-like, multisegmented antennae, and relatively small compound eyes. The first thoracic segment is always larger than the second and third segments, while the posterior thoracic segments are joined immovably to one another. Forewings are rather narrow and somewhat hardened. Hindwings are broader and membranous, and they are folded fan-like under the forewings when at rest. Both pairs of wings are richly veined; in some species, the wings are shortened or absent. With the exception of Locusts (see), members of this order can sustain only brief bursts of flight. In the mole cricket, the forelegs are modified for digging. In most species, the hindlegs are adapted for jumping. Two short, single- or multisegmented cerci are located at the posterior end of the abdomen. Females are recognizable by their usually long ovipositor. Many members possess stridulating and auditory organs. Most *S.* are plant eaters, inhabiting dry, warm regions. Larval development respresents a form of incomplete metamorphosis (paurometabolous development).

Eggs. Eggs (usually elongate or ovoid) are laid with the aid of an ovipositor, either in the earth (3 to 8 cm deep) or in plant tissues. Eggs of the *Phaneropterinae* have a flattened shape. Eggs may be laid separately *(Ensifera),* but more usually 40 to 500 eggs are laid in batches in a mass of frothy material, which hardens to form a cylindrical egg case. In temperate climates, eggs laid in late summer or autumn remain dormant through the winter.

Larvae. The head of young larvae is large in comparison with the body. Wing primordia appear early in development, so that the larva generally closely resembles the imago. Larva and imago also have a similar mode of existence. The adult insect appears after 5 to 6 molts (short-horned grasshopper and locusts) or after 6 to 9 molts (long-horned grasshopper, katydids and crickets). In temperate regions the total life cycle is usually about 9 months, the larval stage occupying 3 to 4 months.

Economic importance. In central Asia and North America, cereal crops may be severely attacked by several species of locusts and long-horned grasshoppers, but the resulting damage is relatively small, compared with the devastation caused by locusts in Africa and other warm areas. In central Europe, these insects cause hardly any damage to crops.

Classification.

1. Suborder: *Ensifera.* Antennae are multisegmented, and as long as, or longer than, the body. Stridulation is by rubbing together of the forewings. Auditory (tympanal) organs are present on the tibiae of the forelegs.

1. Superfamily: *Tettigonioidea* [Katydids (Bush crickets), Long-horned grasshoppers], e.g. *Leptophyes punctatissima* ***(Speckled bush cricket),*** *Decticus verrucivorus* L. ***(Wart biter),*** *Tettigonia viridissima* L. ***(Great green bush cricket).***

2. Superfamily: *Grylloidea* (True crickets), e.g. *Gryllus campestris* L. ***(Field cricket),*** *Acheta domesticus* L. ***(House cricket),*** *Gryllotalpa gryllotalpa* L. ***(Mole cricket).***

2. Suborder: *Caelifera.* Antennae are shorter than the body, with fewer than 30 segments. Stridulation is by rubbing of the inner side of each hind femur (bearing small projections) against the hardened radial vein of the closed forewing. Auditory organs are situated at the base of the abdomen on the sides of the first segment.

Superfamily: *Acridioidea* ***(Short-horned grasshoppers*** and ***Locusts),*** e.g. *Omocestus viridulus* ***(Common green grasshopper),*** *Stenobothrus lineatus* ***(Stripe-winged grasshopper).*** Locust is a collective term for more than a dozen species which display periodic population increases, followed by mass migration, often in gigantic swarms. In Africa and Asia this behavior is shown by species of *Locusta* and *Schistocerca.* Swarms may be a 100 km wide, extending over distances of 1 000 or even 2 000 km. When they descend on an area, the entire vegetation is usually consumed. In earlier centuries, swarms of *Locusta migratoria* sometimes even reached Europe, causing extensive crop damage, e.g. in 1889 in Brandenburg.

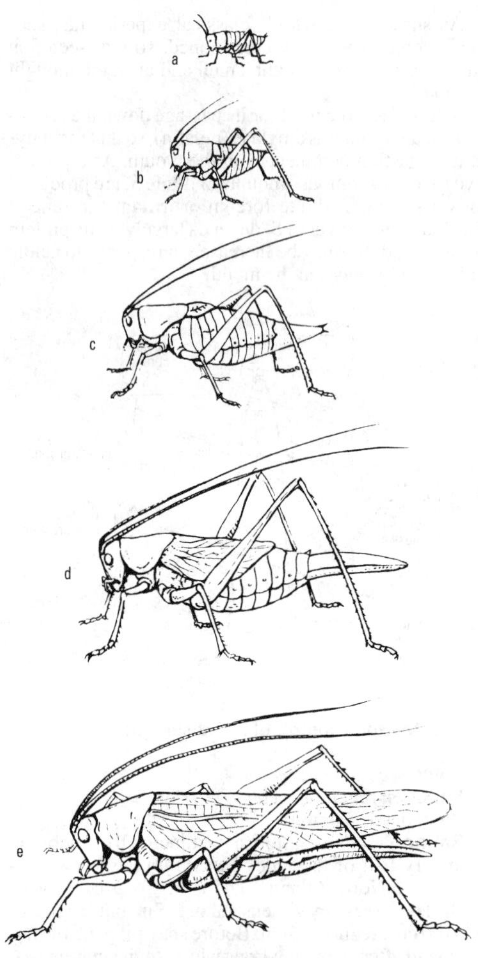

Fig.1. *Saltatoptera.* Metamorphosis of the great green bush cricket *(Tettigonia viridissima* L.). *a-d* Larval satges. *e* Adult (female).

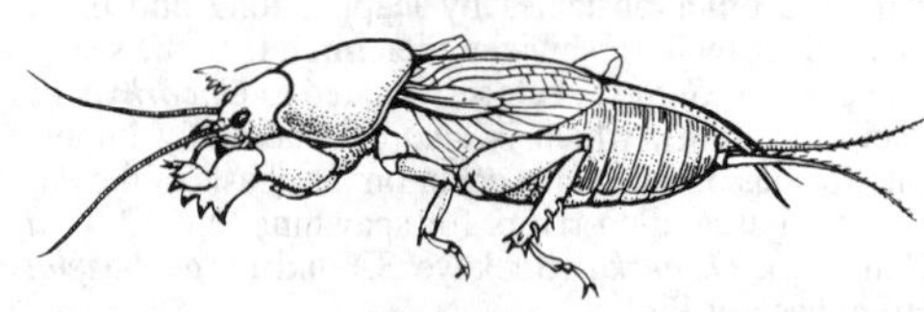

Fig.2. *Saltatoptera.* Mole cricket *(Gryllotalpa* L.).

Saltatorial animals: see Movement.

Saltatory conduction: see Neuron.

Salticicidae, ***jumping spiders:*** the largest family of the *Aranaceae* (see), containing about 4 000, small to medium-sized, mostly tropical species, with a square-fronted carapace. They hunt during the day, and jump on their prey (e.g. flies) from a distance of a few centimeters. The prey is turned on its back, grasped with the chelicerae and injected with venom. Before mating, the male performs a highly stereotyped, ritual dance in front of the female, which suspends her hunting drive. They have great visual acuity, and are possibly sensitive to color and the plane of polarized light.

Salt toad: see *Bufonidae.*

Salvelinus, ***charrs*** and ***brook trout:*** a genus of the family *Salmonidae,* order *Salmoniformes.* They are brightly spotted, economically important predatory fishes with tasty flesh. *S. alpinus* ***(Charr)*** is gray-blue to greenish brown with yellow spots and a pinkish underside. It occurs in the Arctic waters of the Atlantic, in the Pacific and Arctic Oceans, and sporadically in the Alpine lakes of Europe and in Scandanavian lakes. *S. fontinalis* ***(Brook trout)*** resembles the Charr in coloration and patterning, but is brighter. Its presence in Europe, South America, Asia and New Zealand is due to its introduction from North America, where it is found in the Great Lakes and the upper reaches of the Mississippi.

Salvia: see *Lamiaceae.*

Saltwater crocodile, *Crocodylus porosus:* a member of the *Crocodylidae* (see), about 6 m in length and found from India along the entire coast of Southeast Asia to northern Australia and to Fiji. It has 2 nasal ridges that converge anteriorly. An excellent swimmer, it prefers salt water, and ventures hundreds of kilometers out to sea. It has a reputation as a maneater, and is indeed the species responsible for most deaths due to crocodiles.

Samandaridine: see Salamander alkaloids.

Samandarin: see Salamander alkaloids.

Samandarone: see Salamander alkaloids.

Sambar, *Cervus unicolor, Cervus equinus:* a large black-brown deer with conspicuously large preorbital glands. It is found from India to Southeast Asia, and has been introduced in Australia and New Zealand. The branched antlers have a long brow tine at an acute angle to the beam, and the terminal forks point forward. Two other species are also known as Ss: *Cervus timorensis* ***(Sunda sambar, Rusa deer*** or ***Timor deer;*** Java, Celebes, Timor, Moluccas); *Cervus mariannus* ***(Philippine sambar;*** Philippines). See Deer.

Sambucus: see *Caprifoliaceae.*

Sand boas, *Eryx:* snakes of the subfamily *Erycinae,* family *Boidae* (see). Brownish in color with dark patches, and hardly 1 m in length, S.b. are mainly subterranean in habit, burrowing in earth or sand. Head and body are continuous with no obvious neck; the tail is blunt and the scales are smooth. The diet consists mainly of small mammals and lizards. Like larger members of the *Boidae,* S.b. kill their prey by constriction. Most species are native to Africa and Asia. In Europe, the genus is represented by the ***Western sand boa*** *(E. jaculus),* which ranges as far as the Balkan peninsula, and the 40 cm-long ***Dwarf sand boa*** *(E. miliaris),* which occurs from the north coast of the Caspian Sea to Central Asia. All are viviparous.

Sand culture: culturing of higher plants in quartz sand with the addition of macro- and micronutrients. Development of the method is particularly associated with the name of Hellriegel (1883). The nutrient salts are either mixed with the sand before starting the culture, or solutions of nutrient salts are added periodically or continually during the growth of the plants. This combination of sand and water culture is sometimes referred to as *sand-water culture.* Like pure Water culture (see), S.c. has proved particularly useful as an exactly reproducible method for investigations of plant physiology, especially for the study of mineral metabolism. Specific experimental conditions, such as vessel size and shape, rate of watering, different type of sand (e.g. different particle size, sea sand or sharp sand) and pretreatments (e.g. washing with acid, heating to red heat), can be chosen according to the requirements of the investigation.

[Hugh G. Gauch, *Inorganic Plant Nutrition,* Dowden, Hutchinson & Ross, Inc. Stroudsburg Pa. 1972]

Sand dollars: see *Echinoidea.*

Sand eels: see *Ammodytidae.*

Sandersiella: see *Cephalocarida.*

Sandfish: see Common skink.

Sand flea: see *Amphipoda.*

Sandgaper: see *Bivalvia.*

Sandgrouse: see *Columbiformes.*

Sand lances: see *Ammodytidae.*

Sand lizard, *Lacerta agilis:* a 20-cm-long, blunt-nosed, relatively common lacertid (see *Lacertidae)* of central and eastern Europe. The upper surface is green in males and brown in females, and in both sexes is patterned with dark-rimmed, light-centered eyespots. It is a diurnal animal, favoring warm lowland and hilly habitats. The diet consists of insects worms and snails. Mating occurs from April to July, and the female lays 10–15 parchment-shelled eggs, which it buries in the ground. Its numbers have decreased due to human destruction of its habitat.

Sand lizards: see *Lacertidae, Psammodromas.*

Sandpipers: see *Scolopacidae.*

Sand plovers: see *Charadriidae.*

Sand shrimps: see Shrimps.

Sand stargazers: see *Blennioidei.*

Sand tigers: see *Selachimorpha.*

Sand viper:

1) ***Long-nosed viper,*** *Vipera ammodytes:* a member of the family *Viperidae* (see), existing as several subspecies, which are distributed from southern Europe to the Caucasus. It is sometimes locally quite common in dry, stony areas of Jugoslavia and Bulgaria. This relatively stout viper attains lengths of up to 90 cm, has a broad, dark zig-zag band along its back, and possesses a small scaly horn on the tip of the snout. It gives birth to 5–18 live young. Its bite is more dangerous than that of the Common viper. The venom is used pharmaceutically to prepare remedies against rheumatism, since it promotes blood flow.

2) ***Common sand viper,*** *Cerastes vipera,* see Horned vipers.

Sansevieria: see *Agavaceae.*

Santa Cruz long-toed salamander: see *Ambystomatidae.*

Saperda: see Long-horned beetles.

Sap-feeding beetles: see *Coleoptera.*

Sapogenin: see Saponins.

Saponification: hydrolytic cleavage of esters to carboxylic acids and alcohols. In the narrow sense, S. is the hydrolysis of neutral fats by treatment with aqueous NaOH or KOH, with the production of glycerol and the sodium or potassium salts of fatty acids (soaps).

Saponins: a large and widely distributed group of secondary plant products, named for their ability to form strongly foaming, soap-like solutions with water. They are all glycosides, and are classified according to the nature of the aglycon (also known as a *sapogenin),* e.g. steroid, triterpene and steroid-alkaloid S.; the latter are also known as glycoalkaloids. Common sugar components are D-glucose, D-galactose, D-glucuronic acid, D-xylose, as well as L-arabinose and L-rhamnose. S. are powerful surfactants and they cause hemolysis. Although they are potent plasma toxins and have long been used as fish poisons, they have no toxic effects when ingested by humans, because they are not absorbed. Many S. have antibiotic activity, mainly against lower fungi. With sterols, S. form poorly soluble 1:1 molecular compounds, which can be used for the analytical separation of S. or sterols, e.g. Digitonin (see) is used for the precipitation, isolation and quantitative determination for cholesterol and other sterols. S. have been found in more than 100 plant families; they occur, e.g. in soapwort, rape, soybean, foxglove, horse chestnuts and sycamore.

Sapphirine gunard: see *Triglidae.*

Saprobe: see Saprotroph.

Saprolegniales: see *Phycomycetes.*

Sapropel: see Humus.

Saprophytes: heterotrophic organisms which obtain their organic nutrients from dead substrates (i.e. the dead remains of previously living organisms) (see Heterotrophism). Many bacteria and fungi and a few flowering plants (e.g. yellow bird's-nest, *Monotropa hypopithys)* live saprophytically.

The actual growth substrate requirement varies widely according to the species of saprophyte. In addition to organic materials, a carbon source is also often necessary, e.g. carbohydrate, and in certain cases there may also be a requirement for alcohols, fats, organic acids, or even hydrocarbons, like paraffins. Proteins may also be utilized as a carbon source. Often carbon sources serve not only as substrates for the synthesis of cell material, but they are also respired to obtain energy. S. often secrete exoenzymes; these degrade extracellular materials, like lignin, cellulose or protein into absorbable products, which are taken up and incorporated into the pathways of basal metabolism.

S. also use a wide variety of nitrogen sources. Some molds and yeasts utilize inorganic nitrogen compounds, like nitrates and ammonium salts, i.e. they are N-autotrophic. Certain soil bacteria even assimilate atmospheric nitrogen (see Nitrogen fixation). Other S. need organic sources of nitrogen, like amino acids, peptides or proteins.

S. display varying degrees of specialization with respect to their feeding requirements. Omnivorous forms are not very selective, whereas highly specialized forms can live at the expense of only certain compounds. For example, if the mold, *Penicillium glaucum,* is presented with racemic tartaric acid, it first utilizes only the (+)-form. Other S. have an absolute requirement for individual vitamins or amino acids. Microorganisms with such requirements may be used in the laboratory for the microbiological determination of the compound in question.

S. are responsible for the important natural processes of putrifaction and decay. Usually groups of different organisms contribute jointly to these processes, so that the cleavage and waste products of one species serve as growth substrates for another, which in turn produces materials that support the growth of yet another species, until finally all the organic material or dead animals and plants or their parts (e.g. leaf litter) is converted into inorganic compounds (i.e. it is remineralized). S. therefore perform an important function in the cycling of material is the biosphere. The biological self-purification of polluted water also depends on the activity of S. In the active sludge method for industrial water purification, communities of S. are responsible for the processing of organic sewage material.

Many organisms display feeding behavior that is intermediate between a saprophytic mode of existence and parasitism. *Holosaprophytes* obtain their entire carbon supply from dead organic material, whereas hemisaprophytes utilize some inorganic materials. Some hemisaprophytes develop via a holosaprophytic young stage. Facultative S. are parasites which can live saprophytically for a certain period. In contrast, facultative parasites (see Parasites) are S. that under certain circumstances may exist at the expense of living tissue. *Perthophytes* invade cells and tissues, kill them, then live on the dead organic remains.

Saprophytic system: the classification of freshwater organisms according to their tolerance of pollution, in particular their ability to withstand oxygen deficiency and the toxic effects of putrifaction products. The S.s. is used to determine the degree of pollution of water by recording the presence and absence of certain organisms (see Biological water analysis), and it avoids the time-consuming and highly technical chemical analysis of the water. It is based either on the quantitative presence of certain index species, or on the analysis of entire biocommunities. See Saprophobic index.

Polysaprophytic organisms inhabit waters containing high concentrations of putrifaction products, e.g. sewage. They are the earliest in a chain of organisms that contribute to the self purification of water. Their habitat contains little or no oxygen, and organic material is degraded mainly by reduction (putrifaction). The bottom is covered with black, putrid mud, and hydrogen sulfide is usually present. Only a few species are adapted to these extreme conditions, but the absence of competition and the high availability of nutrients leads to the presence of large numbers of individual organisms. A massive development of bacteria (more than 1 million cells per cm^3) is characteristic, most of which are facultatively anaerobic. The fauna consists of protozoa that feed on bacteria, some rotifers, worms and insect larvae. Index forms are bacterial species, such as *Sphaerotilus natans, Zoogloea ramigera* and *Beggiatoa alba,* as well as numerous flagellates and ciliates, like *Euglena, Paramecium* and *Colpidium.* Multicellular organisms of polysaprophobic waters are, e.g. *Tubifex rivulorum* and larvae of the alderfly, *Eristalis tenax.*

As the process of water self-purification progresses, polysaprophytic organisms are replaced by ***α-mesosaprophytic organisms.*** Examples of α-mesosaprophytic waters are village ponds, pools and fertilized carp ponds. The oxygen concentration of the water may be high, and reductive processes are replaced by oxidative degradation. Large quantities of organic material are still present, but hydrogen sulfide is absent. In addition to bacteria, there are often large numbers of algae. There are also some snails, bivalves, lower crustaceans and insect larvae; the fish population includes carp, tench, crucian carp and eel. During the night, respiration (dissimilation) by the large algal population may lead to oxygen depletion and the death of fish. In contrast, during the day, the photosynthetic activity of the algae may lead to over-saturation of the water with oxygen. Index forms of the α-mesosaprophytic zone include several species of cyanobacteria of the genus *Oscillatoria,* the diatoms, *Navicula cryptocephala* and *Nitzschia palea,* the green alga, *Closterium acerosum* (order *Conjugales),* the waste water fungus, *Leptomitus lacteus,* the flagellates, *Chilomonas paramecium* and *Anthophysa vegetans,* the ciliates, *Stentor coeruleus* and *Spirostomum ambiguum,* the mud leech, *Herpobdella atomaria,* larvae of the soldier fly, *Stratiomys chamaeleon,* and the bivalve known as the horny orb shell *(Sphaerium corneum).*

β-Mesosaprophytic organisms inhabit waters in which the concentration of organic material has been greatly decreased by mineralization. The oxygen concentration is high, and it is no longer subject to wide variation. Bacteria are less numerous than in α-mesosaprophytic waters, and there is a rich variety of animals and plants. Many species of diatoms, green algae, cyanobacteria, higher plants, sponges, bryozoans, worms, crustaceans, snails, bivalves, insect larvae, amphibians and fishes are present. All inhabitants are very sensitive to variations in oxygen concentration and to putrifaction products like hydrogen sulfide. Many large lowland lakes, e.g. those of the Havel river in North Germany, as well as the ouflow of well managed sewage works, are β-mesosaprophytic. Index forms are the cyanobacteria, *Microcystis flos-aquae* and *Aphanizomenon flos aquae,* the diatoms, *Asterionella formosa* and *Fragillaria crotonensis,* the green algae, *Spirogyra crassa, Closterium moniliferum, Scendesmus quadricauda, Pediastrum boryanum* and *Cladophora crispata,* the heliozoan, *Actinosphaerium eichhorni,* the flagellates, *Synura uvella* and *Uroglena volvox,* the ciliates, *Vorticella campanula* and *Didinium nasutum,* the polychaete worm, *Stylaria lacustris,* and the larva of the mayfly, *Cloëon dipterum.*

Oligosaprophytic organisms inhabit the clear waters of mountain lakes and streams. Oxidation of organic material in the mud and in the water is now complete. The oxygen concentration of the water is constantly high, and the bacterial count is less than 100 cells/cm^3. Flagellates, ciliates and algae are also less numerous. Larvae of aquatic insects are quite common. All inhabitants are very sensitive to organic pollution and putrifaction toxins, and to variations in the oxygen concentration. Index forms are the diatom, *Cyclotella bodanica,* the green algae, *Closterium dianae, Ulothrix zonata* and *Cladophora glomerata,* the red algae, *Lemanea annulata* and *Batrachiosperma vagum,* the aquatic bryophyte, *Fontinalis antipyretica,* the flagellate, *Diplosiga socialis,* the ciliate, *Halteria cirrifera,* the rotifer, *Notholca longispina,* the turbellarian, *Planaria alpina,* the freshwater mussel, *Margaritana margaritifera,* the crustacean, *Holopedium gibberum,* the larva of the stonefly, *Perla bipunctata,* the larva of the mayfly, *Rhitrogena semicolorata,* and the larva of the caddisfly, *Molanna angustata.*

The division into oligo-, meso- and poly-saprophytic conditions is based on the system of Kolkwitz and Marsson (1909), extended by Liebmann (1947, 1969). Other systems have also been devised. The system of Sladecek (1973) is presented below.

Catharosaprophytic conditions (unpolluted water)		
Limnosaprophytic conditions (polluted surface and ground waters)	Oligosaprophytic β-Mesosaprophytic α-Mesosaprophytic Polysaprophytic	Analogous to the system of Kolkwitz and Marsson
Eusaprophytic conditions (domestic and industrial waste water with bacterial decomposition)	Isosaprophytic Metasaprophytic	Ciliate zone Zone of colorless flagellates
	Hypersaprophytic Ultrasaprophytic	Bacterial zone Abiotic, but nontoxic zone
Trans-saprophytic conditions (no bacterial decomposition)	Antisaprophytic Radiosaprophytic	Toxic zone Radioactive zone

Saprotroph, *saprobe, saprovore:* an organism that lives by absorbing nutrients from decaying material, in particular an inhabitant of water polluted with excretory material, including feces, and products from the decay of dead organisms. Plant Ss. are known as Saprophytes (see), animal Ss. as saprozoites.

Saprotrophic index: in the biological investigation of water, a value representing water quality, determined from the frequency of living organisms and their sensitivity as biological indicators. See indicator organisms.

Saprovore: see Saprotroph.

Saprozoite: see Saprotroph.

Sap wood: see Wood.

Sararanga: see *Pandanaceae.*

Sarcina: a genus of anaerobic Cocci (see), which forms packets of 8 cells (see Fig. to Cocci). In the morphological sense, the name is also applied to all cocci that form a packet-like arrangement of cells, irrespective of their taxonomy. *Sarcina lutea,* characterized by its yellow colonies, is used in the assay of antibiotics, notably penicillin.

Sarcocystis: see *Sarcosporidia, Toxoplasmida.*

Sarcoma: see Cancer cell.

Sarcoplasmic reticulum: see Endoplasmic reticulum.

Sarcoptes: see Scabies mite.

Sarcoptic mange: see Scabies mite.

Sarcoptidae: see Scabies mite.

Sarcorhamphus: see *Falconiformes.*

Sarcosine, *Sar, N-methylglycine:* CH_3—NH—CH_2—COOH, a nonproteogenic amino acid, which is biosynthesized by the N-methylation of glycine. It serves as an intermediate in the metabolism of choline in liver and kidney mitochondria, where it is also formed by the demethylation of dimethylglycine. It is present in mammalian muscle, and it has also been isolated from starfish and sea urchins, where it appears to be a major metabolite. Certain peptides contain Sar.

Sarcosomes: see Muscle.

Sarcosporidia: small coccidia (phylum *Protozoa)* which parasitize mammals and birds. In the vegetative stage they are curved in the shape of a banana or half moon, with a pointed anterior end and a rounded posterior end. Multiplication occurs by binary fission, intracellularly or in cysts. The cysts of *Sarcocystis,* which develop in parasitized muscle, may attain a diameter of 1 cm. In humans, *Toxoplasma gondii* can infect the fetus via the placenta, so that the newborn infant suffers from *Toxoplasmosis* (see) (hydrocephaly, inflammation of arteries and retina), or it is miscarried or stillborn.

Sarda: see Bonito.

Sardina: see Pilchard.

Sardine: see Pilchard.

Sargassum: see *Phaeophyceae.*

Sarkastodon: see *Creodonta.*

Sarothamnus: see *Papilionaceae.*

Sarracenia: see Carnivorous plants.

Sarraceniaceae: see Carnivorous plants.

Sarsia: see *Hydroida.*

Sasa: see *Graminae.*

Sassabies: see *Alcelaphinae.*

Satan's mushroom: see *Basidiomycetes (Agaricales).*

Satellite chromosomes, ***Sat-chromosomes:*** small distal sections of chromosomes, separated from the main chromosome in the interphase nucleus by secondary, nonstainable constrictions in the nucleolus organizer region. S.c. display a different buoyant density from the remainder of the chromosome, and they often contain highly repetitive DNA.

Satellite virus: see Deficient virus.

Satiation: see Hunger.

Satiety: see Hunger.

Satureja: see *Lamiaceae.*

Saturniidae: see *Lepidoptera.* See *Bombycidae.*

Satyridae: see *Lepidoptera.*

Sauria: see *Lacertilia.*

Saurischia: see Dinosaurs.

Sauromatum: see *Araceae.*

Sauropods: see Dinosaurs.

Savanna monkey: see Guenons.

Savin: see *Cupressaceae.*

Savory: see *Lamiaceae.*

Sawbills: see *Anseriformes.*

Sawflies (plates 1, 2 & 45): members of the superfamily, *Tenthredinoidea,* which comprises the families *Tenthredinidae, Cimbicidae, Diprionidae* and *Pergidae* (see *Hymenoptera).* Like all plant wasps, the S. lack a "wasp waste", but wing venation is well developed. Females possess a saber-like or saw-like ovipositor, which is used to place the eggs beneath the epidermis of the larval food plant. The caterpillar-like larvae (see *Hymenoptera,* section Larvae) often live in large groups. Some S. are occasionally pests of orchard trees and bushes, of members of the *Cruciferae,* and of ornamental plants, e.g. the ***Apple and plum sawfly*** *(Hoplocampa),* ***Gooseberry sawfly*** *(Pterodinea)* and ***Brassica sawfly*** *(Athalia rosae* L.).

Saw-scaled viper: see Carpet viper.

Sawtooth termites: see *Isoptera.*

Saxaul: see *Chenopodiaceae.*

Saxicola: see *Turdinae.*

Saxicoloides: see *Turdinae.*

Saxifragaceae, ***saxifrage family:*** a family of dicotyledonous plants. containing about 700 species, found mainly in mountainous regions of the Northern Hemisphere. Leaves are usually alternate, exstipulate or with small sheathing stipules. Flowers are radially symmetrical, rarely slightly zygomorphic, and nearly always 5-merous. There are 2 carpels (rarely 3–5), and pollination is by insects. The fruit is a capsule.

Saxifragaceae. a London pride *(Saxifraga umbrosa). b* Bergenia *(Bergenia* sp.). *c Saxifraga rosaceae* Moench.

Many ornamental plants belong to the large genus, *Saxifraga* ***(Saxifrages),*** whose members occur almost exclusively in mountainous and rocky areas, and are grown in rockeries. *Saxifraga umbrosa* ***(London pride)*** is a much cultivated garden plant, occurring in several different varieties, and often naturalized in subalpine localities; it is truly wild only in the Pyrenees, Corsica, northern Spain, Portugal and southwestern Ireland, preferring limestone woodland up to 1 850 m. It has rosette ground leaves and white flowers in loosely branching clusters; each petal has a few pale red spots and 2 yellow spots, and the sepals turn downward. The easy-to-grow, undemanding ***Bergenias*** (various species of *Bergenia),* with their large, leathery, glossy leaves and large white or reddish flowers, are also popular garden plants. The ***Garden astilbes*** *(Astilbe* x *Arendsii* hybrids) have pinnatifid leaves, and their wild progenitor species occur in eastern Asia.

Saxifrage: see *Saxifragaceae.*

Scab: see Scabies mite.

Scab disease: see *Ascomycetes.*

Scabies mite, *Sarcoptes scabiei:* a mite (see *Acari)* that parasitizes humans and various domestic animals. *S. scabiei* var. *hominis* is specifically associated with humans. About 0.4 mm in length, it is just visible to the naked eye. It bores into thin and tender parts of the skin, especially between the fingers, forming tunnels up to 5 cm in length, containing excrement and eggs. Other favored sites are the skin of the wrists, the outside of the elbows, feet and ankles, and male genitalia. Women tend to be attacked on the palms of their hands, while the palms of the hands and feet of young children are particularly vulnerable. The larvae bore to the surface and are transferred to another host (in humans primarily by hand shaking), or they bore a new tunnel in the old host. This boring activity is responsible for an unbearable itching, often accompanied by a general feeling of ill health. Close bodily contact is necessary for transfer between hosts; for example, using the same bed linen as a scabies patient, but in the absence of the patient, is insufficient for transfer. *S. scabiei* belongs to the family *Sarcoptidae,* generally known as "itch mites". Related forms are responsible for *sarcoptic mange* or *scab* on domestic animals. Thus, the genus *Notoedres* has been found on a wide range of mammals, where it causes "notoedric mange".

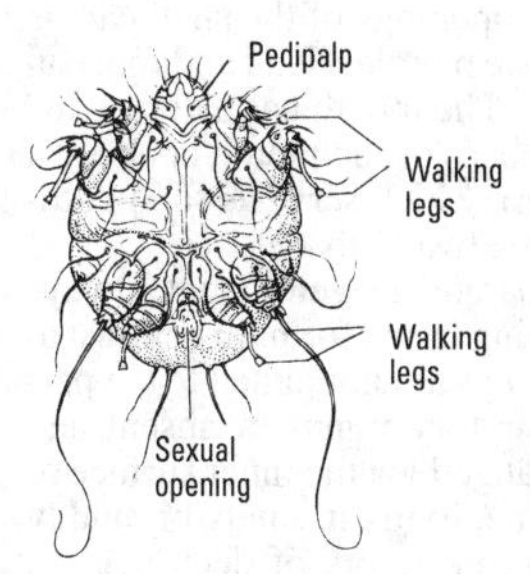

Scabies mite (Sarcoptes scabiei) (male).

Scad, ***horse mackerel,*** *Trachurus trachurus:* a predatory marine fish of the family *Carangidae* (jacks and pompanos), order *Perciformes.* Its curved lateral line is marked by keeled bony plates. Up to 40 cm in length, and predominantly gray in color, it occurs in the eastern Atlantic (Norway to South Africa) and western Atlantic (Argentinian and Brazilian coasts), as well as in the Mediterranean and Black Seas. A commercially important fish with tasty flesh, it forms large shoals in open water over the continental shelf.

Scaldfish: see *Pleuronectiformes.*

Scale insect: see *Homoptera.*

Scaleless black dragonfishes: see *Stomiiformes.*

Scales: a botanical and zoological term for a variety of morphological structures.

1) ***Plant S.***

a) Epidermal structures (modified hairs) adpressed to the plant surface.

b) Colorless or green scalelike lower leaves, which precede the development of true leaves in many seedlings; they are also present on the annual shoots of many trees (bud S.), and on the innovations of perennial herbs. Such scalelike leaves are also present on subterranean shoots, developing into foliage leaves when exposed to light.

c) *Bract S.* and *seed S.* of the strobilate flowers of *Pinaceae* (instead of sporophylls, the axis of the female flower carries spirally arranged sterile bract S. with axillary fertile ovuliferous S.), as well as the hard *cone S.* of the cones (modified flowers) of coniferous trees.

2) ***Animal S.***

a) Epidermal structures that partly or totally cover the body surface of animals, and which display a variety of structures and functions. The S. of insects are flattened and composed of chitin; in the fully developed state they are hollow, air-filled true hairs. The uniform scale covering of the *Lepidoptera* essentially determines the general outward appearance of this order of insects; S. similar to those of the *Lepidoptera* also occurs in the *Thysanura,* some beetle species and the *Diptera.* By virtue of their fine structure, *iridescent S.,* which are responsible for the colors of many butterflies, produce color by optical interference, rather than by pigmentaation. Some male butterflies also possess *gland S. Sensory S.* are also present on some invertebrates. In vertebrates, different skin layers give rise to different types of S. The *placoid S.* or *dermal denticles* of sharks and rays structurally resemble the teeth of higher vertebrates, and are considered to be the evolutionary precursors of the latter. In turn , placoid S. can be interpreted as the evolutionary derivatives of ancient armored skin plates. In the development of a placoid S., a conical thorn of dentine, covered with a layer of enamel, arises centrally from a rhombic basal plate anchored in the corium. The enamel is derived from the enamel organ of the epidermis. Placoid S. give the skin a granular appearance, like shagreen leather. The S. of more advanced fishes are bony structures, which arise in the corium, push through and raise the epidermis, and overlap each other like roof tiles, forming a small pocket between neighboring S. The concentric growth lines of these bony S. are used for determining the ages of fishes. They represent a degenerate evolutionary derivative of the more primitive rhomboid *ganoid S.,* which consist of a superficial layer of enamel-like ganoin, a middle layer of dentine, and a basal layer of vascular bony tissue. Ganoid S. are found in the *Acipenseriformes* (sturgeons) and *Lepisosteiformes* (gars), whereas the bony S. of teleosts fishes are either *cycloid S.* or *ctenoid S.,* the latter carrying small spines or a comb-like structure on their posterior edge. The *horny S.* of quadrupeds are epidermal structures, which arise above the papillae of the corium. In some reptiles, e.g. the blindworm *(Aguis fragilis),* thin bony plates develop in the corium below the horny S. In lizards and snakes, the overlapping S. serve a special protective

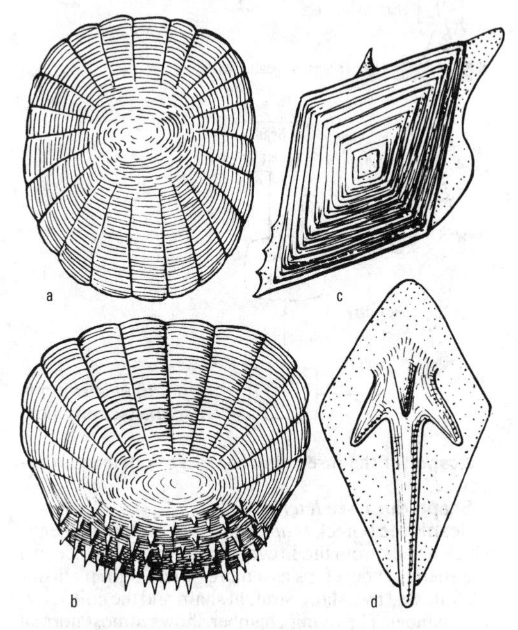

Scales. Different types of fish scales. *a* Cycloid. *b* Ctenoid. *c* Ganoid. *d* Placoid.

function, especially around the head where they are modified to large plates or shields. The shells of tortoises and turtles have the same origin as (i.e. are homologous with) the horny S. of other reptiles. Among the mammals, the *Pholidota* (pangolins) and *Dasypodidae* (armadillos) are largely covered with S. In marsupials, insectivores and rodents, the tail carries small horny S., which, like the S. on the legs of birds, are also evolutionary derivatives of the horny S. of reptiles.

b) Fragments of the cornified skin layer of mammals that are continually lost by sloughing are called S. In this sense, the term is especially common in medicine.

Scallops: see *Bivalvia.*

Scaly anteaters: see *Pholidota.*

Scaly dragonfishes: see *Stomiiformes.*

Scaly polyporus: see *Basidiomycetes (Poriales).*

Scaly weavers: see *Ploceidae*

Scammony: see *Convolvulaceae.*

Scanning electron microscope: a special type of Electron microscope (see) for studying the microscopic structure of surfaces. The electron beam generated at the cathode is concentrated by 2 or 3 electromagnetic lenses to a diameter of 5–20 nm, then used to explore the surface of the specimen. A deflector coil provides a line scanning pattern. When the electron beam irradiates the sample, secondary electrons are emitted, which are registered on an appropriate detector, video-amplified, and fed to a cathode ray tube. A resolution of 5–10 nm is achieved. This is low in comparison with transmission electron microscopy, which has a resolving power of about 1 mm, but the images from scanning electron microscopy are brilliant and rich in contrast. The method has been used by biologists for investigating the surface structure of foraminifera, diatoms, insects, pollen grains, cultured cells, etc.

Scanning electron microscope (schematic)

Scape (of the insect antenna): see Antennae of insects.

Scaphidura: see *Icteridae.*

Scaphites (Greek *scaphe* boat): an ammonite genus which occurs from the Lower to the Upper Cretaceous. The shell first develops as a more or less tight planispiral, followed by a short, straight shaft, and the coiled living chamber. The living chamber shows almost normal coiling with a tendency to curve back upon itself (forming a recurved hook), and it does not extend over the starting coil. The end of the starting coil is closed by a collar-like constriction. Phylogenetically, *S.* represents a degenerate form of ammonite ribbing and shell shape.

Scaphopoda, *Solenoconcha,* ***tooth shells, tusk shells:*** a class of the *Mollusca* (subphylum *Conchifera),* containing about 350 living species, up to 13.5 cm in length (most species 2–5 cm), which are morphologically intermediate between snails and bivalves.

Morphology. All species have an elongate body, enclosed in a shell which is open at both ends, and which is usually slightly curved, tapering and tubular (resembling an elephant tusk), or less often spindle shaped. The anterior and posterior openings of the shell can be closed by ring muscles of the mantle. Head and foot can be extended from the shell. The mouth has a dorsal jaw and radula. Situated at the base of the mouth cone are a number of Captaculae (see), which serve as food-gathering appendages. Eyes are always absent.

Biology. Sexes are separate. Development always proceeds via a free swimming larval form. In contrast to all other mollusks, the kidneys are not joined to the pericardium. Specialized respiratory organs are absent, and respiratory gases are exchanged via the inner surface of the mantle. All *S.* are marine, living in sandy or muddy bottoms, many species on the floors of deep oceans. With the aid of the foot, they burrow at an angle into the substratum, leaving only a few millimeters of the posterior end of the shell above the surface. They feed on microorganisms, e.g. foraminifera, which are detected, captured and delivered to the mouth by the captaculae.

Geological record. Fossils are found from the Ordovician to the present. In the Paleozoic and Lower Mesozoic, relatively large numbers of *S.* are only found locally in certain strata. They increase in the Cretaceous and Tertiary, but they do not attain the abundance of true index fossils.

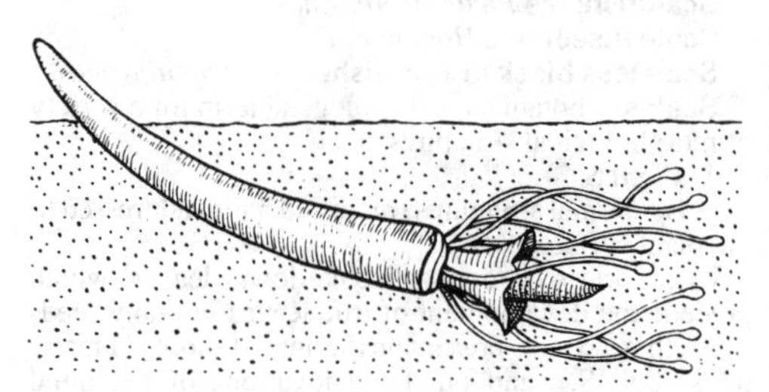

Scaphopoda. Tusk shell *(Dentalium dentale* L.).

Scapula: see Pectoral girdle.

Scarabaeinae: see Lamellicorn beetles.

Scarabaeoidea: see Lamellicorn beetles.

Scardafella: see *Columbiformes.*

Scardinius: see Rudd.

Scaridae, *Callyodontidae,* ***parrotfishes:*** a family (about 68 species in 11 genera) of fishes of the order *Perciformes.* They are vividly colored, marine tropical fishes, with a parrot-like "beak", in which the jaw teeth are fused into sharp edges. With the aid of the beak, they feed by grazing on the dead coral substrate of coral reefs. They change their color and patterning as they grow, and when they change sex. Change of sex is common, and most of the studied species have both primary and secondary males. They are popular fishes of marine aquaria.

Scarlet fire thorn: see *Rosaceae.*
Scarlet pimpernel: see *Primulaceae.*
Scarlet runner: see *Papilionaceae.*
Scatophagidae, ***scats:*** a family of fishes in the order *Perciformes.* The body is deep and laterally compressed, with a maximal length of about 35 cm. They are omnivorous, and occur in large shoals in coastal regions, especially estuaries of the tropical Indopacific. Young fish can be kept for a long period in fresh water, but adults require saline conditions. Two genera are recognized: *Selenotoca* and *Scatophagus,* with about 4 species.
Scats: see *Scatophagidae.*
Scaup: see *Anseriformes.*
Sceloporus: see Spiny lizards.
Scenopoeetes: see *Ptilinorhynchidae.*
Scents: see Odiferous principles.
Schardinger dextrins: see Dextrins.
Schemochromes: see Structural colors.
Schindylesis articulation: see Joints.
Schistoceras (Greek *shistos* slit, *keras* horn): a characteristic goniatite ammonite of the Upper Carboniferous and Permian. The shell is generally disk-shaped and planispirally coiled, with grid-like surface markings. The suture line shows simple folding, with 10–16 lanceolate lobes.
Schistocerca: see *Saltatoptera.*
Schistostegales: see *Musci.*
Schizicephala: see *Diplopoda.*
Schizocarp: see Fruit.
Schizocoel: a primary coelom produced from numerous cleavage cavities in the mesenchyme between the ectoderm (epidermis) and endoderm (intestinal tract). The S. is a reticular, branching system of cavities, filled with liquid and containing migratory cells. It is present, e.g. in *Platyhelminthes* and *Nemertini.*
Schizodont: a hinge type of the *Bivalvia* (see).
Schizodus (Greek *schizein* split, *odus* tooth): an extinct bivalve genus with an elongated triangular or trapezoid, smooth shell. There is a single large, deeply indented, triangular hinge tooth, and 2 smooth lateral teeth, one in front and one behind. Its fossils are found from the Carboniferous to the Permian. *Schizodus obscurans* is an index fossil of the Upper Permian German Zechstein.
Schizogany: see *Telosporidia.*
Schizogenic: a term applied to cavities or spaces produced by the detachment, cleavage, or movement apart of surrounding material. In plant tissue, intercellular spaces are S., i.e. they are formed by the local separation of cell walls.
Schizogenic secretory reservoir: see Excretory tissues of plants.
Schizogeny: in protozoa, asexual multiplation by fission into many pieces. The process is preceded by the appearance in the cell of a corresponding number of nuclei. See Reproduction.
Schizogony: a form of asexual reproduction in the protozoa. The offspring are known as schizozoites or merozoites. The nucleus divides into several copies, so the cell may be temporarily multinucleate. By multiple fission, each nucleus is then assigned to its own cell. See Reproduction, Cell multiplication.
Schizogregarinaria: see *Telosporidia.*
Schizomids: see *Uropygi.*
Schizophyceae: see *Cyanophyceae.*
Schizozoite: see *Telosporidia.*
Schlegel's pit viper: see Lance-head snakes.
Schlotheimia (named after the paleontologist, E.v. Schlotheim, 1764–1832): a characteristic ammonite genus of the Lower Lias (Lias α_2 or Hettangian stage, also known as the Schlotheim layers). The shell is planispirally coiled with a wide umbilicus and a rounded outer edge. The simple ribs curve forward on the ventral side, and are usually interrupted by a smooth and occasionally sunken median zone on the outer edge. The suture line is weakly serrated.
Schlumbergera: see *Cactaceae.*
Schoinbates: see *Phalangeridae.*
Schöngastia: see *Trombiculidae.*
Schrätzer, *Acerina (= Gymnocephalus) schraetzer:* a small perch (family *Percidae)* found in the Danube and its tributaries, and closely related to the Pope. The fusiform body has lemon sides and a pointed head. It lives near the bottom over gravel or sand, and females lay their eggs over a wide area on stones and other solid substrates.
Schwann cell: see Nerve fiber, Neuron.
Sciadopitys: see *Taxodiaceae.*
Sciaenidae, ***drums, croakers*** or ***meagres:*** a family (about 160 species) of large, elongate fishes of the order *Perciformes,* found mainly in shallow marine waters, but also in brackish and freshwater habitats, throughout the tropics and subtropics. They produce rhythmic, species-specific sounds (knocking, croaking, grunting, snoring or drumming) by using the swim bladder as a resonating chamber. There are two dorsal fins and an anal fin with two spines. Many species are important food fishes, and some are popular sporting fishes.
Scimitar babblers: see *Timaliinae.*
Scincidae, ***skinks:*** a family of the *Lacertilia* (see), containing more than 600 species in about 50 genera, distributed in all tropical and many subtropical regions, and most numerous in Southeast Asia. Most species have a pointed head and a cylindrical body, with smooth glossy scales which are often underlain by small bony plates (osteoderms). The teeth are pleurodont, i.e. attached by one side to the inner surface of the jawbone With few exceptions, skinks are ground-dwelling animals. Many species excavate subterranean passages. Eyes and ear openings are often regressed. Some species have well developed, if not particularly powerful limbs, others have greatly reduced limbs, and there are snake-like species without external limbs. Skinks feed predominantly on insects, and large species also eat vegetation. Most species lay eggs, some are vivi-oviparous, and a few are viviparous. Some examples are *Tiliqua* (Blue-tongued skinks) (see), *Scincus scincus* (Sandfish) (see), *Chalcides* (see), *Mabuya* (see), *Ablepharus* (see), *Ergenia* (see), *Typhlosaurus* (Blind skinks) (see).
Sciuridae, ***sciurids, squirrels and allies:*** a family of Rodents (see) containing 262 recent species in 49 genera, distributed worldwide, except in the Australian region, southern South America, North Africa and Madagascar. A unique rodent feature displayed by the sciurids is the presence of a postorbital process on the frontal bone. Four toes are present on the front feet, and each of the 5 toes on the hindfeet has a sharp claw. The tail is usually long, the eyes large and protruding, and the eyesight is sharp with a wide field of vision. Sciurid thumbs are usually underdeveloped, often carrying an unusually large claw (e.g. Prairie dogs), or a flat nail (Mar-

mots). The upper lip is split and the nostrils are divided by a furrow. Dental formula: 1/1, 0/0, 1–2/1, 3/3 = 20–22. They are basically diurnal herbivores, but other food is also taken (e.g. tree squirrels sometimes take birds' eggs, and insects feature in the diets of some sciurids) and flying squirrels are nocturnal. Sizes range from that of the ***Giant Asiatic tree squirrels*** *(Ratufa;* head plus body 25.5–50 cm, tail about the same length as the body) to that of the ***African pygmy squirrel*** from Gabon *(Myosciurus pumilo;* head plus body about 7 cm, i.e. the size of a human thumb).

Tree squirrels *(Sciurus;* head plus body 20–31.5 cm, tail 20–31 cm) comprise about 55 species, all of which are arboreal in habit and very agile climbers, with a characteristic, large, bushy tail. They are distributed in most of Europe and Asia south of the northern tree limit, in Japan, and from southern Canada to northern Argentina, in deciduous, coniferous and tropical forests. *Sciurus vulgaris* ***(Red squirrel;*** head plus body 20–25 cm, tail 16.5–20 cm; southern somewhat larger than northern subspecies) is found throughout Europe and Scandanavia and into Asia, and is the only species of *Sciurus* found in Central Europe. With its bushy tail and tufted ears, it is the archetypal squirrel of legend and popular definition. It builds a spherical nest *(drey)* in the branches, and its habit of burying food stores, including fruits and nuts, in the ground, is an aid to tree dispersal. The Red squirrel is named for its red fur, but its winter coat also contains gray hair, and the proportion of gray is greater in the eastern and northeastern populations. In some northern European populations and especially in the Siberian subspecies, gray-white is the predominant color. The white fur of northern subspecies is used by furriers, under the name Siberian squirrel or miniver. Subspecies include the ***Central European red squirrel*** *(Sciurus vulgaris fuscoater),* whose lowland populations have a red coat, while mountain populations are black. The ***British red squirrel*** *(S. v. leucourus),* also red in color, has been largely replaced by the ***Gray squirrel*** *(Sciurus carolinensis)* in most areas of Britain. The Gray squirrel originally inhabited only the oak woodlands of eastern USA, later invading most other types of deciduous woodland in North America, where it damages trees by stripping the bark, and is considered to be a pest when this activity is extended to the sugar maple. In 1876, in the county of Bedfordshire in UK, 350 Gray squirrels were released, and this species is now common in most British woodlands. It is rather larger and more powerful than the Red squirrel, and in the Old World it is restriced entirely to Britain. ***North American red squirrels*** *(Tamiasciurus hudsonicus* and *T. douglasii;* head plus body 16.5–23 cm, tail; 9–16 cm), also known as ***Chickarees,*** are colored reddish brown to olive brown, and also have a typical squirrel appearance. They are true tree squirrels, inhabiting mainly coniferous, but also deciduous forests, and they also spend much time on the ground.

Marmots *(Marmota;* head plus body 55–90 cm, tail 11–16 cm) are found throughout the Northern Hemisphere, across the Arctic Circle to the southern ranges of the temperate zone in North America, Asia and Europe. They are plump and sturdy with a round, broad head, short tail and short legs. Their hair is long and dense, and the tip of the thumb carries a small nail. Thirteen species and subspecies are recognized. The ***Alpine marmot*** *(Marmota marmota)* inhabits the European Alps and the Carpathians. It burrows extensively, and the whole family (up to 15 animals) hibernates underground for 6 months or more in the winter. The thumb has practically disappeared, being represented only by the metacarpal bone, which has an amorphous structure. Marmot fat is greatly valued in folk medicine in the European Alps as a remedy for rheumatism. In many areas of the Alps the marmot became extinct, but these areas have now been successfully restocked. The ***Bobac marmot*** *(Marmota bobak),* which is trapped for its fur, is larger than the Alpine marmot. It also excavates a subterranean system of burrows, and it is found in a continuous belt from eastern Europe across Central Asia to eastern Siberia.

Chipmunks *(Eutamias and Tamias),* comprising about 16 species, are found in both the Old and the New World. ***Western chipmunks*** occur throughout most of North America, from the Yukon and Mackenzie drainages to California and Mexico. The ***Siberian chipmunk*** *(Eutamias sibiricus)* is found across Central and northern Asia from Russia as far as northern China and northern Japan. The back of this small, gray rodent is patterned by 5 black-brown longitudinal stripes. It excavates subterranean burrows with food storage chambers. The ***Eastern American chipmunk*** *(Tamias striatus)* is the only species of its genus. *Tamias* has only 1 premolar on each side of the upper jaw (cf. 2 premolars in *Eutamias*). It is found over most of eastern USA and southeastern Canada, where it prefers deciduous forests, especially when these are broken by open ground. With a head plus body length of 13.5–18.5 cm (tail 7.5–11.5), *Tamias* is larger than the American and Asian species of *Eutamias.* The upper parts of *Tamias* are red-brown, normally with 5 dark back stripes, while the underparts are whitish.

Old World species of the genus *Spermophilus (= Citellus)* are generally known as ***Sousliks*** or ***Susliks,*** whereas those of the New World are called ***Ground squirrels*** or ***Gophers*** . These are slender, terrestrial rodents (head plus body 13–40.5 cm, tail 4–25 cm). For sciurids, the ears are unusually small and almost hidden. The coat is grizzled gray-brown to yellow-brown with irregular waving. They live in social colonies, inhabiting underground burrows. Thirty two species are recognized. The ***European souslik*** *(Spermophilus citellus)* is found in Germany and parts of eastern Europe bordering the west coast of the Black Sea. Although colonial, each individual lives singly in its own den within the colony. The coat is usually unpatterned, but occasionally displays faint spots. It is crepuscular in habit, i.e. it is active in the morning and evening. The back and flanks of the North American ***Thirteen-lined ground squirrel*** *(Spermophilus tridecemlineatus)* are patterned with 13 light-colored, broken stripes. It is a typical prairie inhabitant from Canada to the central USA.

The combined ranges of the 5 species of *Ammospermophilus* cover the sparsely vegetated plains and lower mountain slopes of Southwest USA, extending to western Texas and northern Mexico. These are true ground squirrels, living in subterranean burrows. The small (about 90 g) ***Antelope ground squirrel*** *(Ammospermophilus leucurus)* displays exceptional tolerance to high temperatures (70 °C in the sun) in the deserts of Southwest USA, even withstanding a body temperature of 43.6 °C without perspiring or panting. It is preyed upon by snakes, bobcats, foxes and hawks.

Prairie dogs *(Cynomys)* (plate 38) are heavily built, short tailed, short legged squirrels (head plus body 28–33 cm, tail 3–11.5 cm). All species are similar in appearance, with grizzled yellow-gray or brown-black coats, slightly paler on the underparts. Their mode of existence is very similar to that of marmots, and they possess a high-pitched barking voice. They excavate a very extensive network of interconnected burrows, which are shared by all members of the social group or *coterie.* Many coteries form a "town", usually covering an area of about 100 hectares. They inhabit open plains and plateaus. The ***Black-tailed prairie dog*** *(Cynomys ludovicianus)* is a typical inhabitant of the buffalo grass prairies, and its numbers have decreased due to control measures by farmers. The ***White-tailed prairie dog*** *(Cynomys gunnisoni)* is found in more mountainous areas. Hibernation occurs but appears to be very fitful. Prairie dogs are commonly preyed upon by hawks and eagles, coyotes, bobcats and bull snakes.

Flying squirrels are represented by 34 species in 13 genera, most of them in the forests of the Old World (Finland to southern Asia, especially Southeast Asia). One genus *(Glaucomys,* ***New World flying squirrels****)* is found in the New World *(G. volans* in southeastern Canada, eastern USA, Mexico and Honduras; *G. sabrinus* in Alaska, Canada and adjoining northern and western USA, and in the Appalachians). All possess a broad fold of skin (the *patagium)* between each forefoot and hindfoot, which is extended by a rod of cartilage at the outer edge of the wrist. The two patagia act as gliding surfaces. True flight does not occur, but after an initial jump the extended patagia enable these rodents to glide over considerable distances from high to lower sites among the trees. The bushy tail also assists and acts as a rudder. Thus, it is claimed that the ***Giant flying squirrels*** *(Petaurista)* of Southeast Asia can glide up to 450 meters, also turning in midair, but a gliding distance of about 100 m is more usual. Two species of *Pteromys* ***(Old World flying squirrels)*** are known. Both are silvery to pale brown with gray underparts and whitish patagia (head plus body 12–23 cm, tail 11–13 cm). *Pteromys volans* inhabits the entire evergreen forest zone of Eurasia from Finland and the Baltic to eastern Siberia. *P. momonga* is restricted to the Japanese islands of Honshu and Kyushu. They are preyed upon by the eagle owl *(Bubo bubo)* and by cats.

Some other sciurids are: *Rhinosciurus laticaudatus* ***(Long-nosed squirrel,*** a monospecific genus found in Malaysia, Sumatra and Borneo); *Lariscus insignis* ***(Three-striped ground squirrel*** or ***Malayan black-striped squirrel);*** *Callosciurus* (Indo-Malaysian ***Beautiful squirrels,*** notable for their gold, brown and red colors with black and white markings); *Hyosciurus heinrichi* ***(Long-snouted squirrel*** from Celebes); *Protoxerus* ***(Oil palm squirrels,*** Africa); *Epixerus* ***(Ebien squirrels*** or ***African palm squirrels);*** *Heliosciurus* ***(Sun squirrels,*** Africa).

Sclera: see Light sensory organs.

Scleractinian corals: see *Madreporaria.*

Sclereids: see Mechanical tissues.

Sclerenchyma: see Mechanical tissues.

Sclerites:

1) Components of the sclerotized exoskeleton of invertebrate; in arthropods, the exoskeletal plates. Dorsal Ss. are called a *tergites,* ventral Ss. are *sternites,* and the lateral Ss. are *pleurites.* Individual Ss. are connected to each other by flexible articulations (cuticular membranes).

2) Skeletal elements of sponges, secreted by sclerocytes, and more commonly known as spicules. They consist of either calcium carbonate or silicic acid. See *Porifera.*

3) Skeletal elements in the mesoglea of the *Octocorallia.*

Sclerocyte: see *Porifera.*

Scleroderma: see *Basidiomycetes (Gasteromycetales).*

Sclerophylls: evergreen plants with stiff, coriaceous (leathery) leaves, e.g. *Laurus, Myrtus, Olea.* They are components of Sclerophylous vegetation (see). See Xerophytes.

Sclerophylous vegetation: vegetation type of regions with a Mediterranean-type climate (mild, moist winters and long, dry, hot summers). Areas of S.v. are found in South Africa (southwest cape), California, central Chile and southwest Australia. Characteristic tree species of Mediterranean sclerophylous forests are *Quercus ilex, Quercus suber, Olea eropaea, Erica arborea, Ceratonia siliqua, Arbutus unedo, Pinus pinea,* various other pines *(Pinus halepensis, P. maritima), Cupressus sempervirens,* etc. Sclerophylous forest is often modified (degraded), taking the form of open xerophytic thickets of evergreen, often thorny, mostly small-leaved bushes and scrubby trees, with spring herbaceous plants (usually developing from bulbs), annuals and geophytes. It is then subject to fire and to degradation by grazing animals, with subsequent erosion. Throughout the Mediterranean region, this particular type of biome of evergreen shrubs is known as *Maquis* or *Macchia.* It is also known as *Garigue* (Italy and France), *Tormillares, Encinar or Jarales* (Spain), *Phrygana* (Balkans), *Fynbos* (South Africa), *Mallee* (South Australia), *Matorral* (Chile), *Monte* or *Chaparral* (Southwest USA and Mexico). Typical vegetation consists of evergreen bushes of the genera *Erica, Cistus, Pitacia, Arbutus* and *Juniperus,* evergreen shrubs of the genera *Ulex* and *Calicotome,* together with herbaceous *Lamiaceae, Liliaceae* and *Orchidaceae.*

Scleropodium: see *Musci.*

Sclerospongiae: *see Porifera.*

Scerotium: spherical or ellipsoid resting organ of certain basidiomycetes. It consists of a hard, tuberous mass of hyphae, which develop transverse walls, forming a pseudoparenchyma of binucleate cells. After a lengthy resting period, the S. produces fruiting bodies. See *Ascomycetes* (order *Clavicipitales),* Ergot alkaloids.

Sclerotome: see Axial skeleton.

Sclerurinae: a subfamily of the *Furnariidae* (see).

Scolecodonts (Greek *skolex* worm, *odons* tooth): fossil jaw parts and jaw apparatus of polychaete worms. The complete jaw apparatus consists of several paired structures; starting with the most posterior elements, these are: 1) carriers of various shapes, 2) denticulate forceps (also known as maxilla I; relatively large compared with other elements), 3) an adjacent basal plate (sometimes absent), and 4) between 2 and 6 denticulate maxilla. Other structures include a single intercalary tooth above the basal plate, lateral teeth marginal to maxilla I, and anterior, unpaired teeth. The mandibles (a pair of noneversible jaw pieces, with a long posterior shaft capped by a frontal plate) are situated outside the pharynx. S. are black, consisting mainly of silicic acid, with smaller quanities of chitin and calcium carbonate, and they range in size from 0.1 to 5 mm. S. first appear in the Cambrian, and are particularly abundant from the Ordovician to the Devonian (with a few less abundant periods in the Jurassic and Cretaceous). They have a worldwide distribution, and their potential as an aid to stratigraphic correlation has not yet been fully explored.

Scolecomorphidae: see *Apoda.*

Scolecophidia: see Snakes.

Scolex: see *Cestoda.*

Scolites (Greek *skolos* post, palisade, *lithos* stone): stone casts of closely packed, parallel, vertical tubes, 0.2 to 1 cm in diameter, which are thought to be fossilized burrows of marine worms. They are particularly common in early Paleozoic sandstone and in Cambrian and Ordovician quartzite (scolite quartzite).

Scolopacidae, ***long-billed waders, sandpipers and allies:*** an avian family in the order *Charadriiformes* (see), consisting of 81 species (in 24 genera) of small to medium-sized (13–61 cm), ground-dwelling wading birds. The slender, straight, decurved, or recurved bill is always as long as the head and sometimes much longer. There is a short hallux (absent in one species), and the 3 front toes are always long. Plumage is cryptic, the upper parts patterned with mottled browns and grays. The family is represented widely in both Hemispheres, but the majority of species breed in the Northern Hemisphere, often at high latitudes, migrating to winter feeding grounds deep in the Southern Hemisphere. Four subfamilies are recognized.

1. Subfamily: *Tringinae* (***Curlews*** and ***Tringine sandpipers***). Curlews, the largest of the *Scolopacidae,* constitute the genus *Numenius.* Of the 8 known species, the ***Eskimo curlew*** *(N. borealis)* is probably extinct. The ***Whimbrel*** *(N. phaeopus;* 38–41 cm, bill 9 cm) has a discontinuous distribution, breeding on moorland and marshy tundra in Iceland, Scotland and northern Europe, northeastern Siberia, Alaska and Hudson Bay, overwintering on sandy and muddy shores at the other end of the globe in Tierra del Fuego, South Africa, Tasmania and New Zealand. The ***Curlew*** *(N. arquata;* 56–63 cm, bill about 12.5 cm) breeds on moorland, heaths and grassland, as well as on dunes and cultivated land across Europe to central Asia, overwintering further south within its breeding range, or undertaking migrations to sub-Saharan Africa and the Oriental region, where it feeds on muddy coasts or inland marshes.

The 9 species of *Tringa* (***Tringine sandpipers*** or ***Tattlers***) breed in temperate latitudes. The ***Redshank*** *(T. totanus;* 28 cm) breeds in various habitats (salt marshes, dunes, lowland grassland, as well as steppes up to 4 500 m) in a broad band from northern Europe and Scandanavia to southern China and Siberia. In winter, it moves to milder coasts in the Northern Hemisphere, or migrates to South Africa or Southeast Asia. The ***Greenshank*** *(T. nebularia;* 30.5 cm) breeds in a band across Scotland, Scandanavia and northern Asia to the Pacific, migrating to shores and swamps in the tropics, and sometimes travelling as far as Australia. The ***Common sandpiper*** *(T. hypoleucos = Actitis hypoleucos;* 19.5 cm) breeds near to fresh or salt water in almost the entire Palaearctic region between 40 ° and 65 ° north, as well as in an isolated area in the highlands of East Africa. It overwinters mainly in the tropics of the Southern Hemisphere, and is often seen on the backs of wallowing hippopotamus. ***Godwits*** and ***Willets*** are also placed in the *Tringinae.*

2. Subfamily: *Scolopacinae* (***Snipes*** and ***Woodcock;*** 25 species). These have relatively short legs and very long bills, which are used for probing the mud. ***Snipes*** (12 species) constitute the genus *Capella* or *Gallinago.* They breed on all continents except Australia, which receives a winter visit from the migrating ***Japanese snipe*** (known in Australia as the Australian snipe). The ***Common snipe*** *(Gallinago gallinago;* 27 cm) has a circumpolar breeding range in the Northern Hemisphere, overwintering further south within the breeding range, or migrating below the equator. Some authorities consider the African snipe *(Gallinago nigripennis;* sub-Saharan Africa) to be a race of the Common snipe; both breed on high moorland. The ***Eurasian woodcock*** *(Scolopax rusticola;* 35.5 cm) inhabits open wet, swampy forests, probing the soft ground for worms. Also placed in the *Scolopacinae* are the North American ***Dowitchers,*** which breed in the muskeg region from Alaska to Hudson Bay, overwintering further south on coastal mudflats and beaches; they have bright red breasts and are sometimes called red-breasted snipes or robin snipes.

3. Subfamily: *Eroliinae* (***Sandpipers and allies;*** 23 species). The sandpipers of this subfamily (including the Eurasian ***Stints***) are also known as ***Calidritine sandpipers*** (from an alternative classification, where they are placed with the Godwits, Dowitchers and Ruff in the subfamily *Calidritinae).* They are small, running, beach birds, all rather similar in appearance, especially in winter plumage, which is brown or gray and paler below. Examples are: ***Dunlin*** or ***Red-breasted sandpiper*** *(Calidris alpina;* 17–19 cm; circumpolar in temperate to Arctic zones; differs from the other species of the genus in having a slightly decurved bill); ***Knot*** *(Calidris canutus;* 25.5 cm; discontinuous distribution in the Arctic, overwintering on northern temperate estuaries and coasts, also reaching Argentina, South Africa and New Zealand); ***Curlew sandpiper*** *(Calidris ferrugina;* 19 cm; breeds in Arctic Siberia, overwintering in Oriental region, sub-Saharan Africa and Australasia); ***Purple sandpiper*** *(Calidris maritima;* 20.5 cm; discontinuous breeding distribution in sub-Arctic and Arctic, and south to Kuriles in the North Pacific). Most aberrant and largest member of this sandpiper subfamily is the ***Ruff*** *(Philomachus pugnax;* 28–30.5 cm). Sexual dimorphism is very pronounced, and only the females incubate the eggs and care for the young. In breeding plumage, the male is brilliantly colored with a head tuft and an elaborate collar or ruff of long feathers, which are expanded in the courtship display. No two males have exactly the same color and pattern of feathers, which consist of various combinations of brown, white, black and buff. In contrast, the females are cryptically colored and well camouflaged. On account of this marked difference in appearance of the sexes, the female is sometimes called a ***reeve,*** while the name, ruff, is retained for the male. Males gather on display sites or ***leks,*** which are visited by females for copulation. The Ruff breeds in northwestern Europe and northern Asia, overwintering in Africa, Southwest Asia and in western and southern Europe.

4. Subfamily: *Arenariinae* (***Turnstones*** and ***Surfbird;*** 3 species). Although the two plump, short-necked Turnstones are the least typical of the family, they are probably the most primitive and most closely allied to

the original ancestral forms. The bill is short, thin and flattened, and slightly, upturned at the tip. Plumage is broadly patterned with contrasting light and dark patches. The ***Turnstone*** *(Arenaria interpres;* 23 cm) breeds in the Arctic tundra around the entire North Pole, migrating down the coastlines of America, Asia and Europe to Tierra del Fuego, the Cape of Good Hope, Australia and New Zealand. The rather darker colored ***Black turnstone*** is very similar in habits and appearance, breeding only on the west and south coasts of Alaska, and migrating along the Pacific coast of America to Cape Horn. The ***Surfbird*** *(Aphriza vergata;* 24 cm) breeds on rocky tundra above 1000 m in the Alaskan mountains, overwintering all along the Pacific coast of North and South America as far as Tierra del Fuego. Out of the breeding season it is a bird of rocky shores and the surfline.

Scolopale: see Scolopidium.

Scolopax: see *Scolopacidae.*

Scolopendra: see *Chilopoda.*

Scolopendrellidae: see *Symphyla.*

Scolopendridae: see *Chilopoda.*

Scolopendromorpha: see *Chilopoda.*

Scolophore: see Scolopidium.

Scolo(po)phorous sensillum: see Scolopidium.

Scolo(po)phorus organ: see Chordotonal organ.

Scolopidium, *scolophore, chordotonal sensillum, scolophorous sensillum:* an insect proprioceptor, which occurs singly, but more commonly in groups, in Chordotonal organs (see). It consists of a linear arrangement of 3 cells, extended between two points of attachment: an elongated sensory neurone, a sheath cell (envelope cell), and a cap cell (attachment cell). Within the cap cell, the elongated process of the sensory neurone (largely ensheathed by the sheath cell) terminates in a sensory rod (a ribbed cuticular structure known as the *scolops* or *scolopale).* In the *amphinematic S.,* the scolopale is joined to the chitin cuticle by a terminal strand that passes through the cap cell. In the *mononematic S.,* the terminal strand is lacking and even the cap cell may no longer be attached to the cuticle. The mononematic scolopidia are no longer extended like a string, and are ofen found in Tympanal organs (see). Many other types of chordotonal organ contain bundles of amphinematic scolopidia.

Scolops: see Scolopidium.

Scolytidae: see Engraver beetles.

Scomber: see Mackerel.

Scombridae: see *Scombroidei.*

Scombroidei: a suborder of the *Perciformes,* containing about 100 predatory species in 45 genera and 7 families, many of which have considerable economic and commercial importance. The upper jaw is nonprotrusible, except in the genera *Scombrolabrax* and *Luvarus.* They are oceanic fishes of the open sea, with fusiform bodies and a powerful, deeply notched caudal fin. Some species are arguably the fastest swimming fishes known. Many species have several small fins behind the dorsal and anal fins. Seven families are recognized: 1) *Scombrolabracidae* (single species: *Scombrolobrax heterolepis);* 2) *Gempylidae* ***(Snake mackerels;*** about 22 species in 15 genera; marine, tropical and subtropical); 3) *Trichiuridae* ***(Cutlassfishes;*** about 17 species in 9 genera; marine, Atlantic and Indo-Pacific); 4) *Scombridae* ***(Mackerels*** and ***Tunas;*** 48 species in 15 genera; marine, tropical and subtropical; see Mackerel, Tunny); 5) *Xiphiidae* [single species: ***Swordfish*** (see)]; 6) *Luvaridae* (single species: *Luvarus,* ***Louvar);*** 7) *Istiophoridae* or *Histiophoridae* ***[Billfishes*** (about 10 species in 3 genera), ***Sailfishes*** (2 species), ***Spearfishes*** (6 species), ***Marlins*** (2 species)].

Scombrolabracidae: see *Scombroidei.*

Scombrolabrax: see *Scombroidei.*

Scophthalmus: see *Pleuronectiformes.*

Scopidae: see *Ciconiiformes.*

Scopolamine: a highly toxic, anticholinergic tropane alkaloid, which occurs in various members of the *Solanaceae,* especially *Datura stramonium* (thorn-apple), *Hyoscymus niger* (henbane) and *Atropa belladonna* (deadly nightshade). S. is levorotatory, and it differs structurally from atropine only in the possession of an epoxy group. Like atropine, S. suppresses the secretion of gastric juice and perspiration, and causes expansion of the pupils. Furthermore, it paralyses the central nervous system, and behaves as a narcotic and intoxicant. The hydrobromide has been used to sedate mental patients, as a preanesthetic, and to control motion sickness.

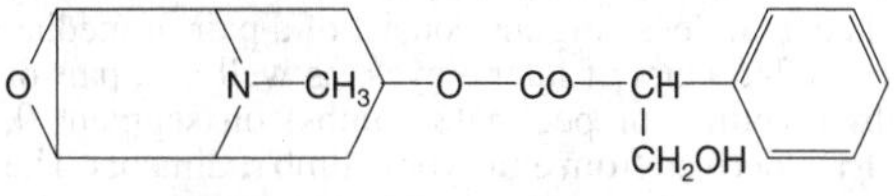

O N–CH_3 —O—CO—CH— CH_2OH

Scopolamine

Scopoletin: a derivative of coumarin. S. is widely distributed in higher plants, e.g. deadly nightshade *(Atropa belladonna),* tobacco, and oat seedlings.

Scops: see *Strigiformes.*

Scopus: see *Ciconiiformes.*

Scorpaena: see Red scorpion fish.

Scorpaenidae, *rockfishes, scorpionfishes:* a marine family of the order *Scorpaeniformes* (see), containing about 330 species, distributed worldwide in tropical and temperate seas. They have a large, broad, often bizarre head with prominent skin appendages, and a wide mouth. The body is laterally compressed. Many are strikingly patterned and colored, various shades of red being very common. Spiny fins are common, and poison glands are present at the base of the dorsal fin. Many are bottom dwellers, e.g. Red scorpion fish (see), while others live in open water, e.g. Redfish (see), Turkeyfish.

Scorpaeniformes: an order of small to medium-sized teleost fishes, mostly marine in all the seas of the world, but with some freshwater species. The head is large and partly or entirely covered with bony armor. All species possess a bony element across the cheeks, and are therefore sometimes called *mail-cheeked fishes.* The body is completely or partly covered with ctenoid or cycloid scales. Spines are common, and the gill covers have two bony processes that terminate in spines. The swim bladder is poorly developed or absent; when present, it is often modified as a sound-producing organ. Representative families are *Scorpaenidae* (scorpionfishes) (see), *Triglidae* (searobins) (see), *Cottidae* (sculpins and bullheads) (see), and the *Cyclopteridae* (lumpfishes) (see). The order also includes the extremely venomous stonefishes *(Synancejidae).*

Scorpionfishes: see *Scorpaenidae.*

Scorpion flies: see *Mecoptera.*

Scorpions, *Scorpiones:* an order of the *Arachnida*

(see) containing about 600 species, which occur in the tropics and subtropics, especially in steppes and deserts. Some occur in less dry, wooded habitats, and in buildings. Only a few species have penetrated into the southern part of central Europe. S. are solitary, nocturnal animals, concealing themselves during the day in fissures and under stones, or burying themselves in the sand.

The elongated body (length 4–9 cm) consists of a short prosoma with 6 pairs of limbs, and a long opisthosoma (divided into mesosoma and metasoma). The posterior part of the opisthosoma (segments 15–19) can be moved in all directions. At the terminus of the opisthosoma is a telson, which is modified to a venom gland with a sting. There are small, pincer-like chelicera, and large powerful pedipalps. The last segment of the pedipalps is usually enlarged into large pincers, which are held horizontally in front of the body. Prey (any arthropod that can be overcome) is held in the pedipalp pincers and killed by crushing. Powerful prey are dispatched with the venomous sting, which is brought forward by arching of the abdomen. The effect of the toxin is usually instantaneous, and the prey is then dismembered by the chelicera. Sense organs consist of a pair of median eyes and 2–5 pairs of lateral eyes, as well as a pair of combs (pectines or pectinate combs) on segment 9, which are derived from embryonic limb rudiments. The pectines are sensitive to mechanical stimulation, and it is thought that they assist the male in selecting a suitable mating ground. External respiration is served by a pair of lung books on each of segments 10, 11, 12 and 13 (segments 3–6 of the mesosoma). Male and female S. can usually be distinguished by their external appearance. Mating is preceded by a complicated dance in which male and female face one another; the male, walking backward, pulls the female with him. Finally the male deposits a spermatophore on the ground, and as he drags the female over it she picks it up with her sex opening. Many species are viviparous. The new-born young climb onto the mother and obtain food by sharing in her feeding activities.

Many species stridulate by drawing the fluted border of an appendage over stiff bristles. Venom toxicity toward humans varies with the species. Desert species in particular deliver an extremely painful sting, and its toxic symptoms persist for a relatively long period. S. use their sting against humans only if they feel threatened or attacked.

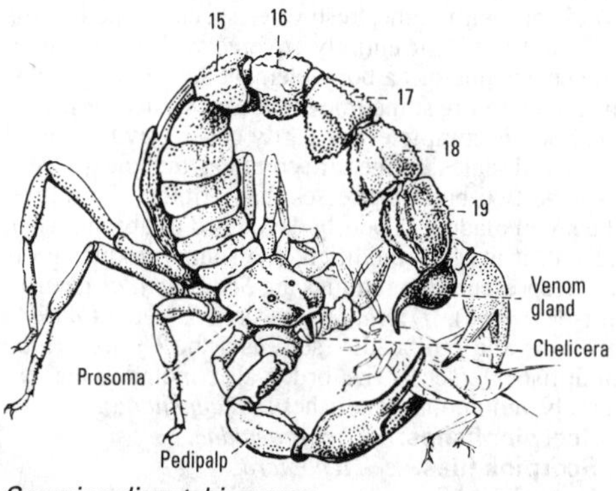

Scorpion dispatching prey

Scorzonera: see *Compositae.*

Scoters: see *Anseriformes.*

Scrapie: see Prion.

Screamers: see *Anseriformes.*

Screw pine: see *Pandanaceae.*

Scrobicularia: see Modes of nutrition.

Scrophulariaceae, ***figwort family:*** a cosmopolitan family of dicotyledonous plants, containing nearly 3 000 species. They are usually herbaceous plants, sometimes shrubs, rarely trees, with undivided, alternate or opposite leaves. The calyx is 5-merous (sometimes 4-merous, the top lobe being absent), and the flowers are pollinated by insects. There are basically 5 stamens which alternate with the corolla lobes, but more frequently 4 stamens are present, the fifth upper one being absent or represented by a staminode. The superior 2-celled ovary develops into a multiseeded capsule, rarely a berry. The family is exceptional in the variety of forms displayed by its members, e.g. various stages in the development of flower structure are observed from actinomorphism to bilateral symmetry; and different modes of existence are represented, including autotrophic, semiparasitic and parasitic. Important medicinal plants are, e.g. the ***European common foxglove*** or ***Purple foxglove*** *(Digitalis purpurea)* and the southeastern European ***Grecian foxglove*** *(Digitalis lanata),* which contain cardiac glycosides. The ***Yellow foxglove*** *(Digitalis grandiflora),* found in warm deciduous forests, on forest edges and in copses and clearings in France, Belgium and the Balkans, is a protected species.

Certain species of ***Mullein*** *(Verbascum)* from western Asia, North Africa and Europe are also official. *Verbascum* species are exceptional among the *S.* in possessing 5 stamens. Common European summer garden flowers are the ***Snapdragon*** *(Antirrhinum majus),* originally from the Mediterranean region; the South African ***Nemesia*** *(Nemesia strumosa)* with a wide range of flower colors; and the Chilean ***Monkey flower*** *(Mimulus luteus),* usually with brown-spotted flowers. Several species of the South American ***Slipperwort*** *(Calceolaria)* are grown as decorative pot plants. Other large genera of *S.* that provide ornamentals are European and Asian ***Toadflax*** *(Linaria);* ***Speedwell*** *(Veronica)* from temperate and cold regions; and the North American ***Pentstemon*** or ***Beardtongue*** *(Pen(t)stemon).*

Semiparasitic *S.* are ***Cow-wheat*** *(Melampyrum* spp.), ***Eyebright*** *(Euphrasia* spp.) and ***Lousewort*** or ***Wood betony*** *(Pedicularis* spp.). A fully parasitic species is the chlorophyll-free ***Toothwort*** *(Lathraea squamaria),* which parasitizes the roots of various woody plants. The fully parasitic genera, *Lathraea* and *Orobanche* ***(Broomrapes),*** are usually placed in the separate family, *Orobanchaceae* (see), which is closely related to the *S.,* but distinguished by the absence of chlorophyll and the parasitic existence of its members.

Scrub fowl: see *Megapodiidae.*

Scrub robins: see *Turdinae.*

Scrub typhus: see *Trombiculidae.*

Sculpins: see *Cottidae.*

Scutigerella: see *Symphyla.*

Scutigerellidae: see *Symphyla.*

Scutigeridae: see *Chilopoda.*

Scutigeromorpha: see *Chilopoda.*

Scutum: see *Ixodides.*

Scyliorhininae: see *Selachimorpha.*

Scyliorhinidae: see *Selachimorpha.*

Scrophulariaceae. a Antirrhinum majus (snapdragon). *b Linaria vulgaris* (toadflax). *c Euphrasia officinalis* (eyebright). *d Veronica chamaedrys* (germander speedwell). *e Lathraea squamaria* (toothwort).

Scyllaridae: see Spiny lobsters.
Scyllarus: see Spiny lobsters.
Scyphistoma: see *Scyphozoa.*
Scyphomedusa: see *Scyphozoa.*
Scyphopolyp: see *Scyphozoa.*

Scyphozoa: a class of the phylum, *Cnidaria.* Scyphozoans occur in 2 main forms, i.e. polyps and medusae.

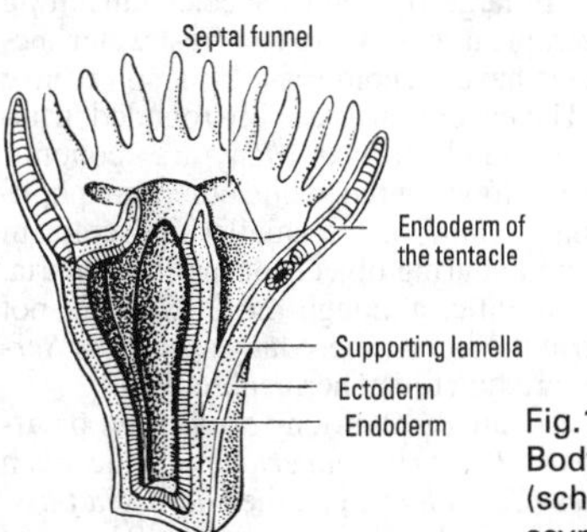

Fig.1. *Scyphozoa.* Body structure (schematic) of a scyphopolyp.

The *scyphopolyp* (or *scyphistoma)* is solitary, and only 1–7 mm in length. It consists of a body and an oral disk, which carries a central mouth and a peripheral circle of tentacles. The gastrovascular space is always divided into 4 compartments by septa. In each septum, the ectoderm sinks inward as a funnel-like structure, infiltrating the cell-containing supporting lamella (mesoglea) and forming a muscular extension to the basal disk. The polyps live in the litttoral and sublittoral zones of the sea, and they represent the small and inconspicuous larval stage of the larger scyphomedusae, which are popularly called "jelly fish".

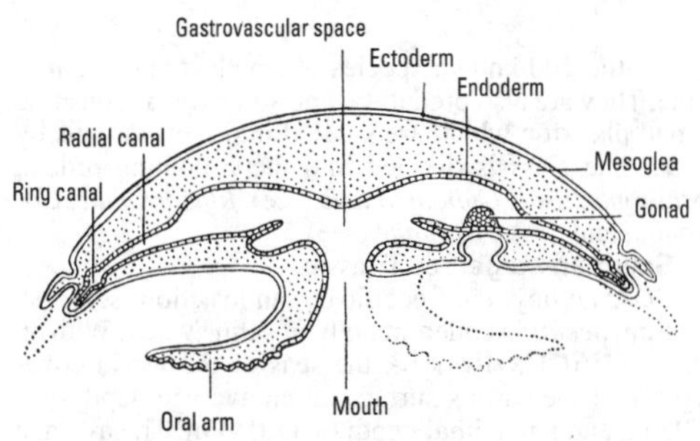

Fig.2. *Scyphozoa.* Body structure (schematic) of a scyphomedusa.

In contrast to the polyps, *scyphomedusae* are relatively large, some species attaining a diameter of 1 m. They are usually shaped like a shallow umbrella, often with numerous tentacles and sense organs around the periphery. Unlike the hydromedusae, the scyphomedusae do not possess a velum. A layer of mesoglea is present between the ectoderm and endoderm. A square-shaped manubrium hangs from the underside of the umbrella (subumbrella), and this is drawn out into 4 or 8 often frilly arms, which aid in the capture and ingestion of prey. The central gastrovascular space above the manubrium is strongly folded into finger-like gastral filaments, which correspond to the mesenterial filaments of the *Anthozoa,* and which secrete enzymes for external digestion. Four radial canals extend from the gastrovascular space, and these are interconnected by a ring canal in the periphery of the medusa. Numerous branched secondary radial canals are also present. The nematocysts of the medusa are often sufficiently powerful to harm humans. Scyphomedusae are usually unisexual, and the gonads are formed in the endoderm of the wall of the gastrovascular space.

Reproduction normally occurs with an alternation of generations. The fertilized egg develops via a Planula (see) into a polypoid larva called a *scyphistoma.* The scyphistoma can multiply asexually by budding, a process in which the daughter polyps always separate from the mother. Alternatively, the tentacles of the scyphistoma may regress, followed by division in the transverse plane at the oral end, a process called Strobilation (see). Many transverse divisions may form at the same time, forming a stack of immature medusae (Fig.3). The single tranverse bud is called an Ephyra (see). The released, free simming ephyrae develop into medusae, which reproduce sexually. Sometimes a polyp is transformed directly into a medusa, whereas in a few species there is no polyp generation, and the planula larva develops directly into a medusa.

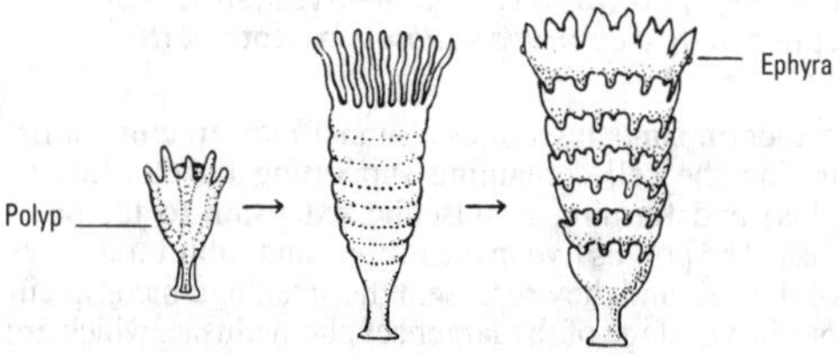

Fig.3. *Scyphozoa.* Strobilation of a scyphopolyp.

All the 200 known species of scyphozoans are marine. They are also predatory, and some forms consume small plankton which are swirled into their mouths by cilia. The class is divided into the following orders: *Stauromedusae, Cubomedusae* (see), *Rhizostomae, Coronata* and *Semaeostomae* (see).

Scythian stage: see Triassic.

SDA: acronym of Specific dynamic action (see).

Sea, *ocean:* a coherent body of saline water. With an area of 361.1 x 10^6 km^2, the seas of the world cover 70.8% of the earth's surface, at an average depth of 3 795 m, and a maximal depth of 11 034 m. The average salt concentration is 3.5%, with considerable variations in certain smaller, partially landlocked seas (0.2% in the northern part of the Baltic Sea, and 4.1% in the Red Sea). The sea is subdivided into different regions, according to the distance from the coast and the depth. The coastal strip down to a depth of 200 m covers the *continental shelf,* after which the bottom drops away relatively steeply as the *continental slope,* followed by the less steep *continental rise,* and finally the *abyssal plain* between 4 000 and 6 000 m below the surface. 75.9% of the total area of the sea lies above the abyssal plain, which is sporadically criss-crossed by deep sea trenches. The sea is divided vertically (Fig.) as follows: 1) the deep sea trenches form the *hadopelagic zone;* 2) water at the depth of the abyssal plain is the *abyssopelagic zone;* 3) the *bathypelagic zone* lies between the depths of 1 000 and 3 000; 4) the *mesopelagic zone* (also called the *transition zone)* lies between 200 and 1 000 m; 5) the *epipelagic zone* lies at the depth of the continental shelf. Biologically, the surface of the sea bottom is divided into: 1) the *littoral zone* (continental shelf), which is subdivided into the *supralittoral* (spray and splash zone of the beach), *littoral* in the strict sense (intertidal) and *sublittoral* (subtidal) zones; 2) the *archibenthal zone* (continental slope); 3) the *bathylal zone* (continental rise); 4) the *abyssal zone* (abyssal plain); 5) the *ultra-abyssal* or *hadal zone* (sides and bottoms of deep sea trenches). The entire sea bottom below the continental shelf is known as the *Profundal* (see). The total sea bed is called the *benthic zone* or *benthal,* and the sum of all vertical zones is the *pelagic zone* or *pelagial.*

The sea flora consists mainly of thallophytes, in particular red and brown algae. Flowering plants, like mangroves, glasswort and seagrass are restricted to the beach *(supralittoral),* the intertidal zone *(littoral)* and the upper part of the continental shelf *(sublittoral).* Phytoplankton develop in the illuminated *epipelagial.*

All animal phyla are represented in the sea fauna, and animals are found in all regions of the sea. Animal organisms of the open water (pelagic animals) form the zooplankton and the necton, while bottom-dwelling animals form the zoobenthos. The variety of species, as well as total population numbers, decrease with increasing depth. Thus, pelagic and benthic animals are distributed from the surface waters to the greatest depths, but the greatest numbers of individuals and of species occur in the sublittoral and epipelagial zones.

See Benthal, Benthos, Littoral, Pelagial, Profundal.

Sea anemones: see *Actiniaria.*

Sea butterflies: see *Gastropoda.*

Sea basses, *Serranidae:* a large family of fishes (about 370 species in about 35 genera) in the order *Perciformes.* They are large fish, most species inhabiting tropical and subtropical seas, with a few freshwater species. Most species have ctenoid scales, but some have cycloid scales. The caudal fin is usually rounded, truncate or lunate, and rarely forked. Many are economically important as a food source, while others are popular sporting fishes. Some of the smaller and colorful species of S.b. are enduring objects of marine aquaria. S.b. are hermaphroditic, although the 2 sexes do not usually develop at the same time. Many species of *Serranus,* however, are functional hermaphrodites.

Examples of S.b. are the 2 m-long ***Stone bass*** or ***Atlantic wreck fish*** *(Polyprion americanus),* the even larger ***Jew-fishes*** *(Stereolepis,* e.g. the ***Giant sea bass,*** *S. gigas;* length up to 200 cm, weight up to 300 kg), and the up to 25 cm-long *Serranellus (= Serranus) scriba.* The carnivorous ***Groupers*** *(Promicrops* and *Epinephelus),* which are often encountered around coral reefs, have sharp, basally linked teeth that can be pointed inward, while the front part of the jaws is often supplied with large grasping teeth. The bodies of the ***Hamlets***

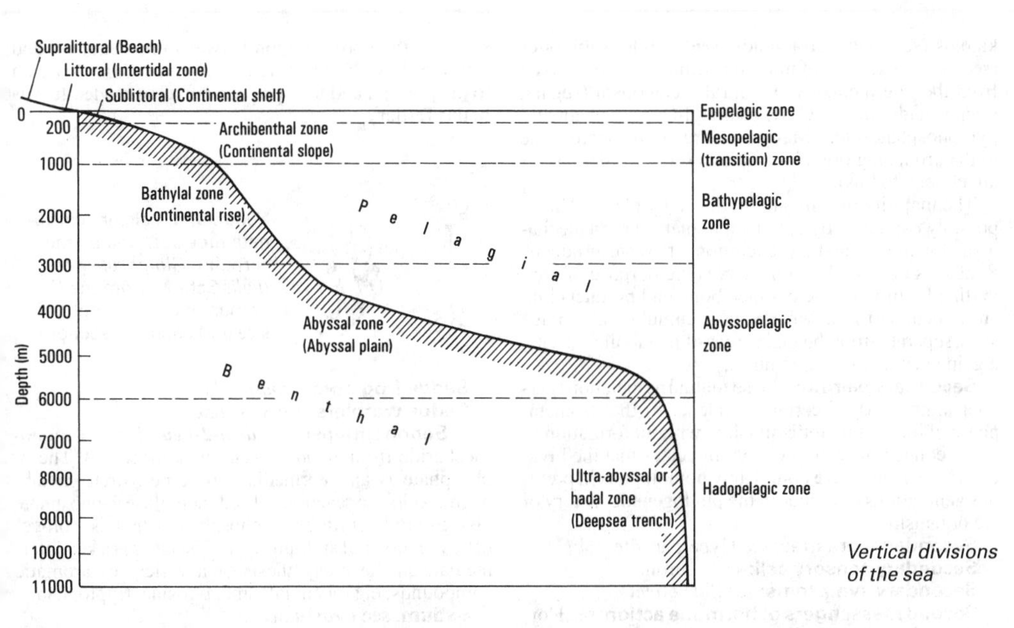

Vertical divisions of the sea

(*Hypoplectrus;* length about 12 cm; east coast of America) are more high set and compressed than those of any other S.b. Apart from their differences in color (blue, yellow, indigo, brown, etc.), the various species of *Hypoplectrus* are remarkably similar, and classed by some authorities as subspecies.

Sea bat: see Batfish.

Sea belt: see *Phaeophyceae.*

Sea-buckthorn: see *Elaeagnacee.*

Sea cow: see *Sirenia.*

Sea cucumbers: see *Holothuroidea.*

Sea ducks: see *Anseriformes.*

Sea eagles: see *Falconiformes.*

Sea elephants: see *Phocidae.*

Sea fans: see *Gorgonaria.*

Sea hares: see *Gastropoda.*

Sea holly: see *Umbelliferae.*

Sea horses: see *Syngnathoidei.*

Sea laces: see *Phaeophyceae.*

Sealions: see *Otariidae.*

Seals: see *Phocidae, Otariidae.*

Sea mats: see *Bryozoa.*

Sea mouse: see *Errantia.*

Sea otter, *Enhydra lutris:* a large member (head plus body 1–1.2 m, tail 25–37 cm) and monotypic genus of the *Mustelidae* (see). Adapted to an amphibious existence, it is found in coastal waters from southern California, north along the west coast of North America and the offshore islands, to the Aleutian Islands, among the Kurile and Commander Islands, and into Asian waters along the coast of Kamchatka. It is usually observed off rocky mainland coasts or island shores, and it rarely ventures more than 1 km out to sea. Head, throat and chest are gray to cream in color, while the rest of the fur is black to dark brown. The large, blunt head is carried on a short, thick neck. The hindfeet are webbed and flattened into broad flippers, while the forefeet are small. Unusually for a marine mammal, there is no subcutaneous insulating layer of blubber, and insulation is conferred by a layer of air trapped in the long, soft fur. It is the only carnivore with 4 incisor teeth in the lower jaw; the cheek teeth are also unique, the molars being broad and flat and adapted for crushing the shells of crustaceans, mussels, etc. At one time, the S.o. was overhunted for its fur, and the species was threatened with extinction, but conservation measures have aided its recovery.

Sea pansy: see *Pennatularia.*

Sea pens: see *Pennatularia.*

Sea pikes: see *Sphyraenidae.*

Searching movements, ***seeking movements:*** helical nutational movements of the growing tips of plant organs, especially rendrils. See Tendril movements, Cyclonasty, Circumnutation.

Searobins: see *Triglidae.*

Sea scorpion: see *Cottidae, Gigantostraca.*

Sea slugs: see *Gastropoda.*

Sea smail: see *Cyclopteridae.*

Sea snail: see *Cyclopteridae.*

Sea snakes: see *Hydrophiidae.*

Sea spiders: see *Pycnogonida.*

Sea stars: see *Asteroidea.*

Sea turtles: see *Chelonidae.*

Seatworm: see Enterobiasis.

Sea urchins: see *Echinoidea.*

Sea wasps: see *Cubomedusae.*

Sebastes: see Redfish.

Secale: see *Graminae.*

Secernentea: see *Nematoda.*

Secondary association center: see Association areas of the cerebrum.

Secondary meristems: see Meristems.

Secondary metabolites: collective term for compounds produced by the *secondary metabolism* of microorganisms, plants and animals. S.m. are often notable for their color, smell, taste or conspicuous physiological action. S.m. of animals are e.g. Toad poisons (see) and Ommochromes (see). Some plant S.m. are Alkaloids (see) and Carotenoids (see), while Antibiotics (see) are an example of microbial S.m. They are derived from the intermediates of central metabolism (sugars, amino acids, acetyl-CoA, malonyl-CoA, isopentenyl pyrophosphate, etc.). Many S.m. are of no apparent use to the producing organism. Others serve for defense or attack (e.g. toxins).

The majority of S.m. are produced by plants. This is possibly explained by the fact that plants lack a mechanism for the immediate excretion of waste products. Synthesis of S.m. therefore serves the purpose of converting harmful or useless metabolic end products into an innocuous form. Most S.m. accumulate at storage sites, separate from the main areas of metabolic activity, e.g. in vacuoles and resin ducts.

Secondary pairing: a loose mutual association (parallel alignment) of certain bivalents in the prometaphase of the first meiotic division, without formation of a true connexion. S.p. probably indicates that the bivalents in question were completely homologous in previous generations or earlier in the phylogenetic history of the organism.

Secondary parasites: see Hyperparasites.

Secondary sensory cell: see Neuron.

Secondary symptoms: see Side effects.

Second messengers of hormone action: see Hormones, Calcium, Cyclic adenosine 3′,5′-monophosphate.

Secretary bird: see *Falconiformes.*

Secretin: a polypeptide hormone, M_r 3 050, containing 27 amino acid residues. S. shows considerable sequence homology with glucagon, vasoactive intestinal peptide and gastric inhibitory peptide, and it is thought that these four hormones evolved from a common ancestral protein by a process of gene multiplication. Production of S. by the duodenal mucosa is stimulated by the acidic pH of the chyle and by large peptides ("secretogogues") from the incomplete hydrolysis of dietary protein, fat and alcohol, i.e. it is generally stimulated by food with the exception of carbohydrate. S. is secreted into the blood. It stimulates the formation and secretion of $NaHCO_3$-rich pancreatic juice and $NaHCO_3$-rich bile, and it inhibits HCl production by the stomach.

As the first proteohormone discovered by Baylis and Starling in 1902, S. led to the development of the hormone concept.

Secretory tissues of plants: see Excretory tissues of plants.

Sedentaria: an order or subclass of the *Polychaeta* (see), including the bamboo worms, coneworms, fanworms and lugworms. They are entirely marine, and the segments and parapodia display considerable variation along the length of the animal. All are burrowers or tube dwellers.

The common Lugworm *(Arenicola marina,* family *Arenicolidae)* attains a length of 30 cm. It does not secrete a tube, but burrows in the sands of shallow and tidal waters, and is used widely as an angling bait.

The serpulids (family *Serpulidae)* secrete a tube of calcium carbonate in which they live. This is usually cemented to the substratum; it may form an arch, a knotted structure, or a spiral, and in transverse section it is circular or oval. Serpulid tubes are known as fossils, and they have contributed to the formation of rock strata, e.g. the serpulite of the German malm, and the serpulid sands in the border region between the Cenoman and Turon of Saxony. They are known from the Ordovician to the present, and they provide numerous index fossils in the Tertiary.

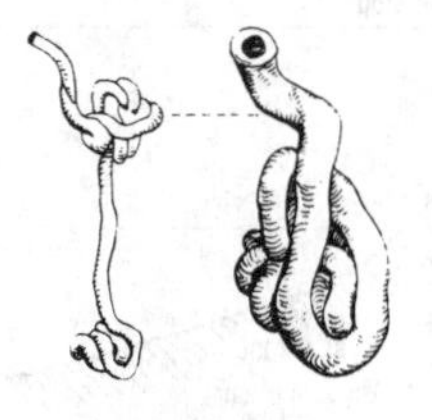

Sedentaria. Calcium carbonate tube secreted by the serpulid, *Glomerula gordialis* Schlot.; from the Cenoman of Saxony (natural size and enlarged section).

Sedge frogs: see *Hyperoliidae.*

Sedge warblers: see *Sylviinae.*

D-Sedoheptulose, *D-altro-2-heptulose:* a monosaccharide from *Sedum spectabile* (stonecrop). The 7-phosphate is an intermediate in carbohydrate metabolism. Aldolase reactions of sedoheptulose 7-phosphate give rise to D-erythrose 4-phosphate, which is a biogenetic precursor of shikimic acid. The latter is a key intermediate in the biosynthesis of a variety of aromatic compounds, e.g. phenylalanine, tyrosine, tryptophan.

Sedum: see *Crassulaceae.*

Seed: a dispersal organ of higher plants, which arises from an ovule (see Flower). The S. contains an *embryo* in a state of temporary dormancy. Special *nutrient storage tissue (endosperm)* is also usually present, and the whole is enveloped by the *seedcoat (testa).* Ss. vary greatly in size, ranging from the giant double coconut of the Seychelles Coco de Mer *(Lodoicea maldivica,* family *Aceraceae),* which weighs several kilograms, to the minute, dust-like Ss. of orchids, which weigh only a few thousanths of a milligram.

Very early in its development, the ***embryo*** can be seen to contain the primordia of the three basic cormophyte organs, i.e. the *radicle, hypocotyl* and *cotyledons.* These usually surround the *plumula* (shoot primordium) of the embryo, which subsequently gives rise to the plant shoot. Some embryos also possess an *epicotyl,* which is a section of stem between the first foliage leaves and the cotyledons. According to the number of cotyledons, a distinction is drawn between monocotyledonous plants (monocotyledons) and dicotyledonous plants (monocotyledons).

The ***nutritive storage tissue*** consists of cells filled with reserve material, especially starch or other storage carbohydrates, storage protein (in particular aleurone granules), or fat, as well as phosphate-containing materials such as phytin. During the early stages of S. germination, the developing and growing embryo relies upon these food reserves. The reserves of some Ss. are also sufficient to support the early stages of the growth of the young plant. The storage tissue either surrounds the embryo or lies to one side of it. In Ss. of the *Gymnospermae,* the nutritive storage tissue consists of haploid primary endosperm, derived from the prothallus which is already present in the embryo sac before fertilization. The Ss. of angiosperms characteristically possess a triploid secondary endosperm, formed by the fusion of two embryo sac cells and a sperm cell. Nutritive tissue may also consist of diploid *nucellus* tissue *(perisperm)* [The nucellus is the central tissue of the ovule, containing the embryo sac and surrounded by integument(s); see

Flower]. According to the nature of the predominant storage material, Ss. can be classified as starchy or starch-rich Ss. (e.g. cereals, buckwheat), oil-Ss. or oil-rich Ss. (e.g. rape, sunflower, poppy, groundnut), and high-protein Ss. (e.g. seeds of the *Papilionaceae).* The minute Ss. of orchids contain no storage material, and they depend upon fungi for their germination.

The ***seedcoat (testa)*** is derived from one (advanced groups), or both (more primitive spermatophytes) integuments of the ovule. Thus, in the most primitive angiosperms and the cycads it consists of an inner, lignified *sclerotesta* and an outer, fleshy, usually brightly colored *sarcotesta.* Ss. of dehiscent fruits have a robust and hard testa, which protects the embryo. In indehiscent fruits, the S. testa is thin and membranous. A testa is absent from the Ss. of certain parasitic plants. The surface of the testa is often sculptured with reticulate ridges and tubercles, etc. Often appendages are also present that aid dispersal, e.g. hairs (cotton), membranous aerofoil borders (willow), and appendages known as elaiosomes containing food or other substances attractive to ants (violet), which therefore gather and disperse the Ss.

The testa often becomes mucilaginous (e.g. flax, quince, tomato), and is then known as a *myxotesta.* The *hilum* is usually visible and prominent on the testa, representing the site at which the S. was formerly attached to the funicle or to the placenta. Often the *micropyle* is also recognizable as a small pore in the testa; this represents the former integument-free entrance to the ovule (in many cases the site of access to the ovule of the pollen tube); in some mature Ss. it serves as a site for the entry of water when the S. germinates. Some Ss. possess a *caruncle,* an outgrowth of the testa which overlaps the micropyle. Thus, in the castor oil S. *(Ricinus communis)* the caruncle is a fleshy, 2-lobed structure, and the micropyle is not an open pore. Testas of anatropous Ss. (see Flower) characteristically possess a *raphe,* a narrow funicular strand supplied with vascular bundles, representing the line of fusion of the funiculus with the integument. In some Ss. (in particular those of the *Papilionaceae),* the entry of water is controlled by a small opening in the testa, known as the *strophiole* or *strophiolar cleft,* which is lined with suberized cells; removal or damage of these cells permits the entry of water. Some Ss. are more or less enveloped by a tissue called the *aril,* which is formed after fertilization from the sarcotesta, e.g. in yew *(Taxus)* and water lily *(Nymphaea).*

S. development commences directly after fertilization of the ovum. First the zygote (I in Fig.) enlarges and divides into a small embryo cell (II E) and a large suspensor cell (II SZ). Further divisions of the latter produce a chain suspensor cells (S in III and IV), terminating with a greatly enlarged *basal cell* (BZ). Basal cell and suspensor cells together form the *suspensor.* The entire structure of suspensor plus embryo cell is the *proembryo,* which is at first a thread-like structure. The embryo cell soon divides into quadrants then octants, forming an early spherical embryo. As the suspensor elongates, it pushes the embryo into the developing endosperm in the embryo sac. Cotyledons (VI Co) and a plumule then become differentiated at the free end of the embryo, the hypocotyl (VI Hy) differentiates more centrally, while the radicle (VI W) is formed near the site of attachment of the embryo to the suspensor (with its root cap directed toward the suspensor, i.e. toward the micropyle). Endosperm formation is completed during embryonal development. Immediately after fertilization, the *endosperm nucleus* is stimulated to divide vigorously. The resulting triploid nuclei continue to divide, and become arranged on the wall of the expanding embryo sac, where they form a *nuclear tapetum.* This is converted into a *cell tapetum* by the synthesis of tranverse cell walls between the free nuclei. Further cell division of the cell tapetum produces the endosperm, which normally fills the entire embryo sac. In the coconut and a few other plants, however, endosperm formation ends prematurely, leaving a large vacuole inside the greatly expanded embryo sac. This vacuole contains a fluid enriched with emulsified fat (e.g. coconut milk) . Degradation of the formed endosperm provides nutrients which are utilized and sometimes also stored by the embryo. In many Ss., the secondary endosperm is totally resorbed during embryo growth, so that the S. contains no separate nutrient store, e.g. in the *Cruciferae.* In other Ss., e.g. beans and peas, endosperm is never formed in the first place, and storage material is usually laid down in the embryo itself e.g. in the cotyledons (Ss. of the *Papilionaceae,* walnut, horse chestnut), or in the hypocotyl (Brazil nut). During endosperm formation, the nucellus of the ovule usually disappears, but in a number of cases, e.g. *Piperaceae* and *Caryophyllidae,* the nucellus serves as a food reserve and becomes the perisperm. During embryogenesis and endosperm formation, the testa is formed from the integuments of the ovule.

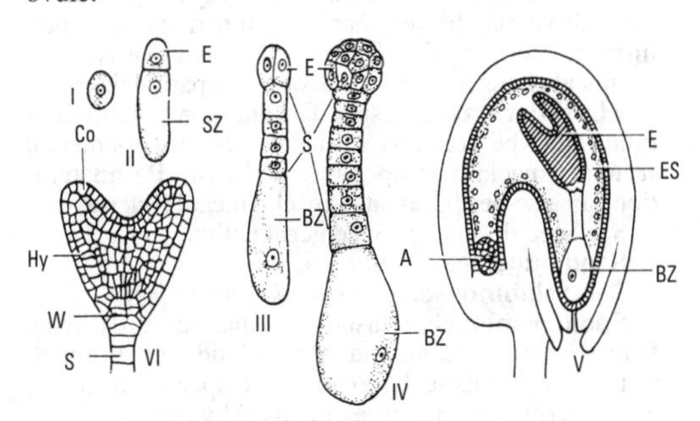

Seed. Embryonal development in the seed of shepherd's purse *(Capsella bursa-pastoris). ES* Embryo sac with nuclear tapetum. *A* Antipodal cells. See text for further details.

In certain plants, sexual reproduction has been lost and replaced by asexual reproduction, so that embryos are formed without fertilization (see Apomixis). Several embryos may also be present in a single S. (see Polyembryony).

S. maturation. This marks the completion of S. development. In the course of S. maturation, growth processes inside the S. come to an end, and all cells lose water and pass from an active into a dormant state (see Seed dormancy). In certain cases, the S. may become dormant before the completion of embryonal development. The more or less continuous process of S. maturation can, for practical (agricultural) purposes, be divided into different stages, e.g. the milk development, dought development, and ripening of cereal grains. After the conclusion of S. maturation and the onset of S.

dormancy, the Ss. of many plant species are potentially capable of germination, i.e. they will germinate if placed under the necessary favorable conditions. In other species, however, germination is possible only after the completion of a period of postmaturation, and these Ss. therefore lie for a long time in the soil. Thus, the Ss. of apple, pear and cherry require a postmaturation period of about 90 days at relatively low temperatures. The postmaturation period for the Ss. of hornbeam, ash and Swiss stone pine is at least a year. A period of postmaturation is also necessary for the Ss. of many spring plants, e.g. winter aconite *(Eranthis hyemalis)* and larkspur *(Delphinium ajacis)*. Postmaturation is initiated endogenously (i.e. not by external factors) and it is demanded for different reasons. For example, in Ss. of pilewort *(Ranunculus ficaria)*, orchids and some species of ash, the embryo is *morphologically immature*, i.e. its morphological development is incomplete and the postmaturation period is required for its completion. In other Ss., e.g. those of pine, cherry or juniper, the embryo is physiologically immature. In these Ss., the excised embryo is found to be morphologically fully developed, but is nevertheless unable to germinate. The reasons for this are incompletely understood, but it has been shown in many cases that various genes are blocked, whose products are necessary for germination. In other cases, germination inhibitors are present (see Inhibitors), and these must be washed or soaked out, or degraded. Possibly the gradual synthesis or activation of germination-stimulating substances is also involved. In addition, the testa, or other covering layers of the embryo may be very hard and impermeable ("coat-imposed dormancy"). Dormancy is then broken by chemical changes in the testa, or by its partial degradation by soil microorganisms. The duration of postmaturation may be crucially influenced by environmental factors, especially temperature and light. Postmaturation can also be abbreviated by chemical influences.

See Seed dormancy, Seed germination.

Seed bugs: see *Heteroptera.*

Seed clubmosses: see *Lycopodiatae.*

Seed dispersal: transport of mature seeds away from the mother plant, as a means of aiding multiplication and geographical spread of the species. Seeds of angiosperm fruits are either released by opening of the fruit wall, or they are distributed together with the fruit wall as an indehiscent fruit.

Active ejection of seeds *(self-dispersal* or *autochory)* by plants *(autochorous plants)* is relatively uncommon. Autochorous mechanisms rely either on turgor (e.g. explosive capsules of *Impatiens,* squidging action of *Oxalis,* squirting action of *Ecballium)*, or on hygroscopic movements (e.g. torsion in legumes and *Dictamnus,* catapulting capsules of *Geranium,* squidging action of various *Viola* species). In the majority of cases, S.d. is passive and due to external forces *(allochory),* like wind, water and animals. In wind dispersal *(anemochory),* the wind shakes the seeds from the open capsule and scatters or blows them away. The minute, dust-like seeds of the *Orchidaceae* are easily blown over large distances, but most wind-distributed seeds and fruits possess hairs or other appendages or outgrowths of the seed wall or fruit coat, which increase the surface area and aid spinning or gliding flight. Animal dispersal *(zoochory)* occurs mainly as *endozoochory,* i.e. an animal eats a fruit, often because of its color or taste, but the hard testas of the seeds resist the digestive processes, and the viable seeds are excreted with the feces. In *epizoochory,* seeds or fruits become attached to animals. Thus, mistletoe berries adhere by their sticky surface, whereas other seeds and fruits possess glandular hairs or barbed hooks. Also, seeds and fruits of many water plants are carried considerable distances in the mud attached to wild water fowl. Many birds and rodents also pick up seeds in the process of feeding and food collection, without actually ingesting them, and letting them fall to the ground again at some distant site.

Dispersal by ants *(myrmecochory)* occurs when a fruit or seed has appendages (elaiosomes) containing food or material attractive to ants, which then gather and disperse them. Elaiosomes occur on the seeds of species of, e.g. *Viola, Cyclamen, Melampyrum, Allium, Galanthus.*

Anthopochory is the dispersal of seeds by humans. In many parts of the world, anthropochores (intentionally introduced agricultural plants and unintentionally introduced arable weeds) have displaced local flora.

Water dispersal *(hydrochory)* occurs in water and shore plants, whose fruits and seeds are made buoyant by air sacs or are unwettable (e.g. seeds of *Nymphaea,* or the utricles of *Carex),* or they possess specialized flotation tissue (e.g. coconut, and swamp and water plants like *Iris pseudacorus* and *Potamogeton).* Water dispersal also occurs in *Sedum* and in members of the *Aizoa-*

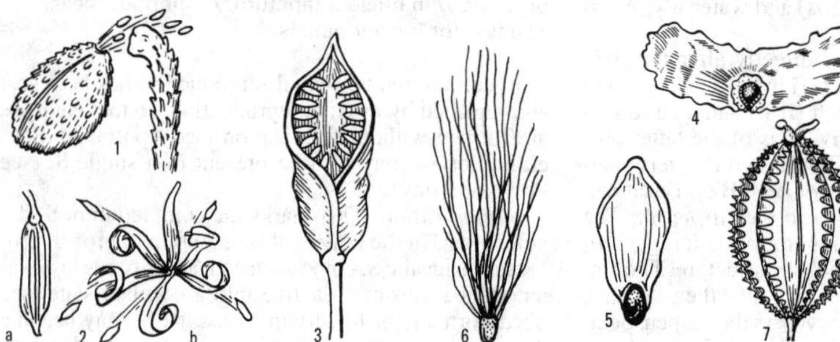

Seed dispersal. 1 Squirting mechanism due to turgor pressure (squirting cucumber, *Ecballium*). *2* Catapulting mechanism due to turgor pressure (water balsam, *Impatiens*), *a* closed fruit, *b* exploded fruit. *3* Multiseeded follicle which behaves as a scatter fruit (larkspur, *Delphinium*). *4* Winged gliding seed *(Macrozanonia,* family *Cucurbitaceae*). *5* Winged gyrating seed (spruce, *Picea*). *6* Seed with flight hairs (willow-herb, *Epilobium*). *7* Burr fruit (carrot, *Daucus*).

ceae, which have hydrochastic capsules, i.e. the capsules open only when wet, so that the seeds are washed out by rainwater.

Seed dormancy: the period of dormancy in seeds between maturation and the onset of the ability to germinate. In the majority of plants, seed maturation is accompanied by an extensive loss of water, and is followed by a period of *total dormancy.* In this latent state, seeds are very resistant to adverse external conditions, such as cold, heat and the effect of chemicals. Depending on the plant species, seeds remain viable for varying lengths of time. Poplar and willow seeds remain dormant for only a few weeks, after which they must germinate or die. Under suitable conditions, seeds of the *Papilionaceae* may remain dormant and viable for 50–150 years, while seeds of the Indian lotus flower may remain dormant and viable for at least 400 years.

The ***duration, causes*** and ***localization*** of S.d. are very varied. Seeds of certain plants, especially agricultural and horticultural plants, are capabale of germination immediately after maturation, i.e. they are only in a state of *imposed dormancy,* which can be abolished at any time by favorable external conditions. Seeds of many other species, however, display *endonomic* or *true dormancy,* which persists, even when conditions are otherwise favorable for germination. Some cases of endonomic dormancy, are due to blockage of the imbibition of water and of gas exchange by the seedcoat. In other cases, growth and development of the embryo is inhibited. Factors inhibiting the ability to germinate are gradually abolished only by a more or less lengthy period of maturation, a process that usually requires specific environmental conditions. On the completion of this maturation process, germination is possible, but if water is not available, the seed remains in a dormant condition, known as *aitionomic dormancy.*

There are different types of embryo dormancy. One cause of S.d. is the *morphological immaturity* of the embryo, i.e. the embryo is incompletely developed, e.g. in seeds of the pilewort *(Ranunculus ficaria),* orchids, and some species of ash. Alternatively, the embryo may be *physiologically immature,* i.e. despite complete morphological development, the embryo still does not grow, e.g. the seeds of cherry, pine and juniper. Physiological immaturity has not been fully explained. It appears to be primarily the result of the blockage of various genes, whose products are necessary for germination. It has been shown that physiologically immature embryos possess all the enzymes of protein biosynthesis, but they do not synthesize mRNA, and therefore do not synthesize proteins. On the other hand, mRNA is synthesized by nondormant, dry seeds. Furthermore, embryo dormancy is enforced by *germination inhibitors,* which may be present in the embryo itself, in the endosperm, in the seedcoat, in the fruit flesh, or in the fruit rind. These inhibitors are mostly derivatives of benzoic acid, cinnamic acid or coumarin, and abscisic acid is also such an inhibitor; they act by inhibiting enzymatic reactions. A quantitative relationship has been demonstrated between S.d. and the concentration of abscisic acid in seeds and fruits.

Termination of S.d. Even more varied than the causes of S.d. are the mechanisms of its termination. Further development of morphologically or physiologically immature embryos to the stage where they are fully formed and capable of germination is known as *post-maturation.* It occurs spontaneously, i.e. without the influence of external factors. A progressive decrease in the histone content of the DNA, which may be associated with the abolition of gene blockage, has been observed during physiological post-maturation of embryos. Germination inhibitors may be removed by elution with water, e.g. the seeds of some desert plants are stimulated to germinate by the washing out of inhibitors by rain. After elution, many germination inhibitors are adsorbed by the soil, and they can also be also adsorbed by artifical germination media, active charcoal or filter paper. If such seeds are sown too densely, the local concentrations of inhibitors remain too high for complete adsorption by the surrounding medium, and germination is still inhibited. In other cases, germination inhibitors appear to be oxidized by atmospheric oxygen. Inhibitors are also removed by enzymatic degradation. Finally, illumination and cold conditions can lead to the disappearance of germination inhibitors. Gibberellins and occasionally cytokinins are often synthesized or released from inactive combination toward the end of dormancy; these compounds can alter the concentrations of inhibitors or act as antagonists of inhibitor action. In some plants, S.d. can be terminated by ethylene, thiourea, potassium nitrate and other substances. Termination of dormancy in the seeds of certain parasites, e.g. of the genus *Orobanche,* requires the presence of the host plant. Similarly, the seeds of certain symbiotic plants require the presence of a symbiotic partner, e.g. orchid seeds require the presence of a suitable fungus. In these cases, germination is stimulated (dormancy is broken) by excreted compounds, e.g. nicotinamide in the case of orchid seeds. In certain plants, e.g. *Nicotiana tabacum* or *Lactuca sativa,* in order to break seed dormancy, the seeds must be illuminated after they have become swollen by the imbibition of water. In contrast to these *light germinators,* the seeds of *dark germinators,* e.g. *Phacelia tanacetifolia,* will germinate only in continual darkness. In *long-* and *short-day germinators,* the termination of S.d. is regulated by photoperiodicity. This action of light on S.d. is due to a reversible photoreaction, controlled by the Phytochrome (see) system, which is also involved in other light-dependent growth and developmental processes in plants. In many cases, low temperatures lead to the termination of S.d., e.g. the seeds of the lime (linden) tree, sycamore, oak and apple. Generally, more or less brief exposure to 0 °C (sometimes only +10 °C) is sufficient. Temperatures below 0 °C are required by *frost germinators,* many of which are alpine plants. In other cases, S.d. is broken by high temperatures, e.g. seeds of soybean, cotton and millet.

S.d. caused by the presence of the seedcoat can easily be abolished experimentally by its complete or partial removal, by its abrasion, or by treatment with sulfuric acid, etc. High and low temperature treatment, as well as illumination or maintenance in the dark are also easily applied in the laboratory.

Seed dressing, ***chemical treatment of seeds:*** application of chemical Plant protection agents (see) to seeds, tubers, bulbs, etc. to protect them from damage by bacteria, fungi, insects, rodents, etc. Dressing is intented to kill disease organisms that may already be present on or in the seeds, and to counteract attack from outside during storage or prior to germination in the soil. There are various dressing techniques, depending on the type of protection required and the type of agent applied. 1. *Steep treatment:* seeds are soaked for a predetermined period in a solution or suspension of the chemical agent, or in the undiluted agent if it is a liquid, then drained and dried. 2. *Sprinkle treatment:* seeds are sprinkled with a liquid agent, or with a solution or suspension, left damp for a definite period, then dried. 3. *Dusting:* seeds are mixed thoroughly with the dry agent in dust form until the seeds are well coated (usually about 2 g of mercurial fungicide per kg seeds). 4. *Slurry treatment:* seeds are mixed with the dry agent in dust form, together with small calibrated quantities of liquid (5–20 ml/kg seeds), to form a slurry, which insures coating without undue wetting. 5. *Wet treatment:* seeds are mixed with relatively small quantities of concentrated liquid agent (100–300 ml/kg seeds); no excess liquid remains, but drying is necessary. 6. *Quick-wet* (or *short-wet) treatment:* seeds are mixed with a small quantity of liquid or concentrated dissolved fungicide (20–40 ml/kg seeds) to insure a good coating. No drying is necessary, provided the seeds are sown within a few days. 7. *Oil-fungicide (Panogen) treatment:* seeds are mixed with extremely small quantities of the slightly volatile mercurial fungicide, Panogen, in oily suspension (1–3 ml/ kg seeds). Subsequent drying is not necessary. The method is also used with highly volatile pesticides such as chlorinated hydrocarbon insecticides. 8. *Fumigation:* seeds are treated in air-tight containers for a definite period with volatile fungicide, nematocide or insecticide. 9. *Pelleting:* seeds are first coated with methocel (a cellulose acetate adhesive agent or "sticker") in dilute solution, then agitated with the pesticide or fertilizer, etc. in dust form, so that each seed becomes a pellet with an outer layer of protectant. Before pelleting, seeds are sometimes treated with fungicide, e.g. sugar beet seeds are steeped in ethylmercuric phosphate before pelleting.

Seed ferns: see *Lyginopteridatae.*

Seed germination: in spermatophytes, the recommencement of embryonal development, after being halted by seed maturation. During S.g., nutrient stores in the cotyledons and endosperm are mobilized, i.e. they are degraded into soluble, diffusable materials that serve the nutritional requirements of the growing embryo. Storage starch and cellulose, as well as other solid carbohydrates are converted by diastase and similar enzymes into soluble, diffusable dextrin or sugars. Stored fatty oils are first metabolized to carbohydrates. Insoluble proteins (aleurone, gluten) are converted into soluble albumins. Provided water is available and the seed is no longer in a state of total dormancy (see Seed dormancy), S.g. starts with the imbibition of water, which leads to swelling. The end of S.g. is marked by the point at which the seedling becomes independent of food reserves in the seed. For practical purposes, easily established criteria are required for the existence of germination and its termination. These are: a) visible emergence of the radical (primary seedling root) from the testa; b) the unfolding of those parts of the embryo that are important for the formation of a normal plant; c) the appearance of aerial plant parts at the soil surface. The term, *seed germination rate,* is often used to indicate the percentage of a batch of seeds that germinate under favorable conditions, rather than the speed at which the germination process occurs.

In *epigeal germination,* the hypocotyl elongates and lifts the cotyledons out of the ground, where they turn green (Fig.). In *hypogeal germination,* the hypocotyl remains short, the cotyledons remain in the ground, and stretching growth of the epicotyl raises the first foliage leaves above the ground, where they turn green (Fig.). Monocotyledonous plants represent a special case. The single cotyledon is usually modified to a scutellum, and the plumule and radicle are enclosed in sheaths (coleoptile and coleorhiza, respectively), which are broken through during germination.

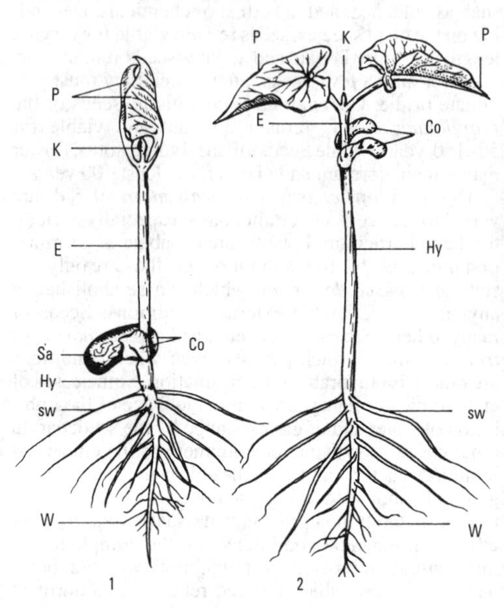

Seed germination. 1 Hypogeal germination (scarlet runner bean, *Phaseolus coccinus). 2* Epigeal germination (French or kidney bean, *Phaseolus vulgaris). Sa* Seed coat. *Co* Cotyledon. *E* Epicotyl. *P* Primary leaf. *K* Shoot apex. *Hy* Hypocotyl. *W* Main root. *sw* Lateral roots.

A series of Inhibitors (see) (blastocolins) are known to act as natural regulators of S.g., as well as certain phytohormones, in particular gibberellins and phytokinins. In barley caryopses and other starch-rich seeds, the stimulation of germination by gibberellins is due to the activation and de novo synthesis of starch-degrading enzymes. Gibberellins accelerate and generally stimulate germination of the seeds of many plants. The response of some species is especially striking, and the effects of gibberellins, temperature and light are clearly interrelated. S.g. can be influenced in many ways by these external factors, although the precise action of temperature and light depends crucially on the degree and type of seed dormancy.

S.g. in certain plant species is promoted by sunlight (light germinators), while in others it is inhibited (dark germinators). Only parts of the spectrum are active. Thus, "bright-red" light (630–680 nm) stimulates, whereas "dark-red" light (720–750 nm) inhibits S.g. Sunlight contains more bright-red than dark-red light, so that light germinators *(e.g. Lepidium sativum, Nicotiana tabacum)* automatically receive a stimulatory light dose. The light receptor systems of dark germinators (e.g. *Am-*

aranthus, Phacelia, Lamium amplexicaule) are more sensitive to the dark-red region of the spectrum, and these seeds are accordingly inhibited by sunlight. This action of light on S.g. is due to a reversible photoreaction, controlled by the Phytochrome (see) system, which is also involved in other light-dependent growth and developmental processes. In addition to this low-energy reaction to red light, the blue region of the spectrum (400–500 nm) also has a certain influence on S.g. (high energy reaction), which is not yet understood in detail. Light sensitivity depends strongly on the state of swelling of the seeds; when seeds are in an air-dried, dormant condition, light has no effect. As swelling progresses, there is sometimes a change in the relative sensitivity to different regions of the spectrum. Many light germinators can be induced to germinate by a single, brief flash of light; germination will then proceed in total darkness. For example, exposure to 15 min of moonlight is sufficient to initiate the germination of tobacco seeds. In other species, the optimal conditions for the initiation of S.g. are provided by several alternating periods of light and dark. A distinction is drawn between *short-day germinators* [e.g. garden cress *(Lepidium)* and western hemlock *(Tsuga)]* and long-day germinators [e.g. birch *(Betula), Kalanchoe* and *Begonia].* The light effect on germination can be modified in various ways by temperature. Thus, certain species of *Amaranthus* and *Physalis* behave as light germinators at high temperatures and as dark germinators at low temperatures. The extent of maturation and post-maturation is also important. The seeds of many light germinators gradually lose their light sensitivity and become indifferent to light with prolonged storage. Many light germinators germinate in the dark when stimulated by gibberellin, and in certain cases by kinetin In certain light-dependent seeds, germination is stimulated by nitrate (in particular potassium nitrate) or thiourea. See Seed dormancy.

Seedsnipes: see *Thinocoridae.*

Seeking movements: see Searching movements.

Segment: see Mesoblast, Metamerism.

Segmental duct: see Excretory organs, Nephridia.

Segmental organs: see Nephridia.

Segmental polyploidy: see Allosyndesis, Alloploidy.

Segmentation: see Mesoblast, Metamerism.

Segmentation of the intestine: see Intestinal motor activity.

Segmented latex duct: see Excretory tissues of plants.

Segregation: separation of homologous pairs of dissimilar alleles and their distribution to different cells. S. normally occurs during meiosis, but it can also arise by somatic recombination in heterozygous cells, and during somatic nuclear and cell division.

Segregational load: see Genetic load.

Segregation generation: offspring (F_2 generation) from a cross within an F_1 generation, where the F_1 generation was the result of a hybridization. The S.g. comprises geno- and phenotypically distinct individuals, resulting from the Mendelian segregation of the heterozygous allele pairs.

Seicercus: see *Sylviinae.*

Seismonasty: a nondirectional movement of a plant or plant part in response to a mechanical shock or vibration (see Nasty). It is due to sudden, reversible changes of turgor in specific regions of tissue (see Variational movements), so that cell contraction due to loss of turgor occurs on only one side of an organ (e.g. in a pulvinus), leading to a hinge-like movement.

Seismonastic movements are usually all-or-nothing reactions, e.g. certain movements of stamens and stigmata (see Flower movements) and leaf movements in some carnivorous plants, e.g. venus' fly trap *(Dionaea muscipula),* and in particular the movements of the leaf pinnules of tropical mimosas ("sensitive plants"), e.g. *Mimosa pudica.* Many different kinds of mechanical stimuli and even wounding or electrical stimuli cause the pinnules of *M. pudica* fold obliquely upward in pairs until their surfaces touch, and finally the entire leaf collapses downward. All these movements are due to changes in the turgor of pulvini, which form the movable joints. Powerful stimuli are even transmitted to neighboring leaves, which then respond in the same way. The stimulus is conducted at a rate of 4–30 mm/s, and in extreme cases (especially after physical injury to the leaf) it travels at 100 mm/s, which corresponds to the slowest impulse conduction in animal nerves. The stimulus can be propagated from cell to cell, each cell stimulating an adjacent cell to develop its own action potential. In addition, stimulated tissues release an excitation substance, which appears to be a nitrogenous hydroxy acid, and which can be transported throughout the entire plant via the phloem, parenchyma and apparently also via the xylem After completion of the seismonastic movement, the process of restitution begins. In the cells that received or conducted the stimulus, restitution proceeds relatively quickly, whereas the cells that responded require more time for recovery, so that leaves do not attain their normal position until 10 to 20 minutes later.

Sei whale: see Whales.

Selachimorpha, *Pleurotremata,* ***sharks:*** a superorder of fishes of the class *Chondrichthyes.* The body is mostly spindle- to torpedo-shaped. The snout extends anteriorly, so that the mouth (well armed with sharp teeth) is ventrally situated. The upper lobe of the heterocercal caudal fin is elongated. The surface area of the short intestine is increased by the presence of a spiral valve. Sharks occur mainly in warm seas, and many representatives bear live young, e.g. *Mustelus mustelus* ***(Smooth hound;*** family *Carcharhinidae), Sphyrna zygaena* ***(Hammerhead;*** family *Sphyrnidae),* and *Squalus acanthus* ***(Spiny dogfish;*** family *Squalidae).* They have little economic or commercial importance, although a vitamin-rich oil is obtained from the liver, and strips of smoked ventral flesh of the Spiny dogfish are sold as "rock salmon", and its dorsal flesh is sold in aspic as "sea-eel". Several species have been reported to occasionally attack humans, but the ***Hammerhead,*** the ***Blue shark*** *(Prionace glauca;* family *Carcharhinidae)* , the 6–7 m-long ***White shark*** *(Carcharodon carcharius;* subfamily *Lamninae,* family *Lamnidae),* and the ***Porbeagle*** *(Lamna nasus;* subfamily *Lamninae,* family *Lamnidae)* are recognized as the most dangerous. Six orders are recognized:

1. Order: *Ctenacanthiformes.* Extinct.
2. Order: *Hybodontiformes.* Extinct.
3. Order: *Hexanchiformes (Notidanoidei),* containing the 2 families *Chlamydoselachidae* (comprising a single species, the ***Frill shark,*** *Chlamydoselachus anguineus)* and *Hexanchidae* ***(Cow sharks;*** 3 genera and 4 species).

4. Order: *Heterodontiformes,* represented by the single family *Heterodontidae* ***(Bullhead, Horn*** or ***Port Jackson sharks;*** 1 genus, 8 species).

5. Order: *Lamniformes (Galeoidea),* with 7 families, about 65 genera and 239 species.

Families: 1) *Rhincodontidae* (represented by a single species, the plankton-feeding ***Whale shark,*** *Rhincodon typus).* 2) *Orectolobidae* ***(Carpet*** or ***Nurse sharks;*** about 11 genera and 28 species). 3) *Odontaspididae (Carchariidae)* ***(Sand tigers),*** with the subfamilies *Odontaspidinae* ***(Sand sharks)*** and *Scapanorhynchinae* ***(Goblin sharks).*** 4) *Lamnidae,* with the subfamilies *Alopiinae* ***(Thresher sharks),*** *Cetorhininae* ***(Basking sharks)*** *and Lamninae* ***(Mackerel sharks).*** 5) *Scyliorhinidae* ***(Cat sharks),*** with the subfamilies *Pseudotriakinae, Scyliorhininae,* and *Proscylliinae.* 6) *Carcharhinidae,* with the subfamilies *Triakinae* ***(Smooth dogfishes)*** and *Carcharhininae* ***(Requiem sharks).*** 7) *Sphyrnidae* ***(Hammerhead sharks).***

6. Order: *Squaliformes (Tectospondyli),* containing 3 families: *Squalidae* [with the subfamilies *Echinorhininae* ***(Bramble sharks),*** *Squalinae* ***(Dogfish sharks)*** and *Dalatiinae* ***(Sleeper sharks)],*** *Pristiophoridae* ***(Saw sharks)*** and *Sqatinidae* ***(Angel sharks).***

Seladang: see *Bovinae.*

Selaginellales: see *Lycopodiatae.*

Selection: the variable contribution of different genotypes to subsequent generations. This may occur naturally due to differences in the Fitness (see) of different genotypes, or it may be due to intentional selection by a breeder. 1) *Stabilizing S.* eliminates deviant individuals from the population, including not only genetically abnormal organisms, but also healthy organisms with certain extremely developed characters. Thus, stabilizing S. maintains the appropriate level of adaptation of the population to its environment, once this has been attained. 2) *Directed S.* eliminates individuals of one extreme, so that the average value of the character in question shifts within the population. Directed S. improves the degree of adaptation or the efficiency of existence of a population, or enables it to adapt to new environmental conditions. It is an essential evolutionary factor (see Natural selection), and its operation can be demonstrated in natural populations. Well investigated examples are the development of resistance to insecticides and antibiotics, as well as Industrial melanism (see). It can also be observed in laboratory populations. 3) *Disruptive S.* eliminates the average individuals of a population, leading either to a greater variability of characters or to Polymorphism (see). 4) see *Frequency-dependent S.*

A special form of S. in natural populations operates in the choice of sexual partner for breeding. This has led to the evolution, inter alia, of patterning and coloration, ornamentation and vocalization, which the male uses as an advertisement to the female or to deter rivals.

Artificial S. is used by breeders to produce efficient races of agricultural plants or farm animals.

Other types of S. include Group selection (see), k-Selection (see), r-Selection (see), Kin selection (see).

Selection coefficient, ***s:*** an index of the advantage or disadvantage of a genotype containing a particular gene or gene combination. It is expressed within the limits of ±1, where +1 represents completely positive selection, 0 is selective neutrality, and –1 represents total contraselection. For example, if only 999 gametes with allele a survive in a generation, compared with 1 000 gametes with allele a^+, then the S.c. of a^+ has a value + 0.001. See Fitness

Selection disadvantage: see Fitness.

Selectionist-neutralist controversy: see Non-Darwinian evolution.

Selection pressure: the rate of change of gene frequencies in a population living under defined environmental conditions, as a measure of the intensity of Natural selection (see). See Mutation pressure.

Selective growth media: Growth media (see), which support the growth of only certain microorganisms. They are used for the enrichment or isolation of the latter. S.g.m. may contain special nutrients which can be utilized by only a few organisms, e.g. chitin-containing media are used for the selective culture of streptomycetes. Other S.g.m. contain inhibitors to prevent the growth of other organisms. For example, S.g.m. for tubercle bacilli contain malachite green, which inhibits other bacteria.

Selective ion uptake: see Mineral metabolism (section on plants).

Selenicereus: see *Cactaceae.*

Selenium, ***Se:*** an element toxic in large quantities, but an essential micronutrient for mammals, birds, many bacteria, probably fish and other animals. A requirement by higher plants is uncertain. Se is an essential component of the enzyme, glutathione peroxidase, which is important in the protection of red cell membranes and other tissues from damage by peroxides. Normal sheep muscle contains a low M_r selenoprotein of unknown function, which is absent from the muscle of Se-deficient sheep suffering from dystrophic white muscle disease. The amino acid, selenocysteine, is present in bacterial formate dehydrogenase and glycine reductase.

Selenium anions compete with sulfate ions for uptake by plant roots, and Se is even incorporated into some proteins in place of sulfur. In low concentrations, Se has a favorable effect on the growth of some plants, but larger quantities cause, inter alia, chlorosis and stunted growth. Sunflowers, maize, wheat, rye and barley display a high tolerance of Se. Soils with high Se contents are found mainly in arid climatic zones, and the high Se content of animal fodder grown on such soils can cause loss of hair and feathers, tooth disease and hoof malformation. These problems can be prevented by the application of sulfate fertilizers, which suppress the uptake of Se by plant roots.

Self assembly, ***cooperative self assembly:*** the generation of complex biological structures by the spontaneous association of their macromolecular components. The occurrence of S.a. indicates that the information required for generation of the correct structure is inherent in the macromolecules. Examples of S.a. are the assembly of actin filaments, microtubules, ribosomes and phages, as well as entire membrane transport systems. The principle of S.a. is important in morphogenesis. It is difficult to establish whether any final structure is determined thermodynamically or kinetically (i.e. dependent on the reaction pathway).

Self domestication: the biological alteration of humans resulting from their own behavior (e.g. manufacture of tools, living in settlements, development of agriculture and animal husbandry). Similarities in the mode of existence and in the resulting biological changes that occur in domestic animals and humans have led some authors to regard human development as a S.d. process. Both live under conditions that are very different from those in the wild, with restricted freedom of movement, a certain degree of control over climatic influences, and planned or voluntary management of the food supply and reproduction. Changes in the conditions of sexual selection and natural selection has led to an alteration of many characters in both humans and domestic animals. The relaxation of instinct-dependent behavior as observed in domestic animals, is an important precondition for the freedom of action displayed by humans. Nevertheless, it is difficult to determine whether these are parallel phenomena, or whether domestication is a critical causal factor.

Self fertilization, ***autogamy:*** fertilization (with subsequent seed formation) of a plant ovule by pollen from the same flower or from a flower on the same plant. *Obligatory S.f.* occurs in peas, soybeans, French beans, runner beans, potatoes. *Facultative S.f.* (S.f. is the norm, but not obligatory) occurs in wheat, oats, barley *(Hordeum distichon* and *H. vulgare),* rice, lentils, common vetch *(Vicia sativa),* birdsfoot *(Ornithopus),* flax, certain lupins *(Lupinus albus* and *L. angustifolius),* yellow clover *(Trifolium campestre),* tomato, lettuce, tobacco, walnut, and all varieties of European grape vines.

S.f. provides a means of sexual reproduction for isolated plants. The process is therefore common among pioneer plants and weeds, as well as plants of island floras, which were first established from single propagules. S.f. is often also the only method of sexual reproduction in extreme habitats with a scarcity of animal pollinators (e.g. Arctic tundra and deserts).

Selfing: forced self pollination; a horticultural Inbreeding (see) technique sometimes used on cross pollinators by preventing access of pollen from other sources.

Self-pollination: see Pollination.

Self replication: the characteristic ability of genetic material to replicate itself. S.r. is a property of the DNA of all DNA-containing organisms, and of the RNA of RNA-viruses. As early as 1922, Muller recognized that the genetic material of the cell is represented by those structures that are capable of S.r. The S.r. of DNA and viral RNA depends on the double stranded structure of these nucleic acids (a temporary condition for viral RNA), which is determined by base complementarity. In addition, it requires specific enzymes, i.e. DNA polymerases and RNA-dependent RNA-polymerases. mRNA, tRNA and rRNA are synthesized by transcription from a DNA (or viral RNA) template; they are not capable of S.r., and therefore do not represent genetic material. See DNA replication.

Self sterility: the inability of pollen to successfully fertilize the ovule of the same plant or the ovule of a genotypically identical plant, i.e. pollination does not lead to seed formation. If different varieties of the same species display S.s., the condition is called Intersterility (see). S.s. is genetically determined. See Incompatibility.

Self stimulation: a technique of brain physiology, introduced by Olds and Milner in 1954. Stimulatory electrodes of fine wire are implanted into specific regions of an animal brain, so that the animal can electrically stimulate neuronal populations of its own brain by pressing a lever with its paw and completing an electrical circuit. Experimental rats then perform S.s. until they are physically exhausted, i.e. they stimulate themselves up to a thousand times an hour. In all investigated mammalian species S.s. occurs when the stimulatory electrodes are placed in certain positions, mainly in parts of the limbic system. S.s. in rats can be demonstrated as early as three days after birth. The animal never reaches a state of satiation, although the rate of S.s. does depend on quantities such as blood pressure. Injection of certain substances into the same brain structures causes an increase (e.g. morphine) or a decrease (e.g. vasopressin) in the rate of S.s. Self injection experiments have also been performed, in which animals self administer solutions of enkephalin or morphine to themselves. It is assumed that stimulation of certain populations of neurons causes the experimental animal to experience positive emotions, in that the stimulated part of the brain houses a reward system. Stimulation of certain other neuronal populations causes negative emotions, thus revealing the existence of a punishment system. The total system is known as the *pleasure and pain system* or the *reward and punishment system,* but it is sometimes difficult to clearly distinguish between pain and negative emotions.

Semaeostomae: an order of the *Scyphozoa* with bowl-shaped bells with scalloped margins. Radial canals extend from the central gastrovascular cavity to the bell margin. The corners of the mouth of the medusa are drawn out into 4 long frilled lips, with ciliated grooves on their inside surfaces. The order includes *Pelagia noctiluca* which generates patches of light on the sea surface, and *Aurelia* (see).

Semaphoront: an organism which for a limited period characterizes a stage of an ecological cycle, e.g. egg, larva, pupa and imago are all insect Ss. The different Ss. of a cycle can belong to totally different biocoenoses, and fulfil different functions in the ecosystem. The same Ss. of one species form a S. population, e.g. the tadpoles of *Rana temporaria* in a pond.

Semicircular canals: see Auditory organ.

Semilunar valves: see Heart valves.

Seminal vesicles, ***vesiculae seminalis:*** glands in male mammals, whose secretion aids sperm motility. They empty into the vas deferens, which then continues as the ejaculatory duct.

Semipermeability: see Osmosis.

Semipalmated goose: see *Anseriformes.*

Semipermanent modification: see Permanent modification.

Semiplume: see Feathers.

Semisterility: a condition of heterozygous hybrids, in which only about half of all gametes produced are functional.

Sempervivum: see *Crassulaceae.*

Semul tree: see *Bombacaceae.*

Senegal kassina: see *Hyperoliidae.*

Senescence: see Lifespan, Ontogenesis, Development.

Senescence factors: see Lifespan, Abscission, Correlations.

Senility: see Age diagnosis.

Senna: see *Caesalpiniaceae.*

Sensation, ***stimulus intensity evaluation:*** the conscious perception of impulses generated in sensory systems in response to specific stimuli from the environment and from various parts of the body. The intensity of S. is correlated with the discharge rate of afferent fibers.

The relation between differences in S. and changes in the magnitude of the stimulus depend on the absolute magnitude of the stimulus. Thus, a weight of 1 kg is easily perceived as different from 2 kg, whereas it is difficult to distinguish between 50 kg and 51 kg, although in both cases the difference is 1 kg. In 1834, Ernst Weber expressed this relationship in the formula: $\Delta S = K \times S$ (Weber's law), where ΔS is the minimal perceivable intensity difference, relative to a reference stimulus or background stimulus *(S); K* is a constant. In 1860, Gustav Fechner refined Weber's law, and expressed the relationship as: $I = K \log (S/S_0)$ (Fechner's law), where I is the subjectively experienced intensity, S is the suprathreshold stimulus used in the estimation of stimulus intensity, and S_0 is the threshold value. In 1953, S.S. Stevens found that the subjective magnitude of a sensation is not proportional to the logarithm of the intensity of the suprathreshold stimulus, but to its *n*th power, expressed as the formula: $I = K \times (S - S_0)^n$, (Stevens' law).

Sense of touch: see Tactile stimulus reception.

Sense organs, *receptor organs:* structures of multicellular organisms, which are supplied with sensory neurons, and which receive sensory stimuli.

In animals, the functional unit of the sense organ is the Sensory cell (see), which responds to specific energy forms or stimuli.

In the simplest case, sensory perception is performed by individual sensory cells or free nerve endings scattered in the epidermis of the external body surface. These may be *indifferent,* i.e. responsive to a variety of stimuli. More or less structurally complicated S.o. (sensory buds, sensory epithelia, etc.) are formed by the aggregation of sensory cells. These are usually associated with auxillary structures, which specify the type of stimulus to which the system will respond (the adequate stimulus), and prevent the system from responding to other types of stimuli (inadequate stimuli). According to the nature of the stimulus and the type of reaction that it initiates, S.o. are classified into the following groups: chemical S.o. *(olfactory* and *gustatory* S.o.); light S.o. *(visual organs);* temperature S.o. *(thermoreceptors, heat-sensitive organs);* mechanical S.o. [organs of touch *(haptoreceptors),* flow, equilibrium *(statoreceptors)* and sound *(auditory organs)].* In arthropods, Chordotonal organs (see) serve as proprioceptive organs.

Plants also possess special anatomical structures for sensory perception, which are usually much simpler in construction than those of animals. Examples are the haptosensitive and chemosensitive glandular heads of the tentacles of *Drosera,* haptosensitive sensory pits in the outer walls of the epidermal cells of some tendrils, and the photosensitive coleoptile tip of grasses. In plants, however, the susception of stimuli is not restricted to S.o.

Sensilla ampullacea: see *Solifugae,* Olfactory organ.

Sensilla (basiconica, coelonconica, placodea, styloconica, trichoid olfactoria): see Olfactory organ.

Sensilli: see Sensory cells.

Sensitive phase: a limited period in the development of an individual for the obligatory learning of environmentlly oriented behavior. The critical period (see Imprinting) is a special case or phase of a S.p. The S.p. for the acquisition of the mother language by human children is approximately the age range 9 to 36 months.

Sensitive plant: see *Mimosaceae.*

Sensitivity: see Irritability.

Sensory cell, *receptor cell, sensilla:* a specialized cell of the ectodermal epithelial tissue. S.cs. may occur individually, or they may be associated to form the sensory epithelium of a sense organ. Three types are recognized (Fig.). 1) A *Primary S.c.* conducts stimuli via its own axon to the axon of a neuron. 2) Stimuli received by a *Secondary S.c.* are conducted directly from the cell by the dendrites of another neuron. 3) *Sensory neurons,* also called *free nerve endings,* are the sensitive, profusely branched dendrites of neurons, with no associated sensory cell. Sensory neurons can also be interpreted as the apical poles of deeply buried S.cs.

S.cs. often possess filiform dendrites, which serve for the reception of stimuli, and are variously known, e.g. as visual rods, auditory hairs, olfactory microvilli, etc. According to their function, the respective S.cs. are called visual, auditory, olfactory or touch receptors, etc.

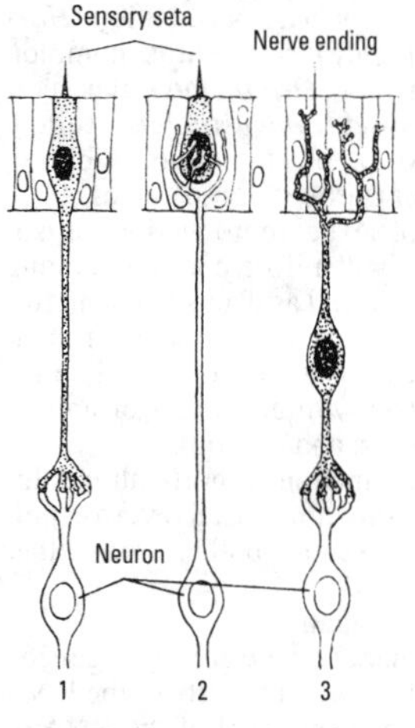

Sensory cell. Types of sensory cells and their associated nerve cells. *1* Primary sensory cell. *2* Secondary sensory cell. *3* Sensory neuron (free nerve ending).

Sensory epithelium: see Epithelium, Sensory cells.

Sensory nerve cell (Sensory neuron): see Neuron, Sensory cell.

Sensory physiology: a branch of Physiology (see) concerned with the functions and mechanisms of animal sense organs.

Sensory transduction: conversion of an external physical chemical stimulus into the electrical response of a sensory cell (the primary electrical response or receptor potential). Stimulus intensity is encoded in analog form by the potential and current of the receptor potential. The receptor current represents an alteration of the Resting potential (see) of the receptor cell, and it can initiate an Action potential (see). Through regulated amplification processes, the relatively low energy of the stimulus can trigger the release of much of the energy available from the metabolism of the sensory cell. Thus, the absorption of a single light quantum with an energy of about 5×10^{-19} J can trigger a reaction with an energy content that is 1 000 to 1000 000 times greater, i.e. the amplification value is 10^3–10^6.

Examples of S.t. are Mechanoreception (see), Chemoreception (see), Photoreception (see) and Electroreception (see Electroreceptors).

Senilis: see Age diagnosis.

Sepia: see *Cephalopoda.*
Sepioidea: see *Cephalopoda.*
Sepiola: see *Cephalopoda.*
Sepioteuthis: see *Cephalopoda.*
Septibranchia: see *Bivalvia.*
Sequoia: see *Taxodiaceae.*
Sequoiadendron: see *Taxodiaceae.*
Seradella: see *Papilionaceae.*
Sergestes: see Shrimps.
Seriata: see Planarians, *Turbellaria.*
Sericulus: see *Ptilinorhynchidae.*
Seriemas: see *Gruiformes.*

L-Serine, *Ser:* L-α-amino-β-hydroxypropionic acid, HO—CH_2—$CH(NH_2)$—COOH, a proteogenic, nonessential, glucoplastic amino acid. It occurs widely in many proteins, and is a notable constituent of silk fibroin. The β-C atom is an extremely important source of one carbon units in intermediary metabolism. Ser is synthesized metabolically from glycine or from 3-phosphoglycerate. In animals and *Escherichia coli:* phosphoglycerate → phosphohydroxypyruvate → phosphoserine → serine (reduction, transamination, dephosphorylation). In plants: phosphoglycerate → glyceric acid → hydroxypyruvate → serine (dephosphorylation, reduction, transamination).

Serin finches: see *Fringillidae.*
Serinus: see *Fringillidae.*

Serology: a branch of Immunology (see); the study of antigen-antibody reactions and the role of complement in vitro, the investigation of the physical chemical basis of these reactions, and the development of test systems for the specific detection of antigens or antibodies. Serological methods (immunoassays) are extremely specific and sensitive, and they permit the detection of antigens down to a few ng or even pg. In the most sensitive techniques, labels (e.g. radioactive isotopes, fluorescent compounds or enzymes) are coupled with the specific antibody, so that even smaller quantities of antigen can be detected (see Immunofluorescence technique). Radioimmunoassays and Enzymeimmunoassays are used primarily for the quantitation of hormones, e.g. in blood or cerebrospinal fluid. Serological methods are so named because the antibodies that they employ are prepared from serum; moreover, in certain assays or detection methods, the relatively impure serum is used. Many assays now employ monoclonal antibodies, which are prepared from hybridoma cultures, and therefore have no association with serum. The term "immunoassay" is therefore more appropriate for all procedures using antibodies. "Serological method" is best reserved for methods that actually use serum.

Serosa: see Serous membrane.

Serous membrane, *serosa:*

1) ***Amniogenic chorion, amniochorion:*** a thin, transparent, fragile membrane, which is formed in the embryonic development of certain vertebrates. It merges into the extraembryonic ectoderm and mesoderm, and surrounds the yolk sac and exocoelom (provided the exocoelom exists and is not filled in by the growing allantois). The S.m. may partially fuse with the outer wall of the allantois to form an embryonic respiratory organ called the allantochorion. See Extraembryonic membranes.

2) A transparent membrane of pavement epithelium and underlying connective tissue which lines certain large internal cavities of the body, e.g. the abdominal peritoneum, thoracic pleura and pericardium are S.ms.

Serows, *Capricornis:* horned ungulates (see *Bovidae)* of the subfamily *Caprinae* (see). Head plus body length is 140–180 cm. Both sexes have horns, which are slightly curved. Upper parts are gray or black, and the underparts are whitish. They favor mountains and high ridges covered with brush or trees at elevations up to 2 700 m. Two species are recognized: *C. sumatraënsis* (central and southern China, Assam, Burma, Thailand, Indochina, Peninsula Malaysia, Sumatra); *C. crispus* (Honshu, Shikoku, Kyushu, Taiwan).

Serpent eagles: see *Falconiformes.*
Serpentes: see Snakes.
Serpent stars: see *Ophiuroidea.*
Serpula: see *Basidiomycetes (Poriales).*
Serpulidae: see *Sedentaria.*
Serpulids: see *Sedentaria.*
Serranellus: see Sea basses.
Serranidae: see Sea basses.
Serranus: see Sea basses.
Serrasalminae: see *Characiformes.*
Serrasalmus: see *Characiformes.*
Serrate antennae: see Antennae of insects.

Serratia: a genus of peritrichously flagellated Enterobacteria (see), represented by the type genus *S. marcescens.* Most strains form red colonies, due to the presence of the intracellular red pigment, prodigiosin. Especially large amounts of the pigment are produced when the organism is grown aerobically on starchy media at room temperature. *S.* is widely distributed in water, soil and foodstuffs, but it is not found in the intestine. In the bacteriological laboratory, *S.* is employed as an oxygen scavenger in closed culture vessels, in order to lower the oxygen tension sufficiently for the growth of anaerobes.

Sertularia cupressina: a thecate cnidarian species of the order, *Hydroida.* The delicately branched, usually spiral, 45 cm-high polyp colonies have a creeping hydrorhiza. They are found in the North Sea and the western Baltic. Although the medusoid stage is degenerate and remains attached to the colony, it is still capable of sexual reproduction. The polyps can completely withdraw into the large hydrothecae, which are arranged in pairs along the stem; each hydrotheca is covered by a lid. Dried and green-dyed colonies are used as a room decoration.

Serum sickness: a clinical syndrome caused by intolerance of foreign protein, in particular foreign serum protein. It is rarely due to the intolerance of other biological substances like penicillin, sulfonamides, etc. It can arise from the injection of large quantities of foreign serum, e.g. antitoxic antisera. Antibodies are produced in the body in response to the injected protein. These antibodies bind to the remaining antigen, and the resulting soluble antigen-antibody complex activates the Complement system (see), which in turn gives rise to inflammatory reactions. Because relatively large quantities of foreign protein are injected, sensitization by, and reaction to, the antigen can occur during a single exposure. As in all other forms of Allergy (see), however, sensitization often occurs during the first exposure, and the reaction is seen during subsequent exposure. Clinical symptoms are vasculitis, uticaria, arthritis, glomerulitis, lymphadenitis, and other life-threatening conditions.

Serval: see *Felidae.*

Sesamoid bone: a small rounded nodule of bone

within the tendon and ligaments of mammalian limbs. The most well developed S.bs. occur in the hands and feet of the less advanced mammals. They arise from a primordium of hyaline cartilage. The patella is an "oversized" S.b.

Sesquiterpenes: aliphatic, mono, di or tricyclic terpenes, formed from three isoprene units (and therefore containing 15 carbon atoms). About 100 structural types are known and about 1 000 natural representatives, forming the largest class of terpenes. Many are found in the volatile oils of plants and can be classified as secondary metabolites with no apparent physiological significance. Some are isolated technically for use in perfumery. On the other hand, the plant hormone, abscisic acid, is also a S.

Sessile oak: see *Fagaceae.*

Seston: the totality of suspended material in water. The live component is called the *bioseston,* e.g. plankton, neuston, pleuston and necton. The nonliving component or detritus is known as the *abioseston.*

Seta: see *Musci.*

Setaceous antennae: see Antennae of insects.

Setae: stiff hairs with thick shafts present in some mammals (e.g. pigs).

Setipalpia: see *Plecoptera.*

Setophaga: see *Parulidae.*

Seville orange: see *Rutaceae.*

Sewage farms: see Biological waste water treatment.

Sewage irrigation: see Biological waste water treatment.

Sewage lands: see Biological waste water treatment.

Sewage ponds: see Biological waste water treatment.

Sewage treatment: see Biological waste water treatment.

Sewall Wright distribution: see Wright distribution.

Sewall Wright effect: the evolutionary influence of random genetic drift as emphasized in the mathematical theory of Sewall Wright. Accordingly, population structure is critical in evolution, and optimal evolutionary opportunity arises when a population is divided into numerous subgroups, which are almost but not completely separated. Some random drift of gene frequency occurs in each subgroup, and a particularly favorable gene combination might arise, which then spreads through the entire population. Selection therefore occurs not only between individuals, but also between subpopulations.

An alternative view, held especially by R.A. Fisher, is that evolution occurs optimally in very large populations. Potential genetic variablilty is then at its greatest, random fluctuations are minimized, and a genotype with only a very small selective advantage will ultimately succeed.

Seychelles frogs: see *Sooglossidae.*

Seychelles tortoise: see *Testudinidae.*

Sex chromatin: sex chromosomes that are stainable with basic or fluorescent dyes in mammalian cells during interphase. One of the two X-chromosomes (see Sex chromosomes) of the female mammal is heterochromatic during interphase. When stained with basic dyes, it can be seen as a densely stained mass at the nuclear membrane or associated with the nucleolus in many cells (e.g. leukocytes), where it is known as a Barr body (see). In granulocytes, this takes the form of a drumstick or a nodule (facultative heterochromatin; see Heterochromatin). In male humans, part of the Y-chromosome chromatin is revealed by fluorescent labeling with quinacrine. Investigation of human S.c. is an important procedure for the early diagnosis of sex, and for the determination of chromosomal sex (e.g. in hermaphrodites). Barr body, heterochromatic X-chromosome and S.c. are all loosely used as synonyms.

Sex chromosomes, *heterochromosomes, heterosomes, gonosomes, allosomes:* chromosomes which contain, inter alia, sex determinants, and are responsible for genotypic Sex determination (see). S.c. differ structurally and functionally from the other chromosomes (autosomes). In particular, all homologous autosomes are similar, whereas homologous S.c. do not always conform to this rule: the females of most plant and animal species, including humans, possess two structurally identical S.c. (X-chromosomes), but females have one X- and one Y-chromosome (male heterogamy). In birds, butterflies, moths, caddis flies and some other insects, the female is heterogametic (female XY; male XX).

An X-chromosome is defined as the sex chromosome that occurs twice in the homogametic sex, but only once in the heterogametic sex. In the heterogametic sex, the partner chromosome is structurally different from the X-chromosome, and is designated the Y-chromosome (XY-type). A partner chromosome may also be absent (XO-type, e.g. in the heteropteran *Protenor belfragei).* Investigation of individuals with abnormal sex chromsomes has led to the discovery of two different sex determinants. The mammalian Y-chromosome carries a dominant M (male) determinant, so that individuals lacking Y (XO, XX, XXX, XXXX) are phenotypic females. The female fruit fly *(Drosophila)* has two X-chromosomes, whereas the male has one X- and one Y-chromosomes. In *Drosophila,* however, the M determinant is not localized on the Y-chromosome, but on the autosomes. In *Drosophila,* the Y-chromosome therefore has no importance in sex determination, and XO individuals are male (not female as in mammals). The F (female) determinant of *Drosophila* consists of several genes and is located on the X-chromosome. Both male and female individuals of *Drosophila* therefore carry determinants for both sexes. Sex is determined by the numerical ratio X:A in the genome (A = autosomal chromosomes carrying M determinants): 2A + 1X = MMF = male; 2A + 2X = MMFF = female, indicating that F is more strongly expressed than M.

In female mammals during interphase, only one X-chromosome is active in transcription. The other X-chromosome becomes heterochromatic and is visible as a Barr body (see Sex chromatin). In males, part of the chromatin of the Y-chromosome fluoresces after treatment with a fluorochrome.

S.c. also contain genes with no direct involvement in sexual differentiation. For example, the X-chromosomes of human, horse, hare and house mouse carry a gene for glucose 6-phosphate dehydrogenase. The X-chromosome of *Drosophila* carries a gene for eye color. Human X-chromosomes carry genes for blood coagulation and for part of the color vision system. Their inheritance is therefore sex-linked, and their mutations can give rise to hemopathies and red-green color blindness, respectively. Aneuploidy of the S.c. in humans leads to *Ulrich-Turner* or to *Klinefelter syndrome.* In Ulrich-

Turner syndrome, one of the two X-chromosomes is missing. These female XO-types (22 autosomal pairs + XO) show stunted growth and marked underdevelopment of the female sex organs. Patients with Klinefelter syndrome are male with one X-chromosome too many (22 autosomal pairs + XXY); they show a marked reduction in the development of the male sex organs and often display a rather female body form.

X- and Y-chromosomes are also present in lower organisms (thallophytes, bryophytes and pteridophytes). In some cases they are morphologically different, and they contain sex-determinant genes.

Sex-controlled inheritance: inheritance of characters, which are expressed differently (quantitatively and/or qualitatively) in males and females.

Sex determinants: see Sex chromosomes, Sex determination.

Sex determination: an early and usually irreversible stage in the development of a bisexual organism, which determines whether subsequent development will lead to predominantly male or female characteristics. Actual *sex differentiation* occurs later. The totality of genes containing information for development of male characteristics is known as the *A-complex,* that for female characteristics being known as the *G-complex.* Under certain conditions a bisexual organism may also express the characters of both sexes *(bisexual potency).* The phenotype, however, usually displays the characters of only one sex. Bisexual potency has been demonstrated in many organisms, e.g. treatment of male tadpoles of the clawed toad *(Xenopus)* with the female sex hormone, estradiol, causes them to develop into females that produce normal fertile eggs. Probably both the A- and the G-complex are present in bisexual organisms, and sex is determined by activation of one and inactivation of the other. S.d. is usually genetic ***(genotypic S.d.),*** but in exceptional cases it occurs independently of the genome and is due to environmental factors ***(phenotypic S.d.).*** The factors responsible for genotypic S.d. are usually present in the Sex chromosomes (see), but in exceptional cases they may also be present in autosomes (e.g. the male determinants in *Drosophila).* In organisms with cytologically differentiated sex chromosomes, S.d. is based on their inheritance. If the homogametic sex is female, then female is determined by the inheritance of XX, whereas individuals carrying XY or X become males. The converse is true in male homogamy.

In metazoan animals and in flowering plants both the diploid and the haploid phase are sexually differentiated. Sex in the diploid phase is determined by the diploid chromosome complement ***(diplogenotypic S.d.),*** and the haploid phase assumes the sex of the diploid phase *(predetermination).* In this process, no part is played by the determinants allocated during meiosis, e.g. females always produce eggs, and males always produce sperm, irrespective of whether these gametes contain determinants of the opposite sex, e.g. a sperm may contain an X-chromosome.

In flowering plants, most metazoa and humans, the female is homogametic, i.e. it is homozygous for the sex determinant (FF, where F = female). The male is heterozygous (FM, where M = male; M is relatively more strongly expressed than F). Therefore, only F-type eggs are formed during meiosis, but 50% of the sperms are F-type and 50% M-type. In this form of female homogamy, S.d. occurs at the moment of gamete fusion: F + F = female; M + F = male (M is expressed more strongly than F). In birds and certain insects such as butterflies, moths, caddis flies and others, the female sex is heterogametic (MM = male; FM = female; F is expressed more strongly than M). In this form of female heterogamy, S.d. occurs during meiosis. The mature egg carries either F or M. In both female homogamy and female heterogamy, the sex ratio is about 1:1. In some animals (e.g. various insects) deviations from this ratio occur due to peculiarities in sex chromosome inheritance, or linkage of S.d. with facultative parthenogenesis (in bees and other hymenopterans). The honey bee carries a sex-determining gene (x) which exists as many alleles. Heterozygous individuals (e.g. x_ax_b) are female. Homozygous individuals (e.g. x_ax_a) are male, and they are eaten by the nurse bees while they are still larvae. Hemizygous individuals (e.g. x_a from an unfertilized egg) are male drones. S.d. is controlled by the queen bee; she can allow her eggs to be fertilized by releasing sperms from her store in the seminal receptacle (these eggs then develop into females); alternatively, she can keep the sphincter of the seminal receptacle closed, so that the eggs remain unfertilized and develop into male drones. As a rule, S.d. is due to the presence of a determinant gene or a closely linked group of such genes ***(monofactorial genotypic S.d.).*** In some cases this determinant function is shared by several or even all the chromosomes of the genome ***(polyfactorial genotypic S.d.).*** Sex is then determined by the relative numbers of alleles for male and female. In these species the sex ratio is usually very variable, due to strain differences in the numbers of male- and female-determinant alleles.

In some species, male- and female-determinant genes are in equilibrium, each cancelling the effect of the other, so that genotypic S.d. is not possible. In these cases, sex is determined by specific external factors which induce phenotypic sex differences. Genuine phenotypic S.d. can only occur among the members of a clone (e.g. in the sporozoan, *Eucoccidium),* because only then is it certain that all members have exactly the same genetic constitution. Sporozoites produced by mitosis from the macrogametes are genetically identical, but under suitable conditions some develop into female and others into male gamonts.

In the echiurid worm, *Bonellia viridis,* the obviously phenotypic S.d. is possibly also influenced by polyfactorial genes, since the sexes are separate and all the offspring cannot be genetically identical. After 4 days attached to the prostomium of an adult female, the sexually undifferentiated larvae of this species differentiate into males [S.d. is influenced by active substances (termones) in the prostomium]. Larvae that make no contact with adult females develop into females. If they are treated for 4 days with an extract of the adult female prostomium, they develop into males. Shorter periods of treatment lead to intersexual forms.

Sex diagnosis: establishment of sex, in particular of fetuses, individuals with abnormal sexual development, parts of corpses, and complete and incomplete skeletal remains. See Sexual dimorphism.

Sex differentiation: see Sex chromosomes, Sex determination.

Sex hormones: steroid hormones produced in the testes or ovaries, and in the adrenal glands of both sexes. According to their biological action, they are classified as male S.h. (see Androgens) and female S.h. (see Estrogens, Gestagens). All types of S.h. are produced by both sexes, but in different quantities. Ovaries secrete larger quantities of estrogens than of androgens, whereas the testes produce large quantities of androgens and only small quantities of estrogens. Biosynthesis and secretion of S.h. by the testes and ovaries is regulated by Follicle stimulating hormone (see) and Luteinizing hormone (see), both produced in the hypophysis. S.h. are responsible for the growth and function of sex organs and for the development and maintenance of secondary male and female characters.

Sex linkage: behavior of genes (sex-linked genes) belonging to linkage groups situated on the sex chromosomes. Sex-linked genes become segregated in meiosis in the same way as the sex chromosomes that carry them. S.l. is manifested as sex-linked inheritance. See Holandric genes, Hologynous genes.

Sex-linked inheritance, ***allosomal inheritance:*** inheritance of genes located on the Sex chromosomes (see) rather than on the autosomes.

Sex ratio, ***sexual index:*** numerical ratio of males to females in a population in which the sexes are separate. The S.r. is influenced by the mechnaism of sex determination in the population. For example, female honey bees (queens and workers) develop from fertilized eggs, whereas males (drones) develop from unfertilized eggs. The colony contains a few hundred males, but thousands of females; usually only one of the females (the queen) is sexually active.

The primary S.r. is that immediately following fertilization; the secondary S.r. is that at birth (or hatching); the tertiary ratio is that at sexual maturity. A statistically constant ratio of males to females is expected from the genotypic sex determination (x/y-, x/o-type) found in many organisms. As a result of differences in the activity and mortality of individual gametes, however, the primary S.r. can deviate from the predicted value even at the stage of fertilization. Further alterations of the S.r. may occur during the period leading to sexual maturity, due to differences in the viability of the sexes. In fish and bird populations, males often exceed females, whereas females are in the majority in populations of housemice, sheep and horses. The human S.r. for newly born children is 105 males: 100 females. With increasing age, this ratio shifts in favor of women, due to their longer life expectancy. There are also considerable regional differences in the human S.r.

The S.r. of animal populations is an important ecological factor. It is subject to seasonal variations; frequently the number of females, i.e. the rate of offspring production, is optimized, so that males become predominant in a pessimum (see Ecological valency).

Sex-restricted inheritance: existence of certain genes in both sexes, which are expressed phenotypically only in one sex. For example, the defective gene for blood coagulation factor VIII, which is responsible for hemophilia A, is carried by females but only expressed in males.

Sexual dimorphism: occurrence of morphological differences, other than primary sexual characters, between the sexes of a species. In animals as in humans, S.d. is determined primarily by the sex chromosomes. Normally, the somatic cells of the female carry two X-chromosomes, those of the male one X- and one Y-chromosome. Even in the embryo and during childhood, a number of sexual differences arise, in addition to the obvious primary differences in the sexual organs, e.g. the time of appearance of centers of ossification, the timing of tooth eruption, shape of the head and pelvis, general body proportions, etc. Sexual characters may be primary, secondary or tertiary. Primary characters are the sex organs and their associated glands. Secondary characters include hair growth, voice quality, development of mammary glands, whereas tertiary characters include body size, bone structure, cardiac and respiratory performance, etc. Sexual differentiation is completed in puberty, mainly under the influence of pituitary and gonadal hormones; it includes both physical and psychological changes, and the psychological development is greatly influenced by social relationships.

S.d. is qualitatively similar in all human populations, but its intensity of expression may differ. Female body mass is always 90–96% of male body mass. Women have somewhat different body proportions, expressed mainly in a smaller shoulder width compared with the pelvic width, relatively long trunk, relatively short extremities with small hands and feet, and particularly short forearms. The female skeleton is more slender and delicate with a less pronounced muscle relief; it accordingly supports a less developed musculature, but there is more subcutaneous fat. In men, skeletal muscle and skeleton account for about 60% of body mass, while fat accounts for 20%. In women, these values are 50% and 30%, respectively. S.d. is particularly pronounced in the pelvis. The hip bones are spread more widely in women, and the outlet of the pelvis is larger and is transverse oval in shape, whereas in men it is rather heart-shaped; the angle between the two pubic bones is larger and the sacrum is broader in women than in men. As a rule, the S.d. of the body skeleton is further emphasized by the soft tissues that cover it. The volume of the female skull is on average about 100 cm^3 less, the frontal bone rises steeply above the nasal bone, the crown of the head is flattened, and the tuberosities of the frontal and parietal bones are usually pronounced. The muscle relief of the male skull is powerfully developed, the forehead lightly receeding, the crown of the skull well arched, and the tuberosities of the frontal and parietal bones hardly palpable. S.d. of the skull and skeleton is important for sex diagnosis of skeletons and skeletal parts in paleoanthropology and forensic medicine.

All possible degrees of expression of S.d. occur in humans. Truly intermediate forms represent a condition known as Intersex (see). With an incidence of 0.1–0.2%, the present world population contains several million intersexual individuals.

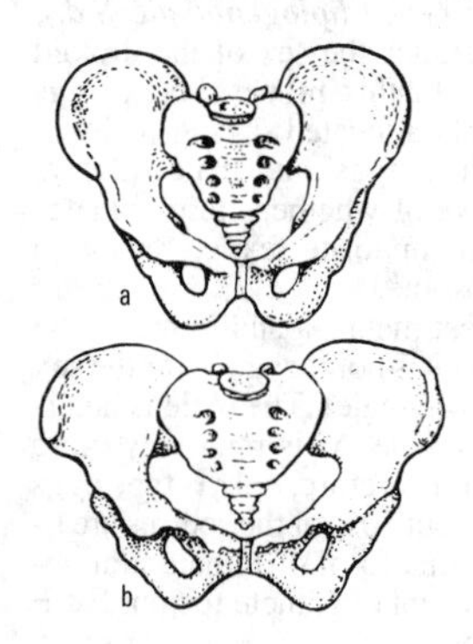

Sexual dimorphism. Human pelvis. *a* Male. *b* Female. (After Bach)

Sexual maturity: developmental stage and age at which the sex glands become functional, the sex drive and ability to copulate are manifest, and secondary sex characters develop. In humans this stage of development is known as *puberty.* S.m. is a precondition for subsequent normal sexual activity. Time of onset of S.m. varies according to species and race, and is subject to wide variations. It is genetically determined, and is influenced by environmental conditions, e.g. it may be delayed by poor nutrition.

Sexual partnerships: in the animal kingdom, types of companionship and interaction between sexes, characterized by their duration and intimacy. S.p. may represent the first stage in the formation of societies (see Sociology of animals, Colony formation).

Absence of pair bonding. Sexual partners meet only briefly for the transfer of gametes, then quickly separate. This type of union is often arbitrary and forceful (e.g. in many worms), but in other cases it may be linked to complex pairing behavior. (e.g. in scorpions). It is displayed by most invertebrates, many fishes, all amphibians, most reptiles, and some birds and mammals. If such a union is largely unregulated and the choice of partners is random, the mating is said to be *promiscuous.* Termites, dung beetles, some crustaceans and some diplopods appear to be exceptional, in that thay remain together for a relatively long time. However, the close, lifelong bodily contact between the sexual partners of some trematodes (e.g. *Schistosoma haematobium)* and nematodes cannot be classified as pair bonding in the present sense.

Pair bonding in the animal kingdom is characterized by an internal drive to remain together after copulation. Birds may remain bonded until one brood is reared (site bonding), for the entire breeding period (annual bonding), or for several years or for life (permanent bonding).

Among mammals, the orang-utan and gorilla probably pair for life. Both live in family groups. Dolphins, rhinoceros and pigs display annual or permanent bonding. Many herd animals remain together only during the mating season, e.g. red deer. Insectivores, bats and rodents are entirely promiscuous.

Special forms of sexual partnership are *polygamy* (e.g. the harems of male hooved animals and of male baboons) and polyandry (e.g. painted snipe and other bird species, in which the females possess the secondary sexual characters and fight one another for possession of males).

Sexual drive, an instinct for brood care, and social drive are the most significant factors in the establishment of the more stable types of pair bonding. These same factors therefore represent essential preconditions for the establishment of societies by mammals and colonial animals.

Sexual reproduction: see Reproduction.

Shad, *Alosa:* a genus of fishes of the family *Clupeidae,* order *Clupeiformes,* distinguished from other genera of the *Clupeidae* by very long scales at the base of the caudal fin. All species are anadromous, i.e. they migrate into fresh water to spawn. ***Allis shad*** *(Alosa alosa = Clupea alosa)* is found in the eastern Atlantic coastal waters of Europe and Scandanavia, the North Sea and the Mediterranean. Its flesh is very tasty, but on account of its scarcity it is no longer economically important. In particular, it has practically disappeared from the rivers of northern Europe. ***Finta shad*** or ***Twaite shad*** *(Alosa fallax = Clupea fallax)* has a similar geographical distribution to the Allis shad, but has little economic importance. ***Black-headed shad*** *(Alosa kessleri = Caspialosa kessleri)* is found in the Black and Caspian Seas; its flesh is of high quality, and it has local economic importance. ***Caspian shad*** *(Alosa caspia = Caspialosa caspia),* also an inhabitant of the Black and Caspian Seas, exists as several subspecies; it is good to eat and commercially quite important.

Shaddock: see *Rutaceae.*

Shade leaves: foliage leaves on the shady north side and in the interior of the crown of a tree, e.g. beech *(Fagus sylvatica).* They are generally much thinner and less differentiated than Sun leaves (see) on the same plant, and they usually lack typical palisade parenchyma. These anatomical differences are the result of the relatively low intensity of illumination (see Light as an environmental factor), as well as the accompanying alterations of transpiration, nutrient supply, and the ratio of potassium and calcium ions. The leaf type is often determined during the preceding vegetative period, i.e. by the light intensity during development of the buds containing the leaf primordia.

Shade plants: see Sun plants.

Shadflies: see *Ephemeroptera.*

Shags: see *Pelicaniformes.*

Shaketails: see *Furnariidae.*

Shaking platform: a motorized horizontal platform or rack, which performs a reciprocating circular motion (rotary shaker) or oscillates. It carries holders for culture vessels, and is used for growing shake cultures of microorganisms and plant tissues.

Shamas: see *Turdinae.*

Shannon-Wiener diversity index, ***H:*** an index of diversity calculated as

$$H = \sum_{i=1}^{s} (Pi)(\log_2 p_i),$$

where p_i (the probability that one individual belongs to species i) can be derived from $p_i = N_i/N$, where N_i is the number of individuals of species i and N is the total number of individuals. The minimal value of H is attained when all individuals belong to the same species, and the maximal value when each individual belongs to a different species.

Shantung silk: see *Bombycidae.*

Shaoyao: see *Paeoniaceae.*

Shared settlement: see Relationships between living organisms.

Sharks: see *Selachimorpha.*

Sharp-tailed streamcreeper: see *Furnariidae.*

Shchistomordnik: see Halys viper.

Shearwaters: see *Procellariiformes.*

Sheathbills: see *Chionididae.*

Sheath-forming bacteria: see *Chlamydobacteriaceae.*

Sheep, *Ovis:* medium-sized ungulates of the subfamily *Caprinae* (see), family *Bovidae* (see), with curved, often spirally twisted horns. All S. have lachrymal grooves and interdigital glands, which are absent in goats. Various species and subspecies of wild S. inhabit hilly and mountainous areas in Eurasia and North America, and they have the widest distribution of all horned ungulates. Like goats, which also have mountainous habitats, the distribution of S. is insular, so that the exact number of species and subspecies is a matter of debate; the approximately 37 different forms that are recognized may be classified into 8 or 2 different species, depending on the authority. ***Domestic sheep*** *(Ovis ammon v. aries)* are descended mainly from Near Eastern wild forms, in particular the ***Mouflon*** *(Ovis ammon musiman).* The Mouflon, the smallest subspecies of wild sheep, now exists as small, poorly protected populations on Sardinia and Corsica, and it has occasionally been reintroduced into parts of Europe. At the beginning of the Late Stoneage, it was distributed from eastern and southern Europe to the Mediterranean.

Shelducks: see *Anseriformes.*

Shell: the hard exterior structure housing many animals, e.g. bivalves, snails, etc. It consists of chitin or calcareous material (mostly calcium carbonate).

Shellac: a resinous exudate encrusting the body of the female East Asian scale insect *(Tachardia lacca).* Principal trees visitied by by the lac insects are *Aleurites laccifera (Euphorbiaceae), Ficus religiosa (Urticaceae), Schleichera trijuga (Sapindaceae), Butea frondosa (Papilionaceae), Acacia arabica (Papilionaceae)* and *Cajanus indica (Papilionaceae),* the last two species being cultivated for the purpose of S. production. The raw product consists of the broken twigs of trees, which are coated with up to 7 mm of brown to red S. containing the bodies of the insects. This crude S. is removed, crushed, and extracted with water, leaving partially decolorized S. still containing insect debris. To prepare the S. of commerce, this last product is melted by heating, filtered through cloth, then allowed to set. S. is largely used in the manufacture of varnishes.

Shelled amebas: see *Arcellinida.*

Shell gland:

1) In branchiopods, an excretory organ, the duct of which is visible through the wall of the carapace.

2) A gland of the female reproductive apparatus, which secretes material involved in the formation of the egg shell, e.g. in platyhelminths and elasmobranch fishes.

Sheltopusik ("yellow belly" in slavonic languages), ***European glass lizard, European glass snake,*** *Ophisaurus apodus:* the largest member of the *Anguidae* (see). Chestnut brown in color with a yellowish underside, it is as broad as a child's arm and attains a length of 1.25 m. It is limbless, with only scale-like rudiments of the hindlimbs. The diamond-shaped scales are underlain by tough bony plates, and arranged in rings around the body There is a deep, longitudinal, lateral groove on either side of the body. It inhabits warm, dry areas of the Balkans, southern Europe and western Asia, and is active during the day In addition to snails, it eats small lizards, mice and young birds. The female lays 6–10 elongated white eggs.

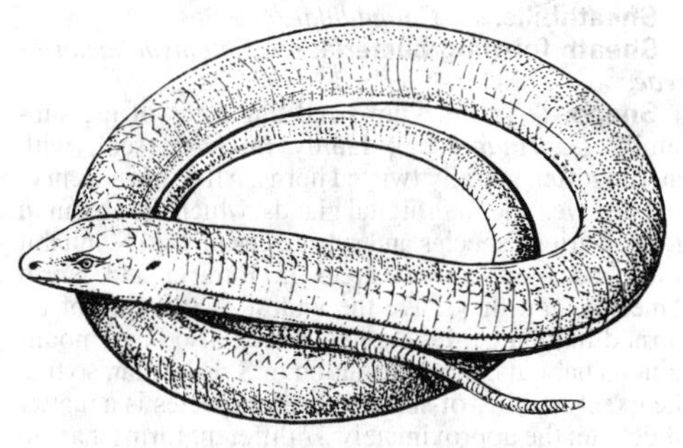

Sheltopusik (Ophisaurus apodus)

Shield bugs: see *Heteroptera.*

Shigella: a genus of nonmotile, rod-shaped Enterobacteria (see), which occur in the intestinal tract of humans and apes, and which are responsible for diseases that vary in severity from mild diarrhea to acute bacillary dysentery. Dysentery is common throughout the world; it is characterized by a sudden onset of abdominal pain, pyrexia and prostration, with frequent stools that eventually become composed of blood and mucus. The type species is *S. dysenteriae.* The genus is named after its discoverer, the Japanese bacteriologist, U. Shiga.

Shikimic acid: a hydroxylated, alicyclic carboxylic acid, isolated, inter alia, from Japanese star-anise *(Illicium anisatum),* ginkgo leaves, figs and pine needles. It is an intermediate in the pathway of aromatic biosynthesis in plants and microorganisms (generally known as the "shikimic acid pathway"), which produces the benzene ring system of phenylalanine, tyrosine, tryptophan and *p*-aminobenzoic acid, and the ring systems of ubiquinone and vitamin K_2.

COOH
HO OH
OH

Shikimic acid

Shining leaf chafers: see Lamellicorn beetles.

Shinisaurus: see *Xenosauridae.*

Shipworms: see *Bivalvia,* Boring bivalves.

Shivering: muscular movements of mammals, which serve to increase heat production. S. arises from excitation of the heat-promoting center of the posterior hypothalamus (see Temperature regulation of the body). Strong impulses are transmitted from the hypothalamus into the bulboreticular formation and red nucleus of the hindbrain. Signals from these regions increase muscular tone and therefore muscular heat production. In addition, neurons of the spinal cord promote the stretch reflex, and facilitation of these neuronal circuits causes oscillation, i.e. contraction of a muscle in response to the stretch reflex stretches the antagonistic muscle, which then displays its own stretch reflex, etc., setting in motion visible and conspicuous oscillations known as S., which have a frequency of 10–20 Herz. The resulting muscular activity may increase body heat production by 300–400%. S. is abolished by curare and other drugs that act on motor endplates (muscle relaxants).

Shock organ: the organ most affected in Anaphylaxis (see). Mammals are affected differently by the pharmacologically active substances released in anaphylaxis. The guinea pig dies from anaphylactic shock by suffocation, whereas the rabbit dies from heart failure.

Shoebill: see *Ciconiiformes.*

Shoe flowers: see *Malvaceae.*

Shoemaker frog: see *Myobatrachidae.*

Shoot extension growth: see Extension growth.

Shoot mutant: see Chimera.

Shore bugs: see *Heteroptera.*

Shore crab: see Swimming crabs.

Shore lark: see *Alaudidae.*

Shoreweed: see *Plantaginaceae.*

Short-day plants: plants that are sensitive to the photoperiodicity of their environment, and which under natural conditions produce flowers only when the period of darkness is not less than a certain critical minimum. Although they are called short-day plants, they are really "long-night" plants, the crucial factor in the initiation of flowering being the length of the period of uninterrupted darkness. Typical short-day plants are often native to the tropics, e.g. chrysanthemums, poinsettias, certain species of soybean and tobacco, *Amaranthus caudatus, Kalanchoe blossfeldiana, Perilla frutescens, Salvia occidentalis* and *Xanthium pensylvanicum.* In some plants, flower formation is favored by, but is not necessarily dependent on, short days (long nights), e.g. pineapple, cotton, coffee, rice, sugar cane, *Cosmos bipinnatus, Primula malacoides* and *Salvia splendens.*

See Photoperiodicity, Flower development.

Short horned toad: see Horned toad.

Short-long-day plants: plants that are sensitive to the photoperiodicity of their environment, and which do not flower until they have experienced first short-day conditions and later long-day conditions (see Short-day plants, Long-day plants, Short-long-day plants). Examples are canterbury bell *(Campanula media),* white clover *(Trifolium repens),* devil's bit scabious *(Succisa pratensis,* or *Scabiosa succisa),* and *Echeveria harmsii.* Under natural conditions, they flower in the late spring or early summer. See Photoperiodicity, Flower development.

Shortwings: see *Turdinae.*

Shou: see Red deer.

Shoulder girdle: see Pectoral girdle.

Shovelers: see *Anseriformes.*

Shovelfoot frogs: see *Myobatrachidae.*

Shovel-nosed frogs: see *Hemisinae.*

Shovel nose lobsters: see Spiny lobsters.

Shovel-snouted legless lizards: see *Anniellidae.*

Shrikes: see *Laniidae.*

Shrimps: a collective name for the members of 3 groups of macrurous crustaceans of the order *Decapoda* (see). They have large, fringed pleopods, which are adapted for swimming. The legs are long and slender. Chelipeds may or may not be present. The body is laterally compressed, and in many species the cephalothorax has a keel-shaped, serrated rostrum. Most are intermittently swimming benthic forms, but some species are pelagic. Many (e.g. the cleaning S.) are commensal with other animals. Larger species are often called ***prawns.*** Several species of S. and prawns are caught commercially as a food source. Most species are marine and they are ecologically important in the food chain as the food of fish.

Classification. Formerly, all shrimps were assigned to the suborder, *Natantia,* with the 4 families: *Penaeidae, Pandalidae, Palaemonidae* and *Crangonidae.* This older classification system is replaced by the following.

1. Suborder: *Dendrobranchiata.*

Infraorder: *Penaeidea* (primitive S.). These are also the most primitive decapods. Pleura of the second abdominal segment do not overlap those on the first. The rostrum is well developed. Gills are dendrobranchiate, i.e. the central axis of the gill bears a series of paired main branches, which give rise to numerous side branches. The first 3 pairs of legs are chelate, but not with enlarged chelae. Females do not carry their eggs, and the first larval stage is a nauplius. Genera are *Penaeus, Sergestes, Acetes, Lucifer (Leucifer).*

Along the Atlantic coast of USA, in the Gulf of Mexico, and elsewhere in the world, species of *Penaeus* are the most important shrimps that are caught commercially. Lucifer is the most aberrant of the penaeids, possessing a very slender macrurous body, with an extremely elongated head, long eyestalks, no chelae and no gills.

2. Suborder: *Pleocyemata.*

1. Infraorder: *Stenopodidea.* These S. have a more or less cylindrical cephalothorax. Pleura of the second abdominal segment do not overlap those of the first. Gills are trichobranchiate (filamentous), i.e. along the main axis are several series of filamentous (but not subbranched) branches. The first 3 pairs of legs are chelate, and one member of the third pair is enlarged. Female carry the eggs on their pleopods, and the first larval form is a zoea. Genus: *Stenopus.*

2. Infrorder: *Caridea.* These S. have a more or less cylindrical cephalothorax. Pleura of the second abdominal segment overlap those of the first and third. Gills are phyllobranchiate, i.e. along the main gill axis are arranged 2 (usually) series of flattened branches or lamellae. The third pair of legs are chelate, and first 2 pairs are chelate or subchelate. In many species, either the first or the second pair of legs are longer or heavier than the others. Females carry the eggs on their pleopods, and the first larval stage is a zoea.

This infraorder contains the largest number of species and genera of S. *Crago franciscorum* is an important commercial shrimp on the west coast of North America. Species of the genus, *Crangon,* are known as ***Sand S.*** The ***North sea shrimp*** *(Crangon crangon),* which occurs in large swarms in the North sea, is caught commercially, and the ***Baltic shrimp*** *(Palaemon squilla = P. adspersus)* is also sometimes harvested and canned. *P. longirostris* (Fig.) also occurs in the Baltic Sea. *Leander* ***(Common prawn)*** is also caught commercially. ***Pistol*** or ***Snapping S.*** (e.g. *Alpheus, Synalpheus;* family *Alphaeidae)* possess one greatly enlarged cheliped, which can be cocked, then closed with sufficient force to generate a snapping sound. The *Atyidae* (e.g. *Atya)* are a freshwater family. Members of the family, *Olophoridae,* are pelagic. The cleaning shrimp, *Periclimenes,* also belongs to the *Caridea.*

Cleaning S., such as *Periclimenes* and *Stenopus,* clean the skin surface of reef fish by removing small parasites and other animal or plant growths. The fish display a surprising tolerance of the S., which even insert their chelipeds into the fish's gills.

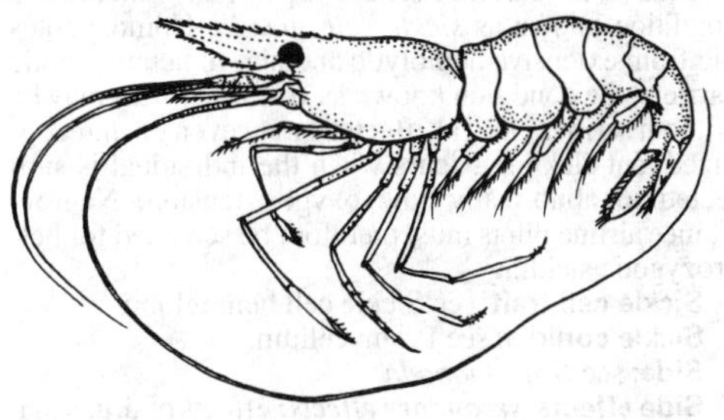

Shrimps. Palaemon longirostris (Baltic sea).

Shrub: a perennial, woody plant, less than 10 m high, which branches below or near ground level (see Basitony). A single main axis is usually absent, and replaced by several main stems of equal status. In deciduous Ss., foliage is lost at the end of the growing season, but there is otherwise no die-back of the aerial parts. *Dwarf shrubs* are less than 0.5 m high. *Subshrubs* tend to be smaller than Ss.; only the basal parts are woody, and these give rise to abundant, branching, nonwoody growth, which dies back at the end of the growing season.

Sialia: see *Turdinae.*

Sialoidea: see *Megaloptera.*

Siamang: see Gibbons.

Siamese fighting fishes: see Gouramies.

Siberian moccasin: see Halys viper.

Siberian ruby throat: see *Turdinae.*

Siberian salamander: see *Hynobiidae.*

Siberian squirrel: see *Sciuridae.*

Sibon: see Snail-eating snakes.

Sibynomorphus: see Snail-eating snakes.

Sicalis: see *Fringillidae.*

Sicista: see *Zapodidae.*

Sicistinae: see *Zapodidae.*

Sicklebill: a name applied to various birds that possess a sickle-shaped bill, e.g. *Falculea palliata* (family *Vangidae), Hemignathus procerus* (family *Drepaninidae),* species of *Drepanornis* and *Epimachus* (family *Paradisaeidae),* and species of *Eutoxeres* (family *Trochilidae).* It is also sometimes applied to species of *Campylorhynchus* (family *Dendrocolaptidae),* but the term "scythebill" is more common in this case. In Australia, the White Ibis, *Threskiornis molucca* (family *Threskiornithidae)* is also sometimes called a sicklebill.

Sicklebird: see *Vangidae.*

Sickle cell anemia: see Sickle cell hemoglobin.

Sickle cell hemoglobin, ***HbS:*** one of the most frequently occurring abnormal hemoglobins, especially in negroids. As a result of a point mutation, the glutamic acid residue at position 6 in the β-chain (normal hemoglobin) is replaced by valine. The α-chain is normal. There are no marked differences in the conformations of HbS and normal hemoglobin. Deoxy-HbS undergoes self association and forms a liquid crystalline phase, which distorts the erythrocytes into a sickle shape. In this phase, deoxy-HbS monomers are in equilibrium with polymers. The polymers consist of 6–8 helical deoxy-HbS strands lying side by side to form a tubular shape of diameter 140–148Å. Deformation of the erythrocytes into sickle shapes ("sickling") leads to their aggregation and to a decrease in blood circulation. Clinical symptoms are anemia and acute ischemia, tissue infarction and chronic failure of organ function, a condition known as *sickle cell anemia.* Homozygotes die from extensive hemolytic anemia. In heterozygotic carriers, the condition known as *sickle cell trait* may be without serious clinical effects; it may even go unrecognized, but sickling occurs when the individual is subjected to abnormally low oxygen tension. Negroid trainee airline pilots must therefore be screened for heterozygotic sickling.

Sickle cell trait: see Sickle cell hemoglobin.

Sickle conidia: see Promycelium.

Sida: see *Branchiopoda.*

Side effects, ***secondary effects:*** effects of drugs that are additional to their intended therapeutic effects. Usually, S.e. are undesirable, e.g. the antimalarial drug, chloroquine, may have adverse effects on vision; the powerful fungicidal drug, griseofulvin, was withdrawn from use, because it was found to cause liver damage.

The term is also applied to disease symptoms that are not directly related to the disease process, but which may be useful for diagnosis, e.g. plants lacking an essential nutrient or diseased in some other way may show an increased susceptibility to infestation by pests, so that the presence of colonies of aphids or their secretions is diagnostic. In such cases, terms such as "supplementary symptoms", or "secondary symptoms" are also appropriate.

Sideramines: see Siderochromes.

Siderochromes: iron-containing, red-brown, water-soluble secondary metaboliotes produced by microorganisms. They include a series of antibiotics, the ***sideromycins*** (albomycin, ferrimycin, danomycin, grisein, etc.) and a class of compounds with growth factor properties for certain microorganisms, the ***sideramines*** (ferrichrome, coprogen, ferrioxmaine, ferrichrysin, ferrirubin, etc.). The sideramines competitively inhibit the antibiotic activity of the sideromycins.

S. are specific ligands for iron. Their synthesis and secretion by microorganisms are increased under conditions of iron deficiency. S. of the catechol type are produced by anaerobic organisms, whereas the hydroxamate type are produced by aerobic microorganisms. They typically contain a central iron(III)-trihydroxamate complex, and specifically bind Fe^{3+}. As metal chelating agents, sideramines fulfil two functions: they transport iron into the microbial cell and/or make the iron available for heme synthesis. In analogy with animal transferrin, the iron of microbial ferrichrome is transferred enzymatically into the porphyrin molecule during heme synthesis.

Other microbial chelating agents, e.g. mycobactin, aspergillic acid and schizokinen, are sometimes classified with the sideramines.

Sideromycins: see Siderochromes.

Siderophilins: nonheme, iron-binding, single chain animal glycoproteins, M_r about 77 000, carbohydrate content about 6%. On the basis of their occurrence, they are classified as transferrin (vertebrate blood), lactoferrin (mammalian milk and other body secretions) and conalbumin or ovotransferrin (avian blood and avian egg white). S. differ in their physical, chemical and physiological properties, but each possesses two binding sites for iron(III). Transferrin (a β-globulin) is the best studied S.; its main function is the transport of absorbed dietary iron(III) to iron depots (liver and spleen), or to the reticulocytes and their precursors in the bone marrow. Transferrin becomes bound to surface receptors of the reticulocytes, enters the cell and releases its iron, then is returned to the blood as iron-free apotransferrin.

By virtue of their ability to chelate iron, all S. inhibit bacterial growth; this property is important in avian eggs and milk.

Siderophores: collective term for substances secreted by microorganisms, which form tight complexes with iron, and transport iron into the cell via specific receptors. Plant compounds with a similar function are called *phytosiderophores.* It is unclear whether the plant constituent, nicotianamine, is a phytosiderophore, but it certainly plays an important part in iron transport in plants.

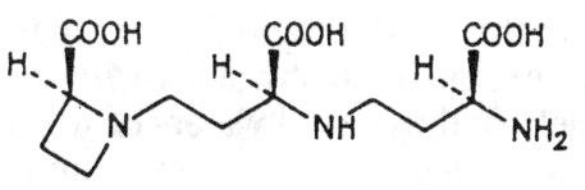

COOH COOH COOH H N NH NH$_2$

Nicotianamine

Siegennian stage: see Devonian.

Sievert: see Dose.

Sieve tube: see Conducting tissue, Shoot.

Sifaka: see Lemurs.

Sign: see Symptom.

Signal behavior: production of signals for the purpose of Biocommunication (see). S.b. has generally evolved from Maintenance activity (see) The process leading to the establishment of S.b. is known as Ritualization (see).

Sign stimulus, ***recognition stimulus, key stimulus:*** an external stimulus or combination of external stimuli, which triggers an inherited behavioral response. If the stimulus has evolved for communication (e.g. an odor, sound or coloration), it is also called a *releaser*. Thus, for a male stickleback *(Pungitius)* guarding a nest, the color red is a S.s. for fighting, since under natural conditions the only source of the color red is the ventral surface of another male stickleback. Carefully constructed models of male sticklebacks lacking red ventral color do not elicit the fighting response, whereas a patch of red color on practically any object acts as a S.s. to a nest-guarding male stickleback. The term *Schlüsselreiz* ("key stimulus") was coined by Konrad Lorenz, who regarded the external stimulus as analogous to a key, which "unlocked" a releasing mechanism. The concept also involves Filtering (see). Complicated patterns of behavior, e.g. courtship, often arise through a chain of sequential S.ss. Experimentally, it has been possible to construct models that are more effective than natural stimuli in releasing fixed patterns of behavior, e.g. gray lag geese direct egg-rolling retrieval movements toward very large dummy eggs in preference to their own eggs; plovers also perform egg-rolling movements more readily when they are given outsized dummy eggs. Such S.ss. are said to be "supernormal". Sex hormones can shift the threshold for a R.s., e.g. male sticklebacks injected with testosterone perform courtship displays to immature females.

Sika deer, ***Japanese deer,*** *Cervus nippon:* a medium-sized deer (up to 150 cm in length, shoulder height 65–109 cm) distributed from southeastern Siberia to Manchuria, Korea and eastern China, and southward to Vietnam. The summer coat is chestnut brown with lines of diffuse white spots on the sides, while the winter coat is grayer with less distinct spots. There is a characteristic black-bordered, white disk on the rump. This forest-dwelling animal is solitary or lives in small groups, and it has been introduced into Europe, America and New Zealand. It is kept in Russia and the former USSR for its young bast antlers, which are used in a pharmaceutical preparation. See Deer.

Silage: animal feed prepared and preserved by a lactic acid fermentation. It is prepared by densely packing high carbohydrate plant material (maize chaff, maize stalks, alfalfa, turnip leaves, steamed potatoes, etc.) in closed silos (tall cylindrical tanks). These anaerobic conditions encourage development of lactic acid bacteria and support an active lactic acid fermentation. Inappropriate preparation and management of the silo contents leads to development of undesirable microorganism, such as butyric acid bacteria and molds. Good S. must contain at least 1% lactic acid to suppress the development of putrifactive organisms.

Silent mutation: see Gene mutation.

Silesian subperiod: see Carboniferous.

Silicon, ***Si:*** a nonmetallic element. The Si content of plants varies considerably with species, organ, stage of development and habitat. Ferns, horsetails *(Equisetatae),* grasses, and the needles of conifers contain particularly high concentrations of Si. On account of their high Si content, horsetails were formerly used for polishing metal vessels (the ash of these plants can contain more than 90% SiO_2). In horsetails and in diatoms (which form complicated shells of SiO_2), Si clearly serves a structural function. Monocotyledonous plants, e.g. grasses, store SiO_2 in their cell walls, usually in an amorphous form. The structural stability of cereal crops is apparently not affected by this stored SiO_2, but it is thought to have a protective function, e.g. against mildew diseases.

In most plants, Si is primarily a ballast material. Deficiency studies indicate, however, that it may also be a micronutrient. Thus Si promotes the growth of a number of plants, such as barley, maize, tobacco, beans and squash. Symptoms of Si deficiency have even been induced experimentally in oats and rice. The function of Si in the plant is, however, unknown. There appears to be a certain metabolic relationship between Si and phosphorus, and to certain extent these two elements may even have interchangeable functions in plant physiology. The availability of phosphate to the plant is increased by the presence of silicate in the soil. Si is taken up and transported by plants predominantly as the silicate anion. Organic compounds of Si also occur in plants, e.g. galactose silicate.

For the role of Si in animals, see Mineral metabolism.

Siliqua: see Fruit.

Silk: see *Bombycidae.*

Silk cotton family: see *Bombacaceae.*

Silk rubber: see *Apocyanaceae.*

Silky flycatchers: see *Bombycillidae.*

Silurian (after the celtic tribe, the Silures): a period of the Paleozoic between the Ordovician and the Devonian (see Geological time scale). It lasted for 35 million years, and is divided into four stages: Llandovery, Wenlock, Ludlow and Pridoli. Biostratigraphic classification of the S. is based on freshwater graptoliths, and on the marine corals, brachiopods, nautiloids and trilobites of the littoral regions of continental shelves. A fine division into 21 zones is based on the continual evolutionary simplification of the graptoliths, leading to the evolutionary peak of the monograptids. The sandstone and calcareous deposits are classified from their constituent trilobites, nautiloids, brachiopods and reef-forming corals. Coastal zones of warm Silurian seas are characterized by reef complexes of sponges, corals, bryozoa, brachiopods and echinoderms. Fish, which were previously scarce, became abundant for the first time in the S.; they are represented by more than 100 species, most of them armored forms of the *Agnatha.* In contrast to the other fauna, the fishes inhabited mainly freshwater lagoons. The S. is considered as part of the Eophytic (see), although the first terrestrial vascular plants (primitive psilophytes) appeared in the flooded coastal plains, marshes and bogs of the Upper S.

Siluriformes, *Nematognathi,* ***catfishes:*** an order of mostly bottom-feeding fish, related to the *Cypriniformes,* and comprising 25–30 familes and about 2000 species. Like the *Cypriniformes,* they also possess a Weberian apparatus (see). True scales are lacking. In some species the body is naked, while in others it carries bony plates. The latter are sometimes arranged in rows, while in certain species they form an armored covering. The bony plates may also carry dermal denticles. Barbels, usually several in number, arise from the borders of the mouth. The barbels contain taste organs, and in some species they also have a supportive cartilage. Several species are known to be venomous; usually the spines of the pectoral fin inject a poison produced by glandular cells in the epidermal tissue covering the spines. Stinging appears to be passive, but *Heteropneustes fossilis* (see *Heteropneustidae)* is known to show aggression. Some members display brood care, and the small number South American species comprising the *Trychomycteridae* are pseudoparasites. Catfishes occur mainly in freshwater, but some species occur in the oceans. They are economically important, some species being highly prized for their flavor. The ***Wels*** *(Siluris glanis),* a member of the European catfish family *(Siluridae),* can attain a weight of 130 kg; it is found in freshwater throughout Europe, and the largest specimens are reputed to be in the Danube. ***Dwarf catfishes*** *(Mystus,* family *Bagridae)* are popular aquarium species, e.g. the ***Striped dwarf catfish*** *(Mystus tengara).* The ***Electric catfish*** *(Malapterurus electricus),* which occurs in tropical Africa, possesses a well developed electric organ, which can deliver pulses of up to 400 volts. Members of the genus *Ameiurus* are generally known as ***Bullheads.***

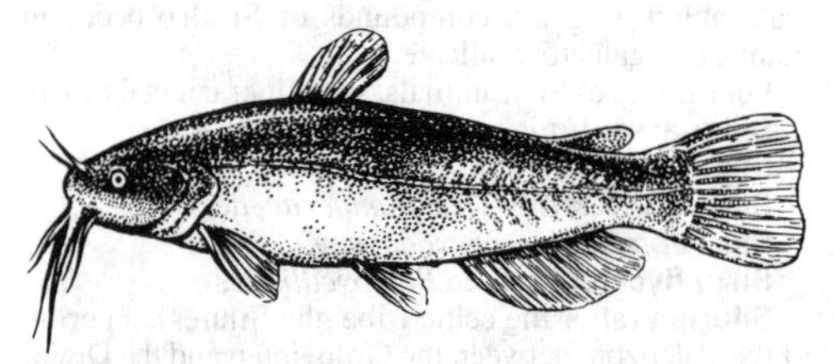

Siluriformes. Dwarf wels or bullhead *(Ameiurus nebulosus).*

Siluris: see *Siluriformes.*

Silvea: see Biocoenosis.

Silver birch: see *Betulaceae.*

Silver bream, *Blicca bjoerkna:* a high-backed fish of the family *Cyprinidae,* found in lakes and rivers of central and northern Europe. It grows up to 30 cm in length and resembles the common bream *(Abramis brama),* but is distinguished from the latter by its strikingly large eyes and by the orange or red bases of the paired fins. The flesh is very bony, and the fish has no economic importance as food.

Silver dollars: see *Characiformes.*

Silverfish: see *Zygentoma.*

Silver leaf disease: see *Basidiomycetes (Poriales).*

Silver line system: lines or reticulated structures demonstrable by silver staining in the pellicle of different protozoa, particularly ciliates.

Silver orfe: see Orfe.

Silversides, *Atherinidae:* a family (suborder *Antherinoidei)* of small, slender schooling fish with two dorsal fins, occurring in the coastal regions of temperate seas, and to some extent in freshwater. *Atherina presbyter* is found widely in the coastal waters of western Europe. The closely related rainbow fishes (family *Melanotaeniidae,* suborder *Antherinoidei)* are often kept in freshwater aquaria.

Silver-Y moth, *Autographa gamma* L.: a noctuid moth (see *Noctuidae)* with a marking on each forewing resembling a γ or a Y. It is one of the commonest migratory species in the Old World, migrating in varying numbers annually from the south into Central Europe (see Migratory Lepidoptera). Adults are active in the day as well as at night. The caterpillars eat root crops, vegetables, clover and other plants, and during occasional periods of marked population increase (a female can lay up to 1000 eggs), they are responsible for widespread damage.

Simians, ***anthropoids,*** *Anthropoidea:* a suborder of the primates (see), comprising monkeys, apes and man. S. possess chisel-shaped teeth in an even dental arcade, with no diastemic intervals. The cerebrum is especially well developed. S. are mostly diurnal, whereas Prosimians (see) are mostly nocturnal. The young are carried around by the mother for some considerable time after birth. S. are divided into two infraorders, New World monkeys (see) and Old World monkeys (see). For the fossil record, see Primates.

Similarity, ***Similarity coefficient:*** a measure of the degree of similarity between two specimens or taxa, based on any of a range of values, conditions, or expressions of taxonomic characters that are homologous, i.e. that are phylogenetically related in the compared organisms or groups (see Homology). Alternatively, the comparison may based on the functional similarity of phylogenetically unrelated characters (see Analogy). Only homologous characters are appropriate for the elucidation and evaluation of phylogenetic relationships.

Taxonomy is a system of ranked similarities, in which members of the same species possess numerous similar characters and display the highest possible S.c., and in which the S.c. decreases as the degree of relatedness decreases in the higher taxonomic categories, i.e. in the order, species → genus → family → order, etc.

Similarity index: a measure of the similarity in the species compositions of two biological communities. Various indexes may be used, e.g. Jaccard's coefficient (see), Kulezinski's coefficient (see), Sørensen's coefficient (see).

Similarity quotient: see Quotient of similarity.

Similitude: see Model system.

Simple latex duct: see Excretory tissues of plants.

Simpson index: an index of diversity (D), based on the probability of selecting two organisms at random that are different species:

$D = 1 - \sum_{i=1}^{i} (P_i)^s$, where P_i is the proportion of individuals of species i in a community containing s different species.

Simulation analysis: see Model system.

Simulium: see Onchoceriasis.

Sinai spiny mouse: see *Muridae.*

Sinew: see Connective tissues

Single cell culture: a tissue culture or microbiological culture, produced by the growth and multiplation of a single original cell. By starting a culture from a single cell, the purity of the final culture is guaranteed. Such cultures are usually started by stepwise dilution of a microorganism suspension until a single cell can be separated from it (based on microscopic observation, or plating out techniques).

Single cell protein: see *Ascomycetes.*

Sinupalliate: a term applied to members of the *Bivalavia,* in which the pallial line (line of attachment of the shell to the mantle) is interrupted by an indentation toward the rear of the shell. See Integripalliate.

Sinus:

1) A space in an organ or body part, e.g. the air-filled sinuses in mammalian facial bones, which connect with the nasal cavity.

2) A venous blood space in the brain, e.g. the cranial venous sinuses which receive blood from the terminal cerebral veins, and which are situated between the outer and inner layers of the dura mater

3) A recess or depression in the palial line of certain bivalves (usually those with a burrowing habit).

Sinus gland: the main Neurohemal gland (see) of crustaceans. It is usually situated on the eye stalk, between the medulla interna and and the lamina ganglionaris (see Brain), and in close proximity to a blood sinus. In the simplest case (in *Mysidaceae* and *Euphausiacea)* it is a thickening of the neuropilem of the medulla terminalis. In the *Decapoda* it is cup-shaped (e.g. in palaemonid shrimps) or branched (e.g. in crabs). In the *Isopoda* it is a more or less spherical structure on the ventral side of the optic lobe. Histologically, the S.g. of advanced crustaceans consists of branched, distended nerve endings surrounded by connective tissue. Specialized nerve tracts conduct basophilic and acidophilic neurosecretions from different regions of the optic lobe and brain to the S.g., and from there they are released into the blood system.

Siphonaptera, *Aphaniptera, **fleas:*** an insect order of about 1 500 species. The earliest fossil *S.* are found in Baltic amber.

Imagoes. Adults are between 1 and 7 mm in length, with piercing-sucking mouthparts. The laterally compressed, heavily sclerotized body usually carries an armature of posteriorly directed spines and bristles. Wings are absent. The coxae are very large, and the last tarsal segment of each leg is provided with a pair of strong, pointed, hooked claws. The middle legs and especially the hindlegs are adapted for jumping. Compound eyes are absent, but two lateral ocelli may be present. All *S.* are ectoparasites of birds and mammals. Metamorphosis is complete (holometabolous development).

Eggs. Oocytes mature only after the female has fed on blood. Eggs are about 0.5 mm long with an adhesive surface, so that they adhere to the host or some part of its habitation. Between 8 and 10 eggs are laid at intervals, during which the female requires a blood meal. The so-called ***Human flea*** *(Pulex irritans* L.), whose normal host is the pig, lays up to 500 eggs. The reproductive cycle of the ***Rabbit flea*** *(Spilopsyllus cuniculi)* is controlled by the level of gonadal hormones in the host's blood.

Larvae. The whitish, vermiform, eyeless and legless larvae possess biting-chewing mouthparts; they feed on organic debris in the habitations (burrows, nests, etc.) of their future hosts. There are 3 larval stages.

Pupae. Before pupation, the larva spends a few days in a resting state, during which it is curled into a U-shape within a cocoon manufactured from its salivary secretion. It then changes into a pupa libera (see Insects).

Economic importance. Fleas are an irritant nuisance of humans and domestic animals, but more importantly they can carry disease. A few species are monoxenous (restricted to a single host species), but most are polyxenous. Some hosts may carry more than one flea species. Any such lack of host specificity means that the flea is a potential interspecies disease vector. Thus, the ***Rat flea*** *(Xenopsylla cheopis)* carries bubonic plague and typhus to humans. The ***Sand flea*** *(Tunga penetrans* L.) is one of the most unpleasant pests of the tropics; the female burrows into the skin of domestic animals or humans, where it is known as a jigger or chigoe flea. Feet and hands are preferred, especially under nails and between toes or fingers; the abdomen of the flea is left exposed, awaiting insemination, after which the jigger becomes swollen to the size of a pea. Fleas are intermediate hosts of some dog and rodent tapeworms, which can also affect humans.

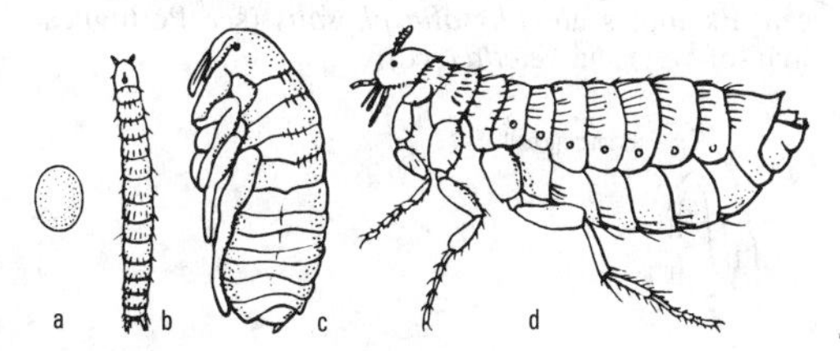

Siphonaptera. Human flea *(Pulex irritans* L.). *a* Egg. *b* Larva. *c* Pupa. *d* Imago (male).

Classification. There are 3 suborders (sometimes regarded as superfamilies).

1. Suborder: *Pulicoidea* (2 families).

1. Family: *Pulicidae.* This family contains many cosmopolitan fleas that attack humans and animals, e.g. *Xenopsilla, Pulex, Ctenocephalides canis* **(Dog flea),** *Ctenocephalides felis* **(Cat flea),** *Echidnophaga gallinacea* ***(Sticktight flea*** of chickens, which anchors itself in the skin by its mouthpoarts).

2. Family: *Tungidae,* e.g. the jigger.

2. Suborder: *Malacopsylloidea* (2 families).

1. Family: *Rhopalopsyllidae.* These are fleas of sea birds and mammals. Most are neotropical with some representatives in Australia and North America.

2. Family: *Malacopsyllidae.* Found only on armadilloes in South America.

3. Suborder: *Ceratophylloidea* (12 families). The family, *Pygiopsyllidae,* is mainly Australian, found on marsupials, monotremes, rodents, passerine birds and sea birds. The *Ischnopsyllidae* are found only on bats. The *Hystrichopsyllidae* are mainly Holarctic, parasitizing rodents and shrews, with some carnivores and marsupials. Many cosmopoliotan species are found in the *Ceratophyllidae,* parasitizing rodents and birds, and some are hosts of the tapeworm, *Hymenolepis diminuta.*

Siphon of sea urchins: see Digestive system.

Siphonogamy: fertilization with the aid of a pollen tube. S. evolved in the gymnosperms, when motile spermatozoids (still encountered in the cycads) lost their flagella and became nonmotile sperm cells.

Siphonoglyph: see *Actiniaria.*

Siphonophora: an order of the class, *Hydrozoa.* Members are free swimming, pelagic polyp colonies, in which the individual polyps display a wide range of different forms. The upper pole of the colony is formed by a sterile medusa, which lacks a mouth and tentacles, and which serves as a swimming bell for locomotion of the colony; several swimming bells are often present. From the bell hangs a long thread with side branches, which carry feeding polyps, and from which sexual medusae are formed by budding. In some species, the terminal, sterile medusa is modified as a gas-filled buoyancy aid, known as a Pneumatophore (see), which by storing or releasing gas enables the colony to move vertically in the water. The pneumatophore may also project above the surface of the water like a sail. Feeding is performed by the feeding polyps (gastrozooids), which have a terminal mouth and a single, long, contractile fishing tentacle, armed with batteries of stinging nematocysts. These sometimes extremely long fishing tentacles capture prey such as small fish, copepods, etc.

The 150 brightly colored species all live in the open sea. Examples are *Physalia physalis* (see Portuguese man-of-war) and *Velella* (see).

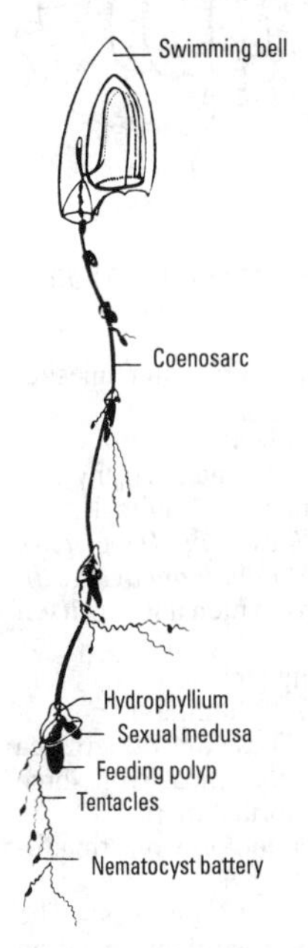

Siphonophora. Muggiaea kochii (length about 5 mm), which is found among Mediterranian plankton.

Siphonopoda: see *Cephalpoda.*

Siphonostomatoida: see *Copepoda.*

Siphuncle: a tubular organ present in various mollusks. The term is applied in particular to a long, narrow, calcified tube that runs internally through all the chambers of the cephalopod shell. It contains the siphuncular cord of tissue, which extends from the visceral mass. Maked structural differentiations of the nautiloid S. are used taxonomically for the determination of genera. See *Cephalopoda.*

Sipunculida, ***peanut worms:*** a phylum of the *Protostomia,* containing about 320 known species in about 17 genera. Four families are recognized.

Morphology. *S.* are 1 to 66 cm in length, almost cylindrical, and nonsegmented. The body consists of a thick trunk (which is often longitudinally and transversely furrowed, or covered with tubercles) and a slender, protrusible trunk or *introvert.* When protruded, the introvert accounts for 1/3 to 2/3 of the body length. The mouth is situated at the end of the introvert, and in almost all species it is surrounded by a wreath of small tentacles. There is a long, U-shaped gut, and the anus is located on the anterior dorsal surface. The body cavity is a uniform, spacious, undivided coelom. There is no vascular system. Sexes are separate, and the testes or ovaries lie in strips on the coelom wall. Sperms and eggs leave via the exit ducts of the metanephridia (1 or 2 metanephridia are present).

Biology. They are exclusively marine, living in the sand or mud bottoms of mainly tropical seas, often in their own tubes, rock clefts or the shells of other animals. Most species feed on animal and plant detritus, which is ingested with the sand and mud, and some species are carnivorous. Development proceeds via a free-swimming larval stage that closely resembles the polychaete trochophore larva.

Collection and preservation. *S.* are dug with a spade from practically level sand bottoms, or they may be excavated from greater depths with a large, toothed dredge. *S.* are also often found in secondarily hardened bottoms, which must be fragmented to obtain specimens. Animals are narcotized by placing in sea water, to which 70% ethanol or cocaine hydrochloride is added dropwise. Protrusion of the introvert usually occurs during narcotization. Specimens are preserved in 70% ethanol or 2% formaldehyde.

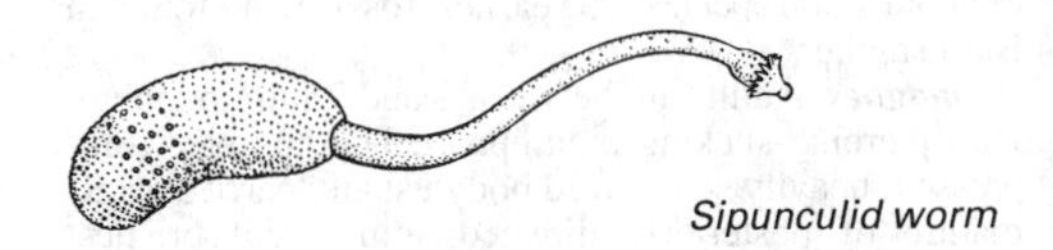

Sipunculid worm

Sirenia, ***sirenians, manatees*** and ***dugongs:*** a rather anomalous mammalian order containing 4 living species in 2 genera. Heavy-bodied, slow-moving animals, they are the only completely aquatic mammals that are also herbivorous. The body is fusiform, plump and hairless, except for bristles on the snout. Hindlimbs are absent, and the pelvis is vestigial. The forelimbs are modified to flippers, and the tail is a horizontal fluke. Skull structure and dentition suggest an affinity to the ungulates. The lungs are unlobed and unusually long, oriented horizontally, and separated from the massive gut by a long, horizontal diaphragm. Sirenians graze underwater on water plants, and remain submerged for up to 15 minutes.

The ***Dugong*** or ***Sea cow*** *(Dugong dugong)* inhabits the coastal waters of the Indian Ocean, Red Sea and the Malay Archipelago. Its tail is notched like that of a whale. Skeletal characteristics include robust neural spines and ribs, small premaxillaries, the presence of nasal bones, and a long nasal cavity. Its teeth are large and columnar, covered with cement and lacking enamel; they have open roots, and the occlusal surfaces are wrinkled. Dental formula; 1/0, 0/0, 0/0, 2–3/2–3 = 10–14.

There are 3 species of ***Manatees*** *(Trichechus),* all found in marine bays and/or sluggish rivers, usually in turbid water. *T. manatus* inhabits the coast and coastal rivers of southeastern USA, west to Texas and the West Indies, and the adjacent mainland from Vera Cruz to northern South America. *T. inunguis* is found in the Amazon and Orinoco drainages of northeastern South America. *T senegalensis* is found along the West African coast from the Senegal River to the Cuanaga River, as well in Lake Chad and its associated drainages. The manatee tail is not notched, but rounded like a spoon. Manatees have slender neural spines and ribs, large premaxillaries and a short nasal cavity; nasal bones are absent. They have an indefinite large number of enamel-covered, cementless teeth, each with 2 cross ridges and closed roots. There are no functional incisors, and as the teeth at the front wear out, they are replaced by posterior teeth moving forward.

A fifth member of the order, ***Steller's sea cow*** *(Hydrodamalis stelleri = gigas),* became extinct in 1769. About 8 m in length and probably weighing in excess of 6 000 kg, it inhabited the shallow areas of the Bering Sea, where it fed on seaweed.

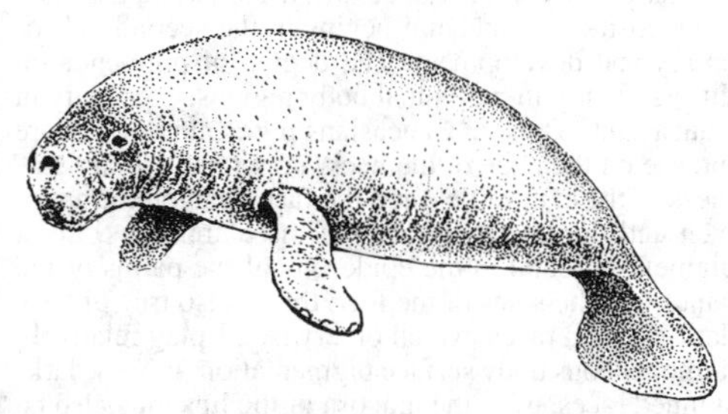

Sirenia. African manatee *(Trichechus senegalensis).*

Sirenidae, *sirens:* a North American family of the *Urodela* (see) comprising 3 species in 2 genera, which live permanently in muddy pools. Sirens grow up to 1 m in length and have eel-like bodies, with small forelimbs close to the gills. Hindlimbs and pelvic girdle are absent. The eyes are very small, and the body is olive-green or dark gray with pale spots. The jaws contain horny ridges and no teeth. Adults are neotenic "permanent larvae", their several pedomorphic features including the absence of eyelids, the presence of gill slits, and the posession of external gills which are retained throughout life. Lungs, however, are also present. Metamorphosis has never been observed, even experimentally. In dry periods, they burrow into the mud and their gills degenerate. Their phylogenetic relationships are uncertain. The 3 species are: *Pseudobranchus striatus* ***(Dwarf siren;*** 3 fingers on forefoot), *Siren lacertina* ***(Greater siren;*** 4 fingers), *Siren intermedia* ***(Lesser siren;*** 4 fingers).

Sirenin: see Gamones.

Sirens: see *Sirenidae.*

Siricidae: see *Hymenoptera.*

Sisal: see *Agavaceae.*

Siskins: see *Fringillidae.*

Sister chromatids: the two genetically identical chromatids that make up each chromosome at the end of mitosis, and during the subsequent period from the interphase nucleus to anaphase. Each sister chromatid is a conservative copy of the original chromatid. As a result of Recombination (see), genetically nonidentical S.c. can occur temporarily between two mitoses and after meiosis I. In the heterozygous condition, S.c. and nonsister chromatids (of homologous chromosomes) carry different alleles of the same genes. In the homozygous condition, both S.c. and nonsister chromatids carry the same alleles of all their genes.

Sitatunga: see *Tragelaphinae.*

Sitosterols: a group of phytosterols structurally related to the parent hydrocarbon, stigmastane (see Steroids). They are widely distributed in higher plants, and usually occur in mixtures that are difficult to resolve. They sometimes occur as glycosides. Chief representative is β-sitosterol (stigmast-5-ene-3β-ol), which ocurs, e.g. in cotton seed, germinating wheat and rye, sugar cane wax, potatoes, tobacco and pine bark.

HO H H

β-*Sitosterol*

Sitta: see *Sittidae.*

Sittidae, *nuthatches:* an avian family of 31 small (9.5–19 cm) species, belonging to the order *Passeriformes* (see). With the aid of very sturdy toes and claws, they climb on the vertical surfaces of tree trunks and rocks. The short tail is not used as a support when climbing (cf. woodpeckers and treecreepers, which use their tail as a climbing support). They climb obliquely, hanging from one foot and placing the other lower for support. The long, symmetrically tapered bill is used for prizing insects, spiders and other small animals from crevices. Some species also eat nuts and seeds, and are able to break open hazel nuts by hammering them into cracks (hence the name, nuthatch, which is derived from "nut hack"). They tend to nest in hollows and holes, often narrowing the entrance with a mud wall. The upper surface of the body is usually blue-gray (but blue-green in 3 southern Asian species: ***Velvet-fronted nuthatch,*** *S. frontalis;* ***Azure nuthatch,*** *S. azurea;* ***Beautiful nuthatch,*** *S. formosa),* while the underparts vary from white through gray to chestnut. There is hardly any sexual dimorphism. The most widely distibuted species is the ***Nuthatch*** *(Sitta europaea)* (plate 6) which is found throughout Europe and northern Asia as far as the treeline; it also enters a small area of North Africa across the straits of Gibraltar. *Sitta canadensis* is widely distributed in North America, and some individuals of this species are migratory; otherwise, nuthatches are not migratory.

The family also includes the genera, *Neositta* ***(Treerunners;*** 6 species restricted to Australia and New Guinea), *Hypositta* (confined to Madagascar), *Trichodroma* ***(Wallcreepers;*** in the high mountains of Europe and Asia), *Salpornis* (Africa and India), and *Rhabdornis* (Philippines).

Sivatherium: see *Giraffidae.*

Six-lined grass lizard: see *Takydromas.*

Six-lined long-tailed lizard: see *Takydromas.*

Skatole, ***3-methylindole:*** an unpleasantly smelling compound, contributing to the odor of feces. It is formed from the amino acid tryptophan during the putrifaction of proteins.

Skimmers: see *Rynchopidae.*

Skin, ***integument:*** outer body covering of multicellular animals. As the boundary layer that separates the organism from the environment, the S. and its various associated structures have a wide variety of functions. It protects the internal organs from mechanical damage, prevents invasion of the body by microorganisms, and acts as a first defense against physical and chemical influences of the environment. In addition to these passive functions, the S. also actively participates in respiration and excretion, and has a regulatory role in the water and salt economy of the body. Through the responses of its specialized sensory cells, which register temperature, pressure and pain, the S. also serves to mediate contact between the organism and its external environment. In warm-blooded animals, the S. is important in the regulation of body temperature.

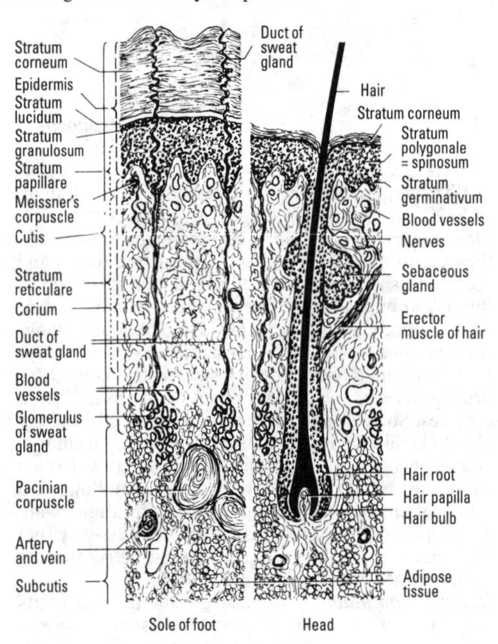

Skin. Section of human skin.

Invertebrate S. usually consists of a single layer of ectodermal cells, the epidermis, which is covered on its exterior surface by a secreted more or less thick protective Cuticle (see). Vertebrate S. consists of two distinct layers, the multilayered outer *epidermis,* and the inner *corium* or *dermis* of connective tissue. Epidermis and dermis together form the *cutis.* The *subcutis* (connective tissue layer below the cutis) stores subcutaneous fat, which acts as heat insulation for the body, as well as serving as a metabolic fat reserve. The subcutis is also the movable and displaceable layer that intervenes between the S. and the musculature. The epidermis is generally subdivided into a basal germinal layer and an outer cornified layer. New cells are continually synthesized in the germinal layer; as these move outward they lose their nucleus and cytoplasmic contents, and become flattened and keratinized. Aquatic vertebrates display only slight cornification of the epidermis, whereas the cornified zone is highly developed in terrestrial vertebrates. The cornified layer is continually sloughed off as small scales, or periodically shed in its entirety (e.g. in snakes). The S. may contain lacteal glands, sweat glands and sebaceous glands, as well as pigments. Other structures within, or derived from, the S. are hairs, feathers, nails, claws, hooves, horns and teeth.

Skin beetles: see *Coleoptera.*

Skin color: in humans, one of a graded series of brown tones between very pale and very dark brown (almost black). These different shades of brown are due mainly to differences in the quantity of Melanin (see), i.e. the number of melanocytes in the germinal layer of the epidermis and in the connective tissue cells of the dermis. S.c. is also influenced by the red color of the blood, which shows through the skin capillaries and the sometimes very thin epidermis, and by the presence of carotene in the epidermal cells and subdermal tissues.

Deposition of melanin begins in the second half of embryonal development. The degree of pigmentation differs greatly in individual body regions, especially in Caucasians. Thus, in Caucasians pigmentation is more intense on the nape of the neck, upper back, outer surfaces of the extremities, and the face, than on the abdomen and the inside surfaces of the extremities. Least pigment is found in the epidermis of the palms of the hands and the soles of the feet; this is also true of very dark-skinned races, which otherwise display relatively homogeneous body surface pigmentation. In very dark-skinned races, even the mucosa of the lips may also be pigmented, so that the lips appear brown or brownish red to violet. In the most heavily pigmented groups, pigment is also present in the mucosa of the cheeks and in the gums. For anthropological purposes, skin color is classified by direct comparision with a chart of 36 standard tones, which are available as color-standardized glass squares for direct physical comparison.

Skin pigmentation has an important protective function, in particular against short-wave, ultraviolet light, which causes burning of the skin. The occurrence of different degrees of pigmentation in the world populations must be seen primarily as an adaptation to regional differences in irradiation by the sun.

Skin musculature: see Dermal muscles.

Skinner box, ***operant box:*** a box or cage for the investigation of learning in animals, in particular for the study of operant conditioning (instrumental conditioning). The experimental arrangement is designed primarily for rats. In principle, in order to release food or water, the animal must learn to perform a particular activity (e.g. press a bar or lever) in response to a certain signal. Depending on the conditioning process under investigation, there are numerous variations in the design of the S.b., including electronic control and recording devices, so that the entire operation and data processing can be automated.

Skippers: butterflies of the family *Hesperiidae;* see *Lepidoptera.*

Sklerea: see Biocoenosis.

Skuas: see *Stercorariidae.*

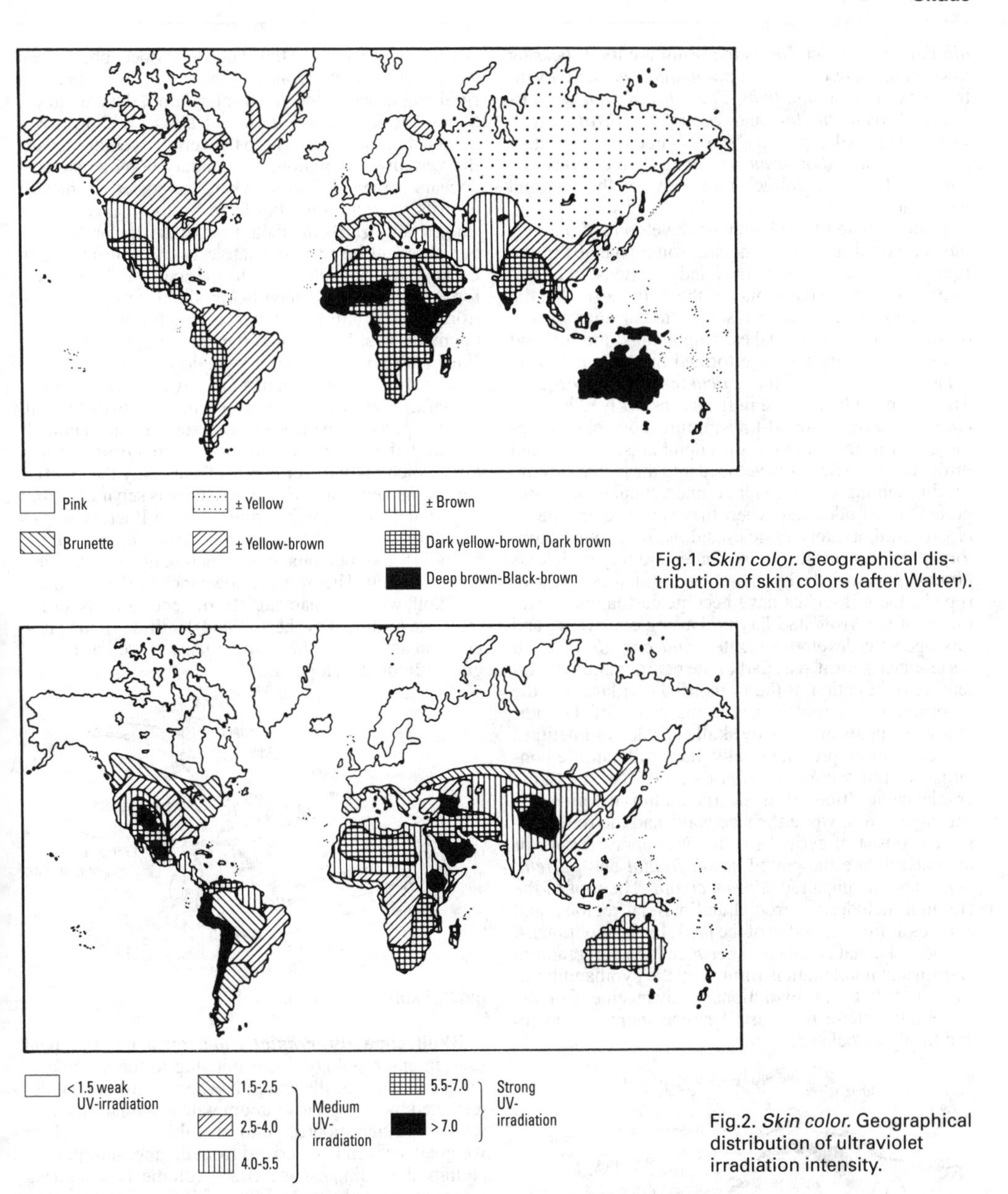

Fig. 1. *Skin color.* Geographical distribution of skin colors (after Walter).

Fig. 2. *Skin color.* Geographical distribution of ultraviolet irradiation intensity.

Skull, ***cranium:*** the totality of the head skeleton of vertebrates. It consists of two distinct regions, the *neurocranium* which surrounds the brain and sensory organs, and the *splanchnocranium (branchiocranium* or *viscerocranium)* which is associated with the visceral or branchial arches (jaws, gills and derived structures). There are two basic structural elements: 1) the bony *dermatocranium* or *exocranium* (consisting of dermal bones), which arises without a cartilaginous precursor, and 2) the cartilaginous neural and visceral *endocranium,* which may be secondarily bony. Elasmobranchs (sharks, rays) do not have a dermatocranium, and the endocranium is not secondarily ossified; the secondary formation of bony tissue does not occur in these animals. In primitive fishes (lungfishes and *Crossopterygii),* the cartilage of the endocranium is almost entirely replaced by secondary bony tissue. In teleosts and quadrupeds, extensive regions of the dermatocranium are always fused with parts of the endocranium to form a new type of unitary structure. In order from front to back, the neurocranium includes three pairs of capsules: the olfactory, optic and auditory (otic) capsules, while the most posterior part of the neurocranium is the occiput or occipital region. The splanchnocranium evolved from a gill skeleton which formerly provided bar-like supports for the gill openings. It consists of: 1) the primary *man-*

dibular arch, which forms the primitive tooth-bearing jaws, consisting of a dorsal *palatoquadrate* and a ventral lower jaw or *mandible;* 2) the *hyoid arch,* with its dorsal *hyomandibular* and central *basihyal,* which forms the hyoid apparatus, and 3) three to seven (usually five) *visceral* or *branchial* arches, which consitute the gills. There is a Spiracle (see) between the jaws and the hyoid arch.

In the course of phylogenetic development, numerous structural modifications and some changes in the functions of certain parts have led to extensive alterations of the above basic plan of the S. In particular, the splanchnocranium has been subject to considerable evolutionary change. In fishes, amphibians, reptiles and birds, the primary jaws are formed from the *quadrate* (inner lower jaw) and the *articular* (inner upper jaw). The hyomandibular bone in fishes has evolved into the *columella auris,* a sound-transmitting, bony or cartilaginous rod in the inner ear of amphibians, reptiles and birds. The spiracle has developed into the *tympanic cavity.* In mammals, the quadrate and articular have migrated to the middle ear, where they have become the *incus* (central auditory ossicle) and *malleus* (outer auditory ossicle), while the *stapes* (inner auditory ossicle) is homologous with the hyomandibular of fishes. In quadrupeds, the gill arches have become cartilaginous elements of the *hyoid* and larynx. During embryonic and phylogenetic development, the *skull capsule* of adult vertebrates is formed as part of the neurocranium by intensive ossification of the primordial cartilage and the incorporation of roof bones. Demarcation of the individual elements of the bony skull capsule is sometimes difficult to interpret. Internally, the skull capsule contains the brain within the *cranial cavity.* The base of the cranial cavity (floor of the neurocranium) is formed by cartilage bones, while the side walls and the roof *(calvaria)* consist of dermal (membrane) bones. Bones of the calvaria are the paired *nasal, frontal* and *parietal bones* and the unpaired *supraoccipital.* The roof of the cranium includes the frontal and parietal regions, and extends as far as the edge of the jaws. In elasmobranchs, the only skeletal connection between the neurocranium and splanchnocranium is formed by the hyomandibular cartilages, but with evolutionary advancement of the vertebrates, these two parts become more intimately and firmly associated.

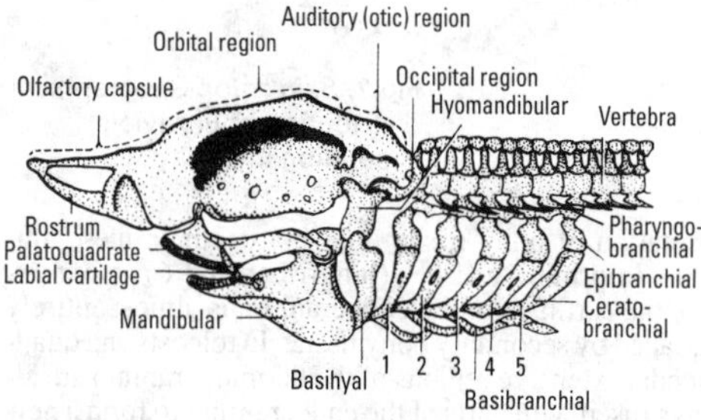

Fig.1. *Skull.* Cartilaginous cranium of an elasmobranch fish.

In fishes the skull capsule is relatively small. Behind the jaw is a gill cover composed of 4 main parts: opercular, subopercular, interopercular and the branchiostegals. The 5th pair of gill arches rarely carries gills, and is often toothed (e.g. the well developed pharyngeal teeth of the *Cypriniformes).* Whereas in fishes the S. is rigidly attached to the vertebral column, the amphibian S. has 2 *occipital condyles,* which allow nodding movements of the skull relative to the vertebral column. During vertebrate evolution, a secondary bony palate first appears in the reptilian S., where the posterior, internal nostrils *(choanae)* are bordered by the *pterygoids.* In snakes and lizards the palato-maxillary apparatus (upper jaw) can be moved relative to the neurocranium. Linkage of the reptilian S. to the axial skeleton is afforded by a single condyl below the occipital foramen (foramen magnum), the latter serving for the passage of the medulla oblongata into the cranial cavity. The bird skull also possesses only a single occipital condyle. Modern birds have no teeth. On the whole, the bird skull is characterized by the firm and sutureless fushion of all its bony parts, even in the juvenile state. In the mammalian skull, the lower jaw is formed entirely from the *dentary,* which in other vertebrates forms only the anterior part of the lower jaw. The nasal cavity is separated from the cranial cavity by the *ethmoid* (the olfactory nerves pass through perforations in the *cribriform plate* of the ethmoid). Two occipital condyles link the S. to the vertebral column. The overall appearance of the mammalian skull, which is characteristic of genus, family, order, etc., is determined by the size and dentition of the jaws, position and size of the sense organs, as well as the degree of brain development.

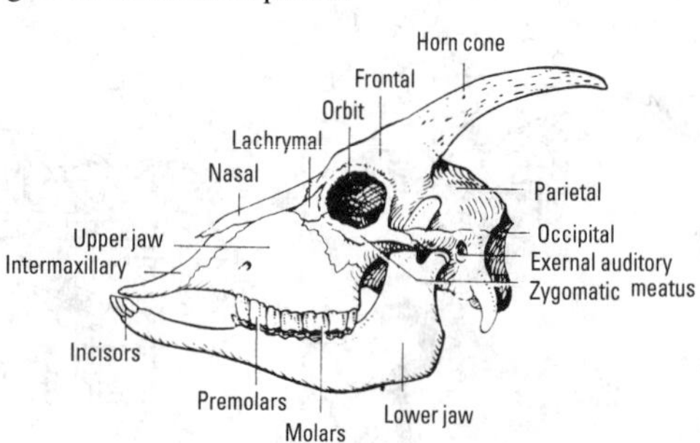

Fig.2. *Skull.* Mammalian skull.

Skull capacity, *cranial capacity:* a measurement used in anthropology, corresponding to the volume of the inner space of the cranium. The volume can be determined by filling the cranium with granular material, which is then poured into a calibrated vessel or weighed, or it can be derived from the measurement of certain skull dimensions. Brain volume is somewhat less than S.c. In humans, a S.c. of 100 cm^3 represents a brain mass of 91–95 g. The great apes have an average S.c. of 450 cm^3, the minimal value for the female orangutan being 290 cm^3, and the maximal value for the male gorilla 685 cm^3. In *Homo sapiens sapiens,* the European average is 1 300 cm^3 for men and 1450 cm^3 for men. Some Neandertalers had a conspicuously large S.c. (e.g. 1723 cm^3 for Spy II)

The performance of the brain is not determined by size alone, but also depends on the relative sizes and fine structure of different brain parts.

Skull planes: planes of orientation defined by selected points on the skull, which allow an exact comparison of different skulls with respect to the heights and angles of their components. When describing or measuring the human skull, it is assumed to be in its natural position, i.e. as carried on the body. This position is attained when the upper margins of the two ear holes and the lowest point of the margin of one orbit are all in a horizontal plane. The horizontal ear-eye plane defined by these three points is called the *Frankfort plane.* As viewed from points in the Frankfort plane, there are four true profile views of the skull: *norma facialis* (full face view), *norma lateralis sinistra* (left side view), *norma lateralis dextra* (right side view), and the *norma occipitalis* (back view). At right angles to these, and viewed from points on the median sagittal plane, are the *norma verticalis* (view from above the skull) and the *norma basalis* (view from below the skull) Of particular importance is the *median sagittal plane,* which divides the skull into two mirror halves, and which serves as a reference for several useful parallel sagittal planes (Fig.).

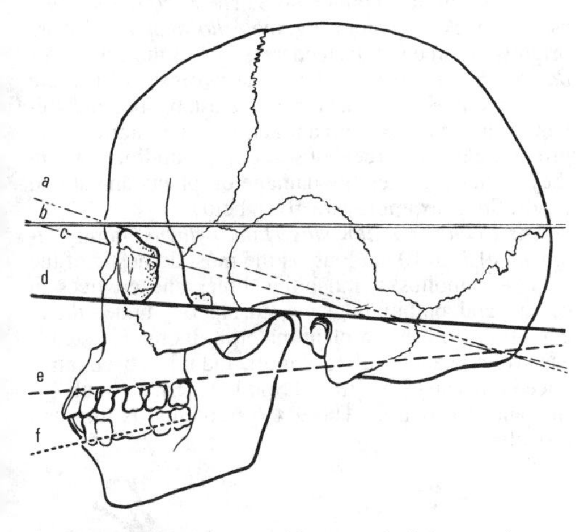

Skull planes. Side view of human skull (norma lateralis sinistra). showing skull planes (after Martin). *a* Glabello-inion line. *b* Glabello- lambda line. *c* Nasion-inion line. *d* Frankfort plane. *e* Prosthion-condyle plane. *f* Alveolar line of the mandible. [*Glabella:* the most anterior point in the midline at the level of the supraorbital ridges, when skull is in the Frankfort plane. *Inion:* the most prominent point of the external occipital protuberance. *Lambda:* junction of the sagittal and lambdoid sutures. *Nasion:* junction of the fronto-nasal and internasal sutures. *Prosthion:* the lowest point of the alveolar margin of the maxilla between the upper incisors]

Skunks: species of several genera, all belonging to the *Mustelidae* (see). As part of their defense strategy, all S. turn their back to any perceived source of threat or danger and eject a repulsively smelling liquid over a distance of several meters from their well developed anal glands. Examples are *Mephitis mephitis* ***(Striped skunk;*** southern Canada and USA to northern Mexico), *Mephitis macroura* ***(Hooded skunk;*** southwest USA to Central America), *Spilogale putorius* ***(Spotted skunk;*** British Columbia and most of USA south to Mexico and Central America), *Spilogale pygmaea* (Pacific coast of Mexico; known only from 4 specimens), *Conepatus* (6 species of ***Hog-nosed skunks*** distributed from southwest USA to the Strait of Magellan). *Mephitis* occupy a variety of habitats including woods, plains and deserts. Both species have black and white color patterns, with considerable individual variation. *M. mephitis* usually has white on top of the head and nape, extending posteriorly in 2 stripes (Fig.). *S. putorius* has a white forehead patch, and the body is patterned with white stripes and spots, with no large single areas of white. *Conepatus* are the only S. in South America. Nocturnal and largely solitary, they have the coarsest fur of all S., and like all other S. they are patterned black and white.

Skunks. Striped skunk (*Mephitis mephitis).*

Skylark: see *Alaudidae.*

Slave-making ants: see Ants.

Sleep: a state of decreased activity with greatly reduced sensitivity to external stimuli and greatly reduced motor activity. Nervous activity is considerably reduced, but the autonomic system continues to function. Onset and duration of S. are often regulated by exogenous Zeitgeber (e.g. alternation of light and dark). S. is a metabolically determined state, which is essential for the subsequent continuation and maintenance of physiological function. See Hibernation, Sleeping behavior.

Sleeping behavior: behavior leeding to, and insuring the state of Sleep (see). In animals that display typical sleeping phases during the course of each 24 hours, S. is determined by a behavioral predisposition. Orienting appetitive behavior consists of seeking a suitable resting or sleeping site. If a suitable site is found, behavior is concentrated on selecting the actual place where the animal will lie, and possibly on its preparation (e.g. chimpanzees construct a simple sleeping nest). The behavioral pattern ends with the adoption of the species-specific sleeping position, which may alter during sleeping, e.g. by changing the side of the body next to the substratum. After a species- and situation-specific time, the predisposition for sleeping is inhibited, and the animal awakes, i.e. a new phase of motor activity is introduced. See Hibernation, Sleep.

Sleeping sickness: see Trypanosomes.

Slender blind snakes: see *Leptotyphlopidae.*

Sliding tubule mechanism: see Cilia and Flagella.

Slime tubes: see Cuvierian organs.

Slipper orchid: see *Orchidaceae.*

Slipper shell: see *Gastropoda.*

Sloe: see *Rosaceae.*

Sloped culture, ***culture slope:*** a culture of microorganisms on the surface of a sloping solid culture medium, inside a tube. Tubes for this purpose are similar to ordinary test tubes, but they lack a lip. One quarter to one third of a culture tube is filled with culture medium containing agar (more rarely gelatine). The tube is plugged with nonabsorbent cotton wool or heat-resistant plastic foam, or provided with an aluminium cap, then heat sterilized (autoclave). While the medium is still hot and liquid, the tubes are laid at an angle on their sides (i.e. they are "sloped"), so that the medium solidifies at an angle to the axis of the tube. Such media are suitable for surface cultures, and for the storage of living cultures for long periods.

Sloth bear, *Melurus ursinus:* a shaggy black bear native to India and Sri Lanka, with a strongly protruding, mobile gray-white snout and a white, yellow or chestnut-brown V- or Y-shaped chest marking. The lips are protrusible, mobile and naked, and the nostrils can be closed at will. The inner pair of upper incisors is absent, leaving a gap, and the palate is hollowed so that the mouth and snout are adapted for sucking and blowing. It feeds mainly on insects and fruits. Using its strong, sickle-shaped claws, it opens termite nests and sucks out the termites, making a noise that is audible over a considerable distance.

Sloths, *Bradypodidae:* a family of the order *Edentata.* S. have a rounded head and shaggy coat. With the aid of long, hooked claws on each of the 4 limbs, they hang motionless for long periods from branches. They spend virtually their entire existence in trees, and progress on the ground only with difficulty.

Unais or ***Two-toed sloths*** (2 species, *Choleopus hoffmanni* and *C. didactylus)* grow up to 68 cm long, and have light gray-brown fur, which becomes greenish from the growth of algae on the hairs; the tail is absent or vestigial. The ***Ais*** or ***Three-toed sloths*** (about 6 species, e.g. *Bradypus tridactylus* and the ***Maned sloth,*** *B. torquatus)* are even more lethargic than the unais; their tails are up to 7 cm long, and they attain head plus body lengths of up to 60 cm.

Whereas other mammals possess 7 cervical vertebrae, the ai has 9, and the unai has 6 or 7; both are able to turn their heads through 270 °. S. live in tropical South America, and their food consists of leaves and fruits.

Although now extinct, some ***Ground sloths*** are considered to be recent mammals, e.g. *Parocnus, Acratocnus* and *Mylodon.* They existed in the Americas after the establishment of humans, and their skeletal remains and even fur have been found.

Two-toed sloth or unai (Choleopus hoffmanni)

Sludge activation systems: see Biological waste water treatment.

Slugs: terrestrial members of the subclass *Pulmonata,* class *Gastropoda,* lacking a visible outer shell, which is either greatly reduced and represented by a thin plate almost covered by the mantle, or it is totally absent. The reduced anteriodorsal mantle forms a mantle shield with a lateral pneumostome. No single criterion precisely defines the nature of S. Whereas the most obvious feature is shell reduction or absence, biologically more significant is the incorporation of visceral organs into the head-foot. Several recent species show all intermediate stages between truly shelled snails and naked S. Their evolution is random and their origin polyphyletic. Thus, 7 pulmonate families contain S.: *Veronicellidae, Rathouisiidae, Arionidae, Philomycidae, Limacidae, Testacellidae, Athoracophoridae.* On the other hand, the orders, *Orthurethra* (9 families) and *Mesurethra* (4 families) contain no S. The S. comprise a variety of sizes and colors, e.g. the *Athoracophoridae* are brightly colored fungus feeders, whereas the *Testacellidae* are more soberly colored carnivores. Most S. are herbivorous. Several have become established in habitats influenced by human activity, i.e. agricultural ground, gardens, greenhouses, damp buildings, where they inflict considerable damage on plants and stored foods. Some examples are given below.

The ***Cellar*** or ***Yellow slug*** *(Limax flavus;* family *Limacidae),* 7.5–10 cm long, is the most domestic of the European mollusks, inhabiting holes and crevices in cellars and outbuildings, venturing out under damp conditions to feed on plant material. Its range extends east to Syria and south to Tripoli, and it has been introduced into Japan, North and South America, South Africa and Australasia. The 9 x 6 mm shell is thin and broadly oval.

Slugs. a Agriolimax (= Deroceras) agreste (field slug) and damaged horticultural products (x 0.5). *b Arion rufus* (black slug) (x 0.4).

Field slugs *(Agriomilax agreste* and *A. reticulatum;* family *Limacidae),* 5–6 cm long, are colored variously from whitish to mottled, or reticulated brown with black. They extend all over Europe and through China, Japan and North Africa; they have been introduced into North and South America, the West Indies, South Africa and Australasia. Green plants are the chief source of food, and they cause significant damage in gardens and greenhouses. The thin, oval-oblong shell measures 5 x 3 mm.

The ***Black slug*** *(Arion rufus;* family *Arionidae),* up to 20 cm long, is typically black, often reddish with many other color varieties. It inhabits woods, marshes, fields, hedgerows and gardens throughout Europe, and has been introduced into New Zealand. When touched, it contracts into a hemispherical shape. The shell is represented by a soft calcareous secretion, which occasionally forms a hard mass.

Small-angle scattering: scattering of X-rays and neutron radiation in monodisperse solutions of macromolecules. Many biological molecules cannot be crystallized and submitted to X-ray diffraction analysis, but information on their molecular properties can be obtained from their small angle scattering curves in monodisperse solution. Since there is no macroscopic order in solution, and the molecules are very large relative to the wavelength, the scatter due to the entire molecule has only a small angle. A small angle scattering curve is constructed by plotting the intensity of the scattered radiation against the angle of scatter. This provides information on conformation changes, reactions with small molecules, solvent interactions, etc.

Small circuit theory: see Excitation conductance.

Small gerbils: see *Gerbillinae.*

Small-leaved lime: see *Tiliaceae.*

Small winter moth: see Winter moths.

Smear:

1) A microbiological technique, in which microorganisms in a drop of fluid are spread in a very thin layer across a microscope slide, using an inoculation loop, a cover slip, or another slide. The smear is dried in very gentle heat, immersed in fixative, followed by appropriate stains, then examined under the microscope. Blood may be prepared for examination in the same way. Blood smears are particularly appropriate for confirming the presence of blood parasites, e.g. malaria parasites.

2) A sample of wound exudate, mucosal coating, etc., taken with an inoculation loop or sterile swab, for the purpose of microscopical examination, or for the culture of microorganisms. Smears are used for identifying the causative agents of disease, and for the detection of malignant cells (e.g. in cervical smears).

Smectite: see Sorption.

Smelt, *Osmerus eperlanus:* a small (up to 30 cm in length), economically important, slender, pelagic marine fish, found in European coastal waters from Spain to Scandanavia (family *Osmeridae,* order *Salmoniformes).* Spawning occurs in the brackish waters of river estuaries. Freshwater races are also established in European rivers and lakes. The dorsal surface is dark metallic blue, and the sides and belly are silvery gray. The mouth is armed with conspicuous teeth. Although described as delicious, the taste of the flesh is acquired, being somewhat musty, sometimes described as resembling cucumber.

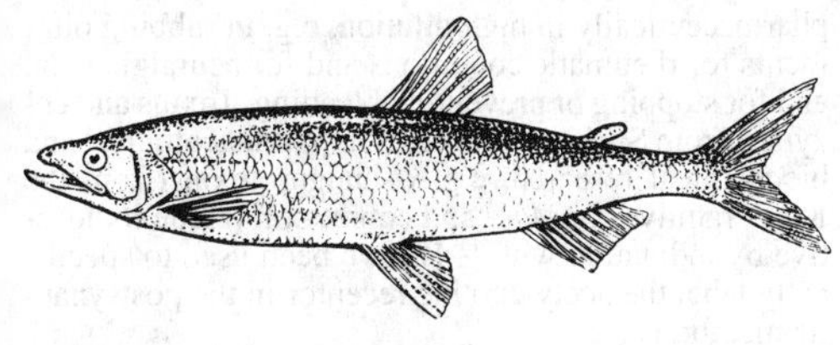

Smelt (Osmerus eperlanus)

Smew: see *Anseriformes.*

Smoke tree: see *Anacardiaceae.*

Smooth newt: see *Salamandridae.*

Smooth snake: see European smooth snake.

Smut fungi: see *Basidiomycetes (Ustilaginales).*

Snaggletooths: see *Stomiiformes.*

Snail-eating snakes: nocturnal, tree-dwelling members of the subfamilies *Dipsadinae* and *Pareinae* of the family *Colubridae* (see). They are specialized for feeding on snails. In the Southeast Asian genus, *Dipsas,* the chin scutes are partially fused, there is no throat groove, and the floor of the mouth is very firm. The lower jaw, which has elongated teeth, is pushed into the opening of the snail shell and turned from side to side to gain a purchase in the flesh of the snail. The snail is then pulled out and devoured. The American genera, *Sibon* and *Sibynomorphus,* still possess a chin fold, and their dentition is not so highly modified.

Snails: see *Gastropoda.*

Snake bird: see *Pelicaniformes.*

Snake eagles: see *Falconiformes.*

Snake flies: see *Raphidioptera.*

Snake lizards: see *Cordylidae, Pygopodidae.*

Snake mackerels: see *Scombroidei.*

Snake-root: see *Polygonaceae.*

Snakes, *Ophidia, Serpentes:* the most specialized and most recent suborder of the *Squamata* (see). S. evolved from lizard-like ancestors. The fossil record shows a marked increase in diversity at the beginning of the Tertiary, coinciding with the appearance of small rodents, which are the preferred prey of many S. The most obvious characteristic of S. is their elongated body, which in large species may attain lengths of 10 m or more. Elongation of the body is accompanied by a corresponding elongation of the internal organs. Thus, the heart is long and slender, in some species representing one quarter of the length of the body. The right lobe of the lung is highly developed, whereas the left lobe and left pulmonary artery are regressed or absent. Members of the *Boidae,* however, do not display such a marked reduction of the left lung (the lungs of boids are nearly equal in length, while the left lung of the pythons is shortened). There are 4 main types of locomotion. 1) Horizontal forward undulations of the body, i.e. "serpentine" movement. 2) Sidewinding movement, in which loops of the body are thrown sideways, and the snake travels obliquely. 3) Forward movement, in which the vertebral column remains straight, and alternate lifting and lowering of groups of ribs produces a harmonic wave along the body; the belly skin with its large scales (scutes) is pulled across the ribs, then pressed down again, so that the snake appears to glide over the surface. 4) Alternate constriction and expansion of the body, so that the vertebral column does not remain horizontal, but displays a wavelike movement. Other forms of locomotion include the swimming of aquatic S. and the ability of some tree-dwelling species to throw themselves from one branch to another. Possibly the swimming movements of the earliest S., which evolved from amphibian lacertilian ancestors, gave rise to the horizontal undulatory locomotion of modern S. The number of vertebrae is greatly increased, being on average 200, maximally 435. With the exception of the first neck vertebra (the atlas vertebra), all vertebrae can carry ribs. There is no sternum, and the free ribs are movable by special joints, facilitating a forward gliding movement (above). A pectoral girdle is totally absent. Some less advanced familes possess remnants of pelvic girdle. The *Boidae* also possess tiny rudiments of hindlimbs, which are evident as anal spurs. There is no urinary bladder, and the cloacal opening lies across the long axis of the body. In the highly specialized skull, the jaw, palatine and sphenoid bones have additional joints and are loosely attached to one another by ligaments, making it possible to engulf and swallow large animal prey. The eyes are covered by a transparent capsule, which is formed by fusion of the degenerate upper lid with the lower lid; this capsule is shed with the skin during molting. External auditory openings are absent. The bifid (forked) tongue, which can be extended through the rostral opening when the mouth is closed, is both tactile and olfactory, transferring scent material from the exterior to Jacobson's organ (see) in the roof of the mouth. Some species also possess a temperature-sensitive organ, and this is especially well developed in the pit vipers (see *Crotalidae).*

About 12% of all snakes are venomous, and the salivary glands of these species are modified to venom glands (see Snake venoms). In most species the front teeth of the upper jaw are elongated. There are four main types of dentition: *aglyphic* (teeth solid and ungrooved), *opisthoglyphic* (back teeth of the upper jaw are grooved), *proteroglyphic* (front teeth of the upper jaw are grooved), *solenoglyphic* (front teeth of the upper jaw are hollow, i.e. they contain a longitudinal duct). If the teeth are associated with venom glands, then the venom is delivered via the groove or canal into the wound made by the teeth. Boids are aglyphic. Members of the *Colubridae* may be aglyphic or opisthoglyphic. The *Elapidae* and *Hydrophiidae* are proteroglyphic, whereas the *Viperidae* and *Crotalidae* are solenoglyphic. S. feed exclusively on living animals, which they swallow whole, undivided and without mastication. The stomach capacity is appropriately large. Large S. may go for a year without feeding. Most S. lay elongated, parchment-shelled eggs. Some species, like pythons and the king cobra, display brood care. Other species, e.g. boas, rattlesnakes, most vipers and some sea snakes are ovo-viviparous, i.e. they give birth to live young enveloped in a gelatinous egg covering, which is penetrated before or very shortly after birth. Very few species are truly viviparous (i.e. the young are nourished via a placenta).

The S. are subclassified into 3 infraorders and 12 families. Each of the following families is listed separately.

1. Infraorder: *Scolecophidia* (Blind snakes). Families: *Typhlopidae, Leptotyphlopidae.*

2. Infraorder: *Henophidia* (Primitive snakes). Families: *Aniliidae, Uropeltidae, Xenopeltidae, Acrochrodidae, Boidae.*

3. Infraorder: *Xenophidia* or *Caenophidia* (Advanced snakes). Families: *Colubridae, Elapidae, Hydrophiidae, Viperidae, Crotalidae.*

Snake teiids, *Ophiognomon:* a genus of the lizard family, *Teiidae* (see). All 3 species are found in the upper Amazon basin of tropical South America. They have a burrowing habit, and their limbs are degenerate.

Snake venoms: mixtures of toxins produced by venomous snakes, which vary in composition according to the genus and species of the snake. They consist of antigenic polypeptides and proteins (some of them with enzymatic activity), which are produced and stored in the venom glands (parotid glands of the upper jaw). Even "nonvenomous" aglyphic snakes may possess venom glands, which release proteolyic enzymes into the alimentary canal to facilitate the digestion of the undivided prey. In opisthoglyphic, proteroglyphic and solenoglyphic snakes (see Snakes), the venom gland is associated with grooved or canalized teeth.

Some protein components of S.v. are true toxins, causing paralysis and death of prey (or aggressors), while others facilitate the spread of the toxins (e.g. hyaluronidase). ***Neurotoxins*** (nerve poisons) show curare-like activity, preventing neuromuscular transmission by blocking receptors for transmitters at the synapses of autonomic nerve endings and at the motor end plate of skeletal muscle; the victim dies from cessation of breathing. Certain ***protease inhibitors*** also act on the nervous system by inhibiting acetylcholine esterase, thereby preventing nervous transmission and causing paralysis (the catalytic mechanism of acetylcholinesterase closely resembles that of chymotrypsin, to the extent that is also inhibited by certain protease inhibitors). ***Cardiotoxins*** (heart muscle poisons) cause irreversible depolarization of the cell membranes of heart muscle and nerve cells, causing heart failure. ***Hemotoxic activity*** is due phospholipase, which causes hemolysis, and to certain proteases that cause blood coagulation and tissue necrosis.

Neurotoxic proteins from the cobra, *Naja naja atra,* are linear peptides, of which there are 2 types: short chain toxins (containing 60–62 amino acid residues, M_r about 7 000) or long chain toxins (71–74 amino acids, M_r about 8 000). Cobra neurotoxins are structurally related to components of scorpion venom. Neurotoxins from the *Viperidae* have higher M_r values, e.g. M_r of rattlesnake crotoxin is 30 000. Some neurotoxins show subunit structure, e.g. taipoxin from the Taipan *(Elapidae)* consists of 2 nonidentical polypeptide chains. Owing to the presence of disulfide bridges the neurotoxins are extremely stable, suffering no loss of activity when treated with 8 mol/l urea for 24 hours at 25 °C, or for 30 min at 100 °C. Cardiotoxins and protease inhibitors consist of 60 amino acid residues. Despite their pathophysiological differences, the cardio- and neurotoxins of the *Elapidae* (cobras, mambas, etc.) show sequence homologies.

The symptoms following snake bite vary according to the snake species, and to a certain extent are dependent on other factors such as the health and age of the bitten individual. The only more or less reliable remedy for the poisonous effects of snake bite is the injection of antiserum. Antisera are prepared by immunization of large animals, in particular horses, with the venom itself. Monovalent sera act specifically against the venom of one species, while polyvalent sera act against the venoms of several different species. Some S.v. are used pharmaceutically in high dilution, e.g. in rubbing ointments for rheumatic conditions and for neuralgic pain, and for stopping or preventing bleeding. Toxins and enzymes from S.v. are useful aids in pharmacological and biochemical research, e.g. α-bungarotoxin (from the Krait, family *Elapidae)* and cobratoxin (made radioactive by iodination with ^{125}I) have been used to specifically label the acetylcholine receptor in the postsynaptic membrane.

Snapping shrimps: see Shrimps.

Snapping turtles, *Chelydridae:* a family of freshwater turtles containing only 2 living species, distributed from Canada to northwest South America. The plastron is much reduced and cross-shaped. The widely separated abdominal scutes are confined mainly to the narrow bridge, which links the plastron to the strongly vaulted carapace. Even in very old specimens, the knobbly carapace is never entirely ossified. The tail is almost as long as the shell, and has a double row of sharp serrations. The head is large with powerful hooked jaws. They are predators, and they spend most of the time in the water, walking on the bottom and rarely swimming. Adults can even represent a danger to humans. Females dig holes in the ground, in which they lay their eggs. *Macroclemys temminckii* ***(Alligator snapping turtle)*** inhabits the Mississippi region. With a maximal weight of 100 kg and a shell length of 1 m, it is the largest living freshwater turtle, although most specimens are smaller than this. The carapace has 3 keels, and it is usually densely overgrown with algae. *Macroclemys* catches fish by lying in wait on the bottom with open mouth, displaying a red, mobile tongue appendage which acts as bait.

Chelydra serpentina ***(Snapping turtle)*** of North America attains a weight of up to 30 kg with a shell length of 50 cm. Its powerful legs cannot be withdrawn inside the shell. It lies in wait on the bottom, and in addition to aquatic vertebrates, it also feeds on vegetation and carrion.

Sneezewort: see *Compositae.*

Snipe eels: see *Nemichthyidae.*

Snipes: see *Scolopacidae.*

Snout beetles: see Weevils.

Snout moth: see *Pyralididae.*

Snowball tree: see *Caprifoliaceae.*

Snowbell: see *Primulaceae.*

Snowberry: see *Caprifoliaceae.*

Snow bunting: see *Fringillidae.*

Snowcocks: see Pheasants.

Snowdrop: see *Amaryllidaceae.*

Snow finches: see *Ploceidae.*

Snow leopard: see *Felidae.*

Snow vole: see *Arvicolinae.*

Snub-nosed monkeys, *Rhinopithecus:* a genus of the *Colobinae* (see Colobine monkeys). *R. roxellana* ***(Chinese snub-nosed monkey)*** inhabits coniferous forest, bamboo jungle and rhododendron thickets in mountainous central and western China. It is a large, robust animal (head plus body 750 mm, tail 725 mm) with a dark gray to gray-brown, dense coat, which forms a mane over the shoulders. In *R.r. roxellana* ***(Golden monkey)*** the mane contains golden strands, the underparts of the hands and feet are bright orange, and the skin around the eyes is naked and bright blue.

R. avunculus ***(Tonkin snub-nosed monkey)*** inhabits the monsoon forests of northern Vietnam. It is somewhat smaller than *R. roxellana,* and its coat is black with yellowish white underparts.

Soa-soa: see Sail fin lizard.

Sociability: an analytical character for characterizing the distribution of individuals of one species in a stand of vegetation. According to Braun-Blanquet, it is expressed on a 5 point scale, where "1" indicates that each plant grows alone, "2" indicates groups or thickets, "3" indicates bunches, patches or cushions, "4" is for relatively large patches, and "5" indicates extensive massed growth. The sociability number is quoted after the number for cover abundance.

Social communication: information exchange or signalling between individuals of a social sysem (see Sociolgy of animals). It serves for partner recognition, space-time coordination of action, and the generation and maintenance of the existing social structure. Optical, acoustic, chemical and tactile signals are used, often in combination. The associated behavior is collectively known as Communicative display (see), and this often has the property of a releaser stimulus. Communicative display is essentially inherited, but can be extended by learning. The following examples of communicative display also serve for S.c. 1) body movements, e.g. the supplicatory movements of many birds and mammals, the intimidation, threat, attack and submission gestures of social vertebrates for the preservation of ranking order, recognition and marking of territory, and pairing. The dance of honeybees (see Colony formation) also employs movement sequences for communicating the location of pollen sources. 2) Movements of body parts, e.g. the tail, ears or extremities, have a high signal content. Finely differentiated facial expressions have high signalling status in primate social behavior. 3) Movements that display color and shape are widely used in the courtship behavior of birds. 4) Movements that involve bodily contact, like mutual nibbling by horses, or the social grooming of primates, the naso-genital or naso-anal contact of many carnivores, as well as muzzle and beak contact. Establishment of contact is often also associated with chemical signals. 5) The use of chemical signals (usually olfactory, less often gustatory factors) is widespread, particularly in social insects (see Pheromones) and mammals. The active materials either remain on the body or they are released into the environment. They serve for mutual recognition, group coherence, territory marking, orientation, reproduction and brood care. As alarm substances, they trigger flight and defense behavior, and therefore have a protective function. 6) Acoustic signals are used by certain insects (stridulation), but especially by birds and mammals. Sound signals (vocalizations) are used for calling to (searching for) other animals, such as a mate or offspring. Other sounds are used for begging, warning, threatening, attracting and for communicating humility or surrender.

S.c. also serves for Social stimulation (see).

Social hierarchy: see Rank order.

Socialia: see *Isoptera.*

Social regulation: see Colony formation.

Social stimulation: generation, by signal behavior, of behavioral motivation in other members of the same social group (see Social communication). Both *group effects* and *partner effects* contribute to S.s. The former are manifested as increased feeding by each animal in the group, in the calming of individuals after separation from the group, and in the general conformity of many actions and aspects of behavior. In many animals, the partner effect leads to the synchronization of sexual behavior.

Sociation: a vegetative unit that is characterized by one or a few dominant species in each layer. If only one layer of a multilayered vegetation unit contains a dominant species, this unit is called a *consociation.* In complex plant communities, particularly those with a large variety of species, practically only consociations can be recognized. Ss. and consociations are usually determined by the frequency method (see Frequency). Ss. are used only by the Scandanavian school of plant sociology, which considers the Ss. to be fundamental vegetative units. Ss. are very effective for describing the finer quantitative differences in vegetation, especially in species-poor plant societies, or in floristically poor regions, such as northern Europe with its wide stretches of uniform vegetation. See Sociology of plants.

Sociology of animals, ***animal sociology, zoosociology:*** an ecological discipline concerned with social systems (societies) in the animal kingdom. Animals form groups in which the different functional relationships transcend individual survival strategies. Important criteria for the operation and coherence of an animal society are its internal structures (age and sex distribution, group size, synchronization, rank order), its functions (behavioral inhibition, behavioral activation, feeding, reproduction, mutual protection, migration, resting and hiberantion strategies), and its system of information exchange or signalling (see Social communication). Such groups may be temporary or permanent. They are formed and maintained by social attraction between individuals and typified by the communal performance of certain functions or tasks. The main aspects of animal sociology are subjects of ethological research.

1) Groups of individuals of one species (intraspecific animal collectives) are called *homotypic societies.* These are known as homomorphic if they consist of a single class of the species (e.g. the same sex, or the same stage of development), and heteromorphic when several classes are involved, as in animal colonies. *Resting* or *sleeping* groups sometimes consist of a communal gathering of animals of the same sex *(Hymenoptera,* some *Diptera,* crows, starlings, bats and flying foxes). *Overwintering* or *hibernating communities* are formed by solitary bees, and by many amphibians, reptiles and bats. *Molting groups* and some *feeding societies* are homomorphic (molting groups of black arches *(Lymantria monacha),* feeding groups of lepidopteran larvae). Often there is a Social stimulation (see) effect, leading to higher individual rates of feeding. *Communal acquisition of prey* is practised by some birds (feeding groups of cormorants and pelicans) and mammals (hunting packs of wolves, Cape hunting dogs and lions). *Migratory groups* (see Animal migration) are formed by locusts, larvae of fungus gnats *(Mycetophilidae)* and migratory lepidoptera, as well as invading and migratory birds, lemmings and many hoofed grazing mammals. *Colonial groups* or *colonies* are sometimes difficult to distinguish from family groups. A *brood colony* is recognizable from the temporary accumulation of individual nests (e.g. cormorants), whereas an *animal colony* in the normal sense is formed by the permanent cohabitation of several families (e.g. marmots). Breeding colonies are temporary gatherings at particular sites for the purpose of pairing (mating grounds of fur seals, the rut of many hoofed grazing animals, and the mass spawning accumulations of marine invertebrates, fish and amphibians).

Within all the above-mentioned homotypic groupings, individuals display a high degree of anonymity, and are freely interchangeable. In *family groups,* pair formation leads to relatively long lasting relationships within the group, which may extend to close personal bonds and ranking within a hierarchy. This results in parent families (cichlids, song birds, birds of prey), families of female parents (scorpions, ruffs), and families of male parents (sticklebacks, scrubfowl). If the young animals remain in the family group and reproduce within that group, the result is an extended family or kinship group, as in the case of the brown rat *(Rattus norvegicus)* and some spiders. Individualized closed groups of several males, females and young animals are formed by many monkeys and apes, and by some predators (Cape hunting dog). The most highly developed homotypic society is the animal colony (see Colony formation).

2) *Heterotypic societies* (interspecific animal collectives). Associations of different species may be loose and temporary, or they may be permanent interactions. Transitional forms between these two extremes are also known. Mixed migratory and feeding communities are found among insects (dragonflies and butterflies), birds (sandpipers and titmice), and hoofed grazing animals. Particularly varied interspecies relationships occur in probiosis and symbiosis (see Relationships between living organisms).

Sociology of plants, ***plant sociology, phytosociology:*** the study of plant communities and their relationships with the environment. It is a branch of plant geography, and it studies the basis for the repeated occurrence of particular groups of plant species in specific plant communities or societies. The plant cover of an area can be characterized by many different properties (floristic composition, appearance, nature of the substratum and other environmental factors, etc.), so that plant societies can be delineated from different standpoints.

1) Vegetation can be classified *floristically* according to its Characteristic species combination (see) or its Characteristic species group combination (see) *(Halle school),* according to one or a few species *(Nordic* or *Scandanavian school),* or according to statistically determined groups, whose constituent species are known as character or differential species *(Alpine school);* in parts of southern, western and central Europe, the methods of the alpine school are followed in modified form. Classification can also be based on species of the same geographical distribution (see Areal, Chorological species), species with the same ecological behavior (see Ecological species group), or species of the same dynamic behavior (see Succession). Succession is also of general importance in plant sociology.

2) Vegetation classification according to *physiognomic* criteria. This type of classification is based on life form and growth form, often in combination with some floristic characters. Many authors use a combination of floristic and physiognomic classifications.

Plant sociology may be *descriptive,* in that it investigates the composition of plant societies, and is therefore concerned with clearly delineated plant sociological units. Alternatively, *synecology* studies the energy and material balance (budget) of plant societies and environmental factors that determine the composition and behavior of plant societies. *Syndynamics* studies the development of plant societies with time. *Synchorology* is concerned with the geographical distribution of plant societies. The aim of *Syntaxonomy* (classification of plant societies) is to elaborate a clear natural system or taxonomy of plant societies. Finally, *synsociology* is devoted to the mapping of the phytosociological complexes of individual land areas.

Plant sociology developed in the 1920s mainly from the study of formations. At the turn of the 20th century, the major types of vegetative cover in central Europe were well defined as physiognomic descriptions of their formations. This approach proved to be less appropriate for the study of the variety of vegetation in smaller areas. More objective methods were therefore developed for the study of relatively small vegetative units, otherwise known as plant communities, plant societies, etc.

Sodium, *Na:* an alkali metal. All living cells accumulate potassium ions almost to the exclusion of Na^+. Na^+ does not appear to be essential for nonmarine plants, but it is required by some C4 salt-tolerant plants, and it plays an essential part in crassulaceaen acid metabolism. Na is a component of felspars, whose weathering produces soluble forms of Na in the soil. Different mineral fertilizers contain Na as ballast material in the form of sodium chloride (NaCl). Na^+ ions cause silting and incrustation of soils, so that NaCl fertilizers must be used with caution on heavy soils. In contrast to potassium, Na is not taken up actively by plants, and plants are unable to control or prevent Na uptake. Some plants tolerate high Na concentrations, and respond favorably to Na-containing mineral fertilizers. Such *natrophilic plants* (e.g. spinach, beet, barley, celery, various species of cabbage) often have a high Na content of about 15–25 mg/g dry mass. In contrast, the aerial parts of *natrophobic plants* (e.g. beans, sunflower, maize) usually contain less than 1 mg/g dry mass, although their roots may contain relatively high Na concentrations, suggesting that Na transport from the roots is inhibited.

Na^+ is required by multicellular animals that regulate their internal body fluids, maintaining high extracellular and low intracellular Na^+ concentrations (see Ion pumps). With potassium, it is involved in the excitation of muscle and nervous tissue. The majority of body Na is contained in the blood plasma, where, as NaCl, it is partially responsible for regulation of the osmotic pressure of the blood. The dietary source of Na is mainly NaCl.

Sodium channel, ***Na^+channel, voltage-gated Na^+channel:*** a localized site within the plasma membranes of electrically excitable cells (mainly nerve and muscle cells), which accounts for their selective and regulable permeability to Na^+ ions. The Na^+ channel consists of a protein (M_r approx. 230 000, containing about 2000 amino acid residues), which spans the plasma membrane. cDNA for this protein has been cloned from the membranes of the electric organ (electroplax) of the electric eel *(Electrophorus electricus)* and from two different Na^+ channels in rat brain. Striking similarities in the structures of all three cDNAs (and therefore in the structures of the corresponding Na^+ channel proteins) indicate that the structure of Na^+ channel proteins has been greatly conserved during evolution. Structural conservation is especially apparent in 4 repeated regions (which probably arose by gene duplication) of the protein. Each region contains 6 helices, which span the membrane and form the Na^+ channel (Fig.). One of these helices contains a relatively large number of basic amino acid residues (arginyl and lysyl residues), and all 4 copies of this helix are probably important in voltage gating (see below). The purified protein can be incorporated experimentally into phospholipid bilayers, and the behavior of such reconstituted systems is very similar to that of native plasma membranes, i.e. they exhibit Na^+ transport dependent on a threshold voltage.

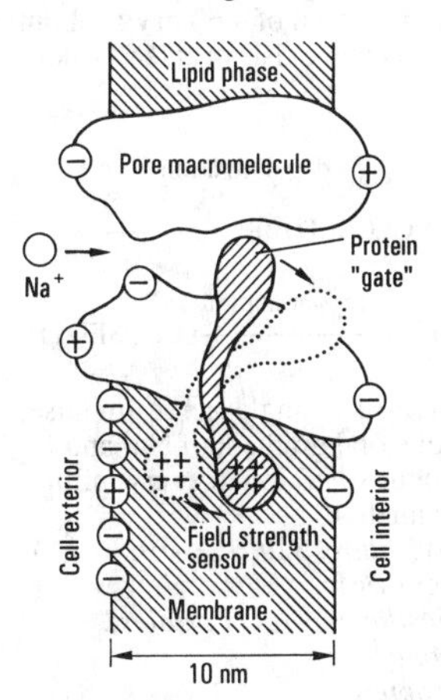

Model of a sodium channel

If the rate of Na^+ transport via Na^+ channels is titrated against pH, it reveals a p*K* of 5.2. This suggests that a negatively charged carboxylate ion is present in the channel, probably enabling Na^+ transport by specifically hydrogen bonding with the hydrated Na^+ ion. Selectivity is also determined by the size of the hydrated ion, radii greater than 5 Å being excluded. Na^+ (ionic radius 0.95 Å) is transported 11 times more quickly than K^+ (ionic radius 1.33 Å). Tetrodotoxin (see) exerts its toxicity by specifically blocking Na^+ channels in nerve and muscle cells (K^+ channels are unaffected). Saxitoxin (see) has a similar mode of action. From the kinetics of the inhibition of Na^+ transport by tetrodotoxin (dissociation constant for binding of tetrodotoxin to Na^+ channel = 10^{-9}M), it is calculated that about 20 Na^+ channels are present per μm^2 of the membrane surface of unmyelinated nerve fibers, i.e. the average distance between any two Na^+ channels is about 2000 Å. In contrast, Na^+ channels are present at a density of about $10^4/\mu m^2$ in the nodes of Ranvier of myelinated nerve fibers. Batrachotoxin (see) exerts its toxicity by causing a selective and irreversible opening of Na^+ channels, and its action is antagonized by tetrodotoxin.

The Na^+ channel is said to be *gated by the membrane voltage,* i.e. the "gate" of a Na^+ channel is opened during an action potential (nerve impulse), only when the depolarization voltage reaches a critical value. Na^+ then enters the cell, and this influx of positive charge causes further depolarization, leading to the opening of more Na^+ channels. A single, open Na^+ channel allows the passage of 10^7 Na^+ ions per second. The resulting cascade effect (positive feedback) results in a rapid, local change in membrane potential (from about – 60 mV to about +30 mV in 1 millisecond), but it does not result in the permanent opening of all Na^+ channels. Since the Na^+ equilibrium potential lies at +30 mV, the flow of Na^+ stops when the potential of the membrane reaches +30 mV. An independent and apparently automatic mechanism then inactivates (closes) the Na^+ channel. As the Na^+ channels are closing, the K^+ channels begin to open, resulting in an outward flow of K^+, which restores the membrane potential to its former negative value. Thus, about 2 milliseconds after the initial impulse, the membrane potential reaches – 75 mV (the K^+ equilibrium potential), followed by a slight rise, attaining the membrane resting potential of – 60 mV at about 3 milliseconds. The Na^+ channel exists in at least 4 conformational states:

$$\text{CLOSED (1)} \xrightarrow{\text{depolarization}} \text{CLOSED (2)} \xrightarrow{\text{depolarization}}$$
inactivation

$$\text{OPEN} \xrightarrow{\text{spontaneous}} \text{INACTIVE} \xrightarrow{\text{repolarization}} \text{CLOSED (1)}$$

During the transmission of a single nerve impulse, only a small fraction of the total gradient of Na^+ and K^+ is dissipated, e.g. only about 1 Na^+ ion in every million passes through a Na^+ channel.

[M. Noda et al. (1986) *Nature*, ***320***, 188–192]

Sodoku: Rat bite fever (see).

Soft amadou: see *Basidiomycetes (Poriales).*

Soft corals: see *Alcyonaria.*

Soft scales: see *Homoptera.*

Soft-shelled clams: see *Bivalvia.*

Softshell turtles: see *Trionychidae.*

Soft ticks: see *Ixodides.*

Soil bacteria: see Soil organisms.

Soil biology, *pedobiology:* a branch of soil ecology, which deals with the composition and activity of biotic communities and populations in the soil.

Major areas of S.g. research are the taxonomy, morphological and physiological adaptations, quantitative and qualititive composition and distribition of soil biotic communities, as well as their influence on the formation, breakdown and fertility of soils. Soil zoology (pedozoology) has developed as a branch of zoology, while Soil microbiology (see) is an important branch of soil science. Functional S.b. investigates the role of soil organisms in the total soil ecosystem, with particular regard to the processes of Decomposition (see), soil productivity, cycling of mineral nutrients, and the formation and maintenance of soil structure. Diagnostic S.b. uses various biological indicators for assessing current and historical anthropogenic influences on the soil (agriculture, fertilizer application, action of pesticides and environmental pollutants), as well as for the classification of soils. Applied S.b. is concerned with: the exploitation of soil biological processes for agriculture and forestry (e.g. plowless agriculture; the management and biological control of soil pests, especially when new agricultural methods are tested (e.g. single seed sowing); the capacity of soil biotic communities to survive the stress of pollution, to purify their own environemnt (i.e. to act as depolutant systems), and to serve for the treatment of sewage; the regulation of anti- and symbiosis, especially in the rhizosphere (e.g. in order to increase nitrogen fixation); manipulation of humus production and mineralization; and the re-establishment of biotic communities in semisterile soils, e.g. in areas affected by toxic waste, mine spoil heaps and slag heaps.

Historical. Nineteenth century pioneers were Pasteur and Winogradsky (soil microbiology), Charles Darwin and Hensen (soil zoology). S.b. was established as an independent discipline around 1950 by the work of Gilarov, Kühnelt, Franz and Delamare-Deboutteville, and by the flounding of international journals (*Pedobiologia* was launched in 1961).

Soil microbiology: a branch of microbiology and soil biology, which studies living microorganisms in the soil. In addition to aquatic environments, the soil is a major habitat for microorganisms. One gram of good agricultural soil can contain up to ten thousand million (10^{10}) live bacterial cells, several hundred thousand fungal cells, 100 000 algae and 10 000 primitive fauna. Many of the activities of soil microorganisms are essential for soil fertilitiy and structure, e.g. production of nutrients by mineralization, fixation of atmospheric nitrogen, and improverment of crumb structure by formation of colloidal material. See Soil organisms, Soil biology, Humus.

Soil microfauna and microflora: see Soil organisms.

Soil organisms, *edaphic organisms, edaphobionts:* plants and animals living permanently or temporarily in the pores and cavities of the soil interior, or in the crevices of the soil surface. The totality of S.o. is known as the *edaphon.* S.o. are abundant in the humus layer and the surface layer (A-horizon), and far less numerous in the nutrient-poor subsoil (B-horizon).

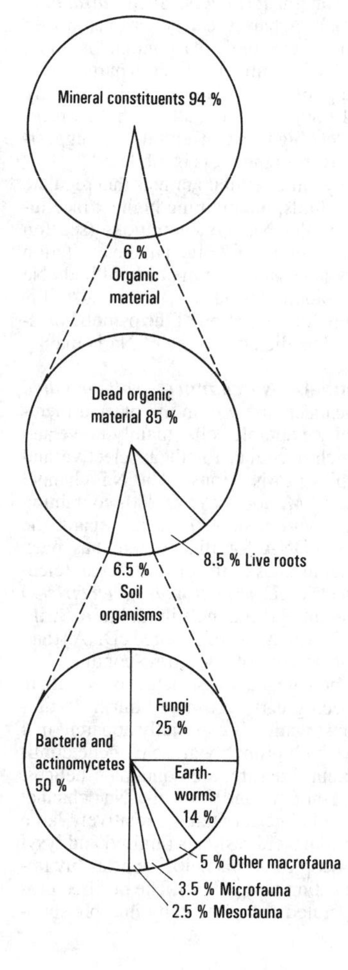

Fig. 1. *Soil organisms.* Relative masses of soil organisms in the top soil of a deciduous forest.

Significant numbers are not found in the soil parent material (C-horizon), except where abnormal accumulations of organic material occur, e.g. in graveyards and after deep plowing. S.o. are crucially important for soil fertility. Soil microflora perform the essential function of degrading organic material and converting it into humin (see Humus). Soil fauna consume vegetable debris, and their tissues are protein-rich and easily decomposed; since many soil animals are short-lived, they make a large contribution to the decay process, but animals do not themselves form humus. In addition, animals help to create and maintain an open, porous soil structure.

In the soil of a temperate deciduous forest, organic material contributes 1–10% of the mass of the humus-rich surface soil layer, while the roots of living plants contribute 10–15% (Fig.1). Microflora constitute about 75% of the S.o. mass, while the soil fauna constitute about 25%. The population density and mass of the main groups of S.o. are listed in the table.

Group	Numbers beneath 1 m^2 of soil surface	Mass (g)
Bacteria	10^{14}–10^{16}	100–700
Actinomycetes	10^{13}–10^{15}	100–500
Fungi	10^{11}–10^{14}	100–1 000
Green algae and blue-green bacteria	10^{8}–10^{11}	20–150
Protozoa	10^{8}–10^{10}	5–150
Nematodes	10^{6}–10^{8}	5–50
Enchytraeids	10^{4}–10^{5}	5–50
Earthworms	10^{1}–10^{2}	30–200
Mites	10^{4}–10^{5}	0.6–4
Apterygotes	10^{4}–10^{5}	0.5–4
Others	10^{2}–10^{4}	5–100

Soil microflora. The most numerous and diverse components of the microflora are the bacteria, representing an average mass of 500 kg/hectare. Most are heterotrophic, obtaining energy and carbon from organic materials. Nitrogen fixation in the soil is economically very important; soil bacteria capable of fixing atmospheric nitrogen include free-living (e.g. *Azotabacter),* symbiotic (e.g. *Rhizobium)* and photosynthetic *(Cyanophyceae)* species. Nitrifying bactera are important soil autotrophs. These various types of soil bacteria all contribute in different ways to the cycling of organic materials and nutrients, and all have different ecological requirements. High oxygen concentrations usually favor nitrifying and nitrogen-fixing species, but inhibit the growth of reducing bacteria. In natural, biologically active soils and in agricultural soils, bacteria make a major contribution to the formation of mineral plant nutrients, the synthesis and degradation of humus, and the stabilization of soil crumb structure.

Actinomycetes are relatively numerous in humus-rich, adequately drained, nonacidic (pH 6–8) soils, attaining population densities of 10^5 to 10^7 organisms/g soil, and a mass of 100 to 700 kg/ hectare. Actinomycetes are mainly aerobic, but like many bacteria, some of them also live as symbionts in the intestines of soil fauna.

Fungi are more numerous in adequately aerated, acidic, humus-rich or heavily manured soils; they are mainly aerobic, rarely partially anaerobic. At population densities of 10^3 to 10^7 organisms/g soil, fungi can attain a mass of 500 to 10 000 kg/hectare. They occur in a wide variety of morphologically and physiologically different forms, the most important being yeasts, molds and basidiomycetes. Molds and basidiomycetes make a very large contribution to the biological activity of the soil, especially as mycorrhizal fungi.

Since algae are photosynthetic, they require light and live mainly in the upper soil layer. Some species, however, can live in deeper soil layers, by utilizing long-wave irradiation, or by obtaining energy heterotrophically by decomposition of organic compounds. Although they can withstand long periods of drought, algae require high moisture levels for growth and development. Soil algae are mostly members of the *Chlorophyceae.* The soil also contains high concentrations of *Cyanophyceae* (blue-green bacteria), which are prokaryotic, photosynthetic, nitrogen-fixing organisms. Green algae (both unicellular and filamentous) and blue-green bacteria are the predominant photosynthetic soil microorganisms, but diatoms are often also common in humus-rich soils. Under optimal conditions for their development, soil photosynthetic organisms can attain a biomass of more than 1 000 kg/hectare, making a considerable contribution to the biological activity of the soil.

There is a close relationship between the type of Humus (see) and its microflora. Bacteria and actinomycetes predominate in mull humus, whereas fungi are more prevalent in mor humus and leaf litter. Soil microorganisms have a worldwide distribution, but the composition of a local microfloral population depends on the climate, soil type and the nature of the material undergoing decomposition. In any soil, there may be nonactive microorganisms (known as potential microflora), which are incapable of competing under the prevailing conditions. High temperatures increase the activity (respiration) and often the biomass of soil microorganisms. Moisture-dependent bacteria and actinomycetes attain maximal population densities usually in the spring and autumn, whereas fungi tend to be more dominant in the summer. Acidic conditions (pH 3–5) favor fungal development, but these low pH-values are especially inhibitory to nitrogen fixation and nitrification. The region of maximal population density and biological activity of microorganisms lies immediately below the surface layer of the soil, where temperatures are still high, the moisture content is relatively constant, aeration is efficient, and there is a high concentration of organic material. Also within this layer, the effects of environmental desiccation are less than on the surface, and illumination from the sun is considerably attenuated. Deeper soil layers contain relatively larger populations of anaerobic organisms. Digging and plowing, changes in the type of plant cover, and the application of organic and inorganic fertilizers, herbicides or pesticides, all selectively and profoundly affect the soil microflora, but this effect is often only short-term.

The species composition of the microorganism population is also greatly influenced by biotic interactions. Various types of antagonism are common, e.g. competition for nutrients, production of antibiotics (e.g. penicillin, aureomycin), parasitism (e.g. certain bacteria parasitize fungi), heterotrophic nutrition at the expense of other microorganisms (e.g. myxobacteria and slime molds feed selectively on certain bacteria). Positive interactions (synergism) occur by cross-utilization of metabolic products (e.g. between nitrogen-fixing and cellulose-degrading bacteria), by production of growth factors (e.g. vitamin B_{12}) and by true symbiosis (e.g. between fungi and algae in lichens).

Relationships between soil microorganisms and soil fauna are primarily trophic. Soil fauna feed on microorganisms and distribute them in the soil. Spores and other resistant stages often pass undigested through the animal, and are still viable in the feces. Grazing of microflora by soil fauna (e.g. protozoa) has been observed to "rejuvenate" the microfloral population (see Protozoan theory). Soil fauna also produce physiologically active substances that stimulate or inhibit soil microorganisms. Various characteristic interactions of microflora and plant roots, e.g. micorrhiza and root nodules, occur in the rhizosphere.

Soil fauna are classified as *permanent* or *true soil fauna,* when all stages of their life cycle are spent entirely in the soil. Other types are: *temporary soil fauna,* which spend only one stage of their life cycle in the soil (e.g. insect larvae); *periodic soil fauna,* which frequently leave the soil and return to it (e.g. certain mammals); *partial soil fauna,* i.e. temporary soil fauna, which also spent part of their "open air" adult phase in the soil (e.g. dung beetles); *alternating soil fauna,* in which one or more soil-living generations alternate with generations above ground (e.g. vine phylloxera); and *transitory soil fauna,* whose presence in the soil is restricted to inactive forms, like eggs and pupae.

The following classification of soil fauna, based on size, is accepted internationally: *microfauna* (up to 0.2 mm long; protozoa); *mesofauna* or *meiofauna* (0.2–4 mm long; rotifers, nematodes, tardigrades, mites, apterygotes); *macrofauna* (4–80 mm long; enchytraeids, earthworms, snails, myriapods, isopods, beetle larvae, dipteran larvae); *megafauna* (more than 800 mm long; vertebrates). The limits of these size ranges cannot be applied strictly, and there is considerable overlap in the lengths of certain organisms.

Soil fauna can also be separated ecologically, i.e. according to the layers of the soil profile which they inhabit. *Euedaphic* organisms occupy the pores and channels of the soil interior. These include aquatic fauna in water-filled capillaries and in the water film around soil particles, consisting of sessile (protozoa) and motile (rotifers and nematodes) forms. The air-filled pores contain nonburrowing, small cavity dwellers; these are mainly microarthropods, like mites and springtails, which are usually small and wormlike, being nonpigmented and blind, with only short limbs and few or no surface hairs. *Hemiedaphic fauna* occupy the upper soil layer and the surface debris, and they are relatively adaptabe to atmospheric changes. Morphologically and physiologically, hemiedaphic fauna represent a transition to the *epedaphic soil fauna,* which depend primarily on the upper soil layer for food, but usually spend the day in cavities in the superficial humus layer. Epedaphic fauna are mostly large, well pigmented and actively motile. Large, burrowing soil fauna *(fodients)* are not classified as euedaphic, since they create an extensive soil-air interface, which extends into the soil interior (earthworms, small rodents). The *atmobiotic soil fauna* consist of species at risk of desiccation, e.g. springtails, which feed in the vegetation layer, but must regularly (daily) enter the soil to balance their water economy.

Soil fauna are equipped in various ways to meet the physiological dangers of changing soil water content. Some possess complex sensory organs for monitoring humidity (e.g. the postantennal organ) and some can adapt physiologically to desiccation. Conversely, a capability for increased osmoregulation is important when soils are flooded. S.o. often live in an atmosphere different from that above ground. Thus, by facultative anaerobiosis, many euedaphic fauna are able to live in high concentrations of carbon dioxide (up to 3%) and decreased oxygen tensions. For small arthropods with a thin integument, the physiologically critical soil moisture content lies at or below that which results in permanent plant wilting. Most species have a low, narrow optimal temperature range (cool-stenothermic organisms). Xerothermophyllic soil fauna are unknown. The most important food sources are dead organic material (humus) and other S.o., so that soil fauna make an important contribution to Decomposition (see). Relatively few groups of soil fauna regularly feed on and damage plant roots and germinating seeds, e.g. nematodes and some insect larvae. Many more species are facultative pests, becoming phytopathologically important when other food and water supplies are scarce. Such soil pests are best controlled by appropriate husbandry, such as crop rotation. Soil fauna also play a part in the maintenance of soil texture and structure by soil crumb formation, and by mixing and loosening the soil, which improves aeration and drainage. In temperate climates, earthworms are very largely responsible for these processes, but in warm, dry soils, arthropods have more influence on the physical structure of the soil. For example, termites are particularly active in tropical soils. In European soils, the formation of soil crumbs (fecal casts) can amount to 100 tonnes/hectare/year, equivalent to a surface layer of 5 mm. In tropical soils, this value can be as high as 240 tonnes/hectare/year. The richest soil fauna is found in fresh, humus-rich mull soils, whereas humus-deficient sandy soils have a poor soil fauna. Tropical rainforest soils contain less soil fauna that deeper, humus-rich temperate soils.

Methods of investigation. The total activity of soil microorganisms can be determined indirectly from the respiration (carbon dioxide production) of a soil sample, directly by counting organisms under the microscope, or by colony counting of plated extracts. Even when living and dead microorganisms are differentiated by staining, a direct microscopic count gives a value of 10 to 100 times greater than that obtained by plating. Soil flora can be differentiated and identified by plating out in the presence of different nutrients and antibiotics. Special aids, however, are usually needed for the isolation of soil fauna. Earthworms can be driven to the surface by applying 0.5% formaldehyde solution (5 l per 1/4 m^2). On account of their small weight-to-volume ratio, small arthropods can be separated from mineral soils by flotation, by shaking with concentrated sodium chloride solution. Nematodes and enchytraeids actively abandon a damp soil sample if it is warmed, and they can be collected in a water funnel placed beneath (Baermann's funnel, Fig.2).

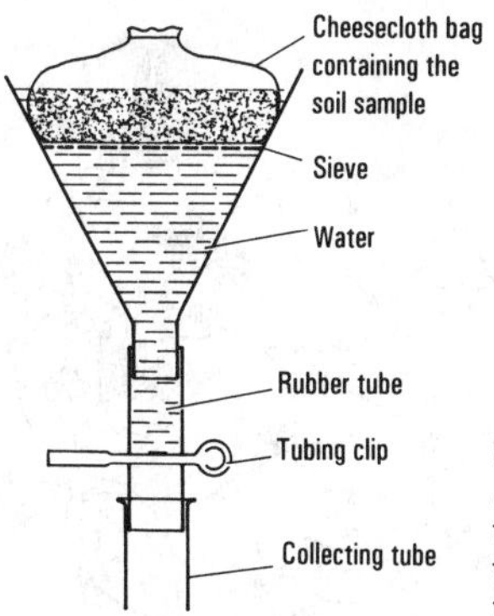

Fig.2. *Soil organisms.* Baerman's funnel for the isolation of nematodes from soil.

Small arthropods (mites, springtails) can be driven from a soil sample by warming, drying, and illumination from above. They emerge from the bottom of the sample, and can be collected in a tube of ethanol (Berlese's or Tullgren's apparatus, Fig.3). Animals of the soil surface can be caught in traps (Barber's trap, Fig.4).

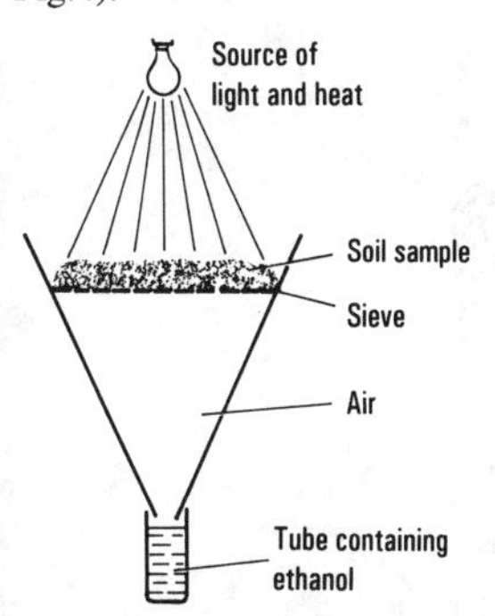

Fig.3. *Soil organisms.* Berlese's apparatus for the isolation of small soil arthropods.

The more familiar nonsoil organisms can usually be quickly identified on the basis of easily recognizable taxonomic features. By comparison, the species determination of practically all S.o. is more involved, and often very difficult.

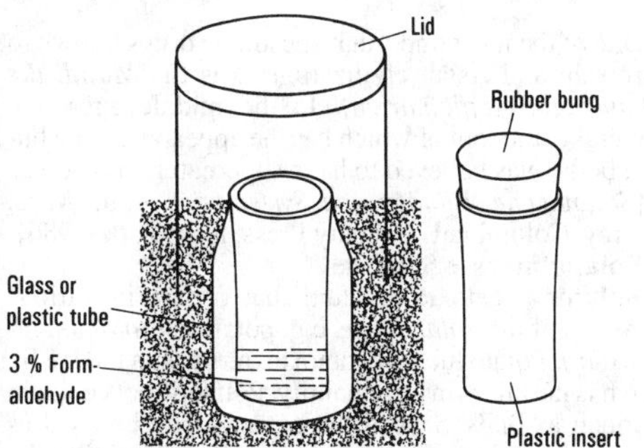

Fig.4. *Soil organisms.* Barber's trap for the collection of soil surface organisms.

Soil pests: soil animals that decrease the economic efficiency of agricultural or forestry land. Four types are recognized: 1) transitory soil inhabitants that damage the aerial parts of plants, and which can be controlled by attacking their resting stages in the soil; 2) true soil inhabitants that feed obligatorily on subterranean plant parts, e.g. certain nematodes, wire worms, beetle larvae, cutworms, slugs; 3) animals that behave as semiparasites, i.e. they normally feed on humus, but under abnormal conditions will consume parts of living plants to the extent that they constitute a pest, e.g. many nematodes, springtails, diplopods, dipteran larvae; 4) soil animals that are considered as pests, because they serve as vectors of parasites of domestic animals and of plant diseases (e.g. viral diseases).

Soil sickness, ***soil tiredness:*** a phenomenon which leads to a decrease in agricultural yield, when the same crop is repeatedly planted in the same soil. To the extent that S.s. is understood, it is due to: 1. accumulation of bacterial, fungal or animal pathogens and parasites, which persist in the soil as resisitant cysts, spores or eggs (e.g. potato eel worm in potato fields); 2. accumulation of growth-inhibiting root secretions; 3. depletion of specific nutrients; 4. the spread of certain weeds.

Soil sterilization, ***soil disinfection:*** biological, physical or chemical treatment of horticultural and agricultural soils, in order to kill pests and disease organiosms. For research purposes, S.s. may be used to kill useful organisms, in order to investigate their contribution to soil enzyme production, humus formation and humus breakdown, etc.

Physical methods of S.s. include X-, γ- or electron irradiation, steaming and autoclaving. Examples of chemical methods (largely for control of nematodes and fungi) are fumigation with formaldehyde or methyl bromide. Other chemicals are added as washes, but are active by vaporization, e.g. methyl isothiocyanate, carbon disulfide. Such chemicals are also harmful to plants, and time (weeks) must be allowed for them to disappear from the soil before a crop is planted. 1,2-Dibromo-3-chloropropane is of sufficiently low toxicity to standing plants to be used for nematode control in citrus orchards.

Solanaceae, ***nightshade family:*** a family of dicotyledonous plants containing 90 genera and 2 300 species, distributed mainly in the tropics and subtropics. They are herbaceous, rarely woody plants, with simple or pinnate, alternate, exstipulate leaves, which are sometimes paired by adnation in the upper parts of the plant. The 5-merous, conspicuous, insect-pollinated flowers are usually actinomorphic, sometimes weakly zygomorphic. The superior, 2-locular ovary develops into a multiseeded capsule or berry. Many species of the genus *Solanum* are andromonoecious, i.e. they produce both hermapohrodite and staminate flowers.

Alkaloids occur widely and abundantly in the *S.*, especially tropane alkaloids (e.g. hyoscyamine, atropine, scopolamine), nicotine alkaloids (e.g. nicotine, nornicotine, anabasine) and steroid alkaloids. Many species are therefore pharmaceutically important or poisonous. Poisonous species include: ***Bittersweet*** or ***Woody nightshade*** *(Solanum dulcamara)* with violet flowers and red berries; ***Black nightshade*** *(Solanum nigrum)* with white flowers and black berries; ***Dwale or Deadly nightshade*** *(Atropa belladonna,* Fig.1a), which grows on calcareous soils and has very poisonous black berries; ***Henbane*** *(Hyoscyamus niger,* Fig.1b), which grows in ruderal habitats; and the ***Thorn apple*** *(Datura stramonium,* Fig.1c).

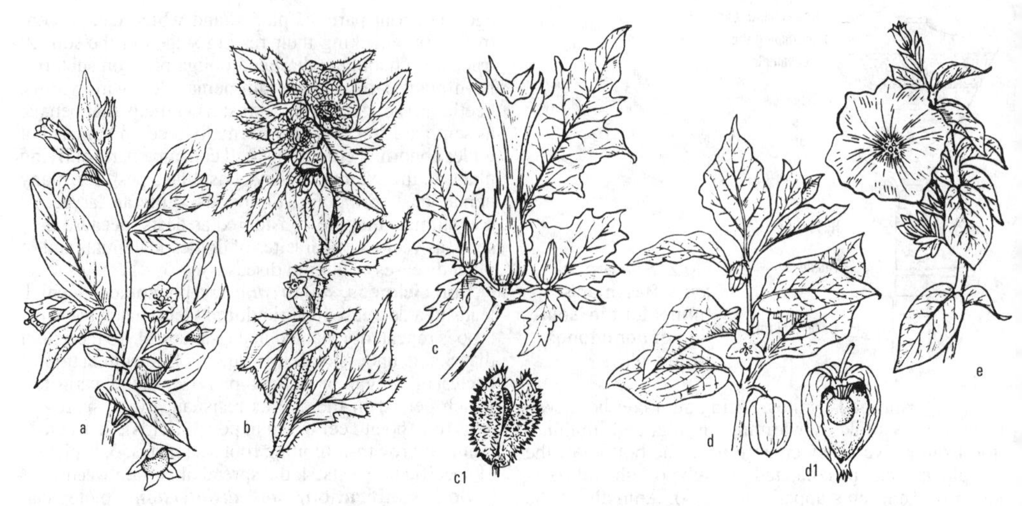

Fig.1. *Solanaceae. a* Deadly nightshade *(Atropa belladonna). b* Henbane *(Hyoscyamus niger). c* Thorn apple *(Datura stramonium) (c1* dehisced fruit). *d* Chinese lantern *(Physalis alkekengi) (d1* calyx cut away to show fruit). *e* Petunia *(Petunia hybrida).*

One of the world's important food crops is the ***Potato*** *(Solanum tuberosum)*, which originated in the foothills of the Andes, and was introduced into Europe in the 16th century. On hundred and sixty tuber-forming species of *Solanum* are known. These are all exploited for breeding cultivars that are resistant to various fungal, bacterial and viral pests of cultivated potatoes, and for breeding varieties with improved storage properties.

Also originating from South and Central America is the ***Tomato*** *(Lycopersicum esculentum),* many varieties of which are now grown throughout the world as a vegetable. The large fruit of ***Paprika*** or ***Sweet pepper*** *(Capsicum annum),* indigenous to South America, is a popular vegetable with a very low content of the hot-tasting principle, capsaicin. In contrast, the smaller fruits of ***Cayenne pepper, Red pepper*** or ***Chilli pepper*** *(Capsicum frutescens),* cultivated in the tropics, are especially rich in capsaicin.

Another well known vegetable is the fruit of the ***Egg plant*** or ***Aubergine*** *(Solanum melongena,* Fig.2), which originated in the Indian subcontinent. *Physalis peruviana* ***(Cape gooseberry)*** is also cultivated for its fruits. Two species of ***Tobacco*** are cultivated, *Nicotiana tabacum* and *N. rustica* ***(Turkish tocacco),*** neither species being known with certainty in the wild. The commercially less important *N. rustica* probably originated in Peru. *N. tabacum* is allotetraploid, and it probably originated in Northwest Argentina from the diploid wild *N. silvestris* and *N. otophora.* Smoking tobacco is the cured leaf of *N. tabacum* or *N. rustica.* In addition, tobacco serves as a commercial source of the alkaloid, nicotine, which is used as an insecticide.

Physalis alkekengi ***(Chinese lantern,*** Fig.1d) is a popular ornamental plant with conspicuous, balloon-like inflated red calyces. Many varieties of ***Petunia*** *(Petunia hybrida,* Fig.1e) are grown in gardens and window boxes. ***Jerusalem*** or ***Winter cherries*** *(Solanum capsicastrum* and *S. pseudocapsicum)* are grown for their attractive, cherry-sized, orange-colored, inedible fruits.

Fig.2. *Solanaceae.* Aubergine *(Solanum melongena).*

One of the most important anesthetic drugs known to European and Asian civilizations was the ***Mandrake*** *(Mandragora officinarum).* In the middle ages, the mandrake, the root of which has the appearance of a human body, was believed to have supernatural powers.

[*Solanaceae, Biology and Systematics,* edit. W. G. D'Arcy, Columbia University Press, New York, 1986]

Solanidine: see Solanine.

Solanine: a toxic alkaloid that occurs in various members of the *Solanaceae,* e.g. potatoes *(Solanum tuberosum).* Potato tubers contain no more than 0.01% S., which is harmless in this quantity. Potato shoots contain as much as 0.5% S., and therefore cannot be used as food. S. is a glycoside (glycoalkaloid), containing the aglycon *solanidine* (solanid-5-ene-3β-ol) and a carbohydrate residue. Various solanines are known, which differ in their carbohydrate residues. The commonest is α-S., in which the solanidine is linked to the trisaccharide β-solatriose (a trisaccharide of D-galactose, D-glucose and L-rhamnose) (Fig.).

D—Glucose
D—Galactose—O
L—Rhamnose

α-Solanine

Solanum: see *Solanaceae.*

Solasodine: see Solasonine.

Solasonine: a steroid alkaloid occurring in many *Solanum* species, e.g. *S. sodomaeum, S. aviculare, S. laciniatum, S. nigrum.* It is a glycoside (glycoalkaloid) containing the aglycon *solasodine* [(22R:25R)-spirosol-5-ene-3β-ol] and a branched trisaccharide of L-rhamnose, D-glucose and D-galactose. Like tomatidine (see Tomatine), solasodine is a steroid with a secondary amino group, and it is important as a starting material for the laboratory synthesis of other pharmacologically active steroids, e.g. adrenal cortical hormones and sex hormones.

Solaster: see Sunstar.

Solatriose: see Solanine.

Soldier beetles: see *Coleoptera.*

Sole: see *Pleuronectiformes, Solea.*

Solea, *soles:* a genus of marine flatfish (family *Soleidae,* order *Pleuronectiformes)* inhabiting sandy and clay sea bottoms at depths of 10–100 m off the coasts of Europe and Africa, and in the Mediterranean. They are important commercially, and have fine–tasting flesh. The small, rounded head extends beyond the crooked mouth, so that the mouth lies at the side (not at the end, as in the *Pleuronectidae).* Examples are the ***Dover sole*** *(S. solea)* and ***Sand sole*** *(S. lascaris = Pegusa lascaris).*

Solenium: see *Helioporida.*

Solenoconcha: see *Scaphopoda.*

Solenocyte: see Protonephridia.

Solenodontidae, *solenodons:* a family of the *Insectivora* (see), represented by a single living genus containing only 2 species. Solenodons are nocturnal in habit and rat-like in appearance with a markedly long, pointed, cartilaginous snout. *Solenodon paradoxus* ***(Hispaniola solenodon)*** is found in remote regions of Haiti. The cartilaginous skeleton of its snout articulates with the skull by a ball and socket joint. *S. cabanus* ***(Cuban solenodon)*** is similar to *S. paradoxus,* but its coat is finer and its tail is slightly shorter, and the snout does not articulate with the skull. The submaxillary glands of *S. paradoxus* (and presumably also those of *S. cabanus)* produce a toxic saliva. Solenodons feed on beetles and other invertebrates in the soil and in tree litter. Dental formula: 3/3, 1/1, 3/3, 3/3 = 40. Both are now threatened with extinction. The already extinct genus, *Nesophontes,* inhabited the Caribbean islands at least until the arrival of the Spaniards.

Solenodon

Solenogastres: see *Aplacophora.*

Solenoglyphic dentition: see Snakes.

Solenostomidae: see *Syngnathoidei.*

Solidago: see *Compositae.*

Solifugae, *windscorpions:* an order of the *Arachnida* (see). Some species attain a body length of 7 cm. The prosoma consists of a proterosoma and two free segments, the opisthosoma is segmented, and the carapace carries a pair of large median eyes. Long sensory hairs occur over the entire body surface. Chelicerae are very large, chelate and armed with teeth, but they do not contain venom glands. Pedipalps are leg-like and provided at the tip with an eversible attachment vesicle. The first pair of walking legs is often short and weak, and their tarsi contain sensilla ampullacea (pits with short sensory hairs). The terminal segments of the last pair of walking legs carry racquet organs on their ventral surfaces; a racquet organ is a sensory organ, resembling a tennis racquet in shape and containing sensory nerve endings.

About 800 species are known, inhabiting desert and steppe areas, mainly nocturnal or crepescular, and distributed mainly in India, West Indies, eastern USA and northern Mexico. There are 6 European species. When faced with an adversary, they immediately adopt a defense posture, facing the enemy with open chelicerae and emitting a hissing sound. Prey (other arthropods) is caught by chase or ambush and dispatched mechanically with the chelicerae; poison is never used.

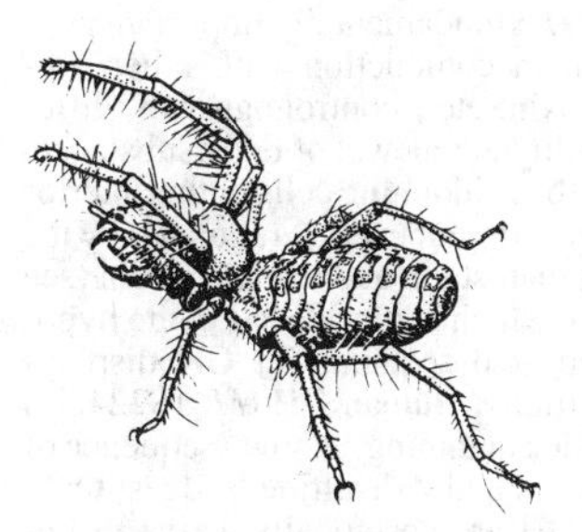

Solifugae. The windscorpion, *Galeodes arabs,* in a defensive posture.

Solitaire: a bird belonging to the extinct family, *Raphidae* (see). In North America, it is also the name given to species of *Myadestes* (see *Passeriformes).*

Solitary bees: nonsocial bees, comprising 9 families: *Colletidae* (membrane bees), *Andrenidae* (digger bees), *Halictidae* (sweat bees), *Megachilidae* (leaf cutter and mason bees), *Anthophoridae* (carpenter and miner bees), *Melittidae, Oxaeidae, Fideliidae* (3 small families with no common names), *Apidae* (solitary honey, stingless, solitary bumble and orchid bees). Each female independently mates, makes her own nest of about 10 brood cells, stocks the cells with food, lays an egg in each cell, then dies before the eggs hatch. S.b. play an important role in pollinating crops and wild plants, some of which (e.g. alfalfa) are not effectively pollinated by social honey bees. See bees.

Solitary oocyte nutrition: see Oogenesis.

Soma: see Neuron.

Somaclonal variation: genetic variation in clones of somatic cells, due to changes in karyotype, chromosome rearrangement, or changes in the extrachromsomal (plastids, mitochondria) genetic complement.

Somateriini: see *Anseriformes.*

Somatic: an adjective applied to those parts, functions and processes of an organism that are not directly associated with sexual reproduction.

Somatic disjunction: see Chimera.

Somatic inconstancy: an unusual phenomenon, in which the number of chromosomes in the somatic cells of an organism display more or less marked variations from cell to cell. S.i. is a type of somatic aneuploidy, and it is oberved in pathological cells (e.g. cancer cells).

Somatic segregation: see Chimera.

Somatocoel: the hind, third section of the paired secondary coelom of echinoderms. Unlike the two anterior sections of the coelom, the S. remains paired, forming the adult coelom, as well as two canal systems. The two canal systems are an aboral circular genital canal, and an oral hyponeural canal. The latter consists of a ring with 5 radial canals, which encircles the foregut, and runs parallel to the water vascular system. See Mesoblast.

Somatogamy: see Reproduction.

Somatology: in anthropology, the study of the living human body by somatometry and somatoscopy.

Somatolysis: see Protective adaptations.

Somatometry: measurement of an animal or human body. See Anthropometry.

Somatopleure: see Mesoblast, Extraembryonal membranes.

Somatoscopy: morphological description of the living human or animal body.

Somatotropin, ***somatotropic hormone, STH, growth hormone, GH:*** a fundamentally important peptide hormone, which in conjunction with other hormones (insulin, thyroxin, etc.) controls growth, differentiation and the continual renewal of body substance. GH is synthesized in the acidophilic cells of the anterior pituitary. Its secretion and synthesis are regulated by the peptide hormones, somatostatin and somatoliberin (see Releasing hormones), which are produced in the hypothalamus. The amino acid sequence of GH displays marked species specificity. Human GH (M_r 18224.52) is a monomeric peptide containing a known sequence of 191 amino acid residues and 2 disulfide bridges. Only human and primate GH are hormonally active in humans.

The growth stimulating activity of GH is especially apparent in pituitary dwarfism, in which the production of GH is defective or absent. Treatment with GH results in increased growth of muscle, skeleton and organs, and an increased functional capacity of the kidneys and liver. GH acts directly on cellular metabolism (stimulation of nucleic acid and protein biosynthesis, as well as cellular carbohydrate and fat metabolism) without the intermediate participation of other glandular hormones, as well as stimulating the liver to produce somatomedin-1, which in turn causes growth of muscle and bone.

Somite: see Mesoblast

Song: a term applied to complicated sequence of sounds in animal vocalizations, which can normally be analysed as a basic pattern of strophes. In the biological sense, "song" can therefore usually be defined as a sequence of strophes (see Strophe). The term is used primarily for the vocalizations of song birds.

Song babblers: see *Timaliinae.*

Song birds: all members of the suborder, *Oscines,* of the *Passeriformes* (see). Unlike the other passerine birds, the *Oscines* have 5–9 pairs of song muscles, and the syrinx is of the trachobronchial type.

Song thrush: see *Turdinae.*

Sonneratia: see Mangrove swamp.

Sonoran spadefoot: see *Pelobatidae.*

Sooglossidae, ***sooglossids, Seychelles frogs:*** a family of the *Anura* (see) containing 3 species in 2 genera, restricted to the islands of Mahé and Silhouette in the Seychelles.

The digits of sooglossids characteristically terminate in small, pointed disks. Eight presacral vertebrae with persistent notochord and free intervertebral bodies are present in sub-adults. In adults the intercentral bodies are fused to form procoelous centra. The neural arches are nonimbricate (nonoverlapping). Presacrals I and II are not fused. The omosternum is cartilaginous and the sternum is ossified. The sacrum has dilated diapophyses and it displays a monocondylar articulation with the coccyx. Small, proximal transverse processes are present on the coccyx. The pectoral girdle is arciferal, but the epicoracoid cartilages are fused at their anterior and posterior extremities. Maxillary and premaxillary teeth are present. Ribs are absent. The pupil is horizontal. Most adult *Anura* possess a complete cricoid cartilage (cartilaginous ring of the larynx), formed by the ventral and lateral fusion of paired cartilaginous structures. In the sooglossids (also in the myobatrachins), these paired cartilages are not fused ventrally, and the cricoid cartilage is therefore incomplete.

Nesomantes has a compact, toad-like appearance, whereas species of *Sooglossus* superficially resemble the African long-fingered frogs. These small (snout to vent 40 mm) frogs are independent of standing water at all stages of their life history; the only water in their mountainous forest habitats consists of steep rushing streams. The large-yolked eggs are laid in groups of 10–15 in small gelatinous heaps on the ground, where they are guarded by the male. In *Sooglossus gardineri* ***(Gardiner's Seychelles frog)*** the eggs undergo direct development. In *S. se(y)chellensis* ***(Seychelles frog)*** and *Nesomantis thomasseti* ***(Thomasset's Seychelles frog),*** the eggs hatch as nonfeeding tadpoles, which lack internal gills and mouthparts (i.e. no horny jaws or teeth are present). Tadpoles are carried on the back of the male until their development to the miniature adult is complete, during which time they are sustained by the yolk reserve of the egg. Although they are not rare, all 3 species are listed as endangered in the IUCN red book, because only small changes in their habitat would lead to their extinction.

The phylogenetic relationships of the *Sooglossidae* are unclear; they have also been variously classified as a subfamily of the *Ranidae,* and as a sister group of the myobatrachins.

Sooty chats: see *Turdinae.*

Sooty molds: see *Ascomycetes.*

D-Sorbitol: a sweet-tasting C6-sugar alcohol, which is present in many fruits, e.g. rowan berries, plums, pears and apples. It can be prepared by the catalytic or electrochemical reduction of the configurationally related sugars, D-glucose, D-fructose or L-sorbose. Preparation by the catalytic reduction of D-glucose is technically important.

It is used as a sweetening agent for diabetics, as a

starting material for the technical synthesis of ascorbic acid, as a preservative in the food industry, and as a softening agent for sweets and candy.

By heating in the presence of acid catalysts, D-S. undergoes an intramolecular loss of water with the formation of internal ethers. One such ether, 1,4-sorbitan, is converted commercially into dispersants and emulsifying agents by partial esterification with fatty acids and reaction with ethylene oxide.

$$\begin{array}{c} CH_2OH \\ | \\ H—C—OH \\ | \\ HO—C—H \\ | \\ H—C—OH \\ | \\ H—C—OH \\ | \\ CH_2OH \end{array}$$

D-Sorbitol

L-Sorbose: a monosaccharide hexulose, present in certain plant juices, e.g. rowan berries, and biosynthesized by the dehydrogenation of D-sorbitol. It is an intermediate in the biosynthesis of ascorbic acid.

Sorensen's quotient of similarity: see Quotient of similarity.

Sorex: see *Soricidae.*

Sorghum: see *Graminae.*

Soricidae, ***shrews*** (plate 7): a family of the order *Insectivora* (see) containing about 290 species in 20 genera, distributed throughout the world, except in Australia, most of South America and the polar regions. They are small and mouse-like in appearance. *Suncus etruscus* ***(Savi's pygmy shrew)*** is the smallest living mammal, weighing 1.5–2.5 g. Larger species of shrews may weigh up to 180 g. The snout is extended and the teeth are pointed. The first teeth are shed in the embryo stage, so that they are born with permanent teeth, ranging in number from 26 to 32. The first upper incisor is typically enlarged with a hook at the end, and with a cusp projecting ventrally at the base. The first lower incisor is enlarged and projects forward and slightly upward. Examples of dental formulas are: ***Long-tailed shrews*** *(Sorex):* 3/1, 1/1, 3/1, 3/3 = 32; ***Water shrews*** *(Neomys):* 3/2, 1/0, 2/1, 3/3 = 30; ***White-toothed shrews*** *(Crocidura):* 3/1, 1/1, 1/1, 3/3 = 28; ***Savi's pygmy shrew:*** 3/1, 1/1, 2/1, 3/3 = 30. They require a high food intake in relation to their size, and they eat any creature that they can overpower. A powerful musk scent is used for territorial marking and for defense against predators

Sorption: the property of small soil particles, especially colloids, to bind molecules and ions. *Adsorption* refers to attraction and binding on an exposed surface, whereas *absorption* refers to attraction and penetration. Sorption is used to include both adsorption and absorption. None of these terms specifies whether true adsorption (i.e. binding due to surface geometry, which obeys a Langmuir adsorption isotherm) or ion exchange phenomena are responsible for the attraction and binding. In the complex system of soil colloids, several different forces are probably involved, but cations and anions are retained and exchanged primarily by ion exchange.

By sorption, numerous materials in soil water are retained by the soil, and therefore avoid loss by drainage. The most important soil fractions for ion exchange are the organic and mineral components with effective particle diameters of less than 20 mm. This includes a portion of the silt and all the clay fraction (< 2 µm), as well as colloidal organic matter. In the organic fraction, negative charges arise mainly (80–95%) from the dissociation of H^+ from carboxylic (—COOH) and phenolic (—C_6H_4OH) groups. Negative charges in the inorganic fraction arise from isomorphous substitution (e.g. Mg^{2+} for Al^{3+}, or Al^{3+} for Si^{4+}) in layer silicate minerals such as montmorillonite, and from exposed AlOH groups in layer silicates. There are 3 general types of layer silicate clay minerals in soil: 2:1 (each layer consists of a sheet of alumina between 2 sheets of silica, e.g. montmorillonite, illite, vermiculite); 2:1:1 (the 2:1 structure plus an interlayer hydroxide sheet, e.g. chlorites);1:1 (series of layers, each containing 1 silica sheet and 1 alumina sheet, e.g. kaolinite, halloysite).

The greater the clay and humus content of the soil, the more plant nutrients it can bind. Thus, montmorillonite (smectite) has a very high binding capacity, whereas pure quartz sand has none at all, and sandy soils have only a low binding capacity. Different ions have different binding affinities, and some ions may replace other ions from sorption complexes in the soil, i.e. ion exchange occurs. *Cation exchange* (formerly called *base exchange)* is more closely related to soil fertility than anion exchange. The sum of all exchangeable cations is known as the *cation exchange capacity* (CEC). It is quoted in milliequivalents of cations per 100 g soil dried at 105 °C.

CEC is often measured by extracting a dried soil sample with neutral 1 mol/l ammonium acetate. All the exchangeable cations are replaced by ammonium ions, so that the CEC becomes saturated with ammonium ions. This ammonium-saturated soil is then extracted with a solution of a different salt, e.g. 1 mol/l KCl, and the suspension of soil and potassium chloride is filtered. The quantity of ammonium ions in the filtrate (displaced by K^+, and equivalent to the quanitiy of original exchangeable cations) is a measure of the CEC. Some authors prefer to extract with unbuffered salt, so that the CEC is measured at the normal pH of the soil; the resulting value is then the *effective CEC.*

In addition to the physical processes of sorption, the biological fixation and/or retention of certain materials (e.g. nitrogen) in the cell material of soil organisms is also important for the localization of nutrients in the upper soil layers.

Sorrel: see *Polygonaceae.*

Sotalia: see Whales.

Sousa: see Whales.

Souslik: see *Sciuridae.*

South African kingdom: see Cape region.

South Island wren: see *Xenicidae.*

South American bullfrog: see *Leptodactylidae.*

Southeast Asian rice frog: see *Raninae.*

Southern armyworm: see *Cutworms.*

Southern beech: see *Fagaceae.*

Southern spadefoot: see *Pelobatidae.*

Sowbread: see *Primulaceae.*

Sow bug: see *Isopoda.*

Soybean (Soya bean): see *Papilionaceae.*

Spadefoot toads: see *Pelobatidae.*

Spadicidae: see *Monocotyledoneae.*

Spadix: see Flower, section 5) Inflorescence.

Spalacidae, ***mole rats:*** a family of rodents (see), comprising a single genus *(Spalax)* with 3 species, restricted the eastern Mediterranean region and southern Europe. Mole rats have no tail; their eyes are functionless, and the pinnae reduced. They are fossorial, and the incisors (which are even more enlarged than in other rodents) rather than the forefeet are used for digging. Dental formula: 1/1, 0/0, 0/0, 3/3 = 16. The supraorbital region of the skull slopes forward, and the narrow zygomatic plate turns downward. As in most fossorial mammals, the pelage is short and reversible.

Spalacopus: see *Octodontidae.*

Spalax: see *Spalacidae.*

Spanish artichoke: see *Compositae.*

Spanish chestnut: see *Fagaceae.*

Spanish desman: see Desmans.

Spanish fly: see *Coleoptera,* Cantharidin.

Spanish lobsters: see Spiny lobsters.

Sparidae, ***porgies:*** a family (about 100 species in 29 genera) of fishes of the order *Perciformes,* found predominantly in warm seas (rarely in brackish or fresh water). They are high-backed fishes with tasty flesh, and they have local importance as a food source. There is a single continuous dorsal fin, the anterior part of which is supported by 10–13 spines, the posterior part consisiting of 10–15 soft rays The anal fin has 3 spines and 8–14 soft rays.

Sparmannia: see *Tiliaceae.*

Sparrow hawk: see *Falconiformes.*

Sparrows: see *Ploceidae, Fringillidae.*

Sparrow weavers: see *Ploceidae.*

Spatangus: see *Echinoidea.*

Spathebothridea: see *Cestoda.*

Spathicarpa: see *Araceae.*

Spathiphyllum: see *Araceae.*

Spea: see *Pelobatidae.*

Spearfishes: see *Scombroidei.*

Spearmint: see *Lamiaceae.*

Speciation: the development of genetic differences and reproductive Isolation (see) between different populations of a single species. The most frequent type of S. is *geographical* or *allopatric S.,* in which selection and genetic drift lead to considerable differences between spatially separated populations, and in which these differences also affect the reproductive system (see Genetic revolution). The flow of genes between the populations is then difficult or impossible. If contact is re-established between two populations that have become separated by this type of postzygotic isolation mechanism, then prezygotic isolation mechanisms often develop in addition.

The importance of geographical isolation in S. is demonstrated by the fact that the lizard species, *Lacerta bocagei,* has formed only 3 subspecies on the entire Iberian Peninsula, whereas the related *Lacerta pityusensis,* which inhabits the western Balearic Islands (Ibiza, Formentera and many smaller, rocky islands) has split into 37 subspecies, while another relative, *Lacerta lilfordi,* which inhabits the eastern Balearic Islands (Mallorca, Menorca and several smaller islands) has split into 13 subspecies.

It is fundamental to the definition of a new species that reproduction is no longer possible between it and the species from which it developed, i.e. there is complete reproductive isolation. The necessary time span for S. is very variable. The island of Fernando Po became separated from the African mainland 12–15 000 years ago. Thirty percent of the birds and mammals on Fernando Po have formed races that differ somewhat from those of the mainland, but they have not yet evolved into separate species. On the other hand, if females of *Drosophila pseudoobscura* from Colombia are crossed with males from USA, the F2 males are nonfertile, despite the fact that the Colombian population of this fruit fly did not arrive from the USA until 1960.

Sympatric S. is the gradual accumulation of genetic differences and the development of genetic isolation between individuals living in the same territory. Whether sympatric S. is at all possible is a matter of dispute. At least, it is very rare.

In the plant kingdom, S. frequently occurs by *polyploidy.* About 50% of all plant species are polyploids. Almost all of these are allopolyploids, produced either directly by the crossing of species, or arising later from the progeny of such crosses.

In both plants and animals, chromosome mutations can spread rapidly through a small population by drift and become homozygotic, despite the fact that the heterozygous carriers may have a relatively low potential for survival. The fit and competitive homozygous individuals thus represent a new species, which is reproductively isolated from the ancestral form, due to the reduced vitality of the heterozygotes. This process was discovered in Australian locusts, and named *stasipatric S.*

Even when species occupy a large, self-contained area, they change gradually, so that after a certain time they meet the criteria for a new species. This process is called *phyletic S.*

Species: the most important taxonomic category in the systematic classification of plants and animals. Speciation was originally based on shared morphological and typological characters, and differences of these characters from those of other species. The untenability of this separation is apparent from differences in the appearances of geographically separated populations, from the existence of Subspecies (see), and from the joint occurrence of Twin species (see), which are sometimes distinguishable only by biochemical or karyotype analysis.

A species is now defined biologically as an evolutionary and potentially propagative unit, which is isolated reproductively from other species. Thus, reproduction between different species does not normally occur. If it does occur, the resulting offspring is not viable, or if it does survive it is sterile and incapable of reproduction. A species is therefore an objective reality, and as such it has a special taxonomic status. Higher taxa, such as genus, family, order, etc., cannot be objectivized in this way.

Although this definition is normally adequate, at least for higher animals and plants, new definitions of species are continually being proposed. This is because: a) the speciation of plants differs in certain respects from that of animals (e.g. plants are more easily hybridized, and polyploidy can exist in plants; b) in many cases, the presence and effectiveness of reproductive barriers cannot be tested; and c) intermediate forms between subspecies and species are theoretically very predictable and are known to exist.

In well researched animal groups, like birds and mammals, there is a predominance of *polytypic* or *polymorphic* species which can be divided into 2 or more subspecies, in contrast to *monotypic* or *monomorphic* species which display no differences that would justify further subdivision.

Whether closely related forms are grouped as species or divided into subspecies is usually determined by their geographical distribution, Sympatry (see) being characteristic of species, while Allopatry (see) is characteristic of subspecies. There are, however, exceptions. Thus, the existence of very closly related species, separated geographically like subspecies, caused Rensch to add to his concept of species as a family of races the concept of a family of species; both cases are included in the older term, Formenkreis (see). Such groupings are now called *superspecies,* the individual members of which are more or less reproductively isolated and are called *semispecies.* The concept of superspecies is also justified when several subspecies show an overlapping distribution, in which neighboring subspecies along the chain are able to interbreed, but the members most distantly separated at opposite ends of the chain are genetically isolated. A well known example is the herring gull, whose separated populations underwent different degrees of differentiation in their Refugia (see) during the Ice Age. After re-extension of their territory, they behaved partly like separate species (e.g. D in Fig.), and partly like subspecies. Breeding may have occurred between gulls that crossed the North Atlantic and gulls on the coasts of the Baltic and the Atlantic coasts of Northwest Europe (A + M), but this was exceptional.

Normally, this type of reproductive isolation goes unnoticed, because the territories of the members at the extremities of the distribution do not overlap (sympatry is absent). In recent times, insect populations from widely separated areas have been used, with a certain amount of success, for genetic pest control, instead of artificially sterilized animals.

In naming and characterizing a species, the essential characters must be determined that qualify for the species name, irrespective of alterations in the demarkation of the species or other corrections of its description. Since Linnaeus, the species name has consisted of 2 parts (known as the Latin binomial), i.e. the genus and the actual species name. The complete name also includes the author and year of description, e.g. *Pierris brassica* Linnaeus, 1758. An undetermined or unspecified species of the genus is abbreviated to spec. or sp. (plural spp.), e.g. *Mantis* sp. means a single undetermined species of *Mantis; Epilobium* spp. means 2 or more undetermined species of *Epilobium.*

Species activity density: see Activity density.

Species-area curve: see Minimal area.

Species combination: a group of species that occur repeatedly and often in association with one another at a particular site or in a particular habitat. The characteristic S.c. of a habitat is determined by the specific interaction of various Environmental factors (see). Affinity (see) provides an index of the extent to which these environmental factors influence the joint occurrence of species.

Species pairs: different species of plants of the same genus that display only minor morphological differences, and which occupy similar ecological niches in different parts of the world, especially between the New World and the Old World. Examples (the first member of the pair is European, the second American): the Yellow water lilly *(Nuphar luteum* and *N. americanum);* May lily *(Maianthemum bifolium* and *M. canadense);* Way-fairing tree *(Viburnum lantana* and *V. alnifolium);* Mountain ash *(Sorbus aucuparia* and *S. americana).*

Species protection, ***species conservation:*** measures for the preservation of animal and plant species that are threatened with extinction (see Nature conservancy). The term, animal protection, means both the protection of animals from mistreatment, and the conservation of threatened animal species. Similarly, plant

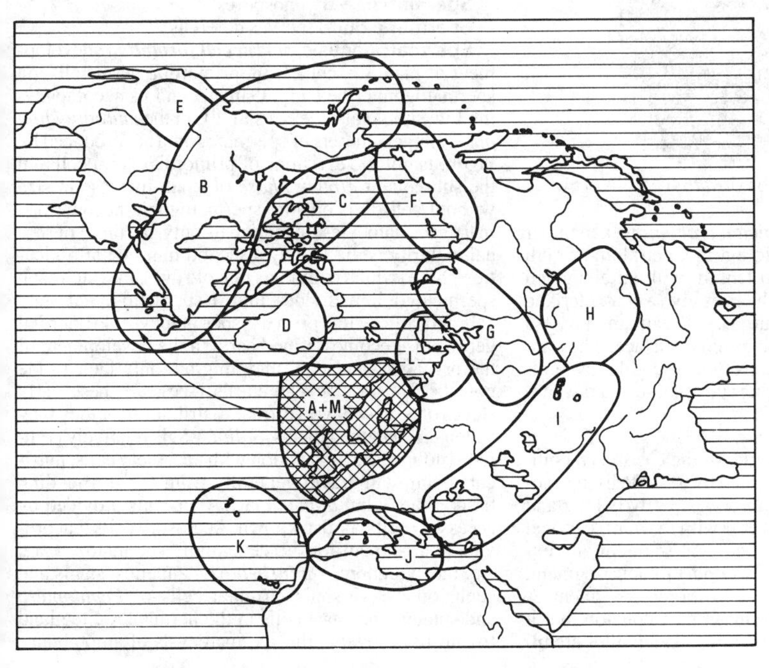

Species. Geographical distribution of different forms of the herring gull *(Larus argentatus)* (after Mayr).

protection refers to the measures of agricultural husbandry for maintaining healthy, disease-free crops, as well as the conservation of threatened species.

Specific dynamic action, ***SDA, calorigenic effect:*** an increase in body heat production immediately after a meal. SDA is not due to the oxidation of absorbed nutrients, and it cannot be attributed entirely to digestion and absorption. As a percentage of calorific values, the SDA of protein is 30%, carbohydrate 6% and fat 4%. The SDA must be allowed for when calculating the calorific equivalents of foods. Thus, the total energy value of 20 g protein is 20 x 4 kcal/g = 80 kcal; the corresponding SDA is 0.3 x 80 kcal = 24 kcal. These 24 kcal would be wasted as heat, so that 80 – 24 kcal = 56 kcal of utilizable energy is available. However, the SDA of a mixed diet is not the sum of the SDA of its components, because the SDA of each component is decreased in the presence of the others, and especially by fat. Therefore, between 6 and 10% is normally added to the calculated energy equivalent of a mixed diet to account for SDA. [1 calorie = 4.184 Joules]

Specific vigilance reaction: see Vigilance.

Spectacled bear, *Tremarctos ornatus:* a black or blackish brown bear, with large circles or semicircles of white around the eyes and a white semicircle on the lower side of the neck. These head and chest markings are variable and may sometimes be lacking. Body length is 1.5. to 1.8 m, shoulder height about 760 cm. It inhabits the Andes at 1 500 to 2 000 m, and is the only South American bear.

Spectacled bear (Tremarctos ornatus)

Spectacled caiman, *Caiman crocodilus:* a member of the *Alligatoridae* (see) found in Central and South America. It is about 2 m in length, with an elongated snout. Bony ridges around the eyes give a characteristic spectacled appearance (plate 5). All caimans possess hard bony plates on their abdominal surface. They are therefore less endangered than other crocodiles by hunting, because their ventral skin is unsuitable as a raw material for "crocodile leather".

Speirops: see *Zosteropidae.*

Spelaeogriphacea: an order of the *Crustacea* (subclass *Malacostraca),* represented by a single living species, *Spelaeogriphicus lepidops.* The body is long and cylindrical. A short carapace (with a faint cervical groove) is fused to the first thoracic segment and extends dorsally to cover the second thoracic segment, also overlapping part of the third thoracic segment. A movable ocular lobe is present on the cephalon, but it lacks pigments or optical structures. Antennules are biramous and well developed. Antennae carry a small scaphocerite. The mandible has a lacinia mobilis and an unjointed palp. Thoracopods 2 to 4 have flap-like, setose exopods, while 5 to 7 (and sometimes 8) have branchial exopods. Pleopods 1 to 4 are well developed, while 5 is vestigial. The uropods are biramous and flap-like. *S. lepidops* is found only in a freshwater stream that flows through Bat Cave in Table Mountain, South Africa. A single fossil form is also known: *Acadiocaris novascotica* from the Lower Carboniferous (Mississippian) of Canada.

Speothos: see *Canidae.*

Speotyto: see *Strigiformes.*

Spermateleosis: see Spermatogenesis.

Spermaceti: a solid animal wax, m.p. 45–50 °C, obtained from the head of the sperm whale, *Physeter macrocephalus.* An oily, liquid, crude sperm oil is secreted in a special large cylindrical organ in the upper region of the jaw above the right nostril. After capture, this cavity is emptied of its oil, which on cooling deposits crystalline S. This is separated by pressure and purified by re-melting and washing with dilute NaOH to remove the last traces of oil. Up to 5 000 kg of S. are obtained from a single sperm whale. S. consists chiefly of cetyl palmitate, accompanied by smaller quantities of cetyl laurate and myristate. It is used in the pharmaceutical and cosmetic industries as a base for creams. Part of the S. of commerce is also obtained from the bottle-nosed whale, *Hyperoodon rostratus.* See Whales.

Spermatheca, ***receptaculum seminis:*** a pouch or sac of female or hermaphrodite organisms, which serves for the reception and storage of spermatozoa. Spermathecae are found in mollusks, annelids and arthropods. In many insects, for example, pairing occurs only once, and since egg maturation may extend over a prolonged period, the spermatozoa stored in the S. are used to fertilize eggs from time to time.

Spermatia: see Pycnospores

Spermatic duct: see Vas deferens.

Spermatogenesis, ***sperm cell formation:*** development of male sex cells from primordial germ cells via spermatogonia (the entry, Gamete, and its accompanying Fig. should also be consulted). In the *multiplication phase,* large numbers of spermatogonia are produced by vigorous mitotic divisions of primordial germ cells. In the subsequent *growth phase* (a small increase in size without cell division) the spermatogonia become primary spermatocytes. In mammals only, the onset of sexual maturity is marked by a second mitotic multiplication phase (which continues into old age), in which each spermatogonium divides into a spermatogonial stem cell and a growing primary spermatocyte. Primordial germ cells become defined early in the development of the organism, and they may migrate only later to the gonads. Thus, as in Oogenesis (see), all these early stages of S. may have a diffuse distribution, or they may be localized in the gonads. *Solitary* S. is relatively common (no specific association with accessory cells, nutrients being withdrawn generally from the surrounding tissue fluids), but *alimentary* S. (nutrients provided by accessory cells) is also known. Accessory cells that provide nourishment for the developing spermatozoa are, e.g. the cytophores of *Turbellaria,* annelids, snails and cephalopods, Verson's terminal cells of *Lepidoptera* and spiders, the basal cells in the hermaphrodite gland (ovotestis) of *Helix,* the spermatocysts of amphibians, and the Sertoli cells of mammalian testes. The third or *maturation* phase of S. involves Meiosis (see), and it gives rise to haploid spermatids. Differentiation of spermatids into active spermatozoa (motile and capable of penetrating an egg) is known as *spermiohistogenesis, spermiogenesis* or *spermateleosis,* a process which differs somewhat from species to species, depending on the final shape and structure of the Spermatozoa (see).

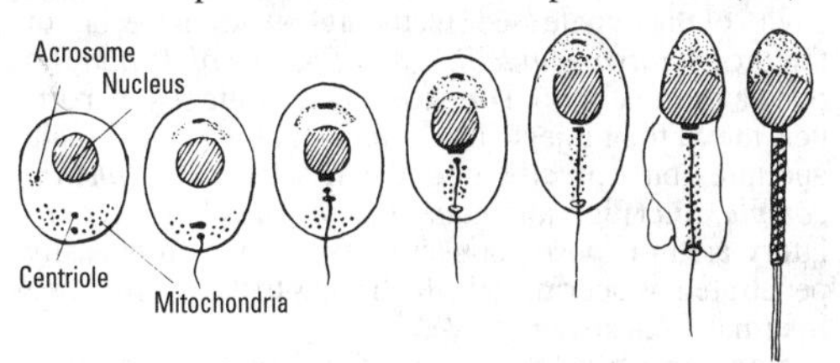

Fig. 1. *Spermatogenesis.* Spermiohistogenesis (spermiogenesis, spermateleosis) in the human (schematic).

The generalized chordate spermatozoon consists of head, middle piece (midpiece) and tail. Alternatively, the entire extended region behind the head is known as the tail, which is then divided into the middle piece and principal piece. In the present account, the former nomenclature is used (i.e. "tail" does not include the middle piece). The sperm head arises from the spermatid nucleus by shrinkage, chromatin condensation, stretching and changes of shape caused by the action of contractile fibrils; in its final form the nucleus/sperm head may be spherical (carp), elongate (chameleon), hook-shaped (rat, mouse), scimitar-shaped (domestic fowl), helical (snails, sharks and rays, spadefoot toad, birds), ellipsoid (bull, stallion), pear-shaped (human), shovel-shaped (guinea pig). The posterior part of the head contains a higher proportion of DNA than the anterior region, whereas the anterior region is richer in protein. In mammals, the anterior region also contains hyaluronidase. There is also an anterior caplike structure, known as the Acrosome (see), which varies in size and shape. Within the Golgi apparatus of the spermatid, small granules (proacrosomal granules) accumulate then coalesce to form a single acrosomal granule within an acrosomal vesicle. This continues to grow by acquisition of other Golgi vesicles, then spreads over the anterior surface of the nucleus, eventually forming the acrosome (or acrosomal vesicle), which consists of an inner and an outer layer, each presumably of mucopolysaccharide. After treatment of spermatozoa with acridine, the acrosome can be clearly distinguished under the fluorescence microscope as a bright red-fluorescing structure capping the green-fluorescing sperm head.

The middle piece, joined to the head by a neck region, is usually short and spherical to rod-shaped. In humans it is 5–7 µm in length with a cross section diameter of about 1 mm, extending from a slender connecting piece in the neck to its junction (marked by the annulus) with the tail. The middle piece is formed during spermiohistogenesis by elongation of the bulk of the spermatid cytoplasm that is displaced behind the caudal pole of the nucleus. At the same time, the proximal centriole of the original spermatid becomes localized at the posterior pole of the nucleus, while the distal centriole acts as a basal corpuscle for the flagellum. An axial filament complex or axoneme (2 central tubules surrounded by 9 peripheral pairs of tubules; for detailed structure, see Cilia and Flagella) extends from the base of the head, through the middle piece and through the length of the flagellum. In addition, 9 coarse fibers or "outer" fibers form a fibrous sheath around the axoneme. At the centrioles, these outer fibers fuse to form the wall of the neck. Within the middle piece the mitochondria are arranged in a ring, sphere or helix (in human sperm with 8–9 turns) around the outer fibers, forming a structure known as the mitochondrial sheath.

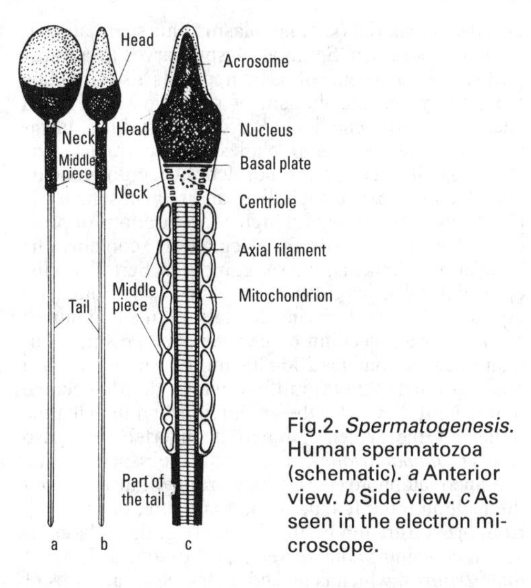

Fig. 2. *Spermatogenesis.* Human spermatozoa (schematic). *a* Anterior view. *b* Side view. *c* As seen in the electron microscope.

The tail is variable in length, but is usually about 10 times longer than the head. It consists of a protein sheath around the axial filament and its fibrous sheath (all nine of the outer fibers may not continue from the middle piece into the tail, e.g. in rodents only 7 outer fibers continue into the tail). A small, naked end piece of the axial filament, surrounded only by plasma membrane, projects from the end of the tail.

With the completion of spermiohistogenesis, most of the excess spermatid cytoplasm is shed as a droplet, which can be seen still adhering to immature spermatozoa. The remainder of the cytoplasm forms a thin layer over the head and middle piece, and extends into the tail.

Spermatozoa of some animals are totally different in construction from the flagellated type described above (see Spermatozoa). Flagellated spermatozoa are formed by coelenterates, annelids, insects, echinoderms and chordates. They are usually the smallest cells of the body, e.g. in humans they are 0.05–0.06 mm in length, but in rare cases they can attain an enormous size, e.g. 7 mm in ostracods. As a rule, each species produces spermatozoa of uniform shape and size, and variations from this norm usually have a pathological cause. Occasionally, however, sperm dimorphism occurs normally (see Spermatozoa). In insects and mammals, the small forms of dimorphic spermatozoa give rise to male offspring, the large forms to females.

Spermatozoa are always ejaculated in large numbers, suspended in seminal plasma (secretions of the testes and epididymis and accessory sex glands). The total ejaculated material (seminal plasma plus spermatozoa) is known as semen. Seminal plasma serves as a transport vehicle, and some of its components also stimulate the activity and metabolism of spermatozoa. Typical chemical constituents include fructose, sorbitol, inositol, citric acid, lactic acid, cholesterol, spermine, spermidine, amino acids, phosphorylcholine, ergothioneine and glycerylphosphorylcholine, and human semen has also been found to contain high concentrations of prostaglandins. In mammals, the ejaculate also contains various formed elements, such as giant cells, Sertoli's cells, epithelial cells, leukocytes, prostate cells, Böttcher's crystals, Charcot-Neumann crystals, pigment granules, fat droplets and lecithin bodies. An average single human ejaculate contains 200–300 million spermatozoa, a horse ejaculate 22 000 million, which do not become motile until they enter the epididymis and the alkaline milieu of the semen. Numerous invertebrates [also newts *(Salamandridae)]* incapsulate their spermatozoa in a Spermatophore (see), which is either introduced by the male into the female genital aperture, or taken up from the substratum by the female (e.g. the gelatinous sperm-containing cone secreted by the cloacal glands of male *Triturus,* which is picked up by the cloacal lips of the female).

Spermatozoa have a lifespan of a few to several days, e.g. when stored in the cold, human spermatozoa survive for about 9 days. By storage in a specialized structure of the female (spermatheca or *Receptaculum seminis),* spermatozoa remain viable for years, e.g. in pulmonate snails and the queen bee.

Spermatophore: an aggregation of sperm held together by gelatinous material *(Urodela),* or a gelatinous packet of sperm (many insects), which is inserted into, or attached to, the female as part of reproductive behavior. The gelatinous material is produced by accessory sex glands of the male. Ss. serve for the transfer of sperm in many worms, arthropods, mollusks and *Urodela.* They display a variety of shapes: bottle-shaped, tubular, spherical, etc. In many arthropods and urodeles the S. is attached to the substratum by a complicated carrier structure (e.g. ***a*** in Fig.) then taken up later by the female which places itself over it. Cephalopods employ a complicated transfer mechanism, involving a specialized tentacle (hectocotylized arm) which serves as an intromittent organ for transfer of the S. into the mantle cavity of the female (see Hectocotylus). Most leeches of the orders *Rhynchobdellida* and *Pharyngobdellida* deposit a spermatophore on the surface of the sexual partner; the S. then injects its contents, not into the female aperture, but directly into the tissues. *Glossiphonia complanata* (Fig.) and many *Piscicolidae* have a copulatory area of loose connective tissue, which is easily penetrated by sperm, and which leads to the ovaries. See Spermatogenesis.

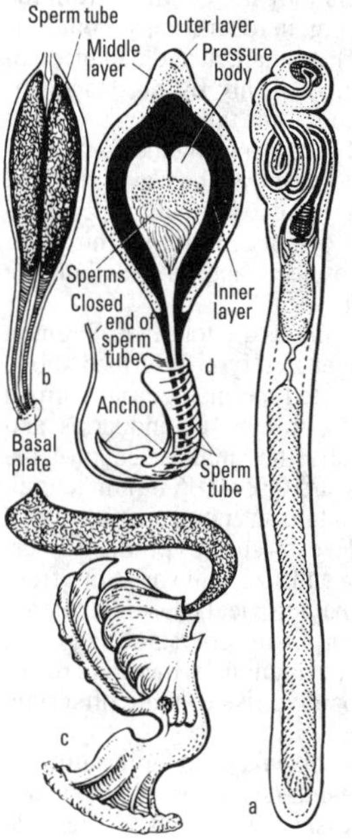

Different types of spermatophores. *a* Spermatophore of *Sepia* (cuttlefish) with elastic sperm container and explosive apparatus for ejecting the sperm. *b* Spermatophore of the leech, *Glossiphonia complanata* (see text). *c* Spermatophore of *Triturus vulgaris* (smooth newt) with attachment structure. *d* Spermatophore of *Liogryllus campestris* (field cricket).

Spermatophyta, *Anthophyta,* ***phanerogams, spermatophytes, flowering plants, seed plants:*** the most highly evolved division of the plant kingdom, containing 227 000 species. They are cormophytes, characterized especially by the possession of flowers and the production of seeds. *S.* have a heteromorphic alternation of generations, coupled with an alternation of the nuclear phase. The haploid gametophyte, which always remains enclosed by the sporophyte, is always strongly reduced. Archegonia occur only in primitive *S.* In more advanced *S.,* the gametophyte consists of only a few cells, and these give rise to the female and male sex cells (egg and sperm cells). The sporophyte is derived from the diploid zygote, which develops into the embryo within the seed. The seed serves as the dispersal unit of the *S.* The diploid plant is finally produced by germination of the seed. The mega- and microsporophylls of the flowers, which are known as carpels and stamens, carry the ovules (megasporangia) and pollen sacs (microsporangia). Ovules may lie exposed on the surface of the carpels (gymnosperms) or they may be enclosed by overgrowth of the carpels (angiosperms).

Pollination (transfer of pollen to the female flower parts) is usually performed by insects, more rarely by birds or other animals, frequently by the wind, and in special cases by water. Fertilization involves the fusion of sperm and egg cells to form a zygote; this may occur within a few days or hours after pollination (angiosperms), or it can take more than a year (gymnosperms).

Since pollination, fertilization and the dispersal of seeds (i.e. dormant young plants) do not require moisture, the *S.* are much better adapted to a terrestrial existence than the pteridophytes, and they have been able to invade practically every type of terrestrial habitat. The oldest known *S.* (their fossils occur in the Upper Devonian) probably evolved from the *Progymnospermae,* a group of iso- or heterosporous Ferns (see) that represent a link between the *Psilophytatae* and the primitive *S.* Evolutionary transition from spore-scattering to seed-bearing forms probably occurred several times in parallel, and this is taken into account in the systematic classification of the *S.* Thus, the gymnosperms represent the earliest evolutionary stage within the *S.,* but they are not a homogeneous phylogenetic group. Rather the *S.* consist of 3 subdivisions (Fig.): fork- and needle-leaved gymnosperms *(Coniferophytina),* pinnate-leaved gymnosperms *(Cycadophytina),* and Angiosperms *(Angiospermae, Magnoliophytina).* See Gymnosperms, Angiosperms.

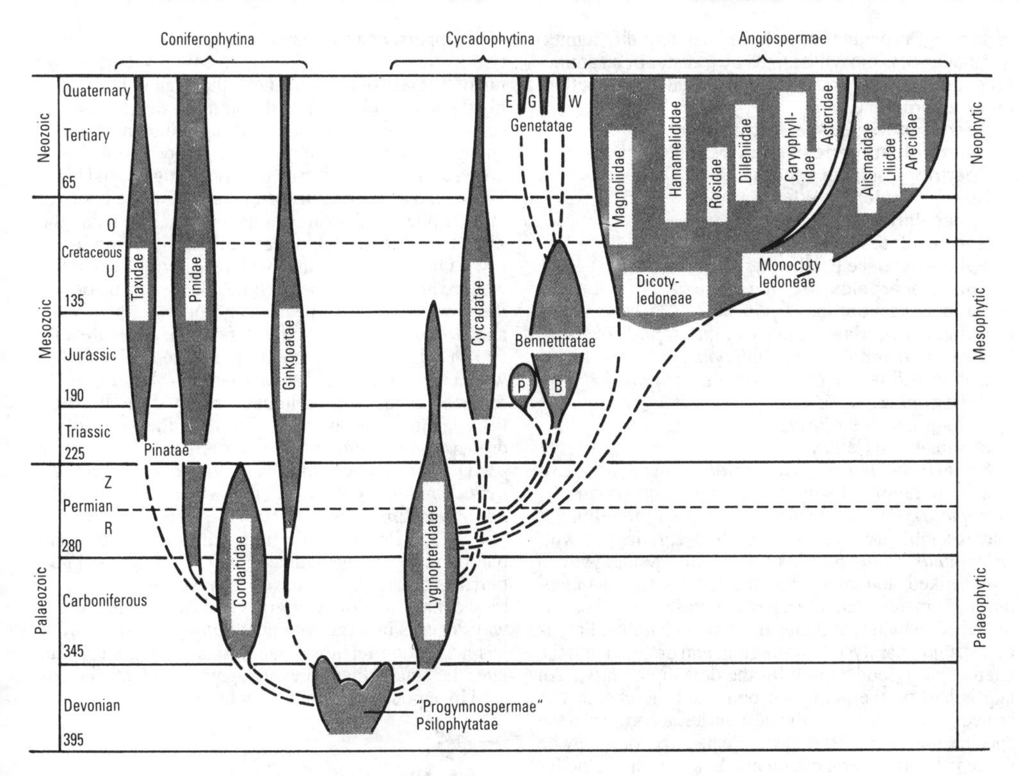

Spermatophyta. Hypothetical phylogenetic relationships between groups of seed-bearing plants, and their evolution in geological time. Numbers at the beginning of geological formations indicate millions of years. Relationships not documented by fossil evidence are shown as dotted lines and are left white. *B Bennettitidae. P Pentoxylidae. E Ephedridae. G Gnetidae. W Welwitschiidae.*

Spermatozoa (sing. **Spermatozoon),** *sperms* (plates 22 &23): mature male Gametes (see) possessing a simple haploid chromosome number, and produced in the animal testis (the entry, Spermatogenesis, should also be consulted). The shape and structure of the S. is characteristic for each animal species. Most species produce flagellated, filiform S., consisting of a *head* (which contains the nucleus and only a small quantity of cytoplasm), a *middle piece* (or *midpiece)* (containing mitochondria) attached to the head via a neck (which contains a centrosome), and a *tail* which serves as a locomotory organelle. The tip of the head often displays a specialized structure, known as the rod or perforatorium, which is thought to aid penetration of the egg, e.g. it may be dagger-like (frog), hooked (mouse) or helical (spadefoot toad). In analogy with certain flagellates, the tail sometimes carries an undulating membrane (e.g. in urodeles). The S. of some animal groups do not conform to the filiform structure. Many lower crustaceans and numerous arachnids produce spherical S. Prawns and *Ascaris* produce spike-like S. containing a refractive body of specific proteins. The S. of mites, aschelminths and some lower crustaceans display ameboid movement. Decapods *(Astacus fluviatilis, Pagurus striatus, Galathea strigosa)* produce explosive S., consisting of a double-walled chitinous capsule with rigid buoyancy processes, containing a spring mechanism derived by modification of the centriole. The latter opens the capsule explosively and actually drives the spermatozoon into the egg.

Occasionally, two forms of S. occur together in the same species *(sperm dimorphism),* e.g. in some snails and *Lepidoptera.* Here, the typical (eupyrenous) S. are accompanied by atypical (oligo and apyrenous) S., which have an incomplete complement of chromosomes and often exhibit abnormal morphology. It is thought that the atypical S. carry Gamones (see) which activate the typical S.

Between species there is a very wide variation in the size of S., e.g. human S. are about 0.05 mm in length, while the S. of ostracods attain 7 mm.

Spermatozoid: see Reproduction, Gametes.

Spermidine, $H_2N—(CH_2)_3—NH—(CH_2)_4—NH_2$: an aliphatic triamine (see Amines), which together with spermine is responsible for the alkaline reaction of semen. Derivatives of S. have been found in numerous plants. S. is biosynthesized by transfer of an aminopropyl group from decarboxylated *S*-adenosyl methionine to putrescine (which in turn is derived from the decarboxylation of ornithine).

Spermine, $H_2N—(CH_2)_3—NH—(CH_2)_4—NH—(CH_2)_3—NH_2$: a strongly basic aliphatic tetra-amine (see Amines) found in yeast and semen, which together with spermidine is responsible for the alkaline reaction of semen. Derivatives of S. have been found in numerous plants. S. is biosynthesized by transfer of an aminopropyl group from decarboxylated *S*-adenosyl methionine to spermidine.

Spermiogenesis: see Spermatogenesis.

Spermiohistogenesis: see Spermatogenesis.

Spermogonia: see Pycnospores.

Sperm whale: see Whales.

Sphacelariales: see *Phaeophyceae.*

Sphaeriales: see *Ascomycetes.*

Sphaerium: see Freshwater mussels.

Sphaerocarpales: see *Hepaticae.*

Sphaerocarpus: see *Hepaticae.*

Sphaeromatidae: a family of the *Isopoda* (see).

Sphaerotheriidae: see *Diplopoda.*

Sphaerotilus: see *Chlamydobacteriaceae.*

Sphagnidae: see *Musci.*

Sphagnum: see *Musci.*

S-phase: see Cell cycle.

Sphecidae, *digger wasps, sand wasps:* a large cosmopolitan family of solitary hunting wasps (order *Hymenoptera),* containing about 7 600 species. Imagos may resemble true wasps *(Vespidae),* e.g. the bee-wolf *(Philanthus triangulum),* or they are slender with a long-stalked abdomen, e.g. the sand wasps *(Ammophila).* Females excavate a passage in sandy soil or rotten wood, which terminates in a small chamber. Prey is then caught, paralysed by stinging, and dragged into the chamber as a food supply for the developing larva. An egg is laid on the paralysed body and the chamber is sealed. The intricacy of the nest varies according to the species; it may consist of a simple passage, or it may be branched with several chambers. In a few cases, the female continues to care for her larvae and tops up the food supply. Different species of sphecids are prey-specific, and there is tendency for primitive sphecids to hunt primitive insects (cockroaches, crickets), while more advanced species hunt flies and beetles. Some even hunt spiders.

They are closely related to bees (the superfamily *Sphecoidea* consists of sphecids and bees). Body hairs of sphecids are unbranched (cf. bees, in which they are branched), and the hind basitarsus has the same width as the subsequent segments (cf. bees, in which it is wider).

Spheciospongia: see *Porifera.*

Sphecotheres: see *Oriolidae.*

Sphenisciformes, *penguins:* an order of birds with 16 species in 6 genera, distributed in the Southern Hemisphere from the Antarctic to the equator. The greatest numbers of penguins are found on Antarctic coasts and sub-Antarctic islands, but the greatest species diversity occurs in the New Zealand area and the Falkland islands. Although some species are found in warm climates (larva shores of equatorial islands and sandy subtropical beaches), they are all basically adapted to cold conditions, and their tropical habitats are always in the vicinity of cold water currents, e.g. the Humboldt current along the Pacific coast of South America, and the Benguela and Agulhas currents off South Africa. They feed on fish, squid and crustaceans. They are excellent swimmers, many species "porpoising" through the water, using their modified wings for propulsion and their feet and tail for steering. Swimming speeds up to 60 km/h are claimed, but 5–10 km/h is more usual. The wing bones are flattened and expanded, and the wings do not fold. The wings thus function as strong, narrow, stiff flippers, and they are covered with strongly modified feathers which act like scales. With a dense covering of 3 layers of short feathers, the streamlined body is highly waterproof. Below the skin is a well defined, insulating fat layer. Flippers and legs contain a heat exchange system, whereby venous blood from the exposed extremities is warmed by outgoing arterial blood. To avoid overheating, tropical penguins have relatively large flippers and an area of naked facial skin, and they often escape from the direct sunlight by living in burrows. On land, the normal posture is upright. Some species are also agile runners, while others can hop or waddle. Some Antarctic species can flop onto their bellies and "toboggan" down snow slopes. The sexes are alike, and in all species the coloration is generally similar, with a white breast and a black or blue-black back. Differences occur in the plumage color of the head and neck. Some species have crests on the sides of the crown (e.g. the ***Macaroni penguin,*** *Eudyptes chrysolophus)* . The smallest species ***(Little blue penguin,*** *Eudyptula minor)* is 30–40 cm tall, while the largest ***(Emperor penguin,*** *Aptenodytes forsteri)* can attain a height of 150 cm. Penguins are monogamous, and the same pair bond is usually maintained for life. Eggs are laid in burrows or caves, or in the open. When nests are made, they consist of stones or grass and twigs. Breeding always occurs in large colonies. *Aptenodytes* (Antarctic) has no nesting territory, but carries its single egg (and later the small chick) pressed against the tarsi and covered by a pouch-like fold of abdominal skin.

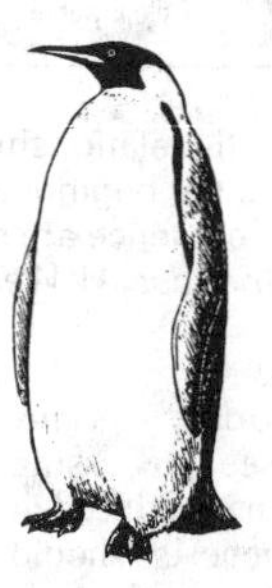

Sphenisciformes. Emperor penguin *(Aptenodytes forsteri).*

Sphenodon: see *Rhynchocephalia.*

Sphenodontidae: see *Rhynchocephalia.*

Sphenophyllales: see *Equisetatae.*

Sphenopsida: see *Equisetatae.*

Sphenopsidales: see *Equisetatae.*

Spherosomes: see Oleosomes.

Sphiggurus: see *Erethizontidae.*

Sphincter: a circular muscle that closes the entrance to a hollow organ, e.g. the cardiac and pyloric Ss. of the stomach, and the S. of the urinary bladder.

Sphingidae, *hawk moths, sphinx moths:* a family of the *Lepidoptera* containing about 1 000 species, most of which are found in the tropics, but with many representatives in temperate regions. The imago has long, slender forewings (wingspan up to 19 cm) and short hindwings, and a stout, spindle-shaped body. Most S. fly at night, and many species feed on nectar with the aid of a long proboscis while hovering on the wing. Some species are migratory (see Migratory Lepidoptera). S. are among the fastest flying insects; speeds up to 50 km/h having been recorded. Caterpillars are almost hairless, and carry a curved, posterior horn. The favored food plant of the caterpillar is often indicated in the popular name of the insect. With the exception of the ***Pine hawk moth*** *(Sphinx pinastri),* all European species are protected. Other examples are: ***Eyed hawk moth*** *(Smerinthus ocellatus;* Europe, temperate Asia excluding Japan; food plants: *Salix, Malus, Ligustrum, Poplus);* ***Lime hawk moth*** *(Mimas tiliae;* Europe, through Asia to Siberia and Japan); ***Spurge hawk moth*** (Fig.) *(Hyles euphorbiae;* Canary Islands to northern Iran and northwest India; widely distributed in Europe from Spain to Caucasus; introduced into western Canada in an attempt to control *Euphorbia* weed species); ***Death's head hawk moth*** *[Acherontia atropos;* distributed throughout Europe, migrating to the British Isles; also found in the Middle East, Africa, Madagascar and Seychelles; feeds on various species of *Solanaceae* as well as ash *(Fraxinus excelsior)];* ***Privet hawk moth*** *(Sphinx ligustri;* most of Europe, through Asia to China and Japan); ***Humming bird hawk moth*** *(Macroglossum stellatarum;* southern Europe, North Africa, temperate Asia to Japan). In Europe, *Macroglossum* is often more common in gardens than in the countryside; the caterpillars favor bedstraws *(Galium)* and the imago is fond of jasmine nectar. The ***Bee hawk moths*** [e.g. the ***Broad bordered bee hawk*** *(Hemaris fuciformis);* Europe, northern and central Asia, North Africa] are day-flying species, which take nectar from various flowers, especially honeysuckle and rhodendendron, most commonly in the morning. Their buzzing, flight habit resembles that of bees, and this resemblance is increased by the yellowish, furry body and transparent areas on the wings resulting from an early loss of scales. The ***Tobacco hornworm*** or ***Carolina sphinx*** *(Manduca sexta;* northern USA to tropical South America) and the ***Tomato hornworm*** *(Manduca quinquemaculata;* northeastern Canada to USA and Hawaii) are the most common tomato pests in North America. The caterpillars attack both the foliage and the green fruits, while those of *M. sexta* also attack peppers, potatoes, tobacco and other members of the *Solanaceae.* Most widely distributed of all the hawk moths is the ***Striped hawk moth*** or ***White-lined sphinx*** *(Hyles lineata),* which is found from Canada to Central America, in Africa and Asia, and as a migrant in most of Europe from its breeding areas in the Mediterranean region and North Africa.

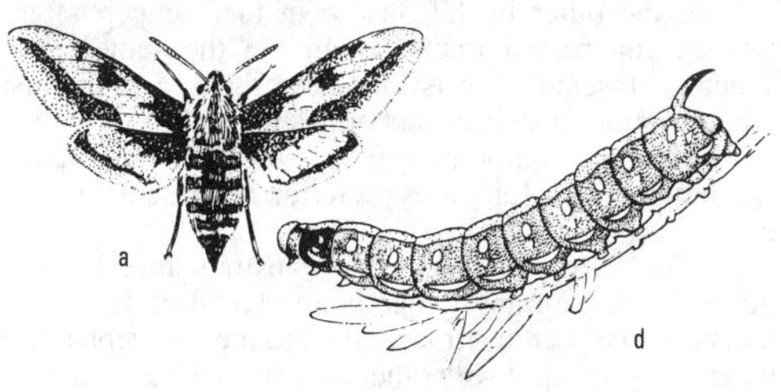

Sphingidae. Spurge hawk-moth *[Hyles (= Celerio) euphorbiae]. a* Imago. *b Caterpillar.*

Sphingomyelins, ***sphingophospholipids:*** phospholipids, in which the long-chain, unsaturated aminoalcohol, *sphingosine,* is combined with a higher fatty acid, phosphoric acid and choline. All P. therefore consist of a strongly hydrophobic portion (fatty acid in amide linkage with the sphingosine chain) and a highly hydrophilic portion (phosphorylcholine linked to the primary acohol group of the sphingosine). They therefore form bilayers in aqueous media and have excellent emulsifying properties. As indicated by their name, S. occur in the myelin sheaths of nerves. The fatty acid residue is often nervonic acid. Rich sources for the isolation of S. are brain, spleen , liver and lungs.

Sphingosine: see Glycolipids.

Sphyraena: see *Sphyraenidae.*

Sphyraenidae, ***barracudas:*** a family (1 genus, 18 species) of pelagic, predatory, slim, elongate, marine teleost fishes of the order *Perciformes.* They posses a jutting lower jaw with strong fanglike teeth, and the body is covered with small cycloid scales. Superficially, they resemble pikes, and in many languages they are known as "sea pikes". Inhabitants of tropical and subtropical coastal waters at depths of up to 100 m, they usually hunt in shoals and can also be dangerous to humans. In some areas they are even more feared than sharks. The lateral line is well developed. There are 2 widely separated dorsal fins, the first with 5 spines, the second with a single spine and 9 soft rays. Specimens longer than about 1.8 m are rarely encountered, but *Sphyraena barracuda* has been reported to attain 2–3 m. The ***Mediterranean barracuda*** *(Sphyraena sphyraena)* has a maximal length of 1 m. Many species are economically important as food.

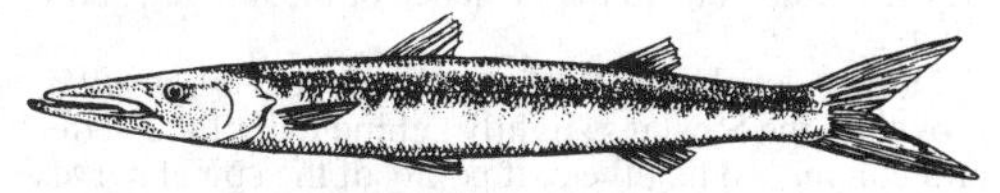

Sphyraenidae. Mediterranean barracuda *(Sphyraena sphyraena).*

Sphyrnidae: see *Selachimorpha.*

Spider crabs, *Majidae:* a family of true crabs (see *Brachyura),* which are exclusively marine. The triangular carapace (longer than it is wide) and long legs usually carry hooked setae for the attachment of foreign bodies (pieces of seaweed, sponges, etc.) for camouflage purposes. The use of camouflage is more common among juveniles than adults. The walking legs (pereopods) and the chelipeds and chelae are slender and elongated.

Spiderhunters: see *Nectariniidae.*

Spider mites: see *Tetranychidae.*

Spider monkeys, *Ateles:* a genus of the subfamily *Atelinae* (see Cebid monkeys). They are extremely well adapted to their arboreal existence, inhabiting forests from southern Mexico to the Amazon basin, and feeding mainly on fruits with some leaves. They are rather pot-bellied, with a long prehensile tail and very long slender limbs. The tail is used as a fifth limb, for locomotion and for hanging from branches. The metacarpal of the thumb is virtually absent. Four species have been described: *A. paniscus* ***(Black spider monkey),*** *A. belzebuth* ***(Long-haired spider monkey),*** *A. fuscipeps* ***(Brown-headed spider monkey)*** and *A. geoffroyi* ***(Black-handed spider monkey).*** The Woolly spider monkey (see) belongs to a separate genus.

Spiders: see *Araneae.*

Spider wasps: see *Hymenoptera.*

Spike:

1) An Action potential (see), especially its graphical representation.

2) One of a group of glycoproteins in the envelope of

enveloped viruses (e.g. influenza virus, herpes virus, arbovirus). Also known as *peplomers,* they protrude from the envelope, forming a fringe over the lipoprotein coat of the virion. E.g. influenza virus has two kinds of peplomer, a hemagglutinin and a neuramidase.

3) See Flower, section 5) Inflorescence.

Spikelet: see *Graminae.*

Spike oil: see *Lamiaceae.*

Spilogale: see Skunks.

Spinach: see *Chenopodiaceae.*

Spinachia: see *Gasterosteidae.*

Spinach leaf miner: *see Anthomyiids.*

Spinacia: see *Chenopodiaceae.*

Spinal bulb: see Brain.

Spinal canal: see Axial skeleton.

Spinal cord, *medulla spinalis:* part of the vertebrate central nervous system enclosed in the spinal canal of the spinal column. It extends from the medulla oblongata of the brain almost to the end of the vertebral column, where it terminates in a tapering extremity called the conus medullaris. A slender prolongation of the conus medullaris (the filum terminale) anchors the S.c. to the back of the coccyx. In transverse section, the S.c. displays a characteristic H- or butterfly-shaped central region of *gray matter,* which consists mainly of ganglion cells, and which is surrounded by *white matter,* whose color is due to the presence of medullated nerve fibers.

During development of the *Anura* and *Mammalia,* growth of the S.c. lags slightly behind that of the vertebral column, so that the exit points of the spinal nerves (each connected to the S.c. by a dorsal and a ventral root) are shifted posteriorly (caudally). This leads to a particularly dense concentration of spinal nerves within the spinal canal in the lumbar and sacral regions. In limbless vertebrates, the S.c. is about the same diameter along its entire length. In quadrupeds, the regions of the S.c. associated with nerves from the extremites are markedly swollen (intumescent). Before the dorsal and ventral roots merge to form the mixed nerve, the dorsal root contains a *spinal ganglion,* containing the ganglion cells (cell bodies) of receptor neurons. The ventral roots contain mainly motor fibers. The S.c. is ensheathed by one (fishes) or several (quadrupeds) membranes called *meninges.* The internal central canal (derived from the original neural canal) contains cerebrospinal fluid (liquor cerebrospinalis), which is also present in the brain ventricles. In mammals, cerebrospinal fluid is also present in the subarachnoid space, whose lateral extensions communicate with the roof of the floor of the fourth ventricle.

The vertebrate S.c. serves two functions. The gray matter contains the centers for spinal reflexes, so that the entire reflex activity is an independent function of the S.c. The S.c. also contains nervous pathways that link the peripheral nerves of the body with the brain, so that the S.c. also has a conducting and routing function. With vertebrate evolution, the independent reflex function of the S.c. has decreased, while the conducting pathways to and from higher centers in the brain have become more developed.

Conducting pathways are located in the white matter. In mammals (as distinct from other quadrupeds), the white matter contains numerous ascending and descending nerve tracts. Most functionally similar nerve fibers travel together in the same tract. *Ascending* or *afferent tracts* conduct impulses to higher brain centers. *Descending* or *efferent* tracts conduct impulses from different brain regions into the S.c. *Long tracts* or *projection tracts* travel between the brain and the S.c., whereas *short* or *intersegmental tracts* link the individual segments of the S.c. *Crossed tracts* cross from one side to the other in the S.c. or in the diencephalon, whereas uncrossed tracts remain on the same side. Tracts or fasciculi consist of nerve fibers. A funiculus contains more than one tract or fasciculus.

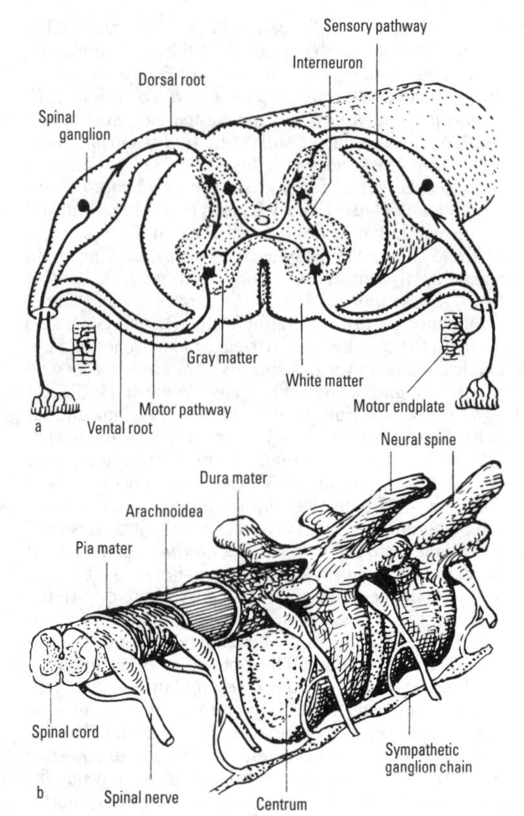

Spinal cord. a Structure. *b* Location within the vertebral column.

1) The most important *afferent tracts* are the tractus spinothalamicus, funiculus posterior, and tractus spinocerebellaris.

a) The *spinothalamic tract* transmits mainly impulses from pain and temperature receptors. In mammals the fibers of this tract extend to the diencephalon, where they synapse with other neurons whose dendrites travel to the brain.

b) The *funiculus posterior (dorsal funiculus)* carries impulses from pressure and touch receptors in the skin and from the muscles. The fibers of this tract travel uncrossed to the brainstem. After synapsing with other ganglion cells, they cross to the opposite side, travel to the diencephalon and then to the cerebral cortex. The funiculus posterior system is present in reptiles and more advanced vertebrates, but not in amphibians and fishes. It constitues 12% of the white matter in reptiles,

21% in the rabbit, and 39% in humans. The spinothalamic tract is phylogenetically older and is also present in fishes

c) The *spinocerebellar tract (spinocerebellar fasciculus)* conducts impulses from receptors in skeletal muscle, tendons and joints to the cerebellum, thereby relaying information about muscle tone and the position of limbs.

2) The *efferent (motor)* system is divided into pyrimidal tracts and extrapyrimidal tracts

a) *Pyrimidal tracts* (see) are found only in mammals. In humans, this conducting system is particularly well developed, constituting 20% of the white matter of the S.c. Most of the pyrimidal fibers (80–90%) constitute the *crossed pyrimidal tract (lateral corticospinal tract, lateral cerebrospinal fasciculus)* and they cross to the other side in the lower section of the brain stem. The remaining fibers travel uncrossed in the *direct pyrimidal tract (fasciculus cerebrospinalis anterior, ventral corticospinal tract, ventral cerebrospinal fasciculus)*.

b) The *extrapyrimidal tract* has various origins. Only some of the fibers arise in the cerebral cortex, while others originate in the cerebellum and brainstem. In the advanced warm-blooded vertebrates and in humans, these tracts are often interrupted by synapses. The fibers of this tract function mainly in the initiation and coordination of involuntary movement and in the control of muscle tone. Furthermore, they conduct impulses from different brain regions to the Motor cells of the anterior horn (see). The extrapyramidal system is present in fishes, and it is particularly well developed in birds.

Spinasterols: a group of very similar phytosterols found in higher plants. Chief representative is α-spinasterol (5α-stigma-7,22-diene-3β-ol) isolated, e.g. from spinach, senaga root and lucerne. α-, β-, γ- and δ-S. differ in the position of the side chain double bond.

Spindle, ***spindle apparatus:*** a spindle-shaped structure, which is formed in eukaryotic cells at the beginning of mitosis and meiosis. The "S. fibers", which are visible under the light microscope, consist largely of bundles of Microtubules (see). Chromosomes become aligned on the equator of the S., which is also the equatorial plane of the cell. This alignment or orientation appears to arise from interactions between the kinetochores (granular regions associated with the centromeres of the chromatids) and special microtubules of the S. known as *kinetochore microtubules* or *chromosome fibers*. The chromatids or chromosomes then move to the cell poles. The motive force for this migration is possibly derived from the shortening of microtubules attached to the kinetochores. Alternatively, a force may be generated by the sliding action between the microtubules attached to the kinetochore and those that extend between the poles of the cell; the true mechanism is uncertain. The structure of the S. varies according to whether it is formed with or without the agency of Centrioles (see). In centriole-containing cells, a pair of centrioles is present at each pole. Each centriole pair generates an aster, which serves for the orientation of the chromosome fibers, and the S. is said to be *amphiastral*. Microtubules grow outward from opposite centrosomes, forming a star-shaped arrangement of short microtubules radiating from the centrosome; this is known as an *aster,* and it consists of *astral microtubules*. On one side of the aster, the microtubules continue to grow toward the opposite pole, eventually overlapping with similar microtubules extending from the other aster. These microtubules are then known as *polar microtubules*. In their region of overlap, they become cross-linked, mutually stabilizing one another and giving the appearance of single fibers *(polar fibers)* extending from pole to pole. In contrast, kinetochore microtubules (see above) extend only from the pole to the equator, where they are attached to a chromosome. The number of microtubules linked to a kinetochore (i.e. the number of microtubules constituting a single chromosome S. fiber) varies from one (e.g. yeasts and certain other microorganisms) to between 20 and 40 (e.g. human cells). In the absence of centrioles (most spermatophytes, many fungi, vegetative cells of some pteridophytes, mosses and algae) no aster is observed, and the S. is said to be *anastral*. Thus, in yeast there are no centrosomes, and the nuclear membrane remains intact during S. formation. Integrated constituents of the nuclear membrane, known as *S. pole bodies,* serve as *microtubule organizing centers*. Microtubules extend from opposite S. pole bodies into the nucleoplasm, eventually interdigitating to form a complete S. Chromosome fibers also extend from each S. pole body to the equator of the S. where they are attached to the centromeres of chromatids [in yeast, a morphologically conspicuous kinetochore is not present, but the core of the centromere contains specific kinetochore proteins to which the single microtubule of the chromosome fiber (kinetochore microtubule) becomes attached]. During Mitosis (see) the S. undergoes characteristic changes, which depend on the rapid dissociation and resynthesis of microtubules. Formation of the S., and therefore the distribution of chromatids or chromosomes to the daughter nuclei are prevented by the alkaloid, colchicine (for the action of colchicine, see Microtubules). On the other hand, colchicine does not prevent the separation of chromosomes into chromatids, so that polyploidy can be induced by colchicine treatment. In many unicellular eukaryotes and certain fungi (e.g. yeasts and the myxomycete, *Physarium polycephalum),* mitosis is intranuclear, i.e. the nuclear membrane does not disintegrate, and the S. is formed inside the intact nuclear membrane. "Primitive" extranuclear Ss. have been observed in dinoflagellates; these form outside the nuclear membrane, which remains intact during cell and nuclear division.

Much relevant information will also be found under Microtubules.

Spindle poisons: mitotic poisons which interfere with the formation and function of the mitotic spindle, and therefore decrease or prevent chromosome migration during nuclear division. S.p. are used experimentally to induce gene mutations (primarily polyploidy). The most familiar of the S.p. is Colchicine (see). See Microtubules.

Spindle pole body: see Spindle.

Spindle spores: spindle-shaped fungal spores (see Conidia) that occur inter alia in the dermatophytes, *Achorion* and *Trichophyton*. They are usually thick-walled and often colored.

Spined loach, *Cobitis taenia:* an elongate, flat-sided, freshwater bottom fish of the family *Cobitidae* (order *Cypriniformes),* found throughout the European mainland and across Central Asia. Mainly nocturnal in habit, it favors shallow water with a sandy bottom, often burying itself during the day with only the head protruding. It is gray-brown with rows of dark spots, and a pale yellow to orange belly. The mouth is surrounded by

3 pairs of barbels, and below each eye is an erectile, fork-tipped bony spine, which is capable of inflicting a painful wound.

Spine-tailed swifts: see *Apodiformes*.

Spinetails: see *Furnariidae*.

Spinochromes: derivatives of 1,4-naphthoquinone (see Benzoquinones), responsible for the red or orange color of sea urchin shells. More than 20 different hydroxylated S. are known. In the native state they are present as calcium and magnesium salts. They differ from the echinochromes, which occur in the eggs, perivisceral fluid and internal organs of the sea urchin, e.g. echinochrome A is a pentahydroxy-1,4-naphthoquinone with an ethyl group at C6.

Echinochrome A

Spiny anteaters: see *Monotremata*.

Spiny dormice: see *Platacanthomyidae*.

Spiny eels: see *Mastacembeloidei*.

Spiny-headed worms: see *Acanthocephala*.

Spiny lizards, ***swifts, fence lizards***, *Sceloporus:* medium-sized, ground living iguanas (see *Iguanidae)*. An especially large number of species is found in Mexico and 14 species are found in North America. The scales on the back are very rough to the touch, each carrying a keel ending in a sharp spine. On the upper surface the body is usually gray or brown, decorated with transversely arranged spots or longitudinal stripes. Certain species have a dark neck band. The throat and underside of males is usually radiant blue. They favor agricultural and rural areas, hence the name, fence lizard, from their habit of spending long periods sunning themselves on fences. They also occupy desert, brush, grasslands, woods and mountains up to 4 000 m. In many areas of North America they are extremely common. The diet consists of insects and spiders; the Mexican species, *S. poinsetti*, also eats leaves and buds. Most species lay eggs, but a few montane species are viviparous. Familiar species are *S. undulatus* ***(Eastern fence lizard)*** and *S. occidentalis* ***(Western fence lizard)***.

Spiny lobsters, ***crawfish:*** crustaceans of the family *Palinuridae,* infraorder *Palinura,* order *Decapoda* (see). In contrast to other Lobsters (see), they lack chelipeds, and have large spiny antennae. Like other lobsters, they are caught for their meat. *Palinurus vulgaris* ***(European rock lobster*** or ***Langouste)*** attains a length of 45 cm and a weight of 8 kg. *Jasus lalandei* ***(Cape spiny lobster)*** is caught off the coasts of Australia and South Africa. Close relatives are the ***Spanish, Shovel nose*** or ***Locust lobsters*** *(Scyllarus,* family *Scyllaridae),* which have short, flattened antennae, and are also caught for their meat.

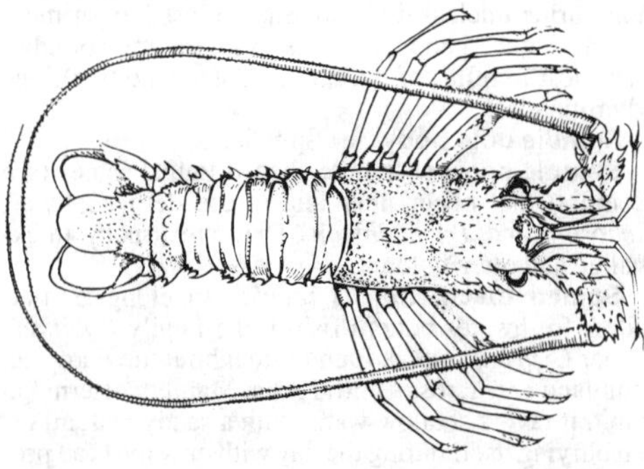

Spiny lobsters. Langouste or European rock lobster *(Palinurus vulgaris).*

Spiny mice: see *Muridae*.

Spiny rats: see *Echimyidae*.

Spiny sharks: see Acanthodians.

Spiny-tailed lizards, *Uromastyx:* agamids (see *Agamidae)* native to North Africa and Southwest and South Asia. They have a plump, cylindrical body and a short, broad tail. The latter is covered with spiny scales, and is used for defense. As an adaptation to the vegetarian diet, the back teeth have wide grinding surfaces, and the incisors of the lower jaw are fused. These agamids readily conceal themselves in rock clefts, where they swell up and become firmly anchored if threatened by predators.

Spiny-tailed skinks: see *Ergenia*.

Spiracle:

1) A small, vestigial gill aperture of elasmobranchs and some primitive ray-finned fishes, situated before the gill clefts and between the mandibular and hyoid arches. Rays *(Rajidae)* and angel sharks *(Squatinidae)* have large Ss., which serve for the inflow of respiratory water. In the *Anura* and evolutionary higher vertebrates, the S. has become modified to the tympanic chamber.

2) See Stigmata.

Spiral-horned antelopes: see *Tragelaphinae*.

Spiralia: a group of protostomian animal phyla, whose embryonal development characteristically involves spiral cleavage. In those members of the *S.* that pass through a larval stage, this is always a trochophore. The *S.* comprises the following phyla: *Platyhelminthes, Nemertini, Entoprocta, Mollusca, Sipunculida, Echiurida* and *Annelida*.

Spirea: see *Rosaceae*.

Spiriferida (Latin *spira* coil, *ferre* carry): an order of variously shaped, hinged brachiopods. Most members have radically ribbed and conspicuously winged shells. The brachial valve carries a spiralium or brachial apparatus, consisting of two lateral coiled arms. The apex of each coil extending into the outer corners of the straight hinge (Fig.). *S.* are distributed worldwide from the Middle Ordovician to the Permian, and sporadically from the Permian to the Lower Jurassic. Some species are useful index fossils of the Devonian and Lower Carboniferous.

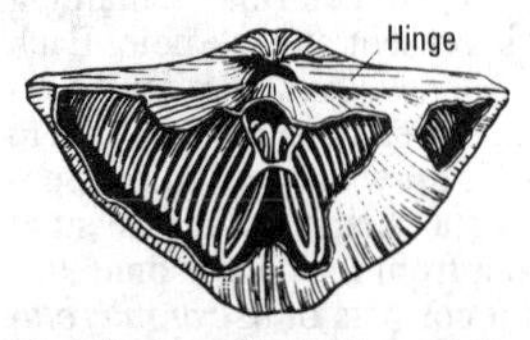

Spiriferida. Shell of a spiriferid brachiopod *(Spirifera striatus).* Part of the brachial valve is broken away to reveal the spiralium (x 0.75).

Spirillum: a genus of bacteria with spirally twisted cells, which may be up to 60 μm in length but only 0.5–3 μm in diameter. They occur in both fresh and salt water, and they display rapid, darting, nonundulatory movement. Between 1 and 7 flagella are present, usually at each pole of the cell. Metachromatic storage granules of polyhydroxybutyrate (also known as volutin) are often present, and these are especially prominent the large species, *S. volutans.* Most species are harmless saprophytes living in stagnant or polluted water, and in putrifying materials. Isolation in pure culture is difficult. *S. minus* (sometimes described as a spirochete) is, however, a causative agent of Rat bite fever (see). It occurs naturally in the blood of wild rats and other wild rodents.

Spirobolidae: see *Diplopoda.*

Spirochetes: slender bacteria, which are capable of active flexing of the cell body. They are motile, although flagella are absent. There is no rigid cell wall, and the cell is wound spirally around an axial filament, which lies beneath the exterior cell envelope. S. include free-living as well as parasitic bacteria.

Treponema grows up to 18 μm in length and is anaerobic. It has a typical corkscrew form, with coils of wavelength 1.0–1.5 μm. Some species are parasitic in humans or animals, where they cause disease. Best known is *Treponema pallidum,* the causative agent of *syphilis. Treponema pertenue* is the causative agent of the tropical disease, *yaws,* while *Treponema carateum* is the responsible for *pinta,* a disease of the West Indies, Central and South America. Both yaws and pinta may be transmitted by the fly, *Hippolates pallipes.*

Borrelia has large, open coils (wavelength 2–3 μm), grows up to 20 μm in length and has an elastic cell envelope. It is an obligate anaerobe, and its metabolism is fermentative. Different species of *Borrelia* are responsible for various diseases in humans and animals, and they are transmitted by insects and ticks. *Borrelia recurrentis,* which causes *European relapsing fever* in humans, is transmitted by lice.

Species of *Leptospira* are known as leptospires; they include blood parasites and free living forms. In humans and animals, they cause febrile diseases, known as *leptospiroses.* One of the more virulent forms is *Leptospira icterohaemorrhagiae,* which is carried by rats, and is the most frequent cause of Weil's disease (hemorrhagic jaundice). The coils are very fine and close (wavelength about 0.5 μm), and hardly visible by dark-ground light microscopy, but electron microscopy clearly reveals the coils around a single axial filament.

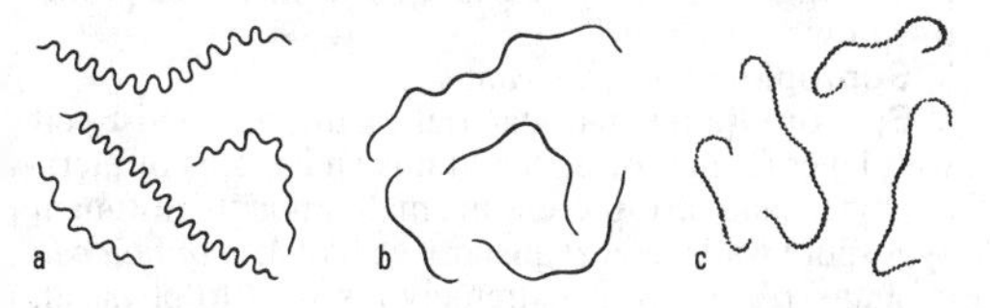

Spirochetes. a Treponema. b Borrelia. c Leptospira.

Spirocyst: see Nematocyst.

Spirodela: see *Lemnaceae.*

Spiroplasma: a genus of the Mycoplasmas (see), with a characteristic spiral structure. Some species are responsible for plant diseases, e.g. *Spiroplasma citri,* the causative agent of "citrus stubborn" disease in citrus plants.

Spirostreptidae: see *Diplopoda.*

Spirotricha: see *Ciliophora.*

Spirula: see *Cephalopoda.*

Spirura: see *Nematoda.*

Spirurida: see *Nematoda.*

Spittle bug: see *Homoptera.*

Spizixos: see *Pycnonotidae.*

Splanchnocranium: see Skull.

Splanchnopleure: see Mesoblast, Extraembryonic membranes.

Spleen: a richly vascularized organ of vertebrates, located near to the stomach. In cyclostomes it is represented by a network of tissue around the intestine, whereas in other vertebrates it is a separate, well defined organ with its own blood supply. Internally it is differentiated into regions of *white pulp,* which are filled with white blood cells, and regions of *red pulp* with accumulations of red blood cells. In the embryo and in nonmammals, the M. serves for the production of both white and red blood cells, whereas in adult mammals, it is the site of synthesis of only lymphocytes. In addition, it degrades aged red blood cells. It is also a storage site for red blood cells, which it releases into the circulation when needed.

Spodoptera: see Cutworms.

Spogel seeds: see *Plantaginaceae.*

Sponge-berry tree: see Chaulmoogra oil.

Sponges: see *Porifera.*

Spongiaria: see *Porifera.*

Spongillidae: see *Porifera.*

Spong(i)ocoel: see Digestive system, *Porifera.*

Spongocyte: see *Porifera.*

Spontaneous heating: the rise in temperature that occurs in densely packed, damp organic material, like hay, tobacco, animal dung. The process occurs in several stages. By the action of mesophilic organisms, the temperature is raised to about 40 °C, followed by a further increase to about 80 °C due to the activities of thermophilic organisms like *Bacillus stearothermophilus* and thermophilic actinomycetes. Sudden admission of oxygen may then lead to *spontaneous combusion,* which is a purely chemical process. S.h. and spontaneous combustion, which can cause considerable agricultural losses, are avoided by adequate drying of hay and straw.

Spoonbills: see *Ciconiiformes.*

Spoon worms: see *Echiurida.*

Sporangiolum: a small daughter sporangium formed on the parent sporangium. It contains relatively few spores, and spore production occurs later than in the parent sporangium. Sporangolia are produced by molds of the family *Cephalidae* (order *Mucorales),* e.g. by the genus *Mucor.*

Sporangium (plural **Sporangia):** the site of formation of endospores. In algae and fungi, it is a single cell, differing only in shape from other cells of the vegetative organism. In mosses and ferns, sporangia are generally multicellular structures, in which external, sterile cell layers surround the spore-forming (sporogenous) tissue. The interior often contains a columnar structure, called a *columella.*

Sporeformers: bacteria that form endospores. Only 2 genera of bacteria are S.f., i.e. *Bacillus* (aerobic) and *Clostridium* (anaerobic). The spores are extremely resistant to high temperatures and other influences, so that these bacteria present special problems of sterilization in medicine and industry. The spore may be located centrally, terminally or subterminally without distortion of the sporangium wall (1 & 2 in Fig.), or the oval or round spore may cause a terminal bulge, so that the sporangium resembles a drumstick (plectridium type) (3 & 4 in Fig.), or the spore may cause a subterminal bulge, so that the sporangium resembles a spindle (Clostridium type) (5 in Fig.). The shape and location of spores is taxonomically important, e.g. primary subdivisions of *Bacillus* are based on spore characteristics:

Group I *(B. subtilis, B. licheniformis, B. megaterium, B. cereus, B. anthracis, B. thuringensis):* spores oval, thin-walled, central or subterminal; sporangium not swollen.

Group II *(B. polymyxa, B. macerans, B. circulans):* spores oval, thick-walled, central or subterminal; sporangium swollen.

Group III *(B. pasteurii):* spores spherical, thick-walled, terminal; sporangium swollen.

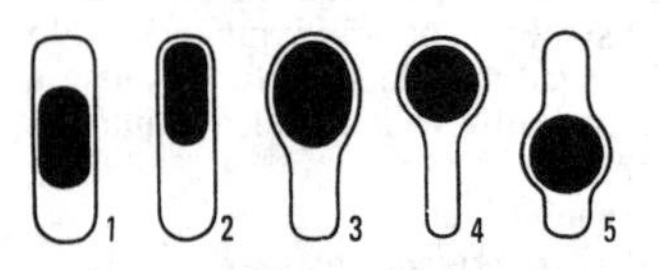

Sporeformers. Bacterial cells, showing different shapes and positions of spores. *1* and *2* Cell shape unchanged. *3* and *4* Terminal spore with bulging of bacterial cell wall (drumstick form). *5* Subterminal spore with swelling of bacterial cell wall (Clostridium form).

Spores, ***agametes:*** unicellular (rarely multicellular) reproductive bodies, which separate from the organism, and without fertilization develop directly or indirectly into a new individual.

Ss. abstricted at the ends of fungal hyphae are called *conidia.* Ss. formed inside specific cells or organs (sporangia, see Sporangium) are called *endospores* or *sporangiospores.* The latter include Ascospores (see) and Bacterial spores (see). These are either ciliated, motile *Zoospores* (see) or *aplanospores (akinetes).* Aplanospores are distributed by air currents after rupture of the sporangium wall, and some are propelled in a mechanism involving liquid secretion (see Ballistospores). All these types of spores are formed by plants as an evolutionary adaptation to a terrestrial existence. They usually have a tough cell wall that protects against damage and desiccation (see Resting spores). Cylindrical Arthrosporess (see) formed by lower fungi are known as Oidia (see).

Ss. produced by normal mitosis in the course of vegetative reproduction are called *mitospores.* They may be produced during the haplophase or the diplophase, and are then known respectively as *haplo-mitospores* or *diplo-mitospores.*

If S. formation is preceded by meiosis, the products are *meiospores* or *gonospores.* There is a causal relationship with the marked nuclear phase change in plants, which is frequently linked to an alternation of generations. The generation that is formed from meispores is known as the *sporophyte,* whereas the generation arising from haploid spores is the *gametophyte.* The sex cells (see Gamete) are carried by the gametophyte in variously structured gametangia. The appearance and contribution of the sporophyte (and gametophyte) to the total developmental cycle of the plant varies widely in different plant groups. In lower plants, algae, fungi, etc., the gametophyte is usually predominant, whereas the sporophyte is the predominant in higher plants. For example, in flowering plants the entire plant body can be considered as the sporophyte, whereas the haploid gametophyte is restricted to the cells in the pollen grains or ovules. Mosses and ferns occupy an intermediate position. A moss plant is a gametophyte, but the sporophyte grows upon it in the form of a stalked spore capsule. The gametophyte of ferns consists of a small prothallium, and the actual fern plant is the sporophyte, which carries various types of sporangia. In many fern species, as in the mosses, all Ss. are identical *(isospory, homospory).* Some groups, however, produce two types of spores: large *macrospores (megaspores gynospores)* in Macrosporangia (see) and small *microspores (androspores)* in Microsporangia (see). Macrospores develop into female prothallia, whereas microspores develop into male prothallia. The formation of differently sized, sexually differentiated Ss. is termed *heterospory.*

The term, S., is also applied to a number of other cells, such as various thick-walled resting cells which may be gametes, zygotes or vegetative cells. To avoid confusion, these are more appropriately called Cysts (see).

Ss. of the rust fungi (see Aecidiospores, Basidiospores, Pycnospores, Teleutospores, Uredospores) and smut fungi (brand spores) are important agents in the transmission of plant fungal diseases.

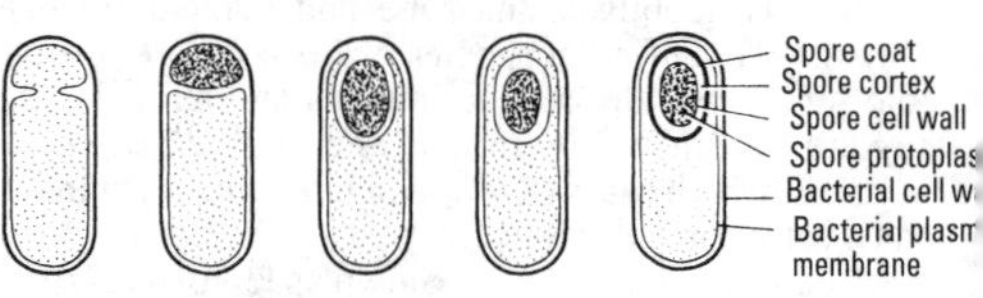

Spores. Spore formation in bacteria (schematic).

Sporidia: see Promycelium.

Sporocyst: see Cyst, Liver flukes.

Sporogamy: Reproduction by Spores (see).

Sporophyll: a leaf that carries sporangia. The Ss. of ferns are often simpler in structure than the assimilatory leaves (trophophylls). The term is usually applied to the small scale-like leaves in the cones of *Lycopodium* and *Selaginella,* which carry sporangia on their upper surfaces. Aggregations of Ss. can be considered as primitive flowers.

Sporopipes: see *Ploceidae.*

Sporopollenin: the material of the outermost cell wall layer (exine) of pollen grains and spores of pteridophytes, and also present in small amounts in fungal zygospore walls (e.g. zygospore wall of *Mucor mucedo* contains 1% S.). S. is extremely resistant to physical, chemical and biological degradation. Pollen grains are therefore well preserved in geological strata, and have proved archeologically useful as quantitative and qualitative markers of previous plant life and agriculture. P. is an intimate mixture or complex of 10–15% cellulose, 10% of an ill-defined xylan fraction, 10–15% of a lignin-like material, and 55–65% of a lipid. It is now generally accepted that the lipid material is formed by oxidative polymerization of carotenoids; a virtually identical material can be synthesized in the laboratory by catalytic oxidation of β-carotene.

Sporozoa, ***sporozoans:*** a class of the *Protozoa.* All members live parasitically, and they are characterized by asexual multiplication of the zygote (sporogony). The resulting sporozoites or sporozoids invade a new host. Sporogony alternates with sexual reproduction (gamogony). One or several stages of asexual multiplication (schizogony) may result in the inundation of the host with parasites (e.g. *Plasmodium)* before the occurrence of gamete formation. The life cycle of the more advanced members includes an intermediate host, which is a blood-sucking arthropod. Under the electron microscope, the anterior pole of the vegetative stage can be seen to contain an organelle complex, which aids penetration of the host cells. Feeding or locomotary organelles are absent. There are 2 main subclasses: *Telosporidia* (see) and *Cnidosporidia* (see). See also Piroplasmids, *Toxoplasmida.*

Recent authorities raise the sporozoans to the status of a phylum, the *Apicomplexa,* named for the organelle complex at the anterior pole.

Sporozoite: see *Telosporidia.*

Sport: see Chimera.

Spotted cutworm: see Cutworms.

Spotted deer: see Chital.

Spotted flycatcher: see *Muscicapinae.*

Spotted grass frog: see *Myobatrachidae.*

Spotted salamander: see *Ambystomatidae.*

Sprat, *Sprattus sprattus:* a shoaling marine fish of the family *Clupeidae* (order *Clupeiformes),* found in European coastal waters, the Mediterranean, Baltic and Black Sea, being especially common in brackish water near river mouths. The belly is keel-like with a row of sharp, saw-like scales. It has a metallic blue or greenish back with a silvery white belly and sides, with an additional golden lateral streak at spawning time. It is economically important, especially as a tinned fish.

Sprattus: see Sprat.

Spraying: a method for the application of solutions, suspensions or emulsions of Plant protection agents (see). In coarse sprays the average droplet size is greater than 0.15 mm, whereas in fine sprays the droplet size varies between 0.05 and 0.15 mm. In misting the droplet size is less than 0.05 mm.

Spotted snake millipede: see *Diplopoda.*

Spring: an ouflow of groundwater. A *pond S.* consists of a basin-like depression, fed with groundwater, whose overflow forms a stream. The substratum is usually sandy and overgrown with higher plants. A *torrent S.* flows directly from a sloping overhang; there is no source pond, and there is often a long drop to the valley below. A *marsh S.* or *ooze* converts the source area into a marsh, which feeds an overflow stream.

According to the temperature and chemical composition of the water, a S. is classified as acratopegic, thermal or mineral.

The water temperature of an acratopegic S. is constantly low, corresponding roughly to the average yearly temperature of the S. site. Thus, in summer the water is cooler, in winter warmer, than the prevailing air temperature. The exiting ground water has a low oxygen content, but this increases rapidly on contact with the atmosphere. The content of mineral salts is determined by the chemical composition of the local rocks.

Due to its constant low temperature regime, an acratopegic S. provides an environment in which aquatic plants do not die in the winter (because the temperature does not fall too low), and reproduce vegetatively (because the low summer temperature prevents flowering). The relatively still or very slow flowing water of a pond S. often supports stands of higher plants, such as *Nasturtium officinale* (watercress) and *Ranunculus trichophyllus* (dark hair crowfoot). Torrent Ss. are colonized primarily by algae and liverworts and mosses, e.g. *Marchantia polymorpha* and *Fontinalis antipyretica.* S. fauna consist of true S. animals (crenobionts), as well as species that choose to inhabit the S., but originate from neighboring habitats (crenophiles). Other species that occur sporadically and accidentally in a S. are known as crenoxenic organisms. *Crenobionts (crenon)* are mostly cold-loving, pure water species with a low oxygen requirement, e.g. the turbellarian, *Polycladodes alba,* the copepods, *Canthocamptus echinatus* and *Canthocamptus zschokkei,* several ostracods, snails of the genus *Bythinella* (in particular *B. dunkeri* and *B. alta),* as well as many aquatic mites and insects. *Crenophilic species* originate from the uppermost region of the water outflow, e.g. *Planaria alpina* and *Gammarus pulex,* or from the ground water, e.g. the tiny, colorless waterflea, *Niphargus puteanus* (a gammarid, and a close relative of *Eriopisa* and *Eriopisella,* which are found in sea shore gravel), the annelid, *Haplotaxis gordioides* (sometimes even found in tap water), and the white *Planaria vita.*

The average yearly water temperature of a *thermal S.* is higher than that of the climatic temperature at the S. site.

Thermal Ss. are usually of volcanic origin, and they are classified as hypothermal (below 18 °C), hilarothermal (18–30 °C), euthermal (30–50 °C), acrothermal (50–70 °C) and hyperthermal (above 70 °C). Up to 40 °C, the flora and fauna of a thermal S. are still diverse and contain many species that also flourish at normal temperatures. Between 40 and 45 °C there is a sharp decrease in the number of species. Only a few specially adapted thermal species are able to live in waters warmer than 45 °C. Upper temperature limits for organisms found in thermal waters are (in °C): bacteria (88.0), blue-green algae (85.2), diatoms (40.0), green algae (38.0), protozoa (55.0), rotifers (45.0), snails (53.0), insects (50.0), fish (37.5), flowering plants (40).

The water of *mineral Ss.* contains at least 1 g of dissolved solid inorganic material per kg water, or trace elements or gases in excess of established limits. According to the chemical content of the water, mineral Ss. are known, e.g. as acid-, iron-, salt-, iodine- and sulfur-Ss., etc.

Springbuck: see *Antilopinae.*

Spring frog: see *Raninae.*

Spring hare: see *Pedetidae.*

Spring snowflake: see *Amaryllidaceae.*

Springtails: see *Collembola.*

Spring trap: see Carnivorous plants.

Spring wood, ***early wood:*** the wood of trees and bushes, formed in the spring and consisting largely of wide vessels.

Spruce: see *Pinaceae.*

Spruce bark beetle: see Engraver beetles.

Spurfowl: see *Phasianidae.*
Spurge: see *Euphorbiaceae.*
Spurge laurel: see *Thymelaeaceae.*
Spur-tailed Mediterranean tortoise: see *Testudinidae.*
Spur-thighed Mediterranean tortoise: see *Testudinidae.*
Spurwinged goose: see *Anseriformes.*
Squalene: a triterpene hydrocarbon containing six double bonds. S. is a colorless oil, first isolated from shark liver, and subsequently found to be present in small quantities in many animal and vegetable oils. It is biosynthesized from acetyl-CoA via isopentenylpyrophosphate, and it is an intermediate in the biosynthesis of all cyclic triterpenes, e.g. steroids.

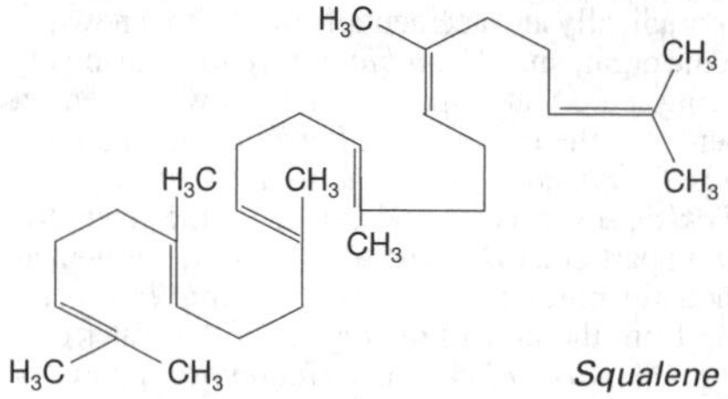

Squalene

Squalidae: see *Selachimorpha.*
Squaliformes: see *Selachimorpha.*
Squalinae: see *Selachimorpha.*
Squamata, *lizards, snakes and allies:* an order of terrestrial and aquatic reptiles, with well developed, regressed, or totally absent limbs. The dry skin carries horny epidermal scales or scutes, and its uppermost horny layers are shed at intervals. The teeth are not sunk into distinct sockets, and in some species they display strong functional modifications. With a few exceptions, chemical stimuli in the surrounding air are perceived with the aid of the forked or notched tongue, which is continually darted in and out, transferring olfacory principles to the Vomeronasal organ (see) in the roof of the buccal cavity. The cloacal opening is tranverse, and the reproductive organs are paired. Most species lay "parchment-shelled" eggs. Some are vivioviparous, i.e. the eggs remain in the body of the female during embryonal development, and the young leave the gelatinous envelope shortly before or after discharge. In a very few species the embryo is nourished via a placenta-like organ, i.e. they are truly viviparous.

The order contains about 5 700 living species, accounting for about 95% of all recent reptiles. Three suborders are recognized: *Lacertilia* (lizards and allies), *Amphisbaenia* (worm lizards; considered to be a family of the *Lacertilia* by some authorities), *Ophidia* (snakes). Members of the *S.* display varying degrees of modification of the primitive reptilian skull. Whereas the boundaries of the upper and lower temporal vacuities are retained in the primitive skull (e.g. as in *Sphenodon,* see *Rhynchocephalia),* the lower temporal arch of the *S.* is incomplete due to reduction of the jugal and squamosal processes, thereby allowing mobility of the quadrate and an increased gape. The embryonic history of all *S.* is also very similar. Members of this very successful order display great evolutionary diversity, and they have invaded every type of habitat. With the exception of the polar regions, the order is represented in all the climatic zones of the world.

Squamous epithelium: see Epithelium.
Squash: see *Cucurbitaceae.*
Squatinidae: see *Selachimorpha.*
Squids: see *Cephalopoda.*
Squirrel monkeys, *Saimiri:* the smallest of the Cebid monkeys (see) (head plus body 320 mm, tail 410 mm), and the only members of the subfamily, *Saimirinae.* The ***Common squirrel monkey (S. sciureus)*** is found in most rainforest areas of South America, whereas the ***Red-backed squirrel monkey (S. oerstedii)*** is restricted the Costa Rica and Panama. S.m. are white-faced with dark hair on the forehead. The body coat is short-haired and grayish, and the legs are yellow. In view of the great variety of coat shade and shape of the facial mask, it is thought that there may be more than 2 species. The diet is mainly fruits with some insects.
Squirrels: see *Sciuridae.*
Squirting cucumber: see *Cucurbitaceae.*
Squirting mechanisms: a special type of Catapulting mechanism (see) in plants , which serves for the distribution of spores and seeds. Squirting is due to the bursting of cells that have a high turgor pressure, e.g. ascospores of ascomycetes or the spores of *Pilobolus* are ejected by S.m. In the squirting cucumber, *Ecballium,* touching the ripe fruit releases a turgor pressure of 15 MPa, which ejects the contents over a distance of a few meters.
Stab culture: see Culture of microorganisms.
Stag beetles: see Lamellicorn beetles.
Stagemaker: see *Ptilinorhynchidae.*
Stagnicole: Organisms and associations of organisms living in stagnant bodies of water. The converse of S. is Torrenticole (see).
Stag's horn coral: see *Madreporaria,* Coral reef.
Stag's horn moss: see *Lycopodiatae.*
Staining: methods for increasing the visibility and contrast of cell structures and cellular material in light microscopy. Biological structures almost always show poor contrast under the light microscope, but this can be increased by exploiting the solubility of dyes in cellular material, the electrostatic and colloidal binding of dyes to cellular materials and structures, and the specific reaction of dyes with certain cellular compounds.
Stamen: see Flower.
Stamen sensitivity: seismonastic sensitivity (see Seismonasty) of the stamens of certain plant species. See Flower movements.
Staminodes: nonfertile, often rudimentary stamens, which lack anthers, or have anthers with no pollen. S. are often more or less modified to petals, and they are responsible for the "doubling" of flowers. See Flower.
Standard bicarbonate: a measure of the acid-binding capacity of blood plasma. It is determined under standardized conditions by measuring the carbon dioxide released by adding strong acid to a known quantity of blood plasma. Blood bicarbonate has a concentration of 0.02–0.03 M (20–30 mEq/l), and it constitutes the main buffering system of the blood.
Standard petal: see *Papilionaceae.*
Standard type: in genetics, a particular genotype, used as a reference base for genetic studies. The genes of the S.t., which may also be a wild type, are generally symbolized by "+" in genetic formulas.
Standing crop, *biomass:* the total quantity of living

organic material at a given time in a particular biome, usually expressed as fresh or dry weight per unit area or volume, e.g. kg/m^2. The corresponding quanitity of dead organic material is known as the Necromass (see).

Standing water: permanent freshwater habitats like Lakes (see), artificial lakes and Ponds (see), and temporary accumulations of water, such as puddles.

St. Anthony's fire: see *Ascomycetes.*

Stapelia: see *Asclepiadaceae.*

Stapeliads: see *Asclepiadacea.*

Stapes of the middle ear: see Auditory organ.

Staphylococci: facultative anaerobic Micrococci (see), up to 1.5 μm in diameter, which grow in clusters. They can be generally defined as cocci which form relatively large, opaque, white or colored, smooth colonies on solid media, and which ferment glucose both aerobically and anaerobically, and may liquify gelatin. They often occur on the skin, and some species are disease agents. *Staphylcoccus aureus* is a common human commensal, found on the skin; it is responsible for numerous superficial inflammatory lesions, wound suppuration and boils, and sometimes for food poisoning. They are of ubiquitous occurrence, and apart from human and animal sources, S. can be isolated from air, water, soil and dust.

Staple diet: the basic necessary diet. For an adult human, the recommended S.d. consists of 12–15% protein, 20–30% fat and more than 50% carbohydrate.

Starch: a macromolecular polysaccharide, empirical formula $(C_6H_{10}O_5)_n$, consisting of up to 20% water-soluble Amylose (see) and up to 80% water-insoluble Amylopectin (see). S. is biologically very important as the reserve carbohydrate of most higher plants, particular storage sites being roots, tubers and parenchyma. In plant metabolism, S. first appears as an assimilation product in the chloroplasts. It is then degraded, the products of degradation are translocated, and the S. is resynthesized as storage S. in the form of grains or granules. The size, shape and stratification of these S. grains (simple, compound, centric, acentric) is characteristic of the plant species, so that flour from different sources (e.g.maize, rice, wheat, rye) can be identified microscopically. S. is biosynthesized from adenosine-diphosphate-glucose. During animal digestion, S. is hydrolysed by amylases. α-Amylase hydrolyses α1,4 linkages at random, forming a mixture of glucose and maltose. Maltose is hydrolysed to glucose by α-D-glucosidase (maltase). Amylose is thus completely degraded to glucose. α-Amylase cannot hydrolyse α1,6 linkages at the branch points of amylopectin, so that the product of α-amylase action on amylopectin is a large, highly branched core dextrin or limit dextrin (see Dextrins). The 1,6 linkages are subsequently hydrolysed by an oligo-α1,6-glucosidase in the small intestine. β-Amylases are found especially in germinating seeds (e.g. malt); they act on the nonreducing end of the polysaccharide chain, removing successive maltose units. In plant cells, storage S. is remobilized by phosphorolysis with inorganic phosphate to glucose 1-phosphate.

S. is very important in human nutrition, supplying most of the 500 g of dietary carbohydrate required per day. Potatoes, wheat, rice, maize and bananas are particularly rich in S. Metabolism of 1 g S. supplies 16.75 kJ (4kcal).

Starch gum: see Dextrins.

Starfishes: see *Asteroidea.*

Starflower: see *Primulaceae.*

Starfruit: see *Oxalidaceae.*

Star-gazer, *Uranoscopus scaber:* a sluggish bottom fish, which grows up to 30 cm in length; family *Uranoscopidae,* order *Perciformes.* The eyes are situated on the flat top of the massive head, and the gape of the mouth is large and vertical. A branched fleshy outgrowth of the lower jaw is extruded as a lure for prey. It is found in the shallow coastal waters of the Mediterranean, the Black Sea and the Eastern Atlantic. The flesh makes good eating (especially in soup). The fish is used if caught, but it is not economically important.

Starlings: see *Sturnidae.*

Starling's law of the heart: a law discovered by Starling, which decribes the ability of the isolated heart to adapt to changes in the pressure and volume of venous inflow. Based on results obtained with his heart-lung preparation, Starling made the following generalization, commonly called Starling's law of the heart, which represents the reaction of the heart to increasing distension: *the energy of contraction is a function of the length of the muscle fibers.* Alternatively, the Frank-Starling law (Starling based some of his calculations on the results of Frank) can be expressed as: *the heart will pump whatever amount of blood enters the right atrium within the physiological limits of the heart's pumping capacity, and without a significant build-up of back pressure in the right atrium.*

Heart-lung preparation. An experimental animal is anesthetized and artificial respiration is maintained with a pump, and a cannula inserted in the trachea. The chest is opened and a cannula is placed in a branch of the aorta, the other branches and the rest of the descending aorta being tied off. The superior vena cava is cannulated and connected to a reservoir of defibrinated blood, and the inferior vena cava is attached to a water manometer. Blood can therefore flow from the left ventricle only via the coronary arteries and the aortic cannula, the latter leading to a manometer which records the mean arterial pressure, and from there into a vessel containing air. Compression of this air during ventricular systole imitates the elasticity of the arterial walls and provides the pressure to drive the blood through an artificial resistance (representing arterial resistance) during diastole. The artificial resistance consists of a rubber tube that passes through a sealed glass container. The air pressure within the glass container can be varied to restrict or relax the rubber tube, thereby controlling "arterial resistance". Blood is then conducted to a venous reservoir, and from there to the superior vena cava. After passing into the right atrium, the blood is pumped through the lungs by the right ventricle, finally returning via the pulmonary veins to the left side of the heart. Provision is also made to maintain the blood at body temperature.

This heart-lung preparation has yielded valuable information about the performance of the heart as a pump, but Starling's law has only limited application to the heart of the intact animal. Moreover, unless the coronary sinus is also cannulated, no allowance is made for blood flow through the coronary vessels, which accounts for about 5% of the total output of the heart-lung preparation.

Change of volume. An isolated heart produces a larger Stroke volume (see) from the left ventricle, if it receives an increased volume of blood during diastole.

More blood then also accumulates in the right side of the heart, which in turn pumps out the excess. This autonomous mechanism of the heart insures that there are no differences in performance between the two sides of the heart, i.e. that the left and right sides of the heart deliver the same quantity of blood to the aorta and pulmonary artery, respectively. In the whole organism, the operation of this regulatory mechanism is restricted, because reflex control by the cardiac sympathetic system prevents over-extension of the heart.

Change of pressure. If the volume of blood delivered to the isolated heart is kept constant, but the aortic pressure is increased, the stroke volume at first decreases. In the body, a decrease in stroke volume is accompanied by a decreased venous blood flow, so that vessels are less extended and Starling's law does not fully apply.

Consider pressure-volume diagrams 1 and 2 in the Fig. If all the valves of the left ventricle are closed experimentally during systole, there is an isovolumetric rise of pressure to a maximum (in the diagram, this maximum is directly above the starting volume on the abscissa). By starting with different volumes, other pressure maxima are attained. The line connecting all these maxima is the curve of the *isovolumetric pressure maxima.* Alternatively, at any time during systole the intraventricular pressure may be made constant, so that less blood is ejected and the quantity of blood remaining in the ventricle is increased (since this quantity has a lower rather than an upper limit, it is in fact a minimum). A line joining these minima is the curve of the *isobaric volume minima.* During the pre-ejection period the intraventricular pressure increases isovolumetrically from A to B on the pressure-volume curve of the relaxed ventricle. At B, the intraventricular pressure just exceeds the aortic pressure and the ejection period begins. As ejection progresses, the pressure continues to rise, attains its systolic peak, then decreases. The intraventricular volume decreases during the ejection period. Point C signifies the end of the ejection period, with the closing of the aortic valve. Point C is located on the *U curve,* which lies between the curve of isovolumetric pressure maxima and the curve of isobaric volume minima. [U curves show the endsystolic pressures and volumes of the ventricle under natural conditions.] The first phase of diastole is characterized by the drop from C to D, which indicates isovolumetric relaxation. Point D represents the start of the filling phase, during which the ventricular pressure at first continues to decrease. When the ventricle is completely relaxed, the pressure increases again until it reaches the end-diastolic filling pressure; under steady state conditions, the latter is the same as that indicated by point A. Area ABCD is an index of the mechanical work performed by the left ventricle. The distance between abscissa values B and C corresponds to the *stroke volume* (the ejected volume). The abscissa value of point C corresponds to the *residual volume,* which under normal resting conditions is about the same as the stroke volume.

Pressure-volume diagrams 3 and 4 show the effects of volume changes and pressure changes, respectively. Diagram 5 shows how excitation of the cardiac sympathetic system increases the force of contraction and decreases the residual volume.

Star-nosed mole, *Condylura cristata:* a long-tailed, semi-aquatic mole, inhabiting eastern and central North America from Labrador to Lake Winnipeg as far south as Georgia. It is the only species of the *Condylura.* Its characteristic muzzle is ringed in front with 22 pink fleshy appendages, known as rays or tentacles. The eyes are very small but visible, and the ears are much reduced and hardly discernable externally. Like other

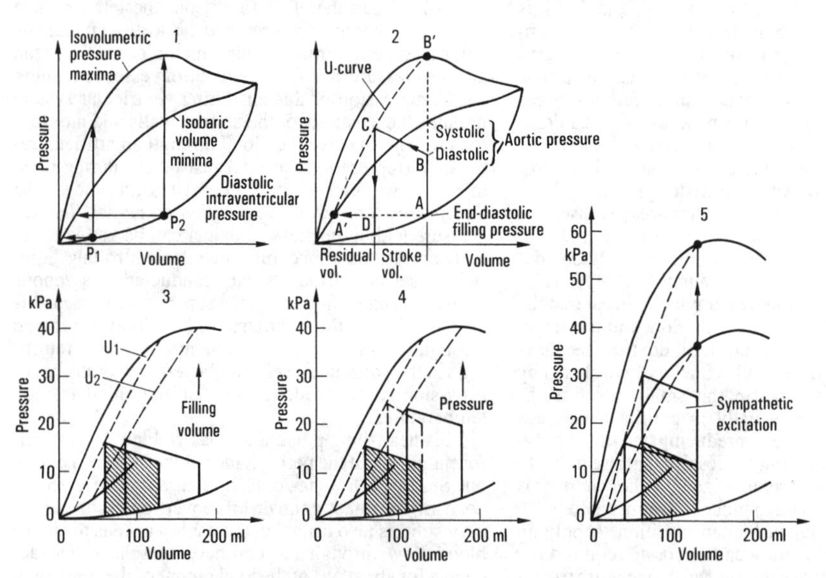

Starling's law of the heart. Pressure-volume diagrams of the left ventricle (see text for explanation). Diagrams 1 and 2 are recorded for normal filling pressure and volume. Diagrams 3 and 4 show the effects of increasing the filling volume and the filling pressure, respectively. Diagram 5 shows the effect of sympathetic stimulation.

Moles (see), *Condylura* excavates underground passages, in which it lives and breeds. It is also an expert swimmer and diver, possessing a dense, almost black, waterproof coat. It feeds on aquatic insects, small fish, earthworms and crustaceans.

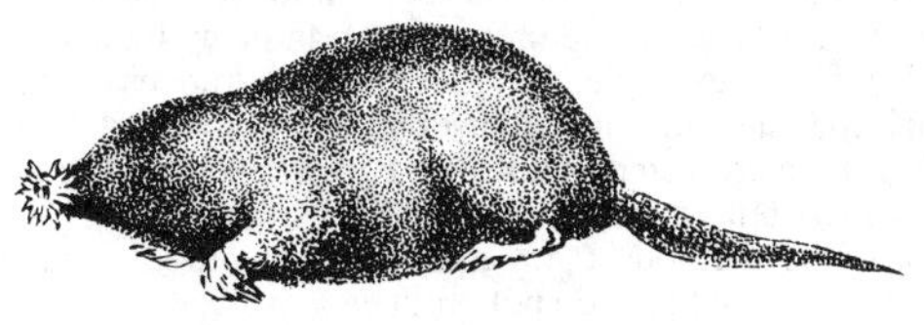

Star-nosed mole

Starter cultures, ***starters:*** lactic acid bacteria used for the souring of milk or cream, or for the production of aroma substances, in the industrial preparation of cheese, sour milk products and butter. They are commercially produced, mixed cultures of selected acid-producing and aroma-forming strains of Lactobacilli and Streptococci.

Stasipatric speciation: see Speciation.

Statins: see Releasing hormones.

Statoblast: see *Bryozoa.*

Statocyst: an equilibrium organ present in many coelenterates, platyhelminths, aschelminths, mollusks, crustaceans, etc. The S. is a fluid-filled vesicle containing one or several crystalline bodies lying on the sensory hairs of a sensory epithelium.

Status-quo hormone: see Juvenile hormone.

Status system: see Rank order.

Staunching of blood flow, ***arrest of hemorrhage:*** a complex process that occurs between the formation of a wound and its closure. In humans, the bleeding time from a wound is 1–3 minutes, and the blood clotting time is 3–5 minutes. Thus blood flow is staunched by preliminary wound constriction or closure, and finally by Blood coagulation (see).

Bleeding eventually stops of its own accord only in medium-sized and small arteries, and in arterioles veins and venules. In large arteries, any plugs of clumped platelets or small clots are washed away by the high-pressure blood flow. Seepage bleeding may occur from the damaged capillaries of rigid tissue (bone) or from especially large wound surfaces of parenchymatous organs (liver); otherwise, damaged capillaries do not bleed, because they become sealed by collapsing and adhering to one another.

Physical damage to arterial vessels acts as a mechanical stimulus, which causes the smooth musculature of the vessel walls to constrict. Blood loss and the accompanying drop in pressure favor vessel closure, because the walls of the vessel are then less distended. Wounding also brings the blood platelets into contact with the collagen fibers of the vessel wall. This causes the platelets to disintegrate, with the release of: 1) the vasoconstrictor substances, serotonin, adrenalin and noradrenalin, 2) ATP and ADP, 3) coagulation-promoting substances, and 4) phospholipids, which initiate coagulation. Depolarization of the damaged cells of the vessel wall also results in the release of ATP, which is converted enzymatically into ADP, which in turn causes reversible clumping of blood platelets. Within a few seconds, the wound is sealed by a white plug of clumped platelets. Formation of the final, red-colored, permanent clot by true blood coagulation is initiated by contact with the damaged tissue, and by the release of factor(s) in tissue fluids from the damaged area, which contribute to the activation of proconvertin (factor VII).

Stauromedusae: see *Scyphozoa.*

Staurotypidae, ***Mexican musk turtles:*** a family of the order *Chelonia* (see), containing 3 species in 2 genera, distributed from southern Mexico to Honduras. They are sometimes considered as a subfamily of the *Kinosternidae.* The carapace has 3 longitudinal ridges, and the plastron is reduced and cross-shaped.

Steady state: a stationary state of an open system, characterized by a minimal entropy increase and the action of forces that remain constant with time. Reaction rates and flow rates are equal in both directions, so that all macroscopic processes and quantities remain constant. A chain of chemical reactions is in a steady state when the concentrations of all intermediates remain constant, despite a net flow of material through the system, i.e. the concentration of intermediates remains constant, while a product is formed at the expense of a substrate. An ecosystem represents a S.s., but it is also subject to forces that are not constant with time. The S.s. of an ecosystem is therefore an idealized state. See Ecological equilibrium.

Steamer ducks: see *Anseriformes.*

Steaming: see *Anseriformes.*

Steam steriliser: a heatable cylindrical metal vessel, which possesses a lid, and which is used for killing microorganisms with hot steam (see Sterilization). Steam is generated by boiling water in the base of the vessel, and the material to be sterilized is placed on a grid above the boiling water. This form of sterilization is used primarily for unstable culture media and for other materials that cannot withstand autoclave temperatures, i.e. temperatures greater than 100 °C. Since resistant bacterial spores are not killed at 100 °C, heat treatment in the S.s. is repeated on 3 subsequent days. Between treatments, the spores start to germinate and therefore become vulnerable. The procedure was introduced into microbiology by R. Koch.

Stearic acid, *n-octadecanoic acid:* CH_3—$(CH_2)_{16}$—COOH, a higher fatty acid, which, when purified in the laboratory, forms a white, water-insoluble, crystalline mass. Esterified with glycerol, S.a. is one of the most abundant and most widely distributed fatty acids, occurring in practically all animal and plant oils and fats, e.g. 34% in cocoa butter, 30% in mutton fat, 18% in beef fat, 5–15% in milk fat. It is used commercially in the manufacture of candles, soaps, detergents, antifoams, pharmaceuticals and cosmetics. Pure stearin is the triglyceride (triacylglycerol) of S.a., but commercial stearin also contains some palmitin.

Steatopygia: the accumulation of large quantities of fat in the buttocks. It is common in female hottentots and bushmen, occurring primarily in the region of the coccyx, and to a lesser extent in the lateral parts of the thighs. Protrusion of the seat is increased by a marked lordosis in the lumber region of the vertebral column. S. commences in young girls, increases considerably during puberty, and becomes excessive during pregnancy. To a certain extent, it is influenced by the nutritional state.

Steatornis: see *Caprimulgiformes.*

Steenbock: see *Neotraginae.*

Steeple head: see Acrocephaly.

Steeple skull: see Acrocephaly.

Steinbock: see *Neotraginae.*

Steinheim: an archeological site on the River Murr, north of Stuttgart, Germany, known for the discovery of a human skull dating from the Mindel-Riss Interglacial period. See Anthropogenesis.

Steganopodes: see *Pelicaniformes.*

Steganopus: see *Phalaropodidae.*

Steganura: see *Ploceidae.*

Stegocephalia (Greek *stegos* roof, *kephale* head): old term for all the quadrupeds of the Paleozoic and Triassic. Since all forms were included without consideration for phylogenetic relationships, the *S.* is not a systematic unit. Many members of the *S.* are Labyrinthodonts (see). All S. share the characteristics of a strongly ossified skull roof, and the absence of temporal fossae. See Triassic.

Stegosaurus (Stegosaurs): see Dinosaurs.

Stele: the central vascular cylinder of stems and roots, consisting of pericycle, xylem, phloem and sometimes pith and medullary rays. See Stem, Telome theory.

Stellaroidea: see *Echinodermata.*

Steller's sea cow: see *Sirenia.*

Stem, ***axis:*** a cylindrical, rod-like structure of higher plants (see Cormus), which carries the leaves, transports nutrients and metabolites between roots and leaves, and often contains storage materials. The aerial stems of herbaceous plants possess chloroplasts and are able to perform photosynthesis.

Development. The stem originates from the *apical meristem* (vegetative apex, growing point of the shoot), which is derived by successive divisions of cells in the embryo, and is therefore a primary or primitive meristem. In aquatic plants, the apical meristem is often a conspicuous spherical structure, whereas in most land plants it is a conical or flattened dome. The outer cells of the apical meristem grow in such a way that all the resulting cell walls are perpendicular to the surface (anticlinical division, producing anticlinical walls). One or several layers of cells displaying only anticlinical division constitute the *tunica.* The outer layer of the tunica is called the *protoderm* or *dermatogen,* since it eventually becomes the primary boundary layer or epidermis. The tunica forms a sheath around the inner ground mass of the apical meristem, which is known as the *corpus.* Cells of the corpus divide both periclinically (i.e. walls are formed parallel to the surface) and anticlinically, resulting in an enlargement of the apical meristem, and the loss of the initial tubular shape of the tunica. The apical meristem merges without a visible boundary into the *zone of determination* or *zone of organogenesis* (0.02–0.08 mm behind the apex of the meristem). Cells in the zone of determination are already programmed for their subsequent transformation into permanent tissues, e.g. cortex, vascular and medullary tissue, a fact which to some extent can be demonstrated by special staining methods and specific histological tests for enzymes. At the same time, certain cells in the tunica, especially in its second layer, undergo periclinical division to form leaf primordia, i.e. protuberances of the tunica, which develop into leaves. Development of leaves from these primordia is often rapid, so that they form a protective envelope around the apical meristem, i.e. a Bud (see) is formed. The positions on the stem at which leaves are inserted are called *nodes,* whereas the intervening parts of the stem, which are free of leaves and which undergo extension growth during the course of development, are called *internodes.*

During differentiation, some promeristematic cells remain undivided but retain a latent capacity for division and differentiation. These are known as *residual meristems* (see meristematic tissue). In many monocotyledons, for example, the bases of the internodes retain their meristematic character for a long period, and they form an interpolated (intercalary) growth zone. Intercalary meristems enable a rapid and considerable elongation of the internodes; they contribute to rapid extension of the stem, but they do not produce leaves.

Anatomy. After the completion of differentiation, the stem possesses its primary structure. The cortex (with no vascular tissue, and bounded by the epidermis) surrounds the central cylinder which contains the vascular tissue. The cortex consists mainly of parenchyma. In aerial shoots, the cells in the outer layers of the parenchyma contain chloroplasts, whereas the inner layers may be colorless and often contain storage materials. Cortical parenchyma usually contains mechanical tissue in the form of collenchyma and sclerenchyma, which give the plant shoot tensile strength and resistance to bending. As a rule, the innermost cortex layer has no intercellular spaces. In many water plants and subterranean shoots, this inner cortex layer resembles the endodermis of the root, whereas in aerial shoots it contains large, easily mobilized starch grains, and can be looked upon as a starch sheath. Often, however, the cortex is not clearly delineated from the central cylinder. Roots that emerge from the shoot originate from the outer layer (pericycle) of the central cylinder. In dicotyledons and gymnosperms the vascular bundles of the pericycle are arranged in a circle and separated from each other by radial sectors of parenchyma tissue, known as *medullary rays.* These transport material between the medulla and cortex. In the vascular bundles, the phloem faces outward, the xylem inward. In dicotyledons the xylem and phloem are often separated by a zone of meristematic tissue, known as *cambium,* which becomes active again during secondary growth. Vascular bundles with this type of structure are called *open collateral bundles.* Within the ring of vascular bundles there is a central zone of large-celled parenchyma, known as the *medulla* or *pith.* Vascular bundles of monocotyledons have no cambium, so that the phloem and xylem are in contact with each other. These *closed collateral bundles* of the monocotyledon stem are not arranged in a circle. Instead they are more or less scattered, so that a transverse section of the monocotyledon stem shows numerous small bundles toward the periphery, and larger and less numerous bundles toward the interior. Medulla and medullary rays are absent. In the central cylinder of some subterranean shoots, e.g. the rhizome of lily of the valley *(Convalaria majalis),* as well as some monocotyledons, the xylem of some central bundles surrounds the central phloem. On the other hand, the central cylinder of most pteridophytes contains concentric bundles in which a central strand of xylem is surrounded by phloem. The central cylinder or *protostele* of the stem and root of some species of *Lycopodium* consists of *radial vascular bundles.* In transverse section, the xylem (containing vessels) of the radial vascular bundle appears as a star-shaped mass. Between the rays of the star lies the phloem (containing sieve tubes). In other species of *Lycopodium*, the xylem and phloem are rather intermixed, and in yet other species the xylem and phloem occur (in transverse section) as alternating, transverse, parallel bands. A star-shaped protostele is known as a *haplostele,* while separate, parallel bands of xylem are called a *plectostele.*

According to the *stelar theory,* the lignified vascular and mechanical elements (arranged at the periphery of the axis, thereby conferring resistance to bending) represent functional units which have evolved from each other and ultimately from the central column of tracheids or *protostele* of primitive land plants. The inner parenchyma of the protostele was surrounded by relatively undifferentiated *prosenchymatous* cells (elongated cells connected with one another by the overlaying of their pointed ends), which are considered to represent a primitive phloem. The juvenile stages of many living ferns display a typical protostele. Multiple longitudinal division and spreading of the protostele leads to a *polystele* or *actinostele,* consisting of numerous vascular bundles, which show a general distribution in a transverse section of the stem, e.g. in bracken *(Pteridium aquilinum).* Evolution of the polystele into the *eustele* of spermatophytes is then conceivable, i.e. thickening of the axis is accompanied by migration of the vascular bundles to a peripheral position. A central, cylindrical arrangement of vascular strands enclosing a parenchymatous pith is known as a *siphonostele;* both pith and cortex contain reserve material, and the leaf traces arise in the same way as those of the polystele (e.g. *Gleicheniaceae, Schizaeaceae and Lepidodendron).* In the *dictylostele,* the branching leaf traces leave parenchyma-filled gaps (pith rays) in the central vascular cylinder, which serve as connections between the pith and cortex (e.g. *Osmundia* and many other ferns). According to the Telome theory (see), the eustele may have arisen in a somewhat different way. Thus, thickening of the stem during evolution would have displaced the individual bundles of the polystele to the periphery to give mechanical strength, followed by their fusion into a siphonostele. Development of transverse anastomoses would then have led to the modern eustele. With regard to stelar evolution and structure, the *atactostele* of modern monocotyledons occupies an isolated position. Although the bundles are distributed over the entire transverse section (like those of the polystele), the stele does not consist of parallel cauline bundles, but of numerous individual leaf traces. An extreme example of the atactostele is found in the *Palmae,* in which numerous bundles extend from the base of the leaf, which folds closely around the axis. Each median bundle runs almost to the center of the central cylinder, while lateral bundles penetrate less deeply.

Stem thickening. As the plant grows the axis must thicken, in order to provide an increasing number of conducting elements for the transport of water, mineral substances and organic materials between the roots and leaves, as well as generating more mechanical tissue for the increasing weight of the aerial parts of the plant. Thickening occurs by *primary thickening* and *secondary thickening.* Primary thickening occurs in the direct vicinity of the meristem, and it continues for only a limited period. Plants that display exclusively primary thickening (the majority of monocotyledons and certain other plants) therefore have a slender shape. Primary thickening in monocotyledons occurs by the periclinical division of a mantle-like meristem between the tunica and corpus. In many monocotyledons (e.g. cereal species) this meristem is not very active, and the stem remains thin. In palm trees, on the other hand, large quantities of tissue are formed by the meristem mantle. A short distance from the apical meristem, the meristem mantle does not divide or undergoes very limited cell division, so that those actively dividing parts furthest from the apex extend upwards, and the actual meristem lies in a crater-like depression or groove. The sometimes considerable diameter attained at this stage is then retained without further increase throughout the life of the monocotyledonous plant. In dicotyledons, primary thickening does not involve a mantle-like meristem; it occurs rather by the irregular division of parenchymatous pith cells (medullary thickening, e.g. the thickening of kohlrabi, celeriac and potato tubers) or of parenchyma cells in the primary cortex (cortical thickening, e.g. in *Cactaceae* and to a certain extent in potato tubers). As in monocotyledons, primary thickening in dicotyledons can be so extensive that an apical depression is formed, e.g. the "eyes" of a potato represent such a depression.

Secondary thickening is a further type of thickening that occurs in dicotyledons and gymnosperms, and is not initiated until primary thickening is near completion. It occurs by division of a peripheral meristem, the *cambium,* which is active throughout the life of the plant. Cambium consists of slender elongated cells, and it is derived from the original meristem, i.e. it is the residue of the procambium of the apical region, which has become transformed into *fascicular cambium.* It lies between the xylem and phloem of the open collateral bundles. Before secondary thickening commences, however, cells in the pith rays between the vascular bundles are transformed into cambium (i.e. they become meristematic) so that a continuous cylinder of cambium is formed. These newly formed "gap-filling" cambial cells represent a secondary meristem, known as *interfascicular cambium.*

The resulting closed cylinder of cambium separates the central wood from the surrounding mantle of phloem and cortex, and it divides to form new cells on both its outer and inner surfaces. In particular, this cell division and differentiation leads to the production of new vascular tissue. New xylem is added to the existing xylem so that the vascular bundles become wider. In the region of the interfascicular cambium, parenchymatous pith ray tissue is formed. The primary medullary rays linking the pith to the cortex become elongated in accordance with the increased girth of the stem, but they do not become wider. At certain sites the fascicular cambium forms parenchyma cells instead of xylem or phloem; these form the secondary medullary rays, which end blindly in xylem or phloem. All tissues elements that divide off from the inner surface of the cambium, irrespective of their type and the nature of their differentiation, are known as Wood (see). All those produced from the outer surface of the cambium are known as Bast (see). Continued formation of wood pushes the cambium ring ever further outward. The cambium (as well as the outer layer of bast) must therefore undergo *dilatation,* i.e. it must continually increase its circumference by anticlinal cell division. As a result of strong secondary growth the epidermis and the primary cortex are usually split and often destroyed. The primary boundary layer must then be replaced by a secondary boundary layer, known as cork or periderm (see Boundary tissue), which arises from a special secondary meristem, the *cork cambium* or *phellogen.*

Stem branching. Some plants, especially annuals, may remain unbranched. When branching does occur, it is of two main types: *dichotomous branching (Lycopodiatae* and some related pteridophytes) and *lateral* or *axillary branching* (the remaining pteridophytes and all spermatophytes). In dichotomous branching the shoot apex divides into two new apices; these may be equal *(isotomous* dichotomous branching) or unequal *(anisotomous* dichotomous branching) in size and status. In anisotomous branching, if the growth of the smaller apex is less than that of the larger apex, the plant superficially appears to display lateral branching.

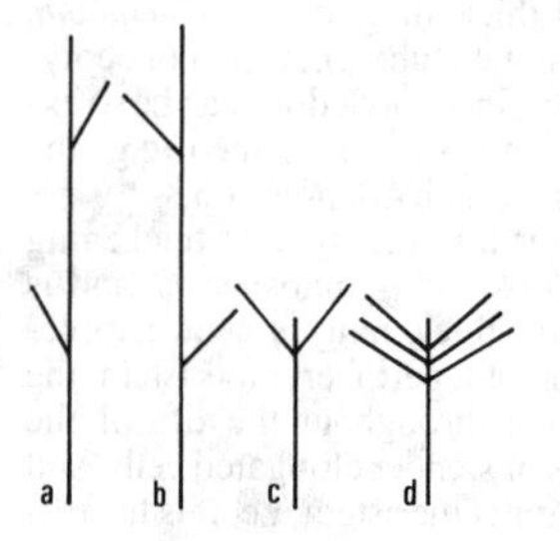

Stem. Types of branching. *a* Monopodium. *b* Sympodium. *c* Dichasium. *d* Pleiochasium.

Lateral or axillary branching occurs from lateral (axillary) buds, which normally arise as peripheral (exogenous) outgrowths of the parent axis at the bases of leaf primordia; like the latter, they form a sequence toward the apex, i.e. an acropetal sequence. The leaf in whose axil the side shoots occur (axillary shoots) is known as the subtending leaf or bract. By intercalary growth in the basal tissue, the individual buds may come to lie much higher on the stem than their subtending leaves (concaulescence). Alternatively, the base of the subtending leaf may grow beneath the axillary bud so that the bud is carried out with the leaf, and the axillary shoot is actually on the subtending leaf (recaulescence).

If the side axes remain subordinate to the main axis, the branching system is *monopodial.* This type of branching system possesses a true, uniform main axis, known as a *monopodium* (e.g. poplar and various conifers such as spruce and fir). If the lateral shoot grows more strongly than the main axis, the branching system is *sympodial.* In *monochasial branching*, the main axis is pushed aside by a single lateral shoot, which then becomes dominant and starts to branch, forming a *monochasium.* The side branches of the monochasium overtop their parent axes, and a main axis *(sympodium)* appears to be present, which actually consists of the side branches of different orders. If this apparent axis is straight, the result is a sympodial system with a supficial resemblance to a monopodial system [e.g. grape vine *(Vitis vinifera),* lime tree *(Tilia),* hornbeam *(Carpinus betulus)].* If branching occurs by two lateral branches of the same order (situated more or less opposite each other), the system becomes a *dichasium,* which superficially resembles dichotomous branching [e.g. mistletoe *(Viscum),* buckthorn *(Rhamnus cathartica),* lilac *(Syringa vulgaris)].* If three or more axillary buds produce mutually equivalent side shoots while the extension of the main axis is suppressed, the result is a pleiochasium (e.g. the cyathia of *Euphorbia* spp.). Lateral branches vary greatly in their development. Thus, *long shoots* develop toward the apex of fruit trees; these are lateral shoots with extended internodes and usually unlimited capacity for growth. *Short shoots* develop in the intermediate region of the tree; these are lateral shoots with compressed internodes and a limited capacity for extension growth. In many fruit trees (e.g. apple, pear, cherry) flowering is restricted to the short shoots. *Renewal shoots* or *innovations* arise if the primary shoot stops growing due to flower formation or exhaustion.

Modifications of the stem. The stems of some plants display morphological and functional adaptations. For example, in dry climates, leaves may be absent and photosynthesis is performed entirely by the chlorophyll-containing stem. *Cladodes* or *cladophylls* are a further adaptation to dry habitats; these are modified lateral shoots, which take over the role of photosynthesis. They may have a needle-like shape (e.g. asparagus), or they may be flattened and leaf-like *(platycladodes).* Succulents (e.g. *Cactaceae)* have fleshy, thickened stems, which store water.

The *Rhizome* or *rootstock* is a modified stem, which serves for the storage of reserve materials. It is usually a persistent, subterranean, sometimes climbing shoot axis with short, thickened internodes (e.g. asparagus, iris). *Stem tubers,* found on certain annuals (e.g. kohlrabi), are formed as a result of strong internodal compression and robust thickening. In the tuber of the cyclamen, the hypocotyl is also involved in the thickening. Purely *hypocotyl tubers* are those of the beetroot and certain varieties of radish.

Tuberous swellings at the ends of runners are known as *stoloniferous tubers,* e.g. potato tubers and the tubers of chufa *(Cyperus esculentus)* and Chinese artichoke *(Stachys sieboldi).* The storage organs known as *bulbs* (e.g. onion, hyacinth) consist of a disk-like flattened axis with strongly compressed internodes (corresponding to a corm or bulbotuber), into which are inserted fleshy leaves. *Stem thorns* are short, modified shoots [e.g. sloe or blackthorn *(Prunus spinosa),* sea buckthorn *(Hippophae),* furze *(Genista),* spiny restharrow or thorny ononis *(Ononis spinosa)],* and like normal branches they are borne in the axils of leaves. Branched, three-pointed thorns occur on the honey locust *(Gledits(ch)ia).* Occasionally, stem thorns develop small, usually deciduous leaves, thus providing further evidence that they are modified stems. Thorns may also have an assimilatory function. *Tendrils* are modified lateral shoots, which are sensitive to contact stimulus and enable the climbing shoot to become attached to supporting structures (e.g. grape vine and passion flower).

The stems of parasitic plants and xerophytes are often greatly reduced, and foliage leaves may also be greatly reduced or absent.

Stem mother: see Fundatrix, *Homoptera.*

Stem sawflies, *Cephidae:* a family of the insect order *Hymenoptera* (see), of the suborder *Symphyta* (plant wasps). Adult insects are 6 to 15 mm in length, and usually patterned black and yellow. The larvae live in grass halms, in the pith of shrubs, and occasionally in woody twigs. Larvae of the European ***Wheat stem sawfly*** *(Cephus pygmaeus* L.) attacks especially rye and wheat, as well as barley and wild grasses; it is widely distributed in Europe, North Africa and the Near East, and as an introduced species it has caused increasing damage in the USA, east of the Mississippi, since 1887. The ***American wheat stem sawfly*** *(Cephus cinctus)* is a severe pest in grain-growing areas of Canada.

Stenella: see Whales.

Stenidae: see Whales.

Steno: see Whales.

Steno-: a prefix denoting low tolerance of changes in the environment, i.e. a close dependence on the constancy of certain environmental factors. Thus, stenohaline species tolerate only a narrowly defined salt concentration. See Eury-.

Stenodelphinidae: see Whales.

Stenoecious conditions: see Ecological valency.

Stenolaemata: see *Bryozoa.*

Stenopodidea: Shrimps (see), an infraorder of the *Decapoda* (see).

Stenopodium: see Biramous limb.

Stenopus: see Shrimps.

Stenostomata: see *Bryozoa.*

Stenothermal: see Temperature factor.

Stenotopic species: see Ecological valency.

Stenson's duct: see Vomeronasal organ.

Stentor: see *Ciliophora.*

Stephanian stage: see Carboniferous.

Stephen Island wren: see *Xenicidae.*

Stephens wren: see *Xenicidae.*

Steppe: a treeles or almost treeless vegetation formation dominated by grasses, which together with tall herbaceous perennials, tuber and bulb geophytes and annuals, form a more or less continuous cover. It occurs in continental and subcontinental regions with dry summers and low annual precipitation (400–500 mm). In the Eurasian Ss. which have formed on substrata of loess, the soil has developed into a highly fertile black earth. The main growth period is spring and early summer. Late summer and autumn are dry and dormant. Ss. occur on the mountain plains of Asia *(mountain S.)* and on the plains and rolling hills of southern Siberia, Mongolia and Ukraine. extending into central Europe. The corresponding formations in North America are known as *prairies,* and in South America (in particular Argentina) as *pampas.* With increasing rainfall and increasing temperature from the north to south of eastern Europe, the vegetation is zoned, starting with *forest steppe* in the north (islands of forest, consisting of pendunculate oak, elm and wild fruit trees, separated by areas of S.) which merges into a brightly spring-flowering, species-rich *meadow S.* This is bordered in the south, where conditions are drier, by *feather grass S.,* consisting of grass thickets of *Stipa* and *Festuca,* very few herbaceous perennials, but abundant annuals. Even further south are the *short grass Ss.* with still fewer species, consisting of xerophytic fescues *(Festuca)* with spring annuals, mosses and lichens between the grass thickets. This finally merges into *desert S.* (semidesert). As islands of extrazonal vegetation (see Zonal vegetation), the eastern European Ss. extend far into central Europe (see Steppe heath). The mountain Ss. of the Near East and central Asia lie at altitudes of 3 000 m and higher, and contain a rich variety of species.

Steppe cat: see *Felidae.*

Stepped series: see Form series.

Steppe heath: a practically treeless vegetation formation in dry areas of central Europe, consisting mostly of continentally and subcontinentally distributed grasses and herbaceous plants. Flat, stony regions colonized by species of Mediterranean/sub-Mediterranean origin are known as *stone heaths.* The floristically richest dry habitat plant communities of central Europe are found on S.hs. See Steppe.

Steppe heath theory: the theory that during the post glacial climatic optimum (see Pollen analysis) predominantly treeless areas of steppe heath or lightly wooded steppe provided the first opportunity for permanent settlement by neolithic humans. This is based on the observation that the distribution of Early Stone Age and Bronze Age settlements largely corresponds to the occurrence of steppe heath. As the climate became moister and more favorable for tree growth, invasion of these open areas by forest was actively prevented by humans.

Stepping stone model: a model of population genetics for investigating the results of Gene flow (see). It is assumed that individuals of each generation migrate backward and forward only between neighboring populations. The one dimensional S.s.m., in which each population exchanges members only with two neighboring populations, is particularly easy to interpret.

Sterane: see Steroids.

Stercobilin: see Bile pigments.

Stercobilinogen: see Bile pigments.

Stercoraridae, ***skuas*** or ***jaegers:*** a small avian family in the order *Charadriiformes* (see), consisting of 4 species of strong-flying, largely pelagic birds. Closely related to gulls, they are distinguished from these by possession of a fleshy cere across the base of the upper mandible, which houses the opening of the nostrils. The ***Great skua*** *(Stercorarius (= Catharacta) skua;* 58 cm; uniformly dark brown with white patch at the base of the flight feathers) has a circumpolar distribution on the fringes of Antarctica and on sub-Antarctic islands, as well as the coasts of southern Chile and New Zealand. It also occurs, but is less numerous, at the other end of the globe around Iceland, the Faroes and the Shetland Islands, overwintering at sea further south down to latitude 35 ° north. Populations in the Southern Hemisphere move north in their winter in all the southern oceans, also crossing the equator into the North Pacific. Northern and southern populations display only slight differences in color and size, and there is no evidence that they meet or interbreed. Much food is obtained by harassing other sea birds to disgorge their catch, and during the breeding season they feed themselves and their brood on the eggs and nestlings of other sea birds such as petrels, shearwaters and penguins. They also scavenge the shore line for dead animal remains, and they are fond of offal thrown overboard from fishing boats.

The remaining 3 species are: ***Pomatorhine skua*** or ***Pomarine jaeger*** *(Stercorarius pomarinus;* 51 cm), ***Arctic skua*** or ***Parasitic jaeger*** *(Stercorarius parasiticus;* 45.5. cm), and ***Long-tailed skua*** or ***Long-tailed jaeger*** *(Stercorarius longicaudis;* 56 cm), all with a circumpolar distribution on Arctic coasts and islands, overwintering further south in the Atlantic and the Pacific, often well into the Southern Hemisphere. Easily recognizable from their elongated central tail feathers, all are powerful, fast-flying predators, with scavenging and piratical dietary habits similar to those of the Great skua. All have 2 phases of plumage color: uniformly gray-brown with some pale yellow around the cheeks and neck (dark phase), or with the addition of white underparts (light phase).

Stercorarius: see *Stercoraridae*.

Sterculiaceae, *cocoa family:* a family of dicotyledonous plants, containing about 1 000 exclusively tropical species. They are mostly trees and bushes (usually with soft wood), rarely herbaceous plants, with simple or digitately compound leaves. The flowers are usually actinomorphic, hermaphrodite and 5-merous, with numerous stamens fused into a tube. The ovary contains 2–5 (rarely 10–12) united or free carpels (rarely reduced to 1 carpel), developing into a dry or rarely fleshy indehiscent or variously dehiscent fruit.

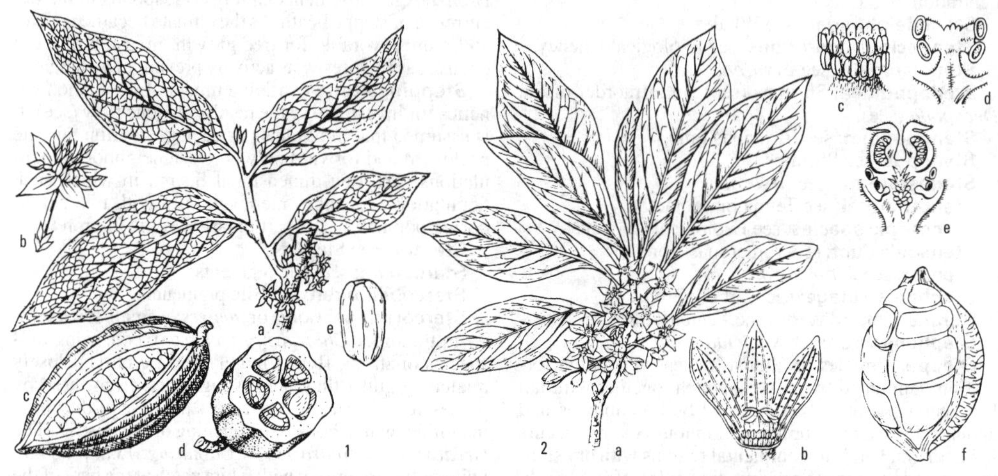

Sterculaceae. 1 Cocoa tree, *a* flowering shoot, *b* flower, *c* opened fruit, *d* cross section of fruit, *e* seed. *2* Cola tree, *a* flowering shoot, *b* female flower, *c* stamen tube of male flower, *d* longitudinal section of male flower, *e* longitudinal section of female flower, *f* fruit with seeds (cola nut).

Economically the most important member is the ***Cocoa tree (Theobroma cacao),*** with stem-borne flowers and large, melon-sized, indehiscent fruits, containing numerous seeds (cocoa beans). The large cotyledons of cocoa beans contain, inter alia, a fat known as cocoa butter, and the stimulatory alkaloid, theobromine. The fat is partially expressed from the beans, and the residue yields cocoa powder. Indigenous to Central and South America, the Cocoa tree is now cultivated in many parts of the tropics, especially West Africa.

Also economically important are the native West African ***Cola trees,*** *Cola acuminata* and *Cola nitida,* which are also cultivated in other tropical countries, such as the West Indies, Brazil, Java, etc. Their seeds (cola nuts) contain 1.5–2.5% of the alkaloid, caffeine. Cola nuts are chewed fresh by African natives. They are also used to some extent pharmaceutically as a nerve stimulant, but their main use is in the preparation of refreshing drinks.

Stereolepis: see Sea basses.

Stereomicroscope: a microscope that uses two objectives, two tubes and two eyepieces. Prisms between the objectives and eyepieces reverse the image. The distance between the objectives and specimen is relatively large, the specimen is illuminated form the side or from above, and the image is three-dimensional. The S. provides low magnification (about 100 x), and is used for the observation and manipulation of small animal and plant specimens.

Stereotyped movements: periodic uniform patterns of movement, observed in higher vertebrates, which are interpreted as substitute movements. They occur as a result of the limitation of movement (e.g. in caged animals), or the absence of environmental factors that are required for normal movement (e.g. the absence of earth for species that normally burrow). S.m. therefore display a strong linkage to the environment of the animal. Common S.m. are head movements, body movements (e.g. rocking), movements of the extremities without locomotion (turning round repeatedly on one spot), stereotyped walking backward and forward on the same stretch of ground, often with dragging of the hindquarters.

Stereum: see *Basidiomycetes (Poriales).*

Sterigma (plural **Sterigmata):** in certain fungi, a cell extension or stalk from which conidia are abstricted, e.g. in many *Phycomycetes* and in the *Basidiomycetes* (see). As distinct from a Phialide (see), an S. is not an entire cell, but only part of a cell. Nevertheless, the two terms are often used synonymously.

Sterile:

1) Free of living microorganisms.

2) Infertile and therefore incapable of producing offspring.

Sterility, *infertility:* the partial or complete inability of an individual to produce functional gametes, and in the wider sense viable zygotes, under the existing environmental conditions. Genetically determined S. can be due to sterility genes *(genic S.)* or the inadequate homology (or total absence of homology) of the respective sets of chromosomes *(chromosomal S.).* The *somatoplastic S.* observed in plants is due to incompatibility between endosperm and embryo, which is expressed as abnormal seed development and embryo abortion. For further information on S., see Intersterility, Hybrid sterility, Self sterility.

Sterilization: removing or killing microorganisms and their spores and the inactivation or removal of viruses, e.g. on medical instruments, medical dressings, culture media for microorganisms and canned foods. S. can be performed by different physical and chemical methods. These can be broadly classified as either heat or cold S.

1) *Heat sterilization* is performed: a) by dry heating for two hours at 160–180 °C in a hot air sterilizer, e.g. for S. of glass apparatus; b) steam S. (20–60 min at 100 °C in a steamer). Since this does not kill all bacterial spores, the procedure is often perfomed in stages (Tyndallization), i.e. 20 min at 100 °C on three consecutive days, e.g. for the S. of unstable nutrient media; c) by autoclaving (see Autoclave) (10–20 min in superheated steam at 120–135 °C) for the S. of relatively heat-stable materials, e.g. canned meat. Ultrahigh heating (see) is carried out at even higher temperatures; d) by heating in a flame until the article glows, or by dipping in ethanol and burning off, e.g. inoculation loops, inoculation needles, scissors, knives, etc.

2) Cold sterilization serves primarily for the S. of heat-sensitive apparatus and materials, as well as for the S. of gases. It is performed: a) chemically by the action of ethylene oxide, formaldehyde, ethanol, β-propiolactone, mercuric chloride and hydrohalides, for S. of rooms, apparatus, plastic articles, living organisms, tissues, etc.; b) by physical methods by irradiation with UV-light, X-rays and γ-rays, for S. of apparatus, sterile cabinets and foods; c) by filtration (see Filtration sterilization). See Disinfection.

Sterlet: see *Acipenseridae.*

Sterna: see *Laridae.*

Sterninae (terns): *see Laridae.*

Sternoptychidae: see *Stomiiformes.*

Sternopygidae: see *Gymnotiformes.*

Sternorrhynchi: see *Homoptera.*

Sternotherus: see *Kinosternidae.*

Steroid hormones: an extensive group of animal Steroids (see), which are derived biosynthetically from cholesterol, and which function as hormones. According to the number of C-atoms in their molecule, the S.h. are classified as: 1) C_{27}-steroids (see Ecdysone, Ecdysterone); 2) C_{21}-steroids (see Progesterone, Adrenal corticosteroids); 3) C_{19}-steroids (see Androgens); 4) C_{18}-steroids (see Estrogens).

According to their biological activity they are grouped as mammalian Glucocorticoids (see), Mineralocorticoids (see), Androgens (see), Estrogens (see) and Gestagens (see), or they are insect molting hormones (see Insect hormones).

Human S.h. are biosynthesized mainly in the adrenal cortex and the gonads, a process that starts with the oxidative side chain cleavage of cholesterol. Subsequent key intermediates are pregnenolone and progesterone. Unlike peptide hormones, S.h. are not stored in the organism, but released as they are synthesized. They are transported in the circulation, largely bound to carrier protein. The hormone-specific receptors of S.h. are not present in the cell membrane of the target organ (cf. adrenalin, insulin, etc.), but within the cell. These receptors appear to be in the cytosol, but there is now strong evidence that they are in fact localized in the nucleus and readily released into the cytosol by the laboratory methods (cell homogenization and high speed centrifugation) used for their preparation. S.h. are degraded in the liver, and their metabolites are excreted in the urine.

Artificially synthesized S.h., which represent structural modifications of the natural S.h., are used as pharmacologically as ovulation inhibitors, anabolic steroids and anti-inflammatory agents.

Steroids: a large group of terpenoids, including many biologically important compounds, e.g. sterols, bile acids, cardiac glycosides, steroid saponins, steroid alkaloids and steroid hormones, as well as pharmacologically important synthetic S., such as ovulation inhibitors, anaboloic steroids and various structurally modified steroid hormones.

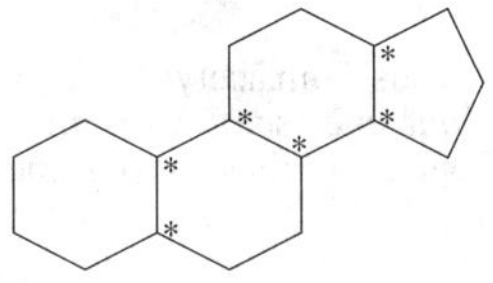

Fig.1. *Steroids.* Cyclopentanoperhydrophenanthrene, the parent ring structure of the steroids.

Structurally, S. are derivatives of cyclopentanoperhydrophenanthrene, which contains 6 asymmetrical centers (marked with asterisks in Fig.1). There are thus $2^6 = 64$ theoretically possible stereoisomers, but relatively few of these are known among the S. The nomenclature of the tetracyclic ring structure and the numbering of the carbon atoms are shown in Fig.2. A number of parent hydrocarbons are defined by the substituents at C-atoms 10, 13 and 17 (Table). The simplest member is gonane (formerly called sterane) in which rings B/C and C/D are *trans* to one another, as in the majority of natural S. The configuration at C-5 depends on the *cis-trans* relationship of rings A and B; thus 5α-gonane is A/B *trans,* and 5β-gonane is A/B *cis.* With substitution at C-13, the nomenclature becomes 5α- or 5β-estrane. Introduction of a further methyl group at C-10 gives rise to 5α-androstane (formerly called testane) or 5β-androstane (formerly called etiocholane), etc. Both of the angular methyl groups are present in most natural S. Sterically, they are above the fixed plane of the ring system (Fig.3); all such atoms or groups above the plane of the molecule are designated by the prefix β. In drawn or printed formulas, the bond linking a β-methyl group to the ring carbon is shown as a solid line. Bonds to α-groups or atoms (i.e. below the plane of the ring) are shown as dotted lines. Structural differences among S. arise from the type, number, position and configuration of the substituents, as well as the number and position of double bonds.

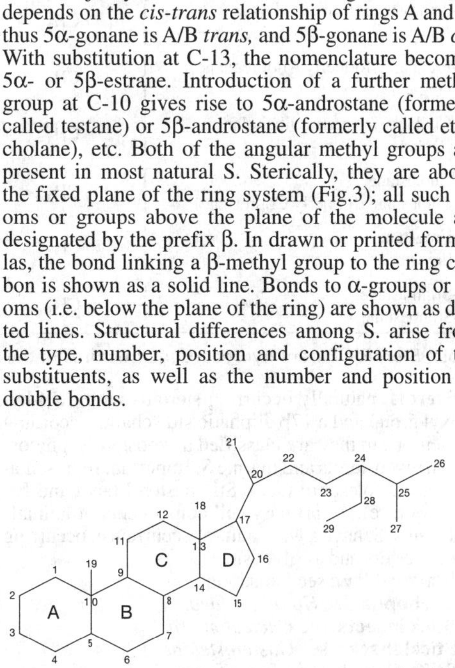

Fig.2. *Steroids.* Ring nomenclature and numbering of carbon atoms in the steroid ring system and its side chains.

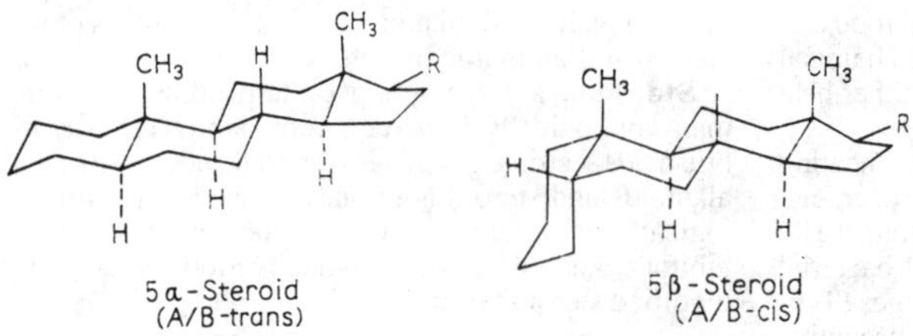

5 α-Steroid (A/B-trans)

5 β-Steroid (A/B-cis)

Fig.3. *Steroids.* Conformations of the steroid molecule.

The biosynthesis of S. proceeds from acetyl-CoA via the open chain triterpene, Squalene (see). More than 20 000 S. are known, and about 2% of these have some medical significance.

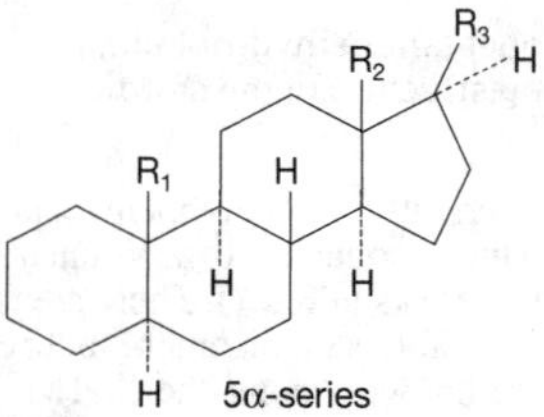

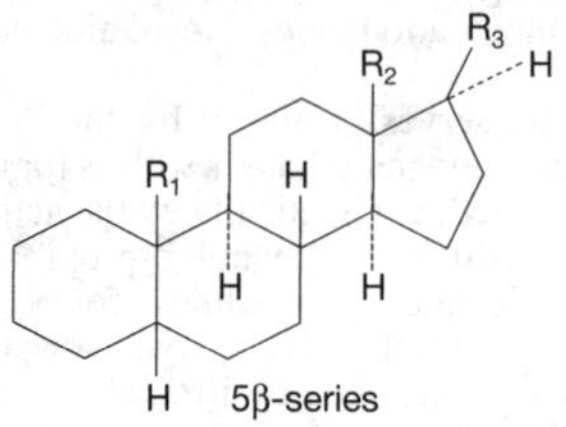

5α-series

5β-series

Name	R_1	R_2	R_3
5α-Gonane (formerly Sterane) 5β-Gonane	H	H	H
5α-Estrane 5β-Estrane	H	CH_3	H
5α-Androstane (formerly Testane) 5β-Androstane (formerly Etiocholane)	CH_3	CH_3	H
5α-Pregnane (formerly Allopregnane) 5β-Pregnane	CH_3	CH_3	C_2H_5
5α-Cholane (formerly Allocholane) 5β-Cholane	CH_3	CH_3	$CH(CH_3)CH_2CH_2CH_3$
5α-Cholestane 5β-Cholestane (formerly Coprostane)	CH_3	CH_3	$CH(CH_3)CH_2CH_2CH_2CH(CH_3)_2$
5α-Erogstane 5β-Ergostane	CH_3	CH_3	$CH(CH_3)CH_2CH_2CH(CH_3)CH(CH_3)_2$
5α-Stigmastane 5β-Stigmastane	CH_3	CH_3	$CH(CH_3)CH_2CH_2CH(C_2H_5)CH(CH_3)_2$

Table. *Steroids.* Parent hydrocarbon structures of the steroids.

Sterols: naturally occurring steroids with a 3β-hydroxyl group and a 17β aliphatic side chain. According to their origin they are classified as zoosterols, phytosterols, mycosterols and marine S. Important representatives are Cholesterol (see), Stigmasterol (see) and Ergosterol (see). As primary cell constituents of animals and plants, S. have a wide natural occurrence, occurring free, as esters and as glycosides.

Stevens' law: see Sensation.

Stichopus: see *Holothuroidea.*

Stick insects: see *Cheleutoptera.*

Sticklebacks: see *Gasterosteidae.*

Stifftail ducks: see *Anseriformes.*

Stifftails: see *Anseriformes.*

Stigmastane: see Steroids.

Stigmasterol: 5α-stigmasta-5,22-diene-3β-ol, a widely distributed phytosterol, first isolated from calabar beans and later from many other source, e.g. soybeans, sugar cane wax, carrots and coconut oil. It is used as a starting material in the technical synthesis of steroid hormones.

Stigmasterol

Stigmata (singular **Stigmatum**), *spiracles:* the openings of the tracheae that are generally situated in the lateral body integument of onychophorans, arachnids, myriapods and insects. The simplest S. lead directly into the trachea. In most cases, however, this primary opening is below the surface of the integument, and preceded by a chitinous atrium or tracheal chamber. The atrium may be highly sclerotized, and in most insects it is furnished with a closure mechanism which prevents the entry of foreign material. Practically all S. are supplied with special muscles, which control the rate of ventilation. In the *Orthoptera,* a directional air stream can be generated in the tracheal system by the controlled opening and closing of S. In the primary condition, S. are arranged segmentally, every body segment possessing one pair (except in onychophorans). Thus most myriapods have one pair of S. on each leg-bearing segment, except the first and the last. In the basic body plan of insects, thoracic segments 2 and 3, and the first 8 abdominal segments each carry a pair of S. (holopneustians). Frequently, however, individual pairs of S. are closed or reduced (hemi- and hypopneustians). In aquatic insect larvae, all the S. may be closed (branchiopneustians with tracheal gill respiration). In yet other insects, the S. are located at the ends of specialized body appendages *(siphons),* which enable respiratory gas exchange at the water surface, e.g. mosquito and hoverfly larvae, and water bugs. Larvae of the sedge beetle have specially modified S., with which they can bore into the air-filled intercellular spaces of water plants. In arachnids the arrangement and number of S. is variable. They occur as paired openings mainly near to the roots of the legs and mouthparts, or as paired and unpaired openings on the underside of the abdomen, e.g. in spiders. Harvestmen *(Opiliones)* have secondarily evolved S. on their long legs.

Stilts: see *Recurvirostridae.*

Stimulus (plural **Stimuli**): an event in the environment of a living system, which leads to a change in the state of the system. Stimuli cause excitation in nerve cells, or change existing forms of excitation. Traditionally stimuli are classified as chemical, osmotic, thermal, mechanical, electrical, acoustic or optical. Receptors (see) are generally specialized for the detection of a specific type of stimulus or signal, e.g. photoreceptors in the eye. Under certain circumstances, specific receptors may be stimulated by an inappropriate S., such as pressure. A S. is characterized by certain quantities, e.g. intensity, duration, rate of increase in intensity, changes of intensity during stimulation of the receptor, polarity of an electrical S. In order to stimulate a receptor, the stimulus must attain a threshold intensity. A sudden pulse of direct current has a much lower threshold value than a slowly increasing direct current.

Stimulus summation rule: a rule of behavioral biology that the actions of Sign stimuli (see) are summated in the initiation of species-specific behavior. In the simplest case, the activities are simply additive, but more frequently the summation is more complex. In voles it has been shown that the inhibitory action of visual and acoustic stimuli on reconnaissance activity is almost purely additive.

Stimulus threshold: the minimal intensity of a stimulus that must be applied to a receptor, in order to cause an excitation. The *absolute S.t. (threshold intensity)* is the minimal threshold intensity per unit time that must be applied to achieve excitation. In the dark-adapted human eye, the absolute S.t. for blue light (507 nm) is 5.6×10^{-17} *J*/s. For excitation to occur, this threshold intensity must have a minimal duration, or *effective period,* which in the above example is 0.5 s. From this it can be calculated that the threshold energy of the stimulation is about 3×10^{-17} *J*. The calculated absolute S.t. of single sensory cells sometimes approaches the limit of physical possibility. Thus, one quantum of light is sufficient to excite a single rod, and one molecule of an odiferous principle will excite a single olfactory cell.

Stinging hairs:

1) see Boundary tissue, Plant hairs, *Urticaceae.*

2) Glandular hairs of lepidopteran larvae. They contain an irritant liquid that causes an inflammatory reaction when the hairs penetrate human or animal skin.

Stinging nettle: see *Urticaceae.*

Stink bird: see *Opisthocomiformes.*

Stink flies: see *Neuroptera.*

Stinkhorn: see *Basidiomycetes (Gasteromycetales).*

Stinking smut: see *Basidiomycetes (Ustilaginales).*

Stints: see *Scolopacidae.*

Stipiturus: see *Sylviinae.*

Stirrup of the middle ear: see Auditory organ.

Stizostedion: see Pike-perch.

Stoat: see Ermine.

Stochastic variable: see Variate.

Stoichactis: see *Actiniaria.*

Stolephoridae: see *Engraulidae.*

Stolidobranchiata: see *Ascidiacea.*

Stolons:

1) Aerial hyphae that normally grow horizontally over the growth substrate, and at their points of contact form rhizoids and tufts of sporangiophores. S. are characteristric of the fungal genus, *Rhizopus.*

2) Plagiotropic (horizontal) *lateral shoots* with greatly extended internodes and often markedly reduced leaves. They become rooted at some distance from the parent plant, leaving a new and independent plant when the intermediate stolon dies. They therefore serve for vegetative reproduction. Aerial S. are formed by strawberry plants, white clover and foxglove. Subterranean S. are formed by potato plants, couch grass, valerian and many perennial meadow grasses.

3) A rootlike structure that attaches colonial invertebrates to the substratum, and from which new individuals may be produced by budding, e.g. in bryozoans and ascidians. See Stolo prolifer.

Stolo prolifer: the stolon of salpids (see *Salpida)* and sea squirts (see *Ascidiacea),* which develops as an outgrowth of the oozoid. Buds are formed asexually from the S.p. This type of reproduction is usually one stage in a complicated alternation of generations.

Stoma: singular of Stomata (see).

Stomach: see Digestive system.

Stomachus: see Digestive system.

Stomata (singular **Stoma**): closable openings in the epidermis of plants. The opening (cleft, pore or stoma) is bordered by two chloroplast-containing guard cells, which are often bean-shaped. S. serve for the exchange of gases between the outer atmosphere and the intercellular system of the plant, and for the passage of water vapor from the leaf tissue (see Transpiration). Usually, the two guard cells are surrounded by subsidiary cells, which are distinct from the other epidermal cells. The adjoining walls of guard cells and subsidiary cells are particularly thin and elastic, so that the guard cells are effectively "hinged" to the subsidiary cells. Guard cells and subsidiary cells comprise the *stomatal apparatus.*

The appearance and mechanism of the stomatal apparatus varies according to the shape of the guard cells and the relative thicknesses of their cell walls. There are three main types. a) The *Amaryllidaceae-type* or *Helleborus-type* is found in many monocotyledonous and dicotyledonous plants. On the side adjacent to the stoma (ventral wall) each guard cell has an upper and lower ridge of thickening, whereas the opposite side (dorsal wall) is unthickened and elastic. An increase of turgor pressure within the guard cell pushes the thin dorsal wall back against the subsidiary cells. The nonelastic ventral wall is thereby drawn back and the stoma opens. The junction of the guard cells and subsidiary cells usually behaves as an attenuated hinge, so that as the stoma opens it also moves slightly above or below the level of the epidermal surface. b) The *Graminae-type* occurs primarily in members of the *Graminae (Poaceae)* and *Cyperaceae.* In this type, the guard cells are dumb-bell shaped. The widened ends are thin-walled, whereas the upper and lower walls of the narrow intervening region are powerfully thickened. As turgor pressure increases within the guard cells, the thin-walled ends expand, causing the thickened central regions to move apart. c) *Mnium-type* S. are characteristic of mosses and ferns; the ventral walls of the bean-shaped guard cells are always thin, whereas the dorsal, inner and outer walls are thickened. As the turgor pressure within the guard cells increases, the outer and inner walls move away from each other, the cells rise toward the surface of the epidermis and become less convex around the stoma, and the stoma opens. A decrease in turgor pressure causes the reverse sequence of events, and the stoma closes. Throughout all these changes, the dorsal wall remains in more or less the same position.

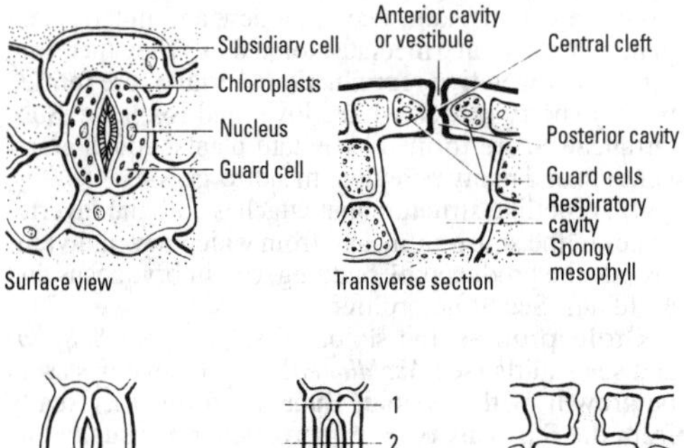

Stomata. Stomatal apparatus (above) and types of stomata (below).

Opening and closing of S. is a nastic processes (see Stomatal movements) due primarily to changes in turgor within the guard cells. Changes in turgor are caused by several different stimuli, such as light, water tension, temperature changes and chemical influences. Loss of water vapor from the plant (transpiration) and the rate of diffusion of CO_2 into the plant for photosynthesis are both regulated by adjustment of stomatal apertures, and this adjustment occurs largely in response to water tension (hydronastic or hygronastic effect) and light (photonastic effects).

Stomatal apparatus: see Stomata.

Stomatal movements: nastic movements of the guard cells of stomata, which are brought about by changes in turgor pressure within the guard cells, and which are responsible for the opening and closing of stomata. The opening and closing of stomata is extremely important in the regulation of transpiration and photosynthesis. Increased turgor pressure within the guard cells leads to stomatal opening, whereas a decrease leads to closure. Various stimuli are capable of activating S.m. 1) Light (photonastic S.m.): photonastic opening in response to moderate illumination; closure in the dark. 2) Water supply (hygronastic S.m.): even a small water deficit in the leaf causes hygronastic stomatal closure, whereas water saturation causes stomatal opening. 3) Temperature. 4) Chemical stimuli. Thermonastic and chemonastic reactions are less important. Under natural conditions, the size of the stomatal aperture is determined by the interplay of the effects of light and water tension. According to the classical concept, the immediate cause of stomatal opening is the decrease in CO_2 concentration in the assimilatory parenchyma and guard cells, as well as in the intercellular spaces, or the accompanying increase in pH of the guard cells, which occurs when photosynthesis begins in response to illumination. By affecting the equilibrium position of starch phosphorylase (which is very pH-dependent between pH 6.4 and 7.3), high pH values are thought to favor an increase in the degradation of starch to glucose 1-phosphate, which in turn causes an increased turgor pressure. Indeed, starch grains that accumulate in the guard cells during the night, disappear as the guard cells subsequently reopen as a result of being illuminated on the next day. It is doubtful, however, whether the removal of CO_2 can sufficiently alter the pH of the well buffered cells. Since open guard cells contain a much higher concentration of potassium ions than closed guard cells, it is now widely accepted that the changes in turgor pressure are caused by a light-induced activation of potassium ion pumps. Since transport of potassium ions from neighboring cells into the guard cells decreases the potassium ion concentration of the former, the differences in turgor pressure required for stomatal opening could occur rapidly in this way. In the dark, the above series of events of would be reversed. In addition, it is also thought that guard cells detect the CO_2 concentration of the adjoining intercellular space, closing when this is above a certain threshold value (about 0.05%) and opening when it falls below this value. In fact, stomata can be made to open experimentally in the dark by removing CO_2, and made to close in the light by introducing CO_2. Among the phytohormones, S.m. are influenced in particular by abscisic acid; an increase in abscisic acid concentration causes stomata to close.

If too much water vapor is being lost from the plant, hygronastic stomatal movements lead to closure of stomata, even in the presence of light; the water potential of the guard cells falls below a certain threshold value, so that the turgor pressure falls, even though the osmotic value may remain unchanged. Such stomatal movements, in which the amount or nature of dissolved material in the guard cells does not change, are known as *hydropassive (hygropassive)* movements. Usually, however, hydronastic stomatal movements are *hydroactive (hygroactive),* i.e. due to quantitative or qualitative changes in the concentration of dissolved material in the guard cells, e.g. the closure of S. during drought stress, due to a rapid increase in the concentration of abscisic acid. Phaseic acid, *all-trans*-farnesol and certain other substances have the same effect as abscisic acid, but this is possibly because they are converted into abscisic acid.

Photonastic and hygronastic S.m. are not due simply to alterations of water content or turgor pressure, but to turgor differences between the guard cells and the neighboring (subsidiary) cells of the epidermis. The large number of ectodesmata in the outer walls of guard cells no doubt contribute to the establishment of these differences.

Stomatopoda, ***mantis shrimps, stomatopods:*** an order of the *Crustacea,* subclass *Malacostraca.* They possess a dorsoventrally flattened, elongated body with a large and powerful abdomen, and only a small, shield-like carapace, which is united with the first 2 thoracic segments. The first 5 pairs of thoracic appendages are uniramous and subchelate. The second pair, however, is extremely well developed for raptorial feeding; the inner edge of the movable finger carries long spines or has the form of a blade, so that the terminal joints can be brought together like a pocket knife. Blows from this raptorial appendage are so powerful that specimens have been known to crack the glass walls of aquaria. It is used to stun prey and to smash the hard exoskeletons of crabs and lobsters. Most of the recognized 300 (approx.) species are tropical, and they vary in length from 5 to 36 cm. They are bottom dwellers, mostly in coastal waters. Some dig tubular burrows in the sand, while others conceal themselved in rock fissures. Many are brilliantly colored (green, blue, red with mottling, or striped). Among the few temperate species, *Squilla empusa* (length about 18 cm) is common in the Atlantic coastal waters of North America. In many coastal regions of the tropics mantis shrimps are eaten by local people.

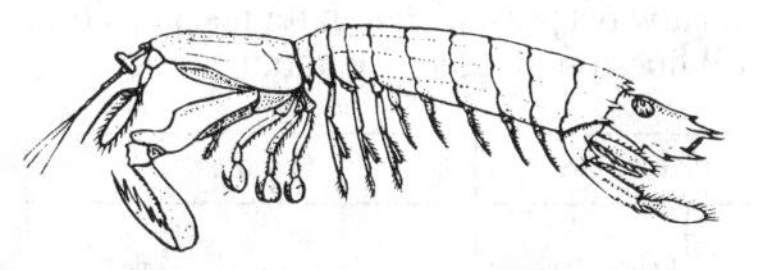

Stomatopoda. Mantis shrimp *(Squilla mantis).*

Stomiidae: see *Stomiiformes.*

Stomiiformes, *Stomiatiformes:* an order of deep sea fishes, monstrous in appearance, often with very large gapes and luminescent organs. Some species have dwarf males, and the larval stages of some species have stalked eyes.

Nine families are recognized: *Gonostomatidae* ***(Lightfishes*** or ***Bristlemouths;*** 6 genera, 12 species), *Sternoptychidae* (10 genera, 41 species, including the ***Marine hatchetfishes),*** *Photichthyidae* (7 genera, 18 species), *Chauliodontidae* ***(Viperfishes;*** 1 genus, 6 species), *Stomiidae* ***(Scaly dragonfishes;*** 2 genera, 8 species), *Astronesthidae* ***(Snaggletooths;*** 6 genera, 27 species), *Melanostomiidae* ***(Scaleless black dragonfishes;*** 16 genera, about 104 species), *Malacosteidae* ***(Loosejaws;*** 4 genera, 13 species), *Idiacanthidae* ***(Black dragonfishes;*** 1 genus, about 4 species).

Stomochord: a structure present in the *Hemichordata,* consisting of an unpaired, blind, tubular extension of the foregut into the anterior body region (protosome). It functions as a supporting rod and has a similar structure to the notochord of the chordates. It is, however, not an evolutionary precursor of the chordate notochord, and it is neither homologous nor analogous to this structure.

Stomodeum: see Digestive system.

Stone basse: see Sea basses.

Stone cells: see Mechanical tissues.

Stonechat: see *Turdinae.*

Stone crabs, lithode crabs, *Lithodidae:* a family in the order *Decapoda* (see). Except in the primitive subfamily *Haplogastrinae,* the abdomen, as in the true crabs, is tightly compressed under the cephalothorax. The abdominal tergites consist of several pieces, rather than uniform plates. Gills are phyllobranchiate. Uropods are absent. Adults and larvae (especially in the *Haplogastrinae)* are morphologically similar to those of the *Paguridae,* suggesting that these two groups share a common origin. S.c. resemble true crabs, but are in fact relatives of the hermit crabs. Notable differences between between S.c. and the true crabs (see *Brachyura)* are: 1) the S.c. possess an asymmetrical abdomen; 2) in S.c. the scales on the second article of the second antennae and fifth pleopods are short, and may be hidden in the branchial chamber; 3) all S.c. are cold water animals. In the primitive subfamily *Haplogastrinae,* the abdomen is not folded beneath the carapace, so that the dorsal plate of the second pleomere is visible from above. The dorsal plates of the third, fourth and fifth pleomeres are soft and unsegmented. In contrast, the subfamily *Lithodinae* contains crablike forms, e.g. *Lithodes,* which has a triangular, heavy, mineralized, sculptured carapace with dorsal tubercles and spines. In Maine and further north, *Lithodes maia* is frequently caught in lobster traps. On the West Coast of the USA, *Lithodes aequispina* (up to 9 kg in weight) is caught and canned. The ***King crab*** *(Paralithodes camtschatica)* of the northern Pacific is caught and canned in large quantities for food.

Stonecrop: see *Crassulaceae.*

Stoneflies: see *Plecoptera.*

Stone flower: see *Aizoaceae.*

Stone loach, *Noemacheilus barbatulus:* a freshwater fish in the order *Cypriformes,* superfamily *Catostomoidea,* family *Cobitidae* **(loaches),** subfamily *Noemacheilinae.* The genus *Noemacheilus* contains about 115 species. All loaches are wormlike to fusiform in shape, with a ventral mouth and 3–6 pairs of barbels. The body of the stone loach is elongated, anteriorly cylindrical and posteriorly slightly compressed. It normally attains a length of up to 12 cm, but some specimens may be up to 18 cm in length. The skin is very slimy. It has 3 pairs of barbels, 2 of which are rostral, while the other pair (often longer) are maxillary and arise in the corners of the mouth. The stone loach is a bottom fish of oxygen-rich, clear, rapidly flowing streams and rivers with pebbly bottoms, and inshore parts of clear lakes. It is also occasionally found in the brackish water of the Baltic. Throughout its very extensive distribution range (the whole of Eurasia) there are numerous subspecies and different color varieties; Central European forms are yellow-brown.

Stone marten, *Martes foina:* a 70 cm-long mustelid (see *Mustelidae),* with a yellow-brown coat and a forked, white throat patch. It is distributed throughout Europe with the exception of the British Isles and Scandanavia. It inhabits low mountainous regions and lowlands, preferring areas that provides quiet hiding places.

Stony corals: see *Madreporaria.*

Stool shoot: see Adventitious shoots.

Storage materials, ***reserve materials:*** substances that are temporarily removed from active metabolism, and stored at various sites in plants and animals. After a latent or rest period, or at times of inadequate nutrient availability or prolonged absence of nutrients, S.m. are subsequently reincorporated into metabolism for the purposes of growth and energy provision.

The main S.m. of plants are carbohydrates, proteins and fats. A very common and abundant reserve carbohydrate is Starch (see), which is a primary constituent of all basic human foods, like bread, potatoes, rice and millet. It consists of 2 different polysaccharides: α-amylose (consisting of α1,4-linked glucose) and amylopectin (α1,4-linked glucose with α1,6 branches). Starch is produced in chloroplasts during photosynthesis in the form of small, lens-shaped granules *(primary* or *assimilation starch).* During the night, primary starch is degraded to its sugar components, which are transported to the storage organs, then reconverted to *reserve starch* in the colorless amyloplasts. The size and shape of starch granules vary widely in different plants: rice 3–10 µm, maize 10–15 µm, beans 25–60 µm, potatoes 50–100 µm. Starch granules usually reveal a distinctly layered structure under the microscope.

Animal *Glycogen* (see) has a similar structure to that of amylopectin, but with a greater proportion of α1,6 branch-points in the glucose chain; it also occurs in fungi, bacteria and cyanobacteria. Water-soluble *Inulin* (see), a β2,l-linked polymer of fructose, is the reserve polysaccharide of Jerusalem artichoke tubers *(Helianthus tuberosus),* and is generally found in members of the *Compositae. Reserve celluloses,* like arabans, xylans, mannans and galactoarabans, are laid down as cell wall material, particularly in seeds and fruits, a notable example being the seed endosperm of the palm, *Phytelephas macrocarpa,* which is the source of "vegetable ivory". Date stones (seeds) also consist of reserve cellulose. The disaccharide, *Sucrose* (see), is stored in sugar beets and in the stems of sugar cane. The monosacccharides, *Glucose* (see) and *D-Fructose* (see) are widely distributed in plants, and present in particularly high concentrations in fruits. *Reserve proteins,* which are structurally different from cytoplasmic and structural proteins, occur dissolved in the cell sap or as *protein* or *aleurone granules;* the latter represent the solid residue from the dehydration of protein-containing vacuoles. In addition to proteins (some of which are present in crystalline form), the rounded aleurone granules often contain globules of phytin. In cereal grains, the cells of the aleurone layer (the outer cell layer of the endosperm) are completely filled with small aleurone granules. In the manufacture of white flour, this protein layer remains mainly in the bran. The cotyledons of legume seeds are very rich in aleurone granules. Oily fats are stored in the seed endosperm of many plants. They are usually present in small amounts, but extremely large quantities are stored in the seeds of oil plants. These oils may form an emulsion of minute droplets in the cell cytoplasm, or they may coalesce to form relatively large fat vacuoles.

In the germination of reserve-rich seeds and vegetative reproductive organs (tubers, rhizomes, buds, etc.), S.m. are mobilized by enzymatic cleavage, then used as synthetic precursors for growth, or respired as a source of energy.

In animals, S.m. are stored as adipose tissue beneath the skin and in the mesenteries. Liver and muscle are sites of glycogen storage.

Storks: see *Ciconiiformes.*

Storksbill: see *Geranaceae.*

Storm flies: see *Thysanoptera.*

Storm-petrels: see *Procellariiformes.*

St. Peter's fish: see John Dory.

Strain: a term meaning the same as race, i.e. the smallest systematic unit. The different strains of a species usually display very small physiological or morphological differences, and they are usually named by placing specific letters and numbers after the Latin binomial of the species. Antibiotic synthesis by the commercially exploited *high production strains* of microorganisms is several fold greater than that of the corresponding wild strain. Such strains are obtained by mutation and selection. Similarly, high yielding strains of agricultural crop plants are obtained by breeding and selection.

Strangling fig: see *Moraceae.*

Strap floret: see *Compositae.*

Stratification: see Relevé.

Stratified epithelium: see Epithelium.

Stratigraphic dating: see Dating methods.

Stratum (plural **Strata),** ***layer, horizon:*** a zone in a vertical zonation of soil and vegetation. The following zones are distinguished: soil, litter, herbage, bushes, tree stems, tree cowns, each of which can be further subdivided (Fig.). The living populations of these strata are known as *stratocoenoses,* each consisting of *choriocoenoses,* which are habitat-specific, recurring species combinations.

The soil surface horizon is particularly important (in forests, this is a typical litter layer). It is particularly suitable for the global comparisons, because (unlike the other strata) it is always present, and it represents a link between the soil horizon and the vegetation.

	Coenotypes	Coenoses		
Atmobial	Canopy layer	Atmobios	Euatmobios	Epiphytos Epizoobios
	Trunk layer			
Aerial Zoal Phytal		Aerobios Zoobios Phyrobios		Endophytos Endozoobios
	Bush layer		Pseudo-atmobios	Mesophytos Mesozoobios
Soil surface	Herbaceous layer			
	Moss layer		Epiedaphon	Epigaion
Edaphal	A- L-Horizon		Hemiedaphon	Hypogaion
	A- F-Horizon			
	A- H-Horizon			
	B-Horizon	Edophon	Euedaphon	Endogaion Mesogaion
	C-Horizon			

Stratum. Vertical zonation of terrestrial coenotypes and coenoses.

Strawberry: see *Rosaceae.*

Strawberry crown mite, *Tarsonemus pallidus fragariae* Zimmermann: an almost microscopic, soft-skinned mite, white to light brown in color, which feeds on young, unfolded leaves in the crowns of strawberry plants. Growth is stunted, the plants are dwarfed, the leaves form rosettes due to the failure of the stems to elongate and the plant dies. It is often considered identical to the Cyclamen mite, *Tarsonemus pallidus* Banks, but individuals on strawberry plants fail to breed normally when transferred to other host plants. Likewise, *T. pallidus* taken from cyclamen (or aster or ivy) does not breed normally on strawberry. Clearly, there is a difference at the subspecies level, but specific status is not confirmed.

Strawberry tree: see *Ericaceae.*

Streaking: a microbiological technique for isolating colonies derived from single cells. Using an inoculation loop (usually of platinum or nichrome wire), a small area of the solid culture substrate in a Petri dish is inoculated with a drop of bacterial suspension, preferably toward the edge of the dish. The loop is sterilized (flamed), cooled by immersion in a sterile area of the substrate, then placed lightly on the inoculated site and drawn ("streaked") once across the surface of the substrate. The loop is again sterilized and drawn at right angles across the first streak, sterilized again and drawn across the second streak, etc. At each stage the cell mass carried in the path of the streak is diluted, until separate individual cells are present. After incubation, the culture plate displays heavy confluent growth in the area of the first inoculum, but this becomes progressively less dense in the subsequent streaks, until separate colonies are seen, which are derived from single cells.

Stream frog: see *Dendrobatidae.*

Streber: see Zingel.

Strepsiptera, ***stylopids, stylopoids:*** an insect order containing only about 400 known species, mainly in the Holarctic region, but also represented in all other zoogeographical regions. The adults display marked sexual dimorphism.

Adult males. These are free-living, 1 to 7 mm in length, with very reduced, vein-less forewings, and enlarged, fan-like, folding hindwings with few veins. The head carries protruding compound eyes, which appear to be on stalks (stylops = "stalk eye"). In the original condition, the antennae were 7-segmented, but in many species fusion has occurred, and the third segment has a lateral projection (flabellate antennae). The biting mouthparts are often fused and reduced. Forelegs and middle legs lack a trochanter; the legs are generally weak and are used for holding onto the female during copulation. Males survive for only a few hours after emergence, while they actively seek a female.

Adult females. These are invariably wingless and 1.5 to 3 mm in length. In the family, *Mengeidae,* the adult female is larviform and free-living. All other female *S.* are endoparasitic on certain wasps, cicadas and bugs. The abdomen of the parasitic *S.* remains within the last larval cuticle or puparium, which itself is buried in the abdomen of the host. Only the reduced head and thorax (fused to form a cephalothorax) of the parasite protrude from the host's abdomen. Between the puparium and the ventral surface of the female is a cavity, called the brood passage, which is open to the exterior. The brood passage is used by the male for insemination, and as an exit for first instar larvae.

In adults of both sexes, the reduced gut ends blindly, and Malpighian tubules are absent.

Larvae. First instar larvae are called triungulins. Usually several thousand are produced by a single female, and these escape via the brood passage. They are active, and in some species capable of jumping 2–3 cm. Their legs are well developed, but the mouthparts are reduced and antennae are lacking. An immature stage of the host is attacked, and when such an opportunity occurs the triungulin enters the host's abdomen, apparently aided by a secretion from the mouth which softens the host's cuticle. Inside the host, the triungulin molts to a grub-like, apodous form. Further development occurs within the host.

Metamorphosis is complete (holometabolous development). Male pupae are adecticous and exarate (pupa libera), but female pupae are suppressed (except in the *Mengeidae).*

Parasitic attack by *S.* leads to atrophy of the gonads and other internal organs of the host, which is then said to be "stylopized".

Classification. Recent stylopids are grouped into 5 families: *Mengeidae* (the most primitive; the female is free-living; larvae are parasitic on bristletails); *Stylopidae* (the largest family; parasites of bees and wasps; cosmopolitan); *Halictophagidae* (parasites of *Homoptera, Auchenorrhyncha* and *Tridactylidae*; cosmopolitan); *Corioxenidae* (*Callipharixenidae*) (parasites of *Heteroptera*; Ethiopia, Japan, New Guinea, Australia); *Myrmecolacidae* (male larvae develop in ants, female larvae in *Orthoptera*; East Asia, Australia, Africa, South America).

Streptobacillus moniliformis: see Rat bite fever.

Streptococci: in the wider sense a family of nonmotile Cocci (see) that occur in chains or pairs, and which possess a fermentative metabolism. They include the genera *Leuconostoc* (see) and *Streptococcus.* In the restricted sense, S. are members of the genus *Streptococcus,* found in sour milk, on plants, in silage and in the intestinal tract. Some species are used industrially, e.g. *Streptococcus lactis* is the common milk-souring organism, several varieties being used in cheese making. *Streptococcus cremoris* is used as a starter culture in butter and cheese production. Other species are disease agents. Most of those associated with primary streptococcal infections in man and animals produce a soluble hemolysin, and they are further classified according to the chemical nature of the carbohydrate of the cell (C antigen). Skin infections such as impetigo and erysipelas, scarlet fever (scarlatina) and tonsilitis are all caused by *Streptococcus pyogenes.* Bovine mastitis is caused by *Streptococcus agalactiae.*

Streptomyces: see *Actinomycetales.*

Streptomycetaceae: see *Actinomycetales.*

Streptomycin: a clinically important antibiotic synthesized by *Streptomyces griseus,* and produced commercially on a large scale by industrial fermentation. It is active against mycobacteria, many Gram-negative bacteria and some staphylococci. Chemically, it is an aminoglucoside, in which *streptidine* (an inositol derivative) is linked glycosidically to the disaccharide, *streptobiosamine* (consisting of streptose and *N*-methyl-L-glucosamine). S. inhibits protein biosynthesis on 70S ribosomes; it binds to the 23S core protein of the 30S ribosomal unit, thereby preventing the binding of mRNA. It was discovered in 1944 by Waksmann, and it is used against tuberculosis.

Streptoneura: see *Gastropoda.*

Streptoneury: see *Gastropoda.*

Streptopelia: see *Columbiformes.*

Stress: a term introduced by Seyle for the physiological condition caused by excessive unspecific and/or unusual and environmental or psychological pressures on an organism. It is often accompanied by behavioral changes.

Stress compounds: see Phytoalexins.

Stretching syndrome: a complex of movements initiated by proprioceptive stimuli, and involving relatively slow muscle movements with considerable muscle tonus and tension. Typical components of the syndrome are stretching and straightening of the body, craning and yawning. The pattern of movements depends on the species and its physiological state.

Stretch reflex: the reflex contraction of a skeletal muscle in response to the sudden stretching of the same skeletal muscle. Some S.rs. are exploited clinically, e.g. the knee jerk reflex (patellar reflex).

Stretch receptors: see Baroceptors.

Stridulation: sound generation by animals with the aid of S. organs. A S. organ consists of a pars stridens (with a comb-like roughened surface) and a plectrum or file, which are rubbed against each other to produce a sound. In arthropods, these structures are present on different body parts. If only one of these parts is moved while the other remains stationary, the organ is named after both parts and the moving part is quoted first, e.g. the S. organ of short-horned grasshoppers is called the *organum stridens tegmine metafemorale,* i.e. the metafemur is rubbed against the forewing (tegmen). At least 30 different types of S. organs are known in insects. In fishes, S. is performed by rubbing teeth, jaws, pharyngeal bones, fin rays and other skeletal parts against each other.

The sounds generated by S. consists of pulses (the smallest signalling element) and elementary oscillations. Pulses can be organized into phrases, phrases can be combined into verses, and a sequence of verses is a 1st order sequence or strophe. A sequences of strophes is a 2nd order sequence or "song". Combined with postures and body movements, the sounds produced by S. constitute species-specific signals (e.g. in numerous members of the *Orthoptera),* usually in the search for, and choice of, a sexual partner. They are also produced during the mating process, and they are used to deter sexual rivals.

Strigidae: see *Strigiformes.*

Strigiformes, ***owls:*** an order of birds (see *Avies)* with a worldwide distribution. They are all birds of prey (raptors), with a strongly hooked beak and powerful, sharp talons. Their superficial resemblance to the *Falconiformes* is the result of parallel evolution. Structurally, the owls are most closely allied to the *Caprimulgiformes,* a phylogenetic relationship more recently confirmed by the analysis of the egg white proteins of these two orders. To distinguish them from the largely diurnal *Falconiformes,* owls are sometimes called *nocturnal birds of prey* or *nocturnal raptors,* although some owls do hunt by day. Obvious characteristics are a large head, forward-facing eyes, a facial disk, and a hooked bill that is partly hidden in the plumage between the 2 halves of the facial disk. Owls usually swallow their prey whole. Bones, fur, feathers, insect integuments, etc., remaining after digestion, are then regurgitated in the form of a neat pellet, commonly known as an owl pellet. Some owls (see below) have 2 tufts of feathers on the head, which resemble ears or horns ("eared owls" and "horned owls"), but these are in no way associated with the true ears. The true ears are large, partly covered by flaps of skin, and completely hidden under the head plumage.

Two families are recognized.

1. Family: *Tytonidae* ***(Barn owls*** and ***Bay owls).*** Barn and Bay owls are recognizable by their heart-shaped facial disk. The middle and inner toes are of equal length (the inner is shorter in the strigids), the claw of the middle toe has a serrated comb (otherwise found only in herons and some nightjars), and the furcula (wishbone) is fused to the sternum. The ***Common barn owl*** *(Tyto alba;* 34 cm) is one of the most widely distributed birds, occurring in North, Central and South America, Europe, Arabia, India, Southeast Asia, Australia, and North and sub-Saharan Africa. The male is golden buff with a white face and underparts, and the female tends to be a little grayer. Most owls appear to have a keen sense of hearing and the ability to accurately locate prey by sound. Controlled experiments have shown that this ability is extraordinarily well developed in *Tyto alba,* which is able to locate a moving mouse with pinpoint accuracy in the total darkness of a blacked out room. It feeds almost entirely on rats and mice. The 9 other members of the family are confined to the Old World (Africa, Madagascar, Southeast Asia and Australia). Several inhabit open country and are known as ***Grass owls.*** The African Bay owl is known only from a single specimen collected in Zaire in 1951.

2. Family: *Strigidae* ***(Typical owls).*** Typical owls range in size from the American ***Elf owl*** *(Micrathene whitneyi;* 14 cm) to the ***Great gray owl*** *(Strix nebulosa;* 61–84 cm; female larger than the male) of northern North America and western Eurasia. Most of the 125 species (25 genera) are sedentary and tree-living. Depending on the species, they hunt all forms of animal prey from rabbits to fish and insects, and their habitats range from the Arctic tundra to tropical grasslands. The large eyes are not capable of independent movement, and direction of sight is controlled by movement of the

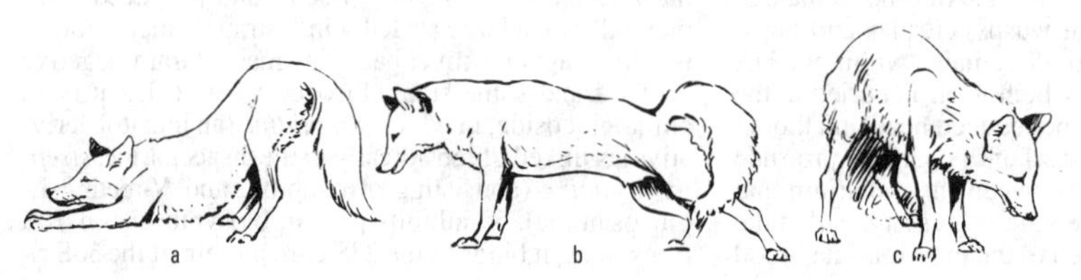

Stretching syndrome (red fox). *a* and *b* Successive stretching movements *(a* stretching of the forebody, *b* stretching of the hindbody). *c* Simultaneous stretching. (After Tembrock)

very flexible neck. The ***Snowy owl*** *(Nyctea scandiaca;* 63.5 cm) has a circumpolar distribution in the Arctic tundra, while the ***Great horned owl*** *(Bubo virginianus;* 63.5 cm) is found in the New World from the northern tree line to the southern tip of South America. Two examples of ***Pygmy owls*** are the ***Eurasian pygmy owl*** *(Glaucidium passerinum;* 16.5 cm) and the ***Ferruginous pygmy owl*** *(Glaucidium brasilianum;* 18 cm), which occurs from southwestern USA to Patagonia. ***Spectacled owl*** is the name given to the 3 species of *Pulsatrix,* which are found in the jungles of the new World. The 11 species of *Strix* ***(Wood owls)*** include the large *S. nebulosa* (see above) and the ***Tawny owl*** *(S. aluco;* 38 cm) (plate 6), which is found throughout Europe into North Africa, to Iran and almost to the Persian Gulf, and in western Siberia. Separate populations exist in the Himalayas and the mountains from Burma to China. Plumage is mottled brown, with a grayish brown facial disk, and dark brown to bluish black eyes. European populations also pass through phases when the plumage is gray. It is well known for its distinctive "kee wick" call and hooting song. Analysis of pellets in Europe reveals that the main diet is small rodents and beetles, but frogs, crabs, fish, snails and earthworms are also taken, depending on the habitat. The neck is exceptionally flexible and can be turned almost full circle. Various nesting sites are used, such as a tree hole, a deserted animal burrow, a cliff ledge, the space among tree roots or thick foliage, a squirrel drey, and ledges and spaces in buildings. It hunts by night, and during the day it is often mobbed by smaller birds, but remains passively on its perch apparently unperturbed by their attentions. The 11 species of *Bubo* have two conspicuous "ear tufts". They also have large, yellow, glaring eyes. ***Verreaux's eagle owl*** or the ***Milky owl*** *(Bubo lacteus;* 66 cm) is a dark blackish gray species found in the depths of the African forests. The genus *Otus* contains 36 species of small to medium-sized owls known as ***Scops*** or ***Screech owls,*** which are widely distributed in the New World. *Asio flammens* ***(Short-eared owl;*** 42 cm female, 35.5 cm male) is found throughout Eurasia, including Iceland, northeastern Siberia, most of North America, from Peru to Patagonia and the Falkland Islands, as well as in Hawaii, the Carolines and the Galapagos. Fairly open habitats are preferred, e.g. marshes, grassland, moorland, steppes and forest clearings, and it is found up to 300 m in the mountains of South America. Populations in the extreme north and south migrate during their winters. Its upper parts are richly marked buff and dark brown, and the wings (relatively long) and tail are barred. The well defined facial disk is darker around the golden yellow eyes. The ear tufts are hardly visible. Short-eared owls hunt by quartering the ground like harriers, searching for small rodents, as well as birds and beetles. They are especially fond of voles, and their breeding numbers increase during vole plagues. Since conifer plantations provide an ideal habitat for voles, the range of the Short-eared owl in England and Scotland is gradually extending southward with the tree-planting activities of the forestry commission. *Speotyto cunicularia* ***(Burrowing owl;*** 23 cm) is found in the prairie states of western North America, through Florida and all the way south to Tierra del Fuego. Its upperparts are brown, polka-dotted with white, with a white throat, black collar and yellow eyes. Insects, especially *Orthoptera,* and occasionally small rodents are hunted at dusk. During the day *Speotyto* is commonly seen resting on the ground or sitting on a fence post. Unusually for an owl, *Speotyto* breeds in a burrow. It may occupy an abandoned animal burrow, such as that of a prairie dog, or it may excavate a nesting burrow up to 2 m or more in length. Both parents incubate and feed the young.

Strigopinae: see *Psittaciformes.*

Strigops: see *Psittaciformes.*

Stringocephalus (Greek: *strinx* owl, *kephale* head): a genus of the *Brachiopoda* with a looped brachial skeleton, and hinged, bioconcave, usually smooth valves. The ventral valve is strongly arched with a curved, posterior beak (umbo), resembling an owl. The dorsal valve is less strongly arched. The genus is distributed worldwide in the Middle Devonian, and it is an index fossil for the marine Upper Devonian limestone, known as the Calcaire de Givet, or the Givetian stage, which extends from Belgium into the calcareous flagstones of the Rhineland west of Bonn in Germany.

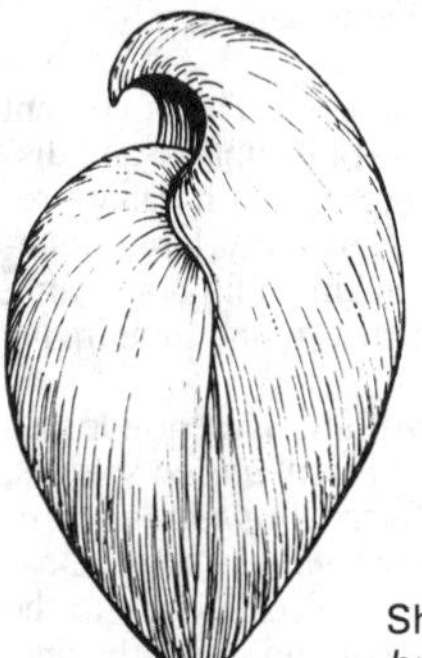

Shell of *Stringocephalus burtini* Defr. (x 0.5)

Striped field mouse: see *Muridae.*

Striped hairy-footed hamster: see *Cricetinae.*

Strip spraying: herbicide spraying along rows of cultivated plants, in contrast to blanket spraying of the entire area.

Strix: see *Strigiformes.*

Strobila: see *Cestoda.*

Strobilation (Greek *strobilos* pine cone): a form of asexual reproduction in the *Scyphozoa* (see). The tentacles of the sessile polypoid larva regress, and small

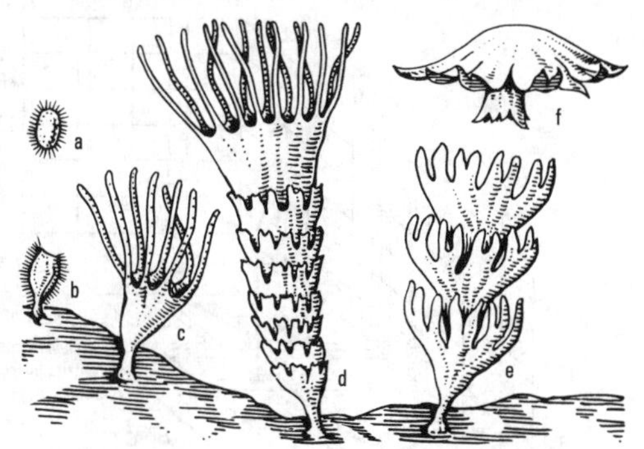

Strobilation. Stages in the strobilation of the jellyfish, *Aurelia aurita. a* Free-swimming ciliated planula larva. *b* Attached planula larva. *c* Young polyp at the scyphistoma stage. *d* Strobila stage. *e* Abstriction of ephyrae. *f* Free-swimming ephyra.

disk-shaped medusal larvae are formed by transverse cleavage at the oral (anterior) end of the polyp. Each larva, known as an Ephyra (see), may be released as it is formed, or a number of such structures may become stacked on top of each other like a pile of plates. The resulting structure then bears a certain resemblance to a pine cone. The released ephyrae live as free swimming, immature medusae.

Stroke volume: the volume of blood ejected into the aorta by one heart beat. In diastole, the left ventricle fills with blood (expelled from the left atrium during atrial systole). During the isometric phase, the contents of the ventricle undergo a rapid rise in pressure, followed by ventricular systole, in which part of the contents (the S.v.) is ejected. The remaining part, the *residual volume* or *end-systolic volume,* remains in the ventricle (Fig.). During the heart beat of an adult human at rest, the S.v. is 70 ml and the residual volume 50 ml.

Strongylida: see *Nematoda.*

Strongylocentrotus: see *Echinoidea.*

Strongylus: see *Nematoda.*

Strontium, ***Sr:*** an alkali earth metal. The Sr content of plants is 1–10 mg % dry mass. In its uptake and distribution, Sr resembles calcium (see), but it cannot replace calcium physiologically. Relatively large quantities of Sr cause damage to plants, including browning and death of lower leaves. Sr is also found in animals, but its significance is unknown.

Strophanthins, ***strophanthosides:*** cardenolide cardiac glycosides from the seeds of *Strophanthus* species, e.g. g-strophanthin (ouabain) from *Strophanthus gratus,* and k-strophanthin from *Strophanthus kombé.* Ouabain consists of rhamnose linked glycosidically to the aglycon, *ouabagenin.* In k-strophanthin, the aglycone, *strophanthidin,* is linked glycosidically to the sugars, cymarose and β-glucose. Strophanthidin is the aglycon of other cardiac glycosides, e.g. convallatoxin of *Convallaria majalis* (lily of the valley).

S. are highly potent toxins, which have been used in Africa since antiquity as arrow poisons. In medicine they are therapeutically very important as cardiac stimulants that act more rapidly than the Digitalis glycosides.

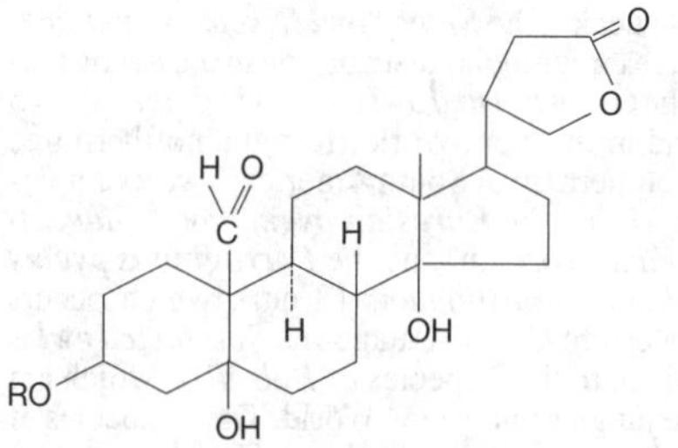

k-Strophanthin: R = -cymarose-β-glucose-β-glucose
k-Strophanthin-b: R = -cymarose-β-glucose
Convallatoxin: R = -rhamnose

Ouabain is a specific inhibitor of the membrane-bound ($Na^+ + K^+$)-ATPase, which is responsible for maintaining high intracellular concentrations of K^+ and low intracellular concentrations of Na^+. Ouabain therefore inhibits sodium ion transport out of the cell and potassium ion transport into the cell, and simultaneously inhibits entry of glucose and amino acids.

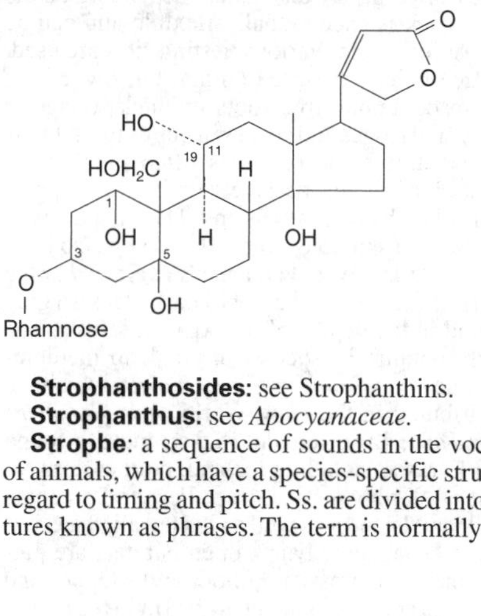

Oubain

Strophanthosides: see Strophanthins.

Strophanthus: see *Apocyanaceae.*

Strophe: a sequence of sounds in the vocalizations of animals, which have a species-specific structure with regard to timing and pitch. Ss. are divided into substructures known as phrases. The term is normally used only

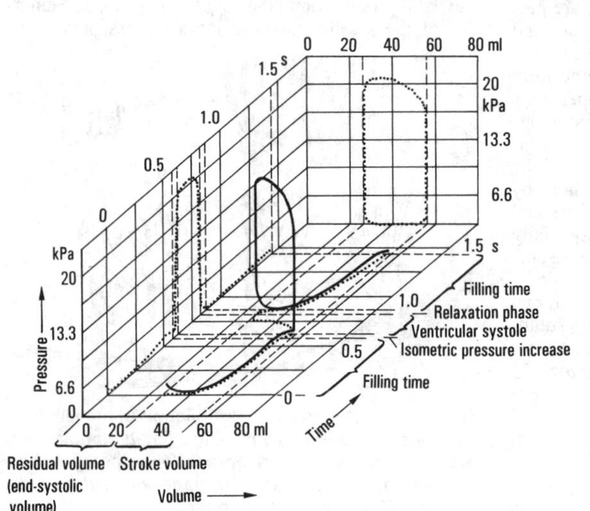

Stroke volume. Changes of pressure and volume in the left ventricle of a dog during a single heart beat.

for those Ss. that occur as the recognizable sequences known as Song (see).

Structural colors, ***schemochromes:*** colors seen in, e.g. bird feathers and butterfly wings, which are due to the physical nature of the illuminated surface, and not to the deposition of pigments. Thus, interference or refraction produce iridescent colors that change with the angle of view. Noniridescent colors are produced by light scattering, e.g. the Tyndall scattering of shorter light wavelengths is responsible for most blue colors, such as the blue of the domestic budgerigar. In the latter case, light is scattered by air-filled cavities within the keratin of the feather barbs.

Structural genes: genes that encode the synthesis and molecular structure (amino acid sequence) of enzymes or structural proteins, and are therefore directly involved in the expression of the phenotype of an organism, rather than serving a regulatory role. The encoded information is transcribed as messenger RNA, which is then translated into the polypeptide (see Operon, Regulator genes).

Structural heterozygosity: a condition in which a cell, tissue or individual is heterozygotic for one or more chromosome mutations. Meiosis of structural heterozygotes is characterized by the occurrence of special pairing configurations (see Inversion, Translocation). *Cryptic S.h.* is due to very small chromosomal structural alterations, which have little or no effect on pairing phenomena, but result in more or less marked sterility.

Structural hybrids: hybrids whose heterozygotic state is due to structural differences in homologous chromosomes, and in which these structural differences arise by modification of chromosome structure.

Structural type: the essential structural characteristics of an organism. Comparative anatomy shows that the organisms within each higher systematic unit (e.g. insects, birds, mammals, etc.) do not differ fundamentally in structure, but rather represent variations of a common structural type. The varying degrees of expression of each character within the basic structural plan are the result of adaptation. However, every organism in a systematic unit displays adaptations, so that the basic structural type is actually a theoretical type. The realization that the polymorphism of organisms can be rationalized by assigning each organism to a structural type was orginally seen as the expression of supernatural concepts, which delayed its causal interpretation. It is now accepted that common structural types and homologous organs constitute important evidence for the theory of evolution, which provides the only satisfactory explanation of their existence.

Type specimens and *type organisms* of a systematic unit do not display the basic structural plan or type, but are typical species chosen to represent the group, and against which all other species of the group are described. See Life form.

Struggle for life: an expression coined by Charles Darwin (1809–1882) for the fact that under natural conditions many young organisms are produced, but normally only a few survive. Each individual must therefore struggle for survival. This struggle leads to natural Selection (see).

Struthioniformes, ***ostriches:*** an order of flightless, running birds, represented by the single species *Struthio camelus* ***(Ostrich).*** Five or 6 subspecies or races are recognized. *S.c. massaicus* ***(Masai ostrich),*** which has pink neck, is found in eastern Kenya south of the River Tana. *S.c. molybdophanes* ***(Somali ostrich),*** which has a blue neck, occurs in the Horn of Africa. *S.c. spatzi,* from Rio de Oro, was described in 1926 on the basis of its small egg and its shell structure; its existence in this remote area remains unconfirmed. *S.c. camelus* ***(North African ostrich),*** which has a pink neck, is no longer found in the North African desert, but it occurs south of the Atlas mountains from Senegal and Niger to Sudan and central Ethiopia. The ***Syrian ostrich*** *(S.c. syriacus)* originally inhabited the entire Syrian and Arabian deserts to the banks of the Euphrates, later becoming restricted to the Arabian desert, and finally becoming extinct in 1941. *S.c. australis* ***(Southern ostrich),*** which has a blue neck, occupies open savanna and veld between the Zambesi and Cunene rivers, and it was formerly farmed commercially for its plumes in the Little Karoo area of Cape Province. This early commercial importance of the ostrich led to attempts at breeding and domestication, and introduction of the bird into Australia. Consequently, small feral populations of ostriches exist in southern Australia, derived from crosses between *S.c. australis* and *S.c. camelus.*

The body plumage of the male is black, with white, plume-like wing quills and tail, whereas the female plumage is uniformly brownish with pale edges to the feathers. In both sexes, the face, the legs, and about two thirds of the long, muscular, mobile neck are naked. The black eyelashes are very prominent. Their powerful legs enable ostriches to run at high speeds (up to 50 km/h), often outpacing antelopes. They also run with great agility, being able to swerve, change direction abruptly, and double back. Their eyesight is exceptionally acute, so that in open country it is difficult to approach ostriches closer than about 100 meters. Lions prey upon young birds, and the eggs are taken by jackals, but the extinction of the ostrich in certain areas is due entirely to profligate hunting by man.

Ostriches inhabit open, low rainfall areas of Africa, often associating with herds of zebras, antelopes and giraffes. In these big game areas, ostriches may form groups of 40–50, but groups of 5 or 6 are more usual. Single birds are also encountered. Food consists mainly of vegetable matter, e.g. flowering shrubs, parasitic creepers, ground-creeping gourds, fallen wild figs, but small animals sich as lizards are also taken.

During the courtship display, the male arches his wings in a characteristic posture, while the naked skin of his front legs and neck undergoes a momentary color change to fiery red. Each cock associates with 3–5 hens, and all the hens belonging to one cock lay their eggs in a common nest, which is a shallow excavation in the sandy soil. The number of eggs incubated in one nest varies according to the subspecies (15–20 for *S.c. camelus*; 50–60 for *S.c. massaicus).* Incubation, which lasts from 5 to 6 weeks, is shared by the male and females.

Ostriches differ from all other birds by the presence of only 2 toes. The inner toe, which has a stout nail, corresponds to the middle anterior toe of other birds, while the smaller, outer toe has no nail. Adult males weigh up to 156 kg, and they measure about 242 cm to the top of the head, and 137 cm to the top of the back.

The "American ostrich" is not an ostrich, but a Rhea (see *Rheiformes).* Members of the *Struthioniformes, Rheiformes, Casuariiformes* and *Apterygiformes* were formerly grouped together as the "Ratitae"; this close relationship is now doubted, but the members of these 4 orders are still collectively known as Ratites. In all ratites, the wing skeleton and pectoral girdle show varying degress of reduction, and the sternum is flat and raftlike with no keel. See *Avies.*

Strychnine: a highly toxic, very bitter-tasting alkaloid from the seeds of *Strychnos nux-vomica* and other *Strychnos* species. It is accompanied by lesser quantities of other Strychnos alkaloids. It forms colorless crystals, and possesses a complicated chemical structure of seven fused rings. It causes a blockade of central inhibition by selectively antagonizing the effect of glycine. (Glycine is an inhibitory transmitter of 4-aminobutyrate receptors in the central nervous system of vertebrates, particularly in the spinal cord). S. is thought to interact with receptor sites on postsynaptic membranes, thus preventing access of glycine. 5 mg in children and 30–100 mg in adults are sufficient to cause death from muscle rigor, due to the increased reflex excitability of the spinal cord. Low doses of S. nitrate are used medically to stimulate respiration and blood circulation, but the high toxicity of S. precludes its extensive use in medicine. It is also used as a rodent poison (mixed with red-dyed wheat grains).

Strychnos alkaloids: see Strychnine.

Stump-tailed skink: see Blue-tongued skinks.

Sturgeon: see *Acipenseridae.*

Sturnidae, ***starlings, mynahs*** and ***oxpeckers:*** an Old World avian family containing 111 species (18–43 cm in length), in the order *Passeriformes* (see). They breed in pairs, but tend to form flocks when seeking food, which consists of insects and fruit. They usually build open or cavity nests, but many other types of nest are also bulit, e.g. the South pacific ***Glossy starling,*** *Apolonis metallica,* builds a large, woven hanging nest. The family is most numerous in India and Southeast Asia, and is also well represented in Africa. All but 3 species are placed in the subfamily, *Sturninae.*

Of the few Palearctic species, the ***Common starling*** *(Sturnus vulgaris;* fresh plumage dark, glossy and spangled) is well known. Like many successful members of the family, it has established a close relationship with humans. It has been taken from Europe to many countries throughout the world (e.g. South Africa, Australia, New Zealand, Jamaica, North America), where it has usually become an agricultural pest, as well as being a serious competitor of native species, e.g. of bluebirds *(Sialia)* in North America. In the winter it forms large foraging flocks containing thousands or even hundreds of thousands, which have evolved the habit of invading cities in the evening and roosting on roofs and balconies. Its successful invasion of new habitats is a clear demonstration of the adaptability and aggressiveness of the Common starling, which are also characters of most other members of the family.

In its native habitat in the Middle East, the ***Rosy pastor*** *(Sturnus roseus;* head and wings glossy black with a purple sheen; body pale pink) occasionally undergoes a population explosion and enters eastern Europe in large numbers. Like the African ***Wattled starling*** *(Creatophora cinerea),* the Rosy pastor is a useful predator on locust swarms. The ***Common mynah*** *(Acridothera tristis)* of southern Asia appears to be the tropical counterpart of the Common starling; it has been successfully introduced in many parts of the world, where it is often a pest. Many members of the family display an ability to imitate sounds, especially the mynahs, and in particular the ***Hill mynah*** *(Gracula religiosa)* of India and the East Indies.

The 2 African ***Oxpeckers*** or ***Tick-birds*** *(Buphagus)* form the subfamily *Buphaginae.* They search for ticks on large game animals and livestock, and they sip the blood that oozes from the tick wounds; they serve as lookouts for their hosts, flying up with a rattling cry at the approach of danger.

Sturninae: see *Sturnidae.*

Sturnus: see *Sturnidae.*

Stygobiont: see Ground water fauna.

Stylaster: see *Millepora.*

Stylasterina: see *Millepora.*

Style, ***stylus:*** unsegmented, rodlike appendage on the coxae of all or some leg pairs of certain apterygote insects and of the larvae of some pterygote insects.

Stylet barbule: see Feathers.

Styliconic sensilla: see Olfactory organ.

Stylomecon: see *Papaveraceae.*

Stylommatophora: see *Gastropoda.*

Stylopidae: see *Strepsiptera.*

Stylopids and **Stylopoids:** see *Strepsiptera.*

Styrene, ***monostyrene:*** the starting material for synthesis of polystryene. In the pure state, S. is a colorless liquid with an odor resembling that of benzene. It is very unstable, polymerizung even at room temparature to liquid distyrene and solid polystyrene.

C_6H_5–CH = CH_2

Styrene

Subalpine zone: see Altitudinal zonation.

Subassociation, ***subcommunity:*** a unit of vegetation within an association. It differs from similar Ss. of the same association only by its content of differential species. According to Tüxen, a typical S. does not possess its own differential species. See Character species.

Subatlantic: a post-Ice Age vegetative and climatic period, lasting from about 500 BC to the present. The early S. is characterized by a marked spread of beech and hornbeam, and its later stages by anthropogenic influences on the vegetation, leading to "cultivated woodlands" and "cultivated grasslands". See Pollen analysis.

Subboreal: a post-Ice Age vegetative and climatic period, which lasted from about 5 000 to 3 000 BC, beginning with the development of mixed oak forests and alder stands, followed by the oak-beech transition period. The climate was more or less continental and becoming cooler. See Pollen analysis.

Subculture: propagation of microorganisms or cultured tissue by transfer of cells into fresh growth medium. By regular S., cell cultures can be maintained indefinitiely.

Subcutis: see Skin.

Subdominance: the status below that of dominance in the biosocial hierarchy of a ranked society. In certain nonanonymous biosocial animal societies, group stabilization is achieved by a functional dominance hierarchy, in which individual dominance rankings are seen as functional roles. The existence of individuals with subdominant status can greatly increase the total fitness of the dominant animals or the dominant individual, which is important in groups whose members are interrelated.

Suberic acid, ***octanedioic acid:*** HOOC—$(CH_2)_6$—COOH, a saturated dicarboxylic acid formed by the oxidation of cork or ricinus oil with nitric acid.

Subfossil: an organism that has become extinct in historical time. The remains of these previous life forms represent a phylogenetic and chronological link between true fossil forms and the modern flora and fauna.

Subgenual organ: see Chordotonal organ.

Subimago: the last larval stage of a mayfly *(Ephemeroptera).* It has fully developed wings and is capable of flight, but it must molt in order to become a sexually mature imago.

Submontane zone: see Altitudinal zonation.

Suboscines: in the narrow sense an alternative name for the suborder, *Menurae,* of the *Passeriformes.* In the broad sense the 3 suborders, *Eurylaimi, Tyranni* and *Menurae,* of the *Passeriformes* (see).

Subshrub: see Shrub.

Subsidiary cell: see Stomata.

Subspecies, ***geographical race*** (abb. ssp., plural sspp.): an infraspecific category, i.e. a further subdivision of Species (see), which is given a Latinized trinomial. S. differ genetically, and they occur (at least in the reproductive period) in different geographical regions (see Allopatry). Since they can cross with other S. of the same species and produce fertile offspring, their S. status would be abolished under conditions of Sympatry (see). A good example is the carrion crow (species name *Corvus corone),* whose two Ss. (hooded crow, *Corvus corone cornix;* carrion crow, *Cornus corone corone)* meet in central Europe more or less along the River Elbe. In the contact area the population is mixed, due to occasional mating between Ss. The same phenomenon is observed everywhere in the world that Ss. come into contact on the boundaries of their ranges.

Substance P, ***SP:*** an undecapeptide: Arg-Pro-Lys-Pro-Gln-Gln-Phe-Phe-Gly-Leu-Met-NH_2, found in the brain and intestine of mammals (including man), birds, reptiles and fishes. The highest concentrations of SP measured by radioimmunoassay (SP-like immunoreactivity) in brain are found in the substantia nigra, habenula, hypothalamus, spinal cord sensory ganglia, dorsal roots and substantia gelatinosa. It is now established that SP acts as a neurotransmitter, probably throughout the central nervous system, and certainly in the sensory neurons of the spinal cord. It stimulates contraction of isolated rabbit jejunum, and causes transient hypotension when injected intravenously into anesthetized rabbits.

Substitute behavior: see Displacement activity

Substitute plant community, ***replacement plant community:*** a plant community that has replaced a natural community, not by natural succession, but as a result of human intervention. For example, pasture and meadow communities have been established after clearing alder carrs and woods on the flood plains of large rivers. Also, stands of pine and spruce have been planted in place of deciduous woods to produce a new type of forest community. Whether a community is original or substitute can be determined by comparing the present vegetation with that depicted on old maps, by pollen analysis, and by the study of *contact communities,* i.e. communities directly bordering the community of interest, and whose constitution can be used as an index of whether neighboring vegetation is primary or secondary.

Substitutional load: see Genetic load.

Substrate feeders: see Modes of nutrition.

Substrate hyphae: fungal hyphae that grow in or on a substrate, as distinct from Aerial hyphae (see).

Substrate mycelium, ***nutritive mycelium:*** a fungal lawn consisting of vegetative hyphae, as distinct from reproductive aerial hyphae.

Substrate specificity: see Enzymes.

Subtilin: an antibiotic synthesized by *Bacillus subtilis.* It is a basic polypeptide, and it inhibits the development of some Gram-positive and some Gram-negative bacteria.

Subunguis: see Claws, hoofs and nails.

Succession: the chronological succession of different organism collectives resulting from directional changes in habitat conditions. In a S., the organism composition continues to diverge and is never recompensated, as distinct from, e.g. seasonal variations in the quantitative composition of biocoenoses. S. is not necessarily continuous, and it may pass through relatively stable intermediate phases when the species composition is relatively stable for a number of years. S. continues until a new Dynamic equibrium (see) is attained in the energy and substance economy of the biocoenosis. This represents the climax, after which no further changes in composition occur over a long period of time. Ss. are a useful aid for the classification and ranking of plant societies.

S. can be *autogenic* (induced by the biocoenosis itself, e.g. the silting of a lake), or *allogenic* (caused by external influences, such as human disturbance of the habitat, or a change of climate, e.g. over hundreds of years as the climate warms after an Ice Age). If the interfering factor is removed (e.g. a field is no longer used agriculturally), a *secondary progressive S.* commences, and over a period of years a climax society develops which is appropriate to the soil and climatic conditions.

Primary S. is the development of a completely new ecological structure on an initially organism-free substratum, e.g. the colonization of a newly formed volcanic island, or a habitat produced artificially by human activity. *Secondary S.* starts from an established biocoenosis, which has been subjected to change (fire clearance, melioration, etc.). The end result of S. can, however, be greatly influenced by the random and fortuitous incorporation of particular species in certain phases of the S. Thus, the regrowth of the cleared areas of open cast mines, and their development into useful forests is very dependent on the chance development of an earthworm population.

Succinic acid, ***ethane dicarboxylic acid:*** an organic dicarboxylic acid (see Carboxylic acids), which occurs in esterified form in amber and other natural resins, and as the free acid in plant juices and animal fluids. It is an intermediate in the Tricarboxylic acid cycle (see), it is generated in the Glyoxylate cycle (see), and it serves as a precursor in the biosynthesis of porphyrins.

Succinite: see Amber.

Succinum: see Amber.

Succory: see *Compositae.*

Succulents: see Xerophytes.

Succus: incorrect spelling of Sucus (juice).

Succus juniperi: see *Cupressaceae.*

Sucking behavior: see Suckling.

Sucking lice: see *Phthiraptera.*

Suckling: the feeding of mammalian infants with

their mother's milk during the period immediately after birth. Milk is usually taken by sucking from the nipples or teats of the mammary glands, but monotremes feed on milk secreted from specialized sweat glands that do not possess central nipples. The period of suckling varies according to the species. In wild animals, milk production (lactation) ceases naturally after a given period, when the young offspring have developed sufficiently and are able to rely entirely on other food. It may also stop in response to a new pregnancy, or the loss of the suckled offspring. Regular removal of milk is essential to maintain the flow. In domestic animals, lactation ceases if the young are removed from the mother, and in dairy cattle more frequent milking produces more milk. The onset and continuation of lactation are under hormonal control. S. causes reflex stimulation of the hypothalamus, which in turn stimulates the anterior pituitary to produce more lactogenic hormone.

The S. reflex of the infant occurs in response to contact with, and chemical stimuli from, the mother. Sucking rhythms are species-specific (Fig.). In infants that grow and develop within a litter or brood, the constant number of teats may lead to a S. order, e.g. in cats and pigs. Before S., some species of infants display instinctive head swaying movements while searching for a teat, followed by pressing of the head against the teat while feeding. Altricial infants characteristically perform foot treadling movements in search of the milk supply.

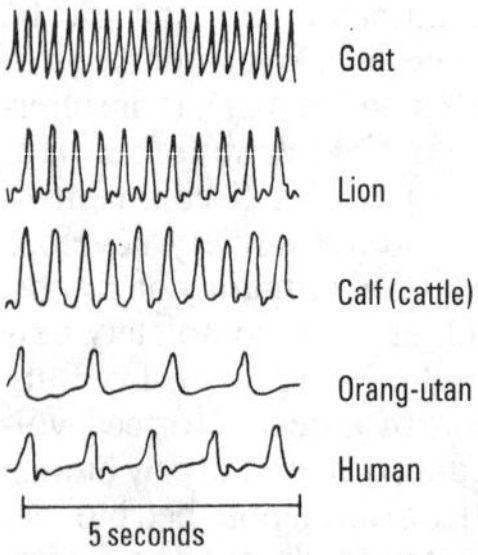

Goat
Lion
Calf (cattle)
Orang-utan
Human
5 seconds

Suckling. Sucking rhythms of infant mammals.

Sucrose, ***cane sugar, beet sugar:*** α-D-glucopyranosyl-β-D-fructofuranoside, a nonreducing disaccharide, consisting of β-1,4 glycosidically linked D-fructose and D-glucose. There are two important commercial sources: the expressed juice of sugar cane *(Saccharum officinarum),* which grows in the tropics, contains 14–21% S., while the expressed juice of sugar beet *(Beta vulgaris),* grown in temperate climates, contains 12–20% S. It is also obtained in small quantities from sugar maple.

Hydrolytic cleavage by dilute acid, or by enzymes like maltase or invertase yields a mixture of equal parts of D-fructose and D-glucose, known as Invert sugar (see). S. is not fermented directly, but supports fermentation after enzymatic cleavage by the fermenting organism to the fermentable sugars, fructose and glucose. In high concentrations, however, it inhibits growth of microorganisms, and is therefore used as a food preservative, as well as a food sweetener. S. serves as a substrate in the industrial fermentative production of ethanol, butanol, glycerol, citric and levulinic acids, and as a preservative, sweetener and demulcent in pharmaceutical preparations. Esters, ethers and other chemical modification products of S. are used in the preparation of detergents, soaps and plastics.

As a weak acid, S. (and saccharides derived from S.) forms salts (saccharates) with alkali and alkaline earth metal hydroxides. S. is widely distributed in plants, where it serves as a transport form of soluble carbohydrate. It is an important metabolic product in all chlorophyll-containing plants, and cannot be synthesized by animals.

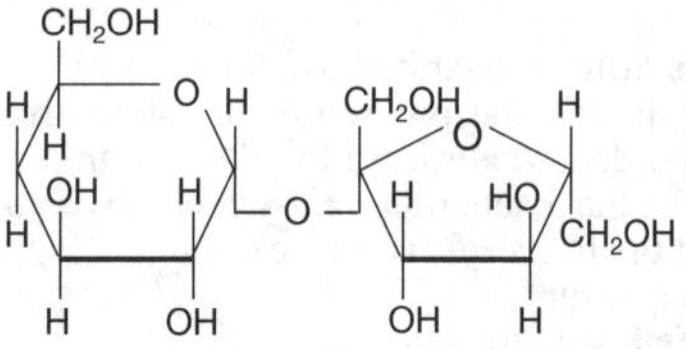

Sucrose

Suction organs: see Adhesion organs.

Suction trap: see Carnivorous plants.

Suctoria: see *Ciliophora.*

Sucus entericus: see Intestinal juice.

Sugar: in the broad scientific sense, any mono- or oligosaccharide. Blood S. is specifically glucose. In the kitchen and the grocery store, the term means sucrose.

Sugarbirds: see *Meliphagidae.*

Sugar cane: see *Graminae.*

Sugar thief: see *Zygentoma.*

Sugar gliders: see *Phalangeridae.*

Suidae, ***pigs, swine:*** a family of the order *Artiodactyla* (see), represented by 8 or 9 recent species in 5 genera. Pigs have relatively well developed hindtoes and a stout body. The snout is extended, mobile, and terminally truncated with disk-like cartilage at the tip. In all pigs the snout is strengthened by an unusual prenasal bone, situated below the tip of the nasal bones of the skull. In both sexes, the upper and lower canines form tusks, which curve outward, forward and upward. They are more developed in the male, and in some species they represent a formidable weapon. Pigs are omnivorous, grubbing for roots and tubers (aided by the tusks and the flat, sensitive nose), as well as eating carrion and small animals (even poisonous snakes). It is exclusively an Old World family, with representatives in Eurasia, the Orient and sub-Saharan Africa. *Sus scrofa* (wild pig and domestic derivatives) has been introduced and is well established in the USA. Fossils of early ancestors are found in the Oligocene; these were massive piglike animals *(Entelodontidae)* with skulls up to 1 m in length.

The following members are described under separate entries: African bush pig, Babirusa, Wart hog, Wild pig. The domestic pig, *Sus scrofa* var. *domestica,* is derived from the Eurasian Wild pig.

Suint: see Yolk.

Sula: see *Pelicaniformes.*

Sulawesi black ape: see Macaques.

Sulawesi pig: see Babirusa.

Sulfatases: esterases that catalyse the hydrolytic cleavage of sulfate esters with release of the sulfate anion. S. are found in animal tissues.

Sulfate: see Sulfates.

Sulfate activation: see Sulfur (section Assimilation).

Sulfate assimilation: see Sulfur.

Sulfate reduction: see Sulfur.

Sulfates: salts or esters of sulfuric acid (H_2SO_4). Dissociation of S. produces the sulfate anion (SO_4^{2-}). The sulfate anion is also called sulfate.

Sulfatides: see Glycolipids.

Sulfides: salts of hydrogen sulfide (H_2S). Dissociation of S. produces the sulfide anion (S^{2-}). The sulfide anion is also called sulfide. See Sulfur.

Sulfur, *S:* a component of many biomolecules, e.g. cysteine and methionine (and proteins and peptides containing these amino acids), as well as vitamins (biotin and thiamin) and various complex lipids, etc.

S is present in the soil as sulfate (mainly gypsum $CaSO_4 \cdot 2H_2O$ and anhydrite $CaSO_4$), and as sulfide (iron pyrites FeS_2, and iron(II) sulfide FeS). The soil also contains a large quantity of organic S. Most plants derive their S from inorganic sulfates, in particular gypsum, which is sufficiently water-soluble to be utilizable. In the soil, sulfides can be oxidized to S by purely chemical processes in the soil; the S is then oxidized to sulfate by S bacteria. Soil bacteria convert organic S compounds into reduced forms of S, which are oxidized to free S and then to sulphate by colorless S bacteria (e.g. *Beggiatoa, Thiobacillus, Thiothrix),* thereby playing an important role in the circulation of S in the biosphere (Fig.). In stagnant water and rice paddies, etc., hydrogen sulfide (H_2S) produced by the degradation of organic material is converted into elemental S by green S bacteria, and the S is oxidized to sulfate by purple S bacteria.

Under anaerobic conditions, e.g. in dense, nonaerated soils, sulfate can be reduced to H_2S, leading to a decrease in the availability of S to plants *(desulfurication).* Bacteria largely responsible for this conversion belong to the genera *Desulfovibrio* and *Desulfotomaculatum,* which use the sulfate ion as an electron acceptor for respiration in place of molecular oxygen (sulfate respiration). The released hydrogen sulfide also decreases the availability of iron, by forming iron(II) sulfide (FeS) and iron(II) disulfide (FeS_2).

Relatively large quantities of S are needed by plants (see Plant nutrients). Higher plants take in sulfate ions through their roots. To a certain extent they can also absorb atmospheric SO_2 through their leaves and oxidize it to sulfate, but this is insufficient to satisfy the total requirement, especially since high concentratrions of atmospheric SO_2 cause plant damage. Injurious levels of atmospheric SO_2 are found especially in areas near to chemical industry, where high SO_2 emission is due to the burning of coal with a high S content, e.g. in the industrial areas of the former German Democratic Republic, and other parts of industrialized eastern Europe.

For its assimilation into organic cell compounds, S (like nitrogen) must be present in a reduced form Only bacteria, fungi and green plants are able to convert sulfate to sulfide. Animals, in contrast, must receive reduced S compounds in their diet. In green plants, reduction of sulfate occurs primarily in the chloroplasts, but it can also occur in the roots. Like biological nitrate reduction, sulfate reduction occurs in two stages:

$$SO_4^{2-} \xrightarrow{2e^-} SO_3^{2-} \xrightarrow{6e^-} S^{2-}$$

The process begins with the formation of *phosphoadenosine phosphosulfate* (PAPS; "active sulfate"). This is transferred to a low molecular mass protein *(M_r* about 5 000) and the sulfate is reduced via protein-bound sulfite to the oxido-reduction level of hydrogen sulfide. Reduced ferredoxin serves as the reducing agent. During the conversion of sulfate to sulfide, all intermediates remain bound to the protein surface, and no intermediate is detectable in the free state. The resulting hydrogen sulfide is transferred to O-acetyl-L-serine to form cysteine. The SH-group of cysteine is then transferred in the synthesis of other S-containing compounds, e.g. methionine, coenzyme A, lipoic acid, thiamin, mustard oils, etc.

With regard to its function in living organisms, S cannot be replaced by any other element. Although Selenium (see) can replace S in organic structures, it cannot functionally replace S in the living cell.

The primary effects of *S deficiency* are seen in protein metabolism. In plants, S deficiency symptoms first appear in the youngest leaves, and are rather similar to those of nitrogen deficiency. The leaves become light green to yellow, with areas of reddish coloration. Members of the *Cruciferae* remain stunted with slender leaves. In Europe, S deficiency is practically unknown, since the soils are treated with adequate quantities of S-containing mineral fertilizers, like ammonium sulfate, potassium sulfate, superphosphate, etc., and in addition considerable quantities of SO_2 are washed into the soil by rain from the atmosphere. Weathering of minerals also relases sulfate into the soil. S deficiency has economic importance in the cultivation of tea.

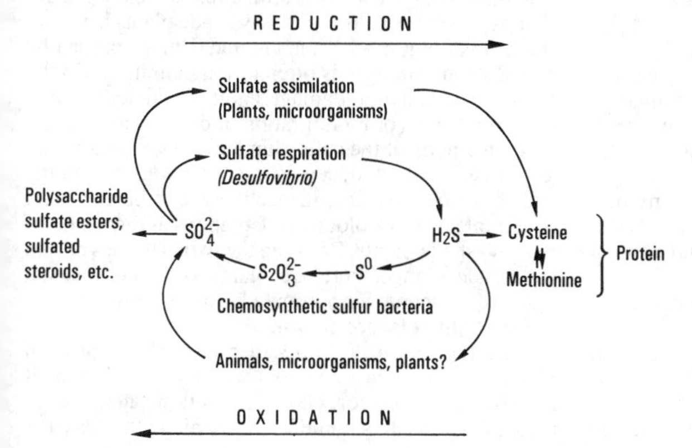

Sulfur cycle

The role of S in animal metabolism is described under Mineral metabolism (see).

Sulfur bacteria: a collective term for autotrophic bacteria that oxidize hydrogen sulfide or other reduced sulfur compounds to elemental sulfur. The sulfur is either deposited intracellularly, or released to the exterior. Colorless S.b., e.g. *Beggiatoa* (see) utilize hydrogen sulfide as an energy source for carbon dioxide assimilation (see Chemosynthesis). Purple sulfur bacteria (see Rhodobacteria, *Chromatium)* and Chlorobacteria (see) utilize hydrogen sulfide as a source of reducing equivalents for photosynthetic carbon dioxide assimilation (see Photosynthesis).

Sulfur-bottomed whale: see Whales.

Sulfur cycle: see Sulfur.

Sulidae: see *Pelicaniformes.*

Sultanas: see *Vitaceae.*

Sumach: see *Anacardiaceae.*

Summation: a phenomenon in which a single stimulus in the sensory nerve is insufficient to produce a propagated action potential in the postsynaptic neuron, whereas a propagated action potential is produced in response to 2 or more stimuli. S. may be spatial or temporal. In *spatial S.,* more than one presynapse is stimulated in spatially separated areas of a neuron, and each presynapse releases transmitter simultaneously. Thus, the presynapses facilitate one another by jointly producing sufficient transmitter to attain the firing level of the neuron. In *temporal S.,* the postsynapse becomes depolarized (i.e. an action potential is propagated in the postsynaptic neuron) in response to 2 (or more) nonsimultaneous impulses arriving at the presynapse. Each of the single impulses alone is insufficient to cause postsynaptic depolarization. Temporal S. therefore occurs when repeated afferent stimuli cause new excitatory postsynaptic potentials (see Synaptic potential) before previous ones have decayed. Ss. are important for the spread of excitation in coupled neurons (see Neuronal circuit), and they have been implicated in theories of nerve reflexes.

Summer cypress: see *Chenopodiaceae.*

Sun animalcules: see Heliozoans.

Sunbeam snakes: see *Xenopeltidae.*

Sun bear, *Malayan sun bear,* *Helarctos malayanus:* a relatively short-haired, black-brown bear with yellowish chest markings, native to Southeast Asia. It feeds mainly on vegetables and fruit, and also takes insects, ground birds, small vertebrates and the nests of wild bees. Its legs are turned inward when walking.

Sunbirds: see *Nectariniidae.*

Sunbittern: see *Gruiformes.*

Sun compass orientation: orientation in space, using the sun as a reference point. In order to maintain a constant position, or a constant direction of movement, compensation must be made for the change in the relative position of the sun during the day. In the S.c.o. of honey bees and other animal species, this compensation is achieved by internal programming, which is based largely on local changes in the horizontal and vertical components of the sun's azimuth.

Sundew: see Carnivorous plants, *Droseraceae.*

Sunfish, *Mola mola* (plate 2): a very large (up to 2 m long, and 1 000 kg in weight), disk-shaped, truncated fish of the high seas (family *Molidae,* order *Tetraodontiformes).* It has long dorsal and anal fins, both jointed to the vestigial caudal fin. The flesh has little value and is virtually inedible. The brownish gray, scaleless body is covered with thick, elastic skin.

Sunfishes, *Centrarchidae:* a family of North American freshwater fishes in the order *Perciformes.* The family also includes the basses, and it is closely related to the true perches. Most are nest builders, and they display brood care. The body is commonly deep and strongly compressed, except in the genera *Aplites, Pomoxis, Elassoma* and *Micropterus,* which are lower and rather more elongated. The lateral line is absent in *Elassoma,* but present in all other genera, where it is complete or almost so. Some species, e.g. the ***Pumpkinseed sunfish*** *(Lepomis gibbosus)* are popular aquarium fishes, while the ***Bluegill*** *(Lepomis macrochirus)* has been used in experimental physiology and ecology. The largest member is the ***Largemouth bass*** *[Micropterus (=Aplites) salmoides],* which attains lengths of up to 83 cm.

Sunflower: see *Compositae.*

Sungazer: see *Cordylidae.*

Sungrebes: see *Gruiformes.*

Suni: see *Neotraginae.*

Sun leaves: external foliage leaves on the south side of a tree, e.g. beech *(Fagus sylvatica).* S.l. are generally much thicker than Shade leaves (see) on the same plant, and they possess well developed palisade parenchyma.

Sun pitcher: see Carnivorous plants.

Sun plants, *heliophytes:* plant species adapted to habitats with high light intensities, e.g. inhabitants of deserts, steppes and stony heaths. Leaves of S.p. are often aligned in such a way that they do not receive the maximal possible solar irradiation. Other common protective adaptations against irradiation include a thick epidermis, a thick cuticle, and surface hairs, as well as reduced or narrow leaves which have a relatively small surface area. Often the protoplasm displays a marked resistance to the damaging effects of radiation, but the nature of this resistance is not understood. In contrast, *shade plants* thrive by utilizing 1/25 (in extreme cases up to 1/400) of the light falling on the surface of their habitat. They perform photosynthesis most efficiently with a radiant energy consumption that is greatly suboptimal for S.p., and they lead a satisfactory existence under light conditions that are insufficient for S.p. to attain a positive metabolic balance. Important factors for shade plants are an increased CO_2 concentration in the air layers near the ground, and certain anatomical features, such as thin leaves and large intercellular spaces, which allow efficient light utilization and easy gas exchange. The protoplasm of many shade plants is not resistant to strong irradiation, and the damage caused by direct, strong sunlight is often of long duration. On the other hand, radiation damage is often counteracted by shedding leaves or by migration of chloroplasts to less irradiated parts of the cells. Sun leaves (see) and Shade leaves (see), which occur on the outer edge and in the interior of tree crowns, respectively, exhibit the same structural and physiological adaptations as the leaves of S.p. and shade plants. Between the two extremes of S.p. and shade plants, there are plants with intermediate light requirements for optimal photosynthesis.

Sun squirrels: see *Sciuridae.*

Sunstar, *Solaster papposus:* a starfish (phylum *Echinodermata,* class *Asteroidea)* with 8 to 14 arms. It is browish red in color with bright yellow lateral bands across the arms. It is found in the sea along the coasts of

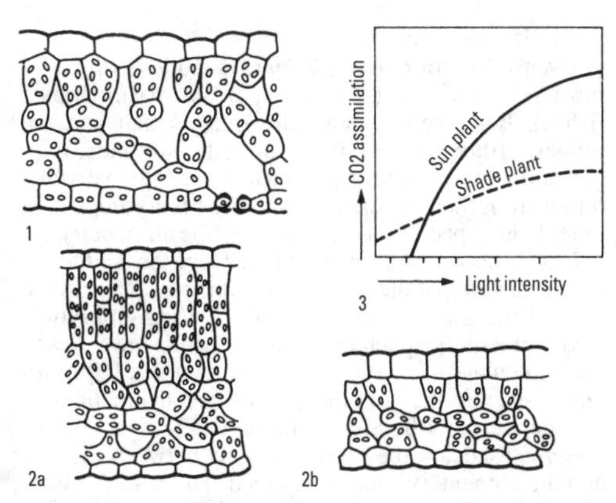

Sun plants. Adaptations of plants to the light intensity of their habitat. *1* Transverse section of the leaf of a shade plant *(Impatiens)*. *2* Transverse section of a sun leaf *(a)* and a shade leaf *(b)* of beech. *3* Intensity of photoassimilation of a sun plant and a shade plant in dependence on the light intensity.

the Northern Hemisphere, including the North Sea and parts of the Baltic.

Superdominance: a situation in which a heterozygous allele combination (+a) surpasses both homozygous types (++ and aa) with respect to certain characters.

Supernumerary limbs: see Regeneration.

Superposition: superimposition of the elements of signal behavior that belong to two different behavioral patterns. Most frequently encountered is the combination of attack and flight motivation. By Ritualization (see), the pattern of superimposed behavior can itself become true signal behavior. In chronobiology, the superimposition of different periods is also called S.

Superposition image: see Light sensory organs.

Superspecies: see Formenkreis.

Support tissues: see Mechanical tissues.

Suppressor mutation: a mutation that functionally but not genetically compensates or reverses the effect of another mutation. It has the effect of converting the mutant phenotype into the wildtype phenotype, and therefore has the appearance of a true back mutation. An *intragenic S.m.* occurs in the same functional gene (cistron) as the first mutation. An *extragenic S.m.* occurs in a different functional region of the genome. As a rule, an extragenic S.m. only partially compensates the effect of the first mutation, which enables it to be easily distinguished from a true back mutation.

Supraesophageal ganglion: see Brain.

Suprapharyngeal ganglion: see Brain.

Suprarenal gland: see Adrenal gland.

Supraphonation: emission of sounds by more than one individual, with a regular ordering of the individual sound components. This may be performed as a "duo" by two animals. If the sounds alternate, it is known as *antiphone*. S. in unison for all or part of the time is known as *antiphonic song* or a "duet". More than two animals take part in *group phonation,* e.g. the roaring calls of lions and the chorous howling of wolves and coyotes. S. has many different functions, and it may have several functions in a single species. It can stabilize partner relationships, serve as an identifying signal of a pair or group, serve for territorial marking of a pair or group, regulate the space-time distribution of partners, signal a particular status, and probably serve many other functions.

Surface epithelium: see Epithelium.

Surface potential: the electrical potential difference between the surface and the bulk of an aqueous solution. Dissociable groups are often localized on biological membranes, so that the membrane surface carries a net charge. The counter-ions present in the solution are subject to thermal agitation. This results in a partial separation of charges, leading to a potential difference between the surface and the total liquid volume, i.e. an Electrical double layer (see) is formed. The concentration of counter-ions is increased while that of co-ions is decreased, a fact which must taken into account in the investigation of membrane transport. The influence of S.p. must also be considered in cell-cell interaction.

Surfbird: see *Scolopacidae.*

Surgeonfishes: see *Acanthuridae.*

Surinam toad, *Pipa pipa:* a member of the *Pipidae* (see), up to 20 cm in length, living in the jungle rivers of Brazil and Guayana. The flattened body is colored uniformly gray-brown, and is well camouflaged against the black river mud. The head is triangular with a pointed nose and flaps of skin around the jaws. On the tip of each of the flexible fingers is a star-shaped pad of sensitive glandular hairs, which is used for locating prey in mud and opaque silt waters. When mating, the male clasps the female in front of her hindlegs. With characteristic, ritualized, body rotations, the female drops a few eggs at a time onto the belly of the male. After a somersaulting movement of the male, the eggs slide onto the back of the female, where they are fertilized. The soft dorsal skin of the female is honeycombed with special pits, each with a horny lid. Each fertilized egg (up to 60 in number) develops inside one of these pits. After 3 to 4 months, the fully metamorphosed young toads emerge from the pits, which then regress.

Surra: see Trypanosomes.

Susceptibility: the opposite of Resistance (see). S. may be horizontal (not specific for one race; not necessarily present in genetically related species), or vertical (race-specific; associated with a particular gene or genetic complement).

Susception: a term from sensory physiology, meaning the purely physical stage of stimulus reception. See, e.g. Geotropism.

Suslik: see *Sciuridae.*

Suspension feeders: see Modes of nutrition.

Sus scrofa: see Wild pig, *Suidae.*

Susu: see Whales.

Sutural bones: see Wormian bones.

Suture:

1) See Ammonites.

2) See Joints.

Swallowers: see *Saccopharyngoidei.*

Swallowing, ***deglutition:*** reflex transport of food from the buccal cavity to the stomach. The mouth is closed and breathing is temporarily suspended. Voluntary raising of the tongue then forces food backward toward the pharynx. As food enters the pharynx, the S. reflex is triggered, in which the larynx is closed by the epiglottis and the nasal passages are closed by the soft palate. Contact of food with the mucosa stimulates a reflex wave of contraction, known as peristalsis, which transports food toward the stomach. As part of this reflex, the cardiac sphincter is also relaxed, admitting food to the stomach. The reflex center is located in the medulla oblongata; it is stimulated by the afferent sensory glossopharyngeal nerve, and the reflex response is due to the action of four motor nerves: hypoglossus, trigeminus, glossopharyngeal and vagus.

Swallows: see *Hirudinidae.*

Swallowtail butterflies: butterflies of the family *Papilionidae;* see *Lepidoptera.*

Swallow-tailed kite: a name applied to 2 members of the avian order, *Falconiformes* (see), family *Accipitridae:* the African swallow-tailed kite *(Chelictina,* subfamily *Elaninae)* and the American swallow-tailed kite *(Elanoides,* subfamily *Perninae).*

Swallow-wort: see *Asclepiadaceae.*

Swamp cypress: see *Taxodiaceae.*

Swamp deer: see Barasingha deer.

Swamp eels: see *Synbranchiformes.*

Swamp warblers: see *Sylviinae.*

Swan goose: see *Anseriformes.*

Swan mussel: see Freshwater mussel.

Swans: see *Anseriformes.*

Swarming: behavior leading to the formation and maintenance of a loose community of a large numbers of individuals, known as a *swarm.* S. is a form of semibiosocial behavior, involving mutual attraction and coordination of behavior, as seen in the movements of fish shoals and bird flocks. In a *polarized swarm,* the bodies of all members are aligned in parallel, whereas in an *unpolarized swarm,* they lie in different directions. Large herds of ungulates are an example of the latter. S. by large herbivores is usually referred to a *herding.* S. is a special form of environmental adaptation, associated with, e.g. migration and protection against predators. S. is more specifically applied to the behavior of bees, when a young queen leaves the hive, taking large numbers of workers with her.

Sweat: an opalescent, colorless liquid, actively secreted by S. glands. Average S. production by humans is 0.5–0.7 liters per day, but with physical exertion, this may increase to 4 liters, and in the tropics to 10–15 liters. S. is an aqueous solution of inorganic salts (sodium and potassium chloride, calcium and magnesium phosphate, sulfates), protein, urea, uric acid, creatine, amino acids, ammonia, fatty acids, acetone and lactic acid. Volatile fatty acids are responsible for the odor of S. The most important function of S. is its contribution to Temperature regulation of the body (see): moistening with S. greatly increases the heat conductance of the skin, and evaporation of S. extracts considerable quantities of heat.

Horses and donkeys also secrete large quantities of S. Cattle and sheep produce very little S. Many animals (birds, monotremes and many rodents) possess no S. glands and produce no S. See Perspiration.

Sweet chestnut: see *Fagaceae.*

Sweet flag: see *Araceae.*

Sweet gale: see *Myricaceae.*

Sweet pea: see *Papilionaceae.*

Sweet pepper: see *Solanaceae.*

Sweet potato: see *Convolvulaceae.*

Swiftlets: see *Apodiformes.*

Swift moths: moths of the family *Hepialidae;* see *Lepidoptera.*

Swifts: see *Apodiformes.*

Swim bladder, ***gas bladder:*** a hollow organ of most teleost fishes (it is absent from elasmobranch fishes), lying between the alimentary canal and the kidneys, filled with a mixture of carbon dioxide, oxygen and nitrogen, whose proportions are often very different from those of the atmosphere. Embryologically, the S.b. is a specialized derivative of the alimentary canal; it develops as a diverticulum from the middorsal or lateral walls of the gut, and in some larval fishes the wall of the S.b. has peptic glands. During embryonic development, the pneumatic duct between the S.b. and the alimentary canal may be lost or retained. In *physoclistic* species, the proximal part of the pneumatic duct degenerates and the S.b. is closed, whereas in *physostomatous* species, the pneumatic duct between the S.b. and the alimentary canal is retained. The closed S.b. of physoclistic species is differentiated into two morphologically different regions: an anterior secretory region containing the gas gland and associated *retia mirabilia* or *rete mirabile* ("wonder net"), and a posterior resorbent region called the *oval.* The oval develops from the distal end of the degenerating pneumatic duct. In *euphysoclistic* fishes a diaphragm separates the posterior, gas-resorbing part of the S.b. from the anterior gas-producing section, whereas in *paraphysoclistic* fishes the gas-producing and gas-secreting areas are not well separated. The S.b. is innervated by the sympathetic system via a branch from the coeliacomesenteric ganglion and by branches from the left and right intestinal vagus nerves.

The S.b. may function as a hydrostatic organ, or it may contribute to respiratory gas exchange. It can also serve as a sense organ, or act as a resonator in sound production (see Weberian ossicles).

For the mechanism of gas production in the S.b., see Hairpin loop countercurrent principle.

Swimming: active locomotion of a living organism in a liquid medium (see Movement). Swimming organs generate propulsive forces which overcome the resistance of the surrounding medium. Biomechanically, a distinction must be made between organisms in which the generation of forward motion and the generation of drag can be separated, and those organisms in which a single unit is responsible for propulsion and the generation (and minimization of) drag. The former organisms are those with rigid bodies, such as turtles and water beetles. In contrast, propulsion and drag generation cannot be separated in, e.g. fishes, whales, flagellates, spermatozoa and penguins. S. is fundamentally influenced by the density of the organism in relation to that of the surrounding medium. In many cases, the density of the organism is decreased by the presence of a hydrostatic organ, such as the swim bladder of fishes or the air-filled chambers of the cuttle fish phragmocone, or in smaller organisms by the inclusion of oil droplets. On the other hand, some animals, e.g. sharks, must generate buoyancy and lift by the development of a suitable body shape and by continuous S.

The relationship between drag and speed is given by the Reynolds number (see). S. organisms display many different adaptations for the optimization (minimization) of their drag. Thus, the body shape of the Great diving beetle *(Dytiscus marginalis)* is a compromise between low drag and high S. stability. Organisms of the open sea tend to have the shape of a rather wide spindle (elongated elipsoid), which provides optimal mass transport for a given velocity with the lowest possible drag. For example, dolphins, porpoises and tuna fish are elongated elipsoids, pointed at the rear, and with the largest cross section well to the rear. The same principle is used in e.g. the design of aeroplanes. Many fishes (e.g. mackerel, swordfish) have keeled scales in the region of their greatest width which act as vortex generators. By making the boundary layer turbulent in this region, it is prevented from separating and causing a large pressure drag; it rather separates further back where the body cross section is smaller and the resulting drag smaller.

In S. organisms it is extremely important to suppress turbulence at the front of the body. In dolphins this is achieved by the presence of elastic skin structures, which damp out the oscillations that would otherwise develop into turbulence. Even traces of dissolved fish slime can greatly reduce the frictional drag of the surrounding medium. In extreme situations, such as fleeing from a predator or capturing prey, where high speed is required, boundary layer oscillations (caused by high velocity) agitate the slime and cause it to dissolve in the boundary layer just where separation is about to occur, thereby preventing separation and reducing drag.

Swimming crabs, *Portunidae:* a family of true crabs (see *Brachyura),* with representatives in the coastal regions of all seas. The last pair of legs have a broad, flattened terminal segment, which serves as a swimming aid. Members include the ***Shore crab*** *(Carcinus maenus)* (Fig.), the commonest crab of the North Sea, which has also penetrated the western Baltic. It lives in the intertidal zone, and buries itself in the sand during the ebb tide. The ***True swimming crabs*** *(Portunus),* capture prey by darting rapidly from a resting position on the coastal sea bottom, e.g. the ***Common swimming crab*** or ***Velvet crab*** *(Portunus holsatus)* of the Atlantic. ***Blue crabs*** *(Callinecta)* are common in the warm seas around Central America. *Podophthalmus vigil* of the Indo-Pacific is notable for its enormous eye stalks.

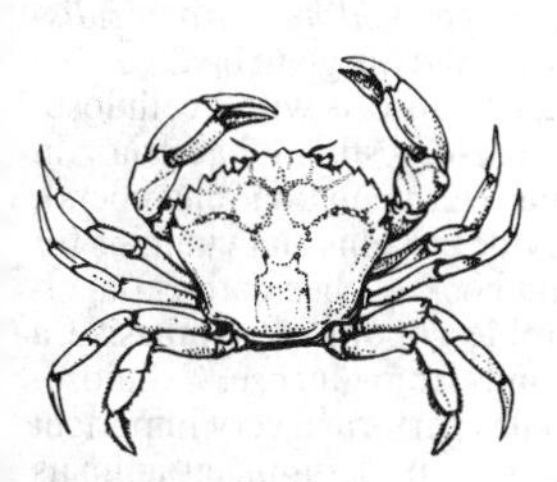

Swimming crabs. Shore crab *(Carcinus maenas).*

Swine: see *Suidae.*

Swiss cheese plant: see *Araceae.*

Swordfish, *Xiphias gladius:* a 4 m-long, economically important, tasty marine fish, distributed worldwide, and the only living representative of the family *Xiphiidae,* suborder *Scombrodei* (see), order *Perciformes.* The upper jaw is flattened, tapered and very elongated. Adults lack teeth and have no scales. Ventral fins are lacking. It hunts shoal fish, such as herring or mackerel, as well as feeding on squid. It is a popular sporting fish, and can be dangerous to humans.

Swordtail butterflies: butterflies of the family *Papilionidae;* see *Lepidoptera.*

Sycamore: see *Aceraceae.*

Sycon: see *Porifera,* Digestive system.

Sylvia: see *Sylviinae.*

Sylvicapra: see *Cephalophinae.*

Sylvietta: see *Sylviinae.*

Sylviinae, *Old World warblers:* an avian subfamily of the *Muscicapidae* (see) in the order *Passeriformes* (see). Most of the 300 species (in about 30 genera) are found in Eurasia or Africa, northern members being migratory. "True warblers" belong to the genus *Sylvia,* e.g. the ***Blackcap*** *(S. atricapilla;* found throughout Europe, extending into northwest Asia and Asia Minor, migrating to southern Mediterranean and equatorial Africa in the winter). The ***Dartford warbler*** *(S. undata)* is found in southern Europe and a small part of North Africa, and leads a precarious existence in southern England, where it is in danger from severe winters (it is nonmigratory) and the exploitation of its natural habitat (gorsy heathland) for farming and suburban expansion. The European ***Whitethroat*** *(S. communis)* (plate 6), found in open country and woodland edges, overwinters in Africa and Arabia. ***Reed*** and ***Sedge warblers*** *(Acrocephalus)* are common in marshy habitats throughout the Old World. In Africa similar habitats are occupied by ***Swamp warblers*** *(Calamocichla).* The genera, *Cettia and Bradypterus,* are found in marshy habitats and forest undergrowth in subtropical and tropical Asia and Africa, e.g. ***Cetti's warbler*** *(Cettia cetti)* of marshes and riversides in southern Europe. *Cettia diphone* ***(Japanese bush warbler)*** features in Japanese folk lore as a melodious song bird. ***Tit warblers*** or ***Tit babblers*** *(Leptopoecile)* are represented by only 2 species, found in western China and Central Asia; they have fluffy, multicolored plumage, which are rather aberrant characters for the *Sylviinae.* Most of the 70 species of ***Grass warblers*** *(Cisticola)* are found in Africa; they build exquisite globular or bottle-shaped nests, using plant fibers and cobwebs. The ***Canegrass warblers*** *(Megalurus;* southern Asia, Australia) are similar to, but larger than the grass warblers. ***Australian wrens*** or ***Wren warblers*** *(Malurus;* Australia, New Guinea, New Zealand) have brilliant plumage and long tails. Together with the Australian ***Emu wrens*** *(Stipiturus), Malurus* is placed by many authorities in a separate subfamily, *Malurinae.* Emu wrens are largely brown in color and possess only 6 tail feathers, which are long and loosely barbed like emu feathers. ***Grasshopper warblers*** *(Locustella)* are found widely throughout Europe and Asia, migrating to southern Asia in the winter. The ***Leaf warblers*** (including the ***Willow warbler*** , ***Wood warbler and Chiffchaff)*** *(Phylloscopus)* are small greenish and yellowish birds, mainly distributed in the Palaearctic and Oriental regions. In the tropical regions of Asia and Africa *Phylloscopus* gives way to genera such as *Seicercus, Abroscopus, Apalis, Sylvietta* and *Eremomela,* which are are more brightly colored and patterned than *Phylloscopus,* with longer and often fan-shaped tails, and shorter, rounder wings. In many characteristics they are intermediate between *Phylloscopus* and the ***Wren-warblers*** *(Prinia)* and the ***Tailor birds*** *(Orthotomus).* Tailor birds, found from Sri-Lanka to southern China and Java, are famous for their nests, which are built in a cradle formed by sewing together the edges of one large leaf, or 2 or more leaves. The leaf is pierced with the bill, and separate stitches are made, using threads of wool, cotton, spider's web, etc. The nest itself is a soft cup of plant down lined with hair and grass. The New Zealand

***Fernbird** (Bowdleria punctata)* is also included in the *Sylviinae.*

The ***Kinglets** (Regulus)* have a fragmented range, including Europe and various parts of Asia, as well as North America. The ***Firecrest** (R. ignicapillus)* is mainly European, while *R. goodfellowi* occurs only in mountain pine forests of Formosa. The ***Goldcrest** (R. regulus;* 9 cm), Europe's smallest bird, has yellow-green plumage and a double wing bar; the male has an orange crest, and the female's is more yellow, both with a black border at the base. It inhabits mainly pine forests, where it hunts small insects and especially favors spiders' eggs. Its globular nests are sometimes suspended from the fine ends of branches, sometimes at considerable heights. *Regulus* is represented in North America by the ***Golden-crowned kinglet** (R. sapata)* and the ***Ruby-crowned kinglet** (R. calendula).* The fiery crown colors and fluffy plumage of *Regulus* set the genus apart as a rather aberrant genus, sometimes allocated to a separate subfamily, *Regulinae.*

In some taxonomies, ***Gnatcatchers** (Polioptila,* 8 species,Canada to Mexico and the Bahamas) and ***Gnatwrens** (Microbates* and *Ramphocaenus,* 4 Neotropical species) are also included in the *Sylviinae.* More usually, however, they are assigned to the subfamily, *Polioptilinae.* With the exclusion of the gnatcatchers, gnatwrens, and kinglets, the *Sylviinae* becomes an exclusively Old World group.

Sylvilagus: see *Leporidae.*

Symbiosis: temporary or permanent association of different species, with marked mutual dependence, and to the benefit of both partners. In the broad sense, S. includes alliance and mutualism (see Relationships between living organisms). The smaller and more simply organized of the 2 partners is called the *symbiont.*

Lichens (see) consist of a symbiotic association of an alga and a fungus, in which the fungus receives some of the carbohydrate synthesized by the alga, and the alga receives water and nutrient salts from the fungus. Important symbiotic associations of bacteria and higher plants also involve metabolic cooperation. Thus, root nodule bacteria fix atmospheric nitrogen and make it available to the plant in a utilizable form; 1 hectare of lupins is capable of fixing 200 kg of atmospheric nitrogen during the growth season. In turn, the bacteria receive carbohydrate from the plant. The root nodule system may be considered as a rather one-sided S., largely to the advantage of the plant, since the bacteria are eventually destroyed.

A symbiotic association between a higher plant and a fungus is called a Mycorrhiza (see).

Animals enter into many different symbiotic relationships. In *ectosymbiosis,* one partner remains outside the body of the other. In *endosymbiosis,* the smaller partner (symbiont) lives inside the body of its larger partner.

Pollination S. between flowering plants and pollinators (insects, birds, bats) is more correctly considered as a form of mutualism. The same applies to the distribution of seeds and fruits by animals (zoochory).

Cleaning S. between large animals and birds that eat insect grubs, and between fishes and their smaller cleaning fish serves as a protection against ectoparasites and skin diseases. It has been shown that the absence of cleaning fishes has a definite deleterious effect on the health of the larger partner.

S. between coral fishes and stinging cnidaria serves as a protection against predators. A well known example is the S. between the hermit crab, *Eupagurus bernhardus* and the sea anemone, *Calliactis parasitica.* Without the stinging protection of the sea anemone, the crab is far more likely to be consumed by a cuttlefish; the sea anemone profits from fragments of the crab's prey and from the wider distribution afforded by transport.

Feeding S. occurs as trophobiosis between ants and insects that suck plant juices (aphids, scale insects and cicadas); the ants consume the sugary excrement (honeydew) of the juice sucker, and in turn protect it from natural enemies. Leaf-cutter ants, termites, some beetles and some wood wasps cultivate fungi in special chambers in the colony nest or in chewed out tunnels, and these fungi serve as a valuable food source. The fungal spores are often transferred in the digestive tract, or in vaginal or intersegmental pouches.

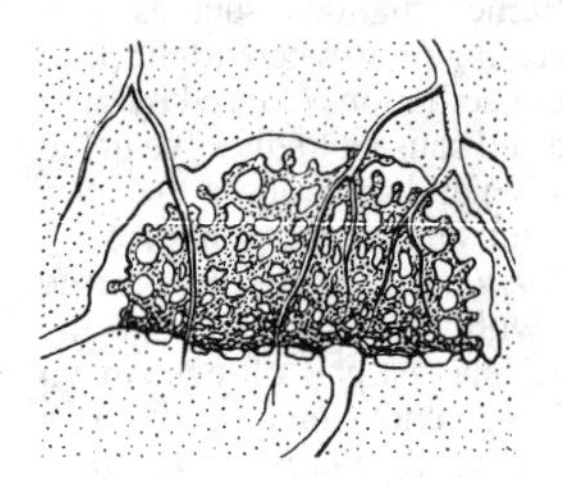

Symbiosis. Section though a fungus chamber of the leaf-cutting ant, *Atta.*

Endosymbiotic bacteria, algae, fungi and protozoa reside in natural cavities or specialized organs inside the body of their partner, and sophisticated mechanisms often exist for their transfer to the next generation. Most of these systems represent feeding S. Thus, protozoa, sponges, cnidarians, turbellarians and mollusks utilize the photoassimilate and the oxygen produced by green algae and cyanobacteria that live in their cells or tissues, e.g. the thecate ameba, *Paulinella chromatophora,* contains cyanobacteria, the polyp, *Chlorohydra viridissima,* contains zoochlorellae, and the giant bivalve, *Tridacna,* contains zooxanthellae. Insects with a cellulose-rich diet carry cellulose-digesting bacteria in evaginations of their gut (fermentation chambers); these and yeasts often provide proteins and vitamins for their partner. Termites and cockroaches harbor flagellates, which under anaerobic conditions synthesize a variety of organic compounds from glucose. Cellulose-digesting bacteria and ciliates perform a very important symbiotic function in the rumen of ruminant animals and in the cecum of many rodents. By their continual growth and death, the bizarre-shaped rumen ciliates and rumen bacteria provide a valuable source of proteins and vitamins for their host. On gram of rumen contents may contain 13 thousand million (13^9) bacteria and 1 million ciliates. Rodents obtain vitamins by eating their own cecal contents (cecotrophy). The human infant depends on *Lactobacillus bifidus* for the degradation of oligosaccharides in the maternal milk; the symbiont enters the mouth of the neonate from the vaginal flora during birth.

Symbranchii: see *Synbranchiformes.*

Symmetry: the ordered repetition of identical or similar components. Uniform distribution of similar elements in a straight line is called *longitudinal S.* or *trans-*

lational S. In animals, especially in many arthropods (e.g. myriapods), longitudinal S. is expressed mainly as body segmentation. The successive, longitudinally arranged segments are called metameres. In algae, rhythmic alteration of the alignment of the axis of the mitotic spindle of the apical cells leads to dichotomy. Alternatively, unequal division of the subapical cells leads to lateral branching and differentiation of the thallus into nodes and internodes. In seed plants, the stem is also organized into nodes and internodes. In all cases, therefore, the plant body is divided into similar segments which are repeated periodically along the longitudinal axis, i.e. there is a condition of longitudinal S. Transverse repetition of similar structural elements results in *lateral S.,* which can be either rotation S. or mirror S. In *rotation S.,* similar elements coincide if they are rotated around a central axis of S. *Mirror S.* (the strict meaning of S.) means that the two halves of the bisected organism or organ are exact mirror images of each another. Often mirror S. is combined with rotation S. If the bisection plane can be rotated around the central axis to other positions of mirror S., the condition is known as *radial S., multilateral S.* or *polysymmetry.* This occurs in radial or *actinomorphic* structures, which are common among coelenterates and echinoderms (e.g. starfish). In spermatophyes, radial structure often arises from uniform branching, e.g. in conifers and in numerous flowers. If only two intersecting planes of S. are present, the condition is that of *disymmetry,* also known in botany as *bilateral S.* [e.g. the flattened shoots of opuntia and the flowers of *Cruciferae* and *Dicentra spectabilis* (bleeding heart)]. If there is only one possible mirror plane, e.g. the dorsal and ventral surfaces are dissimilar, the condition is called *monosymmetry* or *dorsiventrality* by botanists, and also called *bilateral S.* by zoologists. Dorsiventrality occurs widely in animals, including humans. Most leaves display dorsiventrality. Dorsiventral flowers (e.g. pansy) are also frequently called *zygomorphic.* In the absence of any plane of S., i.e. there is no lateral S. (e.g. *Canna* flowers) the condition is one of *asymmetry.*

Sympathetic nervous system: see Autonomic nervous system.

Sympathetic tone: see Autonomic nervous system.

Sympatric speciation: see Speciation.

Sympatry: the common occurrence of 2 taxa in the same geographical area. On the basis of the Competition-exclusion principle (see), the sympatric occurrence of 2 closely related forms is taken as important evidence for their genetic isolation, and therefore for their status as separate species. See Polymorphism. The converse of S. is Allopatry (see).

Sympedium: an association of young animals, often siblings. Sympedia are formed by young spiders and by butterfly larvae. As young spiders develop to the stage where they hunt or catch their own prey, they disperse, whereas caterpillars of the oak processionary moth remain together for feeding, resting, molting and migrating.

Young fish also form sympedia. Shoals of young minnows only admit new members whose body lengths differ by no more than 1.2 cm from the group average. Groups of young birds, known as crèches, are formed by penguins, pelicans and the common shelduck. Young walruses also sometimes form crèches. Such crèches, however, are not pure sympedia, because they are supervised by one or more older animals.

Within an S. there are varying degrees of bonding. Possible reasons for S. formation are an early tendency for social bonding, communal feeding or protection, and bonding to a common locality.

Sympetalae tetracyclicae: see *Dicotyledoneae.*

Symphile: see Myrmecophiles; Relationships between living organisms.

Symphily: see Relationships between living organisms.

Symphoricarpos: see *Caprifoliaceae.*

Symphyla: a class of myriapodous arthropods (see *Myriapoda).* They are white, thin-skinned animals with 11–12 pairs of legs and string-like antennae. The anal segment carries a pair of large spinnerets and a pair of long sensory hairs (trichobothria). Most species are between 4 and 5 mm in length, some species attaining maximally 10 mm. The genital openings lie on the ventral side of the fourth trunk segment. Eyes are absent, but there are 2 well developed organs of Tömösvary (disk-shaped organs with a central pore, which receives the endings of subcuticular sensory cells). The tracheal system opens to the exterior via a pair of stigmata below the antennae. Barely 130 species are known, and 3 families are recognized: *Scolopendrellidae, Scutigerellidae,* and *Geophilellidae.* They are found in the damp humus layer of the soil down to 50 cm, in leaf litter, and beneath moss and tree bark. Decaying organic material appears to be the main dietary material; some species, notably *Scutigerella immaculata,* feed on the fine root hairs of higher plants. The resulting agricultural damage may be considerable, with population densities reaching 1000–20 000 individuals below each square meter of soil surface. Eggs are laid in earth cavities, attached to a short stalk. Emergent larvae possess 6 pairs of legs. It has been estimated that the *S.* live for about 7 years.

The mouthparts of the *S.* bear are structurally similar to those of insects, but functionally different. The suggested affinity of the *S.* to insects is not generally accepted.

Symphyla. Symphylella vulgaris.

Symphysis: see Joints.

Symphysodon: see Cichlids, Discus fish.

Symphyta: see *Hymenoptera.*

Symptom: a characteristic sign or indication of a particular disease or injury. In phytopathology, Ss. are the totality of the external, visible changes in the organism resulting from a disease or injury. In clinical medicine, the strict definition of S. is an indication of a disease or disorder noticed by the patient, a *presenting S.* being a S. that leads a patient to consult a physician. In contrast, the indication of a disorder recognized by the physician, but not by the patient, is strictly speaking a *sign.*

Symptomology: the totality of externally recognizable primary and secondary symptoms or signs of a disease, disorder or injury. It has a similar meaning to Clinical picture (see), but may be applied to nonclinical situations, e.g. in plant pathology.

Synallaxinae: a subfamily of the *Furnariidae* (see).

Synalpheus: see Shrimps.

Synangium: a structure formed by fusion of sporangia. It occurs in some psilophytes, ferns and lyginopterids (seed ferns).

Synanthropy: intimate association between a living organism and humans, human dwellings or a habitat that exists through human agency. Such organisms (e.g. housefly, brown rat) are also known as hemerophilic species. The number of synanthopic organisms increases as the climatic temperature decreases (synanthropy rule).

Synanthropy rule: see Ecological principles, rules and laws.

Synapse: a site for the transfer of excitation from one cell to another. Ss. consist of three morphologically separate regions: 1) the ending of a nerve fiber or sensory cell (the *presynaptic region* or *presynapse),* 2) the *synaptic cleft* (a functional space between the cells), and 3) the membrane region of the effector cell (the *postsynaptic region* or *postsynapse)* (Fig.). The latter may be a nerve cell *(neuroneuronal S.),* a gland cell *(neuroglandular S.),* or a muscle fiber *(neuromuscular S.).* Neuroneuronal Ss. are further subdivided into *axodendritic Ss., axosomatic Ss., axoaxonal Ss.,* and *dendrodendritic Ss.,* according to which parts of the neurons form the pre- and postsynapse. A distinction is drawn between electrical and chemical Ss. *Electrical Ss.* occur in invertebrates: each action potential arriving at the presynapse is transferred directly to the postsynaptic membrane, where it gives rise to a similar action potential. In the typical *chemical S.,* found in both vertebrates and invertebrates, impulses arriving in the presynapse cause the release of Neurotransmitters (see) by exocytosis into the synaptic cleft. Molecules of the neurotransmitter then become bound to specific Receptors (see) on the postsynaptic membrane, resulting in changes of electrical potential across this membrane. In the presynapse, neurotransmitter substances are contained mainly or exclusively in synaptic vesicles, i.e. the neurotransmitter molecules are enclosed within a membrane. Different types of presynaptic vesicle can be distinguished according to their diameter (40–200 nm) and other characters observable under the electron microscope, e.g. vesicles containing noradrenalin are electron-dense, whereas those containing acetylcholine are clear. According to Dale's principle, it was formerly thought that each nerve ending produces only one neurotransmitter, but more recent work suggests that at least two chemically different neurotransmitters may be operative in any synapse (see Neuropeptides). Release of neurotransmitter into the 15–150 nm-wide synaptic cleft is quantized. It is generally accepted that excitation can by propagated in only one direction, i.e. from the pre- to the postsynapse, but there is evidence that in a few special cases, even in chemical Ss., excitatory transfer may be possible in the reverse direction. By serving as junctions for the unidirectional propagation of excitatory stimuli, Ss. facilitate the organization of neurons into Neuronal circuits (see). The number of presynapses associated with an effector cell varies from 1 to about 120 000, e.g. the Purkinje neurons of the human cerebellum receive inputs from more than 100 000 other neurons. In the human brain each effector nerve cell receives inputs from about 1 000 presynaptic endings. Thus for 10^{11} neurons, the total number of Ss. is about 10^{14}.

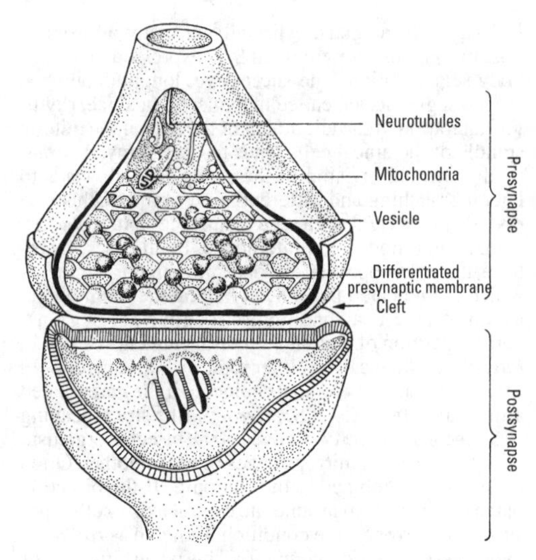

Synapse. Schematic representation of a neuro-neuronal synapse (modified from Akert).

Synaptic cleft: see Neuron, Synapse.

Synaptic potential: alterations of postsynaptic potential, which occur when transmitters occupy receptors on the postsynaptic membrane. The membrane may be depolarized. The initial depolarizing response produced by a single stimulus begins about 0.5 ms after the afferent impulse enters the spinal cord; it reaches a peak 1–1.5 ms later, then declines exponentially. During this potential, excitability by other stimuli is increased, so that the potential is called an *excitatory postsynaptic potential* (EPSP). Alternatively, hyperpolarization occurs, i.e. an *inhibitory postsynaptic potential* (IPSP) is produced. See Excitation, Endplate potential.

Synaptic vesicle: see Neuron, Synapse.

Synaptonemal complex: a structure, visible under the electron microscope, which forms just before pachytene during Meiosis (see) and disappears immediately afterward. It apparently serves as a scaffolding structure, keeping the homologous chromosomes together and closely aligned with each other. It must be present in order for crossing over to occur. Between 160 and 200 nm in width, it consists of two long lateral protein elements which are closely apposed to the chromosomes (the two homologous chromosomes are aligned linearly on the opposite sides of each element) and connected to a central element by filaments. It therefore has the appearance of a long, double ladderlike structure.

Synarchic interactions: see Calcium.

Synascidiae, *Ascidiae compositae:* colonial ascidians, whose individuals are surrounded by a common envelope and often share a common cloacal cavity. See *Monascidiae.*

Synbranchiformes, *Symbranchii,* ***swamp eels:*** an order of eel-like teleost fishes, represented by a single family, *Synbranchidae.* All are tropical or subtropical freshwater fishes, some species occasionally being found in brackish water. Pectoral and pelvic fins are absent, although pectorals are present during early development in some species. Dorsal and anal fins are reduced to vestigial rayless ridges, while the caudal fin is small, vestigial or absent. Scales are absent, except in the subgenus, *Amphipnous*. All species have small eyes, which may even be nonfunctional and below the skin. The gill membranes are united, and the gills communicate with the exterior via a slit or pore under the head or throat. There is no swim bladder and ribs are absent. Through modification of the gill apparatus most species can breathe air when their swamp water becomes depleted of oxygen. In Asia, they are prized as food fish. Four genera are recognized: *Macrotrema* (1 species in fresh and brackish water; Thailand and Malay Peninsula); *Ophisternon (= Furmastix;* 6 species, 2 in the Americas, 4 in Asia); *Synbranchus* (2 species, Mexico and Central and South America); *Monopterus* (subgenera *Amphipnous* and *Typhlosynbranchus;* 6 species, Liberia and India to Japan).

Synbranchus: see *Synbranchiformes.*

Syncarida: see *Malacostraca.*

Syncerus: see *Bovinae.*

Synchorology: study of the geographical distribution of plant societies, also involving the mapping of higher vegetative units. See Sociology of plants.

Synchrondrosis articulation: see Joints.

Synchronization: temporal coordination between events and processes of an individual and those of the ecosystem, population, or another individual. Many organisms synchronize their pattern of behavior during the day with the cycle of light and dark, or their behavior may undergo seasonal changes during the year. In the *ethological* sense, the near-field behavior of individuals is often mutually synchronized, e.g. the temporal coordination of movement and body contact during copulation, or the synchronization of behavior during courtship display or group hunting. In the *chronobiological* sense, S. is the establishment of a quasi-stationary phase relationship between two or more oscillations.

Synchronous culture: a culture of unicellular organisms or eukaryotic cells in which all cells divide at the same time (synchronously). With the aid of S.c., cells or microorganisms can be investigated at separate stages of their development, e.g. S.c. can be used to analyse different stages of the cell cycle. If a bacterial culture is maintained at a certain suboptimal temperature, all cells grow until they are ready to divide, then remain in this condition until the temperature is raised. Similarly, bacterial cells in the stationary phase of growth are largely in synchrony, so that they divide synchronously after being placed in fresh growth medium. One of the earliest techniques for obtaining a S.c. of mammalian cells was to expose the cells to a drug that interferes with a specific step in the cell cycle and prevents cells from progressing beyond it; eventually, all cells in the culture are halted at the same developmental stage. After removal of the block, all the cells then cycle together. It is also possible to exploit the fact that cells in the M phase become rounded and hardly adhere to the culture dish, so that they are easily removed by gently agitation. Cells obtained by this method of "mitotic shake-off" are then ready to enter the G_1 phase. In addition, since cells become progressively larger as they pass through the cell cycle, sedimentation or filtration can be used to select cells at each stage of the cycle.

In all eukarytoic and prokarytoic cells, the time between cell divisions involves one or more random components, so that a synchronous cell population soon (within a few cell divisions) drifts out of synchrony.

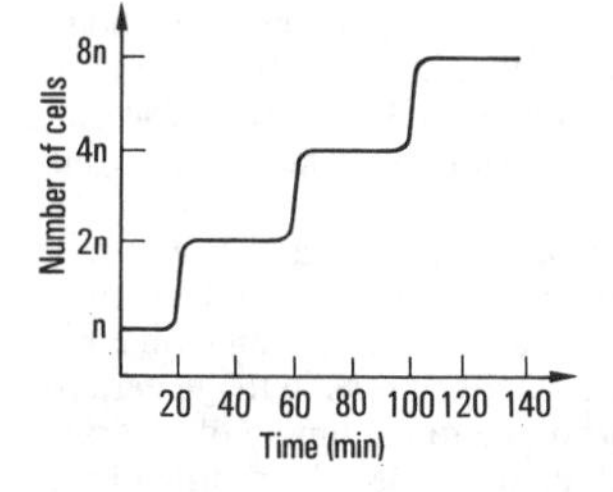

Synchronous culture. Growth curve of a synchronous culture of microorganisms.

Synchytrium: see *Phycomycetes.*

Synchytrium papillatum: see Galls.

Syndesmochorial placenta: see Placenta.

Syndesmosis articulation: see Joints.

Syndrome: a family of different but related symptoms that is characteristic of a specific disease or illness.

Syndynamics: see Sociology of plants.

Synechthran: see Myrmecophiles.

Synecology: an branch of ecology concerned with associations of living organisms (heterotypic organism collectives) and their relationship with the environment. Important areas are plant sociology, biocoenology and production biology, as well as applied sciences, like soil ecology, agrarian ecology, marine ecology and limnology. There are also close connections with biogeography and geomorphology. See Sociology of plants.

Synergids: the two cells which together with the egg cell form the egg apparatus of the embryo sac. See Macrosporogenesis, Flower.

Synergism: cooperative activity of materials or factors, so that the total activity is greater than the sum of the individual activities. Certain ions act synergistically in the mineral metabolism of plants. S. also exists between different phytohormones. Environmental factors act synergistically on the metabolism, growth and development of plants. S. sometimes occurs between pharmacological agents, and between plant protection agents such insecticides. The converse is Antagonism (see).

Synergists:

1) Materials or factors which promote the activity of each other. The combined effect is greater than the sum of the effects when each is acting alone. The phenomenon is called synergism.

2) Muscles that perform the same function and act together.

Syngamus: see *Nematoda.*

Syngnathidae: see *Syngnathoidei.*

Syngnathoidei, *Lophobranchii,* ***ghost pipefishes, pipefishes*** and ***seahorses:*** a suborder of fishes of the order *Sygnathiformes (Dactylopteriformes).* They are small, curiously and characteristically shaped fishes, lacking scales, but possessing an armor consisting of rings of hard dermal plates or scutes. Food (small crustaceans, small fish) is sucked into the small, toothless mouth at the end of an elongated tubular snout. The gills are tufted and lobelike. Some species, especially the sea horse, anchor themselves by coiling their prehensile tail around underwater plants. There is no caudal fin. The suborder contains 2 familes:

1. Family: *Solenstomidae* (ghost pipefishes). Represented by a single genus, *Solenostomus (= Solenichthys = Solenostomatichthys),* of the tropical Indo-Pacific

seas. Each of the 5 species has a short, compressed body (maximal length 16 cm). The first dorsal fin has 5 long spines, while the second has 18–23 soft rays on an elevated base. In the female, the relatively large pelvic fins form a brood pouch.

2. Family: *Syngnathidae* (pipefishes and seahorses). The body is characteristically elongate and tapered. Pectoral and pelvic fins are absent, and the anal fin is rudimentary or absent. Their movements in the water appear ungainly, but appropriate to the tangle of eelgrass and seaweed among which they live. The profile of the distinctive head of the seahorses, with tubular jaws at right angles to the body, resembles that of a horse or the knight of a chess set. Some seahorse species are popular in marine aquaria. In many species (e.g. *Syngnathus typhle),* the male develops a ventral brood pouch at spawning time, in which the female lays her eggs. Examples are: ***Great pipefish*** *(Syngnathus acus,* 30–40 cm, shores of the western Atlantic from southern half of Norwegian coast to northern Spain), ***Broad-nosed pipefish*** *(Syngnathus typhle,* 20–25 cm, European shores of the Atlantic, coasts of the Baltic, Mediterranean and Black seas), ***Straight-nosed pipefish*** *(Nerophis ophidion,* 20–25 cm, female 30 cm, Atlantic coasts from northern Norway to Northwest Africa, coasts of the Baltic, Mediterranean and Black Seas, and river mouths). Most of the approx. 25 species of seahorse are inhabitants of tropical and subtropical seas, only 3 species living in temperate European waters, e.g. ***Spotted sea horse*** *(Hippocampus guttulatus,* 10–15 cm, European shores of Atlantic from UK to Gibraltar, coasts of the Mediterranean and Black Seas).

Syngnathus: see *Syngnathoidei.*

Syngonium: see *Araceae.*

Synkaryon: see Cell fusion.

Syno(e)kete: see Myrmecophiles.

Synrhabdosome: see Graptolites.

Synsociology: see Sociology of plants.

Syntaxonomy: see Sociology of plants.

Synthetic theory of evolution: a modern theory of evolution incorporating Darwin's theory of selection, Mendelian genetics, and genetics at the molecular level. Five mechanisms are assumed to operate: ***1)*** *Mutation* (see). Mutations of genes, chromosomes and genomes continually cause nondirectional changes in the genetic composition, chromosome number and chromosome structure. "Nondirectional" means that the mutations themselves are not oriented toward an improved adaptation of the organism to the environment. The resulting genetic and phenotypic variability within populations are amplified by ***2)*** *Recombination* (see). Recombination accelerates evolutionary processes by uniting mutant alleles that arise in different chromosome sections, chromosomes or individuals. It also enables genes to avoid mutual competition. The complement of genes in a population is altered by ***3)*** *Selection* (see), i.e. the preferential survival and above-average rate of increase of individuals that are best adapted to prevailing environmental conditions. ***4)*** Genetic change in a population is also influenced by *Genetic drift* (see). ***5)*** Geographical *Isolation* (see) is also important as a prerequisite for the formation of new species, i.e. for the occurrence of genetically determined reproductive isolation between populations.

Selection primarily ensures that phylogenetic evolution results in a more appropriate adaptation of the organism to its environment. In contrast, the other four mechanisms occur more or less accidentally and are nondirectional with respect to the form and function of the organism. Both microevolution (differentiation and divergence up to the status of species) and macroevolution (formation of larger and more comprehensive taxa, e.g. at and above the rank of genus) are ultimately due to the operation of these five factors within populations.

The basis of the synthetic theory was first summarized in 1926 by S.S. Cetverikov (1880–1959). The theory was not fully developed and recognized until 1937, with the publication of the book "Genetics and the origin of species" by Th. Dobzhanski (1900–1975).

Syntrophism, ***cross feeding:*** the growth of two different species or strains of microorganisms on certain nutrient media, which do not support the growth of one or both oganisms alone. Growth factors needed by one organism are supplied by the other. S. occurs, e.g. between auxotrophic mutants of bacteria.

Synura: see *Chrysophyceae.*

Synusia:

1) A unit of vegetation consisting of different plant species with the same life form, which flourish under the same habitat conditions. The definition of a S. does not take account of the species composition. In contrast to a formation, the S., is a very small unit, and it is related to only one distinct layer in an area of vegetation. Like a formation, a S. is a physiognomic vegetation unit (see Sociology of plants). It is not widely used by plant sociologists.

2) In animal ecology, a group of different species with similar life forms that occur in a particular habitat.

Syphilis: see Spirochetes.

Syrian brown bear: see Brown bear.

Syringa: see *Oleaceae.*

Syringes: the plural of Syrinx (see).

Syrinx (plural **Syringes):** the vocal organ of birds, located at or near the junction of the trachea and bronchi. Birds possess a larynx, but this lacks vocal cords and is not known to generate vocal sounds. The larynx may, however, influence the intensity and the harmonic spectrum of sounds generated in the S., as well as controlling their onset and termination. Whereas the vibrating structures of the larynx are its internal vocal cords, the S. produces sound by the vibration of the tympaniform membranes that form part of its wall. There are 3 types of S.

1. ***Tracheobronchial S.*** This is the most common type of S. It contains a cylindrical resonating chamber or *tympanum,* which is a direct continuation of the trachea, formed by the fusion (or close apposition) and ossification of the lowest rings of the trachea immediately or very shortly before the division into the bronchi. The incomplete bronchial rings neighboring the tympanum (known as the bronchial syringeal cartilages) form the divided part of the S. Their medial ends are connected by the paired *medial* (or *internal) tympaniform membrane* , each membrane forming most of the medial surface of the divided part of the S. At the angle of the bronchi, these 2 medial tympaniform membranes unite and protrude into the syringeal lumen as a semilunar fold. A wedge-shaped, cartilaginous or bony bar *(pessulus)* lies dorsoventrally across the base of the trachea, separating the 2 medial tympaniform membranes. In species lacking a pessulus, the 2 medial tympaniform membranes are connected by a band of elastic tissue *(bronchides-*

mus). There are also 2 *lateral* (or *external) tympaniform membranes,* one on each side of the S. These may be well defined external membranes stretched between well separated cartilages on the lateral aspect of the S., or they may be reduced to membranous strips of tissue between bronchial syringeal cartilages, or they may be absent (e.g. in some songbirds, i.e. passerines of the subfamily *Oscines).* Vocalization is produced by vibration of the membranes, and the sound frequency is modulated by extrinsic and intrinsic syringeal muscles. For example, increased tension of the sternotracheal muscle (an extrinsic muscle) pulls the trachea down onto the S., causing the syringeal membranes to relax and bulge into the bronchial lumen. In contrast, contraction of intrinsic muscles (e.g. the tracheolateral muscle) increases the tension on the tympanic membranes. Intrinsic muscles are sometimes numerous and complex, their number and modes of insertion having taxonomic significance. Thus, birds in which the intrinsic muscles insert on the lateral or middle parts of the bronchial half rings are assigned to the *Mesomyodi (= Tyranni).* In the *Acromyodi (=Menurae + Oscines),* the muscles are inserted on the ends of the half rings (where the bronchial rings give way to the medial tympaniform membrane). Muscles may also be inserted on the dorsal (anacromyodian type), ventral (catacromyodian type) or both ends (diacromyodian type) of the bronchial half rings. The most complicated syringeal musculature is found in the songbirds (order *Passeriformes,* suborder *Oscines),* which possess from 5 to 9 pairs of intrinsic syringeal muscles.

2. ***Tracheal S.*** This type of S. is found only in the tropical families, *Conopophagidae, Rhinocryptidae, Formicariidae* and *Dendrocolaptidae.* Known as *Tracheophonae* or *Clamatores,* they are all loud voiced birds, which vocalize with a swelling of the throat. The last 6 rings of the trachea are very thin, and the tracheal wall in this region (known as the tracheal membrane) is membranous. Medial and lateral tympaniform membranes are also present.

3. ***Bronchial S.*** This consists of expanded lateral tympaniform membranes between the half rings of the bronchi. A medial membrane may also be present. Examples are found in the Oilbird *(Steatornis caripensis),* Nightjars *(Caprimulgus)* and Cuckoos *(Cuculidae).*

In some species, syringeal structure displays sexual dimorphism, and this is correlated with differences in the vocalization of the sexes. Thus, in most male ducks the base of the trachea is modified into a large bony bulla, which is lacking in the female. In several species (e.g. European blackbird, *Turdus merula;* Zebra finch, *Poephila guttata),* the syringeal skeletal elements and musculature are smaller and less robust in the female than in the male.

Syrrhaptes: see *Columbiformes.*

Systellommatophora: see *Gastropoda.*

System:

1) A distinct unit of interacting elements. Quantities acting on the S. from outside, e.g. stimuli, energy sources, information, are known as the *input,* whereas the response by the S. is the *output.* Removal or malfunction of any element affects the S. as a whole.

2) See Geological time scale.

3) See Taxonomy.

Systemic agents: plant protection agents, whose active constituents penetrate the plant via the roots, leaves or cut regions of the stem, and are stored there for a certain time. During this period they protect against pests. Certain fungicides and organophosphorus insecticides are used systemically.

Systemic herbicides: see Herbicides.

System mutation: see Macromutation.

System theory: a branch of cybernetics which relates the behavior of a system to the interaction of its components, and vice versa. The main concern of S.t. is the analysis of systems. There are basically three different types of problem. 1) The input and the laws governing system behavior are known, and the output must be determined. In practice, this is a common situation, e.g. the need to determine to response of a known ecosystem to environmental changes. 2) The laws governing system behavior are known and a particular reaction of the system is observed. The causes of this reaction must be determined. 3) The input and output of a system are known, and the nature of the system and the laws governing the interaction of its parts must be determined. Experimentally, the reaction of a system to different stimuli can be studied, and the results used to elucidate the nature of the system or "black box". This is the most difficult task of S.t.

The main approach of S.t. is to divide complex systems (e.g. a living organism) into subsystems, and to study each of these separate elements. In addition, however, the interactions between the subsystems must also be elucidated. These interactions can be displayed as flow diagrams. Elucidation of the hierarchy of a system is crucial, since this makes it possible to describe very large and complex systems on the basis of relatively simple laws and interrelationships. In biology, S.t. is intimately involved in the study of evolution, the analysis of growth processes on the biochemical, cellular and organismal level, and analysis of the development of populations and entire ecosystems.

Systole: see Cardiac cycle.

Syzygy: see *Telosporidia.*

T

Table Mountain ghost frog: see *Heliophrynidae.*

Table reef: see Coral reef.

Tabulata, ***tabulate corals:*** an extinct subclass of colonial anthozoans with high, massive, sometimes branched, tubular or prismatic calcareous skeletons, forming colonies up to 2 m in diameter. The skeletal tubes characteristically contain horizontal platforms or tabulae on which the polyps rested, and regressed, thorn-like septa. Examples are *Favosites* (Fig.) and *Halysites.* They existed from the Ordovician to the Permian, and possibly in the Cambrian.

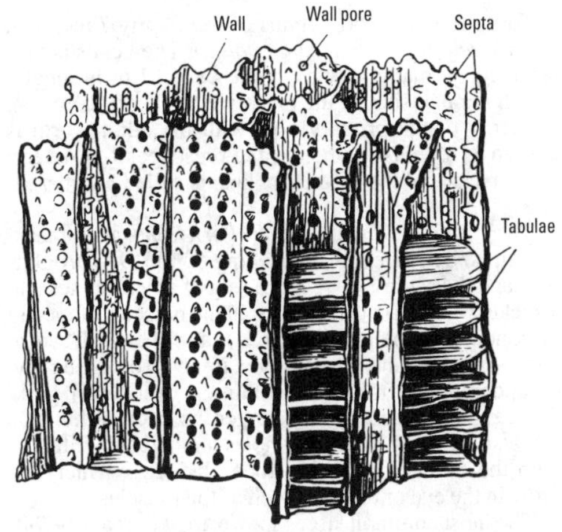

Tabulata. Part of a colony of *Favosites,* showing horizontal platforms (tabulae), wall pores and small, thorn-like, degenerate septa (x 4).

Tabulate corals: see *Tabulata.*

Tachyeres: see *Anseriformes.*

Tachyglossidae: see *Monotremata.*

Tachytely: see Evolutionary rate.

Tactile feeders: see Modes of nutrition.

Tactile stimulus reception, ***thigmoreception, haptoreception:*** reception by, and stimulation of, appropriate receptors by the mechanical stimuli of stroking, pressure and deformation, commonly referred to in humans as a "sense of touch". Receptors are situated mainly in the surface tissues of the body and in the hollow organs (alimentary canal, urogenital system), and to a lesser extent at other sites within the body. Touch receptors are often aggregated on, or associated with, other external body structures [e.g. sensory hairs of many arthropods, vibrissae (whiskers) and muzzle tip of mammals, bill tip of birds, balls of the feet, fingertips]. The various receptors are all derived from two main types: ciliated primary or secondary sensory cells, or free nerve endings. Some ciliates possess stiffened sensory cilia. Tactile-sensitive cells of the *Turbellaria, Nemertini, Chaetognatha* and others often have densely packed cilia that form *tactile chaetae.* The cilia of sensory cells may also be ensheathed by cuticle, e.g. in annelids. In the lateral line organ of fishes and aquatic amphibians, the sense organs *(neuromasts)* are composed of four different types of secondary sensory cells. Each centrally located sense cell has a rounded base and tapers toward the surface where it carries a sensory hair (kinocilium). It is surrounded by three different types of supporting cells, which are also ciliated (stereocilia) (basal; interstitial supporting; and investing mantle cells). The sensory area of the neuromast is covered by a gelatinous *cupula.* Water flow distorts or disturbs the cupula, the kinocilium is bent toward the stereocilia, and a nervous impulse is triggered. The fundamental element of the tactile organs of arthropods is the *hair sensilla, trichoid sensilla* or *sensory seta* (Fig.1). This consists of a trichogenous cell which gives rise to the seta, a hair membrane cell which secretes the fine membrane that articulates the seta with the body wall, and a bipolar nerve cell or receptor cell that lies within the trichogenous cell. A small cuticular sheath (scolops) extends from the base of the seta and covers the tip of the cilium of the receptor cell. Transfer of forces from the scolops to the cilium, e.g. by movement of the sensory hair, causes excitation of the sensory cell. A *trichobothrium* (spiders and insects) (Fig.2) is a long, extended sensory hair, inserted in a flexible membrane in groove in the exoskelton, and linked to a sensory nerve. Trichobothria detect air movements. Other tactile organs of invertebrates include *sensory clefts* on the tarsi and pedipalps of spiders, and the *pectinal combs* of scorpions.

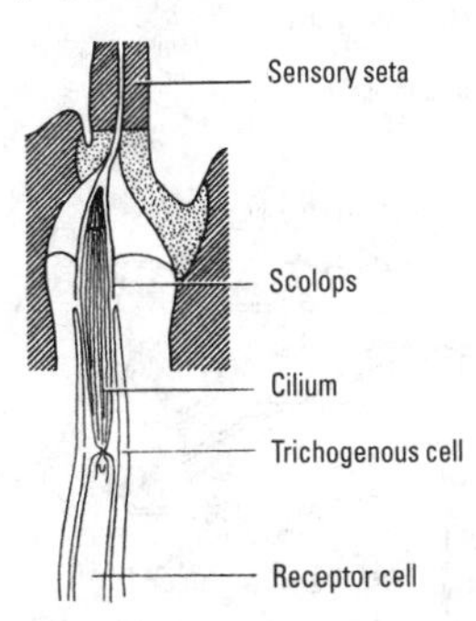

Fig.1. *Tactile stimulus reception.* Sensory hair of a honeybee (after Welsch and Storch, 1973).

Free nerve endings are rarely found in invertebrates. In vertebrates, these are the fine termini of nerves that extend from cell bodies in the spinal ganglia of the dorsal roots of the spinal cord. Free nerve endings lie among the cells of the epidermis or corium. Nerve fiber termini can also be encapsulated, particularly in higher vertebrates. *Pacinian corpuscles* (corpuscula lamel-

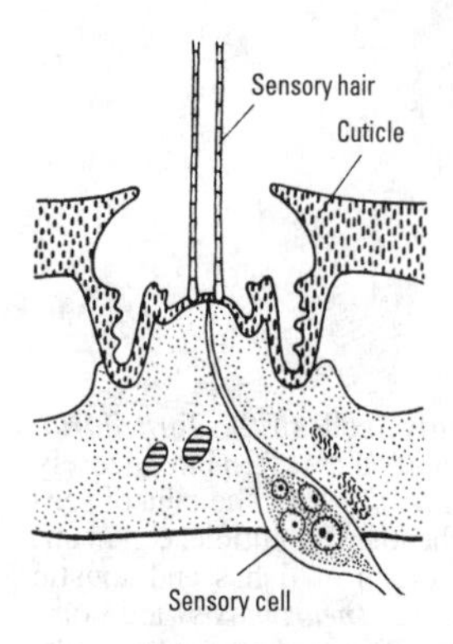

Fig.2. *Tactile stimulus reception.* Section through a trichobothrium.

losa) (Fig.3) are encapsulated nerve fiber termini, about 1 mm in diameter, and they respond to pressure on the body surface (very numerous in finger tips and balls of the feet), or in the body interior, where they are found in the connective tissue of tendons, joints and blood vessels. *Meissner's corpuscles* (corpusculus tactus, Meissneri) (flexor aspect of hands and feet and particularly in the skin of the toe ends and fingertips) (Fig.3) allow the discrimination of fine textures. *Krause's end bulbs* (corpuscula bulboidea, Krausii) are found on the lips, glans penis, etc. and consist of a connective tissue capsule enclosing a core of cells with ramifications of the nerve fibril. *Expanded tip receptors* (Merkel's tactile disks) (Fig.3) are capable of prolonged response to very light pressure. *Ruffini's end-organ* detects stretch in tissues and joint capsules. Mammals also possess *tactile hair receptors,* in which nerve fibers are twined around the root of each hair. Birds possess touch-sensitive end bulbs known as Herbst corpuscles and Grandry corpuscles (Fig.3). Phasic touch receptors adapt rapidly, whereas phasic-tonic touch receptors adapt more slowly. Some tactile receptors, e.g. sensory clefts of spiders, and Pacinian and Herbst corpuscles of vertebrates, respond to vibrations rather than actual tactile stimuli, a phenomenon known as *seismoreception.*

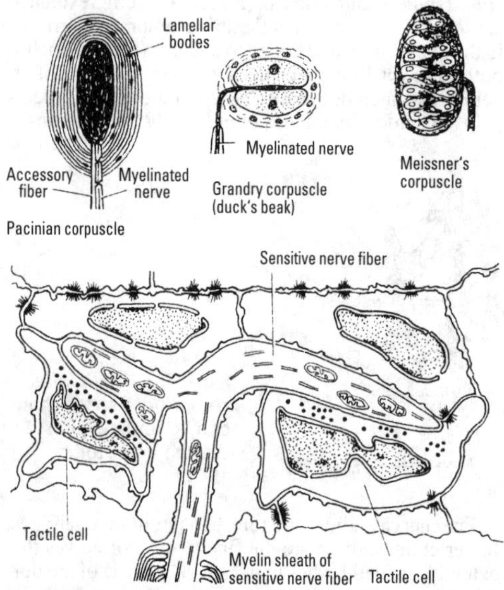

Fig.3. *Tactile stimulus reception.* Different mechanoreceptors.

Tadaria: see *Molossidae.*

Tadorna: see *Anseriformes.*

Tadornini: see *Anseriformes.*

Tadpole: the aquatic, gill-breathing larva of most amphibians, with a fish-like vascular system. Ts. have a streamlined shape with a laterally compressed tail. Anuran Ts. have horny jaws with small abrasive teeth for the grazing of algae. During metamorphosis to the adult amphibian, the gills and gill circulation degenerate and are replaced by lungs and a pulmonary circulation, and the horny jaws are lost; in anurans the tail is digested and absorbed, whereas in urodeles only the transparent border of the tail fin is resorbed. In urodeles, the forelimbs develop first, whereas in anurans these are preceded by the hindlimbs. Many Ts. that inhabit fast flowing water possess an adhesive disk for attachment to the substratum.

Taenia: a tapeworm genus of the family *Taeniidae,* order *Cyclophyllidea (see Cestoda).* The genus is represented by many species, some over 12 m in length, which are parasites in the digestive tracts of mammals. Infection of the primary host with the adult worm is known as *taeniasis,* whereas the presence of cysticerci in the tissues of the secondary host is called *cysticercosis.*

The strobila of *Taenia solum* ***(Pork tapeworm)*** measures up to 8 m in length, consisting of 600–900 proglottids, and the scolex possesses a ring of apical spines and 4 suckers. The cysticercus lives in the muscles of the pig (secondary or intermediate host), and the adult worm parasitizes humans, who become infected by eating raw or undercooked measly (containing cysticerci) pork. *Taenia solum* is particularly dangerous, because humans can also serve as secondary hosts and become infected with the cysticerci (metacestode bladders), which can grow in the eye or brain, as well as the muscles.

The most common large tapeworm of humans is *Taenia saginata* ***(Beef tapeworm).*** This species possesses 4 suckers on a knob-like scolex. Its strobila attains lengths of up to 14 m with up to 2000 proglottids, which are progressively larger and more mature from the neck region backward. The male reproductive system matures before the female system. Each gravid proglottid contains about 80 000 developing embryos. Between 7 and 9 proglottids pass from the body with the feces each day. The main intermediate or secondary host is an animal of the genus *Bos* (especially the domestic cow). Egg capsules are passed to the cow on the hands of infected persons or by flies, or they may be ingested with contaminated vegetation or water. The egg capsule is removed by the combined action of gastric and intestinal enzymes and bile, and the released oncosphere uses its hooks and penetration glands (boring glands) to bore through the intestinal wall and enter the blood or lymph. It then migrates to the muscles of the jaw or heart, and within 2–3 months it grows into a 1 cm-long cysticercus (metacestode or bladder worm). Humans become infected by eating raw or undercooked measly (containing cysticerci) beef.

The adult worm of *Taenia multiceps* (up to 60 cm in length) is a cosmopolitan intestinal parasite of dogs, jackals and foxes. Its coenurid larva is an egg to fist-sized, fluid-filled bladder with multiple invaginated scoleces, which develops in the brain or spinal cord of sheep, cattle and horses, causing blindness and staggering, followed by death.

Taeniasis: see *Taenia.*

Taeniocrada: see *Psilophytatae.*

Tagetes: see *Compositae.*

Tagma (plural **Tagmata):** see Metamerism.

Tahrs, *Hemitragus:* goatlike animals of the subfamily *Caprinae* (see), family *Bovinae* (see), with short, broad, somewhat laterally flattened horns, which are not twisted. Head plus body length is 90–140 cm. The muzzle is naked, and interdigital glands are present. Males have a silky, glossy mane, but no beard. Three species are recognized, all inhabiting mountain slopes and similar rugged high terrain *H. jemlahicus* ***(Tahr;*** Himalayas from Kashmir to Sikkim) has a red-brown to dark brown coat, and a shaggy mane around the neck and shoulders that extends to the knees. *H. jayakari* ***(Arabian tahr;*** Oman) has a gray to tawny brown coat consisting of brittle short hairs. *H. hylocrius* ***(Nilgiri tahr;*** Nilgiri hills and adjoining hills in southern India) is dark yellow-brown with a grizzled area on the back. *H. jemlahicus* has been introduced into New Zealand, and the established population of 20–30 000 is hunted commercially and for sport.

Taiga: the northern European and Siberian conifer forest region of the Boreal floral zone. It consists primarily of pine, spruce, fir and larch, and is characteristic of regions with relatively high summer temperatures but very low winter temperatures, and a short growth season. It also includes parts of northwestern and eastern Siberia with permafrost soils. Some authorities restrict the term to the more open lichen woodlands on the northern fringe of the Boreal forest. See Biocoenosis.

Tail, ***cauda:*** an extension of the vertebrate body, containing a variable number of caudal vertebrae. It is absent from the *Anura (Ecaudata),* great apes and humans. In aquatic animals, the T. is an organ of propulsion, and in birds the T. feathers are used for steering. In mammals the T. has a wide variety of different forms.

Tailed frogs: see *Ascaphidae.*

Tail fan: in many crustaceans, a fan-like organ formed from the last pair of abdominal appendages (uropods) and the telson. It is used primarily as an oar in recoil swimming.

Tailless whipscorpions: see *Amblypygi.*

Tailor birds: see *Sylviinae.*

Taipan, *Oxyuranus scutellatus:* a member of the *Elapidae* (see) found in remote coastal stretches of northern Australia and New Guinea, and on the islands of the Torres Strait. Somber black-brown in color, it attains a length of over 3.5 m, and is one of the largest and most dangerous poisonous snakes in the world. Its bite can kill a human in a few minutes, if a specific antitoxin is not injected. The venom contains both neurotoxic and hemotoxic components (see Snake venoms). It lays eggs, and feeds mainly on rats.

Takahe: see *Gruiformes.*

Takakia: see *Hepaticae.*

Takin, *Budorcas taxicolor:* a sturdy, 1 m-tall horned ungulate (see *Bovidae)* of the subfamily *Caprinae* (see). It lives in the high mountains of Central Asia, and is difficult to place phylogenetically, since it shares characters of both cattle and mountain goats. Three subspecies are recognized: ***Mishmi takin*** *(B.t. taxicolor),* ***Szechwan takin*** *(B.t. tibetana),* ***Shensi takin*** *(B.t. bedfordi).*

Takydromas, ***oriental grass lizards, long-tailed lizards:*** a genus of the family *Lacertidae* (see), with an unusally long, autotomous tail that tapers like a whip, and may be 8 times the length of the head plus trunk. They have a very slender shape, with imbricating scales on the back, and small granular scales on the flanks. Inhabitants of open grassland, they occur on the East Asian mainland, as well as Japan, Formosa and the Ryukyu Islands, and Southeast Asian islands like Borneo, Sumatra and Java.

The very light-weight, agile ***Six-lined long-tailed lizard*** or ***Six-lined grass lizard*** *(T. sexlineatus)* of Indochina, Malaya, Sumatra, Borneo and Java, has an overall length of up to 36 cm, and actually runs along the tops of plants.

Talegalla: see *Megapodiidae.*

Talopins, *Miopithecus:* the smallest of the Old World monkeys (see) (head plus body 350 mm; tail 430 mm). They inhabit lowland forest from Cameroon to Angola, where they feed on leaves, buds, pith and insects. The ***Southern talopin*** *(M. talopoin)* has a southerly range, extending south to Angola, while the ***Northern talopin*** (which has not yet been assigned a species name) is found in Cameroon, Gabon and Rio Muni. T. are accomplished swimmers, noted for stealing cassava placed for steeping in forest pools, and sleeping above water in overhanging low branches.

Talpidae, ***moles, shrew moles*** and ***desmans:*** a family of the order *Insectivora* (see), comprising 27 species in 12 genera, with representatives in Eurasia and North America. Dental formula: 2–3/1–3, 1/0–1, 3–4/3–4, 3/3 = 32–44. The powerful forelimbs of these subterranean mammals are modified for digging and burrowing, and they are used to excavate an extensive system of underground passages. They consume large numbers of insect larvae and earthworms. See Desmans and Star-nosed mole.

Tamarao: see *Bovinae.*

Tamaraw: see *Bovinae.*

Tamarind: see *Caesalpiniaceae.*

Tamarins: see *Marmosets.*

Taming: the abolition of those behavioral and psychological barriers in an animal, which distance it from humans. T. is a complex process, the essential component of which is the elimination of the readiness for flight. The T. of social or gregarious animals can be accelerated by the presence of tame individuals, a phenomenon that is exploited in the familiarization ("settling down") of Asiatic elephants. Familiarization normally precedes T.

Tanagers: see *Thraupidae.*

Tanaidacea: an order of the *Crustacea,* subclass *Malacostraca.* The body (1–5 mm in length) is elongated, and the abdomen is short in relation to the thorax. A very short carapace covers, and is fused to, only the first 2 thoracic segments, and it forms a lateral branchial cavity which functions as a gill. The first pair of thoracopods are modified as maxillipeds, which have a large segmented branchial epipodite. The modified second pair of thoracopods (which are large and chelate and known as gnathopods) represent one of the characteristic features of the *T.* Eggs develop in a brood pouch (marsupium) of the female. Almost all of the 350 (approx.) known species are marine, and most are bottom dwellers of the littoral zone, where they live in rock crevices or burrow into the mud; a few species inhabit the deep ocean. Food particles are filtered by the second maxillae. In addition, large detritus particles may be retained on the brushing setae of the maxillipeds, and the diet may also be supplemented by raptorial feeding. Throughout the order, there is an evolutionary tendency toward the loss of filter feeding, culminating in the family, *Tanaidae,* which are entirely raptorial.

Tangel humus: a humus layer, sometimes as much as 40 cm deep, formed from the highly fibrous, poorly decomposable residues of heather, rhododendron, pine, etc., by the feeding and mixing activities of soil fauna. It accumulates on the top of biologically active mull humus, which in turn is layered on limestone. T.h. occurs mainly on rendzina soils (tangel-rendzina).

Tangerine: see *Rutaceae.*

Tangle: a common name often applied to a variety of large seaweeds (brown, red and some large green algae), but more correctly applied to the single species of brown algae, *Laminaria digitata,* which occurs at and below the low tide mark. The latter has a branched and rootlike holdfast and a smooth cylindrical stipe, and its blade is divided into several straps or ribbons. See *Phaeophyceae.*

Tank farming: see Water culture.

Tannic acid: see Tannins.

Tannins: a widely distributed, chemically heterogeneous and structurally complicated group of plant substances, which have an astringent action on mucosae, taste bitter, inhibit bacterial growth, precipitate proteins and alkaloids from solution, and produce green to black precipitates with ferric salts. A long exploited property of T. is their ability to tan animal hides, i.e. to convert animal skin into leather. During the tanning process, soluble and/or hydratable skin proteins become insoluble and nonhydratable, due to crosslinking of peptide chains by the phenolic hydroxyl groups of the T. There are two classes of vegetable T.

1. ***Hydrolysable T.*** These are depsides which can be hydrolysed to glucose (or sometimes another polyhydric alcohol) and gallic acid (gallotannins) or ellagic acid (ellagitannins). They are present, e.g. in the wood and bark of the sweet chestnut (5–10%), and especially in certain galls (see below). The simplest known gallotannin is β-glucogallin (1-*O*-galloyl-β-D-glucopyranose) (Fig.) from *Rheum officinale* (Chinese rhubarb). In contrast, Chinese gallotannin, found widely in the families *Hamamalidaceae, Paeonaceae, Aceraceae* and *Anacardiaceae,* and sporadically in the *Ericaceae,* may contain up to 8 galloyl groups. Certain members of the *Anacardiaceae* (sumach family) contain particularly large amounts of useful T., e.g. T. from the leaves and shoots of *Rhus coriaria* are used to prepare morocco leather.

Ellegitannins are derivatives of hexahydroxydiphenic acid, which becomes lactonized to ellagic acid during hydrolysis. The simplest known ellagitannin is corilagin from *Caesalpina coriaca* and other plants.

Other phenolic compounds are sometimes found in place of gallic or hexahydroxydiphenic acid, e.g. chebulic acid (in *Myrobalans* T.) and brevifelin carboxylic acid (in *Algarobilla* T.).

The "tannic acid" or "gallotannic acid" of commerce consists of glucosides of gallic acid and the depside of *m*-digallic acid, all of which can be hydrolysed to D-

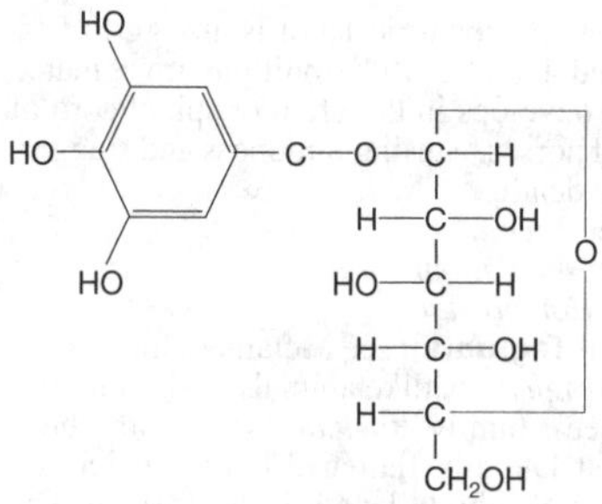

β-*Glucogallin*

glucose and gallic acid with enzymes or acid. In its purest form, this gallic acid is a white powder (usually, however, yellowish white) with a bitter astringent taste. In addition to its technical use in tanning and in the textile industry as a mordant for basic dyes, tannic acid is also used medicinally (against diarrhea, as an antiseptic, and for stopping bleeding), in the manufacture of ink (its aqueous solution is colored dark blue by ferric salts) and in the clarification of wine. Particularly abundant sources are the twig galls of *Quercus infectoria* (Aleppo galls) (50–70%) and the leaf galls of *Rhus semialata* (Chinese or Japanese galls) (about 70%).

2. ***Condensed T., nonhydrolysable T., catechin T.*** These are polymers of phenolic flavans (usually flavan-3-ols, also known as catechins), in which the flavan units are linked by C–C bonds; ester linkages are absent. Many higher oligomers and polymers of proanthocyanidins are therefore condensed vegetable T., e.g. the procyanidin polymer from the seed coat of sorghum. Condensed T. have been synthesized in the laboratory from flavan-3-ols by the action of heat, acid or enzymes. Abundant sources of condensed T. are oak bark (10%) and horse chestnut and spruce bark (10–18%).

Tantalus monkey: see Guenons.

Tapaculos: see *Rhinocryptidae.*

Tapetum: the innermost layer of the sporangium wall of pteridophytes and the pollen sack of seed plants. See Flower.

Tapeworm: see *Cestoda.*

Taphocoenosis (Greek *tafos* grave, *koinoo* society): a recent or extinct society whose members died at the same site. It contains both autochthonous and allochthonous elements. If the remains are prehistoric and completely embedded, the society is rather called a Thanatocoenosis (see).

Taphrinales: see *Ascomycetes*

Tapiridae, ***tapirs:*** a family of odd-toed ungulates (see *Perissodactyla)* containing 4 species in a single genus. Tapirs are relatively large, short bodied mammals, weighing 225–300 kg. Head plus body length is 180 to almost 250 cm, with a short tail of 5–10 cm. Height at the shoulder is 73.5–103 cm. Short bristly hairs are scattered rather sparsely on the body, the thickest coat being found on *Tapirus roulini.* A short, narrow mane is present, but often hardly discernible. Nose and upper lip are elongated to form a short proboscis or trunk, bearing the nostrils near the tip. Four toes are present on the forefeet, 3 on the hindfeet. They are found solitary or in pairs, feeding on leaves and fruit. Fond of bathing, they wear paths to permanent bodies of water. Dental formula: 3/3, 1/1, 4/4, 3/3 = 44; incisors are chisel-shaped, canines conical and well developed. In most species the coloration is dark to reddish brown, often paler below. Young tapirs are dark with yellow and white stripes and spots, and this patterning is usually lost in 6–8 months.

Tapirus terrestris ***(Brazilian tapir)*** is the most widely distributed of all the tapirs. Uniformly brown in color, it is found in Colombia and Venezuela, and south to Paraguay and Brazil. *T. roulini* ***(Woolly, Andean*** or ***Mountain tapir)*** occurs on the forested hill and mountain slopes of the Andes of Colombia and Ecuador. *T. bairdi* ***(Baird's tapir)*** is found west of the Andes in Colombia and Ecuador, northward through Central America to southern Mexico. *T. indicus* ***(Asiatic tapir)*** occurs in the dense jungles of Burma, Thailand, Malaysia, Sumatra and Indochina. This latter species displays exceptional patterning and coloration: the front half of the body and the hindlegs are very dark, almost black, while the central part of the body, including the rear half above the legs, is white.

Fossil record. Fossil tapirs are found from the Oligocene to the present, and their ancestors are found in the Eocene (see *Lophiodon*). Up to the Ice-age, tapirs lived in Europe and North America, but they then withdrew into tropical regions.

Tapirs: see *Tapiridae.*

Tarantulas: a term applied to various arachnids, especially spiders, which according to folk lore are particularly dangerous to humans. In the Mediterranean region a number of Wolf spiders (see) are called T. Although wolf spiders are relatively harmless, there are the most extraordinary tales about their venomous nature. In America, spiders of the family *Aviculariidae* and arachnids of the order *Amblypygi* (whipspiders) are often called T.

Taraxacum: see *Compositae.*

Tardigrada, ***water bears:*** a phylum of the *Protostomia,* whose members are only 0.2 to 1.2 mm in length. The body is short and barrel-shaped, and ventrally flattened. Segmentation is not externally apparent, but 5 segments are recognizable during embryonal development, and the nervous system is distinctly metameric. Head and trunk are not demarcated. Two pointed, chitinous stylets can be protruded from the mouth opening. Locomotion is by 4 pairs if nonjointed, stumpy, retractable appendages (cylindrical extensions of the ventrolateral body wall), which have 4–8 terminal claws, more rarely suctorial pads. The body is covered with a smooth or sculptured cuticle, consisting of chitin and mucopolysaccharide with a laminated ultrastructure. Some species have symmetrically arranged dorsal plates. The underlying epidermis is composed of a constant number of cells. Many other organs also show a constancy of cell number. Most T. possess a pair of cup-shaped red or black pigment cells (simple eye spots), each associated with a single light sensory cell innervated from the brain. All species are dioecious, and only a single, sacular gonad (testis or ovary) is present. In the female, a single oviduct opens via the gonopore to the exterior shortly before the anus, or it opens into the hindgut, which then becomes a cloaca. Males have 2 sperm ducts, which open via a single median gonopore immediately before the anus. Cleavage is holoblastic. During embryonal development, five pairs of coelomic pouches are formed by folding of the archenteron. This type of mesoderm formation (enterocoelous) is also found in brachiopods, but is otherwise characteristic of deuterostomes rather than protostomes. The epithelia of the 4 anterior coelomic pouches later disintegrate and their cells form mainly the body musculature. The fifth pair of coelomic sacs fuse to form the single, unpaired gonads.

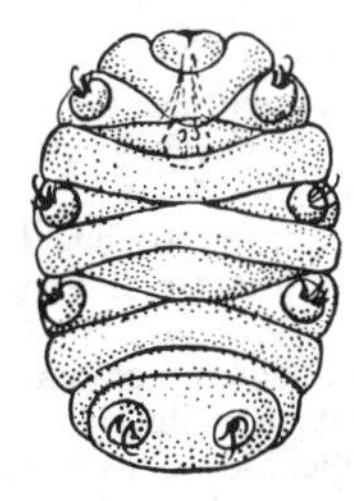

Fig.1. *Tardigrada.* Desiccated resting form of *Hypsibius* (ventral view).

T. occur on all continents. About 400 species are known, many of them cosmopolitan. Most are freshwater species, living in the bottom detritus or on the aquatic plants of ponds and lakes. They are also found in the moisture film of the soil, and are especially abundant in the water films of terrestrial moss and lichen cushions, on roofs, and on trees. A few marine species are known from the bottom mud and sand of both shallow and deep waters. Under dry conditions, many species lose water and shrivel to form a resting stage that can withstand cold and desiccation for several years. On re-exposure to water, they become active again in about 1 hour. Most species feed by piercing plant cells with their oral stylets, and sucking out the contents with the aid of their suctorial pharynx.

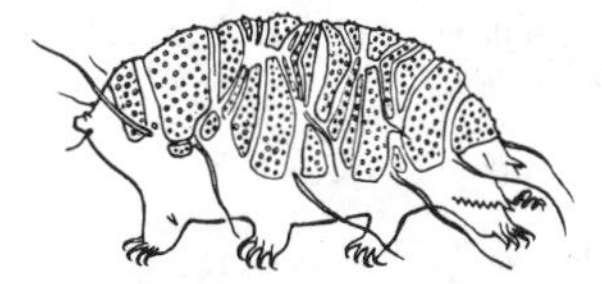

Fig.2. *Tardigrada. Echiniscus scrofa* Richters.

Classification. Three orders are recognized: *Heterotardigrada, Mesotardigrada, Eutardigrada.*

Collection and preservation. Depending on availablilty, 1 or 2 cubic centimeters of collected material are agitated with water, and the animals are picked out visually from the washings. They are then killed and extended by exposure to the vapor from 40% formaldehyde solution. For fixation and storage they are kept in 1% neutral formaldehyde solution.

Tarentola mauritanica: see Common gecko.

Target theory: a model for the quantitative description of the biological effects of radiation. The T.t. postulates the existence of radiation-sensitive targets within a cell or molecule. Within each target, one or more ionizations or excitations are necessary for inactivation of the cell or molecule. The logarithm of a quantifiable radiation effect (e.g. survival rate of an organism or tissue culture, residual activity of an enzyme, the yield of a crop plant) is plotted against the radiation dose. The resulting *dose-effect curve* usually conforms to one of the following four types.

1. *Exponential single hit curve.* This a straight line, and it is interpreted to mean that only a single ionization or excitation (one hit) is necessary for inactivation or killing. It is often observed for the inactivation of viruses or enzymes.

2. *Multiple hit curve.* This is often observed for the ir-

radiation of bacteria or higher organisms. At low doses the effects are small or absent. As the dose increases, the plot becomes exponential and resembles that of the single hit curve. It extrapolates, however, to a value of the effect that is greater than 100%; finally, it curves to form a shoulder and crosses the effect axis at 100%. Two interpretations are possible: a single target must be hit more than once, or more than one target must be hit, in order to effect killing or inactivation.

3. *Biphasic curve.* This is observed if the irradiated population contains individuals of different radiation sensitivity, e.g. different strains of bacteria, or cells at different stages of the cell cycle. Low doses preferentially kill the more sensitive individuals, while higher doses effect both the more sensitive and less sensitive individuals. The plot consists of two staight lines of different gradients, so that the curve appears to "sag".

4. *Stimulation curve.* At low doses, the system is actually stimulated, rather than inactivated, an effect sometimes observed with higher plants, bacteria, fungi and animals. Thus, in comparison with nonirradiated controls, the irradiated organisms may show a higher yield, increased height (plants), increased RNA synthesis, etc., indicating the operation of repair processes that overcompensate for the irradiation damage. The curve therefore rises to a dose effect greater than 100%. At higher doses it then falls, eventually crossing the effect axis at 100%.

The T.t. is based on the premise that irradiation causes damage directly, but indirect effects (e.g. due to the toxicity of reactive free radicals produced by irradiation) also have a very significant effect on the survival of organisms and the stability of proteins and nucleic acids.

Taricha: see *Salamandridae.*

Taro: see *Araceae.*

Tarozzi-bouillon: see Anaerobes.

Tarpan: see Wild horse.

Tarpon: see *Megalopidae.*

Tarragon: see *Compositae.*

Tarsiers (plate 40): species of the genus *Tarsius* of the family *Tarsiidae,* usually placed in its own suborder *(Tarsioidea)* of the Primates (see). T. are rat-sized animals with enormous eyes and a long tail. Tibia and fibula are fused in the lower third of their length, and the tarsal bones are considerably elongated (hence the name of the animal), providing leverage for a powerful leap. T. are predators, eating birds and large insects, and even poisonous snakes. They leap upon their prey and kill by biting.

Superficially T. display the characters of Prosimians (see), but they possess a dry, simian-type nose, and placentation is typically simian. The retina of the eye has a fovea (a structure associated with daylight vision and therefore useless in the nocturnal T.), whereas the rest of the retina is adapted for nocturnal vision. T. therefore appear to have evolved from simian-type ancestors, but they have adopted a prosimian mode of existence. Thus, they are also classified as haplorhines (see Primates).

The genus has three, approximately equal-sized species (weight 120 g; head plus body 125 mm; tail 220 mm), all found in South-East Asia: *Tarsius syrichta* ***(Philippines tarsier***, Philippines), *Tarsius bancanus* ***(Horsfield's tarsier***, Borneo, Sumatra and nearby islands), *Tarsius spectrum* ***(Spectral tarsier***, Sulawesi and nearby islands).

Tartaric acid, ***dihydroxysuccinic acid:*** a colorless, crystalline, dibasic organic acid containing 2 asymmetrical carbon atoms. The L(+) form is distributed widely in the plant kingdom, either as the free acid or as its salts, e.g. in grapes, apricots, rowan berries, rock cherries *(Prunus mahaleb),* hawthorn, dandelion. During wine production, poorly soluble potassium hydrogen tartrate precipitates out as tartar. T.a. has many uses in the food and textile industries.

```
      COOH
       |
     HC—OH
       |
  HO—CH
       |
      COOH
```

L(+)-Tartaric acid

Tarweed: see *Compositae.*

Tasmacetus: see Whales.

Tarsonemus: see Strawberry crown mite.

Tasmanian tiger: see Tasmanian wolf.

Tasmanian wolf, ***Tasmanian tiger, marsupial wolf, thylacine,*** *Thylacinus cynocephalus:* a dog-like marsupial, total length about 1.5 m. Previously a feared predator in Australia and Tasmania, it was thought to have been exterminated, the last known survivor dying in Hobart zoo in 1936. Reports that it may still survive in Western Australia have not been confirmed [S.J. Paramonov (1967) *Western Australian Naturalist,* **10,** 171–172; A.M. Douglas (1986) New Scientist, No. 1505, 44–47]. Some authorities place it in the *Dasyuridae,* but it is usually allocated to a separate family, *Thylacinidae.*

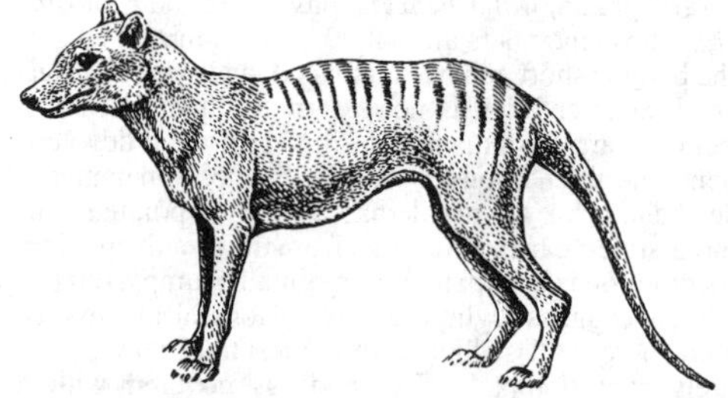

Tasmanian wolf

Taste buds, ***taste organs, gustatory organs:*** structures containing groups of sensory cells that serve for the perception of dissolved chemical substances. The structure of vertebrate T.b. is basically similar in all species. A sheath of supporting cells surrounds a central group of long narrow sensory cells, forming a bud-like structure, which is embedded in the stratified epithelium of the epidermis or buccal cavity. The free ends of the sensory cells carry short hair-like processes, while their basal parts are associated with nerve fibers. Depending on the position of the T.b., the associated nerve fibers contribute to the VII, IX or X cranial nerves. In fish, T.b. commonly occur on the exterior epidermis, especially on lips and barbels. In most vertebrates, however, they are located in the mouth cavity. In man, they are aggregated on the tongue. There are only four primary sensations of taste: sweet, bitter, sour, salty. The variety of flavors actually experienced is due to different combinations of these four, and their integration with olfacory sensations. T.b. on the epiglottis of man, or at the entrance of the swim bladder of some fishes, trigger reflexes that protect against chemicals or particles entering the air passages.

Taste organ: see Taste buds.

Tattlers: see *Scolopacidae.*

Taurine, ***aminoethane sulfonic acid:*** $H_2N—CH_2—CH_2—SO_3H$, a widely distributed compound in living organisms, e.g. in vertebrate bile, where it is a constituent of taurocholic acid (see Bile acids). T. is an endproduct of L-cysteine biodegradation.

Taurocholic acids: see Bile acids.

Taurotragus: see *Tragelaphinae.*

Taurulus: see *Cottidae.*

Taxaceae, ***yew family:*** the only family of the gymnosperm subclass, *Taxidae.* Members of the *Taxidae* differ from those of the *Pinidae,* particularly in their floral structure (see Gymnosperms). Five genera *(Amentotaxus, Austrotaxus, Pseudotaxus, Taxus, Torreya)* and about 15 species are recognized, most of which occur in the Northern Hemisphere. Members are evergreen trees or shrubs, which are usually dioecious. Only the male strobili form cones, which are catkin-like with numerous sporophylls. Female strobili are found in the leaf (needle) axils; they do not form cones, and they possess a single, erect ovule or seed, not borne on a scale but partly or totally surrounded by a fleshy aril when ripe.

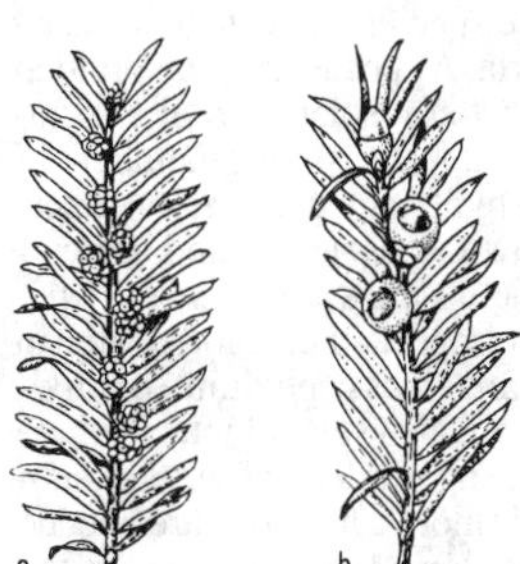

Taxaceae. a Twig of yew *(Taxus baccata)* with male flowers (cones). *b* Twig of yew with 3 seeds; the lower 2 seeds are ripe and each is surrounded by a fleshy cup-shaped aril; the upper seed is unripe.

The genus, *Taxus* ***(Yews),*** is widely distributed in Europe, North America, East Asia and Asia Minor. *Taxus baccata* **(Common yew)** is a densely branching tree with scaly, reddish-brown bark. Its flat, pointed, spirally arranged needles are borne on elongate shoots, and they have a dark, glossy upper surface. The red, fleshy, cup-like aril is the only part of the plant that does not contain the toxic alkaloids known as taxins. Male and female strobili usually occur on different trees. In many countries, the yew is a protected plant.

The genus, *Torreya,* is found in North America, China and Japan, e.g. *Torreya californica* ***(Californian "nutmeg"),*** which produces an edible nutlike seed.

Taxidea taxus: see American badger.

Taxis: movement of a motile organism in response to the direction and intensity of a stimulus. Normally the movement is toward (positive T.) or away from (negative T.) the stimulus source. Taxes are classified according to the type of stimulus: *chemotaxis* (chemical stimuli), *thigmotaxis* (tactile stimuli), *rheotaxis* (water flow), *anemotaxis* (air currents), *phonotaxis* (sound), *geotaxis* (gravitational fields) and *phototaxis* (light stimuli). *Phobotaxis* is applied to nondirectional, random turning in different directions ("fright reactions"), whereas *topotaxis* refers to ordered and directional turning. *Tropotaxis* (see) is orientation on the basis of symmetrical excitation, e.g. via two sense organs, whereas *Telotaxis* (see) can be defined as directly goal-oriented turning movements. Other types of T. are Thermotaxis (see), Hydrotaxis (see), osmotaxis, Aerotaxis (see), Galvanotaxis (see) and Traumatotaxis (see).

Taxodiaceae, ***swamp cypress family:*** a family of needle-leaved gymnosperms containing 15 living species in 10 genera, which show a relict distribution. They are found mainly in the Northern Hemisphere, and their evolutionary peak was in the Cretaceous and the Tertiary. They have predominantly spirally arranged needles and woody cones, and their pollen grains lack air bladders. They include some of the world's tallest and oldest trees, such as the Californian ***Mammoth trees*** or ***Redwoods:*** *(Sequoiadendron giganteum)* from the Sierra Nevada, and *Sequoia sempervirens* from the coastal mountains of western North America, which may live for 1 500 to more than 3 000 years, and attain heights of more than 100 m. In 1944, living specimens of *Metasequoia glyptostroboides,* previously known only from fossil remains in the Mesozoic and Tertiary, were discovered in China. So far, this is the only time in the history of botany that a plant has first been described from its fossil remains, then discovered live. *Metasequoia glyptostroboides* attains a height of 35 m, and it sheds its short, deciduous, needle-bearing shoots in the autumn. This rapidly growing tree can now be seen in many botanical gardens throughout the world, and has been given the trivial name of ***Dawn redwood.*** A further deciduous member of this family is the ***Swamp cypress*** *(Taxodium distichium)* of swamps and river banks in Central and North America, which grows to a height of 50 m; as an aid to respiratory gas exchange, its modified roots project from the water or the swamp mud. Other

Taxodiaceae. Dawn redwood *(Metasequoia glyptostroboides),* branch with cones.

examples are the ***Umbrella fir*** *(Sciadopitys;* East Asia) and the ***Water spruce*** *(Glyptostrobus;* East Asia). All the above-mentioned species were widely distributed in the Northern Hemisphere during the Tertiary.

Taxodium: see *Taxodiaceae.*

Taxodont: a hinge type of the *Bivalvia* (see).

Taxol: see Microtubules.

Taxonomy, *systematics:* a branch of biology which describes, names and groups organisms according to their natural affinities, at the same time serving the practical need for ordering the immense number of living organisms in a manageable classification system. Modern T. is concerned mainly with the phylogenetic relationships of organisms within their evolutionary tree. About 400 000 plant species and almost 2 miliion animal species have been described, and new species are discovered every day.

The older *systems* of the plant and animal kingdoms were very crude, and based on properties such as shape. Such systems were created in ancient times by Aristotle (384–322 BC), Dioscorides (about 30–80 AD) and others. These were still in use in the Middle Ages, and, like the *herbals* that appeared mainly in the 16th century (Brunfels 1530, Fuchs 1539, Bock 1542), they contained only a small fraction of existing species. Linneus (1707–1778) described many new plant and animal species, and in 1735 prepared a system of plants based on the number and type of stamens in the flower. This was therefore known as a *sexual system,* and like all earlier systems, it was *artificial,* because the classification depended on only one or a very small number of characters. In contrast, the founders of *natural systems* used as many characters as possible for classification purposes. Natural plant systems were created, e.g. by A.L. de Jussieu (1789), A.P. De Candolle (1813) and A. Brongniart (1843). At first, these natural systems were not based on evolutionary or phylogenetic relationships, but they provided an objective view of graduated similarities of organisms. Such classifications were later recognized as an expression of natural affinities, and they provided important support for the theory of evolution.

Modern phylogenetic systems are based on evolutionary relationships, and they are subject to continual modification and improvement in the light of new discoveries and research. Recent authors of phylogentic plant systems are, e.g. von Takthajan, Cronquist and Dahlgren.

Nowdays, T. utilizes information from most other branches of biology. In addition to morphological characters (which are still of primary importance for classification), many other important properties are used in an attempt to develop a truly natural system, e.g. research findings from biochemistry, anatomy, histology, cytogenetics, palynology, embryology, chorology, ecology, etc. A complex of characters (the totality of taxonomically relevant charcaters) is established that is shared by all members of a group, or which serves to differentiate between groups. The closer the natural relationship between two organisms or groups of organisms, the greater the number of shared characters or properties. It must be insured that the chosen characters are taxonomically relevant, i.e. that similarities have a common evolutionary origin and are therefore homologous (see Homology). Phylogenetic relationships cannot be inferred from similarities that have arisen by Convergence (see). Homologous characters must also be further evaluated to determine whether thay are primitive *(plesiomorphic)* or derived *(apomorphic).* Differences in the evolutionary rate of characters may mean that mosaics of both primitive and derived characters are present in the same group (heterobathmy).

An attempt is made to construct a system or *evolutionary tree* of a *monophyletic group* of organisms, i.e. all members of the group are derived from a single ancestral species. *Polyphyletic groups* (containing two or more evolutionary lines) and *paraphyletic groups* (which do not contain all the descendants of the ancestral species) therefore represent preliminary stages of classification, still requiring further subdivision and completion. *Hennig's principle* is often used in the reconstruction of phylogenetic trees of recent groups, especially in zoology. This is based on the hypothesis that the monphyletic nature of a group is indicated only by the common possesssion of derived characters that represent conversions of the same starting material, and not by the common possession of primitive characters. This type of character evaluation reveals the existence of sister groups, which can be represented as a dichotomy of the phylogenetic tree.

Otherwise, classification is based on a *hierarchical system,* i.e. an internationally recognized ranked system of taxonomic categories, known as *taxa* (singular *taxon).* The most important category is the *species,* which is the totality of all individuals that share the same essential characters, can interbreed freely, and are restricted to a defined areal. For all sexually reproducing plants and animals, the species is a category based entirely on objective criteria. A species may be divided into *subspecies* or *races* that are isolated from one another geographically. In botanical systems, members of a species showing only slight, but genetically determined differences are known as *varieties,* and agricultural and horticultural varieties are known as *cultivars.* After the species, the most important categories are (in descending order) the genus, class, phylum (or division) and kingdom. There are also several intermediate categories, like subfamily, superorder, infraorder, subclass, etc., which are used if more categories are needed for an objective representation of major steps in the phylogenetic development of the entire group. The main systematic categories (taxa) are presented in the following table for the classification of the edible plum, *Prunus domestica,* and the mosquito, *Culex pipiens.*

Kingdom	Plant *(Eukaryota-Plantae)*
Phylum	*Spermatophyta*
Subphylum	*Angiospermae*
Class	*Dicotyledoneae*
Order	*Rosales*
Family	*Rosaceae*
Genus	*Prunus*
Species	*Prunus domestica*
Subspecies	*Prunus domestica* subsp. *italica*
Kingdom	Animal *(Eukaryota-Animalia)*
Phylum	*Arthropoda*
Subphylum	*Tracheata*
Class	*Insecta*
Order	*Diptera*
Family	*Culicidae*
Genus	*Culex*
Species	*Culex pipiens*

Tayassuidae, *peccaries, javelinas:* a family of the Ungulates (see), comprising a single recent genus and 2 species. *Tayassu (= Dicotyles) tajacu* ***(Collared pec-***

cary) is found in Texas, New Mexico and Arizona in deserts and arid woodlands. *Tayassu (= Dicotyles) pecari* ***(White-lipped peccary)*** ranges from southern Mexico to Paraguay in humid tropical forests. Head plus body length of adults is 75–100 cm, and the height at the shoulder is 45–57 cm. The snout is elongate, cartilaginous and mobile, mainly naked at the end around the nostrils. A rudimentary tail of only 6–9 vertebrae is present. The legs are long and slim with 4 toes on the forefeet and 3 toes on the hindfeet. (Some fossil species have 2 toes on the hindfeet). The third and fourth foot bones are fused at their proximal ends (i.e. as in ruminants), in contrast to the *Suidae* (pigs) in which these bones are separate. Bristly hair covers the body, and the complexity of the stomach approaches that of ruminant. Dental formula: i 2/3, c 1/1, pm 3/3, m 3/3 = 38. The upper canine tusks are not recurved, but directed downward, and they are smaller than those of the *Suidae.* Both species possess a gland, about 20 cm in front of the tail, which produces a musk-scented liquid. This is used by the animals to scent-mark their paths in the undergrowth, and members of a pack often rub their musk glands on each other.

Tayassu: see *Tayassuidae.*

Tea family: see *Theaceae.*

Teal: see *Anseriformes.*

Tectum: see Brain.

Tectum opticum: see Brain.

Teeth, ***dentes:*** hard structures in the buccal cavity of vertebrates. T. are modified epidermal structures, homologous with the placoid scales of elasmobranch fishes. The totality of the T. of an organisms is known as the Dentition (see).

Each mammalian tooth consist of 1) the *root,* which is inserted in the alveolus of the jaw bone, 2) the *neck,* which is surrounded by gum tissue (gingivae), and 3) the *crown,* which extends free into the mouth cavity.

T. are composed of *dentine* (ivory), covered in the crown region with extremely hard *enamel,* and in the neck and root regions with *cement.* The *pulp cavity* inside the tooth is filled with soft connective tissue, blood vessels and nerve fibers (all referred to colloquially as the tooth"nerve"). Dentine is permeated by numerous fine canals containing thin protoplasmic processes from the *odontoblasts* that lie along the periphery of the pulp cavity.

The root is enclosed in a bony socket, the *alveolus,* where it is anchored by cement and by the attachment of gum tissue in the neck region. Between the root and the alveolar wall is a continuation of the bone periosteum which forms the vascular *alveolar* or *peridontal membrane.*

Tooth attachment in the jaw may be *acrodont* (arising from the midline of the top edge of the jaw; amphibians, snakes, some lizards), *pleurodont* (arising from the inner edge of the jaw; many reptiles), or *thecodont* (inserted into an alveolus of the jawbone; mammals).

The simplest T. are conical with a single point. More commonly they have several points or they are somewhat flattened like incisors. Amphibians have small, pointed T. on the palatine and pterygoid. Lizards and snakes have T. on the jaws, as well as on the palatine, pterygoid and vomer.

Extinct, primitive birds *(Archaeopterygiformes)* possessed T. Modern birds and turtles have no T. Mammalian T. may be *bunodont* (the crown possesses blunt protuberances, e.g. pig), *hypselodont* (cylindrical with a high crown, e.g. elephant tusks), or *lophodont* (combined into grinding surfaces, e.g. cattle). Continuous growth resulting from permanently open roots is displayed by the tusks of the elephant (incisors), the gnawing T. of lagomorphs and rodents, and the tusks (canines) of the wild boar. See Dentition, Tooth development.

Teff: see *Graminae.*

Tegmentum: see Brain.

Tegu, *Tupinambis nigropunctatus:* a member of the lizard family *Teiidae* (see). An inhabitant of Central and South America, it is up to 1.2 m in length, with a black, yellow and blue patterned body, powerful limbs and a rounded tail. The eggs are laid in termite nests. The 4 species of the genus *Tupinambis* are all relatively large, the largest being the ***Common tegu*** *(T. teguixin),* which is up to 1.4 m in length, and patterned with bands of transverse yellow spots on a black background.

Teichmann's crystals: see Hemoglobin.

Teiidae, ***teiid lizards, whiptails*** and ***racerunners:*** a family of the *Lacertilia* (see) containing about 200 species in about 45 genera, found in South and Central America with a single genus in North America. They are the New World counterparts of the Old World *Lacertidae.* In addition to large predatory species, the family also contains small forms of normal lizard-like appearance, as well as soil-burrowing, snake-like representatives with degenerate limbs. All members have a characteristically long and fragile tail, moderately large eyes with oval pupils, well developed movable eyelids (usually covered with scales), and a long, narrow tongue, which is deeply notched at the tip and covered on its upper surface with fleshy protuberances. The front teeth are conical, while the side teeth may have 3 points, or are flattened to some extent, or may even form a broad grinding surface, depending on the species. In contrast to the *Lacertidae,* the teeth are not hollow at the base, and some of the plates on the head are not fused to the bones of the skull. Representatives include: Whiptail lizards (see), Jungle runners (see), Tegu (see) and Snake teiids (see).

Teiid lizards: see *Teiidae.*

Tela choroidea: see Brain.

Telencephelon: see Brain.

Teleosaurus (Greek *teleos* complete, *sauros* lizard): an extinct, long-tailed marine genus of crocodiles, whose long snout and sharp teeth indicate that they preyed on fish. Both the dorsal and ventral surfaces were armored. The limbs were adapted as paddles, but each retained 5 toes; the forelimbs were shorter and more strongly modified. Fossils occur in the European Jurassic.

Teleost: see *Osteichthyes.*

Teleutospores, ***teliospores, winter spores:*** single- or multi-celled, diploid, thick-walled perennating organs, formed by rust fungi (see *Basidiomycetes,* Figs. 3.4 and 3.5). When first formed, the cells are binucleate, and the nuclei fuse when the cell matures. After overwintering, each diploid cell puts out a tubular basidium (or promycelium). Meiosis then produces 4 haploid nuclei, separated by transverse septa, and each cell produces a haploid basidiospore (see Basidiospores) carried on a sterigma. The life cycle of the fungus is completed when these basidiospores are carried by the wind to a plant host.

Teliospores: see Teleutospores.
Tellins: see Modes of nutrition.
Telmatoplankton: see Plankton.
Teloblasty: see Mesoblast.
Telome theory: a theory developed by W. Zimmermann to explain the evolutionary origin and further differentiation of cormophyte organs (see *Cormophyta*). The primordial telomic state is seen as consisting of telomes and mesomes (Fig. 1); the basic structure consists of a central strand of prosenchymatous lignified cells (primordial stele or prostele) and a parenchymatous sheath or cortex, with an exterior layer of cutinized epidermis. The peripheral, unbranched parts of shoots are called telomes; these extend from the shoot or organ apex to the junction with other telomes. Parts of shoots between telomes are called mesomes. The conceived growth habit of the primordial telome more or less corresponds to that of *Rhynia*, a psilopsid fern. Telomes may be sterile and are then called phylloids; if the whole or part of a phylloid has a reproductive function (fertile telome), it is called a sporangium. According to the T.t., the typical organs of higher plants (e.g. reproductive organs and especially leaves and roots), as well as the variety of different morphologies that they display, have been generated by the following 5 phylogenetic elementary processes (Fig. 2).

1. *Overtopping.* Originally the telomes have equal status, but this is changed when the growth of one shoot is favored. This shoot becomes dominant, while the other (overtopped) shoot(s) becomes a lateral appendage.

2. *Planation.* Telomes of the lateral branches come to lie in the same plane, i.e. a 2-dimensional arrangement is formed. Planation is, for example, a prerequisite for the volution of flat, planar leaves.

3. *Fusion.* Telomes and mesomes may fuse directly or become united by parenchyma. If planar telomes (resulting from planation) fuse with one another, a typical leaf form is produced (IIIa). Fusion in the axis (i.e. where telomes have a 3-dimensional arrangement) produces a genuine "stem" structure containing several vascular bundles (IIIb).

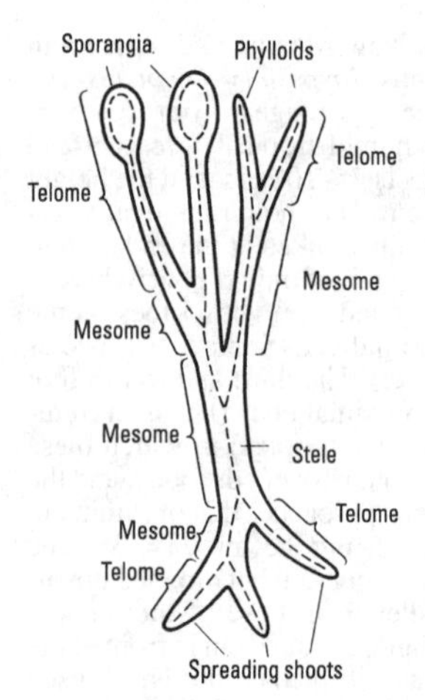

Fig. 1. *Telome theory.* The primitive telome condition.

4. *Reduction.* Individual telomes can become reduced, resulting in simplification. Reduction can occur at different sites. For example, the small, single-veined leaflets (microphylls) of the clubmosses and *Asteroxylon* (a psilopsid fern) could have arisen in this way.

5. *Involution.* This occurs by unequal growth of opposite sides of a telome, and it is thought to be especially important in the formation of reproductive organs.

According to the T.t., these 5 elementary processes have occurred repeatedly independently of each other and in combination, during the evolution of higher plants.

On the basis of these concepts, and in complete agreement with paleobotanical evidence, it is possible to explain the formation and derivation of the main characters of the various classes of pteridophytes. Thus, the *Primofilices* (from the Middle and Upper Devonian) display a range of forms, which can be interpreted to represent continuous evolution (by overtopping, planation, fusion, reduction and involution) from early land plants to the most advanced modern plants.

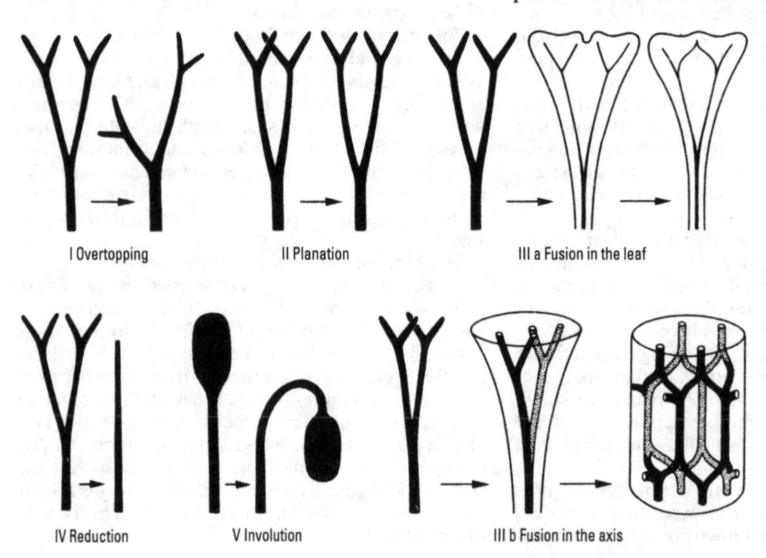

Fig. 2. *Telome theory.* The 5 elementary phylogenetic processes.

Telophase: see Meiosis, Mitosis.

Telosporidia: a subclass of the *Sporozoa* (see). *T.* and the *Sporozoa* are often considered to be synonymous. The adult of the vegetative stage has a single nucleus, and the end of this stage is marked by spore formation. Spore cases, if present, are usually simple in structure, and they usually contain several sporozoites. The vegetative stage *(trophozoite)* may be ameboid (e.g. *Plasmodium),* but in most species it has a definite shape. In those forms displaying trophozoite fission (agamogony), this fission is multiple, and is called *schizogony;* the resulting cells are then called *schizozoites* or *merozoites.* The first division of the zygote is meiotic, so that most of the life cycle is haploid. Two orders are recognized.

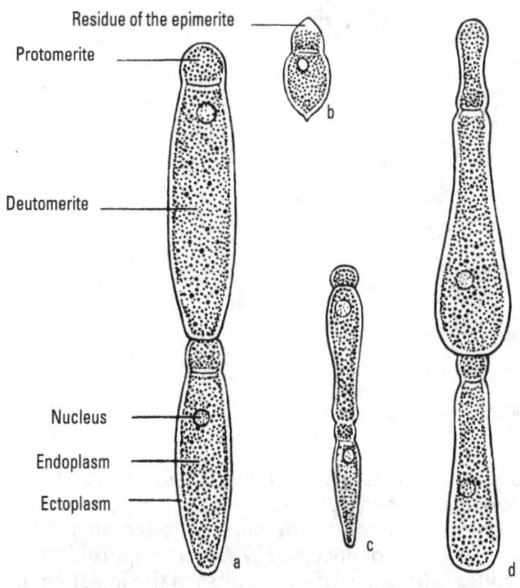

Fig.1. *Telosporidia.* Gregarines from the intestine of the meal worm. *a Gregarina polymorpha* (2 animals attached end to end in syzygy). *b* Young stage of *a. c Gregarina steini* (2 animals in syzygy). *d Gregarina cuneata* (2 animals in syzygy). (All x 100)

1. Order: *Gregarinidea* (Gregarines). Adult trophozoites outgrow their host cell and become extracellular. Male and female gametes are merogametes.

1. Suborder: *Schizogregarinaria.* Schizogony precedes gametogamy. They are parasites of annelids, arthropods, echinoderms and ascidians, usually occurring in the body cavities.

2. Suborder: *Eugregarinaria.* Schizogony does not occur, and the sporozoites develop directly into gamonts. The body is divided into 3 segments: epimerite (fixing organ), protomerite and deutomerite (contains the nucleus). Gamonts tend to adhere in pairs, a phenomenon known as *syzygy.* They encyst, and the separating walls disintegrate so that both gamonts are contained within a single cyst. The gamonts do not fuse, but undergo multiple fission into several gametes. Pairs of gametes then fuse, the members of each pair being derived from a different gamont. Each zygote transforms into a walled body, which is usually called a spore (previously called a *pseudonavicella).* Each spore undergoes 3 nuclear divisions to form 8 sporozoites (sporogony).

2. Order: *Coccidiomorpha.* The trophozoite remains inside a host cell. Schizogony represents a stage of active multiple fission, resulting in the inundation of the host with merozoites. Syngamy is developed to the stage of oogamy, and the zygote is an oocyst. The latter becomes a sporocyst, within which two phases of multiplication occur, i.e. the zygote divides to form sporoblasts, then each sporoblast divides into 2 sporozoites (sporogony).

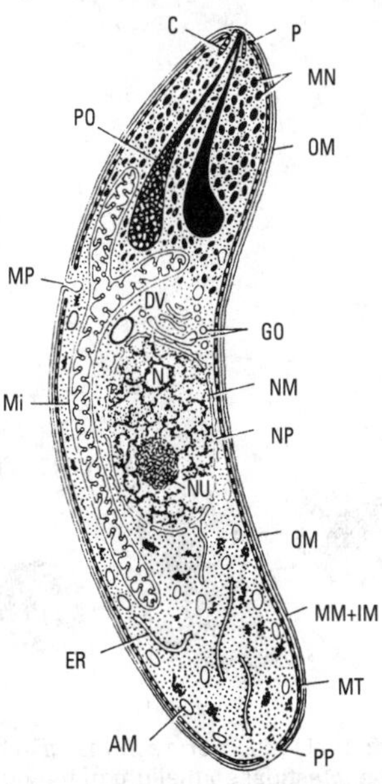

Fig.2. *Telosporidia.* Fine structure of a coccidian merozoite in longitudinal section. *AM* Amylopectin. *C* Conoid. *DV* Thick-walled vesicle. *ER* Endoplasmic reticulum. *GO* Golgi apparatus. *IM* Inner membrane. *Mi* Mitochondria. *MM* Middle membrane. *MN* Micronemes. *MP* Micropore or cytostome. *MT* Microtubules. *N* Nucleus. *NM* Nuclear membrane. *NP* Pore in nuclear membrane. *NU* Nucleous. *OM* Outer membrane. *P* Anterior or apical polar ring. *PO* Rhoptries (paired organelles). *PP* Posterior polar ring.

1. Suborder: *Coccidia.* These are mostly intracellular gut parasites. The zygote is nonmotile and sporozoites are usually encased. Species of *Eimeria* are parasitic in the intestinal epithelium of vertebrates and invertebrates (Fig.3), and are responsible for coccidoses of cattle, rabbits and poultry. Toxoplasmosis of mammals (including humans) and birds is caused by *Toxoplasma gondii. [Toxoplasma* has a doubtful phylogenetic status, so that some authorities place it in a separate order, *Toxoplasmida* (see)].

2. Suborder: *Haemosporidia.* These are always intracellular blood parasites of vertebrates, and they are transmitted by blood-sucking invertebrates. The sporozoite is naked and the zygote (ookinete) is motile. Intraerythrocytic stages in the vertebrate host contain granules of pigment derived from the host hemoglobin. Agamogony and gamete formation occur within the erythrocytes of the vertebrate host, and gamete formation and fertilization occur in the invertebrate vector.

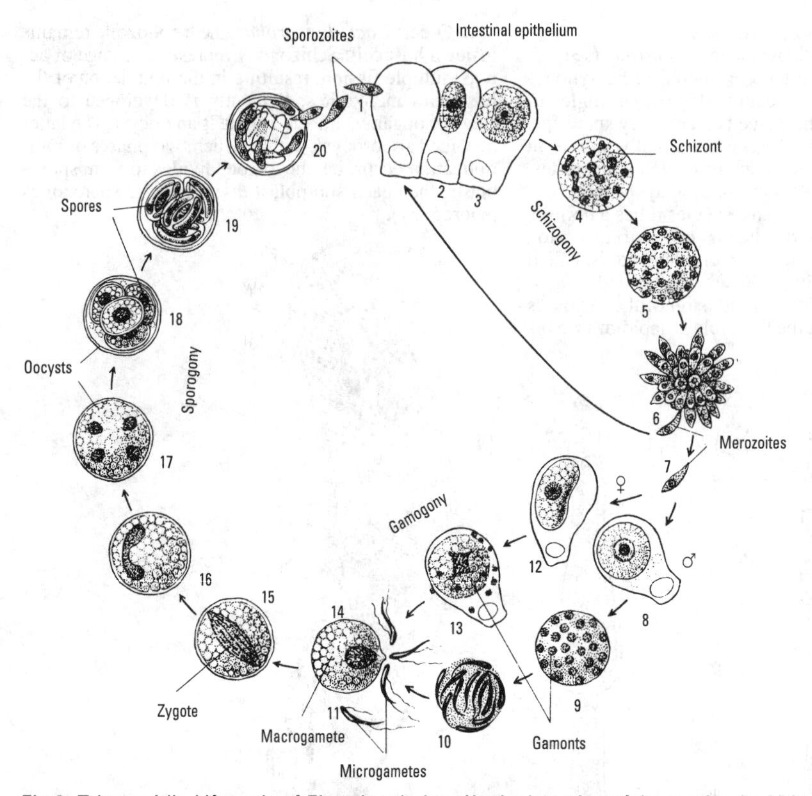

Fig.3. *Telosporidia.* Life cycle of *Eimeria schubergi* in the intestine of the centipede, *Lithobius forficatus. 1, 2, 3* Infection of a cell in the intestinal epithelium of the host, followed by growth of the agamont. *4, 5, 6, 7* Schizogony (agamogony). *8, 9* Male gamogony. *10, 11* Development and release of male microgametes. *12, 13* Female gamogony. *14* Conjugation (syngamy). *15, 16, 17, 18* Division of encysted zygote into sporoblasts. *19* Division of each sporoblast within its cyst into two sporozoites. *20* Escape of sporozoites in the intestine of a new host.

Various species of *Plasmodium,* transmitted by mosquitoes of the genus *Anopheles,* are causative agents of malaria (see plate 43). Piroplasmids (see) are sometimes classified as *Haemosporidia.*

Telotaxis, *goal orientation:* movement (taxis) of a motile organism in response to a single stimulus, without reference to other sources of stimulation, i.e. movement directed to a goal. See Klinotaxis, Tropotaxis.

Telson: the tail plate (final abdominal "segment") of many crustaceans. Unlike a true abdominal segment, it possesses neither a ganglion nor a pair of appendages.

Temperate phages: phages that do not cause lysis of the host cell immediately after infection, i.e. infection is not followed directly by the synthesis of new phage particles. The genetic material of the phage (called the "prophage") is usually integrated into the host genome, where it is reproduced and passed on to each daughter cell. Host cell lysis usually does not occur until the host (and with it the prophage) has replicated several hundred times. Bacteria carrying this type of prophage, and therefore "predestined" to lyse, are called *lysogenic bacteria,* and the condition is called *lysogeny.* Integration of the T.p. genome into the genome of a lysogenic bacterium may confer new characters on the host *(lysogenic conversion).* Thus, diphtheria bacteria do not produce toxins until they are attacked by certain T.p. The abiltiy to form flagella can also be conferred by certain T.p., i.e. infection with a T.p. can convert a nonmotile bacterium into a motile form. There are also systems in which the prophage is not incorporated into the host genome ("extrachromosomal lysogeny", exhibited, e.g. by coliphage P_1).

ABCDEFGHMLKIJ — b_2 — int. $C_{III}NC_I$ x $C_{II}OP$ QR
Phage (head) | Phage (tail) | Integration | Regulation | DNA | Lysozyme Late protein

Fig.1. *Temperate phages.* Simplified linear gene map of λ-phage.

The regulatory processes leading to the establishment of lysogeny have been well studied in the system λ-phage/*Escherichia coli.* The simplified gene map of λ-phage (Fig.1) shows the structural genes that encode the phage head and tail, genes (A–R) that regulate the lytic process, and groups of genes whose activity determines the establishment and maintenance, as well as the loss of lysogeny. The latter include genes C_I, C_{II} and C_{III}, which shortly after penetration of the phage into the host cell suppress development of the lytic process (see

Phages) of λ-phage. C_I gene encodes the λ-repressor. This acidic protein (M_r ⟩⟩ 30 000) has a high affinity for DNA sequences in the operator region, so that it prevents transcription of a large part of the phage genome, i.e. it represses genes X, C_{II} and O (these together form an operon, which is transcribed from X to O), as well as gene N. Some of these genes encode early proteins (see Virus, Phages), which are needed, inter alia, for the replication of phage nucleic acid. As soon as the λ-repressor is formed, further infection with the same type of phage is suppressed, so that the bacterium is immune to superinfection.

In the second phase of lysogeny, the repressed linear genome of the λ-phage (the prophage) is converted into a circular genome. Each end of the open, double-stranded DNA has a short, single-stranded projection of about 15 nucleotides. These represent complementary nucleotide sequences, which can therefore become spliced by base pairing, converting the DNA molecule into a circular structure. Integration of the circular phage genome is then promoted by the action of gene products of the integration zone of the phage genome. Phage DNA and bacterial DNA become aligned with one another in the *homologous region,* i.e. where the nucleotide sequences of virus and host DNA correspond (1, 2, 3, 4 in Fig.2). In the middle of this region, both DNA circles are cleaved by endonucleases, then linked crosswise with each other by the action of enzymes similar to repair enzymes. The viral "chromosome" is therefore integrated into the bacterial chromosome in the same way that plasmids can be integrated into the "main" DNA strand of a bacterium. The prophage is incorporated between the genes for galactose synthesis and biotin synthesis.

Synthesis of new phage particles occurs when lysogeny is reversed. Often, this spontaneous prophage induction does not occur until the host has passed through hundreds or thousands of cycles of cell division. Experimentally, the most potent means of triggering prophage induction are UV-irradiation and treatment with mitomycin C. Prophage induction occurs by inactivation or neutralization of the phage repressors, including the λ-repressor; the subsequent events represent a reversal of the processes leading to lysogeny. The phage DNA is released from the host genome; genes for early proteins are then transcribed, followed by replication and linearization of the phage DNA, production of virus proteins, maturation of phage particles and lysis of the host cell (see Phages).

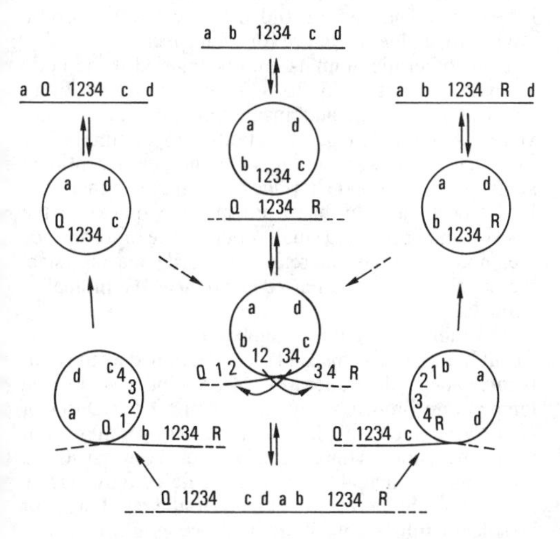

Fig.2. *Temperate phages.* Campbell model for the integration and excision of the genome of λ-phage. After integration, the phage genome (with genetic markers *a, b, c, d)* lies between two homologous regions. Left and right: production of transducing phages, in which the bacterial marker, *Q (gal*-region), is exchanged for phage marker *b* (left), and *R (bio*-region) is exchanged for *c* (right).

Occasionally, errors occur in the excision of the prophage (about 1 in every 10^6 particles). Parts of the phage genome are replaced by host genes from the regions neighboring the insertion site of the prophage *(gal-* and *bio*-regions). As a result of this loss of part of their genome, such phages are *defective,* and cannot complete their replication cycle without assistance from another phage (helper phage) carrying the missing genes, i.e. in a host cell superinfected with both helper and defective phages. Phages carrying part of the host genome can transfer genes from the first host to a second host. Thus, strains of *Escherichia coli* that have lost the ability to utilize galactose regain this ability if they are infected with a phage carrying the *gal*-region from a previous host. Such a transfer of genetic material by T.p. is called *transduction.*

Temperate viruses: animal viruses which complete their replication cycle (i.e. form new virus particles and cause severe cell damage or cell death) in only some infected cells. In this respect, they resemble Temperate phages (see). In most cells, the viral replication cycle is interrupted, and the viral genome becomes incorporated into the host genome. The provirus is then replicated with each host cell division and passed on to each daughter cell. Such a repressed viral genome may confer new properties on the host cell, such as uncontrolled cell growth and division, leading to the production of a tumor. See Tumor viruses.

Temperature coefficient, Q_{10}: in plant physiology, the change in the rate of a process brought about by a temperature change of 10 °C. For example, $Q_{10} = 2$ means that the reaction rate doubles when the temperature is increased by 10 °C.

Temperature factor, ***temperature effect, temperature relationships:***

1) The temperature relationships that influence the structure and performance of a plant in its habitat. Plants and microorganisms can exist and survive within a very wide temperature range. Arctic-Alpine pioneer plants can withstand temperatures as low as –70 °C. Resistant structures of some primitive plants can survive temperatures down to –200 °C for a short period, and the dehydrated spores of some bacteria can survive boiling water for 1 hour. Plants, however, are normally biologically active between 0 °C and +45 °C, with species differences within this range. Plants often differ with respect to their *temperature minimum* and/or *temperature maximum,* i.e. the lowest and/or highest temperature at which the species in question can live. Species also differ from one another with respect to their *temperature optima,* i.e. the temperature at which they grow best (this is also usually the temperature of optimal development). The minimum, optimum and maximum are *cardinal values* of the temperature relationship.

Eurythermal plants thrive within a relatively wide temperature range. Furthermore, between their temperature minimum and the critical low temperature that kills them, and between their temperature maximum and the critical, nonsurvivable high temperature, they pass through transition stages in which they are still viable but totally inactive (i.e. there are high and low temperature ranges of suspended function). In contrast, *stenothermal plants* display only a narrow temperature amplitude, and periods of " suspended function" are largely absent. According to whether the optimum of the temperature curve lies at a low or high temperature, plants are classified as warm- or cold-loving, and all intermediate types are also known. Cold-loving plants emerge rapidly from cold dormancy into an actively growing state, and they are capable of metabolic assimilation at low temperatures. They are therefore able to exploit with relative efficiency the usually short growth season of their habitat. For this purpose they often have well developed winter buds. In addition, their foliage is retained all the year round, and is often xeromorphic and therefore resistant to the drying effects of winter frosts. It is these adaptations that enable the sudden sprouting and flowering of high alpine vegetation. Many cold-loving plants are characterized by frost resistance, i.e. resistance of their organs to extremes of low temperature. Frost resistance displays annual periodicity. Thus, it can be shown experimentally that the frost resistance of plants is lower in summer and higher in winter. In cold-loving plants, the temperature optima for photosynthesis and respiration are low, so that at low temperatures sufficient material is assimilated and enough energy is generated for growth and development. In *warm-loving plants,* assimilation and energy production are often totally prevented by those temperatures at which cold-loving plants are optimally active. Many warm-loving plants "freeze" at temperatures above the freezing point of water, and in extreme cases this may occur at temperatures as high as +10 °C to +20 °C. The temperature optimum of warm-loving plants is relatively high, and the temperature limit for vegetative tissues may also be high, sometimes even higher than the coagulation temperature of protein. Some cyanobacteria and other bacteria live in thermal springs at temperatures up to +75 °C, and the basis of their resistance to these temperatures is still not fully explained. In terrestrial plants, the danger of overheating of organs is avoided in various ways, e.g. by high transpiration, alignment of leaves to the sun (see Compass plants), columnar habit of the axis (as in many cacti and succulents).

2) For animals, temperature is the most important abiotic environmental factor. The most significant source of heat energy is irradiation from the sun. Metabolic activity enables the animal to produce its own heat. In *poikilothermic* ("cold-blooded") animals, endogenous heat production is only small and easily lost to the surroundings; their body temperature therefore changes with that of the environment. A certain degree of temperature regulation is, however, possible; e.g. in mass aggregations of bees or weevils, body temperature production leads to an increase in the temperature of the aggregate. Changes of body position and posture, water evaporation from the body surface (e.g. from the tongue) and oxidative metabolism are all employed to some extent by poikilothermic animals as means of temperature regulation.

In *homeothermic* ("warm-blooded") animals (birds and mammals), temperature regulation is much more highly developed, employing heat production and efficient insulation. Constant body temperature is maintained by the efficient coordination of heat production at the center and heat loss at the periphery (see Temperature regulation of the body). In fact, only the temperature at the center of the body is truly constant, the peripheral tissue temperature varying with (but not necessarily corresponding to) that of the environment. The tolerable or favorable temperature range varies considerably according to the animal species. *Eurythermal animals* tolerate a wide range of temperature, whereas *stenothermal animals* tolerate only a relatively narrow temperature range. *Cold-stenothermal animals* have a narrow tolerable temperature range in the cold region, and *warm-stenothermal animals* have a narrow tolerable temperature range in the warm region.

The lower temperature limit for poikilothermic animals is usually below 0 °C, exceptions being found among tropical species which can die above 0 °C. In contrast, roundworms and rotifers in a state of anabiosis can withstand extreme temperatures below freezing. Cold sensitivity is influenced by water content, salt concentration, stage of development, sex, and the possession of additional means of protection. The duration of exposure to low temperatures and the rate of cooling also effect the extent of damage caused by cold temperatures.

At the lower temperature limit, there is frequently a lack of movement, i.e. Cold torpor (see). This is associated with an extensive decrease of metabolism, and in poikilothermic animals it is an essential component of hibernation. The onset of cold torpor occurs at a specific environmental temperature for each species.

Homeothermic animals are endangered if the body temperature falls to 15–20 °C (except in hibernation). By heat generation and insulation, this can be prevented, but maintenance of a body temperature higher than that of the surroundings depends on a continual supply of food. Food shortages in winter can therefore lead to death, and this happens especially quickly in the case of small birds and small rodents. The ratio, surface area/body volume, increases with decreasing size, thereby favoring an increased rate of heat loss in smaller animals.

The range of the upper temperature limit is narrow for all living organisms, and it is determined by the temperature-dependent alteration of proteins. Usually, the temperature limit is not greater than 50 °C. Onset of *heat torpor* indicates that the upper temperature limit has been attained. This generally occurs between 40 and 50 °C, and the actual temperature is defined by narrow limits. A slight increase then leads to the rapid onset of heat death. Inhabitants of thermal springs are characterized by a relatively high heat death temperture. Cold-stenotherm animals, like young trout, are killed by temperatures only as low as 15 °C. In addition, the degree of heat damage is influenced by many internal and external factors. Within the tolerance range, each species displays a *preferred temperature,* which may be altered by other factors, e.g. time of day, time of year, stage of development, etc. The preferred temperature and *tem-*

perature tolerance range can be changed within certain limits by acclimatization and habituation.

Temperature can affect physical body characters, and when several generations are produced in one year, they may display seasonal dimorphism (e.g. in *Lepidoptera).* Water fleas also display different body shapes during the course of the year, a phenomenon known as Cyclomorphosis (see). Striking alterations in coloration and patterning may also occur through seasonal dimorphism. Thus, low temperatures lead to increased melanin production in insects. Both temperature extremes often have the same effect, thereby accounting for seasonal color variations in free-living insects.

Practically all the life functions of poikilothermic animals are temperature-dependent, especially their rate of metabolism. Bodily activity is therefore also strongly dependent on temperature. Their length of life is also related to environmental temperature. Within the tolerable temperature range, low temperatures prolong life. However, the resulting delay in the onset of sexual maturity, and decrease in the rate of development of offspring counteract the apparently positive effect of the increased lifespan. In the development of poikilothermic animals, there is an optimal temperature at which the greatest proportion of individuals complete their development in the shortest time. Since a certain amount of heat is necessary for the development of a species, the time required for development is related to temperature.

See Cold death, Cold injury, Cold torpor, Temperature regulation of the body.

Temperature preference, ***preferred temperature, thermal indifference zone:*** in humans, also known as the range of temperature comfort, a temperature range within which the individual experiences no temperature sensation, i.e. feels neither warm nor cold. The T.p. differs according to the animal species, and varies within each species in dependence on the physiological state. In thermotaxis (orientation within a temperature gradient), the thermotactic optimum is also known as the T.p. or thermal indifference zone.

Temperature receptors: see Thermoreceptors.

Temperature regulation of the body: in homeothermic animals (birds and mammals), maintenance of a constant body temperature by nervous and hormonal control. Temperature receptors in the hypothalamus respond to temperatures that are above or below the nominal or target value for blood temperature. This value is normally constant, although it may vary slightly during each 24 hours (see Body temperature), and it changes more dramatically in Fever (see) and during Hibernation (see). The central nervous system also constantly receives signals from peripheral Thermoreceptors (see), which monitor body temperature. As a result, autonomic and motor nerves from the regulatory systems of the hypothalamus change their impulse frequencies, thereby activating a wide range of temperature adjustment processes, such as central heat production, alteration of peripheral blood flow (vasoconstriction and vasodilation), and perspiration (Fig.1).

Role of the hypothalamus. The most anterior part of the hypothalamus contains the *preoptic center,* a nucleus or group of neurons that responds directly to temperature. An increase in blood temperature causes an increase in the rate of discharge of preoptic neurons. Conversely, a decrease in blood temperature is accompanied by a decrease in the rate of discharge of preoptic

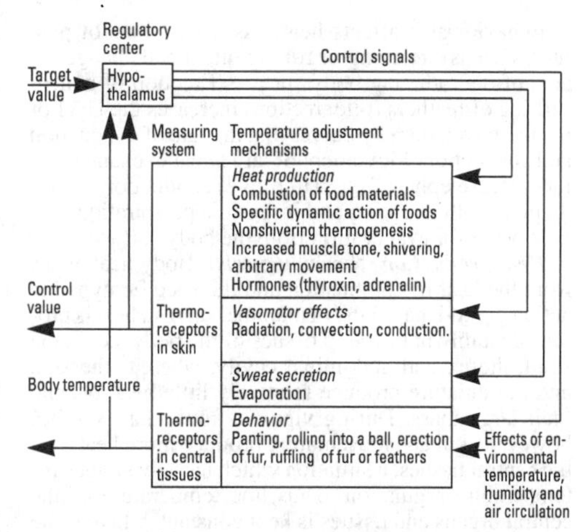

Fig.1. *Temperature regulation of the body*

neurons. Signals from the preoptic center are transmitted to other parts of the hypothalamus, i.e. to 1) the anterior *heat-losing center* (consisting mainly of hypothalamic parasympathetic centers, which, when stimulated, activate various mechanisms for decreasing body heat), and to 2) the posterior *heat-promoting center* (which operates mainly via the sympathetic nervous system, and when stimulated activates mechanisms for increasing body heat).

Heat loss. Heat transported by the blood from the deep tissues of the body is then lost from the body surface by radiation and conduction (convection), and by the evaporation of sweat (Fig.2). Physical factors that influence heat loss are the temperature difference between the body and environment (which determines the intensity of heat loss by radiation), as well as atmospheric pressure, humidity and air movement, which determine the extent of conduction and convection. Physiologically, heat loss is controlled by the heat-losing center of the hypothalamus, which reduces body heat by promoting vasodilation (increased blood flow in the periphery of the body), promoting sweating, and by causing decreased muscle tone. If the body becomes overheated, fibers of the sympathetic system are inhibited, the peripheral blood vessels dilate, and the skin radiates more heat. Heat stimulates cholinergic fibers of the sympathetic nerve, which promote sweat secretion.

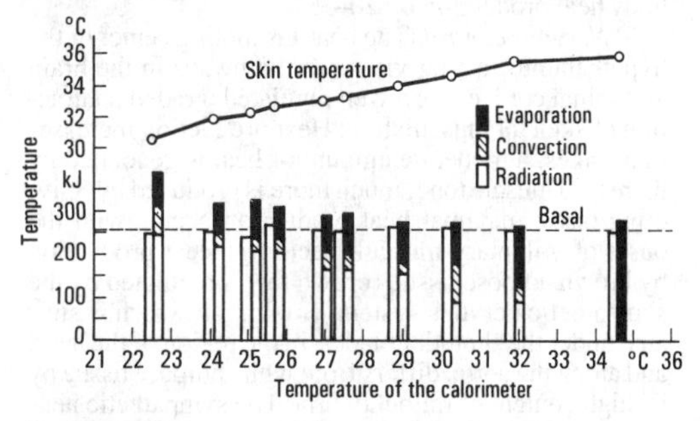

Fig.2. *Temperature regulation of the body.* Processes of heat loss at different skin temperatures in the human.

Behavior also affects heat loss. Alterations of posture, such as stretching or rolling into a ball change the area of the radiating body surface. Erection of hair or ruffling of feathers (piloerection) increases the layer of insulating air, thereby decreasing the rate of conduction and convection. Movement of large heat exchange surfaces (e.g. elephants' ears) improves conduction and radiation. Both sweating and insensible perspiration provide water for evaporation from the body surface.

Heat production (thermogenesis). Body heat arises from the heat of combustion and the specific dynamic action of food materials. In mammals at rest, heat is produced mainly in the deep tissues of the body, i.e. in the skull, thorax and abdominal cavity, whereas the skin and musculature produce relatively little heat, despite their large mass. During vigorous muscular exercise, however, skin and musculature produce more heat than the central tissues, a situation which has advantages for temperature regulation. Thus, the temperature of the central organs and tissues is kept constant, whereas the temperature of the peripheral tissues is allowed to vary. At rest, the heat necessary for the maintenance of body temperature is produced directly in the central tissues. Additional heat from muscle activity is easily lost, because it follows a very short route to the exterior. Long term processes of heat production are regulated hormonally, whereas short term heat production is largely under nervous control.

1) *Hormonal control.* If the body is exposed to cold for several weeks, e.g. with the onset of winter, the thyroid enlarges and produces more thyroxin and triiodothyronine. Cold stimulation of thermoreceptors leads to increased production of a neurosecretory hormone in the preoptic region of the hypothalamus. This, in turn, stimulates the release of thyrotropin (thyroid stimulating hormone; TSH) from the anterior hypophysis, which then stimulates production of thyroxin and triiodothyronine by the thyroid. Slightly reduced temperatures lead to increased storage of inactive thyreoglobulin in the thyroid, but during cold stress, thyroxin and triiodothyronine are released from the thyreoglobulin by proteolysis. Cold stress also increases the production of adrenalin and noradrenalin, which promote the further release of thyroxin and triiodothyronine. Thyroid hormones increase the rate of combustion of carbohydrates, fats and proteins, with a resulting increase in body heat production; they also decrease the rate of protein synthesis in the body. In the long term, under very cold conditions, enhanced thyroid activity may increase body heat production by 20–30%.

2) *Nervous control.* The heat-promoting center of the hypothalamus, acting via motor pathways in the brain and spinal cord, cause a well regulated, graded contraction of skeletal musculature. Heat production increases in 3 stages: a moderate amount of heat is generated by increased muscle tone; much more is produced by Shivering (see); maximal heat production occurs with the onset of voluntary muscular activity. Heat production by brown adipose tissue (brown fat) is regulated by the sympathetic nervous system. Brown fat, which is situated under the shoulder blades, in the region of the heart and along the aorta, differs from white adipose tissue by its high content of mitochondria. The sympathetic neurotransmitter, noradrenalin, causes the mobilization of free fatty acids in brown fat. Since the mitochondria of brown fat are largely uncoupled, this results in high heat production, known as *nonshivering thermogenesis.* The musculature, heart and liver also make a contribution to nonshivering thermogenesis.

See Body temperature, Thermoreceptors.

Temperature tolerance: see Temperature factor.

Temple: a region on the head of humans and other vertebrates, lying between the eye and ear and above the cheek.

Tenancy: see Relationships between living organisms.

Tench, *Tinca tinca, Tinca vulgaris:* a freshwater fish of the family *Cyprinidae* (minnows and carps) (order *Cypriniformes).* It is native to Europe and northern Asia, and has been introduced into other continents. Maximal length is 60 cm, but 30–45 cm is more usual, with a weight of about 5 kg. It inhabits lakes with mud bottoms, and is sometimes found in marshy river estuaries. It is one of the most important commercial freshwater fishes, and the flesh is white and fine-tasting. The small scales are inserted deeply into the skin. It has a blue-green dorsal surface, greenish sides with a gold luster, and a light yellowish belly. Gold and red specimens are also encountered. Spawning occurs from the end of May to the end of July. It spends most of the time on the bottom, feeding on insect larvae and crustaceans, and spends the winter in shoals, sometimes buried in the bottom mud.

Tendon: see Connective tissues.

Tendril movements: movements of Tendrils (see) leading to the grasping of a support. Many tendrils perform autonomic cycling or spiraling movements (see Cyclonasty). As soon as a tendril makes contact with a potential support, a haplonastic bending movement is initiated (see Haplonasty). Turgor decreases on the surface against the support, apparently due to loss of the semipermeable properties of the cell membranes in that region. At the same time, there is an increase in the turgor of the cells on the opposite side of the tendril, so that the tendril begins to bend around the support. This first phase of curvature is accompanied by the utilization of adenosine 5′-triphosphate (ATP). Specific motility proteins are also probably involved. The second phase consists of auxin-dependent growth on the side of the tendril not in contact with the support, so that the tendril often forms several coils around the support.

Tendrils: contact-sensitive, branched or unbranched, filamentous modifications of lateral shoots, leaves or (more rarely) roots, which serve to attach climbing plants to their supports. *Shoot T.* (metamorphosed shoot axes) are found e.g. in the grape vine and other members of the *Vitaceae.* The *leaf T.* of the *Cucurbitaceae* consist of modified midribs, whereas the leaf T. of the *Papilionaceae* are modified leaf laminas (e.g. modified upper leaflets of the pinnate leaf of pea and vetch). The simple tendril of *Lathyrus aphaca* is derived from the lamina, and the two large stipules are developed into photosynthetic organs. The elongated petioles of *Tropaeolum* ("nasturtium", family *Tropaeolaceae)* also display contact sensitivity and behave as T. *Root T.* are found in certain tropical climbers, e.g. the vanilla plant *(Orchidaceae),* where they become entwined in the neighboring vegetation.

Tenrecidae, *Tenrecs* and ***Madagascar hedgehogs:*** a primitive family (34 species in 12 genera) of the mammalian order, *Insectivora* (see). Dental formula: 2/3, 1/1, 3/3, 4/3 = 40. The family is native only to Madagascar and the Comoro islands, but members have been introduced to other islands in the Indian Ocean. The coat hairs of these animals are modified to bristles and/or spines. Body temperature regulation is incomplete, e.g. the body temperature of the 40 cm-long ***Tenrec*** *[Centetes (= Tenrec) ecaudatus]* varies between 24 ° and 35 °C. Another well known member of the family is the ***Small Madagascar "hedgehog"*** *(Echinops telfairi).*

Tensile strength: mechanical resistance to tension, which is particularly well developed in the roots and rootstock of terrestrial plants, and in the stems of aquatic plants that live in fast flowing water. T.s. is conferred mainly by sclerenchyma fibers, which, in accordance with the laws of mechanics, are aligned longitudinally in the central core of the organ.

Tentacle sensor: see *Apoda.*

Tentaculata, *Micropharyngea:* a class of aquatic, mostly sessile, bottom-dwelling organisms, with three body sections: protosoma, mesosoma, metasoma, the posterior section containing paired coelomic cavities. In all *T.* the mesosome carries a tentacular apparatus or lophophore which surrounds the mouth opening. The 4 phyla of the *T.* are listed separately: *Phoronida, Bryozoa, Brachiopoda, Entoprocta.*

Tentaculitida, ***tentaculites*** (Latin *tentaculum* feeler): an extinct group of mollusks in the class *Cricoconarida.* The millimeter- to centimeter-long, pointed conical shells consist of mother-of-pearl covered on both sides with a layer of prismatic material. There is a smooth, pointed protoconch (embryonic chamber). Some forms are smooth, while others have transverse or longitudinal ribbing. Fossils are found from the Lower Ordovician to the Upper Devonian. Between the Upper Silurian and Middle Devonian, *T.* were responsible for the formation of sedimentary rocks (tentaculite chalk and slate of the Devonian).

Tent caterpillar moths: moths of the family *Lasiocampidae* (see), whose caterpillars spin tent-like shelters.

Tenthredinidae: see Sawflies.

Tenuis larva: see *Carapidae.*

Tepal: see Flower.

Terathopius: see *Falconiformes.*

Teratocarcinoma: see Chimera.

Teratology: the study of Malformations (see). About 3% of all full term human pregnancies result in malformed or deformed offspring. The study of the genetic origin of malformations and their developmental physiology is therefore important in human genetics (see Human genetic counselling). Such studies are also important for understanding normal developmental processes.

Teratornis: see *Falconiformes.*

Terebrantes: see *Hymenoptera.*

Terebrantia: see *Thysanoptera.*

Terebratula (Latin *terebratus* bored through): a genus of the order *Terebratulida,* phylum *Brachiopoda,* with a plump, rounded to oval, smooth or concentrically lined shell with a folded anterior border. The edge of the hinge is short and curved. The beak-like posterior umbo of the ventral valve (which has a large pedicle foramen) extends well beyond the dorsal valve, and the looped brachial skeleton is alkylopegmatic. The order provides numerous index fossils from the upper Silurian to the present, but the family *Terebratulae in sensuo stricto* is restricted to the Miocene and Pliocene.

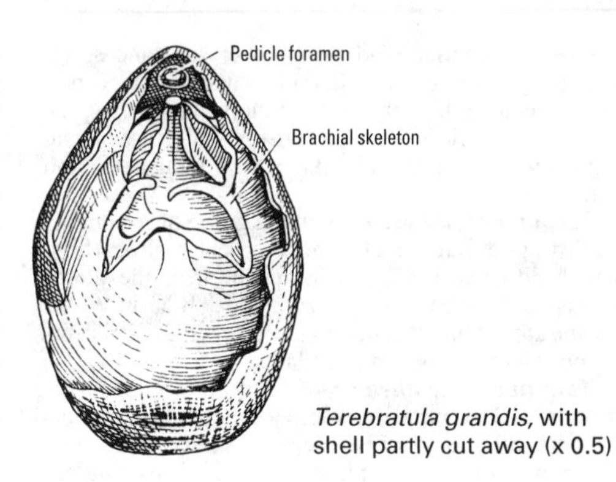

Terebratula grandis, with shell partly cut away (x 0.5)

Tergomya: see *Monoplacophora.*

Terminal chiasma: in meiosis, the single remaining chiasma between paired chromosomes at the beginning of anaphase I. During the late stage of prophase I in meiosis, many bivalents contain more than one chiasma. The chiasmata then decrease in number, starting from the kinetochores, until the chromosomes are held together by only one chiasma, i.e. the T.c.

Terminal filament: an unpaired extension of the 11th abdominal segment of insects, found predominantly in lower insects (e.g. *Archaeognatha, Zygentoma),* and sometimes leaflike and modified to a tracheal gill (caudal gill) as in dragonfly larvae.

Terminalization: see Meiosis.

Terminalization of chiasmata: see Chiasma terminalization.

Terminal knob: the terminal, knob-like expansion of a presynaptic nerve ending, i.e. T.k. is synonymous with the presynaptic terminal of a neuron or axon collateral. In the narrow sense, the term is applied to "Held-Auerbach knobs", i.e. the presynaptic terminals of motor neurons, which may attain a diameter of about 1 µm. See Synapse.

Termites: see *Isoptera.*

Termitida: see *Isoptera.*

Termitina: see *Isoptera.*

Termitophile: see Inquiline.

Terns: see *Laridae.*

Terpenes, ***terpenoids, isoprenes, isoprenoids:*** an extensive group of natural products, with the empirical formula $(C_5H_8)_n$. Inspection of the formulas of T. shows that they can be built up from a C_5 unit, which Wallach suggested might be isoprene. Ruzicka put forward the *isoprene rule,* in which the hydrocarbon skeleton of many open chain and cyclic terpenes is constructed from isoprene units arranged head-to-tail. The rule has proved useful in the assignment of structure, although there are exceptions, e.g. squalene contains a tail-to-tail arrangement. Free isoprene has not been demonstrated as a naturally occurring substance, but it is now known that T. are biosyntheized from the active 5-carbon unit, isopentenylpyrophosphate ("active isoprene"), thus confirming the isoprene rule. Isopentenylpyrophosphate is in turn biosynthesized from acetyl-CoA via mevalonic acid. According to their molecular size and/or the number of linked isoprene units, T. are classified as mono-, sesqui-, tri-, tetra- and polyterpenes. They include numerous olfactory substances, camphor, abscisic acid, gibberellins, resin acids, steroids, carotenoids, caoutchouc, and many other substances. The structures of more than 5 000 plant and animal T. have been elucidated, including alcohols, aldehydes, ketones, carboxylic acids, etc., some of which are physiologically important, while others have industrial applications.

Terpinenes, ***dihydrocymene:*** a group of 3 isomeric, doubly unsaturated, cyclic, monoterpene hydrocarbons, which differ according to the position of the double bonds. α-T. occurs in *Elettaria* spp., β-T. in *Pittosporum* spp., while γ-T. has a generally wide distribution in the essential oils of many plants.

Terpineol, ***menthene-8-ol:*** a naturally occurring mixture of 3 isomeric monoterpene alcohols (α-, β- and γ-T.). An oily liquid with a pleasant smell of lilac, it is a component of essential plant oils, and it is used in perfumery.

Terramycin: see Tetracyclins.

Terrapene: see *Emydidae.*

Terrapins: see *Emydidae.*

Terrestrial bugs: see *Heteroptera.*

Territorial behavior, ***territoriality:*** animal behavior that serves to establish, demarcate and defend an area. The area in question may be the entire range within which an animal or group of animals live and hunt or forage. Alternatively, it may be a relatively small area, which is chosen and defended for a specific purpose, e.g. a nesting site. In principle, T. is a population-specific, special case of Competition (see). In *stationary T.b.,* the territorial boundary is marked, e.g. by scent markings, or by visual markers, such as tree bark stripping by male deer (the smooth tree surface is then anointed with a secretion from glands beneath the eyes), or the signal pyramids of some fiddler crabs. *Nonstationary T.b.* is generally displayed by individual animals, e.g. acoustic signals, or visual body displays. Some species display in pairs, especially when they are bonded long-term or for life.

Establishment of territory insures freedom from disturbance and a provision of sufficient food for subsequent offspring. In many vertebrates, establishment of territory is a prerequisite for breeding.

Territory: an area or space within the habitat of a particular species, which is claimed and defended against intruders by an individual animal, a family, or a larger grouping. Within its T., an animal attains behavioral superiority over would-be intruders or aggressors, and the establishmen and defense of T. is a biological imperative. Association with a T. may be temporary, e.g. birds establish T. only during the breeding season, and T. is important to insects and bats only for the purposes of sleeping and hibernation. Permanent association with T. is observed, e.g. in some fishes, the field mouse, the brown rat and marmots.

In higher animals, T. is selected and claimed by *marking.* Mammals mark their T. with odiferous substances, birds and howler monkeys by vocalization, fiddler crabs by visual signals (waving of the large cheliped), and vertebrates in general employ a variety of imposing and threatening attitudes.

Strong association with a single T. is frequently accompanied by a highly developed homing instinct, especially with respect to breeding T., e.g. carrier pigeons, many migratory birds, salmon and amphibians travel large distances and return with great precision to their original breeding grounds. Homing ability is based on optical and chemical stimuli and on various other incompletely understood mechanisms (see Bird migration).

Tertiary: the oldest system of Cenozoic Era, The name refers to the third phase in the evolution of living organisms, and was formerly applied to the entire Cenozoic. The T. lasted 70 million years and is divided into 5 stages. In addition to marine deposits, there is an increasing frequency and quantity of limnic/continental formations. Important index fossils of the marine sediments are foraminifera, nannoplankton and mollusks. Nummulites (up to 11 cm in diameter) occur primarily in alpine deposits, where they make a local contribution to rock formation. Bivalves and snails attained a notable evolutionary peak, with large species and wide individual diversity; some members are important index fossils. Snails of the T. are closely related to modern living forms, and terrestrial snails occur in large numbers. Sea urchins also serve as index fossils. Sponges, corals, bryozoans and brachiopods are also present in large numbers, but have essentially no stratigraphic importance. A rich insect fauna is also present, sometimes preserved in amber. Mammals make a characteristic contribution to the continental deposits, sometimes revealing phylogenetic evolutionary series *(Equidae, Proboscidea).* All orders of mammals are represented, and during the T. they undergo explosive evolution with multiple and widely varied phylogenetic differentiation.

In terms of plant evolution, the T. belongs to the Cenophytic (see), containing both gymnosperms and angiosperms. The fossil record of central Europe indicates that the climate during the Lower T. was partly subtropical,and partly tropical, followed by a gradual cooling until the end of the Upper (Late) T. See Geological time scale.

Tessellated snake, *Natrix tessellata:* a slender, olive-gray colubrid snake (see *Colubridae)* distributed from western China to western Europe, and especially common in southeastern Europe. Its body is patterned with cube-shaped spots. It lives mainly in water, and feeds mostly on fish. In the eastern part of its range it attains a length of 1.5 m. Central European specimens are found only in warmer areas (e.g. Rhine-Main area) and are 0.75 m-long.

Testacea: see *Arcellinida.*

Testacida: see *Arcellinida.*

Testane: see Steroids.

Testate amebas: see *Arcellinida.*

Testis, ***testicle, orchis:*** the male sex gland of animals, in which the male gametes (spermatozoa, sperms) are produced.

Sponges have individual sperm-forming cells in their dermal layer, but they do not possess true testes.

In *coelenterates,* male glands are localized in various body sections. Thus *freshwater polyps* (e.g. *Hydra)* possess nipplelike prominences in the region of the oral cone, the testes of *anthozoans* are developed from groups of endodermal cells in the wall of the gastrovascular cavity, while in *ctenophorans* the testes are saclike appendages of the subcostal canals.

The paired testes of *platyhelminths* consist of branched, racemose tubes, or small, hollow vesicles distributed in large numbers in the body parenchyma. The exit ducts of the testes (vasa efferentia) unite to form a vas deferens, which, together with the female reproductive duct, empties into the genital atrium. The end of the vas deferens may expand to form a seminal vesicle with various accessory glands. *Nematodes* usually have a single, unpaired, tubular T.

In the primitive condition, the testes of *mollusks* are paired, but in advanced forms (snails, cephalopods) they are unpaired. They are hollow organs, and their cavities consist of the residue of the secondary coelom, the gonocoelom. In hermaphrodite snails, the T. may be united with the ovary to form the hermaphrodite gland.

Arthropod testes are paired in the primitive condition, but they may be secondarily fused into more or less single, uniform structures. They consist of a variable number of blind sperm tubes (testicular follicles), which unite to form a common duct (vas deferens).

Each of the paired testes of *insects* usually consists of a group of short testicular follicles (multifollicular T.), or more rarely of a single testicular follicle (unifollicular T.). Testicular follicles may be arranged like a comb or crest, in a raceme, or clustered in a terminal tuft (Fig.). In higher insects, the testicular follicles are often contained in a common envelope (peritoneal envelope), giving the external appearance of a single smooth structure. Internally, each testicular follicle is divided into zones: 1) the germarium or formative zone (at the blind tip of the follicle), containing primordial germ cells (spermatogonia) in the process of multiplication; 2) a growth zone, in which the spermatogonia increase in size, undergo repeated mitotoic division, and develop into spermatocytes; 3) maturation zone, in which spermatocytes develop into spermatids by meiosis; 4) transformation zone, in which spermatids become mature spermatozoa. From the spermatocyte stage onward, the developing cells are separated into groups, each group enveloped by an epithelial-like arrangement of testicular cyst cells. Growth and development occur within these structures, and the surrounding envelope is not lost until the spermatozoa are fully mature and capable of fertilization. The abandoned cyst cells degenerate after transfer of the seminal liquid to the female. From each follicle, spermatozoa are led into the vas deferens via a vas efferens. The two vasa deferentia combine to form the unpaired, medial ectodermal ejaculatory duct, which usually opens at the tip of a tubular copulatory organ (penis or phallus). As a rule, each vas deferens dilates into a seminal vesicle immediately before it unites with its counterpart, and there are usually accessory glands arising as blind tubes from the seminal vesicles and the vasa deferentia.

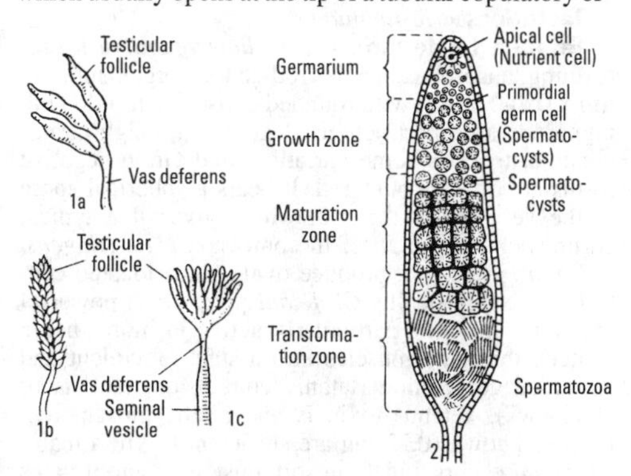

Testis. The insect testis. *1* Multifollicular testes. *1a* Comb-shaped testis of *Tetrophthalmus* sp. (a biting louse, *Mallacostraca). 1b* Racemose testis of a grasshopper *(Tetrix* sp.). *1c* Testis of a mayfly *(Leuctra* sp.) (the testicular follicles form a terminal tuft). *2* Schematic longitudinal section through a testicular follicle.

Echinoderms have 5 pairs of radially arranged testes in the body wall. In nonmammalian *vertebrates,* the testes remain at their site of development in the abdominal cavity, but in most *mammals* they migrate through the inguinal canal into an exterior *scrotum* (see Descent of testes). Several vasa efferentia arise from each T., which unite to form the much convoluted epididymis. The latter eventually continues as the vas deferens, which empties into the urethra or cloaca.

Testosterone, ***17β-hydroxy-androst-4-en-3-one:*** an important Androgen (see), m.p. 155 °C, synthesized in the interstitial cells of the testes. Although T. is recognized as the actual circulating male hormone, its metabolically active form within the cell is 5α-dehydrotestosterone. In the testes, T. is biosynthesized from cholesterol via progesterone, 17α-hydroxyprogesterone and androstenedione. In the adrenal cortex, it is biosynthesized from cholesterol via pregnenolone, 17α-hydroxypregnenolone, androstenolone and androstenedione. The adult human male secretes up to 12 mg T. daily. During its transport in the blood, T. is largely bound to a carrier protein. It is metabolically inactivated by reduction of the A-ring. Major excretory metabolites of T. are androsterone and 3α-hydroxy-5β-androstan-17-one.

T. stimulates the growth and development of male reproductive organs, is responsible for the development of the secondary male sex characteristics, and promotes the maturation of sperm. In addition to androgenic activity, T. is also an active anabolic hormone, stimulating protein synthesis and nitrogen retention. It was first isolated in 1935 from bull testes.

Certain structural modifications, e.g. 17-nortestosterone, have weaker androgenic potency but stronger anabolic activity then T. itself., and they are used therapeutically as "anabolic steroids".

OH
H_3C
H_3C
O
Testosterone

Testudinata: see *Chelonia.*

Testudines: see *Chelonia.*

Testudinidae, ***tortoises:*** a family of the order *Chelonia* (see), containing 41 species in 10 genera. The largest genus, *Testudo,* is divided into several subgenera. All members are terrestrial, most preferring dry habitats. With few exceptions, the heavy shell is highly vaulted, the feet are columnar and clublike, and the toes (but not the claws) are fused. These plump animals progress more slowly than members of other chelonian families. The head can be fully retracted beneath the carapace. The legs are angled, so that when they are withdrawn under threat only the outer surface (covered with large scales) is visible. Almost all tortoises are herbivorous, in contrast to the pond and river turtles, which are largely carnivorous. Eggs are laid in self-excavated holes. Tortoises are represented in all parts of the tropics and subtropics, except Australia. The habitats of all species have been greatly altered by human activity, and many species are threatened with extinction.

Testudo hermanni ***(Spur-tailed Mediterranean tortoise)*** inhabits dry sunny areas and undergrowth on plains and hills in eastern Spain, southern France, the Mediterranean islands, southern Italy, and the Balkan Peninsula south of the Danube. It has a high vaulted, yellow-brown, black-spotted shell, and attains a length of 25 cm. Above the tail, the posterior marginal scute of the carapace is divided into two, there is a spur on the end of the tail, and the thighs do not carry spurs (cf. *Testudo graeca,* which is otherwise very similar in appearance to *T. hermanni).* It buries itself in the ground for winter hibernation. The male has an indented plastron, while the plastron of the female is smooth. As a prelude to mating, the male pursues the female and repeatedly rams her with his shell. Ten to twelve eggs are buried in the ground, and the newly hatched young are about the size of a walnut. It feeds mainly on vegetation, with some worms and snails. In captivity, it thrives on lettuce, dandelion, endive and cabbage, and shows a great liking for soft fruit; it will also eat raw mincemeat, liver, boiled rice and white bread soaked in milk.

Testudo graeca ***(Greek tortoise*** or ***Spur-thighed Mediterranean tortoise)*** exists as several subspecies from southern Spain via southeastern Europe and North Africa to the Near East. It attains a length of 30 cm, and bears a close external resemblance to *Testudo hermanni,* but differs from the latter in that the posterior marginal scute is not divided, each upper thigh carries a spur, and there is no spur on the tail. In captivity, it needs warmer conditions than *T. hermanni.* It feeds mainly on vegetation and fruit, but also takes worms, caterpillars, snails, etc.

Testudo (Agrionemys) horsfieldi, found on the steppes and deserts of Central Asia from the Caspian Sea to eastern Pakistan, has a rather flat and rounded carapace. It attains a length of 20 cm, and has only 4 claws on each foot. Long periods of dormancy are spent in both summer and winter in earth burrows up to 2 m in length.

Testudo = Asterochelys = Geochelone radiata ***(Radiated tortoise)*** is a warmth-loving species found in dry areas of Madagascar. It attains a length of 45 cm, and is characterized by bright radiating patterns on a highly vaulted dark carapace. The diet consists mainly of the shoots and roots of succulent plants.

Testudo = Chelonoidis = Geochelone elephantopus ***(Galapagos giant tortoise)*** exists as several subspecies distributed on the separate islands of the Galapagos. Up to 1.2 m in length, it weighs over 100 kg. The carapace is black, and in contrast to the Seychelles tortoise, it does not have a cervical scute at the nape of the neck. In the 17th and 18th centuries they were very numeorus, and were killed in large numbers for food by sailors. The eggs were also taken for their oil. In addition to this exploitation by humans, young and eggs were also taken by introduced animals that had reverted to the wild, with the result that the species was nearly exterminated. Like most Galapagos animals, they are now under strict international protection, and numbers have stabilized and even increased, although these measures were too late to save some subspecies. They live mainly in the hot, dry larva fields on the lowlands of the islands, from where they regularly migrate to the wet highlands to bathe and drink in ponds.

Testudinidae. Testudo elephantopus (Galapagos giant tortoise).

Testudoi = Aldabrachelys = Geochelone gigantea ***(Seychelles tortoise)*** was formerly widely distributed on the Seychelles and the neigboring islands of the Indian Ocean, but now occurs in reasonable numbers only on Aldabra. Up to 1.3 m in length, it has a black carapace, and differs from the very similar Galapagos tortoise in possessing a small cervical scute at the nape of the neck. Its greatest proven age is 180 years, the greatest of any living reptile.

Testuda = Chelonoidis denticulata ***(Red-legged tortoise)*** is found in tropical South America east of the Andes. The carapace, up to 0.5 m in length, is dark with swirled and wavy markings.

The genus *Kinixys* (3 species) comprises the ***Hinge-backed tortoises,*** found in central and southern Africa. The posterior third of the carapace in divided from the rest of the carapace, and can be folded downward for protection. They attain a length of 20–30 cm, and unlike other tortoises they often seek out water.

Testudo: see *Testudinidae.*

Tetanus bacterium, *Clostridium tetani:* a spore-forming, anaerobic, slender rod-shaped organism (2–5 μm x 0.4–0.5 μm) with rounded ends. It is motile with numerous long peritrichous flagella. Although Gram-positive, there is some variation, and Gram-negative forms are also encountered. It bears a spherical spore at the very tip of the rod, which gives it a typical "drumstick" appearance. In contrast, *Cl. perfringens, Cl. sporogenes,* etc. produce oval spores located centrally or excentrically. *Cl. tetani* produces a powerful exotoxin, which is particularly active in motor nerve centers, throwing muscles into a state of violent and sustained contraction (tetanic convulsion, tetanus or "lockjaw"). Tetanus toxin is one of the most potent poisons known (0.25 mg are sufficient to kill a man). *Cl. tetani* occurs widely in soil, dust, etc., and it gains access to the body through wounds. Since it is a strict anaerobe, it cannot invade the body, and it can only live saprophytically in dead, anaerobic tissues of deep wounds. Tetanus antitoxin was the first antitoxin to be used for conferring passive immunity (von Behring and Frankel, 1890).

Tethys sea, ***Tethyan seaway*** *(Tethys* was the wife of the Greek sea god, *Oceanus):* a central, east-west oriented, shallow sea, separating the Mesozoic supercontinents of Laurasia and Gondwana (see). It existed from the Paleozoic to the Tertiary, and for much of the Cenozoic it separated Europe from Africa. Many sediments and deposits of the T.s. are now exposed at high altitudes due to tectonic movements, e.g. Triassic reefs now form part of the Alps. The dispersal of animals was crucially influenced by this belt of water, which provided a barrier to land fauna and a migration route for marine fauna, although the cosmopolitan character of dinosaur faunas indicates that land bridges probably existed between the two supercontinents during the Mesozoic. The present Mediterranean represents a small residue of the T.s.

Tetrabranchiata: see *Cephalopoda.*

Tetracerini: see *Tragelaphinae.*

Tetracerus: see *Tragelaphinae.*

Tetracorallia, *Tetracoralla, Rugosa:* an extinct order of mostly solitary corals, the fossils of which are found from the Cambrian to the Permian. They possess a system of major and minor radial sclerosepta, e.g. *Zaphrentis.*

Tetracyclins: a family of antibiotics from various species of *Streptomyces,* and now produced on a large scale by industrial fermentation. T. contain 4 linearly fused 6-membered rings, and individual T. differ according to the ring substituents (see Fig. and Table). T. are active against many Gram-positive and Gram-negative bacteria, viruses, rickettsias and certain spirochetes. ***Chlortetracyclin*** (from *S. aureofaciens)* has been used successfully to treat bronchitis, peritonitis, whooping cough, scarlet fever, syphilis and typhus. ***Oxytetracyclin*** (from *S. rimosus)* has a similar action spectrum, and it is indicated for septicemia, bronchitis, urinary tract infections and typhus. Both are also widely used as additives to animal feedstuffs, resulting in increased growth and decreased infant mortality. On account of side reactions and increasing resistance of bacteria to T., their use in medicine and as food additives is now decreasing. T. inhibit protein synthesis by preventing the binding of aminoacyl-tRNA to ribosomes.

The structures of some tetracyclins

Name	R_1	R_2	R_3	R_4	R_5
Chlortetracyclin (aureomycin)	H	H	OH	CH_3	Cl
Oxytetracyclin (terramycin)	H	OH	OH	CH_3	H
Tetracyclin (achromycin, ambramycin)	H	H	OH	CH_3	H
Methacyclin (rondomycin)	H	OH	$CH_2=$		H
Doxycyclin (vibramycin)	H	OH	H	CH_3	H

Tetrad:

1) A bundle of 4 homologous chromatids during the first meiotic division (see Meiosis).

2) The 4 haploid cells resulting from a single meiosis. With regard to their genetic composition, these 4 cells may constitute a di-T. or a tetra-T. If an organism is produced by the fusion of gametes ab^+ and a^+b, (i.e. it is heterozygous for 2 allele pairs), then meiosis of any one of its cells results in one of the following T. types: ab^+, ab^+, a^+b, a^+b *(parent di-type T.);* a^+b^+, a^+b^+, ab, ab *(nonparent di-type T.);* and ab^+, a^+b, a^+b, ab (tetra-type T.). If the gene pairs in question belong to different linkage groups, the three T. types are produced with equal average frequency, with 50% recombination types (a^+b^+, ab). If the genes are linked, the percentage Recombination (see) is always less than 50%, i.e. the genes are not free, and their recombination by crossing over is restricted.

Tetrad analysis is used to study the behavior of chromosomes and genes during crossing over during meiosis. For this purpose, the products of meiosis must remain together so that they can be counted as units, e.g. meiospores in the asci of certain fungi like *Aspergillus, Neurospora* and *Saccharomyces.*

Tetrahymena: a genus of ciliate protozoa (suborder *Tetrahymenina,* order *Hymenostomatida)* with conspicuous oral cilia, consisting of 3 membranelles and an undulating membrane. Species of *T.* are easily grown in sterile, defined nutrient media, and are therefore widely used for laboratory studies. See *Ciliophora.*

Tetranychidae, ***spider mites:*** a cosmopolitan family of mites, suborder *Trombidiformes,* order *Acari* (see), which live parasitically on plants. They have spinnerets that open anteriorly, and their secretion can be drawn out into a thread. They often live in large colonies on the underside of leaves under a feltlike covering of spun silk that largely protects them from the weather. With their specialized mouthparts, they pierce leaves and suck out the cell contents, causing white patches to form on the leaf surface. Heavy infestations cause the leaves to dry out and die, causing considerable damage to fruit trees and soft fruit bushes. Almost all major food crops and ornamental plants are vulnerable to attack by one or more species. Colors range from green to yellow-green, red and orange.

Red species are generally and indiscrimately referred to as red spider mites. The ***Fruit tree red spider mite*** *(Metatetranychus ulmi* Koch = *Paratetranychus pilosus* Canestrini & Fanzago = *Panonychus ulmi)* lays bright red overwintering eggs. In the spring, the newly hatched larva is brown to orange in color, becoming red as it matures. Summer eggs are also laid, and several generations are produced each year. Mass population increases are favored by dry conditions, when the finely spun compartments on leaf undersides contain the living animals, eggs, cast larval skin and excrement, all visible to the naked eye. The leaf responds by producing light patches, which then discolor, and finally the leaf dries out. Natural predators of the Fruit tree red spider mite are, e.g. ladybirds, as well as predatory bugs and other mites. Commercial acaricides are available for its control. *Tetranychus urticae (= Tetranychus telarius)* is well known in glass houses, and is responsible for considerable crop destruction worldwide. Variously known as the ***Carmine mite, Green two-spotted mite, Glasshouse red spider*** and ***Hop red spider mite,*** it has two summer feeding forms, one red and one green, the green form turning red when starved.

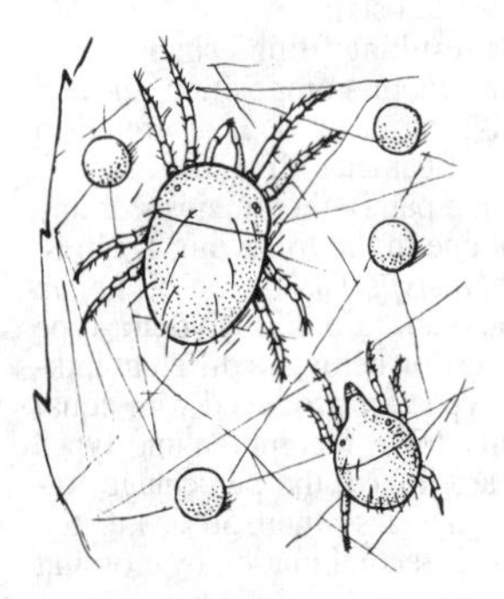

Tetranychidae. Glasshouse red spider *(Tetranychus urticae);* mature mite, larva and eggs under a covering of spun silk on the underside of a leaf (enlarged).

Tetrao: see *Tetraonidae.*

Tetraodontidae, ***puffers:*** a family of tropical and subtropical marine teleost fishes (about 118 species in 16 genera) of the order *Tetraodontiformes.* They are closely related to the porcupinefishes *(Diodontidae).* Many species of *Tetraodon* are found in fresh water, mainly in the Zaire River and in southern Asia. When danger is perceived they can inflate their stomach by taking in water or air. Some species contain the powerful neurotoxin, tetraodontoxin. Many species are popular aquarium fishes. The body is naked, but erectile spines are present in the skin.

Tetraodontiformes, *Plectognathi:* an order of variously shaped marine teleost fishes with a relatively small mouth. The external opening of the operculum is usually a hole. Teeth are separate, or fused into grinding plates which are grouped into a beaklike structure. Scales are usually modified as spines, shields or plates. Eight families are recognized, with about 92 genera and 329 species: *Triacanthodidae* (Spikefishes), *Triacanthidae* (Triplespines), *Balistidae* (Leatherjackets) (see), *Ostraciidae (Ostraciontidae,* Boxfishes) (see), *Triodontidae* (Three-toothed puffer), *Tetraodontidae* (Puffers) (see), *Diodontidae* (Porcupinefishes), *Molidae* (Molas).

Tetraonidae, ***grouse*** and ***ptarmigan:*** a small family (18 species) of the avian order *Galliformes* (see), all distributed in the Northern Hemisphere. Unlike other gallinaceous birds, their tarsi and sometimes their feet are completely or partly feathered. Other characteristics are feathered nostrils, and a rather short neck with an inflatable area of bare skin or erectile feathers. A patch of brightly colored bare skin is often present over each eye. The hindtoe is raised, but there is no spur. Largest of the family is the turkey-sized ***Capercailie*** (or ***Caperkaizie)*** *(Tetrao urogallus),* which inhabits pine forests from northern Spain and central and northern Europe to central Siberia and Mongolia. The cock weighs up to 7.7 kg and attains a length of up to 90 cm, but the hen is only two thirds of this size. In courtship the male spreads his tail into a fan, stretches his neck vertically upward, ruffles his wattles and drops his wings. The display takes place first on strong, horizontal branches, later on the ground. The male of the ***Black grouse*** *(Lyrurus tetrix;* 53 cm) is glossy blue-black with a lyre-shaped tail, and a white wing bar and undertail coverts. Patches of ground called "leks" are cleared by the male for his courtship display. The female (43 cm) is mottled brown and black with a forked tail. The ***Prairie chicken*** *(Tympanuchus cupida;* 45 cm) is a grouse that originally inhabited the tall grass prairies of North America, but is now more localized in central and southern USA, where it is also found on scrub grass and cultivated fields. Its plumage is barred white on shades of brown, with a blackish rounded tail, and a dark brown head and crest. Males display on "booming grounds", which are slight rises or open ridges. Bare air sacs at the sides of the neck are inflated during courtship, and they add resonance to the booming voice of the courtship ritual. The male also has orange-yellow wattles, as well as tufts of long, erectile feathers on the sides of the neck. ***Ptarmigan*** are small, primitive grouse found only in northerly latitudes, the ***Willow grouse*** *(Lagopus lagopus lagopus)* and the ***Rock ptarmigan*** *(Lagopus mutus)* having a circumpolar distribution. *Lagopus leucurus* ***(White-tailed ptarmigan;*** 32 cm) is found in the Rocky Mountains from New Mexico to Central Alaska, while the ***Red grouse*** *(Lagopus lagopus scoticus)* inhabits moorlands and heaths in Britain and Ireland, and has also been introduced to Belgium. The dark rufous brown plumage of the Red grouse is retained throughout the year (the female is somewhat lighter in the spring), whereas all 3 of the other ptarmigans have characteristic white winter plumage.

Tetrapanax: see *Araliaceae.*

Tetraparental animals: see Chimera.

Tetraphyllidea: see *Cestoda.*

Tetraploidy: polyploidy, in which cells, tissues or individual organisms have 4 sets of chromosomes, i.e. they are tetraploid.

Tetrapodili, ***gall mites:*** a suborder of the *Acari* (see), containing a single family, the *Eriophyidae.* Other names are blister mites, rust mites, bud mites and eriophyid mites. Their average length is only 0.15–0.2 mm, and they are some of the smallest arthropods. They pierce plants cells with their chelicera, and secrete enzymes to liquify the contents, which they then suck out. The plant reacts by forming felt-like hairs, spherical galls, or sacklike galls, by extruding lateral buds, or by the twisting of axes, etc. The *T.* display considerable host specificity. The ***Blackcurrant gall mite*** *(Eriophyes ribis)* primarily attacks blackcurrant bushes, causing a condition known as "big bud". The mites feed inside the buds, which swell into a large spherical shape before drying out. The mite may also transfer a debilitating virus disease to blackcurrants, which persists after the mites have been destroyed. Control is effected by removal of the swollen buds, and by the application of appropriate commercial acaricides, but it is usually necessary to remove and burn affected plants and to restock. The ***Vineleaf gall mite*** or ***Grape erineum mite*** *(Eriophyes vitis)* attacks vine leaves, causing the formation of vesicular growths on the upper surface, which are differently colored according to the variety of the grape. In heavy infestations, a dirty white to red felted layer appears on the underside of the leaf. Control is effected by extensive cutting back of the plant, and by the application of appropriate commercial acaricides.

Tetrapteryx: see *Gruiformes.*

Tetrasomy: a form of aneuploidy, in which cells or individuals possess four copies of a homologous chromosome in their otherwise diploid chromosomal complement (2n + 2). More than one homologous chromosome may be represented four times, e.g. (2n + 2 + 2) represents *double T.* Spontaneous T. is rare. It is more likely to occur from the crossing of trisomic individuals (see Trisomy), when the extra chromosome is transmitted via the female or male gamete. See Aneuploidy, Disomy, Monosomy, Nullisomy, Polysomy, Trisomy.

Tetrazolium salts: salts of 2,3,5-trisubstituted tetrazol. The colorless, water-soluble 2,3,5-triphenyltetrazolium chloride is particularly important as a biological reagent. T. are reduced to red-colored (usually), water-insoluble formazans. Enzymes that catalyse biological reductions also convert T.s. to their colored, insoluble reduction products. T.s. are therefore useful biological indicators for determining the viability of microorganisms, the germination potential of seeds, or the histochemical demonstration of reducing enzymes in different tissues.

[N—N—C_6H_5 / H_5C_6— / N=$N^{\oplus}$—C_6H_5] Cl^-

2,3,5-Triphenyltetrazolium chloride

Tetroses: monosaccharides with the empirical formula, $C_4H_8O_4$, e.g. erythrose and threose. They occur as intermediates of carbohydrate metabolism.

Teuthidae: see *Acanthuridae.*

Texas fever: see Piroplasmids.

Texas horned toad: see Horned toad.

Textularia: see *Foraminiferida.*

TGN: acronym of Trans Golgi network; see Golgi apparatus.

Thalamencephalon: see Brain.

Thalamus: see Brain.

Thalassinoidea: marine burrowing shrimps, a superfamily of the *Decapoda* (see).

Thalattosuchia: a fossil suborder of the *Crocodylia,* which existed from about the end of the Jurassic to the beginning of the Cretaceous. They were predators with formidibly large teeth in tapering jaws. Ideally adapted to a marine existence, they possessed a fusiform body, a powerful laterally compressed tail with a vertical tail fin, and paddlelike modified limbs. Fossils reveal no traces of body armor. Well preserved skeletons with outlines of the body surface have been found in the Solnhofen shale. *Geosaurus giganteus* was described as early as 1816 by the Munich anatomist, Sömmering.

Thaliacea: a class of pelagic tunicates (see *Tunicata)* containing about 40 species and divided into 3 orders: *Doliolida* (see), *Pyrosomida* (see) and *Salpida* (see).

Sizes vary from a few millimeters to several centimeters. Some forms grow as aggregates, forming chains many meters in length. Individual animals are broad for their length, and more than half of the body is taken up by the pharynx and wide mouth opening. Gill clefts and anus empty into a large caudal (posterior) cloacal chamber. Beneath the mantle, the bodies of all *T.* are encircled by broad muscle bands, resembling the hoops of a barrel. The intestine, with its sac-like stomach, is generally short, and the vascular system is weakly developed. Only the *Doliolida* have a tailed larval stage. In all *T.* there is an alternation of generations between an asexual solitary form (oozoid) and a sexual aggregate form (blastozooid).

Thallo-chamaephyte: see Life form.

Thallo-cryptophyte: see Life form.

Thallo-epiphyte: see Life form.

Thallo-geophyte: see Life form.

Thallo-hemicryptophyte: see Life form.

Thallo-therophyte: see Life form.

Thallophyta, *thallophytes:* a nonsystematic term for plants that are not differentiated into root, stem and leaves. Their vegetative body is called a thallus. Algae *(Phycophyta),* fungi *(Mycophyta)* and lichens *(Lichenes)* are T. The *Bryophyta* are transitional between T. and Cormophytes (see), since they often possess a shoot axis and leaves, but have rhizoids rather than true roots.

Thallus (plural **Thalli**)**:** a weakly differentiated, multicellular vegetative plant body, which is not differentiated into stems, leaves and roots. The term is normally applied to nonvascular plants, such as algae, fungi, lichens and liverworts. A T. may consist of spherical or filamentous cell associations, or it may display highly differentiated external shapes, but unlike a cormus it possesses no specialized tissues. It may be an *aggregation T.,* which arises by the aggregation of previously free-living single cells, e.g. the green algae, *Pediastrum* and *Hydrodictyon,* in which a large number of freely motile single cells aggregate to form a uniform, multicellular organism prior to vegetative reproduction. Most T., however, are *true T.*, and these arise by incomplete separation of daughter cells after cell division, accompanied by development of common cell membranes, i.e. they are true multicellular organisms. In the simplest case, all the individual cells of the T. are equivalent and capable of division. More highly evolved forms show some degree of polar differentiation, growing by continuous division of a cap cell, and often fixed to the substratum by an attachment organ (see Rhizoids) consisting of modified cells. Forked or dichotomous branching arises from an alteration of the alignment of the axis of the mitotic spindle in the cap cell, so that the ensuing longitudinal division produces two cap cells. Lateral branching occurs when the ability to divide is restored to a cell some distance from the cap cell, and the axis of the mitotic spindle is also altered.

The T. of the highly evolved brown algae consist of true tissues. Many species have leaflike assimilatory organs *(phylloids),* stemlike central structures *(cauloids)* and rootlike attachment organs *(rhizoids).*

The T. of red algae and the fruiting bodies of higher fungi consist largely of plectenchyma, which can be very similar to true tissue. The actual T. of fungi consists of threadlike, profusely branched hyphae, known collectively as a mycelium, which permeates the fungal substratum

Thamnophis: see Garter snakes.

Thanatocoenosis (Greek *thanatos* death, *koinos* common)**:** all the embedded fossils at a single discovery site. They often originate from different habitats, having been brought together by water currents. An *autochthonous T.* consists of organisms that lived, died and became embedded at the same site. If the death and embedding site are different, the T. is said to be *allochthonous.*

Thanatosis: see Protective adaptations.

Thaumatin: a sweet-tasting, strongly basic, histidine- and carbohydrate-free single chain protein (M_r 21 000), isolated from the fruits of *Thaumatococcus danielli* (family *Marantaceae).* It is 750–1 600 times (mass basis) or 30 000–100 000 times (molar basis) sweeter than sucrose. It shows considerable sequence homology with, and is immunologically releated to, the B-chain of another sweet protein, Monellin (see).

Thaumetopoeidae: see *Lepidoptera.*

Theaceae, *tea family:* a dicotyledonous plant family of 550 species in 18 genera, found mainly in the tropics and subtropics. They are trees or shrubs with simple, alternate, leathery, exstipulate leaves. Flowers are actinomorphic, usually hermaphrodite, 5-7-merous, with numerous stamens. The ovary is superior, rarely half-inferior *(Annesleya, Visnea)* or inferior *(Symplocarpon),* with 2–10 united carpels and as many free or united styles; placentation is axile, and ovules (2 to numerous in each loculus) are anatropous. The fruit is a capsule or dry drupe, and the usually curved embryo possesses little or no endosperm.

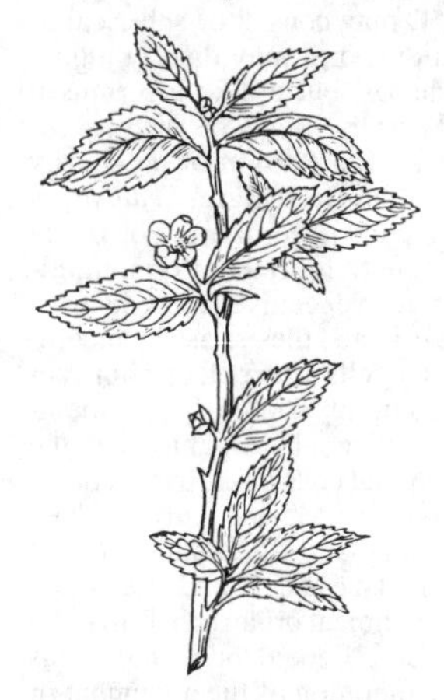

Theaceae. Flowering branch of the tea plant *(Camelia sinensis).*

Camellias (varieties of *Camellia japonica),* originating from East Asia, are grown in the west as ornamental plants; they have dark green, glossy leaves and white, pink or red flowers. In its native habitat, the camellia is a 15 m-high tree, and the flowers are pollinated by birds; oil from the seeds is used in East Asia. The economically important ***Tea plant*** *(Camellia sinensis = Thea sinensis)* has been cultivated in China and other countries of southern and eastern Asia for centuries; its leaves are processed to produce commercial green or black tea, which contains the stimulatory alkaloids, caffeine, theobromine and theophylline. The stimulatory action of the caffeine (sometimes present in very high concentration) is decreased by the presence of the antagonist, adenine.

Theca:

1) The chitinous cover of the ectoderm of members of the *Hydrozoa.* In the *Athecata* (see *Hydroidea)* it covers the body of the individual animal as far as the hydranth head. In the *Thecophora* it forms a small bell (hydrotheca).

2) The calcareous wall (98–99.7% $CaCO_3$ crystallized as fibrous aragonite) secreted around the periphery of the pedal disk of stony corals *(Madreporaria),* and which joins the outer edges of the sclerosepta.

3) The chitinous cup of graptoliths, in which the zooid of a colony lives.

4) The bony outer shell of tortoises and turtles.

5) Part of a plant serving as a receptacle, e.g. pollen sac, spore case, capsule, etc.

Theca externa: see Oogenesis.

Theca folliculi: see Oogenesis.

Theca interna: see Oogenesis.

Thecate amebas: see *Arcellinida.*

Thecate hydroids: see *Hydroida.*

Thecodont: see Teeth.

Thecodontia (Greek *theke* container, *odontes* teeth): an order of extinct animals, which were the evolutionary ancestors of the dinosaurs, crocodilians and pterosaurs. *T.* superficially resembled the *Crocodylia.* They were small, carnivorous predators with shortened forelimbs, progressing bipedally on their hindlegs in a semi-upright posture, using their powerfully developed tail as a balancing organ. Their teeth were housed in characteristic deep sockets in the jaws. Fossil *T.* are found in the Triassic.

Thecosomata (shell pteropods): see *Gastropoda.*

Thecurus: see *Hystricidae.*

Theilerian: see Piroplasmids.

Theileridae: see Piroplasmids.

Thein: see Caffeine.

Thelazia: see *Nematoda.*

Thelotornis: see Twig snake.

Thelytoky: production of only female progeny; see Parthenogenesis.

Theobroma: see *Sterculiaceae.*

Theobroma, oil of: see Cocoa butter.

Theobromine, *3,7-dimethylxanthine:* a bitter tasting purine alkaloid, which constitutes up to 1.8% of cocoa beans *(Theobroma cacao).* It is poorly soluble in water, and is responsible for the pharmacological activity of cocoa. It acts as a diuretic, smooth muscle relaxant, cardiac stimulant and vasodilator. The 1-hexylderivative of T., called pentifylline, is used as a vasodilator; it has increased lipid solubility which favors absorption. Other preparations include molecular compounds with diethanolamine or isopropanolamine, which have improved water solubility and are used as diuretics. T. is structurally related to Theophylline (see) and Caffeine (see).

Theophylline, *1,3-dimethylxanthine:* a purine alkaloid present in small quantities in tea leaves. It has similar pharmacological properties to Caffeine (see), but has a relatively greater stimulatory activity on the central nervous system. It is used as a diuretic.

Theria: a subclass of the mammals, containing 3 infraclasses: *Pantotheria* (extinct ancestral forms), *Metatheria* (marsupials), and *Eutheria* (placental mammals). They are separated from the *Prototheria* mainly by dentitiion, and by the presence in the *T.* of large squamosal and alisphenoid bones of the skull.

Theriodontia (greek *therion* animal, *odontes* teeth), ***mammal-toothed reptiles:*** a suborder of the *Therapsida* (subclass *Synapsida* or mammal-like reptiles). The dentition of these tiger-sized predators was functionally and structurally differentiated into incisors, canines and molars. The T. represent an important stage in the evolution of pelycosaurs via the therapsids to the ictidosaurs, which are considered to be the immediate ancestors of the mammals.

Fossil *T.* are found from the Permian to the Triassic, and are particularly abundant in South Africa.

Thermal waters: see Spring.

Thermoacidophiles: see Archaebacteria.

Thermogenesis: see Temperature regulation of the body.

Thermomorphosis: a temperature-controlled change in the growth morphology of a plant. See Morphoses.

Thermonasty: a nondirectional movement of a plant organ or part in response to a change in temperature. It is due to differences of gowth rate on opposite sides of the organ concerned (see Nasty). Striking and familiar examples are the opening and closing movements of flowers (see Flower movements). In some species the flower stalks also show a thermonastic response [anemones *(Anemone),* wood sorrel *(Oxalis),* cranesbill *(Geranium)* and others]. Foliage leaves generally do not display T., although thermonastic reactions may be involved in the movements of stomata. Some tendrils coil thermonastically in response to variations in temperature.

Thermoperiodism: in plants, a diurnal variation in the optimal temperature for growth and development. Optimal development only occurs if the temperature does undergo a day/night fluctuation. The tomato plant displays T.

Thermophilic organisms, *thermophiles:* microorganisms whose optimal growth temperature is between 45 ° and 75 °C. T.o. are primarily spore-forming bacteria and actinomycetes. Some are present in thermal sulfur springs, while others are involved in the spontaneous heating of damp hay, animal manure, etc. Sometimes, T.o. cause spoilage of canned foods, since some species can survive temperatures up to 80 °C.

Thermoreceptors: sensory nerve endings which are stimulated by body temperature. *Central T.* are located centrally in the body, whereas *peripheral T.* are located in the skin. They are also classified as *cold* or *warm* receptors, in accordance with their response to elevated or decreased temperatures. Peripheral T. are distributed irregularly over the body. In humans they are aggregated in the head region and on the extremities; cold receptors are about 10 times more numerous than warm receptors. Peripheral T. are responsible for the sensation of skin temperature, and are involved in Temperature regulation of the body. Central T. are all warm receptors, located in the hypothalamus and spinal cord. They do not provide temperature sensation, but serve as receptors for the regulation of body temperature. Peripheral cold receptors are stimulated by chilling. Thus leads to constriction of peripheral blood vessels and a marked increase in heat production by the body. Usually, heat production is greater than required, resulting in stimulation of central T. and cessation of heat production. Central T. show a similar response when physical exertion leads to excess heat production. Thus, peripheral T. are responsible for the rapid onset of temperature regulation in response to chilling, whereas central T. are responsible for the rapid cessation of heat production when the body is in danger of overheating. *Heat stress* causes stimulation of both the central and peripheral warm receptors. Regulatory centers of the hypothalamus then promote active processes of heat loss, e.g. increased sweating in humans. See Temperature regulation of the body.

Thermoregulatory centers: see Temperature regulation of the body.

Thermosbaenacea: an order of the *Crustacea,* subclass *Malacostraca.* It is a small order (8 known species) and the only order of the superorder *Pancarida.* The carapace is reduced in size (especially in males) and fused to the first thoracic segment. In females, the carapace remains free above the other thoracic segments, forming a dorsal brood pouch. Body length is no greater than 4 mm. They are very similar to the *Peracarida,* but differ from these in possessing a dorsal brood pouch. The first pair of thoracopods are modified as maxillipeds, while thoracopods 2 to 8 are biramous and lack epipodites. Pleopods are markedly reduced or absent. The uropods and telson form a tailfin. Some inhabit the interstitial water of the wet sand of thermal springs, others are found in warm coastal ground waters and others in cave pools. Those in the sand space system of hot springs thrive at a temperature of 45 °C and become stiff and immobile at 30 °C. The order displays a disjunct geographical distribution, being found near to and around the Mediterranean region and in the Caribbean. *Mondella texana* is found in several localities some distance from the coast in Texas; *M. sanctaecrucis* occurs in the West Indies; and several species of *Monodella* occur in the Mediterranean region. *Halosbaena acanthura* from Curaçao is a close relative of *Limnobaena finki* of Yugoslavia. The ancestors of this group probably performed an early migration from the sea into coastal ground water, and were later separated into different populations by plate tectonics.

Thermostat: an electronic, electrical or mechanical control device for stabilizing the chosen temperature attained by a heating system. Ts. are used e.g. for keeping incubation cabinets or water baths at constant temperature.

Thermotaxis: directional movement (see Taxis) of motile organisms or cell organelles in response to a temperature difference. T. is shown by spirillae, flagellates and diatoms, and by the plasmodia of certain myxomycetes. At room temperatures or somewhat higher, in a temperature gradient of only 0.05 °C/cm, mixomycetes move up the gradient to regions of higher temperature.

Thermotropism: directional growth of plant organs in relation to a unilateral heat source. The orientation of the leaves of compass plants is the result of T. combined with phototropism. Negative thermotropic bending is shown by the sporangiophores of certain fungi (e.g. *Phycomyces)* in response to local warming.

Theromorphs, *Theromorpha* (Greek *therion* animal, *morphe* form or shape): an extinct order of the most primitive reptiles, which combined amphibian and reptilian characters (see *Cotylosauria).* They also contain a group known as *Pelycosauria* which display some primitive mammalian characters. The roof of the skull was perforated on each side by a temporal fossa. The teeth, which were fused to the base of a groove, show the first signs of differentiation. The T. progressed by striding. Fossils occur from the Upper Carboniferous to the Triassic.

Therophytes: rooted annual plants that survive winter cold or periodic drought usually as seeds. This ability to survive unfavorable conditions, and their easy dispersal as seeds has led to the establishment of T. in otherwise hostile environments. T. can be classified as *summer annuals,* which die before the winter after producing seeds, and *winter annuals,* which overwinter, e.g. as rosette plants, but still only live for one year. Many desert plants and weeds are T. See Life form.

Theropithecus gelada: see Gelada.

Theropods: see Dinosaurs.

Thick-billed lark: see *Alaudidae.*

Thickening of plant tissues: a form of plant Growth (see) in which the diameter of an organ is increased. For *primary* and *secondary* thickening see Shoot, Root.

Thick filaments: see Myosin.

Thickheads: see *Pachycephalinae.*

Thick-skinned honeycreepers: see *Drepanididae.*

Thienemann's rules: see Ecological principles, rules and laws.

Thigmomorphosis: see Mechanomorphosis.

Thigmonasty: see Haptonasty.

Thigmoreaction: see Haptoreaction.

Thigmoreception: see Tactile stimulus reception.

Thigmostimulus: see Contact stimulus, Haptonasty, Haptotropism.

Thigmotricha: see *Ciliophora.*

Thigmotropism: see Haptotropism.

Thinocoridae, ***seedsnipes:*** a small South American avian family (4 species) in the order *Charadriiformes* (see). Seedsnipes are plump little birds with short legs, long pointed wings, and sharp conical bills. Their slit-shaped nostrils are covered by a unique opercular flap. They run rapidly with head forward, and when threatened they attempt to conceal themselves on the ground with head and neck extended. Flight is rapid and erratic. Unusually for members of the *Charadriiformes,* the diet is mainly vegetarian (seeds, buds, etc), although some insects are also taken. Two species are placed in the genus *Thinocorus,* e.g. *T. rumicivorus* ***(Patagonian seedsnipe;*** 18 cm), which is found from Ecuador and Bolivia to Tierra del Fuego. The other 2 species belong to the genus *Attagis,* e.g. *A. malouinus* ***(White-bellied seedsnipe;*** 28 cm), which occurs from the Rio Negro in Argentina to Tierra del Fuego and the Falkland Islands.

Thinocorus: see *Thinocoridae.*

Thiobacillus: a genus of rod-shaped, Gram-negative, usually aerobic bacteria, which may be nonmotile, or may possess polar flagella. According to their autotrophic mode of existence, they can be classified as Sulfur bacteria (see), which obtain energy by oxidizing hydrogen sulfide, sulfur, etc. They occur in sea water, fresh water, seepage water from mines, industrial effluents, and soil.

Thiocyanic acid: HSCN, an acid with a penetrating, pungent odor. It is present in saliva, where it functions as a disinfectant.

Thirst: a sensation which is the expression of a general need for water by the body. It is the result of a complex physiological process which responds to water loss from the body, generates the sensation of T. (a desire for water), and thereby leads to the restoration of normal water balance throught the drinking of water.

T. arises from the drying of the buccal and pharyngeal mucosae, the decrease in blood volume, and the hyperosmolarity of body fluids. Drying of the mucosae is recorded by sensitive nerve fibers in the buccopharyngeal cavity. Decreases in blood volume are registered by stretch receptors of the vagus nerve in the heart atria. Hyperosmolarity stimulates arterial receptors, as well as leading to the loss of water from osmoreceptors in the hypothalamus. Electric or osmotic stimulation of the hypothalamus generates a desire to drink, whereas its destruction leads to death from water deprivation due permanent loss of the desire to drink. The actual sensation of thirst is first felt in the mouth, presumably because one of the earliest effects of water deprivation is a reduction in the output of saliva. Thus, administration of atropine stops salivary secretion and generates the sensation of thirst, even when the body does not need water.

Water loss of up to 8% of the body mass can be survived, whereas deficiencies of 15–20% are life-threatening. In humans, a water loss of 3% leads to a decrease of salivary secretion and decreased urine production (antidiuretic hormone is produced in response to the hypertonicity of the blood). A loss of 5% leads to acceleration of the respiration rate and an increase in the rate of the heart beat; in addition, the blood pressure sinks, since the percentage water loss from the blood is three times greater than the percentage loss from the body as a whole. In addition, the rectal temperature increases. A water deficiency of 10% leads to a rapid decline of physical and mental performance, and to the cessation of salivary secretion. Death from T. is mainly due to circulatory failure, arising from the marked decrease in blood volume.

T. is satisfied or quenched by drinking. In some animals, T. is reflexly quenched by the intake of water, e.g. in cats and dogs by stimulation of the pharyngeal nerves. In other animals, like the rat, guinea pig and human, it is reflexly quenched by stimulation of the stretch receptors of the stomach. Reflex cessation of T. occurs quickly, but is of only short duration. Complete removal of T. occurs after water has been transported through the stomach and intestinal walls, and has resulted in a sufficient decrease in the osmolarity of the blood for this to be registered by the osmoreceptors of the hypothalamus.

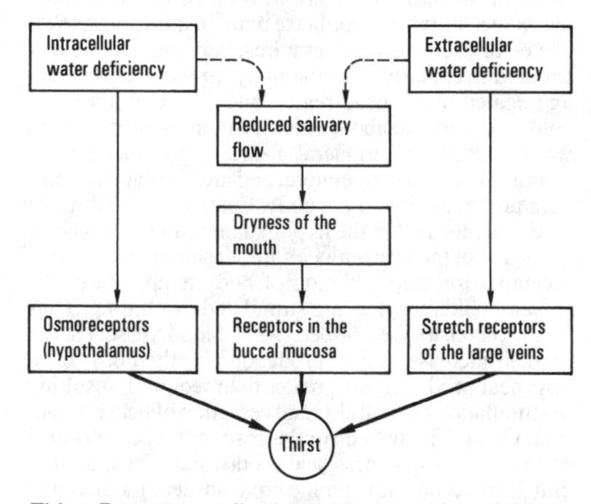

Thirst. Factors contributing to the sensation of thirst.

Thistle: see *Compositae.*

Thoma's counting chamber: see Counting chamber.

Thoracica: see *Cirripedia.*

Thoracic respiration: see Respiratory mechanics.

Thoracomeres: segments of the custacean thorax. See *Crustacea.*

Thoracopods: appendages of the crustacean thorax. See *Crustacea.*

Thorax: a skeletal structure of vertebrates formed by the ribs, thoracic vertebrae, sternum, cartilage and ligaments, which contains the heart and lungs, and which performs respiratory movements. The *ribs* are rod-like, arched, cartilaginous or bony elements, attached to the thoracic vertebrae (see Axial skeleton). In reptiles, birds and mammals, the *true ribs (vertebro-sternal ribs)* articulate directly with the *sternum* (which lies in the median line at the front of the T.), the *false ribs (vertebrochondral ribs)* are united by their cartilages to the cartilage of the preceding rib (and therefore only indirectly to the sternum), and the *floating ribs (vertebral ribs)* are free at their anterior extremities, so that they lie free in the trunk musculature with no connection to the sternum. By articulating with the clavicle (collar bone), the sternum (breastbone) with its attached ribs represents a linkage of the pectoral girdle to the axial skeleton. In birds with efficient flight, the sternum supports an extension of the anterior body wall, and it possesses a keel *(carina, crista sterni),* which serves as an enlarged surface for the attachment of the flight muscles. In fishes, limbless reptiles, snakes and turtles, a sternum is absent. The space enclosed by the T. has a superior aperture or inlet formed by the body of the first thoracic vertebra behind, the arches of the first rib on either side, and the upper part of the sternum in front. The lower aperture, bounded by the last thoracic vertebra and ribs, is usually larger than the superior aperture. In transverse section, the shape of the T. depends on sex, age, body structural type and function. In burrowing and swimming species of higher vertebrates, e.g. mole and beaver, it is barrel-shaped, while in efficient running and walking species, e.g. dog and horse, it is heart-shaped.

Thorium-uranium dating: see Dating methods.

Thorn apple: see *Solanaceae.*

Thorny devil, *moloch,* *Moloch horridus:* a small agamid (see *Agamidae)* inhabiting the deserts and plains of central and southern Australia. Body and tail are covered with large spiny scales. There are two especially large spines on the head, and a spiny hump of fat on the neck. Coloration is a camouflaging pattern of yellow, red and brown. It feeds almost exclusively on black ants.

Thorny-headed worms: see *Acanthocephala.*

Thr: standard abbreviation of L-threonine.

Thrashers: see *Mimidae.*

Thraupidae, *Emberizidae,* ***tanagers:*** in some classifications, a family of the *Passeriformes* (see), but here subsumed in the *Fringillidae* (see). The 222 species of this family are confined almost entirely to the tropics and subtropics of the Americas. All are nonmigratory, except for 4 species that occur in North America. Some of the Central and South American tanagers move to different altitudes with the seasons. Most members are brightly colored, including both males and females. Small to medium in size, most tanagers are less than 20 cm in length. The usually short and rounded wings have 9 primaries. There are conspicuous rictal bristles, and tangers appear to be more closely allied with the cardinal finches than with the icterids. They differ from buntings mainly in their adaptation to a fruit and nectar diet.

Threadworm: see Enterobiasis.

Threefin blennies: see *Blennioidei.*

Three-strand crossing over, *three-strand exchange:* a reciprocal exchange of segments between chromosomes during meiosis. Two crossing over events are involved. The second crossing over occurs between a chromatid that was involved in the first event and a third chromatid that was not involved in the first event (Fig.). See Crossing over, Four-strand crossing over, Two-strand crossing over, Recombination.

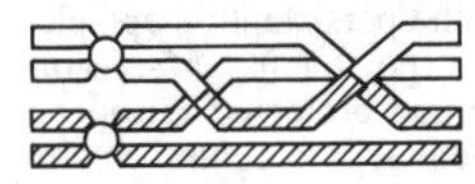

Three strand crossing over

Three-striped ground squirrel: see *Sciuridae.*

Three-toed amphiuma: see *Amphiumidae.*

Three-toothed puffer: see *Tetraodontiformes.*

L-Threonine, *Thr, L-threo-α-amino-β-hydroxybutyric acid:* $H_3C—CH(OH)—CH(NH_2)—COOH$, a proteogenic, essential, glucoplastic amino acid, with two asymmetric carbon atoms. In most organisms, Thr is degraded by deamination to α-ketobutyric acid (2-oxobutyric acid) by a pyridoxal phosphate-dependent enzyme, threonine dehydrase. In plants and microorganisms, Thr is biosynthesized from phosphohomoserine (derived from aspartate) by γ-elimination of phosphate followed by β-replacement with an OH-group, the total rection being catalysed by the pyridoxal phosphate-dependent enzyme, threonine synthase.

Threskiornis: see *Ciconiiformes.*

Threskiornithidae: see *Ciconiiformes.*

Thripidae: see *Thysanoptera.*

Thrips: see *Thysanoptera.*

Thrombocytes: see Blood, Blood coagulation, Hemocytes.

Thrombus: a blood clot in the lumen of a blood vessel. It may hinder or prevent blood flow (thrombosis). Thrombosis in an artery obstructs blood flow to the tissue it supplies, e.g. thrombosis of an artery to the brain may cause a stroke, and thrombosis of coronary arteries causes a myocardial infarction (heart attack). Thrombosis in a vein results in inflammation (phlebitis). A T. may become detached and lodge in another part of the vascular system, causing an *embolism,* e.g. pulmonary embolism in which a T. lodges in the pulmonary artery.

Thrum-eyed flower: see Pollination.

Thrushes: see *Turdinae.*

Thrush nightingale: see *Turdinae.*

Thuidium: see *Musci.*

Thuja: see *Cupressaceae.*

Thujone: a cyclic, monoterpene ketone, present in the essential oils of *Thuja* spp. (arbor vitae), *Artemisia absinthium* (wormwood), *Chrysanthemum (= Tanacetum) vulgare* (tansy) and *Salvia* (sage). It is a colorless oil which smells like camphor.

Thunnus: see Tunny.

Thuringium: see Zechstein.

Thylacininae: see *Dasyuridae,* Tasmanian wolf.

Thylacine: see Tasmanian wolf.

Thylakoid: see Chloroplasts, Membrane, Plastids.

Thylogale: see Kangaroos.

Thymallus: see Grayling.

Thyme: see *Lamiaceae.*

Thymelaeaceae, *daphne family:* a family of dicotyledonous plants, with about 650 species in 50 genera. The family has a cosmopolitan distribution in temperate and tropical regions, and is especially well represented in Africa. They are mostly shrubs (more rarely trees, lianes or herbs), with entire, alternate (rarely opposite), exstipulate leaves. Flowers are 4-6-merous, with a colored petaloid calyx of united sepals, the corolla being reduced to scales or absent. After pollination by insects, the superior ovary develops into an achene, berry, drupe or capsule, often enclosed in the persistent receptacle. The early spring-flowering, poisonous ***Mezereon*** *(Daphne mezereum)* is a rare and protected European species, occurring from Scandanavia to central Spain, northern Greece and temperate Asia (to the Altai Mountains). On account of its attractive purple calyx and bright red berries, it is sometimes grown in cottage gardens. In contrast, the ***Spurge laurel*** *(Daphne laureola)* is a widespread and rather common European species, occurring in England, from Germany to Spain, Corsica and Macedonia, in Asia Minor and the Azores, becoming rare in North Africa.

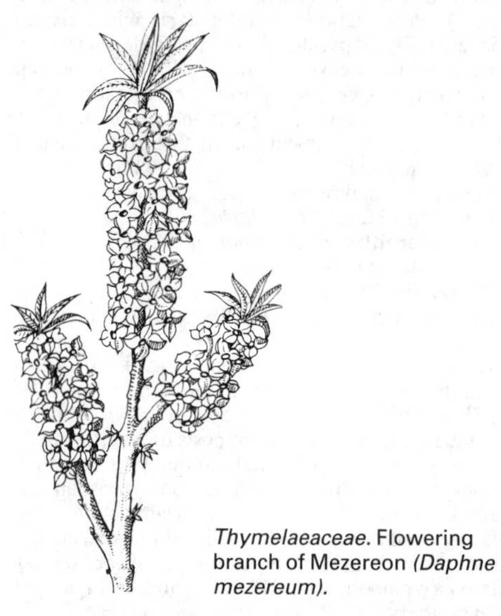

Thymelaeaceae. Flowering branch of Mezereon *(Daphne mezereum).*

Thymol, ***3-hydroxy-4-isopropyltoluene:*** a compound present in the essential oil of thyme *(Thymus* spp.), and largely responsible for the typical smell of these plants. When purified, it forms large, colorless crystals. On account of its antiseptic properties it is used medicinally, e.g. in the treatment of bronchitis, whooping cough and digestive disturbances.

Thymus: in vertebrate quadrupeds, a paired organ located behind the sternum in the lower neck. It is part of the lymphoid system of the body, and the functional counterpart of the Bursa fabricii (see) of birds. Substances synthesized and secreted by epithelial cells of the T. (e.g. a family of largely identical monomeric polypeptide hormones, known as thymopoietin) induce the differentiation of lymphocytes to *T-lymphocytes,* which enter the circulation or peripheral lymphatic organs. T-lymphocytes play a crucial role in cell-mediated immune reactions, e.g. in the rejection of transplants, defense against tumors and cell-mediated allergic reactions. Most of the preprocessing of T-lymphocytes in the T., however, occurs before and shortly after birth. Removal of the T. some months after birth therefore does not seriously impair the T-lymphocyte immune system.

Absence or underdevelopment of the T. in the fetus results in severe defects in the body's immune system, which may be life-threatening.

Embryonically, the T. arises from the gill pouches or gill clefts.

Thyroid cartilage: see Larynx.

Thyroid gland, ***glandula thyreoidea:*** an endocrine organ present in various forms in all vertebrates. Phylogenetically, the T.g. is derived from the endostyle of primitive chordate animals. Ontogenetically, it arises as an evagination of the floor of the pharynx (from the 5th gill pocket of the embryonic gut). Situated in the neck region near to the trachea, the T.g. is a paired structure in amphibians and birds, but usually unpaired in reptiles and mammals. Histologically the gland is relatively simple, consisting of small vesicles or follicles enclosed in connective tissue. The walls of the follicles secrete the iodine-containing hormones, Thyroxin (see) and triiodothyronine. Thyroxin is an important regulatory hormone of metamorphosis; e.g. it is essential for the development of eel larvae and tadpoles into their adult forms. There are also close relationships between thyroid function and the sexual cycle, pregnancy and parturition. Thyroxin and triiodothyronine increase the basal metabolic rate, and have the general effect of accelerating most bodily activities.

Calcitonin (see) is a hormone produced by the parafollicular cells of the T.g. of mammals and by the ultimobranchial gland of nonmammalian species. Calcitonin causes a rapid but short-lived drop in the level of calcium and phosphate in the blood by promoting the incorporation of these ions into the bones.

Hyperthyroidism is caused by overactivity of the T.g. Hypothyroidism results from decreased hormone production, and may be due to iodine deficiency, administration of goitrogens, defects in the enzymes of hormone synthesis, autoimmune thyroiditis (antibodies formed against the body's own thyroid tissue), etc. Prolonged hypothyroidism may result in dwarfism, mental deficiency, goiter and myxedema.

The rate of hormone synthesis by the T.g. is stimulated by Thyrotropin (see).

Thyroid tissue located outside the T.g. is known as accessory T.g.

Thyrostatic agents, ***antithyroid compounds:*** collective term for substances that inhibit thyroid function, usually by inhibition of iodine peroxidase (conversion of iodide to "active iodine": $H_2O_2 + 2I^- + 2H^+ \rightarrow 2\text{"I"} + 2H_2O$), the iodination of tyrosine residues, or the coupling of monoiodotyrosine and diiodotyrosine residues to form thyroxin (T_4) and triiodothyronine (T_3). The resulting low plasma levels of T_3 and T_4 stimulate the release of thyrotropin from the anterior pituitary, which in turn causes hypertrophy of the thyroid gland. Such an enlargement without inflammation or malignancy is called a goiter. T.a. that also lead to goiter development are called goitrogens (e.g. 2-thiouracil, thiourea, sulfaguanidine, propylthiourea, 2-mercaptoimidazole, 5-vinyl-2-thiooxazolidone, allylthiourea). Other T.a. are, e.g. pantothenic acid, reserpine, *p*-aminosalicylic acid, many sulfonamides, phenols, etc. Synthetic T.a. are used to control hyperthyroidism. See Thyroxin.

Thyrotropin, ***thyroid stimulating hormone, TSH:*** a glandular hormone produced by the basophilic cells of the anterior pituitary. TSH generally stimulates the thyroid gland. It is a glycoprotein, M_r 25 000 (bovine), containing 15–20% carbohydrate. The protein contains 201 amino acid residues, and consists of two peptide chains (α- and β-chains). For hormone activity, the two chains must be associated, and each is inactive in the absence of the other. The α-chain is identical to that of luteinizing hormone. FSH acts by stimulating the adenylate cyclase system (see Hormones). It stimulates growth of the thyroid gland, and increases its the metabolic activity, leading to increased iodine uptake, and increased biosynthesis and secretion of thyroxin and triiodothyronine. Both synthesis and secretion of TSH are stimulated by thyrotropin releasing hormone (thyroliberin) from the hypothalamus, and inhibited by thyroxin. TSH is inactivated in the liver.

Thyroxin, ***3,5,3′,5′-tetraiodothyronine, T_4:*** a hormone produced by the thyroid gland and absolutely essential for growth and development. T_4 and another thyroid hormone, 3,5,3′-triiodothyronine (T_3), are synthesized from L-tyrosine residues in thyroglobulin, a dimeric glycoprotein that constitutes the bulk of the thyroid follicle. Specific tyrosine residues in thyroglobulin become iodinated, so that the protein contains several mono- and diiodotyrosine residues. In the subsequent coupling reaction, iodinated rings are transfered from some iodotyrosine residues to form ether linkages by reaction with the hydroxyl function of other iodinated tyrosine residues. T_4 and T_3 are then released by proteolysis of thyroglobulin. T_4 and T_3 synthesis depend on an adequate dietary supply of iodine. Synthesis and secretion of T_4 and T_3 are regulated by thyrotropin from the anterior pituitary. Both hormones are carried in the blood to all body cells, partly in the free form and partly bound to prealbumin and glycoprotein, T_4 being more tightly bound than T_3.

Both hormones generally cause in increase in the basal metabolic rate. Thus, they cause increased oxygen uptake by mitochondria and increased heat production (calorigenic effect). In physiological concentrations, they both increase RNA and protein synthesis. In higher doses, they act catabolically, causing negative nitrogen balance and mobilization of fat depots. Independently of their calorigenic effect, they increase the rate of cell differentiation and metamorphosis, e.g. development of tadpoles into frogs. Biological half-life of T_4 is 7–12 days. Degradation consists of removal of iodine (reused by the thyroid gland), deamination and coupling with glucuronic acid or sulfate in the liver, followed by urinary excretion.

Hyperthyroidism is caused by overactivity of the thyroid gland, leading to excess production of T_4 and T_3. Hypothyroidism results from decreased hormone production; this may be caused by iodine deficiency, administration of goitrogens, defective enzymes in hormone synthesis, autoimmune thyroiditis, etc. Prolonged hypothyroidism may result in dwarfism, mental deficiency, goiter and myxedema. See Thyrostatic agents.

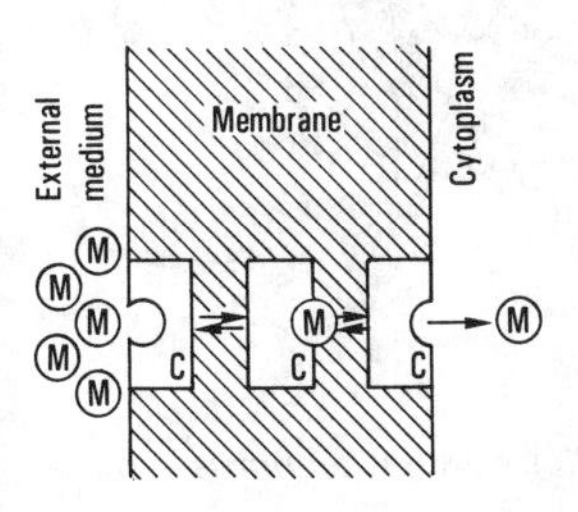

Thyroxin

Thyrsus: see Flower, section 5) Inflorescence.

Thysanoptera, *Physopoda,* ***thrips:*** an insect order of worldwide distribution, containing about 4 500 species, over 600 of which occur in the USA and 220 in Central Europe. The earliest fossil *T.* are found in the Permian.

Imagoes. The adult body is flattened and elongate, ranging in length from 0.5 to 10 mm, or somewhat longer in some tropical species. The squarish head carries small, prominent compound eyes, and ocelli may be absent or present. Mouthparts are adapted for piercing and sucking. In some species the wings are reduced or absent, but otherwise there are 2 pairs of slender wings fringed with long setae. The simple legs are short and carry a 2-jointed foot (tarsus). Between the weakly developed claws of the terminal tarsal segment is an adhesive bladder (arolium), which enables the insect to cling to and walk on smooth surfaces; this bladder or lobe can be folded, but contrary to earlier reports, it apparently cannot be reversibly everted and invaginated. *T.* are found on plants, and most of them feed on plant juices. Some species, however, are predatory, feeding e.g. on mites and aphids. Certain species, in particular the widespread ***European grain thrips*** *(Limnothrips cerealium* and *Limnothrips denticornis),* swarm in enormous numbers on humid days in early summer, thereby acquiring the name of "storm flies". Metamorphosis is partly complete and partly incomplete (remetabolous development).

Eggs. Some species are viviparous. Egg laying species lay 20 to 200 eggs, which are cemented to plant parts *(Tubulifera)* or injected into plant tissue with a saw-like ovipositor *(Terebrantia).* In temperate climates, the eggs hatch after 7 to 10 days.

Larvae. Young larvae resemble the adult, but with shorter antennae and no wings. Wing primordia do not appear until the third instar. The third instar is a nonfeeding, resting stage, known as prepupa, propupa or pronymph, which undergoes some degree of metamorphosis. The final instar (or final 2 instars in the *Tubulifera)* is also a nonfeeding form, resembling (but not necessarily homologous with) the pupa of endopterygotes.

Economic importance. Several species living on plant sap are horticultural and agricultural pests, especially of cereals, legume crops, tobacco, cotton, etc. They cause damage by weakening the host plant, and by acting as vectors of disease-causing viruses.

Classification.

1. Suborder: *Terebrantia* ***(Boring thrips).*** The suborder contains 4 families: *Aeolothripidae* (most members are predators of other insects), *Heterothripidae* and *Merothripidae* (small families, mainly of the neotropics) and *Thripidae* (the largest family, containing most of the pest species). The female possesses an ovipositor adapted for drillng, and the eggs are inserted into plant tissue. Males are often much smaller than females, and with a rounded abdomen. The wing surface is hairy. Examples: ***Onion thrips*** *(Thrips tabaci* Lind., which prefers onions, but also feeds on tomatoes, tobacco, beans, etc., and transmits the virus of tomato spotted wilt), flax fly *(Thrips lini* Ladureau), ***Greenhourse thrips*** *(Heliothrips haemorrhoidalis* Bouche).

2. Suborder: *Tubulifera* ***(Banded thrips).*** All are members of the single family, *Phlaeothripidae.* The female lacks an ovipositor, and in both sexes the end of the abdomen is tubular. The wing surface is not hairy.

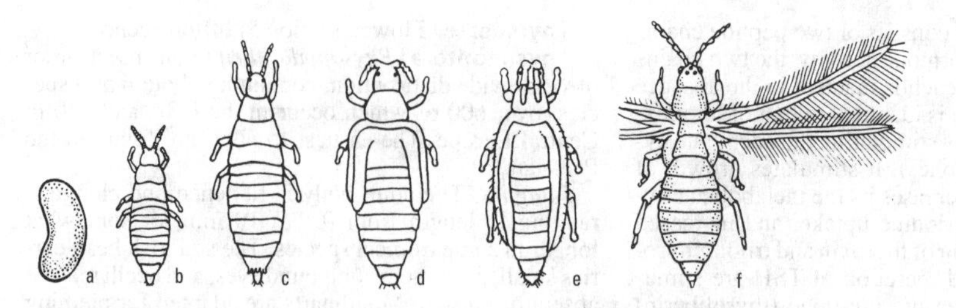

Thysanoptera. Metamorphosis of the pear thrips, *Taeniothrips inconsequens. a* Egg. *b* First larval stage. *c* Second larval stage. *d* Third larval stage (pronymph, propupa or prepupa). *e* Fourth larval stage (nymph). *f* Imago.

Examples: ***Wheat thrips*** *(Haplothrips tritici* Kurdj.), ***Banded greenhouse thrips*** *(Hercinothrips femoralis),* ***Mullein thrips*** *(Haplothrips verbasci).* This suborder includes the largest thrips, e.g. the 13 mm-long *Mecynothrips wallacei* of New Guinea.

Thysanura: a collective, obsolete term for the bristletails and silverfish, which are ectognathous, primarily flightless insects (see *Apteryogota).*

Tiaris: see *Fringillidae.*

Tibetan antelope: see *Saiginae.*

Tibetan brown bear: see Brown bear.

Ticinosuchus: see *Chirotherium.*

Tick-birds: see *Sturnidae.*

Ticks: see *Ixodides.*

Tidal volume of the lungs: see Lung volume.

Tiger: see *Felidae.*

Tiger beetles: see *Coleoptera.*

Tiger finch: see *Ploceidae.*

Tiger moths: see *Arctiidae.*

Tiger salamander: see *Ambystomatidae.*

Tight junction: see Epithelium.

Tiglic acid: see Angelic acid.

Tiliaceae, *linden family:* a cosmopolitan family (but occurring mainly in the tropics) of dicotyledonous plants, containing about 450 species in 35 genera. They are almost exclusively trees or bushes, with spirally arranged or alternate (rarely opposite), simple, dentate or lobed leaves. Stipules are usually small and caducous (falling off at an early stage), often functioning as bud scales and sometimes never present. The actinomorphic, hermaphrodite, 5-merous flowers are pollinated by insects. Numerous stamens and a multilocular ovary are usually present, the latter developing into a capsule, drupe or nut (rarely a berry or druplets) with 1–5 seeds. The flowers form a cymose inflorescence; in some species this is fused with a prominent bract, which later serves as a flight organ of the infructescence.

Small-leaved lime *(Tilia cordata* Mill.) has heart-shaped leaves, 3–6 cm in length, with tufts of rusty hairs at the axils of the veins on the undersurface, and an upper surface that is dark green and more or less glossy. ***Large-leaved lime*** *(Tilia platyphyllos* Scop., *Tilia grandiflora* Ehrh. ex Hoffm.) has broadly ovate leaves, 8–12 cm in length with simple hairs covering the pale green undersurface, together with whitish tufts of hairs at the axils of the veins; the upper surface being dark dull green with practically no hairs. Both species are found in mixed deciduous woodland on fertile soils, and they are popular avenue and parkland trees. The small-leaved lime always flowers profusely and is a valuable bee food plant.

The ***African sparmannia*** *(Sparmannia africana)* is a well known and widely planted indoor ornamental plant; the leaves are large, pale green and covered with white hairs, and the flowers have yellow, sensitive stamens. In its native South Africa, it grows as a large bush. Important fiber plants in this family are *Corchorus capsularis* and *C. olitorius* from Southeast Asia, whose pericycle fibers provide jute for the manufacture of coarse sacking (burlap).

Tiliqua (Tiliquine skinks): see Blue-tongued skinks.

Tillering: the formation of numerous lateral shoots with many adventitious roots, close to, or beneath the ground. T. usually occurs from damaged parts of the shoot of herbaceous plants, mainly grasses and especially cereal species.

Tilletiaceae: see *Basidiomycetes.*

Timaliinae, *babblers* and ***wren tit:*** an avian subfamily of the *Muscicapidae* (see) in the order *Passeriformes* (see). The systematics of the more than 280 species of this group are much debated. Some members look like crows, others like shrikes or finches, and others like pittas or titmice. Members tend to have fairly bright plumage, and to spend much time in cover on or near the

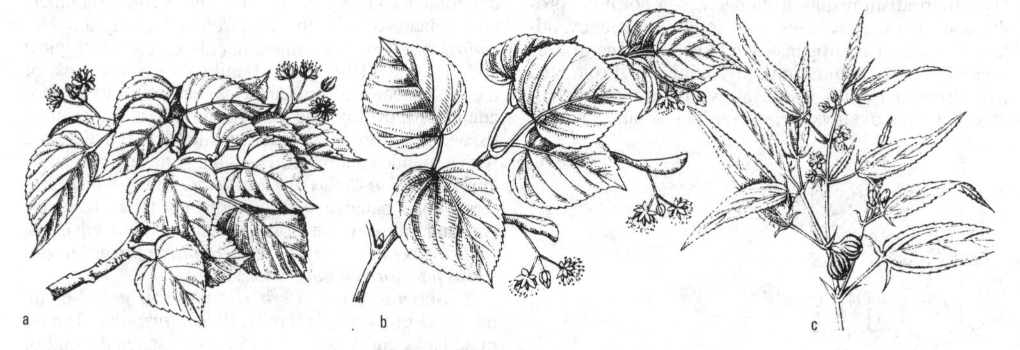

Tiliaceae. a Small-leaved lime *(Tilia cordata). b* Large-leaved lime *(Tilia platyphylla). c* Jute *(Corchorus).*

ground. With the exception of the monotypic North American ***Wren tit*** *(Chaemaea fasciata;* mainly brown coloration; Pacific coast from Oregon to California in low scrub), the subfamily is found only in the Old World, where it occupies the same ecological niche as the Old World Antbirds *(Formicariidae).* Representatives are found throughout the Oriental region, including China and the Philippines, extending to Celebes, New Guinea, Australia, Africa and Madagascar. The ***Bearded tit*** or ***Reedling*** *(Panurus biarmicus)* is a monotypic Palearctic species, but the genus *Panurus* is assigned by some authorities to the subfamily, *Paradoxornithinae* (Parrotbills), which also contains 20 species of *Paradoxornis* and the monotypic *Conostoma* . Most closely related to the wren tit are the 3 species of *Chrysomma:* the ***Rufous crowned babbler*** *(C. poecilotis;* western China), the Oriental ***Golden-eyed babbler*** *(C. sinense)* and ***Jerdon's babbler*** *(C. altirostre;* India). The 4 species of ***Jerys*** *(Neomixis),* which resemble small tree babblers, are found only on Madagascar, e.g. the ***Green Jery*** *(N. viridis).* Other representatives are the ***Jungle babblers, Scimitar babblers, Wren babblers, Tit babblers, Song babblers, Laughing*** or ***Jay thrushes,*** and 2 large, rather aberrant West African ***Rock fowl*** *(Picathartes): P. gymnocephalus* ***(Gray-necked rockfowl,*** with a bright yellow bald head) and *P. oreas* ***(White-necked rockfowl,*** with a naked pink head).

Timber line: see Forest limit.

Timberman beetle: see Long-horned beetles.

Timor deer: see Sambar.

Timothy: see *Graminae.*

Tinamiformes, *Crypturi, tinamous:* an order of birds containing 46 species in 9 genera, distributed in Central, and South America. They are ground birds, bearing a superficial resemblance to game birds, especially guinea fowl, but they are not closely related to the *Galliformes.* On the basis of the analysis of bone structure, DNA and egg white proteins, they are more closely related to the *Rheiformes.* They also resemble ratites, in that incubation of the eggs and care of the young are performed entirely by the male. The sternum is, however, keeled (carinate) and it serves for the attachment of strong flight muscles. Surprisingly, the tinamous are poor and clumsy fliers, soon becoming exhausted, possibly on account of their relatively small heart. They are equally poor and clumsy runners, so that they rely heavily on concealment by their protective gray-brown and mottled coloration. The bill is slender, elongate and slightly decurved. Examples are: *Tinamus major* ***(Great tinamou),*** *Crypturellus saltuarius* ***(Magdalena tinamou),*** *Rhynchotus rufescens* ***(Red-winged tinamou),*** *Crypturellus tataupa* ***(Tataupa tinamou),*** *Eudromia elegans* ***(Elegant crested tinamou),*** *Crypturellus undulatus* ***(Undulated tinamou).***

Tinca: see Tench.

Tinder fungus: see *Basidiomycetes (Poriales).*

Tineidae: see *Lepidoptera* for classification. See European grain moth, Clothes moth.

Tinkerbirds: see *Capitonidae.*

Tiredness: see Fatigue.

Tissue: an association of more or less similarly differentiated, functionally coordinated cells, bound together by their secreted Intercellular substance (see). Most of the T. types mentioned below are listed separately.

1) The following main types of T. are found in animals: Epithelial T., Connective T., Support T., Muscle T., Nerve T.

2) The main types of T. in lower plants are Meristematic T., Parenchyma and Reproductive T., while Plectenchyme is a highly specialized T. type. Higher plants contain additional T. types, which have evolved in association with a terrestrial existence and an increasing division of labor between different plant parts: Boundary T., Absorption T. (in particular water-absorbing T.), Storage T., Secretory T., Permanent T., Vascular T. and Skeletal T.

Tissue hormones: a group of hormones produced in specialized paracrine cells, as distinct from endocrine hormones which are produced in glands. Paracrine cells occur individually or in very small groups in tissues. T.h. reach their target tissue in the blood or by diffusion. Examples are glucagon, gastrin, secretin, pancreozymin, bradykinin, melatonin and others.

Locally acting T.h. that diffuse to, and act upon, cells in their immediate vicinity are also known as *mediators;* they are closely related to both hormones and neurotransmitters, e.g. prostaglandins, serotonin, histamine and others.

Tissue tension: a high grade structural principle, which relies on the different expansibility of the internal and external tissues of a plant organ. The fully turgid, extensible cells of the central tissue are pressed against the less elastic external tissues, such as the epidermis. This tension becomes evident when the different types of tissue are separated, e.g. when the peduncle of a dandelion or a rhubard stem is severed by a transverse cut, then divided toward the summit and placed in water, the separated segments curve outward and roll up; the inner layers are no longer restrained by the outer tissues, and are able to expand under their turgor pressure. Maintenance of T.t. depends on an adequate water supply.

Tit babblers: see *Timaliinae.*

Titicaca grebe: see *Podicipediformes.*

Titis, *Callicebus:* small New World monkeys (see) of the family *Cebidae,* subfamily *Callicebinae* (see Cebid monkeys), which live in family groups of one male, one female and young. The genus is represented by three species. *C. personatus* ***(Masked titi)*** is found in southeastern Brazil. The smallest species, *C. moloch* ***(Dusky titi;*** brown and gray), and the largest species, *C. torquatus* ***(Widow monkey*** or ***White-handed titi;*** dark red with white collar and white hands), occur in the tropical rainforests of Colombia, Venezuela, the southern Amazon basin and Bolivia. All species have long bushy coats. The diet is mainly fruits and may also include leaves *(C. moloch)* or invertebrates *(C. torquatus).* Male titis possibly display the most marked paternal behavior of all primate fathers, carrying the young except when they are suckling.

Titmice: see *Paridae.*

Tits: see *Paridae.*

Tit shrike: see *Vangidae.*

Tit warblers: see *Sylviinae.*

Tityras: see *Cotingidae.*

Tmeripteris: see *Psilophytatae.*

Tmetothylacus: see *Motacillidae.*

Toad-headed lizards, *Phrynocephalus:* small agamids (see *Agamidae),* about 10 cm in length, which inhabit the steppes and semideserts of Central Asia.

They have a round head and flat body, and species living in sandy soil bury themselves by sideway body movements when threatened. On the other hand, *P. mystaceus* responds to threat by extending its lateral cheek flaps. The front surfaces of these flaps are blood red, and they make the mouth look twice its normal size, giving the animal the appearance of a small"dragon". Some montane species are viviparous.

Toads: see *Anura.*

Toadstool: see *Basidiomycetes (Agaricales).*

Toad toxins: compounds in the venoms (secretions of the skin glands) of toads *(Bufonidae),* which are extremely poisonous to mammals, as well as to other vertebrates, including the anurans that produce them. T.t. comprise 2 different types of compounds:

1. ***Bufadienolides.*** These contain a steroid ring system, and they may be present as free sterols (bufogenins), or as esters of suberic acid or suberoylarginine. Bofotoxin was first isolated in 1922 by Wieland and Alles from the common European toad *(Bufo bufo vulgaris).* "Ch'an Su", the dried venom of the East Asian toad *(Bufo bufo gargarizans),* which is used in traditional Chinese medicine for the treatment of dropsy, has yielded several bufodienolides, e.g. marinobufagin, resibufagin, cinobufagin and bufalin. The physiological action of the bufodienolides is similar to that of the cardiac glycosides, i.e. they have a digitalis-like effect on the heart; small doses strengthen and slow the heart beat. They are present in toad blood at a dilution of 1:5000 to 1:20 000, where they are necessary for normal heart function.

2. ***Bufotenines.*** These are basic, nitrogen-containing toxins, i.e. animal alkaloids, which contain an indole ring system and are biosynthesized from tryptophan, e.g. bufotenine, dehydrobufotenine and *O*-methylbufotenine. The venoms of some toad species also contain adrenalin and related compounds. Bufotenines increase the blood pressure and have a paralysing effect on the motor centers of the brain and spinal column.

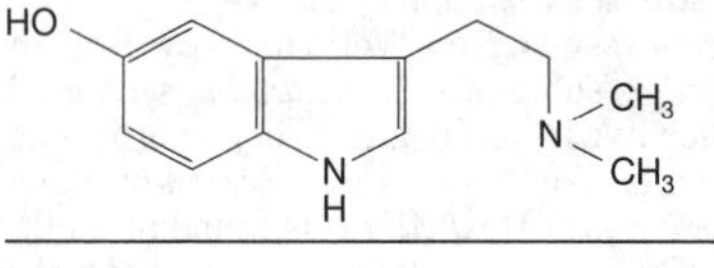

T.t. also have an anesthetic effect which is several times more potent than that of cocaine.

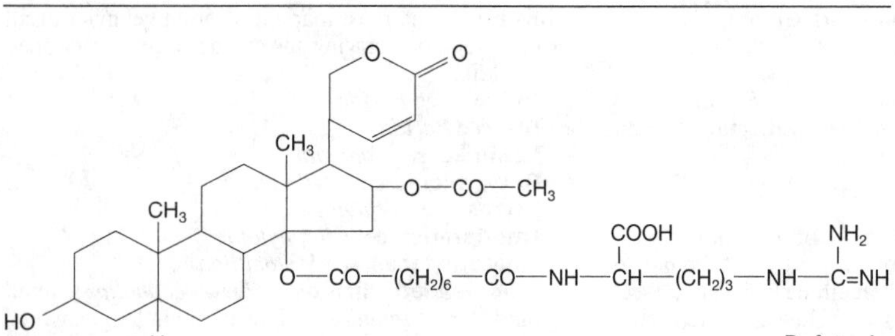

[Y. Kamano et al., *Tetrahedron letters* (1968) 5669–5672 & 5673–5676; H.O. Linde-Tempel, *Helv. Chim. Acta* **53** (1970) 2188–2196]

Tobacco: see *Solanaceae.*

Tobacco alkaloids: a family of alkaloids occurring in the tobacco plant, comprising nicotine, nornicotine, nicotyrine, anabasine and several other related minor alkaloids.

Tobacco mosaic virus, *TMV* (plate 19): a worldwide distributed helical virus with a rod-shaped virion, 300 nm long and 18 nm diameter. As a rule, an infected host cell contains between 1 million and 10 million virus particles. It is readily transmitted mechanically, attacking primarily cultivated members of the *Solanaceae,* in particular, tobacco, tomato and paprika. It has also been found in low concentrations in fruit trees and grape vine. There are numerous strains of the virus, which cause a mosaic of light and dark green, a yellow mosaic, or marked leaf deformations. Infection with TMV also often leads to stunted growth. Normally (e.g. in *Nicotiana tabacum,* which includes most of the commercially grown varieties of tobacco), the virus spreads throughout the entire plant, only the meristems remaining virus-free. The infection is then said to be *systemic,* since the virus is transported in the vascular tissue, usually in the phloem. In other species, e.g. *Nicotiana glutinosa,* the plant reacts strongly against the infection, and the virus is restricted to (encapsulated in) certain cells. *Necrotic lesions* are formed at the foci of infection; these are easily counted, and under experimental conditions, the number of necrotic lesions is an index of the concentration or virulence of the virus particles in an inoculum.

TMV is not only extremely infectious, but also very resistant to desiccation, heating and numerous chemical agents. This is one of the main reasons that it became a model for plant virus research, leading to the elucida-

Bufotenine

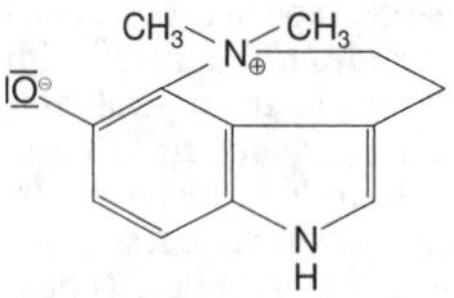

Dehydrobufotenine

Bufotoxin

tion of many fundamental problems of molecular biology. The replication system of TMV was studied in detail, using virus-infected cultures of protoplasts (plants cells, whose cell walls have been removed by enzymatic treatment), as well as synchronous virus replication. It also serves as a model RNA virus. After virus particles enter a cell (usually through a wound in the plant tissue), part of the protein of the nucleocapsid (see Viruses) is stripped off, exposing a cistron that encodes early proteins. One of these early proteins induces or accelerates the synthesis of RNA-dependent RNA polymerase, i.e.an enzyme that catalyses RNA synthesis on a template of RNA. This enzyme is encoded in the genome of the host plant (earlier, it was erroneously thought to be encoded by the viral RNA). Replication of the TMV-RNA then commences with the aid of the RNA-dependent RNA polymerase. At first, mainly double stranded RNA is synthesized (known as the RF form or replicative form). At the same time, heterogeneous (with respect to size), replicative intermediates are detectable (RI forms). RI forms consist of infective template strands with attached, complementary, growing daughter stands, the two held together by replicases. Very soon, increasing amounts of replicated complete viral RNA become detectable. The ensuing exponential phase of viral RNA replication continues for about 10 hours in the host protoplasm. Translation of the coat protein starts early in the exponential phase of RNA synthesis. Very soon there are more coat protein subunits than replicase molecules. This is probably made possible by translation of the coat protein on a separate monocistronic mRNA, which is produced either by the processing of newly synthesized viral RNA or by transcription of part of the viral RNA. Production of such large amounts of coat protein leads to the incorporation of ever increasing quantities of newly synthesized TMV-RNA into nucleocapsids (each RNA strand is coated with 2 130 coat protein subunits). About 20 hours after infection of the protoplasts, practically all synthesized viral RNA strands are coated with protein. It is not clear why newly synthesized virus particles do not become uncoated like infecting particles; they are possibly protected from the relevant enzymes by inclusion in membrane vesicles. Like the particles of other viruses, TMV particles often accumulate as relatively large, light microscopically visible *inclusion bodies,* which in plants are also known as *X-bodies.*

Tobacco rattle virus: see Virus groups.

Tobamoviruses: see Virus groups, Tobacco mosaic virus.

Tobraviruses: see Virus groups.

Toddy cat: see Palm civets.

Todidae: see *Coraciiformes*

Todies: see *Coraciiformes*

Todus: see *Coraciiformes.*

Tokay gecko, ***great house gecko,*** *Gekko gekko:* a large, aggressive gecko (see *Gekkonidae),* common in Southeast Asia. One of the largest geckos, it attains a length of 30 cm or more, and feeds on small reptiles, birds and mammals, as well as insects. It has a very broad head, and the body has reddish markings on an olive-brown background. It has a loud call of "to-kay".

Tolu balsam: a brown-yellow, viscous balsam with a pleasant odor of vanilla, produced mainly in Colombia and Venezuela by making incisions in the trunk of *Myroxylon balsamum* (= *M. touifera)* (family *Papilionaceae).* It consists of cinnamic and benzoic esters, vanillin, terpenes and resins, and it is used medicinally in cough mixtures. Tolu is a small town on the northern coast of Colombia. See Balsams.

α-Tomatine, ***tomatine:*** the main alkaloid of the tomato plant *(Lycopersicon esculentum).* It is a glycoside (glycoalkaloid) of the aglycon, tomatidine [a steroid, (22S:25S)-5α-spirosolane-3β-ol], and the tetrasaccharide, β-lycotetraose (consisting of D-galactose, D-xylose and 2 molecules of D-glucose). T. protects the tomato plant from attack by the Colorado beetle, and it possesses antibiotic activity against the causative agent of tomato wilt and other pathogenic fungi.

Tomato: see *Solanaceae.*

Tomato bushy stunt virus: see Virus groups.

Tomato spotted wilt virus: see *Thysanoptera.*

Tombusviruses: see Virus groups.

Tomistoma: see False gharial.

Tömösvary's organ: see Postantennal organ, *Symphyla.*

Tongue, ***lingua, glossa:*** a muscular organ arising from the floor of the buccopharyngeal cavity of vertebrates, and attached to the hyoid bone (os hyoideum or os linguae). Fishes do not possess a T., or it is represented by a nonprotrusible pad of tissue on the mouth cavity floor. Some reptiles lack a T., but in the majority of species that have a T., the organ is slightly divided at the front, and attached anteriorly to the floor of the mouth cavity; it is curled outward and around captured prey. In snakes and lizards, the T. is frequently forked at the tip and very mobile. In many birds, the end of the T. is strongly keratinized. In ducks it acts jointly with the keratinized upper bill to form a sieving apparatus. The T. of touracos (plantain eaters) is finely divided like a brush. Woodpeckers have a long protrusible T. (with backward directed barbs at the tip), which in most species can be extruded to great lengths beyond the bill; this is made possible by massive development of the horns of the hyoid bone, which when at rest lie arched around the eyes. Mammals characteristically have a muscular and very glandular T., which is used to grasp food, and to manipulate and transport it within the mouth cavity. Anteaters and bats have a protrusible T. The mammalian T. is used as an aid to vocalization, and for body washing and grooming. It is also the site of the Taste buds (see).

Tonna: see *Gastropoda.*

Tonofibrils: fine fibers (0.7–0.8 nm diameter), which cause double diffraction of polarized light, and which are particularly abundant in epidermal cells, but are also found in other vertebrate cell types. T. consist of bundles of noncontractile cytoplasmic filaments (tonofilaments, 10 nm diameter). They are frequently inserted into desmosomes or zonulae adhaerens, and appear to anchor these to the cytoplasm. With their relatively high tensile strength, T. contribute to the mechanical stability of cells and cell associations.

Tonsils, ***palatine tonsils:*** two masses of lymphoid tissue, one on either side of the triangular space between the glossopalatine and the pharyngopalatine arches. The surface of the T. contains openings called crypts that communicate with channels throughout the tissue. Blood is supplied by the lingual and internal maxillary arteries, and the nerve supply is derived from both divisions of the autonomic nervous system. The T. function like lymph nodes, contributing to the formation of leukocytes and serving as filters to prevent the entry of microorganisms. Although they protect against infection, they may in severe conditions become foci of infection (tonsillitis); the resulting swelling may interfere with the passage of air to the lungs.

Other masses of lymphoid tissue beneath the tongue are called the *lingual tonsils.*

Tooth-billed pigeon: see *Columbiformes.*

Tooth development, *eruption of the teeth, dentition:* Most mammals have two different generations of teeth, i.e. the milk teeth (dentes decidui), followed by the permanent dentition (dentes permanentes). The milk teeth are usually vertically displaced and forced out by the growing permanent teeth. In elephants, sea cows and kangaroos, the teeth grow horizontally and migrate forward. In the embryo, the skin covering the jaw bears a longitudinal groove or furrow *(dental groove),* which marks the site of a vertical intucking of the epidermis to form a *dental lamina.* At intervals along this lamina, dental papillae are formed as outgrowths or evaginations, and these constitute the tooth rudiments. In the human embryo, tooth rudiments begin to form in both jaws early in the second month. Tooth crowns appear first, and development proceeds toward the roots. After birth, the teeth erupt through the gum tissue, normally in a regular sequence. The middle incisors appear in the 6–9th month after birth, and the lateral incisors appear in the 8–11th month, followed by the premolars (12–16 months), canines (16–20 months) and the molars (20–24 months). Even some of the permanent teeth start their development in the embryo. During growth, the food supply to the milk teeth becomes increasingly restricted, resulting in degeneration which proceeds from root to crown. Finally, only a thin enamel cap remains, which is pushed away by the emerging permanent tooth. Replacement of milk teeth by the permanent dentition also occurs in an ordered sequence.

The milk dentition of males usually erupts somewhat earlier than that of females, whereas eruption of the permanent dentition is somewhat earlier in females. The programmed eruption of different teeth at different ages is used in anthropology and forensic medicine for age diagnosis.

In addition to the process of tooth development and eruption, the term, Dentition (see), also refers to the number and type of teeth present in the jaws of any species, i.e. it is also almost synonymous with the dental formula.

In elasmobranch fishes, the teeth are replaced continuously.

Toothed whales: see Whales.

Tooth fungus: see *Basidiomycetes (Poriales).*

Tooth shells: see *Scaphopoda.*

Topaz: see *Apodiformes.*

Topoclimate: the climate of the air layers at the soil surface, largely determined by the topography of the land.

Top yeasts: see Brewer's yeasts.

Torgos: see *Falconiformes.*

Tormillares: see Sclerophylous vegetation.

Toro: see *Echimyidae.*

Torrenticole: the population of organisms that live in fast flowing streams. Torrenticolous organisms have a high requirement for oxygen, and therefore inhabit only powerfully agitated water, such as mountain streams and sea surf. In quiet or stagnant water, they usually rapidly suffocate. In animals, the main adaptation to life in fast flowing water is flattening of the body shape. A shield-like arched or vaulted shape is an adaptation to an existence on the bottom or bed of the stream (e.g. the larvae of beetles and mayflies, and snails with flattened shallow shells). Suction disks are common, enabling organisms to attach themselves to solid structures in the current. Larvae of the neuropteran fly, *Liponeura,* have several suction disks. Certain genera of the *Sisordidae* (Sucker catfishes) have developed various suction devices, e.g. the suction apparatus of *Glyptothorax* consists of skin folds on the lower surface of the forebody and head, while species of *Oreoglanis* can adhere to solid surfaces with the aid of the first rays of the pelvic and pectoral fins, which are greatly broadened and diagonally striped, as well as using their wide and flattened lips as an auxillary suction organ. Tadpoles of the toad, *Bufo penangensis,* possess a single suction disk. Many insect larvae attach themselves to the substratum with silk threads, which also often serve as a net for trapping food, e.g. larvae of the caddis fly, *Hydropsyche.* In addition to structural adaptations, torrenticolous organisms also exhibit behavioral adaptations. Thus, freely motile forms are positively thigmotropic, i.e. they strive to maintain their bodies in constant contact with the substratum; for external respiration they do not rise to the water surface, but respire via gills or by gas exchange over the entire body surface.

Torsion: movement of plant organs or organ parts, in which the direction of growth is maintained, but the organ turns or twists on its own axis, while the base of the organ is unaffected. T. may occur autonomously, or it may be induced by external stimuli (see Geotorsion), e.g. in flower stalks, or it may be a Hygroscopic movement (see), e.g. in the awn of the one-seeded storkbill fruit.

Tortoises: see *Testudinidae.*

Tortricidae, *tortricid moths, fruit moths, bell moths, leaf roller moths:* a worldwide family of the *Lepidoptera,* containing many species (500 species in central Europe, well over 300 in Britain, 600 in North America, 800 in Australia, etc.). The wingspan varies between 10 and 25 mm. Most of the moths have a typical "tortricid" or bell shape (also described as a tent shape) when the wings are at rest. The relatively broad forewings are usually patterned with various shades of brown or gray, whereas the hindwings are mostly uniform in color. Caterpillars live in shelters formed by rolling leaves or binding several leaves together, or they bore into fruits, stems or roots. Many are plant pests, e.g. Apple codlin moth (see), Appleskin worm (see), Pea moth (see), Bud moths (see), Plum fruit moth (see), Vine moths (see).

Tortricid moths: see *Tortricidae.*

Torutein: see *Ascomycetes.*

Total body water: see Water economy.

Totipotency: the ability of a single somatic cell to differentiate into any other type of cell of the same organism, and to give rise to a whole new organism. All the living somatic cells of an organism have the same complement of genetic information as the original fertilized egg cell, because a complete replica of the DNA (and therefore the genome of the cell) is passed to each daughter cell during mitosis. T. is suppressed but not lost by cell differentiation. This is demonstrated dramatically by the fact that, under appropriate conditions of nutrient supply and hormonal supplementation, a single somatic plant cell (from leaf, stem, root, petal etc.) will divide to form an undifferentiated callus (i.e. a tissue culture), which can then be induced to develop (differentiate) into a whole new plant. Similarly, a begonia leaf placed on moist sand will produce an adventitious shoot from a single epidermal cell, especially at the site of a cut leaf vein, leading eventually to the generation of a new plant. Formation of wound callus occurs by the dedifferentiation of parenchyma cells, representing a reactivation of suppressed potency (reactivation of genes with a temporary restoration of T.). T. of differentiated cells can also be shown experimentally in animals. For example, if cell nuclei from differentiated muscle or intestinal epithelium of larvae of the clawed toad are transplanted to enucleated, activated egg cells, these subsequently develop into entirely normal larvae.

Toucanets: see *Ramphastidae.*

Toucans: see *Ramphastidae.*

Touch, sense and **stimulus:** see Contact stimulus, Haptonasy, Haptotropism, Tactile stimulus reception

Touchwood: see *Basidiomycetes (Poriales).*

Touraco: see *Cuculiformes.*

Tournaisian stage: see Carboniferous.

Tower head: see Acrocephaly.

Tower skull: see Acrocephaly.

Towhees: see *Fringillidae.*

Toxicodendron: see *Anacardiaceae.*

Toxoglossa: see *Gastropoda.*

Toxoplasma: see *Telosporidia.*

Toxoplasmida: an order of the *Sporozoa,* recognized by some authorities, and containing organisms otherwise assigned to the *Telosporidia* (see). A well known member is *Sarcocystis muri,* which infects rats and mice. Ingested spores hatch in the intestine and liberate amebulae that penetrate the cells of the intestinal epithelium. Within the epithelial cells, trophozoites grow and multiply by schizogony. Merozoites then migrate to the muscles, where they grow to form a multinucleate plasmodium, which divides by plasmotony. Eventually, masses of parasites form long, slender, cylindrical structures with pointed ends, known as "Miescher's tubes", which contain sickle-shaped spores.

Toxoplasma gondii is responsible for toxoplasmosis (see plate 42), a cosmopolitan disease of man and many other animals. The final host, in which sexual reproduction occurs, is the cat. There may be several, serially infected intermediate hosts, including numerous mammals, humans and birds. Intermediate hosts support a special form of asexual reproduction, known as *endodyogeny,* which is responsible for the acute phase of the disease. This is followed by the chronic stage, characterized by the appearance of large resting stages in the musculature and brain. Symptoms are unspecific, with headache, fever and lymphopathies. Since *Toxoplasma* easily crosses the placenta, it can also damage the fetus, leading to miscarriage and premature birth. Humans are infected by unhygienic practices, and by eating the undercooked meat of another intermediate host. Most human infections are subclinical, and about 50% of all humans are infected during their lifetime.

Toxoplasmosis: see *Toxoplasmida.*

Toxostoma: see *Mimidae.*

Toxotidae, ***archerfishes:*** a family (6 species in 1 genus) of fishes of the order *Perciformes.* They occur in southern Asia, from India, through the Malay Archipelago as far as Australia and the New Hebrides. They are found in marine water, well inland in fresh water, and particularly in the brackish water of coastal mangroves. The deeply compressed body has a high back. They feed on insects, which they wash down from overhanging stems and branches with a forceful and accurate ejection of water from their pointed mouths. Most species are less than 16 cm in length, but *Toxotes chatareus* attains 40 cm. *Toxotes jaculatrix* is occasionally kept in aquaria.

TPN: see Nicotinamide-adenine-dinucleotide phosphate.

Trabecular bone: see Bones.

Trace elements: chemical elements required in extremely small quantities for the normal functioning of living organisms. Notable T.e. are boron, copper, iron, manganese, molybdenum and zinc. In humans, animals and plants, deficiency of T.e. leads to severe physiological disturbances. See the entries for individual elements.

Trachea:

1) In air-breathing vertebrates, a tubular structure lined with ciliated epithelium and supported by cartilaginous rings. It leads from the larynx, then divides into 2 bronchii, which enter the lungs.

2) See Tracheae.

Tracheae: air-filled, narrow, tubular or saclike invaginations of the body integument of terrestrial *Onychophora* and *Arthropoda,* which serve for the conduct of respiratory gases to and from the body tissues and organs. The tracheal wall consists of an outer cellular matrix, a tracheal epithelium and an internal chitinous intima The intima, which is stiffened by annular or spiral thickenings (tanidia), is a continuation of the exoskeleton; like the latter it is renewed at each molt. T. communicate with the exterior via paired apertures called *spiracles* or *stigmata,* which are arranged segmentally. In insects the spiracles are furnished with a closing mechanism which regulates the air flow and prevents entry of foreign bodies. Each spiracle opens into an *atrium* or *tracheal chamber,* which in turn gives rise to tracheal trunks. The main trunks divide into dorsal and ventral systems of T., which branch, anastomose and extend to all parts of the body.

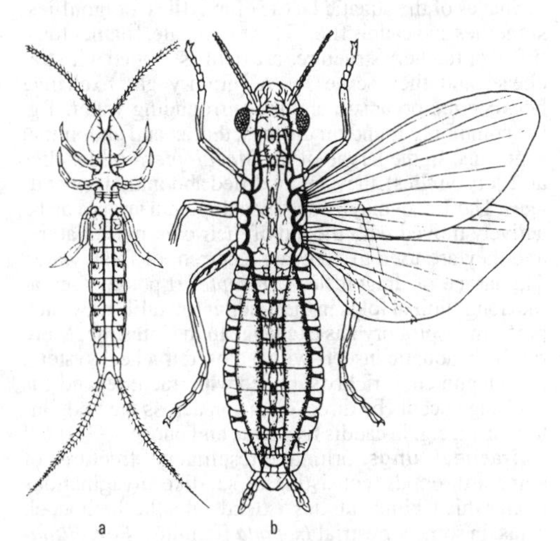

Fig.1. *Tracheae.* Basic plan of the insect tracheal system. *a* Primary wingless insect *(Campodea),* ventral view. *b* Winged insect.

In many myriapods and primitive insects, each segment has a separate tracheal system with its associated spiracles. In advanced insects, T. may communicate with the Tracheoles (see) of neighboring segments, so that tracheal trunks run the entire length of the body, giving rise to segmental branches. Oxygen in the T. and tracheoles diffuses into the surrounding tissue or into the hemolymph. In many myriapods, insect larvae and insect pupae, the respiratory gases enter and leave the T. by diffusion via the spiracles, whereas in winged insects the movement of respiratory gases is accelerated by the extension and contraction (usually dorso-ventral) of the abdomen. This "pumping action" extends and compresses the thicker trunks of the T. and their less thickened expansions (see Tracheal vesicles). In insects that are secondarily adapted to an aquatic existence, all the spiracles are closed, and frequently the T. are totally or partly regressed. In such insects, respiratory gases are exchanged via Tracheal gills (see) or via the integument (surface respiration). Surface respiration also occurs in the very small larvae of ichneumon flies that parasitize other insects. In addition to normal tubular T., many insect larvae possess Tracheal lungs (see). T. also generally serve to support the body organs and hold them in position.

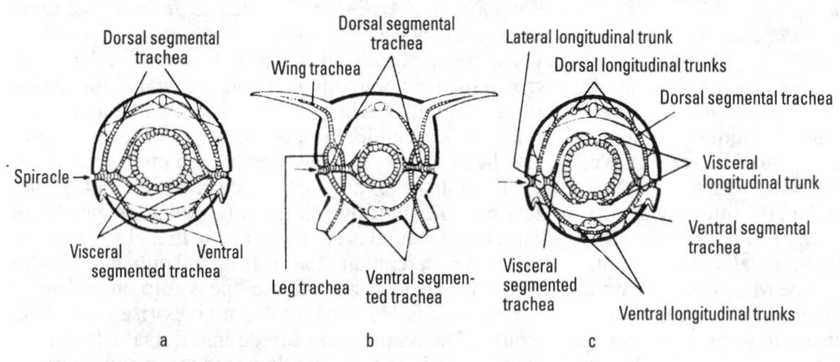

Fig.2. *Tracheae.* Transverse section of the insect tracheal system. *a* Basic plan of an abdominal segment. *b* Basic plan of a thoracic segment. *c* Secondary modifications, showing longitudinal stems and transverse connections.

Tracheal gills, ***pseudobranchi:*** thin-walled body appendages of the aquatic larvae of mayflies, dragonflies, stoneflies and caddis flies. These laminate, filamentous, tufted or feathery structures are richly supplied with tracheae, and they serve for respiratory gas exchange between the organism and the surrounding water. T.g. are commonly found on the legs, thorax and abdominal segments. In the larvae of the *Megaloptera* (alder flies and dobson flies), they are modified abdominal appendages. The T.g. of mayfly and megalopteran larvae can be actively moved with the aid of their own musculature, and they are used to generate a current of fresh water. The larvae of dragonflies *(Anisoptera)* possess septa-like longitudinal folds in the hindgut (rectal T.g.), which perform respiratory gas exchange in the intestine. More rarely, in aquatic insects with a closed tracheal system, the integument is richly supplied with tracheae, and gas exchange occurs by direct diffusion across the body integument, e.g. in caddis fly larvae and pupae.

Tracheal lungs: primitive respiratory structures of some arthropods, consisting of sac-like invaginations from which blind tubules extend into the hemocoel. Thus, in some terrestrial *Isopoda* [families: *Porcellionidae* and *Armadillidiidae* (woodlice)], thin-walled invaginations of the expodites of the abdominal legs give rise to blind tubules, which extend into the hemocoel of the expodite. The *Scutigeromorpha* (subphylum *Myriapoda)* have 7 unpaired T.l. located mid-dorsally near the posterior margins of the tergal plates covering the leg-bearing segments, and opening to the exterior via a spiracle. Each lung consists of an atrium, which gives rise to two large fans of short tracheal capillaries (up to 600 per fan), the latter extending into the pericardium. T.l. differ from the more advanced tracheal system (see Tracheae), which penetrates deeper into the body, and terminates in fluid-filled cavities adjacent to the tissues. See Book lungs, Book gills.

Tracheal membrane: see Syrinx.

Tracheal organ: see *Isopoda.*

Tracheal vesicles: vesicular expansions of tracheae, which serve as air storage organs, swim bladders or sound-amplifying organs. The chitinous intima is not spirally thickened (cf. tracheae). In some cases fine branches extend from T.v. into the surrounding tissues. T.v. that adjoin or are very close to the body surface (i.e. communicating more or less directly with the exterior via spiracles) are called tracheal chambers or atria. Other T.v. are expansions of the tracheal roots; these are particularly abundant and well developed in flying insects, e.g. honeybee and housefly, where they act as air storage organs and serve as reservoirs for the equalization pressure within the tracheal system during flight. In the aquatic larvae of the *Chironomidae,* the T.v. function as a hydrostatic apparatus.

Tracheal gills. a Plate-like gills of a mayfly larva *(Siphlurus). b* Schematic representation of tracheal gills associated with a closed tracheal system.

Tracheata, *Antennata:* a subdivision of the *Mandibulata.* The latter is no longer regarded as a systematic taxon. Tracheates respire with the aid of tracheae and possess a single pair of antennae. About 860 000 species are known, and 850 000 of these are insects. All species are primarily terrestrial, although the larvae or imagoes of a few species are secondarily adapted to an aquatic existence. In certain groups the number of spiracles is reduced. The antennae are homologous with the antennulae of crustaceans and they are innervated from the deuterocerebrum. The following segment (which in crustaceans carries a second pair of antennae) also contains a pair of ganglia (tritocerebrum), but never carries appendages.

Tracheids: see Conducting tissue, Wood.

Tracheoles, ***tracheal capillaries:*** blind, filiform or ramose, extremely fine terminal branches of tracheae. It is in the T. that respiratory gas exchange occurs between the air of the tracheal system and the O_2-consuming/CO_2-producing tissues. In insects, T. have a diameter of 1 µm or less. They are tubular and thin-walled, and the spiral thickening of their chitinous intima can only be observed by electron microscopy. The chitinous intima is not shed during molting. T. develop from terminal cells of the tracheae, then ramify between and into the tissue cells. At rest, the T. are filled with fluid, but during organ activity the fluid is resorbed by the tissue. Specialized aggregations of T., known as Tracheal lungs (see) are present in some fly and lepidopteran larvae and in the thorax of water bugs *(Hydrocorisae).*

Tracheophonae: see Syrinx.

Trachinidae, ***weeverfishes, weevers:*** a family (1 genus, at least 4 species) of marine benthic (bottom) teleost fishes (order *Perciformes)* found in the eastern Atlantic as far north as Norway, especially common in the Mediterranean, and also found in the Black Sea and off the coast of Chile. There is a large, superior mouth, and the eyes are situated on top of the head. The anterior dorsal fin is short and spiny, the posterior dorsal fin being much larger and composed of soft rays.

The ***Greater*** or ***Dragon weever*** *(Trachinus draco;* 20–39 cm, max. 45 cm) occurs on the continental shelf of the Atlantic from Norway to South Africa, as well as in the Mediterranean and the Black Sea, spending most of the day buried in the sand with just its eyes protruding. It has venom glands at the base of the spiny pectoral fins, the first dorsal fin and the spiny processes of the opercula; all of these secrete highly potent, dangerous toxins. It is gray-brown with several narrow, oblique, lateral dark stripes, and a yellowish belly.

The ***Lesser weever*** *(T. vipera;* 8–15 cm, max. 20 cm) has a rather more southerly distribution and is less common in the Mediterranean. It is arguably the most venomous of all fishes, carrying an extremely potent toxin in the venom glands at the base of the first dorsal fin and the spiny processes of the opercula. It also spends much time buried in the sand.

Up to 50 cm in length, *T. araneus* is the largest species. Found mostly in the Mediterranean, it is characterized by 6 or 7 lateral dark spots, and it is reported to have very tasty flesh.

Trachinus: see *Trachinidae.*

Trachurus: see Scad.

Trachylina: see *Hydrozoa.*

Trachyceras (Greek *trachys* crude): an index ammonite genus of the Middle and Upper Alpine Triassic. The discoid, planispiral shell has a rounded outer edge with a medial groove, which is bordered on each side by a row of small nodules. The shell surface is decorated with simple to forked ribs, which are coved with numerous nodules. The suture line is totally serrated.

Track, ***trail, spoor:*** a sequence of footprints revealing the presence of animals (Fig.), including nocturnal species that are rarely seen. The T. of an animal provides information on its identity, and often also indicates whether the animal was walking (wandering, roaming) or running (fleeing). The hardness of the ground or depth of snow influence the appearance of a T. Thus, the dew claw of a deer forms an imprint only when the foot sinks deeply into the substratum. See Fig. on p. 1202.

Tradition, ***biotraditional behavior:*** acquisition of individual behavior by learning processes that are based mainly on imitation or following an example, sometimes in association with biosocial stimulation. The imitation may be motor or sensory. In the latter case, biotraditional behavior is observed (heard or seen) then incorporated into the behavior of the observer. In the UK, bluetits have learned to pierce the aluminium tops of milk bottles for a drink of cream, and this behavior has become a tradition of the species in certain areas. In recent years, numerous examples of biotraditional behavior have been described in primates, including the use of tools for obtaining food (chimpanzees insert a grass stalk or stick into an ants' nest, withdraw the stick, then lick off and eat the ants). The feeding and bathing behavior of certain Japanese macaques is an interesting example of recently acquired biotradional behavior (see Macaques).

Tragacanth: see *Papilionaceae.*

Tragelaphinae: a subfamily of the *Bovidae* (see). The subfamily is divided into 3 tribes.

1. Tribe: *Tragelaphini;* genera: *Tragelaphus* ***(Spiral-horned antelopes)*** and *Taurotragus* ***(Eland).***

2. Tribe: *Boselaphini;* genus: *Boselaphus* ***(Nilgai*** or ***Bluebuck).***

3. Tribe: *Tetracerini;* genus: *Tetracerus* ***(Four-horned antelope).***

The ***Nilgai*** *(Boselaphus tragocamelus)* is a monotypic species, head plus body length 180–210 cm, found in eastern Pakistan and India, where it prefers forest and low jungle, but is sometimes seen in more open areas. Males are about 20% larger and heavier than females. Horns are short and only present on the male. Upper parts of adult males are blue-gray in color, with white underparts, white rings on the fetlocks and white stripes in the ears. Females are lighter in color. Both sexes have a neck mane, and the male also develops a tuft of hair on the throat.

Six species of *Tragelaphus* are recognized. All are medium-sized to large, with patterns of white spots and stripes on the face and body. Only the males possess horns, which characteristically have the shape of an open spiral. *T. angasi* ***(Nyala;*** plains and mountains of southern Malawi, Mozambique, Zimbabwe, eastern South Africa) has a head plus body length of 135–155 cm, and a record horn length of 835 cm. It is rather grayer than most other species, and adult males are characterized by a long curtain of hairs, which hangs from the lower neck, lower shoulders, belly and thighs. *T. strepsiceros* ***(Greater kudu;*** woodland thickets; southern Chad to Somalia, south to South Africa) has a head plus body length of 195–245 cm, and very long,

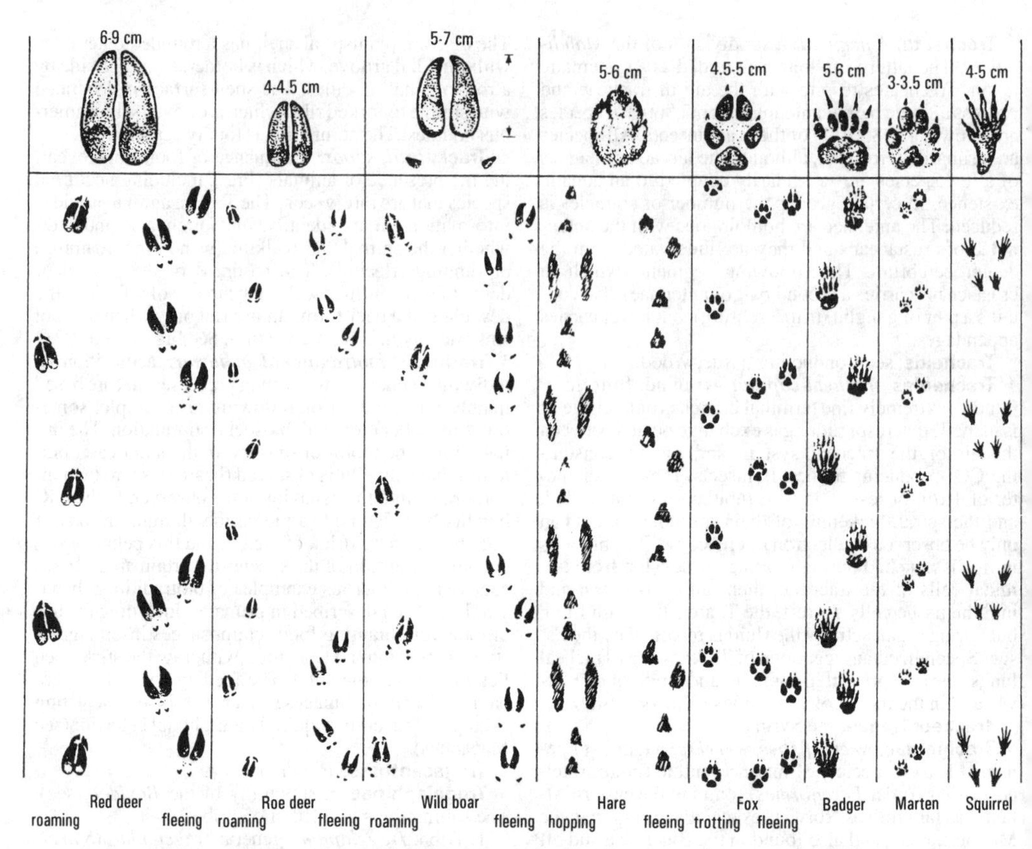

Track. Tracks and footprints of some European animals. Measurements in cm refer to the length of a single foot impression from the back of the heel to the end of the most extended digit.

twisted horns (about 102 cm in a straight line and 132 cm along the curves). It leaps with ease over obstacles (bushes) up to 2.5 m in height. *T. imberbis* ***(Lesser kudu;*** dense vegetation; South Yemen, Northeast Africa through Uganda to Kenya and eastern Tanzania) has a head plus body length of 110–140 cm, and horns of 60–90 cm. Males tend to be deep yellowish gray, although some individuals may be darker; females are pale brown. Other species are:*T. buxtoni* ***(Mountain nyala;*** highlands of southern Ethiopia), *T. spekei* ***(Sitatunga;*** swamplands; Gambia to southern Sudan, south to northern Botswana), *T. scriptus* ***(Bushbuck;*** vegetation near to water in most of sub-Saharan Africa).

Three species of *Taurotragus* are recognized. *T. euryceros* ***(Bongo)*** inhabits the forest zone from Sierra Leone to Kenya. *T. oryx* ***(Eland),*** found on the savannas of eastern and southern Africa, attains a height of 1.9 m and a head plus body length of 180–345 cm. It is yellow-brown in color, and posesses a neck dewlap; the horns are relatively straight with a spiral keel. *T. derbianus* ***(Derby*** or ***Giant Eland;*** savanna from Senegal to southern Sudan) (sometimes considered to be a subspecies of *T. oryx)* is colored rich fawn, and its black neck has a white base. For antelopes, both the Eland and Derby have a rather ox-like massiveness and appear to be phylogenetically close to cattle.

Tragelaphini: see *Tragelaphinae.*

Tragelaphus: see *Tragelaphinae.*

Tragulidae, ***chevrotains:*** a family of primitive, rabbit- to hare-sized ungulates, represented by only 4 living species in 2 genera: *Tragulus* (Asian) and *Hyemoschus* (African). It is the only surviving family of the superfamily, *Traguloidea* (see). They possess large upper canines, which in males are extended as tusks. Upper incisors are absent, and both sexes lack horns. The subdivisions of the stomach are less complex than in other ruminants. Each limb carries four digits. The cannon bone (formed by fusion of the middle metapodials) is completely formed in the hindlimbs, only partially formed in the forelimbs of *Tragulus,* and absent from the forelimbs of *Hyemoschus.* Tibia and fibula are fused.

Traguloidea: a superfamily comprising the *Tragulidae* (see) and the extinct families, *Hypertragulidae, Protoceratidae* and *Gelocidae,* whose fossils are found from the Eocene to the Pliocene. Hypertragulids were the ancestors of the tragulids; their fossils are found mainly in North America, and they characteristically possesss modified lower premolars that functioned as canines. The North American protoceratids shared this dental feature, but they also had horns and their faces were longer. The Old World gelacids of the Oligocene may represent a transition to modern ruminants.

Tragulus: see *Tragulidae.*

Training: a learning process in animals, which is designed and controlled by humans. In *positive T.,* the correct choice of behavior is rewarded (reinforced). In *negative T.,* the incorrect choice of behavior is punished. *Differential T.* combines both the positive and negative approaches. In training for performance, as in a circus act, or in the *dressage* of horses, an attempt is made to totally adapt the behavior of the animal to a particular task, if possible by exploiting insight learning (see Learning).

Transamination: an important reaction of amino acid metabolism, catalysed by pyridoxal phosphate-dependent transaminases (aminotransferases) (see Transferases). The amino group of an amino acid is transferred reversibly to an α-ketoacid, such as α-ketoglutarate or oxaloacetate: amino acid + α-ketoglutarate $\rightleftharpoons$ α-ketoacid + glutamate. T. is important in mammals because it is mechanism for the removal of amino nitrogen from excess dietary amino acids and from amino acids during protein turnover; the amino-nitrogen of glutamate and aspartate can enter the urea cycle and become incorporated into urea which is excreted. T. is generally important in all organisms as a means of synthesizing amino acids. Thus, in heterotrophic green plants, the carbohydrate skeletons of the amino acids (i.e. the corresponding ketoacids) are first synthesized, then transaminated with glutamate to produce amino acids. In animals, some nonessential amino acids can be synthesized by transamination of the appropriate α-ketoacids (some amino acid biosyntheses are rather more complicated, e.g. tryptophan and arginine are not produced simply by transamination of the ketoacid).

Trans-configuration: a term applied to the configuration of the alleles of two coupled genes (see Coupling), or the Heteroalleles (see) of a cistron. In heterozygous or heterogenic cells, each coupling structure (chromosome or genophore) carries a mutant and a normal allele, so that the genotype can be expressed as +b/a+. See Cis-configuration, Cis-trans test.

Transcription: the DNA-dependent synthesis of RNA (messenger RNA, transfer RNA, ribosomal RNA). T. is the first step in the expression of the genetic information encoded in DNA. The process is catalysed by RNA polymerase, of which there is more than one type (e.g. RNA polymerase I, II and III), depending on the site of synthesis (nucleolus or nucleoplasm) and on the type of RNA that is transcribed. For example, RNA polymerase II is present in the nucleoplasm of eukaryotic cells and is mainly responsible for the synthesis of messenger RNA. Protein factors are also involved (sigma factor, psi factor, rho factor). RNA polymerase is also called DNA-dependent RNA polymerase, nucleoside triphosphate:RNA nucleotidyltransferase, and transcriptase.

Transduction: transfer of bacterial genes or genome fragments from one bacterial cell (the donor) to another (the recipient) by a bacteriophage (the vector). The transferred genetic material is carried by the phage DNA. The recipient cell becomes a merozygote. The transduced DNA either 1) remains in the cytoplasm of the recipient and is transferred to only one of the daughter cells at cell division, or 2) becomes integrated into the recipient DNA and is replicated with it.

Transferases: a large class of enzymes of great importance in the metabolism of all living cells. They catalyse the reversible transfer of a chemical group (R) between two substrates (S and S′): R—S + S′ $\rightleftharpoons$ R—S′ + S. The following classes of T. are particularly important.

1) ***C_1-transferases*** catalyse the transfer of C_1-groups, such as methyl, hydroxymethyl, formyl and carboxyl. Methyltransferases transfer methyl groups (—CH_3) to the C, N or O atoms of suitable acceptors, usually from the methyldonor, *S*-adenosylmethionine. Tetrahydrofolic acid (see Vitamins) is the coenzyme for transfer of hydroxymethyl (—CH_2OH) and formyl (—CHO) groups. Biotin (see Vitamins) is the cofactor of T. that catalyse the transfer of carboxyl (—COOH) groups.

2) ***Acyltransferases (transacylases)*** catalyse the transfer of acyl groups (R—CO—), in particular acetyl groups (CH_3CO—), from acetyl-coenzyme A to acceptors. This is an essential process in the synthesis and degradation of fatty acids, in the synthesis of bile acid conjugates, and many other metabolic reactions.

3) ***Transaminases*** (see Transamination).

4) ***Aldo-*** and ***ketotransferases,*** e.g. transketolase, catalyse reactions in carbohydrate metabolism. Transketolase catalyses the transfer of a C_2-fragment (a glycolaldehyde group) from a ketose to an aldose. Thiamin pyrophosphate (see Vitamins) is the coenzyme.

5) ***Transglycosidases (glycosyltransferases)*** transfer glycosidically bound sugar residues to the hydroxyl groups of suitable acceptors. They are paticularly important in polysaccharide biosynthesis.

6) ***Transamidases*** catalyse the transfer of the amide nitrogen of glutamine, and are necessary for several reactions of intermediary nitrogen metabolism.

7) ***Transphosphatases (phosphokinases)*** catalyse the transfer of phosphate residues to suitable acceptors. The donor is usually adenosine triphosphate (ATP), but some transphosphatases use uridine, cytidine or guanine triphosphate. Transphosphatases play an essential role in carbohydrate and nucleic acid metabolism. See Kinases.

Transferrin: see Siderophilins.

Transfer RNA, *tRNA:* the smallest known functional RNA, present in all living cells and essential for Protein biosynthesis (see), and constituting about 10% of total cell RNA. Different tRNAs contain between 70 and 85 nucleotide residues; the average M_r is 25 000. There is at least one specific tRNA in any living cell for each of the 20 proteogenic amino acids. Often, however, there is a multiplicity of tRNAs (between 50 and 70 tRNAs in one cell), due to organelle specificity, and the fact that there may be two or more different but specific tRNAs for one amino acid. Source and specificity of a tRNA are indicated by a code, e.g. $tRNA^{Val}_{yeast}$ is the valine-specific tRNA from yeast.

Two sites on the tRNA molecule are of particular importance for the specific function of tRNA in translation, i.e. the amino acid recognition region and the codon recognition region. The former determines which amino acid will be esterified at the 3′-terminus of the tRNA (this is always cytosine-cytosine-adenosine, —CCA—OH3′). The latter, contained within the anticodon loop of the folded tRNA structure, consists of three nucleotide residues, known as an anticodon triplet.

tRNA becomes esterified with its specific amino acid by the action of aminoacyl-tRNA synthetase. The resulting aminoacyl-tRNA becomes bound to the acceptor site of the 50S-subunit of a ribosome, where antiparallel base pairing occurs between the anticodon of the tRNA and the complementary codon of the associated mRNA. The specificity of this base pairing insures that the amino acid is incorporated into the correct position in the growing polypeptide chain. During translation, the deacylated tRNA is released from the ribosome and becomes available for recharging with its amino acid.

Transformation: conceptually the simplest form of genetic transfer. "Naked" DNA from a donor cell enters a recipient cell and is incorporated into the recipient DNA by genetic recombination. No other carrier substance or structure is involved; small fragments of donor DNA simply penetrate the membrane (and wall if it is present) of the recipient cell. T. was first decribed in 1944 by Avery (USA) for the T. of R (rough)-type nonpathogenic pneumococci into S (smooth)-type pathogenic pneumococci by treatment with killed S-type cells. The "transforming principle" was eventually shown to be DNA, thus providing the first proof that the genetic material of the cell is DNA. Frequency of T. may be low (< 1%) owing to rapid degradation of donor DNA before genetic recombination can occur.

Trans Golgi network: see Golgi apparatus.

Transgression: the occurrence of genotypes in an F2 generation from an F1 hybrid (see Mendelian segregation), which, with respect to certain characters, surpass their parents or the F1 generation. T. is particularly common for characters controlled by numerous genes (see Polygeny). In contrast to heterosis (hybrid vigor), T. effects can be transmitted to subsequent generations.

Transition: see Gene mutation.

Transitional load: see Genetic load.

Transitional zone: a biogeographical region whose flora and fauna are mixtures of organisms from adjoining regions, and whose actual boundaries cannot be satisfactorily defined, or defined only according to the distribution of perhaps a single class of animals. The main T.zs. are between Central America and southern North America, in the desert belt that extends from West Africa to Central Asia, and between some of the Indoaustralian islands..

Within recorded history, North Africa was home to elephants, giraffes, leopards and lions (representatives of sub-Saharan Africa) and to deer, wild boar and bears (Eurasian animals). For a long time, the position of the boundary between the Oriental region (see) and the Australian region (see) was disputed. Nowdays, the existence of an Indoaustralian transition zone (see Wallacea) is largely accepted.

The T.z. between the Nearctic (see) and Neotropical region (see) is much more difficult to interpret and it is less investigated than other T.zs. In 1887, Heilprin defined the *sonoric T.z.,* which included a large part of the southern Nearctic. In contrast, according to Schmidt (1954), the T.z. is a Caribbean subregion (Central America and Caribbean islands, derived mainly from parts of the Neotropical region), and he considered that this adjoined the Holarctic (see). Compared with the Indoaustralian transition zone, this *Caribbean T.z.* is much more extensive and contains a much richer variety of species.

Translation: the second stage in the expression of the genetic information encoded in DNA. In the wider sense, it is equivalent to protein biosynthesis. In the narower sense, T. is the decoding process whereby each codon in mRNA is translated into one of 20 amino acids during protein synthesis on polysomes. See Transcription, Genetic code, Transfer RNA, Ribosome, Protein biosynthesis.

Translocation:

1) A structural change in a chromosome (see Chromosome mutation), in which a chromosome segment is incorporated into a new position in the same chromosome, or is transferred to a different chromosome, or in which two segments are exchanged between homologous or nonhomologous chromosomes. The first type is intrachromosomal, and is also known as a *transposition* or *shift,* while the second type is interchromosomal and known as *reciprocal T.* A reciprocal T. is asymmetrical, if it results in a dicentric chromosome. Conversely, a T. is symmetrical if the chromosome remains monocentric following the reciprocal displacement of the segment. Asymmetrical T. is often lethal to the cell.

Fig.1. *Translocation.* Chromosome pairing for reciprocal translocation in the heterozygous state.

Ts. may be homozygous or heterozygous. Chromosomes carrying heterozygous Ts. display a number of cytological and genetic peculiarities. When a single chromosome is heterozygous for a reciprocal T., the changed homology results in the formation of a characteristic crossed structure containing 4 chromosomes in the prophase of the first meiotic division. Crossing over and chiasma formation, involving 3 or all 4 of the homologous segments of this structure, produce respectively a ring or chain that remains stable up to anaphase. During anaphase, depending on the orientation of the centromeres, either alternate or neighboring pairs of chromosomes within this ring or chain migrate to the same cell pole. Depending on the type of polar distribution in anaphase, two types of cell tetrad (see Tetrads) are produced by meiosis: 1) some cells contain a complete genome, and some cells contain a deleted genome with missing segments; 2) all cells contain a duplicated genome. Such gametes are often incapable of fertilization, but if fertilization occurs and the genetically unbalanced chromosome complement is not compensated by the other gamete, the resulting zygote usually fails to develop. Many translocation heterozygotes are therefore sterile.

Reciprocal Ts., like inversions, play an important part in the evolution of the karyotype.

2) See Animal migration.

3) see Transport *(2).*

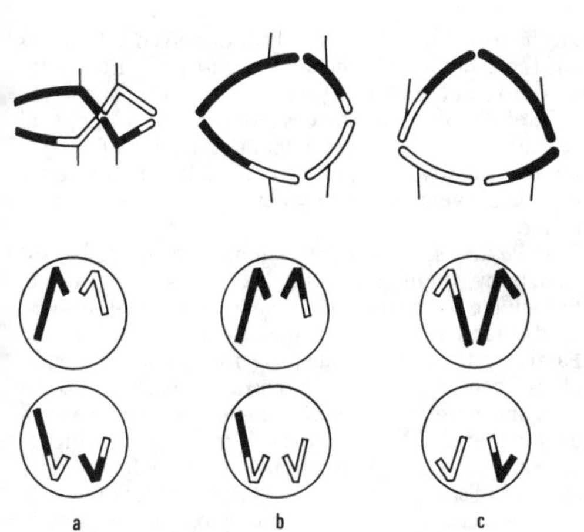

Fig.2. *Translocation.* Possible chromosome configurations during translocation in the heterozygous state, showing the resulting genotypes. *a* Functionally unimpaired gametes. *b* and *c* Unbalanced gametes which are nonfunctional or have a limited function.

Transmission: see Absorbance.

Transpiration:

1) In plants, the release of water vapor to the atmosphere, especially through the leaves. Between 90 and 95% of transpired water is lost by *stomatal T.,* i.e. it passes to the exterior through the Stomata (see), which lie mainly on the undersides of the leaves, and whose aperture size is actively regulated in response to environmental conditions. Cells in the leaf interior lose water by evaporation into the system of intercellular spaces, which communicate with the stomata. Although the total area of all the stomatal apertures (there can be several hundred per mm^2 of leaf surface) account for only 1–2% of the leaf area, the rate of T. via stomata can be as high as 50–70% of the rate of evaporation from a leaf-sized area of open water. This is attributed to an edge effect, i.e. at the edge of a stomatal aperture, exiting water molecules have a lateral field of free diffusion, whereas those in the center are surrounded on all sides by other water molecules and can only diffuse vertically. A small quantity of water (5–10%) is lost by evaporation through the cuticle. This *cuticular T.* is nonregulable; it is more rapid in plants of moist habitats (see Hygrophytes) which have a very thin cuticle, than in plants of dry habitats (see Xerophytes) which are protected by outer waxy layers and a very strongly developed cuticle.

T. is driven by the diffusion pressure of water vapor from the water-saturated atmosphere of the leaf interior into the nonsaturated atmosphere surrounding the leaf (alternatively interpreted as the suction pressure of the external air). Even at 99% relative humidity, the suction pressure of the unsaturated air is equal to that of soil at the Wilt point (see) of most agricultural plants.

Since plants have a relatively high water vapor pressure, the *T. rate* depends on the water vapor concentration gradient between the plant and the external air. All factors that increase this gradient also increase the T. rate. In this connection, it should be noted that an increase in temperature at constant air humidity decreases the relative humidity (and therefore the atmospheric vapor pressure), thereby increasing the rate of diffusion of water vapor from the plant. T. is also increased by wind (which removes the moist air layers from the leaf surface) and by a high water content of the plant (equivalent to a high water vapor pressure within the plant). T. is also increased by all factors that induce stomatal opening, such as illumination, increased temperature and high water content of the plant. In general, under normal weather conditions, the T. of higher plants displays a diurnal rhythm, consisting of a slow increase in the morning to a maximum in the late afternoon (often with a temporary decrease at midday due to stomatal closure), followed by a decrease toward evening. The cooling effect of the evaporating water helps to counteract the heating effect of solar irradiation, but more importantly, the transpiration stream provides a means of transporting nutrients from the soil to the aerial parts of the plant.

On account of the large total leaf area of most plants, considerable quantities of water are continually lost by T. and replaced by root uptake. A sunflower plant on a sunny day can lose 1 liter, and a birch tree 60–70 liters (on very hot days as much as 400 liters). If T. is suppressed by high air humidity, water is lost instead by Guttation (see).

The quantity of water that passes through the plant for each kg of synthesized dry material is known as the *T. coefficient.* For agricultural plants in temperate regions the T. coefficient is 300–900 liters, but this increases in dry years and on light soils, and is generally higher in dry climates. Good fertilization decreases both T. and the T. coefficient. Plant species with low T. coefficients (e.g. maize, millet) can still produce a good harvest with relatively little soil water.

The quantity of water lost by T. can be determined by repeated weighing of a plant, or with a Potometer (see).

2) The term is not usually applied to humans or animals, where the equivalent process is known as Perspiration (see).

Transplantation:

1) In plants, the transfer of a plant part to another plant. Cell T. (transfer of parts of unicellular plants) has been performed mainly with *Acetabularia,* to elucidate the role of the cell nucleus in cell differentiation. T. is also possible with fungal fruiting bodies. In higher plants, T. is called *grafting,* in which a plant part bearing a bud (the scion) is grafted onto a part bearing roots (the stock or rootstock). Grafting is widely used in horticulture and agriculture as an artificial means of propagation.

2) In animals and humans, the transfer and implantation of organs and tissues. T. performed within the same organism is known as autotransplantation, e.g. bone grafting and skin grafting within the same individual to repair or replace accidentally damaged tissue. In this case, the donor and recipient tissues are genetically identical, so there is no immune reaction with tissue rejection.

T. between genetically identical individuals, e.g. between monozygotic twins, or between animals (e.g. mice) of an inbred strain is called *isotransplantation.* In this case, the transplanted material does not promote an immune reaction and it is therefore not rejected.

Allotransplantation (earlier called homotransplantation) is medically important, involving T. between individuals of the same species (humans) that are not genetically identical. The transplanted material is recognized as "foreign" by the immune system of the recipient, and the resulting immune response destroys the transplant, if immunosuppressive therapy is not applied. The long-term successful T. of human organs is essentially restricted to kidney T., although some success has also been achieved with heart transplants, heart and lung transplants, and liver transplants.

For successful T., it is necessary to accurately *type* the donor and recipient. Since practically all transplanted organs originate from corpses, all potential recipients are typed (patients dependent on chronic dialysis and awaiting a kidney transplant, patients with heart disease and awaiting a heart transplant, etc.) and the results held in a computer. Typing is based mainly on the leukocyte antigen system *(HLA-system),* determined with antisera with specific binding properties for the membrane antigens. As soon as an organ becomes available, typing is performed on the donor or donor materal, and the resulting data fed to the computer. If the computer finds an appropriately matching (closely similar) recipient, the organ is transported carefully (with cooling) and quickly to the transplantation unit, where the recipient is waiting.

Since an immune reaction must be anticipated, immunosuppressive therapy is started immediately after the T. After a kidney T., immunosuppressive therapy is continued for several weeks. In the most favorable cases, this leads to development of specific nonreactivity against the transplant. Despite the need for subsequent continuous medical observation, kidney transplant recipients can usually return to a full working life with a good life expectancy. Heart or liver T. also causes an immune reaction in the recipient. These other organs, however, appear to be more sensitive to immune reactions, and moreover they tend to be damaged by immunosuppressive drugs. If a transplanted kidney is rejected, the recipient can be kept alive by dialysis, but there is no corresponding treatment for a heart or liver patient. In this case, management of the immune system and prevention of rejection is crucial.

In animal experiments, it has been possible to generate a specific immunological nonreactivity called *immunotolerance,* a condition of great theoretical and practical significance. For their fundamental experimental work on immunotolerance, the English zoologist, Medawar, and the Australian biologist, Burnet, received the Nobel prize.

T. between individuals of different species is so far only of theoretical interest, e.g. between sheep and goats, or between apes and humans. This is known as *xenotransplantation* (earlier called *heterotransplantation).* The resulting immune reactions are so strong that they are not prevented by intensive immunosuppressive therapy. T. of animal organs to humans is therefore not practicable.

Transport:

1) Passage of material through biological membranes. It is an essential precondition for the existence of living systems that cells are separated from their surroundings by membranes and are internally compartmentalized by membranes. Mechanisms must therefore exist for the selective uptake and excretion of material across membranes. For this purpose, biological membranes contain many different *T. systems.* Different mechanisms operate for the T. of dissolved substances and the T. of colloids, the latter occurring by Endocytosis (see) and Exocytosis (see).

T. of dissolved substances along an electrochemical potential gradient occurs without utilization of energy, i.e. it is *passive.* T. against the gradient of an electrochemical potential is energy-dependent and is said to be *active.*

1) *Passive T.* can occur by simple or facilitated diffusion, or by coupling with the flux of another substance. For simple diffusion, the transported material must be lipid-soluble in order to permeate the membrane lipids. Passive diffusion is inadequate for important metabolites, such as glucose and amino acids, because it is too slow and nonselective. Such metabolites are transported by facilitated diffusion (Figs 1 & 2) by specific T. systems, in which T. is promoted by conformational changes of specific T. proteins. The existence of mobile macromolecular carriers in biological membranes is now doubtful (see Carrier), but certain small molecules (e.g. macrocyclic antibiotics produced as membrane poisons by different bacteria) do operate as mobile carriers. The systems or mechanisms for the facilitated diffusion of ions are also known as pores or channels. Ion T. is influenced not only by the concentration gradient, but also by the difference in electrical potential. Of particular importance are the regulable channels for K^+, Na^+ and Ca^{2+} in excitation processes, e.g. brief permeabilty to Na^+, K^+ and Cl^- ions is induced in the motor endplate of muscle by release of acetylcholine in the synapse.

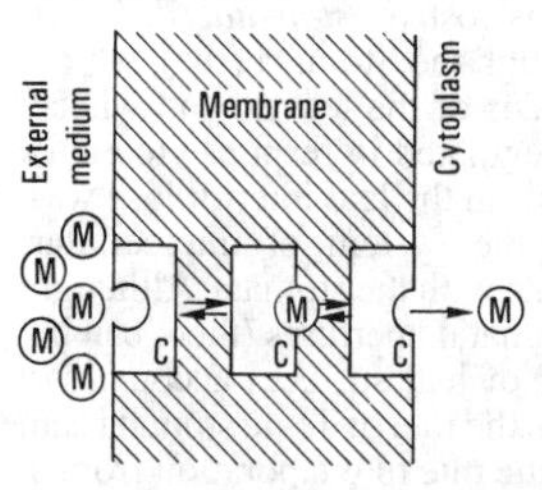

Fig.1. *Transport.* Facilitated diffusion by a carrier mechanism. Mobile carrier (C) binds metabolite (M) on the exterior of the membrane. On the cytoplasmic side of the membrane, the equilibrium is shifted in favor of the release of M.

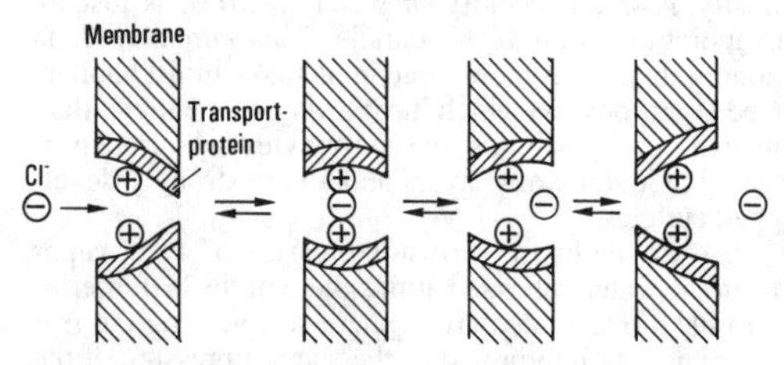

Fig.2. *Transport.* Facilitated diffusion of Cl^- in the erythrocyte membrane. Binding of Cl^- alters the conformation of the transport protein, causing Cl^- to be released on the cytoplasmic side.

In passive T. by flow coupling, the T. of one type of molecule is coupled with the passive T. of another type of molecule. One type of molecule may therefore be transported against a concentration gradient. In *sym-*

port, both molecules are transported in the same direction. In *antiport,* they flow in opposite directions. In many cases, one of the T. partners is water or Na^+. In most cases, however, coupling with Na^+ flow is a type of active T., because generation of the the necessary Na^+ concentration gradient is energy-dependent.

2) *Active T.* All known animal cells and some plant cells possess a T. mechanism which pumps Na^+ ions out of the cell and K^+ ions into the cell. The process is energy-dependent, i.e. it is linked to the cleavage of ATP into ADP and inorganic phosphate. This chemical energy is used to drive the ions across the membrane against their respective concentration gradients. The differing permeability of the membrane to the asymmetrically distributed ions results in a potential difference between the cell interior and the cell exterior (see Membrane potential). Early in evolution, this ionic equilibrium would certainly have prevented the osmotic swelling of cells. The Na^+ gradient was then probably adapted later for flow coupling *(secondary active T.,* see Fig.3). Enzymes that transport ions with simultaneous cleavage of ATP are known as ATPases. The Na^+- and K^+-transporting enzyme is therefore a K-Na-ATPase. The sarcoplasmic reticulum of muscle fibers also contains an important Ca-Mg-ATPase.

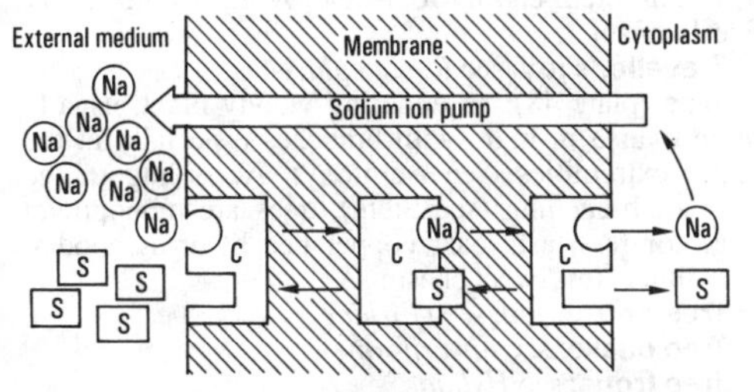

Fig.3. *Transport.* Secondary active transport of a substrate (S) by coupling to Na transport by carrier C.

The inner mitchondrial membrane contains an entirely different type of active T. system which is part of the mechanism of oxidative phosphorylation. Energy from the flow of electrons in the electron T. chain is used to pump protons from the mitchondrial matrix to the exterior. The resulting proton gradient drives the synthesis of ATP. Proton carriers (e.g. dinitrophenol) uncouple ATP synthesis from proton T. by short circuiting the proton gradient. The energy that would have been used for ATP synthesis is dissipated as heat, so that poisoning with e.g. dinitrophenol results in excessively high body temperatures. An analogous situation is found in chloroplasts where the energy for proton T. is also derived from electron T. in the light reaction of photosynthesis. In the halophytic bacterium, *Halobacterium halobium,* proton T. is coupled directly with light absorption.

See Carrier

2) Translocation of material in plants and animals. Higher plants employ two main systems of T.

1) Long distance transport of water and dissolved substances (nutrient transport) from the root into the shoot system occurs in vessels and is a function of the *transpiration stream.* On the other hand, there is no direct relationship between the rate of water uptake and the rate of transpiration, or between the rates of water uptake and ion transport. In fact, different ions are transported within the plant at different rates. Long distance T. of assimilate from synthesis sites (leaves) to utilization sites (roots, fruits, shoot tips) occurs in the *assimilate stream* in the sieve tubes of the phloem at a rate of 0.5–1 m/h. The sieve tube sap is a 10–25% solution of assimilated material, and this concentration is higher in the day than at night. It consists of about 90% carbohydrate, mainly sucrose (in some plants raffinose). In addition to sucrose there are almost always small quantities of raffinose, stachyose and verbascose, i.e. sucrose with an additional 1, 2 or 3 galactose residues; sometimes the assimilate stream also contains mannitol and sorbitol. Amino acids, nucleotides, carboxylic acids, vitamins, phytohormones and ATP are also present. A clear understanding of the forces involved in long distance transport in phloem is lacking. See Pressure flow theory.

2) Short distance T. to and from the xylem or phloem occurs through the cells of nonspecialized tissue. Three separate systems can be recognized. a) The *symplast,* i.e. the protoplasm of all cells, forms a continuum via the plasmodesmata, and transports organic and inorganic substances and water over short distances (see Mineral metabolism). b) The *apoplast* consists of the vacuoles and all the intermicellar and interfibrillar spaces of the cell wall, and it transports water and inorganic ions over short distances. c) The discontinuous system of *vacuoles* is concerned exclusively with water transport.

The long distance and short distance T. (diffusion) of gases, in particular carbon dioxide, oxygen and and water vapor, occurs within a ventilation system of intercellular spaces, which communicates with the external atmosphere via the stomatal apertures and lenticels.

In higher animals, the blood and vascular system are responsible for T.

Transport ATPase: see Carrier.

Transposase: see Transposon.

Transposon: a mobile genetic element that can move its position independently within a chromosome, and which carries genes ("jumping genes"), e.g. genes for resistance to antibiotics in bacteria. As a rule, Ts. are flanked by insertion sequences, which, together with specific enzymes *(transposases,* which are encoded by the T.) are responsible for incorporation of the T. into the DNA molecule. In addition to Ts. that move directly, eukaryotes also contain *retrotransposons,* which transcribe RNA; this RNA is in turn transcribed by reverse transcriptase, and the resulting DNA is inserted at a new site. Ts. can give rise to mutations (e.g. deletions and translocations) at the site of their incorporation.

Small bacterial Ts., containing only the genes necessary for their transfer, are called *Insertional sequences* (see).

Trans-saprophytic conditions: see Saprophytic system.

Trans-specific evolution: see Macroevolution.

Transverse process: see Axial skeleton.

Transverse tropism: see Tropism.

Transversion: see Gene mutation.

Trapa: see *Onagraceae.*

Trapaceae, *Hydrocaryaceae,* ***water-nut family:*** a family of dicotyledonous plants, containing 3 species in a single genus, distributed in Europe and Asia. They are floating plants with alternate leaves and small stipules. Flowers are hermaphrodite and 4-merous, the fruits are formed underwater. The fruit is a drupe-like nut, functionally an anchor fruit, with sharp, hooked thorns derived from the calyx. The floating aquatic *Trapa natans* is native to Central Europe, but it is becoming rarer in its natural habitat of relatively warm, standing water. The family is interesting because it represents a transition between the orders *Myrtales* and *Haloragales.*

Trapezoid body: see Brain.

Trap layers: see Modes of nutrition.

Trapping and collection methods: in animal ecology, methods for the quantitative and qualitative recording of animals in their habitats and natural ranges. The design and rationale of these methods depends on the animal or animals of interest. Thus, various insects can be attracted at night with ultraviolet light, while others are attracted to trapping sites with baits such as fermenting beer, fruit, sugar solution, carrion, or species-specific pheromones. At the capture site they may simply be gathered, or they may be retained in special traps. Important collecting devices for the fauna of vegetation layers and tree canopies are beating trays (canvas stretched on a shallow frame like an inverted umbrella) to collect falling insects dislodged by beating tree branches and bushes, and sweeping nets or bags (fairly strong material on a rigid, heavy frame), which are swept among grasses and soft vegetation. Organisms of the ground litter are captured with a Barber trap (see Soil organisms), a pitfall trap (a container sunk into the ground), or by sieving topsoil and litter. Simple, unbaited tentlike structures exploit the instinct of insects to continually climb the walls rather than escaping by the lower opening. Forceps or an aspirator may be used to collect insects from any surface, such as trees, walls and the soil. Special methods have been developed for the collection of Soil organisms (see). Flying insects are also caught with the aid of light, cotton mesh nets, window traps and adhesive surfaces.

Birds are caught with various types of traps (clap net, Chardonneret trap, Japanese mist net, pigeon cage trap with bob wires, top entry funnel for rooks, ground entry trap for sparrows). Many stationary trapping devices are based on the Reusen principle, which is used primarily for catching fish. Small mammals are caught in various types of live traps and killing traps. Animals caught in live traps can be marked and released; their recapture can provide information on their habits, distribution, population density, size of territory, migration routes,etc. The use of color marking, plastic tags, small radiotransmitters, marking with radioactive materials and ringing (especially birds) are all customary methods.

Large wild animals are captured live for tagging or attachment of radiotransmitters by shooting with anesthetic darts, which inject an appropriate dose of general anesthetic.

For the quantification, the number of captured animals is expressed in relation to the surface area or volume of the habitat (stationary density), or as the number of animals that pass a boundary in a given time (see Activity density). The correct choice and application (number and distribution of collecting sites) of a collecting method is influenced by the Minimal area (see). See Lincoln index.

Traube's cell: a simple osmotic system (see Osmosis) consisting of a semipermeable membrane of copper hexacyanoferrate (II), which is formed when a copper sulfate crystal is immersed in a 5% solution of potassium hexacyanoferrate (II). The concentrated $CuSO_4$ solution within the membrane takes up water by osmosis. Consequently the membrane is stretched and then bursts. A new membrane is then formed outside the ruptured first membrane, etc., so that the T.c. is continually enlarged.

Traumatochorism: the active rejection of plant organs, especially flower parts, in response to wounding.

Traumatodinesis: see Dinesis.

Traumatonasty: a Nasty (see) induced by wounding. Collapse of the leaflets of mimosa is normally a seismonastic response (see Seismonasty), but it may also occur in response to cell damage, in which case it is a T.

Traumatotaxis: taxis as a result of wounding. In plant cells, the nucleus often displays traumatotactic migration. If the nucleus is in the immediate vicinity of the cell damage, it moves away, i.e. it shows *negative T.,* but if it is far removed from the site of injury, it tends to move toward it, i.e. it displays *positive T.* Such movements of the nucleus depend on cytoplasmic streaming.

Traumatotropism: growth curvature of plant organs caused by wounding. Curvature may occur toward the damage (positive T.) or away from it (negative T.). T. is exhibited, e.g. by roots and by grass coleoptiles. See Tropism.

Traveller's joy: see *Ranunculaceae.*

Tree (plate 48): an enduring woody plant, usually more than 3 m high, with branches especially at the upper extremities (see Acrotony). According to the form of branching (see Stem), trees are monopodial (e.g. conifers, ash, oak, copper beech) or sympodial (e.g. lime, elm, sweet chestnut).

Tree dormouse: see *Gliridae.*

Tree ducks: see *Anseriformes.*

Tree frogs: see *Hylidae.*

Tree hopper: see *Homoptera.*

Tree hunters: see *Furnariidae.*

Tree limit: see Forest limit.

Tree mosses: see *Musci.*

Tree-nesting birds: birds that nest in tree trunks and on tree branches. See Cavity-nesting birds.

Tree runners: see *Furnariidae, Sittidae.*

Tree shrews: see *Tupaiidae.*

Tree snakes, *Boiga:* a genus of large, arboreal Boigine vipers (see) found in South and Southeast Asia. They are mildly venomous, and they live mainly on birds and tree lizards. The ***Mangrove snake*** *(Boiga dendrophila)* measures up to 2.5 m in length, and is colored blue-black with bright yellow transverse bands.

Trees of the desert: see *Chenopodiaceae.*

Tree squirrels: see *Sciuridae.*

Trehalose: a nonreducing disaccharide consisting of two glucopyranoside residues. There are three forms of T., depending on the nature of the glycosidic linkage: $\alpha\alpha$-T., $\alpha\beta$-T. (neotrehalose), or $\beta\beta$-T. (isotrehalose). T. is present in fungi, bacteria and algae, and it occurs sporadically in nonphotosynthetic tissues of higher plants. It is the "blood sugar" of insects.

Trematoda, ***flukes:*** a class of flatworms (phylum *Platyhelminthes),* containing more than 6 000 known species, which are always parasitic in the adult stage, living on, or usually in, vertebrates.

Morphology. Lengths range from less than 1 mm to 7 m, but most species are a few centimeters. The more or less flattened, oval to lanceolate body always lacks visible segmentation; cylindrical and threadlike species are also known. In the order *Monogenea,* the anterior end carries a usually paired attachment organ (glands or suckers), and the posterior end carries a large attachment disk, which is often armed with hooklets. Members of the order *Digenea* have an anterior oral sucker and a ventral or posterior suction disk. In the *Aspidocotylea,* the entire or greater part of the ventral surface is covered by a large attachment disk, which may be divided into a number of areolae by longitudinal and transverse ridges. Most flukes are practically colorless, but a few species are somewhat yellow, red or brown. There is a thick cuticle, which often carries spines or hooks, especially at the anterior end. Many members of the *Monogenea* and some free-living larvae of the *Digenea* possess between one and four pairs of simple eyes (pigment spots or cup-shaped pigment cells). The anterior mouth lies at the base of an oral suction disk or in a funnel-shaped depression; it leads into a muscular pharynx, then into a blind alimentary canal, which divides into two lateral branches. As a rule, flukes are hermaphrodite, with the exception of e.g. Schistosomes (see). The hermaphrodite sexual apparatus consists of one or two testes (many testes in the *Monogenea)* and always a single ovary, as well as accessory yolk glands and a shell gland.

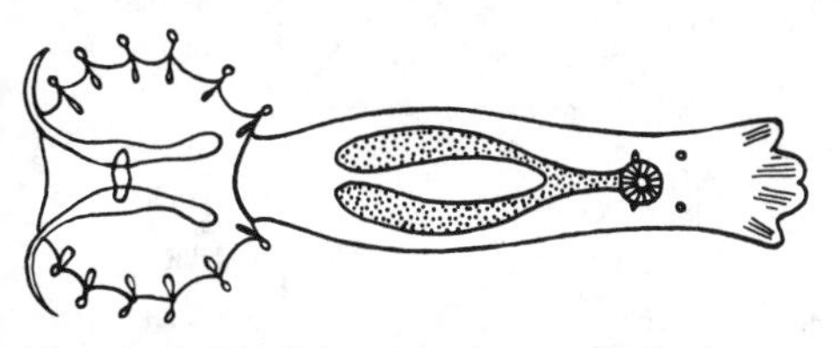

Trematoda. The fluke, *Dactylogyrus* (order *Monogenes).*

Biology. All adult flukes are parasitic. Most members of the *Monogena* live in the skin or gills of fishes, amphibians, reptiles, crustaceans or cephalopods, and more rarely in the nasal and pharyngeal cavities and urinary bladder of fishes, amphibians and reptiles. The *Aspidocotylea* occur primarily in various internal organs of fishes, turtles, gastropods or bivalves. The reproductive generation of the *Digenea* lives in the internal organs (intestine, liver, lungś or blood vessels) of all classes of vertebrates, whereas the nonsexual generation parasitizes gastropods and sometimes bivalves. Flukes feed on tissue fragments, blood, lymph or intestinal juices of their host.

Development. In most of the *Monogenea* and in the *Aspidocotylea,* development is direct, i.e. young flukes, resembling miniature adults, hatch from the eggs, then immediately attach themselves to, or penetrate, a host animal. Members of the *Digenea* pass through different stages of development, and at least one stage of their life history is spent in an invertebrate host, almost invariably a mollusk. This alternation of hosts is decribed in more detail for the Liver fluke (see). See also Schistosomes.

Economic importance. Many flukes are serious parasites of humans, as well as domestic and exploited animals. Some fluke infections are responsible for considerable health problems and economic losses, e.g. the liver fluke *(Fasciola hepatica)* is responsible for liver rot in ruminants, and the dreaded tropical and subtropical human disease, Bilharzia, is caused by species of *Schistosoma.* Entire broods of fish may be lost or become severely depleted by blood loss, due to gill or skin infections by species of *Dactylogyrus* and *Diplozoon.*

Classification. Based on the position and structure of the adhesion organs, as well as the mode of development of the fluke, three orders are recognized:

1. *Monogenea* (ectoparasitic flukes), with, e.g. the genera *Dactylogyrus* and *Diplozoon.*
2. *Digenea* (endoparasitic flukes), with e.g. the greater and lesser liver flukes *(Fasciola),* the schistosomes *(Schistosoma).*
3. *Aspidocotylea* (digenic in general anatomy, but monogenic in development).

Tremblers: see *Mimidae.*

Tremellales: see *Basidiomycetes.*

Trepang: a high-value, easily digested food, containing 50–60% protein. It is prepared from the muscular body wall of holothurians (sea cucumbers). The animal is cut open, and the intestine and other viscera removed. The remaining flesh is boiled, dried and smoked. T. is eaten, particularly in soups, mainly in East Asia and the Mediterranean region. The animal itself is also known as T. or bêche-de-mer.

Treponema: see Spirochetes.

Treroninae: see *Columbiformes.*

Tiacanthidae: see *Tetraodontiformes.*

Triacanthodidae: see *Tetraodontiformes.*

Triacylglycerols: see Fats.

Triakinae: see *Selachimorpha.*

Triassic (Greek *trias* trinity): the oldest system of the Mesozoic (see Geological time scale). The type T. region is central Germany, where two continental sequences are separated by a marine complex; these three (hence the name) sequences are known as the Buntsandstein (sometimes called the "bunter" in English), the Muschelkalk, and the Keuper (the Keuper is the youngest, the bunter the oldest). In the pelagic marine Alpine-Mediterranean region, the T. is represented by a nearly complete sequence of six geosynclinal strata of complex facies relations, consisting of highly fossiliferous limestone and other deposits, with abundant ammonoids, which form the main basis of the T. correlation. These six Alpine-Mediterranean stages are taken as the standard for the correlation of marine T. deposits. From the upper to the lower, they are named as follows: Rhaetian, Norian, Carnian, Ladinian, Anisian, Scythian (Upper T. = Rhaetian + Norian + Carnian; Middle T. = Ladinian + Anisian; Lower T. = Scythian). In the northeastern Alps they correspond respectively to the Kossen formation, Haupt Dolomite, Raibl formation, Wetterstein Dolomite, Dolomites, Werfen formation. T. formations vary greatly throughout the world, several stages often being absent.

The T. lasted 30 million years. Its most important index fossils are Ammonites (see). These became very scarce at the beginning and end of the T., showing a peak in the middle of the system. Ammonites of the early T. tend to be smooth-shelled, but most genera of the middle and late T. are highly ornamented with lines, ridges, nodes, or spines. The suture is ceratitic or ammonite. Some dominant forms display a partial or complete phylloceratid suture (a leaflike pattern in the minor denticulations of the lobes and saddles). The ribbing of ammonites becomes more complex with the advance of the T. On account of this great variety of ornamentation and the relatively short periods of existence of many species, the ammonites are very useful markers of stratigraphic zones and geological time. Distinctive bivalves also contribute to strata classification, the anisomyarians *(Pectinacea, Limacea, Mytilacea)* being more abundant than the heteromyarians *(Unionidae, Trigonidae, Megalodontidae).* Gastropods also attained considerable diversity and ornamentation in the T. Foraminifera, sponges, corals, brachiopods, bryozoans, echinoderms and fishes are all present and sometimes numerous, but have little stratigraphic importance.

Mesozoic amphibians were very numerous in the T. Heavily built stegocephalians, known as labyrinthodonts, did not persist beyond the T. Stegocephalian remains found in the Upper T. of the Germanic basin belonged to a species, which has been named *Mastodonsaurus giganteus;* it attained a length of more than 1 meter, and is the largest known amphibian since their first appearance to the present. Reptiles show wide diversity in the T., which marks the beginning of their great Mesozoic expansion. Crocodile-like aquatic reptiles from the Western American T., known as *phytosaurs,* inhabited sluggish streams and shallow ponds, and were notable for their very long snout. Dinosaurs were abundant, and displayed a wide range of sizes. Of considerable phylogenetic importance is the first appearance of primitive mammals in the Upper T., i.e. the *Eozostrodontidae* of Europe, China and Africa, and the *Kuehneotheriidae* from Wales.

The gymnosperms, which appeared in the Permian, evolved further in the T. Where tropical lands rose above the sea, there were extensive forests, predominantly of conifers closely related to *Auracaria,* which now extends across parts of the Southern Hemisphere. Dramatic remains are the great agatized tree trunks of the petrified forest of Arizona.

Tricarboxylic acid cycle, ***TCA-cycle, citric acid cycle, Krebs cycle:*** the most important route for the terminal oxidation of acetyl-CoA (active acetate, see Coenzyme A), which arises from the degradation of carbohydrates, fatty acids and certain amino acids. Inspection of the reactions of the TCA-cycle (Fig.) shows that CO_2 is produced by decarboxylation of oxoacids (ketoacids), i.e. oxalosuccinate and 2-oxoglutarate. Hydrogen and electrons from the dehydrogenation of substrates are oxidized via the respiratory chain, and this oxidation is coupled to the synthesis of Adenosine 5′-triphosphate (see). Intermediates of the TCA-cycle can also serve as intermediates in the biosynthesis of cell constituents (e.g. 2-oxoglutarate and oxaloacetate may be transaminated to glutamate and aspartate, respectively; succinyl-CoA, an intermediate in the conversion of 2-oxoglutarate to succinate, is a heme precursor). The TCA-cycle was discovered almost simultaneously in 1937 by Krebs, and by Martius and Knoop. In eukaryotes, the cycle operates in the mitochondria, where it is structurally and functionally integrated with the respiratory chain and with the degradation of fatty acids; the enzymes are present in the inner mitochondrial matrix, loosely associated with the inner mitochondrial membrane, or forming an integral part of it (e.g. succinate dehydrogenase). In prokaryotes, the enzymes of the TCA-cycle are localized in the cell cytoplasm.

Acetyl-CoA enters the TCA-cycle by reacting with oxaloacetate to form citrate. Through the action of aconitase, citrate is in reversible equilibrium with isocitrate via the intermediate, *cis*-aconitate. Isocitrate dehydrogenase catalyses the dehydrogenation of isocitrate to oxalosuccinate, which is then decarboxylated to 2-oxoglutarate. By decarboxylation and simultaneous dehydrogenation (i.e. oxidative decarboxylation), 2-oxoglutarate is converted irreversibly to succinate. The latter is dehydrogenated to fumarate by the action of succinate dehydrogenase. Fumarase catalyses addition of water to fumarate to form maleate, which is then oxidized to oxaloacetate by the action of malate dehydrogenase in the final step of the cycle.

In each turn of the cycle there are four reactions involving dehydrogenation. The respiratory chain therefore operates four times. In the respiratory chain, hydrogen from the reduced coenzyme reacts with oxygen to form water in a stepwise series of reactions. In three of the dehydrogenations, the cofactor is NAD, so that the respiratory chain oxidizes NADH + H^+ (oxidation of one molecule of NADH + H^+ by the respiratory chain is capable of providing energy for formation of three molecules of ATP, i.e. the P/O ratio is 3). One dehydrogenation produces $FADH_2$ (oxidation of one molecule of

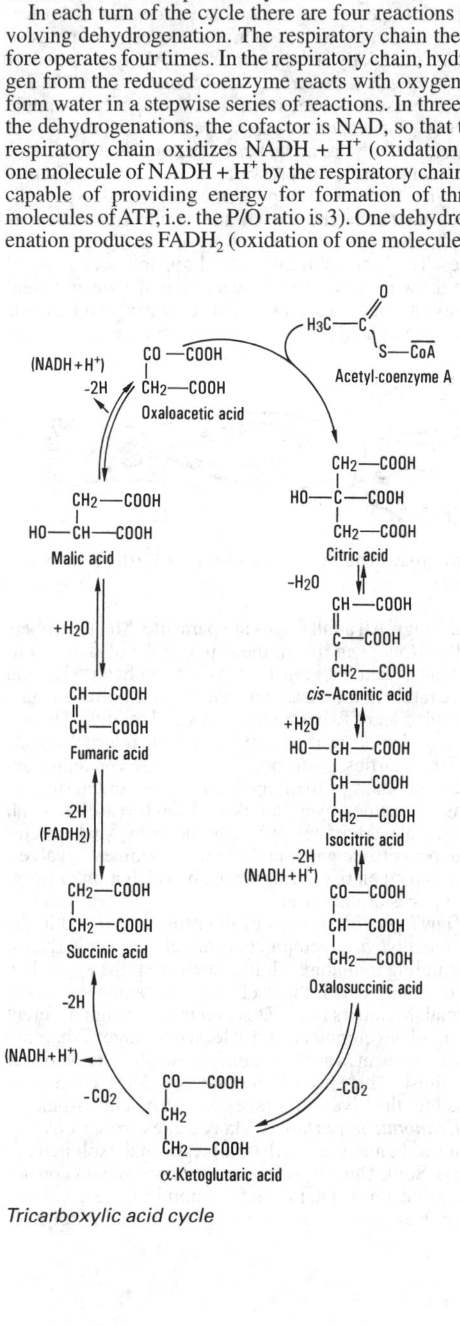

Tricarboxylic acid cycle

$FADH_2$ by the respiratory chain is capable of providing energy for the formation of two molecules of ATP, i.e. the P/O ratio is 2). The full potential for ATP synthesis in the respiratory chain may, however, not be realized. Some of the available energy may instead be released directly as heat, which contributes to the maintenance of body temperature in warm-blooded animals. See Respiratory chain, Oxidative phosphorylation.

Triceps muscle: see Musculature.

Triceratops (Greek *treis* three, *keras* horn, *ops* face): a genus of plant-eating dinosaurs belonging to the order, *Ornithischia.* They were 5–8 m long and 2.5–3 m high, with a beak-like upper jaw. Fore- and hindlimbs were of approximately equal length, and they progressed on all four feet. The robust skull was more than 2 m in length and armed with 3 horns; it was also furnished with a bony shield that protected the neck and nape. Fossils are found in the Upper Cretaceous of North America.

Trichdroma: see *Sittidae.*

Trichechus: see *Sirenia.*

Trichinella: see Trichinosis; *Nematoda.*

Trichinosis: a disease of caused by the parasitic nematode, *Trichinella spiralis,* which has a cosmopolitan distribution. The infection is transmitted by the ingestion of infected ("trichinized") meat, and it is most commonly acquired by humans through eating undercooked, infected pork. *Trichinella* is primarily a rat parasite, and rats acquire it by eating one another, but it is also found in other carnivorous animals, including humans, pigs, dogs, cats, foxes and badgers. After fertilizing the female, the adult male (1.4–1.6 mm) dies. Fertilized adult female worms (3–4 mm in length) burrow deeply into the lymph vessels in the wall of the small intestine. The eggs hatch within the female, and the larvae are released into the lymphatic system. Via the lymphatics and the blood vessels, they migrate to all parts of the body and finally lodge in voluntary muscle, especially in the diaphragm, chest, tongue, throat and eyes. Here they form calcified cysts, which remain viable for at least 10 years. If the infection is high, it may cause pain and stiffness, as well as circulatory disturbances. Symptoms of the intestinal infection are nausea, fever and diarrhea. Meat is now routinely inspected for *Trichinella* cysts, and the incidence of trichinosis has decreased markedly.

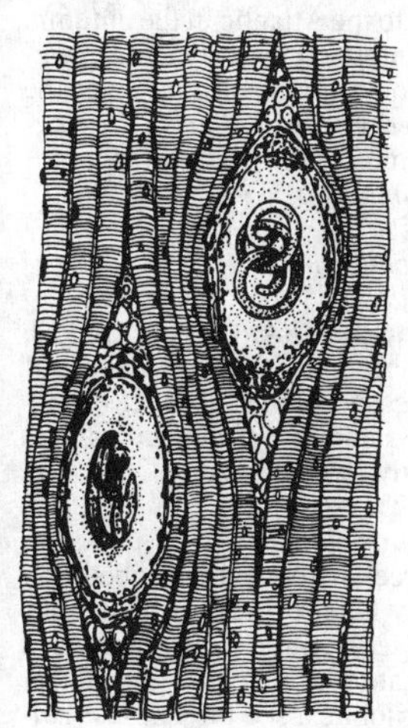

Trichinella spiralis encysted in muscle

Trichiuridae: see *Scombroidei.*

Trichobranchiate gills: see Shrimps.

Trichochromes: see Melanins.

Trichocysts: organelles analogous to nematocysts, found in the pellicle of the *Ciliatea* and some *Flagellata.* Some T. discharge a barbed, threadlike shaft through a pore. Others are filled with fluid and discharge mucus and toxins. Their function is not completely understood, but they serve partly for defense and attack.

Trichodurus: see Eelworms.

Trichogaster: see Gouramies.

Trichogastrinae: see Gouramies.

Trichoglossus: see *Psittaciformes.*

Trichogyne: see *Ascomycetes.*

Trichoid sensilla: see Olfactory organ.

Trichomys: see *Echimyidae.*

Trichome:

1) A threadlike association or chain of vegetative cells, formed by the *Hormogonales (Cyanobacteriaceae)* and certain bacteria. The cells communicate via pores. The Ts. of cyanobacteria are often surrounded by a slimy sheath.

2) A plant hair.

Trichome hydathode: see Excretory tissues of plants.

Trichoniscidae: a family of the *Isopoda* (see).

Trichoptera, ***Phryganoidea,*** ***caddis flies:*** an insect order containing about 7 000 species, distributed worldwide, but mainly in the Northern Hemisphere. The earliest fossil *T.* are found in the Lower Jurassic.

Imagoes. Adult flies are between 1.5 and 40 mm in length; the wingspan of larger species may exceed 60 mm. Fore- and hindwings are coupled in flight, and the adult insect has a mothlike appearance. In some males the compound eyes are very large and almost meet at the vertex; otherwise the compound eyes are usually small. Ocelli are either 3 in number or absent. Mouthparts are reduced, with small maxillae closely associated with the labium, and vestigial mandibles. Labrum and labium are elongated into a proboscis-like rostrum, and the hypopharynx is usually modified for sucking up plant juices. Antennae are very long and mainly setaceous or filiform. The 2 pairs of dull gray, brown or yellowish wings are covered with long hairs, or rarely with scale-like structures. Ten abdominal segments are visible. *T.* live in the vicinity of water, and they are usually found at rest during the day on waterside plants, with wings held rooflike over the body. Metamorphosis is complete (holometabolous development).

Eggs. Eggs are deposited below or above the water level, glued to stones or vegetation. In some species the eggs are embedded in a gelatinous mass, forming a spawn, whose shape (tubular, circular, spiral, or spherical) is species-specific.

Larvae. With few exceptions, the larvae are aquatic in fresh water. A few species are found in brackish or salt water; the larva of *Enoicyla* lives in damp moss. The head is well sclerotized with 2 pairs of very short antennae, mandibulate mouthparts, and 2 lateral clusters of ocelli. There are 2 types of larvae. 1. Eruciform larvae: these resemble lepidopterous caterpillars; the head is inclined at a marked angle to the body, which is cylindrical in shape; papillae are present on the first abdominal segment; a lateral line and tracheal gills are present. 2. Campodeiform (campodeoid) larvae: the head is not inclined at an angle; the body is compressed; the lateral line and abdominal papillae are absent; tracheal gills are usually absent. The majority of species are of the eruciform type, living in tessellated tubes, which are spun from silk with the addition of fragments of vegetation, small snail shells, stones and similar materials. These encased larvae feed mainly on fresh or rotting vegetation and especially algae. Most of the campodeiform species do not build a case, but a silken retreat, which may be used by several larvae; they are mostly carnivorous, feeding on small water animals, which they trap in a silken underwater web, or actively capture by hunting. All *T.* larvae are a popular fishing bait, and the name "caddis" was originally an angling term applied only to the larvae.

Pupae. These are decticous and exarate (pupa libera). Larvae pupate inside their case, which they first anchor to the substrate with silk threads, before closing off each end with a silk wall. Non-case making forms spin a silk cocoon, also attached to an underwater object. Before emergence of the mature adult, the pharate adult leaves the housing and swims to the surface or crawls onto the bank.

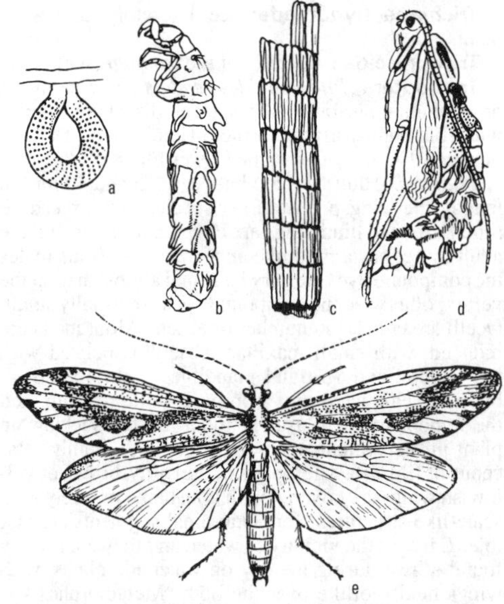

Trichoptera. Stages in the development of caddis flies (family *Phryganeidae). a* Eggs in a suspended gelatinous mass. *b* Larva. *c* Larval case. *d* Pupa. *e* Imago. *(a, b, c, e Phryganea grandis* L.; *d Neuronia ruficrus* Scop.).

Classification.

1. Suborder: *Annulipalpia* ***(Web-making T.).*** Larvae campodeiform; females have distinct cerci; larval anal hooks are long, slender and curved; adults lack a superatentorium; larvae live in silken, nonportable retreats; tracheal gills lacking, but anal blood gills are often present; some species are carnivorous and build silken snares; some species swim freely and forage for prey between stones.

2. Suborder: *Integripalpia.* Both free-living and case-building forms; adults have a supratentorium; females lack cerci; larval anal are hooks short and stout; the suborder includes both eruciform and campodeiform larvae.

1. Superfamily: *Rhyacophiloidea.* This superfamily contains 3 families, all with campodeiform larvae, which are either free-living or build simple cases.

2. Superfamily: *Lymnephiloidea.* This superfamily contains the most specialized of the *T.* All species have eruciform larvae, which build relatively elaborate cases; these may be straight, curved or helical, and they are enlarged by the larva as it grows.

Trichostomata: see *Ciliophora.*

Trichosurus: see *Phalangeridae.*

Trichurida: see *Nematoda.*

Trichuris: see *Nematoda.*

Trichys: see *Hystricidae.*

Trickling filter: see Biological waste water treatment.

Tricladida: see Planarians, *Turbellaria.*

Tricuspid valve: see Heart valves.

Tridacna: see *Bivalvia.*

Trifacial nerve: the V cranial nerve; see Cranial nerves.

Trifolium: see *Papilionaceae.*

Trigeminal nerve: the V cranial nerve; see Cranial nerves.

Triggerfishes: see *Balistidae.*

Triglidae, ***gurnards, searobins:*** a marine family of bottom-dwelling fishes of the order *Scorpaeniformes* (see). About 85 species are recognized, distributed worldwide in temperate and tropical seas. They have a large, armored head, with a broad, forward-projecting mouth. The slender body is laterally compressed, and covered with small scales. The lowest 2 or 3 rays of the pectoral fins are free from the fin membrane and can be moved independently, serving for the detection of food and for locomotion on the sea bottom. Otherwise, the pectoral fins are particularly well developed and prominent; their surfaces reflect an entire spectrum of color, and their propulsive movement resembles a wing beat. Two subfamilies are recognized: *Triglinae* and *Peristediinae,* and some authorities regard them as separate families. The ***Sapphirine gunard*** *(Chelidonichthys lucernus,* subfamily *Triglinae)* occurs from South Africa to the Lofoten Islands, in the Mediterranean and in the Black Sea, migrating far into the Atlantic in the autumn. The *Peristediinae* occur almost exclusively in deep waters, and they have spiny outgrowths of bone over the entire body, e.g. the ***Armed gurnard*** *(Peristedion cataphractus),* which is common in the Mediterranean.

Triglochin: see *Juncaginaceae.*

Triglycerides: see Fats.

Trigonoceps: see *Falconiformes.*

Trigonoida: see *Bivalvia.*

2,3,5-Triiodobenzoic acid, ***TIBA:*** $C_6H_2I_3COOH$, a synthetic auxin which interferes with the transport of natural auxin, gibberellins and other subsances, possibly by competing for the same carrier molecules. It is sometimes used as a selective weedkiller.

Triiodothyronine: see Thyroxin.

Trilling frog: see *Myobatrachidae.*

Trilobita, ***trilobites*** (Greek *trilobos* three-lobed): an extinct class of exclusively Paleozoic arthropods, subphylum *Trilobitomorpha* (see), with a thick exoskelton of chitin, calcium carbonate and calcium phosphate. For diagrams of typical trilobites, see the entries *Asa-*

phus and *Paradoxides.* The body was divided transversely into a head *(cephalon),* segmented trunk *(thorax)* and a tail shield *(pygidium).* On the dorsal surface, the entire length of the organism was divided by two longitudinal furrows, giving a 3-lobed appearance; the median or axial lobe was flanked on either side by lateral lobes. The raised median lobe of the cephalon is known as the *glabella,* flanked on either side by the *cheeks.* Each cheek carried a crescent-shaped to disk-shaped compound eye, which was occasionally stalked. Eyeless forms are also known. A *facial suture* divided the cheeks into an outer area (the *free cheeks)* and inner area (the *fixed cheeks).* When eyes were present, they lay along the facial suture on the free cheeks. The posteriolateral angles of the free cheeks often extended backward as spines. The thorax consisted of 2–42 imbricating segments, which are thought to have been united during life by a thin flexible integument. Each segment was a single rigid plate, although it was differentiated into three regions. The central region of the thorax is known as the *rachis,* while the two flanking regions are the *pleural lobes,* the latter often being extended as spines along the sides of the thorax. The pygidium was a single trilobate shield, movably articulated with the thorax, and composed of a variable number (2–29) of fused segments. In life it covered the abdominal part of the body, and in some species it was prolonged into a *caudal spine.*

Ventrally, there were 5 pairs of appendages on the cephalon and one pair of appendages on each thoracic and pygidial segment. All appendages were biramous, consisting of an endopodite and an exopodite, which arose from the outer extremity of a large basal coxopodite. The endopodite had 6 segments, the outer bearing 3 bristles; in life it was probably used for walking. The exopodite was a flattened shaft with closely spaced bristles along its posterior margin; in life it may have served the dual function of swimming and respiratory gas exchange.

Most trilobites were 50–75 mm in length but some were less than 10 mm, while others were in excess of 50 cm. They were exclusively marine, inhabiting shallow coastal areas, creeping on the sea floor or burrowing in the mud.

The oldest trilobite fossils occur in the Lower Cambrian. Peak development occurred in the Cambrian and Ordovician, and they became rare after the Devonian. Only a few generalized genera are found in the Carboniferous, and extinction occurred in the Permian. They had a wordwide distribution, and many are important index fossils, especially of the Permian. Local formations with particularly well preserved remains are e.g. Bergess shale, Lodi siltstone, Utica shale and Niagaran limestone.

Trilobites: see *Trilobita.*

Trilobitomorpha: the most primitive arthropod subphylum, comprising the major class of the *Trilobita* (see), together with the minor fossil classes of the *Merostomoidea, Marrellomorpha, Pseudocrustacea* and *Arthropleurida.* The T. first appeared in the early Cambrian and became extinct in the Permian, showing their greatest diversity and widest distribution in the Cambrian and Ordovician. More than 4 000 fossil species have been described, and these are placed in about 1 300 genera. They differ from all modern living arthropods by the uniformity of their appendages. With the exception of a single pair of uniramous, multiarticulate antennae, all other appendages were biramous and similar to each other, i.e. there were no structurally specialized mouthparts, mandibles, grasping organs, respiratory appendages or swimming appendages. This uniformity of limb structure, and the complete segmentation of the trunk are considered to be primitive characters.

Trimeresurus: see Lance-head snakes.

Trimmatom: see *Gobiidae.*

Tringa: see *Scolopacidae.*

Tringinae: see *Scolopacidae.*

Trinucleus: an important genus of trilobites occurring from the Ordovician to the Middle Silurian. The cephalon has a broad, regularly pitted border. The glabella widens slightly toward the front, and is divided longitudinally into 3 smooth, bulbous lobes. The free cheeks are almost entirely ventral (the facial suture is marginal), and they carry very long genal spines. Eyes are absent in the adult. There are between five and seven thoracic segments, and the triangular pygidium, which shows clear division into 3 segments, is smaller than the cephalon.

Triodontidae: see *Tetraodontiformes.*

Trionychidae, ***softshell turtles:*** a family of the order *Chelonia* (see), containing 25 species in 7 genera, inhabiting fresh water in America, Africa and Asia; the Giant softshell also commonly enters estuaries and is occasionally observed at sea. Horny scutes are absent, and the bony carapace and plastron are much reduced, very flat, and covered with a thick leathery skin. Plastron and carapace are connected with ligaments or cartilage rather than bone. The snout is extended into a short trunk, the limbs are modified as paddles, and the fingers and toes are webbed. Most species leave the water only to lay eggs.

The commonest representative is *Trionyx sinensis,* found from Indochina to Mongolia. *Trionyx spiniferans* ***(Eastern spiny softshell)*** is found in the Mississippi basin. *Trionyx triunguis* ***(African softshell turtle),*** which has only 3 claws on each limb, occurs in the swamps of equatorial Africa, ranging northward via Egypt to Syria. They are skilfull swimmers and live by predation, mainly on fish. All softshell turtles will readily bite humans, if they are handled.

Some members of the family posses a pair of fleshy cutaneous flaps on the rear of the plastron, which cover the hindlimbs when withdrawn; these species are generally known as ***Flapshells,*** and they are placed in a separate subfamily, the *Cyclanorbinae.*

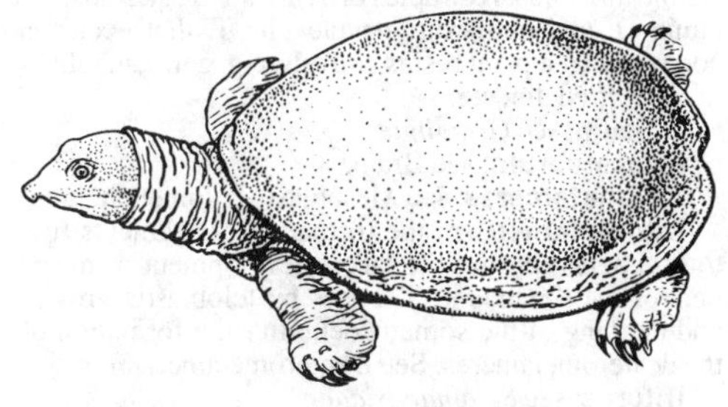

Trionychidae. African softshell turtle *(Trionyx triunguis).*

Trionyx: see *Trionychidae.*

Triops: see *Branchiopoda.*

Triose: the simplest type of monosaccharide. By definition, all Ts.contain a chain of 3 carbon atoms. Important examples are D-glyeraldehyde and dihydroxyacetone. Phosphorylated Ts., such as D-glyceraldehyde 3-phosphate and dihydroxyacetone phosphate, occur as intermediates of carbohydrate metabolism, e.g. in photosynthesis, alcoholic fermentation and glycolysis.

Triphosphopyridine nucleotide: see Nicotinamide-adenine-dinucleotide phosphate.

Triplespines: see *Tetraodontiformes.*

Triplocaulous: an adjective applied to plants that produce flowers on axes of the second order, i.e. on the second lateral branches, and are unable to form reproductive organs on the main axis, or on the first lateral branches. For example, the main axis of *Plantago major* (Great plantain) forms a rosette of leaves; the first lateral branches carry bracts, while the second order branches arise from the axils of the bracts and terminate in flowers. The first order branches are therefore the axes of the inflorescence. See Diplocaulous, Haplocaulous.

Triploidy: a form of polyploidy in which cells, tissues or individual organisms possess 3 sets of chromosomes, i.e. they are triploid.

Tripton: see Detritus.

Trisaccharide: an oligosaccharide consisting of three glycosidically linked monosaccharides. Ts. do not occur widely in living organisms. An important representative is Raffinose (see).

Trisetum: see *Graminae.*

Trisomy: a form of aneuploidy in which otherwise diploid cells, tissues, or individual organisms possess extra chromosomes homologous with existing pairs. If one chromosome pair has an extra homolog, the chromosomal complement can be represented as $2n + 1$. If two chromosome pairs each have an extra homolog, the chromosomal complement is $2n + 1 + 1$, etc. Trisomy arises from nondisjunction. See Aneuploidy, Polysomy.

Tristeza virus: see Virus groups.

Tritanopy: see Color blindness.

Triterpenes: an extensive group of terpenes biosynthesized from six isoprene units. Apart from Squalene (see), which is acyclic, most of this group are tetra- or pentacyclic hydroaromatic compounds, based on the parent hydrocarbon, sterane, which contains 30 carbon atoms. Some natural T., however, possess fewer than 30 carbon atoms, due to subsequent loss of one or more carbon atoms. Addition of extra carbon atoms and incorporation of heteroatoms are also possible, e.g. in the steroid alkaloids. Examples of T. are all the steroids, vitamin D, and numerous pentacyclic T. that occur as components of plant resins and oils, e.g. germanicol, taraxerol, friedelin, etc.

Triticum: see *Graminae.*

Tritocerebrum: see Brain.

Tritometameres: see Mesoblast, Tritometamerism.

Tritometamerism: the formation of segments (tritometameres) during the larval development of members of the *Articulata.* T. arises by teloblastic growth and budding of the somatocoel, after the formation of the deuterometameres. See Deuterometamerism.

Triturus: see *Salamandridae.*

Triungulin: see *Strepsiptera.*

Trivial names: names of living organisms in local languages, i.e. non-Latinized, nonscientific names. A single organism may have several T.n. even within one language area, e.g. *Taraxacum officinale* has a few dozen different German names. T.n. used by the Greeks and Romans were used as a starting point for the scientific naming of plants and animals in the Middle Ages. See Nomenclature.

Trivalent: a group of three homologous chromosomes in synapsis, which exists up to anaphase of the first meiotic division in polyploid or aneuploid cells.

Trochilidae: see *Apodiformes.*

Trochites: (Greek *trochos* wheel, disk): stalk sections of sea lilies, often contributing to the formation of sedimentary rock strata, especially the trochite chalk of the upper Muschelkalk.

Trochlear nerve: the IV cranial nerve; see Cranial nerves.

Trochoides joint: a pivot joint (see Joints).

Trochophore: a larval stage in the development of polychaetes and many marine mollusks (but not cephalpods). It is usually free swimming, but in some mollusks it remains inside the egg membrane. Externally, the T. is separated into two poles or regions by a ciliated band (prototroch) around its equator. The mouth lies just below this equatorial ciliated band. A tuft of cilia is usually present on the crown of the upper pole. The lower pole may possess one or two postoral rings of cilia (e.g. the metatroch), and sometimes a tuft of cilia at its lower end.

The polychaete T. develops into the adult by growth from two primordia: 1) The first quartet of micromeres of the gastrula becomes the apical organ of the T., which develops into the prostomium of the adult, containing the brain, eyes, etc. 2) One cell (the d^2 cell or somatoblast) of the second quartet of the gastrula proliferates to form the ventral plate of the T. Together with a mesoblast cell, this proliferates to form the trunk, so that the lower pole of the T. grows backward into a cylindrical process, which is clearly constricted into segments, and which contains the hindgut. The mollusk T. develops into a Veliger larva (see). See Mesoblast (Fig.2), *Polychaeta* (Fig.2).

Trochus: one of the ciliated bands constituting the "wheel organ" of the *Rotifera* (see).

Troctomorpha: see *Psocoptera.*

Trogiomorpha: see *Psocoptera.*

Troglobionts, *cavernicolus animals:* animals that live permanently in caves or other subterranean cavities. They are adapted to the uniform low temperatures and uniformly high air humidity, and the absence of daily or annual cycles of light and temperature. Except for the region around cave entrances, air movement and changes of illumination are totally absent. The fauna of the entrance region therefore differs from that of the actual cave. Basic nutrients must be provided from outside.

True T. (eutroglobionts) are largely or entirely unpigmented [e.g. cave fishes, cave salamander *(Proteus anguinus)],* and they often have degenerate eyes and/or flight organs, but their organs of touch are very well developed, e.g. the antennae of arthropods. As a result of the uniformity of abiotic factors, activity and reproductive rhythms are often completely lost. A number of relict Ice Age species have survived to the present as T., at low cave temperatures. Absence of competition in caves can lead to high population densities of some species.

Examples of typical T. are the apterygote insect *Schaefferia emucronata,* the beetle *Duvalis hungaricus,* the amphipod *Niphargus aquilex,* and the amphibian *Proteus anguinus.*

The entrance region of caves is often exploited by non-T. for winter hibernation.

Troglodytes: see *Troglodytidae.*

Troglodytidae, ***wrens:*** an avian family of 60 insectivorous species in the order *Passeriformes* (see), and related to the dippers and mockingbirds. They have a generally uniform appearance, ranging in length from 10 to 20 cm, with brown plumage streaked with black, often slightly streaked with white, and with paler underparts. Their thin probing beaks are slightly curved, their legs are strong and long-shanked, and their tails are usually short and uptilted. Some species are polygamous. Nest building is performed mainly by the male, which may build several nests; one is usually better built and better concealed than the others, and this is selected by the female which completes the construction by providing a nest lining. The male often sleeps in one or more of the other more flimsy and less well hidden nests, which may also serve to fool nest parasites such as cowbirds. Most wrens live near to the ground in dense undergrowth, and several are common in garden shrubbery and hedgerows. Many are cavity nesters, exploiting any space among entangled roots, or clefts in rocks Most genera and species are found in tropical Central and South America. Various habitats are exploited, such as tropical forests, marshes, coastal areas, mountains and deserts. Thus, the large ***Cactus wren*** *(Campylorynchus bruneicapillus;* 20 cm) is found in the deserts and arid scrublands of western USA and Texas. The ***Zapata wren*** *(Ferminia cerverai;* 16 cm) is found only in the dense scrub of the Zapata swamp on the south coast of Cuba. The Holarctic ***Wren*** or ***Winter wren*** *(Troglodytes troglodytes;* 10–12.5 cm) is found over much of Eurasia, in North Africa and in North America. The ***House wren*** *(Troglodytes aëdon;* 11.5–14 cm; southern Canada to Tierra del Fuego) is a common user of bird boxes. Similar in geographical distribution, but different in habitat is the ***Short-billed marsh wren*** *(Cistothorus platenis;* 10 cm; southern Canada to Tierra del Fuego).

Trogonidae: see *Trogoniformes.*

Trogoniformes, ***trogons:*** an order of birds containing a single family *(Trogonidae)* of 35–40 species, with a relict distribution in Central and South America (about 20 species in 5 genera), Ethiopian Africa (3 species) and Southeast Asia (11 species in 1 genus, *Harpactes).* Trogons are unique in being zygodactylous with the inner (second) toe pointing backward. The forward pointing toes are joined at the base. Bills are short and broad with rictal bristles, and they are adapted to take insects and pick fruit on the wing. American and African species tend to be metallic green above with bright yellow or red underparts. Asian species are also conspicuously colored. They incubate their eggs in hollow trees and termite mounds.

A notable member is the ***Quezal*** *(Pharomachrus mocinno),* the heraldic bird of Guatemala. The lower breast and undertail coverts of the male are crimson, and the golden green wing coverts hide black flight feathers, while the head, crest, neck and upper breast are an intense irridescent green. Two central tail feathers form a train, and the upper tail coverts are variously colored or glossed with gold, green, blue or violet. The female is also very colorful, but less so than the male, with a blue-gray head and gray belly.

Trogons: see *Trogoniformes.*

Trollius: see *Ranunculaceae.*

Trombiculidae: a family of mites of the suborder *Trombidiformes,* order *Acari* (see). Larvae of the genera *Trombicula* and *Schöngastia* are known as chiggers or red-bugs. In many parts of the world chiggers are human pests, their bite causing varying degrees of reaction from an irritiating dermatitis to a blister or skin eruption measuring several centimeters across. Some individuals, however, do not react to their bite, and appear to be immune. In the Orient, *Trombicula* is the vector of a rickettsial disease called scrub typhus or tsutsugamushi disease.

Tropane, ***8-methyl-8-azabicyclo[3.2.1]octane:*** a bicyclic ring system present in numerous alkaloids, collectively known as tropane alklaoids, e.g. atropine, L-hyoscyamine, littorine, cocaine.

Tropane ring system

Trophic environmental demands: demands placed on the environment by the nutritional requirements of a living organism. See Environmental resisitance.

Trophic factors: see Neurotrophic factors.

Trophic function: see Neurotrophic factors.

Trophic interrelationships: see Nutritional interrelationships.

Trophobiosis: see Relationships between living organisms, Colony formation.

Trophocyte, ***nurse cell, nutritive cell:*** a modified egg cell, often called a yolk cell, which supplies nutrients to the developing egg cell and is consumed in the process (nutrimental egg development). Several Ts. or a single T. may be enclosed with the egg cell in the egg membrane (e.g. in platyhelminths), or they may be located in nutrient chambers in the Ovary (see) (e.g. in insects), where they supply nutrients to the growing egg cell directly or via plasmatic nutritive strands.

Trophogenic layer: the surface layer of a body of water, which is the predominant site of synthesis of organic material, i.e. the site of photoassimilation by littoral plants and phytoplankton. The depth of the T.l. depends on the depth to which the light is able to penetrate.

Tropholytic layer: the deep layer of a body of water, in which the degradation of organic material mainly occurs. Light does not penetrate to this layer, so that assimilatory activity is not possible.

Trophophytes: plants which change their growth rhythm and organ formation so that are always appropriate to the seasonal variations of temperature and water availability.

In the season that is unsuitable for growth (i.e. cold period in temperate regions, dry period in the tropics), woody plants often drop their leaves and protect their apical meristems by forming closed resting buds, usually consisting of overlapping or firmly apposed bracts (bud scales), which are coated with resin or mucilage.

Alternatively, open buds may be formed by the development of a dense layer of hairs on the young leaves before they dry out. Only species with xeromorphic leaves (e.g. many coniferous trees) retain their leaves. Shrubs, perrenial herbs and biannuals survive by means of resting buds. These arise on aerial or subterranean stems, and they occur at ground level or somewhat higher (see Chamaephytes, Hemicryptophytes). Some herbs lose all their aerial parts and survive as subterranean resting buds, which may also serve as storage organs. Resting buds above the soil surface are usually protected by a snow layer. Annual plants survive as seeds, or as rosette plants pressed close to the soil. See Life form.

Trophozoite: see *Telosporidia.*

Tropic acid, ***α-phenyl-β-hydroxypropionic acid:*** an acid which occurs esterified in various alkaloids, from which it can be obtained by hydrolysis. For example, L-hyoscyamine is an ester of L-T.a., and atropine is an ester of DL-T.a.

Tropical land hermit crab: see Coconut crab.

Tropic birds: see *Pelicaniformes.*

Tropism: movement of an organ of a sessile plant or animal in response to a stimulus, which occurs in a direction related to the direction or gradient of that stimulus. The physical asymmetry of the stimulus gradient generates a physiological asymmetry, which is expressed primarily in the distribution of auxin. The concentration of auxin increases on the side of the plant organ that faces away from the direction of the stimulus. Growth therefore increases on that side, causing the organ to bend toward the source of the stimulus. The following types of T., classified according to the nature of the active stimulus, are listed separately: Phototropism, Geotropism, Haptotropism, Traumatotropism, Thermotropism, Chemotropism. If the stimulated plant part bends in the direction of the source of the stimulus, it is known as a *positive T.* If it bends away, it is a *negative T.* Positive and negative Ts. aligned directly toward the stimulus source are also collectively known as *orthotropism.* If bending occurs at an angle to the direction of the stimulus source (between 1 ° and 179 °), it is known as *plagiotropism.* A special case of plagiotropism is *transverse T.,* in which the plant organs become aligned at an angle of 90 ° to a line from the stimulus source. A tropic bending movement may be followed by recurvature that returns the plant part to its original alignment; such compensatory movements are known as Autotropism (see).

Tropomyosin: a fibrillar protein associated with actin in muscle fibers and in the cytoskeleton of other cell types. Striated muscle contains two similar forms: α-T. and β-T., both containing 284 amino acid residues (M_r 33 000). The molecule is a dimer (αα, αβ, or ββ). Dimers polymerize head-to-tail to form a fiber, which lies in a groove of the F-actin helix. T. binds tightly to troponins (proteins which regulate contraction). In nonmuscle cells, T. is associated with some microfilaments, where it probably stabilizes F-actin.

Tropotaxis: movement (taxis) of a motile organism, typically in a straight line, directly toward or away from the source of a stimulus received by symmetrically arranged, bilateral sense organs, e.g. the eyes. [cf. Klinotaxis (see)]

Trough shells: see *Bivalvia.*

Troupials: see *Icteridae.*

Trout: spotted, predatory fishes of the family *Salmonidae* (order *Salmoniformes),* which inhabit clear fresh water. They all make excellent eating.

The ***Rainbow T.*** *(Salmo gaidneri),* originally from the Pacific area of North America, was introduced into Europe in 1880. It is now bred and fattened commercially in ponds.

The ***Sea T.*** *(Salmo trutto),* originally native to Europe, is a migratory (anadromous) species, which begins its upstream migration in the spring, spawning in the upper reaches of rivers from October to January. The fry develop in fresh water, then migrate downstream, reaching the sea between the end of their second year and the beginning of their fourth. They remain in inshore waters for 1–3 years, then migrate inland into fresh water after attaining sexual maturity. There are three main subspecies: *Salmo trutta fario* ***(Brown T.),*** *S.t. trutta* ***(Sea*** or ***Salmon T.),*** and *S.t. lacustris* ***(Lake T.).***

Trout-perches: see *Percopsiformes.*

Trout region: see Running water.

True crabs: see *Brachyura, Decapoda.*

True frogs: see *Raninae.*

True lizards: see *Lacertidae.*

True millers: moths of the family *Tineidae* (see).

True mosses: see *Musci.*

True swimming crabs: see Swimming crabs.

True toads: see *Bufonidae.*

True tree frogs: see *Hylidae.*

True weavers: see *Ploceidae.*

True widowbirds: see *Ploceidae.*

Truffles: see *Ascomycetes.*

Trumpeter bullfinches: see *Fringillidae.*

Trumpeters: see *Gruiformes.*

Trypanorhynchea: see *Cestoda.*

Trypanosomes (plate 42): a genus of parasitic flagellated protozoans [genus *Trypanosoma,* family *Trypanosomidae,* class *Zoomastigophorea* (see)]. Many T. pass through four developmental stages, which differ in size, shape, and development of the flagellum and its associated membrane (Fig.1). All stages possess a Kinetoplast (see), which has two constituent structures: a granule (blepharoplast) from which the axoneme of the flagellum arises, and the parabasal body. The *leishmanial* stage (amastigote) occurs in a vertebrate host; it is an oval body, containing a nucleus, kinetoplast and rhizoplast, but no flagellum. In the *leptomonad* stage (promastigote), which is parasitic in the gut of insects, the flagellum arises from the basal granule at one end of the elongated body. In the *crithidial* stage (epimastigote), which is also parasitic in the gut of insects, the flagellum arises from a basal granule near the middle of the elongated body, and the flagellum is united with the body by an undulating membrane. The *trypanosomal* stage (trypomastigote) occurs in the body fluids of the final vertebrate host; its long flagellum extends backward along the length of the elongated body and beyond, and is united with the body by an undulating membrane.

T. are the causative agents of various diseases of vertebrates known as *trypanosomiases,* and they are usually transmitted to the vertebrate host by blood sucking insects (flies, bugs). Only asexual reproduction has been observed. At the site of entry or injection into the vertebrate host, the parasite divides vigorously and causes local reactions, followed by invasion of the blood system, then the lymphatic system; depending on the T. species, it may finally enter the cerebrospinal fluid (sleeping sickness) or other organs.

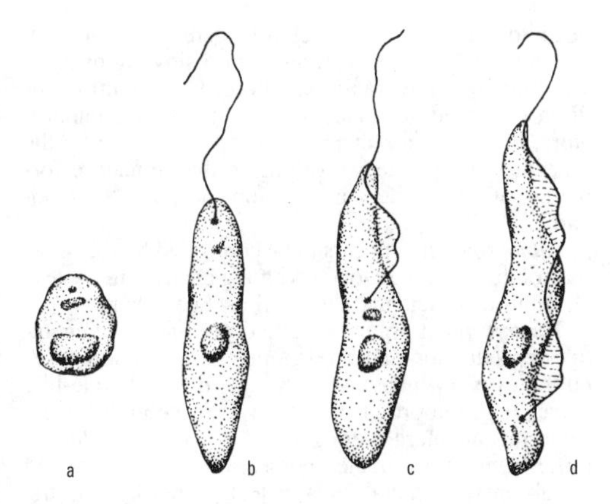

Fig. 1. *Trypanosomes.* The four developmental stages of the *Trypanosomatidae.* *a* Leishmanial stage. *b* Leptomonad stage. *c* Crithidial stage. *d* Trypanosomal stage.

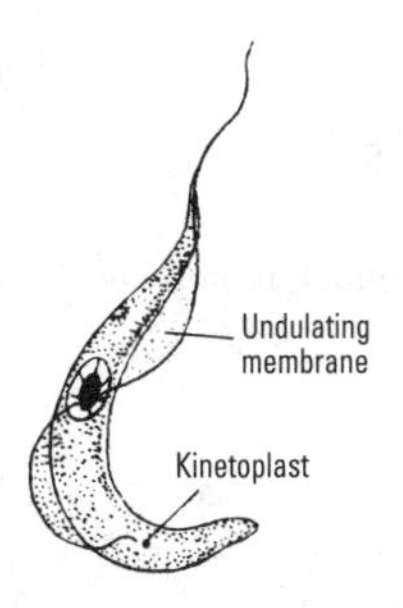

Fig. 2. *Trypanosomes. Trypanosoma gambiense,* one of the causative agents of human sleeping sickness.

Important trypanosomiases are: 1. ***African cattle sickness, Nagana,*** or ***Tstetse fly disease,*** which occurs in horses, donkeys, dogs, cats, rodents and cattle in Central Africa. The causative organism, *Trypanosoma brucei,* is transmitted by the tsetse fly, *Glossina morsitans,* and certain other species of *Glossina.* 2. ***Sleeping sickness*** of humans, caused by *Trypanosoma gambiense* (Fig.2)(West and Central Africa) or *Trypanosoma rhodesiense* (East Africa). *T. rhodesiense* causes a more acute form of the disease. Both forms are transmitted by tsetse flies, especially *Glossina morsitans* and *Glossina palpalis.* A large reservoir of *T. rhodesiense* exists in domestic and wild animals, especially in the bush buck *(Tragelaphus).* 3. ***Chagas' disease,*** a human trypanosomiasis of South and Central America, caused by *Trypanosoma cruzi,* and transmitted by assassin bugs *(Reduviidae).* In particular, it is transmitted by the assassin bug *Triatoma megista,* known as a "barbeiro" or "kissing bug", because it tends to bite the lips of sleeping people. The bug passes infected feces, which are then rubbed into the wound when the bit is scratched. 4. ***Dourine,*** a worldwide trypanosomiasis of horses, caused by *Trypanosoma equiperdum.* There is no intermediate insect host, and the parasite is transmitted directly during coitus. 5. ***Surra,*** a trypanosomiasis of camels and horses, which occurs in North Africa, Southeast Asia, and South and Central America. It is caused by *Trypanosoma evansi,* and transmitted by stable flies and horse flies *(Tabanidae),* and by vampire bats.

Trypanosomiases: see Trypanosomes.

Trypetesa: see *Cirripedia.*

Trypetidae: see Fruit flies.

Trypsin: a prominent proteolytic enzyme of the vertebrate digestive system, which is present as a zymogen or proenzyme (trypsinogen) in the pancreas of all vertebrates. Trypsinogen is released via the pancreatic duct into the duodenum. Conversion of trypsinogen into T. is initiated in the small intestine by enterokinase (enteropeptidase), then accelerated autocatalytically by traces of T. itself. Enterokinase catalyses removal of the *N*-terminal acidic hexapeptide, Val(Asp)$_4$Lys, from trypsinogen. The resulting strongly basic β-T. is cleaved by limited autolysis of the Lys$_{131}$-Ser$_{132}$ bond to give the 2-chain structure of α-T. Further cleavage of Lys$_{176}$-Asp$_{177}$ produces a 3-chain active enzyme called pseudo T., pH optimum 7–9. The 3 chains are united by 6 disulfide bonds. Calcium ions accelerate the activation process and make it more specific, thus avoiding formation of inactive byproducts through other less selective cleavages. T. is stable at pH 2–4 and +4 °C for several weeks, but at pH 9 and 30 °C it undergoes progressive autolysis of its intact Lys- and Arg-bonds, and becomes totally inactive within 24 h. T. displays very high substrate specificity, catalysing hydrolysis of only Lys- and Arg-bonds in dietary proteins; in all the resulting oligopeptides the C-terminal amino acid residue is therefore either lysine or arginine.

T. belongs to a group of proteases known as "serine proteases", i.e. its active center contains a reactive serine residue (Ser$_{183}$), and catalytic activity is inhibited by synthetic inhibitors like diisopropylfluorophosphate, which forms an ester with the hydroxyl group of the serine. Primary and secondary structure of T. are known. Bovine trypsin, M_r 24 000, contains 223 amino acid residues. T. displays a close structural relationship to Chymotrypsin (see).

Enzymes similar to T. have been found in many invertebrates, like crabs and insects, e.g. cocoonase produced by silk moths to attack their proteinaceous cocoon and aid escape.

N.B. All sequence positions quoted above are for the original trypsinogen.

Trypsinogen: the inactive precursor (proenzyme) of Trypsin (see).

Tryptamine, *β-indolyl-(3)-ethylamine:* a Biogenic amine (see), produced by decarboxylation of L-tryptophan. It stimulates contraction of smooth muscle, increases blood pressure, and causes local pain. It occurs as a bacterial degradation product of L-tryptophan, and is widely distributed in small quantitites in plants. Experimentally in plants, it behaves as a weak auxin and promotes growth. It is also found in small quantities in animals.

Trypterygiidae: see *Blennioidei.*

Tryptomastigote: see Trypanosomes.

L-Tryptophan, *Trp:* an essential, proteogenic amino acid. It is an aromatic amino acid, i.e. its structure contains a benzene ring, and it is a product of the Shikimic acid (see) pathway of biosynthesis in plants and bacteria. Trp is the biosynthetic precursor of tryptamine, indole-3-acetic acid, serotonin and melatonin.

Tryptophan

Tryptophyte: a parasite which kills a living tissue then invades it. The tissue is killed by the action of enzymes secreted by the T.

Tsetse fly disease: see Trypanosomes.

Tshiir: see *Coregonus.*

Tsilik marasi: a canker disease of grapevines, caused by *Xanthomonas ampelina.* See *Xanthomonas.*

Tsutsugamushi disease: see *Rickettsiae, Trombiculidae.*

T-tubules: see Muscle tissue.

Tuatara: see *Rhynchocephalia*

Tuba auditiva: see Eustachian tube.

Tuba Eustachii: see Eustachian tube.

Tubeblennies: see *Blennioidei.*

Tube floret: see *Compositae.*

Tube-noses: see *Procellariiformes.*

Tuber: see Tubers.

Tuberales: see *Ascomycetes.*

Tubercula quadrigemina: see Brain.

Tubercle bacilli: see *Actinomycetales.*

Tuberculosis: see *Actinomycetales.*

Tuber development: in certain perennial and biennial herbs, the thickening and metamorphosis of certain parts of the stem or root, resulting in the formation of organs of vegetative reproduction, overwintering and/or food storage. Stem tubers may arise from the leafy epicotyl above the cotyledons (e.g. stem tuber of kohlrabi), from the hypocotyl below the cotyledons (e.g. stem tubers of radish, beetroot and *Cyclamen),* or from the ends of subterranean plagiotropic side branches (stolons) (e.g. stem tubers of potato and Jerusalem artichoke).

Root tubers usually arise by the swelling of adventitious roots [e.g. *Dahlia, Ranunculus ficaria* (Pilewort) and terrestrial orchids], and rarely from primary roots [e.g. *Lathyrus tuberosus* ("Earthnut pea")].

T.d. occurs in 3 stages: induction, tuber growth, and tuber ripening (i.e. transition to dormancy).

Induction involves differentiation of the meristem, loss of polarity and apolar plasmatic and extension growth. In many potato varieties, Jerusalem artichoke and begonias, these changes are induced by short day conditions, the response to the photoperiod being mediated by the Phytochrome (see) system. In some plants, T.d, like flower development, is controlled entirely by the photoperiod. As in the case of the flowering stimulus, the photoperiodic effect is transmitted from foliage leaves to the site of action, and can also be transmitted by grafting appropriately stimulated plant parts onto nonstimulated plants. There is evidence that the stimulus for T.d. is transmitted by a group of unspecific substances. In some species, T.d. appears to be induced by abscisic acid, the production of which is increased under short day conditions. T.d. can also be activated by cytokinins. T.d. induction is very persistent. Thus, tissue cultures of potato plants previously induced for tuber formation produce well-formed tubers, whereas cultures of noninduced plants are hardly able to form tubers.

Tuber growth depends primarily on the photosynthetic activity of the plant. Favorable environmental conditions for tuber growth are those that enable high starch production and accumulation. Various anomalies of T.d. have been observed in the potato plant. For example, aerial tubers may be be formed in the leaf axes above ground, due to damage to the vascular system by mechanical injury or parasites. Dwarf tubers or divided tubers may develop if there is a temporary interruption of the growth phase or a slowing of tuber swelling by unfavorable environmental conditions or disease. If seed tubers are stored at too high a temperature, they have a reduced capacity for growth by the time they are planted, resulting in the premature formation of small tubers, a condition known as "little potato".

Tuber ripening appears to be controlled by the interplay of growth regulators, involving changes in the levels of abscisic acid and gibberellins. See Tubers.

Tubers: plant organs modified as organs of vegetative reproduction and food storage. *Stem T.* still carry buds ("eyes") which arise in the axes of small scale-like leaves, e.g. the orthotropic stem T. of kohlrabi or the subterranean lateral stem T. of the potato plant. *Root T.* differ from stem T. in their possession of a root cap, and the absence of lateral buds or leaf primordia, e.g. the root T. of *Dahlia* and terrestrial orchids. See Tuber development.

Tubifex: see *Oligochaeta.*

Tubificidae: see *Oligochaeta.*

Tubiflorae: see *Compositae.*

Tubilifera: see *Thysanoptera.*

Tubuliflorae: see *Compositae.*

Tubinates: see *Procellariiformes.*

Tuboidea: see Graptolites.

Tubuli: see Mitochondria.

Tubulidentata: an order of mammals represented by a single species, the Aardvark (see).

Tubulin: see Microtubules.

Tuco-tucos: see *Ctenomyidae.*

Tucuxi: see Whales.

Tui: see *Meliphagidae.*

Tulasnellales: see *Basidiomycetes.*

Tulip tree: see *Magnoliaceae.*

Tullgren's apparatus: a device for isolating large soil fauna. See Soil organisms.

Tulu: see Dromedary.

Tumor immunology: a branch of Immunology (see), concerned with the immunological reactions of tumor-bearing organisms to their tumors. This leads to the development of methods for the immunodiagnosis of cancers and tumors, and provides new approaches to the clinical management of cancer.

All tumor cells possess components that are antigenic to the affected organism, and which are different from the components of normal body cells. All cancers therefore evoke an immune reaction, which, however, does not always destroy the cancer. Immunological cancer diagnosis relies on this sensitization of the organism against its cancer.

Immunotherapy of cancer is still in its infancy. Present clinical methods are based only on the unspecific stimulation of the immune system, which supposedly also leads to an increased defense reaction against the cancer. The long-term goal, however, is direct stimulation of the specific immune reaction of the patient against the cancer.

Tumor virsues, *cancer viruses, oncogenic viruses:* viruses which cause malignant tumors in animals. Only a few T.v. are known that cause malignant tumors in humans. Alternatively, the virus infection may represent a "risk factor", which together with other influences (chemical, genetic, hormonal, physical) is involved in the development of a malignant tumor, i.e. the genesis of the tumor may be multifactorial. T.v. are found among retroviruses, as well as nearly all the main groups of animal DNA viruses, in particular papova-, adeno- and herpesviruses (see Virus families). As a rule, T.v. also cause *malignant transformation* of cells cultured in vitro, a phenomenon associated with altered growth and multiplication. After malignant transformation in vitro, some cells develop into malignant growths when transplanted to appropriate host animals, thus providing an in vitro model of carcinogenesis. There are, however, numerous viruses (e.g. human adenoviruses), which transform human cells in vitro, but are rarely carcinogenic in the human organism. On the other hand, some oncogenic viruses (e.g. chronic oncornaviruses, or hepatitis B virus) do not transform human cells in vitro.

The genome of nearly all T.v. can be incorporated into the genome of the host cell by nonhomologous recombination (see Molecular genetics). This is also possible for oncogenic RNA viruses (oncornaviruses), which, like other retroviruses, possess an RNA-dependent DNA polymerase (revertase, reverse transcriptase) for the synthesis of virus-specific double-stranded DNA. Incorporation of viral DNA leads to the regular replication and transmission of the viral genome to all susequent cell generations, i.e. the malignant dedifferentiation becomes genetically based. At least in the case of chronic oncornaviruses, provirus integration also appears to be necessary for the subsequent malignant transformation (see below). The pro-oncoviruses share many similarities with jumping genes. Although the proviruses of all T.v. appear to behave like the prophages of temperate bacteriophages, no analogous repressor activity has yet been demonstrated in cells infected with T.v.

Each tumor virus is oncogenic only in certain animal species, and usually only in specific organs or tissues. As a rule, when a virus is associated with malignant growth, it does not multiply in the affected cells, i.e. the cells are "nonpermissive" for the virus in question. Thus, most T.v. cause, or contribute to the development of, specific cancers, e.g. papilloma virus causes skin tumors, while hepatitis B virus is a factor in the development of liver cancer, etc. The multifactorial nature of carcinogenesis is particularly obvious in the action of the herpesvirus, *Epstein-Barr virus (EBV).* In nonimmune individuals in Europe and North America, this virus causes a benign febrile illness called infectious mononucleosis (Pfeiffer's glandular fever). In Africa and Papua New Guinea, a combination of infection by the malaria parasite, *Plasmodium falciparum,* and a chromosome mutation caused by EBV is reponsible for the development of a malignant tumor, known as *Burkitt's lymphoma* (a cancer of B lymphocytes). In southern China and Greenland (Inuit tribes), EBV, in combination with other unknown factors (histocompatibility genotype? Consumption of salted fish in infancy?), causes malignant tumors of the nasopharyngeal cavity. EBV can therefore be classified as a human tumor virus, or as a contributory factor for the development of human tumors. Other human T.v. are: 1. *Human papilloma virus type 5 (HPV5)* (family *Papovaviruses),* which (in dependence on other factors) causes a very rare, wart-like malignant skin tumor. Certain papillomaviruses are thought to cause cancer of the uterine cervix. Infection by papillomavirus is usually benign, the viral DNA being maintained as plasmids in basal epithelial cells. Very occasionally, however, a fragment of a viral plasmid becomes incorporated into the host genome. The resulting unregulated production of viral replication protein drives the host cell into S-phase, with the eventual development of a cancer. 2. *Hepatitis B virus (HBV),* the causative agent of serum hepatitis. In tropical Africa and Southeast Asia, and in combination with other factors (aflatoxins in food, alcoholism, smoking, other viruses), HBV causes primary liver cancer (hepatocellular carcinoma), which is one of the commonest human cancers.

One retrovirus, *T-cell lymphoma virus (TCLV),* has been identified as a human cancer risk factor in Japan and the West Indies. Other viruses, especially other types of papilloma viruses, as well as herpes simplex virus type II, are suspected of involvement in the development of human cancers, e.g. cancers of the genital system. To the relatively small list of human T.v. must now be added the human immunodeficiency virus (HIV-1; the AIDS virus), a retrovirus which is thought to be a contributory factor in cancer of the endothelial cells of blood vessels (Kaposi's sarcoma) in Central Africa. Although relatively few human cancer-causing T.v. are known, many viruses have been identified (in particular retroviruses) which cause malignant tumors in practically all other groups of vertebrates. Some of these are therefore economically important.

There is no single uniform mechanism of malignant transformation or carcinogenesis by T.v. The process may even differ between members of the same virus genus. For example, a gene product is reponsible for the oncogenic action of the polyomavirus, murine papova virus (plate 19), whereas this gene product is totally absent from simian virus SV40, which belongs to the same genus.

There are at least 3 principal mechanisms of virus-induced carcinogenesis:

1. *Virus-specific genes* and their products are directly responsible for the oncogenic action, e.g. in the case of polyoma- and adenoviruses.

2. *A cellular oncogene (c-onc)* or *proto-oncogene* is activated by the T.v. This is the mode of action of the retroviruses known as chronic oncornoviruses, which do not possess a viral oncogene. Activation occurs when the provirus is incorporated into the host genome close to or even within the proto-oncogene. This activation process of cellular oncogenes is known as *insertional mutagenesis.* Oncogenesis can also be activated by virus-induced chromosome modification. Thus, Epstein-Barr virus (see above) causes chromosome modification, and this is a contributory factor in the development of Burkitt's lymphoma.

3. The virus carries a *viral oncogene (v-onc)* into the cell, e.g. in the case of oncornaviruses. The viral oncogene is homologous with one of 15 to 20 cellular *onc* genes, and it was originally incorporated into the viral genome from a host genome during the evolution of the virus. The prototype of such an acute oncornavirus is the *Rous sarcoma virus,* the first tumor virus to be discovered (Peyton Rous, 1911; Rous received the Nobel prize for this discovery in 1966). The oncogene of the Rous sarcoma virus is the *src* oncogene, which codes for a protein kinase. Conversion of a proto-oncogene

into an oncogene by incorporation into a retrovirus may occur by alteration or truncation of the gene sequence so that it codes for a protein with abnormal activity. Alternatively, within the viral genome, the gene may be placed under the control of powerful promoters and enhancers, resulting in excess production of gene product, or formation of gene product when it is not required. Both effects (gene modification and overproduction of gene product) are often involved.

Tuna: see Tunny.

Tundra: treeless vegetation of the Arctic floristic zone, beyond the northern timber line. Between the true forest limit and the treeless plain of the T., there are transitional zones *(forest T.),* consisting of *Betula tortuosa* in oceanic regions, or *Picea* (spruce) and *Larix* (larch) in continental regions. The vegetation of the T. is characterized by dwarf woody plants, woody ground espalier plants and tortuous creeping undershrubs e.g. willow species *(Salix polaris, S. herbacea* and *S. retusa),* dwarf birch, avens *(Dryas),* together with bog plants like *Eriophorum* (cotton grass), sedges *(Carex* species), rushes *(Juncus* species) and wood rushes *(Luzula* species). Grasses are also usually present.

T. soils are classified as *inceptisols,* i.e. mineral soils with one or more horizons in which minerals have been weathered or removed. Such soils are in an early stage of formation of visible horizons, and display only the beginning of the development of a distinctive soil profile, particularly in slightly raised regions, where the effects of alternate freezing and thawing contribute to profile formation. T. soils are generally poorly drained (mainly on account of permafrost), acidic, and 30–60 cm deep. Since decomposition is inhibited by the low temperatures, there is an accumulation of organic matter on the surface. The permafrost thaws only on the surface during the brief growth season, leading to water flow (solifluction) on hillsides and slopes. *Epilithic* or *Rock T.* is typified by creeping shrubs and cushion plants. *Dwarf woody T.* occurs on drier soils with varying degrees of humus formation, and is characterized by the predominance of bearberry *(Arctostaphylos uva-ursi),* crowberry *(Empetrum nigrum)* and bog whortleberry *(Vaccinium uliginosum). Lichen T.* consists of dense growths of relatively tall branched lichens, like reindeer moss *(Cladonia rangiferina)* and Iceland moss *(Cetraria islandica).* See Biocoenosis.

Tundra vole: see *Arvicolinae.*

Tundu disease: see Eelworms.

Tung oil: see *Euphorbiáceae.*

Tunica: see Meristems.

Tunica muscularia: see Digestive system.

Tunica propria: see Digestive system.

Tunicata, ***tunicates,*** *Urochordata:* a subphylum of the *Chordata* containing about 2 100 species, which possess a notochord and a neural tube, at least in the larval stage. The body is characteristically encased in a mantle or tunic. The mantle contains living cells and usually fibers of a cellulose-like polysaccharide. Calcareous or organic spicules are also sometimes present, as well as blood vessels and nerves.

Morphology. Beneath the mantle lies the body epithelium, then gelatinous connective tissue in which all the organs are embedded. The Pharynx (see), which is typical of all *Chordata,* accounts for a large proportion of the body. *T.* are unsegmented, typical coelom is lacking, and the main body cavity is hemocoelic. The pericardium may, however, represent a coelomic space. In addition, in most species, two epicardial sacs develop from the posterior pharynx, and these may represent enterocoelic pouches. Nephridia are absent. The nervous system, represented by the neural tube, is reduced after metamorphosis to a ganglion or brain (situated above the foregut), which supplies nerve fibers to the organs. Beneath this ganglion is the neural gland, which communicates with the pharynx by a duct, and is homologous with the vertebrate pituitary. The vascular system is lacunar, and the tubular heart continually alternates the direction of its beat.

Biology. *T.* are exclusively marine, attached to the substratum (ascidians) or free swimming, and often aggregating in large numbers. Almost all species are hermaphrodite. Development is direct in the *Appendicularia.* All other forms possess free-swimming larvae, which have a tail containing the notochord; the tail notochord and dorsal nerve cord are resorbed during metamorphosis. Asexual reproduction by budding is also common. A ring of oral cilia generates a current through the pharynx. Food particles filtered out by the pharynx are transferred via the endostyle to the often U-shaped intestine.

Classification. Three classes are recognized: *Appendicularia* (see), *Thaliacea* (see) and *Ascidiacea* (see).

Collection and preservation. Sedentary ascidia (sea squirts) are removed from the substratum with a dredge or rake. To prevent contraction of the animals, 5% cocaine or 7% magnesium chloride must be added slowly to the sea water. Magnesium chloride is particualrly successful with medium-sized to large sea squirts. Storage is most effective in a mixture of 40% formaldehyde and alcohol. Free-swimming salpids and pyrosomes are collected with a net and preserved in neutral 2% formaldehyde solution.

Tunny, ***blue-fin tuna, Atlantic tuna,*** *Thunnus thynnus:* a very important commercial marine fish of the family *Scombridae* (mackerels and tunas), order *Perciformes.* Specimens of 5 m are sometimes caught, but the usual length is about 3 m. It is found in all oceans, but is more common in warm seas. An extremely fast and powerful swimmer, it hunts shoaling fish, and is closely related to the bonito.

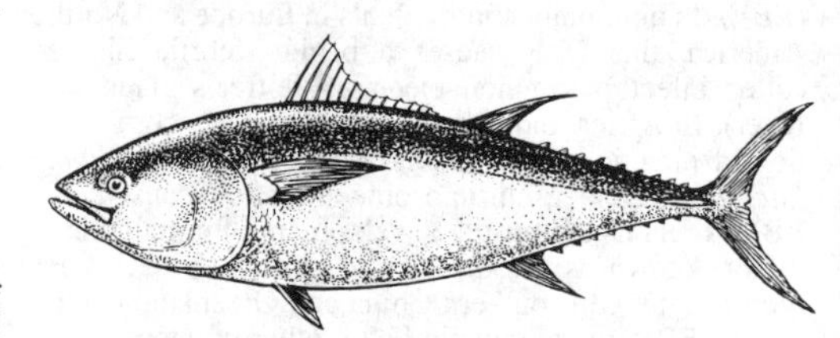

Tunny (Thunnus thynnus)

Tun shell: see *Gastropoda.*

Tupaiidae, ***tree shrews:*** a family of mammals (16 species in 5 genera) with the appearance of long-snouted squirrels. Some authorities consider them to be insectivores, while others place them with the prosimians. Dental formula: 2/3, 1/1, 3/3, 3/3 = 38. Various tree shrews inhabit India and Southeast Asia, including Sumatra, Borneo and the Phillipines.

Tree shrew

Tupinambis: see Tegu.

Turacin: see *Cuculiformes.*

Turacos: see *Cuculiformes.*

Turacoverdin: see *Cuculiformes.*

Turbatrix: see *Nematoda.*

Turbellaria: a class of the *Platyhelminthes.* Nearly all of the 3 000 species of *T.* are free-living in water or wet soil, and only a few are parasitic.

Morphology. Most *T.* do not exceed 20 mm, and only certain terrestrial forms attain lengths greater than 50 cm. Body shapes display all transitions, from markedly flattened forms with an oval leaflike or lanceolate outline, to spindle-shaped or stringlike elongated species with a flattened underside which serves as a creeping sole. Small species are often translucent or whitish in color (sometimes also brown, gray or black), whereas many large marine or terrestrial forms are brightly colored and patterned. Some species display a yellow to green coloration, due to the presence of symbiotic unicellular green algae (zoochlorellae) and unicellular brown algae (zooxanthellae). The exterior surface is completely or partly covered with cilia, which serve for locomotion and for renewing the surrounding layer of water. Scattered cells in or below the ectoderm contain crystalline, rod-shaped bodies known as rhabdites, which when extruded become hydrated, swell, and dissolve. Sometimes sagittocysts are also present; these secretory capsules contain a fine, extrusible spine, and they serve for defense and for the capture of food. Rhabdites may also serve a protective function. Between one pair and three pairs of light-sensitive organs (pigment spots or simple cup-shaped organs) occur at the anterior end, or several to 1 000 may be present on the dorsal surface or the margins of the body. In all species, the intestine is blind, being sacklike or rod-shaped, sometimes with lateral cecae, or multiply branched. Some primitive species *(Acoela)* do not have a hollow gut but a central syncytium of endodermal cells, which serve for digestion. The mouth is usually situated on the ventral surface, near to the anterior end. There is usually a well developed, extrusible pharynx. With the exception of a few parthenogenetically reproducing species, the majority of *T.* are hermaphrodite. The sexual apparatus shows a variety of forms, but always consists of testes, ovaries, and egg-forming organs.

Biology. Most species live free in the sea or fresh water (mainly on the bottom, under stones or on water plants), or in wet terrestrial habitats. A few species are commensals or parasites of aquatic invertebrates, especially crustaceans. With the aid of their cilia, they creep on the surface of the substratum. Swimming forms are less common. They feed mainly on living and dead animal material, many of the larger species feeding by predation on small invertebrates.

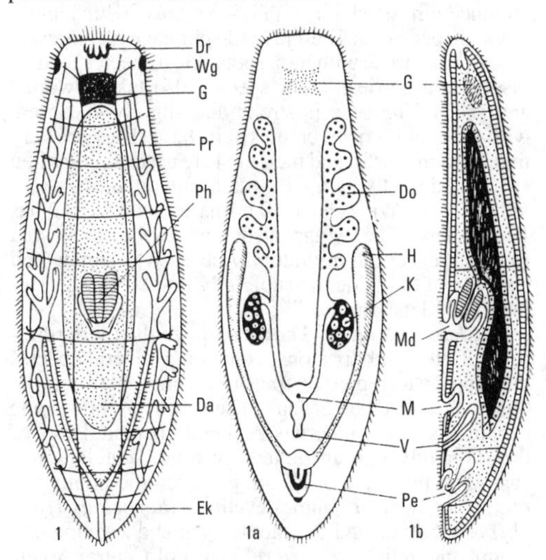

Turbellaria. Structure of a turbellarian worm. *1* and *1a* Dorsal view. *1b* Longitudinal section. *Da* Intestine. *Do* Yolk sac or vitellarium. *Dr* Frontal gland. *Ek* Common duct of the excretory system. *G* Cerebral ganglion (brain). *H* Testis. *K* Ovary. *M* Common oviduct. *Md* Mouth. *Pe* Penis. *Ph* Pharynx. *Pr* Protonephridia. *V* Opening of genital atrium to the exterior. *Wg* Ciliated groove.

Classification.

1. Subclass: *Archoophora*

Orders: *Acoela* (no true gut cavity or pharynx); *Macrostomida; Catenulida; Polycladida* (a free-living larval stage is present)

2. Subclass: *Neoophora*

Orders: *Prolecithophora; Lecithoepitheliata; Seriata* or *Tricladida* (see Planarians); *Neorhabdocoela* or *Rhabdocoela*

Development. Development is usually direct, and the newly hatched animal resembles the mature form. Only certain marine species of the *Polycladida* develop via a larval stage (Müller's larva or Goette's larva). In addition to normal eggs, some species produce resistant eggs with a low yolk content, which are able to survive unfavorable environmental conditions. Members of the orders *Catenulida* and *Macrostomida* can reproduce asexually by transverse fission, followed by regrowth of the lost organs. This results in the formation of chains of individuals (zooids) at different stages of development. Regular alternation between this type of asexual reproduction and a sexual generation is not uncommon. Other *T.*, in particular the planarians, are capable of autotomy, dividing themselves by transverse cleavage in the middle of the body, followed by regeneration from each fragment.

Turbot: see *Pleuronectiformes.*

Turdinae, *thrushes and allies:* a large (300 species in 60 genera) avian subfamily of the *Muscicapidae* (see), in the order *Passeriformes* (see). The subfamily has a worldwide distribution, being absent only from Antarctica and a few isolated islands; formerly absent from New Zealand, it is now represented there by the introduced song thrush and the blackbird. The *Turdinae* are medium sized birds (11.5–33 cm), which build bowl-shaped nests. Their juvenile plumage is characteristically patterned with light spots. Most species feed on insects and berries, larger species also taking worms and snails. The beak is strong and slightly decurved, with a few or no rictal bristles at its base. They are very much ground birds, and their tarsi are accordingly often strong and relatively large, with horny rather than scutellate plates. Wings have 10 primaries, and the tail has 12 feathers. The wings also tend to be somewhat rounded, except in migratory species. It is simple and convenient to divide the subfamily into "Chats and allies" and "True thrushes".

1. ***Chats and allies.*** These comprise the more primitive of the 2 subdivisions, containing several species that appear to represent transition forms between the *Turdinae* and other groups. They also provide some exceptions to the otherwise general characteristics of the subfamily, e.g. unspotted juveniles, scutellate tarsus. Most primitive of all are probably the nonmigratory, short-winged, jungle-dwelling ***Shortwings*** *(Brachypteryx:* India and Southeast Asia; and *Zeledonia,* a monotypic, relict, New World genus of Central America) and the ***Scrub robins*** *(Drymodes:* Australia and New Guinea; and *Erythropygia:* Africa). These primitive groups appear to be intermediate between the *Turdinae* and the *Sylviinae,* and they are probably ancestral to the ***Old World robins*** *(Erithacus)* and ***Nightingales*** *(Luscinia).* Representative of the Old World robins is the ***Eurasian robin*** *(Erithacus rubecula),* which is widespread from Britain to the Canary Islands in a band extending to western Siberia, with related species in Northern Asia and Africa. It favors well undergrown deciduous and coniferous forests up to altitudes of 2 000 m, where it feeds on insects, worms, etc. from the forest litter. In Britain and Europe it is also a common garden visitor, and it is found near human habitations throughout the year. Nightingales are extremely close relatives of the robins, the most familiar European representative being the ***Nightingale*** *(Luscinia megarhynchos),* which occurs from Britain to North Africa, as far as Central Asia. Its range overlaps that of the ***Thrush nightingale*** *(Luscinia luscinia),* which is more common in Northeast Europe and favors moister habitiats. Further east in Asia, especially in the conifer forests of the taiga, is found the ***Siberian ruby throat*** *(Luscinia calliope),* the male of which has a bright red throat with white stripes above the eyes and below the cheeks. The ***Bluethroat*** *(Luscinia svecica)* occurs throughout most of northern Eurasia, overwintering in North Africa and southern Asia; it has a black-bordered, blue (male) or whitish (female) throat, and a red-brown tail. Evolutionary branches of the robins and nightingales are the ***Robin-chats*** *(Cossypha and Alethe),* the ***Magpie robins*** or ***Shamas*** *(Copsychus),* and the ***Redstarts*** *(Phoenicurus).* The ***Firecrest aletha*** *(Alethe diademata),* found in dense jungle undergrowth from Sierra Leone to Uganda, is chestnut brown above and whitish below, with an orange-striped crown. Redstarts are mainly rock-dwelling birds, and all have a red-brown tail. *Phoenicurus phoenicurus* ***(Redstart)*** (plate 6) occurs from western Europe to central Siberia and Afghanistan, and it is one of the commonest European birds. The somewhat darker *Phoenicurus ochrurous* ***(Black redstart)*** is found in western Europe, the Mediterranean region, the Near East, northern China and the Himalayas. Also in this group is the glistening blue *Rhyacornis fluiginosus* (the only species of the genus) which is found from the Himalayas to China. There are 3 species of colorful North American ***Bluebirds*** *(Sialia),* e.g. the ***Eastern bluebird*** *(Sialia sialis),* which occurs from Canada to Central America, is indigo blue above and red-brown below, with a white belly. The ***Stonechat*** *(Saxicola torquata)* is found throughout Europe (not Scandanavia) and Asia Minor, extending through the Caucasus over much of Asia as far as China, Korea and Japan, as well as many areas of sub-Saharan Africa, and in Madagascar. It is nonmigrátory, although the most northerly populations may move further south within their overall range in the winter. It favors open country with rough grassland and scattered trees and bushes. In summer the male has a chestnut, almost red breast shading into whitish below, a black head and back, and white patches on the neck, wings and rump, while the female has streaky brown upper parts and no rump patch; both sexes are duller in the winter. The ***Whinchat*** *(Saxicola rubetra)* favors habitats similar to those of the stonechat, but unlike the stonechat it is highly migratory, overwintering in Africa and moving in the summer into Europe, northeastern Asia, and Scandanavia. Compared with the stonechat, the back of the whinchat is more streaked, and the male has a white eye stripe. Also included with the chats and their allies are the black-and-white ***Forktails*** *(Enicurus)* of Southeast Asia, treated by some authorities as a separate family. ***Wheatears*** *(Oenanthe)* are found in the deserts, tundra and meadows of Eurasia, Africa and India. Closely related are the rock-dwelling ***Sooty chats*** or ***Ant chats*** *(Myrmecocichla)* and the ***Indian robin*** *(Saxicoloides fulicata).*

2. ***True thrushes.*** The dominant and most widely distributed genus of the family is *Turdus,* represented by about 63 species, many of them migratory. Examples are the popular European ***Blackbird*** *(Turdus merula;* 25.5 cm). Once confined to woodland, the blackbird has now adapted to parks and gardens, as well as open heaths and rocky areas. The male is black with a yellow beak, whereas the female is dark brown, lighter below, slightly mottled, and has a brown beak. The blackbird is particularly noted for its beautiful song, being one of the main performers in the dawn chorus. Although nonmigratory, the most northerly breeding birds may move further south in the winter within the overall range. In addition to the whole of Europe and southern Scandanavia, the range extends into North Africa and the Near East, across the Caucasus and the Himalayas to China. The New World counterpart of the Blackbird is the ***American robin*** *(Turdus migratorius;* 25.5 cm). Occupying very similar habitats and ecological niches to *T. merula,* the American robin breeds from northwest Alaska through much of Canada, and all the way south through USA to Mexico and Guatemala. The male has a black head with a white patch around the eye, generally gray upper parts and chestnut red underparts, and white streaks or patches on the throat and under the tail; the female is duller and paler. The ***Song thrush*** *(Turdus philomelos;* 23 cm) ranges over Europe, including the entire British Isles, but excludung southern Spain. It also extends into Asia Minor and western and central Asia. The

sexes are alike, and the breast is notably streaked with chestnut, shading into white on the belly. The more northerly populations migrate to the Mediterranean and southwest Asia in the Winter. The song thrush is well known for its habit of using a stone as an anvil for breaking open snail shells. The range of the ***Mistle thrush*** *(Turdus viscivorus;* 26.5 cm) is similar to that of the song thrush, but it covers the whole of Europe and extends into a small part of North Africa, the eastern Mediterranean, and through the Caucasus to northern India, where it occurs in forests at over 3000 m in the Himalayas. The sexes are alike and bear a close resemblance to the song thrush, but the mistle thrush is larger, has white outer tail feathers, the spotting and streaking of the breast is much bolder, and the stance is very upright. Rather smaller than the song thrush, the ***Redwing*** *(Turdus iliacus = musicus;* 21 cm) occupies a northerly breeding range stretching across Scandanavia and northern Asia, where it is found in the open woodland of the taiga. It overwinters in southern Europe, North Africa and southwest Asia, but appears to be extending its breeding range further south, and some birds now breed, e.g. in Scotland. Red patches are present on the flanks and underwings, the breast and underparts are streaked and speckled with dark brown, and there is a characteristic prominent white eye stripe. Like the redwing, the ***Fieldfare*** *(Turdus pilaris;* 25.5 cm) has a northerly breeding range, covering Scandanavia and northern Asia, but extending also into northern Europe. Southerly migrations are undertaken in the winter into Britain and southern Europe. The ***Ring ouzel*** *(Turdus torquatus;* 24 cm) superficially resembles the blackbird, but it has a white crescent on its breast, and the closed wings are somewhat paler than the rest of the sooty black body; the female is smaller and browner. It overwinters mainly in the Atlas mountains and the Mediterranean region, and visits Europe and Scandanavia during the summer. The ring ouzel is a timid, upland bird, nesting on or near the ground, often near a moorland stream, and rarely below 300 m. Other true thrushes are the ***Rock thrushes*** *(Monticola),* which are chat-like, African and Eurasian birds of open country, often with bright blue and brown in their plumage. The "***Whistling schoolboy***" or ***Whistling thrush*** *(Myiophoneus caeruleus)* is a beautiful dark blue song bird of Southeast Asia. The New World, nonmigratory *Zoothera* thrushes are represented by, e.g. ***White's thrush*** or ***Golden mountain thrush*** *(Zoothera dauma),* the ***Varied thrush*** *(Z. naevia)* and the ***Aztec thrush*** *(Z. pinicola).* Close relatives are the Caribbean ***Forest thrushes*** *(Cichlherminia),* the New World ***Nightingale-like thrushes*** *(Catharus)* (e.g. the ***Hermit thrush,*** *C. guttata;* and the ***Veery,*** *C. fuscescens),* and the North American ***Wood thrush*** *(Hylocichla mustelina).*

Turdus: see *Turdinae.*

Turgor movements: plant movements resulting from changes in the turgor pressure of tissues. See Catapulting mechanisms, Explosive mechanisms, Squirting mechanisms, Variational movements.

Turgor pressure, *turgor:* the internal pressure exerted on plant cell walls, which is due to the osmotic pressure (see Osmosis) of the plant cell sap. If the osmotic value of the cell sap is sufficiently high, the cell withdraws water from the external medium. The vacuole enlarges, and the protoplasm becomes pressed against the partially elastic cell wall. T.p. and Tissue tension (see) both contribute to the rigidity of the plant body. If water is withdrawn from the cells, e.g. by a high transpiration rate with inadequate water uptake, T.p. is decreased, cells become limp, and the plant wilts.

Turions: winter buds of aquatic plants that serve for asexual reproduction. They are produced in the autumn, separate from the parent plant, and sink to the bottom. In the spring they rise to the surface and develop into a new plant. Formation of T. is favored by darkness and falling temperatures. Morphologically, T. resemble the winter buds of terrestrial plants. Their lower leaf-like organs contain reserve materials.

Turkey oak: see *Fagaceae.*

Turkeys: see *Meleagrididae.*

Turkish filbert: see *Corylaceae.*

Turkish gecko: see *Hemidactylus.*

Turmeric: see *Zingiberaceae.*

Turnicidae: see *Gruiformes.*

Turnip dart: see Cutworms.

Turnip moth: see Cutworms.

Turnip yellow mosaic virus: see Virus groups.

Turnix: see *Gruiformes.*

Turnstones: see *Scolopacidae.*

Turpentine: see Balsams.

Turricephaly: see Acrocephaly,

Turrid shells: see *Gastropoda.*

Tursiops: see Whales.

Turtledoves: see *Columbiformes.*

Turtles: see *Chelonia.*

Tusk shells: see *Scaphopoda.*

Tussah silk: see *Bombycidae.*

Tussilago: see *Compositae.*

Tussock moths: see *Arctiidae, Lymantriidae.*

Twig snake, *Thelotornis kirtlandii:* a snake of the family *Colubridae* (see), subfamily *Boiginae* (see). A native of tropical Africa, the T.s. attains a length of 1.5 m. It blows up its neck into a sac when threatened, and its bite can be fatal to humans.

Twining plants: Climbing plants (see) which ascend by winding their slender shoots helically around a vertical support. In most T.p. the direction of twining is constant, and usually the ascending helix rotates to the left (e.g. runner bean). Only a few species spiral to the right, e.g. hop and honeysuckle, or show a variable direction of twining, e.g. black bindweed *(Polygonum convolvulus).*

At first the newly germinated shoot grows orthotropically by negative geotropism. At a later stage of development, the shoot tip becomes plagiogeotropic and displays marked helical, cyclonastic movements (see Cyclonasty), e.g. the shoot tip of the hop describes a rising circle of about 50 cm diameter. The time required for a complete cycle varies between 2 and 9 hours, depending on the plant species. If the "searching" tip encounters a more or less vertical support, it twines around it. Since the twining shoot is not sensitive to touch, thin supports such as wire are climbed by continuation of the cyclonastic movements (circumnutation), whereas progress up thicker supports depends on lateral geotropism (see Geotropism). In plants that lack lateral geotropism, the helix is tightened upon the support by Torsion (see).

Twin research: study of twins, triplets, quadruplets, etc. of plants, animals and humans. *Monozygotic* or *identical twins* ("true twins") arise from the cleavage of a single egg or organ primordium, whereas *nonidentical* or *dizygotic twins* develop separately but simultaneously from different eggs in the same female.

T.r. is primarily an area of Human genetics (see). Dizygotic twins behave genetically as normal siblings, i.e. 50% of their genes are the same, whereas monozygotic twins are genetically identical. Monozygotic twins may, however, display genetic differences, due to differences in chromosome behavior during cell division, and mutation in one sibling.

By T.r. it is possible to differentiate and quantify the genetic and environmnetal influences that contribute to the establishment of a particular phenotype. The identity of monzygotic twins is generally genetically determined (see Concordance), but it may also be due to identical developmental conditions. Differences in phenotype (see Discordance) are usually due to environmental influences. The environmental stability or instability of genes can be studied very conveniently in monozygotic twins, which since early childhood have grown and developed in different environments.

See Multiple births, Twin diagnosis.

Twin species: morphologically very similar species, which inhabit the same geographical region (see Sympatry), and are reproductively isolated from each other. When morphological differentiation is difficult, T.s. can be distinguished by cytological, physiological or ethological methods.

Twinspot: see *Ploceidae*.

Twin spots: see Chimera.

Twitch time: the duration of a single twitch of a muscle (see Muscle contraction). The shortest T.ts. are shown by insect flight muscle (housefly 0.003 s, bee 0.005 s, dragonfly 0.035 s). Mammalian limb muscles have T.ts. between 0.1 and 0.3 s. The T.t. of some smooth muscle is in the order of minutes.

Twite: see *Fringillidae*.

Two-banded monitor: see Water monitor.

Two-pronged bristletails: see *Diplura*.

Two-strand crossing over, *two-strand exchange:* a reciprocal exchange of segments between paired chromosomes during meiosis. Crossing over occurs twice between the same two chromatids (Fig.). See Crossing over, Four-strand crossing over, Three-strand crossing over, Recombination.

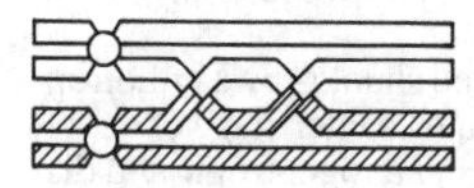

Two strand crossing over

Two-toed amphiuma: see *Amphiumidae*.

Tychoplankton: see Plankton.

Tylenchida: see *Nematoda*.

Tylas: see *Oriolidae*.

Tylonycteris: see *Vespertilionidae*.

Tylopilus: see *Basidiomycetes (Agaricales)*.

Tylosis: see Wood.

Tymoviruses: see Virus groups.

Tympanal organs: Chordotonal organs (see) of insects which serve for the susception of sound waves. They may be located, e.g. on the legs or body of grasshoppers and locusts, and they all have the same basic structure, i.e. a thin cuticular membrane appressed to a trachea, forming a "drum" or tympanum, associated with a group of scolopidia (see Scolopidium).

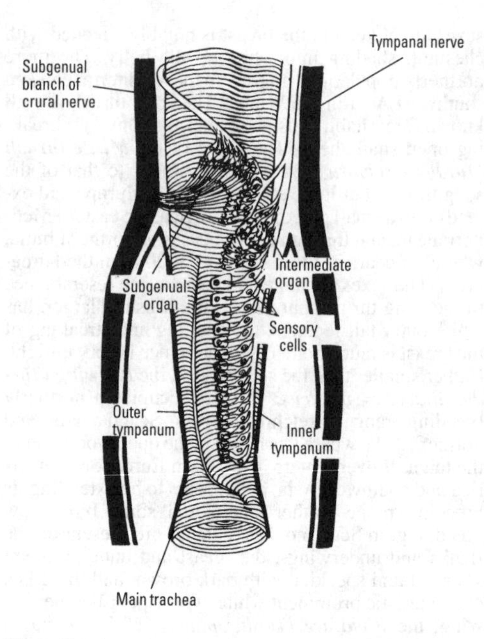

Fig.1. *Tympanal organs.* Tibial organ of a long-horned grasshopper (family *Tettigoniidae).*

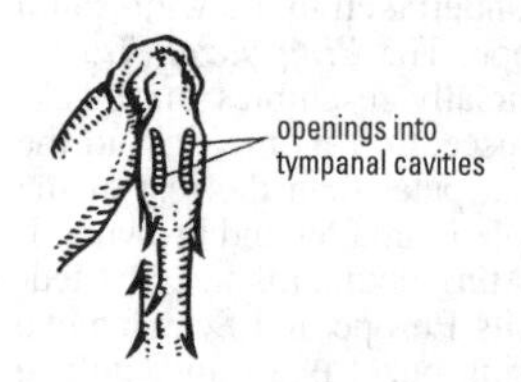

Fig.2. *Tympanal organs. Surface view of the tibial tympanal organ of* a long-horned grasshopper (family *Tettigoniidae),* showing the slit-like openings into the tympanal cavities.

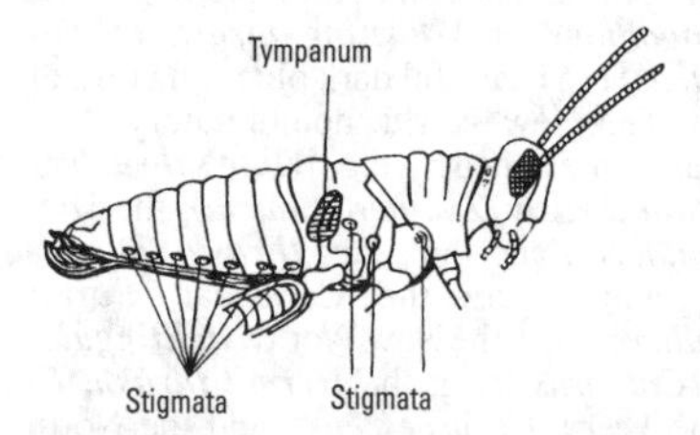

Fig.3. *Tympanal organs.* Side view of a short-horned locust (family *Acrididae),* with wings removed, showing the position of the tympanum on the first abdominal segment.

Tympanic chamber: see Auditory organ.

Tympaniform membranes: see Syrinx.

Tympanuchus: see *Tetraonidae*.

Tympanum: see Syrinx.

Type organism: see Structural type.

Type specimen: see Structural type.

Typhlonectidae: see *Apoda*.

Typhlopidae, *blind snakes:* a family of small, primitive, wormlike Snakes (see), adapted to a subterranean, burrowing existence, and found throughout the tropics and subtropics. The fused skull bones form a container-like skull. Vestigial eyes are hidden beneath the head scales, and the entire body is covered with uniform, smooth, glossy scales. The very short tail often ends in a spine. The upper jaw is almost perpendicular to the longitudinal body axis, and it carries teeth. The lower jaw lacks teeth, except in the genus, *Anomalepis,* which has a single tooth in the lower jaw. Pelvic bones are entirely absent, or may be represented by a small bony bar. Systematically, the position of the *T.* is unclear, and a close relationship with the *Anguinidae* seems unlikely. The family contains more than 200 species, all of which feed on ants and termites. It has been shown that *Ramphotyphlops nigrescens* is able to follow week-old trails left by foraging worker ants, enabling it to track down the ant larvae on which it feeds [J. Webb and R. Shiner, *Animal Behavior,* ***43,*** 941 (1992)]. Some species are viviparous, while others lay eggs. The only European species is the 20 cm-long, egg-laying ***Vermiform blind snake*** *(Typhlops vermicularis),* whose range extends from Southwest Asia into Jugoslavia and Greece, where it lives under stones in dry bushy country with rock debris and scree. ***Peter's blind snake*** *(Typhlops dinga)* of tropical West Africa is the largest species, growing up to 75 cm in length. In Southeast Asia the most common representative is the ***Common blind snake*** *(Typhlops braminus).*

Typhlops: see *Typhlopidae.*

Typhlosole: see Digestive system.

Typhlosynbranchus: see *Synbranchiformes.*

Typhoid: see *Salmonella.*

Typhus: see *Rickettsiae.*

Tyr: standard abbreviation for Tyrosine.

Tyramine: a Biogenic amine (see), formed in animal tissues and by bacteria by decarboxylation of L-tyrosine. It also occurs in many plants, e.g. milk-thistle, broom, shepherd's purse. It has been used medicially on account of its hypertensive action and its ability to cause uterine contraction.

Tyrannidae, ***tyrant flycatchers:*** a family (over 350 species in 115, mainly monotypic genera) in the order *Passeriformes* (see). Tyrant flycatchers are found all over the Americas, mainly in forest habitats, with some species adapted to open country. They range in length from 10 to 40 cm, and they possess hook-tipped bills. The 2 front toes are joined as in the cotingas and manakins. Plumage is inconspicuously colored gray and/or green, but a characteristic bright yellow or orange crest is present, which is erected only when the bird is excited. Members of the genus, *Tyrannus,* have an upright stance and they habitually occupy an observation point from which they dart out to catch insects, thus showing convergence with the Old World flycatchers.

Tyrannosaurus: see Dinosaurs.

Tyrian purple: 6,6′-dibromoindigotin, a violet-purple pigment containing two brominated and oxidized indole ring systems. It is formed from the colorless glandular secretion of marine gastropods known as drills (e.g. *Purpura* and *Murex,* rachiglossate genera of the suborder *Neogastropoda).* The secretion becomes violet-purple when exposed to light. In antiquity and in the Middle Ages it was a much prized and expensive fabric dye, and it was prepared mostly from *Murex brandaris* and *Murex trunculus.* The structural elucidation (Friedlander, 1909–1911) was performed on T.p. prepared from the hypobranchial body of *Murex brandaris.*

Tyroglyphidae: see *Acaridae.*

Tyroglyphus: see *Acaridae.*

Tyrophagus: see *Acaridae.*

L-Tyrosine, ***Tyr:*** an essential proteogenic amino acid, which is both glucoplastic and ketoplastic. It is an aromatic amino acid, i.e. its structure contains a benzene ring, and it is a product the Shikimic acid (see) pathway of biosynthesis in plants and bacteria. In the animal body, it can be synthesized from L-phenylalanine, and is itself the precursor of adrenalin, thyroid hormones and melanin.

HO—C_6H_4—CH_2—CH(NH_2)—COOH

L-Tyrosine

Tyrothricin: a mixture of different polypeptide antibiotics, discovered as early as 1939 in culture filtrates of *Bacillus brevis.* The mixture consists of about 80% basic cyclic decapeptides known as tyrocidins, and about 20% neutral Gramicidins (see).

Tyto: see *Strigiformes.*

Tytonidae: see *Strigiformes.*

U

Uakaris, *Cacajao:* a genus of the subfamily *Pitheciinae.* U. are medium-sized, exclusively vegetarian Cebid monkeys (see) of the infraorder *Platyrrhini* (see New World monkeys), inhabiting the rainforest of the upper Amazon basin, especially in areas subject to seasonal flooding. *C. calvus* ***(Bald*** or ***White uakari)*** and *C. rubicundus* ***(Red uakari)*** have bare crimson faces, bald heads and long shaggy, yellowish white *(calvus)* or rust-red *(rubicundus)* fur. *C. melanocephalus* ***(Black-headed uakari)*** has a less shaggy, yellowish brown coat with red legs and tail, and black arms, hands and feet.

Ubiquinone, ***coenzyme Q:*** a low molecular mass, lipophilic electron transport component of the respiratory chain. Structurally, it is a 2,3-dimethoxy-5-methyl-benzoquinone, carrying an isoprenoid side chain at position 6. According to the number of carbon atoms in the side chain, the different Us. are designated U-30, U-35, U-40, U-45 and U-50. Us. have a wide natural occurrence in animals and plants. They are mitochondrial lipids, structurally related to vitamins K and E; but they are not vitamins, since they can be synthesized in the organism from tyrosine (precursor of the ring system) and from isoprene units (precursors of the side chain).

In the electron transport chain they operate between the flavoproteins and the cytochromes. They are reduced by a specific flavoprotein to ubihydroquinone, which becomes reoxidized by transferring electrons to cytochrome *b.* U./ubihydroquinone is therefore a redox system.

O
H_3CO CH_3
H_3CO H
n
O

Ubiquinone

Ubiquitous species: a euryecious plant or animal species occurring in various habitats. U.ss. do not usually display a wide geographical distribution, whereas Cosmopolitan species (see) are found worldwide. The term, ubiquitous, refers to habitat, whereas the term, cosmopolitan, refers to geographical distribution.

Uca: see Fiddler crabs.

Uegitglanis: see *Clariidae.*

Ulmaceae, ***elm family:*** a family of dicotyledonous plants containing about 150 species in about 13 genera, distributed mainly in the temperate regions of the Northern Hemisphere. They are trees with simple, stipulate, asymmetrical leaves. The unisexual or hermaphrodite, 4- to 6-merous flowers occur singly or in clusters, are wind pollinated, and have an inconspicuous herbaceous perianth, The superior ovary develops into a single-seeded, compressed, dry or slightly fleshy, often winged fruit (nut).

About 30 species of ***Elm*** *(Ulmus)* occur in northern temperate regions and in the mountains of tropical Asia. They are large (> 30 m), deciduous trees, usually with suckers. Their flowers appear before the leaves; the perianth is campanulate, persistent, and usually 4-5-lobed; anthers are reddish and the ovary is usually unilocular. All produce a compressed fruit with broad wings. *U. diversifolia (= U. minor,* ***Lock elm, Small-leaved elm)*** and *Ulmus glabra (= U. montana = U. scabra,* ***Wych elm)*** are both planted as avenue and park trees. They have, however, been decimated in recent years by Dutch elm disease, caused by the fungus *Ceratocystis ulmi,* the spores of which are carried by bark beetles *(Scolytidae),* mainly *Scolytus scolytus* (Elm bark beetle). The fungus prevents water transport by blocking phloem vessels, so that the tree dies rapidly.

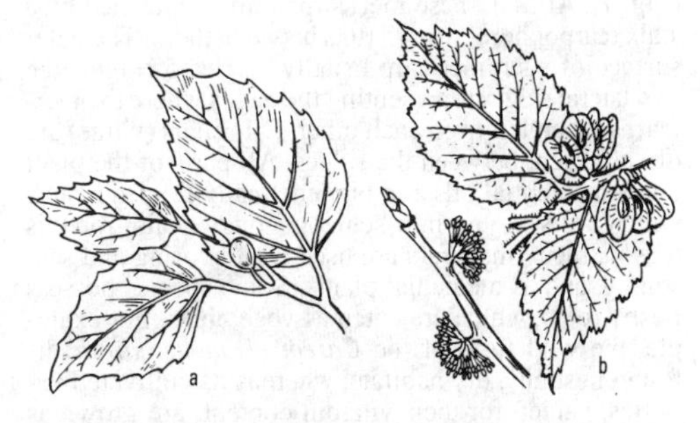

Ulmaceae. a Southern hackberry *(Celtis australis). b* Small leaved elm or lock elm *(Ulmus minor* auct., *Ulmus diversifolia* Melville).

The ***Fluttering elm*** *(U. laevis)* attains 35 m; it is less common than the above two species, although it is widely distributed across most of Europe (not in the Iberian peninsula), and is known in Britain only as a nonindigenous ornamental. Unlike the other two elms, its flowers and fruits have long stalks. The wood is also less valuable, being porous with a very wide area of sapwood.

"Nettle trees" or "Hackberries" (genus *Celtis)* are elegant, fast-growing, medium-sized trees. *Celtis occidentalis* (the North American ***Western hackberry)*** has cherry-like fruit and is occasionally planted as an ornamental. *Celtis australis* ***(Southern hackberry;*** southern Europe, Asia Minor, North Africa) has sloe-like, brown, sweet fruits, and its wood is used for carving and for the manufacture of flutes. Other examples of the genus are: *C. caucasia* ***(Caucasian nettle tree;*** eastern Bulgaria, western Asia); and *C. laevigata (= C. mississippiensis;* ***Mississippi hackberry;*** southeastern USA).

Ulmus: see *Ulmaceae.*

Ultradian rhythm: see Biorhythm.

Ultrahigh heating, ***ultrapasteurization:*** a protective method for the sterilization of food, particularly milk. The material is heated to temperatures between 135 and 150 °C for 1 to 2 seconds.

Ultrapasteurization: see Ultrahigh heating.

Ultrasaprophytic conditions: see Saprophytic system.

Umbel: see Flower, section 5) Inflorescence.

Umbelliferae, *Apiaceae, **umbellifer family***: a family of dicotyledonous plants containing about 200 genera and 2 700 species, which are cosmopolitan but distributed mainly in the Northern Hemisphere. They are mostly herbs, with alternate, exstipulate often multiply divided, pinnate leaves. Petioles usually form well developed sheaths at the leaf bases. Flowers are 5-merous, white or yellow, usually arranged in a compound umbel, sometimes a simple umbel. The compound umbel (often subtended by an involucer of bracts) is formed from partial umbels, each of which is subtended by an involucer of bracteoles, called an involucel. The inferior, 2-celled (rarely 1-celled) ovary carries 2 styles. Each style base is often enlarged to form a nectar-secreting stylopodium, which attracts pollinating insects, especially beetles and flies. The two-seeded fruit is a dry schizocarp with a septum (commissure) between the carpels. A split usually occurs along the septum, dividing the carpels into 2 mericarps, each containing a single seed (Fig. 1). At first, these mericarps remain attached by a stalk (carpophore), which runs between them. The outer surface of each mericarp usually carries 5-ridges, the two lateral ridges representing the edges where the mericarps separated from each other. Oil canals (vittae) are often present between the ridges. All parts of the plant contain essential oils and biogenetically related resins, which are secreted into schizogenous canals; for this reason, many members are used for flavoring and seasoning, and as medicinal plants. Some species possess fleshy roots, which are eaten as vegetables. For example, the wild form of the ***Carrot*** *(Daucus carota),* is found in sunny, dry habitats, whereas its cultivated varieties, valued for their vitamin content, are grown as vegetable and fodder plants in all temperate and subtropical regions; in Europe they have probably been used since the 13th century.

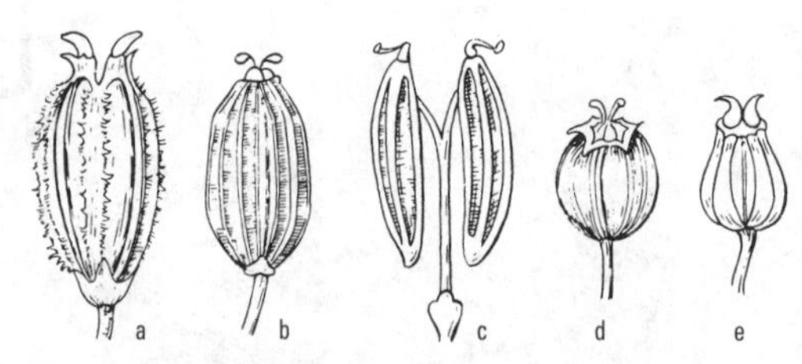

Fig.1. *Umbelliferae.* Fruits. *a* Cummin *(Cuminum cyminum). b* Dill *(Peucidanum graveolens). c* Caraway *(Carum carvi). d* Coriander *(Coriandrum sativum). e* Parsley *(Petroselenium crispum).*

Celery *(Apium graveolens)* is an almost cosmopolitan plant in saline habitats; different varieties are cultivated as seasoning and vegetable plants. Another root vegetable is the ***Parsnip*** *(Pastinaca sativa),* grown widely in well-fertilized soils. ***Parsley*** *(Petroselinum crispum),* a plant native to the Mediterranean region, is used both as a seasoning and as a vegetable; roots as well as leaves are used. The poisonous ***Fool's parsley*** *(Aethusa cynapium),* a weed of nitrogen-rich habitats, differs from true parsley by the possession of 3–4 recurved bracteoles on the outer side only of the partial umbels. Other seasoning plants are ***Chervil*** *(Anthriscus cerefolium)* from southeastern Europe, ***Lovage*** *(Levisticum officinale)* from western Asia, and ***Dill*** *(Anethum graveolens)* from the eastern Mediterranean, whose fruits are used as a pickling spice. Other exploited plants are: the cultivated form of ***Fennel*** *(Foeniculum vulgare),* whose fleshy leaf sheaths are eaten as a vegetable, especially in the Mediterranean region; ***Anise*** *(Pimpinella anisum),* which yields aniseed; ***Coriander*** *(Coriandrum sativum);* common ***Caraway*** *(Carum carvi),* which is also a familiar wild species in Europe; and ***Cumin*** *(Cuminum cyminum)* which is native to the Mediterranean region. Garden ***Angelica*** *(Angelica archangelica)* grows on river banks and in marshy areas, where it attains heights up to 2.5 m; the roots are used for medicinal purposes and for the preparation of a liqueur.

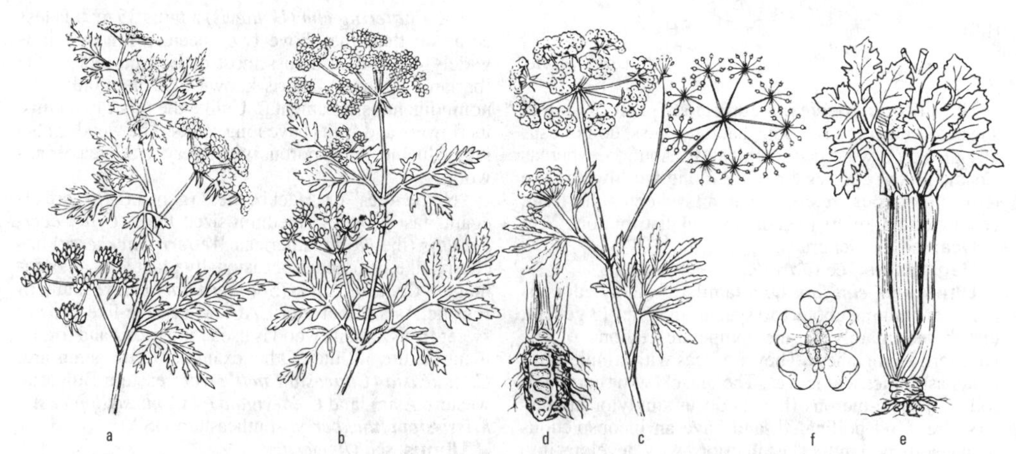

Fig.2. *Umbelliferae. a* Fool's parsley *(Aethusa cynapium). b* Hemlock *(Conium maculatum). c* Cowbane *(Cicuta virosa). d* Root of cowbane. *e* Cultivated celery *(Apium graveolens). f* Single flower of celery.

Some umbellifers are dangerously poisonous, especially ***Hemlock** (Conium maculatum),* which contains the alkaloid coniine; it favors waste dumps and boundary fences, and it possesses a characteristic smooth, red-flecked stem and unpleasant odor. Also poisonous is cowbane or ***Water hemlock** (Cicuta virosa),* which is found on river banks and in marshy areas, and possesses a chambered tuberous rhizome.

In Europe, the only protected umbellifer is the ***Sea holly** (Eryngium maritimum)* of sandy and shingly shores, which has blue-green, glabrous, spiny leaves and bracts, and a simple umbel.

Umbellifer family: see *Umbelliferae.*

Umbelliferone, *7-hydroxycoumarin:* a coumarin derivative, found in the bark of *Daphne mezerium,* released from damaged leaves of *Hydrangea macrophylla,* and obtained by the dry distillation of resin from a number of members of the *Umbelliferae.* It is also present in chamomile oil and carrots.

Umbellula: see *Pennatularia.*

Umbilical circulation: see Allantois.

Umbra: see *Umbridae.*

Umbrellabird: see *Cotingidae.*

Umbrella fir: see *Taxodiaceae.*

Umbridae, *mudminnows:* a family of freshwater fishes found in southern Europe and North America. It belongs to the order *Salmoniformes,* and it is closely related to the *Esocidae* (Pikes). Three genera and 5 species are recognized: *Novumbra* (1 species), *Dallia* (1 species, the ***Alaska blackfish**),* *Umbra* (3 species). Respiratory gas exchange is supplemented by the swimbladder, so that these fishes can survive in oxygen-poor waters. Some species bury themselved temporarily in the bottom mud.

Unconditioned reflex: see Conditioned reflex.

Unconditioned stimulus: see Conditioned reflex.

Underwing moths, *Catocala:* a genus of noctuid moths (see *Noctuidae)* containing more than 200 species, distributed mostly in the northern temperate zone, with some species in Central America, Southeast Asia and India. They all possess cryptically colored forewings, which camouflage the resting moth against various backgrounds such as tree bark. The hindwings, however, carry bands of bright color. The ***Blue underwing*** or ***Clifden nonpareil** (Catocala fraxini* L.; Europe, temperate Asia to eastern Russia and Japan) is a large species (wingspan 10 cm), the caterpillar of which feeds on species of *Populus* (poplar and aspen); the hindwings are pale brown with a transverse blue band. The ***Red underwing** (C. nupta* L.; wingspan 8.0 cm) has a similar distribution to *C. fraxini,* and its larvae feed on poplar and willow; the hindwings are red with two transverse dark bands and a pale margin. New World species include the ***White underwing** (C. relicta* Walker), found from Canada to Arizona; its hindwings are actually black with a white transverse band parallel to the outer margin, as well as a white marginal band; the larvae feed on poplar and willow. The ***Bronze underwing** (C. cara* Guenée), with a wingspan of almost 9 cm, is found from Canada to Florida; its hindwings are pale red to orange with a black base, a black central band and black marginal band, while the outer margin is pale yellow.

All European species are protected.

Unfolding movements, *opening movements:* the opening of leaf and flower buds due to increased unilateral growth. In the bud, leaves that are folded upward are opened by the sudden onset of growth on their upper surface. Those that are folded downward are opened by growth on their underside. Illumination usually causes flowers to open, although night flowering species (e.g. *Silene nutans)* show the opposite behavior and open in the dark. Hyponasty (see) is a typical U.m.

Unguis: see Claws, hoofs and nails.

Ungula: hoof; see Claws, hoofs and nails.

Ungulata: collective term for the orders of hoofed mammals, i.e. the modern *Perissodactyla* (see) and *Artiodactyla* (see), and the extinct *Condylartha* from which they evolved. Fossil ungulates occur from the Tertiary to the present. Evolution of the *U.* in the Tertiary is characterized by the following important developments: increase in the size of the animal, reduction of the number of digits, elongation of the metatarsus, and modification of the teeth for grazing a vegetarian diet (high-crowned, cement-covered teeth with a diastema between front and rear). The *Perissodactyla* appeared in the Eocene and reached their evolutionary peak in the Middle Tertiary, while the *Artiodactyla* underwent a burst of adaptive radiation in the Eocene and early Oligocene.

Unguligrade: walking on the tips of digits which have hooves. Ungulates are therefore U. animals. In the *Perissodactyla* (odd-toed ungulates; horse, tapir, rhinoceros) the weight-bearing axis of the foot lies along the third toe. In the *Artiodactyla* (even-toed ungulates; pig, hippopotamus, camel, deer, giraffe, sheep, antelope, cattle) the weight bearing axis lies between the third and fourth toes. See Digitigrade, Plantigrade.

Unica: see *Felidae.*

Uniformitarianism: see Actualism.

Unilocular hydatid disease: see Dog tapeworm.

Unio: see Freshwater mussels.

Unionoida: see *Bivalvia.*

Units of enzyme activity: see Enzymes.

Unken reflex: display of the brightly colored underside by amphibia when threatened. Newts, for example, hold themselves rigidly immobile with tail and chin elevated, displaying their brighly colored undersides. These animals are often also poisonous and inedible (e.g. European Alpine newt), and predators, birds in particular, learn to recognize the color display and the avoid them. A similar reflex is shown by Fire bellied toads, which are known in German as "Unken".

Unscheduled DNA replication: see DNA replication.

Unsegmented latex duct: see Excretory tissues of plants.

Upas tree: see *Moraceae.*

Upper montane forest zone: see Altitudinal zonation.

Upupa: see *Coraciiformes.*

Upupidae: see *Coraciiformes.*

Uraeginthus: see *Ploceidae.*

Uraeotyphlidae: see *Apoda.*

Uranoscopus: see Star-gazer.

Urate oxidase: see Uricase.

Urban ecology: an ecological discipline devoted to the study of cities (usually large cities) as a living environment for humans, plants and animals. It is characterized by extreme microclimatic conditions, specific environmental rhythms, high pollutant concentrations in the air and water, etc. Thermophilic organisms tend to accumulate in permanently warm areas, rock inhabitants colonize walls of buildings, various coelenterates gather in cellars and beneath rubble, halophilic species accumulate in chemical and processing plants, and many pests adapt physiologically and thrive in the anthropogenically altered environment. As a measure of the effects of urbanization (e.g. of pollution), the presence (or absence) and numbers of certain indicator species is often more informative than the use of physical and chemical measurements.

Urea: main nitrogenous excretory product of mammals. It is synthesized in the liver by the Urea cycle (see), and excreted in the urine. It is colorless, odorless, crystallizable, and highly water-soluble. Between 80 and 90% of the human dietary nitrogen intake is excreted as U. (25–35 g daily). Some organisms possess Urease (see), which catalyses the hydrolytic cleavage of U. to ammonia and carbon dioxide.

Urea cycle, ***arginine-urea cycle, ornithine cycle, Krebs-Henseleit cycle:*** a metabolic cycle present in mammals and other ureotelic animals (e.g. adult amphibians), which results in the synthesis of urea from carbon dioxide, ammonia and the α-amino nitrogen of aspartic acid (Fig.). The process is energy-dependent: synthesis of one molecule of urea or L-arginine requires 3 molecules of ATP, and involves the expenditure of 4 high energy bonds (2 molecules of ATP are cleaved to ADP and inorganic phosphate and one is cleaved to AMP and pyrophosphate, the latter being further hydrolysed to inorganic phosphate). The U.c. is catalytic, and is based on the recycling of the catalytic molecule, L-ornithine. The primary function of the U.c. is to convert waste nitrogen into nontoxic, soluble urea, which can be excreted. The cycle may also be completed, not by hydrolysis of L-arginine to urea plus ornithine, but by transfer of the amidine group to glycine, to form L-ornithine and guanidinoacetic acid (the precursor of creatine; see Phosphagen). A further function is synthesis of the proteogenic amino acid, L-arginine. The U.c. has, in fact, evolved from an original pathway for L-arginine synthesis. In animals, the U.c. is more or less primed by the synthesis of L-ornithine from L-glutamate, and to some extent by syntheis of L-ornithine from the products of degradation of L-proline. Dietary L-arginine may serve to supplement the cycle, or the synthesis of L-ornithine and its conversion into L-arginine may supplement the dietary requirement for L-arginine.; the balance of these two processes depends on the animal species, its physiological state and its diet, e.g. many young, growing animals have a dietary requirement for L-arginine, whereas the adult appears to be able to synthesize its total requirement.

The chief site of the U.c. is the liver. Conversion of L-ornithine to L-citrulline and the synthesis of carbamoyl phosphate occur in the mitochondrial matrix, and all the other reactions of the U.c. occur in the cytoplasm. Kidney cytoplasm contains the enzymes for conversion of L-citrulline to L-ornithine, but kidney mitochondria lack the necessary enzymes for converting L-ornithine to L-

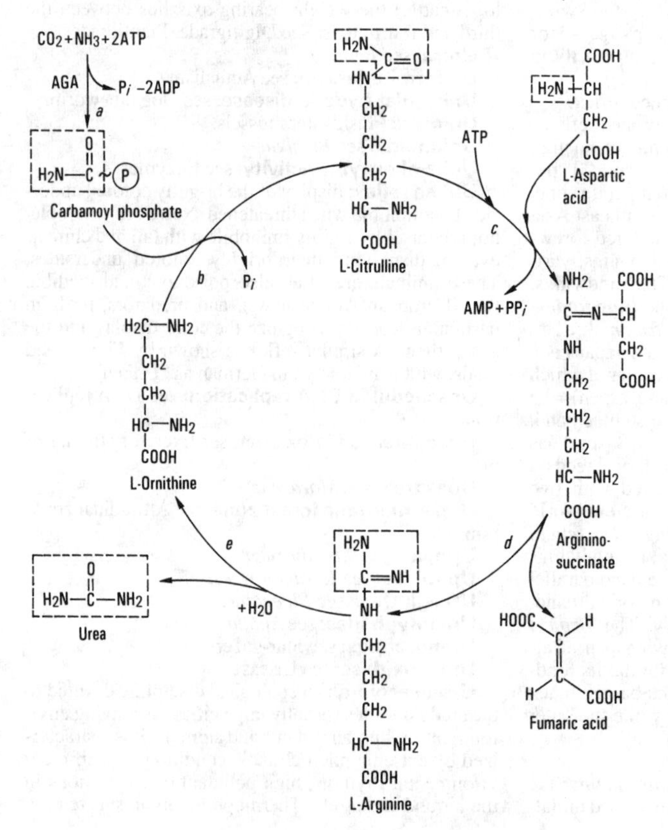

Urea cycle. a Carbamoyl phosphate synthetase (ammonia). *b* Ornithine carbamoyltransferase. *c* Argininosuccinate synthetase. *d* Argininosuccinate lyase. *e* Arginase. *AGA* = *N*-Acetylglutamic acid, a stimulatory allosteric effector of carbamoyl phosphate synthetase.

citrulline, and for synthesizing carbamoyl phosphate. Some L-citrulline is tranported from the liver to the kidneys, where it is converted to L-ornithine and urea.

Fumarate produced in the U.a. (Fig.) can enter the mitochondria and be converted to oxaloacetate (by reactions of the tricarboxylic acid cycle). Transamination of the latter to aspartate, which then participates in the U.a., represents a means of chanelling waste nitrogen into urea synthesis.

Urease: a hydrolase enzyme with high catalytic activity, which catalyses the hydrolysis of urea (the main nitrogenous excretory product of mammals) into carbonic acid and ammonia: $O{=}C(NH_2)_2 + 2H_2O \rightarrow H_2CO_2 + 2NH_3$. This cleavage of urea is an important part of the nitrogen cycle of the biosphere, since the ammonia (as ammonium ions, or after oxidation to nitrate) can then be used by bactera, fungi and higher plants. U. is found especially in plant seeds and microorganisms, as well as invertebrates (crabs, mussels). It is not present in vertebrates. Apart from urea, it only attacks hydroxy- and dihydroxyurea, which also act as noncompetitive inhibitors. The U. of soybean is a protein of M_r 489 000, which at neutral pH in 0.1% sodium dodecyl sulfate dissociates into 8 subunits, each of which consists of two covalently bound chains (each chain M_r 30 000). U. is remarkably resistant to denaturation by its own substrate, urea, which at 8 M concentration denatures numerous proteins. In 8–9 M urea, U. dissociates into M_r 60 000-subunits, which still possess catalytic activity. U. was the first enzyme to be crystallized (Sumner 1926).

Uredinales: see *Basidiomycetes.*

Uredospores: single-celled "summer spores" of the rust fungi (see *Basidiomycetes,* Fig. 3.3). Each uredospore arises from the swollen end of a hypha, and it contains a pair of nuclei. U. occur in groups in small, linear, rust-colored sori, which rupture the epidermis of the host. As soon as they are ripe, they produce a simple germination tube, which is capable of entering a stoma and infecting a new host plant. Infected plants produce millions of U., which serve to spread the fungal disease during the summer.

Urera: see *Urticaceae.*

Urethra: canal for the conduct of urine to the exterior. In teleost fishes, the U. usually exits via a separate papilla. In most other nonmammalian vertebrates it empties into the cloaca. In female mammals it extends from the bladder to its external orifice, which is embedded in the anterior wall of the vagina. In male mammals the entire tubular structure from the bladder to the tip of the penis is called the U. That portion between the entry of the vas deferens and the exterior (which therefore serves for the conduct of both urine and seminal fluid) is also called the *urogenital duct.*

Uria: see *Alcidae.*

Uric acid, ***2,6,8-trihydroxypurine:*** a colorless, weakly acidic, odorless and tasteless purine derivative. In reptiles, birds and many invertebrates (notably insects), it is the main nitrogenous excretory product, i.e. in these organisms waste nitrogen (as ammonia and as amido and amino nitrogen) is chanelled into purine synthesis; only a small proportion of the product is incorporated into nucleic acids, and the majority is directly oxidized to U.a. and excreted (purine biosynthesis → adenine → hypoxanthine → xanthine → U.a.). Thus, bird excrement (guano) consists of up to 90% of the ammonium salt of U.a. In mammals, adult amphibia and some fishes, the major nitrogenous excretory product is urea, which is produced by an entirely different route and not from the degradation of U.a. (see Urea cycle), but small quantities of U.a. are produced from the turnover of nucleic acids. In most mammals, this U.a. is degraded to allantoin, but not in humans and higher apes, whose urine contains low concentrations of U.a. In fishes, amphibians and many invertebrates (notably the *Echinodermata)* the U.a. from nucleic acid turnover is degraded to urea (U.a. → allantoin → allatoic acid → urea + glyoxylic acid).

Gout is the result of deposition of U.a. in synovial capsules of the joints. Certain types of bladder and kidney stones also consist largely of U.a.

Uricase, ***urate oxidase:*** a copper-containing aerobic oxidase which, in the presence of molecular oxygen, catalyses the oxidation of poorly soluble uric acid into soluble allantoin, with the formation of hydrogen peroxide. It occurs in all vertebrates and invertebrates with the exception of many insects (only flies and related groups possess U.). U. is particularly active in the liver, where it is stored in uricase-rich microbodies called uricosomes, The enzyme from porcine liver has M_r 125 000 and consists of 4 subunits.

Urinary bladder, ***vesica urinalis:*** collection and storage site for urine produced in the kidneys. The urine in released from the U.b. via the urethra at intervals. In elasmobranch fishes and some primitive teleosts, the U.b. or urino-genital sinus is an expansion at the junction of the two kidney ducts. In all other organisms, the U.b. is derived from the stalk of the allantois during embryological development, and is therefore sometimes called the *allantoic bladder.* There is considerable evidence that the U.b. of terrestrial amphibians serves as a water storage organ to prevent desiccation. Some adult lizards and all adult snakes, crocodiles and birds (with the exception of the ostrich) do not have a U.b.

Urinary pigments: compounds reponsible for the yellowish to brown color of urine. Commonest U.p. are urochrome, uroerythrin, stercobilin, indican and urobilin. Urine of nonmammals is usually colorless.

Urine: liquid produced in the kidney (see Kidney function) and temporarily stored in the bladder. Mammalian U. is colored yellowish to brown by pigments.

U. composition varies between species and is very dependent on the diet. The most abundant compound and main nitrogenous excretory product of mammalian U. is urea. Birds, however, excrete nitrogenous waste mainly in the form of crystalline uric acid. In fishes, amphibians and reptiles, the nitrogenous excretory product, guanine, is deposited in the skin, where it is reponsible for a silvery sheen. Herbivores excrete more hippuric acid and potassium than do carnivores.

Formed components of human U. include epithelial cells from the urethra. Changes in urinary composition are important in the diagnosis of many diseases.

Urine production: see Kidney functions.

Uriniferous tubule: see Excretory organs.

Urnatellidae: see *Entoprocta.*

Urobilin: see Bile pigments.

Urobilinogen: see Bile pigments.

Urodela, *Caudata,* ***urodeles, newts*** and ***salamanders:*** an amphibian order, whose adult members have an elongated body and tail, i.e. the tail (which is cylindrical or laterally compressed) is retained by the adult after metamorphosis. Usually 4, relatively nonrobust limbs are present (the *Sirenidae* lack hindlimbs). There are generally 4 fingers and 5 toes. Depending on the species, the body is 5–150 cm in length and it is always naked. The gilled larvae and the adults of aquatic species have a lateral line system (see Fishes), which is sensitive to changes of pressure and to chemical stimuli. In terrestrial forms this lateral line has degenerated. The jaws contain teeth, and ear drums are absent. The eyes are small and sometimes degenerate, but in contrast the sense of smell is highly developed. Urodeles live either permanently in moist terrestrial environments, perhaps seeking out water only for mating and egg laying, or they live permanently in water. In most species the male deposits a spermatophore in the water or on the ground, which the female takes up into her cloaca (sometimes it is transmitted actively into her cloaca), i.e. fertilization is internal. External fertilization of the eggs by the male occurs only in 2 families: *Hynobiidae* and *Cryptobranchidae*. Some species spawn in water, and their gill-bearing tadpoles do not leave the water until they have metamorphosed to the lung-breathing stage (e.g. *Triturus).* In contrast, viviparous species give birth to live, lung-breathing offspring on the land. A range of intermediate forms exists between these 2 extremes. As the tadpole develops the forelimbs appear first (cf. the *Anura,* in which the hindlimbs emerge first). The aquatic larvae display a wide variety of adaptations to the environment, which have no taxonomic significance, e.g. various types of gills, fins on the body and tail, and attachment organs. In several species the female displays brood care; in a very small number of species this behavior is also found in the male. Like anuran tadpoles (see *Anura),* urodele larvae undergo extensive reorganization and differentiation, finally emerging as tetrapods capable of a terrestrial existence. During this process the gills are lost, the branchial clefts fuse, and the larval branchial skeleton is replaced by the adult hypobranchial. In addition, the eyes bulge and acquire lids, and the palate, jaws and skull bones undergo profound changes. In contrast to anuran metamorphosis, however, the spike teeth of the urodele larvae are usually replaced by bicuspid teeth, and the tail is retained. Not all of these changes occur in all species of urodeles, and varying degrees of Neoteny (see) are observed (see entries for individual families).

With few exceptions, the approximately 300 species of urodeles inhabit wet areas of the temperate regions of the Northern hemisphere, extending across the Equator only in South America.

Classification. Nine families in 3 suborders (or superfamilies) are recognized, based mainly on anatomical features, structure of the vertebral column, relative positions of the skull bones, and the distribution of teeth in the skull bones. See separate entry for each family.

1. Suborder (superfamily): *Cryptobranchoidea*

Families: *Cryptobranchidae* (giant salamanders); *Hynobiidae* (hynobiids).

2. Suborder (superfamily): *Salamandroidea.*

Families: *Salamandridae* (salamandrids, newts and salamanders); *Amphiumidae* (amphiumids, congo eels); *Proteidae* (*Necturidae*) (proteids, necturids, olm, mudpuppy and waterdogs); *Ambystomatidae* (mole salamanders); *Dicamptodontidae* (Pacific mole salamanders); *Plethodontidae* (lungless salamanders).

3. Suborder (superfamily): *Sirenoidea*

Family: *Sirenidae* (sirens).

On account of their skull structure and primitive mode of reproduction (external fertilization), the *Hynobiidae* and *Cryptobranchidae* are also known as the "lower urodeles", whereas the remaining 7 families are called "higher urodeles".

Fossil remains are found from the Cretaceous to the present, and well preserved fossils are especially abundant in the Tertiary layers. Phylogenetic separation of urodeles from primitive amphibians probably occurred in the Permian, but ancestral forms have not been found in the Triassic or Jurassic. Compared with ancestral forms, the modern urodeles show many adaptations as well as degenerate features. It is assumed that all ancestral forms were aquatic.

Urogenital system: collective term for the excretory and genital systems of vertebrates, based on their common embryological and phylogenetic origin.

Urogenital duct: see Urethra.

Uroglena: see *Chrysophyceae.*

Uromastyx: see Spiny-tailed lizards.

Uronic acids: aldehyde carboxylic acids formed by oxidation of the terminal primary alcohol group (to a COOH group) of aldoses. Biological conversion of aldoses into U.a. proceeds via the nucleotide diphosphate sugar, e.g. UDP-glucose (uridine diphosphate glucose) is oxidized enzymatically to UDP-glucuronic acid, coupled with the reduction of NAD^+ to $NADH + H^+$, a reaction catalysed by UDP-glucose dehydrogenase. Direct oxidation of free aldoses to free U.a. may be possible in animals, but is uncertain. U.a. are named by adding the ending "-uronic acid" to the stem of the parent monosaccharide, e.g. D-glucuronic, D-galacturonic and D-mannuronic acid are derived from glucose, galactose and mannose, respectively. U.a. tend to form lactones, usually γ-lactones. They give the usual chemical reactions for sugars and are widely distributed in nature as components of glycosides, polyuronides, polysaccharides and mucopolysaccharides (in particular the mucopolysaccharide, hyaluronic acid, an important constituent of animal connective tissue and synovial fluid). Many hydroxylated compounds (alcohols and phenols) are excreted as conjugates of glucuronic acid (glucuronides). Thus, hydroxylated metabolic products of drugs and other xenobiotics, as well as endogenous products of steroid metabolism are often excreted in the urine as their glucuronides, e.g. estriol 17-glucuronide, estrone 3-glucuronide. Glucuronides are synthesized in the liver by reaction of the hydroxylated compound with UDP-glucuronic acid, catalysed by the enzyme UDP-glucuronyltransferase. In general, glucuronidation may be considered to be a detoxication reaction, since it renders xenobiotics more soluble and more readily excretable.

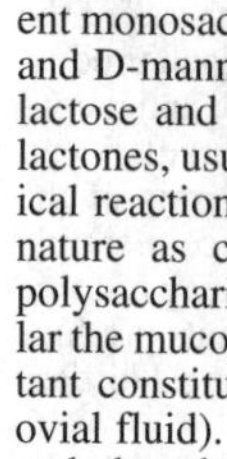

```
   C(=O)H
   |
H—C—OH
   |
HO—C—H
   |
H—C—OH
   |
H—C—OH
   |
   C(=O)OH
```

Glucuronic acid

In most animals, but not in primates, D-glucuronic acid is a biosynthetic precursor of ascorbic acid (vitamin C).

Uropeltidae, *uropeltid snakes* (Greek *oura* tail, *pelte* shield): a relict group of 43, rough-tailed, burrowing snakes restricted to the mountains and foothills of Central India and Sri Lanka. Uropeltids are small (20–80 cm in length, up to 2 cm in diameter), often brightly colored and/or iridescent snakes, with tiny pointed heads. The eyes are diminutive and the lower jaw is countersunk, and the tail is very short and blunt, ending in an enlarged shield, often with spines, and supported internally by a bony plate. These primitive snakes seem to be phylogenetically close to the *Aniliidae.* The original preferred habitat is the humus layer of dense forests, but some have survived commercial destruction of forests by forming deep tunnels in exposed laterite soils, gaining access to crevices in underlying rocks. They have also been found in compost heaps and well irrigated, overgrown agricultural areas.

Uropeltid snakes: see *Uropeltidae.*

Uroporphyrin, *uroporphyrin III:* the first described natural porphyrin. It is formed by the oxidation of uroporphyrinogen III and excreted in the urine. (Uroporphyrinogen III is derived from 5-aminolevulinate via porphobilinogen, and it serves as an intermediate in the biosynthesis of hemoglobin, cytochromes and chlorophylls.)

Uropygi, *schizomids* and *whipscorpions:* a suborder of the order *Pedipalpi,* class *Arachnida* (see). The suborder contains about 130 species (75 whipscorpions and more than 50 schizomids) with many resemblances to scorpions. Whipscorpions may attain a length of 7.5 cm, whereas schizomids are much smaller with lengths of up to 7 mm. The body consists of a short prosoma and a relatively long, segmented opisthosoma, the last segment carrying a whiplike, multisegmented telson appendage (flagellum), which is long in whipscorpions and much shorter in schizomids. The large, long and leg-like pedipalps are carried flexed and extended horizontally in front of the animal. Often provided with terminal pincers, these pedipalps are used for catching prey, and they are fused at their bases to form a trough, in which food is predigested. The very small chelicerae carry terminal pincers. The first pair of walking legs serve as feelers. An anal spray gland is present in the last abdominal segment, and its secretion (acetic acid with some caprylic acid) is accurately squirted at adversaries, but not at prey. Lung books serve for respiratory gas exchange.

Most uropygids are inhabitants of the tropics and subtropics. They live beneath stones, fallen leaves, loose bark, etc., and are nocturnal in habit. Many species excavate passages in sandy soils. Reproduction is similar to that of scorpions, and females carry the eggs in a secreted egg case at the sexual opening. See Fig.1 of *Arachnida.*

Uropygial gland, *preen gland, oil gland:* a bilobed gland present in most birds, relatively large in some aquatic species, and absent in the ostrich, emu, cassowaries, bustards, frogmouths, and many pigeons, woodpeckers and parrots. It is situated between the dorsal skin and muscles at the dorsal base of the tail. The number of ducts opening to the surface varies according to the species, e.g. a single duct in the hoopoe, 2 in the domestic fowl, and 18 in penguins. Duct papillae are usually bare, except for a tuft of down feathers, known as the uropygial wick. The sometimes unpleasantly smelling, waxy/oily sebaceous secretion of the U.g. has remarkable spreading properties, and it protects the feathers against wetting by water; it consists predominantly of waxes, i.e. esters of long-chain fatty acids and long-chain alcohols. The U.g. is the only true epidermal gland of birds. Other glands include those of the external ear, which secrete waxy material, and those at and around the vent, which secrete mucus. Individual cells of the epidermis also produce lipid sebaceous secretions.

Ursidae, *bears:* a family of very powerful, large, somewhat plump terrestrial mammals in the order *Carnivora* (see), with a short or very short tail and non-prominent canine teeth. There are 6 extant genera and probably only 7 or 8 distinct living species. Eight fossil genera are known from the Middle Miocene of Europe and the Lower Pliocene of North America. Bears are plantigrade animals, i.e. they place the entire area of the sole of the foot against the ground when walking. They are omnivorous, favoring insects and fruit, but also eating flesh, fish and carrion. They occur only in Eurasia and the Americas. The following examples are listed separately: Brown bear, Black bear, Asiatic black bear, Polar bear, Sloth bear, Sun bear, Spectacled bear, Giant panda. See also Climatic laws.

Ursolic acid: an unsaturated pentacyclic triterpene carboxylic acid ($C_{30}H_{48}O_3$), which occurs widely in plants as the free acid, esterified, or as the aglycon of triterpene saponins, e.g. in the wax layer on the surface of apples, pears and cherries, in the skin of bilberries and cranberries, and in the leaves of many members of the *Rosaceae, Oleaceae* and *Lamiaceae.*

Urostyle: see Axial skeleton.

Urtica: see *Urticaceae.*

Urticaceae, *stinging nettle family:* a family of dicotyledonous plants, containing about 600 species in about 41 genera, distributed mainly in the tropics. They are herbaceous plants, rarely soft-wooded trees, with simple, alternate or opposite, usually stipulate leaves. Cystoliths are usually present in the epidermal cells. The 4-5-merous flowers are usually unisexual, mono- or dioecious, small and inconspicuous, and wind-pollinated. Male flowers have 4–5 stamens inserted opposite the perianth segments, and a rudimentary ovary is usually present. In male flower buds, the stamens are inflexed, springing outward when the flower opens and scattering pollen. Female flowers have a 1-celled ovary with a solitary orthotropous ovule; small staminodes are often also present. The fruit is an achene or drupe.

The ***Stinging nettle*** *(Urtica dioica)* and the ***Small nettle*** *(Urtica urens)* are both widely distributed ruderal plants of temperate regions, and both possess stinging hairs. Stinging hairs are also present on species of the tropical genera, *Laportea* and *Urera;* the sting of *Laportea* is particularly painful and its effects last for weeks. Other species lack stinging hairs, e.g. ***Ramie*** *(Boehmeria nivea),* which has been cultivated since antiquity in southern Asia as an economically important fiber plant. Some species of the genus *Pilea* are used horticulturally as leaf plants.

Ustinales: see *Basidiomycetes.*

Uterus, *womb:* an internal female sex organ which houses the developing fertilized egg (embryo), and which communicates with the Vagina (see). In vivaparous animals the U. incorporates various mechanisms for the nutrition of the embryo (see {Placenta).

In invertebrates (in particular the *Arthropoda)* part of the oviduct or vagina may be expanded to form a U. For example, in viviparous *Onychophora,* the embryo is temporarily connected to the U. wall by a nutritional organ resembling a placenta. In viviparopus insects, e.g. aphids and tsetse flies *(Glossinidae),* the anterior section of the vagina is developed into a U., where the larvae hatch and develop, in some cases to the stage of pupation; in tsetse flies the larvae are nourished by a milky secretion from well developed uterine glands.

In vertebrates, the U. is a differentiated part of the Müllerian duct, arising either as a fold of the Wolffian duct (urodeles and elasmobranchs), or as an independent lateral structure of the urinogential tract (higher vertebrates). There are several different types of mammalian U. In the *U. duplex,* two uteri are entirely separate. Partial fusion with a common uterine canal is seen in the *U. biparitus.* The *U. bicornis* is a single U. with two horns, while the *U. simplex* of the higher primates is single hollow structure with no differentition into horns or canal (Fig.1).

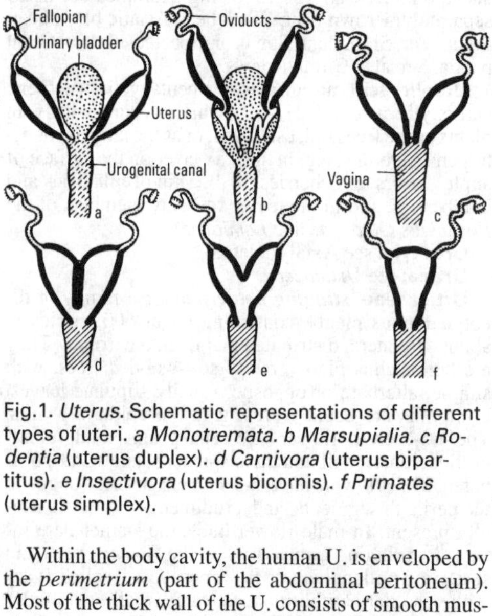

Fig.1. *Uterus.* Schematic representations of different types of uteri. *a Monotremata. b Marsupialia. c Rodentia* (uterus duplex). *d Carnivora* (uterus bipartitus). *e Insectivora* (uterus bicornis). *f Primates* (uterus simplex).

Within the body cavity, the human U. is enveloped by the *perimetrium* (part of the abdominal peritoneum). Most of the thick wall of the U. consists of smooth muscle *(myometrium),* and the inner layer of uterine mucosa is known as the *endometrium.* The endometrium consists of a finely fibrous connective tissue, covered by a single layer of ciliated epithelium, and containing uterine glands (glandulae uterinae), which extend from the surface to the muscle layer beneath. During pregnancy, the U. undergoes considerable expansion.

During the 28 days of the menstrual cycle, the endometrium displays periodic changes; these are regulated by hormonal changes in the ovary, which in turn are under the hormonal control of the pituitary. During the maturation phase of a Graafian follicle, under the influence of follicle stimulating hormone, the endometrium develops to a depth of 7 mm *(proliferation phase).* After rupture of the follicle *(ovulation),* and under the influence of progesterone from the corpus luteum (see Placenta), the secretory activity of the uterine glands increases, and the egg is received by the endometrium. This marks the onset of the *secretory phase,* which lasts from the 14th to the 28th day. If, during this phase, the egg is not fertilized and pregnancy is not initiated, the secretion of progesterone ceases, the corpus luteum undergoes fatty degeneration, and a large proportion of the endometrium (the *decidua menstruationis* or the *funcionalis)* is shed together with the egg (menstrual blood flow). This *desquamation phase* lasts from 3 to 6 days. The basal layer of endometrium remains and serves for renewed development during the subsequent proliferation phase. If the egg is fertilized, a trophoblast is formed and the endometrium is retained (see Placenta).

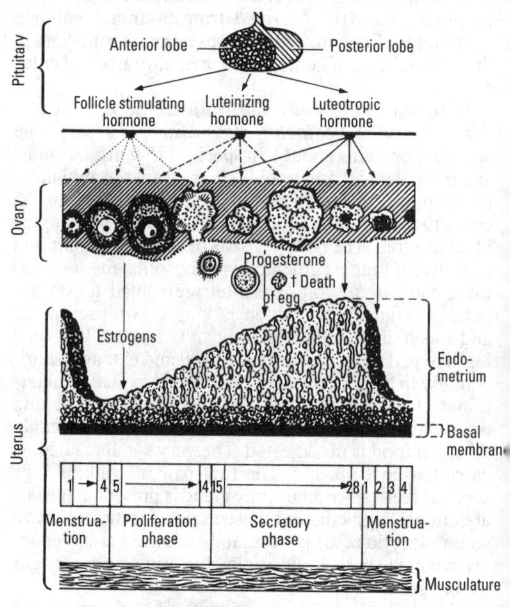

Fig.2. *Uterus.* Cycle of processes occurring in the human uterine mucosa and ovary in the absence of egg fertilization.

Utricularia: see Carnivorous plants.

Utriculus (utricle) of the inner ear: see Auditory organ.

V

Vaccine: a term coined by Edward Jenner for the preparation of cowpox virus, which confers immunity to smallpox when scratched into the arm of a human. The preparation consists of the serum or lymph from the pustules on the udder of a cow with cowpox. In humans, this preparation ("vaccine") causes a very mild disease (vaccinia), with formation of a single poxlike lesion at the site of application. Either the cowpox and smallpox viruses are distinct species, but sufficiently related so that one immunizes against the other, or the cowpox virus is derived from the smallpox virus by passage through cows. V. is now used as a general term for any material injected for the purpose of conferring protective immunity. See Immunity, Inoculation.

Vaccinium: see *Ericaceae.*

Vagina: the final section of the internal female genital tract, which receives the male copulatory organ (see Cirrus, Penis), and which serves to conduct eggs or live young to the exterior. The invertebrate V. often carries a dorsal, unpaired Seminal receptacle (see). Dorsal, paired glands are also frequently present, and these usually produce a cement material. Ventrally, the invertebrate V. often carries a copulatory pouch (bursa copulatrix), which receives sperm during copulation, before it is transferred to the seminal receptacle. In some invertebrates, this copulatory pouch communicates separately with the body surface, e.g. in the *Lepidoptera.*

In humans, the entrance to the V. is protected by the membranous *hymen.* This usually perforates spontaneously before puberty, and is completely torn *(defloration)* during the first sexual intercourse. The cyclic phenomena that are regulated by follicle stimulating hormone can be monitored by the microscopic examination of a *vaginal smear* (mouse, rat, guinea pig) for keratinized epithelial cells (desquamation phase).

Vagus nerve, *nervus vagus:* the 10th cranial nerve. It is a composite nerve. In the *Craniata,* four branches of the V.n. each fork into a pre- and post-trematic branch around each of the last four gill clefts. A further branch (ramus lateralis or lateral line nerve) innervates the lateral line sense organs, while another branch (ramus visceralis) innervates the viscera of the body cavity.

In the human, the V.n. passes through the neck and thorax to the abdomen. Its motor fibers (arising from the nucleus ambiguus) supply the muscles of the pharynx, larynx, trachea, heart, mouths of the large arteries and veins, aortic arch, esophagus, stomach, small intestine, pancreas, liver, spleen, ascending colon, kidneys and visceral blood vessels. It also supplies inhibitory fibers to the heart, and secretory fibers to the gastric and pancreatic glands. Sensory fibers of the V.n. (arising from the jugular ganglion and from the ganglion nodosum) are distributed to the mucous membranes of the larynx, trachea, lungs, esophagus, stomach, intestines and gall bladder.

Vagus tone: see Autonomic nervous system.

Val: the standard abbreviation of Valine (see).

Valerianaceae, *valerian family:* a family of dicotyledonous plants, comprising 400 species in 13 genera, distributed throughout the world, except Australia, and mainly in the temperate zones of the Northern Hemisphere and in the Andes, with some representatives in South Africa and tropical North America. They are mostly herbs, rarely shrubs, with opposite, exstipulate and often pinnatifid leaves, and often strong-smelling rhizomes. The insect-pollinated, weakly zygomorphic flowers are hermaphrodite or unisexual, and they occur in cymose inflorescences. The corolla is funnel-shaped, usually consisting of 5 united petals, often saccate or spurred at the base, with imbricate lobes. Stamens are 1–4 in number, alternating with the corolla lobes, and the anthers are introrse. The inferior ovary consists of 3 united carpels, but usually only one of these is fertile, containing a single pendulous, anatropous ovule. The fruit is a single dry, indehiscent achene, often with persistent winged or plumose pappus formed from the calyx.

Valerianaceae. Common valerian *(Valeriana officinalis).* Flowering stalk with leaf.

Valeriana officinalis ***(Common valerian)*** is an erect (1–1.5 m), nearly glabrous, herbaceous perennial found from Europe to Japan, and common in moist situations. Its leaves are unpaired and pinnate. The small white to pink flowers contain 3 stamens, and they occur in terminal inflorescences. The rhizome (official Valerian root) is used pharmaceutically; it has a characteristic, disagreeble odor (becoming more powerful and penetrating as the rhizome is dried), due to the presence of essential oils which have sedative properties.

Valerianella locusta ***(Lamb's lettuce*** or ***Corn salad)*** is sometimes cultivated in Europe for its edible leaves.

Centranthus ruber ***(Red valerian),*** native to the Mediterranean region, has been cultivated widely as an ornamental, and is now naturalized as an escape in many parts of Europe and elswhere.

The fragrant rhizomes of *Nardostachys jatamansi* ***(Spikenard),*** which is native in the Himalayas, are a source of oil used in perfumery.

n*-Valeric acid,** ***pentanoic acid: $CH_3(CH_2)_3COOH$, a liquid, colorless, unpleasant smelling fatty acid, which occurs naturally in esters, e.g. in apples.

L-Valine, ***Val,*** ***L-α-aminoisovaleric acid:*** $(CH_3)_2CH—CH(NH_2)COOH$, an aliphatic, neutral, essential, glucogenic, proteogenic amino acid, which is biosynthesized from pyruvate, and degraded via propionic acid to succinyl-CoA. The intact molecule of Val is incorporated in the biosynthesis of penicillin. It is present in most poroteins, and constitutes 17.4% of the elastin from bovine tendons.

Valvifera: a suborder of the *Isopoda* (see).

Valvula pylorica: see Digestive system.

Vampire bats: see *Desmodontidae.*

Vampire squids: see *Cephalopoda.*

Vampyromorpha: see *Cephalopoda.*

Vanadium, ***V:*** a metal that is present in very small quantities in the ash of numerous plants. It is a trace element required for normal growth by animals; most studies have been performed on rats and chickens. A reliable estimate of human V requirement is lacking; most dietary items contain less than 100 ng V/g. V is taken up as V^{5+} (present in the environment as the vanadate ion VO_4^{3-}) and reduced to V^{3+} in the cell. V deficiency causes an increase in plasma cholesterol and triacylglycerols. It stimulates the oxidation of phospholipids and decreases cholesterol synthesis by inhibiting squalene synthase (a liver microsomal enzyme system); it also stimulates acetoacetyl-CoA deacylase in liver mitochondria. V deficiency also leads to abnormal bone growth; injected radioactive V shows a high incorporation into regions of active mineralization in dentine and bone. V. is required for optimal growth of some green algae, and it inhibits growth of *Mycobacterium tuberculosis.* Nitrogen fixation by *Azotobacter* is increased by V. Certain ascidians show a remarkable ability to concentrate V from the surrounding sea water.

Vanda: see *Orchidaceae.*

Van der Waals forces, ***London forces:*** weak interatomic or intermolecular, attractive, electrodynamic forces (1–2 kJ/mol) arising from the interaction of induced dipole moments of molecules that do not possess permanent dipole moments. Their existence is due to the mutual influence of fluctuations in the electrical fields of neighboring molecules or atoms. As a crude simplification, these interactions generate antiparallel, fluctuating dipoles, which since they are antiparallel show mutual attraction, resulting in a net attractive force. Between ions, this force varies as the seventh power of the separatory distance. Since a theoretical derivation is very complicated and based on quantum mechanics, an effective interaction constant (Hamaker constant) is determined experimentally, and this serves as a very adequate description of the V.d.W.f. between different bodies.

Vanellus: see *Charadriidae.*

Vanga shrikes: see *Vangidae.*

Vangidae, ***vanga shrikes:*** a small (12 species) avian family in the order *Passeriformes* (see). All 12 species are restricted to the forests of Madagascar. They show a close anatomical resemblance to shrikes *(Laniidae),* and they presumably evolved from an immigrant shrike ancestor. Sizes range from 13 to 30 cm. The smallest species is the short-billed ***Red-tailed vanga*** or ***Tit shrike*** *(Calicalicus madagascariensis).* Typically, they have stout, strong, hooked bills, especially the black and white ***Hook-billed vanga*** *(Vanga curvirostris),* whose bill is particularly heavy and sharply hooked. The bill of the ***Helmetbird*** *(Aerocharis = Euryceros prevostii;* 30 cm) carries a casque, so that the bill appears large and inflated, while the ***Sicklebird*** or ***Sicklebill*** *(Falculea palliata)* has a thin, decurved bill. Although the variations of bill shape and size are an obvious example of adaptive radiation, corresponding specializations in the use of the bill are not apparent. All are arboreal, occupying different niches in the forest canopy. They are largely insectivorous, and they also prey on small vertebrates such as tree frogs and lizards. Most are metallic black above and white below, or marked with chestnut and gray. An exception is the ***Blue vanga shrike*** *(Leptopterus madagascarinus;* 15 cm) which has blue upper parts, wings and tail, and a white breast and belly.

Vanilla: see *Orchidaceae.*

Vanillin: a pleasantly smelling aromatic aldehyde, which crystallizes as colorless needles. It is the characteristic odiferous principle of vanilla pods, which contain about 2% V., and is also present in clove oil, as well as several flowers such as *Scorzonera* and *Spiraea.* The widely used vanilla flavor of the food industry is derived either from vanilla pods or from laboratory-synthesized V.

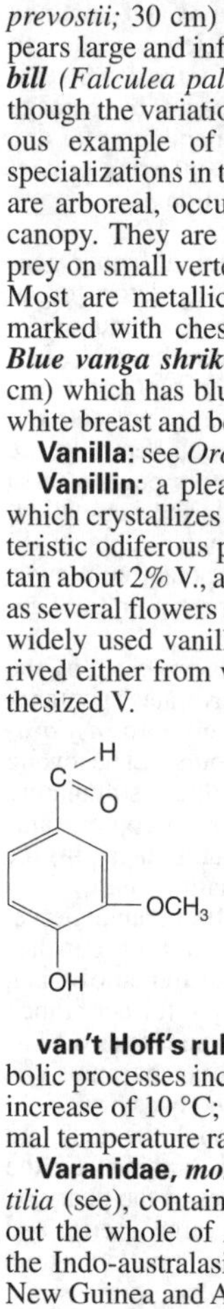

H C O OCH3 OH

Vanillin

van't Hoff's rule: a rule stating that the rate of metabolic processes increases 2–3 fold for each temperature increase of 10 °C; this only valid, however, in the optimal temperature range of any species.

Varanidae, ***monitor lizards:*** a family of the *Lacertilia* (see), containing 31 species, distributed throughout the whole of Africa, extending to southern Asia, the Indo-australasian Archipelago, Philippines, Papua New Guinea and Australia. They are large reptiles with small granular scales, a long head and neck, conspicuous ear openings, powerful legs with long sharp claws, and a long, muscular tail, which is used as both a rudder and a weapon. The long, slender, deeply forked tongue is very extensible. Monitor lizards are the largest living lizards, most species being 1–2 m in length. The smallest species is the Australian ***Short-tailed***

monitor *(Varanus brevicauda)* with a length of 20 cm, while the largest is the Komodo dragon (see). Both monitors and snakes evolved directly from the same ancestors, and there are close phylogenetic relationships between these two groups. They are active in the day and live by predation. In addition to small mammals, birds and reptiles, they also take eggs and carrion. After tearing with teeth and claws, pieces of prey are swallowed whole without chewing (snakes also swallow prey whole without chewing). Most adult monitors are brown, gray or black, but the young are more colorful. Earth cavities are excavated by most species, but a few green species are adapted to an arboreal existence. Many are fond of water, and possess a laterally compressed tail. All species lay eggs, which are buried in the soil, or concealed in tree cavities. On account of their edible flesh and the value of their skin for leather manufacture, they are extensively hunted in some countries. The following selected species are listed separately: Komodo dragon, Water monitor, Nile monitor, Variegated lizard.

Varanus salvator: see Water monitor.

Variable lizards: see Changeable lizards.

Variability: changes in organisms in the course of a generation or within a population, due to noninherited modifications, and due to inherited variations resulting from recombination and mutation. Types that deviate from the norm (standard type) are called *variants.* Phenotypically, V. may expressed as a smooth gradation of characters between variants *(continual, fluctuating* or *quantitative V.)* or it may be manifested by extreme variants that are not linked by intermediate forms *(discontinuous, alternative* or *qualitative V.).*

In the case of genotypic (inherited) V., it is useful to differentiate between free and potential V. *Free V.* is the appearance of new phenotypes by Mendelian segregation and recombination. In a population, the *potential* or *cryptic V.* refers to the total reservoir of V., which has not yet been phenotypically realized, and which will not be phenotypically realized until appropriate Mendelian segregation and recombination has occurred. The significance of potential V. is that it represents material which by selection can lead to the establishment of new geno- and phenotypes in a population. Conversion of potential into free V. (and vice versa) within a population by crossing and segregation is known as *V. flow.*

Variant:

1) A unit of vegetation characterized by a frequently recurring combination of species, without the presence of actual differential species. According to the definition of Character species (see), a V. is the smallest distinguishable vegetative unit that can be defined on the basis of differential species.

2) see Variability.

Variate, *random variable, stochastic variable:* a variable that takes on any one of a specific set of values (determined by measuring or counting) with a specified relative frequency or probability.

Variational movements: movements of plant organs caused by reversible changes of turgor. Organs that display V.m. often possess tissues under tension or capable of swelling. Thus, tissues consisting of highly vacuolated cells (and therefore capable of generating considerable turgor) are present and often clearly visible in leaf bases, or (e.g. in bean leaves) as joints or pulvini at the bases of leaf pinnae. Well known *induced V.m.* are nasties such as Seismonasty (see) amd Stomatal movements (see). *Autonomous* or *endogenous V.m.* include rhythmic leaf movements observed in different leguminous plants. For example, in the Malaysian *Desmodium gyrans,* two small lateral leaflets move so rapidly when the temperature is sufficiently high (35 °C) that the tip describes an ellipse in about 30 seconds. The leaves of *Trifolium pratense* move more slowly, swinging up and down every 2–4 hours in the dark. See Bending movements.

Varied thrush: see *Turdinae.*

Variegated cutworm: see Cutworms.

Variegated lizard, *Varanus varius:* a common Australian monitor lizard (see *Varanidae),* up to 2 m in length, and vividly patterned with broad, transverse black-brown and yellow-white bands. In addition to small mammals, it feeds mostly on birds' eggs.

Variegation: collective term for patterns of broken leaf color, due to a deficiency or absence of chlorophyll, which arises by various mechanisms, e.g. demixing of genetically different plastids, gene instability, positional effects, somatic crossing over, and different kinds of gene or plasmid mutation. Thus, in some cases it is inherited. It can also be caused by virus diseases. In *marginal V.* only the leaf margins are white, a condition known in ivy, ash, elder and other plants. In *sectorial V.,* white and green stripes radiate from the midrib to the leaf margin, and in parallel-veined leaves these stripes can extend from the leaf insertion to the leaf tip. Sectorial V. is very common, occurring, inter alia, in meadow saffron, iris, maple, beech, clover, blackberry, roses and pines. The occurrence of scattered, more or less large, green-white spots on the leaf surface is known as *marbled V.* If the spots are smaller and less extensive the condition is known as *pulverulent V.* The latter two types are encountered in elm, elder, roses, clover and species of *Polygonum.* Several types of V. may occur on the same leaf. Many ornamental plants, e.g. garden pelargoniums, have variegated leaves.

Variety: a systematic category below that of species (see Taxonomy).

Varve chronology: see Dating methods.

Vasa recta: see Hairpin countercurrent principle, Kidney functions.

Vascular bundle: see Conducting tissue, Shoot, Root.

Vascularization of organs: see Blood supply to organs.

Vascular system: see Blood circulation.

Vascular water flow: see Water economy.

Vas deferens, *spermatic duct:* a duct that carries spermatozoa from the male gonad. In mammals, it empties into the urethra. See Testicle.

Vasoactive intestinal polypeptide: see Gastrointestinal hormones.

Vasomotor center: the most important center for Circulatory regulation (see). It lies in the Medulla oblongata, and it is responsible for the nervous regulation of the internal diameter of peripheral blood vessels, cardiac activity, and the secretion of adrenalin by the adrenal medulla (Fig.). Impulses are constantly arriving at the V.c. from the baroceptors in the carotid sinus and aortic arch, and from the cardiac and respiratory centers in the medulla itself.

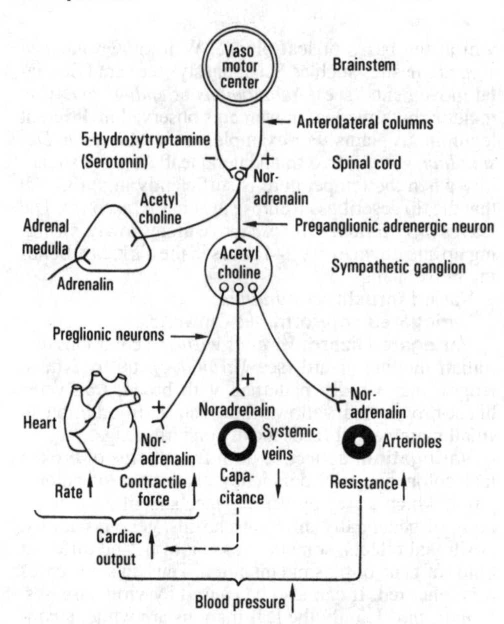

Vasomotor center. Regulation of the sympathetic outflow of the cardiovascular system by the vasomotor center in the brainstem. Pressor neurons release 5-hydroxytryptamine, and depressor neurons release noradrenalin. The preganglionic neurons liberate acetylcholine at their endings in the sympathetic ganglia and the adrenal medulla. This in turn activates postganglionic neurons which control the response of the heart, resistance blood vessels (arterioles) and capacitance blood vessels (veins); the magnitude of each response depends on the amount of liberated noradrenalin (norepinephrin). In the adrenal medulla, acetylcholine causes the release of catecholamines, mainly adrenalin (epinephrin).

Vasopressin, *antidiuretic hormone, antidiuretin, pitressin:* a mammalian neurosecretory peptide hormone with antidiuretic activity on the kidneys, as well as causing vasoconstriction of peripheral vessels, with slowing of the heart beat and increase of blood pressure. Structure: Cys-Tyr-Phe-Gln-Asp-Cys-Pro-Arg-Gly-NH_2 (the two cysteine residues form a disulfide bridge), M_r 1084. The amino acid residues in positions 3,4 and 8 are variable, e.g. [8-arginine]vasopressin (V. in which the tyrosyl residue is replaced by an arginyl residue) is physiologically and pharmacologically one of the most active forms of the hormone; 2 ng are sufficient to cause pronounced antidiuresis in the adult human.

V. is synthesized in the supraoptic nucleus of the hypothalamus by protelytic cleavage of precursor prohormones. In combination with the carrier protein, neurophysin II, it is transported as granules down nerve fibers (supraoptico-hypophysis) to the posterior pituitary (neurohypophysis), where it is stored. Release of V. into the blood circulation is stimulated by thirst and lack of water (i.e. its release is regulated by the degree of hydration of the organism). It activates the adenyl cyclase system in the distal tubule of the kidney, resulting in increased resorption of water and increased excretion of Na^+. V. is structurally related to Oxytocin (see).

Vasotocin: Cys-Lys-Phe-Gln-Asp-Cys-Ile-Arg-Gly-NH_2 (the two cysteine residues form a disulfide bridge), a peptide hormone secreted by the posterior pituitary of birds, reptiles and some amphibians. It is structurally and functionally related to the mammalian hormone, Vasopressin (see).

Veery: see *Turdinae.*

Vegetable sponge: see *Cucurbitaceae.*

Vegetation, *plant cover:* the totality of plants and plant communities covering an area. A distinction is drawn between V. and Flora (see). The *natural V.* consists of the V. that is not influenced directly or indirectly by human activity, and which is in equilibrium with its environment. Modern agricultural landscapes contain only residues of natural V., e.g. plant communities in undisturbed lakes and rivers, reedbeds, salt marshes, dunes, moors, cliff faces, inaccessible high mountain meadows, etc. Without human intervention, many parts of the world, would be covered with forest and woodland. Much of the apparently natural V. is in fact only seminatural V.; although it exists independently of human activity and is in equilibrium with the environment, its establishment was originally made possible by human activity, e.g. the clearing of woodlands. Truly natural V. is better described as *original V.,* meaning the V. that was established before any changes occurred due to human agency. Historical research is often necessary to identify original V. In large areas of Central Europe, it seems that the first appreciable influence of human activity on the V. occurred in the early Neolithic. Since the climate has changed considerably since the early Neolithic, it is very difficult to compare the original V. of that time with the present natural V. The *potential natural V.* is the natural V. that would become established if human activity were to cease. In contrast to the natural V., the potential natural V. is (would be) determined by the irreversible alterations of habitat caused by human activity. Since the natural V. is an indicator of the productive capacity, the determination of the potential natural V. is of economic value for predicting the current and future agricultual potential of a particular location or site. See also Sclerophyllous vegetation, Zonal vegetation.

Vegetation layering: see Relevé.

Vegetation stratification: see Relevé.

Vegetative nervous system: see Autonomic nervous system.

Vegetative reproduction: see Reproduction.

Veil community: general term for a plant community frequently found interwoven like a veil within other plant communities. It consists largely of rambling and climbing plants, and is found primarily in reed beds, water meadows and willow thickets.

Veins:

1) Structures present in leaves, consisting of bundles of vascular tissue enclosed in sclerenchyma, often in relief on the undersurface of the leaf, and lighter in color than the rest of the leaf surface (see Leaf).

2) Blood vessels that return blood from the organs and peripheral tissues to the heart (see Blood circulation).

Velamen: see Absorption tissue, Epiphytes, *Orchidaceae.*

Velella: a colonial, pelagic, floating hydroid of the order, *Siphonophora.* Each colony is polypoid, consisting of a large gastrozooid, the aboral surface of which is modified as a pneumatophore. A raised vertical ridge on the pneumatophore protrudes from the water and enables the colony to exploit the wind and truly sail across the ocean, some colonies sailing on a starboard reach, others to port. Several circles of tentacles (dactylozooids) hang down from the periphery of the float. Between the tentacles and the central mouth of the gastrozooid are gonozooids, which bud off free, sexually reproductive medusae. *Velella velella* occurs in the Mediterranean and the Atlantic. Related genera are *Porpita* and *Porpema.*

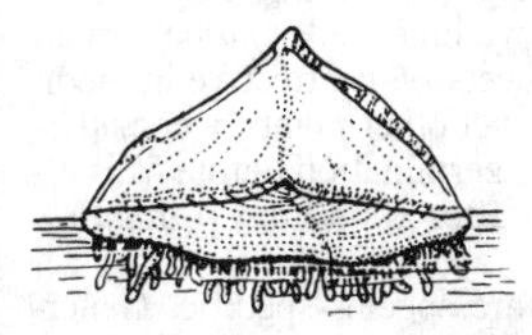

Velella velella ("little sail"), a member of the order, *Siphonophora,* and common in warm seas.

Veliger larva: a free-swimming larva of marine *Lamellibranchiata* (including *Dreissena)* and many snails. It develops from a trochophore larva, and it possesses conspicuous rudiments of the adult foot, mantle and shell. Anteriorly, the prototroch is developed into a 2- or multi-banded peripheral ciliated sail or *velum.* It is a typical plankton feeder, living planktonically for different periods (according to the species), before finally metamorphosing into a young snail or bivalve. In the V.l. of gastropods, the foot increases in size, and the shell may become quite prominent and coiled, before transformation to the young snail.

Velum:

1) An ectodermal lamella extending inward from the umbrella margin of a hydromedusa (see *Hydrozoa).*

2) A ciliated, sail-like appendage of a veliger larva.

3) A double belt of cilia (sometimes called the "wheel organ") surrounding the mouth of rotifers (see *Rotifera).*

4) An integument present in the young fruiting bodies of members of the *Agaricales,* which covers only the underside of the parasol *(Velum partiale),* or completely covers the parasol and stipe *(Velum universale).* During rapid growth of the fruiting body the edge of the Velum partiale is ripped, so that a hanging sleeve *(Armilla)* or a ring *(Anulus)* is left on the stipe, or the Velum may disappear completely. In fully grown fruiting bodies, the remains of the Velum universale often form a cup-like structure *(Volva)* at the bottom of the stipe (e.g. in Amanita), or they may be left as small pieces on the top of the cap (e.g. in the Fly agaric).

Velvet ants: see Ants, *Hymenoptera.*

Velvet crab: see Swimming crabs.

Velvet sponge: see Bath sponges.

Velvet water bugs: see *Heteroptera.*

Venation:

1) The pattern of leaf veins (see Leaf).

2) The pattern of wing veins in insects. These veins are longitudinal and transverse vessels, sometimes containing tracheae, which transport nutrients to the developing wing during metamorphosis, and act as support structures to the unfolding wing after emergence from the pupa. Insect wing V. patterns are taxonomically useful.

Vendace: see *Coregonus.*

Veneroida: see *Bivalvia.*

Venomous animals: animals that employ toxins for defense or immobilization of prey. Toxins afford individual protection if they repel an aggressor or cause it to reject the toxin-producer as a source of food. If the toxin becomes effective only after its producer has been eaten (passive effect), it provides no active individual protection, but it confers a selective protection on the species, provided the predator is capable of learning from experience. There are many chemically different toxins, and their physiological effects range from the generation of a feeling of nausea to rapid death. A dog can recognize the injurious nature of a toad simply by sniffing at it; experienced farmyard fowls do not peck at earthworms on accouint of the bad taste of their coelomic fluids; the inky fluid of dibranchiate cephalopods appears to paralyse the olfactory organs of their main enemy, the conger eel. The polecat uses its stink glands for repulsion of aggressors.

Active use of obnoxious or toxic substances for protection is widespread among insects, which release glandular secretions or evil-tasting body fluids. Obnoxious substances are sometimes propelled at the perceived aggressor, e.g. the secretion of the anal glands of the skunk can be sprayed over a distance of 2 m. Alternatively, the toxin is injected into the body of the aggressor, e.g. by explosive injection of the nematocyts of the *Cnidaria,* injection by a sting (scorpions, bees, wasps), introduction of the toxin by the mouthparts (spiders, venomous snakes). Ants make a wound by biting, then introduce the toxin (formic acid) into the wound from their abdominal gland.

Venomous fishes: fishes that possess venom glands. The glands normally open to the exterior on spinous fin rays. When a spine pierces the flesh of another animal, the venom enters the wound directly. Some examples are *Trachinus draco* ***(Dragon weaver), Pterois (Turkeyfish),*** *Dasyatis pastinaca* ***(Stingray),*** and members of the families *Synanceidae* ***(Stonefishes)*** and *Scorpaenidae* ***(Scorpionfishes*** or ***Rockfishes).*** "Poisonous fishes", on the other hand, are species whose edible body parts contain toxins, either permanently or at certain times, e.g. the Puffer fish *(Tetrodon).*

Venomous lizards: see *Helodermatidae.*

Venomous snails: collective term for various species and genera of marine snails which paralyse or kill their prey with the aid of toxic glandular secretions, e.g. Tun shells, Cone shells, Turrids and Auger shells. The secretions contain inorganic acids, in some cases 2–4% sulfuric acid, and can even be dangerous to humans. See *Gastropoda.*

Ventilation tissue: see Parenchyma.

Ventral glands: paired, independent molting glands of hemimetabolous insects. They resemble the prothoracic glands of holometabolous insects in their position, shape, origin and function. The V.g. also degenerate in the imago (exceptions being the workers and soldiers of termites). In the larval stages, V.g. are responsible for the production of molting hormone (see Ecdysone).

Ventral plate: see Mesoblast.

Ventricle:

1) Ventricles of the brain. During the embryonic and phylogenetic development of vertebrates, the neural canal becomes partially divided into communicating cavities by thickening of the wall. These divisions persist in the adult, where they constitutes the cavities or ventricles of the brain, i.e. lateral ventricles (ventricles I and

II) within the cerebral hemispheres, ventricle III in the thalamencephalon and ventricle IV in the medulla oblongata, all communicating with one another and filled with cerebrospional fluid. Ventricle III lies behind the lateral ventricles, and is connected to each via small openings called the foramina of Munro. Ventricle IV is in the front of the cerebellum behind the pons Varolii and the medulla. Ventricle III communicates with ventricle IV via a narrow canal called the *aqueduct of Sylvius* or *iter* (abbreviation of "iter a tertio ad quartum ventriculum"). The roof of ventricle IV contains 3 openings, the median one being the foramen of Magendie. By means of these 3 openings, the ventricle communicates with the subarachnoid cavity, and cerebrospinal fluid can circulate from one to the other. The so-called fifth ventricle is not a true ventricle, but a narrow space in front of ventricle III, with no connections with other ventricles. Sometimes the optic lobes are hollow, each containing an *optocoel* and opening out from the iter. If the olfactory lobes are well formed, large structures, they are also frequently hollow, enclosing a rhinocoel, each communicating with a lateral ventricle in the cerebral hemisphere.

2) Ventricles of the heart: see Heart.

Ventriculus: see Digestive system.

Venus' flowerbasket: see *Porifera.*

Venus' fly trap: see Carnivorous plants, *Droseraceae.*

Veratric acid, ***3,4-dimethoxybenzoic acid, dimethylprotocatechuic acid:*** an aromatic carboxylic acid present in various species of hellebore (in particular *Veratrum sabadilla)* and in seeds of *Schoenocaulon officinale.*

Veratrum alkaloids: a group of steroid alkaloids (more than 40 are known), which occur in liliaceous plants of the suborder *Melanthaceae,* in particular in the hellebores *(Veratrum).* Important representatives are jervine, veratramine, cevine, veracevine, rubijervine and isorubijervine. V.a. have positive ionotropic action on the heart, and cause a decrease of blood pressure by reflex inhibition of the vasomotor centers. They were used for treatment of hypertension, but have been replaced by Rauwolfia alkaloids and other drugs. They have also been used as insecticides.

Verdin: see *Remizidae.*

Verdoglobin: see Bile pigments.

Vermiculite: see Sorption.

Vermiform appendix: see Cecum.

Vernalin: see Flower development.

Vernalization: induction of flower development (i.e. the change from the vegetative to the reproductive phase of plant development) by low temperature. V. is necessary for flower development in many biannuals and winter annuals (plants that germinate in autumn and spend the winter as resting seedlings). Numerous perennials, especially shrubs, also require low temperature for induction of flower development. Some plants display an *absolute requirement* for low temperature, and spend the entire year in the vegetative phase if they are not vernalized, e.g. beets, turnips, cabbage, celery, as well as biannual races of *Hyoscyamus niger.* Plants with a *relative requirement* for cold conditions, e.g. lettuce, radishes, winter cereals, produce flowers without V., but flower developoment is then delayed.

Almost all cold-dependent plants also require long day conditions for induction of flower development, and they must be exposed to this second environmental factor after V. has occurred. Accordingly, V. does not actually induce flower development, but renders the vegetative meristem responsive to the flower-inducing effects of long day conditions.

The low temperature stimulus is perceived only by tissues that contain dividing cells, i.e. usually the shoot meristem. In certain plants, in which cell division occurs in the leaves (e.g. the garden species of honesty, *Lunaria rediviva),* the low temperature stimulus is also perceived by the leaves. Some plants can even become vernalized in the embryo, e.g. the swelling caryopsis of winter cereals and other swelling seeds can be vernalized. In other plants, e.g. beets, V. is effective immediately after germination. In yet other plants, a considerable time must elapse after germination before V. is effective, e.g. *Hyoscyamus* (after 10–30 days), *Lunaria* (after 8 weeks).

Optimal V. conditions are species-specific. Generally the temperature must lie between 1 and 7 °C for 2–10 weeks. Winter cereals whose seeds have been vernalized before sowing (allowed to swell by imbibing water then cooled to a low temperature), can be sown in the spring, and they still produce a harvest in the same year. For the same result without prior V. the seeds must be sown in the previous autumn. Development of many varieties of summer cereals is accelerated by V. Cereal grains can be swelled and vernalized, then dried again to await sowing.

V. proceeds in several stages. In the first days of cold treatment, the effect can be reversed by warming the plant or seed to a higher temperature (about 35 °C), a process known as *devernalization.* V. of devernalized plants or seeds *(revernalization)* requires the full period of V. Several days after cold treatment, a stable state is attained, in which devernalization is no longer possible. V. may then be interrupted by warming the plant to 7–22 °C, but V. continues without loss of time when the plant is cooled again.

V. only occurs in the presence of oxygen. Sufficient respiratory substrate (carbohydrates) must also be available. V. appears to operate at a genetic level, by highly stable differential gene activation (stable determination). Thus, changes in the histone fractions have been demonstrated following the V. of winter cereals. Once V. has been established by cooling, its effect is retained, i.e. it is passed from cell to cell for weeks, months or even years, until flower development is induced by long day conditions.

Verrucosa amebas: see Amebas.

Versene: see Ethylenediaminetetraacetic acid.

Verson's glands: see Molting glands.

Vertebra: see Axial skeleton.

Vertebrata: see Vertebrates.

Vertebrates, *Vertebrata,* ***craniates,*** *Craniata:* the most extensive subphylum of the *Chordata.* V. possess a conspicuous cartilaginous or bony axial endoskeleton (derived from the Chorda dorsalis), consisting of the cranium (skull) and the vertebral column. The central nervous system is highly evolved, the neural tube expanding anteriorly to form a definite brain, which is enclosed in the cranium. The basic body plan of head, trunk and caudal region has been widely modified in the course of evolution, but there is always a high degree of cephalization with a well differentiated head region. In the primitive state, all V. have two pairs of extremities, but in some groups these have undergone secondary loss (e.g. in snakes).

Specialized, complex sense organs are prominently developed. A closed vascular system is present. The blood is usually red, due to the presence of hemoglobin-containing corpuscles, and it is driven at relatively high pressure by a chambered muscular heart with valves, which arises by differentiation of part of the ventral vessel. At least one venous portal system is always present. Excretion of nitrogenous waste is performed by the kidneys, which require a high blood pressure for their operation. The digestive system always contains a stomach (sometimes more than one), a liver and a pancreas. Larval stages of some primitive V. possess an endostyle, but in all adult V. this is replaced by the thyroid gland. There is a well developed system of endocrine organs. In the primitive condition, respiratory gas exchange occurs via gills (derived from the visceral clefts), which have been retained only by the cyclostomes and the fishes, and by the larval stages of amphibians. In all other V., gills have been lost in the course of phylogenetic development, and replaced by lungs. The great adaptability of V. has enabled them to colonize a variety of habitats. Like the insects before them, they have evolved into aquatic, terrestrial and aerial forms. V. comprise the *Agnatha* (with the single living class, *Cyclostomata),* fishes in the wide sense (classes: *Placodermi, Choanichthyes, Actinopterygii, Chondrichthyes),* amphibiams, reptiles, birds and mammals. Excluding the *Agnatha* and the fishes, the remaining groups are known collectively as *tetrapods* or *quadrupeds.*

Geological record. Fossils are known from the Upper Cambrian *(Agnatha* and fishes) to the present, but they do not become a regular feature of the fossil record until the Silurian. Of these earliest forms, only parts of the endoskeleton have survived as fossils. Such remains, however, are easily assigned to their sites in the original skeleton, which in turn bears a close relationship to the internal organization of the animal, so that the interpretation of phylogenetic changes is possible. Because of their relatively infrequent occurrence (compared with the abundance of certain invertebrate fossils), fossil V. are largely unsuitable for stratigraphic identification and dating. Exceptions are the *Agnatha* and the fishes in the continental Devonian (Old Red), amphibians and reptiles of the South African Karru formation, and the Tertiary and diluvial mammals.

Vertebral column: see Axial skeleton.

Vervet monkey: see Guenons.

Vesica urinalis: see Urinary bladder.

Vespertilionidae, *vespertilionid bats:* a family of bats (see *Chiroptera),* comprising about 275 species in 38 genera, distributed worldwide in temperate and tropical regions. *Myotis* ***(Little brown bats,*** or ***Mouse-eared bats)*** has the widest distribution of any bat genus. Most species inhabit caves, and nearly all species feed on insects. The dental formula varies from 1/2, 1/1, 1/2, 3/3 = 28 to 2/3, 1/1, 3/3, 3/3 = 38. The molars have a "W" pattern of cusps and ridges, and the incisors are small and separated medially. A nose leaf is lacking. Some members have large bulges or folds on the snout, caused by glands beneath the skin. All species of *Glischropus, Eudiscopus* and *Tylonycteris,* and some species of *Pipistrellus* and *Hesperoptenus* have suction pads on the soles and/or wrists. *Pizonyx* and *Cistugo* have wing glands. Some other members are: *Myotis daubentoni* ***(Water bat),*** *Myotis dasycneme* ***(Pond bat),*** *Myotis myotis* ***(Brown bat),*** *Plecotus auritus* ***(Long-eared bat),*** and *Pipistrellus pipistrellus* ***(Common pipistrelle,*** which is quite common in cities), all of which are found in both Europe and Asia. *Pipistrellus nanus* ***(African banana bat)*** is found in many parts of Africa, where it spends the day concealed inside young, unopened banana leaves. The ***Noctule bat*** *(Nyctalus noctula)* is a very widely distributed, powerfully built species, dark to yellowish brown above and paler below; it flies fast with sudden turns, and starts its feeding flights in the early evening.

Vespidae: see Wasps.

Vessel tone: the constant tension of blood vessel walls. If this tension decreases, the vessels dilate, the blood pressure falls, and the amount of blood carried by the vessels (and therefore the amount of blood flowing through the vascularized tissue) is increased. An increase in wall tension causes constriction, with an increase in blood pressure and a decrease in the quantity of blood flowing through the tissue or organ in question. Vessel diameter is regulated both locally and by the nerves to the vessel. Local effectors that cause vessel dilation are heat, oxygen deficiency, excess carbon dioxide, and acids (usually lactic acid), as well as regulatory substances produced in the body, i.e. bradykinin, histamine and prostaglandins. Vessel constriction is promoted by adenalin and noradrenalin from the sympathetic nervous system or adrenal medulla, and by angiotensin and serotonin.

Nervous control of V.t. is exerted on the arterioles (rather than the venules), via the vasoconstrictor fibers of the sympathetic nerve. Increased sympathetic activity leads to an increased secretion of adrenalin and noradrenalin at the nerve endings; these catecholamines cause the release of Ca^{2+}, which promotes contraction of the circular muscles of the arterioles. The arterioles constrict, their flow resistance increases, the blood pressure rises, and quantity of blood flowing through the tissue is decreased. When this sympathetic activity ceases, V.t. decreases and the vessels are dilated passively by

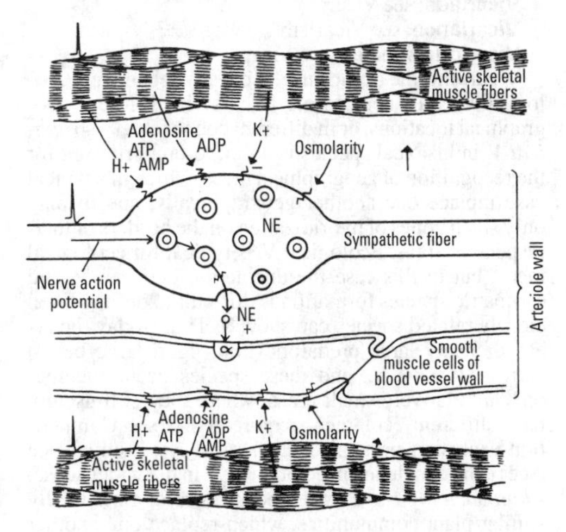

Vessel tone. Nervous and local regulation of arteriole diameter. *NE* = Noradrenalin (norepinephrin).

the pressure of the blood. Nervous constriction decreases blood flow in the innervated region. The decreased blood supply leads to an oxygen deficit and to an accumulation of carbon dioxide and lactic acid, but since these factors cause dilation, the V.t. of the innervated vessels is decreased and they return to their original diameter. See Vasomotor center.

Vesica fellea: see Gall bladder.

Vestibule of the inner ear: see Auditory organ.

Vestigial organs: see Rudimentary organs.

Vetch: see *Papilionaceae.*

Vetiver grass: see *Graminae.*

Vetiveria: see *Graminae.*

Vexillifer larva: see *Carapidae.*

Vexillum: see Feathers.

Vibraculum: see *Bryozoa.*

Vibramycin: see Tetracyclins.

Vibrio: a genus of Gram-negative, mainly aerobic bacteria, which are motile by means of one or more polar flagella. They characteristically have the shape of short rods, about 1.5–3 µm x 0.5 µm, which are always curved or shaped like a comma. In liquid culture, *V. cholerae* occurs singly, in pairs, or in chains end to end with the curves alternating. Species of *V.* live saprophytically in stagnant fresh or saltwater ponds and polluted rivers, or they are disease-causing organisms in humans and animals. *V. cholerae (= V. comma)* was discovered by R. Koch as the causative agent of Asiatic cholera.

A curved or comma-shaped bacterium is commonly known as a "vibrio", irrespective of genus. By this definition, certain other bacteria can also be called vibrios, e.g. *Photobacterium* (see Bioluminescence).

Vibrios

Vibrissae: see Feathers.

Viburnum: see *Caprifoliaceae.*

Vicariads: see Vicars.

Vicariance: see Vicarism.

Vicarious species: see Vicars.

Vicarism: the occurrence of closely related and ecologically equivalent forms (vicars) in different geographical locations, or in different coenoses. *Geographical V.* in bisexual species is an important criterion for the recognition of geographical races, which in a typical case replace one another geographically, and display only small zones of hybridization on the borders of their respective areas. Ecological V. is typical for ecological races, but in this case the distinction between V. and sympatric species formation is difficult to define. Even closely related species can show V. Thus, today the areas of many large predators (lion, tiger, large bears) show extensive V., and these species occur together only in relatively small areas, where they still usually have different ecological requirements (see Competition-exclusion principle). Ecological site equivalence (see) can be included in ecological V. in the wider sense. *Plant sociological V.* is the occurrence of floristically similar plant communities, which replace one another geographically.

Vicariance is often used as a synonym of V., but is more appropriately defined as the splitting of taxa by the development of a natural geographical barrier.

Vicars, *vicariads, vicarious species:* related and ecologically equivalent forms that are mutually exclusive, and therefore occupy disjunct geographical areas. See Vicarism.

Vicia: see *Papilionaceae.*

Victoria amazonica and cruziana: see *Nymphaeaceae.*

Victoriornis: see *Raphidae.*

Vicuña, *Lama vicugna:* a dainty and relatively short-headed, nonhumped member of the *Camelidae* (see), which inhabits the high mountain regions of the Andes. It is much sought for its soft coat, and for this reason is extinct in many parts of its orginal distribution range. See Lama.

Viduinae: see *Ploceidae.*

Vigilance: the average excitation level of the entire behavioral system, scaled along a continuum from deep dreamless sleep to the extremes of alertness and responsiveness. Terms with similar or closely overlapping meanings are *specific vigilance reaction, arousal, psychological arousal, activation, degree of activation,* and *general drive.* V. of the waking phase is often divided into three states: 1) relaxed waking state,

2) state of wakeful attentiveness, 3) extreme alertness. In humans, these different states are associated with the development of different electric potentials in the brain, and they can be differentiated with the aid of an electroencephalogram. More recently, comparable states have been shown to exist in invertebrates, such as mollusks and arthropods, and they are apparently linked to the existence of a centralized nervous system. For investigatative purposes, V. is determined primarily by measuring reaction times to specific stimuli.

Villikinin: see Gastrointestinal hormones.

Vinblastine: see Microtubules.

Vinca: see *Apocyanaceae.*

Vincetoxicum: see *Asclepiadaceae.*

Vincristine: see Microtubules.

Vinegar: see Acetic acid fermentation.

Vinegar eelworm: see *Nematoda.*

Vineleaf gall mite: see *Tetrapodili.*

Vine moths: two pests of the grape vine in central Europe. *Eupoecilia ambiguella* Hb. is a member of the family, *Phaloniidae,* whereas *Lobesia botrana* D. & S. (L'Eudemis de La Vigne) belongs to the *Tortricidae.* During the year there are two generations of caterpillars (called grape worms); the first generation ("Heuwurm") feeds on the flower buds, and the second generation ("Sauerwurm") attacks the unripe berries of the grape vine.

Vine phylloxera: see *Homoptera.*

Viola: see *Violaceae.*

Violaceae, *violet family:* a cosmopolitan family of dicotyledonous plants containing about 850 species in about 16 genera. They include some shrubs and climbing plants, but most are herbs with alternate, simple, stipulate leaves. With the exception of the gynaecium, flowers are 5-merous (5 stamens alternate with 5 petals), hermaphrodite, and usually zygomorphic with a spurred corolla. The unilocular, superior ovary develops into a multiseeded, usually 3-valved dehiscent capsule. The small seeds contain a fatty endosperm. Pollination is usually by insects, and cleistogamy (self-pollination in the closed flower) is common. The most widely distributed genus is *Viola* ***(Violet, Pansy),*** in which the anterior petal is produced as a spur (into which nectar is secreted by processes of the two anterior stamens). Most of the European species of *Viola* are blue, and they are distributed in a variety of habitats (deciduous and coniferous woodland, hedge bottoms, pastures, meadows, cultivated fields). The ***Common violet*** *(Viola riviniana),* usually blue-violet in color, inhabits hedgebanks, heaths, pastures and mountain rocks in Europe, Scandanavia and Morocco (Atlas), and is found on any type of soil, provided it is not too wet. The ***Wild pansy*** *(Viola tricolor),* which has yellow-white or multicolored flowers, is an agricultural weed; it is also used medicinally on account of its mucilaginous leaves and flowers. Many varieties of the ***Garden pansy*** *(Viola wittrockiana)* are used for spring planting in parks, cemetries and gardens.

Violet family: see *Violaceae.*

Violoxanthin, ***3(S),3'(S)-dihydroxy-β-carotene-(5R,6S,5'R,6'S)-5,6,5',6'-diepoxide, zeaxanthin diepoxide:*** an abundant and widespread xanthophyll, present as an orange or brown-yellow pigment in all green leaves, and especially plentiful in flowers and fruits of *Viola tricolor, Taraxacum, Tagetes, Citrus, Cytisus, Laburnum,* etc.

VIP: see Gastrointestinal hormones.

Viperfishes: see *Stomiiformes.*

Viperidae, ***vipers:*** a family of venomous Snakes (see) distributed in Europe, Asia and Africa. Vipers are characterized by their usually plump body, short tail and a triangular head, which is well demarcated from the neck. In contrast to the *Colubridae* (see) and *Elapidae* (see), the large head shields of most vipers consists of many small scales. The short, mobile upper jaw carries 2 foldable venom fangs, which are solenoglyphic (see Snakes). When the mouth is closed, the fangs are folded upward. As the jaws open, the fangs are erected by a highly evolved lever mechanism, consisting of several articulated joints to the upper jaw bone. Most vipers are nocturnal, and have vertical pupils. They feed mainly on small rodents, which are killed or paralysed by a venomous bite, then seized again and swallowed. The venom, unlike that of the *Elapidae,* is mainly hemotoxic (attacking the blood and blood vessels) (see Snake venoms). Several species are among the most dangerous of the venomous snakes. The following examples are listed separately: Common viper, Sand viper, Russell's viper, Puff adders, Horned vipers, and the Carpet viper.

Virescence: deposition of chlorophyll in plant parts that are normally not green, in particular in flowers. V. is often caused by viruses and mycoplasms, sometimes by phsiological disturbances.

Virescence period: a phase of rapid appearance of new species, families or systematic units of higher order within a group of organisms; e.g. 12 new families of mussels appeared in the Ordovician, whereas only 2 appeared in the preceding Cambrian, and only one in the subsequent Silurian. Various other animal groups also underwent rapid evolutionary diversification during the Ordovician. Certain systematic units showed a V.p. directly after their first appearance. For other groups, the V.p. occurred after a preliminary period of slower evolution. For example, the mammals appeared in the Trias, but did not attain their modern abundance of forms until the late Tertiary, when a short, explosive phase of evolution gave rise to 17 mammalian orders containing many families and genera. Many groups, e.g. the ammonites, experienced several V.ps.

V.ps. are the result of the appearance of superior organisational types, or the absence of competitors. V.ps. therefore often occur after the extinction of competitors. Alternatively, a group of organisms may become successful by entering a new and favorable environment, which is free from competitors, or occupied by inferior forms. Often the V.p. of one group leads to the V.p. of another, e.g. the number of parasite species usually increases with the number of host species.

V.p. is one of several terms used to express the idea that evolution is periodically accelerated. See also Adaptive radiation, Punctuated evolution, Quantum evolution, Typostrophism.

Virginia creeper: see *Vitaceae.*

Virgula: see Graptolites.

Virion: see Viruses.

Viroids: virus-like group of plant disease agents, consisting of small, naked RNA circles, only 150–400 nucleotides in length, with no protein coat. Extensive hydrogen bonding between complementary bases causes the molecule to compress and form a rod with laterally projecting hydrogen-bonded loops. They replicate in suitable live host cells, and no protein coat is formed at any stage of replication. First of these infective agents to be called a viroid was the *potato spindle tuber viroid* (PSTV), which causes potatoes to become spindle-shaped, but also attacks 128 other plant species in 11 families. Other V. are listed in the Table. For their replication, V. are almost totally dependent on the replication systems of the host cell. It has been shown that the DNA of noninfected PSTV hosts contains nucleotide sequences complementary to those of the viroid RNA, a situation analogous to that in RNA tumor viruses. The M_r of PSTV is only about 120 000, and it is therefore much smaller than any known viral RNA. Also, it is much more infectious than any plant virus, 10 molecules being sufficient to infect a potato plant.

Viroids (after Gross, 1980)

Viroid	Acronym	Disease
Potato spindle tuber viroid	PSTV	Spindle tuber disease of potatoes
Citrus exocortis viroid	CEV	Exocortis disease of citrus plants
Cucumber pale fruit viroid	CPFV	Pale fruit disease of cucumber
Chrysanthemum stunt viroid	CSV	Stunting of chrysamthemums
Chrysanthemum chlorotic mottle viroid	CCMV	Chlorotic leaf mottle of chrysanthemums
Coconut cadang-cadang viroid	CCCV	Cadang-cadang disease of coconut palms
Hop stunt viroid	HSV	Stunting of hops
Columnea erythrophae viroid	CV	
Avacado sunblotch viroid	ASV	

Virola: see *Myristicaceae.*

Virology: study of Viruses (see) and Virus diseases (see). V. is concerned with the shape, size, structure and chemical composition of viruses, their interaction with animal, plant and bacterial cells, methods of culture, replication and transmission (especially with respect to virus diseases). V. also attempts to classify viruses on the basis of their phylogentic relationships, independently of any disease symptoms that they cause. Experimental exploitation of viruses (especially bacteriophages) as "mobile genetic elements" has enabled significant advances in molecular biology and molecular genetics.

Viroses: see Virus diseases.

Virulence: a measure of the rapidity and/or severity with which a parasite or pathogen attacks its host. V. (e.g. of bacteria and viruses) can be expressed quantitatively as the 50% lethal dose (LD50) or minimal lethal dose (MLD). See Lethal dose.

Virus coat proteins: proteins with M_r values as high as 40 x 10^6, and therefore of greater molecular mass than all other known proteins. Most V.c.p. are composed of identical subunits.

Virus diseases, ***viroses:*** functional disturbances of an organism, due to infection with a virus. These disturbances are mainly due to the effects of virus replication on host metabolism.

Symptoms. V.d. are manifested as a variety of external and internal symptoms, but the exact chain of events leading to specific symptoms is not always clearly understood. In seed-forming plants many processes are accelerated by virus infections, e.g. increased protein degradation, chlorophyll degradation, changes in the activities of certain enzymes, leading to premature aging of leaves. This is manifested as chlorosis, which may be expressed as a mosaic of chlorotic areas or as a generalized yellowing. V.d. often cause tissue death or necrosis, affecting more or less large areas of tissue. Disturbances of phytohormone metabolism lead to growth malformations, e.g. curling, vesicular expansions of leaves, swelling of shoots, excessive branching (witches broom), or growth inhibition leading to regression of laminas to a few middle and side ribs (threadlike leaves) or to dwarfing (stunting). Some V.d. cause suppression of flower formation, or development of cancerous growths (e.g. wound tumor virus). Petals may display spotted or striped color changes (e.g. the color breaks of tulip petals). In other cases, flower parts develop into structures resembling foliage leaves (folification). Fruits may carry spots, rings, raised vesicles, warty outgrowths or necrotic areas, or they may have an abnormal shape. The fruit flesh is often leathery and hard. V.d. markedly decrease the total plant mass of fodder crops, as well as reducing fruit and seed yield. V.d. of potatoes are responsible for an annual crop loss of 20%. Considerable losses are caused to fruit trees by the plum pox (scharka) virus, which causes premature ripening and fruit drop. On account of the widespread occurrence of this virus disease, a hundred thousand infected trees in western Bulgaria and 16 million infected trees in former Yugoslavia had to be destroyed and replaced with healthy trees. In an attempt to eradicate cocoa swollen shoot disease, which caused serious losses in West Africa, especially Ghana, more than 40 million cocoa trees were cleared in the Gold Coast (pre-independence name of Ghana) in the period 1945–55. By the end of 1982, in Ghana, 186 million trees had been removed. Cocoa swollen shoot virus is transmitted by a mealy bug and is as yet unclassified; it may belong with the maize chlorotic dwarf virus group.

In warm blooded (homeothermic) animals, virus infections frequently cause nonsuppurative inflammations, e.g. of the mucosae. Such inflammations can give rise to red spots, e.g. in measles. Other viruses may cause cell necrosis and death of infected cells. If the nervous system is affected, the resulting damage or loss of nerve cells can cause severe symptoms, e.g. in rabies or poliomyelitis. Damage and necrosis of liver cells are manifested as jaundice. Other symptoms include cell exudations in lung tissue (alveoli and bronchioles) or in the trachea, which are typical of various cold infections. Many infections, e.g. by influenza virus, are accompanied by the formation of toxins which greatly exacerbate the disease condition. Other viruses stimulate cell division, leading to formation of excrescences, e.g. the pocks of chickenpox and smallpox. Many viruses, notably the agent of German measles (rubella), can cause damage to a growing embryo, if they infect the pregnant mother. In homeothermic animals, V.d. are often characterized by a biphasic fever curve. After initial multiplication in the region of the infection site, the virus becomes distributed throughout the entire body, causing generalized disturbances (1st phase). In the 2nd phase, the disease becomes localized in an organ or organs. In humans and homeothermic animals, the initial entry of most viruses causes a powerful but relatively short-lived reaction, during which antibody formation against the virus is initiated.

In insects, many viruses cause only slight or no symptoms, and the virus remains in an inactive state during the entire life of the insect. For this reason, insects are often dangerous reservoirs and vectors of V.d. of plants, animals and humans. In addition, some viruses are powerfully pathogenic for insects (see Insect viruses)

Control of V.d. is more difficult than the control of diseases caused by other groups of organisms, owing to the close integration of virus metabolism with that of the host. Many control measures are therefore directed to the *prevention of infection,* such as quarantine procedures and removal of diseased organisms. Thus, virus-diseased potato plants are rejected for the culture of seed material, virus-diseased fruit trees are removed and destroyed, and diseased animals (e.g. with rinderpest) are culled. Disinfection and sterilization are also important preventative measures, e.g. the knife used for removing the side shoots of tomatoes should be sterilized before being used on a different plant, and strict disinfection and isolation are necessary to prevent the spread of bovine foot and mouth disease. Other measures include the control of virus vectors. Specially cultured, virus-free seed potato plants are often protected from infection by the use of insecticides to prevent attack by virus-carrying aphids. Similarly, the insecticidal control of mosquitoes protects humans against yellow fever. Transmission of different plant viruses, e.g. cucumber mosaic virus, potato viruses A and Y, can be prevented by spraying with skim-milk or emulsified oil.

Provided a plant is only slightly affected by high temperatures, and the infecting virus is temperature-sensitive, the plant can be made virus-free by keeping it at a high temperature for a long period. Starting with totally virus-infected plant material, healthy clones of vegetatively propagated plants can be prepared by culturing samples of meristematic tissue. Plant viruses usually do not penetrate as far as the meristematic tip; the latter is therefore separated and induced to form an undifferentiated tissue culture in a liquid or solid (agar) growth medium. The new virus-free plant is obtained by inducing or allowing the callus material redifferentiate to form shoot and roots. In combination with heat therapy, this method is widely used to prepare virus-free potatoes, chrysanthemeums and carnations.

Many V.d. of homeothermic animals can be controlled by *passive* or *active immunization (vaccination)*. Passive immunization consists of the injection of antisera taken from organisms shortly after they have overcome the same V.d., e.g. infective hepatitis or rubella. Active immunization is performed by injecting attenuated live viruses, or viruses killed by treatment with formaldehyde, phenol or other chemicals. This induces the formation of specific antibodies, which eliminate the virus. Once formed, these antibodies remain in the circulation for relatively long periods, protecting the organism against infection with virulent and strongly pathogenic forms of the same virus. Smallpox was controlled and eventually eliminated (the disease is now extinct) by active immunization (vaccination) with less virulent virus strains. The poliomyelitis vaccine (Sabin vaccine) is administered orally. The measles vaccine consists of an appropriately treated measles virus. Inactivated virus preparations are also used to control foot and mouth disease, fowlpest and other animal viroses.

Chemotherapy of V.d. is a more recent development. Particularly herpesviruses (see Virus families), but also poxviruses and some myxo- and paramyxo-, as well as adenoviruses can be controlled with adamantin or thiosemicarbazone. Considerable success has also been achieved with analogs of pyrimidine and purine bases or their ribosides, in particular, 5-bromo-2′-deoxyuridine and 5-iodo-2′-deoxyuridine. Virazol (1-β-D-ribofuranosyl-1,2,4-triazole-3-carboxaminde), an analog of an intermediate of purine biosynthesis, has been widely tested clinically, and found to be active against numerous human viruses, as well as many plant viruses.

Viruses (plates 19 & 28): very small, infective particles, which are not retained by ultrafine filters that hold back the smallest bacteria. They consist of RNA or DNA enclosed in a protein coat, and they can replicate (multiply) only within a suitable live host cell. They suppress the genetic information of the host cell, and exploit the ribosomes, energy-producing mechanisms and various enzymes of the host in support of their own replication. There is a wide variety of different virus types, differing in the type of nucleic acid (RNA or DNA, single-stranded or double-stranded, linear or circular), structure of the protein coat, mode of infection, and mechanism of replication. Virus infections often cause functional disturbances of the host organism, known as Virus diseases (see) or viroses.

Occurrence and distribution. Viruses are found in practically all groups of organisms. The numerous viruses of the *Schizophyta* (bacteria and cyanobacteria) are called Phages (see). Many viruses are known which attack fungi (Mycoviruses), and a smaller number are known to infect green algae. No virus disease of mosses or liverworts has been identified with certainty, but several pteridophyte viruses are known. Spermatophytes (seed-bearing plants) are attacked by hundreds of different viruses. Virus-like infective agents have also been identified as disease organisms of protozoa. Among invertebrates, viruses have been isolated in particular from nematodes and insects, some insect viruses being responsible for severe epidemics (see Insect viruses), while others hardly affect their hosts. In view of the growing list of newly discovered invertebrate viruses, it can be assumed that viruses are abundant and widely distributed among most groups of invertebrates. Virus diseases of fishes and reptiles are also well known. Of particular importance, however, are the viruses of homeothermic animals, especially those of humans, which cause hundreds of different, sometimes serious, diseases.

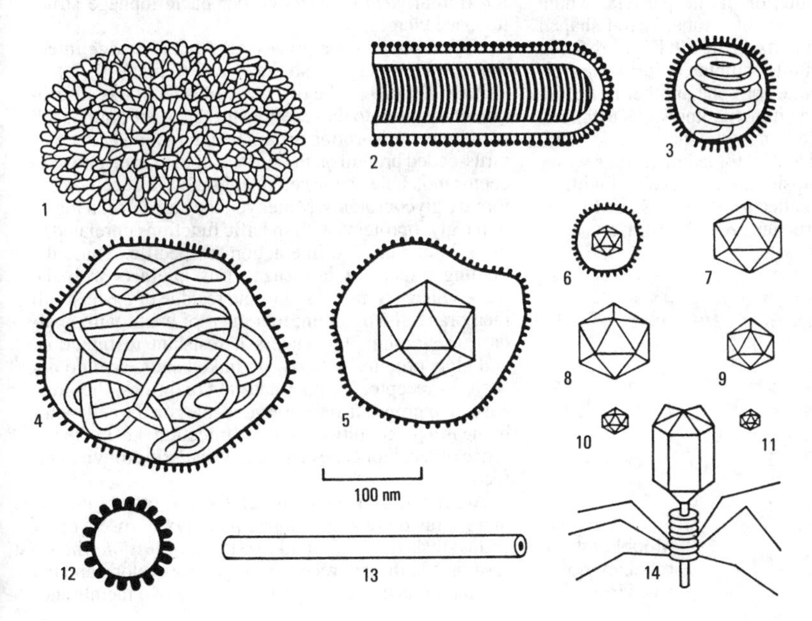

Fig.1. *Viruses.* Schematic representation of virus particles, all drawn to scale. The type of genome (RNA or DNA) is shown in brackets. *Enveloped viruses: 1* Pox virus (DNA). *2* Rabies virus (RNA). *3* Influenza virus (RNA). *4* Measles virus (RNA). *5* Chickenpox virus (DNA). *Naked or unenveloped viruses: 6* Yellow fever virus (RNA). *7* Adenovirus (DNA). *8* Reovirus (RNA). *9* Wart-papilloma virus (DNA). *10* Poliomyelitis virus (RNA). *11* Parvovirus (RNA). *12* Corona virus (RNA). *13* Tobacco mosaic virus (RNA). *14* Bacteriophage T2 (DNA).

Structure. Virus particles display a wide range of different forms, but the particle size and shape of each species are usually constant and characteristic (Fig.1). The diameter of isometric species ranges between 15 and 300 nm, while rod-shaped and helical particles vary in length between 180 and 2 000 nm.

Most virus particles contain a single nucleic acid molecule, but in some species (e.g. wound tumor virus) the genome is distributed among several separate nucleic acid molecules. Such viruses are known as *divided genome viruses.* In *Multipartite viruses* (see), the genome is divided among different particles. In most plant viruses (e.g. tobacco mosaic virus), many animal viruses and some bacterial viruses, the genome consists of single-stranded RNA. In many RNA-viruses, however, the genome is linear, helical and double-stranded, e.g. rice dwarf virus, the reoviruses and some insect viruses. Other animal and bacterial viruses, as well as a few plant viruses, contain circular single-stranded DNA (e.g. bacteriophage ΦX 174), while some very small viruses (e.g. parvoviruses) contain single-stranded DNA. Double-stranded linear DNA is found, e.g. in bacteriophage T4 and herpesvirus, while double-stranded circular DNA forms the genomes of, e.g. SV40 and polyoma virus. Adenoviruses contain double-stranded linear DNA with covalently linked terminal protein, while the linear double-stranded DNA of poxviruses is covalently sealed at both ends.

In most viruses, the nucleic acid is protected (e.g. from the action of nucleases) by a protein coat. The only exception to this structural feature is provided by the Viroids (see), which lack any proteins of their own. The *coat protein* consists of many identical subunits. For example, the subunit of the coat protein of Tobacco mosaic virus (see) (TMV) contains 158 amino acid residues of known primary sequence. In the mature virus, between 2 100 and 2 700 of these subunits are arranged like the steps of a spiral staircase, and the spirally wound nucleic acid lies in a groove in each subunit. A cavity remains in the interior of the particle, which therefore has the appearance of a tube, is rod-shaped and displays helical symmetry (plate 19; Figs 1 & 2). In other viruses, 2 or 4 subunits combine to form *capsomers,* which in turn associate with one another to form a hollow body with cubic symmetry, known as a *capsid.* The capsid encloses and protects the nucleic acid, which is spirally wound and/or folded in various ways, in order to fit into the capsid interior. A capsid with its enclosed nucleic acid is called a *nucleocapsid.* Most viruses with this type of structure are polyhedra, e.g. icosahedra, and are therefore referred to as isodiametric or isometric, and sometimes incorrectly as spherical (plate 19; Fig.3). In some cases, e.g. the adenoviruses, the capsomers forming the corners of the polyhedron possess a fibrillar appendage (a glycoprotein) which aids attachment of the virus to the host cell (Fig.3 of plate 19).

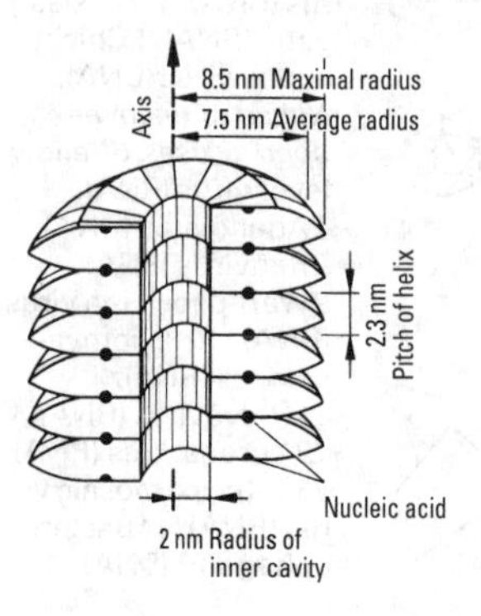

Fig.2. *Viruses.* Model of the tobacco mosaic virus.

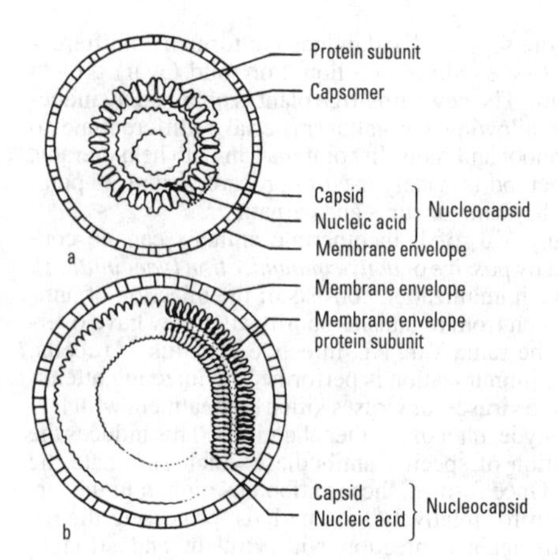

Fig.3. *Viruses.* Schematic representation of enveloped isodiametric *(a)* and helical *(b)* viruses.

A number of viruses have a secondary envelope (the *membrane envelope)* surrounding the nucleocapsid. In addition to protein, this secondary envelope usually contains carbohydrates and lipids, and many of its components are derived from the host cell membrane. The nucleocapsid of helical viruses is frequently spirally wound within the membrane envelope (Fig.5 of plate 19; Fig.3). Viruses with a membrane envelope are called *enveloped* viruses, whereas those with only a nucleocapsid are referred to as *naked* or *unenveloped.* The complete virus particle of a species is also called a *virion* (plural *viria* or *virions).* For bacteriophage structure, see Phages.

Infection. For the process of *bacteriophage* infection, see Phages (see also Figs 6 & 7 of plate 19).

Animal viruses. The first stage of infection is adsorption of the virus to the exterior surface of the animal cell membrane. Adsorption occurs by interaction between a virus-coded protein on the surface of the virion and a receptor molecule on the cell membrane. Most cell receptors are glycoproteins. Moreover, they are normal membrane glycoproteins with specific functions unrelated to virus infection. The interaction is specific. Thus, the binding protein of influenza virus interacts with the $\alpha 2 \rightarrow 3$ linked terminal sialic acid residue of the host cell membrane glycoprotein; treatment of a cell with sialidase (neuraminidae) renders it resistant to infection, and glycoproteins with $\alpha 1 \rightarrow 6$ linked sialic acid do not serve as receptors. Under natural conditions, the presence of appropriate receptor molecules on the cell membrane is a precondition for virus infection, i.e. in the absence of receptors, the cell is not *permissive* for virus infection

Animal viruses pass through the cell membrane, either by fusion (e.g. enveloped paramyxoviruses) or by endocytosis (e.g. Semliki Forest virus). *Fusion:* the virus binds to the cell receptor, the viral envelope and the cell membrane become perforated, the two membranes seal together, and the capsid enters the cell. *Endocytosis:* following adsorption of the virus, a depression forms on the cell surface in the receptor region (adsorption site). Continued invagination forms a coated pit (lined on the cytoplasmic side with clathrin), eventually enclosing the virus in a vesicle, which is finally released into the interior of the host cell by abstriction, still surrounded by the vesicle membrane; the latter loses its clathrin coat and fuses with other cellular vesicles to form vesicles called endosomes; viral and endosomal membranes then fuse and the nucleocapsid is relased into the host cytoplasm.

Plant viruses enter cells through wounds, or they are injected by biting insects, e.g. *Homoptera* and *Heteroptera.* Passage from cell to cell within the plant is then possible via plasmodesmata.

Within the host cell, the virus nucleic acid is released from both the nucleocapsid and the membrane envelope (if present). This process, known an *uncoating,* involves, inter alia, the action of specific proteases. The duration of uncoating increases with the complexity of the virus and with the tightness of the packing of the nucleic acid. Uncoating of small viruses, e.g. picornaviruses is complete in about 2 h., whereas the uncoating of poxviruses lasts 10–12 h.

Replication. For replication of bacteriophages, see Phages.

Part of the uncoated nucleic acid is transcribed and translated (DNA viruses) or simply translated (RNA viruses), with the aid of host enzymes, to form *early proteins,* which are generally enzymes or regulatory proteins. The early proteins of many RNA viruses include RNA-dependent RNA-polymerase (replicase or transcriptase), which is necessary for replication of the genome of RNA viruses, and which is absent from most hosts. In DNA viruses, other early proteins promote transcription of viral DNA rather than host DNA to form early viral mRNA, while in RNA viruses early proteins determine the balance of messenger and template function. Early proteins also slow down or stop the biosynthesis of host nucleic acids and proteins. The appearance of early proteins is followed by replication of the viral genome.

Double-stranded viral DNA genomes are replicated and transcribed by mechanisms already well documented for prokaryotic and eukaryotic DNA, using the enzymic apparatus of the host cell. Single stranded DNA genomes consist of the noncodogenic DNA strand (– strand); alternatively, the codogenic strand (+ strand) and – strand may be packaged in separate virions. Noncodogenic strands quickly become double-stranded by synthesis of a complementary strand, using host enzymes. In fact, the entire process of transcription and replication of a single-stranded DNA viral genome depends on the enzymatic apparatus of the host, the only proteins encoded by the viral genome being capsid proteins.

RNA genomes depend on different mechanisms for their replication. The single-stranded RNA of certain animal viruses functions directly as mRNA, and is therefore called a *positive* or *plus* (+) *strand* genome; such (+)RNA always specifies enzymes for the replication of RNA via its complementary (–) strand. Other single-stranded RNA viruses contain the nontranslatable strand of RNA, known as a *negative* or *minus* (–) *strand* genome. In this case, a plus strand or mRNA must be made from the minus strand, using virus-coded transcriptase, which is packaged in the virion, and released together with the (–)RNA during infection. Double-stranded RNA viruses also carry a packaged transcriptase for the synthesis of mRNA from this type of template. Replication occurs in the host nucleus (primarily the nucleolus), in the cytoplasm, or in other organelles.

As viral genome replication slows down, *late proteins* are synthesized in increasing quantities; these include certain soluble proteins (eventually packaged with the nucleic acid in the mature phage), maturation proteins and structural proteins (e.g. capsid and envelope proteins). The phase of late protein synthesis is followed by the maturation phase, in which new virus particles are assembled from the newly synthesized nucleic acid and proteins.

Finally, in infections of bacterial and animal cells, virus progeny are released. Especially in animal enveloped viruses, this can occur continually by cell budding or secretion of the virus (plate 19), in which the host cell remains intact, and the virus acquires its membrane envelope from the host plasma membrane. Bacteriophages are released by lysis, and therefore death, of the host cell. Newly formed plant viruses often remain for months or years in their host cells, until they are transferred to a new host by wound contact or by insects (see Virus vectors).

The replication cycle varies considerably, depending on the virus, host and type of viral nucleic acid (see Phages, Tobacco mosaic virus). Occasionally, the replication cycle of certain viruses cannot be completed. Thus, if the genes for early phage proteins are repressed, the viral nucleic acid becomes incorporated into the host bacterial genome (see Temperate phages). Alternatively, certain enzymes required for replication or for maturation may be absent, due to mutation (see Defective viruses, Deficient viruses). Under certain circumstances, the genome of a deficient virus may be integrated into the host genome, where it promotes tumor formation (see Tumor viruses).

Classification and nomenclature. See Virus families, Virus groups, Virus taxonomy.

Virus families: taxonomic groups of viruses, each consisting of well studied, chemically, physically and biologically similar viruses, and accorded familial status under the rules of the International Committee on the Taxonomy of Viruses (ICTV). Most families have a latinized name ending in *-idea.* Genera are also recognized. The following important families are largely composed of species that infect vertebrates, with a few that attack plants and insects. ***Poxviridae*** (poxviruses) are large, cuboid, DNA viruses with several membrane envelopes, e.g. viruses of smallpox , myxomatosis. ***Herpesviridae*** (herpesviruses) are large enveloped particles, about 200 nm in diameter, with an isometric nucleocapsid, e.g. viruses of herpes simplex, shingles, chickenpox. Other species cause illnesses in monkeys, rabbits, cattle and pigs. Epstein-Barr virus causes infectious mononucleosis, and is also implicated in the development of certain malignant tumors (Burkitt's lymphoma in Africa, and nasopharyngeal carcinoma in Southeast Asia). Herpesviruses typically from aggregates in the host cell nucleus. ***Adenoviridae*** (adenoviruses) consist of medium-sized, DNA-containing, naked isometric particles, often found in adenoidal, glandular-like tissues such as the tonsils. They replicate in the host nucleus, where they often appear as crystal-like aggregates. Many forms are responsible for cold infections. Deficient forms are known which cause animal tumors. ***Papoviridae*** (papoviruses) are small, naked, ether-resistant, DNA-containing, isometric virions, which replicate in the host cell nucleus. Two genera are recognized, papillomavirus and polyomavirus. Wart-papilloma virus causes various warts in humans and animals, and a few papillomaviruses, e.g. human papillomavirus type 5 (HPV5), are clearly implicated as risk factors in the development of malignant tumors. Polyomaviruses include two human papoviruses, BKV and JCV, the latter thought to be wholly or partly responsible for a chronic degenerative disease of the nervous system, known as progressive multifunctional leucoencephalopathy (PML). Other polyomaviruses are the monkey virus, SV40 (oncogenic in hamsters), mouse tumor virus and others. Papoviruses, in particular SV40 and polyomaviruses, are among the best studied animal viruses. SV40 and papillomavirus are used extensively as molecular vehicles in gene technology. ***Retroviridae*** (retroviruses) are RNA viruses with a lipid-containing membrane envelope. The virion contains two identical single-stranded RNA molecules, consisting of between 8 and 10 thousand nucleotide residues, as well as RNA-dependent DNA polymerase (revertase, reverse transcriptase). The retroviruses are divided into 3 subfamilies: *Oncovirinae* (oncornaviruses; oncogenic RNA viruses), which include all RNA tumor viruses, as well as some related, nononcogenic viruses; *Lentivirinae* or "slow viruses", which cause chronic degenerative diseases of the nervous system; *Spumavirinae,* which cause persistent infections without clinical symptoms. The oncornaviruses, which cause malignant tumors in animals, are especially well studied. An intensive search for human RNA tumor viruses led, in 1981, to the discovery of the T-cell lymphoma virus (TCLV), which is responsible for T-cell lymphoma mainly in Southeast Asia. The RNA of oncornaviruses serves as a template for synthesis of double-stranded DNA, a process catalysed by reverse transcriptase. This DNA is integrated as a provirus into the host genome by a process of nonhomologous recombination, then replicated and transcribed like host DNA. Acute oncornaviruses possess an onc gene (v-*onc),* which encodes a tyrosine-specific protein kinase involved in malignant (neoplastic) transformation. Normal cells contain a homologous sequence known as a proto-oncogene (c-*onc),* which is activated by rearrangment of the chromosome, by mutation, or by insertion next to it of a segment of viral DNA. ***Paramyxoviridae*** (paramyxoviruses) are enveloped viruses. The usually isometric envelope surrounds a helical, RNA-containing nucleocapsid, which is often wound into multiple spirals. Paramixoviruses cause clumping of erythrocytes (hemagglutination). Both the nucleocapsid and the membrane envelope are formed in the host cell nucleus. Important members of the family are the causative agents of parainfluenza, measles, mumps, rinderpest (cattle plague), atypical fowlpest and canine distemper. ***Orthomyxoviridae*** (orthomyxoviruses) are enveloped viruses, similar to, but somewhat smaller than, the paramyxoviruses. The genome is often divided among several helical nucleocapsids, all within the same membrane envelope. Both rounded and filamentous orthomyxoviruses are known. The helical nucleocapsid is synthesized in the host cell nucleus, the envelope in the cytoplasm. Maturation of the virion occurs on the host cell membrane. Important representatives are the agents of human influenza and swine influenza. ***Rhabdoviridae*** (rhabdoviruses) are short, rod-shaped enveloped viruses, often with one flattened end (i.e. bullet-shaped, bacilliform particles), and containing a helical nucleocapsid. They include the causative agents of rabies, vesicular stomatitis, hemorrhagic septicemia of the rainbow trout, yellow dwarf disease of the potato, and beet leaf curl. ***Togaviridae*** (togaviruses) consist of several genera of small RNA viruses. *A-* and *B-arboviruses* are transmitted by arthropods (arthropod-borne), mostly by biting insects. More than 160 species are known, which multiply in humans, horses, bats, birds, snakes, mosquitoes, ticks and other organisms. The B-group of arboviruses includes the feared yellow fever virus. The genus, *Rubiovirus,* contains the agent of German measles (rubella), while the genus, *Pestivirus,* contains the agent of swine cholera. ***Reoviridae*** (reoviruses) have a naked nucleocapsid consisting of two protein coats. The genome consists of several molecules of double-stranded RNA, as well as shorter single-stranded RNA. Some forms infect homeothermic animals, including humans in which they cause mild diseases of the respiratory and alimentary tracts (respiratory, enteric, orphan viruses). Other forms are responsible for insect polyhedroses (see Insect viruses). Others, e.g. wound tumor virus, cause plant diseases, and they also multiply in the insect vector. ***Picornaviridae*** (picornaviruses) are very small (pico), RNA-containing, isometric particles. Members of the genus, *Enterovirus,* are acid-resistant; they include agents of influenza-like infections, often associated with diarrhea and other gastrointestinal disturbances, as well as the virus of infantile paralysis (poliomyelitis). Members of the genus, *Rhinovirus,* are acid-labile, and include more than 60 known species, many being responsible for cold infections of humans. Of the animal rhinoviruses, the agent of foot and mouth disease is well known and much feared.

Virus groups: taxonomic groups, constructed according to the same principles as Virus families (see), but not yet accorded familial status by the International Committee on the Taxonomy of Viruses. Names of plant virus groups are usually constructed from an abbreviation or acronym of the English name of a typical group member. The most important groups of plant viruses are as follows. ***Caulimoviruses*** (cauliflower mosaic virus group) are the only plant viruses whose genetic material is DNA, and they occur as isometric particles. Familiar members are cauliflower mosaic virus and carnation etched ring virus. ***Cucumoviruses*** (cucumber mosaic virus group), like all the following groups of plant viruses, possess an RNA genome. They are isometric particles with a diameter of 30–40 nm. Mechanical transmission readily occurs, as well as nonpersistent transmission by aphids. The type representative is the cucumber mosaic virus, which infects hundreds of different plant species, and has an unusually wide distribution in the temperate zones of the world. ***Luteoviruses*** (luteo = yellow; barley yellow dwarf virus group; leaf roll virus group) consist of particles of 25–30 nm. They are generally transmitted by insects, in which the infection is persistent. Since they are confined to the plant phloem they cause mainly dwarfing and yellowing. Important representatives are barley yellow dwarf virus and potato leaf roll virus. ***Nepoviruses*** (nematode portable viruses) consist of isometric particles 25–30 nm in diameter. All are transmitted by nematodes, but they are also easily transmitted mechanically. A high proportion of infections is due to transfer in the seed. Examples are the agents of tobacco ringspot, Arabis mosaic, tomato black ring, strawberry latent ringspot, tomato ringspot. Disease symptoms, for the most part ringspots, generally regress in young plants and finally disappear completely, a phenomenon known as *recovery*. ***Tombusviruses*** (tomato bushy stunt virus group) consist of isometric particles of 30 nm diameter. Chief representative is the tomato bushy stunt virus, which attacks both tomatoes and geraniums. ***Ilarviruses*** (isometric labile ringspot virus group) are isometric particles 26–35 nm in diameter. Four different RNA components are packaged in at least 3 different particles (see Multipartite viruses). Known representatives are the viruses of tobacco streak, Prunus necrotic ringspot, apple mosaic and rose mosaic. They can be transmitted mechanically, and some are transmitted with pollen. ***Tymoviruses*** (turnip yellow mosaic virus group) are isometric particles with 32 capsomers per virion. Diseased plants generally show yellow mosaic mottling. Natural vectors are beetles. In addition to the yellow turnip mosaic virus, the group includes the cucumber (wild) mosaic virus and the cocoa yellow mosaic virus. ***Comoviruses*** (cowpea mosaic virus group) are Mulitpartite viruses (see) with isometric particles. Known representatives include the viruses of cowpea mosaic, squash mosaic, radish mosaic and broad bean true mosaic. Transmission occurs both mechanically and due to feeding by beetles. ***Bromoviruses*** (brome mosaic virus group) are isometric particles of 25 nm diameter. Brome mosaic virus attacks a wide variety of grasses. Another member is the broad bean mottle virus. Routes of transmission are unclear. ***Hordeiviruses*** (*Hordeum* = barley; barley stripe mosaic group) consist of elongate particles with helical symmetry, length 110–160 nm, diameter 20–25 nm. They are Multipartite viruses (see), consisting of 2–4 particles, and 2 or 3 of which are required for infection. Best known member is the barley stripe mosaic virus, which is transmissible by vectors, as well as by pollen and seeds. ***Tobraviruses*** (tobacco rattle virus group) is a Multipartite virus (see), and its genome is distributed among rod-shaped particles of different lengths. Best known member of this group is tobacco rattle virus, which, in addition to tobacco, attacks many other woody and herbaceous plants from numerous families, causing necrotic spots, patterns and rings, as well as leaf curling. The viruses are transmitted mechanically, as well as in seeds and by nematodes. ***Tobamoviruses*** (tobacco mosaic virus group) consist of stiff helical rods, length about 300 nm, diameter 18 nm. They are transmitted mechanically, and they attain high concentrations in infected plants. ***Potexviruses*** (potato-X-virus group) are flexible helical rods, length 480–580 nm, diameter 15 nm. They are transmitted mechanically and to some extent by aphids, and can attain high concentrations in infected plants. Members include potato virus X, potato aucuba virus, narcissus mosaic virus, hydrangea ringspot virus, and cactus virus X. ***Carlaviruses*** (carnation latent virus group) are flexible rods (e.g. carnation latent virus, lily symptomless virus). ***Potyviruses*** (potato-Y-virus group) consist of flexible helical particles, normally between 720 and 800 nm long, diameter 15 nm. Transmission is mechanical or by aphids. Considerable damage is caused to the plant by very low virus concentrations. The group contains many species, some widely distributed, e.g. the viruses of potato-Y disease (leaf drop streak), potato-A disease, beet mosaic, pea mosaic, bean mosaic, soybean mosaic, sugarcane mosaic, lettuce mosaic, plum pox (sharka). ***Closteroviruses*** (Greek *kloster* spindle, thread; beet yellow virus group) are very flexible rods, normal length 1 250 nm–2 000 nm, diameter about 15 nm. These are the longest of all plant viruses. Mechanical transmission is very difficult, but infection occurs very readily by aphid transmission. Members include beet yellows virus, Festuca necrosis virus and the tristeza virus of citrus fruits. Infections by members of this group typically involve yellowing and necrosis, in particular of the phloem.

Virus interference: mutual inhibition or promotion of the replication of two different viruses in the same host. V.i. between certain bacteriophages and between certain plant viruses is so pronounced that replication of both species is totally blocked. Thus, preinfection with a very mild strain of tobacco mosaic virus, which causes hardly any damage to the plant, can completely suppress the deleterious effects of a subsequent infection with an aggressive strain of the same species. This phenomenon of *preimmunity* is used to protect crops (e.g. tomatoes) from aggressive viruses. The mechanism of V.i. is unclear. V.i. between animal viruses may be due to Interferons (see).

Virus taxonomy: a branch of virology concerned with the description, naming and classification of viruses. There is as yet no universal system for virus classification, since viruses display no clear evolutionary relationships. Some viruses may have arisen by regression of parasitic organisms, others from cell organelles. Bacterial viruses [in particular Temperate phages (see)] and Tumor viruses (see) may have evolved from plasmids. Some of the difficulties associated with devising a comprehensive classification and nomenclature have been bypassed by constructing groups of viruses that display chemical, structural, biochemical and serological similarities. Many of these groups have been accorded familial status according to the rules of the International Committee on the Taxonomy of Viruses (ICTV) (see Virus families). Some families contain distinct subdivisions, which are given the status of genera. The taxonomic status of other similar viruses, primarily plant viruses, still awaits more detailed investigation, and they are placed as a preliminary measure in Virus groups (see).

Virus transmission: spread of a virus infection from one host to another. The mode of V.t. varies greatly for different viruses. Resistant viruses can be transferred mechanically, especially by contact between virus-diseased and healthy individuals. Human and animal viruses are transmitted mechanically in saliva droplets, food, dirt and contaminated water. Rabies is transmitted by the bite of rabid individuals. Plant viruses are often transmitted by machines and implements used in cultivation, e.g. by knives for removing side shoots. Other viruses are transmitted by vectors (see Virus vectors). Most virus vectors are animals (mainly piercing and sucking insects, and in some cases nematodes), but lower fungi and parasitic plants [e.g.

dodder *(Cuscuta)* which forms a physical connection between different individuals] are also known to transmit certain viruses. Some plant viruses are transmitted by seeds and pollen, while others are transmitted in the soil.

Virus vectors: organisms which on account of their motility, feeding activity or other behavior carry virus infections between host organisms. A very large number of virus infections, especially those of plants, is transmitted by piercing and sucking insects. Vectors of animal and human viruses are primarily lice, fleas, mosquitoes and ticks. Plant viruses are carried by cicadas, leaf bugs, thrips, and in particular by hundreds of different species of aphids. A very important vector is the green peach aphid, *Myzus persicae,* which attacks hundreds of plant species and transmits more than 50 virus species. The beet aphid, *Aphis fabae,* is also responsible for transmitting a range of different viral diseases. Some viruses multiply inside their vector, as well as in the main host, e.g. the yellow fever virus replicates inside the mosquito, Aedes *aegypti.*

According to their relationship with the vector, virus species can be divided into 2 groups. *Persistent viruses* are often not easily transmitted mechanically. Persistent viruses remain in their vector for a long time, often until death of the host, and during this time they remain viable and capable of infecting another host. After entering their vector, persistent viruses become latent. Depending on the virus and the insect species, the latent period lasts for 1 hour to several days; during this time the insect does not pass on the infection when it bites another virus host organism. During the latent period, the virus travels from the intestinal wall of the V.v. into the circulation, and finally to the salivary gland. It may also multiply within the vector. It is finally introduced into a new plant host when the vector pierces the host tissues and injects saliva, prior to sucking the plant sap. *Nonpersistent viruses* are usually easily transmitted by mechanical means, and they can be transmitted by the vector immediately after uptake by the insect, i.e. there is no latent period; but they lose their infectivity relatively quickly within the vector. Spread of persistent viruses is effectively reduced by controlling the vectors with insecticides. This may, however, have the opposite effect on the spread of nonpersistent viruses, because for a brief period after contact with the insecticide many insect vectors become excessively active, fly fom plant to plant, and cause a rapid spread of disease.

In addition to arthropods, nematodes are also well known V.v. Some parasitic, soil fungi also serve as V.v., in particular the primitive fungus, *Olpidium brassicae,* as well as other species of *Olpidium,* and species of *Synchytrium, Polymyxa* and *Spongospora.* Finally, parasitic plants, e.g. dodder *(Cuscuta),* which make physical contact with different individual host plants, can act as vectors of plant viruses.

Viscacha: see *Chinchillidae.*

Viscacha rat: see *Octodontidae.*

Viscera: in the narrow sense, the organs present in the abdominal cavity. In the broad sense, all the organs of the body.

Visceral nervous system: see Autonomic nervous system.

Visceral skeleton: collective term for all the elements of the pharyngeal skeleton of vertebrates. In the primitive condition, these consist of the cartilaginous and bony supports of the gill apparatus. In most embryos it consists of visceral arches, one behind the mouth and one behind each of the visceral clefts. Each arch encircles the pharynx, but the two halves of the arch do not neet on the dorsal side. Phylogenetically, it is a very adaptable part of the skeleton, having been modified for respiration and for feeding. In fishes, the V.s. includes the jaws and the skeleton of the pharynx. In mammals, the anterior visceral arches have evolved into the jaw bones, palate and allied structures, while the remaining arches contribute to the cartilages of the larynx. See Skull.

Visceroceptor: see Proprioceptors.

Viscerocranium: see Skull.

Viscidium: see *Orchidaceae.*

Viscosity: a property that opposes the relative movement of adjacent layers of a fluid, generally regarded as the internal friction of liquids and gases. The reciprocal value of V. is fluidity. *Dynamic V.* is the force (in Pascal · seconds; Pa · s) per m^2 required to maintain a velocity difference of 1 m/s between two parallel layers of fluid, 1 m apart. *Kinematic V.* is the ratio of dynamic V. to density. The *relative V.* of a suspension is the ratio of the V. of the suspension to the V. of the suspension medium. In the ideal case, V. is independent of the rate of sheer (Newtonian liquids). Due to the orientation and deformation of proteins and other colloidal components, biological fluids usually display nonNewtonian behavior. Increase of V. with the rate of sheer is known as *dilatancy.* Fluids whose V. decreases with the rate of sheer are said to be *thixotropic.*

V. is measured with viscometers, e.g. rotation, capillary and falling sphere viscometers. In the *rotation viscometer,* a torsion device is used to measure the force on a rotor; the resulting value includes the effects of defined sheering forces. The *capillary viscometer* measures the time taken for a certain quantity of fluid to flow through a narrow tube, and V. is then calculated from Poiseuille's equation: $h = pr^4tp/8vL$, where h is the coefficient of V., and v is the volume of fluid that flows through a tube of length L cm and radius r cm in t sec, under a driving pressure of p dynes/cm^2. In the *falling sphere viscometer,* the time taken for a small steel sphere to fall between two marks in a tall cylinder of the liquid under test is measured. The measurement is repeated with the same sphere in another liquid of known V., and the V. is calculated from the equation: $h_1/h_2 = t_1(d' - d_1)/t_2(d' - d_2)$, where d' is the density of the steel ball, d_1 and d_2 the respective densities of the two liquids, and t_1 and t_2 the respective times taken for the sphere to fall.

V. has been measured in vivo by introducing small magnetic particles and measuring their velocity in a magnetic field. V. can also be determined from Brownian molecular movement, by making use of Stoke's law which describes the movement of small spherical particles in a viscous medium; this is valid only when the Reynolds number (see) is sufficiently small.

Visean stage: see Carboniferous.

Visual depth: the water depth at which the outline of a white disk *(Secchi disk,* 20–30 cm diameter) is just still visible. The V.d. depends on the absorption of light by the water and its dissolved materials, and on the scattering of light by turbidity. The greatest V.d. (66.5 m) is found in the Sargasso sea. The V.d. of oligotrophic lakes is between 16 and 20 m. In eutrophic lakes it depends strongly on the time of year, e.g. 10 m in winter, but only a few cm in summer, due to the presence of algal water bloom. The greatest reported V.d. of a freshwater lake (40 m) is that of Lake Baikal.

Visual organs: see Light sensory organs.

Vitaceae, *grape vine family:* a family of dicotyledonous plants containing about 700 species in about 11 genera. Most species are restricted to the tropics, but some species are distributed in the Northern Hemisphere. They are mostly woody, climbing plants, which climb with the aid of shoot tendrils. They have alternate, variously lobed, digitate, simple or divided leaves, sometimes with stipules. The usually inconspicuous 4-5-merous flowers are actinomorphic, insect-pollinated, and almost always hermaphrodite, and they are aggregated into corymbose or paniculate inflorescenses. In most species, the 2-celled ovary develops into a 4-seeded juicy berry.

Economically the most important species is the ***Grape vine** (Vitis vinifera),* which was cultviated by the ancients. Its ripe berries (grapes) are a popular fruit. Large, ripe, seedless grapes grown in the Balkans are dried to make the raisins and sultanas of commerce. Small, seedless, blue-skinned grapes, originally grown in the area of Corinth, are dried to make currants. The expressed juice of specially selected and propagated varieties of grapes is fermented to make wine.

Vitaceae. Virginia creeper *(Parthenocissus* sp.)

Species of *Parthenocissus* ***(Virginia creeper)*** are cultivated widely as climbing plants with decorative foliage, for covering walls and buildings. Most commonly planted is *P. tricuspidata,* whose tendrils end in adhesive disks.

Vital capacity of the lungs: see Lung volume.

Vitalism: a theory that the function, behavior and evolution of living organisms is subject to a vital force. Old theories of V. were influenced by the church, as well as existing at a time when many life phenomena were unexplained and therefore attributable to mysterious forces. The advances of reductionist experimental science, especially in the latter half of the 20th century, have revealed much of the molecular mechanisms of living organisms, and progress has also been made in understanding the integration of separate systems within the organism, and of the organism in its environment. In view of these advances, the mechanist is confident that subsequent research will explain the organization of the living organism in its entirety, and denies the existence of, or the need for, a vital force. *Neovitalism* is usually attributed to the interpretation of results from developmental mechanics and developmental physiology, in particular the work of H. Driesch, who obtained a dwarf but otherwise normal larva from individual cleavage cells of the sea urchin egg. Driesch interpreted his findings as evidence for the existence of a nonmaterial causal factor, or a "prospective potency" of the embryonic cell. Although much is still unclear regarding the mechanism of embryonic determination, it seems highly likely that this subject will also eventually by explained in purely mechanistic terms. True neovitalism is more closely related to the cosmological theory that the universe has a directional purpose (i.e. since the Big Bang it has evolved according to a fixed pattern). Neovitalists recognize this directional purpose in the exquisite organization of living organisms. Mechanists and vitalists can agree that the idea of human consciousness existed as a possibility at the time of the Big Bang, but the mechanist believes that the realisation of this possibility was due to random selection, whereas the vitalist claims that is was preordained and inevitable.

[F. Cramer, *Chaos and Order – The Complex Structure of Living Systems,* VCH Publishers, Weinhein, New York, Cambridge, Tokyo, 1993]

Vitamins (Latin *vita* + amine): substances present in the animal diet in only small amounts, and indispensable for growth and maintenance of the organism. A dietary requirement is implicit in the definition of a V. Most of the substances that are V. in animals are essential for the metabolism of all living organisms, but plants and microorganisms can synthesize them (some fat-soluble V., however, may have metabolic roles unique to animals). The dietary requirement in the animal results from the evolutionary loss of this biosynthetic ability. Animals differ in their ability to synthesize certain V., and they therefore display different dietary requirements for V. For example, ascorbic acid (V.C) is a V. only for primates and a few other animals (e.g. guinea pig); most animals can synthesize it, and for them it is therefore not a V. Some V. can be synthesized from provitamins in the diet. In addition some the V. requirement of humans and higher animals is supplied by the intestinal flora, e.g. most of the V.K required by humans is supplied in this way.

The role of V. is largely catalytic, most V. serving as coenzymes and prosthetic groups of enzymes. V.D, however, acts as a regulator of bone metabolism, and therefore resembles a hormone. As a component of the visual pigments, V.A acts as a prosthetic group, but it is not known whether it is associated with catalytic proteins in its other functions. Nicotinamide and riboflavin are constituents of hydrogen-transferring coenzymes. Biotin, folic acid, pantothenic acid, pyridoxine, cobalamin and thiamin are (or are precursors of) coenzymes of group transfer reactions. The low daily requirement for

V. reflects their catalytic and/or regulatory roles. Thus, V. are nutritionally quite different from fat, carbohydrate or protein, which are required in the diet in considerable quantities as substrates of tissue synthesis and energy metabolism.

The biological activity of a pure V. can be expressed in International Units (IU). Thus, 0.3 μg V.A (retinol), 8 μg thiamin hydrochloride, 0.18 μm biotin, 50 μg L-ascorbic acid, 0.025 μg ergocalciferol or 1 mg DL-α-tocopherol acetate each corresponds to 1 IU. The system of IU is retained, even though the structures of all V. are known, because in most cases a V. is a family of closely related compounds all with the same action, but with different activities.

Lack or deficiency of a V., as a result of unbalanced nutrition, leads to characteristic metabolic disturbances. Complete absence of a V. leads to avitaminosis, with typical clinical symptoms. Relative deficiency of a V. causes hypovitaminosis. Such conditions are reversible by administration of the appropriate V. Excessive intake of certain V., e.g. V.A or V.D, can lead to hypervitaminosis.

Formerly, V. were named after the diseases they cured, e.g. antiscorbutic V., antirachitic V., antiberiberi factor. Not all V., however, have such a pronounced specificity, and the clinical pictures of many avitaminoses and hypovitaminoses are complex and variable. A nomenclature based on letters of the alphabet was developed simultaneously; the designations A, B, C, D and E were applied in the historical order of discovery. Subscripts were applied as more refined chemical analysis revealed that the originally isolated substances were in fact complex mixtures. This was especially true for the B vitamins. Partly because of confusion over the "B complex", trivial names which give an indication of the chemical structure of the V. (e.g. pyridoxine or pyridoxol for V.B_6) are now preferred. The V. represent a heterogeneous group of substances, which are classified into 2 main groups: fat-soluble and water-soluble, depending on whether they can be extracted from foodstuffs with organic solvents or water.

Retinol (V.A; axerophthol; xerophthol) (obsolete names: epithelial protection V.; growth V.) is a fat-soluble V. with polyisoprenoid structure. The alcohol, retinol, is also known as V.A_1. 3-Dehydroretinol (V.A_2) has an additional double bond between C3 and C4 in the ring. V.A is essential to the visual process, as well as for growth, skeletal development, normal reproductive function, and the maintenance and differentiation of tissues.

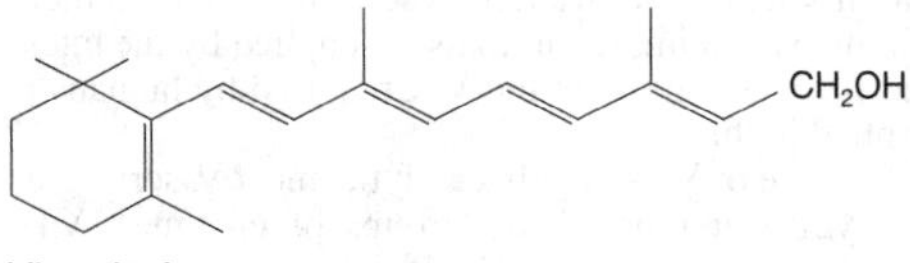

Vitamin A_1

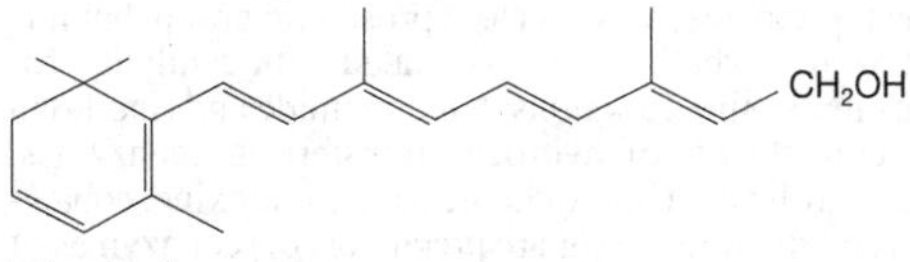

Vitamin A_2

It occurs predominantly in animal products, such as milk, butter, egg yolk, cod liver oil and the body fat of many animals. All the carotenes, which are abundant in green plants and fruits, have provitamin A activity. Conversion of carotenes to V.A occurs in the small intestine, but other organs, such as muscle, lungs and serum also have a limited ability to perform the same conversion. β-Carotene is oxidatively cleaved by the intestinal mucosa into 2 molecules of retinal (α- and γ-carotenes yield only 1 molecule of V.A), which are then reduced to all-*trans*-retinol, and esterified with a fatty acid, mainly palmitate. This V.A palmitate is transported in the lymph to the liver for storage. Free retinol is released by hydrolysis and transported from the liver by a retinol-binding plasma protein. Retinol is removed from the plasma by the cells of the retina, where it is oxidized to all-*trans*-retinal (retinaldehyde). This is isomerized to 11-*cis*-retinal, which is a component of visual purple (rhodopsin).

An early symptom of V.A deficiency in humans is night blindness, caused by deficient regeneration of rhodopsin. Later, the deficiency leads to hyperkeratosis of the epithelia of the eye (xerophthalmia), skin follicles, respiratory tract and digestive tract. In children, V.A deficiency also leads to growth arrest, and in adults to resorption of fetuses, stillbirths and birth defects. The daily requirement for adults is 1.5–2.0 mg.

Thiamin (V.B_1; aneurin; antiberiberi factor; antineuritic V.) is a water-soluble V. of widespread occurrence in natural materials, especially in yeasts and germinating cereal grains. Chemically, it consists of a pyrimidine and a thiazole ring. Both ring systems are synthesized separately as phosphorylated derivatives, which then become linked via a quaternary nitrogen. The pyrophosphorylated form of V.B_1, thiamin pyrophosphate, is a coenzyme of decarboxylases, transketolases and 2-oxoacid dehydrogenases. Deficiency of V.B_1 results in disturbances of carbohydrate metabolism, accompanied by an increase in the concentration of blood oxoacids (mostly pyruvate), which reflects the role of thiamin pyrophosphate as a coenzyme of pyruvate dehydrogenase. The typical deficiency disease, beriberi, results from a diet exclusively of polished rice. It is characterized by disturbances of the central and peripheral nervous system (polyneuritis) and of cardiac function. The daily requirement for thiamin is about 1 mg.

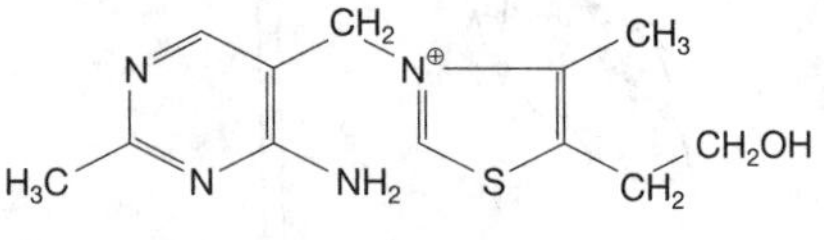

Vitamin B_1

Riboflavin (V.B_2; lactoflavin; 6,7-dimethyl-9-(D-1′-ribityl)-isoalloxazine) is a water-soluble yellow flavin derivative, occurring chiefly in a bound form in flavin nucleotides or flavoproteins in yeasts, animal products and legume seeds. Milk contains free riboflavin. It is required as a precursor of flavin mononucleotide and flavin-adenine-dinucleotide, which are coenzymes of the flavin enzymes. In rats, experimental riboflavin deficiency causes growth failure and dermatitis around the nostrils and eyes. In humans, riboflavin deficiency (ariboflavinosis) is characterized by lip lesions, a seborrheic dermatitis about the nose, ears and eyelids, and loss of hair. Angular stomatitis, glossitis, cheilosis, and ocular changes such as photobia, indistinct vision and corneal vascularization have also been reported as typical of human ariboflavinosis.

The biosynthetic precursors of riboflavin are a purine (probably guanine), ribitol and diacetyl. Most human diets contain adequate riboflavin. The daily requirement is about 1 mg.

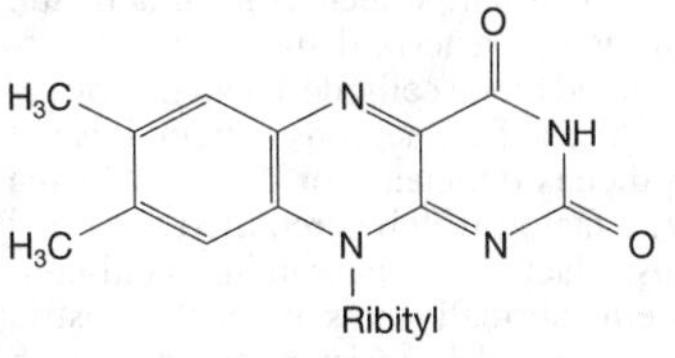

Riboflavin

Folic acid (pteroylglutamic acid) is a pteridine derivative, especially plentiful in liver, yeast and green plants. Chemically, it consists of 3 moieties: 2-amino-4-hydroxypteridine, *p*-aminobenzoic acid and one or more glutamic acid residues, linked by peptide bonds via their γ-carboxyl groups. Folic acid is a growth factor for some bacteria.

The biochemically active form of folic acid is 5,6,7,8-tetrahydrofolic acid (FH_4), a coenzyme in the metabolism of one carbon units. Hydroxymethyl and formyl groups are carried on atoms N^5 and N^{10} of FH_4. For example, when the amino acid, L-serine, is converted metabolically to glycine, its hydroxymethyl group is transferred to FH_4 to form $N^{5,10}$-methylene-FH_4. Once attached to FH_4, one carbon units can be oxidized or reduced. Thus $N^{5,10}$-methylene-FH_4 can be reduced to N^5-methyl-FH_4, which serves as the methyl donor in methylation reactions.

In humans, folic acid avitaminosis is more often caused by faulty uptake and/or utilization, than by dietary deficiency. It usually results in blood abnormalities, e.g. megablastic anemia and thrombocytopenia. Antimetabolites of folic acid are aminopterin and methopterin, which are used therapeutically in the treatment of leukemia. The sulfonamides are antimetabolites of *p*-aminobenzoic acid, and therefore act as inhibitors of bacterial folic acid synthesis.

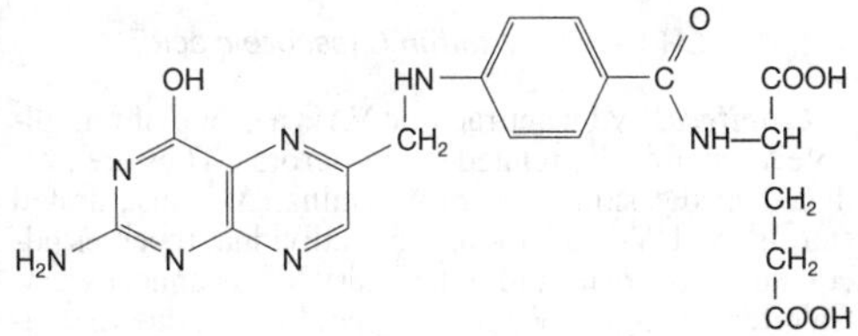

Folic acid

Nicotinic acid (niacin) and ***Nicotinamide*** (niacinamide) (pellagra preventative factor, V.PP) are metabolically interconvertible, water-soluble, simple pyridine derivatives with equal V. activity. They are widely distributed in nature, and are especially plentiful in liver, fish, yeast and germinating cereal grains. Nicotinic acid and nicotinamide are nutritionally equivalent, since both can be assimilated for the purposes of NAD(P) synthesis. For therapeutic purposes, however, nicotinamide is preferred, because large doses of nicotinic acid may have undesirable side effects. In many mammals and in fungi, the nicotinamide moiety of NAD(P) can be derived by the oxidative degradation of L-tryptophan. The extent to which the dietary nicotinamide requirement of animals can be spared by dietary tryptophan varies according to the species. Thus, if the definition of a V. implies a dietary requirement, nicotinamide is not a V. for the rat, which can satisfy its total requirement by the degradation of tryptophan. Synthesis of the nicotinamide moiety of NAD(P) in bacteria and plants occurs by a different pathway, in which aspartic acid and dihydroxyacetone phosphate act as precursors.

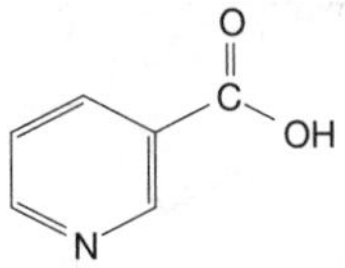

Nicotinic acid 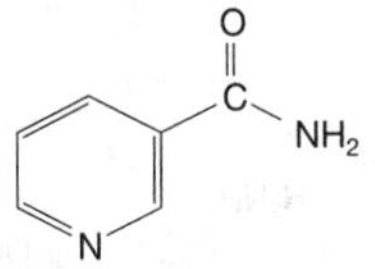　　Nicotinamide

Under certain nutritional conditions, e.g. when maize forms the bulk of the human diet, the deficiency of niacin leads to pellagra. Treatment of maize with lime, as practised by Central Americans, releases niacin precursors. Pellagra is therefore not common among them, in spite of their monotonous diet, but was common among poor Europeans who did not treat their maize in this way. This deficiency state affects the skin (brown coloration), the digestive system (diarrhea) and the nervous system (dementia). Pellagra can be cured by feeding tryptophan; or nicotinamide may be administered therapeutically. The daily requirement is 1–2 mg nicotinic acid or nicotinamide.

Pantothenic acid is a widely distributed compound in animals and plants, consisting of 2,4-dihydroxy-3,3-dimethylbutyric acid (pantoic acid) linked to β-alanine by an amide bond. Most organisms have the ability to synthesize pantoic acid from valine, and β-alanine from asparate, but humans lack the enzyme, pantothenate synthetase, which catalyses the condensation of β-alanine and pantoic acid to form pantothenic acid. Only the D(+)-form of pantothenic acid is biologically active. It is required for the synthesis of Coenzyme A (see). Nonexperimental human deficiency states have not been observed, so pantothenic acid is presumably present in sufficient quantity in all diets.

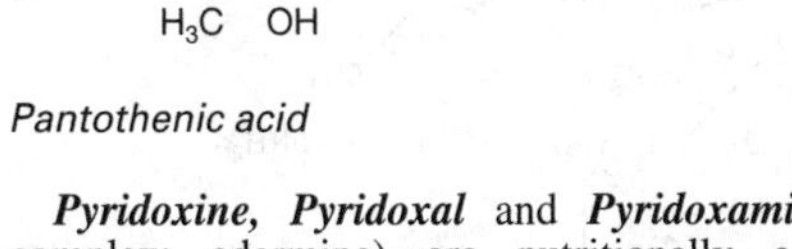

Pantothenic acid

Pyridoxine, Pyridoxal and ***Pyridoxamine*** ($V.B_6$ complex; adermine) are nutritionally equivalent, widely distributed, water-soluble V., found, e.g. in yeast, wheat, maize, liver, potatoes, vegetables, etc. All forms are metabolically interconvertible. Pyridoxal phosphate is one of the most important coenzymes of amino acid metabolism, taking part in transamination, decarboxylation and elimination reactions. Sufficient $V.B_6$ is present in all basic foods, and no typical deficiency state is known in humans. Xanthurenic acid excretion is used as an index of V.B_6 deficiency, since V.B_6 deficiency inhibits the degradation of L-tryptophan, with a resulting increase in the excretion of xanthurenic acid. In the rat, V.B_6 deficiency causes a pellagra-like condition, with loss of hair, edema and red scaly skin. The daily human requirement is 1.5-2 mg.

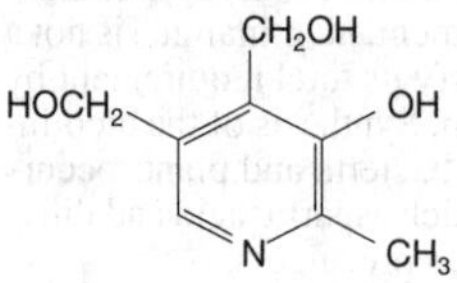

Pyridoxol (Pyridoxine)

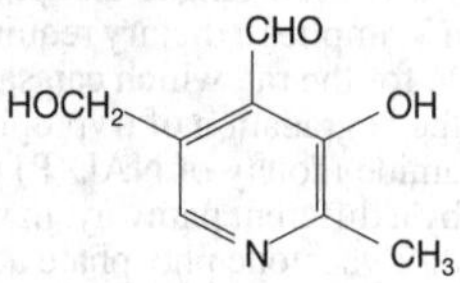

Pyridoxal

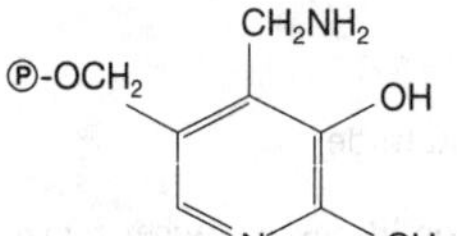

Pyridoxamine-phosphate

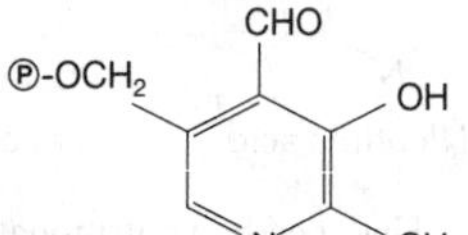

Pyridoxal-phosphate

Cobalamin (V.B_{12}; extrinsic factor; animal protein factor) represents a group of water-soluble corrinoids required in very small amounts (daily human requirement 0.005 mg V.B_{12}).

The corrinoid structure consists of a complex corrin ring system with a centrally bound trivalent cobalt atom, a base-nucleotide moiety, and a monovalent group (called the cobalt ligand) bound to the cobalt (X in the Fig.). In biological systems, the cobalt ligand is —OH, H_2O, —CH_3 or deoxyadenosine, the last two being found in cobalamin coenzymes. Extraction usually yields V.B_{12} in the form of cyanocobalamin. 5′-Deoxyadenosyl- and methylcobalamin are important coenzymes of rearrangement reactions.

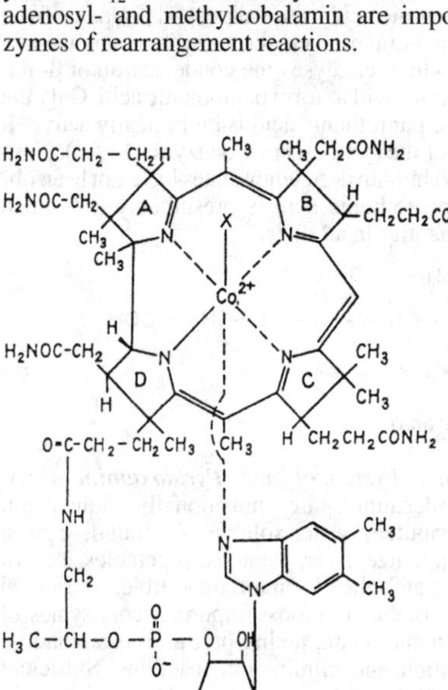

Vitamin B_{12}

V.B_{12} occurs predominantly in animal tissues and animal products. It is synthesized principally by bacteria. Green plants contain little or no V.B_{12}. Deficiency symptoms are sometimes observed in strict vegetarians, most often in breast-fed infants whose mothers consume no animal products. The body reserves of cobalamin are usually so large that an adult can survive for many years on them in the absence of a dietary intake.

V.B_{12} is also known as antipernicious anemia factor. Pernicious anemia is characterized by a severely reduced production of red blood cells, deficient gastric secretion and disturbances of the nervous system. It is not usually caused by dietary deficiency of V.B_{12}, but by the absence of intrinsic factor, which is required for V.B_{12} absorption. Intrinsic factor is a neuraminic acid-containing glycoprotein, normally present in the gastric mucosa, which forms a pepsin-resistant complex with V.B_{12}, and enables V.B_{12} absorption in the lower part of the intestinal tract.

Ascorbic acid (V.C; antiscorbutic V.) is a heat-sensitive, water-soluble V. with a wide natural distribution, especially in fresh vegetables and fruit. V.C is the γ-lactone of 2-oxo-L-gulonic acid, derived from carbohydrate metabolism. In most mammals it is synthesized from D-glucuronate. Higher primates (including humans) and guinea pigs cannot synthesize V.C, because they lack the enzyme, L-gulonolactone oxidase. For these animals, it is therefore a true V. and must be supplied in the diet. V.C is a powerful reducing agent, and by its reversible conversion to dehydroascorbic acid it serves as a biochemical redox system. It is involved in several metabolic hydroxylation reactions, e.g. hydroxylation of the proline in collagen. Deficiency results in scurvy, a long-known avitaminosis, characterized by rupture of blood capillaries, hemorrhage of the skin and mucosas, inflammation of the gums, loosening of the teeth and painful swellings of the joints. Resistance to infectious diseases is also reduced. The recommended daily intake is 75 mg, which is considerably higher than for other V.

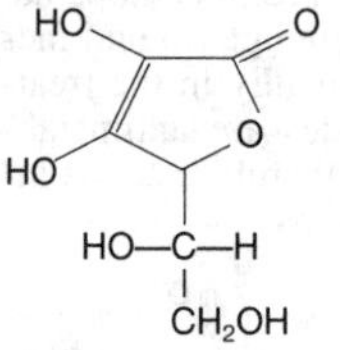

Vitamin C (ascorbic acid)

Calciferol (V.D; antirachitic V.) is a group of fat-soluble V. chemically related to the steroids. They are produced in the skin from provitamins ($\Delta^{5,7}$-unsaturated sterols) by UV-irradiation. If an individual receives adequate exposure to sunlight, dietary V.D is unnecessary. Calciferol is thus a V. only for people who, due to confinement indoors or highly pigmented skin at higher latitudes, are unable to synthesize a sufficient amount. In the conversion of sterols to calciferol, ring B of the steroid is opened between C9 and C10, forming precalciferol, which acts as a precursor of the V.D group.

V.D_2 (ergocalciferol) is derived from ergosterol in the skin by UV irradiation (via the intermediate precalciferol); tachysterol and lumisterol are also formed from precalciferol.

V.D_3 (cholecalciferol) is derived from 7-dehydrocholesterol in the skin by UV-irradiation. V.D_3 is con-

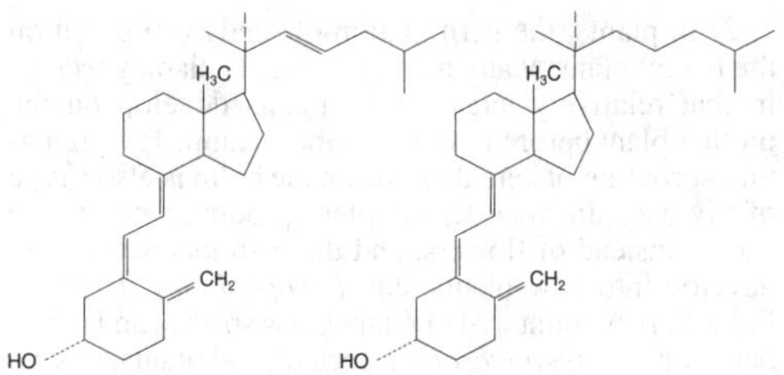

Vitamin D_2 Vitamin D_3

verted enzymatically in the liver and kidneys to 25-hydroxy-cholecalciferol, followed by hydroxylation to the highly active 1α,25-dihydroxycholecalciferol (1α,25-dihydroxy-V.D_3); this compound represents the active form of the V. in humans, and its behavior resembles that of a hormone, rather than a biocatalyst. V.D_3 is present in cod liver oil in particularly large quantities.

V.D_1 is a molecular compound of lumisterol and ergocalciferol.

V.D_4 is 22-dihydroergocalciferol, produced from 22-dihydroergosterol by UV-irradiation.

The V.D complex is also present in e.g. herrings, egg yolk, butter, cheese, milk, pig liver and edible fungi.

V.D is required for calcium absorption and the mineralization of bone. The V.D deficiency disease of children, known as rickets, is characterized by a softening and malformation of the bones. It results poor absorption of calcium, coupled with deficient incorporation of calcium into bone tissue. Rickets can be cured by exposure to sunlight and by administration of synthetic V.D. Because V.D is stored in the liver, overdosage can lead to hypervitaminosis, with disturbances of calcium and phosphate metabolism and withdrawal of calcium from the bones. Exposure to sunlight never leads to hypervitaminosis D. The daily human requirement for V.D is 0.1 mg during growth, and 0.02 mg for adults.

Tocopherol (V.E; antisterility factor) is a group of fat-soluble V. containing a chromane ring with a polyisoprenoid side chain. Eight compounds with V.E activity are known: α-, β-, γ-tocopherol, etc., which differ in the number and positions of the methyl groups in the aromatic ring. Biologically, the most important member is α-tocopherol. Since tocopherol is easily oxidized to a quinone, V.E acts as a naturally occurring antioxidant. It prevents the spontaneous oxidation of highly unsaturated material, e.g. unsaturated fatty acids. It has other biological functions which have not been elucidated in detail. In animal experiments, V.E deficiency results in the death of the embryo in pregnant females. In the male there is atrophy of the gonads and muscle dystrophy. Neither deficiency states nor V.E-hypervitaminosis have been described in humans. V.E occurs in wheat seedlings, and has been isolated from wheat germ oil. It is also present in lettuce, celery, cabbage, maize, palm oil, ground nuts, castor oil and butter. The daily requirement is estimated at about 5 mg.

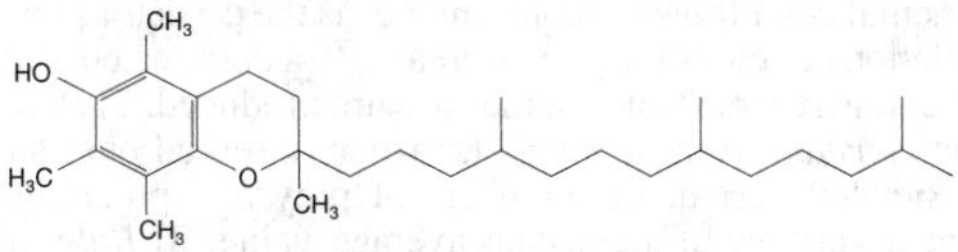

α-Tocopherol

Essential fatty acids (V.F). These are unsaturated fatty acids (in particular, linoleic, linolenic and arachidonic acid) which cannot be synthesized in the body. They serve as components of membrane lipids and as precursors of prostaglandins.

Biotin (V.H; bios II; coenzyme R) is a sulfur-containing, water-soluble V. Chemically, it contains 2 condensed 5-membered rings, and it is a cyclic urea derivative: 2´-oxo-3,4-imidazoline-2-tetrahydrothiophene-*n*-valeric acid. It was discovered as a yeast growth factor, and has been isolated from liver extracts and egg yolks. Biotin is biosynthesized from cysteine, pimelic acid and carbamoyl phosphate. There are 8 stereoisomers, the most important biological isomer being D-biotin. In animal tissues. Biotin functions as a coenzyme of many important carboxylation reactions, during which it becomes covalently bound (via an amide bond) to a lysyl residue in the carboxylation enzyme. The amino acid conjugate, ε-N-biotinyllysine, is also present in animal tissues, and is known as biocytin. In animals, biotin deficiency causes skin disorders [hence V."H" for the German Haut (skin)] and loss of hair. Excessive intake of raw eggs causes avitaminosis in humans; egg whites contain the glycoprotein, avidin, which specifically and tightly binds biotin and prevents its absorption. Biotin is present in practically all foods, and the daily requirement (0.25 mg) is in any case generally provided by gut bacteria. Biotin deficiency is therefore practically unknown.

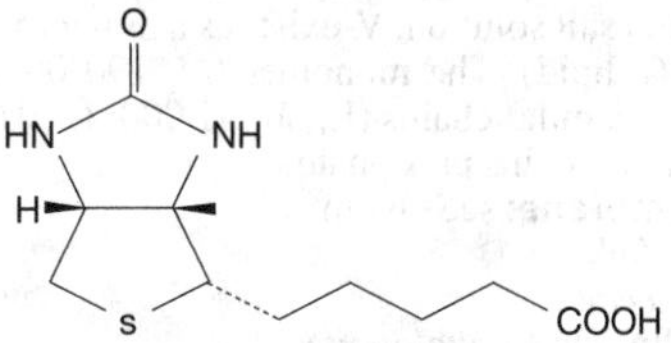

Biotin

Phylloquinone (V.K; antihemorrhagic V.; coagulation V.) is a group of fat-soluble naphthoquinone compounds, with varying sizes of isoprenoid side chain. Mammals can synthesize the side chain but not the naphthoquinone moiety. V.K_1 is plentiful in green plants. V.K_2 (farnoquinone; menaquinoine-6; 2-methyl-1,4-naphthoquinone) is found chiefly in bacteria. V.K_3 (menadione; 2-methyl-1,4-naphthoquinone) is actually a provitamin.

In many bacteria, V.K is a component of the respiratory chain, in place of ubiquinone. In animals, V.K deficiency leads to deficient production of blood coagulation factors, in particular prothrombin, leading to abnormally long clotting times and a marked tendency to hemorrhage. V.K serves as a cofactor in the carboxylation of glutamic acid residues during post-translational modification of blood clotting factors II (prothrombin), VII, IX and X. Avitaminosis is rare in human children and adults, because sufficient V.K is provided by the gut flora. However, it does not cross the placenta well, so that neonates are at risk of avitaminosis. Fatal hemorrhage sometimes occurs in breast-fed infants whose mother's milk does not contain adequate amounts of the vitamin. V.K_3 is used in the treatment of hemorrhages and liver diseases.

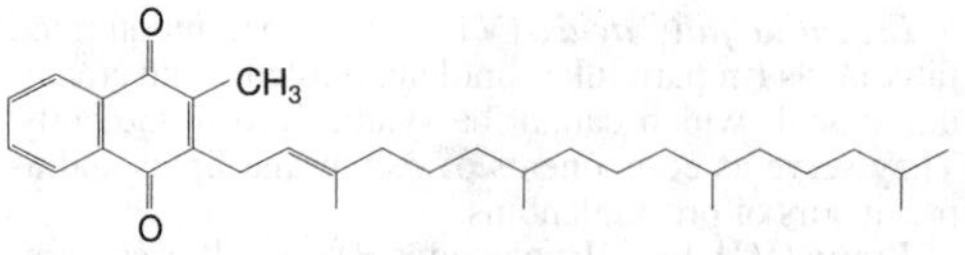

Vitamin K_1

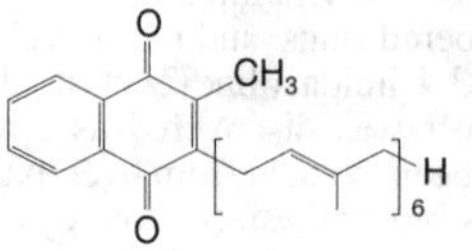

Vitamin K_2

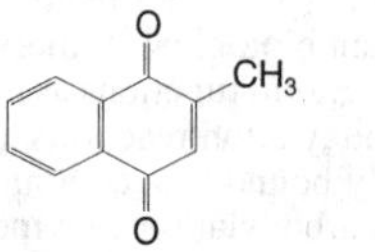

Vitamin K_3 (Menadione)

Warfarin is a V.K antagonist used as a rodenticide; it causes death by hemorrhage after the animals have fed on it repeatedly. It is also used clinically as an anticoagulant. Dicoumarol is an important antagonist of V.K; it is present in moldy clover hay, and is responsible for hemorrhage in cattle.

Vitellarium: see Yolk gland.

Vitellin: a lipophosphoprotein present in egg yolk together with phosphovitin. It is present in higher concentration than phosphovitin, but in contrast to this phosphorus-rich protein, V. contains only 1% phosphate. In yolk and in neutral salt solution, V. exists as a dimer (M_r 380 000; 16-22% lipid). The monomer (M_r 190 000) consists of two dissimilar chains (L_1, M_r 31 000; L_2, M_r 130 000); only L_1 contains phosphate.

Vitelline membrane: see Ovum.

Vitellus: see Yolk.

Vitis: see *Vitaceae.*

Vitreledonella: see *Cephalopoda.*

Vitta: see *Umbelliferae.*

Viverra: see Civets.

Viverridae, ***viverrids:*** a family of terrestrial carnivores (see *Fissipedia).* They are small to fox-sized mammals, generally elongate in shape with a long tail and pointed snout, and bearing a close superficial resemblance to to martens and/or cats. As nimble, nocturnal predators, they feed on small vertebrates, insects and birds' eggs, as well as occasional fruits. They are distributed in Southwest Europe, Africa and Asia. The following representatives are listed separately: Civets, Genets, Palm civets, Masked palm civet, Binturong, Mongooses.

Viviparidae: see *Gastropoda.*

Viviparous seedling: see Mangrove swamp.

Viviparus: see *Gastropoda* (subclass *Streptoneura,* order *Stylommatophora).*

Vivipary:

1) In animals, the birth of live young, which have completed their embryonic development and sometimes also the early stages of postembryonic development within the female parent. The egg membranes are ruptured before or during birth. All mammals and some fishes and reptiles are viviparous. There is no sharp division between V. and Ovovivipary (see). Thus, the tsetse fly, *Glossina* (vector of sleeping sickness), "gives birth" to larvae which pupate immediately. The converse of V. is Ovipary (see).

2) In plants, the germination of seeds while still on the parent plant. Many mangrove plants display *true V.,* in that relatively large young plants develop on the mother plant before falling into the swamp. The premature sprouting of cereal grains on the halm is also a type of vivipary. In *false V.,* the plant produces vegetative shoots instead of flowers, and these shoots detach and develop into new plants, e.g. *Bryophylum calycinum.* False V. is encountered in many grass species and members of the *Polygonaceae;* in agricultural plants, it is responsible for decreased seed production.

Vizcacha: see *Chinchillidae.*

Vocal cords: see Larynx.

Volantia: see Movement.

Volatinia: see Galapagos finches.

Voles: see *Arvicolinae.*

Vollhering: see Herring.

Voltage clamp technique: a method for determining electric currents at the membrane of excitable cells, by holding the membrane potential at a constant value. The membrane potential changes very rapidly during an action potential, so that the kinetics of current components can be analysed only if the membrane potential is kept constant experimentally. The time course of the current is then measured at a defined constant membrane potential. By introducing inhibitors of individual ion fluxes (e.g. blockage of Na^+ currents with tetrodoxin), the currents associated with particular ionic species can be determined separately. With modern electronic techniques it is possible to measure the current through a single ion channel.

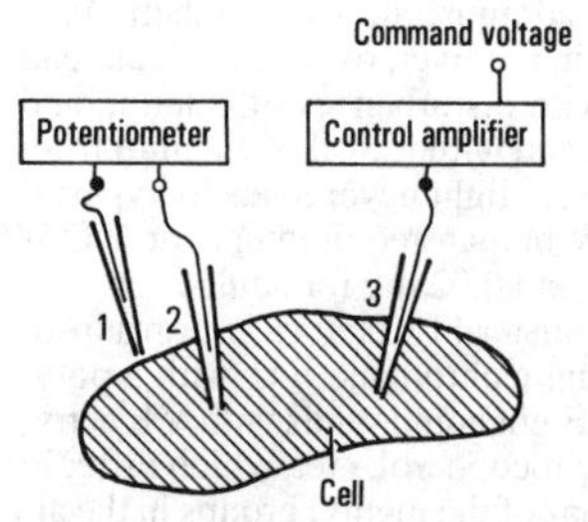

Voltage clamp technique. The membrane potential (difference between the intra- and extracellular potentials) is measured between electrodes 1 and 2, and compared with a command voltage. The control amplifier compensates the differences between the measured and control voltages by passing current into the cell via electrode 3. This compensating clamp current, which has the same value (but the opposite sign) as the membrane current, is recorded.

Volterra's laws: laws governing the population dynamics of predators and prey (parasite and host), based on simplified mathematical models, derived independently by V. Volterra and A. Lotka in 1926. 1) *Rule of periodicity:* population cycles of predators and prey display variations (oscillations), even when environmental conditions are constant, i.e. as the predator population increases the prey decreases to a point where the trend is reversed and oscillations are produced. 2) *Rule of constant average numbers:* when observed over an extended period, the numbers of prey and predators each vary about a constant average value. 3) *Rule of perturbation of the average value:* if both predators and prey are subjected to negative influences, the number of predators at first decreases and that of the prey increases.

As shown by May (1980) V.l. are only valid in a restricted cases. Depending on the relation between the specific rates of increase of predators and prey, the population cycles may be stable, or they may oscillate periodically around 2,4,8,16 or more points, or they may vary chaotically.

Volutin: see *Spirillum.*

Vombatidae: see Wombats.

Vomeronasal organ, *Jacobson's organ:* an organ housed in a specialized section of the anterior nasal cavity. It serves for the perception of olfactory substances from the buccal cavity, and is connected to the latter via Stenson's duct. In lizards and snakes it is completely separate from the nasal cavity, and the forked tongue of these reptiles is placed into the pouch of the V.o. It is also functional in a number of mammals, in particular those with an acute sense of smell, but it is degenerate in whales, Old World monkeys, humans and many bats. In mammals, it serves as a receptor for sexual olfactory stimuli. The V.o. is thought to play a role in the behavior known as Flehmen (see).

Vomiting reflex: the reflex expulsion of the stomach contents. In humans, the V.r. is initiated by many different stimuli (Fig.), including: revolting sights, rotary motion, nauseating smells, physical irritation of the pharynx, over-extension of the stomach, residues of tainted food in the intestine, chest pains, excitation of chemoreceptors in the brain, and pregnancy (morning sickness, pregnancy sickness). The reflex center in the medulla oblongata is responsible for the coordinated activation of the following nerves: the chorda tympani and the glossopharyngeal nerve to the pharynx and palate, the vagus fibers to the stomach, and the phrenic nerve and spinal nerves to the thoracic and abdominal musculature. First signs of the V.r. are an increased salivary secretion, a decrease in the breathing rate, decreased stomach muscle tone, and a choking and retching sensation due to uncoordinated movements of the respiratory apparatus. Vomiting occurs after a deep inspiration, closure of the nasopharynx, and relaxation of the cardiac sphincter of the stomach. The negative pressure generated in the thoracic cavity by the initial deep inspiration causes the esophagus to widen, and the stomach contents are expelled by powerful contractions of the diaphragm and abdominal muscles.

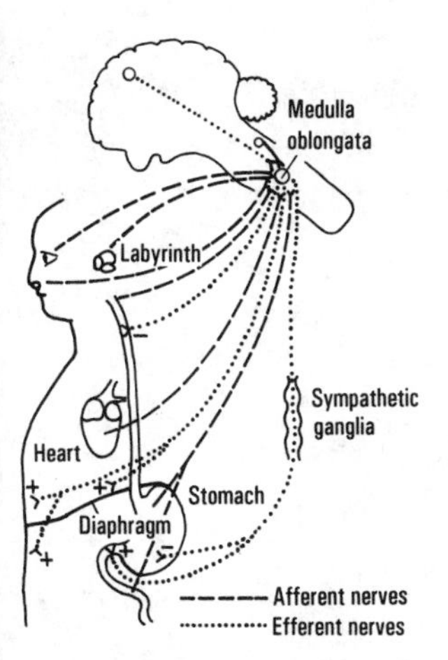

Vomiting reflex

Vorticella: see *Ciliophora.*

Votsotsa: see *Nesomyinae.*

Vulpes: see Foxes.

Vultur: see *Falconiformes.*

Vultures: see *Falconiformes.*

Vulva, *vestibulum vaginae:* in female mammals, a cleft-like depression of the urogenital cavity (sinus urogenitalis) housing at its base the openings of the vagina and urethra. The sides of the V. are bordered by paired folds of skin known as the labia majora and labia minora. V. is also used as a general term for the external genitalia of the human female.

Vu Quang ox, *Pseudoryx nghetinensis:* a member of the family *Bovidae,* described and reported for the first time in 1993 [Vu Van Dung et al. (1993) *Nature, **363,*** 443–445] during a joint survey of the Vu Quang Nature Reserve, Ha tinh province, Vietnam, by the Ministry of Forestry and the World Wide Fund for Nature. Diagnostic features of the V.Q.o. are long, smooth, almost straight slender horns (almost circular in cross section), elongated premolars, large facial glands, and distinctive color patterning on the face, neck and rump. The length from nose to anus is about 1.5 m, and the height at the shoulder is 80–90 cm. There is a short tail (about 13 cm) with a fluffy black tassle. The skull (32–36.5 cm in length) has a highly bridged nasal area. DNA analysis places the V.Q.o. with the oxen *(Bovinae),* and it is provisionally placed in the tribe *Boselaphini* on account of the facial glands and coat patterning. It lives in the montane forests of northern Vietnam, occurring at all altitudes according to the season. In the 1993 report, the animal was described on the basis of skins, skulls and skeletons supplied by hunters. The first live specimen, a female calf, was caught in 1994 (*WWF News,* Autumn 1994).

Discovery of a new (to science) large mammal is now unusual. The last such event was the discovery, more than 50 years previously, of the kouprey *(Bos = Novibos sauveli),* also in Indochina. Until the discovery of the V.Q.o., only two native bovid genera were known in this part of Asia, i.e. *Bos* and *Naemorhedus (= Capricornis).*

W

Wagner's gerbil: see *Gerbillinae.*
Wagtails: see *Motacillidae.*
Waldrapp: see *Ciconiiformes.*
Wallabia: see Kangaroos.
Wallabies: see Kangaroos.
Wallacea, ***Indoaustralian transition zone:*** the transition zone between the Oriental and Australian zoogeographical regions. The fauna is very impoverished [from Wallace's point of view (see Wallace's line), these were exceptionally poor collecting grounds] and individual taxa display very variable mixtures of Oriental and Australian animals. During the lowering of the sea level in the Ice Age, land connections were not formed between the islands of W. (see Fig.) and the enlarged Asiatic and Australasian continents. For mammals, the Bali strait represents a very distinct barrier (see Wallace's line); the relatively small size of Bali has also restricted the number of Oriental mammals that it can carry. Lombok has 3 monkeys and 3 carnivore species, while Sumbawa has 2 species of each, and Flores has 1 species of each. Sulawesi is relatively well endowed with Oriental mammals (e.g. 4 viverrids, 3 primates, the babyroussa and the pygmy buffalo). It is very difficult, however, to reconstruct the natural distribution of mammals, because particularly in W. many animals have been introduced. Owing to the marked depletion of mammalian species from west to east, and only slight penetration of Australian mammals, there is little mixing of the two types. Mixing occurs, however, on Sulawesi, which is home to 2 climbing marsupials. Mixing of Oriental and Australasian fauna is much more evident among the birds, reptiles and amphibians, although Oriental bird and reptile species are more numerous than their Australasian counterparts. One notable reptile is the Komodo dragon. Amphibians are well represented, and they contain a high proportion of Australasian species.

Weber's line is an imaginary line separating the Oriental and Australasian zoogeographical regions, and it represents the boundary at which Oriental and Australian vertebrates are in equilibrium, i.e. the limit of the relatively successful invasion of W. by Oriental animals. The eastern boundary of W. may be considered as Weber's line or Lydekker's line (Fig.). W. is bounded in the west by Wallace's line (see).

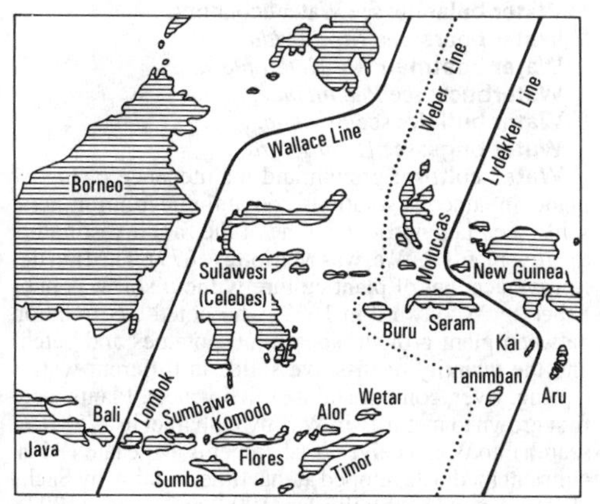

Wallacea. The Indoaustralian transition zone, bounded in the west by Wallace's line and in the east by Lydekker's line.

Wallace's line (British biologist, Alfred Russel Wallace, 1823–1913): an imaginary line passing between the Pilippines and the Moluccas in the north, then southwest between Sulawesi and Borneo, and continuing south between Lombok and Bali. During his travels in the Malay Archipelago between 1854 and 1862, Wallace was struck by the fact that the small islands of Lombok and Bali, only 15–20 miles (24–32 km) apart, carried strongly contrasting faunas. The original Wallace line was, in fact a line drawn through the narrow strait between these two islands. Even winged species generally keep to their own side of the narrow strait, so that on Bali Wallace found many of the same birds that he had collected on the Malay Peninsula and Borneo (e.g. green woodpeckers and barbets). On Lombok, however, the prominent birds were typical Australasians, e.g. sulfur-crested cockatoos, honey suckers and scrub fowl *(Megapodius).* On later journeys, Wallace studied local fauna in the light of his Bali-Lombok experience and classified islands as either predominantly Oriental or Australasian. See Wallacea.
Wallaroo: see Kangaroos.
Wallcreepers: see *Sittidae.*
Wallerian degeneration: see Degeneration.
Wall lizard, *Lacerta (Podarcis) muralis:* a slender, sharp-snouted, 20 cm-long lacertid (see *Lacertidae),* whose many subspecies are found in dry, sunny areas of southern Europe. It has also been introduced into the Channel Islands and Mainland Britain. The back is gray to red-brown with dark reticulate markings; the center of the back is usually spotted. It is diurnal in habit and requires a warm habitat. It associates itself with human settlements, and climbs very skilfully on smooth walls. Eggs are buried in the sand.
Wall lizards: see *Lacertidae.*
Wall pepper: see *Crassulaceae.*
Walnut: see *Juglandaceae.*
Walrus: see *Odobenidae.*
Wapiti: see Red deer.
Warblers: see *Sylviinae.*
Warburg's respiratory enzyme: see Cytochromes.
Warm-stenothermal: see Temperature factor.
Warning coloration: see Protective adaptations.
Wart biter: see *Saltatoptera.*
Wart hog, *Phacochoerus aethiopicus:* a monotypic

species of pig (see *Suidae)* found in most of Africa, and most numerous in the savanna and light forest regions of the east and south. There is a sparse mane of coarse hair, and the body is otherwise covered in bristles; both skin and hair are dark gray in color, sometimes with a tinge of brown. Males have large upper tusks, which may be more than 63 cm in length. Only males have prominent warts, which are skin growths on the sides of the head and in front of the eyes. They are usually diurnal in habit, but retreat into shade during the heat of the day, sometimes into abandoned aardvark burrows.

Wart snakes: see *Acrochordidae.*

Washington palm: see *Arecaceae.*

Wasmannian mimicry: see Protective adaptations.

Wasps (plates 1 & 2): in the broad sense, all *Hymenoptera,* with the exception of ants and bees. In the narrow sense, W. are members of the family *Vespidae.* The latter include the black and yellow-patterned, social W., which are attracted particularly by fruit and and sugary food material in late summer, and which build their familiar "paper" nests on roof beams and similar sites. Other examples are the 35 mm-long brown-yellow hornet *(Vespa crabo* L.) and solitary species like field and wall W.

Waste water fish ponds: see Biological waste water treatment.

Waste water ponds: see Biological waste water treatment.

Waste water treatment: see Biological waste water treatment.

Water, H_2O: a compound of hydrogen and oxygen; a crucial environmental factor which is essential for life.

In their natural state, proteins of living organisms are hydrated. In living systems water contains and transports many dissolved polar and nonpolar substances, as well as serving as the medium in which the numerous extracellular and intracellular enzyme-catalysed reactions occur. Water is also important for temperature regulation (see) in homeothermic animals. Water loss and water deficiency in higher animals leads to severe physiological stress and very quickly to death. Most animals acquire water directly, either by drinking, or as the water that is present in practically all food materials. In addition, some animals (e.g. snails and insects) can take up water or water vapor through their body surface. In all organisms, water is a metabolic product from the oxidation of food and reserve materials (in particular fat, but also carbohydrate and protein), and this source is important in desert inhabitants and in insects that feed on dry wood.

Animals continually lose water through their excretory processes. Water-conserving animals produce relatively dry feces and concentrated urine (e.g. rabbits, flour beetles), whereas at the other extreme some animals produce brei-like feces and voluminous urine (e.g. cattle, blood- and sap-sucking parasites, and aquatic animals). Unless they live permanently in a moisture-saturated environment, terrestrial animals also lose water by evaporation, a process influenced by air humidity, temperature and air movement.

Animals of dry environments therefore have numerous protective adaptations against water loss, such as chitinous or calcareous shells, closure of body apertures, and behavioral patterns appropriate to water conservation. Mammals die if they lose 15% of their body water, whereas slugs can lose up to 80%, and aschelminths as much as 86%. Tolerance of water loss varies greatly with the stage of development or metamorphosis; e.g. low air humidity severely hinders molting and ecdysis in terrestrial arthropods.

The duration of development or metamorphosis and the lifespan of poikilothermic animals is affected by water availability. In general, an adequate water supply shortens the time of developoment and increases the lifespan. On the other hand (e.g. in some insect groups), the lifespan is considerably extended if periods of drought enforce the development of resting stages. In response to drought, protozoa, rotifers, aschelminths and tardigrades can survive for many years in an anabiotic state.

Excess water can injure animals whose skin is waterpermeable, e.g. soil-dwelling scarabaeiform larvae become distended through water uptake when the soil is flooded. On the other hand, many beetles can survive for a long time in water. The ability of a terrestrial arthropod to withstand water exposure depends on the permeability of the integument, its degree of waterproofing, and the osmotic value of the body fluid.

Adaptation to moist environments is known as *hydrophily,* while adaptation to dry conditions is *xerophily,* but between these two extremes many transitional states are displayed by a wide range of organisms. Each species is optimally adapted to a particular level of water availability and water exposure, and this specific water preference of a species exerts a considerable influence on its natural behavior. Even relatively neutral species will actively seek water, if their water balance is disturbed by withdrawing or withholding water from them.

Many terrestrial animals are stimulated by moisture; in wet weather, snails, amphibians and earthworms become very active, and biting midges become very aggressive. In isopods and some insects, increasing dryness of the environment causes unrest and searching movements (hygrokinesis). Water also acts as an orientation factor in hygrotaxis, e.g in midges, bees, wireworms and isopods.

Water analysis, biological: see Biological water analysis.

Water balance: see Water economy.

Water bears: see *Tardigrada.*

Water boatman: see *Heteroptera.*

Waterbuck: see *Reduncinae.*

Water buffalo: see *Bovinae.*

Water bugs: see *Heteroptera.*

Water culture: growth and maintenance of higher plants in aqueous solutions containing certain mineral salts (see Plant nutrients). One of the first investigators to grow plants in W.c. was Mariotte (1679). The first definitive account of plant culture without soil was published by Woodward in 1699, who noted a correlation between plant growth (spearmint, potatoes and vetch) and the quantity of dissolve solids in different waters (spring, river, conduit and distilled water). Plants were first grown to maturity in W.c. by Duhamel in 1758. Research into W.c. received fresh impetus in the late 1850s. Nutrient media developed at that time, notably by Sachs (1860), Knop (1865), Pfeffer (1900) and Crone (1902), are still used today, with the exception that trace elements are now usually added in chelated form. W.c. is a widely used technique, enabling the detailed investigation of plant mineral metabolism, and helping to solve numerous problems of agricultural chemistry. As an exactly reproducible culture method for experimental plants, it has made a considerable contribution to plant physiology, phytopathology and plant breeding.

Depending on the problem under investigation, the culture medium is either a complete nutrient medium containing all macro- and micronutrients, or it may lack one or more essential elements. In a *standing culture,* the medium is not changed or changed only periodically during the experiment. In *flow culture,* the medium is continually renewed, so that its composition and properties remain constant. Some investigations are performed under sterile conditions, necessitating more elaborate apparatus and extra laboratory facilities. In the special technique of *air culture,* the plant roots grow in moist air and are sprayed intermittently with nutrient medium. In *Sand culture* (see), the roots grow in pure quartz sand saturated with nutrient medium.

W.c. is usually performed in cylinders or wide-necked flasks of various sizes, constructed of glass (quartz glass may be used for special investigations), enamelled iron or synthetic polymers. The neck of the vessel is covered with a layer of cork or board containing a hole for the plant, the latter being held in position with nonabsorbent cotton wool, foam rubber or other suitable material. To avoid possible effects of illumination on root physiology, to prevent undue warming by illumination, and to discourage algal growth, glass and plastic vessels are covered or painted to make them opaque. The medium may also be aerated. W.c. can be started with very young seedlings, but the plants must possess a shoot and roots before they are transferred to the nutrient medium.

W.c. may be modified and simplified in various ways, in particular for horticultural purposes. This modified W.c., known as *hydroponic culture* or *hydroponics,* can be performed in various ways. In *tank farming,* the medium is held in a large container of wood, iron or concrete, covered with a wire mesh or tray of hardware cloth, which supports a layer of peat, straw, wood shavings or similar relatively inert material. Seeds or plants are placed in the upper layer, which is kept moist until the roots grow through the wire or cloth into the medium. In *aggregate culture,* tanks, basins or pots contain a layer of coarse sand, quartz gravel, cinders, clinker, pummice stone, broken building brick or similar material. Medium is either periodically washed in and through from above, supplied by continual overhead irrigation from a piped supply, or pumped in from below with an appropriate run-off.

Nutriculture is a general term for several soilless methods of growing plants in artificial media, such as W.c., all forms of hydroponics, sand culture, etc.

[Hugh G. Gauch, *Inorganic Plant Nutrition,* Dowden, Hutchinson & Ross, Inc. Stroudsburg Pa. 1972]

Water dog: see *Proteidae.*

Water dragons, *Physignathus:* agamids (see *Agamidae)* native to Australia and the mainland of Southeast Asia. They are up to 80 cm in length, and usually greenish with dark transverse bands. The body is laterally compressed, carrying distinct crests of long scales on the neck and back, and in some species on the tail. Like the South American Green iguana (see), they live in trees by water. They are good swimmers, using the tail as an oar. The clutch of eggs is buried in the river bank.

Water economy: the uptake, transport and loss of water, as well as its metabolic utilization and production. Water is a fundamental component in cell metabolism, and it is essential for normal life processes. All plants and animals must therefore have an adequate supply of water and a correctly balanced water economy.

1) W.e. of plants

a) Water uptake. Higher terrestrial plants take up water through their roots from the surrounding soil, mainly by Osmosis (see), with some contribution from Hydration (see). In both mechanisms, the energy for the movement of water from soil to plant arises from the accompanying decrease in the chemical potential of the water. Water uptake usually commences with the simple diffusion of water and dissolved nutrients, largely in the region of the root hairs and especially via the root hairs themselves. The epidermis of the root hairs and neighboring cells is normally neither cutinized nor covered by a cuticle (or the cuticle is extremely thin).

For water uptake to occur, the total water absorption power of the root must be greater than that of the soil; since the soil water is usually a dilute salt solution of low osmolarity, this requirement is generally satisfied. In saline soils, however, the soil water can attain an osmotic value of about 100 atmospheres. In addition, part of the soil water, even in normal soils, is very tightly bound to soil particles by adsorption and hydration, and in clay soils these forces are extremely powerful. As an adaptation to such conditions, plant root cells may have a high osmotic pressure and an appropriate pressure-resistant construction. To a certain extent, higher terrestrial plants can take up dew and rain through their leaves. Some Epiphytes (see) take up water predominantly from the air with the aid of specialized absorption hairs; others absorb rainwater which collects in the cistern-like container formed by the leaf rosette. Epiphytic orchids take up water via specialized aerial roots. Many filmy-ferns *(Hymenophyllaceae),* which can exist only in habitats of high atmospheric humidity, are also dependent on an atmospheric water supply. Mosses take up water exclusively through their leaves. Water plants (hydrophytes) absorb water by osmosis, either through their total surface (totally submerged water plants), or with the aid of root-like modified leaves (e.g. floating aquatic ferns of the order *Salviniales).*

b) Water transport. Water entering root hair or epidermal cells is first transported via the root cortex to the vascular tissues. In this extravascular transport, water passes from vacuole to vacuole of adjacent cells by osmosis. Extravascular water flow also occurs via the capillary systems formed by the intermicellar and interfibrillar spaces of the cell walls (apoplastic route), until it reaches the endodermis, where further progress is prevented by the water-impermeable Casparian strip (see); at this point the water must enter the endodermal cells. From the endodermal cells, water passes (active transport may be involved directly or indirectly) into the vessels of the vascular bundles in the central cylinder. Within the vascular bundles an unbroken column of water (vascular water flow) passes upwards from the roots to the leaves, and replaces water lost by transpiration. With the aid of this vascular water movement, water in trees is transported to heights of 30 to 50 m, and even higher than 100 m in Eucalyptus species and some lianas. In lianas, flow rates greater than 100 m/h have been measured. Herbaceous plants attain a vascular flow rate of 60 m/h, while the rates of trees with wide-pored vessels (e.g. oak) are 20 to 45 m/h. Trees with narrow-pored vessels (e.g. beech) have flow rates of 1 to 4 m/h, while pine trees, which have simple vascular vessels, have vascular flow rates of about 1 m/h.

Various forces are involved in raising the water against gravity. These include root pressure, which in temperate climates is important in the spring before leaf emergence. After leaf emergence and the resulting increase in transpiration, however, root pressure alone is insufficient to supply and raise all the water lost by transpiration. In any case, root pressure is normally no higher than 1 atmosphere which can raise water only to a height of 10 m. Another contributing force is the negative pressure generated by transpiration from the leaf surfaces, which pulls water upward through the vessels of the vascular bundles. For this suction to be effective, however, the columns of water in the conducting vessels must be unbroken. Provided that gas is excluded and there is no possibility of gas embolism within the vessels, the unbroken continuity of the water column is guaranteed by the extremely high cohesive forces (about 250 atmospheres) holding the water molecules together. Thus, the "cohesion theory" has been put forward to explain the upward movement of water in plants, although strictly speaking, cohesion is not a driving force, but a prerequisite for water movement. The actual lifting force corresponds to the difference in vapor pressure between the soil surface and the atmosphere, which is maintained by the thermal energy of the sun. The plant is simply a system linking the two extremes of this energy gradient, and apart from the root pressure, the plant itself does not supply energy for water movement.

After the water leaves the vascular system, its transport through the leaf parenchyma is mainly apoplastic.

c) Water loss by plants occurs mainly by Transpiration (see). When the transpiration rate is inadequate, water is also lost by Guttation (see). After injury, water is lost by Bleeding (see).

The critical factor determining water loss by plants is the water balance, i.e. the ratio of water taken up to that lost by transpiration. Although water balance may become temporarily negative, it must be equalized over a period of time. If it remains negative for too long, the plant begins to wilt, first reversibly, then irreversibly.

Plants, which maintain a daily water balance by means of a well developed root system, by sensitive regulation of stomatal opening, and often by the possession of water storage tissues, are called *hydrostable species;* these include trees, shade plants and succulents, and some grasses. In contrast, *hydrolabile species* show much wider variations in water balance, but their cells are able to withstand such conditions; these include many plants of warm habitats, and many grasses.

Water balance is determined experimentally by measurement of the maximal or saturated water content (W_s) and the actual water content (W_a) of a plant organ. The water saturation deficit (WSD) is then derived from the equation: $WSD(\%) = (W_s–W_a/W_s) \times 100$. Water balance of an organ is also directly related to the Water potential (see).

2. W.e. of animals and humans. Terrestrial vertebrates contain 60–75% water, and this value may fall to 50% in fattened domestic animals. Medusae consist of 97–99% water. In most animals, the loss of a few percent of their water leads to severe physiological disturbances. Thus, for vertebrates, a 10–15% water loss is fatal. Invertebrates, however, are better able to survive desiccation.

Total body water in vertebrates is distributed in three compartments, which are in contact with one another via semipermeable membranes: 1) water of the circulating blood (4% of body mass in humans); 2) water of the interstitial fluid of the body cells (16% of body mass); 3) intracellular water (40% of body mass).

Wide fluctuations in the salt:water ratio of the blood can be buffered in such a way that blood osmolarity and blood volume change only slightly (see Osmoregulation, Volume regulation).

Water uptake. In all aquatic invertebrates and many terrestrial invertebrates of moist habitats, in amphibians, and in mouthless parasites living inside other animals, water is taken in through the skin. Many terrestrial invertebrates, reptiles, birds and mammals drink water or obtain it from the water present in food. Mammals require an especially large quantity of water, since soluble waste products (especially urea) are excreted in aqueous solution. The main nitrogenous excretory product of birds and reptiles is the relatively insoluble uric acid, which is voided as a semi-solid mass, so that less water is required. In addition to water from the environment (eating and drinking), water is also formed internally by the cellular metabolism of respiratory substrates (metabolic or respiratory water). This is important for animals that have no dietary water, like wood beetles and clothes moths, or for camels, zebu cattle and fat-tailed sheep, which can survive for long periods without water. The following table lists the quantity of water produced by the metabolism of different chemical classes of foods:

100 g protein	41.3 g water
100 g carbohydrate	55.9 g water
100 g fat	107.0 g water

Water removal. Excess water is excreted. A large proportion of this is removed by the kidneys (see Kidney function) and voided by Micturition (see). Evaporation of water from the outer skin surface or from mucosae contributes to Temperature regulation (see) in warm-blooded animals. An adult human normally loses 2.5 l of water per 24 h, via the kidneys, lungs and skin, and in the feces, and this must be replaced by food, drink and metabolic water.

Water fleas: see *Amphipoda, Branchiopoda.*

Waterfowl: see *Anseriformes.*

Water frogs: see *Raninae.*

Water hemlock: see *Umbelliferae.*

Water hog: see African bush pig, *Hydrochoeridae.*

Watering pot shells: see *Bivalvia.*

Water lily family: see *Nymphaeaceae.*

Water loss: see Water economy.

Water lungs: see Respiratory trees.

Water measurers: see *Heteroptera.*

Water mint: see *Lamiaceae.*

Water mites: see *Hydrachnella.*

Water moccasin, *cottonmouth*, *Agkristrodon piscivorus:* a common pit viper (see *Crotalidae)* of North America, especially the swamps of Florida and neighboring regions, and the rice fields of the southern States. It attains lengths of 1.5 m and more, and adult animals are colored blue-black. Young animlas are vividly patterned with red-brown, white and black bands. It prefers damp habitats and is usually found near water. It is, however, not a truly aquatic snake, and the fish and frogs of its diet are caught from the shore. Females give birth to 5–15 live young.

Water monitor, *two-banded monitor,* *Varanus salvator:* a monitor lizard found in various habitats that provide plentiful water (e.g. humid forests, sea coasts and river shores) in Southeast Asia from eastern Bengal to southern China, in Sri-Lanka, and on the Sunda Islands, but not in New Guinea. It attains a length of almost 3 m. The coloration and patterning of body and tail (dark backgound with transverse stripes of lighter spots or rosettes) is most conspicuous in young specimens. The W.m. is an efficient climber, and it swims readily in both fresh water and the sea. It consumes practically any prey it can catch, including other reptiles. It also climbs trees to lie in wait for birds, and is known to steal poultry. Females lay 15–30 eggs, which are about 7 x 4 cm in size.

Water-nut family: see *Trapaceae.*

Water plants: see Hydrophytes.

Water potential, ψ_w: the difference between the chemical potential of water at a given site (μ_w), e.g. a plant vacuole, and that of pure water at atmospheric pressure (μ_{ow}). Taking into account the partial molar volume of water (V_w), the W.p. is given by: $\psi_w = \mu_w - \mu_{ow} / V_w$. The W.p. is applicable to osmotic phenomena, as well as imbibition.

Water slater: see *Isopoda.*

Water snakes: the name given in the USA to snakes of the genera *Natrix, Regina, Clonophis, Thamnophis* (Garter nakes) and *Helicops,* all of which belong to the subfamily *Natricinae* of the family *Colubridae* (see).

Water spider: see *Argyroneta aquatica.*

Water spruce: see *Taxodiaceae.*

Water saturation deficit: see Water economy.

Water scorpions: see *Heteroptera.*

Water starwort: see *Callitrichaceae.*

Water striders: see *Heteroptera.*

Water transport: see Conducting tissue, Water economy.

Water treaders: see *Heteroptera.*

Water uptake: see Water economy.

Water vole: see *Arvicolinae.*

Waterwheel plant: see Carnivorous plants, *Droseraceae.*

Watson's rule: see Mosaic evolution.

Wattle: see *Mimosaceae.*

Wattlebirds: see *Callaeidae.*

Wattled crow: see *Callaeidae.*

Wattle eyes: see *Muscicapinae.*

Watussi ox: see *Bovinae.*

Waxbills: see *Ploceidae.*

Waxes: esters of long chain, even numbered fatty acids and monohydric, straight chain, aliphatic alcohols or sterols. Examples of wax acids are: lauric (C_{12}), myristic (C_{14}), palmitic (C_{16}), carnaubic (C_{24}), cerotic (C_{26}), montanic (C_{28}) and melissic (C_{30}) acids. Some wax alcohols are: cetyl (C16), carnaubyl (C_{24}), ceryl (C_{26}) and myricyl (C_{30}) alcohols. W. are very hydrophobic, forming a waterproof layer on the aerial parts of plants, e.g. leaves and fruits, which prevents water loss by evaporation. In animals, W. are found on skin and feathers, and are used by bees to build the honeycomb. Some animal W. are spermaceti, shellac, beeswax and wool wax (lanolin). Carnauba wax is obtained from the leaves of the palm tree, *Copernica cerifera.*

Wax moths, *honeycomb moths, bee moths:* two inconspicuous gray-brown species of pyralid moths, whose caterpillars feed on wax in beehives. The ***Greater wax moth*** (*Galleria melonella* L.) (Europe, North America, Africa, Australia) has a wingspan of about 30 mm. Its caterpillars riddle the honeycombs with silk-lined tunnels, and they pupate in durable, papery cocoons. The ***Honey moth*** or ***Lesser wax moth*** (*Achroia grisella* F.) occurs worldwide. Its has a wingspan of about 20 mm, and the wings are yellow-brown and rather plain. The head of the moth is glossy yellow. In addition to beeswax, the caterpillar feeds on other stored food products. Old beehives are especially susceptible to attack.

Wax plant: see *Asclepiadaceae.*

Wax scales: see *Homoptera.*

Waxwing: see *Bombycillidae.*

Weasel, *common weasel,* *Mustela nivalis:* the smallest of the native European mustelids (see *Mustelidae).* In coloration, shape, etc., it resembles a small ermine, but the tail is relatively shorter and does not have a black tip. It ranges over wide areas of Eurasia. In the winter in the colder parts of its areal its coat color changes to white.

Weatherfish, *Misgurnis fossilis:* a fish of the family *Cobitidae* (loaches), order *Cypriniformes.* The very elongate body (maximal length 35 cm; more commonly 20–25 cm) is cylindrical in cross section, and there are 10 barbels on the rim of the mouth. It inhabits muddy fresh water in Europe, and is also found in lagoons off the Baltic Sea, usually buried in the bottom mud. It performs intestinal respiration by swallowing air, and it can survive briefly in the mud if the water dries out. The sides and belly are yellow or reddish, with 3 dark bands (1 wide, 2 narrow) from head to tail on each side.

Weaver ants: see Ants.

Weavers: see *Ploceidae.*

Webbing clothes moth: see Clothes moth.

Weber-Fechner law: see Weber's law.

Weberian apparatus, *Weberian ossicles:* A small row of bones discovered by E.H. Weber, which unites the swim bladder with the auditory organ in fishes of the orders *Cypriniformes* and *Siluriformes.* The swim bladder functions as a resonance chamber, transmitting vibrations to the inner ear via the W.a. Embryologically, the W.a. is derived from anterior vertebrae and ribs. It is analogous to the auditory ossicles of terrestrial vertebrates, and these respective structures represent an example of an evolutionary convergence of function. See *Cypriniformes.*

Weberian ossicles: see Weberian apparatus, *Cypriniformes.*

Weber's law, *Weber-Fechner law:* the relation dI/I = constant, established in 1834 by Weber, in which I is the intensity of an active stimulus, and dI is the smallest appreciable difference between two stimuli that can be conciously appreciated by a test person. Thus, the smallest appreciable difference between two stimuli depends on the ratio of that difference to their magnitudes, and not on the absolute difference between their magnitudes. Thus, a modest light will brighten a dark room, but it is unnoticable in sunlight. It may be possible to sense the difference in mass between 10 g and

11 g, i.e. a difference of 10%, but not between 100 g and 101 g; again a difference of 10% is required, so that the difference does not become appreciable until the second mass is at least 110 g. This limiting factor, usually expressed as a percentage, is constant within certain limits, and is known as the *differential threshold.* From this relation, Fechner (1859) formulated the Weber-Fechner law: the intensity of a Sensation (see) = $K \log I$, where K is the proportionality constant for a particular species of receptor, and I is the stimulus intensity, i.e. sensation varies with the natural logarithm of the stimulus. W.l. is applicable only between limits which vary with each sense organ, and it is invalid for very weak or very strong stimuli. In the case of temperature receptors and certain mechanoreceptors, sensation and stimulus are approximately directly proportional, rather than the sensation being proportional to the logarithm of stimulus intensity. W.l. is fundamentally important in the evaluation of the sensitivity of sensory cells, and for determining sensitivity to differences (rather than absolute values) between stimulus intensities. It also shows that conscious experiences can be subject to physical and mathematical laws.

W.l. also appears to be valid for numerous phenomena in plant physiology, although on the whole it must be applied tentatively. For example, if an oat coleoptile is illuminated on opposite sides by two light sources, it bends toward the stronger source only if this exceeds the weaker by at least 3%; the differential threshold is therefore 3%.

Weber's line: see Wallacea.

Webspinners: see *Embioptera.*

Weeping merulius: see *Basidiomycetes (Poriales).*

Weeverfishes: see *Trachinidae.*

Weevils, *snout beetles,* *Curculionidae:* the largest family of the *Coleoptera*, and indeed the most numerous of all animal families, containing some 50 000 species, the majority in warm climates. Most species are extremely hard and difficult to crush and their shape is markedly convex. The head is characteristically extended into a short, long or curved rostrum, which carries the mouthparts at its apex. Antennae are usually geniculate (elbowed), and they issue from a groove in the rostrum. Most of the 1 000 European species are less than 20 mm in length. The majority of W. lack membraneous hindwings, and are incapable of flight. Adults are herbivorous, often boring into stems, seeds and leaves. Larvae are legless and grub-like, and they feed on both lignified and nonlignified tissues inside a wide variety of different plant organs. Some are leaf-miners or gall-formers. Many W. are pests, e.g. the ***Beet weevil*** *(Bothynoderes punctiventris* Germ.), ***Apple-blossom weevil*** *(Anthonomous pomorum* L., Fig. 1), and ***Granary weevil*** *(Calandra = Sitophilus granaria* L., Fig. 2).

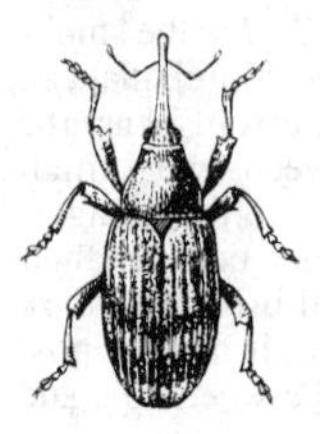

Fig.1. *Weevils.* Apple blossom weevil *(Anthonomus pomorum* L.).

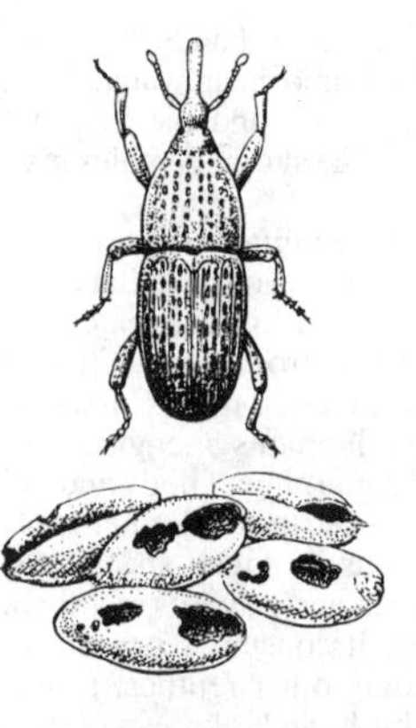

Fig.2. *Weevils.* Granary weevil *(Calandra granaria* L.) with damaged wheat grains.

Weichselian glaciation: see Würm.

Weigela: see *Caprifoliaceae.*

Weil's disease: see Spirochetes.

Weka: see *Gruiformes.*

Welwitschia: see *Gnetatae.*

Weissmann's ring: see Corpora allata, Corpora cardiaca.

Wenlock (Wenlock edge, a mountain in Shropshire, UK): a stage of the Silurian (see).

Western house mouse: see *Muridae.*

Western red cedar: see *Cupressaceae.*

Western spadefoot: see *Pelobatidae.*

Western spinebill: see *Meliphagidae.*

Western toad: see *Bufonidae.*

Westfalian stage: see Carboniferous.

West Indian rock iguanas, *West Indian ground iguanas,* *Cyclura:* large, powerful iguanas (see *Iguanadae*), native to and found only on the islands of the Caribbean. They are phylogenetically old forms of ground iguanas, and their island existence has resulted in the evolution of numerous subspecies. The back of the head is greatly broadened, especially in old males, and they possess a low dorsal crest. Normally they are dusky gray or olive green with cross bands that are often barely discernable. The tail is clothed in rings of keeled scales and it serves as a striking weapon. *Cyclura* inhabit open landscapes and rocky areas, and they do not climb trees. Their diet consists mainly of plant material and fruits; large insects and young rodents are also taken. These iguanas are now threatened with extinction, having suffered the effects of mongooses, feral dogs, cats, goats and cattle. *Cyclura cornuta* ***(Rhinoceros iguana)*** lives on Haiti in areas of dry thorn bush and cactus; the male carries 3 prominent horny protuberances on the tip of the snout, which are rather inconspicuous in the female.

Whalebone: see Whales.

Whale lice: see *Amphipoda.*

Whales, *cetaceans,* *Cetacea:* an order of large to gigantic, aquatic (mostly marine) mammals, completely adapted to life in water and unable to survive on land. The largest whale, and indeed the largest living animal, is the Blue whale (see below). The forelimbs of W. are modified to form balancing flippers with little or no locomotory function [see Limbs: Fig. (e)]. Hindlimbs are totally absent. Only vestiges of the pelvic girdle remain, i.e. two small rods of bone representing part of the ischial region, which serve for attachment of the corpora cavernosa of the penis or clitoris. In some Right W. and occasionally in Sperm W., small, highly regressed remnants of the femurs are also found. Clearly, the skeletal structure of W. is greatly modified in comparison with that of terrestrial vertebrates; in particular the skull shows many extreme modifications. As in terrestrial mammals, the anterior part of the skull (rostrum) is formed from the premaxillae and maxillae, but in W. it is greatly elongated. In the skulls of Toothed W., the maxillae and premaxillae also form an anterior facial plate, so that other structures, like the frontals and parietals, are pushed to the rear and the sides. Also, the skulls of Toothed W. are asymmetrical, being larger on the right side than the left; this asymmetry is correlated with the reduction of one of the nasal passages, leaving the other as a single tube, which opens to the exterior through a single nasal aperture or "blow hole". In contrast, the Baleen W. possess 2 nasal passages and paired blow holes. In all W. the blow holes are situated on top of the head. W. must regularly come to the surface to breathe. Air is exhaled through the blow hole(s) under considerable pressure, forming a visible cloud of condensation. A large whale may exhale up to 2000 liters of air, then inhale a corresponding volume, all in the space of 2 seconds or less, before submerging again. Variable periods are are spent under water, but the maximal period of submersion is 50–90 minutes. The eyes are situated rather low, near the corners of the mouth. Like other marine mammals, (e.g. walrus), W. also possess an insulating layer of fibrous, fatty tissue known as blubber, which varies in thickness from less than 2 cm to 30 cm or more, depending on the species. The blubber, which lies beneath the dermis, can be peeled away as a separate layer, being only loosely attached to the skin muscles on its inner surface. The skin is smooth and naked; in cerain species a few small hairs around the mouth are probably equivalent to the vibrissae or "whiskers" of land mammals. Forward motion is achieved by the action of a pair of horizontal flukes (the tail), which consist of tough, dense, fibrous connective tissue attached to the caudal vertebrae and enveloped by a ligament layer; the outer cutaneous layer of the flukes is continuous with that of the body, but it is free from blubber. Gestation lasts 10 to 13 months. The single calf is suckled under water, the fat-rich milk being actively squirted into its mouth in response to sucking.

W. produce sounds of very high frequency, which enable them to determine the distance and direction of surrounding objects from the received echo. Most studies on this sonar or echo-location system have been performed on dolphins, but there is much evidence to support the use of ultrasonic vibrations by all W. for finding food and forming a picture of the environment. In addition, W. have a vocal repertoire for conveying anger, distress, alarm and sexual arousal, and generally for keeping contact between members of a group. Thus, Toothed W. (odontocetes) produce short pulses ("clicks") of variable frequency and duration, and at variable rates, depending on the species, e.g. Orcas emit 6–18 clicks per second; each click contains frequencies between 250 and 500 Herz, with additional higher frequencies of weaker amplitude, and it lasts 10–25 milliseconds. In contrast, the 24 millisecond click of the Sperm whale consists of 9 shorter pulses of 0.1–2 milliseconds each. Odontocetes also produce squeals and whistles limited to a narrow frequency band, or even consisting of a single frequency. The sounds of Baleen W. (mysticetes) are typically low frequency moans and screams, which are generally not as complex as those of odontocetes, e.g. 20 Herz in the Fin Whale to 1000 Herz in the Humpback. Many sites have been suggested for the origin of these sounds: the larynx, the lips of the arytenoid cartilage, the membrane across the posterior face of the dorsal narial opening, the lateral lips of the nasal plugs, or even the lips of the blow hole; possibly more than one site is involved. A fatty body called the "melon" lies in front of the blow hole, and it has been suggested that this is an acoustic lens, reflecting sound waves forward as a narrow beam. In addition, the lower jaw, which contains a column of fatty tissue (intramandibular fat body, containing lipids similar to those in the melon) may act as an acoustic probe, conducting sound waves to the inner ear. [K.S. Norris, The echolocation of marine mammals in *The Biology of Marine Mammals*, edit. H.T. Anderson, pp. 391–423, Academic press (1969), New York and London].

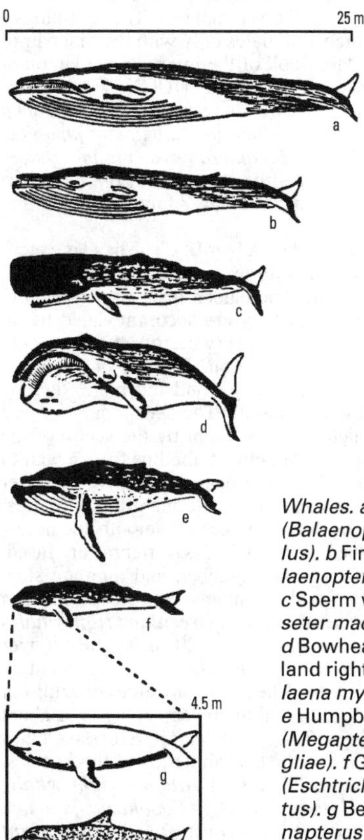

Whales. a Blue whale *(Balaenoptera musculus). b* Fin whale *(Balaenoptera physalus). c* Sperm whale *(Physeter macrocephalus). d* Bowhead or Greenland right whale *(Balaena mysticetus). e* Humpback whale *(Megaptera novaeangliae). f* Gray whale *(Eschtrichtius robustus). g* Beluga *(Delphinapterus leucas). h* Bottle-nosed dolphin *(Tursiops truncatus).*

Classification. The W. are divided into 2 suborders.

1. Suborder: *Mystacoceti, Mysticeti* ***(mysticetes, Baleen W., Whalebone W.).*** These W. lack teeth and possess a large filtration apparatus consisting of baleen plates, which enables them to filter out large quantities of crustaceans (krill) and other small animals from the

water. They therefore interact at a relatively low level in the marine feeding pyramid (see Nutritional interrelationships). In Antarctic waters the main source of food is the planktonic shrimp, *Euphausia superba,* whereas in northern waters *Boreophausia* (a smaller shrimp) is important. In the far north pteropods (planktonic mollusks known as sea butterflies) form a significant part of the Baleen Whale diet. Smaller Baleen W. may prefer smaller planktonic animals, e.g. cocepods. Each baleen plate consists of a series of tubes with an outer covering (this gives it the strength and elasticity valued and exploited in the 18th and 19th centuries for making "whalebone" corsets). There are 2 rows of these winglike, horny plates hanging from the palate across the mouth cavity (i.e. at 90 ° to the long axis of the body), with a channel between the inner borders of the two rows. These inner borders are provided with fimbriae or fringes, which are the exposed ends of the tubular skeleton of the plate. The lower floor of this filter bed is formed by the tongue. Mysticetes also have a symmetrical skull, paired external nasal apertures (blow holes), single headed ribs, and no sternal ribs. The sternum is a single bone, which articulates only with the first pair of ribs. The Red Data Book of the International Union of Conservationists and Naturalists (IUCN) lists the following 7 mysticetes as endangered species: *Balaena mysticetus, Eubalaena glacialis, Eubalaena japonica, Eubalaena australis, Megaptera novaeangliae, Balaenoptera physalis,* and both subspecies of *Balaenoptera musculus.* This suborder contains 3 families with living representatives.

1. Family: *Balaenidae* ***(Right W.).*** The head is exceptionally large, comprising up to two fifths of the total length of the animal. The baleen plates are especially long and flexible, and they are accommodated by an upper jaw (rostrum) that is very narrow and massively arched in the fore-aft direction. The lips of the lower jaws continue upward far beyond the bone, thereby closing the sides of the mouth. The open mouth forms a cavern into which water is swept by the scoop of the lower lips. As the mouth closes, the lips form a wall to the sides of the baleen; the rising tongue forces water between the plates and out of the margins of the lips. The surface of the throat and neck is smooth and unfurrowed (cf. Finback W.). All 7 neck vertebrae are fused. Right W. have very thick blubber, and they are slow swimmers. The family contains 3 species. *Balaena mysticetus* ***(Bowhead whale*** or ***Greenland right whale)*** (see Fig.) attains lengths up to 20 m (slightly longer specimens have been claimed), and is restricted to northern polar waters. The head constitutes one third of the total body length, and the baleen consists of about 300 pairs of plates, 3–4.5 m in length. Excessive hunting has almost led to the extermination of this species. *Eubalaena glacialis* ***(Black*** or ***Atlantic right whale, Nordkapper,*** or ***Biscay whale),*** *Eubalaena japonica* ***(Right whale)*** and *Eubalaena australis* ***(Southern right whale)*** are probably more correctly regarded as subspecies, genetically isolated by their restriction to different oceans. *E. glacialis* (male 18.3 m, female slightly longer) is widely distributed in all temperate waters of the world. It has 230 pairs of baleen plates, about 2.5 m in length. The head carries a number of callosities, known as the "bonnet". The skin is black, but many specimens also have some white coloration along the center of the belly. *Caperea marginata* ***(Pygmy right whale,*** about 6.4 m) occurs only in the waters of the Southern Hemisphere, and its range does not extend to the tropics. It has more ribs (17 pairs) than any other cetacean, and 230–250 pairs of baleen plates, the longest about 70 cm.

2. Family: *Eschrictiidae* ***(Gray W.).*** This family is represented by a single species, *Eschrictius robustus* (= *E. gibbosus*) ***(Gray whale,*** male 15 m, female 13 m, North Pacific) (see Fig.). There is no dorsal fin, but a line of small dorsal humps (up to 10) is present, resembling those of the Sperm Whale. The throat carries 2 (sometimes 4) short furrows, reminiscent of the numerous long furrows of the Finback W. Each autumn, Gray W. leave their summer feeding grounds in the Bering and Chukchi Seas and migrate 5000 miles southward in the Pacific coastal waters of North America to the sheltered bays and lagoons of Baja California; here they mate and give birth to young conceived during the previous winter. The return migration northward starts in late January, thereby completing a 10 000 mile round trip in 6–9 months. Between the 1840s and 1940s, intensive whaling reduced the population of eastern Pacific Gray W. from an estimated 24 000 to possibly less than 1000. They were finally accorded full protection in 1946, and by 1987 the population had recovered to an estimated 18 000.

3. Family: *Balaenopteridae* ***(Finners, Finback W.*** or ***Rorquals).*** These W. characteristically possess a dorsal fin (set rather far aft, consisting of connective tissue) and pointed flippers, in contrast to the smooth backs and rounded flippers of the Right W. Very characteristic is the series of furrows or pleats running longitudinally on the underside of the throat and chest, which allow the throat to be distended during feeding. The family is represented by 2 genera. *Balaenoptera musculus* ***(Blue whale*** or ***Sibbald's rorqual)*** (see Fig.) can exceed 30 m in length and 130 tonnes in weight in the Southern Hemisphere, but specimens from the north are generally smaller. Although this whale is naturally slaty blue, it can acquire a film of diatoms which may be colored, hence the old name, ***Sulfur-bottomed whale.*** It is the largest animal that has ever lived, and it is found in all the oceans of the world. In addition to meat, a single animal yields about 9000 liters of oil and 250 kg of baleen. Other species of *Balaenoptera* are as follows. *B. musculus brevicauda* ***(Pygmy blue whale)*** is now rare, but has been reported in the Indian Ocean near the Kerguelen Archipelago. *B. physalus* ***(Fin whale, Common rorqual*** or ***Razorback,*** 24 m, second in size to the Blue whale) (see Fig.) is found especially in the colder zones of all the oceans. Its large dorsal fin attains a height of about 61 cm, and its head is relatively smaller and its snout more pointed than those of the Blue whale. *B. borealis* ***(Sei whale*** or ***Rudolphi's rorqual,*** 18.5 m) also has a pointed snout and relatively very large dorsal fin. *B. edeni* ***(Bryde's whale,*** 14 m) is similar in appearance to the Sei whale. *B. acutorostrata* ***(Minke whale, Lesser rorqual,*** or ***Piked rorqual,*** 9.1 m) is the smallest Finback whale, and it has a very pointed snout. *Megaptera novaeangliae* ***(Humpback whale)*** (see Fig.), the only species of its genus, is an especially wide-bodied Finback whale with very elongated flippers, which are nearly one third of the body length. It occurs in all the large oceans, especially near to coasts.

2. Suborder: *Odontoceti* ***(odontocetes, Toothed W.).*** These W. possess teeth [see Dentition: Fig.(e)], al-

though in some species the teeth do not erupt through the gums. In each species the teeth are all of one type, unlike those of the extinct suborder, *Archaeoceti,* which were differentiated into incisors, canines and molars. The skull is asymmetrical and there is a single external nasal opening (blow hole). Double headed anterior ribs are present, as well as sternal ribs; the latter may be cartilaginous in certain familes. The sternum consists of 3 or more sternebrae, which articulate with 3 or more pairs of ribs. The suborder contains between 7 and 9 families (depending whether the *Iniidae* and *Stenodelphinidae* are recognized as separate families) with living representatives.

1. Family: *Platanistidae* ***(River dolphins).*** These small dolphins (1.5–2.7 m) are considered to be the most primitive of living W. They live in small groups, feeding on fish and small water animals. *Platanista gangetica* ***(Ganges susu*** or ***Gangetic dolphin)*** is found in the waters of the Ganges and associated coastal waters of India, while *Platanista indi* similarly inhabits the Indus and its coastal waters; they may be varieties of the same species. The eyes of the susu have no lens and are largely nonfunctional, probably serving only to differentiate between light and dark. *Inia geoffrensis* ***(Boutu*** or ***Amazon dolphin)*** from the Amazon and Orinoco rivers of South America is similar to the susu, but its eyes are more functional; it is sometimes placed in a separate family *(Iniidae). Lipotes vexillifer* ***(White-flag dolphin*** or ***Chinese river dolphin),*** which occurs only in Lake Tung Ting and the Yangtze Kiang River in China, is one of the least known W.; its eyes are reduced and practically nonfunctional, and its snout curves upward. *Pontoporia (= Stenodelphis) blainvillei* ***(La Plata dolphin)*** is a coastal and esturine species, found only along the Atlantic coast of South America, and sometimes placed in a separate family *(Stenodelphinidae).*

2. Family: *Ziphiidae* ***(Beaked W.).*** These are medium sized W. with an elongated snout, known as a beak. Dentition is usually reduced, with only 1 or 2 pairs of erupted teeth in the lower jaw. Between 2 and 7 of the cervical vertebrae are fused. The throat carries 2 longitudinal grooves. Most familiar is *Hyperoodon* ***(Bottle-nosed whale,*** male 9 m, female 7.5 m), which occurs as 2 closely related species, *H. ampullatus* ***(Northern bottle-nosed whale)*** and *H. planifrons* ***(Southern bottle-nosed whale).*** *Hyperoodon* carries an oil reserve in its upper lip, which gives rise to a very characteristic bulbous "forehead"; in old males this character is further exaggerated by the presence of maxillary crests on the rostrum. Other members of this family are: *Ziphius cavirostris* ***(Cuvier's beaked whale,*** male 5.4 m, female 6.1 m, no bulbous forehead, widely distributed in all the oceans); *Berardius bairdii* ***(Baird's beaked whale,*** male 11.8 m, female 12.8 m, North Pacific, bulbous forehead not as pronounced as in *Hyperoodon); Berardius arnouxii* ***(Arnoux's beaked whale,*** similar in size and shape to *B. bairdii,* southern oceans); *Mesoplodon densirostris* ***(Blainville's beaked whale,*** worldwide in tropical and temperate waters); *Mesoplodon stejnegeri* ***(Stejneger's beaked whale,*** North Pacific); *Mesoplodon bidens* ***(Sowerby's whale,*** Atlantic); *Mesoplodon mirus* ***(True's beaked whale,*** Atlantic, very rare) (species of *Mesoplodon* are 4.9–5.2 m in length);*Tasmacetus shepherdii* ***(Shepherd's beaked whale,*** southern oceans, 52 teeth in lower jaw, 38 teeth in upper jaw, very rare).

3. Family: *Physeteridae* ***(Sperm W.).*** Represented by 3 species in 2 genera. Most familiar is *Physeter macrocephalus (P. catodon* ; ***Sperm whale, Great sperm whale,*** 17–20 m; in all the warmer oceans) (see Fig.). About one third of the bulk of the animal is accounted for by the head, the large size of which is due to the presence of an enormous internal cavity filled with a waxy mass of spermaceti. To accommodate the spermaceti, the skull has a high transverse crest with a concave basin. The dorsal fin is practically absent, but between its vestige and the tail are 4–5 small humps. The flippers are rounded. Only the lower jaw has teeth (conical in shape), which fit into sockets in the upper jaw. The remaining 2 species, *Kogia breviceps* ***(Pygmy sperm whale)*** and *Kogia simus* ***(Dwarf sperm whale)*** are not well studied; they both have well defined dorsal fins, bluntly pointed flippers, and teeth (backward curving) only in the lower jaw.

4. Family: *Monodontidae* ***(White W.*** and ***Narwhals).*** The family is represented by only 2 members. *Monodon monoceros* ***(Narwhal)*** is a mottled gray whale well known for the 3 m long, spirally twisted tusk of the male, which is derived from the left (rarely the right) incisor. The other incisor usually develops normally and there are no other teeth. Occasionally the female also shows slight development of a tusk. The tusk appears to have no function, and it may be a structure that has become specialized beyond usefulness. The Narwhal is usually found only in the high Arctic. Its forehead is bulbous, the flippers are relatively short and rounded, and the mouth does not form a beak. *Delphinapterus leucas* ***(White whale*** or ***Beluga)*** (see Fig.) favors shallow and estuarine waters in the Arctic and sub-Arctic. Calves are gray, but the mature whale (2.8–4.4 m in length) is white to cream in color. The forehead is high and swollen, and the mouth (which is not formed into a beak) usually contains 10 pairs of teeth in the upper jaw and 8 pairs in the lower jaw.

The following 3 closely related families of porpoises and dolphins comprise the superfamily, *Delphinoidea.* There are few fundamental differences between the *Phocoenidae* and the *Delphinidae.* Strictly speaking, the *Phocoenidae* do not possess a beak and their teeth are spade-shaped. The terms, porpoise and dolphin, are, however, applied freely to both families.

5. Family: *Stenidae* ***(Long-snouted dolphins, Rough-toothed*** and ***Humpbacked dolphins).*** The family contains 3 genera, and is separated from the *Phocoenidae* and *Delphinidae* by the specialized air sinus system of the skull. *Steno bredanensis* ***(Rough-toothed dolphin,*** 2.2–2.75 m) occurs in tropical and subtropical waters of the Atlantic and western Pacific. The crowns of the teeth are characteristically furrowed. *Sotalia fluviatilis* ***(Tucuxi)*** occurs in large rivers on the Atlantic coast of South America. Mature females in the Amazon may be only 146 cm in length, but a larger size (1.6 m) is more normal. *Sousa chinensis* ***(Pacific humpbacked dolphin*** or ***Chinese white dolphin,*** coastal waters of China) and *Sousa teuszii* ***(Atlantic humpbacked dolphin*** or ***West African white dolphin,*** coastal waters and river deltas of West Africa) are both found in warmer waters; they are milk white with reddish flippers, they have a characteristic hump which carries the dorsal fin, the beak is well developed, and each jaw contains 32 pairs of teeth.

6. Family: *Phocoenidae* ***(Porpoises).*** The family is represented by 6 closely related species. *Phocoena pho-*

coena (**Harbor porpoise,** male 1.6 m, female 1.8 m) occurs in the coastal waters and river estuaries of the North Atlantic. *Phocoena sinus* (**Cochito**) is found only in the Gulf of California. *Phocoena spinipinnis* (**Burmeister's porpoise**) occurs in relatively large numbers in the Atlantic and Pacific coastal waters of South America. *Phocoena dioptrica* (**Spectacled porpoise**) is an uncommon species found in the Atlantic coastal waters of South America. All *Phocoena* are relatively small, rarely exceeding 2 m in length. *Phocoenoides dalli* (**Dall's porpoise,** 2.5 m) favors the open waters of the North Pacific. It is captured off the northeast coast of Japan for food. *Neophocoena phocoenoides* (**Finless porpoise,** 184 cm) is widely but thinly distributed in the coastal waters and rivers of India, Southeast Asia, China, Japan and Korea. It has a low dorsal ridge and no dorsal fin.

7. Family: *Delphinidae* (**True Dolphins**), comprising 4 subfamilies.

1. Subfamily: *Orcininae* (**Killer W., Orcas**). Orcas feed on and actively hunt other marine mammals like seals and small dolphins; they are therefore at the very peak of the marine feeding pyramid (see Nutritional interrelationships). Marine mammals, however, comprise only up to one third of the Orcas' food, the major portion being fishes and cephalopods. Although most dolphins are more patterned than other W., the coloration of *Orcinus orca* (**Killer whale** or **Orca,** male over 9 m, female up to 6 m) is especially striking. The white of the belly extends over the flanks between the flukes and the sharply pointed dorsal fin (triangular in mature males, sickle shaped in mature females). The back is otherwise very black, with a gray saddle behind the dorsal fin, and there is a white patch behind the eye. *Pseudorca crassidens* (**False killer whale,** male 5.6 m, female 5.0 m) is uniformly black in color, except for white spots on the leading edges of the tapered, elongated flippers. The dorsal fin is sickle shaped. *Feresa attenuata* (**Pygmy killer whale,** male 244 cm, female 227 cm) occurs in tropical and warm temperate waters, but it is rare and not well studied. *Orcaella brevirostris* (**Irrawaddy dolphin**) is a gray to white dolphin occurring in the coastal waters and large coastal rivers of eastern India, Southeast Asia and northern Australia. *Peponocephala electra* (**Electra dolphin** or **Melon-headed whale,** 2.5–3.0 m) is a poorly known species, generally dark in color, with a white patch between jaw and flippers and a long white ventral patch. The genus *Globicephala* (**Pilot W.**) is represented by 2 species: *G. melaena* (**Long-finned pilot whale;** 2 separate stocks in the colder waters of the 2 hemispheres; the southern form is also called *G. edwardii* or *G. sieboldii* , and the northern form is sometimes called the **Atlantic blackfish**) and *G. macrorhynchus* (**Short finned** or **Indian pilot whale,** in warmer waters between the 2 stocks of *G. melaena*). These are long, slender dolphins (male 8 m, female 5 m) with bulging foreheads, very long flippers, and a sickle-shaped dorsal fin. There is a white throat patch extending toward the tail, and the body is otherwise black.

2. Subfamily: *Lissodelphininae* (**Right whale dolphins**). *Lissodelphis borealis* (**Northern right whale dolphin,** 2.7 m, North Pacific) and *Lissodelphis peronii* (**Southern right whale dolphin,** 1.8 m, subtemperate waters above the Antarctic) are the only 2 living members of this subfamily. Like the Right W., they lack a dorsal fin. The beak is slender with about 40 pairs of teeth in each jaw. A white line runs from the vent to the flukes, which have white undersurfaces, and ventral white patches are seen between the flippers and beneath the chin.

3. Subfamily: *Cephalorhynchinae.* Represented by 4 species in one genus: *Cephalorhynchus heavisidii* (**Heaviside's dolphin,** coastal waters of South Africa); *C. eutropia* (**Black dolphin,** coastal waters of Chile); *C. hectori* (**Hector's dolphin,** coastal waters of New Zealand); *C. commersonii* (**Commerson's dolphin,** western South Atlantic and South Atlantic islands). The beak is short, with about 30 pairs of teeth in each jaw. All species are small (about 1.8 m), and variously patterned black and white.

4. Subfamily: *Delphininae* . Contains 6 genera. *Delphinus* is represented by a single species, *D. delphis* (**Common dolphin,** 2.6 m), which is distributed worldwide in temperate and tropical waters. The dorsal surface is black or very dark brown and the ventral surface is white. On the flanks is an intricate pattern of alternate light and dark bands, while 2 lines of yellow and white meet at the level of the dorsal fin. Patterning also includes a dark circle around each eye and tapering light bands from the bases of the flippers to the sides of the lower jaw. It is one of the commonest and most familiar of all marine mammals. *Lagenorhynchus* is represented by 6 species, e.g. *L. albirostris* (**White-beaked dolphin,** 3.1 m, North Atlantic) is of stout build and has a short beak with 22–25 pairs of teeth in each jaw. Others are: *L. acutus* (**White-sided dolphin,** North Atlantic), *L. obscurus* (**Dusky dolphin,** southern oceans), *L. cruciger* (**Hourglass dolphin,** southern oceans), *L. australis* (**Peale's dolphin,** coastal waters of South America), *L. obliquidens* (**Pacific white-sided dolphin,** North Pacific). *Lagenodelphis hosei* (**Fraser's dolphin**) is found in the tropical waters of the eastern and central Pacific. *Tursiops truncatus* (**Bottle-nosed dolphin,** 3.7 m) (see Fig.) is found in all temperate and tropical waters. It is very amenable to training, and it is the most familiar dolphin seen in captivity. Both jaws contain 22–25 pairs of teeth. *Stenella* is represented by at least 3 species, but the great variety of forms in this genus probably includes other separate species. *S. caeruleoalba* (**Blue-white, Striped** or **Euphrosyne dolphin,** 2.5 m) is found in most warm and tropical seas. It has a prominent beak with 43–50 pairs of teeth in each jaw. In general appearance it resembles the Common Dolphin, but it lacks the lateral yellow coloration. Characteristic narrow stripes from the eyes to the vent, with branches to the bases of the flippers, give rise to the name, Striped Dolphin. *S. longirostris* (**Spinner dolphin,** separate populations in all tropical seas) is named for its habit of leaping from the water and spinning on its axis in mid flight. *S. dubia* (= *S. frontalis* = *S. attenuata* = *S. palgiodon* ; **Spotted dolphin**) occurs in coastal waters throughout the tropics and in certain warm temperate regions of the North and South Atlantic. The beak is especially long, and the body is patterned with pale spots on a darker ground. *Grampus griseus* (**Risso's dolphin,** all tropical and temperate waters) differs from all other *Delphininae* in not possessing a beak. The head is bulbous, teeth are absent from the upper jaw, and in general appearance it resembles a Pilot whale.

[*Whales* by E.J. Slijper, English translation 1962 (Hutchinson: London); *Whales* by W. Nigel Bonner, 1980 (Blandford Press: Poole, Dorset, UK); *Cetacean*

Behavior by Louis M. Herman, 1980 (John Wiley & Sons: New York, Chichester, Brisbane, Toronto); *Sea Guide to Whales of the World* by Lyall Watson & Tom Ritchie, 1981 (Hutchinson: London, Sydney, Auckland, Johannesburg); *The Sierra Club Handbook of Whales and Dolphins* by Stephen Leatherwood, Randall R. Reeves & Larry Foster, 1983 (Sierra Club Books: San Francisco); *The Sei Whale* by Joseph Horwood, 1987 (Croom Helm: London, New York, Sydney)]

Wharton's jelly: see Connective tissues.

Wheat: see *Graminae.*

Wheat bulb fly: see Anthomyiids.

Wheatears: see *Turdinae.*

Wheat gall: see Eelworms.

Wheat midge: see Gall midges.

Wheel animalcules: see *Rotifera.*

Wheel organ: see *Rotifera.*

Whelk: see *Gastropoda.*

Whidas: see *Ploceidae.*

Whimbrel: see *Scolopacidae.*

Whinchat: see *Turdinae.*

Whip corals: see *Gorgonaria.*

Whip-poor-will: see *Caprimulgiformes.*

Whipscorpions: see *Uropygi.*

Whip snakes, *Coluber:* a numerous genus of colubrid snakes (see *Colubridae),* widely distributed in America, Europe, North Africa and Asia. They are vey fast, agile, slender snakes, with a strongly demarcated head and a long tail, and smooth or keeled scales. Lizards, small snakes and rodents form the main part of the diet. Some species have venom glands, but these are not associated with fangs or teeth. All lay eggs. The ***European whip snake*** *(C. gemonensis)* lives in southern Europe and West Africa. Since some specimens attain a length of over 270 cm, it is the longest European snake. Although not venomous, it is very aggressive and much given to biting. *C. constrictor* of North America attains 1.5 m, and on account of its speed is known as the ***Racer.***

Whipspiders: see *Amblypygi.*

Whiptail lizards, *racerunners,* *Cnemidophorus:* a genus of the lizard family, *Teiidae* (see), found in South and Central America. Twenty to 45 cm in length, with a tapered head, most species are brown-gray in color with longitudinal stripes. Bands of imbricating or granular scales run longitudinally, transversely or diagonally. The belly is often covered with a regular pattern of very large scales. They are quick, agile, diurnal inhabitants of open ground. The male often has a brightly colored throat and belly. *Cnemidophora sexlineatus* ***(Six-lined racerunner)*** also occurs in North America. Several species reproduce parthenogenetically, and only the female is known, e.g. *C. cozumela* of Yucatan.

Whipworm: see *Nematoda.*

Whirligig beetles: see *Coleoptera.*

Whisk fern: see *Psilophytatae.*

Whistlers: see *Pachycephalinae.*

Whistling ducks: see *Anseriformes.*

Whistling hares: see *Ochotonidae.*

Whistling schoolboy: see *Turdinae.*

Whistling thrush: see *Turdinae.*

White ants: see *Isoptera.*

White body: see *Isopoda.*

White dodo: see *Raphidae.*

White eyes: see *Zosteropidae.*

Whitefishes: see *Coregonus.*

Whitefly: see *Homoptera.*

White-footed mouse: see *Cricetinae.*

White-line dart: see Cutworms.

White-lipped peccary: see *Tayassuidae.*

White-lipped tree viper: see Lance-head snakes.

White oak: see *Fagaceae.*

White pine: see *Araucariaceae.*

Whites and sulphurs (butterflies): see *Pieridae.*

White's thrush: see *Turdinae.*

White-tailed deer, *Odocoileus virginianus:* a medium-sized deer (head plus body 85–210 cm, shoulder height 55–110 cm) distributed from southern Canada to Peru and northeastern Brazil, with basket-like antlers consisting of a single main beam with many branches. The white underside of the tail serves as a species-specific following signal, which is revealed when the tail is held high during flight. There are only 2 species of *Odocoileus,* the other being the closely related *O. hemionus* (see Mule deer). Indian buckskin leather was made from the hide of both species, and both yield venison of excellent flavor. See Deer.

White-tailed kites: see *Falconiformes.*

White-tailed rat: see *Cricetinae.*

Whitethroat: see *Sylviinae.* See plate 6.

White whale: see Whales.

Whiting, *Merlangius merlangus:* A fish of the order *Gadiformes,* found in the eastern Atlantic (North pole to Atlantic coast of Spain) , the Mediterranean, and the Black sea. Up to 70 cm in length, it is light red-brown in color, somewhat resembling a haddock but with smaller lateral black spots. It is fished commercially, and is popular as a fried fish.

Whortleberry: see *Ericaceae.*

Whydas: see *Ploceidae.*

Wide-mouthed toads: see *Leptodactylidae.*

Widow weavers: see *Ploceidae.*

Wielandiella: a fossil plant group of the *Bennettitatae* (see).

Wigeon: see *Anseriformes.*

Wild ass, *Equus asinus:* a member of the family *Equidae* (see), order *Perissodactyla* (see) with relatively slender hooves, long ears, and a tail with hairs only on the distal half. The gray to rufous gray coat carries a dark dorsal stripe and usually a dark shoulder stripe. In its North African native habitat the W.a. is extremely rare. It is the ancestor of the domestic ass or donkey.

Wild boar: see Wild Pig.

Wildebeests: see *Alcelaphinae.*

Wild hog: see Wild pig.

Wild horse, *Equus przewalskii:* a rufous to yellow-brown member of the family *Equidae* (see), order *Perissodactyla* (see) with a black dorsal stripe, black mane, long hairy tail and whitish tip to the snout. Living in small herds led by a stallion, it originally inhabited the steppes or forests of Asia. It is the main ancestor of the domestic horse, *Equus przewalskii caballus.* The last free-living population of the *Equus przewalskii przewalskii* has disappeared from Mongolia, and the species exists only in zoos. The ***Forest tarpan*** *(Equus przewalskii silvaticus)* and the ***South Russian plains tarpan*** *(Equus przewalskii gmelini),* both now extinct, were very similar to, but rather less heavily built than *Equus przewalskii przewalskii.*

Wild pig, *wild hog, wild boar,* *Sus scrofa:* a species of pig (see *Suidae),* which exists as numerous subspecies, distributed throughout large areas of Europe, Asia and North Africa, and now also established in the USA. Although, strictly speaking, male pigs are called boars and the females sows, the W.p. is often generally referred to as a wild boar. The body has stiff bristles, and usually some black to brown finer hair, but the body covering is usually very scant. Many individuals have side whiskers and a neck mane. The 6–12 young, born after a gestation of 16–20 weeks, have protective brown-yellow striping. There are 4 continually growing tusks, one in each jaw; dental formula: 3/3, 1/1, 4/4, 3/3 = 44. *Sus scrofa* is the ancestor of the domestic pig.

Wild type: the phenotype that characterizes most of the wild forms of a race or species in a natural environment. In genetic notation, genes responsible for individual characters of a wild phenotype are symbolized by a plus sign, e.g. A $tryp^+$ strain of *Escherichia coli* is able to synthesize tryptophan, and this property is also present in the W.t.

Willets: see *Scolopacidae.*

Williamsonia (English gelogist, Williamson): a group of fossil plants belonging to the *Bennettitatae* (see).

Williamsoniella: see *Bennettitatae.*

Willie wagtail: see *Muscicapinae.*

Willow-herb: see *Onagraceae.*

Willow moss: see *Musci.*

Willows: see *Salicaceae.*

Willow tit: see *Paridae.*

Willow warbler: see *Sylviinae.*

Willow weed: see *Polygonaceae.*

Wilting: reversible or irreversible decrease in the rigidity or tautness of plant organs (i.e. they become limp), due to decreased turgor. W. occurs when the rate of water loss from the plant exceeds the rate at which it is replaced. This can arise from decreased water uptake, blocked water transport, or increased water loss. *Physiological W.* is caused by environmental factors, such as drought, frost, or nutrient deficiency. W. is particularly symptomatic of potassium deficiency. *Pathological W.* is caused by pathogenic bacteria and fungi, and, in contrast to physiological W., it is usually irreversible. Pathological W. is due to the action of Wilt toxins (see), the action of certain enzymes secreted by the invading organism, toxicity of other metabolic products of the pathogen, as well as physical blockage of vessels by the invading organism and its secreted products. The extent of pathological W. is generally influenced by environmental factors, especially water availability and temperature.

Wilting coefficient: see Wilt point.

Wilt point, ***wilting coefficient:*** the maximal water content of the soil that still leads to the wilting of plants, i.e. the highest water content that cannot be exploited by plants. It is measured in g water per 100 g dry soil mass, and it is determined by placing plants in soil in a vessel, with the surface of the soil sealed against evaporation. The plants are allowed to grow until wilting occurs, and until the degree of wilting is such that the plants do not recover after 24 hours exposure to moist air. The soil is weighed, then dried by heating or by exposure to low pressure in the presence of a powerful desiccant, and weighed again. The W.p. varies according to the plant and the soil, e.g. the W.p. for rye in fine sand is 3.1 g/100 g, whereas the W.p. for rye in marl is 15.5 g/100 g.

Wilt toxins, ***wilting agents, marasmines:*** toxins produced by plant pathogenic microorganisms, which cause irreversible wilting in higher plants. The metabolic products of these pathogens also usually cause further types of injury, so that a clear distinction does not always exist between W.t. and other toxic agents. Some W.t. are also antibiotics.

Well studied W.t. are Fusaric acid (see) and Lycomarasmine (see). Both are produced by fungi of the genus *Fusarium,* and they have a similar physiological mode of action. Thus, they cause irreversible wilting, due primarily to toxic effects on the protoplasm of parenchyma cells, leading to increased permeability and loss of turgor. These W.t. also chelate heavy metals, in particular iron, causing severe inhibition of certain enzymes.

Fusarium species also produce W.t. known as *enniatines,* cyclic peptides with antibiotic activity against mycobacteria.

Several other metabolic products of microorganisms cause wilting. Some polysaccharides formed by plant pathogenic bacteria and fungi cause wilting by blocking the water-conducting tissues of the host plant (e.g. in bacterial infections of tomato plants). The condition of the infected plant is probably made worse by the action of pectin-cleaving enzymes, secreted by the pathogenic organism. Furthermore, organisms that cause vessel blockage also generally produce W.t. (e.g. *Corynebacterium michiganense).*

Wind factor: effects on plants and animals due to wind. In plants, wind serves directly for pollination and seed distribution, while trees and smaller plants may be damaged or uprooted by strong winds. By promoting transpiration, wind has a considerable influence on the water balance of plants. When the soil is frozen and water uptake thereby greatly decreased, the drying effect of wind can lead to desiccation damage. If the drying effects of wind are counteracted by stomatal closure, this also leads to decreased gas exchange with an inhibition of carbon dioxide assimilation. Experimentally, it can be shown that plants continually exposed to wind remain much smaller and develop a smaller leaf mass than control plants grown in relatively still air. Since wind velocity is lower near the soil surface (due to drag), increasing markedly with height (e.g. from 0.01 m/s at 2 cm above ground level to 6 m/s at 2 m), trees and parts of tall plants are particularly influenced by wind.

In animals, wind is often important in orientation. Movement or establishment of a position in relation to wind direction is known as anemotaxis. Thus, locusts move against the wind, while many birds and mammals adopt a head-to-wind resting position. Some bugs and beetles fly with the wind. Insects of high mountain environments and wind-exposed coastal regions cease flying and creep when it is windy, or they may display adaptive wing reduction (e.g. flies on Kerguelen island).

By diverting animals from their flight path, by increasing or decreasing their flying efficiency and by passive wind transport, wind also influences the active and passive distribution of flying animals. Passive transport occurs when the flying performance of a small animal is transcended by the power of the wind. By anemochoric transport, birds and insects may be carried over large distances, often greater than 1 000 km. Sometimes auxillary aids are developed, e.g. some insects produce spun threads, wax threads (aphids), or long hairs (e.g. eggs of the gypsy moth are covered with hairs from the adult). Thermals (upward air flow) enable many birds to glide, and they carry many small insects to great heights where they become part of the aerial plankton. Strong air movements, however, may then move such organisms away from land, with the catastrophic result that mass deposits of insects are washed up later on the coast.

To counteract the drying and cooling effect of wind, trees and hedges are often planted as wind breaks, and the same protection is afforded on a smaller scale by walled gardens. Protection from the wind is beneficial to all types of plant and animal agriculture. It increases the average temperature, decreases water loss, in some cases increases rainfall, and is particularly effective in coastal regions, where it also decreases soil erosion.

See Wing reduction.

Wind-pollination: see Pollination.

Windscorpions: see *Solifugae.*

Wine yeasts: selected varieties of *Saccharomyces cerevisiae,* which are used for the fermentation of grape juice for wine production. They may also be accorded the status of an independent species, called *Saccharomyces ellipsoideus.* The different varieties are usually named according to their origin, e.g. Bordeaux yeast, Tokay yeast, etc. The cells of W.y. are oval to elongated. Individual varieties differ mainly with respect to the type and quantity of aroma substances that they produce, the activity of their fermentative metabolism, the final ethanol concentration that they produce, and the sugar concentration to which they are most suited.

Wing degeneration: see Wing reduction.

Wing petals: see *Papilionaceae.*

Wing reduction, ***wing regression, wing degeneration:*** the degeneration or total loss of flight organs in certain animals (see Life form), in accordance with their mode of existence. W.r. occurs in cave-dwelling species, in occupants of habitats strongly influenced by wind (mountains, islands, deserts, areas of drifting sand and sea coasts), in occupants of isolated habitats free from predators (e.g. islands) and in numerous parasites. Thus, all cave-dwelling insects are flightless. The extinct gigantic flightless pigeons, known as dodos and solitaires, were island dwellers of the Mascarenes. A particularly large number of short-winged or wingless bird species occurs in New Zealand. The flightless cormorant *(Nannopterus harrisi)* nests along the shores of two of the main Galapagos Islands (see Fig. in *Pelecaniformes)* . Steppe regions are home to a large number of running and walking birds. Many crickets and beetles of high mountain regions (e.g. 54% of crickets in the high regions of the Kärtner Alps) have reduced wings. Flies that inhabit coastal beaches make only short jumping flights. Most bembid beetles (family *Carabidae,* subfamily *Harpalinae)* and rove beetles (family *Staphylinidae)* found in coastal spray zones are flightless. In the drifting sand regions of Finnish coasts, 14.3% of the stenotopic insects display wing regression. Such wing reductions confer a selective advantage, because the insects thereby avoid being blown about by the wind.

The mode of existence of a large number of ectoparasites has rendered their ability to fly superfluous, so that wings are reduced or absent, e.g. in fleas, animal lice and bed bugs. Some louse-flies *(Hippoboscidae)* (ectoparasites of birds and large mammals) are wingless or have shortened wings, while others develop wings then shed them.

Wing nut: see *Juglandaceae.*

Wing regression: see Wing reduction.

Wings:

1) The two lateral petals of flowers of the family *Papilionaceae.*

2) Membranous appendages of fruits or seeds, which aid wind dispersal. E.g. fruits of maple, elm and ash, and seeds of tulip, lily and iris.

3) Animal organs which serve for active flight. The W. of insects are evaginations of the exoskeleton, The W. of vertebrates (birds, bats) are modified forelimbs (see Extremities).

Winter cherry: see *Solanaceae.*

Wintergreen: see *Pyrolaceae.*

Wintergreen oil: see Salcylic acid.

Winter killing: a term for the injury suffered by agricultural field plants, attributable mainly to the following causes: ***freezing*** by very low temperatures in the absence of a protective snow covering; ***desiccation*** due to water loss (transpiration) under the influence of wind and/or sun, combined with the inability of the roots to take up water from the frozen soil; ***suffocation*** and ***putrifaction*** beneath encrusted snow or in melt water puddles, where the oxygen tension is low and the carbon dioxide tension is high; ***freezing out*** by frequent alternate freezing and thawing of the soil, leading to tearing of the roots and lifting of plants with weakened or poorly developed root systems from the ground. The results of W.k. usually become apparent in the spring. Weakened plants are then more likely to fall victim to various fungal diseases, e.g. clover stemrod caused by *Scerotina trifoliorum,* snow mold of cereal crops caused by *Calonectra graminicola (= Fusarium nivale),* and typhula blight (or speckled snow mold) caused by *Typhula incarinata (= T. intoana).* Animal pests also exacerbate the effect of winter killing, e.g. field mice, wire worms, garden march fly *(Bibio hortulana),* wheat bulb fly *(Phorbia = Hylemya coarctata)* and the flea beetle, *Psylliodes chrysocephala,* whose larvae feed on cabbage, turnip and other *Cruciferae.*

Winter moths (plate 45): moths of the family, *Geometridae,* which appear during mild spells from October to December in temperate climates. The ***Small winter moth*** *(Operophtera brumata* L.), a pest of apple orchards, occurs in Europe, Asia, Japan, Nova Scotia, New Brunswick and Prince Edward Island; its caterpillars also feed on oak and hawthorn. The ***Mottled umber*** *(Erannis defoliaria* Cl.), a pest of fruit orchards, occurs in Europe and most of northern and temperate Asia; the caterpillars feed on apple, oak, birch and other wood-

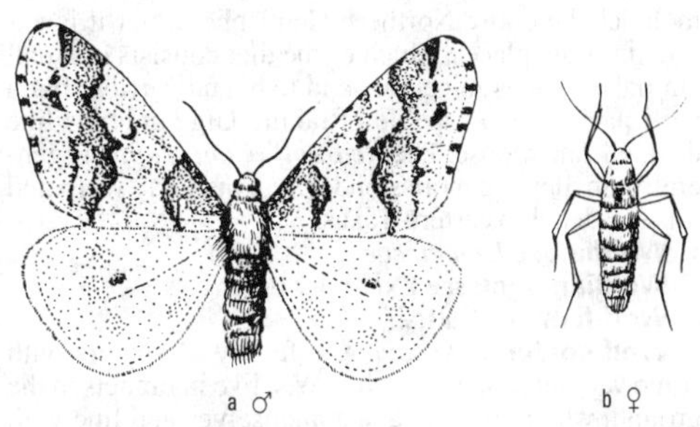

Winter moths. Mottled umber *(Erannis defoliaria* Cl.). *a* Male. *b* Female.

land trees, sometimes causing total defoliation. Caterpillars of the ***Northern winter moth*** *(Operophtera fagata;* northern Europe) feed on birch. In all these species, the female is wingless, possessing only vestigial wing stumps. Eggs are laid in crevices in the bark or buds of the larval food tree. Larvae hatch early in the spring, and eat the buds and leaves.

Winter spores: fungal spores, usually with thick walls, which survive during the winter, e.g. Teleutospores (see), Perennating spores (see).

Wireworms: see Click beetles.

Wisconsin glaciation: the last glacial advance in North America, which occurred between 70 000 and 10 000 years ago. The corresponding period in northern Europe is known as the Würm glaciation.

Wisent: see *Bovinae.*

Wistar rat: see *Muridae.*

Wistaria: see *Papilionaceae.*

Wisteria: see *Papilionaceae.*

Witches' broom: a malformation (see Galls) in the crown of an otherwise normal tree, consisting of a densely packed proliferation of thin shoots, often with reduced leaves and deformed leaf lamina. The onset of W.b. formation is marked by an abnormally rapid sprouting of buds, leading to an irregular accumulation of shoots, which disturb and inhibit the development of one another. The resulting proliferation of branches or twigs resembles a broom, and from a distance may be mistaken for a bird's nest. W.bs. are usually caused by infections, although some are of nonparasitic origin. The W.bs. of cherry and plum trees, which cause considerable loss of fruit production, are caused by rust fungi of the genus *Taphrina.* During the flowering period, flowers are absent from the affected branches, and the W.bs. appear as green bushes in a flowering tree. *Taphrina deformans* and *T. turgida* cause W.bs. on the birch species, *Betula pubescens* and *B. pendula,* respectively. Similar malformations also occur in forest trees, e.g. in the silver fir, where it is caused by the rust fungus *Melampsorella caryophyllacearum.* See *Ascomycetes.*

Withers: in domestic animals, the foremost (and often highest), gradually backward sloping part of the back, which is formed by the strongly developed neural spines of the 3rd to the 13th thoracic vertebrae. The height of the W. of a horse is measured at the 5th thoracic vertebra.

Wolf, ***gray wolf, timber wolf,*** *Canis lupus:* a large (head plus body 107–137 cm), powerful member of the *Canidae* (see), which in summer is solitary or lives in family groups, and in winter lives in large packs and often undertakes long migrations. Its natural distribution includes the entire Northern Hemisphere, but it is extinct in some places. Much of the diet consists of small animals, such as mice, rats and fish, and carrion, but a wolf pack is also capable of taking large animals like deer, elk and moose. The ***Red wolf*** *(Canis rufus)* is generally smaller and more slender than the Gray wolf, and it is restricted to eastern Texas.

Wolffia: see *Lemnaceae.*

Wolffian duct: see Excretory organs

Wolf-fish: see Catfish.

Wolf spiders, *Lycosidae:* a family of spiders with some very large species. Most W.s. live in tunnels in the ground, which they excavate themselves and line with silk threads. They lie in wait for prey at the tunnel entrance, or they actually hunt and chase their prey (Greek *Lycosa,* wolf), especially at night. They may be commonly observed running through grass or lurking under stones, especially in damp situations. In nearly all species, the cuticle is clothed with simple or squamose hairs. All carry their egg sac attached to the spinnerets by a bundle of threads. After hatching, the young are carried on the mother's body for some time. The harmless southern European ***Tarantula*** *(Lycosa tarantula),* whose bite was erroneously thought to cause dancing madness, belongs to this family.

Wolverine, ***gluton,*** *Gulo gulo:* a monotypic genus of the *Mustelidae* (see). Compared with other mustelids, the W. is plump and has a relatively short tail. Head plus body length is 65–87 cm, tail 17–26 cm, and height at the shoulder 36–45 cm. Females are smaller than males. It inhabits the northern taiga and forest tundra of Eurasia and North America, and is also found in the western USA. The long, dense fur is blackish brown in color, with a light brown band from shoulder to rump along each side of the body, joining over the base of the tail. It is usually solitary, does not hibernate, and displays alternating 3-4-hour cycles of activity and rest. For a mammal of its size, it is extremely strong, and can even drive bears and mountain lions from their kill. It is even powerful enough to kill elk and moose if these are hampered by snow. The preferred diet is, however, carrion, as well as the eggs of ground-nesting birds, wasp larvae and berries. Predatory feeding is more common in winter.

Wombats, *Phascolomidae, Vombatidae:* a family of Marsupials (see) containing 2 (possibly 3) species and 2 genera, native to Australia and Tasmania. They are short-legged, stout animals, up to 1 m in length, which display a marked structural convergence toward the rodents. They feed on grasses, roots and fungi, and with the aid of their sickle-like claws, they excavate large subterranean burrows. The marsupium opens posteriorly. Dental formula: 1/1, 0/0, 1/1, 4/4 = 24. The incisors are large and resemble those of rodents, and they are separated from the cheek teeth by a wide diastema.

Wonder net of the fish swim bladder: see Hairpin loop countercurrent principle, Swim bladder.

Wood: the secondary permanent tissues of plants, which develop toward the stem center from the inside the fascicular cambium. Several types of tissue are involved: Conducting tissue (see) (vessels), Mechanical tissue (see) (sclerenchyma fibers) and Parenchyma (see) (storage and conducting parenchyma). Wood conducts water and increases the mechanical strength of the shoot and root system, as well as serving for the storage of organic materials.

The wood of most gymnosperms (gymnospermous or coniferous wood) has a relatively simple structure. Coniferous wood consists predominantly of tracheids, with walls of varying thickness, depending on the time of year in which they were formed. In the spring, tracheids with wide lumina are produced, which serve exclusively for water transport (spring wood), whereas in later summer there is an increased production of thick-walled cells with narrow lumina, which serve mainly as mechanical tissue (late wood). In some conifers (e.g. *Taxus),* wood parenchyma is found only in the pith rays. In others (e.g. *Pinus)* it is found only around the schizogenous resin ducts, which penetrate the wood longitudinally. Neighboring resin ducts are linked by transverse channels, thereby forming a network. Numerous medullary (pith) rays, usually consisting of only a single cell layer, run radially between the tracheids.

Compared with coniferous wood, the wood of dicotyledonous plants displays a progressive differentiation of cells and tissues. In particular, water-conducting elements are differentiated from support material. The most conspicuous elements of dicotyledonous wood are vessels with wide lumina [the result of Intussusception (see) of the walls]; in some tree species, equal-sized, wide-lumen vessels may be distributed uniformly throughout the entire wood, whereas in others they predominate only in the spring, with narrow-lumen vessels in the late wood. Dicotyledonous wood also contains wood fiber cells as supporting elements. Finally, dicotyledonous wood always contains more parenchyma than does coniferous wood. It surrounds the longitudinal vessels as wood parenchyma, as well as being present as radial pith parenchyma (as in conifers). The latter contains one or more layers of parenchyma cells, and forms a coherent system of living cells with the wood parenchyma.

A sharp boundary usually exists between the late wood of one year and the spring wood of the next year. This results in the formation of *annual rings,* which are easily recognized with the naked eye, so that the age of a tree can be determined simply by counting the annual rings in a cross section of the trunk. Many tropical trees, however, which grow continuously and uniformly throughout the year, do not possess annual rings.

In most trees, the youngest, peripheral annual rings form the soft *sap wood,* whose living cells serve for conduction and storage. The older, central annual rings constitute the *heart wood* or *duramen,* which has lost the ability to transport water and serves only for support. Heart wood is often commercially more valuable than sap wood, because in many tree species it is protected from decay by the presence of various chemical compounds, in particular tannins and their oxidation products, which are released from the dying wood parenchyma. Dyestuffs may also be present, e.g. the hematoxylin of *Haematoxylon campechianum* (log wood). In the absence of these substances, the center of the tree is more likely to be attacked by fungi, so that the trunk becomes hollow, e.g. poplar and willow. The darker the heart wood the more durable and commercially valuable it usually is, e.g. teak and ebony. Heart wood may undergo secondary changes, such as the occlusion of vessels by *tyloses,* which are vesicular distensions of the pit membranes.

Wood ants: see Ants.

Wood-boring isopods: see *Isopoda.*

Woodcock: see *Scolopacidae.*

Woodcreepers: see *Dendrocolaptidae.*

Woodfordia: see *Zosteropidae.*

Wood frogs: see *Raninae.*

Wood hedgehog: see *Basidiomycetes* (Fig.4.2).

Woodland: a vegetation community in which mature trees predominate, but are usually more widely spaced and have a more spreading form than in a Forest (see). A closed canopy is not formed. Various communities exist between the trees, e.g. grass, heath or scrub.

Woodland spring plants, ***early woodland plants:*** more or less light-loving ground plants of beech and oak woods, which develop early in the year, and which usually flower before the trees come into leaf.

The rapid development of these plants is made possible by food reserves in their subterranean shoot organs, and by the warming of the upper leaf litter layers before the spring. Many such plants have already produced seeds when the tree foliage begins to unfold, and their seeds are often dispersed by ants (e.g. *Primula, Viola, Asarum,* etc.).

Wood lark: see *Alaudidae.*

Wood lemming: see *Arvicolinae.*

Wood lice: see *Isopoda.*

Woodpecker finch: see Galapagos finches, Adaptive radiation.

Woodpeckers: see *Picidae.*

Wood pigeon: see *Columbiformes.*

Woodruff: see *Rubiaceae.*

Wood sorrel family: see *Oxalidaceae.*

Wood termites: see *Isoptera.*

Wood thrush: see *Turdinae.*

Wood warblers: see *Parulidae, Sylviinae.*

Woody nightshade: see *Solanaceae.*

Woolly aphid: see *Homoptera.*

Woolly monkeys, *Lagothrix:* a genus of the subfamily, *Atelinae* (see Cebid monkeys), inhabiting tropical and montane forests of the Amazon basin. The diet is mainly fruit, with some seeds and leaves. They are round-headed, large, robust animals, with a dense woolly coat and a powerful prehensile tail. The underside of the lower third of the tail is naked and divided into dermatoglyphics, which facilitate a strong grip on branches. There are 2 species: *L. lagothricha* ***(Common woolly monkey*** or ***Humboldt's woolly monkey;*** black, gray or brown) and *L. flavicauda* ***(Yellow-tailed woolly monkey*** or ***Hendee's woolly monkey;*** mahogany red with a yellow stripe outlining the bare skin of the prehensile tail, and a yellow scrotal tuft).

Woolly spider monkey, *Brachyteles arachnoides:* a single species of the subfamily, *Atelinae* (see Cebid monkeys). It has the slender, lanky build and prehensile tail of the Spider monkeys (see), and thumbs are also absent, but the coat is more woolly and it is larger than the latter. The diet is largely vegetarian, consisting of leaves, fruits, seeds and nectar. The species is severely threatened with extinction, due to destruction of its habitat, the eastern coastal rainforest of Brazil.

Woolly rhinoceros: see *Rhinocerotidae.*

Wool sponge: see Bath sponges.

Work of heart: see Cardiac work.

Worldwide Fund for Nature, ***WWF:*** an organization formerly known as the World Wildlife Fund (the name was changed in 1988), founded in 1961 in Switzerland (Gland) for the protection of animals threatened with extinction. As an international organization, it is represented in many countries, and it works closely with the IUCN (see). Together with the IUCN, the WWF releases funds for projects that exceed the means of countries or national nature conservation organizations. The WWF has been particularly active in purchasing threatened habitats. The symbol of the WWF is the giant panda.

Wormian bones, ***sutural bones, epactal bones:*** isolated bones of irregular shape and variable size, which are occasionally encountered along the line of the cranial sutures and in the region of the fontanelles. They are named after the Danish anatomist, Wormius.

Worm lizards: see *Amphisbaenia, Anguidae.*

Worm mollusks: see *Aplacophora.*

Wormwood: see *Compositae.*

Wound healing: in plants, the sealing of regions exposed by damage. In cases of slight damage, the affected cells die and the underlying undamaged cells become suberized to form a cork layer. After extensive wounding, cells bordering the affected area proliferate to produce a callus. In most cases, cork cambium develops on the periphery of the callus tissue, and generates an external layer of cork. After the completion of healing, Restitution (see) may occur.

Wound hormones: substances, whose existence was first postulated in 1914 by Haberland, which were thought to be released from damaged plant cells, and to cause cell division and callus formation in adjacent cells. A substance called traumatic acid (2-dodecenoic acid) was isolated from bean pods, which caused callus formation in young bean pods, but the occurrence and action of this substance is apparently restricted to beans. It is now thought that cytokinins released from damaged cells stimulate neighboring cells to divide.

Wrack: see *Phaeophyceae.*

Wrasses, *Labridae:* a family of fishes (about 500 species in about 57 genera) in the order *Perciformes.* W. are found mainly in the warm seas of the Indian and Pacific Oceans. They are often very colorful, and they display brood care. The mouth is protractile with characteristically thick lips, and the jaw teeth usually project outward. Many species build nests, and most species bury themselves in the sand at night. Some smaller species clean ectoparasites from larger fishes. Nontropical species include *Ctenolabrus rupestris* ***(Gold-sinny)*** and *Crenilabrus melops* ***(Black-eyed wrasse),*** both of the western Baltic.

Wren babblers: see *Timaliinae.*

Wrens: see *Troglodytidae.*

Wren tit: see *Timaliinae.*

Wren-warblers: see *Sylviinae.*

Wright distribution, ***Sewall-Wright distribution:*** the expected frequency of alleles and genotypes in an inbred population of diploid organisms, with different relative frequencies of genes at one locus. If two alleles, A and a, are present at relative frequencies p and q (where $p + q = 1$), then the frequencies of the three genotypes are given by: $p^2(1 - f) + pf$: $2pq(1 - f)$: $q^2(1 - f) + qf$, where f is the inbreeding coefficient. If $f = 0$, the W.d. reduces to the Hardy-Weinberg law. When $f = 1$, the population consists entirely of homozygous individuals. W.d. is a modification of the Hardy-Weinberg law (see).

Wright effect: see Sewall Wright effect.

Wright's inbreeding coefficient: the proportion of the genes of 2 individuals that are identical by descent. It is synonymous with Panmictic index and the Coefficient of relationship. See Inbreeding coefficient.

Wrinkled bark beetles: see *Coleoptera.*

Wrybill plover: see *Charadriidae.*

Wryneck frogs: see *Microhylidae.*

Wrynecks: see *Picidae.*

Wuchereria: see Filariasis, *Nematoda.*

Würm, ***Weichselian, Devensian:*** the last glacial advance in northern Europe, which occurred between 70 000 and 10 000 years ago. The corresponding period in North America is known as the Wisconsin glaciation.

WWF: see Worldwide Fund for Nature.

X

Xanthine, ***2,6-dihydroxypurine:*** a purine discovered in 1817 in renal stones. It crystallizes as colorless platelets, containing one molecule of water of crystallization. It also occurs naturally in e.g. blood, urine, liver, various fruit and vegetable juices and coffee beans. In higher animals it arises by the deamination of guanine and the oxidation of hypoxanthine, and in turn is further oxidized to uric acid by the action of Xanthine oxidase (see). It is therefore an important intermediate in metabolic purine degradation. The purine alkaloids, caffeine, theobromine and theophylline are methylated derivatives of X.

Xanthine oxidase, ***xanthine dehydrogenase, Schardinger enzyme:*** an iron- and molybdenum-containing oxidase, which contains flavin-adenine-dinucleotide as a prosthetic group. It is responsible for the metabolic oxidation of hypoxanthine to xanthine and the further oxidation of xanthine to uric acid, thereby playing an important part in aerobic purine degradation. Its substrate specificity is not very great, and it also catalyses the oxidation of certain aliphatic, aromatic and N-heterocyclic compounds. It is widely distributed and especially abundant in liver and milk.

Xanthocillin, ***1,4-di-(4-hydroxyphenyl)-2,3-diisonitrilobutadiene(1,3):*** a bacteriostatic antibiotic from *Penicillium notatum,* used against local infections caused by Gram-positive and Gram-negative organisms.

Xanthomonas: a genus of rod-shaped (0.2–0.8 μm x 0.6–2.0 μm), motile (single polar flagellum), Gram-negative bacteria, which are usually yellow when grown on agar media. They are strict aerobes, which never reduce nitrates, and never exhibit fermentative metabolism. Species of *X.* are plant pathogens, usually causing the death of the infected tissue. Thus, *X. campestris* causes vascular and parenchymatous disease in *Brassica* species, *X. fragariae* is responsible for leaf spot disease of strawberries, *X. albilineans* causes leaf scald of sugar cane, and *X. ampelina* causes "Tsilik Marasi", a canker disease of grapevines.

Xanthophyceae, ***yellow-green algae:*** a class of algae formerly known as the *Heterocontae,* because their swarming cells and gametes possess two unequally long flagella.

Different levels of thallus organization are displayed by different species, i.e. unicellular, filamentous and tubular. Growth and development usually occur by mitotic cell division. *X.* contain chlorophyll *a*, carotene, and various xanthophylls which are unique to this class of algae. Storage materials are oil and the polysaccharide, chrysolaminarin.

X. occur predominantly on mud and in fresh water, but marine forms are also known.

They are classified according to the level of organization of the thallus. The filamentous members of the genus *Tribonema* are relatively common in fresh water; the walls of each pair of neighboring cells have the appearance of a letter H. Tubular forms are represented by the widely distributed *Vaucheria* species, which consist of multinucleate tubes (not partitioned into cells), and by *Botrydium* species, which possess vesicle-shaped tubes. Both genera are attached to their substrate by rhizoids.

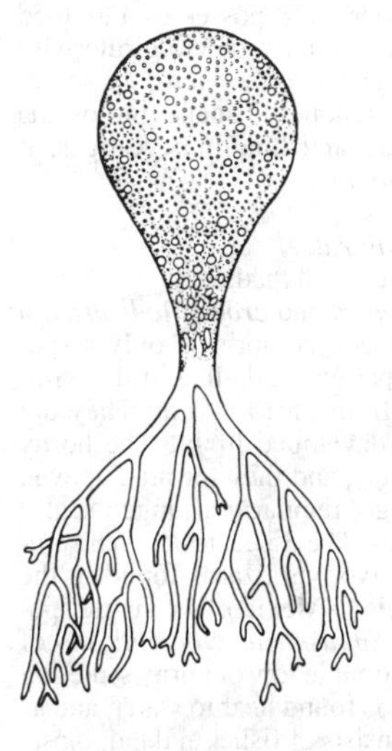

Xanthophyceae. Entire plant of *Botrydium granulatum* (x 30).

Xanthophylls: a group of oxygen-containing Carotenoids (see). Lutein (see) is also known as xanthophyll.

Xanthosoma: see *Araceae.*

Xanthusia: see Night lizards.

Xanthusiidae: see Night lizards.

Xenicidae, *Acanthisittidae,* ***New Zealand wrens:*** an avian family of 3 insectivorous species (order *Passeriformes)* found only on the islands of New Zealand. The ***Rifleman*** *(Acanthisitta chloris* Sparrman 1787) is found in forests, on farmlands and scrubland on the North and South Islands, as well as on Stewart, Great Barrier and Little Barrier Islands. The forest-dwelling ***Bush wren*** *(Xenicus longipes* Gmelin 1789) is rare on the South Island, while on the North Island it is restricted to the Lake Waikaremoana area; it is also found on Stewart Island. The ***Rock wren*** or ***South Island wren*** *(Xenicus gilviventris* Pelzeln 1867) is restricted to the South Island, where it is found in scrub and rocky areas above the timberline on high mountains. A fourth flightless species, the ***Stephen Island wren*** or ***Stephens wren*** *(Xenicus lyalli* Rothschild 1894) was discovered in 1894 and soon afterwards exterminated by a cat.

Xenicus: see *Xenicidae.*

Xenies (sing. **Xeny):** hybrid seeds formed in plants by fertilization by a different species or genus. The hybrid nature of the seeds is expressed in their color, shape and/or size, as well as the type and quantity of materials they contain, giving rise to color-, shape- , size- and chemo-X. Formation of X. is known in maize, rye, millet, sweet peas, garden peas, beans and lupins. Hybrid characters are expressed in the seed, because while one nucleus of the pollen tube fuses with the ovum to form the zygote, the other fuses with the secondary embryo sac nucleus to form the endosperm nucleus.

Xenogamy: see Pollination.

Xenogenous: see Allochthonous.

Xenopeltidae, *sunbeam snakes:* a family of South and Southeast Asian snakes, which is related to the *Boidae* (see). Up to 1 m in length, and round in cross section, they have glossy brown scales with a rainbow-like iridescence. The only definite member of the family is the monotypic *Xenopeltis unicolor,* which is found across India and Southeast Asia, and which feeds on small snakes, frogs and rodents. It possesses a toothed maxillary bone, and a rudimentary pelvic girdle with hindlimb vestiges. The genus, *Loxocemus,* is sometimes also classified as a xenopeltid snake, but on the basis of its skull structure and hindlimb vestiges, it probably belongs in the *Boidae.*

Xenophidia: see Snakes.

Xenopneusta: see *Echiurida.*

Xenopus: see African clawed toad.

Xenosauridae, ***xenosaurs*** and ***crocodile lizards:*** a family of the *Lacertilia* (see), comprising only 4 species, all of them vivi-oviparous, and all skilfull swimmers, varying in length from 25 to 40 cm. They are sturdy lizards with well developed limbs. The horny scales have a bony underlay, and they form an uneven surface, consisting of large, regularly arranged scales interrupted by small scales. The 3 species of *Xenosaurus* are crepescular insectivorous lizards found in the rainforests from Guatemala to Mexico. The single species of ***Crocodile lizard (Shinisaurus crocodilurus)*** of southwestern China has a double row of horny scales on the tail like a crocodile. It is found near to water, and as far as is known its diet consists of fishes and tadpoles.

Xenosaurus: see *Xenosauridae.*

Xenusion: a bilaterally symmetrical, segmented wormlike animal with annulated uniramous appendages, found in the Swedish Xenusion sandstone and in glacial quarzite of Central Europe. Of Precambrian or Early Cambrian origin, its systematic relationships are uncertain It appears to lie between the annelids and the arthropods, and may represent an ancestral form of the *Onychophora.* Practically complete specimens are in the Natural History Museum in Berlin and the Geiseltal Museum in Halle (Saal, Germany).

Xerocomus: see *Basidiomycetes (Agaricales).*

Xerophytes: plants of dry habitats, which survive conditions of prolonged drought without becoming dehydrated. X. possess morphological adaptations which permanently or temporarily prevent water loss by transpiration (see Xeromorphism), and they also often possess very long roots. X. grow in extremely dry areas, especially deserts and steppes, on dry rocks and as epiphytes; but also in cold winter regions as evergreen woody plants which show protective adaptations against frost desiccation.

There are two main types: hard-leaved or leafless *sclerophyllic X.* and soft-leaved, mostly hairy *malacophyllic X.* The small, hard evergreen leaves of the sclerophylls show a marked reduction of spongy mesophyll, only a few intercellular spaces, and a high proportion of palisade cells and sclerenchymous elements (sclereids). The outer epidermal walls, like the cuticle, are often thickened and covered with wax, resin or calcium carbonate. Furthermore, some X. possess a multilayered epidermis and a subepidermal layer of schlerenchyma (mechanical tissue). The leaves are often rolled or folded, or they are aligned vertically to the sun's rays. Stomata are often narrow and sunken, or arched over by a felt of hairs, thereby reducing the evaporation rate. On the other hand, the leaf surface may possess an especially high density of stomata, favoring photosynthesis.

The most effective protection against water loss by transpiration is reduction of the area of the transpiring surfaces; this is achieved by leaf fall, dwarfing the whole plant (nanism), limited branching, decrease in the total quantity of foliage, reduction of the stem and leaf blades and formation of stem and leaf thorns, and more rarely root thorns. Reduction of leaves is often coupled with a flattening and leaf-like appearance of the green stem (cladode), or only the leaf blades are reduced, and the blade-like flattened leaf stalks serve as phyllodes for photosynthesis. Many epiphytic orchids show leaf and stem reduction, and their flattened green aerial roots function as leaves.

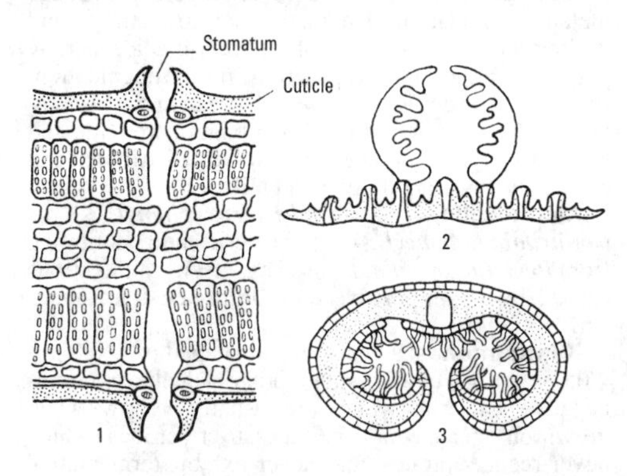

Xerophytes. Anatomical adaptations of xerophytes. *1* Leaf structure of the xerophytic plant, *Calistemon.* *2* Folding leaf of the grass, *Stipa capillata,* in the normal extended state, and rolled during the dry period (transverse section). *3* Rolled leaf of *Tylanthus,* a xerophyte from the Cape region (transverse section).

Sclerophylls (sclerophyllous plants) are predominantly hard-leaved plants of frost-free areas with warm summers (e.g. Mediterranean region, Capensis), including laurel, olive, myrtle, carob, *Erica* species, as well as numerous conifers like pines and firs of temperate and boreal latitudes.

Malacophylls (malacophyllous plants) possess soft leaves, mostly covered with a thick felt of hairs to minimize evaporation. In dry periods, the leaves wilt, and in a prolonged drought they are shed, so that only the youngest leaves remain as buds with a thick covering of hairs, e.g. *Lamiaceae, Scrophulariaceae, Boraginaceae, Compositae.* During a drought, at least a little water is necessary for the survival of these X.

X. also include *succulent plants,* which show a largely xeromorphic structure, but in wet periods they are able to take up and retain water in water storage tissues in leaves, shoots or roots, against subsequent and often prolonged periods of drought. During drought conditions, they take up no water, and their tap roots die off. According to their water storage organs, they are classifed as leaf succulents, stem succulents and root succulents. Leaf succulents have thickened leaves, which may be cylindrical, e.g. stonecrops, aloes, agaves, stoneflowers. The stem succulents are especially well adapted to the dry climates of deserts and semideserts by leaf reduction or by premature leaf shedding, and their columnar or clubbed shape presents the minimal surface area for transpiration. Convergent evolution of stem succulence is seen in cacti and species of spurge, *Stapelia, Kleinia* and *Cissus*. Root succulents possess subterranean thickened storage roots adapted for water storage. While less numerous than the leaf and stem succulents, they include the root crops (turnip, beet, etc.) and some representatives of the genera *Pelargonium* and *Oxalis,* which live in steppes and deserts. The solute concentration in the cell sap of these root succulents is very low, and it remains constant during water loss in long dry periods.

Certain halophytic plants (see Halophytes) are also succulent. They take up salt into their cell sap, and are unable to excrete it (chloride storage). The salt concentration of the cell sap therefore increases until the water potential of the plant exceeds that of the soil, so that during dry periods, succulent halophytic plants are still able to take up water and nutrients. Some perennial halophytic shrubs shed their leaves when the salt concentration of the leaf sap becomes too high.

Plants of dry regions, which complete their development during short wet periods (see Ephemerism) and survive for the remaining time as seeds (see Therophytes) or in the soil (see Geophytes) show no other special adaptation to water deficiency, apart from this seasonal rhythm (see Tropophytes). Ephemeral plants show close similarities to a group of poikilohydric plants, which retain their vegetative organs during drought, but survive complete desiccation without damage and revive in the presence of water, e.g. bacteria, certain lichens and mosses, as well as certain scaly or sclerophyllous ferns.

Xerostomia: deficient production of saliva.

Xerotherm grasslands: a general term for plant communities, predominantly grasses, of hot dry habitats. See Steppe.

Xiphias: see Swordfish.

Xiphiidae: see *Scombroidei,* Swordfish.

Xiphosura: an order or subclass of the *Merostomata.* The large, spade-shaped body consists of an anterior prosoma (cephalothorax) and a posterior opisthosoma (abdomen). The prosoma carries a dorsal, broad, smooth, shovel-like, horseshoe-shaped carapace, which is hinged to the opisthosoma. Six, dorsally fused segments are recognizable in the opisthosoma. In addition, a long caudal spine articulates with the posterior of the opisthosoma; this is usually regarded as a telson, but it does not contain the anal opening, and it possibly incorporates lost opisthosomal segments (similar fossil animals, e.g. *Hemiaspis,* occur in the Paleozoic, which possess 3 additional opisthosomal segments and a shorter caudal spine). The caudal spine is highly mobile; it is not used for defense, but rather for righting the body if it is overturned.

On the underside of the prosoma are six pairs of appendages, comprising a single pair of three-jointed chelicerae, four pairs of chelate walking legs, and a single pair of nonchelate walking legs. The first four pairs of chelate legs are all similar, and the median sides of their coxae are heavily armed with spines (this prominent spiny coxa is known as a *gnathobase).* The coxae of the fifth pair of prosomal legs carry a short spatulate process (known as a flabellum, and used for cleaning the gills), and the first tarsal segments carry four leaflike processes.

On the underside of the opisthosoma are seven pairs of appendages, starting with a pair of chilaria (small flattened processes, which are reduced to a single article armed with hairs and spines). which represent the pregenital segment. These are followed by the genital operculum (a membranous structure formed by median fusion of the paired appendages of the eighth or genital segment). The remaining five pairs of appendages are all flattened and membranous, fused along the midline, and modified as gills (generally known as book gills); the undersurface of each of these flattened structures is formed into about 150 leaflike folds called lamellae, which provide a surface for respiratory gas exchange.

There is a pair of small, simple, median eyes, and a pair of large, lateral, compound eyes. The median eyes are invaginated cups; the interior of the cup forms a lens, which is continuous with the exterior cornea, and surrounded by retinal cells. The compound eyes consist of 8–14 retinular cells grouped around a rhabdome, each unit possessing a lens and cornea. Although each unit is called an ommatidium, they do not form a compact array as in other arthropods, and the cornea and lens are also structually different.

Food (mollusks, worms, algae, etc.) is gathered by the chelate appendages, macerated by the gnathobases, then transported to the mouth, which opens just posterior to the chelicerae. The mouth opens into a sclerotized esophagus, which extends anteriorly to the crop, then into the gizzard. The alimentary canal then changes direction, passing posteriorly through a valve into the nonsclerotized stomach, followed by the intestine. Longitudinal cuticular folds of the gizzard are armed with denticles, and the entire structure is provided with a strong musculature. Food is ground in the gizzard, undigestible material is regurgitated, and useful material passes into the stomach. Ramifications of two large glandular hepatic ceca are found throughout the prosoma; these ceca empty into the stomach via two pairs of ducts. Malpighian tubules are absent. Paired gonads are present in the prosoma, and they communicate with the exterior via paired ducts opening on the genital operculum.

The blood contains hemocyanin, and a single type of amebocyte which contributes to clotting. A dorsal tubular heart (with 8 ostia) pumps blood through three large anterior arteries and four pairs of lateral arteries. Arterial branches finally empty into tissue sinuses, which in turn empty into two large ventral longitudinal sinuses. The latter provide blood to the five book gills on each side of the body.

Four pairs of brick-red coxal glands (two on each side of the gizzard) are responsible for excretion. Waste removed from the blood by each gland collects in a chamber, then passes to the exterior via a pore at the base of the last pair of walking legs.

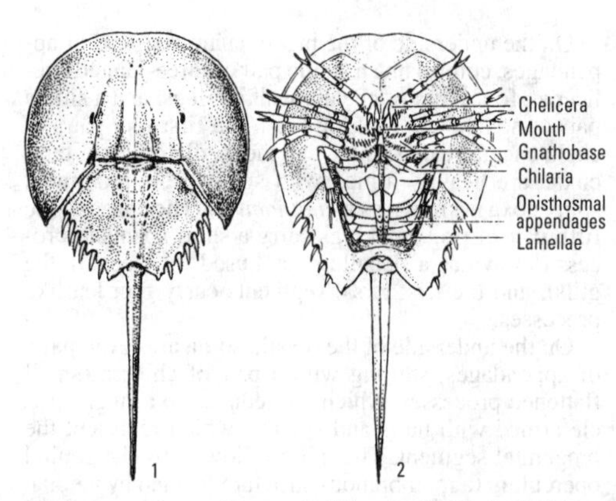

Xiphosura. Horseshoe crab *(Limulus). 1* Dorsal view. *2* Ventral view.

Only five living species (in 3 genera) are known. Most familiar is the ***Horseshoe crab*** *(Limulus polyphemus),* found on the northwestern Atlantic coast of America and the Gulf of Mexico. Other species are found on Asian coasts from Japan and Korea south through the East Indies and the Philippines. They inhabit shallow waters with soft sand or mud bottoms, where they burrow or plough through the upper surface. During mating, the male climbs onto the back of the female and may be carried in this way for weeks. The female then excavates a hole in the sand (about 15 cm deep) and deposits numerous, heavily yolked eggs, which are fertilized by sperm from the male; thus, despite the mating position of the male, copulation does not occur. The eggs hatch into a fully segmented, free-swimming, 6 mm-long juvenile stage (trilobite stage), possessing a very short caudal spine and only two pairs of gill legs on the opisthosoma. During the first year the young organism molts five to six times. In subsequent years there are only one or two molts, and sexual maturity is not attained until the tenth year. Maximal length is about 60 cm.

Collection and preservation. Specimens are most commonly encountered in the spring, and may be caught in the intertidal zone with baskets. They are killed with chloroform vapor, injected with 4% formaldehyde solution, then stored either in 2% formaldehyde solution, or as a dry preparation (especially fully grown specimens).

X-organ: two morphologically distinct structures on the optic lobe of decapod crustaceans. *Hanström's X-organ* is a vesicular structure at the base of the eye papilla; in some crustaceans it consists of two parts: a group of secretory cells (pars ganglioniaris X-organ) and another part (pars distalis X-organ), which has a storage function and has the structure of a neurohemal organ. In addition to Hanström's X-organ, there are certain groups of neurosecretory cells in the optic lobe, whose axons usually lead to the sinus gland, but also to Hanström's X-organ; these clustered neurosecretory cells are known as the *medulla-terminalis-X-organ,* the *medulla-externa-X-organ* and the *medulla-interna-X-organ.* In most decapods, Hanström's X-organ is separate, but in the crabs it forms a single unit by fusion with the medulla-terminalis-X-organ. The neurosecretory cells of the X-organ produce a molt-inhibiting hormone, which is stored in the sinus and released from it into the hemolymph during the premolt and intermolt periods. A decrease in the circulating concentration of this hormone stimulates the Y-organ to initiate molting by producing molting hormone. The X-organ probably also produces a hormone that promotes molting.

X-ray microprobe analysis: see Microprobe analysis.

Xylem: see Conducting tissue, Shoot.

Xylene, ***dimethylbenzene,*** $C_6H_4(CH_3)_2$**:** a flammable liquid with a characteristic smell, which is used as a solvent for paraffin wax, Canada balsam, etc. It is a mixture of the three isomers, *o-, m-* and *p*-dimethylbenzene, and it is immiscible with water.

Xylitol: an optically inactive C5-sugar alcohol. It is formed from the hemicellulose, xylan, during wood saccharification, and it can also be prepared by the catalytic hydrogenation of xylose. It is fully utilized by the human organism, and can therefore be used as a sugar substitute in diabetic diets. Its sweetness is similar to that of sucrose. Since it is not converted to organic acids by dental bacteria, it is also used in caries prophylaxis.

Xylophage: see Modes of nutrition.

Xyloplax: see *Concentricycloidea.*

D-Xylose, ***wood sugar:*** a monosacccharide pentose, which is not fermented by yeast. It is produced by acid hydrolysis of xylans, and it is used as a sugar substitute for diabetics.

Y

Yak: see *Bovinae.*

Yam: see *Convolvulaceae.*

Yard builders: see *Ptilinorhynchidae.*

Yarkand deer: see Red deer.

Yarmouthian interglacial: interglacial period in the middle of the Quaternary Ice Age in North America.

Yarrow: see *Compositae.*

Yaws: see Spirochetes.

Yearly cycles: see Annual cycles.

Yeast: see Baker's yeast, Brewer's yeast, Wine yeast, *Ascomycetes.*

Yeast protein: a protein- and vitamin-rich food material, produced by large-scale culture of certain yeast species or yeast-like fungi, e.g. *Torulopsis utilis* and *Candida tropicalis* (food yeasts, protein yeasts). The commercially available dried product contains 50–60% protein. Since growth of the microorganism requires only carbohydrate and mineral nutrients, excess carbohydrate is used in this way to close the protein gap. Molasses, whey, hydrolysates of cereal wastes, and even waste sulfite liquors from the cellulose industry are used as crude growth substrates, thereby also providing a means of economically processing industrial waste. A further cheap source of substrate is wood hydrolysate.

If the final product is intended for human consumption, selected raw materials are used in the growth medium, and the harvested protein is further purified.

Yellow ear-rot: see Eelworms.

Yellow enzymes: see Flavoproteins.

Yellow fir gall louse: see Galls.

Yellow-green algae: see *Xanthophyceae.*

Yellowhammer: see *Fringillidae.* See plate 6.

Yellow necked termite: see *Isoptera.*

Yellow slug: see Slugs.

Yellow smut: see *Basidiomycetes (Ustilaginales).*

Yellow sponge: see Bath sponges.

Yellow-striped armyworm: see Cutworms.

Yersinia pestis, *Pasteurella pestis:* the causative agent of Oriental plague. It is a nonmotile coccobacillus (about 1.5 x 0.7 μm) and a member of the Enterobacteria (see). It occurs in rats and other rodents, and is transmitted to humans by rat fleas *(Xenopsylla cheopis).* Oriental plague, also known as the black death, is one of the major historical pestilences of the world; for example, it killed 25% of the population of Europe in the 14th century. Severe cases are characterized by sudden onset, high fever, prostration and delirium. The disease may progress by rapid septicemia (high concentration of pathogens found in the spleen at post mortem), or it may be of the bubonic or pneumonic types, both of which may also become septicemic. In *bubonic plague,* the organism becomes localized in regional lymph glands, which, together with surrounding tissue, form a swollen mass known as a bubo. In *pneumonic plague,* the infection is concentrated in the lungs, with fulminating hemorrhagic bronchopneumonia.

Yew family: see *Taxaceae.*

Y-mechanism of DNA replication: see DNA replication.

Yoda's power law: a law that quantitatively describes the process of self-thinning among plant seedlings. Above a certain density of sowing, the number of surviving plants is not related to the initial sowing density, but a constant relationship is observed between the density of survivors and their total biomass. $W = Cr^{-3/2}$, where W is the dry mass of the surviving plants, r is the density of the surviving plants, and C is a constant which is specific for each species.

Yohimbine, ***quebrachine:*** an indole alkaloid, and the main alkaloid in the bark and leaves of *Corynanthe yohimbe* K. Schum. (= *Pausinystalia yohimbe* Pierre), family *Rubiaceae,* an indigenous tree of the Cameroons. It causes vasodilation, and is used for the treatment of arteriosclerosis, and as an aphrodisiac by the natives and in veterinary medicine. It is also the main alkaloid of *Aspidosperma quebrachoblanco* Schlecht (family *Apocyanaceae)* from Argentina.

Yoldia: a genus of the order *Protobranchiata* (class *Bivalvia)* with a posteriorly extended, truncated and somewhat gaping shell. The surface is only weakly sculptured. *Y. arctica,* which still survives in northern regions, is an index fossil for the strata of the Yoldia period, a postglacial development stage of the Baltic Sea (see *Gastropoda).*

Yoldia sea: see *Gastropoda.*

Yolk:

1) deutoplasm, vitellus, lecithus: the totality of reserve material in the egg cytoplasm, such as amino acids, proteins, glycogen, phospholipids (lecithins) and vitamins; it is stored in membrane-bound structures known as *Y. bodies* or *Y. platelets.* Y. may be synthesized autonomously by the egg cell (e.g. in many lower animals, such as echinoderms), where it accumulates in Golgi vesicles, in the endoplasmic reticulum, or even in mitochondria before being packaged in Y. bodies. In many species large amounts of material for Y. synthesis are taken up by the egg cell by pinocytosis; some of this material is derived from follicle cells or nurse cells that surround the growing oocyte. In some organisms, e.g. *Drosophila,* a common progenitor cell divides into the oocyte and nurse cells, leaving connecting bridges of cytoplasm which aid the direct transfer of material to the oocyte (see Oogenesis). A large proportion of material taken up by the oocyte for Y synthesis, however, is derived from the blood. Thus, in the eggs of amphibians and birds the Y. consists largely of proteins and lipids derived from a large precursor molecule, *vitellogenin,* secreted by the liver. Oocyte plasma membranes possess vitellogenin receptors that enable to oocyte to take up vitellogenin from the circulation by endocytosis. Endocytic vesicles then deliver the molecule to Y. bodies, where it is cleaved to the smaller storage molecules, lipovitellin and phosvitin. The quantity of Y. stored in reptile and bird eggs is relatively enormous compared with that stored in other eggs; it represents an extreme condition, in which the bulk of the egg cytoplasm containing the nucleus is restricted to a small germinal disk on the surface of the Y. mass. The term, Y., is also applied in a slightly different sense to this central yellow mass (see *2* below). In addition, the egg white or egg albumin represents a further food supply, which is secreted around the central Y. mass.

The quantity of Y. is related to the number of eggs laid and the mechanism or type of brood care. The more extended the brood care, the fewer the eggs and the less Y. they contain. Exceptions are the Y.-poor eggs of mammals, whose embryos are nourished via the placenta in the uterus of the female parent. See Cleavage, Egg.

2) The large, yellow, spheroidal mass of food material in the egg of a reptile or bird.

3) A greasy material present in sheep wool, also known as *suint.*

Yolk gland, ***vitellarium:*** a gland, usually paired, of the female reproductive system of the *Platyhelminthes* (flatworms). The Y.g. has evolved and separated from the ovary, and has developed its own duct *(vitelline duct).* The vitelline ducts empty into the oviduct, which then continues as the *ductus communis,* leading to the genital atrium. The Y.gs. supply each fertilized, nutrient-poor ovum with yolk cells (up to 60) and liquid shell substance. "Shell glands" at the junction of the oviducts and vitelline ducts produce a secretion which appears to promote hardening of the shell material. In the resulting "compound egg" the yolk cells provide nutrients for development of the embryo.

In other members of the animal kingdom yolk is more commonly formed in the ovary. The separation into ovarium and vitellarium is a peculiarity of the *Platyhelminthes.*

Yolk sac, ***saccus vitellinus, lecithoma:*** a spherical or pear-shaped organ, filled with yolk for the nutrition of the embryo. Yolk sacs are typically associated with the embryos that develop from the extremely telolecithal eggs of fish, reptiles, birds and monotremes, as well as the embryos of mammals, including humans, whose eggs are almost yolk-free. Cephalopods are the only invertebrates that possess a Y.s.

In fishes (Fig. 1), reptiles and birds, the Y.s. arises from the extraembryonic region of the gastrula. In this process of Y.s. formation, first the ectoderm, then the endoderm, and finally the parietal and visceral mesodermal layers begin to spread out on all sides over the yolk mass until they envelop it completely. Thus, the yolk mass, which acts as a barrier to embryonal development, is isolated in a four-layered vesicle or sac. Formation of the extraembryonic coelom separates this layer into the true yolk sac wall and the outer wall. The true yolk sac wall consists of endoderm plus visceral (splanchnic) mesoderm, i.e. it is extraembryonic splanchnopleure, and it is continuous with the intraembryonic gut. The outer wall consists of ectoderm plus parietal (somatic) mesoderm, i.e. it is extraembryonic somatopleure, and it is equivalent to the ventral wall of the embryo. As the embryo increases in size, the Y.s. connection with the mid gut region of the embryo becomes narrower, forming a constricted area of splanchnopleure, called the yolk stalk, which may be short and narrow (teleosts, reptiles, birds), or long and narrow (selachians). The lumen of the yolk stalk is known as the vitellointestinal duct, vitelline duct, omphalomesenteric duct, or umbilical duct. Direct utilization of the yolk by the embryo via the vitelline duct is of minor quantitative importance. Nutrients are mainly transported from the yolk by an extraembryonal system of blood vessels developed from the visceral mesoderm, which communicates with the intraembryonal blood system via two omphalomesenteric veins and two omphalomesenteric arteries. The extraembryonal system of blood vessels is especially dense in birds. As the yolk is utilized, the Y.s. gradually becomes smaller, and its walls are slowly degraded. It is finally absorbed completely at about the same time that the embryo attains its final outer form and has the appearance of a miniature adult. The embryo may hatch from the egg before the Y.s. is completely degraded. Such larval stages occur in the cyclostomes and in some fishes, e.g. selachians, sturgeons, lungfishes and some teleosts (e.g. salmon). In fishes, the Y.s. is entirely nutritional in function, but in reptiles and birds it is also responsible for the exchange of respiratory gases, until this function is taken over by the allantois. The allantois, together with the chorion (see Extraembryonic membranes), is pressed close against the egg membrane, and it takes up oxygen into its vessels by osmosis. Monotremes are the only mammals that possess both yolk and a functional Y.s. Marsupials have less yolk, and the placental mammals have none. In the early developmental stages of

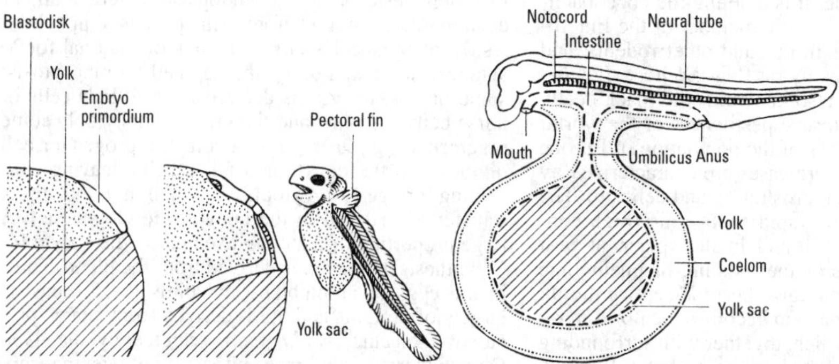

Fig.1. *Yolk sac.* Generation of the external body form and yolk sac of a fish.

mammals the Y.s. is nothing more than the blastocyst lined with endoderm (see Gastrulation). In humans only, the Y.s. cavity (part of the internal space of the blastocyst) is formed beneath the embryonal disk by the spread of mesodermal cells from the original plate of primary endoderm and/or by the organization of cells which become detached from the trophoblast mesoderm (Fig. 2, a to e). The resulting membrane is called Heuser's membrane or the extracoelomic membrane, and the total structure is called the primary Y.s. Endoderm cells then grow from the lower layer of the embryonal disk, and form the endodermal epithelium of the secondary Y.s., which displaces the primary structure; the proximal portion of the latter remains intact and continuous with the embryonic endoderm. The secondary Y.s. takes in fluid and increases in size, attaining a diameter of 5 to 10 mm. In the third week, blood vessels appear in the Y.s. mesoderm, which transfer nutrients from the Y.s. to the embryo (the quantity of available nutrients is, however, very small). The lumen of the Y.s. is soon compressed on all sides by the expanding amniotic cavity, until it atrophies to an insignificant umbilical vesicle. In the region of the allantoic stalk or connecting stalk (the mesodermal component of the allantois) the mesoderm around the umbilical vesicle unites with that around the allantoic vesicle (allantoic diverticulum). The resulting structural complex of amniotic tissue, enteric and allantoic diverticula, and splanchnopleure mesoderm is called the umbilical cord. As a rudimentary embryonal organ, the umbilical vesicle is evidence that evolutionary ancestors of the mammals possessed telolecithal eggs.

Y-organ, ***carapace gland:*** a paired gland of ectodermal origin, present in crustaceans, which lies either in the first maxillary or antennal segment. In decapod crustaceans, it consists of small masses of glandular tissue situated ventrally and laterally in the eye socket, directly above the junction of the branchiostegite with the cuticle of the lateral body wall at the anterior end of the branchial chamber. Functionally controlled by neurohormones, it is the site of synthesis of molting hormone (see 20-Hydroxyecdysone), and is probably involved in the control of other processes besides molting, e.g. calcium carbonate mobilization and deposition.

Yorkshire fog: see *Graminae.*

Yponomeutidae: a worldwide family of small moths, including a number of agricultural and forestry pests. Caterpillars feed on the surface or tunnel into shoots, and they often spin living quarters of a thick, gray-white network of silk, which may cover entire trees. Adult moths of many tropical species are brightly colored.

Members of the genus, *Yponomeuta,* are serious fruit tree pests. The caterpillars, in their gauzy network of silk, strip branches of buds and leaves. Adults of different species are often indistinguishable, with a wingspan of 20–25 mm, and white wings with small black spots

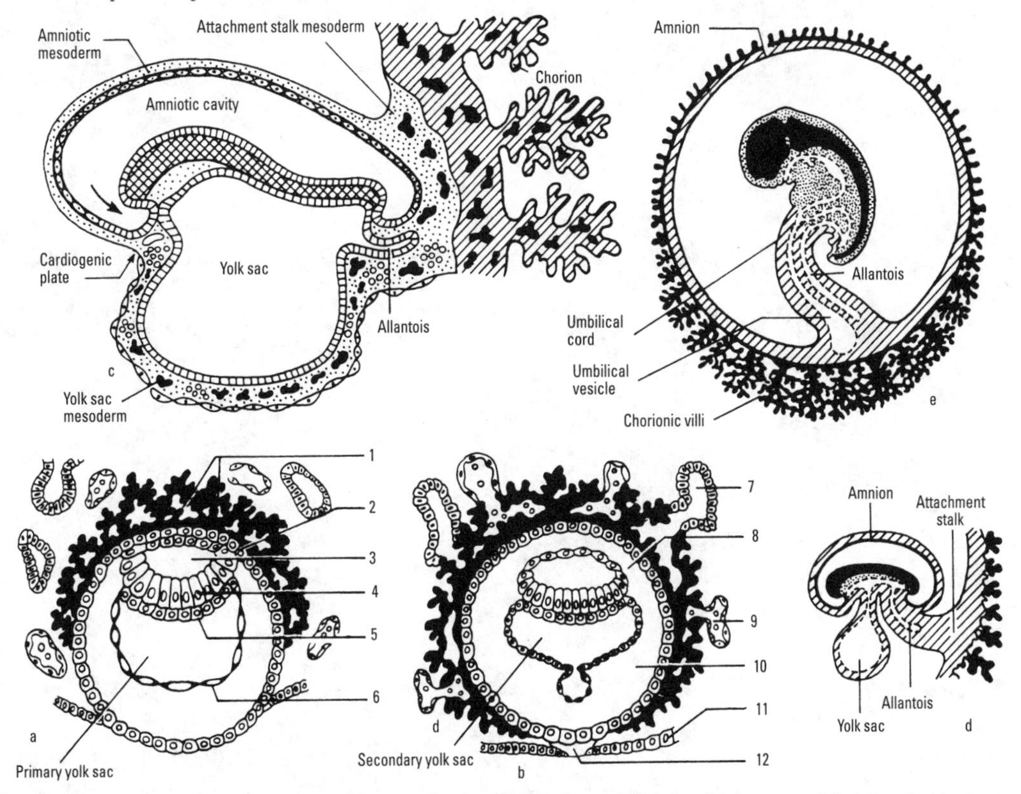

Fig.2. *Yolk sac. a-e* Formation of the human yolk sac. *1* Syncytiotrophoblast with lacunae. *2* Cytotrophoblast. *3* Amniotic cavity. *4* Ectoderm. *5* Endoderm. *6* Heuser's membrane. *7* Mucosal gland. *8* Extraembryonic mesoderm. *9* Maternal blood vessel. *10* Extraembryonic coelom. *11* Endometrial epithelium. *12* Coagulated fibrin.

(hence the name, ***Ermine moth).*** Caterpillars, however, are specific for their food plants, e.g. *Yponomeuta cagnagella* H. ***(Spindle ermine moth),*** a pest of *Euonymous japonicus* and other ornamental spindle trees in Europe, Asia Minor, the Middle East and Siberia; *Y. evonymella* L. ***(Bird cherry ermine),*** a pest of various cherries and plum in the Palearctic region from Europe to Japan; *Y. malinellus* Z. ***(Apple small ermine, Apple ermine),*** an apple pest throughout the Palearctic region where apples are grown; *Y. padella* L. ***(Small ermine, Orchard ermine, Cherry ermine),*** a pest of hawthorn, plum, cherry and blackthorn in Europe, Asia Minor and Siberia; *Y. rorrella* H. ***(Willow ermine),*** a pest of willow and poplar in Europe.

A notorious pest is *Plutella maculipennis* ***(Cabbage leaf miner, Cabbage web moth, Diamond-back moth),*** which is very destructive to members of the *Cruciferae,* and is now one of the most universally distributed *Lepidoptera.* It appears to be readily introduced by commercial routes, and its range continues to increase. *Argyresthia conjugella* Z. ***(Apple fruit moth)*** normally feeds on rowan berries, but is sometimes found on apples. *Argyresthia cornella* F., which is common on apples, has now been introduced into North America, its indigenous range being Europe to Siberia and Japan. *Argyresthia laevigatella* Herrich-Schaffer ***(Larch shoot moth, Larch shoot borer)*** is a larch pest in northern and central Europe. *Argyresthia pruniella* Clerck ***(Cherry fruit moth),*** a cherry pest of temperate Europe, has been introduced into North America. *Prays citri* Millière ***(Citrus flower moth)*** is a pest of lemon, orange, citron, etc. in many citrus growing areas of the world.

Yucca: see *Agavaceae.*

Z

Zalambdodont: a term applied to molars (see Dentition) that are V-shaped, narrow at the front and rear, with a sharp apex, and with two cusps partly or completely fused.

Zalophus: see *Otariidae.*

Zanclinae: see *Acanthuridae.*

Zanclus: see *Acanthuridae.*

Zantedeschia: see *Araceae.*

Zapata wren: see *Troglodytidae.*

Zaphrentis: see *Tetracorallia.*

Zapodidae, ***jumping mice*** and ***birchmice:*** a family of Rodents (see) containing 14 species in 4 genera, divided into 2 subfamilies. In general appearance, they are mouse-like, with a very long tail and relatively long hindlegs. The upper incisors are characteristically grooved. Dental formula: 1/1, 0/0, 1–0/0, 3/3 = 18 or 16. They are not very active burrowers, and generally live on the surface, hiding by day in clumps of vegetation. Nests are sometimes found in hollow logs, and the hibernation nest is often at the end of a burrow in a bank. Hibernation is long and deep, lasting 6–9 months according to the species. With their long hindlegs, all species are able to jump or bound, e.g. the Woodland jumping mouse can bound 1.5–3 m at a time. Most species, however, are more likely to be found crawling among vegetation or progressing by running with short hops.

1. Subfamily: *Zapodinae* ***(Jumping mice):*** 5 species in 3 genera; head plus body 7.6–11 cm, tail 15–16.5 cm. One species is found in China *(Eozapus setchuanus),* all others being native to North America, e.g. *Napaeozapus insignis* ***(Woodland jumping mouse)*** and *Zapus hudsonius* ***(Meadow jumping mouse).***

2. Subfamily: *Sicistinae* ***(Birchmice):*** 9 species in a single genus, all in Eurasia; head plus body 5–9 cm, tail 6.5–10 cm. Examples are: *Sicista betulina* (northern Eurasia, Central Europe), *S. caucasica* (western Caucasus and Armenia), *S. concolor* (China), *S. subtilis* (former USSR and eastern Europe).

Zapodinae: see *Zapodidae.*

Zapus: see *Zapodidae.*

Z-band: see Muscle tissue.

Z-disk: see Muscle tissue.

Zea: see *Graminae.*

Zeatin: see Cytokinins.

Zeaxanthin, ***3,3′-dihydroxy-β-carotene, (3R,3′R)-β,β-carotene-3,3′-diol:*** a xanthophyll isomeric with Lutein (see), and one of the commonest plant pigments (yellow-orange). It is responsible for the yellow color of maize seeds *(Zea mays),* as well as yellow bird feathers. It occurs free and esterified in many flowers and fruits, e.g. crocus, rose hips, paprika, oranges, peaches, as well as algae and some bacteria. *Physalin,* the waxy, deep red dipalmitate of Z., is present in very high concentration in Chinese lanterns *(Physalis alkekengi).*

Zebra mussel, *Dreissena polymorpha:* a mussel with an appearance similar to that of *Mytilus* of the Black Sea. Compared with other bivalves, *Dreissena* displays a high tolerance for different degrees of salinity, enabling it to invade both marine and freshwater habitats. In 1824 it was reported in the Thames in London and at the mouth of the Rhine, and it has now spread throughout all the Central European river systems. It attaches itself firmly by byssus threads to posts and stones, and it is transported over great distances attached to ships. The free-swimming veliger larva enters industrial outflow pipes and filtration systems, leading to the establishment of adult colonies that cause serious and costly blockages.

Zebras: black and white striped members of the family *Equidae,* found only in Africa in large or small herds. Various subspecies of ***Burchell's zebra*** or the ***Common zebra*** *(Equus burchelli)* are found in large herds on the

Zebras. Left to right: Grevy's zebra, Hartmann's zebra, Grant's or Boehm's zebra, Chapman's zebra.

African plains south of the Sahara, e.g. ***Grant's zebra***, ***Boehm's zebra*** or the ***Bontkwagga*** *(Equus burchelli boehmi)* and ***Chapman's zebra*** *(Equus burchelli antiquarum).* The nominate race of Burchell's zebra, which inhabited southern Africa, became extinct in 1910. ***Grevy's zebra*** *(Equus grevyi)* is the largest of the Z.; it has particularly narrow stripes and is found on the plains of Somaliland, Ethiopia and northern Kenya. The original zebra of southern Africa, the ***Mountain zebra*** *(Equus zebra),* is now scarce; it occurs as several subspecies, e.g. ***Hartmann's zebra***, which lives in the mountains of South West Africa. The now extinct ***Quagga*** *(Equus quagga)* of Southern Africa was also a zebra, with brownish, less distinct stripes; it became extinct in the wild before 1860, and the last survivor died in an Amsterdam zoo in 1883.

Zebu: see *Bovinae.*

Zechstein, *Thuringium:* the upper division of the Permian (see) of the German basin. See Geological time scale.

Zeitgeber: see Biorhythm.

Zeledonia: see *Turdinae.*

Zenaidura: see *Columbiformes.*

Zeocephus: see *Muscicapinae.*

Zeta potential: see Electrical double layer.

Zetorchestidae: see *Oribatei.*

Zeus: see John Dory.

Ziege, *Pelecus cultratus:* a fish of the family *Cyprinidae,* found in rivers flowing into the Baltic, Black, Caspian and Aral Seas, and in the brackish waters of these seas. The flat-sided, straight-backed body is shaped like a pocket knife. Up to 50 cm in length, it is silvery to bluish gray on the back with silvery white sides and belly. The lateral line is wavy, and the mouth is upturned and superior. The Z. has some commercial importance around the Sea of Azov, but the flesh is not particularly good to eat.

Ziehl-Neelsen stain: see *Actinomycetales.*

Zimocca sponge: see Bath sponges.

Zingel, ***streber,*** *Aspro zingel, Aspro streber, Zingel streber, Zingel zingel:* a long (up to 17.5 cm) slender, large-headed fish of the family *Percidae,* order *Perciformes,* found in the rivers Danube, Dnester, Vardar and Prut. Its yellowish brown body has 4 or 5 dark oblique stripes and is almost circular in cross section; the anterior dorsal fin is supported entirely by spines. The flesh is good to eat, but the fish is not caught on a commercial scale; it is occasionally used as a bait fish. Its number are declining.

Zingiberaceae, ***ginger family:*** a family of monocotyledonous plants, containing about 1 000 species in 49 genera. They are predominantly tropical species, occurring mainly in Southeast Asia, but also extending though tropical Africa to Central and South America. They are herbaceous perennials, and many species are aromatic. Possession of isodiametric oil cells with suberized walls is characteristic of the family. Most species possess sympodial, fleshy rhizomes, which are covered with distichous scale leaves; tuberous roots are often also present. Aerial shoots are formed mainly by the rolling of leaf sheaths. The hermaphrodite, zygomorphic flowers usually have conspicuous bracts, and they may be solitary or aggregated in racemose or cymose inflorescences. There is a calyx of 3 united sepals, and the corolla consists of 3 united petals. There are basically 2 whorls of 3 stamens. Only one stamen is fertile. The remaining inner stamens form varying numbers of sometimes petaloid staminodes, 2 of which are fused to form a 2- or 3-lobed labellum. The inferior ovary of 3 united carpels develops into a 3-valved, loculicidal capsule, which is sometimes fleshy, indehiscent and berry-like.

Zingiberaceae. Ginger *(Zingiber officinale).* Rootstock with vegetative and flowering shoots.

Since the Z. are rich is esential oils, the rhizomes of several species are used medicinally or as spices. ***Ginger*** is the rhizome of *Zingiber officinale,* native to tropical Asia, but grown commercially elsewhere, in particular in Jamaica. ***Turmeric*** *(Curcuma longa,* Indo-China) is used medicinally, as a coloring agent, and as a curry spice. ***Cardamom,*** used in the prepartion of cakes and pastries, and for flavoring liqueurs, is derived from the fruits of *Elettaria cardamomum* (India) or *Elettaria major* (Sri-Lanka).

Zinc, *Zn:* an essential bioelement for growth and development of plants, animals and microorganisms. It is taken up by plant roots as the cation, Zn^{2+}. In the soil, very little is present in solution, the major proportion being firmly bound in sorption complexes. The soil also contains poorly soluble Zn salts, whose ability to serve as a Zn source for plants decreases with increasing pH. Thus, liming of the soil decreases Zn availability. Zn is also less available in soils with a high phosphate content. High concentrations of Zn are present in the minerals, biotite, augite and hornblende. Plants generally contain between 10 and 100 mg Zn per kg dry mass. The Zn requirement of agricultural plants is variable. Maize, hops, flax, *Phaseolus* beans and fruit trees are particularly sensitive to Zn deficiency. Deficiency states are most often encountered in fruit trees, the symptoms being sparse foliage, with small, lanceolate leaves, which are often also chlorotic and arranged in rosettes. Growth of branches is inhibited. Zn deficiency in other plant species also leads to dwarf growth. Orchards are supplemented with Zn either as a leaf spray, or by addition of Zn chelates to the soil. An absolute Zn deficiency in the soil is rare; rather the Zn is tightly bound and therefore unavailable.

Zn has a high affinity for nitrogen and sulfur ligands, and occurs in the cell in association with many different compounds, e.g. proteins (insulin), amino acids, nucleic acids. It is a tightly bound component of zinc-metalloenzymes, and it stimulates in vitro the activity of zinc-metal-enzyme complexes. About 20 zinc-metalloenzymes are known, e.g. dehydrogenases, phosphatases, carboxypeptidases, carbonic anhydrase. Zn also activates the enzymatic synthesis of tryptophan.

Humans contain 2–4 g Zn, the majority being intracellular. Blood contains only 7–8 mm Zn/ml, of which 85% is in the erythrocytes, where it is required for the activity of carbonic anhydrase. Relatively high quantities of Zn are found in the pancreas.

Zinnia: see *Compositae.*

Ziphiidae: see Whales.

Ziphius: see Whales.

Zoantharia: see *Hexacorallia.*

Zoanthidea: see *Hexacorallia.*

Zoarium: see *Bryozoa.*

Zoea: pelagic larval stage of most decapod crustaceans. It hatches directly from the egg, and is characterized by eight pairs of trunk appendages free of the carapace.

Zoecium: see *Bryozoa.*

Zoidiogamy: in plants, fertilization of the egg cell by spermatozoids, e.g. in *Cycas* species.

Zokors: see *Cricetinae.*

Zonal vegetation: the predominant vegetation of an extended region or zone as determined by the prevailing climate of that zone, e.g. steppe, deciduous forest and pine forest zones. The term, *extrazonal vegetation,* is applied to plants that appear atypically in a zone and belong to another, usually a neighboring zone, e.g. many species of the Russian and Asiatic steppes are also found north and south of the Carpathians in hot, dry areas of Central Europe (e.g. Central Bohemia), which belong to the temperate deciduous forest zone. Such extrazonal occurrences are also called disjunctions, the above example representing a steppe plant disjunction. *Azonal* vegetation does not belong to a single vegetation zone, e.g. reed banks represent communities of water plants whose location is determined by edaphic factors, i.e. environmental conditions determined by the physical, chemical and biological properties of the soil.

Zona pellucida: see Oogenesis.

Zonation community, ***zonation biocoenois:*** see Biocoenosis.

Zonites: external delineations of the cuticle of *Kinorhyncha* (see), which have the appearance of segments.

Zonotrichias: see *Fringillidae.*

Zonula adherens: see Epithelium.

Zonula occludentes: see Epithelium.

Zoobenthos: animal organisms of the Benthos (see).

Zoocecidium: see Galls.

Zoochlorellae: green algae living symbiotically in animal cells.

Zoochory: see Seed dispersal.

Zooflagellata: see *Mastigophorea.*

Zoogamy: see Pollination.

Zoogeographical regions: see Animal geographical regions.

Zoogl(o)ea: a bacterial colony embedded in a gelatinous mass, usually observed as a thin membrane or scum on the surface of water. The gelatinous material is secreted by the bacterial cells. See Biological waste water treament.

Zooid: see *Bryozoa.*

Zoology: a branch of biology; the science of the structure of animals and all phenomena of animal life. Z. is accordingly divided into many areas of scientific interest. *Morphology* (including *cytology), histology, organography* and *anatomy* are concerned with the study of internal and external structure. *Physiology* is the study of the functions and mechanisms of the animal body and its parts. *Ecology* is concerned with the variety of interactions of animals with their abiotic and biotic environment. *Taxonomy* or *systematic* Z. classifies the animal species of the world into naturally related groups, with the aim of constructing a system based on natural (evolutionary) affinities. Taxonomy draws on information from most of the other branches of Z. Further specialized areas of Z. are embryology, animal geography (animal distribution), genetics, psychological basis of behavior (animal psychology, ethology), phylogenetics (evolution) and others. Areas of interest are also classified according to the type of animals studied, e.g. mammalogy (mammals), ornithology (birds), entomology (insects), malacozoology (mollusks), herpetology (reptiles and amphibians).

Zoomastigina: see *Mastigophorea.*

Zoomastigophorea: see *Mastigophorea.*

Zoomimesis: see Protective adaptations.

Zoonosis: a disease that can be transmitted between humans and other vertebrates under natural conditions. The other vertebrates consitute a reservoir of the disease, e.g. rabies, toxoplasmosis, trichinosis, psittacosis.

Zooparasite: see Parasite.

Zoophage: an organism that feeds by Zoophagy (see).

Zoophagus: see Carnivorous plants.

Zoophagy:

1) A mode of nutrition by certain plants, which capture animals, digest their body materials and absorb the products. See Carnivorous plants.

2) See Modes of nutrition

Zoophily: see Pollination.

Zooplankton: see Plankton.

Zoosemiotics, ***animal semiotics:*** study of the use of signs and signals by animals for the purpose of communication and information transfer. See Biocommunication.

Zoosociology: see Sociology of animals.

Zoosporangium: a sporangium of lower fungi, within which zoospores are formed.

Zoospores: flagellated, actively motile spores, produced mainly by aquatic fungi and algae. In terrestrial fungi that produce Z., e.g. *Phytophthera infestans* and other phytopathogenic fungi, Z. dispersal depends on rainwater and dew. The associated fungal infections therefore occur with greater frequency and severity in wet weather.

Zoosterols: sterols biosynthesized by animals. Important examples are: Cholestanol (see), Cholesterol (see), Coprosterol (see), 7–Dehydrocholesterol (see), Lanosterol (see) and Poriferstérol (see).

Zoothera: see *Turdinae.*

Zoraptera: an insect order first described in 1913 by Silvestri. The 22 known species belong to a single genus *(Zorotypus),* which is represented in all geographical regions except Australia. Adults are barely 3 mm in length, and resemble small termites; they also form small colonies beneath tree bark or in soil cavities. A species may have both winged and wingless individuals, which may be of either sex. Fore- and hindwings are dissimilar, membranous and only weakly veined. Wings are often shed at the base, but there is no basal suture as in termites. Winged individuals are darkly pigmented with both compound eyes and ocelli, and the thorax is covered by a hard shield. Wingless individuals are only very weakly pigmented and eyeless, and their thoracic cuticle is soft. Two different types of larvae are also sometimes observed. There appears to be no division of labor between winged and wingless individuals. Metamorphosis is complete (paurometabolous development). The Z. are systematically related to the *Psocoptera.*

Zosterophyllacea: see *Psilophytatae.*

Zosterophyllum: see *Psilophytatae.*

Zosteropidae, ***white eyes:*** an avian family in the order *Passeriformes* (see), suborder *Oscines.* The 85 species are distributed from Africa to Micronesia and Polynesia, including New Zealand, from sea level to the timberline in mountainous areas. Colors are fairly uniform, e.g. yellow-green, gray, brown or white. It is a very homogeneous Old World family of small, insect and fruit-eating birds, with slightly decurved, sharply pointed bills, brush tongues, and rather rounded wings with only 9 primaries. Nearly all species have a white ring around each eye, consisting of minute silky white feathers; in some species, however, it is so narrow that it is hardly visible, whereas in some African species it forms a large patch. Some authorities place all 85 species in the genus *Zosterops,* while others place 23 in 11 other genera. Largest of the alternative genera is *Lophozosterops,* comprising 6 Indonesian species with peculiar patterning on the head. Two large species of *Woodfordia,* which have long bills and no eye ring, are found respectively on the island of Santa Cruz and on the southeastern Solomon Islands. Also in isolated habitats are the 4 species of *Speirops,* found respectively on the islands of São Tomé, Principe, and Fernando Po, and on the top of Cameroon mountain, all large species and 3 lacking the white eye ring.

Zosterops: see *Zosteropidae.*

Zygapophysis: see Axial skeleton.

Zygocactus: see *Cactaceae.*

Zygolophodon: see *Mastodon.*

Zygentoma, ***silverfish, firebrats:*** an order of primarily wingless, primitive insects (see *Apterygota),* previously placed with the bristletails in the obsolete taxon, *Thysanura.* The ectognathous, chewing mouthparts are easily visible, and the mandibles have dicondylic articulation; the Z. therefore appear to be closely related to the pterygote insects. The dorsoventrally flattened body (5 to 25 mm in length) has a thin, delicate integument, which is often scaly, light or silvery in color, rarely darkly pigmented. Compound eyes are small or absent and ocelli are always lacking. The head carries long, filiform, multisegmented antennae. Of the 11 abdominal segments, numbers 7 to 9 usually carry a ventral pair of styli, but in the *Nicoletiidae* these are found on segments 2 to 9. Three long caudal appendages are present (paired cerci and a terminal filament), and they are usually equal in length (Fig.). A tracheal system, with longitudinal and transverse branches, is served by 2 pairs of thoracic and 8 pairs of abdominal spiracles. The alimentary canal has a crop and a proventriculus, and 4 to 8 Malpighian tubules at the junction of the midgut and hindgut.

Development is epimetabolous, i.e. Z. continue to molt several times after attaining sexual maturity. Sperm transfer is indirect and accompanied by a ritualized courtship. The 280 species (in 48 genera) are distributed worldwide, mainly in the tropics and subtropics. Members of the family, *Nicoletiidae,* have no eyes, and some have no scales. In central Europe, this family is represented by a single species, *Atelura formicaria,* which is a commensal in ants' nests. Members of the family, *Lepismatidae,* have eyes, and are the only Z. commonly encountered. In Europe they are found in human dwellings, e.g. the ***Silverfish*** or ***Sugar thief*** *(Lepisma saccharina),* which thrives best at 25 °C; and the ***Oven silverfish*** or ***Firebrat*** *(Thermobia domestica)* with a preferred temperature of 38 °C, which has been introduced into the USA, where it is a pest in bakeries. All Z. feed mainly on plant refuse, together with some animal residues. Firebrats and silverfish, which are pests of flour products, as well as paper and textiles, are relatively easy to control with the usual insecticides.

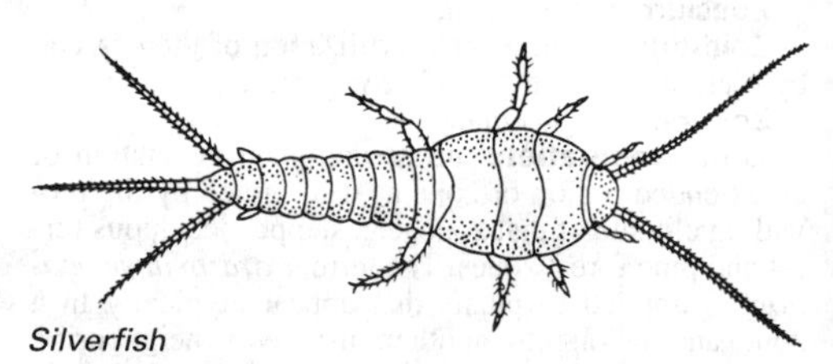
Silverfish

Zygomycetes: see *Phycomycetes.*

Zygoptera: see *Odonatoptera.*

Zygosity analysis: the determination of whether twins have arisen from the same egg (monozygotic), or from two different eggs (dizygotic). In humans, investigation of the embryo membranes and placenta do not always provide a clear result, since monozygotic twins can also have separate membranes and placentas, and dizygotic twins may share the same placenta. Since monozygotic twins are genetically the same, they must also be identical with respect to characters that are strictly inherited. Z.a. is performed mainly on blood cells and serum, using characters whose inheritance is well understood, e.g. ABO blood groups, Rh-factors, HLA-factors, enzyme patterns. A decision can then usually be made on the basis of gene contributions from both parents. In doubtful cases, an anthropological comparison of characters, i.e. a *polysymptomatic similarity analysis* can be performed. Alternatively, the entire analysis can be based on the modern technique of DNA fingerprinting. See Forensic anthropology.

Zygospores: large, thick-walled resting spores which arise by gametangiogamy, and which are produced chiefly by molds. In heterothallic fungi, when mature hyphae of different mating types grow very close to each other or actually meet, club-shaped *copulation branches* develop. By chemotropism, the ends of these copulation branches come together, and each develops a terminal multinucleate gametangium, separated from the rest of the hypha by a transverse wall. These morphologically identical, but physiologically different gametangia fuse, and the fusion product develops into a Z.

Zygote, ***fertilized egg cell:*** diploid eukaryotic cell, produced by the cellular and nuclear fusion of a female haploid gamete and a male haploid gamete. In diplonts, the Z. develops directly into a diploid organism. In haplonts, the Z. undergoes Meiosis (see) to produce four haploid cells, each of which develops into a haploid organism.

Zygotene: see Meiosis.

Zygotony: see Reproduction.

Zymase: a term coined at the end of the 18th century for a mixture of enzymes extracted from yeast, which catalyses the alcoholic fermentation of glucose. Z. activity was found in 1897 by E. and H. Buchner in "Presssaft", a cell-free liquid obtained by submitting yeast to mechanical pressure. At the time, Z. was considered to be a single enzyme. Its discovery marks the beginning of enzymology, and it disproved the theory that only intact cells are capable of alcoholic fermentation.

Zymogens, ***proenzymes:*** catalytically inactive precursors of enzymes, usually proteolytic enzymes, which are converted into a catalytically active form by limited proteolysis. Familiar examples are the Z. of digestive enzymes (pepsinogen, trypsinogen, chymotrypsinogens A, B and C, proelastase, and procarboxypeptidases A and B), which are converted into their active forms in the stomach or duodenum. The blood coagulation cascade operates by the sequential proteolytic activation of proenzymes (e.g. prothrombin and plasminogen). Z. can be stored at their sites of synthesis without causing self digestion of cells or tissues. An example is the synthesis, storage and secretion of the zymogen, trypsinogen, by the secretory cells of the pancreas. Conversion of trypsinogen into proteolytically active trypsin is initiated in the small intestine by enterokinase, and accelerated autocatalytically by trypsin itself. (Certain proteohormones are also produced as inactive precursors, e.g. proinsulin.)

Zymosterol, ***5-α-cholesta-8(9),24-dien-3β-ol:*** an intermediate in the biosynthesis of cholesterol from lanosterol in animals and of ergosterol from lanosterol in yeast. It is a mycosterol, which occurs in relatively large quantities in yeast, from which it can be isolated.

Plate Section

Index to the plates

1 ***Insects.*** ***1*** Hover fly *(Syrphus contuscus)* on a chicory flower. ***2*** Brimstone butterfly *(Gonepteryx rhamni)* sucking nectar from a bean flower. ***3*** Shield bug *(Picromerus bidens)* sucking the body contents of a cabbage white caterpillar. ***4*** A robber fly *(Sturmia sentellata)* has laid eggs of the body surface of a gypsy moth caterpillar. ***5*** A ground beetle *(Carabus auratus)* with a captured cockchafer. ***6*** Larve of a green lacewing *(Chrysopa)* feeding on an aphid. ***7*** Ichneumon fly *(Diaeretus rapae)* injecting its eggs into an aphid (this fly parasitizes in particular *Brevicoryne brassicae).* ***8*** Ichneumon fly *(Pimpla instigator)* pierces a pupa of the moth, *Lymantria monacha* (black arches) with its short ovipositor, and lays a single egg.

Plate 2

2 ***Protective adaptations of insects.*** ***1*** Caterpillar of the dagger moth *(Acronicta aceris)* with long dense hairs. ***2*** Ermine moth caterpillars *(Yponomeuta* sp.) in a woven silk nest. ***3*** Pupation cocoon of a stream-dwelling caddis fly, protected and camouflaged by interwoven foreign bodies (small stones). ***4*** Defensive movements of a sawfly larva *(Croesus septentrionalis).* ***5*** The larva of a green lacewing camouflages itself with food remains and drained teguments of its aphid victims. ***6*** Camouflage by color and shape: the locust, *Oedipoda caerulescens.* ***7*** Color camouflage of a sawfly larva *(Platycampus luridiventris)* on alder. ***8*** Mimicry by the hooded owlet moth caterpillar *(Cucullia artemisiae).* ***9*** Mimicry by the peppered moth caterpillar *(Biston betularia).*

3 ***Brood care in the animal kingdom.*** ***1*** Proboscid leech *(Helobdella stagnalis;* up to 12 mm; cosmopolitan) with eggs on ventral surface. ***2*** Earwig *(Forficula auricularia)* licking its eggs. ***3*** Scorpion with young on its back. ***4*** Male of the midwife toad *(Alytes obstetricans,* Harz Mountains to France) with eggs. ***5*** Mouth brooding cichlid. ***6*** Male of the poison-arrow frog *(Phyllobates bicolor)* with tadpoles on its back. ***7*** Sri-Lankan tailed caecilian *(Ichthyophis glutinosus)* with eggs. ***8*** Reticulated python *(Python reticulatus)* brooding its eggs. ***9*** Female African ostrich *(Struthio camelus)* guarding young. ***10*** Mother and young of the giant anteater *(Myrmecophaga tridactyla).*

4 *Fishes.* ***1*** Common hammerhead shark *(Sphyrna zygaena)*, 4 m. ***2*** Thornback ray *(Raja clavata)*, male up to 70 cm, female up to 1.25 m. ***3*** Common Atlantic sturgeon *(Acipenser sturio)*, 2–4 m. ***4*** Longnose gar *(Lepisosteus osseus)*, 1.5 m. ***5*** Deepsea hatchet fish *(Argyropelecus affinis)*, 9 cm. ***6*** Seahorse *(Hippocampus guttulatus)*, 12 cm. ***7*** Flyingfish *(Exocoëtus* sp.), 20 cm. ***8*** Ocean sunfish *(Mola mola)*, 2.5 m. ***9*** Flounder *(Pleuronectes = Platichthys flesus)*, 30 cm. ***10*** Mudskipper *(Periophthalmus barbarus)*, 15 cm. ***11*** Coelacanth *(Latimeria chalumnae)*, 1.5 m. ***12*** African lungfish *(Protopterus dolloi)*, 85 cm.

5 ***Amphibians and reptiles.*** Amphibians: ***1*** European banded fire salamander *(Salamandra salamandra terrestris,* family *Salamandridae),* 20 cm. ***2*** Surinam toad *(Pipa americana,* family *Pipidae),* 20 cm (jungles of Brazil and Guiana). ***3*** A European toad *(Bufo viridis,* family *Bufonidae),* 8 cm. ***4*** European tree frog *(Hyla arborea,* family *Hylidae),* 4 cm. ***5*** South American horned frog or Brazilian horned toad *(Ceratophrys* sp., family *Leptodactylidae),* 15 cm (the eyelids are extended into long tips or "horns"). ***6*** American bullfrog *(Rana catesbeiana,* family *Ranidae),* 20 cm. Reptiles: ***7*** Tuatara *(Sphenodon punctatus,* family *Sphenodontidae),* 70 cm (New Zealand). ***8*** Central American pseudemid turtle or slider *(Pseudemys scripta elegans,* family *Kinosternidae),* 28 cm. ***9*** Spectacled caiman *(Caiman crocodilus,* family *Alligatoridae),* 2.5 m. ***10*** Jewelled lizard, eyed lizard or ocellated green lizard *(Lacerta lepida,* family *Lacertidae),* 60 cm (Central and South America). ***11*** Komodo dragon *(Varanus komodoensis,* family *Varanidae),* 3 m (Komodo and neighboring islands in the Malay archipelago). ***12*** Anaconda *(Eunectes murinus,* family *Boidae),* ***7*** m (South Amrica). ***13*** Leopard snake or rat snake *(Elaphe situla,* family *Colubridae).* 1.1 m (Europe). ***14*** Gaboon viper *(Bitis gabonica,* family *Viperidae),* 1.5 m (West and Central African rainforest).

Plate 6

6 ***Native European birds. 1*** Black-necked grebe *(Podiceps nigricollis)*. ***2*** Little bittern *(Ixobrychus minutus)*. ***3*** Moorhen *(Gallinula chloropus)*. ***4*** Teal *(Anas crecca)*. ***5*** Sparrowhawk *(Accipiter nisus)*. ***6*** Little ringed plover *(Charadrius dubius)*. ***7*** Common tern *(Sterna hirundo)*. ***8*** Collared dove *(Streptopelia decaocto)*. ***9*** Cuckoo *(Cuculus canorus)*. ***10*** Tawny owl *(Strix aluco)*. ***11*** Swift *(Apus apus)*. ***12*** Green woodpecker *(Picus viridis)*. ***13*** House martin *(Delichon urbica)*. ***14*** Yellow wagtail *(Motacilla flava)*. ***15*** Red backed shrike *(Lanius collurio)*. ***16*** Whitethroat *(Sylvia communis)*. ***17*** Redstart *(Phoenicurus phoenicurus)*. ***18*** Yellowhammer or yellow bunting *(Emberiza citrinella)*. ***19*** Bullfinch *(Pyrrhula pyrrhula)*. ***20*** Nuthatch *(Sitta europaea)*. ***21*** Magpie or black-billed magpie *(Pica pica)*.

7 Small mammals of central Europe. Dormice *(Gliridae)*: ***1*** Hazel mouse *(Muscardinus avellanarius)*; ***2*** Edible dormouse *(Glis glis)*. Microtine rodents *(Arvicolinae)*: ***3*** European water vole *(Arvicola terrestris)*; ***4*** Bank vole *(Clethrionomys glareolus)*. Old World rats and mice *(Muridae)*: ***5*** Harvest mouse *(Micromys minutus)*; ***6*** Yellow-necked field mouse *(Apodemus flavicollis)*; ***7*** House mouse *(Mus musculus)*; ***8*** European long-tailed field mouse *(Apodemus sylvaticus)*. Red-toothed shrews *(Soricidae*, genus *Sorex)*: ***9 Wood shrew*** *(Sorex araneus)*; ***10*** Dwarf shrew *(Sorex minutus)*. White-toothed shrews *(Soricidae*, genus *Crocidura)*: ***11*** House shrew *(Crocidura russula)*. Skulls: ***12*** Vole, ***13*** Red-toothed shrew, ***14a*** and ***14b*** Front teeth of a red-toothed and white-toothed shrew, respectively.

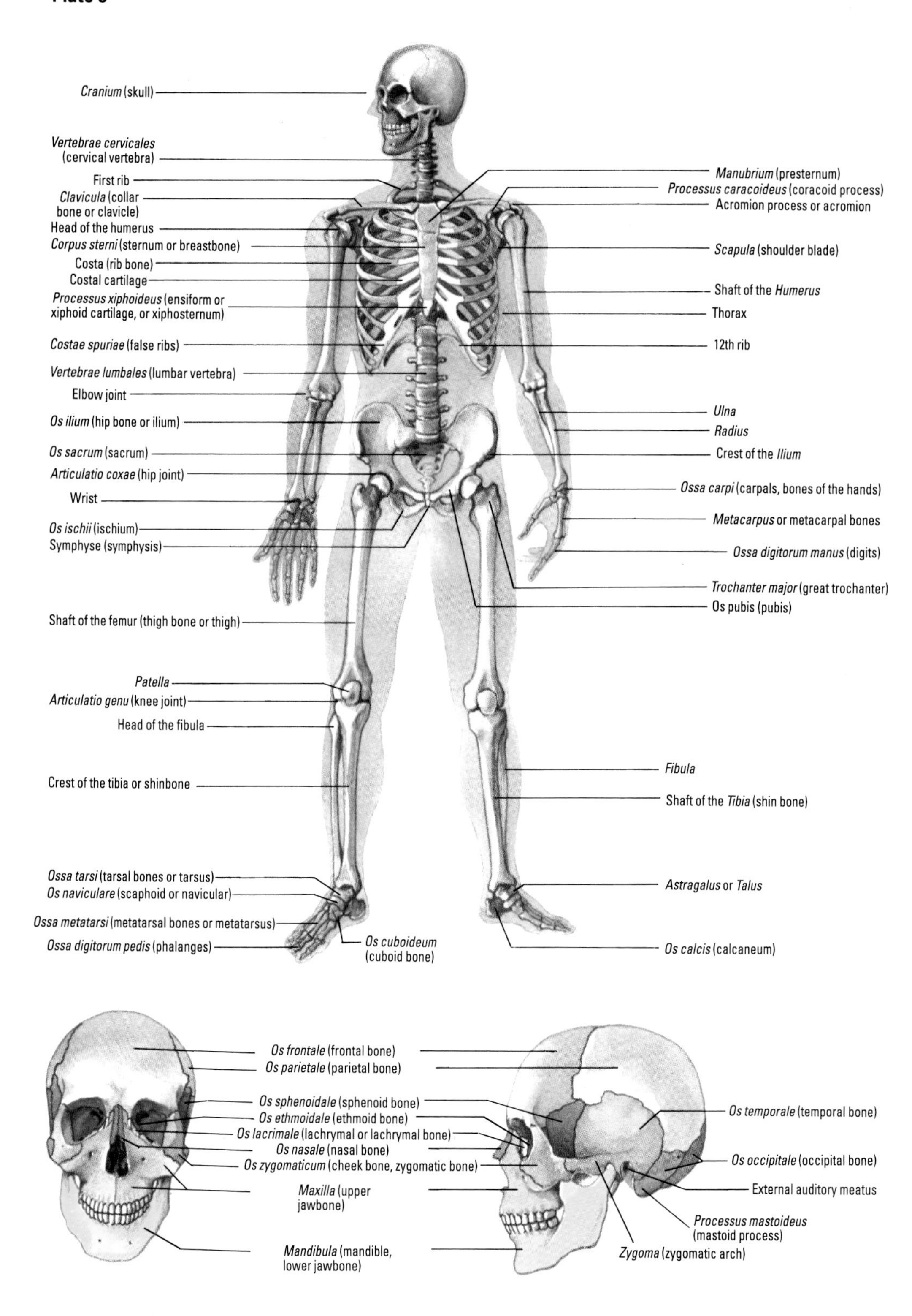
Cranium (skull)
Vertebrae cervicales (cervical vertebra)
First rib
Clavicula (collar bone or clavicle)
Head of the humerus
Corpus sterni (sternum or breastbone)
Costa (rib bone)
Costal cartilage
Processus xiphoideus (ensiform or xiphoid cartilage, or xiphosternum)
Costae spuriae (false ribs)
Vertebrae lumbales (lumbar vertebra)
Elbow joint
Os ilium (hip bone or ilium)
Os sacrum (sacrum)
Articulatio coxae (hip joint)
Wrist
Os ischii (ischium)
Symphyse (symphysis)
Shaft of the femur (thigh bone or thigh)
Patella
Articulatio genu (knee joint)
Head of the fibula
Crest of the tibia or shinbone
Ossa tarsi (tarsal bones or tarsus)
Os naviculare (scaphoid or navicular)
Ossa metatarsi (metatarsal bones or metatarsus)
Ossa digitorum pedis (phalanges)
Os cuboideum (cuboid bone)
Manubrium (presternum)
Processus caracoideus (coracoid process)
Acromion process or acromion
Scapula (shoulder blade)
Shaft of the Humerus
Thorax
12th rib
Ulna
Radius
Crest of the Ilium
Ossa carpi (carpals, bones of the hands)
Metacarpus or metacarpal bones
Ossa digitorum manus (digits)
Trochanter major (great trochanter)
Os pubis (pubis)
Fibula
Shaft of the Tibia (shin bone)
Astragalus or Talus
Os calcis (calcaneum)
Os frontale (frontal bone)
Os parietale (parietal bone)
Os sphenoidale (sphenoid bone)
Os ethmoidale (ethmoid bone)
Os lacrimale (lachrymal or lachrymal bone)
Os nasale (nasal bone)
Os zygomaticum (cheek bone, zygomatic bone)
Maxilla (upper jawbone)
Mandibula (mandible, lower jawbone)
Os temporale (temporal bone)
Os occipitale (occipital bone)
External auditory meatus
Processus mastoideus (mastoid process)
Zygoma (zygomatic arch)

8 ***Human anatomy I.*** *Above:* skeleton. *Below left:* front view of skull. *Below right:* side view of skull.

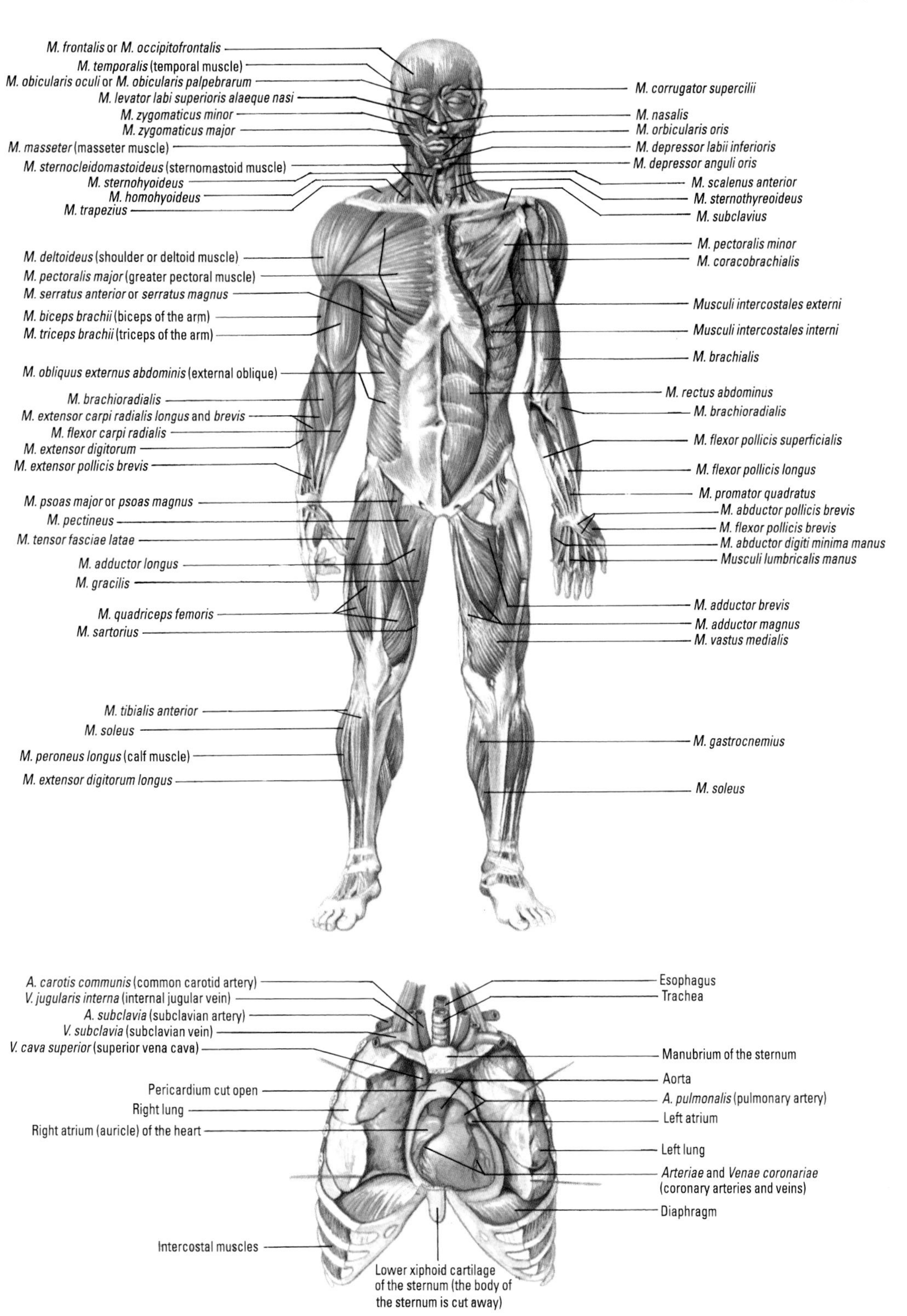
M. frontalis or M. occipitofrontalis
M. temporalis (temporal muscle)
M. obicularis oculi or M. obicularis palpebrarum
M. levator labi superioris alaeque nasi
M. zygomaticus minor
M. zygomaticus major
M. masseter (masseter muscle)
M. sternocleidomastoideus (sternomastoid muscle)
M. sternohyoideus
M. homohyoideus
M. trapezius
M. corrugator supercilii
M. nasalis
M. orbicularis oris
M. depressor labii inferioris
M. depressor anguli oris
M. scalenus anterior
M. sternothyreoideus
M. subclavius
M. deltoideus (shoulder or deltoid muscle)
M. pectoralis major (greater pectoral muscle)
M. serratus anterior or serratus magnus
M. biceps brachii (biceps of the arm)
M. triceps brachii (triceps of the arm)
M. pectoralis minor
M. coracobrachialis
Musculi intercostales externi
Musculi intercostales interni
M. brachialis
M. obliquus externus abdominis (external oblique)
M. brachioradialis
M. extensor carpi radialis longus and brevis
M. flexor carpi radialis
M. extensor digitorum
M. extensor pollicis brevis
M. rectus abdominus
M. brachioradialis
M. flexor pollicis superficialis
M. flexor pollicis longus
M. promator quadratus
M. abductor pollicis brevis
M. flexor pollicis brevis
M. abductor digiti minima manus
Musculi lumbricalis manus
M. psoas major or psoas magnus
M. pectineus
M. tensor fasciae latae
M. adductor longus
M. gracilis
M. quadriceps femoris
M. sartorius
M. adductor brevis
M. adductor magnus
M. vastus medialis
M. tibialis anterior
M. soleus
M. peroneus longus (calf muscle)
M. extensor digitorum longus
M. gastrocnemius
M. soleus
A. carotis communis (common carotid artery)
V. jugularis interna (internal jugular vein)
A. subclavia (subclavian artery)
V. subclavia (subclavian vein)
V. cava superior (superior vena cava)
Pericardium cut open
Right lung
Right atrium (auricle) of the heart
Intercostal muscles
Esophagus
Trachea
Manubrium of the sternum
Aorta
A. pulmonalis (pulmonary artery)
Left atrium
Left lung
Arteriae and Venae coronariae
(coronary arteries and veins)
Diaphragm
Lower xiphoid cartilage
of the sternum (the body of
the sternum is cut away)

9 ***Human anatomy II.*** *Above:* the most important muscles; superficial muscles on the left, deeper muscles on the right. *Below:* thorax.

Plate 10

10 *Biology of flowers.* ***1*** and ***2*** Male and female flowers of the monoecious wind-pollinated walnut *(Juglans regia)*. ***3*** The "flower" (cyathium) of the milkweed *(Euphorbia* sp.), with its conspicuous crescent moon-shaped nectaries, is actually an inflorescence. ***4*** In many flowers (illustrated here by the foxglove, *Digitalis purpurea)* the way into the interior of the flower is marked for pollinators by honey-guides. ***5*** The spadix of *Calla palustris* consists of a number of nectarless flowers. Its unpleasant odor attracts many flies. ***6*** The poppy is one of the most prolific pollen producers; the field poppy *(Papaver rhoeas)* can produce 2.6 million pollen grains. The flower reflects a large proportion of UV-light, and is therefore visible to red-blind bees and flies. ***7*** A honeybee has become powdered with pollen from chicory *(Cichorium intybus)*, which will be brushed together and transferred to the extensive pollen sacs. Chicory produces pollen only from about 7 to 12 am. ***8*** Next to bees, hover flies (illustrated is *Episyrphus balteata)* play a very important role as pollinators. ***9*** A bumble bee has activated the lever mechanism of a sage flower *(Salvia glutinosa)*, which presses its pollen vessels onto the back of the bee.

11 *Plant communities.* ***1*** Ruderal vegetation: Scotch thistle community *(Onopordetum acanthii).* ***2*** Weed vegetation of arable land: Chamomile community *(Aphano-Matricarietum).* ***3*** Aquatic vegetation: Water lily community *(Nymphaeetum albo-candidae).* ***4*** to ***6*** Meadow and pasture vegetation. ***4*** Club awn grass vegetation *(Spergulo-Corynephoretum).* ***5*** Tall oat grass community *(Arrhenatheretum elatius).* ***6*** Mat-grass vegetation *(Nardetum).* ***7*** to ***9*** Woodland and forest vegetation. ***7*** Riparian woodland *(Fraxino-Ulmetum).* ***8*** Sessile oak-hornbeam-linden (lime) woodland *(Tilio-Carpinetum).* ***9*** Pine forest *(Piceetum hercynium).*

Plate 12

12 *Plant galls I* 1 Root nodule of alder *(Alnus)* caused by the actinomycete, *Frankia alni.* ***2*** Pineapple-like gall on spruce, caused by the homopteran, *Sacchiphantes viridis.* ***3*** Leaf galls on willow, caused by the sawfly, *Pontania proxima.* ***4*** Oyster gall on the leaf vein of oak, caused by the gall wasp, *Andricus ostrea;* the other galls are lenticulate "spangle" galls, caused by the gall wasp, *Neuroterus quercus-baccarum* (parthenogenetic generation). ***5*** On oak, the bisexual generation of *Neuroterus quercus-baccarum* causes a spherical (5–8 mm) gall which grows on leaves or, as shown here, on male catkins. ***6*** Oak leaf galls formed by *Neuroterus laeviusculus* fall to the ground in autumn together with their larval tenants. This oak produced an estimated 1 million galls.

13 ***Plant galls II.*** ***1*** Galls on the underside of an oak leaf, caused by the oak gall wasp, *Cynips longiventris.* ***2*** Artichoke gall on oak, caused by the female generation of the gall wasp, *Andricus fecundatrix (= fecundator),* which lays its eggs in a leaf bud. ***3*** Section through an old robin's pin-cushion or bedeguar gall on a rose, caused by the small gall wasp, *Diplolepis rosae* L. ***4*** Galls on an aspen leaf *(Populus tremula),* caused by the gall gnat, *Harmandia tremulae.* ***5*** Willow "rose" gall, caused by the gall gnat, *Rhabdophaga rosaria.* A gall of the sawfly, *Pontania vesicator,* is also present on one leaf. ***6*** Gall of *Didymomyia tiliacea* on lime *(Tilia).* On the left, a gall is being extruded and will fall to the ground. An empty crater is seen on the right.

14 *Monocotyledonous plants.* ***1*** Vriesia *(Vriesia* sp., family *Bromeliaceae)* ***2*** Crown imperial *(Fritillaria imperialis* v. *lutea,* family *Liliaceae).* ***3*** Star tulip *(Tulipa tarda = T. dasystemon,* family *Liliaceae).* ***4*** Fragrant orchid *(Gymnadenia conopsea,* family *Orchidaceae).* ***5*** Slipper orchid *(Paphiopedilum* v. *aladin,* family *Orchidaceae).* ***6*** Large rooted alocasia *(Alocasia macrorrhiza,* family *Araceae).*

15 *Dicotyledonous plants.* *1* Pheasant's eye *(Adonis vernalis,* family *Ranunculaceae).* ***2*** Austrian briar *(Rosa foetida,* family *Rosaceae).* ***3*** Guelder rose *(Viburnum opulus,* family *Caprifoliaceae).* ***4*** Red passion flower *(Passiflora racemosa,* family *Passifloraceae).* ***5*** Japanese rhododendron *(Rhododendron japonicum,* family *Ericaceae).* ***6*** Silver thistle *(Carlina acaulis,* family *Compositae).*

Plate 16

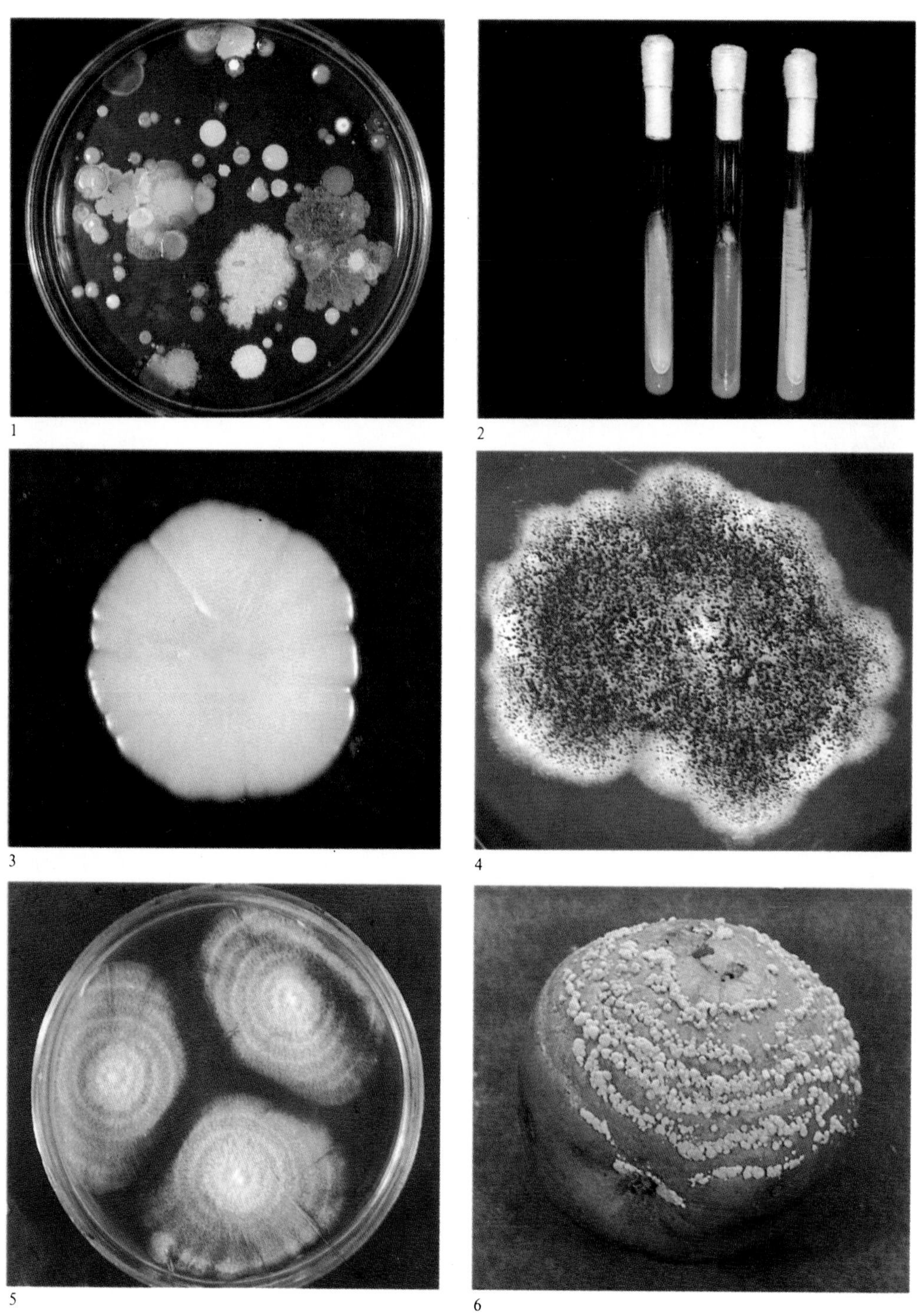

16 *Microbiology I.* ***1*** Colonies of microorganisms (mostly bacteria) developing on nutrient agar in a Petri dish after exposing the medium for some time to the air. ***2*** Nutrient agar slopes with pure bacterial cultures; left to right: *Bacillus subtilis, Serratia marcescens, Micrococcus luteus.* ***3*** Large colony of the yeast, *Saccharomyces cerevisiae.* ***4*** Large colony of the mold, *Aspergillus niger* [the aerial spores (conidia) are black]. ***5*** The rose mold, *Trichothecium (= Cephalothecium),* often found on bread. ***6*** Brown rot or Monilia-rot on an apple, caused by *Scerotinia fructigena.* The concentric circles of fungus are due to the diurnal light-dark cycle.

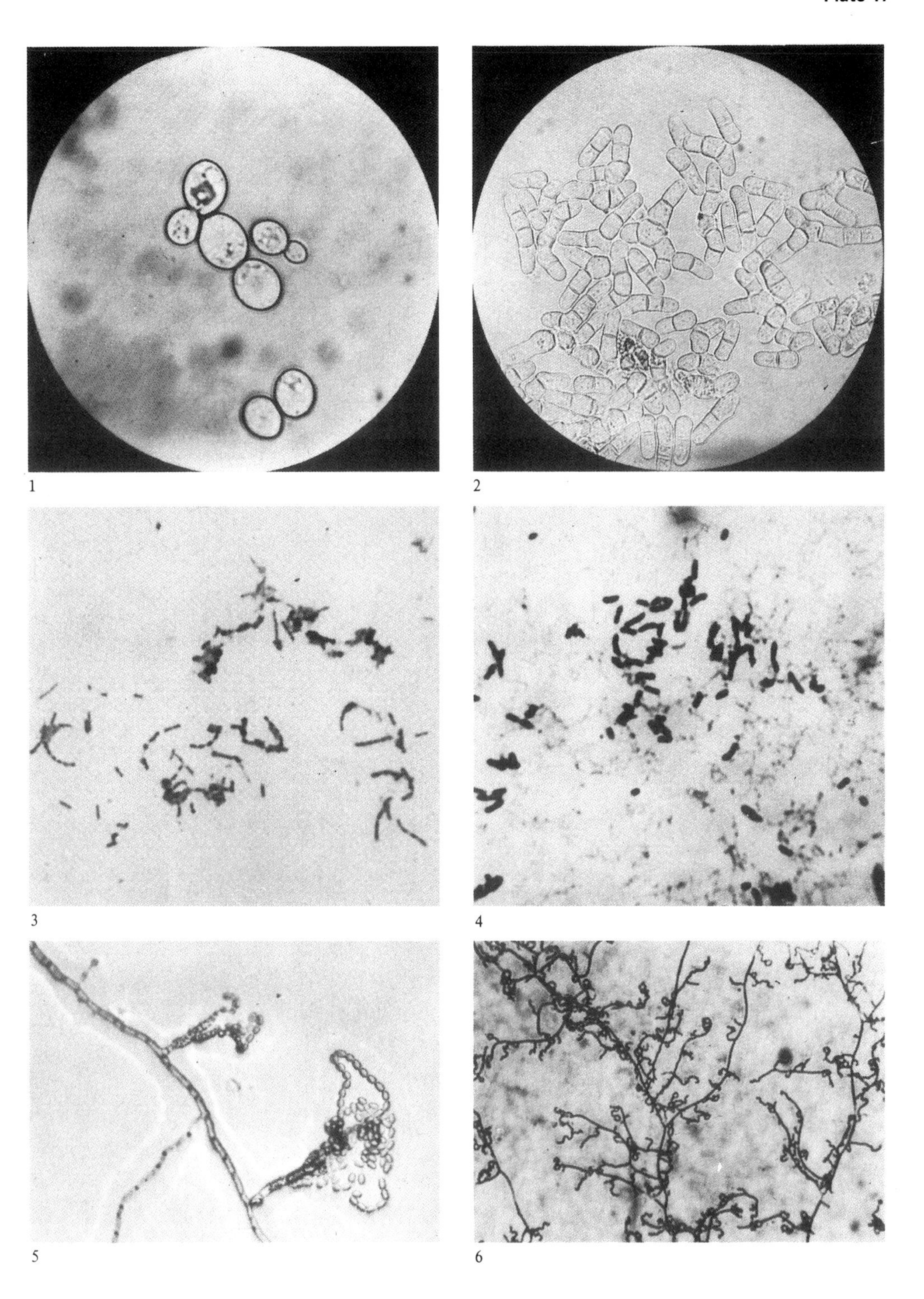

17 ***Microbiology II.*** ***1*** Cultured yeast, *Saccharomyces cerevisiae* (x 600). ***2*** Cultured yeast *Schizosaccharomyces octosporus* (x 600). ***3*** Lactic acid bacterium, *Lactobacillus delbrueckii* (x 400). ***4*** Acetic acid bacterium, *Acetobacter aceti* (x 900). ***5*** Mold, *Scopulariopsis brevicaulis* (x 200). ***6*** Actinomycete, *Streptomyces albus* (x 200).

Plate 18

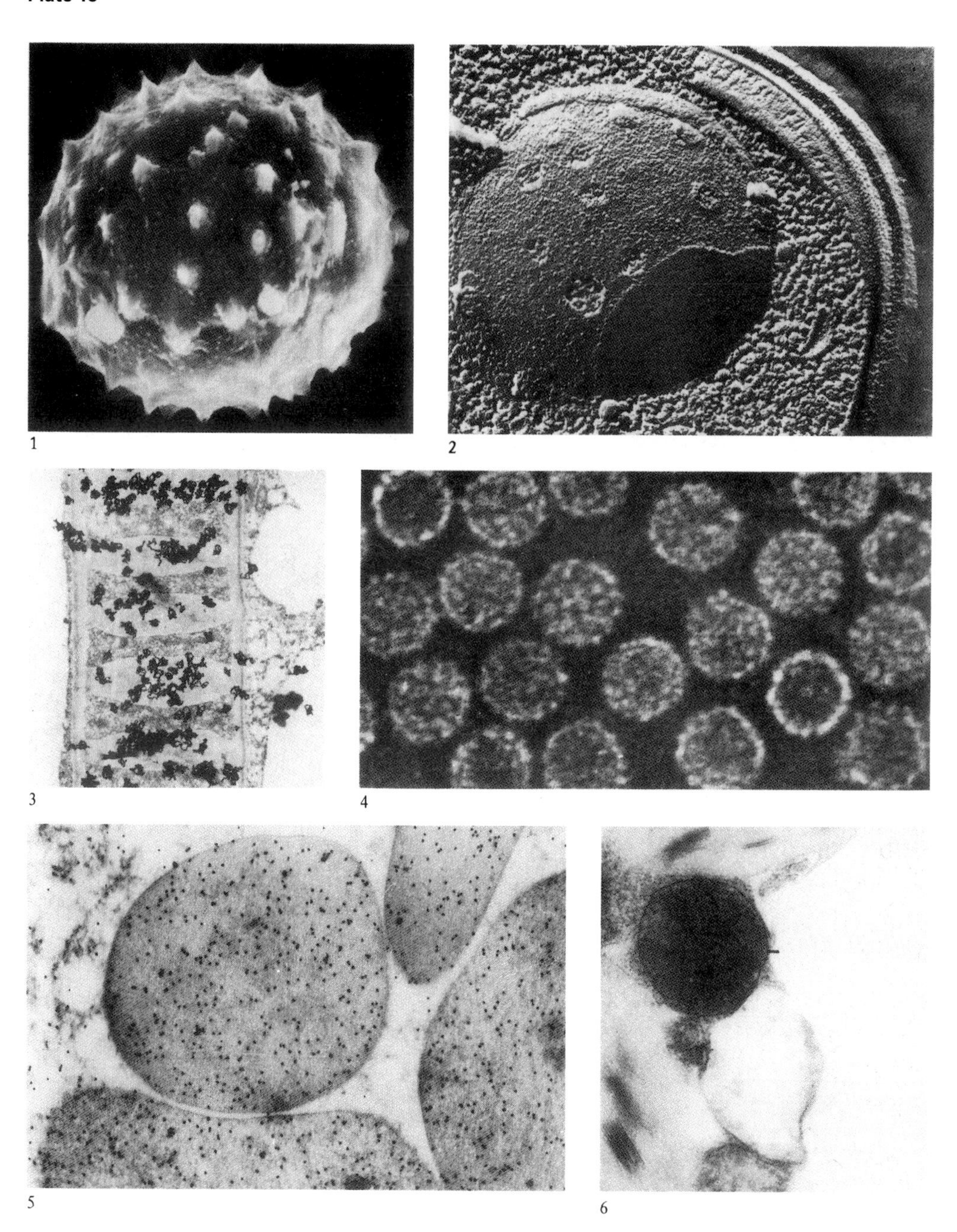

18 ***Electron microscope methods.*** ***1*** Scanning electron micrograph of a pollen grain of *Arbutilon* (x 3 000). ***2*** Freeze-etched preparation of a yeast cell. The freeze fracture has exposed cleavage surfaces of both membranes of the nuclear envelope. At the outer cell boundary, the freeze etching has exposed a narrow edge of the cell surface between the transversely broken cell wall and the ice level (x 50 000). ***3*** Autoradiogram of a young xylem vessel of tobacco, after feeding radioactive nicotine and exposing to photographic emulsion. The black filaments are silver grains, showing the distribution of radioactive material in the specimen (x 5 400). ***4*** Negative contrast (with phosphotungstic acid) of hamster papover virus from skin warts. The proteins of the virus coat are clearly distinguishable by this negative contrast technique. ***5*** Immunohistochemical demonstration of storage proteins in broad bean seeds *(Vicia faba)*. The antibodies are labeled with colloidal gold, which forms points of high contrast in the electron microscope (x 33 600). ***6*** Histochemical demonstration of catalase in a barley leaf cell. The high contrast of cytosomes (–) shows that the enzyme is restricted to these organelles (x 24 000).

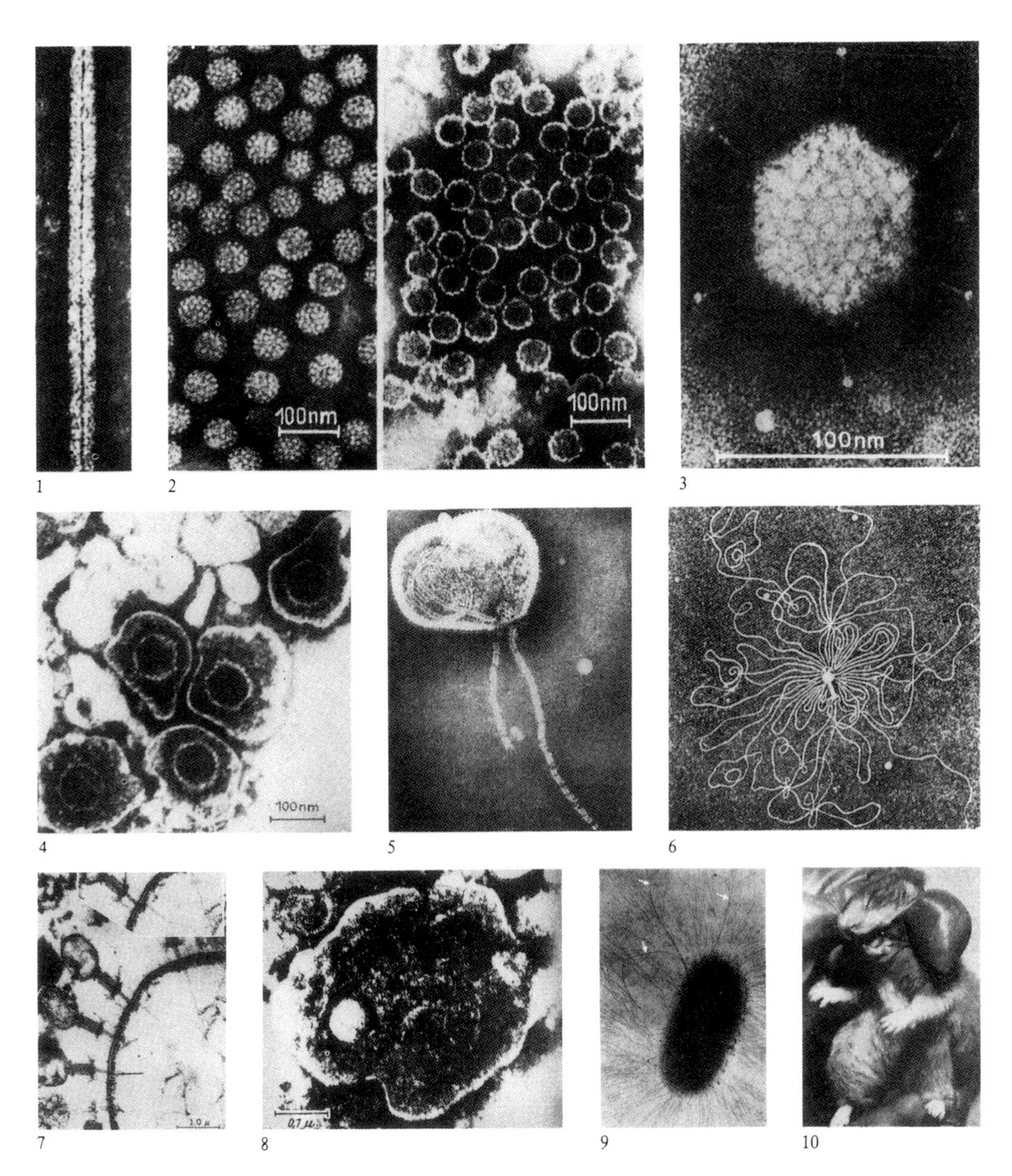

1 2 3 4 5 6 7 8 9 10

19 ***Viruses and phages.*** ***1*** Electron micrograph of a particle of tobacco mosaic virus (negative contrast). The central channel is clearly visible (after Leberman). See Viruses. ***2*** Human wart virus; *left:* full particles (nucleocapsids), *right:* nucleic acid-free "empty" particles (capsids) (electron micrograph with negative contrast) (after Follet). See Viruses. ***3*** Adenovirus. The capsomeres at the corners of the icosahedron have fibrous appendages terminating in knoblike structures (after Valentine and Pereira). See Viruses. ***4*** *Herpes simplex,* an enveloped virus with an isodiametric nucleocapsid (negative contrast) (after Watson and Wildy). See Viruses. ***5*** Measles virus, an enveloped virus with an extended, helical, multiply wound nucleocapsid, which has partly escaped from the envelope during preparation (x 200 000) (after Pereira). See Viruses. ***6*** Exploded particle of T2 phage, showing protein envelope surrounded by extensive strand of escaped DNA (x 40 000) (after Kleinschmidt). See Phages. ***7*** T4 phage adsorbed to the cell wall of *Escherichia coli.* The arrow shows a phage tail core that has penetrated the cell wall. The thin threads (diam. ≈ 2 nm) are probably phage DNA (after Simon and Anderson). See Phages, Viruses. ***8*** Incorporation of influenza virus into a host cell. The particles are adsorbed onto the exterior surface of the host cell, then gradually phagocytosed by the latter (after Hoyle). See Viruses. ***9*** Continuous release of the threadlike *Escherichia coli* phage-M13. The bacterial flagella (shown by arrows) are clearly distinguishable from the phage particles (after Mach). See Phages, Viruses. ***10*** Tumor in the cheek pouch of a hamster, which was injected subcutaneously shortly after birth with polyoma virus (after Eddy). See Tumor viruses.

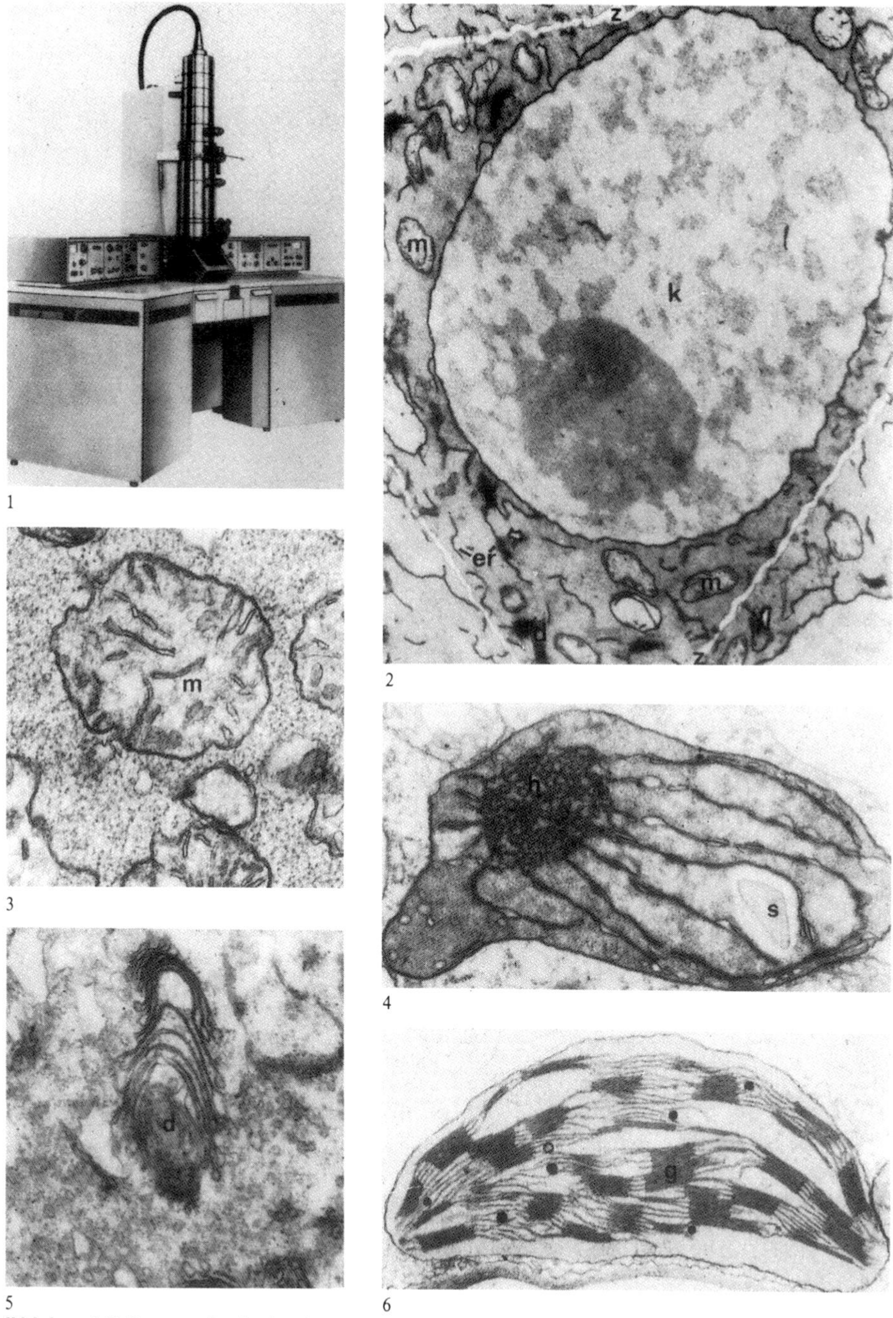

20 *Cell biology I.* Cell organelles in the electron microscope. ***1*** Standard electron microscope BS 500. ***2*** Root tip cell of an onion. The picture of this meristematic cell is dominated by the nucleus (k). The other cell organelles are relatively small and many lack the structural features of fully developed organelles: mitochondria (m), dictyosomes (d), endoplasmic reticulum (er), cell wall (z) (x 9 000). ***3*** Mitochondrion (m) from the phytoflagellate, *Euglena gracilis* (x 40 000). ***4*** Leukoplast from the opium poppy. Leukoplasts represent a stage in the development of chloroplasts. They contain starch (s), a few membranes, and a crystal-like structure (Heitz-Leyon's crystal, h), which regresses than disappears on illumination. ***5*** Dictyosome (d) from the phytoflagellate, *Euglena gracilis*. This organelle consists of stacked membranes, which abstrict small vesicles. Dictyosomes have an excretory function (x 36 000). ***6*** Chloroplast from maize. The membranes, which contain chlorophyll, are regularly stacked at specific intervals to form grana (g) (x 27 000).

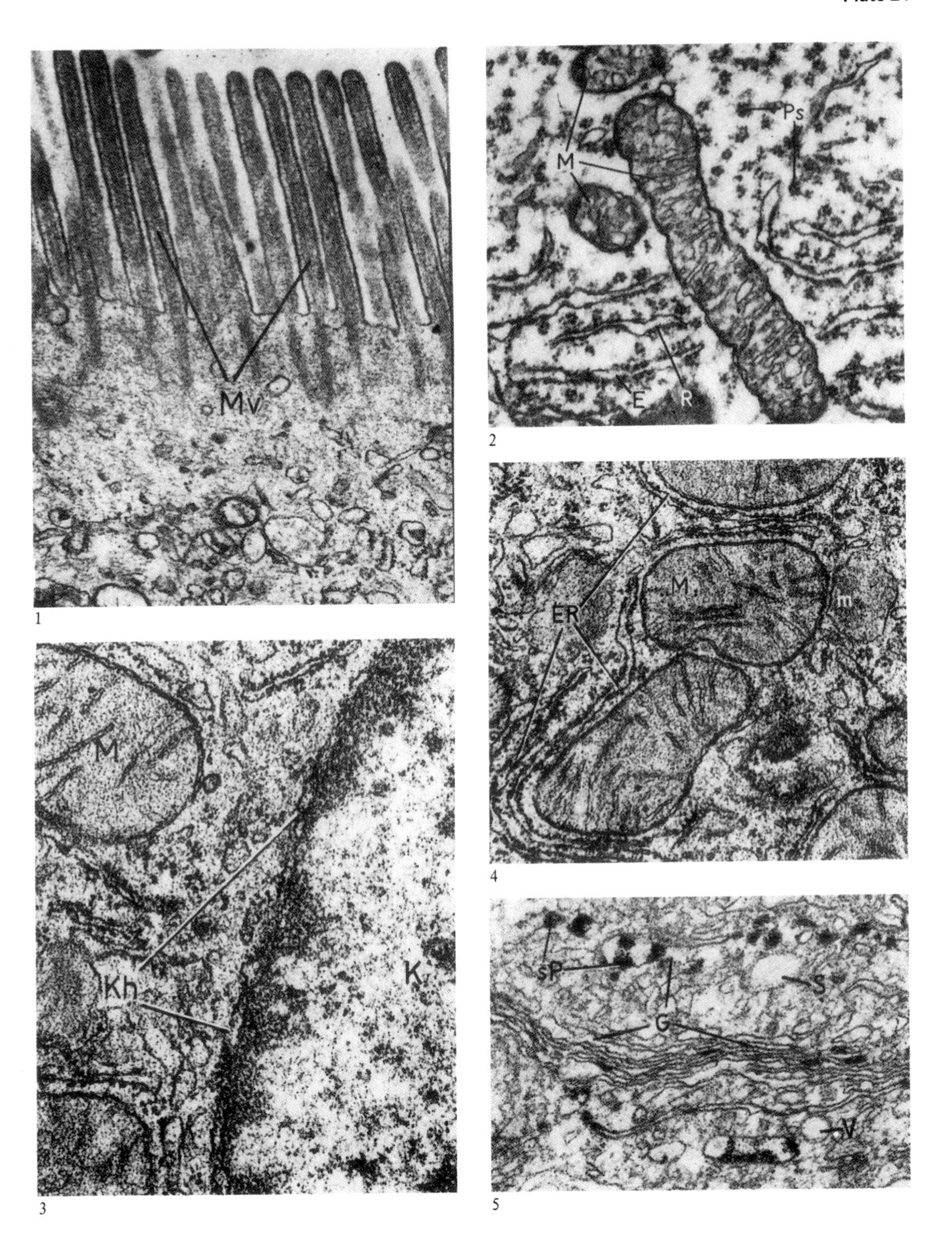

21 *Cell biology II.* Electron micrographs of cell sections. ***1*** Small intestinal epithelium of mouse (x 45 000). ***2*** Neuron of rat brain (x 43 000). ***3*** and ***4*** Rat liver cell (x 62 000). ***5*** Secretory ependymal cell of a frog *(Rana esculenta):* ultrahistochemical demonstration of acid phosphatase (dark reaction product) in the Golgi apparatus (x 61 000). ***ER*** Rough endoplasmic reticulum. ***G*** Golgi cisternae. ***K*** Cell nucleus. ***Kh*** Nuclear membrane. ***M*** Mitochondrion (crista type). ***m*** Microbody. ***Mv*** Microvilli. ***Ps*** Polysome. ***S*** Secretory vesicle. ***sP*** Reaction product showing presence of acid phosphatase. ***V*** Golgi vesicle.

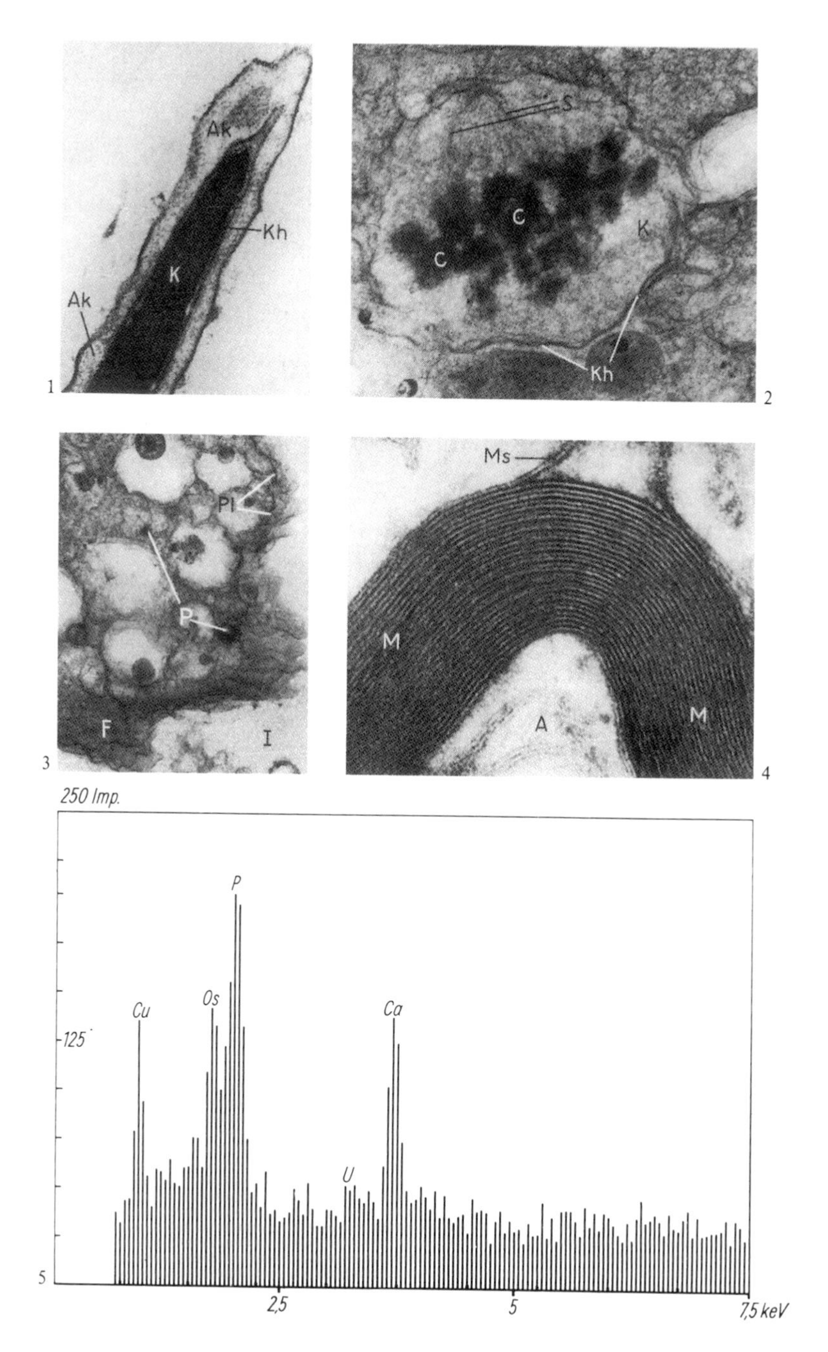

22 *Cell biology III.* 1 Anterior end of a ram sperm (x 86 000). ***2*** Part of the ectoplasm of a plasmodium of the myxomycete, *Physarum polycephalum,* showing intranuclear mitosis; 1 mm section (x 18 000). ***3*** Ectoplasm from the plasmodium of *Physarum polycephalum;* 1 mm section (x 12 000). ***4*** Myelin sheath of a nerve fiber in rat brain; 0.5 mm section (x 170 000). ***5*** X-ray microanalysis of an electron-dense calcium phosphate particle in *Physarum polycephalum,* showing the energy distribution of the emitted X-rays. The analysis in Fig.5 was made with the EDAX-system of the Institute for Solid-state Physics and Electron Microscopy, Academy of Sciences, GDR (former), Halle; Figs. 2, 3 and 4 were taken with the high voltage electron microscope JEM-1000 of the same institute, using 1 000 kV. ***A*** Axon. ***Ak*** Acrosome. ***C*** Chromosome (metaphase). ***F*** Actomyosin fibrils. ***I*** Lumen of the invagination system. ***K*** Cell nucleus. ***Kh*** Nuclear membrane. ***M*** Myelin sheath. ***Ms*** External mesaxon. ***P*** Calcium phosphate particle. ***Pl*** Plasma membrane. ***S*** Spindle.

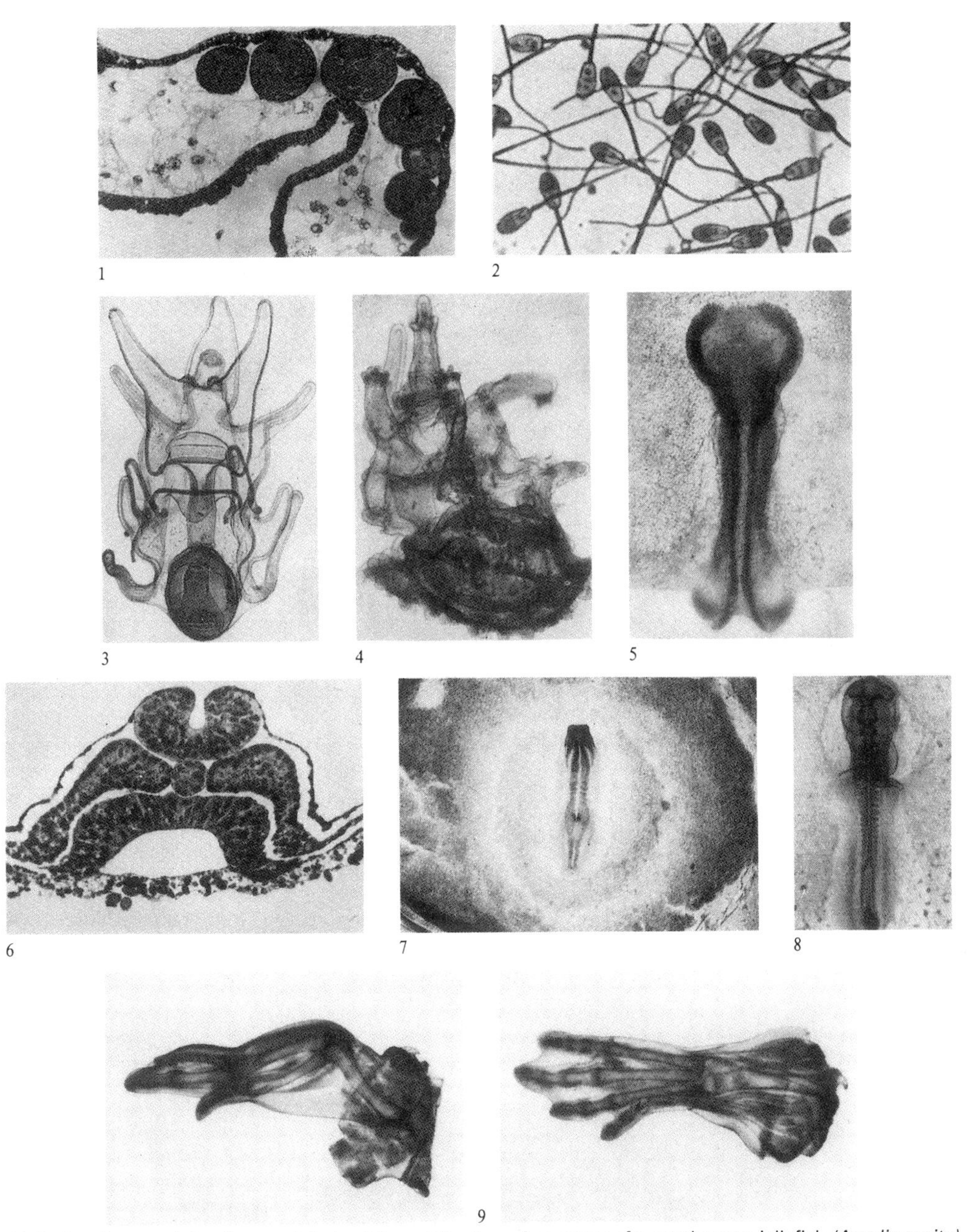

23 ***Embryology and developmental physiology.*** ***1*** Section of an ovary of a scyphozoan jellyfish *(Aurelia aurita)* showing individual eggs. ***2*** Spermatozoa of the domestic goat *(Capra hircus).* ***3*** Brachiolaria larva of a starfish. ***4*** Metamorphosis of a starfish within the brachiolaria larva. ***5*** Early embryo of a dogfish *(Scylliorhinus* sp.) on the blastodisk. The primordial brain is uppermost. ***6*** Section though the middle of the dogfish embryo in ***5.*** The neural groove is uppermost, and the chorda dorsalis is seen below it as a small circle. Right and left of the chorda dorsalis is a pair of slightly curved primordial segments lying over the upper curve of the intestine. The intestinal lumen lies between the intestinal wall and the edge of the yolk sac (not shown). ***7*** Chicken blastodisk after about 25 hours incubation. The anterior (uppermost in the Fig.) of the embryo shows an early stage of head and brain development. Four somites have been formed. The hindmost part of the primitive groove can still be seen behind the end of the neural groove. ***8*** Chicken embryo at the end of the second incubation day. Five brain sections and about twenty pairs of somites have been formed. The neural groove is almost closed. At the end of the anterior third of the body, the tubular heart arches outward to the right. ***9*** Differentiation of bird limbs. Chicken wing and leg on the 8th day of embryonal development.

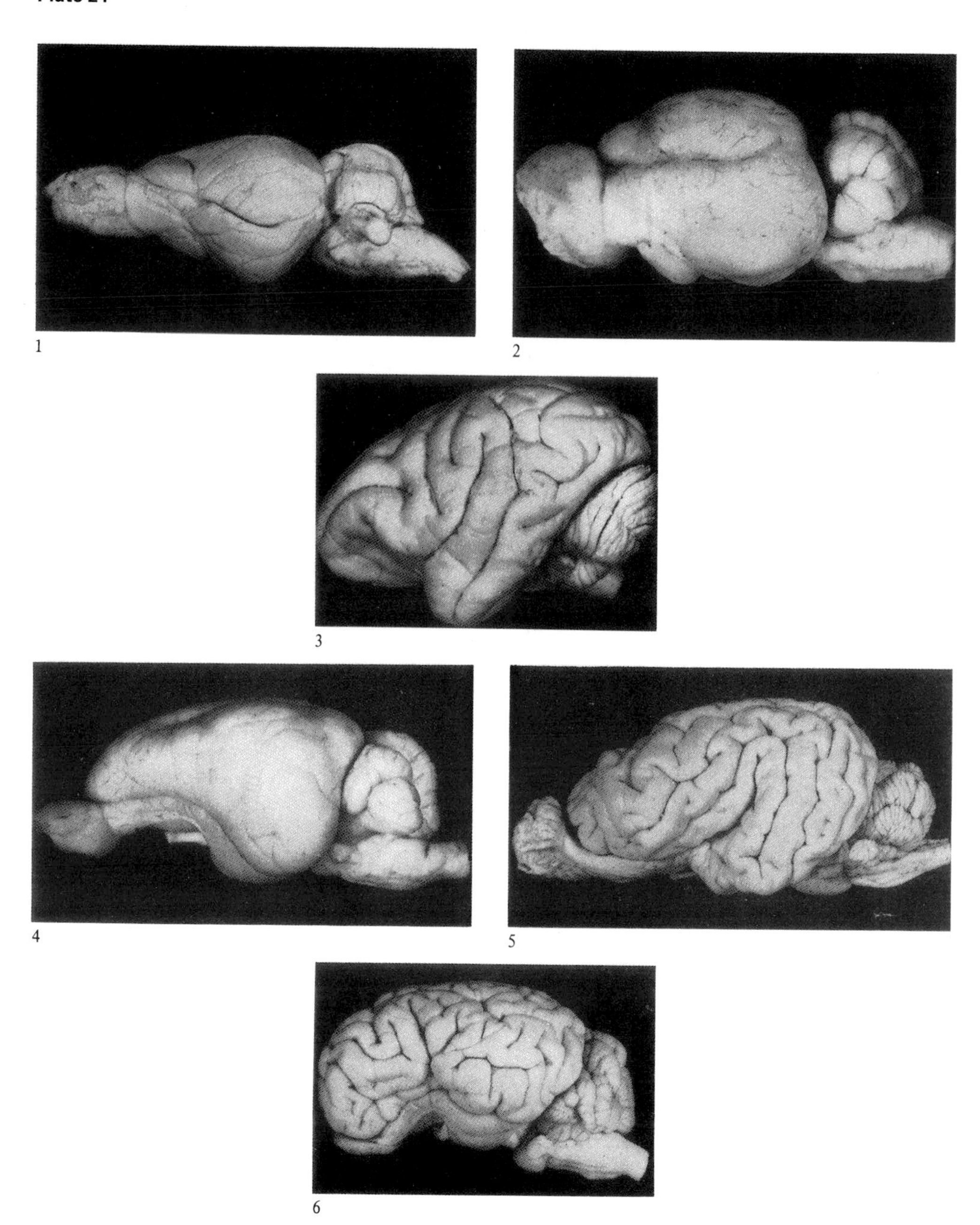

24 ***Neurobiology I.*** Side views of the brains of different mammals. ***1*** Opossum *(Metachirus* sp., order *Marsupalia).* ***2*** Hedgehog *(Erinaceus europaeus,* order *Insectivora).* ***3*** Silvery gibbon *(Hylobates lar,* order *Primates).* ***4*** Hare *(Lepus capensis,* order *Lagomorpha).* ***5*** Polar bear *(Thalarctos maritimus,* order *Carnivora).* ***6*** Barbary sheep *(Ammotragus lervia,* order *Artiodactyla).*

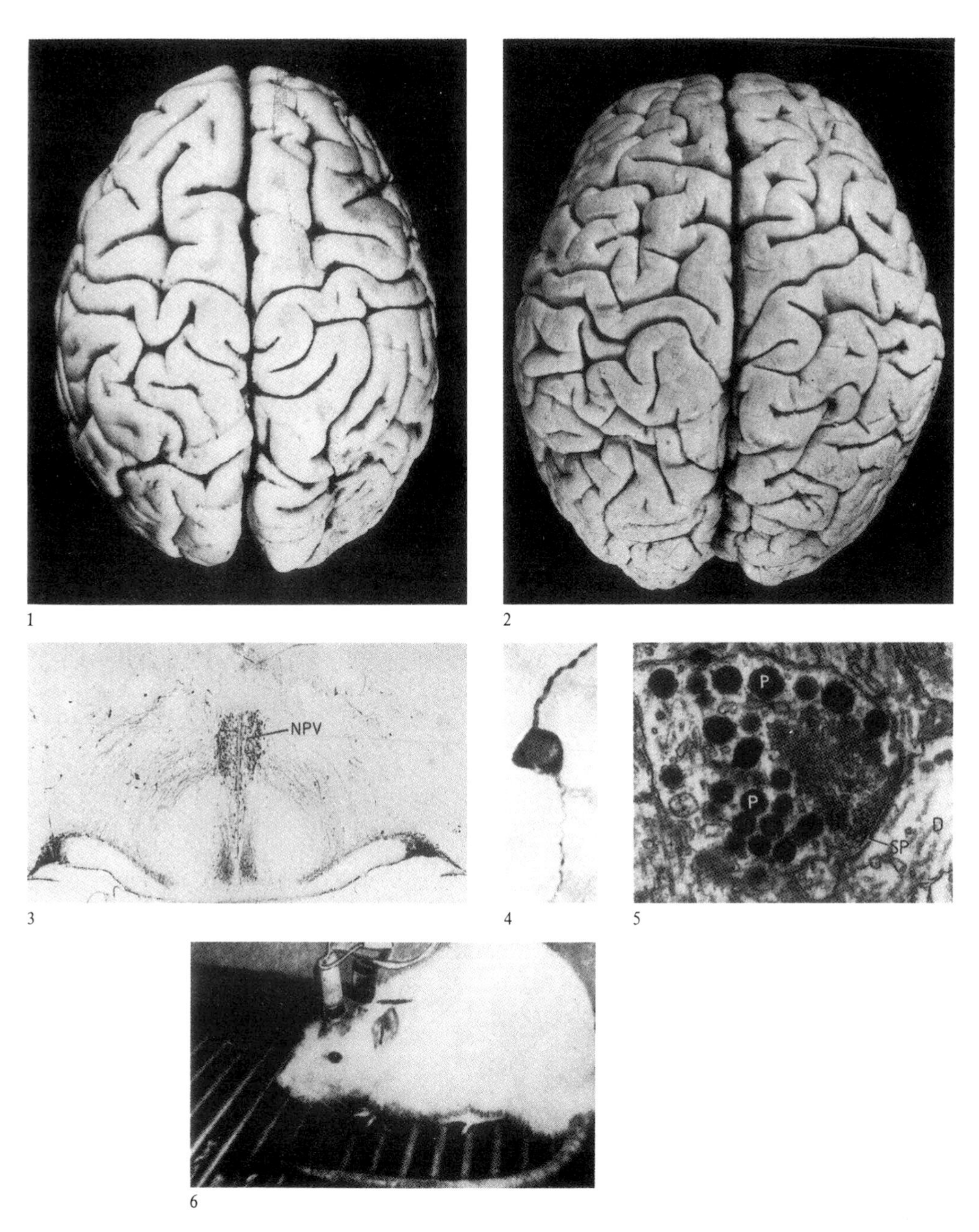

25 ***Neurobiology II. 1*** Surface of chimpanzee brain (mass about 400 g). ***2*** Surface of human brain (mass about 1210 g). Note the different numbers of sulci (grooves or infoldings); original pictures by Dr. Kurt Brauer, Paul Flechsig Institute, Karl-Marx University, Leipzig; from A. Ermisch, 1978. ***3*** Frontal section through a mouse brain. The dark regions are nerve cell bodies and nerve fibers, specifically stained for the peptide, neurophysin, using an immunocytological method. Neurophysin is a carrier for other peptides (oxytocin, vasopressin) (x 50). ***4*** Neuron with dendrites, from the nucleus paraventriculus of ***3*** (x 1 500). ***3*** and ***4*** are original pictures by H. Petter, Biosciences section, Karl-Marx University, Leipzig. ***5*** Synaptic endings of a nerve fiber in the brain of *Pleurodeles waltlii* (salamander). Peptide-containing granules (P) are clearly visible in the presynaptic region, as well as in the synaptic cleft (SP) and part of the dendrite that forms the postsynaptic region (x 4 300); original picture by Dr. R. Wegelin, Biosciences section, Karl-Marx University, Leipzig. ***6*** Rat with electrodes implanted in the brain, linked by a flexible cable to an EEG apparatus; original picture by Dr. F. Klingberg, Paul Flechsig Institute, Karl-Marx University, Leipzig; from A. Ermisch, 1978.

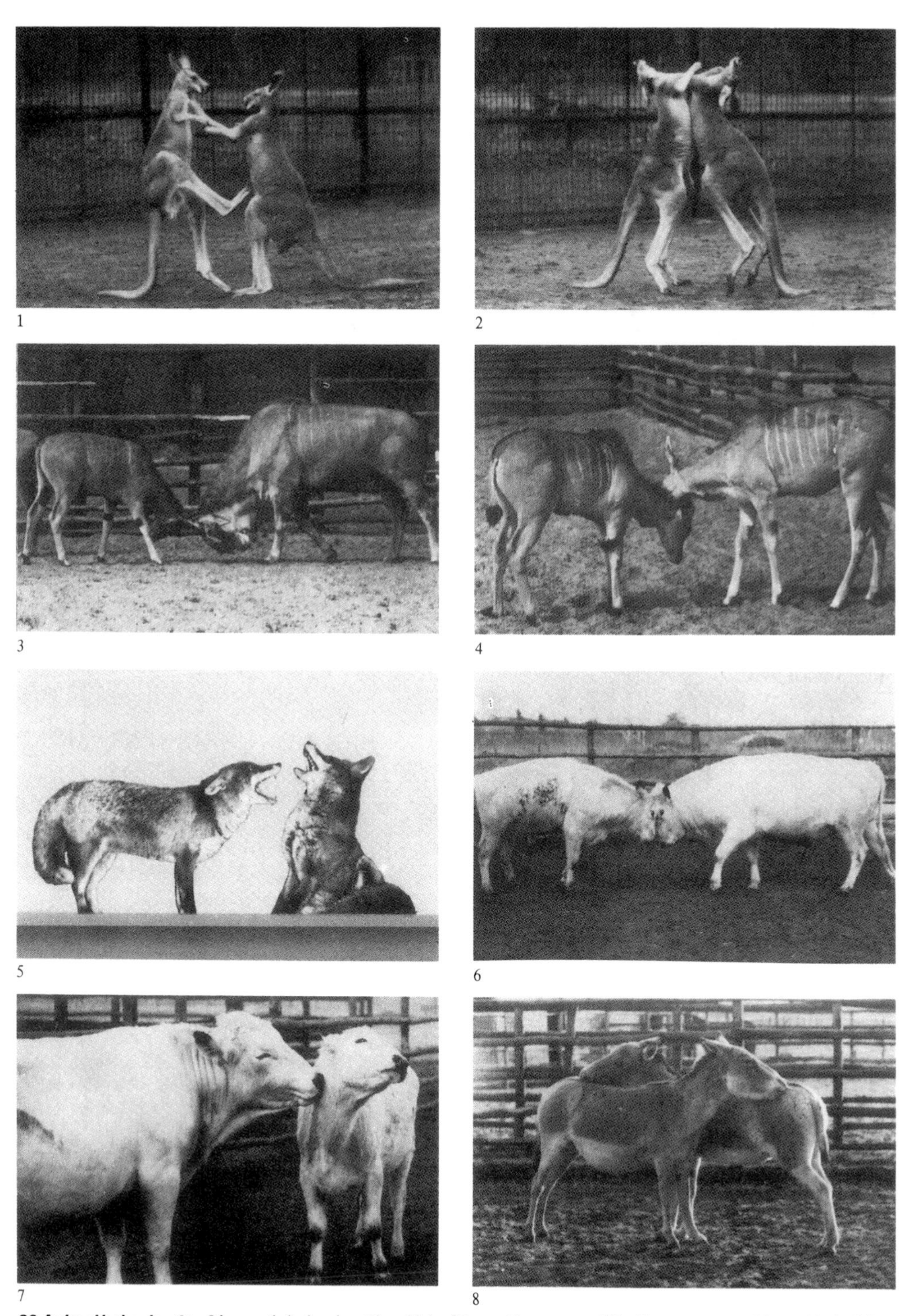

26 *Animal behavior.* *1* to ***6*** Agnostic behavior. ***1*** Leg kick of the red kangaroo. ***2*** Red kangaroos striking with their forepaws, with necks bent backward. ***3*** Playful horn butting of the Eland antelope. ***4*** Neck butting of young Eland antelopes. ***5*** Agonistic elements of ranking (social hierarchy) behavior: right, male red fox offering neck, which inhibits the antagonist. ***6*** Hornless cattle in phase 2 of head butting (pushing and pressing). ***7*** and ***8*** Social behavior (epimiletic behavior). ***7*** Hornless cattle: unritualized licking. ***8*** Onagers: mutual ritualized licking of specific body parts.

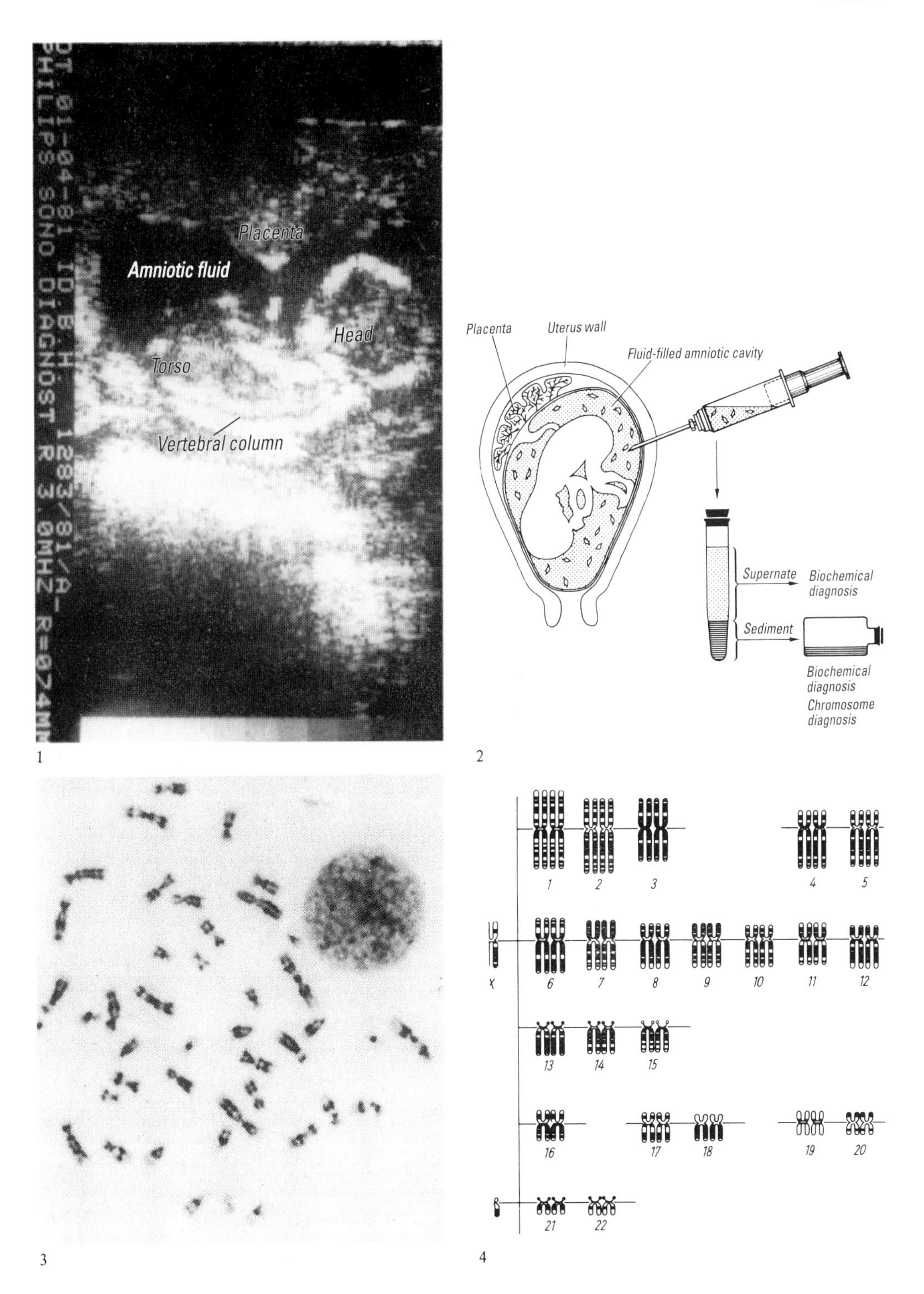

1 2 3 4

27 *Human genetics.* *1* Ultrasound image of a fetus in the 15th week of pregnancy. ***2*** Amniocentesis (sampling of amniotic fluid) in order to test for biochemical or cytogenetic abnormalities in the fetus. ***3*** Human chromosomes in the metaphase of the cell cycle. ***4*** Arrangement of human chromosomes into groups (schematic) to facilitate the search for chromosome aberrations.

Plate 28

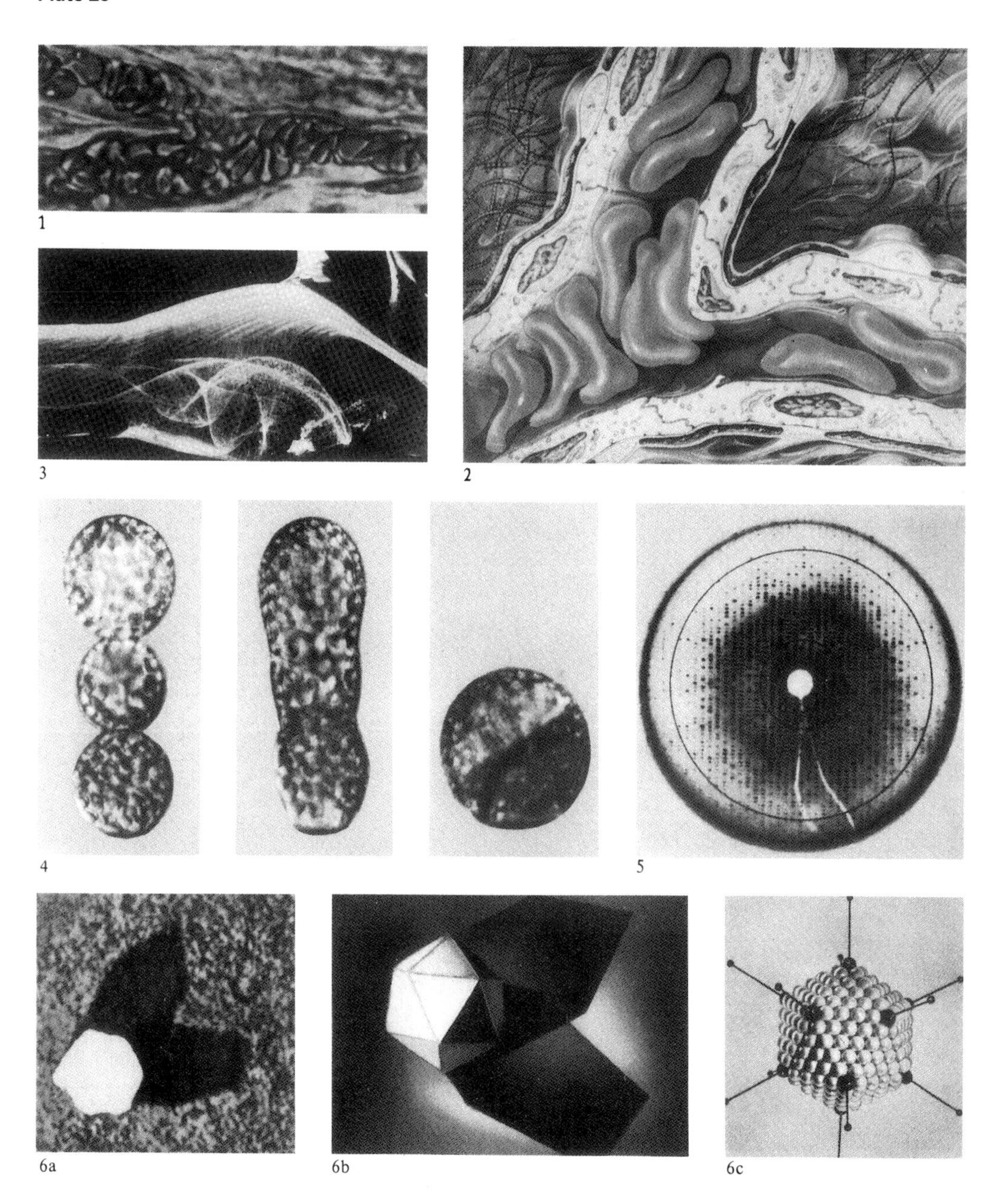

28 ***Biophysics.*** ***1*** to ***3*** are concerned with the study of flow in biological systems (biorheology). Hemorheology is an important branch of this biophysical discipline. Deformation properties of erythrocytes in capillaries, as well as blood flow in large vessels, are medically important, e.g. in conditions of restricted blood flow to organs and tissues. ***1*** Photograph of an exposed capillary. ***2*** Drawing of the situation in ***1.*** ***3*** Schlieren interference photograph of flow in a transparent plastic tube, designed to mimic the aorta. ***4*** A brief voltage pulse can be used to cause cell fusion. Cells are first brought into contact (formation of a cell chain) by a high frequency, low strength, alternating field. Application of a brief impulse (about 2 00 V/ cm) across the touching membranes causes the cells to fuse. The resulting cell hybrid rapidly becomes spherical. This method (shown here for oat protoplasts) is important in gene technology. ***5*** Molecules are so small that they will only cause diffraction of extremely short wavelength electromagnetic radiation, e.g. X-rays. With computer aided calculations, it is possible to evaluate the X-ray diffractogram of a myoglobin molecule, and to reconstruct a picture of the crystallized molecule. ***6*** Special biophysical techniques can be used to optimize the diagnostic power of electron microscopy. Thus, shadowing of a virus with an oblique ray of metal ions from a specific angle produces a picture ***(6a),*** which can be reconstructed as a geometric model ***(6b),*** then used to derive a molecular picture of the virus ***(6c).***

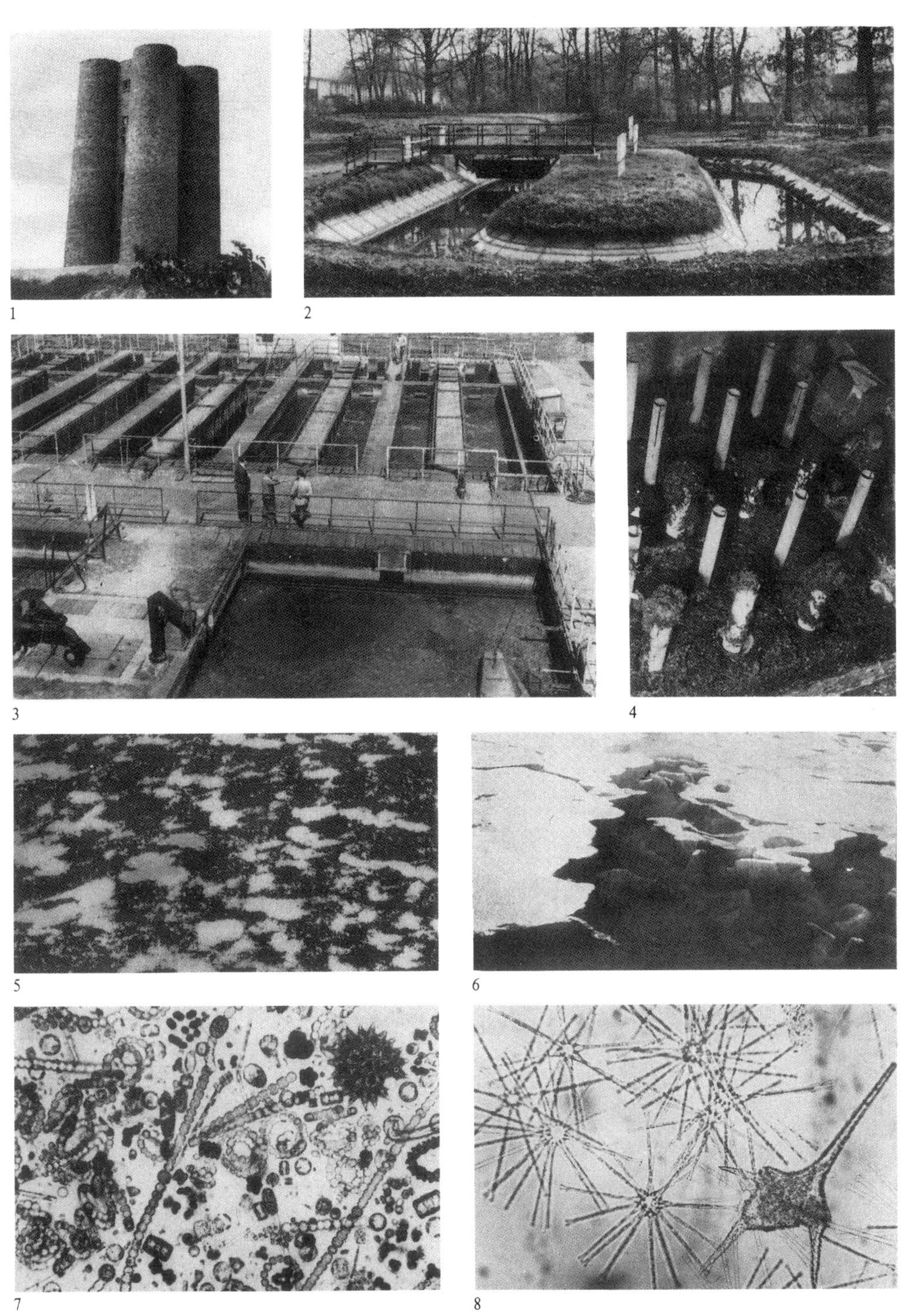

29 ***Biological waste water treatment.*** ***1*** Percolation towers containing trickling filters. ***2*** Oxidation ditch. ***3*** Sludge activation basins of a town sewage works. ***4*** Mass growth of bacteria of the genus *Sphaerotilus* in the water pretreatment tank of a boiler. ***5*** Mass growth of the cyanobacterium, *Anabaena flos-aquae,* in waste water. ***6*** Mass growth of green algae of the genus *Cladophora* in waste water. ***7*** and ***8*** Photomicrographs of phytoplankton.

Plate 30

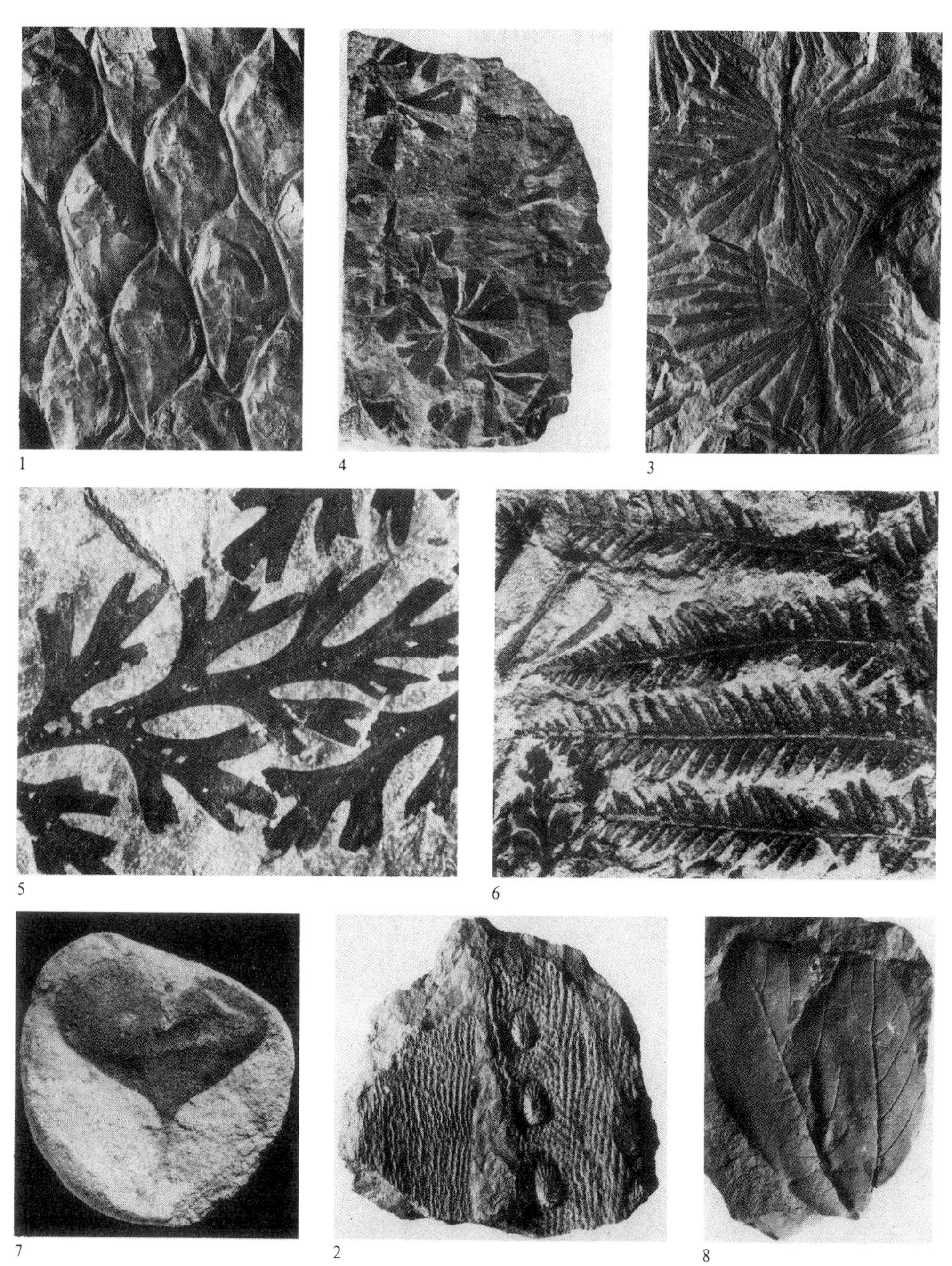

30 ***Paleobotany.*** ***1*** *Lepidodendron aculeatum* (clubmoss tree): Upper Carboniferous of the Saar region. Stem surface with scars of rhombic leaf cushions (x 1.5). ***2*** *Lepidodendron veltheimii* (clubmoss tree): Culm of Dolny Slask. Stem surface with three scars (diam. 2.5–3 cm) left by the abscission of short branches, which carried terminal sporophylls in cones. ***3*** "Leaf" rosettes of *Annularia stellata* (horsetail) (diam. about 3 cm): Upper Carboniferous of Wettin. ***4*** "Leaf" whorls of *Sphenophyllum emarginatum:* Upper Carboniferous of Zwickau. Diam. of whorl 2.5–3 cm. ***5*** Pinnae of a frond of *Sphenopteris divaricata (Pteridospermae):* Upper Carboniferous of Dolny Slask. Length 1.2–1.5 cm. ***6*** Pinnae of a fertile frond *Scolecopteris plumosa (Pteridospermae):* Upper Carboniferous of the Saar region. Length of each pinna 3–4 cm. ***7*** Neocomian sandstone with fertile frond of the fern, *Hausmannia kohlmanni:* Cretaceous of Quedlinburg. Maximal width of frond 5 cm. ***8*** Leaf impression of *Credneria triacuminata (Platanaceae):* Upper Cretaceous of Blankenburg (Harz). Length 18 cm.

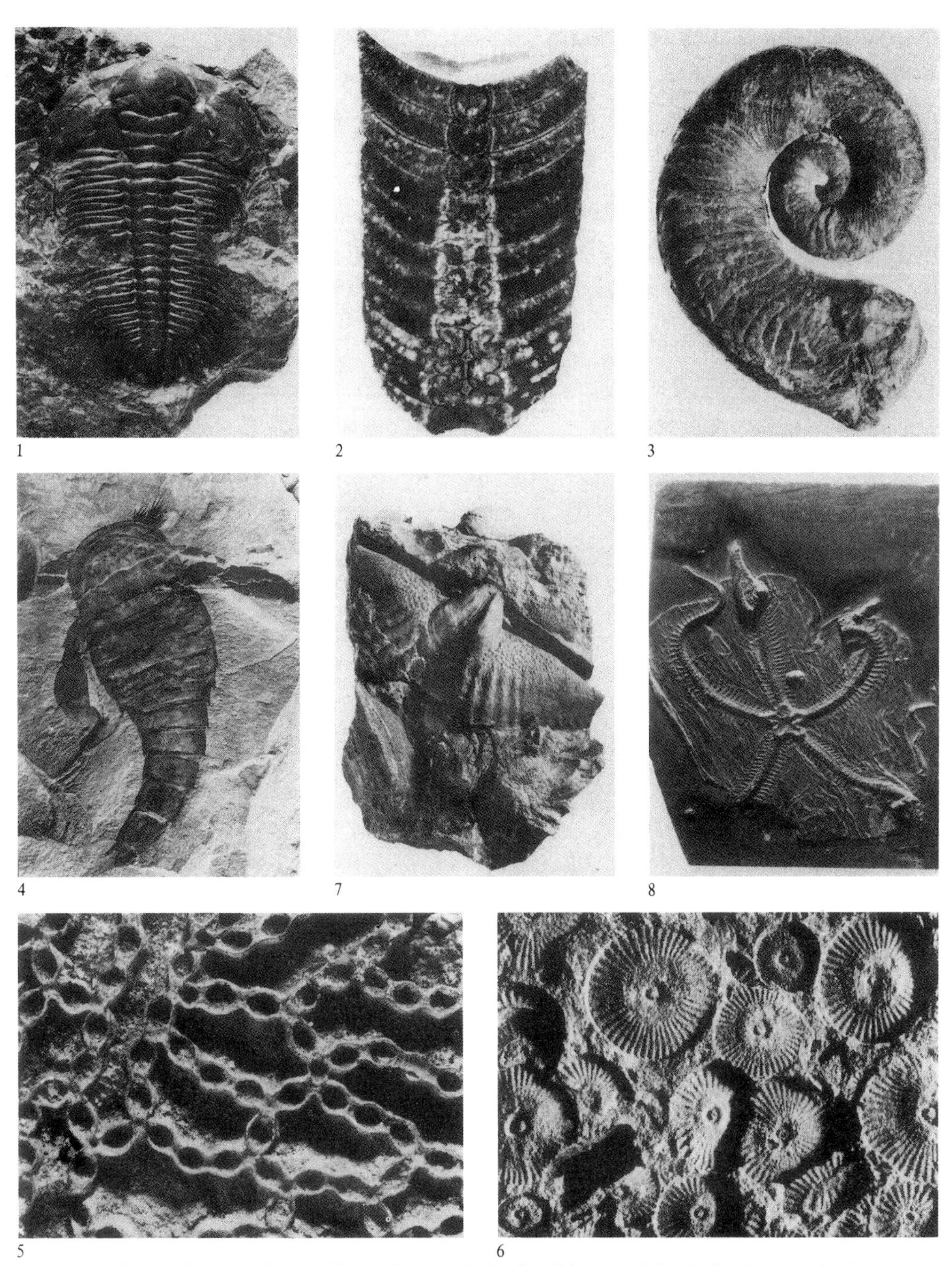

31 *Paleozoology.* *1* *Paradoxides gracilis* (trilobite), an index fossil from the Middle Cambrian of the Barrandium of Bohemia; length 9 cm. ***2*** *Actinoceras* sp. (nautiloid); longitudinal section through part of the chambered shell; from the Silurian of Lochov (Bohemia); length 7.5 cm, width 5 cm. ***3*** *Discoceras antiquissimum* (nautiloid) from the Ordovician of southern Sweden; the growth lines are clearly seen; diameter of spiral 4.2 cm. ***4*** *Eurypterus lacustris* (arthropod, gigantostracan) from the Silurian of North America; total length 24 cm. ***5*** Transverse section through a colony of *Halysites catenularia* (chain coral) from the Ordovician to the Silurian of England; diameter of a single tube 1 mm. ***6*** *Ctenocrinus typus* (crinoid) from the Devonian of Westphalia; diameter 2–4 mm. ***7*** *Euryspirifer paradoxus (Brachiopoda);* stone cast of a hinged lampshell from the Lower Devonian of the Rhineland slate; width 7.6 cm. ***8*** *Euzonosoma tischbeiniana (Ophiuroidea);* a brittle star from the Lower Devonian Bundenbach slate of Eifel; diagonal width 13 cm.

32 *Mosses, liverworts and ferns.* ***1*** *Leucobryum glaucum* (moss). ***2*** *Bryum argenteum* (moss). ***3*** *Conocephalum conicum* (liverwort). ***4*** *Riccardia multifida* (liverwort). ***5*** *Marchantia polymorpha* (liverwort). ***6*** *Rhodobryum roseum* (moss). ***7*** Floating moss *(Salvinia natans)* (fern). ***8*** Hart's tongue fern *(Scolopendrium vulgare = Phyllitis scolopendrium)* (fern).

33 *Ferns and lichens.* 1 Rusty-back fern *(Ceterach officinarum)*. **2** Maidenhair spleenwort *(Asplenium trichomonas)* (fern). **3** Pillwort *(Pilularia globulifera)* (fern). **4** *Cladonia pyxidata* (lichen). **5** *Parmelia* sp. (lichen). **6** *Baeomyces roseus* (lichen). **7** *Peltigera* sp. (lichen). **8** *Leconora* sp. (lichen). **9** *Umbilicaria hirsuta* (lichen).

Plate 34

34 *Flowers and fruits of deciduous trees.* ***1*** Flowers of witch hazel *(Hamamelis virginiana)*. ***2*** Flowers and ***3*** Fruit of the tulip tree *(Liriodendron tulipifera)*. ***4*** Flowers of the sweet chestnut *(Castanea sativa)*. ***5*** Flower and ***6*** Fruit of magnolia *(Magnolia yulan)*. ***7*** Fruits of the plane tree *(Platanus acerifolia)*. ***8*** Fruits of the Oregon grape *(Mahonia aquifolium = Berberis aquifolium)*. ***9*** Flowers of holly *(Ilex aquifolium)*.

35 ***Flowers and fruits of conifers.*** ***1*** Veitch's fir *(Abies veitchii)*. ***2*** Oregon Douglas fir *(Pseudotsuga taxifolia douglasi)*. ***3*** Blue spruce *(Picea pungens glauca)*. ***4*** Serbian spruce *(Picea omorika)*. ***5*** White pine *(Pinus strobus)*. ***6*** Ponderosa pine or Western yellow pine *(Pinus ponderosa* Dougl.). ***7*** Nootka cypress *(Chamaecyparis nootkaensis)*. ***8*** Japanese larch *(Larix leptolepis)*. ***9*** Yew *(Taxus baccata)*.

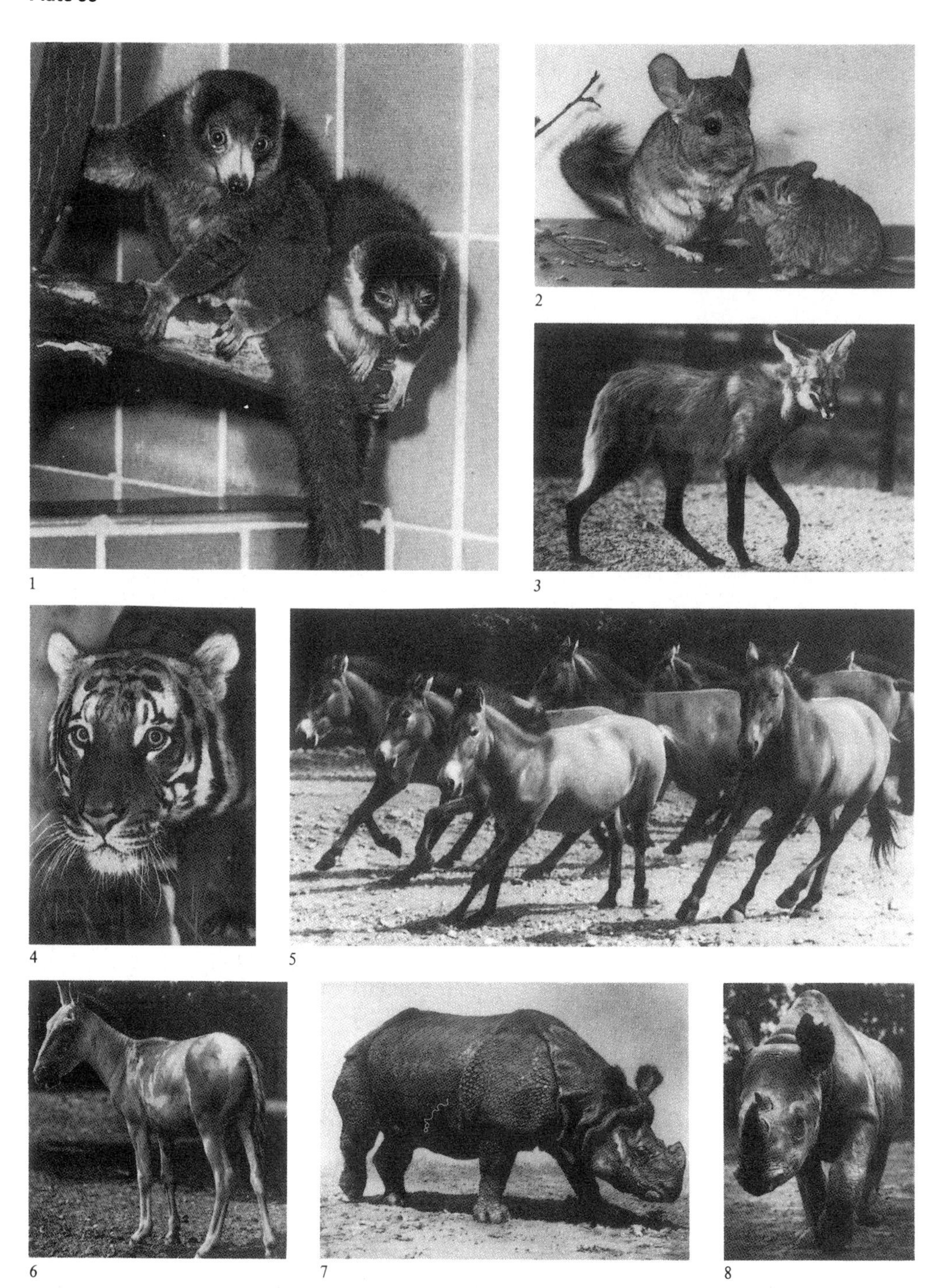

36 *Animals threatened with extinction I.* ***1*** Mongoose lemur *(Lemur mongoz mongoz)*. ***2*** Long-tailed chinchilla *(Chinchilla laniger;* survival insured as a domestic pet). ***3*** Maned wolf *(Chrysocyon brachyurus;* the small residual population is scattered over an immense area. ***4*** Tiger *(Panthera tigris;* some subspecies severely threatened or already extinct). ***5*** Wild horse *(Equus przewalskii)*. ***6*** Onager *(Equus hemionus onager)*. ***7*** Indian one-horned rhinoceros *(Rhinoceros unicornis)*. ***8*** African black rhinoceros *(Diceros bicornis;* rapidly reduced in recent years by poaching).

37 *Animals threatened with extinction II.* 1 Guanaco *(Lama guanicoë;* extinct in many parts of its areal). ***2*** Sambar *(Cervus unicolor;* some subspecies endangered). ***3*** Wisent *(Bison bonasus).* ***4*** Bison *(Bison bison).* ***5*** Pasang *(Capra aegagrus;* pure line endangered by crossing with domestic goats). ***6*** Ostrich *(Struthio camelus;* has disappeared from many areas; Arabian subspecies extinct). ***7*** Peregrine falcon *(Falco peregrinus;* threatened with extinction in certain areas, in particular central Europe). ***8*** Great bustard *(Otis tarda).* Also threatened are: Giant panda, Okapi, Arabian oryx (plates 38 & 39) and many other species.

38 *Mammals I.* ***1*** Brush-tailed possum *(Trichosurus vulpecula;* order *Marsupialia,* family *Phalangeridae).* **2** Tree kangaroo *(Dendrolagus* sp., order *Marsupialia,* family *Phalangeridae).* ***3*** Dwarf armadillo *(Euphractus pichiy,* order *Edentata,* family *Dasypodidae).* ***4*** Prairie dog *(Cynomys ludovicianus,* order *Rodentia,* family *Sciuridae).* ***5*** Giant panda *(Ailuropoda melanoleuca,* order *Carnivora,* family *Ailuropodidae).* ***6*** Swamp mongoose *(Herpestes paludinosus,* order *Carnivora,* family *Viverridae).* ***7*** Cape hunting dog *(Lycaon pictus,* order *Carnivora,* family *Canidae).*

1 2 3 5 6 7 8

39 ***Mammals II.*** ***1*** Serval *(Felis = Leptailurus serval,* order *Carnivora,* family *Felidae).* ***2*** Cheetah *(Acinonyx jubatus,* order *Carnivora,* family *Felidae).* ***3*** Kulan *(Equus hemionus kulan,* order *Perissodactyla,* family *Equidae).* ***4*** Père David's deer *(Elaphurus davidianus,* order *Artiodactyla,* family *Cervidae).* ***5*** Okapi *(Okapia johnstoni,* order *Artiodactyla,* family *Giraffidae).* ***6*** Arabian oryx *(Oryx gazella leucoryx,* order *Artiodactyla,* family *Bovidae).* ***7*** Saiga antelope *(Saiga tatarica,* order *Artiodactyla,* family *Bovidae).* ***8*** Musk ox *(Ovibos moschatus,* order *Artiodactyla,* family *Bovidae).*

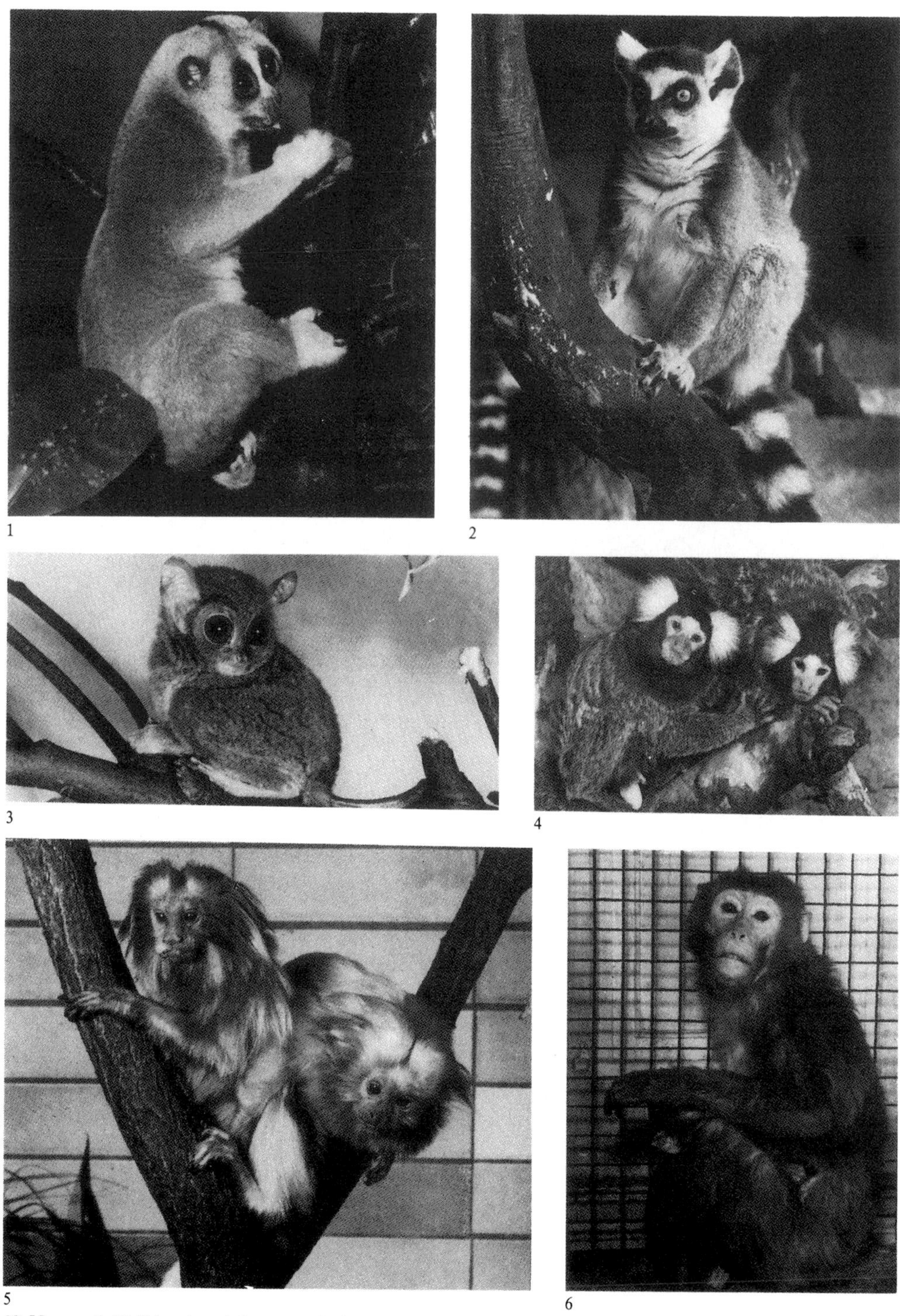

40 ***Mammals III.*** Primates. ***1–3*** are prosimians. ***4*** and ***5*** are New World monkeys. ***6*** is an Old World monkey. ***1*** Slow loris *(Nycticebus coucang,* family *Lorisidae).* ***2*** Ring-tailed lemur *(Lemur catta,* family *Lemuridae).* ***3*** Tarsier *(Tarsius,* family *Tarsiidae).* ***4*** White fronted marmoset or common marmoset *(Callithrix jacchus,* family *Callitrichidae).* ***5*** Lion tamarin *(Leontopithecus rosalia,* family *Callitrichidae).* ***6*** Rhesus monkey *(Macaca mulatta,* family *Cercopithecidae).*

41 *Mammals IV.* Primates. ***1–6*** are all Old World monkeys or apes. ***1*** Red-eared guenon *(Cercopithecus erythrotis,* family *Cercopithecidae).* ***2*** Hamadryas baboon or sacred baboon *(Papio hamadryas,* family *Cercopithecidae).* ***3*** Agile or dark-handed gibbon *(Hylobates lar,* family *Hylobatidae).* ***4*** Orang-utan *(Pongo pygmaeus,* family *Pongidae).* ***5*** Gorilla *(Gorilla gorilla,* family *Pongidae,* alternatively *Hominidae).* ***6*** Chimpanzee *(Pan troglodytes,* family *Pongidae,* alternatively *Hominidae).*

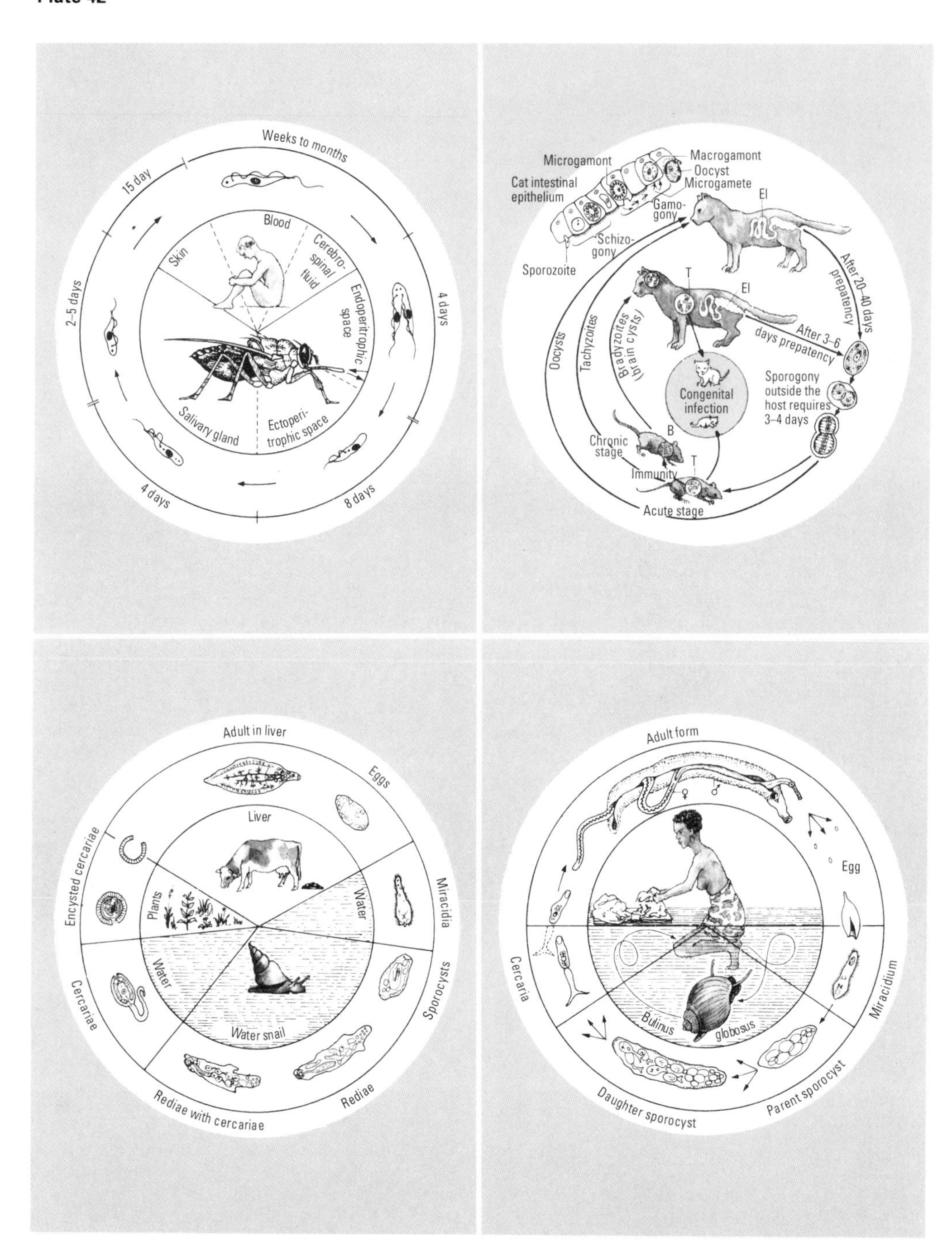

42 Parasite life cycles I. Top left: *Trypanosoma gambiense (Zooflagellata),* causative agent of sleeping sickness. The inner circle shows the hosts, the outer circle the respective stages of the parasite. The times given are for minimal periods of development (after Dönges. 1980). ***Top right:*** *Toxoplasma gondii (Sporozoa),* causative agent of toxoplasmosis. The natural intermediate host is the mouse; many warm blooded animals, including humans, are potential intermediate hosts (after Dönges, 1980). ***B*** Bradyzoites. ***EI*** Enteroepithelial infection (only in the cat). ***T*** Tachyzoites in the lymph system. ***Bottom left:*** *Fasciola hepatica (Trematoda),* Liver fluke (after Kämpfe, 1983). ***Bottom right:*** *Schistosoma haematobium (Trematoda),* causative agent of Bilharzia (after Dönges, 1980).

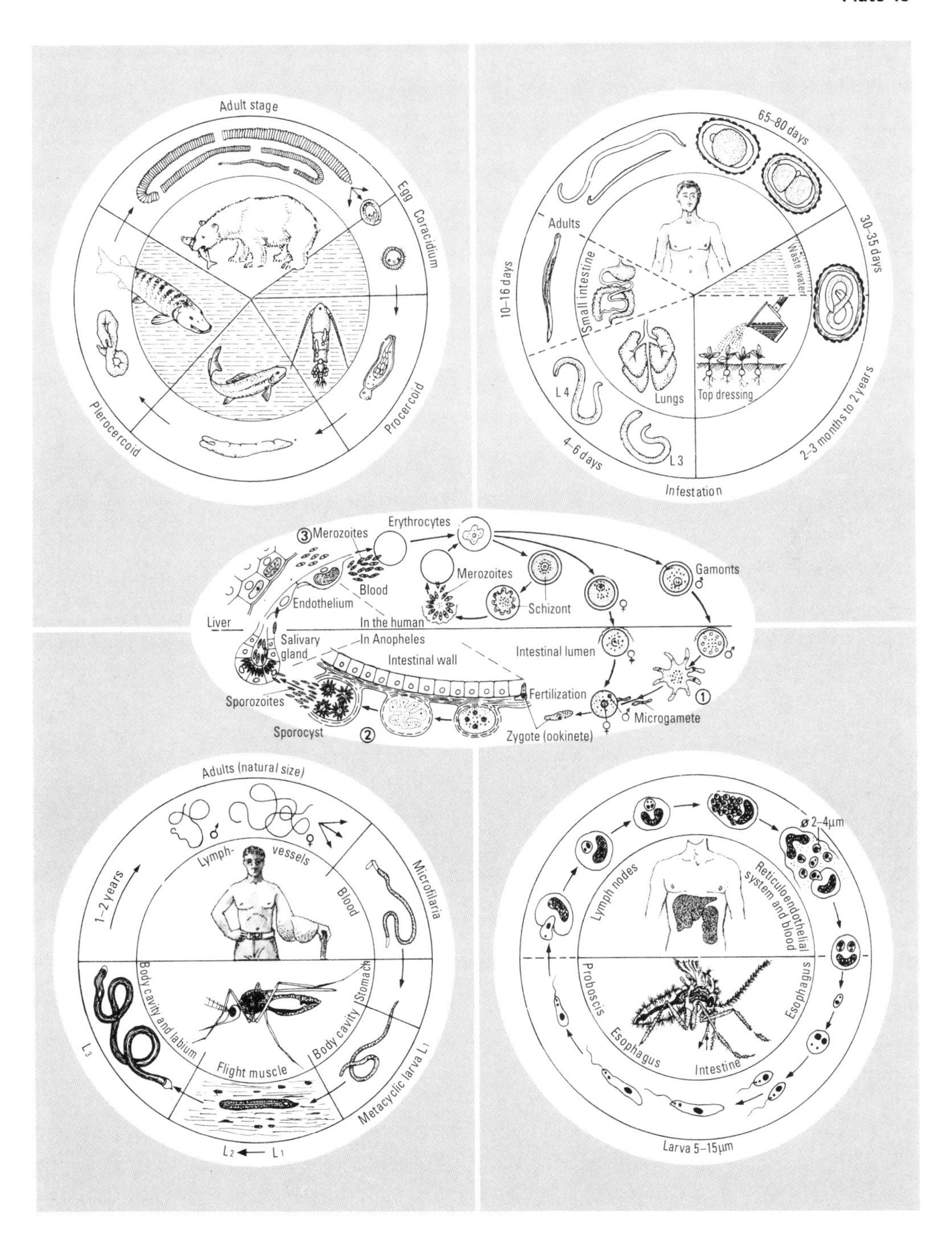

43 ***Parasite life cycle II. Top left:*** *Diphyllobothrium latum (Cestoda),* broad fish tapeworm (after Dönges, 1980). ***Top right:*** *Ascaris lumbricoides (Nematoda),* round worm (after Kämpfe, 1983). ***Middle:*** *Plasmodium,* causative agent of malaria, showing hosts and parasite life cycle (from Remane, Storch and Welsch, 1980). ***1*** Gametogony starts in the human erythrocyte and is concluded in the mosquito intestine. ***2*** Sporogony in the body of the mosquito. ***3*** Schizogony in the human reticuloendothelial system and liver parenchyma, then later in erythrocytes. ***Bottom left:*** *Wuchereria bancrofti (Nematoda),* causative agent of elephantiasis. L1, L2 and L3 are larval stages 1 to 3 (after Dönges, 1980). ***Bottom right:*** *Leishmania donovani (Zooflagellata),* causative agent of leishmaniasis (splenomegaly) (after Dönges, 1980).

Plate 44

44 *Different modes of feeding in animals.* ***1*** Filter feeder: an anostracan *(Chirocephalus grubei)*. ***2*** Tentacle feeder: sea anemone *(Anemone sulcata)*. ***3*** Grazing animal: rabbit *(Oryctolagus cuniculus)*. ***4*** Hunter- predator: crying eagle *(Aquila pomarina)*. ***5*** Hunter-predator: tiger beetle *(Cicindela hybrida)*. ***6*** Licking or lapping: map butterfly *(Araschnia levana)*. ***7*** Ambush hunter: praying mantis *(Mantis religiosa)*. ***8*** Trap-setter: spider *(Argiope* sp.). ***9*** Piercing and sucking parasite: pig louse *(Haematopinus suis)*.

45 ***Animal pests I.*** ***1*** Asparagus fly *(Platyparea poeciloptera)*. ***2*** Larva of the rose sawfly *(Arge rosae)*. ***3*** Larva of the apple leaf psyllid *(Psylla mali)*. ***4*** Larva of the large cabbage white butterfly *(Pieris brassicae)*. ***5*** Female of the mottled umber *(Erannis defoliaria)*. ***6*** Colorado beetle *(Leptinotarsa decemlineata)*. ***7*** Rape plants damaged by the weevil, *Ceuthorhynchus napi*. ***8*** Silver Y moth *(Autographa gamma)*. ***9*** Grub of the summer chafer *(Amphimallon solstitiale)*.

Plate 46

46 ***Animal pests II.*** ***1*** Brown tail moth *(Euproctis chrysorrhoea);* crowns of oak trees stripped of foliage, showing the winter nests of the caterpillars. ***2*** Winter nest of caterpillars of the brown tail moth. ***3*** Galleries of the large elm bark beetle *(Scotylus scotylus).* ***4*** Gypsy moth *(Lymantria dispar)* (female). ***5*** Caterpillars of an yponomeutid micromoth *(Yponomeuta* sp.). ***6*** Black arches moth *(Lymantria monacha).* ***7*** Black arches caterpillar. ***8*** Galleries of the engraver beetle, *Ips (= Tomicus) minor.* ***9*** Oriental cockroach, kitchen cockroach or kakerlak *(Blatta orientalis)* with fresh egg capsule.

47 ***Landscape management and nature conservancy.*** ***1*** Highly polluted, biologically depleted water. The water is dark in color from coal residues, and the foam is due to pollution with phenols. ***2*** Moorland destroyed by sewage pollution. ***3*** Bank erosion due to the absence of trees and bushes. ***4*** Spruce forest in a nature conservancy area. ***5*** Lake formed by renovation and redevelopment of the flooded depression of an old open cast mine. The sides of the mine are largely consolidated by trees and bushes. ***6*** Black-headed gulls over a breeding colony.

48 ***Trees as monuments.*** ***1*** Oak, approximate age 1 000 years. ***2*** Ancient village elm. ***3*** Horse chestnut. ***4*** Lime (linden). ***5*** Beech.